CHILTON'S
DRIVEABILITY MANUAL

Publisher and Editor-in-Chief Kerry A. Freeman, S.A.E
Managing Editors Peter M. Conti, Jr. □ W. Calvin Settle, Jr., S.A.E
Assistant Managing Editor Nick D'Andrea
Senior Editors Debra Gaffney □ Ken Grabowski, A.S.E., S.A.E.
Michael L. Grady □ Richard J. Rivele, S.A.E.
Richard T. Smith □ Jim Taylor □ Ronald T. Webb
Project Managers Benjamin Greisler, S.A.E. □ Martin J. Gunther
Jeffrey M. Hoffman □ Steven Morgan □ James B. Steele
Service Editors Peter B. Bilotta, A.S.E. □ Lawrence C. Braun, S.A.E., A.S.C.
Thomas P. Browne III □ Hugh J. Brulliea □ Dean G. Callahan □ Michael M. Carroll
William C. Cottman, A.S.E. □ Robert B. Day, Jr. □ Paul D. DeGuiseppi, A.S.E.
Robert F. Dougherty, Jr. □ Robert E. Doughten □ Sam Fiorani
Andrew J. Folz, A.S.E. □ Edward J. Giacomucci, A.S.E. □ Jacques Gordon
Neil Leonard, A.S.E. □ Kevin Maher □ Robert McAnally □ Raymond K. Moore
Craig P. Nangle, A.S.E. □ Charles T. Ramsey □ Roy Ripple, A.S.E.
John H. Rutter, A.S.E. □ Don Schnell, A.S.E., S.A.E. □ Larry E. Stiles
Jeff N. Sullivan □ Anthony Tortorici, A.S.E., S.A.E. □ Thom Young

Director of Manufacturing Mike D'Imperio
Assistant Manager of Manufacturing Robin Norman
Assistant Production Manager Andrea Steiger
Production Assistants Marsha Park Herman □ Monica Santa Maria □ Margaret Stoner
Mechanical Artists Lisa Gressen □ Kim Hayes

National Sales Manager Lawrence R. Rufo
National Administrator, Sales Benjamin K. Tatta
Eastern Regional Manager Howie Freeman
Western Regional Manager Larry W. Marshall
Mid West Regional Manager Bruce McCorkle
Special Accounts Manager Herbert Marshall
National Accounts Manager Kevin Mullen

OFFICERS
President, Chilton Enterprises David S. Loewith
Senior Vice President Ronald A. Hoxter

CHILTON BOOK COMPANY
ONE OF THE ABC PUBLISHING COMPANIES,
A PART OF CAPITAL CITIES/ABC, INC.

Manufactured in USA ©1993 Chilton Book Company • Chilton Way, Radnor, Pa. 19089
ISBN 0-8019-8474-2   1234567890 2109876543   ISSN 1053-220X

HOW TO USE THIS MANUAL

For ease of use, this manual is divided into sections:

SECTION 1	Self-Diagnostic Systems
SECTION 2	Electronic Ignition Systems
SECTION 3	Multi-Port Fuel Injection Systems
SECTION 4	Central Port Fuel Injection Systems
SECTION 5	Diesel Fuel Systems

The **CONTENTS**, inside the front cover, summarizes the subjects covered in each section.

To quickly locate the proper service section, use the chart on the following pages. It references applicable **MODELS** and **SERVICE SECTIONS** for major engine performance control systems.

It is recommended that the service technician be familiar with the applicable **GENERAL INFORMATION** and **SERVICE PRECAUTIONS** (if any) before testing or servicing the system. Major service sections are grouped by individual vehicle or component. Each sub-section contains:

- **GENERAL INFORMATION** Information pertaining to the operation of the system, individual components and the overall logic by which components work together.

- **SERVICE PRECAUTIONS (if any)** Precautions of which the service technician should be aware to prevent injury or damage to the vehicle or components.

- **TESTS, ADJUSTMENTS AND COMPONENT R&R** Performance tests and specifications, adjustments and component replacement procedures.

- **FAULT DIAGNOSIS** Complete troubleshooting for the entire system.

SAFETY NOTICE

Proper service and repair procedures are vital to the safe, reliable operation of all motor vehicles, as well as the personal safety of those performing repairs. This manual outlines procedures for servicing and repairing vehicles using safe, effective methods. The procedures contain many NOTES and CAUTIONS which should be followed along with standard safety procedures to eliminate the possibility of personal injury or improper service which could damage the vehicle or compromise its safety.

It is important to note that the repair procedures and techniques, tools and parts for servicing motor vehicles, as well as the skill and experience of the individual performing the work vary widely. It is not possible to anticipate all of the conceivable ways or conditions under which vehicles may be serviced, or to provide cautions as to all of the possible hazards that may result. Standard and accepted safety precautions and equipment should be used when handling toxic or flammable fluids, and safety goggles or other protection should be used during cutting, grinding, chiseling, or any other process that can cause material removal or projectiles.

Some procedures require the use of tools specially designed for a specific purpose. Before substituting another tool or procedure, you must be completely satisfied that neither your personal safety, nor the performance of the vehicle will be endangered.

PART NUMBERS

Part numbers listed in this reference are not recommendations by Chilton for any product by brand name. They are references that can be used with interchange manuals and aftermarket supplier catalogs to locate each brand supplier's discrete part number.

Although information in this manual is based on industry sources and is as complete as possible at the time of publication, the possibility exists that some vehicle manufacturers made later changes which could not be included here. While striving for total accuracy, Chilton Book Company cannot assume responsibility for any errors, changes or omissions that may occur in the compilation of this data.

Portions of materials contained herein have been reprinted with permission of General Motors Corporation, Service Technology Group.

SELF-DIAGNOSTIC
Section 1

ELECTRONIC IGNITION
Section 2

MULTIPORT FUEL INJECTION (MFI)
Section 3

CENTRAL PORT FUEL INJECTION (CPI)
Section 4

DIESEL FUEL SYSTEMS
Section 5

THROTTLE BODY INJECTION (TBI)
Section 6

SEQUENTIAL PORT FUEL INJECTION (SFI)
Section 7

DIAGNOSTIC EQUIPMENT
Section 8

GLOSSARY
Section 9

DRIVEABILITY — APPLICATIONS
BUICK/CADILLAC

BUICK

Model	Body VIN	Engine Liter	Engine VIN	Ignition Type	Fuel Type	Diagnostic Charts
Century	A	2.2	4	DIS	MFI	3-100
		2.5	R	DIS	TBI	6-77
		3.1	T①	C³I	SFI	7-17
		3.3	N	C³I	MFI	3-748
Estate Wagon	B	5.7	7	HEI	TBI	6-160
LeSabre	H	3.8	L	C³I	SFI	7-242
Park Avenue	L	3.8	L	C³I	SFI	7-242
Park Avenue Ultra	L	3.8	1	C³I	SFI	7-331
Regal	W	3.1	T	DIS	MFI	3-614
		3.1	T①	C³I	SFI	7-17
		3.8	L	C³I	SFI	7-157
Riviera	E	3.8	L	C³I	SFI	7-422
Roadmaster	B	5.7	7	HEI	TBI	6-160
Skylark	N	2.3	D	IDI	MFI	3-427
		2.3	3	IDI	MFI	3-427
		3.1	T①	C³I	SFI	7-17
		3.3	N	C³I	MFI	3-749

C³I—Computer Control Command Ignition
DIS—Distributorless Ignition System
HEI—High Energy Ignition
IDI—Integrated Direct Ignition
MFI—Multi-Port Fuel Injection
TBI—Throttle Body Injection
SFI—Sequential Fuel Injection
① California

CADILLAC

Model	Body VIN	Engine Liter	Engine VIN	Ignition Type	Fuel Type	Diagnostic Charts
Allante	V	4.5	8	HEI	SFI	7-498
		4.6	9	DIS	SFI	7-608
Brougham	D	5.0	E	HEI	TBI	6-160
		5.7	7	HEI	TBI	6-160
DeVille	C	4.9	B	HEI	SFI	7-877
Fleetwood	C	4.9	B	HEI	SFI	7-877
Eldorado	E	4.6	Y	DIS	SFI	7-741
		4.6	9	DIS	SFI	7-741
		4.9	B	HEI	SFI	7-982
Seville	K	4.6	9	DIS	SFI	7-741
		4.9	B	HEI	SFI	7-982

DIS—Distributorless Ignition System
HEI—High Energy Ignition
SFI—Sequential Fuel Injection
TBI—Throttle Body Injection

CHEVROLET

Model	Body VIN	Engine Liter	Engine VIN	Ignition Type	Fuel Type	Diagnostic Charts
Beretta	L	2.2	4	DIS	MFI	3-210
		2.3	A	IDI	MFI	3-321
		3.1	T	DIS	MFI	3-550
		3.1	T③	DIS	SFI	7-17
Camaro	F	3.1	T	HEI	MFI	3-549
		3.4	S	DIS	SFI	7-80
		5.0	E	HEI	TBI	6-161
		5.0	F	HEI	MFI	3-826
		5.7	8	HEI	MFI	3-826
		5.7	P①	Opti	MFI	3-876
Caprice	B	4.3	Z	HEI	TBI	6-160
		5.0	E	HEI	TBI	6-160
		5.7	7	HEI	TBI	6-160
Cavalier	J	2.2	4	DIS	MFI	3-144
		3.1	T	DIS	MFI	3-549
Corsica	L	2.2	4	DIS	MFI	3-210
		2.3	A	IDI	MFI	3-321
		3.1	T	DIS	MFI	3-550
		3.1	T③	DIS	SFI	7-17
Corvette	Y	5.7	8	HEI	MFI	3-948
		5.7	J	DIS	SFI	7-1079
		5.7	P①	Opti	MFI	3-949
Lumina	W	2.2	4	DIS	MFI	3-275
		2.5	R	DIS	TBI	6-177
		3.1	T	DIS	MFI	3-614 ②
		3.1	T③	DIS	SFI	7-17
		3.4	X	DIS	MFI	3-615

DIS—Distributorless Ignition System
HEI—High Energy Ignition
IDI—Integrated Direct Ignition
Opti—Opti-Spark Ignition
MFI—Multi-Port Fuel Injection
SFI—Sequential Fuel Injection
TBI—Throttle Body Injection
① 1993 only; in 1994 this engine came equipped
   with Sequential Fuel Injection (SFI)
② VFV—Variable Fuel Vehicle also available:
   3-687
③ California

OLDSMOBILE

Model	Body VIN	Engine Liter	Engine VIN	Ignition Type	Fuel Type	Diagnostic Charts
Achieva	N	2.3	D	IDI	MFI	3-427
		2.3	A	IDI	MFI	3-427
		2.3	3	IDI	MFI	3-427
		3.1	T ①	C^3I	SFI	7-17
		3.3	N	C^3I	MFI	3-749
Cutlass Supreme	W	3.1	T	DIS	MFI	3-614
		3.1	T ①	C^3I	SFI	7-17
		3.4	X	DIS	MFI	3-615
Custom Cruiser	B	5.0	E	HEI	TBI	6-160
		5.7	7	HEI	TBI	6-160
Cutlass Cruiser	A	2.2	4	DIS	MFI	3-100
		2.5	R	DIS	TBI	6-77
		3.1	T ①	C^3I	SFI	7-17
		3.3	N	C^3I	MFI	3-748
Cutlass Ciera	A	2.2	4	DIS	MFI	3-100
		2.5	R	DIS	TBI	6-77
		3.1	T ①	C^3I	SFI	7-17
		3.3	N	C^3I	MFI	3-748
Eighty-Eight	H	3.8	L	C^3I	SFI	7-242
Ninety-Eight	C	3.8	L	C^3I	SFI	7-242
		3.8	1	C^3I	SFI	7-331
Toronado	E	3.8	L	C^3I	SFI	7-423
Toronado Trofeo	E	3.8	L	C^3I	SFI	7-423

C^3I—Computer Control Command Ignition
DIS—Distributorless Ignition System
HEI—High Energy Ignition
IDI—Integrated Direct Ignition
MFI—Multi-Port Fuel Injection
SFI—Sequential Fuel Injection
TBI—Throttle Body Injection
① California

PONTIAC

Model	Body VIN	Engine Liter	Engine VIN	Ignition Type	Fuel Type	Diagnostic Charts
Bonneville	H	3.8	L	C³I	SFI	7-242
		3.8	1	C³I	SFI	7-331
Firebird	F	3.1	T	HEI	MFI	3-549
		3.4	S	DIS	SFI	7-80
		5.0	E	HEI	TBI	6-161
		5.0	F	HEI	MFI	3-827
		5.7	8	HEI	MFI	3-826
		5.7	P①	Opti	MFI	3-876
Grand Am	N	2.3	A	IDI	MFI	3-427
		2.3	3	IDI	MFI	3-427
		2.3	D	IDI	MFI	3-427
		3.1	T②	C³I	SFI	7-17
		3.3	N	C³I	MFI	3-614
Grand Prix	W	3.1	T	DIS	MFI	3-614
		3.1	T②	C³I	SFI	7-17
		3.4	X	DIS	MFI	3-615
LeMans	T	1.6	6	HEI	TBI	6-29
Sunbird	S	2.0	H	DIS	MFI	3-40
		3.1	T	DIS	MFI	3-549

C³I—Computer Control Command Ignition
DIS—Distributorless Ignition System
HEI—High Energy Ignition
IDI—Integrated Direct Ignition
Opti—Opti-Spark Ignition
MFI—Multi-Port Fuel Injection
SFI—Sequential Fuel Injection
TBI—Throttle Body Injection
① 1993 only; in 1994 this engine came equipped
   with Sequential Fuel Injection (SFI)
② California

GEO

Model	Body VIN	Engine Liter	Engine VIN	Ignition Type	Fuel Type	Diagnostic Charts
Metro	M	1.0	6	ESC	TBI	6-290
Prizm	S	1.6	5	ESC	MFI	3-1079
		1.6	6	ESC	MFI	3-1108
		1.6	8	ESC	MFI	3-1174
		1.8	U	ESC	TBI	6-320
Storm	R	1.6	6	ESC	MFI	3-1222
		1.8	8	ESC	MFI	3-1225
Tracker	J	1.6	U	ESC	TBI	6-320

ESC—Electronic Spark Control
MFI—Multi-Port Fuel Injection
TBI—Throttle Body Injection

DRIVEABILITY — APPLICATIONS
SATURN/LIGHT TRUCKS AND VANS

SATURN

Model	Body VIN	Engine Liter	Engine VIN	Ignition Type	Fuel Type	Diagnostic Charts
SL/SL1/SW1/SC1	Z	1.9	9	EI	TBI	6-357
SL2/SW2/SC2	Z	1.9	7	EI	MFI	3-1271

EI—Electronic Ignition
TBI—Throttle Body Injection
MFI—Multi-Port Fuel Injection

GENERAL MOTORS LIGHT TRUCKS, VANS, AND APVS

Model	Body VIN	Engine Liter	Engine VIN	Ignition Type	Fuel Type	Diagnostic Charts
S-Series, Bravada and Astro/Safari	L,M,S,T,	2.5	A	HEI	TBI	6-444
		2.8	R	HEI	TBI	6-445
		4.3	Z[1]	HEI	MFI	3-1319
		4.3	W	HEI	CPI	4-13
Light Trucks and Vans	C,K,R,V,G,P	4.3	Z	HEI	TBI	6-445
		5.0	H	HEI	TBI	6-448
		5.7	K	HEI	TBI	6-448
		6.2	C	Diesel	TBI	5-11
		7.4	N	HEI	TBI	6-449
Lumina APV	U	3.1	D	HEI	TBI	6-405
		3.8	L	C^{3}I	SFI	7-1230
Silhouette	U	3.1	D	HEI	TBI	6-405
		3.8	L	C^{3}I	SFI	7-1230
Trans Sport	U	3.1	D	HEI	TBI	6-405
		3.8	L	C^{3}I	SFI	7-1230

C^{3}I—Computer Control Command Ignition
HEI—High Energy Ignition
TBI—Throttle Body Injection
CPI—Central Port Injection
MFI—Multi-Port Fuel Injection
SFI—Sequential Fuel Injection
[1] MFI with turbocharger

SELF-DIAGNOSTIC SYSTEMS

EXCEPT CADILLAC, RIVIERA, TORONADO, TROFEO, LIGHT TRUCKS, VANS, GEO AND SATURN

General Description

ELECTRONIC CONTROL MODULE

The ECM is the control center of the fuel injection system. It continuously monitors information from a variety of sensors and switches, and controls the systems that affect vehicle performance. The ECM also has self-diagnostic capability, and can detect system problems, alert the driver through the Malfunction Indicator Lamp (MIL) and record the appropriate Diagnostic Trouble Code (DTC). The numerical code will identify the specific circuit to aid the technician in diagnosis. The ECM is capable of storing more than one code, with possibilities ranging from 12 to 999, however not all systems use every code. In addition, similar code numbers can relate to different circuits depending on the vehicle. For example, on the 3.3L (VIN N), code 46 indicates a fault found in the power steering pressure switch circuit. The same code on the 5.7L (VIN F) engine indicates a fault in the VATS anti-theft system.

Learning Ability

The ECM can compensate for minor variations within the fuel system through the long term fuel trim (block learn) and short term fuel trim (fuel integrator) systems. The short term fuel trim monitors the oxygen sensor output voltage, adding or subtracting fuel to drive the mixture rich or lean as needed to reach the ideal air fuel ratio of 14.7:1. The integrator values may be read with a scan tool; the display will range from 0–255 and should center on 128 if the oxygen sensor is seeing a 14.7:1 mixture.

Powertrain Control Module (PCM)

The PCM is used when vehicles are equipped with electronically controlled transmissions. Its function is identical to that of the ECM with the added features of controlling shift points in the transmission, as well as cruise control operation and diagnostics. This unit may be referred to as the PCM, the ECM/PCM or the PCM/TCM (transmission control module). The integrated functions of engine and transmission control allow accurate gear selection and improved fuel economy.

For engine diagnostics, the PCM may be considered identical to an ECM system, although the combined unit will display additional codes relating to transmission function and components.

> **NOTE: The terms Electronic Control Module (ECM), Powertrain Control Module (PCM) and Body Control Module (BCM) may sometimes be used interchangably.**

Quad Drivers (QDM)

Some control modules have eliminated the use of individual transistors to control each output circuit. A single QDM can individually control 4 outputs from the com-

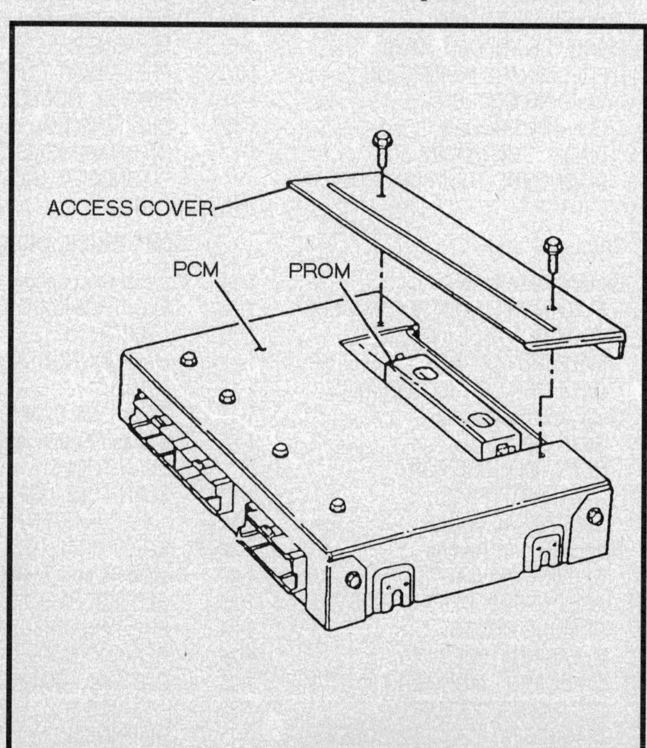

Electronic Control Module (ECM) — Camaro and Firebird shown

Powertrain Control Module (PCM) — Cavalier shown

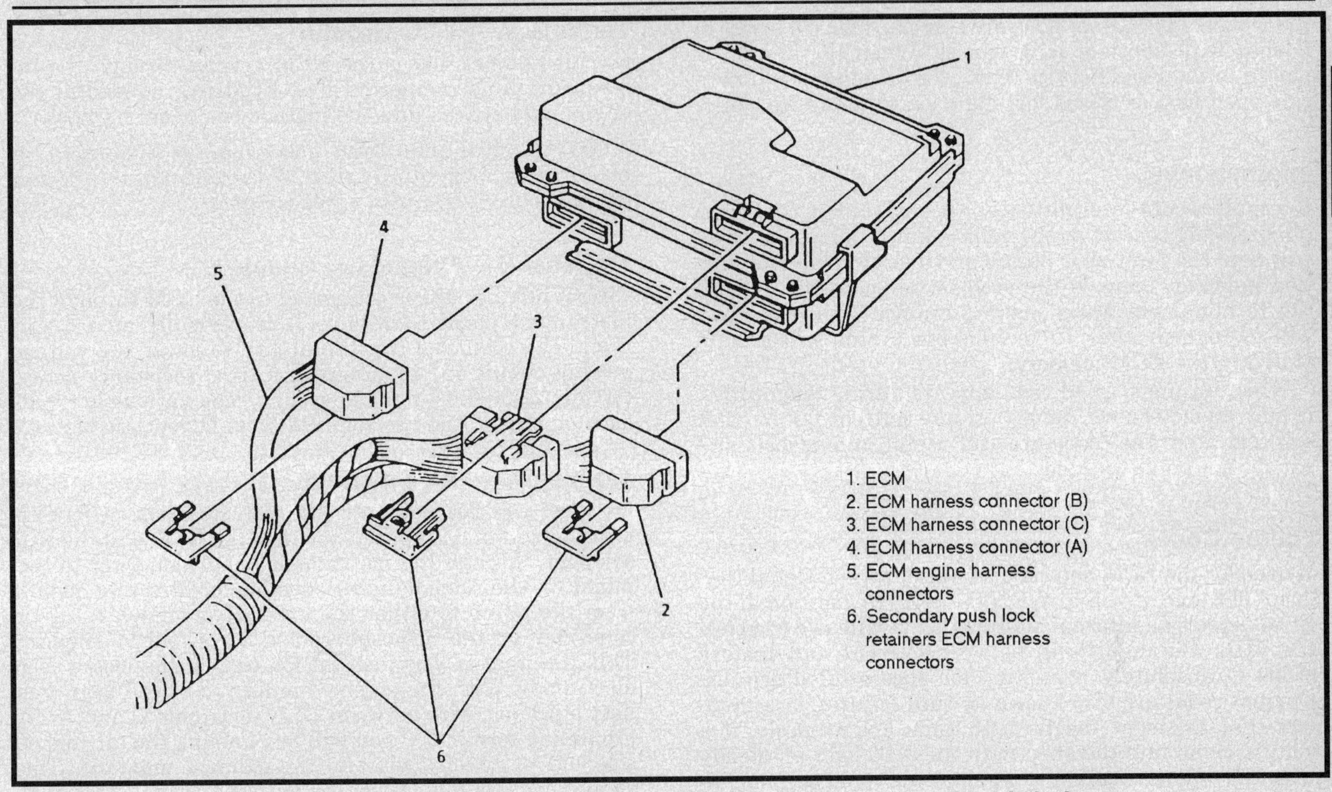

Electronic Control Module with re-programmable EEPROM — Camaro and Firebird shown

puter. Since these control modules are internally fault protected, a single faulty circuit or component may cause the other circuits on that QDM to be ON or OFF all the time. Do not confuse this with a failed ECM; if the Quad Driver itself has failed, the control module output will usually test either open or shorted.

The quad driver check procedure should be used before replacing an expensive ECM/PCM assembly. The check will not test all computer functions but will determine if an external circuit has disabled a QDM. Be particularly suspicious of an ECM/PCM exhibiting any of these characteristics:
- Malfunction indicator lamp ON with no codes stored.
- Malfunction indicator light flickers, intermittent or dim.
- ECM/PCM will not flash Code 12 and/or engine will not start.
- Controlled output, such as TCC, is inoperative or ON all the time.
- Engine stalls, misfires or surges.
- Scan tool erratic or inoperative.

BODY CONTROL MODULE

Body Control Module (BCM) is the communication center for the ECM/PCM, Instrument Panel Cluster (IPC), HVAC (Programmer), Electronic Climate Control Panel (ECCP), and the Supplemental Inflatable Restraint (SIR) module. A combination of inputs from these major components, as well as other sensors and switches, communicate with the BCM either as individual inputs or over the common communications link called the data line. The various inputs to the BCM combine with program instructions within the system memory to provide accurate control over the many subsystems involved. When a subsystem circuit exceeds pre-programmed limits, a system malfunction is indicated and certain backup functions may be provided.

Providing control over the many subsystems from the BCM is done by controlling system outputs. These outputs can be either direct or transmitted along the data line to one of the five other major components. The process of receiving, storing, testing and controlling information is continuous. The data communication gives the BCM control over the component's self-diagnostic capabilities in addition to its own.

In a method similar to that of a telegraph system, the BCM's internal circuitry rapidly switches a circuit between 0 and 5 volts like a telegraph key. This process is used to convert information into a series of pulses which represents coded data messages understood by the other components. Also, much like a telegraph system, each major component has its own recognition code or address. When messages are transmitted along the data line, only the component that matches the assigned recognition code will accept the message, while the other components will ignore it.

MALFUNCTION INDICATOR LAMP

The primary function of the malfunction indicator lamp is to advise the operator and the technician that a fault has been detected, and, in most cases, a code stored. Under normal conditions, the malfunction indicator lamp will illuminate when the ignition is turned **ON**. Once the engine is started and running, the ECM will perform a system check and extinguish the warning lamp if no fault is found.

Additionally, the malfunction indicator lamp can be used to retrieve stored codes after the system is placed in the Diagnostic Mode. Codes are transmitted as a series of flashes with short or long pauses. When the system is placed in the Field Service Mode, the dash lamp will indicate open loop or closed loop function to the technician.

Intermittents

If a fault occurs intermittently, such as a loose connector pin breaking contact as the vehicle hits a bump, the ECM will note the fault as it occurs and energize the malfunction indicator lamp. If the problem self-corrects, as with the terminal pin again making contact, the dash lamp will extinguish after 10 seconds but a code will remain stored in the ECM memory.

When an unexpected code appears during diagnostics, it may have been set during an intermittent failure that self-corrected; the codes are still useful in diagnosis and should not be discounted.

Failure Codes

Whenever the ECM detects a sensor or switch signal that is out of range, it will turn ON the MIL. In addition, if the ECM detects an internal malfunction, it will also turn ON the MIL. Distinguishing between current and history codes is absolutely necessary for successful diagnosis. Current codes are also known as hard failures, or a problem that exists at the present time. For example, if a vehicle came into the shop with the MIL light on steady with the engine running, a trouble code could be immediately retrieved and that particular diagnostic trouble chart could be referenced.

History codes, also known as intermittents, are failures which have taken place at one time, but are not presently causing the MIL light to turn ON. An example of this would be if the ignition timing had been recently adjusted on a 5.0 liter engine with TBI. When the bypass wire was disconnected to make the adjustment, the open circuit would have caused the ECM to revert to base timing. Because of this change, the ECM detected a problem and turned on the MIL while simultaneously recording a trouble code 42. When the timing adjustment was completed and the bypass wire was reconnected, the ECM waited approximately 10 seconds and then turned off the MIL, but retained the code. The ECM would retain this information up to a maximum of 50 ignition cycles. If code retrieval was attempted in the meantime by another technician unaware of the timing adjustment, their only recourse would be to reference the wiring diagram associated with the code 42 chart and physically inspect all connections.

DATA LINK CONNECTOR (DLC)

The data link connector is located underneath the instrument panel and is used to access the on-board microprocessors. These include the ECM, PCM, BCM as well as various other controllers such as the anti-lock brake system module. The use of a scan tool such as GM's Tech 1 will allow the technician to monitor data and control some of the microprocessor's operations. The scan tool will read the serial data signal from the DLC and translate that signal into useful diagnostic information on the display screen. The DLC can also be used without a scan tool for retrieving trouble codes, entering certain diagnostic modes and other valuable testing.

Terminal A — DLC Ground

Circuit number 450 is the ECM ground through the instrument panel connector. The ECM is grounded at the engine and splices into the instrument panel harness.

NOTE: Most scan tools can disguise problems related to ECM grounds due to the alternate ground path through the tool's power cord.

Terminal B — Diagnostic Enable

Circuit number 451 is connected to the ECM through the instrument panel connector. It is normally at a 5 volt level. The scan tool has the ability to lower the voltage level of circuit 451 by dropping a fixed resistance across the circuit or shorting it to ground. The various low voltages on circuit 451 are referred to as Diagnostic Modes.

Terminal C — Air Solenoid

On most vehicles with Air Injection Reaction (AIR) Systems, circuit 436 is spliced between the ECM and the AIR solenoid, through the instrument panel connector to terminal C. The solenoid is responsible for directing airflow from the pump to either the exhaust manifold, catalytic converter or the atmosphere via an attached silencer. This terminal is very helpful for testing the closed loop electronics. With the engine running in closed loop, connect a jumper wire between DLC terminals C and A. By grounding terminal C you will be enabling the air control solenoid to direct airflow to the exhaust manifold. This should cause a low O_2 sensor output voltage and a high (greater than 128) short term fuel trim number. If connected long enough, this action will trigger a code 44, lean exhaust.

In Open Loop operation, this solenoid will be enabled to quickly bring the O_2 sensor up to operating temperature. Once the system goes to Closed Loop the solenoid will become disabled causing airflow to be directed to either the catalytic converter or the atmosphere depending on the type of system. These actions can be verified using a digital voltmeter by probing terminal C with the engine running. In open loop the voltmeter should read 0 volts (solenoid energized) and in closed loop the reading should be battery voltage (solenoid de-energized).

Terminal D — Malfunction Indicator Lamp

Circuit 419 is spliced between the ECM and the MIL light to terminal D. With the key **ON**, the MIL light is turned ON through ECM terminal A5. If during a normal bulb check the MIL fails to turn on, a quick test can be performed before condemning the ECM. Insert a jumper between terminals D and A of the DLC with the key **ON**. This action should turn ON the MIL. If it does not, a defective bulb is the most likely problem.

Terminal E — Serial Data Line

Circuit 461 is connected through the instrument panel connector to the ECM. This is a discreet data line used on fuel injection systems with a slow data transmission rate (baud rate).

Terminal F — Torque Converter Clutch Solenoid

Circuit 422 is connected through the instrument panel connector and spliced into the ground-controlled TCC solenoid circuit. The ECM will turn on the TCC solenoid based on a variety of sensed inputs. By inserting a

DIAGNOSTIC MODE	VOLTAGE AT TERMINAL B	RESISTANCE ACROSS A AND B
Road Test or Open	5V	Infinite
Diagnostic or 10K mode	2.5V	10K ohms
Back Up Fuel	1.5V	3.9K ohms
Field Service	0V	0 ohms

Diagnostic modes

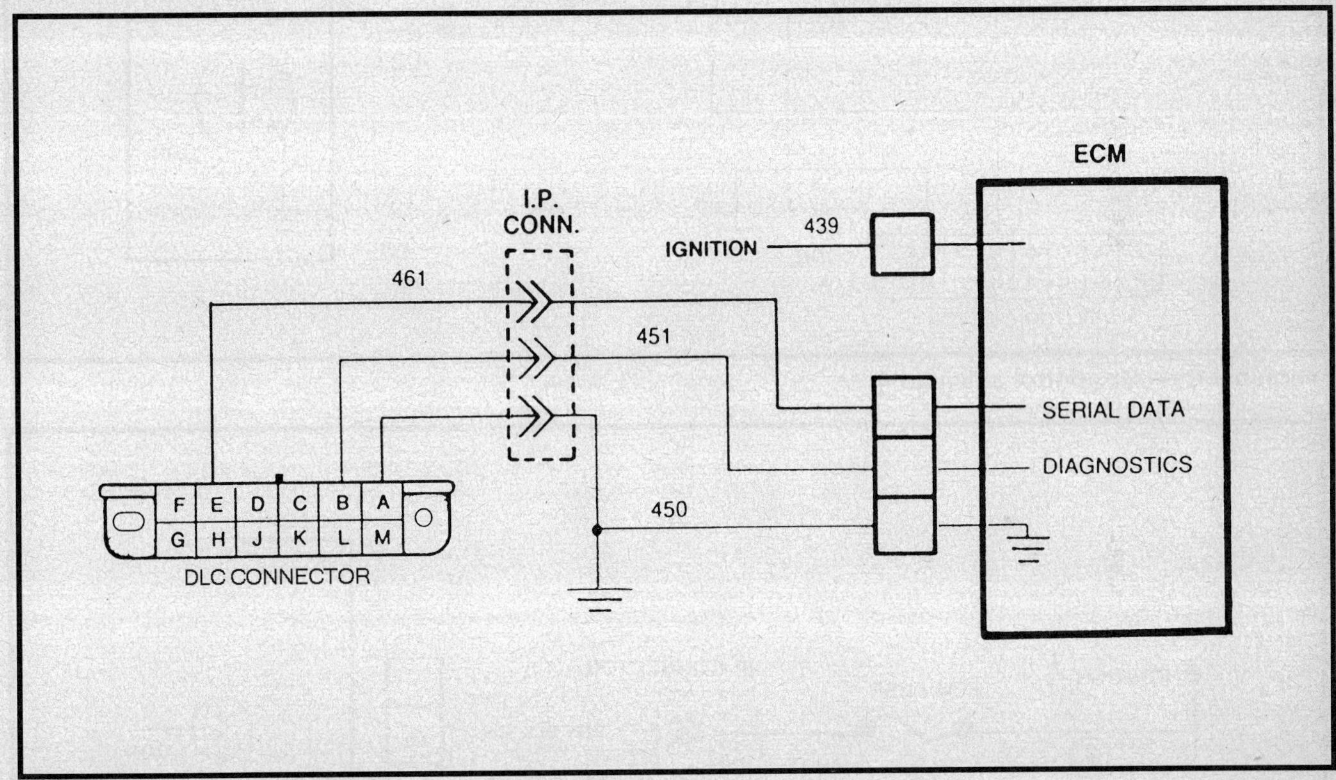

Terminals A, B and E — DLC

testlight between DLC terminals F and A, TCC command status can be observed. With the brake switch closed (brakes not applied) and the vehicle in 3rd gear (3rd gear switch closed), the test light should be ON. As soon as the ECM commands TCC engagement, the test light should turn OFF.

Terminal G — Fuel Pump Test Point

On 1992 Camaro and Firebird, circuit 490 is spliced into the electric fuel pump feed through the de-energized fuel pump relay. Most other models have a fuel pump test connector located in the engine compartment. Applying B+ voltage through a fused jumper wire will energize the in-tank fuel pump. This is helpful when performing fuel pressure testing and is more convenient than cycling the

key to turn ON the pump. On other models that use terminal G in the DLC, it is for other purposes than fuel pump. For example, it is used for keyless entry on 1993–94 Camaro and Firebird. When terminal G is either vacant or used for another purpose, a fuel pump test lead will be located in the engine compartment, usually near the battery.

Terminal M — Serial Data Line

Circuit 461 is the serial data line for fast baud rate ECMs such as the 3.1L (Vin T) engine. Terminal M may also include serial data transmission from other on-board microprocessors such as an anti-lock brake system controller.

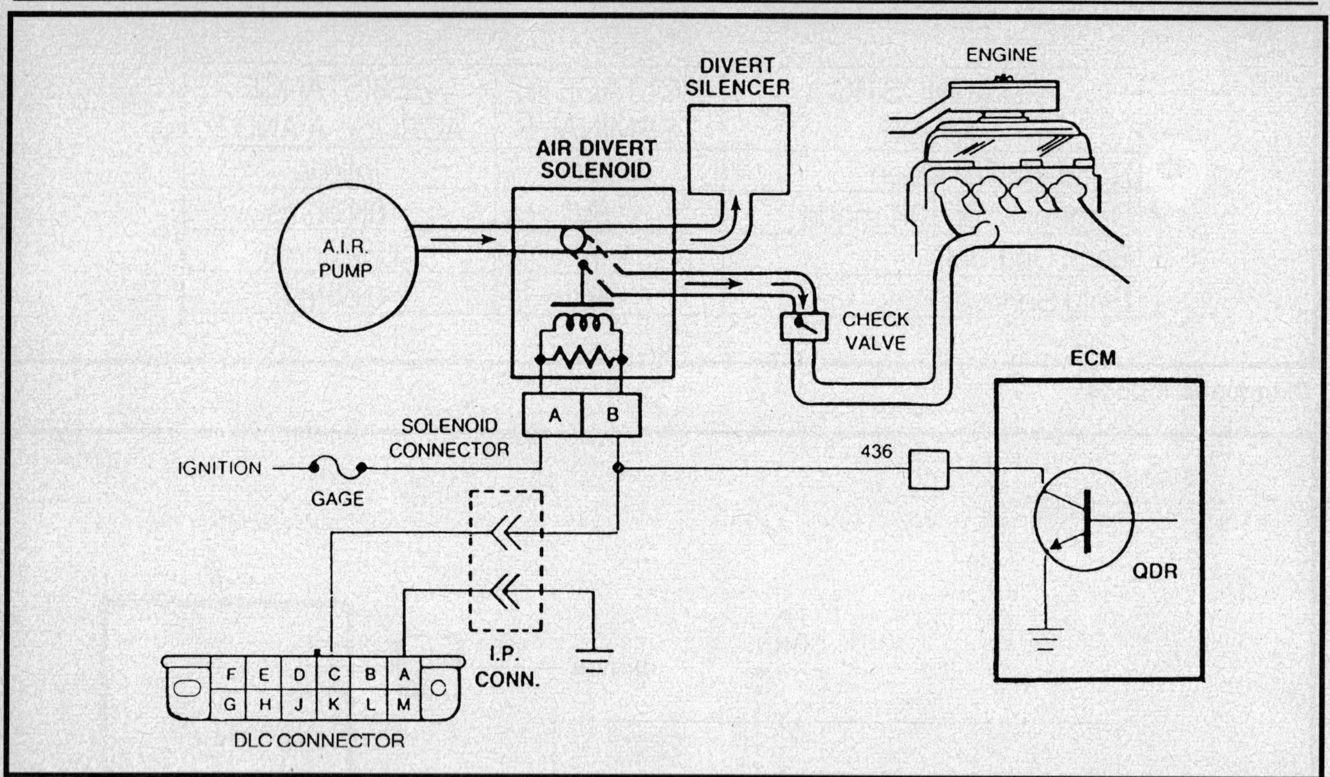

Terminal C — Air control solenoid

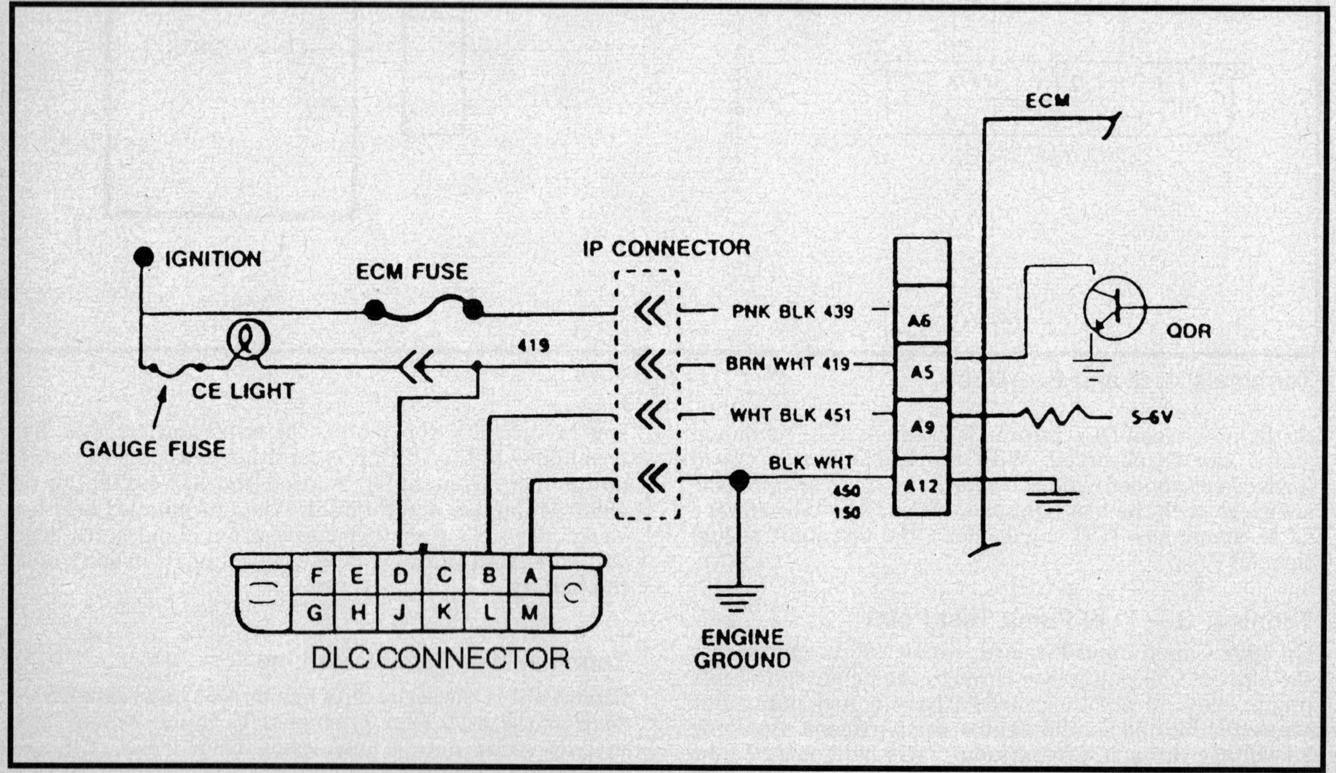

Terminal D — Malfunction indicator lamp

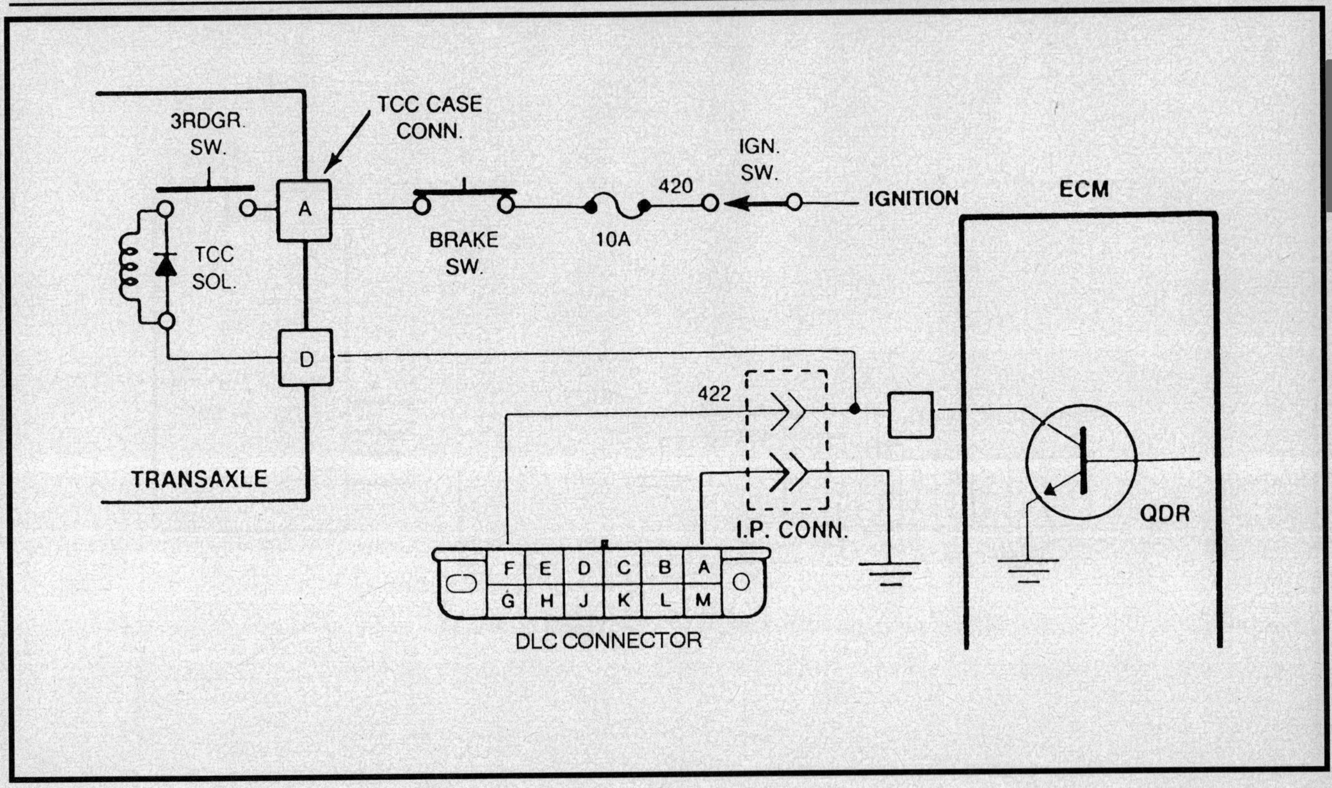

Terminal F — Torque converter clutch solenoid

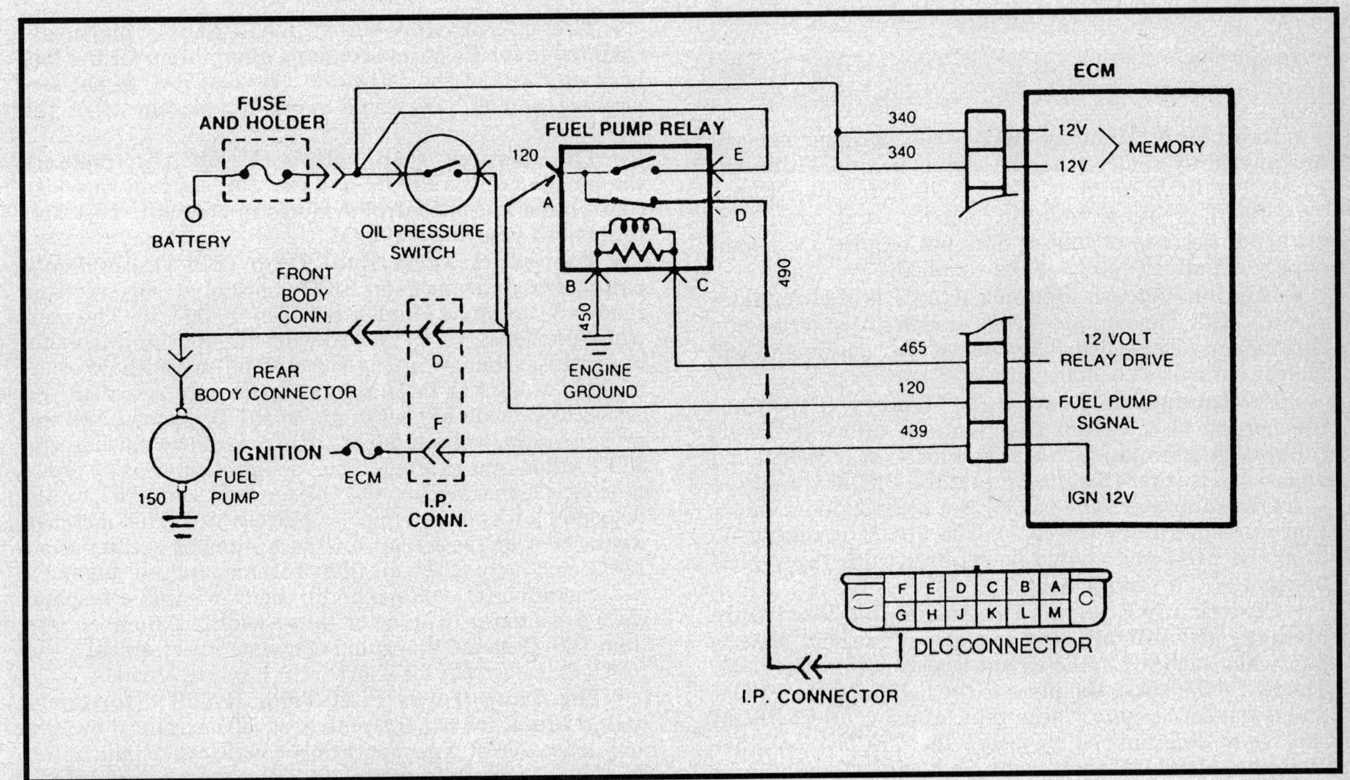

Terminal G — Fuel pump test point

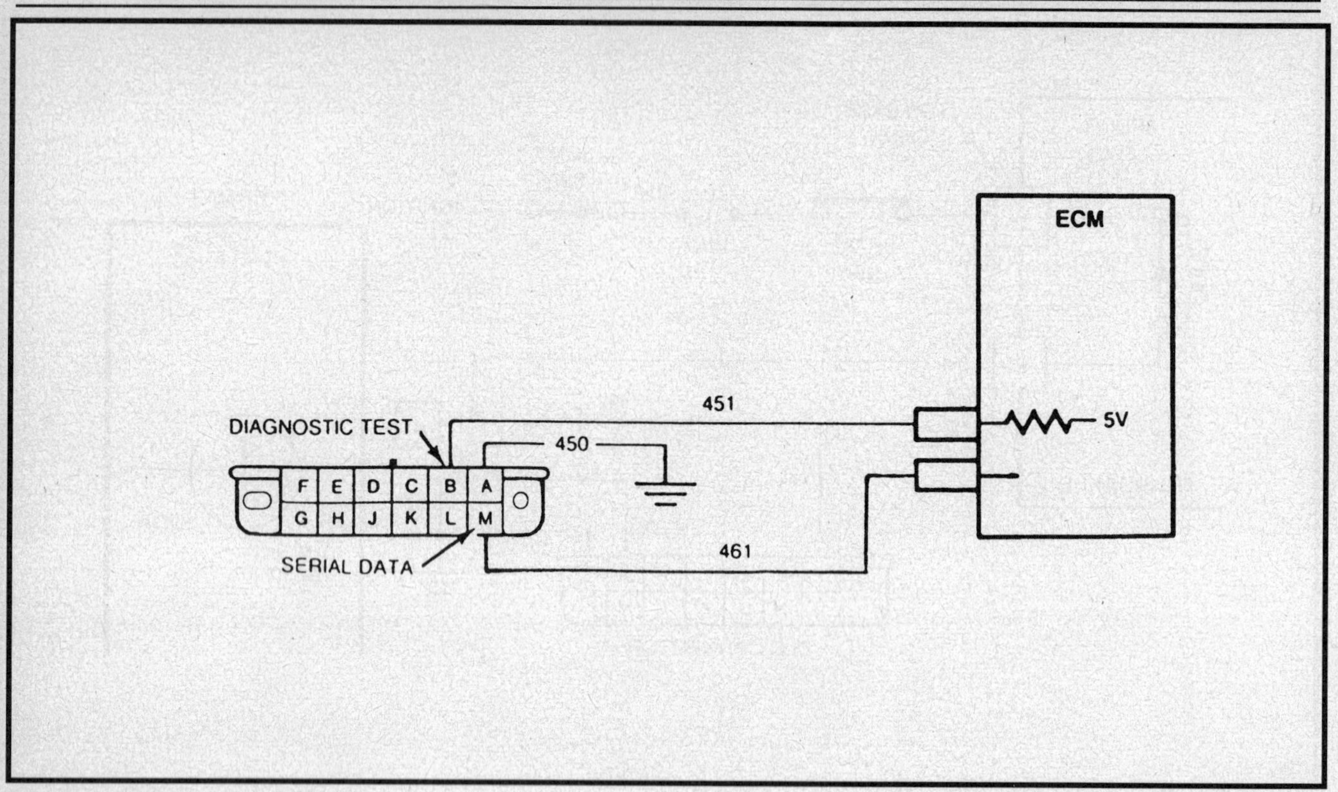

Terminal M — Serial data line

MEMORY

NOTE: Not every type of memory control is used in each vehicle application.

- **Read Only Memory (ROM)** is a permanent memory physically soldered to the circuit boards within the ECM. The ROM contains the overall control algorithms. The ROM is a non-volatile memory and cannot be changed. Because of this, it does not require a constant supply of battery voltage to be retained.
- **Random Access Memory (RAM)** is the computer scratch pad. The computer can read from or write into this memory as needed. This memory is volatile and will be lost without a constant supply of battery voltage.
- **Programmable Read Only Memory (PROM)** is the portion of the ECM that contains different engine calibration information that is specific to year, model and emissions. It is for this reason that the PROM is referred to as the calibrator. The PROM is a non-volatile memory that is read only by the ECM. The PROM is removable from the ECM and should be retained with the vehicle following ECM replacement.
- **Electrically Erasable Programmable Read Only Memory (EEPROM)** is a permanent memory that is physically soldered to the circuit boards within the ECM. The EEPROM takes the place of the ROM and the PROM and is not serviceable. Those vehicles using an EEPROM can be re-programmed by using the TECH-1 or other Techline terminal equipment and equivalents. Some ECM's using an EEPROM use a separate knock sensor module underneath the computer's access cover. This chip must be removed before replacing the old ECM and reinstalled into the new computer.

- **The Calibration Pack (CALPAC)** is physically soldered to the ECM and contains information for the fuel back-up mode of the computer. The CALPAC is not serviceable and failure would require replacement of the ECM.
- **The Memory Calibration (MEMCAL)** combines the function of the PROM and CALPAC as well as cruise control and knock control modules in one unit. This unit is serviced just like a PROM.
- **The Short Term Fuel Trim (STFT) (formerly called integrator)** is an ECM volatile memory register that will contain a number between 0 and 255. The neutral value for STFT is 128; any deviation from this value indicates a change in the injector pulse width as commanded from STFT. This function is only active in the closed loop mode of operation. As the ECM monitors the oxygen sensor voltage input, it is constantly varying the STFT value. For example, if a vacuum leak were to occur causing a lean condition (O_2 voltage low), the STFT would respond by increasing injector pulse width. This increase would be seen on a scan tool as a number greater than 128. Conversely, if the air filter became plugged causing a rich condition (O_2 voltage high), the STFT would respond with a decrease in injector pulse width; a number less than 128. Because this value is updated very quickly, the STFT only corrects for short term mixture trends.
- **The Long Term Fuel Trim (LTFT) (formerly called block learn)** is a matrix of cells arranged by rpm and load. As engine operating conditions change, the ECM will switch from cell to cell to determine which LTFT factor is appropriate. While in any given block, the ECM also monitors the STFT. If the STFT is far enough from 128 in either direction, the ECM will change the LTFT value for that given cell. Once the LTFT value is

changed the STFT value should revert back to 128, this represents a neutral condition. If the mixture is still not correct (as judged by the O_2 sensor), the process of short and long term fuel correction will continue until they reach the limits of their control. These limits are programmed into the PROM and would be different for each vehicle. At this point of maximum correction a trouble code 44 or 45 would be registered in the computer's RAM. As with STFT the long term fuel trim is only active in closed loop operation. The LTFT represents the computer's learning capability as these values can be tailored to the specific driving habits of the operator. Once battery power is lost however, the LTFT values will revert back to default settings, and a loss of performance may be noticed. Original performance should return after the vehicle is operated for a period of time in closed loop.

Tools and Equipment

SCAN TOOLS

The system can communicate a wide variety of information through the Data Link Connector (DLC), previously known as the ALDL. Depending on the application, serial data is transmitted to terminals E or M of the DLC. This data is transmitted at a high frequency which requires a quality scan tool for interpretation. Although stored codes may be read with only the use of a small jumper wire, the use of a hand-held scan tool such as GM's TECH 1 or equivalent, is recommended. There are many manufacturers of these tools; a purchaser must be certain that the tool is proper for the intended use.

The scan tool allows any stored codes to be read from the ECM memory. The tool also allows the operator to view the data being sent to the ECM while the engine is running. This ability has obvious diagnostic advantages; the use of the scan tool is frequently required by the diagnostic charts.

With an understanding of the data stream that the tool will display and knowledge of the circuits involved, the tool can be be an extremely valuable tool when examining the system. Without scan tools, diagnostic information would be difficult to obtain in some circuits and impossible to retrieve in others. Scan tools do not make the use of trouble code charts unnecessary, nor do they pinpoint the exact area in a circuit where the problem exists. However, this type of equipment will give more specific information, such as sensor input readings. An example of this would be the coolant sensor circuit. While the ECM looks at specific voltage levels to determine coolant temperature, the scan tool will display this information as an actual temperature reading. This information is not only useful in determining system malfunctions, but is also very helpful in diagnosing mechanical problems. For example, if the engine is overheating, the scan tool can be used to look at the actual engine temperature. This is helpful when checking for electric cooling fan enable switch problems, sticking thermostats, etc.

The ECM is capable of communicating with a scan tool in 3 modes: Normal, DCL and Factory Test.

Normal or Open Mode

This mode is not applicable to all engines. When engaged, certain engine data can be observed on the scanner without affecting engine operating characteristics. The number of items readable in this mode varies with engine

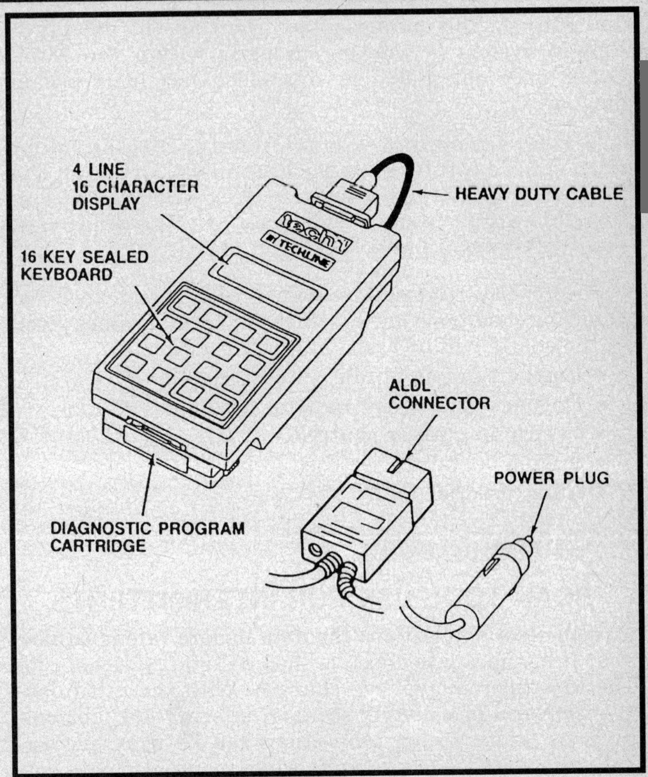

Typical scan tool — GM's Tech 1 shown

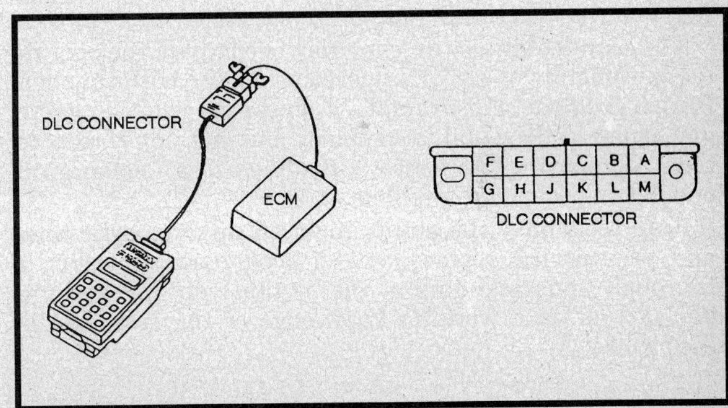

DLC connection to scan tool

family. Most scan tools are designed to change automatically to the DLC mode if this mode is not available.

DLC Mode

Also referred to as the ALDL, 10K or SPECIAL mode, the scanner will present all readable data as available. Certain operating characteristics of the engine are changed or controlled when this mode is engaged. The closed loop timers are bypassed, the spark (EST) is advanced and the PARK/NEUTRAL restriction is bypassed. If applicable, the IAC controls the engine speed to 1000 rpm ± 50, and, on some engines, the canister purge solenoid is energized.

Factory Test

Sometimes referred to as BACK-UP mode, this level of communication is primarily used during vehicle assembly and testing. This mode will confirm that the default or limp-in system is working properly within the ECM. Other data obtainable in this mode has little use in diagnosis.

> **NOTE: A scan tool that is known to display faulty data should not be used for diagnosis. Although the fault may be believed to be in only one area, it can possibly affect many other areas during diagnosis, leading to errors and incorrect repair.**

To properly read system values with a scan tool, the following conditions must be met. All normal values given in the charts will be based on these conditions:

- Engine running at idle, throttle closed
- Engine warm, upper radiator hose hot
- Vehicle in park or neutral
- System operating in closed loop
- All accessories **OFF**

SCAN TOOL USE FOR INTERMITTENTS

In some tool applications the data update rate is so slow that it becomes less effective than a digital voltmeter for checking sporadic voltage changes. With the rapid data transmission of a quality scanner, intermittent problems such as faulty wiring connections can be more successfully diagnosed. For example, while manipulating a suspect electrical circuit or component, observe the scan tool display screen. A rapid voltage change or erratic reading will uncover the trouble area.

The scan tool is also an easy way to compare the operating parameters of a poorly operating engine with that of a known good one. For example, a sensor may shift in value and cause a driveability problem, but not set a trouble code. Comparing the sensor's readings to a known good one may uncover the problem.

Scan tools have the ability to speed up diagnostic time and prevent the replacement of good parts, however a thorough understanding of the system you are working on, as well as a working knowledge of the scan tool is essential.

ELECTRICAL TOOLS

The most commonly required electrical diagnostic tool is the digital multimeter, allowing voltage, resistance and amperage to be read by one instrument. The multimeter must be a high-impedance unit, with 10 megohms of impedance in the voltmeter. This type of meter will not place an additional load on the circuit it is testing; this is extremely important in low voltage circuits. The multimeter must be of high quality in all respects. It should be handled carefully and protected from impact or damage. Replace batteries frequently in the unit.

Other necessary tools include an unpowered test light, a quality tachometer with inductive (clip-on) pick up and the proper tools for releasing GM's Metri-Pack, Weather Pack and Micro-Pack terminals as necessary. The Micro-Pack connectors are used at the ECM connector. A vacuum pump/gauge may also be required for checking sensors, solenoids and valves.

Diagnosis and Testing

SERVICE PRECAUTIONS

- To prevent internal ECM damage, the ignition must be **OFF** when disconnecting or reconnecting power to the computer.
- When handling a PROM, CALPAC or MEMCAL, do not touch the component leads. Also, do not remove the integrated circuit from the carrier.
- Never allow welding cables to lie on, near or across any vehicle electrical wiring.
- Leave new components and modules in the shipping package until ready to install them.
- When performing electrical tests on the system, use a high impedance multimeter, digital voltmeter (DVM) J-34029-A or equivalent.
- To prevent possible electrostatic discharge damage to the ECM, do not touch the connector pins or soldered components on the circuit board.

PRELIMINARY INSPECTION

In order to successfully locate and repair problems in complex computer controlled systems, the technician needs to adhere to a comprehensive, sequential method of diagnostics. Many times a vehicle has a malfunction and the worst is suspected. Premature conclusions often result in the replacement of good parts and wasted time. A proper sequence of steps needs to be established and strictly adhered to when troubleshooting a vehicle. These are as follows:

1. If possible, try to speak directly to the customer, especially if the complaint is not associated with an intermittent or steady MIL.
2. Road test the vehicle, if necessary, to verify the complaint.
3. Perform a thorough visual inspection. This step can very often eliminate the need for further testing. Loose wires, especially ground circuits, can create mysterious driveability problems. However, a complete visual inspection can uncover these problems quickly, and prevent the replacing of good parts and reduce vehicle downtime.
4. Perform the Diagnostic Circuit Check. This test confirms that the diagnostic system has not failed and is able to communicate through the malfunction indicator lamp.
5. After locating and repairing the problem area, road test the vehicle to confirm the repair and to verify no additional problem areas exist.

Visual Underhood Inspection

This is possibly the most critical step of diagnosis. A detailed examination of connectors, wiring and vacuum hoses can often lead to a repair without further diagnosis. Performance of this step relies on the skill of the technician performing it; a careful inspector will check:

- The undersides of hoses as well as the integrity of hard-to-reach hoses blocked by the air cleaner or other components.
- The wiring carefully for any sign of strain, burning, crimping, or terminal pull-out from a connector.
- The connectors at components or in harnesses as required; usually, pushing them together will reveal a loose fit.

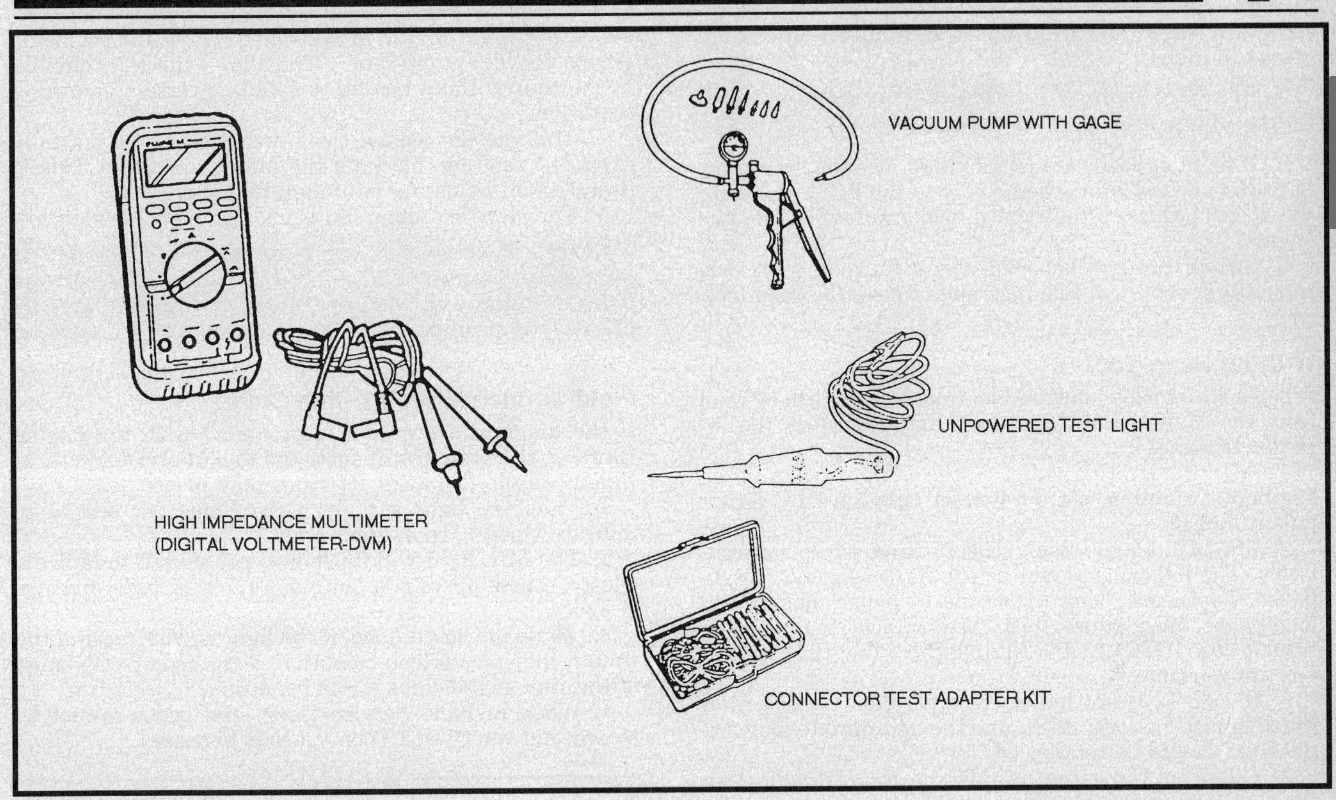

Electrical system diagnostic tools

Diagnostic Circuit Check

This step is used to check that the on-board diagnostic system is working correctly. A system which is faulty or shorted may not yield correct codes when placed in the Diagnostic Mode.

If the diagnostic system is not operating correctly, or if a problem exists without the malfunction indicator lamp being lit, refer to the vehicle's A-Charts. These charts cover such conditions as Engine Cranks but Will Not Run or No Service Engine Soon Light.

1. Turn the ignition switch **ON** but do not start the engine.

2. The MIL should illuminate and and stay lit until the engine is started.

3. If the MIL does not light, check the bulb and circuits for an open or short.

4. Connect jumper wire from terminal A to B in the DLC. The MIL should flash code 12 at least 3 times. If not, turn the ignition switch **OFF** and disconnect the ECM.

5. Turn the ignition switch **ON** but do not start the engine. If the MIL is ON, look for short in circuit 419.

6. If the MIL does not light, check the MEMCAL PROM for proper installation. If no light, check ECM or look for open or short in circuit.

Circuit/Component Diagnosis and Repair

Using the appropriate chart(s) based on the Diagnostic Circuit Check, the fault codes and the scan tool data will lead to diagnosis and checking of a particular circuit or component. It is important to note that the fault code indicates a fault or loss of signal in an ECM-controlled system, not necessarily in the specific component. Detailed procedures to isolate the problem are included in each code chart; these procedures must be followed accurately to insure timely and correct repair. Following the procedure will also insure that only truly faulty components are replaced.

READING CODES

With Scan Tool

NOTE: Scan tool functions and procedures may vary between manufacturers. Consult owners manual for proper connection and operation of each scan tool.

Once the visual inspection and diagnostic circuit check has been completed, enter the Diagnostic Mode and read any stored codes. To enter the diagnostic mode:

1. Turn the ignition switch **OFF**. Locate the Data Link Connector (DLC) also called ALDL, usually under the instrument panel. It may be within a plastic cover or housing labeled DIAGNOSTIC CONNECTOR. This link is used to communicate with the ECM.

2. Connect the scan tool correctly to the DLC.

3. Turn the ignition switch to the **ON** position but DO NOT start the engine. A Code 12 may be displayed. Code 12 is not a fault code. It is used as a system acknowledgment or handshake code; its presence indicates that the ECM can communicate as requested. Code 12 is used to begin every diagnostic sequence. Some vehicles also use Code 12 after all diagnostic codes have been sent.

4. After Code 12 has been transmitted 3 times, the fault codes, if any, will each be transmitted 3 times. The

codes are stored and transmitted in numeric order from lowest to highest.

NOTE: The order of codes in the memory does not indicate the order of occurrence.

5. If no fault codes are transmitted, use the scan functions to view the values being sent to the ECM. Compare the actual values to the typical or normal values for the engine.

6. Switch the ignition **OFF** when finished with code retrieval or scan tool readings and remove the scan tool.

Without Scan Tool

When a scan tool is unavailable, trouble codes can be read from the Malfunction Indicator Lamp (MIL) on the instrument panel.

1. With the ignition switch in the **ON** position and the engine not running, place a jumper between DLC terminals A and B.

2. The MIL should flash code 12 three times consecutively. The following would be the flash sequence: flash, pause, flash-flash, long pause, flash, pause, flash-flash, long pause, flash, pause, flash-flash, indicating diagnostic trouble code (DTC) 12. DTC 12 indicates that the diagnostics are working.

3. If code 12 is not indicated, a problem exists within the diagnostic system itself, and the appropriate diagnostic chart should be referenced.

4. Following the output of DTC 12, the MIL will flash the next highest trouble code (e.g. DTC 13) if one exists. It will output this 3 consecutive times before moving on to the next code in sequence. When all codes have been displayed, the cycle will repeat starting with code 12.

5. If no codes are stored in the system's memory it will continue to flash code 12.

6. To exit the diagnostic mode, turn the ignition switch **OFF** and remove the jumper wire or scan tool.

CLEARING CODES

To clear diagnostic trouble codes from the memory by disconnecting the power feed for at least 30 seconds. Depending on how the vehicle is equipped, this can be accomplished by:

- Removing the power fuse or positive battery pigtail.
- Disconnect the negative battery terminal, however, other on-board memory data such as the radio and clock settings will be lost.
- On some applications, clearing DTC's can also be done through the scan tool.

DIAGNOSTIC MODES

Manual Diagnostic Mode (0 Ohm State Field Service)

When the diagnostic terminal in the data link connector (DLC) is grounded with the key **ON** and the engine OFF, the system will enter the diagnostic mode. In this mode the following activities take place:

1. Displays a DTC 12 indicating the system is operational.

2. Displays any stored fault codes 3 times consecutively.

3. Energizes all system controlled relays and solenoids except the fuel pump relay. This allows checking circuits in the shop without having to simulate certain operating conditions.

4. The idle air control (IAC) valve moves to it's fully extended position, blocking the idle air passage. This is useful when adjusting minimum idle speed.

5. The electric cooling fan is enabled on some vehicles.

NOTE: Battery voltage may fall considerably if left in this mode for extended periods. This voltage drop could cause false readings. Connect a battery charger if time exceeds 30 minutes.

Field Service Mode (0 Ohm State)

If the diagnostic terminal is grounded with the engine running, the system will enter the Field Service Mode. In this mode the following activities take place:

1. The MIL light will flash 2.5 times per second to indicate Open Loop operation.

2. The MIL light will flash once per second to indicate Closed Loop operation and a 14.7-to-1 fuel mixture delivery.

3. While in Closed Loop, if the light is OFF most of the time it indicates a lean condition, if the light is ON most of the time it indicates a rich condition.

4. When in Field Service Mode, new codes cannot be stored and the Closed Loop timer is bypassed.

WARNING

The vehicle should not be driven while in field service mode. Damage to the catalytic converter could result.

Road Test Mode (Open Mode)

Road test mode can be accessed with the appropriate scan tool. This mode allows the ECM/PCM to function normally, without affecting fuel control or spark timing. This is the preferred mode when testing with a scan tool. On some vehicles, only a few data parameters may be displayed on the scan tool while a NO DATA message may appear on others. In these situations, the Diagnostic Mode (10K Mode) must be selected to access the data stream. This mode is very helpful on those vehicles that use multiple microprocessors. On these models the data stream is active all the time because it is constantly transmitting information to the BCM, IPC, etc..

Diagnostic Mode (10K Mode)

Diagnostic mode can be accessed with the appropriate scan tool. In this mode all available serial data is displayed on the scan tool. In addition, the following activities take place:

- ECM timers are bypassed.
- On systems equipped with a knock sensor, spark timing is advanced.
- IAC valve is controlled to maintain a fixed idle speed, around 1000 RPMs.
- Park/Neutral restrict functions are disabled.
- On some vehicles the charcoal canister purge solenoid will be commanded to disable purge.

NOTE: When in diagnostic mode the idle speed will be abnormally high. With some scan tools this will occur as soon as the tool is connected.

Back Up Fuel Mode (3.9K Mode)

This mode verifies the operation of the fuel back up circuitry inside the computer, namely the CALPAC. Engine operation is very erratic in this mode due to the limited amount of sensor information being used. While in back up fuel mode, electronic spark timing is not functional and fuel calculations are based on the following:

- RPM (distributor reference signal)
- TPS (throttle position signal)
- ETC (engine coolant temperature signal)

DIAGNOSTIC TROUBLE CODES (DTC)

- DTC **12** — Diagnostic check only (flash code)
- DTC **13** — Oxygen (O_2) sensor circuit open
- DTC **14** — Engine Coolant Temperature (ECT) sensor circuit out of range (high)
- DTC **15** — Engine Coolant Temperature (ECT) sensor circuit out of range (low)
- DTC **16** — System voltage high or low — **OR** — Direct/electronic ignition system fault line circuit error
- DTC **17** — Camshaft position sensor/spark reference circuit error
- DTC **18** — Crankshaft/camshaft error
- DTC **19** — Ignition control module signal circuit error
- DTC **21** — Throttle Position Sensor (TPS) circuit voltage out of range
- DTC **22** — Throttle Position Sensor (TPS) circuit voltage out of range
- DTC **23** — Intake/manifold air temperature sensor circuit voltage out of range
- DTC **24** — Vehicle speed sensor circuit signal error
- DTC **25** — Intake/manifold air temperature sensor circuit voltage out of range
- DTC **26** — Quad-driver module circuit error
- DTC **27** — Gear switch circuits error
- DTC **28** — Gear switch circuits error — **OR** — Quad-driver module control circuit
- DTC **29** — Gear switch circuits error
- DTC **31** — PRNDL/transmission neutral start/park-neutral position switch circuit error — **OR** — Turbo wastegate overboost — **OR** — Camshaft sensor circuit error
- DTC **32** — Exhaust Gas Recirculation (EGR) system fault
- DTC **33** — Mass air flow/manifold absolute pressure sensor circuit voltage out of range
- DTC **34** — Mass air flow/manifold absolute pressure sensor circuit voltage out of range
- DTC **35** — Idle speed/air/idle air control error
- DTC **36** — Transaxle shift problem — **OR** — Ignition control signal circuit error
- DTC **38** — Brake switch circuit error
- DTC **39** — Torque converter/transaxle clutch switch circuit error
- DTC **41** — Cam position sensor/cylinder select error
- DTC **42** — Electronic spark timing/ignition control circuit open or shorted
- DTC **43** — Knock sensor/electronic spark control circuit error
- DTC **44** — Oxygen (O_2) sensor circuit indicates system lean
- DTC **45** — Oxygen (O_2) sensor circuit indicates system rich

- DTC **46** — Power steering pressure circuit error — **OR** — Vehicle anti-theft system circuit
- DTC **48** — Misfire diagnosis
- DTC **51** — PCM prom memory error
- DTC **52** — PCM prom memory error — **OR** — Engine oil temperature sensor circuit indicates low
- DTC **53** — System over voltage — **OR** — Exhaust Gas Recirculation (EGR) system malfunction — **OR** — Vehicle anti-theft system circuit error
- DTC **54** — Fuel pump circuit voltage low — **OR** — Exhaust Gas Recirculation (EGR) system fault
- DTC **55** — PCM error — **OR** — Mixture control solenoid/fuel lean monitor circuit error — **OR** — Exhaust Gas Recirculation (EGR) system fault
- DTC **56** — Quad-driver module circuit error — **OR** — Vacuum sensor circuit error — **OR** — Secondary air inlet valve actuator vacuum sensor circuit error
- DTC **57** — Boost control solenoid circuit error
- DTC **58** — Personal automotive security system (PASS-key®II) fuel enable circuit error
- DTC **61** — Degraded oxygen (O_2) sensor — **OR** — Cruise control vent solenoid circuit error — **OR** — A/C system circuit error — **OR** — Secondary port throttle valve system error
- DTC **62** — Transaxle gear switch signal circuit error — **OR** — Cruise control vacuum solenoid circuit error — **OR** — Engine oil temperature sensor circuit
- DTC **63** — Cruise control system error — **OR** — Exhaust Gas Recirculation (EGR) flow check — **OR** — Oxygen (O_2) sensor (right bank) circuit open
- DTC **64** — Exhaust Gas Recirculation (EGR) flow check — **OR** — Oxygen (O_2) sensor (right bank) circuit indicates lean
- DTC **65** — Cruise control pressure sensor circuit error — **OR** — Exhaust Gas Recirculation (EGR) flow check — **OR** — Oxygen (O_2) sensor (right bank) circuit indicates rich — **OR** — Fuel injector circuit error
- DTC **66** — A/C refrigerant pressure sensor circuit — **OR** — Engine power switch circuit error
- DTC **67** — Cruise control switches circuit error
- DTC **68** — A/C compressor relay circuit shorted — **OR** — Cruise control system problem
- DTC **69** — A/C compressor relay circuit open — **OR** — A/C head pressure switch circuit error
- DTC **70** — A/C refrigerant pressure sensor circuit indicates high
- DTC **71** — A/C evaporator temperature sensor circuit indicates low
- DTC **72** — Gear selector switch circuit error
- DTC **73** — A/C evaporator temperature sensor circuit indicates high
- DTC **75** — Exhaust Gas Recirculation (EGR) No. 1 solenoid circuit error
- DTC **76** — Exhaust Gas Recirculation (EGR) No. 2 solenoid circuit error
- DTC **77** — Exhaust Gas Recirculation (EGR) No. 3 solenoid circuit error
- DTC **81** — Brake switch error
- DTC **82** — Ignition control signal error
- DTC **85** — PCM prom error
- DTC **86** — A/C multiplexer chip error
- DTC **87** — Electronically Erasable Programmable Read Only Memory (EEPROM) error

CADILLAC

General Description

These systems differ from the on-board diagnostic systems found on Buick and Oldsmobile models, in that the DLC on Cadillac (FWD) has a limited role in driveability diagnosis, and is typically used for the connection of a scan tool to troubleshoot other vehicle systems. Similarities, however, due exist in terms of utilizing a BCM as the central communication point for all other vehicle computer systems. Additionally, the basic format for using the Electronic Climate Control Panel (ECCP), as well as the Instrument Panel Cluster (IPC) for the operation of the self-diagnostic system, is very closely related.

A striking difference between Cadillac (FWD) and ALL other models is the operation of the Malfunction Indicator Lamp (MIL). These systems use two service telltale lights, which function differently from the standard MIL used on all other systems.

ELECTRONIC CONTROL MODULE

The ECM is the control center of the fuel injection system. It continuously monitors information from a variety of sensors and switches, and controls the systems that affect vehicle performance. The ECM also has self-diagnostic capability, and can detect system problems, alert the driver through the Malfunction Indicator Lamp (MIL) and record the appropriate Diagnostic Trouble Code (DTC). The numerical code will identify the specific circuit to aid the technician in diagnosis. The ECM is capable of storing more than one code, with possibilities ranging from 12 to 999, however not all systems use every code. In addition, similar code numbers can relate to different circuits depending on the vehicle. For example, on the 3.3L (VIN N), code 46 indicates a fault found in the power steering pressure switch circuit. The same code on the 5.7L (VIN F) engine indicates a fault in the VATS anti-theft system.

Learning Ability

The ECM can compensate for minor variations within the fuel system through the long term fuel trim (block learn) and short term fuel trim (fuel integrator) systems. The short term fuel trim monitors the oxygen sensor output voltage, adding or subtracting fuel to drive the mixture rich or lean as needed to reach the ideal air fuel ratio of 14.7:1. The integrator values may be read with a scan tool; the display will range from 0–255 and should center on 128 if the oxygen sensor is seeing a 14.7:1 mixture.

Powertrain Control Module (PCM)

The PCM is used when vehicles are equipped with electronically controlled transmissions. Its function is identical to that of the ECM with the added features of controlling shift points in the transmission, as well as cruise control operation and diagnostics. This unit may be referred to as the PCM, the ECM/PCM or the PCM/TCM (transmission control module). The integrated functions of engine and transmission control allow accurate gear selection and improved fuel economy.

For engine diagnostics, the PCM may be considered identical to an ECM system, although the combined unit will display additional codes relating to transmission function and components.

> **NOTE: The terms Electronic Control Module (ECM), Powertrain Control Module (PCM) and Body Control Module (BCM) may sometimes be used interchangably.**

Quad Drivers (QDM)

Some control modules have eliminated the use of individual transistors to control each output circuit. A single QDM can individually control 4 outputs from the computer. Since these control modules are internally fault protected, a single faulty circuit or component may cause the other circuits on that QDM to be ON or OFF all the time. Do not confuse this with a failed ECM; if the Quad Driver itself has failed, the control module output will usually test either open or shorted.

The quad driver check procedure should be used before replacing an expensive ECM/PCM assembly. The check will not test all computer functions but will determine if an external circuit has disabled a QDM. Be particularly suspicious of an ECM/PCM exhibiting any of these characteristics:

- Malfunction indicator lamp ON with no codes stored.
- Malfunction indicator light flickers, intermittent or dim.
- ECM/PCM will not flash Code 12 and/or engine will not start.

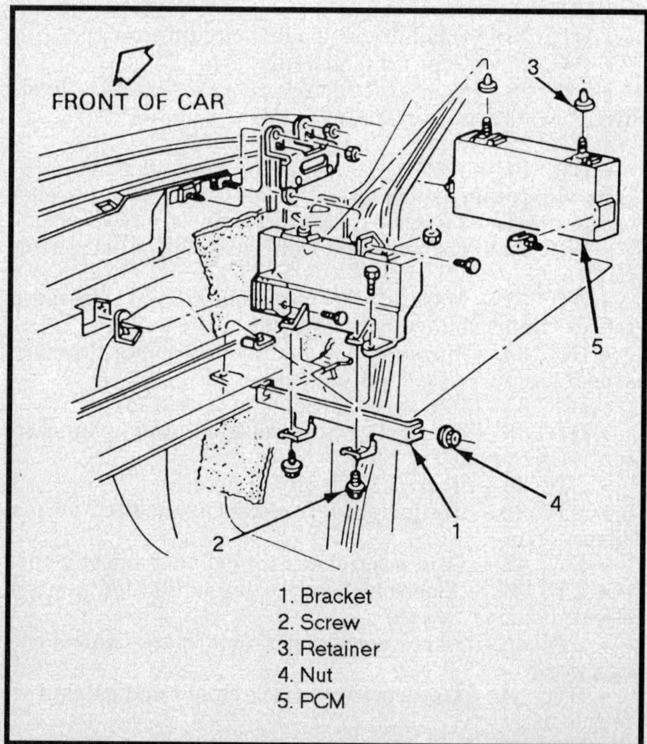

Powertrain Control Module (PCM) — Allante shown

- Controlled output, such as TCC, is inoperative or ON all the time.
- Engine stalls, misfires or surges.
- Scan tool erratic or inoperative.

BODY CONTROL MODULE

Body Control Module (BCM) is the communication center for the ECM/PCM, Instrument Panel Cluster (IPC), HVAC (Programmer), Electronic Climate Control Panel (ECCP), and the Supplemental Inflatable Restraint (SIR) module. A combination of inputs from these major components, as well as other sensors and switches, communicate with the BCM either as individual inputs or over the common communications link called the data line. The various inputs to the BCM combine with program instructions within the system memory to provide accurate control over the many subsystems involved. When a subsystem circuit exceeds pre-programmed limits, a system malfunction is indicated and certain backup functions may be provided.

Providing control over the many sub systems from the BCM is done by controlling system outputs. These outputs can be either direct or transmitted along the data line to one of the five other major components. The process of receiving, storing, testing and controlling information is continuous. The data communication gives the BCM control over the component's self-diagnostic capabilities in addition to its own.

In a method similar to that of a telegraph system, the BCM's internal circuitry rapidly switches a circuit between 0 and 5 volts like a telegraph key. This process is used to convert information into a series of pulses which represents coded data messages understood by the other components. Also, much like a telegraph system, each major component has its own recognition code or address. When messages are transmitted along the data line, only the component that matches the assigned recognition code will accept the message, while the other components will ignore it.

MALFUNCTION INDICATOR LAMP

The primary function of the malfunction indicator lamp is to advise the operator and the technician that a fault has been detected, and, in most cases, a code stored. Under normal conditions, the malfunction indicator lamp will illuminate when the ignition is turned **ON**. Once the engine is started and running, the ECM will perform a system check and extinguish the warning lamp if no fault is found.

Additionally, the malfunction indicator lamp can be used to retrieve stored codes after the system is placed in the Diagnostic Mode. Codes are transmitted as a series of flashes with short or long pauses. When the system is placed in the Field Service Mode, the dash lamp will indicate open loop or closed loop function to the technician.

The Deville and Fleetwood use two service telltale lights located on the instrument cluster. These are; SERVICE ENGINE SOON and SERVICE VEHICLE SOON.

On the Eldorado, Seville, and Allante, the SERVICE ENGINE SOON telltale light is used on the instrument cluster, while SERVICE VEHICLE SOON light appears as a message on the Driver Information Center (DIC) display.

Except Allante

The SES telltale will turn **ON** whenever the ECM detects a fault in the engine control system. The SVS telltale will turn **ON** whenever the ECM detects a fault in the following circuits only: Generator, Fuel Pump, VCC Brake Switch, PRNDL Switch, and Heated Windshield (if equipped). A display message on the Driver Information Center of Eldorado and Seville, which reads SERVICE VEHICLE SOON, replaces the function of the SVS telltale. The following lists the operation of the telltale status lights.

> **NOTE: Even though the SERVICE VEHICLE SOON message on Eldorado and Seville has the same function as the SVS telltale, it does not operate during the following sequence unless a specific code is set, see number 5.**

1. Key **ON**: SES and SVS lights will be **ON** for two seconds.
2. Cranking: Both indicator lights turn **OFF**.
3. Start: Both indicator lights will will turn **ON** for two seconds after the engine starts.
4. Diagnostic mode: Both indicator lights will turn **ON**.
5. Code set: Depending on the malfunctioning circuit(s) either one or both indicator lights will turn **ON**.

Allante

The SES telltale will turn **ON** whenever the ECM detects a fault in the engine control system. A SERVICE VEHICLE SOON message will be displayed on the Driver Information Center whenever the ECM detects a fault in the following circuits only: Generator, Fuel Pump, VCC Brake Switch or PRNDL Switch.

1. Key **ON**: The SES light will turn **ON** for a bulb check.
2. Cranking: The SES light will be illuminated steadily.
3. Start: The SES light will turn **OFF**.
4. Diagnostic mode: The SES light will turn **ON**.
5. Code set: Depending on the malfunctioning circuit(s) either the SES telltale will turn **ON**, or the SERVICE VEHICLE SOON message will appear on the Driver Information Center.

Intermittents

If a fault occurs intermittently, such as a loose connector pin breaking contact as the vehicle hits a bump, the ECM will note the fault as it occurs and energize the malfunction indicator lamp. If the problem self-corrects, as with

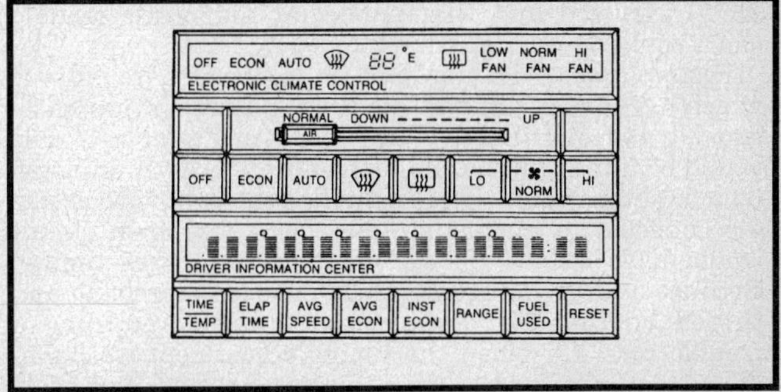

Climate control driver information center — 1992 Allante, Eldorado and Seville

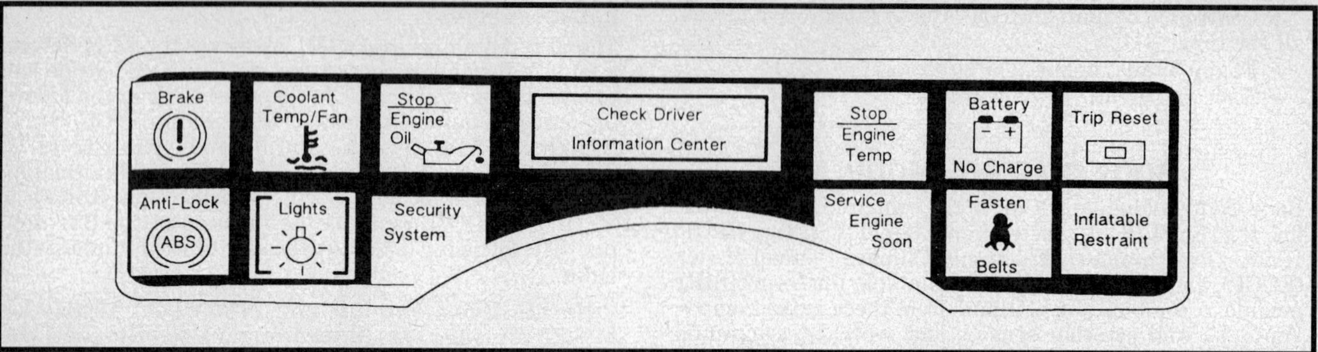

Telltale warning lights — Eldorado and Seville

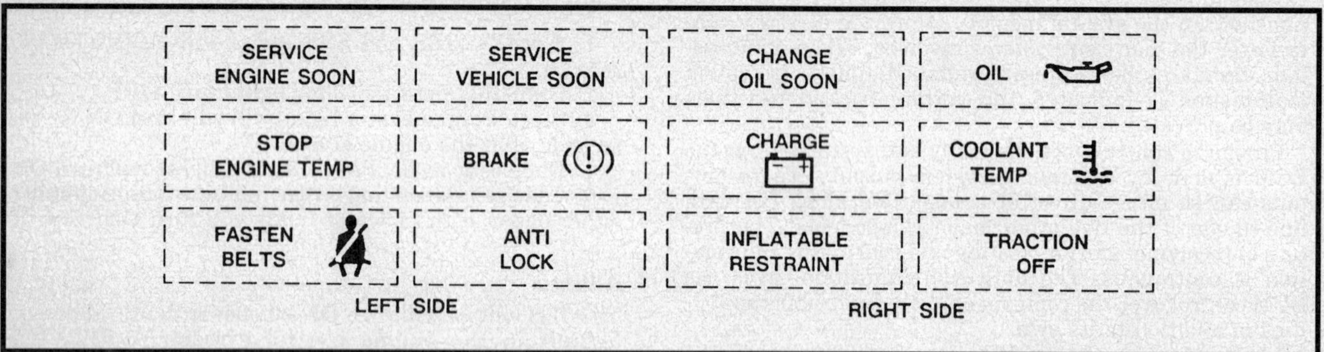

Telltale warning lights — Fleetwood and Deville

the terminal pin again making contact, the dash lamp will extinguish after 10 seconds but a code will remain stored in the ECM memory.

When an unexpected code appears during diagnostics, it may have been set during an intermittent failure that self-corrected; the codes are still useful in diagnosis and should not be discounted.

Failure Codes

Whenever the ECM detects a sensor or switch signal that is out of range, it will turn ON the MIL. In addition, if the ECM detects an internal malfunction, it will also turn ON the MIL. Distinguishing between current and history codes is absolutely necessary for successful diagnosis. Current codes are also known as hard failures, or a problem that exists at the present time. For example, if a vehicle came into the shop with the MIL light on steady with the engine running, a trouble code could be immediately retrieved and that particular diagnostic trouble chart could be referenced.

History codes, also known as intermittents, are failures which have taken place at one time, but are not presently causing the MIL light to turn ON. An example of this would be if the ignition timing had been recently adjusted on a 5.0 liter engine with TBI. When the bypass wire was disconnected to make the adjustment, the open circuit would have caused the ECM to revert to base timing. Because of this change, the ECM detected a problem and turned on the MIL while simultaneously recording a trouble code 42. When the timing adjustment was completed and the bypass wire was reconnected, the ECM

waited approximately 10 seconds and then turned off the MIL, but retained the code. The ECM would retain this information up to a maximum of 50 ignition cycles. If code retrieval was attempted in the meantime by another technician unaware of the timing adjustment, their only recourse would be to reference the wiring diagram associated with the code 42 chart and physically inspect all connections.

DATA LINK CONNECTOR (DLC)

The data link connector is located underneath the instrument panel and is used to access the on-board microprocessors. These include the ECM, PCM, BCM as well as various other controllers such as the anti-lock brake system module. The use of a scan tool such as GM's Tech 1 will allow the technician to monitor data and control some of the microprocessor's operations. The scan tool will read the serial data signal from the DLC and translate that signal into useful diagnostic information on the display screen. The DLC can also be used without a scan tool for retrieving trouble codes, entering certain diagnostic modes and other valuable testing.

Terminal A — DLC Ground

Circuit number 450 is the ECM ground through the instrument panel connector. The ECM is grounded at the engine and splices into the instrument panel harness.

> **NOTE: Most scan tools can disguise problems related to ECM grounds due to the alternate ground path through the tool's power cord.**

Terminal B — Diagnostic Enable

Circuit number 451 is connected to the ECM through the instrument panel connector. It is normally at a 5 volt level. The scan tool has the ability to lower the voltage level of circuit 451 by dropping a fixed resistance across the circuit or shorting it to ground. The various low voltages on circuit 451 are referred to as Diagnostic Modes.

Terminal C — Air Solenoid

On most vehicles with Air Injection Reaction (AIR) Systems, circuit 436 is spliced between the ECM and the AIR solenoid, through the instrument panel connector to terminal C. The solenoid is responsible for directing airflow from the pump to either the exhaust manifold, catalytic converter or the atmosphere via an attached silencer. This terminal is very helpful for testing the closed loop electronics. With the engine running in closed loop, connect a jumper wire between DLC terminals C and A. By grounding terminal C you will be enabling the air control solenoid to direct airflow to the exhaust manifold. This should cause a low O_2 sensor output voltage and a high (greater than 128) short term fuel trim number. If connected long enough, this action will trigger a code 44, lean exhaust.

In Open Loop operation, this solenoid will be enabled to quickly bring the O_2 sensor up to operating temperature. Once the system goes to Closed Loop the solenoid will become disabled causing airflow to be directed to either the catalytic converter or the atmosphere depending on the type of system. These actions can be verified using a digital voltmeter by probing terminal C with the engine running. In open loop the voltmeter should read 0 volts

DIAGNOSTIC MODE	VOLTAGE AT TERMINAL B	RESISTANCE ACROSS A AND B
Road Test or Open	5V	Infinite
Diagnostic or 10K mode	2.5V	10K ohms
Back Up Fuel	1.5V	3.9K ohms
Field Service	0V	0 ohms

Diagnostic modes

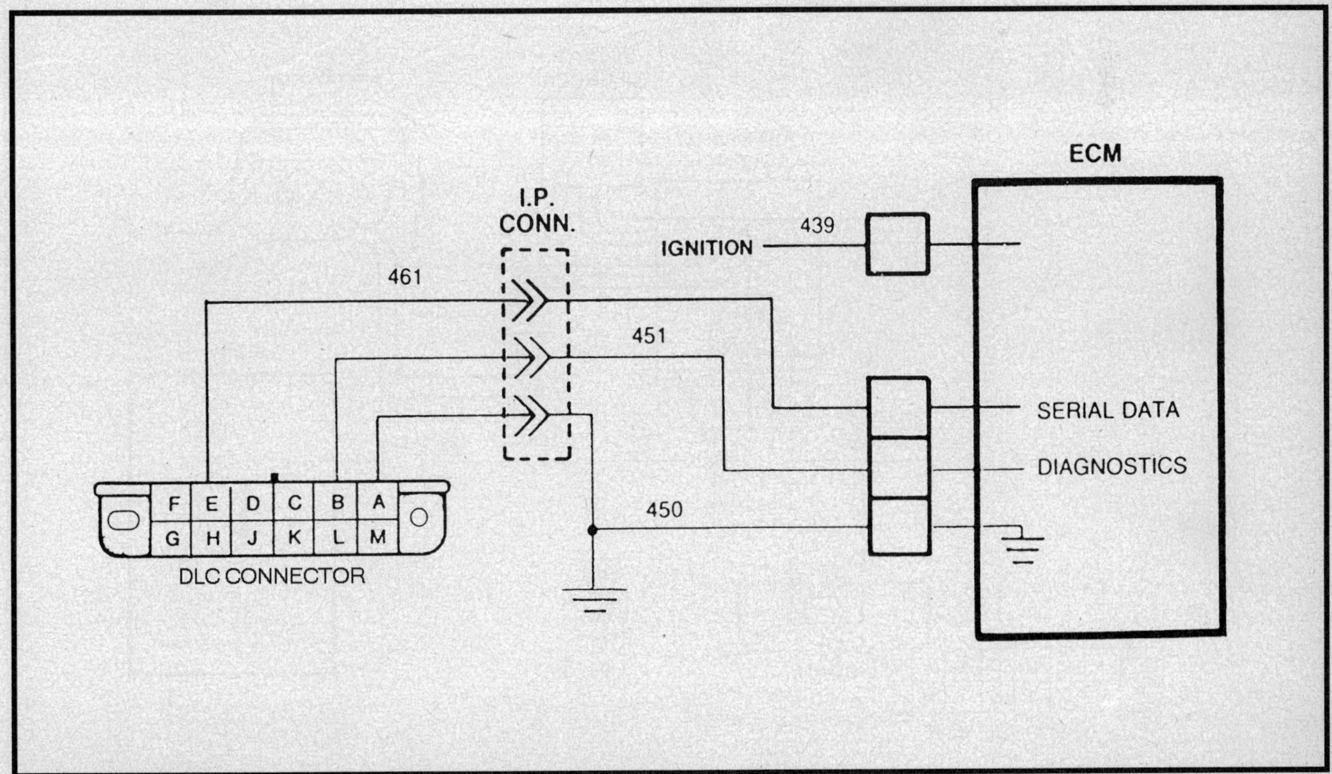

Terminals A, B and E — DLC

(solenoid energized) and in closed loop the reading should be battery voltage (solenoid de-energized).

Terminal D — Malfunction Indicator Lamp

Circuit 419 is spliced between the ECM and the MIL light to terminal D. With the key **ON**, the MIL light is turned ON through ECM terminal A5. If during a normal bulb check the MIL fails to turn on, a quick test can be performed before condemning the ECM. Insert a jumper between terminals D and A of the DLC with the key **ON**. This action should turn ON the MIL. If it does not, a defective bulb is the most likely problem.

Terminal E — Serial Data Line

Circuit 461 is connected through the instrument panel connector to the ECM. This is a discreet data line used on fuel injection systems with a slow data transmission rate (baud rate).

Terminal F — Torque Converter Clutch Solenoid

Circuit 422 is connected through the instrument panel connector and spliced into the ground-controlled TCC solenoid circuit. The ECM will turn on the TCC solenoid based on a variety of sensed inputs. By inserting a testlight between DLC terminals F and A, TCC command status can be observed. With the brake switch closed (brakes not applied) and the vehicle in 3rd gear (3rd gear switch closed), the test light should be ON. As soon as the ECM commands TCC engagement, the test light should turn OFF.

Terminal G — Fuel Pump Test Point

Most models have a fuel pump test connector located in the engine compartment. Applying B+ voltage through a fused jumper wire will energize the in-tank fuel pump. This is helpful when performing fuel pressure testing and is more convenient than cycling the key to turn ON the pump. On other models that use terminal G in the DLC, it is for other purposes than fuel pump. When terminal G is either vacant or used for another purpose, a fuel pump test lead will be located in the engine compartment, usually near the battery.

Terminal M — Serial Data Line

Circuit 461 is the serial data line for fast baud rate ECMs on some models. Terminal M may also include serial data transmission from other on-board microprocessors such as an anti-lock brake system controller.

MEMORY

NOTE: Not every type of memory control is used in each vehicle application.

- **Read Only Memory (ROM)** is a permanent memory physically soldered to the circuit boards within the ECM. The ROM contains the overall control algorithms. The ROM is a non-volatile memory and cannot be changed. Because of this, it does not require a constant supply of battery voltage to be retained.
- **Random Access Memory (RAM)** is the computer scratch pad. The computer can read from or write into this memory as needed. This memory is volatile and will be lost without a constant supply of battery voltage.

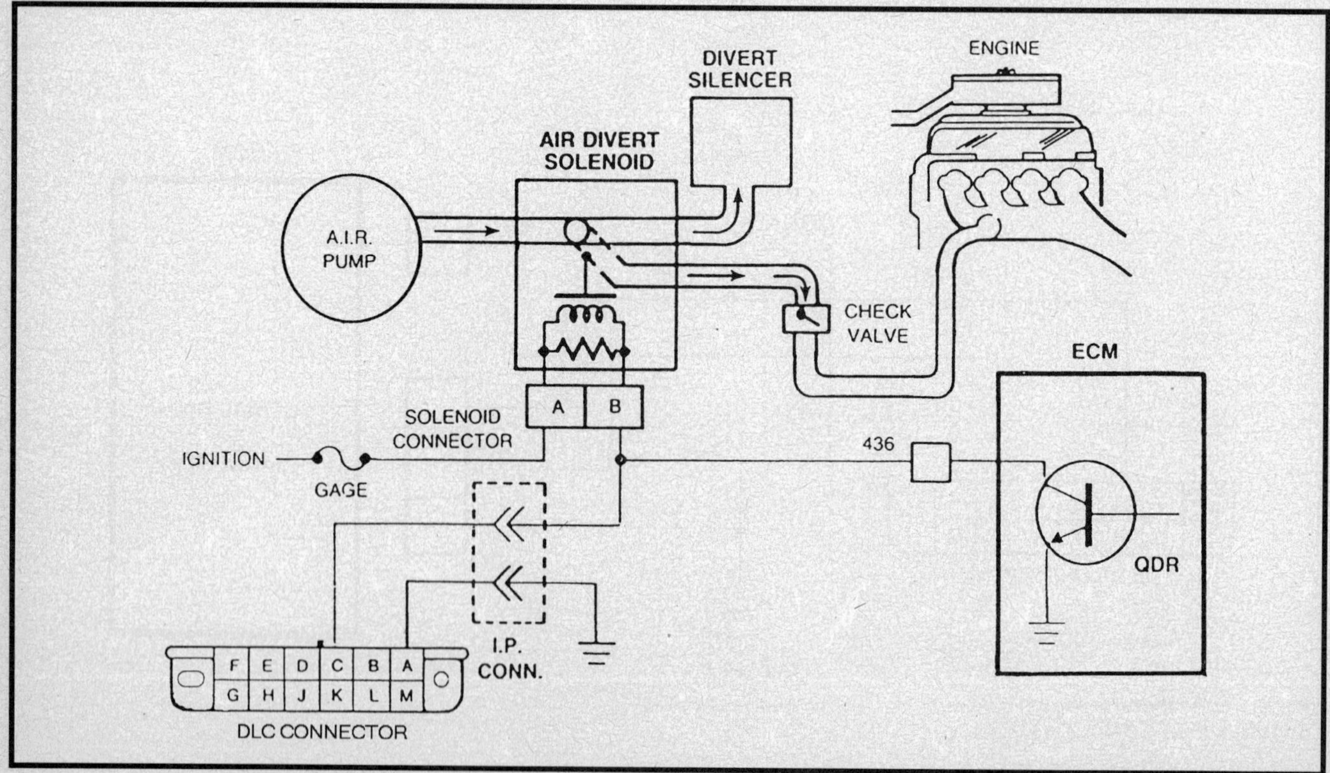

Terminal C — Air control solenoid

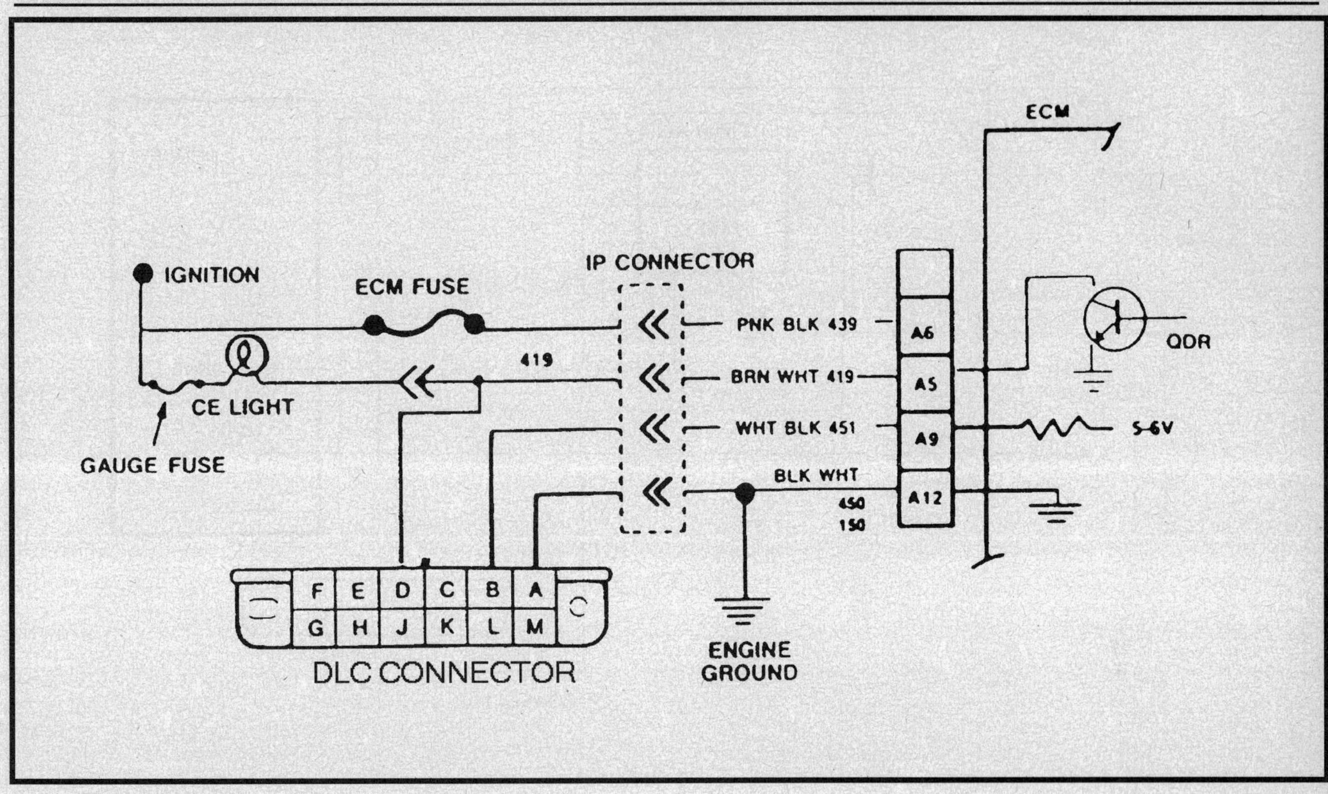

Terminal D — Malfunction indicator lamp

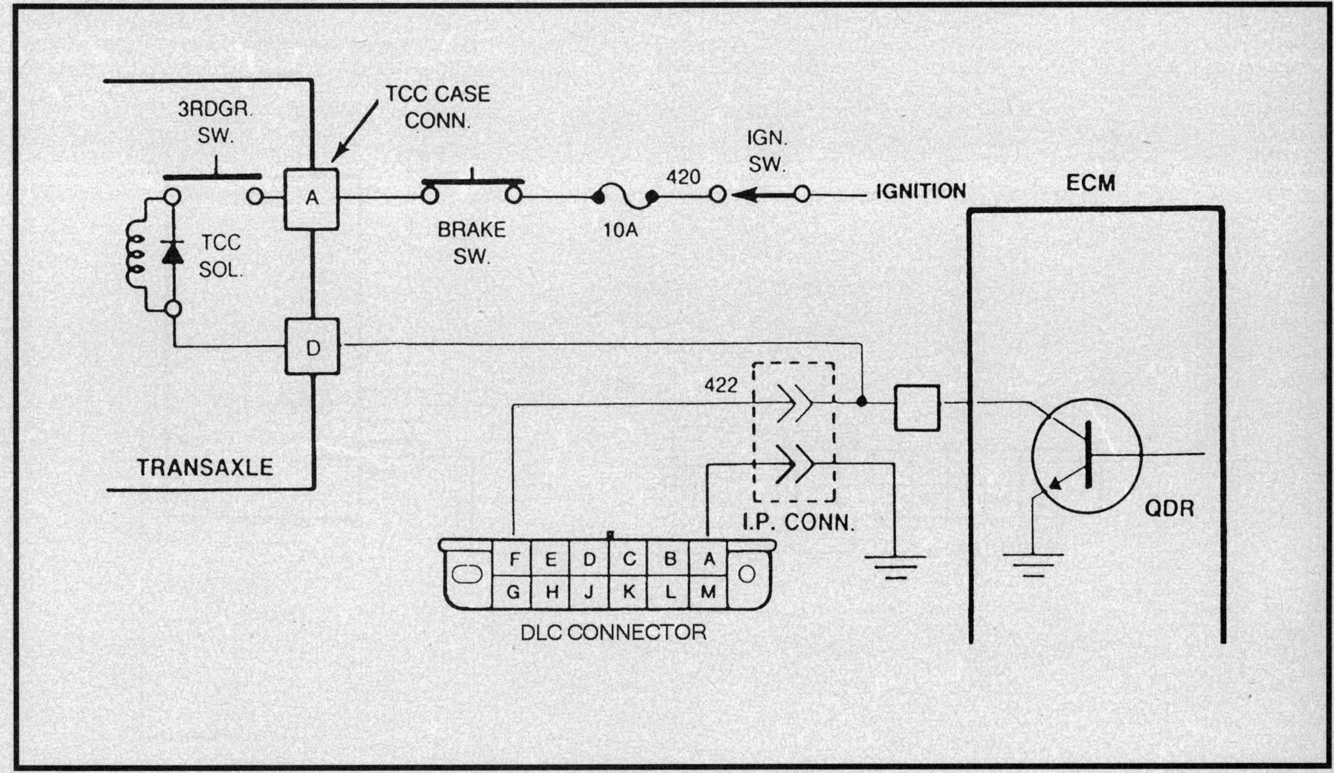

Terminal F — Torque converter clutch solenoid

SELF-DIAGNOSTIC SYSTEMS
CADILLAC

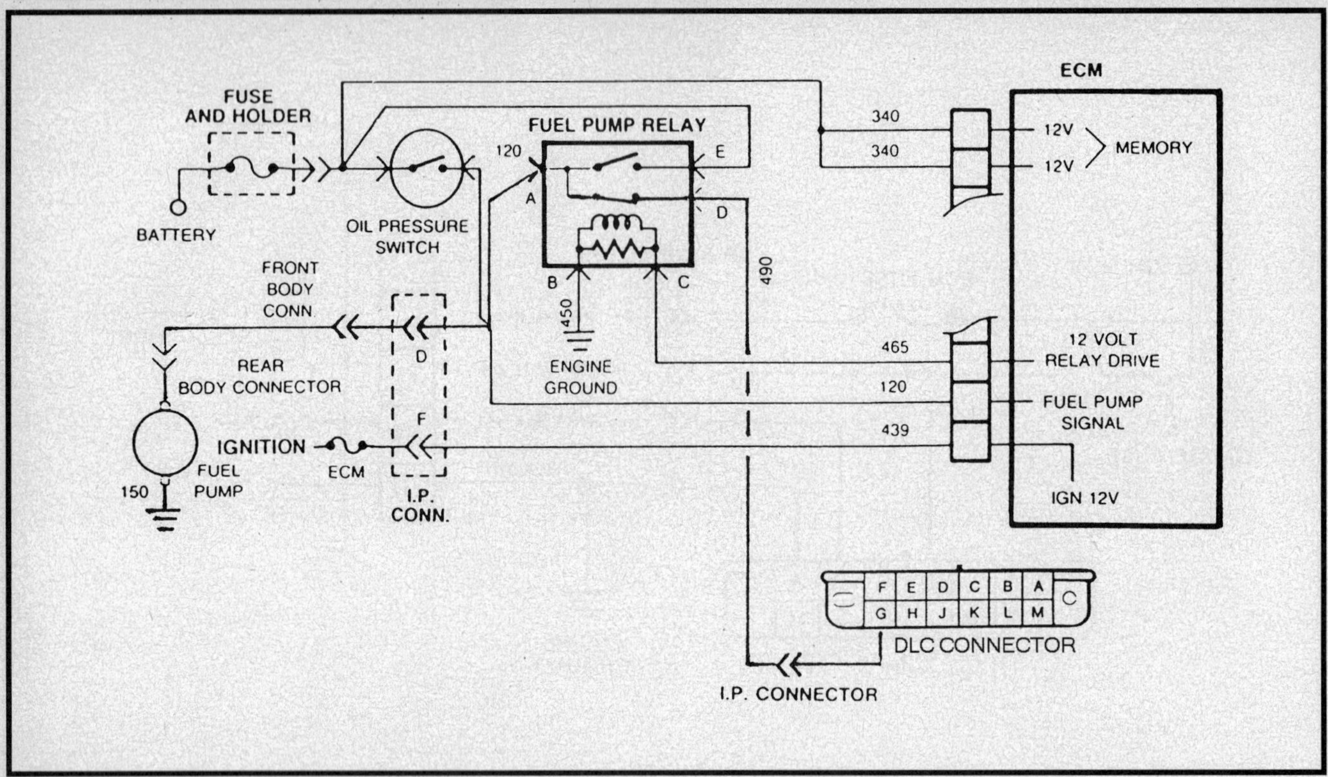

Terminal G — Fuel pump test point

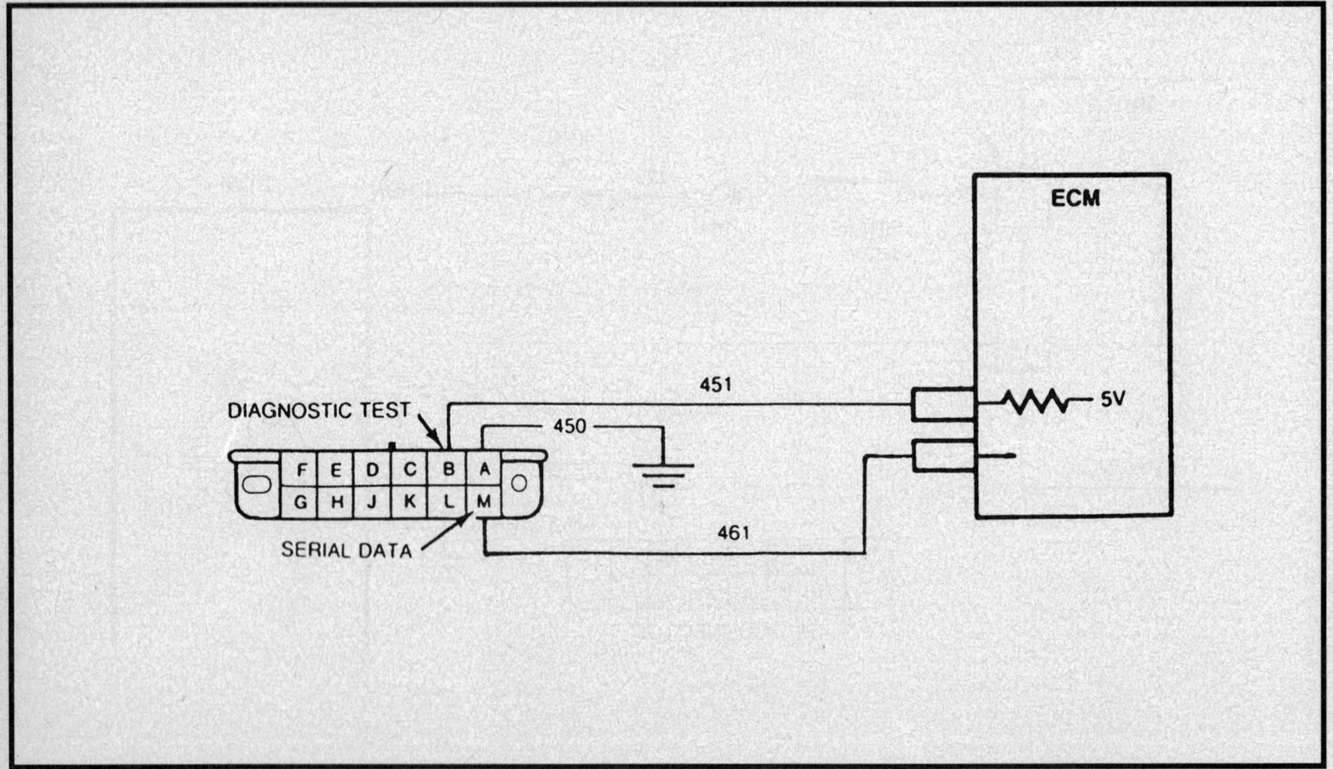

Terminal M — Serial data line

• **Programmable Read Only Memory (PROM)** is the portion of the ECM that contains different engine calibration information that is specific to year, model and emissions. It is for this reason that the PROM is referred to as the calibrator. The PROM is a non-volatile memory that is read only by the ECM. The PROM is removable from the ECM and should be retained with the vehicle following ECM replacement.

• **Electrically Erasable Programmable Read Only Memory (EEPROM)** is a permanent memory that is physically soldered to the circuit boards within the ECM. The EEPROM takes the place of the ROM and the PROM and is not serviceable. Those vehicles using an EEPROM can be re-programmed by using the TECH-1 or other Techline terminal equipment and equivalents. Some ECM's using an EEPROM use a separate knock sensor module underneath the computer's access cover. This chip must be removed before replacing the old ECM and reinstalled into the new computer.

• **The Calibration Pack (CALPAC)** is physically soldered to the ECM and contains information for the fuel back-up mode of the computer. The CALPAC is not serviceable and failure would require replacement of the ECM.

• **The Memory Calibration (MEMCAL)** combines the function of the PROM and CALPAC as well as cruise control and knock control modules in one unit. This unit is serviced just like a PROM.

• **The Short Term Fuel Trim (STFT) (formerly called integrator)** is an ECM volatile memory register that will contain a number between 0 and 255. The neutral value for STFT is 128; any deviation from this value indicates a change in the injector pulse width as commanded from STFT. This function is only active in the closed loop mode of operation. As the ECM monitors the oxygen sensor voltage input, it is constantly varying the STFT value. For example, if a vacuum leak were to occur causing a lean condition (O_2 voltage low), the STFT would respond by increasing injector pulse width. This increase would be seen on a scan tool as a number greater than 128. Conversely, if the air filter became plugged causing a rich condition (O_2 voltage high), the STFT would respond with a decrease in injector pulse width; a number less than 128. Because this value is updated very quickly, the STFT only corrects for short term mixture trends.

• **The Long Term Fuel Trim (LTFT) (formerly called block learn)** is a matrix of cells arranged by rpm and load. As engine operating conditions change, the ECM will switch from cell to cell to determine which LTFT factor is appropriate. While in any given block, the ECM also monitors the STFT. If the STFT is far enough from 128 in either direction, the ECM will change the LTFT value for that given cell. Once the LTFT value is changed the STFT value should revert back to 128, this represents a neutral condition. If the mixture is still not correct (as judged by the O_2 sensor), the process of short and long term fuel correction will continue until they reach the limits of their control. These limits are programmed into the PROM and would be different for each vehicle. At this point of maximum correction a trouble code 44 or 45 would be registered in the computer's RAM. As with STFT the long term fuel trim is only active in closed loop operation. The LTFT represents the computer's learning capability as these values can be tailored to the specific driving habits of the operator.

Once battery power is lost however, the LTFT values will revert back to default settings, and a loss of performance may be noticed. Original performance should return after the vehicle is operated for a period of time in closed loop.

Tools and Equipment

SCAN TOOLS

The system can communicate a wide variety of information through the Data Link Connector (DLC), previously known as the ALDL. Depending on the application, serial data is transmitted to terminals E or M of the DLC. This data is transmitted at a high frequency which requires a quality scan tool for interpretation. Although stored codes may be read with only the use of a small jumper wire, the use of a hand-held scan tool such as GM's TECH 1 or equivalent, is recommended. There are many manufacturers of these tools; a purchaser must be certain that the tool is proper for the intended use.

The scan tool allows any stored codes to be read from the ECM memory. The tool also allows the operator to view the data being sent to the ECM while the engine is running. This ability has obvious diagnostic advantages; the use of the scan tool is frequently required by the diagnostic charts.

With an understanding of the data stream that the tool will display and knowledge of the circuits involved, the tool can be an extremely valuable tool when examining the system. Without scan tools, diagnostic information would be difficult to obtain in some circuits and impossible to retrieve in others. Scan tools do not make the use of trouble code charts unnecessary, nor do they pinpoint the exact area in a circuit where the problem exists. However, this type of equipment will give more specific information, such as sensor input readings. An example of this would be the coolant sensor circuit. While the ECM looks at specific voltage levels to determine coolant temperature, the scan tool will display this information as an actual temperature reading. This information is not only useful in determining system malfunctions, but is also very helpful in diagnosing mechanical problems. For example, if the engine is overheating, the scan tool can be used to look at the actual engine temperature. This is helpful when checking for electric cooling fan enable switch problems, sticking thermostats, etc.

The ECM is capable of communicating with a scan tool in 3 modes: Normal, DCL and Factory Test.

Normal or Open Mode

This mode is not applicable to all engines. When engaged, certain engine data can be observed on the scanner without affecting engine operating characteristics. The number of items readable in this mode varies with engine family. Most scan tools are designed to change automatically to the DLC mode if this mode is not available.

DLC Mode

Also referred to as the ALDL, 10K or SPECIAL mode, the scanner will present all readable data as available. Certain operating characteristics of the engine are changed or controlled when this mode is engaged. The closed loop timers are bypassed, the spark (EST) is advanced and the PARK/NEUTRAL restriction is bypassed. If applicable, the IAC controls the engine speed to 1000 rpm ± 50, and, on some engines, the canister purge solenoid is energized.

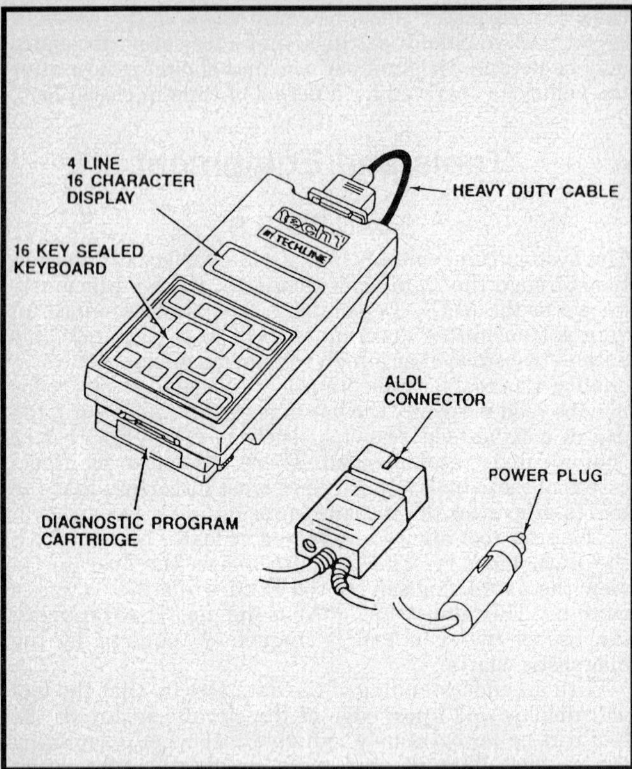

Typical scan tool — GM's Tech 1 shown

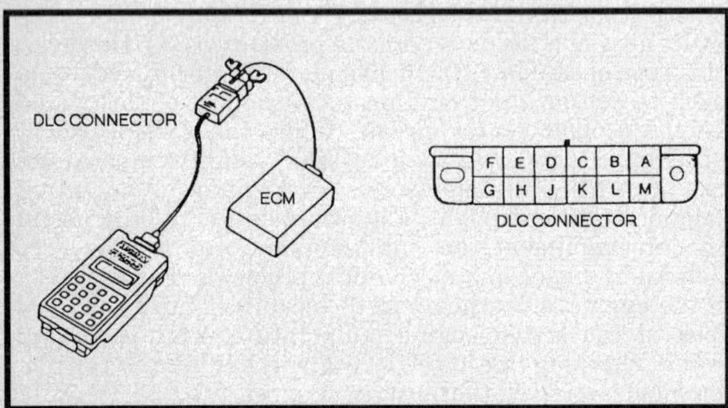

DLC connection to scan tool

Factory Test

Sometimes referred to as BACK-UP mode, this level of communication is primarily used during vehicle assembly and testing. This mode will confirm that the default or limp-in system is working properly within the ECM. Other data obtainable in this mode has little use in diagnosis.

> **NOTE: A scan tool that is known to display faulty data should not be used for diagnosis. Although the fault may be believed to be in only one area, it can possibly affect many other areas during diagnosis, leading to errors and incorrect repair.**

To properly read system values with a scan tool, the following conditions must be met. All normal values given in the charts will be based on these conditions:

- Engine running at idle, throttle closed
- Engine warm, upper radiator hose hot
- Vehicle in park or neutral
- System operating in closed loop
- All accessories **OFF**

SCAN TOOL USE FOR INTERMITTENTS

In some tool applications the data update rate is so slow that it becomes less effective than a digital voltmeter for checking sporadic voltage changes. With the rapid data transmission of a quality scanner, intermittent problems such as faulty wiring connections can be more successfully diagnosed. For example, while manipulating a suspect electrical circuit or component, observe the scan tool display screen. A rapid voltage change or erratic reading will uncover the trouble area.

The scan tool is also an easy way to compare the operating parameters of a poorly operating engine with that of a known good one. For example, a sensor may shift in value and cause a driveability problem, but not set a trouble code. Comparing the sensor's readings to a known good one may uncover the problem.

Scan tools have the ability to speed up diagnostic time and prevent the replacement of good parts, however a thorough understanding of the system you are working on, as well as a working knowledge of the scan tool is essential.

ELECTRICAL TOOLS

The most commonly required electrical diagnostic tool is the digital multimeter, allowing voltage, resistance and amperage to be read by one instrument. The multimeter must be a high-impedance unit, with 10 megohms of impedance in the voltmeter. This type of meter will not place an additional load on the circuit it is testing; this is extremely important in low voltage circuits. The multimeter must be of high quality in all respects. It should be handled carefully and protected from impact or damage. Replace batteries frequently in the unit.

Other necessary tools include an unpowered test light, a quality tachometer with inductive (clip-on) pick up and the proper tools for releasing GM's Metri-Pack, Weather Pack and Micro-Pack terminals as necessary. The Micro-Pack connectors are used at the ECM connector. A vacuum pump/gauge may also be required for checking sensors, solenoids and valves.

Diagnosis and Testing

SERVICE PRECAUTIONS

- To prevent internal ECM damage, the ignition must be **OFF** when disconnecting or reconnecting power to the computer.
- When handling a PROM, CALPAC or MEMCAL, do not touch the component leads. Also, do not remove the integrated circuit from the carrier.
- Never allow welding cables to lie on, near or across any vehicle electrical wiring.
- Leave new components and modules in the shipping package until ready to install them.

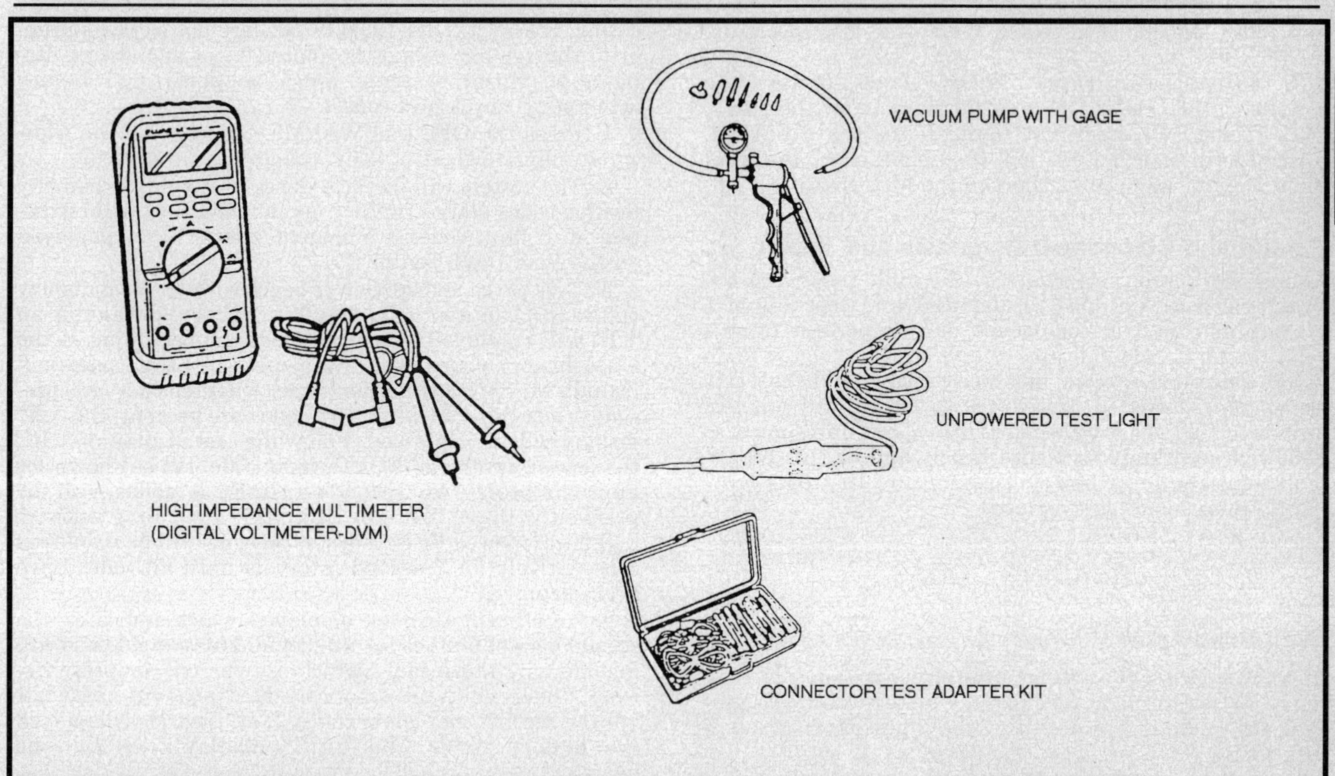

Electrical system diagnostic tools

• When performing electrical tests on the system, use a high impedance multimeter, digital voltmeter (DVM) J-34029-A or equivalent.

• To prevent possible electrostatic discharge damage to the ECM, do not touch the connector pins or soldered components on the circuit board.

PRELIMINARY INSPECTIONS

In order to successfully locate and repair problems in complex computer controlled systems, the technician needs to adhere to a comprehensive, sequential method of diagnostics. Many times a vehicle has a malfunction and the worst is suspected. Premature conclusions often result in the replacement of good parts and wasted time. A proper sequence of steps needs to be established and strictly adhered to when troubleshooting a vehicle. These are as follows:

1. If possible, try to speak directly to the customer, especially if the complaint is not associated with an intermittent or steady MIL.

2. Road test the vehicle, if necessary, to verify the complaint.

3. Perform a thorough visual inspection. This step can very often eliminate the need for further testing. Loose wires, especially ground circuits, can create mysterious driveability problems. However, a complete visual inspection can uncover these problems quickly, and prevent the replacing of good parts and reduce vehicle downtime.

4. Perform the Diagnostic Circuit Check. This test confirms that the diagnostic system has not failed and is able to communicate through the malfunction indicator lamp.

5. After locating and repairing the problem area, road test the vehicle to confirm the repair and to verify no additional problem areas exist.

Visual Underhood Inspection

This is possibly the most critical step of diagnosis. A detailed examination of connectors, wiring and vacuum hoses can often lead to a repair without further diagnosis. Performance of this step relies on the skill of the technician performing it; a careful inspector will check:

• The undersides of hoses as well as the integrity of hard-to-reach hoses blocked by the air cleaner or other components.

• The wiring carefully for any sign of strain, burning, crimping, or terminal pull-out from a connector.

• The connectors at components or in harnesses as required; usually, pushing them together will reveal a loose fit.

Diagnostic Circuit Check

This step is used to check that the on-board diagnostic system is working correctly. A system which is faulty or shorted may not yield correct codes when placed in the Diagnostic Mode.

If the diagnostic system is not operating correctly, or if a problem exists without the malfunction indicator lamp being lit, refer to the vehicle's A-Charts. These charts cover such conditions as Engine Cranks but Will Not Run or No Service Engine Soon Light.

1. Turn the ignition switch **ON** but do not start the engine.

2. The MIL should illuminate and and stay lit until the engine is started.

3. If the MIL does not light, check the bulb and circuits for an open or short.

4. Connect jumper wire from terminal A to B in the DLC. The MIL should flash code 12 at least 3 times. If

not, turn the ignition switch **OFF** and disconnect the ECM.

5. Turn the ignition switch **ON** but do not start the engine. If the MIL is ON, look for short in circuit 419.

6. If the MIL does not light, check the MEMCAL PROM for proper installation. If no light, check the ECM then look for an open or short in the MIL circuit.

Circuit and Component Diagnosis and Repair

Using the appropriate chart(s), the fault codes and the scan tool data will lead to diagnosis and checking of a particular circuit or component. It is important to note that the fault code indicates a fault or loss of signal in an ECM-controlled system, not necessarily in the specific component. Detailed procedures to isolate the problem are included in each code chart; these procedures must be followed accurately to insure timely and correct repair. Following the procedure will also insure that only truly faulty components are replaced.

READING CODES

With Scan Tool

NOTE: Scan tool functions and procedures may vary between manufacturers. Consult owners manual for proper connection and operation of each scan tool.

Once the integrity of the system is confirmed, enter the Diagnostic Mode and read any stored codes. To enter the diagnostic mode:

1. Turn the ignition switch **OFF**. Locate the Data Link Connector (DLC) also called ALDL, usually under the instrument panel. It may be within a plastic cover or housing labeled DIAGNOSTIC CONNECTOR. This link is used to communicate with the ECM.

2. Connect the scan tool correctly to the DLC.

3. Turn the ignition switch to the **ON** position but DO NOT start the engine. A Code 12 may be displayed. Code 12 is not a fault code. It is used as a system acknowledgment or handshake code; its presence indicates that the ECM can communicate as requested. Code 12 is used to begin every diagnostic sequence. Some vehicles also use Code 12 after all diagnostic codes have been sent.

4. After Code 12 has been transmitted 3 times, the fault codes, if any, will each be transmitted 3 times. The codes are stored and transmitted in numeric order from lowest to highest.

NOTE: The order of codes in the memory does not indicate the order of occurrence.

5. Switch the ignition **OFF** when finished with code retrieval or scan tool readings and remove the scan tool.

Without Scan Tool — Deville and Fleetwood

The self-diagnostic capabilities may be used to retrieve stored codes. The codes will be displayed on the instrument panel in the odometer window below the tachometer.

Before entering the diagnostic mode, the visual underhood inspection must be performed. The Diagnostic Circuit check must also be performed. Before codes can be read, the SERVICE ENGINE SOON warning lamp must be operative and the system must be capable of communi-

cating. If the CC/DIC, BCM or IPC are not working properly, the system cannot be placed into diagnostics. Repairs to these systems must be performed before attempting to retrieve codes.

1. Press the OFF and WARMER buttons on the Electronic Climate Control (ECC) panel simultaneously.

2. The system will perform the segment check then any trouble codes stored in the computer memory will be displayed. If fault codes are present, they will be displayed on the Fuel Data Center (FDC).

3. Trouble code display will begin with an 8.8.8 display on the FDC for approximately 1 second. Following this an "..E" will displayed which indicates the beginning of the ECM history codes. The initial pass of ECM codes will include all detected malfunctions whether they are present or not. If no ECM history codes are present, the "..E" display will be bypassed. Following the display of "..E", the lowest numbered ECM history code will be shown for approximately two seconds. All ECM codes will be prefixed with an "E" (i.e. E13, E14, etc.). Progressively higher numbered ECM codes, if present, will be displayed consecutively for 2 second intervals until all codes have been displayed.

4. ".E.E" will then be displayed which indicates the second pass of fault codes will begin. This second pass will include only those codes which are current, or presently exist. Codes which are displayed during the first pass but not the second are history codes. If all the codes displayed are history codes, the ".E.E" display will then be bypassed.

5. At the conclusion of the ECM code display, the BCM fault codes will be shown in a similar fashion. The differences between ECM and BCM code display are as follows:

   a. BCM codes are prefixed with an "F".

   b. "..F" precedes the first pass.

   c. ".F.F" precedes the second pass.

6. After all ECM and BCM codes have been displayed, or if no fault codes are present, .7.0 will be displayed indicating system readiness for the selection of the next diagnostic feature.

Without Scan Tool — Eldorado, Seville and Allante

1. Turn the ignition switch **ON**.

2. On the Climate Control panel, press the OFF and WARM controls simultaneously.

3. Hold the buttons depressed until all display segments on the CC/DIC and Instrument Panel Cluster (IPC) illuminate. On 1992 models, a separate combination panel houses both the ECC and DIC. On 1993–94 models, the DIC is part of the instrument panel cluster.

4. After the segment check, the diagnostic trouble codes will be displayed. Each trouble code consists of the system identifier, a three digit code identifier and the current or history identifier. If no codes are present for a system, a NO X CODE or NO X DATA message will be displayed.

- **E** — ECM trouble code.
- **P** — PCM trouble code.
- **C** — Current trouble code.
- **H** — History trouble code.
- **NO X CODE** — No code present (X represents the system).
- **NO X DATA** — Communication line not operating (X representing the system).

CLEARING CODES

Deville and Fleetwood

To erase stored codes in the ECM memory, enter diagnostics and then press the OFF and HI buttons simultaneously and hold until "E.0.0" is displayed. To remove BCM trouble codes, press the OFF and LO buttons simultaneously and hold until "F.0.0" appears on the display. After either display, a ".7.0" will appear. At this point turn the ignition **OFF** for at least 10 seconds before reentering the diagnostic mode.

Eldorado, Seville and Allante

Selection of the CLEAR CODES? test display the message CODES CLEAR. This action will result in all codes being erased from that system's memory. The CODES CLEAR message will appear for 3 seconds.

ON BOARD DIAGNOSTIC SYSTEM

Operation — Deville and Fleetwood

TO ENTER DIAGNOSTICS:

Press the OFF and WARMER buttons on the Electronic Climate Control (ECC) panel simultaneously and hold until all display panel segments illuminate. This indicates the beginning of the diagnostic readout. The purpose of this panel illumination is to verify that all segments of the displays are working. Do not attempt diagnosis if all segments do not appear as this will lead to misdiagnosis because the readout will be difficult to interpret. If any of the segments are inoperative, the affected display panel will need to be replaced.

> **NOTE: Upon entering the diagnostic mode, the ECC will operate in whichever mode was being commanded prior to pressing the OFF and WARMER buttons.**

TROUBLE CODE DISPLAY

At the conclusion of the segment check, any trouble codes stored in the computer memory will be displayed. If fault codes are present, they will be displayed on the Fuel Data Center (FDC) as follows:

1. Trouble code display will begin with an 8.8.8 display on the FDC for approximately 1 second. Following this an "..E" will displayed which indicates the beginning of the ECM history codes. The initial pass of ECM codes will include all detected malfunctions whether they are present or not. If no ECM history codes are present, the "..E" display will be bypassed. Following the display of "..E", the lowest numbered ECM history code will be shown for approximately two seconds. All ECM codes will be prefixed with an "E" (i.e. E13, E14, etc.). Progressively higher numbered ECM codes, if present, will be displayed consecutively for 2 second intervals until all codes have been displayed.

2. "..E.E" will then be displayed which indicates the second pass of fault codes will begin. This second pass will include only those codes which are current, or presently exist. Codes which are displayed during the first pass but not the second are history codes. If all the codes displayed are history codes, the "..E.E" display will then be bypassed.

3. At the conclusion of the ECM code display, the BCM fault codes will be shown in a similar fashion. The differences between ECM and BCM code display are as follows:
   a. BCM codes are prefixed with an "F".
   b. "..F" precedes the first pass.
   c. ".F.F" precedes the second pass.

4. After all ECM and BCM codes have been displayed, or if no fault codes are present, .7.0 will be displayed indicating system readiness for the selection of the next diagnostic feature.

EXITING THE DIAGNOSTIC MODE

To exit the diagnostic mode, press the AUTO button, or turn the ignition switch **OFF** for 10 seconds. The temperature setting will reappear on the ECC panel. The fault codes will not be erased from this procedure.

STATUS LIGHT DISPLAY

While in the diagnostic mode, the ECC mode indicator lights are used to determine the activity of certain operating circuits within the system. These lights will either be turned ON or OFF. For example, loop status will be shown by the AUTO indicator on the ECC. The AUTO indicator will be turned ON to indicate closed loop, and turned OFF to indicate open loop.

CODE .7.0

Code .7.0 is a decision point. When .7.0 is displayed, the technician selects a diagnostic feature. The following choices are available:

ECM switch tests — Press cruise control switch to **ON**
ECM data display — Press LO then HI
ECM code snapshot — Press LO then ECON and WARMER
ECM snapshot — Press LO then ECON and COOLER
ECM output cycling — Press HI then HI
ECM output overrides — Press HI then ECON and WARMER
BCM data display — Press OUTSIDE TEMPERATURE then HI
ECC program override — Press OUTSIDE TEMPERATURE
Clear Codes ECM/BCM — Press OFF and HI
Exit Diagnostics — Press AUTO

ECM SWITCH TEST SERIES

As long as .7.0 is displayed on the FDC panel, the switch tests can begin. To initiate the switch test sequence, turn the cruise control OFF/ON switch to **ON**. The switch tests begin as the display changes from .7.0 to E.7.0. If the display does not advance to E.7.0 refer to the diagnostic chart for switch test fault code E.7.0. This will help determine the cause of the ECM not receiving or processing the cruise control ON/OFF signal.

The switch test number is displayed on the FDC panel. Individual test scan be selected in the series by using the HI button to move to the next test number, or the LO button to move to the previous test number. When the switch test input has been recognized by the ECM, the display will alternate between E.0.0 and the test number. The switch test sequence is initiated as follows:

> **NOTE: Upon the completion of each step, the display will change to the next switch test number in sequence.**

1. With E.7.0 displayed, depress the brake pedal to test the cruise control brake switch.

2. With E.7.1 displayed, press and release the brake pedal again to test the VCC brake switch circuit.

3. With E.7.2 displayed, press the accelerator pedal from idle to wide open slowly, then release the throttle slowly. This checks the operation of the throttle switch circuit.

4. With E.7.3 displayed, move the gear selector from park to drive. This action checks the P/N switch.

5. With E.7.5 displayed, switch the cruise control ON/OFF switch from **ON** to **OFF**, and then back to **ON**. This action checks the operation of the switch.

6. With E.7.6 displayed and the cruise control switch still in the **ON** position, press and release the set/coast button to check the operation of this switch.

7. With E.7.7 displayed and the cruise control switch still in the **ON** position, press and release the resume/accel switch to check its function.

8. With E.7.8 displayed and the engine running, turn wheels from straight ahead to full left or right lock. This checks the PSPS signal to the ECM. Upon the completion of the last test, .7.0 will again be displayed on the FDC awaiting the next command prompt from the technician. Any tests that have failed will be displayed on the FDC, however a failed cruise control ON/OFF switch, or cruise control brake switch, will prevent the test sequence from being initiated.

ECM DATA DISPLAY PARAMETERS

To enter the ECM data display mode, Press the LO button then the HI button. The ECM data parameters are similar to the data information that is available on other vehicles through the use of a scan tool. The following is a list of parameters with a brief explanation of each.

- P.0.1 — The throttle angle is displayed in degrees with a range from -9 to +90.0. The indicated angle is what the ECM sees and not a corrected value. A decimal point will appear before the last digit.
- P.0.2 — Map sensor values are expressed in kilopascals (kpa), with a range from 14 to 109. To convert kpa to psi, divide kpa by 6.895.
- P.0.3 — The computed BARO value is displayed in kpa, with a range from 60 to 102. The value is derived from sampling MAP readings at W.O.T, which eliminates the need for a separate BARO sensor.
- P.0.4 — Engine coolant temperature is expressed in degrees Celsius, with a display range of -40 to 151 degrees Celsius. To convert to Fahrenheit, multiply Celsius by 1.8 then add 32.
- P.0.5 — The intake air temperature is displayed in Celsius, with a display range of -40 to 151 degrees Celsius.
- P.0.6 — The spark advance value is displayed in degrees, and will agree with a timing light within ±2 degrees as long as the base setting is correctly adjusted. Display range is from 0 to 90.
- P.0.7 — Battery voltage is displayed with a range of 0 to 25.5 volts. A decimal point will appear before the last digit.
- P.0.8 — Engine speed is displayed in RPM's, with a range from 0 to 637. Multiply the reading by 10 to get the actual RPM number. For example, 120 will indicate 1200 RPM's.
- P.0.9 — Vehicle speed is displayed in MPH, with a range from 0 to 255.
- P.1.2 — The injector on time is expressed in milliseconds, with a display range from 0 to 99.6. A decimal point will appear before the last digit.

- P.1.4 — The oxygen sensor reading is displayed in volts, with a display range from 0 to 1.14. A decimal point will appear before the last two digits.
- P.1.6 — Oxygen sensor cross counts are expressed 0 to 255. This number will be increased by one each time the O_2 voltage crosses the threshold line in 1 second.
- P.1.8 — Fuel integrator counts are expressed 0 to 255. See Short Term Fuel Trim (STFT) for a detailed explanation.
- P.2.0 — Block learn counts are expressed 0 to 255. See Long Term Fuel Trim (LTFT) for a detailed explanation.
- P.2.1 — The cruise control servo position is expressed in percentage, with a range of 0 to 100%. 0% indicates servo position is fully extended, while 100% indicates full retraction.
- P.2.2 — PRNDL switch status on the B and C output lines from the ECM. 00 indicates switch closed state, while 11 indicates switch open state. See DTC E91 for detailed information.
- P.2.3 — PRNDL switch status on the A and P output lines from the ECM. 00 indicates switch closed state, while 11 indicates switch open state. See DTC E91 for detailed information.
- P.2.4 — The ignition cycle counter display represents the number of times the ignition has bee cycled OFF since an ECM fault code was last detected. After 50 cycles without fault detection all previous codes are cleared.
- P.2.5 — The ECM PROM identification (ID) is a number up to three digits long which can be used to verify correct prom usage. When the ECM data display is initiated, the FDC will display the parameter number for one second, followed by the parameter value for nine seconds. For example, P.0.2 (MAP) will display for 1 second, then 14 (kpa) will display for 9 seconds. This sequence will continue to repeat until another parameter selection is made.

ECM SNAPSHOT MODES

The ECM snapshot mode allows parameters to be viewed at any instant; either when a code sets, or at any other time chosen.

ECM Code Snapshot

When selecting ECM snapshot mode, the parameter displayed are those stored in memory when the last code set. This last code and the parameter values relating to it will be prefixed by an L (L.0.1, L.0.2 etc.). To enter ECM code snapshot press the ECON and WARMER buttons with E.9.0 on the display. When ECM codes are cleared this information will be lost.

ECM Instant Snapshot

The instant snapshot is similar to the code snapshot except the technician can determine the exact moment the data will be recorded. With E.9.0 on the display, press the ECON and COOLER buttons. Once the snapshot is recorded, the ECC will display 5.9.0. Pressing the HI button will begin a review of the recorded data. Data parameters will be prefixed by an S (S.0.1, S.0.2 etc.).

NOTE: The letter S and the number 5 look similar on the display panel.

ECM Output Cycling Mode

After .7.0 is displayed on the FDC and the engine is OFF, press the HI button. The FDC will display E.9.5 to indicate the start of the test sequence. At this time the HI and

LO buttons can be used to select the specific output test. Once the test has been selected, the display will alternate between E.9.6 and the code number for the specific test. E.9.6 indicates that the solenoid is de-energized, and the test code number indicates that the solenoid is energized. Devices will be turned ON and OFF at 3 second intervals.

ECM Overrides

After .7.0 is displayed on the FDC, press the HI button to advance the display to E.9.5. Press the ECON and WARMER to change the display to E.5.0, indicating the start of the override tests. The HI and LO buttons can be used to select the next higher or lower test. The COOLER and WARMER buttons are used to activate the selected override. The FDC will display the override as a percentage of apply with 00 equaling OFF and 99 equaling ON. Whenever an override test is active the FDC display will alternate between the override test number and either 00 or 99. This action indicates the specific override and its present status, either OFF or ON. When neither the WARMER nor COOLER buttons are held down to activate or deactivate a particular override, the ECM will resume command of that circuit.

- E.5.0 — No Overrides Commanded
- E.5.1 — VCC Solenoid — With the WARMER button held down, the solenoid will be turned ON and the VCC engaged. As long as the COOLER button is held down, the solenoid will be turned OFF and the VCC disengaged.
- E.5.2 — EGR Solenoid — With the WARMER button held down, the EGR valve will receive a vacuum signal and the vacuum signal will cease as long as the COOLER button is held down. This test will only verify the integrity of the vacuum signal to the EGR and not its mechanical operation. The reason for this is the use of positive backpressure valves.
- E.5.3 — ISC Motor — While in this mode the ISC motor can be commanded to either a fully extended position with the WARMER button held down, or a fully retracted position with the COOLER button down. For this function to be enabled, the vehicle must be standing still with the transmission in **P** or **N**. During this test, A/C and EGR are commanded OFF, and spark advance is fixed.
- E.5.4 — Fuel Injector Cutout — With the engine running and the transmission in **P** or **N** depress the accelerator to the desired speed to begin the power balance test. ISC motor and A/C clutch are commanded OFF during this override which makes it necessary to hold the throttle at a selected position. The WARMER button will be used to select the specific injector, and when held down, will increment to each injector at a rate of one selection per second. The COOLER button will be used to disable the selected injector for as long as the button is held down.
- E.5.5 — Fuel Pump Relay — With the vehicle running in **P** or **N**, and the backup circuit disconnected (oil pressure switch), press and hold the COOLER button to disable the relay. As long as the backup circuit has been disconnected, and the fuel pump relay is good, the engine should stop. The WARMER button does apply to this test.
- E.5.6 — No Valid Test
- E.5.7 — Cruise Servo — The engine should be run before doing this test in order to charge the vacuum reservoir. With the engine OFF and the vehicle in **P**, press and hold the WARMER button to retract the servo. With the COOLER button held down the servo will extend.

- E.5.8 — Engine Cooling Fan — While in this mode the COOLER button is depressed to invert the state of the low speed fan relay. 00 will be displayed with the fans not running. With the fans running at low speed and the low speed relay energized, the FDC will display 10. The WARMER button is pressed to invert the operation of the high speed fans. With the high fan relay energized, 01 will be displayed, while 11 will be displayed during the operation of both fan relays.
- E.5.9 — Fixed Spark — While in this mode, spark advance is controlled manually and displayed in degrees on the FDC. The first time the COOLER button is pressed and released, spark timing will be fixed at 10 degrees BTDC. Each time the COOLER button is pressed and released, an additional 1 to 2 degrees of spark is retarded up until 0 degrees is reached. Each time the WARMER button is pressed and released, 1 to 2 degrees of spark timing is added until the initial setting that the ECM was commanding before the override is reached. Spark cannot be advanced beyond this point and the FDC will display 8.8.8.
- E.6.0 — Injector Flow — Each injector can only be tested once during this override to prevent engine flooding. With the transmission in **P** or **N**, press the WARMER button to open the injector. The injector will now remain open for 50 milliseconds. Press and hold the COOLER button down to select the specific injector at a rate of 1 choice per second.
- E.6.1 — Transmission Shift — In this mode the transaxle can be downshifted by depressing the COOLER button, and upshifted by depressing the WARMER button. The transaxle will not shift down from 2nd to 1st above 30 mph, or from 3rd to 2nd above 60 mph.

Operation — Eldorado, Seville and Allante

TO ENTER DIAGNOSTICS:
1. Turn the ignition switch **ON**.
2. On the Climate Control panel, press the OFF and WARM controls simultaneously.
3. Hold the buttons depressed until all display segments on the CC/DIC and Instrument Panel Cluster (IPC) illuminate. On 1992 models, a separate combination panel houses both the ECC and DIC. On 1993–94 models, the DIC is part of the instrument panel cluster.

SEGMENT CHECK
The segment check is the illumination of the IPC, CCP and DIC to verify that all segments of the liquid crystal displays are working. Only the turn signal indicators do not light during this check. Diagnosis should not be attempted unless all segments appear, as this could lead to misdiagnosis. If any portions or segments are inoperative, the faulty component should be replaced.

STATUS LIGHTS
When the system is placed into the Diagnostic Mode, the lighting elements on the climate control panel are used as status lights for various components. The mode of operation is determined by the light being ON or OFF. The use of each lighting element is different in ECM or BCM mode. For example, in the ECM mode the word OFF on the CCP will light to indicate a rich condition sensed by the left oxygen sensor input. In BCM mode, the same

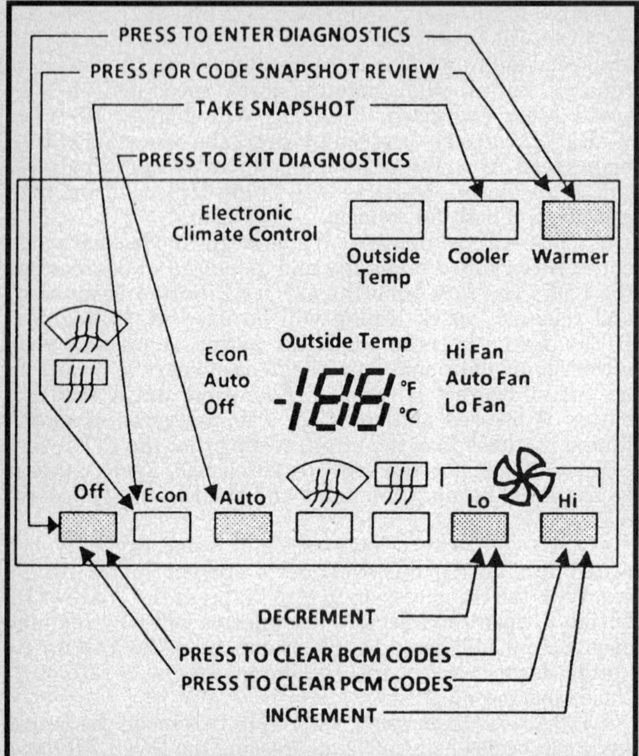

Climate control panel — Deville and Fleetwood

light is ON when the BCM commands the air inlet valve to the recirculate position.

> **NOTE: Although the use of the heater control panel is changed during diagnostics, the heating/cooling system remembers and operates at the last setting chosen before entering diagnostics. Once placed into diagnostics, the Climate Control Panel cannot be used to control the heating/cooling system. The system will respond to normal control after exiting diagnostics. The only exception to this would be on 1993–94 models, when ACP overrides are used to modify system parameters.**

TROUBLE CODE DISPLAY

After diagnostics is entered, any fault codes stored in computer memory will be displayed. For 1992 systems, codes may be stored for the ECM, BCM, or SIR systems. For 1993–94 the system has been expanded to include codes for the PCM, IPC, ACP, SIR, TCS and RTD systems. The Instrument Panel Cluster (IPC) replaces the BCM for 1993–94 models, but serves the same purpose. ACP, SIR, TCS and RTD stand for Air Conditioning Programmer, Supplemental Inflatable Restraint, Traction Control System and Real Time Dampening, respectively. We will only deal with engine control malfunctions here.

Each trouble code consists of the system identifier (E for ECM or P for PCM), a three digit code identifier and the letter C or H to indicate current or history codes. If no codes are present for a system, a NO X CODE message will be displayed, with X representing the system. If the communication line to a component is not operating a NO X DATA message will appear indicating that the BCM/IPC could not communicate with that system.

On 1992 models, the RESET/RECALL button on the DIC will terminate diagnostics and return the vehicle to normal operation. For 1993–94 models, depress the AUTO button on the ECC panel to return to normal operations.

SELECTING THE SYSTEM

Following the display of trouble codes, the first available system will be displayed for testing. The first choice will be ECM on 1992 models or PCM on 1993–94 models. When selecting a system to test, the following actions may be taken to control the display:

1. Depressing the HI button on the ECC will select the displayed system for testing.

2. Depressing the LO button on the ECC will display the next available system selection.

3. Depressing the OFF button will stop the system selection process and return the display to the beginning of the trouble code sequence.

SELECTING THE TEST TYPE

Having selected a system, the first available test type will be displayed (i.e. ECM DATA?). When selecting a test type, the following actions may be taken to control the display:

1. Depressing the HI button on the ECC will select the displayed test type. At this point, the first of several test type choices will appear.

2. Depressing the LO button on the ECC will display the next available test type for the selected system.

3. Depressing the OFF button on the ECC will stop the test type selection process and return the display to the next available system choice.

SELECTING THE TEST

Selection of the DATA?, INPUTS?, OUTPUTS? or OVERRIDES? will result in the first available test being displayed. Four characters will appear on the display which indicate the system, test type and test. For example, a display of PD01 would indicate the following:

- P=PCM (system selection)
- D=DATA (test type selection)
- 01=THROTTLE POSITION (test)

When selecting a specific test, any of the following actions may be taken:

1. Depressing the HI or LO buttons on the ECC will incrementally select the desired numbered test. Use HI to test up and LO to test down.

2. Depressing the OFF button will stop the test selection process and return the display to the next available test type for the selected system.

Snapshot

Selection of the SNAPSHOT? test type will allow the recall of all data and input values for the selected system from a specific point in time. These values may be retrieved from either a manual snapshot, or a snapshot that was triggered when a code was set. If more than one code was set, the data values shown will apply to the last code that was set.

Selecting Codeset Snapshot

If a snapshot for a set code is available, the display will read X###Snapshot?. X represents the system selected such as E for ECM, or P for PCM. ### represents the 3 digit diagnostic code that recorded the snapshot. This snapshot may be selected by pressing HI or bypassed by pressing LO.

DIAGNOSTIC FLOW CHART — DEVILLE AND FLEETWOOD

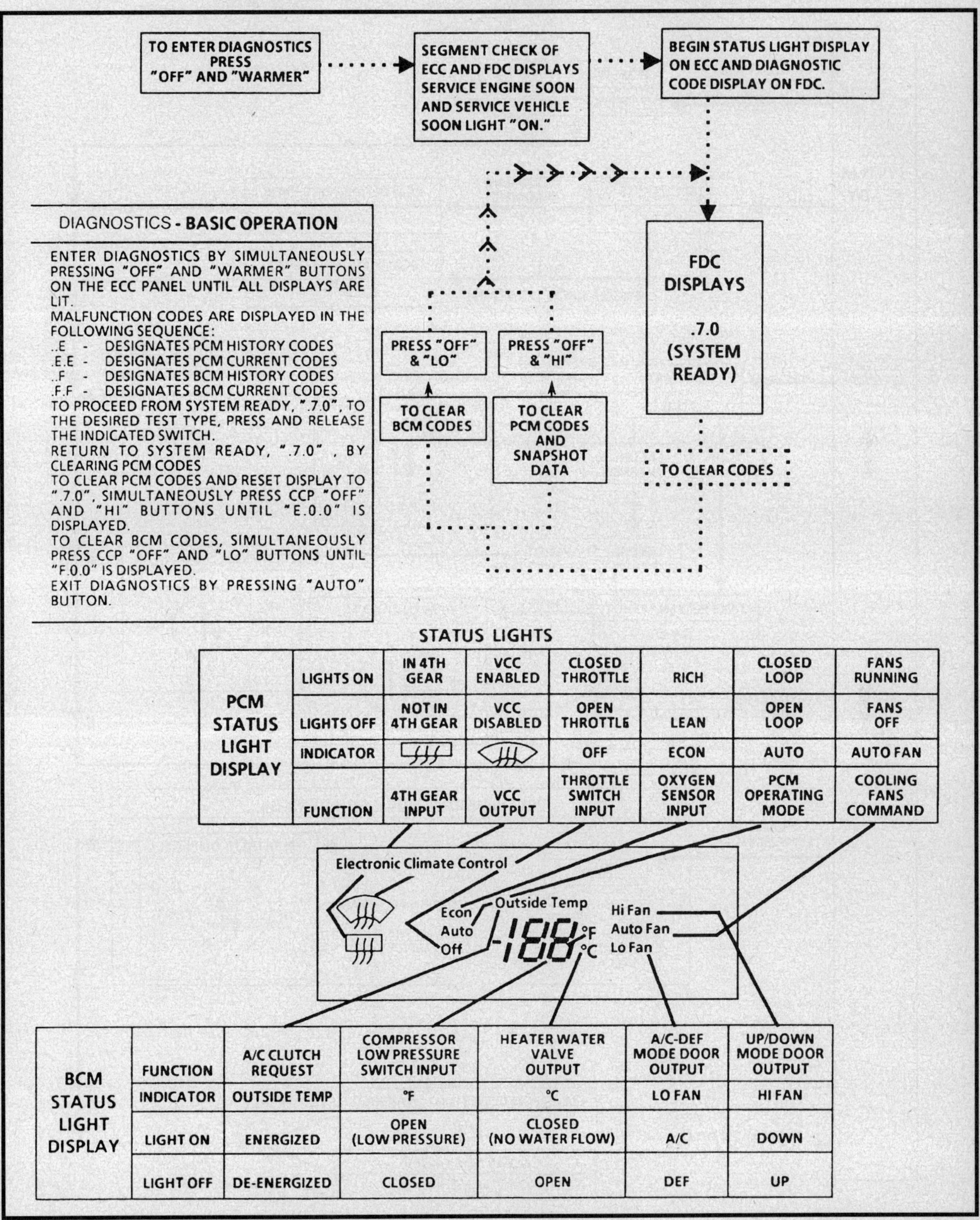

DIAGNOSTICS - BASIC OPERATION

ENTER DIAGNOSTICS BY SIMULTANEOUSLY PRESSING "OFF" AND "WARMER" BUTTONS ON THE ECC PANEL UNTIL ALL DISPLAYS ARE LIT.
MALFUNCTION CODES ARE DISPLAYED IN THE FOLLOWING SEQUENCE:
..E - DESIGNATES PCM HISTORY CODES
.E.E - DESIGNATES PCM CURRENT CODES
..F - DESIGNATES BCM HISTORY CODES
.F.F - DESIGNATES BCM CURRENT CODES
TO PROCEED FROM SYSTEM READY, ".7.0", TO THE DESIRED TEST TYPE, PRESS AND RELEASE THE INDICATED SWITCH.
RETURN TO SYSTEM READY, ".7.0" , BY CLEARING PCM CODES
TO CLEAR PCM CODES AND RESET DISPLAY TO ".7.0", SIMULTANEOUSLY PRESS CCP "OFF" AND "HI" BUTTONS UNTIL "E.0.0" IS DISPLAYED.
TO CLEAR BCM CODES, SIMULTANEOUSLY PRESS CCP "OFF" AND "LO" BUTTONS UNTIL "F.0.0" IS DISPLAYED.
EXIT DIAGNOSTICS BY PRESSING "AUTO" BUTTON.

STATUS LIGHTS

PCM STATUS LIGHT DISPLAY	LIGHTS ON	IN 4TH GEAR	VCC ENABLED	CLOSED THROTTLE	RICH	CLOSED LOOP	FANS RUNNING
	LIGHTS OFF	NOT IN 4TH GEAR	VCC DISABLED	OPEN THROTTLE	LEAN	OPEN LOOP	FANS OFF
	INDICATOR			OFF	ECON	AUTO	AUTO FAN
	FUNCTION	4TH GEAR INPUT	VCC OUTPUT	THROTTLE SWITCH INPUT	OXYGEN SENSOR INPUT	PCM OPERATING MODE	COOLING FANS COMMAND

BCM STATUS LIGHT DISPLAY	FUNCTION	A/C CLUTCH REQUEST	COMPRESSOR LOW PRESSURE SWITCH INPUT	HEATER WATER VALVE OUTPUT	A/C-DEF MODE DOOR OUTPUT	UP/DOWN MODE DOOR OUTPUT
	INDICATOR	OUTSIDE TEMP	°F	°C	LO FAN	HI FAN
	LIGHT ON	ENERGIZED	OPEN (LOW PRESSURE)	CLOSED (NO WATER FLOW)	A/C	DOWN
	LIGHT OFF	DE-ENERGIZED	CLOSED	OPEN	DEF	UP

DIAGNOSTIC FLOW CHART — DEVILLE AND FLEETWOOD

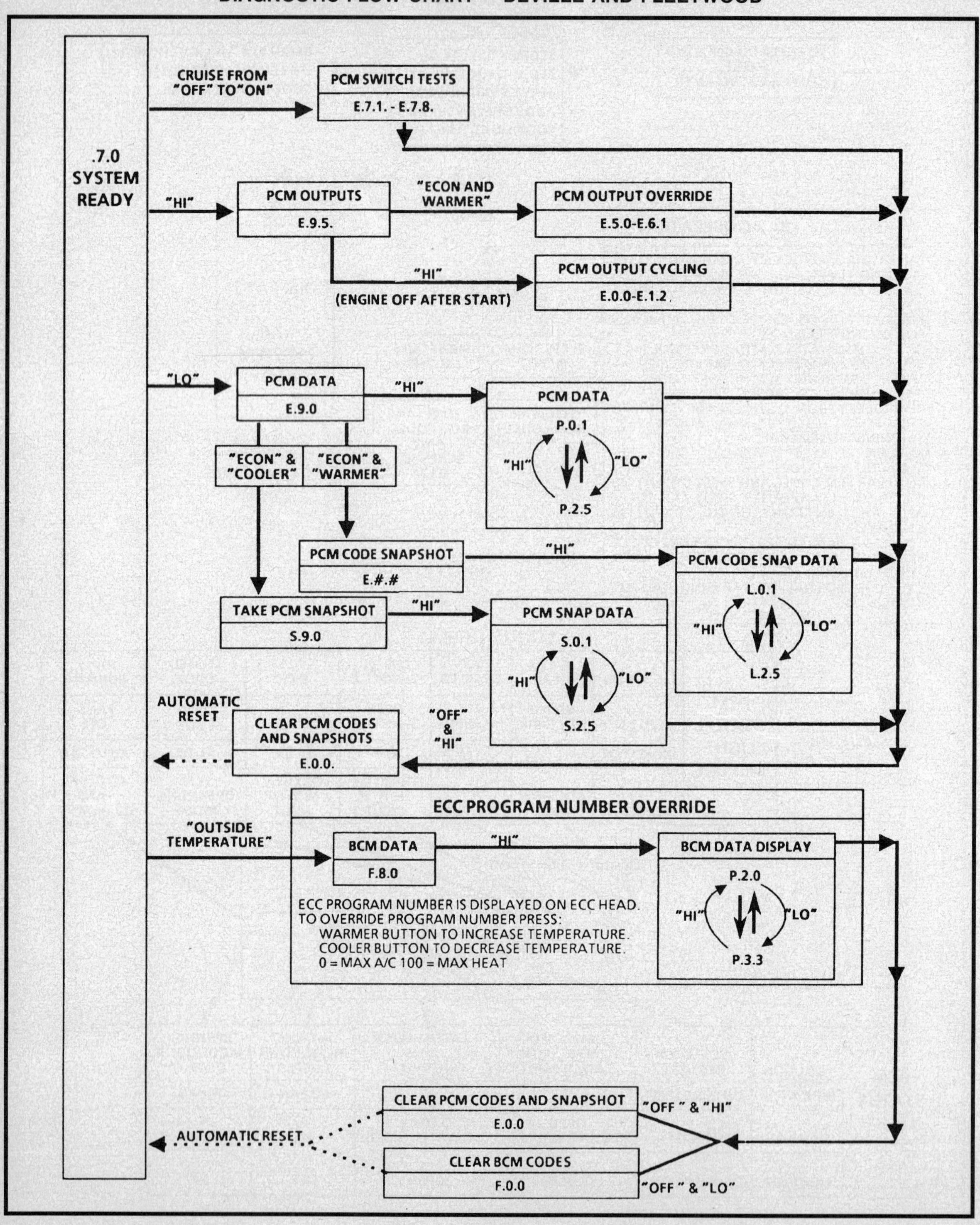

DIAGNOSTIC PARAMETERS — DEVILLE AND FLEETWOOD

PCM SWITCH TESTS

DIAGNOSTIC DISPLAY	CIRCUIT TEST
SWITCH CRUISE "ON" BEFORE TESTING.	
E.7.0	CRUISE BRAKE SWITCH
E.7.1	VCC BRAKE SWITCH
E.7.2	THROTTLE SWITCH
E.7.3	PARK/NEUTRAL
E.7.5	CRUISE "ON/OFF"
E.7.6	CRUISE "SET/COAST"
E.7.7	CRUISE "RESUME/ACCEL."
E.7.8	POWER STEERING PRESSURE SWITCH (ENGINE RUNNING)

PCM OUTPUT OVERRIDES

DIAGNOSTIC DISPLAY	OVERRIDE
E.5.0	NO OVERRIDE
E.5.1	VCC
E.5.2	EGR
E.5.3	ISC
E.5.4	INJECTORS (1-8)
E.5.5	FUEL PUMP RELAY
E.5.7	CRUISE SERVO
E.5.8	COOLANT FANS
E.5.9	FIXED SPARK
E.6.0	INJECTOR FLOW
E.6.1	TRANSAXLE SHIFTING

PRESSING "WARMER" WILL INCREASE OR ACTIVATE.
PRESSING "COOLER" WILL DECREASE OR DEACTIVATE.

FDC DISPLAY OF PCM DATA

PARAMETER NUMBER	PARAMETER	RANGE		UNITS
P.0.1	THROTTLE POSITION	-10.0 TO	90.0	DEGREES
P.0.2	MAP	14 TO	109	KPA
P.0.3	COMPUTED BARO	60 TO	102	KPA
P.0.4	COOLANT TEMPERATURE	-40 TO	151	°C
P.0.5	IAT	-40 TO	151	°C
P.0.6	SPARK ADVANCE	0 TO	90	DEGREES
P.0.7	BATTERY VOLTAGE	0 TO	25	VOLTS
P.0.8	ENGINE SPEED	0 TO	637	RPM/10
P.0.9	VEHICLE SPEED	0 TO	255	MPH
P.1.2	INJECTOR PULSE WIDTH	0 TO	99.6	MILLISECONDS
P.1.4	O₂ SENSOR VOLTAGE	0 TO	1.14	VOLTS
P.1.6	O₂ CROSS COUNTS	0 TO	255	COUNTS/SEC
P.1.8	FUEL INTEGRATOR	0 TO	255	COUNTS
P.2.0	BLOCK LEARN	0 TO	255	COUNTS
P.2.1	CRUISE SERVO POSITION	0 TO	100	PERCENT
P.2.2	PRNDL (CKT 772, 773)	00 TO	11	0 = SW. CLOSED
P.2.3	PRNDL (CKT 771, 776)	00 TO	11	0 = SW. CLOSED
P.2.4	IGNITION CYCLES	0 TO	255	KEY CYCLES
P.2.5	PROM ID	0 TO	999	CODE

PCM OUTPUT CYCLING

PARAMETER NUMBER	PARAMETER
E.0.0	CYCLE NONE
E.0.1	CANISTER PURGE SOLENOID
E.0.2	VCC SOLENOID
E.0.3	EGR SOLENOID
E.0.6	ISC MOTOR
E.0.7	CRUISE VENT SOLENOID
E.0.8	CRUISE VAC SOLENOID
E.0.9	SHIFT SOLENOID "A"
E.1.0	SHIFT SOLENOID "B"
E.1.1	A/C CLUTCH RELAY
E.1.2	CYCLE ALL

FDC DISPLAY OF BCM DATA

PARAMETER NUMBER	PARAMETER	RANGE		UNITS
P.2.0	COMMANDED BLOWER VOLTAGE	-3.3 TO	18.0	VOLTS
P.2.1	COOLANT TEMPERATURE	-40 TO	151	°C
P.2.2	COMMANDED AIR MIX DOOR POSITION	0 TO	100	%
P.2.3	ACTUAL AIR MIX DOOR POSITION	0 TO	100	%
P.2.4	AIR DELIVERY MODE	0 TO	8	CODE □
P.2.5	IN-VEHICLE TEMPERATURE	-39 TO	102	°C
P.2.6	ACTUAL OUTSIDE TEMPERATURE	-40 TO	93	°C
P.2.7	HIGH SIDE TEMPERATURE (CONDENSER OUT)	-15 TO	215	°C
P.2.8	LOW SIDE TEMPERATURE (EVAPORATOR IN)	-40 TO	93	°C
P.2.9	ACTUAL FUEL LEVEL	.8 TO	18.1	GALLONS
P.3.0	IGNITION CYCLE COUNTER	0 TO	99	KEY CYCLES
P.3.1	OIL LIFE FORWARD ODOMETER	0 TO	70	(x 100) MILES
P.3.2	SUN LOAD SENSOR READING	0 TO	255	COUNTS
P.3.3	BCM PROM I.D.	0 TO	255	CODE

□ AIR DELIVERY MODE

BCM AIR DELIVERY MODE IS PARAMETER (P.2.4) OF BCM DATA AND IS DISPLAYED AS A NUMERICAL CODE AS FOLLOWS:

CODE NO.	MODE	CODE NO.	MODE
0	MAX. A/C	4	HEATER
1	A/C	5	OFF
2	BI-LEVEL	6	NORMAL PURGE
3	HEATER/DEFROST	7	COLD PURGE
		8	FRONT DEFOG

Selecting Manual Snapshot

If no codeset snapshots are available or this function is bypassed by pressing LO, the display will then read TAKE SNAPSHOT?. A manually triggered snapshot may be taken at this point by pressing HI. The message SNAPSHOT TAKEN will be taken. The display will change to SNAPDATA? within several seconds.

Viewing Snapshot Information

After a snapshot has been taken or a codeset snapshot selected, the display will read SNAP DATA?, or X###DATA?. At this point the following actions may be taken:

1. Depressing the HI button on the ECC will select the data values for the snapshot.

2. Depressing the LO button on the ECC will select the input values for the snapshot to be selected.

3. Depressing the OFF button will return to the original SNAPSHOT? screen. Pressing this button a second time will return the display to the next available test selection.

Data Display

Serial data is displayed on the Driver Information Center (DIC) in much the same way that a scan tool displays data on other vehicles. This information can be used to monitor sensor circuits with or without the engine running.

Input Displays

When diagnosing a malfunction, the input displays can be helpful in determining if the switched inputs have been properly interpreted. When one of the input tests has been selected, the input status is shown on the display as either HI or LO. Cycling status is displayed as a 0 to indicate that the ECM has not yet seen the component cycle. The 0 will change to an X once the ECM recognizes the change of state, either HI/LO or LO/HI.

Output Displays

The output function allows the technician to cycle ECM controlled devices. The DIC will identify the selected solenoid or relay and the state that the ECM is commanding. HI indicates that the device is energized, and LO indicates that the device has been de-energized. The output function can only be performed with the key **ON** and engine **OFF**.

Override Displays

The override function allows the technician to test specific circuit components regardless of normal program instructions. Upon choosing a test, the current operation of that function will be displayed as a percentage of its full range on the ECC display (0 to 99). The status of the selection will be displayed on the DIC (i.e. ISC Motor Extend). The ECC display will alternate between — for 1 second followed by the normal program command for 10 seconds. This alternating display indicates that the selection is not yet be overridden.

Pressing the WARMER or COOLER buttons on the ECC will initiate the override at which time the ECC display will no longer toggle to —. Press the WARMER button to increase the override value and the COOLER button to decrease the value. Upon release of either button, the display may either remain at the overridden value, or automatically return to normal program control. This action depends on the specific override choice. If the display remains at the override value, normal program control can be restored in 1 of 3 ways:

1. Selection of another override test (HI and LO buttons) will cancel the current override.

2. Selection of another system (OFF button) will cancel the current override.

3. On some system overrides, overriding the value beyond either extreme (0 or 99) will display '—' momentarily on the ECC and jump to the opposite extreme. If the button is released while '—' is displayed, normal program control will be restored and the display will once again toggle.

EXITING DIAGNOSTICS

On 1992 models depress the RESET/RECALL button on the DIC or turn the ignition **OFF**. Trouble codes are not erased when this is done. On 1993–94 models, press the AUTO or DEFOG button on the ECC panel.

SET TIMING MODE/DLC

The Data Link Connector (DLC) on these vehicles is used for adjusting the base timing setting. The "Set Timing Mode" is requested by following the procedure.

1. Allow engine to reach operating temperature, approximately 85 degrees Celsius.
2. Verify that the RPM is less than 900.
3. Make sure the transmission is in **P**.
4. If in the diagnostic mode, exit diagnostics.
5. Insert a jumper between terminals A and B of the DLC. With this procedure completed the ECM will command timing at 10 degrees BTDC. This setting can now be checked with a standard timing light. After checking/adjusting timing, remove the jumper in the DLC. On 1993–94 Northstar systems timing adjustment is not possible.

DIAGNOSTIC TROUBLE CODES (DTC)

Deville and Fleetwood

PCM

A — "SERVICE ENGINE SOON" light ON
B — "SERVICE VEHICLE SOON" light ON
C — No telltale light ON
F — Disengages VCC for ignition cycle
P — Enables canister purge
Q — Disables cruise for ignition cycle
{} — Bracketed systems are disabled when code is present

- DTC **E12** — A — No distributor signal
- DTC **E13** — A/P — Oxygen (O_2) sensor not ready
- DTC **E14** — A — Shorted Engine Coolant Temperature (ECT) sensor circuit signal
- DTC **E15** — A — Open Engine Coolant Temperature (ECT) sensor circuit signal
- DTC **E16** — B — Generator voltage out of range {all solenoids}
- DTC **E19** — B — Shorted fuel pump circuit feedback signal
- DTC **E20** — B — Open fuel pump circuit feedback signal
- DTC **E21** — A — Shorted Throttle Position Sensor (TPS) circuit signal {viscous converter clutch}
- DTC **E22** — A — Open Throttle Position Sensor (TPS) circuit signal {viscous converter clutch}
- DTC **E23** — A — Electronic spark timing circuit signal {Exhaust Gas Recirculation (EGR) system}

DIAGNOSTIC FLOW CHART — 1992–93 ALLANTE, 1992 ELDORADO AND SEVILLE

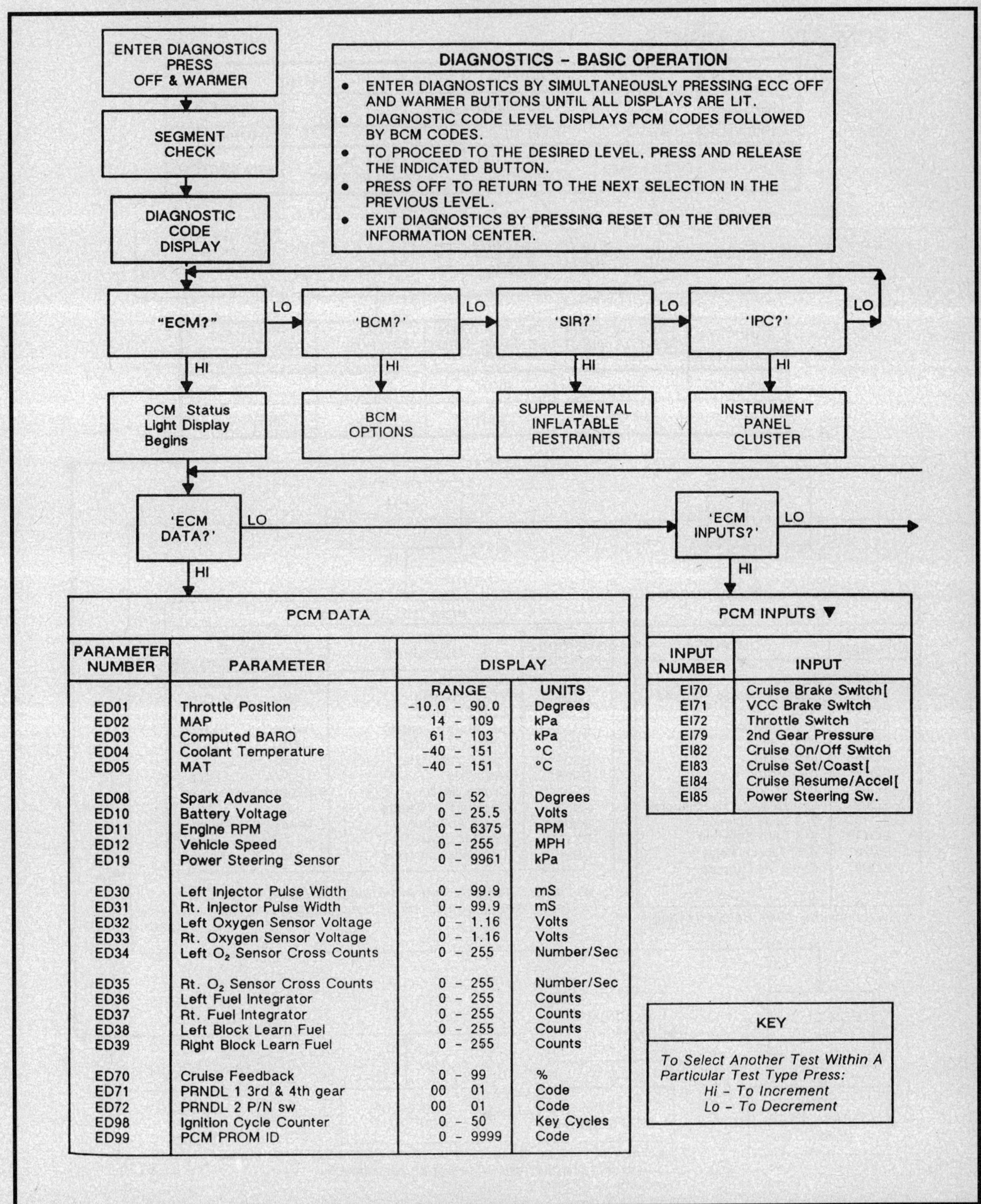

PCM DATA			
PARAMETER NUMBER	PARAMETER	DISPLAY	
		RANGE	UNITS
ED01	Throttle Position	-10.0 – 90.0	Degrees
ED02	MAP	14 – 109	kPa
ED03	Computed BARO	61 – 103	kPa
ED04	Coolant Temperature	-40 – 151	°C
ED05	MAT	-40 – 151	°C
ED08	Spark Advance	0 – 52	Degrees
ED10	Battery Voltage	0 – 25.5	Volts
ED11	Engine RPM	0 – 6375	RPM
ED12	Vehicle Speed	0 – 255	MPH
ED19	Power Steering Sensor	0 – 9961	kPa
ED30	Left Injector Pulse Width	0 – 99.9	mS
ED31	Rt. Injector Pulse Width	0 – 99.9	mS
ED32	Left Oxygen Sensor Voltage	0 – 1.16	Volts
ED33	Rt. Oxygen Sensor Voltage	0 – 1.16	Volts
ED34	Left O_2 Sensor Cross Counts	0 – 255	Number/Sec
ED35	Rt. O_2 Sensor Cross Counts	0 – 255	Number/Sec
ED36	Left Fuel Integrator	0 – 255	Counts
ED37	Rt. Fuel Integrator	0 – 255	Counts
ED38	Left Block Learn Fuel	0 – 255	Counts
ED39	Right Block Learn Fuel	0 – 255	Counts
ED70	Cruise Feedback	0 – 99	%
ED71	PRNDL 1 3rd & 4th gear	00 01	Code
ED72	PRNDL 2 P/N sw	00 01	Code
ED98	Ignition Cycle Counter	0 – 50	Key Cycles
ED99	PCM PROM ID	0 – 9999	Code

PCM INPUTS ▼	
INPUT NUMBER	INPUT
EI70	Cruise Brake Switch[
EI71	VCC Brake Switch
EI72	Throttle Switch
EI79	2nd Gear Pressure
EI82	Cruise On/Off Switch
EI83	Cruise Set/Coast [
EI84	Cruise Resume/Accel [
EI85	Power Steering Sw.

KEY
To Select Another Test Within A Particular Test Type Press: *Hi – To Increment*   *Lo – To Decrement*

SELF-DIAGNOSTIC SYSTEMS
CADILLAC

DIAGNOSTIC FLOW CHART — 1992–93 ALLANTE, 1992 ELDORADO AND SEVILLE

PCM STATUS LIGHTS

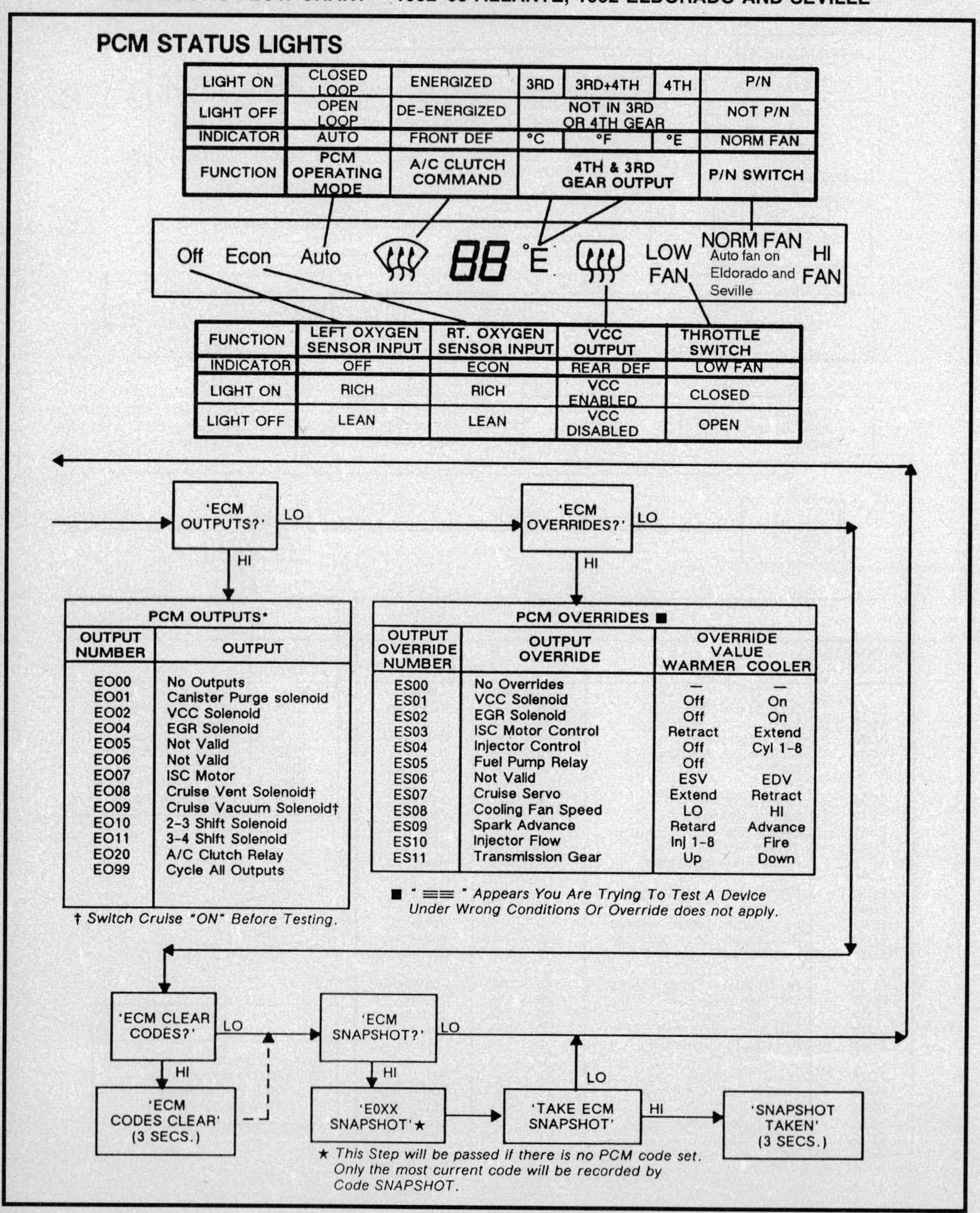

PCM STATUS LIGHTS — 1993–94 ELDORADO AND SEVILLE WITH 4.9L ENGINE

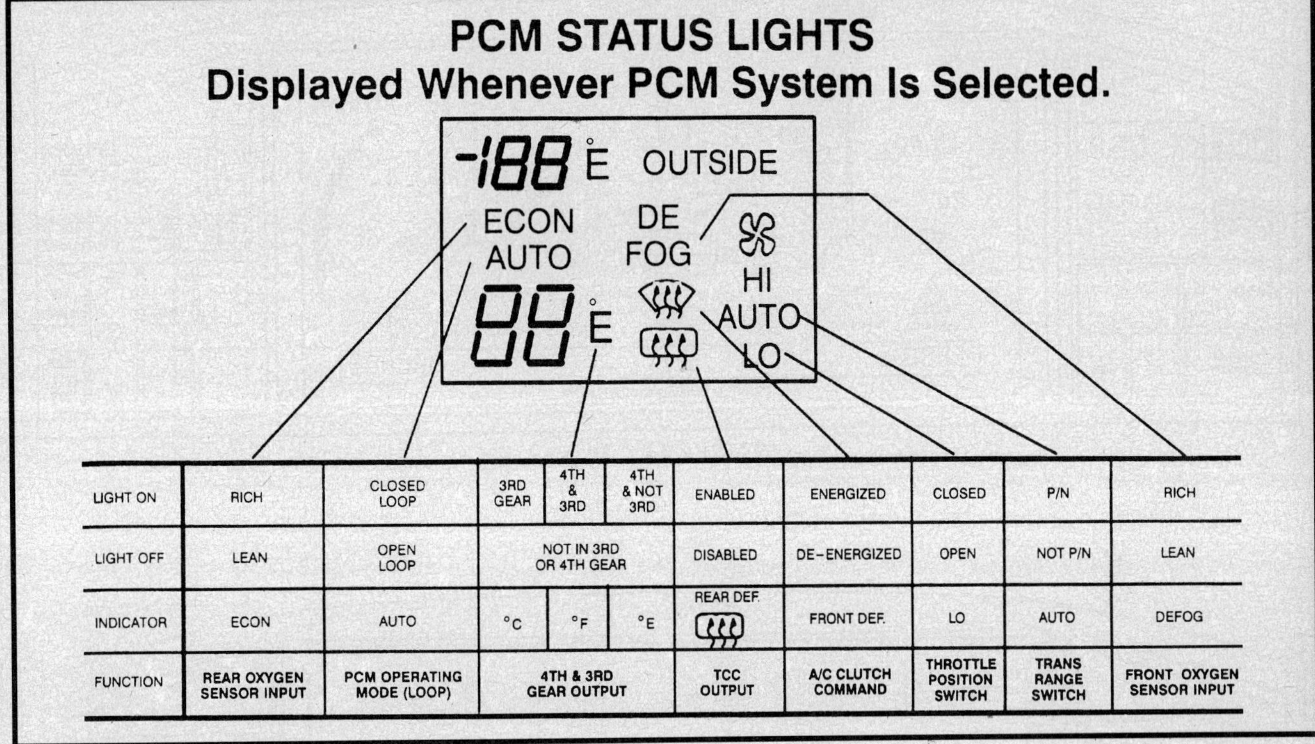

PCM STATUS LIGHTS
Displayed Whenever PCM System Is Selected.

LIGHT ON	RICH	CLOSED LOOP	3RD GEAR	4TH & 3RD	4TH & NOT 3RD	ENABLED	ENERGIZED	CLOSED	P/N	RICH
LIGHT OFF	LEAN	OPEN LOOP	NOT IN 3RD OR 4TH GEAR			DISABLED	DE–ENERGIZED	OPEN	NOT P/N	LEAN
INDICATOR	ECON	AUTO	°C	°F	°E	REAR DEF.	FRONT DEF.	LO	AUTO	DEFOG
FUNCTION	REAR OXYGEN SENSOR INPUT	PCM OPERATING MODE (LOOP)	4TH & 3RD GEAR OUTPUT			TCC OUTPUT	A/C CLUTCH COMMAND	THROTTLE POSITION SWITCH	TRANS RANGE SWITCH	FRONT OXYGEN SENSOR INPUT

- DTC **E24** — A/Q — Viscous converter clutch circuit signal {cruise control, viscous converter clutch}
- DTC **E26** — A — Shorted throttle signal switch circuit {Exhaust Gas Recirculation (EGR) system}
- DTC **E27** — A — Open throttle signal switch circuit {Exhaust Gas Recirculation (EGR) system}
- DTC **E28** — A — Shorted 3rd or 4th gear switch
- DTC **E30** — A — Idle speed control rpm out of range
- DTC **E31** — A — Shorted Manifold Absolute Pressure (MAP) sensor signal
- DTC **E32** — A — Open Manifold Absolute Pressure (MAP) sensor signal
- DTC **E34** — A — Manifold Absolute Pressure (MAP) sensor signal out of range
- DTC **E37** — A — Shorted manifold air temperature sensor signal
- DTC **E38** — A — Open manifold air temperature sensor signal
- DTC **E39** — A/F — Viscous converter clutch engagement problem {viscous converter clutch}
- DTC **E40** — A — Open power steering pressure switch signal
- DTC **E41** — A — Cam sensor signal problem
- DTC **E44** — A/P — Oxygen (O₂) sensor signal lean
- DTC **E45** — A/P — Oxygen (O₂) sensor signal rich
- DTC **E47** — A — BCM/PCM data problem
- DTC **E48** — A — Exhaust Gas Recirculation (EGR) problem
- DTC **E49** — A — Air control problem
- DTC **E52** — C — PCM memory test
- DTC **E53** — C — Distributor signal interrupt
- DTC **E55** — C — Closed throttle angle out of range
- DTC **E58** — C — Pass-key® control problem

- DTC **E60** — C — Cruise-transaxle not in drive {cruise control}
- DTC **E61** — C — Cruise vent solenoid problem {cruise control}
- DTC **E62** — C — Cruise vacuum solenoid problem {cruise control}
- DTC **E63** — C — Cruise setting vs. vehicle speed too high {cruise control}
- DTC **E64** — C — Vehicle acceleration too high {cruise control}
- DTC **E65** — C — Cruise servo position sensor signal problem {cruise control}
- DTC **E66** — C — Cruise-engine rpm too high {cruise control}
- DTC **E67** — C — Cruise control set/coast or resume/accel switch shorted {cruise control}
- DTC **E68** — C/Q — Cruise control command problem {cruise control}
- DTC **E70** — C — Intermittent throttle position signal
- DTC **E71** — C — Intermittent Manifold Absolute Pressure (MAP) sensor signal
- DTC **E73** — C — Intermittent Engine Coolant Temperature (ECT) sensor signal
- DTC **E74** — C — Intermittent manifold air temperature sensor signal
- DTC **E75** — C — Intermittent vehicle speed sensor signal
- DTC **E80** — A — Fuel system rich
- DTC **E85** — A — Idle throttle angle too high
- DTC **E90** — B — Viscous converter clutch brake switch signal problem {cruise control}
- DTC **E91** — B — PRNDL switch problem {cruise control}
- DTC **E92** — B — Heated windshield request problem

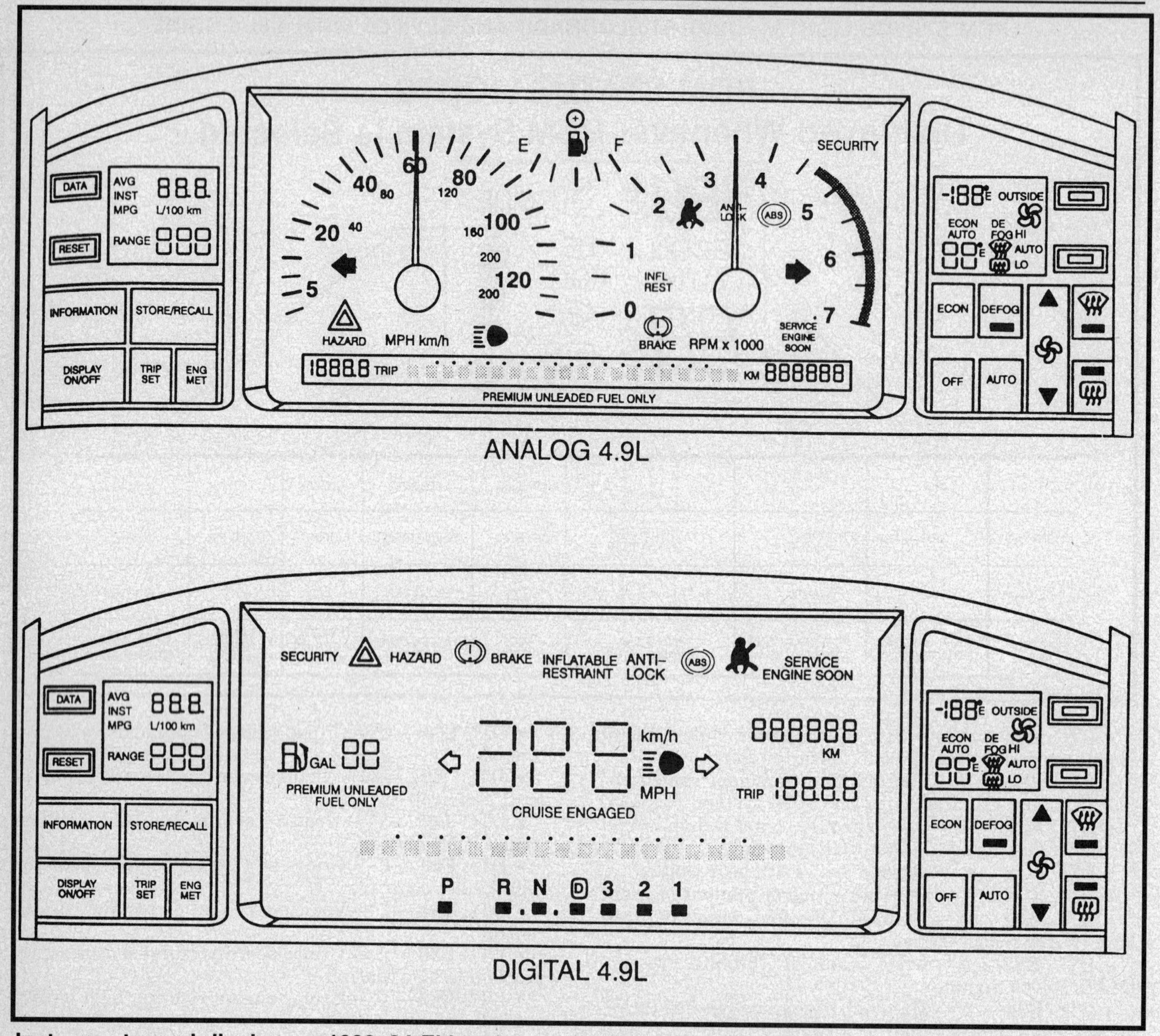

ANALOG 4.9L

DIGITAL 4.9L

Instrument panel displays — 1993–94 Eldorado and Seville with 4.9L engine

- DTC **E96** — A — Torque converter overstress
- DTC **E97** — C — P/N-D/R at high throttle angle
- DTC **E98** — C — High rpm P/N-D/R in idle speed control range
- DTC **E99** — C — Cruise servo apply problem {cruise control}

BCM

- DTC **F10** — Outside air temperature sensor problem
- DTC **F11** — A/C high side temperature sensor circuit problem
- DTC **F12** — A/C low side temperature sensor circuit problem
- DTC **F13** — In-vehicle temperature sensor circuit problem
- DTC **F15** — Sunload sensor circuit failure
- DTC **F30** — Display panels-to-BCM data problem
- DTC **F31** — Display panels-to-BCM data problem
- DTC **F32** — PCM-BCM data problem
- DTC **F40** — Air mix door problem
- DTC **F46** — Low refrigerant charge
- DTC **F47** — Low refrigerant charge
- DTC **F48** — Refrigerant system problem
- DTC **F49** — High A/C temperature clutch disengage
- DTC **F51** — BCM prom error indicator

Allante, Eldorado and Seville

PCM

A — SERVICE ENGINE SOON light ON
B — SERVICE VEHICLE SOON light ON
C — No telltale or message
D — THEFT-SYSTEM PROBLEM-CAR MAY NOT RESTART message

DIAGNOSTIC CHART — 1993–94 ELDORADO AND SEVILLE WITH 4.9L ENGINE

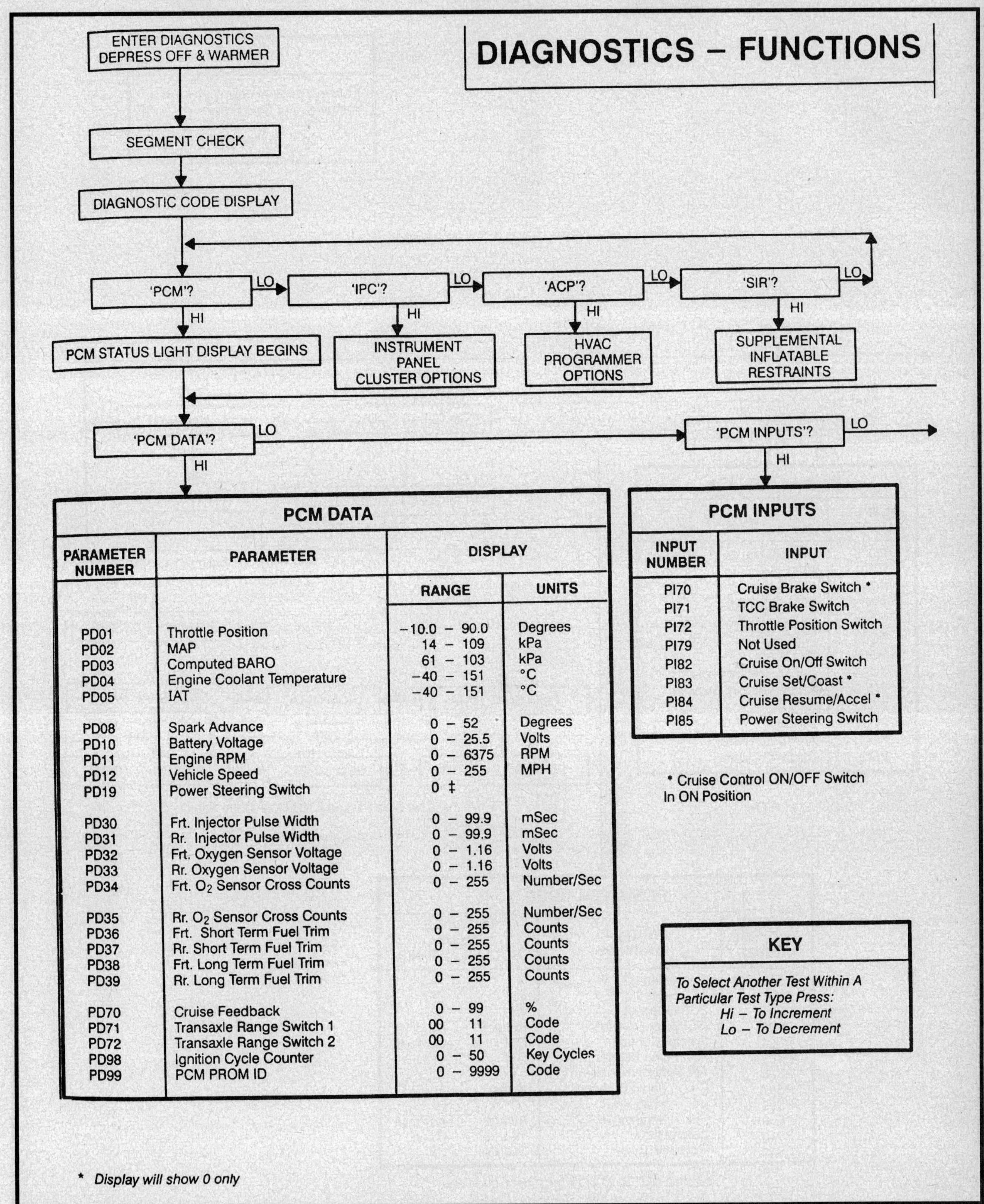

PCM DATA			
PARAMETER NUMBER	PARAMETER	DISPLAY	
		RANGE	UNITS
PD01	Throttle Position	−10.0 − 90.0	Degrees
PD02	MAP	14 − 109	kPa
PD03	Computed BARO	61 − 103	kPa
PD04	Engine Coolant Temperature	−40 − 151	°C
PD05	IAT	−40 − 151	°C
PD08	Spark Advance	0 − 52	Degrees
PD10	Battery Voltage	0 − 25.5	Volts
PD11	Engine RPM	0 − 6375	RPM
PD12	Vehicle Speed	0 − 255	MPH
PD19	Power Steering Switch	0 ‡	
PD30	Frt. Injector Pulse Width	0 − 99.9	mSec
PD31	Rr. Injector Pulse Width	0 − 99.9	mSec
PD32	Frt. Oxygen Sensor Voltage	0 − 1.16	Volts
PD33	Rr. Oxygen Sensor Voltage	0 − 1.16	Volts
PD34	Frt. O_2 Sensor Cross Counts	0 − 255	Number/Sec
PD35	Rr. O_2 Sensor Cross Counts	0 − 255	Number/Sec
PD36	Frt. Short Term Fuel Trim	0 − 255	Counts
PD37	Rr. Short Term Fuel Trim	0 − 255	Counts
PD38	Frt. Long Term Fuel Trim	0 − 255	Counts
PD39	Rr. Long Term Fuel Trim	0 − 255	Counts
PD70	Cruise Feedback	0 − 99	%
PD71	Transaxle Range Switch 1	00 11	Code
PD72	Transaxle Range Switch 2	00 11	Code
PD98	Ignition Cycle Counter	0 − 50	Key Cycles
PD99	PCM PROM ID	0 − 9999	Code

PCM INPUTS	
INPUT NUMBER	INPUT
PI70	Cruise Brake Switch *
PI71	TCC Brake Switch
PI72	Throttle Position Switch
PI79	Not Used
PI82	Cruise On/Off Switch
PI83	Cruise Set/Coast *
PI84	Cruise Resume/Accel *
PI85	Power Steering Switch

* Cruise Control ON/OFF Switch In ON Position

KEY
To Select Another Test Within A Particular Test Type Press: Hi – To Increment Lo – To Decrement

*   Display will show 0 only

DIAGNOSTIC CHART — 1993–94 ELDORADO AND SEVILLE WITH 4.9L ENGINE

KEY

To Select Another Test Within A Particular Test Type Press:
Hi − To Increment
Lo − To Decrement

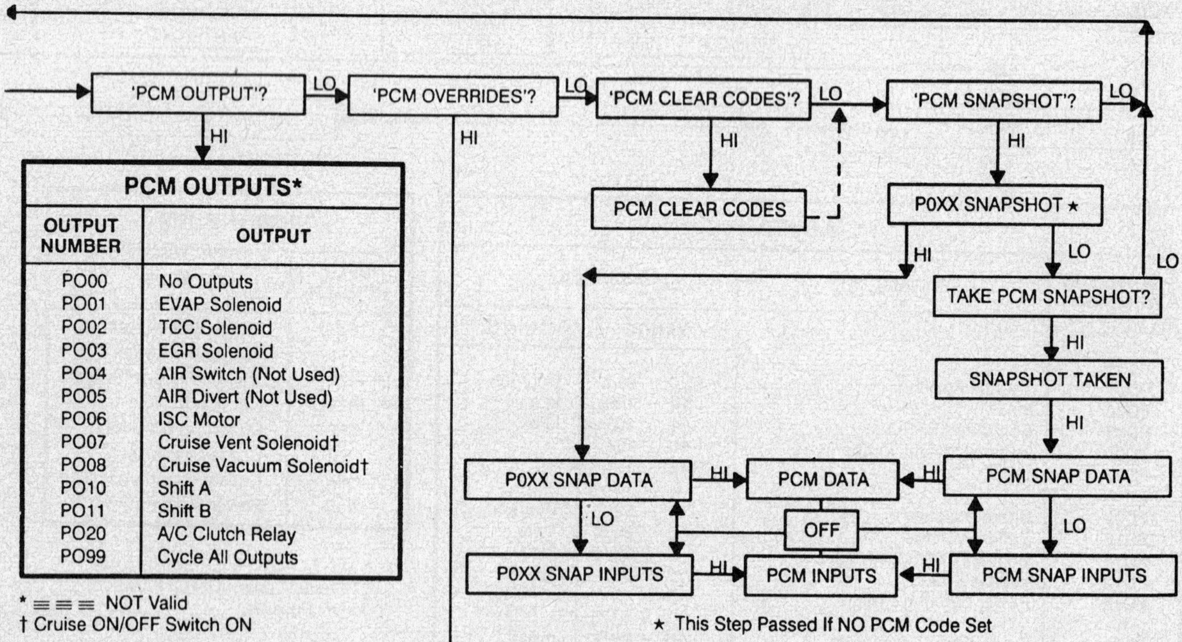

Flowchart: 'PCM OUTPUT'? → (LO) → 'PCM OVERRIDES'? → (LO) → 'PCM CLEAR CODES'? → (LO) → 'PCM SNAPSHOT'? (with LO loop back to top)

'PCM OUTPUT'? — HI →

PCM OUTPUTS*

OUTPUT NUMBER	OUTPUT
PO00	No Outputs
PO01	EVAP Solenoid
PO02	TCC Solenoid
PO03	EGR Solenoid
PO04	AIR Switch (Not Used)
PO05	AIR Divert (Not Used)
PO06	ISC Motor
PO07	Cruise Vent Solenoid†
PO08	Cruise Vacuum Solenoid†
PO10	Shift A
PO11	Shift B
PO20	A/C Clutch Relay
PO99	Cycle All Outputs

* ≡≡≡ NOT Valid
† Cruise ON/OFF Switch ON

'PCM CLEAR CODES'? — HI → PCM CLEAR CODES

'PCM SNAPSHOT'? — HI → P0XX SNAPSHOT ★

P0XX SNAPSHOT ★ — HI / LO → TAKE PCM SNAPSHOT? — HI → SNAPSHOT TAKEN — HI → PCM SNAP DATA

P0XX SNAP DATA — (HI) → PCM DATA ← (HI) — PCM SNAP DATA
P0XX SNAP DATA — LO → P0XX SNAP INPUTS — (HI) → PCM INPUTS ← (HI) — PCM SNAP INPUTS
OFF

★ This Step Passed If NO PCM Code Set

PCM OVERRIDES

OUTPUT OVERRIDE NUMBER	OUTPUT OVERRIDE	OVERRIDE VALUE COOLER	WARMER
PS00	No Overrides	—	
PS01	TCC Solenoid	ON	OFF
PS02	EGR Solenoid	OFF	ON
PS03	ISC Motor Control*	Retract	Exnd
PS04	Injector Control*	OFF	Cyl 1 − 8
PS05	Fuel Pump Relay	OFF	
PS06	AIR System (Not Used)		
PS07	Cruise Servo Command	Relax	Apply
PS08	Cooling Fan Relay	LO	HI
PS09	Spark Advance	Retard	Advance
PS10	Injector Flow*	Inj 1 − 8	Open
PS11	Transaxle Gear*	DOWN	UP

* ≡≡≡ Test NOT Valid or Wrong Test Conditions

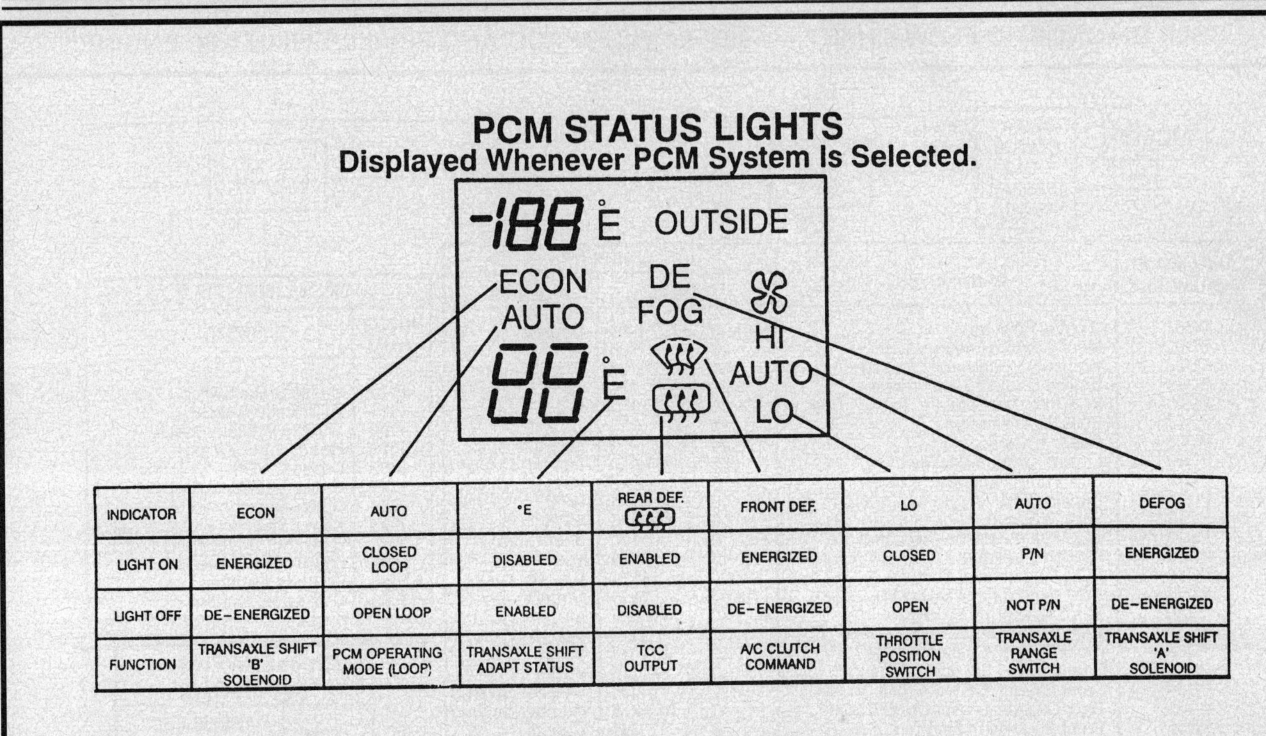

INDICATOR	ECON	AUTO	°E	REAR DEF.	FRONT DEF.	LO	AUTO	DEFOG
LIGHT ON	ENERGIZED	CLOSED LOOP	DISABLED	ENABLED	ENERGIZED	CLOSED	P/N	ENERGIZED
LIGHT OFF	DE-ENERGIZED	OPEN LOOP	ENABLED	DISABLED	DE-ENERGIZED	OPEN	NOT P/N	DE-ENERGIZED
FUNCTION	TRANSAXLE SHIFT 'B' SOLENOID	PCM OPERATING MODE (LOOP)	TRANSAXLE SHIFT ADAPT STATUS	TCC OUTPUT	A/C CLUTCH COMMAND	THROTTLE POSITION SWITCH	TRANSAXLE RANGE SWITCH	TRANSAXLE SHIFT 'A' SOLENOID

PCM status lights — 1993–94 Eldorado and Seville with 4.6L engine

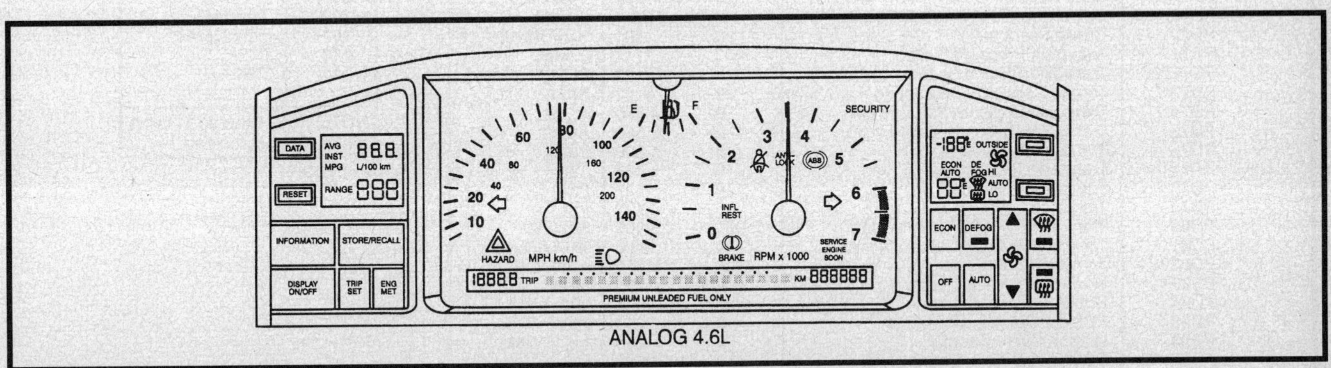

Instrument panel display — 1993–94 Eldorado and Seville with 4.6L engine

E — REDUCED ENGINE POWER message
F — BATTERY NO CHARGE message
{} — Bracketed systems are disabled when code is present
- DTC **P12** — A — No reference signal from ignition control module
- DTC **P13** — A — Rear Oxygen (O_2) sensor not ready
- DTC **P14** — A — Shorted Engine Coolant Temperature (ECT) sensor circuit signal
- DTC **P15** — A — Open Engine Coolant Temperature (ECT) sensor circuit signal
- DTC **P16** — B — Generator voltage out of range {all solenoids}
- DTC **P17** — A — Front Oxygen (O_2) sensor not ready
- DTC **P19** — B — Shorted fuel pump circuit
- DTC **P20** — B — Open fuel pump circuit
- DTC **P21** — A — Shorted Throttle Position Sensor (TPS) circuit signal {TCC, transaxle pressure control}
- DTC **P22** — A — Open Throttle Position Sensor (TPS) circuit signal {TCC, Exhaust Gas Recirculation (EGR)}
- DTC **P23** — A — Ignition control circuit problem {ignition control}
- DTC **P24** — A — Vehicle speed sensor circuit problem {cruise control}
- DTC **P25** — B — Circuit reference signal problem
- DTC **P26** — A — Shorted throttle signal switch circuit {Exhaust Gas Recirculation (EGR) system}
- DTC **P27** — A — Open throttle signal switch circuit {Exhaust Gas Recirculation (EGR) system}
- DTC **P28** — A — Transaxle pressure switch circuit problem

SELF DIAGNOSTIC FLOW CHART — 1993–94 ELDORADO AND SEVILLE WITH 4.6L ENGINE

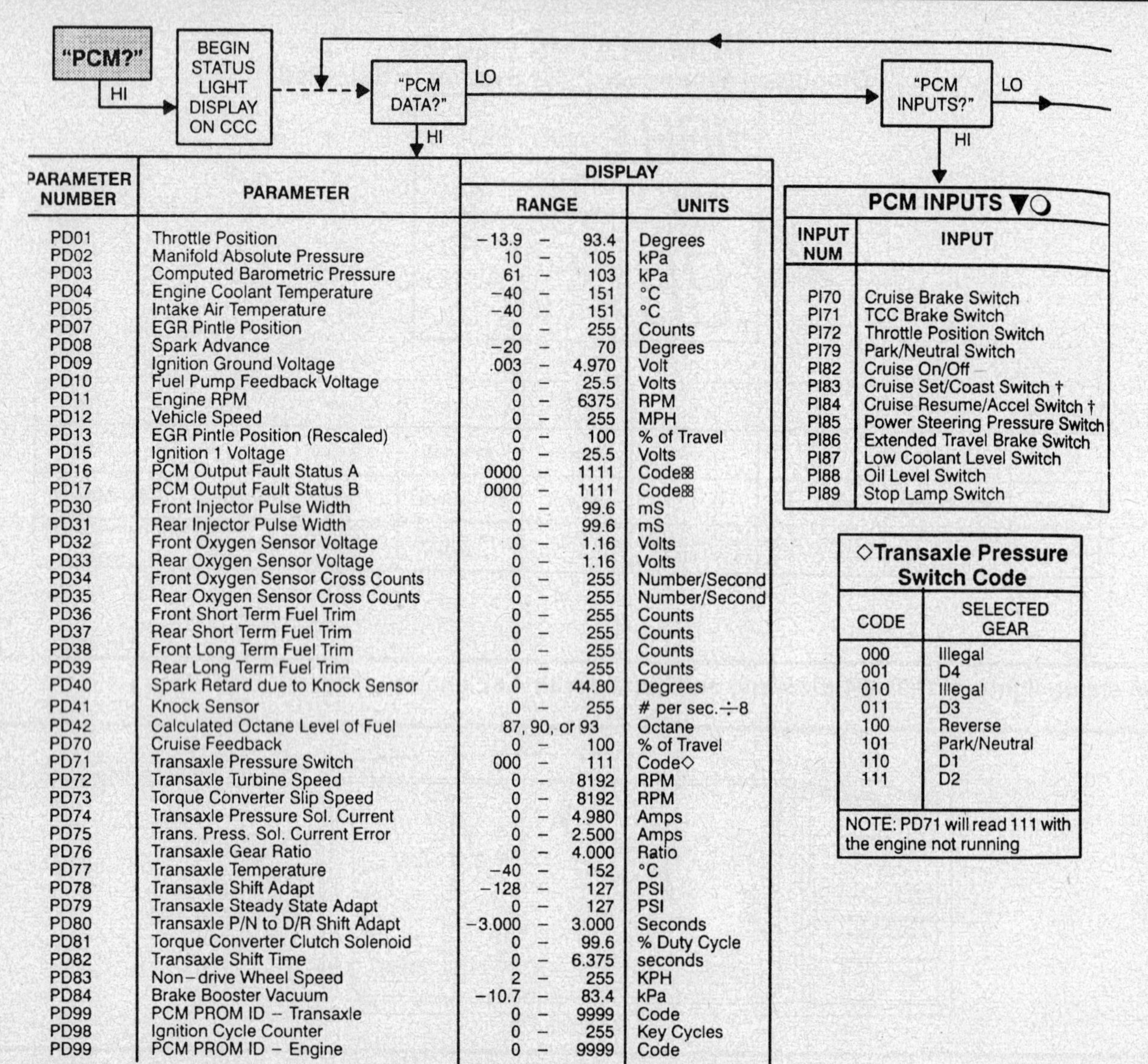

PARAMETER NUMBER	PARAMETER	DISPLAY RANGE		UNITS
PD01	Throttle Position	−13.9	93.4	Degrees
PD02	Manifold Absolute Pressure	10	105	kPa
PD03	Computed Barometric Pressure	61	103	kPa
PD04	Engine Coolant Temperature	−40	151	°C
PD05	Intake Air Temperature	−40	151	°C
PD07	EGR Pintle Position	0	255	Counts
PD08	Spark Advance	−20	70	Degrees
PD09	Ignition Ground Voltage	.003	4.970	Volt
PD10	Fuel Pump Feedback Voltage	0	25.5	Volts
PD11	Engine RPM	0	6375	RPM
PD12	Vehicle Speed	0	255	MPH
PD13	EGR Pintle Position (Rescaled)	0	100	% of Travel
PD15	Ignition 1 Voltage	0	25.5	Volts
PD16	PCM Output Fault Status A	0000	1111	Code⊞
PD17	PCM Output Fault Status B	0000	1111	Code⊞
PD30	Front Injector Pulse Width	0	99.6	mS
PD31	Rear Injector Pulse Width	0	99.6	mS
PD32	Front Oxygen Sensor Voltage	0	1.16	Volts
PD33	Rear Oxygen Sensor Voltage	0	1.16	Volts
PD34	Front Oxygen Sensor Cross Counts	0	255	Number/Second
PD35	Rear Oxygen Sensor Cross Counts	0	255	Number/Second
PD36	Front Short Term Fuel Trim	0	255	Counts
PD37	Rear Short Term Fuel Trim	0	255	Counts
PD38	Front Long Term Fuel Trim	0	255	Counts
PD39	Rear Long Term Fuel Trim	0	255	Counts
PD40	Spark Retard due to Knock Sensor	0	44.80	Degrees
PD41	Knock Sensor	0	255	# per sec.÷8
PD42	Calculated Octane Level of Fuel	87, 90, or 93		Octane
PD70	Cruise Feedback	0	100	% of Travel
PD71	Transaxle Pressure Switch	000	111	Code◇
PD72	Transaxle Turbine Speed	0	8192	RPM
PD73	Torque Converter Slip Speed	0	8192	RPM
PD74	Transaxle Pressure Sol. Current	0	4.980	Amps
PD75	Trans. Press. Sol. Current Error	0	2.500	Amps
PD76	Transaxle Gear Ratio	0	4.000	Ratio
PD77	Transaxle Temperature	−40	152	°C
PD78	Transaxle Shift Adapt	−128	127	PSI
PD79	Transaxle Steady State Adapt	0	127	PSI
PD80	Transaxle P/N to D/R Shift Adapt	−3.000	3.000	Seconds
PD81	Torque Converter Clutch Solenoid	0	99.6	% Duty Cycle
PD82	Transaxle Shift Time	0	6.375	seconds
PD83	Non−drive Wheel Speed	2	255	KPH
PD84	Brake Booster Vacuum	−10.7	83.4	kPa
PD99	PCM PROM ID — Transaxle	0	9999	Code
PD98	Ignition Cycle Counter	0	255	Key Cycles
PD99	PCM PROM ID − Engine	0	9999	Code

PCM INPUTS ▼○

INPUT NUM	INPUT
PI70	Cruise Brake Switch
PI71	TCC Brake Switch
PI72	Throttle Position Switch
PI79	Park/Neutral Switch
PI82	Cruise On/Off
PI83	Cruise Set/Coast Switch †
PI84	Cruise Resume/Accel Switch †
PI85	Power Steering Pressure Switch
PI86	Extended Travel Brake Switch
PI87	Low Coolant Level Switch
PI88	Oil Level Switch
PI89	Stop Lamp Switch

◇Transaxle Pressure Switch Code

CODE	SELECTED GEAR
000	Illegal
001	D4
010	Illegal
011	D3
100	Reverse
101	Park/Neutral
110	D1
111	D2

NOTE: PD71 will read 111 with the engine not running

⊞ PCM Output Fault Status

PARAM.	STATUS A	STATUS B
1st Digit	TCC Sol/Engine Temp Lite	Shift 'A'/Shift 'B'
2nd Digit	EGR/EVAP/Starter Inhibit	Injectors
3rd Digit	Cooling Fans/Service Engine Soon MIL	Generator
4th Digit	A/C Clutch/Delivered Torque/RSS Lift/Dive	Not Used

KEY

To Select Another Test Within A Particular Test Type Press:
Hi − To Increment
Lo − To Decrement

▼ "HI" = High Signal Voltage
"LO" = Low Signal Voltage

○ "O" = Input Same Since Displayed
"X" = Input Changed Since Displayed

† Switch Cruise ON Before Testing

SELF DIAGNOSTIC FLOW CHART — 1993–94 ELDORADO AND SEVILLE WITH 4.6L ENGINE

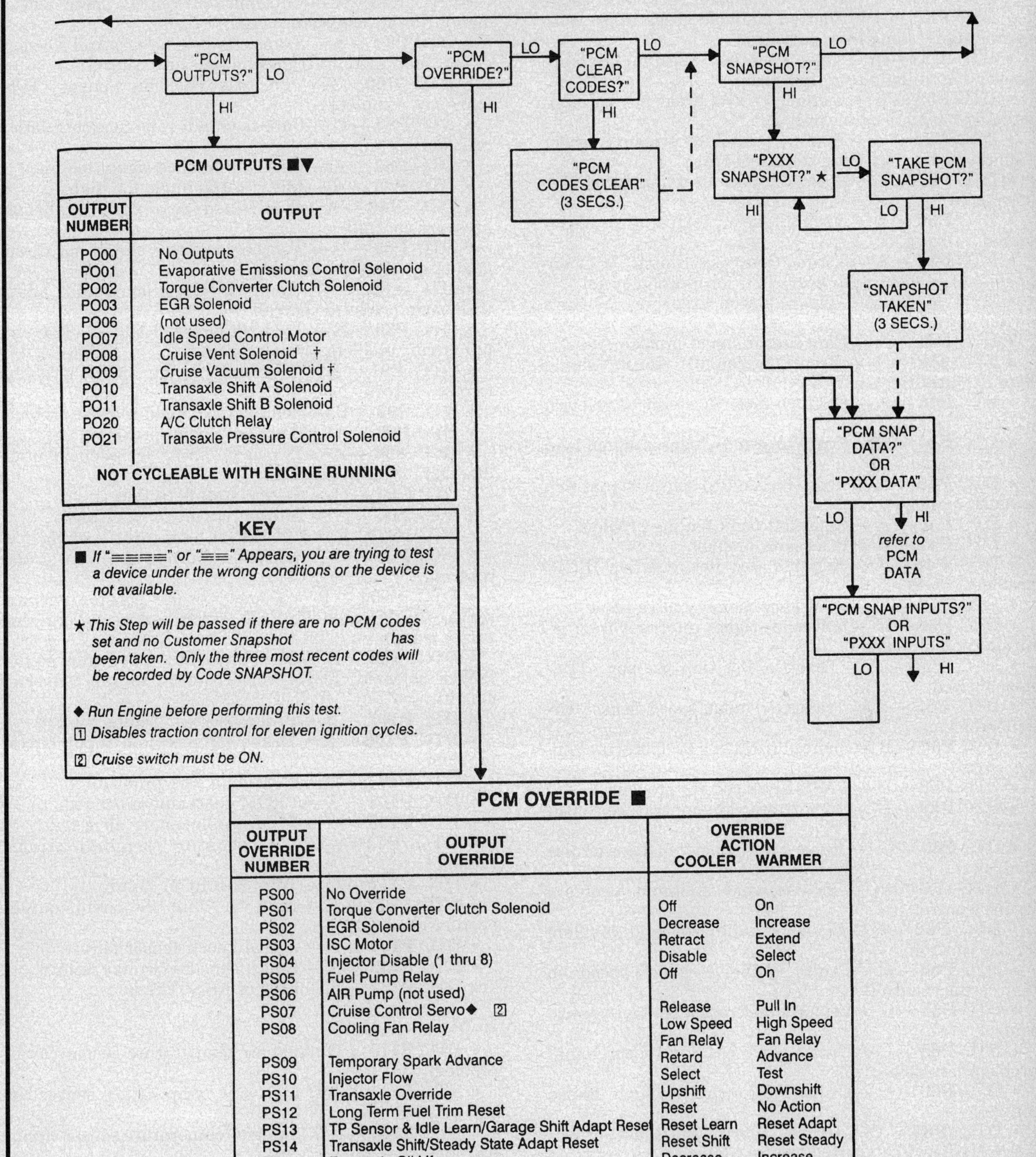

OUTPUT OVERRIDE NUMBER	OUTPUT OVERRIDE	OVERRIDE ACTION COOLER	WARMER
PS00	No Override		
PS01	Torque Converter Clutch Solenoid	Off	On
PS02	EGR Solenoid	Decrease	Increase
PS03	ISC Motor	Retract	Extend
PS04	Injector Disable (1 thru 8)	Disable	Select
PS05	Fuel Pump Relay	Off	On
PS06	AIR Pump (not used)		
PS07	Cruise Control Servo ◆ ②	Release	Pull In
PS08	Cooling Fan Relay	Low Speed Fan Relay	High Speed Fan Relay
PS09	Temporary Spark Advance	Retard	Advance
PS10	Injector Flow	Select	Test
PS11	Transaxle Override	Upshift	Downshift
PS12	Long Term Fuel Trim Reset	Reset	No Action
PS13	TP Sensor & Idle Learn/Garage Shift Adapt Reset	Reset Learn	Reset Adapt
PS14	Transaxle Shift/Steady State Adapt Reset	Reset Shift	Reset Steady
PS15	Transaxle Oil Life	Decrease	Increase
PS20	Transaxle Pressure Control	Decrease	Increase
PS21	Knock Sensor Test	Advance	No Action
PS22	Permanent Spark Retard	Retard	Advance
PS23	Generator	Off	On
PS24	Traction Control Disable ①	Enable	Disable

- DTC **P29** — A — Transaxle shift "B" solenoid problem {1st, 3rd and 4th gear}
- DTC **P30** — A — Idle speed control rpm out of range
- DTC **P31** — A — Shorted manifold absolute pressure sensor signal {long term fuel trim}
- DTC **P32** — A — Open manifold absolute pressure sensor signal {long term fuel trim}
- DTC **P33** — B — Extended travel brake switch input circuit problem {cruise control}
- DTC **P34** — A — Manifold absolute pressure sensor signal out of range {long term fuel trim}
- DTC **P37** — A — Shorted intake air temperature sensor signal
- DTC **P38** — A — Open intake air temperature sensor signal
- DTC **P39** — A — Torque Converter Clutch (TCC) engagement problem {4th gear, TCC for ignition cycle}
- DTC **P40** — A — Open power steering pressure switch open
- DTC **P41** — A — Cam sensor signal problem
- DTC **P42** — A — Front Oxygen (O_2) sensor signal lean exhaust signal
- DTC **P43** — A — Front Oxygen (O_2) sensor signal rich exhaust signal
- DTC **P44** — A — Rear Oxygen (O_2) sensor signal lean exhaust signal
- DTC **P45** — A — Rear Oxygen (O_2) sensor signal rich exhaust signal
- DTC **P46** — A — Right-left bank fueling problem
- DTC **P47** — B — PCM data problem
- DTC **P48** — A — Exhaust Gas Recirculation (EGR) problem {EGR}
- DTC **P52** — C — PCM keep memory alive reset
- DTC **P53** — C — Reference signal interrupt from ignition control module
- DTC **P55** — A — Throttle Position Sensor (TPS) misadjusted
- DTC **P56** — A — Transaxle input speed sensor circuit problem
- DTC **P57** — B — Shorted transaxle temperature sensor circuit
- DTC **P58** — D — PASS-Key® control problem
- DTC **P59** — B — Open transaxle temperature sensor circuit
- DTC **P60** — C — Cruise-transaxle not in drive {cruise control}
- DTC **P61** — C — Cruise vent solenoid problem {cruise control}
- DTC **P62** — C — Cruise vacuum solenoid problem {cruise control}
- DTC **P63** — C — Cruise setting vs. vehicle speed too high {cruise control}
- DTC **P64** — C — Vehicle acceleration too high {cruise control}
- DTC **P65** — C — Cruise servo position sensor signal problem {cruise control}
- DTC **P66** — C — Cruise-engine rpm too high {cruise control}
- DTC **P67** — C — Cruise control set/coast or resume/Accel switch shorted {cruise control}
- DTC **P68** — C — Cruise control servo position out of range {cruise control}
- DTC **P70** — C — Intermittent throttle position signal
- DTC **P71** — C — Intermittent Manifold Absolute Pressure (MAP) sensor signal
- DTC **P73** — C — Intermittent Engine Coolant Temperature (ECT) sensor signal

- DTC **P74** — C — Intermittent intake air temperature sensor signal
- DTC **P75** — C — Intermittent vehicle speed sensor signal {torque converter clutch}
- DTC **P76** — A — Transaxle pressure control solenoid circuit malfunction {transaxle pressure control}
- DTC **P80** — A — Throttle Position Sensor (TPS) learn not complete
- DTC **P81** — C — Cam-to-circuit reference correlation problem
- DTC **P83** — A — Circuit reference signal too high
- DTC **P85** — A — Idle throttle angle too high
- DTC **P86** — A — Undefined gear ratio {transaxle pressure control}
- DTC **P88** — A — Torque converter clutch not disengaging {transaxle adapts}
- DTC **P89** — A — Long shift and maximum adapt {transaxle pressure control}
- DTC **P90** — E — Torque Converter Clutch (TCC) input circuit problem {cruise control}
- DTC **P91** — B — Transaxle range switch problem {cruise control}
- DTC **P92** — B — Heated windshield request problem
- DTC **P93** — A — Traction control problem
- DTC **P94** — A — Transaxle shift "A" solenoid problem {1st, 3rd, 4th gear}
- DTC **P95** — C — Engine stall detection
- DTC **P96** — A — Torque converter overstress
- DTC **P97** — E — P/N-D/R at high throttle angle
- DTC **P99** — E — Cruise servo apply problem {cruise control}
- DTC **P102** — B — Shorted brake booster vacuum sensor {EVAP solenoid, EGR, cruise control, torque converter clutch}
- DTC **P103** — B — Open brake booster vacuum sensor {EVAP solenoid, EGR, cruise control, torque converter clutch}
- DTC **P105** — B — Brake booster vacuum too low
- DTC **P106** — E — Brake lamp switch input circuit problem
- DTC **P107** — C — PCM data link problem
- DTC **P108** — A — PROM checksum mismatch
- DTC **P109** — C — PCM keep memory alive reset
- DTC **P110** — F — Alternator terminal circuit problem
- DTC **P112** — C — Total EEPROM failure
- DTC **P117** — C — Shift "A"/Shift "B" circuit output open or shorted
- DTC **P131** — A — Active knock sensor failure
- DTC **P132** — A — Knock sensor circuitry failure
- DTC **P137** — E — Loss of ABS/TCS data

BCM

- DTC **B110** — Outside air temperature sensor circuit problem
- DTC **B111** — A/C high side temperature sensor circuit problem
- DTC **B112** — A/C low side temperature sensor circuit problem
- DTC **B113** — In-vehicle temperature sensor circuit problem
- DTC **B115** — Sunload sensor circuit failure
- DTC **B119** — Twilight photocell circuit problem
- DTC **B120** — Twilight delay switch pot circuit problem
- DTC **B121** — Twilight enable switch circuit problem

- DTC **B122** — Twilight delay & panel dimming problem
- DTC **B123** — Courtesy lamps panel switch circuit problem
- DTC **B333** — Loss of Supplemental Inflatable Restraint (SIR) Diagnostic Energy Reserve Module (DERM) data
- DTC **B334** — Loss of PCM data
- DTC **B335** — Loss of Climate Control/Driver Information Center (CCDIC) data
- DTC **B336** — Loss of Instrument Panel Cluster (IPC) data
- DTC **B337** — Loss of Heating, Ventilation and Air Conditioning (HVAC) programmer data
- DTC **B410** — Charging system circuit problem
- DTC **B411** — Battery voltage too low or too high
- DTC **B412** — Battery voltage too low or too high
- DTC **B420** — Relay circuits problem
- DTC **B440** — Heating, Ventilation and Air Conditioning (HVAC)air mix door circuit problem
- DTC **B446** — Low refrigerant charge
- DTC **B447** — Low refrigerant charge
- DTC **B448** — Low refrigerant problem
- DTC **B450** — Heating, Ventilation and Air Conditioning (HVAC) coolant temperature too high
- DTC **B552** — Keep alive memory error
- DTC **B556** — Odometer prom error

RIVIERA, TORONADO AND TROFEO

General Description

ELECTRONIC CONTROL MODULE

The ECM is the control center of the fuel injection system. It continuously monitors information from a variety of sensors and switches, and controls the systems that affect vehicle performance. The ECM also has self-diagnostic capability, and can detect system problems, alert the driver through the Malfunction Indicator Lamp (MIL) and record the appropriate Diagnostic Trouble Code (DTC). The numerical code will identify the specific circuit to aid the technician in diagnosis. The ECM is capable of storing more than one code, with possibilities ranging from 12 to 999, however not all systems use every code. In addition, similar code numbers can relate to different circuits depending on the vehicle.

Learning Ability

The ECM can compensate for minor variations within the fuel system through the long term fuel trim (block learn) and short term fuel trim (fuel integrator) systems. The short term fuel trim monitors the oxygen sensor output voltage, adding or subtracting fuel to drive the mixture rich or lean as needed to reach the ideal air fuel ratio of 14.7:1. The integrator values may be read with a scan tool; the display will range from 0–255 and should center on 128 if the oxygen sensor is seeing a 14.7:1 mixture.

Powertrain Control Module (PCM)

The PCM is used when vehicles are equipped with electronically controlled transmissions. Its function is identical to that of the ECM with the added features of controlling shift points in the transmission, as well as cruise control operation and diagnostics. This unit may be referred to as the PCM, the ECM/PCM or the PCM/TCM (transmission control module). The integrated functions of engine and transmission control allow accurate gear selection and improved fuel economy.

For engine diagnostics, the PCM may be considered identical to an ECM system, although the combined unit will display additional codes relating to transmission function and components.

> **NOTE: The terms Electronic Control Module (ECM), Powertrain Control Module (PCM) and Body Control Module (BCM) may sometimes be used interchangably.**

Quad Drivers (QDM)

Some control modules have eliminated the use of individual transistors to control each output circuit. A single QDM can individually control 4 outputs from the computer. Since these control modules are internally fault

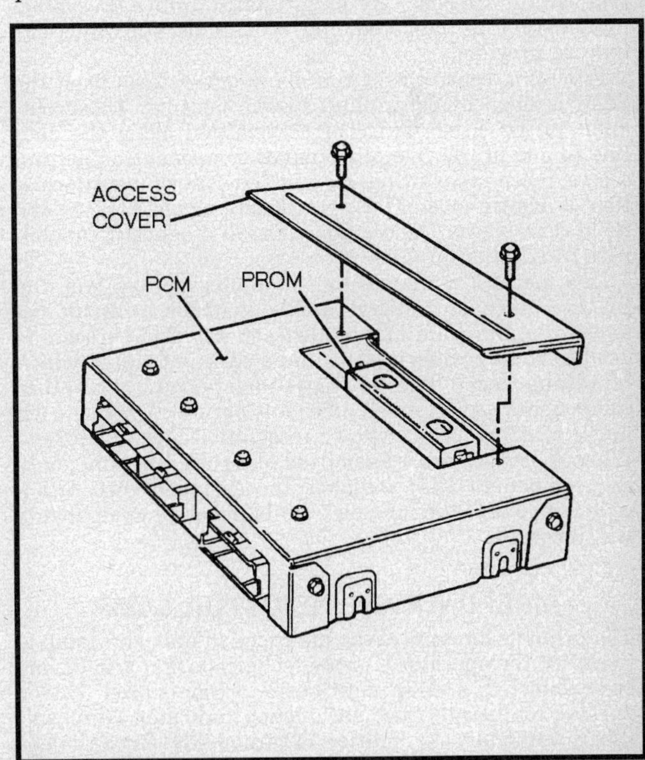

Powertrain Control Module (PCM) — Riviera shown

protected, a single faulty circuit or component may cause the other circuits on that QDM to be ON or OFF all the time. Do not confuse this with a failed ECM; if the Quad Driver itself has failed, the control module output will usually test either open or shorted.

The quad driver check procedure should be used before replacing an expensive ECM/PCM assembly. The check will not test all computer functions but will determine if an external circuit has disabled a QDM. Be particularly suspicious of an ECM/PCM exhibiting any of these characteristics:

- Malfunction indicator lamp ON with no codes stored.
- Malfunction indicator light flickers, intermittent or dim.
- ECM/PCM will not flash Code 12 and/or engine will not start.
- Controlled output, such as TCC, is inoperative or ON all the time.
- Engine stalls, misfires or surges.
- Scan tool erratic or inoperative.

BODY CONTROL MODULE

Body Control Module (BCM) is the communication center for the ECM/PCM, Instrument Panel Cluster (IPC), HVAC (Programmer), Electronic Climate Control Panel (ECCP), and the Supplemental Inflatable Restraint (SIR) module. A combination of inputs from these major components, as well as other sensors and switches, communicate with the BCM either as individual inputs or over the common communications link called the data line. The various inputs to the BCM combine with program instructions within the system memory to provide accurate control over the many subsystems involved. When a subsystem circuit exceeds pre-programmed limits, a system malfunction is indicated and certain backup functions may be provided.

Providing control over the many subsystems from the BCM is done by controlling system outputs. These outputs can be either direct or transmitted along the data line to one of the five other major components. The process of receiving, storing, testing and controlling information is continuous. The data communication gives the BCM control over the component's self-diagnostic capabilities in addition to its own.

In a method similar to that of a telegraph system, the BCM's internal circuitry rapidly switches a circuit between 0 and 5 volts like a telegraph key. This process is used to convert information into a series of pulses which represents coded data messages understood by the other components. Also, much like a telegraph system, each major component has its own recognition code or address. When messages are transmitted along the data line, only the component that matches the assigned recognition code will accept the message, while the other components will ignore it.

MALFUNCTION INDICATOR LAMP

The primary function of the malfunction indicator lamp is to advise the operator and the technician that a fault has been detected, and, in most cases, a code stored. Under normal conditions, the malfunction indicator lamp will illuminate when the ignition is turned **ON**. Once the engine is started and running, the ECM will perform a system check and extinguish the warning lamp if no fault is found.

Additionally, the malfunction indicator lamp can be used to retrieve stored codes after the system is placed in the Diagnostic Mode. Codes are transmitted as a series of flashes with short or long pauses. When the system is placed in the Field Service Mode, the dash lamp will indicate open loop or closed loop function to the technician.

Intermittents

If a fault occurs intermittently, such as a loose connector pin breaking contact as the vehicle hits a bump, the ECM will note the fault as it occurs and energize the malfunction indicator lamp. If the problem self-corrects, as with the terminal pin again making contact, the dash lamp will extinguish after 10 seconds but a code will remain stored in the ECM memory.

When an unexpected code appears during diagnostics, it may have been set during an intermittent failure that self-corrected; the codes are still useful in diagnosis and should not be discounted.

Failure Codes

Whenever the ECM detects a sensor or switch signal that is out of range, it will turn ON the MIL. In addition, if the ECM detects an internal malfunction, it will also turn ON the MIL. Distinguishing between current and history codes is absolutely necessary for successful diagnosis. Current codes are also known as hard failures, or a problem that exists at the present time. For example, if a vehicle came into the shop with the MIL light on steady with the engine running, a trouble code could be immediately retrieved and that particular diagnostic trouble chart could be referenced.

History codes, also known as intermittents, are failures which have taken place at one time, but are not presently causing the MIL light to turn ON. An example of this would be if the ignition timing had been recently adjusted on a 5.0 liter engine with TBI. When the bypass wire was disconnected to make the adjustment, the open circuit would have caused the ECM to revert to base timing. Because of this change, the ECM detected a problem and turned on the MIL while simultaneously recording a trouble code 42. When the timing adjustment was completed and the bypass wire was reconnected, the ECM waited approximately 10 seconds and then turned off the MIL, but retained the code. The ECM would retain this information up to a maximum of 50 ignition cycles. If code retrieval was attempted in the meantime by another technician unaware of the timing adjustment, their only recourse would be to reference the wiring diagram associated with the code 42 chart and physically inspect all connections.

NOTE: On models with a climate control diagnostic display, stored codes are displayed with either a "C" or an "H" to designate between current and history code.

DATA LINK CONNECTOR (DLC)

The data link connector is located underneath the instrument panel and is used to access the on-board microprocessors. These include the ECM, PCM, BCM as well as various other controllers such as the anti-lock brake system module. The use of a scan tool such as GM's Tech 1 will allow the technician to monitor data and control some of the microprocessor's operations. The scan tool will read the serial data signal from the DLC and translate that signal into useful diagnostic information on the display screen. The DLC can also be used without a scan tool for retrieving trouble codes, entering certain diagnostic modes and other valuable testing.

Terminal A — DLC Ground

Circuit number 450 is the ECM ground through the instrument panel connector. The ECM is grounded at the engine and splices into the instrument panel harness.

NOTE: Most scan tools can disguise problems related to ECM grounds due to the alternate ground path through the tool's power cord.

Terminal B — Diagnostic Enable

Circuit number 451 is connected to the ECM through the instrument panel connector. It is normally at a 5 volt level. The scan tool has the ability to lower the voltage level of circuit 451 by dropping a fixed resistance across the circuit or shorting it to ground. The various low voltages on circuit 451 are referred to as Diagnostic Modes.

Terminal C — Air Solenoid

On most vehicles with Air Injection Reaction (AIR) Systems, circuit 436 is spliced between the ECM and the AIR solenoid, through the instrument panel connector to terminal C. The solenoid is responsible for directing airflow from the pump to either the exhaust manifold, catalytic converter or the atmosphere via an attached silencer. This terminal is very helpful for testing the closed loop electronics. With the engine running in closed loop, connect a jumper wire between DLC terminals C and A. By grounding terminal C you will be enabling the air control solenoid to direct airflow to the exhaust manifold. This should cause a low O_2 sensor output voltage and a high (greater than 128) short term fuel trim number. If connected long enough, this action will trigger a code 44, lean exhaust.

In Open Loop operation, this solenoid will be enabled to quickly bring the O_2 sensor up to operating temperature. Once the system goes to Closed Loop the solenoid will become disabled causing airflow to be directed to either the catalytic converter or the atmosphere depending on the type of system. These actions can be verified using a digital voltmeter by probing terminal C with the engine running. In open loop the voltmeter should read 0 volts (solenoid energized) and in closed loop the reading should be battery voltage (solenoid de-energized).

Terminal D — Malfunction Indicator Lamp

Circuit 419 is spliced between the ECM and the MIL light to terminal D. With the key **ON**, the MIL light is turned ON through ECM terminal A5. If during a normal bulb check the MIL fails to turn on, a quick test can be performed before condemning the ECM. Insert a jumper between terminals D and A of the DLC with the key **ON**. This action should turn ON the MIL. If it does not, a defective bulb is the most likely problem.

Terminal E — Serial Data Line

Circuit 461 is connected through the instrument panel connector to the ECM. This is a discreet data line used on fuel injection systems with a slow data transmission rate (baud rate).

Terminal F — Torque Converter Clutch Solenoid

Circuit 422 is connected through the instrument panel connector and spliced into the ground-controlled TCC solenoid circuit. The ECM will turn on the TCC solenoid based on a variety of sensed inputs. By inserting a testlight between DLC terminals F and A, TCC command status can be observed. With the brake switch closed (brakes not applied) and the vehicle in 3rd gear (3rd gear switch closed), the test light should be ON. As soon as the ECM commands TCC engagement, the test light should turn OFF.

Terminal G — Fuel Pump Test Point

Most models have a fuel pump test connector located in the engine compartment. Applying B+ voltage through a fused jumper wire will energize the in-tank fuel pump. This is helpful when performing fuel pressure testing and is more convenient than cycling the key to turn ON the pump. On other models that use terminal G in the DLC, it is for other purposes than fuel pump. When terminal G is either vacant or used for another purpose, a fuel pump test lead will be located in the engine compartment, usually near the battery.

Terminal M — Serial Data Line

Circuit 461 is the serial data line for fast baud rate ECMs on some vehicles. Terminal M may also include serial data transmission from other on-board microprocessors such as an anti-lock brake system controller.

DIAGNOSTIC MODE	VOLTAGE AT TERMINAL B	RESISTANCE ACROSS A AND B
Road Test or Open	5V	Infinite
Diagnostic or 10K mode	2.5V	10K ohms
Back Up Fuel	1.5V	3.9K ohms
Field Service	0V	0 ohms

Diagnostic modes

SELF-DIAGNOSTIC SYSTEMS
RIVIERA, TORONADO AND TROFEO

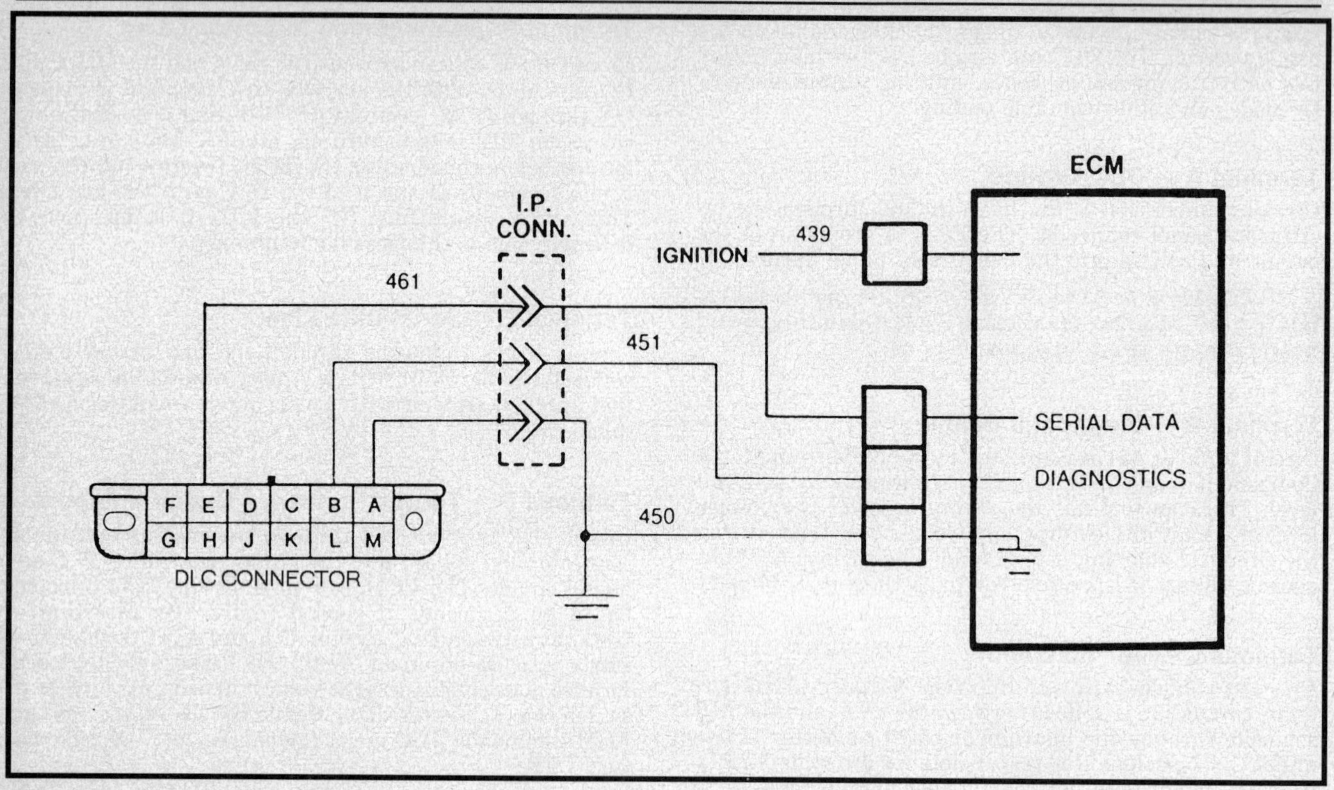

Terminals A, B and E — DLC

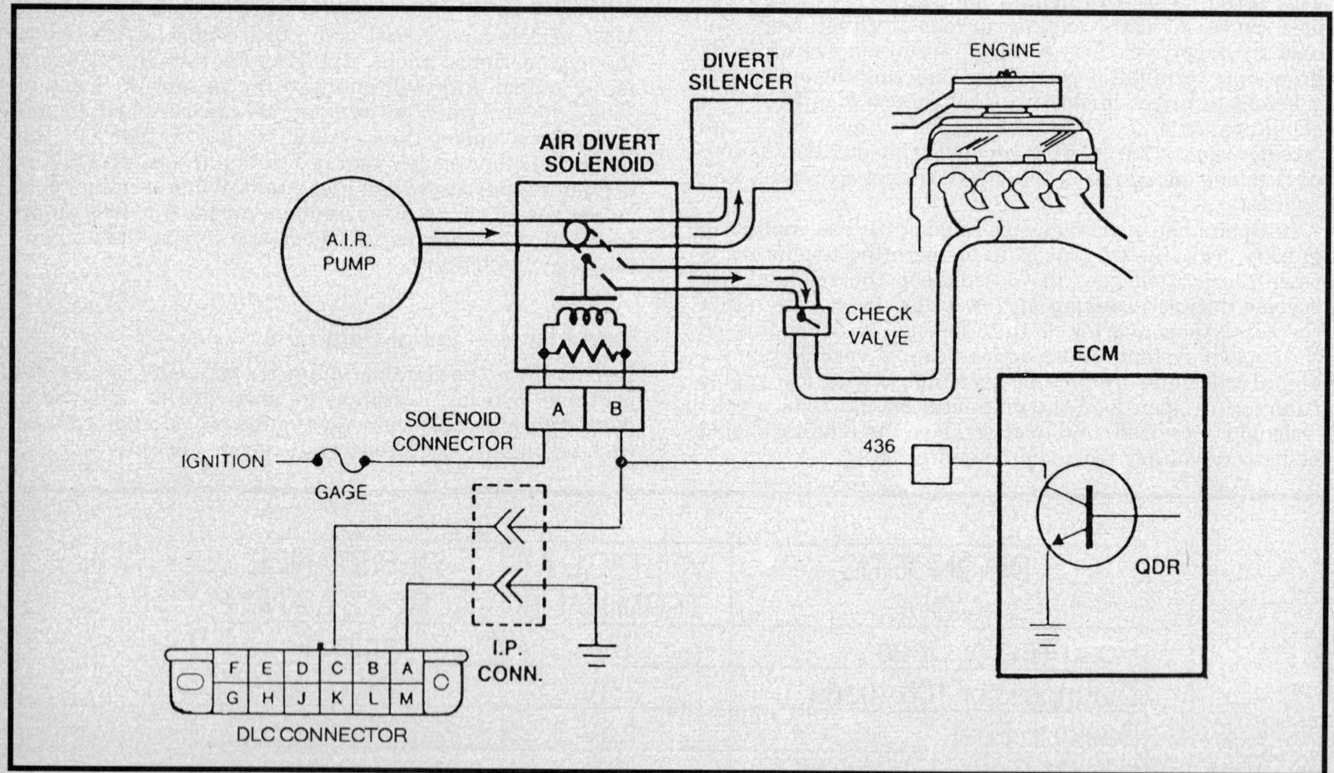

Terminal C — Air control solenoid

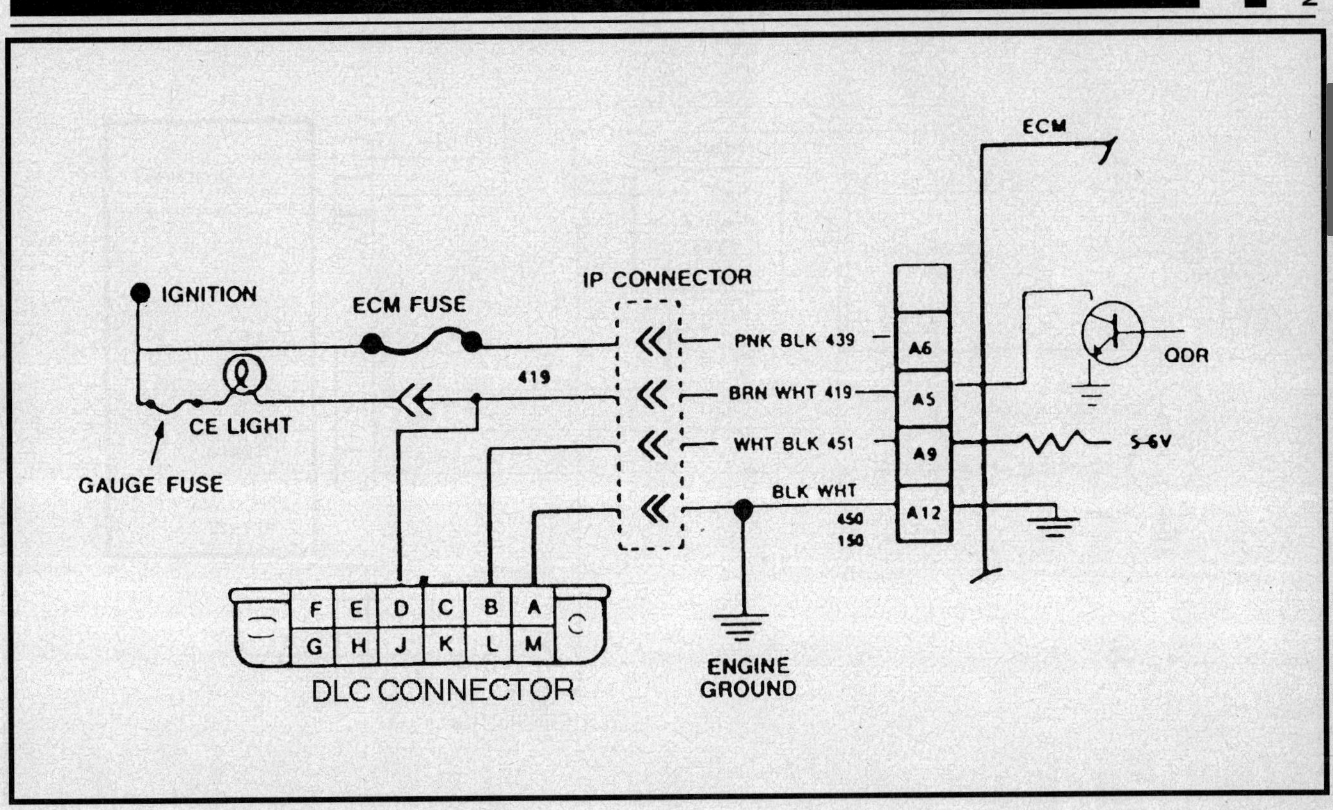

Terminal D — Malfunction indicator lamp

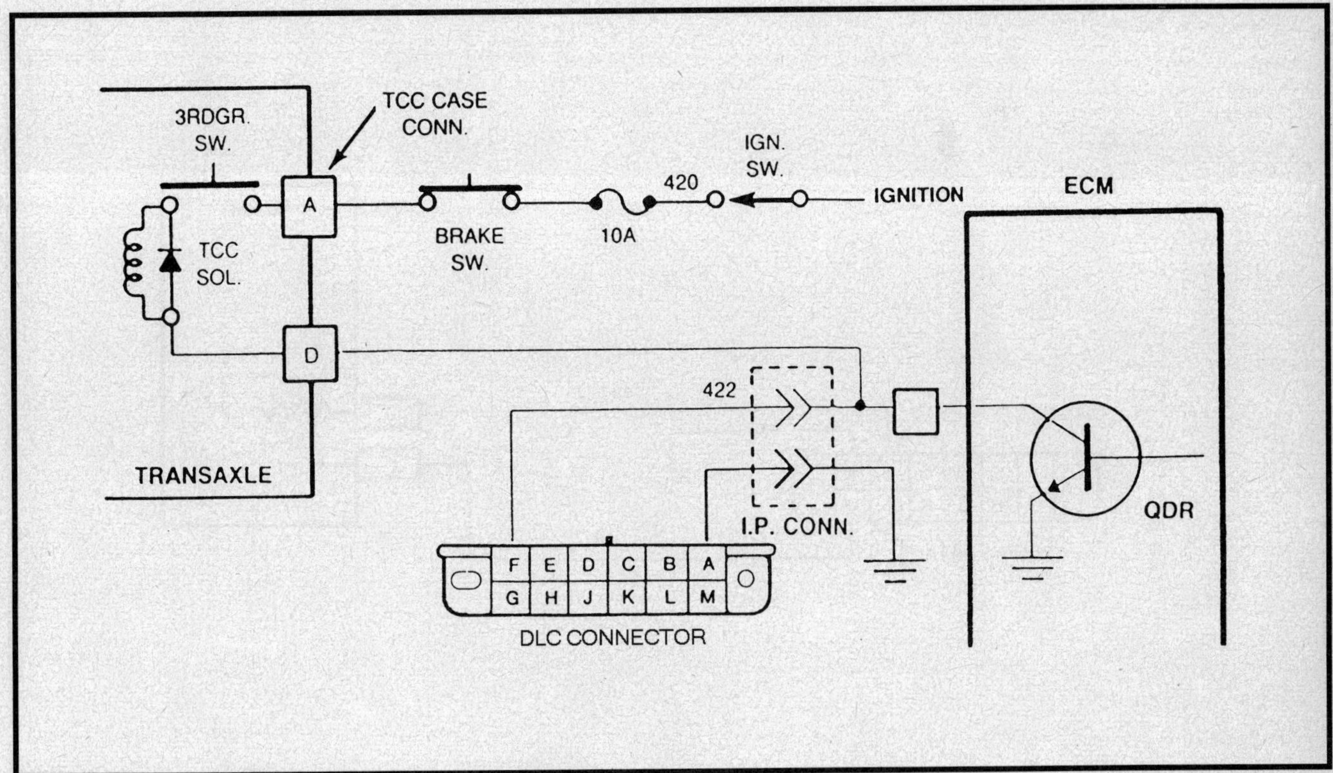

Terminal F — Torque converter clutch solenoid

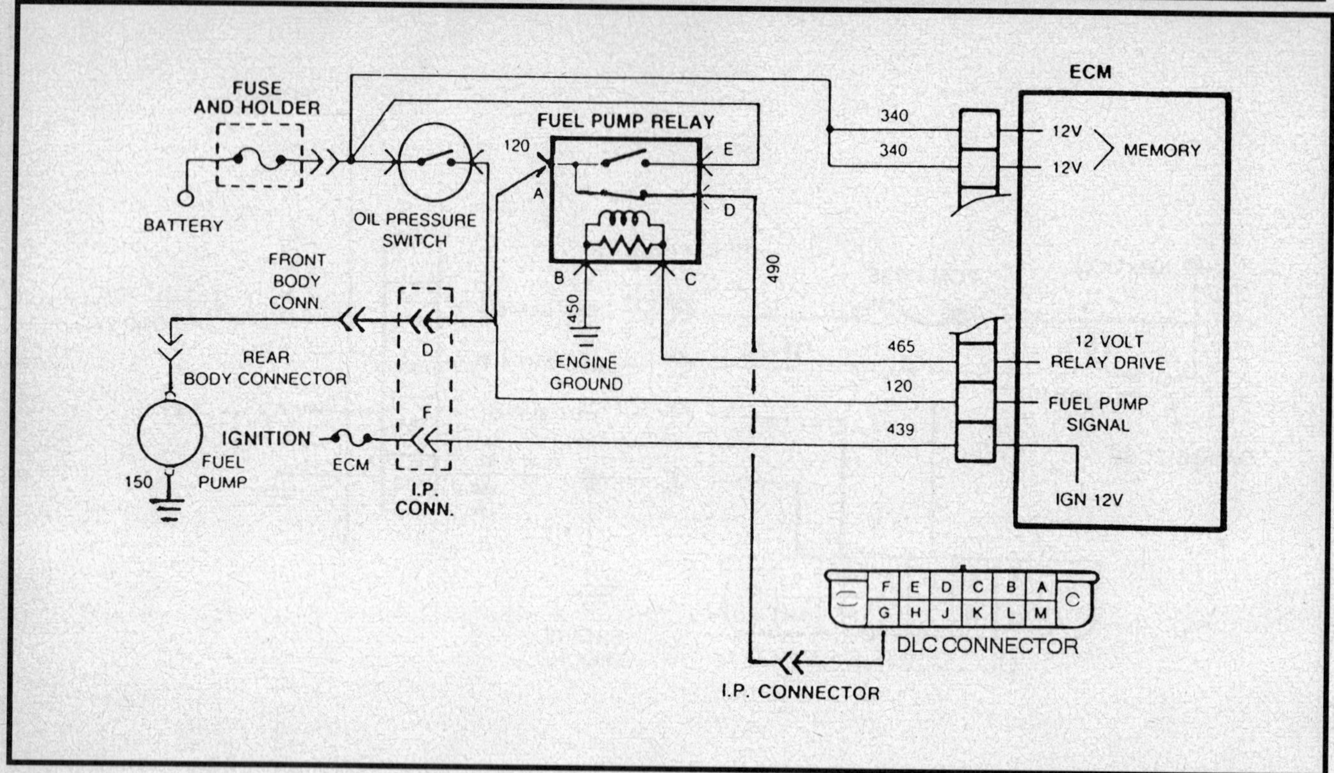

Terminal G — Fuel pump test point

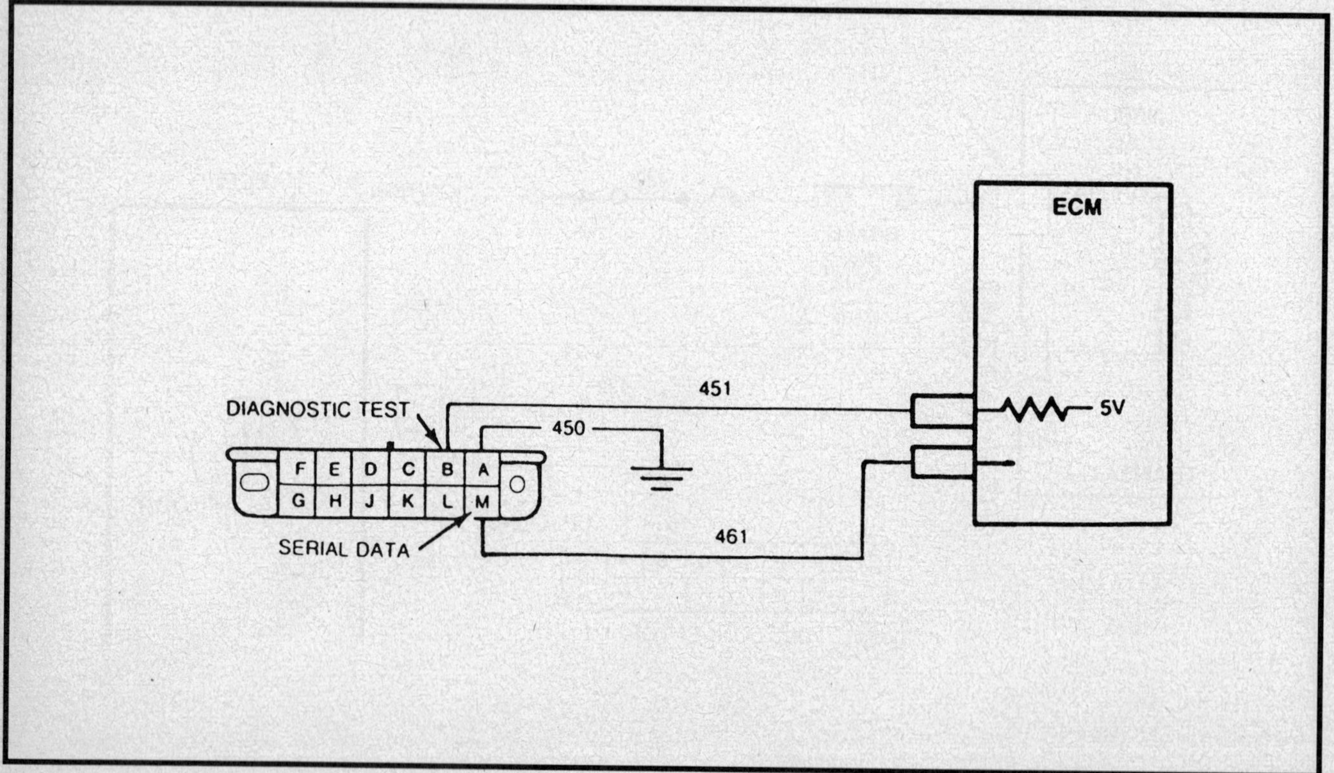

Terminal M — Serial data line

MEMORY

> **NOTE: Not every type of memory control is used in each vehicle application.**

- **Read Only Memory (ROM)** is a permanent memory physically soldered to the circuit boards within the ECM. The ROM contains the overall control algorithms. The ROM is a non-volatile memory and cannot be changed. Because of this, it does not require a constant supply of battery voltage to be retained.
- **Random Access Memory (RAM)** is the computer scratch pad. The computer can read from or write into this memory as needed. This memory is volatile and will be lost without a constant supply of battery voltage.
- **Programmable Read Only Memory (PROM)** is the portion of the ECM that contains different engine calibration information that is specific to year, model and emissions. It is for this reason that the PROM is referred to as the calibrator. The PROM is a non-volatile memory that is read only by the ECM. The PROM is removable from the ECM and should be retained with the vehicle following ECM replacement.
- **Electrically Erasable Programmable Read Only Memory (EEPROM)** is a permanent memory that is physically soldered to the circuit boards within the ECM. The EEPROM takes the place of the ROM and the PROM and is not serviceable. Those vehicles using an EEPROM can be re-programmed by using the TECH-1 or other Techline terminal equipment. Some ECM's using an EEPROM use a separate knock sensor module underneath the computer's access cover. This chip must be removed before replacing the old ECM and reinstalled into the new computer.
- **The Calibration Pack (CALPAC)** is physically soldered to the ECM and contains information for the fuel back-up mode of the computer. The CALPAC is not serviceable and failure would require replacement of the ECM.
- **The Memory Calibration (MEMCAL)** combines the function of the PROM and CALPAC as well as cruise control and knock control modules in one unit. This unit is serviced just like a PROM.
- **The Short Term Fuel Trim (STFT) (formerly called integrator)** is an ECM volatile memory register that will contain a number between 0 and 255. The neutral value for STFT is 128; any deviation from this value indicates a change in the injector pulse width as commanded from STFT. This function is only active in the closed loop mode of operation. As the ECM monitors the oxygen sensor voltage input, it is constantly varying the STFT value. For example, if a vacuum leak were to occur causing a lean condition (O_2 voltage low), the STFT would respond by increasing injector pulse width. This increase would be seen on a scan tool as a number greater than 128. Conversely, if the air filter became plugged causing a rich condition (O_2 voltage high), the STFT would respond with a decrease in injector pulse width; a number less than 128. Because this value is updated very quickly, the STFT only corrects for short term mixture trends.
- **The Long Term Fuel Trim (LTFT) (formerly called block learn)** is a matrix of cells arranged by rpm and load. As engine operating conditions change, the ECM will switch from cell to cell to determine which LTFT factor is appropriate. While in any given block, the ECM also monitors the STFT. If the STFT is far enough from 128 in either direction, the ECM will change the LTFT value for that given cell. Once the LTFT value is changed the STFT value should revert back to 128, this represents a neutral condition. If the mixture is still not correct (as judged by the O_2 sensor), the process of short and long term fuel correction will continue until they reach the limits of their control. These limits are programmed into the PROM and would be different for each vehicle. At this point of maximum correction a trouble code 44 or 45 would be registered in the computer's RAM. As with STFT the long term fuel trim is only active in closed loop operation. The LTFT represents the computer's learning capability as these values can be tailored to the specific driving habits of the operator. Once battery power is lost however, the LTFT values will revert back to default settings, and a loss of performance may be noticed. Original performance should return after the vehicle is operated for a period of time in closed loop.

SELF-DIAGNOSTIC SYSTEM FEATURES

Segment Check

The segment check causes a total illumination of all segments and bulbs on the IPC and ECCP to occur. This check will determine if any of the bulbs or segments of the vacuum florescent display are burned out or are always on. The segment check is initiated whenever diagnostics is first entered.

Diagnostic Trouble Codes (DTC)

In the process of controlling the various subsystems, the ECM and BCM continually monitor operating conditions for possible system malfunctions. By comparing system conditions against standard operating limits, certain circuit and component faults can be detected. A numerical diagnostic trouble code is stored in computer memory when a malfunction is detected by this self-diagnostic system.

The occurrence of certain system problems requires that the vehicle operator be alerted. This helps in prolonging vehicle operation under degraded conditions. Computer controlled telltale lights will illuminate under these conditions indicating service is required as soon as is reasonably possible.

If a specific system malfunction results in unacceptable vehicle operation, the self-diagnostics will attempt to minimize the effect by initiating failsoft action. A typical failsoft action would be the substitution of a fixed input value when a sensor is identified as being open or shorted.

Status Lights

While in Service Diagnostics, the mode indicators on the ECCP are used to determine the status of certain operating modes. The status light indicates specific system operation by either being turned ON or turned OFF.

System Tests

System Tests allows for the display of values as actually seen or commanded by the various systems. An example of this would be throttle position sensor readings or coolant temperature. These readings are displayed either on the Electronic Climate Control Panel (ECCP), the Cathode Ray Tube (CRT) screen or the Instrument Panel Cluster (IPC) in much the same way as a scan tool. The test types available are dependent on the system selected, but may include the following:

- **Data** (analog inputs) — displays analog values as seen by the system.
- **Inputs** (digital inputs) — displays digital values as seen by the system and provides an indication of input cycling.
- **Outputs** (digital outputs) — cycles digital outputs controlled by the system off/on for 3 seconds per cycle.
- **Overrides** (analog outputs) — allows for analog outputs of the system to be set at a desired value.
- **Clear Codes** — clear codes will remove all stored codes for the system currently selected.

NOTE: If the fault is present, the code may immediately reset.

- **Snapshot** — stores all inputs whether analog or digital, to a component at a particular moment; either when a code is set or manually triggered. These values may then be reviewed for diagnosis.

Tools and Equipment

SCAN TOOLS

The system can communicate a wide variety of information through the Data Link Connector (DLC), previously known as the ALDL. Depending on the application, serial data is transmitted to terminals E or M of the DLC. This data is transmitted at a high frequency which requires a quality scan tool for interpretation. Although stored codes may be read with only the use of a small jumper wire, the use of a hand-held scan tool such as GM's TECH 1 or equivalent, is recommended. There are many manufacturers of these tools; a purchaser must be certain that the tool is proper for the intended use.

The scan tool allows any stored codes to be read from the ECM memory. The tool also allows the operator to view the data being sent to the ECM while the engine is running. This ability has obvious diagnostic advantages; the use of the scan tool is frequently required by the diagnostic charts.

With an understanding of the data stream that the tool will display and knowledge of the circuits involved, the tool can be an extremely valuable tool when examining the system. Without scan tools, diagnostic information would be difficult to obtain in some circuits and impossible to retrieve in others. Scan tools do not make the use of trouble code charts unnecessary, nor do they pinpoint the exact area in a circuit where the problem exists. However, this type of equipment will give more specific information, such as sensor input readings. An example of this would be the coolant sensor circuit. While the ECM looks at specific voltage levels to determine coolant temperature, the scan tool will display this information as an actual temperature reading. This information is not only useful in determining system malfunctions, but is also very helpful in diagnosing mechanical problems. For example, if the engine is overheating, the scan tool can be used to look at the actual engine temperature. This is helpful when checking for electric cooling fan enable switch problems, sticking thermostats, etc.

The ECM is capable of communicating with a scan tool in 3 modes: Normal, DCL and Factory Test.

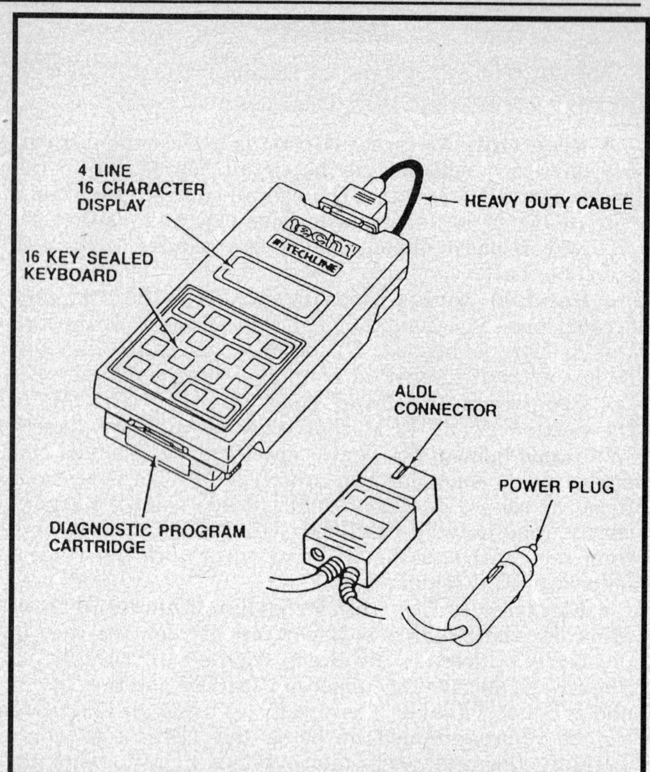

Typical scan tool — GM's Tech 1 shown

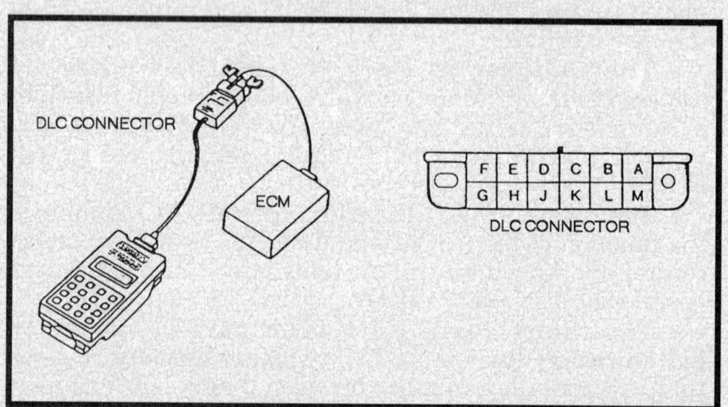

DLC connection to scan tool

Normal or Open Mode

This mode is not applicable to all engines. When engaged, certain engine data can be observed on the scanner without affecting engine operating characteristics. The number of items readable in this mode varies with engine family. Most scan tools are designed to change automatically to the DLC mode if this mode is not available.

DLC Mode

Also referred to as the ALDL, 10K or SPECIAL mode, the scanner will present all readable data as available. Certain operating characteristics of the engine are changed or controlled when this mode is engaged. The closed loop timers are bypassed, the spark (EST) is advanced and the PARK/NEUTRAL restriction is bypassed. If applicable, the IAC controls the engine speed to 1000 rpm ± 50, and, on some engines, the canister purge solenoid is energized.

Factory Test

Sometimes referred to as BACK-UP mode, this level of communication is primarily used during vehicle assembly and testing. This mode will confirm that the default or limp-in system is working properly within the ECM. Other data obtainable in this mode has little use in diagnosis.

> **NOTE: A scan tool that is known to display faulty data should not be used for diagnosis. Although the fault may be believed to be in only one area, it can possibly affect many other areas during diagnosis, leading to errors and incorrect repair.**

To properly read system values with a scan tool, the following conditions must be met. All normal values given in the charts will be based on these conditions:
- Engine running at idle, throttle closed
- Engine warm, upper radiator hose hot
- Vehicle in park or neutral
- System operating in closed loop
- All accessories **OFF**

SCAN TOOL USE FOR INTERMITTENTS

In some tool applications the data update rate is so slow that it becomes less effective than a digital voltmeter for checking sporadic voltage changes. With the rapid data transmission of a quality scanner, intermittent problems such as faulty wiring connections can be more successfully diagnosed. For example, while manipulating a suspect electrical circuit or component, observe the scan tool display screen. A rapid voltage change or erratic reading will uncover the trouble area.

The scan tool is also an easy way to compare the operating parameters of a poorly operating engine with that of a known good one. For example, a sensor may shift in value and cause a driveability problem, but not set a trouble code. Comparing the sensor's readings to a known good one may uncover the problem.

Scan tools have the ability to speed up diagnostic time and prevent the replacement of good parts, however a thorough understanding of the system you are working on, as well as a working knowledge of the scan tool is essential.

ELECTRICAL TOOLS

The most commonly required electrical diagnostic tool is the digital multimeter, allowing voltage, resistance and amperage to be read by one instrument. The multimeter must be a high-impedance unit, with 10 megohms of impedance in the voltmeter. This type of meter will not place an additional load on the circuit it is testing; this is extremely important in low voltage circuits. The multimeter must be of high quality in all respects. It should be handled carefully and protected from impact or damage. Replace batteries frequently in the unit.

Other necessary tools include an unpowered test light, a quality tachometer with inductive (clip-on) pick up and the proper tools for releasing GM's Metri-Pack, Weather Pack and Micro-Pack terminals as necessary. The Micro-Pack connectors are used at the ECM connector. A vacuum pump/gauge may also be required for checking sensors, solenoids and valves.

Diagnosis and Testing

SERVICE PRECAUTIONS

- To prevent internal ECM damage, the ignition must be **OFF** when disconnecting or reconnecting power to the computer.
- When handling a PROM, CALPAC or MEMCAL, do not touch the component leads. Also, do not remove the integrated circuit from the carrier.
- Never allow welding cables to lie on, near or across any vehicle electrical wiring.
- Leave new components and modules in the shipping package until ready to install them.
- When performing electrical tests on the system, use a high impedance multimeter, digital voltmeter (DVM) J-34029-A or equivalent.
- To prevent possible electrostatic discharge damage to the ECM, do not touch the connector pins or soldered components on the circuit board.

PRELIMINARY INSPECTION

In order to successfully locate and repair problems in complex computer controlled systems, the technician needs to adhere to a comprehensive, sequential method of diagnostics. Many times a vehicle has a malfunction and the worst is suspected. Premature conclusions often result in the replacement of good parts and wasted time. A proper sequence of steps needs to be established and strictly adhered to when troubleshooting a vehicle. These are as follows:

1. If possible, try to speak directly to the customer, especially if the complaint is not associated with an intermittent or steady MIL.

2. Road test the vehicle, if necessary, to verify the complaint.

3. Perform a thorough visual inspection. This step can very often eliminate the need for further testing. Loose wires, especially ground circuits, can create mysterious driveability problems. However, a complete visual inspection can uncover these problems quickly, and prevent the replacing of good parts and reduce vehicle downtime.

4. Perform the Diagnostic Circuit Check. This test confirms that the diagnostic system has not failed and is able to communicate through the malfunction indicator lamp.

5. After locating and repairing the problem area, road test the vehicle to confirm the repair and to verify no additional problem areas exist.

Visual Underhood Inspection

This is possibly the most critical step of diagnosis. A detailed examination of connectors, wiring and vacuum hoses can often lead to a repair without further diagnosis. Performance of this step relies on the skill of the technician performing it; a careful inspector will check:
- The undersides of hoses as well as the integrity of hard-to-reach hoses blocked by the air cleaner or other components.
- The wiring carefully for any sign of strain, burning, crimping, or terminal pull-out from a connector.
- The connectors at components or in harnesses as required; usually, pushing them together will reveal a loose fit.

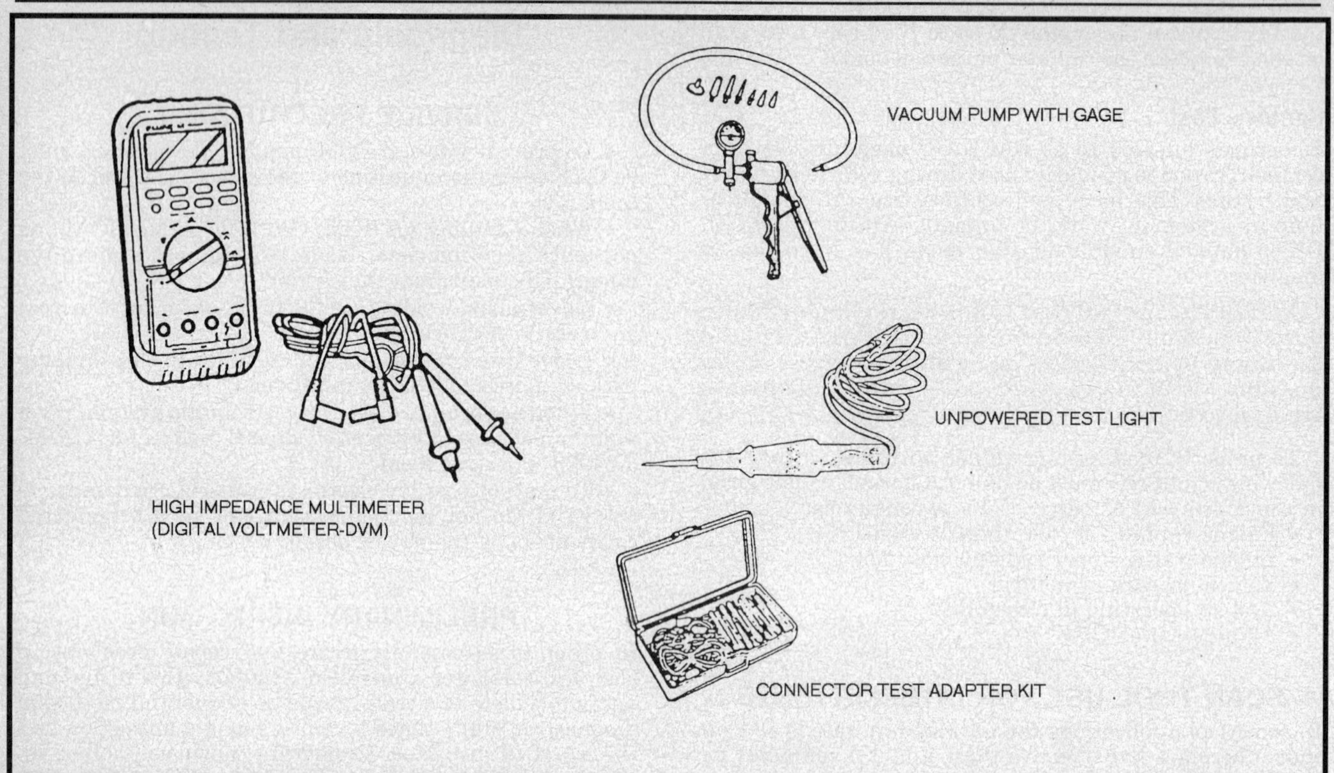

Electrical system diagnostic tools

Diagnostic Circuit Check

This step is used to check that the on-board diagnostic system is working correctly. A system which is faulty or shorted may not yield correct codes when placed in the Diagnostic Mode.

If the diagnostic system is not operating correctly, or if a problem exists without the malfunction indicator lamp being lit, refer to the vehicle's A-Charts. These charts cover such conditions as Engine Cranks but Will Not Run or No Service Engine Soon Light.

1. Turn the ignition switch **ON** but do not start the engine.

2. The MIL should illuminate and and stay lit until the engine is started.

3. If the MIL does not light, check the bulb and circuits for an open or short.

4. Connect jumper wire from terminal A to B in the DLC. The MIL should flash code 12 at least 3 times. If not, turn the ignition switch **OFF** and disconnect the ECM.

5. Turn the ignition switch **ON** but do not start the engine. If the MIL is ON, look for short in circuit 419.

6. If the MIL does not light, check the MEMCAL PROM for proper installation. If no light, check the ECM then look for an open or short in the MIL circuit.

Circuit and Component Diagnosis and Repair

Using the appropriate chart(s), the fault codes and the scan tool data will lead to diagnosis and checking of a particular circuit or component. It is important to note that the fault code indicates a fault or loss of signal in an ECM-controlled system, not necessarily in the specific component. Detailed procedures to isolate the problem are included in each code chart; these procedures must be followed accurately to insure timely and correct repair. Following the procedure will also insure that only truly faulty components are replaced.

READING CODES

With Scan Tool

NOTE: Scan tool functions and procedures may vary between manufacturers. Consult owners manual for proper connection and operation of each scan tool.

Once the diagnostic circuit check has been performed, enter the Diagnostic Mode and read any stored codes. To enter the diagnostic mode:

1. Turn the ignition switch **OFF**. Locate the Data Link Connector (DLC) also called ALDL, usually under the instrument panel. It may be within a plastic cover or housing labeled DIAGNOSTIC CONNECTOR. This link is used to communicate with the ECM.

2. Connect the scan tool correctly to the DLC.

3. Turn the ignition switch to the **ON** position but DO NOT start the engine. A Code 12 may be displayed. Code 12 is not a fault code. It is used as a system acknowledgment or handshake code; its presence indicates that the ECM can communicate as requested. Code 12 is used to begin every diagnostic sequence. Some vehicles also use Code 12 after all diagnostic codes have been sent.

4. After Code 12 has been transmitted 3 times, the fault codes, if any, will each be transmitted 3 times. The

codes are stored and transmitted in numeric order from lowest to highest.

NOTE: The order of codes in the memory does not indicate the order of occurrence.

5. Switch the ignition **OFF** when finished with code retrieval or scan tool readings and remove the scan tool.

Without Scan Tool — Non-CRT/DID Vehicles

1. Turn the ignition to the **ON** position.
2. On Riviera, depress the OFF and TEMP buttons on the ECCP simultaneously and hold until all display segments light. This will initiate the segment check.
3. On Toronado and Trofeo, depress the OFF and WARMER buttons on the ECCP.
4. After diagnostics is entered, any DTC's stored in computer memory will be displayed. Each code consists of the system abbreviation (E, B or R), a three digit code identifier and the letter C (current codes) or H (history codes). All codes for each system are displayed together in sequence, with the lowest number first (i.e. E013, E014 etc.).
- E — ECM trouble code.
- B — BCM trouble code.
- R — SIR trouble code.
- C — Current trouble code.
- H — History trouble code.
- NO X CODE — No code present (X represents the system).
- NO X DATA — Communication line not operating (X representing the system).

NOTE: Any time during DTC display that the FAN DOWN button on the ECCP is depressed, code display will be bypassed.

Without Scan Tool — CRT/DID Vehicles

1. Turn the ignition to the **ON** position.
2. Depress the OFF hardkey and WARM softkey on the CRT/DID simultaneously and hold until all display segments light. This is the Segment Check.
3. After the segment check, any DTC's stored in computer memory will be displayed. Each code consists of the system abbreviation (E, B, I or R), a three digit code identifier and the letter C (current codes) or H (history codes). All codes for each system are displayed together in sequence, with the lowest number first (i.e. E013, E014 etc.).
- E — ECM trouble code.
- B — BCM trouble code.
- I — IPC trouble code.
- R — SIR trouble code.
- C — Current trouble code.
- H — History trouble code.
- NO X CODE — No code present (X represents the system).
- NO X DATA — Communication line not operating (X representing the system).

NOTE: Any time during DTC display that the RTN softkey is depressed, code display will be bypassed.

CLEARING CODES

Selection of the CLEAR CODES? test type displays either the message CODES CLEAR or CODES NOT CLEAR, indicating whether or not the codes were successfully cleared. The message will appear for 3 seconds. After 3 seconds the display will automatically return to the next available test type for the selected system. Cycle the ignition once or test drive the vehicle to ensure the code(s) do not reset.

ON BOARD DIAGNOSTIC SYSTEM

Operating — Non CRT/DTD Vehicles

ENTERING SERVICE MODE

To enter diagnostics, turn the ignition to the **ON** position. On Riviera, depress the OFF and TEMP buttons on the ECCP simultaneously and hold until all display segments light. This will initiate the segment check. On Toronado and Trofeo, depress the OFF and WARMER buttons on the ECCP.

NOTE: Operating the vehicle in the service mode for periods longer than 30 minutes will require the use of a battery charger. This will prevent excessive battery drain which could result in false diagnostic information or a no-start condition.

DIAGNOSTIC DISPLAY

On Riviera, all diagnostic information will be displayed on the Season and Trip Odometers of the IPC unless otherwise noted. On Toronado and Trofeo, all diagnostic information will be displayed on the Driver Information Center (DIC) in the IPC unless otherwise noted. Because of the limited space available, single letter identifiers are often used for each of the major computer systems. These are E for ECM, B for BCM, I for IPC and R for SIR.

NOTE: Upon selecting the diagnostic mode, the climate control will operate in whichever mode was being commanded just prior to entering diagnostics. Even though the ECCP display changes when entering diagnostics, the prior operating mode will be remembered and will resume afterward.

DIAGNOSTIC TROUBLE CODE DISPLAY (DTC)

After diagnostics is entered, any DTC's stored in computer memory will be displayed. Codes may be stored for the ECM, BCM or SIR systems.

NOTE: SIR information refers to the Supplemental Inflatable Restraint System.

Each code consists of the system abbreviation (E, B or R), a three digit code identifier and the letter C (current codes) or H (history codes). All codes for each system are displayed together in sequence, with the lowest number first (i.e. EO13, EO14 etc.). The final character of the code will be a C or an H. Current (C) means the fault was still present the last time the test was run, while History (H) means the failure occurred within the last 50 ignition cycles.

If no codes are present for a system, a NO X CODE message will be displayed, with X representing the particular system letter abbreviation. If the communication line to a component is not operating, a NO X DATA message will be displayed, indicating the BCM could not communicate with that system.

Any time during DTC display that the FAN DOWN button on the ECCP is depressed, code display will be bypassed.

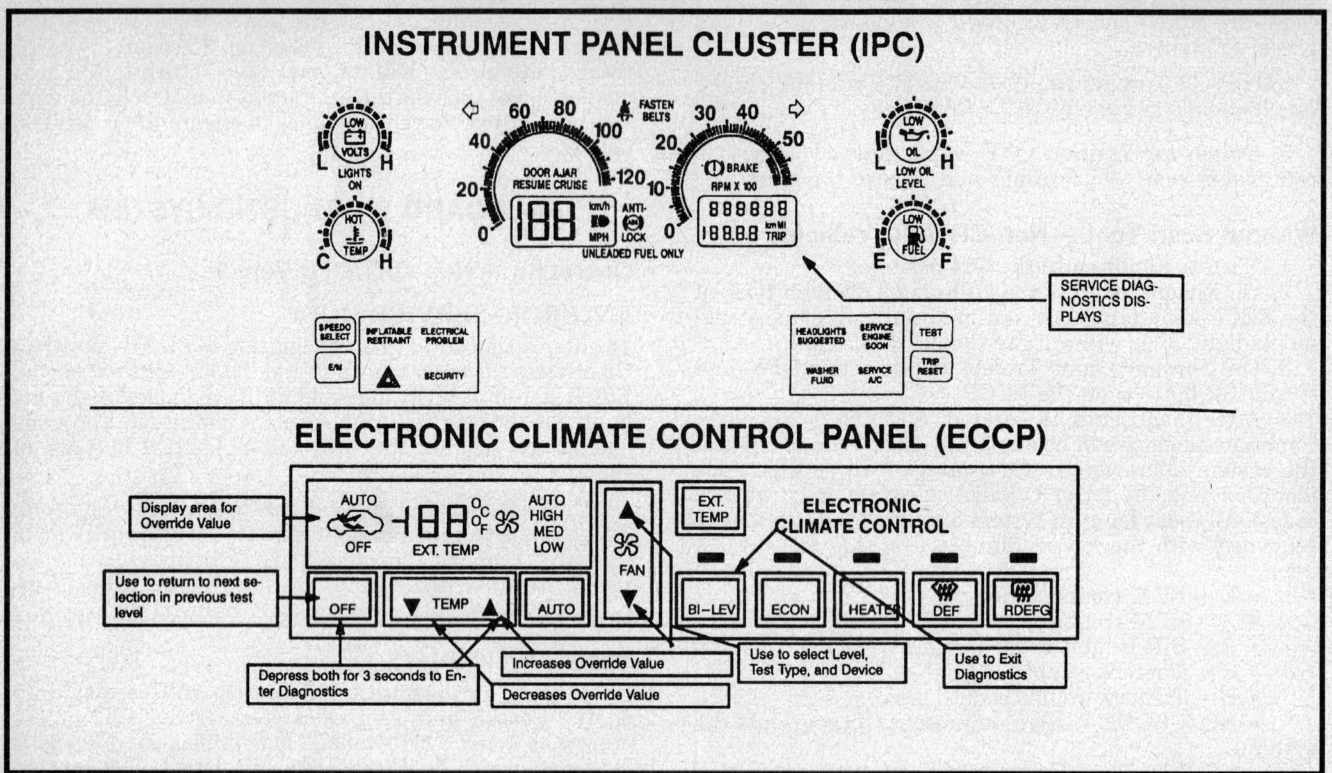

Instrument panel — Riviera

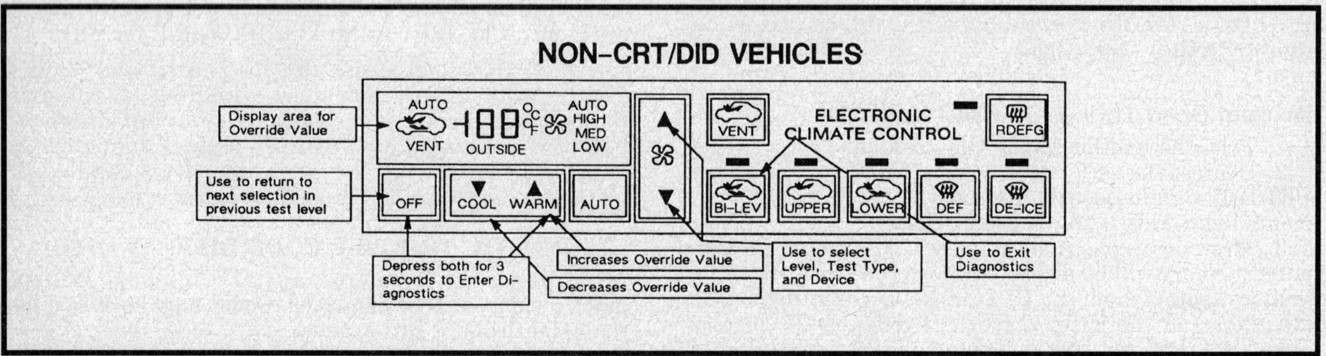

Electronic climate control panel — Toronado and Trofeo without CRT/DID

SELECTING THE SYSTEM

Following the display of DTC's, the first available system for testing will be displayed. For example, EC? would be displayed on Riviera for ECM testing, while on Toronado and Trofeo the message ECM? will appear. To control the display while selecting a system to test, the depress the following buttons on the ECCP:

• **OFF** — terminates the system selection process and return the display to the beginning of the trouble code sequence.

• **FAN DOWN** — displays the next available system selection. This allows the display to be stepped through system choices, as well as repeated following the last system choice.

• **FAN UP** — selects the displayed system for testing.

• **BI-LEVEL** — terminates diagnostics and return the IPC and the ECCP to normal operation.

SELECTING THE TEST TYPE

After selecting a particular system, the first available test type will be displayed (i.e. DATA EC?). On Toronado and Trofeo, the display will read ECM DATA?. The message is clearer on these models due to increased character space in the IPC display area. To control the display while se-

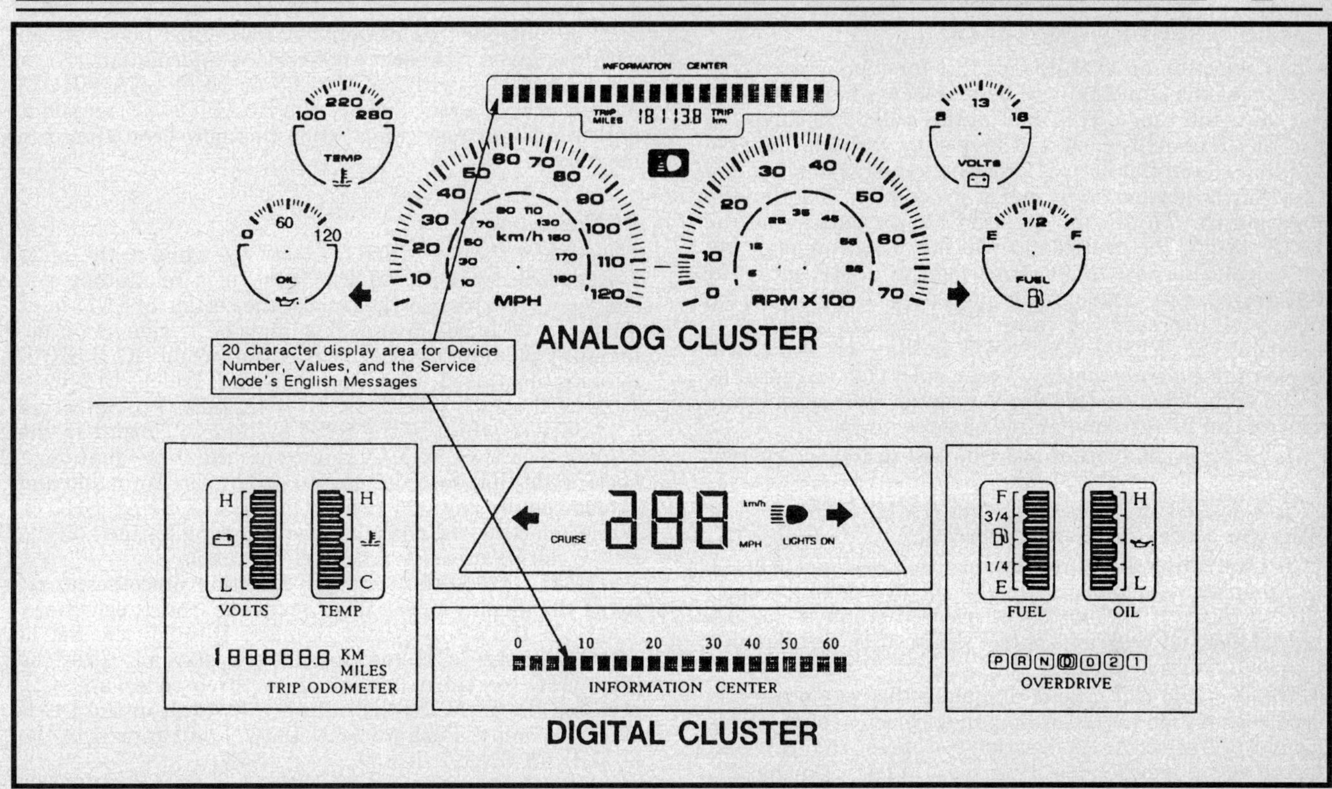

Instrument panel — Toronado and Trofeo

lecting a test type, depress the following buttons on the ECCP:

- **OFF** — returns to the next available system.
- **FAN DOWN** — displays the next available test type for the selected system. This allows the display to be stepped through all available test type choices, as well as repeated following the last test type choice.
- **FAN UP** — selects the displayed test type. At this point the first of several specific tests will appear.
- **BI-LEVEL** — terminates the diagnostics.

SELECTING THE TEST (DATA, INPUTS, OUTPUTS AND OVERRIDES)

On Riviera, four characters of the Season Odometer display on the IPC will contain a test code to identify the selection. The first two characters are letters which identify the system and test type, while the last two characters numerically identify the test. The trip odometer on the IPC will display the selected value as seen by the computer.

On Toronado and Trofeo, information is displayed on the IPC in the Driver Information Center. The first 2 characters are letters which identify the system and test type, while the last 2 characters numerically identify the test. Next will be an English prompt, followed by the value as seen by the computer. While selecting a specific test, any of the following actions may be taken to control the display by pressing the following buttons:

- **OFF** — stops the test selection process and returns the display to the next available test type for the selected system.
- **FAN DOWN** — displays the next lower test number for the selected test type. If this button is pressed with the lowest test number displayed, the highest test number will then appear.
- **FAN UP** — displays the next highest test number for the selected test type. If this button is pressed with the highest test number displayed, the lowest number will then appear.

Specific Data (ECM, BCM)

Selection of the DATA? test type will result in the first available test being displayed. On Riviera, the trip odometer will then display the data value the system computer sees for the selected test. On Toronado and Trofeo, this information will appear in the Driver Information Center on the IPC.

Specific Inputs (ECM, BCM, IPC)

Selection of the INPUT? test type will result in the first available test being displayed. The trip odometer (DIC on the Toronado and Trofeo) will then display the voltage level, HI or LO, of the selected input to the system. Additionally, the third character of the display will be a 0 or an X. A 0 indicates that the input has NOT changed since it was selected, while an X indicates that the input HAS changed since it was selected. This indicator may be reset by selecting a new test and then returning to the original test (i.e. press FAN UP and then FAN DOWN).

Specific Outputs (ECM, BCM)

Selection of the OUPUT? test type will result in the first available test being displayed. The trip odometer (DIC on the Toronado and Trofeo) will then display the voltage level, HI or LO, of the selected output of the system. This voltage level will then be cycled between HI and LO every 3 seconds.

Specific Overrides (ECM, BCM)

Upon selecting an OVERRIDE test function, current operation of that function will be represented as a percentage of its full range. This information will be displayed on the ECCP on ALL models. The display will alternate between '—' and the normal program value as long as that function is not currently being overridden. On Riviera, pressing the TEMP up and TEMP down buttons on the ECCP begins the override at which time the display will no longer alternate to '—'. Pressing the TEMP button up will increase the value, while pressing the TEMP button down will decrease the value. On Toronado and Trofeo, pressing the WARM and COOL buttons on the ECCP begins the override display. Press WARM to increase the value and COOL to decrease the value. Normal program control can be resumed in one of three ways:

1. Selection of another override will cancel the current override.

2. Selection of another system (ECM, BCM, IPC or SIR) will cancel the current override.

3. Overriding the value beyond either extreme (0 or 99) will display '—' momentarily and then jump to the opposite extreme. If the button is released while '—' is displayed, normal program control will resume and the display will again alternate.

The override test type is unique in that any other test type within the selected system may be active at the same time. After selecting an override test, press the OFF button to select another test type (i.e. DATA?). During this time, the ECCP will continue to display the selected override. By selecting another test type and test while simultaneously controlling the override with the TEMP button or WARM and COOL buttons, it is possible to monitor the effect of the override on different vehicle parameters. An example of this would be to select engine RPM in ECM DATA while at the same time manipulating the IAC Motor override and viewing the RPM display on the IPC for the changing value.

SNAPSHOT

Snapshot is a test type that will record all BCM or ECM data and inputs at one instant for review at a later time. The ECM and BCM each have different types of snapshots.

ECM Snapshot

Selection of the SNAPSHOT? test type while in the ECM system level will result in the message SNAP DONE on the display. This message will appear for 3 seconds to indicating all ECM data and inputs have been stored in memory. Following the 3 second interval, the display will automatically proceed to the first available snapshot test type (i.e. SNAP DATA). While selecting a snapshot test type the following buttons on the ECCP can be pressed to control the display:

• **OFF** — stops the test type selection process and returns the display to the next available system selection.

• **FAN DOWN** — display the next available snapshot test type. This allows the display to be toggled between SNAP DATA and SNAP INPUTS.

• **FAN UP** — with SNAP DATA? or SNAP INPUTS? displayed, will select the test type. At this point, the display is controlled as it would normally be, however, all data displayed represents memorized information.

• **FAN UP** — with SNAP EC? or ECM SNAPSHOT? displayed will again display the SNAP DONE message, indicating that new information has now been stored in memory.

BCM Snapshot

Selection of the SNAPSHOT? test type while in the BCM system level will permit the recall of a maximum of 3 snapshots recorded at the time of the setting of BCM fault codes. In addition, 1 snapshot may be triggered on demand by pushing the FAN UP button while DO B SNAP is being displayed. (On Toronado and Trofeo, this message will read TAKE BCM SNAPSHOT). Selecting SNAPSHOT while in the BCM system will result in the display BXXX where XXX represents the three digit diagnostic code that recorded the snapshot. On Toronado and Trofeo the display will read CODE BXXX SNAP SHOT). While selecting the snapshot the following buttons on the ECCP can be pressed to control the display:

• **OFF** — stops the test type selection process and returns the display to the next available system selection.

• **FAN DOWN** — allows scrolling through the list of DTC's that the BCM has stored a snapshot for. After the final DTC (or third if more than three codes are set), pressing the FAN DOWN button will result in the DO B SNAP? display. Pushing FAN DOWN will return to the first BXXX SNAP? display.

• **FAN UP** — with SNAP DATA? or SNAP INPUTS? displayed will select that test type. At this point, the display is controlled as it would normally be, however, all data displayed represents memorized information.

• **FAN UP** — with DO B SNAP? displayed will display the SNAP DONE message indicating that new information has now been stored in memory.

> **NOTE: All Snapshot on codeset information will be lost when the code is cleared.**

EXITING DIAGNOSTICS

To exit diagnostics, depress the BI-LEVEL button on the ECCP or turn the ignition switch **OFF**. This will NOT erase stored trouble codes.

Operating — CRT/DID Vehicles

ENTERING SERVICE MODE

To enter diagnostics, turn the ignition to the **ON** position. Depress the OFF hardkey and WARM softkey on the CRT/DID simultaneously and hold until all display segments light. This is the Segment Check.

> **NOTE: Hardkey refers to those buttons outside of the CRT screen. Softkey refers to those buttons on the CRT display screen.**

DIAGNOSTIC DISPLAY

During diagnostic operation, all information will be displayed on the Driver Information Center (DIC) located in the Instrument Panel Cluster (IPC). Because of the limited space available, single letter identifiers are often used for each of the major computer systems. These are: E for ECM, B for BCM, I for IPC, and R for SIR.

ECM DIAGNOSTIC FLOW CHART — RIVIERA

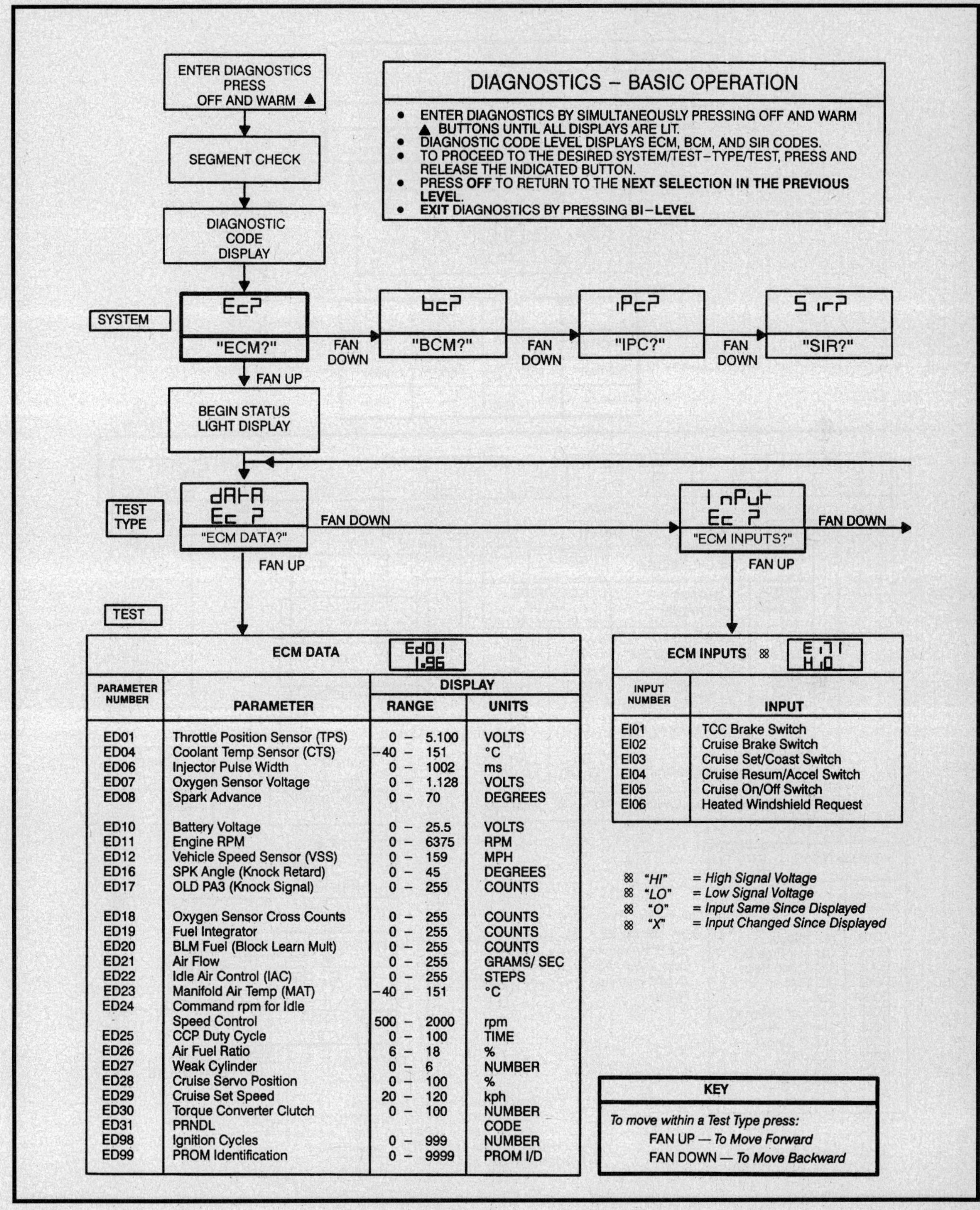

ECM DATA   EdD 1 / 1.96

PARAMETER NUMBER	PARAMETER	DISPLAY	
		RANGE	UNITS
ED01	Throttle Position Sensor (TPS)	0 – 5.100	VOLTS
ED04	Coolant Temp Sensor (CTS)	−40 – 151	°C
ED06	Injector Pulse Width	0 – 1002	ms
ED07	Oxygen Sensor Voltage	0 – 1.128	VOLTS
ED08	Spark Advance	0 – 70	DEGREES
ED10	Battery Voltage	0 – 25.5	VOLTS
ED11	Engine RPM	0 – 6375	RPM
ED12	Vehicle Speed Sensor (VSS)	0 – 159	MPH
ED16	SPK Angle (Knock Retard)	0 – 45	DEGREES
ED17	OLD PA3 (Knock Signal)	0 – 255	COUNTS
ED18	Oxygen Sensor Cross Counts	0 – 255	COUNTS
ED19	Fuel Integrator	0 – 255	COUNTS
ED20	BLM Fuel (Block Learn Mult)	0 – 255	COUNTS
ED21	Air Flow	0 – 255	GRAMS/ SEC
ED22	Idle Air Control (IAC)	0 – 255	STEPS
ED23	Manifold Air Temp (MAT)	−40 – 151	°C
ED24	Command rpm for Idle Speed Control	500 – 2000	rpm
ED25	CCP Duty Cycle	0 – 100	TIME
ED26	Air Fuel Ratio	6 – 18	%
ED27	Weak Cylinder	0 – 6	NUMBER
ED28	Cruise Servo Position	0 – 100	%
ED29	Cruise Set Speed	20 – 120	kph
ED30	Torque Converter Clutch	0 – 100	NUMBER
ED31	PRNDL		CODE
ED98	Ignition Cycles	0 – 999	NUMBER
ED99	PROM Identification	0 – 9999	PROM I/D

ECM INPUTS ⊗   E i71 / H.D

INPUT NUMBER	INPUT
EI01	TCC Brake Switch
EI02	Cruise Brake Switch
EI03	Cruise Set/Coast Switch
EI04	Cruise Resum/Accel Switch
EI05	Cruise On/Off Switch
EI06	Heated Windshield Request

ECM DIAGNOSTIC FLOW CHART — RIVIERA

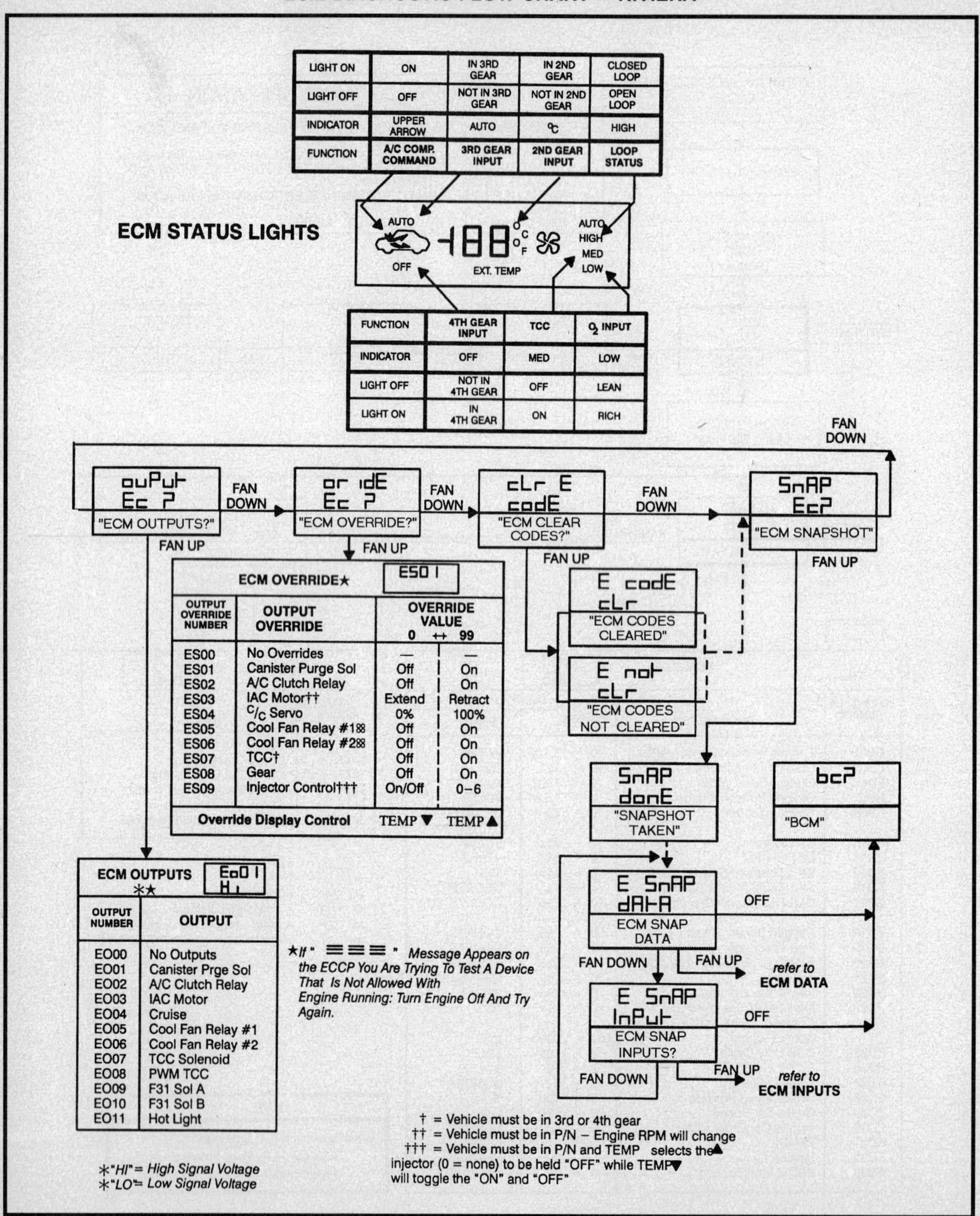

BCM DIAGNOSTIC FLOW CHART — RIVIERA

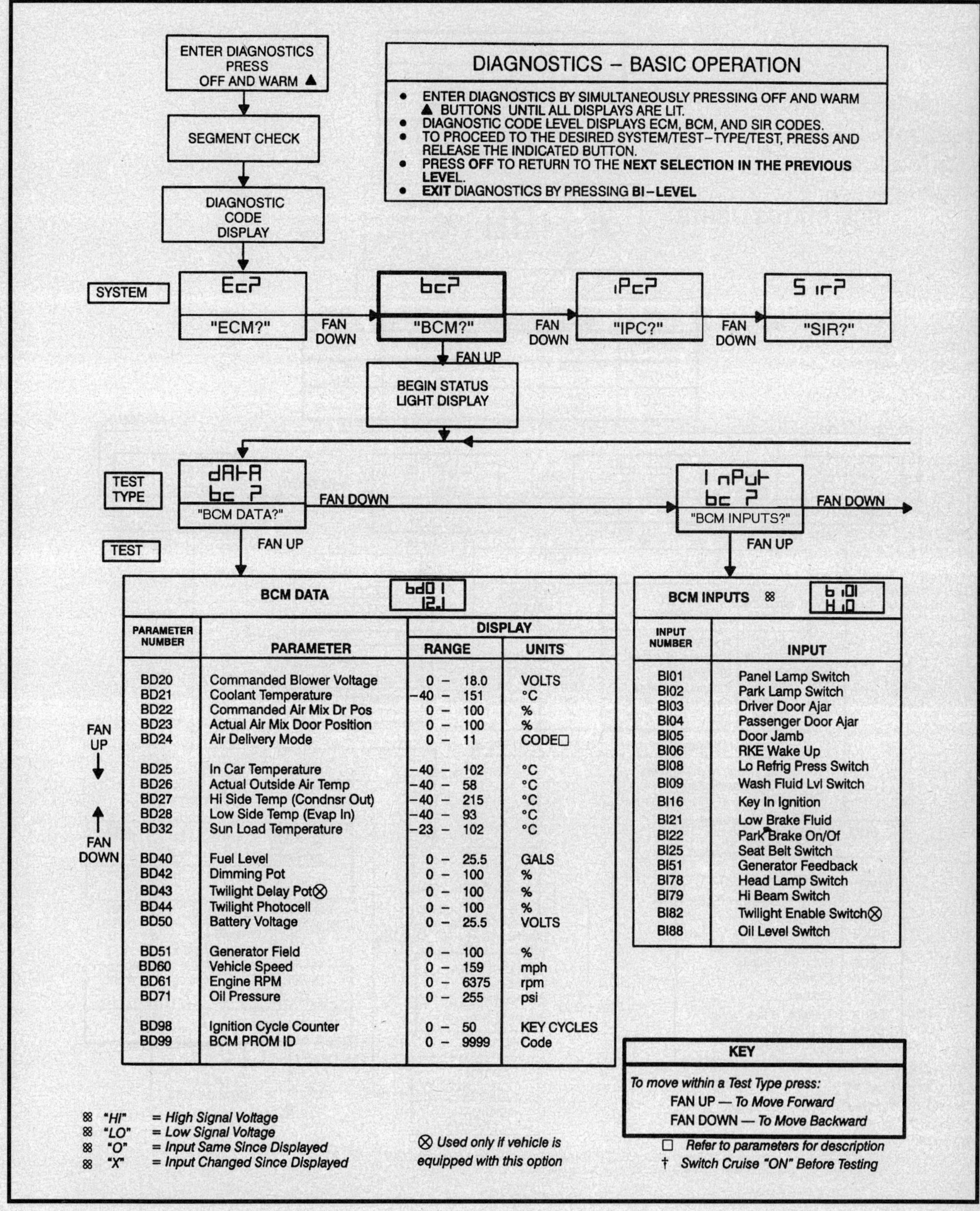

PARAMETER NUMBER	PARAMETER	DISPLAY	
		RANGE	UNITS
BD20	Commanded Blower Voltage	0 — 18.0	VOLTS
BD21	Coolant Temperature	-40 — 151	°C
BD22	Commanded Air Mix Dr Pos	0 — 100	%
BD23	Actual Air Mix Door Position	0 — 100	%
BD24	Air Delivery Mode	0 — 11	CODE□
BD25	In Car Temperature	-40 — 102	°C
BD26	Actual Outside Air Temp	-40 — 58	°C
BD27	Hi Side Temp (Condnsr Out)	-40 — 215	°C
BD28	Low Side Temp (Evap In)	-40 — 93	°C
BD32	Sun Load Temperature	-23 — 102	°C
BD40	Fuel Level	0 — 25.5	GALS
BD42	Dimming Pot	0 — 100	%
BD43	Twilight Delay Pot⊗	0 — 100	%
BD44	Twilight Photocell	0 — 100	%
BD50	Battery Voltage	0 — 25.5	VOLTS
BD51	Generator Field	0 — 100	%
BD60	Vehicle Speed	0 — 159	mph
BD61	Engine RPM	0 — 6375	rpm
BD71	Oil Pressure	0 — 255	psi
BD98	Ignition Cycle Counter	0 — 50	KEY CYCLES
BD99	BCM PROM ID	0 — 9999	Code

BCM DATA   bd01 / 12.1

BCM INPUTS ⊗   bi01 / Hi0

INPUT NUMBER	INPUT
BI01	Panel Lamp Switch
BI02	Park Lamp Switch
BI03	Driver Door Ajar
BI04	Passenger Door Ajar
BI05	Door Jamb
BI06	RKE Wake Up
BI08	Lo Refrig Press Switch
BI09	Wash Fluid Lvl Switch
BI16	Key In Ignition
BI21	Low Brake Fluid
BI22	Park Brake On/Of
BI25	Seat Belt Switch
BI51	Generator Feedback
BI78	Head Lamp Switch
BI79	Hi Beam Switch
BI82	Twilight Enable Switch⊗
BI88	Oil Level Switch

⊠ "HI" = High Signal Voltage
⊠ "LO" = Low Signal Voltage
⊠ "O" = Input Same Since Displayed
⊠ "X" = Input Changed Since Displayed

⊗ Used only if vehicle is equipped with this option

□ Refer to parameters for description
† Switch Cruise "ON" Before Testing

SELF-DIAGNOSTIC SYSTEMS
RIVIERA, TORONADO AND TROFEO

BCM DIAGNOSTIC FLOW CHART — RIVIERA

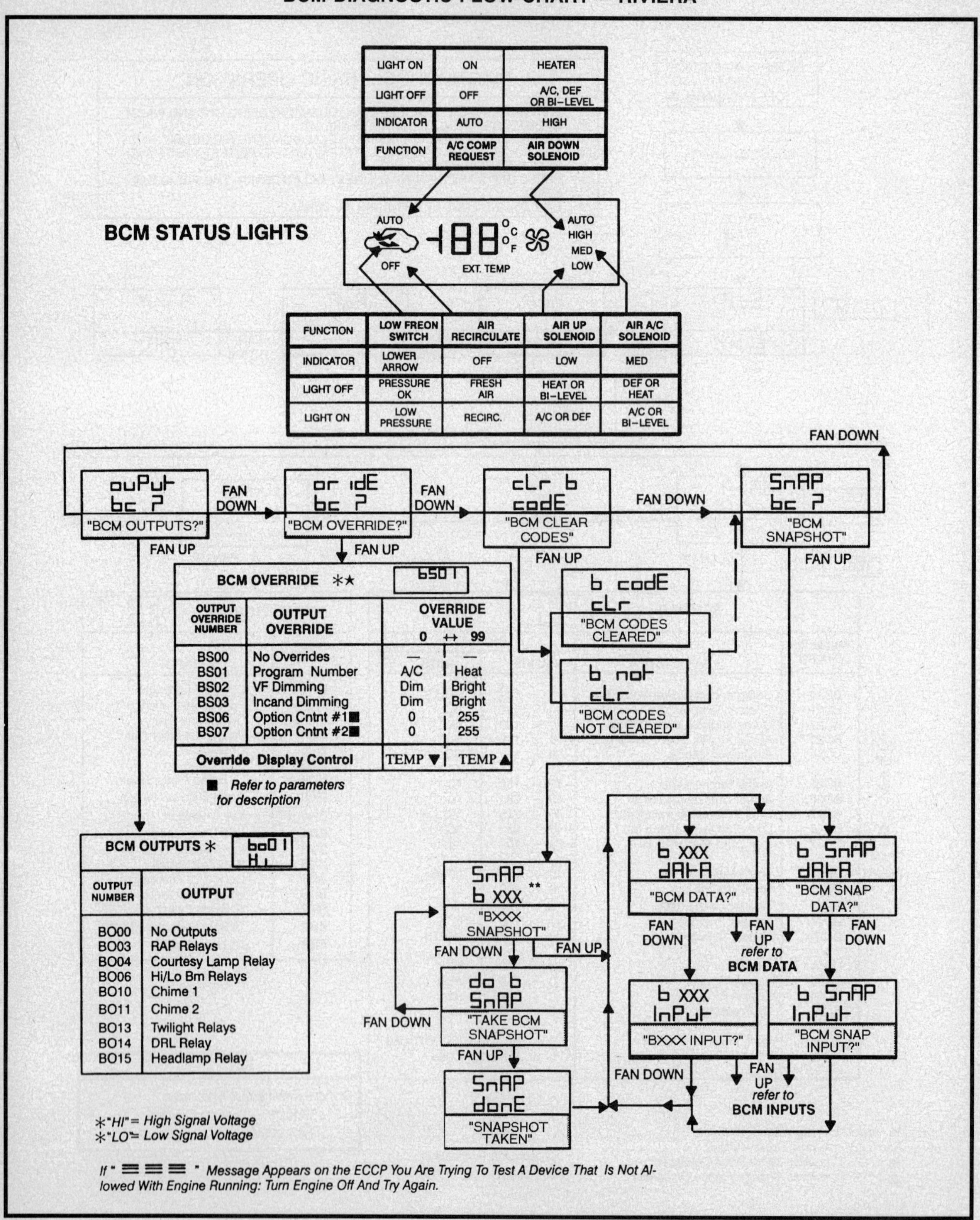

IPC DIAGNOSTIC FLOW CHART — RIVIERA

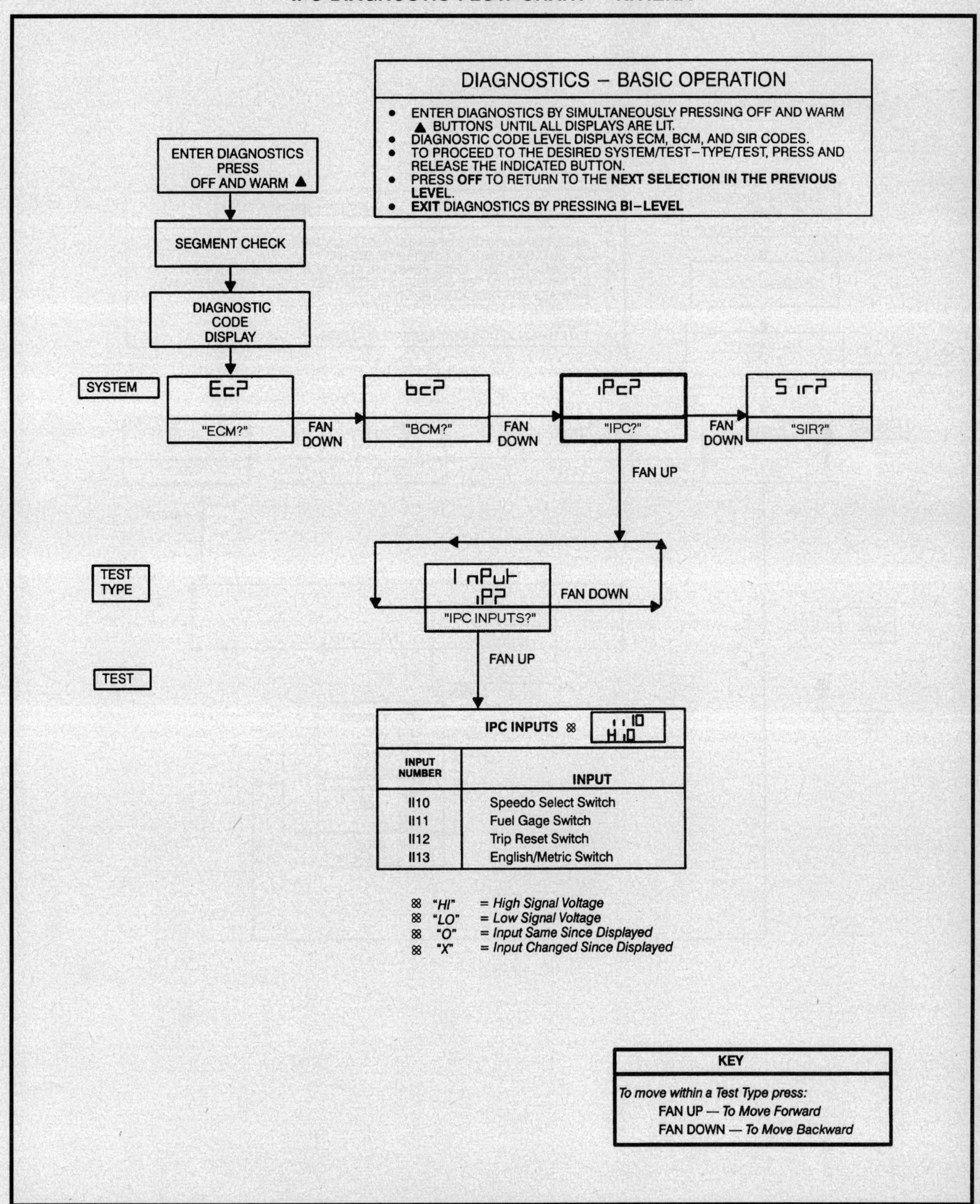

SIR DIAGNOSTIC FLOW CHART — RIVIERA

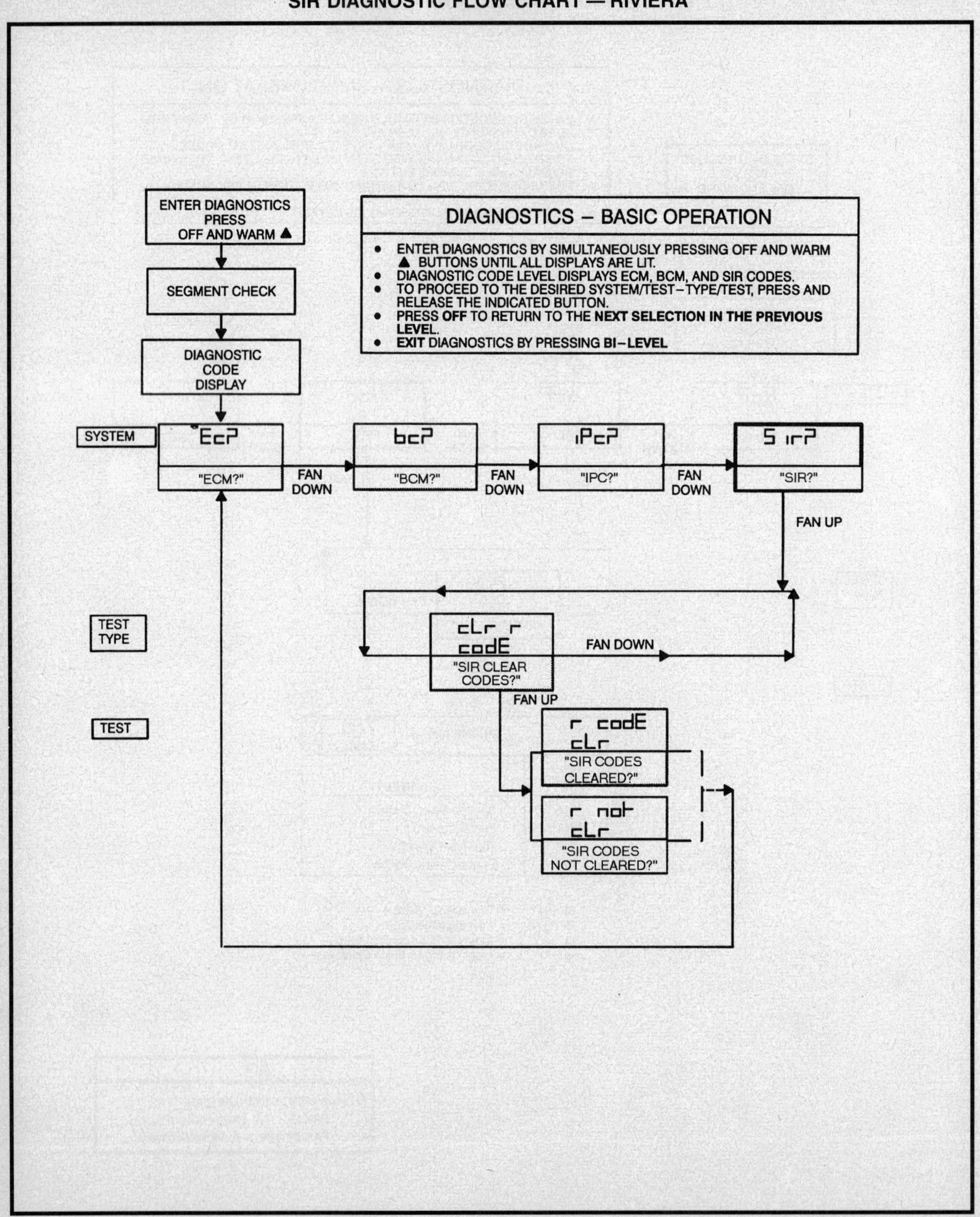

ECM DIAGNOSTIC FLOW CHART — TORONADO AND TROFEO WITHOUT CRT/DID

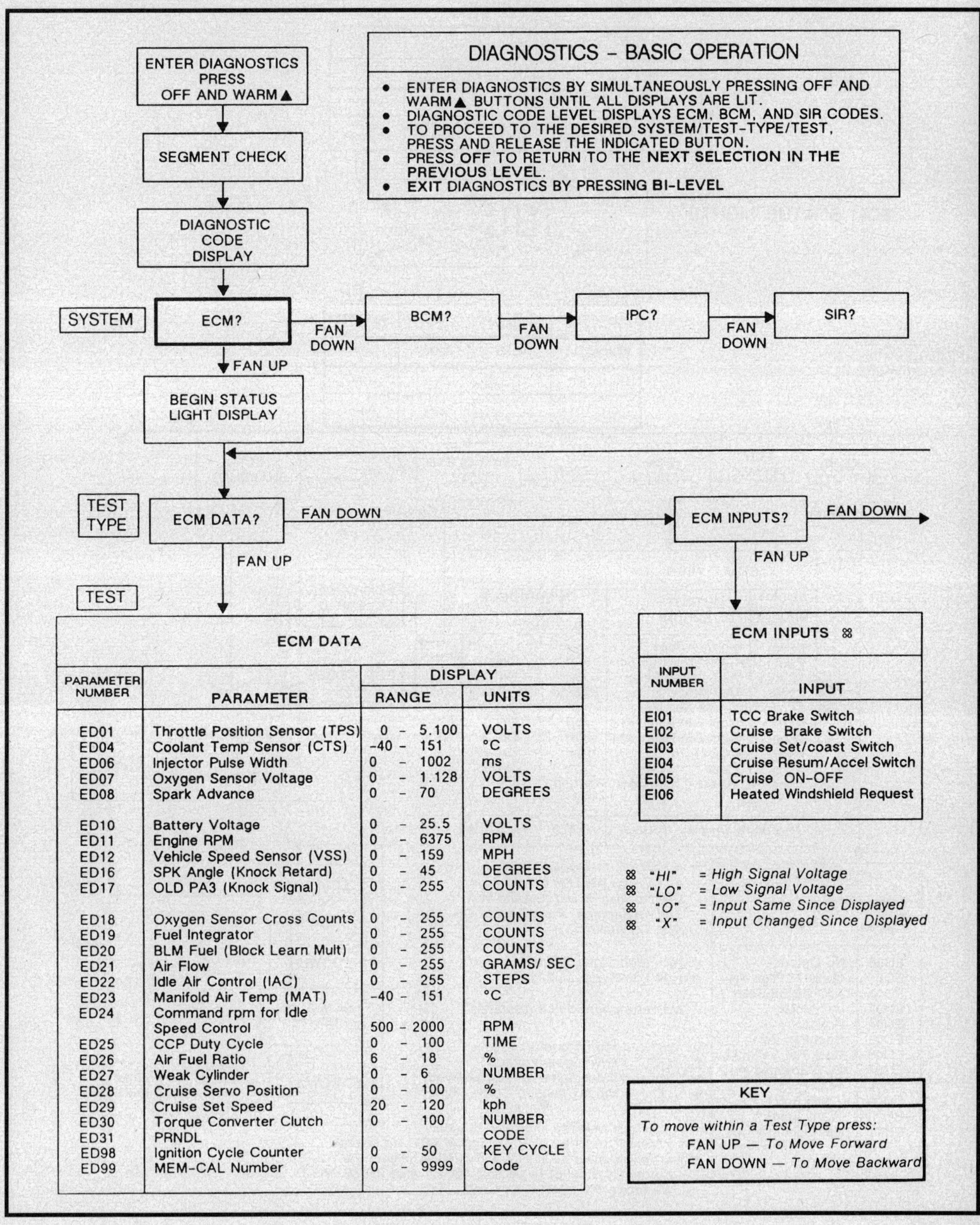

ECM DATA

PARAMETER NUMBER	PARAMETER	DISPLAY	
		RANGE	UNITS
ED01	Throttle Position Sensor (TPS)	0 – 5.100	VOLTS
ED04	Coolant Temp Sensor (CTS)	–40 – 151	°C
ED06	Injector Pulse Width	0 – 1002	ms
ED07	Oxygen Sensor Voltage	0 – 1.128	VOLTS
ED08	Spark Advance	0 – 70	DEGREES
ED10	Battery Voltage	0 – 25.5	VOLTS
ED11	Engine RPM	0 – 6375	RPM
ED12	Vehicle Speed Sensor (VSS)	0 – 159	MPH
ED16	SPK Angle (Knock Retard)	0 – 45	DEGREES
ED17	OLD PA3 (Knock Signal)	0 – 255	COUNTS
ED18	Oxygen Sensor Cross Counts	0 – 255	COUNTS
ED19	Fuel Integrator	0 – 255	COUNTS
ED20	BLM Fuel (Block Learn Mult)	0 – 255	COUNTS
ED21	Air Flow	0 – 255	GRAMS/ SEC
ED22	Idle Air Control (IAC)	0 – 255	STEPS
ED23	Manifold Air Temp (MAT)	–40 – 151	°C
ED24	Command rpm for Idle Speed Control	500 – 2000	RPM
ED25	CCP Duty Cycle	0 – 100	TIME
ED26	Air Fuel Ratio	6 – 18	%
ED27	Weak Cylinder	0 – 6	NUMBER
ED28	Cruise Servo Position	0 – 100	%
ED29	Cruise Set Speed	20 – 120	kph
ED30	Torque Converter Clutch	0 – 100	NUMBER
ED31	PRNDL		CODE
ED98	Ignition Cycle Counter	0 – 50	KEY CYCLE
ED99	MEM-CAL Number	0 – 9999	Code

ECM INPUTS ⊠

INPUT NUMBER	INPUT
EI01	TCC Brake Switch
EI02	Cruise Brake Switch
EI03	Cruise Set/coast Switch
EI04	Cruise Resum/Accel Switch
EI05	Cruise ON-OFF
EI06	Heated Windshield Request

⊠ "HI" = High Signal Voltage
⊠ "LO" = Low Signal Voltage
⊠ "O" = Input Same Since Displayed
⊠ "X" = Input Changed Since Displayed

KEY

To move within a Test Type press:
FAN UP — To Move Forward
FAN DOWN — To Move Backward

ECM DIAGNOSTIC FLOW CHART — TORONADO AND TROFEO WITHOUT CRT/DID

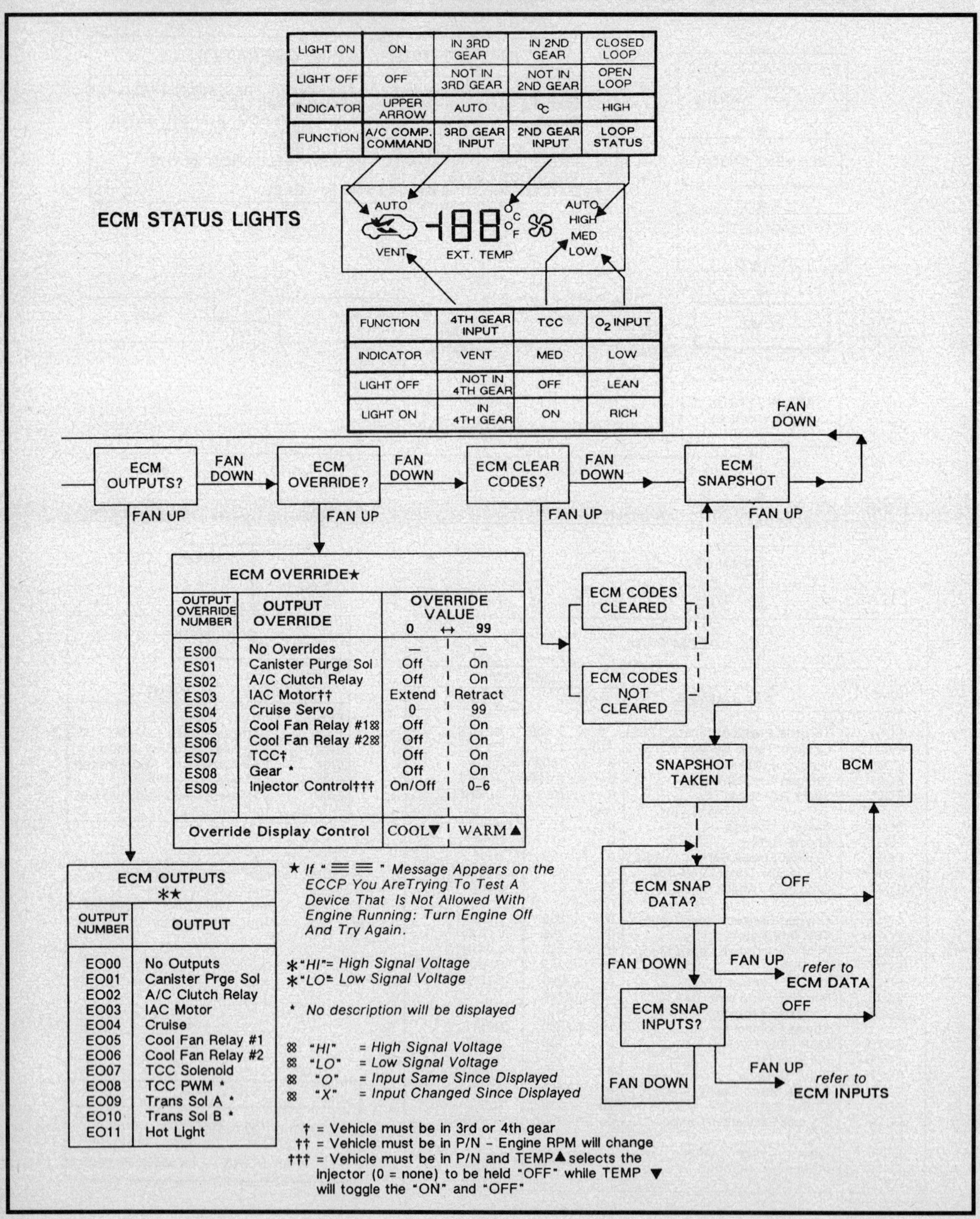

BCM DIAGNOSTIC FLOW CHART — TORONADO AND TROFEO WITHOUT CRT/DID

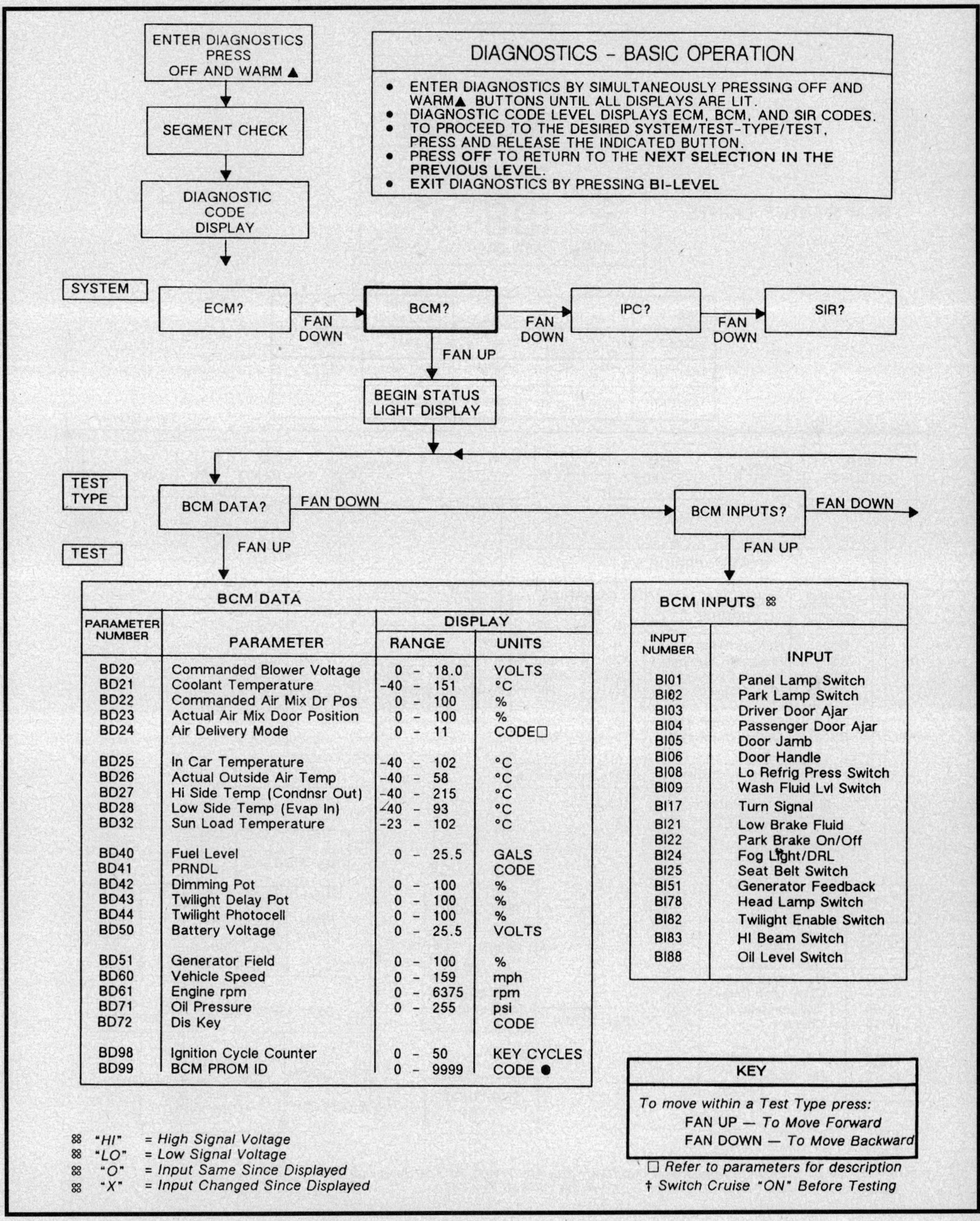

DIAGNOSTICS - BASIC OPERATION

- ENTER DIAGNOSTICS BY SIMULTANEOUSLY PRESSING OFF AND WARM▲ BUTTONS UNTIL ALL DISPLAYS ARE LIT.
- DIAGNOSTIC CODE LEVEL DISPLAYS ECM, BCM, AND SIR CODES.
- TO PROCEED TO THE DESIRED SYSTEM/TEST-TYPE/TEST, PRESS AND RELEASE THE INDICATED BUTTON.
- PRESS OFF TO RETURN TO THE NEXT SELECTION IN THE PREVIOUS LEVEL.
- EXIT DIAGNOSTICS BY PRESSING BI-LEVEL

BCM DATA

PARAMETER NUMBER	PARAMETER	DISPLAY RANGE	DISPLAY UNITS
BD20	Commanded Blower Voltage	0 – 18.0	VOLTS
BD21	Coolant Temperature	-40 – 151	°C
BD22	Commanded Air Mix Dr Pos	0 – 100	%
BD23	Actual Air Mix Door Position	0 – 100	%
BD24	Air Delivery Mode	0 – 11	CODE□
BD25	In Car Temperature	-40 – 102	°C
BD26	Actual Outside Air Temp	-40 – 58	°C
BD27	Hi Side Temp (Condnsr Out)	-40 – 215	°C
BD28	Low Side Temp (Evap In)	-40 – 93	°C
BD32	Sun Load Temperature	-23 – 102	°C
BD40	Fuel Level	0 – 25.5	GALS
BD41	PRNDL		CODE
BD42	Dimming Pot	0 – 100	%
BD43	Twilight Delay Pot	0 – 100	%
BD44	Twilight Photocell	0 – 100	%
BD50	Battery Voltage	0 – 25.5	VOLTS
BD51	Generator Field	0 – 100	%
BD60	Vehicle Speed	0 – 159	mph
BD61	Engine rpm	0 – 6375	rpm
BD71	Oil Pressure	0 – 255	psi
BD72	Dis Key		CODE
BD98	Ignition Cycle Counter	0 – 50	KEY CYCLES
BD99	BCM PROM ID	0 – 9999	CODE ●

BCM INPUTS ※

INPUT NUMBER	INPUT
BI01	Panel Lamp Switch
BI02	Park Lamp Switch
BI03	Driver Door Ajar
BI04	Passenger Door Ajar
BI05	Door Jamb
BI06	Door Handle
BI08	Lo Refrig Press Switch
BI09	Wash Fluid Lvl Switch
BI17	Turn Signal
BI21	Low Brake Fluid
BI22	Park Brake On/Off
BI24	Fog Light/DRL
BI25	Seat Belt Switch
BI51	Generator Feedback
BI78	Head Lamp Switch
BI82	Twilight Enable Switch
BI83	HI Beam Switch
BI88	Oil Level Switch

KEY

To move within a Test Type press:
FAN UP — To Move Forward
FAN DOWN — To Move Backward

□ *Refer to parameters for description*
† *Switch Cruise "ON" Before Testing*

※ *"HI"* = High Signal Voltage
※ *"LO"* = Low Signal Voltage
※ *"O"* = Input Same Since Displayed
※ *"X"* = Input Changed Since Displayed

SELF-DIAGNOSTIC SYSTEMS
RIVIERA, TORONADO AND TROFEO

BCM DIAGNOSTIC FLOW CHART — TORONADO AND TROFEO WITHOUT CRT/DID

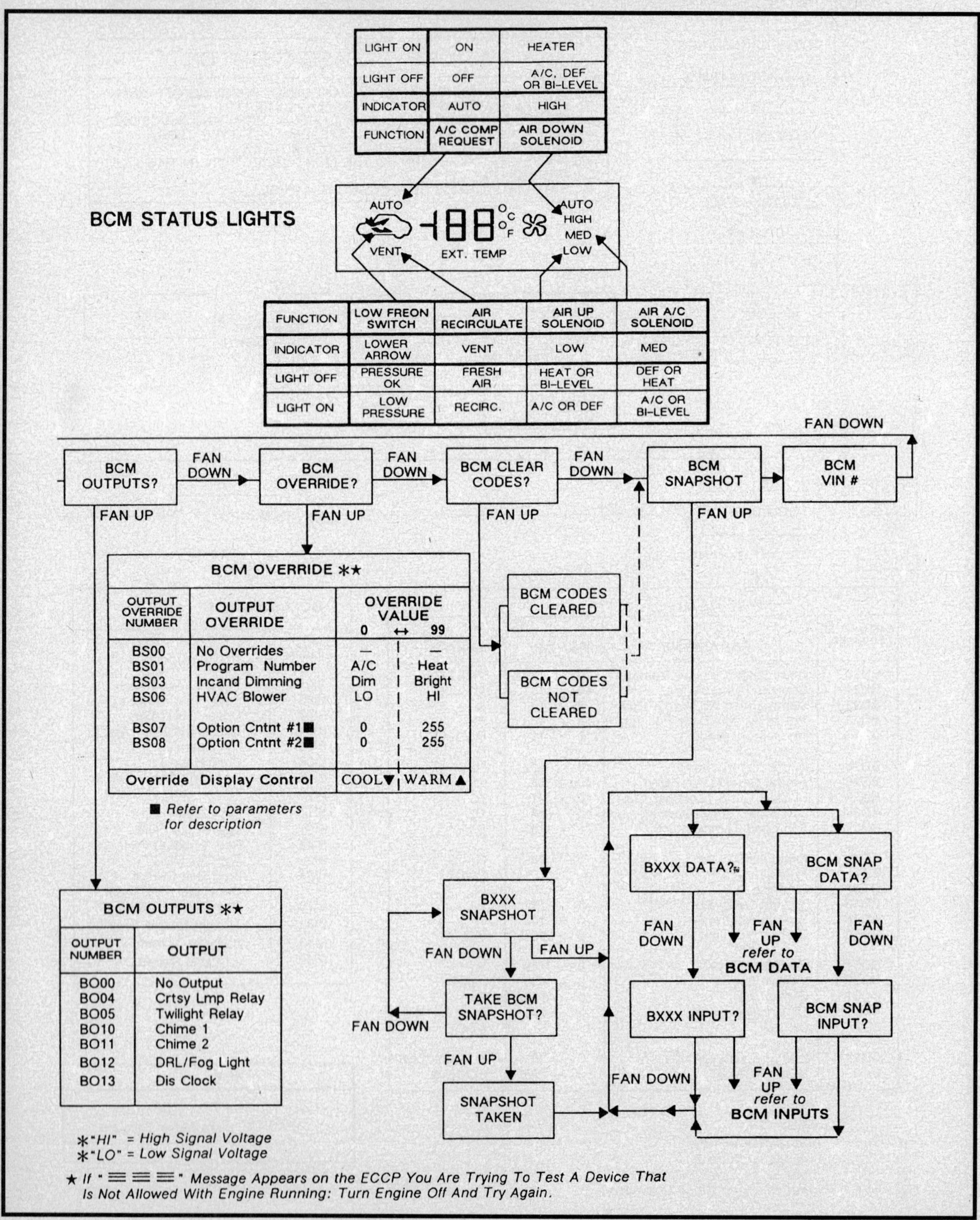

IPC DIAGNOSTIC FLOW CHART — TORONADO AND TROFEO WITHOUT CRT/DID

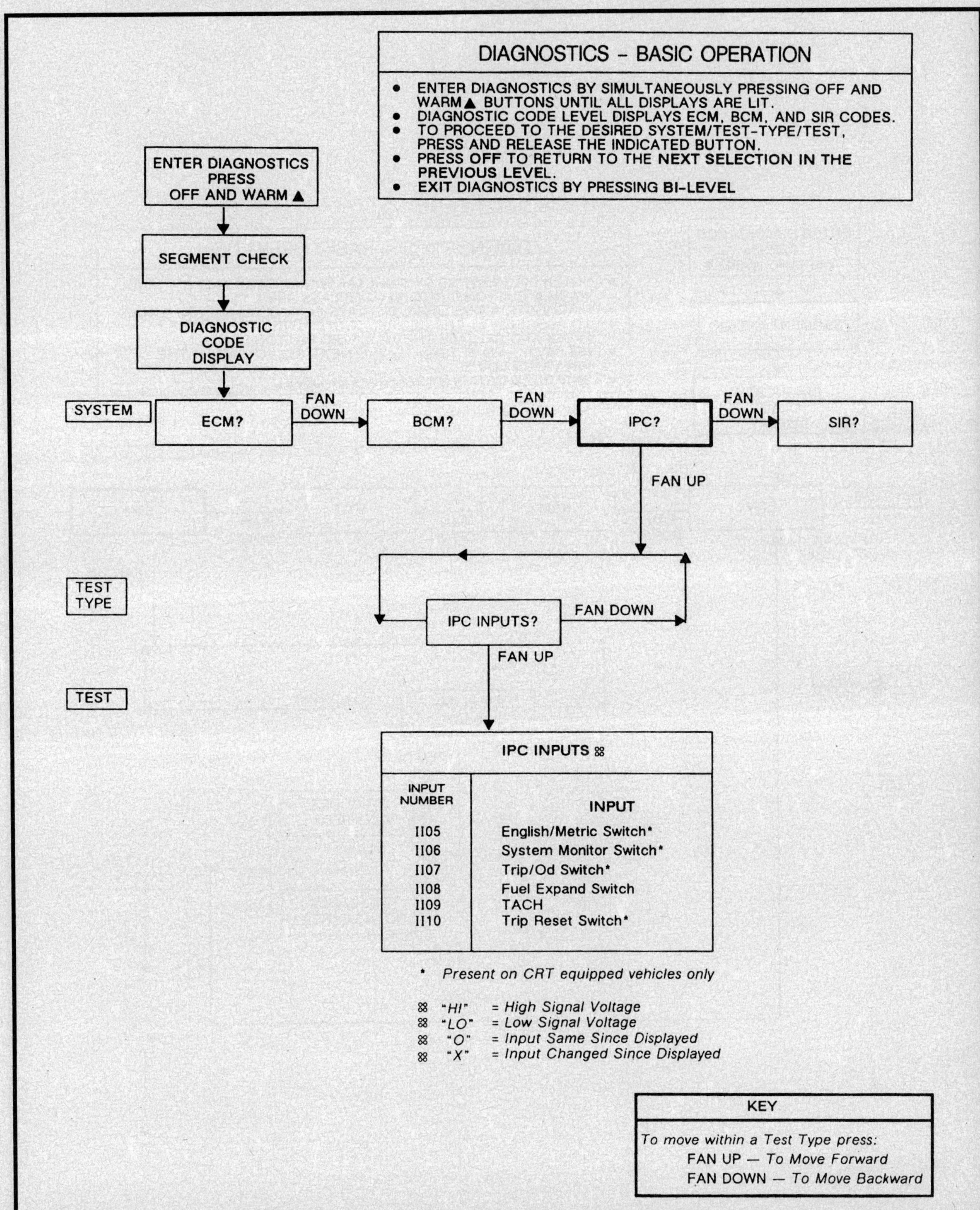

SIR DIAGNOSTIC FLOW CHART — TORONADO AND TROFEO WITHOUT CRT/DID

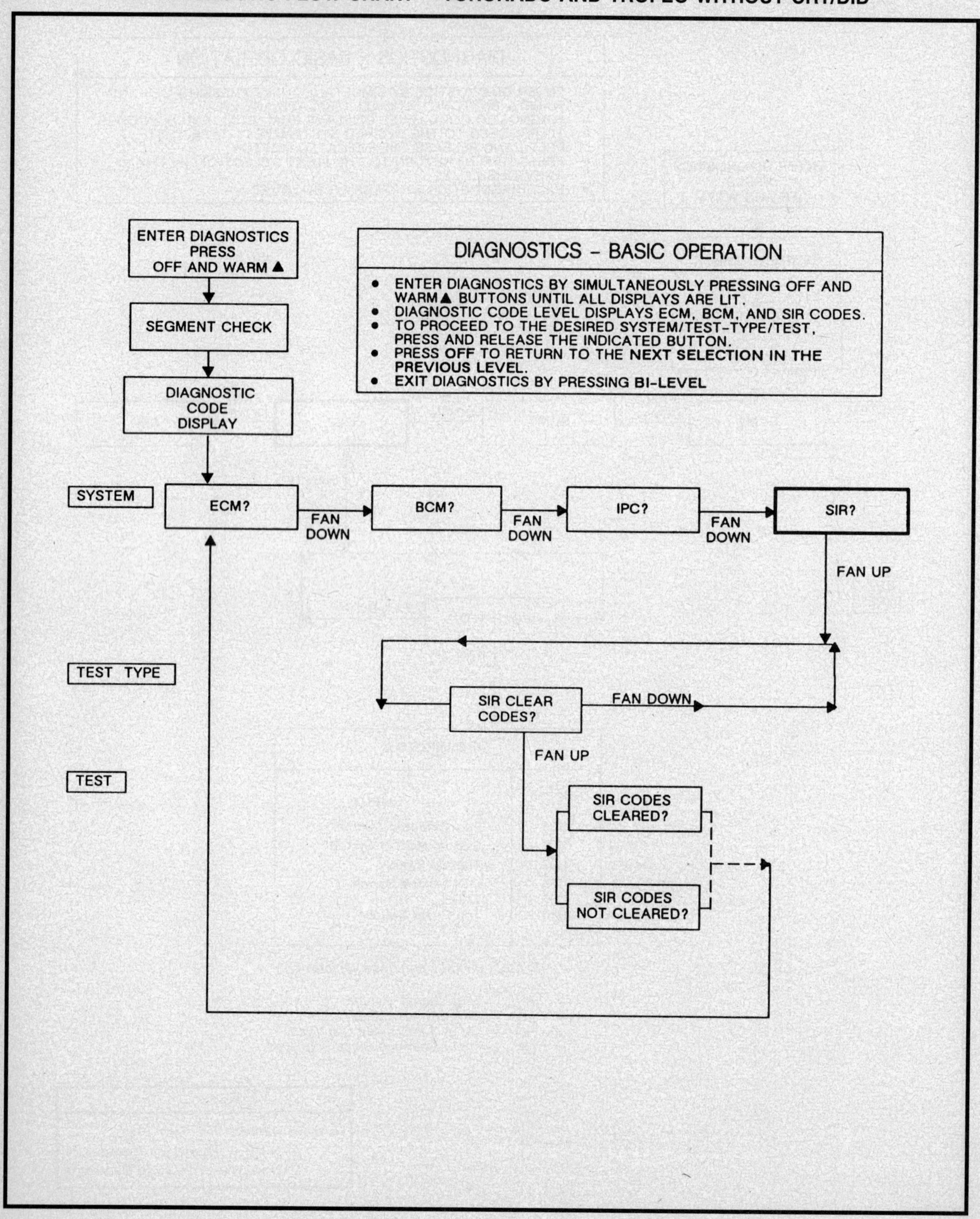

DIAGNOSTIC TROUBLE CODE DISPLAY (DTC)

After diagnostics are entered, any DTC's stored in computer memory will be displayed. Codes may be stored for the ECM, BCM, IPC or SIR systems.

NOTE: SIR refers to the Supplemental Inflatable Restraint System.

Each code consists of the system abbreviation (E, B or R), a three digit code identifier and the letter C (Current codes) or H (History codes). All codes for each system are displayed together in sequence, with the lowest number first (i.e. EO13, EO14 etc.). The final character of the code will be a C or an H. Current (C) means the fault was still present the last time the test was run, while History (H) means the failure occurred within the last 50 ignition cycles.

If no codes are present for a system, a NO X CODE message will be displayed with X representing the particular system letter abbreviation. If the communication line to a component is not operating, a NO X DATA message will be displayed, indicating that the BCM could not communicate with that system. Anytime during DTC display that the RTN softkey is depressed, code display will be bypassed.

SELECTING THE SYSTEM

Following the display of DTC's, the first available system for testing will be displayed (i.e. ECM?). To control the display while selecting a system to test, depress the following softkeys:

- **LEVL** — terminates the system selection process and returns the display to the beginning of the trouble code sequence.
- **NO** — displays the next available system selection. This allows the display to be stepped through system choices, as well as repeated, following the last system choice.
- **YES** — selects the displayed system for testing.
- **RTN** — exits diagnostics and returns the IPC, CRT/DID to normal operation.

SELECTING THE TEST TYPE

After selecting a particular system, the first available test type will be displayed (i.e. ECM DATA?). To control the display depress the following softkeys:

- **LEVL** — returns to the next available system.
- **NO** — displays the next available test type for the selected system. This allows the display to be stepped through all available test type choices, as well as repeated, following the last test type choice.
- **YES** — selects the displayed test type. At this point, the first of several specific tests will appear.
- **RTN** — exits the diagnostics.

SELECTING THE TEST (DATA, INPUTS, OUTPUTS, OVERRIDES)

Four characters of the Season Odometer display will contain a test code to identify the selection. The first 2 characters are letters which identify the system and test type, while the last 2 characters numerically identify the test. The trip odometer on the IPC will display the selected

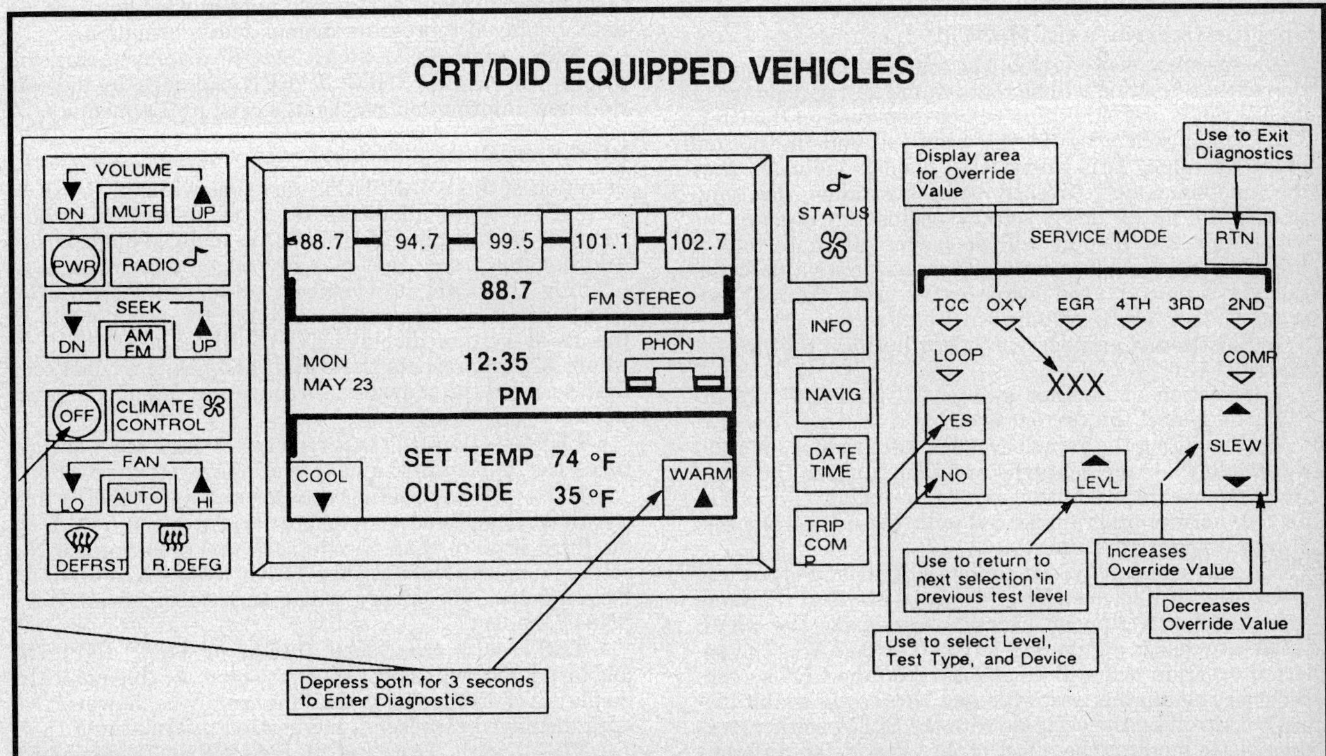

Electronic climate control panel — Toronado and Trofeo with CRT/DID

value as seen by the computer. To control the display depress the following softkeys:

• **LEVL** — stops the test selection process and returns the display to the next available test type for the selected system.

• **NO** — displays the next lower test number for the selected test type. If this button is pressed with the lowest test number displayed, the highest test number will then appear.

• **YES** — displays the next highest number for the selected test type. If this button is pressed with the highest number displayed, the lowest test number will then appear.

• **RTN** — exits the diagnostics.

Specific Data (ECM, BCM)

Selection of the DATA? test type will display the first available test. The DIC will then display the data value the system computer sees for the selected test.

Specific Inputs (ECM, BCM, IPC)

Selection of the INPUT? test type displays the first available test. The DIC will then display the voltage level, HI or LO, of the selected input to the system. Additionally, the DIC will display a 0 or an X. An X indicates that the input has cycled since it was selected. A 0 indicates no change. This indicator may be reset by selecting a new test and then returning to the present test (i.e. press FAN up and then FAN down).

Specific Outputs (ECM, BCM)

Selection of the OUTPUT? test type displays the first available test. The DIC will then display the voltage level, HI or LO, of the selected output of the system. This level will then be cycled between HI and LO every 3 seconds.

Specific Overrides (ECM, BCM)

Upon selecting an Override test function, current operation of that function will be represented as a percentage of its full range and this value will be displayed on the CRT. The display will alternate between '—' and the normal program value. This alternating display indicates that the function is not currently being overridden. Pressing the SLEW up or down softkey begins the override at which time the display will no longer alternate to '—'. Pressing the SLEW up button increases the value while the SLEW down button decreases the value. Normal program control can be resumed in 1 of 3 ways:

1. Selection of another override will cancel the current override.

2. Selection of another system (ECM, BCM, IPC or SIR) will cancel the current override.

3. Overriding the value beyond either extreme (0 or 99) will display '—' momentarily and then jump to the opposite extreme. If the button is released while '—' is displayed, normal program control will resume and the display will again alternate.

The override test type is unique in that any other test type within the selected system may be active at the same time. After selecting an override test, press the LEVL softkey to select another test type (i.e. DATA?). The selected override will still be displayed on the CRT screen. By selecting another test type and test, while simultaneously controlling the override with the SLEW softkey, it is possible to monitor the effect of the override on different vehicle parameters. An example would be to select engine RPM in ECM DATA, while at the same time manipulating the IAC Motor override and viewing the RPM value in the DIC on the IPC.

CLEAR CODES (ECM, BCM, SIR)

Selection of the CLEAR CODES? test type displays either the message CODES CLEAR or CODES NOT CLEAR, indicating whether or not the codes were successfully cleared. The message will appear for 3 seconds. After 3 seconds the display will automatically return to the next available test type for the selected system. Cycle the ignition once or test drive the vehicle to ensure the code(s) do not reset.

SNAPSHOT

Snapshot is a test type that will record all ECM or BCM data and inputs at one instant for review at a later time. The ECM and BCM each have different types of snapshot.

ECM Snapshot

Selection of the SNAPSHOT? test type while in the the ECM system level displays the message SNAP DONE. This message will appear for 3 seconds to indicate all ECM data and inputs have been stored in memory. Following the 3 second interval, the display will automatically proceed to the first available snapshot test type (i.e. SNAPDATA). To control the display depress the following softkeys:

• **LEVL** — stops the test type selection process and returns the display to the next available system selection.

• **NO** — displays the next available snapshot test type. This allows the display to be toggled between SNAP DATA and SNAP INPUTS.

• **YES** — with SNAP DATA? or SNAP INPUTS? displayed, this will select that test type. At this point the display is controlled as it would normally be, however, all data displayed represents memorized information.

• **YES** — with ECM SNAP SHOT? displayed this will display the SNAP SHOT TAKEN message to indicate that new information has been stored in the memory.

BCM Snapshot

Selection of the SNAPSHOT? test type while in the BCM system level will allow the recall of up to 3 snapshots recorded at the time of the setting of BCM fault codes. Additionally, 1 snapshot may be triggered on demand by pushing the FAN up hardkey while the TAKE BCM SNAP SHOT is displayed. Selecting Snapshot while in the BCM system displays CODE BXXX SNAP SHOT, where XXX represents the 3 digit diagnostic trouble code that recorded the snapshot. To control the display depress the following softkeys:

• **LEVL** — stops the test type selection process and returns the display to the next available system selection.

• **NO** — allows scrolling through the list of DTC's that the BCM has stored a snapshot for. After the final DTC (or third if more than 3 codes are set) pressing the NO softkey displays the message TAKE BCM SNAP SHOT?. Pressing the NO softkey will return to the first BXXX SNAP? display.

• **YES** — with the SNAP DATA? or SNAP INPUTS? displayed, this will select that test type. At this point the display is controlled as it would normally be, however, all data displayed represents memorized information.

• **YES** — with TAKE BCM SNAP SHOT? displayed, this will display the SNAP DONE message indicating new information has now been stored in memory.

> **NOTE:** All Snapshot on codeset information will be lost when the code is cleared.

EXITING DIAGNOSTICS

To exit diagnostics, depress the RTN softkey or turn the ignition switch **OFF**. This process, however, will not erase stored trouble codes.

CLIMATE CONTROL IN SERVICE MODE

Upon selecting the diagnostic mode, the climate control will operate in whichever mode was being commanded just prior to entering diagnostics. Even though the display may change just as the buttons are pressed, the prior operating mode is remembered and will resume after diagnostics are complete.

DIAGNOSTIC TROUBLE CODES (DTC)

PCM

- DTC **12** — Diagnostic check only (flash code)
- DTC **13** — Oxygen (O_2) sensor circuit open
- DTC **14** — Engine Coolant Temperature (ECT) sensor circuit out of range (high)
- DTC **15** — Engine Coolant Temperature (ECT) sensor circuit out of range (low)
- DTC **16** — System voltage high or low
- DTC **17** — Spark reference circuit error
- DTC **21** — Throttle Position Sensor (TPS) circuit voltage out of range (high)
- DTC **22** — Throttle Position Sensor (TPS) circuit voltage out of range (low)
- DTC **23** — Intake/manifold air temperature sensor circuit — low temp indicated
- DTC **24** — Vehicle speed sensor circuit signal error
- DTC **25** — Intake/manifold air temperature sensor circuit — high temp indicated
- DTC **26** — Quad-driver module circuit error
- DTC **31** — PRNDL/transmission neutral start/park-neutral position switch circuit error
- DTC **34** — Mass air flow/manifold absolute pressure sensor circuit voltage out of range
- DTC **36** — Transaxle shift problem
- DTC **38** — Brake switch circuit error
- DTC **41** — Cam position sensor/cylinder select error
- DTC **42** — Electronic spark timing/ignition control circuit open or shorted
- DTC **43** — Knock sensor/electronic spark control circuit error
- DTC **44** — Oxygen (O_2) sensor circuit indicates system lean
- DTC **45** — Oxygen (O_2) sensor circuit indicates system rich

- DTC **47** — PCM/BCM Data circuit
- DTC **51** — PROM memory error
- DTC **58** — Personal automotive security system (PASS-key®II) fuel enable circuit error
- DTC **61** — Cruise control vent solenoid circuit error
- DTC **62** — Cruise control vacuum solenoid circuit error
- DTC **65** — Cruise control pressure sensor circuit error
- DTC **67** — Cruise control switches circuit error
- DTC **68** — Cruise control system problem

BCM

- DTC **B110** — Outside air temperature sensor circuit problem
- DTC **B111** — A/C high side temperature sensor circuit problem
- DTC **B112** — A/C low side temperature sensor circuit problem
- DTC **B113** — In-vehicle temperature sensor circuit problem
- DTC **B115** — Sunload sensor circuit failure
- DTC **B119** — Twilight photocell circuit problem
- DTC **B120** — Twilight delay switch pot circuit problem
- DTC **B121** — Twilight enable switch circuit problem
- DTC **B122** — Twilight delay & panel dimming problem
- DTC **B123** — Courtesy lamps panel switch circuit problem
- DTC **B333** — Loss of Supplemental Inflatable Restraint (SIR) Diagnostic Energy Reserve Module (DERM) data
- DTC **B334** — Loss of PCM data
- DTC **B335** — Loss of Climate Control/Driver Information Center (CCDIC) data
- DTC **B336** — Loss of Instrument Panel Cluster (IPC) data
- DTC **B337** — Loss of Heating, Ventilation and Air Conditioning (HVAC) programmer data
- DTC **B410** — Charging system circuit problem
- DTC **B411** — Battery voltage too low or too high
- DTC **B412** — Battery voltage too low or too high
- DTC **B420** — Relay circuits problem
- DTC **B440** — Heating, Ventilation and Air Conditioning (HVAC)air mix door circuit problem
- DTC **B446** — Low refrigerant charge
- DTC **B447** — Low refrigerant charge
- DTC **B448** — Low refrigerant problem
- DTC **B450** — Heating, Ventilation and Air Conditioning (HVAC) coolant temperature too high
- DTC **B552** — Keep alive memory error
- DTC **B556** — Odometer prom error

SELF-DIAGNOSTIC SYSTEMS
RIVIERA, TORONADO AND TROFEO

ECM DIAGNOSTIC FLOW CHART — TORONADO AND TROFEO WITH CRT/DID

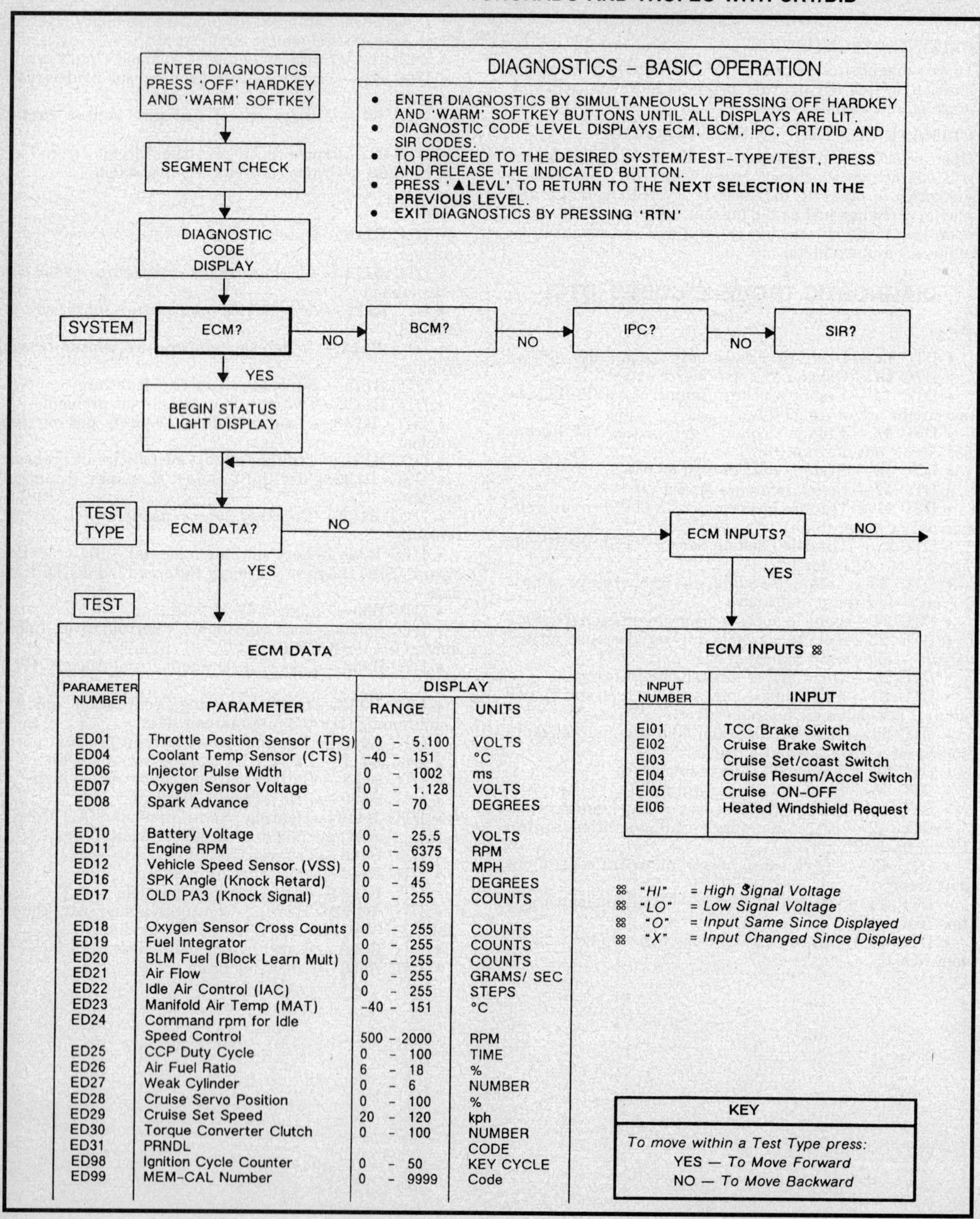

ECM DATA

PARAMETER NUMBER	PARAMETER	DISPLAY	
		RANGE	UNITS
ED01	Throttle Position Sensor (TPS)	0 – 5.100	VOLTS
ED04	Coolant Temp Sensor (CTS)	-40 – 151	°C
ED06	Injector Pulse Width	0 – 1002	ms
ED07	Oxygen Sensor Voltage	0 – 1.128	VOLTS
ED08	Spark Advance	0 – 70	DEGREES
ED10	Battery Voltage	0 – 25.5	VOLTS
ED11	Engine RPM	0 – 6375	RPM
ED12	Vehicle Speed Sensor (VSS)	0 – 159	MPH
ED16	SPK Angle (Knock Retard)	0 – 45	DEGREES
ED17	OLD PA3 (Knock Signal)	0 – 255	COUNTS
ED18	Oxygen Sensor Cross Counts	0 – 255	COUNTS
ED19	Fuel Integrator	0 – 255	COUNTS
ED20	BLM Fuel (Block Learn Mult)	0 – 255	COUNTS
ED21	Air Flow	0 – 255	GRAMS/ SEC
ED22	Idle Air Control (IAC)	0 – 255	STEPS
ED23	Manifold Air Temp (MAT)	-40 – 151	°C
ED24	Command rpm for Idle Speed Control	500 – 2000	RPM
ED25	CCP Duty Cycle	0 – 100	TIME
ED26	Air Fuel Ratio	6 – 18	%
ED27	Weak Cylinder	0 – 6	NUMBER
ED28	Cruise Servo Position	0 – 100	%
ED29	Cruise Set Speed	20 – 120	kph
ED30	Torque Converter Clutch	0 – 100	NUMBER
ED31	PRNDL		CODE
ED98	Ignition Cycle Counter	0 – 50	KEY CYCLE
ED99	MEM–CAL Number	0 – 9999	Code

ECM INPUTS ⊠

INPUT NUMBER	INPUT
EI01	TCC Brake Switch
EI02	Cruise Brake Switch
EI03	Cruise Set/coast Switch
EI04	Cruise Resum/Accel Switch
EI05	Cruise ON-OFF
EI06	Heated Windshield Request

⊠ "HI" = High Signal Voltage
⊠ "LO" = Low Signal Voltage
⊠ "O" = Input Same Since Displayed
⊠ "X" = Input Changed Since Displayed

KEY

To move within a Test Type press:
YES — To Move Forward
NO — To Move Backward

ECM DIAGNOSTIC FLOW CHART — TORONADO AND TROFEO WITH CRT/DID

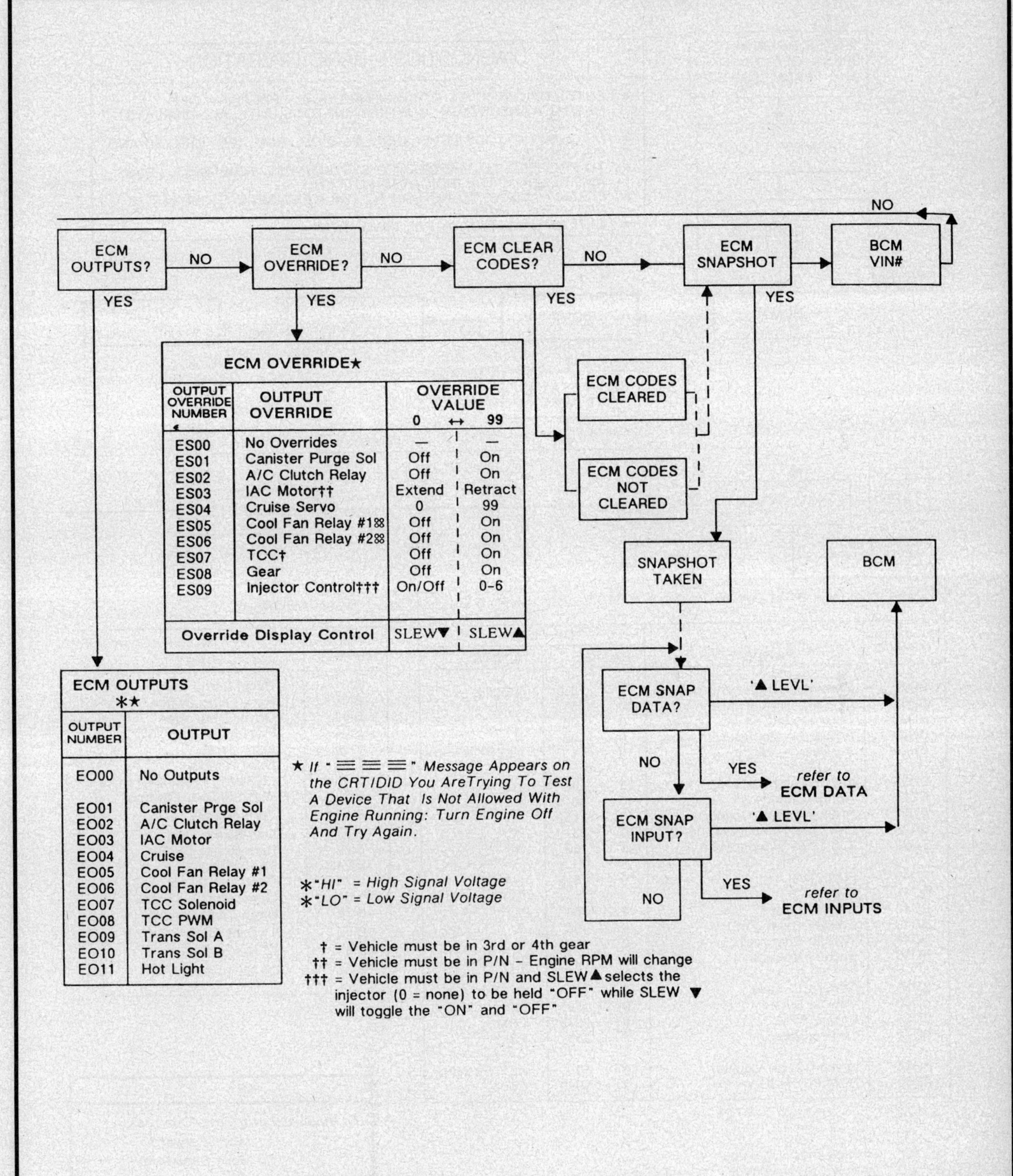

BCM DIAGNOSTIC FLOW CHART — TORONADO AND TROFEO WITH CRT/DID

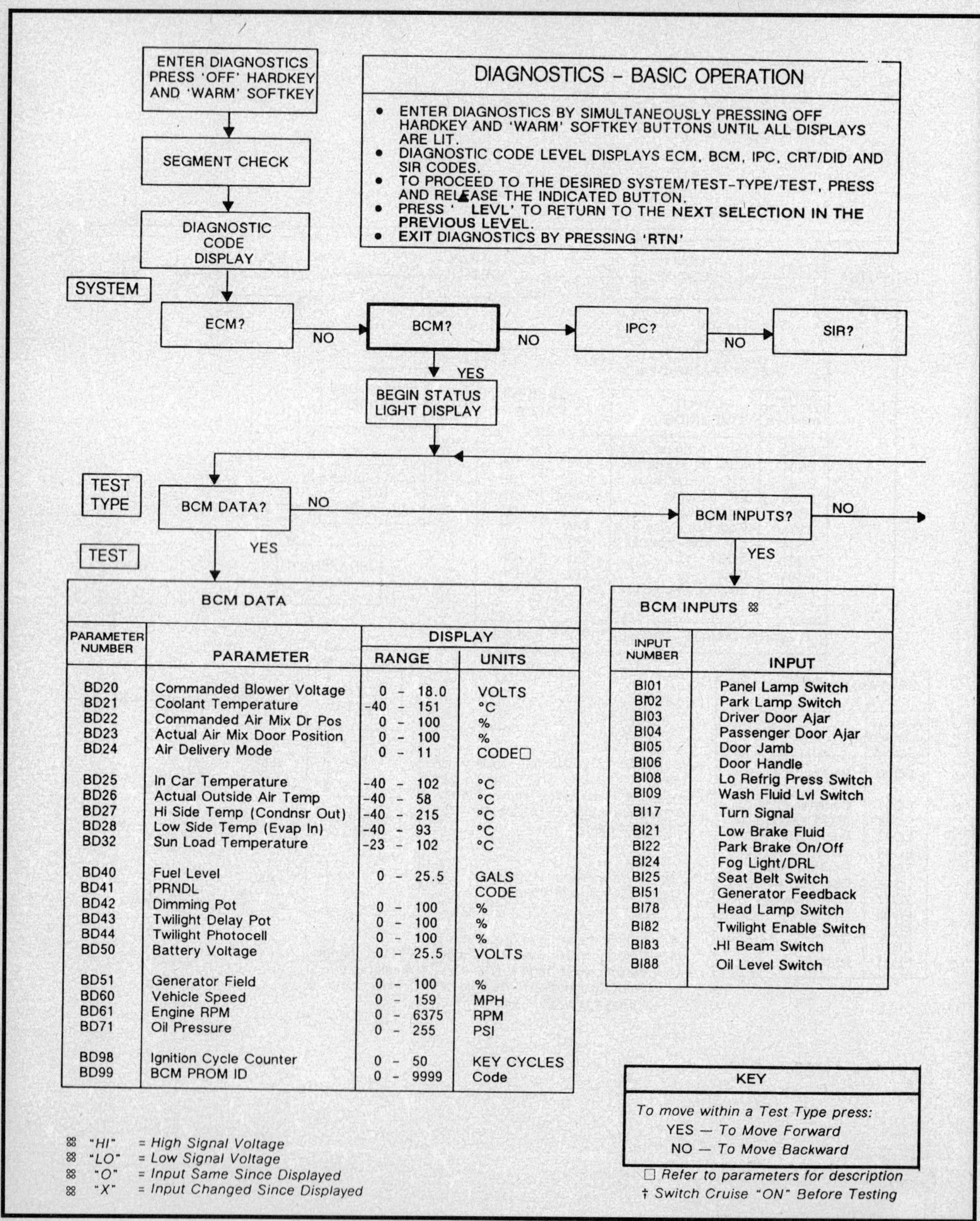

BCM DATA

PARAMETER NUMBER	PARAMETER	DISPLAY	
		RANGE	UNITS
BD20	Commanded Blower Voltage	0 – 18.0	VOLTS
BD21	Coolant Temperature	-40 – 151	°C
BD22	Commanded Air Mix Dr Pos	0 – 100	%
BD23	Actual Air Mix Door Position	0 – 100	%
BD24	Air Delivery Mode	0 – 11	CODE□
BD25	In Car Temperature	-40 – 102	°C
BD26	Actual Outside Air Temp	-40 – 58	°C
BD27	Hi Side Temp (Condnsr Out)	-40 – 215	°C
BD28	Low Side Temp (Evap In)	-40 – 93	°C
BD32	Sun Load Temperature	-23 – 102	°C
BD40	Fuel Level	0 – 25.5	GALS
BD41	PRNDL		CODE
BD42	Dimming Pot	0 – 100	%
BD43	Twilight Delay Pot	0 – 100	%
BD44	Twilight Photocell	0 – 100	%
BD50	Battery Voltage	0 – 25.5	VOLTS
BD51	Generator Field	0 – 100	%
BD60	Vehicle Speed	0 – 159	MPH
BD61	Engine RPM	0 – 6375	RPM
BD71	Oil Pressure	0 – 255	PSI
BD98	Ignition Cycle Counter	0 – 50	KEY CYCLES
BD99	BCM PROM ID	0 – 9999	Code

BCM INPUTS ⊗

INPUT NUMBER	INPUT
BI01	Panel Lamp Switch
BI02	Park Lamp Switch
BI03	Driver Door Ajar
BI04	Passenger Door Ajar
BI05	Door Jamb
BI06	Door Handle
BI08	Lo Refrig Press Switch
BI09	Wash Fluid Lvl Switch
BI17	Turn Signal
BI21	Low Brake Fluid
BI22	Park Brake On/Off
BI24	Fog Light/DRL
BI25	Seat Belt Switch
BI51	Generator Feedback
BI78	Head Lamp Switch
BI82	Twilight Enable Switch
BI83	HI Beam Switch
BI88	Oil Level Switch

KEY

To move within a Test Type press:
YES — To Move Forward
NO — To Move Backward

□ *Refer to parameters for description*
† *Switch Cruise "ON" Before Testing*

⊗ *"HI"* = High Signal Voltage
⊗ *"LO"* = Low Signal Voltage
⊗ *"O"* = Input Same Since Displayed
⊗ *"X"* = Input Changed Since Displayed

BCM DIAGNOSTIC FLOW CHART — TORONADO AND TROFEO WITH CRT/DID

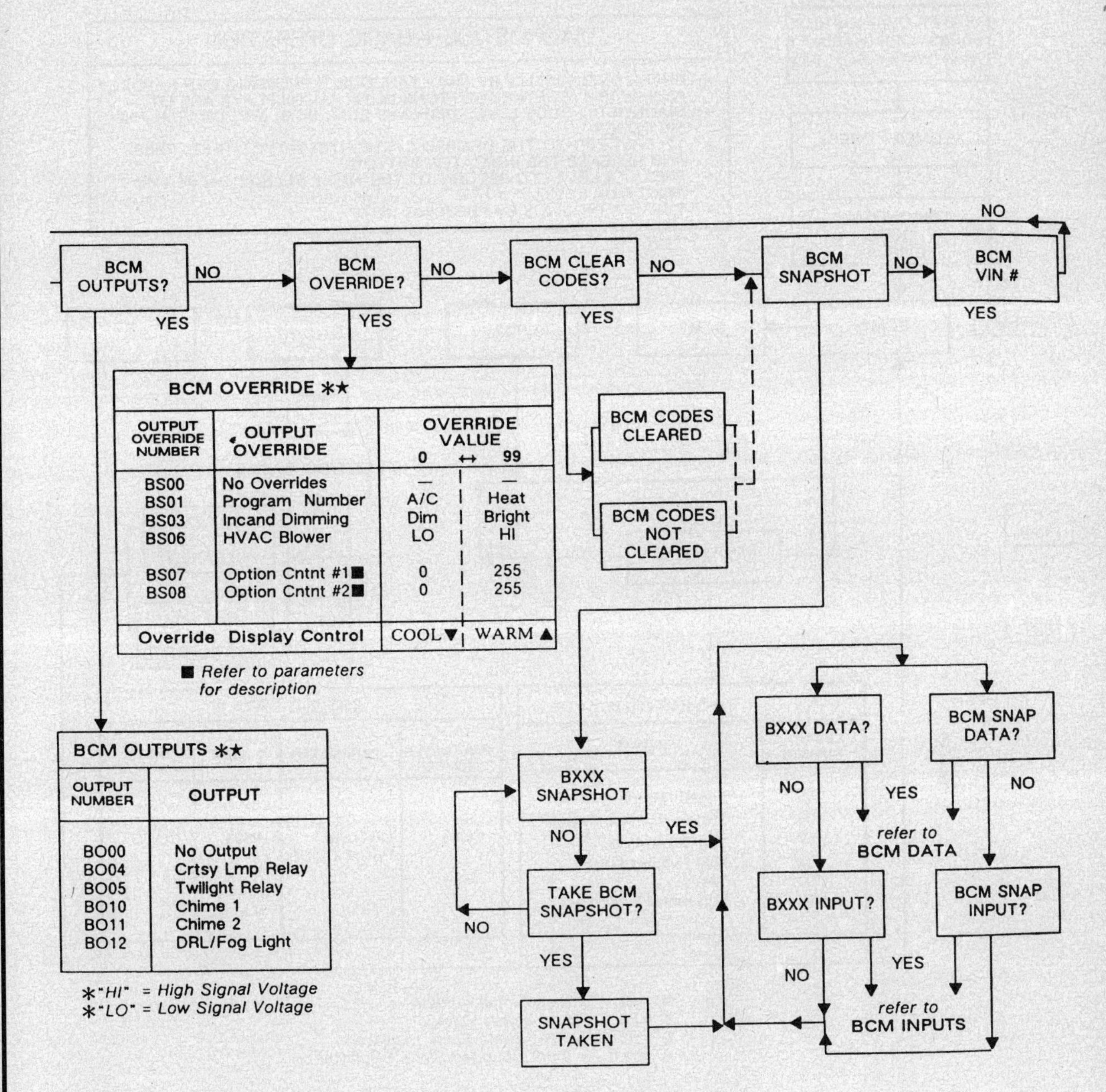

★ If "≡ ≡" Message Appears on the CRT/DID You Are Trying To Test A Device That Is Not Allowed With Engine Running: Turn Engine Off And Try Again.

IPC DIAGNOSTIC FLOW CHART — TORONADO AND TROFEO WITH CRT/DID

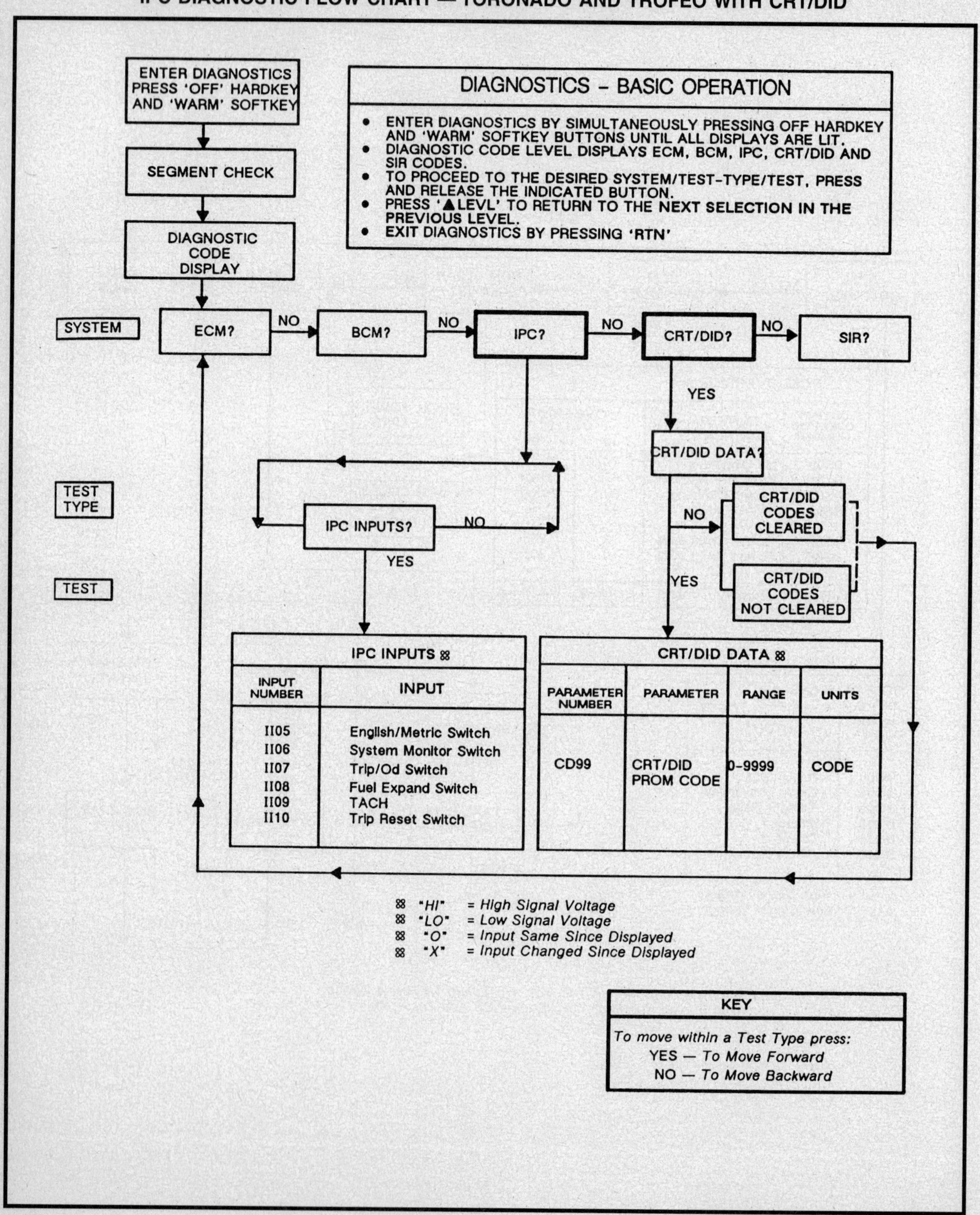

SIR DIAGNOSTIC FLOW CHART — TORONADO AND TROFEO WITH CRT/DID

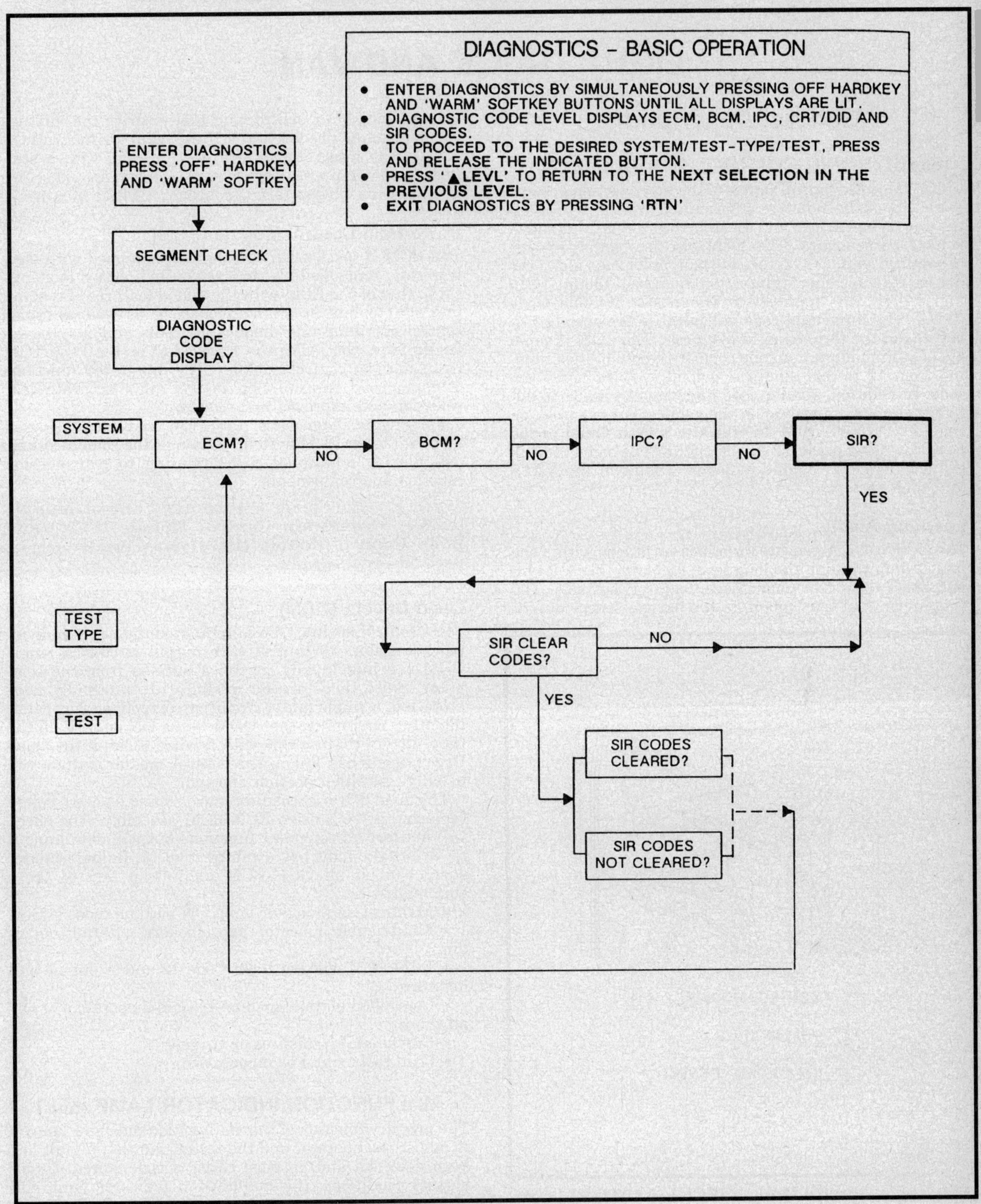

LIGHT TRUCK AND VAN

General Description

ELECTRONIC CONTROL MODULE (ECM)

The ECM is the control center of the fuel injection system. It continuously monitors information from a variety of sensors and switches, and controls the systems that affect vehicle performance. The ECM also has self-diagnostic capability, and can detect system problems, alert the driver through the Malfunction Indicator Lamp (MIL) and record the appropriate Diagnostic Trouble Code (DTC). The numerical code will identify the specific circuit to aid the technician in diagnosis. The ECM is capable of storing more than one code, with possibilities ranging from 12 to 999, however not all systems use every code. In addition, similar code numbers can relate to different circuits depending on the vehicle. For example, on the 3.3L (VIN N), code 46 indicates a fault found in the power steering pressure switch circuit. The same code on the 5.7L (VIN F) engine indicates a fault in the VATS anti-theft system.

Learning Ability

The ECM can compensate for minor variations within the fuel system through the long term fuel trim (block learn) and short term fuel trim (fuel integrator) systems. The short term fuel trim monitors the oxygen sensor output voltage, adding or subtracting fuel to drive the mixture rich or lean as needed to reach the ideal air fuel ratio of 14.7:1. The integrator values may be read with a scan tool; the display will range from 0–255 and should center on 128 if the oxygen sensor is seeing a 14.7:1 mixture.

Powertrain Control Module (PCM)

The PCM is used when vehicles are equipped with electronically controlled transmissions. Its function is identical to that of the ECM with the added features of controlling shift points in the transmission, as well as cruise control operation and diagnostics. This unit may be referred to as the PCM, the ECM/PCM or the PCM/TCM (transmission control module). The integrated functions of engine and transmission control allow accurate gear selection and improved fuel economy.

For engine diagnostics, the PCM may be considered identical to an ECM system, although the combined unit will display additional codes relating to transmission function and components.

NOTE: The terms Electronic Control Module (ECM), Powertrain Control Module (PCM) and Body Control Module (BCM) may sometimes be used interchangably.

Quad Drivers (QDM)

Some control modules have eliminated the use of individual transistors to control each output circuit. A single QDM can individually control 4 outputs from the computer. Since these control modules are internally fault protected, a single faulty circuit or component may cause the other circuits on that QDM to be ON or OFF all the time. Do not confuse this with a failed ECM; if the Quad Driver itself has failed, the control module output will usually test either open or shorted.

The quad driver check procedure should be used before replacing an expensive ECM/PCM assembly. The check will not test all computer functions but will determine if an external circuit has disabled a QDM. Be particularly suspicious of an ECM/PCM exhibiting any of these characteristics:

• Malfunction indicator lamp ON with no codes stored.
• Malfunction indicator light flickers, intermittent or dim.
• ECM/PCM will not flash Code 12 and/or engine will not start.
• Controlled output, such as TCC, is inoperative or ON all the time.
• Engine stalls, misfires or surges.
• Scan tool erratic or inoperative.

MALFUNCTION INDICATOR LAMP (MIL)

The primary function of the malfunction indicator lamp is to advise the operator and the technician that a fault has been detected, and, in most cases, a code stored. Under normal conditions, the malfunction indicator lamp will illuminate when the ignition is turned **ON**. Once the engine is started and running, the ECM will perform a sys-

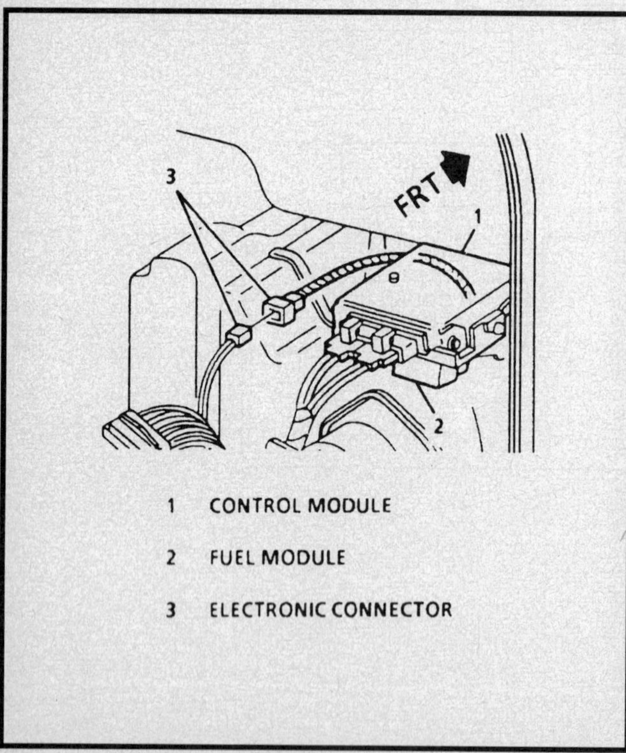

Electronic Control Module (ECM) — 1993–94 light duty truck shown

tem check and extinguish the warning lamp if no fault is found.

Additionally, the malfunction indicator lamp can be used to retrieve stored codes after the system is placed in the Diagnostic Mode. Codes are transmitted as a series of flashes with short or long pauses. When the system is placed in the Field Service Mode, the dash lamp will indicate open loop or closed loop function to the technician.

Intermittents

If a fault occurs intermittently, such as a loose connector pin breaking contact as the vehicle hits a bump, the ECM will note the fault as it occurs and energize the malfunction indicator lamp. If the problem self-corrects, as with the terminal pin again making contact, the dash lamp will extinguish after 10 seconds but a code will remain stored in the ECM memory.

When an unexpected code appears during diagnostics, it may have been set during an intermittent failure that self-corrected; the codes are still useful in diagnosis and should not be discounted.

Failure Codes

Whenever the ECM detects a sensor or switch signal that is out of range, it will turn ON the MIL. In addition, if the ECM detects an internal malfunction, it will also turn ON the MIL. Distinguishing between current and history codes is absolutely necessary for successful diagnosis. Current codes are also known as hard failures, or a problem that exists at the present time. For example, if a vehicle came into the shop with the MIL light on steady with the engine running, a trouble code could be immediately retrieved and that particular diagnostic trouble chart could be referenced.

History codes, also known as intermittents, are failures which have taken place at one time, but are not presently causing the MIL light to turn ON. An example of this would be if the ignition timing had been recently adjusted on a 5.0 liter engine with TBI. When the bypass wire was disconnected to make the adjustment, the open circuit would have caused the ECM to revert to base timing. Because of this change, the ECM detected a problem and turned on the MIL while simultaneously recording a trouble code 42. When the timing adjustment was completed and the bypass wire was reconnected, the ECM waited approximately 10 seconds and then turned off the MIL, but retained the code. The ECM would retain this information up to a maximum of 50 ignition cycles. If code retrieval was attempted in the meantime by another technician unaware of the timing adjustment, their only recourse would be to reference the wiring diagram associated with the code 42 chart and physically inspect all connections.

DATA LINK CONNECTOR (DLC)

The data link connector is located underneath the instrument panel and is used to access the on-board microprocessors. These include the ECM, PCM, BCM as well as various other controllers such as the anti-lock brake system module. The use of a scan tool such as GM's Tech 1, or equivalent, will allow the technician to monitor data and control some of the microprocessor's operations. The scan tool will read the serial data signal from the DLC and translate the signal into useful diagnostic information on the display screen. The DLC can also be used without a scan tool for retrieving trouble codes, entering certain diagnostic modes and other valuable testing.

Terminal A — DLC Ground

Circuit number 450 is the ECM ground through the instrument panel connector. The ECM is grounded at the engine and splices into the instrument panel harness.

> **NOTE: Most scan tools can disguise problems related to ECM grounds due to the alternate ground path through the tool's power cord.**

Terminal B — Diagnostic Enable

Circuit number 451 is connected to the ECM through the instrument panel connector. It is normally at a 5 volt level. The scan tool has the ability to lower the voltage level of circuit 451 by dropping a fixed resistance across the circuit or shorting it to ground. The various low voltages on circuit 451 are referred to as Diagnostic Modes.

Terminal C — Air Solenoid

On most vehicles with Air Injection Reaction (AIR) Systems, circuit 436 is spliced between the ECM and the AIR solenoid, through the instrument panel connector to terminal C. The solenoid is responsible for directing airflow from the pump to either the exhaust manifold, catalytic converter or the atmosphere via an attached silencer. This terminal is very helpful for testing the closed loop electronics. With the engine running in closed loop, connect a jumper wire between DLC terminals C and A. By grounding terminal C you will be enabling the air control solenoid to direct airflow to the exhaust manifold. This should cause a low O_2 sensor output voltage and a high (greater than 128) short term fuel trim number. If connected long enough, this action will trigger a code 44, lean exhaust.

In Open Loop operation, this solenoid will be enabled to quickly bring the O_2 sensor up to operating temperature. Once the system goes to Closed Loop the solenoid will become disabled causing airflow to be directed to either the catalytic converter or the atmosphere depending on the type of system. These actions can be verified using a digital voltmeter by probing terminal C with the engine running. In open loop the voltmeter should read 0 volts (solenoid energized) and in closed loop the reading should be battery voltage (solenoid de-energized).

Terminal D — Malfunction Indicator Lamp

Circuit 419 is spliced between the ECM and the MIL light to terminal D. With the key **ON**, the MIL light is turned ON through ECM terminal A5. If during a normal bulb check the MIL fails to turn on, a quick test can be performed before condemning the ECM. Insert a jumper between terminals D and A of the DLC with the key **ON**. This action should turn ON the MIL. If it does not, a defective bulb is the most likely problem.

Terminal E — Serial Data Line

Circuit 461 is connected through the instrument panel connector to the ECM. This is a discreet data line used on fuel injection systems with a slow data transmission rate (baud rate).

DIAGNOSTIC MODE	VOLTAGE AT TERMINAL B	RESISTANCE ACROSS A AND B
Road Test or Open	5V	Infinite
Diagnostic or 10K mode	2.5V	10K ohms
Back Up Fuel	1.5V	3.9K ohms
Field Service	0V	0 ohms

Diagnostic modes

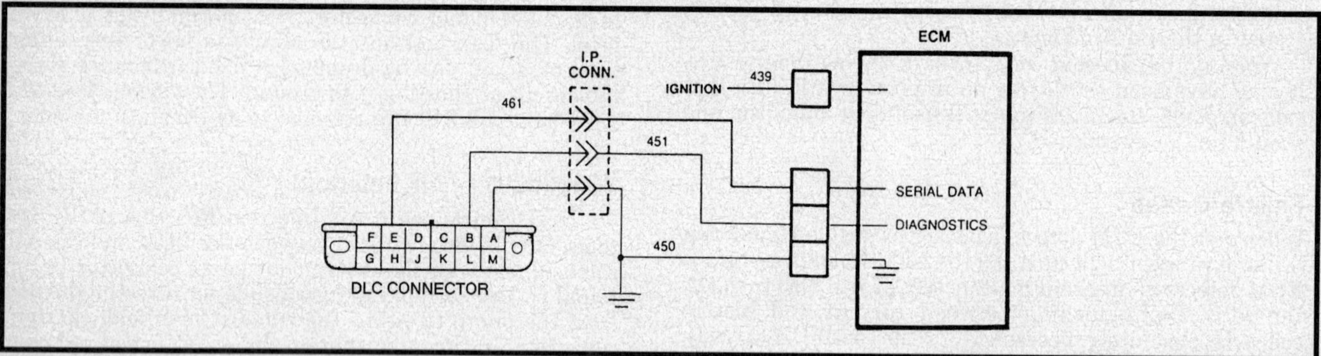

Terminals A, B and E — DLC

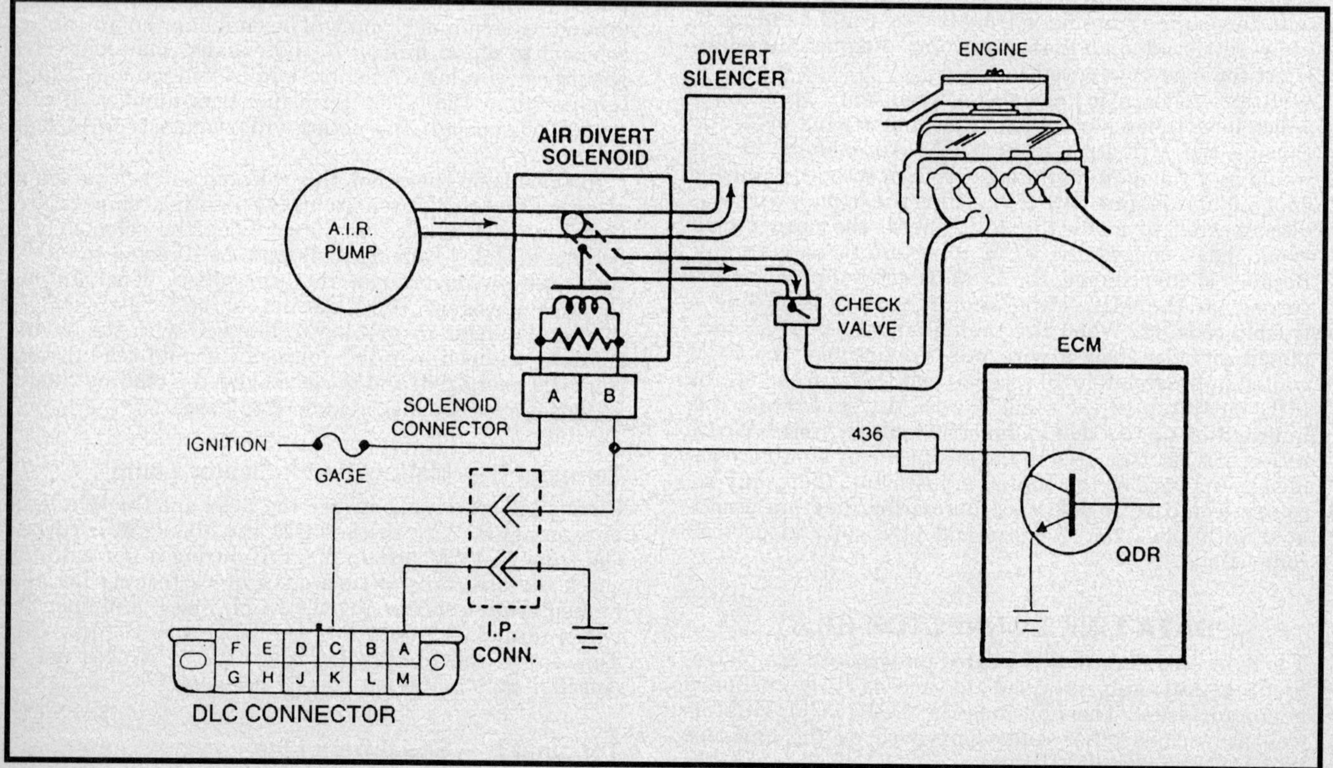

Terminal C — Air control solenoid

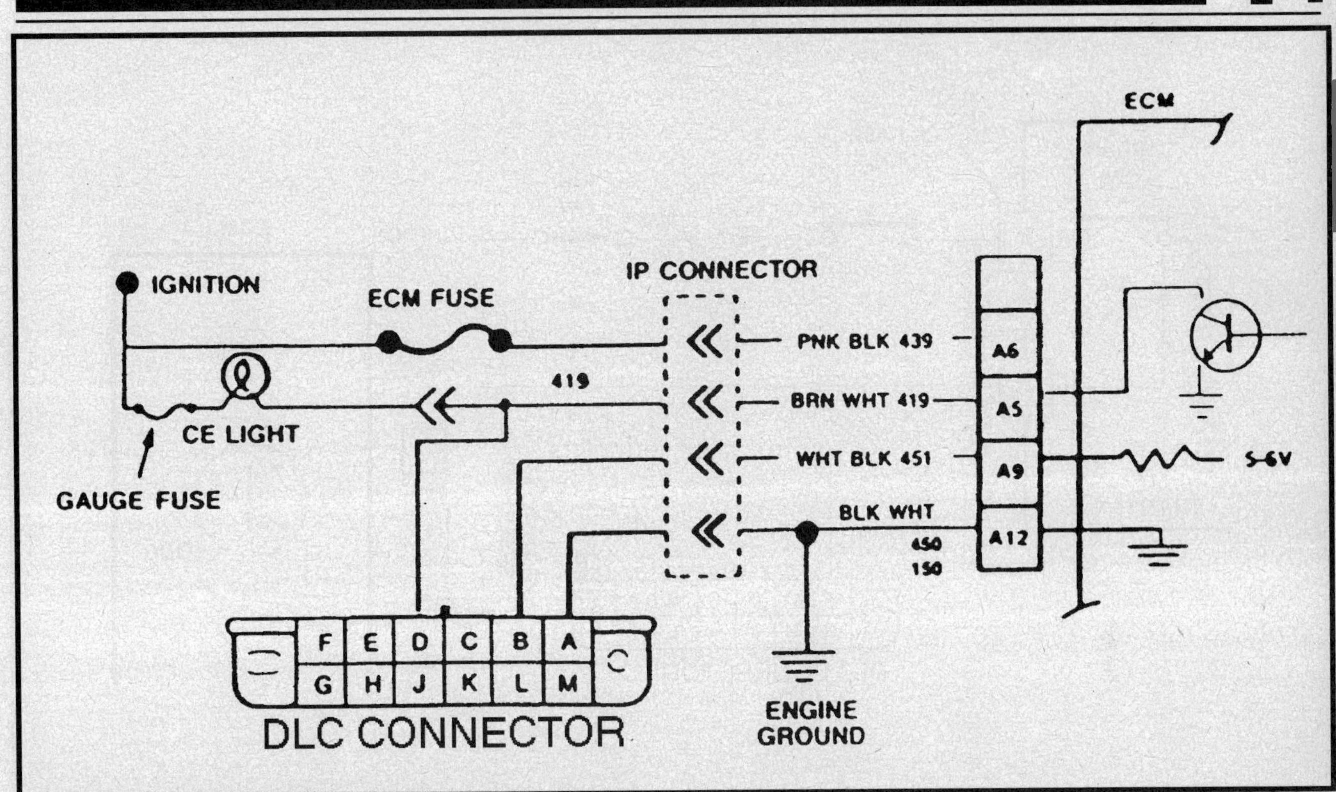

Terminal D — Malfunction indicator lamp

Terminal F — Torque Converter Clutch Solenoid

Circuit 422 is connected through the instrument panel connector and spliced into the ground-controlled TCC solenoid circuit. The ECM will turn on the TCC solenoid based on a variety of sensed inputs. By inserting a test light between DLC terminals F and A, TCC command status can be observed. With the brake switch closed (brakes not applied) and the vehicle in 3rd gear (3rd gear switch closed), the test light should be ON. As soon as the ECM commands TCC engagement, the test light should turn OFF.

Terminal G — Fuel Pump Test Point

Most models have a fuel pump test connector located in the engine compartment. Applying B+ voltage through a fused jumper wire will energize the in-tank fuel pump. This is helpful when performing fuel pressure testing and is more convenient than cycling the key to turn ON the pump. On other models that use terminal G in the DLC, it is for other purposes than fuel pump. When terminal G is either vacant or used for another purpose, a fuel pump test lead will be located in the engine compartment, usually near the battery.

Terminal M — Serial Data Line

Circuit 461 is the serial data line for fast baud rate ECMs such as the 3.1L (Vin T) engine. Terminal M may also include serial data transmission from other on-board microprocessors such as an anti-lock brake system controller.

MEMORY

NOTE: Not every type of memory control is used in each vehicle application.

- **Read Only Memory (ROM)** is a permanent memory physically soldered to the circuit boards within the ECM. The ROM contains the overall control algorithms. The ROM is a non-volatile memory and cannot be changed. Because of this, it does not require a constant supply of battery voltage to be retained.
- **Random Access Memory (RAM)** is the computer scratch pad. The computer can read from or write into this memory as needed. This memory is volatile and will be lost without a constant supply of battery voltage.
- **Programmable Read Only Memory (PROM)** is the portion of the ECM that contains different engine calibration information that is specific to year, model and emissions. It is for this reason that the PROM is referred to as the calibrator. The PROM is a non-volatile memory that is read only by the ECM. The PROM is removable from the ECM and should be retained with the vehicle following ECM replacement.
- **Electrically Erasable Programmable Read Only Memory (EEPROM)** is a permanent memory that is physically soldered to the circuit boards within the ECM. The EEPROM takes the place of the ROM and the PROM and is not serviceable. Those vehicles using an EEPROM can be re-programmed by using the TECH 1 or other equivalent equipment. Some ECM's using an EEPROM use a separate knock sensor module underneath the computer's access cover. This chip must be removed before replacing the old ECM and reinstalled into the new computer.

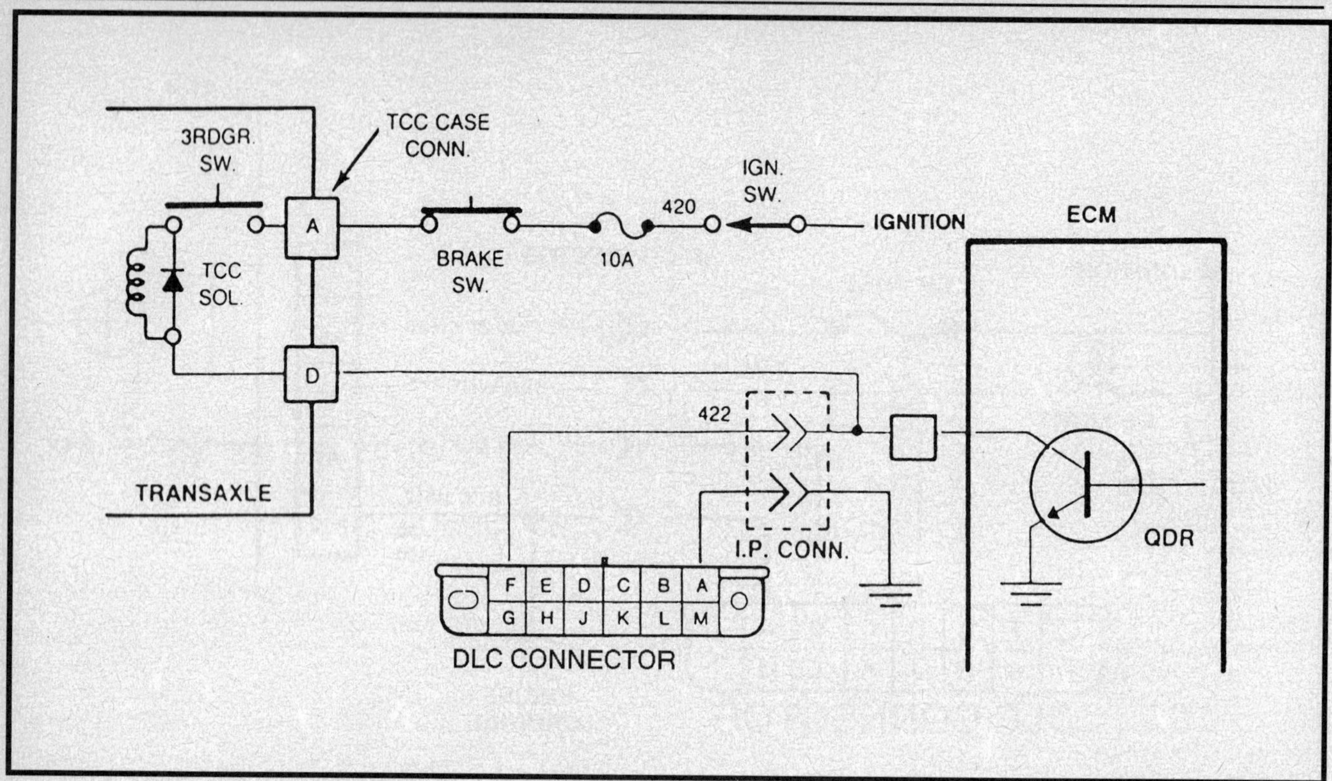

Terminal F — Torque converter clutch solenoid

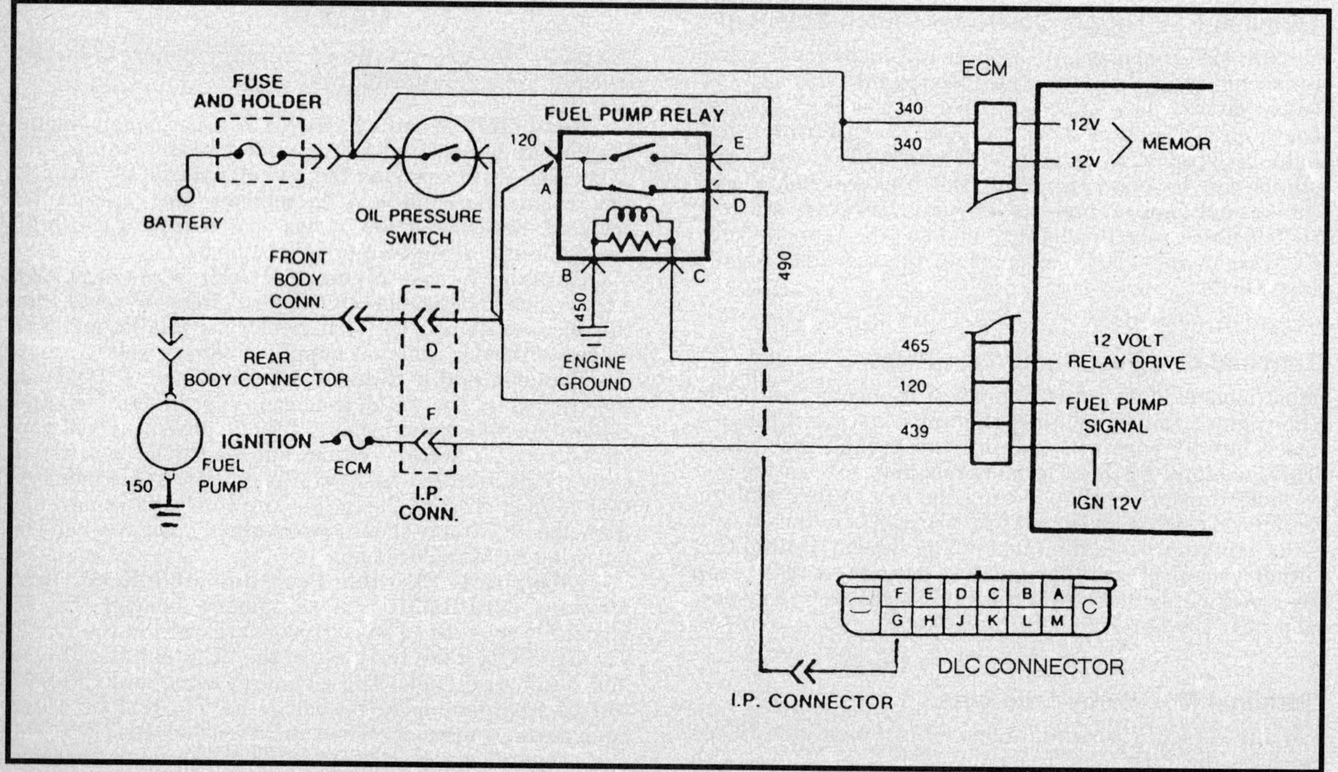

Terminal G — Fuel pump test point

Terminal M — Serial data line

- **The Calibration Pack (CALPAC)** is physically soldered to the ECM and contains information for the fuel back-up mode of the computer. The CALPAC is not serviceable and failure would require replacement of the ECM.
- **The Memory Calibration (MEMCAL)** combines the function of the PROM and CALPAC as well as cruise control and knock control modules in one unit. This unit is serviced just like a PROM.
- **The Short Term Fuel Trim (STFT) (formerly called integrator)** is an ECM volatile memory register that will contain a number between 0 and 255. The neutral value for STFT is 128; any deviation from this value indicates a change in the injector pulse width as commanded from STFT. This function is only active in the closed loop mode of operation. As the ECM monitors the oxygen sensor voltage input, it is constantly varying the STFT value. For example, if a vacuum leak were to occur causing a lean condition (O_2 voltage low), the STFT would respond by increasing injector pulse width. This increase would be seen on a scan tool as a number greater than 128. Conversely, if the air filter became plugged causing a rich condition (O_2 voltage high), the STFT would respond with a decrease in injector pulse width; a number less than 128. Because this value is updated very quickly, the STFT only corrects for short term mixture trends.
- **The Long Term Fuel Trim (LTFT) (formerly called block learn)** is a matrix of cells arranged by RPM and load. As engine operating conditions change, the ECM will switch from cell to cell to determine which LTFT factor is appropriate. While in any given block, the ECM also monitors the STFT. If the STFT is far enough from 128 in either direction, the ECM will change the LTFT value for that given cell. Once the LTFT value is changed the STFT value should revert back to 128, this represents a neutral condition. If the mixture is still not correct (as judged by the O_2 sensor), the process of short and long term fuel correction will continue until they reach the limits of their control. These limits are programmed into the PROM and would be different for each vehicle. At this point of maximum correction a trouble code 44 or 45 would be registered in the computer's RAM. As with STFT the long term fuel trim is only active in closed loop operation. The LTFT represents the computer's learning capability as these values can be tailored to the specific driving habits of the operator. Once battery power is lost however, the LTFT values will revert back to default settings, and a loss of performance may be noticed. Original performance should return after the vehicle is operated for a period of time in closed loop.

Tools and Equipment

SCAN TOOLS

The system can communicate a wide variety of information through the Data Link Connector (DLC), previously known as the ALDL. Depending on the application, serial data is transmitted to terminals E or M of the DLC. This data is transmitted at a high frequency which requires a quality scan tool for interpretation. Although stored codes may be read with only the use of a small jumper wire, the use of a hand-held scan tool such as GM's TECH 1 or equivalent, is recommended. There are many manufacturers of these tools; a purchaser must be certain that the tool is proper for the intended use.

The scan tool allows any stored codes to be read from the ECM memory. The tool also allows the operator to view the data being sent to the ECM while the engine is running. This ability has obvious diagnostic advantages; the use of the scan tool is frequently required by the diagnostic charts.

With an understanding of the data stream that the tool will display and knowledge of the circuits involved, the tool can be an extremely valuable tool when examining the system. Without scan tools, diagnostic information would be difficult to obtain in some circuits and impossible to retrieve in others. Scan tools do not make the use of trouble code charts unnecessary, nor do they pinpoint the exact area in a circuit where the problem exists. However, this type of equipment will give more specific information, such as sensor input readings. An example of this would be the coolant sensor circuit. While the ECM looks at specific voltage levels to determine coolant temperature, the scan tool will display this information as an actual temperature reading. This information is not only useful in determining system malfunctions, but is also very helpful in diagnosing mechanical problems. For example, if the engine is overheating, the scan tool can be used to look at the actual engine temperature. This is helpful when checking for electric cooling fan enable switch problems, sticking thermostats, etc.

To properly read system values with a scan tool, the following conditions must be met. All normal values given in the charts will be based on these conditions:

- Engine running at idle, throttle closed
- Engine warm, upper radiator hose hot
- Vehicle in park or neutral
- System operating in closed loop
- All accessories **OFF**

The ECM is capable of communicating with a scan tool in 4 modes: Manual Diagnostic, Field Service, Road test, Diagnostic and Back Up Fuel.

Manual Diagnostic Mode (0 Ohm State Field Service)

When the diagnostic terminal in the data link connector (DLC) is grounded with the key **ON** and the engine OFF, the system will enter the diagnostic mode. In this mode the following activities take place:

1. Displays a DTC 12 indicating the system is operational.

2. Displays any stored fault codes 3 times consecutively.

3. Energizes all system controlled relays and solenoids except the fuel pump relay. This allows checking circuits in the shop without having to simulate certain operating conditions.

4. The idle air control (IAC) valve moves to it's fully extended position, blocking the idle air passage. This is useful when adjusting minimum idle speed.

5. The electric cooling fan is enabled on some vehicles.

NOTE: Battery voltage may fall considerably if left in this mode for extended periods. This voltage drop could cause false readings. Connect a battery charger if time exceeds 30 minutes.

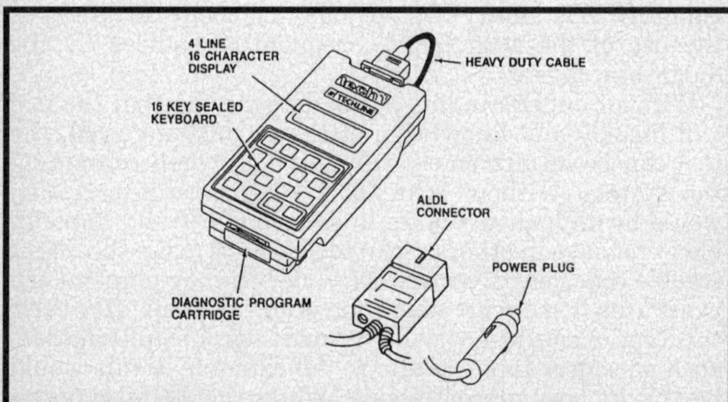

Typical scan tool — GM's Tech 1 shown

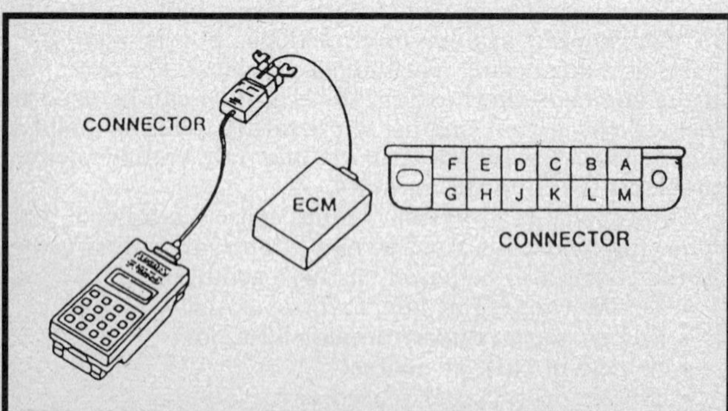

DLC connection to scan tool

Field Service Mode (0 Ohm State)

If the diagnostic terminal is grounded with the engine running, the system will enter the Field Service Mode. In this mode the following activities take place:

1. The MIL light will flash 2.5 times per second to indicate Open Loop operation.

2. The MIL light will flash once per second to indicate Closed Loop operation and a 14.7-to-1 fuel mixture delivery.

3. While in Closed Loop, if the light is OFF most of the time it indicates a lean condition, if the light is ON most of the time it indicates a rich condition.

4. When in Field Service Mode, new codes cannot be stored and the Closed Loop timer is bypassed.

―――――――――――― **WARNING** ――――――――――――
The vehicle should not be driven while in field service mode. Damage to the catalytic converter could result.
―――――――――――――――――――――――――――――――――――――

Road Test Mode (Open Mode)

Road test mode can be accessed with the appropriate scan tool. This mode allows the ECM/PCM to function normally, without affecting fuel control or spark timing. This is the preferred mode when testing with a scan tool. On some vehicles, only a few data parameters may be displayed on the scan tool while a NO DATA message may appear on others. In these situations, the Diagnostic Mode (10K Mode) must be selected to access the data stream. This mode is very helpful on those vehicles that use multiple microprocessors. On these models the data stream is active all the time because it is constantly transmitting information to the BCM, IPC, etc..

Diagnostic Mode (10K Mode or Special Mode)

Diagnostic mode can be accessed with the appropriate scan tool. In this mode all available serial data is displayed on the scan tool. In addition, the following activities take place:

• ECM timers are bypassed.

• On systems equipped with a knock sensor, spark timing is advanced.

• IAC valve is controlled to maintain a fixed idle speed, around 1000 RPMs.

• Park/Neutral restrict functions are disabled.

• On some vehicles the charcoal canister purge solenoid will be commanded to disable purge.

NOTE: When in diagnostic mode the idle speed will be abnormally high. With some scan tools this will occur as soon as the tool is connected.

Back Up Fuel Mode (Factory Test or 3.9K Mode)

This mode verifies the operation of the fuel back up circuitry inside the computer, namely the CALPAC. Engine operation is very erratic in this mode due to the limited amount of sensor information being used. While in back up fuel mode, electronic spark timing is not functional and fuel calculations are based on the following: RPM (distributor reference signal), TPS (throttle position signal) and CTS (engine coolant temperature signal).

SCAN TOOL USE FOR INTERMITTENTS

In some tool applications the data update rate is so slow that it becomes less effective than a digital voltmeter for checking sporadic voltage changes. With the rapid data transmission of a quality scanner, intermittent problems such as faulty wiring connections can be more successfully diagnosed. For example, while manipulating a suspect electrical circuit or component, observe the scan tool display screen. A rapid voltage change or erratic reading will uncover the trouble area.

The scan tool is also an easy way to compare the operating parameters of a poorly operating engine with that of a known good one. For example, a sensor may shift in value and cause a driveability problem, but not set a trouble code. Comparing the sensor's readings to a known good one may uncover the problem.

Scan tools have the ability to speed up diagnostic time and prevent the replacement of good parts, however a thorough understanding of the system you are working on, as well as a working knowledge of the scan tool is essential.

ELECTRICAL TOOLS

The most commonly required electrical diagnostic tool is the digital multimeter, allowing voltage, resistance and amperage to be read by one instrument. The multimeter must be a high-impedance unit, with 10 megohms of impedance in the voltmeter. This type of meter will not place an additional load on the circuit it is testing; this is extremely important in low voltage circuits. The multimeter must be of high quality in all respects. It should be handled carefully and protected from impact or damage. Replace batteries frequently in the unit.

Other necessary tools include an unpowered test light, a quality tachometer with inductive (clip-on) pick up and the proper tools for releasing GM's Metri-Pack, Weather Pack and Micro-Pack terminals as necessary. The Micro-Pack connectors are used at the ECM connector. A vacuum pump/gauge may also be required for checking sensors, solenoids and valves.

Diagnosis and Testing

SERVICE PRECAUTIONS

- To prevent internal ECM damage, the ignition must be **OFF** when disconnecting or reconnecting power to the computer.
- When handling a PROM, CALPAC or MEMCAL, do not touch the component leads. Also, do not remove the integrated circuit from the carrier.
- Never allow welding cables to lie on, near or across any vehicle electrical wiring.
- Leave new components and modules in the shipping package until ready to install them.
- When performing electrical tests on the system, use a high impedance multimeter, digital voltmeter (DVM) J-34029-A or equivalent.
- To prevent possible electrostatic discharge damage to the ECM, do not touch the connector pins or soldered components on the circuit board.

PRELIMINARY INSPECTION

In order to successfully locate and repair problems in complex computer controlled systems, the technician needs to adhere to a comprehensive, sequential method of diagnostics. Many times a vehicle has a malfunction and

the worst is suspected. Premature conclusions often result in the replacement of good parts and wasted time. A proper sequence of steps needs to be established and strictly adhered to when troubleshooting a vehicle. These are as follows:

1. If possible, try to speak directly to the customer, especially if the complaint is not associated with an intermittent or steady MIL.
2. Road test the vehicle, if necessary, to verify the complaint.
3. Perform a thorough visual/physical inspection. This step can very often eliminate the need for further testing. Loose wires, especially ground circuits, can create mysterious driveability problems. However, a complete visual inspection can uncover these problems quickly, and prevent the replacing of good parts and reduce vehicle downtime.
4. Perform the Diagnostic Circuit Check. This test confirms that the diagnostic system has not failed and is able to communicate through the malfunction indicator lamp.
5. After locating and repairing the problem area, road test the vehicle to confirm the repair and to verify no additional problem areas exist.

Visual Underhood Inspection

This is possibly the most critical step of diagnosis. A detailed examination of connectors, wiring and vacuum hoses can often lead to a repair without further diagnosis. Performance of this step relies on the skill of the technician performing it; a careful inspector will check:

- The undersides of hoses as well as the integrity of hard-to-reach hoses blocked by the air cleaner or other components.
- The wiring carefully for any sign of strain, burning, crimping, or terminal pull-out from a connector.
- The connectors at components or in harnesses as required; usually, pushing them together will reveal a loose fit.

Diagnostic Circuit Check

This step is used to check that the on-board diagnostic system is working correctly. A system which is faulty or shorted may not yield correct codes when placed in the Diagnostic Mode.

If the diagnostic system is not operating correctly, or if a problem exists without the malfunction indicator lamp being lit, refer to the vehicle's A-Charts. These charts cover such conditions as Engine Cranks but Will Not Run or No Service Engine Soon Light.

1. Turn the ignition switch **ON** but do not start the engine.
2. The MIL should illuminate and and stay lit until the engine is started.
3. If the MIL does not light, check the bulb, fuses and circuits for an open or short.
4. Connect jumper wire from terminal A to B in the DLC. The MIL should flash code 12 at least 3 times. If not, turn the ignition switch **OFF** and disconnect the ECM.
5. Turn the ignition switch **ON** but do not start the engine. If the MIL is ON, look for short in circuit 419.
6. If the MIL does not light, check the MEMCAL PROM for proper installation. If no light, check the ECM then look for an open or short in the MIL circuit.

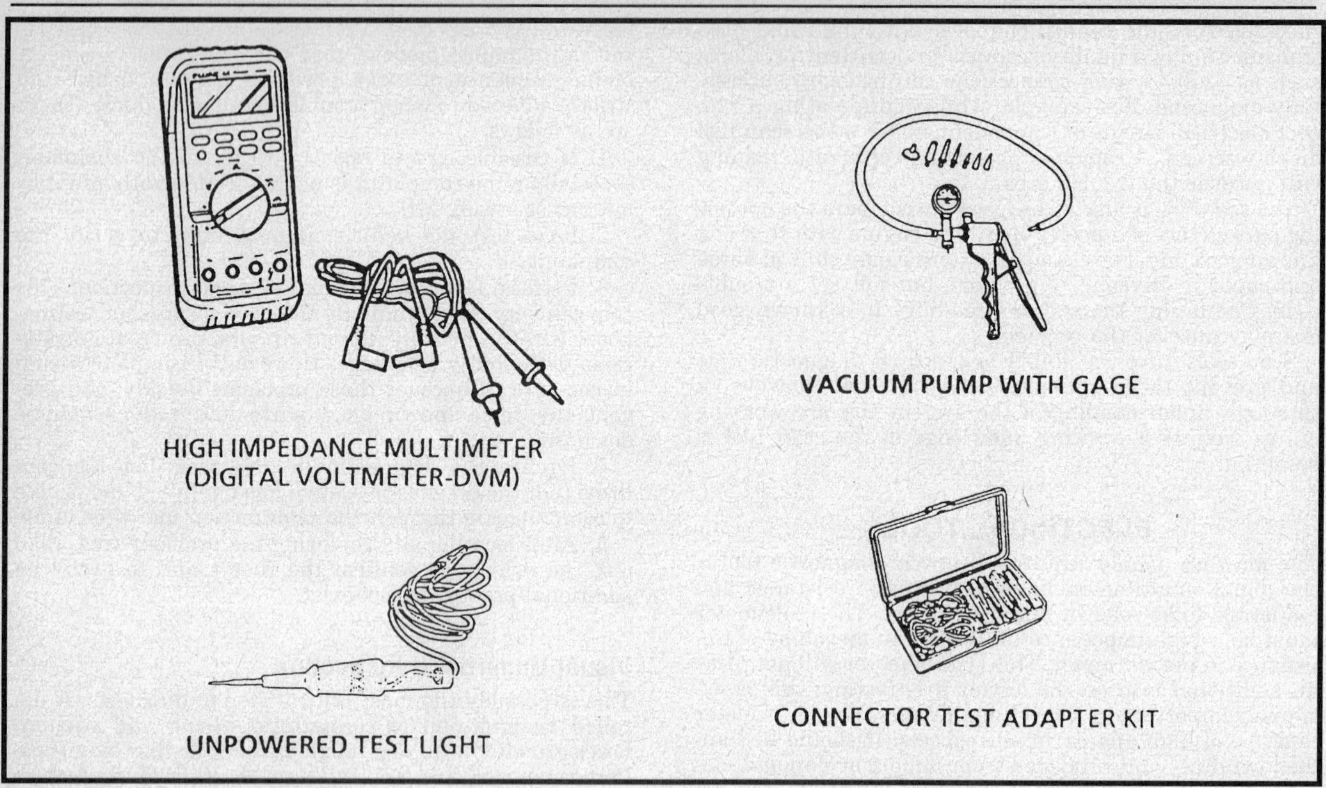

Electrical system diagnostic tools

READING CODES

With Scan Tool

NOTE: Scan tool functions and procedures may vary between manufacturers. Consult owners manual for proper connection and operation of each scan tool.

Once the diagnostic circuit check has been performed, enter the Diagnostic Mode and read any stored codes. To enter the diagnostic mode:

1. Turn the ignition switch **OFF**. Locate the Data Link Connector (DLC) also called ALDL, usually under the instrument panel. It may be within a plastic cover or housing labeled DIAGNOSTIC CONNECTOR. This link is used to communicate with the ECM.

2. Connect the scan tool correctly to the DLC.

3. Turn the ignition switch to the **ON** position but DO NOT start the engine. A Code 12 may be displayed. Code 12 is not a fault code. It is used as a system acknowledgment or handshake code; its presence indicates that the ECM can communicate as requested. Code 12 is used to begin every diagnostic sequence. Some vehicles also use Code 12 after all diagnostic codes have been sent.

4. After Code 12 has been transmitted 3 times, the fault codes, if any, will each be transmitted 3 times. The codes are stored and transmitted in numeric order from lowest to highest.

NOTE: The order of codes in the memory does not indicate the order of occurrence.

5. If no fault codes are transmitted, use the scan functions to view the values being sent to the ECM. Compare the actual values to the typical or normal values for the engine.

6. Switch the ignition **OFF** when finished with code retrieval or scan tool readings and remove the scan tool.

Without Scan Tool

When a scan tool is unavailable, trouble codes can be read from the Malfunction Indicator Lamp (MIL) on the instrument panel.

1. With the ignition switch in the **ON** position and the engine not running, place a jumper between DLC terminals A and B.

2. The MIL should flash code 12 three times consecutively. The following would be the flash sequence: flash, pause, flash-flash, long pause, flash, pause, flash-flash, long pause, flash, pause, flash-flash, indicating diagnostic trouble code (DTC) 12. DTC 12 indicates that the diagnostics are working.

3. If code 12 is not indicated, a problem exists within the diagnostic system itself, and the appropriate diagnostic chart should be referenced.

4. Following the output of DTC 12, the MIL will flash the next highest trouble code (e.g. DTC 13) if one exists. It will output this 3 consecutive times before moving on to the next code in sequence.

5. When all codes have been displayed, the cycle will repeat starting with code 12. If no codes are stored in the system's memory it will continue to flash code 12.

6. To exit the diagnostic mode, turn the ignition switch **OFF** and remove the jumper wire or scan tool.

CLEARING DIAGNOSTIC TROUBLE CODES

To clear diagnostic trouble codes from the memory, the power feed must be disconnected for at least 30 seconds. Depending on how the vehicle is equipped, this can be accomplished by:

- Removing the power fuse or positive battery pigtail.
- Disconnecting negative battery terminal, however, other on-board memory data such as the radio and clock settings will be lost.
- On some applications, clearing DTC's can also be done through the scan tool.

DIAGNOSTIC MODES

Manual Diagnostic Mode (0 Ohm State Field Service)

When the diagnostic terminal B in the data link connector (DLC) is grounded with the key **ON** and the engine **OFF**, the system will enter the diagnostic mode. In this mode the following activities take place:

1. Displays a DTC 12 indicating the system is operational.
2. Displays any stored fault codes 3 times consecutively.
3. Energizes all system controlled relays and solenoids except the fuel pump relay. This allows checking circuits in the shop without having to simulate certain operating conditions.
4. The idle air control (IAC) valve moves to it's fully extended position, blocking the idle air passage. This is useful when adjusting minimum idle speed.
5. The electric cooling fan is enabled on some vehicles.

NOTE: Battery voltage may fall considerably if left in this mode for extended periods. This voltage drop could cause false readings. Connect a battery charger if time exceeds 30 minutes.

Field Service Mode (0 Ohm State)

If the diagnostic terminal is grounded with the engine running, the system will enter the Field Service Mode. In this mode the following activities take place:

1. The MIL light will flash 2.5 times per second to indicate Open Loop operation.
2. The MIL light will flash once per second to indicate Closed Loop operation and a 14.7-to-1 fuel mixture delivery.
3. While in Closed Loop, if the light is OFF most of the time it indicates a lean condition, if the light is ON most of the time it indicates a rich condition.
4. When in Field Service Mode, new codes cannot be stored and the Closed Loop timer is bypassed.

─────────── WARNING ───────────

The vehicle should not be driven while in field service mode. Damage to the catalytic converter could result.

Road Test Mode (Open Mode)

Road test mode can be accessed with the appropriate scan tool. This mode allows the ECM/PCM to function normally, without affecting fuel control or spark timing. This is the preferred mode when testing with a scan tool. On some vehicles, only a few data parameters may be dis-

played on the scan tool while a NO DATA message may appear on others. In these situations, the Diagnostic Mode (10K Mode) must be selected to access the data stream. This mode is very helpful on those vehicles that use multiple microprocessors. On these models the data stream is active all the time because it is constantly transmitting information to the BCM, IPC, etc..

Diagnostic Mode (10K Mode)

Diagnostic mode can be accessed with the appropriate scan tool. In this mode all available serial data is displayed on the scan tool. In addition, the following activities take place:

- ECM timers are bypassed.
- On systems equipped with a knock sensor, spark timing is advanced.
- IAC valve is controlled to maintain a fixed idle speed, around 1000 RPMs.
- Park/Neutral restrict functions are disabled.
- On some vehicles the charcoal canister purge solenoid will be commanded to disable purge.

NOTE: When in diagnostic mode the idle speed will be abnormally high. With some scan tools this will occur as soon as the tool is connected.

Back Up Fuel Mode (3.9K Mode)

This mode verifies the operation of the fuel back up circuitry inside the computer, namely the CALPAC. Engine operation is very erratic in this mode due to the limited amount of sensor information being used. While in back up fuel mode, electronic spark timing is not functional and fuel calculations are based on the following: RPM (distributor reference signal), TPS (throttle position signal) and CTS (engine coolant temperature signal).

DIAGNOSTIC TROUBLE CODES (DTC)

- DTC **12** — Diagnostic check only (flash code)
- DTC **13** — Oxygen (0_2) sensor circuit open
- DTC **14** — Engine Coolant Temperature (ECT) sensor circuit out of range (high)
- DTC **15** — Engine Coolant Temperature (ECT) sensor circuit out of range (low)
- DTC **16** — System voltage high or low — **OR** — Direct/electronic ignition system fault line circuit error
- DTC **17** — Camshaft position sensor/spark reference circuit error
- DTC **18** — Crankshaft/camshaft error
- DTC **19** — Ignition control module signal circuit error
- DTC **21** — Throttle Position Sensor (TPS) circuit voltage out of range
- DTC **22** — Throttle Position Sensor (TPS) circuit voltage out of range
- DTC **23** — Intake/manifold air temperature sensor circuit voltage out of range
- DTC **24** — Vehicle speed sensor circuit signal error
- DTC **25** — Intake/manifold air temperature sensor circuit voltage out of range
- DTC **26** — Quad-driver module circuit error
- DTC **27** — Gear switch circuits error
- DTC **28** — Gear switch circuits error — **OR** — Quad-driver module control circuit
- DTC **29** — Gear switch circuits error
- DTC **31** — PRNDL/transmission neutral start/park-neutral position switch circuit error — **OR** — Turbo wastegate overboost — **OR** — Camshaft sensor circuit error

- DTC **32** — Exhaust Gas Recirculation (EGR) system fault
- DTC **33** — Mass air flow/manifold absolute pressure sensor circuit voltage out of range
- DTC **34** — Mass air flow/manifold absolute pressure sensor circuit voltage out of range
- DTC **35** — Idle speed/air/idle air control error
- DTC **36** — Transaxle shift problem — **OR** — Ignition control signal circuit error
- DTC **38** — Brake switch circuit error
- DTC **39** — Torque converter/transaxle clutch switch circuit error
- DTC **41** — Cam position sensor/cylinder select error
- DTC **42** — Electronic spark timing/ignition control circuit open or shorted
- DTC **43** — Knock sensor/electronic spark control circuit error
- DTC **44** — Oxygen (O_2) sensor circuit indicates system lean
- DTC **45** — Oxygen (O_2) sensor circuit indicates system rich
- DTC **46** — Power steering pressure circuit error — **OR** — Vehicle anti-theft system circuit
- DTC **48** — Misfire diagnosis
- DTC **51** — PCM prom memory error
- DTC **52** — PCM prom memory error — **OR** — Engine oil temperature sensor circuit indicates low
- DTC **53** — System over voltage — **OR** — Exhaust Gas Recirculation (EGR) system malfunction — **OR** — Vehicle anti-theft system circuit error
- DTC **54** — Fuel pump circuit voltage low — **OR** — Exhaust Gas Recirculation (EGR) system fault
- DTC **55** — PCM error — **OR** — Mixture control solenoid/fuel lean monitor circuit error — **OR** — Exhaust Gas Recirculation (EGR) system fault
- DTC **56** — Quad-driver module circuit error — **OR** — Vacuum sensor circuit error — **OR** — Secondary air inlet valve actuator vacuum sensor circuit error
- DTC **57** — Boost control solenoid circuit error
- DTC **58** — Personal automotive security system (PASS-key®II) fuel enable circuit error
- DTC **61** — Degraded oxygen (O_2) sensor — **OR** — Cruise control vent solenoid circuit error — **OR** — A/C

system circuit error — **OR** — Secondary port throttle valve system error
- DTC **62** — Transaxle gear switch signal circuit error — **OR** — Cruise control vacuum solenoid circuit error — **OR** — Engine oil temperature sensor circuit
- DTC **63** — Cruise control system error — **OR** — Exhaust Gas Recirculation (EGR) flow check — **OR** — Oxygen (O_2) sensor (right bank) circuit open
- DTC **64** — Exhaust Gas Recirculation (EGR) flow check — **OR** — Oxygen (O_2) sensor (right bank) circuit indicates lean
- DTC **65** — Cruise control pressure sensor circuit error — **OR** — Exhaust Gas Recirculation (EGR) flow check — **OR** — Oxygen (O_2) sensor (right bank) circuit indicates rich — **OR** — Fuel injector circuit error
- DTC **66** — A/C refrigerant pressure sensor circuit — **OR** — Engine power switch circuit error
- DTC **67** — Cruise control switches circuit error
- DTC **68** — A/C compressor relay circuit shorted — **OR** — Cruise control system problem
- DTC **69** — A/C compressor relay circuit open — **OR** — A/C head pressure switch circuit error
- DTC **70** — A/C refrigerant pressure sensor circuit indicates high
- DTC **71** — A/C evaporator temperature sensor circuit indicates low
- DTC **72** — Gear selector switch circuit error
- DTC **73** — A/C evaporator temperature sensor circuit indicates high
- DTC **75** — Exhaust Gas Recirculation (EGR) No. 1 solenoid circuit error
- DTC **76** — Exhaust Gas Recirculation (EGR) No. 2 solenoid circuit error
- DTC **77** — Exhaust Gas Recirculation (EGR) No. 3 solenoid circuit error
- DTC **81** — Brake switch error
- DTC **82** — Ignition control signal error
- DTC **85** — PCM prom error
- DTC **86** — A/C multiplexer chip error
- DTC **87** — Electronically Erasable Programmable Read Only Memory (EEPROM) error

GEO

General Description

The Geo line — Metro, Tracker, Storm and Prism — contain engines produced by Suzuki, Isuzu and Toyota. Because of the different manufacturers and engine controls, ECM codes and diagnostic procedures will be different for each vehicle.

Although not every engine uses every code, the same code may carry different meanings relative to each engine or engine family. For example, on the 1.6L (VIN U) Suzuki engine, code 21 indicates that the signal voltage on the TPS sensor circuit is too high. On the 1.6L (VIN 5) Toyota engine, the same code identifies a fault in the heated oxygen sensor circuit.

ELECTRONIC CONTROL MODULE (ECM)

The ECM is the control center of the fuel injection system. It continuously monitors information from a variety of sensors and switches, and controls the systems that affect vehicle performance. The ECM also has self-diagnostic capability, and can detect system problems, alert the driver through the Malfunction Indicator Lamp (MIL) and record the appropriate Diagnostic Trouble Code (DTC). The numerical code will identify the specific circuit to aid the technician in diagnosis. The ECM is capable of storing more than one code, with possibilities ranging from 12 to 999, however not all systems use every code. In addition, similar code numbers can relate to different circuits depending on the vehicle.

Powertrain Control Module (PCM)

The PCM is used when vehicles are equipped with electronically controlled transmissions. Its function is identical to that of the ECM with the added features of controlling shift points in the transmission, as well as cruise control operation and diagnostics. This unit may be referred to as the PCM, the ECM/PCM or the PCM/TCM (transmission control module). The integrated functions of engine and transmission control allow accurate gear selection and improved fuel economy.

For engine diagnostics, the PCM may be considered identical to an ECM system, although the combined unit will display additional codes relating to transmission function and components.

NOTE: The terms Electronic Control Module (ECM), Powertrain Control Module (PCM) and Body Control Module (BCM) may sometimes be used interchangably.

Quad Drivers (QDM)

Some control modules have eliminated the use of individual transistors to control each output circuit. A single QDM can individually control 4 outputs from the computer. Since these control modules are internally fault protected, a single faulty circuit or component may cause the other circuits on that QDM to be ON or OFF all the time. Do not confuse this with a failed ECM; if the Quad Driver itself has failed, the control module output will usually test either open or shorted.

The quad driver check procedure should be used before replacing an expensive ECM/PCM assembly. The check will not test all computer functions but will determine if an external circuit has disabled a QDM. Be particularly

1	ECM	4	FUSE BOX
2	BRACKET	5	ECM SCREWS
3	RELAYS		

Electronic Control Module (ECM) — Tracker shown

suspicious of an ECM/PCM exhibiting any of these characteristics:

- Malfunction indicator lamp ON with no codes stored.
- Malfunction indicator light flickers, intermittent or dim.
- ECM/PCM will not flash Code 12 and/or engine will not start.
- Controlled output, such as TCC, is inoperative or ON all the time.
- Engine stalls, misfires or surges.
- Scan tool erratic or inoperative.

MALFUNCTION INDICATOR LAMP (MIL)

The primary function of the malfunction indicator lamp is to advise the operator and the technician that a fault has been detected, and, in most cases, a code stored. Under normal conditions, the malfunction indicator lamp will illuminate when the ignition is turned **ON**. Once the engine is started and running, the ECM will perform a system check and extinguish the warning lamp if no fault is found.

Additionally, the malfunction indicator lamp can be used to retrieve stored codes after the system is placed in the Diagnostic Mode. Codes are transmitted as a series of flashes with short or long pauses. When the system is placed in the Field Service Mode, the dash lamp will indicate open loop or closed loop function to the technician.

Intermittents

If a fault occurs intermittently, such as a loose connector pin breaking contact as the vehicle hits a bump, the ECM will note the fault as it occurs and energize the malfunction indicator lamp. If the problem self-corrects, as with the terminal pin again making contact, the dash lamp will extinguish after 10 seconds but a code will remain stored in the ECM memory.

When an unexpected code appears during diagnostics, it may have been set during an intermittent failure that self-corrected; the codes are still useful in diagnosis and should not be discounted.

Failure Codes

Whenever the ECM detects a sensor or switch signal that is out of range, it will turn ON the MIL. In addition, if the ECM detects an internal malfunction, it will also turn ON the MIL. Distinguishing between current and history codes is absolutely necessary for successful diagnosis. Current codes are also known as hard failures, or a problem that exists at the present time. For example, if a vehicle came into the shop with the MIL light on steady with the engine running, a trouble code could be immediately retrieved and that particular diagnostic trouble chart could be referenced.

History codes, also known as intermittents, are failures which have taken place at one time, but are not presently causing the MIL light to turn ON. An example of this would be if the ignition timing had been recently adjusted on a 5.0 liter engine with TBI. When the bypass wire was disconnected to make the adjustment, the open circuit would have caused the ECM to revert to base timing. Because of this change, the ECM detected a problem and turned on the MIL while simultaneously recording a trouble code 42. When the timing adjustment was completed and the bypass wire was reconnected, the ECM waited approximately 10 seconds and then turned off the MIL, but retained the code. The ECM would retain this information up to a maximum of 50 ignition cycles. If code retrieval was attempted in the meantime by another technician unaware of the timing adjustment, their only recourse would be to reference the wiring diagram associated with the code 42 chart and physically inspect all connections.

MEMORY

NOTE: Not every type of memory control is used in each vehicle application.

• **Read Only Memory (ROM)** is a permanent memory physically soldered to the circuit boards within the ECM. The ROM contains the overall control algorithms. The ROM is a non-volatile memory and cannot be changed. Because of this, it does not require a constant supply of battery voltage to be retained.

• **Random Access Memory (RAM)** is the computer scratch pad. The computer can read from or write into this memory as needed. This memory is volatile and will be lost without a constant supply of battery voltage.

• **Programmable Read Only Memory (PROM)** is the portion of the ECM that contains different engine calibration information that is specific to year, model and emissions. It is for this reason that the PROM is referred to as the calibrator. The PROM is a non-volatile memory that is read only by the ECM. The PROM is removable from the ECM and should be retained with the vehicle following ECM replacement.

• **Electrically Erasable Programmable Read Only Memory (EEPROM)** is a permanent memory that is physically soldered to the circuit boards within the ECM. The EEPROM takes the place of the ROM and the PROM and is not serviceable. Those vehicles using an EEPROM can be re-programmed by using the TECH 1 or other equivalent equipment. Some ECM's using an EEPROM use a separate knock sensor module underneath the computer's access cover. This chip must be removed before replacing the old ECM and reinstalled into the new computer.

Tools and Equipment

SCAN TOOLS

The system can communicate a wide variety of information through the Data Link Connector (DLC), previously known as the ALDL. Depending on the application, serial data is transmitted to terminals E or M of the DLC. This data is transmitted at a high frequency which requires a quality scan tool for interpretation. Although stored codes may be read with only the use of a small jumper wire, the use of a hand-held scan tool such as GM's TECH 1 or equivalent, is recommended. There are many manufacturers of these tools; a purchaser must be certain that the tool is proper for the intended use.

The scan tool allows any stored codes to be read from the ECM memory. The tool also allows the operator to view the data being sent to the ECM while the engine is running. This ability has obvious diagnostic advantages; the use of the scan tool is frequently required by the diagnostic charts.

With an understanding of the data stream that the tool will display and knowledge of the circuits involved, the tool can be an extremely valuable tool when examining the system. Without scan tools, diagnostic information would be difficult to obtain in some circuits and impossible to retrieve in others. Scan tools do not make the use of trouble code charts unnecessary, nor do they pinpoint the exact area in a circuit where the problem exists. However, this type of equipment will give more specific information, such as sensor input readings. An example of this would be the coolant sensor circuit. While the ECM looks at specific voltage levels to determine coolant temperature, the scan tool will display this information as an actual temperature reading. This information is not only useful in determining system malfunctions, but is also very helpful in diagnosing mechanical problems. For example, if the engine is overheating, the scan tool can be used to look at the actual engine temperature. This is helpful when checking for electric cooling fan enable switch problems, sticking thermostats, etc.

To properly read system values with a scan tool, the following conditions must be met. All normal values given in the charts will be based on these conditions:

• Engine running at idle, throttle closed
• Engine warm, upper radiator hose hot
• Vehicle in park or neutral
• System operating in closed loop
• All accessories **OFF**

The ECM is capable of communicating with a scan tool in 2 modes: Diagnostic and Field Service Mode.

Diagnostic Mode

When the diagnostic terminal is grounded with the ignition **ON** and the engine OFF, the system will enter the

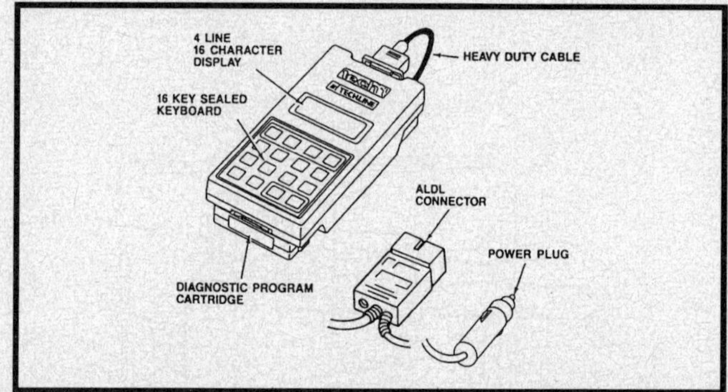

Typical scan tool — GM's Tech 1 shown

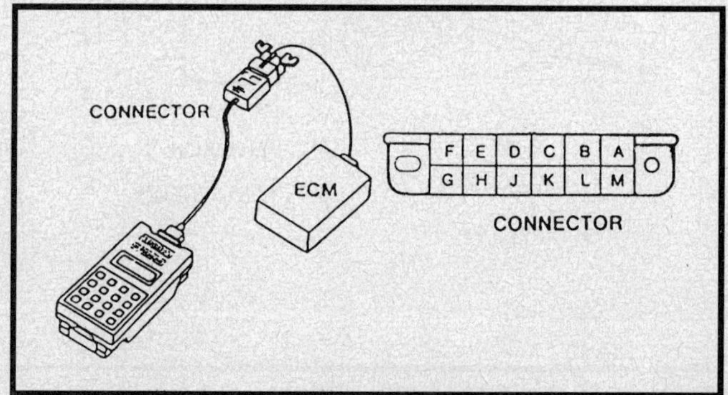

DLC connection to scan tool

diagnostic mode. In this mode the ECM will perform the following:

- Display a Code 12 by flashing the SES telltale light indicating proper diagnostic system operation.
- Display any stored trouble codes by flashing the SES telltale light. Each code will be flashed three times. If no transaxle codes are present the telltale light will repeat the sequence again starting with code 12.
- Energize all ECM controlled solenoids and relays except the fuel pump relay.
- IAC valve is pulsed in and out, with a pause before moving in either direction.

Field Service Mode

If the diagnostic terminal is grounded with the engine running, the system will enter the field service mode. In this mode the SES telltale light will indicate either Open or Closed loop operation by the following flash sequence:

- SES light will flash 2.5 times per second to indicate Open Loop operation.
- SES light will flash once per second to indicate Closed Loop operation.
- In Closed Loop, the SES light will be OFF most of the time to indicate a lean exhaust or the SES light will be ON most of the time to indicate a rich exhaust.

NOTE: While the system is in the field service mode, the Closed Loop timer is by-passed and new trouble codes cannot be stored.

SCAN TOOL USE FOR INTERMITTENTS

In some tool applications the data update rate is so slow that it becomes less effective than a digital voltmeter for checking sporadic voltage changes. With the rapid data transmission of a quality scanner, intermittent problems such as faulty wiring connections can be more successfully diagnosed. For example, while manipulating a suspect electrical circuit or component, observe the scan tool display screen. A rapid voltage change or erratic reading will uncover the trouble area.

The scan tool is also an easy way to compare the operating parameters of a poorly operating engine with that of a known good one. For example, a sensor may shift in value and cause a driveability problem, but not set a trouble code. Comparing the sensor's readings to a known good one may uncover the problem.

Scan tools have the ability to speed up diagnostic time and prevent the replacement of good parts, however a thorough understanding of the system you are working on, as well as a working knowledge of the scan tool is essential.

ELECTRICAL TOOLS

The most commonly required electrical diagnostic tool is the digital multimeter, allowing voltage, resistance and amperage to be read by one instrument. The multimeter must be a high-impedance unit, with 10 megohms of impedance in the voltmeter. This type of meter will not place an additional load on the circuit it is testing; this is extremely important in low voltage circuits. The multimeter must be of high quality in all respects. It should be handled carefully and protected from impact or damage. Replace batteries frequently in the unit.

Other necessary tools include an unpowered test light, a quality tachometer with inductive (clip-on) pick up and the proper tools for releasing GM's Metri-Pack, Weather Pack and Micro-Pack terminals as necessary. The Micro-Pack connectors are used at the ECM connector. A vacuum pump/gauge may also be required for checking sensors, solenoids and valves.

Diagnosis and Testing

SERVICE PRECAUTIONS

- To prevent internal ECM damage, the ignition must be **OFF** when disconnecting or reconnecting power to the computer.
- When handling a PROM, CALPAC or MEMCAL, do not touch the component leads. Also, do not remove the integrated circuit from the carrier.
- Never allow welding cables to lie on, near or across any vehicle electrical wiring.
- Leave new components and modules in the shipping package until ready to install them.
- When performing electrical tests on the system, use a high impedance multimeter, digital voltmeter (DVM) J-34029-A or equivalent.
- To prevent possible electrostatic discharge damage to the ECM, do not touch the connector pins or soldered components on the circuit board.

PRELIMINARY INSPECTION

In order to successfully locate and repair problems in complex computer controlled systems, the technician needs to adhere to a comprehensive, sequential method of diagnostics. Many times a vehicle has a malfunction and the worst is suspected. Premature conclusions often result in the replacement of good parts and wasted time. A proper sequence of steps needs to be established and strictly adhered to when troubleshooting a vehicle. These are as follows:

1. If possible, try to speak directly to the customer, especially if the complaint is not associated with an intermittent or steady MIL.

2. Road test the vehicle, if necessary, to verify the complaint.

3. Perform a thorough visual inspection. This step can very often eliminate the need for further testing. Loose wires, especially ground circuits, can create mysterious driveability problems. However, a complete visual inspection can uncover these problems quickly, and prevent the replacing of good parts and reduce vehicle downtime.

4. Perform the Diagnostic Circuit Check. This test confirms that the diagnostic system has not failed and is able to communicate through the malfunction indicator lamp.

5. After locating and repairing the problem area, road test the vehicle to confirm the repair and to verify no additional problem areas exist.

Visual Underhood Inspection

This is possibly the most critical step of diagnosis. A detailed examination of connectors, wiring and vacuum hoses can often lead to a repair without further diagnosis.

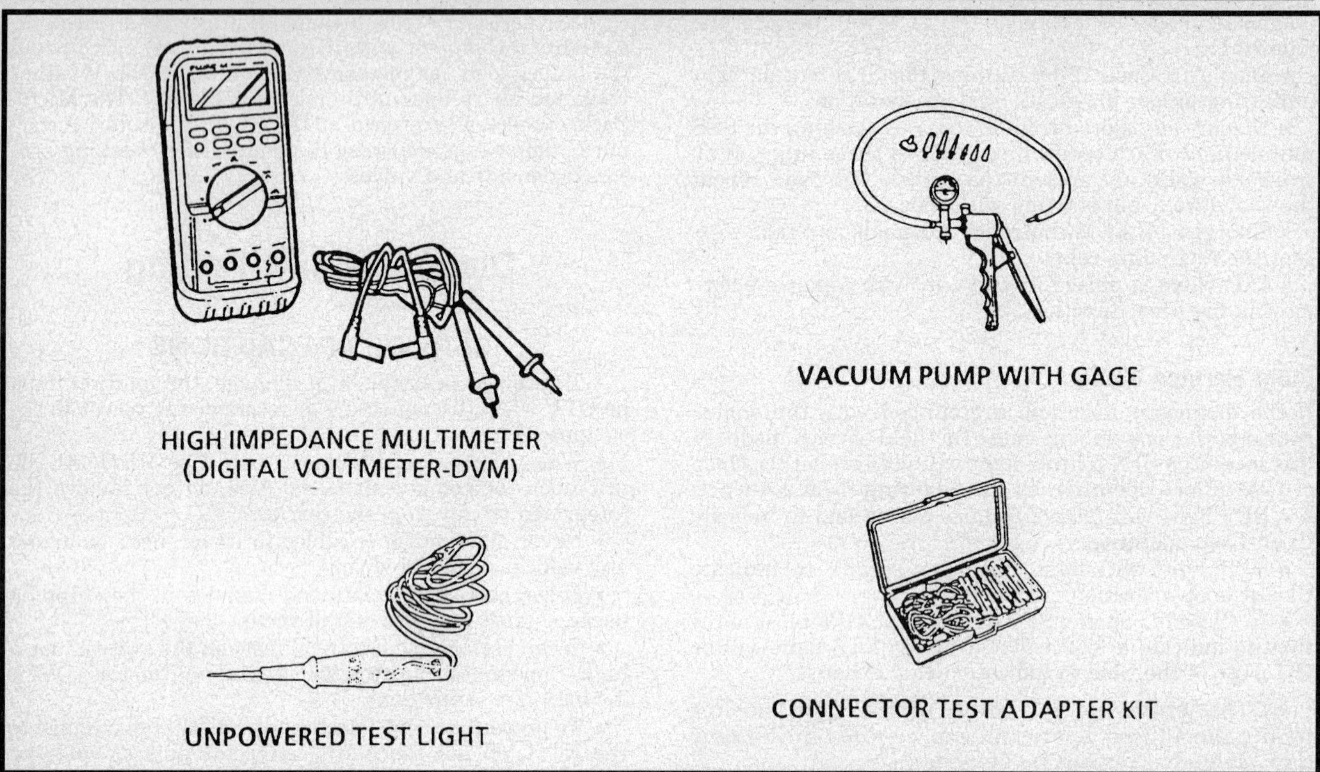

Electrical system diagnostic tools

Performance of this step relies on the skill of the technician performing it; a careful inspector will check:

• The undersides of hoses as well as the integrity of hard-to-reach hoses blocked by the air cleaner or other components.

• The wiring carefully for any sign of strain, burning, crimping, or terminal pull-out from a connector.

• The connectors at components or in harnesses as required; usually, pushing them together will reveal a loose fit.

Diagnostic Circuit Check

This step is used to check that the on-board diagnostic system is working correctly. A system which is faulty or shorted may not yield correct codes when placed in the Diagnostic Mode.

If the diagnostic system is not operating correctly, or if a problem exists without the malfunction indicator lamp being lit, refer to the vehicle's A-Charts. These charts cover such conditions as Engine Cranks but Will Not Run or No Service Engine Soon Light.

1. Turn the ignition switch **ON** but do not start the engine.

2. The MIL should illuminate and and stay lit until the engine is started.

3. If the MIL does not light, check the bulb and circuits for an open or short.

4. Connect jumper wire from terminal A to B in the DLC. The MIL should flash code 12 at least 3 times or flash OFF and ON. If not, turn the ignition switch **OFF** and disconnect the ECM.

5. Turn the ignition switch **ON** but do not start the engine. If the MIL is ON, look for short in the MIL circuit.

6. If the MIL does not light, check for open in the MIL circuit.

Circuit and Component Diagnosis and Repair

Using the appropriate chart(s) based on the Diagnostic Circuit Check, the fault codes and the scan tool data will lead to diagnosis and checking of a particular circuit or component. It is important to note that the fault code indicates a fault or loss of signal in an ECM-controlled system, not necessarily in the specific component. Detailed procedures to isolate the problem are included in each code chart; these procedures must be followed accurately to insure timely and correct repair. Following the procedure will also insure that only truly faulty components are replaced.

DATA LINK CONNECTOR (DLC)

Metro

To communicate with the ECM, access can be gained by the DLC, the DIAG SW connector in the junction block or the diagnostic request terminal in the DUTY CHECK DLC. The DLC and the junction block are located underneath the instrument panel, to the left of the steering column. The DUTY CHECK DLC is located in the left rear of the engine compartment near the strut tower.

Tracker

The DLC is located underneath the instrument panel, to the left of the steering column. The DLC is primarily used for the attachment of a scan tool. In addition, there is a DUTY CHECK DLC located in the right rear of the engine compartment, near the battery.

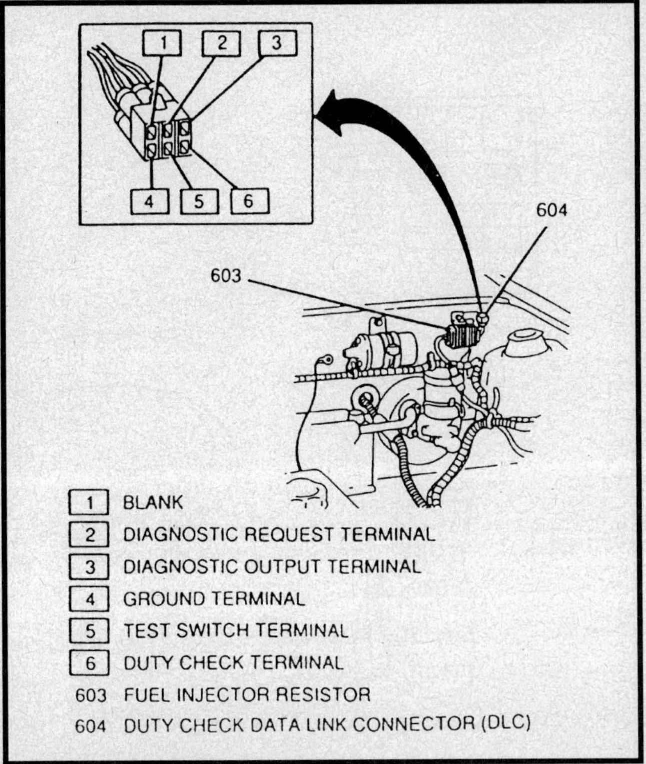

Data Link Connector (DLC) — Metro

1	BLANK
2	DIAGNOSTIC REQUEST TERMINAL
3	DIAGNOSTIC OUTPUT TERMINAL
4	GROUND TERMINAL
5	TEST SWITCH TERMINAL
6	DUTY CHECK TERMINAL
603	FUEL INJECTOR RESISTOR
604	DUTY CHECK DATA LINK CONNECTOR (DLC)

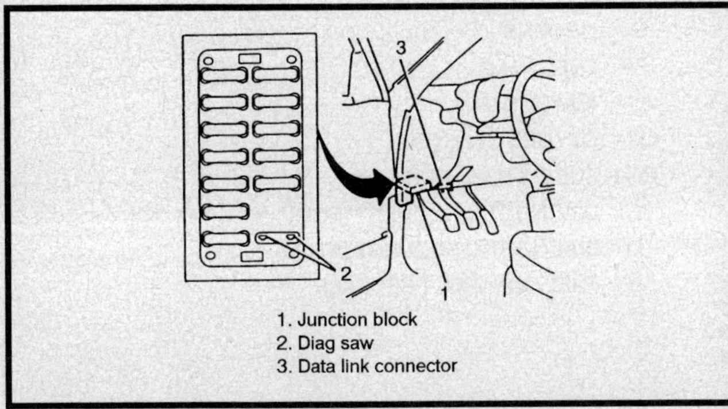

DLC and junction block — Metro

Prizm

The DLC is located in the left rear of the engine compartment, on the strut tower. There is no means provided for the connection of a scan tool. Communication with the ECM is done through the diagnostic terminals TE1 and TE2, 1993–94 models only, of the DLC.

Storm

The DLC is a 3 terminal white connector located behind the right kick panel. Terminal 1 of the DLC is the diagnostic request line, terminal 2 is the serial data line and terminal 3 is the ground.

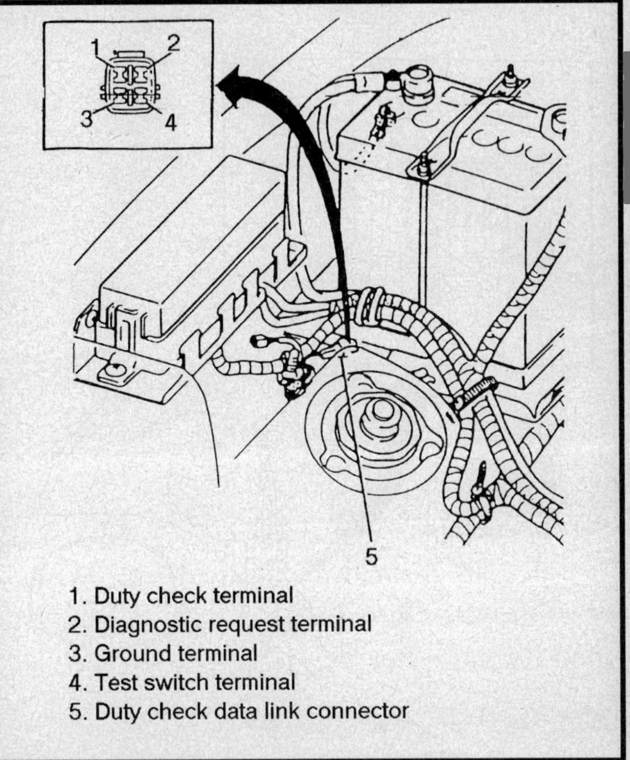

1. Duty check terminal
2. Diagnostic request terminal
3. Ground terminal
4. Test switch terminal
5. Duty check data link connector

Data Link Connector (DLC) — Tracker

READING CODES

With Scan Tool

On all but the Prizm, the DTC's can be retrieved by turning the key to the **OFF**, plugging in a scan tool, then turning the key **ON** and retrieving codes on the scan tool.

Without Scan Tool

METRO — USING MIL

1. Turn the key **ON**. Do not start the engine.
2. Ground either the DIAG SW connector in the junction block or the diagnostic request/switch terminal in the underhood DUTY CHECK DLC.
3. Count the flashes of the MIL (i.e. 12 = flash, pause, flash, flash).

METRO — USING TEST LIGHT

NOTE: The test light method will only work on models with a 6 pin DUTY CHECK DLC.

1. Turn the key **ON**. Do not start the engine.
2. Connect a test light to the diagnostic output terminal.
3. Ground either the DIAG SW connector in the junction block or the diagnostic request/switch terminal in the underhood DUTY CHECK DLC.
4. Count the test light flashes. (i.e. 12 = flash, pause, flash, flash).

TRACKER

1. Turn the key **ON**. Do not start the engine.
2. To retrieve trouble codes on systems using the 4 pin ECM check connector, insert a jumper between terminal

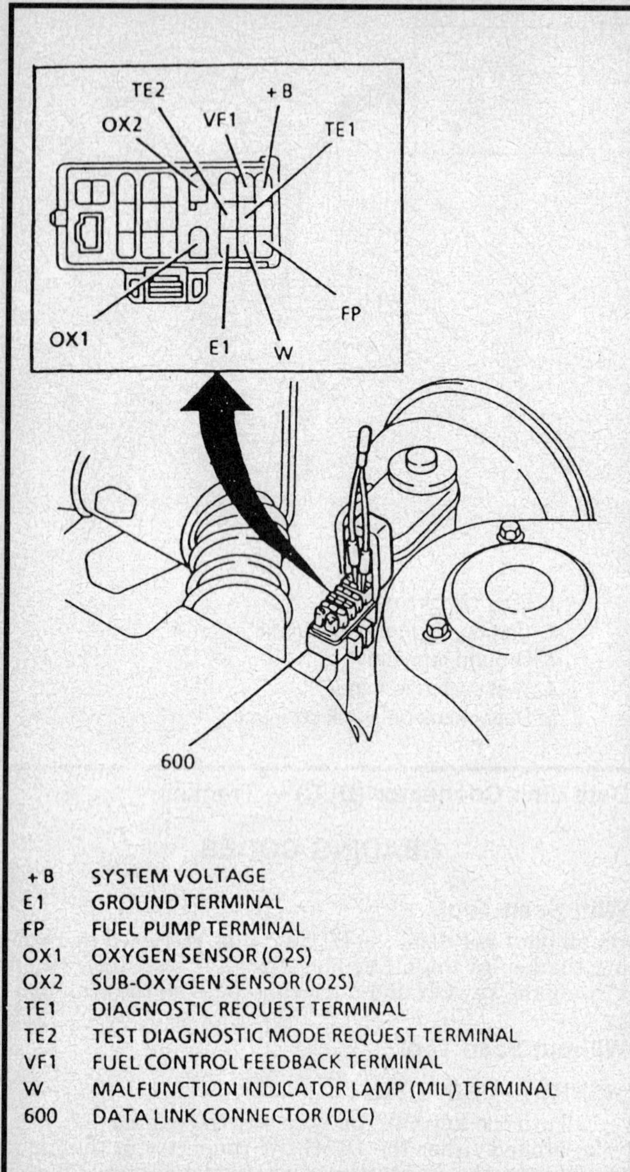

+B	SYSTEM VOLTAGE
E1	GROUND TERMINAL
FP	FUEL PUMP TERMINAL
OX1	OXYGEN SENSOR (O2S)
OX2	SUB-OXYGEN SENSOR (O2S)
TE1	DIAGNOSTIC REQUEST TERMINAL
TE2	TEST DIAGNOSTIC MODE REQUEST TERMINAL
VF1	FUEL CONTROL FEEDBACK TERMINAL
W	MALFUNCTION INDICATOR LAMP (MIL) TERMINAL
600	DATA LINK CONNECTOR (DLC)

Data Link Connector (DLC) — 1993–94 Prizm

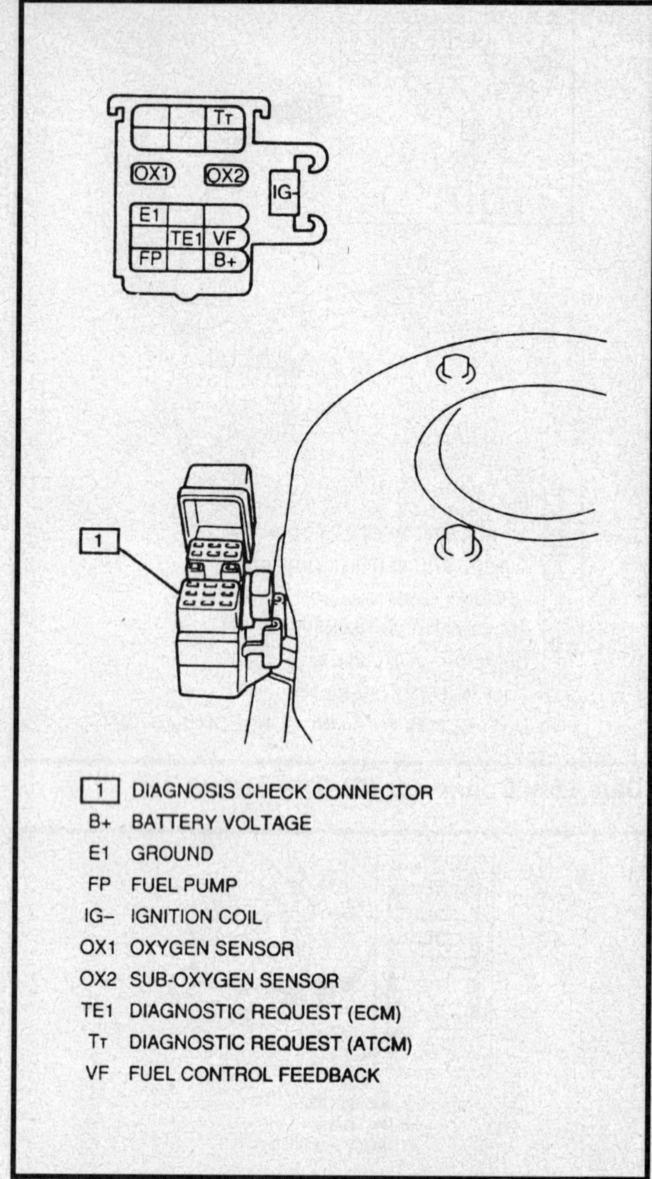

1	DIAGNOSIS CHECK CONNECTOR
B+	BATTERY VOLTAGE
E1	GROUND
FP	FUEL PUMP
IG-	IGNITION COIL
OX1	OXYGEN SENSOR
OX2	SUB-OXYGEN SENSOR
TE1	DIAGNOSTIC REQUEST (ECM)
TT	DIAGNOSTIC REQUEST (ATCM)
VF	FUEL CONTROL FEEDBACK

Data Link Connector (DLC) — 1992 Prizm

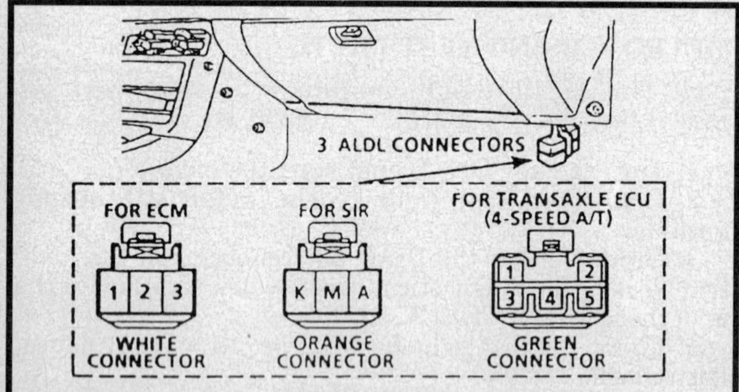

Data Link Connector (DLC) — Storm

3 (diagnostic request) and terminal 4 (ground), then count the MIL flashes (i.e. flash, pause, flash, flash=code 12). Code 12 will only flash if there are no stored trouble codes.

3. Diagnostic information can also be retrieved by grounding the diagnostic request terminal in the DUTY CHECK DLC (terminal 2) and counting the MIL flashes (i.e. 12 =flash, pause, flash, flash).

PRIZM

1. Turn the key **ON**. Do not start the engine.
2. Ground the diagnostic request terminal TE1 in the DLC (terminal E1 is the ground terminal).

NOTE: Because code 12 is used as a malfunction code, a steady series of MIL flashes will be used to indicate a proper functioning system with no stored trouble codes.

3. The flash sequence should be a continuous flash, pause, flash, pause, etc. if there are no stored codes. If a code exists, it will flash in the usual manor. For example, code 12 = flash, pause, flash, flash, long pause.

STORM

1. Turn the key **ON**. Do not start the engine.
2. Insert a jumper wire between terminals 1 and 3 of the DLC.
3. Count the MIL flashes (i.e. 12 = flash, pause, flash, flash).

CLEARING CODES

Metro

To clear the diagnostic trouble codes from the memory, remove the tail light fuse for at least 30 seconds. The fuse is located in the junction block. Make sure the key is in the **OFF** position when removing or re-applying power to the ECM to prevent damage to the ECM. When using the scan tool to retrieve diagnostic data, codes can be cleared from the memory through the scan tool without removing power to the ECM.

Tracker

Remove the 15 amp TAIL/DOME fuse for 30 seconds. Make sure the key is in the **OFF** position when disconnecting or reconnecting power to the ECM to avoid computer damage.

Prizm

To clear stored codes from the ECM on 1992 models, disconnect the power from the ECM by removing the STOP fuse while the ignition is in the **OFF** position. To clear stored codes from the ECM on 1993–94 models, disconnect the power from the ECM by removing the EFI/F-HTR fuse while the ignition is in the **OFF** position.

Storm

DTC's can be cleared from the system memory by removing power to the ECM with the key **OFF** for approximately 30-45 seconds. This can be accomplished by removing the 30 amp main fuse from the relay/fuse center in the engine compartment. The scan tool can also be used to clear codes.

DIAGNOSTIC MODES

Manual Diagnostic Mode

METRO

When the diagnostic terminal in either the junction block or the DUTY CHECK DLC is grounded with the key **ON** and the engine not running, the ECM enters the diagnostic mode. In this mode the following activities will take place:
- Display diagnostic trouble code 12 to indicate the system is operating properly.
- Display any stored codes by flashing the MIL. If any DTC's are contained in memory, DTC 12 will not be flashed.
- Output idle speed control duty through terminal 6 of the DUTY CHECK DLC.
- ON time of idle speed control solenoid valve is kept constant.

TRACKER

When the diagnostic request terminal in the DUTY CHECK DLC is grounded with the key **ON**, and the engine OFF, the system enters the diagnostic mode. In this mode the following activities take place:
- Display diagnostic trouble code 12 to indicate the system is operating properly.
- Display any stored codes by flashing the MIL. If any DTC's are contained in memory, DTC 12 will not be flashed.
- Output idle speed control duty through terminal 1 of the DUTY CHECK DLC.

1992 PRIZM

When the diagnostic request terminal TE1 is grounded with the ignition switch **ON** and the engine OFF, the ECM will enter the Diagnostic mode. In this mode, the ECM will perform the following:
- Display code 51 by flashing the MIL indicating the system is functioning normally, or flash the MIL **ON** and **OFF** continuously indicating that there may be a problem in the idle switch, **P** and **N** circuits or the air conditioner switch.
- Display any stored codes by flashing the MIL. If any codes other than code 51 are set, then code 51 will not be displayed.

1993–94 PRIZM

The ECM operates the system in 2 modes: normal and test. When the diagnostic request terminal TE1 is grounded with the ignition switch **ON** and the engine OFF, the ECM will enter the Normal mode. In this mode, the ECM will flash the MIL continuously or display any stored codes.

The test mode is used to troubleshoot intermittent problems. To enter the test mode, the ignition switch must be **OFF** and then ground terminal TE2 of the DCL. The ECM will lower the threshold for which codes can be set. The vehicle can be test driven under the conditions that originally set the code. After the road test, while the engine is still running, ground terminal TE1 to retrieve any codes which may have been set.

If terminal TE2 in the DLC is grounded with the ignition switch is in any position other than **LOCK**, test mode will not function. A code 42 will set if the vehicle is not driven over 3 mph (a normal condition if the vehicle speed was under 3 mph). Also, a code 51 will set if either the accelerator pedal is pressed off the idle position for at least 3 seconds with automatic transmission in **P** or **N** or if the air conditioner is turned **ON**. Under these conditions a code 51 is normal and indicates the idle and A/C systems are operating properly.

STORM

When the key is **ON** and the diagnostic request terminal grounded (terminal 1), the ECM will be in the diagnostic mode. The following activities will occur:
- DTC 12 will flash 3 times to indicate diagnostics are functioning (i.e. 12 = flash, pause, flash,flash).
- Additional codes (if any) will be displayed in sequence.

Field Service Mode

METRO

With the engine running and the test switch terminal in the DUTY CHECK DLC grounded, the ECM will set the ignition timing to its base setting. The ECM will output the A/F duty through terminal 6, when both the diagnostic terminal in either the junction block or DUTY CHECK DLC and the test switch terminal are grounded simultaneously.

TRACKER

With the engine running and the test switch terminal in the DUTY CHECK DLC grounded, the ECM will adjust ignition timing to its base setting. If both the diagnostic request and test switch terminals are grounded simultaneously, the ECM will output the A/F duty through the duty check terminal of the DUTY CHECK DLC.

PRIZM

When the diagnostic request terminal is grounded with the engine running, the ECM will enter the field service mode. In this mode the ECM will command the ignition at the base setting.

STORM

With the engine running and terminal 1 grounded in the DLC, the ECM will enter the field service mode. In this state, the following activities will occur:
- The MIL will flash 2.5 times per second to indicate Open Loop operation.
- The MIL will flash once per second to indicate Closed Loop operation. While in Closed Loop, the MIL will stay ON most time if the system is rich (O_2 voltage high). The MIL will be OFF most of the time to indicate that the system is lean (O_2 voltage low).
- New trouble codes cannot be stored in the ECM.
- The Closed Loop timer is bypassed.

DIAGNOSTIC TROUBLE CODE DESCRIPTION

Except Prizm
- DTC **12** — Diagnostic check only (flash code)
- DTC **13** — Oxygen (O_2) sensor circuit open
- DTC **14** — Engine Coolant Temperature (ECT) sensor circuit out of range (high)
- DTC **15** — Engine Coolant Temperature (ECT) sensor circuit out of range (low)
- DTC **21** — Throttle Position Sensor (TPS) circuit voltage out of range
- DTC **22** — Throttle Position Sensor (TPS) circuit voltage out of range
- DTC **23** — Intake/manifold air temperature sensor circuit voltage out of range
- DTC **24** — Vehicle speed sensor circuit signal error
- DTC **25** — Intake/manifold air temperature sensor circuit voltage out of range
- DTC **31** — Manifold absolute pressure sensor circuit voltage out of range
- DTC **32** — Storm only — Exhaust Gas Recirculation (EGR) system fault
- DTC **33** — Manifold absolute pressure sensor circuit voltage out of range
- DTC **41** — No ignition signal
- DTC **42** — Storm only — Electronic spark timing/ignition control circuit open or shorted
- DTC **42** — Camshaft position sensor — no signal for 2 seconds
- DTC **44** — Storm only — Oxygen (O_2) sensor circuit indicates system lean
- DTC **44** — Idle switch circuit open
- DTC **45** — Storm only — Oxygen (O_2) sensor circuit indicates system rich
- DTC **45** — Idle switch circuit shorted
- DTC **51** — Storm only — PCM prom memory error
- DTC **51** — Exhaust Gas Recirculation (EGR) circuit
- DTC **53** — Ground circuit open

Prizm
- DTC **12** — No RPM signal for 2 seconds
- DTC **13** — No RPM signal for 2 seconds
- DTC **14** — Ignition control module signal circuit error
- DTC **21** — Oxygen (O_2) sensor circuit open or shorted
- DTC **22** — Engine Coolant Temperature (ECT) sensor circuit open or shorted
- DTC **24** — Intake air temperature sensor circuit voltage out of range
- DTC **25** — Oxygen (O_2) sensor circuit indicates system lean
- DTC **26** — Oxygen (O_2) sensor circuit indicates system rich
- DTC **27** — Sub oxygen sensor circuit open or shorted
- DTC **31** — Mass air flow sensor circuit voltage out of range
- DTC **41** — Throttle position sensor circuit open or shorted
- DTC **42** — Vehicle speed sensor circuit
- DTC **43** — No crank signal from starter
- DTC **51** — Switch condition signal
- DTC **52** — Knock sensor signal open or short
- DTC **53** — Knock control signal, faulty ECM
- DTC **71** — EGR temperature sensor circuit

SATURN

General Description

ELECTRONIC CONTROL MODULE (ECM)

The ECM is the control center of the fuel injection system. It continuously monitors information from a variety of sensors and switches, and controls the systems that affect vehicle performance. The ECM also has self-diagnostic capability, and can detect system problems, alert the driver through the Malfunction Indicator Lamp (MIL) and record the appropriate Diagnostic Trouble Code (DTC). The numerical code will identify the specific circuit to aid the technician in diagnosis. The ECM is capable of storing more than one code, with possibilities ranging from 12 to 999, however not all systems use every code. In addition, similar code numbers can relate to different circuits depending on the vehicle.

Powertrain Control Module (PCM)

The PCM is used when vehicles are equipped with electronically controlled transmissions. Its function is identical to that of the ECM with the added features of controlling shift points in the transmission, as well as cruise control operation and diagnostics. This unit may be referred to as the PCM, the ECM/PCM or the PCM/TCM (transmission control module). The integrated functions of engine and transmission control allow accurate gear selection and improved fuel economy.

For engine diagnostics, the PCM may be considered identical to an ECM system, although the combined unit will display additional codes relating to transmission function and components.

> **NOTE: The terms Electronic Control Module (ECM), Powertrain Control Module (PCM) and Body Control Module (BCM) may sometimes be used interchangably.**

Quad Drivers (QDM)

Some control modules have eliminated the use of individual transistors to control each output circuit. A single QDM can individually control 4 outputs from the computer. Since these control modules are internally fault protected, a single faulty circuit or component may cause the other circuits on that QDM to be ON or OFF all the time. Do not confuse this with a failed ECM; if the Quad Driver itself has failed, the control module output will usually test either open or shorted.

The quad driver check procedure should be used before replacing an expensive ECM/PCM assembly. The check will not test all computer functions but will determine if an external circuit has disabled a QDM. Be particularly

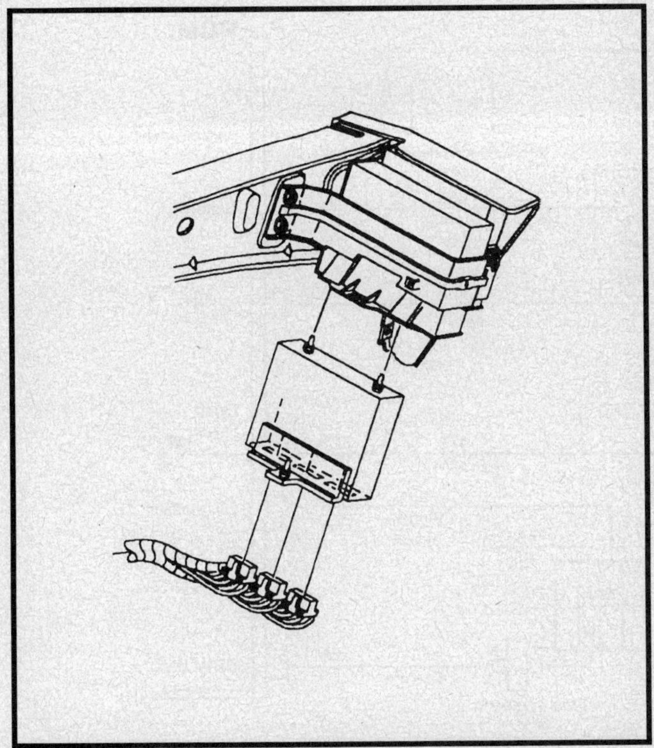

Powertrain Control Module (PCM) — Saturn shown

suspicious of an ECM/PCM exhibiting any of these characteristics:

- Malfunction indicator lamp ON with no codes stored.
- Malfunction indicator light flickers, intermittent or dim.
- ECM/PCM will not flash DTC 12 and/or engine will not start.
- Controlled output, such as TCC, is inoperative or ON all the time.
- Engine stalls, misfires or surges.
- Scan tool erratic or inoperative.

MALFUNCTION INDICATOR LAMP (MIL)

The primary function of the malfunction indicator lamp is to advise the operator and the technician that a fault has been detected, and, in most cases, a code stored. Under normal conditions, the malfunction indicator lamp will illuminate when the ignition is turned **ON**. Once the engine is started and running, the ECM will perform a system check and extinguish the warning lamp if no fault is found.

Additionally, the malfunction indicator lamp can be used to retrieve stored codes after the system is placed in the Diagnostic Mode. DTCs are transmitted as a series of flashes with short or long pauses. When the system is placed in the Field Service Mode, the dash lamp will indicate open loop or closed loop function to the technician.

Intermittents

If a fault occurs intermittently, such as a loose connector pin breaking contact as the vehicle hits a bump, the ECM will note the fault as it occurs and energize the malfunction indicator lamp. If the problem self-corrects, as with the terminal pin again making contact, the dash lamp will extinguish after 10 seconds but a code will remain stored in the ECM memory.

When an unexpected code appears during diagnostics, it may have been set during an intermittent failure that self-corrected; the codes are still useful in diagnosis and should not be discounted.

Failure Codes

Whenever the ECM detects a sensor or switch signal that is out of range, it will turn ON the MIL. In addition, if the ECM detects an internal malfunction, it will also turn ON the MIL. Distinguishing between current and history codes is absolutely necessary for successful diagnosis. Current codes are also known as hard failures, or a problem that exists at the present time. For example, if a vehicle came into the shop with the MIL light on steady with the engine running, a trouble code could be immediately retrieved and that particular diagnostic trouble chart could be referenced.

History codes, also known as intermittents, are failures which have taken place at one time, but are not presently causing the MIL light to turn ON. An example of this would be if the ignition timing had been recently adjusted on a 5.0 liter engine with TBI. When the bypass wire was disconnected to make the adjustment, the open circuit would have caused the ECM to revert to base timing. Because of this change, the ECM detected a problem and turned on the MIL while simultaneously recording a trouble code 42. When the timing adjustment was completed and the bypass wire was reconnected, the ECM waited approximately 10 seconds and then turned off the MIL, but retained the code. The ECM would retain this information up to a maximum of 50 ignition cycles. If code retrieval was attempted in the meantime by another technician unaware of the timing adjustment, their only recourse would be to reference the wiring diagram associated with the code 42 chart and physically inspect all connections.

DATA LINK CONNECTOR (DLC)

The DLC is located underneath the instrument panel and is used to communicate with the ECM. This can be done either manually with a jumper inserted between terminals A and B, or with a scan tool.

Terminal A — DLC Ground

Circuit number 153 the ECM ground through the instrument panel connector. The ECM is grounded at the engine and splices into the instrument panel harness.

NOTE: Most scan tools can disguise problems related to ECM grounds due to the alternate ground path through the tool's power cord.

Terminal B — Diagnostic Enable

Circuit number 451 is connected to the ECM through the instrument panel connector.

Terminal M — Serial Data Line

Circuit 461 is the serial data line for fast baud rate ECMs. Terminal M may also include serial data transmission from other on-board microprocessors such as an anti-lock brake system controller.

MEMORY

NOTE: Not every type of memory control is used in each vehicle application.

- **Read Only Memory (ROM)** is a permanent memory physically soldered to the circuit boards within the ECM. The ROM contains the overall control algorithms. The ROM is a non-volatile memory and cannot be changed. Because of this, it does not require a constant supply of battery voltage to be retained.
- **Random Access Memory (RAM)** is the computer scratch pad. The computer can read from or write into this memory as needed. This memory is volatile and will be lost without a constant supply of battery voltage.
- **Programmable Read Only Memory (PROM)** is the portion of the ECM that contains different engine calibration information that is specific to year, model and emissions. It is for this reason that the PROM is referred to as the calibrator. The PROM is a non-volatile memory that is read only by the ECM. The PROM is removable from the ECM and should be retained with the vehicle following ECM replacement.
- **Electrically Erasable Programmable Read Only Memory (EEPROM)** is a permanent memory that is physically soldered to the circuit boards within the ECM. The EEPROM takes the place of the ROM and the PROM and is not serviceable. Those vehicles using an EEPROM can be re-programmed by using the TECH 1 or other equivalent equipment. Some ECM's using an EEPROM use a separate knock sensor module underneath the computer's access cover. This chip must be removed before replacing the old ECM and reinstalled into the new computer.

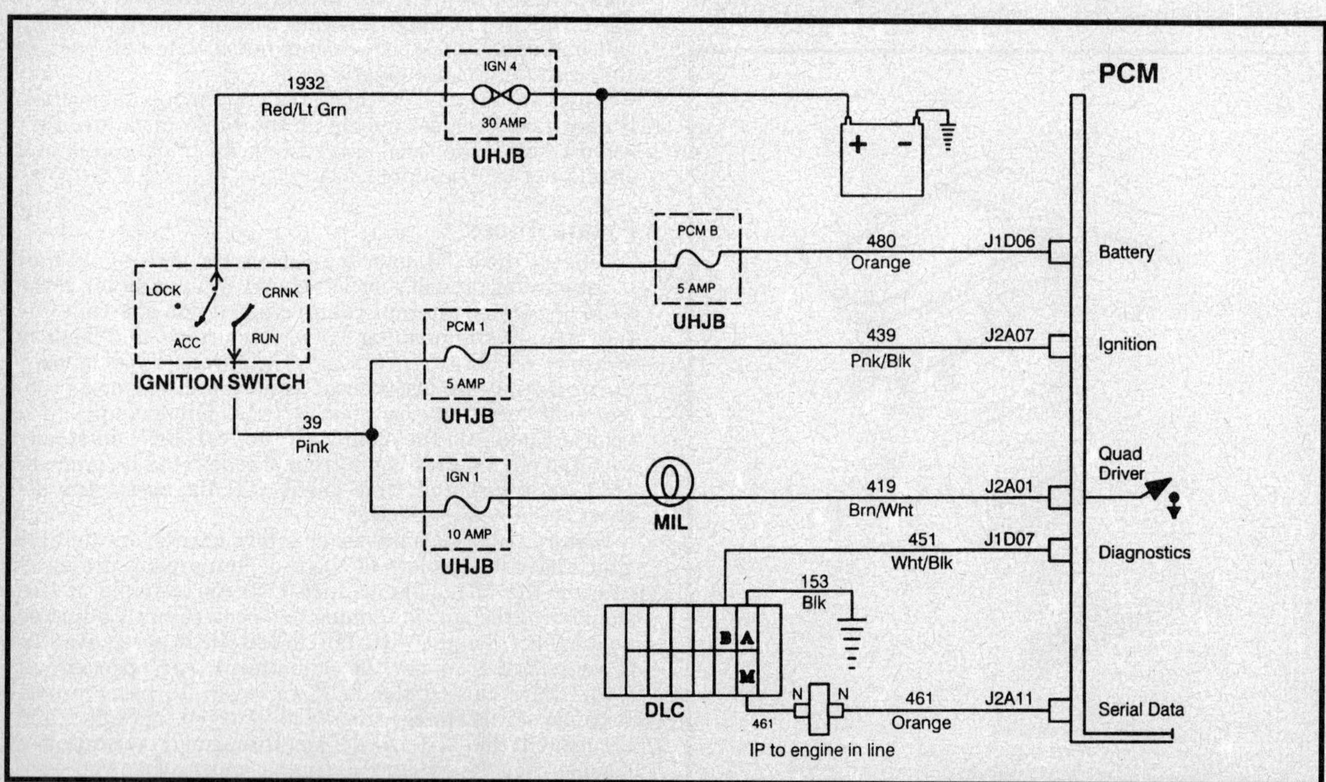

Data Link Connector (DLC) — Saturn

Tools and Equipment

SCAN TOOLS

The system can communicate a wide variety of information through the Data Link Connector (DLC), previously known as the ALDL. Depending on the application, serial data is transmitted to terminals E or M of the DLC. This data is transmitted at a high frequency which requires a quality scan tool for interpretation. Although stored codes may be read with only the use of a small jumper wire, the use of a hand-held scan tool such as GM's TECH 1 or equivalent, is recommended. There are many manufacturers of these tools; a purchaser must be certain that the tool is proper for the intended use.

The scan tool allows any stored codes to be read from the ECM memory. The tool also allows the operator to view the data being sent to the ECM while the engine is running. This ability has obvious diagnostic advantages; the use of the scan tool is frequently required by the diagnostic charts.

With an understanding of the data stream that the tool will display and knowledge of the circuits involved, the tool can be an extremely valuable tool when examining the system. Without scan tools, diagnostic information would be difficult to obtain in some circuits and impossible to retrieve in others. Scan tools do not make the use of trouble code charts unnecessary, nor do they pinpoint the exact area in a circuit where the problem exists. However, this type of equipment will give more specific information, such as sensor input readings. An example of this would be the coolant sensor circuit. While the ECM looks at specific voltage levels to determine coolant temperature, the scan tool will display this information as an actual temperature reading. This information is not only useful in determining system malfunctions, but is also very helpful in diagnosing mechanical problems. For example, if the engine is overheating, the scan tool can be used to look at the actual engine temperature. This is helpful when checking for electric cooling fan enable switch problems, sticking thermostats, etc.

To properly read system values with a scan tool, the following conditions must be met. All normal values given in the charts will be based on these conditions:
- Engine running at idle, throttle closed
- Engine warm, upper radiator hose hot
- Vehicle in park or neutral
- System operating in closed loop
- All accessories **OFF**

The ECM is capable of communicating with a scan tool in 2 modes: Diagnostic and Field Service Mode.

Diagnostic Mode

When the diagnostic terminal is grounded with the ignition **ON** and the engine OFF, the system will enter the diagnostic mode. In this mode the ECM will perform the following:
- Display a DTC 12 by flashing the SES telltale light indicating proper diagnostic system operation.
- Display any stored trouble codes by flashing the SES telltale light. Each code will be flashed three times. If no transaxle codes are present the telltale light will repeat the sequence again starting with code 12.
- Energize all ECM controlled solenoids and relays except the fuel pump relay.
- IAC valve is pulsed in and out, with a pause before moving in either direction.

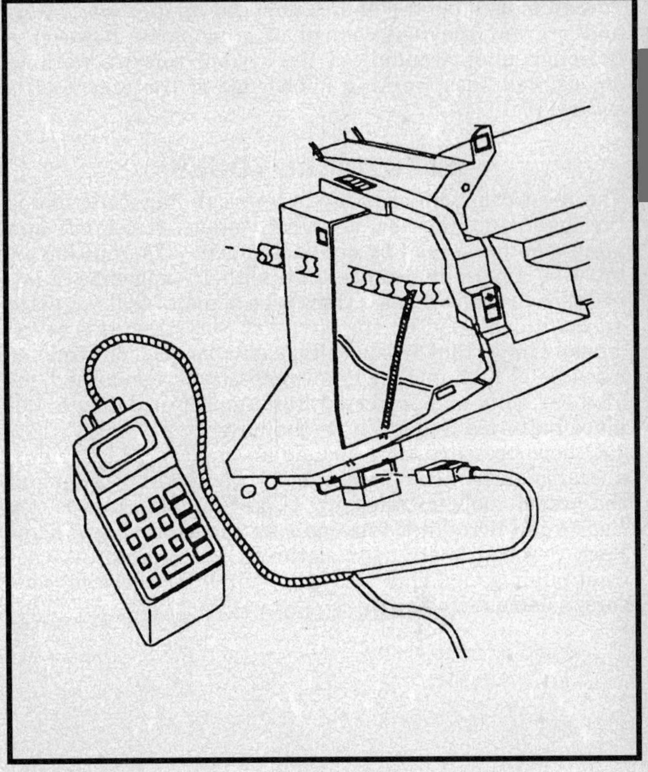

Typical scan tool — Saturn scan tool shown

Field Service Mode

If the diagnostic terminal is grounded with the engine running, the system will enter the field service mode. In this mode the SES telltale light will indicate either Open or Closed loop operation by the following flash sequence:
- SES light will flash 2.5 times per second to indicate Open Loop operation.
- SES light will flash once per second to indicate Closed Loop operation.
- In Closed Loop, the SES light will be OFF most of the time to indicate a lean exhaust or the SES light will be ON most of the time to indicate a rich exhaust.

NOTE: While the system is in the field service mode, the Closed Loop timer is by-passed and new trouble codes cannot be stored.

SCAN TOOL USE FOR INTERMITTENTS

In some tool applications the data update rate is so slow that it becomes less effective than a digital voltmeter for checking sporadic voltage changes. With the rapid data transmission of a quality scanner, intermittent problems such as faulty wiring connections can be more successfully diagnosed. For example, while manipulating a suspect electrical circuit or component, observe the scan tool display screen. A rapid voltage change or erratic reading will uncover the trouble area.

The scan tool is also an easy way to compare the operating parameters of a poorly operating engine with that of a known good one. For example, a sensor may shift in value and cause a driveability problem, but not set a trouble code. Comparing the sensor's readings to a known good one may uncover the problem.

Scan tools have the ability to speed up diagnostic time and prevent the replacement of good parts, however a thorough understanding of the system you are working on, as well as a working knowledge of the scan tool is essential.

ELECTRICAL TOOLS

The most commonly required electrical diagnostic tool is the digital multimeter, allowing voltage, resistance and amperage to be read by one instrument. The multimeter must be a high-impedance unit, with 10 megohms of impedance in the voltmeter. This type of meter will not place an additional load on the circuit it is testing; this is extremely important in low voltage circuits. The multimeter must be of high quality in all respects. It should be handled carefully and protected from impact or damage. Replace batteries frequently in the unit.

Other necessary tools include an unpowered test light, a quality tachometer with inductive (clip-on) pick up and the proper tools for releasing GM's Metri-Pack, Weather Pack and Micro-Pack terminals as necessary. The Micro-Pack connectors are used at the ECM connector. A vacuum pump/gauge may also be required for checking sensors, solenoids and valves.

Diagnosis and Testing

SERVICE PRECAUTIONS

• To prevent internal ECM damage, the ignition must be **OFF** when disconnecting or reconnecting power to the computer.

• When handling a PROM, CALPAC or MEMCAL, do not touch the component leads. Also, do not remove the integrated circuit from the carrier.

• Never allow welding cables to lie on, near or across any vehicle electrical wiring.

• Leave new components and modules in the shipping package until ready to install them.

• When performing electrical tests on the system, use a high impedance multimeter, digital voltmeter (DVM) J-34029-A or equivalent.

• To prevent possible electrostatic discharge damage to the ECM, do not touch the connector pins or soldered components on the circuit board.

PRELIMINARY INSPECTION

In order to successfully locate and repair problems in complex computer controlled systems, the technician needs to adhere to a comprehensive, sequential method of diagnostics. Many times a vehicle has a malfunction and the worst is suspected. Premature conclusions often result in the replacement of good parts and wasted time. A proper sequence of steps needs to be established and strictly adhered to when troubleshooting a vehicle. These are as follows:

1. If possible, try to speak directly to the customer, especially if the complaint is not associated with an intermittent or steady MIL.

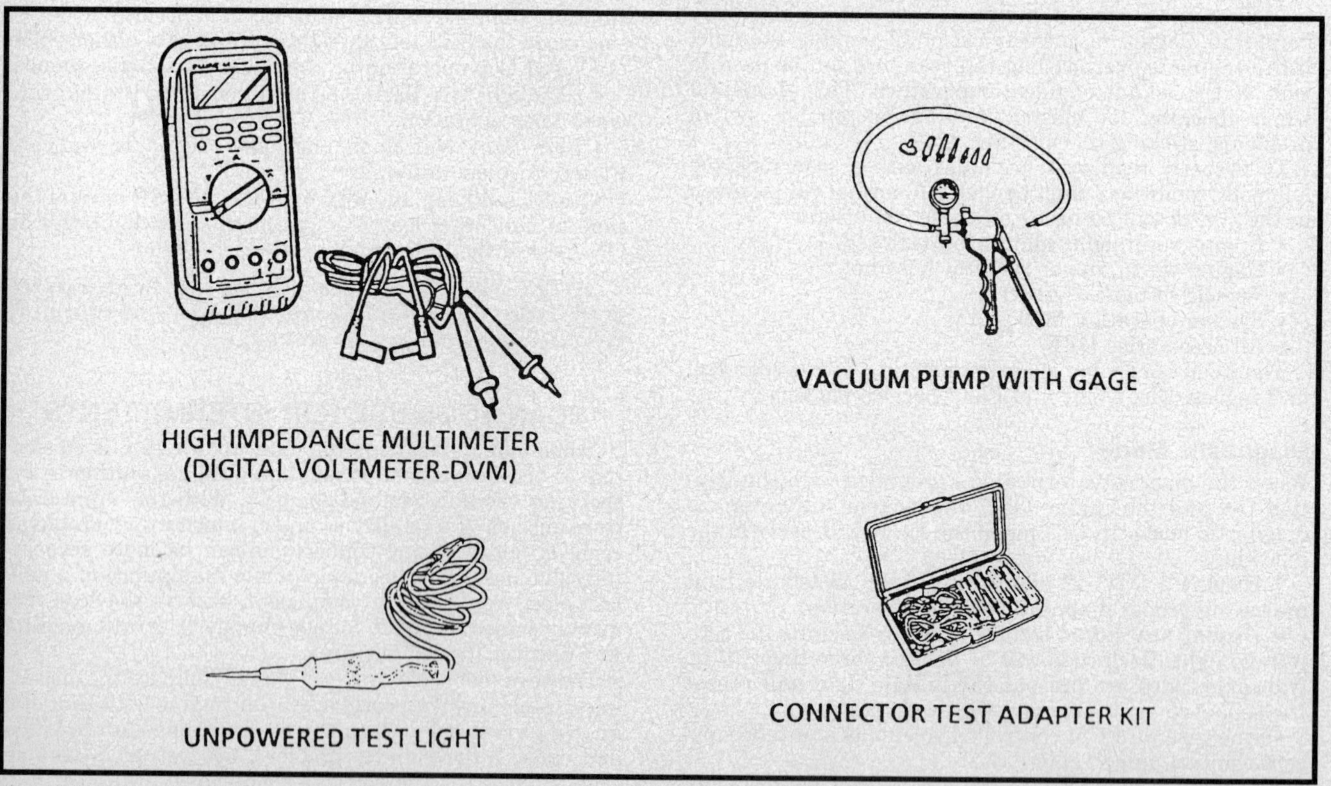

Electrical system diagnostic tools

2. Road test the vehicle, if necessary, to verify the complaint.

3. Perform a thorough visual inspection. This step can very often eliminate the need for further testing. Loose wires, especially ground circuits, can create mysterious driveability problems. However, a complete visual inspection can uncover these problems quickly, and prevent the replacing of good parts and reduce vehicle downtime.

4. Perform the Diagnostic Circuit Check. This test confirms that the diagnostic system has not failed and is able to communicate through the malfunction indicator lamp.

5. After locating and repairing the problem area, road test the vehicle to confirm the repair and to verify no additional problem areas exist.

Visual Underhood Inspection

This is possibly the most critical step of diagnosis. A detailed examination of connectors, wiring and vacuum hoses can often lead to a repair without further diagnosis. Performance of this step relies on the skill of the technician performing it; a careful inspector will check:

• The undersides of hoses as well as the integrity of hard-to-reach hoses blocked by the air cleaner or other components.

• The wiring carefully for any sign of strain, burning, crimping, or terminal pull-out from a connector.

• The connectors at components or in harnesses as required; usually, pushing them together will reveal a loose fit.

Diagnostic Circuit Check

This step is used to check that the on-board diagnostic system is working correctly. A system which is faulty or shorted may not yield correct codes when placed in the Diagnostic Mode.

If the diagnostic system is not operating correctly, or if a problem exists without the malfunction indicator lamp being lit, refer to the vehicle's A-Charts. These charts cover such conditions as Engine Cranks but Will Not Run or No Service Engine Soon Light.

1. Turn the ignition switch **ON** but do not start the engine.

2. The MIL should illuminate and and stay lit until the engine is started.

3. If the MIL does not light, check the bulb, fuses and circuits for an open or short.

4. Connect jumper wire from terminal A to B in the DLC. The MIL should flash code 12 at least 3 times. If not, turn the ignition switch **OFF** and disconnect the ECM.

5. Turn the ignition switch **ON** but do not start the engine. If the MIL is ON, look for short in circuit 419.

6. If the MIL does not light, check for an open in the MIL circuit.

Circuit and Component Diagnosis and Repair

Using the appropriate chart(s) based on the Diagnostic Circuit Check, the fault codes and the scan tool data will lead to diagnosis and checking of a particular circuit or component. It is important to note that the fault code indicates a fault or loss of signal in an ECM-controlled system, not necessarily in the specific component. Detailed procedures to isolate the problem are included in each code chart; these procedures must be followed accurately to insure timely and correct repair. Following the

procedure will also insure that only truly faulty components are replaced.

READING CODES

With Scan Tool

Saturn also uses information flags. These flags do not indicate a failure and will not turn on the SES telltale light. These flags are used as diagnostic aids to the technician, and can only be read with a scan tool.

The PCM has the ability to take a snapshot of 4 pieces of engine data with each trouble code stored. This valuable information can be retrieved with a Saturn scan tool or equivalent.

1. Turn the ignition switch **OFF**.

2. Plug a scan tool into the DLC, then turn the ignition switch **ON** and retrieve the codes with the scan tool.

3. To exit the diagnostic mode, turn the ignition switch **OFF** and remove the scan tool.

Without Scan Tool

1. Turn the ignition switch **OFF**.

2. Install a jumper wire between terminals A and B of the DLC.

3. Turn the key to the **ON** position.

4. Count the flashes of the MIL. To acknowledge diagnostic mode, the MIL will flash code 12 three times consecutively before moving on to stored trouble codes, if any exist. If there are codes in the system memory, they will be flashed in sequence three times consecutively, while the absence of stored codes will cause code 12 to flash continually. The MIL flash sequence occurs as follows: 12 = flash, pause, flash, flash, long pause (three times). Code 12 indicates that the ECM diagnostic system is operating properly, and is not a malfunction code.

> **NOTE: If more than one fault code is stored in memory, the codes will be displayed from the lowest to the highest, with the exception of code 11. Code 11 indicates transaxle codes and will always be flashed last if transaxle codes exist. Transaxle codes are not read with the SES light.**

5. To exit the diagnostic mode, turn the ignition switch **OFF** and remove the jumper wire or scan tool.

CLEARING CODES

When a code is set, the ECM stores it in two tables: General Information and Malfunction History. When removing codes from General Information without a scan tool, turn the ignition **ON** and ground the DLC terminals A and B three times within five seconds. General Information codes will automatically be erased after 50 ignition ON cycles, as long as the problem does not re-occur. Malfunction History can only be cleared with a scan tool. Disconnecting power to the ECM does not effect on Malfunction History, but will clear any codes in General Information.

DIAGNOSTIC TROUBLE CODES (DTC)

• DTC **11** — Transaxle diagnostic code present
• DTC **12** — Diagnostic check only (flash code)
• DTC **13** — Oxygen (O_2) sensor circuit open/not ready
• DTC **14** — Engine Coolant Temperature (ECT) sensor circuit out of range high

- DTC **15** — Engine Coolant Temperature (ECT) sensor circuit out of range low
- DTC **17** — Powertrain Control Module (ECT) fault
- DTC **21** — Throttle Position Sensor (TPS) circuit voltage out of range high
- DTC **22** — Throttle Position Sensor (TPS) circuit voltage out of range low
- DTC **23** — Inlet air temperature sensor circuit voltage out of range high
- DTC **24** — No vehicle speed sensor circuit signal
- DTC **25** — Inlet air temperature sensor circuit voltage out of range low
- DTC **26** — Quad output driver fault
- DTC **32** — Exhaust Gas Recirculation (EGR) system fault
- DTC **33** — Manifold Absolute Pressure (MAP) sensor circuit voltage out of range high
- DTC **34** — Manifold Absolute Pressure (MAP) sensor circuit voltage out of range low
- DTC **35** — Idle Air Control (IAC) rpm out of range
- DTC **41** — Ignition control circuit open or shorted
- DTC **42** — Bypass circuit open or shorted
- DTC **43** — Knock sensor circuit open or shorted
- DTC **44** — Oxygen (O_2) sensor circuit indicates system lean
- DTC **45** — Oxygen (O_2) sensor circuit indicates system rich
- DTC **46** — Power steering pressure circuit open
- DTC **49** — RPM high idle (vacuum leak)
- DTC **51** — Powertrain control module memory error
- DTC **55** — Powertrain control module error
- DTC **81** — Anti-lock braking system (ABS) message fault
- DTC **82** — Powertrain Control Module (PCM) internal communication fault

FLAGS

- Flag **16** — Electronic variable orifice fault
- Flag **48** — RPM signal fault
- Flag **52** — Battery voltage out of range
- Flag **53** — Spark knock present
- Flag **54** — Sensor circuit grounded
- Flag **58** — Battery voltage unstable
- Flag **61** — 6x signal fault
- Flag **65** — Powertrain Control Module (PCM)
- Flag **66** — Powertrain Control Module (PCM) hardware reset
- Flag **67** — Handwheel sensor fault
- Flag **71** — Engine Coolant Temperature (ECT) sensor out of range high
- Flag **72** — Engine Coolant Temperature (ECT) sensor out of range low
- Flag **73** — Engine Coolant Temperature (ECT) sensor unstable
- Flag **74** — Engine Coolant Temperature (ECT) sensor/transaxle ratio error
- Flag **75** — Inlet air temperature sensor unstable
- Flag **76** — Manifold Absolute Pressure (MAP) sensor response fault

ELECTRONIC IGNITION SYSTEMS

HIGH ENERGY IGNITION (HEI) SYSTEM

			BUICK		
Model	**Body VIN**	**Engine Liter**	**Engine VIN**	**Ignition Type**	**Fuel Type**
Century	A	2.2	4	DIS	MFI
		2.5	R	DIS	TBI
		3.1	M	C³I	SFI
		3.3	N	C³I	MFI
Estate Wagon	B	5.7	7	HEI	TBI
		5.7	P	Opti	SFI
LeSabre	H	3.8	L	C³I	SFI
Park Avenue	L	3.8	L	C³I	SFI
Park Avenue Ultra	L	3.8	1	C³I	SFI
Regal	W	3.1	T	DIS	MFI
		3.1	M	C³I	SFI
		3.8	L	C³I	SFI
Riviera	E	3.8	L	C³I	SFI
Roadmaster	B	5.7	7	HEI	TBI
		5.7	P	Opti	SFI
Skylark	N	2.3	D	IDI	MFI
		2.3	3	IDI	MFI
		3.1	M	C³I	SFI
		3.3	N	C³I	MFI

C³I—Computer Control Command Ignition
DIS—Distributorless Ignition System
HEI—High Energy Ignition
IDI—Integrated Direct Ignition
Opti—Opti-Spark Ignition

CADILLAC

Model	Body VIN	Engine Liter	Engine VIN	Ignition Type	Fuel Type
Allante	V	4.5	8	HEI	SFI
		4.6	9	DIS	SFI
Brougham	D	5.0	E	HEI	TBI
		5.7	7	HEI	TBI
		5.7	P	Opti	SFI
DeVille	C	4.9	B	HEI	SFI
Fleetwood	C	4.9	B	HEI	SFI
Eldorado	E	4.6	Y	DIS	SFI
		4.6	9	DIS	SFI
		4.9	B	HEI	SFI
Seville	K	4.6	9	DIS	SFI
		4.9	B	HEI	SFI

C³I—Computer Control Command Ignition
DIS—Distributorless Ignition System
HEI—High Energy Ignition
IDI—Integrated Direct Ignition
Opti—Opti-Spark Ignition

CHEVROLET

Model	Body VIN	Engine Liter	Engine VIN	Ignition Type	Fuel Type
Beretta	L	2.2	4	DIS	MFI
		2.3	A	IDI	MFI
		3.1	T	DIS	MFI
		3.1	M	DIS	SFI
Camaro	F	3.1	T	HEI	MFI
		3.4	S	DIS	SFI
		5.0	E	HEI	TBI
		5.0	F	HEI	MFI
		5.7	8	HEI	MFI
		5.7	P①	Opti	MFI
Caprice	B	4.3	Z	HEI	TBI
		5.0	E	HEI	TBI
		5.7	7	HEI	TBI
		5.7	P	Opti	SFI
Cavalier	J	2.2	4	DIS	MFI
		3.1	T	DIS	MFI
Corsica	L	2.2	4	DIS	MFI
		2.3	A	IDI	MFI
		3.1	T	DIS	MFI
		3.1	M	DIS	SFI
Corvette	Y	5.7	8	HEI	MFI
		5.7	J	DIS	SFI
		5.7	P①	Opti	MFI
Lumina	W	2.2	4	DIS	MFI
		2.5	R	DIS	TBI
		3.1	T②	DIS	MFI
		3.1	M	DIS	SFI
		3.4	X	DIS	MFI

DIS—Distributorless Ignition System
HEI—High Energy Ignition
IDI—Integrated Direct Ignition
Opti—Opti-Spark Ignition
① 1993 only; in 1994 this engine came equipped with Sequential Fuel Injection (SFI)
② VFV—Variable Fuel Vehicle also available

OLDSMOBILE

Model	Body VIN	Engine Liter	Engine VIN	Ignition Type	Fuel Type
Achieva	N	2.3	D	IDI	MFI
		2.3	A	IDI	MFI
		2.3	3	IDI	MFI
		3.1	M	C^3I	SFI
		3.3	N	C^3I	MFI
Cutlass Supreme	W	3.1	T	DIS	MFI
		3.1	M	C^3I	SFI
		3.4	X	DIS	MFI
Custom Cruiser	B	5.0	E	HEI	TBI
		5.7	7	HEI	TBI
Cutlass Cruiser	A	2.2	4	DIS	MFI
		2.5	R	DIS	TBI
		3.1	M	C^3I	SFI
		3.3	N	C^3I	MFI
Cutlass Ciera	A	2.2	4	DIS	MFI
		2.5	R	DIS	TBI
		3.1	M	C^3I	SFI
		3.3	N	C^3I	MFI
Eighty-Eight	H	3.8	L	C^3I	SFI
Ninety-Eight	C	3.8	L	C^3I	SFI
		3.8	1	C^3I	SFI
Toronado	E	3.8	L	C^3I	SFI
Toronado Trofeo	E	3.8	L	C^3I	SFI

C^3I—Computer Control Command Ignition
DIS—Distributorless Ignition System
HEI—High Energy Ignition
IDI—Integrated Direct Ignition
Opti—Opti-Spark Ignition
[1] 1993 only; in 1994 this engine came equipped with Sequential Fuel Injection (SFI)

PONTIAC

Model	Body VIN	Engine Liter	Engine VIN	Ignition Type	Fuel Type
Bonneville	H	3.8	L	C^3I	SFI
		3.8	1	C^3I	SFI
Firebird	F	3.1	T	HEI	MFI
		3.4	S	DIS	MFI
		5.0	E	HEI	TBI
		5.0	F	HEI	MFI
		5.7	8	HEI	MFI
		5.7	P①	Opti	MFI
Grand Am	N	2.3	A	IDI	MFI
		2.3	3	IDI	MFI
		2.3	D	IDI	MFI
		3.1	M	C^3I	SFI
		3.3	N	C^3I	MFI
Grand Prix	W	3.1	T	DIS	MFI
		3.1	M	C^3I	SFI
		3.4	X	DIS	MFI
LeMans	T	1.6	6	HEI	TBI
Sunbird	S	2.0	H	DIS	MFI
		3.1	T	DIS	MFI

C^3I—Computer Control Command Ignition
DIS—Distributorless Ignition System
HEI—High Energy Ignition
IDI—Integrated Direct Ignition
Opti—Opti-Spark Ignition
① 1993 only; in 1994 this engine came equipped with Sequential Fuel Injection (SFI)

ELECTRONIC IGNITION SYSTEMS
HIGH ENERGY IGNITION (HEI) SYSTEM

GENERAL MOTORS LIGHT TRUCKS, VANS, AND APVS

Model	Body VIN	Engine Liter	Engine VIN	Ignition Type	Fuel Type
S-Series, Bravada and Astro/Safari	L,M,S,T	2.2	4	DIS	MFI
		2.5	A	HEI	TBI
		2.8	R	HEI	TBI
		4.3	Z ①	HEI	TBI
		4.3	W	HEI	CPI
Light Trucks and Vans	C,K,R,V,G,P,D	4.3	Z	HEI	TBI
		5.0	H	HEI	TBI
		5.7	K	HEI	TBI
		6.2	C	Diesel	TBI
		7.4	N	HEI	TBI
Lumina APV	U	3.1	D	HEI	TBI
		3.8	L	C³I	SFI
Silhouette	U	3.1	D	HEI	TBI
		3.8	L	C³I	SFI
Trans Sport	U	3.1	D	HEI	TBI
		3.8	L	C³I	SFI

C³I—Computer Control Command Ignition
DIS—Distributorless Ignition System
HEI—High Energy Ignition
ESC—Electronic Spark Control
EI—Electronic Ignition
① MFI with turbocharger

GEO

Model	Body VIN	Engine Liter	Engine VIN	Ignition Type	Fuel Type
Metro	M	1.0	6	ESC	TBI
Prizm	S	1.6	5	ESC	MFI
		1.6	6	ESC	MFI
		1.6	8	ESC	MFI
		1.8	U	ESC	TBI
Storm	R	1.6	5	ESC	MFI
		1.6	6	ESC	MFI
Tracker	J	1.6	U	ESC	TBI

C³I—Computer Control Command Ignition
DIS—Distributorless Ignition System
HEI—High Energy Ignition
ESC—Electronic Spark Control
EI—Electronic Ignition

General Information

The High Energy Ignition (HEI) system controls fuel combustion by providing a spark to ignite the air/fuel mixture at the appropriate time. This system consists of a modified module, which is used in conjunction with the Electronic Spark Timing (EST) timing.

The HEI system can be either equipped with an integral (coil contained in distributor cap) or external mounted ignition coil.

The HEI system features a longer spark duration which is instrumental in firing lean and Exhaust Gas Recirculation EGR diluted fuel/air mixtures. The condenser (capacitor) located within the HEI distributor is provided for noise (static) suppression purposes only and is not a regularly replaced ignition system component. Dwell adjustment is controlled by the ECM and cannot be adjusted.

The HEI distributor is equipped to aid in spark timing changes, necessary for emissions, economy and performance. All spark timing changes in the HEI (EST) distributors are performed electronically by the Electronic Control Module (ECM), which monitors information from the various engine sensors, computes the desired spark timing and signals the distributor to change the timing accordingly. With this distributor, no vacuum or centrifugal advances are used.

To control spark knock and to use maximum spark advance to improve driveability and fuel economy, an Electronic Spark Control (ESC) system is used. This system is consists of a knock sensor and an ESC module (generally part of MEMCAL). The ECM/PCM monitors the ESC signal to determine when engine detonation occurs.

NOTE: The terms Electronic Control Module (ECM), Powertrain Control Module (PCM) and Body Control Module (BCM) may sometimes be used interchangeably.

SYSTEM OPERATION

The HEI distributors use a magnetic pick-up assembly, located inside the distributor containing a permanent magnet, a pole piece with internal teeth and a pick-up coil. When the teeth of the rotating timer core and pole piece align, an induced voltage in the pick-up coil signals the electronic module to open the coil primary circuit. As the primary current decreases, a high voltage is induced in the secondary windings of the ignition coil, directing a spark through the rotor and high voltage leads to fire the spark plugs. The dwell period is automatically controlled by the electronic module and is increased with increasing engine rpm.

To control ignition combustion timing, the ECM needs to know all of the following:
- Engine speed (rpm)
- Crankshaft position
- Engine load (manifold pressure or vacuum)
- Atmospheric (barometric) pressure
- Engine temperature
- Exhaust Gas Recirculation (EGR)

The ESC system is designed to retard spark timing up to 8 degrees to reduce spark knock in the engine. When the knock sensor detects spark knocking in the engine, it sends an A/C voltage signal to the ECM/PCM, which increases with the severity of the knock. The ECM then adjusts the EST to reduce spark knock.

To control EST, the HEI module uses 4 connecting terminals. These terminals provide the following:
- Distributor reference circuit
- Reference ground circuit
- Bypass circuit
- EST circuit

Some vehicles use a capacitor for radio noise suppression. The capacitor is part of the coil wire harness assembly and will seldom need replacement.

SYSTEM COMPONENTS

Electronic Control Module (ECM)

The ECM is the control center of the fuel injection system. It constantly monitors information from various sensors and controls the system. The ECM has 2 replaceable parts, the controller and the MEMCAL.

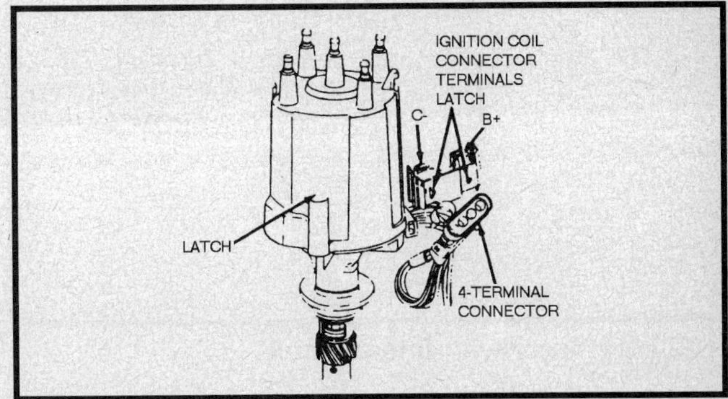

HEI distributor with external coil

SATURN

Model	Body VIN	Engine Liter	Engine VIN	Ignition Type	Fuel Type
SL/SL1/SW1/SC1	Z	1.9	9	EI	TBI
SL2/SW2/SC2	Z	1.9	7	EI	MFI

C³I—Computer Control Command Ignition
DIS—Distributorless Ignition System
HEI—High Energy Ignition
ESC—Electronic Spark Control
EI—Electronic Ignition

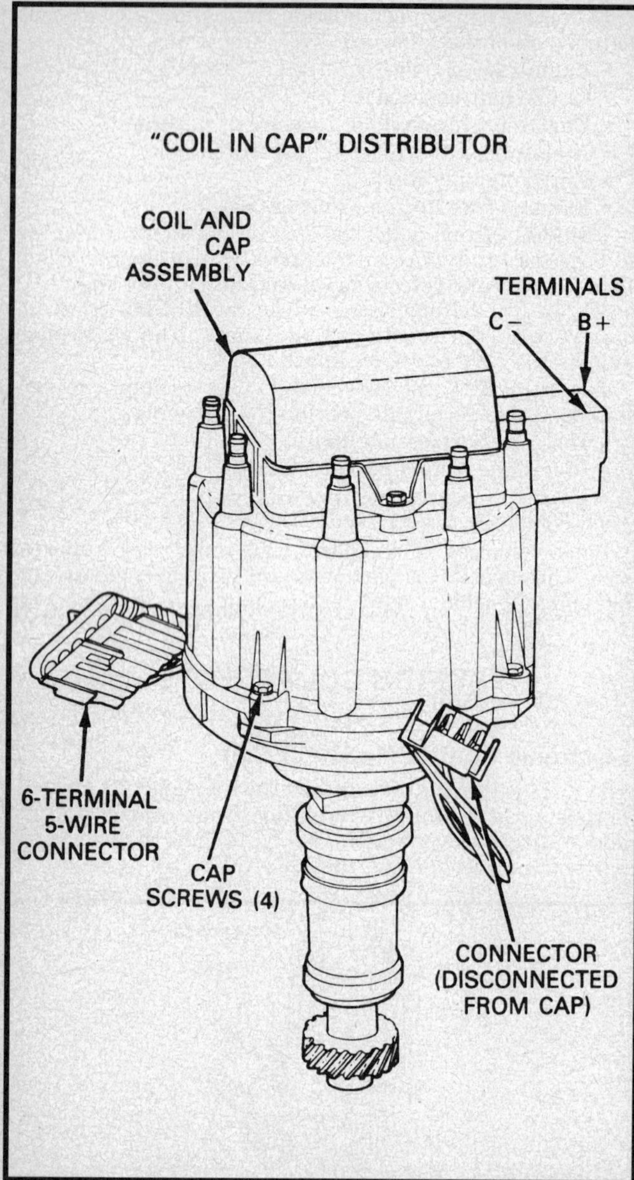

HEI distributor with integral coil

Memory Calibration (MEMCAL) Unit

The MEMCAL, located inside the ECM, allows the ECM to be installed in several different vehicles. It has calibration information based on the vehicle's weight, engine, transmission, axle ratio and several other factors.

Pick-Up Coil

The pick-up coil is a device which generates an alternating current signal to determine crankshaft position.

Knock Sensor

The knock sensor detects spark knock or detonation and send a signal to the ECM to adjust the ignition timing accordingly.

HEI Module

The HEI module is a switching device which operates the primary circuit of the ignition coil.

Ignition Secondary

The ignition secondary consists of the ignition coil, rotor, distributor cap, plugs wires and spark plug. These components supply the high voltage to fire the spark plugs.

Electronic Spark Timing (EST)

The EST system consists of the distributor module, ECM and connecting wires. This system includes the following circuits:

- Distributor reference circuit — provides the ECM with rpm and crankshaft position information.
- Bypass signal — above 500 rpm, the ECM applies 5 volts to this circuit to switch spark timing control from the HEI module to the ECM.
- EST signal — the ECM uses this circuit to trigger the HEI module, after by-pass voltage is applied to the HEI module.
- Reference ground circuit — this wire is grounded through the module and insures that the ground circuit has no voltage drop between the ignition module and the ECM which could affect performance.

Tools and Equipment

SCAN TOOLS

The system can communicate a wide variety of information through the Data Link Connector (DLC), also known as the ALDL. Depending on the application, serial data is

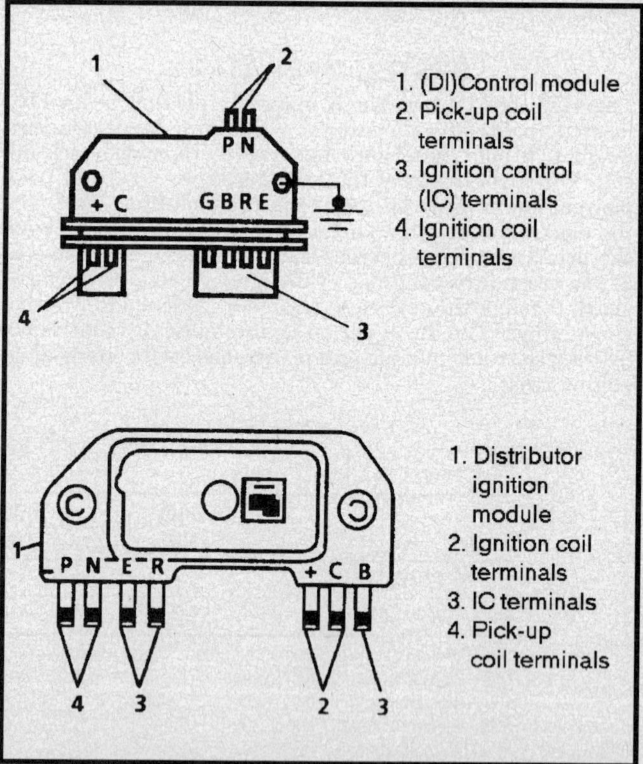

Ignition control modules

transmitted to terminals E or M of the DLC. This data is transmitted at a high frequency which requires a quality scan tool for interpretation. Although stored codes may be read with only the use of a small jumper wire, the use of a hand-held scan tool such as GM's TECH 1 or equivalent, is recommended. There are many manufacturers of these tools; a purchaser must be certain that the tool is proper for the intended use.

The scan tool allows any stored codes to be read from the ECM memory. The tool also allows the operator to view the data being sent to the ECM while the engine is running. This ability has obvious diagnostic advantages; the use of the scan tool is frequently required by the diagnostic charts.

With an understanding of the data stream that the tool will display and knowledge of the circuits involved, the tool can be an extremely valuable tool when examining the system. Without scan tools, diagnostic information would be difficult to obtain in some circuits and impossible to retrieve in others. Scan tools do not make the use of trouble code charts unnecessary, nor do they pinpoint the exact area in a circuit where the problem exists. However, this type of equipment will give more specific information, such as sensor input readings. An example of this would be the coolant sensor circuit. While the ECM looks at specific voltage levels to determine coolant temperature, the scan tool will display this information as an actual temperature reading. This information is not only useful in determining system malfunctions, but is also very helpful in diagnosing mechanical problems. For example, if the engine is overheating, the scan tool can be used to look at the actual engine temperature. This is helpful when checking for electric cooling fan enable switch problems, sticking thermostats, etc.

NOTE: A scan tool that is known to display faulty data should not be used for diagnosis. Although the fault may be believed to be in only one area, it can possibly affect many other areas during diagnosis, leading to errors and incorrect repair.

ELECTRICAL TOOLS

The most commonly required electrical diagnostic tool is the digital multimeter, allowing voltage, resistance and amperage to be read by one instrument. The multimeter must be a high-impedance unit, with 10 megohms of impedance in the voltmeter. This type of meter will not place an additional load on the circuit it is testing; this is extremely important in low voltage circuits. The multimeter must be of high quality in all respects. It should be handled carefully and protected from impact or damage. Replace batteries frequently in the unit.

Other necessary tools include an unpowered test light, a quality tachometer with inductive (clip-on) pick up and the proper tools for releasing GM's Metri-Pack, Weather Pack and Micro-Pack terminals as necessary. The Micro-Pack connectors are used at the ECM connector. A vacuum pump/gauge may also be required for checking sensors, solenoids and valves.

Diagnosis and Testing

SERVICE PRECAUTIONS

> **CAUTION**
>
> *The HEI coil secondary voltage output capabilities can exceed 40,000 volts. Avoid body contact with the HEI high voltage secondary components when the engine is running, or personal injury may result.*

NOTE: To avoid damage to the ECM or other ignition system components, do not use electrical test equipment such as battery or AC powered voltmeter, ohmmeter, etc. or any type of tester other than specified.

- When making compression checks, disconnect the ignition switch feed wire at the distributor.
- Never allow the tachometer terminal to touch ground, as damage to the module and/or ignition coil can result.
- To prevent electrostatic discharge damage, when working with the ECM, do not touch the connector pins or soldered components on the circuit board.
- When handling a PROM, CALPAC or MEMCAL, do not touch the component leads. Also, do not remove the integrated circuit from the carrier.
- Never allow welding cables to lie on, near or across any vehicle electrical wiring.
- Leave new components and modules in the shipping package until ready to install them.
- When performing electrical tests on the system, use a high impedance multimeter, digital voltmeter (DVM) J-34029-A or equivalent.
- Never pierce a high tension lead or boot for any testing purpose; otherwise, future problems are guaranteed.

PRELIMINARY INSPECTION

This is possibly the most critical step of diagnosis. A detailed examination of connectors, wiring and vacuum hoses can often lead to a repair without further diagnosis. Performance of this step relies on the skill of the technician performing it; a careful inspector will check:
- The undersides of hoses as well as the integrity of hard-to-reach hoses blocked by the air cleaner or other components.
- The wiring carefully for any sign of strain, burning, crimping, or terminal pull-out from a connector.
- The connectors at components or in harnesses as required; usually, pushing them together will reveal a loose fit.

READING CODES

The Data Link Connector, also known as the ALDL connector, is used for communicating with the ECM. It is

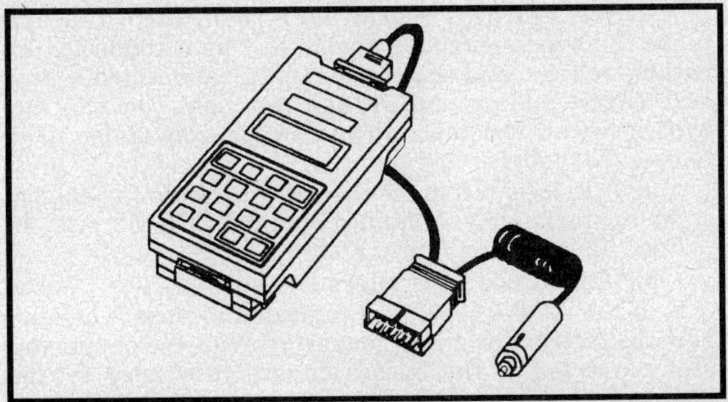

Typical scan tool — GM's Tech 1 shown

usually located under the instrument panel and is sometimes covered by a plastic cover labeled DIAGNOSTIC CONNECTOR. Codes stored in the ECM's memory can be read through a handheld diagnostic scanner plugged into the DLC. Codes can also be read by connecting terminal A to B of the DLC and counting the number of flashes of the SERVICE ENGINE SOON light, with the ignition switch turned **ON**.

CLEARING CODES

To clear codes from the ECM memory, the ECM power feed must be disconnected for at least 30 seconds. Depending on the vehicle, the ECM power feed can be disconnected at the positive battery terminal pigtail, the in-line fuseholder that originates at the positive connection at the battery or the ECM fuse in the fuse block. The negative battery cable may also be disconnected; however,

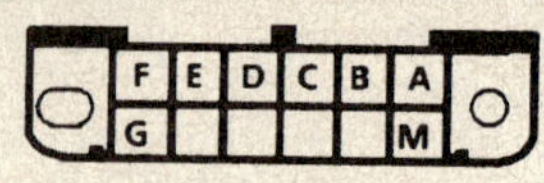

Data link connector (DLC)

other on-board memory data, such as preset radio tuning, will also be lost.

PICKUP COIL PARAMETERS CHART

PICKUP COIL		
System	Test Points	Reading
HEI	Pickup coil lead terminals	500–1500 Ohms

ENGINE CRANKS BUT WILL NOT RUN

Testing

1. Check that the fuel quantity is adequate.
2. Turn the ignition switch **ON**. Verify that the SERVICE ENGINE SOON light is ON.
3. Install the scan tool and check the Throttle Position Sensor (TPS) — if over 2.5 volts, at closed throttle, use Chart Code 21 in the fuel section. Also check the coolant — if less than -22°F (-30°C), use Chart Code 15 in the fuel section.
4. Connect spark checker, J 26792 or equivalent, and check for spark while cranking. Check at least 2 wires.
5. If spark occurs, reconnect the spark plug wires and check for fuel spray at the injector(s) while cranking. If no spark occurs, check for battery voltage to the ignition system. If OK, use the ignition diagnosis chart.

ELECTRONIC SPARK TIMING (EST) CIRCUIT

Testing

1. Clear any previous diagnostic trouble codes.
2. Idle the engine for approximately 1 minute or until Code 42 sets. If Code 42 does not set, Code 42 is intermittent.
3. If Code 42 sets, turn the ignition **OFF** and disconnect the ECM connectors.
   a. Turn the ignition **ON**.

   b. Using an ohmmeter to ground, probe the ECM harness connector, at the EST circuit. Place the ohmmeter selector switch in the 1000–2000 ohms range. The meter should read less than 1000 ohms.
4. If not, check for a faulty connection, open circuit or faulty ignition module.
5. If OK, probe the bypass circuit with a test light to battery voltage.
6. If the test light is come ON, disconnect the ignition 4-way connector and observe the test light.
   a. If the test light goes OFF, its a faulty ignition module.
   b. If the test light stays ON, the bypass circuit is shorted to ground.
7. If the test light remain OFF, from Step 5, again probe the bypass circuit with the test light connected to battery voltage, and the ohmmeter still connected to the EST circuit and ground. As the test light contacts the bypass circuit, resistance should switch from under 1000 to over 2000 ohms.
   a. If it does, reconnect the ECM and idle the engine for approximately 1 minute or until Code 42 sets. If Code 42 sets, its a faulty ECM.
   b. If not, Code 42 is intermittent.
8. If the results are not as indicated in Step 7, disconnect the distributor 4-way connector. With the ohmmeter still connected to the bypass circuit, resistance should have gone high (open circuit).
9. If not, the EST circuit is shorted to ground.

10. If OK, the bypass circuit is open, faulty connections or faulty ignition module.

> **NOTE: When the problem have been corrected, clear codes and confirm Close Loop operation and no SERVICE ENGINE SOON light.**

KNOCK SENSOR

Testing

1. Connect a scan tool and set to knock signal.
2. Start the engine and run at 1500 rpm. Check the knock signal. If a knock signal is present, then sensor is OK.
3. If no signal is present, tap on the engine block near the knock sensor. If tapping produces a knock signal, then system is OK. If tapping does not produce a signal, then sensor is faulty or circuit is shorted.

PICK-UP COIL

Testing

1. Remove the rotor and pick-up coil leads from the module.
2. Connect an ohmmeter between the distributor housing and the pick-up coil lead. Meter should read infinity. Flex the leads by hand while observing the ohmmeter, to check for intermittent opens.
3. Connect the ohmmeter between leads between both of the pick-up coil leads. Meter should read a steady value between 500–1500 ohms.
4. If the readings are not as specified, the pick-up coil is defective and should be replaced.

IGNITION COIL

Testing

EXTERNAL COIL

1. Remove the ignition coil and check the coil with an ohmmeter for an open or ground, as indicated.
2. Connect the meter between C+ and ground (meter on the high scale). Should read very high or infinite. If not, replace the coil.
3. Connect the meter between C+ and C- (meter on the low scale). Should read low or zero ohms. If not, replace the coil.
4. Connect the meter between C+ and the secondary output tower (meter on the high scale). Should not read infinite. If it does, replace the coil.

INTEGRAL COIL

1. Remove the cap from the distributor.
2. With the coil attached to the cap, connect the meter between C- and B+ (meter on the low scale). Should read very low or zero. If not, replace the coil.
3. Connect the meter between GROUND and center cap pickup (meter on the high scale). Should not read infinite. Replace the coil if it does.

EST PERFORMANCE

Testing

EXCEPT 2.5L ENGINE

The ECM will set timing at a specified value when the diagnostic TEST terminal in the DLC is grounded.

1. Run the engine at 2000 rpm with the terminal ungrounded. Note the ignition timing.
2. Ground the TEST terminal and again note the timing. If the EST is operating, there should be a noticeable engine rpm change. A fault in the EST system will set a trouble Code 42.
3. An open or ground in the EST circuit, will set a Code 42 and cause the engine to run on the HEI module timing. This will cause poor performance and poor fuel economy.
4. Loss of the ESC signal, to the ECM, would cause the ECM to constantly retard the EST. This could result in sluggish performance and cause a Code 43 to set.

2.5L ENGINE

1. Run the engine at 2000 rpm. Note the ignition timing.
2. Disconnect the SET TIMING connector and again note the timing.
3. If the EST is operating, there should be a noticeable timing change.

ELECTRONIC CONTROL MODULE

Testing

1. Before replacement of a defective ECM/PCM first check the resistance of each ECM controlled solenoid.
2. Using an ohmmeter and the ECM connector wiring diagram, check the resistance of each solenoid. Any ECM controlled device with low resistance will damage the replacement ECM due to high current flow through the ECM internal circuits.

IGNITION TIMING

Adjustment

If procedures vary from that indicated Vehicle's Emission Control Information (VECI) label set the timing according to the procedure indicated on the VECI label. The VECI label is located in the engine compartment.

> **NOTE: Changes in timing specifications may be made during a production run. Therefore, always check the VECI to verify timing specifications.**

1.6L, 2.5L, 4.5L AND 4.9L ENGINES

1. Turn the ignition switch **OFF**.
2. Connect the pick-up lead of the inductive type timing light to the No.1 spark plug lead.

> **NOTE: Do not pierce the wire or attempt to insert a probe between the boot and the wire. Piercing a wire may create a failure condition that will not be immediately apparent.**

3. Connect the timing light power leads according to the manufacturer's instructions.
4. Start and run the engine until normal operating temperature is reached.

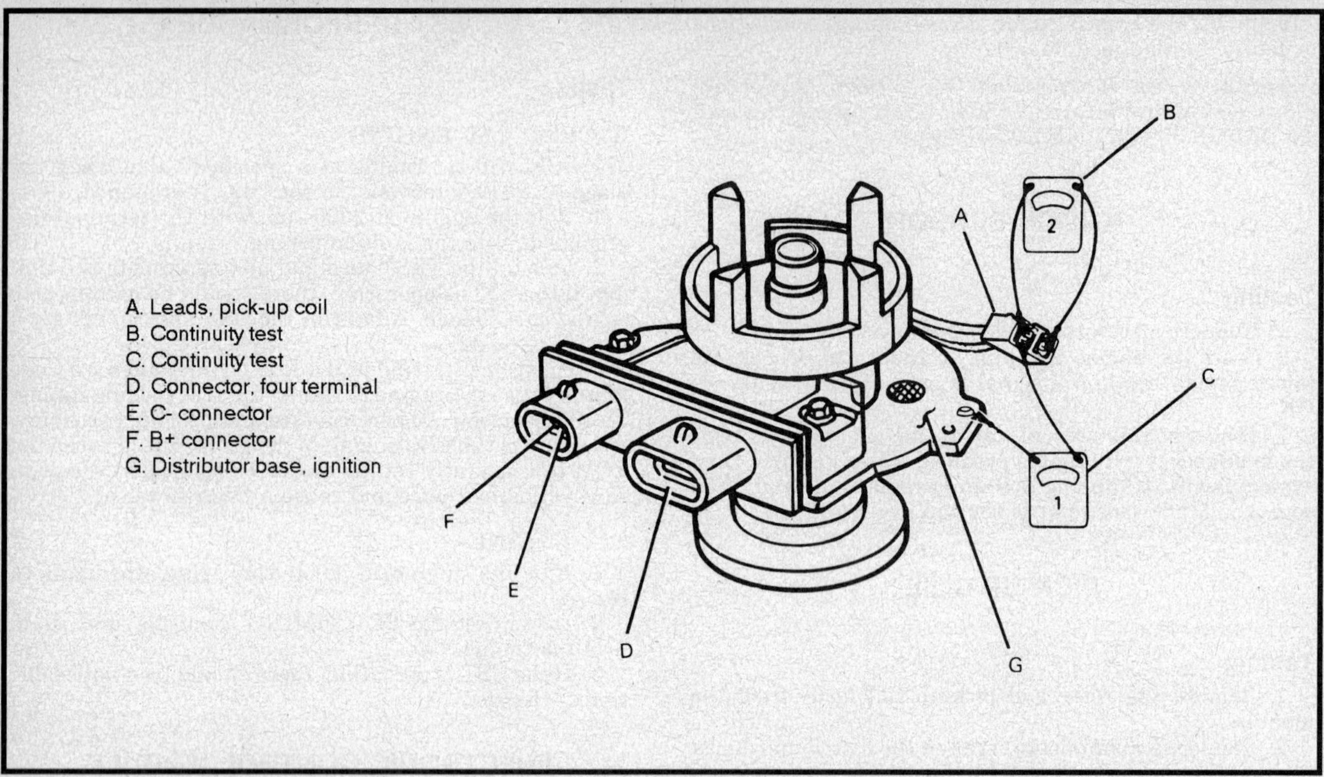

Testing the HEI pick-up coil

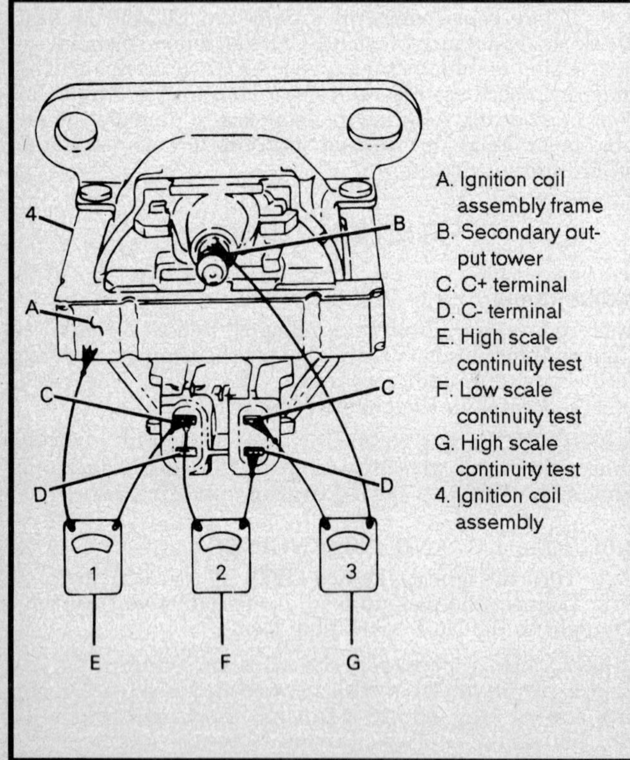

Testing HEI coil — external coil

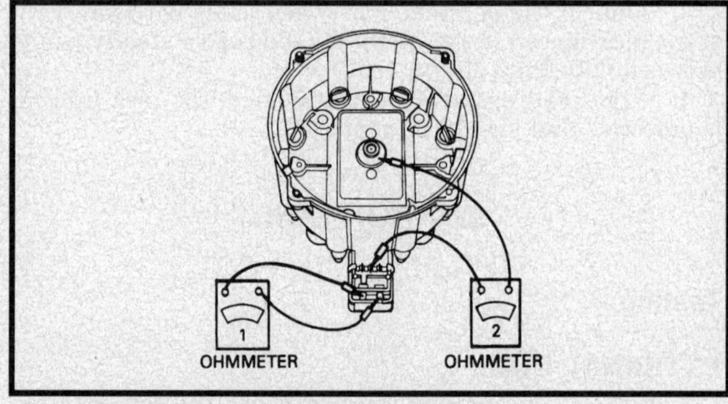

Testing HEI coil — integral coil

5. Put the engine in base timing mode by connecting a jumper wire between terminals A and B of the DLC or follow instructions on the VECI label.

6. Start the engine, then aim the timing light at the timing mark. The line on the pulley should line up at the timing mark. On the 2.5L engine, the timing is set using the average of cylinder No. 1 and No. 4.

7. If adjustment is necessary, loosen the distributor holddown clamp bolt at the base of the distributor. Slightly rotate the distributor until the line indicates the correct timing, according to the specification on the vehicle emission control label in the engine compartment. Then, tighten the distributor holddown clamp bolt. Recheck the timing.

8. Turn the ignition switch **OFF**. Remove the timing light and reconnect the No. 1 spark plug wire, if removed.

3.1L, 4.3L, 5.0L AND 5.7L ENGINES

1. Turn the ignition switch **OFF**.

2. Connect the pick-up lead of the inductive type timing light to the No.1 spark plug lead.

NOTE: Do not pierce the wire or attempt to insert a probe between the boot and the wire. Piercing a wire may create a failure condition that will not be immediately apparent.

3. Connect the timing light power leads according to the manufacturer's instructions.

4. Start and run the engine until normal operating temperature is reached.

5. Put the engine in base timing mode by disconnecting the EST single wire connector. The leads are tan with a black strip and usually break out of the engine wiring harness. On 3.1L (VIN D) engines the connector is in front of the right shock tower. On other engines locations may also be behind the junction block, next to distributor and below heater case in passenger compartment. Exact locations can be found on the VECI label, located in the engine compartment.

NOTE: Do not disconnect the 4 wire connector at the distributor.

6. Start the engine, then aim the timing light at the timing mark. The line on the pulley should line up at the timing mark.

7. If adjustment is necessary, loosen the distributor holddown clamp bolt at the base of the distributor. Slightly rotate the distributor until the line indicates the correct timing, according to the specification on the vehicle emission control label in the engine compartment. Then, tighten the distributor holddown clamp bolt. Recheck the timing.

8. Turn the ignition switch **OFF**. Remove the timing light and reconnect the No. 1 spark plug wire, if removed.

Component Replacement

HEI DISTRIBUTOR

Removal and Installation

1. Disconnect the negative battery cable.

2. Remove the air cleaner assembly.

3. Disconnect the 2 electrical connectors from the side of the distributor.

4. Remove the distributor cover and wire retainer, if equipped. Turn the retaining screws counterclockwise and remove the cap.

5. Mark the relationship of the rotor to the distributor housing and the housing relationship to the engine.

6. Remove the distributor retaining bolt and hold-down clamp.

7. Pull the distributor up until the rotor just stops turning counterclockwise and again note the position of the rotor.

8. Remove the distributor from the engine.

To install:

9. Insert the distributor into the engine, with the rotor aligned to the marks made previously.

10. If the engine was accidentally cranked after the distributor was removed, the following procedure can be used during installation.

  a. Remove the No. 1 spark plug.

  b. Place a finger over the spark plug hole and have a helper crank the engine slowly until compression is felt.

  c. Align the timing mark on the pulley to **0** on the engine timing indicator.

  d. Turn the rotor to point between No. 1 and No. 6 (V-6) or No. 1 and No. 8 (V-8) spark plug towers on the distributor cap.

  e. Install the distributor assembly in the engine.

  f. Install the cap and spark plug wires.

  g. Check and adjust engine timing.

11. Install the distributor hold-down clamp and retaining bolt.

12. If removed, install the wiring harness retainer and secondary wires.

13. Install the distributor cap.

14. Reconnect the 2 connectors to the side of the distributor. Make certain the connectors are fully seated and latched.

15. Reconnect the negative battery cable.

HEI MODULE

Removal and Installation

1. Disconnect the negative battery cable.

2. Remove the air cleaner assembly.

3. Remove the distributor cap and rotor.

4. Remove the module retaining screws, then lift the stamped sheet metal shield and module upwards.

NOTE: Do not wipe the grease from the module or distributor base, if the same module is to be replaced.

5. Disconnect the module leads. Note the color code on the leads, as these cannot be interchanged.

To install:

6. Spread the silicone grease, included in package, on the metal face of the module and on the distributor base where the module seats.

7. Fit the module leads to the module. Make certain the leads are fully seated and latched. Seat the module and metal shield into the distributor and install the retaining screws.

8. Install the rotor and cap.

9. Install the air cleaner assembly.

10. Reconnect the negative battery cable.

PICK-UP COIL

Removal and Installation

1. Disconnect the negative battery cable.

2. Remove the distributor assembly.

3. Support the distributor assembly in a vice and drive the roll pin from the gear. Remove the shaft assembly.

4. To remove the pick-up coil, remove the retainer and shield.

5. Lift the pick-up coil assembly straight up to remove from the distributor.

To install:

6. Assembly the pick-up coil, shield and retainer.

7. Install the shaft.

8. Install the gear and roll pin to the shaft. Make certain the matchmarks are aligned.

9. Spin the shaft and verify that the teeth do not touch the pole piece.

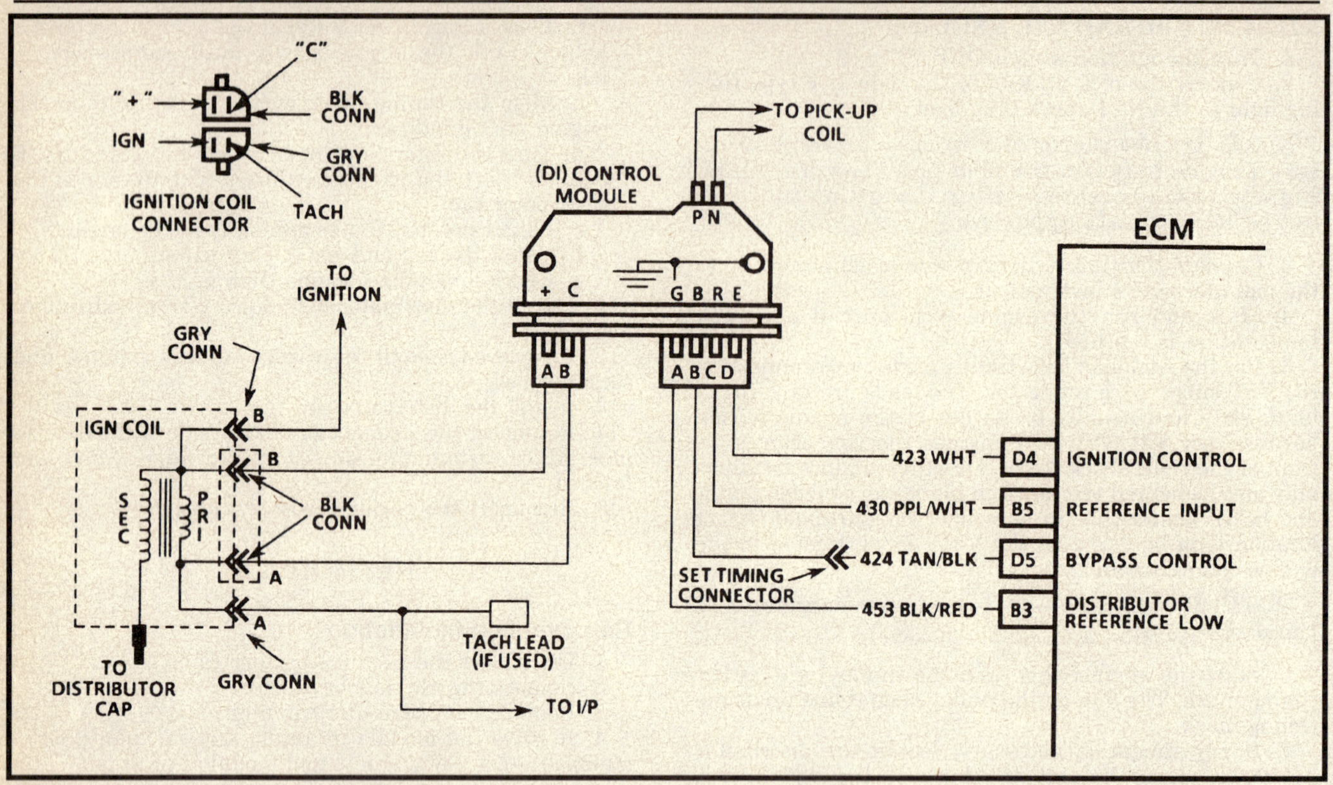

HEI system circuit — External coil

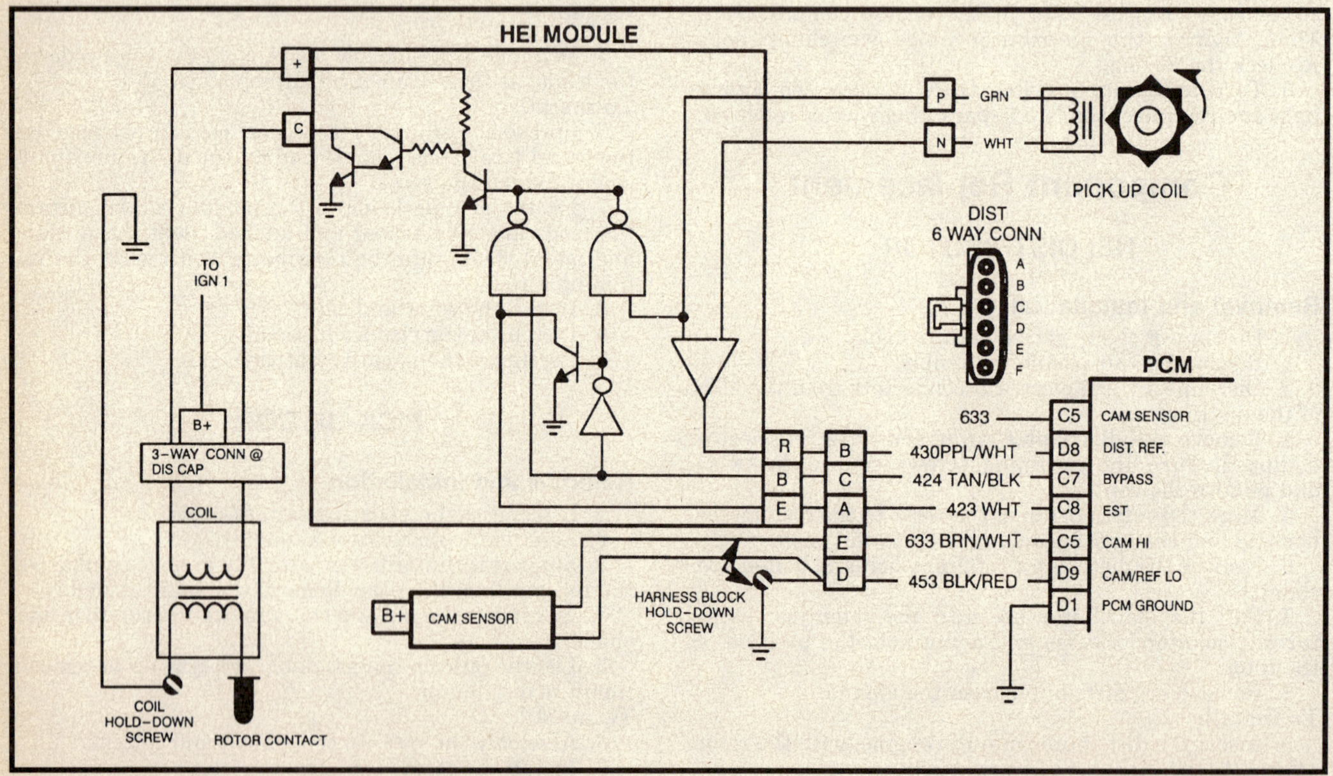

HEI system circuit — Integral coil

HEI SYSTEM DIAGNOSIS — EXTERNAL COIL

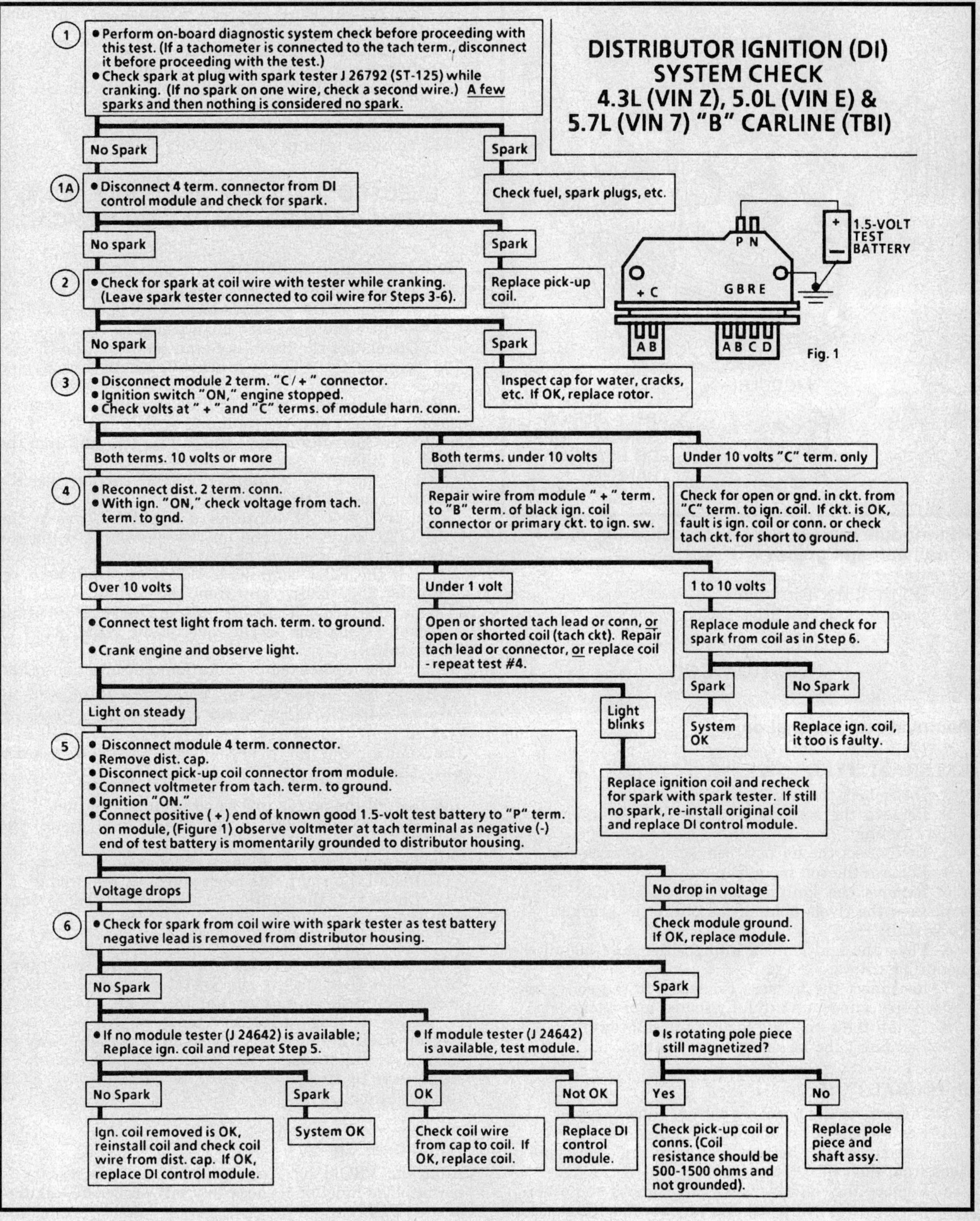

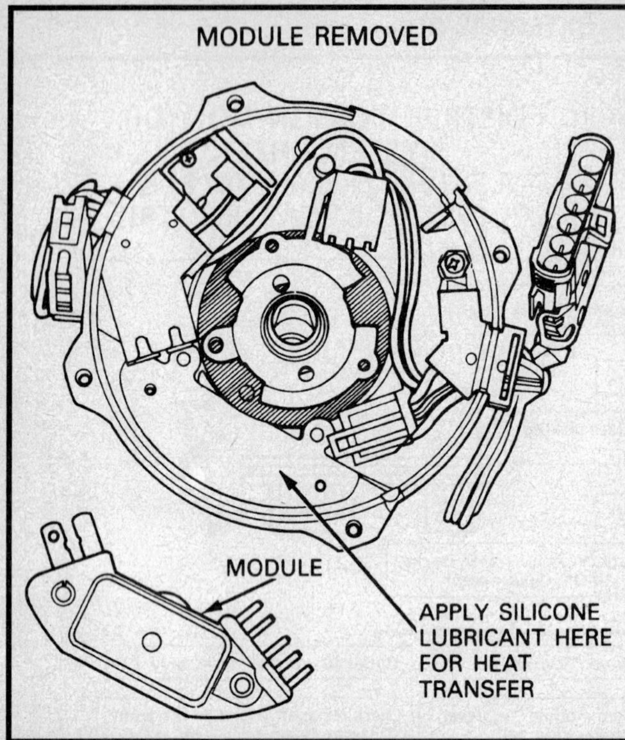

HEI module replacement and application of silicone heat sink grease

10. Reinstall the distributor.
11. Reconnect the negative battery cable.

IGNITION COIL

Removal and Installation

EXTERNAL TYPE

1. Disconnect the negative battery cable.
2. Remove the secondary coil lead. Pull on the boot while twisting.
3. Disconnect the harness connectors from the coil.
4. Remove the coil mounting screws.
5. Remove the ignition coil. If necessary, drill and punch out the rivets holding the coil to the bracket.

To install:

6. Place the ignition coil into position and install the mounting screws.
7. Reconnect the harness connector to the coil. Make certain the connectors are fully seated and latched.
8. Install the secondary lead to the coil tower.
9. Reconnect the negative battery cable.

INTEGRAL COIL

1. Disconnect the negative battery cable.
2. Remove the cover and wire retainer.
3. Disconnect the battery feed wire and coil connections from the cap.
4. Remove the coil cover attaching screws and cover.
5. Remove the coil attaching screws and lift the coil and leads from the cap.

To install:

6. Position the coil and leads into the cap. Be sure the resistor brush, seal and coil grounds are properly positioned.
7. Install the mounting screws.
8. Install the coil cover and retaining screws.
9. Reconnect the feed wire and coil connection to the cap.
10. Install the cover and wire retainer.
11. Reconnect the negative battery cable.

ELECTRONIC CONTROL MODULE (ECM) PROM CARRIER, CALPAC OR MEMCAL

Removal and Installation

1. Turn the ignition switch **OFF**.
2. Disconnect the negative battery cable.
3. Remove the right side hush panel, as required.
4. Disconnect the harness connectors from the ECM.
5. Remove the ECM-to-bracket retaining screws and remove the ECM.
6. If replacement of the calibration unit is required, remove the access cover retaining screws and cover from the ECM. Carefully remove the calibration unit from the ECM, as follows:

   a. If the ECM contains a PROM carrier, use the rocker type PROM removal tool.

   b. If the ECM contains a CALPAC, grasp the CALPAC carrier (at the narrow end only), using the removal tool. Remove the CALPAC carrier.

   c. If the ECM contains a MEMCAL, push both retaining clips back away from the MEMCAL. At the same time, grasp it at both ends and lift it up out of the socket. Do not remove the cover of the MEMCAL.

To install:

7. Fit the replacement calibration unit into the socket.

NOTE: The small notch of the carrier should be aligned with the small notch in the socket. Press on the ends of the carrier until it is firmly seated in the socket. Do not press on the calibration unit, only the carrier.

8. Install the access cover and retaining screws.
9. Position the ECM in the vehicle and install the ECM-to-bracket retaining screws.
10. Reconnect the ECM harness connectors.
11. Install the right side hush panel, as required.
12. Check that the ignition switch is **OFF**. Then reconnect the negative battery cable.
13. Perform the functional check.

NOTE: Before replacement of a defective ECM/PCM first check the resistance of each ECM controlled solenoid. This can be done at the ECM connector, using an ohmmeter and the ECM connector wiring diagram. Any ECM controlled device with low resistance will damage the replacement ECM due to high current flow through the ECM internal circuits.

Functional Check

After the PROM has been replaced, it is necessary to perform the functional check to verify proper installation of the PROM.

1. Turn the ignition switch **ON**.

2. Enter diagnostics, by grounding the appropriate DLC terminals.

a. Allow Code 12 to flash 4 times to verify that no other codes are present. This indicates the PROM is installed properly.

b. If trouble Code 51 is present or if the SERVICE ENGINE SOON light is ON constantly with no codes, the PROM is not fully seated, installed backward, has bent pins or is defective.

——— WARNING ———

Anytime the calibration unit is installed backward and the ignition switch is turned ON, the unit is destroyed.

ESC KNOCK SENSOR

The HEI system uses one or two knock sensors, mounted on the cylinder block.

NOTE: On 2.8L, 4.3L, 5.0L, 5.7L and 7.4L engines the knock sensor is exposed to engine coolant. It will be necessary to drain the cooling system before removing the knock sensor.

Removal and Installation

1. Disconnect the negative battery cable.

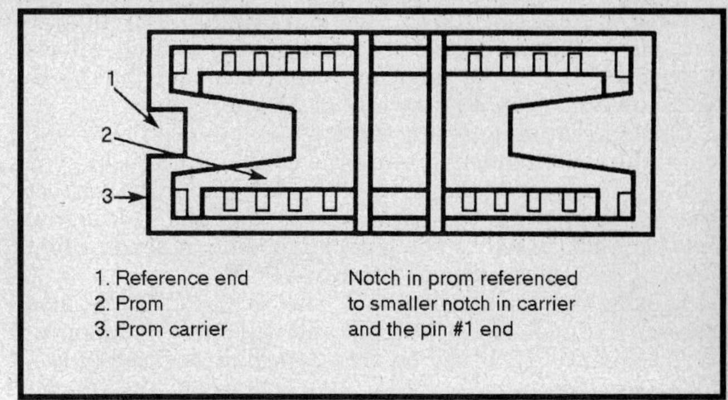

1. Reference end
2. Prom
3. Prom carrier

Notch in prom referenced
to smaller notch in carrier
and the pin #1 end

Proper PROM installation

2. Raise and support the vehicle safely.

3. Disconnect the knock sensor wiring harness connector from the sensor.

4. Remove the sensor from the engine block.

To install:

5. Install the sensor into the engine block. Tighten and torque to 14 ft. lbs. (19 Nm).

6. Reconnect the knock sensor wiring harness connector to the sensor.

7. Lower the vehicle.

8. Reconnect the negative battery cable.

COMPUTER CONTROLLED COIL IGNITION (C³I) SYSTEM

General Information

The computer controlled coil (C³I) ignition system features a distributorless ignition engine. The C³I system consists of 3 ignition coils, a C³I ignition module, a dual crank sensor, camshaft sensor, connecting wires and the Electronic Spark Timing (EST) portion of the Electronic Control Module (ECM)/Powertrain Control Module (PCM).

The C³I system uses the same Electronic Spark Timing (EST) circuits as the distributor-type ignition. The ECM/PCM uses the EST circuit to control spark advance and ignition dwell, when the ignition system is operating in the EST mode. There are 2 modes of ignition system operation. These modes are as follow:

• Module mode — the ignition system operates independently of the ECM/PCM, with module mode spark advance always at 10 degrees BTDC. The ECM/PCM have no control of the ignition system when in this mode.

• EST mode — the ignition spark timing and ignition dwell time is fully controlled by the ECM. EST spark advance and ignition dwell is calculated by the ECM.

To control spark knock, and to use maximum spark advance to improve driveability and fuel economy, an Electronic Spark Control (ESC) system is used. This system consists of a knock sensor and an ESC module (part of MEMCAL). The ECM/PCM monitors the ESC signal to determine when engine detonation occurs.

SYSTEM OPERATION

The C³I ignition system uses a waste spark distribution method. Each cylinder is paired with the cylinder opposite it (Ex: 1–4, 2–5, 3–6). The ends of each coil secondary is attached to a spark plug. These 2 plugs are on companion cylinders, cylinders that are at Top Dead Center (TDC) at the same time. The one that is on the compression stroke is said to be the event cylinder and the one on the exhaust stroke is the waste cylinder. When the coil discharges, both plugs fire at the same time to complete the series circuit.

Since the polarity of the primary and the secondary windings are fixed, one plug always fires in a forward direction and the other in reverse. This is different than a conventional system firing all plugs the same direction each time. Because of the demand for additional energy,

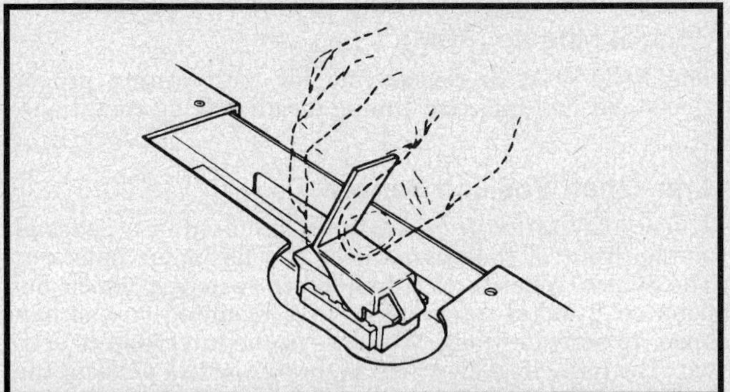

Removing PROM with removal tool

the coil design, saturation time and primary current flow are also different. This redesign of the system allows higher energy to be available from the distributorless coils, greater than 40 kilovolts at all rpm ranges.

During cranking, when the engine speed is below 400 rpm, the C³I module monitors the dual crank sensor sync signal. The sync signal is used to determine the correct pair of cylinders to be sparked first. Once the sync signal has been processed by the ignition module, it sends a fuel control reference pulse to the ECM/PCM.

During the cranking period, the ECM/PCM will also receive a camshaft pulse signal and will operate the injectors sequentially, based on true camshaft position only.

> **NOTE: The terms Electronic Control Module (ECM), Powertrain Control Module (PCM) and Body Control Module (BCM) may sometimes be used interchangeably.**

The sync signal is used only by the ignition module. It is used for spark synchronization at start-up only.

When the engine speed is below 400 rpm (during cranking), the C³I module controls the spark timing. Once the engine speed exceeds 400 rpm (engine running) spark timing is controlled by the EST signal from the ECM. To control EST the ECM/PCM uses the following inputs:

- Crankshaft position
- Engine speed (RPM)
- Engine coolant (Engine Coolant Temperature Sensor — ECT)
- Intake air (Mass Air Flow — MAF)
- Throttle valve position (Throttle Position Sensor — TPS)
- Gear shift lever position (Park/Neutral Switch — P/N)
- Vehicle speed (Vehicle Speed Sensor — VSS)
- ESC signal (Knock Sensor — KS)

The C³I ignition module provides proper ignition coil sequencing during both the module and the EST modes.

The ESC system is designed to retard spark timing up to 10 degrees to reduce spark knock in the engine. When the knock sensor detects spark knocking in the engine, it sends an A/C voltage signal to the ECM, which increases with the severity of the knock. The ECM/PCM then adjusts the EST to reduce spark knock.

SYSTEM COMPONENTS

C³I Module

The C³I module monitors the sync-pulse and the crank signal. During cranking the C³I module monitors the sync-pulse to begin the ignition firing sequence. During this time, each of the 3 coils are fired at a pre-determined interval based on engine speed only. Above 400 rpm, the C³I module is only use as a reference signal.

Ignition Coil

The ignition coil assemblies are mounted on the C³I module. Each coil distributes the spark for 2 plugs simultaneously.

Electronic Spark Control (ESC)

The ESC system incorporates a knock sensor and the ECM. The knock sensor detects engine detonation. When engine detonation occurs, the ECM receives the ESC signal and retards EST to reduce detonation.

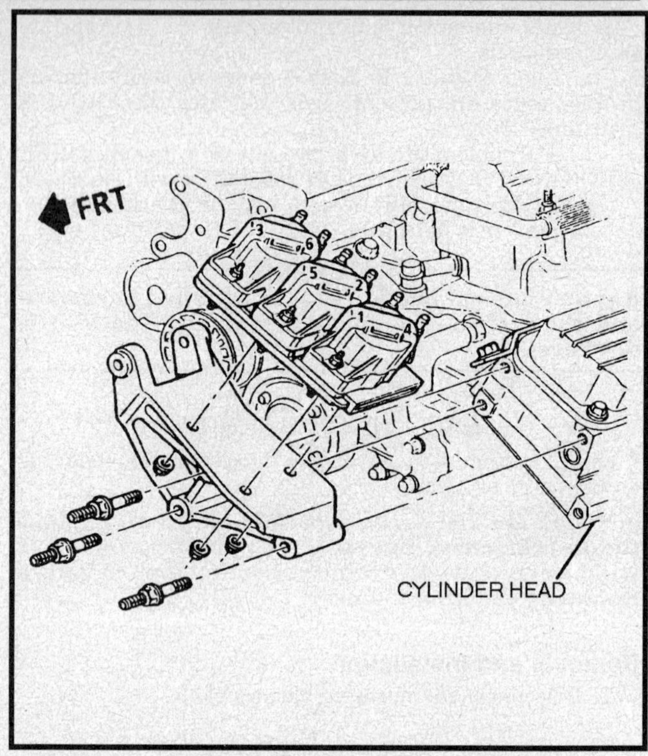

Computer Controlled Coil Ignition (C³I) Ignition coils and module assembly — 3.3L VIN N shown

Electronic Spark Timing (EST)

The EST system is basically the same EST to ECM circuit use on the distributor type ignition systems with EST. This system includes the following circuits:

- Reference circuit — provides the ECM with rpm and crankshaft position information from the C³I module. The C³I module receives this signal from the crank sensor hall-effect switch.
- Bypass signal — above 400 rpm, the ECM applies 5 volts to this circuit to switch spark timing control from the C³I module to the ECM.
- EST signal — reference signal is sent to the ECM via the C³I module during cranking. Under 400 rpm, the C³I module controls the ignition timing. Above 400 rpm, the ECM applies 5 volts to the bypass line to switch the timing to the ECM control.

Electronic Control Module (ECM) or Powertrain Control Module (PCM)

The ECM/PCM is responsible for maintaining proper spark and fuel injection timing for all driving conditions.

Crankshaft Position Sensor

The crankshaft position sensor is mounted in a pedestal on the front of the engine near the harmonic balancer. The sensor consists of 2 hall-effect switches, which depend on 2 metal interrupter rings mounted on the balancer to activate them. Windows in the interrupters activate the hall-effect switches as they provide a path for the magnetic field between the switches transducers and magnets.

Camshaft Sensor

The camshaft sensor sends signal to the ECM/PCM which is used as a sync-pulse to trigger the injectors in the proper sequence.

Tools and Equipment

SCAN TOOLS

The system can communicate a wide variety of information through the Data Link Connector (DLC), also known as the ALDL. Depending on the application, serial data is transmitted to terminals E or M of the DLC. This data is transmitted at a high frequency which requires a quality scan tool for interpretation. Although stored codes may be read with only the use of a small jumper wire, the use of a hand-held scan tool such as GM's TECH 1 or equivalent, is recommended. There are many manufacturers of these tools; a purchaser must be certain that the tool is proper for the intended use.

The scan tool allows any stored codes to be read from the ECM's memory. The tool also allows the operator to view the data being sent to the ECM while the engine is running. This ability has obvious diagnostic advantages; the use of the scan tool is frequently required by the diagnostic charts.

With an understanding of the data stream that the tool will display and knowledge of the circuits involved, the tool can be an extremely valuable tool when examining the system. Without scan tools, diagnostic information would be difficult to obtain in some circuits and impossible to retrieve in others. Scan tools do not make the use of trouble code charts unnecessary, nor do they pinpoint the exact area in a circuit where the problem exists. However, this type of equipment will give more specific information, such as sensor input readings. An example of this would be the coolant sensor circuit. While the ECM looks at specific voltage levels to determine coolant temperature, the scan tool will display this information as an actual temperature reading. This information is not only useful in determining system malfunctions, but is also very helpful in diagnosing mechanical problems. For example, if the engine is overheating, the scan tool can be used to look at the actual engine temperature. This is helpful when checking for electric cooling fan enable switch problems, sticking thermostats, etc.

NOTE: A scan tool that is known to display faulty data should not be used for diagnosis. Although the fault may be believed to be in only one area, it can possibly affect many other areas during diagnosis, leading to errors and incorrect repair.

ELECTRICAL TOOLS

The most commonly required electrical diagnostic tool is the digital multimeter, allowing voltage, resistance and amperage to be read by one instrument. The multimeter must be a high-impedance unit, with 10 megohms of impedance in the voltmeter. This type of meter will not place an additional load on the circuit it is testing; this is extremely important in low voltage circuits. The multimeter must be of high quality in all respects. It should be handled carefully and protected from impact or damage. Replace batteries frequently in the unit.

Other necessary tools include an unpowered test light, a quality tachometer with inductive (clip-on) pick up and the proper tools for releasing GM's Metri-Pack, Weather Pack and Micro-Pack terminals as necessary. The Micro-Pack connectors are used at the ECM connector. A vacuum pump/gauge may also be required for checking sensors, solenoids and valves.

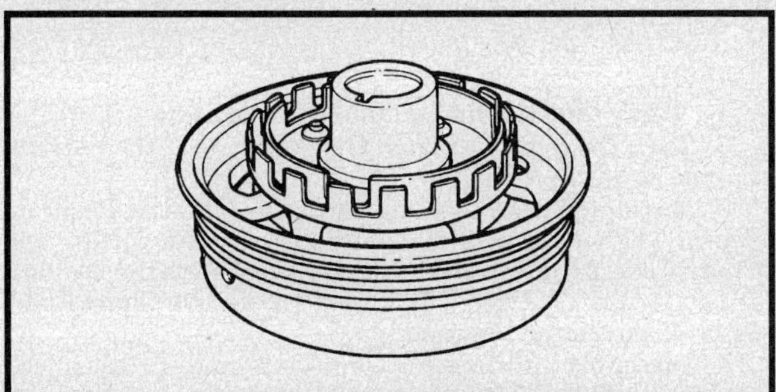

Crankshaft balancer with interrupter rings — C³I system

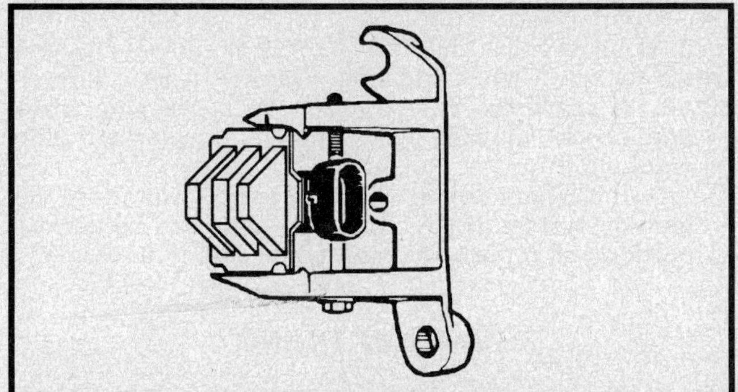

Crankshaft position sensor — C³I system

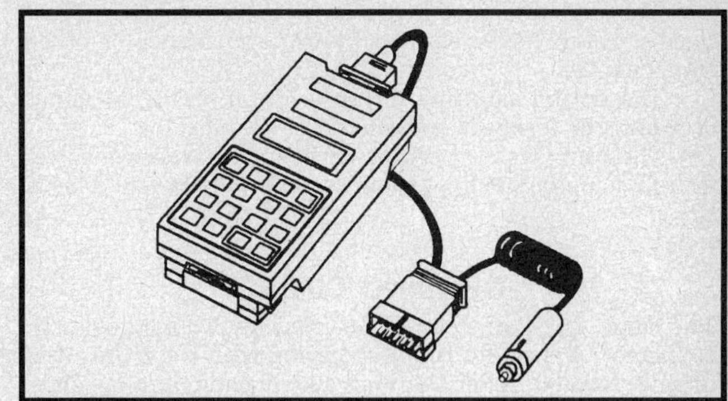

Typical scan tool — GM's Tech 1 shown

Diagnosis and Testing

SERVICE PRECAUTIONS

CAUTION

The ignition coils secondary voltage output capabilities can exceed 40,000 volts. Avoid body contact with the C³I high voltage secondary components when the engine is running, or personal injury may result.

NOTE: To avoid damage to the ECM/PCM or other ignition system components, do not use electrical test equipment such as battery or AC powered voltmeter, ohmmeter, etc. or any type of tester other than specified.

- To properly diagnosis the ignition systems and their problems, it will be necessary to refer to the diagnostic charts in the fuel injection section. Locate the charts that applied to the vehicle in question and follow them from start to stop.
- When performing electrical tests on the system, use a high impedance multimeter, digital voltmeter (DVM) J-34029-A or equivalent.
- To prevent electrostatic discharge damage, when working with the ECM, do not touch the connector pins or soldered components on the circuit board.
- When handling a PROM, CALPAC or MEMCAL, do not touch the component leads. Also, do not remove the integrated circuit from the carrier.
- Never pierce a high tension lead or boot for any testing purpose; otherwise, future problems are guaranteed.
- Do not allow extension cords for power tools or droplights to lie on, near or across any vehicle electrical wiring.
- Leave new components and modules in the shipping package until ready to install them.

PRELIMINARY INSPECTION

This is possibly the most critical step of diagnosis. A detailed examination of connectors, wiring and vacuum hoses can often lead to a repair without further diagnosis. Performance of this step relies on the skill of the technician performing it; a careful inspector will check:

- The undersides of hoses as well as the integrity of hard-to-reach hoses blocked by the air cleaner or other components.
- The wiring carefully for any sign of strain, burning, crimping, or terminal pull-out from a connector.
- The connectors at components or in harnesses as required; usually, pushing them together will reveal a loose fit.

READING CODES

The Data Link Connector (DLC), also known as the ALDL connector, is used for communicating with the ECM. It is usually located under the instrument panel and is sometimes covered by a plastic cover labeled DIAGNOSTIC CONNECTOR. Codes stored in the ECM's memory can be read through a handheld diagnostic scanner plugged into the DLC. Codes can also be read by connecting terminal A to B of the DLC and counting the number of flashes of the Service Engine Soon light, with the ignition switch turned **ON**.

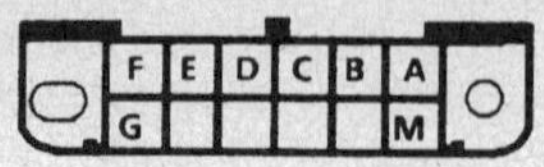

Data link connector (DLC)

CLEARING CODES

To clear codes from the ECM memory, the ECM power feed must be disconnected for at least 30 seconds. Depending on the vehicle, the ECM power feed can be disconnected at the positive battery terminal pigtail, the in-line fuseholder that originates at the positive connection at the battery or the ECM fuse in the fuse block. The negative battery cable may also be disconnected; however, other on-board memory data, such as preset radio tuning, will also be lost.

ENGINE CRANKS BUT WILL NOT RUN

Testing

1. Check that the fuel quantity is adequate.
2. Turn the ignition switch **ON**. Verify that the Service Engine Soon light is ON.
3. Install the scan tool and check the Throttle Position Sensor (TPS) — if over 2.5 volts, at closed throttle, use Chart Code 21 in the fuel section. Also check the coolant sensor — if not between -30°C and 130°C, use Chart Code 14 or 15 in the fuel section.
4. Disconnect all injector connectors and install injector test light (J-34730-2 or equivalent) in injector harness connector. Test light should be OFF.

NOTE: Perform this test on 1 injector from each bank.

5. Connect spark checker, (J 26792 or equivalent), and check for spark while cranking. Check at least 2 wires.

  a. If spark occurs, reconnect the spark plug wires and check for fuel spray at the injector(s) while cranking.

  b. If no spark occurs, check for battery voltage to the ignition system. If OK, check the ignition module connection and replace the ignition module if necessary.

IGNITION TIMING

Adjustment

Because the C³I system uses a crank sensor for EST, ignition timing is not adjustable. However, positioning of the interrupter ring is very important. A clearance of 0.025 inch is required on either side of the interrupter ring.

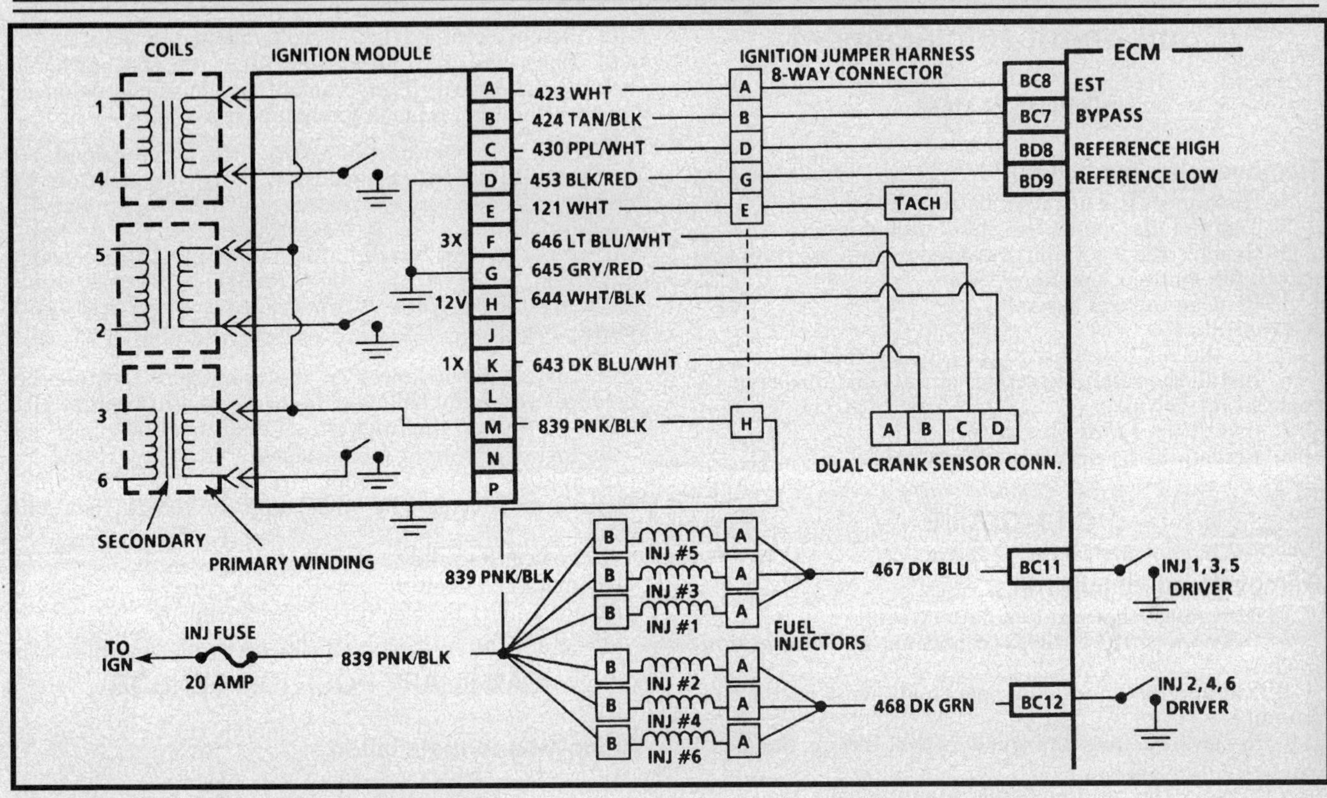

C³I system circuit — 3.3L engine shown

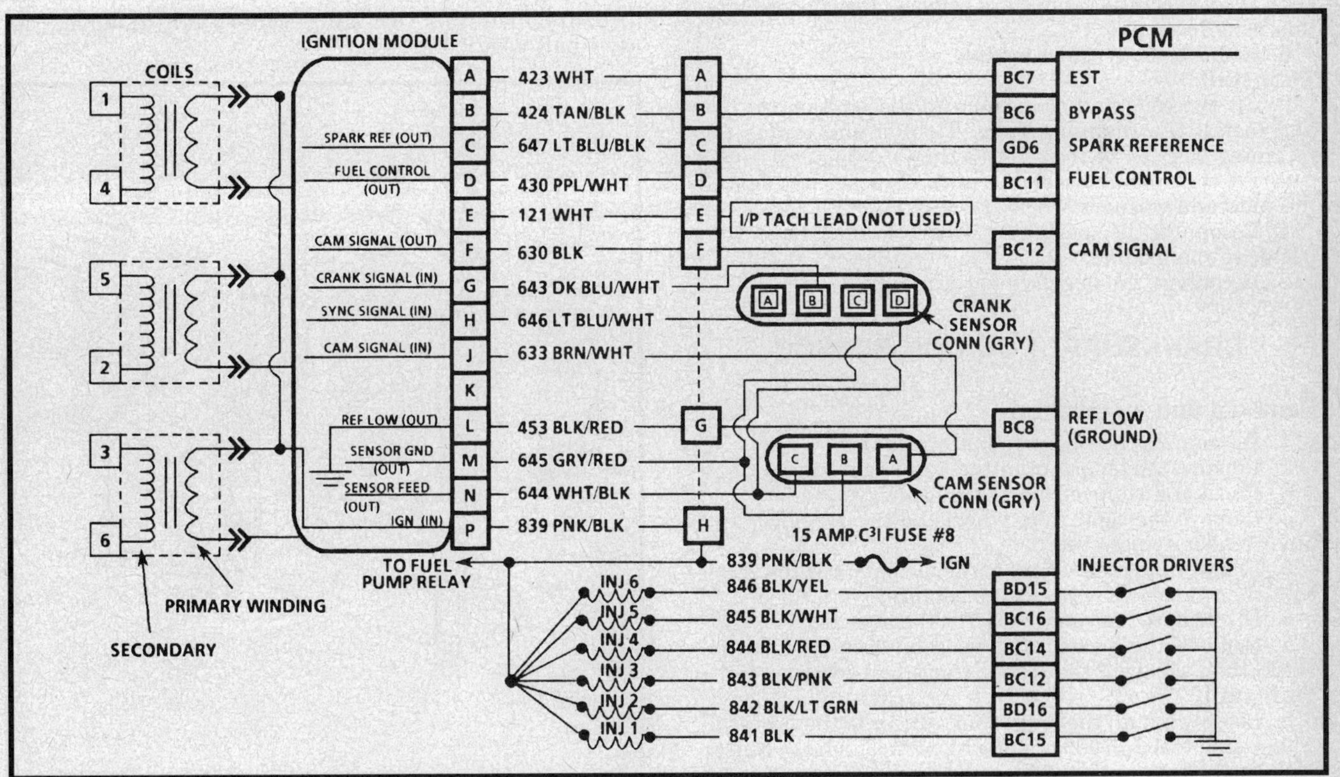

C³I system circuit — 3.8L engine shown

Component Replacement

IGNITION COIL

Removal and Installation

1. Disconnect the negative battery cable.
2. Tag and disconnect the spark plug wires.
3. Remove the 2 retaining screws or nuts securing the coil to the ignition module.
4. Remove the coil assembly.

To install:

5. Fit the coil assembly to the ignition module.
6. Install the retaining screws or nuts and torque to 40 inch lbs. (4–5 Nm).
7. Install the spark plug wires.
8. Reconnect the negative battery cable.

C³I MODULE

Removal and Installation

1. Disconnect the negative battery cable.
2. Disconnect the 14-way connector at the ignition module.
3. Tag and disconnect the spark plug wires at the coil assembly.
4. Remove the nuts and washers that retains the module to the bracket.
5. Remove the coil-to-module retaining bolts.
6. Note the lead colors or mark for reassembly.
7. Disconnect the connectors between the coil and ignition module.
8. Remove the ignition module.

To install:

9. Fit the coils and connectors to the ignition module and install the retaining bolts. Tighten and torque the retaining bolts to 27 inch lbs. (3 Nm).
10. Fit the module assembly to the bracket and install the nuts and washers.
11. Reconnect the spark plug wires and the 14-way connector to the module.
12. Reconnect the negative battery cable.

CRANKSHAFT POSITION SENSOR

Removal and Installation

1. Disconnect the negative battery cable.
2. Remove the belt(s) from the crankshaft pulley.
3. Raise and support the vehicle safely.
4. Remove the right front wheel and tire assembly and inner fender access cover.
5. Remove the crankshaft harmonic balancer retaining bolt, then remove the harmonic balancer.
6. Disconnect the sensor electrical connector.
7. Remove the sensor and pedestal from the engine block, then separate the sensor from the pedestal.

To install:

8. Loosely install the crankshaft sensor to the pedestal.
9. Using tool J-37089 or equivalent, position the sensor with the pedestal attached, on the crankshaft.
10. Install the pedestal-to-block retaining bolts. Tighten and torque to 14–28 ft. lbs. (20–40 Nm).
11. Torque the pedestal pinch bolt 30–35 inch lbs. (20–40 Nm).

12. Remove tool J-37089 or equivalent.
13. Place tool J-37089 or equivalent, on the harmonic balancer and turn. If any vane of the harmonic balancer touches the tool, replace the balancer assembly.

NOTE: A clearance of 0.025 inch is required on either side of the interrupter ring. Be certain to obtain the correct clearance. Failure to do so will damage the sensor. A misadjusted sensor or bent interrupter ring could cause rubbing of the sensor, resulting in potential driveability problems, such as rough idle, poor performance, or a no start condition.

14. Install the balancer on the crankshaft. Install the balancer retaining bolt. Tighten the retaining bolt to 104 ft. lbs. (140 Nm) then tighten an additional 56°.
15. Install the inner fender shield.
16. Install the right front tire and wheel assembly. Tighten and torque the wheel nuts to 104 ft. lbs. (140 Nm).
17. Lower the vehicle.
18. Install the belt(s).
19. Reconnect the negative battery cable.

CAMSHAFT POSITION SENSOR

Removal and Installation

1. Disconnect the negative battery cable.
2. Disconnect the electrical connector from the camshaft position sensor.
3. Remove the retaining screw and remove the camshaft position sensor.

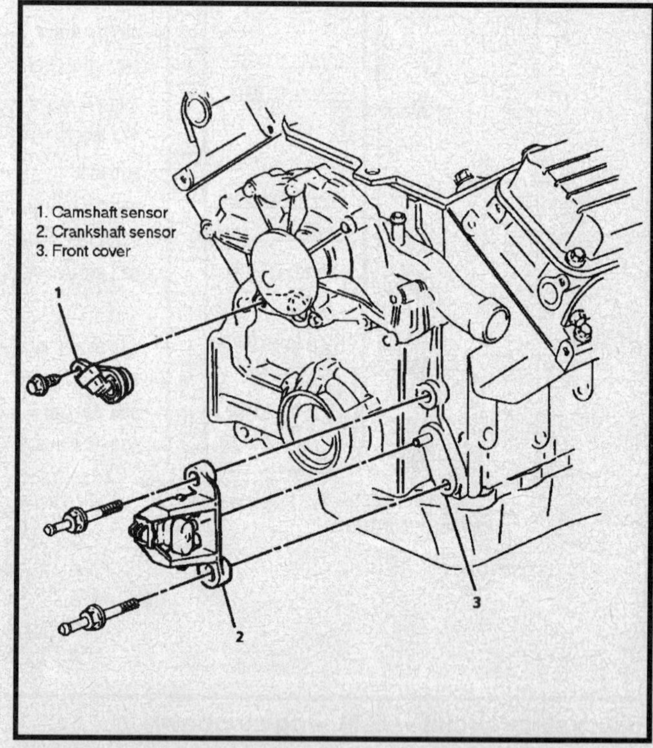

Crankshaft position sensor and camshaft position sensor installation

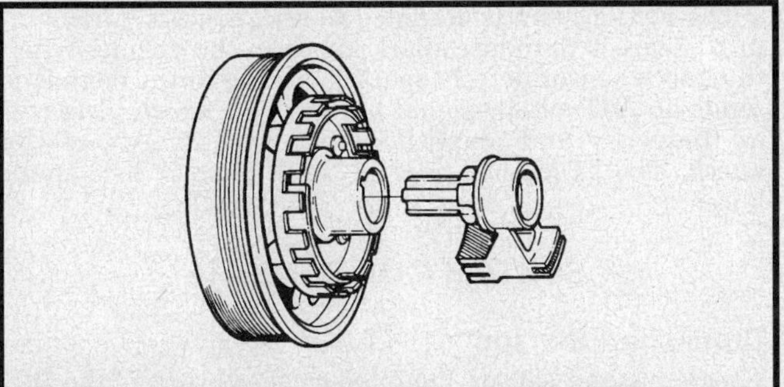

Positioning crankshaft sensor tool on harmonic crankshaft balancer

To install:

4. Install the sensor and tighten the retaining screw to 35–53 inch lbs. (4–6 Nm).

5. Reconnect the electrical connector to the camshaft position sensor.

6. Reconnect the negative battery cable.

ESC KNOCK SENSOR

The C³I system uses one knock sensor mounted on the cylinder block.

Removal and Installation

1. Disconnect the negative battery cable.
2. Raise and support the vehicle safely.
3. Disconnect the harness connector from the knock sensor.
4. Remove the sensor from the engine block.

To install:

5. Clean the threads on the engine block, where the sensor was installed. Install the sensor. Tighten to 11–16 ft. lbs. (15–22 Nm).

6. Reconnect the harness connector to the knock sensor.

7. Lower the vehicle.

8. Reconnect the negative battery cable.

ELECTRONIC CONTROL MODULE (ECM) AND/OR MEMCAL

Removal and Installation

1. Turn the ignition switch **OFF**.
2. Disconnect the negative battery cable.
3. Remove the right side hush panel.
4. Disconnect the harness connectors from the control unit.
5. Remove the control unit-to-bracket retaining screws and remove the control unit.

6. If replacement of the MEMCAL is required, remove the access cover retaining screws and cover from the control unit. Note the position of the MEMCAL for proper installation in the new ECM. Using 2 fingers, carefully push both retaining clips back away from the MEMCAL. At the same time, grasp it at both ends and lift it up out of the socket. Do not remove the cover of the MEMCAL.

To install:

7. Fit the replacement MEMCAL into the socket.

> **NOTE: The small notches in the MEMCAL must be aligned with the small notches in the socket. Press only on the ends of the MEMCAL until the retaining clips snap into the ends of the MEMCAL. Do not press on the middle of the MEMCAL, only the ends.**

8. Install the access cover and retaining screws.
9. Position the control unit in the vehicle and install the control unit-to-bracket retaining screws.
10. Reconnect the control unit harness connectors.
11. Install the hush panel.
12. Check that the ignition switch is **OFF**. Then reconnect the negative battery cable.
13. Perform the functional check.

Functional Check

After the PROM is replaced, it is necessary to perform the functional check to verify proper installation of the PROM.

1. Turn the ignition switch **ON**.
2. Enter diagnostics and allow Code 12 to flash 4 times to verify no other codes are present. This indicates the MEMCAL is installed properly and the control unit is functioning.
3. If trouble Codes 42, 43 or 51 occur, or if the Service Engine Soon light is ON constantly with no codes, the MEMCAL is not fully seated or is defective.
4. If it is not fully seated, turn the ignition switch **OFF**, then press firmly on the ends of the MEMCAL. Make sure all pins are seated in the sockets.

SPARK PLUGS

Removal and Installation

1. Disconnect the negative battery cable.
2. Remove all foreign material from around the spark plugs.
3. Note the position of the plug wires, then twist the spark plug boot ½ turn and remove. Do not pull on spark plug wire.
4. Remove the spark plugs.

To install:

5. Check and adjust the plug gaps.
6. Install the new plugs. Tighten and torque to 11 ft. lbs. (15 Nm).

DIRECT IGNITION SYSTEM (DIS)

General Information

The DIS ignition system features a distributorless ignition engine. The DIS system consists of 2 separate ignition coils (L4), 3 separate coils (V6) or 4 separate coils (V8), a DIS ignition module, a crankshaft sensor, crankshaft reluctor ring, connecting wires and the Electronic Spark Timing (EST) portion of the Electronic Control

Module (ECM). A camshaft sensor is also used on 3.1L VIN T Calif. SFI engines.

The DIS ignition system uses a magnetic crankshaft sensor and a reluctor to determine crankshaft position and engine speed. The reluctor is a special wheel cast into the crankshaft with several machined slots. A specific slot, on the reluctor wheel, is used to generate a sync-pulse.

The camshaft sensor, used 3.1L VIN T Calif. SFI engines, provides a cam signal to identify correct firing sequence. The crankshaft sensor signal triggers each coil at the proper time.

The DIS system uses the same Electronic Spark Timing (EST) circuits as the distributor-type ignition. The ECM uses the EST circuit to control spark advance and ignition dwell, when the ignition system is operating in the EST mode.

NOTE: The terms Electronic Control Module (ECM), Powertrain Control Module (PCM) and Body Control Module (BCM) may sometimes be used interchangeably.

To control spark knock and to use maximum spark advance to improve driveability and fuel economy, an Electronic Spark Control (ESC) system is used on some engines. This system consists of a knock sensor and an ESC module (part of MEMCAL). The ECM monitors the ESC signal to determine when engine detonation occurs.

SYSTEM OPERATION

The DIS ignition system uses a waste spark distribution method. Each cylinder is paired with the cylinder opposite it (Ex: 1–4, 2–3 on a 4-cylinder engine or 1–4, 2–5, 3–6 on a 6-cylinder engine). The ends of each coil secondary wire is attached to a spark plug. These 2 plugs are on companion cylinders, cylinders that are at top dead center at the same time. The one that is on compression is said to be the event cylinder and the one on the exhaust stroke, the waste cylinder. When the coil discharges, both plugs fire at the same time to complete the series circuit.

Since the polarity of the primary and the secondary windings are fixed, one plug always fires in a forward direction and the other in reverse. This is different than a conventional system firing all plugs the same direction each time. Because of the demand for additional energy; the coil design, saturation time and primary current flow are also different. This redesign of the system allows higher energy to be available from the distributorless coils, greater than 40 kilovolts at all rpm ranges.

The DIS ignition system uses a magnetic crankshaft position sensor which protrudes into the engine block at approximately 0.050 inch of the crankshaft reluctor. As the crankshaft rotates, the slots of the reluctor causes a changing magnetic field at the crankshaft sensor, creating an induce voltage pulse. By counting the time between pulses, the ignition module can recognize the specified slot (sync pulse). Based on this sync pulse, the module sends reference signals to the ECM to calculate crankshaft position and engine speed.

To control EST the ECM uses the following inputs:
- Crankshaft position
- Engine Speed (rpm)
- Engine temperature
- Manifold air temperature
- Atmospheric (barometric) pressure
- Engine load (manifold pressure or vacuum)

The ESC system is designed to retard spark timing up to 10 degrees to reduce spark knock in the engine. When the knock sensor detects spark knocking in the engine, it sends an A/C voltage signal to the ECM, which increases in frequency and amplitude with the severity of the knock. The ECM then adjusts the EST to reduce spark knock.

SYSTEM COMPONENTS

Crankshaft Sensor

The crankshaft sensor, mounted on the bottom of the DIS module, is used to determine crankshaft position and engine speed.

Camshaft Position Sensor

The camshaft position sensor is located on the timing cover behind the water pump, near the camshaft sprocket. The camshaft sensor, used 3.1L VIN T Calif. SFI engines, provides a cam signal to identify correct firing sequence.

Ignition Coil

The ignition coil assemblies are mounted on the DIS module. Each coil distributes the spark for 2 plugs simultaneously.

Electronic Spark Timing (EST)

The EST system is basically the same EST to ECM circuit used on the distributor type ignition systems with EST. This system includes the following circuits:
- DIS reference circuit — provides the ECM with rpm and crankshaft position information from the DIS module. The DIS module receives this signal from the crank sensor.
- Bypass signal — above 400 rpm, the ECM applies 5 volts to this circuit to switch spark timing control from the DIS module to the ECM.
- EST signal — reference signal is sent to the ECM via the DIS module during cranking. Under 400 rpm, the DIS module controls the ignition timing. Above 400 rpm, the ECM applies 5 volts to the bypass line to switch the timing to the ECM control.
- Reference ground circuit — this wire is grounded through the module and insures that the ground circuit has no voltage drop between the ignition module and the ECM which could affect performance.

Tools and Equipment

SCAN TOOLS

The system can communicate a wide variety of information through the Data Link Connector (DLC), also known as the ALDL. Depending on the application, serial data is transmitted to terminals E or M of the DLC. This data is transmitted at a high frequency which requires a quality scan tool for interpretation. Although stored codes may be read with only the use of a small jumper wire, the use of a hand-held scan tool such as GM's TECH 1 or equivalent, is recommended. There are many manufacturers of these tools; a purchaser must be certain that the tool is proper for the intended use.

The scan tool allows any stored codes to be read from the ECM memory. The tool also allows the operator to

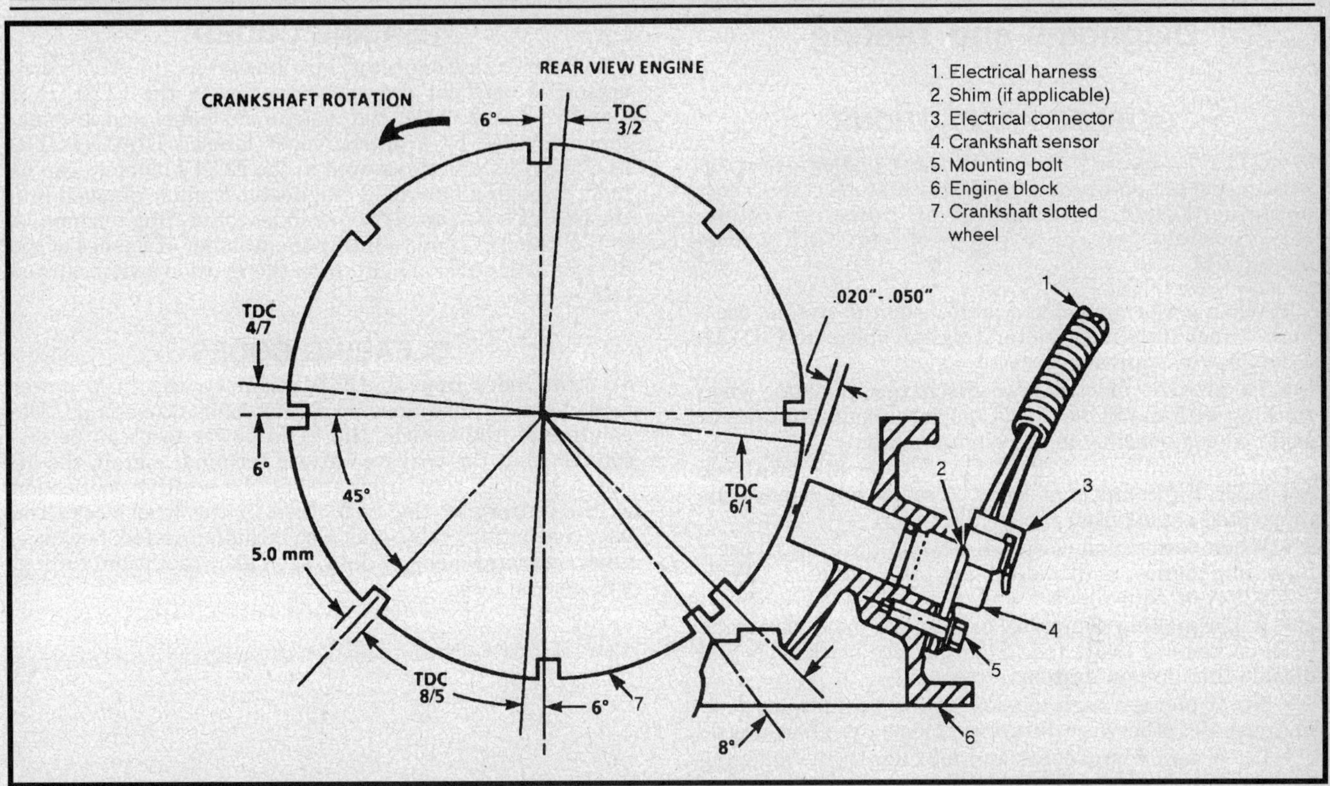

Crankshaft sensor operation — 5.7L engine shown

view the data being sent to the ECM while the engine is running. This ability has obvious diagnostic advantages; the use of the scan tool is frequently required by the diagnostic charts.

With an understanding of the data stream that the tool will display and knowledge of the circuits involved, the tool can be an extremely valuable tool when examining the system. Without scan tools, diagnostic information would be difficult to obtain in some circuits and impossible to retrieve in others. Scan tools do not make the use of trouble code charts unnecessary, nor do they pinpoint the exact area in a circuit where the problem exists. However, this type of equipment will give more specific information, such as sensor input readings. An example of this would be the coolant sensor circuit. While the ECM looks at specific voltage levels to determine coolant temperature, the scan tool will display this information as an actual temperature reading. This information is not only useful in determining system malfunctions, but is also very helpful in diagnosing mechanical problems. For example, if the engine is overheating, the scan tool can be used to look at the actual engine temperature. This is helpful when checking for electric cooling fan' enable switch problems, sticking thermostats, etc.

> **NOTE: A scan tool that is known to display faulty data should not be used for diagnosis. Although the fault may be believed to be in only one area, it can possibly affect many other areas during diagnosis, leading to errors and incorrect repair.**

ELECTRICAL TOOLS

The most commonly required electrical diagnostic tool is the digital multimeter, allowing voltage, resistance and

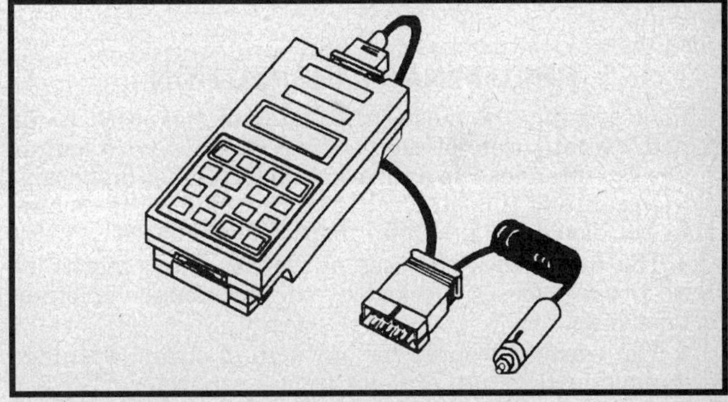

Typical scan tool — GM's Tech 1 shown

amperage to be read by one instrument. The multimeter must be a high-impedance unit, with 10 megohms of impedance in the voltmeter. This type of meter will not place an additional load on the circuit it is testing; this is extremely important in low voltage circuits. The multimeter must be of high quality in all respects. It should be handled carefully and protected from impact or damage. Replace batteries frequently in the unit.

Other necessary tools include an unpowered test light, a quality tachometer with inductive (clip-on) pick up and the proper tools for releasing GM's Metri-Pack, Weather Pack and Micro-Pack terminals as necessary. The Micro-Pack connectors are used at the ECM connector. A vacuum pump/gauge may also be required for checking sensors, solenoids and valves.

Diagnosis and Testing

SERVICE PRECAUTIONS

NOTE: To avoid damage to the ECM or other ignition system components, do not use electrical test equipment such as battery or AC powered voltmeter, ohmmeter, etc. or any type of tester other than specified.

• When performing electrical tests on the system, use a high impedance multimeter, digital voltmeter (DVM) J-34029-A or equivalent.

• To prevent electrostatic discharge damage, when working with the ECM, do not touch the connector pins or soldered components on the circuit board.

• When handling a PROM, CALPAC or MEMCAL, do not touch the component leads. Also, do not remove the integrated circuit from the carrier.

• When performing electrical tests on the system, use a high impedance multimeter, digital voltmeter (DVM) J-34029-A or equivalent.

• When making compression tests on the 5.7L engine (VIN J), remove INJ 1 fuse from the fuse block. This will disable the ignition system.

• Never pierce a high tension lead or boot for any testing purpose; otherwise, future problems are guaranteed.

• Leave new components and modules in the shipping package until ready to install them.

• Never disconnect any electrical connection with the ignition switch **ON** unless instructed to do so in a test.

PRELIMINARY INSPECTION

This is possibly the most critical step of diagnosis. A detailed examination of connectors, wiring and vacuum hoses can often lead to a repair without further diagnosis. Performance of this step relies on the skill of the technician performing it; a careful inspector will check:

• The undersides of hoses as well as the integrity of hard-to-reach hoses blocked by the air cleaner or other components.

• The wiring carefully for any sign of strain, burning, crimping, or terminal pull-out from a connector.

• The connectors at components or in harnesses as required; usually, pushing them together will reveal a loose fit.

READING CODES

The Data Link Connector, also known as the ALDL connector, is used for communicating with the ECM. It is usually located under the instrument panel and is sometimes covered by a plastic cover labeled DIAGNOSTIC CONNECTOR. Codes stored in the ECM's memory can be read through a handheld diagnostic scanner plugged into the DLC. Codes can also be read by connecting terminal A to B of the DLC and counting the number of flashes of the Service Engine Soon light, with the ignition switch turned **ON**.

CLEARING CODES

To clear codes from the ECM memory, the ECM power feed must be disconnected for at least 30 seconds. Depending on the vehicle, the ECM power feed can be disconnected at the positive battery terminal pigtail, the in-line fuseholder that originates at the positive connection at the battery or the ECM fuse in the fuse block. The negative battery cable may also be disconnected; however, other on-board memory data, such as preset radio tuning, will also be lost.

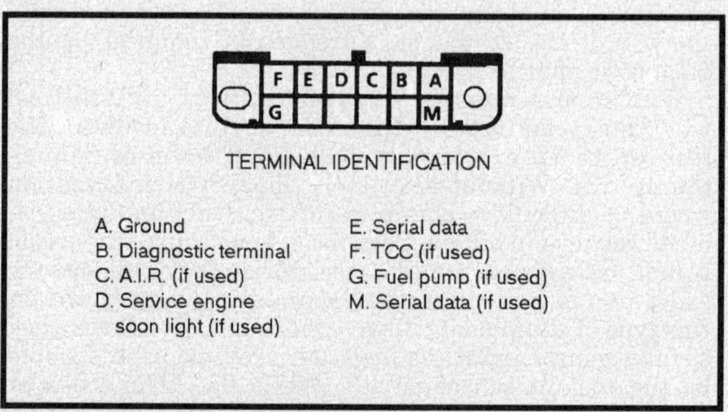

Data link connector (DLC)

CRANKSHAFT SENSOR PARAMETERS CHART

CRANKSHAFT SENSOR

System	Test Points	Reading
DIS	A and C	900–1200 Ohms

ENGINE CRANKS BUT WILL NOT RUN

Testing

1. Check that the fuel quantity is adequate.
2. Turn the ignition switch **ON**. Verify that the SERVICE ENGINE SOON light is ON.
3. Install the scan tool and check the Throttle Position Sensor (TPS) — if over 2.5 volts, at closed throttle, use Chart Code 21 in the fuel section.
4. Connect spark checker, J 26792 or equivalent, and check for spark while cranking. Check at least 2 wires.
5. If spark occurs, reconnect the spark plug wires and check for fuel spray at the injector(s) while cranking.
6. If no spark occurs, check for battery voltage to the ignition system and verify cranking RPM signal. Disconnect the ICM 2 pin connector and connect a test light between the terminals on the harness. If the light is OFF, look for an open or grounded circuit. If the light is ON, disconnect the crankshaft position sensor and measure the resistance between terminals A and C on the harness.
   - Less than 900 Ohms — crankshaft position sensor leads shorted or faulty sensor.
   - Between 900–1200 Ohms — crank the engine over and check the voltage between A and C. If greater than 0.1 volts, replace the ignition control module. If less than 0.1 volts, replace crankshaft sensor.
   - Greater than 1200 Ohms — crankshaft position sensor circuit open or faulty sensor.

IGNITION TIMING

Adjustment

Because the reluctor wheel is an integral part of the crankshaft and the crankshaft sensor is mounted in a fixed position, timing adjustment is not possible. Spark advance is controlled by the ECM.

Spark Advance

The ECM uses information from the MAP and coolant sensor, in addition to rpm to calculate spark advance as follows:
- Low MAP output voltage — more spark advance
- Cold engine — more spark advance
- High MAP output voltage — less spark advance
- Hot engine — less spark advance

Therefore, detonation could be caused by low MAP output or high resistance in the coolant sensor circuit.

Poor performance could be caused by high MAP output or low resistance in the coolant sensor circuit.

Component Replacement

IDLE LEARN PROCEDURE

NOTE: Any time the battery is disconnected, the idle learn procedure must be performed.

This procedure allows the ECM memory to be updated with the correct IAC valve pintle position to provide a stable idle speed. The Idle Learn Procedure must be performed as follows:
1. Place the transaxle in **P** or **N**.
2. Install the Tech 1 scan tool or equivalent.

3. Turn the ignition switch **ON**, engine OFF.
4. Select IAC SYSTEM, then IDLE LEARN in the MISC TEST mode.
5. Proceed with the Idle Learn as directed by the scan tool.

IGNITION COIL

Removal and Installation

1. Disconnect the negative battery cable.
2. Tag and disconnect the spark plug wires.
3. Remove the coil retaining nuts.
4. Separate and remove the coils from the module.

To install:

5. Fit the coils to the module and install the retaining nuts. Tighten and torque the retaining nuts 40 inch lbs. (4.5 Nm).
6. Install the spark plug wires.
7. Reconnect the negative battery cable.

NOTE: Before starting the engine, the Idle Learn procedure must be performed.

DIS ASSEMBLY

Removal and Installation

1. Disconnect the negative battery cable.
2. Disconnect the DIS electrical connectors.
3. Disconnect and tag the spark plugs leads from the coils.
4. Remove the DIS assembly retaining bolts and remove the unit from the engine.

To install:

5. Before installing the DIS assembly, check the crankshaft sensor O-ring for damage or leakage. Replace, if necessary. Lubricate the O-ring with engine oil before installing.
6. Fit the DIS assembly to the engine and install the retaining bolts. Tighten and torque the bolts to 19 ft. lbs. (25 Nm).
7. Reconnect the spark plug wires to the proper coils.
8. Reconnect the DIS electrical connectors.
9. Reconnect the negative battery cable.

NOTE: Before starting the engine, the Idle Learn Procedure must be performed.

IGNITION MODULE

Removal and Installation

1. Disconnect the negative battery cable.
2. Remove the DIS assembly from the engine.
3. Remove the ignition coils.
4. Remove the module from the assembly plate.

To install:

5. Fit the module to the assembly plate. Carefully engage the sensor to the module terminals.
6. Install the ignition coils.
7. Install the DIS assembly to the engine.
8. Reconnect the negative battery cable.

NOTE: Before starting the engine, the Idle Learn procedure must be performed.

ELECTRONIC IGNITION SYSTEMS
DIRECT IGNITION SYSTEM (DIS)

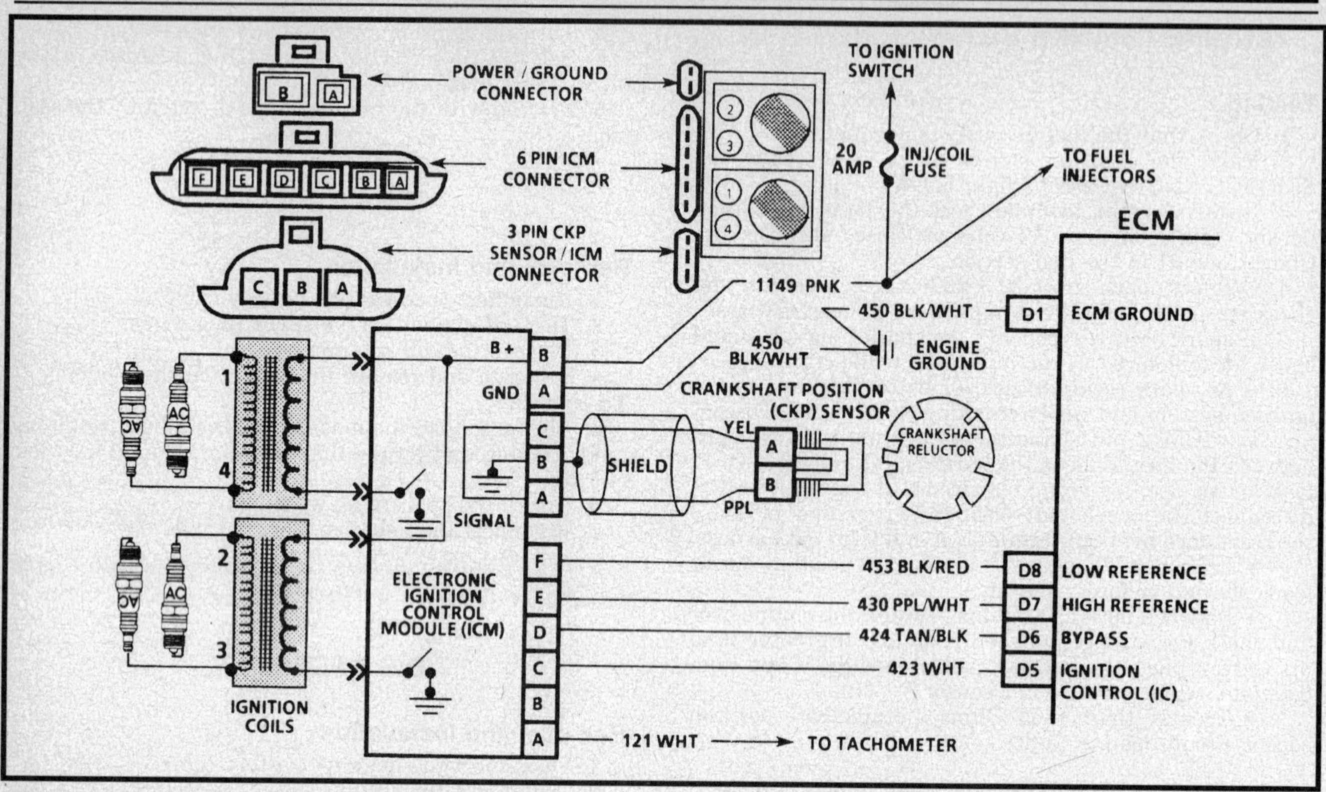

DIS system circuit — 2.2L engine shown

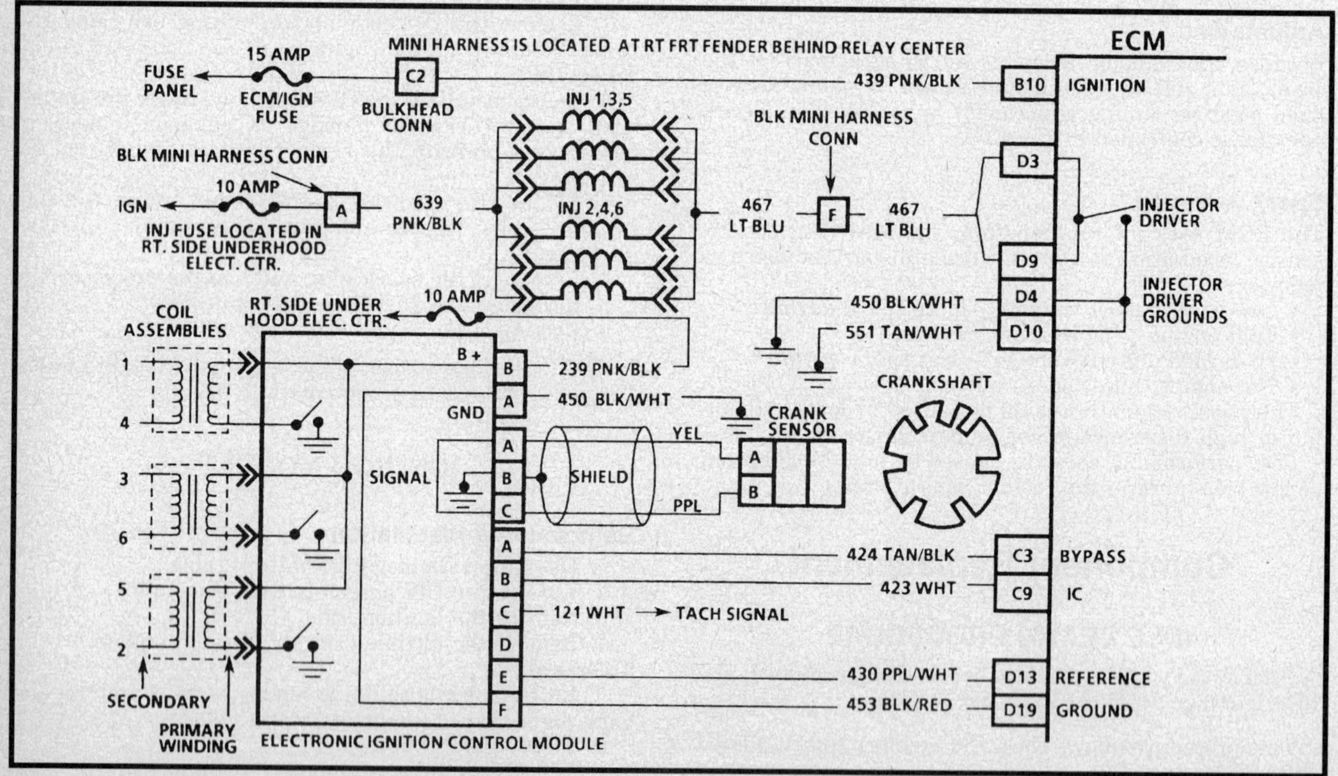

DIS system circuit — 3.1L VIN T Calif. SFI engine

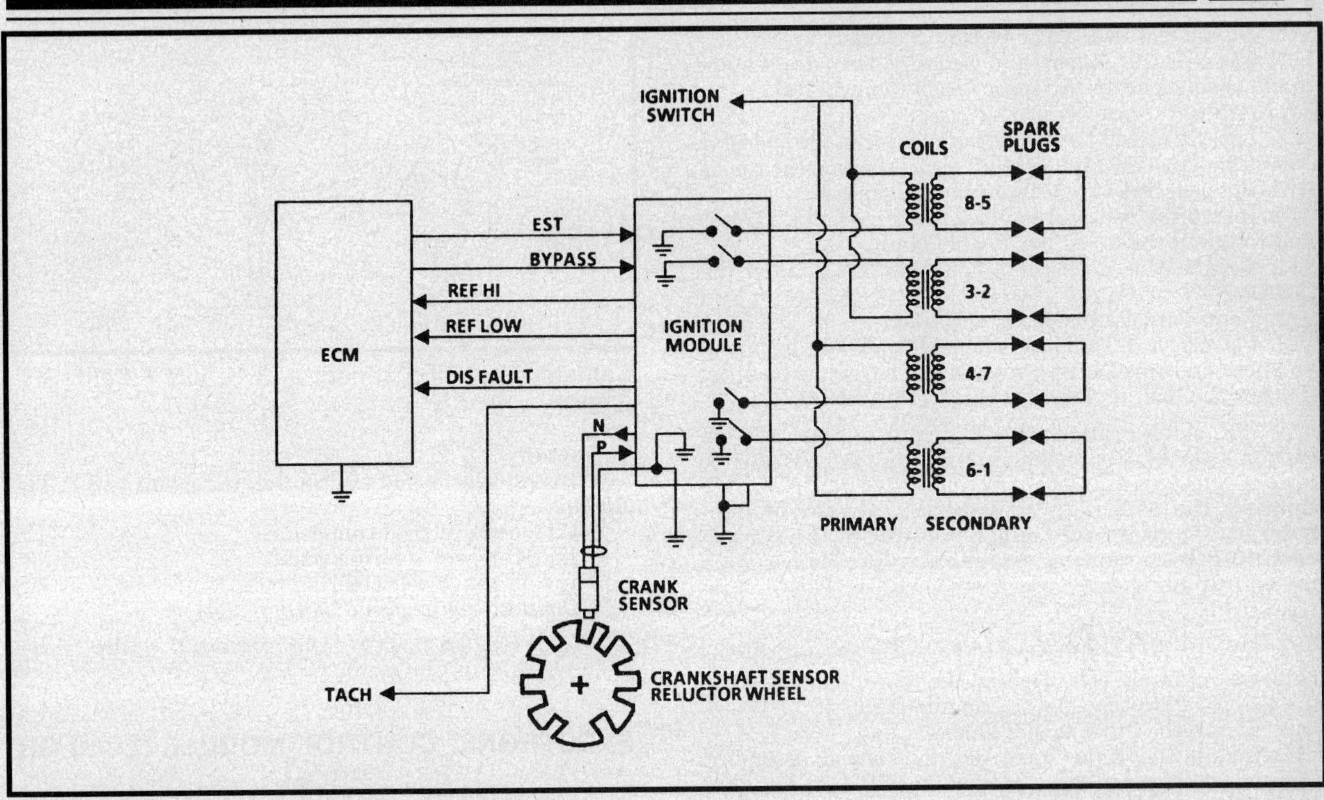

DIS system circuit — 5.7L VIN J engine

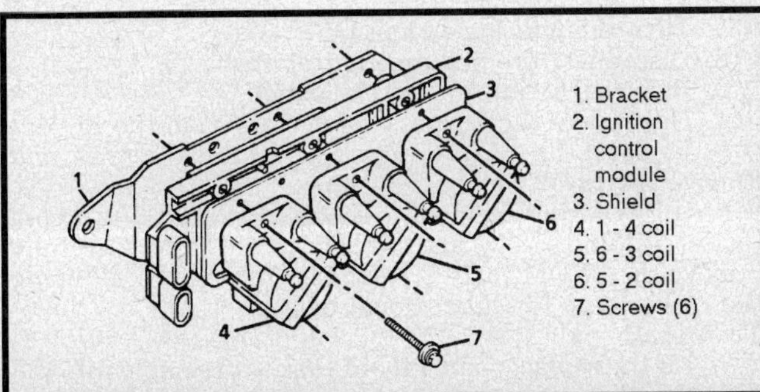

DIS coil and module removal — 3.1L engine shown

CRANKSHAFT SENSOR

Removal and Installation

EXCEPT 3.1L VIN T CALIF. SFI ENGINE

1. Disconnect the negative battery cable.
2. Remove the DIS assembly.
3. Remove the sensor retaining screws and remove the sensor from DIS assembly.

NOTE: Before install the DIS assembly, check the crankshaft sensor O-ring for damage or leakage. Replace, if necessary. Lubricate the O-ring with engine oil before installing.

To install:

4. Fit the sensor to the DIS assembly and install the retaining screws. Tighten and torque the retaining screws 20 inch lbs. (2.3 Nm).
5. Install the DIS assembly to the engine.
6. Reconnect the negative battery cable.

NOTE: Before starting the engine, the Idle Learn procedure must be performed.

3.1L VIN T CALIF. SFI ENGINE

1. Disconnect the negative battery cable.
2. Remove the belt(s) from the crankshaft pulley.
3. Raise and support the vehicle safely.
4. Remove the right front wheel and tire assembly and inner fender access cover.
5. Remove the crankshaft harmonic balancer retaining bolt, then remove the harmonic balancer.

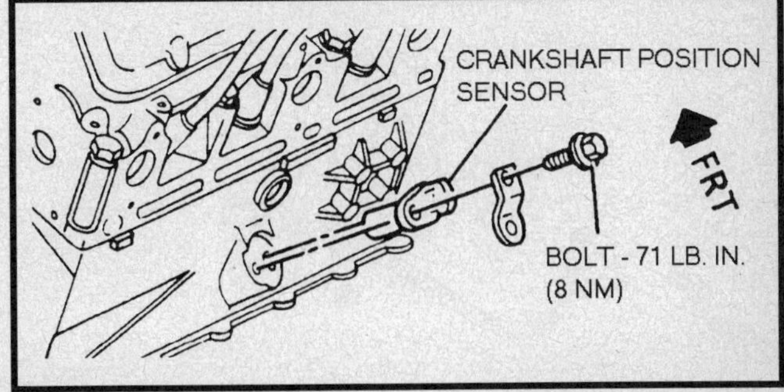

Crankshaft position sensor — Except 3.1L VIN T Calif. SFI engine

6. Disconnect the sensor electrical connector.

7. Remove the sensor and pedestal from the engine block, then separate the sensor from the pedestal.

To install:

8. Loosely install the crankshaft sensor to the pedestal.

9. Using tool J-37089 or equivalent, position the sensor with the pedestal attached, on the crankshaft.

10. Install the pedestal-to-block retaining bolts. Tighten and torque to 14–28 ft. lbs. (20–40 Nm).

11. Torque the pedestal pinch bolt 30–35 inch lbs. (20–40 Nm).

12. Remove tool J-37089 or equivalent.

13. Place tool J-37089 or equivalent, on the harmonic balancer and turn. If any vane of the harmonic balancer touches the tool, replace the balancer assembly.

NOTE: A clearance of 0.025 inch is required on either side of the interrupter ring. Be certain to obtain the correct clearance. Failure to do so will damage the sensor. A misadjusted sensor of bent interrupter ring could cause rubbing of the sensor, resulting in potential driveability problems, such as rough idle, poor performance, or a no start condition.

14. Install the balancer on the crankshaft. Install the balancer retaining bolt. Tighten the retaining bolt to 110 ft. lbs. (150 Nm) then tighten an additional 56°.

15. Install the inner fender shield.

16. Install the right front tire and wheel assembly. Tighten and torque the wheel nuts to 104 ft. lbs. (140 Nm).

17. Lower the vehicle.

18. Install the belt(s).

19. Reconnect the negative battery cable.

NOTE: Before starting the engine, the Idle Learn procedure must be performed.

CAMSHAFT POSITION SENSOR (CMP)

Removal and Installation

NOTE: The camshaft position sensor is only used on the 3.1L VIN T Calif. SFI engine.

1. Disconnect the negative battery cable.

2. Remove the belt from the power steering pulley.

3. Remove the power steering pump.

4. Disconnect the camshaft electrical connector.

5. Remove the retaining bolt and remove the sensor.

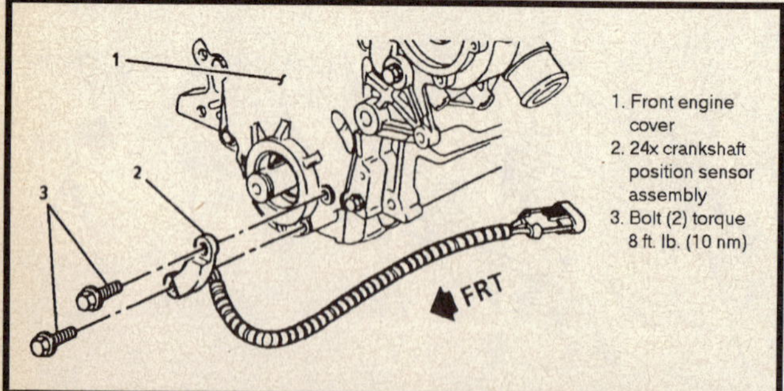

Crankshaft position sensor — 3.1L VIN T Calif. SFI engine

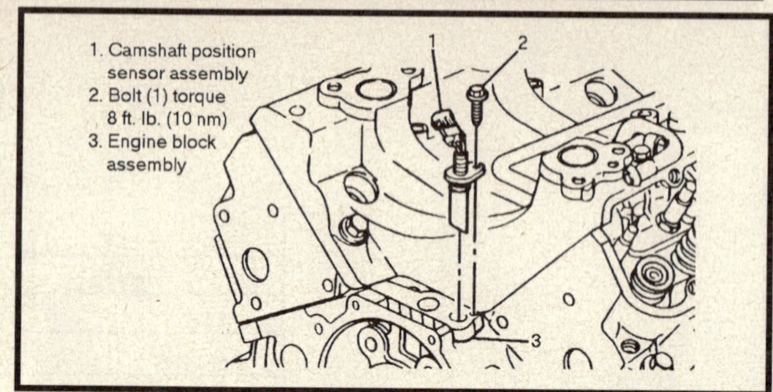

Camshaft position sensor — 3.1L VIN T Calif. SFI engine

To install:

6. Install sensor and tighten retaining bolt to 8 ft. lbs. (10 Nm).

7. Connect electrical connector.

8. Install power steering pump.

9. Install the power steering belt.

10. Connect the negative battery cable.

NOTE: Before starting the engine, the Idle Learn procedure must be performed.

ELECTRONIC CONTROL MODULE (ECM) OR PROM

Removal and Installation

1. Turn the ignition switch **OFF**.

2. Disconnect the negative battery cable.

3. Remove the interior access panel.

4. Disconnect the harness connectors from the ECM.

5. Remove the ECM-to-bracket retaining screws and remove the ECM.

6. If PROM replacement is required, remove the access cover retaining screws and cover from the ECM. Carefully remove the PROM carrier assembly from the ECM, using the rocker type PROM removal tool.

To install:

7. Fit the replacement PROM carrier assembly into the PROM socket.

NOTE: The small notch of the carrier should be aligned with the small notch in the socket. Press on the PROM carrier until if is firmly seated in the socket. Do not press on the PROM, only the carrier.

8. Install the access cover and retaining screws.

9. Position the ECM in the vehicle and install the ECM-to-bracket retaining screws.

10. Reconnect the ECM harness connectors.

11. Install the interior access panel.

NOTE: Before replacement of a defective ECM/PCM first check the resistance of each ECM controlled solenoid. This can be done at the ECM connector, using an ohmmeter and the ECM connector wiring diagram. Any ECM controlled device with low resistance will damage the replacement ECM due to high current flow through the ECM internal circuits.

12. Check that the ignition switch is **OFF**. Then reconnect the negative battery cable.

13. Perform the functional check.

NOTE: Before starting the engine, the Idle Learn procedure must be performed.

Functional Check

After the PROM has been replaced, it is necessary to perform the functional check to verify proper installation of the PROM.

1. Turn the ignition switch **ON**.

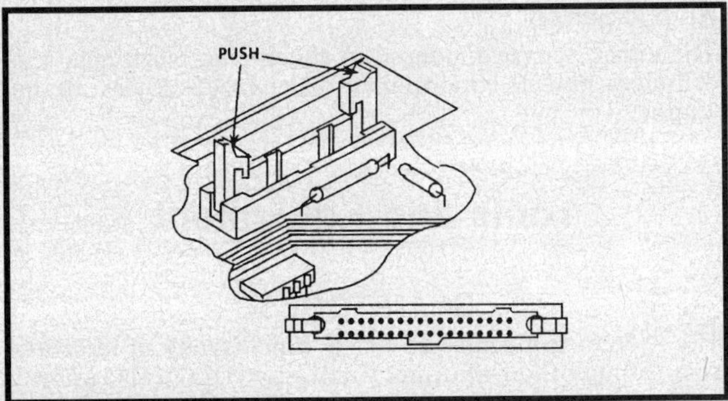

PROM installation

2. Enter diagnostics.

   a. Allow Code 12 to flash 4 times to verify no other codes are present. This indicates the MEMCAL is installed properly and the control unit is functioning.

   b. If trouble Codes 42, 43 or 51 occur, or if the Service Engine Soon light is ON constantly with no codes, the MEMCAL is not fully seated or is defective.

   c. If it is not fully seated, turn the ignition switch **OFF**, then press firmly on the ends of the MEMCAL. Make sure all pins are seated in the sockets.

SPARK PLUGS

Removal and Installation

1. Disconnect the negative battery cable.
2. Remove all foreign material from around the spark plugs.
3. Note the position of the plug wires, then twist the spark plug boot ½ turn and remove. Do not pull on the wires.
4. Remove the spark plugs.

To install:

5. Check and adjust the plug gaps.
6. Install the new plugs. Tighten and torque the plugs to specification.

NOTE: Before starting the engine, the Idle Learn procedure must be performed.

INTEGRATED DIRECT IGNITION SYSTEM (IDI)

General Information

The IDI ignition system features a distributorless ignition engine. The IDI system consists of 2 separate ignition coils, an ignition module and a secondary conductor housing mounted to an aluminum cover plate. The system also consists of a crankshaft sensor and the Electronic Spark Timing (EST) portion of the Electronic Control Module (ECM).

NOTE: The terms Electronic Control Module (ECM), Powertrain Control Module (PCM) and Body Control Module (BCM) may sometimes be used interchangeably.

The IDI ignition system uses a magnetic crankshaft sensor (mounted remotely from the ignition module) and a reluctor to determine crankshaft position and engine speed. The reluctor is a special wheel cast into the crankshaft, with 7 slots machined into it. Six of the slots are equally spaced 60 degrees apart and the seventh slot is spaced 10 degrees from 1 of the other slots. This seventh slot is used to generate a sync-pulse.

The IDI system uses the same Electronic Spark Timing (EST) circuits as the distributor type ignition. The ECM uses the EST circuit to control spark advance and ignition dwell, when the ignition system is operating in the EST mode.

To control spark knock and to use maximum spark advance to improve driveability and fuel economy, an Electronic Spark Control (ESC) system is used. This system is consists of a knock sensor and an ESC module (part of MEMCAL). The ECM monitors the ESC signal to determine when engine detonation occurs.

SYSTEM OPERATION

The IDI ignition system uses a waste spark distribution method. Each cylinder is paired with the cylinder opposite it (Ex: 1–4, 2–3). The ends of each coil secondary is attached to a spark plug. These 2 plugs are on companion cylinders, cylinders that are at top dead center at the same time. The one that is on the compression stroke is said to be the event cylinder and the one on the exhaust stroke, the waste cylinder. When the coil discharges, both plugs fire at the same time to complete the series circuit.

Since the polarity of the primary and the secondary windings are fixed, one plug always fires in a forward direction and the other in reverse. This is different than a conventional system firing all plugs in the same direction each time. Because of the demand for additional energy, the coil design, saturation time and primary current flow are also different. This redesign of the system allows higher energy to be available from the distributorless coils, greater than 40 kilovolts at all rpm ranges.

The IDI ignition system uses a magnetic crankshaft sensor mounted remotely from the ignition module. It protrudes into the block at approximately .050 inch of the crankshaft reluctor. As the crankshaft rotates, the slots of the reluctor causes a changing magnetic field at the crankshaft sensor, creating an induced voltage pulse.

The IDI module sends a reference signal to the ECM, based on the crankshaft sensor pulses, which are used to determine crankshaft position and engine speed. Reference pulses to the ECM occur at a rate of 1 per each 180 degrees of crankshaft rotation. This signal is called the 2X reference because it occurs 2 times per crankshaft revolution.

A second reference signal is sent to the ECM which occurs at the same time as the sync-pulse, from the crankshaft sensor. This signal is called the 1X reference because it occurs 1 time per crankshaft revolution.

By comparing the time between the 1X and 2X reference pulses, the ignition module can recognize the sync-pulse (the seventh slot) which starts the calculation of the ignition coil sequencing. The second crank pulse following the sync-pulse signals the ignition module to fire No. 2–3 ignition coil and the fifth crank pulse signals the module to fire the No. 1–4 ignition coil.

To control EST the ECM uses the following inputs:
- Crankshaft position
- Engine speed (rpm)
- Engine coolant temperature
- Manifold air temperature
- Engine load (manifold pressure or vacuum)

The ESC system is designed to retard spark timing up to 15 degrees to reduce spark knock in the engine. When the knock sensor detects spark knocking in the engine, it sends an A/C voltage signal to the ECM, which increases in frequency and amplitude with the severity of the knock. The ECM then adjusts the EST to reduce spark knock.

On 1992 models, during cranking, the ignition module monitors the sync-pulse to begin the ignition firing sequence and below 700 rpm the module controls spark advance by triggering each of the 2 coils at a pre-determined interval based on engine speed only. Above 700 rpm, the ECM controls the spark timing (EST) and compensates for all driving conditions. The ignition module must receive a sync-pulse and then a crank signal in that order to enable the engine to start.

On 1993–94 models, the ECM controls spark timing and ignition control during crank and run. There is no bypass mode.

SYSTEM COMPONENTS

Crankshaft Sensor

The crankshaft sensor, mounted remotely from the ignition module on an aluminum cover plate, is used determine crankshaft position and engine speed.

Ignition Coil

The ignition coil assemblies are mounted inside the module assembly housing. Each coil distributes the spark for 2 plugs simultaneously.

Electronic Spark Timing (EST)

The EST system is basically the same EST to ECM circuit use on the distributor type ignition systems with EST. This system includes the following circuits:

- Reference circuit — provides the ECM with rpm and crankshaft position information from the IDI module. The IDI module receives this signal from the crank sensor.
- Bypass signal (1992 Only) — above 700 rpm, the ECM applies 5 volts to this circuit to switch spark timing control from the IDI module to the ECM.
- EST signal (1992 Only) — reference signal is sent to the ECM via the DIS module during cranking. Under 700

rpm, the IDI module controls the ignition timing. Above 700 rpm, the ECM applies 5 volts to the bypass line to switch the timing to the ECM control.

- Ignition control circuit (1993–94 Only) — the ECM sends ignition control pulses to the ICM through this circuit. The ECM will then determine which pair of cylinders to fire.
- Reference ground circuit — this wire is grounded through the module and insures that the ground circuit has no voltage drop between the ignition module and the ECM which could affect performance.

Knock Sensor

The knock sensor, mounted in the engine block near the cylinders, detects abnormal vibration (spark knock) in the engine.

Tools and Equipment

SCAN TOOLS

The system can communicate a wide variety of information through the Data Link Connector (DLC), also known as the ALDL. Depending on the application, serial data is transmitted to terminals E or M of the DLC. This data is transmitted at a high frequency which requires a quality scan tool for interpretation. Although stored codes may be read with only the use of a small jumper wire, the use of a hand-held scan tool such as GM's TECH 1 or equivalent, is recommended. There are many manufacturers of these tools; a purchaser must be certain that the tool is proper for the intended use.

The scan tool allows any stored codes to be read from the ECM memory. The tool also allows the operator to view the data being sent to the ECM while the engine is running. This ability has obvious diagnostic advantages; the use of the scan tool is frequently required by the diagnostic charts.

With an understanding of the data stream that the tool will display and knowledge of the circuits involved, the tool can be an extremley valuable tool when examining the system. Without scan tools, diagnostic information would be difficult to obtain in some circuits and impossible to retrieve in others. Scan tools do not make the use of trouble code charts unnecessary, nor do they pinpoint the exact area in a circuit where the problem exists. However, this type of equipment will give more specific information, such as sensor input readings. An example of this would be the coolant sensor circuit. While the ECM looks at specific voltage levels to determine coolant temperature, the scan tool will display this information as an actual temperature reading. This information is not only useful in determining system malfunctions, but is also very helpful in diagnosing mechanical problems. For example, if the engine is overheating, the scan tool can be used to look at the actual engine temperature. This is helpful when checking for electric cooling fan enable switch problems, sticking thermostats, etc.

> **NOTE: A scan tool that is known to display faulty data should not be used for diagnosis. Although the fault may be believed to be in only one area, it can possibly affect many other areas during diagnosis, leading to errors and incorrect repair.**

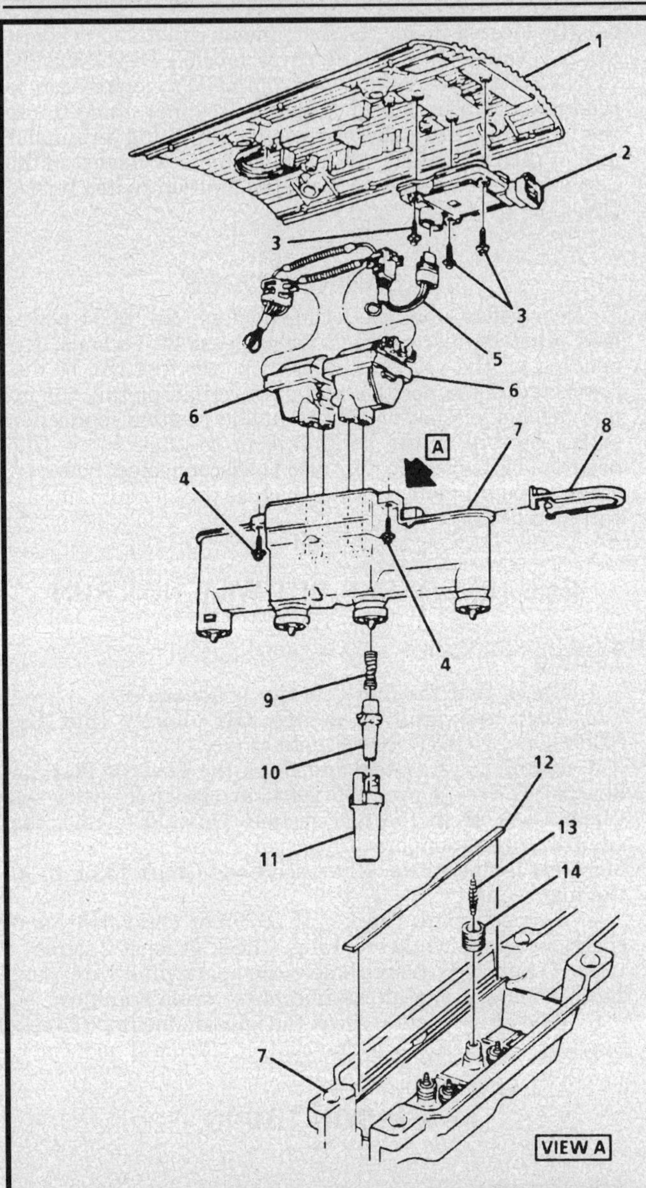

Integrated direct ignition (IDI) assembly

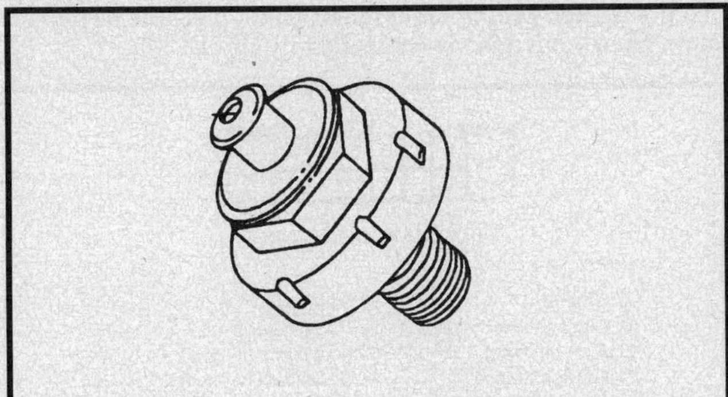

Knock sensor

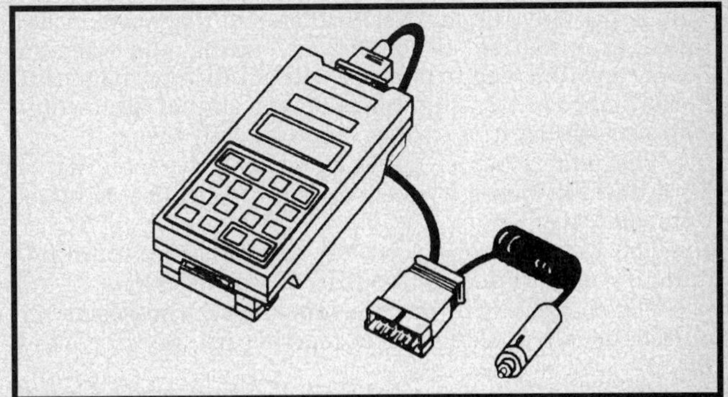

Typical scan tool — GM's Tech 1 shown

ELECTRICAL TOOLS

The most commonly required electrical diagnostic tool is the digital multimeter, allowing voltage, resistance and amperage to be read by one instrument. The multimeter must be a high-impedance unit, with 10 megohms of impedance in the voltmeter. This type of meter will not place an additional load on the circuit it is testing; this is extremely important in low voltage circuits. The multimeter must be of high quality in all respects. It should be handled carefully and protected from impact or damage. Replace batteries frequently in the unit.

Other necessary tools include an unpowered test light, a quality tachometer with inductive (clip-on) pick up and the proper tools for releasing GM's Metri-Pack, Weather Pack and Micro-Pack terminals as necessary. The Micro-Pack connectors are used at the ECM connector. A vacuum pump/gauge may also be required for checking sensors, solenoids and valves.

Diagnosis and Testing

SERVICE PRECAUTIONS

> **NOTE: To avoid damage to the ECM or other ignition system components, do not use electrical test equipment such as battery or AC powered voltmeter, ohmmeter, etc. or any type of tester other than specified.**

- When performing electrical tests on the system, use a high impedance multimeter, digital voltmeter (DVM) J-34029-A or equivalent.
- To prevent electrostatic discharge damage, when working with the ECM, do not touch the connector pins or soldered components on the circuit board.
- When handling a PROM, CALPAC or MEMCAL, do not touch the component leads. Also, do not remove the integrated circuit from the carrier.
- Never pierce a high tension lead or boot for any testing purpose; otherwise, future problems are guaranteed.
- Leave new components and modules in the shipping package until ready to install them.
- Never disconnect any electrical connection with the ignition switch **ON** unless instructed to do so in a test.

PRELIMINARY INSPECTION

This is possibly the most critical step of diagnosis. A detailed examination of connectors, wiring and vacuum hoses can often lead to a repair without further diagnosis. Performance of this step relies on the skill of the technician performing it; a careful inspector will check:

- The undersides of hoses as well as the integrity of hard-to-reach hoses blocked by the air cleaner or other components.
- The wiring carefully for any sign of strain, burning, crimping, or terminal pull-out from a connector.
- The connectors at components or in harnesses as required; usually, pushing them together will reveal a loose fit.

READING CODES

The Data Link Connector (DLC), also known as the ALDL connector, is used for communicating with the ECM. It is usually located under the instrument panel and is sometimes covered by a plastic cover labeled DIAGNOSTIC CONNECTOR. Codes stored in the ECM's memory can be read through a handheld diagnostic scanner plugged into the DLC. Codes can also be read by connecting terminal A to B of the DLC and counting the number of flashes of the Service Engine Soon light, with the ignition switch turned **ON**.

CLEARING CODES

To clear codes from the ECM memory, the ECM power feed must be disconnected for at least 30 seconds. Depending on the vehicle, the ECM power feed can be disconnected at the positive battery terminal pigtail, the in-line fuseholder that originates at the positive connection at the battery or the ECM fuse in the fuse block. The negative battery cable may also be disconnected; however, other on-board memory data, such as preset radio tuning, will also be lost.

ENGINE CRANKS BUT WILL NOT RUN

Testing

1. Check that the fuel quantity is adequate.
2. Turn the ignition switch **ON**. Verify that the SERVICE ENGINE SOON light is ON.
3. Install the scan tool and check the Throttle Position Sensor (TPS) — if over 2.5 volts, at closed throttle, use Chart Code 21 in the fuel section. On cold engine, the engine coolant temperature sensor should scan near ambient air temperature. If not see Code Chart 14 or 15 in the fuel section.
4. Connect spark checker, J 26792 or equivalent, and check for spark while cranking. Check at least 2 wires.
5. If spark occurs, reconnect the spark plug wires and check for fuel spray at the injector(s) while cranking.
6. If no spark occurs, check the ignition module connection and replace the ignition control module if necessary.

IGNITION TIMING

Adjustment

Because the reluctor wheel is an integral part of the crankshaft and the crankshaft sensor is mounted in a fixed position, timing adjustment is not possible or necessary. Spark advance is controlled by the ECM.

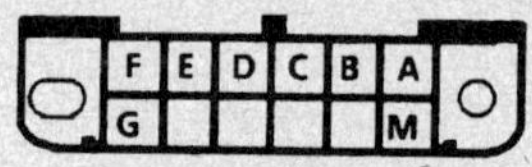

Data link connector (DLC)

Spark Advance

The ECM uses information from the MAP and coolant sensors, in addition to rpm to calculate spark advance as follows:

1. Low MAP output voltage — more spark advance
2. Cold engine — more spark advance
3. High MAP output voltage — less spark advance
4. Hot engine — less spark advance

Therefore, detonation could be caused by low MAP output or high resistance in the coolant sensor circuit.

Poor performance could be caused by high MAP output or low resistance in the coolant sensor circuit.

Component Replacement

IGNITION COIL AND MODULE ASSEMBLY

Removal and Installation

1. Turn the ignition switch **OFF**.
2. Disconnect the negative battery cable.
3. Disconnect the 11-pin IDI ignition harness connector.
4. Remove the ignition system assembly-to-camshaft housing bolts.

NOTE: If the boots are difficult to remove from the spark plugs, use tool J36011 or equivalent, to remove. First twist and then pull upward on the retainers. Reinstall the boots and retainers on the IDI housing secondary terminals. The boots and retainers must be in place on the IDI housing secondary terminals prior to ignition system assembly installation or ignition system damage may result.

5. Remove the ignition system assembly from the engine.

To install:

6. Install the spark plug boots and retainers to the housing.
7. Carefully aligned the boots to the spark plug terminals, while installing the ignition system assembly to the engine.
8. Coat the threads of the retaining bolts with 1052080 or equivalent and install. Tighten and torque to 16 ft. lbs. (22 Nm).
9. Reconnect the 11-pin IDI harness connector.
10. Check that the ignition switch is **OFF**. Then reconnect the negative battery cable.

IGNITION COIL

Removal and Installation

1. Disconnect the negative battery cable.
2. Remove the IDI assembly from the engine.
3. Remove the housing to cover screws.
4. Remove the housing from the cover.
5. Remove the coil harness connector.
6. Remove the coils, contacts and seals from the cover.

To install:

7. Install the coil to the cover.
8. Install the coil harness connectors.
9. Install new seals to the housing.
10. Install the contacts to the housing. Use petroleum jelly to retain the contacts.
11. Install the housing cover and retaining screws. Tighten and torque to 35 inch lbs. (4 Nm).

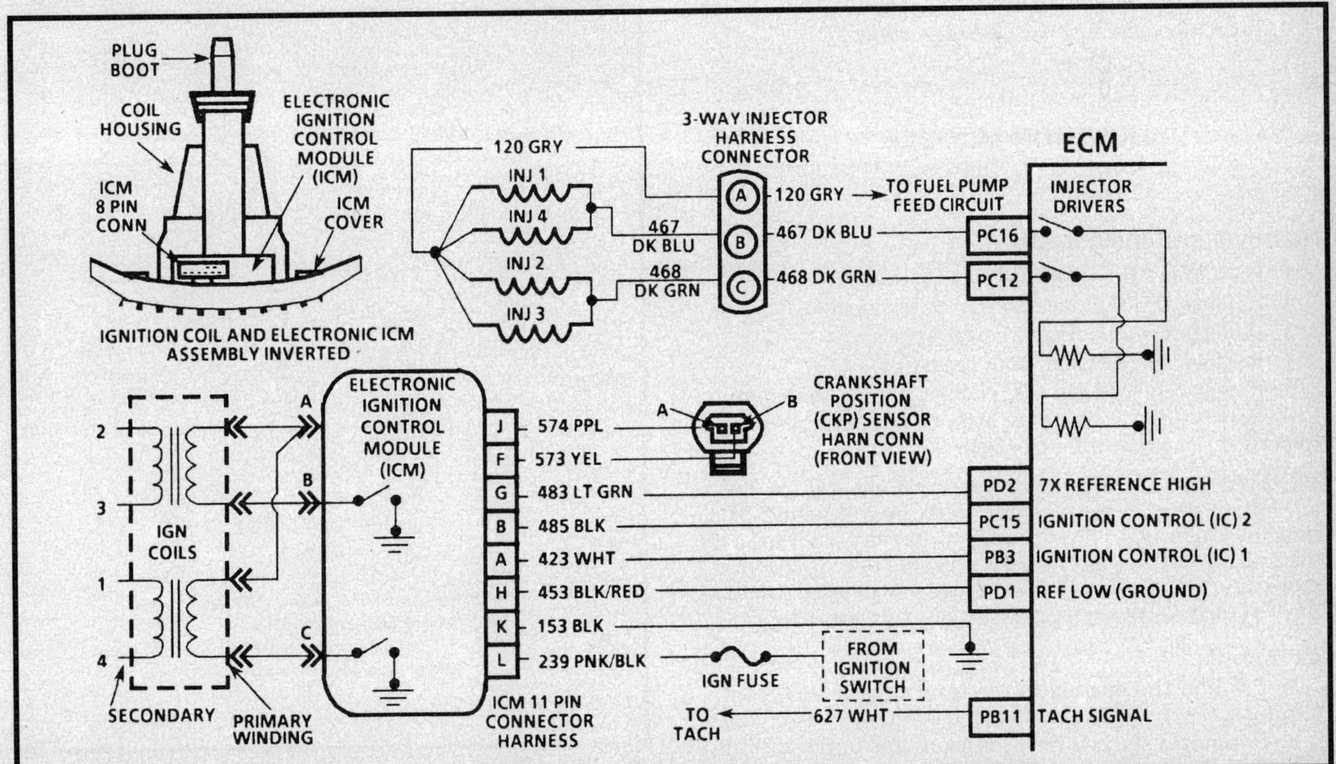

IDI system circuit

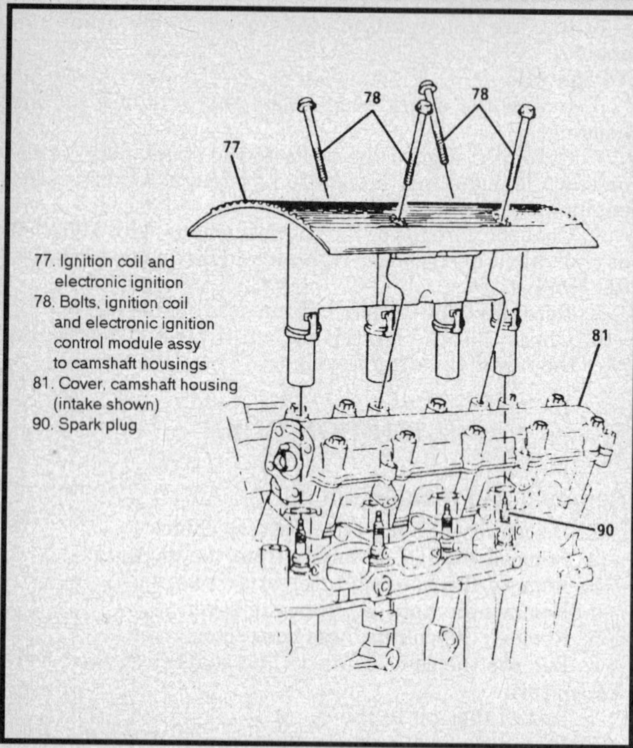

Ignition coil and module assembly removal and installation

12. Fit the spark plug boots and retainers to the housing.

13. Install the IDI assembly to the engine.

14. Reconnect the negative battery cable.

IGNITION MODULE

Removal and Installation

1. Disconnect the negative battery cable.
2. Remove the IDI assembly from the engine.
3. Remove the housing to cover screws.
4. Remove the housing from the cover.
5. Remove the coil harness connector from the module.
6. Remove the module-to-cover retaining screws and remove the module from the cover.

NOTE: Do not wipe the grease from the module or coil, if the same module is to be replaced. If a new module is to be installed, spread the grease (included in package) on the metal face of the new module and on the cover where the module seats. This grease is necessary from module cooling.

To install:

7. Position the module to the cover and install the retaining screws. Tighten and torque to 35 inch lbs. (4 Nm).

8. Reconnect the coil harness connector to the module.

9. Install the housing cover and retaining screws. Tighten and torque to 35 inch lbs. (4 Nm).

10. Fit the spark plug boots and retainers to the housing.

11. Install the IDI assembly to the engine.

12. Reconnect the negative battery cable.

CRANKSHAFT SENSOR

Removal and Installation

1. Disconnect the negative battery cable.
2. Disconnect the sensor harness connector at the sensor.
3. Remove the sensor retaining bolts and pull the sensor from the engine.

To install:

4. Fit a new O-ring to the sensor and lubricate with engine oil. Install the sensor into the engine.
5. Install the sensor retaining bolt. Tighten and torque to 88 inch lbs. (10 Nm).
6. Reconnect the sensor harness connector.
7. Reconnect the negative battery cable.

ELECTRONIC CONTROL MODULE (ECM) OR MEMCAL

Removal and Installation

1. Turn the ignition switch **OFF**.
2. Disconnect the negative battery cable.
3. Remove the right side hush panel.
4. Disconnect the harness connectors from the ECM.

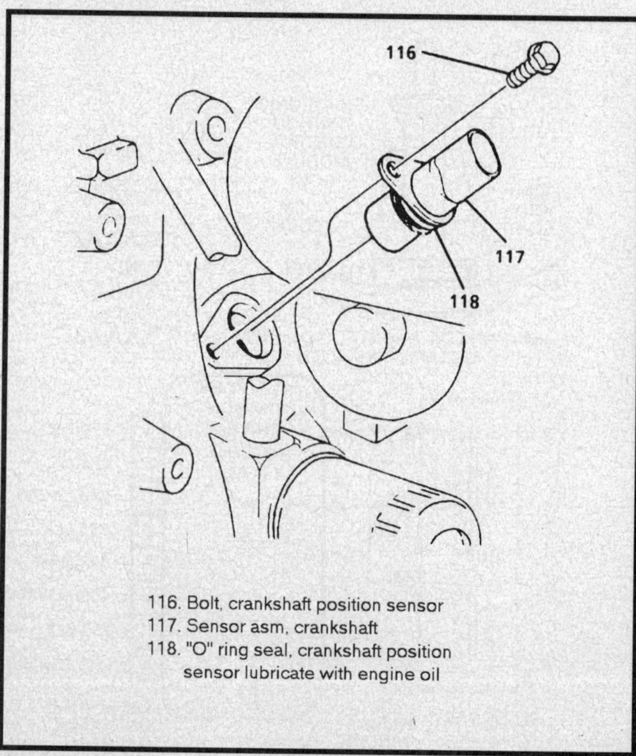

Crankshaft position sensor removal and installation

5. Remove the ECM-to-bracket retaining screws and remove the ECM.

NOTE: Before replacement of a defective ECM/PCM first check the resistance of each ECM controlled solenoid. This can be done at the ECM connector, using an ohmmeter and the ECM connector wiring diagram. Any ECM controlled device with low resistance will damage the replacement ECM due to high current flow through the ECM internal circuits.

6. If replacement of the MEMCAL is required, remove the access cover retaining screws and cover from the ECM. Note the position of the MEMCAL for proper installation in the new ECM. Using 2 fingers, carefully push both retaining clips back away from the MEMCAL. At the same time, grasp it at both ends and lift it up out of the socket. Do not remove the cover of the MEMCAL.

To install:

7. Fit the replacement MEMCAL into the socket.

NOTE: The small notches in the MEMCAL must be aligned with the small notches in the socket. Press only on the ends of the MEMCAL until the retaining clips snap into the ends of the MEMCAL. Do not press on the middle of the MEMCAL, only the ends.

8. Install the access cover and retaining screws.
9. Position the ECM in the vehicle and install the ECM-to-bracket retaining screws.
10. Reconnect the ECM harness connectors.
11. Install the hush panel.
12. Check that the ignition switch is **OFF**. Then reconnect the negative battery cable.
13. Perform the functional check.

Functional Check

After the PROM has been replaced, it is necessary to perform the functional check to verify proper installation of the PROM.

1. Turn the ignition switch **ON**.
2. Enter diagnostics.
   a. Allow Code 12 to flash 4 times to verify no other codes are present. This indicates the MEMCAL is installed properly and the ECM is functioning.
   b. If trouble Codes 19, 43 or 51 occur, or if the Service Engine Soon light is ON constantly with no codes, the MEMCAL is not fully seated or is defective.
   c. If it is not fully seated, turn the ignition switch **OFF**, then press firmly on the ends of the MEMCAL.

KNOCK SENSOR

Removal and Installation

1. Disconnect the negative battery cable.
2. Raise and support the vehicle safely.
3. Disconnect the harness connector from the knock sensor.
4. Remove the sensor from the engine block.

To install:

5. Clean the threads on the engine block, where the sensor was installed. Install the sensor. Tighten to 11–16 ft. lbs. (15–22 Nm).
6. Reconnect the harness connector to the knock sensor.
7. Lower the vehicle.

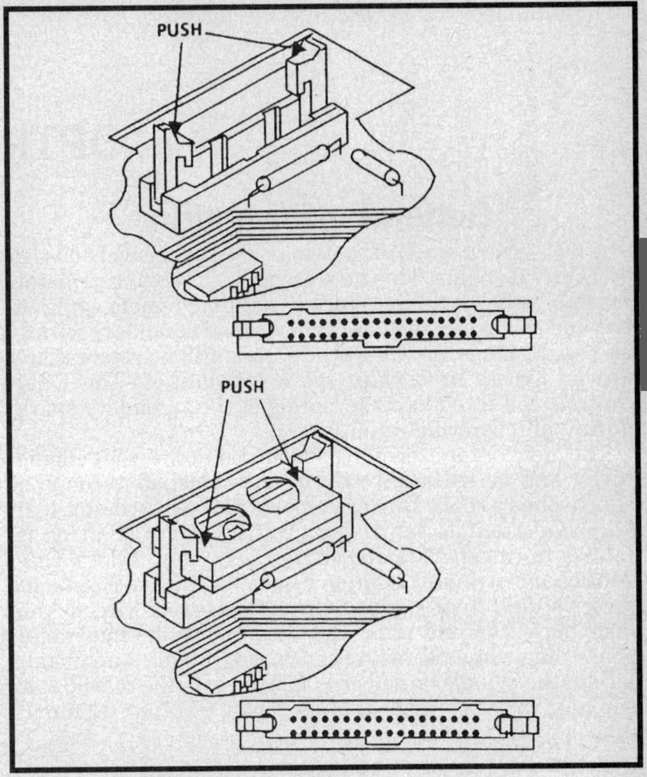

PROM installation

8. Reconnect the negative battery cable.

SPARK PLUGS

Removal and Installation

1. Disconnect the negative battery cable.
2. Remove the Integrated Direct Ignition (IDI) assembly bolts and connector. Pull straight up on the housing and remove.
3. Remove the plug wires from the spark plugs. If the connectors are stuck to the spark plugs, use tool J-36011 or equivalent to remove the connectors.
4. Clean all debris from the spark plugs area.

— WARNING —

The 2.3L engine, VIN D and A, have aluminum cylinder heads. Do not remove the spark plugs while the engine is warm. If the spark plugs are removed while the engine is warm, damage to the cylinder head plug area could result.

5. Remove the spark plugs.

To install:

6. Apply release agent to the porcelain of the spark plugs, in the area where the boot contacts the spark plug. Hand start the spark plug into the cylinder head. Tighten and torque the spark plugs to 17 ft. lbs. (22 Nm).
7. Install the spark plug boot connectors.
8. Fit the IDI assembly over the spark plugs and push straight down.
9. Apply TST Loctite® 592 or equivalent to the IDI assembly to cam housing bolts.
10. Hand start the IDI assembly-to-cam housing bolts. Tighten and torque the bolts 13 ft. lbs. (18 Nm).

11. Reconnect the electrical connector.

12. Reconnect the negative battery cable.

OPTI-SPARK

General Information

The Opti-Spark ignition system is used exclusively on the 5.7L VIN P engine. This new design includes a distributor assembly, ECM timing control circuits, remote ignition coil and coil driver module, primary and secondary wiring and spark plugs. In the Opti-Spark ignition system there are no bypass or backup spark capabilities. The ECM controls coil dwell and the timing of the secondary spark during all operating conditions.

The distributor assembly is mounted to the front engine cover, and is driven directly by the camshaft through a short splined shaft. During removal of the distributor it is very important to mark the position of the shaft as it relates to camshaft-to-distributor alignment. Like a conventional distributor ignition system, it directs the secondary voltage from the coil through the coil wire to the distributor cap and rotor and then out to the individual spark plug wires. To insure correct spark plug wire installation, the cylinder number is molded into the distributor cap next to the corresponding secondary output terminal.

SYSTEM OPERATION

The distributor assembly contains an optical position sensor assembly that along with the rotor is driven by the short splined shaft. The sensor assembly consists of 2 optical diodes and a 2 track slotted disc. The disc has a high resolution and a low resolution track. Each set of slots operates with 1 of the optical sensors to produce 2 separate camshaft position signals to the ECM. One optical sensor senses 360 equally spaced slots in the signal disk which supply 720 switch points per camshaft revolution. Each pulse is equal to 1 degree of crankshaft rotation. This signal is used by the ECM to determine the precise point for fuel delivery and spark control.

The other optical sensor is used to recognize 8 disk slots which provides the ECM with 8 timing pulses per camshaft revolution. These pulses are 45 cam degrees apart. The beginning of each low resolution pulse corresponds to the Top Dead Center (TDC) position of a particular cylinder. Of the 8 slots on the low resolution track, 4 of these have the same length and 4 have variable lengths. The slot that has the longest width is interpreted by the ECM as cylinder No. 2. This is the last cylinder in the engine firing order 1–8–4–3–6–5–7–2. The slot width is 22 degrees between its leading and trailing edges. The next cylinder in the firing order is No. 1, with a slot width of only 2 degrees. Since the leading edge of the 2 degree pulse follows the trailing edge of the 22 degree pulse, the ECM can determine the beginning of the firing order.

The ECM can identify each individual cylinders' TDC position by the leading edges of the low resolution slots. The ECM can further identify which cylinder was in TDC position, by counting the high resolution pulses during the period of time the low resolution signal is high. Both resolution pulses toggle between 0 and 5 volts. By continually comparing the 2 inputs, the ECM can determine when 1 of the signals is missing. An interruption of the high resolution pulse will set a DTC 36. If this failure

occurs the engine will start and run, however, performance and economy will have deteriorated. If the low resolution signal is lost the engine will not run and a DTC 16 will be set. This signal is the reference, or RPM input to the ECM. The high and low resolution signal wires are part of the 4 terminal harness connector that plugs into the top of the distributor housing. These 4 wires are: power feed, ground, high, and low resolution signal wires. The distributor cap and rotor are the only serviceable components within the distributor assembly.

The remote ignition coil and coil driver module assembly are used to provide spark to the distributor. Power is supplied to the ignition coil primary and the coil driver module from the ignition switch. The ECM analyzes inputs from the dual camshaft sensors as well as other sensor inputs to calculate the correct spark advance and coil saturation time. The primary coil winding current is controlled by the ECM via the EST signal line to the coil driver module. An open or short to voltage in this circuit will cause a no-start and set a DTC 41. If the EST line becomes grounded between the distributor and the ECM, the engine will fail to start and a DTC 42 will set. The dual optical sensor in the distributor is grounded by the ECM. An open in this circuit would cause a no-start condition as well, due to the loss of the low resolution signal.

The Knock Sensor system that is used in conjunction with Opti-Spark ignition allows up to 20 degrees of timing retard to reduce detonation in the engine. A malfunction in this circuit should set a DTC 43.

The base timing is preset when the engine is manufactured, for this reason no adjustment is possible. Timing advance and retard are accomplished by the ECM based on sensor input signals.

Tools and Equipment

SCAN TOOLS

The system can communicate a wide variety of information through the Data Link Connector (DLC), also known as the ALDL. Depending on the application, serial data is transmitted to terminals E or M of the DLC. This data is transmitted at a high frequency which requires a quality scan tool for interpretation. Although stored codes may be read with only the use of a small jumper wire, the use of a hand-held scan tool such as GM's TECH 1 or equivalent, is recommended. There are many manufacturers of these tools; a purchaser must be certain that the tool is proper for the intended use.

The scan tool allows any stored codes to be read from the ECM memory. The tool also allows the operator to view the data being sent to the ECM while the engine is running. This ability has obvious diagnostic advantages; the use of the scan tool is frequently required by the diagnostic charts.

With an understanding of the data stream that the tool will display and knowledge of the circuits involved, the tool can be an extremely valuable tool when examining the system. Without scan tools, diagnostic information would be difficult to obtain in some circuits and impossible to retrieve in others. Scan tools do not make the use of trouble code charts unnecessary, nor do they pinpoint the exact area in a circuit where the problem exists. However, this type of equipment will give more specific information, such as sensor input readings. An example of this would be the coolant sensor circuit. While the ECM looks at specific voltage levels to determine coolant temperature, the scan tool will display this information as an actual temperature reading. This information is not only useful in determining system malfunctions, but is also very helpful in diagnosing mechanical problems. For example, if the engine is overheating, the scan tool can be used to look at the actual engine temperature. This is helpful when checking for electric cooling fan enable switch problems, sticking thermostats, etc.

NOTE: A scan tool that is known to display faulty data should not be used for diagnosis. Although the fault may be believed to be in only one area, it can possibly affect many other areas during diagnosis, leading to errors and incorrect repair.

ELECTRICAL TOOLS

The most commonly required electrical diagnostic tool is the digital multimeter, allowing voltage, resistance and amperage to be read by one instrument. The multimeter must be a high-impedance unit, with 10 megohms of impedance in the voltmeter. This type of meter will not place an additional load on the circuit it is testing; this is extremely important in low voltage circuits. The multimeter must be of high quality in all respects. It should be handled carefully and protected from impact or damage. Replace batteries frequently in the unit.

Other necessary tools include an unpowered test light, a quality tachometer with inductive (clip-on) pick up and the proper tools for releasing GM's Metri-Pack, Weather Pack and Micro-Pack terminals as necessary. The Micro-Pack connectors are used at the ECM connector. A vacuum pump/gauge may also be required for checking sensors, solenoids and valves.

Diagnosis and Testing

SERVICE PRECAUTIONS

NOTE: To avoid damage to the ECM or other ignition system components, do not use electrical test equipment such as battery or AC powered voltmeter, ohmmeter, etc. or any type of tester other than specified.

• When performing electrical tests on the system, use a high impedance multimeter, digital voltmeter (DVM) J-34029-A or equivalent.
• To prevent electrostatic discharge damage, when working with the ECM, PROM, CALPAC or MEMCAL, do not touch the connector pins or soldered components on the circuit board.
• Never pierce a high tension lead or boot for any testing purpose; otherwise, future problems are guaranteed.
• Leave new components and modules in the shipping package until ready to install them.
• Never disconnect any electrical connection with the ignition switch **ON** unless instructed to do so in a test.
• No periodic lubrication is required.
• Do not allow the tachometer terminal to touch ground; damage to the ignition module and/or the ignition coil may result.

PRELIMINARY INSPECTION

This is possibly the most critical step of diagnosis. A detailed examination of connectors, wiring and vacuum hoses can often lead to a repair without further diagnosis. Performance of this step relies on the skill of the technician performing it; a careful inspector will check:
• The undersides of hoses as well as the integrity of hard-to-reach hoses blocked by the air cleaner or other components.
• The wiring carefully for any sign of strain, burning, crimping, or terminal pull-out from a connector.
• The connectors at components or in harnesses as required; usually, pushing them together will reveal a loose fit.

READING CODES

The Data Link Connector, also known as the ALDL connector, is used for communicating with the ECM. It is usually located under the instrument panel and is sometimes covered by a plastic cover labeled DIAGNOSTIC CONNECTOR. Codes stored in the ECM's memory can be read through a handheld diagnostic scanner plugged into the DLC. Codes can also be read by connecting terminal A to B of the DLC and counting the number of flashes of the Service Engine Soon light, with the ignition switch turned **ON**.

CLEARING CODES

To clear codes from the ECM memory, the ECM power feed must be disconnected for at least 30 seconds. Depending on the vehicle, the ECM power feed can be disconnected at the positive battery terminal pigtail, the inline fuseholder that originates at the positive connection at the battery or the ECM fuse in the fuse block. The negative battery cable may also be disconnected; however, other on-board memory data, such as preset radio tuning, will also be lost.

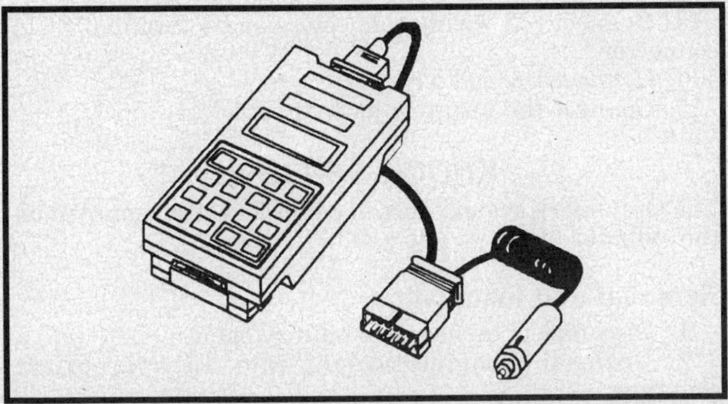

Typical scan tool — GM's Tech 1 shown

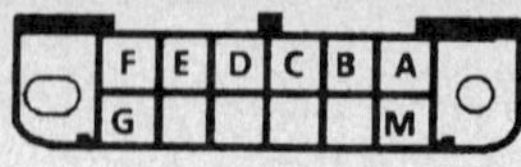

Data link connector (DLC)

ENGINE CRANKS BUT WILL NOT RUN

Testing

1. Check that the fuel quantity is adequate.
2. Turn the ignition switch **ON**. Verify that the SERVICE ENGINE SOON light is ON.
3. Install the scan tool and check the Throttle Position Sensor (TPS) — if over 2.5 volts, at closed throttle, use Chart Code 21 in the fuel section. On cold engine, the engine coolant temperature sensor should scan near ambient air temperature. If not see Code Chart 14 or 15 in the fuel section.
4. Connect spark checker, J 26792 or equivalent, and check for spark while cranking. Check at least 2 wires.
5. If spark occurs, reconnect the spark plug wires and check for fuel spray at the injector(s) while cranking.
6. If no spark occurs, disconnect the ignition control module and turn ignition switch **ON**. Check the voltage from the harness terminals A and D to ground.

• Both terminals 10 volts or more — Crank engine and check voltage between B and ground while cranking engine. If voltage is between 1 and 4 volts, open or short in ignition coil module ground circuit or faulty ignition coil module. If not, check ignition system chart.

• Both terminals under 10 volts — Short or fault in ignition feed circuit or faulty ignition coil.

• Either terminal under 10 volts — Fault in ignition feed circuit or faulty ignition coil or ignition coil connection.

IGNITION TIMING

Adjustment

The base timing is set when the engine is manufactured. Timing advance and retard is controlled by the ECM. Ignition timing is not adjustable.

Component Replacement

DISTRIBUTOR

Removal and Installation

1. Make sure ignition switch is in the **OFF** position.
2. Disconnect negative battery cable.
3. Disconnect intake air temperature sensor harness connector. Remove air intake duct.
4. Remove serpentine drive belt.
5. On Corvette, remove belt tensioner.

6. Drain engine coolant and remove water pump.
7. Remove crankshaft torsional damper assembly.
8. Tag and remove spark plug wires from distributor.
9. Remove the 4 terminal ECM connector from distributor and remove 3 distributor attaching bolts.
10. Pull distributor assembly forward until driveshaft disengages from end of camshaft assembly.
11. Mark the top surface of driveshaft for proper alignment during installation.
12. Remove 4 cap attaching screws and remove cap.
13. Remove 2 rotor retaining screws and remove rotor.
14. Replace rotor and tighten retaining screws to 9 in. lbs. (1 Nm).
15. Replace cap and tighten attaching screws to 25 in. lbs. (2.8 Nm).

To install:

16. Place distributor driveshaft into camshaft assembly keeping previously indexed mark UP.
17. Install 3 attaching bolts and tighten to 97 in. lbs. (11 Nm).
18. Install 4 terminal ECM connector to distributor.
19. Install spark plug wires to distributor.
20. Install crankshaft torsional damper.
21. Install water pump and refill cooling system. Check for leaks.
22. Install serpentine drive belt.
23. Install air intake duct and reconnect IAT sensor.
24. Connect the negative battery cable.

IGNITION COIL OR MODULE ASSEMBLY (ICM)

Removal and Installation

1. Place ignition switch in the **OFF** position.
2. Disconnect the negative battery cable.
3. Remove 4 terminal ECM harness connector at ignition module.
4. Remove ignition coil wire connectors.
5. Remove coil wire.
6. Remove attaching bolts. Remove the unit.
7. Remove coil from bracket by drilling out rivets.
8. Remove 2 module retaining screws and module.
9. If replacing module apply new silicone grease on module bracket and on back of module.
10. Place module on bracket and insert 2 retaining screws and tighten screws to 15 in. lbs. (1.7 Nm).
11. Replace coil on bracket using screws provided with replacement coil.

To install:

12. Install ignition coil/module assembly on cylinder head and tighten retaining bolts to 18 ft. lbs. (25 Nm).
13. Connect coil wiring connectors and 4 terminal ECM connector.
14. Connect the coil wire.
15. Connect the negative battery cable.

KNOCK SENSOR

The Opti-spark system uses one knock sensor mounted on the cylinder block.

Removal and Installation

1. Disconnect the negative battery cable.
2. Drain the engine coolant into an appropriate container.
3. Remove knock sensor wiring connector and sensor.

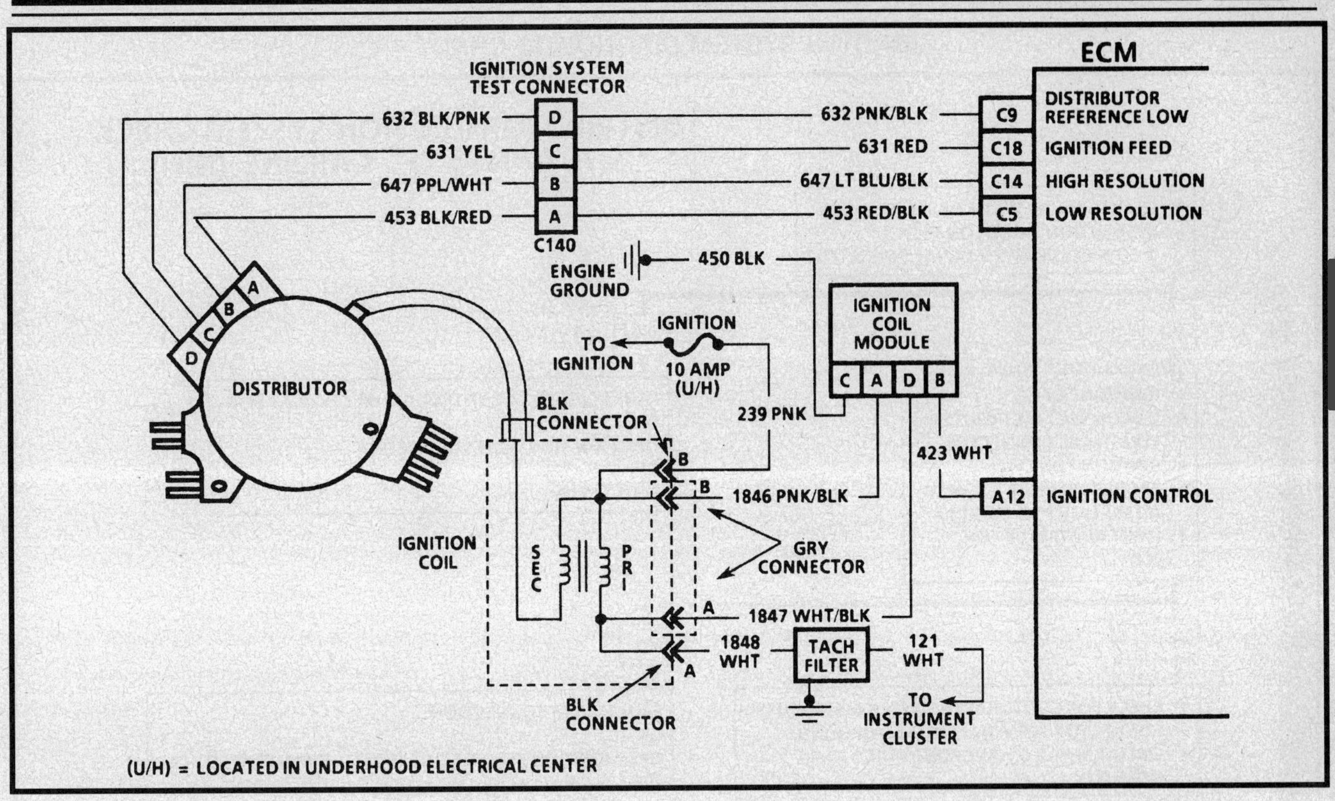

Opti-spark system circuit — 5.7L VIN P engine

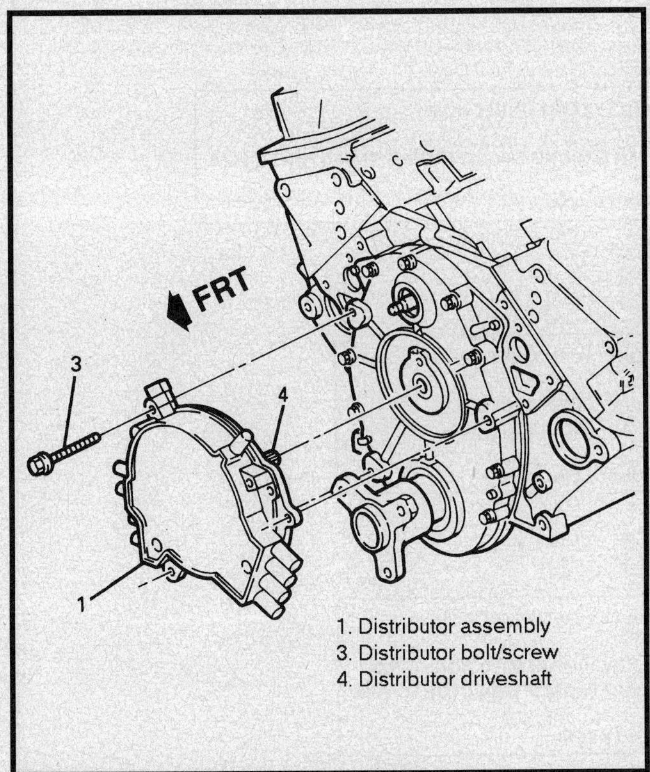

1. Distributor assembly
3. Distributor bolt/screw
4. Distributor driveshaft

Distributor removal and installation

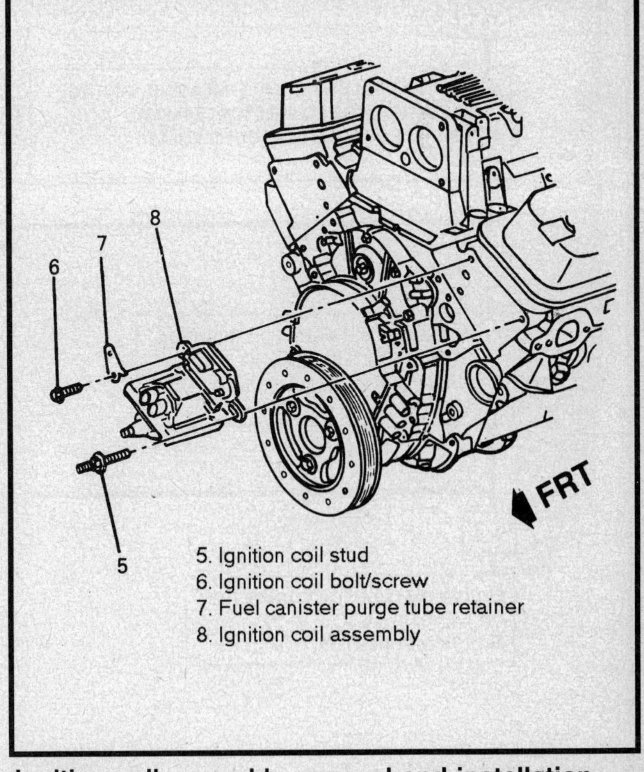

5. Ignition coil stud
6. Ignition coil bolt/screw
7. Fuel canister purge tube retainer
8. Ignition coil assembly

Ignition coil assembly removal and installation

ELECTRONIC IGNITION SYSTEMS
OPTI-SPARK

IGNITION SYSTEM DIAGNOSTIC CHART

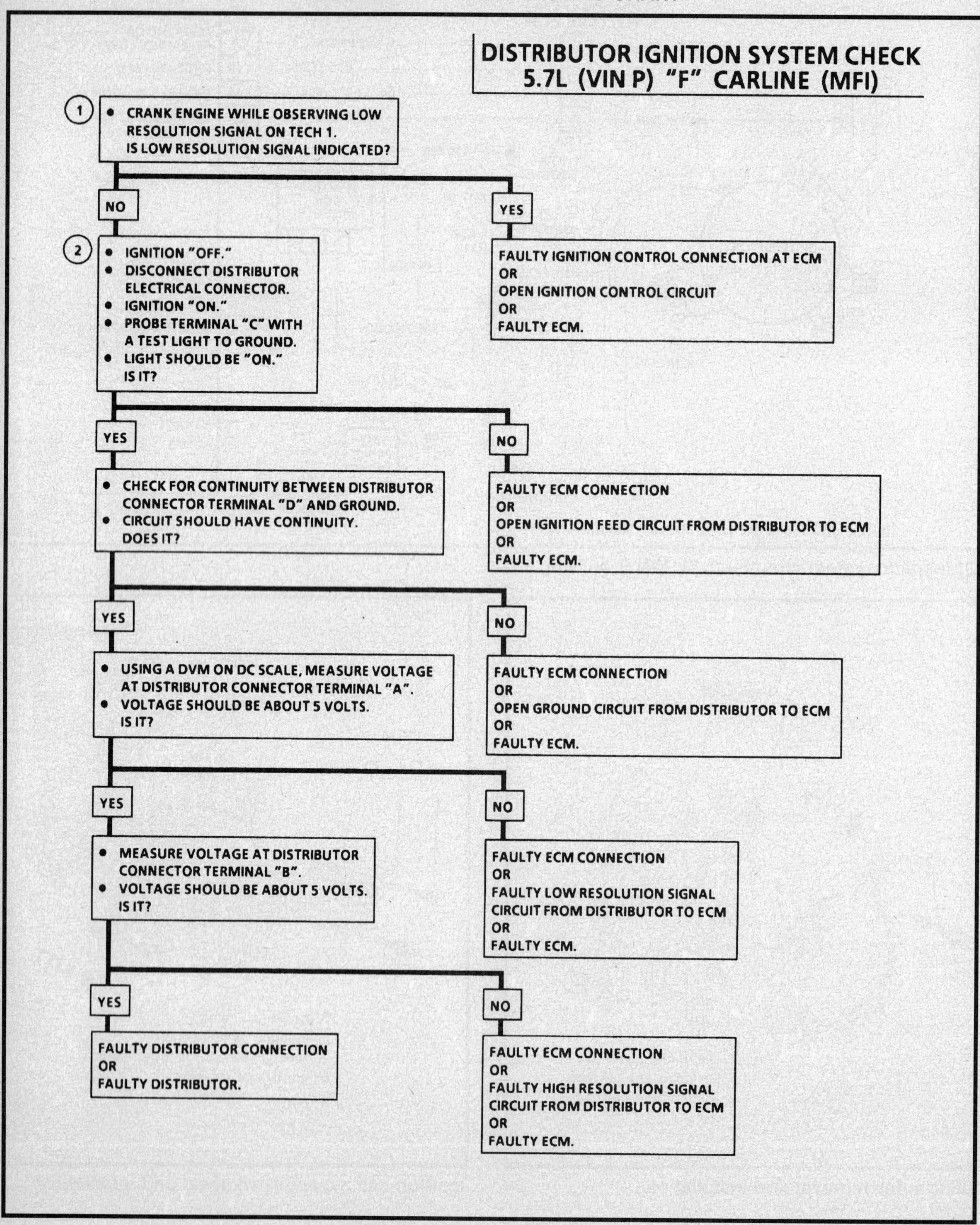

SATURN

General Information

The Electronic Ignition (EI) system on the Saturns provides spark energy for air/fuel combustion in response to timing commands from the Powertrain Control Module (PCM). The system components include an electronic module with 2 coils, secondary wiring and spark plugs.

> **NOTE: The terms Electronic Control Module (ECM), Powertrain Control Module (PCM) and Body Control Module (BCM) may sometimes be used interchangeably.**

SYSTEM OPERATION

When the ignition switch is turned to the **RUN** position, battery voltage (B+) is applied to the EI module. The BYPASS line is held low by the PCM. The REF line is high. No spark occurs at the spark plug until crank begins. Crankshaft rotation initiates crankshaft position sensor (CKP) pulses. With B+ applied to the EI module, any valid CKP pulses will initiate reference pulses. The Bypass line is held low and spark plug firing begins at TDC of the first cylinder following a recognized sync pulse. As engine rotation increases, the Bypass line is held low and the PCM will take control of spark timing.

On 1993–94 vehicles, the EI is designed to accommodate a 10° advance on the crankshaft. This was done to improve startability.

SYSTEM COMPONENTS

Crankshaft Position Sensor

The crankshaft position sensor, mounted on the rear of the engine, is used to determine crankshaft position and engine speed.

Ignition Coil

The ignition coil assemblies are mounted on the EI module. The coil provides a timed signal up to 40,000 volts.

Electronic Spark Timing (EST)

The EST system is an integral part of the PCM. This system includes the following circuits:

- Reference (REF) Line — provides the PCM with rpm and engine position information.
- Bypass Line — The PCM applies 5 volts to this circuit to switch spark timing control from the EI module to the PCM.
- EST Line — reference signal is sent from the PCM to control the ignition timing.
- Reference Low (REF LO) Line — is a low current ground reference supplied by the EI module to the PCM.

Tools and Equipment

SCAN TOOLS

The system can communicate a wide variety of information through the Data Link Connector (DLC), also known as the ALDL. Serial data is transmitted to terminal M of the DLC. This data is transmitted at a high frequency which requires a quality scan tool for interpretation. Although stored codes may be read with only the use of a small jumper wire, the use of a hand-held scan tool such as GM's TECH 1 or equivalent, is recommended. There are many manufacturers of these tools; a purchaser must be certain that the tool is proper for the intended use.

The scan tool allows any stored codes to be read from the ECM memory. The tool also allows the operator to view the data being sent to the ECM while the engine is running. This ability has obvious diagnostic advantages; the use of the scan tool is frequently required by the diagnostic charts.

With an understanding of the data stream that the tool will display and knowledge of the circuits involved, the tool can be an extremely valuable tool when examining the system. Without scan tools, diagnostic information would be difficult to obtain in some circuits and impossible to retrieve in others. Scan tools do not make the use of trouble code charts unnecessary, nor do they pinpoint the exact area in a circuit where the problem exists. However, this type of equipment will give more specific information, such as sensor input readings. An example of this would be the coolant sensor circuit. While the ECM looks at specific voltage levels to determine coolant temperature, the scan tool will display this information as an actual temperature reading. This information is not only useful in determining system malfunctions, but is also very helpful in diagnosing mechanical problems. For example, if the engine is overheating, the scan tool can be used to look at the actual engine temperature. This is helpful when checking for electric cooling fan enable switch problems, sticking thermostats, etc.

> **NOTE: A scan tool that is known to display faulty data should not be used for diagnosis. Although the fault may be believed to be in only one area, it can possibly affect many other areas during diagnosis, leading to errors and incorrect repair.**

ELECTRICAL TOOLS

The most commonly required electrical diagnostic tool is the digital multimeter, allowing voltage, resistance and amperage to be read by one instrument. The multimeter must be a high-impedance unit, with 10 megohms of impedance in the voltmeter. This type of meter will not place an additional load on the circuit it is testing; this is extremely important in low voltage circuits. The multimeter must be of high quality in all respects. It should be handled carefully and protected from impact or damage. Replace batteries frequently in the unit.

Other necessary tools include an unpowered test light, a quality tachometer with inductive (clip-on) pick up and the proper tools for releasing GM's Metri-Pack, Weather Pack and Micro-Pack terminals as necessary. The Micro-Pack connectors are used at the ECM connector. A vacuum pump/gauge may also be required for checking sensors, solenoids and valves.

Diagnosis and Testing

SERVICE PRECAUTIONS

NOTE: To avoid damage to the ECM or other ignition system components, do not use electrical test equipment such as battery or AC powered voltmeter, ohmmeter, etc. or any type of tester other than specified.

- When performing electrical tests on the system, use a high impedance multimeter, digital voltmeter (DVM) J-34029-A or equivalent.
- To prevent electrostatic discharge damage, when working with the ECM, do not touch the connector pins or soldered components on the circuit board.
- When handling a PROM or any electronic semiconductor device, do not touch the component leads. Also, do not remove the integrated circuit from the carrier.
- When performing electrical tests on the system, use a high impedance multimeter, digital voltmeter (DVM) J-34029-A or equivalent.
- Never pierce a high tension lead or boot for any testing purpose; otherwise, future problems are guaranteed.
- Leave new components and modules in the shipping package until ready to install them.
- Never disconnect any electrical connection with the ignition switch ON unless instructed to do so in a test.

PRELIMINARY INSPECTION

This is possibly the most critical step of diagnosis. A detailed examination of connectors, wiring and vacuum hoses can often lead to a repair without further diagnosis. Performance of this step relies on the skill of the technician performing it; a careful inspector will check:
- The undersides of hoses as well as the integrity of hard-to-reach hoses blocked by the air cleaner or other components.
- The wiring carefully for any sign of strain, burning, crimping, or terminal pull-out from a connector.
- The connectors at components or in harnesses as required; usually, pushing them together will reveal a loose fit.

READING CODES

When a scan tool is not available, a jumper can be inserted between terminals A and B of the DLC and the key moved to the ON position, allowing the codes to be retrieved from the system's memory. These codes can be interpreted by counting the flashes of the MIL which will display the message SERVICE ENGINE SOON (SES). To acknowledge diagnostic mode, the SES telltale will flash code 12 three times consecutively before moving on to stored trouble codes, if any exist. If there are codes in the system memory, they will be flashed in sequence three times consecutively, while the absence of stored codes will cause code 12 to flash continually. The MIL flash sequence occurs as follows: 12 = flash, pause, flash, flash, long pause (three times). Code 12 indicates that the ECM

diagnostic system is operating properly, and is not a malfunction code.

If more than one fault code is stored in memory, the codes will be displayed from the lowest to the highest, with the exception of code 11. Code 11 indicates transaxle codes and will always be flashed last if transaxle codes exist. Transaxle codes are not read with the SES light.

Saturn also uses information flags. These flags do not indicate a failure and will not turn on the SES telltale light. These flags are used as diagnostic aids to the technician, and can only be read with a scan tool.

The PCM has the ability to take a snapshot of 4 pieces of engine data with each trouble code stored. This valuable information can be retrieved with a Saturn scan tool or equivalent.

To exit the diagnostic mode, turn the ignition switch OFF and remove the jumper wire or scan tool.

CLEARING CODES

When a code is set, the ECM stores it in two tables: General Information and Malfunction History. When removing codes from General Information without a scan tool, turn the ignition ON and ground the DLC terminals A and B three times within five seconds. General Information codes will automatically be erased after 50 ignition ON cycles, as long as the problem does not re-occur. Malfunction History can only be cleared with a scan tool. Disconnecting power to the ECM does not effect on Malfunction History, but will clear any codes in General Information.

ENGINE CRANKS BUT DOES NOT START

Testing

1. Check for spark at each spark plug with a spark tester. Check at least 2 plug wires.
2. If spark is only detected at some plugs, check for a faulty distributor cap or rotor. Also check the spark plugs and plug wires. Replace, as necessary. If spark is detected on all plugs, check for fuel pressure.
3. If no spark is detected, connect a scan tool, crank the engine and monitor the RPM. If no RPM is indicated, refer to no RPM chart.
4. If RPM is present, Disconnect the 6 pin connector. Turn the key to the ON position. Measure the voltage on circuit 1039 to ground at 6 pin harness connector.
  a. If battery voltage is present, disconnect the 5 pin connector and check the resistance of circuit 450 to ground at EI pin harness connector. If resistance is

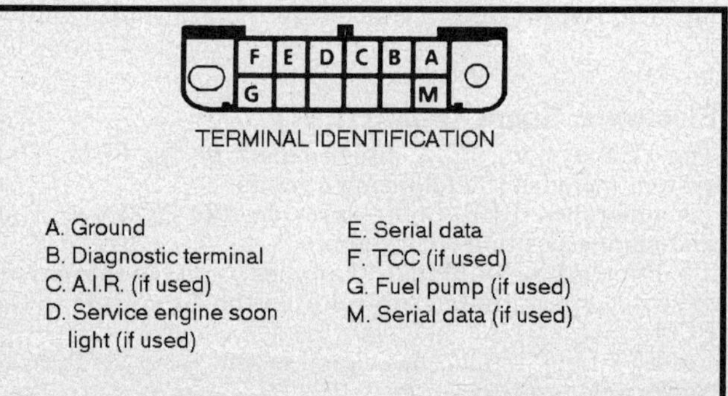

Data link connector (DLC)

below 200 ohms, then terminal is tight or EI module is defective. If resistance is above 200 ohms, there is an open in circuit 450.

b. If battery voltage is not present, check the EI fuse and check circuit 1039 for open or short to ground.

IGNITION COILS

Testing

1. Disconnect the negative battery cable.
2. Measure the resistance from tower to tower on the ignition coil. The resistance should be 7–10K ohms.
3. Connect the negative battery cable.

CRANKSHAFT POSITION SENSOR

Testing

1. Disconnect the negative battery cable.
2. Measure the resistance of the crankshaft position sensor. The resistance should be 700–900 ohms.
3. Connect the negative battery cable.

IGNITION TIMING

Ignition timing is controlled by the PCM and is not adjustable.

Component Replacement

CRANKSHAFT POSITION SENSOR

Removal and Installation

1. Disconnect the negative battery cable.
2. Raise and safely support the vehicle.
3. Disconnect the sensor electrical connector and remove the retaining bolts. Remove the sensor.

To install:

4. Lubricate the O-ring with clean engine oil.
5. Install crankshaft position sensor and retaining bolt. Tighten to 80 inch lbs. (9 Nm).
6. Connect the electrical connector. Lower the vehicle.
7. Connect the negative battery cable.

IGNITION MODULE

Removal and Installation

1. Disconnect the negative battery cable.
2. Tag and disconnect the spark plug wires from the coil towers.
3. Disconnect the connector at the ignition module.
4. Remove the retaining bolts and remove the module.

NOTE: Clean the ignition module mounting holes in the transaxle to remove any thread sealant residue. The tap size is M6 x 1.0 mm.

To install:

5. Install the ignition module using new bolts. Tighten to 71 inch lbs. (8 Nm).
6. Connect the ignition module electrical connector and spark plug wires to the ignition coils.
7. Connect the negative battery cable.

IGNITION COILS

Removal and Installation

1. Disconnect the negative battery cable.
2. Remove the ignition module assembly.
3. Squeeze the retaining tabs together while pulling up. Remove the coils.

To install:

4. Install the coils on the module.
5. Install the ignition module assembly.
6. Connect the negative battery cable.

SPARK PLUGS

Removal and Installation

1. Disconnect the negative battery cable.
2. Remove all foreign material from around the spark plugs.
3. Note the position of the plug wires, then twist the spark plug boot 1/2 turn and remove. Do not pull on the wires.
4. Remove the spark plugs.

To install:

5. Check and adjust the plug gaps.
6. Install the new plugs. Tighten and torque the plugs to 20 ft. lbs. (27 Nm).

NO RPM DETECTED DIAGNOSTIC CHART

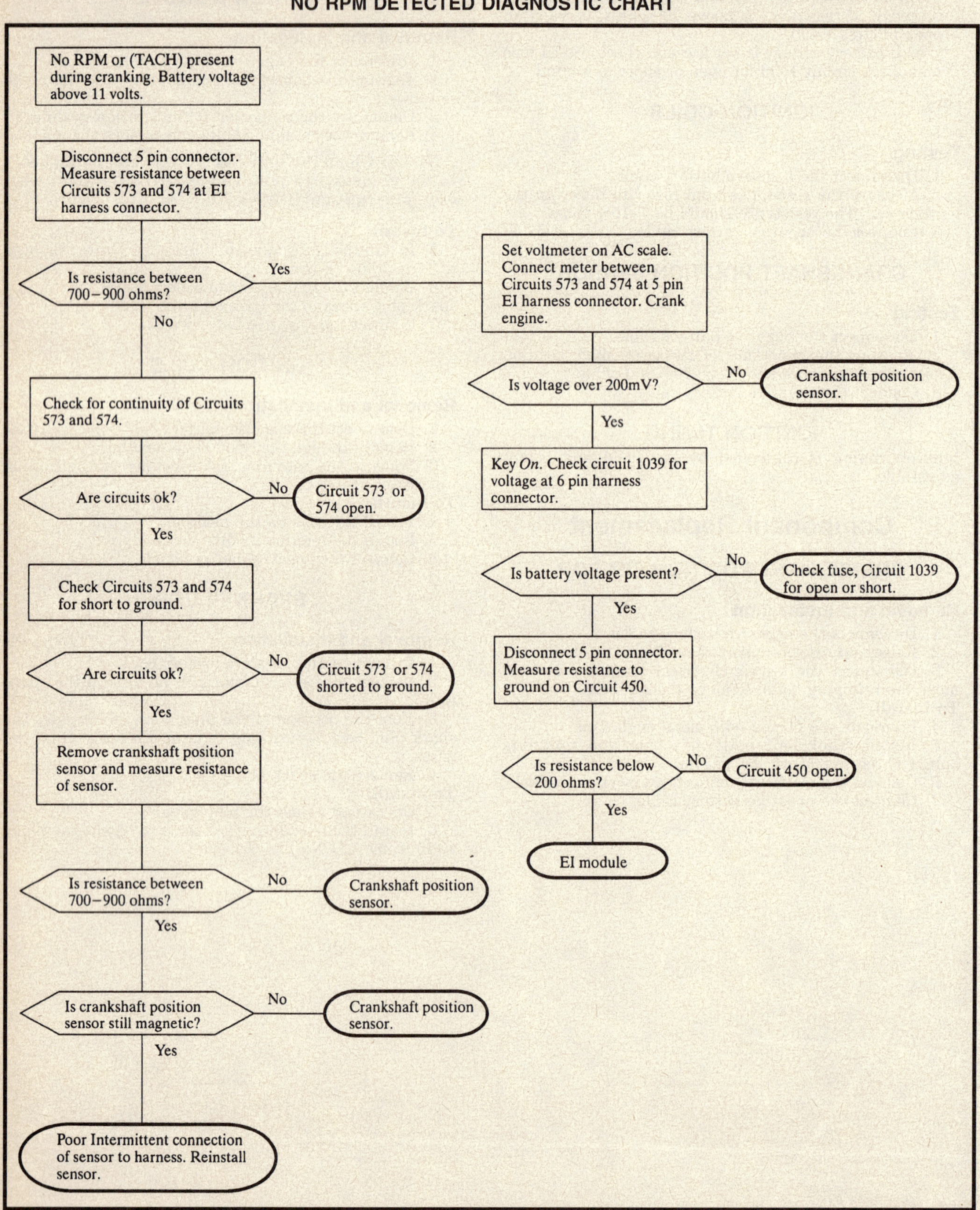

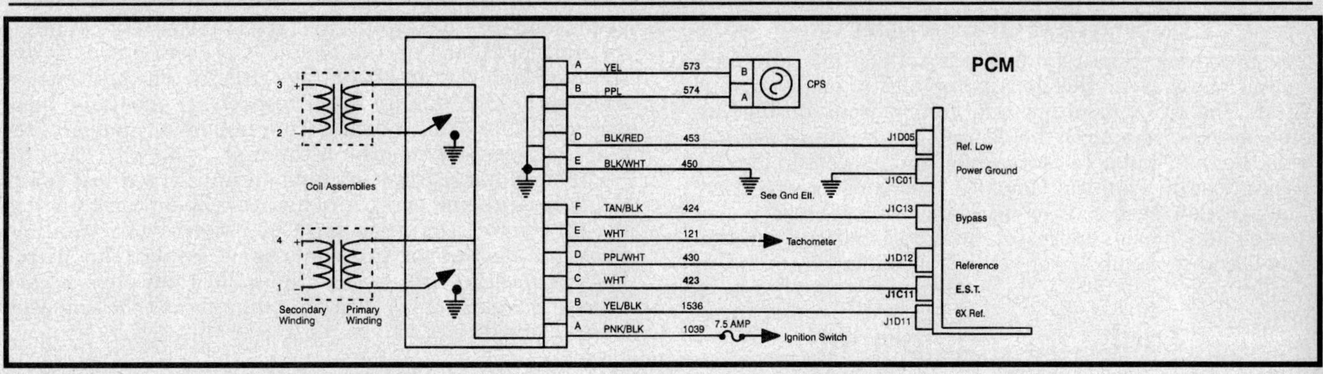

Saturn electronic ignition system

Crankshaft position sensor removal and installation

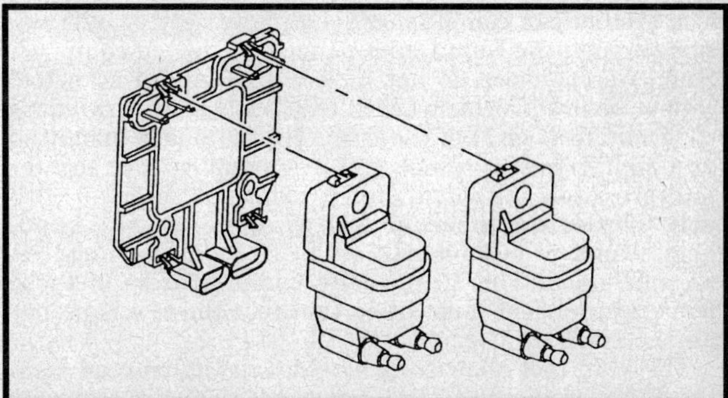

Ignition module removal and installation

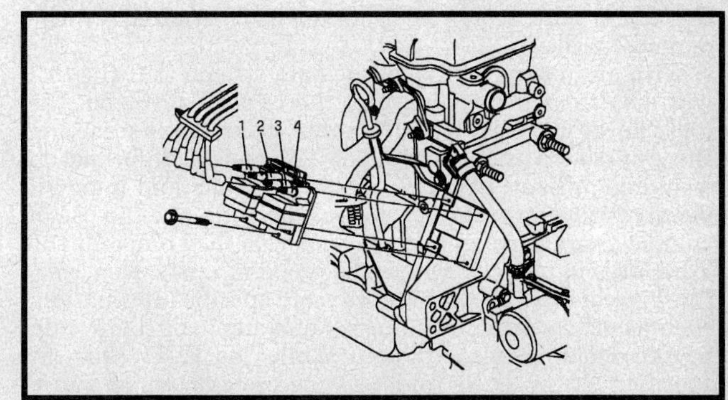

Ignition coil removal and installation

ELECTRONIC SPARK CONTROL (ESC) SYSTEMS

General Description

The Geo automobiles utilize the Electronic Spark Control Ignition System. The system consists of a distributor, igniter, spark plugs, primary and secondary wiring and a camshaft position sensor or pick-up coil. The system is monitored and controlled by the Engine Control Module (ECM).

NOTE: The terms Electronic Control Module (ECM), Powertrain Control Module (PCM) and Body Control Module (BCM) may sometimes be used interchangeably.

SYSTEM OPERATION

The Electronic Spark Control system controls all spark timing changes in the distributor electronically by the ECM. The ECM monitors information from various engine sensors, computes the desired spark timing and signals the distributor to change the timing accordingly. No vacuum or mechanical advance mechanisms are used. The ignition system also uses a noise suppressor condenser and noise suppressor filter to eliminate ignition sound and tachometer reference if so equipped.

Tools and Equipment

SCAN TOOLS

The system can communicate a wide variety of information through the Data Link Connector (DLC), if equipped. Some Geo vehicles do not have a DLC and a scan tool cannot be used. Vehicles with DLC's transmit serial data to terminals E or M of the DLC. This data is transmitted at a high frequency which requires a quality scan tool for interpretation. Although stored codes may be read with only the use of a small jumper wire, the use of a hand-held scan tool such as GM's TECH 1 or equivalent, is recommended. There are many manufacturers of these tools; a purchaser must be certain that the tool is proper for the intended use.

The scan tool allows any stored codes to be read from the ECM memory. The tool also allows the operator to view the data being sent to the ECM while the engine is running. This ability has obvious diagnostic advantages; the use of the scan tool is frequently required by the diagnostic charts.

With an understanding of the data stream that the tool will display and knowledge of the circuits involved, the tool can be an extremely valuable tool when examining the system. Without scan tools, diagnostic information would be difficult to obtain in some circuits and impossible to retrieve in others. Scan tools do not make the use of trouble code charts unnecessary, nor do they pinpoint the exact area in a circuit where the problem exists. However, this type of equipment will give more specific information, such as sensor input readings. An example of this would be the coolant sensor circuit. While the ECM looks at specific voltage levels to determine coolant temperature, the scan tool will display this information as an actual temperature reading. This information is not only useful in determining system malfunctions, but is also very helpful in diagnosing mechanical problems. For example, if the engine is overheating, the scan tool can be used to look at the actual engine temperature. This is helpful when checking for electric cooling fan enable switch problems, sticking thermostats, etc.

> **NOTE: A scan tool that is known to display faulty data should not be used for diagnosis. Although the fault may be believed to be in only one area, it can possibly affect many other areas during diagnosis, leading to errors and incorrect repair.**

ELECTRICAL TOOLS

The most commonly required electrical diagnostic tool is the digital multimeter, allowing voltage, resistance and amperage to be read by one instrument. The multimeter must be a high-impedance unit, with 10 megohms of impedance in the voltmeter. This type of meter will not place an additional load on the circuit it is testing; this is extremely important in low voltage circuits. The multimeter must be of high quality in all respects. It should be handled carefully and protected from impact or damage. Replace batteries frequently in the unit.

Other necessary tools include an unpowered test light, a quality tachometer with inductive (clip-on) pick up and the proper tools for releasing GM's Metri-Pack, Weather Pack and Micro-Pack terminals as necessary. The Micro-Pack connectors are used at the ECM connector. A vacuum pump/gauge may also be required for checking sensors, solenoids and valves.

Diagnosis and Testing

SERVICE PRECAUTIONS

> **NOTE: To avoid damage to the ECM or other ignition system components, do not use electrical test equipment such as battery or AC powered voltmeter, ohmmeter, etc. or any type of tester other than specified.**

- When performing electrical tests on the system, use a high impedance multimeter, digital voltmeter (DVM) J-34029-A or equivalent.
- To prevent electrostatic discharge damage, when working with the ECM, do not touch the connector pins or soldered components on the circuit board.
- When handling a PROM or any electronic semiconductor device, do not touch the component leads. Also, do not remove the integrated circuit from the carrier.
- When performing electrical tests on the system, use a high impedance multimeter, digital voltmeter (DVM) J-34029-A or equivalent.
- Never pierce a high tension lead or boot for any testing purpose; otherwise, future problems are guaranteed.
- Leave new components and modules in the shipping package until ready to install them.
- Never disconnect any electrical connection with the ignition switch **ON** unless instructed to do so in a test.

PRELIMINARY INSPECTION

This is possibly the most critical step of diagnosis. A detailed examination of connectors, wiring and vacuum hoses can often lead to a repair without further diagnosis. Performance of this step relies on the skill of the technician performing it; a careful inspector will check:
- The undersides of hoses as well as the integrity of hard-to-reach hoses blocked by the air cleaner or other components.
- The wiring carefully for any sign of strain, burning, crimping, or terminal pull-out from a connector.
- The connectors at components or in harnesses as required; usually, pushing them together will reveal a loose fit.

READING CODES

With Scan Tool

On all but the Prizm, the DTC's can be retrieved by turning the key to the **OFF**, plugging in a scan tool, then turning the key **ON** and retrieving codes on the scan tool.

Without Scan Tool

METRO — USING MIL

1. Turn the key **ON**. Do not start the engine.
2. Ground either the DIAG SW connector in the junction block or the diagnostic request/switch terminal in the underhood DUTY CHECK DLC.
3. Count the flashes of the MIL (i.e. 12 = flash, pause, flash, flash).

METRO — USING TEST LIGHT

NOTE: The test light method will only work on models with a 6 pin DUTY CHECK DLC.

1. Turn the key **ON**. Do not start the engine.
2. Connect a test light to the diagnostic output terminal.
3. Ground either the DIAG SW connector in the junction block or the diagnostic request/switch terminal in the underhood DUTY CHECK DLC.
4. Count the test light flashes. (i.e. 12 = flash, pause, flash, flash).

TRACKER

1. Turn the key **ON**. Do not start the engine.
2. To retrieve trouble codes on systems equipped with the 4 pin ECM check connector, insert a jumper between terminal 3 (diagnostic request) and terminal 4 (ground), then count the MIL flashes (i.e. flash, pause, flash, flash=code 12). Code 12 will only flash if there are no stored trouble codes.
3. Diagnostic information can also be retrieved by grounding the diagnostic request terminal in the DUTY CHECK DLC (terminal 2) and counting the MIL flashes (i.e. 12 =flash, pause, flash, flash).

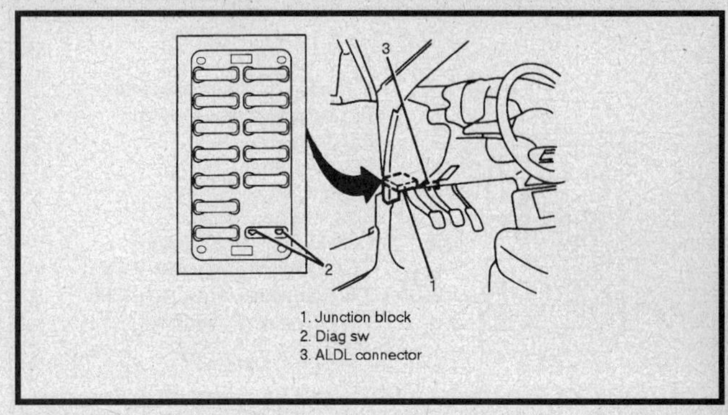

DLC and junction block — Metro

PRIZM

1. Turn the key **ON**. Do not start the engine.
2. Ground the diagnostic request terminal TE1 in the DLC (terminal E1 is the ground terminal).

NOTE: Because code 12 is used as a malfunction code, a steady series of MIL flashes will be used to indicate a proper functioning system with no stored trouble codes.

3. The flash sequence should be a continuous flash, pause, flash, pause, etc. if there are no stored codes. If a code exists, it will flash in the usual manor. For example, code 12 = flash, pause, flash, flash, long pause.

STORM

1. Turn the key **ON**. Do not start the engine.
2. Insert a jumper wire between terminals 1 and 3 of the DLC.

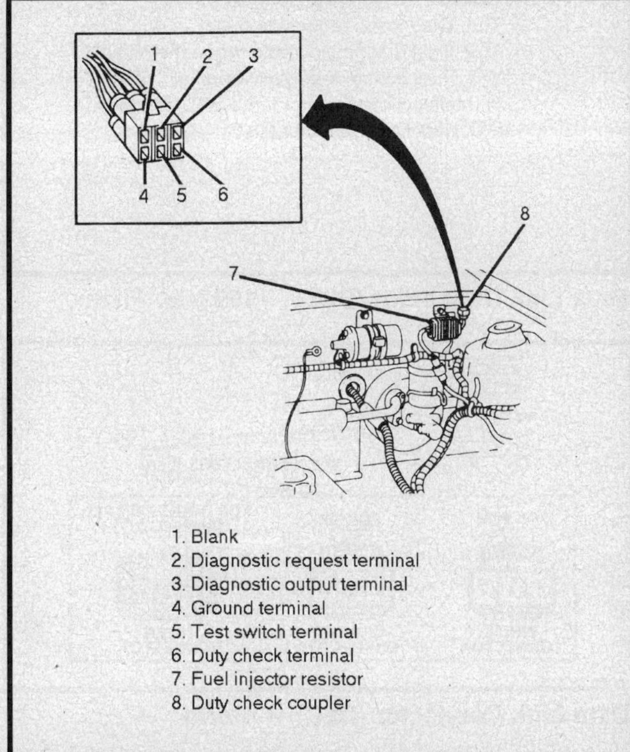

Data Link Connector (DLC) — Metro

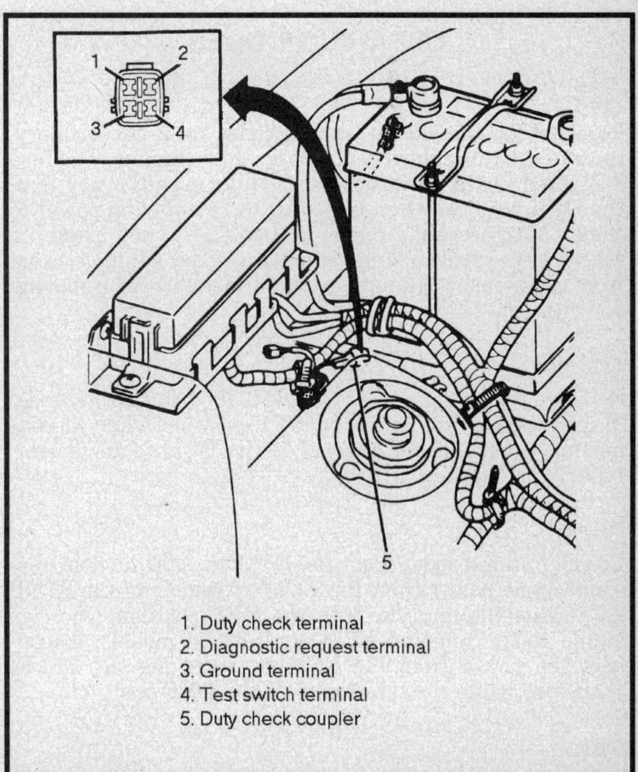

Duty Check Connector — Tracker

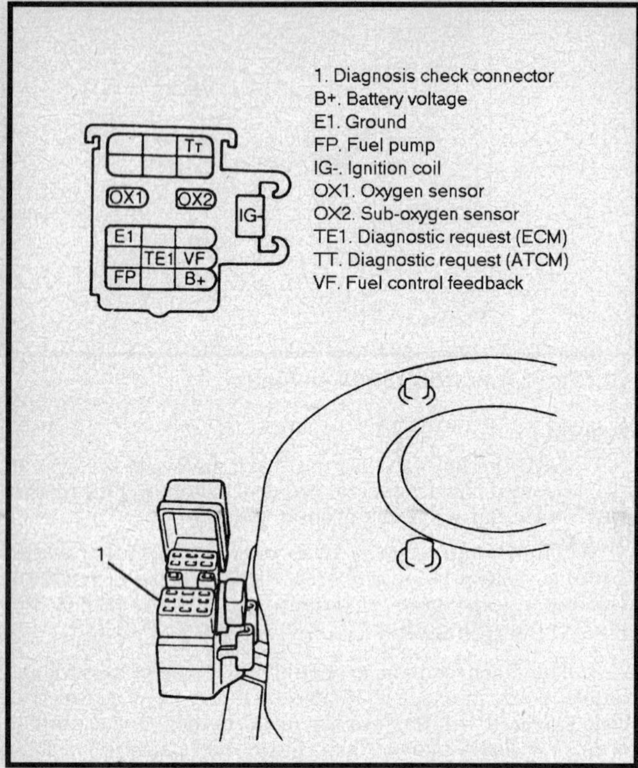

Data Link Connector (DLC) — 1992 Prizm

3. Count the MIL flashes (i.e. 12 = flash, pause, flash, flash).

CLEARING CODES

Metro

To clear the diagnostic trouble codes from the memory, remove the tail light fuse for at least 30 seconds. The fuse is located in the junction block. Make sure the key is in the **OFF** position when removing or re-applying power to the ECM to prevent damage to the ECM. When using the scan tool to retrieve diagnostic data, codes can be cleared from the memory through the scan tool without removing power to the ECM.

Tracker

Remove the 15 amp TAIL/DOME fuse for 30 seconds. Make sure the key is in the **OFF** position when disconnecting or reconnecting power to the ECM to avoid computer damage.

Prizm

To clear stored codes from the ECM on 1992 models, disconnect the power from the ECM by removing the STOP fuse while the ignition is in the **OFF** position. To clear stored codes from the ECM on 1993–94 models, disconnect the power from the ECM by removing the EFI/F-HTR fuse while the ignition is in the **OFF** position.

Storm

DTC's can be cleared from the system memory by removing power to the ECM with the key **OFF** for approxi-

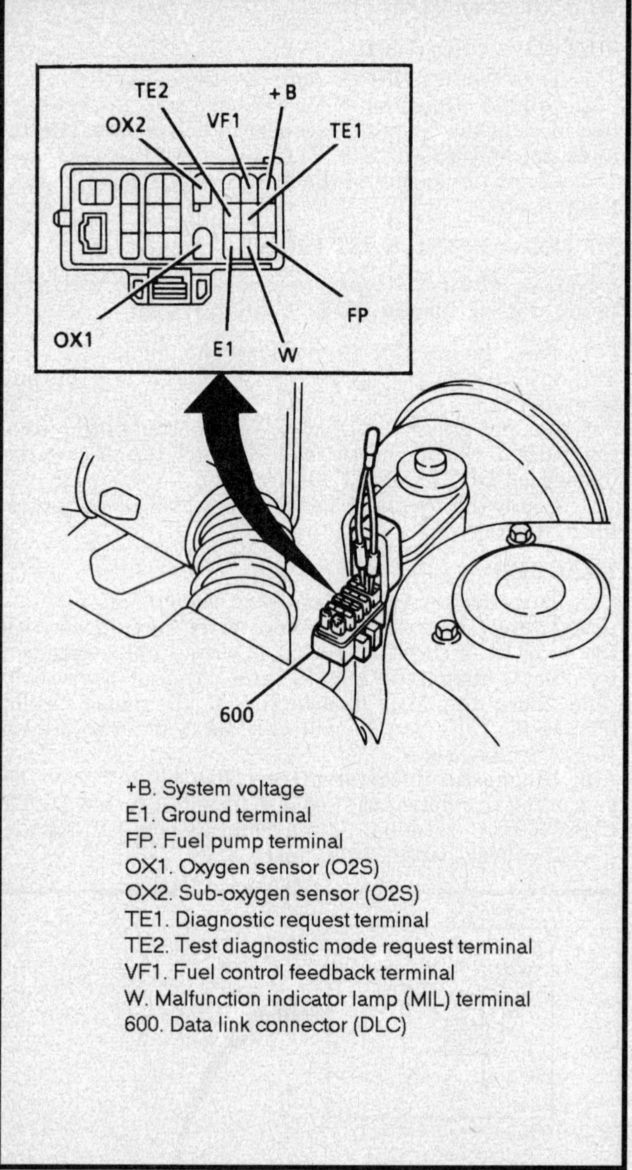

+B. System voltage
E1. Ground terminal
FP. Fuel pump terminal
OX1. Oxygen sensor (O2S)
OX2. Sub-oxygen sensor (O2S)
TE1. Diagnostic request terminal
TE2. Test diagnostic mode request terminal
VF1. Fuel control feedback terminal
W. Malfunction indicator lamp (MIL) terminal
600. Data link connector (DLC)

Data Link Connector (DLC) — 1993–94 Prizm

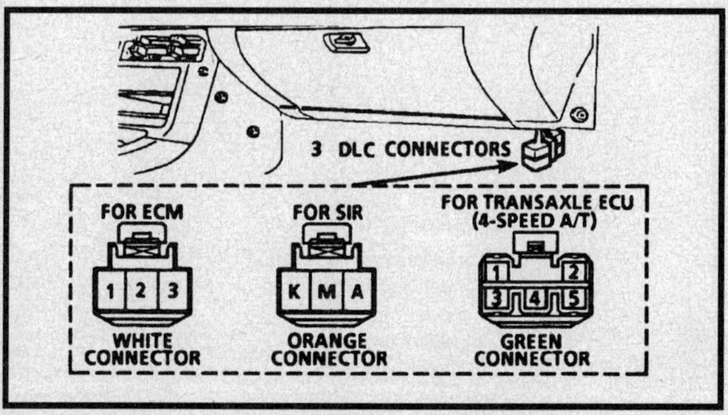

Data Link Connector (DLC) — Storm

mately 30-45 seconds. This can be accomplished by removing the 30 amp main fuse from the relay/fuse center in the engine compartment. The scan tool can also be used to clear codes.

COMPONENT PARAMETER

System	Test Points	Reading
GEO ESC (Except Tracker and 1993–94 Storm)	Primary Windings	1.3–1.6 Ohms
ESC (Tracker)	Primary Windings	0.72–0.88 Ohms
ESC (93–94 Storm)	Primary Windings	0.7–1.4 Ohms
HEI (Distributor Mount Coil)	C – to B + Tower to Ground	0 Ohms Infinite
HEI (External Mount Coil)	C + to Ground C + to C – C + Tower	Infinite 0 Ohms Infinite
Saturn (EZ)	Tower–Tower	7–10K Ohms

NOISE FILTER PARAMETERS

NOISE FILTER

System	Test Points	Reading
GEO (ESC)	Filter to Ground	2.2K Ohms

SPARK PLUG AND COIL WIRES

Testing

1. Remove the distributor cap and spark plug wires from the spark plugs.
2. Invert the cap and measure the resistance of each spark plug wire and the coil wire, using an ohmmeter.
3. If the resistance is not 3–6.5 kilo-ohms/ft. (10–25 kilo-ohms/meter), replace the wire.

IGNITION COIL

Testing

1. Disconnect the negative battery cable.
2. Remove the ignition coil wire.
3. Measure the resistance of the primary winding. Reading should be:
   - Except Tracker and 1993–94 Storm — should be between 1.3–1.6 ohms.
   - Tracker — should be between 0.72–0.88 ohms.
   - 1993–94 Storm — should be between 0.7–1.4 ohms.
4. Measure the resistance of the secondary winding. Reading should be between 10.7–14.5 kilo-ohms.
5. If either readings are not within specification, replace the ignition coil.

NOISE FILTER AND CONDENSER

Testing

NOTE: The Storm is not equipped with a noise filter and condenser.

1. Disconnect the negative battery cable.
2. Disconnect the noise filter/condenser electrical leads.
3. Check the noise resistance between terminals B and C of the noise/filter condenser connector.
4. If the noise filter resistance is not approximately 2.2 kilo-ohms, replace the noise filter.
5. Check for continuity between terminals A and B of the noise filter/condenser connector.
6. If the condenser is conductive, replace the noise filter and condenser.

CAMSHAFT POSITION SENSOR (CMP)

Testing

TRACKER

1. Turn the ignition switch **OFF**.
2. Disconnect the yellow 24 cavity (C2) connector from the ECM.
3. Remove the distributor cap, rotor and signal rotor cover.
4. Connect a digital multimeter between terminal 13 (C2 connector) and terminal 1 of the green 17 cavity (C1) connector. Leave C1 connector connected to the ECM.

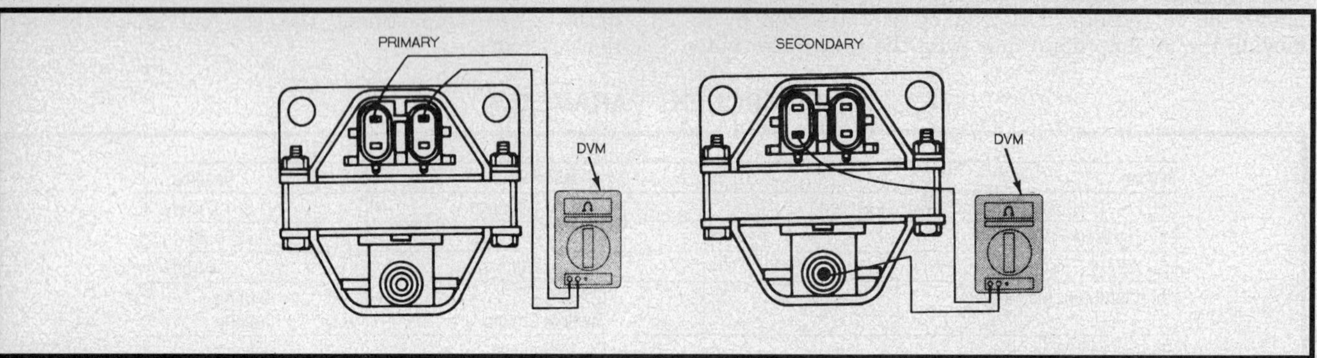

Testing the ignition coil

5. Turn the ignition switch **ON**.

6. Measure the CMP voltage with the signal rotor inserted between the hall element and magnet (magnetic flux cut off). Should indicate 0.0–1.0 volts.

7. If not, check driveability symptoms.

8. If OK, check the CMP voltage without the signal rotor inserted between the hall element and magnet (magnetic flux applied). Should indicate battery voltage.

9. If battery voltage is indicated, the CMP system is operating properly.

10. If battery voltage is not detected, check driveability symptoms.

PRIZM

1. Disconnect the negative battery cable.
2. Remove the distributor cap and rotor.
3. Remove the ignition coil dust cover and gasket.

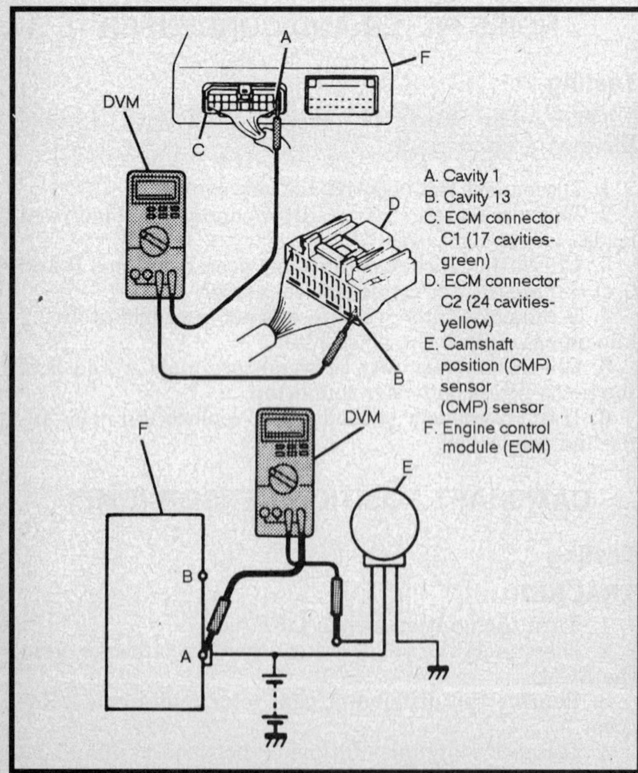

Testing camshaft position sensor — Tracker

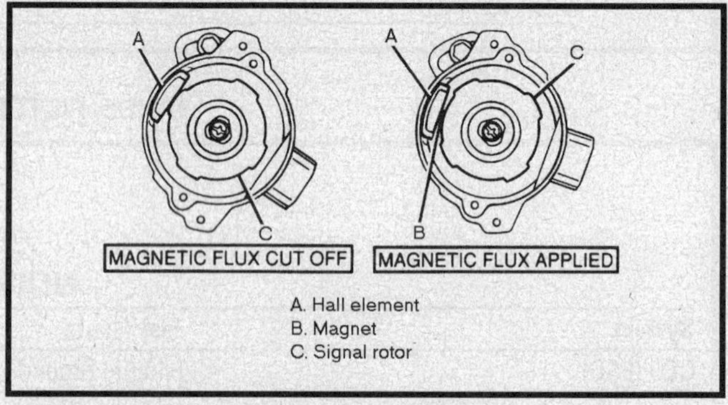

Magnetic flux conditions — Tracker

4. Remove 2 nuts and 4 wires from the ignition coil terminals.

5. Remove 4 ignition coil retaining screws and remove the coil.

6. Remove the 3 screws and 3 wires from the igniter. Remove the 2 igniter retaining screws and remove the igniter.

7. Remove the 2 connectors from the harness retainer and 1 retaining screw.

8. Remove noise suppressor retaining screw and remove suppressor.

9. Check the resistance between terminals 3 and 6 on connector C1. If not within range, replace the distributor housing.

   a. When engine is cold 14°–104° F (-10°–40° C), the resistance should be 185–275 ohms.

   b. When engine is hot 104°–212° F (40°–100° C), the resistance should be 240–325 ohms.

10. Check the resistance between terminals 2 and 5 on connector C1. If not within range, replace the distributor housing.

   a. When engine is cold 14°–104° F (-10°–40° C), the resistance should be 370–550 ohms.

   b. When engine is hot 104°–212° F (40°–100° C), the resistance should be 475–650 ohms.

11. Check the air gap between the signal rotor and the camshaft position sensor. If not within 0.008–0.016 inch (0.2–0.4 mm), Replace the distributor housing.

STORM

1. Disconnect the negative battery cable.
2. Remove the distributor cap and rotor.

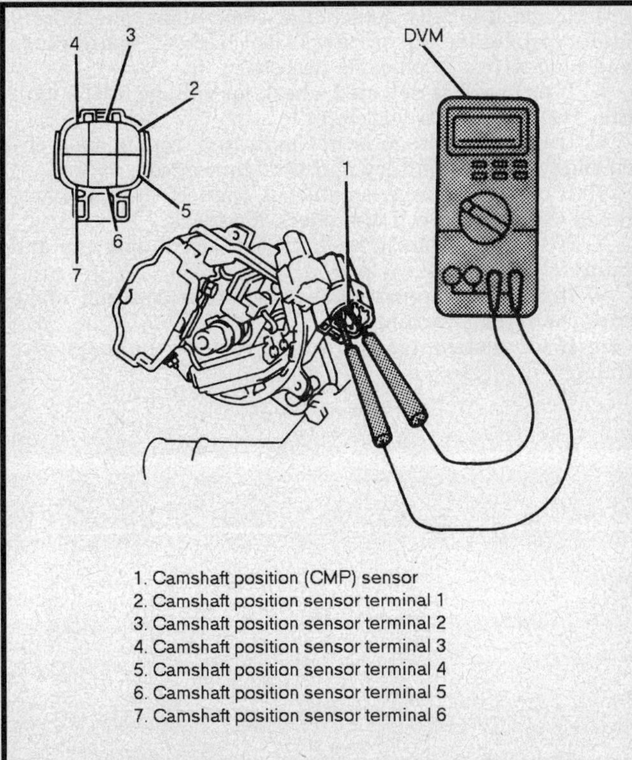

1. Camshaft position (CMP) sensor
2. Camshaft position sensor terminal 1
3. Camshaft position sensor terminal 2
4. Camshaft position sensor terminal 3
5. Camshaft position sensor terminal 4
6. Camshaft position sensor terminal 5
7. Camshaft position sensor terminal 6

Testing camshaft position sensor — Prizm

3. Check the resistance between either camshaft position sensor electrical lead and ground. If the resistance is not infinite, replace the sensor.

4. Check the resistance between the 2 sensor terminals. If the resistance is not between 500 and 1500 ohms, replace the sensor.

METRO

1. Disconnect the negative battery cable.

2. Disconnect the camshaft position sensor connector.

3. Measure the resistance between the 2 sensor terminals.

4. If the resistance is not between 140–180 ohms at approximately 68°F (20°C), replace the sensor.

IGNITION TIMING

Adjustment

Always set the timing according to the procedure indicated in the Vehicle's Emission Control Information label. Follow all instructions on the label.

NOTE: Changes in timing specifications may be made during a production run. Therefore, always check the VECI to verify timing specifications.

1. Start and run the engine until normal operating temperature is reached.

2. Turn all electrical accessories **OFF**.

3. Check that the engine idle speed is 750±50 rpm (vehicles equipped with manual transaxle) or 850±50 rpm (vehicles equipped with automatic transaxle).

4. If the engine idle speed is not within specification, turn the throttle body idle speed adjusting screw until the specified idle speed is obtained.

5. Turn the ignition switch **OFF**. Connect the pick-up lead of the timing light to the No.1 spark plug lead. Use a jumper lead between the wire and plug or an inductive type pick-up.

NOTE: Do not pierce the wire or attempt to insert a probe between the boot and the wire. Piercing a wire may create a failure condition that will not be immediately apparent.

6. Connect the timing light power leads according to the manufacturer's instructions.

7. Install jumper wire as follows:
   • Metro — remove the diagnostic connector cap, located next to the ignition coil. Insert a jumper between connector terminals D and E.
   • Tracker — remove the diagnostic connector cap, located next to the battery. Insert a jumper between connector terminals C and D.
   • Storm — connect a fused jumper wire between terminals 1 and 3 of the DLC connector, located under the right side instrument panel.
   • Prizm — connect a fused jumper wire between terminals E1 and TE1 of the DLC connector, located next to the master cylinder.

8. Start the engine, then aim the timing light at the timing mark. The line on the pulley should line up at the timing mark.

9. If adjustment is necessary, loosen the distributor holddown clamp bolt at the base of the distributor. Slightly rotate the distributor until the line indicates the correct timing, according to the specification on the vehicle emission control label in the engine compartment. Then, tighten the distributor holddown clamp bolt. Recheck the timing.

10. Remove the jumper wire from the diagnostic connector and reinstall the cap. Check to ensure the advanced timing is within specification.

11. Turn the ignition switch **OFF**. Remove the timing light and reconnect the No. 1 spark plug wire, if removed.

ENGINE CRANKS BUT DOES NOT START

Testing

EXCEPT STORM

1. Check for any stored codes. If codes exist, repair necessary circuit.

2. Check for spark at the ignition coil with a spark tester. If no spark, replace the coil.

3. Check for spark at each spark plug with a spark tester. Check at least 2 plug wires.

4. If spark is only detected at some plugs, check the distributor, rotor, cap and plug wires.

5. If no spark is detected, check for a faulty distributor cap or rotor. Also check the coil to distributor wire. Replace, if necessary.

STORM

1. Check for any stored codes. If codes exist, repair necessary circuit.

2. Install the scan tool and check the Throttle Position Sensor (TPS) — if over 2.5 volts, at closed throttle, use Chart Code 21 in the fuel section. Also check the engine coolant — if less than -22°F (-30°C), use Chart Code 14 in the fuel section.

3. Check for spark at each spark plug with a spark tester. Check at least 2 plug wires.

4. If spark is only detected at some plugs, check for a faulty distributor cap or rotor. Also check the spark plugs and plug wires. Replace, is necessary.

5. If no spark is detected, check for voltage at the ignition system using a voltmeter.

6. If battery voltage is not indicated, repair open circuit between the battery and the ignition coil.

7. If battery voltage is detected, connect the spark tester to the ignition coil and check for spark.

8. If spark is detected, Check the distributor cap and rotor.

9. If no spark, remove the distributor cap and make sure the rotor is turning.

10. If the distributor is turning, refer to the diagnostic chart.

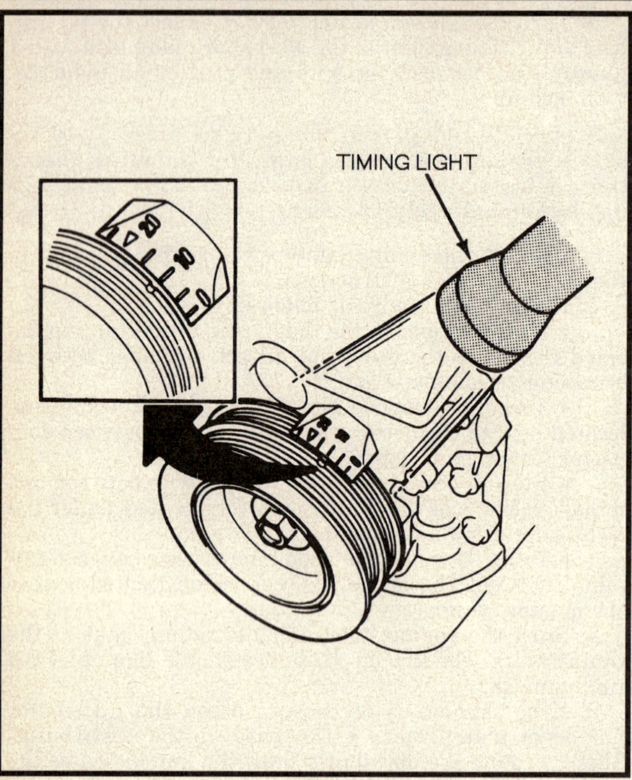

Checking the timing — Prizm

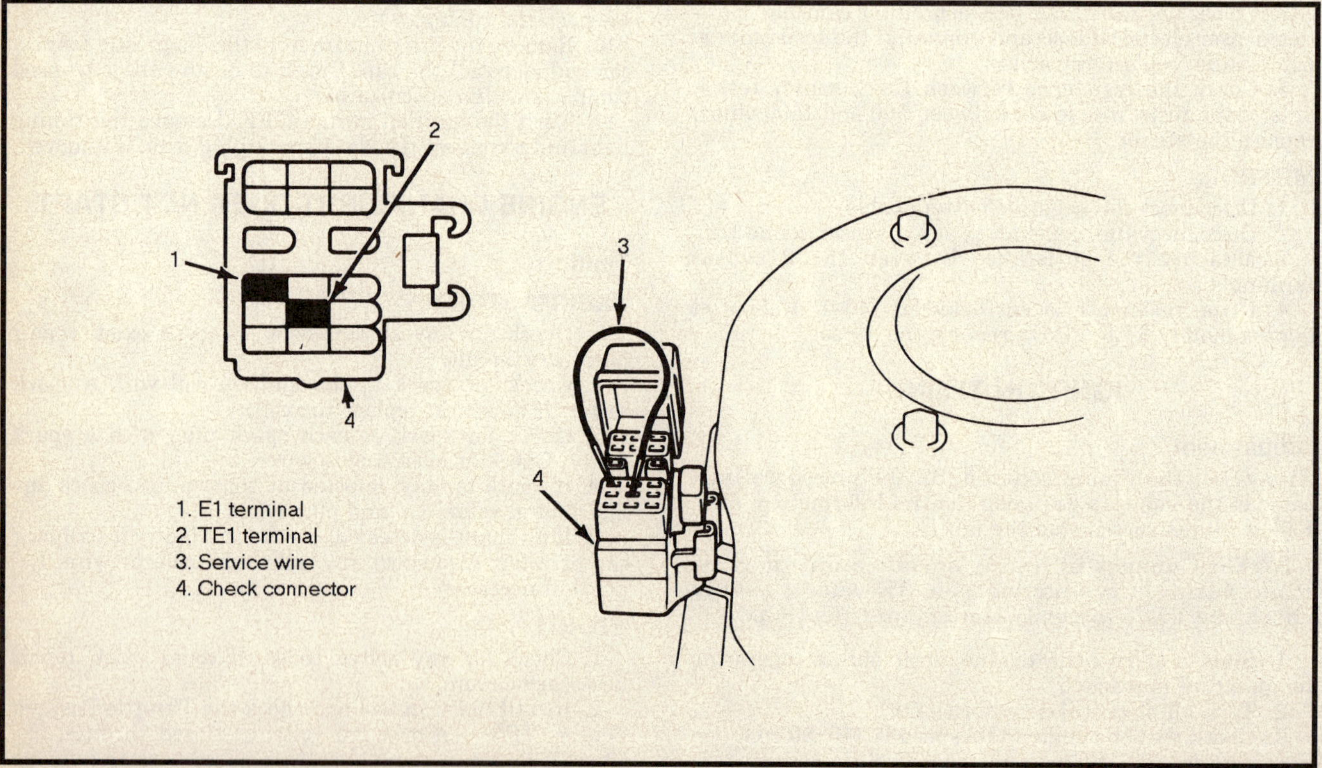

Jumper wire in check connector — Metro

IGNITION DIAGNOSTIC CHART — STORM

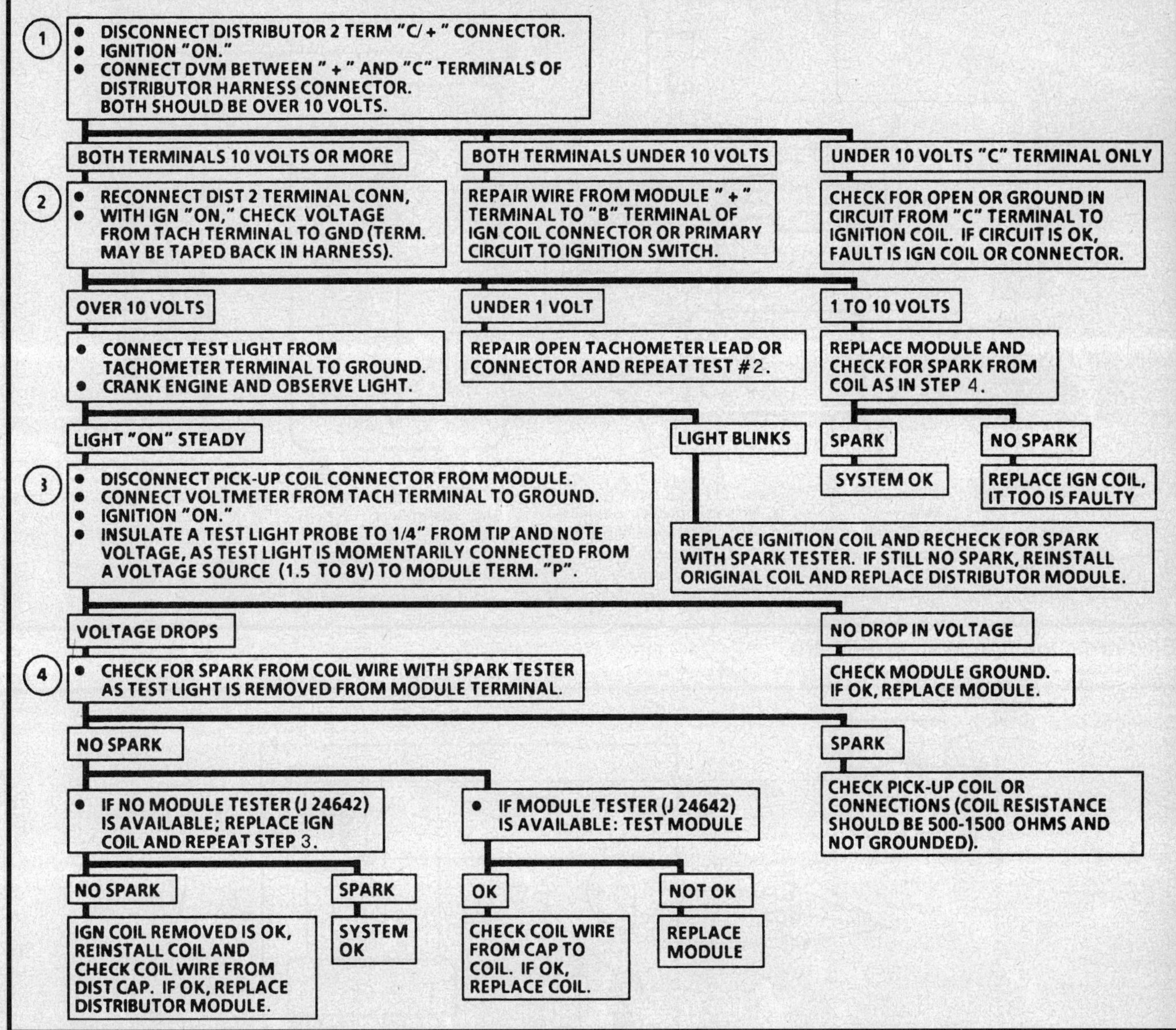

ELECTRONIC IGNITION SYSTEMS
ELECTRONIC SPARK CONTROL (ESC) SYSTEMS

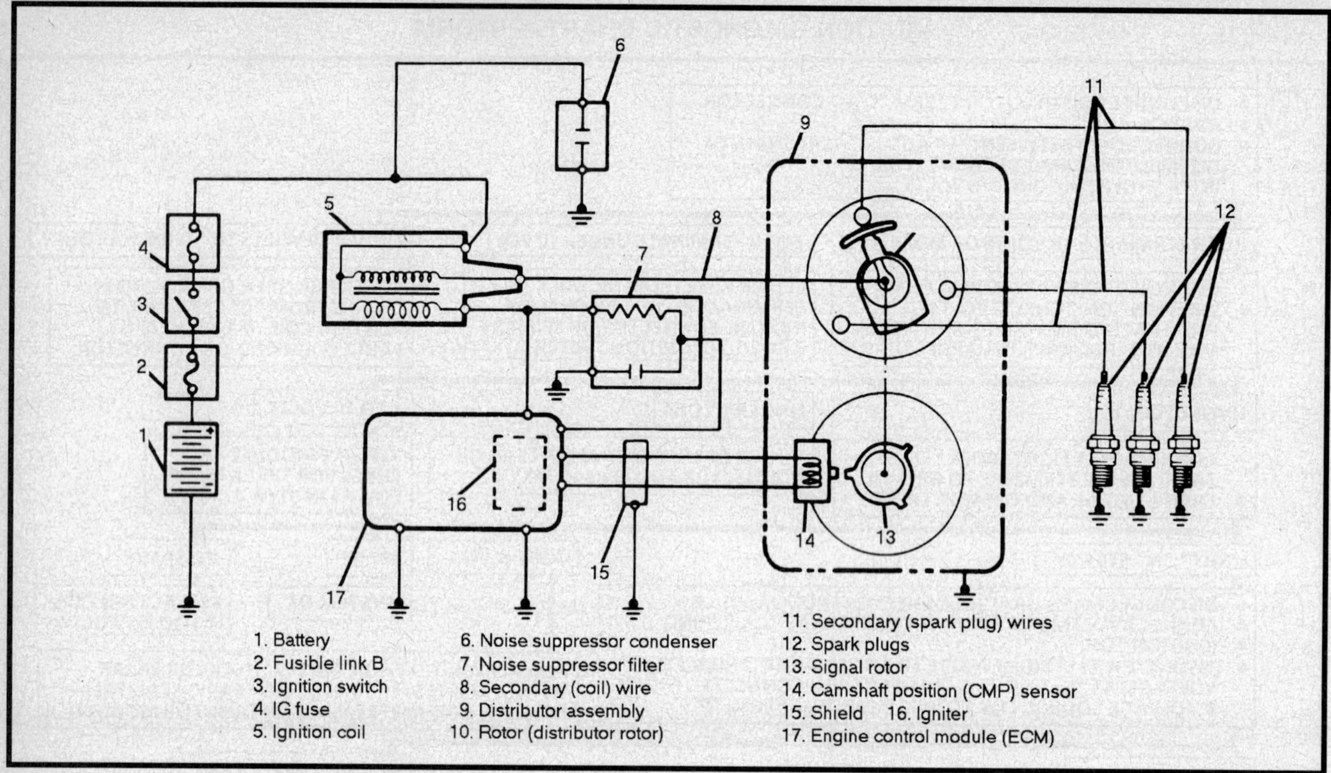

Electronic ignition system — Metro

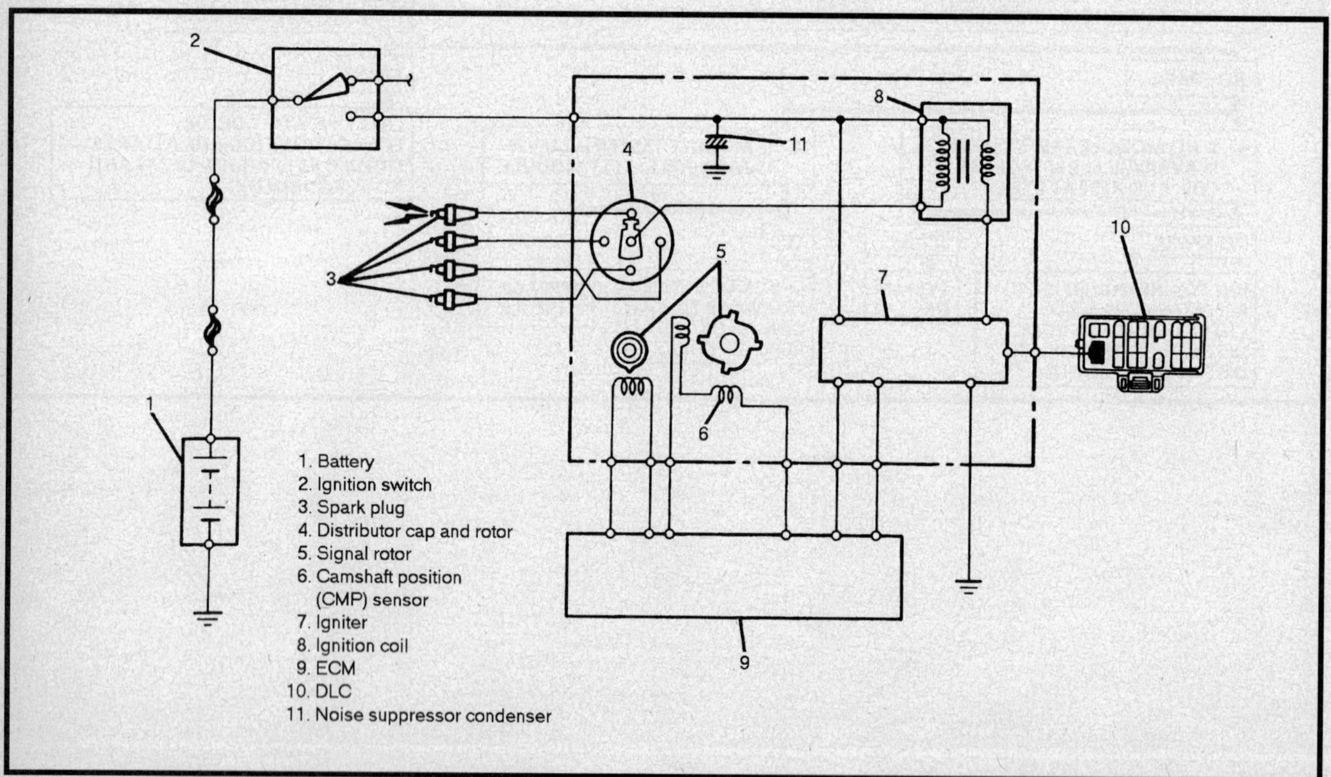

Electronic ignition system — 1993–94 Prizm

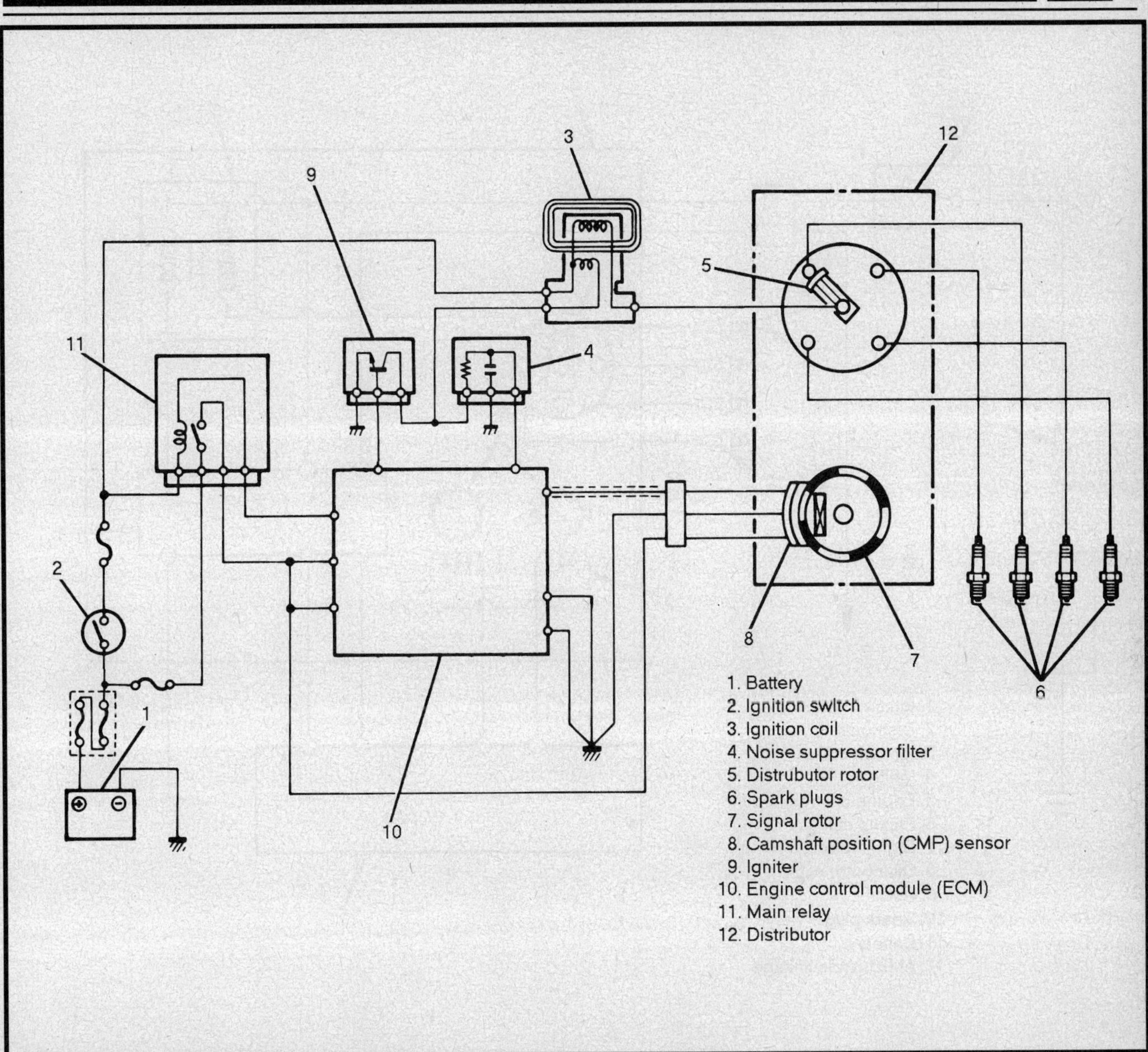

Electronic ignition system — Tracker

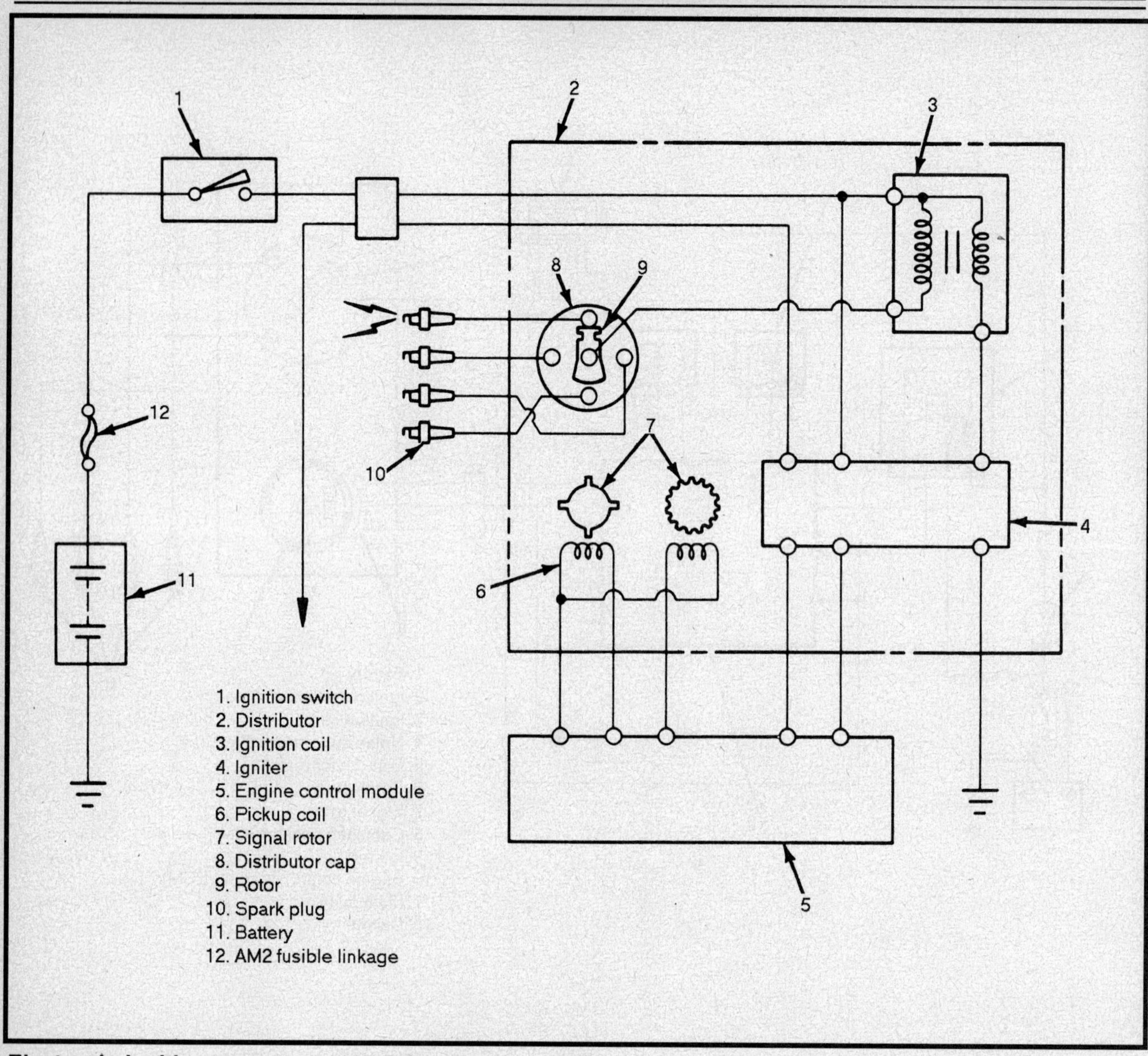

Electronic ignition system — 1992 Prizm — except GSi

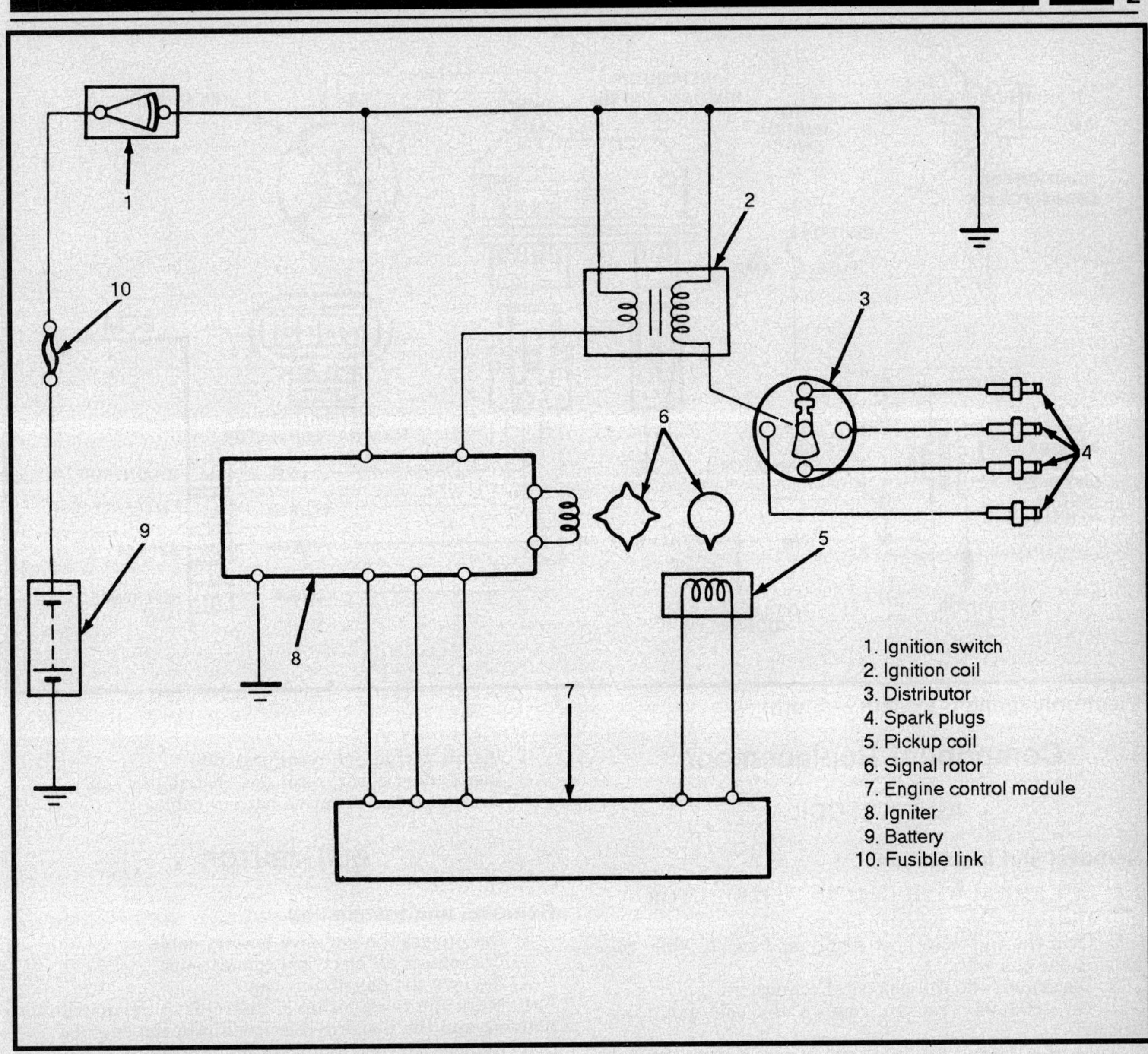

Electronic ignition system — 1992 Prizm GSi

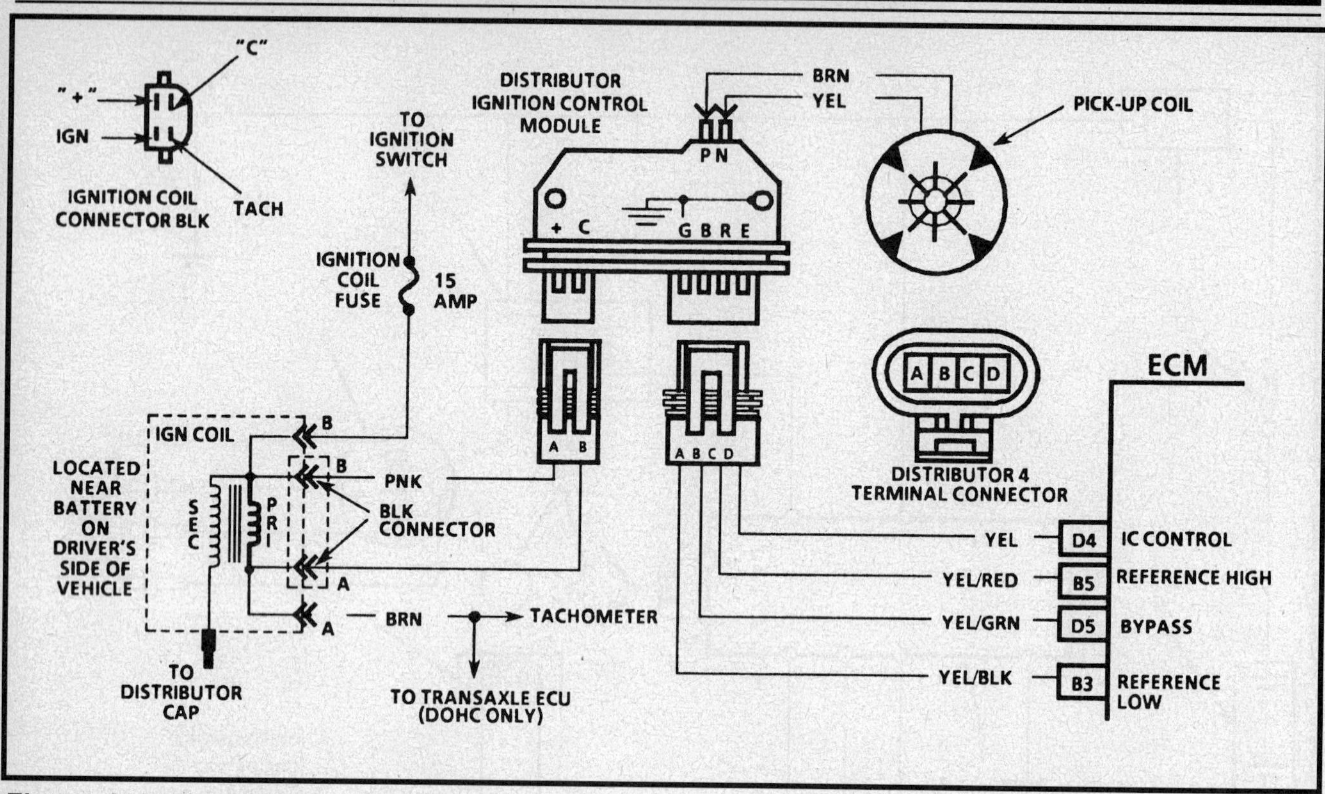

Electronic ignition system — Storm

Component Replacement

IGNITION COIL

Removal and Installation

EXCEPT PRIZM WITH COIL IN DISTRIBUTOR

1. Disconnect the negative battery cable.
2. Grip the coil wire boot firmly and twist, while removing the coil wire.
3. Disconnect the coil electrical connectors.
4. On 1993–94 Tracker, remove the coil mounting bracket.
5. Loosen the coil retaining screw and remove the coil assembly.

To install:

6. Install the ignition coil and tighten the retaining screws.
7. On 1993–94 Tracker, install the coil mounting bracket.
8. Reconnect the coil electrical connectors.
9. Install the coil wire and press the boot firmly.
10. Reconnect the negative battery cable.

PRIZM WITH COIL IN DISTRIBUTOR

1. Disconnect the negative battery cable.
2. Remove the distributor cap and rotor.
3. Remove the ignition coil dust cover.
4. Remove wiring and 2 wire retaining nuts.
5. Remove the coil retaining bolts and remove the ignition coil and gasket.

To install:

6. Install coil and gasket and tighten the retaining screws.

7. Install wiring and retaining nuts.
8. Install dust cover, rotor and distributor cap.
9. Reconnect the negative battery cable.

DISTRIBUTOR

Removal and Installation

1. Disconnect the negative battery cable.
2. Disconnect all electrical connections.
3. Remove the distributor cap.
4. Mark the relationship of the rotor to the distributor housing and the housing relationship to the engine.
5. Remove the distributor retaining bolt and hold-down clamp.
6. Remove the distributor from the engine. Discard the distributor O-ring.

To install:

7. Fit the distributor with a new O-ring. Insert the distributor into the engine, with the marks previously made aligned.
8. Install the distributor hold-down clamp and retaining bolt.
9. Reconnect all electrical connections.
10. Reconnect the negative battery cable.

CAMSHAFT POSITION SENSOR

Removal and Installation

EXCEPT METRO

The camshaft position sensor is an integral part of the distributor housing. The distributor housing assemblies

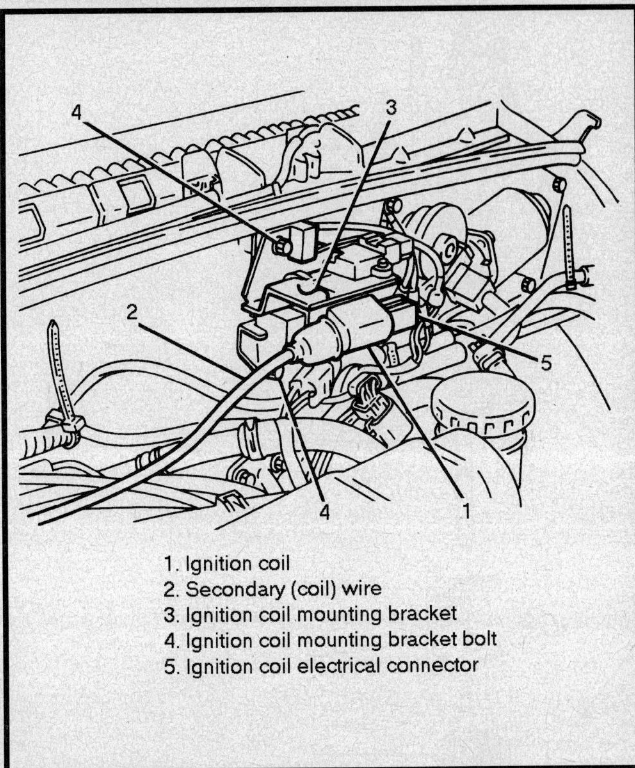

1. Ignition coil
2. Secondary (coil) wire
3. Ignition coil mounting bracket
4. Ignition coil mounting bracket bolt
5. Ignition coil electrical connector

Ignition coil — Tracker

are not serviced separately. The housing assembly is replaced as a unit.

METRO

1. Disconnect the negative battery cable.
2. Remove the distributor cap and rotor.
3. Remove the 2 camshaft position sensor retaining bolts and remove the sensor.

To install:

4. Install the camshaft position sensor to the distributor housing. Then, install but do not tighten the retaining screws.
5. Align a signal rotor ridge with the ridge on the camshaft position sensor. Then, adjust the signal rotor air gap to 0.008–0.016 (0–2 mm). Tighten the retaining screws to 44 inch lbs. (5 Nm).
6. Install the distributor rotor and cap.
7. Reconnect the negative battery cable.

IGNITER

Removal and Installation

TRACKER

1. Disconnect the negative battery cable.
2. Disconnect the electrical connection.
3. Remove the igniter retaining screws and remove the igniter.

To install:

4. Install the igniter and tighten the retaining screws .
5. Reconnect the electrical connection.
6. Reconnect the negative battery cable.

PRIZM

1. Disconnect the negative battery cable.

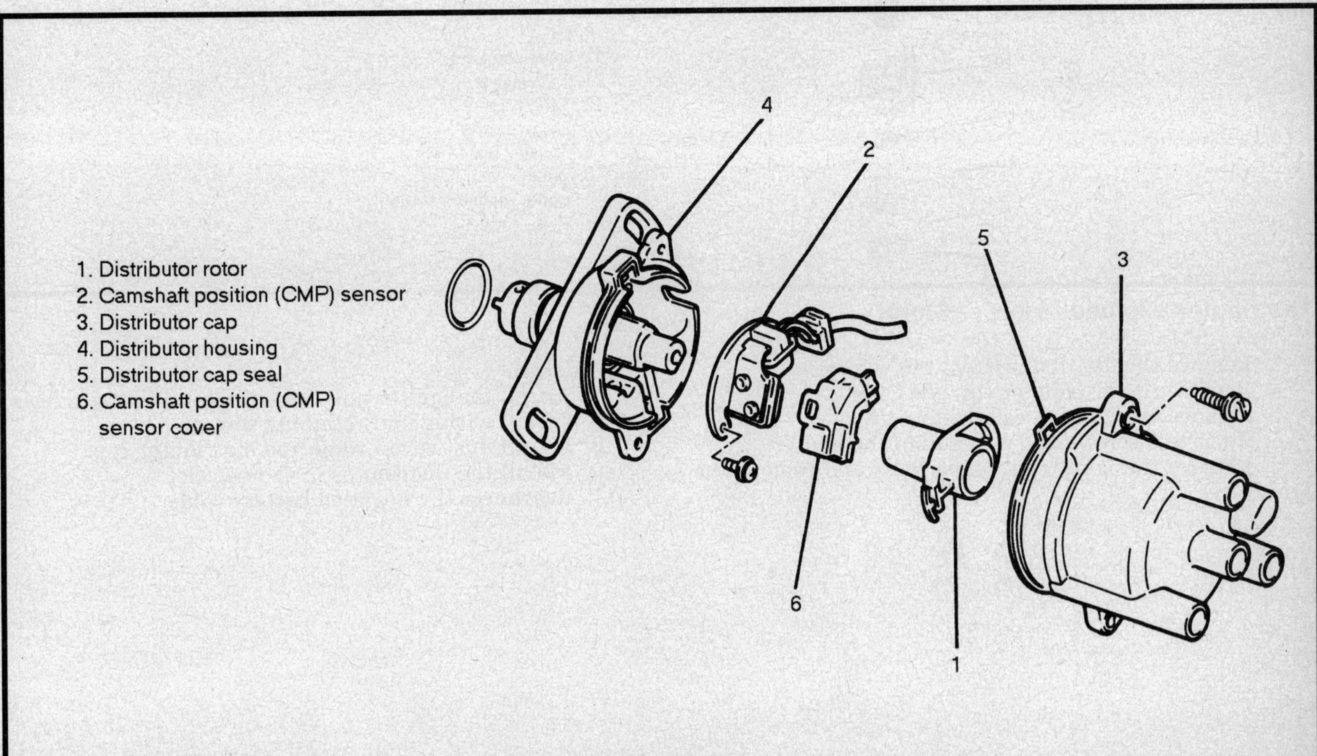

1. Distributor rotor
2. Camshaft position (CMP) sensor
3. Distributor cap
4. Distributor housing
5. Distributor cap seal
6. Camshaft position (CMP) sensor cover

Distributor exploded view — Metro

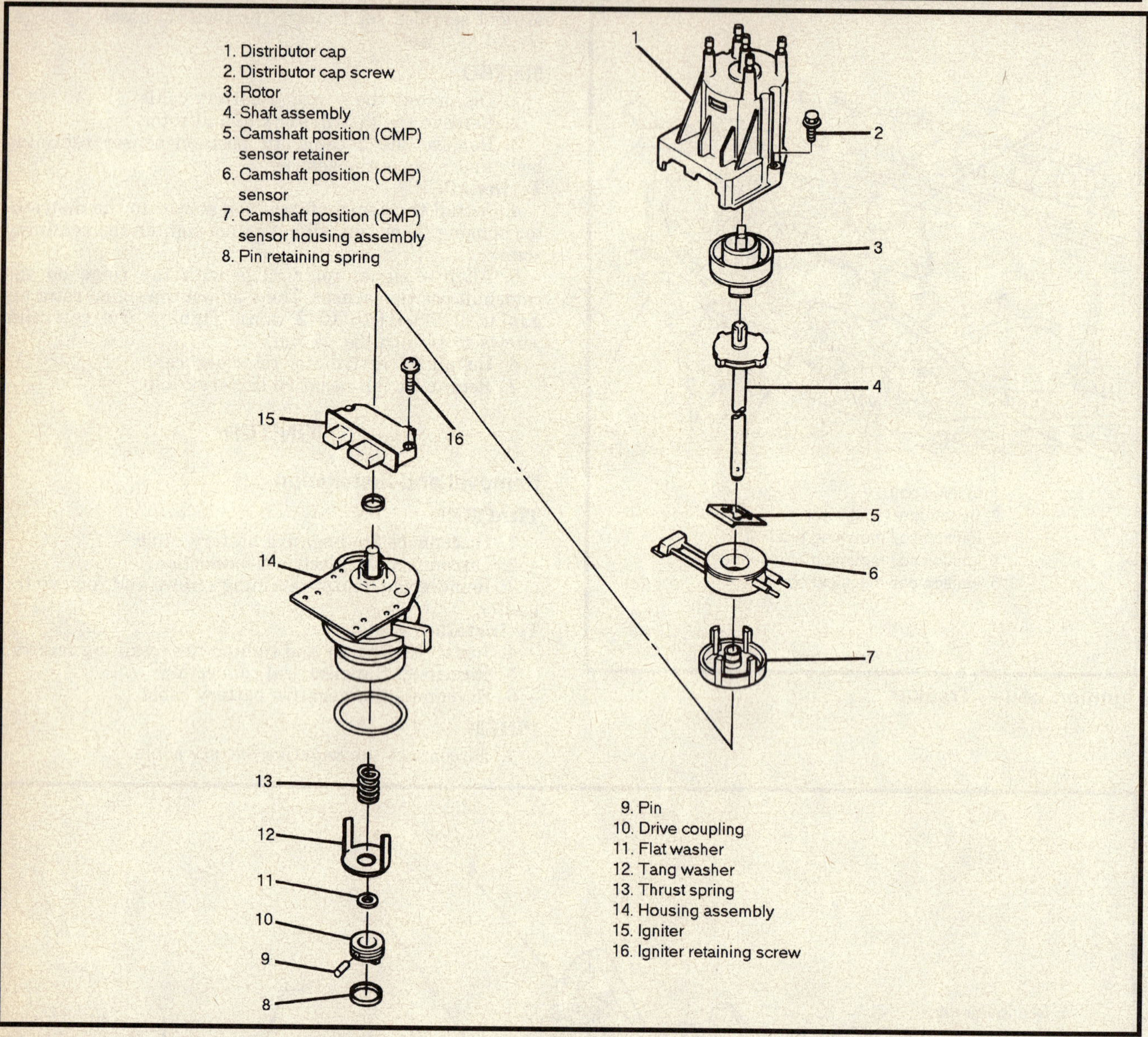

Distributor exploded view — Storm

2. Remove the distributor from the vehicle.
3. Remove the distributor cap and rotor.
4. Remove the ignition coil dust cover.
5. Remove 3 screws and wires from the igniter.
6. Remove the igniter retaining bolts and remove the igniter.

To install:
7. Install the igniter and tighten the retaining screws.
8. Install wiring and retaining screws.
9. Install dust cover, rotor and distributor cap.
10. Install the distributor in the vehicle.
11. Reconnect the negative battery cable.

MULTIPORT FUEL INJECTION (MFI) SYSTEMS

SECTION 3

MULTIPORT FUEL INJECTION (MFI) SYSTEMS

MULTIPORT FUEL INJECTION SYSTEMS

APPLICATION CHART — BUICK

BUICK					
Model	Body VIN	Engine Liter	Engine VIN	Ignition Type①	Fuel System②
Century	A	2.2	4	EI	MFI
		3.3	N	EI	MFI
Regal	W	3.1	T	EI	MFI
Skylark	N	2.3	D	EI	MFI
		2.3	3	EI	MFI
		3.3	N	EI	MFI

① The terms ''EI'' and ''DI'' reflect the new SAE J1930 acronyms used in the automotive market.
   EI—Electronic Ignition; engine does not use distributor
   DI—Distributor Ignition; engine uses a distributor
② The terms ''MFI'' and ''VFV'' reflect the new SAE J1930 acronyms used in the automotive market.
   MFI—Multi-Port Fuel Injection
   VFV—Variable Fuel Vehicle
③ 1993 only; in 1994 this engine came equipped with Sequential Fuel Injection

APPLICATION CHART — CHEVROLET

CHEVROLET					
Model	Body VIN	Engine Liter	Engine VIN	Ignition Type①	Fuel System②
Camaro	F	3.1	T	DI	MFI
		5.0	F	DI	MFI
		5.7	8	DI	MFI
		5.7	P③	DI	MFI
Cavalier	J	2.2	4	EI	MFI
		3.1	T	EI	MFI
Corsica/Beretta	L	2.2	4	EI	MFI
		2.3	A	EI	MFI
		3.1	T	EI	MFI
Corvette	Y	5.7	8	DI	MFI
		5.7	P③	DI	MFI
Lumina	W	2.2	4	EI	MFI
		3.1	T	EI	MFI
		3.1	T	EI	VFV
		3.4	X	EI	MFI

① The terms ''EI'' and ''DI'' reflect the new SAE J1930 acronyms used in the automotive market.
   EI—Electronic Ignition; engine does not use distributor
   DI—Distributor Ignition; engine uses a distributor
② The terms ''MFI'' and ''VFV'' reflect the new SAE J1930 acronyms used in the automotive market.
   MFI—Multi-Port Fuel Injection
   VFV—Variable Fuel Vehicle
③ 1993 only; in 1994 this engine came equipped with Sequential Fuel Injection

APPLICATION CHART — OLDSMOBILE

OLDSMOBILE

Model	Body VIN	Engine Liter	Engine VIN	Ignition Type [1]	Fuel System [2]
Achieva	N	2.3	D	EI	MFI
		2.3	A	EI	MFI
		2.3	3	EI	MFI
		3.3	N	EI	MFI
Ciera/Cutlass Cruiser	A	2.2	4	EI	MFI
		3.3	N	EI	MFI
Cutlass Supreme	W	3.1	T	EI	MFI
		3.4	X	EI	MFI

[1] The terms ''EI'' and ''DI'' reflect the new SAE J1930 acronyms used in the automotive market.
   EI—Electronic Ignition; engine does not use distributor
   DI—Distributor Ignition; engine uses a distributor
[2] The terms ''MFI'' and ''VFV'' reflect the new SAE J1930 acronyms used in the automotive market.
   MFI—Multi-Port Fuel Injection
   VFV—Variable Fuel Vehicle
[3] 1993 only; in 1994 this engine came equipped with Sequential Fuel Injection

APPLICATION CHART — PONTIAC

PONTIAC

Model	Body VIN	Engine Liter	Engine VIN	Ignition Type [1]	Fuel System [2]
Firebird	F	3.1	T	DI	MFI
		5.0	F	DI	MFI
		5.7	8	DI	MFI
		5.7	P [3]	DI	MFI
Grand Am	N	2.3	D	EI	MFI
		2.3	A	EI	MFI
		2.3	3	EI	MFI
		3.3	N	EI	MFI
Grand Prix	W	3.1	T	EI	MFI
		3.4	X	EI	MFI
Sunbird	J	2.0	H	DI	MFI
		3.1	T	EI	MFI

[1] The terms ''EI'' and ''DI'' reflect the new SAE J1930 acronyms used in the automotive market.
   EI—Electronic Ignition; engine does not use distributor
   DI—Distributor Ignition; engine uses a distributor
[2] The terms ''MFI'' and ''VFV'' reflect the new SAE J1930 acronyms used in the automotive market.
   MFI—Multi-Port Fuel Injection
   VFV—Variable Fuel Vehicle
[3] 1993 only; in 1994 this engine came equipped with Sequential Fuel Injection

APPLICATION CHART — GEO

GEO					
Model	Body VIN	Engine Liter	Engine VIN	Ignition Type ①	Fuel System ②
Prizm	S	1.6	5	DI	MFI
		1.6	6	DI	MFI
		1.8	8	DI	MFI
Storm	R	1.6	5	DI	MFI
		1.6	6	DI	MFI

① The terms ''EI'' and ''DI'' reflect the new SAE J1930 acronyms used in the automotive market.
　EI—Electronic Ignition; engine does not use distributor
　DI—Distributor Ignition; engine uses a distributor
② The terms ''MFI'' and ''VFV'' reflect the new SAE J1930 acronyms used in the automotive market.
　MFI—Multi-Port Fuel Injection
　VFV—Variable Fuel Vehicle

APPLICATION CHART — SATURN

SATURN					
Model	Body VIN	Engine Liter	Engine VIN	Ignition Type ①	Fuel System ②
Saturn	Z	1.9	7	EI	MFI

① The terms ''EI'' and ''DI'' reflect the new SAE J1930 acronyms used in the automotive market.
　EI—Electronic Ignition; engine does not use distributor
　DI—Distributor Ignition; engine uses a distributor
② The terms ''MFI'' and ''VFV'' reflect the new SAE J1930 acronyms used in the automotive market.
　MFI—Multi-Port Fuel Injection
　VFV—Variable Fuel Vehicle

APPLICATION CHART — LIGHT TRUCKS, VANS AND APV

GM LIGHT TRUCK, VAN AND APV					
Model	Body VIN	Engine Liter	Engine VIN	Ignition Type ①	Fuel System ②
Compact Pick-up/Utility	S/T	2.2	4	EI	MFI
		4.3	Z	DI	MFI

① The terms ''EI'' and ''DI'' reflect the new SAE J1930 acronyms used in the automotive market.
　EI—Electronic Ignition; engine does not use distributor
　DI—Distributor Ignition; engine uses a distributor
② The terms ''MFI'' and ''VFV'' reflect the new SAE J1930 acronyms used in the automotive market.
　MFI—Multi-Port Fuel Injection
　VFV—Variable Fuel Vehicle

EXCEPT LIGHT TRUCKS, VANS, GEO AND SATURN

General Description

There are 5 variations of multiport fuel injection systems, formerly known as Tuned Port Injection (TPI), Bottom Feed Port Injection (BFPI), Alternating Sychronous Double Firing (ASDF) Fuel Injection and Multi-Port Fuel Injection (MPI or MPFI), all of which the Society of Automotive Engineers (SAE) has now labeled Multiport Fuel Injection (MFI) according to the J-1930 standards which have been designed to universalize automotive terminology among vehicle manufacturers. There is one other new fuel injection system, known as a Variable Fuel Vehicle (VFV), which may use a mixture of gasoline, methanol or ethanol.

All fuel injection systems, except the 2.0L and 2.3L engines, function in same manner by electronically injecting fuel in all (4, 6 or 8) cylinders at the same time, once during each crankshaft revolution. The ASDF system equipped on the 2.0L and 2.3L engines uses 4 injectors, which are alternately pulsed in pairs (cyclinders 1 & 4 and 2 & 3) twice, once during compression and once during the exhaust stroke. The TPI system equipped on the 5.0L (VIN F) and 5.7L (VIN 8) engines use tuned intake runners to increase engine torque and perrformance. The

BFPI system, equipped on the 2.2L (VIN 4) engine, uses a new style injector which is integrally located with the fuel rail inside the lower manifold; instead of fuel only flowing through the injector, fuel flows through and around the entire injector, thus cooling the entire injector which improves combustion and hot restarts because of a denser air/fuel ratio. All other MFI vehicles use an individual injector per cylinder, located in the intake runner.

The 3.1L (VIN T), which is available as a Variable Fuel Vehicle (VFV), is designed to operate on either a 100% concentration of unleaded gasoline or a blend of gasoline and methanol. Fuel produced with 85% methanol and 15% unleaded gasoline is called M85 Fuel.

Another version of the VFV operates on 100% unleaded gasoline or a blend of gasoline and ethanol. Fuel produced with 85% ethanol and 15% unleaded gasoline is known as E85 fuel. Methanol and ethanol fuels cannot be mixed. Each fuel has its own distinct properties which are compatible with the specific VFV fuel system. Any blending of the two fuels will result in poor driveability, and possible component damage.

The fuel supply and fuel metering systems of the VFV are similar to its gasoline designed counterparts in appearance, however, component internal designs are different. In addition the VFV uses a variable fuel sensor, fuel pump speed controller, and remote injector driver, which are unique to the VFV system. The evaporative emission system used on VFV vehicles is designed specifically for use with alcohol/gasoline blended fuels.

SYSTEM OPERATION

MFI Vehicles

All MFI system are controlled by an Electronic Control Module (ECM) which monitors engine operations and generates output signals to provide the correct air/fuel mixture, ignition timing and engine idle speed control. Input to the control unit is provided by an oxygen sensor, coolant temperature sensor, detonation sensor, hot film mass air flow sensor (3.3L engine) or manifold absolute pressure (except 3.3L engine) and throttle position sensor. The ECM also receives information concerning engine rpm, road speed, transmission gear position, power steering and air conditioning.

The injectors are located, one at each intake port, rather than the single injector found on throttle body system. The injectors are mounted on a fuel rail and are activated by a signal from the electronic control module. The injector is a solenoid-operated valve which remains open depending on the width of the electronic pulses (length of the signal) from the ECM; the longer the open time, the more fuel is injected. In this manner, the air/fuel mixture can be precisely controlled for maximum performance with minimum emissions.

Fuel is pumped from the tank by a high pressure fuel pump, located inside the fuel tank. It is a positive displacement roller vane pump. The impeller serves as a vapor separator and pre-charges the high pressure assembly. A pressure regulator maintains 28–38 psi in the fuel line to the injectors and the excess fuel is fed back to the tank. A fuel accumulator is used to dampen the hydraulic line hammer in the system created when all injectors open simultaneously.

The Mass Air Flow (MAF) Sensor (3.3L engine) is used to measure the mass of air that is drawn into the engine cylinders. It is located on the throttle body and consists of a heated film which measures the mass of air, rather than just the volume. A resistor is used to measure the temperature of the film at 75° above ambient temperature. As the ambient (outside) air temperature rises, more energy is required to maintain the heated film at the higher temperature and the control unit uses this difference in required energy to calculate the mass of the incoming air. The control unit uses this information to determine the duration of fuel injection pulse, timing and EGR.

The throttle body incorporates an Idle Air Control (IAC) that provides for a bypass channel through which air can flow. It consists of an orifice and pintle which is controlled by the ECM through a step motor. The IAC provides air flow for idle and allows additional air during cold start until the engine reaches operating temperature. As the engine temperature rises, the opening through which air passes is slowly closed.

The Throttle Position Sensor (TPS) provides the control unit with information on throttle position, in order to determine injector pulse width and hence correct mixture. The TPS is connected to the throttle shaft on the throttle body and consists of a potentiometer with one end connected to a 5 volt source from the ECM and the other to ground. A third wire is connected to the ECM to measure the voltage output from the TPS which changes as the throttle valve angle is changed (accelerator pedal moves). At the closed throttle position, the output is low (approximately 0.2–0.8 volts); as the throttle valve opens, the output increases to a maximum 5 volts at Wide Open Throttle (WOT). The TPS can be misadjusted open, shorted, or loose and if it is out of adjustment, the idle quality or WOT performance may be poor. A loose TPS can cause intermittent bursts of fuel from the injectors and an unstable idle because the ECM thinks the throttle is moving. This should cause a trouble code to be set. Once a trouble code is set, the ECM will use a preset value for TPS and some vehicle performance may return.

Variable Fuel Vehicle (VFV)

Fuel is stored in a stainless steel fuel tank. The fuel tank filler neck is equipped with a threaded type cap which is compatible with alcohol/gasoline blended fuels. On 1992 and later models, the VFV fuel filler cap only vents vacuum, which makes it specific to the application. The in-tank electric pump, which is part of the fuel sender assembly, pumps fuel through an in-line filter, fuel feed pipe, and the variable fuel sensor to the fuel rail. The fuel is delivered at a pressure greater than that which is needed by the injectors. A pressure regulator, which is an integral part of the fuel rail assembly, keeps fuel available to the injectors at a controlled pressure. Fuel in excess of engine requirements is directed back to the tank via the fuel return line. The fuel injectors are controlled by the ECM through a remote injector driver. Due to the low-impedance design of the injectors (approximately 1 ohm), and high output capability, they draw more current than standard MFI injectors. Because the ECM is not capable of dissipating this excessive heat, the VFV system utilizes the remote injector driver. The fuel injectors are pulsed simultaneously, once every crankshaft revolution, thus making the VFV system operate like a standard MFI design.

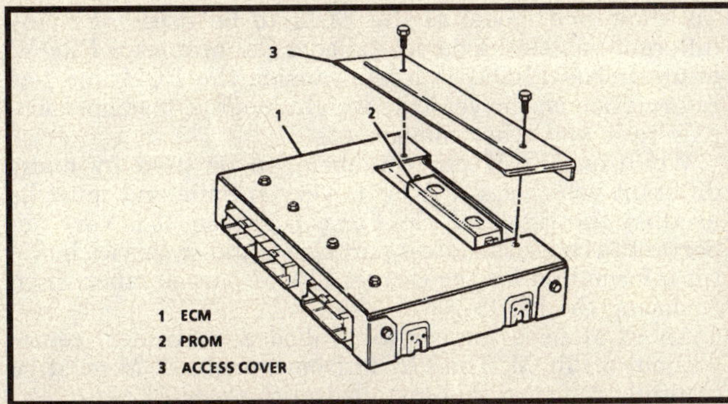

Throttle position sensor

ECM or PCM assembly

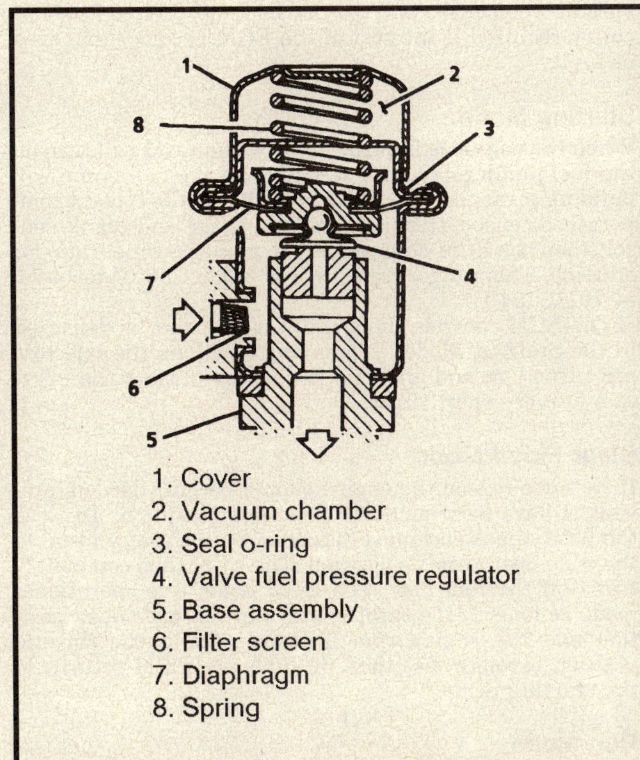

1. Cover
2. Vacuum chamber
3. Seal o-ring
4. Valve fuel pressure regulator
5. Base assembly
6. Filter screen
7. Diaphragm
8. Spring

Fuel pressure regulator assembly

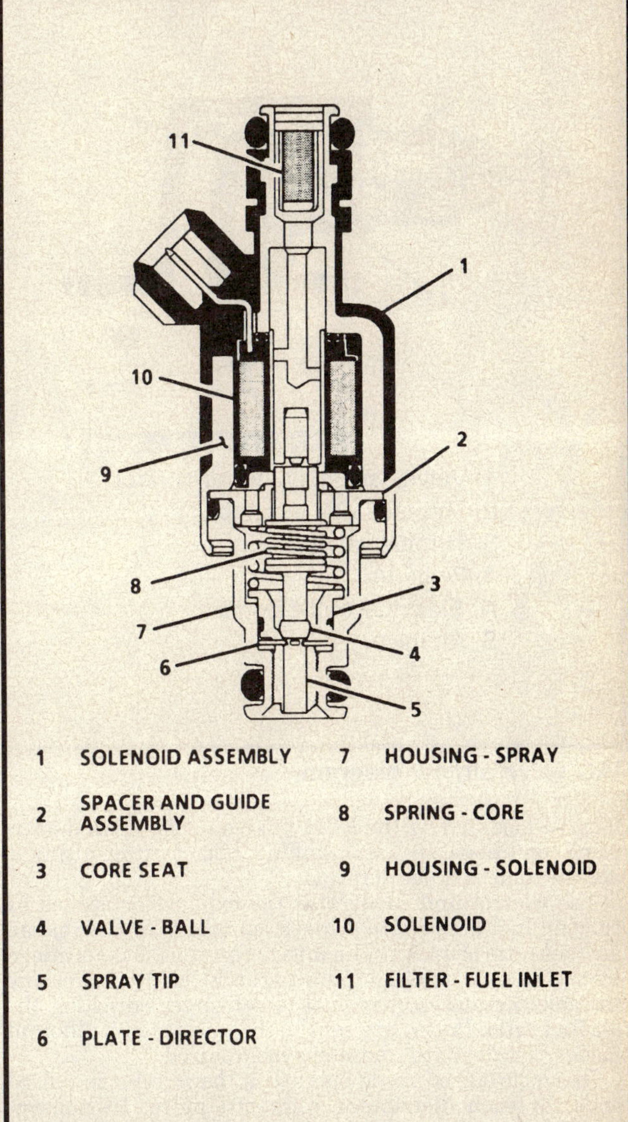

1	SOLENOID ASSEMBLY	7	HOUSING - SPRAY
2	SPACER AND GUIDE ASSEMBLY	8	SPRING - CORE
3	CORE SEAT	9	HOUSING - SOLENOID
4	VALVE - BALL	10	SOLENOID
5	SPRAY TIP	11	FILTER - FUEL INLET
6	PLATE - DIRECTOR		

Fuel injector cut-away view

SYSTEM COMPONENTS

Electronic Control Module (ECM)

All vehicles use an ECM which consists of 2 parts; a Controller (the ECM without the PROM) and a Calibrator called a PROM (Programmable Read Only Memory). The 2.2L engine uses an ECM which is equipped with a Controller and an EEPROM (Electronically Eraseable Programmable Read Only Memory).

CONTROLLER

The fuel injection system is controlled by an on-board computer, the Electronic Control Module (ECM) or controller, usually located in the passenger compartment. The ECM monitors engine operations and environmental conditions (ambient temperature, barometric pressure, etc.) needed to calculate the fuel delivery time (pulse width/injector on-time) of the fuel injector. The fuel pulse

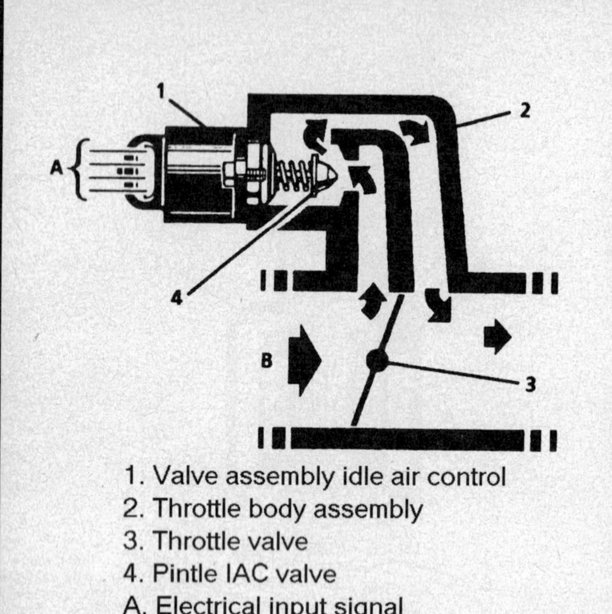

1. Valve assembly idle air control
2. Throttle body assembly
3. Throttle valve
4. Pintle IAC valve
A. Electrical input signal
B. Air inlet

IAC valve airflow diagram

may be modified by the ECM to account for special operating conditions, such as cranking, cold starting, altitude, acceleration and deceleration.

The control unit moderates the exhaust emissions by curving fuel delivery to achieve, as nearly as possible an air/fuel ratio of 14.7:1. The injector on-time is determined by the various sensor inputs to the ECM. By increasing the injector pulse, more fuel is delivered, enriching the air/fuel ratio. Pulses are sent to the injector in 2 different modes, synchonized and non-synchronized.

In synchronized mode operation, the injector is pulsed once for each distributor reference pulse. In nonsynchronized mode operation, the injector is pulsed once every 12.5 milliseconds or 6.25 milliseconds depending on calibration. This pulse time is totally independent of distributor reference.

The ECM constantly monitors the input information, processes this information from various sensors, and generates output commands to the various systems that affect vehicle performance.

The ability of the ECM to recognize and adjust for vehicle variations (engine transmission, vehicle weight, axle ratio, etc.) is provided by a calibration unit (PROM/EEPROM) that is programmed to tailor the ECM for the particular vehicle. There is a specific ECM/PROM combination for each specific vehicle, and the combinations are not interchangeable with those of other vehicles.

The ECM also performs the diagnostic function of the system. It can recognize operational problems, alert the driver through the SERVICE ENGINE SOON light, and store a code or codes which identify the problem areas to aid the technician in making repairs.

NOTE: Instead of an edgeboard connector, newer ECM's have a header connector which attaches solidly to the ECM case. Like the edgeboard connectors, the header connectors have a different pinout identification for different engine designs.

EEPROM

THE EEPROM contains the preset calibrations of the vehicle according to the VIN. The EEPROM is non-repairable and can only be replaced with the controller assembly. The EEPROM will need to be reprogrammed using a GM Techline equipment.

PROM

To allow one model of the ECM to be used for many different vehicles, a device called a Calibrator (or PROM) is used. The PROM is located inside the ECM and has information on the vehicle's weight, engine, transmission, axle ratio and other components.

While one ECM part number can be used by many different vehicles, a PROM is very specific and must be used for the right vehicle. For this reason, it is very important to check the latest parts book and or service bulletin information for the correct PROM part number when replacing the PROM.

An ECM used for service (called a controller) comes without a PROM. The PROM from the old ECM must be carefully removed and installed in the new ECM.

MEM-CAL

This assembly contains the functions of the PROM, Cal-Pak and the ESC module used on other GM applications. Like the PROM, it contains the calibrations needed for a specific vehicle as well as the back-up fuel control circuitry required if the rest of the ECM becomes damaged or faulty.

Starting Mode

When the engine is first turned **ON**, the ECM will turn on the fuel pump relay for 2 seconds and the fuel pump will build up pressure. The ECM then checks the coolant temperature sensor, throttle position sensor and crank sensor, then the ECM determines the proper air/fuel ratio for starting. This ranges from 1.5:1 at -33°F (-36°C) to 14.7:1 at 201°F (94°C).

The ECM controls the amount of fuel that is delivered in the Starting Mode by changing how long the injectors are turned on and off. This is done by pulsing the injectors for very short times.

Clear Flood Mode

If for some reason the engine should become flooded, provisions have been made to clear this condition. To clear the flood, the driver must depress the accelerator pedal to the wide-open throttle position. The ECM then completely turns off the fuel flow. The ECM holds this operational mode as long as the throttle stays in the wide open position and the engine rpm is below 600. If the throttle position becomes less than 62–70%, the ECM returns to the starting mode.

Run Mode

There are 2 different run modes. When the engine is first started and the rpm is above 400, the system goes into open loop operation. In open loop operation, the ECM will ignore the signal from the oxygen (O_2) sensor and calculate the injector on-time based upon inputs from the coolant sensor, MAF or MAP sensor and MAT sensors.

During open loop operation, the ECM analyzes the following items to determine when the system is ready to go to the closed loop mode.

1. The oxygen sensor varying voltage output. (This is dependent on temperature).

2. The coolant sensor must be above specified temperature.

3. A specific amount of time must elapse after starting the engine. These values are stored in the PROM.

When these conditions have been met, the system goes into closed loop operation. In closed loop operation, the ECM will calculate the air/fuel ratio (injector on time) based upon the signal from the oxygen sensor. The ECM will decrease the on-time if the air/fuel ratio is too rich, and will increase the on-time if the air/fuel ratio is too lean.

Acceleration Mode

When the accelerator pedal is depressed, the opening of the throttle valve(s) causes a rapid increase in airflow into the cylinders. The air flow is capable of moving much faster then the flow of fuel and therefore the engine may hesitate; to avoid this situation the ECM increases the fuel injector pulse width according to the TPS and MAF or MAP signals it receives.

Deceleration Mode

Upon deceleration, a leaner fuel mixture is required to reduce emission of hydrocarbons (HC) and carbon monoxide (CO). To adjust the injection on-time, the ECM uses the decrease in MAP and the decrease in throttle position to calculate a decrease in injector on time. To maintain an idle fuel ratio of 14.7:1, fuel output is momentarily reduced. If deceleration is very rapid or for long periods, the ECM can cut off the fuel completely to prevent damage to the catalytic converter.

Battery Voltage Correction Mode

The purpose of battery voltage correction is to compensate for variations in battery voltage to fuel pump and injector response. The ECM compensates by increasing the engine idle rpm.

Battery voltage correction takes place in all operating modes. When battery voltage is low, the spark delivered by the distributor may be low. To correct this low battery voltage problem, the ECM can do any or all of the following:

- Increase injector on time (increase fuel)
- Increase idle rpm
- Increase ignition dwell time

Fuel Cutoff Mode

Depending on the pre-programmed values, the ECM turns **OFF** the fuel flow to the injectors to prevent engine damage.

Converter Protection Mode

In this mode the ECM estimates the temperature of the catalytic converter and then modifies fuel delivery to protect the converter from high temperatures. When the ECM has determined that the converter may overheat, it will cause open loop operation and will enrichen the fuel delivery. A slightly richer mixture will then cause the converter temperature to be reduced.

Data Link Connector (DLC)

Formerly known as the Assembly Line Data Link (ALDL) or Assembly Line Communication Link (ALCL), the Data Link Connector (DLC) was renamed according to the new SAE J-1930 standards in order to universalize automotive terminology among manufacturers. The diagnostic connector located in the passenger compartment, usually under the instrument panel.

The Diagnostic mode or the Field Service Mode may be entered by connecting or jumping terminal B, the diagnostic TEST terminal to terminal A, or ground circuit. This connector is a very useful tool in diagnosing performance and engine related ECM operating conditions. Information from the ECM is available at this terminal and can be read with one of the many popular scanner tools.

FUEL CONTROL SYSTEM

The fuel control system is made up of the following components:

1. Fuel supply system
2. Throttle body assembly
3. Fuel injectors
4. Fuel rail
5. Fuel pressure regulator
6. Idle Air Control (IAC)
7. Fuel pump
8. Fuel pump relay
9. In-line fuel filter

The fuel control system starts at the fuel tank. An electric fuel pump, located in the fuel tank with the fuel gauge sending unit, pumps fuel to the fuel rail through an in-line fuel filter. The pump is designed to provide fuel at a pressure above that needed by the injectors. A pressure regulator in the fuel rail keeps fuel available to the injectors at a constant pressure. Excess fuel pressure is returned to the fuel tank through a separate line.

In order for the fuel injectors to supply a precise amount of fuel at the command of the ECM, the fuel supply system maintains a constant pressure drop across the injectors. As manifold vacuum changes, the fuel system pressure regulator controls the fuel supply pressure to compensate. The fuel pressure accumulator used on select models, isolates fuel line noise. The fuel rail contains a spring loaded pressure tap for testing the fuel system or relieving the fuel system pressure.

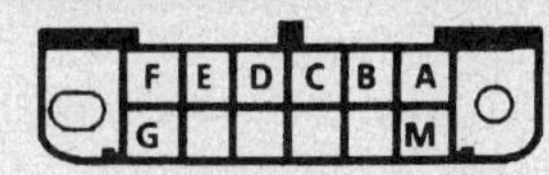

Data Link Connector (DLC) terminal identification for system diagnosis

The injectors are controlled by the ECM. They deliver fuel in one of several modes as previously described. In order to properly control the fuel supply, the fuel pump is operated by the ECM through the fuel pump relay and/or oil pressure switch.

Throttle Body Unit

The throttle body unit has a throttle valve to control the amount of air delivered to the engine. The TPS and IAC valve are also mounted onto the throttle body. The throttle body contains vacuum ports located at, above or below the throttle valve. These vacuum ports generate the vacuum signals needed by various components.

On some vehicles, the engine coolant is directed through the coolant cavity at the bottom of the throttle body to warm the throttle valve and prevent icing.

Fuel Injector

A fuel injector is installed in the intake manifold at each cylinder. The fuel injector is a solenoid operated device controlled by the ECM. The ECM turns on the solenoid, which opens the valve to allow fuel delivery. The fuel, under pressure, is injected in a conical spray pattern at the opening of the intake valve. The regulated fuel, which is not used by the injectors is returned to the fuel tank.

An injector that is partly open, will cause loss of fuel pressure after the engine is shut down, so long crank time would be noticed on some engines. Also dieseling could occur because some fuel could be delivered after the ignition is turned **OFF**.

There are O-rings used to seal the injector at the intake manifold and fuel rail. The O-rings are lubricated and should be replaced whenever the injector is removed. Air leakage at the injector/intake area would create a lean cylinder and a possible driveability problem. The injectors are identified with an ID number cast on the injector near the top side. Injectors manufactured by Rochester® Products have an **RP** positioned near the top side in addition to the ID number.

> **NOTE: The most widely used injector for the MPI units are Bosch injectors, but now there will also be a new injector being used. This new injector will be a Multec MPI injector (a Rochester® product). It is classified as a top feed design because the fuel enters the top of the injector and then flows through the entire length of the injector. It is designed to operate with the system fuel pressures ranging from 36–51 psi (250–350 kPa), and uses a high impedance (12.2 ohms) solenoid coil.**

Fuel Rail

All vehicles except those equipped with the 2.2L engine use a fuel rail. The 2.2L engine uses the lower intake manifold as a replacement to the fuel rail. The standard fuel rail is bolted rigidly to the engine and and used to provide the upper mount for the fuel injectors. It distributes fuel to the individual injectors. Fuel is delivered to the input end of the fuel rail by the fuel lines, goes through the rail, then to the fuel pressure regulator. The regulator keeps the fuel pressure to the injectors steady. The remaining fuel is then returned to the fuel tank. Some fuel rails also contains a spring loaded pressure tap for testing the fuel system or relieving the fuel system pressure.

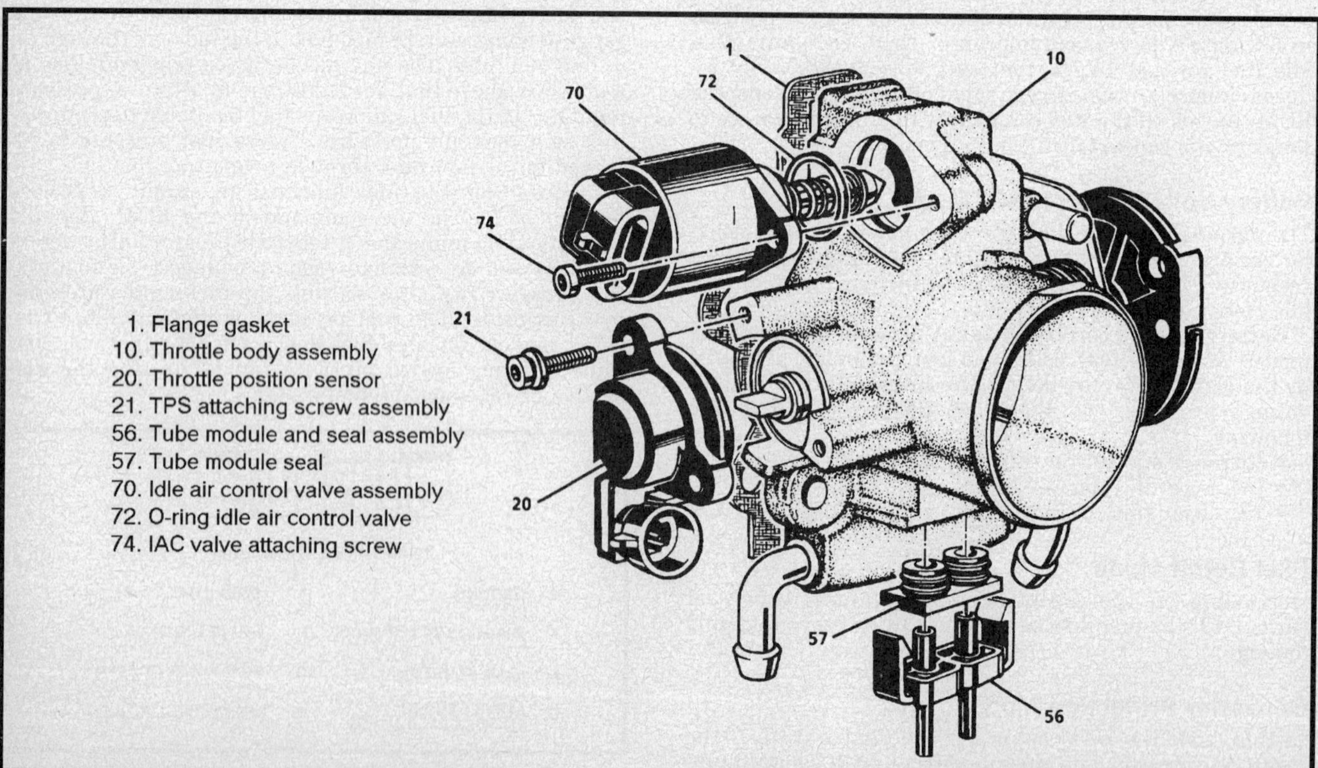

Throttle body assembly — 2.0L engine

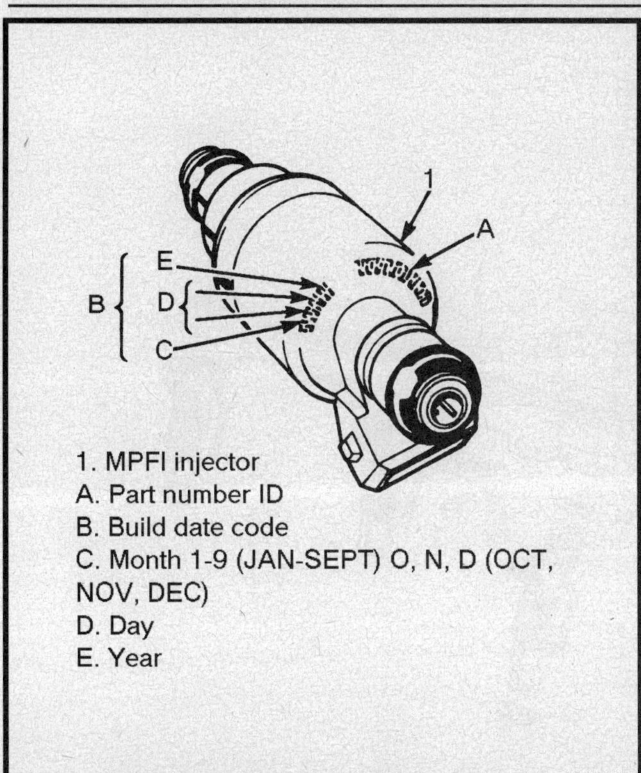

1. MPFI injector
A. Part number ID
B. Build date code
C. Month 1-9 (JAN-SEPT) O, N, D (OCT, NOV, DEC)
D. Day
E. Year

Fuel injector part number location

Pressure Regulator

The fuel pressure regulator contains a pressure chamber separated by a diaphragm relief valve assembly with a calibrated spring in the vacuum chamber side. The fuel pressure is regulated when the pump pressure acting on the bottom spring of the diaphragm overcomes the force of the spring action on the top side.

The diaphragm relief valve moves, opening or closing an orifice in the fuel chamber to control the amount of fuel returned to the fuel tank. Vacuum acting on the top side of the diaphragm along with spring pressure controls the fuel pressure. A decrease in vacuum creates an increase in the fuel pressure. An increase in vacuum creates a decrease in fuel pressure.

An example of this is under heavy load conditions the engine requires more fuel flow. The vacuum decreases under a heavy load condition because of the throttle opening. A decrease in the vacuum allows more fuel pressure to the top side of the pressure relief valve, thus increasing the fuel pressure.

The pressure regulator is mounted on the fuel rail and serviced separately. If the pressure is too low, poor performance could result. If the pressure is too high, excessive odor and a Code 45 may result.

VFV Pressure Regulator Vacuum Solenoid

The Varaiable Fuel Vehicle pressure regulator vacuum valve is a solenoid operated bleed valve located in-line with the manifold vacuum source to the pressure regulator. Its purpose is to increase fuel pressure to prevent vapor lock and improve hot engine restarts. The solenoid is controlled by the ECM, and when energized will bleed off vacuum to the pressure regulator, which results in hot engine restarts. The solenoid is controlled by the ECM, and when energized will bleed off vacuum to the pressure regulator, which results in higher fuel delivery pressure. The ECM looks at coolant temperature, fuel temperature, and alcohol percentage to determine the proper time to turn the solenoid "ON," and vent the vacuum to the atmosphere.

Idle Air Control (IAC)

The purpose of the Idle Air Control (IAC) system is to control engine idle speeds while preventing stalls due to changes in engine load. The IAC assembly, mounted on the throttle body, controls bypass air around the throttle plate. By extending or retracting a conical valve, a controlled amount of air can move around the throttle plate. If rpm is too low, more air is diverted around the throttle plate to increase rpm.

During idle, the proper position of the IAC valve is calculated by the ECM based on battery voltage, coolant temperature, engine load and engine rpm. If the rpm drops below a specified rate and the throttle plate is closed, the ECM will calculate a new valve position.

Three different designs are used for the IAC conical valve. The first design used is single taper while the second design used is a dual taper. The third design is a blunt valve. Care should be taken to ensure use of the correct design when service replacement is required.

The IAC motor has 255 different positions, which are counted in steps. The zero, or reference position, is the fully extended position at which the pintle is seated in the air bypass seat and no air is allowed to bypass the throttle plate. When the motor is fully retracted, maximum air is allowed to bypass the throttle plate. When the motor is fully retracted, maximum air is allowed to bypass the throttle plate. The ECM always monitors how many steps it has extended or retracted the pintle from the zero or reference position; thus, it always calculates the exact position of the motor.

The IAC only affects the engine's idle characteristics. If it is stuck fully open, idle speed is too high (too much air enters the throttle bore) If it is stuck closed, idle speed is too low (not enough air entering). If it is stuck somewhere in the middle, idle may be rough and the engine won't respond to load changes.

If the IAC valve is replaced or disconnected or battery voltage to the ECM is lost for any reason, the valve will need to be reset following the "Idle Learn Procedure".

Fuel Pump

MFI VEHICLES

The fuel is supplied to the system from an in-tank pump. The pump supplies fuel through the in-line fuel filter to the fuel rail assembly. The pump is removed for service along with the fuel gauge sending unit. Once they are removed from the fuel tank, the pump and sending unit can be serviced separately.

The fuel pump delivers more fuel than the engine can consume even under the most extreme conditions. Excess fuel flows through the pressure regulator and back to the tank via the return line. The constant flow of fuel means that the fuel system is always supplied with cool fuel, thereby preventing the formation of fuel-vapor bubbles (vapor lock).

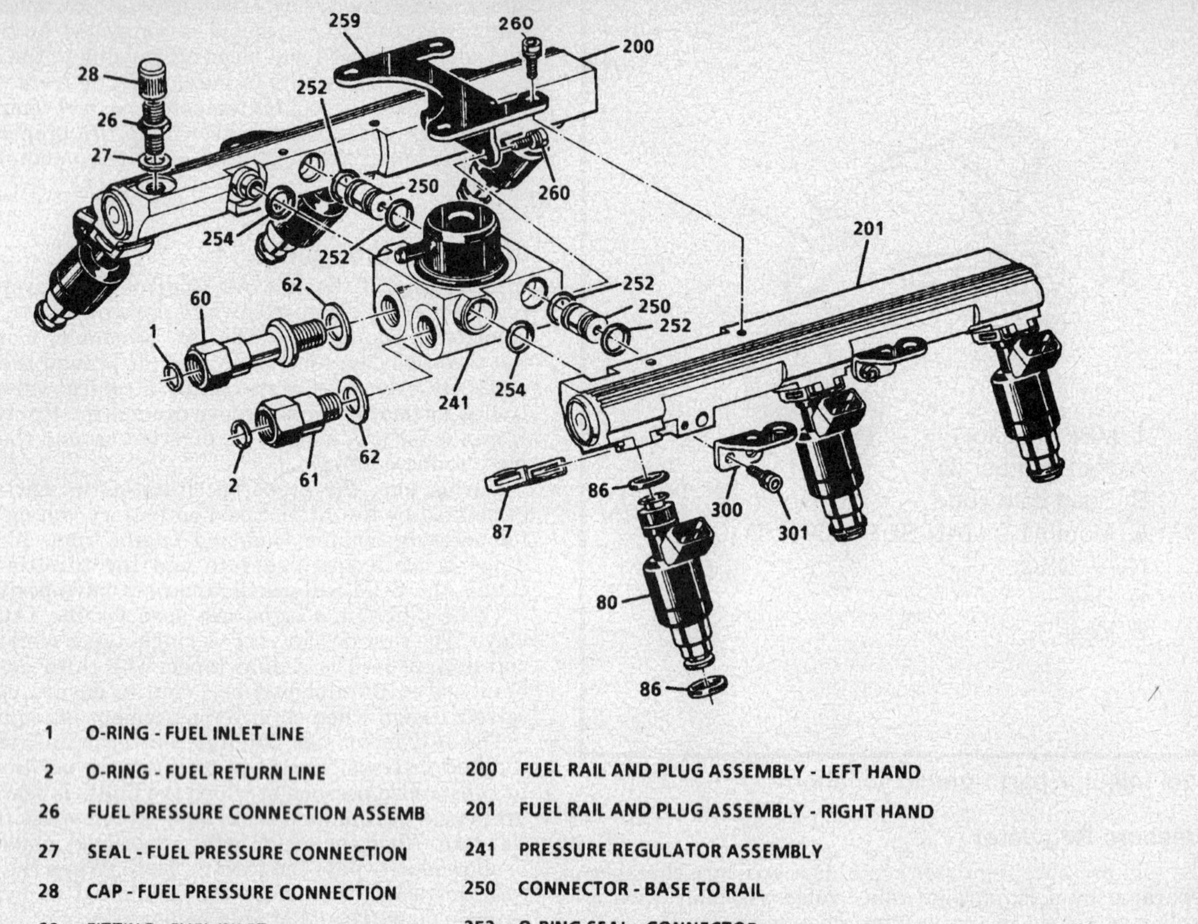

Fuel injector exploded view — Bottom Feed Port Injection

VARIABLE FUEL VEHICLES

The fuel pump is specifically designed for this application. It is an electric high pressure roller vane pump which is mounted to the fuel sender assembly inside the fuel tank. Fuel pump output is variable and is dependent upon the alcohol concentration of the fuel. The more alcohol present, the faster the pump must turn to deliver an adequate supply of fuel for proper engine operation. Fuel pump speed is controlled by the ECM and the fuel pump speed controller. This control is based upon input signals from the variable fuel sensor.

Fuel lines and hoses are constructed of special material that is compatible with the M85 and E85 fuel blends. Similarly, fuel line O-rings are made of a special material that is used specifically for the VFV system. Always replace these O-rings when a fuel line has been disconnected, and never substitute a standard O-ring seal for the original equipment seal as fuel leaks will occur.

Fuel Pump Relay Circuit

The fuel pump relay is usually located on the left or right front inner fenderwell (shock tower) or on the engine side of the firewall (center cowl). The fuel pump electrical system consists of the fuel pump relay, ignition circuit and the ECM circuits are protected by a fuse. The fuel pump relay contact switch is in the normally open position, with no voltage flowing to the pump.

When the ignition is turned **ON** the ECM will for 2 seconds, supply voltage to the fuel pump relay coil, closing the contact switch. The ignition circuit fuse can now supply ignition voltage to the circuit which feeds the relay contact switch. With the relay contacts closed, ignition voltage is supplied to the fuel pump. The ECM will con-

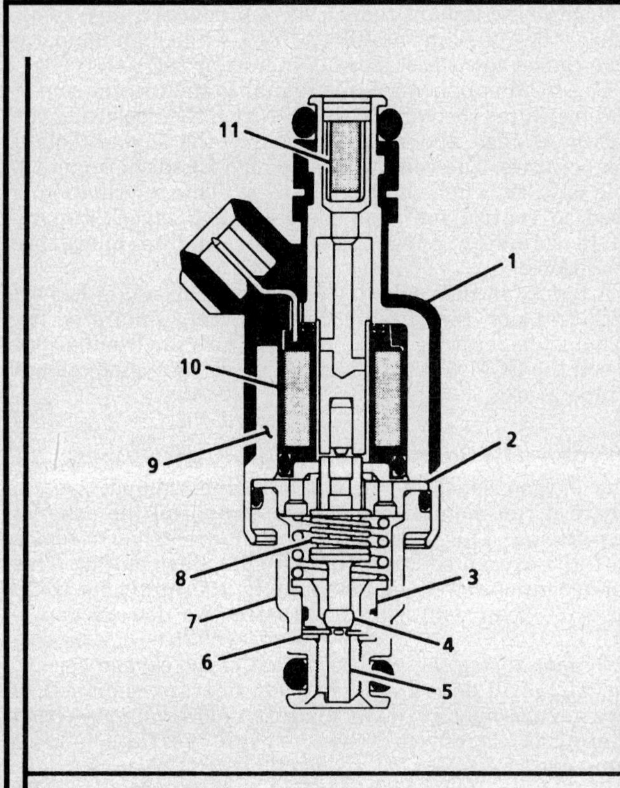

1	SOLENOID ASSEMBLY	7	SPRAY HOUSING
2	SPACER AND GUIDE ASSEMBLY	8	CORE SPRING
3	CORE SEAT	9	SOLENOID HOUSING
4	BALL VALVE	10	SOLENOID
5	SPRAY TIP	11	FUEL INLET FILTER
6	DIRECTOR PLATE		

Exploded view of typical fuel injector

tinue to supply voltage to the relay coil circuit as long as the ECM receives the rpm reference pulses from the ignition module.

The fuel pump control circuit also includes an engine oil pressure switch with a set of normally open contacts. The switch closes at approximately 4–6.5 psi and provides a secondary battery feed path to the fuel pump. If the relay fails, the pump will continue to run using the battery feed supplied by the closed oil pressure switch. A failed fuel pump relay will result in extended engine crank times in order to build up enough oil pressure to close the switch and turn on the fuel pump.

In-Line Fuel Filter

The fuel filter is serviced only as a complete unit. O-rings are used at threaded connections to prevent fuel leakage.

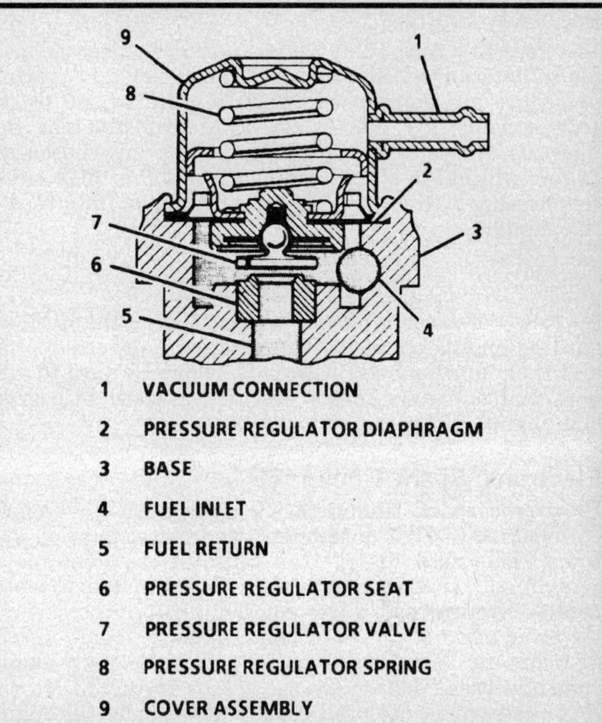

1	VACUUM CONNECTION
2	PRESSURE REGULATOR DIAPHRAGM
3	BASE
4	FUEL INLET
5	FUEL RETURN
6	PRESSURE REGULATOR SEAT
7	PRESSURE REGULATOR VALVE
8	PRESSURE REGULATOR SPRING
9	COVER ASSEMBLY

Cross sectional view of fuel pressure regulator — 3.1L engine

The threaded flex hoses connect the filter to the fuel tank feed line. The filter housing is constructed of steel with either threaded or quick-connect fittings. There is no service interval for the filter, only replace if restricted.

VFV Fuel Sensor

The variable fuel sensor monitors the concentration of alcohol in the fuel feed line. The sensor consists of a variable coaxial capacitor, and electronic circuitry, to generate a signal used by the ECM to adjust engine controls for the energy content of the fuel.

Fuel passing through the sensor causes a voltage to build up within the capacitor, which is translated into a signal sent to the ECM. As the percentage of alcohol increases, the voltage built up by the capacitor increases. When the concentration of alcohol decreases, so does the voltage level of the capacitor.

The variable fuel sensor also compensates for fuel temperature, since temperature will affect the voltage output of the unit. The ECM uses the information supplied from the variable fuel sensor for injection pulse width, and fuel pump output speed control. Because the energy content of these alternative fuels is lower than that of gasoline, the higher the concentration of alcohol in the blend means the greater volume of fuel that needs to be delivered to the engine. Equal amounts of M85 and gasoline for example, would result in an unequal mount of energy available. A malfunction in the variable coaxial capacitor within the variable fuel sensor will cause a DTC 56, 57, or 58 to be stored in the ECM memory. A fault in the temperature sensor portion of the variable fuel sensor will store a 64, or 65.

Pass-Key Signal

The Pass-Key Anti-Theft System employs a decoder module to determine if the correct ignition key and starting procedure are being used. When it recieves the proper code it sends a signal to the ECM to enable the fuel injectors. If this signal is interupted for any reason the engine will fail to start. The scan tool will display either the message "OK" or "FAULT" depending on the status of this signal.

DATA SENSORS

A variety of sensors provide information to the ECM regarding engine operating characteristics. These sensors and their functions are described below. Be sure to take note that not every sensor described is used with every GM engine application.

Electronic Spark Timing (EST)

Electronic spark timing (EST) is used on all engines equipped with HEI distributors and direct ignition systems. The EST distributor contains no vacuum or centrifugal advance and uses a 7-terminal distributor module. It also has 4 wires going to a 4-terminal connector in addition to the connectors normally found on HEI distributors. A reference pulse, indicating both engine rpm and crankshaft position, is sent to the ECM. The ECM determines the proper spark advance for the engine operating conditions and sends an **EST** pulse to the distributor.

The EST system is designed to optimize spark timing for better control of exhaust emissions and for fuel economy improvements. The ECM monitors information from various engine sensors, computes the desired spark timing and changes the timing accordingly. A backup spark advance system is incorporated in the module in case of EST failure.

The basic function of the fuel control system is to control the fuel delivery to the engine. The fuel is delivered to the engine by individual fuel injectors mounted on the intake manifold near each cylinder.

The main control sensor is the oxygen sensor which is located in the exhaust manifold. The oxygen sensor tells the ECM how much oxygen is in the exhaust gas and the ECM changes the air/fuel ratio to the engine by controlling the fuel injectors. The best mixture (ratio) to minimize exhaust emissions is 14.7:1 which allows the catalytic converter to operate the most efficiently. Because of the constant measuring and adjusting of the air/fuel ratio, the fuel injection system is called a CLOSED LOOP system.

Electronic Spark Control (ESC)

When engines are equipped with ESC in conjunction with EST, ESC is used to reduce spark advance under conditions of detonation. A knock sensor signals a separate ESC controller to retard the timing when it senses knock. The ESC controller signals the ECM which reduces spark advance until no more signals are received from the knock sensor.

Engine Coolant Temperature (ECT) Sensor

The coolant sensor is a thermister (an ohms resistor which changes value based on temperature) mounted in the engine coolant stream. As the temperature of the engine coolant changes, the resistance of the coolant sensor changes. Low coolant temperature produces a high resistance (100,700 ohms at -40°C/-40°F), while high temperature causes low resistance (177 ohms at 212°F/100°C).

The ECM supplies a 5 volt signal to the coolant sensor and measures the voltage that returns. By measuring the voltage change, the ECM determines the engine coolant temperature. The voltage will be high when the engine is cold and low when the engine is hot. This information is used to control fuel management, IAC, spark timing, EGR, canister purge and other engine operating conditions.

A failure in the coolant sensor circuit should either set a Code 14 or 15. These codes indicate a failure in the coolant temperature sensor circuit. Once the trouble code is set, the ECM will use a default valve for engine coolant temperature.

Oxygen (O$_2$) Sensor

The oxygen sensor is mounted in the exhaust system where it can monitor the oxygen content of the exhaust gas stream. The oxygen content in the exhaust reacts with the oxygen sensor to produce a voltage output. This voltage ranges from approximately 100 millivolts (high oxygen — lean mixture) to 900 millivolts (low oxygen — rich mixture).

By monitoring the voltage output of the oxygen sensor, the ECM will determine what fuel mixture command to give to the injector (lean mixture — low voltage — rich command, rich mixture — high voltage — lean command).

Remember that oxygen sensor indicates to the ECM what is happening in the exhaust. It does not cause things to happen. It is a type of gauge: high oxygen content = lean mixture; low oxygen content = rich mixture.

The oxygen sensor, if open should set a Code 13. a constant low voltage in the sensor circuit should set a Code 44 while a constant high voltage in the circuit should set a Code 45. Codes 44 and 45 could also be set as a result of fuel system problems.

Manifold Absolute Pressure (MAP) Sensor

The Manifold Absolute Pressure (MAP) sensor measures the changes in the intake manifold pressure which result from engine load and speed changes. The pressure measured by the MAP sensor is the difference between barometric pressure (outside air) and manifold pressure (vacuum). A closed throttle engine coastdown would produce a relatively low MAP value (approximately 20–35 kPa), while wide-open throttle would produce a high value (100 kPa). This high value is produced when the pressure inside the manifold is the same as outside the manifold, and 100% of outside air (or 100 kPa) is being measured. This MAP output is the opposite of what you would measure on a vacuum gauge. The use of this sensor also allows the ECM to adjust automatically for changing altitudes.

The ECM sends a 5 volt reference signal to the MAP sensor. As the MAP changes, the electrical resistance of the sensor also changes. By monitoring the sensor output voltage the ECM can determine the manifold pressure. A higher pressure, lower vacuum (high voltage: 4.0–4.5 volts) requires more fuel, while a lower pressure, higher vacuum (low voltage: 1.0–1.5 volts) requires less fuel. The ECM uses the MAP sensor to control fuel delivery and ignition timing. A failure in the MAP sensor circuit should set a Code 33 or Code 34 and the TPS will then be substituted to control fuel delivery.

Manifold/Intake Air Temperature (MAT/IAT) Sensor

The Manifold/Intake Air Temperature (MAT/IAT) sensor is a thermistor mounted in the intake manifold or air inlet duct. A thermistor is a resistor which changes resistance based on temperature. Low manifold air temperature produces a high resistance (100,700 ohms at -40°F/-40°C), while high temperature cause low resistance (177 ohms at 212°F/100°C).

The ECM supplies a 5 volt signal to the MAT sensor through a resistor in the ECM and monitors the voltage. The voltage will be high when the manifold air is cold and low when the air is hot. By monitoring the voltage, the ECM calculates the air temperature and uses this data to help determine the fuel delivery and spark advance. A failure in the MAT/IAT circuit should set either a Code 23 or Code 25. Once the trouble code is set, the ECM will use an artificial default value for the MAT/IAT and some vehicle performance will return.

Mass Air Flow (MAF) Sensor

The Mass Air Flow (MAF) sensor measures the amount of air which passes through it. The ECM uses this information to determine the operating condition of the engine and to control fuel delivery. A large quantity of air indicates acceleration, while a small quantity indicates deceleration or idle.

A scan tool will display air flow in terms of grams of air per second (gm/sec), with a range from 3 gm/sec to 150 gm/sec. A fully warm engine at idle should produce 4–6 gm/sec. The 3.3L engine is equipped with a MAF sensor, all other engines use a MAP sensor.

Vehicle Speed Sensor (VSS)

> **NOTE: A vehicle equipped with a speed sensor, should not be driven without the speed sensor connected, as idle quality may be affected.**

The vehicle speed sensor (VSS) is mounted behind the speedometer in the instrument cluster or on the transmission/speedometer drive gear. The sensor is a Permanent Magnent (PM), which generates a small AC voltage and transmits the signal to the ECM. The pulses indicate the road speed. The ECM uses this information to operate the IAC, canister purge, and TCC. A failure in the VSS circuit should set a Code 24.

Throttle Position Sensor (TPS)

The Throttle Position Sensor (TPS) is connected to the throttle shaft and is controlled by the throttle mechanism. A 5 volt reference signal is sent to the TPS from the ECM. As the throttle valve angle is changed (accelerator pedal moved), the resistance of the TPS also changes. At a closed throttle position, the resistance of the TPS is high, so the output voltage to the ECM will be low (approximately 0.5 volt). As the throttle plate opens, the resistance decreases so that, at wide open throttle, the output voltage should be approximately 5 volts. At closed throttle position, the voltage at the TPS should be less than 1.25 volts.

By monitoring the output voltage from the TPS, the ECM can determine fuel delivery based on throttle valve angle (driver demand). The TPS can either be misadjusted, shorted, open or loose. Misadjustment might result in poor idle or poor wide-open throttle performance.

An open TPS signals the ECM that the throttle is always closed, resulting in poor performance. This usually sets a Code 22. A shorted TPS gives the ECM a constant wide-open throttle signal and should set a Code 21. A loose TPS indicates to the ECM that the throttle is moving. This causes intermittent bursts of fuel from the injector and an unstable idle. Once the trouble code is set, the ECM will use an artificial default value for the TPS and some vehicle performance will return. The ECM has the ability to auto-zero the TPS voltage when the sensor is indicating a voltage of less than 0.9 volts. This means that any voltage reading less than 0.9 volts is considered 0% throttle angle by the ECM.

Crankshaft Position (CKP) Sensor

All engines equipped with E.I. (Electronic Ignition: engine uses no distributor) use a crank sensor to measure engine rpm, indicate crankshaft position and fire the ignition coils.

All E.I. systems except the 3.3L engine use a magnetic crankshaft position sensor, mounted remotely from the ignition module, which protrudes into the block within approximately 0.050 in. of the crankshaft reluctor. The 3.3L engine uses a Dual Cranks sensor which is mounted externally on the front of the engine near the harmonic balancer.

EXCEPT 3.3L ENGINE

The CKP sensor used on all except the 2.0L engine, have an internal reluctor is a special wheel cast into the crankshaft with 7 slots machined into it, 6 of which are equally spaced (60 degrees apart). A seventh slot is spaced approximately 10 degrees from one of the other slots and severs to generate a **SYNC PULSE** signal. The 2.0L engine also has an internal reluctor on the crankshaft, but the reluctor ring has 58 teeth with 2 teeth missing to allow for absolute crankshaft position referencing. As the reluctor rotates as part of the crankshaft, the slots change the magnetic field of the sensor, creating an induced voltage pulse.

3.3L ENGINE

The dual crank sensor is mounted in a pedestal on the front of the engine near the harmonic balancer. The sensor consists of 2 Hall Effect switches, which depend on 2 metal interrupter rings mounted on the balancer to activate them. Windows in the interrupters activate the hall effect switches as they provide a patch for the magnetic field between the switches transducers and magnets. When one of the hall effect switches is activated, it grounds the signal line to the C^3I module, pulling that signal line's (Sync Pulse or Crank) applied voltage low, which is interpreted as a signal.

Because of the way the signal is created by the dual crank sensor, the signal circuit is always either at a high or low voltage (square wave signal). Three crank signal pulses and one **SYNC PULSE** are created during each crankshaft revolution. The crank signal is used by the C^3I module to create a reference signal which is also a square wave signal similar to the crank signal. The reference signal is used to calculate the engine rpm and crankshaft position by the ECM. The **SYNC PULSE** is used by the C^3I module to begin the ignition coil firing sequence starting with No. 3–6 coil. The firing sequence begins with this coil because either piston No. 3 or piston No. 6 is now at the correct position in compression stroke for the spark plugs to be fired. Both the crank sensor and the **SYNC**

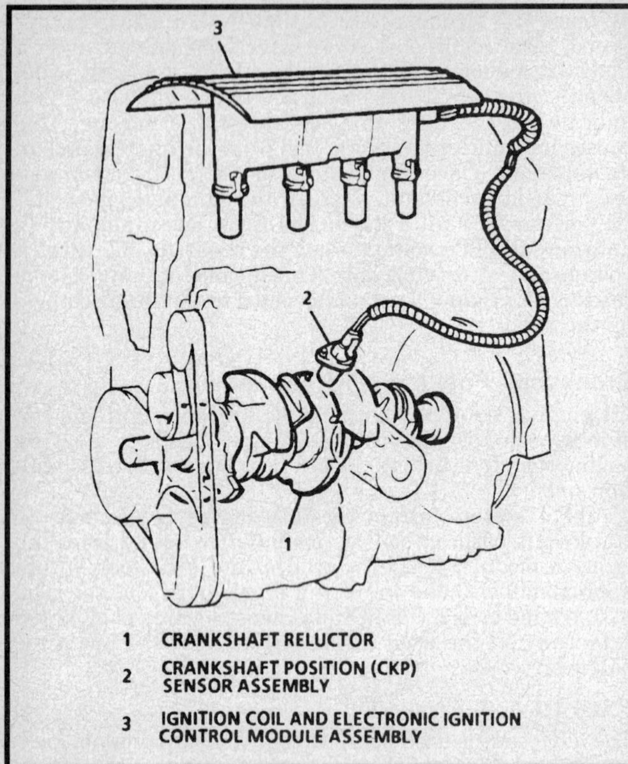

1   CRANKSHAFT RELUCTOR
2   CRANKSHAFT POSITION (CKP) SENSOR ASSEMBLY
3   IGNITION COIL AND ELECTRONIC IGNITION CONTROL MODULE ASSEMBLY

Crankshaft Position (CKP) sensor — 2.3L engine

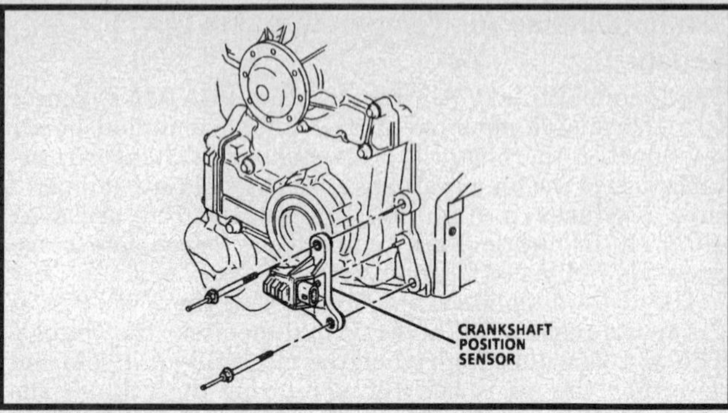

Dual crank sensor — 3.3L engine

PULSE signals must be received by the ignition module for the engine to start. A misadjusted sensor or bent interrupter ring could cause rubbing of the sensor resulting in potential driveability problems, such as rough idle, poor performance, or a nor start condition.

NOTE: **Failure to have the correct clearance will damage the crankshaft sensor.**

The dual crank sensor is not adjustable for ignition timing but positioning of the interrupter ring is very important. A clearance of 0.025 in. is required on either side of the interrupter ring. A dual crank sensor that is damaged, due to mispositioning or a bent interrupter ring, can result in a hesitation, sag stumble or dieseling condition.

To determine if the dual crank sensor could be at fault, scan the engine rpm with a suitable scan tool, while driving the vehicle. An erratic display indicates that a proper reference pulse has not been received by the ECM, which may be the result of a malfunctioning dual crank sensor.

Air Conditioning Pressure Sensor

The air conditioning (A/C) pressure sensor provides a signal to the electronic control module (ECM) which indicates varying high side refrigerant pressure between approximately 0–450 psi. The ECM used this input to the A/C compressor load on the engine to help control the idle speed with the IAC valve.

The A/C pressure sensor electrical circuit consists of a 5 volt reference line and a ground line, both provided by the ECM and a signal line to the ECM. The signal is a voltage that varies approximately 0.1 volt at 0 psi, to 4.9 volts at 450 psi or more. A problem in the A/C pressure circuits or sensor should set a Code 66 and will make the A/C compressor inoperative.

Detonation (Knock) Sensor

This sensor is a piezoelectric sensor located near the back of the engine (transmission end). It generates electrical impulses which are directly proportional to the frequency of the knock which is detected. A buffer then sorts these signals and eliminates all except for those frequency range of detonation. This information is passed to the ESC module and then to the ECM, so that the ignition timing advance can be retarded until the detonation stops.

Park/Neutral Switch

NOTE: **Vehicle should not be driven with the park/neutral switch disconnected as idle quality may be affected in park or neutral and a Code 24 (VSS) may be set.**

This switch indicates to the ECM when the transmission is in **P** or **N**. the information is used by the ECM for control on the torque converter clutch, EGR, and the idle air control valve operation.

Air Conditioning Request Signal

This signal indicates to the ECM that an air conditioning mode is selected at the switch and that the A/C low pressure switch is closed. The ECM controls the A/C and adjusts the idle speed in response to this signal.

Torque Converter Clutch Solenoid

The purpose of the torque converter clutch system is designed to eliminate power loss by the converter (slippage) to increase fuel economy. By locking the converter clutch, a more effective coupling to the flywheel is achieved. The converter clutch is operated by the ECM controlled torque converter clutch solenoid.

Power Steering Pressure Switch

The power steering pressure switch is used so that the power steering oil pressure pump load will not effect the engine idle. Turning the steering wheel increase the power steering oil pressure and pump load on the engine. The power steering pressure switch will close before the load can cause an idle problem. The ECM will also turn the A/C clutch off when high power steering pressure is detected.

Oil Pressure Switch

The oil pressure switch is usually mounted on the back of the engine, just below the intake manifold. Some vehicles use the oil pressure switch as a parallel power supply with the fuel pump relay, which will provide voltage to the fuel pump, after approximately 4 psi (28 kPa) of oil pressure is reached. This switch will also help prevent engine seizure by shutting off the power to the fuel pump and causing the engine to stop when the oil pressure is lower than 4 psi.

EMISSION CONTROL SYSTEMS

Various components are used to control the exhaust emissions from a vehicle. These components are controlled by the ECM based on different engine operating conditions. These components are described in the following paragraphs. Not all components are used on all engines.

Exhaust Gas Recirculation (EGR) System

The EGR is used to lower oxides of nitrogen (NOx) emission levels caused by high combustion temperatures. The EGR meters the exhaust gases into the intake manifold. The amount of exhaust gas admitted is adjusted by a vacuum or electronically controlled valve in response to engine operating conditions. When the valve is opened, the metered amount of exhaust gas is released into the intake manifold and then drawn into the combustion chamber to be reburned, thus lowering the combustion temperature.

There are 2 styles of mechanical EGR valves and 1 style of digitally controlled EGR valve:

The mechanical EGR valves will either have a "P" for Positive backpressure or "N" for Negative backpresure stamped on the top of the valve. The positive backpressure EGR valve has an internal vacuum bleed that will only close when there is a sufficient amount of exhaust backpressure, thus allowing vacuum to open the EGR diaphragm. The negative backpressure EGR valve has an internal vacuum bleed which will only close when there is no backpressure, thus a spring beneath the EGR diaphragm will close off the vacuum bleed and permit the diaphragm to build vacuum and open the EGR valve. The vacuum signal is provided by an EGR vacuum port in the throttle body valve. The ECM controls the EGR vacuum solenoid, permitting the solenoid to open under certain pre-programmed criteria from the TP, ECT and MAP sensors.

The digital EGR valve uses no vacuum lines and is completely controlled by the ECM. The 3 solenoids in the digital EGR are designed with different size orifices whhich permits up to 7 different combinations of exhaust flow into the intake manifold. The ECM will energize each solenoid and then raise the shaft off the seat a pre-determined amount. The ECM uses the TP, MAP and TP sensors to determine when to open the solenoids.

Positive Crankcase Ventilation (PCV) System

A closed Positive Crankcase Ventilation (PCV) system is used to provide more complete scavenging of crankcase vapors. Fresh air from the air cleaner is supplied to the crankcase, mixed with blow-by gases and then passed through a PCV valve into the induction system.

The primary mode of crankcase ventilation control is through the PCV valve which meters the mixture of fresh air and blow-by gases into the induction system at a rate dependent upon manifold vacuum.

To maintain the idle quality, the PCV valve restricts the ventilation system flow whenever intake manifold vacuum is designed to allow excessive amounts of blow-by

gases to backflow through the breather assembly into the air cleaner and through the throttle body to be consumed by normal combustion.

Evaporative Emission Control System (EECS)

The basic evaporative emission control system used on all vehicles uses the carbon canister storage method. This method transfers fuel vapor to an activated carbon storage device for retention when the vehicle is not operating. A ported vacuum signal is used for purging vapors stored in the canister.

CONTROLLED CANISTER PURGE

The ECM controls a solenoid valve which controls vacuum to the purge valve in the charcoal canister. In open loop, before a specified time has expired and below a specified rpm, the solenoid valve is energized and blocks vacuum to the purge valve. When the system is in closed loop, after a specified time and above a specified rpm, the solenoid valve is de-energized and vacuum can be applied to the purge valve. This releases the collected vapors into the intake manifold. On systems not using an ECM controlled solenoid, a Thermo Vacuum Valve (TVV) is used to control purge. See the appropriate vehicle sections for checking procedures.

Air Management Control

The air management system aids in the reduction of exhaust emissions by supplying air to either the catalytic converter, engine exhaust manifold, or to the air cleaner. The ECM controls the air management system by energizing or de-energizing an air switching valve. Operation of the air switching valve is dependent upon such engine operating characteristics as coolant temperature, engine load, and acceleration (or deceleration), all of which are sensed by the ECM.

PULSED SECONDARY AIR INJECTION (PAIR) SYSTEM

The PAIR system utilizes exhaust pressure pulsations to draw air into the exhaust system. Fresh air from the clean side of the air cleaner supplies filtered air to avoid dirt build-up on the check valve seat. The air cleaner also serves as a muffler for noise reduction. The internal mechanism of the Pulsair valve reacts to 3 distinct conditions.

The firing of the engine creates a pulsating flow of exhaust gases which are of positive (+) or negative (-) pressure. This pressure or vacuum is transmitted through external tubes to the Pulsair valve.

If the pressure is positive, the disc is forced to the closed position and no exhaust gas is allowed to flow past the valve and into the air supply line. If there is a negative pressure (vacuum) in the exhaust system at the valve, the disc will open, allowing fresh air to mix with the exhaust gases. Due to the inertia of the system, the disc ceases to follow the pressure pulsations at high engine rpm. At this point, the disc remains closed, preventing any further fresh air flow.

The PAIR shut-off valve also stops airflow to the PAIR system when the ECM determines that it is not necessary, such as deceleration.

AIR INJECTION REACTION (AIR) SYSTEM

The AIR system is capable of injecting air into the exhaust ports, catalytic converter or to atmosphere. The air pump is driven by a belt and mounted on the front of the engine. The Pressure Operated Electric Divert/Electric Switching (PEDES) valve contains 2 solenoidd valves which control the airflow direction. The solenoids are controlled by the ECM.

Catalytic Converter

Of all emission control devices available, the catalytic converter is the most effective in reducing tailpipe emissions. The major tailpipe pollutants are hydrocabons (HC), carbon monoxide (CO), and oxides of nitrogen (NOx).

Diagnosis and Testing

SERVICE PRECAUTIONS

When working around any part of the fuel system, take precautionary steps to prevent fire and/or explosion:

• Disconnect negative terminal from battery (except testing where battery voltage is required).

• Whenever possible, use a flashlight instead of a drop light.

• Keep all open flame and smoking material out of the area.

• Use a shop cloth or similar to absorb fuel when opening a fuel system.

• Relieve fuel system pressure before servicing.

• Use suitable eye protection.

• Always keep a dry chemical (class B) fire extinguisher near the area.

Handling Methanol (M85) or Ethanol (E85) Fuels

• Both methanol and ethanol, like all fuels, are poisonous and should be handled with great care. These fuels can be absorbed into your body through the skin much faster than gasoline can, and can cause headaches, blindness, loss of memory, and even death.

• Avoid spills. If either fuel comes in contact with your skin, wash the exposed area immediately with ample amounts of soap and water. If any fuel contacts your clothing, change clothes right away and wash your skin immediately. Allow contaminated clothes to air dry first, then machine wash.

• Always wear methanol/ethanol resistant gloves, and eye or face protection, and NEVER try to siphon these fuels by mouth. Drinking a mouthful of these fuels can cause death. Additionally, work in a well ventilated area, as these fuels give off harmful vapors which can cause serious bodily harm if ingested over a long period of time. If fumes become prominent, move to an area of fresh air immediately.

• Partial combustion of methanol yields Formaldehyde which is suspected to be a cancer causing agent. A by-product of ethanol fuel combustion is acetaldehyde, which causes burning of the eyes, nose, and throat. When working in an enclosed space with the engine running, it is absolutely essential to use exhaust hoses.

CAUTION

Due to the amount of fuel pressure in the lines, the system must be depressurized prior to performing any work to avoid personal injury or damage to the vehicle.

Electrostatic Discharge Damage

Electronic components used in the control system are often design to carry very low voltage and are very susceptible to damage caused by electrostatic discharge. It is possible for less than 100 volts of static electricity to cause damage to some electronic components. By comparison it takes as much as 4000 volts for a person to even feel the zap of a static discharge.

There are several ways for a person to become statically charged. The most common methods of charging are by friction and induction. An example of charging by friction is a person sliding across a car seat, in which a charge as much as 25000 volts can build up. Charging by induction occurs when a person with well insulated shoes stands near a highly charged object and momentarily touches ground. Charges of the same polarity are drained off, leaving the person highly charged with the opposite polarity. Static charges of either type can cause damage, therefore, it is important to use care when handling and testing electronic components. Follow the guidelines listed below to avoid electrostatic discharge.

• To prevent possible electrostatic discharge damage to the ECM, do not touch the connector pins or soldered components on the circuit board.

• Always touch a known ground before handling any parts.

• When using a voltmeter, always connect the ground lead first.

• Do not remove any parts from there package until it is time to install them.

• Before removing the part from the package, touch the package to a good ground.

ECM LEARNING ABILITY

The ECM has a learning capability. If the battery is disconnected the learning process has to begin all over again. A change may be noted in the vehicle's performance. To teach the ECM, insure the vehicle is at operating temperature and drive at part throttle, with moderate acceleration and idle conditions, until performance returns.

DATA LINK CONNECTOR (DLC)

The DLC, previously known as the Assembly Line Diagnostic Link (ALDL), is a diagnostic connector located in the passenger compartment, usually under the instrument panel.

The manufacturers assembly plant where the vehicle originated used this connector to check the engine for proper operation before leaving. Vehicles with the DLC system, enter the Diagnostic mode or the Field Service Mode by connecting or jumping terminal B (diagnostic TEST terminal) to terminal A (ground circuit). This connector is a very useful tool in diagnosing fuel injected engines. Important information from the ECM is available at this terminal and can be read with one of the many popular scan tools.

READING CODES AND DIAGNOSTIC MODES

This information is able to be read by putting the ECM into 1 of 4 different modes. These modes are entered by inserting a specific amount of resistance between the ALDL connector terminals A and B. The modes and resistances needed to enter these modes are as follows:

Diagnostic Modes — 0 Ohms

When 0 resistance is between terminals A and B of the ALDL connector, the diagnostic mode is entered. There are 2 positions to this mode. One with the engine **OFF**, but the ignition **ON**; the other is when the engine is running called Field Service Mode.

If the diagnostic mode is entered with the engine in the **OFF** position, trouble codes will flash and the idle air control motor will pulsate in and out. Also, the relays and solenoids are energized with the exception of the fuel pump and injector.

As a bulb and system check, the Malfunction Indicator Light (MIL) will come on with the ignition switch **ON** and the engine not running. When the engine is started, the SERVICE ENGINE SOON light will turn off. If the SERVICE ENGINE SOON light remains on, the self-diagnostic system has detected a problem.

If the B terminal is then grounded with the ignition **ON**, engine not running, each trouble code will flash and repeat 3 times. If more than 1 problem has been detected, each trouble code will flash 3 times. Trouble codes will flash in numeric order (lowest number first). The trouble code series will repeat as long as the B terminal is grounded.

A trouble code indicates a problem in a given circuit (Code 14, for example, indicates a problem in the coolant sensor circuit; this includes the coolant sensor, connector harness, and ECM). The procedure for pinpointing the problem can be found in diagnosis. Similar charts are provided for each code.

Also in this mode all ECM controlled relays and solenoids except the fuel pump relay. This allows checking the circuits which may be difficult to energize without driving the vehicle and being under particular operating conditions. The IAC valve will move to its fully extended position on most vehicles, block the idle air passage. This is useful in checking the minimum idle speed.

Field Service Mode — 0 Ohms

When the ALDL connector terminal B is grounded with the engine running, the ECM goes into the field service mode. In this mode, the SERVICE ENGINE SOON light flashes closed or open loop and indicates the rich/lean status of the engine. The ECM runs the engine at a fixed ignition timing advanced above the base setting.

The SERVICE ENGINE SOON light will show whether the system is in Open loop or Closed loop. In Open loop the SERVICE ENGINE SOON light flashes 2 times and one half times per second. In Closed loop the light flashes once per second. Also in closed loop, the light will stay OUT most of the time if the system is too lean. It will stay ON most of the time if the system is too rich. In either case the Field Service mode check, which is part of the Diagnostic circuit check, will lead the technician into choosing the correct diagnostic chart to refer to.

Back-up Mode — 3.9 Kilo-ohms

The backup mode is entered by applying 3.9 kilo-ohms resistance between terminals A and B of the ALDL connector with the ignition switch in the **ON** position. The ALDL scanner tool can now read 5 of the 20 parameters on the data stream. These parameters are as mode status, oxygen sensor voltage, rpm, block learn and idle air control. There are 2 ways to enter the backup mode. Using a scan tool is one way; putting a 3.9 kilo-ohms resistor across terminals A and B of the ALDL is another.

Special Mode — 10 Kilo-ohms

This special mode is entered by applying a 10K ohms resistor across terminals A and B. When this happens the ECM does the following:

1. Allows all of the serial data to be read.
2. Bypasses all timers.
3. Add a calibrated spark advance.
4. Enables the canister purge solenoid on some engines.
5. Idles at 1000 rpm fixed idle air control and fixed base pulse width on the injector.
6. Forces the idle air control to reset at part throttle (approximately 2000 rpm).
7. Disables the park/neutral restrict functions

Open or Road Test Mode — 20 Kilo-ohms

The system is in this mode during normal operation and is used by a scan tool to extract data while driving the vehicle.

Scan Tester Information

A display unit (scanner, monitor, etc), allows a technician to read the engine control system information from the DLC under the instrument panel. It can provide information faster than a digital voltmeter or ohmmeter can. The scan tool does not diagnose the exact location of the problem. The tool supplies information about the ECM, the information that it is receiving and the commands that it is sending plus special information such as integrator and block learn. To use a display tool you should understand throughly how an engine control system operates.

A scanner or monitor puts a fuel injection system into a special test mode. This mode commands an idle speed of 1000 rpm. The idle quality cannot be evaluated with a tester plugged in. Also the test mode commands a fixed spark with no advance. On vehicles with Electronic Spark Control (ESC), there will be a fixed spark, but it will be advanced. On vehicles with ESC, there might be a serious spark knock, this spark knock could be bad enough so as not being able to road test the vehicle in the test mode. Be sure to check the tool manufacturer for instructions on special test modes which should overcome these limitations.

When a tester is used with a fuel injected engine, it bypasses the timer that keeps the system in Open loop for a certain period of time. When all Closed loop conditions are met, the engine will go into Closed loop as soon as the vehicle is started. This means that the air management system will not function properly and air may go directly to the converter as soon as the engine is started.

These tools cannot diagnose everything. They do not tell the technician where a problem is located in a circuit. The diagnostic charts must still be used to pinpoint the problems. These tester's do not let a technician know if a solenoid or relay has been turned on. They only tell the technician the ECM command. To find out if a solenoid has been turned on, check it with a suitable test light or digital voltmeter, or see if vacuum routing through the solenoid changes.

Scan Tool for Intermittents

In some scan tool applications, the data update rate may make the tool less effective than a voltmeter, such as when trying to detect an intermittent problem which lasts for a very short time. Some scan tools have a snapshot function which stores several seconds or even minutes of operation which helps to locate an intermittent problem. Scan tools allow the technisian to manipulate the wiring harness or components under the hood with the engine not running while observing the scan tool's readout.

The scan tool can be plugged in and observed while driving the vehicle under the condition when the MIL turns on momentarily or when the engine driveability is momentarily poor. If the problem seems to be related to certain parameters that can be checked on the scan tool, they should be checked while driving the vehicle. If there does not seem to be any correlation between the problem and any specific circuit, the scan tool can be checked on each position. Watching for a period of time to see if there is any change in the reading that indicates intermittent operation.

The scan tool is also an easy way to compare the operating parameters of a poorly operating engine with typical scan data for the vehicle being serviced or those of a known good engine. For example, a sensor may shift in value but not set a trouble code. Comparing the sensor's reading with those of a known good parameters may uncover the problem.

The scan tool has the ability to save time in diagnosis and prevent the replacement of good parts. The key to using the scan tool successfully for diagnosis lies in the technician's ability to understand the system he is trying to diagnose as well as an understanding of the scan tool's operation and limitations.

CLEARING TROUBLE CODES

When the ECM detects a problem with the system, the MIL will come on and a trouble code will be recorded in the ECM memory. If the problem is intermittent, the MIL will go out after 10 seconds, when the fault goes away. However the trouble code will stay in the ECM memory until the battery voltage to the ECM is removed. Removing the battery voltage for 30 seconds will clear all trouble codes. Do this by disconnecting the ECM harness from the positive battery terminal pigtail for 30 seconds with the key in the **OFF** position, or by removing the ECM fuse for 30 seconds with the key **OFF**.

> **NOTE: To prevent ECM damage, the key must be OFF when disconnecting and reconnecting ECM power.**

DIAGNOSTIC TROUBLE CODE DESCRIPTION

> **NOTE: The list of trouble codes below are a representation of all MFI vehicles, refer to the Troubleshooting Chart for the actual diagnostic code per engine application.**

- DTC **12** — Diagnostic check only (flash code)
- DTC **13** — Oxygen (O_2) sensor circuit open
- DTC **14** — Engine Coolant Temperature (ECT) sensor circuit out of range (high/low)
- DTC **15** — Engine Coolant Temperature (ECT) sensor circuit out of range (low)
- DTC **16** — System voltage high or low — **OR** — Direct/electronic ignition system fault line circuit error
- DTC **17** — Camshaft position sensor/spark reference circuit error
- DTC **18** — Crankshaft/camshaft error

- DTC **19** — Ignition control module signal circuit error — **OR** — Crankshaft position sensor circuit
- DTC **21** — Throttle Position Sensor (TPS) circuit voltage out of range
- DTC **22** — Throttle Position Sensor (TPS) circuit voltage out of range
- DTC **23** — Intake/manifold air temperature sensor circuit voltage out of range
- DTC **24** — Vehicle speed sensor circuit signal error
- DTC **25** — Intake/manifold air temperature sensor circuit voltage out of range
- DTC **26** — Quad-driver module circuit error
- DTC **27** — Gear switch circuits error
- DTC **28** — Gear switch circuits error — **OR** — Quad-driver module control circuit
- DTC **29** — Gear switch circuits error
- DTC **31** — PRNDL/transmission neutral start/park-neutral position switch circuit error — **OR** — Turbo wastegate overboost — **OR** — Camshaft sensor circuit error
- DTC **32** — Exhaust Gas Recirculation (EGR) system fault
- DTC **33** — Mass air flow/manifold absolute pressure sensor circuit voltage out of range
- DTC **34** — Mass air flow/manifold absolute pressure sensor circuit voltage out of range
- DTC **35** — Idle speed/air/idle air control error
- DTC **36** — Transaxle shift problem — **OR** — Ignition control signal circuit error
- DTC **38** — Brake switch circuit error
- DTC **39** — Torque converter/transaxle clutch switch circuit error
- DTC **41** — Cam position sensor/cylinder select error
- DTC **42** — Electronic spark timing/ignition control circuit open or shorted
- DTC **43** — Knock sensor/electronic spark control circuit error
- DTC **44** — Oxygen (O_2) sensor circuit indicates system lean
- DTC **45** — Oxygen (O_2) sensor circuit indicates system rich
- DTC **46** — Power steering pressure circuit error — **OR** — Vehicle anti-theft system circuit
- DTC **48** — Misfire diagnosis
- DTC **51** — ECM, PCM prom memory error
- DTC **52** — PCM prom memory error — **OR** — Engine oil temperature sensor circuit indicates low
- DTC **53** — System over voltage — **OR** — Exhaust Gas Recirculation (EGR) system malfunction — **OR** — Vehicle anti-theft system circuit error
- DTC **54** — Fuel pump circuit voltage low — **OR** — Exhaust Gas Recirculation (EGR) system fault
- DTC **55** — PCM error — **OR** — Mixture control solenoid/fuel lean monitor circuit error — **OR** — Exhaust Gas Recirculation (EGR) system fault
- DTC **56** — Quad-driver module circuit error — **OR** — Vacuum sensor circuit error — **OR** — Secondary air inlet valve actuator vacuum sensor circuit error
- DTC **57** — Boost control solenoid circuit error
- DTC **58** — Personal automotive security system (PASS-key®ii) fuel enable circuit error
- DTC **61** — Degraded oxygen (O_2) sensor — **OR** — Cruise control vent solenoid circuit error — **OR** — A/C system circuit error — **OR** — Secondary port throttle valve system error
- DTC **62** — Transaxle gear switch signal circuit error — **OR** — Cruise control vacuum solenoid circuit error — **OR** — Engine oil temperature sensor circuit

- DTC **63** — Cruise control system error — **OR** — Exhaust Gas Recirculation (EGR) flow check — **OR** — Oxygen (0_2) sensor (right bank) circuit open
- DTC **64** — Exhaust Gas Recirculation (EGR) flow check — **OR** — Oxygen (0_2) sensor (right bank) circuit indicates lean
- DTC **65** — Cruise control pressure sensor circuit error — **OR** — Exhaust Gas Recirculation (EGR) flow check — **OR** — Oxygen (0_2) sensor (right bank) circuit indicates rich — **OR** — Fuel injector circuit error
- DTC **66** — A/C refrigerant pressure sensor circuit — **OR** — Engine power switch circuit error
- DTC **67** — Cruise control switches circuit error
- DTC **68** — A/C compressor relay circuit shorted — **OR** — Cruise control system problem
- DTC **69** — A/C compressor relay circuit open — **OR** — A/C head pressure switch circuit error
- DTC **70** — A/C refrigerant pressure sensor circuit indicates high
- DTC **71** — A/C evaporator temperature sensor circuit indicates low
- DTC **72** — Gear selector switch circuit error
- DTC **73** — A/C evaporator temperature sensor circuit indicates high
- DTC **75** — Exhaust Gas Recirculation (EGR) No. 1 solenoid circuit error
- DTC **76** — Exhaust Gas Recirculation (EGR) No. 2 solenoid circuit error
- DTC **77** — Exhaust Gas Recirculation (EGR) No. 3 solenoid circuit error
- DTC **81** — Brake switch error
- DTC **82** — Ignition control signal error
- DTC **85** — PCM prom error
- DTC **86** — A/C multiplexer chip error
- DTC **87** — Electronically Erasable Programmable Read Only Memory (EEPROM) error

INTEGRATOR AND BLOCK LEARN

The integrator and block learn functions of the ECM are responsible for making minor adjustments to the air/fuel ratio on the fuel injected GM vehicles. These small adjustments are necessary to compensate for pinpoint air leaks and normal wear.

The integrator and block learn are 2 separate ECM memory functions which control fuel delivery. The integrator makes a temporary change and the block learn makes a more permanent change. Both of these functions apply only while the engine is in CLOSED LOOP. They represent the on-time of the injector. Also, integrator and block learn controls fuel delivery on the fuel injected engines as does the MC solenoid dwell on the CCC carbureted engines.

Integrator

Integrator is the term applied to a means of temporary change in fuel delivery. The Integrator value is displayed through the DLC and monitored with a scanner as a number between 0 and 255 with an average of 128. The integrator monitors the oxygen sensor output voltage and adds and subtracts fuel depending on the lean or rich condition of the oxygen sensor. When the integrator is displaying 128, it indicates a neutral condition. This means that the oxygen sensor is seeing results of the 14.7:1 air/fuel mixture burned in the cylinders.

> **NOTE: An air leak in the system (a lean condition) would cause the oxygen sensor voltage to decrease while the integrator would increase (add more fuel) to temporarily correct for the lean condition. If this happened the injector pulse width would increase.**

Block Learn

Although the integrator can correct fuel delivery over a wide range, it is only for a temporary correction. Therefore, another control called block learn was added. Although it cannot make as many corrections as the integrator, it does so for a longer period of time. It gets its name from the fact that the operating range of the engine for any given combinations of rpm and load is divided into 16 cell or blocks.

The computer has a given fuel delivery stored in each block. As the operating range gets into a given block, the fuel delivery will be based on what value is stored in the memory in that block. Again, just like the integrator, the number represents the on-time of the injector. Also, just like the integrator, the number 128 represents no correction to the value that is stored in the cell or block. When the integrator increases or decreases, block learn which is also watching the integrator will make corrections in the same direction. As the block learn makes corrections, the integrator correction will be reduced until finally the integrator will return to 128 if the block learn has corrected the fuel delivery.

BLOCK LEARN MEMORY

Block learn operates on 1 of 2 types of memories depending on application, non-volatile and volatile. The non-volatile memories retain the value in the block learn cells even when the ignition switch is turned **OFF**. When the engine is restarted, the fuel delivery for a given block will be based on information stored in memory.

The volatile memories lose the numbers stored in the block learn cells when the ignition is turned to the **OFF** position. Upon restarting, the block learn starts at 128 in every block and corrects from that point as necessary.

INTEGRATOR/BLOCK LEARN LIMITS

Both the integrator and block learn have limits which will vary from engine to engine. If the mixture is off enough so that the block learn reaches the limit of its control and still cannot correct the condition, the integrator would also go to its limit of control in the same direction and the engine would then begin to run poorly. If the integrators and block learn are close to or at their limits of control, the engine hardware should be checked to determine the cause of the limits being reached, vacuum leaks, sticking injectors, etc.

If the integrator is lied to, for example, if the oxygen sensor lead was grounded (lean signal) the integrator and block learn would add fuel to the engine to cause it to run rich. However, with the oxygen sensor lead grounded, the ECM would continue seeing a lean condition eventually setting a Code 44 and the fuel control system would change to open loop operations.

CLOSED LOOP FUEL CONTROL

The purpose of closed loop fuel control is to precisely maintain an air/fuel mixture 14.7:1. When the air/fuel mixture is maintained at 14.7:1, the catalytic converter is able to operate at maximum efficiency which results in lower emission levels.

Since the ECM controls the air/fuel mixture, it needs to check its output and correct the fuel mixture for deviations from the ideal ratio. The oxygen sensor feeds this output information back to the ECM.

ENGINE PERFORMANCE DIAGNOSIS

Engine performance diagnosis procedures are guides that will lead to the most probable causes of engine performance complaints. They consider the components of the fuel, ignition, and mechanical systems that could cause a particular complaint, and then outline repairs in a logical sequence.

It is important to determine if the MIL light is illuminated or has come on for a short interval while driving. If the MIL has illuminated the computer should be checked for stored **TROUBLE CODES** which may indicate the cause for the performance complaint.

All of the symptoms can be caused by worn out or defective parts such as spark plugs, ignition wiring, etc. If time and/or mileage indicate that parts should be replaced, it is recommended that it be done.

NOTE: Before checking any system controlled by the computer, the Diagnostic Circuit Check must be performed or misdiagnosis may occur. If the complaint involves the MIL, go directly to the Diagnostic Circuit Check.

Basic Troubleshooting

NOTE: The following explains how to activate the trouble code signal light in the instrument cluster. This is not a full system troubleshooting and isolation procedure.

Before suspecting the system or any of its components as faulty, check the ignition system including distributor, timing, spark plugs and wires. Check the engine compression, air cleaner, and emission control components not controlled by the ECM. Also check the intake manifold, vacuum hoses and hose connectors for leaks.

The following symptoms could indicate a possible problem with the system:
1. Detonation
2. Stalls or rough idle-cold
3. Stalls or rough idle-hot
4. Missing
5. Hesitation
6. Surges
7. Poor gasoline mileage
8. Sluggish or spongy performance
9. Hard starting-cold
10. Objectionable exhaust odors (rotten egg smell) or black smoke
11. Cuts out
12. Improper idle speed

As a bulb and system check, the MIL will come on when the ignition switch is turned to the **ON** position but the engine is not started. The MIL will also produce the trouble code or codes by a series of flashes which translate as follows. When the diagnostic test terminal under the dash is grounded, with the ignition in the **ON** position and the engine not running, the MIL will flash once, pause, then flash twice in rapid succession. This is a Code

12, which indicates that the diagnostic system is working. After a long pause, the Code 12 will repeat itself 2 more times. The cycle will then repeat itself until the engine is started or the ignition is turned off.

When the engine is started, the MIL will remain on for a few seconds, then turn off. If the MIL remains on, the self-diagnostic system has detected a problem. If the test terminal is then grounded, the trouble code will flash 3 times. If more than 1 problem is found, each trouble code will flash 3 times. Trouble codes will flash in numerical order (lowest code number to highest). The trouble codes series will repeat as long as the test terminal is grounded.

A trouble code indicates a problem with a given circuit. For example, trouble Code 14 indicates a problem in the cooling sensor circuit. This includes the coolant sensor, its electrical harness, and the ECM. Since the self-diagnostic system cannot diagnose every possible fault in the system, the absence of a trouble code does not mean the system is trouble-free. To determine problems within the system which do not activate a trouble code, a system performance check must be made.

In the case of an intermittent fault in the system, the MIL will go out when the fault goes away, but the trouble code will remain in the memory of the ECM. Therefore, if a trouble code can be obtained even though the MIL is not on, the trouble code must be evaluated. It must be determined if the fault is intermittent or if the engine must be at certain operating conditions (under load, etc.) before the MIL will come on. Some trouble codes will not be recorded in the ECM until the engine has been operated at part throttle for about 5–18 minutes.

INTERMITTENT MIL

An intermittent open in the ground circuit would cause loss of power through the ECM and intermittent MIL operation. When the ECM loses ground, distributor ignition is lost. An intermittent open in the ground circuit would be described as an engine miss.

Therefore, an intermittent MIL, no code stored and a driveability comment described as similar to a miss will require checking the grounding circuit and the Code 12 circuit as it originates at the ignition coil.

UNDERVOLTAGE TO THE ECM

An undervoltage condition below 9 volts will cause the MIL to come on as long as the condition exist.

Therefore, an intermittent MIL, no code stored and a driveability comment described as similar to a miss will require checking the grounding circuit, Code 12 circuit and the ignition feed circuit to terminal C of the ECM. This does nor eliminate the necessity of checking the normal vehicle electrical system for possible cause such as a loose battery cable.

OVERVOLTAGE TO THE ECM

The ECM will also shut off when the power supply rises above 16 volts. The overvoltage condition will also cause the MIL to come illuminate as long as this condition exist.

A momentary voltage surge in a vehicle's electrical system is a common occurrence. These voltage surges have never presented any problems because the entire electrical system acted as a shock absorber until the surge dissipated. Voltage surges or spikes in the vehicle's electrical system have been known, on occasion, to exceed 100 volts.

The system is a low voltage (between 9 and 16 volts) system and will not tolerate these surges. The ECM will be shut off by any surge in excess of 16 volts and will come back on, only after the surge has dissipated sufficiently to bring the voltage under 16 volts.

A surge will usually occur when an accessory requiring a high voltage supply is turned off or down. The voltage regulator in the vehicle's charging system cannot react to the changes in the voltage demands quickly enough and surge occurs. The driver should be questioned to determine which accessory circuit was turned off that caused the MIL to come on.

Therefore, intermittent MIL operation, with no trouble code stored, will require installation of a diode in the appropriate accessory circuit.

BAROMETRIC PRESSURE (BARO) SENSOR PARAMETERS CHART

BAROMETRIC PRESSURE (BARO) SENSOR	
Engine	Key On, Engine Running
2.0L (VIN H)	2.5–5.5 V
2.2L (VIN 4)	varies with barometric pressure
2.3L (VIN A, D, 3)	3.0–5.0 V
3.1L (VIN T)	2.5–5.5 V
3.4L (VIN X)	2.5–5.5 V
5.7L (VIN P)	3.5–4.5 V

NOTE: All specifications reflect those received via a scan tool
V—Volts

ENGINE COOLANT TEMPERATURE (ECT) SENSOR PARAMETERS CHART

ENGINE COOLANT TEMPERATURE (ECT) SENSOR		
Engine	Ohms	Temperature
2.0L (VIN H)	165–286	185–220°F (85–105°C)
2.2L (VIN 4)	165–286	185–220°F (85–105°C)
2.3L (VIN A, D, 3)	150–286	185–239°F (85–115°C)
3.1L (VIN T)	160–286	185–228°F (85–109°C)
3.3L (VIN N)	165–286	185–220°F (85–105°C)
3.4L (VIN X)	160–286	185–228°F (85–109°C)
5.0L (VIN F)	165–286	185–220°F (85–105°C)
5.7L (VIN 8, P)	165–286	185–220°F (85–105°C)

NOTE: All specifications reflect those received via a scan tool

IDLE AIR CONTROL (IAC) MOTOR PARAMETERS CHART

IDLE AIR CONTROL (IAC)

Engine	Key On, Engine Off	Key On, Engine Running
2.0L (VIN H)	above 50 counts	1–50 counts
2.2L (VIN 4)	above 50 counts	1–50 counts
2.3L (VIN A, D, 3)	above 50 counts	5–60 counts
3.1L (DI) (VIN T)	above 50 counts	5–50 counts
3.1L (EI) (VIN T)	above 50 counts	9–20 counts
3.3L (VIN N)	above 50 counts	10–30 counts
3.4L (VIN X)	above 50 counts	5–50 counts
5.0L (VIN F)	above 50 counts	5–50 counts
5.7L (VIN 8, P)	above 50 counts	5–50 counts

NOTE: Values are read via a scan tool

INTAKE/MAINFOLD AIR TEMPERATURE (IAT/MAT) SENSOR PARAMETERS CHART

INTAKE/MANIFOLD AIR TEMPERATURE (IAT/MAT) SENSOR

Engine	Key On, Engine Off	Key On, Engine Running
2.0L (VIN H)	①	50–194°F (10–90°C)
2.2L (VIN 4)	①	varies with air temp
2.3L (VIN A, D, 3)	①	50–194°F (10–90°C)
3.1L (VIN T)	①	50–176°F (10–80°C)
3.4L (VIN X)	①	50–176°F (10–80°C)
5.0L (VIN F)	①	50–194°F (10–90°C)
5.7L (VIN 8, P)	①	50–194°F (10–90°C)

NOTE: All specifications reflect those received via a scan tool
① Either underhood ambient temperature or engine temperature depending where the sensor is located.

MASS AIR FLOW (MAF) SENSOR PARAMETERS CHART

MASS AIR FLOW (MAF) SENSOR

Engine	Key On, Engine Off	Key On, Engine Running
3.3L (VIN N)	0–2 g/s	4–7 g/s

NOTE: All specifications reflect those received via a scan tool
g/s—grams per second

MANIFOLD AIR PRESSURE (MAP) SENSOR PARAMETERS CHART

MANIFOLD AIR PRESSURE (MAP) SENSOR

Engine	Key On, Engine Off	Key On, Engine Running
2.0L (VIN H)	above 4.0 V	1.0–2.0 V
2.2L (VIN 4)	above 4.0 V	1.0–2.0 V
2.3L (VIN A, D, 3)	above 4.0 V	1.0–2.0 V
3.1L (VIN T)	above 4.0 V	1.0–3.0 V
3.3L (VIN N)	above 4.0 V	1.0–3.0 V
3.4L (VIN X)	above 4.0 V	1.0–2.0 V
5.0L (VIN F)	above 4.0 V	1.0–2.0 V
5.7L (VIN 8, P)	above 4.0 V	1.0–2.0 V

NOTE: All specifications reflect those received via a scan tool
V—Volts

OXYGEN (O₂S) SENSOR PARAMETERS CHART

OXYGEN (O_2S) SENSOR

Engine	Key On, Engine Off	Key On, Engine Running
2.0L (VIN H)	350–500 mv	1–1000 mv
2.2L (VIN 4)	350–500 mv	1–1000 mv
2.3L (VIN A, D, 3)	350–500 mv	1–1000 mv
3.1L (VIN T)	350–500 mv	1–1000 mv
3.3L (VIN N)	350–500 mv	1–1000 mv
3.4L (VIN X)	350–500 mv	1–1000 mv
5.0L (VIN F)	350–500 mv	1–1000 mv
5.7L (VIN 8, P)	350–500 mv	1–1000 mv

NOTE: All specifications reflect those received via a scan tool
mv—millivolts

THROTTLE POSITION (TP) SENSOR PARAMETERS CHART

THROTTLE POSITION (TP) SENSOR

Engine	Key On, Engine Running
2.0L (VIN H)	0.33–1.33 V
2.2L (VIN 4)	0.33–1.33 V
2.3L (VIN A, D, 3)	0.20–0.90 V
3.1L (VIN T)	0.29–0.98 V
3.3L (VIN N)	0.20–0.74 V
3.4L (VIN X)	0.29–0.98 V
5.0L (VIN F)	0.36–0.62 V
5.7L (VIN 8)	0.36–0.62 V
5.7L (VIN P)	0.30–0.90 V

NOTE: All specifications reflect those received via a scan tool
V—Volts

THROTTLE POSITION (TP) SENSOR

Testing

Due to the varied application of components, a general procedure is outlined. For the exact procedure for the vehicle being serviced use Code 21 or 22 chart with the appropriate engine application.

WITH SCAN TOOL

1. Use a suitable scan tool to read the TPS voltage.
2. With the ignition switch **ON** and the engine **OFF**, the TPS voltage should be less than 1.25 volts.
3. If the voltage reading is higher than specified, replace the throttle position sensor.

WITHOUT SCAN TOOL

1. Remove air cleaner. Disconnect the TPS harness from the TPS.
2. Using suitable jumper wires, connect a digital voltmeter J–29125–A or equivalent to the correct TPS terminals A and B.
3. With the ignition **ON** and the engine running, The TPS voltage should be 0.3–1.0 volts at base idle to approximately 4.5 volts at wide open throttle.
4. If the reading on the TPS is out of specification, check the minimum idle speed before replacing the TPS.
5. If the voltage reading is correct, remove the voltmeter and jumper wires and reconnect the TPS connector to the sensor.
6. Reinstall the air cleaner.

FUEL SYSTEM

When the ignition switch is turned **ON**, the in-tank fuel pump is energized for as long as the engine is cranking or running and the control unit is receiving signals from the HEI distributor or DIS. If there are no reference pulses, the control unit will shut off the fuel pump within 2 seconds. The pump will deliver fuel to the fuel rail and injectors, then the pressure regulator where the system pressure is controlled to maintain 31–47 psi.

Pump Relay Testing

2.0L AND 2.2L ENGINES

The fuel pump relay is located on the right side of the engine compartment, next to the shock tower. The fuel pump test terminal is located on the left side of the engine compartment. The fuel pump may be tested by using a 10 amp fused jumper wire and applying 12 volts to the fuel pump test terminal while listening for the pump to be running. If the relay is suspected to be malfunctioning, proceed as follows:

1. Disconnect the negative battery cable.
2. Disconnect the fuel pump (5 terminal) relay.
3. Connect an ohmmeter between terminals 30 and 87A of the relay.
4. Verify the resistance is less than 5 ohms.
5. Connect a 12 volt source to terminals 85 and 86.
6. Verify the resistance is greater than 10 kilo-ohms between terminals 30 and 87A.
7. Next, connect an ohmmeter between terminals 30 and 87 of the relay with the 12 volt source still connected between terminals 85 and 86.
8. Verify the resistance is less than 5 ohms.
9. If not as specified, replace the relay.

2.3L AND 3.3L ENGINES

The fuel pump relay is located on the center front of the dash, on the relay bracket. The fuel pump test terminal is located on the left rear of the engine compartment. The fuel pump may be tested by using a 10 amp fused jumper wire and applying 12 volts to the fuel pump test terminal, while listening for the pump to be running.

1. Disconnect the negative battery cable.
2. Disconnect the fuel pump (5 terminal) relay.
3. Connect an ohmmeter between terminals 2 and 5 and then terminals 1 and 3 of the relay.
4. Verify there is continuity.
5. Next, connect a 12 volt source to terminals 2 and 5.
6. Verify the resistance is greater than 10 kilo-ohms between terminals 1 and 3.
7. Then connect an ohmmeter between terminals 1 and 4 of the relay with the 12 volt source still connected.
8. Verify there is continuity.
9. If not as specified, replace the relay.

3.4L ENGINE

The fuel pump relay is located in the right side electrical center. The fuel pump test terminal is located below the left side electrical center. The fuel pump may be tested by using a 10 amp fused jumper wire and applying 12 volts to the fuel pump test terminal, while listening for the pump to be running.

1. Disconnect the negative battery cable.
2. Disconnect the fuel pump (5 terminal) relay.
3. Connect an ohmmeter between terminals 1 and 2 and then terminals 3 and 4 of the relay.
4. Verify there is continuity.
5. Next, connect a 12 volt source to terminals 1 and 2.
6. Verify the resistance is greater than 10 kilo-ohms between terminals 3 and 4.
7. Then connect an ohmmeter between terminals 3 and 5 of the relay with the 12 volt source still connected.
8. Verify there is continuity.
9. If not as specified, replace the relay.

3.1L, 5.0L AND 5.7L (VIN 8) ENGINES

The fuel pump relay is located on the right rear of the engine compartment, behind the strut tower, on the firewall or below the right side instrument panel. The fuel pump test terminal is located on the left rear of the engine compartment on all vehicles. The fuel pump may be tested by using a 10 amp fused jumper wire and applying 12 volts to the fuel pump test terminal, while listening for the pump to be running.

1. Disconnect the negative battery cable.
2. Disconnect the fuel pump (5 terminal) relay.
3. Connect an ohmmeter between terminals 85 (F) and 86 (D) and then terminals 87A (C) and 30 (E) of the relay.
4. Verify there is continuity.
5. Next, connect a 12 volt source to terminals 85 (F) and 86 (D).
6. Verify the resistance is greater than 10 kilo-ohms between terminals 87A (C) and 30 (E).
7. Then connect an ohmmeter between terminals 87 (A) and 30 (E) of the relay with the 12 volt source still connected.
8. Verify there is continuity.
9. If not as specified, replace the relay.

5.7L (VIN P) ENGINE, EXCEPT CORVETTE

The fuel pump relay is located in the passenger compartment, forward of the left kick panel, mounted to the foot rest bracket below the carpet. The fuel pump test terminal is located on the top rear of the right shock tower. The fuel pump may be tested by using a 10 amp fused jumper wire and applying 12 volts to the fuel pump test terminal, while listening for the pump to be running.

1. Disconnect the negative battery cable.
2. Disconnect the fuel pump (6 terminal) relay.
3. Connect an ohmmeter between terminals A1 and C2 of the relay.
4. Verify there is continuity.
5. Next, connect a 12 volt source to terminals A1 and C2.
6. Verify there is continuity between terminals C1 and A2.
7. If not as specified, replace the relay.

5.7L (VIN P) ENGINE, CORVETTE

The fuel pump relay is located below the right side of the instrument panel in the engine compartment. The fuel pump test terminal is located on the top rear of the right shock tower. The fuel pump may be tested by using a scanner, while listening for the pump to be running.

1. Disconnect the negative battery cable.
2. Disconnect the fuel pump (5 terminal) relay.
3. Connect an ohmmeter between terminals 30 and 87A and terminals 85 and 86 of the relay.
4. Verify the resistance is less than 5 ohms between either pair of terminals.
5. Connect a 12 volt source to terminals 85 and 86.
6. Verify the resistance is greater than 10 kilo-ohms between terminals 30 and 87A.
7. Next, connect an ohmmeter between terminals 30 and 87 of the relay with the 12 volt source still connected between terminals 85 and 86.
8. Verify the resistance is less than 5 ohms.
9. If not as specified, replace the relay.

Pressure Testing

1. Connect pressure gauge J–34730–1, or equivalent, to fuel pressure test point on the fuel rail. Wrap a rag around the pressure tap to absorb any leakage that may occur when installing the gauge.
2. Turn the ignition **ON** and check that pump pressure is 41–47 psi. This pressure is controlled by spring pressure within the regulator assembly.
3. Start the engine and allow it to idle. The fuel pressure should drop 10 psi due to the lower manifold pressure.

> **NOTE: The idle pressure will vary somewhat depending on barometric pressure. Check for a drop in pressure indicating regulator control, rather than specific values.**

4. Turn the ignition **OFF** and observe the pressure gauge, verify there is no rapid decrease in fuel line pressure.
5. If the fuel pressure drops, check the operation of the check valve, the pump coupling connection, fuel pressure regulator valve and the injectors. A restricted fuel line or filter may also cause a pressure drop. To check the fuel pump output, restrict the fuel return line and run 12 volts to the pump. The fuel pressure should rise to at least 60 psi with the return line restricted, but do not allow the pressure to go above 60 psi.

—— CAUTION ——

Before attempting to remove or service any fuel system component, it is necessary to relieve the fuel system pressure.

SYMPTOMS CHART — MULTI-PORT INJECTION SYSTEMS

IMPORTANT PRELIMINARY CHECKS

- Before using this section you should have performed the "Diagnostic Circuit Check."
- Verify the customer complaint, and locate the correct SYMPTOM below. Check the items indicated under that symptom.
- If the ENGINE CRANKS BUT WILL NOT RUN, use CHART A-3.
- Several of the following symptom procedures call for a careful visual/physical check.
  The importance of this step cannot be stressed too strongly - it can lead to correcting a problem without further checks and can save valuable time.

BEFORE STARTING

This check should include:
- Vacuum hoses for splits, kinks, and proper connections, as shown on Vehicle Emission Control Information label.
- Air leaks at throttle body mounting and intake manifold.
- Ignition wires for cracking, hardness, proper routing, and carbon tracking.
- Wiring for proper connections, pinches, and cuts.
- The following symptoms cover several engines.
To determine if a particular system or component is used, refer to the ECM wiring diagrams for application.

SYMPTOMS CHART — MULTI-PORT INJECTION SYSTEMS

INTERMITTENTS
(Page 1 of 2)

Definition: Problem may or may not turn "ON" the "Service Engine Soon" light, or store a code.

PRELIMINARY CHECKS

- Perform the careful visual checks as described at start of "Symptoms,"

TROUBLE CODE CHARTS IN "ENGINE COMPONENTS/WIRING DIAGRAMS/DIAGNOSTIC CHARTS," SECTION "6E2-A"

- DO NOT use the Trouble Code Charts in for intermittent problems. The fault must be present to locate the problem. If a fault is intermittent, use of Trouble Code Charts may result in replacement of good parts.

FAULTY ELECTRICAL CONNECTIONS OR WIRING

- Most intermittent problems are caused by faulty electrical connections or wiring. Perform careful check of suspect circuits for:
  - Poor mating of the connector halves, or terminals, not fully seated in the connector body (backed out).
  - Improperly formed or damaged terminals. All connector terminals in problem circuit should be carefully reformed or replaced to insure proper contact tension.
  - Poor terminal to wire connection. This requires removing the terminal from the connector body to check.

ROAD TEST

- If a visual (physical) check does not find the cause of the problem, the vehicle can be driven with a voltmeter connected to a suspected circuit or a "Scan" tool may be used. An abnormal voltage or "Scan" reading, when the problem occurs, indicates the problem may be in that circuit. If the wiring and connectors check OK and a Trouble Code was stored for a circuit having a sensor, except for Codes 43, 44 and 45.

SYMPTOMS CHART — MULTI-PORT INJECTION SYSTEMS

INTERMITTENTS
(Page 2 of 2)

Definition: Problem may or may not turn "ON" the "Service Engine Soon" light, or store a code.

PRELIMINARY CHECKS

- Perform the careful visual checks as described at start of "Symptoms."

INTERMITTENT "SERVICE ENGINE SOON LIGHT"

- An intermittent "Service Engine Soon" light, and No Trouble Codes, may be caused by:
  - Electrical system interference caused by a defective relay, ECM driven solenoid, or switch. They can cause a sharp electrical surge. Normally, the problem will occur when the faulty component is operated.
  - Improper installation of electrical options, such as lights, 2-way radios, cellular phones, etc.
  - EST wires should be routed away from spark plug wires, ignition system components, and generator. Wire for CKT 453 from ECM to ignition system should be a good ground.
  - Ignition secondary shorted to ground.
  - CKT 419 ("Service Engine Soon" light) or CKT 451 (Diagnostic "Test" Terminal) intermittently shorted to ground.
  - ECM power grounds.

LOSS OF TROUBLE CODE MEMORY

- To check, disconnect TPS and idle engine until "Service Engine Soon" light comes "ON." Code 22 should be stored, and kept in memory, when ignition is turned "OFF" for at least 10 seconds. If not, the ECM is faulty.

SYMPTOMS CHART — MULTI-PORT INJECTION SYSTEMS

HARD START
(Page 1 of 2)

Definition: Engine cranks OK, but does not start for a long time. Does eventually run, or may start but immediately dies.

PRELIMINARY CHECKS

- Perform the careful visual checks as described at start of "Symptoms."
- Make sure the driver is using the correct starting procedure.

SENSORS

- **CHECK:** COOLANT TEMPERATURE SENSOR - Using a "Scan" tool compare coolant temperature with ambient temperature on cold engine.
  If coolant temperature reading is 5 degrees greater than or less than ambient air temperature, check for high resistance in coolant sensor circuit or sensor itself. Compare resistance value to Code 15 chart.

- **CHECK:** TPS - If a sticking throttle shaft or binding linkage causes a high TPS voltage (open throttle indication), the ECM will not control idle. Monitor TPS voltage. A "Scan" tool and/or voltmeter should display less than .95 volt with throttle closed.

FUEL SYSTEM

Important

- Fuel pump relay operation - pump should turn "ON" for 2 seconds when ignition is turned "ON." SEE CHART A-7 or Code 54.

- **CHECK:** Fuel pressure, CHART A-7.

- **CHECK:** Water contaminated fuel. For a faulty in-tank fuel pump check valve, which would allow the fuel in the lines to drain back to the tank after the engine is stopped. To check for this condition:
  1. Ignition "OFF."
  2. Disconnect fuel line at the filter.
  3. Remove the tank filler cap.
  4. Connect a radiator test pump to the fuel line and apply 103 kPa (15 psi) pressure. If the pressure will hold for 60 seconds, the check valve is OK.

SYMPTOMS CHART — MULTI-PORT INJECTION SYSTEMS

HARD START
(Page 2 of 2)

Definition: Engine cranks OK, but does not start for a long time. Does eventually run, or may start but immediately dies.

PRELIMINARY CHECKS

- Perform the careful visual checks as described at start of "Symptoms."
- Make sure the driver is using the correct starting procedure.

IGNITION SYSTEM

- **CHECK:** Ignition system for:
  - Proper output with ST-125.
  - Worn distributor shaft.
  - Bare and shorted ignition wires.
  - Pickup coil resistance and connections.
  - Loose ignition coil connections.
  - Moisture in distributor cap.
  - Spark plugs, wet plugs, cracks, wear, improper gap, burned electrodes or heavy deposits.

Important
- If engine starts but then, immediately stalls, disconnect the set timing connector. If engine then starts, and runs OK, replace distributor pickup coil.
- **CHECK:** CKT 423 (EST) for short to ground.

ADDITIONAL CHECKS

- **CHECK:** IAC operation - see CHART C-2C.
- **CHECK:** No crank signal - see CHART C-1B.
- **CHECK:** EGR operation - see CHART C-7.

SYMPTOMS CHART — MULTI-PORT INJECTION SYSTEMS

SURGES AND/OR CHUGGLES

Definition: Engine power variation, under steady throttle or cruise. Feels like the vehicle speeds up and slows down, with no change in the accelerator pedal.

PRELIMINARY CHECKS

- Perform the careful visual checks as described at start of "Symptoms."
- Be sure driver understands transmission/torque converter clutch and A/C compressor operation in owner's manual.
- Use a "Scan" tool to make sure reading of VSS matches vehicle speedometer.

SENSORS

- **CHECK:** Oxygen sensor for silicon contamination from fuel, or use of improper RTV sealant. The sensor may have a white, powdery coating and result in a high but false signal voltage (rich exhaust indication). The ECM will then reduce the amount of fuel delivered to the engine, causing a severe driveability problem.

FUEL SYSTEM

Important
- To determine if the condition is caused by a rich or lean system, the vehicle should be driven at the speed of the complaint. Monitoring block learn will help identify problem.
  Lean - Block learn near 150. Refer to "Diagnostic Aids" on facing page of Code 44.
  Rich - Block learn near 108. Refer to "Diagnostic Aids" on facing page of Code 45.
- **CHECK:** Fuel pressure while condition exists. See CHART A-7.
- **CHECK:** In-line fuel filter. Replace if dirty or plugged.

IGNITION SYSTEM

- **CHECK:** For proper output voltage using spark tester (ST-125) J 26792 or equivalent.
- **CHECK:** Spark plugs. After removing spark plugs, check for wet plugs, cracks, wear, improper gap, burned electrodes, or heavy deposits. Repair or replace as necessary.

ADDITIONAL CHECKS

- **CHECK:** Generator output voltage. Repair if less than 9 or more than 16 volts.
- **CHECK:** Vacuum lines for kinks or leaks.
- **CHECK:** For intermittent EGR at idle, use CHART C-7.
- **CHECK:** TCC operation. Use CHART C-8A.

SYMPTOMS CHART — MULTI-PORT INJECTION SYSTEMS

LACK OF POWER, SLUGGISH, OR SPONGY
(Page 1 of 2)

Definition: Engine delivers less than expected power. Little or no increase in speed, when accelerator pedal is pushed down part way.

PRELIMINARY CHECKS

- Perform the careful visual checks as described at start of "Symptoms."
- Compare customer's vehicle to similar unit. Make sure the customer has an actual problem.
- Remove air filter and check air filter for dirt, or for being plugged, replace as necessary.
- If there is spray from only one injector, then there is a malfunction in the injector assembly, or in the signal to the injector assembly. The malfunction can be isolated, by switching the injector connectors. If the problem remains with the original injector, after switching the connector, the injector is defective. Replace the injector. If the problem moves with the injector connector, the problem is an improper signal in the injector circuits, use CHART A-3.

FUEL SYSTEM

CHECK: For restricted fuel filter, contaminated fuel or improper fuel pressure, use CHART A-7.

IGNITION SYSTEM

- **CHECK:** Proper output voltage with spark tester J 26792 or equivalent (ST-125).
- **CHECK:** Ignition timing. See "Vehicle Emission Control Information" label.
- **CHECK:** Proper operation of EST.

SYMPTOMS CHART — MULTI-PORT INJECTION SYSTEMS

LACK OF POWER, SLUGGISH, OR SPONGY
(Page 2 of 2)

Definition: Engine delivers less than expected power. Little or no increase in speed, when accelerator pedal is pushed down part way.

EXHAUST SYSTEM

- **CHECK:** Exhaust system for possible restriction: See CHART B-1. Inspect exhaust system for damaged or collapsed pipes. Inspect muffler for heat distress or possible internal failure.
1. With engine at normal operating temperature, connect a vacuum gage to any convenient vacuum port on intake manifold.
2. Run engine at 1000 rpm and record vacuum reading.
3. Increase rpm slowly to 2500 rpm. Note vacuum reading at steady 2500 rpm.
4. If vacuum at 2500 rpm decreases more than 3", from reading at 1000 rpm, the exhaust system should be inspected for restrictions.
5. Disconnect exhaust pipe from engine and repeat Steps 3 & 4. If vacuum still drops more than 3", with exhaust disconnected, check valve timing.

ADDITIONAL CHECKS

- **CHECK:** ECM power grounds, see wiring diagrams.
- **CHECK:** Engine valve timing and compression.
- **CHECK:** EGR operation for being open or partly open, all the time. Use CHART C-7.
- **CHECK:** Engine, for proper or worn camshaft.
- **CHECK:** Transmission torque converter operation.
- **CHECK:** Generator output voltage. Repair if less than 9 or more than 16 volts.

SYMPTOMS CHART — MULTI-PORT INJECTION SYSTEMS

DETONATION/SPARK KNOCK
(Page 1 of 2)

Definition: A mild to severe ping, usually worse under acceleration. The engine makes sharp metallic knocks that change with throttle opening.

PRELIMINARY CHECKS

- Perform the careful visual checks as described at start of "Symptoms."

- Make sure the customer has an actual problem.
- Remove air filter and check air filter for dirt, or for being plugged, replace as necessary.
- If there is spray from only one injector, then there is a malfunction in the injector assembly, or in the signal to the injector assembly. The malfunction can be isolated, by switching the injector connectors. If the problem remains with the original injector, after switching the connector, the injector is defective. Replace the injector. If the problem moves with the injector connector, the problem is an improper signal in the injector circuits, use CHART A-3.

COOLING SYSTEM

- **CHECK:** For obvious over heating problems.
- **CHECK:** Low coolant.
- **CHECK:** Loose water pump belt.
- **CHECK:** Restricted air flow to radiator, or restricted water flow thru radiator.
- **CHECK:** Faulty or incorrect thermostat.
- **CHECK:** Correct coolant solution - should be a 50/50 mix of GM #1052753 anti-freeze coolant (or equivalent) and water.

SENSOR

- **CHECK:** Coolant Temperature Sensor (CTS), which has shifted in value.

SYMPTOMS CHART — MULTI-PORT INJECTION SYSTEMS

DETONATION/SPARK KNOCK
(Page 2 of 2)

Definition: A mild to severe ping, usually worse under acceleration. The engine makes sharp metallic knocks that change with throttle opening.

FUEL SYSTEM

⚠ Important

- To determine if the condition is caused by a rich or lean system, the vehicle should be driven at the speed of the complaint. Monitoring block learn will help identify problem.
  Lean - Block learn near 150. Refer to "Diagnostic Aids" on facing page of Code 44.
  Rich - Block learn near 108. Refer to "Diagnostic Aids" on facing page of Code 45.

- **CHECK:** Fuel pressure, CHART A-7.
- **CHECK:** For poor fuel quality, proper octane rating.
- **CHECK:** If "Scan" tool readings are normal (see facing page of "Diagnostic Circuit Check") and there are no engine mechanical faults, fill fuel tank with a premium gasoline that has a minimum octane rating of 92 and re-evaluate vehicle performance.

IGNITION SYSTEM

- **CHECK:** ESC system operation see CHART C-5.
- **CHECK:** Ignition timing. See "Vehicle Emission "Control Information" label.

ENGINE MECHANICAL

- **CHECK:** For incorrect basic engine parts such as cam, heads, pistons, etc.
- **CHECK:** Excessive oil entering combustion chamber.

ADDITIONAL CHECKS

- **CHECK:** For carbon buildup. Remove carbon with top engine cleaner, follow instructions on can.
- **CHECK:** For proper transmission shift points.
- **CHECK:** TCC operation. Use CHART C-8.
- **CHECK:** For incorrect basic engine parts such as cam, heads, pistons, etc.
- **CHECK:** Excessive oil entering combustion chamber.
- **CHECK:** For correct PROM.
- **CHECK:** Spark plugs for correct heat range.
- **CHECK:** Proper operation of THERMAC.

SYMPTOMS CHART — MULTI-PORT INJECTION SYSTEMS

HESITATION, SAG, STUMBLE

Definition: Momentary lack of response as the accelerator is pushed down. Can occur at all vehicle speeds. Usually more severe when first trying to make the vehicle move, as from a stop sign. May cause the engine to stall if severe enough.

PRELIMINARY CHECKS

- Perform the careful visual checks as described at start of "Symptoms."

FUEL SYSTEM

- **CHECK:** TPS - Check TPS for binding or sticking. Voltage should increase at a steady rate as throttle is moved toward Wide Open Throttle (WOT). TPS voltage should be less than .95 volt at idle.
- **CHECK:** MAP SENSOR - See CHART C-1D.

IGNITION SYSTEM

- **CHECK:** Spark plugs for being fouled, worn or cracked.
- **CHECK:** For open ignition system ground, CKT 453.
- **CHECK:** Ignition timing. See "Vehicle Emission Control Information" label.

ADDITIONAL CHECKS

- **CHECK:** Generator output voltage. Repair, if less than 9 or more than 16 volts.
- **CHECK:** EGR valve operation. Use CHART C-7

SYMPTOMS CHART — MULTI-PORT INJECTION SYSTEMS

CUTS OUT, MISSES
(Page 1 of 2)

Definition: Steady pulsation or jerking that follows engine speed, usually more pronounced as engine load increases. The exhaust has a steady spitting sound at idle or low speed.

PRELIMINARY CHECKS

- Perform the careful visual checks as described at start of "Symptoms."
- If there is spray from only one injector, then there is a malfunction in the injector assembly, or in the signal to the injector assembly. The malfunction can be isolated, by switching the injector connectors. If the problem remains with the original injector, after switching the connector, the injector is defective. Replace the injector. If the problem moves with the injector connector, the problem is an improper signal in the injector circuits, use CHART A-3.

FUEL SYSTEM

- **CHECK:** Fuel pressure, CHART A-7.
- **CHECK:** Water contaminated fuel or restricted fuel filter.

IGNITION SYSTEM

- **CHECK:** For missing cylinder by:
1. Start engine, allow engine to stabilize then disconnect IAC motor. Remove one spark plug wire at a time, using insulated pliers.

   CAUTION: Do not perform this test for more than 2 minutes, as this may cause damage to the catalytic converter.

2. If there is an rpm drop, on all cylinders, (equal to within 50 rpm), go to ROUGH, UNSTABLE OR INCORRECT IDLE, STALLING symptom. Reconnect IAC motor.
3. If there is no rpm drop on one or more cylinders, or excessive variation in drop, check for spark, on the suspected cylinder(s) with J 26792 (ST-125) Spark Tester or equivalent. If no spark, use CHART C-4. If there is spark, remove spark plug(s) in these cylinders and check for:
   - Cracks.
   - Wear.
   - Improper Gap.
   - Burned Electrodes.
   - Heavy Deposits.
- **CHECK:** Ignition wire resistance (should not exceed 30,000 ohms), also, check rotor and distributor cap.

[!] **Important**

- If the previous checks did not find the problem:
  Visually inspect ignition system for moisture, dust, cracks, burns, etc. Spray plug wires with fine water mist to check for shorts.

SYMPTOMS CHART — MULTI-PORT INJECTION SYSTEMS

CUTS OUT, MISSES
(Page 2 of 2)

Definition: Steady pulsation or jerking that follows engine speed, usually more pronounced as engine load increases. The exhaust has a steady spitting sound at idle or low speed.

ENGINE MECHANICAL

- **CHECK:** For proper valve timing. Remove rocker covers. Check for bent pushrods, worn rocker arms, broken or weak valve springs, worn camshaft lobes. Repair as necessary.
- **CHECK:** For casting flash in, intake and exhaust manifolds.
- **CHECK:** Low compression. Perform compression check.

SYMPTOMS CHART — MULTI-PORT INJECTION SYSTEMS

POOR FUEL ECONOMY

Definition: Fuel economy, as measured by an actual road test, is noticeably lower than expected. Also, economy is noticeably lower than it was on this vehicle at one time, as previously shown by an actual road test.

PRELIMINARY CHECKS

- Perform the careful visual checks as described at start of "Symptoms."

- Perform "Diagnostic Circuit Check."
- Check owner's driving habits.
  - Is A/C "ON" full time (Defroster mode "ON")?
  - Are tires at correct pressure?
  - Are excessively heavy loads being carried?
  - Is acceleration too much, too often?
- Check air cleaner element (filter) for dirt or being plugged.
- Visually (physically) check: Vacuum hoses for splits, kinks, and proper connections as shown on "Vehicle Emission Control Information" label.

IGNITION SYSTEM

- **CHECK:** ESC operation.

- **CHECK:** Spark plugs. After removing spark plugs, check for wet plugs, cracks, wear, improper gap, burned electrodes, or heavy deposits. Repair or replace as necessary.
- **CHECK:** Ignition wires for cracking, hardness, and proper connections.
- **CHECK:** Ignition timing. See "Vehicle Emission Control Information" label.

COOLING SYSTEM

- **CHECK:** Engine thermostat for faulty part or for wrong heat range.

ADDITIONAL CHECKS

- **CHECK:** TCC operation - Check for proper operation. See CHART C-8. A "Scan" should indicate an rpm drop, when the TCC is commanded "ON."
- **CHECK:** For exhaust system restriction. See CHART B-1.
- **CHECK:** Fuel pressure. See CHART A-7
- **CHECK:** Compression.
- **CHECK:** For proper calibration of speedometer.
- **CHECK:** For dragging brakes.
- Suggest owner fill fuel tank and recheck fuel economy.
- Suggest driver read "Important Facts on Fuel Economy" in Owner's Manual.

SYMPTOMS CHART — MULTI-PORT INJECTION SYSTEMS

STALLING, ROUGH, UNSTABLE OR INCORRECT IDLE
(Page 1 of 2)

Definition: The engine runs unevenly at idle. If bad enough, the vehicle may shake. Also, the idle may vary in rpm (called "hunting"). Either condition may be severe enough to cause stalling. Engine idles at incorrect speed.

PRELIMINARY CHECKS

- Perform the careful visual checks as described at start of "Symptoms,"

SENSORS

- **CHECK:** OXYGEN SENSOR - Inspect sensor for silicon contamination from fuel, or use of improper RTV sealant. The sensor will have a white, powdery coating, and will result in a high but false signal voltage (rich exhaust indication). The ECM will then reduce the amount of fuel delivered to the engine, causing a severe driveability problem.
- **CHECK:** TPS - If a sticking throttle shaft or binding linkage causes a high TPS voltage (open throttle indication), the ECM will not control idle. Monitor TPS voltage. A "Scan" tool and/or voltmeter should display less than .95 volt with throttle closed.
- **CHECK:** COOLANT TEMPERATURE SENSOR (CTS) - Using a "Scan" tool compare coolant temperature with ambient temperature on cold engine. If coolant temperature reading is 5 degrees greater than or less than ambient air temperature. Check for high resistance in coolant sensor circuit or sensor itself. Compare resistance value to Code 15 chart.
- **CHECK:** MAP SENSOR - Refer to CHART C-1D MAP voltage output check.

FUEL SYSTEM

Important
- To determine if the condition is caused by a rich or lean system, the vehicle should be driven at the speed of the complaint. Monitoring block learn will help identify problem.
Lean - Block learn near 150. Refer to "Diagnostic Aids" on facing page of Code 44.
Rich - Block learn near 108. Refer to "Diagnostic Aids" on facing page of Code 45.

- **CHECK:** For fuel in pressure regulator hose. If present, replace regulator assembly.
- **CHECK:** Evaporative emission control system. CHART C-3.
- **CHECK:** Run a cylinder compression check.
- **CHECK:** For injector(s) leaking. Check fuel pressure, see CHART A-7.

SYMPTOMS CHART — MULTI-PORT INJECTION SYSTEMS

STALLING, ROUGH, UNSTABLE OR INCORRECT IDLE
(Page 2 of 2)

Definition: The engine runs unevenly at idle. If bad enough, the vehicle may shake. Also, the idle may vary in rpm (called "hunting"). Either condition may be severe enough to cause stalling. Engine idles at incorrect speed.

IGNITION SYSTEM

- **CHECK:** Ignition system. Refer to "Ignition System (EST)," Section "6E2-C4".
- **CHECK:** Ignition timing. See "Vehicle Emission Control Information" label.

ADDITIONAL CHECKS

- **CHECK:** IAC valve will not move, if system voltage is below 9 or greater than 16 volts.
- **CHECK:** IAC operation - See CHART C-2C.
- **CHECK:** ECM ground circuits.
- **CHECK:** P/N switch circuit. See CHART C-1A, or use "Scan" tool, and be sure tool indicates vehicle is in drive with gear selector is in drive.

Important
- Use "Scan" tool to determine if ECM is receiving A/C status signal. If problem exists with A/C "ON," check A/C system operation CHART C-10.

- **CHECK:** EGR "ON," while idling, will cause roughness, stalling, and hard starting. Use CHART C-7.
- **CHECK:** Battery cables and ground straps should be clean and secure. Erratic voltage will cause IAC to change its position, resulting in poor idle quality.
- **CHECK:** A/C refrigerant pressure too high.
- **CHECK:** For overcharge or faulty pressure switch.
- **CHECK:** PCV valve for proper operation by placing finger over inlet hole in valve end several times. Valve should snap back. If not, replace valve.

ENGINE MECHANICAL

- **CHECK:** Vacuum leaks can cause higher than normal idle.
- **CHECK:** For broken motor mounts.
- **CHECK:** For low compression.

SYMPTOMS CHART — MULTI-PORT INJECTION SYSTEMS

EXCESSIVE EXHAUST EMISSIONS OR ODORS
(Page 1 of 2)

Definition: Vehicle fails an emission test. Vehicle has excessive "rotten egg" smell. Excessive odors do not necessarily indicate excessive emissions.

PRELIMINARY CHECKS

- Perform "Diagnostic Circuit Check."
- If Emission Test shows excessive CO and HC check items which cause vehicle to run RICH. Make sure engine is at normal operating temperature.
- If Emission Test shows excessive NO_x check items which can cause the vehicle to run LEAN or too hot.

SENSORS

⚠ Important
- If the "Scan" tool indicates a very high coolant temperature and the system is running LEAN:
  - Check the coolant system and coolant fan for proper operation. Also check for vacuum leaks.

FUEL SYSTEM

⚠ Important
- If the system is running rich, (block learn near 108), refer to "Diagnostic Aids" on facing page of Code 45. If the system is running lean, (block learn near 150), refer to "Diagnostic Aids" on facing page of Code 44.

- **CHECK:** Fuel pressure. CHART A-7.

⚠ Important
- If test shows excessive NO_x, check items which cause vehicle to run LEAN or too hot.

- **CHECK:** Canister for fuel loading. See CHART C-3.

SYMPTOMS CHART — MULTI-PORT INJECTION SYSTEMS

EXCESSIVE EXHAUST EMISSIONS OR ODORS
(Page 2 of 2)

Definition: Vehicle fails an emission test. Vehicle has excessive "rotten egg" smell. Excessive odors do not necessarily indicate excessive emissions.

COOLING SYSTEM

- **CHECK:** Coolant system and coolant fan for proper operation.

IGNITION SYSTEM

- **CHECK:** Ignition system.
- **CHECK:** For incorrect timing or excessive ignition timing advance. See "Vehicle Emission Control Information" label.
- **CHECK:** Spark plugs, plug wires, and ignition components.

ADDITIONAL CHECKS

- **CHECK:** For lead contamination of catalytic converter (look for the removal of fuel filler neck restrictor).
- **CHECK:** Remove carbon with top engine cleaner. Follow instructions on can.
- **CHECK:** EGR valve for not opening. See CHART C-7. Vacuum leaks.
- **CHECK:** PCV valve for being plugged, stuck, or blocked PCV hose, or fuel in the crankcase.

ENGINE MECHANICAL

- **CHECK:** EGR valve for not opening. See CHART C-7. Vacuum leaks.
- **CHECK:** PCV valve for being plugged, stuck, or blocked PCV hose, or fuel in the crankcase.

SYMPTOMS CHART — MULTI-PORT INJECTION SYSTEMS

DIESELING, RUN-ON

Definition: Engine continues to run after key is turned "OFF," but runs very roughly. If engine runs smoothly, check ignition switch and adjustment.

PRELIMINARY CHECKS

- Perform the careful visual checks as described at start of "Symptoms." B".

FUEL SYSTEM

- **CHECK:** Injector(s) for leaking. Apply 12 volts to fuel pump "test" terminal (located underhood) to turn "ON" fuel pump and pressurize fuel system. Visually check injector(s) and TBI assembly for fuel leakage. Refer to CHART A-7, "Fuel System Diagnosis."

SYMPTOMS CHART — MULTI-PORT INJECTION SYSTEMS

BACKFIRE

Definition: Fuel ignites in intake manifold, or in exhaust system, making a loud popping noise.

PRELIMINARY CHECKS

- Perform the careful visual checks as described at start of "Symptoms."

FUEL SYSTEM

- **CHECK:** Perform fuel system diagnosis check. Use CHART A-7.
- **CHECK:** For plugged fuel filter.

IGNITION SYSTEM

- **CHECK:** Proper output with spark tester J 26792 or equivalent (ST-125).
- **CHECK:** Spark plugs. After removing spark plugs, check for wet plugs, cracks, wear, improper gap, burned electrodes, or heavy deposits. Repair or replace as necessary.
- **CHECK:** Ignition system.
- **CHECK:** For crossfire between spark plugs (distributor cap, spark plug wires, and proper routing of plug wires.)
- **CHECK:** Ignition timing. See "Vehicle Emission Control Information" label.

ENGINE MECHANICAL

- **CHECK:** Compression. Perform a compression check - look for sticking or leaking valves.
  - For proper valve timing.
  - Broken or worn valve train parts.
- **CHECK:** Valve timing.
- **CHECK:** Intake manifold gasket for vacuum leaks.
- **CHECK:** Intake and exhaust manifolds for casting flash.
- **CHECK:** Faulty A.I.R. check valve. Use CHART C-6.
- **CHECK:** EGR operation for being open all the time. Use CHART C-7.

2.0L (VIN H) ENGINE — ENGINE COMPONENT LOCATION CHART — SUNBIRD

2.0L (VIN H) ENGINE — ECM WIRING SCHEMATIC — 1992 SUNBIRD

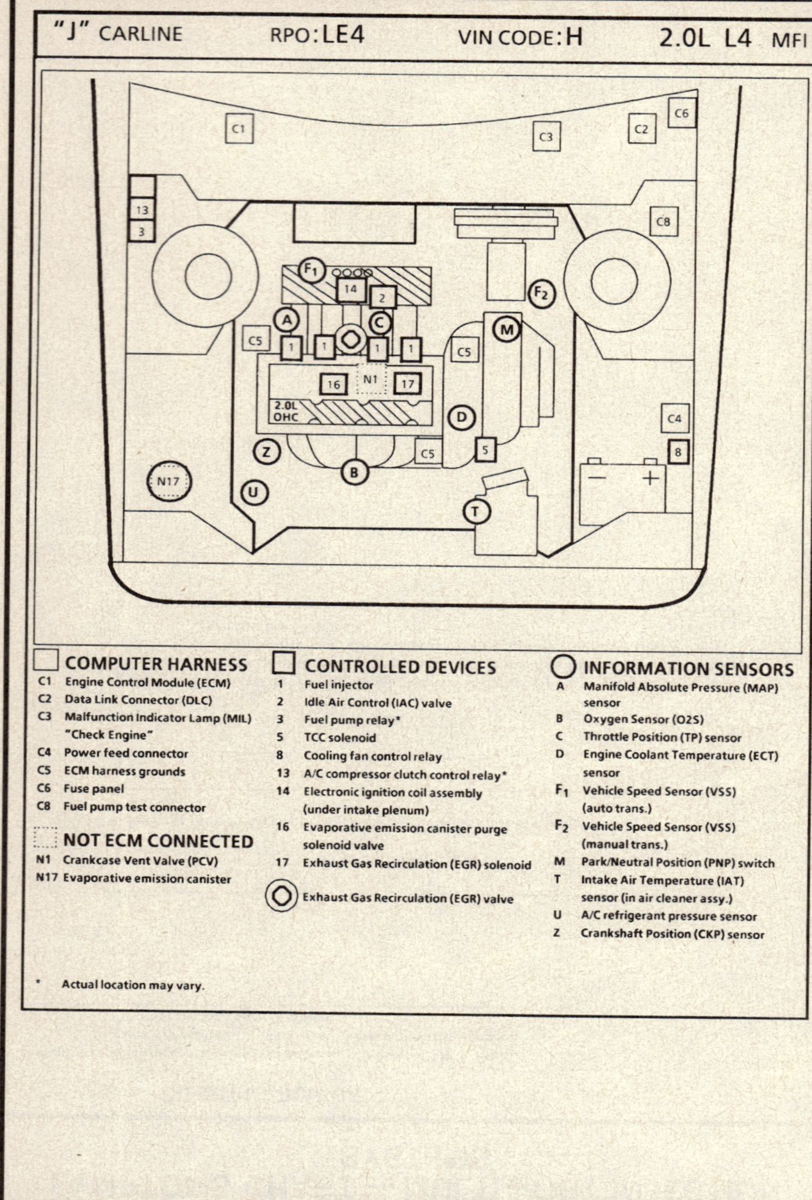

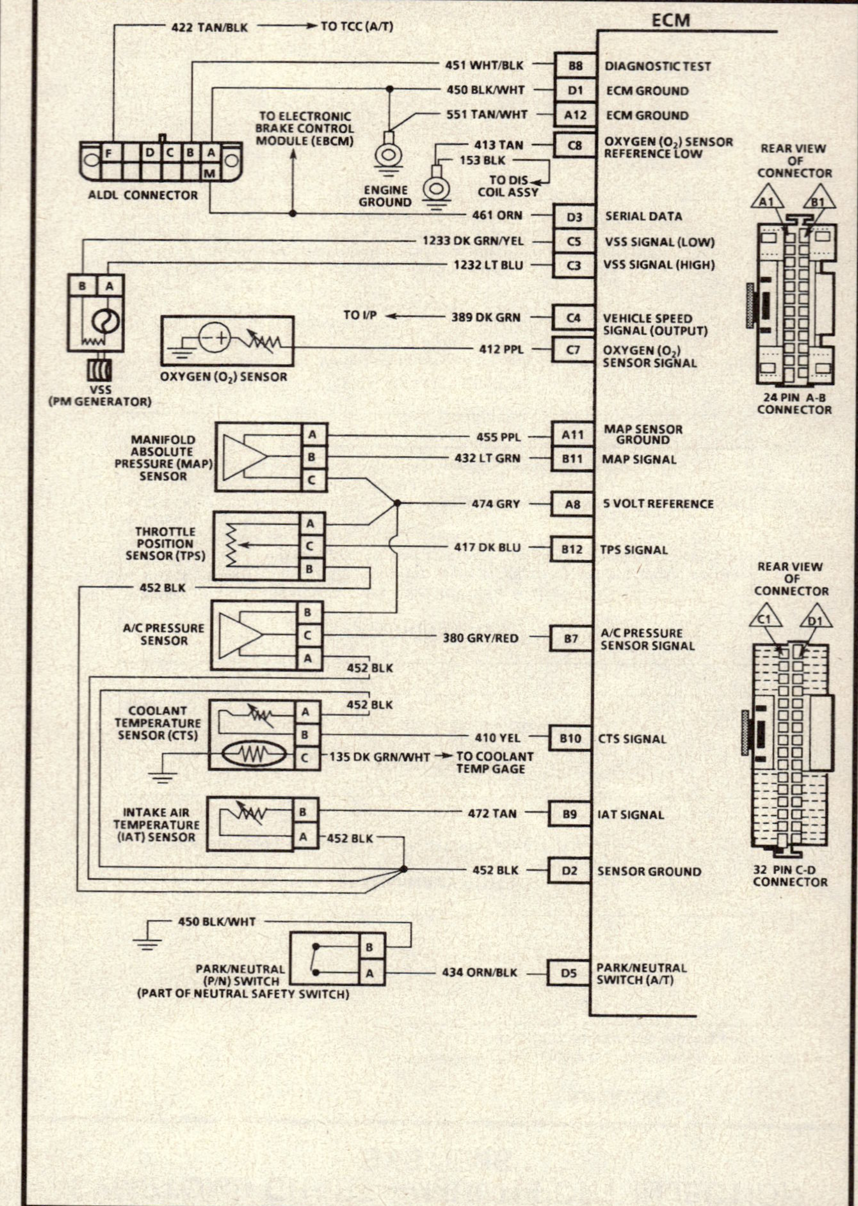

2.0L (VIN H) ENGINE — ECM WIRING SCHEMATIC — 1992 SUNBIRD

2.0L (VIN H) ENGINE — ECM WIRING SCHEMATIC — 1992 SUNBIRD

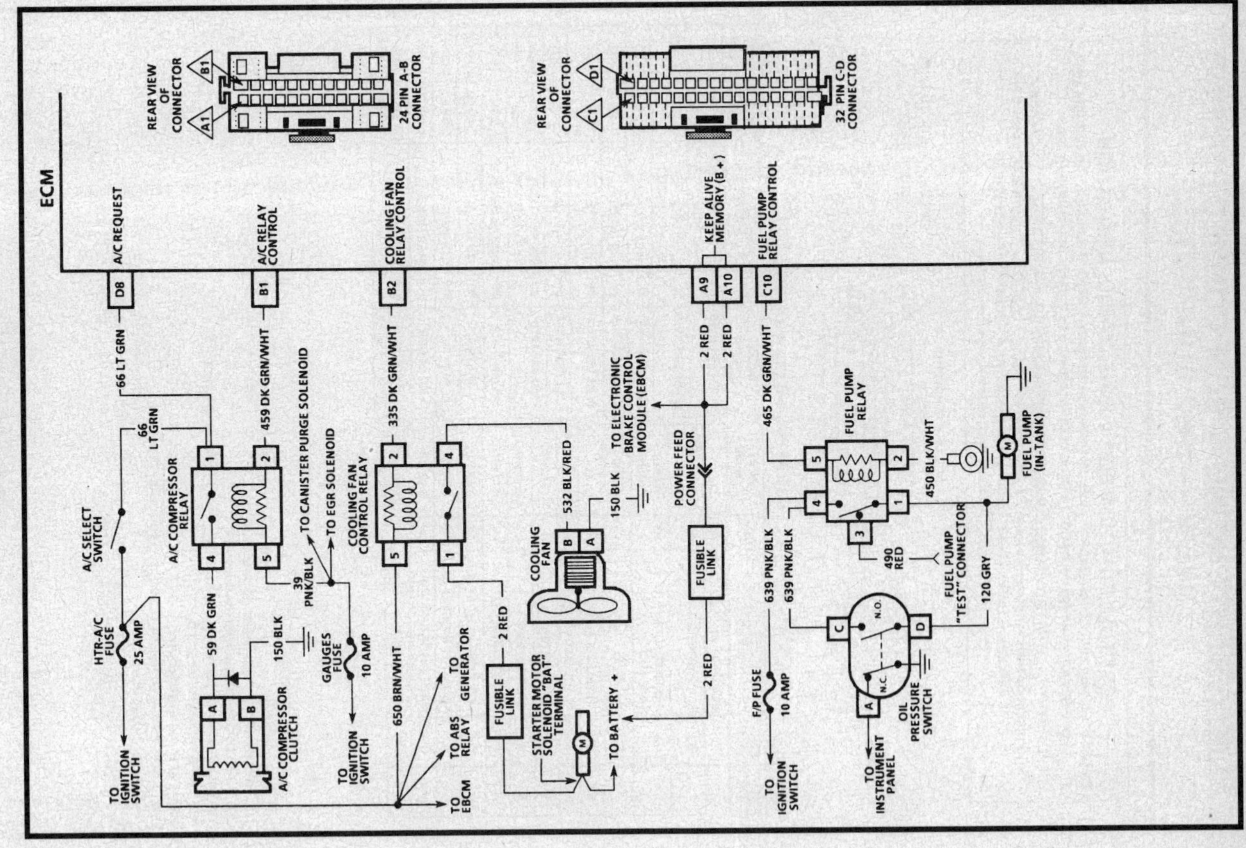

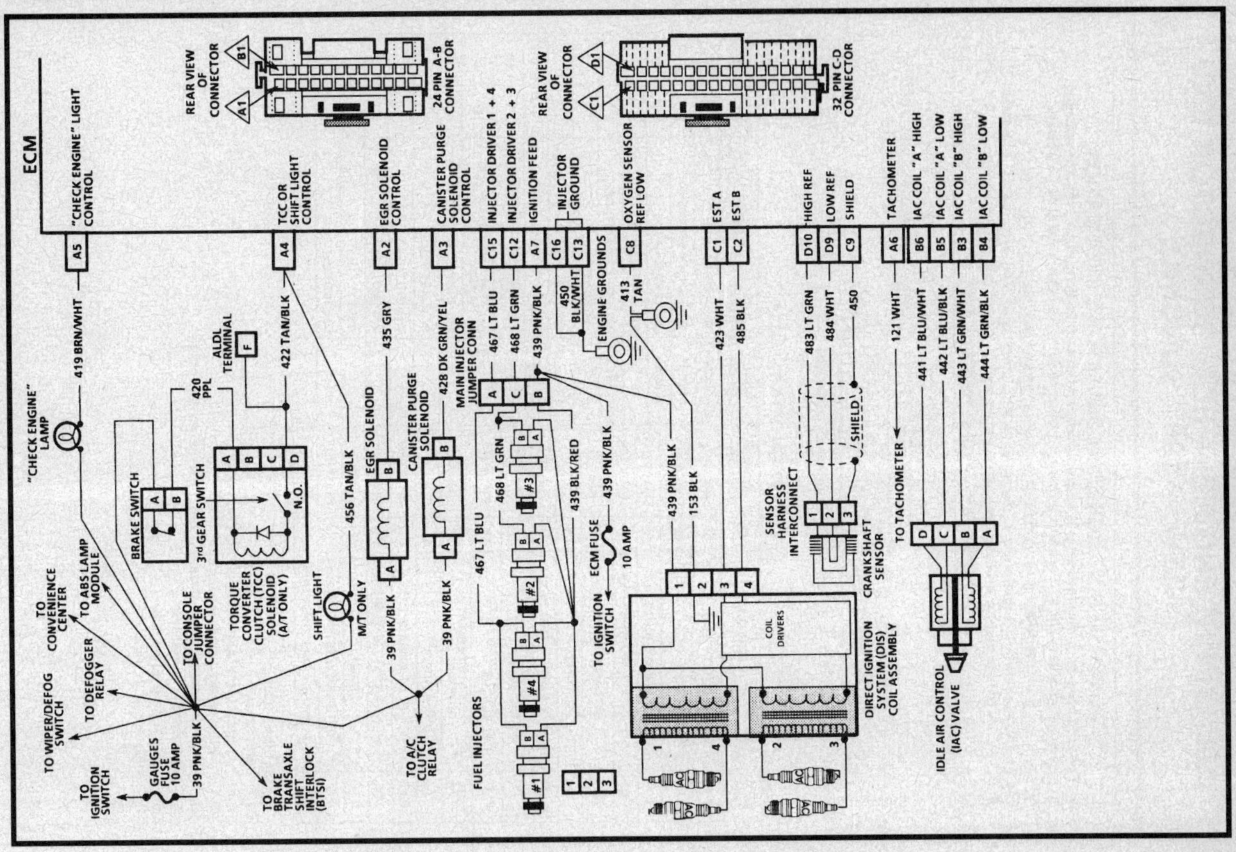

2.0L (VIN H) ENGINE — ECM CONNECTOR END VIEW — 1992 SUNBIRD

PORT FUEL INJECTION ECM CONNECTOR IDENTIFICATION

This ECM voltage chart is for use with a digital voltmeter to further aid in diagnosis. The voltages you get may vary due to low battery charge or other reasons, but they should be very close.

THE FOLLOWING CONDITIONS MUST BE MET BEFORE TESTING:

- Engine at operating temperature • Engine idling in "Closed Loop" (For "Engine Run" column) in park or neutral • "Test" terminal not grounded • "Scan" tool not installed
- B + indicates battery or charging system voltage

24 PIN A-B CONNECTOR — A1 — B1

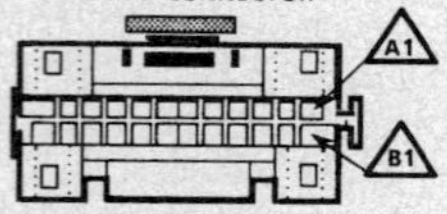

REAR VIEW OF CONNECTOR (LT BLUE)

ENGINE: 2.0L (VIN H)
RPO: LE4

LT BLUE 24 PIN A-B CONNECTOR

VOLTAGE KEY "ON"	ENG. RUN	CIRCUIT	PIN	WIRE COLOR
		NOT USED	A1	
B+	B+	EGR SOLENOID	A2	435 GRY
B+	B+	CANISTER PURGE SOLENOID	A3	428 DK GRN/YEL
0*V	0*V	TCC (A/T) OR		
-----	-----	-----	A4	422 -TAN/BLK- 456
B+	B+	SHIFT LIGHT (M/T)		
0*V	B+	"CHECK ENGINE" LIGHT	A5	419 BRN/WHT
(6)	(6)	TACHOMETER	A6	121 WHT
B+	B+	IGNITION FEED	A7	439 PNK/BLK
5.0V	5.0V	5 VOLTS REFERENCE	A8	474 GRY
B+	B+	KEEP ALIVE MEMORY (B+)	A9	2 RED
B+	B+	KEEP ALIVE MEMORY (B+)	A10	2 RED
0*V	0*V	MAP SENSOR GROUND	A11	455 PPL
0*V	0*V	ECM GROUND	A12	551 TAN/WHT

WIRE COLOR	PIN	CIRCUIT	VOLTAGE KEY "ON"	ENG. RUN
459 DK GRN/WHT	B1	A/C CLUTCH RELAY CONTROL	B + (2)	B + (2)
335 DK GRN/WHT	B2	COOLING FAN RELAY CONTROL	B + (2)	B + (2)
443 LT GRN/WHT	B3	IAC "B" HIGH	(1)	(1)
444 LT GRN/BLK	B4	IAC "B" LOW	(1)	(1)
442 LT BLU/BLK	B5	IAC "A" LOW	(1)	(1)
441 LT BLU/WHT	B6	IAC "A" HIGH	(1)	(1)
380 GRY/RED	B7	A/C PRESSURE SENSOR SIGNAL	VARIES (8)	VARIES (8)
451 WHT/BLK	B8	ALDL DIAGNOSTIC	5.0V	5.0V
472 TAN	B9	IAT SIGNAL	1.3V (3)	1.3V (3)
410 YEL	B10	CTS SIGNAL	2.0V (3)	2.0V (3)
432 LT GRN	B11	MAP SIGNAL	4.75V (7)	1.6V (7)
417 DK BLU	B12	TPS SIGNAL	.33 - 1.33V	.33 - 1.33V

* LESS THAN .5 VOLT.
(1) NOT USABLE.
(2) A/C SELECT "OFF" AND ENGINE COOLING FAN "OFF."
(3) VARIES DEPENDING ON TEMPERATURE.
(4) READS B + FOR 2 SECONDS AFTER KEY "ON," THEN SHOULD READ 0 VOLT.
(5) VARIES WITH VEHICLE SPEED.
(6) VARIES WITH ENGINE RPM.
(7) VARIES WITH ALTITUDE. REFER TO CODE 33 OR CHART C1-D.
(8) REFER TO CODE 66.

2.0L (VIN H) ENGINE — ECM CONNECTOR END VIEW — 1992 SUNBIRD

PORT FUEL INJECTION ECM CONNECTOR IDENTIFICATION

This ECM voltage chart is for use with a digital voltmeter to further aid in diagnosis. The voltages you get may vary due to low battery charge or other reasons, but they should be very close.

THE FOLLOWING CONDITIONS MUST BE MET BEFORE TESTING:

- Engine at operating temperature • Engine idling in "Closed Loop" (For "Engine Run" column) in park or neutral • "Test" terminal not grounded • "Scan" tool not installed
- B + indicates battery or charging system voltage

32 PIN C-D CONNECTOR — C1 — D1

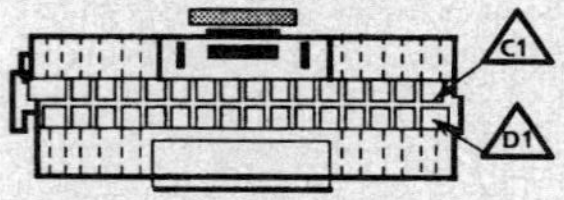

REAR VIEW OF CONNECTOR (LT BLUE)

ENGINE: 2.0L (VIN H)
RPO: LE4

LT BLUE 32 PIN C-D CONNECTOR

VOLTAGE KEY "ON"	ENG. RUN	CIRCUIT	PIN	WIRE COLOR
B+	1.3 - 1.6V (6)	EST "A"	C1	423 WHT
B+	1.3 - 1.6V (6)	EST "B"	C2	485 BLK
(5)	(5)	VSS SIGNAL (HIGH)	C3	1232 LT BLU
(5)	(5)	VSS OUTPUT (4000 PPM)	C4	389 DK GRN
(5)	(5)	VSS SIGNAL (LOW)	C5	1233 DK GRN/YEL
		NOT USED	C6	
.1-.55V	.1-.9V	OXYGEN SENSOR SIGNAL	C7	412 PPL
0*V	0*V	OXYGEN (O$_2$) SENSOR REFERENCE LOW	C8	413 TAN
0*V	0*V	IGNITION REFERENCE SHIELD	C9	450 (BARE)
(4)	B+	FUEL PUMP RELAY CONTROL	C10	465 DK GRN/WHT
		NOT USED	C11	
B+	B + (6)	#2 & 3 INJECTOR DRIVER	C12	468 LT GRN
0*V	0*V	#2 & 3 INJECTOR GROUND	C13	450 BLK/WHT
		NOT USED	C14	
B+	B + (6)	#1 & 4 INJECTOR DRIVER	C15	467 LT BLU
0*V	0*V	#1 & 4 INJECTOR GROUND	C16	450 BLK/WHT

WIRE COLOR	PIN	CIRCUIT	VOLTAGE KEY "ON"	ENG. RUN
450 BLK/WHT	D1	ECM GROUND	0*V	0*V
452 BLK	D2	SENSOR GROUND	0*V	0*V
461 ORN	D3	SERIAL DATA	4.7V	4.7V
	D4	NOT USED		
434 ORN/BLK	D5	PARK/NEUTRAL SW	0*V	0*V
	D6	NOT USED		
	D7	NOT USED		
66 LT GRN	D8	A/C REQUEST	0*V (2)	0*V (2)
484 WHT	D9	IGNITION REFERENCE LOW	(7)	(7)
483 LT GRN	D10	IGNITION REFERENCE HIGH	(7)	(7)
	D11	NOT USED		
	D12	NOT USED		
	D13	NOT USED		
	D14	NOT USED		
	D15	NOT USED		
	D16	NOT USED		

* LESS THAN .5 VOLT.
(1) NOT USABLE.
(2) A/C SELECT "OFF" AND ENGINE COOLING FAN "OFF."
(3) VARIES DEPENDING ON TEMPERATURE.
(4) READS B + FOR 2 SECONDS AFTER KEY "ON," THEN SHOULD READ 0 VOLT.
(5) VARIES WITH VEHICLE SPEED.
(6) VARIES WITH ENGINE RPM.
(7) REFER TO CODE 19 CHART.

2.0L (VIN H) ENGINE — ECM WIRING SCHEMATIC — 1993–94 SUNBIRD

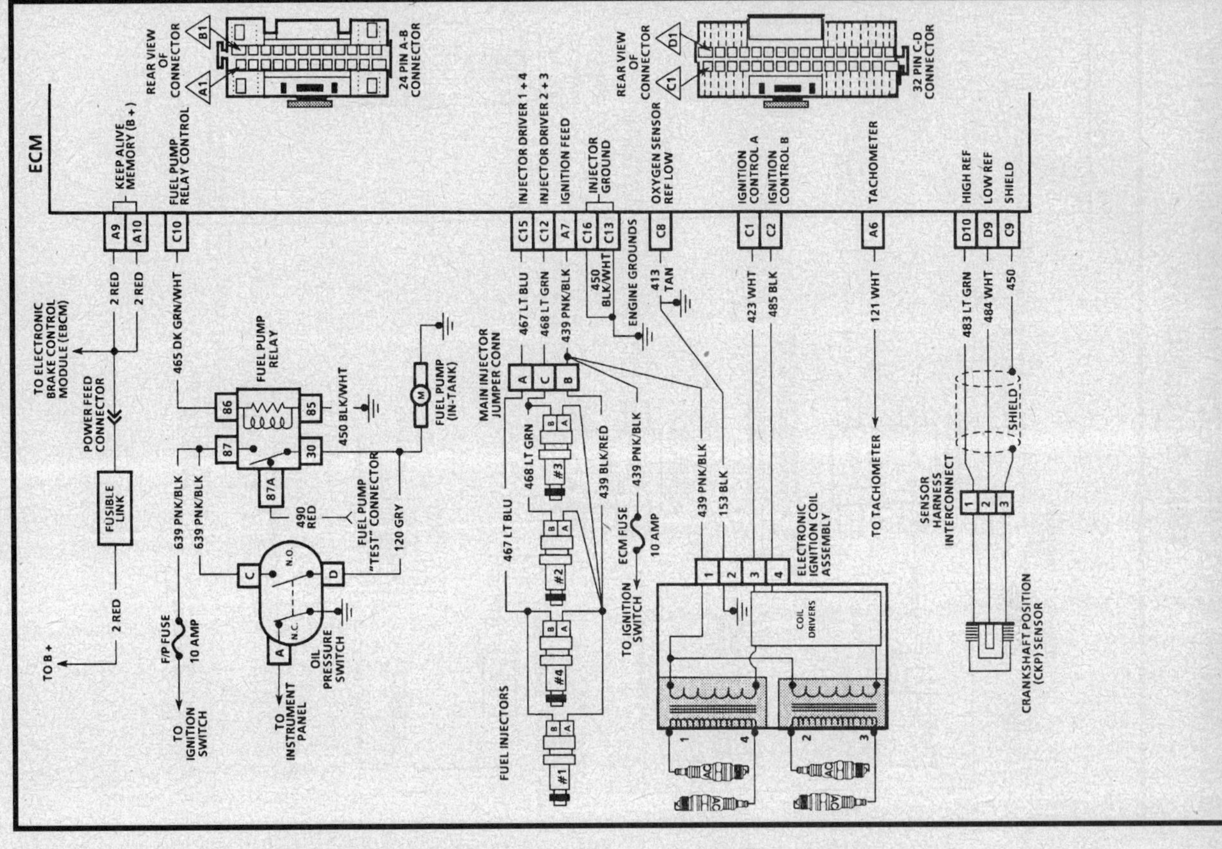

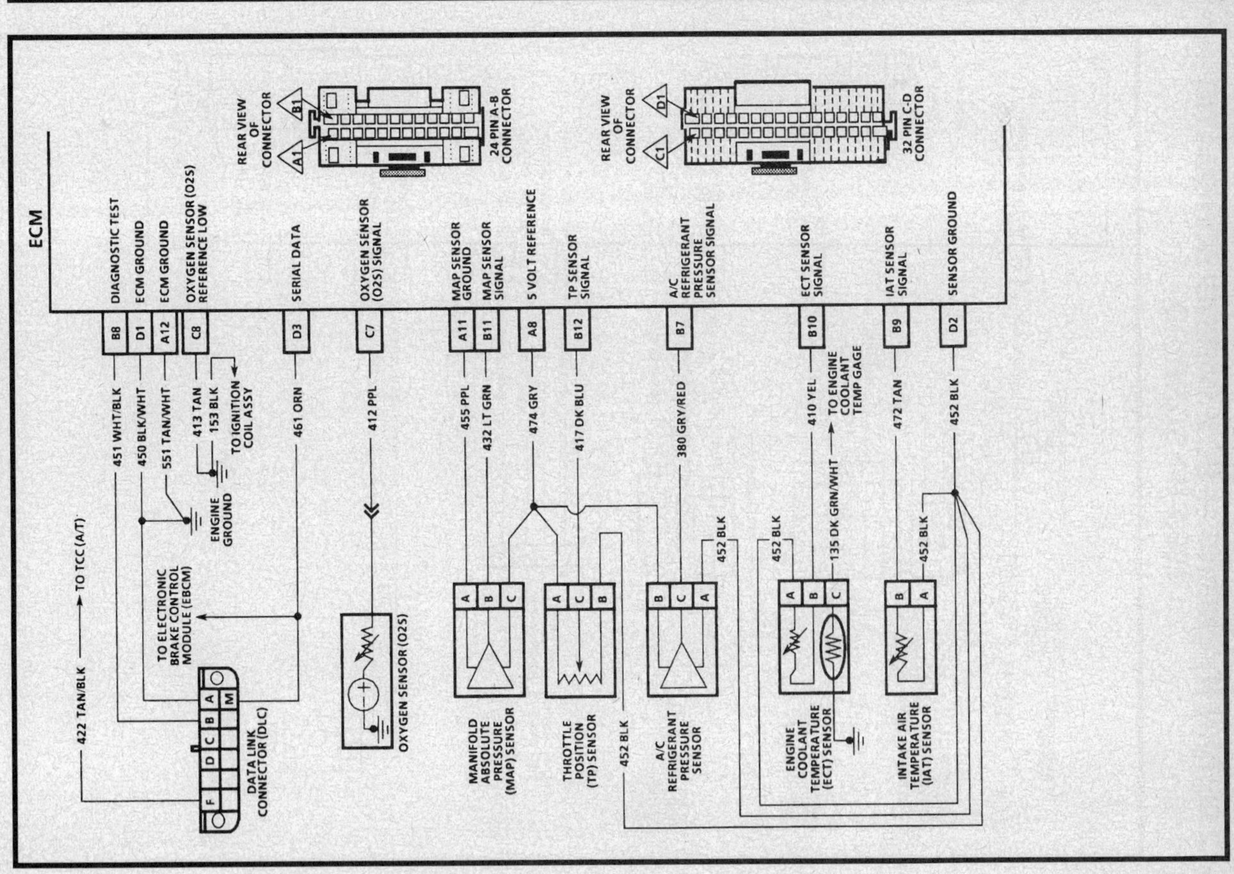

MULTIPORT FUEL INJECTION (MFI) SYSTEMS
EXCEPT LIGHT TRUCKS, VANS, GEO AND SATURN

2.0L (VIN H) ENGINE — ECM WIRING SCHEMATIC — 1993–94 SUNBIRD

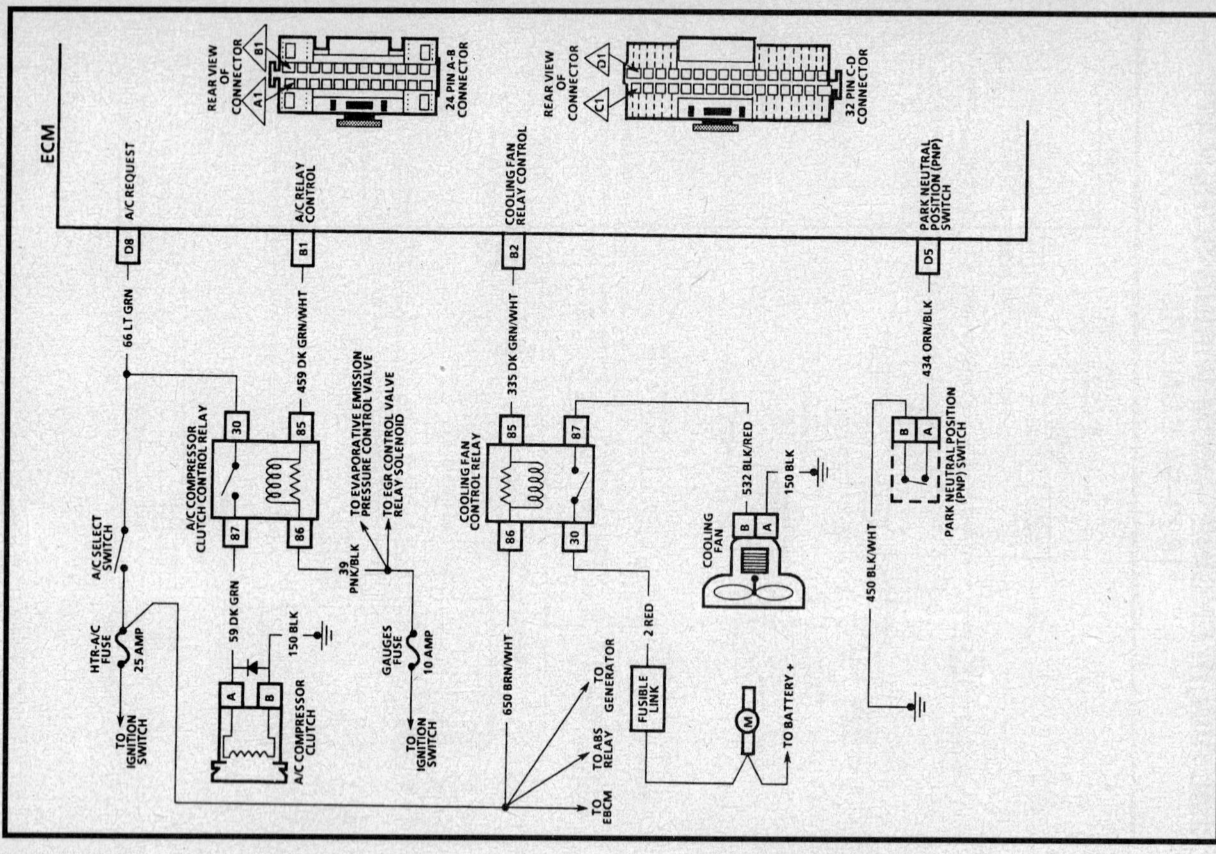

2.0L (VIN H) ENGINE — ECM WIRING SCHEMATIC — 1993–94 SUNBIRD

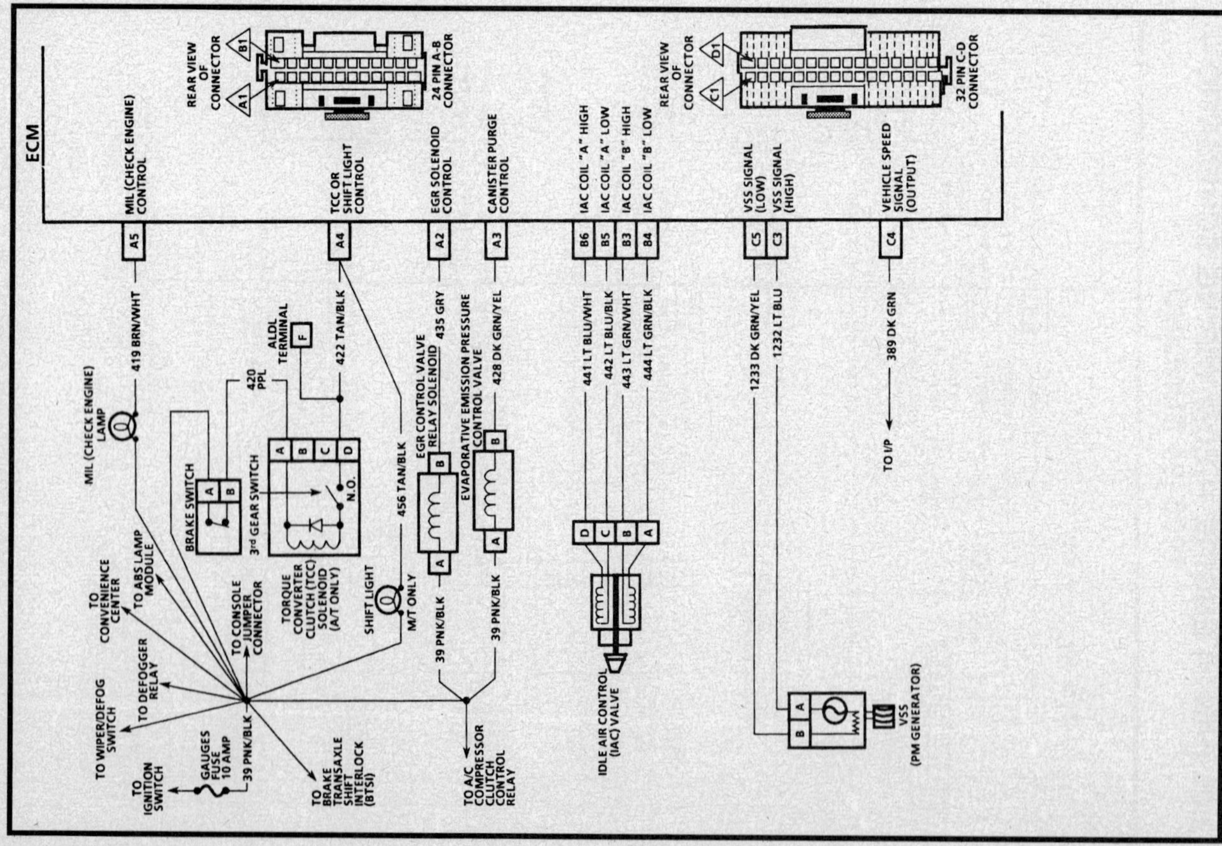

2.0L (VIN H) ENGINE — ECM CONNECTOR END VIEW — 1993–94 SUNBIRD

MULTIPORT FUEL INJECTION ECM CONNECTOR IDENTIFICATION

This ECM voltage chart is for use with J 39200 digital voltmeter to further aid in diagnosis. The voltages you get may vary due to low battery charge or other reasons, but they should be very close.

THE FOLLOWING CONDITIONS MUST BE MET BEFORE TESTING:

- Engine at operating temperature • Engine idling in "Closed Loop" (For "Engine Run" column) in park or neutral • "Test" terminal not grounded • Scan tool not installed • Brake not applied
- B + indicates battery or charging system voltage

24 PIN A-B CONNECTOR

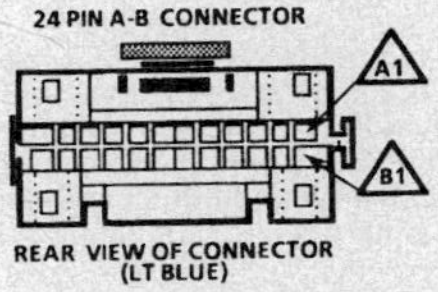

ENGINE: 2.0L (VIN H)
RPO: LE4

REAR VIEW OF CONNECTOR (LT BLUE)

LT BLUE 24 PIN A-B CONNECTOR

KEY "ON"	ENG. RUN	CIRCUIT	PIN	WIRE COLOR
		NOT USED	A1	
B+	B+	EGR SOLENOID CONTROL	A2	435 GRY
B+	B+	CANISTER PURGE SOLENOID CONTROL	A3	428 DK GRN/YEL
0*V	0*V	TCC (A/T) OR	A4	422 -TAN/BLK- 456
B+	B+	SHIFT LIGHT (M/T)	A4	
0*V	B+	MIL (CHECK ENGINE)	A5	419 BRN/WHT
(6)	(6)	TACHOMETER	A6	121 WHT
B + (9)	B + (9)	IGNITION FEED	A7	439 PNK/BLK
5.0V	5.0V	5 VOLTS REFERENCE	A8	474 GRY
B + (9)	B + (9)	KEEP ALIVE MEMORY (B +)	A9	2 RED
B + (9)	B + (9)	KEEP ALIVE MEMORY (B +)	A10	2 RED
0*V	0*V	MAP SENSOR GROUND	A11	455 PPL
0*V (9)	0*V (9)	ECM GROUND	A12	551 TAN/WHT

WIRE COLOR	PIN	CIRCUIT	KEY "ON"	ENG. RUN
459 DK GRN/WHT	B1	A/C COMPRESSOR CLUTCH RELAY CONTROL	B + (2)	B + (2)
335 DK GRN/WHT	B2	COOLING FAN RELAY CONTROL	B + (2)	B + (2)
443 LT GRN/WHT	B3	IAC "B" HIGH	(1)	(1)
444 LT GRN/BLK	B4	IAC "B" LOW	(1)	(1)
442 LT BLU/BLK	B5	IAC "A" LOW	(1)	(1)
441 LT BLU/WHT	B6	IAC "A" HIGH	(1)	(1)
380 GRY/RED	B7	A/C REFRIGERANT PRESSURE SENSOR SIGNAL	VARIES (8)	VARIES (8)
451 WHT/BLK	B8	DIAGNOSTIC TEST	5.0V	5.0V
472 TAN	B9	IAT SENSOR SIGNAL	1.3V (3)	1.3V (3)
410 YEL	B10	ECT SENSOR SIGNAL	2.0V (3)	2.0V (3)
432 LT GRN	B11	MAP SIGNAL	4.75V (7)	1.6V (7)
417 DK BLU	B12	TP SENSOR SIGNAL	.33 - 1.33V	.33 - 1.33V

* LESS THAN .5 VOLT.
(1) NOT USABLE.
(2) A/C SELECT "OFF" AND ENGINE COOLING FAN "OFF."
(3) VARIES DEPENDING ON TEMPERATURE.
(4) READS B + FOR 2 SECONDS AFTER KEY "ON,"
(5) VARIES WITH VEHICLE SPEED.
(6) VARIES WITH ENGINE RPM.
(7) VARIES WITH ALTITUDE.
(8) REFER TO DTC 66.
(9) THIS TYPE OF CIRCUIT SHOULD ALSO BE CHECKED USING J 34142-B TEST LIGHT.

2.0L (VIN H) ENGINE — ECM CONNECTOR END VIEW — 1993–94 SUNBIRD

MULTIPORT FUEL INJECTION ECM CONNECTOR IDENTIFICATION

This ECM voltage chart is for use with J 39200 digital voltmeter to further aid in diagnosis. The voltages you get may vary due to low battery charge or other reasons, but they should be very close.

THE FOLLOWING CONDITIONS MUST BE MET BEFORE TESTING:

- Engine at operating temperature • Engine idling in "Closed Loop" (For "Engine Run" column) in park or neutral • "Test" terminal not grounded • "Scan" tool not installed • Brake not applied
- B + indicates battery or charging system voltage

32 PIN C-D CONNECTOR

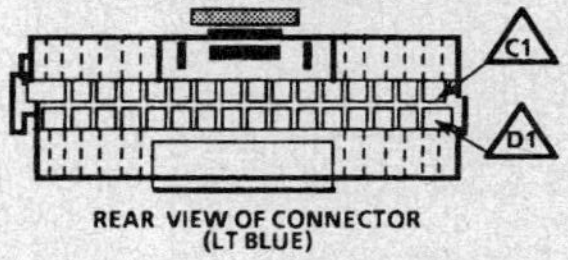

ENGINE: 2.0L (VIN H)
RPO: LE4

REAR VIEW OF CONNECTOR (LT BLUE)

LT BLUE 32 PIN C-D CONNECTOR

KEY "ON"	ENG. RUN	CIRCUIT	PIN	WIRE COLOR
B+	1.3 - 1.6V (6)	IGNITION CONTROL "A"	C1	423 WHT
B+	1.3 - 1.6V (6)	IGNITION CONTROL "B"	C2	485 BLK
(5)	(5)	VSS SIGNAL (HIGH)	C3	1232 LT BLU
(5)	(5)	VSS OUTPUT (4000 PPM)	C4	389 DK GRN
(5)	(5)	VSS SIGNAL (LOW)	C5	1233 DK GRN/YEL
		NOT USED	C6	
.1-.55V	.1-.9V	OXYGEN SENSOR SIGNAL	C7	412 PPL
0*V	0*V	OXYGEN SENSOR REFERENCE LOW	C8	413 TAN
0*V	0*V	IGNITION REFERENCE SHIELD	C9	450 (BARE)
(4)	B+	FUEL PUMP RELAY CONTROL	C10	465 DK GRN/WHT
		NOT USED	C11	
B+	B + (6)	#2 & 3 INJECTOR DRIVER	C12	468 LT GRN
0*V	0*V	#2 & 3 INJECTOR GROUND	C13	450 BLK/WHT
		NOT USED	C14	
B+	B + (6)	#1 & 4 INJECTOR DRIVER	C15	467 LT BLU
0*V	0*V	#1 & 4 INJECTOR GROUND	C16	450 BLK/WHT

WIRE COLOR	PIN	CIRCUIT	KEY "ON"	ENG. RUN
450 BLK/WHT	D1	ECM GROUND	0*V (9)	0*V (9)
452 BLK	D2	SENSOR GROUND	0*V	0*V
461 ORN	D3	SERIAL DATA	4.7V	4.7V
	D4	NOT USED		
434 ORN/BLK	D5	PARK/NEUTRAL POSITION (PNP) SWITCH	0*V	0*V
	D6	NOT USED		
	D7	NOT USED		
66 LT GRN	D8	A/C REQUEST	0*V (2)	0*V (2)
484 WHT	D9	IGNITION REFERENCE LOW	(7)	(7)
483 LT GRN	D10	IGNITION REFERENCE HIGH	(7)	(7)
	D11	NOT USED		
	D12	NOT USED		
	D13	NOT USED		
	D14	NOT USED		
	D15	NOT USED		
	D16	NOT USED		

* LESS THAN .5 VOLT.
(1) NOT USABLE.
(2) A/C SELECT "OFF" AND ENGINE COOLING FAN "OFF."
(3) VARIES DEPENDING ON TEMPERATURE.
(4) READS B + FOR 2 SECONDS AFTER KEY "ON,"
(5) VARIES WITH VEHICLE SPEED.
(6) VARIES WITH ENGINE RPM.
(7) REFER TO DTC 19 CHART.
(9) THIS TYPE OF CIRCUIT SHOULD ALSO BE CHECKED USING J 34142-B TEST LIGHT.

2.0L (VIN H) ENGINE — ON-BOARD DIAGNOSTIC SYSTEM CHART — SUNBIRD

ON-BOARD DIAGNOSTIC (OBD) SYSTEM CHECK
2.0L (VIN H) "J" CARLINE

Circuit Description:

The OBD system check is an organized approach to identifying a problem created by an electronic engine control system malfunction. It must be the starting point for any driveability complaint diagnosis, because it directs the service technician to the next logical step in diagnosing the complaint. Understanding the chart and using it correctly will reduce diagnostic time and prevent the unnecessary replacement of good parts.

Test Description: Number(s) below refer to circled number(s) on the diagnostic chart.

1. This step is a check for the proper operation of the Malfunction Indicator Lamp (MIL) "Check Engine." The MIL (Check Engine) should be "ON" steady.
2. No MIL (Check Engine) at this point indicates that there is a problem with the MIL (Check Engine) circuit or the ECM control of that circuit.
3. This test checks the ability of the ECM to control the MIL (Check Engine). With the diagnostic terminal grounded, the MIL (Check Engine) should flash a Diagnostic Trouble Code (DTC) 12 three times, followed by any DTC stored in memory. Depending upon the type of ECM, a PROM error may result in the inability to flash DTC 12.
4. Most of the procedures use a Tech 1 to aid diagnosis, therefore, serial data must be available. If a PROM error is present, the ECM may have been able to flash DTC 12/51, but not enable serial data.
5. Although the ECM is powered up, a "Cranks But Will Not Run" symptom could exist because of an ECM or system problem.
6. This step will isolate if the customer complaint is a MIL (Check Engine) or a driveability problem with no MIL (Check Engine). Refer to "ECM Diagnostic Trouble Codes" for valid DTC(s). An invalid DTC may be the result of a faulty scan tool, PROM or ECM/PCM.
7. Comparison of actual control system data with the typical values is a quick check to determine if any parameter is not within limits. Keep in mind that a base engine problem (i.e. advanced cam timing) may substantially alter sensor values.
8. Installation of a scan tool will provide a good ground path for the ECM and may hide a driveability complaint due to poor ECM grounds.
9. If the actual data is not within the typical values established. the charts in "Component Systems," will provide a functional check of the suspect component or system.

2.0L (VIN H) ENGINE — ON-BOARD DIAGNOSTIC SYSTEM CHART — SUNBIRD

ON-BOARD DIAGNOSTIC (OBD) SYSTEM CHECK
2.0L (VIN H) "J" CARLINE

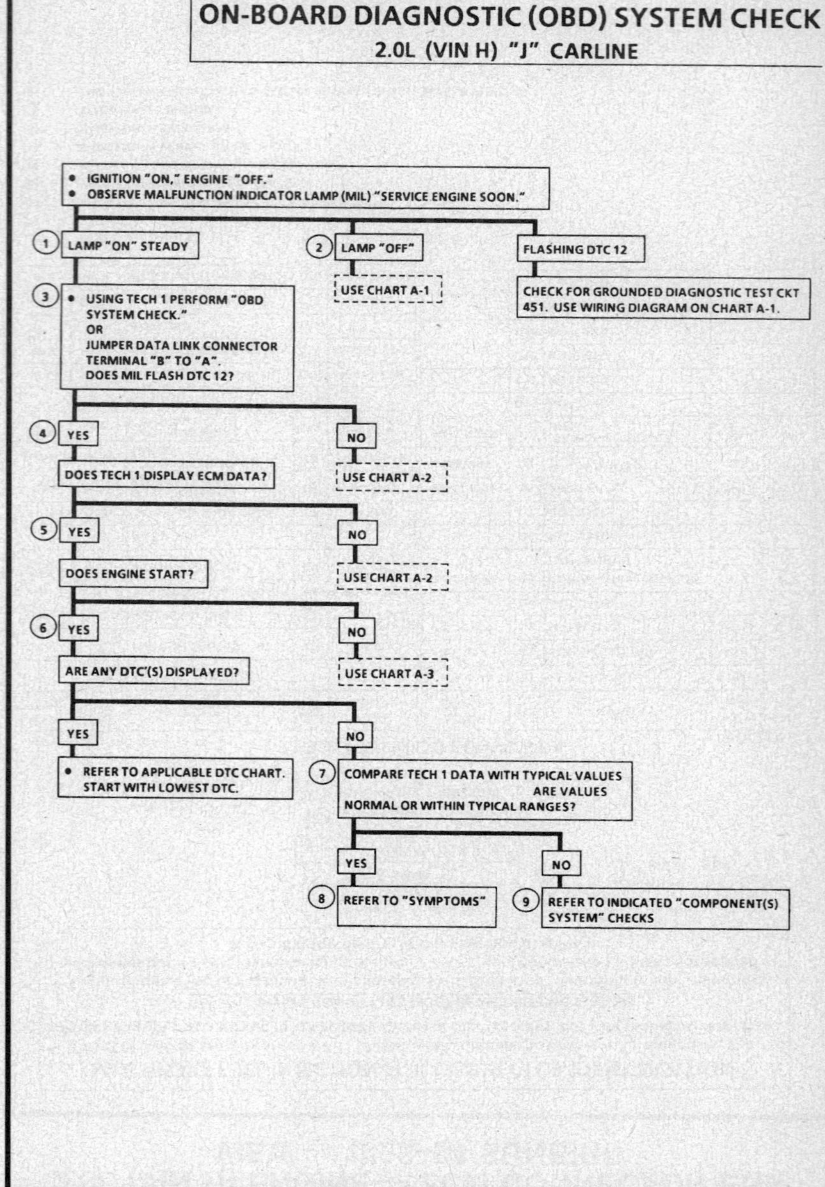

2.0L (VIN H) ENGINE — ECM DIAGNOSTIC TROUBLE CODES — SUNBIRD

ECM DIAGNOSTIC TROUBLE CODES (DTCs)		
DTC	DESCRIPTION	ILLUMINATE MIL (CHECK ENGINE)
13	Oxygen Sensor (O2S) Circuit - open circuit	YES
14	Engine Coolant Temperature (ECT) Sensor Circuit - high/low temperature	YES
19	Crankshaft Position (CKP) Sensor Circuit - incorrect 58X signal - disconnected sensor	YES
21	Throttle Position (TP) Sensor Circuit - signal voltage high/low	YES
23	Intake Air Temperature (IAT) Sensor Circuit - high/low temperature	YES
24	Vehicle Speed Sensor (VSS) Circuit	YES
32	Exhaust Gas Recirculation (EGR) - system failure	YES
33	Manifold Absolute Pressure (MAP) Sensor Circuit - signal voltage high/low - low/high vacuum	YES
44	Oxygen Sensor (O2S) Circuit - lean exhaust indicated	YES
45	Oxygen Sensor (O2S) Circuit - rich exhaust indicated	YES
51	ECM failure (ECM failed or EPROM failure)	YES
--	A/C Refrigerant Pressure Sensor Circuit	NO

If a DTC not listed above appears on Tech 1, ground Data Link Terminal "B" and observe flashed DTCs. If DTC does not reappear, Tech 1 data may be faulty.
If DTC does reappear, check for incorrect or faulty PROM.

2.0L (VIN H) ENGINE — SYSTEM DIAGNOSTIC CHARTS — SUNBIRD

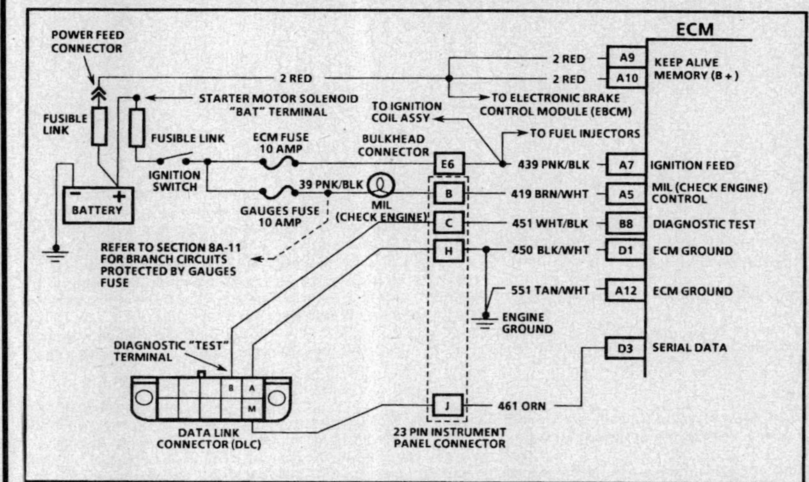

CHART A-1
NO MALFUNCTION INDICATOR LAMP (MIL) "CHECK ENGINE"
2.0L (VIN H) "J" CARLINE

Circuit Description:

There should always be a steady MIL (Check Engine) when the ignition is "ON" and engine stopped. Battery voltage is supplied directly to the light bulb. The Engine Control Module (ECM) will control the lamp and turn it "ON" by providing a ground path through CKT 419 to the ECM.

Test Description: Number(s) below refer to circled number(s) on the diagnostic chart.
1. This test will provide a ground path for the MIL (Check Engine) circuit. If the MIL (Check Engine) is "ON," then the external circuit is OK.
2. If CKTs 439 and 2 have voltage, the ECM grounds or the ECM is faulty.
3. Using a test light connected to B +, probe each of the system ground circuits to be sure a good ground is present. Refer to "ECM Connector End View" in front of this section for ECM pin locations of ground circuits.
4. If a fusible link or fuse is blown, be sure to consider branch circuits and splices to other components.

Diagnostic Aids:

Engine runs OK, check:
- Faulty light bulb.
- CKT 419 open.
- Gauges fuse blown. This will result in no oil or generator lights, seat belt reminder, etc.

Engine cranks, but will not run:
- ECM fuse open.
- Battery CKT 2 to ECM open.
- Ignition CKT 439 to ECM open.
- Poor connection to ECM.
- Fusible link open.

2.0L (VIN H) ENGINE — SYSTEM DIAGNOSTIC CHARTS — SUNBIRD

CHART A-1

NO MALFUNCTION INDICATOR LAMP (MIL) "CHECK ENGINE" 2.0L (VIN H) "J" CARLINE

DOES THE ENGINE START?

YES

(1)
- IGNITION "OFF."
- DISCONNECT ECM CONNECTORS.
- IGNITION "ON."
- PROBE CKT 419 WITH TEST LIGHT TO GROUND. IS THE MALFUNCTION INDICATOR LAMP (MIL) "CHECK ENGINE" "ON"?

- YES: FAULTY ECM CONNECTION OR ECM.
- NO: CHECK:
  - OPEN GAUGES FUSE.*
  - FAULTY CONNECTION.
  - FAULTY BULB.
  - OPEN CKT 419.
  - CKT 419 SHORTED TO VOLTAGE.
  - OPEN IGNITION FEED TO BULB.

NO

(2)
- IGNITION "OFF."
- DISCONNECT ECM CONNECTORS.
- IGNITION "ON."
- PROBE CKTs 2 AND 439 WITH TEST LIGHT TO GROUND. IS THE LIGHT "ON" ON BOTH CIRCUITS?

- YES: (3) FAULTY ECM GROUNDS OR ECM.
- NO: (4) REPAIR OPEN IN CIRCUIT THAT DID NOT LIGHT THE TEST LIGHT. IF A FUSIBLE LINK OR FUSE WAS BLOWN, LOCATE AND CORRECT THE SHORT TO GROUND ON THAT CIRCUIT.

*SEVERAL BRANCH CIRCUITS ARE PROTECTED BY THIS FUSE. REFER TO OTHER BRANCH CIRCUITS THAT COULD CAUSE A BLOWN FUSE.

"AFTER REPAIRS," CONFIRM "CLOSED LOOP" OPERATION AND NO MIL (CHECK ENGINE).

2.0L (VIN H) ENGINE — SYSTEM DIAGNOSTIC CHARTS — SUNBIRD

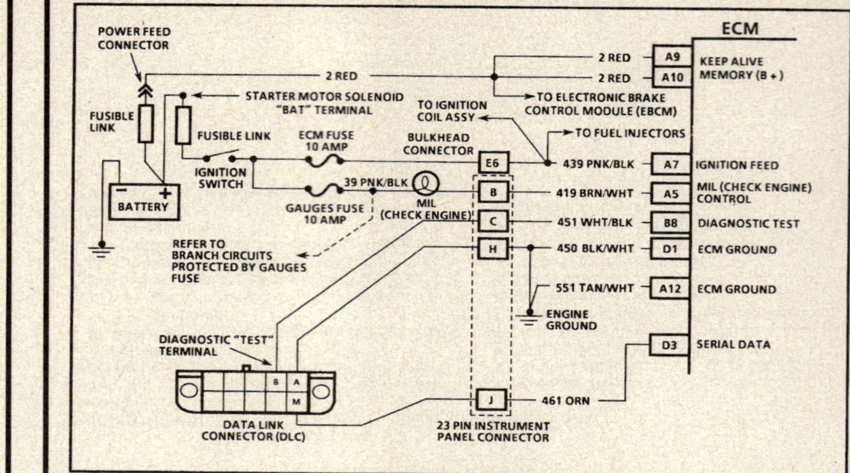

CHART A-2

NO DATA LINK CONNECTOR (DLC) DATA OR WON'T FLASH DTC 12 MALFUNCTION INDICATOR LAMP (MIL) "CHECK ENGINE" "ON" STEADY 2.0L (VIN H) "J" CARLINE

Circuit Description:

There should always be a steady MIL (Check Engine) when the ignition is "ON" and engine the engine is "OFF." Battery voltage is supplied directly to the lamp. The Engine Control Module (ECM) will control the lamp and turn it "ON" by providing a ground path through CKT 419 to the ECM.

With the diagnostic terminal grounded, the lamp should flash a Diagnostic Trouble Code (DTC) 12, followed by any DTC(s) stored in memory. A steady lamp suggests a short to ground in the lamp control CKT 419, or an open in diagnostic CKT 451.

Test Description: Number(s) below refer to circled number(s) on the diagnostic chart.

1. If there is a problem with the ECM that causes a scan tool to not read data from the ECM, then the ECM should not flash a DTC 12. If DTC 12 does flash, be sure that the scan tool is working properly on another vehicle. If the scan is functioning properly and CKT 461 is OK, the EEPROM or ECM may be at fault for the "No Data Link Connector Data" symptom.
2. If the lamp turns "OFF" when the ECM connector is disconnected, then CKT 419 is not shorted to ground.
3. This step will check for an open diagnostic CKT 451.
4. At this point, the MIL (Check Engine) wiring is OK. The problem is a faulty ECM. If DTC 12 does not flash, the ECM should be replaced.

2.0L (VIN H) ENGINE — SYSTEM DIAGNOSTIC CHARTS — SUNBIRD

CHART A-2

**NO DATA LINK CONNECTOR (DLC) DATA OR WON'T FLASH DTC 12
MALFUNCTION INDICATOR LAMP (MIL) "CHECK ENGINE" "ON" STEADY
2.0L (VIN H) "J" CARLINE**

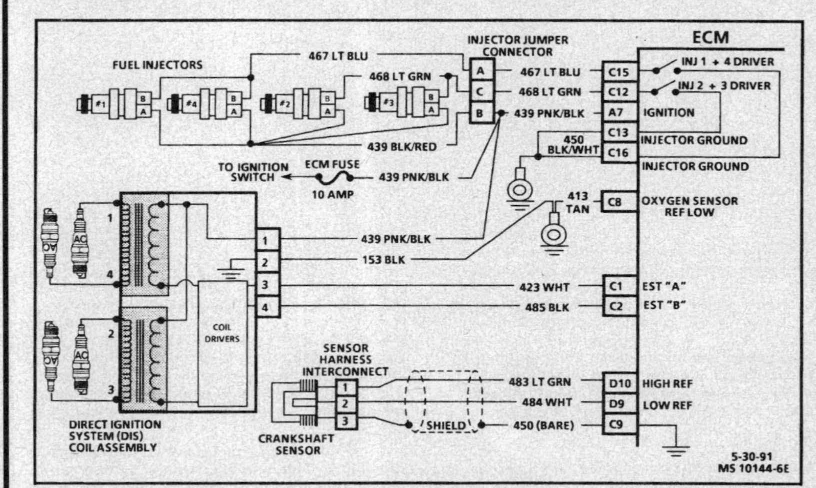

2.0L (VIN H) ENGINE — SYSTEM DIAGNOSTIC CHARTS — 1992 SUNBIRD

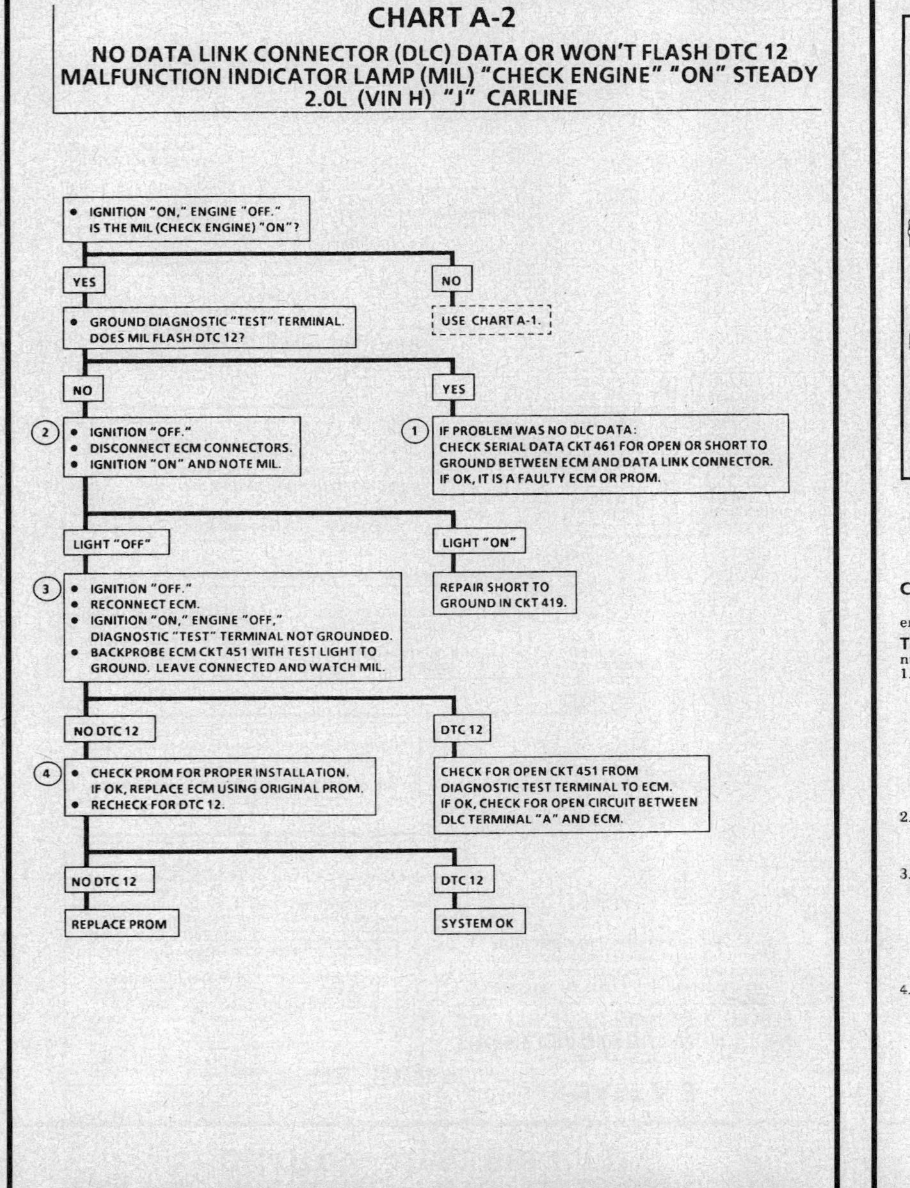

CHART A-3

**(Page 1 of 3)
ENGINE CRANKS BUT WON'T RUN
2.0L (VIN H) "J" CARLINE (PORT)**

Condition:

Engine cranks but will not run, or engine may start, but immediately stops running. Battery condition and engine cranking speed are OK, and there is adequate fuel in the tank.

Test Description: Number(s) below refer to circled number(s) on the diagnostic chart.

1. A "Check Engine" light "ON" is a basic test to determine if there is battery and ignition voltage at the ECM. No ALDL may be the result of an ECM problem, and CHART A-2 will diagnose an ECM. If TPS is less than .2 volt, the TPS 5 volt reference circuit could be shorted to ground. If TPS is over 2.5 volts, the ECM could be in the "Clear Flood Mode" which may cause the engine to not start. The "Scan" tool should display rpm during cranking.
2. Because the Direct Ignition System (DIS) uses two spark plugs and cables to complete the circuit of each coil, the opposite spark plug cable should be left connected.
3. This step determines if injectors or harness are cause of the problem. Nominal resistance of each injector is 11.8 to 12.6 ohms at 20°C (68°F). Resistance will increase slightly at higher temperatures. This test is performed with injectors 1 & 4 then 2 & 3 in parallel, so measurement will be half of the nominal resistance for a single injector.
4. The test light should flash, indicating that the ECM is controlling the injectors. How bright the light flashes is not important.

5. Ignition may have to be cycled "ON" several times to obtain maximum fuel pressure.
6. Damage to the ECM injector driver may occur if any injector resistance measures less than 11.8 ohms (internal short to ignition CKT 439).

Diagnostic Aids:

- Water or foreign material in fuel system can cause a no start condition during freezing weather. The engine may start after approximately 5 minutes in a heated shop. The problem may not recur until an overnight park in very cold or freezing temperatures.
- An EGR valve sticking open can cause a low air/fuel ratio during cranking. Unless the ECM enters "Clear Flood Mode" at the first indication of a flooding condition, it may result in a no start condition.
- Low fuel pressure can result in a very lean air/fuel ratio. See CHART A-7.
- An A/C pressure sensor with a internal short to ground can cause a no start condition. Disconnect the A/C pressure sensor. If vehicle starts, replace faulty sensor.
- A MAP sensor stuck between .5 and 2.5 volts can cause a no start condition. Disconnect the MAP sensor. If vehicle starts, replace faulty sensor

2.0L (VIN H) ENGINE — SYSTEM DIAGNOSTIC CHARTS — 1992 SUNBIRD

CHART A-3
(Page 1 of 3)
ENGINE CRANKS BUT WON'T RUN
2.0L (VIN H) "J" CARLINE (PORT)

NOTICE: PFI SYSTEM UNDER PRESSURE. TO AVOID FUEL SPILLAGE, REFER TO FIELD SERVICE PROCEDURES FOR TESTING OR MAKING REPAIRS REQUIRING DISASSEMBLY OF FUEL LINES OR FITTINGS.

1. • IGNITION "ON." IF "CHECK ENGINE" LIGHT IS "OFF," USE CHART A-1.
   • INSTALL TECH 1 "SCAN" TOOL. IF NO DATA, USE CHART A-2.
   • CHECK THE FOLLOWING:
     - TPS - IF LESS THAN .2 VOLT OR OVER 2.5 VOLTS AT CLOSED THROTTLE, USE CODE 21 CHART.
     - COOLANT - IF BELOW -38°C (-36°F), USE CODE 14 CHART.
     - RPM - IF NO RPM WHILE CRANKING,

• PROBE FUEL PUMP "TEST" TERMINAL WITH A TEST LIGHT TO B +.
• IGNITION "OFF" FOR 10 SECONDS. LIGHT SHOULD BE "ON."
• IGNITION "ON."
TEST LIGHT SHOULD GO OUT FOR ABOUT 2 SECONDS AND THEN COMES BACK "ON."
DOES IT?

— NO → USE FUEL PUMP RELAY CIRCUIT CHART A-5.

— YES →

2. • CRANK ENGINE AND CHECK FOR SPARK WITH ST-125 ON SPARK PLUG CABLES 1 & 2 OR 3 & 4.
   • CHECK ONE CABLE AT A TIME. LEAVE THE OTHER CABLES CONNECTED TO THE PLUGS DURING CRANKING. IS THERE SPARK ON BOTH CABLES?

— NO → USE CHART A-3 (PAGE 2 OF 3).

— YES →

3. • DISCONNECT INJECTOR JUMPER HARNESS AT 3-PIN CONNECTOR.
   • USING DVM, MEASURE RESISTANCE BETWEEN TERMINALS "A" AND "B" OF INJECTOR SIDE OF HARNESS.
   • REPEAT RESISTANCE MEASUREMENT BETWEEN TERMINALS "C" AND "B". RESISTANCE SHOULD BE ABOUT 5.9 TO 6.3 OHMS FOR BOTH TESTS. ARE THEY?

— NO →
  LESS THAN 5.9 OHMS
  GREATER THAN 6.3 OHMS

6. REPAIR SHORT IN HARNESS OR REPLACE ANY INJECTOR(S) THAT MEASURE LESS THAN 11.8 OHMS.

   REPAIR OPEN IN HARNESS OR REPLACE ANY INJECTOR(S) THAT MEASURE GREATER THAN 12.6 OHMS.

— YES →

4. • RECONNECT INJECTOR JUMPER HARNESS CONNECTOR.
   • CONNECT INJECTOR TEST LIGHT J 34730-2B TO INJECTOR #1 HARNESS CONNECTOR.
   • CRANK ENGINE AND OBSERVE TEST LIGHT (SHOULD BLINK).
   • PERFORM TEST AGAIN WITH TEST LIGHT CONNECTED TO INJECTOR #3 HARNESS. LIGHT SHOULD BLINK ON BOTH TESTS. DOES IT?

— NO → USE CHART A-3 (PAGE 3 OF 3)

— YES →

5. • IGNITION "OFF."
   • INSTALL FUEL PRESSURE GAGE (SEE CHART A-7 PAGE 3 OF 3).
   • IGNITION "ON," FUEL PRESSURE SHOULD BE 284-325 kPa (41-47 psi). IS IT?

— NO → USE FUEL SYSTEM DIAGNOSIS CHART A-7.

— YES →

CHECK FOR FOULED SPARK PLUGS
OR
EGR VALVE STUCK OPEN
OR
A/C PRESSURE SENSOR SHORTED.
SEE "DIAGNOSTIC AIDS"

"AFTER REPAIRS," CONFIRM "CLOSED LOOP" OPERATION AND NO "CHECK ENGINE" LIGHT.

2.0L (VIN H) ENGINE — SYSTEM DIAGNOSTIC CHARTS — 1992 SUNBIRD

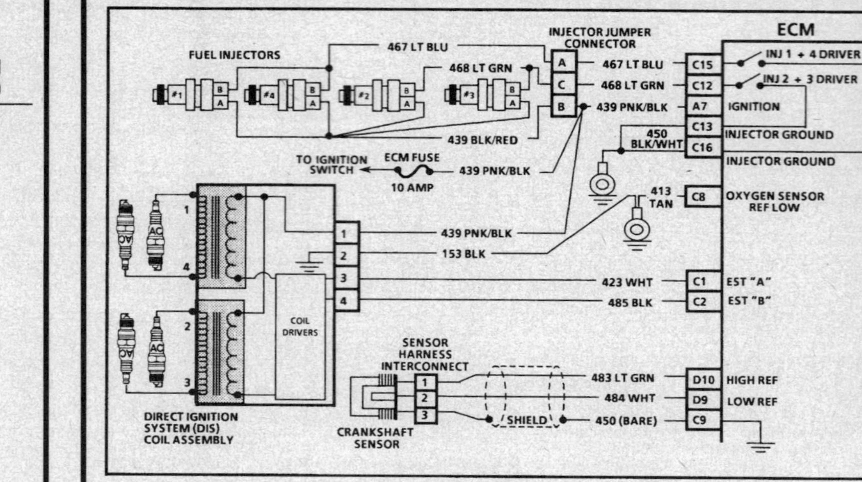

CHART A-3
(Page 2 of 3)
ENGINE CRANKS BUT WON'T RUN
2.0L (VIN H) "J" CARLINE (PORT)

Circuit Description:

This engine is equipped with a distributorless ignition system called Direct Ignition System (DIS) system. The primary circuit of the DIS consists of one ignition coil module, the ECM, and crankshaft sensor as well as the related connecting wires. The secondary circuit consists of the secondary windings of the coil module, the spark plug boot/connector assemblies, and spark plugs.

Test Description: Number(s) below refer to circled number(s) on the diagnostic chart.

1. The Direct Ignition System uses two plugs and cables to complete the circuit of each coil. The other spark plug cable in the circuit must be left connected to create a spark.

2. This test will determine if the 12 volts supply and a good ground is available at coil module.

3. The spark plug cables should be measured with a ohmmeter for correct resistance (each wire should be less than 30000 ohms). This test will determine if the ECM is not triggering the coil module or if the coil module is faulty.

4. This test will determine if the ECM is not triggering the coil module or if the coil module is faulty.

5. This test will determine if the ECM is at fault or the harness from ECM to coil module is open/shorted.

2.0L (VIN H) ENGINE — SYSTEM DIAGNOSTIC CHARTS — 1992 SUNBIRD

NOTICE: WHEN PROBING "DIS" CIRCUITS, AVOID CONTACT WITH ROTATING ENGINE BELTS AND PULLEYS.

CHART A-3
(Page 2 of 3)
ENGINE CRANKS BUT WON'T RUN
2.0L (VIN H) "J" CARLINE (PORT)

1.
- CRANK ENGINE AND CHECK FOR SPARK WITH ST-125 ON SPARK PLUG CABLES 1 & 2 OR 3 & 4.
- CHECK ONE CABLE AT A TIME, LEAVING THE OTHER CABLES CONNECTED TO THE SPARK PLUGS DURING CRANKING.

NO SPARK

SPARK ON ONE WIRE

2.
- IGNITION "OFF."
- DISCONNECT 4-PIN CONNECTOR AT IGNITION MODULE.
- IGNITION "ON."
- PROBE HARNESS TERMINAL "1" WITH A TEST LIGHT TO GROUND. IS THE TEST LIGHT "ON"?

3.
- IGNITION "OFF."
- CHECK SPARK PLUG CABLES. IF OK, DISCONNECT 4-PIN CONNECTOR AT IGNITION MODULE.
- SET VOLTMETER ON DC VOLTS POSITION.
- PROBE HARNESS TERMINALS "3" THEN "4" TO GROUND WHILE CRANKING ENGINE AND OBSERVE VOLTAGE READINGS. BOTH READINGS SHOULD BE GREATER THAN 1 VOLT. ARE THEY?

YES →

PROBE HARNESS TERMINAL "2" (GROUND) WITH A TEST LIGHT TO BATTERY VOLTAGE. IS TEST LIGHT "ON"?

NO → OPEN CKT 439.

NO →

4.
- CHECK CKTs 423 AND 485 FOR OPEN/SHORT.
- CHECK FOR FAULTY CONNECTIONS. ARE CIRCUITS AND CONNECTIONS OK?

YES → FAULTY COIL MODULE.

YES →

4.
- SET VOLTMETER ON DC VOLTS POSITION.
- PROBE HARNESS TERMINALS "3" THEN "4" TO GROUND, WHILE CRANKING ENGINE AND OBSERVE VOLTAGE READINGS. BOTH READINGS SHOULD BE GREATER THAN 1 VOLT. ARE THEY?

NO → OPEN CKT 153.

YES → FAULTY ECM.

NO → REPAIR FAULTY CIRCUITS OR CONNECTIONS.

NO →

5.
- CHECK CKTs 423 AND 485 FOR OPEN/SHORT.
- CHECK FOR FAULTY CONNECTIONS. ARE CIRCUITS AND CONNECTIONS OK?

YES → FAULTY COIL MODULE.

YES → FAULTY ECM.

NO → REPAIR FAULTY CIRCUITS OR CONNECTIONS.

"AFTER REPAIRS," CONFIRM "CLOSED LOOP" OPERATION AND NO "CHECK ENGINE" LIGHT.

2.0L (VIN H) ENGINE — SYSTEM DIAGNOSTIC CHARTS — 1992 SUNBIRD

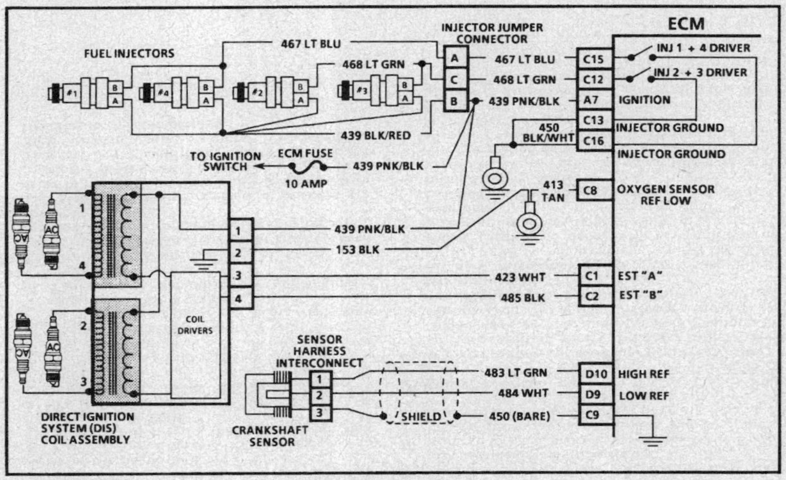

CHART A-3
(Page 3 of 3)
ENGINE CRANKS BUT WON'T RUN
2.0L (VIN H) "J" CARLINE (PORT)

Circuit Description:
Ignition voltage is supplied to the fuel injectors on CKT 439. The paired injectors will be pulsed (turned "ON" and "OFF,") when the ECM opens and grounds injector drive CKTs 467 and 468.

Test Description: Number(s) below refer to circled number(s) on the diagnostic chart.
1. This test checks for ignition voltage at the injector jumper harness connector.
2. Checks for open in CKT 467 or CKT 468 from ECM to injector(s). The test light has a path to voltage through the injector windings to ignition CKT 439.
3. Checks for short to voltage on CKT 467 or CKT 468 from ECM to injectors. Be sure all injectors are disconnected from injector harness connectors.
4. Damage to ECM injector driver may occur if either injector driver CKT 467 or CKT 468 shorts to ignition CKT 439.

2.0L (VIN H) ENGINE — SYSTEM DIAGNOSTIC CHARTS — 1992 SUNBIRD

CHART A-3

(Page 3 of 3)
ENGINE CRANKS BUT WON'T RUN
2.0L (VIN H) "J" CARLINE (PORT)

FROM CHART A-3 PAGE 1

NO LIGHT "ON" ONE OR BOTH

① • IGNITION "ON," ENGINE "OFF."
• PROBE HARNESS CONNECTOR TERMINAL "A" (CKT 439) OF INJECTOR 1 OR 3 WITH A TEST LIGHT TO GROUND.
• IS THE TEST LIGHT "ON"?

STEADY LIGHT "ON" ONE OR BOTH

• DISCONNECT ECM CONNECTORS, AND ALL INJECTOR CONNECTORS.
• CHECK INJECTOR DRIVER CIRCUIT WITH TEST LIGHT FOR SHORT TO GROUND. IS CIRCUIT SHORTED TO GROUND?

YES — REPAIR SHORT TO GROUND ON INJECTOR CIRCUIT.

NO — FAULTY ECM CONNECTIONS, OR FAULTY ECM.

YES

② • IGNITION "OFF," RECONNECT ALL INJECTOR HARNESS CONNECTORS.
• DISCONNECT ECM CONNECTORS.
• TURN IGNITION "ON."
• PROBE ECM TERMINAL "C15" THEN "C12" WITH A TEST LIGHT TO GROUND. IS LIGHT "ON" AT BOTH TERMINALS?

NO — IGNITION CKT 439 IS OPEN.

YES

③ • TURN IGNITION "OFF," DISCONNECT ALL INJECTOR ELECTRICAL CONNECTORS.
• TURN IGNITION "ON."
• PROBE ECM TERMINAL "C15" THEN "C12" WITH A TEST LIGHT TO GROUND. IS LIGHT "ON" AT EITHER TERMINAL?

NO — INJECTOR DRIVER CKT 467 OR 468 IS OPEN.

YES

④ CKT 467 OR 468 IS SHORTED TO VOLTAGE.

NO — FAULTY ECM CONNECTIONS, OR FAULTY ECM.

"AFTER REPAIRS," CONFIRM "CLOSED LOOP" OPERATION AND NO "CHECK ENGINE" LIGHT.

2.0L (VIN H) ENGINE — SYSTEM DIAGNOSTIC CHARTS — 1993–94 SUNBIRD

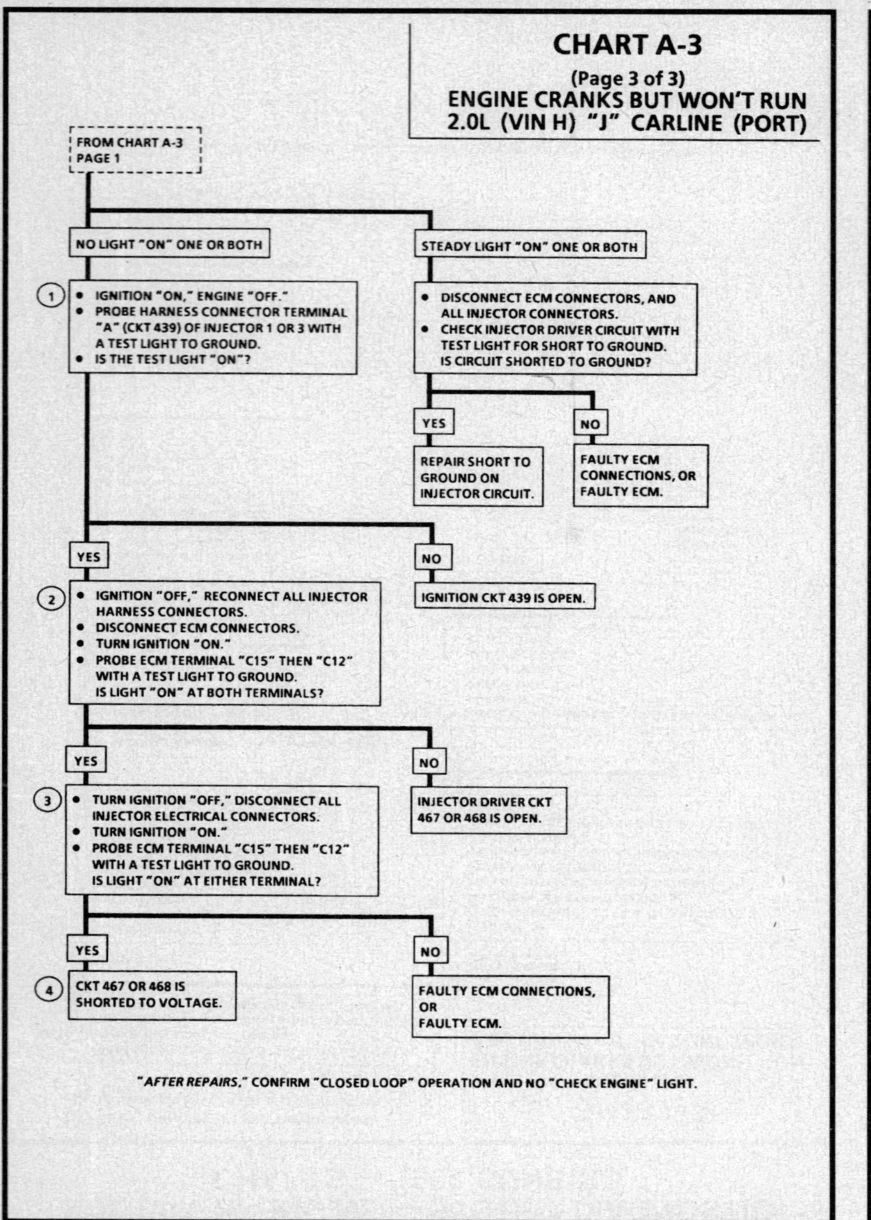

CHART A-3 (Page 1 of 3)

ENGINE CRANKS BUT WON'T RUN
2.0L (VIN H) "J" CARLINE

Condition:
Engine cranks but will not run, or engine may start, but immediately stops running. Battery condition and engine cranking speed are OK, and there is adequate fuel in the tank.

Test Description: Number(s) below refer to circled number(s) on the diagnostic chart.

1. • A MIL (Check Engine) "ON" is a basic test to determine if there is battery and ignition voltage at the ECM.
   • No DLC data may be the result of an ECM problem, and CHART A-2 will diagnose an ECM problem.
   • If TP sensor is less than .2 volt, the TP sensor 5 volt reference circuit could be shorted to ground. If TP sensor is over 2.5 volts, the ECM could be in the "Clear Flood Mode" which may cause the engine to not start.
   • The scan tool should display RPM during cranking.
2. Because the Electronic Ignition (EI) system uses two plugs and cables to complete the circuit of each ignition coil, the companion spark plug cable should be grounded.
3. This step determines if injectors or harness are cause of the problem. This test is performed with injectors 1 & 4 then 2 & 3 in parallel.
4. This test light should flash, indicating that the ECM is controlling the injectors. How bright the light flashes is not important.
5. Ignition may have to be cycled "ON" several times to obtain maximum fuel pressure.
6. Damage to ECM injector driver may occur if any injector resistance measures less than 11.8 ohms (internal short to ignition CKT 439).

Diagnostic Aids:

• Water or contamination in fuel system can cause a no start condition during freezing weather. The engine may start after approximately 5 minutes in a heated shop. The problem may not recur until an overnight park in very cold or freezing temperatures.
• An EGR valve sticking open can cause a low air/fuel ratio during cranking. Unless the ECM enters "Clear Flood Mode" at the first indication of a flooding condition, it may result in a no start condition.
• Low fuel pressure can result in a very lean air/fuel ratio. Refer to CHART A-7.
• An A/C refrigerant pressure sensor with a internal short to ground can cause a no start condition. Disconnect the A/C refrigerant pressure sensor. If vehicle starts, replace faulty sensor.
• A MAP sensor stuck between .5 and 2.5 volts can cause a no start condition. Disconnect the MAP sensor. If vehicle starts, replace faulty sensor.

2.0L (VIN H) ENGINE — SYSTEM DIAGNOSTIC CHARTS — 1993–94 SUNBIRD

NOTICE: MFI SYSTEM UNDER PRESSURE. TO AVOID FUEL SPILLAGE, REFER TO FIELD SERVICE PROCEDURES FOR TESTING OR MAKING REPAIRS REQUIRING DISASSEMBLY OF FUEL LINES OR FITTINGS.

CHART A-3 (Page 1 of 3)
ENGINE CRANKS BUT WON'T RUN
2.0L (VIN H) "J" CARLINE

1.
- IGNITION "ON." IF "CHECK ENGINE" LIGHT IS "OFF," USE CHART A-1.
- INSTALL TECH 1 SCAN TOOL. IF NO DATA, USE CHART A-2.
- CHECK THE FOLLOWING:
  - TP SENSOR - IF LESS THAN .2 VOLT OR OVER 2.5 VOLTS AT CLOSED THROTTLE, USE DTC 21 CHART.
  - COOLANT - IF BELOW -38°C (-36°F), USE DTC 14 CHART.
  - RPM - IF NO RPM WHILE CRANKING, USE DTC 19 CHART.

- PROBE FUEL PUMP "TEST" TERMINAL WITH A TEST LIGHT TO B + .
- IGNITION "OFF" FOR 10 SECONDS. LIGHT SHOULD BE "ON."
- IGNITION "ON."
  TEST LIGHT SHOULD GO OUT FOR ABOUT 2 SECONDS (FUEL PUMP WILL RUN FOR 2 SECONDS) AND THEN COME BACK "ON."
  DOES IT?

YES →

NO → USE FUEL PUMP RELAY CIRCUIT CHART A-5.

2.
- CRANK ENGINE AND CHECK FOR SPARK WITH ST-125 ON SPARK PLUG CABLES.
- CHECK ONE CABLE AT A TIME. CONNECT THE COMPANION CYLINDER SPARK PLUG CABLE TO A GOOD ENGINE GROUND (AWAY FROM ANY ELECTRONIC COMPONENTS). LEAVE THE OTHER CABLES CONNECTED TO THE PLUGS DURING CRANKING.
  IS THERE SPARK ON ALL CABLES?

YES →

NO → USE CHART A-3 (PAGE 2 OF 3).

3.
- DISCONNECT INJECTOR JUMPER HARNESS AT 3-PIN CONNECTOR.
- USING DVM, MEASURE RESISTANCE BETWEEN TERMINALS "A" AND "B" OF INJECTOR SIDE OF HARNESS.
- REPEAT RESISTANCE MEASUREMENT BETWEEN TERMINALS "C" AND "B". RESISTANCE SHOULD BE ABOUT 5.9 TO 7.3 OHMS FOR BOTH TESTS. ARE THEY?

YES →

NO →

4.
- RECONNECT INJECTOR JUMPER HARNESS CONNECTOR.
- CONNECT INJECTOR TEST LIGHT J 34730-2B TO INJECTOR #1 HARNESS CONNECTOR.
- CRANK ENGINE AND OBSERVE TEST LIGHT (SHOULD BLINK).
- PERFORM TEST AGAIN WITH TEST LIGHT CONNECTED TO INJECTOR #3 HARNESS.
  LIGHT SHOULD BLINK ON BOTH TESTS.
  DOES IT?

6.
- MEASURE INJECTOR RESISTANCE OF EACH INJECTOR. NOMINAL RESISTANCE OF EACH INJECTOR SHOULD BE 11.8 TO 12.6 OHMS AT 20°C (68°F). RESISTANCE WILL INCREASE SLIGHTLY AT HIGHER TEMPERATURES.

YES →

NO → USE CHART A-3 (PAGE 3 OF 3).

YES → REPLACE FAULTY INJECTOR.

NO → REPAIR OPEN/SHORT IN INJECTOR HARNESS.

5.
- IGNITION "OFF."
- INSTALL FUEL PRESSURE GAGE (SEE CHART A-7).
- IGNITION "ON," FUEL PRESSURE SHOULD BE 284-325 kPa (41-47 psi).
  IS IT?

YES →

NO → USE FUEL SYSTEM DIAGNOSIS CHART A-7.

CHECK FOR FOULED SPARK PLUGS
OR
EGR VALVE STUCK OPEN
OR
A/C REFRIGERANT PRESSURE SENSOR SHORTED.
REFER TO "DIAGNOSTIC AIDS"

"AFTER REPAIRS," CONFIRM "CLOSED LOOP" OPERATION AND NO MIL (CHECK ENGINE).

2.0L (VIN H) ENGINE — SYSTEM DIAGNOSTIC CHARTS — 1993–94 SUNBIRD

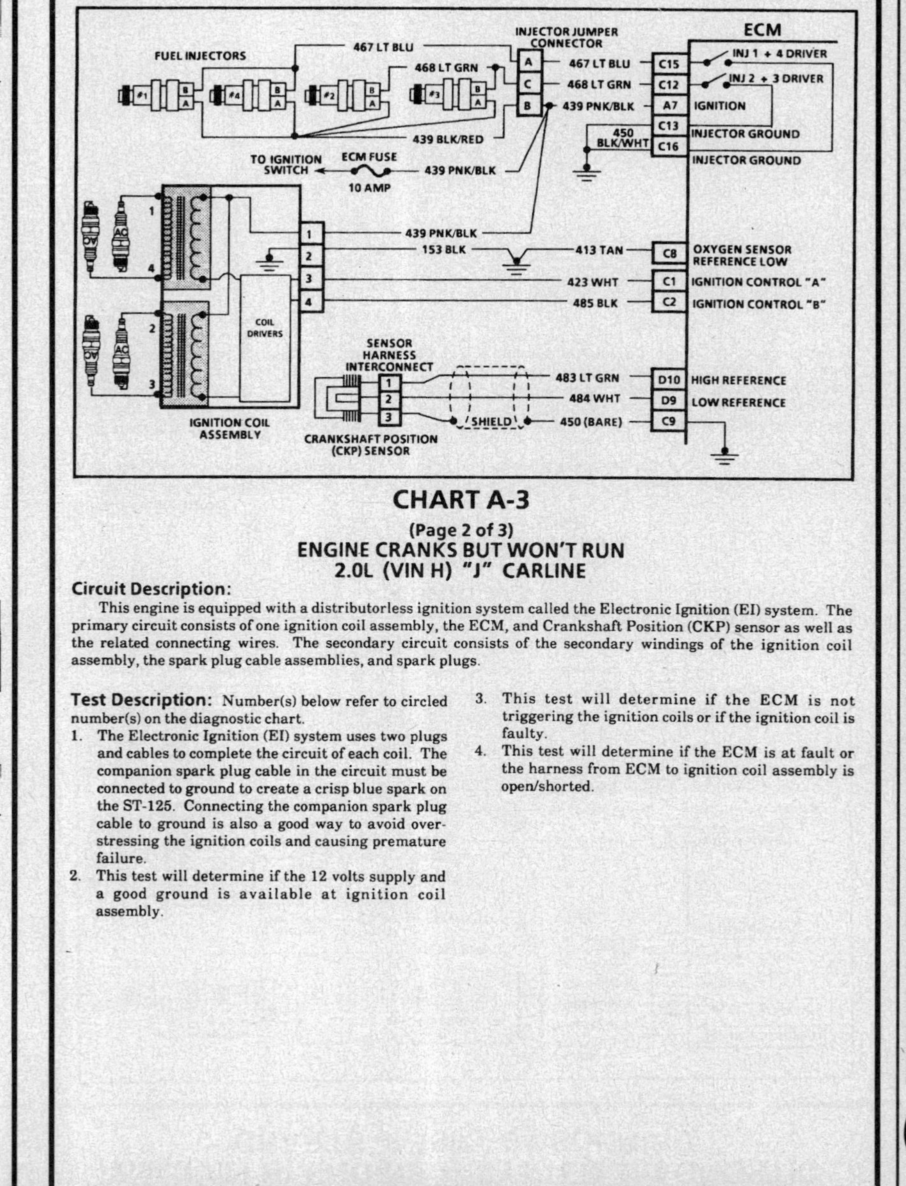

CHART A-3
(Page 2 of 3)
ENGINE CRANKS BUT WON'T RUN
2.0L (VIN H) "J" CARLINE

Circuit Description:

This engine is equipped with a distributorless ignition system called the Electronic Ignition (EI) system. The primary circuit consists of one ignition coil assembly, the ECM, and Crankshaft Position (CKP) sensor as well as the related connecting wires. The secondary circuit consists of the secondary windings of the ignition coil assembly, the spark plug cable assemblies, and spark plugs.

Test Description: Number(s) below refer to circled number(s) on the diagnostic chart.

1. The Electronic Ignition (EI) system uses two plugs and cables to complete the circuit of each coil. The companion spark plug cable in the circuit must be connected to ground to create a crisp blue spark on the ST-125. Connecting the companion spark plug cable to ground is also a good way to avoid over-stressing the ignition coils and causing premature failure.

2. This test will determine if the 12 volts supply and a good ground is available at ignition coil assembly.

3. This test will determine if the ECM is not triggering the ignition coils or if the ignition coil is faulty.

4. This test will determine if the ECM is at fault or the harness from ECM to ignition coil assembly is open/shorted.

2.0L (VIN H) ENGINE — SYSTEM DIAGNOSTIC CHARTS — 1993–94 SUNBIRD

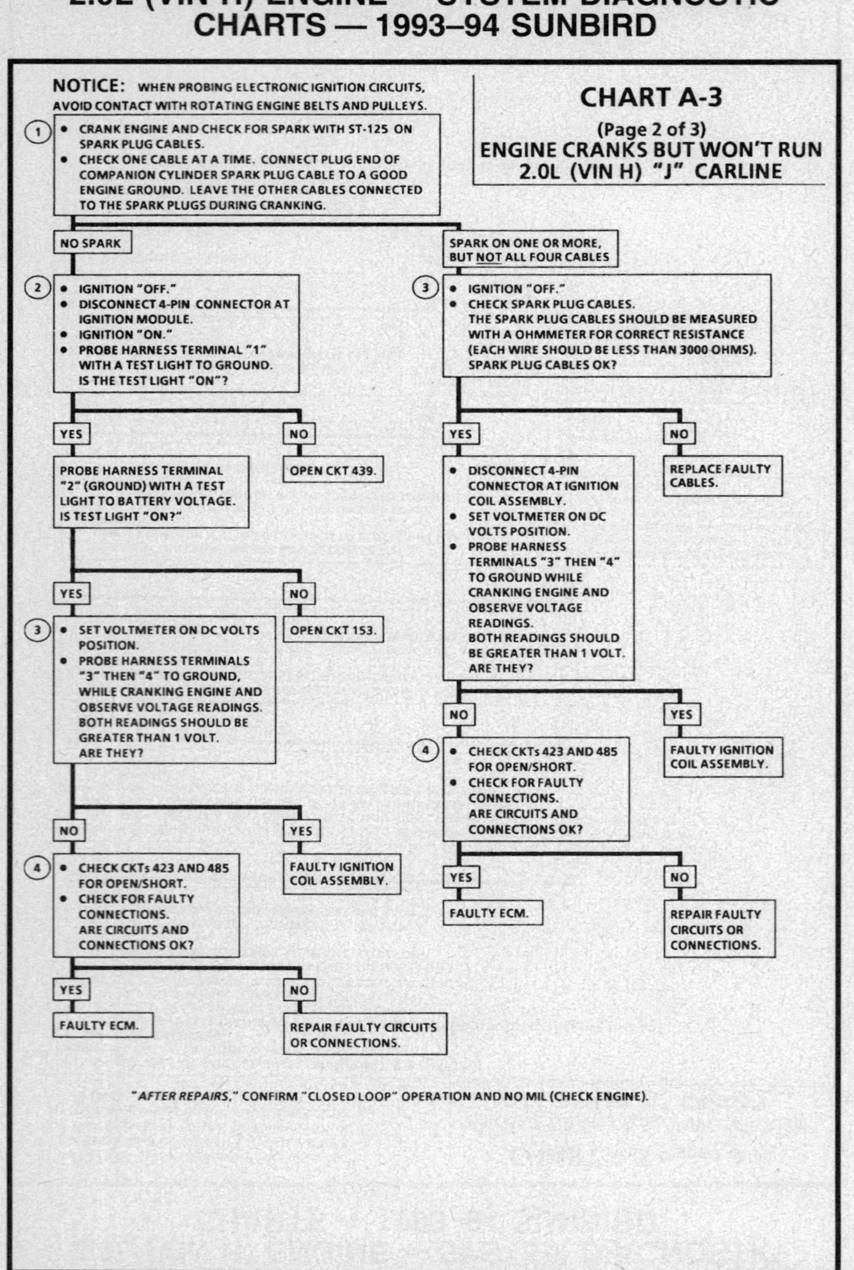

2.0L (VIN H) ENGINE — SYSTEM DIAGNOSTIC CHARTS — 1993–94 SUNBIRD

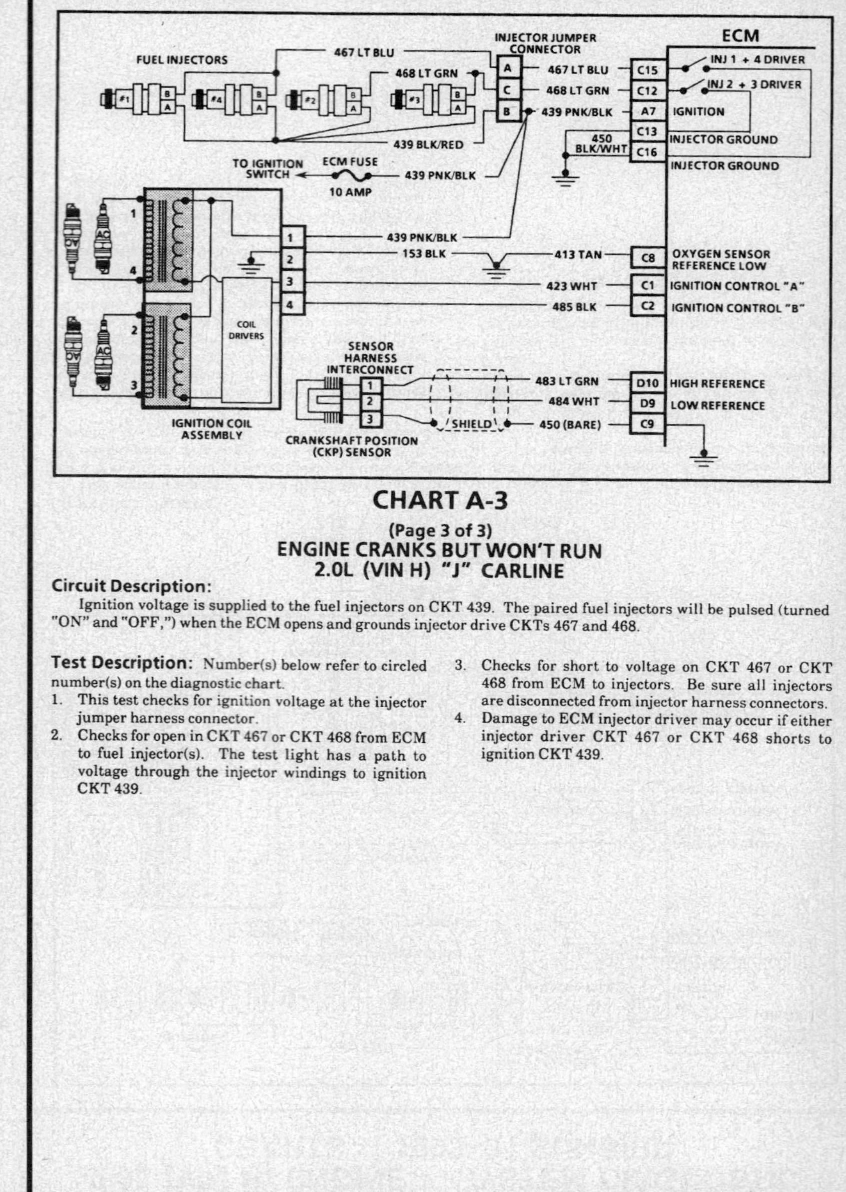

CHART A-3
(Page 3 of 3)
ENGINE CRANKS BUT WON'T RUN
2.0L (VIN H) "J" CARLINE

Circuit Description:

Ignition voltage is supplied to the fuel injectors on CKT 439. The paired fuel injectors will be pulsed (turned "ON" and "OFF,") when the ECM opens and grounds injector drive CKTs 467 and 468.

Test Description: Number(s) below refer to circled number(s) on the diagnostic chart.

1. This test checks for ignition voltage at the injector jumper harness connector.
2. Checks for open in CKT 467 or CKT 468 from ECM to fuel injector(s). The test light has a path to voltage through the injector windings to ignition CKT 439.
3. Checks for short to voltage on CKT 467 or CKT 468 from ECM to injectors. Be sure all injectors are disconnected from injector harness connectors.
4. Damage to ECM injector driver may occur if either injector driver CKT 467 or CKT 468 shorts to ignition CKT 439.

2.0L (VIN H) ENGINE — SYSTEM DIAGNOSTIC CHARTS — 1993–94 SUNBIRD

CHART A-3
(Page 3 of 3)
ENGINE CRANKS BUT WON'T RUN
2.0L (VIN H) "J" CARLINE

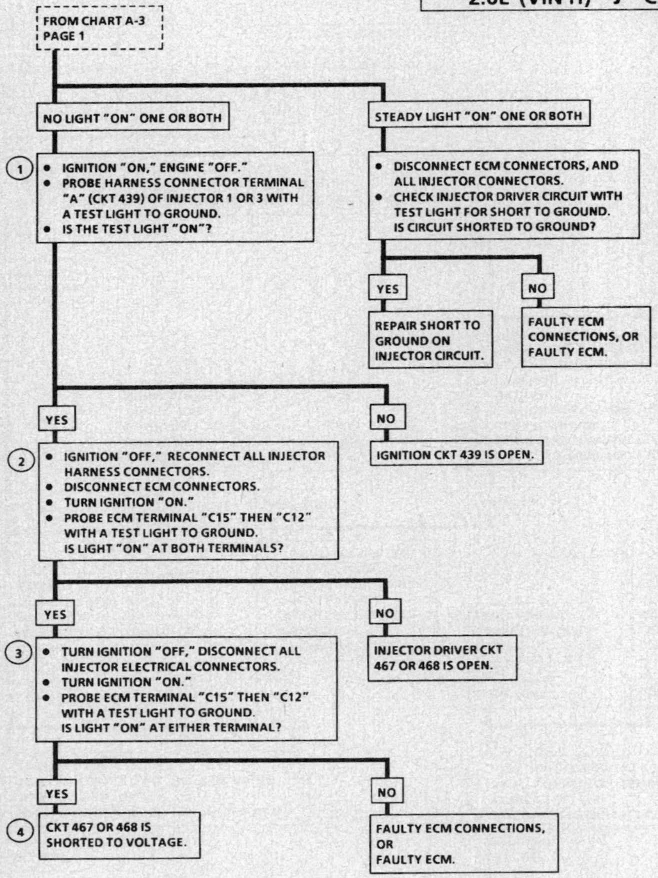

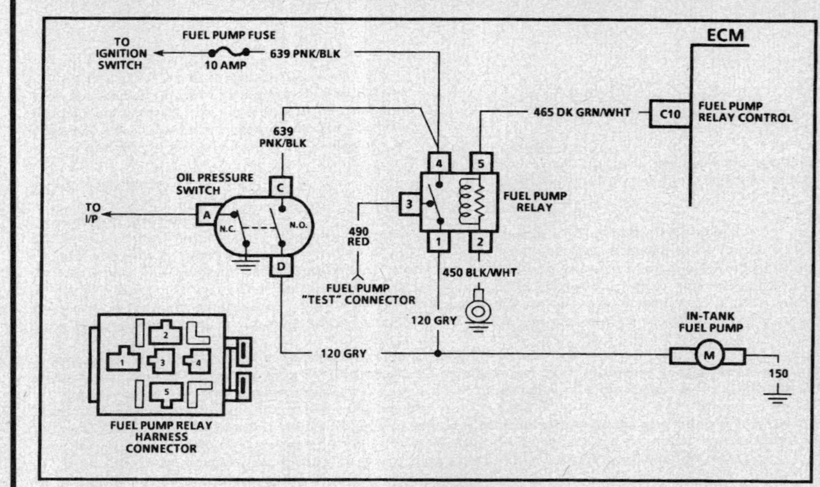

2.0L (VIN H) ENGINE — SYSTEM DIAGNOSTIC CHARTS — 1992 SUNBIRD

CHART A-5
FUEL PUMP RELAY CIRCUIT
2.0L (VIN H) "J" CARLINE (PORT)

Circuit Description:

When the ignition switch is turned "ON," the Electronic Control Module (ECM) will activate the fuel pump relay and run the in-tank fuel pump. The fuel pump will operate as long as the engine is cranking or running and the ECM is receiving ignition reference pulses.

If there are no reference pulses, the ECM will shut "OFF" the fuel pump within 2 seconds after key "ON."

Should the fuel pump relay or the 12 volts relay drive from the ECM fail, the fuel pump will continue to function by way of the oil pressure switch back-up circuit. Oil pressure will close the fuel pump side of the oil pressure switch, completing the circuit to the fuel pump.

The fuel pump test terminal is located in the driver's side of the engine compartment. When the engine is stopped, the pump can be turned "ON" by applying battery voltage to the test terminal. The fuel pump relay is located in the passenger side engine compartment next to the shock tower.

Test Description: Number(s) below refer to circled number(s) on the diagnostic chart.
1. At this point, the fuel pump relay is operating correctly. The back-up circuit through the oil pressure switch is now tested.
2. After the fuel pump relay is replaced, continue with "Oil Pressure Switch Test."

Diagnostic Aids:

An inoperative fuel pump relay can result in long cranking times. The extended crank period is caused by the time necessary for oil pressure to build enough to close the oil pressure switch and turn "ON" the fuel pump. The oil pressure switch closes at about 28 kPa (4 psi).

2.0L (VIN H) ENGINE — SYSTEM DIAGNOSTIC CHARTS — 1993-94 SUNBIRD

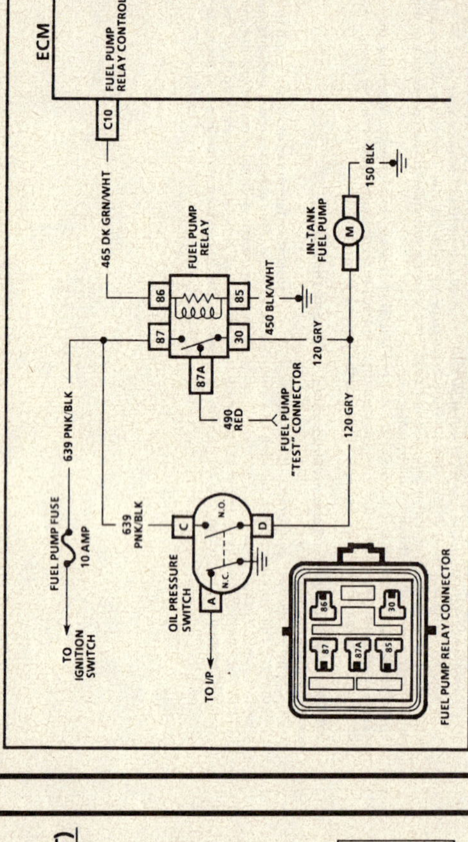

CHART A-5
FUEL PUMP RELAY CIRCUIT
2.0L (VIN H) "J" CARLINE

Circuit Description:

When the ignition switch is turned "ON," the Engine Control Module (ECM) will activate the fuel pump relay and run the in-tank fuel pump. The fuel pump will operate as long as the engine is cranking or running and the ECM is receiving ignition reference pulses.

If there are no reference pulses, the ECM will shut "OFF" the fuel pump within 2 seconds after key "ON."

Should the fuel pump relay or the 12 volts relay drive from the ECM fail, the fuel pump will continue to function by way of the oil pressure switch back-up circuit. Oil pressure will close the fuel pump side of the oil pressure switch, completing the circuit to the fuel pump.

The fuel pump "test" terminal is located in the driver's side of the engine compartment. When the engine is stopped, the pump can be turned "ON" by applying battery voltage to the "test" terminal. The fuel pump relay is located in the passenger side engine compartment next to the shock tower.

Test Description: Number(s) below refer to circled number(s) on the diagnostic chart.

1. At this point, the fuel pump relay is operating correctly. The back-up circuit through the oil pressure switch is now tested.

2. After the fuel pump relay is replaced, continue with "Oil Pressure Switch Test."

Diagnostic Aids:

An inoperative fuel pump relay can result in long cranking times. The extended crank period is caused by the time necessary for oil pressure to reach the pressure required to close the oil pressure switch and supply the necessary current for the fuel pump.

Excess fuel may also cause long cranking times. This would usually be accompanied by a start that is not as fast as normal (once fired, the engine does not build up speed as fast) and a puff of black smoke at the tailpipe. An improperly connected or faulty EVAP canister control valve can cause this problem.

One or more leaking fuel injectors may also extend the cranking time. Perform the "Injector Balance Test" in CHART C2-A.

2.0L (VIN H) ENGINE — SYSTEM DIAGNOSTIC CHARTS — 1992 SUNBIRD

CHART A-5
FUEL PUMP RELAY CIRCUIT
2.0L (VIN H) "J" CARLINE (PORT)

- IGNITION "OFF."
- BACKPROBE FUEL PUMP RELAY TERMINAL "1" WITH A TEST LIGHT TO BATTERY VOLTAGE.

LIGHT "ON"

- DISCONNECT FUEL PUMP RELAY HARNESS CONNECTOR.
- PROBE CONNECTOR TERMINAL "1" WITH A TEST LIGHT TO BATTERY VOLTAGE.

LIGHT "OFF" → OPEN CKT 120, OPEN CKT 150 TO FUEL PUMP, FAULTY CONNECTIONS, OR FAULTY IN-TANK FUEL PUMP.

LIGHT "ON" → (2) FAULTY CONNECTIONS OR FAULTY RELAY.

- IGNITION MUST BE "OFF" FOR TEN SECONDS.
- TURN IGNITION "ON."
- TEST LIGHT SHOULD GO "OFF" FOR ABOUT TWO SECONDS. DOES IT?

NO / YES

- DISCONNECT FUEL PUMP RELAY HARNESS CONNECTOR.
- PROBE CONNECTOR TERMINAL "4" WITH A TEST LIGHT TO GROUND.

LIGHT "OFF" → OPEN CKT 639 (CHECK FUEL PUMP FUSE, WIRING, AND CONNECTIONS).

LIGHT "ON"

- DISCONNECT FUEL PUMP RELAY HARNESS CONNECTOR.
- PROBE CONNECTOR TERMINAL "2" WITH A TEST LIGHT TO BATTERY VOLTAGE.

LIGHT "OFF" → OPEN CKT 450.

LIGHT "ON"

- PROBE FUEL PUMP RELAY HARNESS CONNECTOR TERMINAL "5" WITH A TEST LIGHT TO GROUND.
- IGNITION MUST BE "OFF" FOR TEN SECONDS.
- TURN IGNITION "ON."
- TEST LIGHT SHOULD LIGHT FOR ABOUT TWO SECONDS. DOES IT?

NO → OPEN OR GROUNDED CKT 465 OR FAULTY ECM.

YES → (2) REPLACE FUEL PUMP RELAY.

- OIL PRESSURE SWITCH TEST.
- ALLOW ENGINE TO RUN AT NORMAL TEMPERATURE AND OIL PRESSURE.
- DISCONNECT FUEL PUMP RELAY HARNESS CONNECTOR.
- ENGINE SHOULD CONTINUE TO RUN. DOES IT?

NO → CHECK CKTS 639 AND 120 TO OIL PRESSURE SWITCH. IF OK, REPLACE SWITCH.

YES → NO TROUBLE FOUND.

"AFTER REPAIRS," CONFIRM "CLOSED LOOP" OPERATION AND NO "CHECK ENGINE" LIGHT.

2.0L (VIN H) ENGINE — SYSTEM DIAGNOSTIC CHARTS — 1993–94 SUNBIRD

CHART A-5
FUEL PUMP RELAY CIRCUIT
2.0L (VIN H) "J" CARLINE

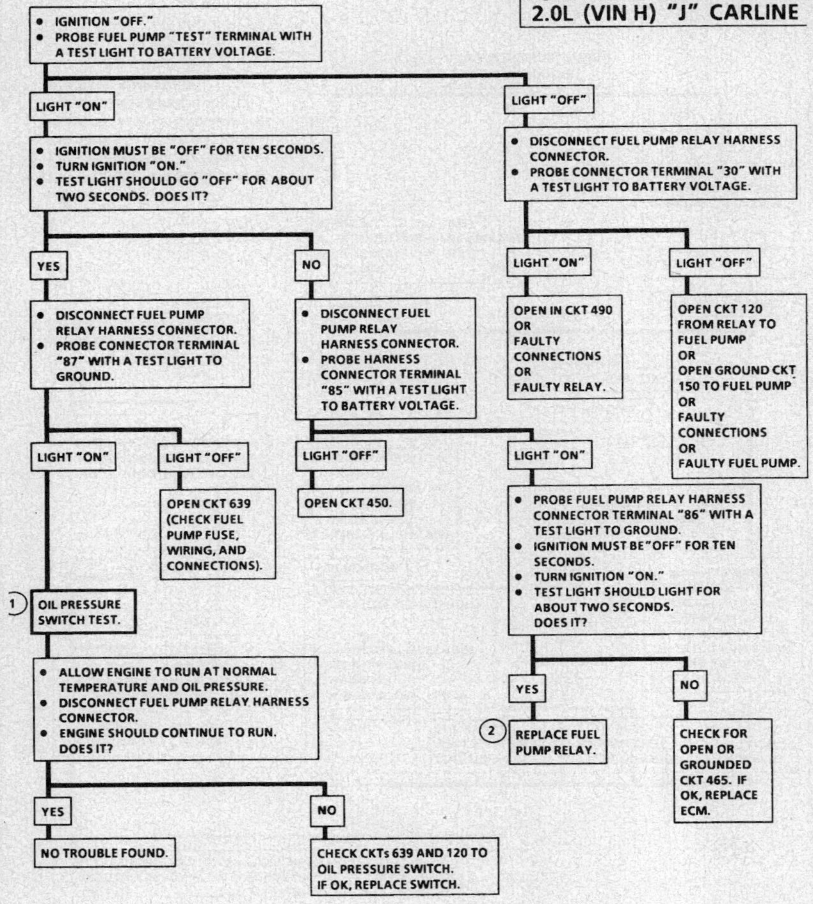

• IGNITION "OFF."
• PROBE FUEL PUMP "TEST" TERMINAL WITH A TEST LIGHT TO BATTERY VOLTAGE.
LIGHT "ON"
LIGHT "OFF"
• IGNITION MUST BE "OFF" FOR TEN SECONDS.
• TURN IGNITION "ON."
• TEST LIGHT SHOULD GO "OFF" FOR ABOUT TWO SECONDS. DOES IT?
• DISCONNECT FUEL PUMP RELAY HARNESS CONNECTOR.
• PROBE CONNECTOR TERMINAL "30" WITH A TEST LIGHT TO BATTERY VOLTAGE.
YES
NO
LIGHT "ON"
LIGHT "OFF"
• DISCONNECT FUEL PUMP RELAY HARNESS CONNECTOR.
• PROBE CONNECTOR TERMINAL "87" WITH A TEST LIGHT TO GROUND.
• DISCONNECT FUEL PUMP RELAY HARNESS CONNECTOR.
• PROBE HARNESS CONNECTOR TERMINAL "85" WITH A TEST LIGHT TO BATTERY VOLTAGE.
OPEN IN CKT 490 OR FAULTY CONNECTIONS OR FAULTY RELAY.
OPEN CKT 120 FROM RELAY TO FUEL PUMP OR OPEN GROUND CKT 150 TO FUEL PUMP OR FAULTY CONNECTIONS OR FAULTY FUEL PUMP.
LIGHT "ON"
LIGHT "OFF"
LIGHT "OFF"
LIGHT "ON"
OPEN CKT 639 (CHECK FUEL PUMP FUSE, WIRING, AND CONNECTIONS).
OPEN CKT 450.
• PROBE FUEL PUMP RELAY HARNESS CONNECTOR TERMINAL "86" WITH A TEST LIGHT TO GROUND.
• IGNITION MUST BE "OFF" FOR TEN SECONDS.
• TURN IGNITION "ON."
• TEST LIGHT SHOULD LIGHT FOR ABOUT TWO SECONDS. DOES IT?
1
OIL PRESSURE SWITCH TEST.
• ALLOW ENGINE TO RUN AT NORMAL TEMPERATURE AND OIL PRESSURE.
• DISCONNECT FUEL PUMP RELAY HARNESS CONNECTOR.
• ENGINE SHOULD CONTINUE TO RUN. DOES IT?
YES
NO
2
REPLACE FUEL PUMP RELAY.
CHECK FOR OPEN OR GROUNDED CKT 465. IF OK, REPLACE ECM.
YES
NO
NO TROUBLE FOUND.
CHECK CKTs 639 AND 120 TO OIL PRESSURE SWITCH. IF OK, REPLACE SWITCH.

"AFTER REPAIRS," CONFIRM "CLOSED LOOP" OPERATION AND NO MIL (CHECK ENGINE).

2.0L (VIN H) ENGINE — SYSTEM DIAGNOSTIC CHARTS — 1992 SUNBIRD

MANIFOLD VACUUM HOSE CONNECTION
PRESSURE REGULATOR
FLEXIBLE HOSE
PINCH HOSE HERE
J 29658-89 QUICK-CONNECT ADAPTERS
ELECTRICAL CONNECTOR
RETURN LINE
J 29658-100 ADAPTER KIT
J 29658-150 RELIEF VALVE
BLEED HOSE
PRESSURE LINE
FUEL FILTER
PINCH HOSE HERE
J 29658-D FUEL PRESSURE GAGE KIT
FUEL PUMP STRAINER
MODULAR SENDER (INCLUDES IN-TANK PUMP)

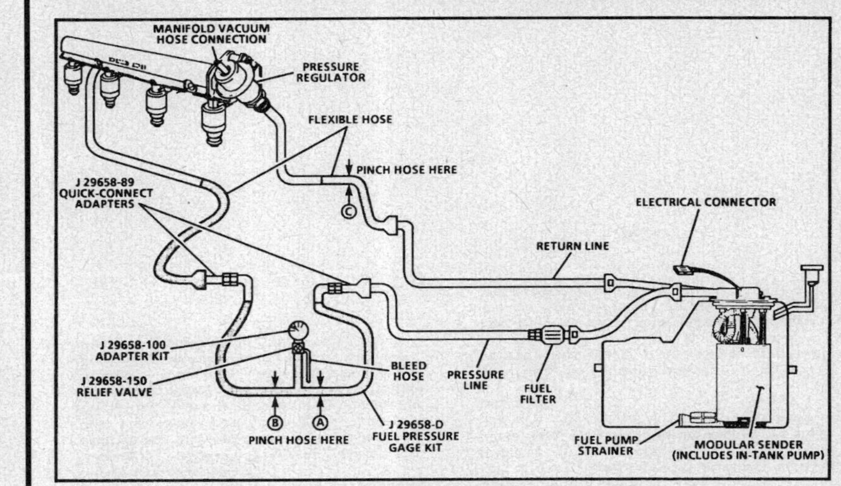

CHART A-7
(Page 1 of 3)
FUEL SYSTEM DIAGNOSIS
2.0L (VIN H) "J" CARLINE (PORT)

Circuit Description:

When the ignition switch is turned "ON," the Electronic Control Module (ECM) will turn "ON" the in-tank fuel pump. It will remain "ON" as long as the engine is cranking or running, and the ECM is receiving reference pulses. If there are no reference pulses, the ECM will shut "OFF" the fuel pump in about 2 seconds after ignition "ON" or 10 seconds after reference pulses stop.

An electric fuel pump, part of the modular fuel sender and located inside the fuel tank, pumps fuel through an in-line filter to the fuel rail assembly. The pump is designed to provide fuel at a pressure above the regulated pressure needed by the injectors. A pressure regulator, attached to the fuel rail, keeps fuel available to the injectors at a regulated pressure. Unused fuel is returned to the fuel tank by a separate line.

Test Description: Number(s) below refer to circled number(s) on the diagnostic chart.

1. Install fuel pressure gage per instructions on Page 3 of 3. Ignition "ON," pump pressure should be 284 to 325 kPa (41-47 psi). This pressure is controlled by spring pressure within the regulator assembly. Ignition may have to be cycled "ON" several times to obtain maximum pressure.
2. When the engine is idling, the manifold pressure is low (high vacuum) and is applied to the fuel regulator diaphragm. This will offset the spring and result in a lower fuel pressure. This idle pressure will vary somewhat depending on barometric pressure, however, the pressure idling should be less indicating pressure regulator control.
3. Pressure that continues to fall quickly is caused by one of the following:
   - In-tank fuel pump check valve not holding.
   - Fuel pressure regulator valve leaking.
   - Injector(s) sticking open.
4. An injector sticking open can best be determined by checking for a fouled or saturated spark plug(s). If a leaking injector cannot be determined by a fouled or saturated spark plug, the following procedure should be used.
   - Remove fuel rail bolts, but leave fuel lines connected.
   - Lift fuel rail out just enough to leave injector nozzles in the ports.

CAUTION: Be sure injector(s) are not allowed to spray on engine and that injector retaining clips are intact. This should be carefully followed to prevent fuel spray on engine which would cause a fire hazard.

   - Pressurize the fuel system and observe for injector(s) leaking.

CHART A-7

(Page 1 of 3)
FUEL SYSTEM DIAGNOSIS
2.0L (VIN H) "J" CARLINE (PORT)

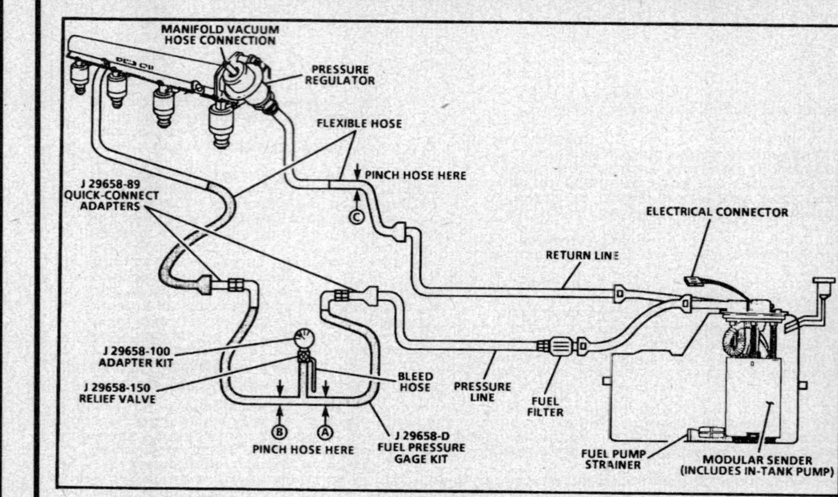

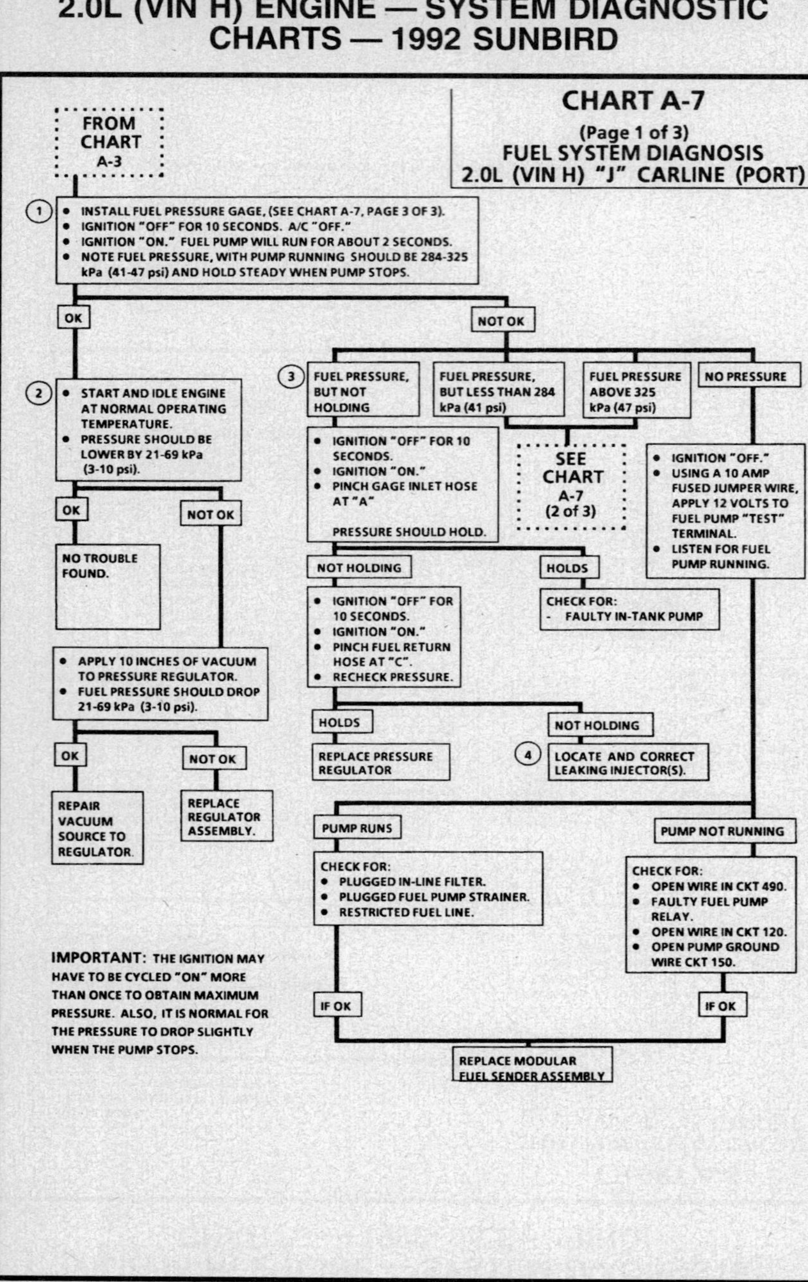

CHART A-7

(Page 2 of 3)
FUEL SYSTEM DIAGNOSIS
2.0L (VIN H) "J" CARLINE (PORT)

Test Description: Number(s) below refer to circled number(s) on the diagnostic chart.

1. Pressure below 284 kPa (41 psi) may cause a lean condition and may set a Code 44 or Code 32. It could also cause hard starting cold and poor driveability. Low enough pressure will cause the engine not to run at all. Restricted flow may allow the engine to run at idle or low speeds, but may cause a surge and stall when more fuel is required, as when accelerating or driving at high speeds.

2. Restricting fuel flow at the fuel pressure gage (at "B") causes fuel pressure to build above regulated pressure. With battery voltage applied to the pump "test" terminal, pressure should rise above 325 kPa (47 psi) as the gage outlet hose is restricted.

   NOTICE: Do not allow pressure to exceed 414 kPa (60 psi) as damage to the regulator may result.

3. This test determines if the high fuel pressure is due to a restricted fuel return line or a faulty fuel pressure regulator.

2.0L (VIN H) ENGINE — SYSTEM DIAGNOSTIC CHARTS — 1992 SUNBIRD

2.0L (VIN H) ENGINE — SYSTEM DIAGNOSTIC CHARTS — 1992 SUNBIRD

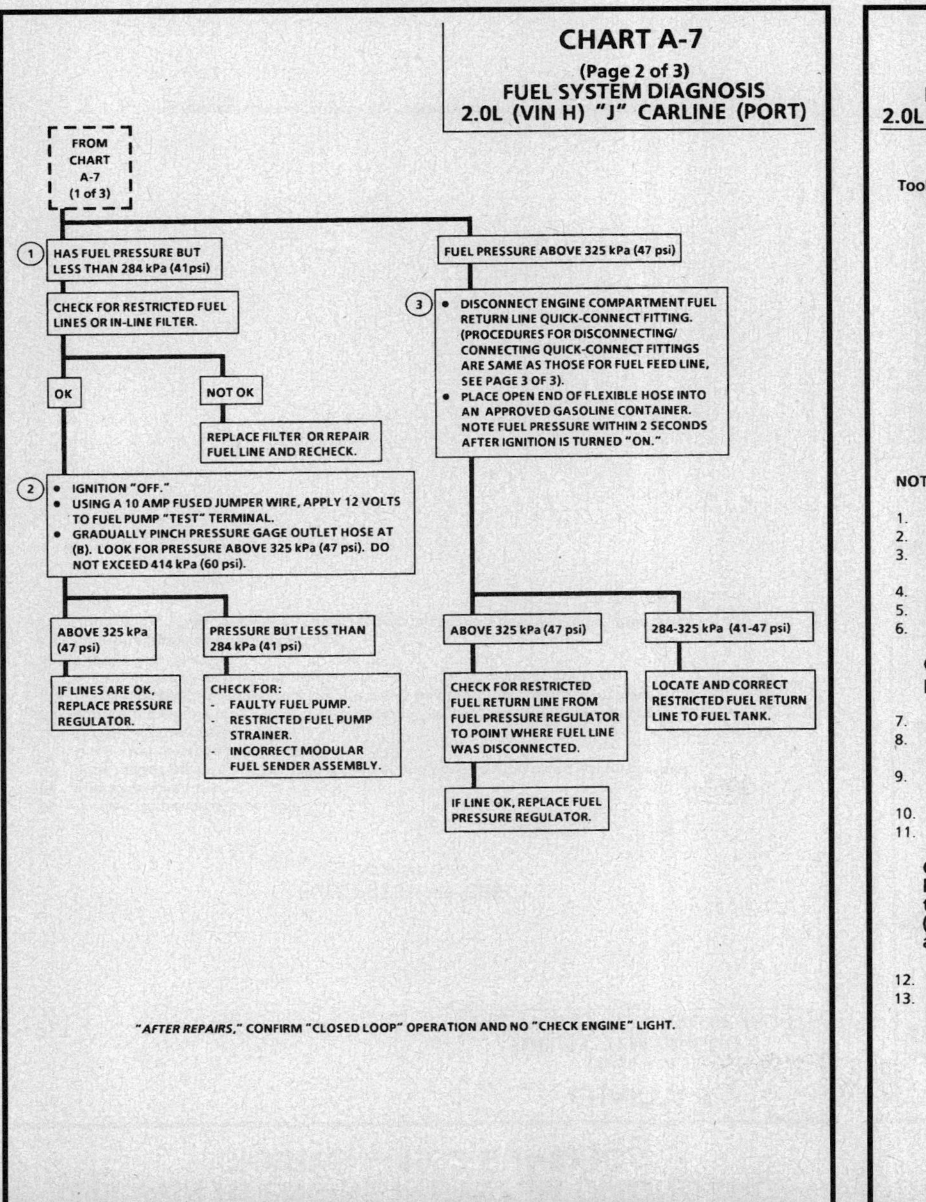

CHART A-7

(Page 3 of 3)
FUEL SYSTEM DIAGNOSIS
2.0L (VIN H) "J" CARLINE (PORT)

FUEL PRESSURE CHECK
-continued-

14. Connect negative battery cable.
15. Check fuel pressure.
16. Place bleed hose into an approved container and open valve to bleed system pressure.
17. Disconnect negative battery cable.
18. Disconnect fuel pressure gage.
19. Lubricate the male tube end of the fuel line, and reconnect quick-connect fitting.
    - Push connector together to cause the retaining tabs/fingers to snap into place
    - Once installed, pull on both ends of connection to make sure it is secure
20. Tighten fuel filler cap.
21. Connect negative battery cable.
22. Cycle ignition "ON" and "OFF" twice, waiting ten seconds between cycles, then check for fuel leaks.

"AFTER REPAIRS," CONFIRM "CLOSED LOOP" OPERATION AND NO "CHECK ENGINE" LIGHT.

CHART A-7 (Page 1 of 3)

FUEL SYSTEM DIAGNOSIS
2.0L (VIN H) "J" CARLINE

Circuit Description:

When the ignition switch is turned "ON," the Engine Control Module (ECM) will turn "ON" the in-tank fuel pump. It will remain "ON" as long as the engine is cranking or running, and the ECM is receiving reference pulses. If there are no reference pulses, the ECM will shut "OFF" the fuel pump in about 2 seconds after ignition "ON" or 10 seconds after reference pulses stop.

An electric fuel pump, part of the modular fuel sender and located inside the fuel tank, pumps fuel through an in-line filter to the fuel rail assembly. The pump is designed to provide fuel at a pressure above the regulated pressure needed by the injectors. A pressure regulator, attached to the fuel rail, keeps fuel available to the injectors at a regulated pressure. Unused fuel is returned to the fuel tank by a separate line.

Test Description: Number(s) below refer to circled number(s) on the diagnostic chart.

1. Install fuel pressure gage as shown in illustration. Refer to Page 3 of 3 for "Fuel Pressure Relief Procedure" and for "Servicing Quick-Connect Fittings." Ignition "ON," pump pressure should be 284 to 325 kPa (41-47 psi). This pressure is controlled by spring pressure within the regulator assembly. Ignition may have to be cycled "ON" several times to obtain maximum pressure.
2. When the engine is idling, the manifold pressure is low (high vacuum) and is applied to the fuel regulator diaphragm. This will offset the spring and result in a lower fuel pressure. This idle pressure will vary somewhat depending on barometric pressure, however, the pressure idling should be less indicating pressure regulator control.
3. Pressure that continues to fall quickly is caused by one of the following:
   - Leaking fuel pump feed hose.
   - Leaking check valve within fuel pump feed hose.
   - Fuel pressure regulator valve leaking.
   - Injector(s) sticking open.
4. An injector sticking open can best be determined by checking for a fouled or saturated spark plug(s). If a leaking injector cannot be determined by a fouled or saturated spark plug, the following procedure should be used.
   - Remove fuel rail bolts, but leave fuel lines connected.
   - Lift fuel rail out just enough to leave injector nozzles in the ports.

CAUTION: Be sure injector(s) are not allowed to spray on engine and that injector retaining clips are intact. This should be carefully followed to prevent fuel spray on engine which would cause a fire hazard.

   - Pressurize the fuel system and observe for injector(s) leaking.

2.0L (VIN H) ENGINE — SYSTEM DIAGNOSTIC CHARTS — 1993–94 SUNBIRD

CHART A-7
(Page 1 of 3)
FUEL SYSTEM DIAGNOSIS
2.0L (VIN H) "J" CARLINE

FROM CHART A-3
1
• INSTALL FUEL PRESSURE GAGE, (SEE CHART A-7, PAGE 3 OF 3).
• IGNITION "OFF" FOR 10 SECONDS. A/C "OFF."
• IGNITION "ON." FUEL PUMP WILL RUN FOR ABOUT 2 SECONDS. THE IGNITION MAY HAVE TO BE CYCLED "ON" MORE THAN ONCE TO OBTAIN MAXIMUM PRESSURE.
• NOTE FUEL PRESSURE, WITH PUMP RUNNING, PRESSURE SHOULD BE 284-325 kPa (41-47 psi). WHEN PUMP STOPS, PRESSURE WILL DROP SLIGHTLY THEN SHOULD HOLD STEADY.
OK
NOT OK
IF FUEL PRESSURE IS WITHIN NORMAL RANGE, BUT IS SUSPECTED OF DROPPING OFF DURING ACCELERATION, CRUISE OR HARD CORNERING SEE CHART A-7 (2 OF 3).
2
• START AND IDLE ENGINE AT NORMAL OPERATING TEMPERATURE.
• PRESSURE SHOULD BE LOWER BY 21-69 kPa (3-10 psi).
OK
NOT OK
NO TROUBLE FOUND.
• APPLY 10 INCHES OF VACUUM TO PRESSURE REGULATOR.
• FUEL PRESSURE SHOULD DROP 21-69 kPa (3-10 psi).
PRESSURE DROPS WITHIN RANGE
PRESSURE DOES NOT DROP
REPAIR VACUUM SOURCE TO PRESSURE REGULATOR.
REPLACE PRESSURE REGULATOR.
3
FUEL PRESSURE NOT HOLDING
FUEL PRESSURE LESS THAN 284 kPa (41 psi)
FUEL PRESSURE ABOVE 325 kPa (47 psi)
NO PRESSURE
• IGNITION "OFF" FOR 10 SECONDS.
• IGNITION "ON."
• PINCH GAGE INLET HOSE AT "A"
PRESSURE SHOULD HOLD.
SEE CHART A-7 (2 OF 3)
• IGNITION "OFF."
• USING A 10 AMP FUSED JUMPER WIRE, APPLY B + TO FUEL PUMP "TEST" TERMINAL.
• LISTEN FOR FUEL PUMP RUNNING.
NOT HOLDING
HOLDS
CHECK FORFAULTY IN-TANK PUMP.
• IGNITION "OFF" FOR 10 SECONDS.
• IGNITION "ON."
• PINCH FUEL RETURN HOSE AT "C".
• RECHECK PRESSURE.
HOLDS
NOT HOLDING
REPLACE PRESSURE REGULATOR.
4
LOCATE AND CORRECT LEAKING INJECTOR(S).
PUMP RUNS
PUMP NOT RUNNING
CHECK FOR:
• PLUGGED IN-LINE FILTER.
• PLUGGED FUEL PUMP STRAINERS.
• RESTRICTED FUEL LINE.
CHECK FOR:
• OPEN WIRE IN CKT 490.
• FAULTY FUEL PUMP RELAY.
• OPEN WIRE IN CKT 120.
• OPEN PUMP GROUND WIRE CKT 150.
IF OK
IF OK
REPLACE FUEL PUMP

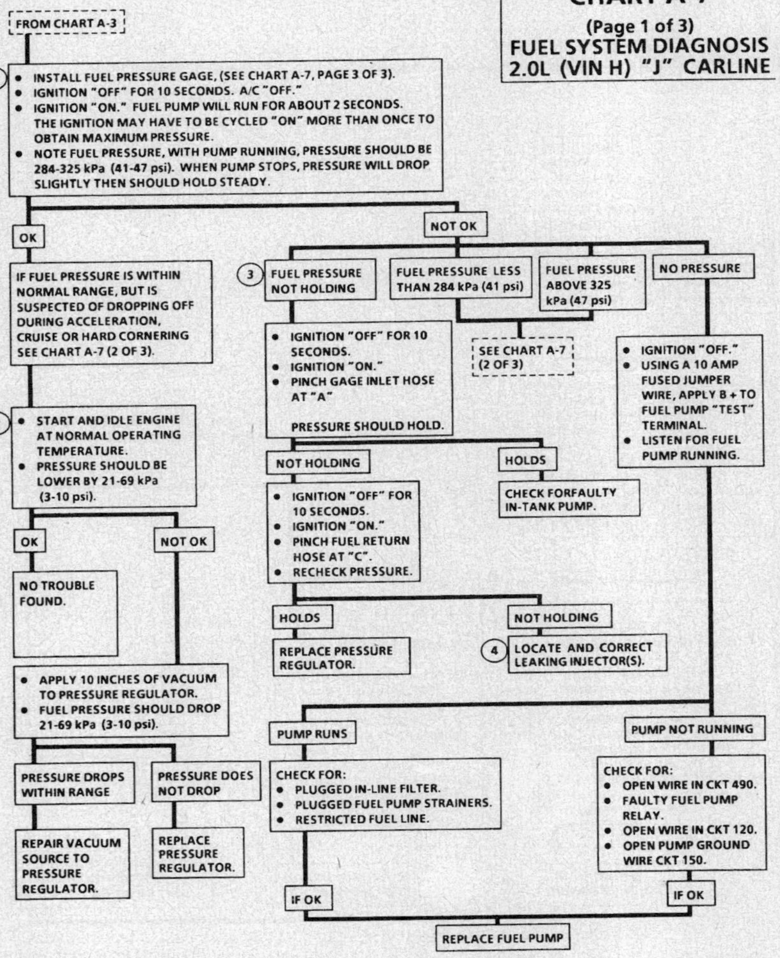

"AFTER REPAIRS." CONFIRM "CLOSED LOOP" OPERATION AND NO MIL (CHECK ENGINE).

2.0L (VIN H) ENGINE — SYSTEM DIAGNOSTIC CHARTS — 1993–94 SUNBIRD

MANIFOLD VACUUM HOSE CONNECTION
PRESSURE REGULATOR
FLEXIBLE HOSE
PINCH HOSE HERE
J 29658-89 QUICK-CONNECT ADAPTERS
ELECTRICAL CONNECTOR
RETURN LINE
J 29658-100 ADAPTER KIT
J 29658-150 RELIEF VALVE
BLEED HOSE
PRESSURE LINE
FUEL FILTER
J 29658-D FUEL PRESSURE GAGE KIT
PINCH HOSE HERE
FUEL SENDER STRAINER
MODULAR FUEL SENDER (FUEL PUMP INSIDE RESERVOIR)

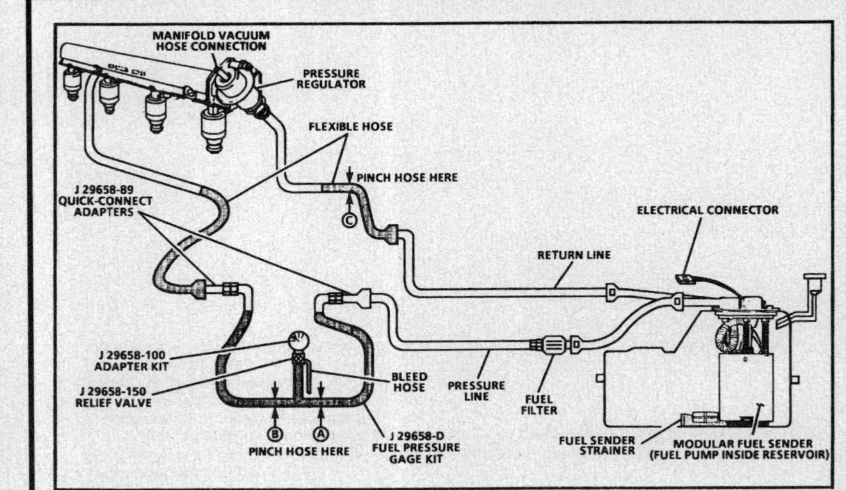

CHART A-7
(Page 2 of 3)
FUEL SYSTEM DIAGNOSIS
2.0L (VIN H) "J" CARLINE

Test Description: Number(s) below refer to circled number(s) on the diagnostic chart.

5. Fuel pressure that drops off during acceleration, cruise or hard cornering may cause a lean condition and result in a loss of power, surging or misfire. This condition can be diagnosed using a Tech 1 scan tool. If the fuel system is very lean, the O2S will stop toggling and output voltage will drop below 500 mV. Also, injector pulse width will increase.

Important
- Make sure system is not operating at "Fuel Cut-Off" which may cause false readings on the scan tool.

6. Fuel pressure below 284 kPa (41 psi) may cause a lean condition and may set a DTC 44 or a DTC 32. Driveability conditions can include hard starting cold, hesitation, poor driveability, lack of power, surging or misfire.
7. Restricting fuel flow at the fuel pressure gage (at "B") causes fuel pressure to build above regulated pressure. With battery voltage applied to the pump "test" connector, pressure should rise above 325 kPa (47 psi) as the gage outlet hose is pinched.

NOTICE: Do not allow pressure to exceed 414 kPa (60 psi) as damage to the regulator may result.

8. Fuel pressure above 325 kPa (47 psi) may cause a rich condition and may set a DTC 45. Driveability conditions can include hard starting (followed by black smoke) and a strong sulphur smell in the exhaust.
9. This test determines if the high fuel pressure is due to a restricted fuel return line or a faulty fuel pressure regulator.

2.0L (VIN H) ENGINE — SYSTEM DIAGNOSTIC CHARTS — 1993–94 SUNBIRD

CHART A-7
(Page 2 of 3)
FUEL SYSTEM DIAGNOSIS
2.0L (VIN H) "J" CARLINE

FROM CHART A-3 PAGE 1
OR
FROM CHART A-7 PAGE 1.
PRESSURE IS SUSPECTED OF DROPPING OFF OR PRESSURE OUT OF SPECIFICATION, 284-325 kPa (41-47 psi).

5 FUEL PRESSURE DROPS OFF DURING ACCELERATION, CRUISE OR HARD CORNERING.

CHECK FOR RESTRICTED FUEL PRESSURE LINE OR IN-LINE FUEL FILTER. IS THERE A RESTRICTION?

NO

YES — REPLACE FILTER OR LOCATE AND CORRECT RESTRICTION IN FUEL LINE AND RECHECK.

CHECK FOR:
- RESTRICTED FUEL SENDER STRAINER.
- RESTRICTED FUEL PUMP STRAINER.
- LEAKING FUEL PUMP FEED HOSE.
- FAULTY FUEL PUMP.
- INCORRECT FUEL PUMP.

6 FUEL PRESSURE LESS THAN 284 kPa (41 psi).

7
- IGNITION "OFF."
- USING A 10 AMP FUSED JUMPER WIRE, APPLY B + TO FUEL PUMP "TEST" CONNECTOR.
- GRADUALLY PINCH FLEXIBLE FUEL RETURN HOSE. PRESSURE SHOULD RISE ABOVE 325 kPa (47 psi). DO NOT EXCEED 414 kPa (60 psi). DOES PRESSURE RISE ABOVE 324 kPa (47 psi)?

NO

YES — REPLACE PRESSURE REGULATOR.

8 FUEL PRESSURE ABOVE 325 kPa (47 psi).

9
- DISCONNECT ENGINE COMPARTMENT FUEL RETURN LINE QUICK-CONNECT FITTING (SEE PAGE 3 OF 3).
- PLACE OPEN END OF FLEXIBLE HOSE INTO AN APPROVED GASOLINE CONTAINER.
- IGNITION "ON." NOTE FUEL PRESSURE WITHIN 2 SECONDS AFTER IGNITION IS TURNED "ON."
- PRESSURE SHOULD BE 284-325 kPa (41-47 psi). IS IT?

NO — CHECK FOR RESTRICTED FUEL RETURN LINE FROM PRESSURE REGULATOR TO POINT WHERE FUEL LINE WAS DISCONNECTED. IS THERE A RESTRICTION?

YES — LOCATE AND CORRECT RESTRICTION IN FUEL RETURN LINE TO FUEL TANK.

NO — REMOVE PRESSURE REGULATOR AND CHECK FOR RESTRICTED FILTER SCREEN (IF EQUIPPED). REPLACE SCREEN IF PLUGGED. IF OK, REPLACE PRESSURE REGULATOR.

YES — LOCATE AND CORRECT RESTRICTION IN FUEL LINE.

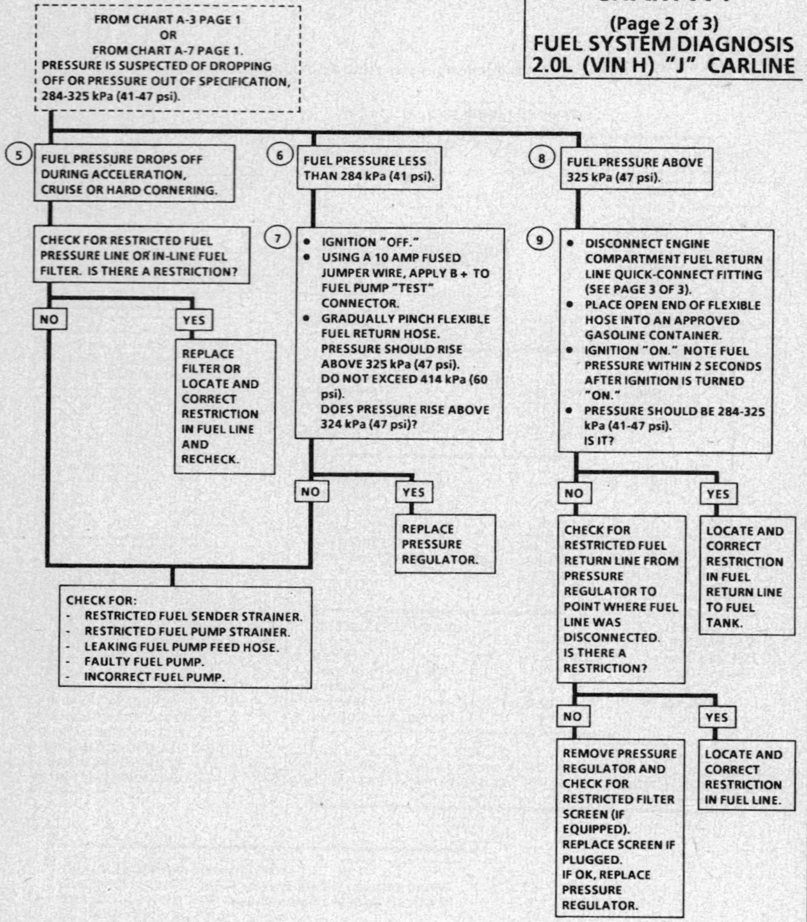

"AFTER REPAIRS," CONFIRM "CLOSED LOOP" OPERATION AND NO MIL (CHECK ENGINE).

CHART A-7
(Page 3 of 3)
FUEL SYSTEM DIAGNOSIS
2.0L (VIN H) "J" CARLINE

FUEL SYSTEM PRESSURE RELIEF PROCEDURE

PFI Engines Without Fuel Pressure Connection (J Car Only)

(Must Be Performed Before Disconnecting Fuel Line Fittings)

CAUTION:
- To reduce the risk of fire and personal injury, it is necessary to relieve fuel system pressure before disconnecting fuel line fittings.
- After relieving system pressure, a small amount of fuel may be released when disconnecting fuel line fittings. In order to reduce the chance of personal injury, cover fuel line fittings with a shop towel before disconnecting, to catch any fuel that may leak out. Place the towel in an approved container when disconnect is completed.

1. Loosen fuel filler cap to relieve tank pressure.
2. Remove fuel pump fuse.
3. Start engine and run until fuel supply remaining in fuel pipes is consumed. Engage starter for 3.0 seconds to assure relief of any remaining pressure.
4. Disconnect negative battery cable to avoid possible fuel discharge if an accidental attempt is made to start the engine.
5. Fuel line fittings are now safe for servicing.
6. Perform service required.
   - If performing fuel pressure check with a gage equipped with a bleed hose, fuel pressure can be relieved through the gage following test. Place bleed hose into approved gasoline container and open valve to bleed system pressure.
7. Tighten fuel filler cap.
8. Connect negative battery cable.
9. Install fuel pump fuse.
10. Cycle ignition "ON" and "OFF" twice, waiting ten seconds between cycles, then check for fuel leaks.

2.0L (VIN H) ENGINE — SYSTEM DIAGNOSTIC CHARTS — 1993–94 SUNBIRD

CHART A-7
(Page 3 of 3)
FUEL SYSTEM DIAGNOSIS
2.0L (VIN H) "J" CARLINE

SERVICING QUICK-CONNECT FITTINGS

Important
- In order to install fuel system diagnostic equipment on vehicles equipped with plastic quick-connect fittings, fuel line separator tools must be used to disconnect the fittings. Use of the separator tools will cause the plastic retainer to remain inside the female connector allowing diagnostic equipment to be connected.

Tools required:
J 37088-A tool set, fuel line quick-connect separator;
J 39504 tool set, fuel line quick-connect separator (restricted access).

Remove or Disconnect
1. Grasp both sides of fitting. Twist female connector 1/4 turn in each direction to loosen any dirt within fitting.

CAUTION: Safety glasses must be worn when using compressed air, as flying dirt particles may cause eye injury.

2. Using compressed air, blow dirt out of fitting.
3. Choose correct tool from J 37088-A or J 39504 tool set for size of fitting. Insert tool into female connector, then push/pull inward to release locking tabs.
4. Pull connection apart.

Clean and Inspect

NOTICE: If it is necessary to remove rust or burrs from fuel pipe, use emery cloth in a radial motion with the pipe end to prevent damage to O-ring sealing surface.

- Using a clean shop towel, wipe off male pipe end.
- Inspect both ends of fitting for dirt and burrs. Clean or replace components/assemblies as required.

Install or Connect

CAUTION: To Reduce the Risk of Fire and Personal Injury:
- Before connecting fitting, always apply a few drops of clean engine oil to the male pipe end of engine fuel pipe, pressure gage adapter or fuel line shut-off adapter. This will ensure proper reconnection and prevent a possible fuel leak. (During normal operation, the O-rings located in the female connector will swell and may prevent proper reconnection if not lubricated.)

1. Apply a few drops of clean engine oil to the male pipe end of engine fuel pipe, pressure gage adapter or fuel line shut-off adapter.
2. Push both sides of fitting together to cause the retaining tabs/fingers to snap into place.
3. Once installed, pull on both sides of fitting to make sure connection is secure.

2.0L (VIN H) ENGINE — DIAGNOSTIC TROUBLE CODE CHART — SUNBIRD

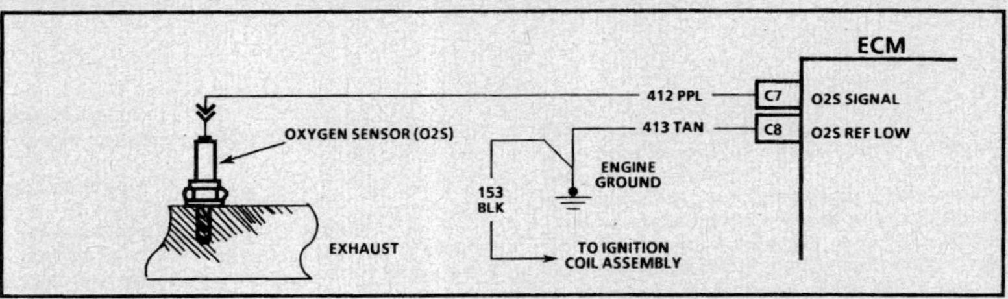

DTC 13
OXYGEN SENSOR (O2S) CIRCUIT
(OPEN CIRCUIT)
2.0L (VIN H) "J" CARLINE

Circuit Description:
The Engine Control Module (ECM) supplies a voltage of about .45 volt between terminals "C7" and "C8". (If measured with a 10 megohm digital voltmeter, this may read as low as .32 volt.)

When the Oxygen Sensor (O2S) reaches operating temperature, it varies this voltage from about .1 volt (exhaust is lean) to about .9 volt (exhaust is rich).

The sensor is like an open circuit and produces no voltage when it is below 316°C (600°F). An open sensor circuit or cold sensor causes "Open Loop" operation.

Test Description: Number(s) below refer to circled number(s) on the diagnostic chart.
1. DTC 13 will set under the following conditions:
   - Engine at normal operating temperature.
   - At least 40 seconds has elapsed since engine start up.
   - O2S signal voltage is steady between .35 and .55 volt.
   - Throttle angle is greater than 5%.
   - All above conditions are met for about 10 seconds.
   If the conditions for a DTC 13 exist, the system will not operate in "Closed Loop."
2. This test determines if the Oxygen Sensor (O2S) is the problem or if the ECM and wiring are at fault.

3. In doing this test, use only a 10 megohm digital voltmeter. This test checks the continuity of CKT 412 and CKT 413. If CKT 413 is open, the ECM voltage on CKT 412 will be over .6 volt (600 mV).

Diagnostic Aids:

Normal scan tool oxygen sensor voltage varies between 100 mV to 999 mV (.1 and 1.0 volt) while in "Closed Loop."

Verify a clean, tight ground connection for CKT 413. Open CKT 412 or CKT 413 will result in a DTC 13. If DTC 13 is intermittent, refer to "Symptoms,"

2.0L (VIN H) ENGINE — DIAGNOSTIC TROUBLE CODE CHART — SUNBIRD

DTC 13
OXYGEN SENSOR (O2S) CIRCUIT
(OPEN CIRCUIT)
2.0L (VIN H) "J" CARLINE

1. • ENGINE AT NORMAL OPERATING TEMPERATURE (ABOVE 80°C/176°F).
   • RUN ENGINE ABOVE 1200 RPM FOR TWO MINUTES.
   • DOES SCAN TOOL INDICATE "CLOSED LOOP"?

NO →

2. • DISCONNECT O2S CONNECTOR.
   • JUMPER HARNESS CKT 412 (ECM SIDE) TO GROUND.
   • SCAN TOOL SHOULD DISPLAY OXYGEN SENSOR VOLTAGE BELOW .2 VOLT (200 mv) WITH ENGINE RUNNING.
   DOES IT?

YES →

DTC 13 IS INTERMITTENT. IF NO ADDITIONAL DTC(s) WERE STORED, REFER TO "DIAGNOSTIC AIDS"

NO →

3. • REMOVE JUMPER.
   • IGNITION "ON," ENGINE "OFF."
   • CHECK VOLTAGE OF CKT 412 (ECM SIDE) AT O2S HARNESS CONNECTOR USING A DVM.

YES →

FAULTY O2S CONNECTION OR SENSOR.

.3-.6 VOLT (300 - 600 mV)

FAULTY ECM.

GREATER THAN .6 VOLT (600 mV)

OPEN CKT 413 OR FAULTY CONNECTION OR FAULTY ECM.

LESS THAN .3 VOLT (300 mV)

OPEN CKT 412 OR FAULTY ECM CONNECTION OR FAULTY ECM.

"AFTER REPAIRS," REFER TO DTC CRITERIA AND CONFIRM DTC DOES NOT RESET.

2.0L (VIN H) ENGINE — DIAGNOSTIC TROUBLE CODE CHART — SUNBIRD

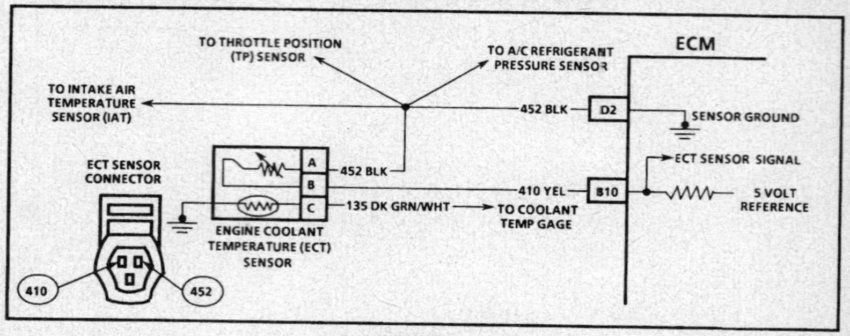

DTC 14
(Page 1 of 2)
ENGINE COOLANT TEMPERATURE (ECT) SENSOR CIRCUIT
(HIGH/LOW TEMPERATURE INDICATED)
2.0L (VIN H) "J" CARLINE

Circuit Description:

The Engine Coolant Temperature (ECT) sensor utilizes a thermistor to control the signal voltage to the Engine Control Module (ECM). The ECM applies a reference voltage on CKT 410 to the sensor. When the engine is cold, the sensor (thermistor) resistance is high. The ECM will then sense a high signal voltage.

As the engine warms up, the sensor resistance decreases and the voltage drops. At normal engine operating temperature, the voltage will measure about 1.5 to 2.0 volts at ECM terminal "B10".

Engine coolant temperature is one of the inputs used to control the following:
- Cooling fan.
- Fuel delivery.
- Ignition Control (IC).
- Idle Air Control (IAC).
- Torque Converter Clutch (TCC).

A separate thermistor within the ECT sensor provides a signal to the engine coolant temperature gage located in the instrument panel.

Test Description: Number(s) below refer to circled number(s) on the diagnostic chart.
1. DTC 14 will set if:
   - The engine has been running for at least 2 minutes.
   - Signal voltage indicates an engine coolant temperature below -30°C (-22°F).
   OR
   - Signal voltage indicates engine coolant temperature above 144°C (291°F).
2. If the ECM recognizes the grounded circuit (low voltage) and displays a high temperature; the ECM and wiring are OK.
3. This test will determine if there is a wiring problem or a faulty ECM. If CKT 452 is open, there may also be a other DTC(s) stored.

Diagnostic Aids:

A scan tool should display engine temperature in degrees celsius.

After the engine is started, the temperature should rise steadily to about 90°C (194°F), then stabilize when the thermostat opens

If the engine has been allowed to cool to an ambient temperature (overnight), engine coolant temperature and intake air temperature may be checked with a scan tool and should read close to each other.

When a DTC 14 is set, the ECM will turn "ON" the engine cooling fan.

If DTC 14 is intermittent, refer to "Symptoms,"

- If DTC 14 is intermittent of a "Hard Start" symptom is present, check engine coolant temperature with a scan tool on a cool engine. Temperature displayed would be within 5 degrees of the ambient. If not, check the ECT sensor using the "Diagnostic Aids"   If sensor is OK, check connections. If connections at sensor and ECM are OK, replace ECM.

2.0L (VIN H) ENGINE — DIAGNOSTIC TROUBLE CODE CHART — SUNBIRD

DTC 14
(Page 1 of 2)
ENGINE COOLANT TEMPERATURE (ECT) SENSOR CIRCUIT
(HIGH/LOW TEMPERATURE INDICATED)
2.0L (VIN H) "J" CARLINE

(1) DOES SCAN TOOL DISPLAY COOLANT TEMPERATURE OF -30℃ (-22°F) OR LESS?

YES → NO

NO → REFER TO DTC 14 (PAGE 2 OF 2)

(2) • DISCONNECT SENSOR CONNECTOR.
• JUMPER HARNESS CKT 410 TO SENSOR GROUND CIRCUIT.
• SCAN TOOL SHOULD DISPLAY 144℃ (291°F) OR HIGHER. DOES IT?

NO → YES

YES → FAULTY SENSOR CONNECTOR OR SENSOR.

(3) • JUMPER CKT 410 TO A KNOWN GOOD GROUND.
• SCAN TOOL SHOULD DISPLAY 144℃ (291°F) OR HIGHER. DOES IT?

YES → NO

YES → OPEN SENSOR GROUND CIRCUIT, FAULTY CONNECTION OR FAULTY ECM.

NO → OPEN CKT 410, FAULTY CONNECTION OR FAULTY ECM.

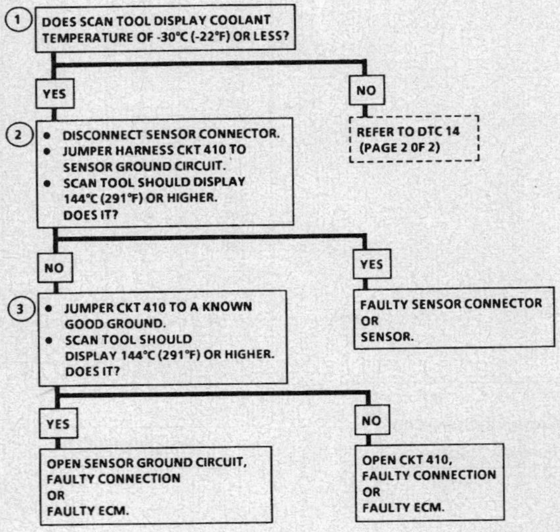

DIAGNOSTIC AID		
ECT SENSOR		
TEMPERATURE VS. RESISTANCE VALUES (APPROXIMATE)		
℃	°F	OHMS
100	212	177
90	194	241
80	176	332
70	158	467
60	140	667
50	122	973
45	113	1188
40	104	1459
35	95	1802
30	86	2238
25	77	2796
20	68	3520
15	59	4450
10	50	5670
5	41	7280
0	32	9420
-5	23	12300
-10	14	16180
-15	5	21450
-20	-4	28680
-30	-22	52700
-40	-40	100700

"AFTER REPAIRS," REFER TO DTC CRITERIA AND CONFIRM DTC DOES NOT RESET.

2.0L (VIN H) ENGINE — DIAGNOSTIC TROUBLE CODE CHART — SUNBIRD

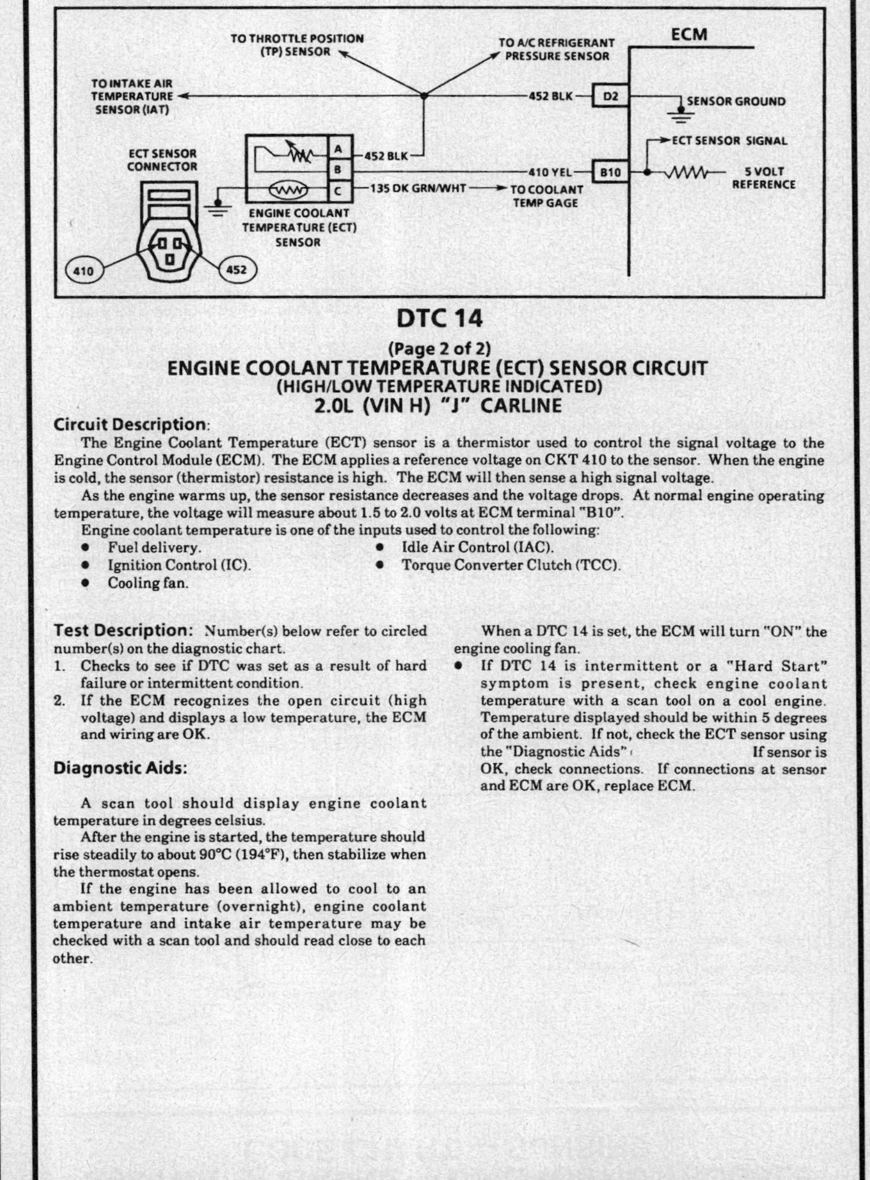

DTC 14
(Page 2 of 2)
ENGINE COOLANT TEMPERATURE (ECT) SENSOR CIRCUIT
(HIGH/LOW TEMPERATURE INDICATED)
2.0L (VIN H) "J" CARLINE

Circuit Description:

The Engine Coolant Temperature (ECT) sensor is a thermistor used to control the signal voltage to the Engine Control Module (ECM). The ECM applies a reference voltage on CKT 410 to the sensor. When the engine is cold, the sensor (thermistor) resistance is high. The ECM will then sense a high signal voltage.

As the engine warms up, the sensor resistance decreases and the voltage drops. At normal engine operating temperature, the voltage will measure about 1.5 to 2.0 volts at ECM terminal "B10".

Engine coolant temperature is one of the inputs used to control the following:
• Fuel delivery.
• Ignition Control (IC).
• Cooling fan.
• Idle Air Control (IAC).
• Torque Converter Clutch (TCC).

Test Description: Number(s) below refer to circled number(s) on the diagnostic chart.
1. Checks to see if DTC was set as a result of hard failure or intermittent condition.
2. If the ECM recognizes the open circuit (high voltage) and displays a low temperature, the ECM and wiring are OK.

Diagnostic Aids:

A scan tool should display engine coolant temperature in degrees celsius.

After the engine is started, the temperature should rise steadily to about 90℃ (194°F), then stabilize when the thermostat opens.

If the engine has been allowed to cool to an ambient temperature (overnight), engine coolant temperature and intake air temperature may be checked with a scan tool and should read close to each other.

When a DTC 14 is set, the ECM will turn "ON" the engine cooling fan.
• If DTC 14 is intermittent or a "Hard Start" symptom is present, check engine coolant temperature with a scan tool on a cool engine. Temperature displayed should be within 5 degrees of the ambient. If not, check the ECT sensor using the "Diagnostic Aids". If sensor is OK, check connections. If connections at sensor and ECM are OK, replace ECM.

DTC 14

(Page 2 of 2)
ENGINE COOLANT TEMPERATURE (ECT) SENSOR CIRCUIT
(HIGH/LOW TEMPERATURE INDICATED)
2.0L (VIN H) "J" CARLINE

FROM DTC 14
PAGE 1 OF 2

(1) DOES SCAN TOOL DISPLAY COOLANT TEMPERATURE OF 144°C (291°F) OR HIGHER?

YES

NO

(2) • DISCONNECT SENSOR CONNECTOR. SCAN TOOL SHOULD DISPLAY TEMPERATURE BELOW -30°C (-22°F) DOES IT?

• DTC 14 IS INTERMITTENT. IF NO ADDITIONAL DTC(s) WERE STORED, REFER TO "DIAGNOSTIC AIDS"

YES

NO

REPLACE COOLANT TEMPERATURE SENSOR.

CKT 410 SHORTED TO GROUND,
OR
CKT 410 SHORTED TO SENSOR GROUND CIRCUIT
OR
FAULTY ECM.

DIAGNOSTIC AID

ECT SENSOR		
TEMPERATURE VS. RESISTANCE VALUES (APPROXIMATE)		
°C	°F	OHMS
100	212	177
90	194	241
80	176	332
70	158	467
60	140	667
50	122	973
45	113	1188
40	104	1459
35	95	1802
30	86	2238
25	77	2796
20	68	3520
15	59	4450
10	50	5670
5	41	7280
0	32	9420
-5	23	12300
-10	14	16180
-15	5	21450
-20	-4	28680
-30	-22	52700
-40	-40	100700

"AFTER REPAIRS," REFER TO DTC CRITERIA AND CONFIRM DTC DOES NOT RESET.

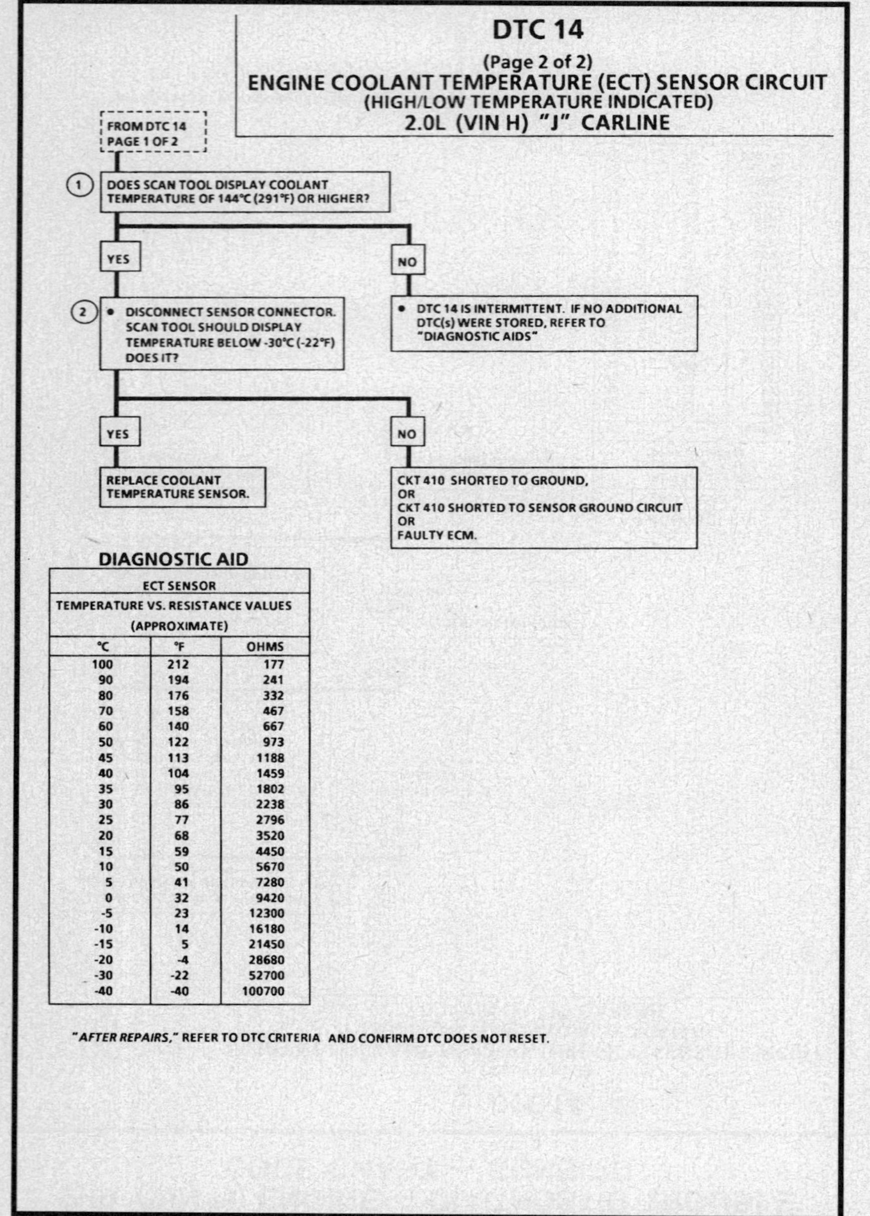

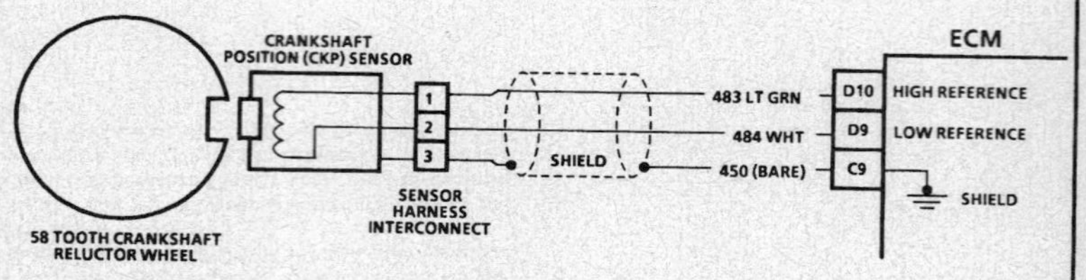

DTC 19

CRANKSHAFT POSITION (CKP) SENSOR CIRCUIT
(INCORRECT 58X SIGNAL, DISCONNECTED SENSOR)
2.0L (VIN H) "J" CARLINE

Circuit Description:

The Crankshaft Position (CKP) sensor (located in the engine block) is positioned directly in line with a 58 tooth crankshaft wheel. As the 58 tooth wheel (crankshaft) rotates, it generates a 58X signal to the ECM. This signal, once converted, is used by the ECM to activate the appropriate ignition coil driver and fuel injectors. The Crankshaft Position (CKP) sensor leads are protected from outside "noise" by a shield which is grounded through the ECM.

Test Description: Number(s) below refer to circled number(s) on the diagnostic chart.

1. DTC 19 will set for the following conditions:
   - No 58X signal to ECM.
     - Engine cranking for at least 10 seconds.
     - At least 4.0 kPa drop in MAP.
     - 1.0 drop in battery voltage.
   - Incorrect 58X signal to ECM.
     - Engine running.
     - ECM reads less than or greater than 58 teeth on crankshaft per revolution of engine.
2. This test checks harness and resistance of Crankshaft Position (CKP) sensor. Resistance may vary slightly with higher temperatures.
3. The Crankshaft Position (CKP) sensor output should measure about 2.5 volts at 340 cranking RPM.

Diagnostic Aids:

- Faulty connections at either the CKP sensor connector or ECM connector can cause an intermittent DTC 19 to set.
- Damage to CKP sensor harness shield may cause an intermittent DTC 19 to set.
- Check for missing teeth, loose or damaged crankshaft reluctor wheel.

2.0L (VIN H) ENGINE — DIAGNOSTIC TROUBLE CODE CHART — SUNBIRD

DTC 19

CRANKSHAFT POSITION (CKP) SENSOR CIRCUIT
(INCORRECT 58X SIGNAL, DISCONNECTED SENSOR)
2.0L (VIN H) "J" CARLINE

(1)
- CLEAR DTC(s).
- CRANK ENGINE FOR 15 SECONDS OR UNTIL ENGINE STARTS. DOES DTC 19 SET?

YES / NO

DTC 19 INTERMITTENT. REFER TO "DIAGNOSTIC AIDS"

(2)
- IGNITION "OFF."
- DISCONNECT HARNESS ECM CONNECTOR "D" FROM ECM.
- MEASURE HARNESS TERMINALS "D10" AND "D9" WITH OHMMETER. DOES IT DISPLAY BETWEEN 480 OHMS AND 680 OHMS?

YES / NO

(3)
- SET VOLTMETER ON AC VOLTAGE SCALE.
- CRANK ENGINE AND OBSERVE VOLTAGE READING. IS READING GREATER THAN .3 VOLT (300 mV)?

480 OHMS OR LESS.

680 OHMS OR GREATER.

CKP SENSOR LEADS SHORTED TOGETHER OR FAULTY CKP SENSOR.

OPEN CKP SENSOR CIRCUIT, FAULTY CONNECTIONS, OR FAULTY CKP SENSOR.

YES / NO

FAULTY CONNECTION OR FAULTY CKP SENSOR.

- RECONNECT HARNESS TO ECM.
- CLEAR DTC(s).
- CRANK ENGINE FOR 15 SECONDS OR UNTIL ENGINE STARTS. IF DTC 19 RESETS, FAULTY CONNECTIONS OR FAULTY CKP SENSOR.

- CLEAR DTC(s) AGAIN.
- CRANK ENGINE FOR 15 SECONDS OR UNTIL ENGINE STARTS. IF DTC 19 RESETS, REPLACE THE ECM.

"AFTER REPAIRS," REFER TO DTC CRITERIA AND CONFIRM DTC DOES NOT RESET.

2.0L (VIN H) ENGINE — DIAGNOSTIC TROUBLE CODE CHART — SUNBIRD

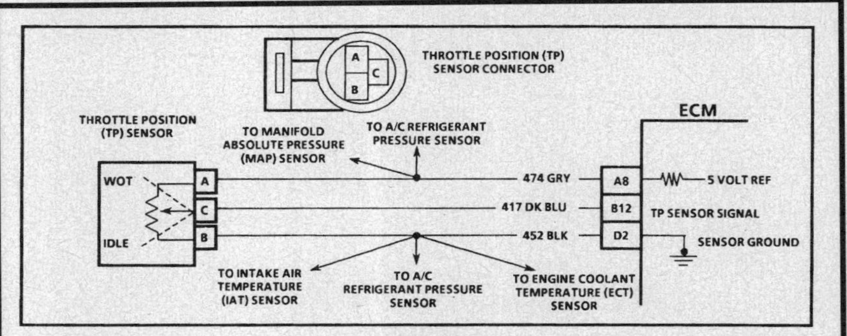

DTC 21

(Page 1 of 2)
THROTTLE POSITION (TP) SENSOR CIRCUIT
(SIGNAL VOLTAGE HIGH/LOW)
2.0L (VIN H) "J" CARLINE

Circuit Description:

The Throttle Position (TP) sensor provides a voltage signal that changes relative to the throttle valve. Signal voltage will vary from .33 to 1.33 volts at idle to about 4.5 volts at Wide Open Throttle (WOT).

The TP sensor signal is one of the most important inputs used by the Engine Control Module (ECM) for fuel control and for many of the ECM controlled outputs.

Test Description: Number(s) below refer to circled number(s) on the diagnostic chart.

1. A DTC 21 will set under the following conditions:
   - Engine idling.
   - TP sensor less than .2 volt or greater than 2.5 volts.
   - MAP reading below 65 kPa.
   - All of the above conditions present for 5 seconds.

   The TP sensor has a auto zeroing feature. If the voltage reading is within the range of about .33 to 1.33 volts, the ECM will use that value as closed throttle. If the voltage reading is outside of the auto zero range at closed throttle, check for a binding throttle cable or damaged linkage. If OK, continue with diagnostics.
2. If the ECM recognizes the change of state, the ECM, CKT 474, and CKT 417 are OK.
3. This step isolates a faulty sensor, ECM, or an open CKT 452. If CKT 452 is open, there may also be other diagnostic trouble codes set.

Diagnostic Aids:

A scan tool displays throttle position in volts. The voltage at closed throttle should be .33 to 1.33 volts. TP sensor voltage should increase at a steady rate as throttle is moved slowly to Wide Open Throttle (WOT) position.

If DTC 21 is intermittent, refer to "Symptoms,"

2.0L (VIN H) ENGINE — DIAGNOSTIC TROUBLE CODE CHART — SUNBIRD

DTC 21

(Page 1 of 2)
THROTTLE POSITION (TP) SENSOR CIRCUIT
(SIGNAL VOLTAGE HIGH/LOW)
2.0L (VIN H) "J" CARLINE

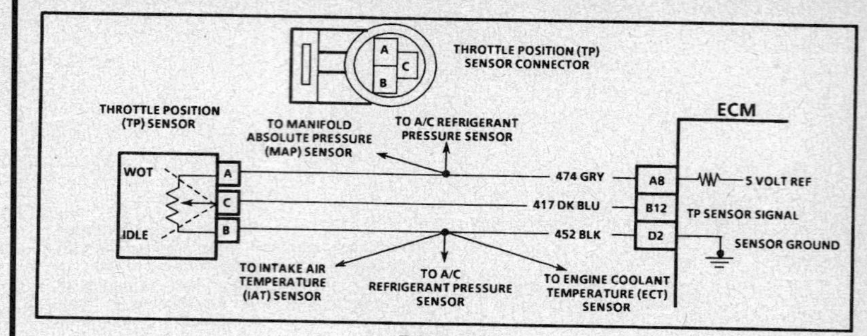

2.0L (VIN H) ENGINE — DIAGNOSTIC TROUBLE CODE CHART — SUNBIRD

DTC 21

(Page 2 of 2)
THROTTLE POSITION (TP) SENSOR CIRCUIT
(SIGNAL VOLTAGE HIGH/LOW)
2.0L (VIN H) "J" CARLINE

Circuit Description:

The Throttle Position (TP) sensor provides a voltage signal that changes relative to the throttle valve. Signal voltage will vary from .33 to 1.33 volts at idle to about 4.5 volts at Wide Open Throttle (WOT).

The TP sensor signal is one of the most important inputs used by the Engine Control Module (ECM) for fuel control and for many of the ECM controlled outputs.

Test Description: Number(s) below refer to circled number(s) on the diagnostic chart.

1. This step checks to see if DTC 21 is the result of a hard failure or an intermittent condition.
2. This step simulates conditions for a DTC 21 to set. If the scan tool displays over 4 volts, the ECM and wiring are OK.
3. This simulates a high signal voltage to check for an open or short in CKT 417. The scan tool should read over 4 volts.
4. With the ignition "ON" (or at idle) and the throttle closed, voltage at ECM terminal "B12" should be .33 to 1.33 volts. If not, replace the TP sensor.
5. If CKT 474 is shorted to ground or sensor ground, there also may be other DTC(s) stored.

Diagnostic Aids:

A scan tool displays throttle position in volts. The voltage at closed throttle should be .33 to 1.33 volts. The TP sensor voltage should increase at a steady rate as throttle is moved to Wide Open Throttle (WOT) position.

If DTC 21 is intermittent, refer to "Symptoms,"

2.0L (VIN H) ENGINE — DIAGNOSTIC TROUBLE CODE CHART — SUNBIRD

DTC 21

(Page 2 of 2)
THROTTLE POSITION (TP) SENSOR CIRCUIT
(SIGNAL VOLTAGE HIGH/LOW)
2.0L (VIN H) "J" CARLINE

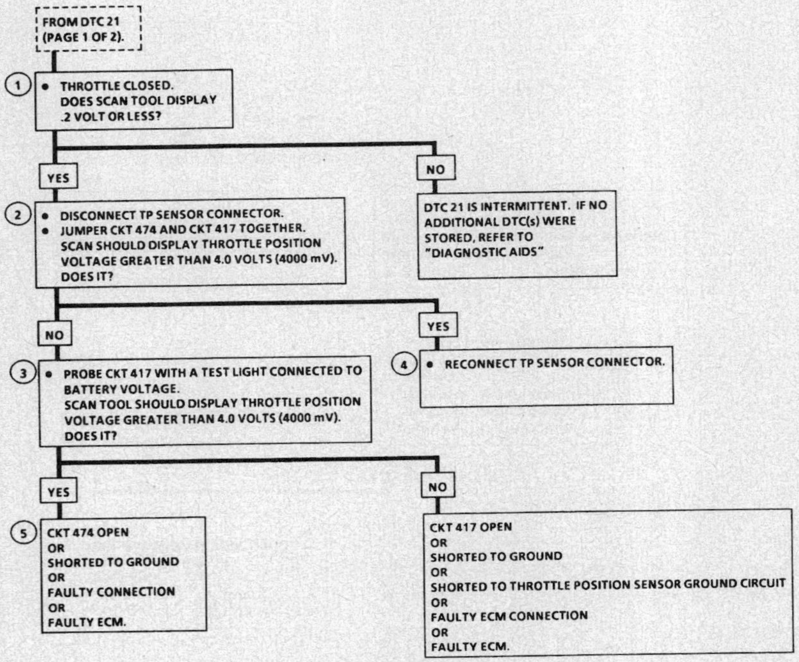

FROM DTC 21
(PAGE 1 OF 2).

(1) • THROTTLE CLOSED.
DOES SCAN TOOL DISPLAY
.2 VOLT OR LESS?

YES

(2) • DISCONNECT TP SENSOR CONNECTOR.
• JUMPER CKT 474 AND CKT 417 TOGETHER.
SCAN SHOULD DISPLAY THROTTLE POSITION
VOLTAGE GREATER THAN 4.0 VOLTS (4000 mV).
DOES IT?

NO → DTC 21 IS INTERMITTENT. IF NO
ADDITIONAL DTC(s) WERE
STORED, REFER TO
"DIAGNOSTIC AIDS"

NO (left path)

(3) • PROBE CKT 417 WITH A TEST LIGHT CONNECTED TO
BATTERY VOLTAGE.
SCAN TOOL SHOULD DISPLAY THROTTLE POSITION
VOLTAGE GREATER THAN 4.0 VOLTS (4000 mV).
DOES IT?

YES → (4) • RECONNECT TP SENSOR CONNECTOR.

YES

(5) CKT 474 OPEN
OR
SHORTED TO GROUND
OR
FAULTY CONNECTION
OR
FAULTY ECM.

NO

CKT 417 OPEN
OR
SHORTED TO GROUND
OR
SHORTED TO THROTTLE POSITION SENSOR GROUND CIRCUIT
OR
FAULTY ECM CONNECTION
OR
FAULTY ECM.

"AFTER REPAIRS," REFER TO DTC CRITERIA AND CONFIRM DTC DOES NOT RESET.

2.0L (VIN H) ENGINE — DIAGNOSTIC TROUBLE CODE CHART — SUNBIRD

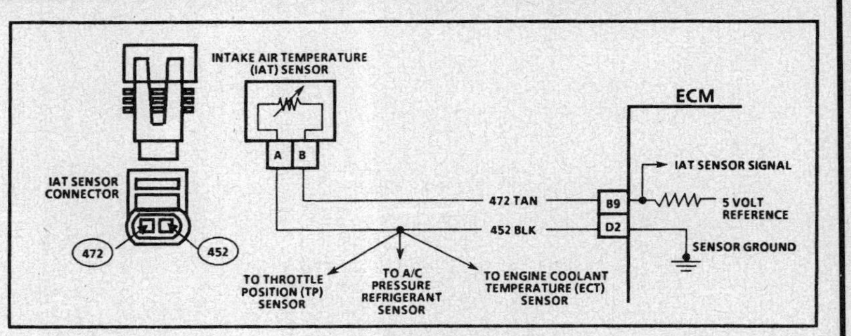

DTC 23

(Page 1 of 2)
INTAKE AIR TEMPERATURE (IAT) SENSOR CIRCUIT
(HIGH/LOW TEMPERATURE)
2.0L (VIN H) "J" CARLINE

Circuit Description:

The Intake Air Temperature (IAT) sensor, located in the air cleaner assembly, uses a thermistor to control the signal voltage to the Engine Control Module (ECM). The ECM applies a reference voltage (5 volts) on CKT 472 to the sensor. When intake air is cold, the sensor (thermistor) resistance is high. The ECM will then sense a high signal voltage. As the air warms, the sensor resistance becomes less and the voltage drops.

Test Description: Number(s) below refer to circled number(s) on the diagnostic chart.

1. A DTC 23 will set under the following conditions:
   • Engine running for 2 minutes or longer.
   • Signal voltage indicates an IAT temperature less than –39°C (–39.4°F).
   OR
   • Signal voltage indicates IAT greater than 146°C (294°F).
2. If the scan tool displays a high temperature, the ECM and wiring are OK.
3. This step checks continuity of CKT 472 and CKT 452. If CKT 452 is open, there may also be other DTC(s) stored.

Diagnostic Aids:

If the engine has been allowed to cool to an ambient temperature (overnight), engine coolant and intake air temperatures may be checked with a Tech 1 scan tool and should display values close to each other.
• If DTC 23 is intermittent or a "Hard Start" symptom is present, check intake air temperature with a scan tool on a cool engine. Temperature displayed should be within 5 degrees of the ambient. If not, check the IAT sensor using the "Diagnostic Aid" If sensor is OK, check connections. If connections at sensor and ECM are OK, replace ECM.

DTC 23
(Page 1 of 2)
INTAKE AIR TEMPERATURE (IAT) SENSOR CIRCUIT
(HIGH/LOW TEMPERATURE)
2.0L (VIN H) "J" CARLINE

1. • DOES SCAN TOOL DISPLAY IAT -39°C (-39.4°F) OR COLDER?

YES / NO

NO → REFER TO DTC 23 (PAGE 2 OF 2)

2. • DISCONNECT IAT SENSOR CONNECTOR.
   • JUMPER HARNESS TERMINALS TOGETHER.
   • SCAN TOOL SHOULD DISPLAY TEMPERATURE OVER 146°C (294°F). DOES IT?

YES → FAULTY CONNECTION OR SENSOR.

NO

3. • JUMPER CKT 472 TO GROUND.
   • SCAN TOOL SHOULD DISPLAY TEMPERATURE OVER 146°C (294°F). DOES IT?

YES → OPEN SENSOR GROUND CIRCUIT, OR FAULTY CONNECTION OR FAULTY ECM.

NO → OPEN CKT 472, OR FAULTY CONNECTION OR FAULTY ECM.

DIAGNOSTIC AID

IAT SENSOR		
TEMPERATURE VS. RESISTANCE VALUES		
(APPROXIMATE)		
°F	°C	OHMS
100	210	185
70	160	450
38	100	1,800
20	70	3,400
4	40	7,500
-7	20	13,500
-18	0	25,000
-40	-40	100,700

"AFTER REPAIRS." REFER TO DTC CRITERIA AND CONFIRM DTC DOES NOT RESET.

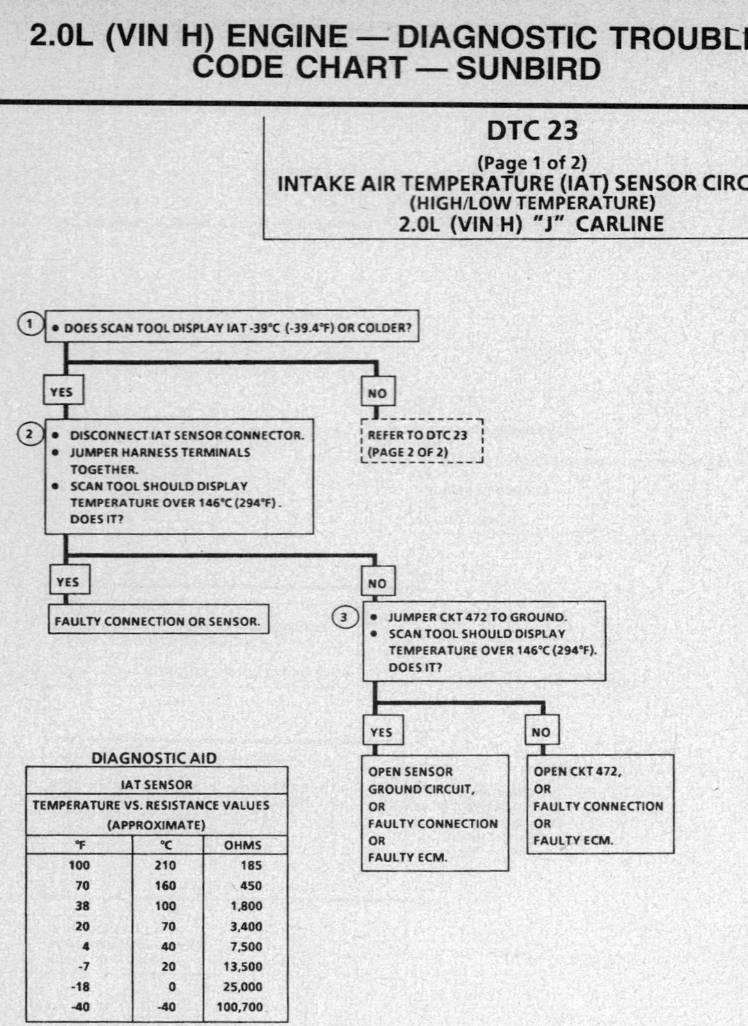

DTC 23
(Page 2 of 2)
INTAKE AIR TEMPERATURE (IAT) SENSOR CIRCUIT
(HIGH/LOW TEMPERATURE)
2.0L (VIN H) "J" CARLINE

Circuit Description:

The Intake Air Temperature (IAT) sensor, located in the air cleaner assembly, uses a thermistor to control the signal voltage to the Engine Control Module (ECM). The ECM applies a reference voltage (5 volts) on CKT 472 to the sensor. When intake air is cold, the sensor (thermistor) resistance is high. Therefore, the ECM will sense a high signal voltage. As the air warms, the sensor resistance becomes less and the voltage drops.

Test Description: Number(s) below refer to circled number(s) on the diagnostic chart.

1. This step determines if DTC 23 is the result of a hard failure or an intermittent condition.
2. If the ECM recognizes the open circuit (high voltage) and displays a low temperature, the ECM and wiring are OK.

Diagnostic Aids:

If the engine has been allowed to cool to an ambient temperature (overnight), engine coolant and intake air temperatures may be checked with a scan tool and should display values close to each other.

• If DTC 23 is intermittent or a "Hard Start" symptom is present, check intake air temperature with a scan tool on a cool engine. Temperature displayed should be within 5 degrees of the ambient. If not, check the IAT sensor using the "Diagnostic Aid" If sensor is OK, check connections. If connections at sensor and ECM are OK, replace ECM.

2.0L (VIN H) ENGINE — DIAGNOSTIC TROUBLE CODE CHART — SUNBIRD

DTC 23
(Page 2 of 2)
INTAKE AIR TEMPERATURE (IAT) SENSOR CIRCUIT
(HIGH/LOW TEMPERATURE)
2.0L (VIN H) "J" CARLINE

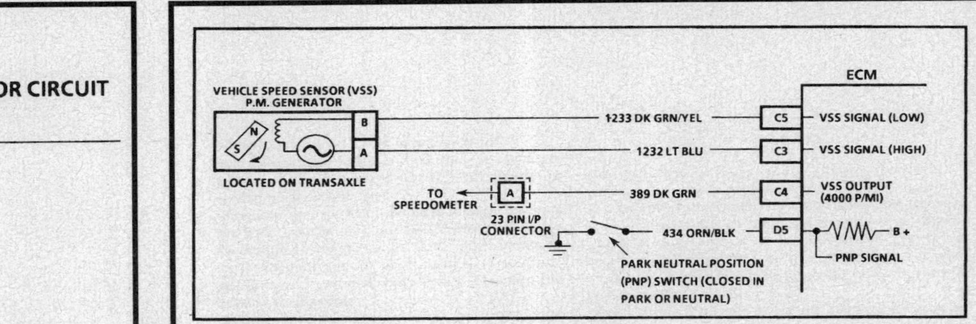

DIAGNOSTIC AID		
IAT SENSOR		
TEMPERATURE VS. RESISTANCE VALUES (APPROXIMATE)		
°C	°F	OHMS
100	210	185
70	160	450
38	100	1,800
20	70	3,400
4	40	7,500
-7	20	13,500
-18	0	25,000
-40	-40	100,700

"AFTER REPAIRS," REFER TO DTC CRITERIA AND CONFIRM DTC DOES NOT RESET.

2.0L (VIN H) ENGINE — DIAGNOSTIC TROUBLE CODE CHART — SUNBIRD

DTC 24
VEHICLE SPEED SENSOR (VSS) CIRCUIT
2.0L (VIN H) "J" CARLINE

Circuit Description:

Vehicle speed information is provided to the Engine Control Module (ECM) by the Vehicle Speed Sensor (VSS), which is a Permanent Magnet (PM) generator, and it is mounted on the transaxle. The PM generator produces a pulsing voltage whenever vehicle speed is over about 3 mph. The AC voltage level and the number of pulses increases with vehicle speed. The ECM then converts the pulsing voltage to mph, which is used for calculations, and the mph/km/h can be displayed with a scan tool.

The function of VSS buffer used in past model years has been incorporated into the ECM. The ECM then supplies the necessary signal to the instrument panel (4000 pulses per mile) for operating the speedometer and the odometer.

Test Description: Number(s) below refer to circled number(s) on the diagnostic chart.

1. DTC 24 will set if vehicle speed is less then 5 mph when:
   - Engine speed is between 1600 and 4400 RPM.
   - Low load condition (low MAP voltage, high manifold vacuum).
   - Transaxle not in park or neutral.
   - All above conditions are met for 4 seconds or more.

   These conditions are met during a road load deceleration.

   Disregard a DTC 24 that sets when the drive wheels are not turning. This can be caused by a faulty PNP switch circuit.

   The PM generator only produces a signal if the drive wheels are turning greater than 3 mph.

2. At this point, the ECM is not sending vehicle speed data to the scan tool. If the scan tool is connected and functioning properly, the ECM is at fault.

Diagnostic Aids:

Scan tool should indicate a vehicle speed whenever the drive wheels are turning greater than 3 mph.

A problem in CKT 389 will not affect the VSS input or the readings on a scan tool.

Check CKTs 1232 and 1233 for proper connections to be sure they are clean and tight and the harness is routed correctly.

- A faulty or misadjusted Park/Neutral Position (PNP) switch can result in a false DTC 24. Use a scan tool and check for the proper signal while in a drive range. Refer to CHART C-1A for the PNP switch check.

2.0L (VIN H) ENGINE — DIAGNOSTIC TROUBLE CODE CHART — SUNBIRD

DTC 24
VEHICLE SPEED SENSOR (VSS) CIRCUIT
2.0L (VIN H) "J" CARLINE

DISREGARD DTC 24 IF SET WHILE DRIVE WHEELS ARE NOT TURNING.

① • RAISE DRIVE WHEELS.

NOTICE: DO NOT PERFORM THIS TEST WITHOUT SUPPORTING THE LOWER CONTROL ARMS SO THAT THE DRIVE AXLES ARE IN A NORMAL HORIZONTAL POSITION. RUNNING THE VEHICLE IN GEAR WITH THE WHEELS HANGING DOWN AT FULL TRAVEL MAY DAMAGE THE DRIVE AXLES.

• WITH ENGINE IDLING IN GEAR, SCAN TOOL SHOULD DISPLAY VEHICLE SPEED ABOVE 0 MPH. DOES IT?

NO → DOES SPEEDOMETER WORK PROPERLY?

YES → DTC 24 IS INTERMITTENT. IF NO ADDITIONAL DTC(S) WERE STORED, REFER TO "DIAGNOSTIC AIDS"

NO →
• IGNITION "OFF."
• DISCONNECT VSS HARNESS CONNECTOR AT TRANSAXLE.
• CONNECT SIGNAL GENERATOR TESTER J 33431-B OR EQUIVALENT TO VSS HARNESS CONNECTOR.
• IGNITION "ON," TOOL "ON" AND SET TO GENERATE A VSS SIGNAL.
• SCAN TOOL SHOULD DISPLAY VEHICLE SPEED ABOVE 0 MPH. DOES IT?

YES → ② REPLACE ECM.

NO →
CKT 1232
OR
1233 OPEN, SHORTED TO GROUND, SHORTED TOGETHER, FAULTY CONNECTIONS, OR
FAULTY ECM.

YES → REPLACE VEHICLE SPEED SENSOR.

"AFTER REPAIRS," REFER TO DTC CRITERIA AND CONFIRM DTC DOES NOT RESET.

2.0L (VIN H) ENGINE — DIAGNOSTIC TROUBLE CODE CHART — SUNBIRD

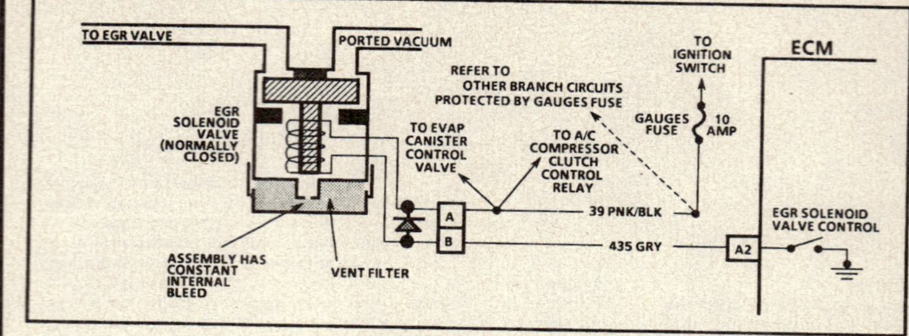

DTC 32
(Page 1 of 2)
EXHAUST GAS RECIRCULATION (EGR) SYSTEM FAILURE
2.0L (VIN H) "J" CARLINE

Circuit Description:

The Exhaust Gas Recirculation (EGR) system is controlled by the ECM. The ECM controls the vacuum being supplied to the valve by energizing and de-energizing a solenoid.

The ECM uses information from various engine sensors to determine when EGR is necessary. Once the ECM has requested EGR by grounding the solenoid circuit, the ECM will monitor engine operating conditions to determine if exhaust gas flow has entered the intake manifold. When the ECM tests for EGR operation and no change in engine operating conditions is indicated, a DTC 32 will set.

Test Description: Number(s) below refer to circled number(s) on the diagnostic chart.

1. **Intake Passage:** Shut "OFF" engine and remove the EGR valve from the manifold. Plug the exhaust side hole with a suitable stopper. Leaving the intake side hole open, attempt to start the engine. If the engine runs at a high idle (up to 3000 RPM is possible) or starts and stalls, the EGR intake passage is not restricted. If the engine starts and idles normally, the EGR intake passage is restricted.

   Exhaust Passage: With EGR valve still removed, plug the intake side hole with a suitable stopper. With the exhaust side hole open, start engine and check for the presence of exhaust gas. If no exhaust gas is present, the EGR exhaust side passage is restricted.

2. By grounding the diagnostic "test" terminal, the EGR solenoid should be energized and allow vacuum to be applied to the gage. The vacuum at the gage may or may not <u>slowly</u> bleed off. It is important that the gage is able to read the amount of vacuum being applied.

3. When the diagnostic "test" terminal is ungrounded, the vacuum gage should bleed off completely through a vent in the solenoid.

The vacuum pump gage may or may not bleed off but this does not indicate a problem.

4. This test will determine if the electrical control part of the system is at fault or if the connector or solenoid is at fault.

5. At this point, it has been determined that the EGR solenoid, the ECM, and the vacuum supply are OK.

Diagnostic Aids:

Vacuum lines should be thoroughly checked for proper routing. Refer to "Vehicle Emission Control Information" label.

Suction from shop exhaust hoses can alter backpressure and may affect the functional check of the EGR valve.

If the EGR system is found to be functioning correctly and DTC 32 is set, low fuel pressure or lean fuel injector(s) could be the cause of setting DTC 32. It may be necessary to monitor fuel pressure while driving the vehicle at various road speeds and/or loads. Refer to "Fuel System Diagnosis," Chart A-7. Perform the "Injector Balance Test" CHART C2-A.

2.0L (VIN H) ENGINE — DIAGNOSTIC TROUBLE CODE CHART — SUNBIRD

DTC 32
(Page 1 of 2)
EXHAUST GAS RECIRCULATION (EGR)
SYSTEM FAILURE
2.0L (VIN H) "J" CARLINE

- WITH ENGINE AT IDLE AND AT NORMAL OPERATING TEMPERATURE, MANUALLY LIFT EGR VALVE DIAPHRAGM.
- RPM SHOULD DECREASE OR ENGINE SHOULD STALL. DOES IT?

YES

- DISCONNECT EGR VACUUM HARNESS AT SOLENOID. INSTALL A VACUUM GAGE TO MANIFOLD SIDE OF HARNESS AND CHECK FOR VACUUM SUPPLY TO SOLENOID.
- RAISE ENGINE TO 2000 RPM AND OBSERVE VACUUM.
- VACUUM SHOULD BE AT LEAST 25 kPa (7" Hg) OF VACUUM AT 2000 RPM. IS IT?

NO

(1) CHECK FOR PLUGGED EGR PASSAGES. IF PASSAGES ARE OK, REPLACE EGR VALVE.

YES

(2)
- ROTATE VACUUM HARNESS AND REINSTALL ONLY THE EGR VALVE SIDE, TO THE SOLENOID.
- INSTALL A VACUUM GAGE IN PLACE OF EGR VALVE.
- INSTALL A HAND HELD VACUUM PUMP TO MANIFOLD SIDE OF EGR SOLENOID.
- IGNITION "ON," ENGINE STOPPED.
- GROUND DIAGNOSTIC TERMINAL.
- APPLY 34 kPa (10" Hg) VACUUM AND OBSERVE GAGE.
- GAGE SHOULD READ VACUUM APPLIED BY PUMP. DOES IT?

NO

PLUGGED VACUUM PORT AT THROTTLE BODY. LEAKING OR RESTRICTED VACUUM SUPPLY LINE TO EGR SOLENOID.

YES

(3)
- APPLY 34 kPa (10" Hg) VACUUM.
- UNGROUND DIAGNOSTIC TERMINAL.
- VACUUM SHOULD BLEED OFF COMPLETELY AT GAGE. DOES IT?

NO

- CONNECT VACUUM PUMP TO EGR VALVE SIDE OF HARNESS.
- APPLY VACUUM AND OBSERVE GAGE.
- GAGE SHOULD READ VACUUM APPLIED BY PUMP. DOES IT?

YES

(5) PROCEED TO DTC 32 (PAGE 2 OF 2).

NO

- DISCONNECT SOLENOID ELECTRICAL CONNECTOR. DOES VACUUM BLEED OFF RAPIDLY AT GAGE?

YES

CKT 435 SHORTED TO GROUND OR FAULTY ECM*.

NO

REPLACE SOLENOID.

YES

(4)
- DISCONNECT EGR ELECTRICAL CONNECTOR.
- CONNECT TEST LIGHT BETWEEN HARNESS CONNECTOR TERMINALS "A" AND "B".
- IGNITION "ON," ENGINE "OFF."
- DIAGNOSTIC "TEST" TERMINAL GROUNDED. TEST LIGHT SHOULD LIGHT. DOES IT?

NO

FAULTY VACUUM SUPPLY LINE TO EGR VALVE.

YES

FAULTY SOLENOID CONNECTION OR FAULTY SOLENOID.

NO

CONNECT TEST LIGHT BETWEEN TERMINAL "A" (CKT 39) AND A KNOWN GOOD GROUND.

LIGHT

OPEN CKT 435 OR FAULTY CONNECTIONS OR FAULTY ECM*.

NO LIGHT

REPAIR OPEN IGNITION FEED CKT 39.

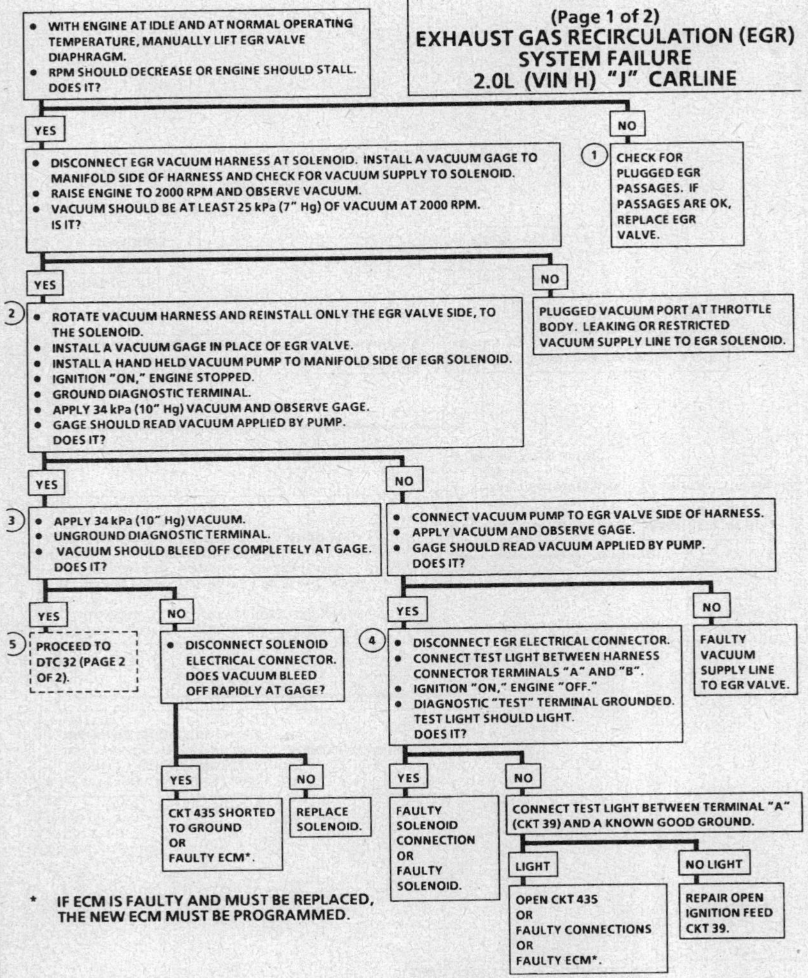

* IF ECM IS FAULTY AND MUST BE REPLACED, THE NEW ECM MUST BE PROGRAMMED.

2.0L (VIN H) ENGINE — DIAGNOSTIC TROUBLE CODE CHART — SUNBIRD

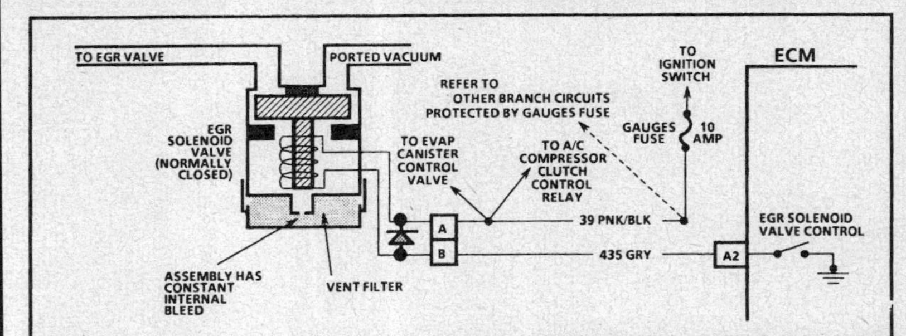

DTC 32
(Page 2 of 2)
EXHAUST GAS RECIRCULATION (EGR) SYSTEM FAILURE
2.0L (VIN H) "J" CARLINE

Circuit Description:

The Exhaust Gas Recirculation (EGR) system is controlled by the ECM. The ECM controls the vacuum being supplied to the valve by energizing and de-energizing a solenoid.

The ECM uses information from various engine sensors to determine when EGR is necessary. Once the ECM has requested EGR by grounding the solenoid circuit, the ECM will monitor engine operating conditions to determine if exhaust gas flow has entered the intake manifold. When the ECM tests for EGR operation and no change in engine operating conditions are indicated, a DTC 32 will set.

Test Description: Number(s) below refer to circled number(s) on the diagnostic chart.

1. The remaining tests check the ability of the EGR valve to interact with the exhaust system. This system uses a positive backpressure EGR valve which will not hold vacuum until sufficient exhaust backpressure is at the base of the valve.

2. The EGR valve diaphragm should move when sufficient exhaust backpressure is present at the base of the valve and when vacuum is being supplied to the valve. Rapidly "snapping" the throttle from idle should provide sufficient exhaust backpressure to the base of the valve which will close an internal vacuum bleed. With the EGR valve's internal vacuum bleed closed, the "jumpered" vacuum supply can now lift the valve off its seat. The amount of valve pintle movement will be small. It is important that the valve pintle moves at the proper time.

3. Excessive exhaust backpressure from bent or restricted exhaust system components could provide enough backpressure at the base of the EGR valve to close the valve's internal bleed and allow undesired EGR valve operation at idle.

4. Plugged EGR exhaust passages can block exhaust backpressure from reaching the EGR valve. With no EGR exhaust backpressure at the base of the valve, the valve's internal bleed will remain open and prevent vacuum from operating the valve.

Diagnostic Aids:

Suction from shop exhaust hoses can alter backpressure and may affect the functional check of the EGR valve.

If the EGR system is found to be functioning correctly and DTC 32 is set, low fuel pressure or lean fuel injector(s) could be the cause of setting DTC 32. It may be necessary to monitor fuel pressure while driving the vehicle at various road speeds and/or loads. Refer to "Fuel System Diagnosis," CHART A-7. Perform the "Injector Balance Test"
CHART C2-A.

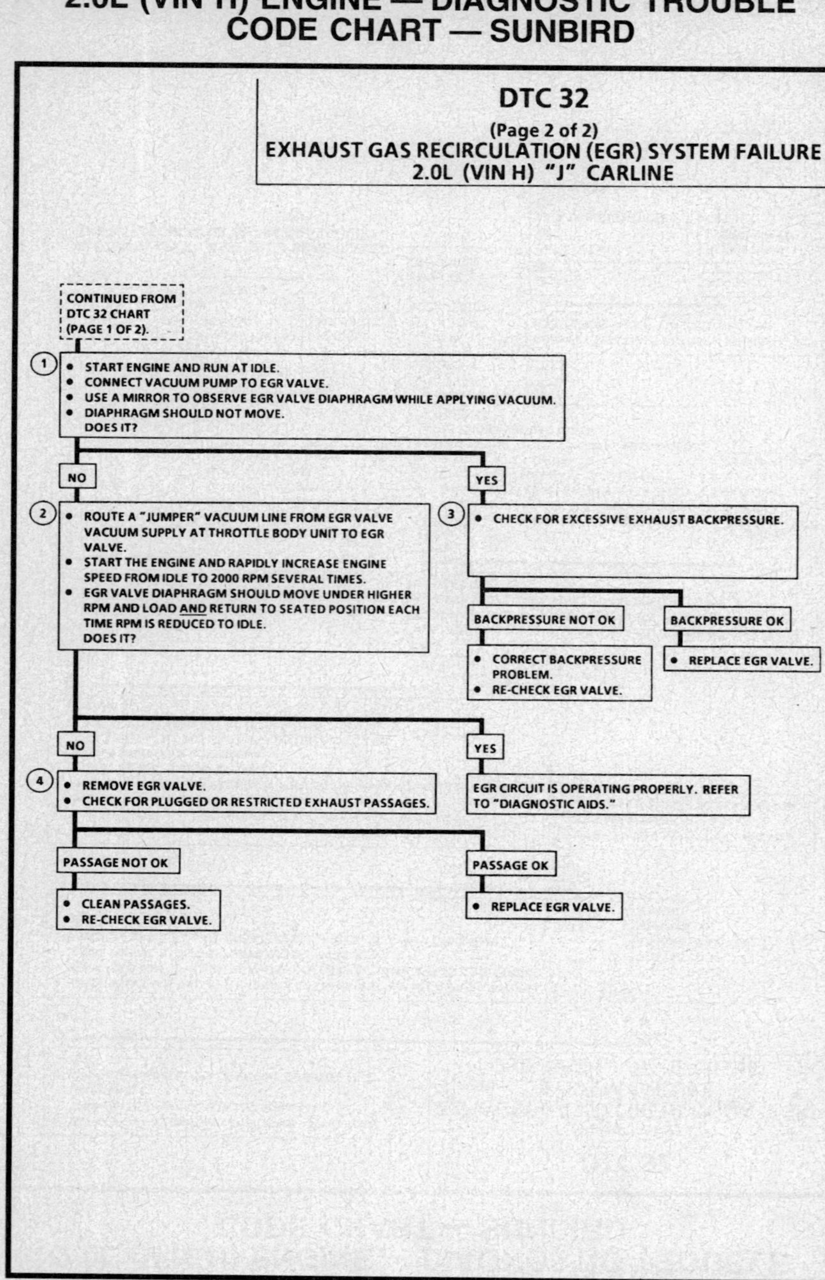

DTC 32

(Page 2 of 2)
EXHAUST GAS RECIRCULATION (EGR) SYSTEM FAILURE
2.0L (VIN H) "J" CARLINE

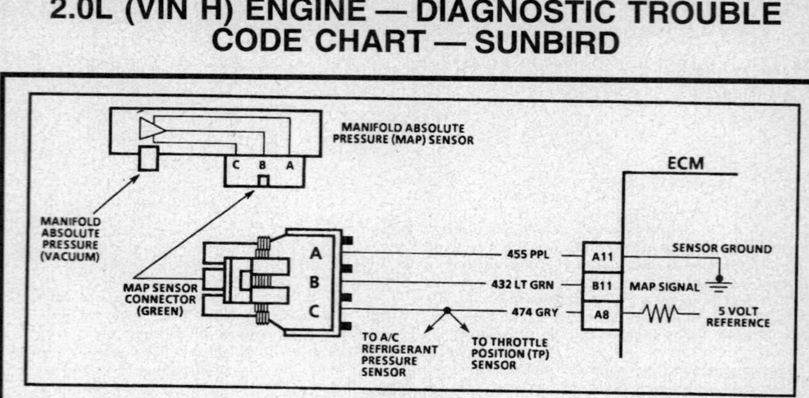

DTC 33

(Page 1 of 2)
MANIFOLD ABSOLUTE PRESSURE (MAP) SENSOR CIRCUIT
(SIGNAL VOLTAGE HIGH/LOW - LOW/HIGH VACUUM)
2.0L (VIN H) "J" CARLINE

Circuit Description:

The Manifold Absolute Pressure (MAP) sensor responds to changes in manifold pressure (vacuum). The ECM receives this information as a signal voltage that will vary from about 1 - 1.5 volts at closed throttle (idle), to 4 - 4.5 volts at Wide Open Throttle (WOT) (low vacuum).

If the MAP sensor fails, the Engine Control Module (ECM) will substitute a fixed MAP value and use the Throttle Position (TP) sensor to control fuel delivery.

Test Description: Number(s) below refer to circled number(s) on the diagnostic chart.

1. A DTC 33 will set under the following conditions:
    *   MAP signal less then .25 volt.
    *   RPM less than 1200 or greater than 1200 with TP sensor greater than 15%.
        OR
    *   MAP signal greater than 3.7 volts.
    *   TP sensor less than 5%.
    *   These conditions exist for 5 seconds or more.
2. If the ECM recognizes the change, the ECM and CKTs 474 and 432 are OK. If CKT 452 is open, there may also be other DTC(s) set.

Diagnostic Aids:

With the ignition "ON" and the engine not running, the manifold pressure is equal to atmospheric pressure, and the signal voltage will be high. This information is used by the ECM as an indication of vehicle altitude and is referred to as BARO. Comparison of this BARO reading with a known good vehicle with the same sensor is a good way to check accuracy of a "suspect" MAP sensor. Reading should be within ± .4 volt.

If CKT 474 is shorted to ground or to CKT 455, other DTC(s) may also be set.

If DTC 33 is intermittent, refer to "Symptoms,"

2.0L (VIN H) ENGINE — DIAGNOSTIC TROUBLE CODE CHART — SUNBIRD

DTC 33
(Page 1 of 2)
MANIFOLD ABSOLUTE PRESSURE (MAP) SENSOR CIRCUIT
(SIGNAL VOLTAGE HIGH/LOW - LOW/HIGH VACUUM)
2.0L (VIN H) "J" CARLINE

(1)
- IF ENGINE IDLE IS ROUGH, UNSTABLE, OR INCORRECT, CORRECT CONDITION BEFORE USING CHART.
- ENGINE IDLING.
- DOES SCAN TOOL DISPLAY A MAP VOLTAGE OF 3.7 VOLTS OR OVER?

YES / NO

(2)
- IGNITION "OFF."
- DISCONNECT MAP SENSOR ELECTRICAL CONNECTOR.
- IGNITION "ON."
- SCAN TOOL SHOULD READ A VOLTAGE OF 1 VOLT OR LESS. DOES IT?

REFER TO DTC 33 (PAGE 2 OF 2).

YES / NO

- PROBE SENSOR GROUND CIRCUIT WITH A TEST LIGHT TO BATTERY VOLTAGE.
- TEST LIGHT SHOULD LIGHT. DOES IT?

CKT 432 SHORTED TO VOLTAGE, SHORTED TO CKT 474 OR FAULTY ECM.

YES / NO

PLUGGED OR LEAKING SENSOR VACUUM HOSE OR FAULTY MAP SENSOR.

OPEN SENSOR GROUND CIRCUIT.

"AFTER REPAIRS," REFER TO DTC CRITERIA AND CONFIRM DTC DOES NOT RESET.

2.0L (VIN H) ENGINE — DIAGNOSTIC TROUBLE CODE CHART — SUNBIRD

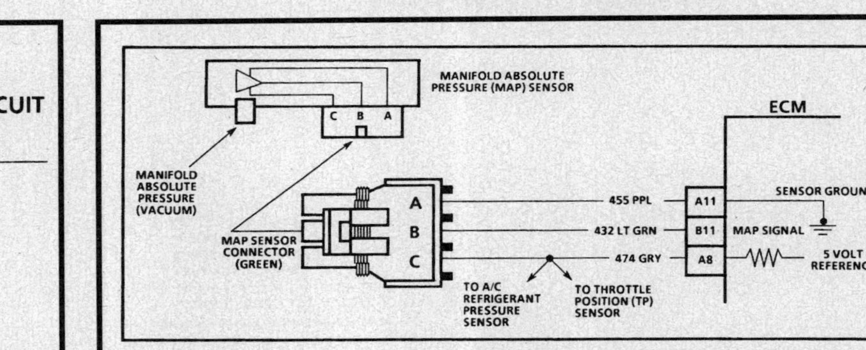

DTC 33
(Page 2 of 2)
MANIFOLD ABSOLUTE PRESSURE (MAP) SENSOR CIRCUIT
(SIGNAL VOLTAGE HIGH/LOW - LOW/HIGH VACUUM)
2.0L (VIN H) "J" CARLINE

Circuit Description:
The Manifold Absolute Pressure (MAP) sensor responds to changes in manifold pressure (vacuum). The ECM receives this information as a signal voltage that will vary from about 1 to 1.5 volts at closed throttle (idle), to 4 to 4.5 volts at Wide Open Throttle (WOT) (low vacuum).
If the MAP sensor fails, the Engine Control Module (ECM) will substitute a fixed MAP value based on RPM and use the Throttle Position (TP) sensor to control fuel delivery.

Test Description: Number(s) below refer to circled number(s) on the diagnostic chart.
1. This step checks to see if DTC 33 is the result of a hard failure or an intermittent condition.
2. Jumpering harness terminals "B" to "C" (5 volts to signal circuit) will determine if the sensor is at fault or if there is a problem with the ECM or wiring.
3. The scan tool may not display 12 volts. The important thing is that the ECM recognizes the voltage as more than 4 volts, indicating that the ECM and CKT 432 are OK.

Diagnostic Aids:
With the ignition "ON" and the engine not running, the manifold pressure is equal to atmospheric pressure, and the signal voltage will be high. This information is used by the ECM as an indication of vehicle altitude and is referred to as BARO. Comparison of this BARO reading with a known good vehicle with the same sensor is a good way to check accuracy of a "suspect" MAP sensor. Reading should be within ± .4 volt.
If DTC 33 is intermittent, refer to "Symptoms,"

2.0L (VIN H) ENGINE — DIAGNOSTIC TROUBLE CODE CHART — SUNBIRD

DTC 33
(Page 2 of 2)
MANIFOLD ABSOLUTE PRESSURE (MAP) SENSOR CIRCUIT
(SIGNAL VOLTAGE HIGH/LOW - LOW/HIGH VACUUM)
2.0L (VIN H) "J" CARLINE

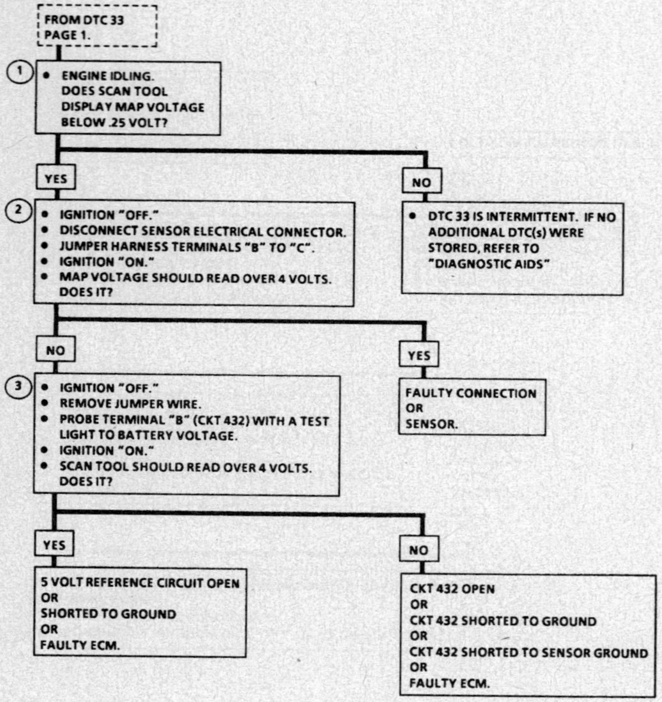
FROM DTC 33 PAGE 1.
1
• ENGINE IDLING. DOES SCAN TOOL DISPLAY MAP VOLTAGE BELOW .25 VOLT?
YES
NO
2
• IGNITION "OFF."
• DISCONNECT SENSOR ELECTRICAL CONNECTOR.
• JUMPER HARNESS TERMINALS "B" TO "C".
• IGNITION "ON."
• MAP VOLTAGE SHOULD READ OVER 4 VOLTS. DOES IT?
• DTC 33 IS INTERMITTENT. IF NO ADDITIONAL DTC(s) WERE STORED, REFER TO "DIAGNOSTIC AIDS"
NO
YES
3
• IGNITION "OFF."
• REMOVE JUMPER WIRE.
• PROBE TERMINAL "B" (CKT 432) WITH A TEST LIGHT TO BATTERY VOLTAGE.
• IGNITION "ON."
• SCAN TOOL SHOULD READ OVER 4 VOLTS. DOES IT?
FAULTY CONNECTION OR SENSOR.
YES
NO
5 VOLT REFERENCE CIRCUIT OPEN OR SHORTED TO GROUND OR FAULTY ECM.
CKT 432 OPEN OR CKT 432 SHORTED TO GROUND OR CKT 432 SHORTED TO SENSOR GROUND OR FAULTY ECM.

"AFTER REPAIRS," REFER TO DTC CRITERIA AND CONFIRM DTC DOES NOT RESET.

2.0L (VIN H) ENGINE — DIAGNOSTIC TROUBLE CODE CHART — SUNBIRD

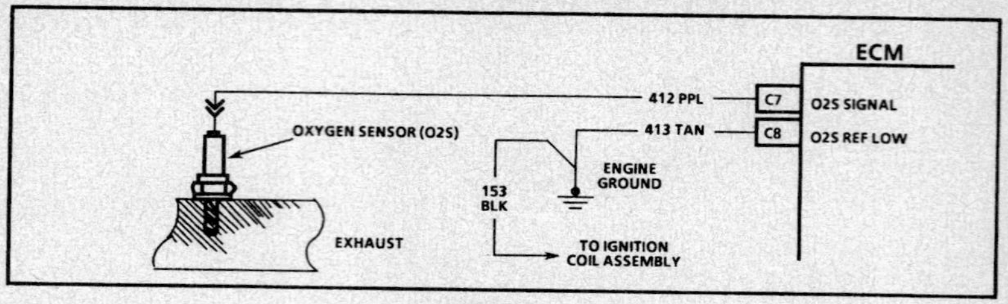
ECM
412 PPL
C7
O2S SIGNAL
413 TAN
C8
O2S REF LOW
OXYGEN SENSOR (O2S)
ENGINE GROUND
153 BLK
EXHAUST
TO IGNITION COIL ASSEMBLY

DTC 44
OXYGEN SENSOR (O2S) CIRCUIT
(LEAN EXHAUST INDICATED)
2.0L (VIN H) "J" CARLINE

Circuit Description:

The Engine Control Module (ECM) supplies a voltage of about .45 volt between terminals "C7" and "C8". (If measured with a 10 megohm digital voltmeter, this may read as low as .32 volt.)

When the Oxygen Sensor (O2S) reaches operating temperature, it varies this voltage from about .1 volt (exhaust is lean) to about .9 volt (exhaust is rich).

The sensor is like an open circuit and produces no voltage when it is below 316°C (600°F). An open sensor circuit or cold sensor causes "Open Loop" operation.

Test Description: Number(s) below refer to circled number(s) on the diagnostic chart.

1. DTC 44 is set when the O2S signal voltage on CKT 412 remains below .35 volt under the following conditions:
   - Engine run time after start is 2 minutes or more.
   - Throttle angle is greater than 5%.
   - These conditions exist for 25 seconds or more.

Diagnostic Aids:

Using the scan tool, observe the long term fuel trim value at different engine speeds. The scan tool also displays the fuel trim cells so the long term fuel trim values can be checked in each of the cells to determine when the DTC 44 may have been set. If the conditions for DTC 44 exists, the long term fuel trim values will probably be around 150 or higher.

Check the following possible causes:

- Oxygen Sensor (O2S) Wire - Sensor pigtail may be routed incorrectly and contacting the exhaust manifold.
  Check for ground in wire between connector and sensor.
- Fuel Contamination - Water, even in small amounts, near the in-tank fuel pump inlet can be delivered to the injector. The water causes a lean exhaust and can set a DTC 44.
- Fuel Pressure - System will be lean if fuel pressure is too low. It may be necessary to monitor fuel pressure while driving the vehicle at various road speeds and/or loads to confirm. Refer to "Fuel System Diagnosis," CHART A-7.
- Exhaust Leaks - If there is an exhaust leak, the engine can cause outside air to be pulled into the exhaust and past the sensor. Vacuum or crankcase leaks can cause a lean condition.
- If DTC 44 is intermittent, refer to "Symptoms,"
- A cracked or otherwise damaged O2S may set an intermittent DTC 44.

2.0L (VIN H) ENGINE — DIAGNOSTIC TROUBLE CODE CHART — SUNBIRD

DTC 44

OXYGEN SENSOR (O2S) CIRCUIT
(LEAN EXHAUST INDICATED)
2.0L (VIN H) "J" CARLINE

(1)
- RUN WARM ENGINE (75°C/167°F TO 95°C/203°F) AT 1200 RPM.
- DOES SCAN TOOL INDICATE OXYGEN SENSOR VOLTAGE FIXED BELOW .35 VOLT (350 mV)?

YES

- DISCONNECT OXYGEN SENSOR CONNECTOR.
- WITH ENGINE IDLING, SCAN TOOL SHOULD DISPLAY OXYGEN SENSOR VOLTAGE BETWEEN .35 VOLT AND .55 VOLT (350 mV AND 550 mV). DOES IT?

YES

REFER TO "DIAGNOSTIC AIDS" ON FACING PAGE.

NO

DTC 44 IS INTERMITTENT. IF NO ADDITIONAL DTC(s) WERE STORED, REFER TO "DIAGNOSTIC AIDS"

NO

CKT 412 SHORTED TO GROUND OR FAULTY ECM.

2.0L (VIN H) ENGINE — DIAGNOSTIC TROUBLE CODE CHART — SUNBIRD

DTC 45

OXYGEN SENSOR (O2S) CIRCUIT
(RICH EXHAUST INDICATED)
2.0L (VIN H) "J" CARLINE

Circuit Description:

The Engine Control Module (ECM) supplies a voltage of about .45 volt between terminals "C7" and "C8". (If measured with a 10 megohm digital voltmeter, this may read as low as .32 volt.)

When the O2S reaches operating temperature, it varies this voltage from about .1 volt (exhaust is lean) to about .9 volt (exhaust is rich).

The sensor is like an open circuit and produces no voltage when it is below 316°C (600°F). An open sensor circuit or cold sensor causes "Open Loop" operation.

Test Description: Number(s) below refer to circled number(s) on the diagnostic chart.
1. DTC 45 is set when the Oxygen Sensor (O2S) signal voltage on CKT 412 remains above .55 volt under the following conditions:
- System is operating in "Closed Loop."
- Engine run time after start is 2 minutes or more.
- Throttle angle is greater than 5%.
- These conditions exist for 51 seconds or more.

Diagnostic Aids:

DTC 45, or rich exhaust, is most likely caused by one of the following:
- Fuel Pressure - System will go rich, if pressure is too high. The ECM can compensate for some increase. However, if it gets too high, a DTC 45 will be set. Refer to "Fuel System Diagnosis" CHART A-7.
- Leaking Injector - Refer to CHART A-7.
- An Open Ground CKT 153 - May result in induced electrical "noise." The ECM interprets this "noise" as reference pulses. The additional pulses result in a higher than actual engine speed signal. The ECM then delivers too much fuel causing the system to go rich. The engine tachometer will also show higher than actual engine speed, which can help in diagnosing this problem.
- Canister Purge - Check for fuel saturation. If EVAP canister is full of fuel, check EVAP canister control valve and hoses.

- MAP Sensor - An output that causes the ECM to sense a higher than normal manifold pressure (low vacuum) can cause the system to go rich. Disconnecting the Manifold Absolute Pressure (MAP) sensor will allow the ECM to set a fixed value for the MAP sensor. Substitute a different MAP sensor if the rich condition is gone, while the sensor is disconnected.
- TP Sensor - An intermittent Throttle Position (TP) sensor output will cause the system to operate rich due to a false indication of the engine accelerating.
- Oxygen Sensor (O2S) Contamination - Inspect O2S for silicone contamination from fuel or use of improper RTV sealant. The sensor may have a white, powdery coating and result in a high but false signal voltage (rich exhaust indication). The ECM will then reduce the amount of fuel delivered to the engine causing a severe surge driveability problem.
- EGR Valve - Exhaust Gas Recirculation (EGR) sticking open at idle is usually accompanied by a rough idle and/or stall condition.
If DTC 45 is intermittent, refer to "Symptoms,"

- Engine Oil Contamination - Fuel fouled engine oil could cause the O2S to sense a rich air/fuel mixture and set a DTC 45.
- Engine Coolant Temperature (ECT) Sensor - Check coolant sensor using "Diagnostic Aid" on DTC 14 chart.

2.0L (VIN H) ENGINE — DIAGNOSTIC TROUBLE CODE CHART — SUNBIRD

DTC 45
OXYGEN SENSOR (O2S) CIRCUIT
(RICH EXHAUST INDICATED)
2.0L (VIN H) "J" CARLINE

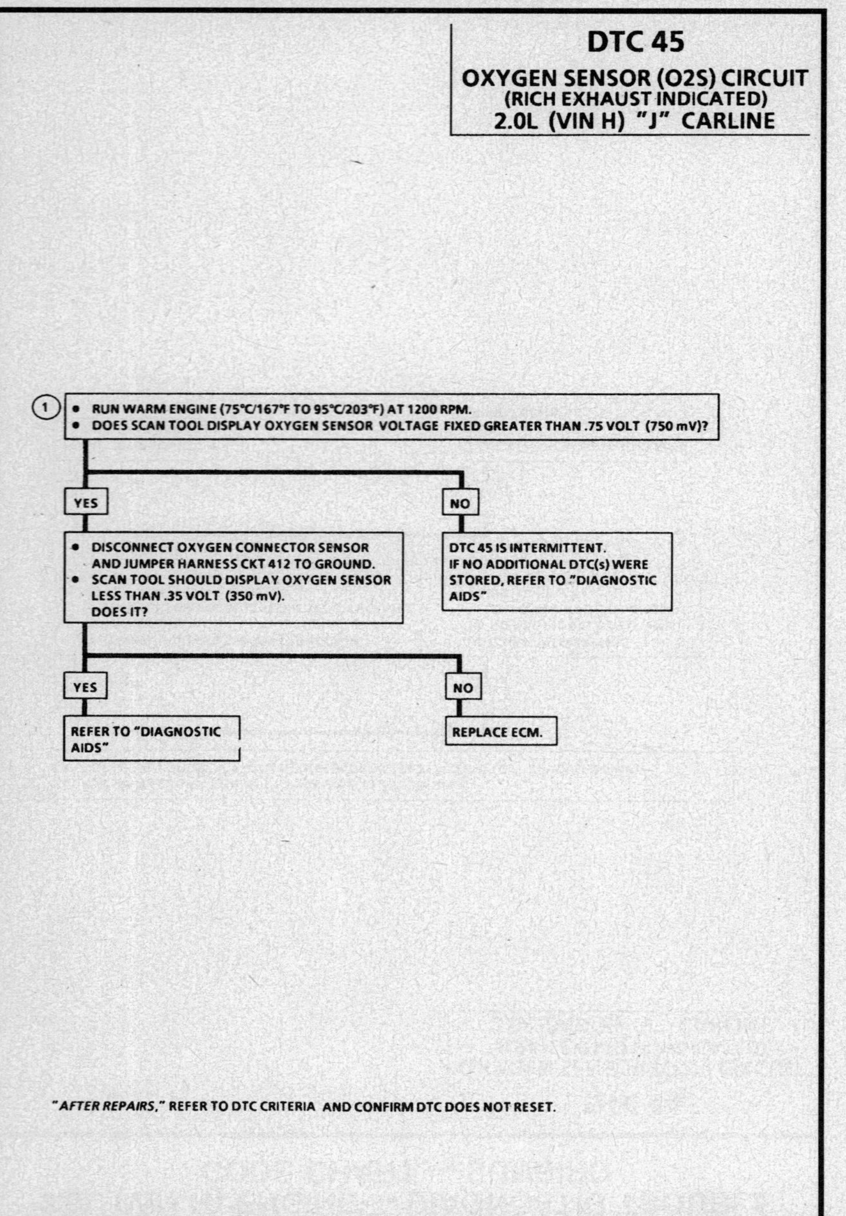

"AFTER REPAIRS," REFER TO DTC CRITERIA AND CONFIRM DTC DOES NOT RESET.

2.0L (VIN H) ENGINE — DIAGNOSTIC TROUBLE CODE CHART — SUNBIRD

DTC 51
2.0L (VIN H) "J" CARLINE

DTC 51

ECM FAILURE
(ECM FAILED OR PROM FAILURE)

CHECK THAT ALL ECM CONNECTIONS ARE GOOD.
IF OK, CLEAR MEMORY AND RECHECK ECM.
IF DTC 51 REAPPEARS, REPLACE ECM.

"AFTER REPAIRS," CONFIRM "CLOSED LOOP" OPERATION AND NO MIL (CHECK ENGINE).

2.0L (VIN H) ENGINE — DIAGNOSTIC TROUBLE CODE CHART — 1992 SUNBIRD

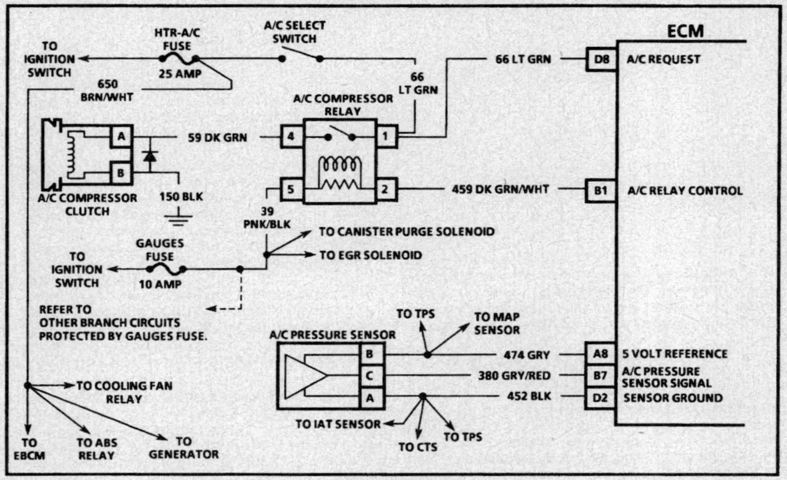

CODE 66
A/C PRESSURE SENSOR CIRCUIT
2.0L (VIN H) "J" CARLINE (PORT)

ECM

Terminal	Circuit	Function
D8	66 LT GRN	A/C REQUEST
B1	459 DK GRN/WHT	A/C RELAY CONTROL
A8	474 GRY	5 VOLT REFERENCE
B7	380 GRY/RED	A/C PRESSURE SENSOR SIGNAL
D2	452 BLK	SENSOR GROUND

Diagram labels: TO IGNITION SWITCH — 650 BRN/WHT — HTR-A/C FUSE 25 AMP — A/C SELECT SWITCH — 66 LT GRN — A/C COMPRESSOR RELAY — 59 DK GRN — A/C COMPRESSOR CLUTCH — 150 BLK — 39 PNK/BLK — TO CANISTER PURGE SOLENOID — TO EGR SOLENOID — GAUGES FUSE 10 AMP — TO IGNITION SWITCH — REFER TO OTHER BRANCH CIRCUITS PROTECTED BY GAUGES FUSE. — TO COOLING FAN RELAY — TO EBCM — TO ABS RELAY — TO GENERATOR — A/C PRESSURE SENSOR (B, C, A) — TO TPS — TO MAP SENSOR — TO IAT SENSOR — TO CTS — TO TPS

Circuit Description:

The A/C pressure sensor responds to changes in A/C refrigerant system high side pressure. This input indicates how much load the A/C compressor is putting on the engine and is one of the factors used by the ECM to determine IAC valve position for idle speed control. The circuit consists of a 5 volt reference and a ground, both provided by the ECM, and a signal line to the ECM. The signal is a voltage which is proportional to the pressure. The sensor's range of operation is 0 to 450 psi. At 0 psi, the signal will be about .1 volt, varying up to about 4.9 volts at 450 psi or above. Code 66 sets if the voltage is above 4.9 volts or below .1 volt for 5 seconds or more. The A/C compressor is disabled by the ECM if the fault is presently occurring, or if pressure is above or below calibrated values

Test Description: Number(s) below refer to circled number(s) on the diagnostic chart.

1. This step checks the voltage signal being received by the ECM from the A/C pressure sensor. The normal operating range is between .1 volt and 4.9 volts.
2. Checks to see if the high voltage signal is from a shorted sensor or a short to voltage in the circuit. Normally, disconnecting the sensor would make a normal circuit go to near zero volts.
3. Checks to see if low voltage signal is from the sensor or the circuit. Jumpering the sensor signal CKT 380 to 5 volts, checks the circuit, connections, and ECM.
4. This step checks to see if the low voltage signal was due to an open in the sensor circuit or the 5 volts reference circuit since the prior step eliminated the pressure sensor.

Diagnostic Aids:

Code 66 sets when signal voltage falls outside the normal possible range of the sensor and is not due to a refrigerant system problem. If problem is intermittent, check for opens or shorts in harness or poor connections. If OK, replace A/C pressure sensor. If Code 66 resets, replace ECM.

2.0L (VIN H) ENGINE — DIAGNOSTIC TROUBLE CODE CHART — 1992 SUNBIRD

CODE 66
A/C PRESSURE SENSOR CIRCUIT
2.0L (VIN H) "J" CARLINE (PORT)

(1)
- KEY "ON," ENGINE NOT RUNNING.
- A/C NOT REQUESTED.
- NOTE TECH 1 SCAN VOLTAGE FOR A/C PRESSURE SENSOR.

Branches: **ABOVE 3.9 VOLTS** | **BELOW .28 VOLT** | **BETWEEN .28 VOLTS AND 3.9 VOLTS**

ABOVE 3.9 VOLTS →

(2)
- DISCONNECT A/C PRESSURE SENSOR ELECTRICAL CONNECTOR.
- DOES "SCAN" DISPLAY LESS THAN 1 VOLT?

NO → CHECK FOR OPEN/SHORT TO VOLTAGE IN CKT 452. OR SHORT TO VOLTAGE IN CKT 380. IF OK, REPLACE ECM.

YES → CHECK FOR POOR SENSOR TERMINAL CONNECTIONS. IF OK, REPLACE A/C PRESSURE SENSOR.

BELOW .28 VOLT →

(3)
- DISCONNECT A/C PRESSURE SENSOR CONNECTOR.
- JUMPER TERMINALS "B" AND "C".
- DOES "SCAN" DISPLAY ABOVE 4.6 VOLTS?

NO →

(4)
- REMOVE JUMPER.
- CONNECT VOLTMETER FROM TERMINAL "A" TO "B".
- IS VOLTAGE ABOUT 5 VOLTS?

  NO → BACK PROBE ECM TERMINAL "A8" WITH VOLTMETER TO GROUND. IS VOLTAGE ABOUT 5 VOLTS?

    NO → CHECK FOR POOR CONNECTION AT ECM TERMINAL "A8" OR SHORT TO GROUND IN CKT 474. IF OK, ECM IS FAULTY.

    YES → REPAIR OPEN IN CKT 474.

  YES → CHECK FOR OPEN/SHORT TO GROUND IN CKT 380. CHECK FOR POOR CONNECTION AT ECM TERMINAL "B7". IF OK, REPLACE ECM.

YES → CHECK SENSOR TERMINAL CONNECTIONS. IF OK, REPLACE A/C PRESSURE SENSOR.

BETWEEN .28 VOLTS AND 3.9 VOLTS → FAULT IS NOT PRESENT AT THIS TIME. SEE "DIAGNOSTIC AIDS."

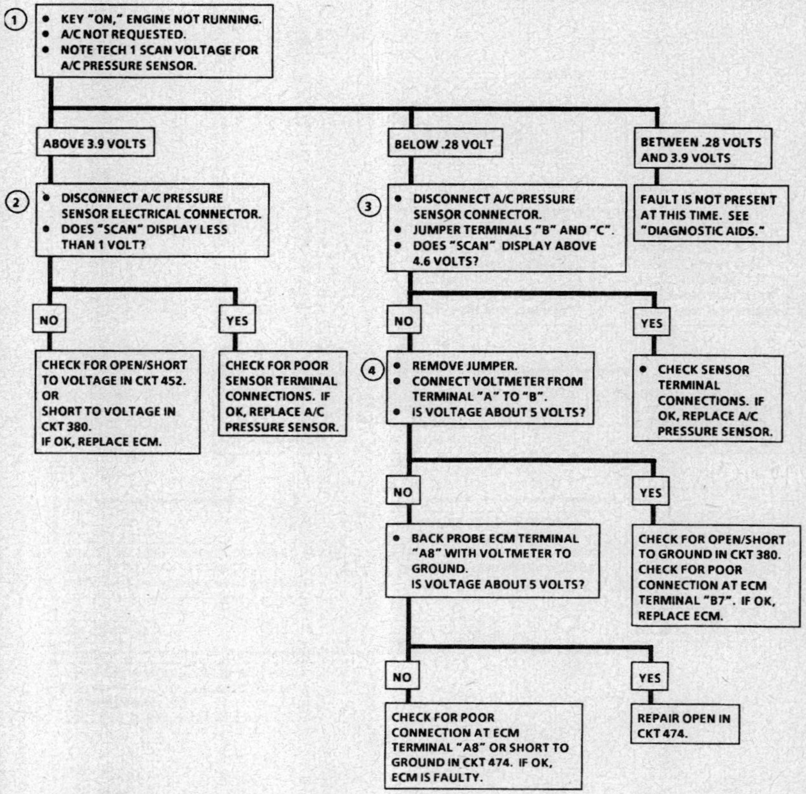

"AFTER REPAIRS," REFER TO CODE CRITERIA AND CONFIRM CODE DOES NOT RESET.

2.0L (VIN H) ENGINE — DIAGNOSTIC TROUBLE CODE CHART — 1993–94 SUNBIRD

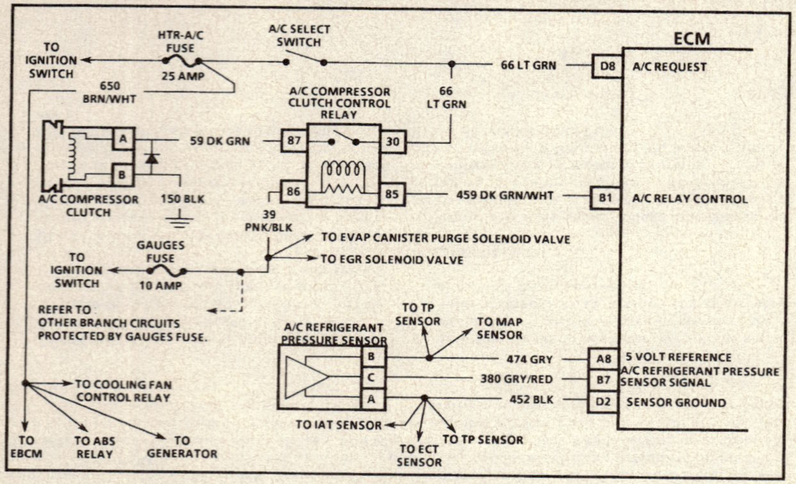

DTC 66

A/C REFRIGERANT PRESSURE SENSOR CIRCUIT 2.0L (VIN H) "J" CARLINE

Circuit Description:

The A/C refrigerant pressure sensor responds to changes in A/C refrigerant system high side pressure. This input indicates how much load the A/C compressor is putting on the engine and is one of the factors used by the ECM to determine IAC valve position for idle speed control. The circuit consists of a 5 volt reference and a ground, both provided by the ECM, and a signal line to the ECM. The signal is a voltage which is proportional to the pressure. The sensor's range of operation is 0 to 454 psi. At 0 psi, the signal will be about .1 volt, varying up to about 4.9 volts at 454 psi or above. DTC 66 sets if the voltage is above 4.9 volts (454 psi) or below .1 volt for 5 seconds or more. The A/C compressor clutch is disabled by the ECM if the fault is presently occurring, or if pressure is above or below calibrated values

Test Description: Number(s) below refer to circled number(s) on the diagnostic chart.

1. This step checks the voltage signal being received by the ECM from the A/C refrigerant pressure sensor. The normal operating range is between .1 volt and 4.9 volts.
2. Checks to see if the high voltage signal is from a shorted sensor or a short to voltage in the circuit. Normally, disconnecting the sensor would make a normal circuit go to near zero volts.
3. Checks to see if low voltage signal is from the sensor or the circuit. Jumpering the sensor signal CKT 380 to 5 volts, checks the circuit, connections, and ECM.
4. This step checks to see if the low voltage signal was due to an open in the sensor circuit or the 5 volts reference circuit since the prior step eliminated the pressure sensor.

Diagnostic Aids:

DTC 66 sets when the signal voltage falls outside the normal possible range of the sensor. If signal voltage is greater than .1 volt, but less than .3 volt, there could be a low refrigerant pressure problem with the A/C system.

2.0L (VIN H) ENGINE — DIAGNOSTIC TROUBLE CODE CHART — 1993–94 SUNBIRD

DTC 66

A/C REFRIGERANT PRESSURE SENSOR CIRCUIT 2.0L (VIN H) "J" CARLINE

NOTICE: TO INSURE PROPER DIAGNOSIS OF THE DTC 66 CIRCUIT, THE A/C PRESSURE SHOULD BE WITHIN LIMITS

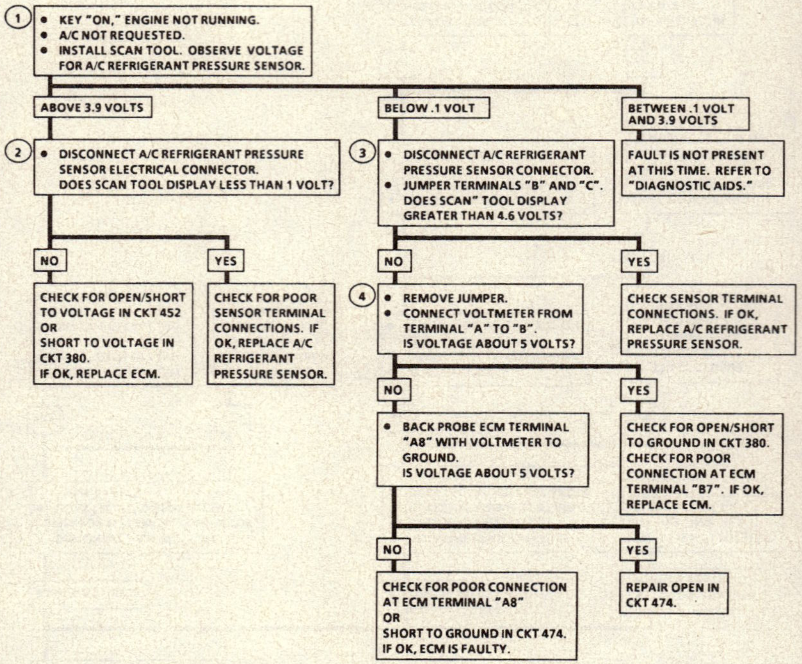

"AFTER REPAIRS," REFER TO DTC CRITERIA AND CONFIRM DTC DOES NOT RESET.

2.0L (VIN H) ENGINE — ECM SYMPTOM CHART — 1992 SUNBIRD

LT BLUE A-B 24 PIN ECM CONNECTOR

ECM PIN FUNCTION	CKT #	WIRE COLOR	COMPONENT/ CONNECTOR CAVITY	NORMAL VOLTAGE KEY "ON"	NORMAL VOLTAGE ENG RUN	CODES AFFECT.	POSSIBLE SYMPTOMS FROM FAULTY CIRCUIT
A2 EGR SOLENOID CONTROL	435	GRY	EGR SOLENOID "B"	B+	B+	32	(3) POSSIBLE SPARK KNOCK. (4) DEGRADED PERFORMANCE.
A3 CANISTER PURGE SOLENOID	428	DK GRN/YEL	CANISTER PURGE SOLENOID "B"	B+	B+		(3) POSSIBLE CANISTER LOADING (RAW FUEL ODOR).
A4 (A/T) TORQUE CONVERTER CLUTCH (TCC) SOLENOID	422	TAN/BLK	TCC SOLENOID "D"	(8)	(8)		(3) TCC INOPERATIVE, POOR FUEL ECONOMY. (4) TCC ENGAGES TOO SOON, POOR PERFORMANCE
A4 (M/T) SHIFT LIGHT CONTROL	456	TAN/BLK	23 PIN I/P CONNECTOR "F"	B+	B+		(3) SHIFT LIGHT INOPERATIVE. (4) SHIFT LIGHT "ON" AT ALL TIMES.
A5 "CHECK ENGINE" LIGHT CONTROL	419	BRN/WHT	23 PIN I/P CONNECTOR "B"	(8)	B+		(3) "CHECK ENGINE" LIGHT INOPERATIVE. (4) "CHECK ENGINE" LIGHT "ON" OR FLASHING -DEGRADED PERFORMANCE.
A6 TACHOMETER	121	WHT	BULKHEAD CONNECTOR "D6"				(5) TACHOMETER INOPERATIVE.
A7 IGNITION	439	PNK/BLK	BULKHEAD CONNECTOR "E6" & ECM FUSE 10 AMP	B+	B+		(3) ENGINE CRANKS, BUT WILL NOT RUN, NO DATA. (4) SAME AS OPEN, ECM 10 AMP FUSE MAY BE BLOWN.
A8 5 VOLT REF FOR MAP, TPS AND A/C PRESSURE SENSORS	474	GRY	MAP "C", TPS "A", AND A/C PRESSURE SENSOR "B"	5.0V	5.0V	33 (5) 21 (5)	(5) ENGINE CRANKS BUT WON'T START.
A9 & A10 KEEP ALIVE MEMORY (B +)	2	RED	POWER FEED CONNECTOR FROM BATTERY POSITIVE CABLE	B+	B+		(3) NO EFFECT IF A9 ONLY OR A10 ONLY ARE OPEN. CRANKS, BUT WILL NOT RUN, IF BOTH ARE OPEN. (4) IF EITHER A9 OR A10 ARE GROUNDED, FUSIBLE LINK FROM BATTERY POSITIVE CABLE WOULD FAIL, NO START.
A11 MAP SENSOR GROUND	455	PPL	MAP "A"	(8)	(8)		(3) DEGRADED PERFORMANCE.
A12 ECM GROUND	551	TAN/WHT	ENGINE GROUND	(8)	(8)		(3) REDUNDANT GROUND. NO EFFECT UNLESS IF CKT 450 OR ECM PIN D1 IS ALSO OPEN.

NOTICE: Voltage may vary due to low battery charge or other reasons, but should be very close. All voltage shown in the "ENG RUN" column are typical with engine at idle, closed throttle, normal operating temperature, park or neutral, system in "Closed Loop" all accessories "OFF," brake not applied.

(1) Increases with vehicle speed. (Measure on "AC volts" scale).
(2) Varies.
(3) Open circuit.
(4) Grounded circuit.
(5) Open or grounded circuit.
(6) Measures B + for 2 seconds after key "ON," then measures "0" volts.
(7) Varies with temperature.
(8) Less than .5 volt (500 mV).

2.0L (VIN H) ENGINE — ECM SYMPTOM CHART — 1992 SUNBIRD

LT BLUE A-B 24 PIN ECM CONNECTOR

ECM PIN FUNCTION	CKT #	WIRE COLOR	COMPONENT/ CONNECTOR CAVITY	NORMAL VOLTAGE KEY "ON"	NORMAL VOLTAGE ENG RUN	CODES AFFECT.	POSSIBLE SYMPTOMS FROM FAULTY CIRCUIT
B1 A/C COMPRESSOR CLUTCH RELAY CONTROL	459	DK GRN/WHT	A/C COMPRESSOR CLUTCH RELAY "2"	B+	B+		(3) NO A/C COOLING, A/C CLUTCH INOPERATIVE (4) SAME AS OPEN, GAUGES FUSE MAY BE BLOWN
B2 COOLING FAN RELAY CONTROL	335	DK GRN/WHT	COOLING FAN RELAY "2"	B+	B+		(3) ENGINE OVERHEATS, DETONATION (4) COOLING FAN RUNS AT ALL TIMES
B3 IDLE AIR CONTROL (IAC) "B" HIGH	443	LT GRN/WHT	IAC VALVE "B"	NOT USABLE	NOT USABLE		(5) INCORRECT, UNSTABLE IDLE, REFER TO CHART C2-C
B4 IAC "B" LOW	444	LT GRN/BLK	IAC VALVE "A"	NOT USABLE	NOT USABLE		(5) INCORRECT, UNSTABLE IDLE, REFER TO CHART C2-C
B5 IAC "A" LOW	442	LT BLU/BLK	IAC VALVE "C"	NOT USABLE	NOT USABLE		(5) INCORRECT, UNSTABLE IDLE, REFER TO CHART C2-C
B6 IAC "A" HIGH	441	LT BLU/WHT	IAC VALVE "D"	NOT USABLE	NOT USABLE		(5) INCORRECT, UNSTABLE IDLE, REFER TO CHART C2-C
B7 A/C PRESSURE SENSOR SIGNAL	380	GRY/RED	A/C PRESSURE SENSOR "C"	(2)	(2)	66 (5)	(5) NO A/C COOLING, A/C CLUTCH INOPERATIVE
B8 ALDL DIAGNOSTIC TEST TERMINAL	451	WHT/BLK	ALDL CONNECTOR "B"	5.0	5.0		(3) NO "FIELD SERVICE MODE" (4) "CHECK ENGINE" LIGHT FLASHES FIELD SERVICE MODE
B9 INTAKE AIR TEMP (IAT) SENSOR SIGNAL	472	TAN	IAT SENSOR "B"	1.3(7)	1.3(7)	23(5)	
B10 COOLANT TEMP SENSOR (CTS) SIGNAL	410	YEL	CTS "B"	2.0(7)	2.0(7)	14(5)	(5) POOR PERFORMANCE, ROUGH IDLE, HESITATION
B11 MANIFOLD ABSOLUTE PRESSURE (MAP) SENSOR SIGNAL	432	LT GRN	MAP SENSOR "B"	4.75	1.6	33(5)	(5) POOR PERFORMANCE, ROUGH IDLE
B12 THROTTLE POSITION SENSOR (TPS) SIGNAL	417	DK BLU	TPS "C"	6	6	21(5)	(5) INCORRECT IDLE, HESITATION

NOTICE: Voltage may vary due to low battery charge or other reasons, but should be very close. All voltages shown in the "ENG RUN" column are typical with engine at idle, closed throttle, normal operating temperature, park or neutral, system in "Closed Loop" all accessories "OFF," brake not applied.

(1) Increases with vehicle speed (Measure on "AC volts" scale).
(2) Varies
(3) Open circuit
(4) Grounded circuit
(5) Open or grounded circuit
(6) Measures B + for 2 seconds after key "ON," then measures "0" volts.
(7) Varies with temperature.
(8) Less than .5 volt (500mV.)

2.0L (VIN H) ENGINE — ECM SYMPTOM CHART — 1992 SUNBIRD

LT BLUE C-D 32 PIN ECM CONNECTOR							
ECM PIN FUNCTION	CKT #	WIRE COLOR	COMPONENT CONNECTOR CAVITY	NORMAL VOLTAGE KEY "ON"	NORMAL VOLTAGE ENG RUN	CODES AFFECT	POSSIBLE SYMPTOMS FROM FAULTY CIRCUIT
C1 EST "A" CYL'S 1 & 4	423	WHT	IGNITION COIL ASSEMBLY CONNECTOR "3"	B +	1.3-1.6V (2)		(5) POOR PERFORMANCE, CYL'S 1 & 4 NOT FIRING
C2 EST "B" CYL'S 2 & 3	485	BLK	IGNITION COIL ASSEMBLY CONNECTOR "4"	B +	1.3-1.6V (2)		(5) POOR PERFORMANCE, CYL 'S 2 & 3 NOT FIRING
C3 VEHICLE SPEED SENSOR SIGNAL (HIGH)	1232	LT BLU	VSS "A"			24(5)	(5) POOR FUEL ECONOMY, TCC DISENGAGED AT ALL TIMES. SPEEDOMETER INOPERATIVE
C4 VSS OUTPUT (4000 PPM)	389	DK GRN	23 PIN I/P CONNECTOR "A"	(1)	(1)		(5) INACCURATE SPEEDOMETER
C5 VSS SIGNAL (LOW)	1233	DK GRN/YEL	VSS "B"			24(5)	(5) POOR FUEL ECONOMY, TCC DISENGAGED AT ALL TIMES, SPEEDOMETER INOPERATIVE
C7 OXYGEN (O₂) SENSOR SIGNAL	412	PPL	OXYGEN(O₂) SENSOR	.01-.55V	.1-.9V	13 (3)	EXHAUST ODOR, POOR PERFORMANCE
C8 OXYGEN (O₂) SENSOR REF LOW	413	TAN	ENGINE GROUND	(8)	(8)	13(3)	EXHAUST ODOR, POOR PERFORMANCE
C9 IGNITION REFERENCE SHIELD	450	(BARE)	CRANKSHAFT POSITION SENSOR "3"	(8)	(8)	19	(3) POOR PERFORMANCE, MISFIRE
C10 FUEL PUMP RELAY CONTROL	465	DK GRN/WHT	FUEL PUMP RELAY "5"	(6)	B +		(3) LONG CRANKING TIME BEFORE ENGINE STARTS
C12 2 & 3 INJECTOR DRIVER	468	LT GRN	MAIN INJECTOR HARNESS CONNECTOR "C"	B +	(2)		(3) POOR PERFORMANCE (4) POSSIBLE DAMAGED FUEL INJECTOR OR BLOWN ECM FUSE
C13 2 & 3 INJECTOR GROUND	450	BLK/WHT	ENGINE GROUND	(8)	(8)		(3) POOR PERFORMANCE (2 & 3 FUEL INJECTORS NOT ENERGIZING)
C15 1 & 4 INJECTOR DRIVER	467	DK BLU	MAIN INJECTOR HARNESS CONNECTOR "A"	B +	(2)		(3) POOR PERFORMANCE (4) POSSIBLE DAMAGED FUEL INJECTOR OR BLOWN ECM FUSE
C16 1 & 4 INJECTOR GROUND	450	BLK/WHT	ENGINE GROUND	(8)	(8)		(3) POOR PERFORMANCE (1 & 4 FUEL INJECTORS NOT ENERGIZING)

NOTICE: Voltage may vary due to low battery charge or other reasons, but should be very close. All voltages shown in the "ENG RUN" column are typical with engine at idle, closed throttle, normal operating temperature, park or neutral, system in "Closed Loop" all accessories "OFF," brake not applied.

(1) Increases with vehicle speed. (Measure on "AC volts" scale).
(2) Varies
(3) Open circuit
(4) Grounded circuit
(5) Open or grounded circuit.
(6) Measures B + for 2 seconds after key "ON," then measures "0" volts.
(7) Varies with temperature.
(8) Less than .5 volt (500 mV).

2.0L (VIN H) ENGINE — ECM SYMPTOM CHART — 1992 SUNBIRD

LT BLUE C-D 32 PIN ECM CONNECTOR							
ECM PIN/FUNCTION	CKT #	WIRE COLOR	COMPONENT CONNECTOR CAVITY	NORMAL VOLTAGE KEY "ON"	NORMAL VOLTAGE ENG RUN	CODES AFFECT	POSSIBLE SYMPTOMS FROM FAULTY CIRCUIT
D1 ECM GROUND	450	BLK/WHT	ENGINE GROUND	(8)	(8)		(3) REDUNDANT GROUND. NO EFFECT. IF OTHER GROUND(S) ARE ALSO OPEN, CRANKS BUT WON'T RUN.
D2 GROUND FOR TPS, CTS, IAT AND A/C PRESSURE SENSOR	452	BLK	TPS "B", CTS "A", IAT "A", AND A/C PRESSURE SENSOR "A"	(8)	(8)	14 (3)	(3) VERY POOR PERFORMANCE, INCORRECT, UNSTABLE IDLE, EXHAUST ODOR.
D3 SERIAL DATA	461	ORN	ALDL CONNECTOR "M"	4.7V	4.7V		(5) NO DATA.
D5 PARK/NEUTRAL SWITCH	434	BLK/ORN	PARK/NEUTRAL SWITCH "A"	(8)	(8)		(5) INCORRECT IDLE.
D8 A/C REQUEST	66	LT GRN	A/C COMPRESSOR CLUTCH RELAY "1"	(8)	(8)		(3) NO A/C COOLING, A/C CLUTCH INOPERATIVE. (4) SAME AS OPEN, HTR-A/C FUSE 25 AMP COULD BE BLOWN.
D9 IGNITION REFERENCE LOW	484	WHT	CRANKSHAFT POSITION SENSOR "2"	(9)	(9)	19 (5)	(5) DEGRADED PERFORMANCE, CANISTER PURGE DISABLED - (RAW FUEL ODOR).
D10 IGNITION REFERENCE HIGH	483	LT GRN	CRANKSHAFT POSITION SENSOR "1"	(9)	(9)	19 (5)	(5) DEGRADED PERFORMANCE, CANISTER PURGE DISABLED - (RAW FUEL ODOR).

NOTICE: Voltage may vary due to low battery charge or other reasons, but should be very close. All voltages shown in the "ENG RUN" column are typical with engine at idle, closed throttle, normal operating temperature, park or neutral, system in "Closed Loop" all accessories "OFF," brake not applied.

(1) Increases with vehicle speed (Measure on "AC volts" scale).
(2) Varies.
(3) Open circuit.
(4) Grounded circuit.
(5) Open or grounded circuit.
(6) Measures B + for 2 seconds after key "ON," then measure "0" volts.
(7) Varies with temperature.
(8) Less than .5 volt (500 mV).
(9) See Code 19 Chart.

2.0L (VIN H) ENGINE — ECM SYMPTOM CHART — 1993–94 SUNBIRD

ECM Connector and Driveability Symptoms Identification

This ECM voltage chart is for use with a J 39200 to further aid in diagnosis. These voltages were derived from a known good vehicle. The voltages you get may vary due to low battery charge or other reasons, but they should be very close.

THE FOLLOWING CONDITIONS MUST BE MET BEFORE TESTING:
- Engine at operating temperature • "Closed Loop" • Engine idling (for "Engine Run" column)
- Test terminal not grounded • Scan tool not installed • Brake not applied.

ECM PIN FUNCTION	CKT #	WIRE COLOR	COMPONENT/ CONNECTOR CAVITY	NORMAL VOLTAGE KEY "ON"	NORMAL VOLTAGE ENG RUN	DTC(S) AFFECTED	POSSIBLE SYMPTOMS FROM FAULTY CIRCUIT
A2 EGR SOLENOID CONTROL	435	GRY	EGR SOLENOID VALVE "B"	B+	B+	32	(3) POSSIBLE SPARK KNOCK. (4) DEGRADED PERFORMANCE.
A3 CANISTER PURGE SOLENOID	428	DK GRN/YEL	CANISTER PURGE SOLENOID VALVE "B"	B+	B+		(3) POSSIBLE EVAP CANISTER LOADING (RAW FUEL ODOR).
A4 (A/T) TORQUE CONVERTER CLUTCH (TCC) SOLENOID	422	TAN/BLK	TCC SOLENOID "D"	(8)	(8)		(3) TCC INOPERATIVE, POOR FUEL ECONOMY. (4) TCC ENGAGES TOO SOON, POOR PERFORMANCE.
A4 (M/T) SHIFT LIGHT CONTROL	456	TAN/BLK	23 PIN I/P CONNECTOR "F"	B+	B+		(3) SHIFT LIGHT INOPERATIVE. (4) SHIFT LIGHT "ON" AT ALL TIMES.
A5 MIL (CHECK ENGINE) LIGHT CONTROL	419	BRN/WHT	23 PIN I/P CONNECTOR "B"	(8)	B+		(3) MIL (CHECK ENGINE) LIGHT INOPERATIVE. (4) MIL (CHECK ENGINE) LIGHT "ON" OR FLASHING -DEGRADED PERFORMANCE.
A6 TACHOMETER	121	WHT	BULKHEAD CONNECTOR "D6"				(5) TACHOMETER INOPERATIVE.
A7 IGNITION	439	PNK/BLK	BULKHEAD CONNECTOR "E6" & ECM FUSE 10 AMP	B+	B+		(3) ENGINE CRANKS, BUT WILL NOT RUN, NO DATA. (4) SAME AS OPEN, ECM 10 AMP FUSE MAY BE BLOWN.
A8 5 VOLT REF FOR MAP, TPS AND A/C PRESSURE SENSORS	474	GRY	MAP "C", TP SENSOR "A", AND A/C REFRIGERANT PRESSURE SENSOR "B"	5.0V	5.0V	33 (5) 21 (5)	(5) ENGINE CRANKS BUT WON'T START.
A9 & A10 KEEP ALIVE MEMORY (B +)	2	RED	POWER FEED CONNECTOR FROM BATTERY POSITIVE CABLE	B+	B+		(3) NO EFFECT IF "A9" ONLY OR "A10" ONLY ARE OPEN. CRANKS, BUT WILL NOT RUN, IF BOTH ARE OPEN. (4) IF EITHER "A9" OR "A10" ARE GROUNDED, FUSIBLE LINK FROM BATTERY POSITIVE CABLE WOULD FAIL. NO START.
A11 MAP SENSOR GROUND	455	PPL	MAP "A"	(8)	(8)		(3) DEGRADED PERFORMANCE.
A12 ECM GROUND	551	TAN/WHT	ENGINE GROUND	(8)	(8)		(3) REDUNDANT GROUND. NO EFFECT UNLESS IF CKT 450 OR ECM PIN "D1" IS ALSO OPEN.

(1) INCREASES WITH VEHICLE SPEED. (MEASURE ON "AC VOLTS" SCALE).
(2) VARIES.
(3) OPEN CIRCUIT.
(4) GROUNDED CIRCUIT.
(5) OPEN OR GROUNDED CIRCUIT.
(6) MEASURES B + FOR 2 SECONDS AFTER KEY "ON," THEN MEASURES "0" VOLTS.
(7) VARIES WITH TEMPERATURE.
(8) LESS THAN .5 VOLT (500 mV).

24 PIN A-B CONNECTOR 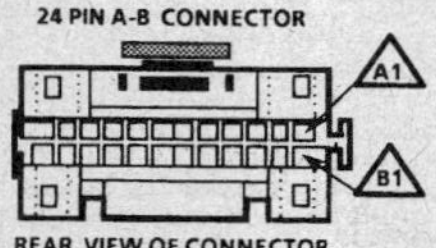
A1
B1
REAR VIEW OF CONNECTOR (LT BLU)

2.0L (VIN H) ENGINE — ECM SYMPTOM CHART — 1993–94 SUNBIRD

ECM Connector and Driveability Symptoms Identification

This ECM voltage chart is for use with a J 39200 to further aid in diagnosis. These voltages were derived from a known good vehicle. The voltages you get may vary due to low battery charge or other reasons, but they should be very close.

THE FOLLOWING CONDITIONS MUST BE MET BEFORE TESTING:
- Engine at operating temperature • Closed Loop • Engine idling (for "Engine Run" column)
- Test terminal not grounded • Scan tool not installed • Brake not applied.

ECM PIN FUNCTION	CKT #	WIRE COLOR	COMPONENT/ CONNECTOR CAVITY	NORMAL VOLTAGE KEY "ON"	NORMAL VOLTAGE ENG RUN	CODES AFFECT.	POSSIBLE SYMPTOMS FROM FAULTY CIRCUIT
B1 A/C COMPRESSOR CLUTCH RELAY CONTROL	459	DK GRN/WHT	A/C COMPRESSOR CLUTCH RELAY "85"	B+	B+		(3) NO A/C COOLING, A/C CLUTCH INOPERATIVE (4) SAME AS OPEN, GAUGES FUSE MAY BE OPEN
B2 COOLING FAN RELAY CONTROL	335	DK GRN/WHT	COOLING FAN RELAY "85"	B+	B+		(3) ENGINE OVERHEATS, DETONATION (4) COOLING FAN RUNS AT ALL TIMES
B3 IDLE AIR CONTROL (IAC) "B" HIGH	443	LT GRN/WHT	IAC VALVE "B"	NOT USABLE	NOT USABLE		(5) INCORRECT, UNSTABLE IDLE, REFER TO CHART C2-C
B4 IAC "B" LOW	444	LT GRN/BLK	IAC VALVE "A"	NOT USABLE	NOT USABLE		(5) INCORRECT, UNSTABLE IDLE, REFER TO CHART C2-C
B5 IAC "A" LOW	442	LT BLU/BLK	IAC VALVE "C"	NOT USABLE	NOT USABLE		(5) INCORRECT, UNSTABLE IDLE, REFER TO CHART C2-C
B6 IAC "A" HIGH	441	LT BLU/WHT	IAC VALVE "D"	NOT USABLE	NOT USABLE		(5) INCORRECT, UNSTABLE IDLE, REFER TO CHART C2-C
B7 A/C REFRIGERANT PRESSURE SENSOR SIGNAL	380	GRY/RED	A/C REFRIGERANT PRESSURE SENSOR "C"	(2)	(2)	66 (5)	(5) NO A/C COOLING, A/C CLUTCH INOPERATIVE
B8 DATA LINK DIAGNOSTIC TEST TERMINAL	451	WHT/BLK	DATA LINK CONNECTOR "B"	5.0	5.0		(3) NO "FIELD SERVICE MODE" (4) MIL (CHECK ENGINE) FLASHES FIELD SERVICE MODE
B9 INTAKE AIR TEMP (IAT) SENSOR SIGNAL	472	TAN	IAT SENSOR "B"	1.3(7)	1.3(7)	23(5)	(5) POOR COLD DRIVEABILITY
B10 ENGINE COOLANT TEMP (ECT) SENSOR SIGNAL	410	YEL	ECT SENSOR "B"	2.0(7)	2.0(7)	14(5)	(5) POOR PERFORMANCE, ROUGH IDLE, HESITATION
B11 MANIFOLD ABSOLUTE PRESSURE (MAP) SENSOR SIGNAL	432	LT GRN	MAP SENSOR "B"	4.75	1.6	33(5)	(5) POOR PERFORMANCE, ROUGH IDLE
B12 THROTTLE POSITION (TP) SENSOR SIGNAL	417	DK BLU	TP SENSOR "C"	.6	.6	21(5)	(5) INCORRECT IDLE, HESITATION

(1) Increases with vehicle speed (Measure on "AC volts" scale).
(2) Varies
(3) Open circuit
(4) Grounded circuit
(5) Open or grounded circuit
(6) Measures B + for 2 seconds after key "ON," then measures "0" volts.
(7) Varies with temperature.,
(8) Less than .5 volt (500mV.)

24 PIN A-B CONNECTOR
32 PIN C-D CONNECTOR 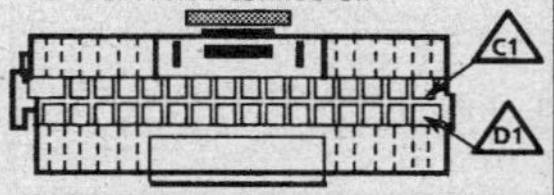
C1
D1
REAR VIEW OF CONNECTOR (LT BLU)

2.0L (VIN H) ENGINE — ECM SYMPTOM CHART — 1993–94 SUNBIRD

ECM Connector and Driveability Symptoms Identification

This ECM voltage chart is for use with a J 39200 to further aid in diagnosis. These voltages were derived from a known good vehicle. The voltages you get may vary due to low battery charge or other reasons, but they should be very close.

THE FOLLOWING CONDITIONS MUST BE MET BEFORE TESTING:

- Engine at operating temperature   • Closed Loop   • Engine idling (for "Engine Run" column)
- Test terminal <u>not</u> grounded   • Scan tool <u>not</u> installed   • Brake <u>not</u> applied.

ECM PIN FUNCTION	CKT #	WIRE COLOR	COMPONENT CONNECTOR CAVITY	NORMAL VOLTAGE KEY "ON"	NORMAL VOLTAGE ENG RUN	DTC(s) AFFECTED	POSSIBLE SYMPTOMS FROM FAULTY CIRCUIT
C1 IGNITION CONTROL "A" CYL'S 1 & 4	423	WHT	IGNITION COIL ASSEMBLY CONNECTOR "3"	B+	1.3-1.6V (2)		(5) POOR PERFORMANCE, CYL'S 1 & 4 NOT FIRING
C2 IGNITION CONTROL "B" CYL'S 2 & 3	485	BLK	IGNITION COIL ASSEMBLY CONNECTOR "4"	B+	1.3-1.6V (2)		(5) POOR PERFORMANCE, CYL 'S 2 & 3 NOT FIRING
C3 VEHICLE SPEED SENSOR SIGNAL (HIGH)	1232	LT BLU	VSS "A"			24(5)	(5) POOR FUEL ECONOMY, TCC DISENGAGED AT ALL TIMES. SPEEDOMETER INOPERATIVE
C4 VSS OUTPUT (4000 PPM)	389	DK GRN	23 PIN I/P CONNECTOR "A"	(1)	(1)		(5) INACCURATE SPEEDOMETER
C5 VSS SIGNAL (LOW)	1233	DK GRN/YEL	VSS "B"			24(5)	(5) POOR FUEL ECONOMY, TCC DISENGAGED AT ALL TIMES, SPEEDOMETER INOPERATIVE
C7 OXYGEN SENSOR SIGNAL	412	PPL	OXYGEN SENSOR	.01-.55V	.1-.9V	13 (3)	EXHAUST ODOR, POOR PERFORMANCE
C8 OXYGEN SENSOR REF LOW	413	TAN	ENGINE GROUND	(8)	(8)	13(3)	EXHAUST ODOR, POOR PERFORMANCE
C9 IGNITION REFERENCE SHIELD	450	(BARE)	CRANKSHAFT POSITION SENSOR "3"	(8)	(8)	19	(3) POOR PERFORMANCE, MISFIRE
C10 FUEL PUMP RELAY CONTROL	465	DK GRN/WHT	FUEL PUMP RELAY "86"	(6)	B+		(3) LONG CRANKING TIME BEFORE ENGINE STARTS
C12 2 & 3 INJECTOR DRIVER	468	LT GRN	MAIN INJECTOR HARNESS CONNECTOR "C"	B+	(2)		(3) POOR PERFORMANCE (4) POSSIBLE DAMAGED FUEL INJECTOR OR BLOWN ECM FUSE
C13 2 & 3 INJECTOR GROUND	450	BLK/WHT	ENGINE GROUND	(8)	(8)		(3) POOR PERFORMANCE (2 & 3 FUEL INJECTORS NOT ENERGIZING)
C15 1 & 4 INJECTOR DRIVER	467	DK BLU	MAIN INJECTOR HARNESS CONNECTOR "A"	B+	(2)		(3) POOR PERFORMANCE (4) POSSIBLE DAMAGED FUEL INJECTOR OR BLOWN ECM FUSE
C16 1 & 4 INJECTOR GROUND	450	BLK/WHT	ENGINE GROUND	(8)	(8)		(3) POOR PERFORMANCE (1 & 4 FUEL INJECTORS NOT ENERGIZING)

(1) Increases with vehicle speed. (Measure on "A/C volts" scale).
(2) Varies
(3) Open circuit
(4) Grounded circuit
(5) Open or grounded circuit.
(6) Measures B+ for 2 seconds after key "ON," then measures "0" volts.
(7) Varies with temperature.
(8) Less than .5 volt (500 mV).

32 PIN C-D CONNECTOR 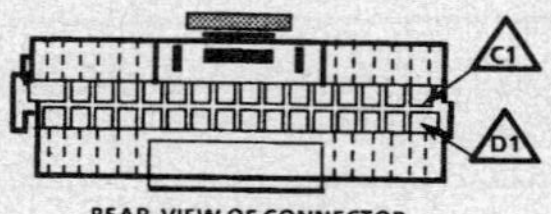
REAR VIEW OF CONNECTOR (LT BLUE)

2.0L (VIN H) ENGINE — ECM SYMPTOM CHART — 1993–94 SUNBIRD

ECM Connector and Driveability Symptoms Identification

This ECM voltage chart is for use with a J 39200 to further aid in diagnosis. These voltages were derived from a known good vehicle. The voltages you get may vary due to low battery charge or other reasons, but they should be very close.

THE FOLLOWING CONDITIONS MUST BE MET BEFORE TESTING:

- Engine at operating temperature   • Closed Loop   • Engine idling (for "Engine Run" column)
- Test terminal <u>not</u> grounded   • Scan tool <u>not</u> installed   • Brake <u>not</u> applied.

ECM PIN/FUNCTION	CKT #	WIRE COLOR	COMPONENT CONNECTOR CAVITY	NORMAL VOLTAGE KEY "ON"	NORMAL VOLTAGE ENG RUN	CODES AFFECT	POSSIBLE SYMPTOMS FROM FAULTY CIRCUIT
D1 ECM GROUND	450	BLK/WHT	ENGINE GROUND	(8)	(8)		(3) REDUNDANT GROUND. NO EFFECT. IF OTHER GROUND(S) ARE ALSO OPEN, CRANKS BUT WON'T RUN.
D2 GROUND FOR TP, CTS, ECT AND A/C REFRIGERANT PRESSURE SENSORS	452	BLK	TP SENSOR "B", CTS "A", ECT SENSOR "A", AND A/C REFRIGERANT PRESSURE SENSOR "A"	(8)	(8)	14(3)	(3) VERY POOR PERFORMANCE, INCORRECT, UNSTABLE IDLE, EXHAUST ODOR.
D3 SERIAL DATA	461	ORN	DATA LINK CONNECTOR "M"	4.7V	4.7V		(5) NO DATA.
D5 PARK/NEUTRAL SWITCH	434	BLK/ORN	PARK/NEUTRAL SWITCH "A"	(8)	(8)		(5) INCORRECT IDLE.
D8 A/C REQUEST	66	LT GRN	A/C COMPRESSOR CLUTCH RELAY "30"	(8)	(8)		(3) NO A/C COOLING, A/C CLUTCH INOPERATIVE. (4) SAME AS OPEN, HTR-A/C FUSE 25 AMP COULD BE BLOWN.
D9 IGNITION REFERENCE LOW	484	WHT	CRANKSHAFT POSITION SENSOR "2"	(9)	(9)	19 (5)	(5) DEGRADED PERFORMANCE, CANISTER PURGE DISABLED - (RAW FUEL ODOR).
D10 IGNITION REFERENCE HIGH	483	LT GRN	CRANKSHAFT POSITION SENSOR "1"	(9)	(9)	19 (5)	(5) DEGRADED PERFORMANCE, CANISTER PURGE DISABLED - (RAW FUEL ODOR).

(1) Increases with vehicle speed (Measure on "A/C volts" scale).
(2) Varies.
(3) Open circuit.
(4) Grounded circuit.
(5) Open or grounded circuit.
(6) Measures B+ for 2 seconds after key "ON," then measure "0" volts.
(7) Varies with temperature.

32 PIN C-D CONNECTOR 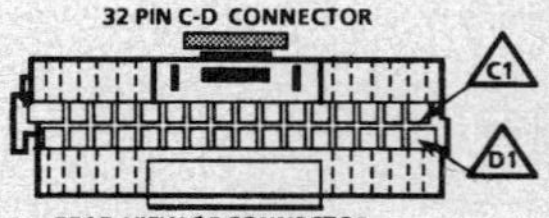
REAR VIEW OF CONNECTOR (LT BLUE)

2.0L (VIN H) ENGINE — COMPONENT DIAGNOSTIC CHART — SUNBIRD

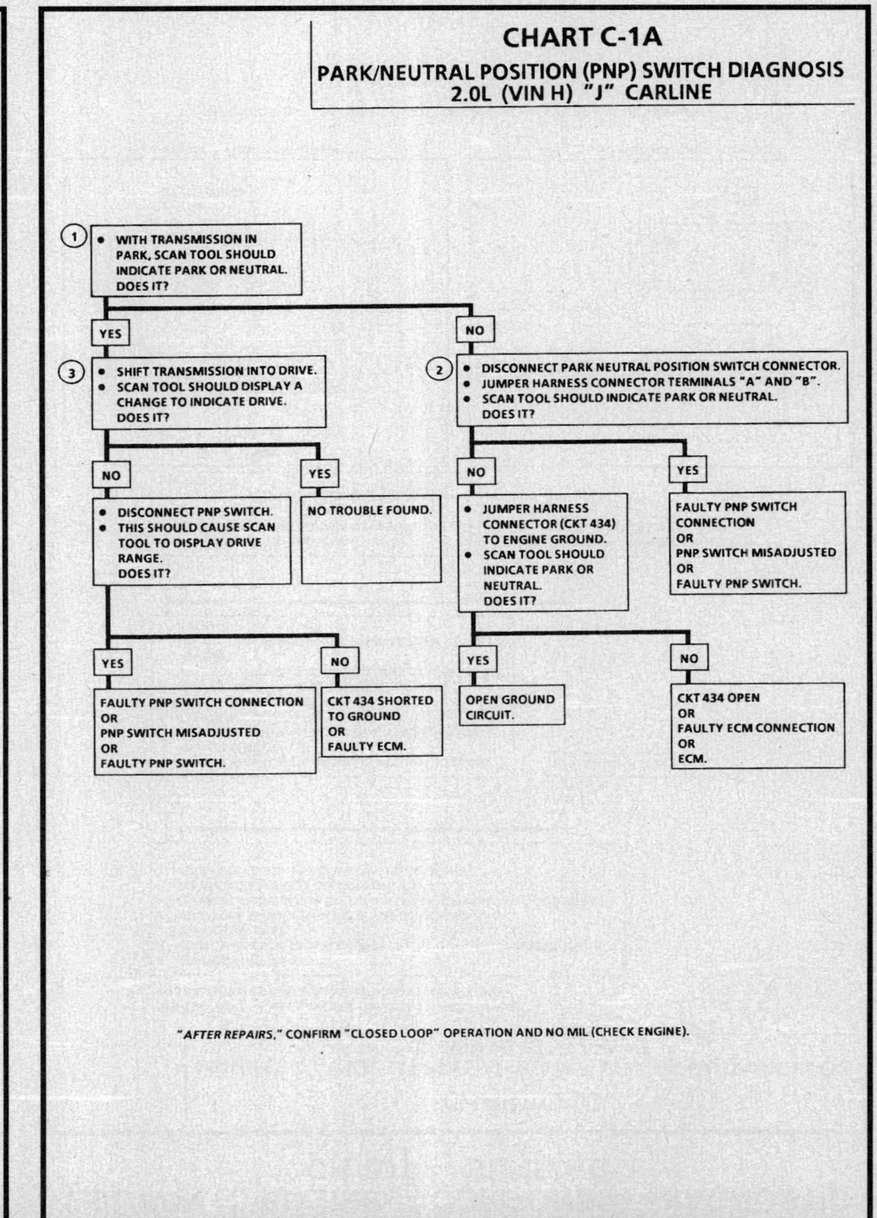

CHART C-1A
PARK/NEUTRAL POSITION (PNP) SWITCH DIAGNOSIS
2.0L (VIN H) "J" CARLINE

Circuit Description:

The Park/Neutral Position (PNP) switch contacts are a part of the neutral start switch and are closed to ground in park or neutral and open in drive ranges.

The ECM supplies ignition voltage through a current limiting resistor to CKT 434 and senses a closed switch when the voltage on CKT 434 drops to less than one volt.

The ECM uses the PNP signal as one of the inputs to control idle air control and VSS diagnostics.

Test Description: Number(s) below refer to circled number(s) on the diagnostic chart.

1. Checks for a closed switch to ground in park position. Different makes of scan tools will read PNP switch differently. Refer to tool operations manual for type of display used.
2. Checks for an open switch in drive range.
3. Be sure scan tool indicates drive, even while wiggling shifter to test for an intermittent or misadjusted switch in drive range.

2.0L (VIN H) ENGINE — COMPONENT DIAGNOSTIC CHART — SUNBIRD

CHART C-1A
PARK/NEUTRAL POSITION (PNP) SWITCH DIAGNOSIS
2.0L (VIN H) "J" CARLINE

"AFTER REPAIRS," CONFIRM "CLOSED LOOP" OPERATION AND NO MIL (CHECK ENGINE).

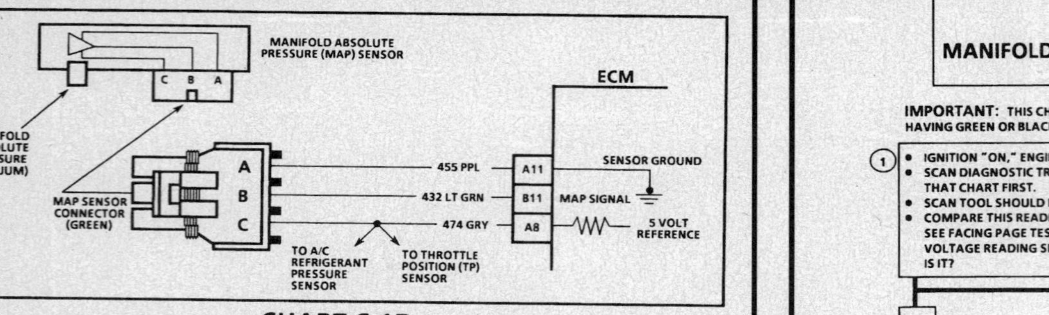

CHART C-1D
MANIFOLD ABSOLUTE PRESSURE (MAP) SENSOR OUTPUT CHECK
2.0L (VIN H) "J" CARLINE

Circuit Description:

The Manifold Absolute Pressure (MAP) sensor measures the changes in the intake manifold pressure which result from engine load (intake manifold vacuum) and RPM changes, and converts these into a voltage output. The ECM sends a 5 volt reference voltage to the MAP sensor. As the manifold pressure changes, the output voltage of the sensor also changes. By monitoring the sensor output voltage, the ECM knows the manifold pressure. A lower pressure (low voltage) output voltage will be about 1 - 2 volts at idle. While higher pressure (high voltage) output voltage will be about 4 - 4.8 at Wide Open Throttle (WOT). The MAP sensor is also used, under certain conditions, to measure barometric pressure, allowing the ECM to make adjustments for altitude changes. The ECM uses the MAP sensor to control fuel delivery and ignition timing.

Test Description: Number(s) below refer to circled number(s) on the diagnostic chart.

Important
- Be sure to use the same diagnostic test equipment for all measurements.

1. When comparing scan readings to a known good vehicle, it is important to compare vehicles that use a MAP sensor having the same color insert and the same "Hot Stamped" number. Refer to Figures 1 and 2
2. Applying 34 kPa (10" Hg) vacuum to the MAP sensor should result in voltage readings of 1.5 to 2.1 volts less than the voltage in Step 1. Upon applying vacuum to the sensor, the change in voltage should be instantaneous. A slow voltage change indicates a faulty sensor.

3. Check vacuum source to sensor for leaking or restriction. Be sure that no other vacuum devices are connected to the MAP vacuum source.

NOTICE: Make sure electrical connector remains securely fastened.

4. Remove sensor from its mounting and twist sensor **(by hand only)** to check for intermittent connection. Output changes greater than .10 volt indicate a faulty sensor or connection. If OK, replace sensor.

CHART C-1D
MANIFOLD ABSOLUTE PRESSURE (MAP) SENSOR OUTPUT CHECK
2.0L (VIN H) "J" CARLINE

IMPORTANT: THIS CHART ONLY APPLIES TO MAP SENSORS HAVING GREEN OR BLACK COLOR KEY INSERT (SEE BELOW).

1
- IGNITION "ON," ENGINE "OFF."
- SCAN DIAGNOSTIC TROUBLE CODES DTC(S). IF DTC 33 IS PRESENT, USE THAT CHART FIRST.
- SCAN TOOL SHOULD INDICATE A MAP SENSOR VOLTAGE.
- COMPARE THIS READING WITH THE READING OF A KNOWN GOOD VEHICLE. SEE FACING PAGE TEST DESCRIPTION, STEP 1. VOLTAGE READING SHOULD BE WITHIN ± .4 VOLT. IS IT?

YES →

NO → REPLACE MAP SENSOR.

2
- REMOVE MAP SENSOR AND PLUG VACUUM PORT ON INTAKE MANIFOLD.
- CONNECT A HAND VACUUM PUMP TO MAP SENSOR.
- START ENGINE.
- NOTE MAP SENSOR VOLTAGE.
- APPLY 34 kPa (10" Hg) OF VACUUM AND NOTE VOLTAGE CHANGE. SUBTRACT SECOND READING FROM THE FIRST. VOLTAGE VALUE SHOULD BE GREATER THAN 1.5 VOLTS. IS IT?

YES →

NO →

3 NO TROUBLE FOUND. CHECK MAP SENSOR VACUUM SOURCE FOR LEAKAGE OR RESTRICTION. OR BE SURE THIS SOURCE SUPPLIES VACUUM TO MAP SENSOR ONLY.

4 CHECK MAP SENSOR CONNECTION. IF OK, REPLACE MAP SENSOR.

Figure 1 - Color Key Insert

Figure 2 - Hot Stamped Number

"AFTER REPAIRS," CONFIRM "CLOSED LOOP" OPERATION AND NO MIL (CHECK ENGINE).

2.0L (VIN H) ENGINE — COMPONENT DIAGNOSTIC CHART — SUNBIRD

CHART C-2A

INJECTOR BALANCE TEST
2.0L (VIN H) "J" CARLINE

The injector balance tester is a tool used to turn the injector on for a precise amount of time, thus spraying a measured amount of fuel into the manifold. This causes a drop in fuel rail pressure that we can record and compare between each injector. All injectors should have the same amount of pressure drop (± 10 kPa). Any injector with a pressure drop that is 10 kPa (or more) greater or less than the average drop of the other injectors should be considered faulty and replaced.

STEP 1

Engine "cool down" period (10 minutes) is necessary to avoid irregular readings due to "Hot Soak" fuel boiling. Relieve fuel pressure in the fuel rail using the "fuel pressure relief procedure" described previously
With ignition "OFF" connect fuel gauge J 29658-C or equivalent (refer to CHART A-7 for connecting fuel pressure gage).

Disconnect harness connectors at all injectors, and connect injector tester J 34730-3, or equivalent, to one injector. Ignition must be "OFF" at least 10 seconds to complete ECM shutdown cycle. Fuel pump should run about 2 seconds after ignition is turned "ON." At this point, insert clear tubing attached to vent valve into a suitable container and bleed air from gauge and hose to insure accurate gauge operation. Repeat this step until all air is bled from gauge.

STEP 2

Turn ignition "OFF" for 10 seconds and then "ON" again to get fuel pressure to its maximum. Record this initial pressure reading. Energize tester one time and note pressure drop at its lowest point. (Disregard any slight pressure increase after drop hits low point.) By subtracting this second pressure reading from the initial pressure, we have the actual amount of injector pressure drop.

STEP 3

Repeat Step 2 on each injector and compare the amount of drop. Usually, good injectors will have virtually the same drop. Retest any injector that has a pressure difference of 10 kPa, either more or less than the average of the other injectors on the engine. Replace any injector that also fails the retest. If the pressure drop of all injectors is within 10 kPa of this average, the injectors appear to be flowing properly.

NOTICE: The entire test should not be repeated more than once without running the engine to prevent flooding. (This includes any retest on faulty injectors.)

2.0L (VIN H) ENGINE — COMPONENT DIAGNOSTIC CHART — SUNBIRD

NOTICE: The entire test should <u>NOT</u> be repeated more than once without running the engine to prevent flooding. (This includes any retest on faulty injectors.)

CHART C-2A

INJECTOR BALANCE TEST
2.0L (VIN H) "J" CARLINE

The fuel pressure test   Chart A-7, should be completed prior to this test.

Step 1. If engine is at operating temperature, allow a 10 minute "cool down" period then connect fuel pressure gage and injector tester.
1. Ignition "OFF."
2. Connect fuel pressure gage and injector tester.
3. Ignition "ON."
4. Bleed off air in gage. Repeat until all air is bled from gauge.

Step 2. Run test:
1. Ignition "OFF" for 10 seconds.
2. Ignition "ON." Record gage pressure. (Pressure must hold steady, if not, see "Fuel System Diagnosis," CHART A-7.
3. Turn injector on, by depressing button on injector tester, and note pressure at the instant the gauge needle stops.

Step 3.
1. Repeat Step 2 on all injectors and record pressure drop on each. Retest injectors that appear faulty (Any injectors that have a 10 kPa (1.5 psi) difference, either more or less, in pressure from the average).

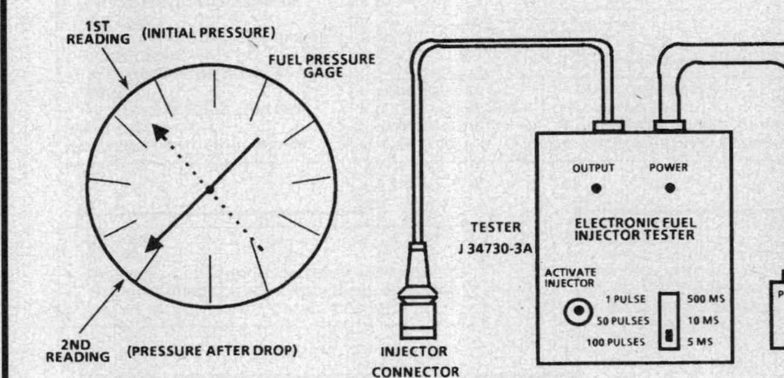

EXAMPLE

CYLINDER	1	2	3
1ST READING	293 kPa (43 psi)	293 kPa (43 psi)	293 kPa (43 psi)
2ND READING	131 kPa (19 psi)	115 kPa (17 psi)	145 kPa (21 psi)
AMOUNT OF DROP	162 kPa (24 psi)	178 kPa (26 psi)	148 kPa (21 psi)
	OK	FAULTY, RICH (TOO MUCH FUEL DROP)	FAULTY, LEAN (TOO LITTLE FUEL DROP)

2.0L (VIN H) ENGINE — COMPONENT DIAGNOSTIC CHART — SUNBIRD

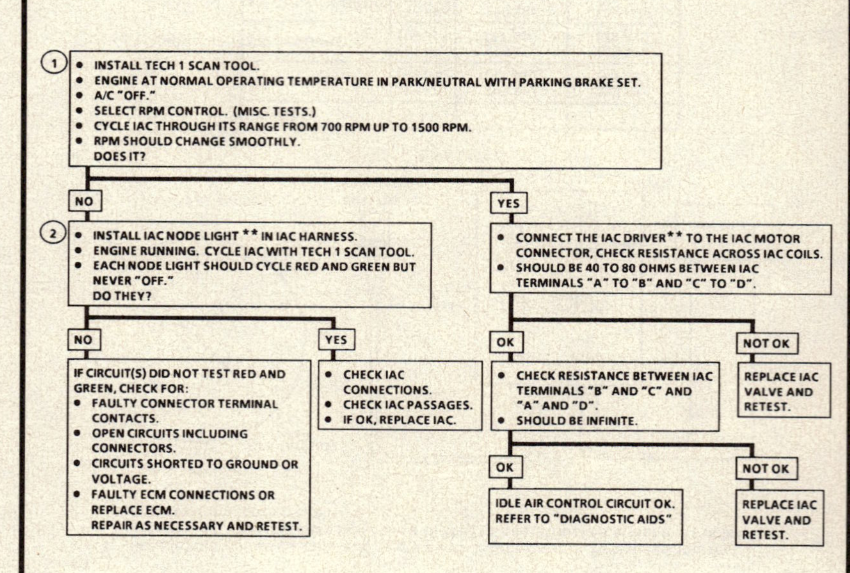

CHART C-2C
IDLE AIR CONTROL (IAC) SYSTEM CHECK
2.0L (VIN H) "J" CARLINE

Circuit Description:

The ECM controls engine idle speed with the IAC valve. To increase idle speed, the ECM retracts the IAC valve pintle away from its seat, allowing more air to bypass the throttle bore. To decrease idle speed, it extends the IAC valve pintle towards its seat, reducing bypass air flow. A Tech 1 scan tool will read the ECM commands to the IAC valve in counts. Higher the counts indicate more air bypass (higher idle). The lower the counts indicate less air is allowed to bypass (lower idle).

Test Description: Number(s) below refer to circled numbers on the diagnostic chart.

1. The Tech 1 RPM control mode is used to extend and retract the IAC valve. The valve should move smoothly within the specified range. If the idle speed is commanded (IAC extended) too low (below 700 RPM), the engine may stall. This may be normal and would not indicate a problem. Retracting the IAC beyond its controlled range (above 1500 RPM) will cause a delay before the RPMs start dropping. This too is normal.

2. This test uses the Tech 1 to command the IAC controlled idle speed. The ECM issues commands to obtain commanded idle speed. The node lights each should flash red and green to indicate a good circuit as the ECM issues commands. While the sequence of color is not important if either light is "OFF" or does not flash red and green, check the circuits for faults, beginning with poor terminal contacts.

Diagnostic Aids:

A slow, unstable, or fast idle may be caused by a non-IAC system problem that cannot be overcome by the IAC valve. Out of control range IAC scan tool counts will be above 60 if idle is too low, and zero counts if idle is too high. The following checks should be made to repair a non-IAC system problem:

- Vacuum Leak (High Idle). - If idle is too high, stop the engine. Fully extend (low) IAC with tester. Start engine. If idle speed is above 800 RPM, locate and correct vacuum leak including crankshaft ventilation system. Also, check for binding of throttle blade or linkage.
- System Too Lean (High Air/Fuel Ratio). - The idle speed may be too high or too low. Engine speed may vary up and down and disconnecting the IAC valve does not help. DTC 44 may be set.
  Scanned Oxygen Sensor (O2S) voltage will be less than 300 mV (.3 volt). Check for low regulated fuel pressure, water in the fuel or a restricted injector.
- System Too Rich (Low Air/Fuel Ratio). - The idle speed will be too low. Scan tool IAC counts will usually be above 80. System is obviously rich and may exhibit black smoke in exhaust. Scanned oxygen sensor voltage will be fixed above 800 mV (.8 volt).
  Check for high fuel pressure, leaking or sticking injector. Silicone contaminated O2S scanned voltage will be slow to respond.
- Throttle Body. - Remove IAC valve and inspect bore for foreign material.
- IAC Valve Electrical Connections. - IAC valve connections should be carefully checked for proper contact.
- Crankcase Ventilation Valve. - An incorrect or faulty crankcase ventilation valve may result in an incorrect idle speed.

- If intermittent poor driveability or idle symptoms are resolved by disconnecting the IAC, carefully recheck connections, valve terminal resistance, or replace IAC.

2.0L (VIN H) ENGINE — COMPONENT DIAGNOSTIC CHART — SUNBIRD

CHART C-2C
IDLE AIR CONTROL (IAC) SYSTEM CHECK
2.0L (VIN H) "J" CARLINE

(1)
- INSTALL TECH 1 SCAN TOOL.
- ENGINE AT NORMAL OPERATING TEMPERATURE IN PARK/NEUTRAL WITH PARKING BRAKE SET.
- A/C "OFF."
- SELECT RPM CONTROL. (MISC. TESTS.)
- CYCLE IAC THROUGH ITS RANGE FROM 700 RPM UP TO 1500 RPM.
- RPM SHOULD CHANGE SMOOTHLY.
- DOES IT?

NO →

(2)
- INSTALL IAC NODE LIGHT ** IN IAC HARNESS.
- ENGINE RUNNING. CYCLE IAC WITH TECH 1 SCAN TOOL.
- EACH NODE LIGHT SHOULD CYCLE RED AND GREEN BUT NEVER "OFF."
- DO THEY?

YES →
- CONNECT THE IAC DRIVER** TO THE IAC MOTOR CONNECTOR, CHECK RESISTANCE ACROSS IAC COILS.
- SHOULD BE 40 TO 80 OHMS BETWEEN IAC TERMINALS "A" TO "B" AND "C" TO "D".

NO →
IF CIRCUIT(S) DID NOT TEST RED AND GREEN, CHECK FOR:
- FAULTY CONNECTOR TERMINAL CONTACTS.
- OPEN CIRCUITS INCLUDING CONNECTORS.
- CIRCUITS SHORTED TO GROUND OR VOLTAGE.
- FAULTY ECM CONNECTIONS OR REPLACE ECM.
REPAIR AS NECESSARY AND RETEST.

YES →
- CHECK IAC CONNECTIONS.
- CHECK IAC PASSAGES.
- IF OK, REPLACE IAC.

OK →
- CHECK RESISTANCE BETWEEN IAC TERMINALS "B" AND "C" AND "A" AND "D".
- SHOULD BE INFINITE.

NOT OK →
REPLACE IAC VALVE AND RETEST.

OK →
IDLE AIR CONTROL CIRCUIT OK. REFER TO "DIAGNOSTIC AIDS"

NOT OK →
REPLACE IAC VALVE AND RETEST.

** IAC DRIVER AND NODE LIGHT REQUIRED KIT
222-L FROM: CONCEPT TECHNOLOGY, INC.
J 37027 FROM: KENT-MOORE, INC.

CLEAR DIAGNOSTIC TROUBLE CODES, CONFIRM "CLOSED LOOP" OPERATION, NO MALFUNCTION INDICATOR LAMP (MIL) "SERVICE ENGINE SOON," PERFORM IAC RESET PROCEDURE AND VERIFY CONTROLLED IDLE SPEED IS CORRECT.

"AFTER REPAIRS," CONFIRM "CLOSED LOOP" OPERATION AND NO MIL (CHECK ENGINE).

2.0L (VIN H) ENGINE — COMPONENT DIAGNOSTIC CHART — SUNBIRD

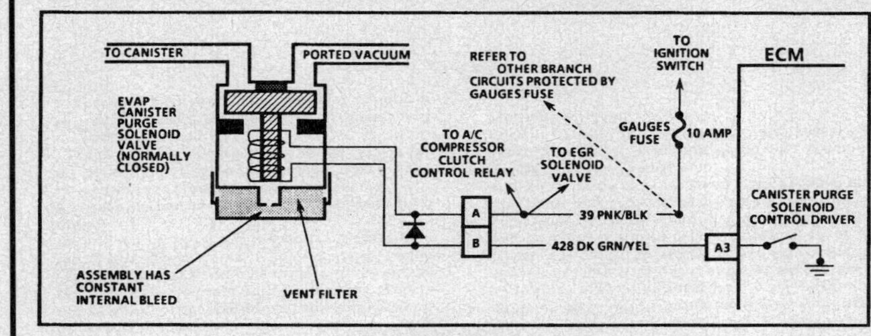

CHART C-3
EVAP CANISTER PURGE SOLENOID VALVE DIAGNOSIS
2.0L (VIN H) "J" CARLINE

Circuit Description:

Canister purge is controlled by a solenoid that allows manifold and/or ported vacuum to purge the EVAP canister when energized. The Engine Control Module (ECM) supplies a ground to energize the solenoid (purge "ON"). The purge solenoid control by the ECM is pulse width modulated (turned "ON" and "OFF" several times a second). The duty cycle (pulse width) is determined by "Closed Loop" feedback from the Oxygen Sensor (O2S). The duty cycle is calculated by the ECM and the output commanded when the following conditions have been met:

- Engine run time after start more than 65 seconds.
- Engine coolant temperature above 56°C (133°F).

Also, if the diagnostic test terminal is grounded with the ignition "ON," engine not running, the canister purge solenoid valve is energized (purge "ON").

Test Description: Number(s) below refer to circled number(s) on the diagnostic chart.
1. Checks to see if the solenoid is opened or closed. The solenoid is normally de-energized in this step, so it should be closed.
2. Checks to determine if solenoid was open due to electrical circuit problem or defective solenoid.
3. Completes functional check by grounding test terminal. This should normally energize the solenoid opening the valve which should allow the vacuum to drop (purge "ON").

Diagnostic Aids:

Make a visual check of vacuum hose(s). Check throttle body for possible cracked, broken, or plugged vacuum block. Check engine for possible mechanical problem.

2.0L (VIN H) ENGINE — COMPONENT DIAGNOSTIC CHART — SUNBIRD

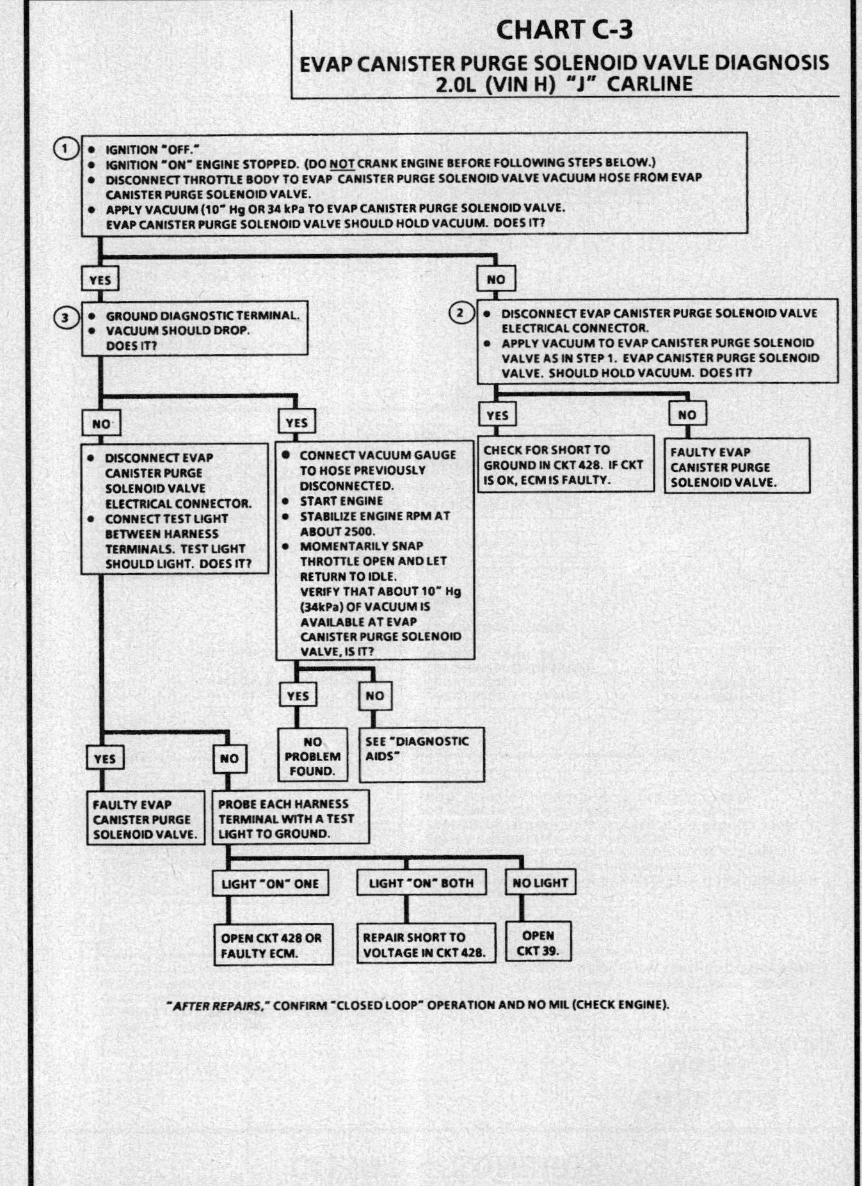

2.0L (VIN H) ENGINE — COMPONENT DIAGNOSTIC CHART — SUNBIRD

CHART C-4

MISFIRE

2.0L (VIN H) "J" CARLINE

Circuit Description:

The Electronic Ignition (EI) system uses a waste spark method of distribution. In this type of system, the ignition coil assembly triggers the #1/4 coil pair resulting in both #1 and #4 spark plugs firing at the same time. #1 cylinder is on the compression stroke at the same time #4 is on the exhaust stroke, resulting in a lower energy requirement to fire #4 spark plug. This leaves the remainder of the high voltage to be used to fire #1 spark plug. On this application, the Crankshaft Position (CKP) sensor is mounted to the engine block and protrudes through the block to within approximately .050 inch of the crankshaft reluctor. Since the reluctor is a machined portion of the crankshaft and the Crankshaft Position (CKP) sensor is mounted in a fixed position on the block, timing adjustments are not possible or necessary.

Test Description: Number(s) below refer to circled numbers on the diagnostic chart.

1. The Electronic Ignition (EI) system uses two plugs and cables to complete the circuit of each coil. The companion cylinder spark plug cable in the circuit must be connected to a good ground to create a spark. Test each spark plug cable with the engine idling (the ignition must be cycled "OFF" when moving the ST-125 tester to a different spark plug cable).
   - Keep disconnected spark plug leads away from sensors and other electronic components.
   - Move quickly through this test. Don't leave any spark plug lead disconnected for longer than 15 seconds.
   - Let the engine run normally for 30 seconds between tests to avoid an excessive buildup of fuel.

2. The test light should flash, indicating that the ECM is controlling the injector(s). How bright the light flashes is not important.

3. Nominal resistance of each injector is 11.8 to 12.6 ohms at 20°C (68°F) (resistance will increase slightly at higher temperatures). All injectors should measure within 0.8 ohms of each other. Damage to the ECM injector driver may occur if any injector resistance measures less than 11.8 ohms (internal short to ignition CKT 439).

4. The spark plug cables should be measured with a ohmmeter for correct resistance (each wire should be less than 30000 ohms).

5. Damage to the ECM injector driver may occur if injector driver CKT 467 or CKT 468 shorts to ignition CKT 439.

2.0L (VIN H) ENGINE — COMPONENT DIAGNOSTIC CHART — SUNBIRD

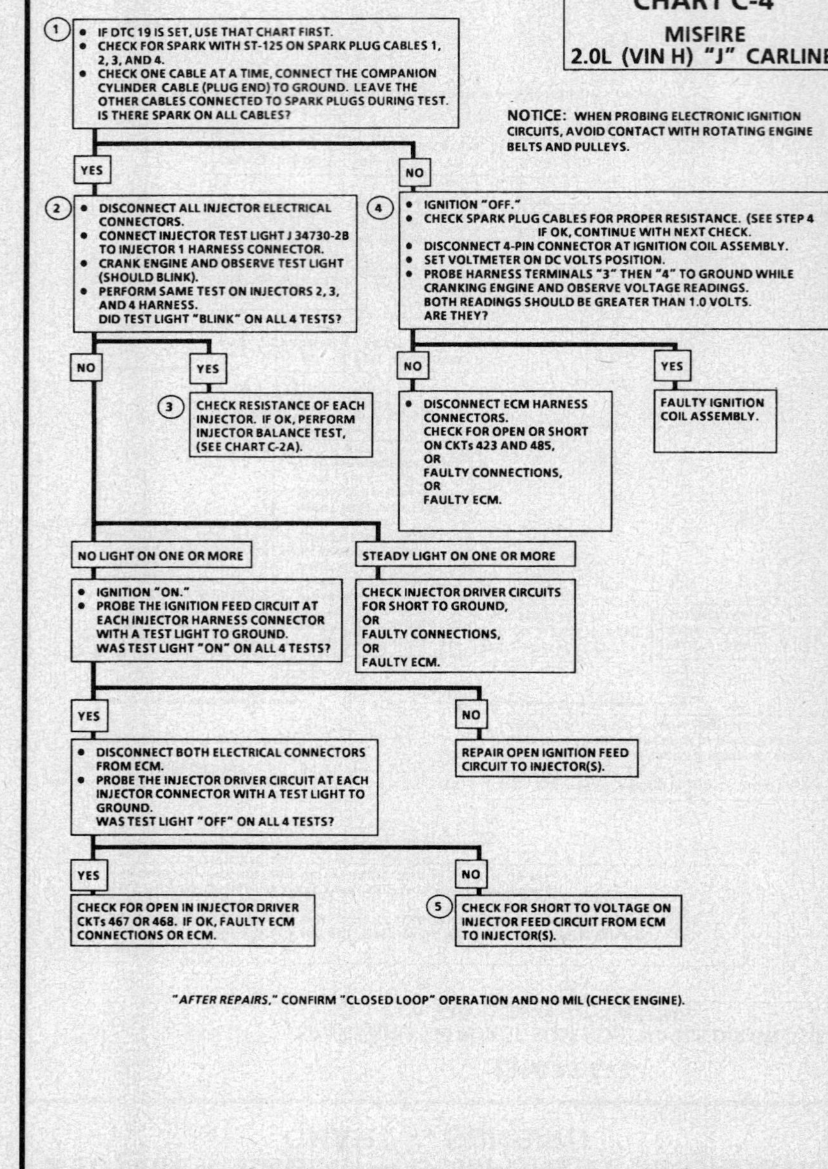

2.0L (VIN H) ENGINE — COMPONENT DIAGNOSTIC CHART — SUNBIRD

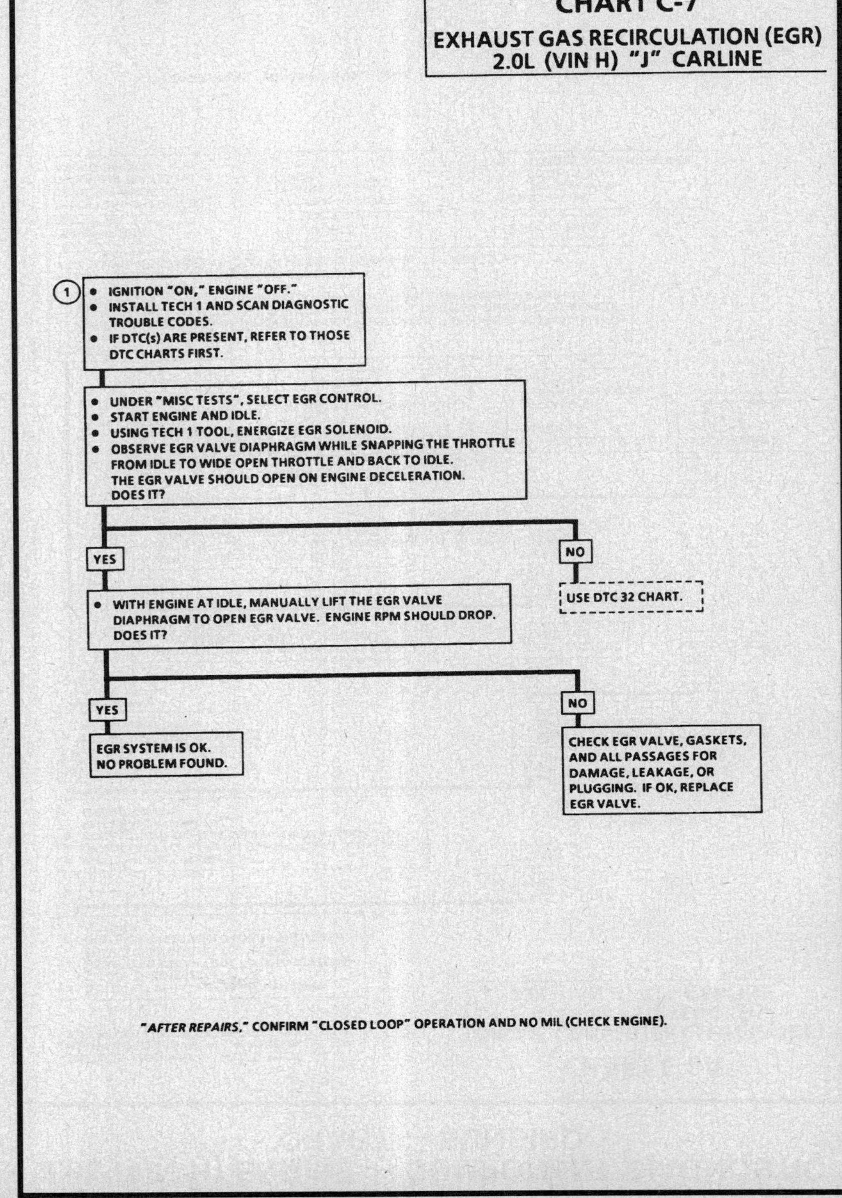

CHART C-7

EXHAUST GAS RECIRCULATION (EGR)
2.0L (VIN H) "J" CARLINE

Circuit Description:

A properly operating EGR system will directly affect the air/fuel mixture requirements of the engine. Since the exhaust gas introduced into the air/fuel mixture cannot be used in combustion (contains very little oxygen), less fuel is required to maintain a correct air/fuel ratio. If the EGR system were to fail in a closed position, the exhaust gas would be replaced with air, and the air/fuel mixture would be leaner. The ECM would compensate for the lean condition by adding fuel, resulting in higher long term fuel trim values.

The fuel control in this engine is conducted within four fuel trim cells. Since EGR is not used at idle, the idle cell would not be affected by EGR system operation. The other fuel trim cells are affected by EGR operation, and when the EGR system is operating properly, the long term fuel trim values in all cells should be close to the same. If the EGR system becomes inoperative, the long term fuel trim values in the open throttle fuel trim cell would change to compensate for the resulting lean mixtures, but the long term fuel trim value in the closed throttle fuel trim cell would not change.

The difference in long term fuel trim values between the idle (closed throttle) fuel trim cell and fuel trim cell 2 is used to monitor EGR system performance. When the difference between the two long term fuel trim values is greater than 12 and the long term fuel trim value in fuel trim cell 2 is greater than 140, DTC 32 is set. The system operates in fuel trim cell 2 during a cruise condition at approximately 55 mph.

Test Description: Number(s) below refer to circled number(s) on the diagnostic chart.
1. DTC(s) should be diagnosed using appropriate chart before performing a functional check. If DTCs 14, 21, 23, 32, or 33 are set, use those charts first.

Diagnostic Aids:

This chart is a functional check of the EGR system. If the EGR system works properly, check other items that could result in high long term fuel trim values in fuel trim cell 2, but not in the closed throttle fuel trim cell.

2.0L (VIN H) ENGINE — COMPONENT DIAGNOSTIC CHART — SUNBIRD

CHART C-7

EXHAUST GAS RECIRCULATION (EGR)
2.0L (VIN H) "J" CARLINE

1.
- IGNITION "ON," ENGINE "OFF."
- INSTALL TECH 1 AND SCAN DIAGNOSTIC TROUBLE CODES.
- IF DTC(s) ARE PRESENT, REFER TO THOSE DTC CHARTS FIRST.

- UNDER "MISC TESTS", SELECT EGR CONTROL.
- START ENGINE AND IDLE.
- USING TECH 1 TOOL, ENERGIZE EGR SOLENOID.
- OBSERVE EGR VALVE DIAPHRAGM WHILE SNAPPING THE THROTTLE FROM IDLE TO WIDE OPEN THROTTLE AND BACK TO IDLE. THE EGR VALVE SHOULD OPEN ON ENGINE DECELERATION. DOES IT?

YES → WITH ENGINE AT IDLE, MANUALLY LIFT THE EGR VALVE DIAPHRAGM TO OPEN EGR VALVE. ENGINE RPM SHOULD DROP. DOES IT?

NO → USE DTC 32 CHART.

YES → EGR SYSTEM IS OK. NO PROBLEM FOUND.

NO → CHECK EGR VALVE, GASKETS, AND ALL PASSAGES FOR DAMAGE, LEAKAGE, OR PLUGGING. IF OK, REPLACE EGR VALVE.

"AFTER REPAIRS," CONFIRM "CLOSED LOOP" OPERATION AND NO MIL (CHECK ENGINE).

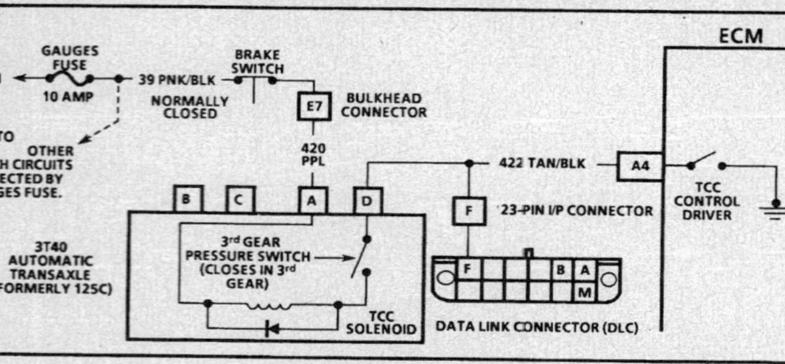

CHART C-8A

TORQUE CONVERTER CLUTCH (TCC)
(ELECTRICAL DIAGNOSIS)
2.0L (VIN H) "J" CARLINE

Circuit Description:

The purpose of the Torque Converter Clutch (TCC) is to eliminate the power loss of the torque converter when the vehicle is in a cruise condition. This allows the convenience of the automatic transaxle and the fuel economy of a manual transaxle.

Voltage is supplied to the TCC solenoid through the brake switch and transaxle third gear apply switch. The ECM will engage TCC by grounding CKT 422 to energize the solenoid.

The TCC will engage under the following conditions:

- Vehicle speed exceeds about 25 mph (40 km/h).
- Engine coolant temperature exceeds 65°C (149°F).
- Throttle position is not changing faster than a calibrated rate (steady throttle).
- Transaxle 3rd gear switch is closed.
- Brake switch is closed.

Test Description: Number(s) below refer to circled number(s) on the diagnostic chart.

1. This test should be performed on a level road with the aid of a helper to operate the Tech 1 and observe the RPM fluctuation.
2. Entering the "field service mode" forces the ECM to close the TCC control driver circuit completing the path for current which should light the test light.
3. Perform this test with the aid of a helper. The 3rd gear switch must be engaged while measuring the TCC solenoid resistance.

Diagnostic Aids:

Fused battery ignition is supplied to the TCC solenoid through the brake switch. The ECM will engage TCC by grounding CKT 422 to energize the solenoid.

An engine coolant thermostat that is stuck open or opens at too low a temperature may result in an inoperative TCC.

CHART C-8A

TORQUE CONVERTER CLUTCH (TCC)
(ELECTRICAL DIAGNOSIS)
2.0L (VIN H) "J" CARLINE

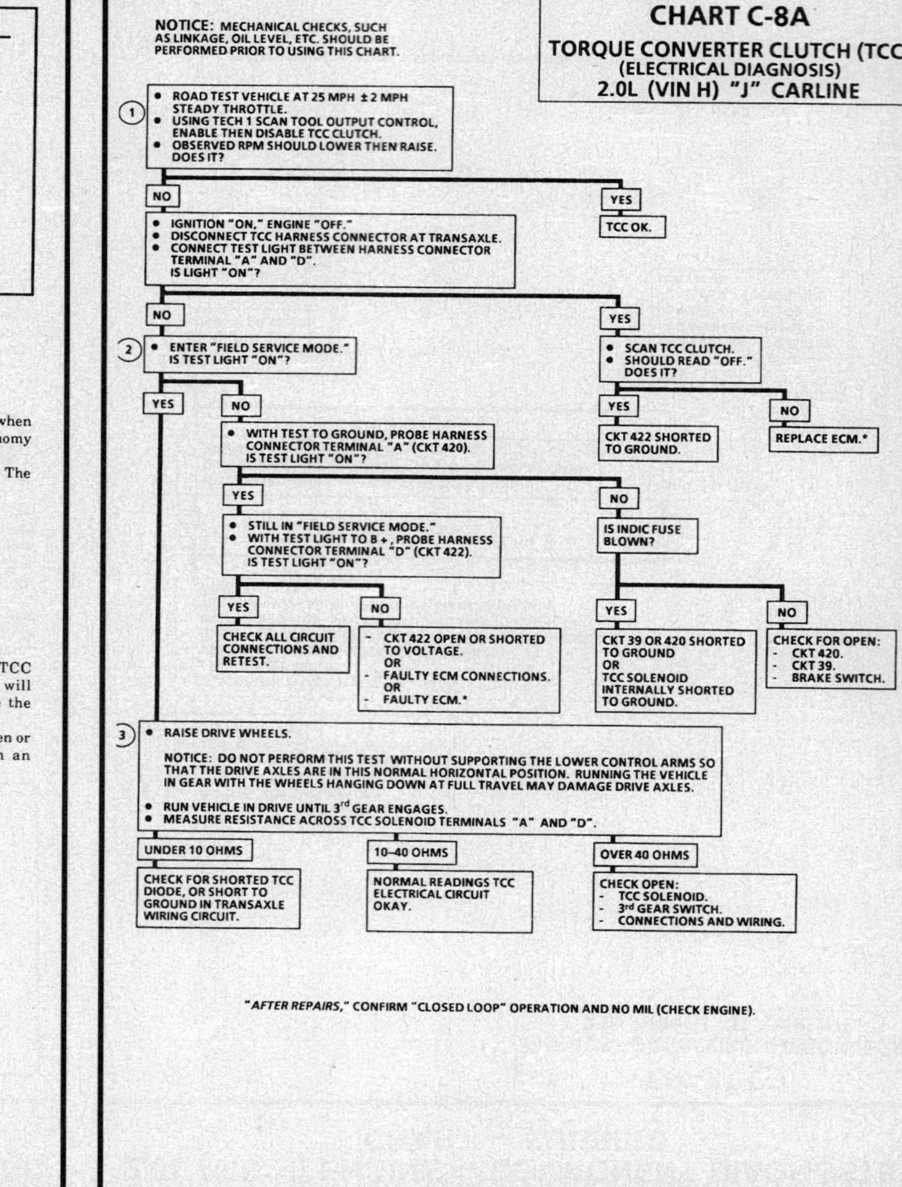

2.0L (VIN H) ENGINE — COMPONENT DIAGNOSTIC CHART — SUNBIRD

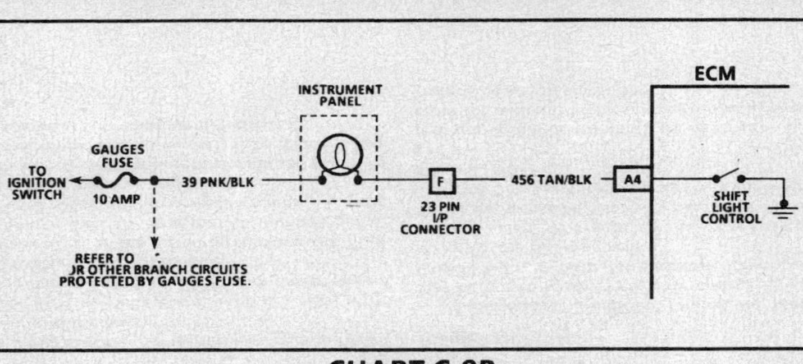

CHART C-8B
MANUAL TRANSAXLE (M/T) SHIFT LIGHT CHECK
2.0L (VIN H) "J" CARLINE

Circuit Description:

The shift light indicates the best transaxle shift point for maximum fuel economy. The light is controlled by the ECM and is turned "ON" by grounding CKT 456.

The ECM uses information from the following inputs to control the shift light:

- Engine Coolant Temperature (ECT) sensor.
- Throttle Position (TP) sensor.
- Vehicle Speed Sensor (VSS).
- Engine speed Crankshaft Position (CKP) sensor.

The ECM uses the measured engine and vehicle speeds to calculate what gear the vehicle is in. It is this calculation which determines when the shift light should be turned "ON."

Test Description: Number(s) below refer to circled number(s) on the diagnostic chart.

1. When the ignition is "ON" and the engine is "OFF," the shift light should be "OFF." If the light is "ON," there is a short to ground in CKT 456 or a fault in the ECM.
2. When the diagnostic terminal is grounded, the ECM should ground CKT 456, and the shift light should be "ON."
3. This checks the shift light circuit up to the ECM connector. If the shift light lights, then the ECM connector is faulty, or the ECM does not have the ability to ground the circuit.

2.0L (VIN H) ENGINE — COMPONENT DIAGNOSTIC CHART — SUNBIRD

CHART C-8B
MANUAL TRANSAXLE (M/T) SHIFT LIGHT CHECK
2.0L (VIN H) "J" CARLINE

"AFTER REPAIRS," CONFIRM "CLOSED LOOP" OPERATION AND NO MIL (CHECK ENGINE).

2.0L (VIN H) ENGINE — COMPONENT DIAGNOSTIC CHART — 1992 SUNBIRD

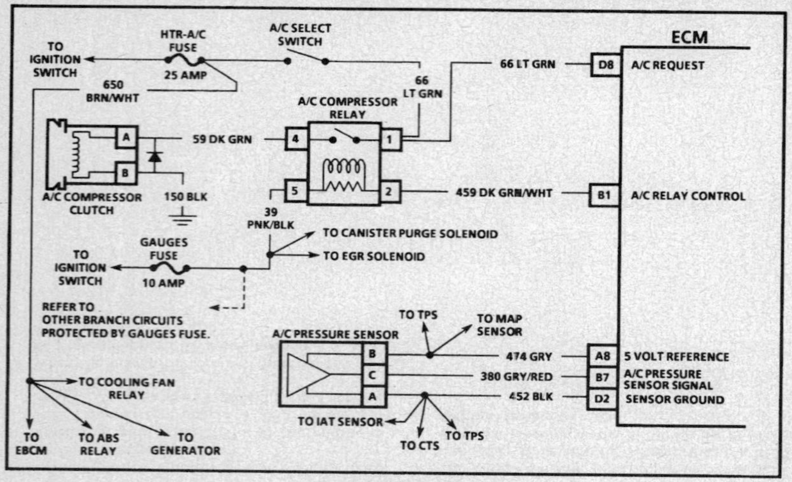

CHART C-10
(Page 1 of 2)
A/C CLUTCH CONTROL CIRCUIT DIAGNOSIS
2.0L (VIN H) "J" CARLINE (PORT)

Circuit Description:

The A/C clutch control relay is energized when the Electronic Control Module (ECM) provides a ground path through CKT 459 and A/C is requested. A/C clutch is delayed about .3 seconds after A/C is requested. This will allow the IAC to adjust engine rpm for the additional load.

The ECM will temporarily disengage the A/C clutch relay for calibrated times for one or more of the following conditions:

- Hot engine restart.
- Wide Open Throttle (WOT) (TPS over 99% disable).
- Engine rpm greater than about 6000 rpm.
- During Idle Air Control (IAC) reset.

The A/C clutch relay will remain disengaged when a Code 66 is present, if pressure is out of range or there is no A/C request signal due to an open A/C select switch or circuit.

Test Description: Number(s) below refer to circled number(s) on the diagnostic chart.

1. The ECM will only energize the A/C relay when the engine is running. This test will determine if the relay or CKT 459 is faulty.
2. Determines if the signal is reaching the ECM through CKT 66 from the A/C control panel. Signal should only be present when an A/C mode or defrost mode has been selected.
3. Be sure to consider branch circuits and splices to other components particularly if the HTR-A/C fuse is blown causing multiple open circuits.

Diagnostic Aids:

If complaint is insufficient cooling, the problem may be caused by an inoperative cooling fan. See CHART C-12 for cooling fan diagnosis. If fan operates correctly, see A/C diagnosis A/C pressure outside of a range of 41 to 426 psi will cause the compressor to be disabled by the ECM. Observe Tech 1 A/C pressure for 2 minutes with engine idling and A/C "ON."

"Scan" pressure should be within 20 psi of actual. If not, check for a circuit problem using Code 66 chart or replace the A/C pressure sensor.

2.0L (VIN H) ENGINE — COMPONENT DIAGNOSTIC CHART — 1992 SUNBIRD

CHART C-10
(Page 1 of 2)
A/C CLUTCH CONTROL CIRCUIT DIAGNOSIS
2.0L (VIN H) "J" CARLINE (PORT)

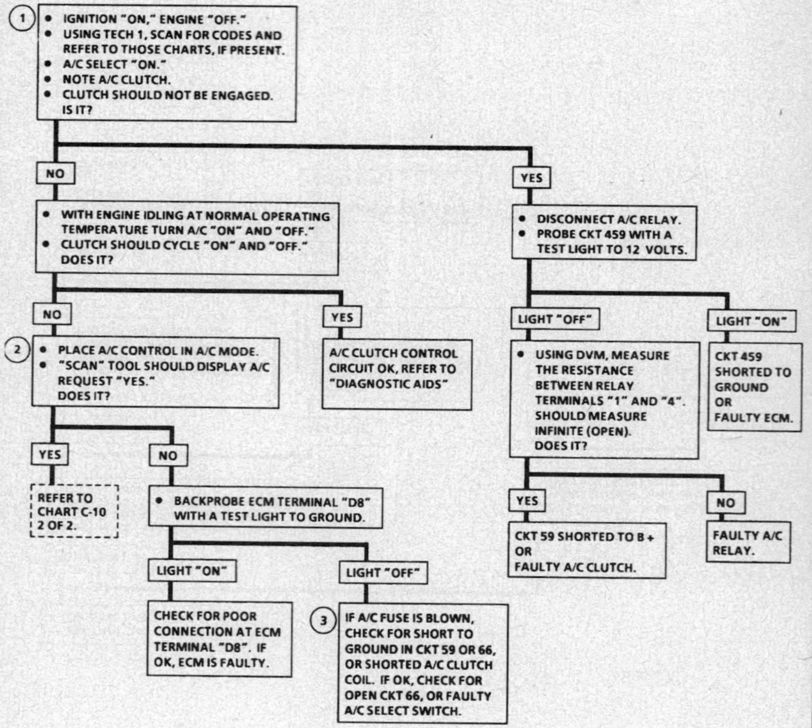

2.0L (VIN H) ENGINE — COMPONENT DIAGNOSTIC CHART — 1992 SUNBIRD

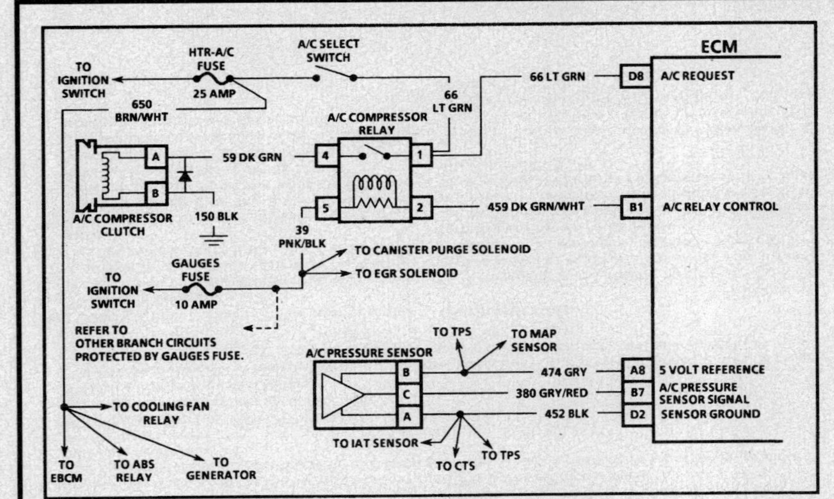

CHART C-10
(Page 2 of 2)
A/C CLUTCH CONTROL CIRCUIT DIAGNOSIS
2.0L (VIN H) "J" CARLINE (PORT)

Circuit Description:
The A/C clutch control relay is energized when the ECM provides a ground path through CKT 459 and A/C is requested. A/C clutch is delayed about .3 seconds after A/C is requested. This will allow the IAC to adjust engine rpm for the additional load.

The ECM will temporarily disengage the A/C clutch relay for calibrated times for one or more of the following conditions:
- Hot engine restart.
- Wide open throttle (TPS over 99% disable).
- Engine rpm greater than about 6000 rpm.
- During IAC reset.

The A/C clutch relay will remain disengaged when a Code 66 is present, if pressure is out of range or there is no A/C request signal due to an open A/C select switch or circuit.

Test Description: Number(s) below refer to circled number(s) on the diagnostic chart.
1. Determines if the pressure transducer is out of range causing the compressor clutch to be disengaged.
2. With the engine stopped and field service mode activated, the ECM should be grounding CKT 459, which should cause the test light to be "ON."
3. Be sure to consider branch circuits and splices to other components, particularly if the gauges fuse is blown causing an open CKT 39.

Diagnostic Aids:
If complaint is insufficient cooling, the problem may be caused by an inoperative cooling fan. See CHART C-12 for cooling fan diagnosis. If fan operates correctly, see A/C diagnosis A/C pressure outside of a range of 41 to 426 psi will cause the compressor to be disabled by the ECM. Observe Tech 1 A/C pressure for 2 minutes with engine idling and A/C "ON."

"Scan" pressure should be within 20 psi of actual. If not, check for a circuit problem using Code 66 chart or replace A/C pressure sensor.

2.0L (VIN H) ENGINE — COMPONENT DIAGNOSTIC CHART — 1992 SUNBIRD

CHART C-10
(Page 2 of 2)
A/C CLUTCH CONTROL CIRCUIT DIAGNOSIS
2.0L (VIN H) "J" CARLINE (PORT)

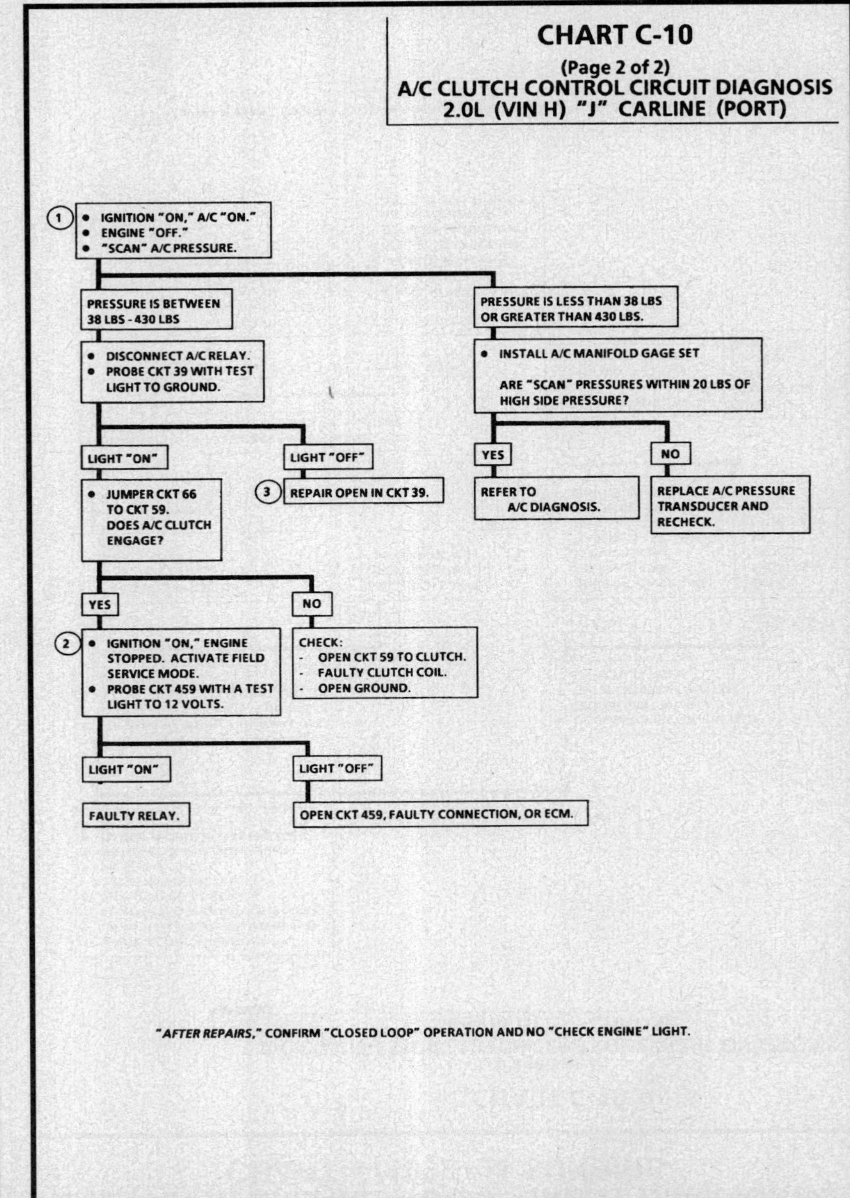

"AFTER REPAIRS," CONFIRM "CLOSED LOOP" OPERATION AND NO "CHECK ENGINE" LIGHT.

2.0L (VIN H) ENGINE — COMPONENT DIAGNOSTIC CHART — 1993–94 SUNBIRD

CHART C-10
(Page 1 of 2)
A/C COMPRESSOR CLUTCH CONTROL CIRCUIT DIAGNOSIS
2.0L (VIN H) "J" CARLINE

Circuit Description:

The A/C compressor clutch control relay is energized when the Engine Control Module (ECM) provides a ground path through CKT 459 and A/C is requested. A/C compressor clutch is delayed about .3 second after A/C is requested. This will allow the IAC valve to adjust engine RPM for the additional load.

The ECM will temporarily disengage the A/C compressor clutch relay for calibrated times for one or more of the following conditions:

- Hot engine restart.
- Wide Open Throttle (WOT) (TP sensor over 99% disable).
- Engine RPM greater than about 6000 RPM.
- During Idle Air Control (IAC) reset.

The A/C compressor clutch relay will remain disengaged when a DTC 66 is present, if pressure is out of range or there is no A/C request signal due to an open A/C select switch or circuit.

Test Description: Number(s) below refer to circled number(s) on the diagnostic chart.

1. The ECM will only energize the A/C compressor clutch control relay when the engine is running. This test will determine if the relay or CKT 459 is faulty.
2. Determines if the signal is reaching the ECM through CKT 66 from the A/C control panel. Signal should only be present when an A/C mode or defrost mode has been selected.
3. Be sure to consider branch circuits and splices to other components particularly if the HTR-A/C fuse is blown causing multiple open circuits.

Diagnostic Aids:

If complaint is insufficient cooling, the problem may be caused by an inoperative cooling fan. Refer to CHART C-12 for cooling fan diagnosis. If fan operates correctly, refer to A/C diagnosis A/C refrigerant pressure outside of a range of 41 to 426 psi will cause the compressor to be disabled by the ECM. Observe Tech 1 A/C refrigerant pressure for 2 minutes with engine idling and A/C "ON."

Scan pressure should be within 20 psi of actual. If not, check for a circuit problem using DTC 66 chart or replace the A/C refrigerant pressure sensor.

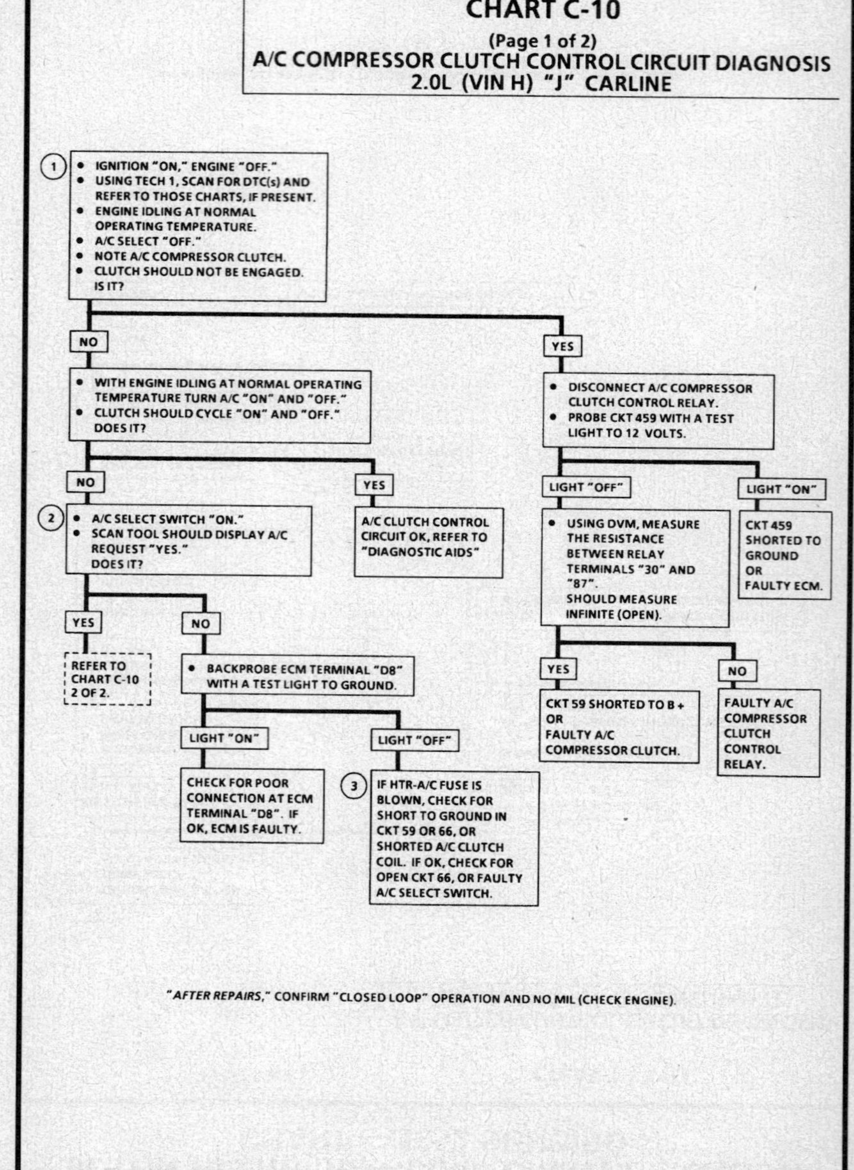

2.0L (VIN H) ENGINE — COMPONENT DIAGNOSTIC CHART — 1993–94 SUNBIRD

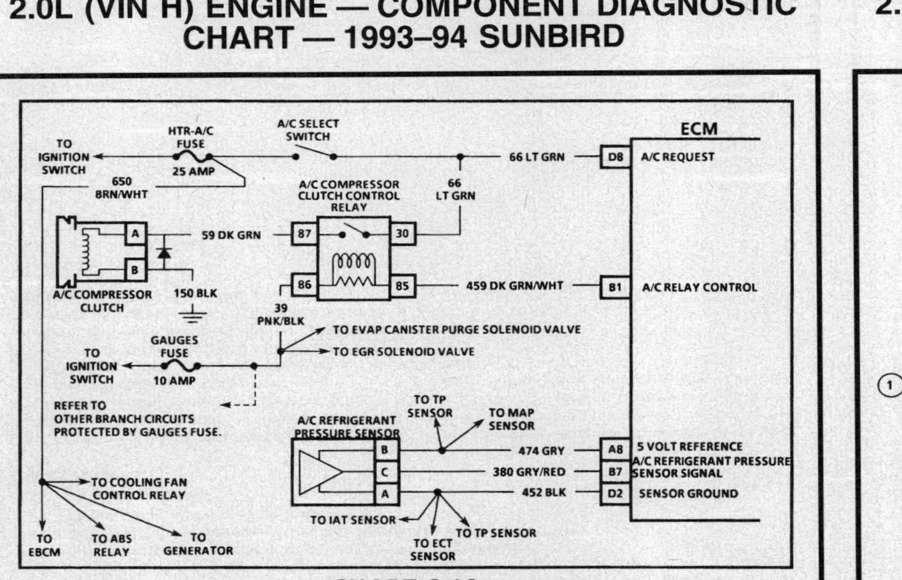

CHART C-10
(Page 2 of 2)
A/C COMPRESSOR CLUTCH CONTROL CIRCUIT DIAGNOSIS
2.0L (VIN H) "J" CARLINE

Circuit Description:

The A/C compressor clutch control relay is energized when the ECM provides a ground path through CKT 459 and A/C is requested. A/C compressor clutch is delayed about .3 second after A/C is requested. This will allow the IAC valve to adjust engine RPM for the additional load.

The ECM will temporarily disengage the A/C clutch relay for calibrated times for one or more of the following conditions:
- Hot engine restart.
- Wide open throttle (TP sensor over 99% disable).
- Engine RPM greater than about 6000 RPM.
- During IAC reset.

The A/C compressor clutch relay will remain disengaged when a DTC 66 is present, if pressure is out of range or there is no A/C request signal due to an open A/C select switch or circuit.

Test Description: Number(s) below refer to circled number(s) on the diagnostic chart.
1. Determines if the A/C refrigerant pressure sensor is out of range causing the compressor clutch to be disengaged.
2. With the engine stopped and field service mode activated, the ECM should be grounding CKT 459, which should cause the test light to be "ON."
3. Be sure to consider branch circuits and splices to other components, particularly if the gauges fuse is blown causing an open CKT 39.

Diagnostic Aids:

If complaint is insufficient cooling, the problem may be caused by an inoperative cooling fan. Refer to CHART C-12 for cooling fan diagnosis. If fan operates correctly, refer to A/C diagnosis. A/C refrigerant pressure outside of a range of 41 to 426 psi will cause the compressor to be disabled by the ECM. Observe Tech 1 A/C refrigerant pressure for 2 minutes with engine idling and A/C "ON."

Scan refrigerant pressure should be within 20 psi of actual. If not, check for a circuit problem using DTC 66 chart or replace A/C refrigerant pressure sensor.

2.0L (VIN H) ENGINE — COMPONENT DIAGNOSTIC CHART — 1993–94 SUNBIRD

CHART C-10
(Page 2 of 2)
A/C COMPRESSOR CLUTCH CONTROL CIRCUIT DIAGNOSIS
2.0L (VIN H) "J" CARLINE

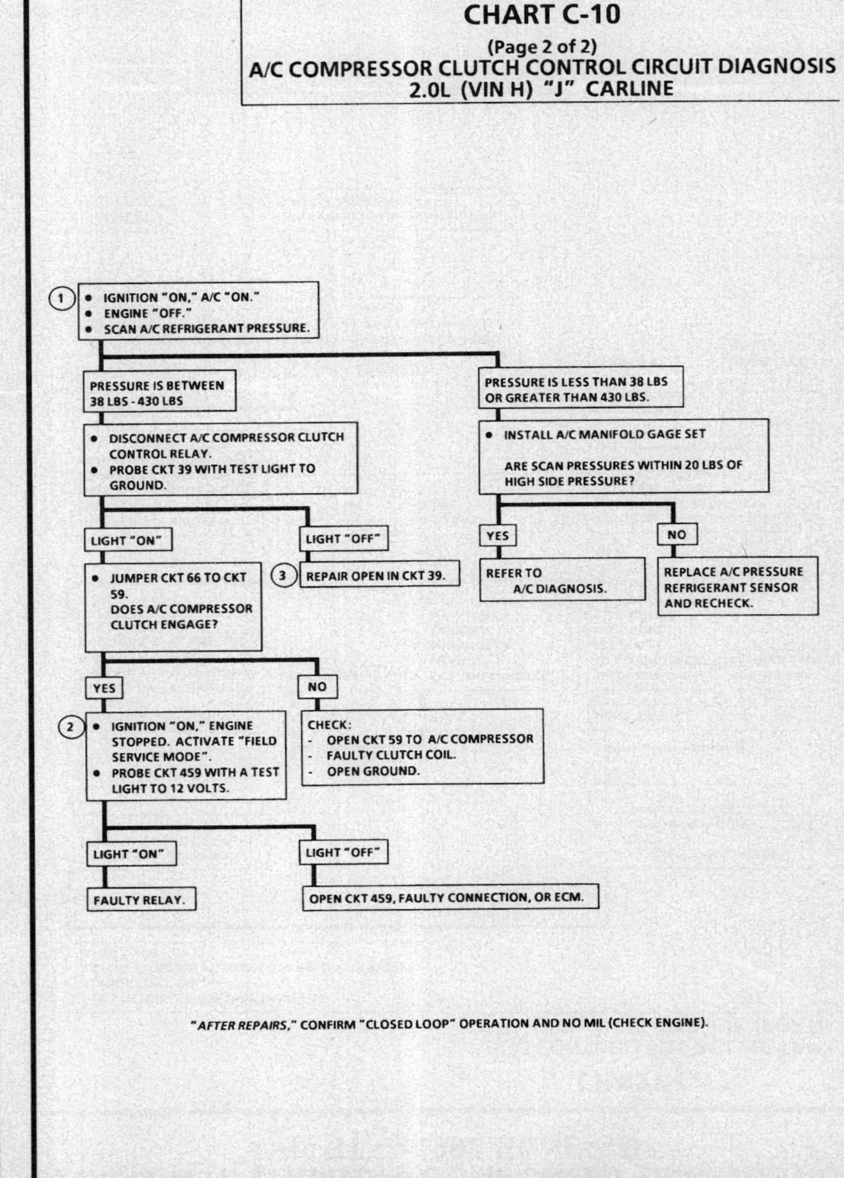

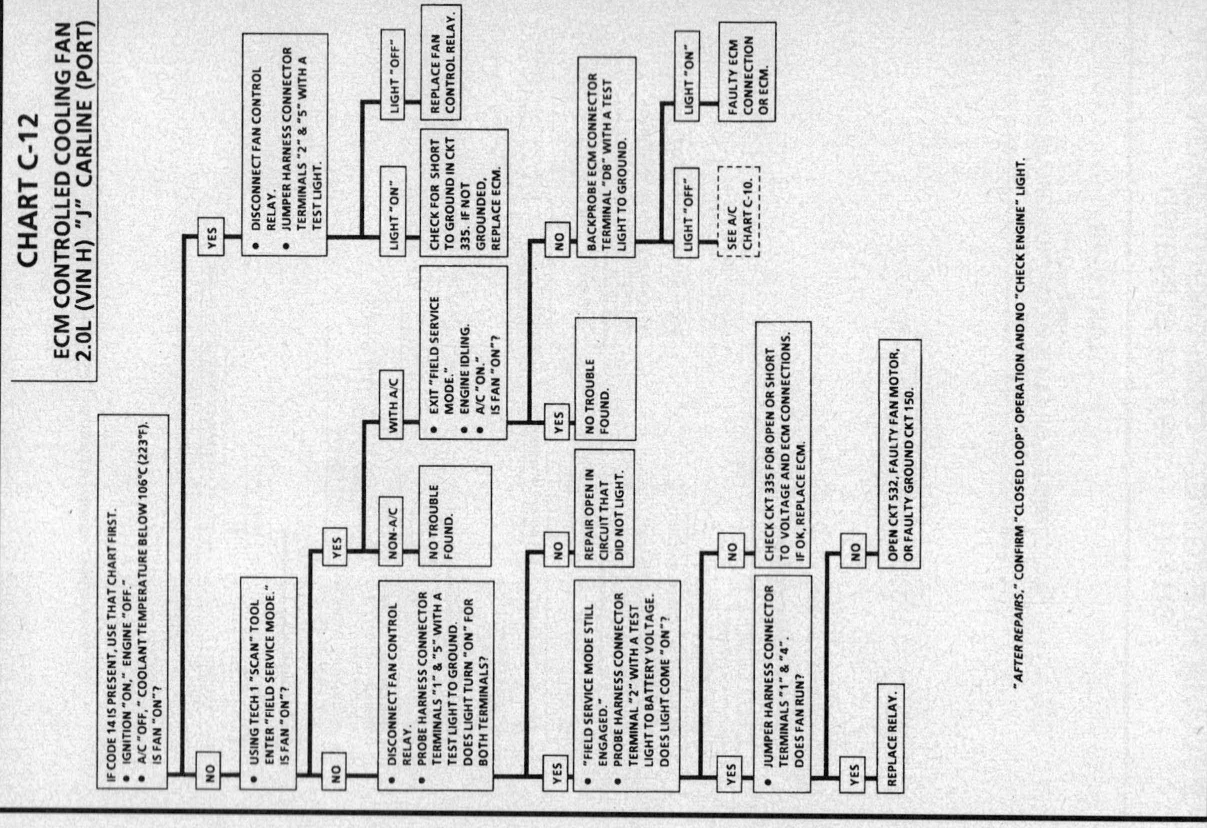

CHART C-12

ECM CONTROLLED COOLING FAN
2.0L (VIN H) "J" CARLINE (PORT)

Circuit Description:

Battery voltage to operate the cooling fan motor is supplied to relay by CKT 2. Ignition voltage to energize the relay is supplied to relay by CKT 650. When the ECM grounds CKT 335, the relay is energized and the cooling fan is turned "ON." When the engine is running, the ECM will energize the cooling fan relay if a coolant temperature sensor (Code 14) has been set, or under the following conditions:

• A/C is "ON" and vehicle speed is less than 70 mph (113 km/h) with A/C "OFF."
• Coolant temperature is greater than 106°C (223°F).

Diagnostic Aids:

If the vehicle has an overheating problem, it must be determined if the complaint was due to an actual boil over, the coolant temperature warning light, or the temperature gage indicated overheating.

If the gage or light indicates overheating but no boil over is detected, the gage circuit should be checked. The gage accuracy can be checked by comparing it to the coolant temperature sensor reading using a Tech 1 "Scan" tool.

If the engine is actually overheating and the gage indicates overheating, but the cooling fan is not turning "ON," the coolant temperature sensor has probably shifted out of calibration and should be replaced.

2.0L (VIN H) ENGINE — COMPONENT DIAGNOSTIC CHART — 1993–94 SUNBIRD

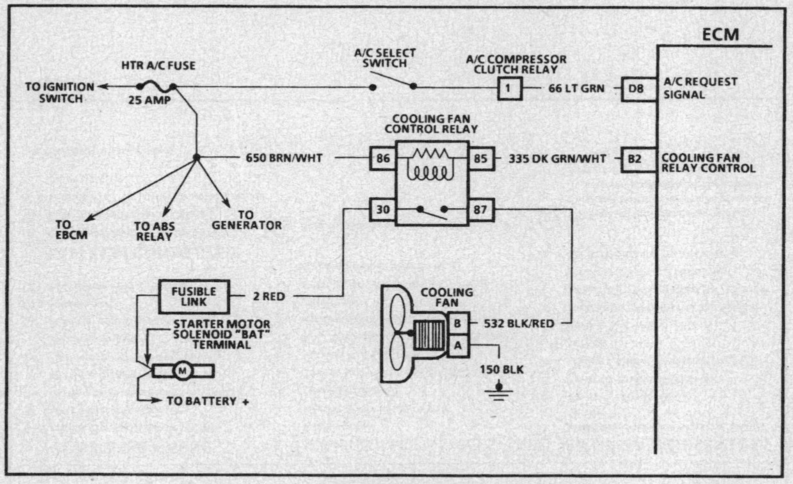

CHART C-12

COOLING FAN CONTROL
2.0L (VIN H) "J" CARLINE

Circuit Description:

Battery voltage to operate the cooling fan motor is supplied to relay by CKT 2. Ignition voltage to energize the relay is supplied to relay by CKT 650. When the ECM grounds CKT 335, the relay is energized and the cooling fan is turned "ON." When the engine is running, the ECM will energize the cooling fan control relay if a coolant temperature sensor (DTC 14) has been set, or under the following conditions:

- Coolant temperature is greater than 106°C (223°F).

  OR

- A/C refrigerant pressure is greater than 51 psi.

Diagnostic Aids:

If the vehicle has an overheating problem, it must be determined if the complaint was due to an actual boil over, the coolant temperature warning light, or the temperature gage indicated overheating.

If the gage or light indicates overheating but no boil over is detected, the gage circuit should be checked. The gage accuracy can be checked by comparing it to the coolant temperature sensor reading using a Tech 1 scan tool.

If the engine is actually overheating and the gage indicates overheating, but the cooling fan is not turning "ON," the coolant temperature sensor has probably shifted out of calibration and should be replaced.

2.0L (VIN H) ENGINE — COMPONENT DIAGNOSTIC CHART — 1993–94 SUNBIRD

CHART C-12

COOLING FAN CONTROL
2.0L (VIN H) "J" CARLINE

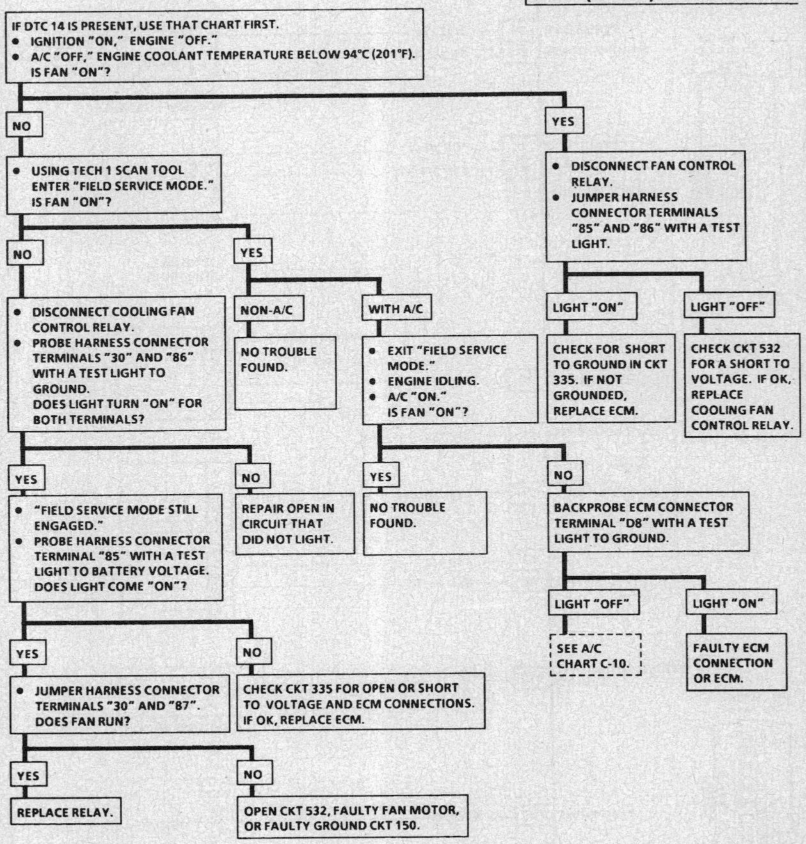

"AFTER REPAIRS," CONFIRM "CLOSED LOOP" OPERATION AND NO MIL (CHECK ENGINE).

2.2L (VIN 4) ENGINE — ECM WIRING SCHEMATIC — CENTURY AND CIERA

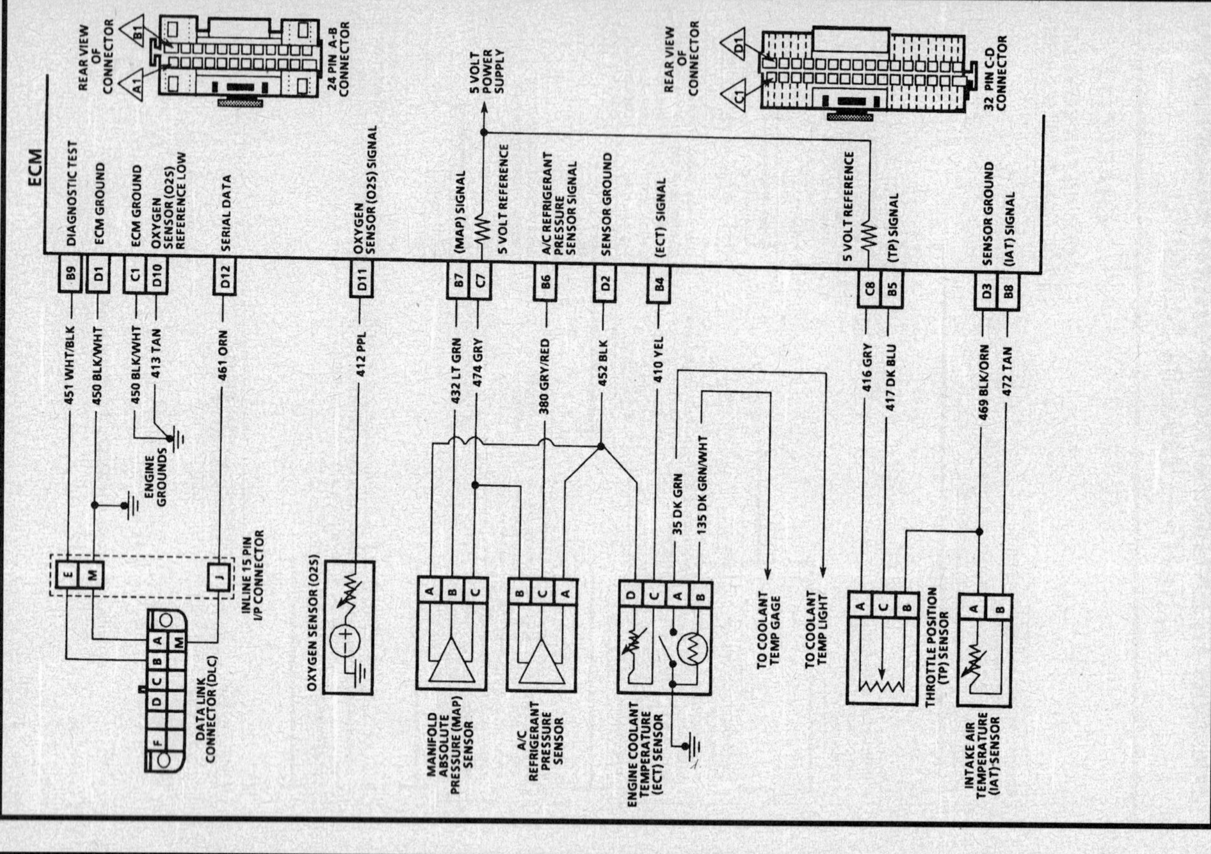

2.2L (VIN 4) ENGINE — ENGINE COMPONENT LOCATION CHART — CENTURY AND CIERA

"A" CARLINE   RPO: LN2   VIN CODE: 4   2.2L L4 MFI

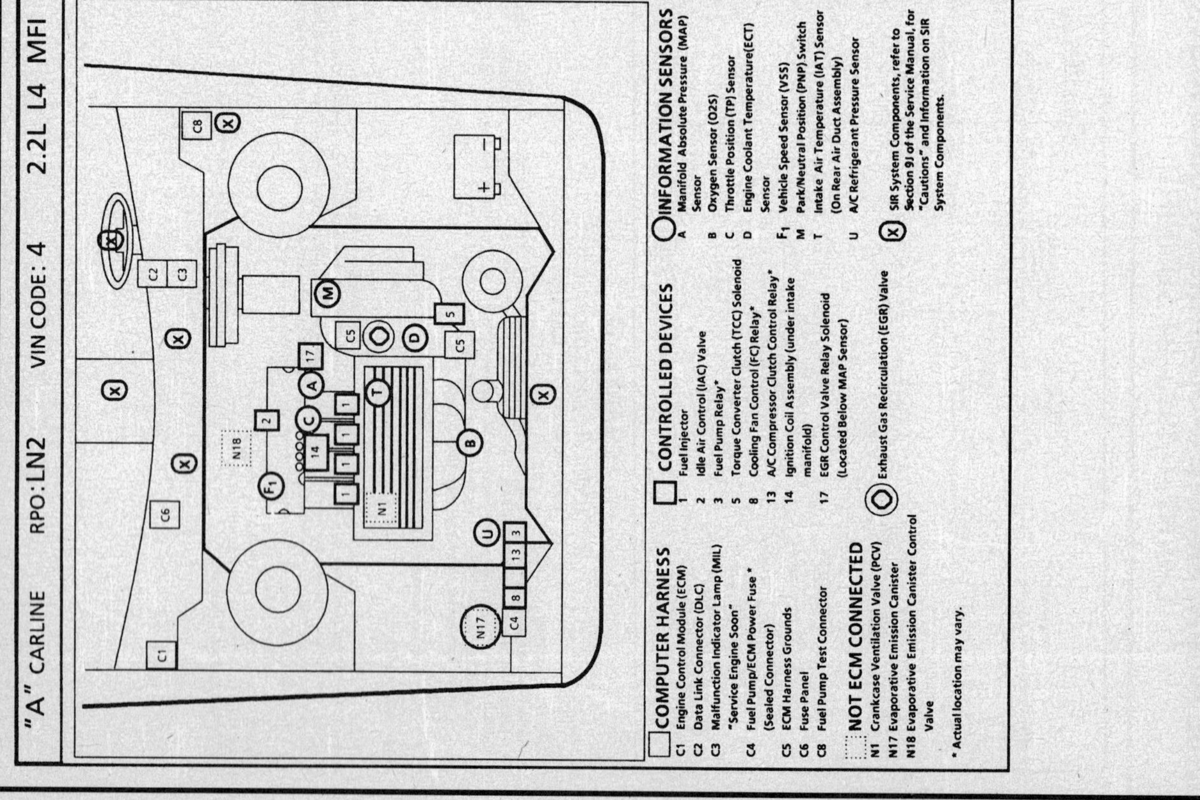

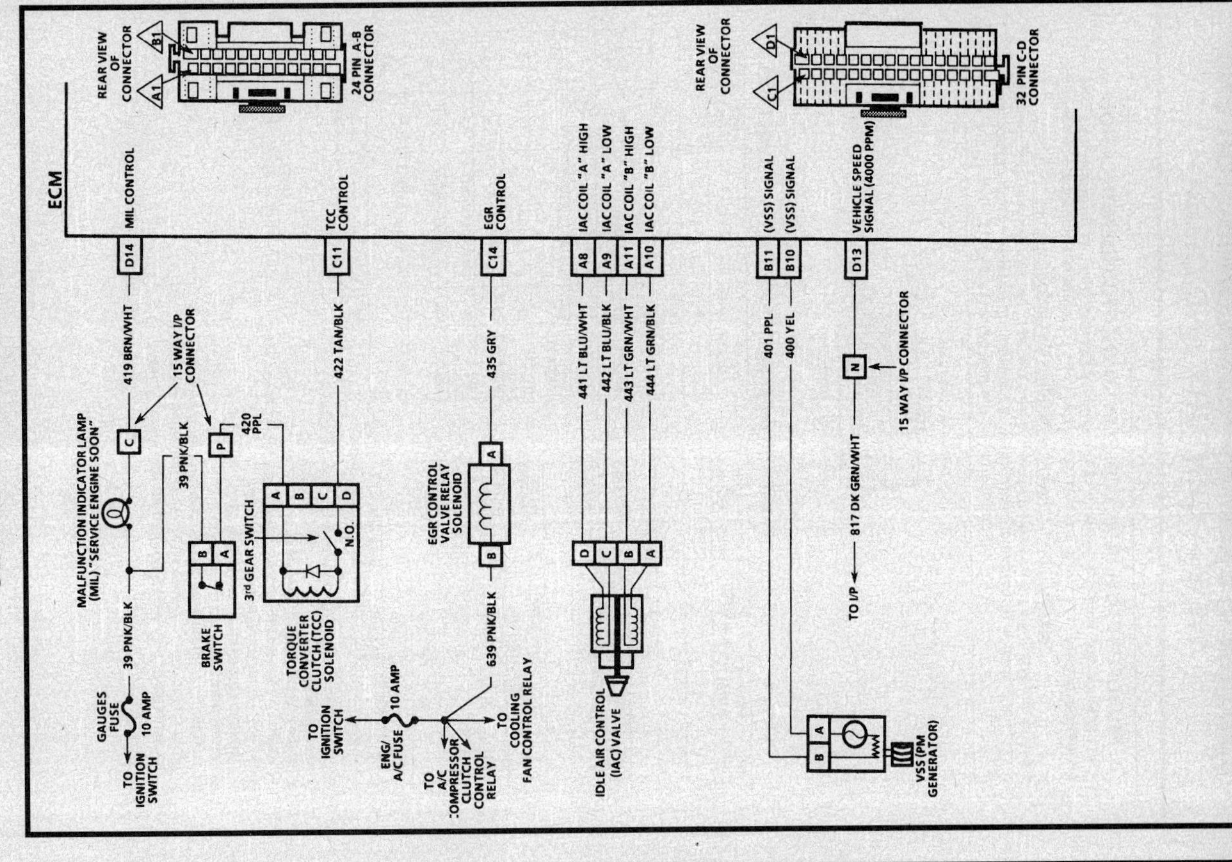

2.2L (VIN 4) ENGINE — ECM WIRING SCHEMATIC — CENTURY AND CIERA
ECM
REAR VIEW OF CONNECTOR
B1
A1
24 PIN A-B CONNECTOR
REAR VIEW OF CONNECTOR
D1
C1
32 PIN C-D CONNECTOR
MIL CONTROL
D14
TCC CONTROL
C11
EGR CONTROL
C14
IAC COIL "A" HIGH
A8
IAC COIL "A" LOW
A9
IAC COIL "B" HIGH
A11
IAC COIL "B" LOW
A10
(VSS) SIGNAL
B11
(VSS) SIGNAL
B10
VEHICLE SPEED SIGNAL (4000 PPM)
D13
419 BRN/WHT
15 WAY I/P CONNECTOR
C
39 PNK/BLK
P
420 PPL
422 TAN/BLK
435 GRY
441 LT BLU/WHT
442 LT BLU/BLK
443 LT GRN/WHT
444 LT GRN/BLK
401 PPL
400 YEL
817 DK GRN/WHT
15 WAY I/P CONNECTOR
N
TO I/P
MALFUNCTION INDICATOR LAMP (MIL) "SERVICE ENGINE SOON"
39 PNK/BLK
GAUGES FUSE 10 AMP
TO IGNITION SWITCH
B A
BRAKE SWITCH
A B C D
B A
3rd GEAR SWITCH
TORQUE CONVERTER CLUTCH (TCC) SOLENOID
TO IGNITION SWITCH
ENG/ A/C FUSE 10 AMP
TO A/C COMPRESSOR CLUTCH CONTROL RELAY
639 PNK/BLK
TO COOLING FAN CONTROL RELAY
A B
EGR CONTROL VALVE RELAY SOLENOID
D C B A
IDLE AIR CONTROL (IAC) VALVE
B A
VSS (PM GENERATOR)

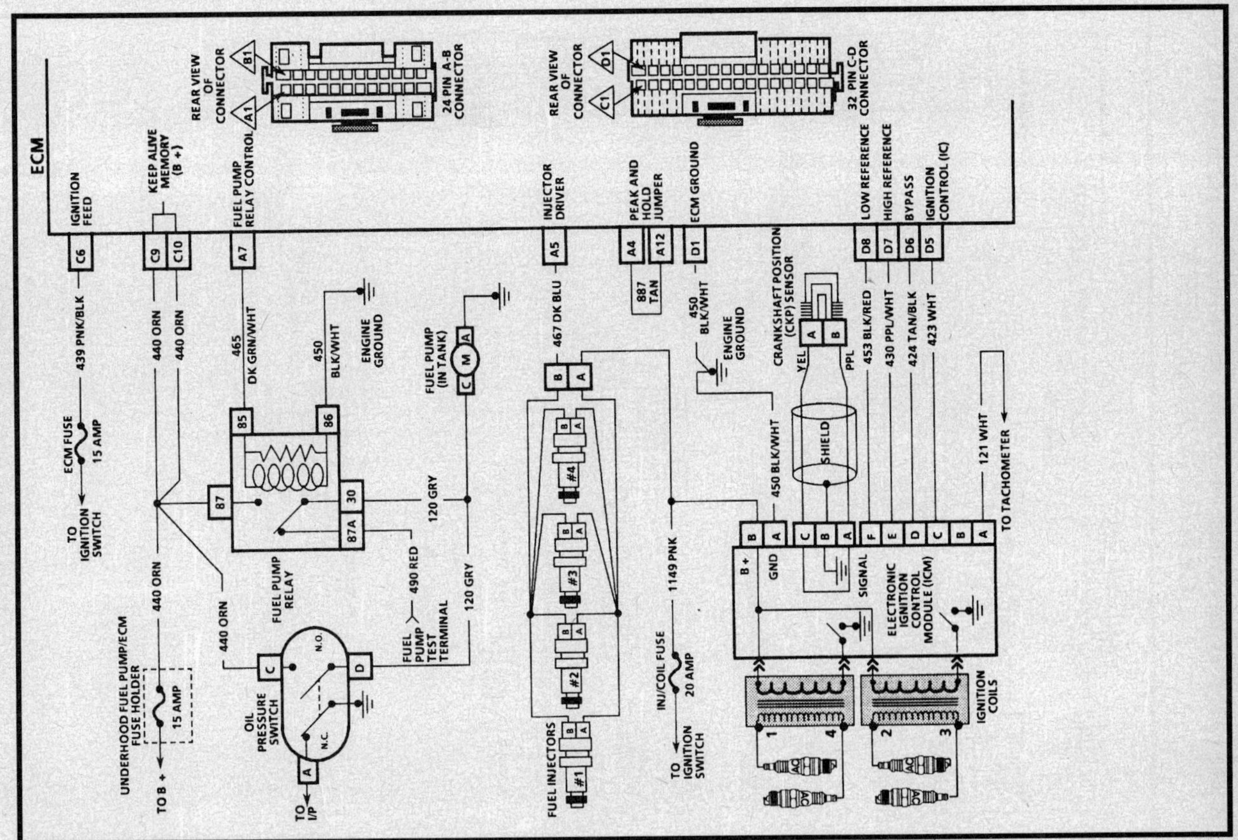

2.2L (VIN 4) ENGINE — ECM WIRING SCHEMATIC — CENTURY AND CIERA
ECM
REAR VIEW OF CONNECTOR
B1
A1
24 PIN A-B CONNECTOR
REAR VIEW OF CONNECTOR
D1
C1
32 PIN C-D CONNECTOR
IGNITION FEED
C6
KEEP ALIVE MEMORY (B+)
C9
C10
FUEL PUMP RELAY CONTROL
A7
INJECTOR DRIVER
A5
PEAK AND HOLD JUMPER
A4
A12
ECM GROUND
D1
LOW REFERENCE
D8
HIGH REFERENCE
D7
BYPASS
D6
IGNITION CONTROL (IC)
D5
439 PNK/BLK
440 ORN
440 ORN
465 DK GRN/WHT
450
BLK/WHT
467 DK BLU
887 TAN
450 BLK/WHT
453 BLK/RED
430 PPL/WHT
424 TAN/BLK
423 WHT
ECM FUSE 15 AMP
TO IGNITION SWITCH
B5
B6
ENGINE GROUND
FUEL PUMP (IN TANK)
M A
ENGINE GROUND
CRANKSHAFT POSITION (CKP) SENSOR
YEL PPL
A B
SHIELD
121 WHT
TO TACHOMETER
B7
87
87A
30
FUEL PUMP RELAY
120 GRY
120 GRY
FUEL PUMP TEST TERMINAL
490 RED
UNDERHOOD FUEL PUMP/ECM FUSE HOLDER
440 ORN
15 AMP
TO B+
OIL PRESSURE SWITCH
N.O.
N.C.
C D
A
TO I/P
440 ORN
#4
#3
#2
#1
FUEL INJECTORS
B A
B A
B A
B A
1149 PNK
INJ/COIL FUSE 20 AMP
TO IGNITION SWITCH
B+ B
GND A
C B
A
SIGNAL
F E D C B
A
450 BLK/WHT
ELECTRONIC IGNITION CONTROL MODULE (ICM)
1 4
2 3
IGNITION COILS

2.2L (VIN 4) ENGINE — ECM WIRING SCHEMATIC — CENTURY AND CIERA

2.2L (VIN 4) ENGINE — ECM CONNECTOR END VIEW — CENTURY AND CIERA

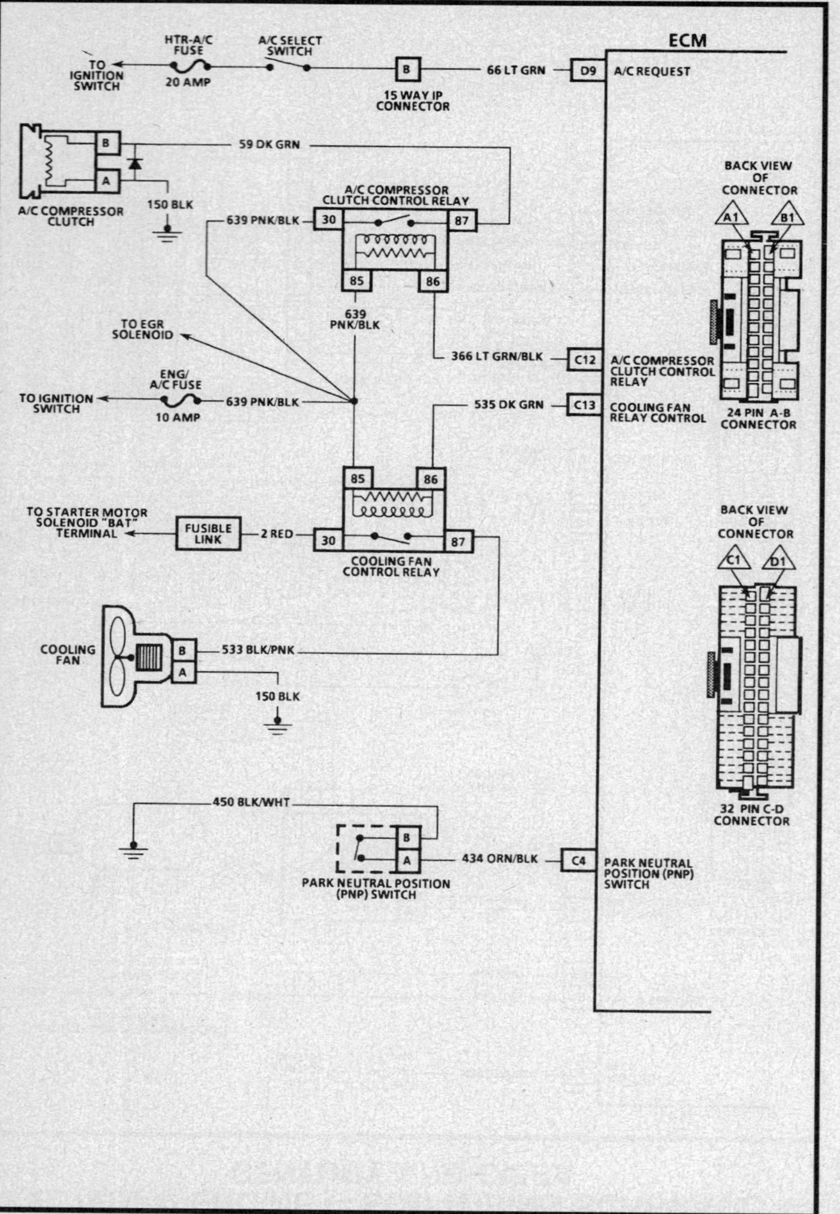

MULTIPORT FUEL INJECTION ECM CONNECTOR IDENTIFICATION

This ECM voltage chart is for use with J 39200 digital voltmeter to further aid in diagnosis. The voltages you get may vary due to low battery charge or other reasons, but they should be very close.

THE FOLLOWING CONDITIONS MUST BE MET BEFORE TESTING:
- Engine at operating temperature ● Engine idling in "Closed Loop" (For "Engine Run" column) in park or neutral ● Test terminal not grounded ● Scan tool not installed ● Brake not applied ● B + indicates battery or charging system voltage

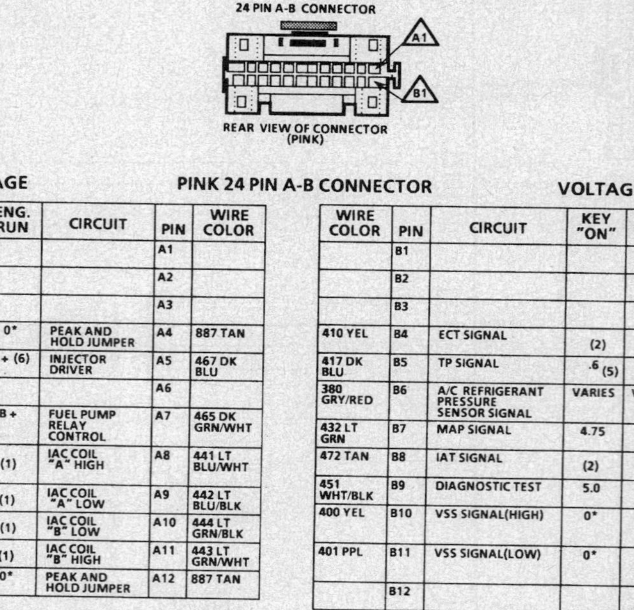

VOLTAGE					PINK 24 PIN A-B CONNECTOR			VOLTAGE	
KEY "ON"	ENG. RUN	CIRCUIT	PIN	WIRE COLOR	WIRE COLOR	PIN	CIRCUIT	KEY "ON"	ENG. RUN
			A1			B1			
			A2			B2			
			A3			B3			
0*	0*	PEAK AND HOLD JUMPER	A4	887 TAN	410 YEL	B4	ECT SIGNAL	(2)	(2)
B+	B+ (6)	INJECTOR DRIVER	A5	467 DK BLU	417 DK BLU	B5	TP SIGNAL	.6 (5)	.6 (5)
			A6		380 GRY/RED	B6	A/C REFRIGERANT PRESSURE SENSOR SIGNAL	VARIES	VARIES
(3)	B+	FUEL PUMP RELAY CONTROL	A7	465 DK GRN/WHT	432 LT GRN	B7	MAP SIGNAL	4.75	1.2
(1)	(1)	IAC COIL "A" HIGH	A8	441 LT BLU/WHT	472 TAN	B8	IAT SIGNAL	(2)	(2)
(1)	(1)	IAC COIL "A" LOW	A9	442 LT BLU/BLK	451 WHT/BLK	B9	DIAGNOSTIC TEST	5.0	5.0
(1)	(1)	IAC COIL "B" LOW	A10	444 LT GRN/BLK	400 YEL	B10	VSS SIGNAL(HIGH)	0*	(4)
(1)	(1)	IAC COIL "B" HIGH	A11	443 LT GRN/WHT	401 PPL	B11	VSS SIGNAL(LOW)	0*	(4)
0*	0*	PEAK AND HOLD JUMPER	A12	887 TAN		B12			

* ALL VOLTAGES SHOWN "0" SHOULD READ LESS THAN .5 VOLT.
(1) NOT USABLE.
(2) VARIES DEPENDING ON TEMPERATURE.
(3) READS B + FOR 2 SECONDS AFTER KEY "ON," THEN SHOULD READ 0 VOLT.
(4) AC VOLTAGE INCREASES WITH VEHICLE SPEED. (REFER TO SECTION 8A-33 FOR DIAGNOSIS.)
(5) WITHIN THE RANGE OF .33 TO 1.33 VOLTS.
(6) ALTERNATE TEST: SET J 39200 DVM TO DC VOLTAGE FREQUENCY SCALE. CONNECT RED LEAD TO IGNITION FEED ("C6" AT ECM) AND BLACK LEAD TO INJECTOR DRIVER ("A5" AT ECM); SHOULD MEASURE ABOUT 9-15 Hz.

ENGINE 2.2L LN2

2.2L (VIN 4) ENGINE — ECM CONNECTOR END VIEW — CENTURY AND CIERA

MULTIPORT FUEL INJECTION ECM CONNECTOR IDENTIFICATION

This ECM voltage chart is for use with J 39200 digital voltmeter to further aid in diagnosis. The voltages you get may vary due to low battery charge or other reasons, but they should be very close.

THE FOLLOWING CONDITIONS MUST BE MET BEFORE TESTING:

- Engine at operating temperature • Engine idling in "Closed Loop" (For "Engine Run" column) in park or neutral • Test terminal not grounded • Scan tool not installed • Brake not applied
- B + indicates battery or charging system voltage

32 PIN C-D CONNECTOR

REAR VIEW OF CONNECTOR (PINK)

VOLTAGE

KEY "ON"	ENG. RUN	CIRCUIT	PIN	WIRE COLOR
0*(5)	0*(5)	ECM GROUND	C1	450 BLK/WHT
			C2	
			C3	
0*	0*	PNP SWITCH	C4	434 ORN/BLK
			C5	
B + (5)	B + (5)	IGNITION FEED	C6	439 PNK/BLK
5.0	5.0	5 V REFERENCE	C7	474 GRY
5.0	5.0	5V REFERENCE	C8	416 GRY
B + (5)	B + (5)	KEEP ALIVE MEMORY (B +)	C9	440 ORN
B + (5)	B + (5)	KEEP ALIVE MEMORY (B +)	C10	440 ORN
0*	0*	TCC	C11	422 TAN/BLK
B + (1)	B + (1)	A/C COMPRESSOR CLUTCH CONTROL	C12	366 LT GRN/BLK
B + (1)	B + (1)	COOLING FAN RELAY CONTROL	C13	535 DK GRN
B +	B +	EGR SOLENOID CONTROL	C14	435 GRY
			C15	
			C16	

VOLTAGE

WIRE COLOR	PIN	CIRCUIT	KEY "ON"	ENG. RUN
450 BLK/WHT	D1	ECM GROUND	0*(5)	0*(5)
452 BLK	D2	SENSOR GROUND	0*	0*
469 BLK/ORN	D3	SENSOR GROUND	0*	0*
	D4			
423 WHT	D5	IGNITION CONTROL (IC)	0*	2.3
424 TAN/BLK	D6	BYPASS	0*	4.8
430 PPL/WHT	D7	REF HIGH	0*	3.2 (4)
453 BLK/RED	D8	REF LOW	0*	0*
66 LT GRN	D9	A/C REQUEST	0*(1)	0*(1)
413 TAN	D10	OXYGEN SENSOR REFERENCE LOW	0*	0*
412 PPL	D11	OXYGEN SENSOR SIGNAL	.01-.55	.1-.9 (3)
461 ORN	D12	SERIAL DATA	4.7	4.7
817 DK GRN/WHT	D13	VSS OUTPUT (4000 PPM)	(2)	(2)
419 BRN/WHT	D14	MIL (SERVICE ENGINE SOON) CONTROL	1.0	B +
	D15			
	D16			

ENGINE   2.2L LN2

* ALL VOLTAGES SHOWN "0" SHOULD READ LESS THAN .5 VOLT.
(1) A/C SELECT "OFF" AND ENGINE COOLING FAN "OFF."
(2) INCREASES WITH VEHICLE SPEED. REFER TO SECTION 8A-33.
(3) VARIES ACTIVELY WITHIN INDICATED RANGE.
(4) ALTERNATE TEST: SET J 39200 DVM TO DC VOLTAGE FREQUENCY SCALE. CONNECT RED LEAD TO REF HIGH ("D7" AT ECM) AND BLACK LEAD TO ECM GROUND ("D1" AT ECM); SHOULD MEASURE ABOUT 25-30 Hz.
(5) THIS TYPE OF CIRCUIT SHOULD ALSO BE CHECKED USING J 34142-B TEST LIGHT.

2.2L (VIN 4) ENGINE — ON-BOARD DIAGNOSTIC SYSTEM CHART — CENTURY AND CIERA

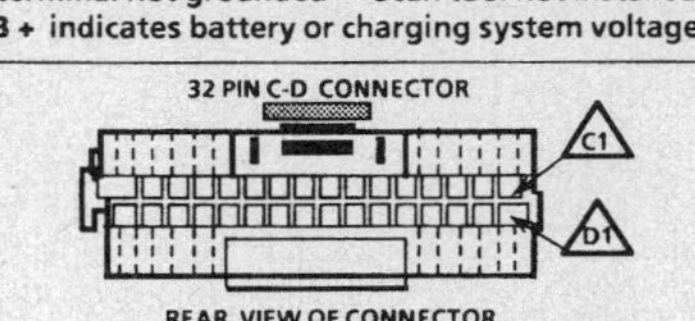

ON-BOARD DIAGNOSTIC (OBD) SYSTEM CHECK
2.2L (VIN 4) "A" CARLINE

Circuit Description:

The OBD system check is an organized approach to identifying a problem created by an electronic engine control system malfunction. It must be the starting point for any driveability complaint diagnosis, because it directs the service technician to the next logical step in diagnosing the complaint. Understanding the chart and using it correctly will reduce diagnostic time and prevent the unnecessary replacement of good parts.

Test Description: Number(s) below refer to circled number(s) on the diagnostic chart.

1. This step is a check for the proper operation of the Malfunction Indicator Lamp (MIL) "Service Engine Soon." The MIL should be "ON" steady.
2. No MIL at this point indicates that there is a problem with the MIL circuit or the ECM control of that circuit.
3. This test checks the ability of the ECM to control the MIL. With the diagnostic terminal grounded, the MIL should flash a DTC 12 three times, followed by any DTC stored in memory. Depending upon the type of ECM, an EEPROM error may result in the inability to flash DTC 12.
4. Most of the 6E procedures use a Tech 1 to aid diagnosis, therefore, serial data must be available. If an EEPROM error is present, the ECM may have been able to flash DTC 12/51, but not enable serial data.
5. Although the ECM is powered up, a "Cranks But Will Not Run" symptom could exist because of an ECM or system problem.
6. This step will isolate if the customer complaint is a MIL or a driveability problem with no MIL. Refer to "ECM Diagnostic Trouble Codes" in this section for a list of valid DTC(s). An invalid DTC may be the result of a faulty scan tool, EEPROM or ECM.
7. Comparison of actual control system data with the typical values is a quick check to determine if any parameter is not within limits. Keep in mind that a base engine problem (i.e. advanced cam timing) may substantially alter sensor values.
8. Installation of a scan tool will provide a good ground path for the ECM and may hide a driveability complaint due to poor ECM grounds.
9. If the actual data is not within the typical values established, the diagnostic charts in "Component Systems,"   will provide a functional check of the suspect component or system.

2.2L (VIN 4) ENGINE — ON-BOARD DIAGNOSTIC SYSTEM CHART — CENTURY AND CIERA

2.2L (VIN 4) ENGINE — ECM DIAGNOSTIC TROUBLE CODES — CENTURY AND CIERA

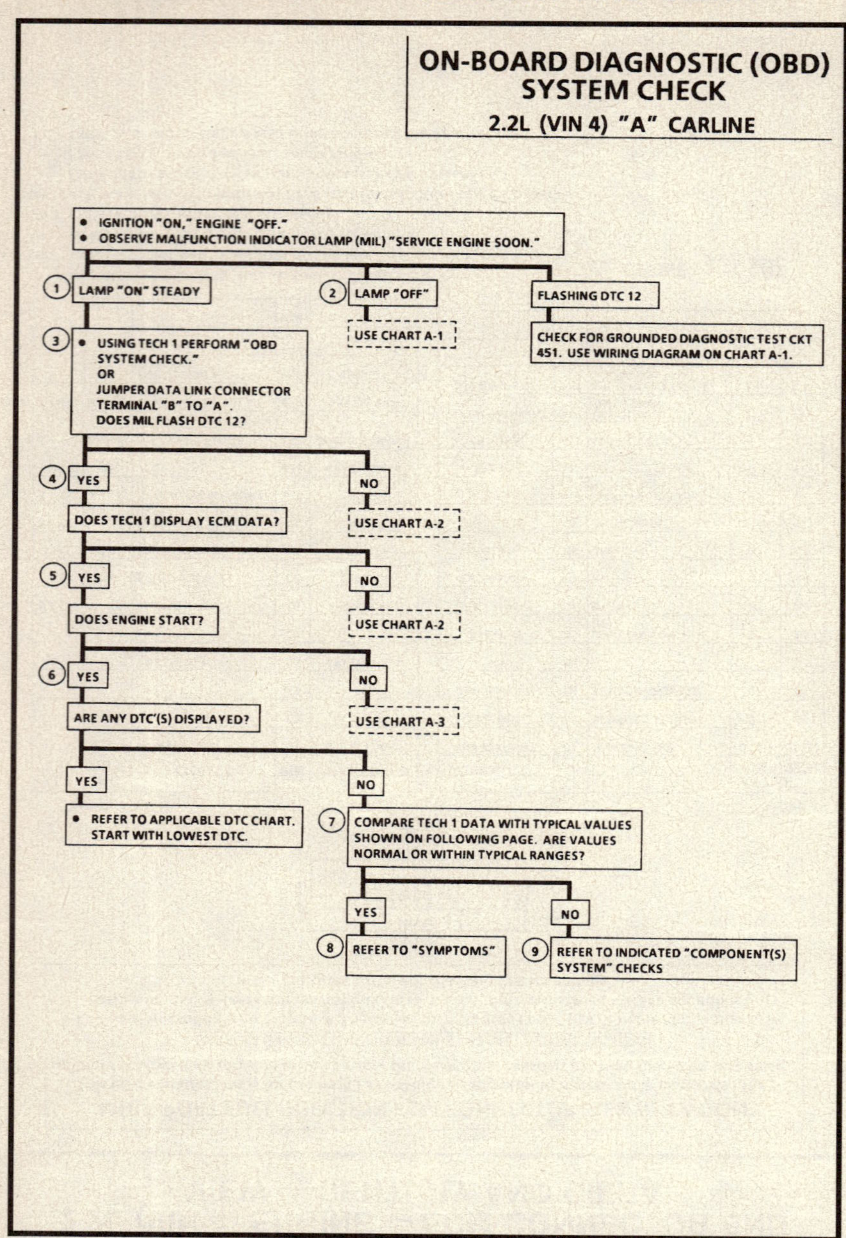

DTC	DESCRIPTION	ILLUMINATE MIL (SERVICE ENGINE SOON)
	ECM DIAGNOSTIC TROUBLE CODES	
13	Oxygen Sensor (O2S) Circuit - open circuit	YES
14	Engine Coolant Temperature (ECT) Sensor Circuit - high/low temperature	YES
21	Throttle Position (TP) Sensor Circuit - Signal voltage high/low	YES
23	Intake Air Temperature (IAT) Sensor Circuit - high/low temperature	YES
24	Vehicle Speed Sensor (VSS) Circuit	YES
32	Exhaust Gas Recirculation (EGR) - System failure	YES
33	Manifold Absolute Pressure (MAP) Sensor Circuit - Signal voltage high/low - low/high vacuum	YES
42	Ignition Control (IC) Circuit	YES
44	Oxygen Sensor (O2S) Circuit - lean exhaust indicated	YES
45	Oxygen Sensor (O2S) Circuit - rich exhaust indicated	YES
51	EEPROM or ECM Failure	YES
66	A/C Refrigerant Pressure Sensor Circuit	NO

If a DTC not listed above appears on Tech 1, ground Data Link Terminal "B" and observe flashed DTC(S).
If DTC does not reappear, Tech 1 data may be faulty.
If DTC does reappear, replace ECM and program. Refer to ECM replacement and programming procedures in Section "6E3-C1".

2.2L (VIN 4) ENGINE — SYSTEM DIAGNOSTIC CHARTS — CENTURY AND CIERA

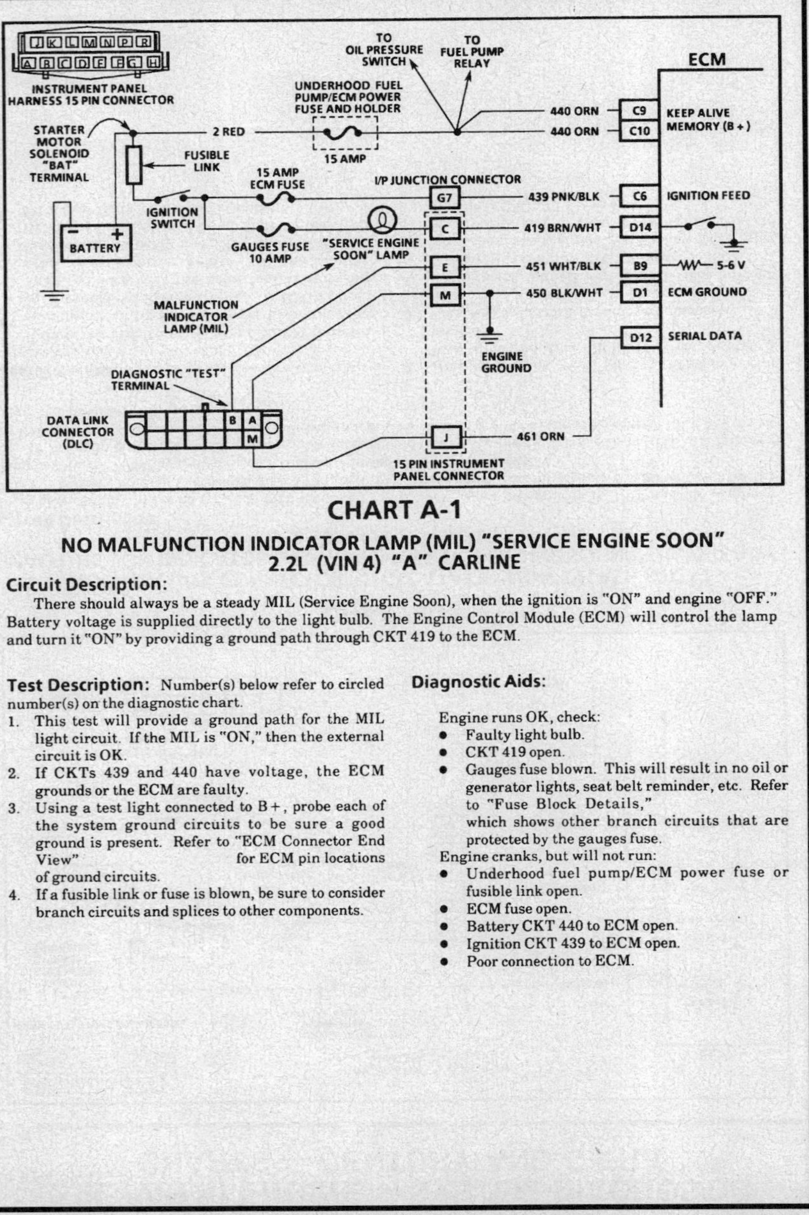

CHART A-1
NO MALFUNCTION INDICATOR LAMP (MIL) "SERVICE ENGINE SOON" 2.2L (VIN 4) "A" CARLINE

Circuit Description:

There should always be a steady MIL (Service Engine Soon), when the ignition is "ON" and engine "OFF." Battery voltage is supplied directly to the light bulb. The Engine Control Module (ECM) will control the lamp and turn it "ON" by providing a ground path through CKT 419 to the ECM.

Test Description: Number(s) below refer to circled number(s) on the diagnostic chart.

1. This test will provide a ground path for the MIL light circuit. If the MIL is "ON," then the external circuit is OK.
2. If CKTs 439 and 440 have voltage, the ECM grounds or the ECM are faulty.
3. Using a test light connected to B+, probe each of the system ground circuits to be sure a good ground is present. Refer to "ECM Connector End View" for ECM pin locations of ground circuits.
4. If a fusible link or fuse is blown, be sure to consider branch circuits and splices to other components.

Diagnostic Aids:

Engine runs OK, check:
- Faulty light bulb.
- CKT 419 open.
- Gauges fuse blown. This will result in no oil or generator lights, seat belt reminder, etc. Refer to "Fuse Block Details," which shows other branch circuits that are protected by the gauges fuse.

Engine cranks, but will not run:
- Underhood fuel pump/ECM power fuse or fusible link open.
- ECM fuse open.
- Battery CKT 440 to ECM open.
- Ignition CKT 439 to ECM open.
- Poor connection to ECM.

2.2L (VIN 4) ENGINE — SYSTEM DIAGNOSTIC CHARTS — CENTURY AND CIERA

CHART A-1
NO MALFUNCTION INDICATOR LAMP (MIL) "SERVICE ENGINE SOON" 2.2L (VIN 4) "A" CARLINE

DOES THE ENGINE START?

YES

1.
- IGNITION "OFF."
- DISCONNECT ECM CONNECTORS.
- IGNITION "ON."
- PROBE CKT 419 WITH TEST LIGHT TO GROUND. IS THE MIL "ON"?

YES → FAULTY ECM CONNECTION OR ECM.*

NO → CHECK:
- BLOWN GAUGES FUSE.**
- FAULTY CONNECTION.
- FAULTY BULB.
- OPEN CKT 419.
- CKT 419 SHORTED TO VOLTAGE.
- OPEN IGNITION FEED TO BULB.

NO

2.
- IGNITION "OFF."
- DISCONNECT ECM CONNECTORS.
- IGNITION "ON."
- PROBE CKT'S 440 & 439 WITH TEST LIGHT TO GROUND. IS THE LIGHT "ON" FOR BOTH CIRCUITS?

YES → 3. FAULTY ECM GROUNDS OR ECM.*

NO → 4. REPAIR OPEN IN CIRCUIT THAT DID NOT LIGHT THE TEST LIGHT. IF A FUSIBLE LINK OR FUSE WAS BLOWN, LOCATE AND CORRECT THE SHORT TO GROUND ON THAT CIRCUIT.

* IF ECM IS FAULTY AND MUST BE REPLACED, THE NEW ECM MUST BE PROGRAMMED. REFER TO ECM REPLACEMENT AND PROGRAMMING PROCEDURES

** SEVERAL BRANCH CIRCUITS ARE PROTECTED BY THIS FUSE.

2.2L (VIN 4) ENGINE — SYSTEM DIAGNOSTIC CHARTS — CENTURY AND CIERA

2.2L (VIN 4) ENGINE — SYSTEM DIAGNOSTIC CHARTS — CENTURY AND CIERA

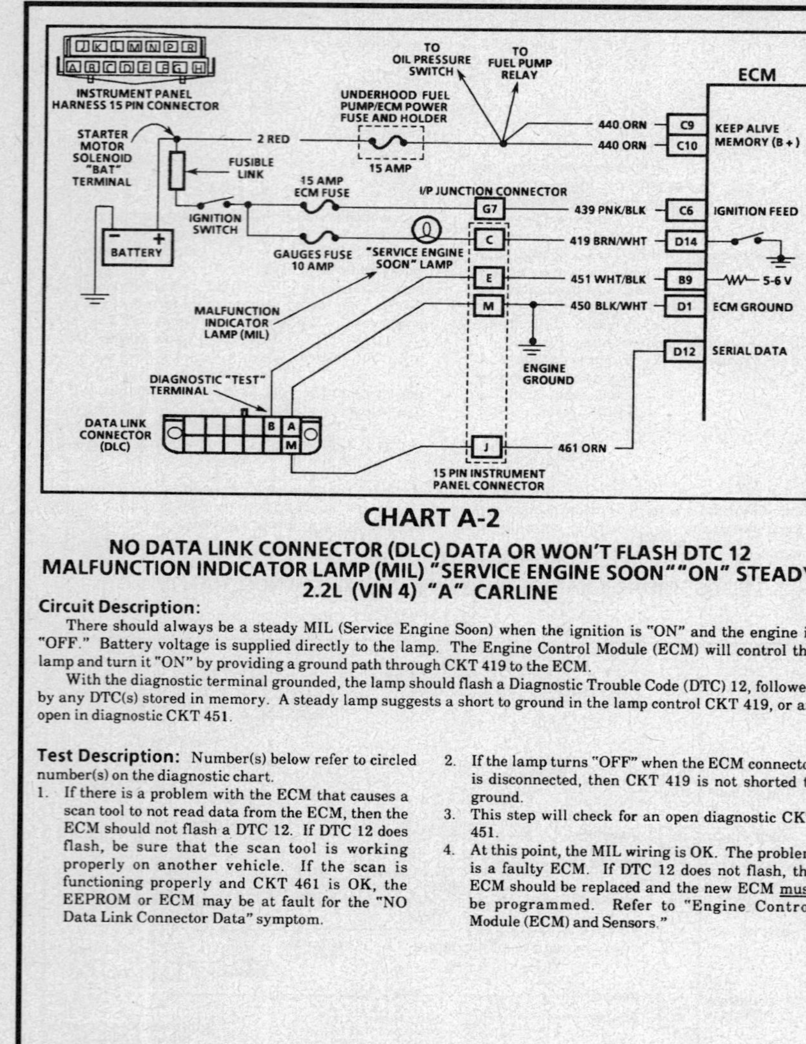

CHART A-2

NO DATA LINK CONNECTOR (DLC) DATA OR WON'T FLASH DTC 12 MALFUNCTION INDICATOR LAMP (MIL) "SERVICE ENGINE SOON" "ON" STEADY 2.2L (VIN 4) "A" CARLINE

Circuit Description:

There should always be a steady MIL (Service Engine Soon) when the ignition is "ON" and the engine is "OFF." Battery voltage is supplied directly to the lamp. The Engine Control Module (ECM) will control the lamp and turn it "ON" by providing a ground path through CKT 419 to the ECM.

With the diagnostic terminal grounded, the lamp should flash a Diagnostic Trouble Code (DTC) 12, followed by any DTC(s) stored in memory. A steady lamp suggests a short to ground in the lamp control CKT 419, or an open in diagnostic CKT 451.

Test Description: Number(s) below refer to circled number(s) on the diagnostic chart.

1. If there is a problem with the ECM that causes a scan tool to not read data from the ECM, then the ECM should not flash a DTC 12. If DTC 12 does flash, be sure that the scan tool is working properly on another vehicle. If the scan is functioning properly and CKT 461 is OK, the EEPROM or ECM may be at fault for the "NO Data Link Connector Data" symptom.

2. If the lamp turns "OFF" when the ECM connector is disconnected, then CKT 419 is not shorted to ground.

3. This step will check for an open diagnostic CKT 451.

4. At this point, the MIL wiring is OK. The problem is a faulty ECM. If DTC 12 does not flash, the ECM should be replaced and the new ECM must be programmed. Refer to "Engine Control Module (ECM) and Sensors."

CHART A-2

NO DATA LINK CONNECTOR (DLC) DATA OR WON'T FLASH DTC 12 MALFUNCTION INDICATOR LAMP (MIL) "SERVICE ENGINE SOON" "ON" STEADY 2.2L (VIN 4) "A" CARLINE

- IGNITION "ON," ENGINE "OFF."
  IS THE MIL (SERVICE ENGINE SOON) "ON"?

YES
- GROUND DIAGNOSTIC "TEST" TERMINAL. DOES LAMP FLASH DTC 12?

NO
USE CHART A-1.

NO

(2)
- IGNITION "OFF."
- DISCONNECT ECM CONNECTORS.
- IGNITION "ON" AND NOTE MIL (SERVICE ENGINE SOON).

YES

(1) IF PROBLEM WAS NO DATA LINK CONNECTOR DATA: CHECK SERIAL DATA CKT 461 FOR OPEN OR SHORT TO GROUND BETWEEN ECM AND DATA LINK CONNECTOR. IF OK, IT IS A FAULTY ECM OR EEPROM.*

LAMP "OFF"

(3)
- IGNITION "OFF."
- RECONNECT ECM.
- IGNITION "ON," ENGINE "OFF." DIAGNOSTIC "TEST" TERMINAL NOT GROUNDED.
- BACKPROBE ECM CKT 451 WITH TEST LIGHT TO GROUND. LEAVE CONNECTED AND WATCH MIL.

LAMP "ON"

REPAIR SHORT TO GROUND IN CKT 419.

NO CODE 12

(4)
- REPLACE ECM AND REPROGRAM.*
- RECHECK FOR DTC 12.

CODE 12

CHECK FOR OPEN CKT 451 TO ECM. IF OK, CHECK FOR OPEN CKT 450 BETWEEN DATA LINK CONNECTOR TERMINAL "A" AND ENGINE GROUND.

* IF ECM IS FAULTY AND MUST BE REPLACED, THEN IT MUST ALSO BE REPROGRAMMED. REFER TO ECM REPLACEMENT AND PROGRAMMING

2.2L (VIN 4) ENGINE — SYSTEM DIAGNOSTIC CHARTS — CENTURY AND CIERA

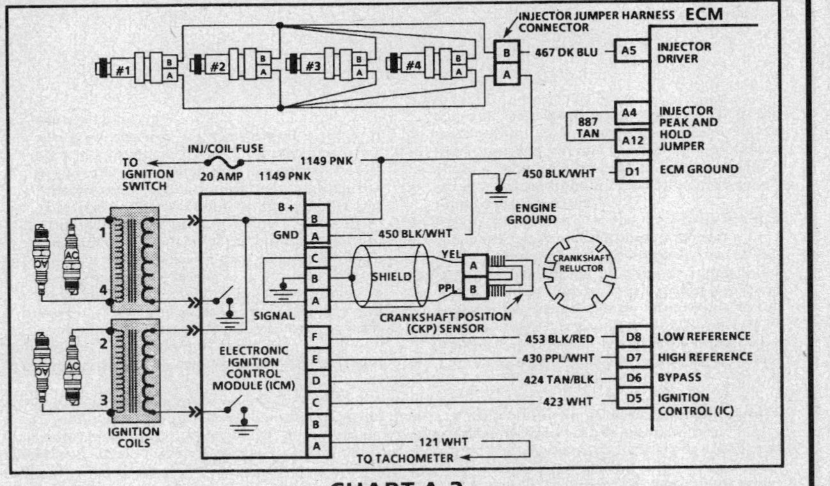

CHART A-3
(Page 1 of 3)
ENGINE CRANKS BUT WON'T RUN
2.2L (VIN 4) "A" CARLINE

Circuit Description:

Before using this chart, battery condition, engine cranking speed, and fuel quantity should be checked and verified as being OK.

Test Description: Number(s) below refer to circled number(s) on the diagnostic chart.

1. - A MIL (Service Engine Soon) "ON" is a basic test to determine if there is battery and ignition voltage at the ECM.
   - No DLC data may be the result of an ECM problem, and CHART A-2 will diagnose an ECM problem.
   - If Throttle Position (TP) sensor is less than .2 volt, the TP sensor 5 volt reference circuit could be shorted to ground. If TP sensor is over 2.5 volts, the ECM could be in the "Clear Flood Mode" which may cause the engine to not start.
   - Compare coolant temperature with intake air temperature when engine is cold. If coolant temperature reading is 10 degrees greater or less than intake air temperature on a cold engine, check resistance of the Engine Coolant Temperature (ECT) sensor circuit or sensor. Compare resistance value to the "Diagnostic Aid" chart found on DTC 14 chart.
   - The scan tool should display RPM during cranking.
2. Because the Electronic Ignition (EI) system uses two spark plugs and cables to complete the circuit of each coil, the companion spark plug cable should be connected to a good ground.
3. This test is performed with injectors 1, 2, 3, and 4 in parallel.
4. The test light should flash, indicating that the ECM is controlling the injectors. How bright the light flashes is not important.
5. Ignition may have to be cycled "ON" several times to obtain maximum fuel pressure.
6. Damage to the ECM injector driver may occur if any injector resistance measures less than 11.6 ohms (internal injector short to CKT 1149).

Diagnostic Aids:

- Water or contamination in fuel system may cause a no start condition during very cold or freezing weather. The engine may start after approximately 5 minutes in a heated shop.
- An EGR valve sticking open can cause a rich air/fuel ratio during cranking. Unless the ECM enters "Clear Flood Mode" at the first indication of a flooding condition, it may result in a no start condition.
- An A/C refrigerant pressure sensor with an internal short to ground can cause a no start condition. Disconnect the A/C refrigerant pressure sensor, if vehicle starts replace faulty sensor.
- A MAP sensor stuck between .5 and 2.5 volts can cause a no start condition. Disconnect the MAP sensor, if vehicle starts replace faulty sensor.

2.2L (VIN 4) ENGINE — SYSTEM DIAGNOSTIC CHARTS — CENTURY AND CIERA

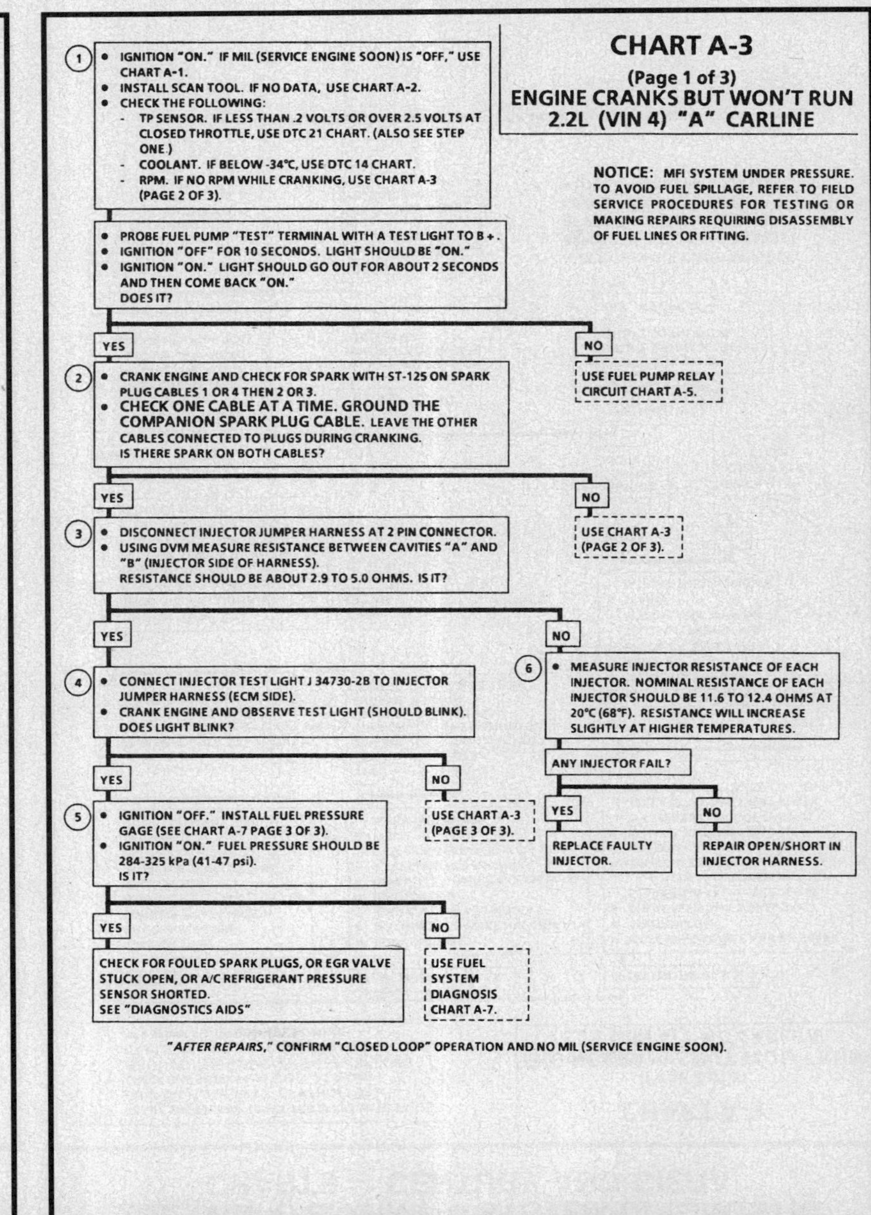

2.2L (VIN 4) ENGINE — SYSTEM DIAGNOSTIC CHARTS — CENTURY AND CIERA

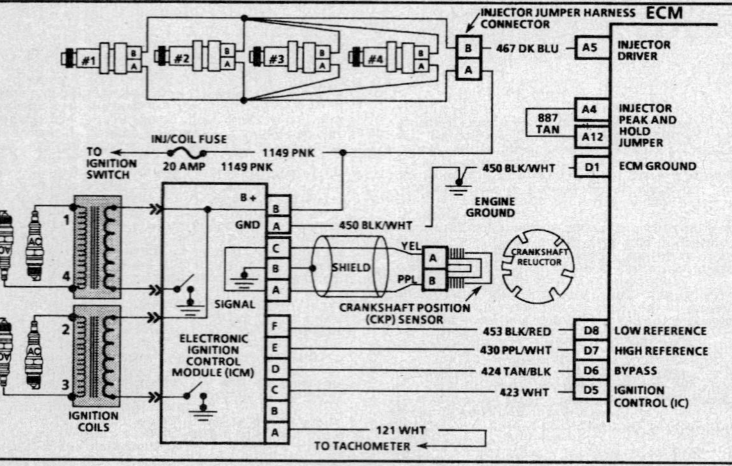

CHART A-3
(Page 2 of 3)
ENGINE CRANKS BUT WON'T RUN
2.2L (VIN 4) "A" CARLINE

Circuit Description:

A magnetic Crankshaft Position (CKP) sensor is used to determine engine crankshaft position, much the same way as the pick-up coil did in Distributor Ignition (DI) type systems. The sensor is mounted in the block, near a slotted wheel on the crankshaft. The rotation of the wheel creates a flux change in the sensor, which produces a voltage signal. The Electronic Ignition Control Module (ICM) processes this signal and creates the reference pulses needed by the Engine Control Module (ECM) to trigger the correct coil at the correct time.

If the scan tool did not indicate cranking RPM, and there is no spark present at the plugs, the problem lies in the ICM, CKP sensor or the power and ground supplies to the ICM.

Test Description: Number(s) below refer to circled number(s) on the diagnostic chart.

1. The Electronic Ignition (EI) system uses two plugs and cables to complete the circuit of each coil. The other spark plug cable in the circuit must be connected to a good ground.
2. This test will determine if the 12 volt supply and a good ground is available at the Electronic Ignition Control Module (ICM).
3. This test will determine if the ICM is not generating the reference pulse, or if the wiring or ECM are at fault. By touching and removing a test light to 12 volts on CKT 430, a reference pulse should be generated. If RPM is indicated, the ECM and wiring are OK.
4. This test will determine if the ICM is not triggering the problem coil, or if the tested coil is at fault. This test could also be performed by substituting a known good coil. The secondary coil winding can be checked with a Digital Voltmeter (DVM). There should be 5,000 to 10,000 ohms across the coil towers. There should not be any continuity from either coil tower to ground.
5. Checks for continuity of the crankshaft position sensor and connections.
6. Normal crankshaft position sensor voltage output range is .8 to 1.4 volts (800 to 1400 mV) with a fully charged battery and engine at room temperature. Minimum output voltage (slow cranking, low battery) can be as low as .3 volt (300 mV).

2.2L (VIN 4) ENGINE — SYSTEM DIAGNOSTIC CHARTS — CENTURY AND CIERA

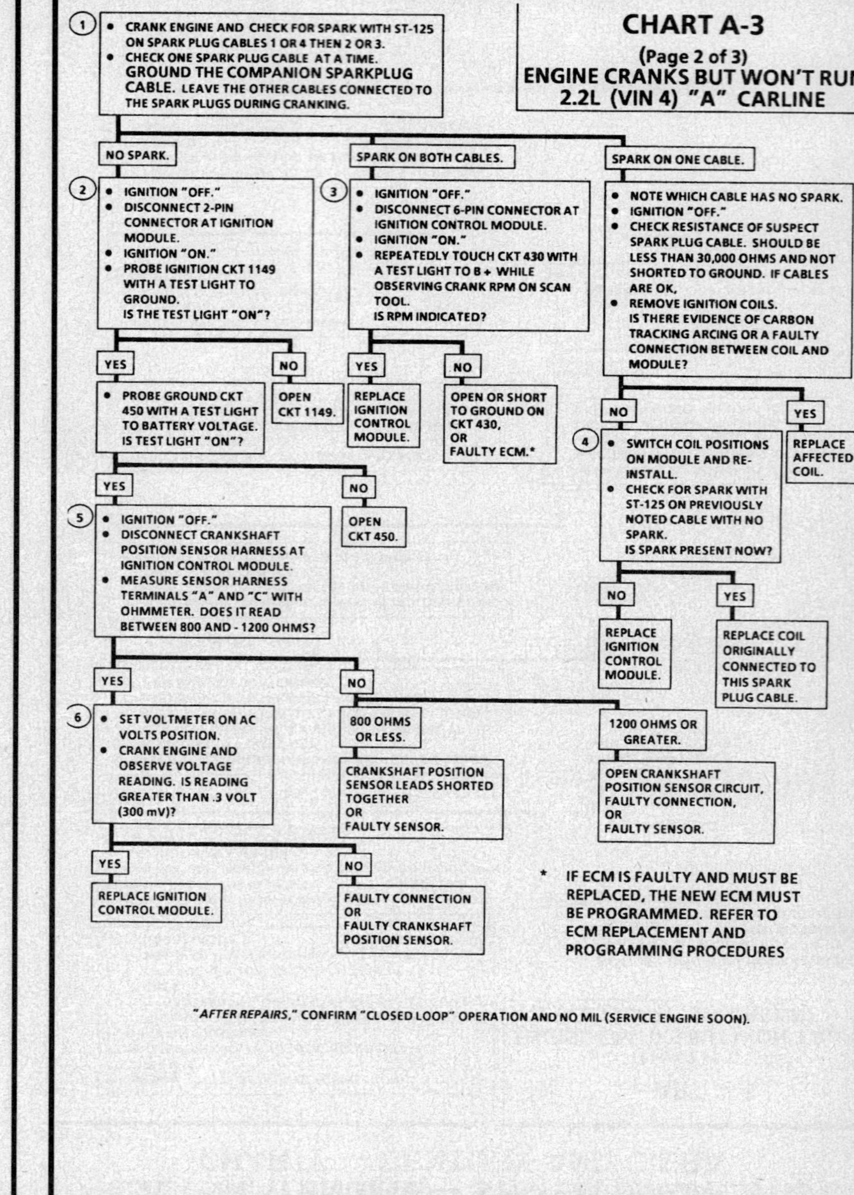

2.2L (VIN 4) ENGINE — SYSTEM DIAGNOSTIC CHARTS — CENTURY AND CIERA

CHART A-3
(Page 3 of 3)
ENGINE CRANKS BUT WON'T RUN
2.2L (VIN 4) "A" CARLINE

2.2L (VIN 4) ENGINE — SYSTEM DIAGNOSTIC CHARTS — CENTURY AND CIERA

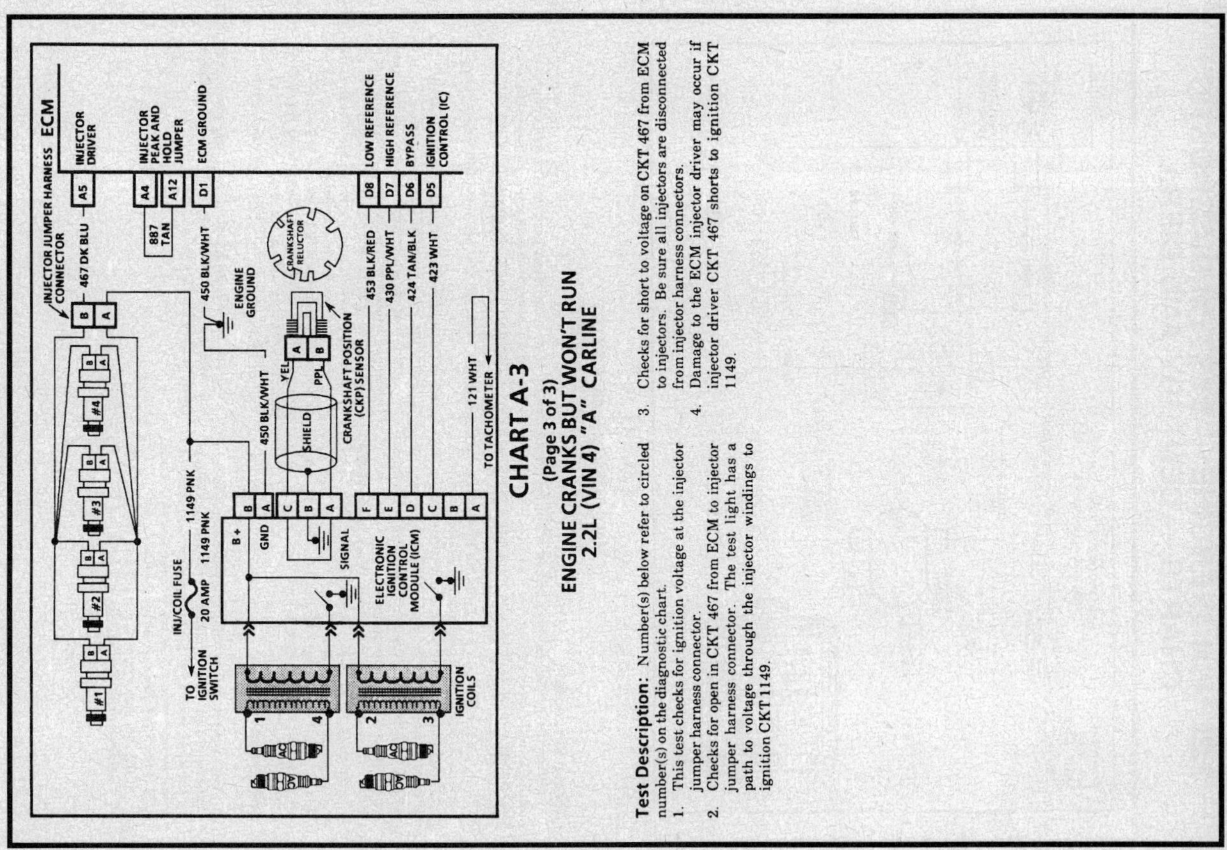

CHART A-3
(Page 3 of 3)
ENGINE CRANKS BUT WON'T RUN
2.2L (VIN 4) "A" CARLINE

Test Description: Number(s) below refer to circled number(s) on the diagnostic chart.

1. This test checks for ignition voltage at the injector jumper harness connector.

2. Checks for open in CKT 467 from ECM to injector jumper harness connector. The test light has a path to voltage through the injector windings to ignition CKT 1149.

3. Checks for short to voltage on CKT 467 from ECM to injectors. Be sure all injectors are disconnected from injector harness connectors.

4. Damage to the ECM injector driver may occur if injector driver CKT 467 shorts to ignition CKT 1149.

2.2L (VIN 4) ENGINE — SYSTEM DIAGNOSTIC CHARTS — CENTURY AND CIERA

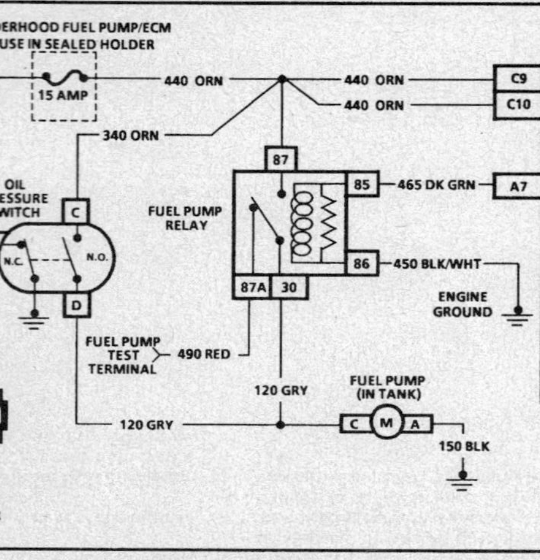

CHART A-5

FUEL PUMP RELAY CIRCUIT
2.2L (VIN 4) "A" CARLINE

Circuit Description:

When the ignition switch is turned "ON," the Engine Control Module (ECM) will activate the fuel pump relay with a 12 volt supply and run the in-tank fuel pump. The fuel pump will operate as long as the engine is cranking or running and the ECM is receiving ignition reference pulses. If there are no ignition reference pulses, the ECM will no longer supply the fuel pump relay signal within 2 seconds after key "ON."

Should the fuel pump relay or the 12 volt relay drive from the ECM fail, the fuel pump will receive supply current through the oil pressure switch back-up circuit.

The fuel pump "test" terminal is located in the driver's side of the engine compartment. When the engine is stopped, the pump can be turned "ON" by applying battery voltage to the "test" terminal.

Test Description: Number(s) below refer to circled number(s) on the diagnostic chart.
1. At this point, the fuel pump relay is operating correctly. The back-up circuit through the oil pressure switch is now tested.
2. After the fuel pump relay is replaced, continue with "Oil Pressure Switch Test."

Diagnostic Aids:

An inoperative fuel pump relay can result in long cranking times. The extended crank period is caused by the time necessary for oil pressure to reach the pressure required to close the oil pressure switch and supply the necessary current for the fuel pump.

If the fuel pump relay circuit checks out OK, refer to "Fuel System Diagnosis," CHART A-7.

Excess fuel may also cause long cranking times. This would usually be accompanied by a start that is not as fast as normal (once fired, the engine does not build up speed as fast) and a puff of black smoke at the tailpipe. An improperly connected or faulty EVAP canister control valve can cause this problem. Disconnect the "EVAP purge hose" from the "EVAP canister control valve" to diagnose. Refer to "Evaporative Emission (EVAP) Control System,"

One or more leaking fuel injectors may also extend the cranking time. Perform the "Injector Balance Test" in CHART C2-A

2.2L (VIN 4) ENGINE — SYSTEM DIAGNOSTIC CHARTS — CENTURY AND CIERA

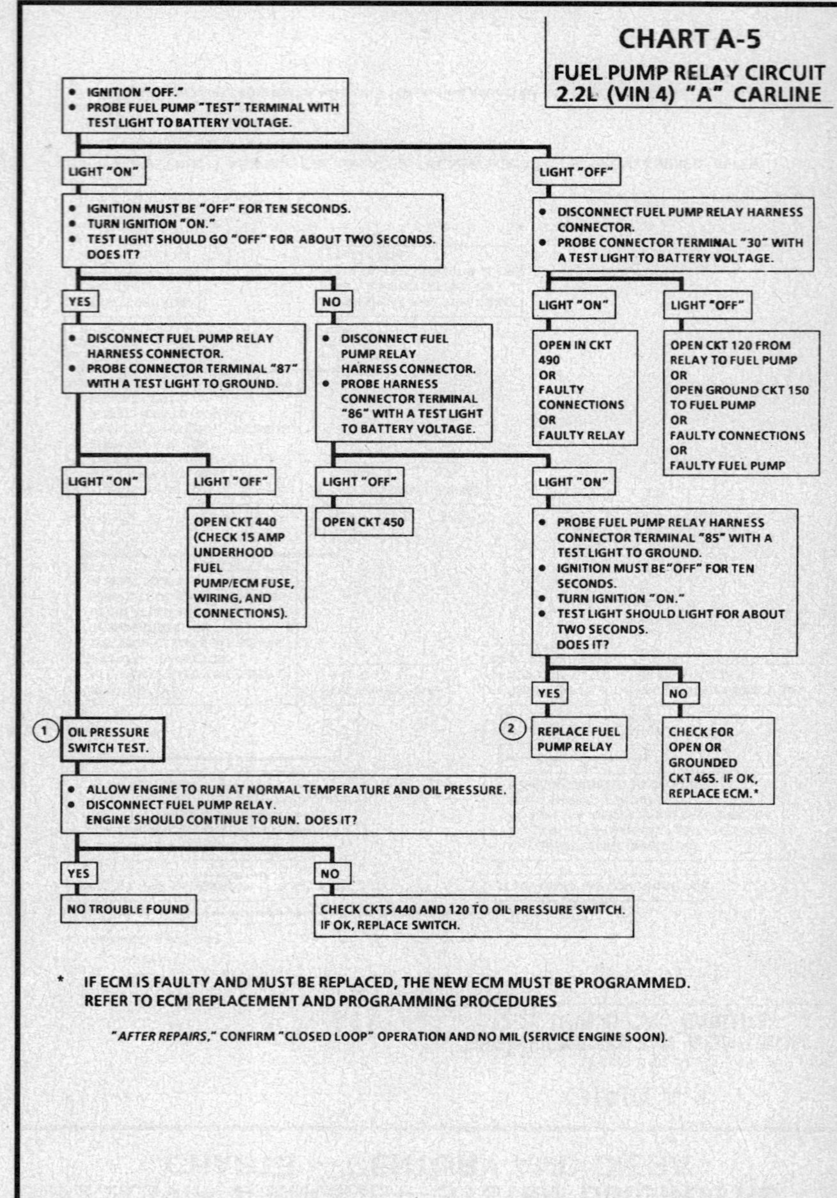

2.2L (VIN 4) ENGINE — SYSTEM DIAGNOSTIC CHARTS — CENTURY AND CIERA

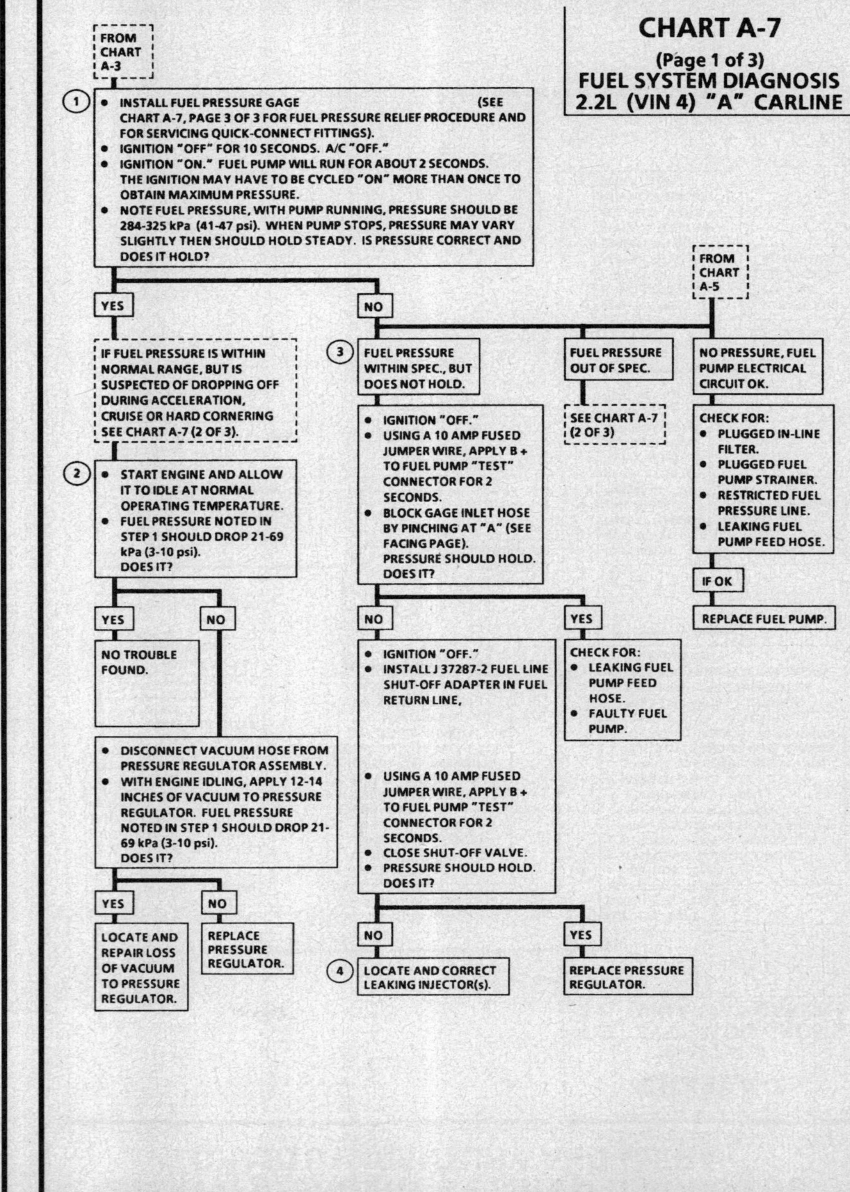

CHART A-7

(Page 1 of 3)
FUEL SYSTEM DIAGNOSIS
2.2L (VIN 4) "A" CARLINE

Circuit Description:

When the ignition switch is turned "ON," the Engine Control Module (ECM) will turn "ON" the in-tank fuel pump. It will remain "ON" as long as the engine is cranking or running, and the ECM is receiving reference pulses. If there are no reference pulses, the ECM will shut "OFF" the fuel pump within 2 seconds after ignition "ON" or engine stops.

An electric fuel pump, attached to the fuel sender assembly (inside the fuel tank), pumps fuel through an in-line filter to the fuel passage within the lower manifold assembly. The pump is designed to provide fuel at a pressure above the regulated pressure needed by the injectors. A pressure regulator, attached to the lower manifold assembly, keeps fuel available to the injectors at a regulated pressure. Unused fuel is returned to the fuel tank by a separate line.

Test Description: Number(s) below refer to circled number(s) on the diagnostic chart.

1. Install fuel pressure gage as shown in illustration. See "Fuel Pressure Relief Procedure" and for "Servicing Quick-Connect Fittings." Ignition "ON" pump pressure should be 284-325 kPa (41-47 psi). This pressure is controlled by spring pressure within the regulator assembly.
2. When the engine is idling, the manifold pressure is low (high vacuum) and is applied to the fuel regulator diaphragm. This will offset the spring and result in a lower fuel pressure.

This idle pressure will vary somewhat depending on barometric pressure, however, the pressure idling should be less, indicating pressure regulator control.

3. Pressure that continues to fall is caused by one of the following:
   - In-tank fuel pump check valve not holding.
   - Leaking fuel pump connecting hose.
   - Fuel pressure regulator valve leaking.
   - Injector(s) sticking open.
4. An injector sticking open can best be determined by checking for a fouled or saturated spark plug(s).

2.2L (VIN 4) ENGINE — SYSTEM DIAGNOSTIC CHARTS — CENTURY AND CIERA

CHART A-7

(Page 1 of 3)
FUEL SYSTEM DIAGNOSIS
2.2L (VIN 4) "A" CARLINE

FROM CHART A-3

1.
   - INSTALL FUEL PRESSURE GAGE (SEE CHART A-7, PAGE 3 OF 3 FOR FUEL PRESSURE RELIEF PROCEDURE AND FOR SERVICING QUICK-CONNECT FITTINGS).
   - IGNITION "OFF" FOR 10 SECONDS. A/C "OFF."
   - IGNITION "ON." FUEL PUMP WILL RUN FOR ABOUT 2 SECONDS. THE IGNITION MAY HAVE TO BE CYCLED "ON" MORE THAN ONCE TO OBTAIN MAXIMUM PRESSURE.
   - NOTE FUEL PRESSURE, WITH PUMP RUNNING, PRESSURE SHOULD BE 284-325 kPa (41-47 psi). WHEN PUMP STOPS, PRESSURE MAY VARY SLIGHTLY THEN SHOULD HOLD STEADY. IS PRESSURE CORRECT AND DOES IT HOLD?

YES → IF FUEL PRESSURE IS WITHIN NORMAL RANGE, BUT IS SUSPECTED OF DROPPING OFF DURING ACCELERATION, CRUISE OR HARD CORNERING SEE CHART A-7 (2 OF 3).

2.
   - START ENGINE AND ALLOW IT TO IDLE AT NORMAL OPERATING TEMPERATURE.
   - FUEL PRESSURE NOTED IN STEP 1 SHOULD DROP 21-69 kPa (3-10 psi). DOES IT?

YES → NO TROUBLE FOUND.

NO →
   - DISCONNECT VACUUM HOSE FROM PRESSURE REGULATOR ASSEMBLY.
   - WITH ENGINE IDLING, APPLY 12-14 INCHES OF VACUUM TO PRESSURE REGULATOR. FUEL PRESSURE NOTED IN STEP 1 SHOULD DROP 21-69 kPa (3-10 psi). DOES IT?

YES → LOCATE AND REPAIR LOSS OF VACUUM TO PRESSURE REGULATOR.

NO → REPLACE PRESSURE REGULATOR.

NO →
3. FUEL PRESSURE WITHIN SPEC., BUT DOES NOT HOLD.
   - IGNITION "OFF."
   - USING A 10 AMP FUSED JUMPER WIRE, APPLY B + TO FUEL PUMP "TEST" CONNECTOR FOR 2 SECONDS.
   - BLOCK GAGE INLET HOSE BY PINCHING AT "A" (SEE FACING PAGE). PRESSURE SHOULD HOLD. DOES IT?

NO →
   - IGNITION "OFF."
   - INSTALL J 37287-2 FUEL LINE SHUT-OFF ADAPTER IN FUEL RETURN LINE,
   - USING A 10 AMP FUSED JUMPER WIRE, APPLY B + TO FUEL PUMP "TEST" CONNECTOR FOR 2 SECONDS.
   - CLOSE SHUT-OFF VALVE.
   - PRESSURE SHOULD HOLD. DOES IT?

NO →
4. LOCATE AND CORRECT LEAKING INJECTOR(s).

YES → REPLACE PRESSURE REGULATOR.

YES → CHECK FOR:
   - LEAKING FUEL PUMP FEED HOSE.
   - FAULTY FUEL PUMP.

FUEL PRESSURE OUT OF SPEC. → SEE CHART A-7 (2 OF 3)

FROM CHART A-5 → NO PRESSURE, FUEL PUMP ELECTRICAL CIRCUIT OK.

CHECK FOR:
   - PLUGGED IN-LINE FILTER.
   - PLUGGED FUEL PUMP STRAINER.
   - RESTRICTED FUEL PRESSURE LINE.
   - LEAKING FUEL PUMP FEED HOSE.

IF OK → REPLACE FUEL PUMP.

2.2L (VIN 4) ENGINE—SYSTEM DIAGNOSTIC
CHARTS—CENTURY AND CIERA

CHART A-7

(Page 2 of 3)
FUEL SYSTEM DIAGNOSIS
2.2L (VIN 4) "A" CARLINE

FROM CHART A-7
(1 OF 3)

(5) FUEL PRESSURE DROPS OFF DURING ACCELERATION, CRUISE OR HARD CORNERING.

CHECK FOR RESTRICTED FUEL PRESSURE LINE OR IN-LINE FUEL FILTER. IS THERE A RESTRICTION?

YES — REPLACE FILTER OR LOCATE AND CORRECT RESTRICTION IN FUEL LINE AND RECHECK.

NO —

CHECK FOR:
• RESTRICTED FUEL PUMP STRAINER.
• LEAKING FUEL PULSE DAMPENER.
• FAULTY FUEL PUMP.
• INCORRECT FUEL PUMP.

(6) FUEL PRESSURE LESS THAN 284 kPa (41 psi)

(7)
• IGNITION "OFF."
• USING A 10 AMP FUSED JUMPER WIRE, APPLY B+ TO FUEL PUMP "TEST" CONNECTOR.
• GRADUALLY PINCH PRESSURE GAGE OUTLET HOSE AT (B). PRESSURE SHOULD RISE ABOVE 325 kPa (47 psi). DO NOT EXCEED 414 kPa (60 psi). DOES PRESSURE RISE ABOVE 325 kPa (47 psi)?

YES — REPLACE PRESSURE REGULATOR.

NO —

(8) FUEL PRESSURE ABOVE 325 kPa (47 psi)

(9)
• IGNITION "OFF."
• DISCONNECT QUICK-CONNECT FITTING AT ENGINE FUEL RETURN PIPE. (PROCEDURES FOR SERVICING QUICK-CONNECT FITTINGS ARE FOUND ON PAGE 3 OF 3.)
• ATTACH 5/16 I.D. FLEX HOSE TO ENGINE SIDE OF FUEL RETURN PIPE. PLACE OPEN END OF HOSE INTO AN APPROVED GASOLINE CONTAINER.
• IGNITION "ON." NOTE FUEL PRESSURE WITHIN 2 SECONDS AFTER IGNITION IS TURNED "ON." PRESSURE SHOULD BE 284-325 kPa (41-47 psi). IS IT?

YES — LOCATE AND CORRECT RESTRICTION IN FUEL RETURN LINE TO FUEL TANK.

NO —

CHECK FOR RESTRICTED FUEL RETURN LINE FROM PRESSURE REGULATOR TO POINT WHERE FUEL LINE WAS DISCONNECTED. IS THERE A RESTRICTION?

YES — LOCATE AND CORRECT RESTRICTION IN FUEL LINE.

NO — REMOVE PRESSURE REGULATOR AND CHECK FOR RESTRICTED FILTER SCREEN (IF EQUIPPED). REPLACE SCREEN IF PLUGGED. IF OK, REPLACE PRESSURE REGULATOR.

"AFTER REPAIRS," CONFIRM "CLOSED LOOP" OPERATION AND NO MIL (SERVICE ENGINE SOON)

2.2L (VIN 4) ENGINE—SYSTEM DIAGNOSTIC
CHARTS—CENTURY AND CIERA

CHART A-7

(Page 2 of 3)
FUEL SYSTEM DIAGNOSIS
2.2L (VIN 4) "A" CARLINE

Test Description: Number(s) below refer to circled number(s) on the diagnostic chart.

5. Fuel pressure that drops off during acceleration, cruise or hard cornering may cause a lean condition and result in a loss of power, surging or misfire. This condition can be diagnosed using a Tech 1 scan tool. If the fuel system is very lean, the oxygen sensor will stop toggling and output voltage will drop below 500 mV. Also, injector pulse width will increase.

Important
• Make sure system is not operating at "Fuel Cut-Off" which may cause false readings on the scan tool.

6. Fuel pressure below 284 kPa (41 psi) may cause a lean condition and may set a DTC 44 or a DTC 32. Driveability conditions can include hard starting cold, hesitation, poor driveability, lack of power, surging or misfire.

7. Restricting fuel flow at the fuel pressure gage (at "B") causes fuel pressure to build above regulated pressure. With battery voltage applied to the pump "test" terminal, pressure should rise above 325 kPa (47 psi) as the gage outlet hose is pinched.

NOTICE: Do not allow pressure to exceed 414 kPa (60 psi) as damage to the regulator may result.

8. Fuel pressure above 325 kPa (47 psi) may cause a rich condition and may set a DTC 45. Driveability conditions can include hard starting (followed by black smoke) and a strong sulphur smell in the exhaust.

9. This test determines if the high fuel pressure is due to a restricted fuel return line or a faulty fuel pressure regulator.

2.2L (VIN 4) ENGINE — SYSTEM DIAGNOSTIC CHARTS — CENTURY AND CIERA

CHART A-7
(Page 3 of 3)
FUEL SYSTEM DIAGNOSIS
2.2L (VIN 4) "A" CARLINE

FUEL SYSTEM PRESSURE RELIEF PROCEDURE

MFI Engines Without Fuel Pressure Connection (Except J Car Applications)

(Must Be Performed Before Disconnecting Fuel Line Fittings)

CAUTION:
- **To reduce the risk of fire and personal injury, it is necessary to relieve fuel system pressure before disconnecting fuel line fittings.**
- **After relieving system pressure, a small amount of fuel may be released when disconnecting fuel line fittings. In order to reduce the chance of personal injury, cover fuel line fittings with a shop towel before disconnecting, to catch any fuel that may leak out. Place the towel in an approved container when disconnect is completed.**

1. Loosen fuel filler cap to relieve tank pressure.
2. Raise vehicle.
3. Disconnect fuel pump electrical connector.
4. Lower vehicle.
5. Start engine and run until fuel supply remaining in fuel pipes is consumed. Engage starter for 3.0 seconds to assure relief of any remaining pressure.
6. Raise vehicle.
7. Connect fuel pump electrical connector.
8. Lower vehicle.
9. Disconnect negative battery cable to avoid possible fuel discharge if an accidental attempt is made to start the engine.
10. Fuel line fittings are now safe for servicing.
11. Perform service required.
    - If performing fuel pressure check with a gage equipped with a bleed hose, fuel pressure can be relieved through the gage following test. Place bleed hose into approved gasoline container and open valve to bleed system pressure.
12. Tighten fuel filler cap.
13. Connect negative battery cable.
14. Cycle ignition "ON" and "OFF" twice, waiting ten seconds between cycles, then check for fuel leaks.

2.2L (VIN 4) ENGINE — SYSTEM DIAGNOSTIC CHARTS — CENTURY AND CIERA

CHART A-7
(Page 3 of 3)
FUEL SYSTEM DIAGNOSIS
2.2L (VIN 4) "A" CARLINE

SERVICING QUICK-CONNECT FITTINGS

Important
- In order to install fuel system diagnostic equipment on vehicles equipped with plastic quick-connect fittings, fuel line separator tools must be used to disconnect the fittings. Use of the separator tools will cause the plastic retainer to remain inside the female connector allowing diagnostic equipment to be connected.

Tools required:
    J 37088-A tool set, fuel line quick-connect separator;
    J 39504 tool set, fuel line quick-connect separator (restricted access).

Remove or Disconnect
1. Grasp both sides of fitting. Twist female connector 1/4 turn in each direction to loosen any dirt within fitting.

CAUTION: Safety glasses must be worn when using compressed air, as flying dirt particles may cause eye injury.

2. Using compressed air, blow dirt out of fitting.
3. Choose correct tool from J 37088-A or J 39504 tool set for size of fitting. Insert tool into female connector, then push/pull inward to release locking tabs.
4. Pull connection apart.

Clean and Inspect

NOTICE: If it is necessary to remove rust or burrs from fuel pipe, use emery cloth in a radial motion with the pipe end to prevent damage to O-ring sealing surface.

- Using a clean shop towel, wipe off male pipe end.
- Inspect both ends of fitting for dirt and burrs. Clean or replace components/assemblies as required.

Install or Connect

CAUTION: To Reduce the Risk of Fire and Personal Injury:
- **Before connecting fitting, always apply a few drops of clean engine oil to the male pipe end of engine fuel pipe, pressure gage adapter or fuel line shut-off adapter. This will ensure proper reconnection and prevent a possible fuel leak. (During normal operation, the O-rings located in the female connector will swell and may prevent proper reconnection if not lubricated.)**

1. Apply a few drops of clean engine oil to the male pipe end of engine fuel pipe, pressure gage adapter or fuel line shut-off adapter.
2. Push both sides of fitting together to cause the retaining tabs/fingers to snap into place.
3. Once installed, pull on both sides of fitting to make sure connection is secure.

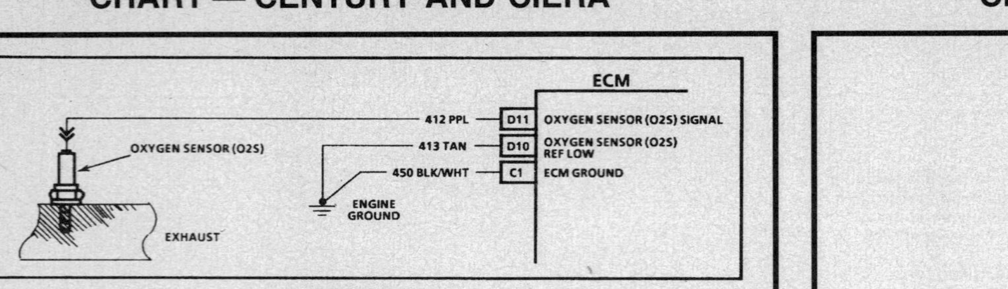

DTC 13

OXYGEN SENSOR (O2S) CIRCUIT
(OPEN CIRCUIT)
2.2L (VIN 4) "A" CARLINE

Circuit Description:

The Engine Control Module (ECM) supplies a voltage of about .45 volt between terminals "D10" and "D11". (If measured with a 10 megohm digital voltmeter, this may read as low as .32 volt.)

When the Oxygen Sensor (O2S) reaches operating temperature, it varies this voltage from about .1 volt (exhaust is lean) to about .9 volt (exhaust is rich).

The sensor is like an open circuit and produces no voltage when it is below 316°C (600°F). An open sensor circuit, or cold sensor, causes "Open Loop" operation.

Test Description: Number(s) below refer to circled number(s) on the diagnostic chart.
1. DTC 13 will set under the following conditions:
   - Engine at normal operating temperature.
   - At least 2 minutes has elapsed since engine start-up.
   - O2S signal voltage is steady between .35 and .55 volt.
   - Throttle angle is above 5%.
   - All above conditions are met for 40.3 seconds or more.
   If the conditions for a DTC 13 exist, the system will not operate in "Closed Loop."
2. This test determines if the Oxygen Sensor (O2S) is the problem or if the ECM and wiring are at fault.

3. In doing this test, use only a 10 megohm digital voltmeter. This test checks the continuity of CKT 412 and CKT 413. If CKT 413 is open, the ECM voltage on CKT 412 will be over .6 volt (600 mV).

Diagnostic Aids:

Normal Tech 1 scan tool Oxygen Sensor (O2S) voltage varies between 100 mV to 999 mV (.1 and 1.0 volt) while in "Closed Loop." DTC 13 sets in about 40 seconds if sensor signal voltage remains between .35 and .55 volt.

Verify a clean, tight ground connection for CKT 413. Open CKT 412 or CKT 413 will result in a DTC 13. If DTC 13 is intermittent, refer to "Symptoms,"

DTC 13

OXYGEN SENSOR (O2S) CIRCUIT
(OPEN CIRCUIT)
2.2L (VIN 4) "A" CARLINE

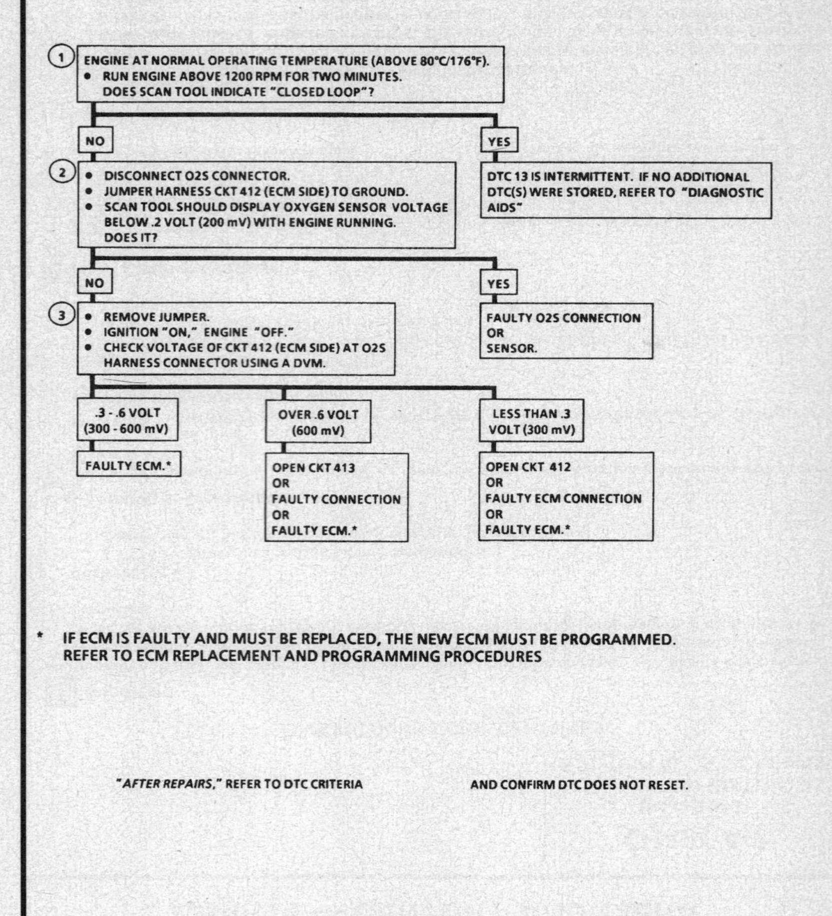

* IF ECM IS FAULTY AND MUST BE REPLACED, THE NEW ECM MUST BE PROGRAMMED. REFER TO ECM REPLACEMENT AND PROGRAMMING PROCEDURES

"AFTER REPAIRS," REFER TO DTC CRITERIA   AND CONFIRM DTC DOES NOT RESET.

2.2L (VIN 4) ENGINE — DIAGNOSTIC TROUBLE CODE CHART — CENTURY AND CIERA

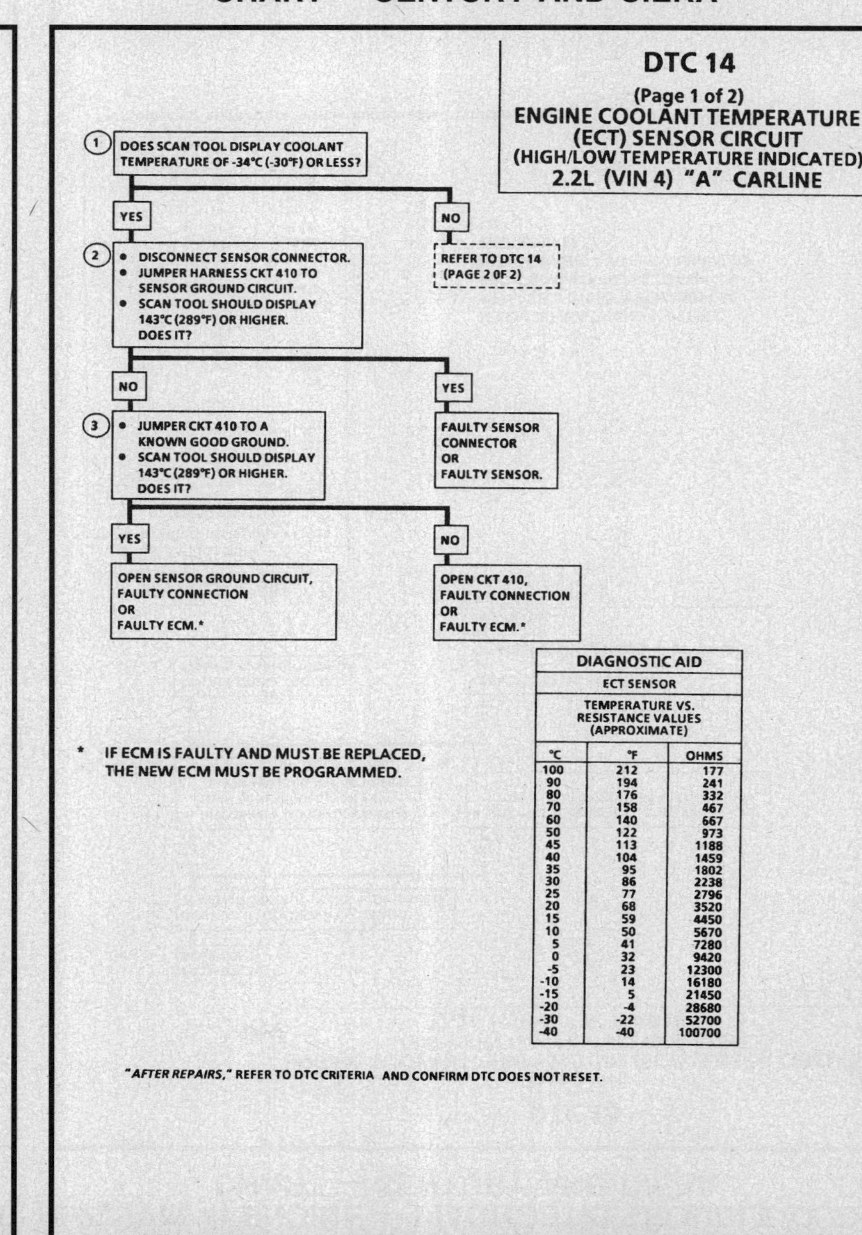

DTC 14
(Page 1 of 2)
ENGINE COOLANT TEMPERATURE (ECT) SENSOR CIRCUIT
(HIGH/LOW TEMPERATURE INDICATED)
2.2L (VIN 4) "A" CARLINE

Circuit Description:

The Engine Coolant Temperature (ECT) sensor utilizes a thermistor to control the signal voltage to the Engine Control Module (ECM). The ECM applies a reference voltage on CKT 410 to the sensor. When the engine is cold, the sensor (thermistor) resistance is high. The ECM will then sense a high signal voltage.

As the engine warms up, the sensor resistance decreases and the voltage drops. At normal engine operating temperature, the voltage will measure about 1.5 to 2.0 volts at ECM terminal "B4".

Coolant temperature is one of the inputs used to control the following:

- Cooling fan.
- Fuel delivery.
- Ignition Control (IC).
- Idle Air Control (IAC).
- Torque Converter Clutch (TCC).

A separate thermistor within the ECT sensor provides a signal to the coolant temperature gage located in the instrument panel.

Test Description: Number(s) below refer to circled number(s) on the diagnostic chart.

1. DTC 14 will set if:
   - The engine has been running for 2 minutes or more.
   - Signal voltage indicates a coolant temperature below -34°C (-30°F).
   OR
   - Signal voltage indicates a coolant temperature above 143°C (289°F) for 3 seconds or more.
2. If the ECM recognizes the grounded circuit (low voltage) and displays a high temperature, the ECM and wiring are OK.
3. This test will determine if there is a wiring problem or a faulty ECM. If CKT 452 is open, there may also be other DTC(s) stored.

Diagnostic Aids:

A scan tool should display engine coolant temperature in degrees Celsius.

After the engine is started, the temperature should rise steadily to about 90°C (194°F), then stabilize when the thermostat opens.

If the engine has been allowed to cool to an ambient temperature (overnight), coolant temperature and Intake Air Temperature (IAT) may be checked with a scan tool and should read close to each other.

When a DTC 14 is set, the ECM will turn "ON" the engine cooling fan.

- If DTC 14 is intermittent or a "Hard Start" symptom is present, check engine coolant temperature with a scan tool on a cool engine. Temperature displayed should be within 5 degrees of the ambient. If not, check the ECT sensor using the "Diagnostic Aid"   If sensor is OK, check connections.

2.2L (VIN 4) ENGINE — DIAGNOSTIC TROUBLE CODE CHART — CENTURY AND CIERA

DTC 14
(Page 1 of 2)
ENGINE COOLANT TEMPERATURE (ECT) SENSOR CIRCUIT
(HIGH/LOW TEMPERATURE INDICATED)
2.2L (VIN 4) "A" CARLINE

1. DOES SCAN TOOL DISPLAY COOLANT TEMPERATURE OF -34°C (-30°F) OR LESS?

   YES →

   2.
   - DISCONNECT SENSOR CONNECTOR.
   - JUMPER HARNESS CKT 410 TO SENSOR GROUND CIRCUIT.
   - SCAN TOOL SHOULD DISPLAY 143°C (289°F) OR HIGHER. DOES IT?

   NO →

   3.
   - JUMPER CKT 410 TO A KNOWN GOOD GROUND.
   - SCAN TOOL SHOULD DISPLAY 143°C (289°F) OR HIGHER. DOES IT?

   YES →
   OPEN SENSOR GROUND CIRCUIT, FAULTY CONNECTION OR FAULTY ECM.*

   NO (from 1) → REFER TO DTC 14 (PAGE 2 OF 2)

   YES (from 2) → FAULTY SENSOR CONNECTOR OR FAULTY SENSOR.

   NO (from 3) → OPEN CKT 410, FAULTY CONNECTION OR FAULTY ECM.*

* IF ECM IS FAULTY AND MUST BE REPLACED, THE NEW ECM MUST BE PROGRAMMED.

DIAGNOSTIC AID		
ECT SENSOR		
TEMPERATURE VS. RESISTANCE VALUES (APPROXIMATE)		
°C	°F	OHMS
100	212	177
90	194	241
80	176	332
70	158	467
60	140	667
50	122	973
45	113	1188
40	104	1459
35	95	1802
30	86	2238
25	77	2796
20	68	3520
15	59	4450
10	50	5670
5	41	7280
0	32	9420
-5	23	12300
-10	14	16180
-15	5	21450
-20	-4	28680
-30	-22	52700
-40	-40	100700

"AFTER REPAIRS," REFER TO DTC CRITERIA AND CONFIRM DTC DOES NOT RESET.

2.2L (VIN 4) ENGINE — DIAGNOSTIC TROUBLE CODE CHART — CENTURY AND CIERA

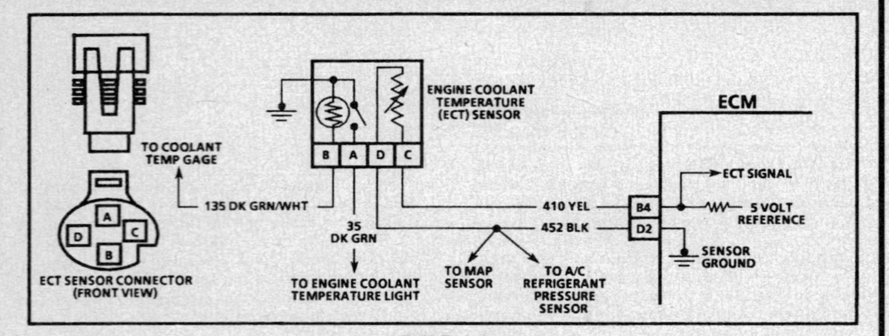

DTC 14

(Page 2 of 2)

ENGINE COOLANT TEMPERATURE (ECT) SENSOR CIRCUIT
(HIGH/LOW TEMPERATURE INDICATED)
2.2L (VIN 4) "A" CARLINE

Circuit Description:

The Engine Coolant Temperature (ECT) sensor utilizes a thermistor to control the signal voltage to the Engine Control Module (ECM). The ECM applies a reference voltage on CKT 410 to the sensor. When the engine is cold, the sensor (thermistor) resistance is high. The ECM will then sense a high signal voltage.

As the engine warms up, the sensor resistance decreases and the voltage drops. At normal engine operating temperature, the voltage will measure about 1.5 to 2.0 volts at ECM terminal "B4".

Coolant temperature is one of the inputs used to control the following:

- Fuel delivery.
- Ignition Control (IC).
- Cooling fan.
- Torque Converter Clutch (TCC).
- Idle Air Control (IAC).

Test Description: Number(s) below refer to circled number(s) on the diagnostic chart.

1. Checks to see if a DTC was set as a result of hard failure or intermittent condition.
2. If the ECM recognizes the open circuit (high voltage), and displays a low temperature, the ECM and wiring are OK.

Diagnostic Aids:

A scan tool should display engine coolant temperature in degrees Celsius.

After the engine is started, the temperature should rise steadily to about 90°C (194°F), then stabilize when the thermostat opens.

If the engine has been allowed to cool to an ambient temperature (overnight), coolant temperature and Intake Air Temperature (IAT) may be checked with a scan tool and should read close to each other.

When a DTC 14 is set, the ECM will turn "ON" the engine cooling fan.

- If DTC 14 is intermittent or a "Hard Start" symptom is present, check engine coolant temperature with a scan tool on a cool engine. Temperature displayed should be within 5 degrees of the ambient. If not, check the ECT sensor using the "Diagnostic Aid" If sensor is OK, check connections.

2.2L (VIN 4) ENGINE — DIAGNOSTIC TROUBLE CODE CHART — CENTURY AND CIERA

DTC 14

(Page 2 of 2)

ENGINE COOLANT TEMPERATURE (ECT) SENSOR CIRCUIT
(HIGH/LOW TEMPERATURE INDICATED)
2.2L (VIN 4) "A" CARLINE

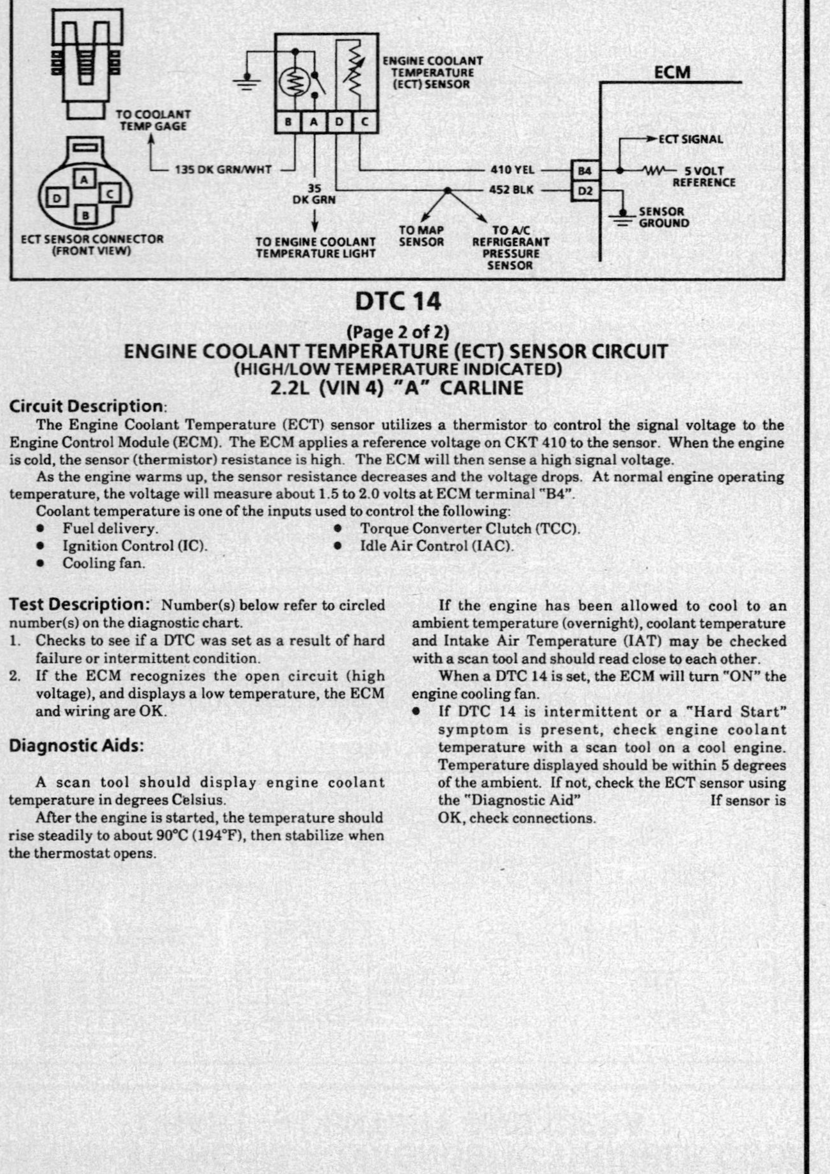

DIAGNOSTIC AID

ECT SENSOR		
TEMPERATURE VS. RESISTANCE VALUES (APPROXIMATE)		
°C	°F	OHMS
100	212	177
90	194	241
80	176	332
70	158	467
60	140	667
50	122	973
45	113	1188
40	104	1459
35	95	1802
30	86	2238
25	77	2796
20	68	3520
15	59	4450
10	50	5670
5	41	7280
0	32	9420
-5	23	12300
-10	14	16180
-15	5	21450
-20	-4	28680
-30	-22	52700
-40	-40	100700

* IF ECM IS FAULTY AND MUST BE REPLACED, THE NEW ECM MUST BE PROGRAMMED. REFER TO ECM REPLACEMENT AND PROGRAMMING PROCEDURES

"AFTER REPAIRS," REFER TO DTC CRITERIA AND CONFIRM DTC DOES NOT RESET.

2.2L (VIN 4) ENGINE — DIAGNOSTIC TROUBLE CODE CHART — CENTURY AND CIERA

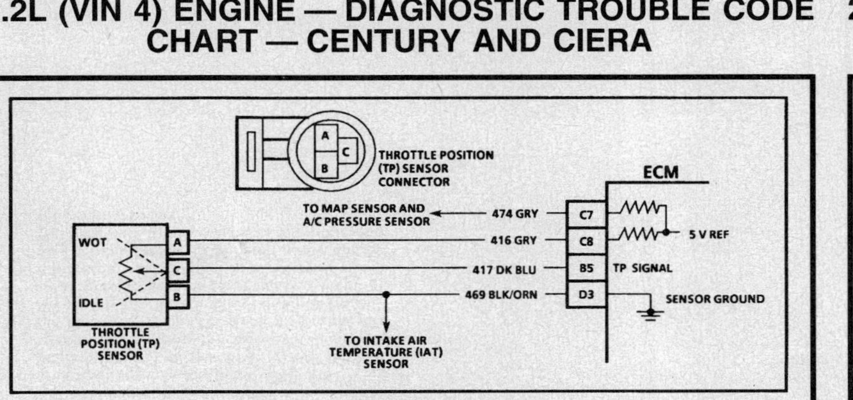

DTC 21
(Page 1 of 2)
THROTTLE POSITION (TP) SENSOR CIRCUIT
(SIGNAL VOLTAGE HIGH/LOW)
2.2L (VIN 4) "A" CARLINE

Circuit Description:

The Throttle Position (TP) sensor provides a voltage signal that changes relative to the throttle valve. Signal voltage will vary from .33 to 1.33 volts at idle to about 4.5 volts at Wide Open Throttle (WOT).

The TP sensor signal is one of the most important inputs used by the Engine Control Module (ECM) for fuel control and for many of the ECM controlled outputs.

Test Description: Number(s) below refer to circled number(s) on the diagnostic chart.
1. A DTC 21 will set under the following conditions:
   - TP sensor reading below 1.9 volts or above 4.9 volts for about 64 seconds or more.
   OR
   - TP sensor reading above 3.9 volts, but below 4.9 volts.
   - MAP reading below 65 kPa.
   - Engine speed less than 1,750 RPM.
   - All of the above conditions present for about 5 seconds.

   The TP sensor has an auto zeroing feature. If the voltage reading is within the range of about .33 to 1.33 volts, the ECM will use that value as closed throttle. If the voltage reading is outside of the auto zero range at closed throttle, check for a binding throttle cable or damaged linkage. If OK, continue with diagnosis.

2. If the ECM recognizes the change of state, the ECM and CKTs 416 and 417 are OK.
3. This step isolates a faulty sensor, ECM, or an open CKT 469. If CKT 469 is open, DTC 23 may also be stored.

Diagnostic Aids:

A scan tool displays throttle position in volts. Closed throttle voltage should be .33 to 1.33 volts. TP sensor voltage should increase at a steady rate as throttle is moved to WOT.

If DTC 21 is intermittent, refer to "Symptoms,"

2.2L (VIN 4) ENGINE — DIAGNOSTIC TROUBLE CODE CHART — CENTURY AND CIERA

DTC 21
(Page 1 of 2)
THROTTLE POSITION (TP) SENSOR CIRCUIT
(SIGNAL VOLTAGE HIGH/LOW)
2.2L (VIN 4) "A" CARLINE

(1) • THROTTLE CLOSED.
DOES SCAN TOOL DISPLAY TP SENSOR OVER 3.9 VOLTS?

YES → (2) • DISCONNECT SENSOR CONNECTOR. SCAN TOOL SHOULD DISPLAY TP SENSOR BELOW .2 VOLT (200mV). DOES IT?

NO → REFER TO DTC 21 (PAGE 2 OF 2)

YES → (3) • PROBE SENSOR GROUND CIRCUIT WITH A TEST LIGHT CONNECTED TO BATTERY VOLTAGE.

NO → TP SENSOR SIGNAL CIRCUIT SHORTED TO VOLTAGE OR FAULTY ECM.*

LIGHT "ON" → FAULTY CONNECTION OR THROTTLE POSITION SENSOR.

LIGHT "OFF" → OPEN SENSOR GROUND CIRCUIT OR FAULTY ECM.*

IF ECM IS FAULTY AND MUST BE REPLACED, THE NEW ECM MUST BE PROGRAMMED. REFER TO ECM REPLACEMENT AND PROGRAMMING PROCEDURES

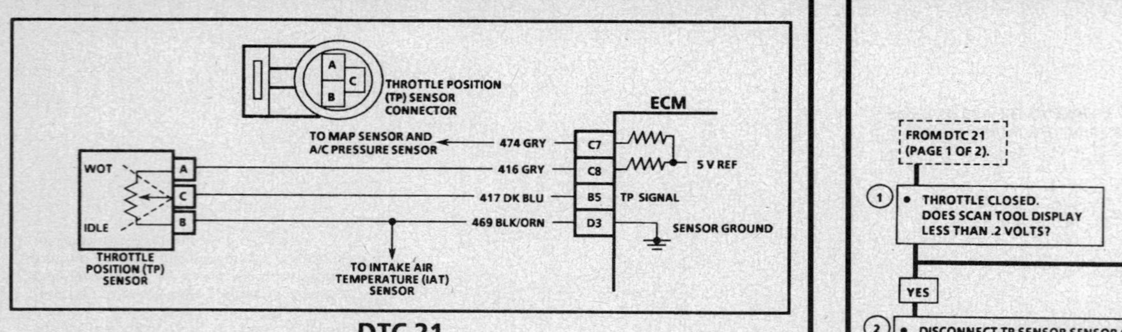

DTC 21
(Page 2 of 2)
THROTTLE POSITION (TP) SENSOR CIRCUIT
(SIGNAL VOLTAGE HIGH/LOW)
2.2L (VIN 4) "A" CARLINE

Circuit Description:

The Throttle Position (TP) sensor provides a voltage signal that changes relative to the throttle valve. Signal voltage will vary from .33 to 1.33 volts at idle to about 4.5 volts at Wide Open Throttle (WOT).

The TP sensor signal is one of the most important inputs used by the Engine Control Module (ECM) for fuel control and for many of the ECM controlled outputs.

Test Description: Number(s) below refer to circled number(s) on the diagnostic chart.

1. This step checks to see if DTC 21 is the result of a hard failure or an intermittent condition.
2. This step simulates conditions for a DTC 21. If the scan tool displays over 4 volts, the ECM and wiring are OK.
3. This simulates a high signal voltage to check for an open or short in CKT 417. The Tech 1 scan tool will not read battery voltage, but the ECM should recognize the signal on CKT 417.
4. Refer to "Fuel Metering System," for TP sensor replacement procedures.
5. The 5 volt reference for the MAP sensor and the A/C refrigerant pressure sensor connect to the same 5 volt power supply in the ECM. If the 5 volt reference circuit to the MAP sensor and A/C refrigerant pressure sensor has a short to ground, the Throttle Position (TP) sensor 5 volt reference circuit will also have a short to ground because they connect to the same power supply in the ECM. A short to ground on either 5 volt reference circuit does not damage the ECM 5 volt power supply. When the short is repaired, the ECM 5 volt supply will return to normal.

Diagnostic Aids:

A scan tool displays throttle position in volts. Closed throttle voltage should be .33 to 1.33 volts. TP sensor voltage should increase at a steady rate as throttle is moved to WOT.

If DTC 21 is intermittent, refer to "Symptoms,"

DTC 21
(Page 2 of 2)
THROTTLE POSITION (TP) SENSOR CIRCUIT
(SIGNAL VOLTAGE HIGH/LOW)
2.2L (VIN 4) "A" CARLINE

FROM DTC 21 (PAGE 1 OF 2).

1. • THROTTLE CLOSED.
   DOES SCAN TOOL DISPLAY LESS THAN .2 VOLTS?

YES →

2. • DISCONNECT TP SENSOR SENSOR CONNECTOR.
   • JUMPER 5 VOLT REFERENCE CIRCUIT AND CKT 417 TOGETHER.
   SCAN SHOULD DISPLAY THROTTLE POSITION OVER 4.0 V (4000 mV).
   DOES IT?

NO → DTC 21 IS INTERMITTENT. IF NO ADDITIONAL DTCS WERE STORED, REFER TO "DIAGNOSTIC AIDS"

NO (from 2) →

3. • PROBE CKT 417 WITH A TEST LIGHT CONNECTED TO BATTERY VOLTAGE.
   SCAN TOOL SHOULD DISPLAY THROTTLE POSITION OVER 4.0V (4000 mV).
   DOES IT?

YES (from 2) →

4. • RECONNECT TP SENSOR CONNECTOR.
   • CHECK FOR FAULTY OR INTERMITTENT CONNECTION. IF OK, TP SENSOR IS FAULTY.

YES (from 3) →

5. 5 VOLT REFERENCE CIRCUIT OPEN
   OR
   SHORTED TO GROUND
   OR
   FAULTY CONNECTION
   OR
   FAULTY ECM.*

NO (from 3) → CKT 417 OPEN
OR
SHORTED TO GROUND
OR
SHORTED TO SENSOR GROUND CIRCUIT
OR
FAULTY ECM CONNECTION
OR
FAULTY ECM.*

* IF ECM IS FAULTY AND MUST BE REPLACED, THE NEW ECM MUST BE PROGRAMMED. REFER TO ECM REPLACEMENT AND PROGRAMMING PROCEDURES

"AFTER REPAIRS," REFER TO DTC CRITERIA AND CONFIRM DTC DOES NOT RESET.

2.2L (VIN 4) ENGINE — DIAGNOSTIC TROUBLE CODE CHART — CENTURY AND CIERA

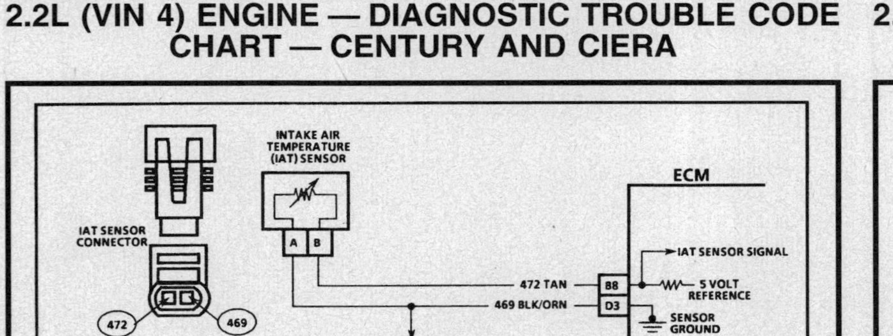

DTC 23
(Page 1 of 2)
INTAKE AIR TEMPERATURE (IAT) SENSOR CIRCUIT
(HIGH/LOW TEMPERATURE INDICATED)
2.2L (VIN 4) "A" CARLINE

Circuit Description:

The Intake Air Temperature (IAT) sensor, located in the rear air duct assembly, uses a thermistor to control the signal voltage to the Engine Control Module (ECM). The ECM applies a reference voltage (4 to 5.5 volts) on CKT 472 to the sensor. When intake air is cold, the sensor (thermistor) resistance is high. The ECM will then sense a high signal voltage. As the air warms, the sensor resistance becomes less and the voltage drops.

Test Description: Number(s) below refer to circled number(s) on the diagnostic chart.

1. A DTC 23 will set under the following conditions:
   - Engine running for 2 minutes or longer.
   - Signal voltage indicates an IAT temperature less than -34°C (-30°F).
     OR
   - Signal voltage indicates an IAT temperature greater than 150°C (302°F) for more than 2 seconds.
2. This test simulates conditions for a DTC 23. If the scan tool displays a high temperature, the ECM and wiring are OK.
3. This step checks continuity of CKT 472 and CKT 469. If CKT 469 is open, there may be a DTC 21 stored.

Diagnostic Aids:

If the engine has been allowed to cool to an ambient temperature (overnight), engine coolant and intake air temperatures may be checked with a Tech 1 scan tool and should read close to each other.
- If DTC 23 is intermittent or a "Hard Start" symptom is present, check intake air temperature with a scan tool on a cool engine. Temperature displayed should be within 5 degrees of the ambient. If not, check the IAT sensor using the "Diagnostic Aid"   If sensor is OK, check connections.

2.2L (VIN 4) ENGINE — DIAGNOSTIC TROUBLE CODE CHART — CENTURY AND CIERA

DTC 23
(Page 1 of 2)
INTAKE AIR TEMPERATURE (IAT) SENSOR CIRCUIT
(HIGH/LOW TEMPERATURE INDICATED)
2.2L (VIN 4) "A" CARLINE

1. • DOES SCAN TOOL DISPLAY IAT -34°C (-30°F) OR COLDER?

   - **YES** →
     2. • DISCONNECT IAT SENSOR CONNECTOR.
        • JUMPER HARNESS TERMINALS TOGETHER.
        • SCAN TOOL SHOULD DISPLAY TEMPERATURE OVER 150°C (302°F). DOES IT?
        - **YES** → FAULTY CONNECTION OR SENSOR.
        - **NO** →
          3. • JUMPER CKT 472 TO GROUND.
             • SCAN TOOL SHOULD DISPLAY TEMPERATURE OVER 150°C (302°F). DOES IT?
             - **YES** → OPEN SENSOR GROUND CIRCUIT, OR FAULTY CONNECTION OR FAULTY ECM.*
             - **NO** → OPEN CKT 472, OR FAULTY CONNECTION OR FAULTY ECM.*

   - **NO** → REFER TO DTC 23 (PAGE 2 OF 2)

DIAGNOSTIC AID		
IAT SENSOR		
TEMPERATURE VS. RESISTANCE VALUES (APPROXIMATE)		
°C	°F	OHMS
100	210	185
70	160	450
38	100	1,800
20	70	3,400
4	40	7,500
-7	20	13,500
-18	0	25,000
-40	-40	100,700

* IF ECM IS FAULTY AND MUST BE REPLACED, THE NEW ECM MUST BE PROGRAMMED. REFER TO ECM REPLACEMENT AND PROGRAMMING PROCEDURES

"AFTER REPAIRS," REFER TO DTC CRITERIA AND CONFIRM DTC DOES NOT RESET.

2.2L (VIN 4) ENGINE — DIAGNOSTIC TROUBLE CODE CHART — CENTURY AND CIERA

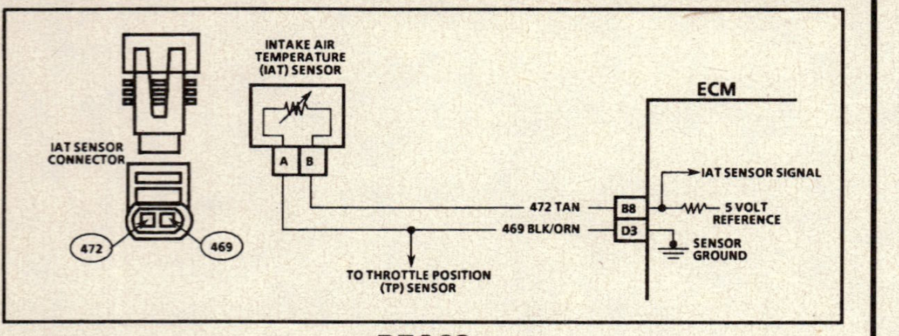

DTC 23

(Page 2 of 2)

INTAKE AIR TEMPERATURE (IAT) SENSOR CIRCUIT
(HIGH/LOW TEMPERATURE INDICATED)
2.2L (VIN 4) "A" CARLINE

Circuit Description:

The Intake Air Temperature (IAT) sensor, located in the rear air duct assembly, uses a thermistor to control the signal voltage to the Engine Control Module (ECM). The ECM applies a reference voltage (4 to 5.5 volts) on CKT 472 to the sensor. When intake air is cold, the sensor (thermistor) resistance is high. Therefore, the ECM will sense a high signal voltage. As the air warms, the sensor resistance becomes less and the voltage drops.

Test Description: Number(s) below refer to circled number(s) on the diagnostic chart.
1. This step determines if DTC 23 is the result of a hard failure or an intermittent condition.
2. If the ECM recognizes the open circuit (high voltage) and displays a low temperature, the ECM and wiring are OK.

Diagnostic Aids:

If the engine has been allowed to cool to an ambient temperature (overnight), engine coolant and intake air temperatures may be checked with a scan tool and should read close to each other.
- If DTC 23 is intermittent or a "Hard Start" symptom is present, check intake air temperature with a scan tool on a cool engine. Temperature displayed should be within 5 degrees of the ambient. If not, check the IAT sensor using the "Diagnostic Aid"   If sensor is OK, check connections.

2.2L (VIN 4) ENGINE — DIAGNOSTIC TROUBLE CODE CHART — CENTURY AND CIERA

DTC 23

(Page 2 of 2)

INTAKE AIR TEMPERATURE (IAT) SENSOR CIRCUIT
(HIGH/LOW TEMPERATURE INDICATED)
2.2L (VIN 4) "A" CARLINE

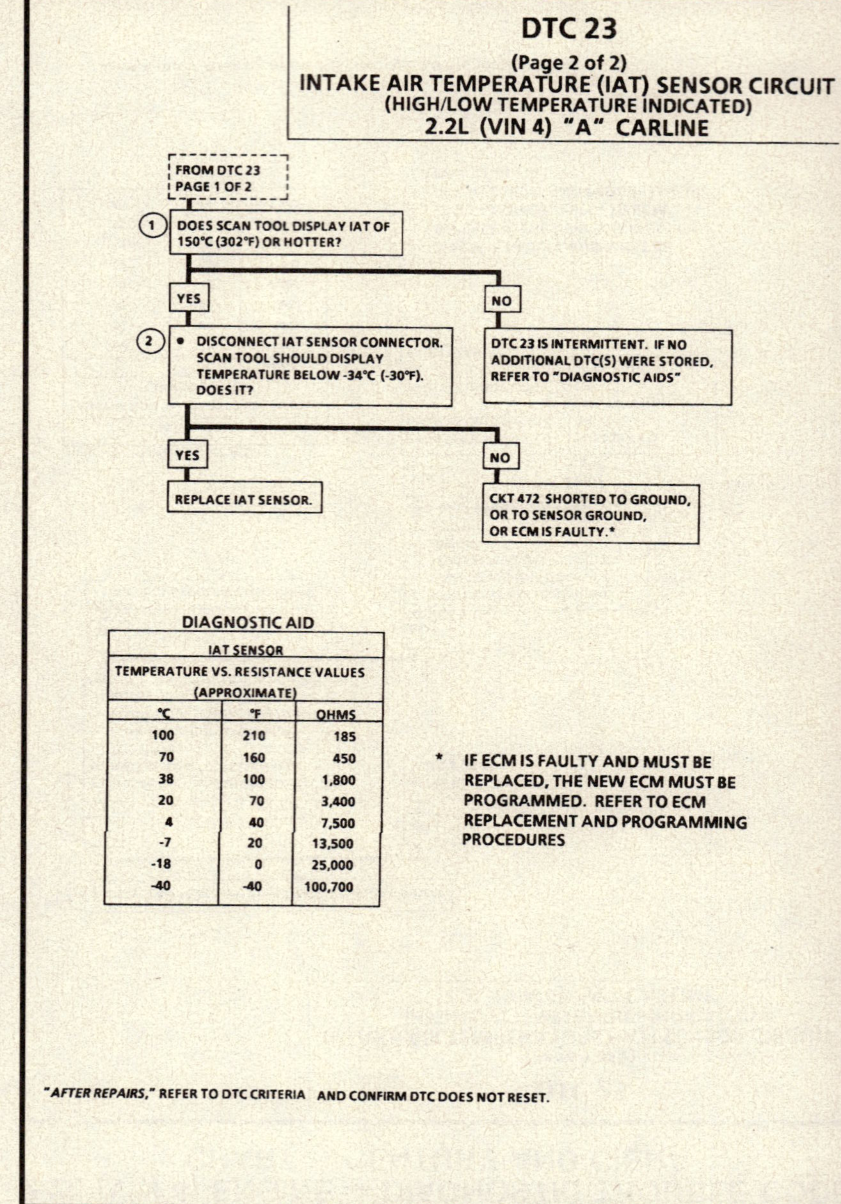

DIAGNOSTIC AID

IAT SENSOR		
TEMPERATURE VS. RESISTANCE VALUES (APPROXIMATE)		
°C	°F	OHMS
100	210	185
70	160	450
38	100	1,800
20	70	3,400
4	40	7,500
-7	20	13,500
-18	0	25,000
-40	-40	100,700

* IF ECM IS FAULTY AND MUST BE REPLACED, THE NEW ECM MUST BE PROGRAMMED. REFER TO ECM REPLACEMENT AND PROGRAMMING PROCEDURES

"AFTER REPAIRS," REFER TO DTC CRITERIA   AND CONFIRM DTC DOES NOT RESET.

2.2L (VIN 4) ENGINE — DIAGNOSTIC TROUBLE CODE CHART — CENTURY AND CIERA

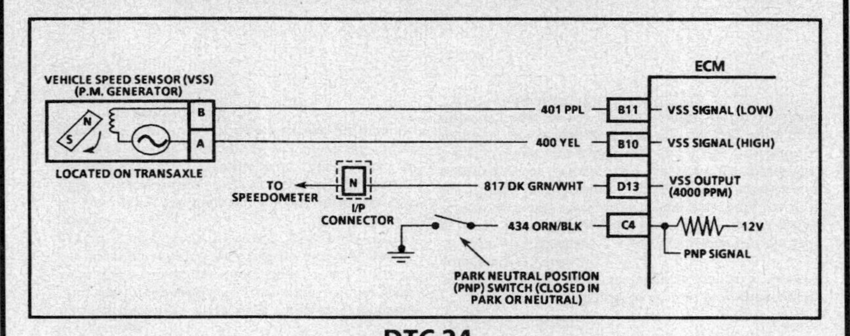

DTC 24
VEHICLE SPEED SENSOR (VSS) CIRCUIT
2.2L (VIN 4) "A" CARLINE

Circuit Description:

Vehicle speed information is provided to the Engine Control Module (ECM) by the Vehicle Speed Sensor (VSS), which is a Permanent Magnet (PM) generator, and it is mounted in the transaxle. The PM generator produces a pulsing voltage, whenever vehicle speed is over about 3 mph. The AC voltage level and the number of pulses increases with vehicle speed. The ECM then converts the pulsing voltage to mph, which is used for calculations, and the mph can be displayed with a scan tool.

The function of VSS buffer used in past model years has been incorporated into the ECM. The ECM then supplies the necessary signal to the instrument panel (4000 pulses per mile) for operating the speedometer and the odometer.

Test Description: Number(s) below refer to circled number(s) on the diagnostic chart.

1. DTC 24 will set if vehicle speed equals 0 mph when:
   - Engine speed is between about 1400 and 3600 RPM.
   - Low load condition (low MAP voltage), high manifold vacuum).
   - Transaxle not in park or neutral.
   - All above conditions are met for 4 seconds.

   These conditions are met during a road load deceleration.

   Disregard a DTC 24 that sets when the drive wheels are not turning. This can be caused by a faulty PNP switch circuit.

   The PM generator only produces a signal if the drive wheels are turning greater than 3 mph.

2. At this point, the ECM is not sending vehicle speed data to the scan tool. If the scan tool is connected and functioning properly, the ECM is at fault.*

Diagnostic Aids:

Scan tool should indicate a vehicle speed whenever the drive wheels are turning greater than 3 mph.

A problem in CKT 817 will not affect the VSS input or the readings on a scan tool.

Check CKTs 400 and 401 for proper connections to be sure they are clean and tight and the harness is routed correctly.

- A faulty or misadjusted Park/Neutral Position (PNP) switch can result in a false DTC 24. Use a scan tool and check for the proper signal while in a drive range. Refer to CHART C-1A for the PNP switch check.

2.2L (VIN 4) ENGINE — DIAGNOSTIC TROUBLE CODE CHART — CENTURY AND CIERA

DTC 24
VEHICLE SPEED SENSOR (VSS) CIRCUIT
2.2L (VIN 4) "A" CARLINE

DISREGARD DTC 24 IF SET WHILE DRIVE WHEELS ARE NOT TURNING.

(1)
- RAISE DRIVE WHEELS.

  NOTICE: DO NOT PERFORM THIS TEST WITHOUT SUPPORTING THE LOWER CONTROL ARMS SO THAT THE DRIVE AXLES ARE IN A NORMAL HORIZONTAL POSITION. RUNNING THE VEHICLE IN GEAR WITH THE WHEELS HANGING DOWN AT FULL TRAVEL MAY DAMAGE THE DRIVE AXLES.

- WITH ENGINE IDLING IN GEAR, SCAN TOOL SHOULD DISPLAY VEHICLE SPEED ABOVE 0 MPH.
  DOES IT?

NO → DOES SPEEDOMETER WORK PROPERLY?

YES → DTC 24 IS INTERMITTENT. IF NO ADDITIONAL DTC(S) WERE STORED. REFER TO "DIAGNOSTIC AIDS"

NO →
- IGNITION "OFF."
- DISCONNECT VSS HARNESS CONNECTOR AT TRANSAXLE.
- CONNECT SIGNAL GENERATOR TESTER J 33431-B OR EQUIVALENT TO VSS HARNESS CONNECTOR.
- IGNITION "ON," TOOL "ON" AND SET TO GENERATE A VSS SIGNAL.
- SCAN TOOL SHOULD DISPLAY VEHICLE SPEED ABOVE 0 MPH.
  DOES IT?

YES → (2) REPLACE ECM.*

NO →
CKT 400
OR
401 OPEN, SHORTED TO GROUND, SHORTED TOGETHER, FAULTY CONNECTIONS
OR
FAULTY ECM.*

YES → REPLACE VEHICLE SPEED SENSOR.

* IF ECM IS FAULTY AND MUST BE REPLACED, THE NEW ECM MUST BE PROGRAMMED. REFER TO ECM REPLACEMENT AND PROGRAMMING PROCEDURES

"AFTER REPAIRS," REFER TO DTC CRITERIA AND CONFIRM DTC DOES NOT RESET.

DTC 32

(Page 1 of 2)
EXHAUST GAS RECIRCULATION (EGR) SYSTEM FAILURE
2.2L (VIN 4) "A" CARLINE

Circuit Description:

The Exhaust Gas Recirculation (EGR) system is controlled by the ECM. The ECM controls the vacuum being supplied to the valve by energizing and de-energizing a solenoid.

The ECM uses information from various engine sensors to determine when EGR is necessary. Once the ECM has requested EGR by grounding the solenoid circuit, the ECM will monitor engine operating conditions to determine if exhaust gas flow has entered the intake manifold. When the ECM tests for EGR operation and no change in engine operating conditions is indicated, a DTC 32 will set.

The difference in long term fuel trim values between the idle (closed throttle) cell and cell 2 is used to monitor EGR system performance. When the difference between the two long term fuel trim values is greater than about 15 and the long term fuel trim value in cell 2 is greater than 135 for 80 seconds or more, DTC 32 is set. The system operates in long term fuel trim cell 2 during a cruise condition at approximately 55 mph.

Test Description: Number(s) below refer to circled number(s) on the diagnostic chart.

1. Intake Passage: Shut "OFF" engine and remove the EGR valve from the manifold. Plug the exhaust side hole with a suitable stopper. Leaving the intake side hole open, attempt to start the engine. If the engine runs at a high idle (up to 3000 RPM is possible) or starts and stalls, the EGR intake passage is not restricted. If the engine starts and idles normally, the EGR intake passage is restricted.

   Exhaust Passage: With EGR valve still removed, plug the intake side hole with a suitable stopper. With the exhaust side hole open, check for the presence of exhaust gas. If no exhaust gas is present, the EGR exhaust side passage is restricted.

2. By grounding the diagnostic "test" terminal, the EGR solenoid should be energized and allow vacuum to be applied to the gage. The vacuum at the gage may or may not <u>slowly</u> bleed off. It is important that the gage is able to read the amount of vacuum being applied.

3. When the diagnostic "test" terminal is ungrounded, the vacuum gage should bleed off completely through a vent in the solenoid. The vacuum pump gage may or may not bleed off but this does not indicate a problem.

4. This test will determine if the electrical control part of the system is at fault or if the connector or solenoid is at fault.

5. At this point, it has been determined that the EGR solenoid, the ECM and the vacuum supply are OK.

Diagnostic Aids:

Vacuum lines should be thoroughly checked for proper routing. Refer to "Vehicle Emission Control Information" label.

The DTC 32 chart is a functional check of the EGR system. If the EGR system works properly but a DTC 32 has been set, check other items that could result in high block learn values during a cruise condition at approximately 55 mph. Low fuel pressure or lean fuel injector(s) may set a DTC 32. It may be necessary to monitor fuel pressure while driving the vehicle at various road speeds and/or loads. Refer to "Fuel System Diagnosis," CHART A-7.

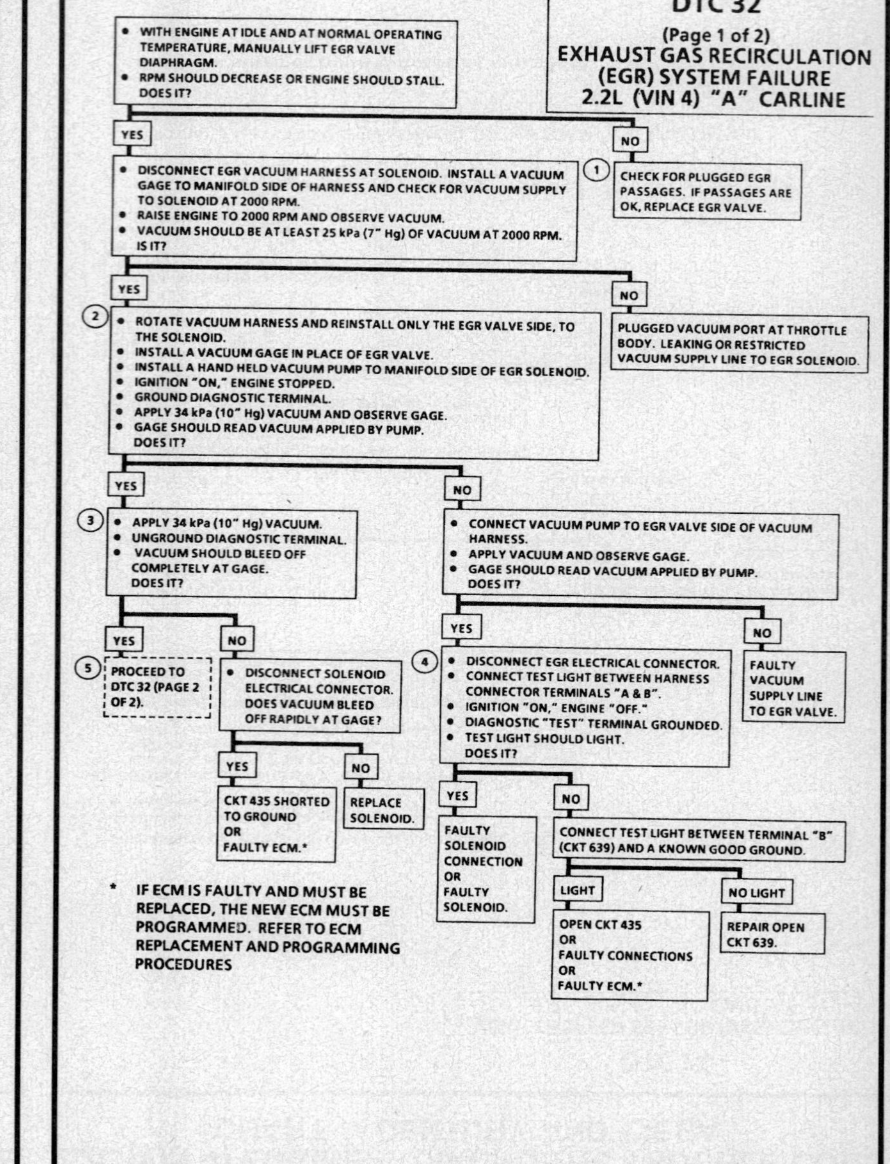

2.2L (VIN 4) ENGINE — DIAGNOSTIC TROUBLE CODE CHART — CENTURY AND CIERA

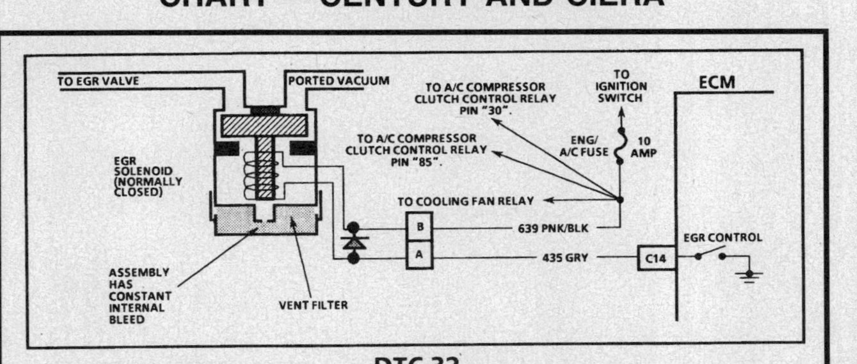

DTC 32
(Page 2 of 2)
EXHAUST GAS RECIRCULATION (EGR) SYSTEM FAILURE
2.2L (VIN 4) "A" CARLINE

Circuit Description:

The Exhaust Gas Recirculation (EGR) system is controlled by the ECM. The ECM controls the vacuum being supplied to the valve by energizing and de-energizing a solenoid.

The ECM uses information from various engine sensors to determine when EGR is necessary. Once the ECM has requested EGR by grounding the solenoid circuit, the ECM will monitor engine operating conditions to determine if exhaust gas flow has entered the intake manifold. When the ECM tests for EGR operation and no change in engine operating conditions is indicated, a DTC 32 will set.

Test Description: Number(s) below refer to circled number(s) on the diagnostic chart.
1. The remaining tests check the ability of the EGR valve to interact with the exhaust system. This system uses a negative backpressure EGR valve which should hold vacuum with engine "OFF." **Be sure shop exhaust hose is not connected during Steps 1 and 2.**
2. When engine is started, exhaust backpressure at the base of the EGR valve should open the valve's internal bleed and vent the applied vacuum allowing the valve to seat. **Because the shop exhaust hose is not installed at this time, do not allow the engine to run longer than 15 seconds.**

Diagnostic Aids:

Low fuel pressure or lean fuel injectors may cause a DTC 32 to set. Use CHART A-7. It may be necessary to monitor fuel pressure while driving the vehicle at various road speeds and/or loads. If fuel pressure is normal, perform the injector balance test, CHART C2-A

2.2L (VIN 4) ENGINE — DIAGNOSTIC TROUBLE CODE CHART — CENTURY AND CIERA

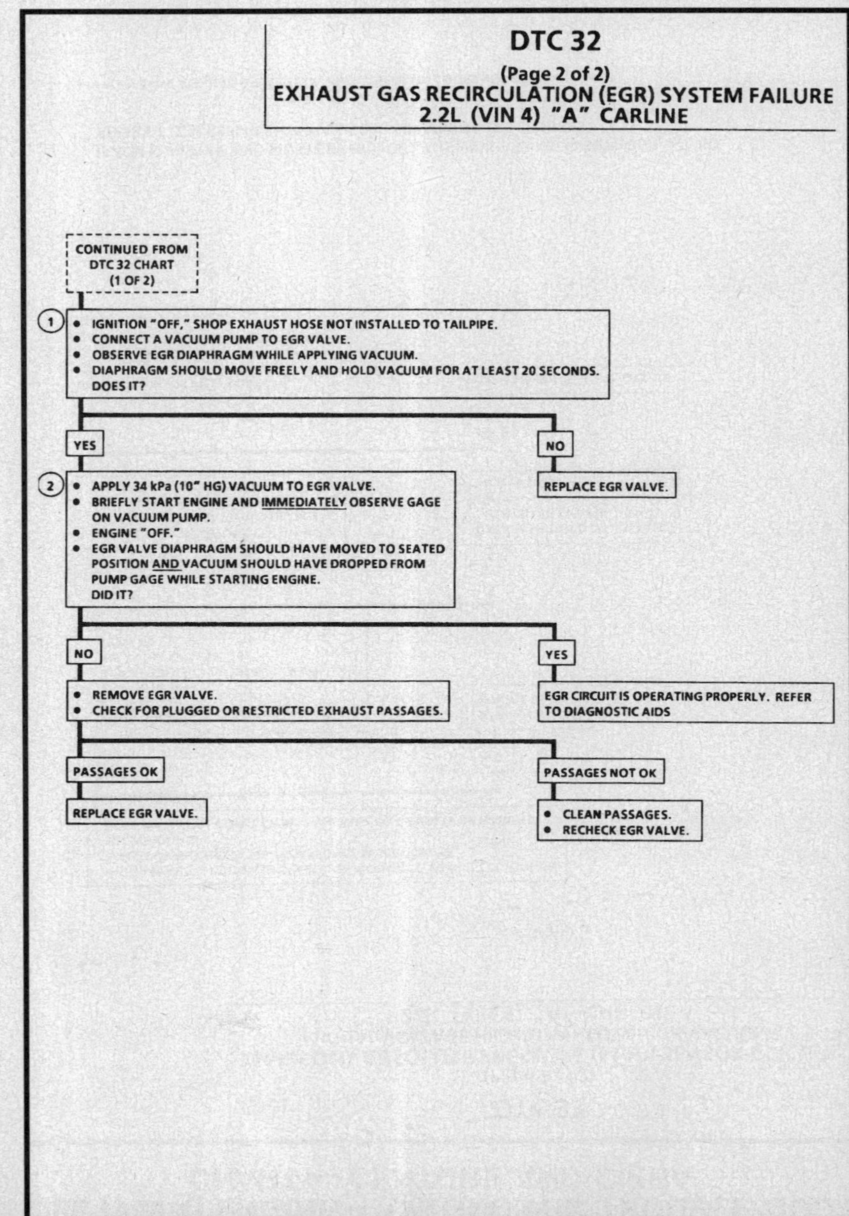

2.2L (VIN 4) ENGINE — DIAGNOSTIC TROUBLE CODE CHART — CENTURY AND CIERA

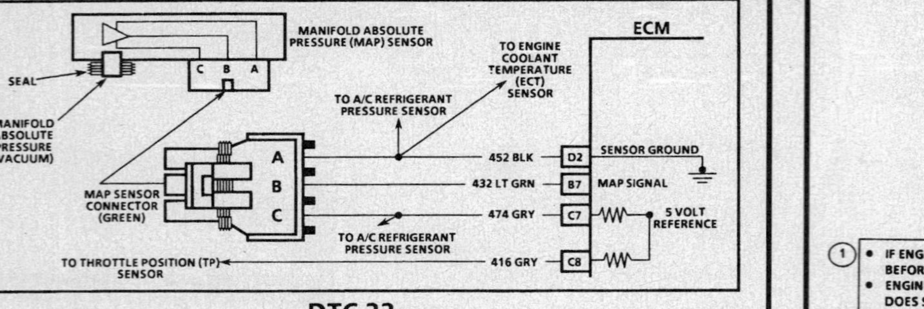

DTC 33
(Page 1 of 2)
MANIFOLD ABSOLUTE PRESSURE (MAP) SENSOR CIRCUIT
(SIGNAL VOLTAGE HIGH/LOW - LOW/HIGH VACUUM)
2.2L (VIN 4) "A" CARLINE

Circuit Description:

The Manifold Absolute Pressure (MAP) sensor responds to changes in manifold pressure (vacuum). The ECM receives this information as a signal voltage that will vary from about 1 to 1.5 volts at closed throttle (idle), to about 4.5 volts at Wide Open Throttle (WOT) (low vacuum).

If the MAP sensor fails, the Engine Control Module (ECM) will substitute a fixed MAP value based on engine RPM.

Test Description: Number(s) below refer to circled number(s) on the diagnostic chart.

1. A DTC 33 will set under the following conditions:
   - MAP signal indicates greater than 3.7 volts (80 kPa).
   - Throttle position is less than 5%.
   - These conditions exist for a time period longer than 5 seconds.
   OR
   - MAP signal indicates less than .3 volt (15 kPa).
2. If the ECM recognizes the change, the ECM and CKTs 474 and 432 are OK. If CKT 452 is open, there may also be other DTC(s) stored.

Diagnostic Aids:

With the ignition "ON" and the engine stopped, the manifold pressure is equal to atmospheric pressure and the signal voltage will be high. This information is used by the ECM as an indication of vehicle altitude and is referred to as BARO. Comparison of this BARO reading with a known good vehicle with the same sensor is a good way to check accuracy of a "suspect" sensor. Reading should be within ± .4 volt.

If DTC 33 is intermittent, refer to "Symptoms,"

2.2L (VIN 4) ENGINE — DIAGNOSTIC TROUBLE CODE CHART — CENTURY AND CIERA

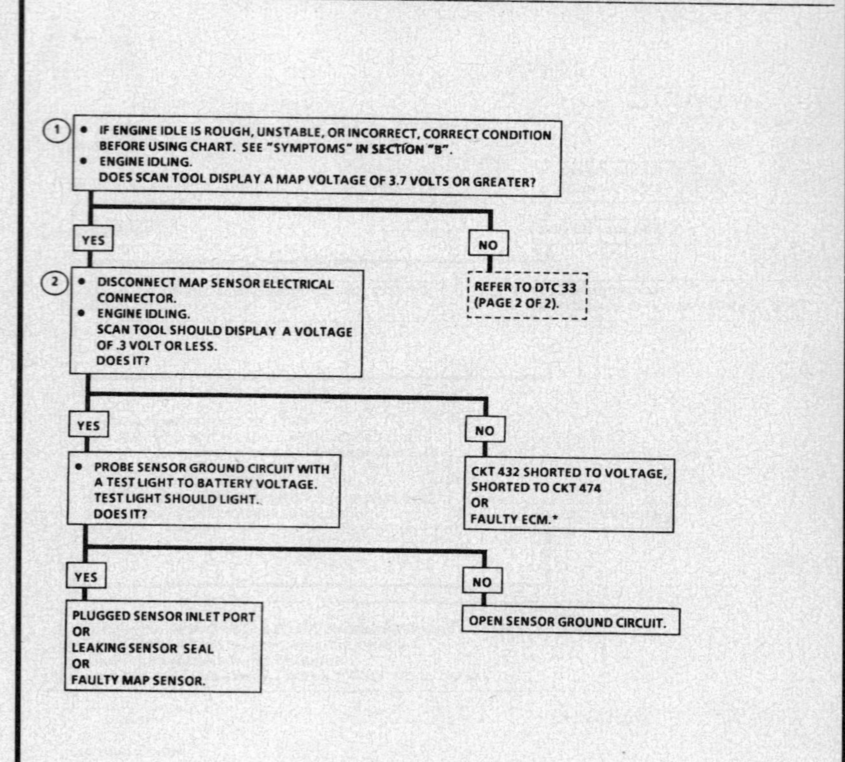

2.2L (VIN 4) ENGINE — DIAGNOSTIC TROUBLE CODE CHART — CENTURY AND CIERA

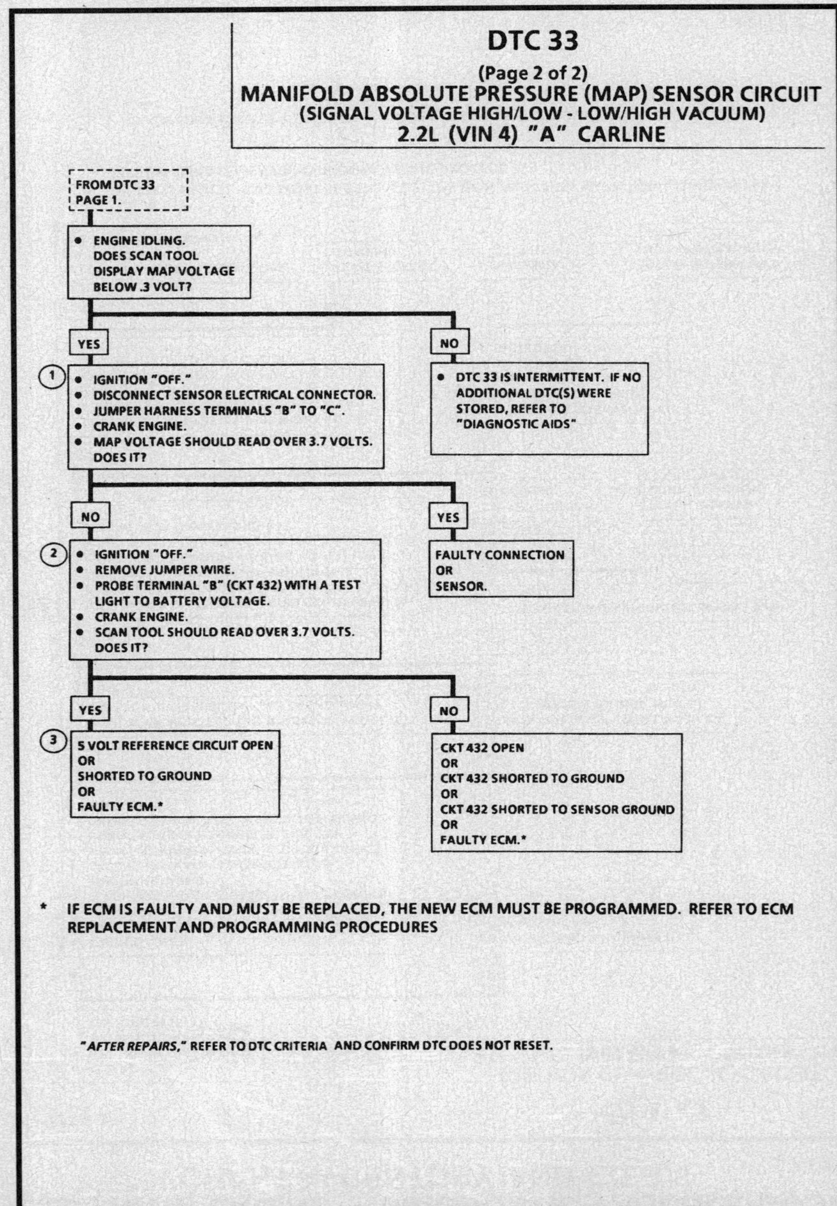

DTC 33

(Page 2 of 2)
MANIFOLD ABSOLUTE PRESSURE (MAP) SENSOR CIRCUIT
(SIGNAL VOLTAGE HIGH/LOW - LOW/HIGH VACUUM)
2.2L (VIN 4) "A" CARLINE

Circuit Description:

The Manifold Absolute Pressure (MAP) sensor responds to changes in manifold pressure (vacuum). The ECM receives this information as a signal voltage that will vary from about 1 to 1.5 volts at closed throttle (idle), to about 4.5 volts at Wide Open Throttle (WOT) (low vacuum).

If the MAP sensor fails, the Engine Control Module (ECM) will substitute a fixed MAP value based on engine RPM and use the Throttle Position (TP) sensor to control fuel delivery.

Test Description: Number(s) below refer to circled number(s) on the diagnostic chart.

1. Jumpering harness terminals "B" to "C" (5 volts to signal circuit), will determine if the sensor is at fault, or if there is a problem with the ECM or wiring.

    The scan tool may not display 5 volts. The important thing is that the ECM recognizes the voltage as more than 4 volts, indicating that the ECM CKTs 432, and 474 are OK.

2. This step determines if CKT 474 or CKT 432 is faulty. The scan tool will not display battery voltage, but should indicate over 4 volts.

3. The 5 volt reference Throttle Position (TP) sensor connects to the same 5 volt power supply in the ECM. If the 5 volt reference circuit to the TP sensor has a short to ground, the MAP sensor and A/C refrigerant pressure sensor 5 volt reference circuit will also have a short to ground because they connect to the same power supply in the ECM. A short to ground on either 5 volt reference circuit does not damage the ECM 5 volt power supply. When the short is repaired, the ECM 5 volt supply will return to normal.

Diagnostic Aids:

With the ignition "ON" and the engine stopped, the manifold pressure is equal to atmospheric pressure and the signal voltage will be high. This information is used by the ECM as an indication of vehicle altitude and is referred to as BARO. Comparison of this BARO reading with a known good vehicle with the same sensor is a good way to check accuracy of a "suspect" sensor. Reading should be within ± .4 volt.

If DTC 33 is intermittent, refer to "Symptoms,"

2.2L (VIN 4) ENGINE — DIAGNOSTIC TROUBLE CODE CHART — CENTURY AND CIERA

DTC 33

(Page 2 of 2)
MANIFOLD ABSOLUTE PRESSURE (MAP) SENSOR CIRCUIT
(SIGNAL VOLTAGE HIGH/LOW - LOW/HIGH VACUUM)
2.2L (VIN 4) "A" CARLINE

FROM DTC 33 PAGE 1.

- ENGINE IDLING.
  DOES SCAN TOOL DISPLAY MAP VOLTAGE BELOW .3 VOLT?

YES

(1)
- IGNITION "OFF."
- DISCONNECT SENSOR ELECTRICAL CONNECTOR.
- JUMPER HARNESS TERMINALS "B" TO "C".
- CRANK ENGINE.
- MAP VOLTAGE SHOULD READ OVER 3.7 VOLTS. DOES IT?

NO
- DTC 33 IS INTERMITTENT. IF NO ADDITIONAL DTC(S) WERE STORED, REFER TO "DIAGNOSTIC AIDS"

NO

(2)
- IGNITION "OFF."
- REMOVE JUMPER WIRE.
- PROBE TERMINAL "B" (CKT 432) WITH A TEST LIGHT TO BATTERY VOLTAGE.
- CRANK ENGINE.
- SCAN TOOL SHOULD READ OVER 3.7 VOLTS. DOES IT?

YES
FAULTY CONNECTION OR SENSOR.

YES

(3)
5 VOLT REFERENCE CIRCUIT OPEN
OR
SHORTED TO GROUND
OR
FAULTY ECM.*

NO
CKT 432 OPEN
OR
CKT 432 SHORTED TO GROUND
OR
CKT 432 SHORTED TO SENSOR GROUND
OR
FAULTY ECM.*

* IF ECM IS FAULTY AND MUST BE REPLACED, THE NEW ECM MUST BE PROGRAMMED. REFER TO ECM REPLACEMENT AND PROGRAMMING PROCEDURES

"AFTER REPAIRS," REFER TO DTC CRITERIA AND CONFIRM DTC DOES NOT RESET.

2.2L (VIN 4) ENGINE — DIAGNOSTIC TROUBLE CODE CHART — CENTURY AND CIERA

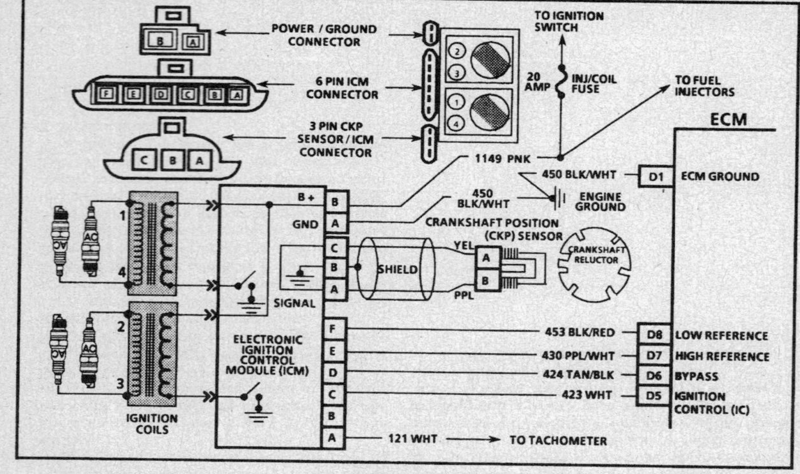

DTC 42
IGNITION CONTROL (IC) CIRCUIT
2.2L (VIN 4) "A" CARLINE

Circuit Description:

The electronic Ignition Control Module (ICM) sends a reference signal to the Engine Control Module (ECM) when the engine is cranking. While the engine speed is under 400 RPM, the ICM controls the ignition timing. When the system is running on the ICM (no voltage on the bypass line), the ICM grounds the Ignition Control (IC) signal. The ECM expects to sense no voltage on the Ignition Control (IC) line during this condition. If it senses a voltage, it sets DTC 42 and will not enter the ignition control mode.

When the engine speed exceeds 400 RPM, the ECM applies 5 volts to the bypass line to switch the timing to ECM control (ignition control). If the bypass line is open or grounded, once the RPM for ignition control is reached, the ICM will not switch to Ignition Control (IC) mode. This results in low Ignition Control (IC) voltage and the setting of DTC 42. If the IC line is grounded, the ICM will switch to Ignition Control (IC), but because the line is grounded, there will be no IC signal. A DTC 42 will be set.

Test Description: Number(s) below refer to circled number(s) on the diagnostic chart.

1. DTC 42 means the ECM has sensed an open or short to ground in the IC or bypass circuits. This test confirms DTC 42 and that the fault causing the DTC 42 is present.
2. Checks for a normal IC ground path through the ICM. An IC CKT 423, shorted to ground, will also read less than 500 ohms, but this will be checked later.
3. As the test light voltage contacts CKT 424, the ICM should switch, causing the ohmmeter to "overrange" if the meter is in the 1000 to 2000 ohms position. Selecting the 10,000 to 20,000 ohms position will indicate a reading above 5000 ohms.

The important thing is that the module "switched."

4. The module did not switch and this step checks for:
   - IC CKT 423 shorted to ground.
   - Bypass CKT 424 open.
   - Faulty ICM connection or ICM.
5. Confirms that DTC 42 is a faulty ECM and not an intermittent in CKTs 423 or 424.

Diagnostic Aids:

If DTC 42 is intermittent, refer to "Symptoms,"

2.2L (VIN 4) ENGINE — DIAGNOSTIC TROUBLE CODE CHART — CENTURY AND CIERA

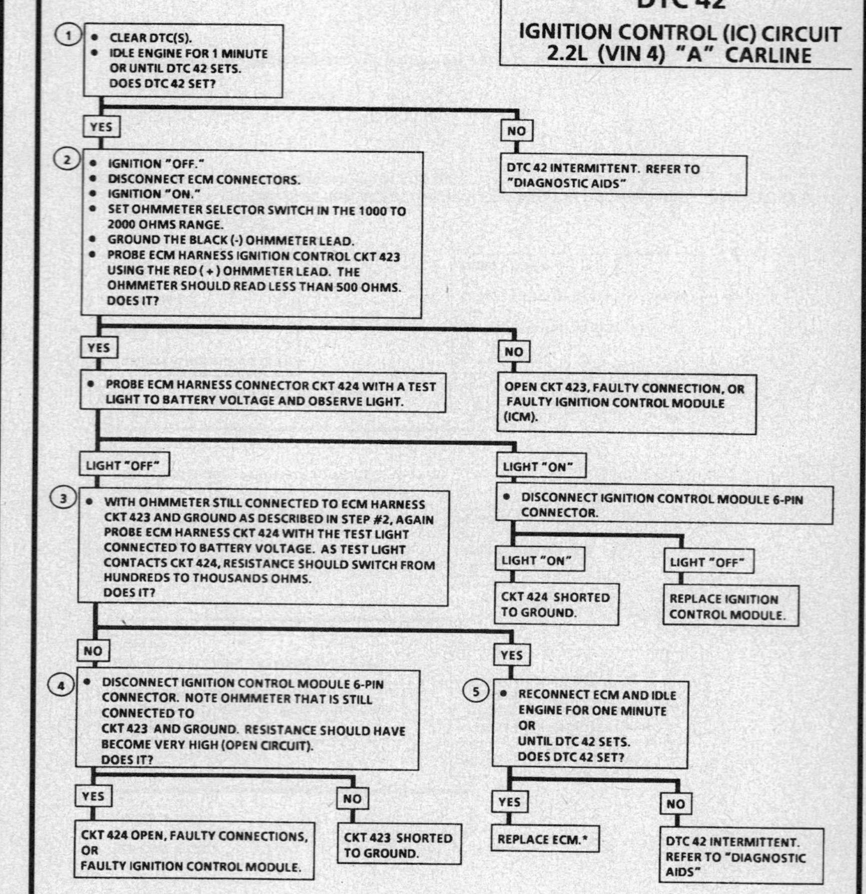

2.2L (VIN 4) ENGINE — DIAGNOSTIC TROUBLE CODE CHART — CENTURY AND CIERA

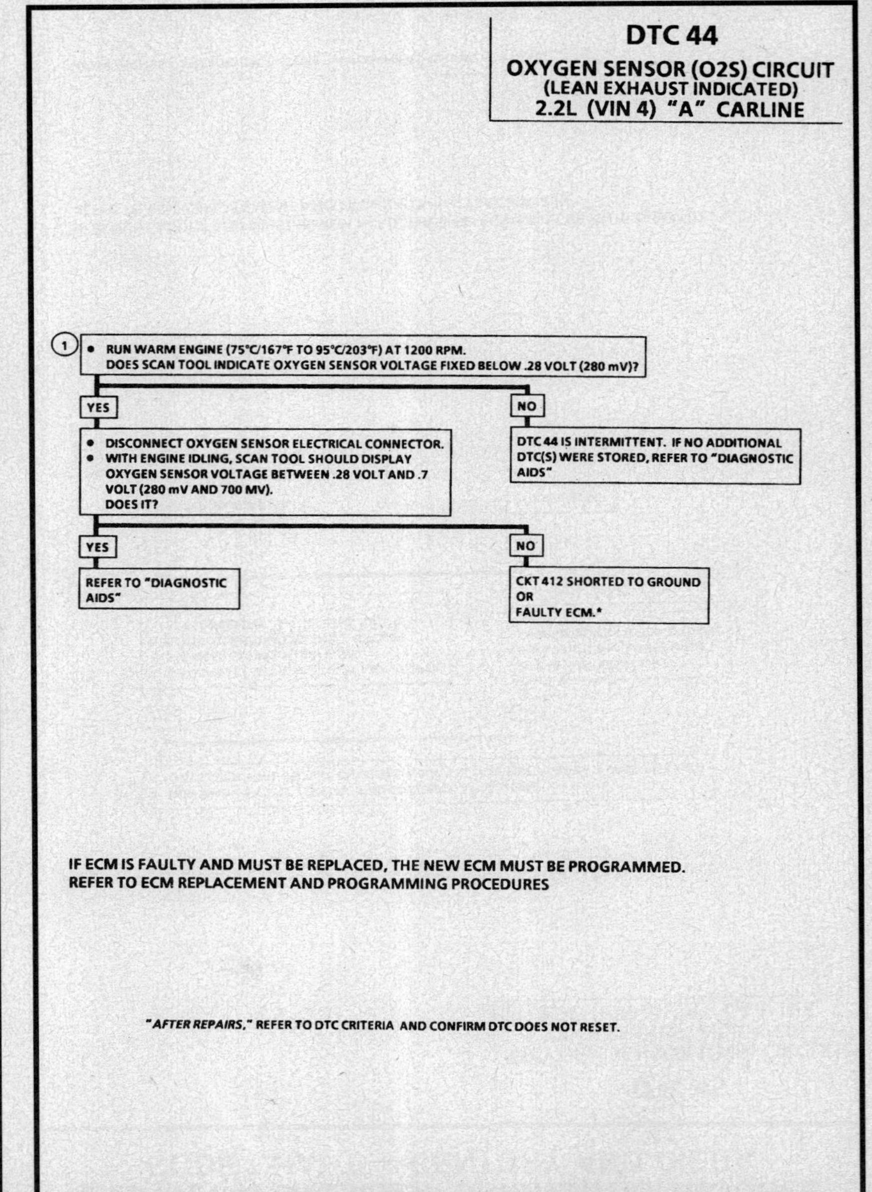

DTC 44
OXYGEN SENSOR (O2S) CIRCUIT
(LEAN EXHAUST INDICATED)
2.2L (VIN 4) "A" CARLINE

Circuit Description:

The Engine Control Module (ECM) supplies a voltage of about .45 volt between terminals "D10" and "D11". (If measured with a 10 megohm digital voltmeter, this may read as low as .32 volt.)

When the O2S reaches operating temperature, it varies this voltage from about .1 volt (exhaust is lean) to about .9 volt (exhaust is rich).

The sensor is like an open circuit and produces no voltage when it is below 316°C (600°F). An open sensor circuit, or cold sensor, causes "Open Loop" operation.

Test Description: Number(s) below refer to circled number(s) on the diagnostic chart.
1. DTC 44 is set when the O2S signal voltage on CKT 412 remains below .28 volt (280 mV) for 25 seconds or more.

Diagnostic Aids:

Using the scan tool, observe the long term fuel trim value at different engine speeds. If the conditions for DTC 44 exists, the long term fuel trim values will be around 160 or higher.
Check the following possible causes:
- Oxygen Sensor Wire - Sensor pigtail may be mispositioned and contacting the exhaust manifold. Check for ground in wire between connector and sensor.

- Fuel Contamination - Water, even in small amounts, near the in-tank fuel pump inlet can be delivered to the injectors. The water causes a lean exhaust and can set a DTC 44.
- Fuel Pressure - System will be lean if fuel pressure is too low. It may be necessary to monitor fuel pressure while driving the vehicle at various road speeds and/or loads to confirm. Refer to "Fuel System Diagnosis," CHART A-7.
- Exhaust Leaks - If there is an exhaust leak, the engine can cause outside air to be pulled into the exhaust and past the sensor. Vacuum or crankcase leaks can cause a lean condition. If DTC 44 is intermittent, refer to "Symptoms,"
- A cracked or otherwise damaged O2S may set an intermittent DTC 44.

2.2L (VIN 4) ENGINE — DIAGNOSTIC TROUBLE CODE CHART — CENTURY AND CIERA

DTC 44
OXYGEN SENSOR (O2S) CIRCUIT
(LEAN EXHAUST INDICATED)
2.2L (VIN 4) "A" CARLINE

(1) • RUN WARM ENGINE (75°C/167°F TO 95°C/203°F) AT 1200 RPM.
DOES SCAN TOOL INDICATE OXYGEN SENSOR VOLTAGE FIXED BELOW .28 VOLT (280 mV)?

YES
- DISCONNECT OXYGEN SENSOR ELECTRICAL CONNECTOR.
- WITH ENGINE IDLING, SCAN TOOL SHOULD DISPLAY OXYGEN SENSOR VOLTAGE BETWEEN .28 VOLT AND .7 VOLT (280 mV AND 700 mV).
DOES IT?

NO
DTC 44 IS INTERMITTENT. IF NO ADDITIONAL DTC(S) WERE STORED, REFER TO "DIAGNOSTIC AIDS"

YES
REFER TO "DIAGNOSTIC AIDS"

NO
CKT 412 SHORTED TO GROUND OR FAULTY ECM.*

IF ECM IS FAULTY AND MUST BE REPLACED, THE NEW ECM MUST BE PROGRAMMED. REFER TO ECM REPLACEMENT AND PROGRAMMING PROCEDURES

"AFTER REPAIRS," REFER TO DTC CRITERIA AND CONFIRM DTC DOES NOT RESET.

2.2L (VIN 4) ENGINE — DIAGNOSTIC TROUBLE CODE CHART — CENTURY AND CIERA

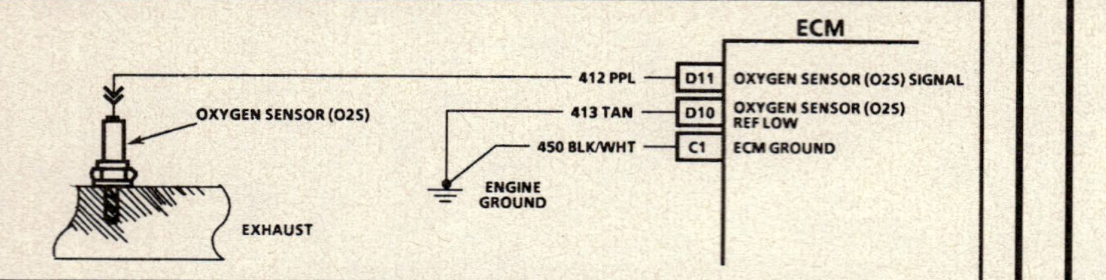

DTC 45
OXYGEN SENSOR (O2S) CIRCUIT
(RICH EXHAUST INDICATED)
2.2L (VIN 4) "A" CARLINE

Circuit Description:

The Engine Control Module (ECM) supplies a voltage of about .45 volt between terminals "D10" and "D11". (If measured with a 10 megohm digital voltmeter, this may read as low as .32 volt.)

When the Oxygen Sensor (O2S) reaches operating temperature, it varies this voltage from about .1 volt (exhaust is lean) to about .9 volt (exhaust is rich).

The sensor is like an open circuit and produces no voltage when it is below 316°C (600°F). An open sensor circuit, or cold sensor, causes "Open Loop" operation.

Test Description: Number(s) below refer to circled number(s) on the diagnostic chart.

1. DTC 45 is set when the O2S signal voltage on CKT 412 remains above .7 volt (700 mV) under the following conditions:
   - Engine run time after start is 2 minutes or more.
   - System is operating in "Closed Loop."
   - Throttle angle is greater than 5%.
   - Above conditions exist for 51 seconds or more.

Diagnostic Aids:

Using the scan tool, observe the long term fuel trim value at different engine speeds. If the conditions for DTC 45 exist, the long term fuel trim values will be around 90 or lower.

Check the following possible causes:
- Fuel Pressure - System will go rich, if pressure is too high. The ECM can compensate for some increase. However, if it gets too high, a DTC 45 will be set. Refer to "Fuel System Diagnosis" CHART A-7.
- Leaking Fuel Injector(s) - Refer to CHART A-7.
- An open ground CKT 453 - May result in induced electrical "noise." The ECM interprets this "noise" as reference pulses. The additional pulses result in a higher than actual engine speed signal. The ECM then delivers too much fuel causing the system to go rich. The engine tachometer will also show higher than actual engine speed, which can help in diagnosing this problem.

- Canister Purge - Check for fuel saturation. If fuel vapor canister is full of fuel, check EVAP canister control valve and hoses. Refer to "Canister Purge," in "Evaporative Emission (EVAP) Control System."
- MAP Sensor - An output that causes the ECM to sense a higher than normal manifold pressure (low vacuum) can cause the system to go rich. Disconnecting the Manifold Absolute Pressure (MAP) sensor will allow the ECM to set a fixed value for the MAP sensor. Substitute a different MAP sensor if the rich condition is gone, while the sensor is disconnected.
- TP Sensor - An intermittent Throttle Position (TP) sensor output will cause the system to operate richly due to a false indication of the engine accelerating.
- O2S Contamination - Inspect Oxygen Sensor (O2S) for silicone contamination from fuel, or use of improper RTV sealant. The sensor may have a white, powdery coating and result in a high but false signal voltage (rich exhaust indication). The ECM will then reduce the amount of fuel delivered to the engine causing a severe surge driveability problem.
- EGR Valve - Exhaust Gas Recirculation (EGR) sticking open at idle is usually accompanied by a rough idle and/or stall condition.
- If DTC 45 is intermittent, refer to "Symptoms,"
- Engine Oil Contamination - Fuel fouled engine oil could cause the O2S to sense a rich air/fuel mixture and set a DTC 45.

2.2L (VIN 4) ENGmdash;DIAGNOSTIC TROUBLE CODE CHART — CENTURY AND CIERA

DTC 45
OXYGEN SENSOR (O2S) CIRCUIT
(RICH EXHAUST INDICATED)
2.2L (VIN 4) "A" CARLINE

(1)
- RUN WARM ENGINE (75°C/167°F TO 95°C/203°F) AT 1200 RPM.
- DOES SCAN TOOL DISPLAY OXYGEN SENSOR VOLTAGE FIXED ABOVE .7 VOLT (700 mV)?

YES
- DISCONNECT OXYGEN SENSOR AND JUMPER HARNESS CKT 412 TO GROUND.
- SCAN TOOL SHOULD DISPLAY OXYGEN SENSOR BELOW .28 VOLT (280 mV). DOES IT?

NO
DTC 45 IS INTERMITTENT. IF NO ADDITIONAL DTCS WERE STORED, REFER TO "DIAGNOSTIC AIDS"

YES
REFER TO "DIAGNOSTIC AIDS"

NO
REPLACE ECM.*

*IF ECM IS FAULTY AND MUST BE REPLACED, THE NEW ECM MUST BE PROGRAMMED. REFER TO ECM REPLACEMENT AND PROGRAMMING PROCEDURES

"AFTER REPAIRS," REFER TO DTC CRITERIA AND CONFIRM DTC DOES NOT RESET.

2.2L (VIN 4) ENGINE — DIAGNOSTIC TROUBLE CODE CHART — CENTURY AND CIERA

2.2L (VIN 4) ENGINE — DIAGNOSTIC TROUBLE CODE CHART — CENTURY AND CIERA

DTC 51

EEPROM OR ECM FAILURE
2.2L (VIN 4) "A" CARLINE

DIAGNOSTIC TROUBLE CODE (DTC) 51

ECM FAILURE
(ECM FAILED OR EEPROM FAILURE)

CHECK THAT ALL ECM CONNECTIONS ARE GOOD.
IF OK, CLEAR MEMORY AND RECHECK ECM.
IF DTC 51 REAPPEARS, REPLACE ECM AND PROGRAM NEW ECM.

* IF ECM IS FAULTY AND MUST BE REPLACED, THE NEW ECM MUST BE PROGRAMMED.
REFER TO ECM REPLACEMENT AND PROGRAMMING PROCEDURES

"AFTER REPAIRS," CONFIRM "CLOSED LOOP" OPERATION, AND NO MIL (SERVICE ENGINE SOON).

TO MANIFOLD ABSOLUTE PRESSURE (MAP) SENSOR
TO THROTTLE POSITION (TP) SENSOR
ECM
A/C REFRIGERANT PRESSURE SENSOR
B
C
A
416 GRY
474 GRY
380 GRY/RED
452 BLK
C8
C7
B6
D2
+5V
A/C REFRIGERANT PRESSURE SIGNAL
SENSOR GROUND
TO MAP SENSOR
TO ENGINE COOLANT TEMPERATURE (ECT) SENSOR
TO IGNITION SWITCH
HTR-A/C FUSE
20 AMP
A/C SELECT SWITCH
66 LT GRN
D9
A/C REQUEST SIGNAL
59 DK GRN
639 PNK/BLK
B
A
A/C COMPRESSOR CLUTCH
150 BLK
30
87
85
86
A/C COMPRESSOR CLUTCH CONTROL RELAY
TO EGR SOLENOID
TO IGNITION SWITCH
ENG/ A/C FUSE
10 AMP
639 PNK/BLK
TO COOLING FAN CONTROL (FC) RELAY
366 LT GRN/BLK
C12
A/C RELAY CONTROL DRIVER

DTC 66

A/C REFRIGERANT PRESSURE SENSOR CIRCUIT
2.2L (VIN 4) "A" CARLINE

Circuit Description:

The A/C refrigerant pressure sensor responds to changes in A/C refrigerant system high side pressure. This input indicates how much load the A/C compressor is putting on the engine and is one of the factors used by the ECM to determine IAC valve position for idle speed control. The circuit consists of a 5 volt reference and a ground, both provided by the ECM, and a signal line to the ECM. The signal is a voltage which is proportional to the pressure. The sensor's range of operation is 0 to 454 psi. At 0 psi, the signal will be about .1 volt, varying up to about 4.9 volts at 454 psi or above. DTC 66 sets if the voltage is above 4.9 volts (454 psi) or below .3 volt (7.5 psi) for 5 seconds or more. The A/C compressor clutch is disabled by the ECM if fault is present.

Test Description: Number(s) below refer to circled number(s) on the diagnostic chart.

1. This step checks the voltage signal being received by the ECM from the A/C refrigerant pressure sensor.
2. Checks to see if the high voltage signal is from a shorted sensor or a short to voltage in the circuit. Normally, disconnecting the sensor would make a normal circuit go to near zero volt.
3. Checks to see if low voltage signal is from the sensor or the circuit. Jumpering the sensor signal CKT 380 to 5 volts, checks the circuit, connections, and ECM.
4. This step checks to see if the low voltage signal was due to an open in the sensor circuit or the 5 volt reference circuit since the prior step eliminated the pressure sensor.
5. The 5 volt reference for the Throttle Position (TP) sensor connects to the same 5 volt power supply in the ECM. If the 5 volt reference circuit to TP sensor has a short to ground, the MAP sensor and A/C refrigerant pressure sensor 5 volt reference circuit will also have a short to ground because they connect to the same power supply in the ECM. A short to ground on either 5 volt reference circuit does not damage the ECM 5 volt power supply. When the short is repaired, the ECM 5 volt supply will return to normal.

Diagnostic Aids:

At temperatures of -20 degrees fahrenheit and below, the A/C pressure can drop below the .3 volt (7.5 psi) threshold. If this happens the DTC 66 will set and disable the A/C system. Clear DTC 66 and check A/C pressure in a heated garage, if pressure is between .3 volt (7.5 psi) and 4.9 volts (454 psi), the A/C pressure transducer is functioning correctly.

DTC 66 sets when the signal voltage falls outside the normal possible range of the sensor. If signal voltage is greater than .1 volt, but less than .3 volt, there could be a low refrigerant pressure problem with the A/C system.

2.2L (VIN 4) ENGINE — DIAGNOSTIC TROUBLE CODE CHART — CENTURY AND CIERA

DTC 66
A/C REFRIGERANT PRESSURE SENSOR CIRCUIT
2.2L (VIN 4) "A" CARLINE

NOTICE: TO ENSURE PROPER DIAGNOSIS OF THE DTC 66 CIRCUIT, THE A/C REFRIGERANT PRESSURE SHOULD BE WITHIN LIMITS

1. • KEY "ON," ENGINE NOT RUNNING.
   • INSTALL SCAN TOOL AND OBSERVE VOLTAGE FOR A/C REFRIGERANT PRESSURE SENSOR.

ABOVE 4.9 VOLTS

2. • DISCONNECT A/C REFRIGERANT PRESSURE SENSOR ELECTRICAL CONNECTOR.
   DOES SCAN TOOL DISPLAY LESS THAN 1 VOLT?

NO → CHECK FOR SHORT TO VOLTAGE IN SENSOR GROUND CIRCUIT. IF NOT SHORTED, REPLACE ECM.*

YES → CHECK FOR OPEN IN SENSOR GROUND CIRCUIT. IF NOT OPEN, CHECK FOR POOR SENSOR TERMINAL CONNECTIONS. IF OK, REPLACE A/C REFRIGERANT PRESSURE SENSOR.

BELOW .3 VOLT

3. • DISCONNECT A/C REFRIGERANT PRESSURE SENSOR CONNECTOR.
   • JUMPER TERMINALS "C" AND "B". DOES SCAN TOOL DISPLAY ABOVE 4.6 VOLTS?

NO →
4. • REMOVE JUMPER.
   • CONNECT VOLTMETER FROM TERMINAL "A" TO "B". IS VOLTAGE ABOUT 5 VOLTS?

YES → CHECK SENSOR TERMINAL CONNECTIONS. IF OK, REPLACE A/C REFRIGERANT PRESSURE SENSOR.

NO → • BACKPROBE ECM TERMINAL "C7" WITH VOLTMETER TO GROUND. IS VOLTAGE ABOUT 5 VOLTS?

YES → CHECK FOR OPEN IN CKT 380. IF OK, CHECK FOR POOR CONNECTION AT ECM TERMINAL "B6". IF OK, REPLACE ECM.*

NO →
5. CHECK FOR POOR CONNECTION AT ECM TERMINAL "C7"
   OR
   SHORT TO GROUND IN CKT 474. IF OK, ECM IS FAULTY.*

YES → REPAIR OPEN IN CKT 474.

BETWEEN .3 VOLT AND 4.9 VOLTS.

FAULT IS NOT PRESENT AT THIS TIME. REFER TO "DIAGNOSTIC AIDS."

* IF ECM IS FAULTY AND MUST BE REPLACED, THE NEW ECM MUST BE PROGRAMMED. REFER TO ECM REPLACEMENT AND PROGRAMMING PROCEDURES

"AFTER REPAIRS," REFER TO DTC CRITERIA AND CONFIRM DTC DOES NOT RESET.

2.2L (VIN 4) ENGINE — ECM SYMPTOM CHART — CENTURY AND CIERA

ECM Connector and Driveability Symptoms Identification

This ECM voltage chart is for use with a J 39200 to further aid in diagnosis. These voltages were derived from a known good vehicle. The voltages you get may vary due to low battery charge or other reasons, but they should be very close.

THE FOLLOWING CONDITIONS MUST BE MET BEFORE TESTING:
• Engine at operating temperature • "Closed Loop" • Engine idling (for "Engine Run" column)
• Test terminal not grounded • Scan tool not installed • Brake not applied.

ECM PIN/FUNCTION	CKT #	WIRE COLOR	COMPONENT CONNECTOR CAVITY	NORMAL VOLTAGES		DTC(S) AFFECTED	POSSIBLE SYMPTOMS FROM FAULTY CIRCUIT
				KEY "ON"	ENG RUN		
A4 PEAK AND HOLD JUMPER	887	TAN	ECM-"A12"	(8)	(8)		(3) CRANKS, BUT WON'T RUN.
A5 INJECTOR DRIVER	467	DK BLU	INJECTOR JUMPER HARNESS CONNECTOR (GRY) "B"	B +	B + (13)		(3) CRANKS, BUT WON'T RUN. (4) POSSIBLE NO START (OPEN 20 AMP INJ/COIL FUSE) OR INJECTORS ENERGIZED AT ALL TIMES WITH KEY "ON."
A7 FUEL PUMP RELAY CONTROL	465	DK GRN/WHT	FUEL PUMP RELAY - "85"	(6)	B +		(3) LONG CRANKING TIME BEFORE ENGINE STARTS.
A8 IAC "A" HIGH	441	LT BLU/WHT	IAC VALVE "D"	NOT USABLE	NOT USABLE		(5) INCORRECT, UNSTABLE IDLE.
A9 IAC "A" LOW	442	LT BLU/BLK	IAC VALVE "C"	NOT USABLE	NOT USABLE		(5) INCORRECT, UNSTABLE IDLE.
A10 IAC "B" LOW	444	LT GRN/BLK	IAC VALVE "A"	NOT USABLE	NOT USABLE		(5) INCORRECT, UNSTABLE IDLE.
A11 IAC "B" HIGH	443	LT GRN/WHT	IAC VALVE "B"	NOT USABLE	NOT USABLE		(5) INCORRECT, UNSTABLE IDLE.
A12 PEAK AND HOLD JUMPER	887	TAN	ECM "A4"	(8)	(8)		(3) CRANKS, BUT WON'T RUN.

(3) OPEN CIRCUIT.
(4) GROUNDED CIRCUIT.
(5) OPEN OR GROUNDED CIRCUIT.
(6) MEASURES B + FOR 2 SECONDS AFTER KEY "ON," THEN MEASURES "0" VOLTS.
(8) LESS THAN .5 VOLT (500 mV).
(13) ALTERNATE TEST: SET J 39200 DVM TO DC VOLTAGE FREQUENCY SCALE. CONNECT RED LEAD TO IGNITION FEED ("C6" AT ECM) AND BLACK LEAD TO INJECTOR DRIVER ("A5" AT ECM); SHOULD MEASURE ABOUT 9-15 Hz.

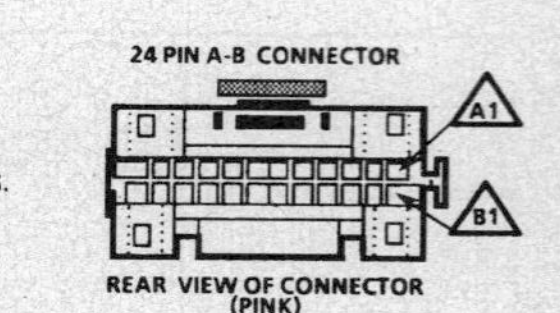

2.2L (VIN 4) ENGINE — ECM SYMPTOM CHART — CENTURY AND CIERA

ECM Connector and Driveability Symptoms Identification

This ECM voltage chart is for use with a J 39200 to further aid in diagnosis. These voltages were derived from a known good vehicle. The voltages you get may vary due to low battery charge or other reasons, but they should be very close.

THE FOLLOWING CONDITIONS MUST BE MET BEFORE TESTING:
- Engine at operating temperature • Closed Loop • Engine idling (for "Engine Run" column)
- Test terminal <u>not</u> grounded • Scan tool <u>not</u> installed • Brake <u>not</u> applied.

ECM PIN/FUNCTION	CKT #	WIRE COLOR	COMPONENT CONNECTOR CAVITY	NORMAL VOLTAGES KEY "ON"	NORMAL VOLTAGES ENG RUN	DTC(S) AFFECT	POSSIBLE SYMPTOMS FROM FAULTY CIRCUIT
B4 ECT SIGNAL	410	YEL	ECT SENSOR "C"	(7)	(7)	14 (5)	(5) INCORRECT IDLE, COOLING FAN RUNS AT ALL TIMES.
B5 TP SIGNAL	417	DK BLU	TP SENSOR "C"	.6V (9)	.6V (9)	21 (5)	(5) POOR PERFORMANCE, HESITATION.
B6 A/C REFRIGERANT PRESSURE SENSOR SIGNAL	380	GRY/RED	A/C REFRIGERANT PRESSURE SENSOR "C"	(2)	(2)	66 (5)	(5) NO A/C COOLING, A/C COMPRESSOR CLUTCH INOPERATIVE, REFER TO DTC 66.
B7 MAP SIGNAL	432	LT GRN	MAP SENSOR "B"	4.75V	(2)	33 (5)	(5) POOR PERFORMANCE, ROUGH IDLE, SURGE REFER TO CHART C-1D.
B8 IAT SIGNAL	472	TAN	IAT SENSOR "B"	(7)	(7)	23 (5)	(5) POOR PERFORMANCE, HARD TO START.
B9 DATA LINK TEST TERMINAL	451	WHT/BLK	DATA LINK CONNECTOR "B"	5.0V	5.0V		(3) NO DATA, WON'T FLASH DTC(S). (4) KEY "ON" - DTC(S) DISPLAYED ENG RUN - FIELD SERVICE MODE "ON."
B10 VSS SIGNAL (HIGH)	400	YEL	VEHICLE SPEED SENSOR (VSS) "A"			24 (5)	(5) NO VSS SIGNAL, INACCURATE OR INOPERATIVE SPEEDOMETER, INOPERATIVE CRUISE CONTROL, TCC INOPERATIVE.
B11 VSS SIGNAL (LOW)	401	PPL	VEHICLE SPEED SENSOR (VSS) "B"			24 (5)	(5) NO VSS SIGNAL, INACCURATE OR INOPERATIVE SPEEDOMETER, INOPERATIVE CRUISE CONTROL, TCC INOPERATIVE.

(2) VARIES.
(3) OPEN CIRCUIT.
(4) GROUNDED CIRCUIT.
(5) OPEN OR GROUNDED CIRCUIT.
(7) VARIES WITH TEMPERATURE.
(9) WITHIN THE RANGE OF .33-1.33 VOLTS.

24 PIN A-B CONNECTOR

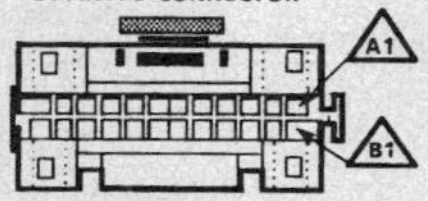

A1

B1

REAR VIEW OF CONNECTOR (PINK)

2.2L (VIN 4) ENGINE — ECM SYMPTOM CHART — CENTURY AND CIERA

ECM Connector and Driveability Symptoms Identification

This ECM voltage chart is for use with a J 39200 to further aid in diagnosis. These voltages were derived from a known good vehicle. The voltages you get may vary due to low battery charge or other reasons, but they should be very close.

THE FOLLOWING CONDITIONS MUST BE MET BEFORE TESTING:
- Engine at operating temperature • Closed Loop • Engine idling (for "Engine Run" column)
- Test terminal <u>not</u> grounded • Scan tool <u>not</u> installed • Brake <u>not</u> applied.

ECM PIN/FUNCTION	CKT #	WIRE COLOR	COMPONENT CONNECTOR CAVITY	NORMAL VOLTAGES KEY "ON"	NORMAL VOLTAGES ENG RUN	DTC(S) AFFECTED	POSSIBLE SYMPTOMS FROM FAULTY CIRCUIT
C1 ECM GROUND	450	BLK/WHT	ENGINE GROUND	(8) (15)	(8) (15)		(3) NO EFFECT UNLESS "D1" IS ALSO OPEN. IF BOTH ARE OPEN, CRANKS, BUT WON'T RUN.
C4 PARK/NEUTRAL POSITION (PNP) SWITCH	434	ORN/BLK	PARK/NEUTRAL POSITION (PNP) SWITCH "A"	(8)	(8)		(5) INCORRECT IDLE.
C6 IGNITION FEED	439	PNK/BLK	I/P JUNCTION CONNECTOR "G7" AND 15 AMP ECM FUSE	B+ (15)	B+ (15)		(5) CRANKS, BUT WON'T RUN. (4) CRANKS, BUT WON'T RUN, OPEN 15 AMP ECM FUSE.
C7 MAP AND A/C REFRIGERANT PRESSURE SENSORS 5V REF	474	GRY	MAP - "C" AND A/C REFRIGERANT PRESSURE SENSOR "B"	5.0V	5.0V	33(5) 66(5)	(5) DEGRADED PERFORMANCE. NO A/C COOLING.
C8 TP SENSOR 5 VOLT REFERENCE	416	GRY	TP SENSOR "A"	5.0V	5.0V	21(5)	(3) DEGRADED PERFORMANCE.
C9 & C10 KEEP ALIVE MEMORY (B +)	440	ORN	FUEL PUMP RELAY "87" AND OIL PRESS SWITCH "C" AND 15 AMP UNDERHOOD FUSE	B+ (15)	B+ (15)		(3) CRANKS, BUT WON'T RUN. (4) CRANKS, BUT WON'T RUN, OPEN UNDERHOOD 15 AMP FUSE.
C11 TCC CONTROL	422	TAN/BLK	TCC SOLENOID "D" AND DATA LINK CONNECTOR "F"	(8)	(8)		(5) TCC INOPERATIVE, POOR FUEL ECONOMY. (4) TCC ENGAGES AT VERY LOW ROAD SPEED.
C12 A/C COMPRESSOR CLUTCH RELAY CONTROL	366	LT GRN/ BLK	A/C COMPRESSOR CLUTCH CONTROL RELAY "86"	B+	B+		(3) NO A/C COOLING. (4) NO A/C COOLING, ENGINE OVERHEATS. OPEN 10 AMP ENG/ A/C FUSE.
C13 COOLING FAN RELAY CONTROL	535	DK GRN	COOLING FAN CONTROL RELAY "86"	B+	B+		(3) ENGINE OVERHEATS. (4) COOLING FAN RUNS AT ALL TEMPERATURES.
C14 EGR SOLENOID CONTROL	435	GRY	EGR SOLENOID "A"	B+	B+	32(5)	(3) EGR VALVE DOESN'T OPEN, HIGH NOx DETONATION, SPARK KNOCK. (4) DEGRADED PERFORMANCE. IMPORTANT: EGR SOLENOID IS DISABLED IF DTC 14, 21, 23, 32 OR 33 IS SET.

(3) OPEN CIRCUIT.
(4) GROUNDED CIRCUIT.
(5) OPEN OR GROUNDED CIRCUIT.
(8) LESS THAN .5 VOLT (500 mV).
(15) THIS TYPE OF CIRCUIT SHOULD ALSO BE CHECKED USING J 34142-B TEST LIGHT.

32 PIN C-D CONNECTOR

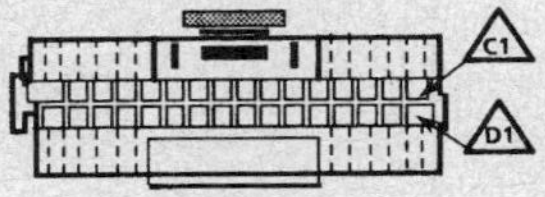

C1

D1

REAR VIEW OF CONNECTOR (PINK)

2.2L (VIN 4) ENGINE — ECM SYMPTOM CHART — CENTURY AND CIERA

ECM Connector and Driveability Symptoms Identification

This ECM voltage chart is for use with a J 39200 to further aid in diagnosis. These voltages were derived from a known good vehicle. The voltages you get may vary due to low battery charge or other reasons, but they should be very close.

THE FOLLOWING CONDITIONS MUST BE MET BEFORE TESTING:

● Engine at operating temperature ● Closed Loop ● Engine idling (for "Engine Run" column)
● Test terminal not grounded ● Scan tool not installed ● Brake not applied.

ECM PIN/FUNCTION	CKT #	WIRE COLOR	COMPONENT CONNECTOR CAVITY	NORMAL VOLTAGES KEY "ON"	NORMAL VOLTAGES ENG RUN	DTC(S) AFFECT	POSSIBLE SYMPTOMS FROM FAULTY CIRCUIT
D1 ECM GROUND	450	BLK/WHT	ENGINE GROUND	(8) (15)	(8) (15)		(3) NO EFFECT UNLESS "C1" IS ALSO OPEN. IF BOTH ARE OPEN, CRANKS, BUT WON'T RUN.
D2 SENSOR GROUND	455	PPL	MAP SENSOR "A", ECT SENSOR "D" AND A/C REFRIGERANT PRESSURE SENSOR "A"	(8)	(8)	14 (3) 33 (3) 66 (3)	(3) HARD START OR CRANKS, BUT WON'T RUN.
D3 SENSOR GROUND	469	BLK/ORN	TP SENSOR "B" AND IAT SENSOR "A"	(8)	(8)	21 (3) 23 (3)	(3) HARD START OR CRANKS, BUT WON'T RUN.
D5 IGNITION CONTROL (IC)	423	WHT	IGNITION CONTROL MODULE 6 PIN CONN. "C"	(8)	2.4V	42 (5)	(5) STUMBLES, UNSTABLE IDLE, DEGRADED PERFORMANCE.
D6 BYPASS	424	TAN/BLK	IGNITION CONTROL MODULE 6 PIN CONN. "D"	(8)	4.8V	42 (5)	(5) DEGRADED PERFORMANCE.
D7 IGNITION REFERENCE HIGH	430	PPL/WHT	IGNITION CONTROL MODULE 6 PIN CONN. "E"	(8)	3.2V (14)		(5) CRANKS, BUT WON'T RUN.
D8 IGNITION REFERENCE LOW	453	BLK/RED	IGNITION CONTROL MODULE 6 PIN CONN. "F"	(8)	(8)		(3) REDUCED PERFORMANCE
D9 A/C REQUEST	66	LT GRN	15 PIN INLINE I/P CONN. "B" AND 5 PIN HVAC CONN. "C"	(8) "OFF" B+ "ON"	(8) "OFF" B+ "ON"		(3) NO A/C COOLING A/C COMPRESSOR CLUTCH WON'T ENGAGE.
D10 OXYGEN SENSOR (O2S) REFERENCE LOW	413	TAN	ENGINE GROUND	(8)	(8)	13 (3)	(3) "OPEN LOOP" OPERATION REDUCED PERFORMANCE.
D11 OXYGEN SENSOR (O2S) SIGNAL	412	PPL	OXYGEN SENSOR	.01-.55V	1.9V (10)	13 44 45	(5) "OPEN LOOP" OPERATION, STRONG EXHAUST ODOR, REDUCED PERFORMANCE.
D12 SERIAL DATA	461	ORN	DATA LINK CONNECTOR "M"	4.7V	4.7V		(5) NO DATA.
D13 VEHICLE SPEED SENSOR (VSS) OUTPUT	817	DK GRN/ WHT	15 PIN INLINE I/P CONNECTOR "N" AND RIGHT CLUSTER CONNECTOR "C4"				(5) INCORRECT SPEEDOMETER READING, CRUISE CONTROL INOPERATIVE.
D14 MIL (SERVICE ENGINE SOON) LAMP CONTROL	419	BRN/WHT	15 PIN INLINE I/P CONNECTOR "C" AND RIGHT CLUSTER CONNECTOR "C14"	.01-1.0V	B +		(3) LAMP NOT "ON" FOR "BULB CHECK," WON'T FLASH DTC(S). (4) LAMP "ON" ALL THE TIME, DOES NOT FLASH, REFER TO "OBD SYSTEM CHECK."

(3) OPEN CIRCUIT.
(4) GROUNDED CIRCUIT.
(8) LESS THAN .5 VOLT (500 mV).
(10) ACTIVELY VARIES WITH INDICATED RANGE.
(14) J 39200 RED LEAD TO REF HIGH ("D7" AT ECM) AND BLACK LEAD TO ECM GROUND ("D1" AT ECM); SHOULD MEASURE ABOUT 25 - 30 Hz IN DC VOLTAGE RANGE.
(15) THIS TYPE OF CIRCUIT SHOULD ALSO BE CHECKED USING J 34142-B TEST LIGHT.

32 PIN C-D CONNECTOR

REAR VIEW OF CONNECTOR (PINK)

2.2L (VIN 4) ENGINE — COMPONENT DIAGNOSTIC CHART — CENTURY AND CIERA

CHART C-1A

PARK/NEUTRAL POSITION (PNP) SWITCH DIAGNOSIS
2.2L (VIN 4) "A" CARLINE

Circuit Description:

The Park/Neutral Position (PNP) switch contacts are a part of the neutral start/back up switch and are closed to ground in park or neutral, and open in drive/reverse ranges.

The ECM supplies ignition voltage through a current limiting resistor to CKT 434 and senses a closed switch, when the voltage on CKT 434 drops to less than one volt.

The ECM uses the PNP signal as one of the inputs to control idle air control and VSS diagnostics.

Test Description: Number(s) below refer to circled number(s) on the diagnostic chart.

1. Checks for a closed switch to ground in park position. Different makes of scan tools will read PNP switch differently. Refer to tool operations manual for type of display used.

2. Checks for an open switch in drive range.

3. Be sure scan tool indicates drive, even while wiggling shifter to test for an intermittent or misadjusted switch in drive range.

2.2L (VIN 4) ENGINE — COMPONENT DIAGNOSTIC CHART — CENTURY AND CIERA

CHART C-1A
PARK/NEUTRAL POSITION (PNP) SWITCH DIAGNOSIS
2.2L (VIN 4) "A" CARLINE

(1) • WITH TRANSAXLE IN PARK, SCAN TOOL SHOULD INDICATE PARK OR NEUTRAL. DOES IT?

YES → (3) • SHIFT TRANSAXLE INTO DRIVE. • SCAN TOOL SHOULD DISPLAY A CHANGE TO INDICATE DRIVE.

NO → • DISCONNECT P/N SWITCH. • THIS SHOULD CAUSE SCAN TOOL TO DISPLAY DRIVE RANGE. DOES IT?

YES → FAULTY PNP SWITCH CONNECTION OR PNP SWITCH MISADJUSTED OR FAULTY PNP SWITCH.

NO → NO TROUBLE FOUND.

NO (from 1) → (2) • DISCONNECT PARK NEUTRAL POSITION (PNP) SWITCH CONNECTOR. • JUMPER HARNESS CONNECTOR TERMINALS "A" AND "B". • SCAN TOOL SHOULD INDICATE PARK OR NEUTRAL. DOES IT?

NO → • JUMPER HARNESS CONNECTOR (CKT 434) TO ENGINE GROUND. • SCAN TOOL SHOULD INDICATE PARK OR NEUTRAL. DOES IT?

YES → OPEN GROUND CIRCUIT.

NO → CKT 434 SHORTED TO GROUND OR FAULTY ECM.*

YES (from 2) → FAULTY PNP SWITCH CONNECTION OR PNP SWITCH MISADJUSTED OR FAULTY PNP SWITCH.

NO → CKT 434 OPEN OR FAULTY ECM CONNECTION OR ECM.*

* IF ECM IS FAULTY AND MUST BE REPLACED, THE NEW ECM MUST BE PROGRAMMED. REFER TO ECM REPLACEMENT AND PROGRAMMING PROCEDURES

"AFTER REPAIRS," CONFIRM "CLOSED LOOP" OPERATION AND NO MIL (SERVICE ENGINE SOON).

2.2L (VIN 4) ENGINE — COMPONENT DIAGNOSTIC CHART — CENTURY AND CIERA

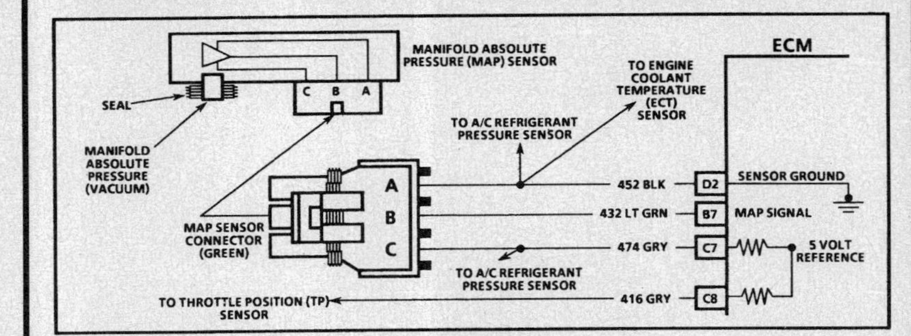

CHART C-1D
MANIFOLD ABSOLUTE PRESSURE (MAP) SENSOR OUTPUT CHECK
2.2L (VIN 4) "A" CARLINE

Circuit Description:

The Manifold Absolute Pressure (MAP) sensor measures the changes in the intake manifold pressure which result from engine load (intake manifold vacuum) and RPM changes; and converts these into a voltage output. The ECM sends a 5 volts reference voltage to the MAP sensor. As the manifold pressure changes, the output voltage of the sensor also changes. By monitoring the sensor output voltage, the ECM calculates the manifold pressure. A lower pressure (low voltage) output voltage will be about 1 to 2 volts at idle. While higher pressure (high voltage) output voltage will be about 4 to 4.8 at Wide Open Throttle (WOT). The MAP sensor is also used, under certain conditions, to measure barometric pressure, allowing the ECM to make adjustments for altitude changes. The ECM uses the MAP sensor to control fuel delivery and ignition timing.

Test Description: Number(s) below refer to circled number(s) on the diagnostic chart.

Important
• Be sure to use the same Diagnostic Test Equipment for all measurements.

1. When comparing scan readings to a known good vehicle, it is important to compare vehicles that use a MAP sensor having the same color insert and the same "Hot Stamped" number. See Figures 1 and 2

2. Applying 34 kPa (10" Hg) vacuum to the MAP sensor should result in voltage readings of 1.5 to 2.1 volts less than the voltage in Step 1. Upon applying vacuum to the sensor, the change in voltage should be instantaneous. A slow voltage change indicates a faulty sensor.

3. Check vacuum seal to sensor for leaking or restriction.

   NOTICE: Make sure electrical connector remains securely fastened.

4. Remove sensor from the intake plenum and twist sensor (by hand only) to check for intermittent connection. Output changes greater than .10 volt indicate a faulty sensor or connection. If OK, replace sensor.

2.2L (VIN 4) ENGINE — COMPONENT DIAGNOSTIC CHART — CENTURY AND CIERA

CHART C-1D
MANIFOLD ABSOLUTE PRESSURE (MAP) SENSOR OUTPUT CHECK
2.2L (VIN 4) "A" CARLINE

IMPORTANT: THIS CHART ONLY APPLIES TO MAP SENSORS HAVING GREEN OR BLACK COLOR KEY INSERT (SEE BELOW).

1.
- IGNITION "ON," ENGINE "OFF."
- SCAN DIAGNOSTIC TROUBLE CODES DTC(S). IF DTC 33 IS PRESENT, USE THAT CHART FIRST.
- SCAN TOOL SHOULD INDICATE A MAP SENSOR VOLTAGE.
- COMPARE THIS READING WITH THE READING OF A KNOWN GOOD VEHICLE. SEE FACING PAGE TEST DESCRIPTION, STEP 1. VOLTAGE READING SHOULD BE WITHIN ± .4 VOLT. IS IT?

YES → NO → REPLACE MAP SENSOR.

2.
- REMOVE MAP SENSOR AND PLUG VACUUM PORT ON INTAKE MANIFOLD.
- CONNECT A HAND VACUUM PUMP TO MAP SENSOR.
- START ENGINE.
- NOTE MAP SENSOR VOLTAGE.
- APPLY 34 kPa (10" Hg) OF VACUUM AND NOTE VOLTAGE CHANGE. SUBTRACT SECOND READING FROM THE FIRST. VOLTAGE VALUE SHOULD BE GREATER THAN 1.5 VOLTS. IS IT?

YES → NO → CHECK MAP SENSOR CONNECTION. IF OK, REPLACE MAP SENSOR.

3. NO TROUBLE FOUND. CHECK SENSOR INLET PORT FOR RESTRICTION OR LEAKING SEAL.

4. CHECK MAP SENSOR CONNECTION. IF OK, REPLACE MAP SENSOR.

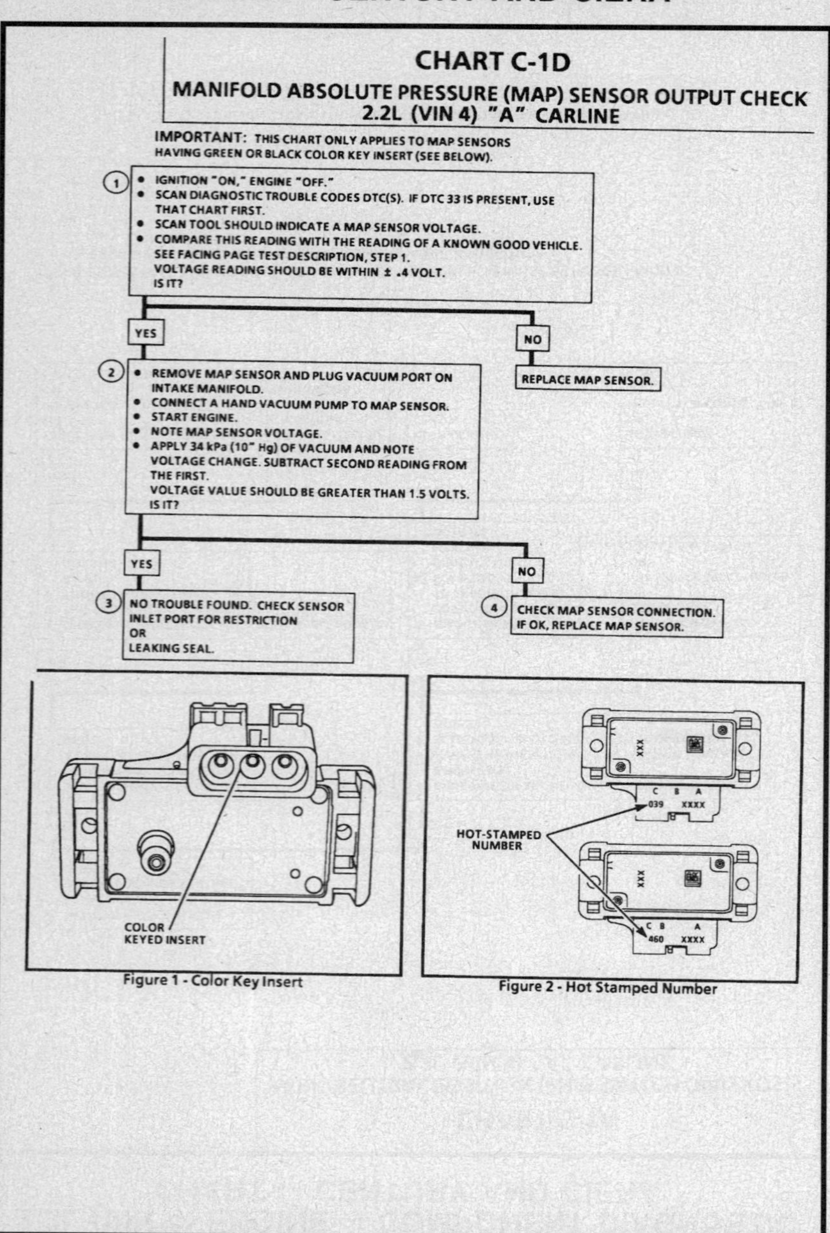

Figure 1 - Color Key Insert

Figure 2 - Hot Stamped Number

2.2L (VIN 4) ENGINE — COMPONENT DIAGNOSTIC CHART — CENTURY AND CIERA

CHART C-2A
INJECTOR BALANCE TEST
2.2L (VIN 4) "A" CARLINE

The injector balance tester is a tool used to turn the injector "ON" for a precise amount of time, thus spraying a measured amount of fuel into the manifold. This causes a drop in fuel rail pressure that we can record and compare between each injector. All injectors should have the same amount of pressure drop (±10 kPa). Any injector with a pressure drop that is 10 kPa (or more) greater or less than the average drop of the other injectors should be considered faulty and replaced.

STEP 1

Engine "cool down" period (10 minutes) is necessary to avoid irregular readings due to "Hot Soak" fuel boiling. Relieve fuel pressure in the fuel rail using the "fuel pressure relief procedure"
With ignition "OFF" connect fuel gage J 29658-D or equivalent. (Refer to CHART A-7 Page 3 of 3 for connecting fuel pressure gage.)
Disconnect electrical connectors at all injectors except #2. Then, connect injector adaptor harness J 34730-200 to the injector jumper harness connector. Now, connect tester J 34730-3 to adaptor harness J 34730-200. Ignition must be "OFF" at least 10 seconds to complete ECM shutdown cycle. Fuel pump should run about 2 seconds after ignition is turned "ON." At this point, insert clear tubing attached to vent valve into a suitable container and bleed air from gauge and hose to insure accurate gauge operation. Repeat this step until all air is bled from gauge.

STEP 2

Turn ignition "OFF" for 10 seconds and then "ON" again to get fuel pressure to its maximum. Record this initial pressure reading. Energize tester one time and note pressure drop at its lowest point. (Disregard any slight pressure increase after drop hits low point.) By subtracting this second pressure reading from the initial pressure, we have the actual amount of injector pressure drop.

STEP 3

Disconnect injector adaptor harness J 34730-200 and reconnect injector jumper harness connector to the engine harness connector. Now connect J 34730-200 directly to each of the remaining injectors. Repeat Step 2 on each injector and compare the amount of drop. Usually, good injectors will have virtually the same drop. Retest any injector that has a pressure difference of 10 kPa, either more or less than the average of the other injectors on the engine. Replace any injector that also fails the retest. If the pressure drop of all injectors is within 10 kPa of this average, the injectors appear to be flowing properly.

NOTICE: The entire test should not be repeated more than once without running the engine to prevent flooding. (This includes any retest on faulty injectors.)

2.2L (VIN 4) ENGINE — COMPONENT DIAGNOSTIC CHART — CENTURY AND CIERA

NOTICE: The entire test should **NOT** be repeated more than once without running the engine to prevent flooding. (This includes any retest on faulty injectors.)

The fuel pressure test in Chart A-7, should be completed prior to this test.

Step 1. If engine is at operating temperature, allow a 10 minute "cool down" period then connect fuel pressure gage and injector tester.
1. Ignition "OFF."
2. Connect fuel pressure gage and injector tester. Refer to Step 1
3. Ignition "ON."
4. Bleed off air in gage. Repeat until all air is bled from gauge.

Step 2. Run test:
1. Ignition "OFF" for 10 seconds.
2. Ignition "ON". Record gage pressure. (Pressure must hold steady, if not see the Fuel System diagnosis, Chart A-7
3. Turn injector on, by depressing button on injector tester, and note pressure at the instant the gauge needle stops.

Step 3. Refer to Step 3

CHART C-2A
INJECTOR BALANCE TEST
2.2L (VIN 4) "A" CARLINE

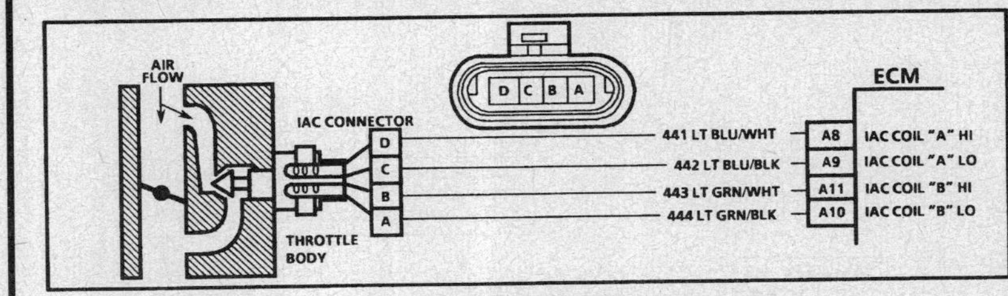

EXAMPLE

CYLINDER	1	2	3
1ST READING	293 kPa (43 psi)	293 kPa (43 psi)	293 kPa (43 psi)
2ND READING	131 kPa (19 psi)	115 kPa (17 psi)	145 kPa (21 psi)
AMOUNT OF DROP	162 kPa (24 psi)	178 kPa (26 psi)	148 kPa (21 psi)
	OK	FAULTY, RICH (TOO MUCH FUEL DROP)	FAULTY, LEAN (TOO LITTLE FUEL DROP)

2.2L (VIN 4) ENGINE — COMPONENT DIAGNOSTIC CHART — CENTURY AND CIERA

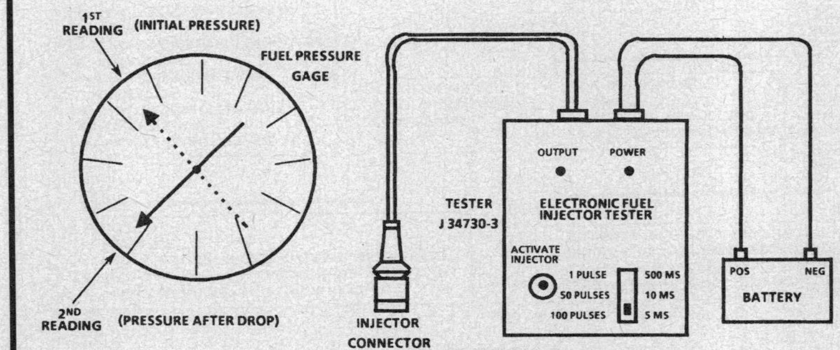

CHART C-2C
IDLE AIR CONTROL (IAC) SYSTEM CHECK
2.2L (VIN 4) "A" CARLINE

Circuit Description:

The ECM controls engine idle speed with the IAC valve. To increase idle speed, the ECM retracts the IAC valve pintle away from its seat, allowing more air to bypass the throttle bore. To decrease idle speed, it extends the IAC valve pintle towards its seat, reducing bypass air flow. A Tech 1 scan tool will read the ECM commands to the IAC valve in counts. Higher the counts indicate more air bypass (higher idle). The lower the counts indicate less air is allowed to bypass (lower idle).

Test Description: Number(s) below refer to circled numbers on the diagnostic chart.
1. The Tech 1 RPM control mode is used to extend and retract the IAC valve. The valve should move smoothly within the specified range. If the idle speed commanded (IAC extended) too low (below 700 RPM), the engine may stall. This may be normal and would not indicate a problem. Retracting the IAC beyond its controlled range (above 1500 RPM) will cause a delay before the RPMs start dropping. This too is normal.
2. This test uses the Tech 1 to command the IAC controlled idle speed. The ECM issues commands to obtain commanded idle speed. The node lights each should flash red and green to indicate a good circuit as the ECM issues commands. While the sequence of color is not important if either light is "OFF" or does not flash red <u>and</u> green, check the circuits for faults, beginning with poor terminal contacts.

Diagnostic Aids:

A slow, unstable, or fast idle may be caused by a non-IAC system problem that cannot be overcome by the IAC valve. Out of control range IAC scan tool counts will be above 50 if idle is too low, and zero counts if idle is too high. The following checks should be made to repair a non-IAC system problem:
- Vacuum Leak (High Idle) - If idle is too high, stop the engine. Fully extend (low) IAC with tester. Start engine. If idle speed is above 800 RPM, locate and correct vacuum leak including crankcase ventilation system. Also check for binding of throttle blade or linkage.

- Crankcase Ventilation Valve - If a high idle condition exists (800 to 1000 RPM), check for vacuum leaks and proper crankcase ventilation valve operation. All throttle bodies are preset at the factory and do not need adjustment. A missing crankcase ventilation valve or grommet or stuck valve can cause this condition.
- System Too Lean (High Air/Fuel Ratio) - The idle speed may be too high or too low. Engine speed may vary up and down and disconnecting the IAC valve does not help. DTC 44 may be set. Scan oxygen voltage will be less than 300 mV (.3 volt). Check for low regulated fuel pressure, water in the fuel or a restricted injector.
- System Too Rich (Low Air/Fuel Ratio) - The idle speed will be too low. Scan tool IAC counts will usually be above 50. System is obviously rich and may exhibit black smoke in exhaust. Scan tool oxygen voltage will be fixed above 800 mV (.8 volt).
Check for high fuel pressure, leaking or sticking injector. Silicone contaminated oxygen sensors scan voltage will be slow to respond.
- Throttle Body - Remove IAC valve and inspect bore for foreign material.
- IAC Valve Electrical Connections - IAC valve connections should be carefully checked for proper contact.

- If intermittent poor driveability or idle symptoms are resolved by disconnecting the IAC, carefully recheck connections, valve terminal resistance, or replace IAC.

2.2L (VIN 4) ENGINE — COMPONENT DIAGNOSTIC CHART — CENTURY AND CIERA

CHART C-2C
IDLE AIR CONTROL (IAC) SYSTEM CHECK
2.2L (VIN 4) "A" CARLINE

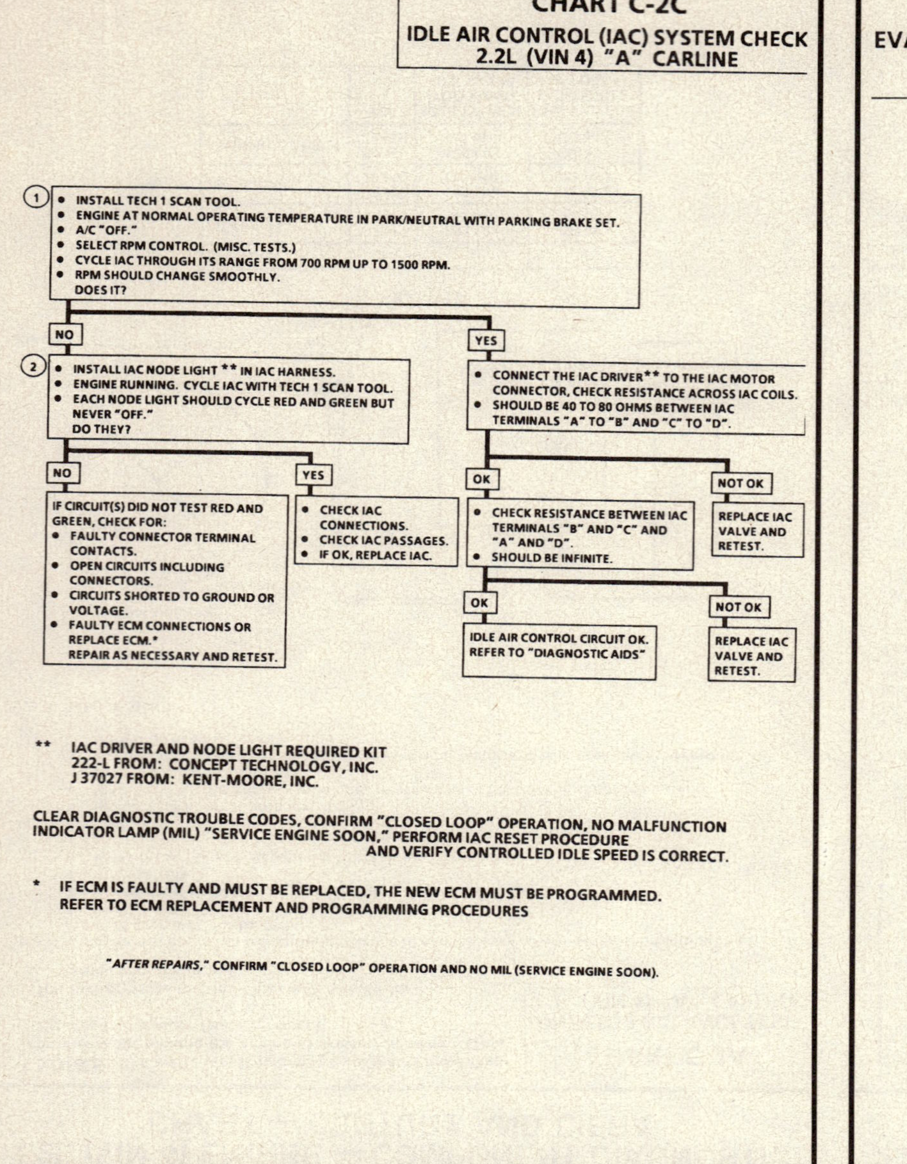

2.2L (VIN 4) ENGINE — COMPONENT DIAGNOSTIC CHART — CENTURY AND CIERA

CHART C-3
EVAPORATIVE EMISSION CANISTER CONTROL VALVE CHECK
(NON-ECM CONTROLLED)
2.2L (VIN 4) "A" CARLINE

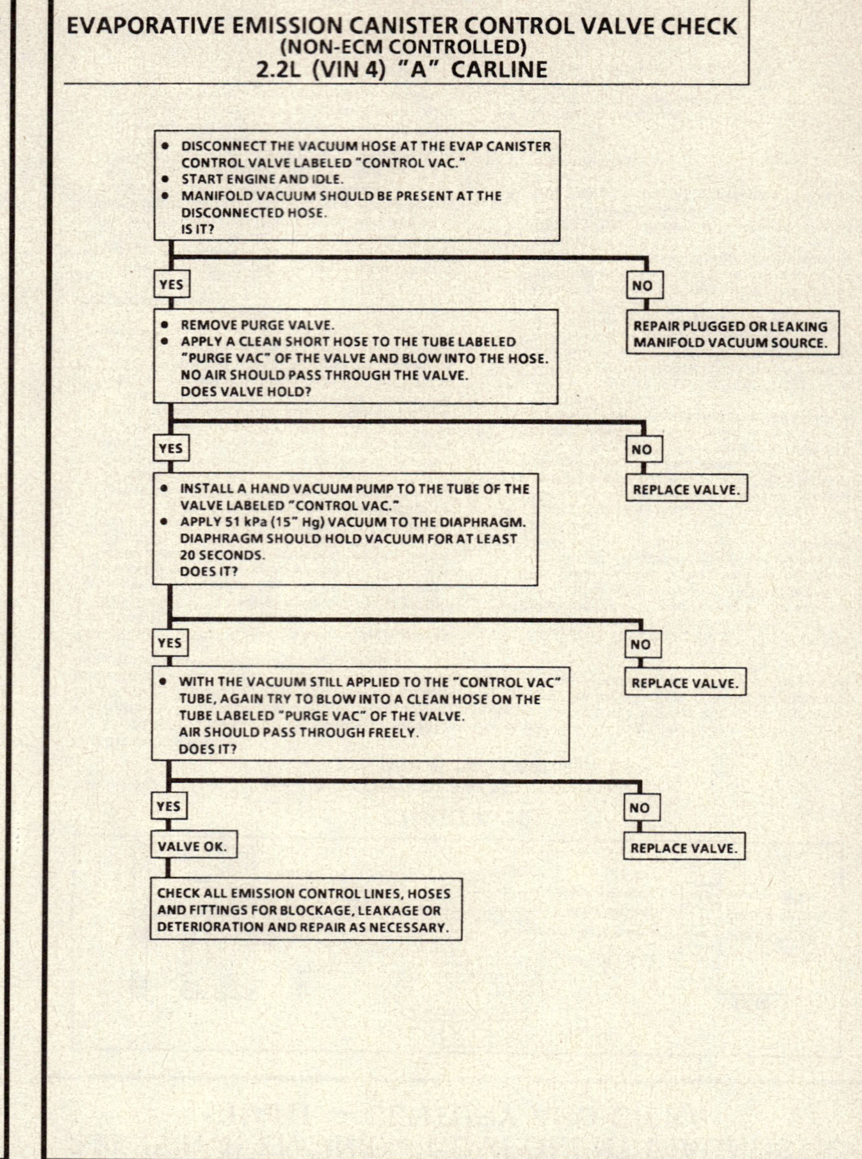

2.2L (VIN 4) ENGINE — COMPONENT DIAGNOSTIC CHART — CENTURY AND CIERA

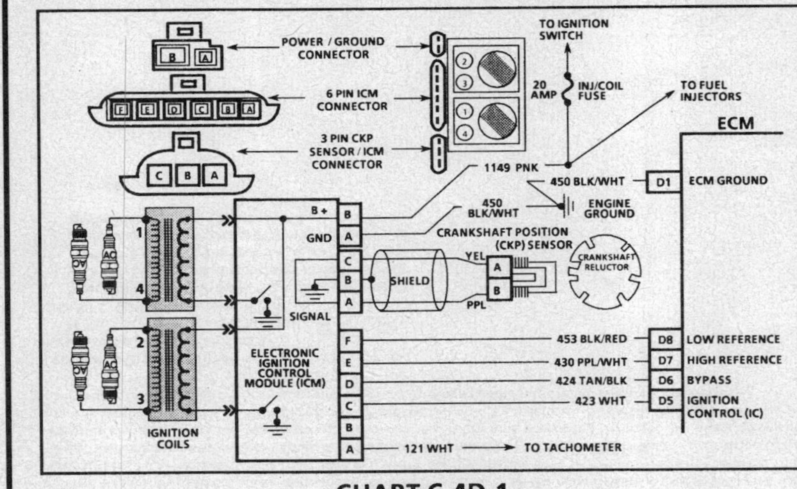

CHART C-4D-1
MISFIRE AT IDLE
2.2L (VIN 4) "A" CARLINE

Circuit Description:

The Electronic Ignition (EI) system uses a waste spark method of distribution. In this type of system, the Electronic Ignition Control Module (ICM) triggers the number 1/4 coil pair resulting in both number 1 and number 4 spark plugs firing at the same time. Number 1 cylinder is on the compression stroke at the same time number 4 is on the exhaust stroke, resulting in a lower energy requirement to fire number 4 spark plug. This leaves the remainder of the high voltage to be used to fire number 1 spark plug. The Crankshaft Position (CKP) sensor is remotely mounted beside the ignition coil assembly and protrudes through the block to within approximately .050" of the crankshaft reluctor. Since the reluctor is a machined portion of the crankshaft and the CKP sensor is mounted in a fixed position on the block, timing adjustments are not possible or necessary.

Test Description: Number(s) below refer to circled number(s) on the diagnostic chart.

1. If the "Misfire" complaint exists under load only, CHART C-4D-2 must be used:
   - Keep disconnected spark plug leads away from sensors and other electronic components.
   - Move quickly through this test. Don't leave any spark plug lead disconnected for longer than 15 seconds.
   - Let the engine run normally for 30 seconds between tests to avoid an excessive buildup of fuel.
2. A spark tester, such as a ST-125, must be used because it is essential to verify adequate available secondary voltage at the spark plug (25,000 volts). If the spark jumps the tester gap after grounding the companion plug cable, it indicates excessive resistance in the plug which was bypassed. A faulty or poor connection at that plug could also result in a misfire condition. Also, check for carbon deposits inside the spark plug boot.
3. If carbon tracking is evident, replace coil and be sure plug cables relating to that coil are clean and tight. Excessive wire resistance or faulty connections could have caused the coil to be damaged.
4. By checking the secondary resistance, a coil with an open secondary may be located.
5. If the no spark condition follows the suspected coil, that coil is faulty. Otherwise, the Electronic Ignition Control Module (ICM) is the cause of no spark. This test could also be performed by substituting a known good coil for the one causing the no spark condition.

2.2L (VIN 4) ENGINE — COMPONENT DIAGNOSTIC CHART — CENTURY AND CIERA

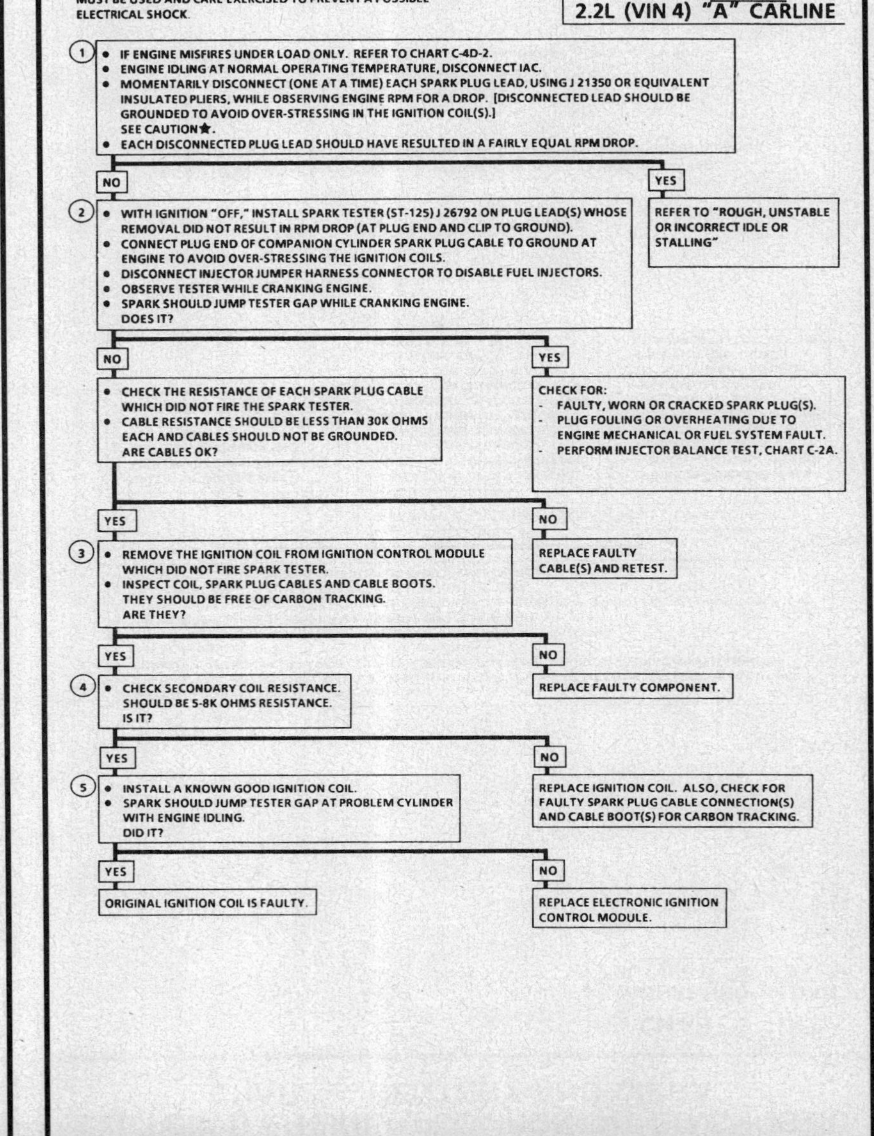

2.2L (VIN 4) ENGINE — COMPONENT DIAGNOSTIC CHART — CENTURY AND CIERA

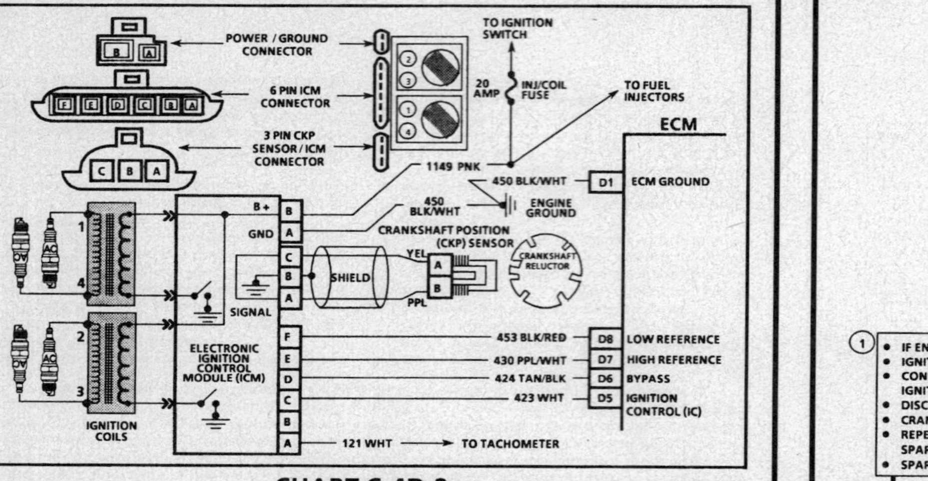

CHART C-4D-2
MISFIRE UNDER LOAD
2.2L (VIN 4) "A" CARLINE

Circuit Description:
The Electronic Ignition (EI) system uses a waste spark method of distribution. In this type of system, the electronic Ignition Control Module (ICM) triggers the number 1/4 coil pair resulting in both number 1 and number 4 spark plugs firing at the same time. Number 1 cylinder is on the compression stroke at the same time number 4 is on the exhaust stroke, resulting in a lower energy requirement to fire number 4 spark plug. This leaves the remainder of the high voltage to be used to fire number 1 spark plug. The Crankshaft Position (CKP) sensor is remotely mounted beside the ignition coil assembly and protrudes through the block to within approximately .050″ of the crankshaft reluctor. Since the reluctor is a machined portion of the crankshaft and the CKP sensor is mounted in a fixed position on the block, timing adjustments are not possible or necessary.

Test Description: Number(s) below refer to circled number(s) on the diagnostic chart.

If the "Misfire" complaint exists at idle only, CHART C-4D-1 must be used.

1. Simulates an engine load. Cranking the engine with the ST-125 installed on any cylinder should produce a crisp, blue spark. A cylinder that produces a spark significantly weaker than the others should be considered as having no spark.

2.2L (VIN 4) ENGINE — COMPONENT DIAGNOSTIC CHART — CENTURY AND CIERA

CHART C-4D-2
MISFIRE UNDER LOAD
2.2L (VIN 4) "A" CARLINE

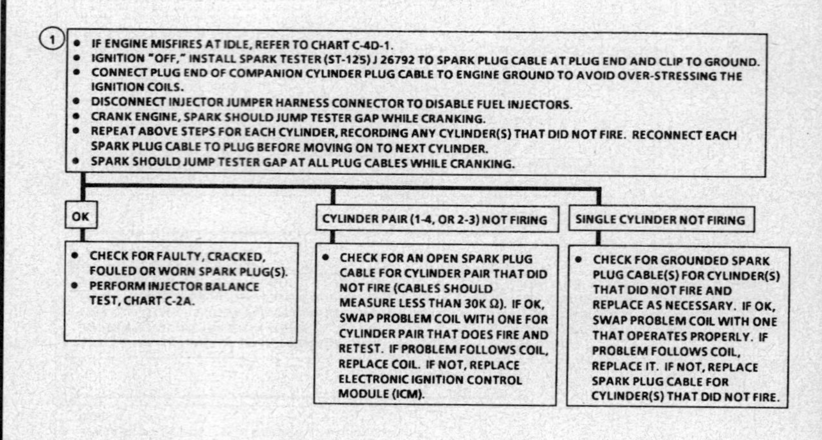

★**CAUTION:** When handling secondary spark plug leads with engine running, insulated pliers must be used and care exercised to prevent a possible electrical shock.

"AFTER REPAIRS," CONFIRM "CLOSED LOOP" OPERATION AND NO MIL (SERVICE ENGINE SOON).

2.2L (VIN 4) ENGINE — COMPONENT DIAGNOSTIC CHART — CENTURY AND CIERA

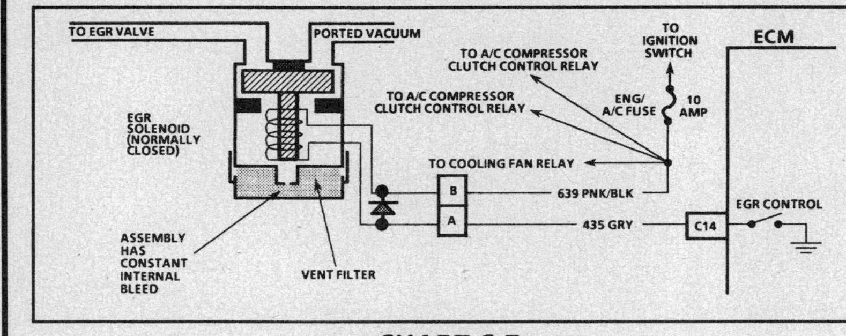

CHART C-7
EXHAUST GAS RECIRCULATION (EGR) FLOW CHECK
2.2L (VIN 4) "A" CARLINE

Circuit Description:

A properly operating EGR system will directly affect the air/fuel mixture requirements of the engine. Since the exhaust gas introduced into the air/fuel mixture cannot be used in combustion (contains very little oxygen), less fuel is required to maintain a correct air/fuel ratio. If the EGR system were to fail in a closed position, the exhaust gas would be replaced with air, and the air/fuel mixture would be leaner. The ECM would compensate for the lean condition by adding fuel, resulting in higher long term fuel trim values.

The fuel control on this engine is conducted within four fuel trim cells. Since EGR is not used at idle, the idle cell would not be affected by EGR system operation. The other fuel trim cells are affected by EGR operation, and, when the EGR system is operating properly, the long term fuel trim values in all cells should be close to the same. If the EGR system becomes inoperative, the long term fuel trim values in the open throttle cells would change to compensate for the resulting lean or rich mixtures, but the long term fuel trim value in the closed throttle cell would not change.

The difference in long term fuel trim values between the idle (closed throttle) cell and cell 2 is used to monitor EGR system performance. When the difference between the two long term fuel trim values is greater than 10, and the long term fuel trim value in cell 2 is greater than 135, Diagnostic Trouble Code (DTC) 32 is set. The system operates in fuel trim cell 2 during a cruise condition at approximately 55 mph.

Test Description: Number(s) below refer to circled number(s) on the diagnostic chart.

1. DTC(s) should be diagnosed using appropriate chart before preparing a functional check. If Diagnostic Trouble Codes 14, 21, 23, 32, or 33 are set, use those charts first.
   Be sure shop exhaust hose is not connected during Steps 2 and 4.
2. The tail pipe ventilation hose must be removed for this test. The ventilation hose can sometimes cause a good EGR valve to fail this test.
3. Intake Passage: Shut "OFF" engine and remove the EGR valve from the manifold. Plug the exhaust side hole with a suitable stopper. Leaving the intake side hole open, attempt to start the engine. If the engine runs at a high idle (up to 3000 RPM is possible) or starts and stalls, the EGR intake passage is not restricted. If the engine starts and idles normally, the EGR intake passage is restricted.

4. Because the shop exhaust hose is not installed at this point, don't allow the engine to run longer than 15 seconds.
   Exhaust Passage: With EGR valve still removed, plug the intake side hole with a suitable stopper. With the exhaust side hole open, check for the presence of exhaust gas. If no exhaust gas is present, the EGR exhaust side passage is restricted.

Diagnostic Aids:

This chart is a functional check of the EGR system. If the EGR system works properly, check other items that result in high long term fuel trim values in fuel trim cell 2, but not in the closed throttle cell.

Incorrect fuel pressure or lean/rich fuel injectors can also cause incorrect long term fuel trim values. Refer to "Fuel Pressure Test" in CHART A-7. It may be necessary to monitor fuel pressure while driving the vehicle at various road speeds and/or loads. If fuel pressure checks out OK, proceed to "Fuel Injector Balance Test" in CHART C-2A.

2.2L (VIN 4) ENGINE — COMPONENT DIAGNOSTIC CHART — CENTURY AND CIERA

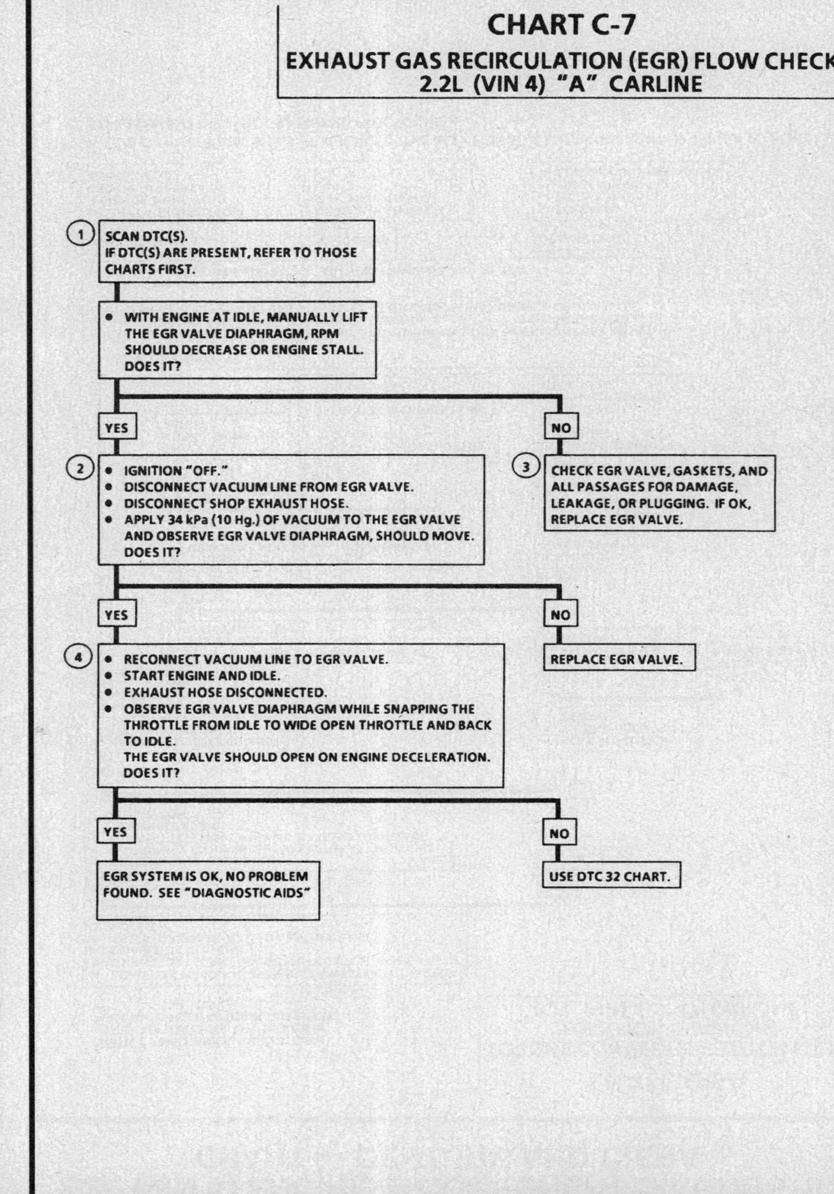

2.2L (VIN 4) ENGINE — COMPONENT DIAGNOSTIC CHART — CENTURY AND CIERA

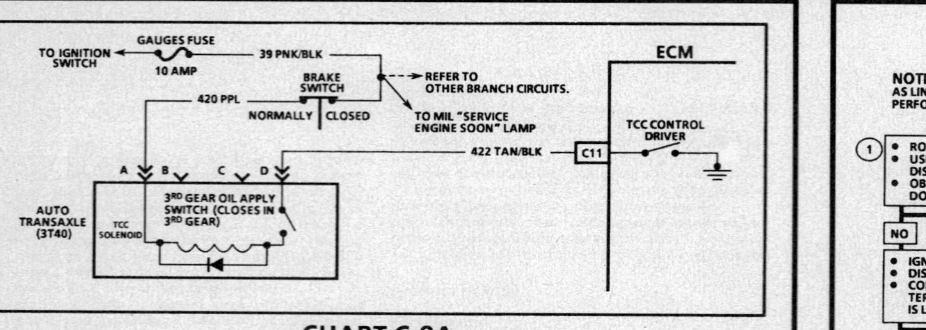

CHART C-8A
TORQUE CONVERTER CLUTCH (TCC)
(ELECTRICAL DIAGNOSIS)
2.2L (VIN 4) "A" CARLINE

Circuit Description:

The purpose of the automatic transaxle Torque Converter Clutch (TCC) feature is to eliminate the power loss of the torque converter stage when the vehicle is in a cruise condition. This allows the convenience of the automatic transaxle and the fuel economy of a manual transaxle.

Test Description: Number(s) below refer to circled number(s) on the diagnostic chart.

1. This test should be performed with the aid of a helper, to operate the Tech 1 and observe the RPM fluctuation.
2. Entering the "Field Service Mode" forces the ECM to close the TCC control driver circuit completing the path for current which should light the test light.
3. Perform this test with the aid of a helper. The 3rd gear switch must be engaged while measuring the TCC solenoid resistance.

Fused battery ignition is supplied to the TCC solenoid through the brake switch. The ECM will engage TCC by grounding CKT 422 to energize the solenoid. TCC will engage when:

- Vehicle speed above about (35 mph with A/C "OFF," 44 mph with A/C "ON").
- Engine at normal operating temperature above about 20°C (68°F).
- Throttle Position (TP) sensor output not changing, indicating a steady road speed.
- Brake switch closed.
- 3rd gear switch applied.

2.2L (VIN 4) ENGINE — COMPONENT DIAGNOSTIC CHART — CENTURY AND CIERA

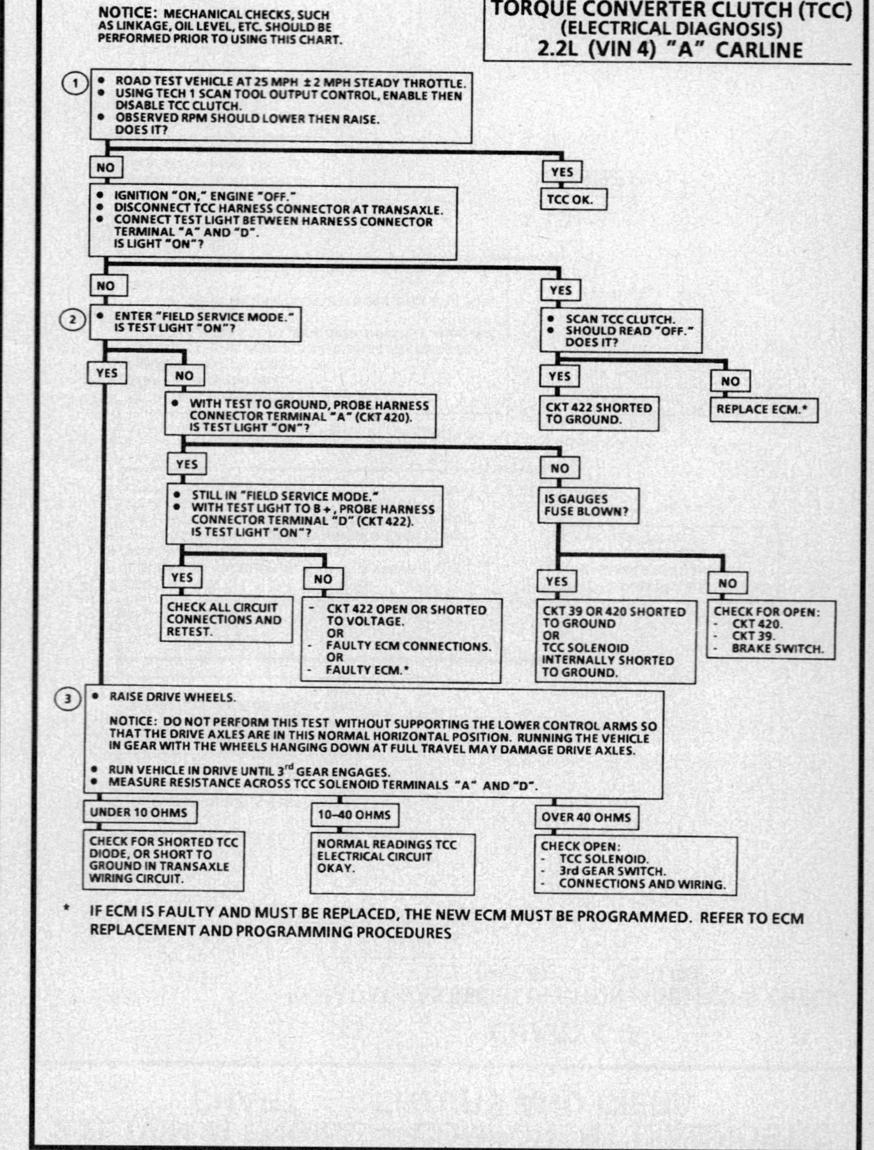

2.2L (VIN 4) ENGINE — COMPONENT DIAGNOSTIC CHART — CENTURY AND CIERA

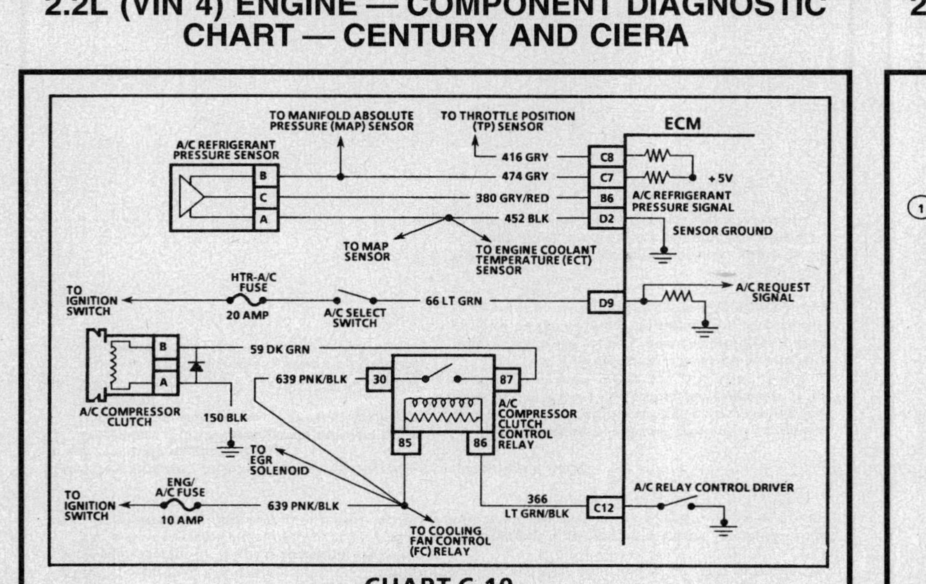

CHART C-10
(Page 1 of 2)
A/C COMPRESSOR CLUTCH CONTROL CIRCUIT DIAGNOSIS
2.2L (VIN 4) "A" CARLINE

Circuit Description:

The A/C compressor clutch control relay is energized when the Engine Control Module (ECM) provides a ground path through CKT 366 and A/C is requested. A/C compressor clutch is delayed about .3 second after A/C is requested. This will allow the IAC to adjust engine RPM for the additional load.

The ECM will disable the A/C compressor clutch for any of the following reasons:

- Engine not running.
- Diagnostic Trouble Code (DTC) 66 is set, and fault is current.
- Throttle position is about 99 percent or greater.
- Engine coolant temperature is about 122°C (251°F) or greater.
- During an IAC motor reset.
- Engine RPM is greater than about 4224 RPM.
- A/C refrigerant pressure is about 429 psi or greater.
- A/C refrigerant pressure is about 38 psi or less.

If the ECM temporarily disabled the A/C compressor clutch, the clutch will be renabled as follows:

- Throttle position drops below about 89 percent.
- Coolant temperature drops below about 119°C (246°F).
- Engine RPM drops below about 3850 RPM.
- A/C refrigerant pressure drops below about 199 psi.
- A/C refrigerant pressure rises above about 47 psi.
- IAC motor reset is complete.

Test Description: Number(s) below refer to circled number(s) on the diagnostic chart.

1. The ECM will only energize the A/C relay when the engine is running. This test will determine if the relay or CKT 366 is faulty.
2. Determines if the signal is reaching the ECM through CKT 66 from the A/C control panel. Signal should only be present when an A/C mode or defrost mode has been selected.

the problem may be caused by an inoperative cooling fan. Refer to CHART C-12 for cooling fan diagnosis. If fan operates correctly, see "A/C Diagnosis"

A/C refrigerant pressure outside of a range of 38 to 429 psi will cause the compressor clutch to be disabled by the ECM. Observe Tech 1 A/C refrigerant pressure for 2 minutes with engine idling and A/C "ON."

Scanned refrigerant pressure should be within 20 psi of actual. If not, check for a circuit problem using DTC 66 chart or replace the A/C refrigerant pressure sensor.

Diagnostic Aids:

Be sure to consider branch circuits and splices to other components. If complaint is insufficient cooling,

2.2L (VIN 4) ENGINE — COMPONENT DIAGNOSTIC CHART — CENTURY AND CIERA

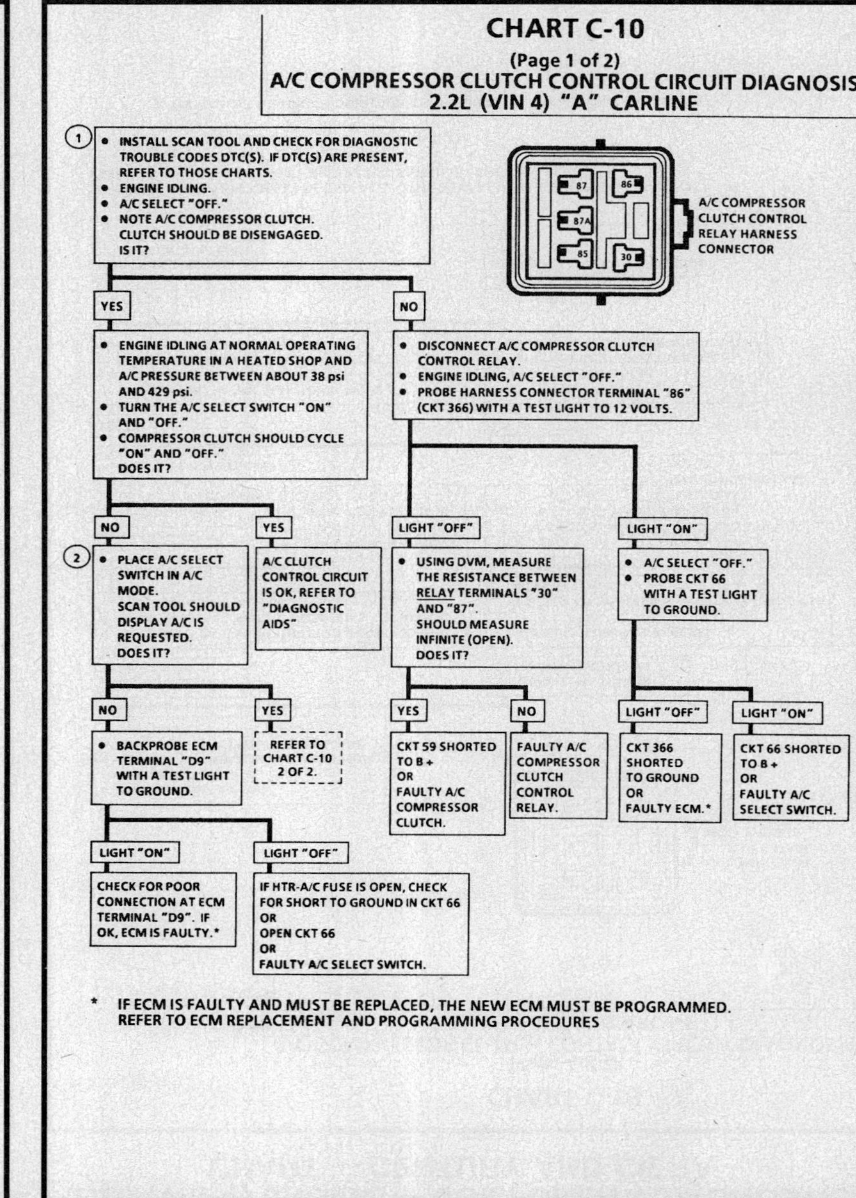

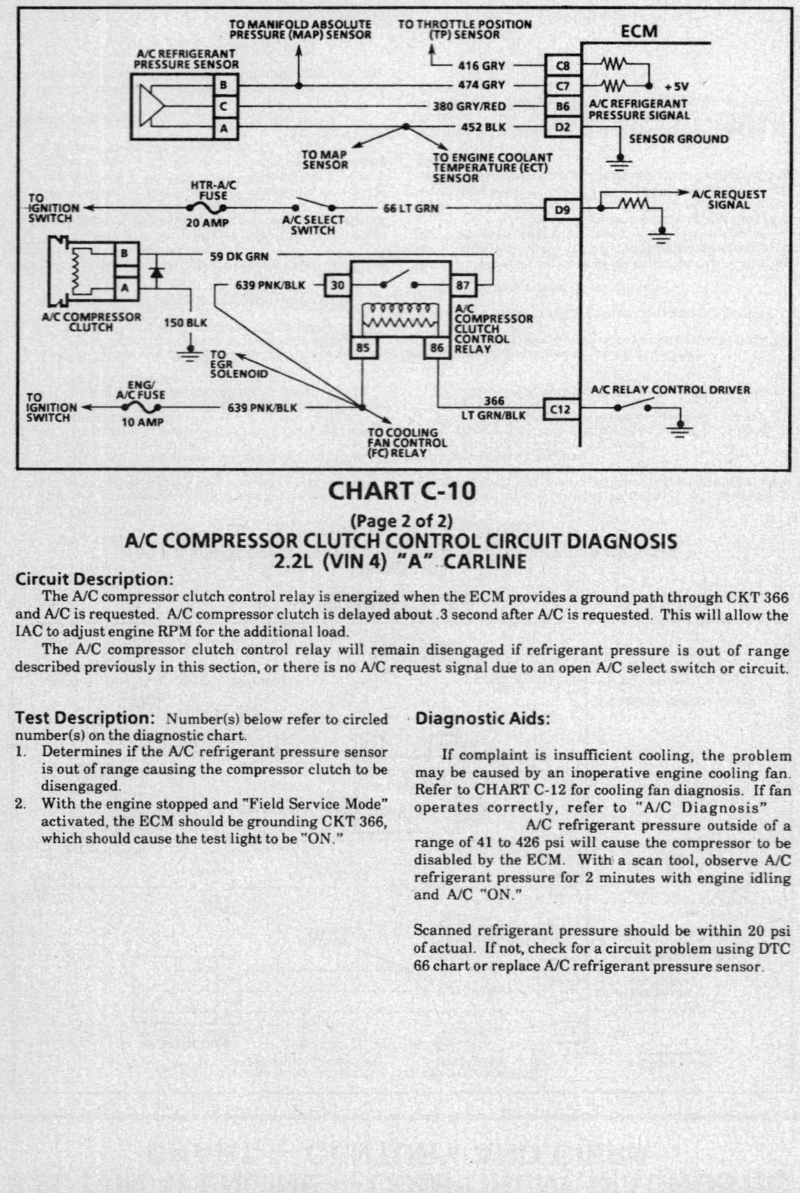

CHART C-10

(Page 2 of 2)
A/C COMPRESSOR CLUTCH CONTROL CIRCUIT DIAGNOSIS
2.2L (VIN 4) "A" CARLINE

Circuit Description:

The A/C compressor clutch control relay is energized when the ECM provides a ground path through CKT 366 and A/C is requested. A/C compressor clutch is delayed about .3 second after A/C is requested. This will allow the IAC to adjust engine RPM for the additional load.

The A/C compressor clutch control relay will remain disengaged if refrigerant pressure is out of range described previously in this section, or there is no A/C request signal due to an open A/C select switch or circuit.

Test Description: Number(s) below refer to circled number(s) on the diagnostic chart.
1. Determines if the A/C refrigerant pressure sensor is out of range causing the compressor clutch to be disengaged.
2. With the engine stopped and "Field Service Mode" activated, the ECM should be grounding CKT 366, which should cause the test light to be "ON."

Diagnostic Aids:

If complaint is insufficient cooling, the problem may be caused by an inoperative engine cooling fan. Refer to CHART C-12 for cooling fan diagnosis. If fan operates correctly, refer to "A/C Diagnosis"

A/C refrigerant pressure outside of a range of 41 to 426 psi will cause the compressor to be disabled by the ECM. With a scan tool, observe A/C refrigerant pressure for 2 minutes with engine idling and A/C "ON."

Scanned refrigerant pressure should be within 20 psi of actual. If not, check for a circuit problem using DTC 66 chart or replace A/C refrigerant pressure sensor.

CHART C-10

(Page 2 of 2)
A/C COMPRESSOR CLUTCH CONTROL CIRCUIT DIAGNOSIS
2.2L (VIN 4) "A" CARLINE

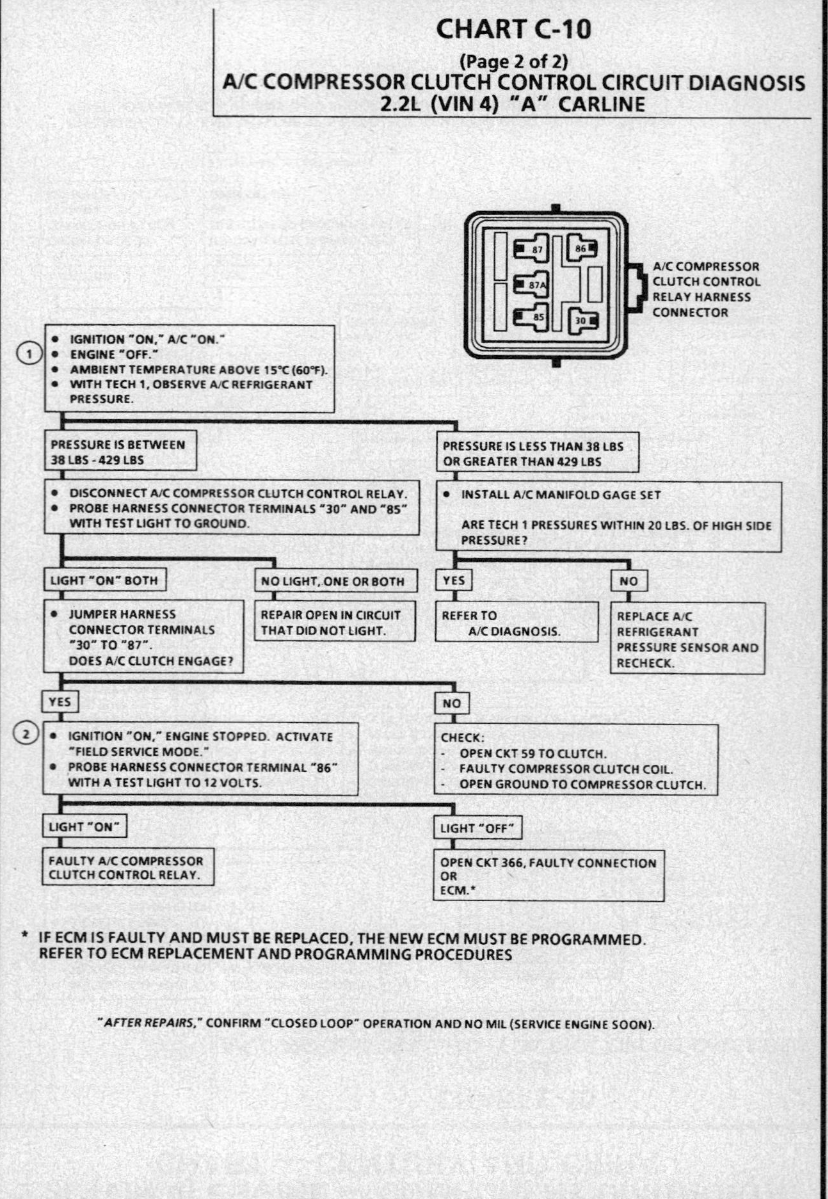

* IF ECM IS FAULTY AND MUST BE REPLACED, THE NEW ECM MUST BE PROGRAMMED. REFER TO ECM REPLACEMENT AND PROGRAMMING PROCEDURES

"AFTER REPAIRS," CONFIRM "CLOSED LOOP" OPERATION AND NO MIL (SERVICE ENGINE SOON).

2.2L (VIN 4) ENGINE — COMPONENT DIAGNOSTIC CHART — CENTURY AND CIERA

CHART C-12

COOLING FAN CONTROL DIAGNOSIS
2.2L (VIN 4) "A" CARLINE

Circuit Description:

Battery voltage to operate the cooling fan motor is supplied to relay by CKT 2. Ignition voltage to energize the FC relay is supplied by CKT 39. When the ECM grounds CKT 535, the relay is energized and the cooling fan is turned "ON."

When the engine is running, the ECM will turn the engine cooling fan "ON" for any of the following reasons:

- Diagnostic Trouble Code (DTC) 14 is set.
- Coolant temperature is 108°C (226°F) or greater.
- A/C refrigerant pressure is 189 psi or greater with A/C selected and a Diagnostic Trouble Code (DTC) 66 is not set.

Once the cooling fan is enabled, the ECM will turn the fan "OFF" when:

- Engine is turned "OFF."
- Coolant temperature drops to 101°C (213°F) or less.
- A/C refrigerant pressure drops to 38 psi or less with A/C selected.

Diagnostic Aids:

If the owner complained of an overheating problem, it must be determined if the complaint was due to an actual boil over, or the hot light, or temperature gage indicated overheating.

If the gage or light indicates overheating, but no boil over is detected, the gage circuit should be checked. The gage accuracy can also be checked by checking the engine coolant temperature reading using a scan tool and comparing its display with the instrument cluster gage.

If the engine is actually overheating and the gage indicates overheating, but the cooling fan is not coming "ON," the ECT sensor has probably shifted out of calibration and should be replaced.

2.2L (VIN 4) ENGINE — COMPONENT DIAGNOSTIC CHART — CENTURY AND CIERA

CHART C-12

COOLING FAN CONTROL DIAGNOSIS
2.2L (VIN 4) "A" CARLINE

IF DTC 14 IS PRESENT, USE THAT CHART FIRST.
- IGNITION "ON," ENGINE "OFF."
- A/C "OFF," COOLANT TEMPERATURE BELOW 101°C (213°F). IS FAN "ON"?

NO
- GROUND DIAGNOSTIC TERMINAL TO ENTER "FIELD SERVICE MODE." IS FAN "ON"?

  NO
  - DISCONNECT COOLING FAN CONTROL (FC) RELAY.
  - PROBE HARNESS CONNECTOR TERMINALS "30" & "85" WITH A TEST LIGHT TO GROUND. DOES LIGHT TURN "ON" FOR BOTH TERMINALS?

    YES
    - "FIELD SERVICE MODE" STILL ENGAGED.
    - PROBE HARNESS CONNECTOR TERMINAL "86" WITH A TEST LIGHT TO BATTERY VOLTAGE. DOES LIGHT COME "ON"?

      YES
      - JUMPER HARNESS CONNECTOR TERMINALS "30" TO "87". DOES FAN RUN?

        YES — REPLACE COOLING FAN CONTROL (FC) RELAY.

        NO — OPEN CKT 533, FAULTY FAN MOTOR, OR FAULTY GROUND CKT 150.

      NO — CHECK CKT 535 FOR OPEN OR SHORT TO VOLTAGE AND ECM CONNECTIONS. IF OK, REPLACE ECM.*

    NO — REPAIR OPEN IN CIRCUIT THAT DID NOT LIGHT.

  YES

  NON-A/C — NO TROUBLE FOUND.

  WITH A/C
  - REMOVE GROUND FROM DIAGNOSTIC TERMINAL TO EXIT "FIELD SERVICE MODE."
  - ENGINE IDLING.
  - A/C "ON" AND A/C PRESSURE 189 psi OR GREATER. IS FAN "ON"?

    YES — NO TROUBLE FOUND.

    NO — BACKPROBE ECM CONNECTOR TERMINAL "D9" WITH A TEST LIGHT TO GROUND.

      LIGHT "OFF" — IF HEATER-A/C FUSE IS BLOWN, CHECK FOR SHORT TO GROUND IN CKT 66. IF OK, CHECK FOR OPEN CKT 66, OR FAULTY A/C SELECT SWITCH.

      LIGHT "ON" — FAULTY ECM CONNECTION OR ECM.*

YES
- DISCONNECT COOLING FAN CONTROL (FC) RELAY.
- JUMPER HARNESS CONNECTOR TERMINALS "85" TO "86" WITH A TEST LIGHT.

  LIGHT "ON" — CHECK FOR SHORT TO GROUND IN CKT 535. IF NOT GROUNDED, REPLACE ECM.*

  LIGHT "OFF"
  - CHECK CKT 533 FOR SHORT TO VOLTAGE. IF OK, REPLACE COOLING FAN CONTROL (FC) RELAY.

IF ECM IS FAULTY AND MUST BE REPLACED, THE NEW ECM MUST BE PROGRAMMED. REFER TO ECM REPLACEMENT AND PROGRAMMING PROCEDURES

COOLING FAN CONTROL RELAY CONNECTOR

"AFTER REPAIRS," CONFIRM "CLOSED LOOP" OPERATION AND NO MIL (SERVICE ENGINE SOON).

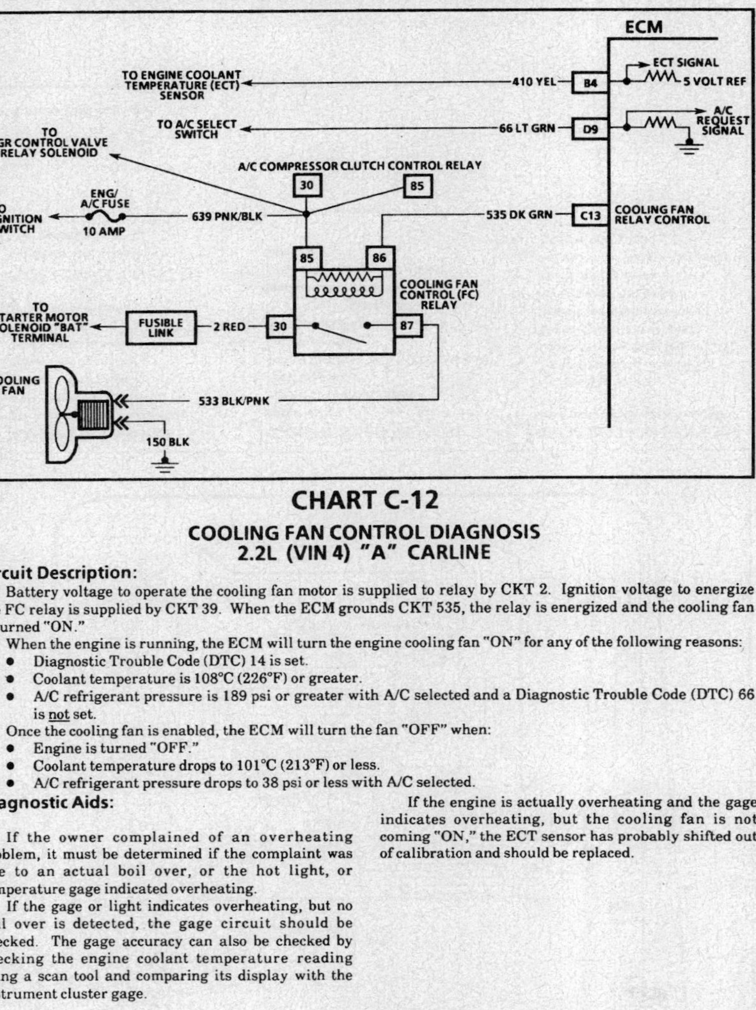

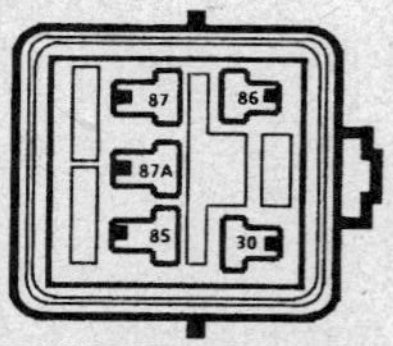

MULTIPORT FUEL INJECTION (MFI) SYSTEMS
EXCEPT LIGHT TRUCKS, VANS, GEO AND SATURN

SECTION 3

2.2L (VIN 4) ENGINE — ENGINE COMPONENT LOCATION CHART — 1992 CAVALIER

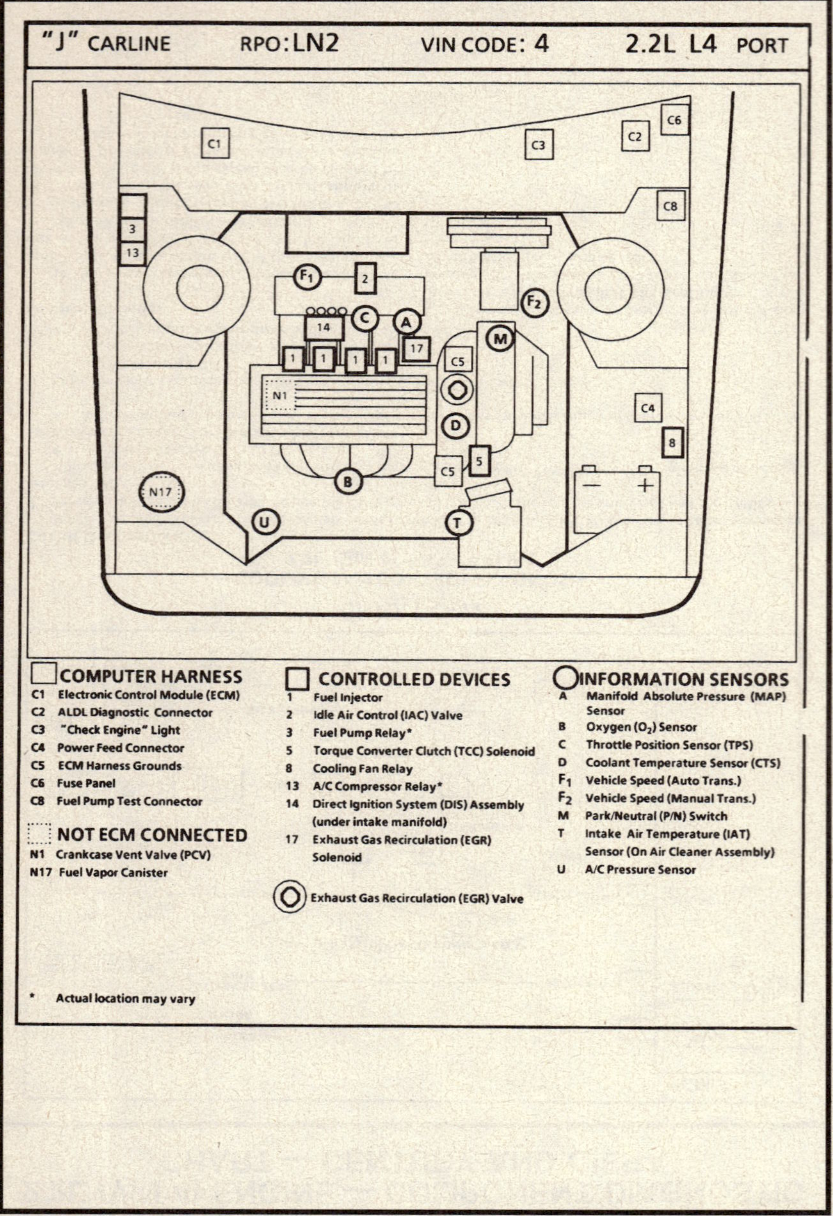

COMPUTER HARNESS
C1 Electronic Control Module (ECM)
C2 ALDL Diagnostic Connector
C3 "Check Engine" Light
C4 Power Feed Connector
C5 ECM Harness Grounds
C6 Fuse Panel
C8 Fuel Pump Test Connector

NOT ECM CONNECTED
N1 Crankcase Vent Valve (PCV)
N17 Fuel Vapor Canister

CONTROLLED DEVICES
1 Fuel Injector
2 Idle Air Control (IAC) Valve
3 Fuel Pump Relay*
5 Torque Converter Clutch (TCC) Solenoid
8 Cooling Fan Relay
13 A/C Compressor Relay*
14 Direct Ignition System (DIS) Assembly (under intake manifold)
17 Exhaust Gas Recirculation (EGR) Solenoid

◎ Exhaust Gas Recirculation (EGR) Valve

INFORMATION SENSORS
A Manifold Absolute Pressure (MAP) Sensor
B Oxygen (O₂) Sensor
C Throttle Position Sensor (TPS)
D Coolant Temperature Sensor (CTS)
F₁ Vehicle Speed (Auto Trans.)
F₂ Vehicle Speed (Manual Trans.)
M Park/Neutral (P/N) Switch
T Intake Air Temperature (IAT) Sensor (On Air Cleaner Assembly)
U A/C Pressure Sensor

* Actual location may vary

2.2L (VIN 4) ENGINE — ENGINE COMPONENT LOCATION CHART — 1993–94 CAVALIER

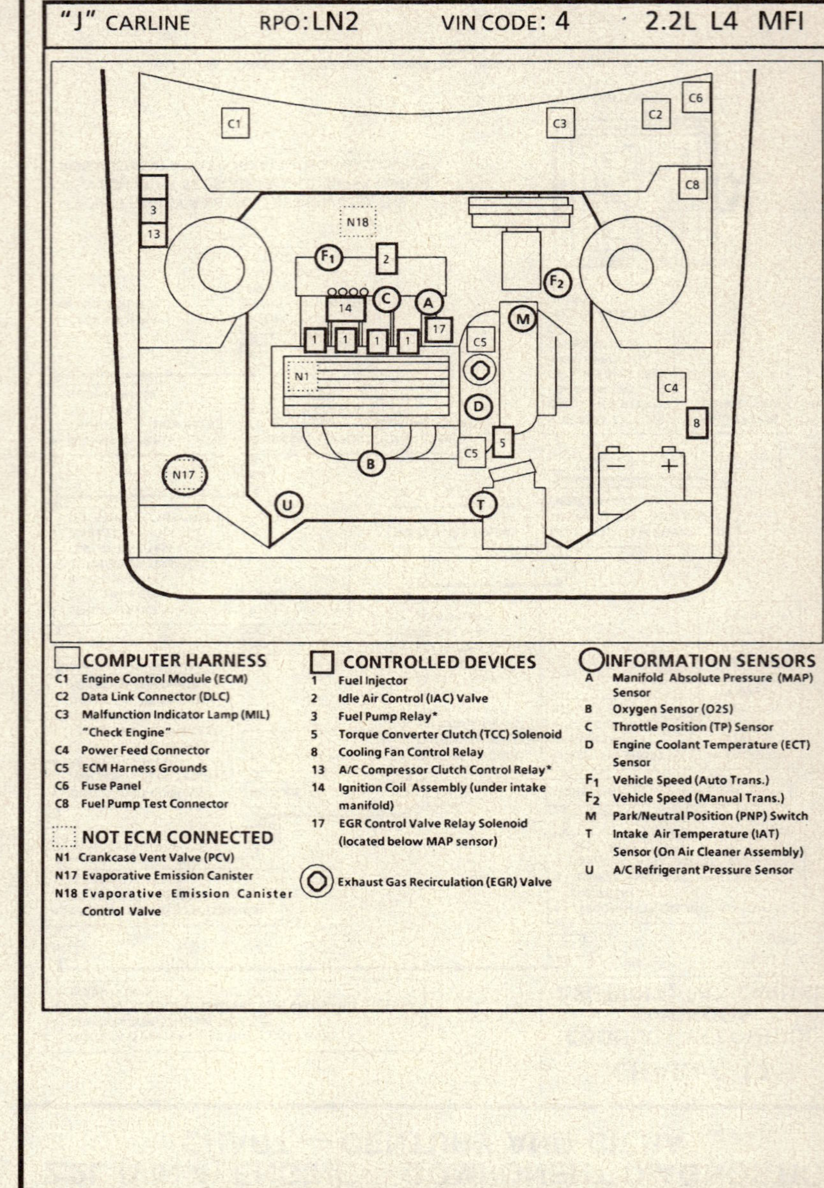

COMPUTER HARNESS
C1 Engine Control Module (ECM)
C2 Data Link Connector (DLC)
C3 Malfunction Indicator Lamp (MIL) "Check Engine"
C4 Power Feed Connector
C5 ECM Harness Grounds
C6 Fuse Panel
C8 Fuel Pump Test Connector

NOT ECM CONNECTED
N1 Crankcase Vent Valve (PCV)
N17 Evaporative Emission Canister
N18 Evaporative Emission Canister Control Valve

CONTROLLED DEVICES
1 Fuel Injector
2 Idle Air Control (IAC) Valve
3 Fuel Pump Relay*
5 Torque Converter Clutch (TCC) Solenoid
8 Cooling Fan Control Relay
13 A/C Compressor Clutch Control Relay*
14 Ignition Coil Assembly (under intake manifold)
17 EGR Control Valve Relay Solenoid (located below MAP sensor)

◎ Exhaust Gas Recirculation (EGR) Valve

INFORMATION SENSORS
A Manifold Absolute Pressure (MAP) Sensor
B Oxygen Sensor (O2S)
C Throttle Position (TP) Sensor
D Engine Coolant Temperature (ECT) Sensor
F₁ Vehicle Speed (Auto Trans.)
F₂ Vehicle Speed (Manual Trans.)
M Park/Neutral Position (PNP) Switch
T Intake Air Temperature (IAT) Sensor (On Air Cleaner Assembly)
U A/C Refrigerant Pressure Sensor

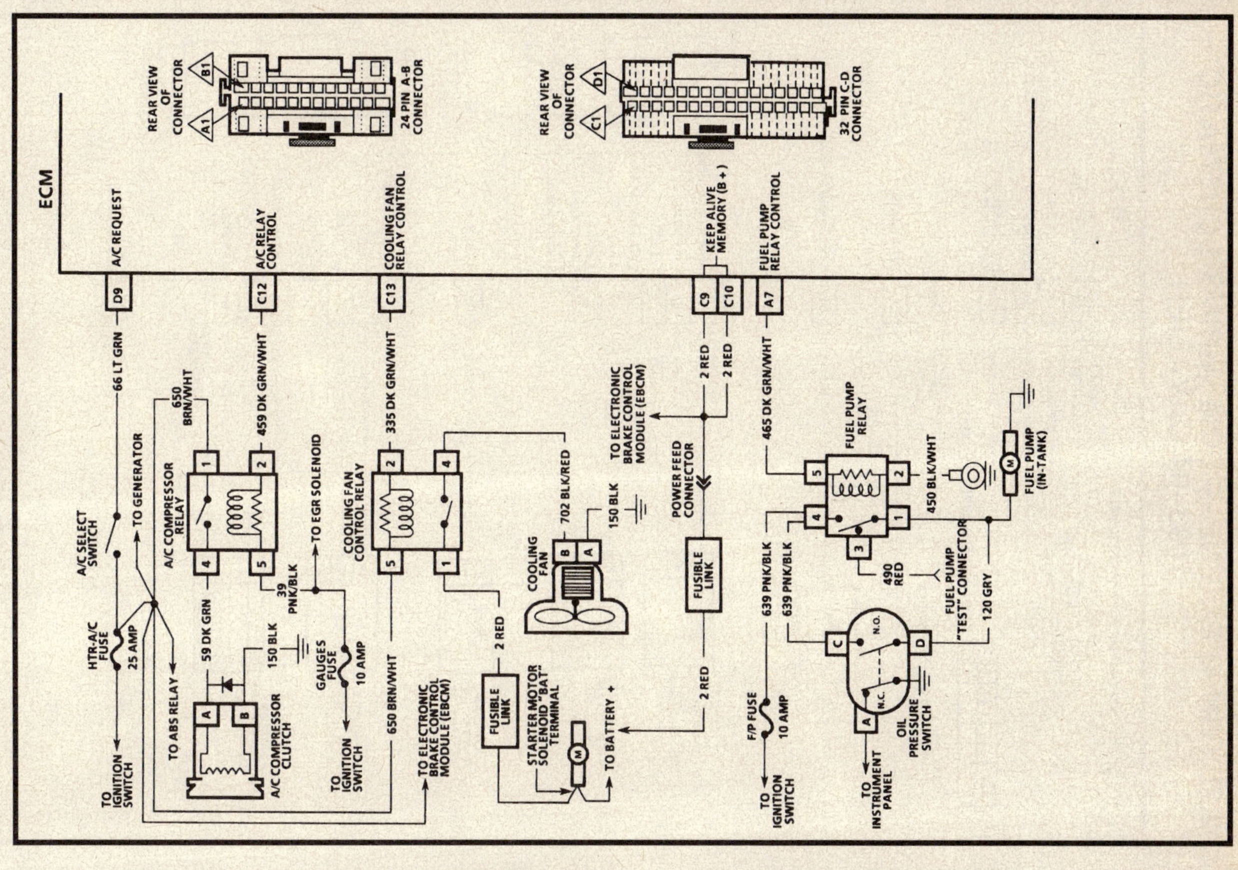
2.2L (VIN 4) ENGINE — ECM WIRING SCHEMATIC — 1992 CAVALIER

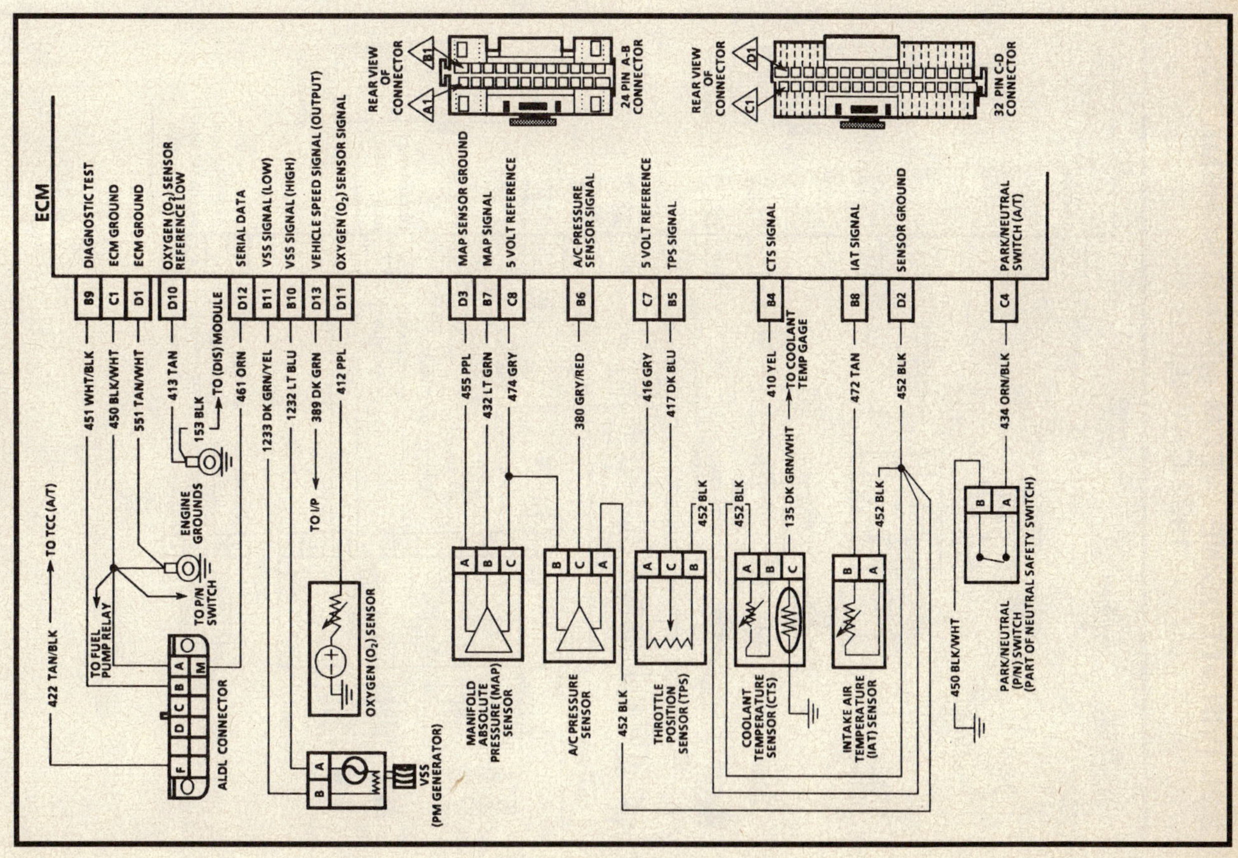
2.2L (VIN 4) ENGINE — ECM WIRING SCHEMATIC — 1992 CAVALIER

MULTIPORT FUEL INJECTION (MFI) SYSTEMS
EXCEPT LIGHT TRUCKS, VANS, GEO AND SATURN

2.2L (VIN 4) ENGINE — ECM WIRING SCHEMATIC — 1993–94 CAVALIER

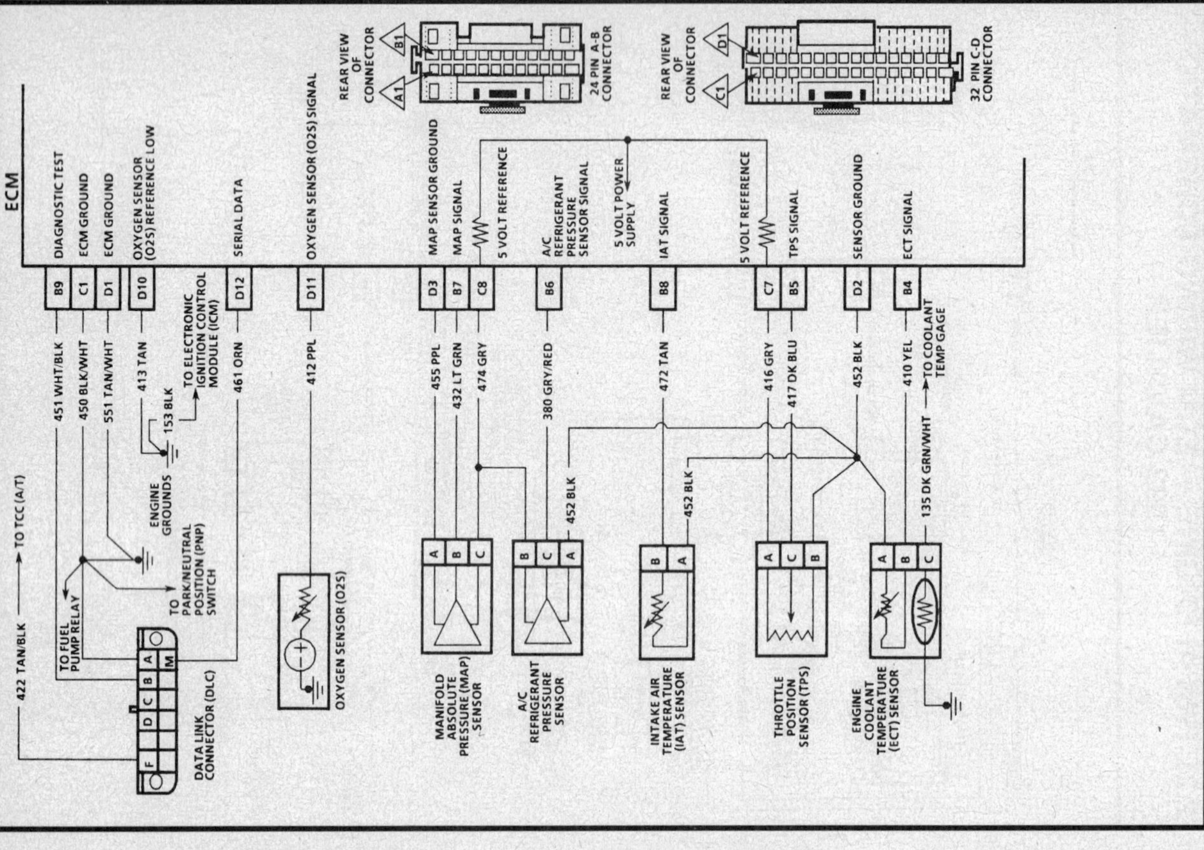

2.2L (VIN 4) ENGINE — ECM WIRING SCHEMATIC — 1992 CAVALIER

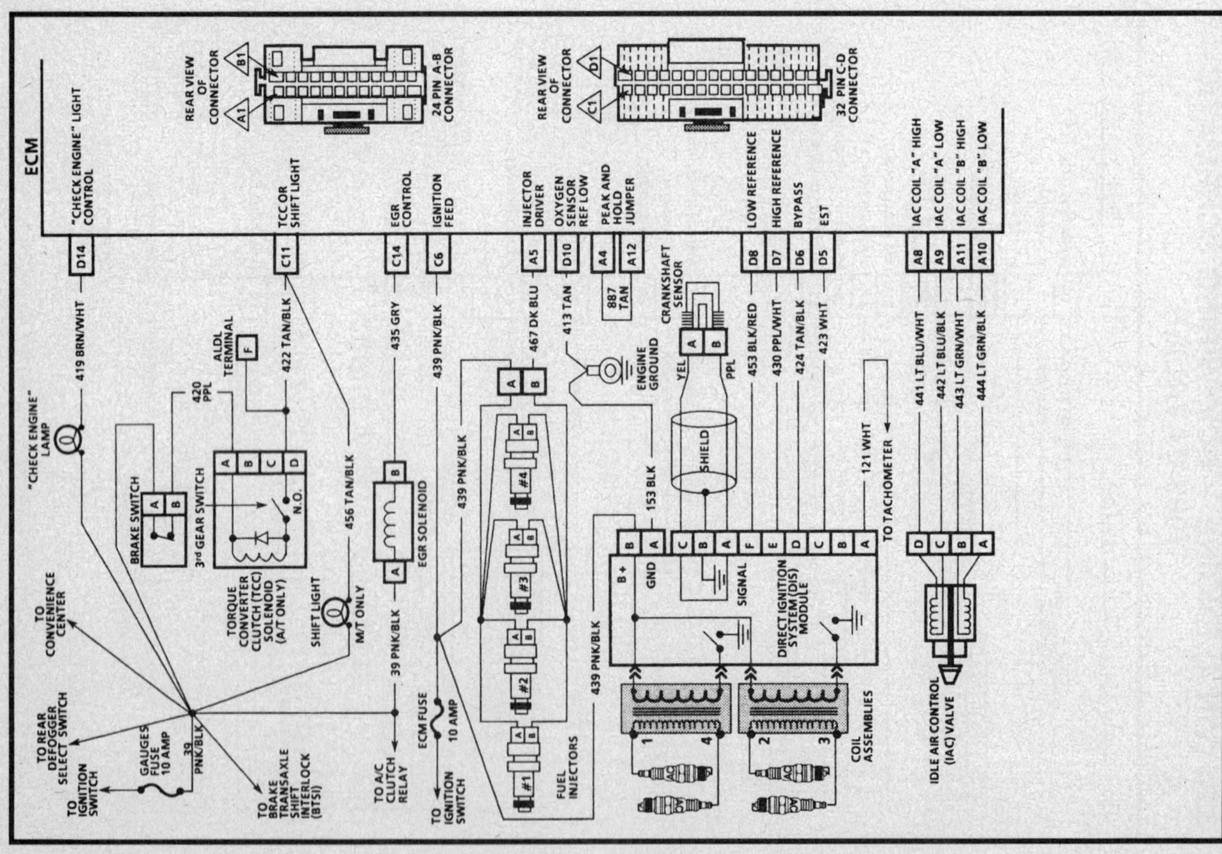

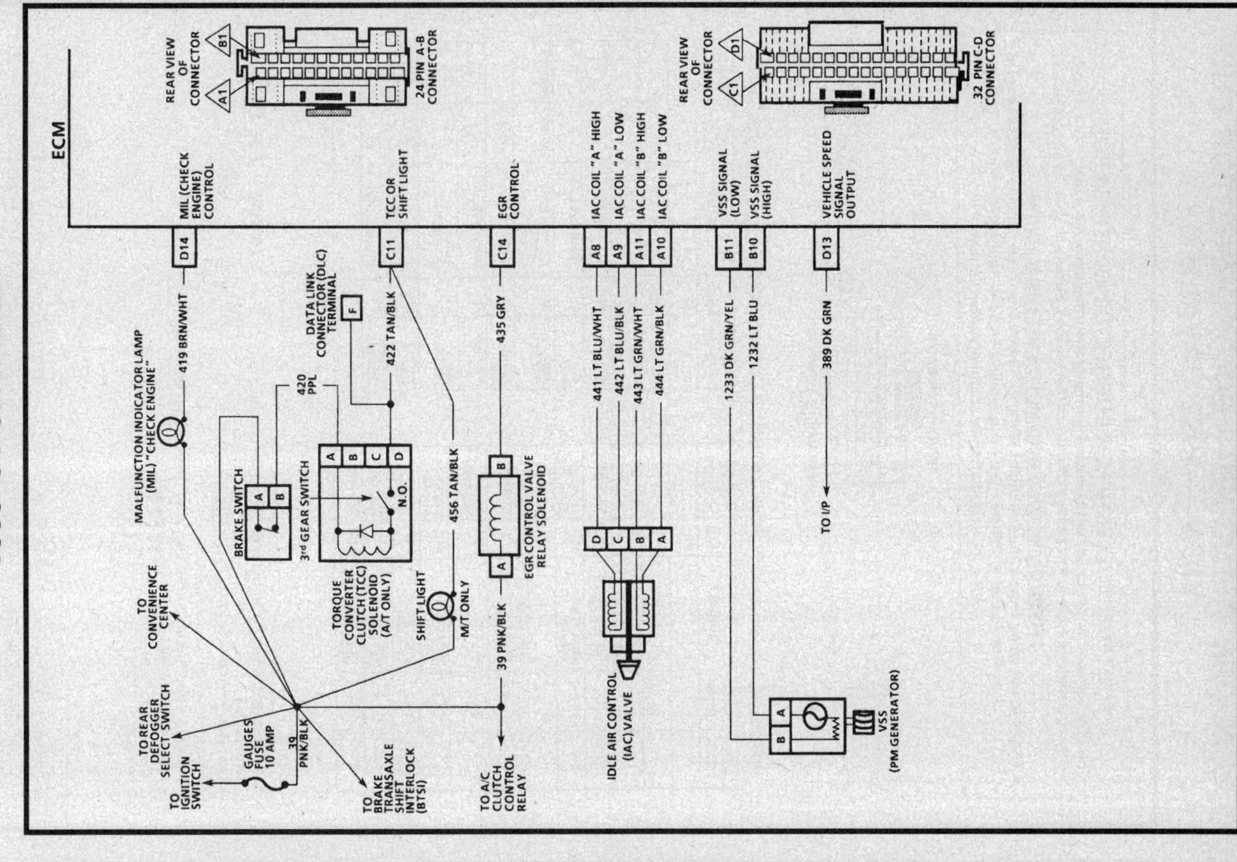

2.2L (VIN 4) ENGINE —ECM WIRING SCHEMATIC — 1993-94 CAVALIER
ECM
REAR VIEW OF CONNECTOR
B1
A1
24 PIN A-B CONNECTOR
REAR VIEW OF CONNECTOR
D1
C1
32 PIN C-D CONNECTOR
MIL (CHECK ENGINE) CONTROL
TCC OR SHIFT LIGHT
EGR CONTROL
IAC COIL "A" HIGH
IAC COIL "A" LOW
IAC COIL "B" HIGH
IAC COIL "B" LOW
VSS SIGNAL (LOW)
VSS SIGNAL (HIGH)
VEHICLE SPEED SIGNAL OUTPUT
D14
C11
C14
A8
A9
A11
A10
B11
B10
D13
419 BRN/WHT
422 TAN/BLK
435 GRY
441 LT BLU/WHT
442 LT BLU/BLK
443 LT GRN/WHT
444 LT GRN/BLK
1233 DK GRN/YEL
1232 LT BLU
389 DK GRN
MALFUNCTION INDICATOR LAMP (MIL) "CHECK ENGINE"
DATA LINK CONNECTOR (DLC) TERMINAL
420 PPL
BRAKE SWITCH
A
B
3rd GEAR SWITCH
A
B
C
D
N.O.
456 TAN/BLK
TORQUE CONVERTER CLUTCH (TCC) SOLENOID (A/T ONLY)
SHIFT LIGHT
M/T ONLY
EGR CONTROL VALVE RELAY SOLENOID
39 PNK/BLK
TO I/P
IDLE AIR CONTROL (IAC) VALVE
VSS (PM GENERATOR)
TO CONVENIENCE CENTER
TO REAR DEFOGGER SELECT SWITCH
TO IGNITION SWITCH
GAUGES FUSE 10 AMP
39 PNK/BLK
TO BRAKE TRANSAXLE SHIFT INTERLOCK (BTSI)
TO A/C CLUTCH CONTROL RELAY

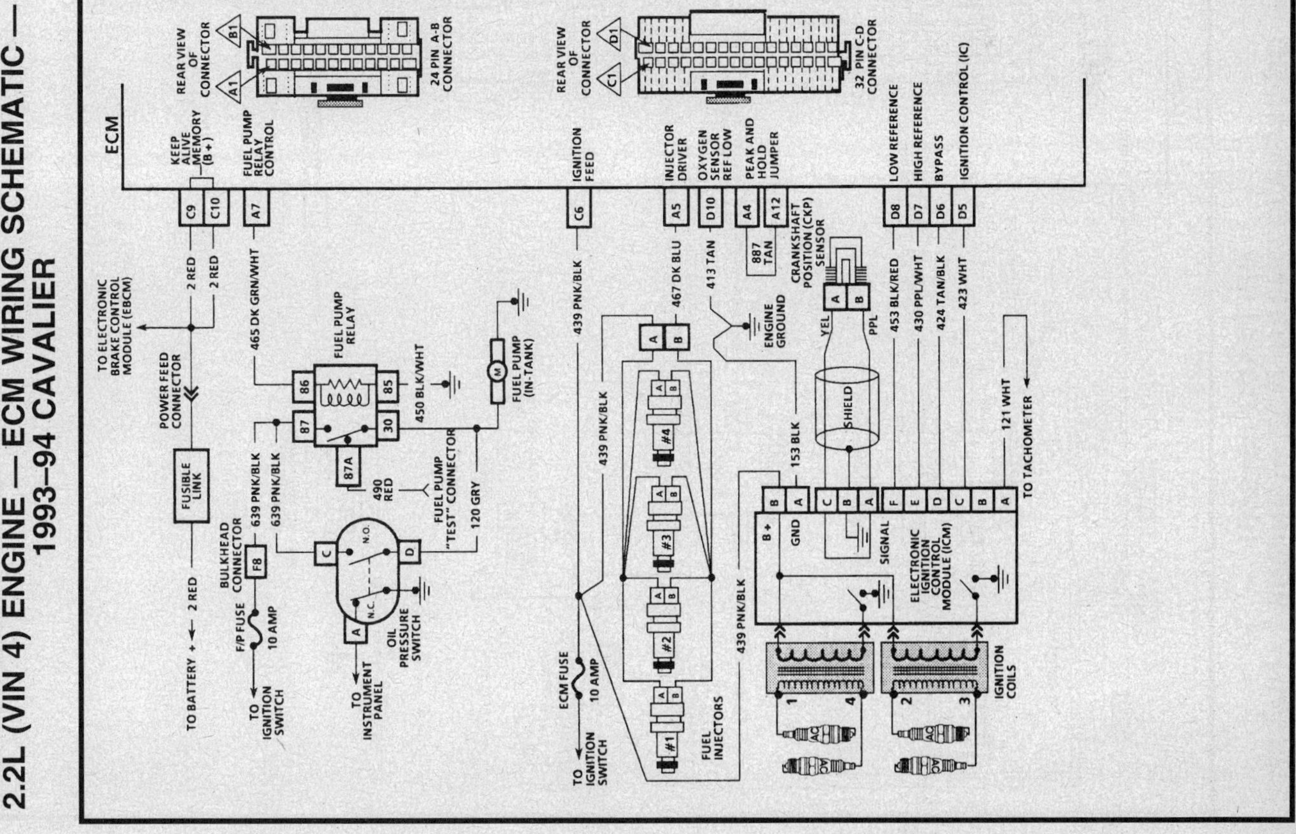

2.2L (VIN 4) ENGINE —ECM WIRING SCHEMATIC — 1993-94 CAVALIER
ECM
REAR VIEW OF CONNECTOR
B1
A1
24 PIN A-B CONNECTOR
REAR VIEW OF CONNECTOR
D1
C1
32 PIN C-D CONNECTOR
KEEP ALIVE MEMORY (B +)
FUEL PUMP RELAY CONTROL
IGNITION FEED
INJECTOR DRIVER
OXYGEN SENSOR REF LOW
PEAK AND HOLD JUMPER
LOW REFERENCE
HIGH REFERENCE
BYPASS
IGNITION CONTROL (IC)
C9
C10
A7
C6
A5
D10
A4
A12
D8
D7
D6
D5
2 RED
2 RED
465 DK GRN/WHT
439 PNK/BLK
467 DK BLU
413 TAN
887 TAN
453 BLK/RED
430 PPL/WHT
424 TAN/BLK
423 WHT
TO ELECTRONIC BRAKE CONTROL MODULE (EBCM)
POWER FEED CONNECTOR
FUSIBLE LINK
BULKHEAD CONNECTOR
F8
F/P FUSE 10 AMP
639 PNK/BLK
639 PNK/BLK
TO BATTERY +
2 RED
TO IGNITION SWITCH
TO INSTRUMENT PANEL
OIL PRESSURE SWITCH
N.O.
N.C.
490 RED
FUEL PUMP RELAY
86
85
87
30
87A
450 BLK/WHT
120 GRY
FUEL PUMP "TEST" CONNECTOR
FUEL PUMP (IN-TANK)
M
439 PNK/BLK
439 PNK/BLK
439 PNK/BLK
ECM FUSE 10 AMP
TO IGNITION SWITCH
#1
#2
#3
#4
FUEL INJECTORS
ENGINE GROUND
CRANKSHAFT POSITION (CKP) SENSOR
YEL
PPL
SHIELD
153 BLK
B +
GND
SIGNAL
ELECTRONIC IGNITION CONTROL MODULE (ICM)
121 WHT
TO TACHOMETER
IGNITION COILS
1
4
2
3

2.2L (VIN 4) ENGINE — ECM WIRING SCHEMATIC — 1993–94 CAVALIER

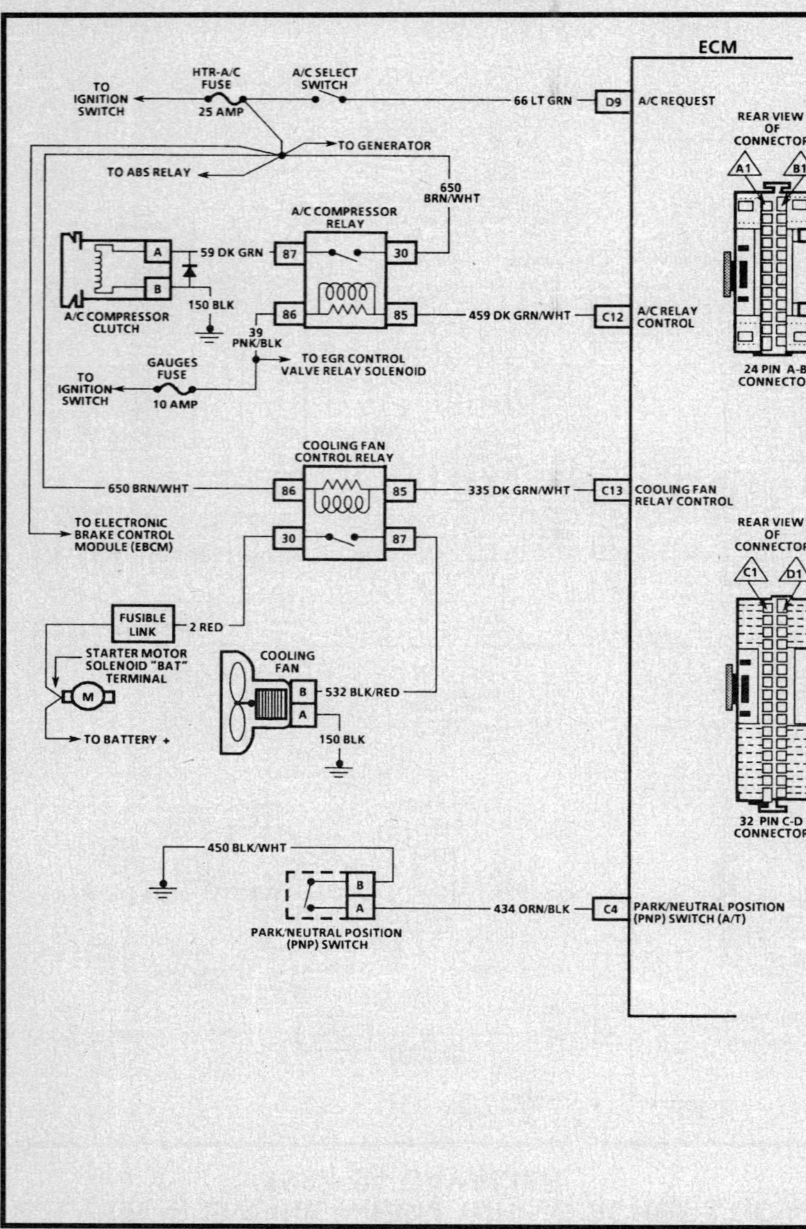

2.2L (VIN 4) ENGINE — ECM CONNECTOR END VIEW — 1992 CAVALIER

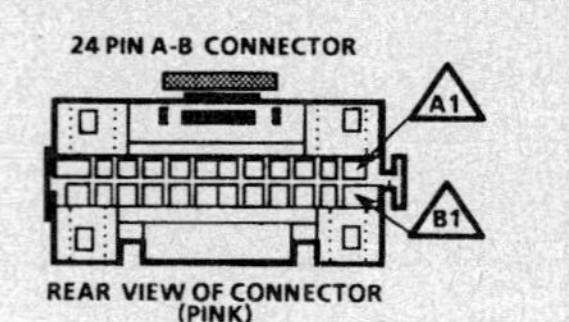

PORT FUEL INJECTION ECM CONNECTOR IDENTIFICATION

This ECM voltage chart is for use with a digital voltmeter to further aid in diagnosis. The voltages you get may vary due to low battery charge or other reasons, but they should be very close.

THE FOLLOWING CONDITIONS MUST BE MET BEFORE TESTING:

- Engine at operating temperature ● Engine idling in "Closed Loop" (For "Engine Run" column) in park or neutral ● Test terminal not grounded ● "Scan" tool not installed
- B + indicates battery or charging system voltage

VOLTAGE — PINK 24 PIN A-B CONNECTOR

KEY "ON"	ENG. RUN	CIRCUIT	PIN	WIRE COLOR
			A1	
			A2	
			A3	
0*	0*	PEAK AND HOLD JUMPER	A4	887 TAN
B+	B+	INJECTOR DRIVER	A5	467 DK BLU
			A6	
(4)	B+	FUEL PUMP RELAY CONTROL	A7	465 DK GRN/WHT
(1)	(1)	IAC COIL "A" HIGH	A8	441 LT BLU/WHT
(1)	(1)	IAC COIL "A" LOW	A9	442 LT BLU/BLK
(1)	(1)	IAC COIL "B" LOW	A10	444 LT GRN/BLK
(1)	(1)	IAC COIL "B" HIGH	A11	443 LT GRN/WHT
0*	0*	PEAK AND HOLD JUMPER	A12	887 TAN

WIRE COLOR	PIN	CIRCUIT	KEY "ON"	ENG. RUN
	B1			
	B2			
	B3			
410 YEL	B4	CTS SIGNAL	(3)	(3)
417 DK BLU	B5	TPS SIGNAL	.6	.6
380 GRY/RED	B6	A/C PRESSURE SENSOR SIGNAL	VARIES	VARIES
432 LT GRN	B7	MAP SIGNAL	4.75	1.5
472 TAN	B8	IAT SIGNAL	(3)	(3)
451 WHT/BLK	B9	ALDL DIAGNOSTIC TEST	5.0	5.0
1232 LT BLU	B10	VSS SIGNAL(HIGH)	0*	(5)
1233 DK GRN/YEL	B11	VSS SIGNAL(LOW)	0*	(5)
	B12			

* All voltages shown "0" should read less than .5 volt.
(1) Not usable
(2) A/C select "OFF" and engine cooling fan "OFF."
(3) Varies depending on temperature
(4) Reads B + for 2 seconds after key "ON," then should read 0 volt.
(5) AC voltage increases with vehicle speed.

ENGINE 2.2L LN2

2.2L (VIN 4) ENGINE — ECM CONNECTOR END VIEW — 1992 CAVALIER

PORT FUEL INJECTION ECM CONNECTOR IDENTIFICATION

This ECM voltage chart is for use with a digital voltmeter to further aid in diagnosis. The voltages you get may vary due to low battery charge or other reasons, but they should be very close.

THE FOLLOWING CONDITIONS MUST BE MET BEFORE TESTING:

- Engine at operating temperature ● Engine idling in "Closed Loop" (For "Engine Run" column) in park or neutral ● Test terminal not grounded ● "Scan" tool not installed
- B + indicates battery or charging system voltage

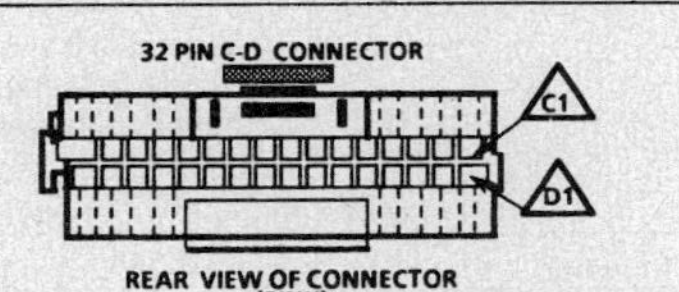

32 PIN C-D CONNECTOR

REAR VIEW OF CONNECTOR (PINK)

KEY "ON"	ENG. RUN	CIRCUIT	PIN	WIRE COLOR
0*	0*	ECM GROUND	C1	450 BLK/WHT
			C2	
			C3	
0*	0*	P/N SWITCH	C4	434 ORN/BLK
			C5	
B+	B+	IGNITION FEED	C6	439 PNK/BLK
5.0	5.0	5 V REFERENCE	C7	416 GRY
5.0	5.0	5V REFERENCE	C8	474 GRY
B+	B+	KEEP ALIVE MEMORY (B+)	C9	2 RED
B+	B+	KEEP ALIVE MEMORY (B+)	C10	2 RED
0*	0*	TCC (A/T) OR	C11	422 --TAN/BLK
B+	B+	SHIFT LIGHT (M/T)		456
B+	B+	A/C RELAY CONTROL (2)	C12	459 DK GRN/WHT
B+	B+	COOLING FAN RELAY CONTROL (2)	C13	335 DK GRN/WHT
B+	B+	EGR SOLENOID CONTROL	C14	435 GRY
			C15	
			C16	

WIRE COLOR	PIN	CIRCUIT	KEY "ON"	ENG. RUN
551 TAN/WHT	D1	ECM GROUND	0*	0*
452 BLK	D2	SENSOR GROUND	0*	0*
455 PPL	D3	MAP SENSOR GROUND	0*	0*
	D4			
423 WHT	D5	EST	0*	2.4
424 TAN/BLK	D6	BYPASS	0*	4.8
430 PPL/WHT	D7	REF HIGH	0*	3.2
453 BLK/RED	D8	REF LOW	0*	0*
66 LT GRN	D9	A/C REQUEST	0*	0* (2)
413 TAN	D10	OXYGEN SENSOR REFERENCE LOW	0*	0*
412 PPL	D11	OXYGEN SENSOR SIGNAL	.01-.55	.1-.9 (6)
461 ORN	D12	SERIAL DATA	4.7	4.7
389 DK GRN	D13	VSS OUTPUT (4000 PPM)	(5)	(5)
419 BRN/WHT	D14	"CHECK ENGINE" LIGHT	0*	B+
	D15			
	D16			

ENGINE **2.2L LN2**

* All voltages shown "0" should read less than .5 volt.
(1) Not usable.
(2) A/C Select "OFF" and engine cooling fan "OFF."
(3) Varies depending on temperature.
(4) Reads B + for 2 seconds after key "ON," then should read 0 volt.
(5) Increases with vehicle speed.
(6) Varies actively within indicated range.

2.2L (VIN 4) ENGINE — ECM CONNECTOR END VIEW — 1993–94 CAVALIER

MULTIPORT FUEL INJECTION ECM CONNECTOR IDENTIFICATION

This ECM voltage chart is for use with J 39200 digital voltmeter to further aid in diagnosis. The voltages you get may vary due to low battery charge or other reasons, but they should be very close.

THE FOLLOWING CONDITIONS MUST BE MET BEFORE TESTING:

- Engine at operating temperature ● Engine idling in "Closed Loop" (For "Engine Run" column) in park or neutral ● Test terminal not grounded ● Scan tool not installed ● Brake <u>not</u> applied
- B + indicates battery or charging system voltage

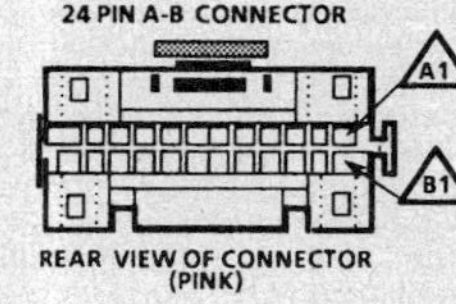

24 PIN A-B CONNECTOR

REAR VIEW OF CONNECTOR (PINK)

PINK 24 PIN A-B CONNECTOR

VOLTAGE KEY "ON"	ENG. RUN	CIRCUIT	PIN	WIRE COLOR
			A1	
			A2	
			A3	
0*	0*	PEAK AND HOLD JUMPER	A4	887 TAN
B+	B+ (6)	INJECTOR DRIVER	A5	467 DK BLU
			A6	
(3)	B+	FUEL PUMP RELAY CONTROL	A7	465 DK GRN/WHT
(1)	(1)	IAC COIL "A" HIGH	A8	441 LT BLU/WHT
(1)	(1)	IAC COIL "A" LOW	A9	442 LT BLU/BLK
(1)	(1)	IAC COIL "B" LOW	A10	444 LT GRN/BLK
(1)	(1)	IAC COIL "B" HIGH	A11	443 LT GRN/WHT
0*	0*	PEAK AND HOLD JUMPER	A12	887 TAN

WIRE COLOR	PIN	CIRCUIT	KEY "ON"	ENG. RUN
	B1			
	B2			
	B3			
410 YEL	B4	ECT SIGNAL	(2)	(2)
417 DK BLU	B5	TP SIGNAL	.6 (5)	.6 (5)
380 GRY/RED	B6	A/C REFRIGERANT PRESSURE SENSOR SIGNAL	VARIES	VARIES
432 LT GRN	B7	MAP SIGNAL	4.75	1.2
472 TAN	B8	IAT SIGNAL	(2)	(2)
451 WHT/BLK	B9	DIAGNOSTIC TEST	5.0	5.0
1232 LT BLU	B10	VSS SIGNAL(HIGH)	0*	(4)
1233 DK GRN/YEL	B11	VSS SIGNAL(LOW)	0*	(4)
	B12			

ENGINE **2.2L LN2**

* ALL VOLTAGES SHOWN "0" SHOULD READ LESS THAN .5 VOLT.
(1) NOT USABLE.
(2) VARIES DEPENDING ON TEMPERATURE.
(3) READS B + FOR 2 SECONDS AFTER KEY "ON," THEN SHOULD READ 0 VOLT.
(4) AC VOLTAGE INCREASES WITH VEHICLE SPEED. (REFER TO SECTION 8A-33 FOR DIAGNOSIS.)
(5) WITHIN THE RANGE OF .33 TO 1.33 VOLTS.
(6) ALTERNATE TEST: SET J 39200 DVM TO DC VOLTAGE FREQUENCY SCALE. CONNECT RED LEAD TO IGNITION FEED ("C6" AT ECM) AND BLACK LEAD TO INJECTOR DRIVER ("A5" AT ECM); SHOULD MEASURE ABOUT 9-15 Hz.

2.2L (VIN 4) ENGINE — ECM CONNECTOR END VIEW — 1993–94 CAVALIER

2.2L (VIN 4) ENGINE — ON-BOARD DIAGNOSTIC SYSTEM CHART — CAVALIER

MULTIPORT FUEL INJECTION ECM CONNECTOR IDENTIFICATION

This ECM voltage chart is for use with J 39200 digital voltmeter to further aid in diagnosis. The voltages you get may vary due to low battery charge or other reasons, but they should be very close.

__THE FOLLOWING CONDITIONS MUST BE MET BEFORE TESTING:__

- Engine at operating temperature ● Engine idling in "Closed Loop" (For "Engine Run" column) in park or neutral ● Test terminal not grounded ● Scan tool not installed ● Brake _not_ applied
- B + indicates battery or charging system voltage

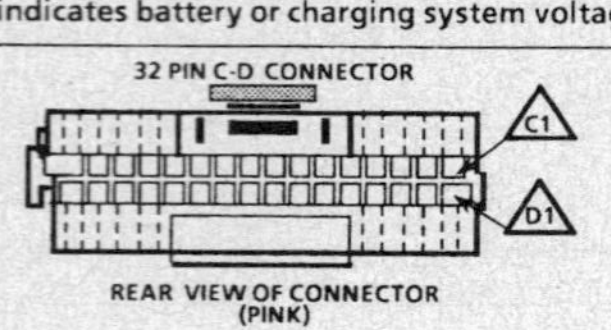

VOLTAGE							VOLTAGE	
KEY "ON"	ENG. RUN	CIRCUIT	PIN	WIRE COLOR	WIRE COLOR	PIN	CIRCUIT	KEY "ON"	ENG. RUN
0* (5)	0* (5)	ECM GROUND	C1	450 BLK/WHT	450 BLK/WHT	D1	ECM GROUND	0* (5)	0* (5)
			C2		452 BLK	D2	SENSOR GROUND	0*	0*
			C3		455 PPL	D3	SENSOR GROUND	0*	0*
0*	0*	PNP SWITCH	C4	434 ORN/BLK		D4			
			C5		423 WHT	D5	IGNITION CONTROL (IC)	0*	2.3
B + (5)	B + (5)	IGNITION FEED	C6	439 PNK/BLK	424 TAN/BLK	D6	BYPASS	0*	4.8
5.0	5.0	5 V REFERENCE	C7	416 GRY	430 PPL/WHT	D7	REF HIGH	0*	3.2 (4)
5.0	5.0	5V REFERENCE	C8	474 GRY	453 BLK/RED	D8	REF LOW	0*	0*
B + (5)	B + (5)	KEEP ALIVE MEMORY (B +)	C9	2 RED	66 LT GRN	D9	A/C REQUEST	0* (1)	0* (1)
B + (5)	B + (5)	KEEP ALIVE MEMORY (B +)	C10	2 RED	413 TAN	D10	OXYGEN SENSOR REFERENCE LOW	0*	0*
0* / B+	0* / B+	TCC (A/T) OR SHIFT LIGHT (M/T)	C11	422 --TAN/BLK 456	412 PPL	D11	OXYGEN SENSOR SIGNAL	.01-.55	.1-.9 (3)
B + (1)	B + (1)	A/C COMPRESSOR CLUTCH RELAY CONTROL	C12	459 DK GRN/WHT	461 ORN	D12	SERIAL DATA	4.7	4.7
B + (1)	B + (1)	COOLING FAN RELAY CONTROL	C13	335 DK GRN/WHT	389 DK GRN	D13	VSS OUTPUT (4000 PPM)	(2)	(2)
B +	B +	EGR SOLENOID CONTROL	C14	435 GRY	419 BRN/WHT	D14	MIL (CHECK ENGINE) CONTROL	0-1.0	B +
			C15			D15			
			C16			D16			

ENGINE   2.2L LN2

- * ALL VOLTAGES SHOWN "0" SHOULD READ LESS THAN .5 VOLT.
- (1) A/C SELECT "OFF" AND ENGINE COOLING FAN "OFF."
- (2) INCREASES WITH VEHICLE SPEED. REFER TO SECTION 8A-33.
- (3) VARIES ACTIVELY WITHIN INDICATED RANGE.
- (4) ALTERNATE TEST: SET J 39200 DVM TO DC VOLTAGE FREQUENCY SCALE. CONNECT RED LEAD TO REF HIGH ("D7" AT ECM) AND BLACK LEAD TO ECM GROUND ("D1" AT ECM); SHOULD MEASURE ABOUT 25-30 Hz.
- (5) THIS TYPE OF CIRCUIT SHOULD ALSO BE CHECKED USING J 34142-B TEST LIGHT.

ON-BOARD DIAGNOSTIC (OBD) SYSTEM CHECK
2.2L (VIN 4) "J" CARLINE

Circuit Description:

The "On-Board Diagnostic (OBD) System Check" is an organized approach to identifying a problem created by an electronic engine control system malfunction. It must be the starting point for any driveability complaint diagnosis, because it directs the service technician to the next logical step in diagnosing the complaint. Understanding the chart and using it correctly will reduce diagnostic time and prevent the unnecessary replacement of good parts.

Test Description: Number(s) below refer to circled number(s) on the diagnostic chart.

1. This step is a check for the proper operation of the Malfunction Indicator Lamp (MIL) "Check Engine." The MIL (Check Engine) should be "ON" steady.
2. No MIL (Check Engine) at this point indicates that there is a problem with the MIL circuit or the ECM control of that circuit.
3. This test checks the ability of the ECM of control the MIL (Check Engine). With the diagnostic terminal grounded, the MIL should flash a DTC 12 three times, followed by any Diagnostic Trouble Code (DTC) stored in memory. Depending upon the type of ECM, a EEPROM error may result in the inability to flash DTC 12.
4. Most of the   procedures use a Tech 1 to aid diagnosis, therefore, serial data must be available. If a EEPROM error is present, the ECM may have been able to flash DTC 12/51, but not enable serial data.
5. Although the ECM is powered up, a "Cranks But Will Not Run" symptom could exist because of an ECM or system problem.
6. This step will isolate if the customer complaint is a MIL (Check Engine) or a driveability problem with no MIL. Refer to "ECM Diagnostic Trouble Codes" for   valid DTC(s) an invalid DTC may be the result of a faulty scan tool, EEPROM or ECM.
7. Comparison of actual control system data with the typical values is a quick check to determine if any parameter is not within limits. Keep in mind that a base engine problem (i.e. advanced cam timing) may substantially alter sensor values.
8. Installation of a scan tool will provide a good ground path for the ECM and may hide a driveability complaint due to poor ECM grounds.
9. If the actual data is not within the typical values established, the charts in "Component Systems," will provide a functional check of the suspect component or system.

2.2L (VIN 4) ENGINE — ON-BOARD DIAGNOSTIC SYSTEM CHART — CAVALIER

ON-BOARD DIAGNOSTIC (OBD) SYSTEM CHECK
2.2L (VIN 4) "J" CARLINE

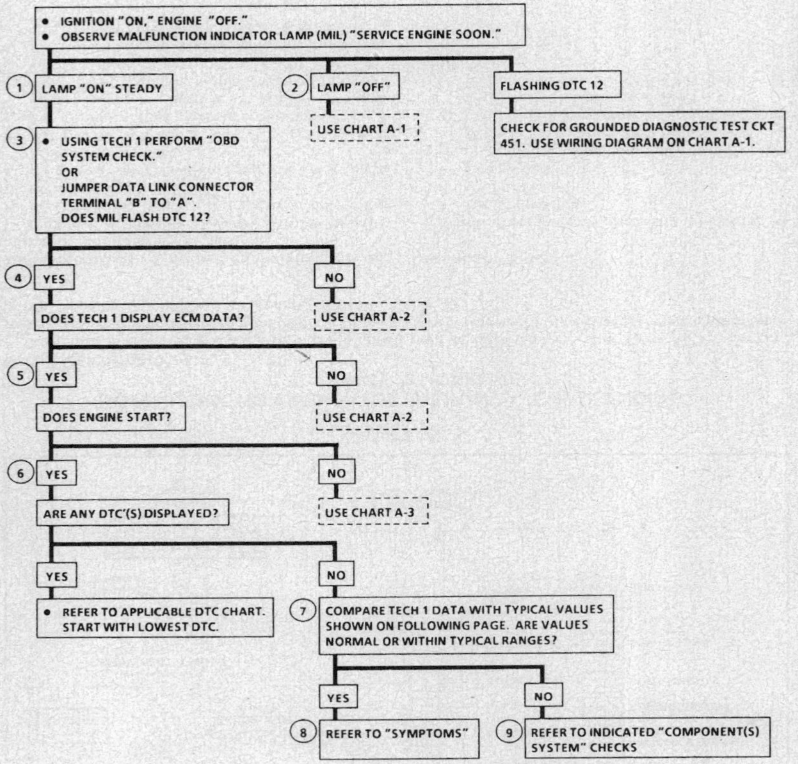

- IGNITION "ON," ENGINE "OFF."
- OBSERVE MALFUNCTION INDICATOR LAMP (MIL) "SERVICE ENGINE SOON."

(1) LAMP "ON" STEADY

(2) LAMP "OFF"
USE CHART A-1

FLASHING DTC 12
CHECK FOR GROUNDED DIAGNOSTIC TEST CKT 451. USE WIRING DIAGRAM ON CHART A-1.

(3) • USING TECH 1 PERFORM "OBD SYSTEM CHECK."
OR
JUMPER DATA LINK CONNECTOR TERMINAL "B" TO "A".
DOES MIL FLASH DTC 12?

(4) YES / NO
DOES TECH 1 DISPLAY ECM DATA?
USE CHART A-2

(5) YES / NO
DOES ENGINE START?
USE CHART A-2

(6) YES / NO
ARE ANY DTC'(S) DISPLAYED?
USE CHART A-3

YES
• REFER TO APPLICABLE DTC CHART. START WITH LOWEST DTC.

(7) COMPARE TECH 1 DATA WITH TYPICAL VALUES SHOWN ON FOLLOWING PAGE. ARE VALUES NORMAL OR WITHIN TYPICAL RANGES?

YES / NO

(8) REFER TO "SYMPTOMS"

(9) REFER TO INDICATED "COMPONENT(S) SYSTEM" CHECKS

2.2L (VIN 4) ENGINE — ON-BOARD DIAGNOSTIC SYSTEM CHART — CAVALIER

If after completing the "On-Board Diagnostic (OBD) System Check" and finding the Tech 1 diagnostics functioning properly and no diagnostic trouble codes displayed, the "Typical Tech 1 Data Values" may be used for comparison with values obtained on the vehicle being diagnosed. The "Typical Tech 1 Data Values," are an average of display values recorded from normally operating vehicles and are intended to represent what a normally functioning system would display.

A SCAN TOOL THAT DISPLAYS FAULTY DATA SHOULD NOT BE USED, AND THE PROBLEM SHOULD BE REPORTED TO THE MANUFACTURER. THE USE OF A FAULTY SCAN TOOL CAN RESULT IN MISDIAGNOSIS AND UNNECESSARY PARTS REPLACEMENT.

Only the parameters listed are used   for diagnosis. If a scan tool reads other parameters, the values are not recommended by General Motors for use in diagnosis. For more description on the values and use of the Tech 1 to diagnosis ECM inputs, refer to the applicable   "Component Systems,"

If all values are within the range illustrated, refer to "Symptoms,"

TYPICAL TECH 1 DATA VALUES
2.2L (VIN 4) "J" CARLINE

Idle / Upper Radiator Hose Hot / Closed Throttle / Park or Neutral / Closed Loop / Acc. OFF/ Brake Not Applied

Tech 1 Parameter	Units Displayed	Typical Data Value
Engine Speed	RPM	875 ± 75 RPM (A/T)
		900 ± 75 RPM (M/T)
Desired Idle	RPM	ECM Commanded (Varies with Temp)
Engine Cool Temp.	°C/ °F	85°-105°C (185°-220°F)
Intake Air Temp.	°C/ °F	Varies with air temp.
MAP	kPa/Volts	29-48 kPa/1-2 volts (varies with manifold and barometric pressures.)
BARO	kPa/Volts	(Varies with barometric pressure.)
Throt Position	Volts	.6 (within the range of .33-1.33)
Throttle Angle	0-100%	0%
Oxygen Sensor	Millivolts	100-999 (actively varying)
Inj. Pulse Width	Milliseconds	.8-3.0
Spark Advance	Degrees	18° (Varies with RPM and engine load)
Short Term (ST) Fuel Trim	Counts	Varies
Long Term (LT) Fuel Trim	Counts	90-160
Idle Air Control	Counts	1-50
Park/Neutral	P-N/R-D-L	P-N
Fuel Trim Cell	Cell Number	0-3 (depends on air flow and RPM)
Fuel Trim Enable	Yes/No	YES/NO (depends on short term fuel trim)
Loop Status	Open/Closed	"Closed Loop"
EGR	"ON/OFF"	"OFF"
Mph / Km/h	0-255	0
TCC/Shift Light	"ON/OFF"	"OFF"
Fan Relay	"ON/OFF"	"OFF" ("ON" when coolant temp. reaches 106°C or when A/C or defrost is selected)
A/C Request	Yes/No	NO
A/C Clutch	"ON/OFF"	"OFF"
A/C Pressure	psi/Volts	(Varies with temp.)
System Voltage	Volts	13.5-14.5
Calibration ID	0-9999	Internal ID only
Time From Start	Hrs/Min	Varies

2.2L (VIN 4) ENGINE — SYSTEM DIAGNOSTIC CHARTS — CAVALIER

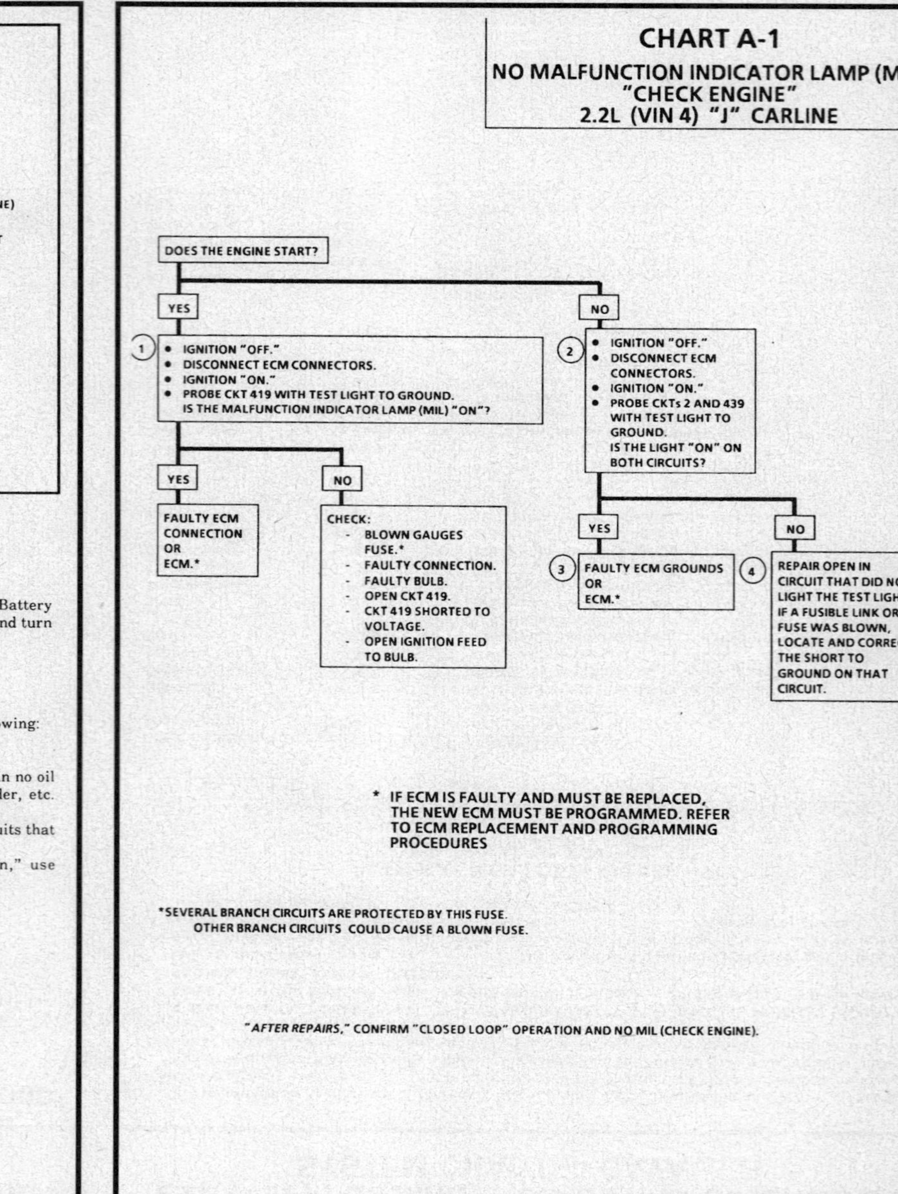

CHART A-1
NO MALFUNCTION INDICATOR LAMP (MIL) "CHECK ENGINE"
2.2L (VIN 4) "J" CARLINE

Circuit Description:

There should always be a steady MIL (Check Engine), when the ignition is "ON" and engine "OFF." Battery voltage is supplied directly to the light bulb. The Engine Control Module (ECM) will control the lamp and turn it "ON" by providing a ground path through CKT 419 to the ECM.

Test Description: Number(s) below refer to circled number(s) on the diagnostic chart.

1. This test will provide a ground path for the MIL (Check Engine) circuit. If the MIL is "ON," then the external circuit is OK.
2. If CKTs 439 and 2 have voltage, the ECM grounds or the ECM are faulty.
3. Using a test light connected to B +, probe each of the system ground circuits to be sure a good ground is present. Refer to "ECM Connector End View" at the beginning of this section for ECM pin locations of ground circuits.
4. If a fusible link or fuse is blown, be sure to consider branch circuits and splices to other components.

Diagnostic Aids:

If engine runs correctly, check for the following:
- Faulty light bulb.
- CKT 419 open.
- Gauges fuse blown. This will result in no oil or generator lights, seat belt reminder, etc. Refer to "Fuse Block Details" which shows other branch circuits that are protected by the gauges fuse.

If "Engine Cranks But Will Not Run," use CHART A-3.

2.2L (VIN 4) ENGINE — SYSTEM DIAGNOSTIC CHARTS — CAVALIER

CHART A-1
NO MALFUNCTION INDICATOR LAMP (MIL) "CHECK ENGINE"
2.2L (VIN 4) "J" CARLINE

DOES THE ENGINE START?

YES

(1)
- IGNITION "OFF."
- DISCONNECT ECM CONNECTORS.
- IGNITION "ON."
- PROBE CKT 419 WITH TEST LIGHT TO GROUND. IS THE MALFUNCTION INDICATOR LAMP (MIL) "ON"?

YES — FAULTY ECM CONNECTION OR ECM.*

NO — CHECK:
- BLOWN GAUGES FUSE.*
- FAULTY CONNECTION.
- FAULTY BULB.
- OPEN CKT 419.
- CKT 419 SHORTED TO VOLTAGE.
- OPEN IGNITION FEED TO BULB.

NO

(2)
- IGNITION "OFF."
- DISCONNECT ECM CONNECTORS.
- IGNITION "ON."
- PROBE CKTs 2 AND 439 WITH TEST LIGHT TO GROUND. IS THE LIGHT "ON" ON BOTH CIRCUITS?

YES — (3) FAULTY ECM GROUNDS OR ECM.*

NO — (4) REPAIR OPEN IN CIRCUIT THAT DID NOT LIGHT THE TEST LIGHT. IF A FUSIBLE LINK OR FUSE WAS BLOWN, LOCATE AND CORRECT THE SHORT TO GROUND ON THAT CIRCUIT.

* IF ECM IS FAULTY AND MUST BE REPLACED, THE NEW ECM MUST BE PROGRAMMED. REFER TO ECM REPLACEMENT AND PROGRAMMING PROCEDURES

*SEVERAL BRANCH CIRCUITS ARE PROTECTED BY THIS FUSE. OTHER BRANCH CIRCUITS COULD CAUSE A BLOWN FUSE.

"AFTER REPAIRS," CONFIRM "CLOSED LOOP" OPERATION AND NO MIL (CHECK ENGINE).

2.2L (VIN 4) ENGINE — SYSTEM DIAGNOSTIC CHARTS — CAVALIER

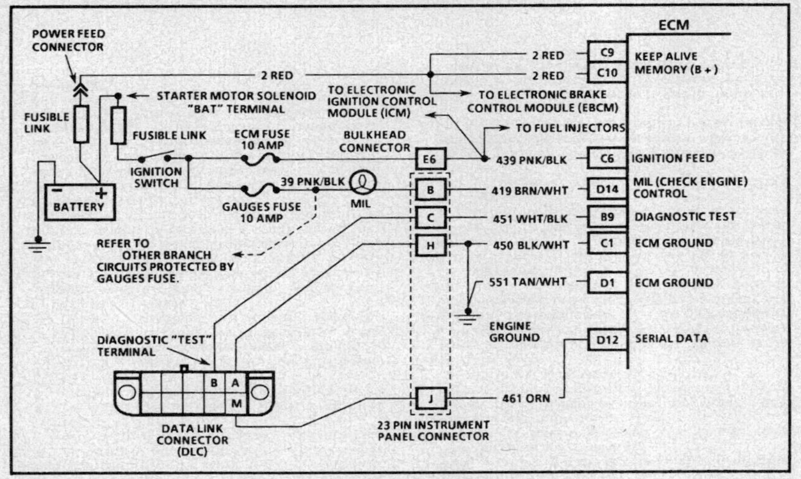

CHART A-2
NO DATA LINK CONNECTOR (DLC) DATA OR WON'T FLASH DTC 12 MIL (CHECK ENGINE) "ON" STEADY 2.2L (VIN 4) "J" CARLINE

Circuit Description:

There should always be a steady MIL (Check Engine) when the ignition is "ON" and the engine is "OFF." Battery voltage is supplied directly to the lamp. The Engine Control Module (ECM) will control the lamp and turn it "ON" by providing a ground path through CKT 419 to the ECM.

With the diagnostic terminal grounded, the light should flash a DTC 12, followed by any DTC(s) stored in memory. A steady lamp suggests a short to ground in the light control CKT 419, or an open in diagnostic CKT 451, or CKT 450 from the DLC to the splice.

Test Description: Number(s) below refer to circled number(s) on the diagnostic chart.

1. If there is a problem with the ECM that causes a scan tool to not read data from the ECM, then the ECM should not flash a DTC 12. If DTC 12 does flash, be sure that the scan tool is working properly on another vehicle. If the scan tool is functioning properly and CKT 461 is OK, the EEPROM or ECM may be at fault for the "NO DLC Data" symptom.

2. If the lamp turns "OFF" when the ECM connector is disconnected, then CKT 419 is not shorted to ground.

3. This step will check for an open diagnostic CKT 451.

4. At this point, the MIL (Check Engine) wiring is OK. The problem is a faulty ECM. If DTC 12 does not flash, the ECM should be replaced and the new ECM must be programmed. Refer to "Engine Control Module (ECM) and Sensors."

2.2L (VIN 4) ENGINE — SYSTEM DIAGNOSTIC CHARTS — CAVALIER

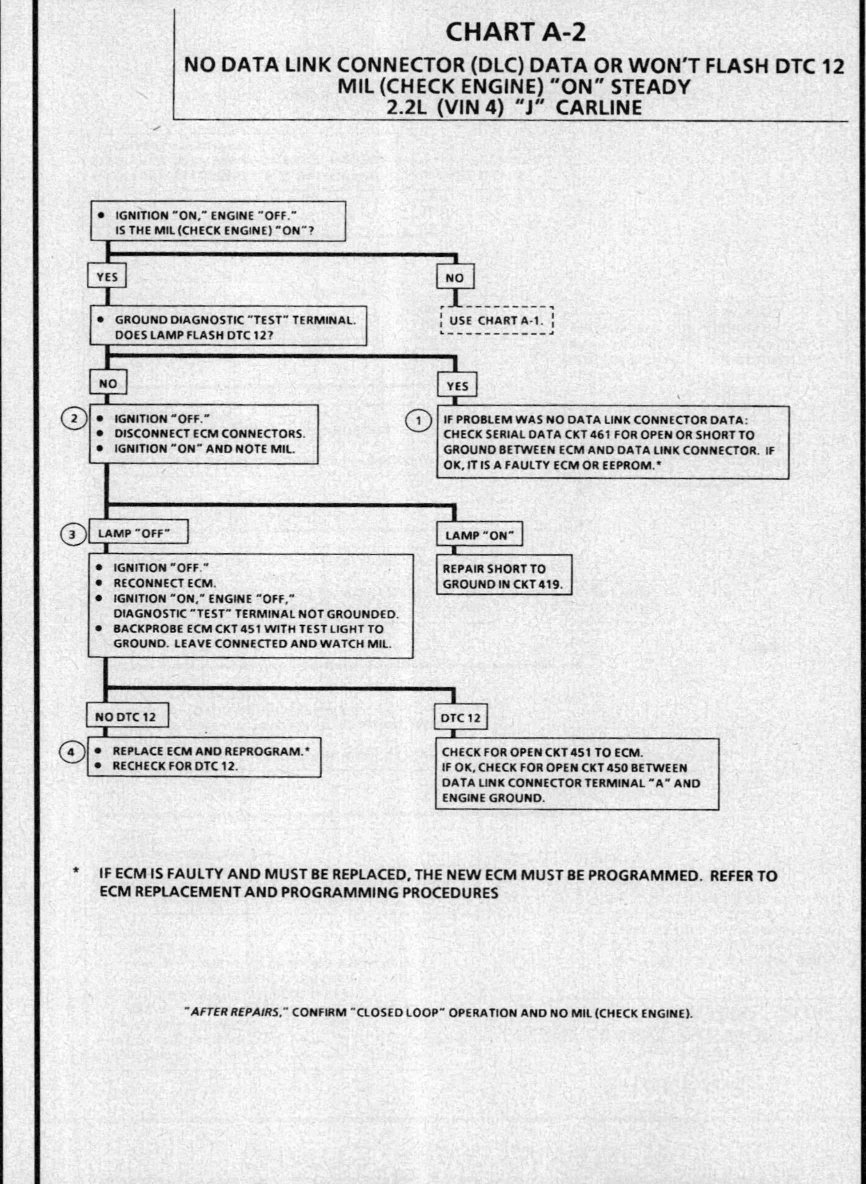

2.2L (VIN 4) ENGINE — SYSTEM DIAGNOSTIC CHARTS — 1992 CAVALIER

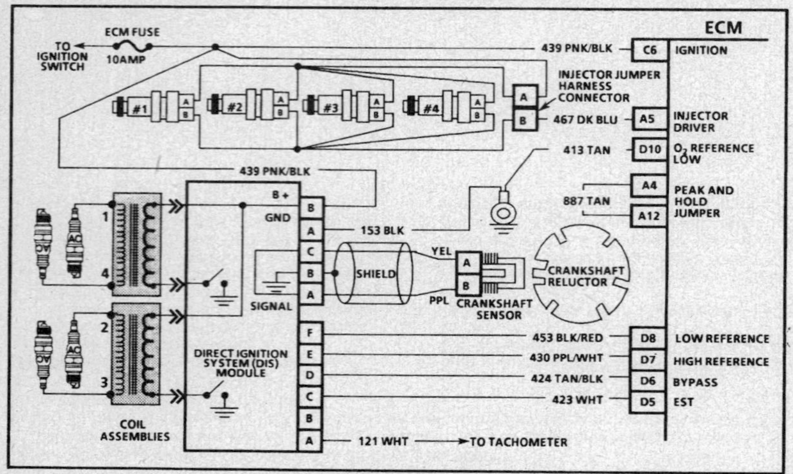

CHART A-3 (Page 1 of 3)

ENGINE CRANKS BUT WON'T RUN
2.2L (VIN 4) "J" CARLINE (PORT)

Circuit Description:

Before using this chart, battery condition, engine cranking speed, and fuel quantity should be checked and verified as being OK.

Test Description: Number(s) below refer to circled number(s) on the diagnostic chart.

1. • A "Check Engine" light "ON" is a basic test to determine if there is battery and ignition voltage at the ECM.
   • No ALDL data may be the result of an ECM problem, and CHART A-2 will diagnose an ECM problem.
   • If TPS is less than .2 volts, the TPS 5 volt reference circuit could be shorted to ground. If TPS is over 2.5 volts, the ECM could be in the "Clear Flood Mode" which may cause the engine to not start.
2. Because the Direct Ignition System (DIS) uses two plugs and wires to complete the circuit of each coil, the opposite spark plug wire should be left connected. If rpm was indicated during crank, the ignition module is receiving a crank signal, but "No Spark" at this test indicates the ignition module is not triggering the coil.
3. This step determines if injectors or harness are cause of the problem. Nominal resistance of each injector is 11.6 to 12.4 ohms at 20°C (68°F). Resistance will increase slightly at higher temperatures. This test is performed with injectors 1, 2, 3, & 4 in parallel, so measurement will be one fourth of the nominal resistance for a single injector.

4. This test light should flash, indicating that the ECM is controlling the injectors. How bright the light flashes is not important.
5. Ignition may have to be cycled "ON" several times to obtain maximum fuel pressure.
6. Damage to ECM injector driver may occur if any injector resistance measures less than 11.6 ohm (internal short to igntion CKT 439).

Diagnostic Aids:

• Water or contamination in fuel system may cause a no start condition during very cold or freezing weather. The engine may start after aproximately 5 minutes in a heated shop. The problem may not recur until an overnight park in very cold or freezing temperatures.
• An EGR valve sticking open can cause a rich air/fuel ratio during cranking. Unless the ECM enters "Clear Flood Mode" at the first indication of a flooding condition, it may result in a no start condition.
• Low fuel pressure can result in a very lean air/fuel ratio, see CHART A-7.
• An A/C pressure sensor with a internal short to ground can cause a no start condition. Disconnect the A/C pressure sensor. In vehicle starts, replace faulty sensor.
• A MAP sensor stuck between .5 and 2.5 volts c, cause a no start condition. Disconnect the MAP sensor. If vehicle starts, replace faulty sensor.

2.2L (VIN 4) ENGINE — SYSTEM DIAGNOSTIC CHARTS — 1992 CAVALIER

CHART A-3

(Page 1 of 3)
ENGINE CRANKS BUT WON'T RUN
2.2L (VIN 4) "J" CARLINE (PORT)

NOTICE: PFI SYSTEM UNDER PRESSURE. TO AVOID FUEL SPILLAGE, REFER TO FIELD SERVICE PROCEDURES FOR TESTING OR MAKING REPAIRS REQUIRING DISASSEMBLY OF FUEL LINES OR FITTING.

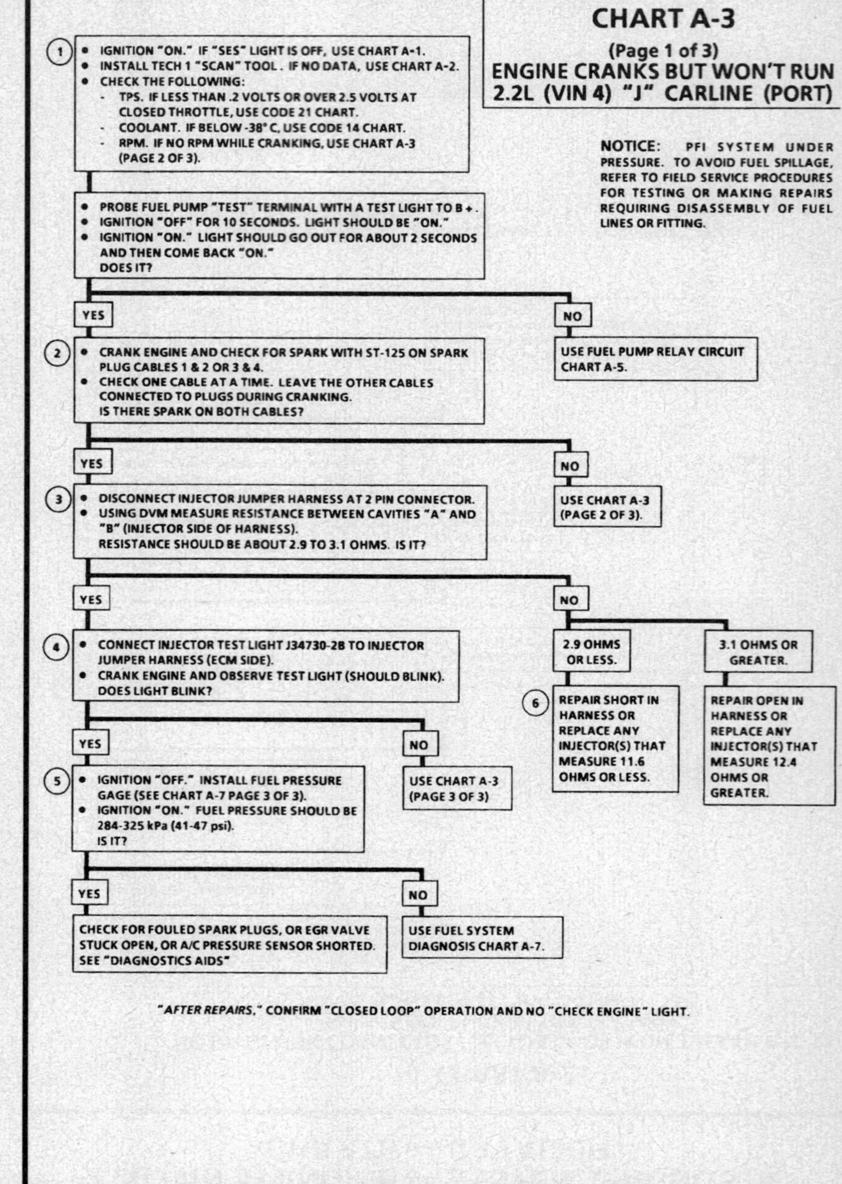

2.2L (VIN 4) ENGINE — SYSTEM DIAGNOSTIC CHARTS — 1992 CAVALIER

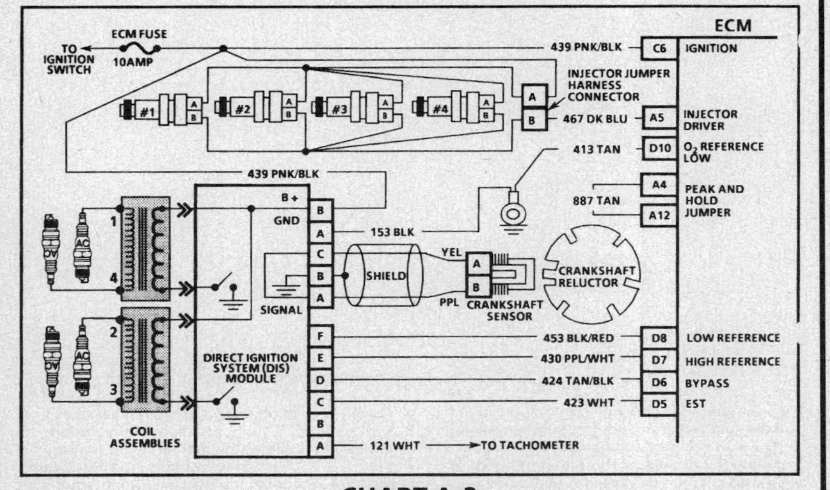

CHART A-3
(Page 2 of 3)
**ENGINE CRANKS BUT WON'T RUN
2.2L (VIN 4) "J" CARLINE (PORT)**

Circuit Description:

A magnetic crank sensor is used to determine engine crankshaft position, much the same way as the pick-up coil did in High Energy Ignition (HEI) type systems. The sensor is mounted in the block, near a slotted wheel on the crankshaft. The rotation of the wheel creates a flux change in the sensor, which produces a voltage signal. The DIS ignition module processes this signal and creates the reference pulses needed by the Electronic Control Module (ECM) to trigger the correct coil at the correct time.

If the "Scan" tool did not indicate cranking rpm, and there is no spark present at the plugs, the problem lies in the Direct Ignition System (DIS) or the power and ground supplies to the module.

Test Description: Number(s) below refer to circled number(s) on the diagnostic chart.

1. The Direct Ignition System (DIS) uses two plugs and cables to complete the circuit of each coil. The other spark plug cable in the circuit must be left connected to create a spark.
2. This test will determine if the 12 volt supply and a good ground is available at the DIS module.
3. This test will determine if the ignition module is not generating the reference pulse, or if the wiring or ECM are at fault. By touching and removing a test light to 12 volts on CKT 430, a reference pulse should be generated. If rpm is indicated, the ECM and wiring are OK.
4. This test will determine if the ignition module is not triggering the problem coil, or if the tested coil is at fault. This test could also be performed by substituting a known good coil. The secondary coil winding can be checked with a Digital Voltmeter (DVM). There should be 5,000 to 10,000 ohms across the coil towers. There should not be any continuity from either coil tower to ground.
5. Checks for continuity of the crank sensor and connections.
6. Normal crank sensor voltage output range is .8 to 1.4 volts (800 to 1400 mV) with a charged battery and engine at room temperature. Minimum output voltage (slow cranking/low battery) can be as low as .3 volt (300 mV).

2.2L (VIN 4) ENGINE — SYSTEM DIAGNOSTIC CHARTS — 1992 CAVALIER

CHART A-3
(Page 2 of 3)
**ENGINE CRANKS BUT WON'T RUN
2.2L (VIN 4) "J" CARLINE (PORT)**

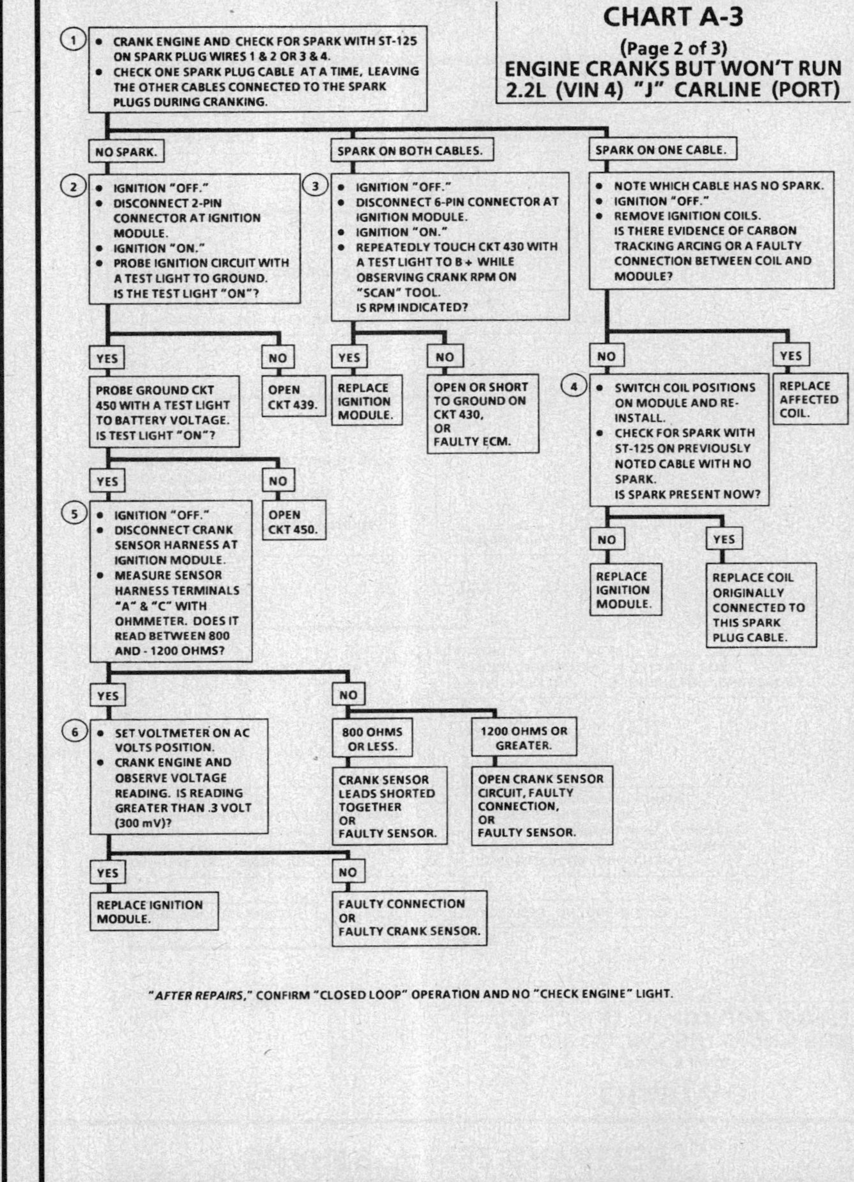

2.2L (VIN 4) ENGINE — SYSTEM DIAGNOSTIC CHARTS — 1992 CAVALIER

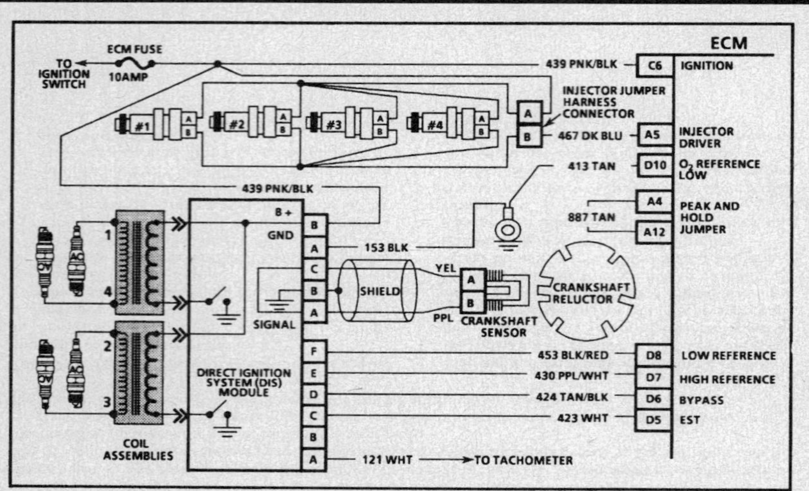

CHART A-3
(Page 3 of 3)
ENGINE CRANKS BUT WON'T RUN
2.2L (VIN 4) "J" CARLINE (PORT)

Test Description: Number(s) below refer to circled number(s) on the diagnostic chart.

1. This test checks for ignition voltage at the injector jumper harness connector.
2. Checks for open in CKT 467 from ECM to injector jumper harness connector.
   The test light has a path to voltage through the injector windings to ignition CKT 439.
3. Checks for short to voltage on CKT 467 from ECM to injectors. Be sure all injectors are disconnected from injector harness connectors.
4. Damage to the ECM injector driver may occur if injector driver CKT 467 shorts to ignition CKT 439.

CHART A-3
(Page 3 of 3)
ENGINE CRANKS BUT WON'T RUN
2.2L (VIN 4) "J" CARLINE (PORT)

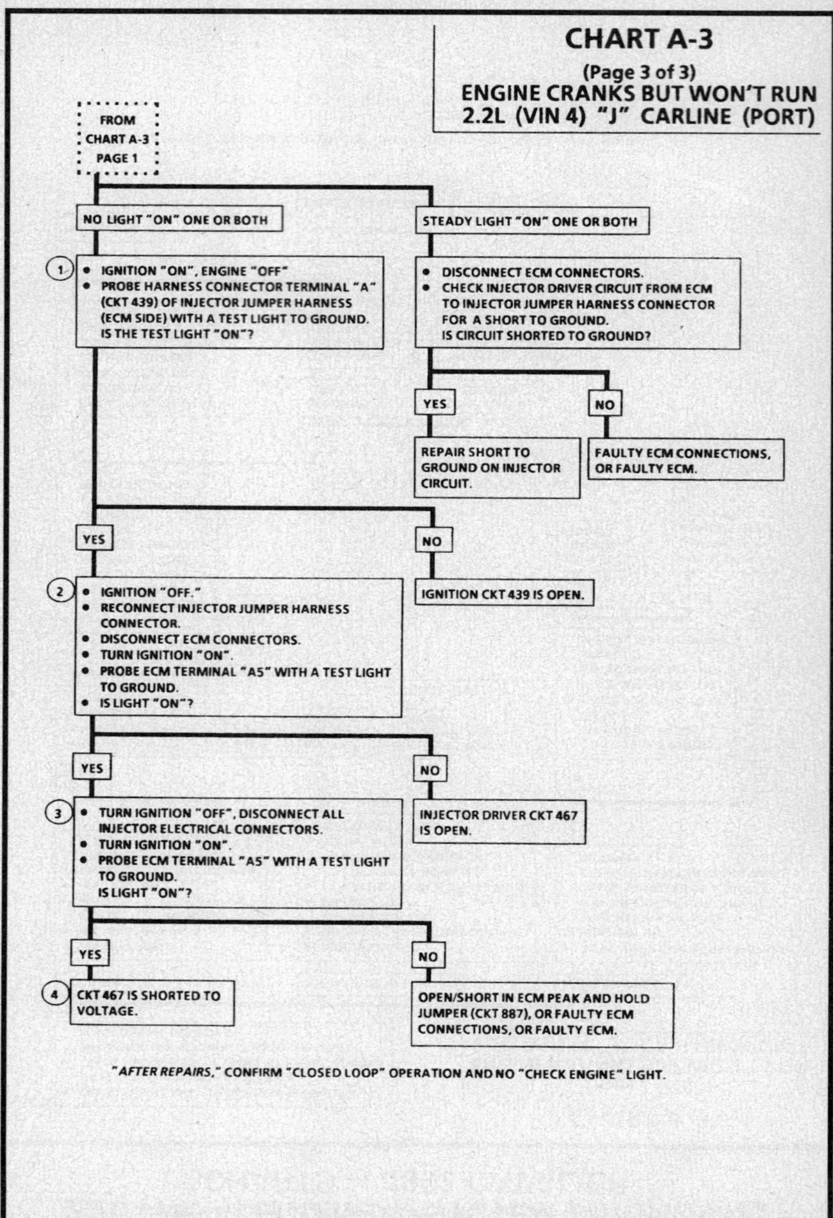

2.2L (VIN 4) ENGINE — SYSTEM DIAGNOSTIC CHARTS — 1993–94 CAVALIER

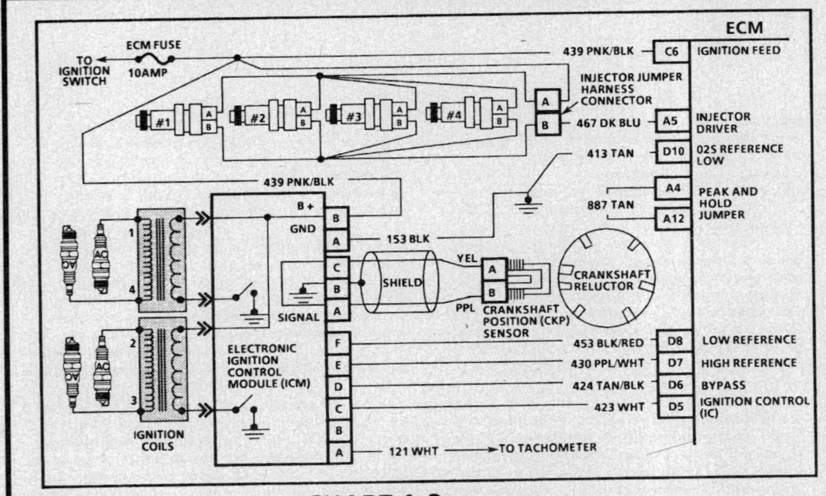

CHART A-3 (Page 1 of 3)
ENGINE CRANKS BUT WON'T RUN
2.2L (VIN 4) "J" CARLINE

Circuit Description:
Before using this chart, battery condition, engine cranking speed, and fuel quantity should be checked and verified as being OK.

Test Description: Number(s) below refer to circled number(s) on the diagnostic chart.

1. A MIL (Check Engine) "ON" is a basic test to determine if there is battery and ignition voltage at the ECM.
- No DLC data may be the result of an ECM problem, and CHART A-2 will diagnose an ECM problem.
- If TP sensor is less than .2 volt, the TP sensor 5 volt reference circuit could be shorted to ground. If TP sensor is over 2.5 volts, the ECM could be in the "Clear Flood Mode" which may cause the engine to not start.
- Compare coolant temperature with intake air temperature when engine is cold. If coolant temperature reading is 10 degrees greater or less than intake air temperature on a cold engine, check resistance of the Engine Coolant Temperature (ECT) sensor circuit or sensor. Compare resistance value to the "Diagnostic Aid" chart found on DTC 14 chart.
2. Because the Electronic Ignition (EI) uses two plugs and wires to complete the circuit of each coil, the opposite spark plug wire should be connected to a good ground. If RPM was indicated during crank, the Electronic Ignition Control Module (ICM) is receiving a crank signal, but "No Spark" at this test indicates the ICM is not triggering the ignition coil.

3. This test is performed with injectors 1, 2, 3, & 4 in parallel.
4. This test light should flash, indicating that the ECM is controlling the injectors. How bright the light flashes is not important.
5. Ignition may have to be cycled "ON" several times to obtain maximum fuel pressure.
6. Damage to ECM injector driver may occur if any injector resistance measures less than 11.6 ohms (internal injector short to ignition CKT 439).

Diagnostic Aids:
- Water or contamination in fuel system may cause a no start condition during very cold or freezing weather. The engine may start after approximately 5 minutes in a heated shop. The problem may not recur until an overnight park in very cold or freezing temperatures.
- Low fuel pressure can result in a very lean air/fuel ratio, refer to CHART A-7.
- An A/C refrigerant pressure sensor with a internal short to ground can cause a no start condition. Disconnect the A/C refrigerant pressure sensor. In vehicle starts, replace faulty sensor.
- A MAP sensor stuck between .5 and 2.5 volts can cause a no start condition. Disconnect the MAP sensor. If vehicle starts, replace faulty sensor.

2.2L (VIN 4) ENGINE — SYSTEM DIAGNOSTIC CHARTS — 1993–94 CAVALIER

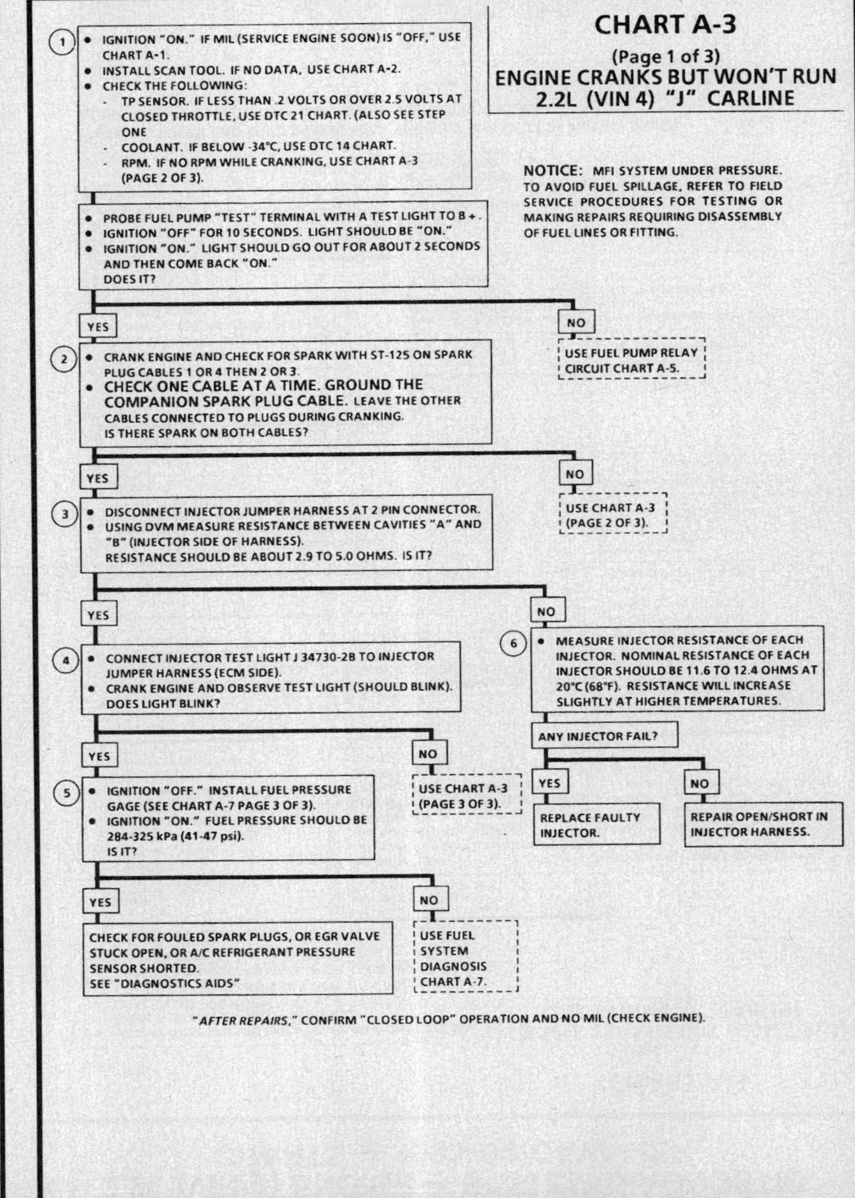

2.2L (VIN 4) ENGINE — SYSTEM DIAGNOSTIC CHARTS — 1993–94 CAVALIER

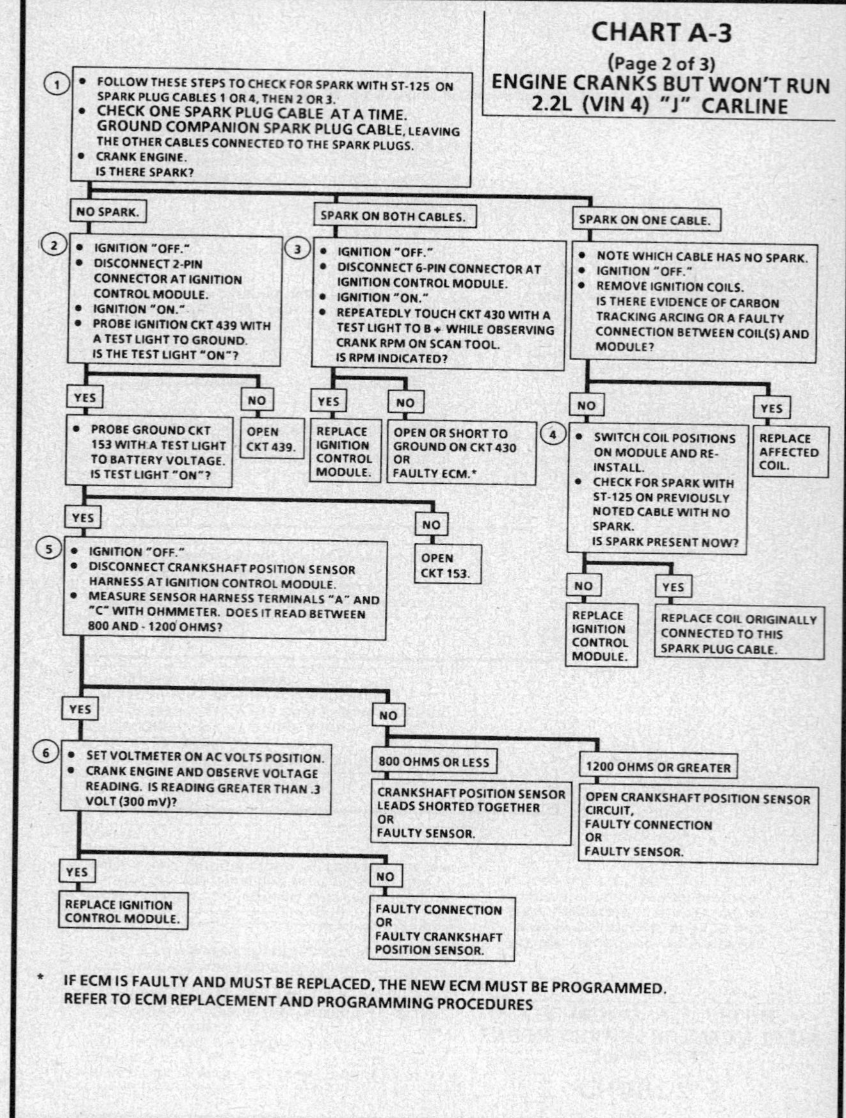

CHART A-3
(Page 2 of 3)
ENGINE CRANKS BUT WON'T RUN
2.2L (VIN 4) "J" CARLINE

Circuit Description:

A magnetic Crankshaft Position (CKP) sensor is used to determine engine crankshaft position, much the same way as the pick-up coil did in Distributor Ignition (DI) type systems. The sensor is mounted in the block, near a slotted wheel on the crankshaft. The rotation of the wheel creates a flux change in the sensor, which produces a voltage signal. The Electronic Ignition Control Module (ICM) processes this signal and creates the reference pulses needed by the Engine Control Module (ECM) to trigger the correct coil at the correct time.

If the scan tool did not indicate cranking RPM, and there is no spark present at the plugs, the problem lies in the ICM, CKP sensor or the power and ground supplies to the ICM.

Test Description: Number(s) below refer to circled number(s) on the diagnostic chart.

1. The Electronic Ignition (EI) system uses two plugs and cables to complete the circuit of each coil. The companion spark plug cable in the circuit must be connected to a good ground.
2. This test will determine if the 12 volt supply and a good ground is available at the Electronic Ignition Control Module (ICM).
3. This test will determine if the ICM is not generating the reference pulse, or if the wiring or ECM are at fault. By touching and removing a test light to 12 volts on CKT 430, a reference pulse should be generated. If RPM is indicated, the ECM and wiring are OK.
4. This test will determine if the ICM is not triggering the problem coil, or if the tested coil is at fault. This test could also be performed by substituting a known good coil. The secondary coil winding can be checked with a Digital Voltmeter (DVM). There should be 5,000 to 10,000 ohms across the coil towers. There should not be any continuity from either coil tower to ground.
5. Checks for continuity of the crankshaft position sensor and connections.
6. Normal crankshaft position sensor voltage output range is .8 to 1.4 volts (800 to 1400 mV) with a fully charged battery and engine at room temperature. Minimum output voltage (slow cranking, low battery) can be as low as .3 volt (300 mV).

2.2L (VIN 4) ENGINE — SYSTEM DIAGNOSTIC CHARTS — 1993–94 CAVALIER

CHART A-3
(Page 2 of 3)
ENGINE CRANKS BUT WON'T RUN
2.2L (VIN 4) "J" CARLINE

1. • FOLLOW THESE STEPS TO CHECK FOR SPARK WITH ST-125 ON SPARK PLUG CABLES 1 OR 4, THEN 2 OR 3.
   • CHECK ONE SPARK PLUG CABLE AT A TIME. GROUND COMPANION SPARK PLUG CABLE, LEAVING THE OTHER CABLES CONNECTED TO THE SPARK PLUGS.
   • CRANK ENGINE. IS THERE SPARK?

NO SPARK.

2. • IGNITION "OFF."
   • DISCONNECT 2-PIN CONNECTOR AT IGNITION CONTROL MODULE.
   • IGNITION "ON."
   • PROBE IGNITION CKT 439 WITH A TEST LIGHT TO GROUND. IS THE TEST LIGHT "ON"?

- YES → • PROBE GROUND CKT 153 WITH A TEST LIGHT TO BATTERY VOLTAGE. IS TEST LIGHT "ON"?
- NO → OPEN CKT 439.

- (Probe ground) YES → (go to 5)
- (Probe ground) NO → OPEN CKT 153.

5. • IGNITION "OFF."
   • DISCONNECT CRANKSHAFT POSITION SENSOR HARNESS AT IGNITION CONTROL MODULE.
   • MEASURE SENSOR HARNESS TERMINALS "A" AND "C" WITH OHMMETER. DOES IT READ BETWEEN 800 AND - 1200 OHMS?

- YES → 6. • SET VOLTMETER ON AC VOLTS POSITION.
   • CRANK ENGINE AND OBSERVE VOLTAGE READING. IS READING GREATER THAN .3 VOLT (300 mV)?
- NO → 800 OHMS OR LESS → CRANKSHAFT POSITION SENSOR LEADS SHORTED TOGETHER OR FAULTY SENSOR. | 1200 OHMS OR GREATER → OPEN CRANKSHAFT POSITION SENSOR CIRCUIT, FAULTY CONNECTION OR FAULTY SENSOR.

- (6) YES → REPLACE IGNITION CONTROL MODULE.
- (6) NO → FAULTY CONNECTION OR FAULTY CRANKSHAFT POSITION SENSOR.

SPARK ON BOTH CABLES.

3. • IGNITION "OFF."
   • DISCONNECT 6-PIN CONNECTOR AT IGNITION CONTROL MODULE.
   • IGNITION "ON."
   • REPEATEDLY TOUCH CKT 430 WITH A TEST LIGHT TO B+ WHILE OBSERVING CRANK RPM ON SCAN TOOL. IS RPM INDICATED?

- YES → REPLACE IGNITION CONTROL MODULE.
- NO → OPEN OR SHORT TO GROUND ON CKT 430 OR FAULTY ECM.*

SPARK ON ONE CABLE.

• NOTE WHICH CABLE HAS NO SPARK.
• IGNITION "OFF."
• REMOVE IGNITION COILS. IS THERE EVIDENCE OF CARBON TRACKING ARCING OR A FAULTY CONNECTION BETWEEN COIL(S) AND MODULE?

- NO → 4. • SWITCH COIL POSITIONS ON MODULE AND RE-INSTALL.
   • CHECK FOR SPARK WITH ST-125 ON PREVIOUSLY NOTED CABLE WITH NO SPARK. IS SPARK PRESENT NOW?
- YES → REPLACE AFFECTED COIL.

- (4) NO → REPLACE IGNITION CONTROL MODULE.
- (4) YES → REPLACE COIL ORIGINALLY CONNECTED TO THIS SPARK PLUG CABLE.

* IF ECM IS FAULTY AND MUST BE REPLACED, THE NEW ECM MUST BE PROGRAMMED. REFER TO ECM REPLACEMENT AND PROGRAMMING PROCEDURES

2.2L (VIN 4) ENGINE — SYSTEM DIAGNOSTIC CHARTS — 1993–94 CAVALIER

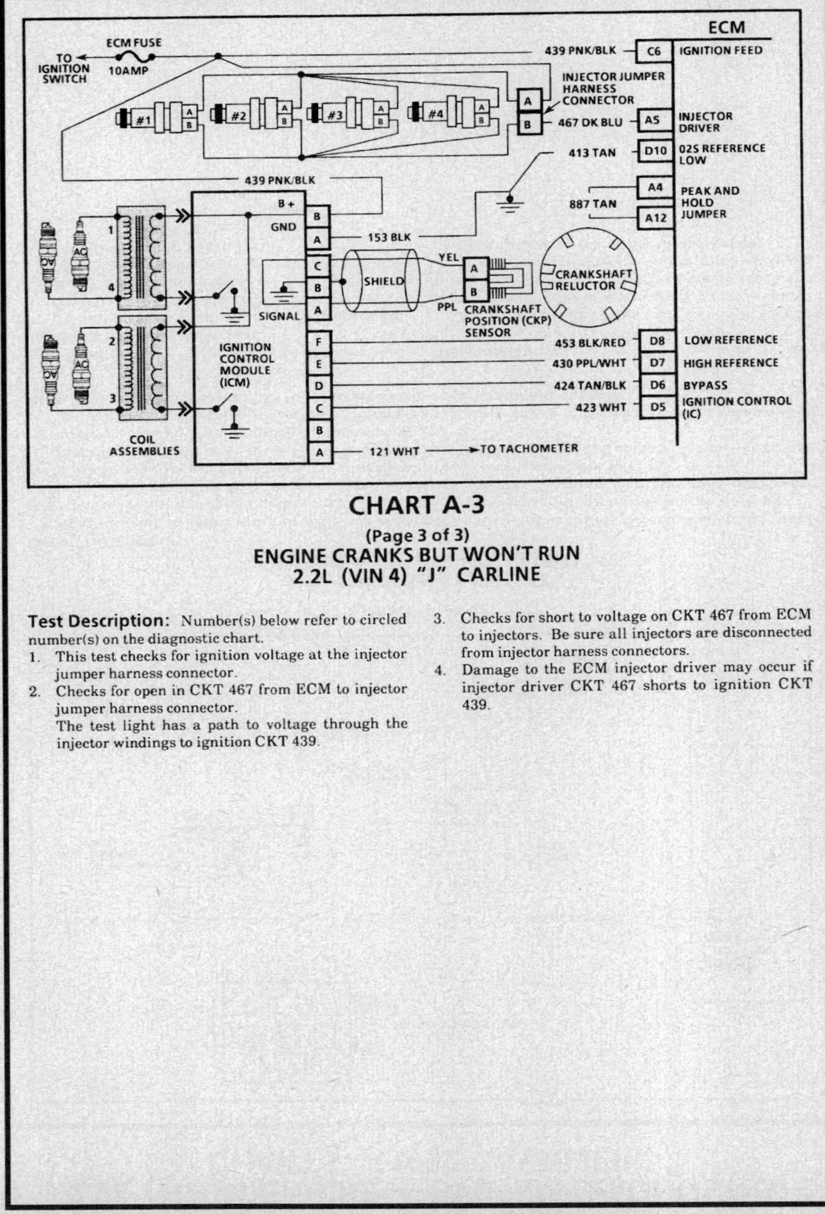

CHART A-3
(Page 3 of 3)
ENGINE CRANKS BUT WON'T RUN
2.2L (VIN 4) "J" CARLINE

Test Description: Number(s) below refer to circled number(s) on the diagnostic chart.

1. This test checks for ignition voltage at the injector jumper harness connector.
2. Checks for open in CKT 467 from ECM to injector jumper harness connector. The test light has a path to voltage through the injector windings to ignition CKT 439.
3. Checks for short to voltage on CKT 467 from ECM to injectors. Be sure all injectors are disconnected from injector harness connectors.
4. Damage to the ECM injector driver may occur if injector driver CKT 467 shorts to ignition CKT 439.

2.2L (VIN 4) ENGINE — SYSTEM DIAGNOSTIC CHARTS — 1993–94 CAVALIER

CHART A-3
(Page 3 of 3)
ENGINE CRANKS BUT WON'T RUN
2.2L (VIN 4) "J" CARLINE

FROM CHART A-3 PAGE 1.

NO LIGHT "ON"

(1) IGNITION "ON," ENGINE "OFF." PROBE HARNESS CONNECTOR TERMINAL "A" (CKT 439) OF INJECTOR JUMPER HARNESS (ECM SIDE) WITH A TEST LIGHT TO GROUND. IS THE TEST LIGHT "ON"?

STEADY LIGHT "ON"

DISCONNECT ECM CONNECTORS. CHECK INJECTOR DRIVER CIRCUIT FROM ECM TO INJECTOR JUMPER HARNESS CONNECTOR FOR A SHORT TO GROUND. IS CIRCUIT SHORTED TO GROUND?

YES — (2) IGNITION "OFF." RECONNECT INJECTOR JUMPER HARNESS CONNECTOR. DISCONNECT ECM CONNECTORS. TURN IGNITION "ON." PROBE ECM TERMINAL "A5" WITH A TEST LIGHT TO GROUND. IS LIGHT "ON"?

NO — IGNITION CKT 439 IS OPEN.

YES — REPAIR SHORT TO GROUND ON INJECTOR CIRCUIT.

NO — FAULTY ECM CONNECTIONS OR FAULTY ECM.*

YES — (3) TURN IGNITION "OFF," DISCONNECT ALL INJECTOR ELECTRICAL CONNECTORS. TURN IGNITION "ON." PROBE ECM TERMINAL "A5" WITH A TEST LIGHT TO GROUND. IS LIGHT "ON"?

NO — INJECTOR DRIVER CKT 467 IS OPEN.

YES — (4) CKT 467 IS SHORTED TO VOLTAGE.

NO — OPEN/SHORT IN ECM PEAK AND HOLD JUMPER (CKT 887) OR FAULTY ECM CONNECTIONS OR FAULTY ECM.*

* IF ECM IS FAULTY AND MUST BE REPLACED, THE NEW ECM MUST BE PROGRAMMED. REFER TO ECM REPLACEMENT AND PROGRAMMING PROCEDURES

"AFTER REPAIRS," CONFIRM "CLOSED LOOP" OPERATION AND NO MIL (CHECK ENGINE).

2.2L (VIN 4) ENGINE — SYSTEM DIAGNOSTIC CHARTS — 1992 CAVALIER

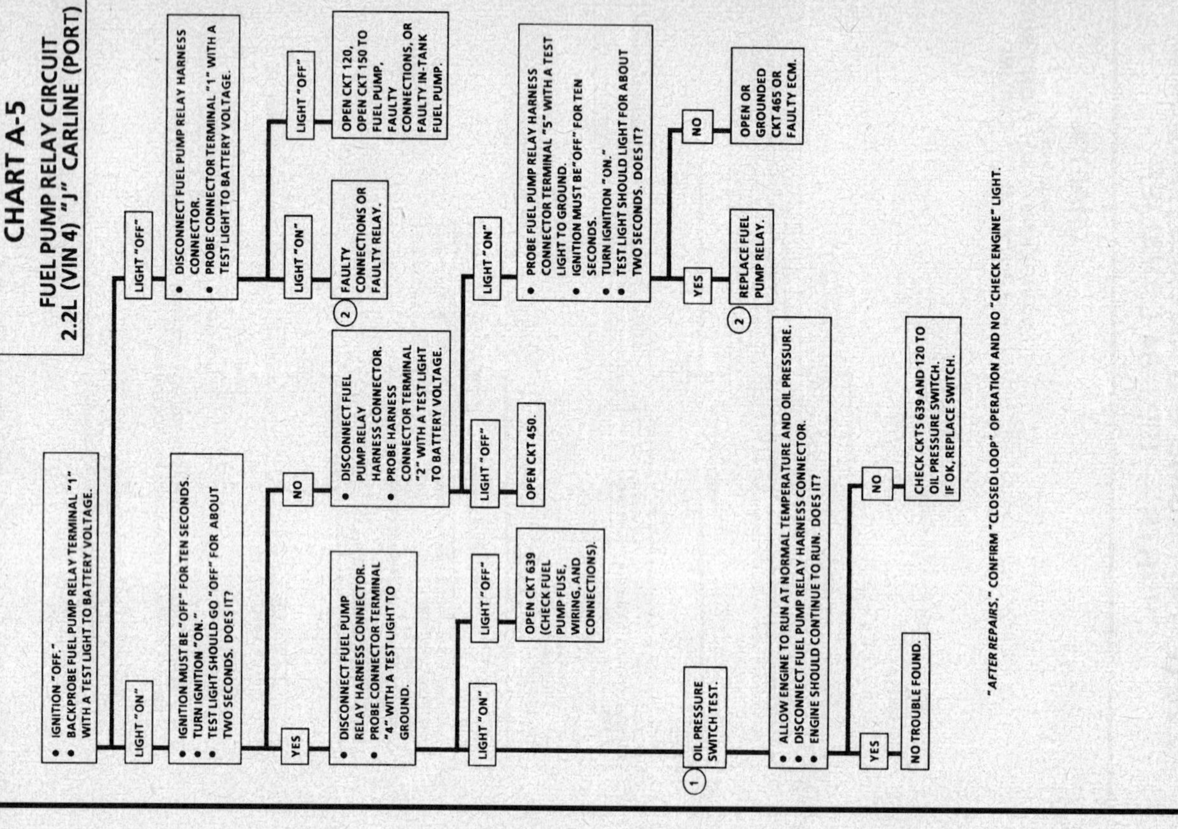

CHART A-5
FUEL PUMP RELAY CIRCUIT
2.2L (VIN 4) "J" CARLINE (PORT)

Circuit Description:

When the ignition switch is turned "ON," the Electronic Control Module (ECM) will activate the fuel pump relay with a 12 volt signal and run the in-tank fuel pump. The fuel pump will operate as long as the engine is cranking or running and the ECM is receiving ignition reference pulses. If there are no ignition reference pulses, the ECM will no longer supply the fuel pump relay signal within 2 seconds after key "ON."

Should the fuel pump relay or the 12 volt relay drive from the ECM fail, the fuel pump will receive electrical current through the oil pressure switch back-up circuit.

The fuel pump test terminal is located in the driver's side of the engine compartment. When the engine is stopped, the pump can be turned "ON" by applying battery voltage to the test terminal.

Test Description: Number(s) below refer to circled number(s) on the diagnostic chart.

1. At this point, the fuel pump relay is operating correctly. The back-up circuit through the oil pressure switch is now tested.
2. After the fuel pump relay is replaced, continue with "Oil Pressure Switch Test."

Diagnostic Aids:

An inoperative fuel pump relay can result in long cranking times. The extended crank period is caused by the time necessary for oil pressure to reach the pressure required to close the oil pressure switch and supply the necessary current for the fuel pump.

2.2L (VIN 4) ENGINE — SYSTEM DIAGNOSTIC CHARTS — 1993–94 CAVALIER

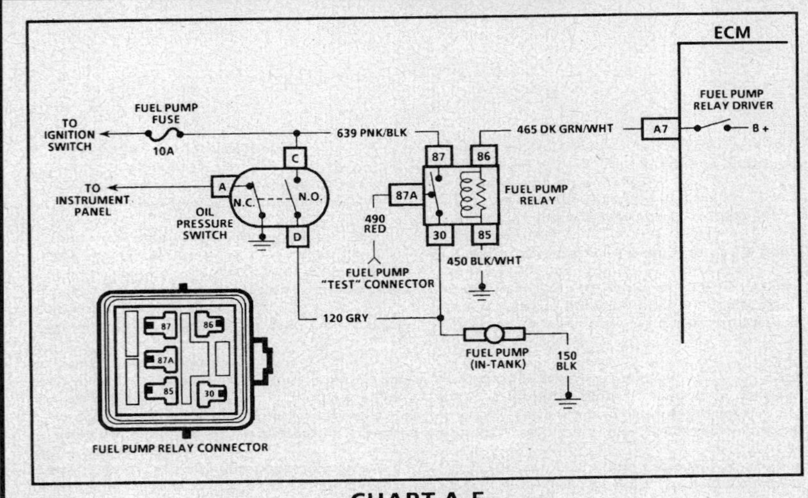

CHART A-5
FUEL PUMP RELAY CIRCUIT
2.2L (VIN 4) "J" CARLINE

Circuit Description:

When the ignition switch is turned "ON," the Engine Control Module (ECM) will activate the fuel pump relay with a 12 volt supply and run the in-tank fuel pump. The fuel pump will operate as long as the engine is cranking or running and the ECM is receiving ignition reference pulses. If there are no ignition reference pulses, the ECM will no longer supply the fuel pump relay signal within 2 seconds after key "ON."

Should the fuel pump relay or the 12 volt relay drive from the ECM fail, the fuel pump will receive supply current through the oil pressure switch back-up circuit.

The fuel pump "test" terminal is located in the driver's side of the engine compartment. When the engine is stopped, the pump can be turned "ON" by applying battery voltage to the "test" terminal.

Test Description: Number(s) below refer to circled number(s) on the diagnostic chart.

1. At this point, the fuel pump relay is operating correctly. The back-up circuit through the oil pressure switch is now tested.
2. After the fuel pump relay is replaced, continue with "Oil Pressure Switch Test."

Diagnostic Aids:

An inoperative fuel pump relay can result in long cranking times. The extended crank period is caused by the time necessary for oil pressure to reach the pressure required to close the oil pressure switch and supply the necessary current for the fuel pump.

If the fuel pump relay circuit checks out OK, refer to "Fuel System Diagnosis," CHART A-7.

Excess fuel may also cause long cranking times. This would usually be accompanied by a start that is not as fast as normal (once fired, the engine does not build up speed as fast) and a puff of black smoke at the tailpipe. An improperly connected or faulty EVAP canister control valve can cause this problem. Disconnect the "EVAP purge hose" from the "EVAP canister control valve" to diagnose. Refer to "Evaporative Emission (EVAP) Control System,"

One or more leaking fuel injectors may also extend the cranking time. Perform the "Injector Balance Test" in CHART C2-A.

2.2L (VIN 4) ENGINE — SYSTEM DIAGNOSTIC CHARTS — 1993–94 CAVALIER

CHART A-5
FUEL PUMP RELAY CIRCUIT
2.2L (VIN 4) "J" CARLINE

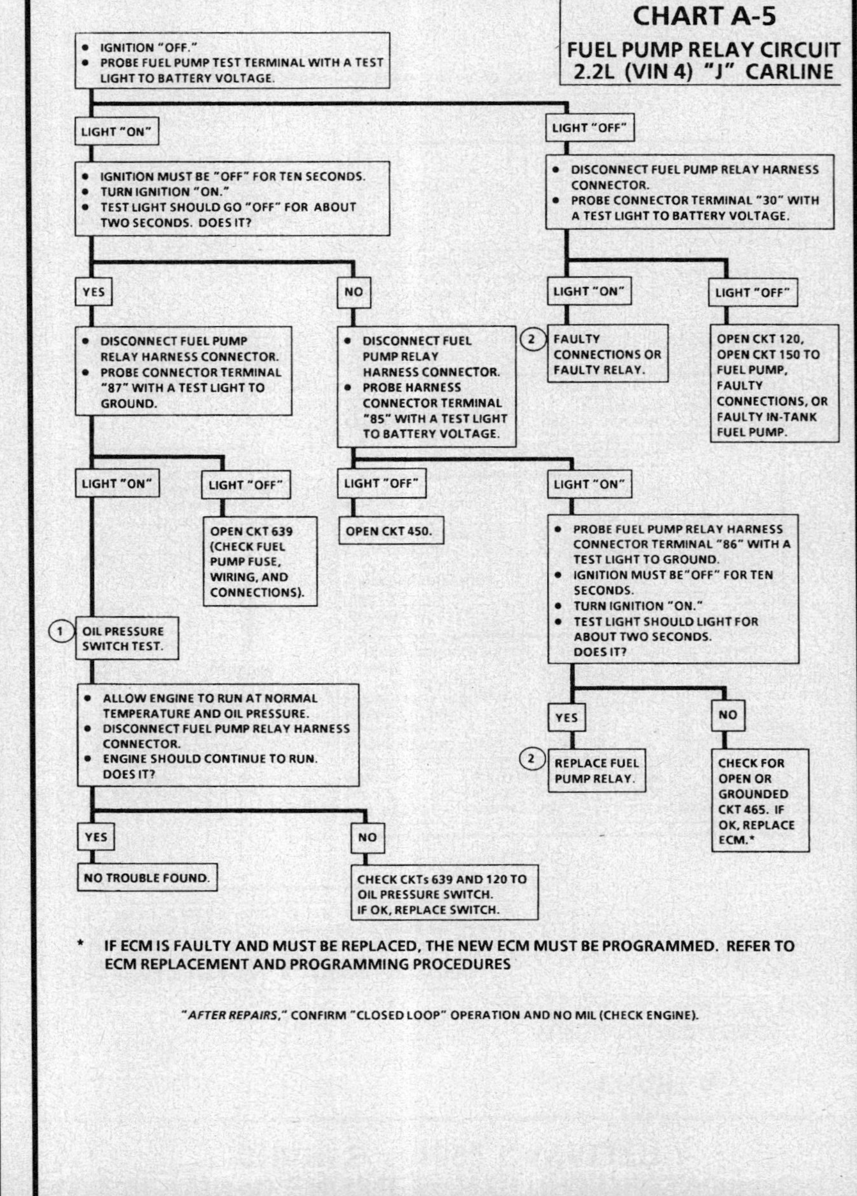

2.2L (VIN 4) ENGINE — SYSTEM DIAGNOSTIC CHARTS — 1992 CAVALIER

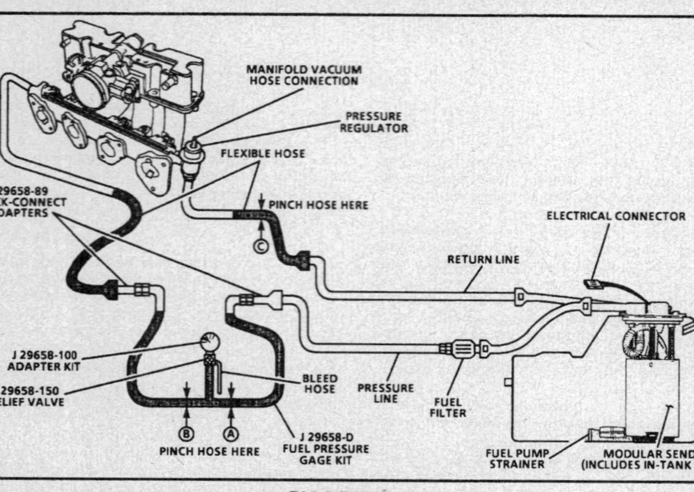

CHART A-7
(Page 1 of 3)
FUEL SYSTEM DIAGNOSIS
2.2L (VIN 4) "J" CARLINE (PORT)

Circuit Description:

When the ignition switch is turned "ON," the Electronic Control Module (ECM) will turn "ON" the in-tank fuel pump. It will remain "ON" as long as the engine is cranking or running, and the ECM is receiving reference pulses. If there are no reference pulses, the ECM will shut "OFF" the fuel pump within 2 seconds after ignition "ON" or engine stops.

An electric fuel pump, part of the modular fuel sender and located inside the fuel tank, pumps fuel through an in-line filter to the fuel passage within the lower manifold assembly. The pump is designed to provide fuel at pressure above the regulated pressure needed by the injectors. A pressure regulator, attached to the lowe. manifold assembly, keeps fuel available to the injectors at a regulated pressure. Unused fuel is returned to the fuel tank by a separate line.

Test Description: Number(s) below refer to circled number(s) on the diagnostic chart.

1. Install fuel pressure gage per instructions on Page 3 of 3. Ignition "ON" pump pressure should be 284-325 kPa to (41-47 psi). This pressure is controlled by spring pressure within the regulator assembly.
2. When the engine is idling, the manifold pressure is low (high vacuum) and is applied to the fuel regulator diaphragm. This will offset the spring and result in a lower fuel pressure.

This idle pressure will vary somewhat depending on barometric pressure, however, the pressure idling should be less, indicating pressure regulator control.

3. Pressure that continues to fall is caused by one of the following:
   - In-tank fuel pump check valve not holding.
   - Fuel pressure regulator valve leaking.
   - Injector(s) sticking open.
4. An injector sticking open can best be determined by checking for a fouled or saturated spark plug(s).

2.2L (VIN 4) ENGINE — SYSTEM DIAGNOSTIC CHARTS — 1992 CAVALIER

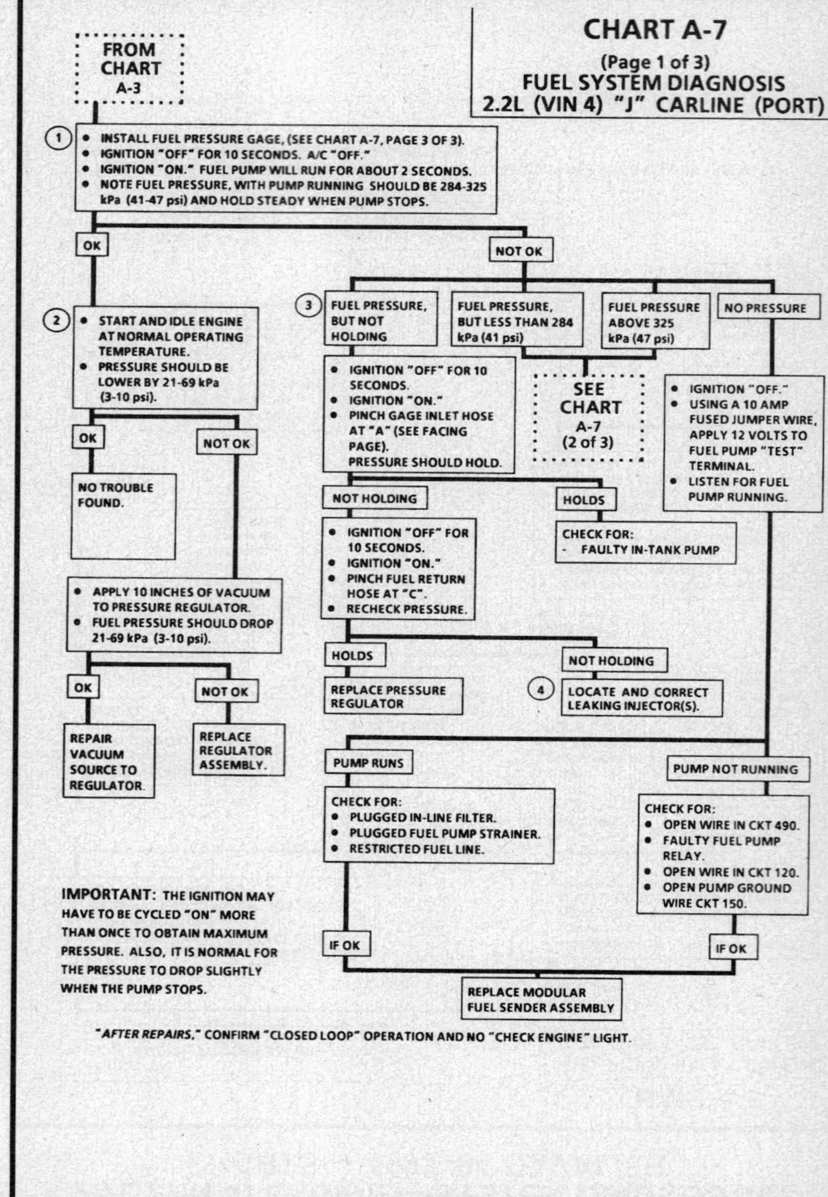

2.2L (VIN 4) ENGINE — SYSTEM DIAGNOSTIC CHARTS — 1992 CAVALIER

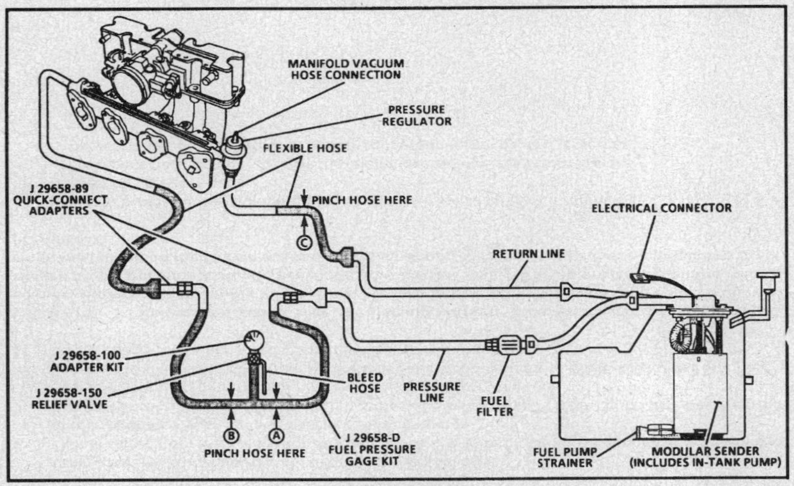

CHART A-7
(Page 2 of 3)
FUEL SYSTEM DIAGNOSIS
2.2L (VIN 4) "J" CARLINE (PORT)

Test Description: Number(s) below refer to circled number(s) on the diagnostic chart.

1. Pressure below 284 kPa (41 psi) may cause a lean condition and may set a Code 44 or Code 32. It could also cause hard starting cold and poor driveability. Low enough pressure will cause the engine not to run at all. Restricted flow may allow the engine to run at idle, or low speeds, but may cause a surge and stall when more fuel is required, as when accelerating or driving at high speeds.

2. Restricting fuel flow at the fuel pressure gage (at B) causes fuel pressure to build above regulated pressure. With battery voltage applied to the pump "test" terminal, pressure should rise above 325 kPa (47 psi) as the gage outlet hose is restricted.

   NOTICE: Do not allow pressure to exceed 414 kPa (60 psi), as damage to the regulator may result.

3. This test determines if the high fuel pressure is due to a restricted fuel return line, or a faulty pressure regulator.

2.2L (VIN 4) ENGINE — SYSTEM DIAGNOSTIC CHARTS — 1992 CAVALIER

CHART A-7
(Page 2 of 3)
FUEL SYSTEM DIAGNOSIS
2.2L (VIN 4) "J" CARLINE (PORT)

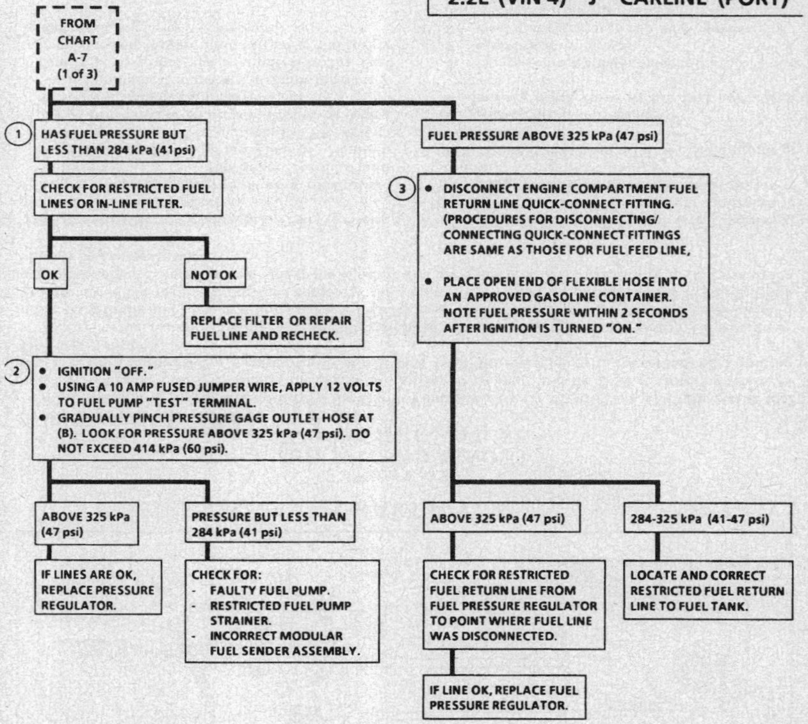

"AFTER REPAIRS," CONFIRM "CLOSED LOOP" OPERATION AND NO "CHECK ENGINE" LIGHT.

CHART A-7

(Page 3 of 3)
FUEL SYSTEM DIAGNOSIS
2.2L (VIN 4) "J" CARLINE (PORT)

FUEL PRESSURE CHECK

Tools Required: J 29658-D - Fuel Pressure Gage Kit
J 29658-150 - Fuel Pressure Gage Relief Valve
J 29658-100 - TBI Pressure Gage Modification Kit
J 29658-89 - Fuel Pressure Quick Connect Adapters
J 37088 - A - Fuel Line Quick-Connect Separators

CAUTION: To Reduce the Risk of Fire and Personal Injury:
- It is necessary to relieve fuel system pressure before connecting a fuel pressure gage.
- After relieving system pressure, a small amount of fuel may be released when disconnecting the fuel lines. Cover fuel line fittings with a shop towel before disconnecting, to catch any fuel that may leak out. Place towel in approved container when disconnect is completed.
- Do not pinch or restrict nylon fuel lines to avoid severing, which could cause a fuel leak.

NOTICE: • If nylon fuel lines become kinked, and cannot be straightened, they must be replaced.

1. Loosen fuel filler cap to relieve fuel tank pressure. (Do not tighten at this time.)
2. Remove fuel pump fuse.
3. Start and run engine until fuel supply remaining in fuel pipes is consumed. Engage starter for three seconds to assure relief of any remaining pressure.
4. Disconnect negative battery cable.
5. Locate engine compartment fuel feed quick-connect fitting.
6. Grasp female end of fitting and twist ¼ turn in each direction to loosen any dirt in fitting.

CAUTION: Safety glasses must be worn when using compressed air, as flying dirt particles may cause eye injury.

7. Using compressed air, blow dirt out of quick-connect fitting.
8. Choose correct tool from separator tool set J 37088-A for size of fitting. Insert tool into female end of connector, then push inward to release male connector.
9. If not previously installed, connect Fuel Pressure Gage Relief Valve J 29658-150, to fuel pressure gage hose assembly.
10. Connect 414 kPa (60 psi) gage from TBI Pressure Gage Modification kit J 29658-100 to hose assembly.
11. Connect gage quick-connect adapters J 29658-89 to hose assembly.

CAUTION: To Reduce the Risk of Fire and Personal Injury: Before connecting fuel line quick-connect fittings, always apply a few drops of clean engine oil to the male tube ends. This will ensure proper reconnection and prevent a possible fuel leak. (During normal operation, the O-rings located inside the female connector will swell and may prevent proper reconnection if not lubricated.)

12. Lubricate the male tube end of the fuel line and the gage adapter with engine oil.
13. Connect fuel pressure gage.
    - Push connectors together to cause the retaining tabs/fingers to snap into place.
    - Once installed, pull on both ends of each connection to make sure it is secure.

CHART A-7

(Page 1 of 3)
FUEL SYSTEM DIAGNOSIS
2.2L (VIN 4) "J" CARLINE

Circuit Description:
When the ignition switch is turned "ON," the Engine Control Module (ECM) will turn "ON" the in-tank fuel pump. It will remain "ON" as long as the engine is cranking or running, and the ECM is receiving reference pulses. If there are no reference pulses, the ECM will shut "OFF" the fuel pump within 2 seconds after ignition "ON" or engine stops.

An electric fuel pump, part of the modular fuel sender and located inside the fuel tank, pumps fuel through an in-line filter to the fuel passage within the lower manifold assembly. The pump is designed to provide fuel at a pressure above the regulated pressure needed by the injectors. A pressure regulator, attached to the lower manifold assembly, keeps fuel available to the injectors at a regulated pressure. Unused fuel is returned to the fuel tank by a separate line.

Test Description: Number(s) below refer to circled number(s) on the diagnostic chart.
1. Install fuel pressure gage as shown in illustration. Refer to   "Fuel Relief Procedure" and for "Servicing Quick-Connect Fittings." Ignition "ON," pump pressure should be 284-325 kPa to (41-47 psi). This pressure is controlled by spring pressure within the regulator assembly.
2. When the engine is idling, the manifold pressure is low (high vacuum) and is applied to the fuel regulator diaphragm. This will offset the spring and result in a lower fuel pressure.

This idle pressure will vary somewhat depending on barometric pressure, however, the pressure idling should be less, indicating pressure regulator control.
3. Pressure that continues to fall is caused by one of the following:
   - Leaking fuel pump feed hose.
   - Leaking check valve within fuel pump feed hose.
   - Fuel pressure regulator valve leaking.
   - Injector(s) sticking open.
4. An injector sticking open can best be determined by checking for a fouled or saturated spark plug(s).

2.2L (VIN 4) ENGINE — SYSTEM DIAGNOSTIC CHARTS — 1993–94 CAVALIER

CHART A-7
(Page 1 of 3)
FUEL SYSTEM DIAGNOSIS
2.2L (VIN 4) "J" CARLINE

FROM CHART A-3

1. • INSTALL FUEL PRESSURE GAGE AS SHOWN (SEE CHART A-7, PAGE 3 OF 3 FOR FUEL PRESSURE RELIEF PROCEDURE AND FOR SERVICING QUICK-CONNECT FITTINGS).
• IGNITION "OFF" FOR 10 SECONDS. A/C "OFF."
• IGNITION "ON." FUEL PUMP WILL RUN FOR ABOUT 2 SECONDS (SEE FACING PAGE).
THE IGNITION MAY HAVE TO BE CYCLED "ON" MORE THAN ONCE TO OBTAIN MAXIMUM PRESSURE.
• NOTE FUEL PRESSURE, WITH PUMP RUNNING, PRESSURE SHOULD BE 284-325 kPa (41-47 psi). WHEN PUMP STOPS, PRESSURE MAY VARY SLIGHTLY THEN SHOULD HOLD STEADY. IS PRESSURE CORRECT AND DOES IT HOLD?

YES → IF FUEL PRESSURE IS WITHIN NORMAL RANGE, BUT IS SUSPECTED OF DROPPING OFF DURING ACCELERATION, CRUISE OR HARD CORNERING SEE CHART A-7 (2 OF 3).

2. • START ENGINE AND ALLOW IT TO IDLE AT NORMAL OPERATING TEMPERATURE.
• FUEL PRESSURE NOTED IN STEP 1 SHOULD DROP 21-69 kPa (3-10 psi). DOES IT?

YES → NO TROUBLE FOUND.

NO → • DISCONNECT VACUUM HOSE FROM PRESSURE REGULATOR ASSEMBLY.
• WITH ENGINE IDLING, APPLY 12-14 INCHES OF VACUUM TO PRESSURE REGULATOR. FUEL PRESSURE NOTED IN STEP 1 SHOULD DROP 21-69 kPa (3-10 psi). DOES IT?

YES → LOCATE AND REPAIR LOSS OF VACUUM TO PRESSURE REGULATOR.

NO → REPLACE PRESSURE REGULATOR.

NO →

3. FUEL PRESSURE WITHIN SPEC., BUT DOES NOT HOLD.

• IGNITION "OFF."
• USING A 10 AMP FUSED JUMPER WIRE, APPLY B + TO FUEL PUMP "TEST" CONNECTOR FOR 2 SECONDS.
• BLOCK GAGE INLET HOSE BY PINCHING AT "A" (SEE FACING PAGE). PRESSURE SHOULD HOLD. DOES IT?

NO → • USING A 10 AMP FUSED JUMPER WIRE, APPLY B + TO FUEL PUMP TEST CONNECTOR FOR 2 SECONDS.
• BLOCK FUEL RETURN LINE BY PINCHING FLEXIBLE HOSE AT "C". PRESSURE SHOULD HOLD. DOES IT?

NO → 4. LOCATE AND CORRECT LEAKING INJECTOR(s).

YES → REPLACE PRESSURE REGULATOR.

YES (from step 3 second box) → CHECK FOR:
• LEAKING FUEL PUMP FEED HOSE.
• LEAKING CHECK VALVE WITHIN FUEL PUMP FEED HOSE.

FUEL PRESSURE OUT OF SPEC.
SEE CHART A-7 (2 OF 3).

FROM CHART A-5

NO PRESSURE, FUEL PUMP ELECTRICAL CIRCUIT OK.

CHECK FOR:
• PLUGGED IN-LINE FILTER.
• RESTRICTED FUEL PRESSURE LINE.
• PLUGGED FUEL SENDER STRAINER.
• PLUGGED FUEL PUMP STRAINER.
• LEAKING FUEL PUMP FEED HOSE.

IF OK → REPLACE FUEL PUMP.

"AFTER REPAIRS," CONFIRM "CLOSED LOOP" OPERATION AND NO MIL (SERVICE ENGINE SOON).

2.2L (VIN 4) ENGINE — SYSTEM DIAGNOSTIC CHARTS — 1993–94 CAVALIER

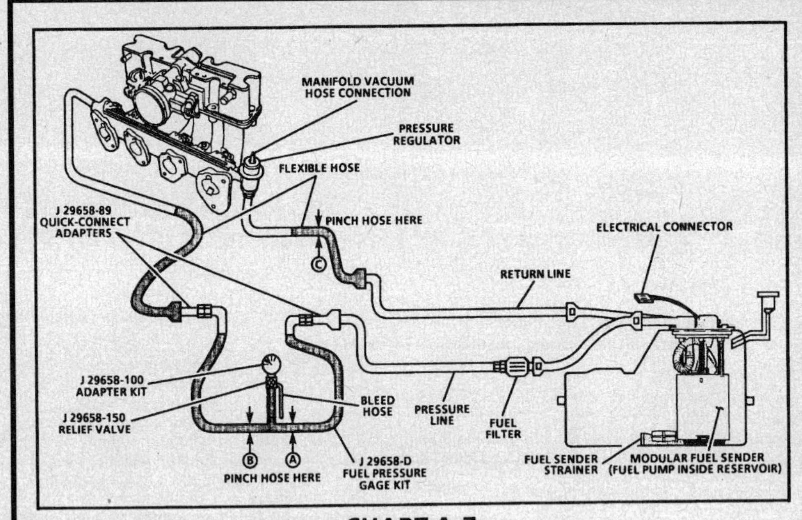

CHART A-7
(Page 2 of 3)
FUEL SYSTEM DIAGNOSIS
2.2L (VIN 4) "J" CARLINE

Test Description: Number(s) below refer to circled number(s) on the diagnostic chart.

5. Fuel pressure that drops off during acceleration, cruise or hard cornering may cause a lean condition and result in a loss of power, surging or misfire. This condition can be diagnosed using a Tech 1 scan tool. If the fuel system is very lean, the O2S will stop toggling and output voltage will drop below 500 mV. Also, injector pulse width will increase.

Important
• Make sure system is not operating at "Fuel Cut-Off" which may cause false readings on the scan tool.

6. Fuel pressure below 284 kPa (41 psi) may cause a lean condition and may set a DTC 44 or a DTC 32. Driveability conditions can include hard starting cold, hesitation, poor driveability, lack of power, surging or misfire.

7. Restricting fuel flow at the fuel pressure gage (at "B") causes fuel pressure to build above regulated pressure. With battery voltage applied the pump "test" connector, pressure should rise above 325 kPa (47 psi) as the gage outlet hose is pinched.

NOTICE: Do not allow pressure to exceed 414 kPa (60 psi), as damage to the regulator may result.

8. Fuel pressure above 325 kPa (47 psi) may cause a rich condition and may set a DTC 45. Driveability conditions can include hard starting (followed by black smoke) and a strong sulphur smell in the exhaust.

9. This test determines if the high fuel pressure is due to a restricted fuel return line or a faulty fuel pressure regulator.

2.2L (VIN 4) ENGINE — SYSTEM DIAGNOSTIC CHARTS — 1993–94 CAVALIER

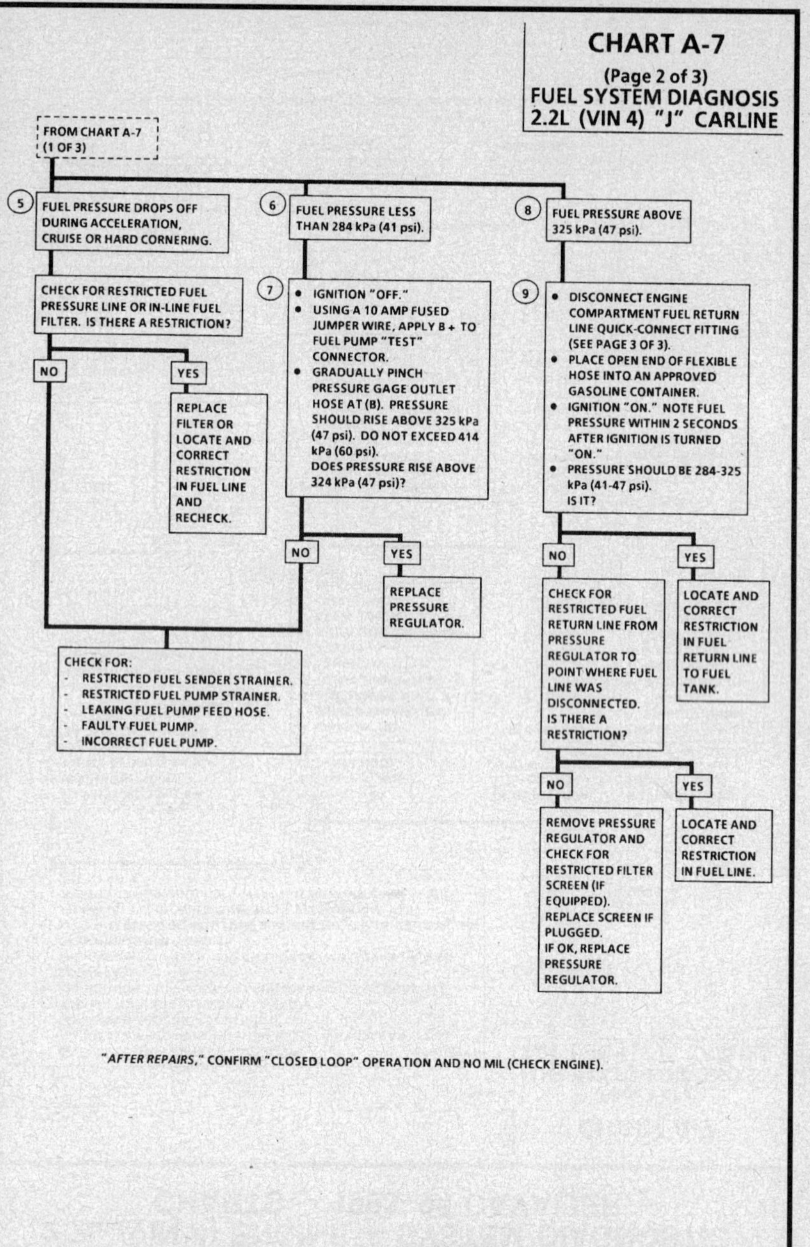

FUEL SYSTEM PRESSURE RELIEF PROCEDURE

MFI Engines Without Fuel Pressure Connection (J Car Only)

(Must Be Performed Before Disconnecting Fuel Line Fittings)

CAUTION:
- To reduce the risk of fire and personal injury, it is necessary to relieve fuel system pressure before disconnecting fuel line fittings.
- After relieving system pressure, a small amount of fuel may be released when disconnecting fuel line fittings. In order to reduce the chance of personal injury, cover fuel line fittings with a shop towel before disconnecting, to catch any fuel that may leak out. Place the towel in an approved container when disconnect is completed.

1. Loosen fuel filler cap to relieve tank pressure.
2. Remove fuel pump fuse.
3. Start engine and run until fuel supply remaining in fuel pipes is consumed. Engage starter for 3.0 seconds to assure relief of any remaining pressure.
4. Disconnect negative battery cable to avoid possible fuel discharge if an accidental attempt is made to start the engine.
5. Fuel line fittings are now safe for servicing.
6. Perform service required.
   - If performing fuel pressure check with a gage equipped with a bleed hose, fuel pressure can be relieved through the gage following test. Place bleed hose into approved gasoline container and open valve to bleed system pressure.
7. Tighten fuel filler cap.
8. Connect negative battery cable.
9. Install fuel pump fuse.
10. Cycle ignition "ON" and "OFF" twice, waiting ten seconds between cycles, then check for fuel leaks.

2.2L (VIN 4) ENGINE — SYSTEM DIAGNOSTIC CHARTS — 1993–94 CAVALIER

CHART A-7
(Page 3 of 3)
FUEL SYSTEM DIAGNOSIS
2.2L (VIN 4) "J" CARLINE

SERVICING QUICK-CONNECT FITTINGS

Tool required for metal quick-connect fittings:
J 37088-A tool set, fuel line quick-connect separator;
J 39504 tool set, fuel line quick-connect separator (restricted access).

Remove or Disconnect

1. Grasp both sides of fitting. Twist female connector 1/4 turn in each direction to loosen any dirt within fitting.

CAUTION: Safety glasses must be worn when using compressed air, as flying dirt particles may cause eye injury.

2. Using compressed air, blow dirt out of fitting.
-*For plastic (Hand Releasable) Fitting-*
3. Squeeze plastic retainer release tabs.
4. Pull connection apart.
OR
-*For Metal Fitting-*
3. Choose correct tool from J 37088-A or J 39504 tool set for size of fitting. Insert tool into female connector, then push/pull inward to release locking tabs.
4. Pull connection apart.

Clean and Inspect

NOTICE: If it is necessary to remove rust or burrs from fuel pipe, use emery cloth in a radial motion with the pipe end to prevent damage to O-ring sealing surface.

- Using a clean shop towel, wipe off male pipe end.
- Inspect both ends of fitting for dirt and burrs. Clean or replace components/assemblies as required.

Install or Connect

CAUTION: To Reduce the Risk of Fire and Personal Injury:
- Before connecting fitting, always apply a few drops of clean engine oil to the male pipe end of engine fuel pipe, pressure gage adapter or fuel line shut-off adapter. This will ensure proper reconnection and prevent a possible fuel leak. (During normal operation, the O-rings located in the female connector will swell and may prevent proper reconnection if not lubricated.)

1. Apply a few drops of clean engine oil to the male pipe end of engine fuel pipe, pressure gage adapter or fuel line shut-off adapter.
2. Push both sides of fitting together to cause the retaining tabs/fingers to snap into place.
3. Once installed, pull on both sides of fitting to make sure connection is secure.

"AFTER REPAIRS," CONFIRM "CLOSED LOOP" OPERATION AND NO MIL (CHECK ENGINE).

2.2L (VIN 4) ENGINE — DIAGNOSTIC TROUBLE CODE CHART — CAVALIER

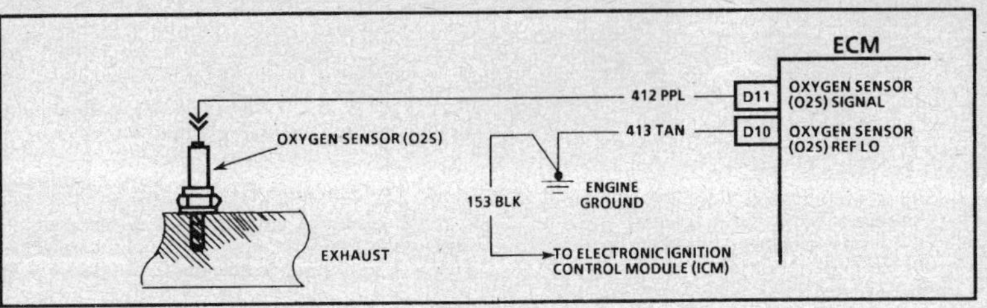

DTC 13
OXYGEN SENSOR (O2S) CIRCUIT
(OPEN CIRCUIT)
2.2L (VIN 4) "J" CARLINE

Circuit Description:

The Engine Control Module (ECM) supplies a voltage of about .45 volt between terminals "D10" and "D11". (If measured with a 10 megohm digital voltmeter, this may read as low as .32 volt.)

When the Oxygen Sensor (O2S) reaches operating temperature, it varies this voltage from about .1 volt (exhaust is lean) to about .9 volt (exhaust is rich).

The sensor is like an open circuit and produces no voltage when it is below 316°C (600°F). An open sensor circuit, or cold sensor, causes "Open Loop" operation.

Test Description: Number(s) below refer to circled number(s) on the diagnostic chart.
1. DTC 13 will set under the following conditions:
   - Engine at normal operating temperature.
   - At least 40 seconds has elapsed since engine start-up.
   - O2S signal voltage is steady between .35 and .55 volt.
   - Throttle angle is above 5%.
   - All above conditions are met for about 40.3 seconds or more.
   If the conditions for a DTC 13 exist, the system will not operate in "Closed Loop."
2. This test determines if the O2S is the problem or if the ECM and wiring are at fault.
3. In doing this test, use only a 10 megohm digital voltmeter. This test checks the continuity of CKT 412 and CKT 413. If CKT 413 is open, the ECM voltage on CKT 412 will be over .6 volt (600 mV).

Diagnostic Aids:

Normal Oxygen Sensor (O2S) voltage varies between 100 mV to 999 mV (.1 and 1.0 volt) while in "Closed Loop." DTC 13 sets in one minute if sensor signal voltage remains between .35 and .55 volt, but the system will go to "Open Loop" in about 15 seconds.

Verify a clean, tight ground connection for CKT 413. Open CKT 412 or CKT 413 will result in a DTC 13. If DTC 13 is intermittent, refer to "Symptoms,"

2.2L (VIN 4) ENGINE — DIAGNOSTIC TROUBLE CODE CHART — CAVALIER

DTC 13
OXYGEN SENSOR (O2S) CIRCUIT
(OPEN CIRCUIT)
2.2L (VIN 4) "J" CARLINE

1. ENGINE AT NORMAL OPERATING TEMPERATURE (ABOVE 80°C/176°F).
   - RUN ENGINE ABOVE 1200 RPM FOR TWO MINUTES.
   DOES SCAN TOOL INDICATE "CLOSED LOOP"?

NO →

2. DISCONNECT O2S CONNECTOR.
   - JUMPER HARNESS CKT 412 (ECM SIDE) TO GROUND.
   - SCAN TOOL SHOULD DISPLAY OXYGEN SENSOR VOLTAGE BELOW .2 VOLT (200 mV) WITH ENGINE RUNNING.
   DOES IT?

YES → DTC 13 IS INTERMITTENT. IF NO ADDITIONAL DTC(S) WERE STORED, REFER TO "DIAGNOSTIC AIDS"

NO →

3. REMOVE JUMPER.
   - IGNITION "ON," ENGINE "OFF."
   - CHECK VOLTAGE OF CKT 412 (ECM SIDE) AT O2S HARNESS CONNECTOR USING A DVM.

YES → FAULTY O2S CONNECTION OR SENSOR.

- .3 - .6 VOLT (300 - 600 mV) → FAULTY ECM.*
- OVER .6 VOLT (600 mV) → OPEN CKT 413 OR FAULTY CONNECTION OR FAULTY ECM.*
- LESS THAN .3 VOLT (300 mV) → OPEN CKT 412 OR FAULTY ECM CONNECTION OR FAULTY ECM.*

* IF ECM IS FAULTY AND MUST BE REPLACED, THE NEW ECM MUST BE PROGRAMMED. REFER TO ECM REPLACEMENT AND PROGRAMMING PROCEDURES

"AFTER REPAIRS," REFER TO DTC CRITERIA AND CONFIRM DTC DOES NOT RESET.

2.2L (VIN 4) ENGINE — DIAGNOSTIC TROUBLE CODE CHART — CAVALIER

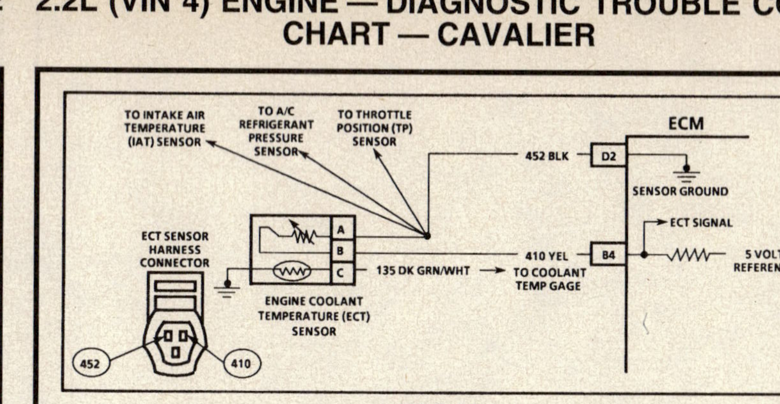

DTC 14
(Page 1 of 2)
ENGINE COOLANT TEMPERATURE (ECT) SENSOR CIRCUIT
(HIGH/LOW TEMPERATURE INDICATED)
2.2L (VIN 4) "J" CARLINE

Circuit Description:

The Engine Coolant Temperature (ECT) sensor utilizes a thermistor to control the signal voltage to the Engine Control Module (ECM). The ECM applies a reference voltage on CKT 410 to the sensor. When the engine is cold, the sensor (thermistor) resistance is high. The ECM will then sense a high signal voltage.

As the engine warms up, the sensor resistance decreases and the voltage drops. At normal engine operating temperature, the voltage will measure about 1.5 to 2.0 volts at ECM terminal "B4".

Coolant temperature is one of the inputs used to control the following:
- Cooling fan.
- Fuel delivery.
- Ignition Control (IC).
- Idle Air Control (IAC).
- Torque Converter Clutch (TCC).

A separate thermistor within the ECT sensor provides a signal to the coolant temperature gage located in the instrument panel.

Test Description: Number(s) below refer to circled number(s) on the diagnostic chart.

1. DTC 14 will set if:
   - The engine has been running for 2 minutes or longer.
   - Signal voltage indicates an engine coolant temperature below -34°C (-30°F).
   OR
   - Signal voltage indicates an engine coolant temperature above 143°C (289°F) for 3 seconds or longer.
2. If the ECM recognizes the grounded circuit (low voltage) and displays a high temperature, the ECM and wiring are OK.
3. This test will determine if there is a wiring problem or a faulty ECM. If CKT 452 is open, there may also be a other DTC(S) stored.

Diagnostic Aids:

A scan tool should display engine coolant temperature in degrees Celsius.

After the engine is started, the temperature should rise steadily to about 90°C (194°F) then stabilize when the thermostat opens.

If the engine has been allowed to cool to an ambient temperature (overnight), Engine Coolant Temperature (ECT) and Intake Air Temperature (IAT) may be checked with a scan tool and should read close to each other.

When a DTC 14 is set, the ECM will turn "ON" the engine cooling fan.
- If DTC 14 is intermittent or a "Hard Start" symptom is present, check engine coolant temperature with a scan tool on a cool engine. Temperature displayed should be within 5 degrees of the ambient. If not, check the ECT sensor using the "Diagnostic Aids"
  If sensor is OK, check connections.

2.2L (VIN 4) ENGINE — DIAGNOSTIC TROUBLE CODE CHART — CAVALIER

DTC 14
(Page 1 of 2)
ENGINE COOLANT TEMPERATURE (ECT) SENSOR CIRCUIT
(HIGH/LOW TEMPERATURE INDICATED)
2.2L (VIN 4) "J" CARLINE

① DOES SCAN TOOL DISPLAY COOLANT TEMPERATURE OF -34°C (-30°F) OR LESS?

YES

NO → REFER TO DTC 14 (PAGE 2 OF 2)

② • DISCONNECT SENSOR CONNECTOR.
• JUMPER HARNESS CKT 410 TO SENSOR GROUND CIRCUIT.
• SCAN TOOL SHOULD DISPLAY 143°C (289°F) OR HIGHER. DOES IT?

NO

YES → FAULTY SENSOR CONNECTOR OR FAULTY SENSOR.

③ • JUMPER CKT 410 TO A KNOWN GOOD GROUND.
• SCAN TOOL SHOULD DISPLAY 143°C (289°F) OR HIGHER. DOES IT?

YES → OPEN SENSOR GROUND CIRCUIT, FAULTY CONNECTION OR FAULTY ECM.*

NO → OPEN CKT 410, FAULTY CONNECTION OR FAULTY ECM.*

DIAGNOSTIC AID		
ECT SENSOR		
TEMPERATURE VS. RESISTANCE VALUES (APPROXIMATE)		
°C	°F	OHMS
100	212	177
90	194	241
80	176	332
70	158	467
60	140	667
50	122	973
45	113	1188
40	104	1459
35	95	1802
30	86	2238
25	77	2796
20	68	3520
15	59	4450
10	50	5670
5	41	7280
0	32	9420
-5	23	12300
-10	14	16180
-15	5	21450
-20	-4	28680
-30	-22	52700
-40	-40	100700

* IF ECM IS FAULTY AND MUST BE REPLACED, THE NEW ECM MUST BE PROGRAMMED. REFER TO ECM REPLACEMENT AND PROGRAMMING PROCEDURES

"AFTER REPAIRS," REFER TO DTC CRITERIA   AND CONFIRM DTC DOES NOT RESET.

2.2L (VIN 4) ENGINE — DIAGNOSTIC TROUBLE CODE CHART — CAVALIER

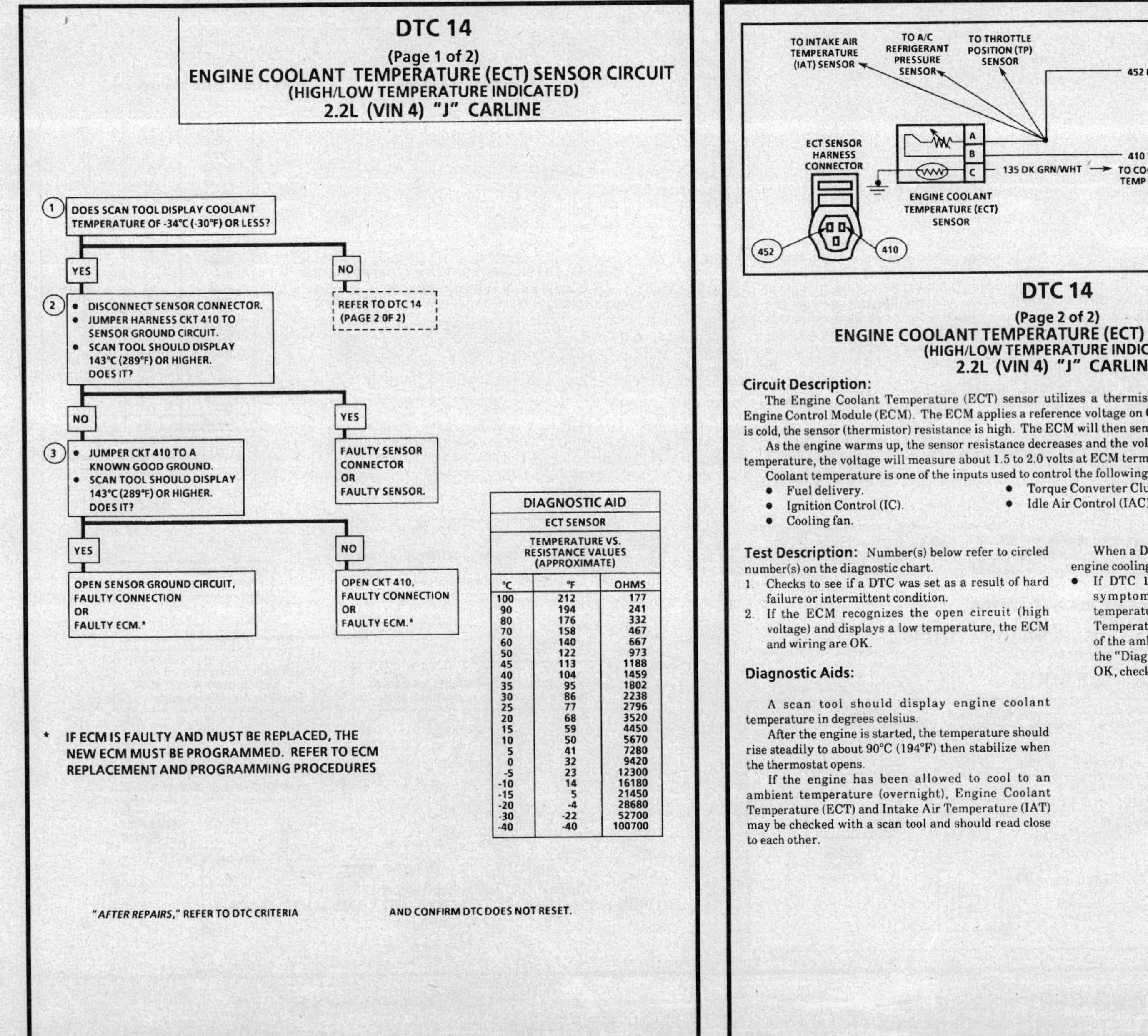

DTC 14
(Page 2 of 2)
ENGINE COOLANT TEMPERATURE (ECT) SENSOR CIRCUIT
(HIGH/LOW TEMPERATURE INDICATED)
2.2L (VIN 4) "J" CARLINE

Circuit Description:

The Engine Coolant Temperature (ECT) sensor utilizes a thermistor to control the signal voltage to the Engine Control Module (ECM). The ECM applies a reference voltage on CKT 410 to the sensor. When the engine is cold, the sensor (thermistor) resistance is high. The ECM will then sense a high signal voltage.

As the engine warms up, the sensor resistance decreases and the voltage drops. At normal engine operating temperature, the voltage will measure about 1.5 to 2.0 volts at ECM terminal "B4".

Coolant temperature is one of the inputs used to control the following:
- Fuel delivery.
- Ignition Control (IC).
- Cooling fan.
- Torque Converter Clutch (TCC).
- Idle Air Control (IAC).

Test Description: Number(s) below refer to circled number(s) on the diagnostic chart.
1. Checks to see if a DTC was set as a result of hard failure or intermittent condition.
2. If the ECM recognizes the open circuit (high voltage) and displays a low temperature, the ECM and wiring are OK.

Diagnostic Aids:

A scan tool should display engine coolant temperature in degrees celsius.

After the engine is started, the temperature should rise steadily to about 90°C (194°F) then stabilize when the thermostat opens.

If the engine has been allowed to cool to an ambient temperature (overnight), Engine Coolant Temperature (ECT) and Intake Air Temperature (IAT) may be checked with a scan tool and should read close to each other.

When a DTC 14 is set, the ECM will turn "ON" the engine cooling fan.
- If DTC 14 is intermittent or a "Hard Start" symptom is present, check engine coolant temperature with a scan tool on a cool engine. Temperature displayed should be within 5 degrees of the ambient. If not, check the ECT sensor using the "Diagnostic Aids"   If sensor is OK, check connections.

2.2L (VIN 4) ENGINE — DIAGNOSTIC TROUBLE CODE CHART — CAVALIER

DTC 14
(Page 2 of 2)
ENGINE COOLANT TEMPERATURE (ECT) SENSOR CIRCUIT
(HIGH/LOW TEMPERATURE INDICATED)
2.2L (VIN 4) "J" CARLINE

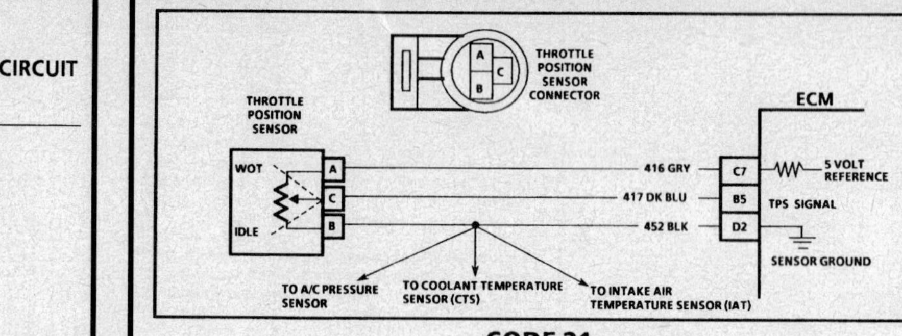

DIAGNOSTIC AID

ECT SENSOR		
TEMPERATURE VS. RESISTANCE VALUES (APPROXIMATE)		
°C	°F	OHMS
100	212	177
90	194	241
80	176	332
70	158	467
60	140	667
50	122	973
45	113	1188
40	104	1459
35	95	1802
30	86	2238
25	77	2796
20	68	3520
15	59	4450
10	50	5670
5	41	7280
0	32	9420
-5	23	12300
-10	14	16180
-15	5	21450
-20	-4	28680
-30	-22	52700
-40	-40	100700

*   IF ECM IS FAULTY AND MUST BE REPLACED, THE NEW ECM MUST BE PROGRAMMED. REFER TO ECM REPLACEMENT AND PROGRAMMING PROCEDURES

"AFTER REPAIRS," REFER TO DTC CRITERIA   AND CONFIRM DTC DOES NOT RESET.

2.2L (VIN 4) ENGINE — DIAGNOSTIC TROUBLE CODE CHART — 1992 CAVALIER

CODE 21
(Page 1 of 2)
THROTTLE POSITION SENSOR (TPS) CIRCUIT
(SIGNAL VOLTAGE HIGH/LOW)
2.2L (VIN 4) "J" CARLINE (PORT)

Circuit Description:

The Throttle Position Sensor (TPS) provides a voltage signal that changes relative to the throttle valve. Signal voltage will vary from .33 to 1.33 volts at idle to about 4.5 volts at Wide Open Throttle (WOT).

The TPS signal is one of the most important inputs used by the Electronic Control Module (ECM) for fuel control and for many of the ECM controlled outputs.

Test Description: Number(s) below refer to circled number(s) on the diagnostic chart.

1. A Code 21 will set under the following conditions:
    *   TPS reading above 3.9 volts.
    *   MAP reading below 65 kPa.
    *   Engine speed less than 1750 rpms.
    *   All of the above conditions present for 5 seconds.

    OR

    *   TPS reading below 1.9 volts for 64 seconds.
    The TPS has an auto zeroing feature. If the voltage reading is within the range of about .33 to 1.33 volts, the ECM will use that value as closed throttle. If the voltage reading is outside of the auto zero range at closed throttle, check for a binding throttle cable or damaged linkage. If OK, continue with diagnosis.

2. If the ECM recognizes the change of state, the ECM and CKTs 474 and 417 are OK.

3. This step isolates a faulty sensor, ECM, or an open CKT 452. If CKT 452 is open, there may also be other codes stored.

Diagnostic Aids:

A "Scan" tool displays throttle position in volt. Closed throttle voltage should be .33 to 1.33 volts. TPS voltage should increase at a steady rate as throttle is moved to WOT.

If Code 21 is intermittent, refer to "Symptoms,"

2.2L (VIN 4) ENGINE — DIAGNOSTIC TROUBLE CODE CHART — 1992 CAVALIER

CODE 21
(Page 1 of 2)
THROTTLE POSITION SENSOR (TPS) CIRCUIT
(SIGNAL VOLTAGE HIGH/LOW)
2.2L (VIN 4) "J" CARLINE (PORT)

1. • THROTTLE CLOSED.
   DOES "SCAN" TOOL DISPLAY TPS OVER 3.9 VOLTS?

YES

2. • DISCONNECT SENSOR CONNECTOR.
   "SCAN" TOOL SHOULD DISPLAY TPS BELOW .2 VOLT (200mV).
   DOES IT?

NO → REFER TO CODE 21 (PAGE 2 OF 2)

YES

3. • PROBE SENSOR GROUND CIRCUIT WITH A TEST LIGHT CONNECTED TO BATTERY VOLTAGE.

NO → TPS SIGNAL CIRCUIT SHORTED TO VOLTAGE OR FAULTY ECM.

LIGHT "ON"

FAULTY CONNECTION OR THROTTLE POSITION SENSOR.

LIGHT "OFF"

OPEN SENSOR GROUND CIRCUIT OR FAULTY ECM.

"AFTER REPAIRS," REFER TO CODE CRITERIA   AND CONFIRM CODE DOES NOT RESET.

2.2L (VIN 4) ENGINE — DIAGNOSTIC TROUBLE CODE CHART — 1992 CAVALIER

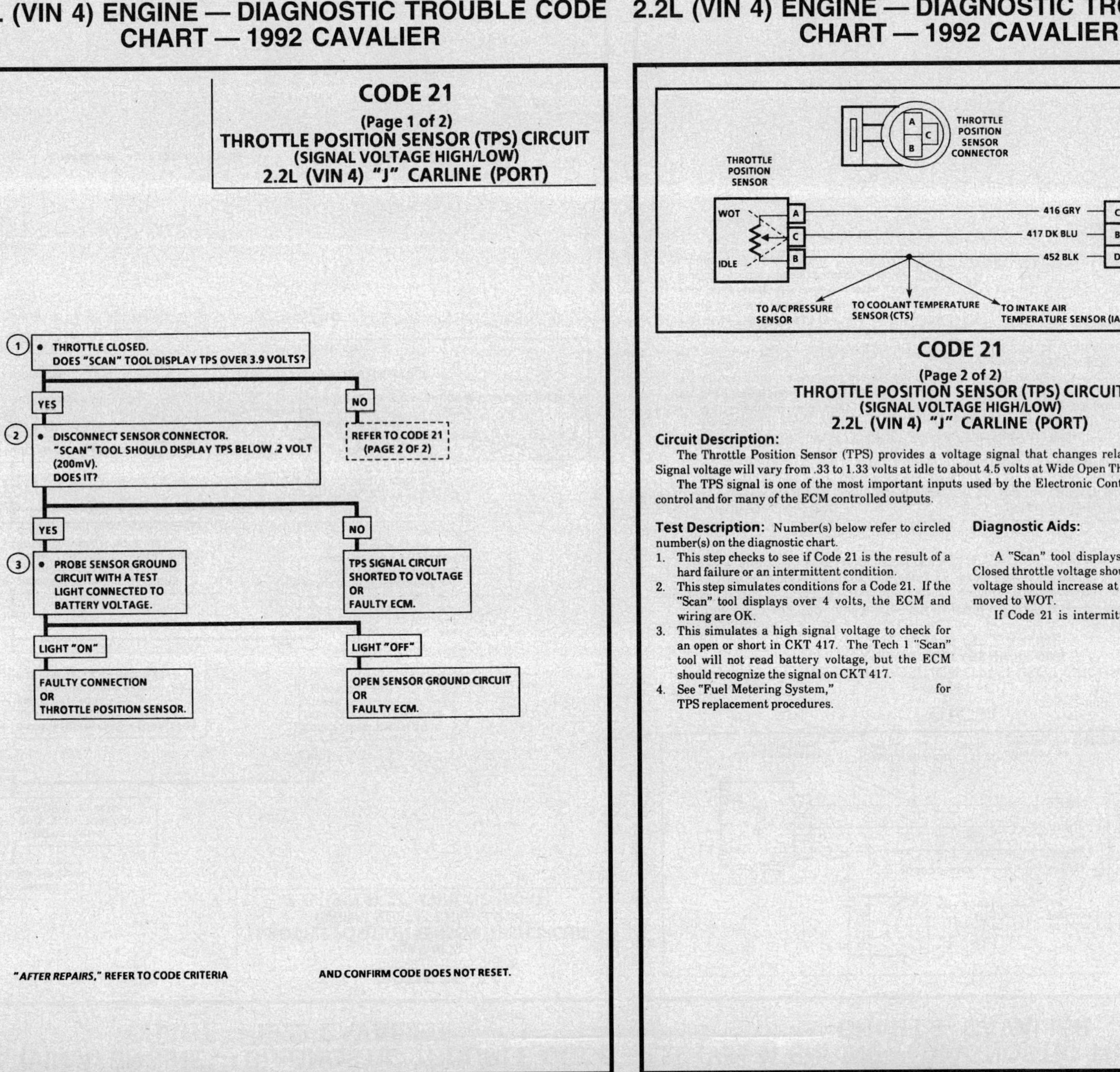

CODE 21
(Page 2 of 2)
THROTTLE POSITION SENSOR (TPS) CIRCUIT
(SIGNAL VOLTAGE HIGH/LOW)
2.2L (VIN 4) "J" CARLINE (PORT)

Circuit Description:

The Throttle Position Sensor (TPS) provides a voltage signal that changes relative to the throttle valve. Signal voltage will vary from .33 to 1.33 volts at idle to about 4.5 volts at Wide Open Throttle (WOT).

The TPS signal is one of the most important inputs used by the Electronic Control Module (ECM) for fuel control and for many of the ECM controlled outputs.

Test Description: Number(s) below refer to circled number(s) on the diagnostic chart.

1. This step checks to see if Code 21 is the result of a hard failure or an intermittent condition.
2. This step simulates conditions for a Code 21. If the "Scan" tool displays over 4 volts, the ECM and wiring are OK.
3. This simulates a high signal voltage to check for an open or short in CKT 417. The Tech 1 "Scan" tool will not read battery voltage, but the ECM should recognize the signal on CKT 417.
4. See "Fuel Metering System," for TPS replacement procedures.

Diagnostic Aids:

A "Scan" tool displays throttle position in volts. Closed throttle voltage should be .33 to 1.33 volts. TPS voltage should increase at a steady rate as throttle is moved to WOT.

If Code 21 is intermittent, refer to "Symptoms,"

CODE 21

(Page 2 of 2)
THROTTLE POSITION SENSOR (TPS) CIRCUIT
(SIGNAL VOLTAGE HIGH/LOW)
2.2L (VIN 4) "J" CARLINE (PORT)

FROM CODE 21 (PAGE 1 OF 2).

(1)
- THROTTLE CLOSED.
  DOES "SCAN" TOOL DISPLAY LESS THAN .19 VOLTS?

YES

(2)
- DISCONNECT TPS SENSOR CONNECTOR.
- JUMPER 5 VOLT REFERENCE CIRCUIT AND CKT 417 TOGETHER. "SCAN" SHOULD DISPLAY THROTTLE POSITION OVER 4.0 V (4000 mV). DOES IT?

NO

CODE 21 IS INTERMITTENT. IF NO ADDITIONAL CODES WERE STORED, REFER TO "DIAGNOSTIC AIDS"

NO

(3)
- PROBE CKT 417 WITH A TEST LIGHT CONNECTED TO BATTERY VOLTAGE. "SCAN" TOOL SHOULD DISPLAY THROTTLE POSITION OVER 4.0V (4000 mV). DOES IT?

YES

(4)
- RECONNECT TPS CONNECTOR.
- CHECK FOR FAULTY OR INTERMITTENT CONNECTION. IF OK, TPS IS FAULTY.

YES

5 VOLT REFERENCE CIRCUIT OPEN
OR
SHORTED TO GROUND
OR
FAULTY CONNECTION
OR
FAULTY ECM.

NO

CKT 417 OPEN
OR
SHORTED TO GROUND
OR
SHORTED TO THROTTLE POSITION SENSOR GROUND CIRCUIT
OR
FAULTY ECM CONNECTION
OR
FAULTY ECM.

"AFTER REPAIRS," REFER TO CODE CRITERIA   AND CONFIRM CODE DOES NOT RESET.

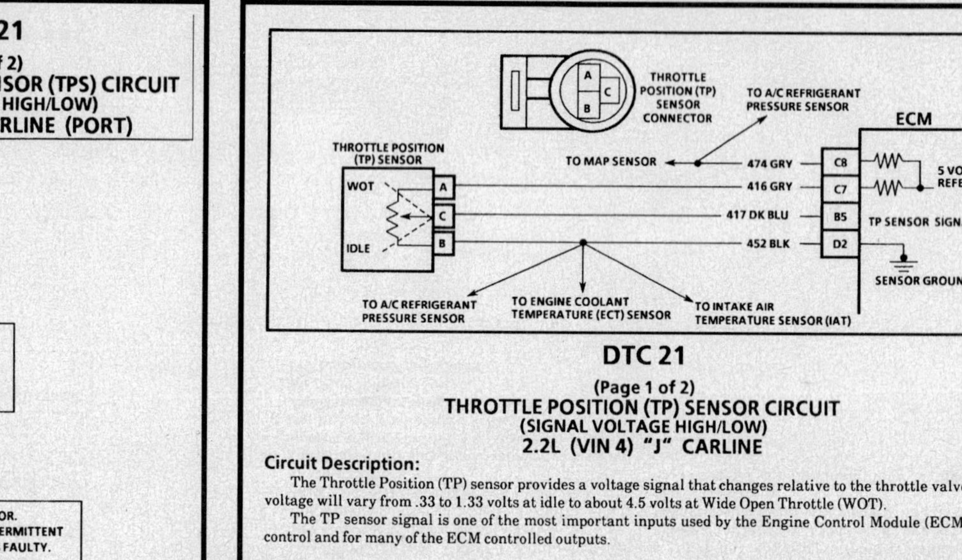

DTC 21

(Page 1 of 2)
THROTTLE POSITION (TP) SENSOR CIRCUIT
(SIGNAL VOLTAGE HIGH/LOW)
2.2L (VIN 4) "J" CARLINE

Circuit Description:

The Throttle Position (TP) sensor provides a voltage signal that changes relative to the throttle valve. Signal voltage will vary from .33 to 1.33 volts at idle to about 4.5 volts at Wide Open Throttle (WOT).

The TP sensor signal is one of the most important inputs used by the Engine Control Module (ECM) for fuel control and for many of the ECM controlled outputs.

Test Description: Number(s) below refer to circled number(s) on the diagnostic chart.

1. A DTC 21 will set under the following conditions:
   - TP sensor reading below 1.9 volts or above 4.9 volts for about 64 seconds.
     OR
   - TP sensor reading above 3.9 volts, but below 4.9 volts.
   - MAP reading below 65 kPa.
   - Engine speed less than 1,750 RPM.
   - All of the above conditions present for about 5 seconds.

   The TP sensor has an auto zeroing feature. If the voltage reading is within the range of about .33 to 1.33 volts, the ECM will use that value as closed throttle. If the voltage reading is outside of the auto zero range at closed throttle, check for a binding throttle cable or damaged linkage. If OK, continue with diagnosis.

2. If the ECM recognizes the change of state, the ECM and CKTs 416 and 417 are OK.
3. This step isolates a faulty sensor, ECM, or an open CKT 452. If CKT 452 is open, there may also be other DTC(S) stored.

Diagnostic Aids:

A scan tool displays throttle position in volts. Closed throttle voltage should be .33 to 1.33 volts. TP sensor voltage should increase at a steady rate as throttle is moved to WOT.

If DTC 21 is intermittent, refer to "Symptoms,"

2.2L (VIN 4) ENGINE — DIAGNOSTIC TROUBLE CODE CHART — CAVALIER

DTC 21

(Page 1 of 2)
THROTTLE POSITION (TP) SENSOR CIRCUIT
(SIGNAL VOLTAGE HIGH/LOW)
2.2L (VIN 4) "J" CARLINE

① • THROTTLE CLOSED.
   DOES SCAN TOOL DISPLAY TP SENSOR OVER 3.9 VOLTS?

YES →

② • DISCONNECT SENSOR CONNECTOR.
   SCAN TOOL SHOULD DISPLAY TP SENSOR BELOW .2 VOLT (200mV).
   DOES IT?

NO → REFER TO DTC 21 (PAGE 2 OF 2)

YES →

③ • PROBE SENSOR GROUND CIRCUIT WITH A TEST LIGHT CONNECTED TO BATTERY VOLTAGE.

NO → TP SENSOR SIGNAL CIRCUIT SHORTED TO VOLTAGE OR FAULTY ECM.*

LIGHT "ON" → FAULTY CONNECTION OR THROTTLE POSITION SENSOR.

LIGHT "OFF" → OPEN SENSOR GROUND CIRCUIT OR FAULTY ECM.*

* IF ECM IS FAULTY AND MUST BE REPLACED, THE NEW ECM MUST BE PROGRAMMED. REFER TO ECM REPLACEMENT AND PROGRAMMING PROCEDURES

"AFTER REPAIRS," REFER TO DTC CRITERIA   AND CONFIRM DTC DOES NOT RESET.

2.2L (VIN 4) ENGINE — DIAGNOSTIC TROUBLE CODE CHART — CAVALIER

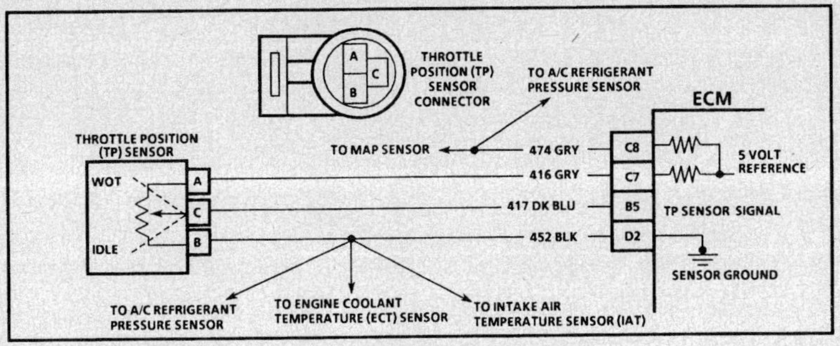

DTC 21

(Page 2 of 2)
THROTTLE POSITION (TP) SENSOR CIRCUIT
(SIGNAL VOLTAGE HIGH/LOW)
2.2L (VIN 4) "J" CARLINE

Circuit Description:

The Throttle Position (TP) sensor provides a voltage signal that changes relative to the throttle valve. Signal voltage will vary from .33 to 1.33 volts at idle to about 4.5 volts at Wide Open Throttle (WOT).

The TP sensor signal is one of the most important inputs used by the Engine Control Module (ECM) for fuel control and for many of the ECM controlled outputs.

Test Description: Number(s) below refer to circled number(s) on the diagnostic chart.

1. This step checks to see if DTC 21 is the result of a hard failure or an intermittent condition.
2. This step simulates conditions for a DTC 21. If the scan tool displays over 4 volts, the ECM and wiring are OK.
3. This simulates a high signal voltage to check for an open or short in CKT 417. The Tech 1 scan tool will not read battery voltage, but the ECM should recognize the signal on CKT 417."
4. Refer to "Fuel Metering System,"
   for TP sensor replacement procedures.
5. The 5 volt reference for the MAP sensor and the A/C refrigerant pressure sensor connect to the same 5 volt power supply in the ECM. If the 5 volt reference circuit to the MAP sensor and A/C refrigerant pressure sensor has a short to ground, the Throttle Position (TP) sensor 5 volt reference circuit will also have a short to ground because they connect to the same power supply in the ECM. A short to ground on either 5 volt reference circuit does not damage the ECM 5 volt power supply. When the short is repaired, the ECM 5 volt supply will return to normal.

Diagnostic Aids:

A scan tool displays throttle position in volts. Closed throttle voltage should be .33 to 1.33 volts. TP sensor voltage should increase at a steady rate as throttle is moved to WOT.

If DTC 21 is intermittent, refer to "Symptoms,"

2.2L (VIN 4) ENGINE — DIAGNOSTIC TROUBLE CODE CHART — CAVALIER

2.2L (VIN 4) ENGINE — DIAGNOSTIC TROUBLE CODE CHART — CAVALIER

DTC 21

(Page 2 of 2)
THROTTLE POSITION (TP) SENSOR CIRCUIT
(SIGNAL VOLTAGE HIGH/LOW)
2.2L (VIN 4) "J" CARLINE

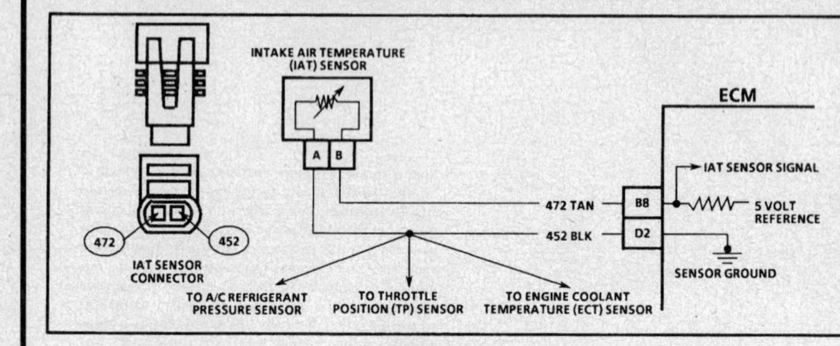

* IF ECM IS FAULTY AND MUST BE REPLACED, THE NEW ECM MUST BE PROGRAMMED. REFER TO ECM REPLACEMENT AND PROGRAMMING PROCEDURES

"AFTER REPAIRS," REFER TO DTC CRITERIA   AND CONFIRM DTC DOES NOT RESET.

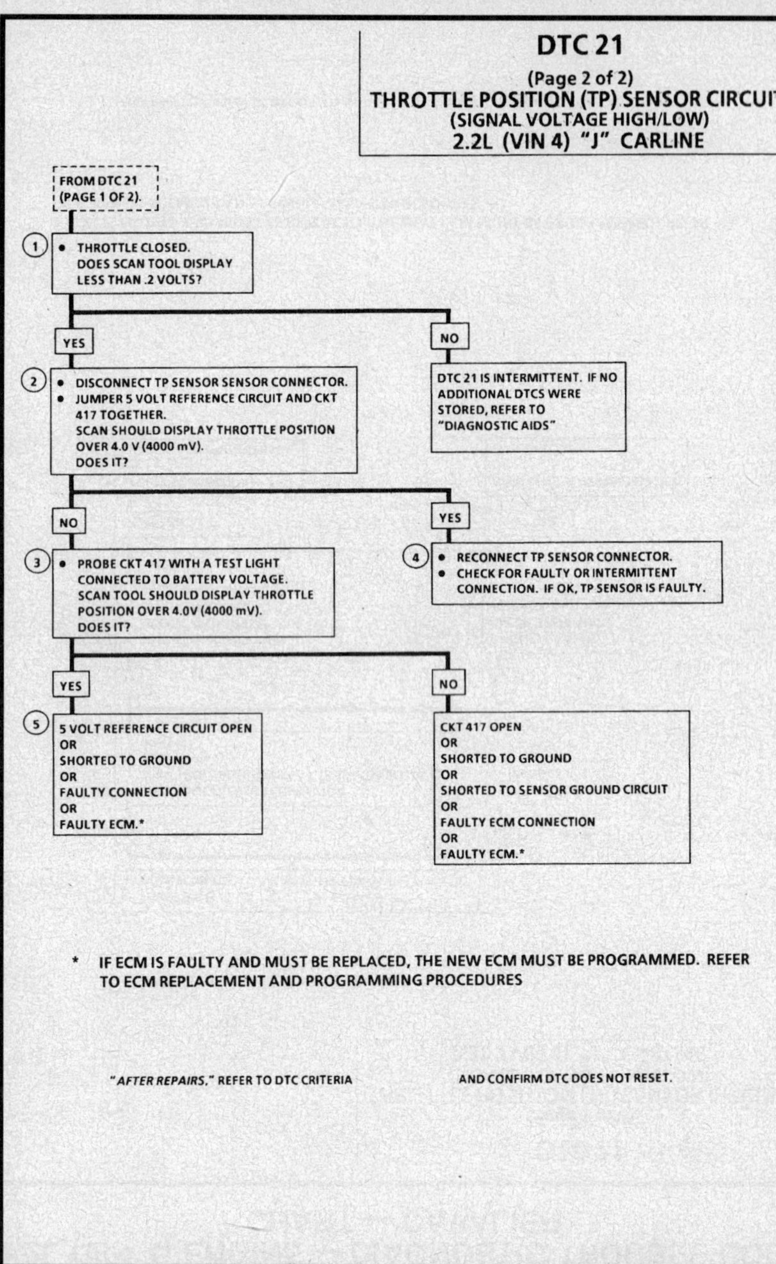

DTC 23

(Page 1 of 2)
INTAKE AIR TEMPERATURE (IAT) SENSOR CIRCUIT
(HIGH/LOW TEMPERATURE INDICATED)
2.2L (VIN 4) "J" CARLINE

Circuit Description:

The Intake Air Temperature (IAT) sensor, located in the air cleaner assembly, uses a thermistor to control the signal voltage to the Engine Control Module (ECM). The ECM applies a reference voltage (4 to 4.5 volts) on CKT 472 to the sensor. When intake air is cold, the sensor (thermistor) resistance is high. The ECM will then sense a high signal voltage. As the air warms, the sensor resistance becomes less and the voltage drops.

Test Description: Number(s) below refer to circled number(s) on the diagnostic chart.

1. A DTC 23 will set under the following conditions:
   - Engine running for 2 minutes or longer.
   - Signal voltage indicates an IAT temperature less than -34°C (-30°F).

   OR
   - Signal voltage indicates an IAT temperature greater than 150°C (302°F) for more than 2 seconds.
2. This test simulates conditions for a DTC 23. If the scan tool displays a high temperature, the ECM and wiring are OK.
3. This step checks continuity of CKT 472 and CKT 452. If CKT 452 is open, there may be other DTC(s) stored.

Diagnostic Aids:

If the engine has been allowed to cool to an ambient temperature (overnight), engine coolant and intake temperatures may be compared with a scan tool and should read close to each other.

- If DTC 23 is intermittent or a "Hard Start" symptom is present, check intake air temperature with a scan tool on a cool engine. Temperature displayed should be within 5 degrees of the ambient. If not, check the IAT sensor using the "Diagnostic Aids" If sensor is OK, check connections.

2.2L (VIN 4) ENGINE — DIAGNOSTIC TROUBLE CODE CHART — CAVALIER

DTC 23
(Page 1 of 2)
INTAKE AIR TEMPERATURE (IAT) SENSOR CIRCUIT
(HIGH/LOW TEMPERATURE INDICATED)
2.2L (VIN 4) "J" CARLINE

① • DOES SCAN TOOL DISPLAY IAT -34°C (-30°F) OR COLDER?

YES

NO → REFER TO DTC 23 (PAGE 2 OF 2)

② • DISCONNECT IAT SENSOR CONNECTOR.
• JUMPER HARNESS TERMINALS TOGETHER.
• SCAN TOOL SHOULD DISPLAY TEMPERATURE OVER 150°C (302°F). DOES IT?

YES → FAULTY CONNECTION OR SENSOR.

NO

③ • JUMPER CKT 472 TO GROUND.
• SCAN TOOL SHOULD DISPLAY TEMPERATURE OVER 150°C (302°F). DOES IT?

YES → OPEN SENSOR GROUND CIRCUIT, OR FAULTY CONNECTION OR FAULTY ECM.*

NO → OPEN CKT 472, OR FAULTY CONNECTION OR FAULTY ECM.*

DIAGNOSTIC AID

IAT SENSOR TEMPERATURE VS. RESISTANCE VALUES (APPROXIMATE)		
°C	°F	OHMS
100	210	185
70	160	450
38	100	1,800
20	70	3,400
4	40	7,500
-7	20	13,500
-18	0	25,000
-40	-40	100,700

* IF ECM IS FAULTY AND MUST BE REPLACED, THE NEW ECM MUST BE PROGRAMMED. REFER TO ECM REPLACEMENT AND PROGRAMMING PROCEDURES

"AFTER REPAIRS." REFER TO DTC CRITERIA AND CONFIRM DTC DOES NOT RESET.

2.2L (VIN 4) ENGINE — DIAGNOSTIC TROUBLE CODE CHART — CAVALIER

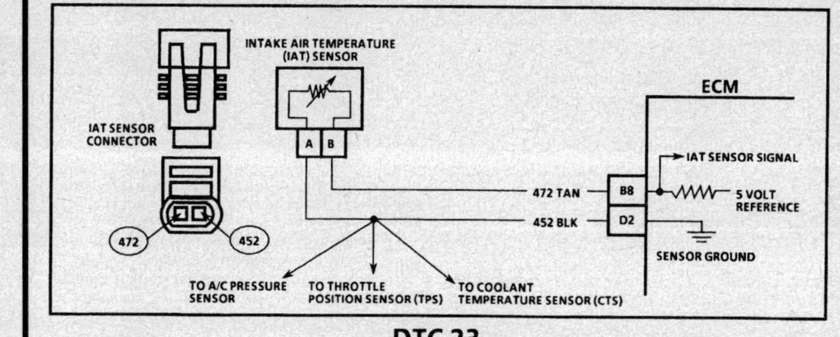

DTC 23
(Page 2 of 2)
INTAKE AIR TEMPERATURE (IAT) SENSOR CIRCUIT
(HIGH/LOW TEMPERATURE INDICATED)
2.2L (VIN 4) "J" CARLINE

Circuit Description:
The Intake Air Temperature (IAT) sensor, located in the air cleaner assembly, uses a thermistor to control the signal voltage to the Engine Control Module (ECM). The ECM applies a reference voltage (4 to 5.5 volts) on CKT 472 to the sensor. When intake air is cold, the sensor (thermistor) resistance is high. Therefore, the ECM will sense a high signal voltage. As the air warms, the sensor resistance becomes less and the voltage drops.

Test Description: Number(s) below refer to circled number(s) on the diagnostic chart.
1. This step determines if DTC 23 is the result of a hard failure or an intermittent condition.
2. If the ECM recognizes the open circuit (high voltage) and displays a low temperature, the ECM and wiring are OK.

Diagnostic Aids:

If the engine has been allowed to cool to an ambient temperature (overnight), engine coolant and intake air temperatures may be checked with a scan tool and should read close to each other.
• If DTC 23 is intermittent or a "Hard Start" symptom is present, check intake air temperature with a scan tool on a cool engine. Temperature displayed should be within 5 degrees of the ambient. If not, check the IAT sensor using the "Diagnostic Aids"

2.2L (VIN 4) ENGINE — DIAGNOSTIC TROUBLE CODE CHART — CAVALIER

2.2L (VIN 4) ENGINE — DIAGNOSTIC TROUBLE CODE CHART — CAVALIER

DTC 23

(Page 2 of 2)
INTAKE AIR TEMPERATURE (IAT) SENSOR CIRCUIT
(HIGH/LOW TEMPERATURE INDICATED)
2.2L (VIN 4) "J" CARLINE

FROM DTC 23 PAGE 1 OF 2

(1) DOES SCAN TOOL DISPLAY IAT OF 150°C (302°F) OR HOTTER?

YES — (2) • DISCONNECT IAT SENSOR CONNECTOR. SCAN TOOL SHOULD DISPLAY TEMPERATURE BELOW -34°C (-30°F). DOES IT?

NO — DTC 23 IS INTERMITTENT. IF NO ADDITIONAL DTC(S) WERE STORED, REFER TO "DIAGNOSTIC AIDS"

YES — REPLACE IAT SENSOR.

NO — CKT 472 SHORTED TO GROUND, OR TO SENSOR GROUND, OR ECM IS FAULTY.*

DIAGNOSTIC AID

IAT SENSOR		
TEMPERATURE VS. RESISTANCE VALUES (APPROXIMATE)		
°C	°F	OHMS
100	210	185
70	160	450
38	100	1,800
20	70	3,400
4	40	7,500
-7	20	13,500
-18	0	25,000
-40	-40	100,700

* IF ECM IS FAULTY AND MUST BE REPLACED, THE NEW ECM MUST BE PROGRAMMED. REFER TO ECM REPLACEMENT AND PROGRAMMING PROCEDURES

"AFTER REPAIRS," REFER TO DTC CRITERIA   AND CONFIRM DTC DOES NOT RESET.

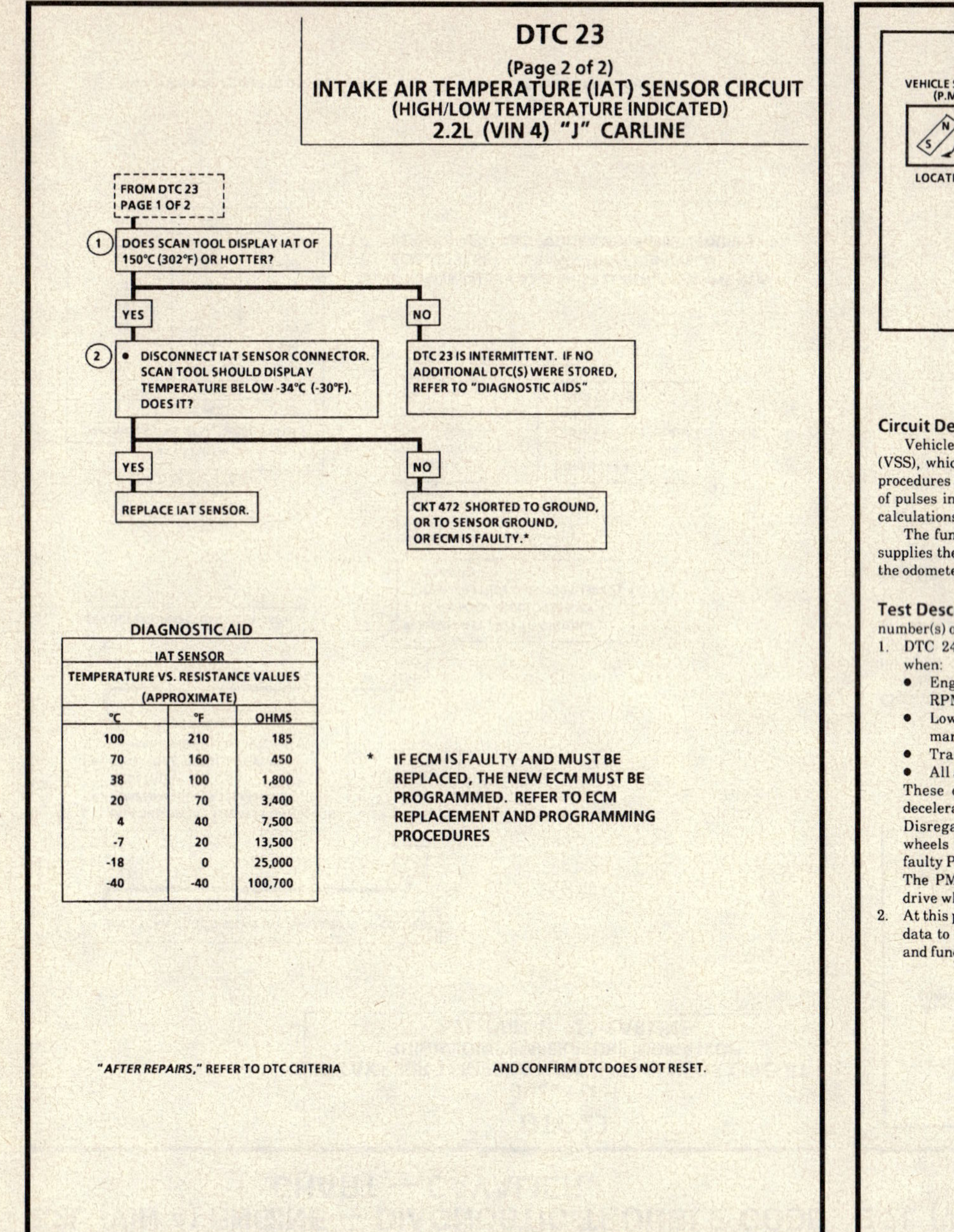

DTC 24

VEHICLE SPEED SENSOR (VSS) CIRCUIT
2.2L (VIN 4) "J" CARLINE

Circuit Description:

Vehicle speed information is provided to the Engine Control Module (ECM) by the Vehicle Speed Sensor (VSS), which is a Permanent Magnet (PM) generator, and it is mounted in the transaxle. The PM generator procedures a pulsing voltage, whenever vehicle speed is over about 3 mph. The AC voltage level and the number of pulses increases with vehicle speed. The ECM then converts the pulsing voltage to mph, which is used for calculations, and the mph can be displayed with a scan tool.

The function of VSS buffer used in past model years has been incorporated into the ECM. The ECM then supplies the necessary signal to the instrument panel (4000 pulses per mile) for operating the speedometer and the odometer.

Test Description: Number(s) below refer to circled number(s) on the diagnostic chart.

1. DTC 24 will set if vehicle speed equals 0 mph when:
   - Engine speed is between about 1400 and 3600 RPM.
   - Low load condition (low MAP voltage, high manifold vacuum).
   - Transaxle not in park or neutral.
   - All above conditions are met for 4 seconds.

   These conditions are met during a road load deceleration.

   Disregard a DTC 24 that sets when the drive wheels are not turning. This can be caused by a faulty PNP switch circuit.

   The PM generator only produces a signal if the drive wheels are turning greater than 3 mph.

2. At this point, the ECM is not sending vehicle speed data to the scan tool. If the scan tool is connected and functioning properly, the ECM is at fault.*

Diagnostic Aids:

Scan tool should indicate a vehicle speed whenever the drive wheels are turning greater than 3 mph.

A problem in CKT 389 will not affect the VSS input or the readings on a scan tool.

Check CKTs 1232 and 1233 for proper connections to be sure they are clean and tight and the harness is routed correctly.

(A/T) - A faulty or misadjusted Park/Neutral Position (PNP) switch can result in a false DTC 24. Use a scan tool and check for the proper signal while in a drive range. Refer to CHART C-1A for the PNP switch check.

DTC 24

**VEHICLE SPEED SENSOR (VSS) CIRCUIT
2.2L (VIN 4) "J" CARLINE**

DISREGARD DTC 24 IF SET WHILE DRIVE WHEELS ARE NOT TURNING.

1. • RAISE DRIVE WHEELS.

   NOTICE: DO NOT PERFORM THIS TEST WITHOUT SUPPORTING THE LOWER CONTROL ARMS SO THAT THE DRIVE AXLES ARE IN A NORMAL HORIZONTAL POSITION. RUNNING THE VEHICLE IN GEAR WITH THE WHEELS HANGING DOWN AT FULL TRAVEL MAY DAMAGE THE DRIVE AXLES.

   • WITH ENGINE IDLING IN GEAR, SCAN TOOL SHOULD DISPLAY VEHICLE SPEED ABOVE 0 MPH. DOES IT?

NO → DOES SPEEDOMETER WORK PROPERLY?

YES → DTC 24 IS INTERMITTENT. IF NO ADDITIONAL DTC(S) WERE STORED, REFER TO "DIAGNOSTIC AIDS"

NO →
• IGNITION "OFF."
• DISCONNECT VSS HARNESS CONNECTOR AT TRANSAXLE.
• CONNECT SIGNAL GENERATOR TESTER J 33431-B OR EQUIVALENT TO VSS HARNESS CONNECTOR.
• IGNITION "ON," TOOL "ON" AND SET TO GENERATE A VSS SIGNAL.
• SCAN TOOL SHOULD DISPLAY VEHICLE SPEED ABOVE 0 MPH. DOES IT?

YES → 2. REPLACE ECM*.

NO → CKT 1232 OR 1233 OPEN, SHORTED TO GROUND, SHORTED TOGETHER, FAULTY CONNECTIONS, OR FAULTY ECM*.

YES → REPLACE VEHICLE SPEED SENSOR.

* IF ECM IS FAULTY AND MUST BE REPLACED, THE NEW ECM MUST BE PROGRAMMED. REFER TO ECM REPLACEMENT AND PROGRAMMING PROCEDURES

"AFTER REPAIRS," REFER TO DTC CRITERIA AND CONFIRM DTC DOES NOT RESET.

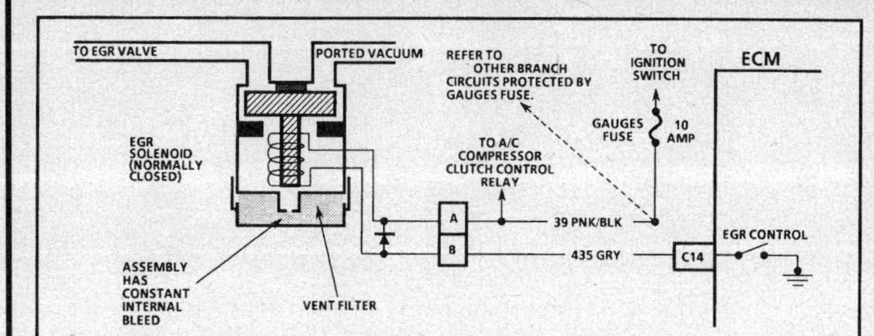

DTC 32

**(Page 1 of 2)
EXHAUST GAS RECIRCULATION (EGR) SYSTEM FAILURE
2.2L (VIN 4) "J" CARLINE**

Circuit Description:

The Exhaust Gas Recirculation (EGR) system is controlled by the ECM. The ECM controls the vacuum being supplied to the valve by energizing and de-energizing a solenoid.

The ECM uses information from various engine sensors to determine when EGR is necessary. Once the ECM has requested EGR by grounding the solenoid circuit, the ECM will monitor engine operating conditions to determine if exhaust gas flow has entered the intake manifold. When the ECM tests for EGR operation and no change in engine operating conditions is indicated, a DTC 32 will set.

The difference in long term fuel trim values between the idle (closed throttle) cell and cell 2 is used to monitor EGR system performance. When the difference between the two long term fuel trim values is greater than about 15 and the long term fuel trim value in cell 2 is greater than 135 for 80 seconds or more, DTC 32 is set. The system operates in long term fuel trim cell 2 during a cruise condition at approximately 55 mph.

Test Description: Number(s) below refer to circled number(s) on the diagnostic chart.

1. Intake Passage: Shut "OFF" engine and remove the EGR valve from the manifold. Plug the exhaust side hole with a suitable stopper. Leaving the intake side hole open, attempt to start the engine. If the engine runs at a high idle (up to 3000 RPM is possible) or starts and stalls, the EGR intake passage is not restricted. If the engine starts and idles normally, the EGR intake passage is restricted.

   Exhaust Passage: With EGR valve still removed, plug the intake side hole with a suitable stopper. With the exhaust side hole open, check for the presence of exhaust gas. If no exhaust gas is present, the EGR exhaust side passage is restricted.

2. By grounding the diagnostic "test" terminal, the EGR solenoid should be energized and allow vacuum to be applied to the gage. The vacuum at the gage may or may not _slowly_ bleed off. It is important that the gage is able to read the amount of vacuum being applied.

3. When the diagnostic "test" terminal is ungrounded, the vacuum gage should bleed off completely through a vent in the solenoid. The vacuum pump gage may or may not bleed off but this does not indicate a problem.

4. This test will determine if the electrical control part of the system is at fault or if the connector or solenoid is at fault.

5. At this point, it has been determined that the EGR solenoid, the ECM and the vacuum supply are OK.

Diagnostic Aids:

Vacuum lines should be thoroughly checked for proper routing. Refer to "Vehicle Emission Control Information" label.

The DTC 32 chart is a functional check of the EGR system. If the EGR system works properly but a DTC 32 has been set, check other items that could result in high block learn values during a cruise condition at approximately 55 mph. Low fuel pressure or lean fuel injector(s) may set a DTC 32. It may be necessary to monitor fuel pressure while driving the vehicle at various road speeds and/or loads. Refer to "Fuel System Diagnosis," CHART A-7

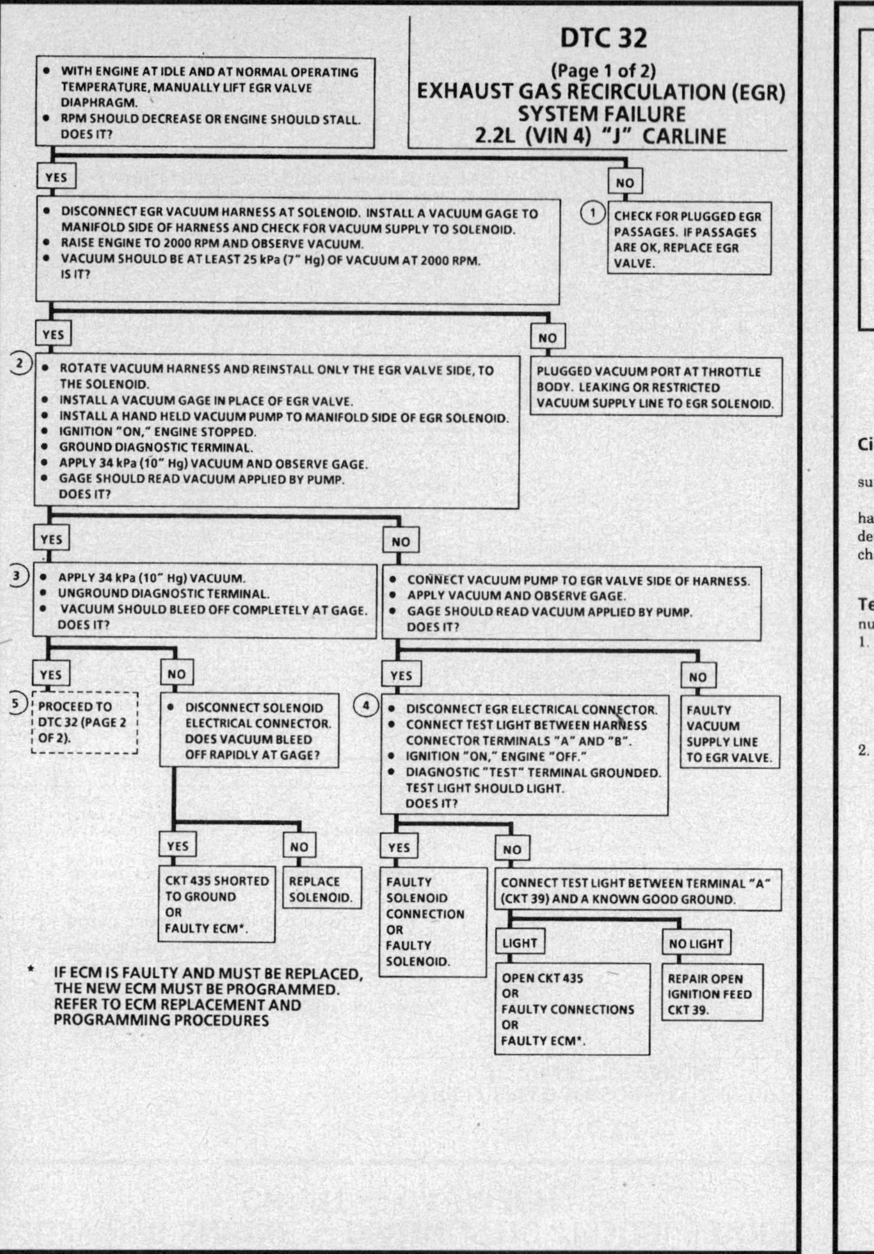

DTC 32

(Page 2 of 2)
EXHAUST GAS RECIRCULATION (EGR) SYSTEM FAILURE
2.2L (VIN 4) "J" CARLINE

Circuit Description:

The Exhaust Gas Recirculation (EGR) system is controlled by the ECM. The ECM controls the vacuum being supplied to the valve by energizing and de-energizing a solenoid.

The ECM uses information from various engine sensors to determine when EGR is necessary. Once the ECM has requested EGR by grounding the solenoid circuit, the ECM will monitor engine operating conditions to determine if exhaust gas flow has entered the intake manifold. When the ECM tests for EGR operation and no change in engine operating conditions is indicated, a DTC 32 will set.

Test Description: Number(s) below refer to circled number(s) on the diagnostic chart.

1. The remaining tests check the ability of the EGR valve to interact with the exhaust system. This system uses a negative backpressure EGR valve which should hold vacuum with engine "OFF." **Be sure shop exhaust hose is not connected during Steps 1 and 2.**

2. When engine is started, exhaust backpressure at the base of the EGR valve should open the valve's internal bleed and vent the applied vacuum allowing the valve to seat. **Because the shop exhaust hose is not installed at this time, do not allow the engine to run longer than 15 seconds.**

Diagnostic Aids:

Low fuel pressure or lean fuel injectors may cause a DTC 32 to set. Use CHART A-7. It may be necessary to monitor fuel pressure while driving the vehicle at various road speeds and/or loads. If fuel pressure is normal, perform the injector balance test, CHART C2-A

2.2L (VIN 4) ENGINE — DIAGNOSTIC TROUBLE CODE CHART — CAVALIER

DTC 32
(Page 2 of 2)
EXHAUST GAS RECIRCULATION (EGR) SYSTEM FAILURE
2.2L (VIN 4) "J" CARLINE

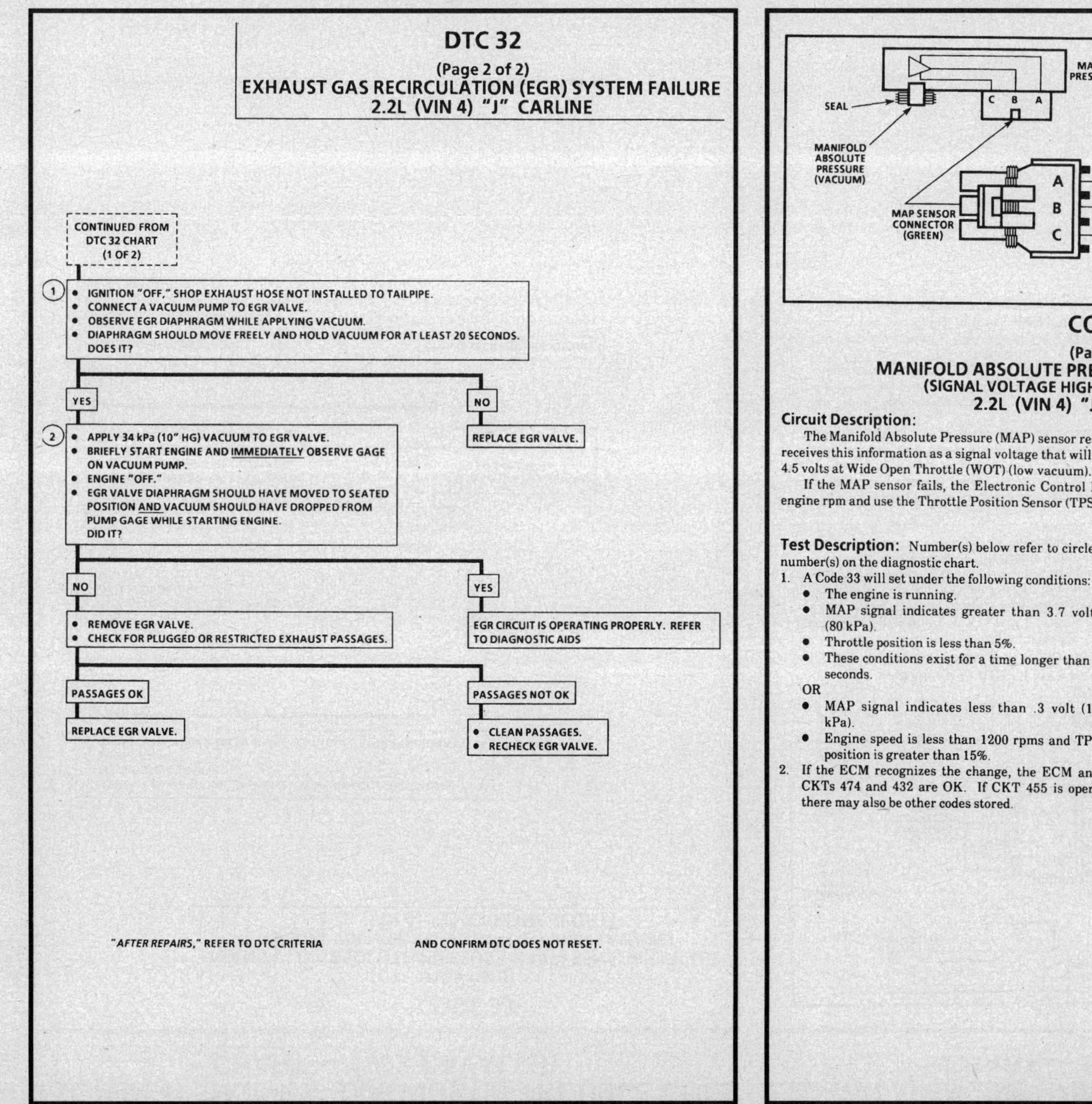

2.2L (VIN 4) ENGINE — DIAGNOSTIC TROUBLE CODE CHART — 1992 CAVALIER

CODE 33
(Page 1 of 2)
MANIFOLD ABSOLUTE PRESSURE (MAP) SENSOR CIRCUIT
(SIGNAL VOLTAGE HIGH/LOW - LOW/HIGH VACUUM)
2.2L (VIN 4) "J" CARLINE (PORT)

Circuit Description:

The Manifold Absolute Pressure (MAP) sensor responds to changes in manifold pressure (vacuum). The ECM receives this information as a signal voltage that will vary from about 1 to 1.5 volts at closed throttle (idle), to 4 to 4.5 volts at Wide Open Throttle (WOT) (low vacuum).

If the MAP sensor fails, the Electronic Control Module (ECM) will substitute a fixed MAP value based on engine rpm and use the Throttle Position Sensor (TPS) to control fuel delivery.

Test Description: Number(s) below refer to circled number(s) on the diagnostic chart.

1. A Code 33 will set under the following conditions:
   • The engine is running.
   • MAP signal indicates greater than 3.7 volts (80 kPa).
   • Throttle position is less than 5%.
   • These conditions exist for a time longer than 5 seconds.
   OR
   • MAP signal indicates less than .3 volt (15 kPa).
   • Engine speed is less than 1200 rpms and TPS position is greater than 15%.
2. If the ECM recognizes the change, the ECM and CKTs 474 and 432 are OK. If CKT 455 is open, there may also be other codes stored.

Diagnostic Aids:

With the ignition "ON" and the engine stopped, the manifold pressure is equal to atmospheric pressure and the signal voltage will be high. This information is used by the ECM as an indication of vehicle altitude and is referred to as BARO. Comparison of this BARO reading with a known good vehicle with the same sensor is a good way to check accuracy of a "suspect" sensor. Reading should be within ± .4 volt.

If Code 33 is intermittent, refer to "Symptoms,"

CODE 33

(Page 1 of 2)
MANIFOLD ABSOLUTE PRESSURE (MAP) SENSOR CIRCUIT
(SIGNAL VOLTAGE HIGH/LOW - LOW/HIGH VACUUM)
2.2L (VIN 4) "J" CARLINE (PORT)

1
- IF ENGINE IDLE IS ROUGH, UNSTABLE, OR INCORRECT, CORRECT CONDITION BEFORE USING CHART.
- ENGINE IDLING.
- DOES "SCAN" TOOL DISPLAY A MAP VOLTAGE OF 3.7 VOLTS OR GREATER?

YES → **2**
- DISCONNECT MAP SENSOR ELECTRICAL CONNECTOR.
- ENGINE IDLING.
- "SCAN" TOOL SHOULD READ A VOLTAGE OF .3 VOLT OR LESS. DOES IT?

NO → REFER TO CODE 33 (PAGE 2 OF 2).

YES →
- PROBE SENSOR GROUND CIRCUIT WITH A TEST LIGHT TO BATTERY VOLTAGE.
- TEST LIGHT SHOULD LIGHT. DOES IT?

NO → CKT 432 SHORTED TO VOLTAGE, SHORTED TO CKT 474 OR FAULTY ECM.

YES → PLUGGED SENSOR INLET PORT OR MISSING OR LEAKING SENSOR SEAL OR FAULTY MAP SENSOR.

NO → OPEN SENSOR GROUND CIRCUIT.

"AFTER REPAIRS," REFER TO CODE CRITERIA AND CONFIRM CODE DOES NOT RESET.

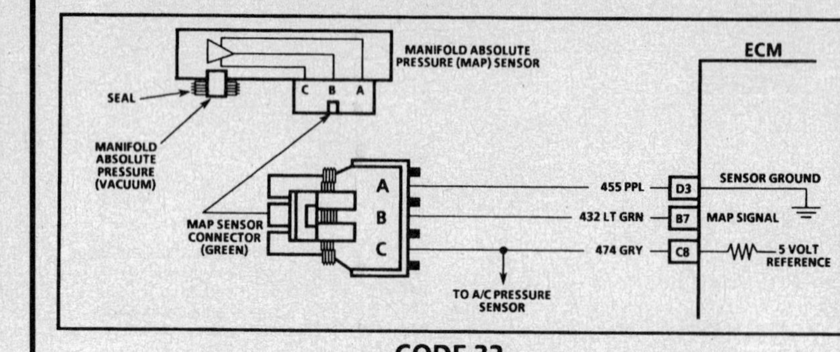

CODE 33

(Page 2 of 2)
MANIFOLD ABSOLUTE PRESSURE (MAP) SENSOR CIRCUIT
(SIGNAL VOLTAGE HIGH/LOW - LOW/HIGH VACUUM)
2.2L (VIN 4) "J" CARLINE (PORT)

Circuit Description:

The Manifold Absolute Pressure (MAP) sensor responds to changes in manifold pressure (vacuum). The ECM receives this information as a signal voltage that will vary from about 1 to 1.5 volts at closed throttle (idle), to 4 to 4.5 volts at Wide Open Throttle (WOT) (low vacuum).

If the MAP sensor fails, the Electronic Control Module (ECM) will substitute a fixed MAP value based on engine rpm and use the Throttle Position Sensor (TPS) to control fuel delivery.

Test Description: Number(s) below refer to circled number(s) on the diagnostic chart.

1. Jumpering harness terminals "B" to "C" (5 volts to signal circuit), will determine if the sensor is at fault, or if there is a problem with the ECM or wiring.
   The "Scan" tool may not display 5 volts. The important thing is that the ECM recognizes the voltage as more than 4 volts, indicating that the ECM, CKT 432, and CKT 474 are OK.
2. This step determines if CKT 474 or CKT 432 is faulty. The "Scan" tool will not display battery voltage but should indicate over 4 volts.

Diagnostic Aids:

With the ignition "ON" and the engine stopped, the manifold pressure is equal to atmospheric pressure and the signal voltage will be high. This information is used by the ECM as an indication of vehicle altitude and is referred to as BARO. Comparison of this BARO reading with a known good vehicle with the sam sensor is a good way to check accuracy of a "suspect" sensor. Reading should be within ± .4 volt.

If Code 33 is intermittent, refer to "Symptoms "

2.2L (VIN 4) ENGINE — DIAGNOSTIC TROUBLE CODE CHART — 1992 CAVALIER

CODE 33
(Page 2 of 2)
MANIFOLD ABSOLUTE PRESSURE (MAP) SENSOR CIRCUIT
(SIGNAL VOLTAGE HIGH/LOW - LOW/HIGH VACUUM)
2.2L (VIN 4) "J" CARLINE (PORT)

FROM CODE 33 PAGE 1.

- ENGINE IDLING.
  DOES "SCAN" TOOL
  DISPLAY MAP VOLTAGE
  BELOW .3 VOLT?

YES

(1)
- IGNITION "OFF."
- DISCONNECT SENSOR ELECTRICAL CONNECTOR.
- JUMPER HARNESS TERMINALS "B" TO "C".
- CRANK ENGINE.
- MAP VOLTAGE SHOULD READ OVER 3.7 VOLTS. DOES IT?

NO

- CODE 33 IS INTERMITTENT. IF NO ADDITIONAL CODES WERE STORED, REFER TO "DIAGNOSTIC AIDS"

NO

(2)
- IGNITION "OFF."
- REMOVE JUMPER WIRE.
- PROBE TERMINAL "B" (CKT 432) WITH A TEST LIGHT TO BATTERY VOLTAGE.
- IGNITION "ON."
- "SCAN" TOOL SHOULD READ OVER 3.7 VOLTS. DOES IT?

YES

FAULTY CONNECTION OR SENSOR.

YES

5 VOLT REFERENCE CIRCUIT OPEN
OR
SHORTED TO GROUND
OR
FAULTY ECM.

NO

CKT 432 OPEN
OR
CKT 432 SHORTED TO GROUND
OR
CKT 432 SHORTED TO SENSOR GROUND
OR
FAULTY ECM.

"AFTER REPAIRS," REFER TO CODE CRITERIA   AND CONFIRM CODE DOES NOT RESET.

2.2L (VIN 4) ENGINE — DIAGNOSTIC TROUBLE CODE CHART — 1993–94 CAVALIER

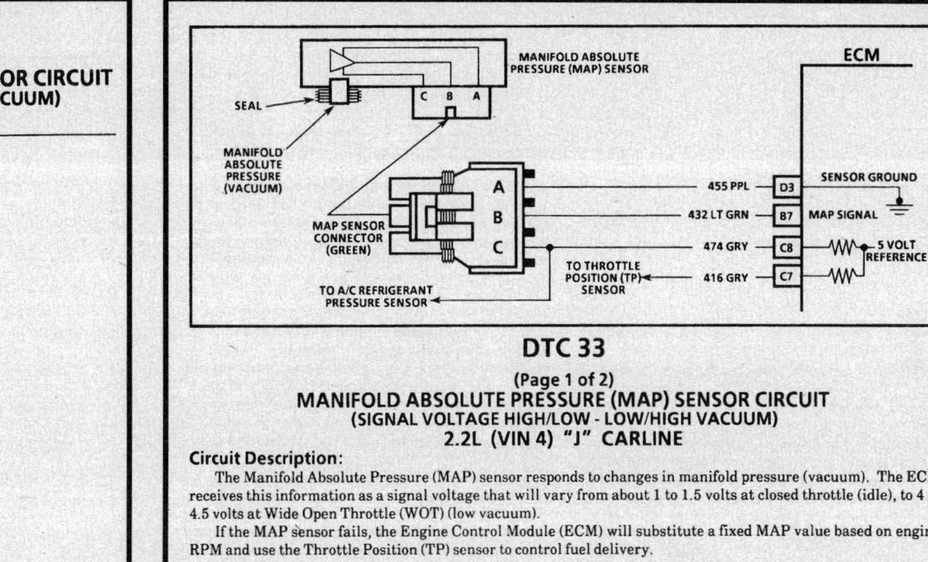

DTC 33
(Page 1 of 2)
MANIFOLD ABSOLUTE PRESSURE (MAP) SENSOR CIRCUIT
(SIGNAL VOLTAGE HIGH/LOW - LOW/HIGH VACUUM)
2.2L (VIN 4) "J" CARLINE

Circuit Description:

The Manifold Absolute Pressure (MAP) sensor responds to changes in manifold pressure (vacuum). The ECM receives this information as a signal voltage that will vary from about 1 to 1.5 volts at closed throttle (idle), to 4 to 4.5 volts at Wide Open Throttle (WOT) (low vacuum).

If the MAP sensor fails, the Engine Control Module (ECM) will substitute a fixed MAP value based on engine RPM and use the Throttle Position (TP) sensor to control fuel delivery.

Test Description: Number(s) below refer to circled number(s) on the diagnostic chart.

1. A DTC 33 will set under the following conditions:
   - MAP signal indicates greater than 3.7 volts (80 kPa).
   - Throttle position is less than 5%.
   - These conditions exist for a time longer than 5 seconds.
   OR
   - MAP signal indicates less than .3 volt (15 kPa).
2. If the ECM recognizes the change, the ECM and CKTs 474 and 432 are OK. If CKT 455 is open, there may also be other DTC(s) stored.

Diagnostic Aids:

With the ignition "ON" and the engine stopped, the manifold pressure is equal to atmospheric pressure and the signal voltage will be high. This information is used by the ECM as an indication of vehicle altitude and is referred to as BARO. Comparison of this BARO reading with a known good vehicle with the same sensor is a good way to check accuracy of a "suspect" sensor. Reading should be within ± .4 volt.

If DTC 33 is intermittent, refer to "Symptoms,"

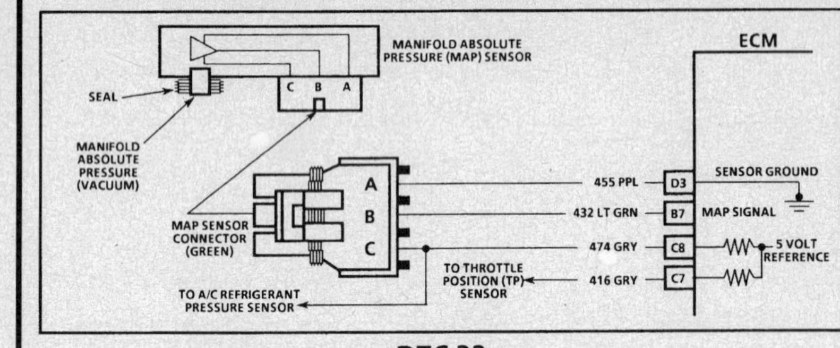

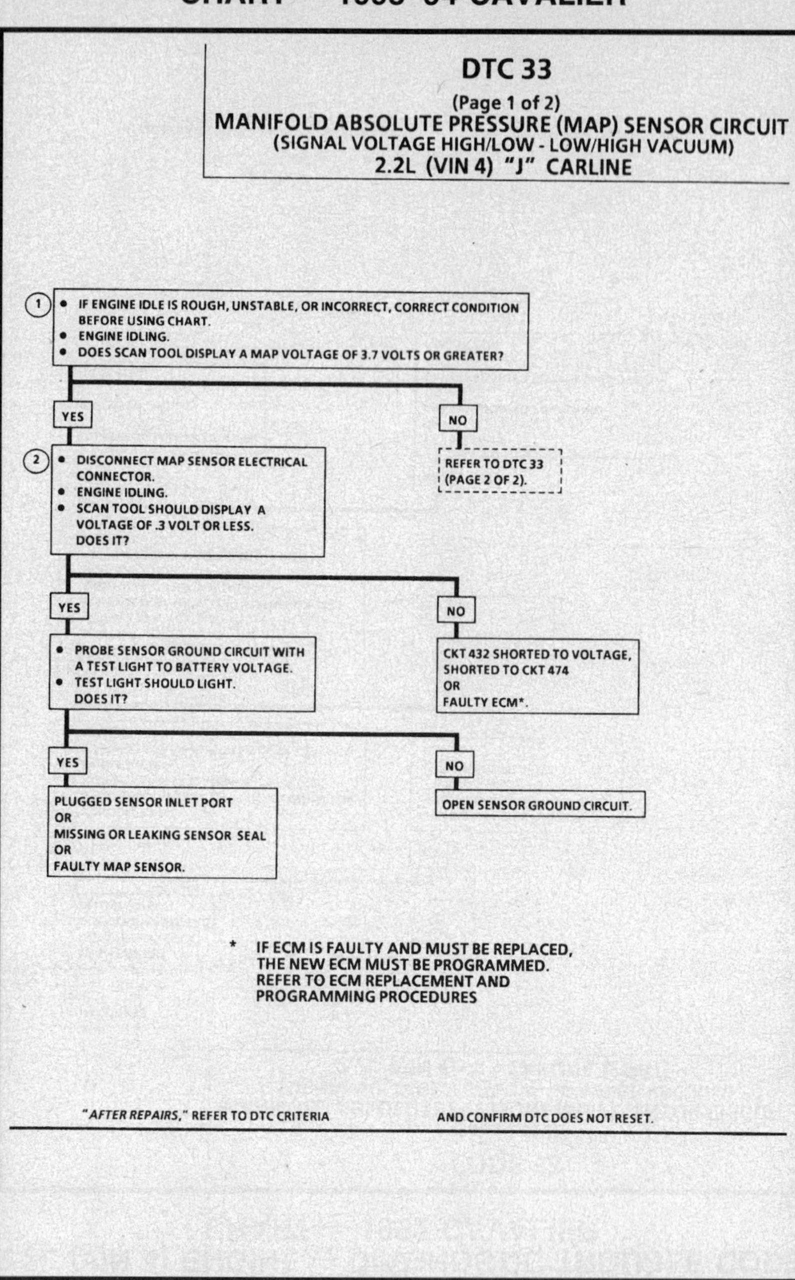

DTC 33

(Page 2 of 2)
MANIFOLD ABSOLUTE PRESSURE (MAP) SENSOR CIRCUIT
(SIGNAL VOLTAGE HIGH/LOW - LOW/HIGH VACUUM)
2.2L (VIN 4) "J" CARLINE

Circuit Description:

The Manifold Absolute Pressure (MAP) sensor responds to changes in manifold pressure (vacuum). The ECM receives this information as a signal voltage that will vary from about 1 to 1.5 volts at closed throttle (idle), to 4 to 4.5 volts at Wide Open Throttle (WOT) (low vacuum).

If the MAP sensor fails, the Engine Control Module (ECM) will substitute a fixed MAP value based on engine RPM and use the Throttle Position (TP) sensor to control fuel delivery.

Test Description: Number(s) below refer to circled number(s) on the diagnostic chart.

1. Jumpering harness terminals "B" to "C" (5 volts to signal circuit), will determine if the sensor is at fault, or if there is a problem with the ECM or wiring.

   The scan tool may not display 5 volts. The important thing is that the ECM recognizes the voltage as more than 4 volts, indicating that the ECM, CKT 432, and CKT 474 are OK.

2. This step determines if CKT 474 or CKT 432 is faulty. The scan tool will not display battery voltage but should indicate over 4 volts.

3. The 5 volt reference Throttle Position (TP) sensor connects to the same 5 volt power supply in the ECM. If the 5 volt reference circuit to the TP sensor has a short to ground, the MAP sensor and A/C refrigerant pressure sensor 5 volt reference circuit will also have a short to ground because they connect to the same power supply in the ECM. A short to ground on either 5 volt reference circuit does not damage the ECM 5 volt power supply. When the short is repaired, the ECM 5 volt supply will return to normal.

Diagnostic Aids:

With the ignition "ON" and the engine stopped, the manifold pressure is equal to atmospheric pressure and the signal voltage will be high. This information is used by the ECM as an indication of vehicle altitude and is referred to as BARO. Comparison of this BARO reading with a known good vehicle with the same sensor is a good way to check accuracy of a "suspect" sensor. Reading should be within $\pm$ .4 volt.

If DTC 33 is intermittent, refer to "Symptoms,"

2.2L (VIN 4) ENGINE — DIAGNOSTIC TROUBLE CODE CHART — 1993–94 CAVALIER

DTC 33

(Page 2 of 2)

MANIFOLD ABSOLUTE PRESSURE (MAP) SENSOR CIRCUIT
(SIGNAL VOLTAGE HIGH/LOW - LOW/HIGH VACUUM)
2.2L (VIN 4) "J" CARLINE

FROM DTC 33 PAGE 1.

- ENGINE IDLING.
  DOES SCAN TOOL
  DISPLAY MAP VOLTAGE
  BELOW .3 VOLT?

YES

(1)
- IGNITION "OFF."
- DISCONNECT SENSOR ELECTRICAL CONNECTOR.
- JUMPER HARNESS TERMINALS "B" TO "C".
- CRANK ENGINE.
- MAP VOLTAGE SHOULD READ OVER 3.7 VOLTS.
  DOES IT?

NO

- DTC 33 IS INTERMITTENT. IF NO
  ADDITIONAL DTC(S) WERE
  STORED, REFER TO
  "DIAGNOSTIC AIDS"

NO

(2)
- IGNITION "OFF."
- REMOVE JUMPER WIRE.
- PROBE TERMINAL "B" (CKT 432) WITH A TEST
  LIGHT TO BATTERY VOLTAGE.
- CRANK ENGINE.
- SCAN TOOL SHOULD READ OVER 3.7 VOLTS.
  DOES IT?

YES

FAULTY CONNECTION
OR
SENSOR.

YES

(3)
5 VOLT REFERENCE CIRCUIT OPEN
OR
SHORTED TO GROUND
OR
FAULTY ECM.*

NO

CKT 432 OPEN
OR
CKT 432 SHORTED TO GROUND
OR
CKT 432 SHORTED TO SENSOR GROUND
OR
FAULTY ECM.*

* IF ECM IS FAULTY AND MUST BE REPLACED, THE NEW ECM MUST BE PROGRAMMED. REFER TO ECM
REPLACEMENT AND PROGRAMMING PROCEDURES

"AFTER REPAIRS," REFER TO DTC CRITERIA   AND CONFIRM DTC DOES NOT RESET.

2.2L (VIN 4) ENGINE — DIAGNOSTIC TROUBLE CODE CHART — CAVALIER

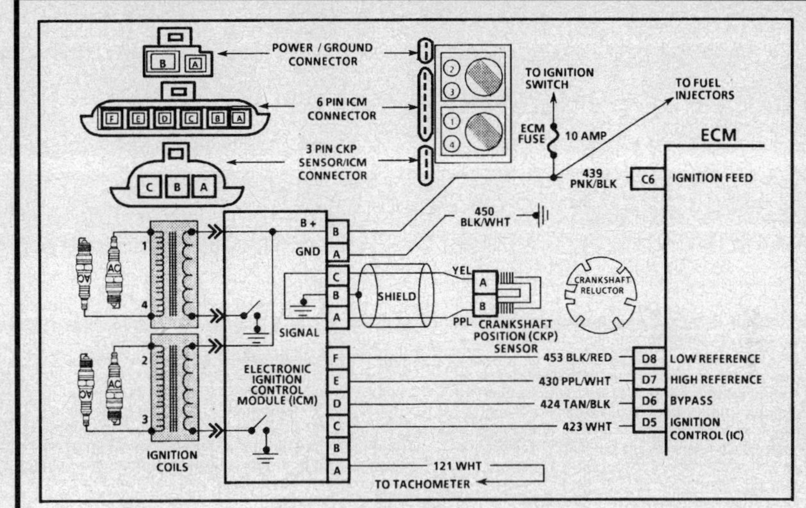

DTC 42

IGNITION CONTROL (IC) CIRCUIT
2.2L (VIN 4) "J" CARLINE

Circuit Description:

The Electronic Ignition Control Module (ICM) sends a reference signal to the Engine Control Module (ECM) when the engine is cranking. While the engine speed is under 400 RPM, the ICM controls the ignition timing. When the system is running on the ICM (no voltage on the bypass line), the ICM grounds the Ignition Control (IC) signal. The ECM expects to sense no voltage on the Ignition Control (IC) line during this condition. If it senses a voltage, it sets DTC 42 and will not enter the ignition control mode.

When the engine speed exceeds 400 RPM, the ECM applies 5 volts to the bypass line to switch the timing to ECM control (ignition control). If the bypass line is open or grounded, once the RPM for ignition control is reached, the ICM will not switch to Ignition Control (IC) mode. This results in low Ignition Control (IC) voltage and the setting of DTC 42. If the IC line is grounded, the ICM will switch to Ignition Control (IC), but because the line is grounded, there will be no IC signal. A DTC 42 will be set.

Test Description: Number(s) below refer to circled number(s) on the diagnostic chart.

1. DTC 42 means the ECM has sensed an open or short to ground in the IC or bypass circuits. This test confirms DTC 42 and that the fault causing the DTC 42 is present.
2. Checks for a normal IC ground path through the ICM. An IC CKT 423, shorted to ground, will also read less than 500 ohms, but this will be checked later.
3. As the test light voltage contacts CKT 424, the ICM should switch, causing the ohmmeter to "overrange" if the meter is in the 1000 to 2000 ohms position. Selecting the 10,000 to 20,000 ohms position will indicate a reading above 5000 ohms. The important thing is that the module "switched."
4. The module did not switch and this step checks for:
   - IC CKT 423 shorted to ground.
   - Bypass CKT 424 open.
   - Faulty ICM connection or ICM.
5. Confirms that DTC 42 is a faulty ECM and not an intermittent in CKT 423 or 424.

Diagnostic Aids:

If DTC 42 is intermittent, refer to "Symptoms,"

DTC 42

IGNITION CONTROL (IC) CIRCUIT
2.2L (VIN 4) "J" CARLINE

1.
- CLEAR DTC(S).
- IDLE ENGINE FOR 1 MINUTE OR UNTIL DTC 42 SETS. DOES DTC 42 SET?

YES →

2.
- IGNITION "OFF."
- DISCONNECT ECM CONNECTORS.
- IGNITION "ON."
- SET OHMMETER SELECTOR SWITCH IN THE 1000 TO 2000 OHMS RANGE.
- GROUND THE BLACK (-) OHMMETER LEAD.
- PROBE ECM HARNESS IGNITION CONTROL CKT 423 USING THE RED (+) OHMMETER LEAD. THE OHMMETER SHOULD READ LESS THAN 500 OHMS. DOES IT?

NO → DTC 42 INTERMITTENT. REFER TO "DIAGNOSTIC AIDS"

YES →
- PROBE ECM HARNESS CONNECTOR CKT 424 WITH A TEST LIGHT TO BATTERY VOLTAGE AND OBSERVE LIGHT.

NO → OPEN CKT 423, FAULTY CONNECTION, OR FAULTY IGNITION CONTROL MODULE (ICM).

LIGHT "OFF" →

3.
- WITH OHMMETER STILL CONNECTED TO ECM HARNESS CKT 423 AND GROUND AS DESCRIBED IN STEP #2, AGAIN PROBE ECM HARNESS CKT 424 WITH THE TEST LIGHT CONNECTED TO BATTERY VOLTAGE. AS TEST LIGHT CONTACTS CKT 424, RESISTANCE SHOULD SWITCH FROM HUNDREDS TO THOUSANDS OHMS. DOES IT?

LIGHT "ON" →
- DISCONNECT IGNITION CONTROL MODULE 6-PIN CONNECTOR.

LIGHT "ON" → CKT 424 SHORTED TO GROUND.

LIGHT "OFF" → REPLACE IGNITION CONTROL MODULE.

NO →

4.
- DISCONNECT IGNITION CONTROL MODULE 6-PIN CONNECTOR. NOTE OHMMETER THAT IS STILL CONNECTED TO CKT 423 AND GROUND. RESISTANCE SHOULD HAVE BECOME VERY HIGH (OPEN CIRCUIT). DOES IT?

YES → CKT 424 OPEN, FAULTY CONNECTIONS, OR FAULTY IGNITION CONTROL MODULE.

NO → CKT 423 SHORTED TO GROUND.

YES (step 3) →

5.
- RECONNECT ECM AND IDLE ENGINE FOR ONE MINUTE OR UNTIL DTC 42 SETS. DOES DTC 42 SET?

YES → REPLACE ECM.*

NO → DTC 42 INTERMITTENT. REFER TO "DIAGNOSTIC AIDS"

*   IF ECM IS FAULTY AND MUST BE REPLACED, THE NEW ECM MUST BE PROGRAMMED. REFER TO ECM REPLACEMENT AND PROGRAMMING PROCEDURES

"AFTER REPAIRS," REFER TO DTC CRITERIA   AND CONFIRM DTC DOES NOT RESET.

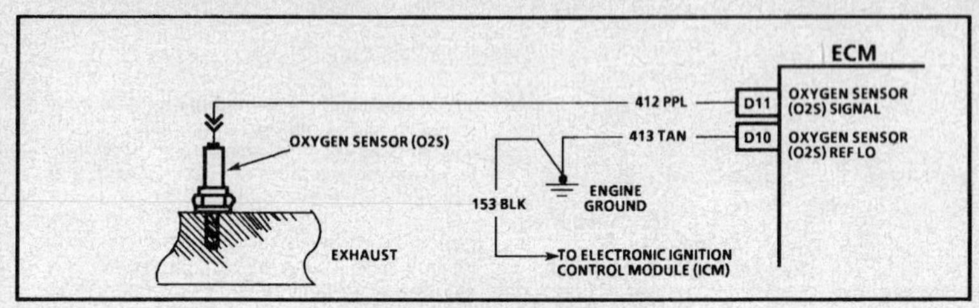

DTC 44

OXYGEN SENSOR (O2S) CIRCUIT
(LEAN EXHAUST INDICATED)
2.2L (VIN 4) "J" CARLINE

Circuit Description:

The Engine Control Module (ECM) supplies a voltage of about .45 volt between terminals "D10" and "D11". (If measured with a 10 megohm digital voltmeter, this may read as low as .32 volt.)

When the O2S reaches operating temperature, it varies this voltage from about .1 volt (exhaust is lean) to about .9 volt (exhaust is rich).

The sensor is like an open circuit and produces no voltage when it is below 316° (600°F). An open sensor circuit, or cold sensor, causes "Open Loop" operation.

Test Description: Number(s) below refer to circled number(s) on the diagnostic chart.

1. DTC 44 is set when the O2S signal voltage on CKT 412 remains below .28 (280 mV) volt for 25 seconds or longer.

Diagnostic Aids:

Using the scan tool, observe the long term fuel trim value at different engine speeds. If the conditions for DTC 44 exist, the long term fuel trim values will be around 160 or higher.

Check the following possible causes:

- Oxygen Sensor Wire - Sensor pigtail may be mispositioned and contacting the exhaust manifold. Check for ground in wire between connector and sensor.

- Fuel Contamination - Water, even in small amounts, near the in-tank fuel pump inlet can be delivered to the injectors. The water causes a lean exhaust and can set a DTC 44.

- Fuel Pressure - System will be lean if fuel pressure is too low. It may be necessary to monitor fuel pressure while driving the vehicle at various road speeds and/or loads to confirm. Refer to "Fuel System Diagnosis," CHART A-7.

- Exhaust Leaks - If there is an exhaust leak, the engine can cause outside air to be pulled into the exhaust and past the sensor. Vacuum or crankcase leaks can cause a lean condition.

- If DTC 44 is intermittent, refer to "Symptoms,"

- A cracked or otherwise damaged O2S may set an intermittent DTC 44.

2.2L (VIN 4) ENGINE — DIAGNOSTIC TROUBLE CODE CHART — CAVALIER

DTC 44
OXYGEN SENSOR (O2S) CIRCUIT
(LEAN EXHAUST INDICATED)
2.2L (VIN 4) "J" CARLINE

① • RUN WARM ENGINE (75°C/167°F TO 95°C/203°F) AT 1200 RPM.
   DOES SCAN TOOL INDICATE OXYGEN SENSOR VOLTAGE FIXED BELOW .28 VOLT (280 mV)?

YES
• DISCONNECT OXYGEN SENSOR ELECTRICAL CONNECTOR.
• WITH ENGINE IDLING, SCAN TOOL SHOULD DISPLAY OXYGEN SENSOR VOLTAGE BETWEEN .28 VOLT AND .7 VOLT (280 mV AND 700 MV). DOES IT?

NO
DTC 44 IS INTERMITTENT. IF NO ADDITIONAL DTC(S) WERE STORED, REFER TO "DIAGNOSTIC AIDS"

YES
REFER TO "DIAGNOSTIC AIDS"

NO
CKT 412 SHORTED TO GROUND OR FAULTY ECM.*

* IF ECM IS FAULTY AND MUST BE REPLACED, THE NEW ECM MUST BE PROGRAMMED. REFER TO ECM REPLACEMENT AND PROGRAMMING PROCEDURES

"AFTER REPAIRS," REFER TO DTC CRITERIA   AND CONFIRM DTC DOES NOT RESET.

2.2L (VIN 4) ENGINE — DIAGNOSTIC TROUBLE CODE CHART — CAVALIER

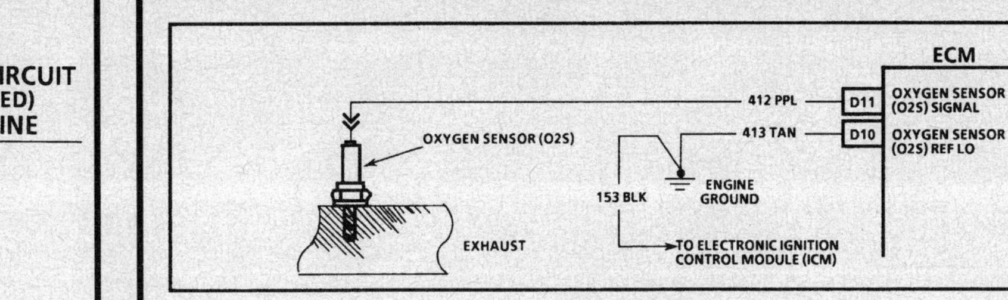

DTC 45
OXYGEN SENSOR (O2S) CIRCUIT
(RICH EXHAUST INDICATED)
2.2L (VIN 4) "J" CARLINE

Circuit Description:

The Engine Control Module (ECM) supplies a voltage of about .45 volt between terminals "D10" and "D11". (If measured with a 10 megohm digital voltmeter, this may read as low as .32 volt.)

When the Oxygen Sensor (O2S) reaches operating temperature, it varies this voltage from about .1 volt (exhaust is lean) to about .9 volt (exhaust is rich).

The sensor is like an open circuit and produces no voltage when it is below 316°C (600°F). An open sensor circuit, or cold sensor, causes "Open Loop" operation.

Test Description: Number(s) below refer to circled number(s) on the diagnostic chart.
1. DTC 45 is set when the O2S signal voltage on CKT 412 remains above .7 (700 mV) volt under the following conditions:
   • Engine run time after start is 2 minutes or more.
   • System is operating in "Closed Loop."
   • Throttle angle is greater than 5%.
   • Above conditions exist for 51 seconds or more.

Diagnostic Aids:

Using the scan tool, observe the long term fuel trim value at different engine speeds. If the conditions for DTC 45 exist, the long term fuel trim values will be around 90 or lower.
   Check the following possible causes:
   • Fuel Pressure - System will go rich, if pressure is too high. The ECM can compensate for some increase. However, if it gets too high, a DTC 45 will be set. Refer to "Fuel System Diagnosis" CHART A-7.
   • Leaking Fuel Injector - Refer to CHART A-7.
   • An open ground CKT 453 - May result in induced electrical "noise." The ECM interprets this "noise" as reference pulses. The additional pulses result in a higher than actual engine speed signal. The ECM then delivers too much fuel causing the system to go rich. The engine tachometer will also show higher than actual engine speed, which can help in diagnosing this problem.

   • Canister Purge - Check for fuel saturation. If evaporative emission vapor canister is full of fuel, check EVAP canister control valve and hoses. Refer to "Evaporative Emission (EVAP) Control System,"
   • MAP Sensor - An output that causes the ECM to sense a higher than normal manifold pressure (low vacuum) can cause the system to go rich. Disconnecting the Manifold Absolute Pressure (MAP) sensor will allow the ECM to set a fixed value for the MAP sensor. Substitute a different MAP sensor if the rich condition is gone, while the sensor is disconnected.
   • TP Sensor - An intermittent Throttle Position (TP) sensor output will cause the system to operate richly due to a false indication of the engine accelerating.
   • O2S Contamination - Inspect Oxygen Sensor (O2S) for silicone contamination from fuel, or use of improper RTV sealant. The sensor may have a white, powdery coating and result in a high but false signal voltage (rich exhaust indication). The ECM will then reduce the amount of fuel delivered to the engine causing a severe surge driveability problem.
   • EGR Valve - Exhaust Gas Recirculation (EGR) sticking open at idle is usually accompanied by a rough idle and/or stall condition.
   • If DTC 45 is intermittent, refer to "Symptoms,"

   • Engine Oil Contamination - Fuel fouled engine oil could cause the O2S to sense a rich air/fuel mixture and set a DTC 45.

2.2L (VIN 4) ENGINE — DIAGNOSTIC TROUBLE CODE CHART — CAVALIER

DTC 45

OXYGEN SENSOR (O2S) CIRCUIT
(RICH EXHAUST INDICATED)
2.2L (VIN 4) "J" CARLINE

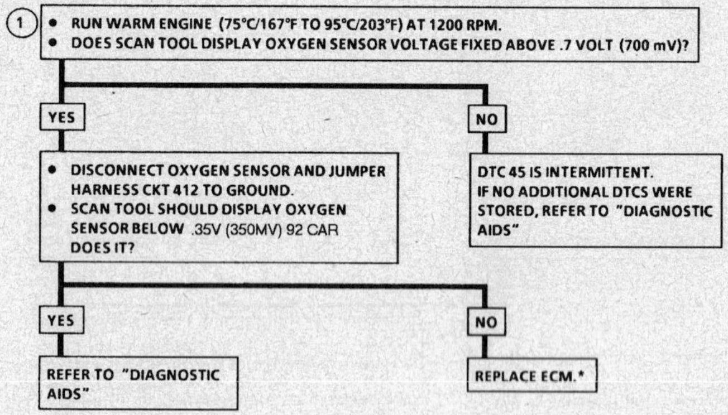

1.
- RUN WARM ENGINE (75°C/167°F TO 95°C/203°F) AT 1200 RPM.
- DOES SCAN TOOL DISPLAY OXYGEN SENSOR VOLTAGE FIXED ABOVE .7 VOLT (700 mV)?

YES
- DISCONNECT OXYGEN SENSOR AND JUMPER HARNESS CKT 412 TO GROUND.
- SCAN TOOL SHOULD DISPLAY OXYGEN SENSOR BELOW .35V (350MV) 92 CAR DOES IT?

NO
DTC 45 IS INTERMITTENT.
IF NO ADDITIONAL DTCS WERE STORED, REFER TO "DIAGNOSTIC AIDS"

YES
REFER TO "DIAGNOSTIC AIDS"

NO
REPLACE ECM.*

*IF ECM IS FAULTY AND MUST BE REPLACED, THE NEW ECM MUST BE PROGRAMMED.
REFER TO ECM REPLACEMENT AND PROGRAMMING PROCEDURES

"AFTER REPAIRS," REFER TO DTC CRITERIA AND CONFIRM DTC DOES NOT RESET.

2.2L (VIN 4) ENGINE — DIAGNOSTIC TROUBLE CODE CHART — CAVALIER

DTC 51

EEPROM OR ECM FAILURE
2.2L (VIN 4) "J" CARLINE

DIAGNOSTIC TROUBLE CODE (DTC) 51
ECM FAILURE
(ECM FAILED OR EEPROM FAILURE)

CHECK THAT ALL ECM CONNECTIONS ARE GOOD.
IF OK, CLEAR MEMORY AND RECHECK ECM.
IF DTC 51 REAPPEARS, REPLACE ECM AND PROGRAM NEW ECM.

* IF ECM IS FAULTY AND MUST BE REPLACED, THE NEW ECM MUST BE PROGRAMMED. REFER TO ECM REPLACEMENT AND PROGRAMMING PROCEDURES

"AFTER REPAIRS," CONFIRM "CLOSED LOOP" OPERATION AND NO MIL (CHECK ENGINE).

2.2L (VIN 4) ENGINE — DIAGNOSTIC TROUBLE CODE CHART — 1992 CAVALIER

CODE 66

A/C PRESSURE SENSOR CIRCUIT
2.2L (VIN 4) "J" CARLINE (PORT)

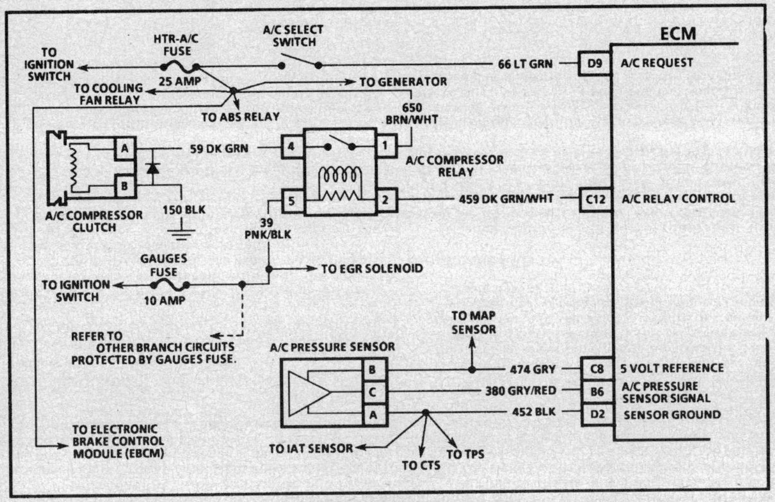

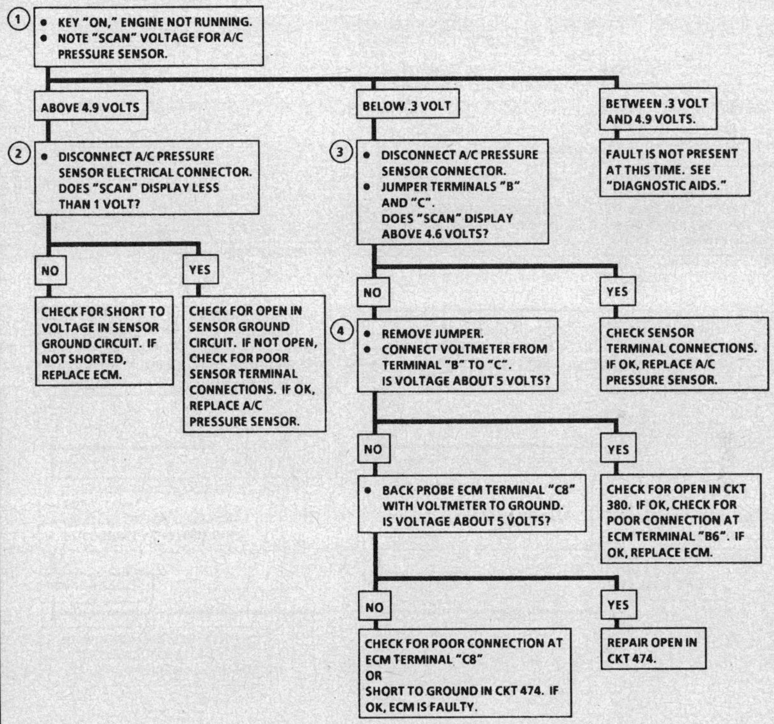

Circuit Description:

The A/C pressure sensor responds to changes in A/C refrigerant system high side pressure. This input indicates how much load the A/C compressor is putting on the engine and is one of the factors used by the ECM to determine IAC valve position for idle speed control. The circuit consists of a 5 volt reference and a ground, both provided by the ECM, and a signal line to the ECM. The signal is a voltage which is proportional to the pressure. The sensor's range of operation is 0 to 450 psi. At 0 psi, the signal will be about .1 volt, varying up to about 4.9 volts at 450 psi or above. Code 66 sets if the voltage is above 4.9 volts (454 psi) or below .3 volt (7.5 psi) for 5 seconds or more. The A/C compressor is disabled by the ECM if fault is currently present.

Test Description: Number(s) below refer to circled number(s) on the diagnostic chart.

1. This step checks the voltage signal being received by the ECM from the A/C pressure sensor. The normal operating range is between .1 volt and 4.9 volts.
2. Checks to see if the high voltage signal is from a shorted sensor or a short to voltage in the circuit. Normally, disconnecting the sensor would make a normal circuit go to near zero volt.
3. Checks to see if low voltage signal is from the sensor or the circuit. Jumpering the sensor signal CKT 380 to 5 volts, checks the circuit, connections, and ECM.
4. This step checks to see if the low voltage signal was due to an open in the sensor circuit or the 5 volt reference circuit since the prior step eliminated the pressure sensor.

Diagnostic Aids:

Code 66 sets when signal voltage falls outside the normal possible range of the sensor and is not due to a refrigerant system problem. If problem is intermittent, check for opens or shorts in harness or poor connections.

2.2L (VIN 4) ENGINE — DIAGNOSTIC TROUBLE CODE CHART — 1992 CAVALIER

CODE 66

A/C PRESSURE SENSOR CIRCUIT
2.2L (VIN 4) "J" CARLINE (PORT)

"AFTER REPAIRS," CONFIRM "CLOSED LOOP" OPERATION AND NO "CHECK ENGINE" LIGHT.

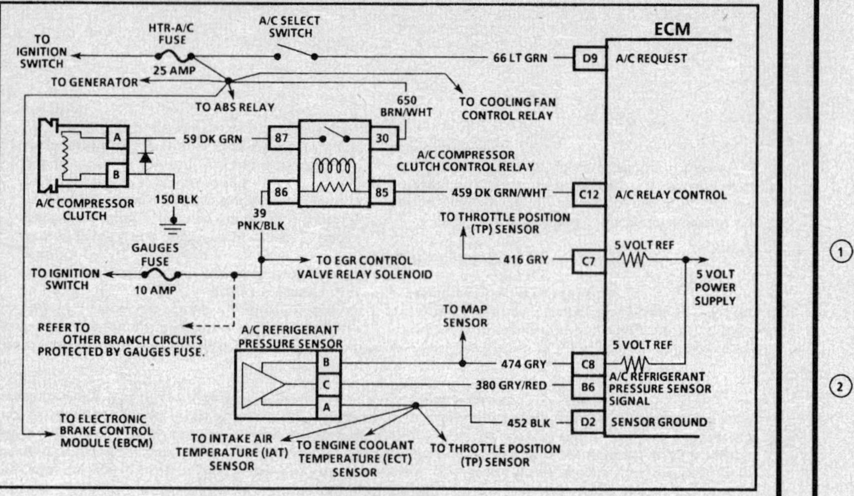

DTC 66
A/C REFRIGERANT PRESSURE SENSOR CIRCUIT
2.2L (VIN 4) "J" CARLINE

Circuit Description:

The A/C refrigerant pressure sensor responds to changes in A/C refrigerant system high side pressure. This input indicates how much load the A/C compressor is putting on the engine and is one of the factors used by the ECM to determine IAC valve position for idle speed control. The circuit consists of a 5 volt reference and a ground, both provided by the ECM, and a signal line to the ECM. The signal is a voltage which is proportional to the pressure. The sensor's range of operation is 0 to 454 psi. At 0 psi, the signal will be about .1 volt, varying up to about 4.9 volts at 454 psi or above. DTC 66 sets if the voltage is above 4.9 volts (454 psi) or below .3 volt (7.5 psi) for 5 seconds or more. The A/C compressor clutch is disabled by the ECM if fault is present.

Test Description: Number(s) below refer to circled number(s) on the diagnostic chart.

1. This step checks the voltage signal being received by the ECM from the A/C refrigerant pressure sensor.
2. Checks to see if the high voltage signal is from a shorted sensor or a short to voltage in the circuit. Normally, disconnecting the sensor would make a normal circuit go to near zero volt.
3. Checks to see if low voltage signal is from the sensor or the circuit. Jumpering the sensor signal CKT 380 to 5 volts, checks the circuit, connections, and ECM.
4. This step checks to see if the low voltage signal was due to an open in the sensor circuit or the 5 volt reference circuit since the prior step eliminated the pressure sensor.
5. The 5 volt reference for the Throttle Position (TP) sensor connects to the same 5 volt power supply in the ECM. If the 5 volt reference circuit to TP sensor has a short to ground, the MAP sensor and A/C refrigerant pressure sensor 5 volt reference circuit will also have a short to ground because they connect to the same power supply in the ECM. A short to ground on either 5 volt reference circuit does not damage the ECM 5 volt power supply. When the short is repaired, the ECM 5 volt supply will return to normal.

Diagnostic Aids:

At temperatures of -20 degrees farenheit and below, the A/C pressure can drop below the .3 volt (7.5 psi) threshold. If this happens the DTC 66 will set and disable the A/C system. Clear DTC 66 and check A/C pressure in a heated garage, if pressure is between .3 volt (7.5 psi) and 4.9 volts (454 psi), the A/C pressure transducer is functioning correctly.

DTC 66 sets when the signal voltage falls outside the normal possible range of the sensor. If signal voltage is greater than .1 volt, but less than .3 volt, there could be a low refrigerant pressure problem with the A/C system.

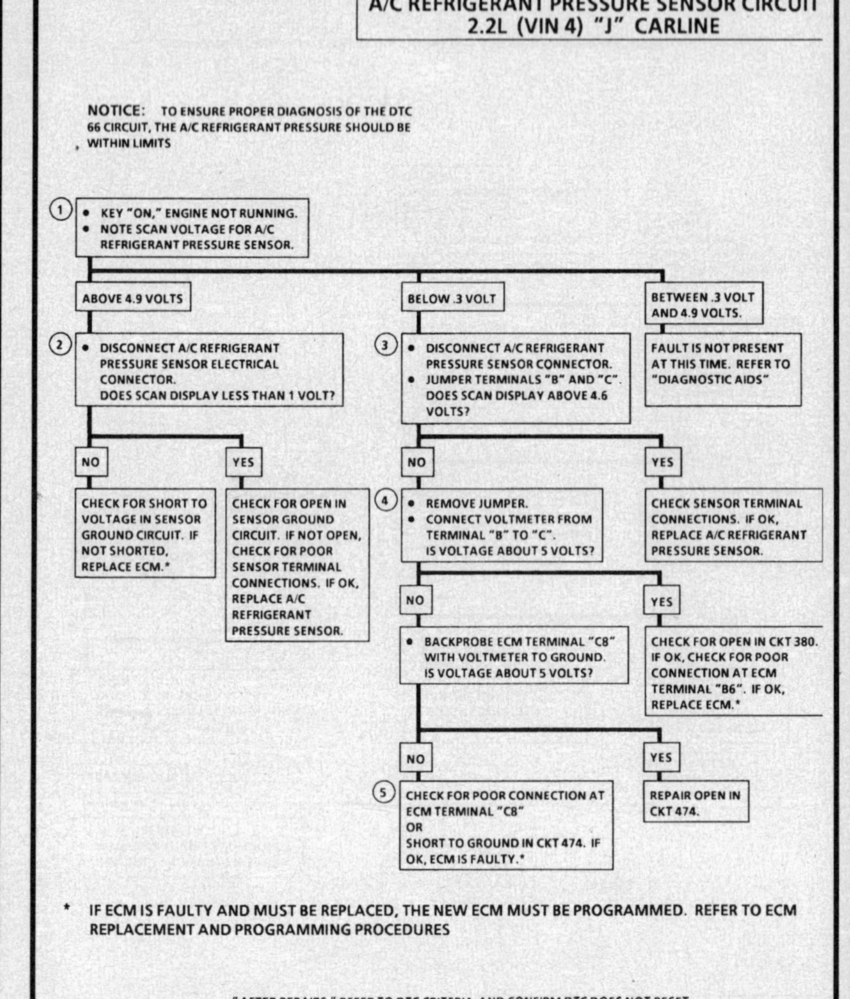

2.2L (VIN 4) ENGINE — ECM SYMPTOM CHART — 1992 CAVALIER

PINK A-B 24 PIN ECM CONNECTOR

ECM PIN/FUNCTION		CKT #	WIRE COLOR	COMPONENT CONNECTOR CAVITY	NORMAL VOLTAGES		CODES AFFECT	POSSIBLE SYMPTOMS FROM FAULTY CIRCUIT
					KEY "ON"	ENG RUN		
A4	PEAK AND HOLD JUMPER	887	TAN	ECM-"A12"	(8)	(8)		(3) CRANKS, BUT WON'T RUN.
A5	INJECTOR DRIVER	467	DK BLU	INJECTOR JUMPER HARNESS CONNECTOR (GRY) "B"	B +	B +		(3) CRANKS, BUT WON'T RUN. (4) POSSIBLE NO START (BLOWN 10 AMP ECM FUSE) OR INJECTORS ENERGIZED AT ALL TIMES WITH KEY "ON."
A7	FUEL PUMP RELAY CONTROL	465	DK GRN/WHT	FUEL PUMP RELAY - "5"	(6)	B +		(3) LONG CRANKING TIME BEFORE ENGINE STARTS.
A8	IAC "A" HIGH	441	LT BLU/WHT	IAC VALVE "D"	NOT USABLE	NOT USABLE		(5) INCORRECT, UNSTABLE IDLE.
A9	IAC "A" LOW	442	LT BLU/BLK	IAC VALVE "C"	NOT USABLE	NOT USABLE		(5) INCORRECT, UNSTABLE IDLE.
A10	IAC "B" LOW	444	LT GRN/BLK	IAC VALVE "A"	NOT USABLE	NOT USABLE		(5) INCORRECT, UNSTABLE IDLE.
A11	IAC "B" HIGH	443	LT GRN/WHT	IAC VALVE "B"	NOT USABLE	NOT USABLE		(5) INCORRECT, UNSTABLE IDLE.
A12	PEAK AND HOLD JUMPER	887	TAN	ECM "A4"	(8)	(8)		(3) CRANKS, BUT WON'T RUN.

NOTICE: Voltage may vary due to low battery charge or other reasons, but should be very close. All voltages shown in the "ENG RUN" column are typical with engine at idle, closed throttle, normal operating temperature, park or neutral, system in "Closed Loop" all accessories "OFF," brake not applied.

(1) Increases with vehicle speed (Measure on "AC volts" scale).
(2) Varies.
(3) Open circuit.
(4) Grounded circuit.
(5) Open or grounded circuit.
(6) Measures B + for 2 seconds after key "ON," then measures "0" volts.
(7) Varies with temperature.
(8) Less than .5 volt (500 mV).

2.2L (VIN 4) ENGINE — ECM SYMPTOM CHART — 1992 CAVALIER

PINK A-B 24 PIN ECM CONNECTOR

ECM PIN/FUNCTION		CKT #	WIRE COLOR	COMPONENT CONNECTOR CAVITY	NORMAL VOLTAGES		CODES AFFECT	POSSIBLE SYMPTOMS FROM FAULTY CIRCUIT
					KEY "ON"	ENG RUN		
B4	CTS SIGNAL	410	YEL	CTS "B"	(7)	(7)	14(5)	(5) INCORRECT IDLE, COOLING FAN RUNS AT ALL TIMES.
B5	TPS SIGNAL	417	DK BLU	TPS "C"	.6V (9)	.6V (9)	21(5)	(5) POOR PERFORMANCE, HESITATION.
B6	A/C PRESSURE SENSOR SIGNAL	380	GRY/RED	A/C PRESSURE SENSOR "C"	(2)	(2)	66(5)	(5) NO A/C COOLING, A/C CLUTCH INOPERATIVE, REFER TO CODE 66.
B7	MAP SIGNAL	432	LT GRN	MAP SENSOR "B"	4.75V	(2)	33(5)	(5) POOR PERFORMANCE, ROUGH IDLE, SURGE REFER TO CHART C-1D.
B8	IAT SIGNAL	472	TAN	IAT SENSOR "B"	(7)	(7)	23(5)	(5) POOR PERFORMANCE, HARD TO START.
B9	ALDL DIAGNOSTIC TEST TERMINAL	451	WHT/BLK	ALDL "B"	5.0V	5.0V	—	(3) NO ALDL DATA, WON'T FLASH CODES. (4) KEY "ON" - CODES DISPLAYED ENG RUN - FIELD SERVICE MODE "ON."
B10	VSS SIGNAL (HIGH)	1232	LT BLU	VEHICLE SPEED SENSOR (VSS) "A"			24(5)	(5) NO VSS SIGNAL, INACCURATE OR INOPERATIVE SPEEDOMETER, INOPERATIVE CRUISE CONTROL, TCC INOPERATIVE.
B11	VSS SIGNAL (LOW)	1233	DK GRN/YEL	VEHICLE SPEED SENSOR (VSS) "B"			24(5)	(5) NO VSS SIGNAL, INACCURATE OR INOPERATIVE SPEEDOMETER, INOPERATIVE CRUISE CONTROL, TCC INOPERATIVE.

NOTICE: Voltages may vary due to low battery charge or other reasons, but should be very close. All voltages shown in the "ENG. RUN" column are typical with engine at idle, closed throttle, normal operating temperature, park or neutral, system in "Closed Loop" all accessories "OFF," brake not applied.

(1) Increases with vehicle speed (Measure on "AC volts" scale).
(2) Varies.
(3) Open circuit.
(4) Grounded circuit.
(5) Open or grounded circuit.
(6) Measures B + for 2 seconds after key "ON," then measures "0" volt.
(7) Varies with temperature.
(8) Less than .5 volt (500 mV).
(9) Within the range of .33-1.33 volts.

2.2L (VIN 4) ENGINE — ECM SYMPTOM CHART — 1992 CAVALIER

PINK C-D 32 PIN ECM CONNECTOR

ECM PIN FUNCTION	CKT #	WIRE COLOR	COMPONENT CONNECTOR CAVITY	NORMAL VOLTAGES KEY "ON"	NORMAL VOLTAGES ENG RUN	CODES AFFECT	POSSIBLE SYMPTOMS FROM FAULTY CIRCUIT
C1 ECM GROUND	450	BLK/WHT	ENGINE GROUND *	(8)	(8)		(3) NO EFFECT UNLESS "D1" IS ALSO OPEN. IF BOTH ARE OPEN, CRANKS, BUT WON'T RUN.
C4 PARK/NEUTRAL SWITCH	434	ORN/BLK	NEUTRAL START BACK-UP SWITCH "A"	(8)	(8)		(5) INCORRECT IDLE.
C6 IGNITION FEED	439	PNK/BLK	"DIS" MODULE 2 PIN CONNECTOR "B", AND FUEL INJECTOR JUMPER HARNESS CONNECTOR "A"	B+	B+		(3) CRANKS, BUT WON'T RUN. (4) CRANKS, BUT WON'T RUN, BLOWN 10 AMP ECM FUSE.
C7 TPS 5 VOLT REFERENCE	416	GRY	TPS-"A"	5.0V	5.0V	21(5)	(5) DEGRADED PERFORMANCE.
C8 MAP & A/C PRESSURE SENSOR 5 VOLT REFERENCE	474	GRY	MAP-"C" A/C PRESSURE SENSOR-"B"	5.0V	5.0V	33(3)	(3) POOR PERFORMANCE.
C9 & C10 KEEP ALIVE MEMORY (B+)	2	RED	POWER FEED CONNECTOR	B+	B+		(3) CRANKS, BUT WON'T RUN. (4) CRANKS, BUT WON'T RUN, BLOWN FUSIBLE LINK.
C11 TCC (A/T) OR SHIFT LIGHT (M/T) CONTROL	422 456	TAN/BLK	TCC SOLENOID "D" ALDL-"F" *	A/T-(8) M/T-B+	A/T-(8) M/T-B+		(3) TCC OR SHIFT LIGHT INOPERATIVE, POOR FUEL ECONOMY. (4)TCC ENGAGES AT VERY LOW ROAD SPEED-(A/T) OR SHIFT LIGHT "ON" AT ALL TIMES-(M/T).
C12 A/C RELAY CONTROL	459	DK GRN/WHT	A/C COMPRESSOR RELAY-"2"	B+	B+		(3) NO A/C COOLING. (4) NO A/C COOLING-"GAUGES" FUSE BLOWN.
C13 COOLING FAN RELAY CONTROL	335	DK GRN/WHT	COOLING FAN RELAY-"2"	B+	B+		(3) ENGINE OVERHEATS, DETONATION, SPARK KNOCK. (4) COOLING FAN RUNS AT ALL TEMPERATURES.
C14 EGR SOLENOID CONTROL	435	GRY	EGR SOLENOID "B"	B+	B+	32(5)	(3) EGR VALVE DOESN'T OPEN. DETONATION, SPARK KNOCK. (4) DEGRADED PERFORMANCE. IMPORTANT: EGR SOLENOID IS DISABLED IF CODE 14, 21, 23, 32 OR 33 IS SET.

NOTICE: Voltage may vary due to low battery charge or other reasons, but should be very close. All voltages shown in the "ENG RUN" column are typical with engine at idle, close throttle, normal operating temperature, park or neutral, system in "Closed Loop" all accessories "OFF," brake not applied.

(2) Varies.
(3) Open circuit.
(4) Grounded circuit.
(5) Open or grounded circuit.
(7) Varies with temperature.
(8) Less than .5 volt (500 mV).
* Refer to wiring diagrams for a complete illustration of components in this circuit.

2.2L (VIN 4) ENGINE — ECM SYMPTOM CHART — 1992 CAVALIER

PINK A-B 24 PIN ECM CONNECTOR

ECM PIN FUNCTION	CKT #	WIRE COLOR	COMPONENT CONNECTOR CAVITY	NORMAL VOLTAGES KEY "ON"	NORMAL VOLTAGES ENG RUN	CODES AFFECT	POSSIBLE SYMPTOMS FROM FAULTY CIRCUIT
D1 ECM GROUND	551	TAN/WHT	ENGINE GROUND *	(8)	(8)		(3) NO EFFECT UNLESS "C1" IS ALSO OPEN. IF BOTH ARE OPEN, CRANKS, BUT WON'T RUN.
D2 SENSOR GROUND	452	BLK	A/C PRESSURE SENSOR-"A", TPS-"B", CTS-"A", AND IAT-"A"	(8)	(8)	14(3) 21(3) 23(3) 66(3)	(3) HARD START OR CRANKS BUT WON'T RUN
D3 SENSOR GROUND	455	PPL	MAP-"A"	(8)	(8)	33(3)	(3) HARD START OR CRANKS BUT WON'T RUN
D5 EST	423	WHT	"DIS" MODULE 6 PIN CONN. "C"	(8)	2.4V	42(5)	(5) STUMBLES, UNSTABLE IDLE. DEGRADED PERFORMANCE
D6 BY PASS	424	TAN/BLK	"DIS" MODULE 6 PIN CONN "D"	(8)	4.8V	42(5)	(5) DEGRADED PERFORMANCE
D7 IGNITION REFERENCE HIGH	430	PPL/WHT	"DIS" MODULE 6 PIN CONN. "E"	(8)	3.2V		(5) CRANKS BUT WON'T RUN.
D8 IGNITION REFERENCE LOW	453	BLK/RED	"DIS" MODULE 6 PIN CONN. "F"	(8)	(8)		(3) REDUCED PERFORMANCE.
D9 A/C REQUEST	66	LT GRN	HVAC MODE SELECT CONN "A"	(8)	(8)		(3) NO A/C COOLING. A/C CLUTCH WON'T ENGAGE
D10 OXYGEN SENSOR REFERENCE LOW	413	TAN	ENGINE GROUND	(8)	(8)	13(3)	(3) OPEN LOOP OPERATION, REDUCED PERFORMANCE
D11 OXYGEN SENSOR SIGNAL	412	PPL	OXYGEN SENSOR	01-55V	1-.9V (10)	13 44 45	(5) OPEN LOOP OPERATION, STRONG EXHAUST ODOR, REDUCED PERFORMANCE.
D12 SERIAL DATA	461	ORN	ALDL CONNECTOR "M"	4.7V	4.7V		(5) NO DATA
D13 VEHICLE SPEED SENSOR OUTPUT	389	DK GRN	I/P CONNECTOR "A", AND I/P CLUSTER CONN "R"				(5) INCORRECT SPEEDOMETER READING. CRUISE CONTROL INOPERATIVE.
D14 "CHECK ENGINE" LIGHT CONTROL	419	BRN/WHT	I/P CONNECTOR "R", AND I/P CLUSTER CONN "I"	(8)	B+		(3) LIGHT NOT "ON" FOR "BULB CHECK," WON'T FLASH CODES (4) LIGHT "ON" ALL THE TIME. DOES NOT FLASH. REFER TO "DIAGNOSTIC CIRCUIT CHECK."

NOTICE: Voltage may vary due to low battery charge or other reasons, but should be very close. All voltages shown in the "ENG RUN" column are typical with engine at idle, closed throttle, normal operating temperature, park or neutral, system in "Closed Loop" all accessories "OFF," brake not applied.

(3) Open circuit.
(4) Grounded circuit.
(5) Open or grounded circuit.
(8) Less than .5 volt (500 mV).
(10) Actively varies within indicated range

2.2L (VIN 4) ENGINE — ECM SYMPTOM CHART — 1993–94 CAVALIER

ECM Connector and Driveability Symptoms Identification

This ECM voltage chart is for use with a J 39200 to further aid in diagnosis. These voltages were derived from a known good vehicle. The voltages you get may vary due to low battery charge or other reasons, but they should be very close.

THE FOLLOWING CONDITIONS MUST BE MET BEFORE TESTING:

- Engine at operating temperature • Closed Loop • Engine idling (for "Engine Run" column)
- Test terminal not grounded • Scan tool not installed • Brake not applied.

ECM PIN/FUNCTION	CKT #	WIRE COLOR	COMPONENT CONNECTOR CAVITY	NORMAL VOLTAGES KEY "ON"	NORMAL VOLTAGES ENG RUN	DTC(S) AFFECTED	POSSIBLE SYMPTOMS FROM FAULTY CIRCUIT
A4 PEAK AND HOLD JUMPER	887	TAN	ECM-"A12"	(8)	(8)		(3) CRANKS, BUT WON'T RUN.
A5 INJECTOR DRIVER	467	DK BLU	INJECTOR JUMPER HARNESS CONNECTOR (GRY) "B"	B+	B+ (13)		(3) CRANKS, BUT WON'T RUN. (4) POSSIBLE NO START (OPEN 10 AMP ECM FUSE) OR INJECTORS ENERGIZED AT ALL TIMES WITH KEY "ON."
A7 FUEL PUMP RELAY CONTROL	465	DK GRN/WHT	FUEL PUMP RELAY - "86"	(6)	B+		(3) LONG CRANKING TIME BEFORE ENGINE STARTS.
A8 IAC "A" HIGH	441	LT BLU/WHT	IAC VALVE "D"	NOT USABLE	NOT USABLE		(5) INCORRECT, UNSTABLE IDLE.
A9 IAC "A" LOW	442	LT BLU/BLK	IAC VALVE "C"	NOT USABLE	NOT USABLE		(5) INCORRECT, UNSTABLE IDLE.
A10 IAC "B" LOW	444	LT GRN/BLK	IAC VALVE "A"	NOT USABLE	NOT USABLE		(5) INCORRECT, UNSTABLE IDLE.
A11 IAC "B" HIGH	443	LT GRN/WHT	IAC VALVE "B"	NOT USABLE	NOT USABLE		(5) INCORRECT, UNSTABLE IDLE.
A12 PEAK AND HOLD JUMPER	887	TAN	ECM "A4"	(8)	(8)		(3) CRANKS, BUT WON'T RUN.

(3) OPEN CIRCUIT.
(4) GROUNDED CIRCUIT.
(5) OPEN OR GROUNDED CIRCUIT.
(6) MEASURES B + FOR 2 SECONDS AFTER KEY "ON," THEN MEASURES "0" VOLTS.
(8) LESS THAN .5 VOLT (500 mV).
(13) ALTERNATE TEST: SET J 39200 DVM TO DC VOLTAGE FREQUENCY SCALE. CONNECT RED LEAD TO IGNITION FEED ("C6" AT ECM) AND BLACK LEAD TO INJECTOR DRIVER ("A5" AT ECM); SHOULD MEASURE ABOUT 9-15 Hz.

24 PIN A-B CONNECTOR

REAR VIEW OF CONNECTOR (PINK)

2.2L (VIN 4) ENGINE — ECM SYMPTOM CHART — 1993–94 CAVALIER

ECM Connector and Driveability Symptoms Identification

This ECM voltage chart is for use with a J 39200 to further aid in diagnosis. These voltages were derived from a known good vehicle. The voltages you get may vary due to low battery charge or other reasons, but they should be very close.

THE FOLLOWING CONDITIONS MUST BE MET BEFORE TESTING:

- Engine at operating temperature • Closed Loop • Engine idling (for "Engine Run" column)
- Test terminal not grounded • Scan tool not installed • Brake not applied.

ECM PIN/FUNCTION	CKT #	WIRE COLOR	COMPONENT CONNECTOR CAVITY	NORMAL VOLTAGES KEY "ON"	NORMAL VOLTAGES ENG RUN	DTC(S) AFFECT	POSSIBLE SYMPTOMS FROM FAULTY CIRCUIT
B4 ECT SIGNAL	410	YEL	ECT SENSOR "B"	(7)	(7)	14 (5)	(5) INCORRECT IDLE, COOLING FAN RUNS AT ALL TIMES.
B5 TP SIGNAL	417	DK BLU	TP SENSOR "C"	.6V (9)	.6V (9)	21 (5)	(5) POOR PERFORMANCE, HESITATION.
B6 A/C REFRIGERANT PRESSURE SENSOR SIGNAL	380	GRY/RED	A/C REFRIGERANT PRESSURE SENSOR "C"	(2)	(2)	66 (5)	(5) NO A/C COOLING, A/C COMPRESSOR CLUTCH INOPERATIVE, REFER TO DTC 66.
B7 MAP SIGNAL	432	LT GRN	MAP SENSOR "B"	4.75V	(2)	33 (5)	(5) POOR PERFORMANCE, ROUGH IDLE, SURGE REFER TO CHART C-1D.
B8 IAT SIGNAL	472	TAN	IAT SENSOR "B"	(7)	(7)	23 (5)	(5) POOR PERFORMANCE, HARD TO START.
B9 DATA LINK TEST TERMINAL	451	WHT/BLK	DATA LINK CONNECTOR "B"	5.0V	5.0V		(3) NO DATA, WON'T FLASH DTC(S). (4) KEY "ON" - DTC(S) DISPLAYED ENG RUN - FIELD SERVICE MODE "ON."
B10 VSS SIGNAL (HIGH)	1232	LT BLU	VEHICLE SPEED SENSOR (VSS) "A"			24 (5)	(5) NO VSS SIGNAL, INACCURATE OR INOPERATIVE SPEEDOMETER, INOPERATIVE CRUISE CONTROL, TCC INOPERATIVE.
B11 VSS SIGNAL (LOW)	1233	DK GRN/ YEL	VEHICLE SPEED SENSOR (VSS) "B"			24 (5)	(5) NO VSS SIGNAL, INACCURATE OR INOPERATIVE SPEEDOMETER, INOPERATIVE CRUISE CONTROL, TCC INOPERATIVE.

(2) VARIES.
(3) OPEN CIRCUIT.
(4) GROUNDED CIRCUIT.
(5) OPEN OR GROUNDED CIRCUIT.
(7) VARIES WITH TEMPERATURE.
(9) WITHIN THE RANGE OF .33-1.33 VOLTS.

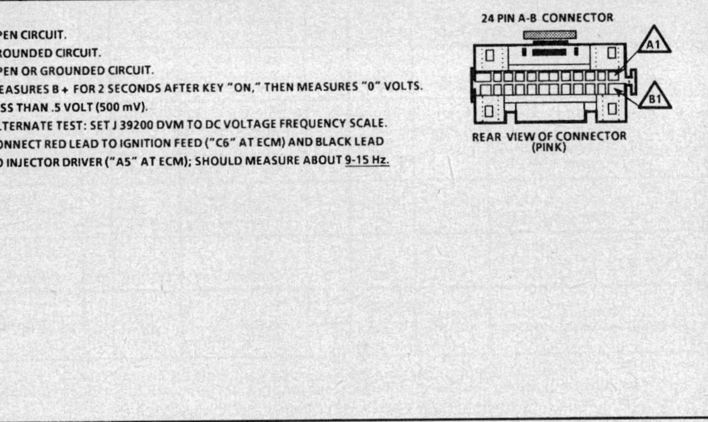

24 PIN A-B CONNECTOR

REAR VIEW OF CONNECTOR (PINK)

2.2L (VIN 4) ENGINE — ECM SYMPTOM CHART — 1993–94 CAVALIER

ECM Connector and Driveability Symptoms Identification

This ECM voltage chart is for use with a J 39200 to further aid in diagnosis. These voltages were derived from a known good vehicle. The voltages you get may vary due to low battery charge or other reasons, but they should be very close.

THE FOLLOWING CONDITIONS MUST BE MET BEFORE TESTING:

- Engine at operating temperature • Closed Loop • Engine idling (for "Engine Run" column)
- Test terminal <u>not</u> grounded • Scan tool <u>not</u> installed • Brake <u>not</u> applied.

ECM PIN/FUNCTION	CKT #	WIRE COLOR	COMPONENT CONNECTOR CAVITY	NORMAL VOLTAGES KEY "ON"	NORMAL VOLTAGES ENG RUN	DTC(S) AFFECTED	POSSIBLE SYMPTOMS FROM FAULTY CIRCUIT
C1 ECM GROUND	450	BLK/WHT	ENGINE GROUND	(8) (15)	(8) (15)		(3) NO EFFECT UNLESS "D1" IS ALSO OPEN. IF BOTH ARE OPEN, CRANKS, BUT WON'T RUN.
C4 PARK/NEUTRAL POSITION (PNP) SWITCH	434	ORN/BLK	PARK/NEUTRAL POSITION (PNP) SWITCH "A"	(8)	(8)		(5) INCORRECT IDLE.
C6 IGNITION FEED	439	PNK/BLK	EI MODULE 2-PIN CONNECTOR - "B" FUEL INJECTOR JUMPER HARNESS - "A"	B + (15)	B + (15)		(5) CRANKS, BUT WON'T RUN. (4) CRANKS, BUT WON'T RUN, OPEN 10 AMP ECM FUSE.
C7 TP SENSOR 5 VOLT REFERENCE	416	GRY	TP SENSOR "A"	5.0V	5.0V	21(5)	(5) DEGRADED PERFORMANCE.
C8 MAP AND A/C REFRIGERANT PRESSURE SENSORS 5 VOLT REF	474	GRY	MAP - "C" A/C REFRIGERANT PRESSURE SENSOR "B"	5.0V	5.0V	33 (5) 66 (5)	(3) DEGRADED PERFORMANCE, NO A/C COOLING.
C9 & C10 KEEP ALIVE MEMORY (B +)	2	RED	POWER FEED CONNECTOR	B + (15)	B + (15)		(3) CRANKS, BUT WON'T RUN. (4) CRANKS, BUT WON'T RUN, OPEN FUSIBLE LINK.
C11 TCC (A/T) / SHIFT LIGHT (M/T) CONTROL	422 / 456	TAN/BLK	TCC SOLENOID "D" DATA LINK CONNECTOR - "F"	(8) (A/T) / B + (M/T)	(8) (A/T) / B + (M/T)		(5) TCC INOPERATIVE, POOR FUEL ECONOMY. (4) TCC ENGAGES AT VERY LOW ROAD SPEED, OR SHIFT LIGHT "ON" AT ALL TIMES-(M/T).
C12 A/C COMPRESSOR CLUTCH RELAY CONTROL	459	DK GRN/ WHT	A/C COMPRESSOR CLUTCH CONTROL RELAY "85"	B +	B +		(3) NO A/C COOLING. (4) NO A/C COOLING, ENGINE OVERHEATS. OPEN 10 AMP GAUGES FUSE.
C13 COOLING FAN RELAY CONTROL	335	DK GRN/ WHT	COOLING FAN CONTROL RELAY "85"	B +	B +		(3) ENGINE OVERHEATS. (4) COOLING FAN RUNS AT ALL TEMPERATURES.
C14 EGR SOLENOID CONTROL	435	GRY	EGR SOLENOID "B"	B +	B +	32(5)	(3) EGR VALVE DOESN'T OPEN, HIGH NOx DETONATION, SPARK KNOCK. (4) DEGRADED PERFORMANCE. IMPORTANT: EGR SOLENOID IS DISABLED IF DTC 14, 21, 23, 32 OR 33 IS SET.

(3) OPEN CIRCUIT.
(4) GROUNDED CIRCUIT.(5) OPEN OR GROUNDED CIRCUIT.
(8) LESS THAN .5 VOLT (500 mV).
(15) THIS TYPE OF CIRCUIT SHOULD ALSO BE CHECKED USING J 34142-B TEST LIGHT.

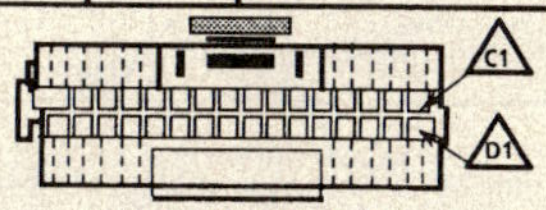

32 PIN C-D CONNECTOR
REAR VIEW OF CONNECTOR (PINK)

2.2L (VIN 4) ENGINE — ECM SYMPTOM CHART — 1993–94 CAVALIER

ECM Connector and Driveability Symptoms Identification

This ECM voltage chart is for use with a J 39200 to further aid in diagnosis. These voltages were derived from a known good vehicle. The voltages you get may vary due to low battery charge or other reasons, but they should be very close.

THE FOLLOWING CONDITIONS MUST BE MET BEFORE TESTING:

- Engine at operating temperature • Closed Loop • Engine idling (for "Engine Run" column)
- Test terminal <u>not</u> grounded • Scan tool <u>not</u> installed • Brake <u>not</u> applied.

ECM PIN/FUNCTION	CKT #	WIRE COLOR	COMPONENT CONNECTOR CAVITY	NORMAL VOLTAGES KEY "ON"	NORMAL VOLTAGES ENG RUN	DTC(S) AFFECT	POSSIBLE SYMPTOMS FROM FAULTY CIRCUIT
D1 ECM GROUND	551	TAN/WHT	ENGINE GROUND	(8) (15)	(8) (15)		(3) NO EFFECT UNLESS "C1" IS ALSO OPEN. IF BOTH ARE OPEN, CRANKS, BUT WON'T RUN.
D2 SENSOR GROUND	452	BLK	TPS - "B", ECT - "A", IAT - "A" AND A/C REFRIGERANT PRESSURE SENSOR "A"	(8)	(8)	14 (3) 21 (3) 33 (3) 66 (3)	(3) HARD START OR CRANKS, BUT WON'T RUN.
D3 SENSOR GROUND	455	PPL	MAP "A"	(8)	(8)	33 (3)	(3) HARD START OR CRANKS, BUT WON'T RUN.
D5 IGNITION CONTROL (IC)	423	WHT	ELECTRONIC IGNITION MODULE 6 PIN CONN. "C"	(8)	2.4V	42 (5)	(5) STUMBLES, UNSTABLE IDLE, DEGRADED PERFORMANCE.
D6 BYPASS	424	TAN/BLK	ELECTRONIC IGNITION MODULE 6 PIN CONN. "D"	(8)	4.8V	42 (5)	(5) DEGRADED PERFORMANCE.
D7 IGNITION REFERENCE HIGH	430	PPL/WHT	ELECTRONIC IGNITION MODULE 6 PIN CONN. "E"	(8)	3.2V (14)		(5) CRANKS, BUT WON'T RUN.
D8 IGNITION REFERENCE LOW	453	BLK/RED	ELECTRONIC IGNITION MODULE 6 PIN CONN. "F"	(8)	(8)		(3) REDUCED PERFORMANCE
D9 A/C REQUEST	66	LT GRN	HVAC MODE SELECT CONNECTOR "A"	(8) "OFF" B + "ON"	(8) "OFF" B + "ON"		(3) NO A/C COOLING A/C COMPRESSOR CLUTCH WON'T ENGAGE.
D10 OXYGEN SENSOR (O2S) REFERENCE LOW	413	TAN	ENGINE GROUND	(8)	(8)	13 (3)	(3) "OPEN LOOP" OPERATION REDUCED PERFORMANCE.
D11 OXYGEN SENSOR (O2S) SIGNAL	412	PPL	OXYGEN SENSOR	.01-.55V	1.9V (10)	13 44 45	(5) "OPEN LOOP" OPERATION, STRONG EXHAUST ODOR, REDUCED PERFORMANCE.
D12 SERIAL DATA	461	ORN	DATA LINK CONNECTOR "M"	4.7V	4.7V		(5) NO DATA.
D13 VEHICLE SPEED SENSOR (VSS) OUTPUT	389	DK GRN	I/P CONNECTOR "A" AND I/P CLUSTER CONN. "R"				(5) INCORRECT SPEEDOMETER READING, CRUISE CONTROL INOPERATIVE.
D14 MIL (CHECK ENGINE) LAMP CONTROL	419	BRN/WHT	I/P CONNECTOR "R" AND I/P CLUSTER CONN. "I"	.01-1.0V	B +		(3) LAMP NOT "ON" FOR "BULB CHECK," WON'T FLASH DTC(S). (4) LAMP "ON" ALL THE TIME, DOES NOT FLASH, REFER TO "OBD SYSTEM CHECK."

(3) OPEN CIRCUIT.
(4) GROUNDED CIRCUIT.
(8) LESS THAN .5 VOLT (500 mV).
(10) ACTIVELY VARIES WITH INDICATED RANGE.
(14) J 39200 RED LEAD TO REF HIGH ("D7" AT ECM) AND BLACK LEAD TO ECM GROUND ("D1" AT ECM); SHOULD MEASURE ABOUT 25 - 30 Hz IN DC VOLTAGE RANGE.
(15) THIS TYPE OF CIRCUIT SHOULD ALSO BE CHECKED USING J 34142-B TEST LIGHT.

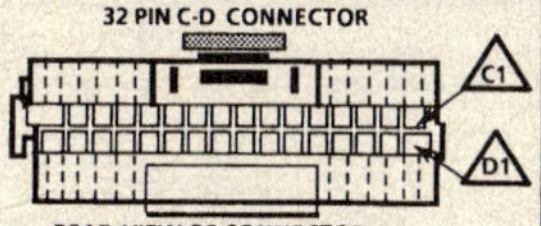

32 PIN C-D CONNECTOR
REAR VIEW OF CONNECTOR (PINK)

2.2L (VIN 4) ENGINE — COMPONENT DIAGNOSTIC CHART — CAVALIER

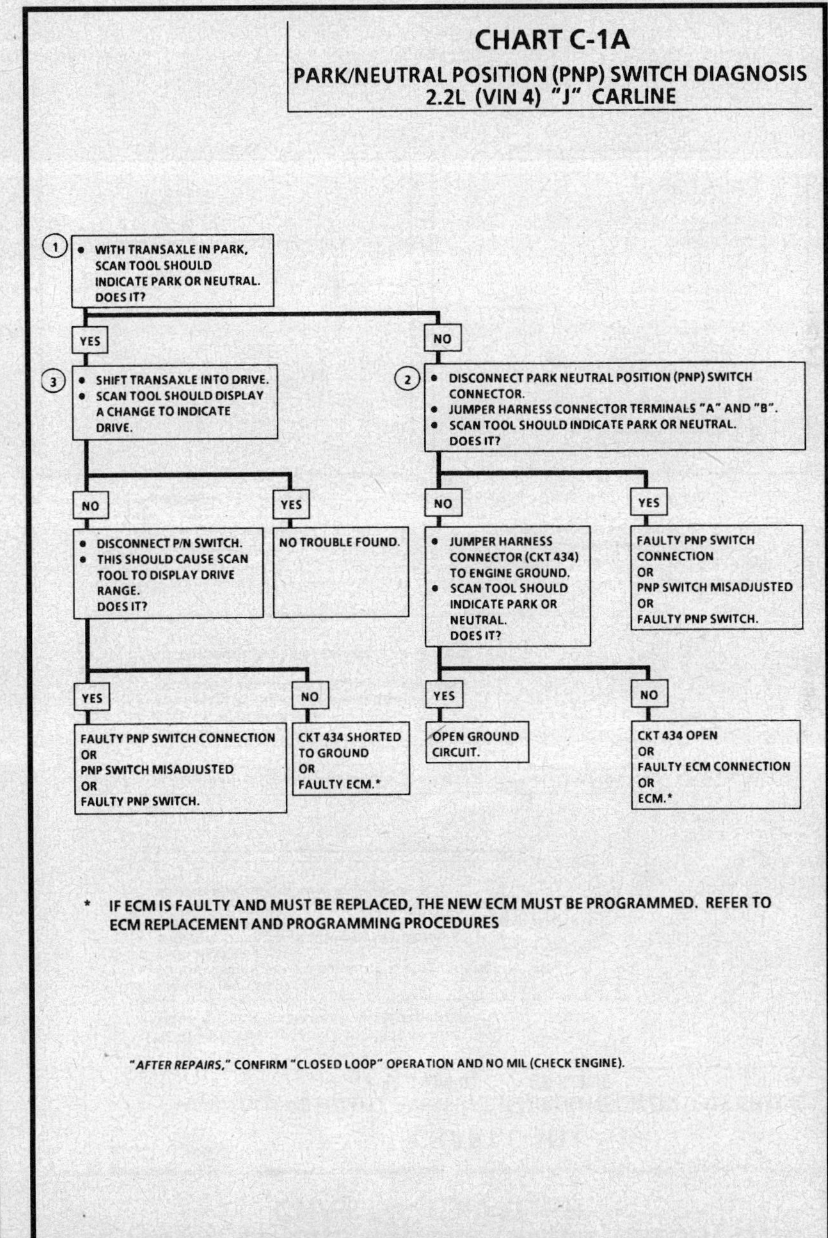

CHART C-1A
PARK/NEUTRAL POSITION (PNP) SWITCH DIAGNOSIS
2.2L (VIN 4) "J" CARLINE

Circuit Description:

The Park/Neutral Position (PNP) switch contacts are a part of the neutral start switch and are closed to ground in park or neutral, and open in drive ranges.

The ECM supplies ignition voltage through a current limiting resistor to CKT 434 and senses a closed switch, when the voltage on CKT 434 drops to less than one volt.

The ECM uses the PNP signal as one of the inputs to control idle air control and VSS diagnostics.

Test Description: Number(s) below refer to circled number(s) on the diagnostic chart.

1. Checks for a closed switch to ground in park position. Different makes of scan tools will read PNP switch differently. Refer to tool operations manual for type of display used.

2. Checks for an open switch in drive range.

3. Be sure scan tool indicates drive, even while wiggling shifter to test for an intermittent or misadjusted switch in drive range.

2.2L (VIN 4) ENGINE — COMPONENT DIAGNOSTIC CHART — CAVALIER

CHART C-1A
PARK/NEUTRAL POSITION (PNP) SWITCH DIAGNOSIS
2.2L (VIN 4) "J" CARLINE

1. • WITH TRANSAXLE IN PARK, SCAN TOOL SHOULD INDICATE PARK OR NEUTRAL. DOES IT?

YES → 3. • SHIFT TRANSAXLE INTO DRIVE. • SCAN TOOL SHOULD DISPLAY A CHANGE TO INDICATE DRIVE.

NO → 2. • DISCONNECT PARK NEUTRAL POSITION (PNP) SWITCH CONNECTOR. • JUMPER HARNESS CONNECTOR TERMINALS "A" AND "B". • SCAN TOOL SHOULD INDICATE PARK OR NEUTRAL. DOES IT?

NO → • DISCONNECT P/N SWITCH. • THIS SHOULD CAUSE SCAN TOOL TO DISPLAY DRIVE RANGE. DOES IT?

YES → NO TROUBLE FOUND.

NO → • JUMPER HARNESS CONNECTOR (CKT 434) TO ENGINE GROUND. • SCAN TOOL SHOULD INDICATE PARK OR NEUTRAL. DOES IT?

YES → FAULTY PNP SWITCH CONNECTION OR PNP SWITCH MISADJUSTED OR FAULTY PNP SWITCH.

YES (from DISCONNECT P/N SWITCH) → FAULTY PNP SWITCH CONNECTION OR PNP SWITCH MISADJUSTED OR FAULTY PNP SWITCH.

NO → CKT 434 SHORTED TO GROUND OR FAULTY ECM.*

YES → OPEN GROUND CIRCUIT.

NO → CKT 434 OPEN OR FAULTY ECM CONNECTION OR ECM.*

* IF ECM IS FAULTY AND MUST BE REPLACED, THE NEW ECM MUST BE PROGRAMMED. REFER TO ECM REPLACEMENT AND PROGRAMMING PROCEDURES

"AFTER REPAIRS," CONFIRM "CLOSED LOOP" OPERATION AND NO MIL (CHECK ENGINE).

2.2L (VIN 4) ENGINE — COMPONENT DIAGNOSTIC CHART — CAVALIER

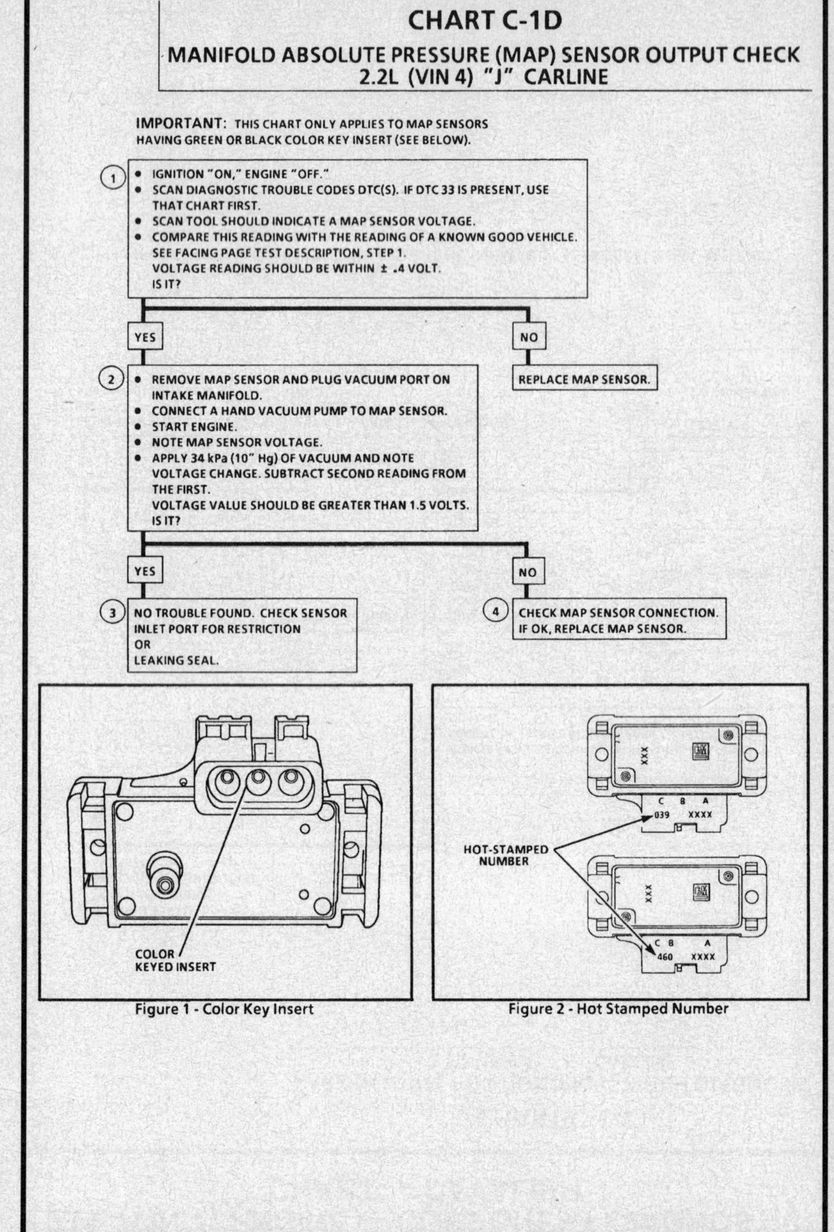

CHART C-1D
MANIFOLD ABSOLUTE PRESSURE (MAP) SENSOR OUTPUT CHECK
2.2L (VIN 4) "J" CARLINE

Circuit Description:

The Manifold Absolute Pressure (MAP) sensor measures the changes in the intake manifold pressure which result from engine load (intake manifold vacuum) and rpm changes; and converts these into a voltage output. The ECM sends a 5 volts reference voltage to the MAP sensor. As the manifold pressure changes, the output voltage of the sensor also changes. By monitoring the sensor output voltage, the ECM knows the manifold pressure. A lower pressure (low voltage) output voltage will be about 1 to 2 volts at idle. While higher pressure (high voltage) output voltage will be about 4 to 4.8 at Wide Open Throttle (WOT). The MAP sensor is also used, under certain conditions, to measure barometric pressure, allowing the ECM to make adjustments for altitude changes. The ECM uses the MAP sensor to control fuel delivery and ignition timing.

Test Description: Number(s) below refer to circled number(s) on the diagnostic chart.

Important

- Be sure to use the same diagnostic test equipment for all measurements.

1. When comparing scan readings to a known good vehicle, it is important to compare vehicles that use a MAP sensor having the same color insert and the same "Hot Stamped" number. Refer to Figures 1 and 2

2. Applying 34 kPa (10" Hg) vacuum to the MAP sensor should result in voltage readings of 1.5 to 2.1 volts less than the voltage in Step 1. Upon applying vacuum to the sensor, the change in voltage should be instantaneous. A slow voltage change indicates a faulty sensor.

3. Check vacuum seal to sensor for leaking or restriction.

   NOTICE: Make sure electrical connector remains securely fastened.

4. Remove sensor from the intake plenum and twist sensor **(by hand only)** to check for intermittent connection. Output changes greater than .10 volt indicate a faulty sensor or connection. If OK, replace sensor.

2.2L (VIN 4) ENGINE — COMPONENT DIAGNOSTIC CHART — CAVALIER

CHART C-1D
MANIFOLD ABSOLUTE PRESSURE (MAP) SENSOR OUTPUT CHECK
2.2L (VIN 4) "J" CARLINE

IMPORTANT: THIS CHART ONLY APPLIES TO MAP SENSORS HAVING GREEN OR BLACK COLOR KEY INSERT (SEE BELOW).

1.
- IGNITION "ON," ENGINE "OFF."
- SCAN DIAGNOSTIC TROUBLE CODES DTC(S). IF DTC 33 IS PRESENT, USE THAT CHART FIRST.
- SCAN TOOL SHOULD INDICATE A MAP SENSOR VOLTAGE.
- COMPARE THIS READING WITH THE READING OF A KNOWN GOOD VEHICLE. SEE FACING PAGE TEST DESCRIPTION, STEP 1. VOLTAGE READING SHOULD BE WITHIN ± .4 VOLT. IS IT?

→ YES

→ NO → REPLACE MAP SENSOR.

2.
- REMOVE MAP SENSOR AND PLUG VACUUM PORT ON INTAKE MANIFOLD.
- CONNECT A HAND VACUUM PUMP TO MAP SENSOR.
- START ENGINE.
- NOTE MAP SENSOR VOLTAGE.
- APPLY 34 kPa (10" Hg) OF VACUUM AND NOTE VOLTAGE CHANGE. SUBTRACT SECOND READING FROM THE FIRST. VOLTAGE VALUE SHOULD BE GREATER THAN 1.5 VOLTS. IS IT?

→ YES

→ NO

3. NO TROUBLE FOUND. CHECK SENSOR INLET PORT FOR RESTRICTION OR LEAKING SEAL.

4. CHECK MAP SENSOR CONNECTION. IF OK, REPLACE MAP SENSOR.

Figure 1 - Color Key Insert

Figure 2 - Hot Stamped Number

2.2L (VIN 4) ENGINE — COMPONENT DIAGNOSTIC CHART — CAVALIER

CHART C-2A

INJECTOR BALANCE TEST
2.2L (VIN 4) "J" CARLINE (PORT)

The injector balance tester is a tool used to turn the injector "ON" for a precise amount of time, thus spraying a measured amount of fuel into the manifold. This causes a drop in fuel rail pressure that we can record and compare between each injector. All injectors should have the same amount of pressure drop (±10 kPa). Any injector with a pressure drop that is 10 kPa (or more) greater or less than the average drop of the other injectors should be considered faulty and replaced.

STEP 1

Engine "cool down" period (10 minutes) is necessary to avoid irregular readings due to "Hot Soak" fuel boiling. Relieve fuel pressure in the fuel rail using the "fuel pressure relief procedure" described previously in this section. With ignition "OFF" connect fuel gage J 29658-D or equivalent. (Refer to CHART A-7 Page 3 of 3 for connecting fuel pressure gage).

Disconnect electrical connectors at all injectors except #2. Then, connect injector adaptor harness J 34730 200 to the injector jumper harness connector. Now, connect tester J 34730-3 to adaptor harness J 34730-200. Ignition must be "OFF" at least 10 seconds to complete ECM shutdown cycle. Fuel pump should run about 2 seconds after ignition is turned "ON." At this point, insert clear tubing attached to vent valve into a suitable container and bleed air from gauge and hose to insure accurate gauge operation. Repeat this step until all air is bled from gauge.

STEP 2

Turn ignition "OFF" for 10 seconds and then "ON" again to get fuel pressure to its maximum. Record this initial pressure reading. Energize tester one time and note pressure drop at its lowest point. (Disregard any slight pressure increase after drop hits low point.) By subtracting this second pressure reading from the initial pressure, we have the actual amount of injector pressure drop.

STEP 3

Disconnect injector adaptor harness J 34730-200 and reconnect injector jumper harness connector to the engine harness connector. Now connect J 34730-200 directly to each of the remaining injectors. Repeat Step 2 on each injector and compare the amount of drop. Usually, good injectors will have virtually the same drop. Retest any injector that has a pressure difference of 10 kPa, either more or less than the average of the other injectors on the engine. Replace any injector that also fails the retest. If the pressure drop of all injectors is within 10 kPa of this average, the injectors appear to be flowing properly.
B".

NOTICE: The entire test should not be repeated more than once without running the engine to prevent flooding. (This includes any retest on faulty injectors).

2.2L (VIN 4) ENGINE — COMPONENT DIAGNOSTIC CHART — CAVALIER

NOTICE: The entire test should <u>NOT</u> be repeated more than once without running the engine to prevent flooding. (This includes any retest on faulty injectors.)

The fuel pressure test in Chart A-7, should be completed prior to this test.

CHART C-2A

INJECTOR BALANCE TEST
2.2L (VIN 4) "J" CARLINE

Step 1. If engine is at operating temperature, allow a 10 minute "cool down" period then connect fuel pressure gage and injector tester.
1. Ignition "OFF."
2. Connect fuel pressure gage and injector tester. Refer to Step 1
3. Ignition "ON."
4. Bleed off air in gage. Repeat until all air is bled from gauge.

Step 2. Run test:
1. Ignition "OFF" for 10 seconds.
2. Ignition "ON". Record gage pressure. (Pressure must hold steady, if not see the Fuel System diagnosis, Chart A-7
3. Turn injector on, by depressing button on injector tester, and note pressure at the instant the gauge needle stops.

Step 3. Refer to Step 3

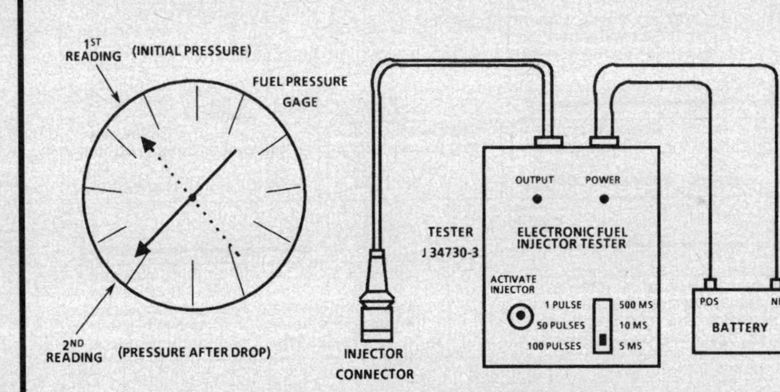

EXAMPLE

CYLINDER	1	2	3
1ST READING	293 kPa (43 psi)	293 kPa (43 psi)	293 kPa (43 psi)
2ND READING	131 kPa (19 psi)	115 kPa (17 psi)	145 kPa (21 psi)
AMOUNT OF DROP	162 kPa (24 psi)	178 kPa (26 psi)	148 kPa (21 psi)
	OK	FAULTY, RICH (TOO MUCH FUEL DROP)	FAULTY, LEAN (TOO LITTLE FUEL DROP)

2.2L (VIN 4) ENGINE — COMPONENT DIAGNOSTIC CHART — CAVALIER

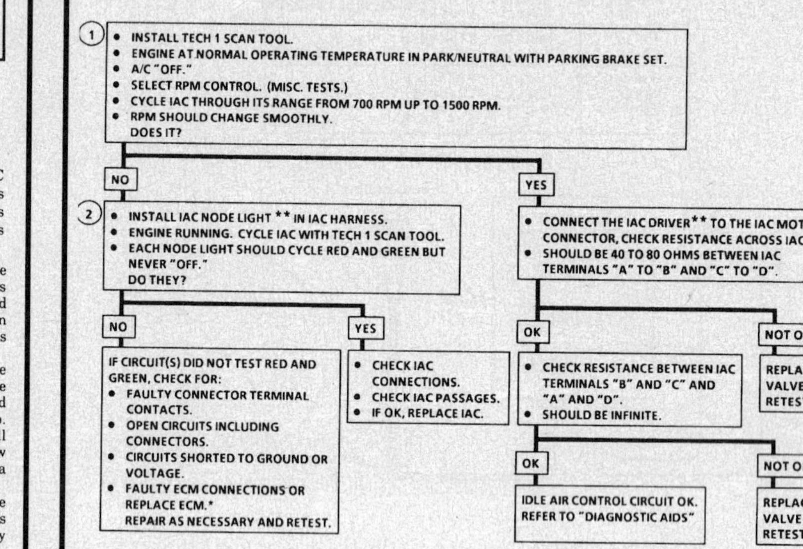

CHART C-2C
IDLE AIR CONTROL (IAC) SYSTEM CHECK
2.2L (VIN 4) "J" CARLINE

Circuit Description:

The ECM controls engine idle speed with the IAC valve. To increase idle speed, the ECM retracts the IAC valve pintle away from its seat, allowing more air to bypass the throttle bore. To decrease idle speed, it extends the IAC valve pintle towards its seat, reducing bypass air flow. A Tech 1 scan tool will read the ECM commands to the IAC valve in counts. Higher the counts indicate more air bypass (higher idle). The lower the counts indicate less air is allowed to bypass (lower idle).

Test Description: Number(s) below refer to circled numbers on the diagnostic chart.

1. The Tech 1 RPM control mode is used to extend and retract the IAC valve. The valve should move smoothly within the specified range. If the idle speed commanded (IAC extended) too low (below 700 RPM), the engine may stall. This may be normal and would not indicate a problem. Retracting the IAC beyond its controlled range (above 1500 RPM) will cause a delay before the RPMs start dropping. This too is normal.
2. This test uses the Tech 1 to command the IAC controlled idle speed. The ECM issues commands to obtain commanded idle speed. The node lights each should flash red and green to indicate a good circuit as the ECM issues commands. While the sequence of color is not important if either light is "OFF" or does not flash red and green, check the circuits for faults, beginning with poor terminal contacts.

Diagnostic Aids:

A slow, unstable, or fast idle may be caused by a non-IAC system problem that cannot be overcome by the IAC valve. Out of control range IAC scan tool counts will be above 50 if idle is too low, and zero counts if idle is too high. The following checks should be made to repair a non-IAC system problem:

- Vacuum Leak (High Idle) - If idle is too high, stop the engine. Fully extend (low) IAC with tester. Start engine. If idle speed is above 800 RPM, locate and correct vacuum leak including crankcase ventilation system. Also check for binding of throttle blade or linkage.
- Crankcase Ventilation Valve - If a high idle condition exists (800 to 1000 RPM), check for vacuum leaks and proper crankcase ventilation valve operation. All throttle bodies are preset at the factory and do not need adjustment. A missing crankcase ventilation valve or grommet or stuck valve can cause this condition.
- System Too Lean (High Air/Fuel Ratio) - The idle speed may be too high or too low. Engine speed may vary up and down and disconnecting the IAC valve does not help. DTC 44 may be set. Scan oxygen voltage will be less than 300 mV (.3 volt). Check for low regulated fuel pressure, water in the fuel or a restricted injector.
- System Too Rich (Low Air/Fuel Ratio) - The idle speed will be too low. Scan tool IAC counts will usually be above 50. System is obviously rich and may exhibit black smoke in exhaust. Scan tool oxygen voltage will be fixed above 800 mV (.8 volt). Check for high fuel pressure, leaking or sticking injector. Silicone contaminated oxygen sensors scan voltage will be slow to respond.
- Throttle Body - Remove IAC valve and inspect bore for foreign material.
- IAC Valve Electrical Connections - IAC valve connections should be carefully checked for proper contact.

- If intermittent poor driveability or idle symptoms are resolved by disconnecting the IAC, carefully recheck connections, valve terminal resistance, or replace IAC.

2.2L (VIN 4) ENGINE — COMPONENT DIAGNOSTIC CHART — CAVALIER

CHART C-2C
IDLE AIR CONTROL (IAC) SYSTEM CHECK
2.2L (VIN 4) "J" CARLINE

1.
- INSTALL TECH 1 SCAN TOOL.
- ENGINE AT NORMAL OPERATING TEMPERATURE IN PARK/NEUTRAL WITH PARKING BRAKE SET.
- A/C "OFF."
- SELECT RPM CONTROL. (MISC. TESTS.)
- CYCLE IAC THROUGH ITS RANGE FROM 700 RPM UP TO 1500 RPM.
- RPM SHOULD CHANGE SMOOTHLY.
- DOES IT?

NO →

2.
- INSTALL IAC NODE LIGHT ** IN IAC HARNESS.
- ENGINE RUNNING. CYCLE IAC WITH TECH 1 SCAN TOOL.
- EACH NODE LIGHT SHOULD CYCLE RED AND GREEN BUT NEVER "OFF."
- DO THEY?

YES (from 1) →
- CONNECT THE IAC DRIVER ** TO THE IAC MOTOR CONNECTOR, CHECK RESISTANCE ACROSS IAC COILS.
- SHOULD BE 40 TO 80 OHMS BETWEEN IAC TERMINALS "A" TO "B" AND "C" TO "D".

NO (from 2):
IF CIRCUIT(S) DID NOT TEST RED AND GREEN, CHECK FOR:
- FAULTY CONNECTOR TERMINAL CONTACTS.
- OPEN CIRCUITS INCLUDING CONNECTORS.
- CIRCUITS SHORTED TO GROUND OR VOLTAGE.
- FAULTY ECM CONNECTIONS OR REPLACE ECM.*
REPAIR AS NECESSARY AND RETEST.

YES (from 2):
- CHECK IAC CONNECTIONS.
- CHECK IAC PASSAGES.
- IF OK, REPLACE IAC.

OK:
- CHECK RESISTANCE BETWEEN IAC TERMINALS "B" AND "C" AND "A" AND "D".
- SHOULD BE INFINITE.

NOT OK:
REPLACE IAC VALVE AND RETEST.

OK:
IDLE AIR CONTROL CIRCUIT OK. REFER TO "DIAGNOSTIC AIDS"

NOT OK:
REPLACE IAC VALVE AND RETEST.

** IAC DRIVER AND NODE LIGHT REQUIRED KIT 222-L FROM: CONCEPT TECHNOLOGY, INC. J 37027 FROM: KENT-MOORE, INC.

CLEAR DIAGNOSTIC TROUBLE CODES, CONFIRM "CLOSED LOOP" OPERATION, NO MALFUNCTION INDICATOR LAMP (MIL) "SERVICE ENGINE SOON," PERFORM IAC RESET PROCEDURE AND VERIFY CONTROLLED IDLE SPEED IS CORRECT.

* IF ECM IS FAULTY AND MUST BE REPLACED, THE NEW ECM MUST BE PROGRAMMED. REFER TO ECM REPLACEMENT AND PROGRAMMING PROCEDURES

"AFTER REPAIRS," CONFIRM "CLOSED LOOP" OPERATION AND NO MIL (CHECK ENGINE).

2.2L (VIN 4) ENGINE — COMPONENT DIAGNOSTIC CHART — CAVALIER

CHART C-3
EVAPORATIVE EMISSION CANISTER CONTROL VALVE CHECK
(NON-ECM CONTROLLED)
2.2L (VIN 4) "J" CARLINE

- DISCONNECT THE VACUUM HOSE AT THE EVAP CANISTER CONTROL VALVE LABELED "CONTROL VAC."
- START ENGINE AND IDLE.
- MANIFOLD VACUUM SHOULD BE PRESENT AT THE DISCONNECTED HOSE.
  IS IT?

YES →

- REMOVE PURGE VALVE.
- APPLY A CLEAN SHORT HOSE TO THE TUBE LABELED "PURGE VAC" OF THE VALVE AND BLOW INTO THE HOSE. NO AIR SHOULD PASS THROUGH THE VALVE.
  DOES VALVE HOLD?

NO → REPAIR PLUGGED OR LEAKING MANIFOLD VACUUM SOURCE.

YES →

- INSTALL A HAND VACUUM PUMP TO THE TUBE OF THE VALVE LABELED "CONTROL VAC."
- APPLY 51 kPa (15" Hg) VACUUM TO THE DIAPHRAGM. DIAPHRAGM SHOULD HOLD VACUUM FOR AT LEAST 20 SECONDS.
  DOES IT?

NO → REPLACE VALVE.

YES →

- WITH THE VACUUM STILL APPLIED TO THE "CONTROL VAC" TUBE, AGAIN TRY TO BLOW INTO A CLEAN HOSE ON THE TUBE LABELED "PURGE VAC" OF THE VALVE. AIR SHOULD PASS THROUGH FREELY.
  DOES IT?

NO → REPLACE VALVE.

YES →

VALVE OK.

NO → REPLACE VALVE.

CHECK ALL EMISSION CONTROL LINES, HOSES AND FITTINGS FOR BLOCKAGE, LEAKAGE OR DETERIORATION AND REPAIR AS NECESSARY.

2.2L (VIN 4) ENGINE — COMPONENT DIAGNOSTIC CHART — 1992 CAVALIER

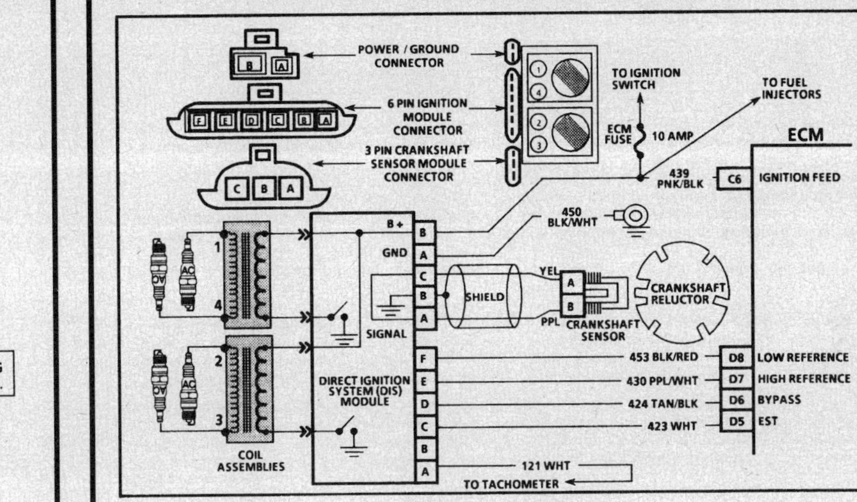

CHART C-4D-1
"DIS" MISFIRE AT IDLE
2.2L (VIN 4) "J" CARLINE (PORT)

Circuit Description:

The Direct Ignition System (DIS) uses a waste spark method of distribution. In this type of system, the ignition module triggers the number 1/4 coil pair resulting in both number 1 and number 4 spark plugs firing at the same time. Number 1 cylinder is on the compression stroke at the same time number 4 is on the exhaust stroke, resulting in a lower energy requirement to fire number 4 spark plug. This leaves the remainder of the high voltage to be used to fire 1 spark plug. The crank sensor is remotely mounted beside the module/coil assembly and protrudes through the block to within approximately .050" of the crankshaft reluctor. Since the reluctor is a machined portion of the crankshaft and the crankshaft sensor is mounted in a fixed position on th block, timing adjustments are not possible or necessary.

Test Description: Number(s) below refer to circled number(s) on the diagnostic chart.

1. If the "Misfire" complaint exists under load only, that the diagnostic chart must be used. Engine rpm should drop approximately equally on all plug leads.
2. A spark tester, such as a ST-125, must be used because it is essential to verify adequate available secondary voltage at the spark plug (25,000 volts).
3. If the spark jumps the tester gap after grounding the opposite plug cable, it indicates excessive resistance in the plug which was bypassed. A faulty or poor connection at that plug could also result in a miss condition. Also, check for carbon deposits inside the spark plug boot.
4. If carbon tracking is evident, replace coil and be sure plug cables relating to that coil are clean and tight. Excessive wire resistance or faulty connections could have caused the coil to be damaged.
5. If the no spark condition follows the suspected coil, that coil is faulty. Otherwise, the ignition module is the cause of no spark. This test could also be performed by substituting a known good coil for the one causing the no spark condition.

CHART C-4D-1
"DIS" MISFIRE AT IDLE
2.2L (VIN 4) "J" CARLINE (PORT)

1. • IF ENGINE MISFIRES UNDER LOAD ONLY, SEE CHART C-4D-2.
   • ENGINE IDLING AT NORMAL OPERATING TEMPERATURE, DISCONNECT IAC.
   • MOMENTARILY DISCONNECT EACH SPARK PLUG LEAD, USING INSULATED PLIERS, WHILE OBSERVING ENGINE RPM.   SEE CAUTION★.
   • ALL PLUG LEAD(S) SHOULD RESULT IN AN RPM DROP.
   DID THEY?

   — YES → SEE "ROUGH, UNSTABLE OR INCORRECT IDLE OR STALLING" IN "SYMPTOMS"

2. (NO) • WITH IGNITION "OFF," INSTALL SPARK TESTER (ST-125) J 26792 OR EQUIVALENT ON PLUG LEAD(S) WHOSE REMOVAL DID NOT RESULT IN RPM DROP.
   • SPARK SHOULD JUMP TESTER GAP WHILE CRANKING ENGINE. DOES IT?

3. (NO) • WITH IGNITION "OFF," GROUND THE OPPOSITE PLUG LEAD OF THE AFFECTED COIL AT SPARK PLUG.
   • SPARK SHOULD JUMP TESTER GAP WHILE CRANKING ENGINE. DOES IT?

   — YES → CHECK FOR:
   − FAULTY, WORN OR CRACKED SPARK PLUG(S).
   − PLUG FOULING DUE TO ENGINE MECHANICAL FAULT. IF SPARK PLUGS CHECK OUT OK, SEE "CUTS OUT, MISSES" IN "SYMPTOMS"

   (from step 2 — YES) → REPLACE THE SPARK PLUG FOR THE LEAD WHICH WAS JUMPERED TO GROUND. IF MISFIRE IS STILL PRESENT, START MISFIRE TEST AGAIN AT STEP #1.

   (NO) • CHECK THE RESISTANCE OF EACH PLUG CABLE OF THE COIL WHICH DID NOT FIRE THE SPARK TESTER.
   • CABLE RESISTANCE SHOULD BE LESS THAN 30,000 OHMS EACH AND CABLES SHOULD NOT BE GROUNDED. ARE CABLES OK?

   — NO → REPLACE FAULTY CABLE(S).

4. (YES) • REMOVE COIL RETAINING NUTS AND REMOVE COILS.
   • COILS SHOULD BE FREE OF CARBON TRACKING. ARE THEY?

   — NO → REPLACE IGNITION COIL. ALSO CHECK FOR FAULTY PLUG CABLE CONNECTION(S) AND CABLE NIPPLE(S) FOR CARBON TRACKING.

5. (YES) • SWITCH A NORMALLY OPERATING COIL WITH THE COIL FROM PROBLEM CYLINDER.
   • SPARK SHOULD JUMP TESTER GAP AT PROBLEM CYLINDER WHILE CRANKING ENGINE. DID IT?

   (YES) → ORIGINAL IGNITION COIL IS FAULTY.

   (NO) → REPLACE DIS MODULE.

★CAUTION: When handling secondary spark plug leads with engine running, insulated pliers must be used and care exercised to prevent a possible electrical shock.

"AFTER REPAIRS," CONFIRM "CLOSED LOOP" OPERATION AND NO "CHECK ENGINE" LIGHT.

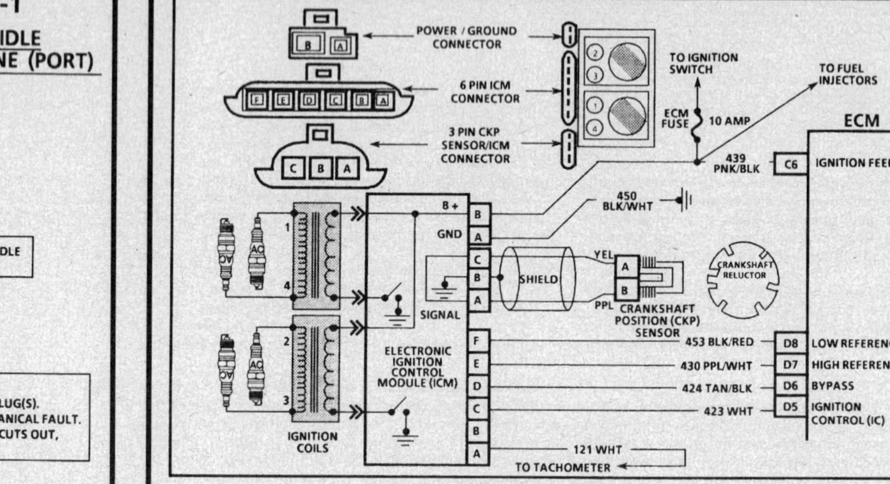

CHART C-4D-1
MISFIRE AT IDLE
2.2L (VIN 4) "J" CARLINE

Circuit Description:

The Electronic Ignition (EI) system uses a waste spark method of distribution. In this type of system, the Electronic Ignition Control Module (ICM) triggers the number 1/4 coil pair resulting in both number 1 and number 4 spark plugs firing at the same time. Number 1 cylinder is on the compression stroke at the same time number 4 is on the exhaust stroke, resulting in a lower energy requirement to fire number 4 spark plug. This leaves the remainder of the high voltage to be used to fire 1 spark plug. The Crankshaft Position (CKP) sensor is remotely mounted beside the ignition coil assembly and protrudes through the block to within approximately .050" of the crankshaft reluctor. Since the reluctor is a machined portion of the crankshaft and the crankshaft sensor is mounted in a fixed position on the block, timing adjustments are not possible or necessary.

Test Description: Number(s) below refer to circled number(s) on the diagnostic chart.

1. If the "Misfire" complaint exists <u>under load only</u>, CHART C-4D-2 must be used.
   • Keep disconnected spark plug leads away from sensors and other electronic components.
   • Move quickly through this test. Don't leave any spark plug lead disconnected for longer than 15 seconds.
   • Let the engine run normally for 30 seconds between tests to avoid an excessive buildup of fuel.
2. A spark tester, such as a ST-125, must be used because it is essential to verify adequate available secondary voltage at the spark plug (25,000 volts). If the spark jumps the tester gap after grounding the companion plug cable, it indicates excessive resistance in the plug which was bypassed. A faulty or poor connection at that plug could also result in a miss condition. Also, check for carbon deposits inside the spark plug boot.
3. If carbon tracking is evident, replace coil and be sure plug cables relating to that coil are clean and tight. Excessive wire resistance or faulty connections could have caused the coil to be damaged.
4. By checking the secondary resistance, a coil with an open secondary may be located.
5. If the no spark condition follows the suspected coil, that coil is faulty. Otherwise, the Ignition Control Module (ICM) is the cause of no spark. This test could also be performed by substituting a known good coil for the one causing the no spark condition.

2.2L (VIN 4) ENGINE — COMPONENT DIAGNOSTIC CHART — 1993–94 CAVALIER

CHART C-4D-1
MISFIRE AT IDLE
2.2L (VIN 4) "J" CARLINE

★CAUTION: WHEN HANDLING SECONDARY SPARK PLUG LEADS WITH ENGINE RUNNING, J 21350 OR EQUIVALENT INSULATED PLIERS MUST BE USED AND CARE EXERCISED TO PREVENT A POSSIBLE ELECTRICAL SHOCK.

1.
- IF ENGINE MISFIRES UNDER LOAD ONLY. REFER TO CHART C-4D-2.
- ENGINE IDLING AT NORMAL OPERATING TEMPERATURE, DISCONNECT IAC.
- MOMENTARILY DISCONNECT (ONE AT A TIME) EACH SPARK PLUG LEAD, USING J 21350 OR EQUIVALENT INSULATED PLIERS, WHILE OBSERVING ENGINE RPM FOR A DROP. [DISCONNECTED LEAD SHOULD BE GROUNDED TO AVOID OVER-STRESSING IN THE IGNITION COIL(S).] SEE CAUTION★.
- EACH DISCONNECTED PLUG LEAD SHOULD HAVE RESULTED IN A FAIRLY EQUAL RPM DROP.

YES → REFER TO "ROUGH, UNSTABLE OR INCORRECT IDLE OR STALLING" IN "SYMPTOMS"

NO ↓

2.
- WITH IGNITION "OFF," INSTALL SPARK TESTER (ST-125) J 26792 ON PLUG LEAD(S) WHOSE REMOVAL DID NOT RESULT IN RPM DROP (AT PLUG END AND CLIP TO GROUND).
- CONNECT PLUG END OF COMPANION CYLINDER SPARK PLUG CABLE TO GROUND AT ENGINE TO AVOID OVER-STRESSING THE IGNITION COILS.
- DISCONNECT INJECTOR JUMPER HARNESS CONNECTOR TO DISABLE FUEL INJECTORS.
- OBSERVE TESTER WHILE CRANKING ENGINE.
- SPARK SHOULD JUMP TESTER GAP WHILE CRANKING ENGINE. DOES IT?

YES → CHECK FOR:
- FAULTY, WORN OR CRACKED SPARK PLUG(S).
- PLUG FOULING OR OVERHEATING DUE TO ENGINE MECHANICAL OR FUEL SYSTEM FAULT.
- PERFORM INJECTOR BALANCE TEST, CHART C-2A.
- REFER TO "CUTS OUT, MISSES" IN "SYMPTOMS,"

NO → CHECK THE RESISTANCE OF EACH SPARK PLUG CABLE WHICH DID NOT FIRE THE SPARK TESTER. CABLE RESISTANCE SHOULD BE LESS THAN 30K OHMS EACH AND CABLES SHOULD NOT BE GROUNDED. ARE CABLES OK?

NO → REPLACE FAULTY CABLE(S) AND RETEST.

YES ↓

3.
- REMOVE THE IGNITION COIL FROM IGNITION CONTROL MODULE WHICH DID NOT FIRE SPARK TESTER.
- INSPECT COIL, SPARK PLUG CABLES AND CABLE BOOTS. THEY SHOULD BE FREE OF CARBON TRACKING. ARE THEY?

NO → REPLACE FAULTY COMPONENT.

YES ↓

4.
- CHECK SECONDARY COIL RESISTANCE. SHOULD BE 5-8K OHMS RESISTANCE. IS IT?

NO → REPLACE IGNITION COIL. ALSO, CHECK FOR FAULTY SPARK PLUG CABLE CONNECTION(S) AND CABLE BOOT(S) FOR CARBON TRACKING.

YES ↓

5.
- INSTALL A KNOWN GOOD IGNITION COIL.
- SPARK SHOULD JUMP TESTER GAP AT PROBLEM CYLINDER WITH ENGINE IDLING. DID IT?

NO → REPLACE ELECTRONIC IGNITION CONTROL MODULE.

YES → ORIGINAL IGNITION COIL IS FAULTY.

2.2L (VIN 4) ENGINE — COMPONENT DIAGNOSTIC CHART — 1993–94 CAVALIER

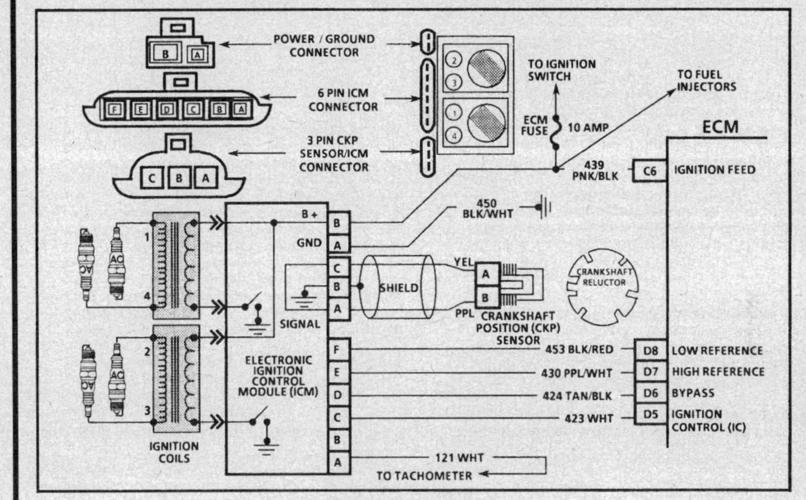

CHART C-4D-2
MISFIRE UNDER LOAD
2.2L (VIN 4) "J" CARLINE

Circuit Description:
The Electronic Ignition (EI) system uses a waste spark method of distribution. In this type of system, the Electronic Ignition Control Module (ICM) triggers the number 1/4 coil pair resulting in both number 1 and number 4 spark plugs firing at the same time. Number 1 cylinder is on the compression stroke at the same time number 4 cylinder is on the exhaust stoke. This leaves the remainder of the high voltage to be used to fire number 1 spark plug. The Crankshaft Position (CKP) sensor is remotely mounted beside the ignition coil assembly and protrudes through the block to within approximately .050" of the crankshaft reluctor is a machined portion of the crankshaft, and the CKP sensor is mounted in a fixed position on the block, timing adjustments are not possible or necessary.

Test Description: Number(s) below refer to circled number(s) on the diagnostic chart.
If the "Misfire" complaint exists at idle only CHART C-4D-1 must be used.
1. Simulates an engine load. Cranking the engine with the ST-125 installed on any cylinder should produce a crisp, blue spark. A cylinder that produces a spark significantly weaker than the other should be considered as having no spark.

2.2L (VIN 4) ENGINE — COMPONENT DIAGNOSTIC CHART — 1993–94 CAVALIER

CHART C-4D-2

MISFIRE UNDER LOAD
2.2L (VIN 4) "J" CARLINE

① • IF ENGINE MISFIRES AT IDLE, REFER TO CHART C-4D-1.
• IGNITION "OFF," INSTALL SPARK TESTER (ST-125) J 26792 TO SPARK PLUG CABLE AT PLUG END AND CLIP TO GROUND.
• CONNECT PLUG END OF COMPANION CYLINDER PLUG CABLE TO ENGINE GROUND TO AVOID OVER-STRESSING THE IGNITION COILS.
• DISCONNECT INJECTOR JUMPER HARNESS CONNECTOR TO DISABLE FUEL INJECTORS.
• CRANK ENGINE, SPARK SHOULD JUMP TESTER GAP WHILE CRANKING.
• REPEAT ABOVE STEPS FOR EACH CYLINDER, RECORDING ANY CYLINDER(S) THAT DID NOT FIRE. RECONNECT EACH SPARK PLUG CABLE TO PLUG BEFORE MOVING ON TO NEXT CYLINDER.
• SPARK SHOULD JUMP TESTER GAP AT ALL PLUG CABLES WHILE CRANKING.

OK
• CHECK FOR FAULTY, CRACKED, FOULED OR WORN SPARK PLUG(S).
• PERFORM INJECTOR BALANCE TEST, CHART C-2A.
• REFER TO "CUTS OUT, MISSES" IN "SYMPTOMS,"

CYLINDER PAIR (1-4, OR 2-3) NOT FIRING
• CHECK FOR AN OPEN SPARK PLUG CABLE FOR CYLINDER PAIR THAT DID NOT FIRE (CABLES SHOULD MEASURE LESS THAN 30K Ω). IF OK, SWAP PROBLEM COIL WITH ONE FOR CYLINDER PAIR THAT DOES FIRE AND RETEST. IF PROBLEM FOLLOWS COIL, REPLACE COIL. IF NOT, REPLACE ELECTRONIC IGNITION CONTROL MODULE (ICM).

SINGLE CYLINDER NOT FIRING
• CHECK FOR GROUNDED SPARK PLUG CABLE(S) FOR CYLINDER(S) THAT DID NOT FIRE AND REPLACE AS NECESSARY. IF OK, SWAP PROBLEM COIL WITH ONE THAT OPERATES PROPERLY. IF PROBLEM FOLLOWS COIL, REPLACE IT. IF NOT, REPLACE SPARK PLUG CABLE FOR CYLINDER(S) THAT DID NOT FIRE.

"AFTER REPAIRS," CONFIRM "CLOSED LOOP" OPERATION AND NO MIL (CHECK ENGINE).

2.2L (VIN 4) ENGINE — COMPONENT DIAGNOSTIC CHART — CAVALIER

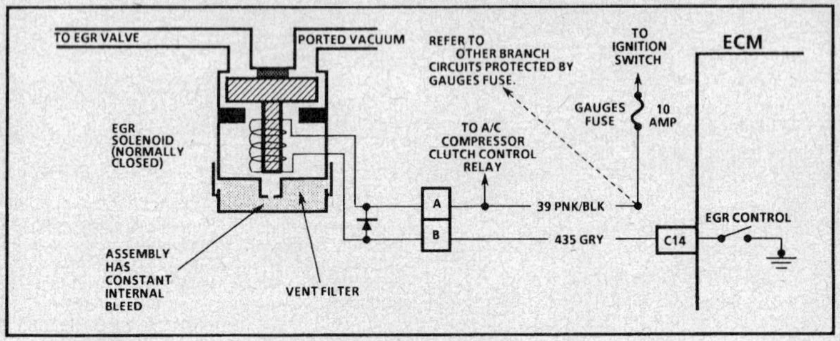

CHART C-7

EXHAUST GAS RECIRCULATION (EGR) FLOW CHECK
2.2L (VIN 4) "J" CARLINE

Circuit Description:

A properly operating EGR system will directly affect the air/fuel mixture requirements of the engine. Since the exhaust gas introduced into the air/fuel mixture cannot be used in combustion (contains very little oxygen), less fuel is required to maintain a correct air/fuel ratio. If the EGR system were to fail in a closed position, the exhaust gas would be replaced with air, and the air/fuel mixture would be leaner. The ECM would compensate for the lean condition by adding fuel, resulting in higher long term fuel trim values.

The fuel control on this engine is conducted within 4 fuel trim cells. Since EGR is not used at idle, the idle cell would not be affected by EGR system operation. The other fuel trim cells are affected by EGR operation, and, when the EGR system is operating properly, the long term fuel trim values in all cells should be close to the same. If the EGR system becomes inoperative, the long term fuel trim values in the open throttle cells would change to compensate for the resulting lean or rich mixtures, but the long term fuel trim value in the closed throttle cell would not change.

The difference in long term fuel trim values between the idle (closed throttle) cell and cell 2 is used to monitor EGR system performance. When the difference between the two long term fuel trim values is greater than 10, and the long term fuel trim value in cell 2 is greater than 135, Diagnostic Trouble Code (DTC) 32 is set. The system operates in fuel trim cell 2 during a cruise condition at approximately 55 mph.

Test Description: Number(s) below refer to circled number(s) on the diagnostic chart.

1. Codes should be diagnosed using appropriate chart before preparing a functional check. If Diagnostic Trouble Codes (DTCs) 14, 21, 23, 32, or 33 are set, use those charts first.
Be sure shop exhaust hose is not connected during Steps 2 and 4.
2. The tail pipe ventilation hose must be removed for this test. The ventilation hose can sometimes cause a good EGR valve to fail this test.
3. Intake Passage: Shut "OFF" engine and remove the EGR valve from the manifold. Plug the exhaust side hole with a suitable stopper. Leaving the intake side hole open, attempt to start the engine. If the engine runs at a high idle (up to 3000 RPM is possible) or starts and stalls, the EGR intake passage is not restricted. If the engine starts and idles normally, the EGR intake passage is restricted.
4. Because the shop exhaust hose is not installed at this point, don't allow the engine to run longer than 15 seconds.
Exhaust passage: With EGR valve still removed, plug the intake side hole with a suitable stopper. With the exhaust side hole open, check for the presence of exhaust gas. If no exhaust gas is present, the EGR exhaust side passage is restricted.

Diagnostic Aids:

This chart is a functional check of the EGR system. If the EGR system works properly, check other items that result in high long term fuel trim values in fuel trim cell 2, but not in the closed throttle cell.

Incorrect fuel pressure or lean/rich fuel injectors can also cause incorrect long term fuel trim values. Refer to "Fuel Pressure Test" in CHART A-7. If fuel pressure checks out OK, proceed to "Fuel Injector Balance Test" in CHART C-2A. It may be necessary to monitor fuel pressure while driving the vehicle at various road speeds and/or loads.

2.2L (VIN 4) ENGINE — COMPONENT DIAGNOSTIC CHART — CAVALIER

2.2L (VIN 4) ENGINE — COMPONENT DIAGNOSTIC CHART — 1992 CAVALIER

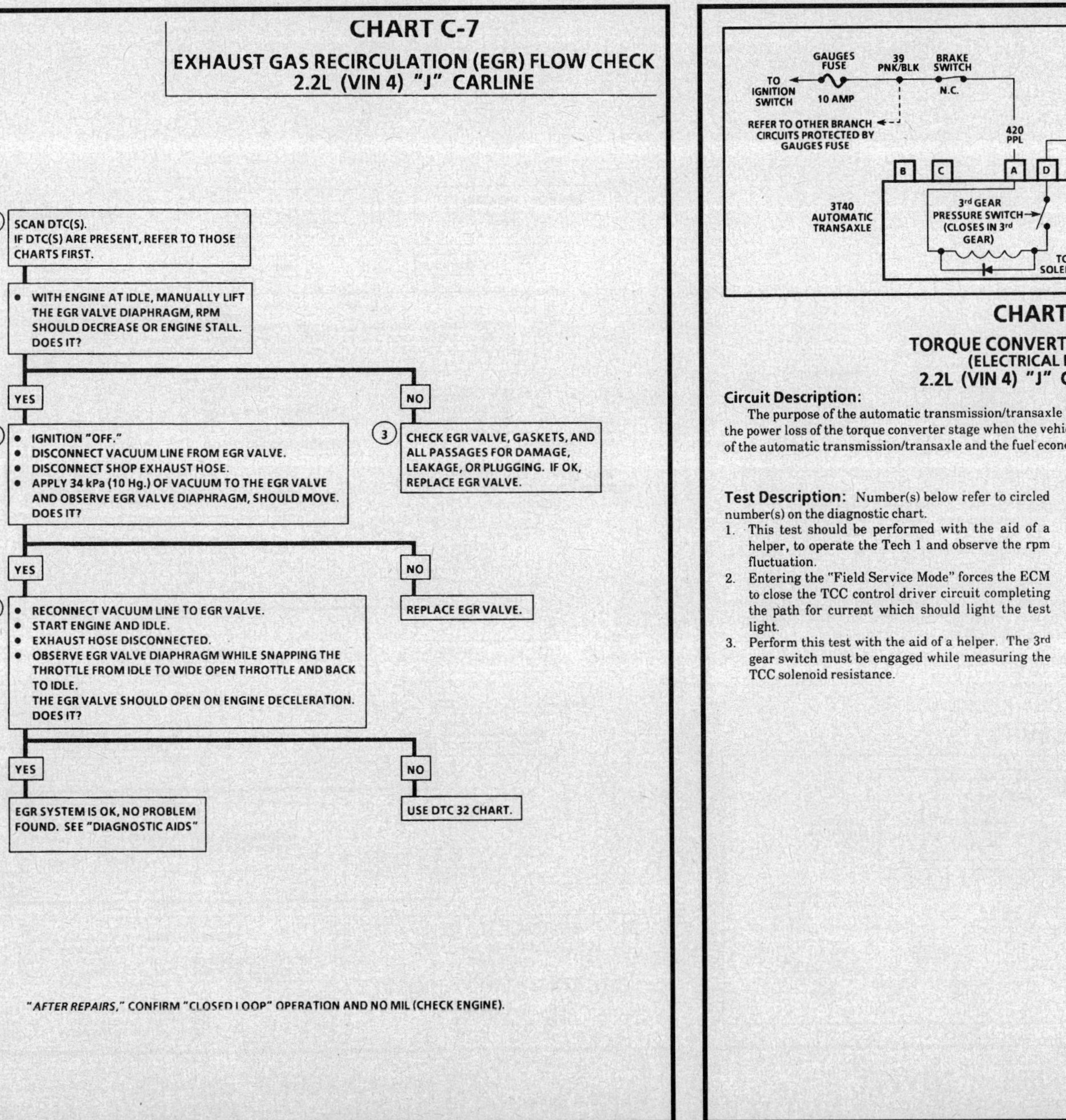

CHART C-8A

TORQUE CONVERTER CLUTCH (TCC)
(ELECTRICAL DIAGNOSIS)
2.2L (VIN 4) "J" CARLINE (PORT)

Circuit Description:

The purpose of the automatic transmission/transaxle Torque Converter Clutch (TCC) feature is to eliminate the power loss of the torque converter stage when the vehicle is in a cruise condition. This allows the convenience of the automatic transmission/transaxle and the fuel economy of a manual transmission.

Test Description: Number(s) below refer to circled number(s) on the diagnostic chart.

1. This test should be performed with the aid of a helper, to operate the Tech 1 and observe the rpm fluctuation.
2. Entering the "Field Service Mode" forces the ECM to close the TCC control driver circuit completing the path for current which should light the test light.
3. Perform this test with the aid of a helper. The 3rd gear switch must be engaged while measuring the TCC solenoid resistance.

Fused battery ignition is supplied to the TCC solenoid through the brake switch. The ECM will engage TCC by grounding CKT 422 to energize the solenoid.

TCC will engage when:

• Vehicle speed above 35 km/h (22 mph).
• Engine at normal operating temperature (above 70°C, 158°F).
• Throttle position sensor output not changing, indicating a steady road speed.
• Brake switch closed.

2.2L (VIN 4) ENGINE — COMPONENT DIAGNOSTIC CHART — 1992 CAVALIER

NOTICE: MECHANICAL CHECKS, SUCH AS LINKAGE, OIL LEVEL, ETC. SHOULD BE PERFORMED PRIOR TO USING THIS CHART.

CHART C-8A

TORQUE CONVERTER CLUTCH (TCC)
(ELECTRICAL DIAGNOSIS)
2.2L (VIN 4) "J" CARLINE (PORT)

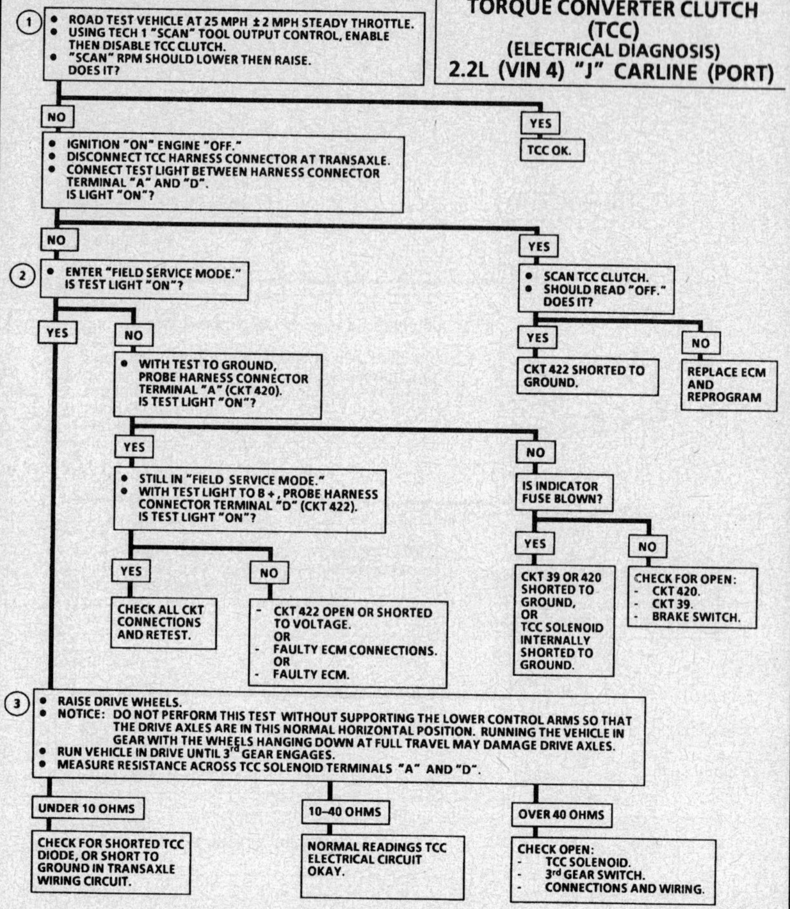

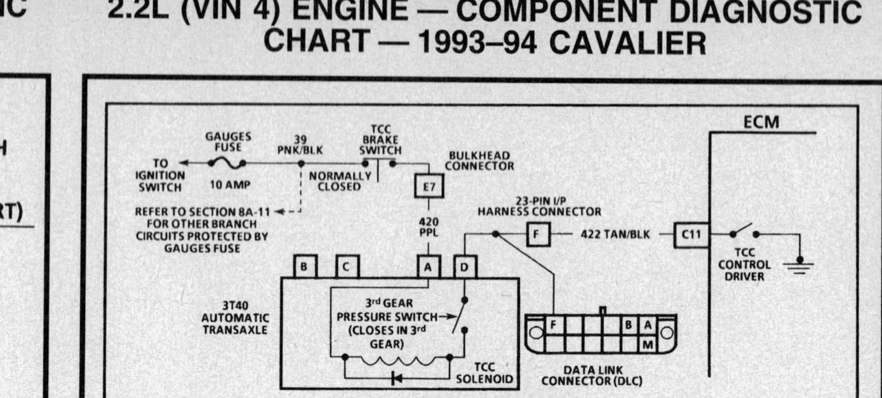

2.2L (VIN 4) ENGINE — COMPONENT DIAGNOSTIC CHART — 1993–94 CAVALIER

CHART C-8A

TORQUE CONVERTER CLUTCH (TCC)
(ELECTRICAL DIAGNOSIS)
2.2L (VIN 4) "J" CARLINE

Circuit Description:

The purpose of the automatic transaxle Torque Converter Clutch (TCC) feature is to eliminate the power loss of the torque converter stage when the vehicle is in a cruise condition. This allows the convenience of the automatic transaxle and the fuel economy of a manual transaxle.

Test Description: Number(s) below refer to circled number(s) on the diagnostic chart.

1. This test should be performed with the aid of a helper, to operate the Tech 1 and observe the RPM fluctuation.
2. Entering the "Field Service Mode" forces the ECM to close the TCC control driver circuit completing the path for current which should light the test light.
3. Perform this test with the aid of a helper. The 3rd gear switch must be engaged while measuring the TCC solenoid resistance.

Fused battery ignition is supplied to the TCC solenoid through the brake switch. The ECM will engage TCC by grounding CKT 422 to energize the solenoid.

TCC will engage when:

- Vehicle speed above about (35 mph with A/C "OFF," 44 mph with A/C "ON").
- Engine at normal operating temperature above about 20°C (68°F).
- Throttle Position (TP) sensor output not changing, indicating a steady road speed.
- Brake switch closed.
- 3rd gear switch applied.

2.2L (VIN 4) ENGINE — COMPONENT DIAGNOSTIC CHART — 1993–94 CAVALIER

CHART C-8A
TORQUE CONVERTER CLUTCH (TCC)
(ELECTRICAL DIAGNOSIS)
2.2L (VIN 4) "J" CARLINE

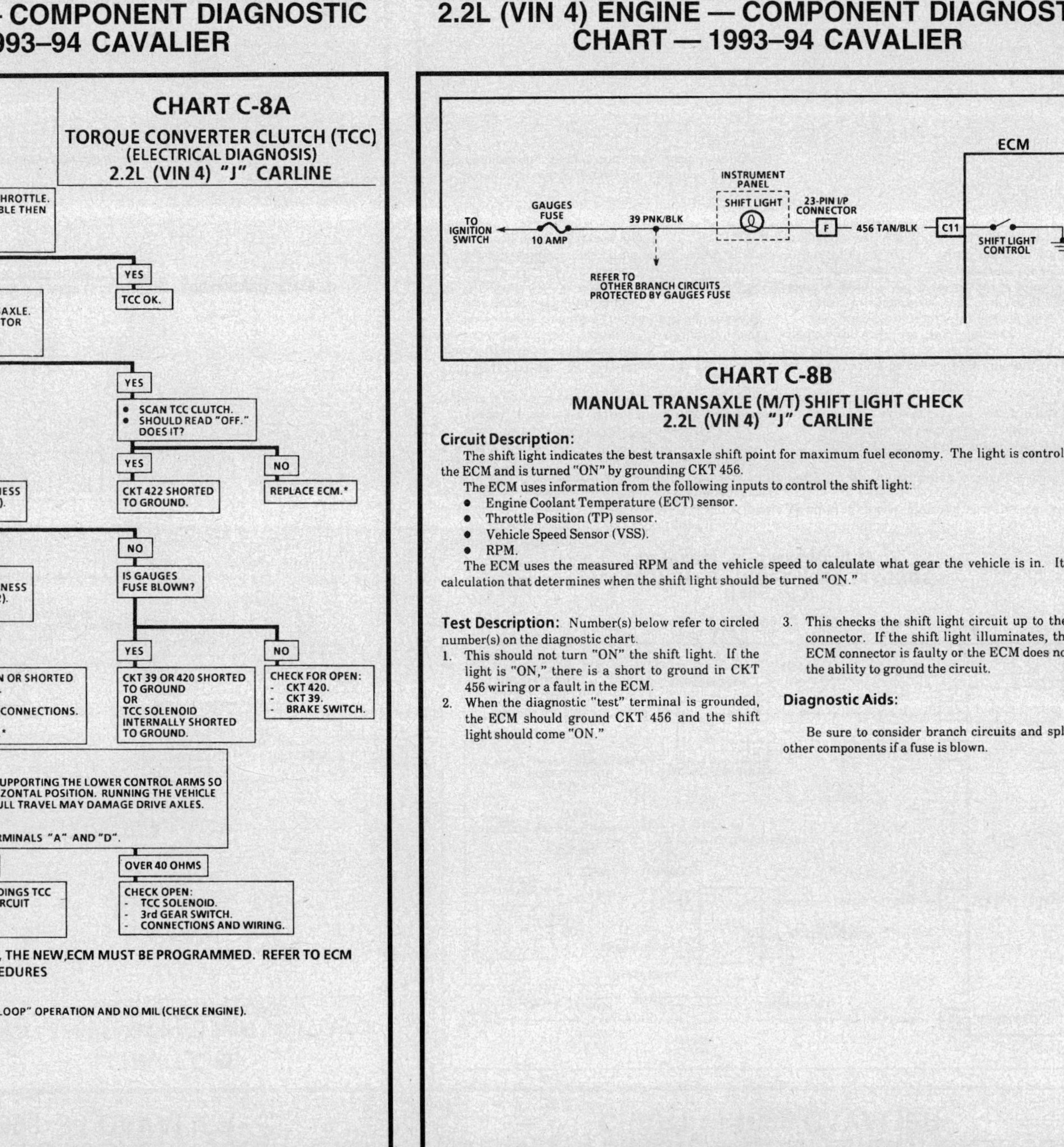

CHART C-8B
MANUAL TRANSAXLE (M/T) SHIFT LIGHT CHECK
2.2L (VIN 4) "J" CARLINE

Circuit Description:

The shift light indicates the best transaxle shift point for maximum fuel economy. The light is controlled by the ECM and is turned "ON" by grounding CKT 456.

The ECM uses information from the following inputs to control the shift light:

● Engine Coolant Temperature (ECT) sensor.
● Throttle Position (TP) sensor.
● Vehicle Speed Sensor (VSS).
● RPM.

The ECM uses the measured RPM and the vehicle speed to calculate what gear the vehicle is in. It's this calculation that determines when the shift light should be turned "ON."

Test Description: Number(s) below refer to circled number(s) on the diagnostic chart.

1. This should not turn "ON" the shift light. If the light is "ON," there is a short to ground in CKT 456 wiring or a fault in the ECM.
2. When the diagnostic "test" terminal is grounded, the ECM should ground CKT 456 and the shift light should come "ON."
3. This checks the shift light circuit up to the ECM connector. If the shift light illuminates, then the ECM connector is faulty or the ECM does not have the ability to ground the circuit.

Diagnostic Aids:

Be sure to consider branch circuits and splices to other components if a fuse is blown.

2.2L (VIN 4) ENGINE — COMPONENT DIAGNOSTIC CHART — 1993–94 CAVALIER

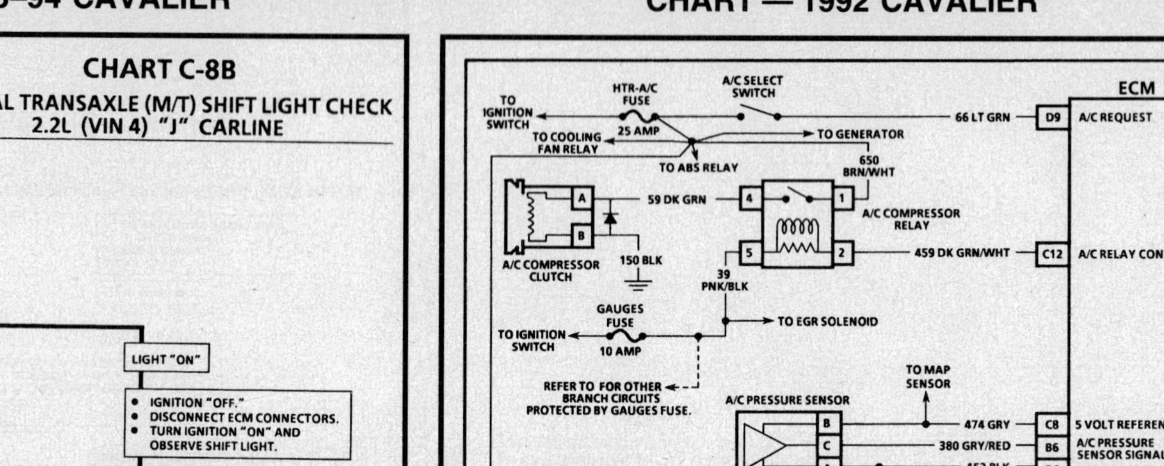

* IF ECM IS FAULTY AND MUST BE REPLACED, THE NEW ECM MUST BE PROGRAMMED. REFER TO ECM REPLACEMENT AND PROGRAMMING PROCEDURES

"AFTER REPAIRS," CONFIRM "CLOSED LOOP" OPERATION AND NO MIL (CHECK ENGINE).

2.2L (VIN 4) ENGINE — COMPONENT DIAGNOSTIC CHART — 1992 CAVALIER

CHART C-10
(Page 1 of 2)
A/C CLUTCH CONTROL CIRCUIT DIAGNOSIS
2.2L (VIN 4) "J" CARLINE (PORT)

Circuit Description:

The A/C clutch control relay is energized when the Electronic Control Module (ECM) provides a ground path through CKT 459 and A/C is requested. A/C clutch is delayed about .3 seconds after A/C is requested. This will allow the IAC to adjust engine rpm for the additional load.

The ECM will temporarily disengage the A/C clutch relay for calibrated times for one or more of the following:

- Hot engine restart.
- Wide Open Throttle (WOT) (TPS over 99%).
- Engine rpm greater than about 5025 rpm.
- During Idle Air Control (IAC) reset.

The A/C clutch relay will remain disengaged if pressure is out of range as described previously in this section or there is no A/C request signal due to an open A/C select switch or circuit.

Test Description: Number(s) below refer to circled number(s) on the diagnostic chart.

1. The ECM will only energize the A/C relay when the engine is running. This test will determine if the relay or CKT 459 is faulty.
2. Determines if the signal is reaching the ECM through CKT 66 from the A/C control panel. Signal should only be present when an A/C mode or defrost mode has been selected.

Diagnostic Aids:

Be sure to consider branch circuits and splices to other components. If complaint is insufficient, cooling, the problem may be caused by an inoperative cooling fan. See CHART C-12 for cooling fan diagnosis. If fan operates correctly, see "A/C Diagnosis".

A/C pressure outside of a range of 41 to 426 psi will cause the compressor to be disabled by the ECM. Observe Tech 1 A/C pressure for 2 minutes with engine idling and A/C "ON" pressure should be within .20 psi of actual. If not, check for a circuit problem using Code 66 chart or replace the A/C pressure sensor

2.2L (VIN 4) ENGINE — COMPONENT DIAGNOSTIC CHART — 1992CAVALIER

CHART C-10
(Page 1 of 2)
**A/C CLUTCH CONTROL CIRCUIT DIAGNOSIS
2.2L (VIN 4) "J" CARLINE (PORT)**

(1)
- IGNITION "ON," ENGINE "OFF."
- USING TECH 1, SCAN FOR CODES AND REFER TO THOSE CHARTS, IF PRESENT.
- A/C SELECT "ON."
- NOTE A/C CLUTCH.
- CLUTCH SHOULD NOT BE ENGAGED. IS IT?

NO
- WITH ENGINE IDLING AT NORMAL OPERATING TEMPERATURE TURN A/C "ON" AND "OFF."
- CLUTCH SHOULD CYCLE "ON" AND "OFF." DOES IT?

NO (2)
- PLACE A/C CONTROL IN A/C MODE.
- "SCAN" TOOL SHOULD DISPLAY A/C REQUEST "YES." DOES IT?

YES
REFER TO CHART C-10 2 OF 2.

NO
- BACKPROBE ECM TERMINAL "D9" WITH A TEST LIGHT TO GROUND.

LIGHT "ON"
- CHECK FOR POOR CONNECTION AT ECM TERMINAL "D9". IF OK, ECM IS FAULTY.

LIGHT "OFF"
- IF A/C FUSE IS BLOWN, CHECK FOR SHORT TO GROUND IN CKT 59 OR 66, OR SHORTED A/C CLUTCH COIL.
- IF OK, CHECK FOR OPEN CKT 66, OR FAULTY A/C SELECT SWITCH. IF OK, REFER TO "DIAGNOSTIC AIDS"

YES
A/C CLUTCH CONTROL CIRCUIT OK, REFER TO "DIAGNOSTIC AIDS"

YES
- DISCONNECT A/C RELAY.
- PROBE CKT 459 WITH A TEST LIGHT TO 12 VOLTS.

LIGHT "OFF"
- USING DVM, MEASURE THE RESISTANCE BETWEEN RELAY TERMINALS "1" AND "4". SHOULD MEASURE INFINITE (OPEN). DOES IT?

YES
CKT 59 SHORTED TO B+ OR FAULTY A/C CLUTCH.

NO
FAULTY A/C RELAY.

LIGHT "ON"
CKT 459 SHORTED TO GROUND OR FAULTY ECM.

2.2L (VIN 4) ENGINE — COMPONENT DIAGNOSTIC CHART — 1992 CAVALIER

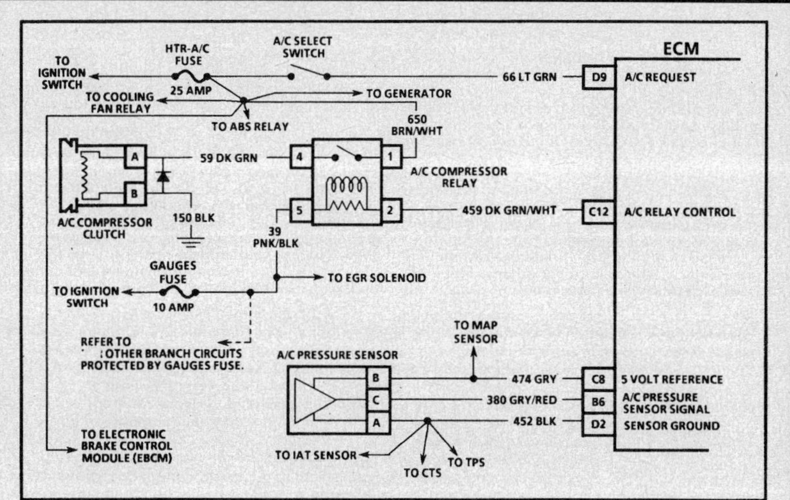

CHART C-10
(Page 2 of 2)
**A/C CLUTCH CONTROL CIRCUIT DIAGNOSIS
2.2L (VIN 4) "J" CARLINE (PORT)**

Circuit Description:

The A/C clutch control relay is energized when the ECM provides a ground path through CKT 459 and A/C is requested. A/C clutch is delayed about .3 seconds after A/C is requested. This will allow the IAC to adjust engine rpm for the additional load.

The ECM will temporarily disengage the A/C clutch relay for calibrated times for one or more of the following:
- Hot engine restart.
- Wide Open Throttle (WOT) (TPS over 99%).
- Engine rpm greater than about 5025 rpm.
- During IAC reset.

The A/C clutch relay will remain disengaged when a Code 66 is present, if pressure is out of range as described previously in this section, or there is no A/C request signal due to an open A/C select switch or circuit.

Test Description: Number(s) below refer to circled number(s) on the diagnostic chart.
1. Determines if the pressure transducer is out of range causing the compressor clutch to be disengaged.
2. With the engine stopped and field service mode activated, the ECM should be grounding CKT 459, which should cause the test light to be "ON."

Diagnostic Aids:

If complaint is insufficient cooling, the problem may be caused by an inoperative cooling fan. See CHART C-12 for cooling fan diagnosis. If fan operates correctly, see A/C diagnosis.   A/C pressure outside of a range of 41 to 426 psi will cause the compressor to be disabled by the ECM. Observe Tech 1 A/C pressure for 2 minutes with engine idling and A/C "ON."

"Scan" pressure should be within 20 psi of actual. If n check for a circuit problem using Code 66 chart. replace sensor.

2.2L (VIN 4) ENGINE — COMPONENT DIAGNOSTIC CHART — 1992 CAVALIER

CHART C-10
(Page 2 of 2)
A/C CLUTCH CONTROL CIRCUIT DIAGNOSIS
2.2L (VIN 4) "J" CARLINE (PORT)

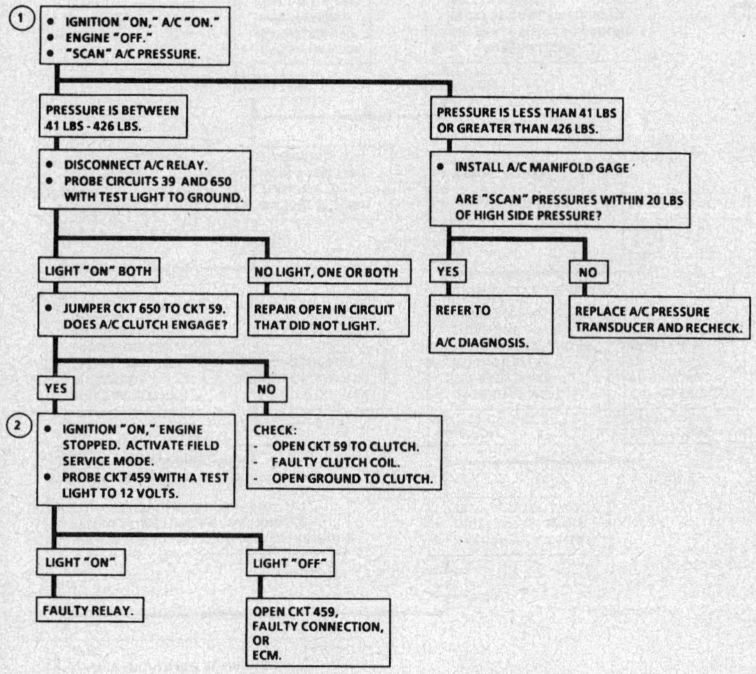

1.
- IGNITION "ON," A/C "ON."
- ENGINE "OFF."
- "SCAN" A/C PRESSURE.

PRESSURE IS BETWEEN 41 LBS - 426 LBS.

- DISCONNECT A/C RELAY.
- PROBE CIRCUITS 39 AND 650 WITH TEST LIGHT TO GROUND.

LIGHT "ON" BOTH

- JUMPER CKT 650 TO CKT 59. DOES A/C CLUTCH ENGAGE?

YES

2.
- IGNITION "ON," ENGINE STOPPED. ACTIVATE FIELD SERVICE MODE.
- PROBE CKT 459 WITH A TEST LIGHT TO 12 VOLTS.

LIGHT "ON"

FAULTY RELAY.

NO LIGHT, ONE OR BOTH

REPAIR OPEN IN CIRCUIT THAT DID NOT LIGHT.

NO

CHECK:
- OPEN CKT 59 TO CLUTCH.
- FAULTY CLUTCH COIL.
- OPEN GROUND TO CLUTCH.

LIGHT "OFF"

OPEN CKT 459, FAULTY CONNECTION, OR ECM.

PRESSURE IS LESS THAN 41 LBS OR GREATER THAN 426 LBS.

- INSTALL A/C MANIFOLD GAGE'

ARE "SCAN" PRESSURES WITHIN 20 LBS OF HIGH SIDE PRESSURE?

YES

REFER TO A/C DIAGNOSIS.

NO

REPLACE A/C PRESSURE TRANSDUCER AND RECHECK.

"AFTER REPAIRS," CONFIRM "CLOSED LOOP" OPERATION AND NO "CHECK ENGINE" LIGHT.

2.2L (VIN 4) ENGINE — COMPONENT DIAGNOSTIC CHART — 1993–94 CAVALIER

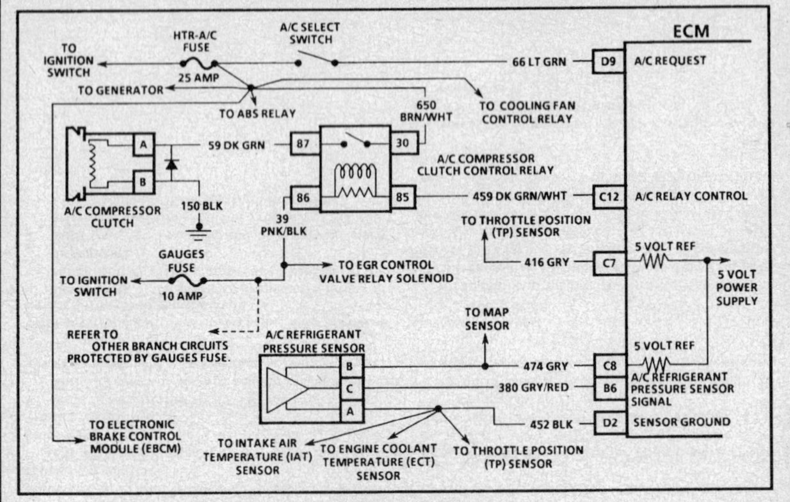

CHART C-10
(Page 1 of 2)
A/C COMPRESSOR CLUTCH CONTROL CIRCUIT DIAGNOSIS
2.2L (VIN 4) "J" CARLINE

Circuit Description:
The A/C compressor clutch control relay is energized when the Engine Control Module (ECM) provides a ground path through CKT 459 and A/C is requested. A/C compressor clutch is delayed about .3 seconds after A/C is requested. This will allow the IAC to adjust engine RPM for the additional load.

The ECM will disable the A/C compressor clutch for any of the following reasons:
- Engine not running.
- DTC 66 is set and fault is current.
- Throttle position is about 99 percent or greater.
- ECT is about 122°C (251°F) or greater.
- During an IAC motor reset.
- Engine RPM is greater than about 4224 RPM.
- A/C refrigerant pressure is about 429 psi or greater.
- A/C refrigerant pressure is about 38 psi or less.

If the ECM temporarily disabled the A/C compressor clutch, the clutch will be renabled as follows:
- Throttle position drops below about 89 percent.
- ECT drops below about 119°C (246°F).
- Engine RPM drops below about 3850 RPM.
- A/C refrigerant pressure drops below about 199 psi.
- A/C refrigerant pressure rises above about 47 psi.
- IAC motor reset is complete.

Test Description: Number(s) below refer to circled number(s) on the diagnostic chart.
1. The ECM will only energize the A/C relay when the engine is running. This test will determine if the relay or CKT 459 is faulty.
2. Determines if the signal is reaching the ECM through CKT 66 from the A/C control panel. Signal should only be present when an A/C mode or defrost mode has been selected.

Diagnostic Aids:
Be sure to consider branch circuits and splices to other components. If complaint is insufficient cooling, the problem may be caused by an inoperative cooling fan. Refer to CHART C-12 for cooling fan diagnosis. If fan operates correctly, refer to "A/C Diagnosis"

A/C refrigerant pressure outside of a range of 38 to 429 psi will cause the compressor clutch to be disabled by the ECM. Observe Tech 1 A/C refrigerant pressure for 2 minutes with engine idling and A/C "ON."

Scanned refrigerant pressure should be within 20 psi of actual. If not, check for a circuit problem using DTC 66 chart or replace the A/C refrigerant pressure sensor.

2.2L (VIN 4) ENGINE — COMPONENT DIAGNOSTIC CHART — 1993–94 CAVALIER

CHART C-10

(Page 1 of 2)

A/C COMPRESSOR CLUTCH CONTROL CIRCUIT DIAGNOSIS
2.2L (VIN 4) "J" CARLINE

(1)
- INSTALL SCAN TOOL AND CHECK FOR DIAGNOSTIC TROUBLE CODES DTC(S). IF DTC(S) ARE PRESENT, REFER TO THOSE CHARTS.
- ENGINE IDLING.
- A/C SELECT "OFF."
- NOTE A/C COMPRESSOR CLUTCH. CLUTCH SHOULD BE DISENGAGED. IS IT?

YES
- ENGINE IDLING AT NORMAL OPERATING TEMPERATURE IN A HEATED SHOP AND A/C PRESSURE BETWEEN ABOUT 38 psi AND 429 psi.
- TURN THE A/C SELECT SWITCH "ON" AND "OFF."
- COMPRESSOR CLUTCH SHOULD CYCLE "ON" AND "OFF." DOES IT?

NO

(2)
- PLACE A/C SELECT SWITCH IN A/C MODE. SCAN TOOL SHOULD DISPLAY A/C IS REQUESTED. DOES IT?

NO
- BACKPROBE ECM TERMINAL "D9" WITH A TEST LIGHT TO GROUND.

LIGHT "ON"
CHECK FOR POOR CONNECTION AT ECM TERMINAL "D9". IF OK, ECM IS FAULTY.*

LIGHT "OFF"
IF HTR-A/C FUSE IS OPEN, CHECK FOR SHORT TO GROUND IN CKT 66
OR
OPEN CKT 66
OR
SHORT TO GROUND IN BRANCH CIRCUITS (CKT 650) REFER TO

FAULTY A/C SELECT SWITCH.

YES
REFER TO CHART C-10 2 OF 2.

YES
A/C CLUTCH CONTROL CIRCUIT IS OK, REFER TO "DIAGNOSTIC AIDS"

NO
- DISCONNECT A/C COMPRESSOR CLUTCH CONTROL RELAY.
- ENGINE IDLING, A/C SELECT "OFF."
- PROBE HARNESS CONNECTOR TERMINAL "85" (CKT 459) WITH A TEST LIGHT TO 12 VOLTS.

LIGHT "OFF"
- USING DVM, MEASURE THE RESISTANCE BETWEEN RELAY TERMINALS "30" AND "87". SHOULD MEASURE INFINITE (OPEN). DOES IT?

YES
CKT 59 SHORTED TO B+
OR
FAULTY A/C COMPRESSOR CLUTCH.

NO
FAULTY A/C COMPRESSOR CLUTCH CONTROL RELAY.

LIGHT "ON"
- A/C SELECT "OFF."
- BACKPROBE ECM TERMINAL "D9" (CKT 66) WITH A TEST LIGHT TO GROUND.

LIGHT "OFF"
CKT 459 SHORTED TO GROUND
OR
FAULTY ECM.*

LIGHT "ON"
CKT 66 SHORTED TO B+
OR
FAULTY A/C SELECT SWITCH.

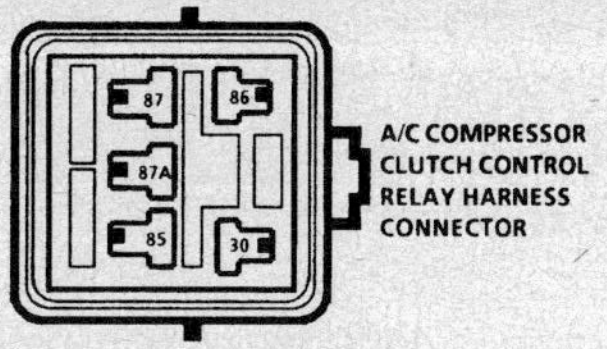

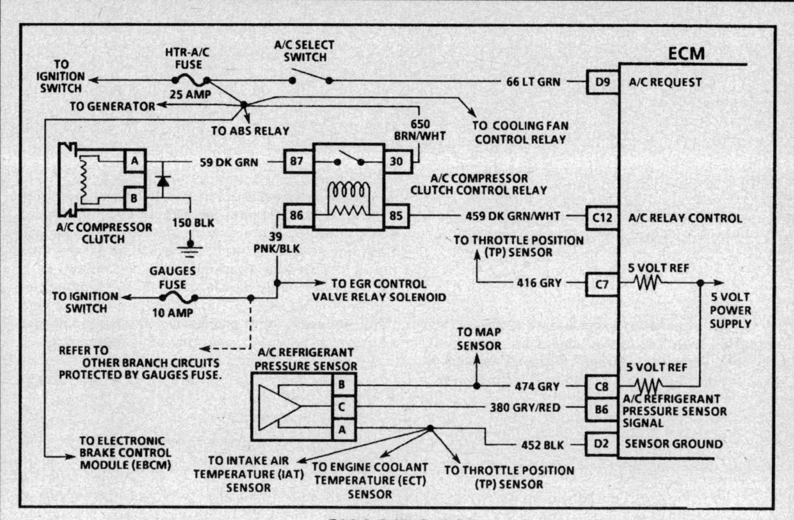

* IF ECM IS FAULTY AND MUST BE REPLACED, THE NEW ECM MUST BE PROGRAMMED. REFER TO ECM REPLACEMENT AND PROGRAMMING PROCEDURES

2.2L (VIN 4) ENGINE — COMPONENT DIAGNOSTIC CHART — 1993–94 CAVALIER

CHART C-10

(Page 2 of 2)

A/C COMPRESSOR CLUTCH CONTROL CIRCUIT DIAGNOSIS
2.2L (VIN 4) "J" CARLINE

Circuit Description:

The A/C compressor clutch control relay is energized when the ECM provides a ground path through CKT 459 and A/C is requested. A/C compressor clutch is delayed about .3 second after A/C is requested. This will allow the IAC to adjust engine RPM for the additional load.

The A/C compressor clutch control relay will remain disengaged if refrigerant pressure is out of range described previously in this section, or there is no A/C request signal due to an open A/C select switch or circuit. Refer to SECTION 1B for more information on A/C refrigerant systems.

Test Description: Number(s) below refer to circled number(s) on the diagnostic chart.

1. Determines if the A/C refrigerant pressure sensor is out of range causing the compressor clutch to be disengaged.
2. With the engine stopped and "Field Service Mode" activated, the ECM should be grounding CKT 459, which should cause the test light to be "ON."

Diagnostic Aids:

If complaint is insufficient cooling, the problem may be caused by an inoperative cooling fan. Refer to CHART C-12 for cooling fan diagnosis. If fan operates correctly, refer to A/C diagnosis A/C refrigerant pressure outside of a range of 41 to 426 psi will cause the compressor to be disabled by the ECM. With a scan tool, observe A/C refrigerant pressure for 2 minutes with engine idling and A/C "ON."

Scan pressure should be within 20 psi of actual. If not, check for a circuit problem using DTC 66 chart or replace A/C refrigerant pressure sensor.

2.2L (VIN 4) ENGINE — COMPONENT DIAGNOSTIC CHART — 1993–94 CAVALIER

CHART C-10

(Page 2 of 2)
A/C COMPRESSOR CLUTCH CONTROL CIRCUIT DIAGNOSIS
2.2L (VIN 4) "J" CARLINE

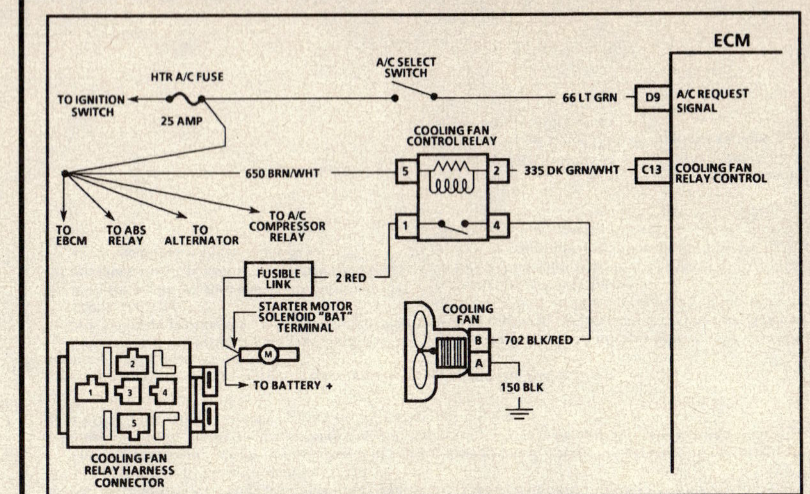

1.
- IGNITION "ON," A/C "ON."
- ENGINE "OFF."
- AMBIENT TEMPERATURE ABOVE 15°C (60°F).
- WITH TECH 1, OBSERVE A/C REFRIGERANT PRESSURE.

PRESSURE IS BETWEEN 38 LBS - 429 LBS

PRESSURE IS LESS THAN 38 LBS OR GREATER THAN 429 LBS

- DISCONNECT A/C COMPRESSOR CLUTCH CONTROL RELAY.
- PROBE HARNESS CONNECTOR TERMINALS "30" AND "85" WITH TEST LIGHT TO GROUND.

- INSTALL A/C MANIFOLD GAGE SET

ARE TECH 1 PRESSURES WITHIN 20 LBS. OF HIGH SIDE PRESSURE?

LIGHT "ON" BOTH

NO LIGHT, ONE OR BOTH

YES

NO

- JUMPER HARNESS CONNECTOR TERMINALS "30" TO "87". DOES A/C CLUTCH ENGAGE?

REPAIR OPEN IN CIRCUIT THAT DID NOT LIGHT.

REFER TO A/C DIAGNOSIS.

REPLACE A/C REFRIGERANT PRESSURE SENSOR AND RECHECK.

YES

NO

2.
- IGNITION "ON," ENGINE STOPPED. ACTIVATE "FIELD SERVICE MODE."
- PROBE HARNESS CONNECTOR TERMINAL "85" WITH A TEST LIGHT TO 12 VOLTS.

CHECK:
- OPEN CKT 59 TO CLUTCH.
- FAULTY COMPRESSOR CLUTCH COIL.
- OPEN GROUND TO COMPRESSOR CLUTCH.

LIGHT "ON"

LIGHT "OFF"

FAULTY A/C COMPRESSOR CLUTCH CONTROL RELAY.

OPEN CKT 459, FAULTY CONNECTION OR ECM.*

* IF ECM IS FAULTY AND MUST BE REPLACED, THE NEW ECM MUST BE PROGRAMMED. REFER TO ECM REPLACEMENT AND PROGRAMMING PROCEDURES

"AFTER REPAIRS," CONFIRM "CLOSED LOOP" OPERATION AND NO MIL (CHECK ENGINE).

2.2L (VIN 4) ENGINE — COMPONENT DIAGNOSTIC CHART — 1992 CAVALIER

CHART C-12

ECM CONTROLLED COOLING FAN
2.2L (VIN 4) "J" CARLINE (PORT)

Circuit Description:

Battery voltage to operate the cooling fan motor is supplied to relay by CKT 2. Ignition voltage to energize the relay is supplied to relay by CKT 650. When the ECM grounds CKT 335, the relay is energized and the cooling fan is turned "ON." When the engine is running, the ECM will turn the cooling fan "ON" if:

- A/C is "ON,"
  or
- Coolant temperature greater than 105°C (221°F)
  or
- Code 14 is set.

Diagnostic Aids:

If the owner complained of an overheating problem, it must be determined if the complaint was due to an actual boil over, or the hot light, or temperature gage indicated overheating.

If the gage or light indicates overheating, but no boil over is detected, the gage circuit should be checked. The gage accuracy can also be checked by comparing the coolant sensor reading using a "Scan" tool and comparing its reading with the gage reading.

If the engine is actually overheating and the gage indicates overheating, but the cooling fan is not coming "ON," the coolant sensor has probably shifted out of calibration and should be replaced.

2.2L (VIN 4) ENGINE — COMPONENT DIAGNOSTIC CHART — 1992 CAVALIER

CHART C-12

ECM CONTROLLED COOLING FAN
2.2L (VIN 4) "J" CARLINE (PORT)

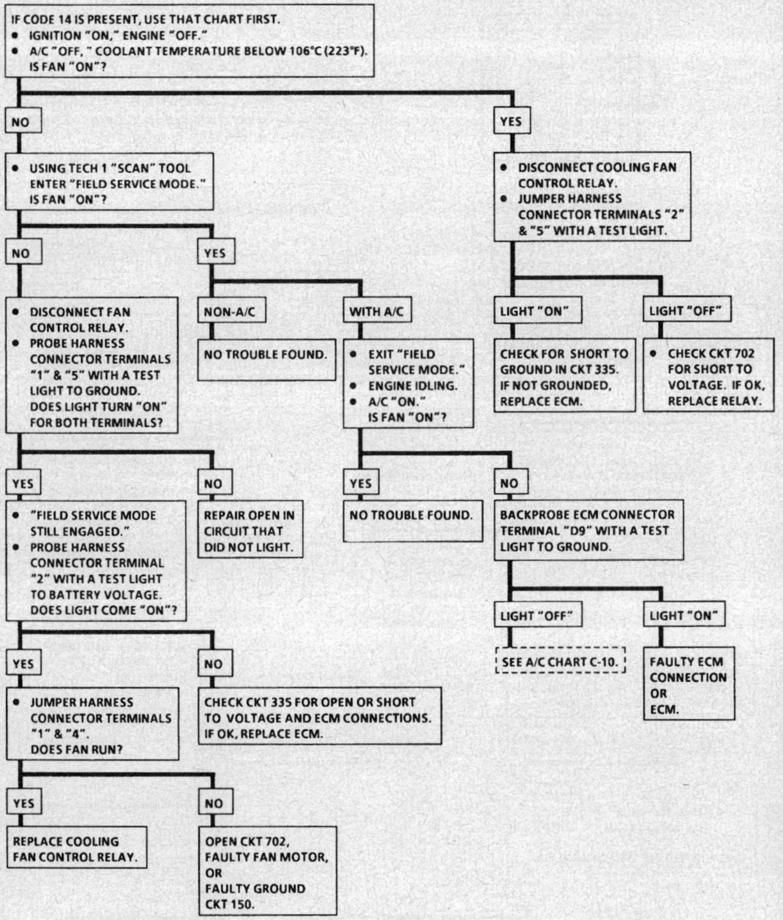

IF CODE 14 IS PRESENT, USE THAT CHART FIRST.
• IGNITION "ON," ENGINE "OFF."
• A/C "OFF," COOLANT TEMPERATURE BELOW 106°C (223°F).
IS FAN "ON"?
NO
YES
• USING TECH 1 "SCAN" TOOL ENTER "FIELD SERVICE MODE." IS FAN "ON"?
• DISCONNECT COOLING FAN CONTROL RELAY.
• JUMPER HARNESS CONNECTOR TERMINALS "2" & "5" WITH A TEST LIGHT.
NO
YES
NON-A/C
WITH A/C
LIGHT "ON"
LIGHT "OFF"
• DISCONNECT FAN CONTROL RELAY.
• PROBE HARNESS CONNECTOR TERMINALS "1" & "5" WITH A TEST LIGHT TO GROUND. DOES LIGHT TURN "ON" FOR BOTH TERMINALS?
NO TROUBLE FOUND.
• EXIT "FIELD SERVICE MODE."
• ENGINE IDLING.
• A/C "ON." IS FAN "ON"?
CHECK FOR SHORT TO GROUND IN CKT 335. IF NOT GROUNDED, REPLACE ECM.
• CHECK CKT 702 FOR SHORT TO VOLTAGE. IF OK, REPLACE RELAY.
YES
NO
YES
NO
• "FIELD SERVICE MODE STILL ENGAGED."
• PROBE HARNESS CONNECTOR TERMINAL "2" WITH A TEST LIGHT TO BATTERY VOLTAGE. DOES LIGHT COME "ON"?
REPAIR OPEN IN CIRCUIT THAT DID NOT LIGHT.
NO TROUBLE FOUND.
BACKPROBE ECM CONNECTOR TERMINAL "D9" WITH A TEST LIGHT TO GROUND.
LIGHT "OFF"
LIGHT "ON"
SEE A/C CHART C-10.
FAULTY ECM CONNECTION OR ECM.
YES
NO
• JUMPER HARNESS CONNECTOR TERMINALS "1" & "4". DOES FAN RUN?
CHECK CKT 335 FOR OPEN OR SHORT TO VOLTAGE AND ECM CONNECTIONS. IF OK, REPLACE ECM.
YES
NO
REPLACE COOLING FAN CONTROL RELAY.
OPEN CKT 702, FAULTY FAN MOTOR, OR FAULTY GROUND CKT 150.

2.2L (VIN 4) ENGINE — COMPONENT DIAGNOSTIC CHART — 1993–94 CAVALIER

ECM
TO IGNITION SWITCH
HTR A/C FUSE
25 AMP
A/C SELECT SWITCH
66 LT GRN
D9
A/C REQUEST SIGNAL
COOLING FAN CONTROL RELAY
650 BRN/WHT
86
85
335 DK GRN/WHT
C13
COOLING FAN RELAY CONTROL
TO EBCM
TO ABS RELAY
TO GENERATOR
TO A/C COMPRESSOR CLUTCH CONTROL RELAY
30
87
FUSIBLE LINK
2 RED
COOLING FAN
B
A
532 BLK/RED
150 BLK
STARTER MOTOR SOLENOID "BAT" TERMINAL
M
TO BATTERY +

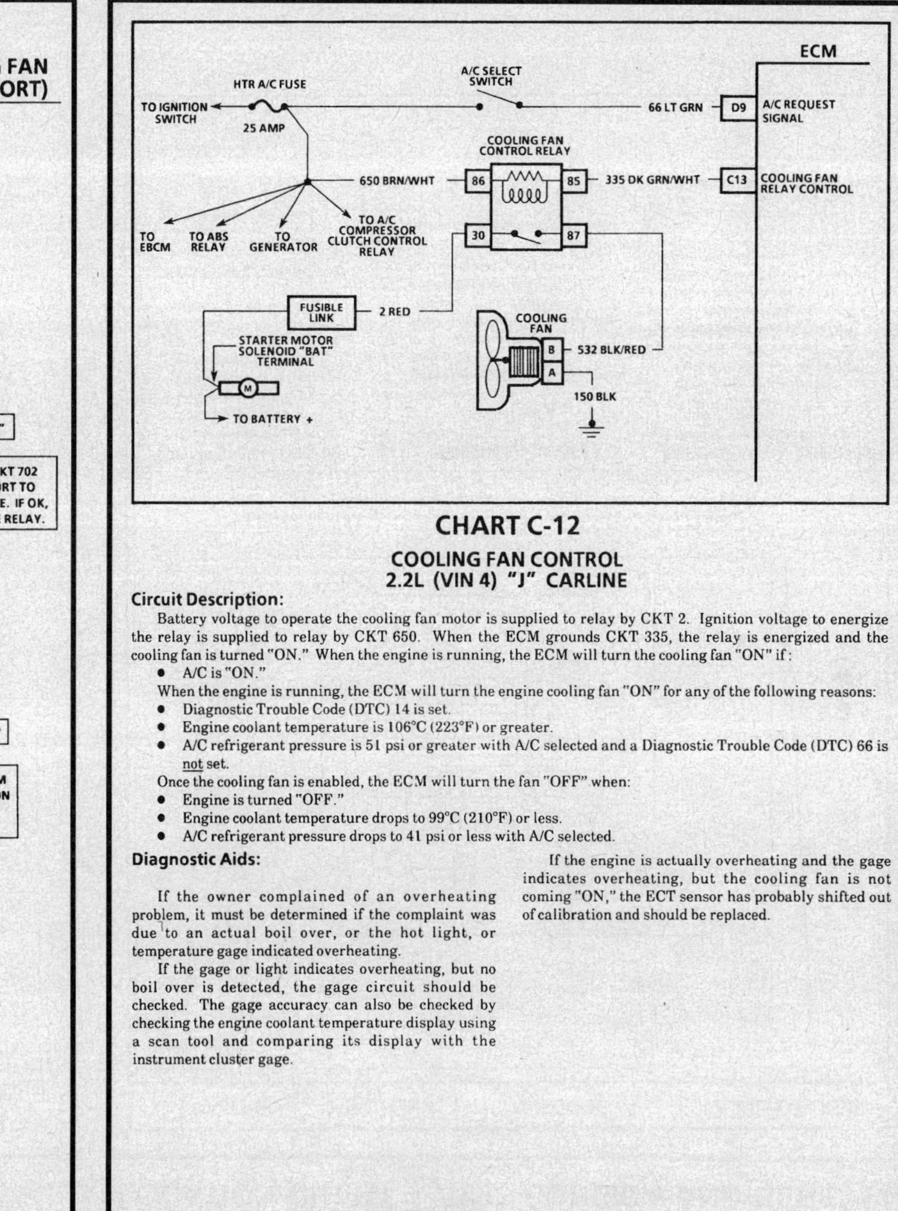

CHART C-12

COOLING FAN CONTROL
2.2L (VIN 4) "J" CARLINE

Circuit Description:

Battery voltage to operate the cooling fan motor is supplied to relay by CKT 2. Ignition voltage to energize the relay is supplied to relay by CKT 650. When the ECM grounds CKT 335, the relay is energized and the cooling fan is turned "ON." When the engine is running, the ECM will turn the cooling fan "ON" if:

- A/C is "ON."

When the engine is running, the ECM will turn the engine cooling fan "ON" for any of the following reasons:

- Diagnostic Trouble Code (DTC) 14 is set.
- Engine coolant temperature is 106°C (223°F) or greater.
- A/C refrigerant pressure is 51 psi or greater with A/C selected and a Diagnostic Trouble Code (DTC) 66 is not set.

Once the cooling fan is enabled, the ECM will turn the fan "OFF" when:

- Engine is turned "OFF."
- Engine coolant temperature drops to 99°C (210°F) or less.
- A/C refrigerant pressure drops to 41 psi or less with A/C selected.

Diagnostic Aids:

If the owner complained of an overheating problem, it must be determined if the complaint was due to an actual boil over, or the hot light, or temperature gage indicated overheating.

If the gage or light indicates overheating, but no boil over is detected, the gage circuit should be checked. The gage accuracy can also be checked by checking the engine coolant temperature display using a scan tool and comparing its display with the instrument cluster gage.

If the engine is actually overheating and the gage indicates overheating, but the cooling fan is not coming "ON," the ECT sensor has probably shifted out of calibration and should be replaced.

2.2L (VIN 4) ENGINE — COMPONENT DIAGNOSTIC CHART — 1993–94 CAVALIER

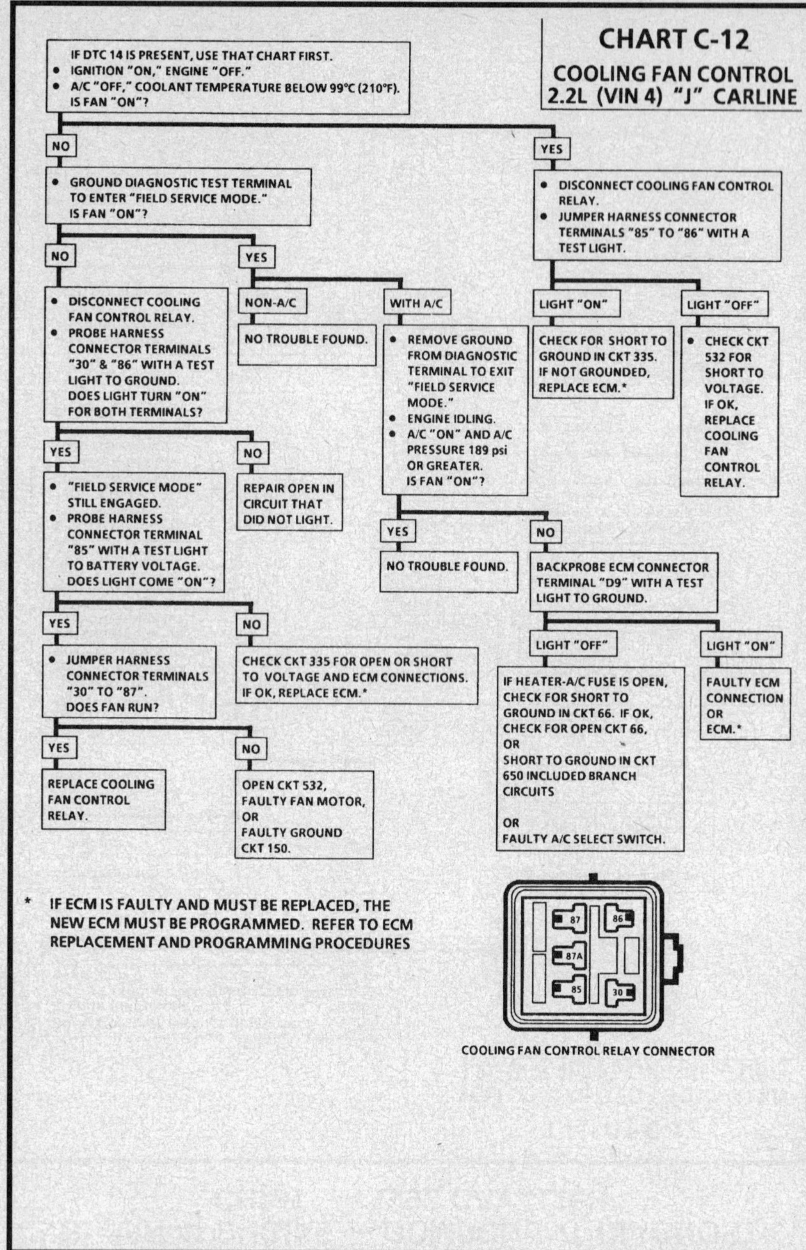

2.2L (VIN 4) ENGINE — ENGINE COMPONENT LOCATION CHART — 1992 CORSICA AND BERETTA

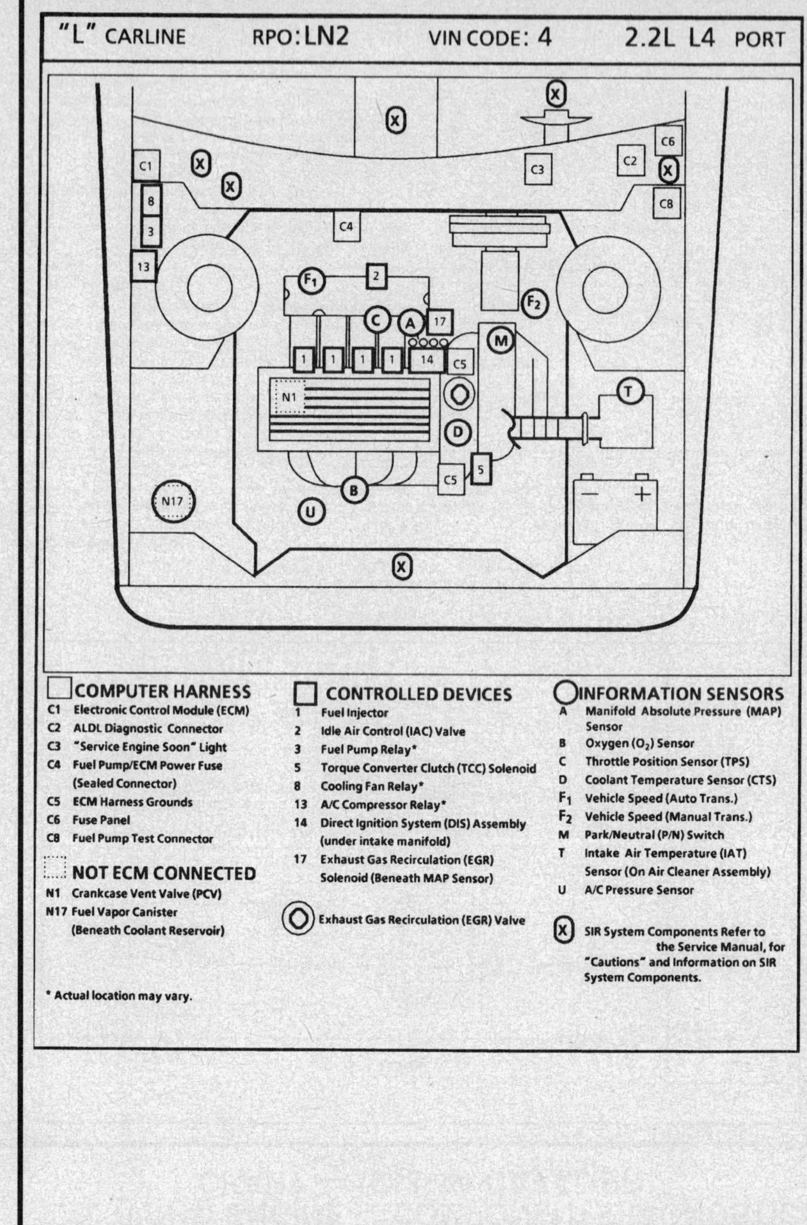

2.2L (VIN 4) ENGINE — ECM WIRING SCHEMATIC — 1992 CORSICA AND BERETTA

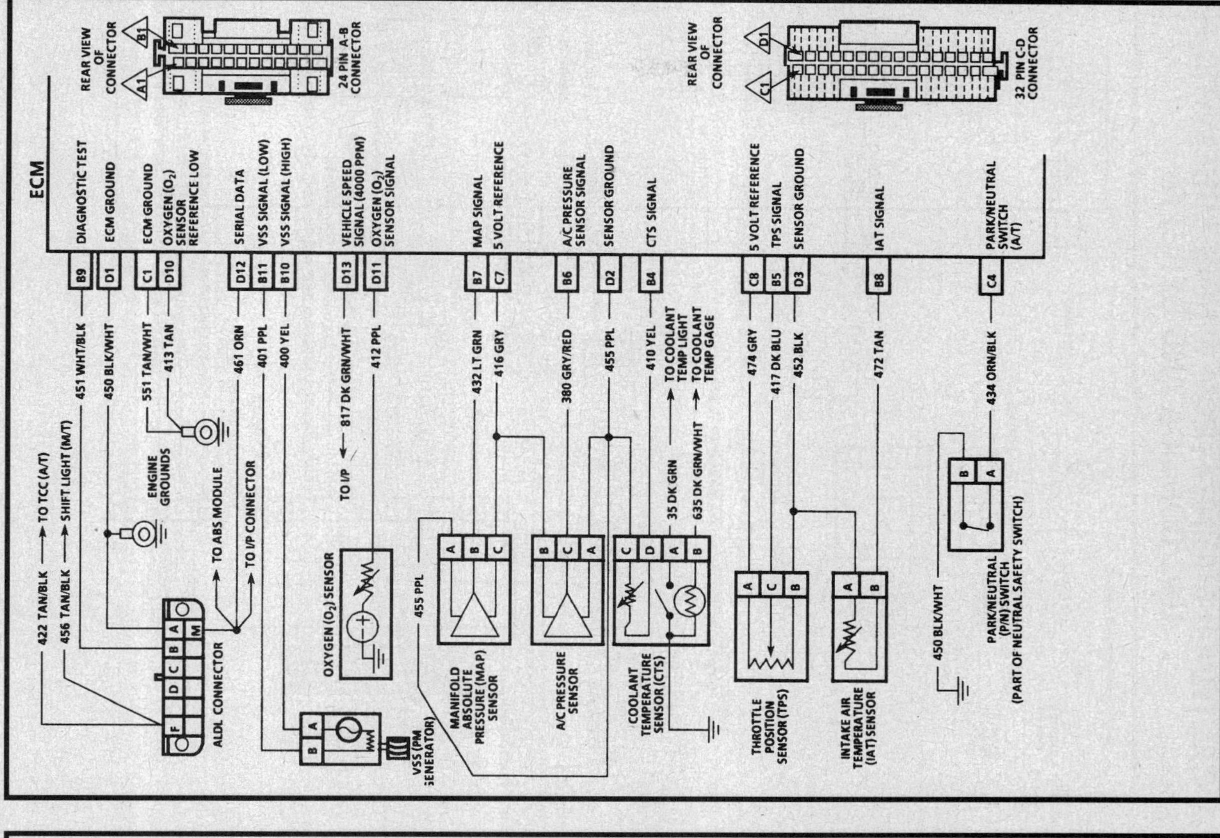

2.2L (VIN 4) ENGINE — ENGINE COMPONENT LOCATION CHART — 1993–94 CORSICA AND BERETTA

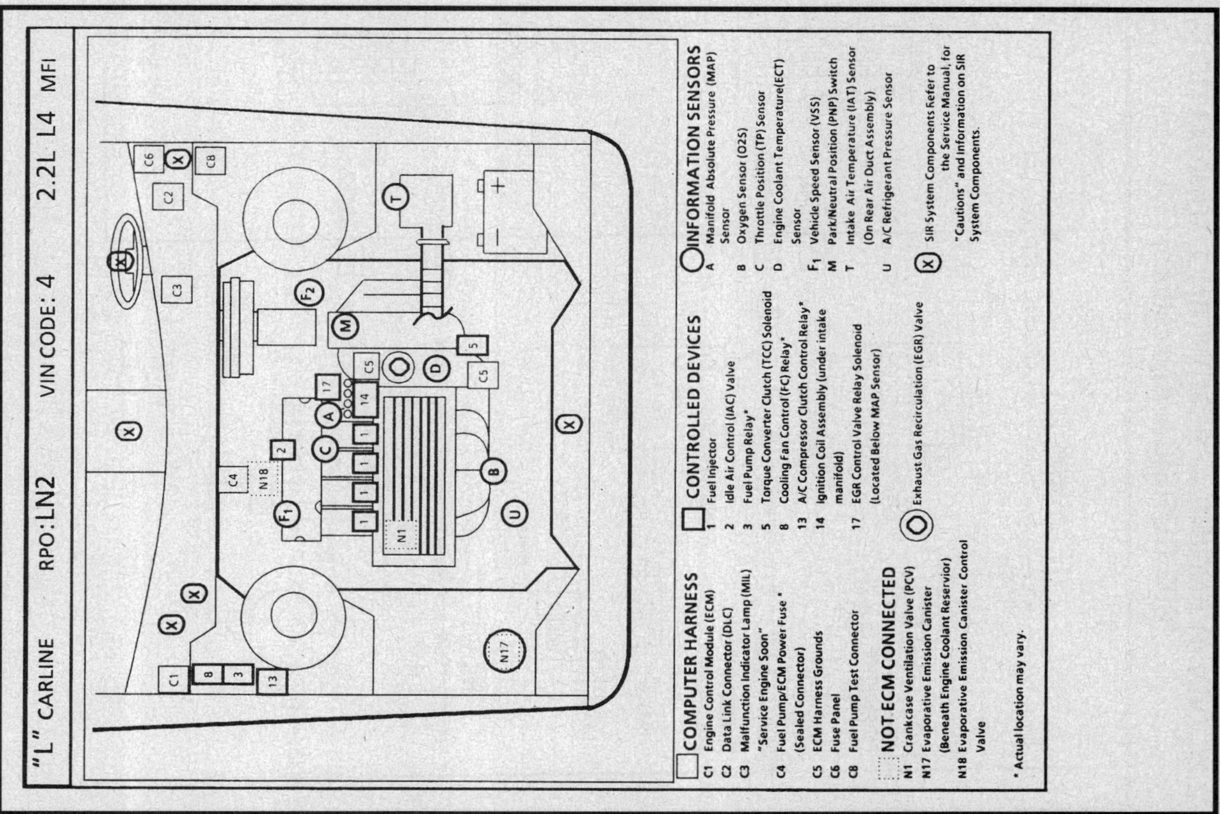

2.2L (VIN 4) ENGINE — ECM WIRING SCHEMATIC — 1992 CORSICA AND BERETTA

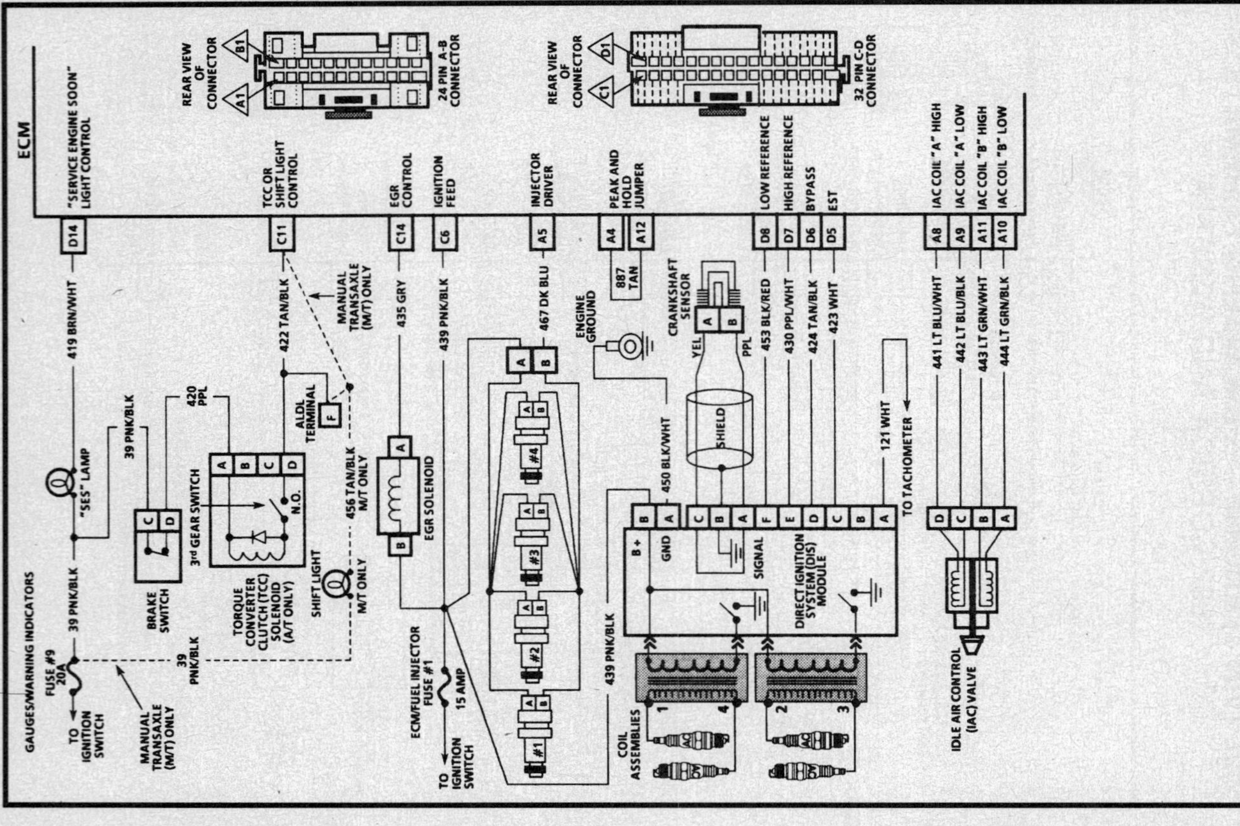

2.2L (VIN 4) ENGINE — ECM WIRING SCHEMATIC — 1992 CORSICA AND BERETTA

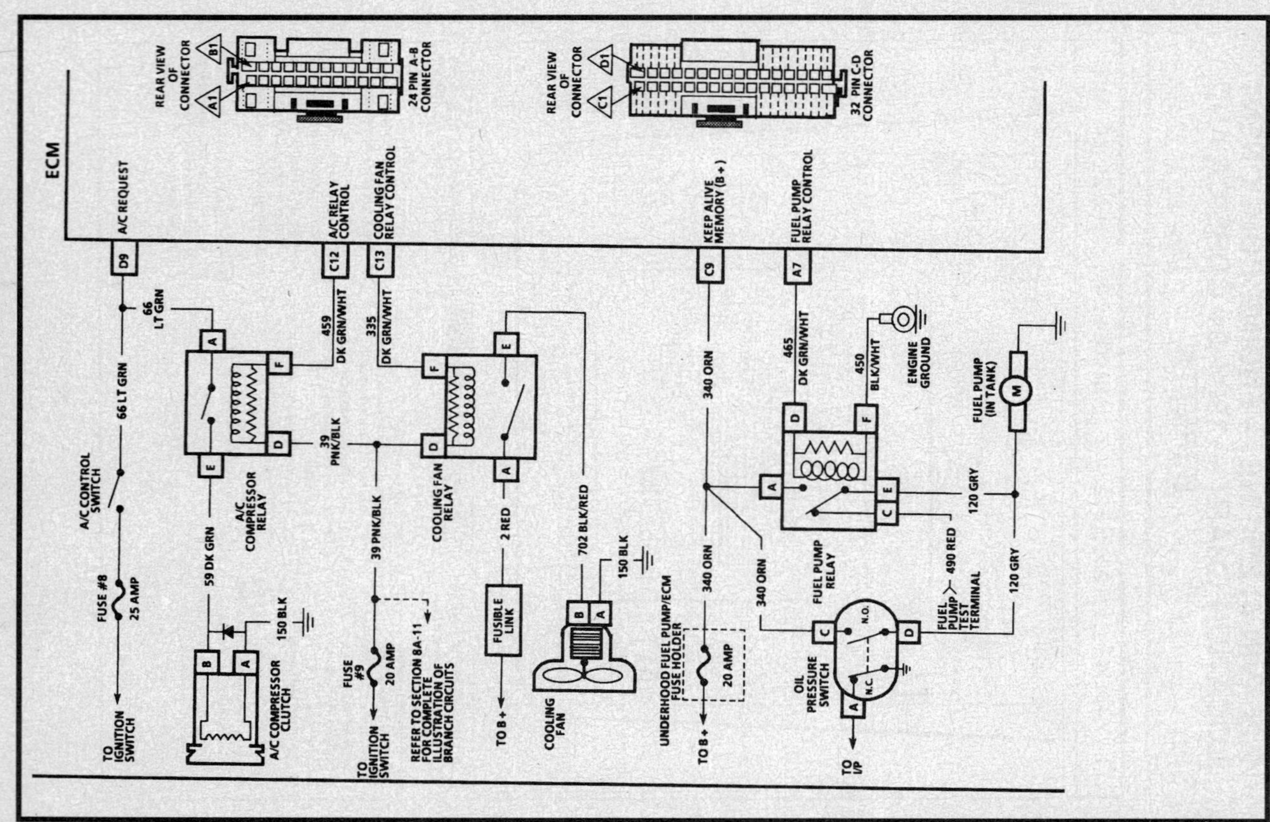

2.2L (VIN 4) ENGINE — ECM WIRING SCHEMATIC — 1993–94 CORSICA AND BERETTA

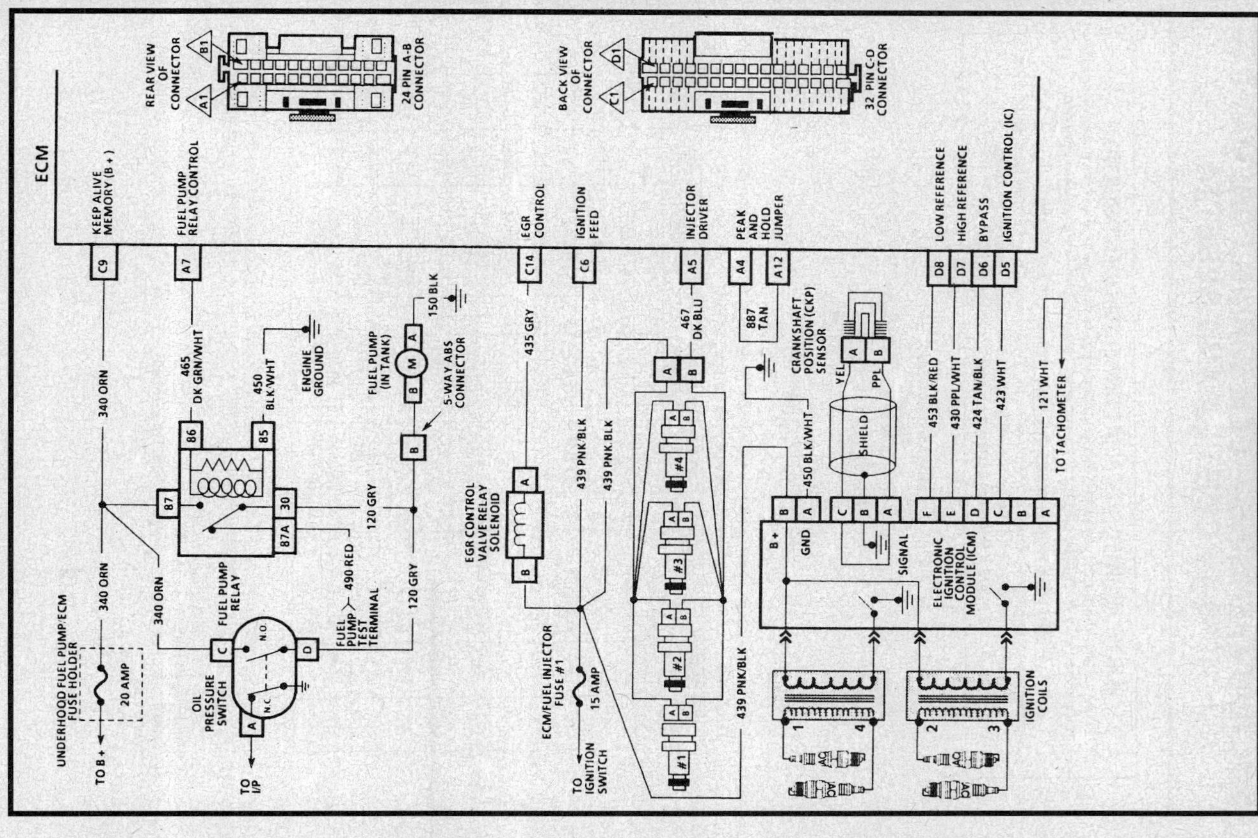
2.2L (VIN 4) ENGINE — ECM WIRING SCHEMATIC — 1993–94 CORSICA AND BERETTA
REAR VIEW OF CONNECTOR
24 PIN A-B CONNECTOR
BACK VIEW OF CONNECTOR
32 PIN C-D CONNECTOR
ECM
KEEP ALIVE MEMORY (B+)
FUEL PUMP RELAY CONTROL
EGR CONTROL
IGNITION FEED
INJECTOR DRIVER
PEAK AND HOLD JUMPER
LOW REFERENCE
HIGH REFERENCE
BYPASS
IGNITION CONTROL (IC)
340 ORN
465 DK GRN/WHT
450 BLK/WHT
ENGINE GROUND
150 BLK
FUEL PUMP (IN TANK)
5-WAY ABS CONNECTOR
435 GRY
467 DK BLU
887 TAN
CRANKSHAFT POSITION (CKP) SENSOR
453 BLK/RED
430 PPL/WHT
424 TAN/BLK
423 WHT
121 WHT
TO TACHOMETER
340 ORN
120 GRY
439 PNK/BLK
#4
#3
#2
#1
439 PNK/BLK
YEL
PPL
SHIELD
B+ B A
GND
SIGNAL
ELECTRONIC IGNITION CONTROL MODULE (ICM)
UNDERHOOD FUEL PUMP/ECM FUSE HOLDER
20 AMP
TO B+
340 ORN
340 ORN
FUEL PUMP RELAY
OIL PRESSURE SWITCH
FUEL PUMP TEST TERMINAL
490 RED
120 GRY
EGR CONTROL VALVE RELAY SOLENOID
ECM/FUEL INJECTOR FUSE #1
15 AMP
TO IGNITION SWITCH
IGNITION COILS
1 4
2 3
TO I/P

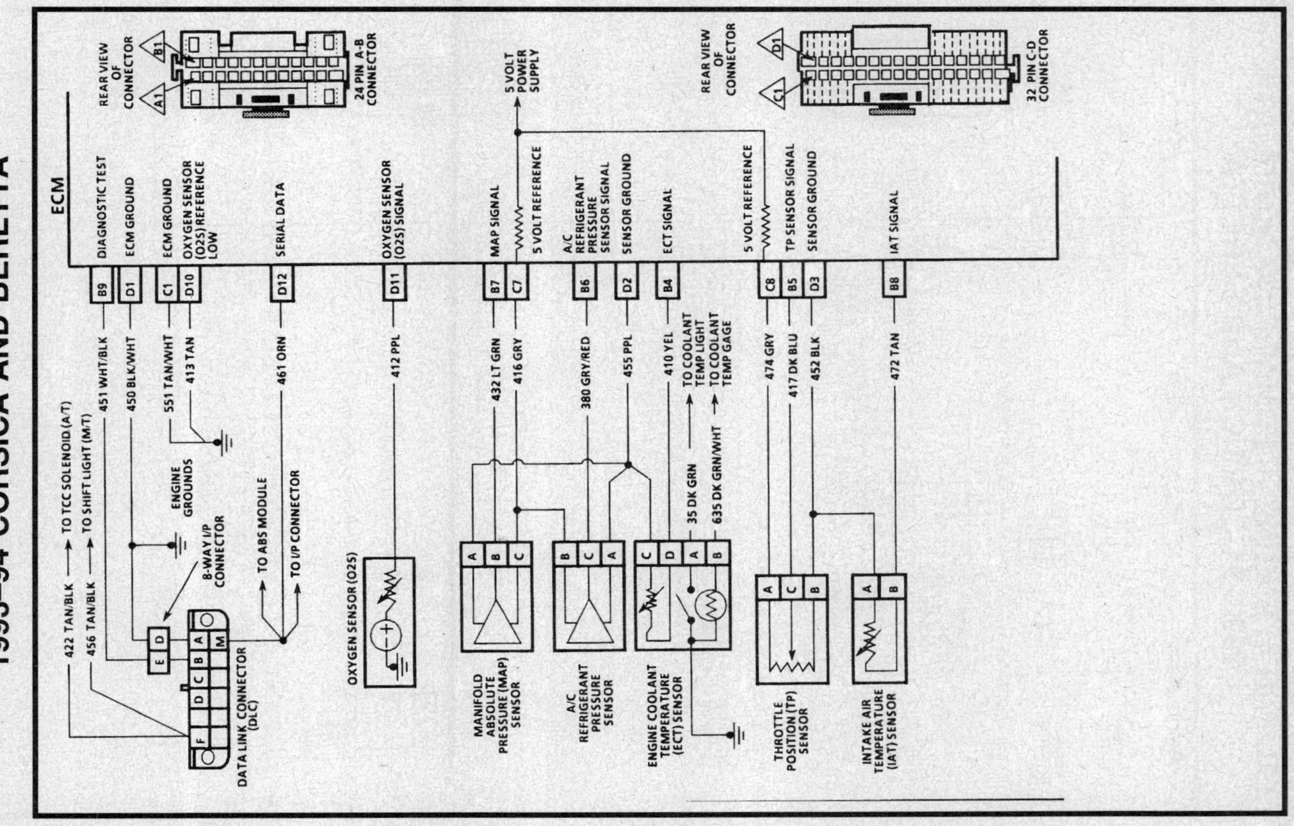
REAR VIEW OF CONNECTOR
24 PIN A-B CONNECTOR
5 VOLT POWER SUPPLY
REAR VIEW OF CONNECTOR
32 PIN C-D CONNECTOR
ECM
DIAGNOSTIC TEST
ECM GROUND
ECM GROUND
OXYGEN SENSOR (O2S) REFERENCE LOW
SERIAL DATA
OXYGEN SENSOR (O2S) SIGNAL
MAP SIGNAL
5 VOLT REFERENCE
A/C REFRIGERANT PRESSURE SENSOR SIGNAL
SENSOR GROUND
ECT SIGNAL
5 VOLT REFERENCE
TP SENSOR SIGNAL
SENSOR GROUND
IAT SIGNAL
451 WHT/BLK
450 BLK/WHT
551 TAN/WHT
413 TAN
461 ORN
412 PPL
432 LT GRN
416 GRY
380 GRY/RED
455 PPL
410 YEL
TO COOLANT TEMP LIGHT
TO COOLANT TEMP GAGE
474 GRY
417 DK BLU
452 BLK
472 TAN
422 TAN/BLK
456 TAN/BLK
TO TCC SOLENOID (A/T)
TO SHIFT LIGHT (M/T)
ENGINE GROUNDS
8-WAY I/P CONNECTOR
TO ABS MODULE
TO I/P CONNECTOR
DATA LINK CONNECTOR (DLC)
OXYGEN SENSOR (O2S)
MANIFOLD ABSOLUTE PRESSURE (MAP) SENSOR
A/C REFRIGERANT PRESSURE SENSOR
35 DK GRN
635 DK GRN/WHT
ENGINE COOLANT TEMPERATURE (ECT) SENSOR
THROTTLE POSITION (TP) SENSOR
INTAKE AIR TEMPERATURE (IAT) SENSOR

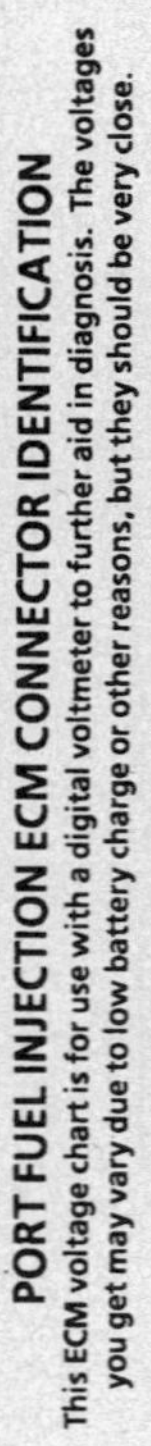

2.2L (VIN 4) ENGINE — ECM WIRING SCHEMATIC — 1993–94 CORSICA AND BERETTA

PORT FUEL INJECTION ECM CONNECTOR IDENTIFICATION

This ECM voltage chart is for use with a digital voltmeter to further aid in diagnosis. The voltages you get may vary due to low battery charge or other reasons, but they should be very close.

THE FOLLOWING CONDITIONS MUST BE MET BEFORE TESTING:

• Engine at operating temperature • Engine idling in "Closed Loop" (For "Engine Run" column) in park or neutral • Test terminal not grounded • "Scan" tool not installed
• B + indicates battery or charging system voltage

PINK 24 PIN A-B CONNECTOR

VOLTAGE KEY "ON"	VOLTAGE ENG. RUN	CIRCUIT	PIN	WIRE COLOR
		NOT USED	A1	
		NOT USED	A2	
		NOT USED	A3	
0*	0*	PEAK AND HOLD JUMPER	A4	887 TAN
B+	B+	INJECTOR DRIVER	A5	467 DK BLU
		NOT USED	A6	
④	B+	FUEL PUMP RELAY CONTROL	A7	465 DK GRN/WHT
①	①	IAC COIL "A" HIGH	A8	441 LT BLU/WHT
①	①	IAC COIL "A" LOW	A9	442 LT BLU/BLK
①	①	IAC COIL "B" LOW	A10	444 LT GRN/BLK
①	①	IAC COIL "B" HIGH	A11	443 LT GRN/WHT
0*	0*	PEAK AND HOLD JUMPER	A12	887 TAN

WIRE COLOR	PIN	CIRCUIT	VOLTAGE KEY "ON"	VOLTAGE ENG. RUN
	B1	NOT USED		
	B2	NOT USED		
	B3	NOT USED		
410 YEL	B4	CTS SIGNAL	2.0	2.0 ③
417 DK BLU	B5	TPS SIGNAL	.6	.6
380 GRY/RED	B6	A/C PRESSURE SENSOR SIGNAL	VARIES	VARIES
432 LT GRN	B7	MAP SIGNAL	4.75	1.6
472 TAN	B8	IAT SIGNAL	1.3	1.3
451 WHT/BLK	B9	ALDL DIAGNOSTIC TEST	5.0	5.0
400 YEL	B10	VSS SIGNAL(HIGH)	VARIES	VARIES ③
401 PPL	B11	VSS SIGNAL(LOW)	VARIES	VARIES
	B12	NOT USED		

ENGINE 2.2L LN2

* All voltages shown "0" should read less than .5 volt.
① Not usable
② A/C select "OFF" and engine cooling fan "OFF."
③ Varies depending on temperature
④ Reads B + for 2 seconds after ignition "ON," then should read 0 volt.

2.2L (VIN 4) ENGINE — ECM WIRING SCHEMATIC — 1993–94 CORSICA AND BERETTA

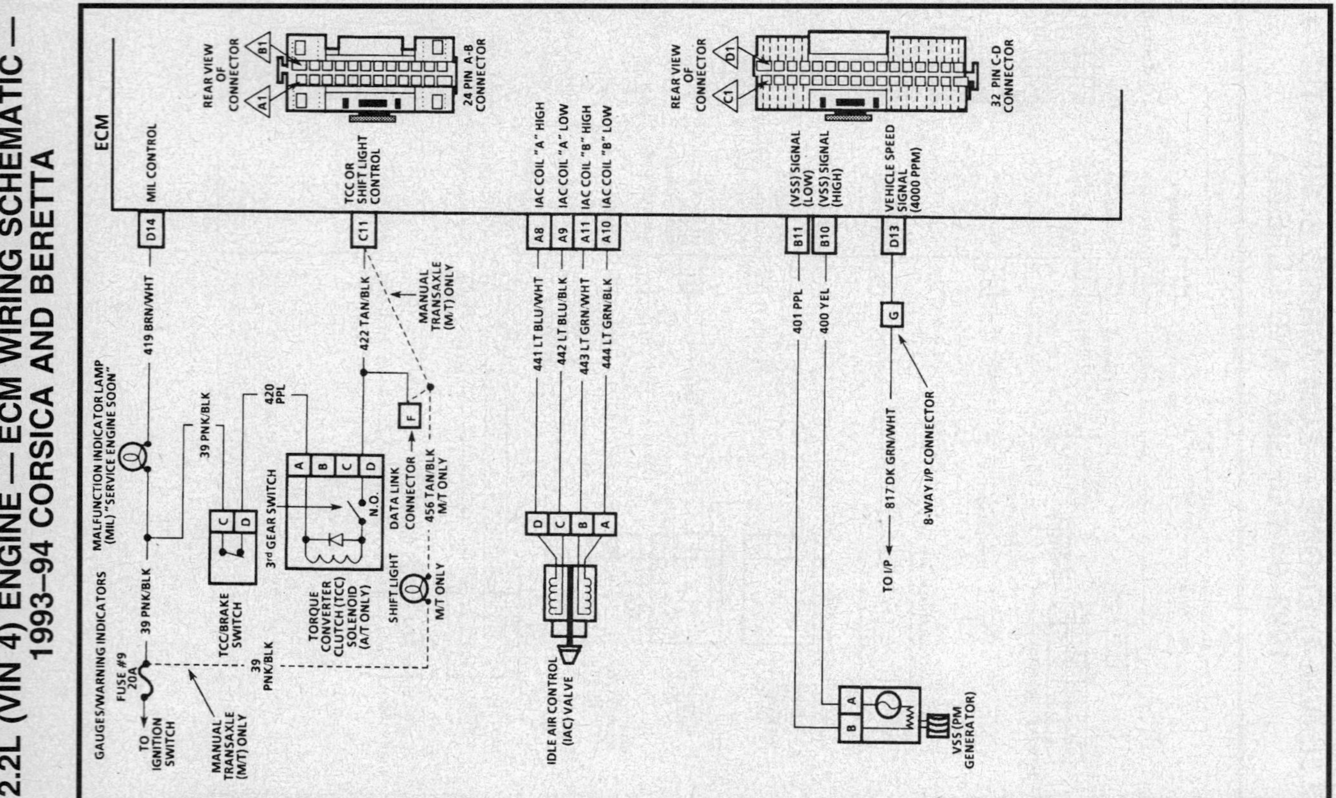

2.2L (VIN 4) ENGINE — ECM CONNECTOR END VIEW — 1992 CORSICA AND BERETTA

PORT FUEL INJECTION ECM CONNECTOR IDENTIFICATION

This ECM voltage chart is for use with a digital voltmeter to further aid in diagnosis. The voltages you get may vary due to low battery charge or other reasons, but they should be very close.

THE FOLLOWING CONDITIONS MUST BE MET BEFORE TESTING:

- Engine at operating temperature ● Engine idling in "Closed Loop" (For "Engine Run" column) in park or neutral ● Test terminal not grounded ● "Scan" tool not installed ● B + indicates battery or charging system voltage

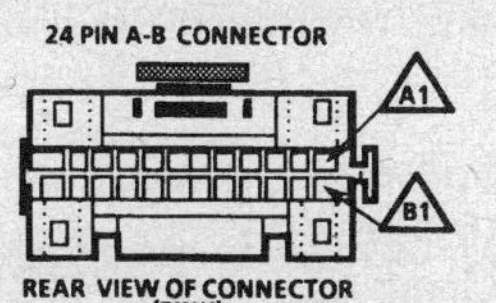

VOLTAGE — PINK 24 PIN A-B CONNECTOR

KEY "ON"	ENG. RUN	CIRCUIT	PIN	WIRE COLOR
		NOT USED	A1	
		NOT USED	A2	
		NOT USED	A3	
0*	0*	PEAK AND HOLD JUMPER	A4	887 TAN
B+	B+	INJECTOR DRIVER	A5	467 DK BLU
		NOT USED	A6	
(4)	B+	FUEL PUMP RELAY CONTROL	A7	465 DK GRN/WHT
(1)	(1)	IAC COIL "A" HIGH	A8	441 LT BLU/WHT
(1)	(1)	IAC COIL "A" LOW	A9	442 LT BLU/BLK
(1)	(1)	IAC COIL "B" LOW	A10	444 LT GRN/BLK
(1)	(1)	IAC COIL "B" HIGH	A11	443 LT GRN/WHT
0*	0*	PEAK AND HOLD JUMPER	A12	887 TAN

WIRE COLOR	PIN	CIRCUIT	KEY "ON"	ENG. RUN
	B1	NOT USED		
	B2	NOT USED		
	B3	NOT USED		
410 YEL	B4	CTS SIGNAL	2.0	2.0 (3)
417 DK BLU	B5	TPS SIGNAL	.6	.6
380 GRY/RED	B6	A/C PRESSURE SENSOR SIGNAL	VARIES	VARIES
432 LT GRN	B7	MAP SIGNAL	4.75	1.6
472 TAN	B8	IAT SIGNAL	1.3	1.3 (3)
451 WHT/BLK	B9	ALDL DIAGNOSTIC TEST	5.0	5.0
400 YEL	B10	VSS SIGNAL(HIGH)	VARIES	VARIES
401 PPL	B11	VSS SIGNAL(LOW)	VARIES	VARIES
	B12	NOT USED		

ENGINE 2.2L LN2

* All voltages shown "0" should read less than .5 volt.
(1) Not usable
(2) A/C select "OFF" and engine cooling fan "OFF."
(3) Varies depending on temperature
(4) Reads B + for 2 seconds after ignition "ON," then should read 0 volt.

2.2L (VIN 4) ENGINE — ECM CONNECTOR END VIEW — 1992 CORSICA AND BERETTA

PORT FUEL INJECTION ECM CONNECTOR IDENTIFICATION

This ECM voltage chart is for use with a digital voltmeter to further aid in diagnosis. The voltages you get may vary due to low battery charge or other reasons, but they should be very close.

THE FOLLOWING CONDITIONS MUST BE MET BEFORE TESTING:

- Engine at operating temperature ● Engine idling in "Closed Loop" (For "Engine Run" column) in park or neutral ● Test terminal not grounded ● "Scan" tool not installed ● B + indicates battery or charging system voltage

KEY "ON"	ENG. RUN	CIRCUIT	PIN	WIRE COLOR
0*	0*	ECM GROUND	C1	551 TAN/WHT
		NOT USED	C2	
		NOT USED	C3	
0*	0*	P/N SWITCH	C4	434 ORN/BLK
		NOT USED	C5	
B+	B+	IGNITION FEED	C6	439 PNK/BLK
5.0	5.0	5 V REFERENCE	C7	416 GRY
5.0	5.0	5 V REFERENCE	C8	474 GRY
B+	B+	KEEP ALIVE MEMORY (B +)	C9	340 ORN
		NOT USED	C10	
0* / B+	0* / B+	TCC (A/T) OR / SHIFT LIGHT (M/T)	C11	422 —TAN/BLK 456
(2) B+	B+	A/C RELAY CONTROL	C12	459 DK GRN/WHT
(2) B+	B+	COOLING FAN RELAY CONTROL	C13	335 DK GRN/WHT
B+	B+	EGR SOLENOID CONTROL	C14	435 GRY
		NOT USED	C15	
		NOT USED	C16	

WIRE COLOR	PIN	CIRCUIT	KEY "ON"	ENG. RUN
450 BLK/WHT	D1	ECM GROUND	0*	0*
455 PPL	D2	SENSOR GROUND	0*	0*
452 BLK	D3	SENSOR GROUND	0*	0*
	D4	NOT USED		
423 WHT	D5	EST	0*	2.4
424 TAN/BLK	D6	BYPASS	0*	4.8
430 PPL/WHT	D7	REF HIGH	0*	3.2
453 BLK/RED	D8	REF LOW	0*	0*
66 LT GRN	D9	A/C REQUEST	0*	0* (2)
413 TAN	D10	OXYGEN SENSOR GROUND	0*	0*
412 PPL	D11	OXYGEN SENSOR SIGNAL	.01-.55	.1-.9
461 ORN	D12	SERIAL DATA	4.7	4.7
817 DK GRN/WHT	D13	VSS OUTPUT (4000 PPM)	(5)	(5)
419 BRN/WHT	D14	SERVICE ENGINE SOON LIGHT	0*	B+
	D15	NOT USED		
	D16	NOT USED		

ENGINE 2.2L LN2

* All voltages shown "0" should read less than .5 volt.
(1) Not useable.
(2) A/C select "OFF" and engine cooling fan "OFF."
(3) Varies depending on temperature.
(4) Reads B + for 2 seconds after ignition "ON," then should read 0 volt.

2.2L (VIN 4) ENGINE — ECM CONNECTOR END VIEW — 1993–94 CORSICA AND BERETTA

MULTIPORT FUEL INJECTION ECM CONNECTOR IDENTIFICATION

This ECM voltage chart is for use with J 39200 digital voltmeter to further aid in diagnosis. The voltages you get may vary due to low battery charge or other reasons, but they should be very close.

THE FOLLOWING CONDITIONS MUST BE MET BEFORE TESTING:

- Engine at operating temperature ● Engine idling in "Closed Loop" (For "Engine Run" column) in park or neutral ● Test terminal not grounded ● Scan tool not installed ● Brake not applied ● B + indicates battery or charging system voltage

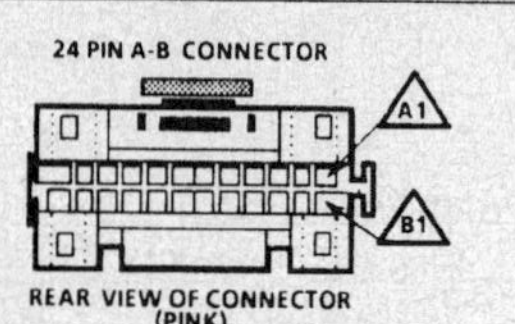

PINK 24 PIN A-B CONNECTOR

VOLTAGE

KEY "ON"	ENG. RUN	CIRCUIT	PIN	WIRE COLOR
			A1	
			A2	
			A3	
0*	0*	PEAK AND HOLD JUMPER	A4	887 TAN
B +	B + (6)	INJECTOR DRIVER	A5	467 DK BLU
			A6	
(3)	B +	FUEL PUMP RELAY CONTROL	A7	465 DK GRN/WHT
(1)	(1)	IAC COIL "A" HIGH	A8	441 LT BLU/WHT
(1)	(1)	IAC COIL "A" LOW	A9	442 LT BLU/BLK
(1)	(1)	IAC COIL "B" LOW	A10	444 LT GRN/BLK
(1)	(1)	IAC COIL "B" HIGH	A11	443 LT GRN/WHT
0*	0*	PEAK AND HOLD JUMPER	A12	8B7 TAN

VOLTAGE

WIRE COLOR	PIN	CIRCUIT	KEY "ON"	ENG. RUN
	B1			
	B2			
	B3			
410 YEL	B4	ECT SIGNAL	(2)	(2)
417 DK BLU	B5	TP SIGNAL	.6 (5)	.6 (5)
380 GRY/RED	B6	A/C REFRIGERANT PRESSURE SENSOR SIGNAL	VARIES	VARIES
432 LT GRN	B7	MAP SIGNAL	4.75	1.2
472 TAN	B8	IAT SIGNAL	(2)	(2)
451 WHT/BLK	B9	DIAGNOSTIC TEST	5.0	5.0
400 YEL	B10	VSS SIGNAL(HIGH)	0*	(4)
401 PPL	B11	VSS SIGNAL(LOW)	0*	(4)
	B12			

* ALL VOLTAGES SHOWN "0" SHOULD READ LESS THAN .5 VOLT.
(1) NOT USABLE.
(2) VARIES DEPENDING ON TEMPERATURE.
(3) READS B + FOR 2 SECONDS AFTER KEY "ON," THEN SHOULD READ 0 VOLT.
(4) AC VOLTAGE INCREASES WITH VEHICLE SPEED.
(5) WITHIN THE RANGE OF .33 TO 1.33 VOLTS.
(6) ALTERNATE TEST: SET J 39200 DVM TO DC VOLTAGE FREQUENCY SCALE. CONNECT RED LEAD TO IGNITION FEED ("C6" AT ECM) AND BLACK LEAD TO INJECTOR DRIVER ("A5" AT ECM); SHOULD MEASURE ABOUT 9-15 Hz.

ENGINE 2.2L LN2

2.2L (VIN 4) ENGINE — ECM CONNECTOR END VIEW — 1993–94 CORSICA AND BERETTA

MULTIPORT FUEL INJECTION ECM CONNECTOR IDENTIFICATION

This ECM voltage chart is for use with J 39200 digital voltmeter to further aid in diagnosis. The voltages you get may vary due to low battery charge or other reasons, but they should be very close.

THE FOLLOWING CONDITIONS MUST BE MET BEFORE TESTING:

- Engine at operating temperature ● Engine idling in "Closed Loop" (For "Engine Run" column) in park or neutral ● Test terminal not grounded ● Scan tool not installed ● Brake not applied ● B + indicates battery or charging system voltage

VOLTAGE

KEY "ON"	ENG. RUN	CIRCUIT	PIN	WIRE COLOR
0* (5)	0* (5)	ECM GROUND	C1	450 BLK/WHT
			C2	
			C3	
0*	0*	PNP SWITCH	C4	434 ORN/BLK
			C5	
B + (5)	B + (5)	IGNITION FEED	C6	439 PNK/BLK
5.0	5.0	5 V REFERENCE	C7	416 GRY
5.0	5.0	5V REFERENCE	C8	474 GRY
B + (5)	B + (5)	KEEP ALIVE MEMORY (B+)	C9	340 ORN
0** / B +	0** / B +	TCC(A/T) OR / SHIFT LIGHT (M/T)	C11	422 --TAN/BLK / 456
B + (1)	B + (1)	A/C COMPRESSOR CLUTCH RELAY CONTROL	C12	459 DK GRN/WHT
B + (1)	B + (1)	COOLING FAN RELAY CONTROL	C13	335 DK GRN/WHT
B +	B +	EGR SOLENOID CONTROL	C14	435 GRY
			C15	
			C16	

VOLTAGE

WIRE COLOR	PIN	CIRCUIT	KEY "ON"	ENG. RUN
450 BLK/WHT	D1	ECM GROUND	0* (5)	0* (5)
455 PPL	D2	SENSOR GROUND	0*	0*
452 BLK	D3	SENSOR GROUND	0*	0*
	D4			
423 WHT	D5	IGNITION CONTROL (IC)	0*	2.3
424 TAN/BLK	D6	BYPASS	0*	4.8
430 PPL/WHT	D7	REF HIGH	0*	3.2 (4)
453 BLK/RED	D8	REF LOW	0*	0*
66 LT GRN	D9	A/C REQUEST	0* (1)	0* (1)
413 TAN	D10	OXYGEN SENSOR REFERENCE LOW	0*	0*
412 PPL	D11	OXYGEN SENSOR SIGNAL	.01-.55	.1-.9 (3)
461 ORN	D12	SERIAL DATA	4.7	4.7
817 DK GRN/WHT	D13	VSS OUTPUT (4000 PPM)	(2)	(2)
419 BRN/WHT	D14	MIL (SERVICE ENGINE SOON) CONTROL	1.0	B +
	D15			
	D16			

* ALL VOLTAGES SHOWN "0" SHOULD READ LESS THAN .5 VOLT.
(1) A/C SELECT "OFF" AND ENGINE COOLING FAN "OFF."
(2) INCREASES WITH VEHICLE SPEED.
(3) VARIES ACTIVELY WITHIN INDICATED RANGE.
(4) ALTERNATE TEST: SET J 39200 DVM TO DC VOLTAGE FREQUENCY SCALE. CONNECT RED LEAD TO REF HIGH ("D7" AT ECM) AND BLACK LEAD TO ECM GROUND ("D1" AT ECM); SHOULD MEASURE ABOUT 25-30 Hz.
(5) THIS TYPE OF CIRCUIT SHOULD ALSO BE CHECKED USING J 34142-B TEST LIGHT.

ENGINE 2.2L LN2

2.2L (VIN 4) ENGINE — ON-BOARD DIAGNOSTIC SYSTEM CHART — CORSICA AND BERETTA

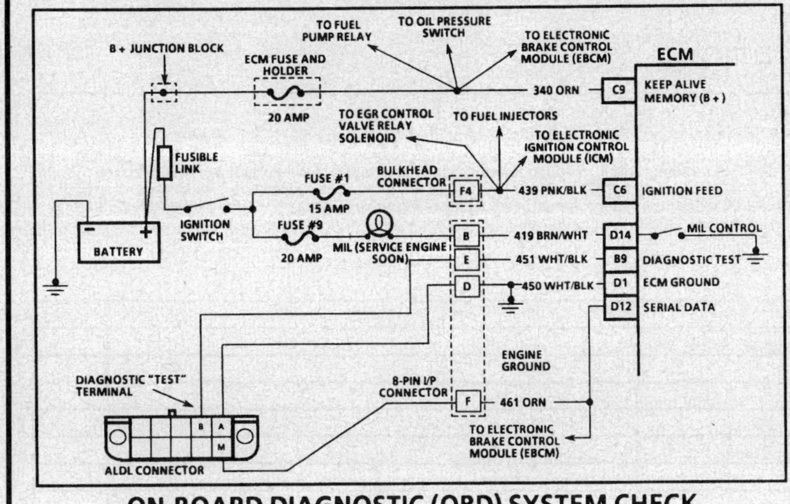

ON-BOARD DIAGNOSTIC (OBD) SYSTEM CHECK
2.2L (VIN 4) "L" CARLINE

Circuit Description:

The OBD system check is an organized approach to identifying a problem created by an electronic engine control system malfunction. It must be the starting point for any driveability complaint diagnosis, because it directs the service technician to the next logical step in diagnosing the complaint. Understanding the chart and using it correctly will reduce diagnostic time and prevent the unnecessary replacement of good parts.

Test Description: Number(s) below refer to circled number(s) on the diagnostic chart.

1. This step is a check for the proper operation of the Malfunction Indicator Lamp (MIL) "Service Engine Soon." The MIL should be "ON" steady.
2. No MIL at this point indicates that there is a problem with the MIL circuit or the ECM control of that circuit.
3. This test checks the ability of the ECM to control the MIL. With the diagnostic terminal grounded, the MIL should flash a DTC 12 three times, followed by any DTC stored in memory. Depending upon the type of ECM, an EEPROM error may result in the inability to flash DTC 12.
4. Most of the 6E procedures use a Tech 1 to aid diagnosis, therefore, serial data must be available. If an EEPROM error is present, the ECM may have been able to flash DTC 12/51, but not enable serial data.

5. Although the ECM is powered up, a "Cranks But Will Not Run" symptom could exist because of an ECM or system problem.
6. This step will isolate if the customer complaint is a MIL or a driveability problem with no MIL. Refer to "ECM Diagnostic Trouble Codes" for valid DTC(s). An invalid DTC may be the result of a faulty scan tool, EEPROM or ECM.
7. Comparison of actual control system data with the typical values is a quick check to determine if any parameter is not within limits. Keep in mind that a base engine problem (i.e. advanced cam timing) may substantially alter sensor values.
8. Installation of a scan tool will provide a good ground path for the ECM and may hide a driveability complaint due to poor ECM grounds.
9. If the actual data is not within the typical values established, the diagnostic charts in "Component Systems," will provide a functional check of the suspect component or system.

2.2L (VIN 4) ENGINE — ON-BOARD DIAGNOSTIC SYSTEM CHART — CORSICA AND BERETTA

ON-BOARD DIAGNOSTIC (OBD) SYSTEM CHECK
2.2L (VIN 4) "L" CARLINE

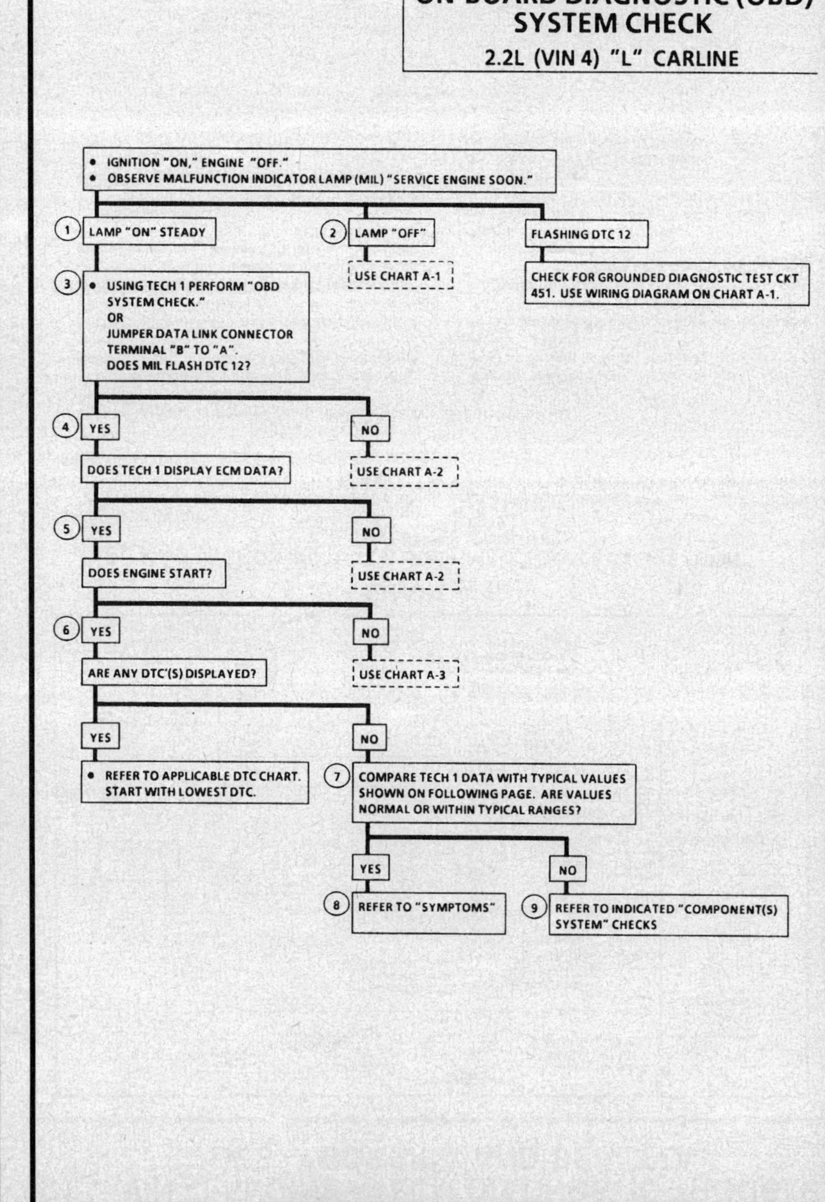

2.2L (VIN 4) ENGINE — ON-BOARD DIAGNOSTIC SYSTEM CHART — CORSICA AND BERETTA

ECM DIAGNOSTIC TROUBLE CODES		ILLUMINATE MIL (SERVICE ENGINE SOON)
DTC	DESCRIPTION	
13	Oxygen Sensor (O2S) Circuit - open circuit	YES
14	Engine Coolant Temperature (ECT) Sensor Circuit - high/low temperature	YES
21	Throttle Position (TP) Sensor Circuit - Signal voltage high/low	YES
23	Intake Air Temperature (IAT) Sensor Circuit - high/low temperature	YES
24	Vehicle Speed Sensor (VSS) Circuit	YES
32	Exhaust Gas Recirculation (EGR) - System failure	YES
33	Manifold Absolute Pressure (MAP) Sensor Circuit - Signal voltage high/low - low/high vacuum	YES
42	Ignition Control (IC) Circuit	YES
44	Oxygen Sensor (O2S) Circuit - lean exhaust indicated	YES
45	Oxygen Sensor (O2S) Circuit - rich exhaust indicated	YES
51	EEPROM or ECM Failure	YES
66	A/C Refrigerant Pressure Sensor Circuit	NO

If a DTC not listed above appears on Tech 1, ground Data Link Terminal "B" and observe flashed DTC(S).
If DTC does not reappear, Tech 1 data may be faulty.
If DTC does reappear, replace ECM and program. Refer to ECM replacement and programming procedures

2.2L (VIN 4) ENGINE — ECM DIAGNOSTIC TROUBLE CODES — CORSICA AND BERETTA

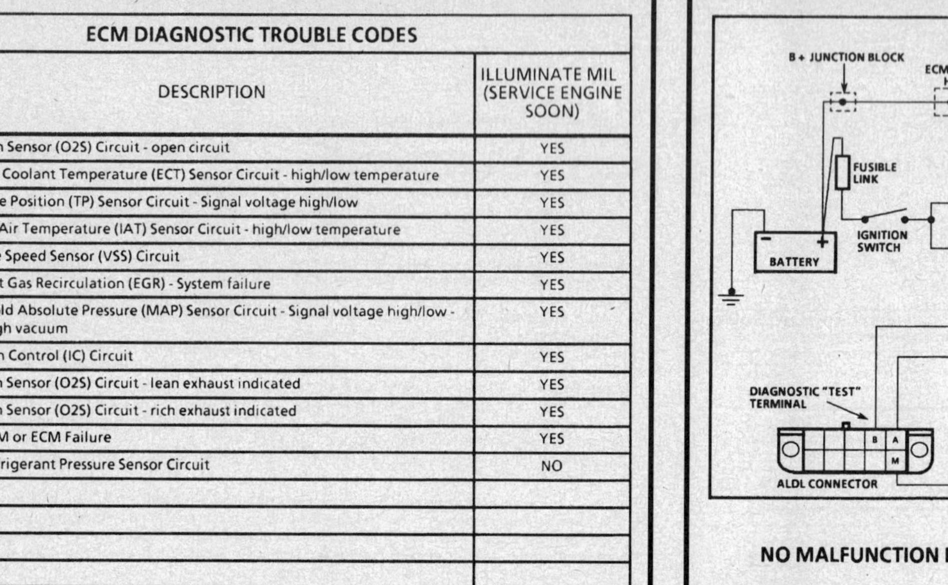

CHART A-1
NO MALFUNCTION INDICATOR LAMP (MIL) "SERVICE ENGINE SOON"
2.2L (VIN 4) "L" CARLINE

Circuit Description:

There should always be a steady MIL (Service Engine Soon), when the ignition is "ON" and engine "OFF." Battery voltage is supplied directly to the light bulb. The Engine Control Module (ECM) will control the lamp and turn it "ON" by providing a ground path through CKT 419 to the ECM.

Test Description: Number(s) below refer to circled number(s) on the diagnostic chart.

1. This test will provide a ground path for the MIL light circuit. If the MIL is "ON," then the external circuit is OK.
2. If CKTs 439 and 340 have voltage, the ECM grounds or the ECM are faulty.
3. Using a test light connected to B+, probe each of the system ground circuits to be sure a good ground is present. Refer to "ECM Connector End View" for ECM pin locations of ground circuits.
4. If a fusible link or fuse is blown, be sure to consider branch circuits and splices to other components.

Diagnostic Aids:

Engine runs OK, check:
- Faulty light bulb.
- CKT 419 open.
- Gauges fuse #9 blown. This will result in no oil or generator lights, seat belt reminder, etc. Refer to "Fuse Block Details," which shows additional branch circuits that are protected by the gauges fuse.

Engine cranks, but will not run:
- Underhood ECM power fuse or fusible link open.
- Fuse #1 open.
- Battery CKT 340 to ECM open.
- Ignition CKT 439 to ECM open.
- Poor connection to ECM.

2.2L (VIN 4) ENGINE — SYSTEM DIAGNOSTIC CHARTS — CORSICA AND BERETTA

CHART A-1

NO MALFUNCTION INDICATOR LAMP (MIL) "SERVICE ENGINE SOON" 2.2L (VIN 4) "L" CARLINE

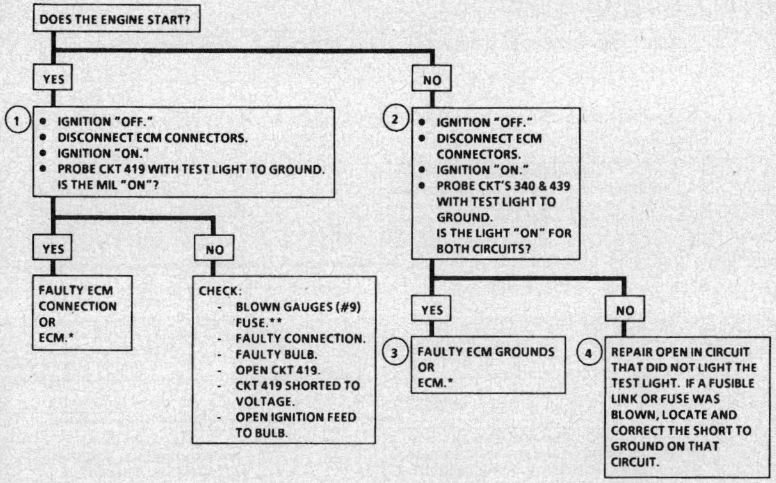

DOES THE ENGINE START?

YES

(1)
- IGNITION "OFF."
- DISCONNECT ECM CONNECTORS.
- IGNITION "ON."
- PROBE CKT 419 WITH TEST LIGHT TO GROUND. IS THE MIL "ON"?

YES

FAULTY ECM CONNECTION OR ECM.*

NO

CHECK:
- BLOWN GAUGES (#9) FUSE.**
- FAULTY CONNECTION.
- FAULTY BULB.
- OPEN CKT 419.
- CKT 419 SHORTED TO VOLTAGE.
- OPEN IGNITION FEED TO BULB.

NO

(2)
- IGNITION "OFF."
- DISCONNECT ECM CONNECTORS.
- IGNITION "ON."
- PROBE CKT'S 340 & 439 WITH TEST LIGHT TO GROUND. IS THE LIGHT "ON" FOR BOTH CIRCUITS?

YES

(3) FAULTY ECM GROUNDS OR ECM.*

NO

(4) REPAIR OPEN IN CIRCUIT THAT DID NOT LIGHT THE TEST LIGHT. IF A FUSIBLE LINK OR FUSE WAS BLOWN, LOCATE AND CORRECT THE SHORT TO GROUND ON THAT CIRCUIT.

* IF ECM IS FAULTY AND MUST BE REPLACED, THE NEW ECM MUST BE PROGRAMMED. REFER TO ECM REPLACEMENT AND PROGRAMMING PROCEDURES

** SEVERAL BRANCH CIRCUITS ARE PROTECTED BY THIS FUSE. OTHER BRANCH CIRCUITS COULD CAUSE A BLOWN FUSE.

2.2L (VIN 4) ENGINE — SYSTEM DIAGNOSTIC CHARTS — CORSICA AND BERETTA

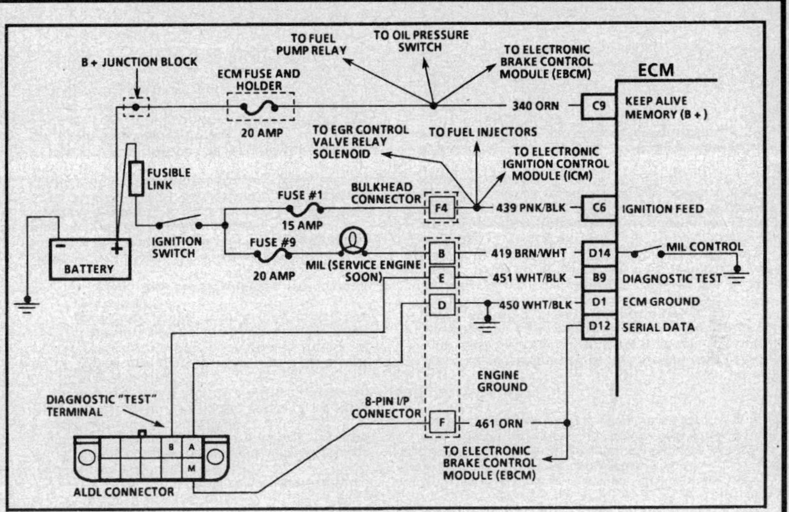

CHART A-2

NO DATA LINK CONNECTOR (DLC) DATA OR WON'T FLASH DTC 12 MALFUNCTION INDICATOR LAMP (MIL) "SERVICE ENGINE SOON" "ON" STEADY 2.2L (VIN 4) "L" CARLINE

Circuit Description:

There should always be a steady MIL (Service Engine Soon) when the ignition is "ON" and the engine is "OFF." Battery voltage is supplied directly to the lamp. The Engine Control Module (ECM) will control the lamp and turn it "ON" by providing a ground path through CKT 419 to the ECM.

With the diagnostic terminal grounded, the lamp should flash a Diagnostic Trouble Code (DTC) 12, followed by any DTC(s) stored in memory. A steady lamp suggests a short to ground in the lamp control CKT 419, or an open in diagnostic CKT 451.

Test Description: Number(s) below refer to circled number(s) on the diagnostic chart.

1. If there is a problem with the ECM that causes a scan tool to not read data from the ECM, then the ECM should not flash a DTC 12. If DTC 12 does flash, be sure that the scan tool is working properly on another vehicle. If the scan is functioning properly and CKT 461 is OK, the EEPROM or ECM may be at fault for the "No Data Link Connector Data" symptom.

2. If the lamp turns "OFF" when the ECM connector is disconnected, then CKT 419 is not shorted to ground.

3. This step will check for an open diagnostic CKT 451.

4. At this point, the MIL wiring is OK. The problem is a faulty ECM. If DTC 12 does not flash, the ECM should be replaced and the new ECM <u>must</u> be programmed.

2.2L (VIN 4) ENGINE — SYSTEM DIAGNOSTIC CHARTS — CORSICA AND BERETTA

CHART A-2

NO DATA LINK CONNECTOR (DLC) DATA
OR WON'T FLASH DTC 12
MALFUNCTION INDICATOR LAMP (MIL)
"SERVICE ENGINE SOON" "ON" STEADY
2.2L (VIN 4) "L" CARLINE

- IGNITION "ON," ENGINE "OFF."
  IS THE MIL (SERVICE ENGINE SOON) "ON"?

YES → NO

NO → USE CHART A-1.

- GROUND DIAGNOSTIC "TEST" TERMINAL.
  DOES LAMP FLASH DTC 12?

NO / YES

(2)
- IGNITION "OFF."
- DISCONNECT ECM CONNECTORS.
- IGNITION "ON" AND NOTE MIL
  (SERVICE ENGINE SOON).

(1)
IF PROBLEM WAS NO DATA LINK CONNECTOR DATA:
CHECK SERIAL DATA CKT 461 FOR OPEN OR SHORT TO
GROUND BETWEEN ECM AND DATA LINK CONNECTOR.
IF OK, IT IS A FAULTY ECM OR EEPROM.*

LAMP "OFF" / LAMP "ON"

(3)
- IGNITION "OFF."
- RECONNECT ECM.
- IGNITION "ON," ENGINE "OFF." DIAGNOSTIC
  "TEST" TERMINAL NOT GROUNDED.
- BACKPROBE ECM CKT 451 WITH TEST LIGHT TO
  GROUND. LEAVE CONNECTED AND WATCH MIL.

REPAIR SHORT TO
GROUND IN CKT 419.

NO CODE 12 / CODE 12

(4)
- REPLACE ECM AND REPROGRAM.*
- RECHECK FOR DTC 12.

CHECK FOR OPEN CKT 451 TO ECM. IF OK, CHECK FOR
OPEN CKT 450 BETWEEN DATA LINK CONNECTOR
TERMINAL "A" AND ENGINE GROUND.

*   IF ECM IS FAULTY AND MUST BE REPLACED, THE NEW ECM MUST BE PROGRAMMED. REFER
    TO ECM REPLACEMENT AND PROGRAMMING PROCEDURES

2.2L (VIN 4) ENGINE — SYSTEM DIAGNOSTIC CHARTS — 1992 CORSICA AND BERETTA

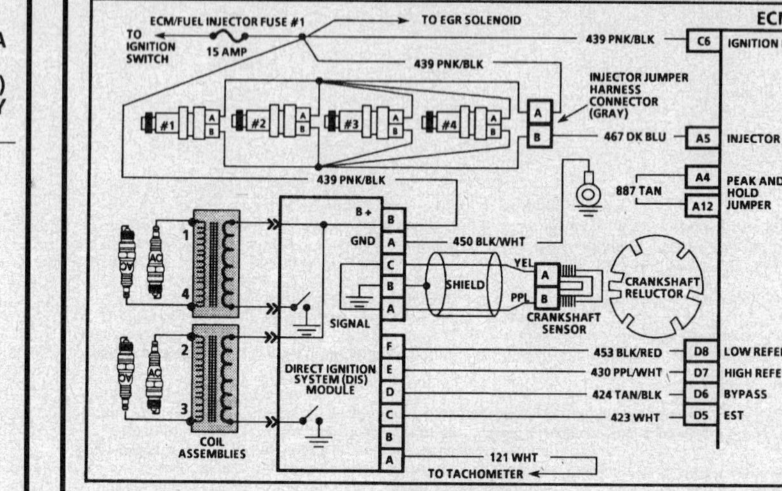

CHART A-3

(Page 1 of 3)
ENGINE CRANKS BUT WON'T RUN
2.2L (VIN 4) "L" CARLINE (PORT)

Circuit Description:

Before using this chart, battery condition, engine cranking speed, and fuel quantity should be checked and verified as being OK.

Test Description: Number(s) below refer to circled number(s) on the diagnostic chart.

1.
   - A "Service Engine Soon" light "ON" is a basic test to determine if there is battery and ignition voltage at the ECM.
   - No ALDL data may be the result of an ECM problem, and CHART A-2 will diagnose an ECM problem.
   - If TPS is less than .2 volts, the TPS 5 volt reference circuit could be shorted to ground. If TPS is over 2.5 volts, the ECM could be in the "Clear Flood Mode" which may cause the engine to not start.
   - The "Scan" tool should display rpm during cranking.
2. Because the DIS uses two spark plugs and cables to complete the circuit of each coil, the opposite spark plug cable should be left connected.
3. Nominal resistance of each injector is 11.6 to 12.4 ohms at 20°C (68°F). Resistance will increase slightly at higher temperatures. This test is performed with injectors 1, 2, 3, and 4 in parallel, so measurement will be one forth of the nominal resistance for a single injector.
4. The test light should flash, indicating that the ECM is controlling the injectors. How bright the light flashes is not important.

5. Ignition may have to be cycled "ON" several times to obtain maximum fuel pressure.
6. Damage to the ECM injector driver may occur if any injector resistance measures less than 11.6 ohms (internal short to ignition CKT 439).

Diagnostic Aids:

- Water or contamination in fuel system may cause a no start condition during very cold or freezing weather. The engine may start after approximately 5 minutes in a heated shop.
- An EGR valve sticking open can cause a rich air/fuel ratio during cranking. Unless the ECM enters "Clear Flood Mode" at the first indication of a flooding condition, it may result in a no start condition.
- An A/C pressure sensor with an internal short to ground can cause a no start condition. Disconnect the A/C pressure sensor, if vehicle starts replace faulty sensor.
- A MAP sensor stuck between .5 and 2.5 volts can cause a no start condition. Disconnect the MAP sensor, if vehicle starts replace faulty sensor.

2.2L (VIN 4) ENGINE — SYSTEM DIAGNOSTIC CHARTS — 1992 CORSICA AND BERETTA

CHART A-3
(Page 1 of 3)
ENGINE CRANKS BUT WON'T RUN
2.2L (VIN 4) "L" CARLINE (PORT)

1
- IGNITION "ON." IF "SES" LIGHT IS OFF, USE CHART A-1.
- INSTALL TECH 1 "SCAN" TOOL. IF NO DATA, USE CHART A-2.
- CHECK THE FOLLOWING:
  - TPS. IF LESS THAN .2 VOLTS OR OVER 2.5 VOLTS AT CLOSED THROTTLE, USE CODE 21 CHART.
  - COOLANT. IF BELOW -38°C, USE CODE 14 CHART.
  - RPM. IF NO RPM WHILE CRANKING, USE CHART A-3 (PAGE 2 OF 3).

NOTICE: PFI SYSTEM UNDER PRESSURE. TO AVOID FUEL SPILLAGE, REFER TO FIELD SERVICE PROCEDURES FOR TESTING OR MAKING REPAIRS REQUIRING DISASSEMBLY OF FUEL LINES OR FITTING.

- PROBE FUEL PUMP "TEST" TERMINAL WITH A TEST LIGHT TO B +.
- IGNITION "OFF" FOR 10 SECONDS. LIGHT SHOULD BE "ON."
- IGNITION "ON." LIGHT SHOULD GO OUT FOR ABOUT 2 SECONDS AND THEN COME BACK "ON."
  DOES IT?

YES · NO → USE FUEL PUMP RELAY CIRCUIT CHART A-5.

2
- CRANK ENGINE AND CHECK FOR SPARK WITH ST-125 ON SPARK PLUG CABLES 1 & 2 OR 3 & 4.
- CHECK ONE CABLE AT A TIME. LEAVE THE OTHER CABLES CONNECTED TO PLUGS DURING CRANKING.
  IS THERE SPARK ON BOTH CABLES?

YES · NO → USE CHART A-3 (PAGE 2 OF 3).

3
- DISCONNECT INJECTOR JUMPER HARNESS AT 2 PIN CONNECTOR.
- USING DVM MEASURE RESISTANCE BETWEEN CAVITIES "A" AND "B" (INJECTOR SIDE OF HARNESS).
  RESISTANCE SHOULD BE ABOUT 2.9 TO 3.1 OHMS. IS IT?

YES · NO → 2.9 OHMS OR LESS. | 3.1 OHMS OR GREATER.

6 → REPAIR SHORT IN HARNESS OR REPLACE ANY INJECTOR(S) THAT MEASURE 11.6 OHMS OR LESS. | REPAIR OPEN IN HARNESS OR REPLACE ANY INJECTOR(S) THAT MEASURE 12.4 OHMS OR GREATER.

4
- CONNECT INJECTOR TEST LIGHT J34730-2B TO INJECTOR JUMPER HARNESS (ECM SIDE).
- CRANK ENGINE AND OBSERVE TEST LIGHT (SHOULD BLINK). DOES LIGHT BLINK?

YES · NO → USE CHART A-3 (PAGE 3 OF 3)

5
- IGNITION "OFF." INSTALL FUEL PRESSURE GAGE (SEE CHART A-7 PAGE 3 OF 3).
- IGNITION "ON." FUEL PRESSURE SHOULD BE 284-325 kPa (41-47 psi). IS IT?

YES · NO → USE FUEL SYSTEM DIAGNOSIS CHART A-7.

CHECK FOR FOULED SPARK PLUGS, OR EGR VALVE STUCK OPEN, OR A/C PRESSURE SENSOR SHORTED. SEE "DIAGNOSTICS AIDS"

2.2L (VIN 4) ENGINE — SYSTEM DIAGNOSTIC CHARTS — 1992 CORSICA AND BERETTA

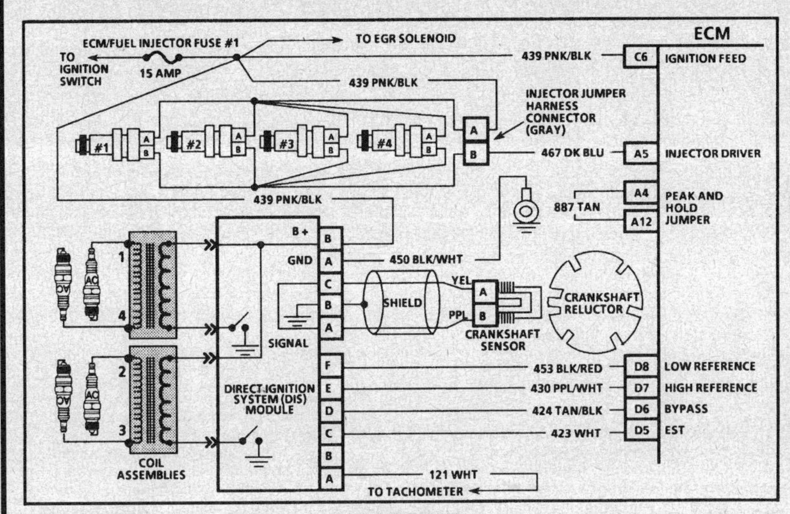

CHART A-3
(Page 2 of 3)
ENGINE CRANKS BUT WON'T RUN
2.2L (VIN 4) "L" CARLINE (PORT)

Circuit Description:

A magnetic crank sensor is used to determine engine crankshaft position, much the same way as the pick-up coil did in High Energy Ignition (HEI) type systems. The sensor is mounted in the block, near a slotted wheel on the crankshaft. The rotation of the wheel creates a flux change in the sensor, which produces a voltage signal. The DIS ignition module processes this signal and creates the reference pulses needed by the Electronic Control Module (ECM) to trigger the correct coil at the correct time.

If the "Scan" tool did not indicate cranking rpm, and there is no spark present at the plugs, the problem lies in the Direct Ignition System or the power and ground supplies to the module.

Test Description: Number(s) below refer to circled number(s) on the diagnostic chart.

1. The Direct Ignition System (DIS) uses two plugs and cables to complete the circuit of each coil. The other spark plug cable in the circuit must be left connected to create a spark.
2. This test will determine if the 12 volt supply and a good ground is available at the DIS module.
3. This test will determine if the ignition module is not generating the reference pulse, or if the wiring or ECM are at fault. By touching and removing a test light to 12 volts on CKT 430, a reference pulse should be generated. If rpm is indicated, the ECM and wiring are OK.

4. This test will determine if the ignition module is not triggering the problem coil, or if the tested coil is at fault. This test could also be performed by substituting a known good coil. The secondary coil winding can be checked with a Digital Voltmeter (DVM). There should be 5,000 to 10,000 ohms across the coil towers. There should not be any continuity from either coil tower to ground.
5. Checks for continuity of the crank sensor and connections.
6. Normal crank sensor voltage output range is .8 to 1.4 volts (800 to 1400 mV) with a fully charged battery and engine at room temperature. Minimum output voltage (slow cranking, low battery) can be as low as .3 volt (300 mV).

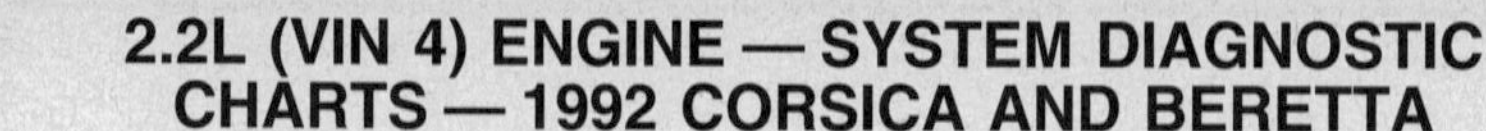

2.2L (VIN 4) ENGINE — SYSTEM DIAGNOSTIC CHARTS — 1992 CORSICA AND BERETTA

CHART A-3
(Page 2 of 3)
ENGINE CRANKS BUT WON'T RUN
2.2L (VIN 4) "L" CARLINE (PORT)

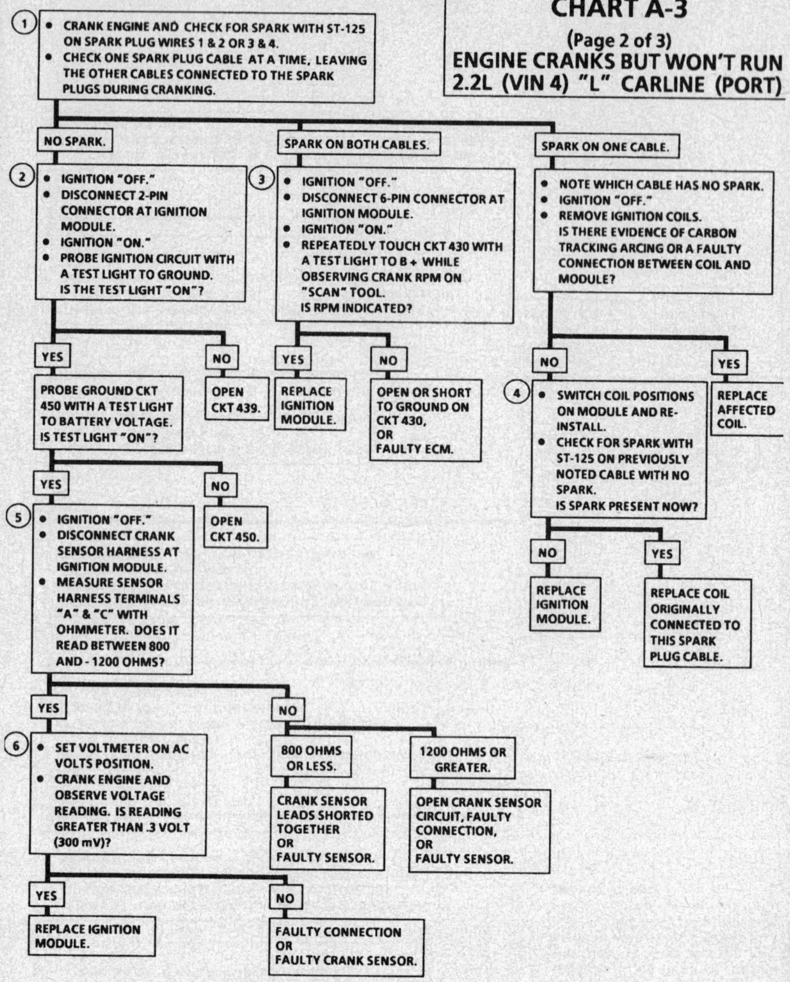

2.2L (VIN 4) ENGINE — SYSTEM DIAGNOSTIC CHARTS — 1992 CORSICA AND BERETTA

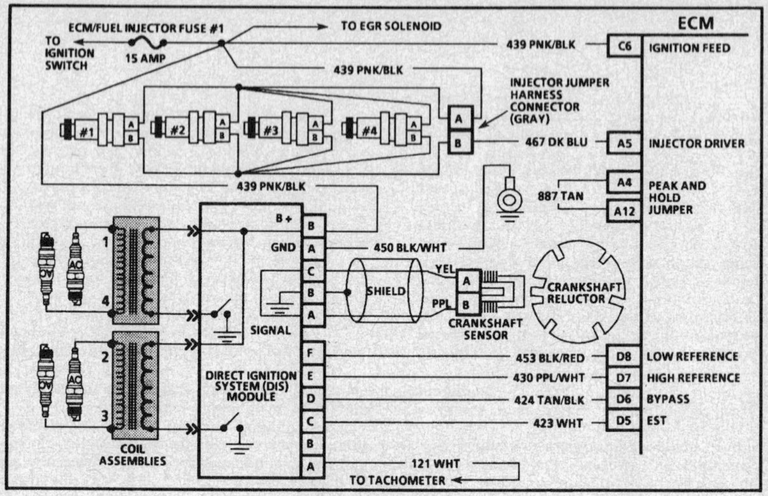

CHART A-3
(Page 3 of 3)
ENGINE CRANKS BUT WON'T RUN
2.2L (VIN 4) "L" CARLINE (PORT)

Test Description: Number(s) below refer to circled number(s) on the diagnostic chart.

1. This test checks for ignition voltage at the injector jumper harness connector.
2. Checks for open in CKT 467 from ECM to injector jumper harness connector. The test light has a path to voltage through the injector windings to ignition CKT 439.
3. Checks for short to voltage on CKT 467 from ECM to injectors. Be sure all injectors are disconnected from injector harness connectors.
4. Damage to the ECM injector driver may occur if injector driver CKT 467 shorts to ignition CKT 439.

2.2L (VIN 4) ENGINE — SYSTEM DIAGNOSTIC CHARTS — 1992 CORSICA AND BERETTA

CHART A-3
(Page 3 of 3)
ENGINE CRANKS BUT WON'T RUN
2.2L (VIN 4) "L" CARLINE (PORT)

FROM CHART A-3 PAGE 1

NO LIGHT "ON" ONE OR BOTH

(1)
- IGNITION "ON", ENGINE "OFF"
- PROBE HARNESS CONNECTOR TERMINAL "A" (CKT 439) OF INJECTOR JUMPER HARNESS (ECM SIDE) WITH A TEST LIGHT TO GROUND. IS THE TEST LIGHT "ON"?

YES →

(2)
- IGNITION "OFF."
- RECONNECT INJECTOR JUMPER HARNESS CONNECTOR.
- DISCONNECT ECM CONNECTORS.
- TURN IGNITION "ON".
- PROBE ECM TERMINAL "A5" WITH A TEST LIGHT TO GROUND.
- IS LIGHT "ON"?

YES →

(3)
- TURN IGNITION "OFF", DISCONNECT ALL INJECTOR ELECTRICAL CONNECTORS.
- TURN IGNITION "ON".
- PROBE ECM TERMINAL "A5" WITH A TEST LIGHT TO GROUND.
- IS LIGHT "ON"?

YES →

(4) CKT 467 IS SHORTED TO VOLTAGE.

NO → IGNITION CKT 439 IS OPEN.

NO → INJECTOR DRIVER CKT 467 IS OPEN.

NO → OPEN/SHORT IN ECM PEAK AND HOLD JUMPER (CKT 887), OR FAULTY ECM CONNECTIONS, OR FAULTY ECM.

STEADY LIGHT "ON" ONE OR BOTH

- DISCONNECT ECM CONNECTORS.
- CHECK INJECTOR DRIVER CIRCUIT FROM ECM TO INJECTOR JUMPER HARNESS CONNECTOR FOR A SHORT TO GROUND. IS CIRCUIT SHORTED TO GROUND?

YES → REPAIR SHORT TO GROUND ON INJECTOR CIRCUIT.

NO → FAULTY ECM CONNECTIONS, OR FAULTY ECM.

"AFTER REPAIRS," CONFIRM "CLOSED LOOP" OPERATION AND NO "SERVICE ENGINE SOON" LIGHT.

2.2L (VIN 4) ENGINE — SYSTEM DIAGNOSTIC CHARTS — 1993–94 CORSICA AND BERETTA

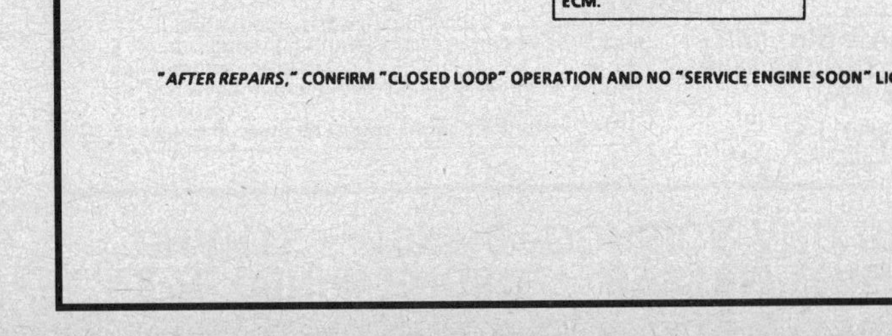

CHART A-3 (Page 1 of 3)
ENGINE CRANKS BUT WON'T RUN
2.2L (VIN 4) "L" CARLINE

Circuit Description:
Before using this chart, battery condition, engine cranking speed, and fuel quantity should be checked and verified as being OK.

Test Description: Number(s) below refer to circled number(s) on the diagnostic chart.

1.
- A MIL (Service Engine Soon) "ON" is a basic test to determine if there is battery and ignition voltage at the ECM.
- No DLC data may be the result of an ECM problem, and CHART A-2 will diagnose an ECM problem.
- If Throttle Position (TP) sensor is less than .2 volt, the TP sensor 5 volt reference circuit could be shorted to ground. If TP sensor is over 2.5 volts, the ECM could be in the "Clear Flood Mode" which may cause the engine to not start.
- Compare Engine Coolant Temperature (ECT) with intake air temperature when engine is cold. If coolant temperature reading is 10 degrees greater or less than intake air temperature on a cold engine, check resistance of the Engine Coolant Temperature (ECT) sensor circuit or sensor. Compare resistance value to the "Diagnostic Aid" chart found on DTC 14 chart.
- The scan tool should display RPM during cranking.

2. Because the Electronic Ignition (EI) system uses two spark plugs and cables to complete the circuit of each coil, the companion spark plug cable should be connected to a good ground.

3. This test is performed with injectors 1, 2, 3, and 4 in parallel.

4. The test light should flash, indicating that the ECM is controlling the injectors. How bright the light flashes is not important.

5. Ignition may have to be cycled "ON" several times to obtain maximum fuel pressure.

6. Damage to the ECM injector driver may occur if any injector resistance measures less than 11.6 ohms (internal injector short to CKT 439).

Diagnostic Aids:
- Water or contamination in fuel system may cause a no start condition during very cold or freezing weather. The engine may start after approximately 5 minutes in a heated shop.
- An EGR valve sticking open can cause a rich air/fuel ratio during cranking. Unless the ECM enters "Clear Flood Mode" at the first indication of a flooding condition, it may result in a no start condition.
- An A/C refrigerant pressure sensor with an internal short to ground can cause a no start condition. Disconnect the A/C refrigerant pressure sensor, if vehicle starts replace faulty sensor.
- A MAP sensor stuck between .5 and 2.5 volts can cause a no start condition. Disconnect the MAP sensor, if vehicle starts replace faulty sensor.

2.2L (VIN 4) ENGINE — SYSTEM DIAGNOSTIC CHARTS — 1993–94 CORSICA AND BERETTA

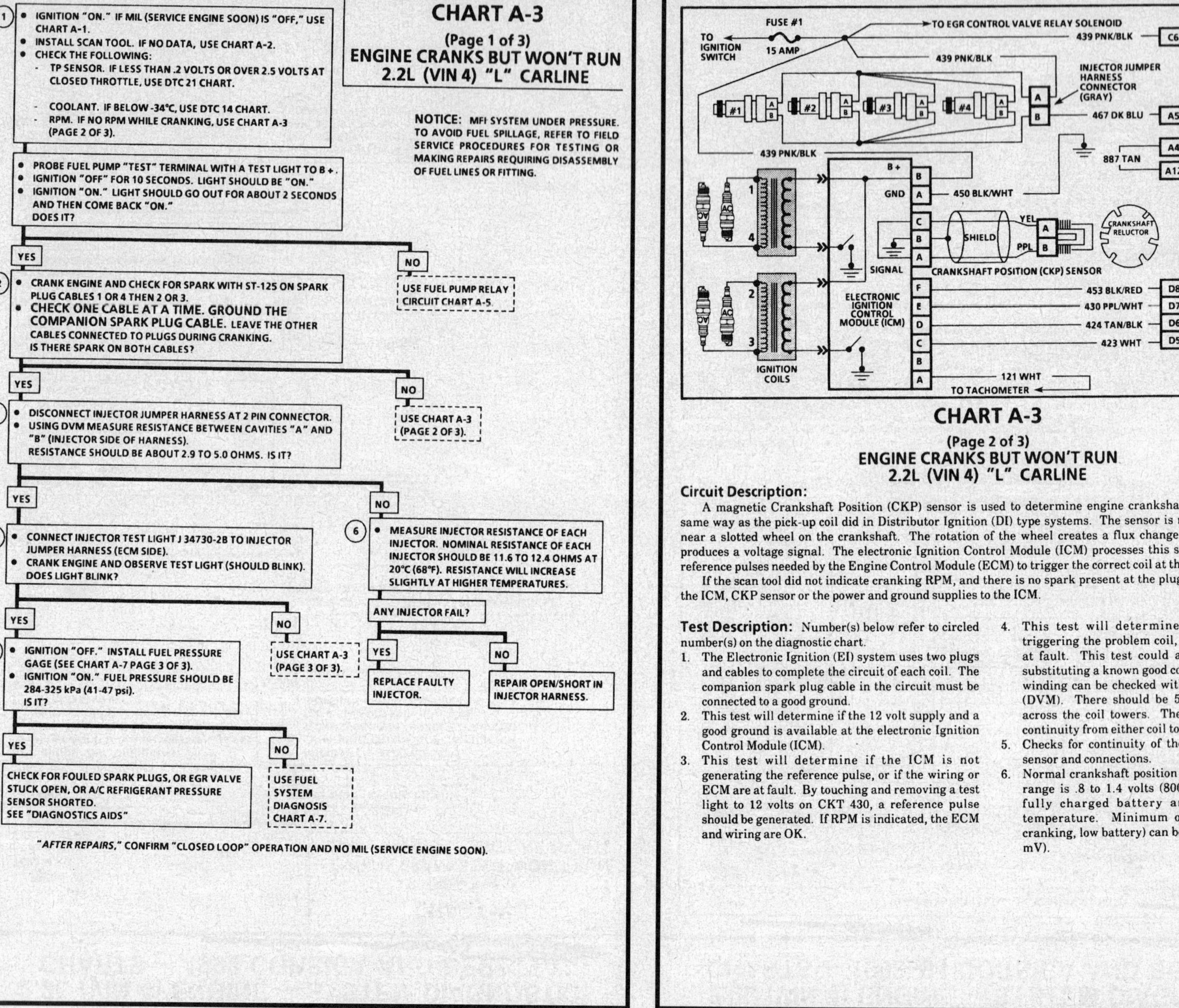

Circuit Description:

A magnetic Crankshaft Position (CKP) sensor is used to determine engine crankshaft position, much the same way as the pick-up coil did in Distributor Ignition (DI) type systems. The sensor is mounted in the block, near a slotted wheel on the crankshaft. The rotation of the wheel creates a flux change in the sensor, which produces a voltage signal. The electronic Ignition Control Module (ICM) processes this signal and creates the reference pulses needed by the Engine Control Module (ECM) to trigger the correct coil at the correct time.

If the scan tool did not indicate cranking RPM, and there is no spark present at the plugs, the problem lies in the ICM, CKP sensor or the power and ground supplies to the ICM.

Test Description: Number(s) below refer to circled number(s) on the diagnostic chart.

1. The Electronic Ignition (EI) system uses two plugs and cables to complete the circuit of each coil. The companion spark plug cable in the circuit must be connected to a good ground.

2. This test will determine if the 12 volt supply and a good ground is available at the electronic Ignition Control Module (ICM).

3. This test will determine if the ICM is not generating the reference pulse, or if the wiring or ECM are at fault. By touching and removing a test light to 12 volts on CKT 430, a reference pulse should be generated. If RPM is indicated, the ECM and wiring are OK.

4. This test will determine if the ICM is not triggering the problem coil, or if the tested coil is at fault. This test could also be performed by substituting a known good coil. The secondary coil winding can be checked with a Digital Voltmeter (DVM). There should be 5,000 to 10,000 ohms across the coil towers. There should not be any continuity from either coil tower to ground.

5. Checks for continuity of the crankshaft position sensor and connections.

6. Normal crankshaft position sensor voltage output range is .8 to 1.4 volts (800 to 1400 mV) with a fully charged battery and engine at room temperature. Minimum output voltage (slow cranking, low battery) can be as low as .3 volt (300 mV).

2.2L (VIN 4) ENGINE — SYSTEM DIAGNOSTIC CHARTS — 1993–94 CORSICA AND BERETTA

CHART A-3
(Page 2 of 3)
ENGINE CRANKS BUT WON'T RUN
2.2L (VIN 4) "L" CARLINE

①
- FOLLOW THESE STEPS TO CHECK FOR SPARK WITH ST-125 ON SPARK PLUG CABLES 1, THEN 2 OR 3, THEN 4.
- CHECK ONE SPARK PLUG CABLE AT A TIME. GROUND COMPANION SPARK PLUG CABLE, LEAVING THE OTHER CABLES CONNECTED TO THE SPARK PLUGS.
- CRANK ENGINE. IS THERE SPARK?

NO SPARK.

②
- IGNITION "OFF."
- DISCONNECT 2-PIN CONNECTOR AT IGNITION CONTROL MODULE.
- IGNITION "ON."
- PROBE IGNITION CKT 439 WITH A TEST LIGHT TO GROUND. IS THE TEST LIGHT "ON"?

YES →
- PROBE GROUND CKT 450 WITH A TEST LIGHT TO BATTERY VOLTAGE. IS TEST LIGHT "ON"?

NO → OPEN CKT 439.

YES →
⑤
- IGNITION "OFF."
- DISCONNECT CRANKSHAFT POSITION SENSOR HARNESS AT IGNITION CONTROL MODULE.
- MEASURE SENSOR HARNESS TERMINALS "A" AND "C" WITH OHMMETER. DOES IT READ BETWEEN 800 AND - 1200 OHMS?

NO → OPEN CKT 450.

YES →
⑥
- SET VOLTMETER ON AC VOLTS POSITION.
- CRANK ENGINE AND OBSERVE VOLTAGE READING. IS READING GREATER THAN .3 VOLT (300 mV)?

NO → 800 OHMS OR LESS. → CRANKSHAFT POSITION SENSOR LEADS SHORTED TOGETHER OR FAULTY SENSOR.

1200 OHMS OR GREATER. → OPEN CRANKSHAFT POSITION SENSOR CIRCUIT, FAULTY CONNECTION, OR FAULTY SENSOR.

YES → REPLACE IGNITION CONTROL MODULE.

NO → FAULTY CONNECTION OR FAULTY CRANKSHAFT POSITION SENSOR.

SPARK ON BOTH CABLES.

③
- IGNITION "OFF."
- DISCONNECT 6-PIN CONNECTOR AT IGNITION CONTROL MODULE.
- IGNITION "ON."
- REPEATEDLY TOUCH CKT 430 WITH A TEST LIGHT TO B + WHILE OBSERVING CRANK RPM ON SCAN TOOL. IS RPM INDICATED?

YES → REPLACE IGNITION CONTROL MODULE.

NO → OPEN OR SHORT TO GROUND ON CKT 430, OR FAULTY ECM.*

SPARK ON ONE CABLE.
- NOTE WHICH CABLE HAS NO SPARK.
- IGNITION "OFF."
- REMOVE IGNITION COILS. IS THERE EVIDENCE OF CARBON TRACKING ARCING OR A FAULTY CONNECTION BETWEEN COIL(S) AND MODULE?

NO →
④
- SWITCH COIL POSITIONS ON MODULE AND RE-INSTALL.
- CHECK FOR SPARK WITH ST-125 ON PREVIOUSLY NOTED CABLE WITH NO SPARK. IS SPARK PRESENT NOW?

NO → REPLACE IGNITION CONTROL MODULE.

YES → REPLACE COIL ORIGINALLY CONNECTED TO THIS SPARK PLUG CABLE.

YES → REPLACE AFFECTED COIL.

* IF ECM IS FAULTY AND MUST BE REPLACED, THE NEW ECM MUST BE PROGRAMMED. REFER TO ECM REPLACEMENT AND PROGRAMMING PROCEDURES

"AFTER REPAIRS," CONFIRM "CLOSED LOOP" OPERATION AND NO MIL (SERVICE ENGINE SOON).

2.2L (VIN 4) ENGINE — SYSTEM DIAGNOSTIC CHARTS — 1993–94 CORSICA AND BERETTA

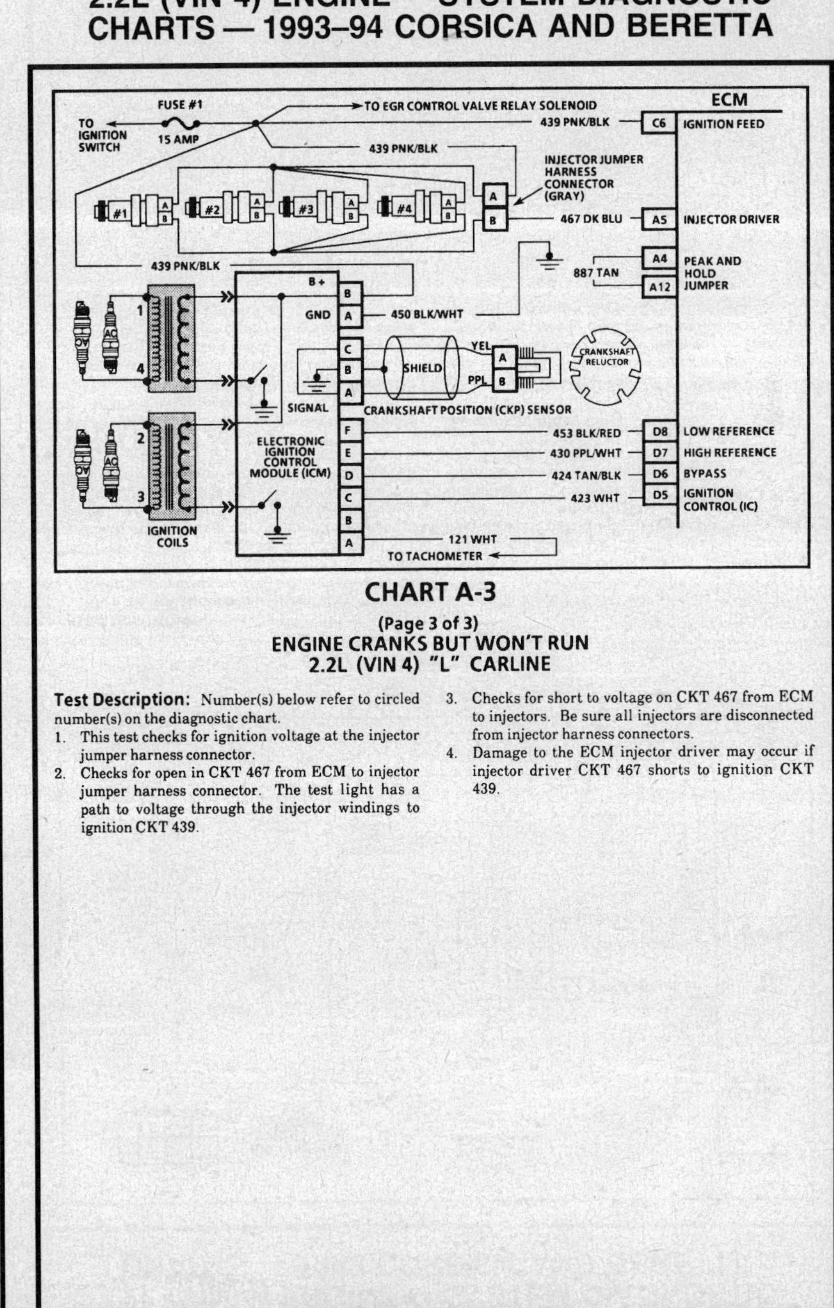

CHART A-3
(Page 3 of 3)
ENGINE CRANKS BUT WON'T RUN
2.2L (VIN 4) "L" CARLINE

Test Description: Number(s) below refer to circled number(s) on the diagnostic chart.

1. This test checks for ignition voltage at the injector jumper harness connector.
2. Checks for open in CKT 467 from ECM to injector jumper harness connector. The test light has a path to voltage through the injector windings to ignition CKT 439.
3. Checks for short to voltage on CKT 467 from ECM to injectors. Be sure all injectors are disconnected from injector harness connectors.
4. Damage to the ECM injector driver may occur if injector driver CKT 467 shorts to ignition CKT 439.

2.2L (VIN 4) ENGINE — SYSTEM DIAGNOSTIC CHARTS — 1992 CORSICA AND BERETTA

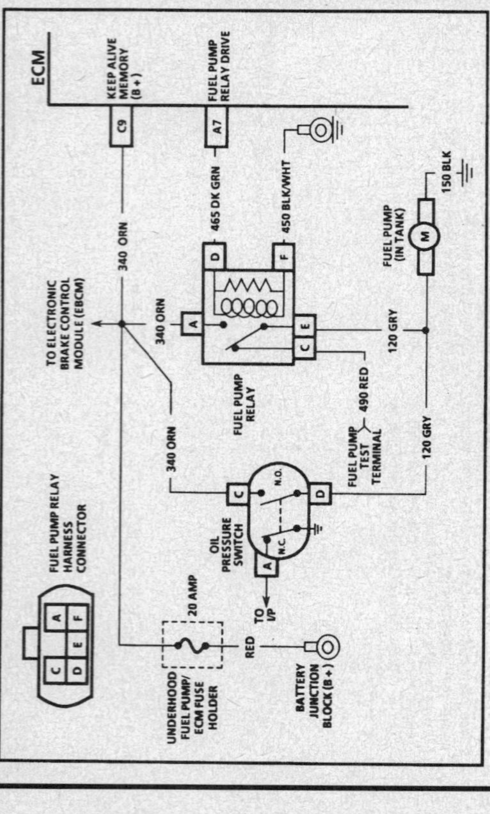

CHART A-5
FUEL PUMP RELAY CIRCUIT
2.2L (VIN 4) "L" CARLINE (PORT)

Circuit Description:

When the ignition switch is turned "ON," the Electronic Control Module (ECM) will activate the fuel pump relay with a 12 volt supply and run the in-tank fuel pump. The fuel pump will operate as long as the engine is cranking or running and the ECM is receiving ignition reference pulses. If there are no ignition reference pulses, the ECM will no longer supply the fuel pump relay signal within 2 seconds after key "ON."

Should the fuel pump relay or the 12 volt relay drive from the ECM fail, the fuel pump will receive supply current through the oil pressure switch back-up circuit.

The fuel pump test terminal is located in the driver's side of the engine compartment. When the engine is stopped, the pump can be turned "ON" by applying battery voltage to the test terminal.

Test Description: Number(s) below refer to circled number(s) on the diagnostic chart.
1. At this point, the fuel pump relay is operating correctly. The back-up circuit through the oil pressure switch is now tested.
2. After the fuel pump relay is replaced, continue with "Oil Pressure Switch Test."

Diagnostic Aids:

An inoperative fuel pump relay can result in long cranking times. The extended crank period is caused by the time necessary for oil pressure to reach the pressure required to close the oil pressure switch and supply the necessary current for the fuel pump.

2.2L (VIN 4) ENGINE — SYSTEM DIAGNOSTIC CHARTS — 1993–94 CORSICA AND BERETTA

CHART A-3
(Page 3 of 3)
ENGINE CRANKS BUT WON'T RUN
2.2L (VIN 4) "L" CARLINE

FROM CHART A-3 PAGE 1 OR CHART C-4D-1

1. IGNITION "ON", ENGINE "OFF."
   - PROBE HARNESS CONNECTOR TERMINAL "A" (IGNITION FEED CIRCUIT) OF INJECTOR JUMPER HARNESS (ECM SIDE) WITH A TEST LIGHT TO GROUND.
   - IS THE TEST LIGHT "ON"?

   STEADY LIGHT "ON"
   - DISCONNECT ECM CONNECTORS.
   - CHECK INJECTOR DRIVER CIRCUIT FROM ECM TO INJECTOR JUMPER HARNESS CONNECTOR FOR A SHORT TO GROUND. IS CIRCUIT SHORTED TO GROUND?
     - NO → FAULTY ECM CONNECTIONS, OR FAULTY ECM.*
     - YES → REPAIR SHORT TO GROUND ON INJECTOR CIRCUIT.

   NO LIGHT "ON" →

2. IGNITION "OFF."
   - RECONNECT INJECTOR JUMPER HARNESS CONNECTOR.
   - DISCONNECT ECM CONNECTORS.
   - TURN IGNITION "ON."
   - PROBE ECM TERMINAL "A5" WITH A TEST LIGHT TO GROUND.
   - IS LIGHT "ON"?
     - NO → IGNITION CKT 1149 IS OPEN.
     - YES →

3. TURN IGNITION "OFF," DISCONNECT ALL INJECTOR ELECTRICAL CONNECTORS.
   - TURN IGNITION "ON."
   - PROBE ECM TERMINAL "A5" WITH A TEST LIGHT TO GROUND.
   - IS LIGHT "ON"?
     - NO → INJECTOR DRIVER CKT 467 IS OPEN.
     - YES →

4. CKT 467 IS SHORTED TO VOLTAGE.
   - NO → OPEN/SHORT IN ECM PEAK AND HOLD JUMPER (CKT 887), OR FAULTY ECM CONNECTIONS, OR FAULTY ECM.*
   - YES →

* IF ECM IS FAULTY AND MUST BE REPLACED, THE NEW ECM MUST BE PROGRAMMED. REFER TO ECM REPLACEMENT AND PROGRAMMING PROCEDURES.

"AFTER REPAIRS," CONFIRM "CLOSED LOOP" OPERATION AND NO MIL (SERVICE ENGINE SOON).

2.2L (VIN 4) ENGINE — SYSTEM DIAGNOSTIC CHARTS — 1992 CORSICA AND BERETTA

CHART A-5
FUEL PUMP RELAY CIRCUIT
2.2L (VIN 4) "L" CARLINE (PORT)

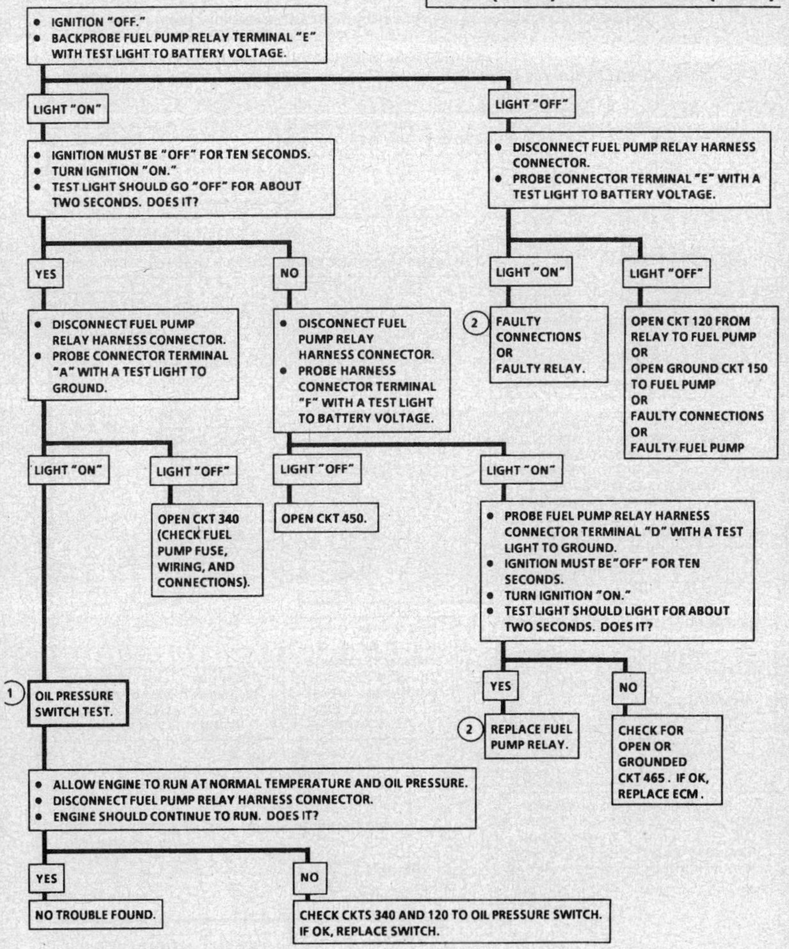

- IGNITION "OFF."
- BACKPROBE FUEL PUMP RELAY TERMINAL "E" WITH TEST LIGHT TO BATTERY VOLTAGE.

LIGHT "ON"

- IGNITION MUST BE "OFF" FOR TEN SECONDS.
- TURN IGNITION "ON."
- TEST LIGHT SHOULD GO "OFF" FOR ABOUT TWO SECONDS. DOES IT?

YES

- DISCONNECT FUEL PUMP RELAY HARNESS CONNECTOR.
- PROBE CONNECTOR TERMINAL "A" WITH A TEST LIGHT TO GROUND.

LIGHT "ON" → **LIGHT "OFF"** → OPEN CKT 340 (CHECK FUEL PUMP FUSE, WIRING, AND CONNECTIONS).

NO

- DISCONNECT FUEL PUMP RELAY HARNESS CONNECTOR.
- PROBE HARNESS CONNECTOR TERMINAL "F" WITH A TEST LIGHT TO BATTERY VOLTAGE.

LIGHT "OFF" → OPEN CKT 450.

LIGHT "ON"

- PROBE FUEL PUMP RELAY HARNESS CONNECTOR TERMINAL "D" WITH A TEST LIGHT TO GROUND.
- IGNITION MUST BE "OFF" FOR TEN SECONDS.
- TURN IGNITION "ON."
- TEST LIGHT SHOULD LIGHT FOR ABOUT TWO SECONDS. DOES IT?

YES → (2) REPLACE FUEL PUMP RELAY.

NO → CHECK FOR OPEN OR GROUNDED CKT 465 . IF OK, REPLACE ECM.

LIGHT "OFF"

- DISCONNECT FUEL PUMP RELAY HARNESS CONNECTOR.
- PROBE CONNECTOR TERMINAL "E" WITH A TEST LIGHT TO BATTERY VOLTAGE.

LIGHT "ON" → (2) FAULTY CONNECTIONS OR FAULTY RELAY.

LIGHT "OFF" → OPEN CKT 120 FROM RELAY TO FUEL PUMP OR OPEN GROUND CKT 150 TO FUEL PUMP OR FAULTY CONNECTIONS OR FAULTY FUEL PUMP

(1) OIL PRESSURE SWITCH TEST.

- ALLOW ENGINE TO RUN AT NORMAL TEMPERATURE AND OIL PRESSURE.
- DISCONNECT FUEL PUMP RELAY HARNESS CONNECTOR.
- ENGINE SHOULD CONTINUE TO RUN. DOES IT?

YES → NO TROUBLE FOUND.

NO → CHECK CKTS 340 AND 120 TO OIL PRESSURE SWITCH. IF OK, REPLACE SWITCH.

"AFTER REPAIRS," CONFIRM "CLOSED LOOP" OPERATION AND NO "SERVICE ENGINE SOON" LIGHT.

2.2L (VIN 4) ENGINE — SYSTEM DIAGNOSTIC CHARTS — 1993–94 CORSICA AND BERETTA

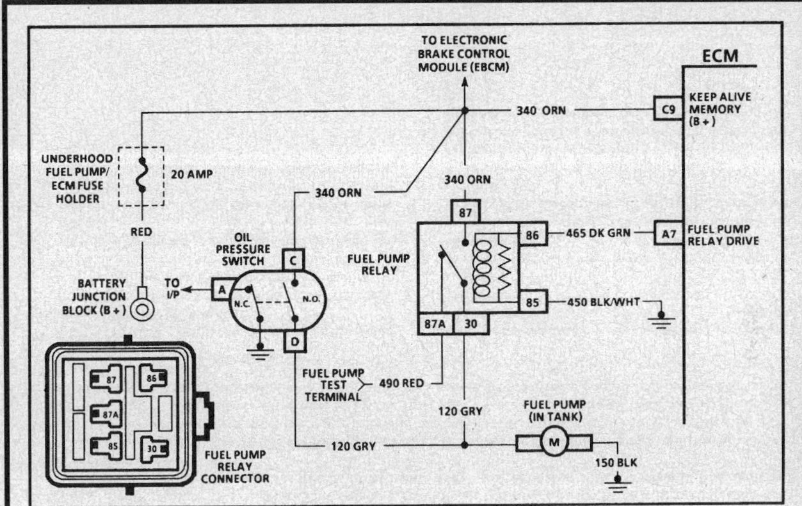

CHART A-5
FUEL PUMP RELAY CIRCUIT
2.2L (VIN 4) "L" CARLINE

Circuit Description:

When the ignition switch is turned "ON," the Engine Control Module (ECM) will activate the fuel pump relay with a 12 volt supply and run the in-tank fuel pump. The fuel pump will operate as long as the engine is cranking or running and the ECM is receiving ignition reference pulses. If there are no ignition reference pulses, the ECM will no longer supply the fuel pump relay signal within 2 seconds after key "ON."

Should the fuel pump relay or the 12 volt relay drive from the ECM fail, the fuel pump will receive supply current through the oil pressure switch back-up circuit.

The fuel pump "test" terminal is located in the driver's side of the engine compartment. When the engine is stopped, the pump can be turned "ON" by applying battery voltage to the "test" terminal.

Test Description: Number(s) below refer to circled number(s) on the diagnostic chart.

1. At this point, the fuel pump relay is operating correctly. The back-up circuit through the oil pressure switch is now tested.
2. After the fuel pump relay is replaced, continue with "Oil Pressure Switch Test."

Diagnostic Aids:

An inoperative fuel pump relay can result in long cranking times. The extended crank period is caused by the time necessary for oil pressure to reach the pressure required to close the oil pressure switch and supply the necessary current for the fuel pump.

If the fuel pump relay circuit checks out OK, refer to "Fuel System Diagnosis," CHART A-7.

Excess fuel may also cause long cranking times. This would usually be accompanied by a start that is not as fast as normal (once fired, the engine does not build up speed as fast) and a puff of black smoke at the tailpipe. An improperly connected or faulty EVAP canister control valve can cause this problem. Disconnect the "EVAP purge hose" from the "EVAP canister control valve" to diagnose. Refer to "Evaporative Emission (EVAP) Control System,"

One or more leaking fuel injectors may also extend the cranking time. Perform the "Injector Balance Test" in CHART C2-A

2.2L (VIN 4) ENGINE — SYSTEM DIAGNOSTIC CHARTS — 1993–94 CORSICA AND BERETTA

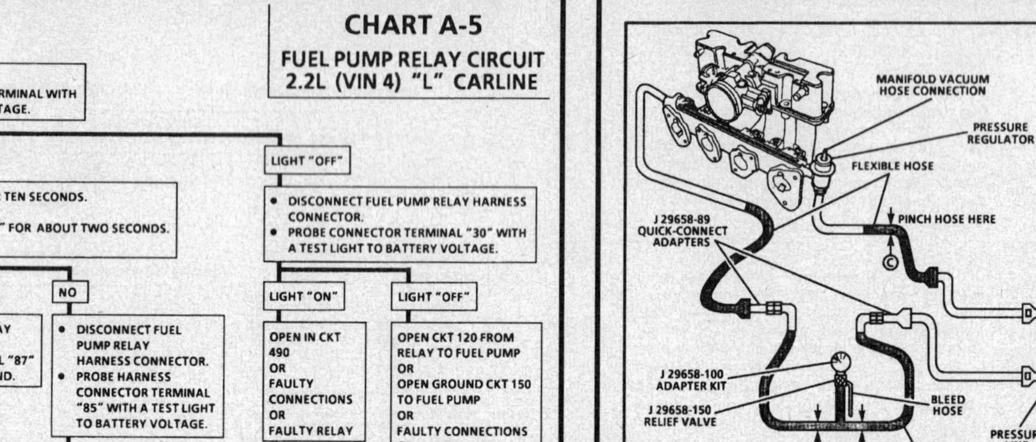

2.2L (VIN 4) ENGINE — SYSTEM DIAGNOSTIC CHARTS — 1992 CORSICA AND BERETTA

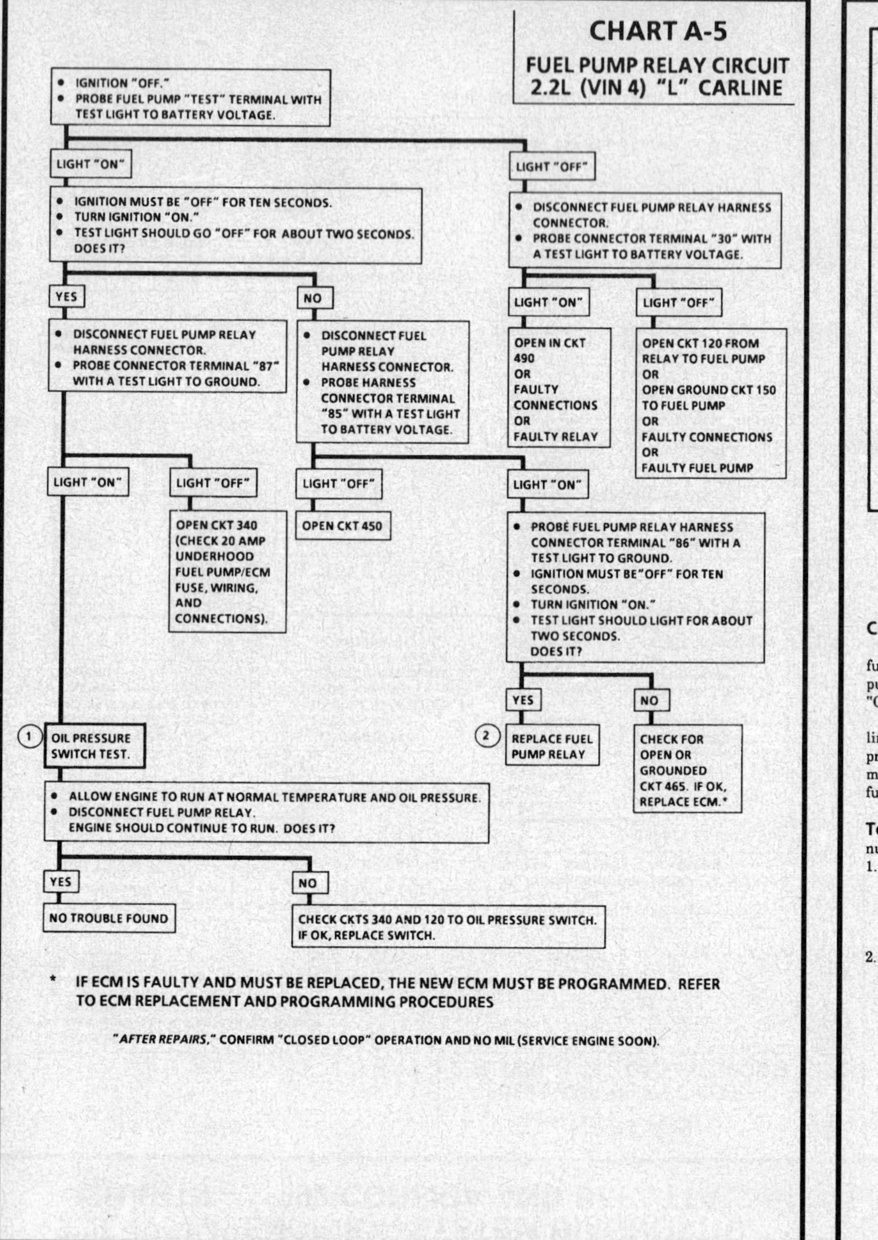

CHART A-7

(Page 1 of 3)
FUEL SYSTEM DIAGNOSIS
2.2L (VIN 4) "L" CARLINE (PORT)

Circuit Description:

When the ignition switch is turned "ON," the Electronic Control Module (ECM) will turn "ON" the in-tank fuel pump. It will remain "ON" as long as the engine is cranking or running, and the ECM is receiving reference pulses. If there are no reference pulses, the ECM will shut "OFF" the fuel pump within 2 seconds after ignition "ON" or engine stops.

An electric fuel pump, attached to the fuel sender assembly (inside the fuel tank), pumps fuel through an in-line filter to the fuel passage within the lower manifold assembly. The pump is designed to provide fuel at a pressure above the regulated pressure needed by the injectors. A pressure regulator, attached to the lower manifold assembly, keeps fuel available to the injectors at a regulated pressure. Unused fuel is returned to the fuel tank by a separate line.

Test Description: Number(s) below refer to circled number(s) on the diagnostic chart.

1. Install fuel pressure gage per instructions on page 3 of 3. Ignition "ON" pump pressure should be 284-325 kPa to (41-47 psi). This pressure is controlled by spring pressure within the regulator assembly.
2. When the engine is idling, the manifold pressure is low (high vacuum) and is applied to the fuel regulator diaphragm. This will offset the spring and result in a lower fuel pressure.

This idle pressure will vary somewhat depending on barometric pressure, however, the pressure idling should be less, indicating pressure regulator control.

3. Pressure that continues to fall is caused by one of the following:
   • In-tank fuel pump check valve not holding.
   • Partially disconnected fuel pulse dampener (pulsator).
   • Fuel pressure regulator valve leaking.
   • Injector(s) sticking open.
4. An injector sticking open can best be determined by checking for a fouled or saturated spark plug(s).

2.2L (VIN 4) ENGINE — SYSTEM DIAGNOSTIC CHARTS — 1992 CORSICA AND BERETTA

CHART A-7
(Page 1 of 3)
FUEL SYSTEM DIAGNOSIS
2.2L (VIN 4) "L" CARLINE (PORT)

FROM CHART A-3
1 • INSTALL FUEL PRESSURE GAGE, (SEE CHART A-7, PAGE 3 OF 3).
• IGNITION "OFF" FOR 10 SECONDS. A/C "OFF."
• IGNITION "ON." FUEL PUMP WILL RUN FOR ABOUT 2 SECONDS.
• NOTE FUEL PRESSURE, WITH PUMP RUNNING SHOULD BE 284-325 kPa (41-47 psi) AND HOLD STEADY WHEN PUMP STOPS.
OK
NOT OK
2 • START AND IDLE ENGINE AT NORMAL OPERATING TEMPERATURE.
• PRESSURE SHOULD BE LOWER BY 21-69 kPa (3-10 psi).
OK
NOT OK
NO TROUBLE FOUND.
• APPLY 10 INCHES OF VACUUM TO PRESSURE REGULATOR.
• FUEL PRESSURE SHOULD DROP 21-69 kPa (3-10 psi).
OK
NOT OK
REPAIR VACUUM SOURCE TO REGULATOR.
REPLACE REGULATOR ASSEMBLY.
3 FUEL PRESSURE, BUT NOT HOLDING
FUEL PRESSURE, BUT LESS THAN 284 kPa (41 psi)
FUEL PRESSURE ABOVE 325 kPa (47 psi)
NO PRESSURE
• IGNITION "OFF" FOR 10 SECONDS.
• IGNITION "ON."
• PINCH GAGE INLET HOSE AT "A"
PRESSURE SHOULD HOLD.
SEE CHART A-7 (2 of 3)
• IGNITION "OFF."
• USING A 10 AMP FUSED JUMPER WIRE, APPLY 12 VOLTS TO FUEL PUMP "TEST" TERMINAL.
• LISTEN FOR FUEL PUMP RUNNING.
NOT HOLDING
HOLDS
• IGNITION "OFF" FOR 10 SECONDS.
• IGNITION "ON."
• PINCH FUEL RETURN HOSE AT "C".
• RECHECK PRESSURE.
CHECK FOR:
- PARTIALLY DISCONNECTED FUEL PULSE DAMPENER (PULSATOR).
- FAULTY IN-TANK PUMP.
HOLDS
NOT HOLDING
REPLACE PRESSURE REGULATOR.
4 LOCATE AND CORRECT LEAKING INJECTOR(S).
PUMP RUNS
PUMP NOT RUNNING
CHECK FOR:
• PLUGGED IN-LINE FILTER.
• PLUGGED FUEL PUMP STRAINER.
• RESTRICTED FUEL LINE.
• DISCONNECTED FUEL PULSE DAMPENER (PULSATOR).
CHECK FOR:
• OPEN WIRE IN CKT 490.
• FAULTY FUEL PUMP RELAY.
• OPEN WIRE IN CKT 120.
• OPEN PUMP GROUND WIRE CKT 150.
IF OK
IF OK
REPLACE IN-TANK PUMP.
IMPORTANT: THE IGNITION MAY HAVE TO BE CYCLED "ON" MORE THAN ONCE TO OBTAIN MAXIMUM PRESSURE. ALSO, IT IS NORMAL FOR THE PRESSURE TO DROP SLIGHTLY WHEN THE PUMP STOPS.

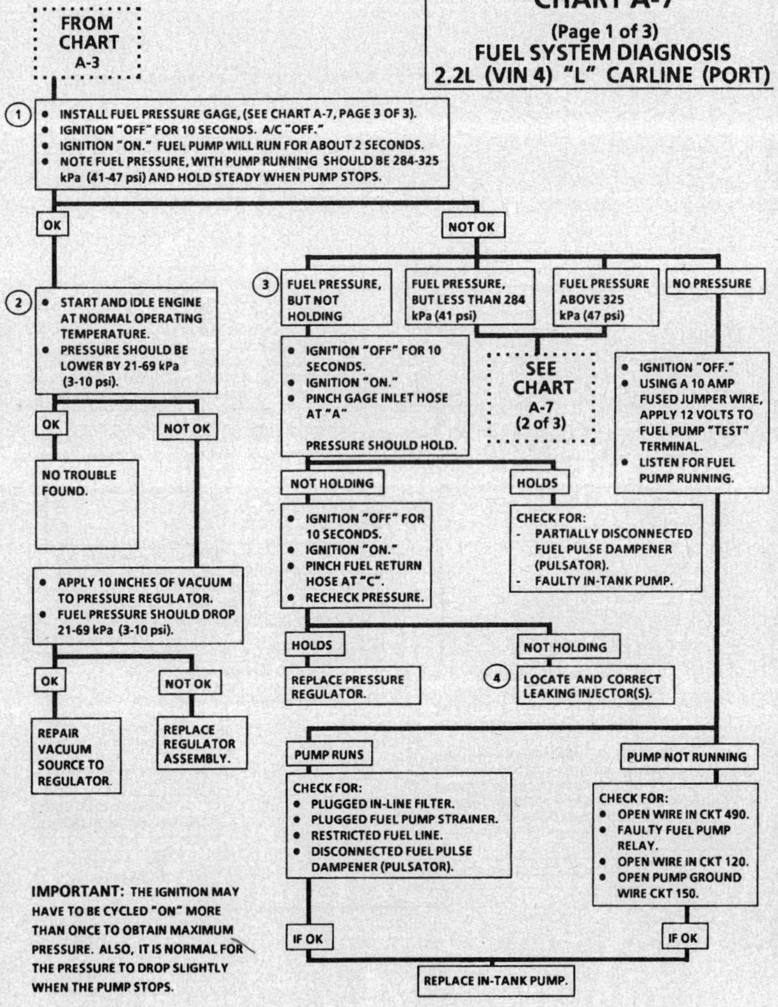

"AFTER REPAIRS," CONFIRM "CLOSED LOOP" OPERATION AND NO "SERVICE ENGINE SOON" LIGHT.

2.2L (VIN 4) ENGINE — SYSTEM DIAGNOSTIC CHARTS — 1992 CORSICA AND BERETTA

MANIFOLD VACUUM HOSE CONNECTION
PRESSURE REGULATOR
FLEXIBLE HOSE
PINCH HOSE HERE
J 29658-89 QUICK-CONNECT ADAPTERS
ELECTRICAL CONNECTOR
RETURN LINE
J 29658-100 ADAPTER KIT
J 29658-150 RELIEF VALVE
BLEED HOSE
FUEL FILTER
PRESSURE LINE
FUEL PULSE DAMPENER (PULSATOR)
IN-TANK PUMP
PINCH HOSE HERE
J 29658-D FUEL PRESSURE GAGE KIT
FUEL PUMP STRAINER

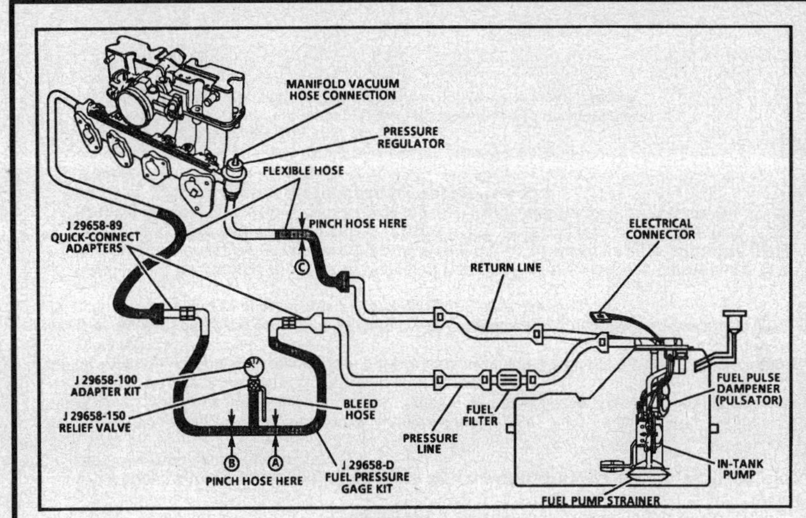

CHART A-7
(Page 2 of 3)
FUEL SYSTEM DIAGNOSIS
2.2L (VIN 4) "L" CARLINE (PORT)

Test Description: Number(s) below refer to circled number(s) on the diagnostic chart.

1. Pressure below 284 kPa (41 psi) may cause a lean condition and may set a Code 44 or Code 32. It could also cause hard starting cold and poor driveability. Low enough pressure will cause the engine not to run at all. Restricted flow may allow the engine to run at idle, or low speeds, but may cause a surge and stall when more fuel is required, as when accelerating or driving at high speeds.
2. Restricting fuel flow at the fuel pressure gage (at B) causes fuel pressure to build above regulated pressure. With battery voltage applied to the pump "test" terminal, pressure should rise above 325 kPa (47 psi) as the gage outlet hose is restricted.

   NOTICE: Do not allow pressure to exceed 414 kPa (60 psi), as damage to the regulator may result.

3. This test determines if the high fuel pressure is due to a restricted fuel return line, or a faulty fuel pressure regulator. High fuel pressure may cause a rich condition and may set a Code 45.

2.2L (VIN 4) ENGINE — SYSTEM DIAGNOSTIC CHARTS — 1992 CORSICA AND BERETTA

CHART A-7
(Page 2 of 3)
FUEL SYSTEM DIAGNOSIS
2.2L (VIN 4) "L" CARLINE (PORT)

FROM CHART A-7 (1 of 3)

① HAS PRESSURE BUT LESS THAN 284 kPa (41 psi)

CHECK FOR RESTRICTED FUEL LINES OR IN-LINE FILTER.

OK

NOT OK

REPLACE FILTER OR REPAIR FUEL LINE AND RECHECK.

② • IGNITION "OFF."
• USING A 10 AMP FUSED JUMPER WIRE, APPLY 12 VOLTS TO FUEL PUMP "TEST" TERMINAL.
• GRADUALLY PINCH PRESSURE GAGE OUTLET HOSE AT (B). LOOK FOR PRESSURE ABOVE 325 kPa (47 psi). DO NOT EXCEED 414 kPa (60 psi).

ABOVE 325 kPa (47 psi)

PRESSURE BUT LESS THAN 284 kPa (41 psi)

IF LINES ARE OK, REPLACE PRESSURE REGULATOR.

CHECK FOR:
- FAULTY FUEL PUMP.
- PARTIALLY DISCONNECTED FUEL PULSE DAMPENER (PULSATOR).
- RESTRICTED FUEL PUMP STRAINER.
- INCORRECT FUEL PUMP.

FUEL PRESSURE ABOVE 325 kPa (47 psi)

③ • DISCONNECT ENGINE COMPARTMENT FUEL RETURN LINE QUICK-CONNECT FITTING. (PROCEDURES FOR DISCONNECTING/ CONNECTING QUICK-CONNECT FITTINGS ARE SAME AS THOSE FOR FUEL FEED LINE,
• PLACE OPEN END OF FLEXIBLE HOSE INTO AN APPROVED GASOLINE CONTAINER. NOTE FUEL PRESSURE WITHIN 2 SECONDS AFTER IGNITION IS TURNED "ON."

ABOVE 325 kPa (47 psi)

284-325 kPa (41-47 psi)

CHECK FOR RESTRICTED FUEL RETURN LINE FROM FUEL PRESSURE REGULATOR TO POINT WHERE FUEL LINE WAS DISCONNECTED.

LOCATE AND CORRECT RESTRICTED FUEL RETURN LINE TO FUEL TANK.

IF LINE OK, REPLACE FUEL PRESSURE REGULATOR.

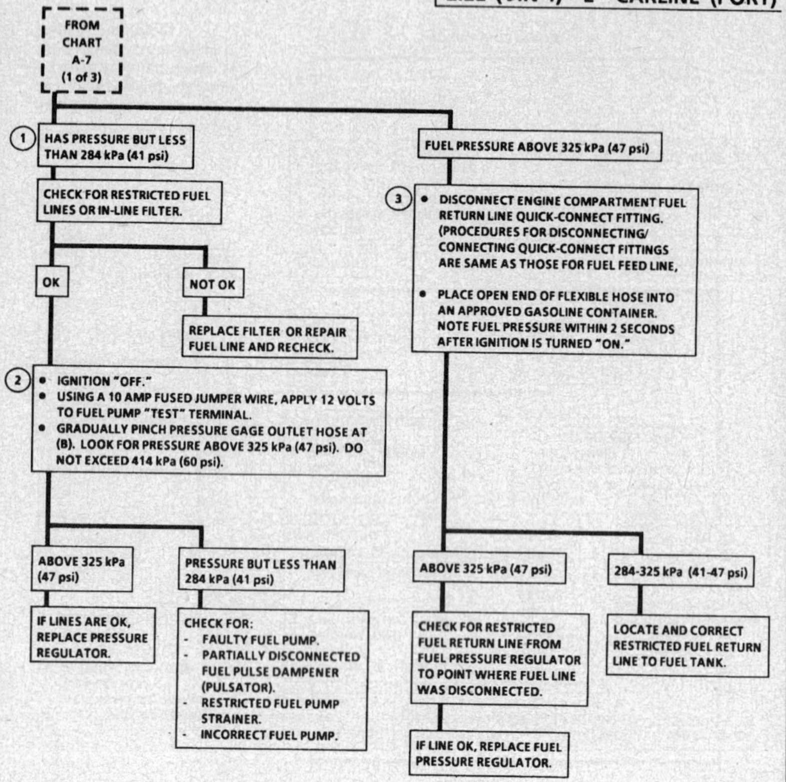

"AFTER REPAIRS," CONFIRM "CLOSED LOOP" OPERATION AND NO "SERVICE ENGINE SOON" LIGHT.

2.2L (VIN 4) ENGINE — SYSTEM DIAGNOSTIC CHARTS — 1992 CORSICA AND BERETTA

CHART A-7
(Page 3 of 3)
FUEL SYSTEM DIAGNOSIS 2.2L (VIN 4) "L" CARLINE (PORT)

FUEL PRESSURE CHECK

Tools Required:
J 29658-D - Fuel Pressure Gage Kit
J 29658-150 - Fuel Pressure Gage Relief Valve
J 29658-100 - TBI Pressure Gage Modification Kit
J 29658-89 - Fuel Pressure Quick Connect Adapters
J 37088 - A - Fuel Line Quick-Connect Separators

CAUTION: To Reduce the Risk of Fire and Personal Injury:
• It is necessary to relieve fuel system pressure before connecting a fuel pressure gage.
• After relieving system pressure, a small amount of fuel may be released when disconnecting the fuel lines. Cover fuel line fittings with a shop towel before disconnecting, to catch any fuel that may leak out. Place towel in approved container when disconnect is completed.
• Do not pinch or restrict nylon fuel lines to avoid severing, which could cause a fuel leak.

NOTICE: If nylon fuel lines become kinked, and cannot be straightened, they must be replaced.

1. Loosen fuel filler cap to relieve fuel tank pressure. (Do not tighten at this time.)
2. Raise vehicle.
3. Disconnect fuel pump electrical connector.
4. Lower vehicle.
5. Start and run engine until fuel supply remaining in fuel pipes is consumed. Engage starter for three seconds to assure relief of any remaining pressure.
6. Disconnect negative battery cable.
7. Locate engine compartment fuel feed quick-connect fitting.
8. Grasp both ends of fitting, twist female end ¼ turn in each direction to loosen any dirt in fitting.

CAUTION: Safety glasses must be worn when using compressed air, as flying dirt particles may cause eye injury.

9. Using compressed air, blow dirt out of quick-connect fitting.
10. Choose correct tool from separator tool set J 37088-A for size of fitting. Insert tool into female end of connector, then push inward to release male connector.
11. If not previously installed, connect Fuel Pressure Gage Relief Valve J 29658-150, to fuel pressure gage hose assembly.
12. Connect 414 kPa (60 psi) gage from TBI Pressure Gage Modification kit J 29658-100 to hose assembly.
13. Connect gage quick-connect adapters J 29658-89 to hose assembly.

CAUTION: To Reduce the Risk of Fire and Personal Injury: Before connecting fuel line quick-connect fittings, always apply a few drops of clean engine oil to the male tube ends. This will ensure proper reconnection and prevent a possible fuel leak. (During normal operation, the O-rings located inside the female connector will swell and may prevent proper reconnection if not lubricated.)

14. Lubricate the male tube end of the fuel line and the gage adapter with engine oil.
15. Connect fuel pressure gage.
   - Push connectors together to cause the retaining tabs/fingers to snap into place.
   - Once installed, pull on both ends of each connection to make sure it is secure.

2.2L (VIN 4) ENGINE — SYSTEM DIAGNOSTIC CHARTS — 1992 CORSICA AND BERETTA

CHART A-7

(Page 3 of 3)
FUEL SYSTEM DIAGNOSIS
2.2L (VIN 4) "L" CARLINE (PORT)

FUEL PRESSURE CHECK
-continued

16. Connect negative battery cable.
17. Check fuel pressure.
18. Place bleed hose into an approved container and open valve to bleed system pressure.
19. Disconnect negative battery cable.
20. Disconnect fuel pressure gage.
21. Lubricate the male tube end of the fuel line, and reconnect quick-connect fitting.
    - Push connector together to cause the retaining tabs/fingers to snap into place.
    - Once installed, pull on both ends of connection to make sure it is secure.
22. Tighten fuel filler cap.
23. Connect negative battery cable.
24. Cycle ignition "ON" and "OFF" twice, waiting ten seconds between cycles, then check for fuel leaks.

2.2L (VIN 4) ENGINE — SYSTEM DIAGNOSTIC CHARTS — 1993–94 CORSICA AND BERETTA

CHART A-7

(Page 1 of 3)
FUEL SYSTEM DIAGNOSIS
2.2L (VIN 4) "L" CARLINE

Circuit Description:
When the ignition switch is turned "ON," the Engine Control Module (ECM) will turn "ON" the in-tank fuel pump. It will remain "ON" as long as the engine is cranking or running, and the ECM is receiving reference pulses. If there are no reference pulses, the ECM will shut "OFF" the fuel pump within 2 seconds after ignition "ON" or engine stops.

An electric fuel pump, part of the modular fuel sender and located inside the fuel tank, pumps fuel through an in-line filter to the fuel passage within the lower manifold assembly. The pump is designed to provide fuel at a pressure above the regulated pressure needed by the injectors. A pressure regulator, attached to the lower manifold assembly, keeps fuel available to the injectors at a regulated pressure. Unused fuel is returned to the fuel tank by a separate line.

Test Description: Number(s) below refer to circled number(s) on the diagnostic chart.

1. Install fuel pressure gage as shown in illustration. Refer to "Fuel Pressure Relief Procedure" and for "Servicing Quick-Connect Fittings." Ignition "ON" pump pressure should be 284-325 kPa to (41-47 psi). This pressure is controlled by spring pressure within the regulator assembly.
2. When the engine is idling, the manifold pressure is low (high vacuum) and is applied to the fuel regulator diaphragm. This will offset the spring and result in a lower fuel pressure.

This idle pressure will vary somewhat depending on barometric pressure, however, the pressure idling should be less, indicating pressure regulator control.

3. Pressure that continues to fall is caused by one of the following:
   - Leaking fuel pump feed hose.
   - Leaking check valve within fuel pump feed hose.
   - Fuel pressure regulator valve leaking.
   - Injector(s) sticking open.
4. An injector sticking open can best be determined by checking for a fouled or saturated spark plug(s).

2.2L (VIN 4) ENGINE — SYSTEM DIAGNOSTIC CHARTS — 1993–94 CORSICA AND BERETTA

CHART A-7
(Page 1 of 3)
FUEL SYSTEM DIAGNOSIS
2.2L (VIN 4) "L" CARLINE

FROM CHART A-3

1. • INSTALL FUEL PRESSURE GAGE (SEE CHART A-7, PAGE 3 OF 3 FOR FUEL PRESSURE RELIEF PROCEDURE AND FOR SERVICING QUICK-CONNECT FITTINGS).
   • IGNITION "OFF" FOR 10 SECONDS. A/C "OFF."
   • IGNITION "ON." FUEL PUMP WILL RUN FOR ABOUT 2 SECONDS

   THE IGNITION MAY HAVE TO BE CYCLED "ON" MORE THAN ONCE TO OBTAIN MAXIMUM PRESSURE.
   • NOTE FUEL PRESSURE, WITH PUMP RUNNING, PRESSURE SHOULD BE 284-325 kPa (41-47 psi). WHEN PUMP STOPS, PRESSURE MAY VARY SLIGHTLY THEN SHOULD HOLD STEADY. IS PRESSURE CORRECT AND DOES IT HOLD?

YES →

IF FUEL PRESSURE IS WITHIN NORMAL RANGE, BUT IS SUSPECTED OF DROPPING OFF DURING ACCELERATION, CRUISE OR HARD CORNERING SEE CHART A-7 (2 OF 3).

2. • START ENGINE AND ALLOW IT TO IDLE AT NORMAL OPERATING TEMPERATURE.
   • FUEL PRESSURE NOTED IN STEP 1 SHOULD DROP 21-69 kPa (3-10 psi). DOES IT?

YES → NO TROUBLE FOUND.

NO →
• DISCONNECT VACUUM HOSE FROM PRESSURE REGULATOR ASSEMBLY.
• WITH ENGINE IDLING, APPLY 12-14 INCHES OF VACUUM TO PRESSURE REGULATOR. FUEL PRESSURE NOTED IN STEP 1 SHOULD DROP 21-69 kPa (3-10 psi). DOES IT?

YES → LOCATE AND REPAIR LOSS OF VACUUM TO PRESSURE REGULATOR.

NO → REPLACE PRESSURE REGULATOR.

NO →

3. FUEL PRESSURE WITHIN SPEC., BUT DOES NOT HOLD.

• IGNITION "OFF."
• USING A 10 AMP FUSED JUMPER WIRE, APPLY B + TO FUEL PUMP "TEST" CONNECTOR FOR 2 SECONDS.
• BLOCK GAGE INLET HOSE BY PINCHING AT "A'

PRESSURE SHOULD HOLD. DOES IT?

NO →
• USING A 10 AMP FUSED JUMPER WIRE, APPLY B + TO FUEL PUMP TEST CONNECTOR FOR 2 SECONDS.
• BLOCK FUEL RETURN LINE BY PINCHING FLEXIBLE HOSE AT "C". PRESSURE SHOULD HOLD. DOES IT?

NO →
4. LOCATE AND CORRECT LEAKING INJECTOR(s).

YES → REPLACE PRESSURE REGULATOR.

YES →
CHECK FOR:
• LEAKING FUEL PUMP FEED HOSE.
• LEAKING CHECK VALVE WITHIN FUEL PUMP FEED HOSE.

FUEL PRESSURE OUT OF SPEC.

SEE CHART A-7 (2 OF 3)

FROM CHART A-5

NO PRESSURE, FUEL PUMP ELECTRICAL CIRCUIT OK.

CHECK FOR:
• PLUGGED IN-LINE FILTER.
• RESTRICTED FUEL PRESSURE LINE.
• PLUGGED FUEL SENDER STRAINER.
• PLUGGED FUEL PUMP STRAINER.
• LEAKING FUEL PUMP FEED HOSE.

IF OK → REPLACE FUEL PUMP.

"AFTER REPAIRS," CONFIRM "CLOSED LOOP" OPERATION AND NO MIL (SERVICE ENGINE SOON).

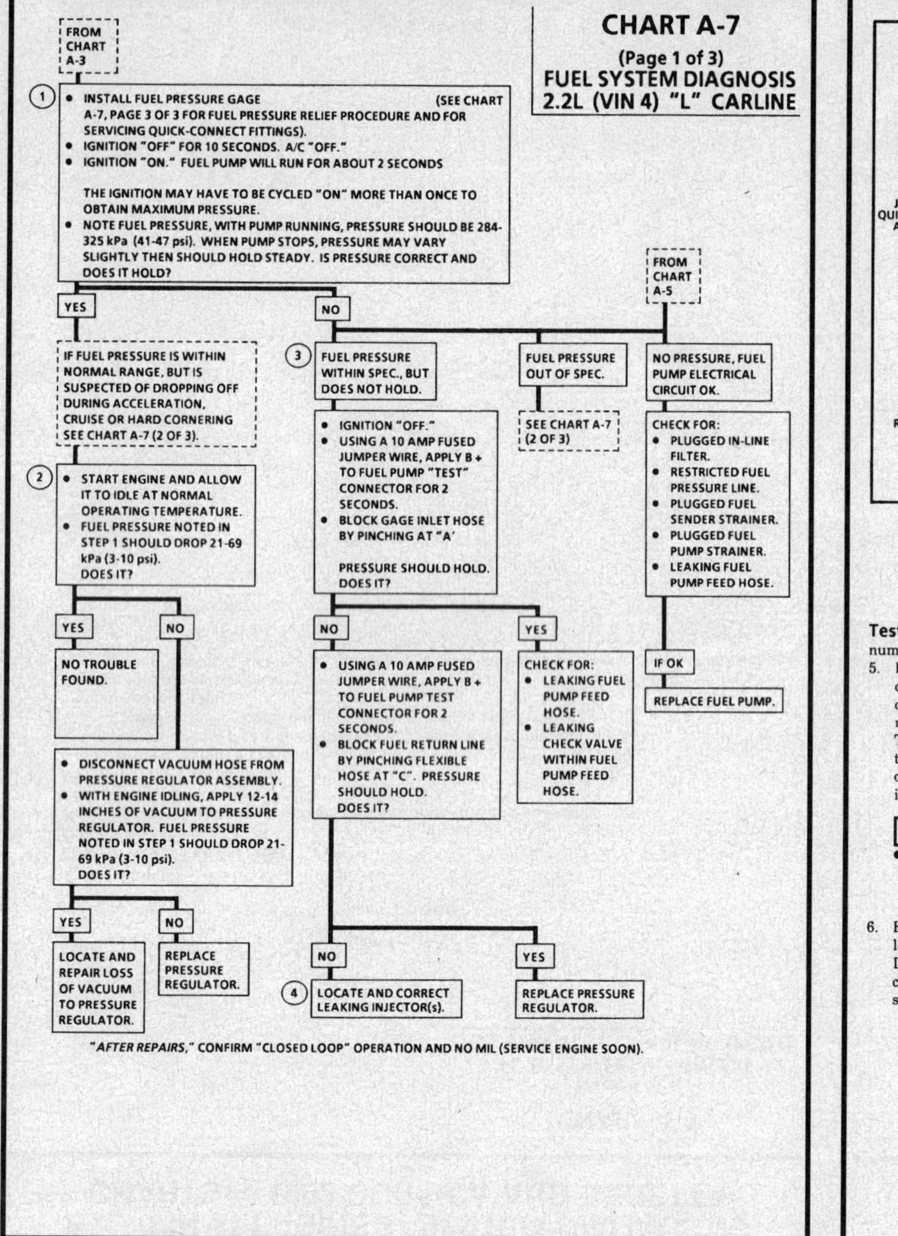

CHART A-7
(Page 2 of 3)
FUEL SYSTEM DIAGNOSIS
2.2L (VIN 4) "L" CARLINE

Test Description: Number(s) below refer to circled number(s) on the diagnostic chart.

5. Fuel pressure that drops off during acceleration, cruise or hard cornering may cause a lean condition and result in a loss of power, surging or misfire. This condition can be diagnosed using a Tech 1 scan tool. If the fuel system is very lean, the Oxygen Sensor (O2S) will stop toggling and output voltage will drop below 500 mV. Also, injector pulse width will increase.

Important
• Make sure system is not operating at "Fuel Cut-Off" which may cause false readings on the scan tool.

6. Fuel pressure below 284 kPa (41 psi) may cause a lean condition and may set a DTC 44 or a DTC 32. Driveability conditions can include hard starting cold, hesitation, poor driveability, lack of power, surging or misfire.

7. Restricting fuel flow at the fuel pressure gage (at "B") causes fuel pressure to build above regulated pressure. With battery voltage applied to the pump "test" connector, pressure should rise above 325 kPa (47 psi) as the gage outlet hose is pinched.

NOTICE: Do not allow pressure to exceed 414 kPa (60 psi), as damage to the regulator may result.

8. Fuel pressure above 325 kPa (47 psi) may cause a rich condition and may set a DTC 45. Driveability conditions can include hard starting (followed by black smoke) and a strong sulphur smell in the exhaust.

9. This test determines if the high fuel pressure is due to a restricted fuel return line or a faulty fuel pressure regulator.

2.2L (VIN 4) ENGINE — SYSTEM DIAGNOSTIC CHARTS — 1993–94 CORSICA AND BERETTA

CHART A-7
(Page 2 of 3)
FUEL SYSTEM DIAGNOSIS
2.2L (VIN 4) "L" CARLINE

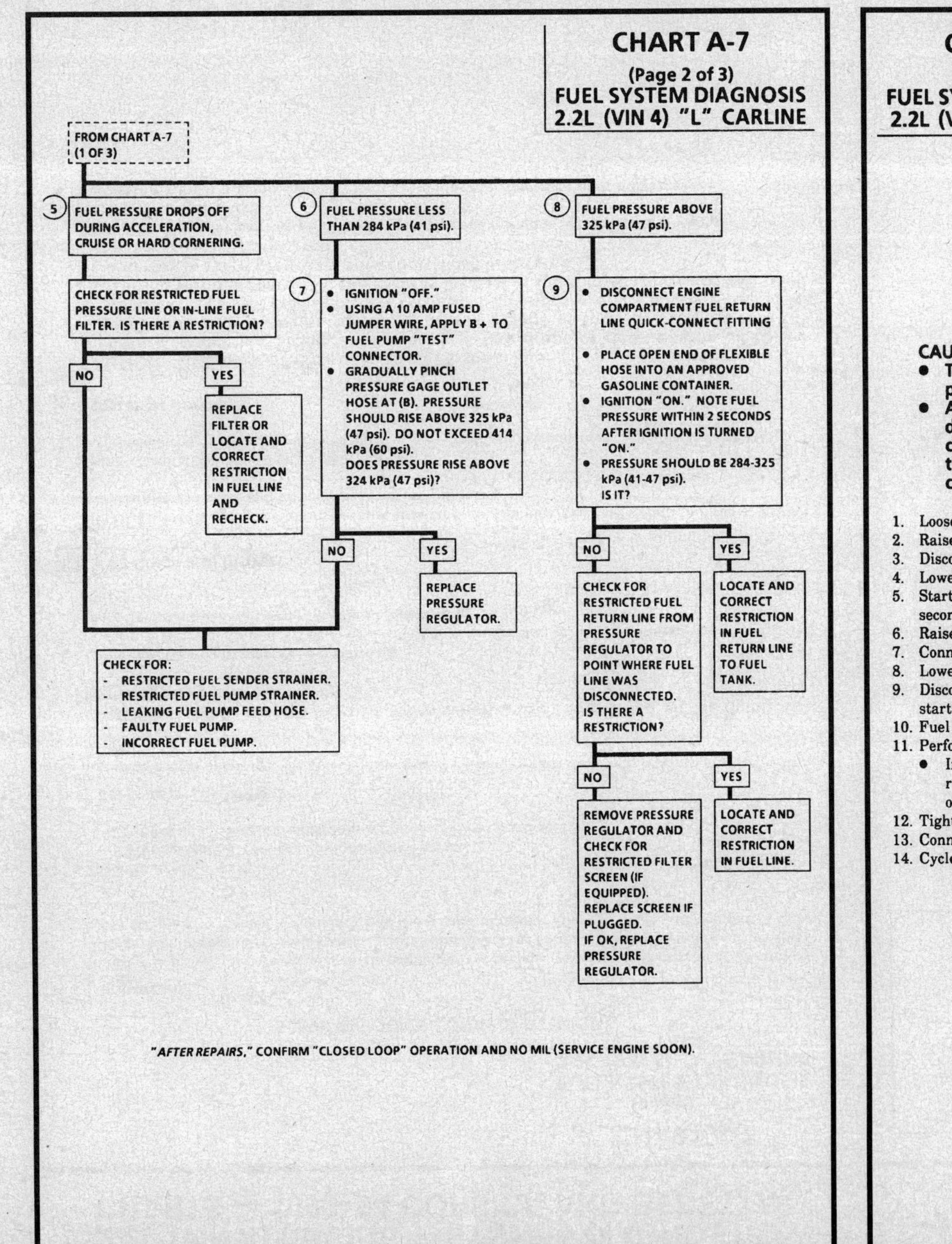

"AFTER REPAIRS," CONFIRM "CLOSED LOOP" OPERATION AND NO MIL (SERVICE ENGINE SOON).

2.2L (VIN 4) ENGINE — SYSTEM DIAGNOSTIC CHARTS — 1993–94 CORSICA AND BERETTA

CHART A-7
(Page 3 of 3)
FUEL SYSTEM DIAGNOSIS
2.2L (VIN 4) "L" CARLINE

FUEL SYSTEM PRESSURE RELIEF PROCEDURE

Engines Without Fuel Pressure Connection (Except J Car Applications)

(Must Be Performed Before Disconnecting Fuel Line Fittings)

CAUTION:
- To reduce the risk of fire and personal injury, it is necessary to relieve fuel system pressure before disconnecting fuel line fittings.
- After relieving system pressure, a small amount of fuel may be released when disconnecting fuel line fittings. In order to reduce the chance of personal injury, cover fuel line fittings with a shop towel before disconnecting, to catch any fuel that may leak out. Place the towel in an approved container when disconnect is completed.

1. Loosen fuel filler cap to relieve tank pressure.
2. Raise vehicle.
3. Disconnect fuel pump electrical connector.
4. Lower vehicle.
5. Start engine and run until fuel supply remaining in fuel pipes is consumed. Engage starter for 3.0 seconds to assure relief of any remaining pressure.
6. Raise vehicle.
7. Connect fuel pump electrical connector.
8. Lower vehicle.
9. Disconnect negative battery cable to avoid possible fuel discharge if an accidental attempt is made to start the engine.
10. Fuel line fittings are now safe for servicing.
11. Perform service required.
   - If performing fuel pressure check with a gage equipped with a bleed hose, fuel pressure can be relieved through the gage following test. Place bleed hose into approved gasoline container and open valve to bleed system pressure.
12. Tighten fuel filler cap.
13. Connect negative battery cable.
14. Cycle ignition "ON" and "OFF" twice, waiting ten seconds between cycles, then check for fuel leaks.

2.2L (VIN 4) ENGINE — SYSTEM DIAGNOSTIC CHARTS — 1993–94 CORSICA AND BERETTA

2.2L (VIN 4) ENGINE — DIAGNOSTIC TROUBLE CODE CHART — CORSICA AND BERETTA

CHART A-7

(Page 3 of 3)
FUEL SYSTEM DIAGNOSIS
2.2L (VIN 4) "L" CARLINE

SERVICING QUICK-CONNECT FITTINGS

Important
- In order to install fuel system diagnostic equipment on vehicles equipped with plastic quick-connect fittings, fuel line separator tools must be used to disconnect the fittings. Use of the separator tools will cause the plastic retainer to remain inside the female connector allowing diagnostic equipment to be connected.

Tools required:
  J 37088-A tool set, fuel line quick-connect separator;
  J 39504 tool set, fuel line quick-connect separator (restricted access).

Remove or Disconnect
1. Grasp both sides of fitting. Twist female connector 1/4 turn in each direction to loosen any dirt within fitting.

CAUTION: **Safety glasses must be worn when using compressed air, as flying dirt particles may cause eye injury.**

2. Using compressed air, blow dirt out of fitting.
3. Choose correct tool from J 37088-A or J 39504 tool set for size of fitting. Insert tool into female connector, then push/pull inward to release locking tabs.
4. Pull connection apart.

Clean and Inspect

NOTICE: If it is necessary to remove rust or burrs from fuel pipe, use emery cloth in a radial motion with the pipe end to prevent damage to O-ring sealing surface.

- Using a clean shop towel, wipe off male pipe end.
- Inspect both ends of fitting for dirt and burrs. Clean or replace components/assemblies as required.

Install or Connect

CAUTION: **To Reduce the Risk of Fire and Personal Injury:**
- **Before connecting fitting, always apply a few drops of clean engine oil to the male pipe end of engine fuel pipe, pressure gage adapter or fuel line shut-off adapter. This will ensure proper reconnection and prevent a possible fuel leak. (During normal operation, the O-rings located in the female connector will swell and may prevent proper reconnection if not lubricated.)**

1. Apply a few drops of clean engine oil to the male pipe end of engine fuel pipe, pressure gage adapter or fuel line shut-off adapter.
2. Push both sides of fitting together to cause the retaining tabs/fingers to snap into place.
3. Once installed, pull on both sides of fitting to make sure connection is secure.

DTC 13

OXYGEN SENSOR (O2S) CIRCUIT
(OPEN CIRCUIT)
2.2L (VIN 4) "L" CARLINE

Circuit Description:
The Engine Control Module (ECM) supplies a voltage of about .45 volt between terminals "D10" and "D11". (If measured with a 10 megohm digital voltmeter, this may read as low as .32 volt.)

When the Oxygen Sensor (O2S) reaches operating temperature, it varies this voltage from about .1 volt (exhaust is lean) to about .9 volt (exhaust is rich).

The sensor is like an open circuit and produces no voltage when it is below 316°C (600°F). An open sensor circuit, or cold sensor, causes "Open Loop" operation.

Test Description: Number(s) below refer to circled number(s) on the diagnostic chart.
1. DTC 13 will set under the following conditions:
   - Engine at normal operating temperature.
   - At least 2 minutes has elapsed since engine start-up.
   - O2S signal voltage is steady between .35 and .55 volt.
   - Throttle angle is above 5%.
   - All above conditions are met for 40.3 seconds or more.
   If the conditions for a DTC 13 exist, the system will not operate in "Closed Loop."
2. This test determines if the Oxygen Sensor (O2S) is the problem or if the ECM and wiring are at fault.

3. In doing this test, use only a 10 megohm digital voltmeter. This test checks the continuity of CKT 412 and CKT 413. If CKT 413 is open, the ECM voltage on CKT 412 will be over .6 volt (600 mV).

Diagnostic Aids:

Normal Tech 1 scan tool Oxygen Sensor (O2S) voltage varies between 100 mV to 999 mV (.1 and 1.0 volt) while in "Closed Loop." DTC 13 sets in about 40 seconds if sensor signal voltage remains between .35 and .55 volt.

Verify a clean, tight ground connection for CKT 413. Open CKT 412 or CKT 413 will result in a DTC 13. If DTC 13 is intermittent, refer to "Symptoms "

2.2L (VIN 4) ENGINE — DIAGNOSTIC TROUBLE CODE CHART — CORSICA AND BERETTA

DTC 13
OXYGEN SENSOR (O2S) CIRCUIT
(OPEN CIRCUIT)
2.2L (VIN 4) "L" CARLINE

1. ENGINE AT NORMAL OPERATING TEMPERATURE (ABOVE 80°C/176°F).
   - RUN ENGINE ABOVE 1200 RPM FOR TWO MINUTES.
     DOES SCAN TOOL INDICATE "CLOSED LOOP"?

NO →

YES → DTC 13 IS INTERMITTENT. IF NO ADDITIONAL DTC(S) WERE STORED, REFER TO "DIAGNOSTIC AIDS"

2. DISCONNECT O2S CONNECTOR.
   - JUMPER HARNESS CKT 412 (ECM SIDE) TO GROUND.
   - SCAN TOOL SHOULD DISPLAY OXYGEN SENSOR VOLTAGE BELOW .2 VOLT (200 mV) WITH ENGINE RUNNING.
     DOES IT?

NO →

YES → FAULTY O2S CONNECTION OR SENSOR.

3. REMOVE JUMPER.
   - IGNITION "ON," ENGINE "OFF."
   - CHECK VOLTAGE OF CKT 412 (ECM SIDE) AT O2S HARNESS CONNECTOR USING A DVM.

.3 - .6 VOLT (300 - 600 mV) → FAULTY ECM.*

OVER .6 VOLT (600 mV) → OPEN CKT 413 OR FAULTY CONNECTION OR FAULTY ECM.*

LESS THAN .3 VOLT (300 mV) → OPEN CKT 412 OR FAULTY ECM CONNECTION OR FAULTY ECM.*

IF ECM IS FAULTY AND MUST BE REPLACED, THE NEW ECM MUST BE PROGRAMMED. REFER TO ECM REPLACEMENT AND PROGRAMMING PROCEDURES

*"AFTER REPAIRS," REFER TO DTC CRITERIA   AND CONFIRM DTC DOES NOT RESET.

2.2L (VIN 4) ENGINE — DIAGNOSTIC TROUBLE CODE CHART — CORSICA AND BERETTA

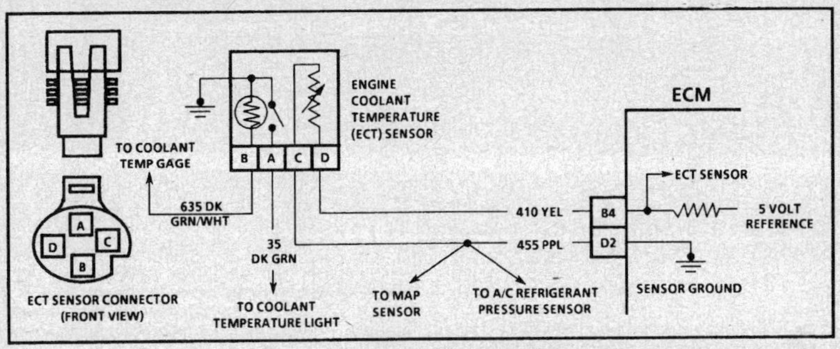

DTC 14
(Page 1 of 2)
ENGINE COOLANT TEMPERATURE (ECT) SENSOR CIRCUIT
(HIGH/LOW TEMPERATURE INDICATED)
2.2L (VIN 4) "L" CARLINE

Circuit Description:

The Engine Coolant Temperature (ECT) sensor utilizes a thermistor to control the signal voltage to the Engine Control Module (ECM). The ECM applies a reference voltage on CKT 410 to the sensor. When the engine is cold, the sensor (thermistor) resistance is high. The ECM will then sense a high signal voltage.

As the engine warms up, the sensor resistance decreases and the voltage drops. At normal engine operating temperature, the voltage will measure about 1.5 to 2.0 volts at ECM terminal "B4".

Coolant temperature is one of the inputs used to control the following:
- Cooling fan.
- Fuel delivery.
- Ignition Control (IC).
- Idle Air Control (IAC).
- Torque Converter Clutch (TCC).

A separate thermistor within the ECT sensor provides a signal to the coolant temperature gage located in the instrument panel.

Test Description: Number(s) below refer to circled number(s) on the diagnostic chart.

1. DTC 14 will set if:
   - The engine has been running for 2 minutes or longer.
   - Signal voltage indicates a coolant temperature below -34°C (-30°F).
     OR
   - Signal voltage indicates a coolant temperature above 143°C (284°F) for 3 seconds or longer.
2. If the ECM recognizes the grounded circuit (low voltage) and displays a high temperature, the ECM and wiring are OK.
3. This test will determine if there is a wiring problem or a faulty ECM. If CKT 455 is open, there may also be other DTC(s) stored.

After the engine is started, the temperature should rise steadily to about 90°C (194°F), then stabilize when the thermostat opens.

If the engine has been allowed to cool to an ambient temperature (overnight), coolant temperature and Intake Air Temperature (IAT) may be checked with a scan tool and should read close to each other.

When a DTC 14 is set, the ECM will turn "ON" the engine cooling fan.

- If DTC 14 is intermittent or a "Hard Start" symptom is present, check engine coolant temperature with a scan tool on a cool engine. Temperature displayed should be within 5 degrees of the ambient. If not, check the ECT sensor using the "Diagnostic Aid" If sensor is OK, check connections.

Diagnostic Aids:

A scan tool should display engine coolant temperature in degrees Celsius.

2.2L (VIN 4) ENGINE — DIAGNOSTIC TROUBLE CODE CHART — CORSICA AND BERETTA

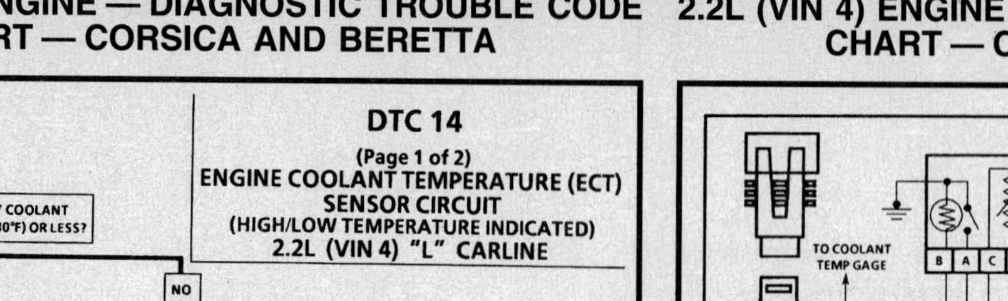

DIAGNOSTIC AID		
ECT SENSOR		
TEMPERATURE VS. RESISTANCE VALUES (APPROXIMATE)		
°C	°F	OHMS
100	212	177
90	194	241
80	176	332
70	158	467
60	140	667
50	122	973
45	113	1188
40	104	1459
35	95	1802
30	86	2238
25	77	2796
20	68	3520
15	59	4450
10	50	5670
5	41	7280
0	32	9420
-5	23	12300
-10	14	16180
-15	5	21450
-20	-4	28680
-30	-22	52700
-40	-40	100700

2.2L (VIN 4) ENGINE — DIAGNOSTIC TROUBLE CODE CHART — CORSICA AND BERETTA

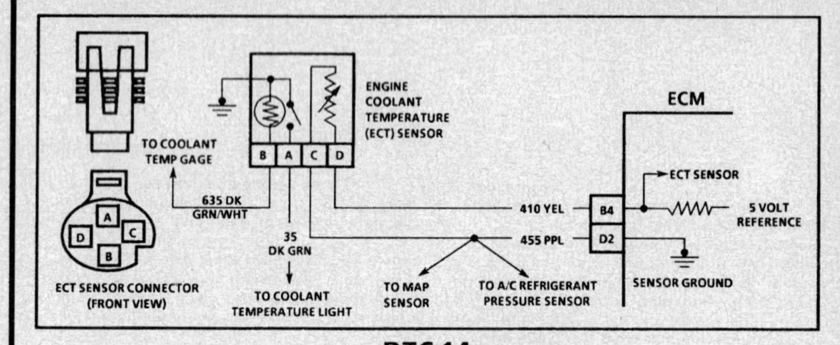

DTC 14
(Page 2 of 2)
ENGINE COOLANT TEMPERATURE (ECT) SENSOR CIRCUIT
(HIGH/LOW TEMPERATURE INDICATED)
2.2L (VIN 4) "L" CARLINE

Circuit Description:

The Engine Coolant Temperature (ECT) sensor utilizes a thermistor to control the signal voltage to the Engine Control Module (ECM). The ECM applies a reference voltage on CKT 410 to the sensor. When the engine is cold, the sensor (thermistor) resistance is high. The ECM will then sense a high signal voltage.

As the engine warms up, the sensor resistance decreases and the voltage drops. At normal engine operating temperature, the voltage will measure about 1.5 to 2.0 volts at ECM terminal "B4".

Coolant temperature is one of the inputs used to control the following:
- Fuel delivery.
- Ignition Control (IC).
- Cooling fan.
- Torque Converter Clutch (TCC).
- Idle Air Control (IAC).

Test Description: Number(s) below refer to circled number(s) on the diagnostic chart.
1. Checks to see if a DTC was set as a result of hard failure or intermittent condition.
2. If the ECM recognizes the open circuit (high voltage), and displays a low temperature, the ECM and wiring are OK.

Diagnostic Aids:

A scan tool should display engine coolant temperature in degrees Celsius.

After the engine is started, the temperature should rise steadily to about 90°C (194°F), then stabilize when the thermostat opens.

If the engine has been allowed to cool to an ambient temperature (overnight), coolant temperature and Intake Air Temperature (IAT) may be checked with a scan tool and should read close to each other.

When a DTC 14 is set, the ECM will turn "ON" the engine cooling fan.
- If DTC 14 is intermittent or a "Hard Start" symptom is present, check engine coolant temperature with a scan tool on a cool engine. Temperature displayed should be within 5 degrees of the ambient. If not, check the ECT sensor using the "Diagnostic Aid"   If sensor is OK, check connections.

2.2L (VIN 4) ENGINE — DIAGNOSTIC TROUBLE CODE CHART — CORSICA AND BERETTA

DTC 14

(Page 2 of 2)
ENGINE COOLANT TEMPERATURE (ECT)
SENSOR CIRCUIT
(HIGH/LOW TEMPERATURE INDICATED)
2.2L (VIN 4) "L" CARLINE

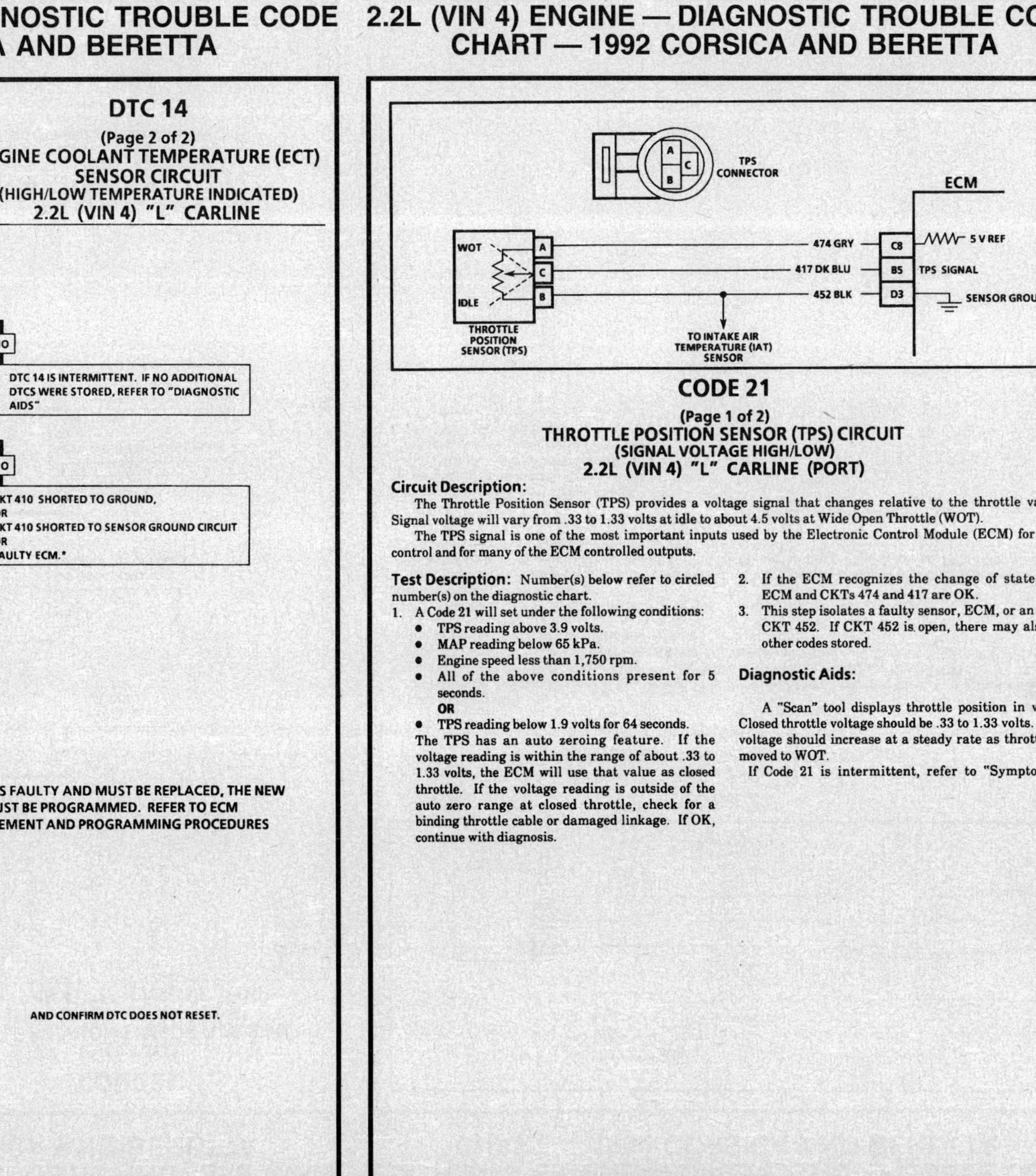

DIAGNOSTIC AID

ECT SENSOR		
TEMPERATURE VS. RESISTANCE VALUES (APPROXIMATE)		
°C	°F	OHMS
100	212	177
90	194	241
80	176	332
70	158	467
60	140	667
50	122	973
45	113	1188
40	104	1459
35	95	1802
30	86	2238
25	77	2796
20	68	3520
15	59	4450
10	50	5670
5	41	7280
0	32	9420
-5	23	12300
-10	14	16180
-15	5	21450
-20	-4	28680
-30	-22	52700
-40	-40	100700

* IF ECM IS FAULTY AND MUST BE REPLACED, THE NEW ECM MUST BE PROGRAMMED. REFER TO ECM REPLACEMENT AND PROGRAMMING PROCEDURES

AFTER REPAIRS," REFER TO DTC CRITERIA

AND CONFIRM DTC DOES NOT RESET.

2.2L (VIN 4) ENGINE — DIAGNOSTIC TROUBLE CODE CHART — 1992 CORSICA AND BERETTA

CODE 21

(Page 1 of 2)
THROTTLE POSITION SENSOR (TPS) CIRCUIT
(SIGNAL VOLTAGE HIGH/LOW)
2.2L (VIN 4) "L" CARLINE (PORT)

Circuit Description:

The Throttle Position Sensor (TPS) provides a voltage signal that changes relative to the throttle valve. Signal voltage will vary from .33 to 1.33 volts at idle to about 4.5 volts at Wide Open Throttle (WOT).

The TPS signal is one of the most important inputs used by the Electronic Control Module (ECM) for fuel control and for many of the ECM controlled outputs.

Test Description: Number(s) below refer to circled number(s) on the diagnostic chart.

1. A Code 21 will set under the following conditions:
   - TPS reading above 3.9 volts.
   - MAP reading below 65 kPa.
   - Engine speed less than 1,750 rpm.
   - All of the above conditions present for 5 seconds.

   OR
   - TPS reading below 1.9 volts for 64 seconds.

   The TPS has an auto zeroing feature. If the voltage reading is within the range of about .33 to 1.33 volts, the ECM will use that value as closed throttle. If the voltage reading is outside of the auto zero range at closed throttle, check for a binding throttle cable or damaged linkage. If OK, continue with diagnosis.

2. If the ECM recognizes the change of state, the ECM and CKTs 474 and 417 are OK.

3. This step isolates a faulty sensor, ECM, or an open CKT 452. If CKT 452 is open, there may also be other codes stored.

Diagnostic Aids:

A "Scan" tool displays throttle position in volts. Closed throttle voltage should be .33 to 1.33 volts. TPS voltage should increase at a steady rate as throttle is moved to WOT.

If Code 21 is intermittent, refer to "Symptoms,"

CODE 21

(Page 1 of 2)
THROTTLE POSITION SENSOR (TPS) CIRCUIT
(SIGNAL VOLTAGE HIGH/LOW)
2.2L (VIN 4) "L" CARLINE (PORT)

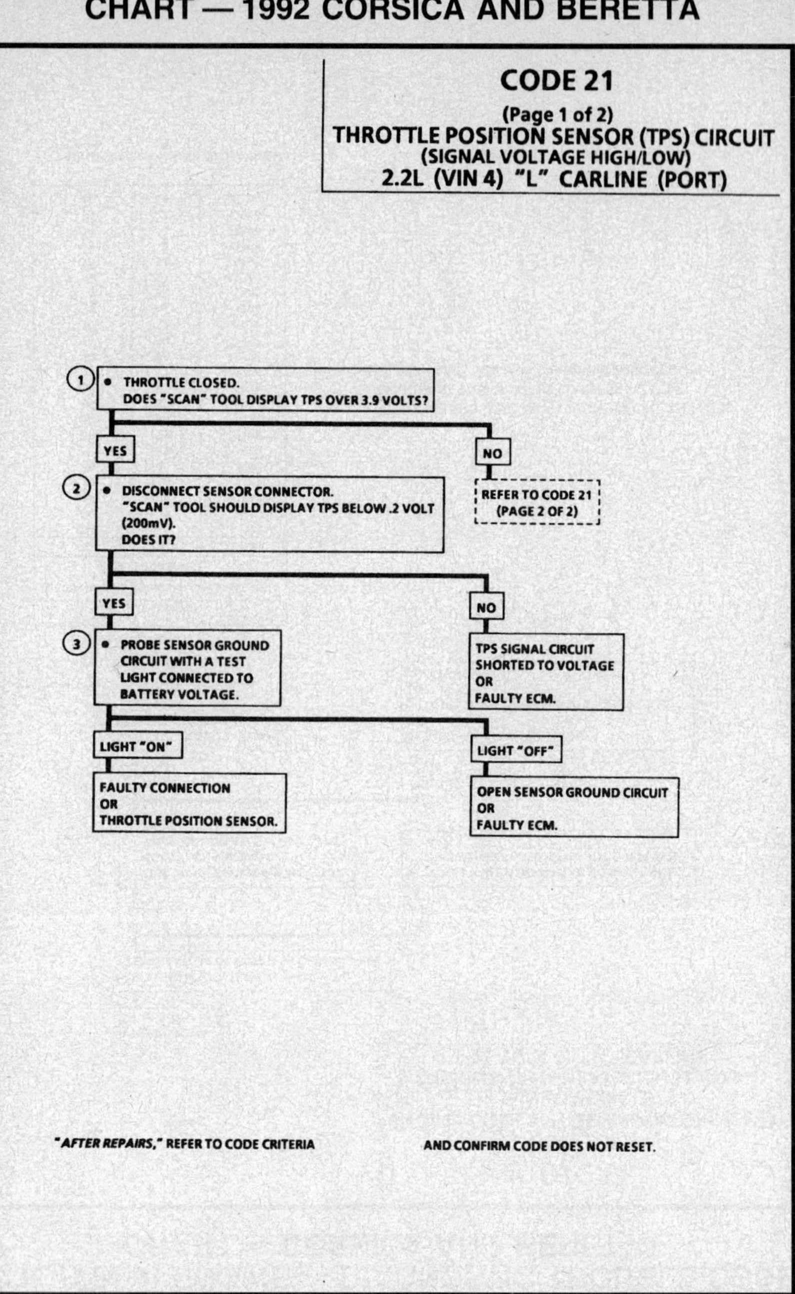

"*AFTER REPAIRS,*" REFER TO CODE CRITERIA AND CONFIRM CODE DOES NOT RESET.

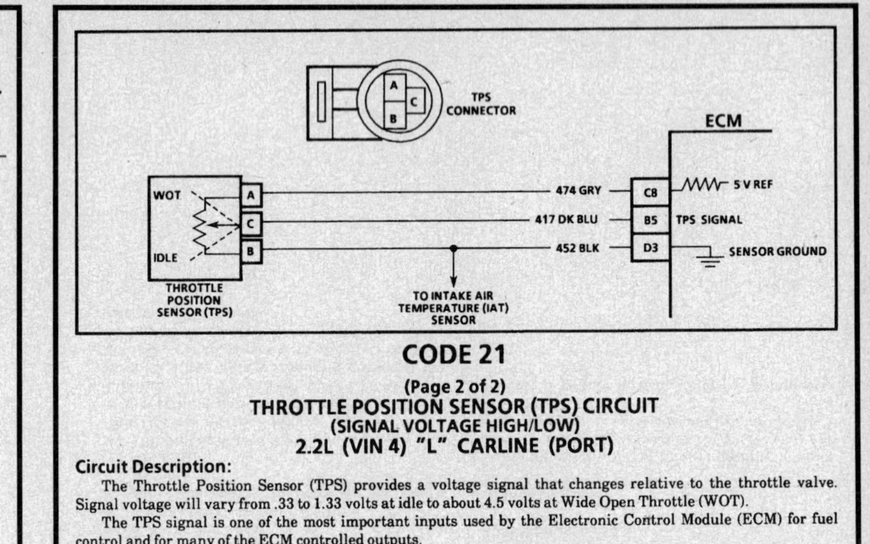

CODE 21

(Page 2 of 2)
THROTTLE POSITION SENSOR (TPS) CIRCUIT
(SIGNAL VOLTAGE HIGH/LOW)
2.2L (VIN 4) "L" CARLINE (PORT)

Circuit Description:

The Throttle Position Sensor (TPS) provides a voltage signal that changes relative to the throttle valve. Signal voltage will vary from .33 to 1.33 volts at idle to about 4.5 volts at Wide Open Throttle (WOT).

The TPS signal is one of the most important inputs used by the Electronic Control Module (ECM) for fuel control and for many of the ECM controlled outputs.

Test Description: Number(s) below refer to circled number(s) on the diagnostic chart.

1. This step checks to see if Code 21 is the result of a hard failure or an intermittent condition.
2. This step simulates conditions for a Code 21. If the "Scan" tool displays over 4 volts, the ECM and wiring are OK.
3. This simulates a high signal voltage to check for an open or short in CKT 417. The Tech 1 "Scan" tool will not read battery voltage, but the ECM should recognize the signal on CKT 417.
4. See "Fuel Metering System," TPS replacement procedures.

Diagnostic Aids:

A "Scan" tool displays throttle position in volts. Closed throttle voltage should be .33 to 1.33 volts. TPS voltage should increase at a steady rate as throttle is moved to WOT.

If Code 21 is intermittent, refer to "Symptoms,"

2.2L (VIN 4) ENGINE — DIAGNOSTIC TROUBLE CODE CHART — 1992 CORSICA AND BERETTA

CODE 21
(Page 2 of 2)
THROTTLE POSITION SENSOR (TPS) CIRCUIT
(SIGNAL VOLTAGE HIGH/LOW)
2.2L (VIN 4) "L" CARLINE (PORT)

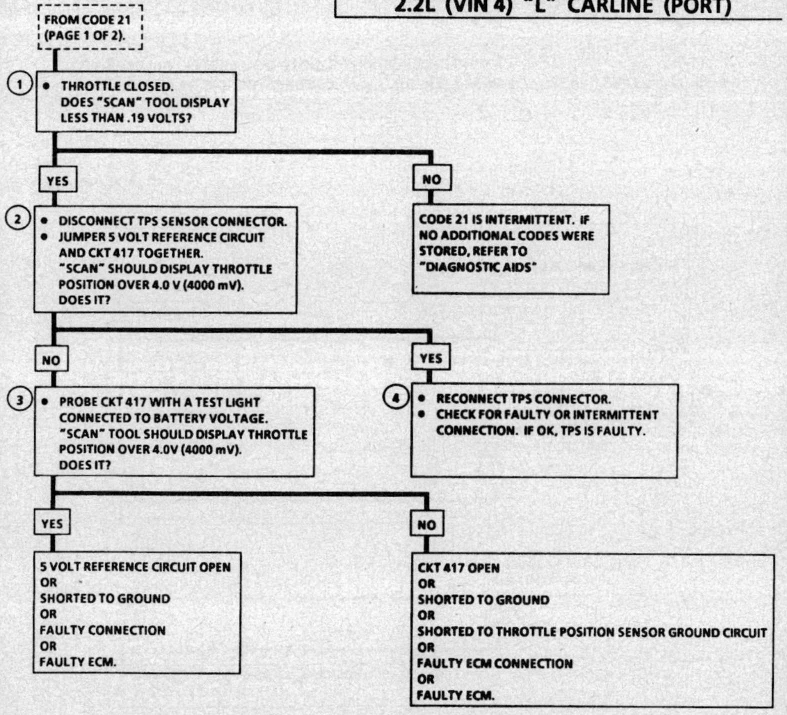

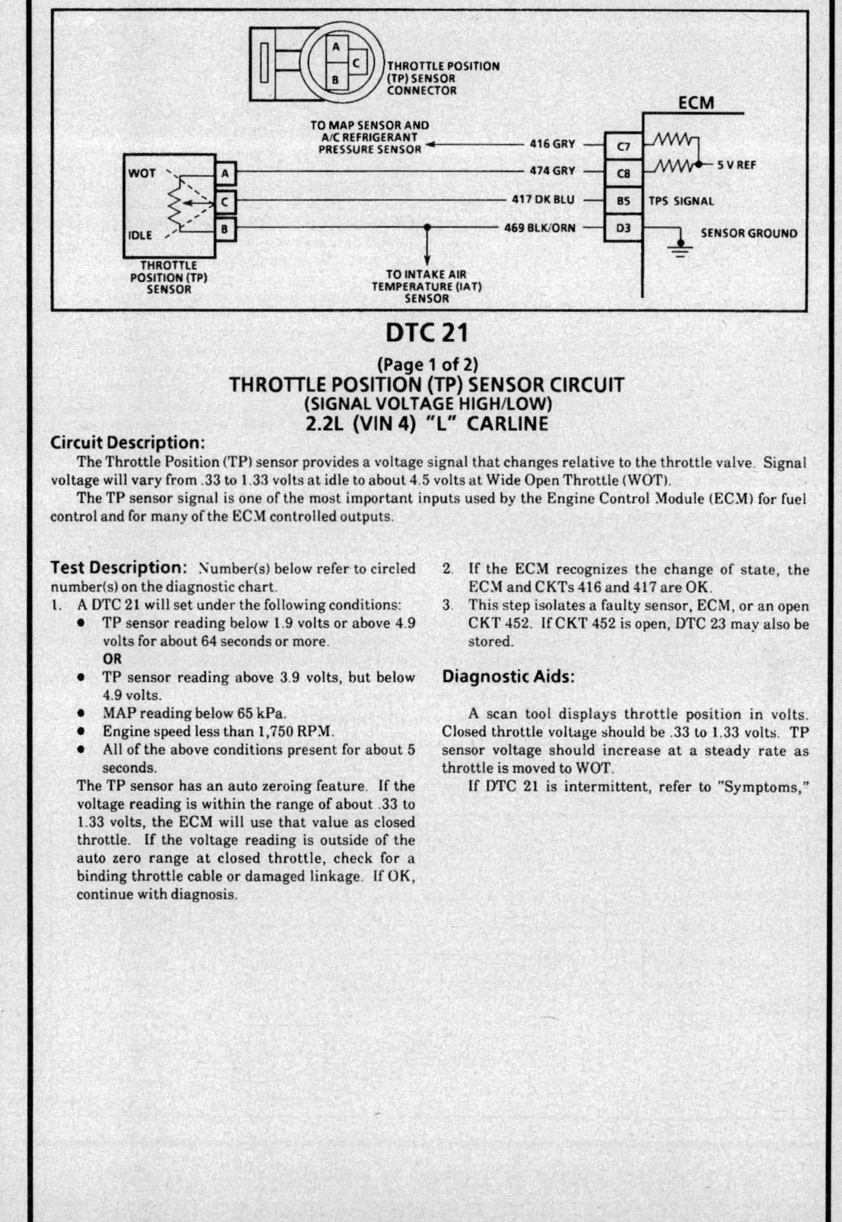

2.2L (VIN 4) ENGINE — DIAGNOSTIC TROUBLE CODE CHART — 1993–94 CORSICA AND BERETTA

DTC 21
(Page 1 of 2)
THROTTLE POSITION (TP) SENSOR CIRCUIT
(SIGNAL VOLTAGE HIGH/LOW)
2.2L (VIN 4) "L" CARLINE

Circuit Description:

The Throttle Position (TP) sensor provides a voltage signal that changes relative to the throttle valve. Signal voltage will vary from .33 to 1.33 volts at idle to about 4.5 volts at Wide Open Throttle (WOT).

The TP sensor signal is one of the most important inputs used by the Engine Control Module (ECM) for fuel control and for many of the ECM controlled outputs.

Test Description: Number(s) below refer to circled number(s) on the diagnostic chart.

1. A DTC 21 will set under the following conditions:
   - TP sensor reading below 1.9 volts or above 4.9 volts for about 64 seconds or more.
   OR
   - TP sensor reading above 3.9 volts, but below 4.9 volts.
   - MAP reading below 65 kPa.
   - Engine speed less than 1,750 RPM.
   - All of the above conditions present for about 5 seconds.

   The TP sensor has an auto zeroing feature. If the voltage reading is within the range of about .33 to 1.33 volts, the ECM will use that value as closed throttle. If the voltage reading is outside of the auto zero range at closed throttle, check for a binding throttle cable or damaged linkage. If OK, continue with diagnosis.

2. If the ECM recognizes the change of state, the ECM and CKTs 416 and 417 are OK.

3. This step isolates a faulty sensor, ECM, or an open CKT 452. If CKT 452 is open, DTC 23 may also be stored.

Diagnostic Aids:

A scan tool displays throttle position in volts. Closed throttle voltage should be .33 to 1.33 volts. TP sensor voltage should increase at a steady rate as throttle is moved to WOT.

If DTC 21 is intermittent, refer to "Symptoms,"

DTC 21

(Page 1 of 2)
THROTTLE POSITION (TP) SENSOR CIRCUIT
(SIGNAL VOLTAGE HIGH/LOW)
2.2L (VIN 4) "L" CARLINE

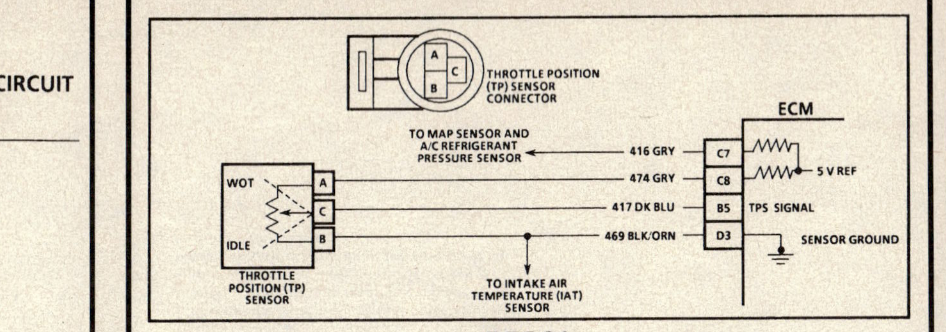

* IF ECM IS FAULTY AND MUST BE REPLACED, THE NEW ECM MUST BE PROGRAMMED. REFER TO ECM REPLACEMENT AND PROGRAMMING PROCEDURES

"AFTER REPAIRS," REFER TO DTC CRITERIA   AND CONFIRM DTC DOES NOT RESET.

DTC 21

(Page 2 of 2)
THROTTLE POSITION (TP) SENSOR CIRCUIT
(SIGNAL VOLTAGE HIGH/LOW)
2.2L (VIN 4) "L" CARLINE

Circuit Description:

The Throttle Position (TP) sensor provides a voltage signal that changes relative to the throttle valve. Signal voltage will vary from .33 to 1.33 volts at idle to about 4.5 volts at Wide Open Throttle (WOT).

The TP sensor signal is one of the most important inputs used by the Engine Control Module (ECM) for fuel control and for many of the ECM controlled outputs.

Test Description: Number(s) below refer to circled number(s) on the diagnostic chart.

1. This step checks to see if DTC 21 is the result of a hard failure or an intermittent condition.
2. This step simulates conditions for a DTC 21. If the scan tool displays over 4 volts, the ECM and wiring are OK.
3. This simulates a high signal voltage to check for an open or short in CKT 417. The Tech 1 scan tool will not read battery voltage, but the ECM should recognize the signal on CKT 417.
4. Refer to Fuel Metering System, for TP sensor replacement procedures.
5. The 5 volt reference for the MAP sensor and the A/C refrigerant pressure sensor connect to the same 5 volt power supply in the ECM. If the 5 volt reference circuit to the MAP sensor and A/C refrigerant pressure sensor has a short to ground, the Throttle Position (TP) sensor 5 volt reference circuit will also have a short to ground because they connect to the same power supply in the ECM. A short to ground on either 5 volt reference circuit does not damage the ECM 5 volt power supply. When the short is repaired, the ECM 5 volt supply will return to normal.

Diagnostic Aids:

A scan tool displays throttle position in volts. Closed throttle voltage should be .33 to 1.33 volts. TP sensor voltage should increase at a steady rate as throttle is moved to WOT.

If DTC 21 is intermittent, refer to "Symptoms,"

2.2L (VIN 4) ENGINE — DIAGNOSTIC TROUBLE CODE CHART — 1993–94 CORSICA AND BERETTA

DTC 21

(Page 2 of 2)
THROTTLE POSITION (TP) SENSOR CIRCUIT
(SIGNAL VOLTAGE HIGH/LOW)
2.2L (VIN 4) "L" CARLINE

FROM DTC 21
(PAGE 1 OF 2).

1. • THROTTLE CLOSED.
   DOES SCAN TOOL DISPLAY
   LESS THAN .2 VOLTS?

   YES

2. • DISCONNECT TP SENSOR SENSOR CONNECTOR.
   • JUMPER 5 VOLT REFERENCE CIRCUIT AND CKT 417 TOGETHER.
   SCAN SHOULD DISPLAY THROTTLE POSITION OVER 4.0 V (4000 mV).
   DOES IT?

   NO

   DTC 21 IS INTERMITTENT. IF NO ADDITIONAL DTCS WERE STORED, REFER TO "DIAGNOSTIC AIDS"

3. • PROBE CKT 417 WITH A TEST LIGHT CONNECTED TO BATTERY VOLTAGE.
   SCAN TOOL SHOULD DISPLAY THROTTLE POSITION OVER 4.0V (4000 mV).
   DOES IT?

   YES

4. • RECONNECT TP SENSOR CONNECTOR.
   • CHECK FOR FAULTY OR INTERMITTENT CONNECTION. IF OK, TP SENSOR IS FAULTY.

   YES

5. 5 VOLT REFERENCE CIRCUIT OPEN
   OR
   SHORTED TO GROUND
   OR
   FAULTY CONNECTION
   OR
   FAULTY ECM.*

   NO

   CKT 417 OPEN
   OR
   SHORTED TO GROUND
   OR
   SHORTED TO SENSOR GROUND CIRCUIT
   OR
   FAULTY ECM CONNECTION
   OR
   FAULTY ECM.*

* IF ECM IS FAULTY AND MUST BE REPLACED, THE NEW ECM MUST BE PROGRAMMED. REFER TO ECM REPLACEMENT AND PROGRAMMING PROCEDURES

"AFTER REPAIRS," REFER TO DTC CRITERIA   AND CONFIRM DTC DOES NOT RESET.

2.2L (VIN 4) ENGINE — DIAGNOSTIC TROUBLE CODE CHART — CORSICA AND BERETTA

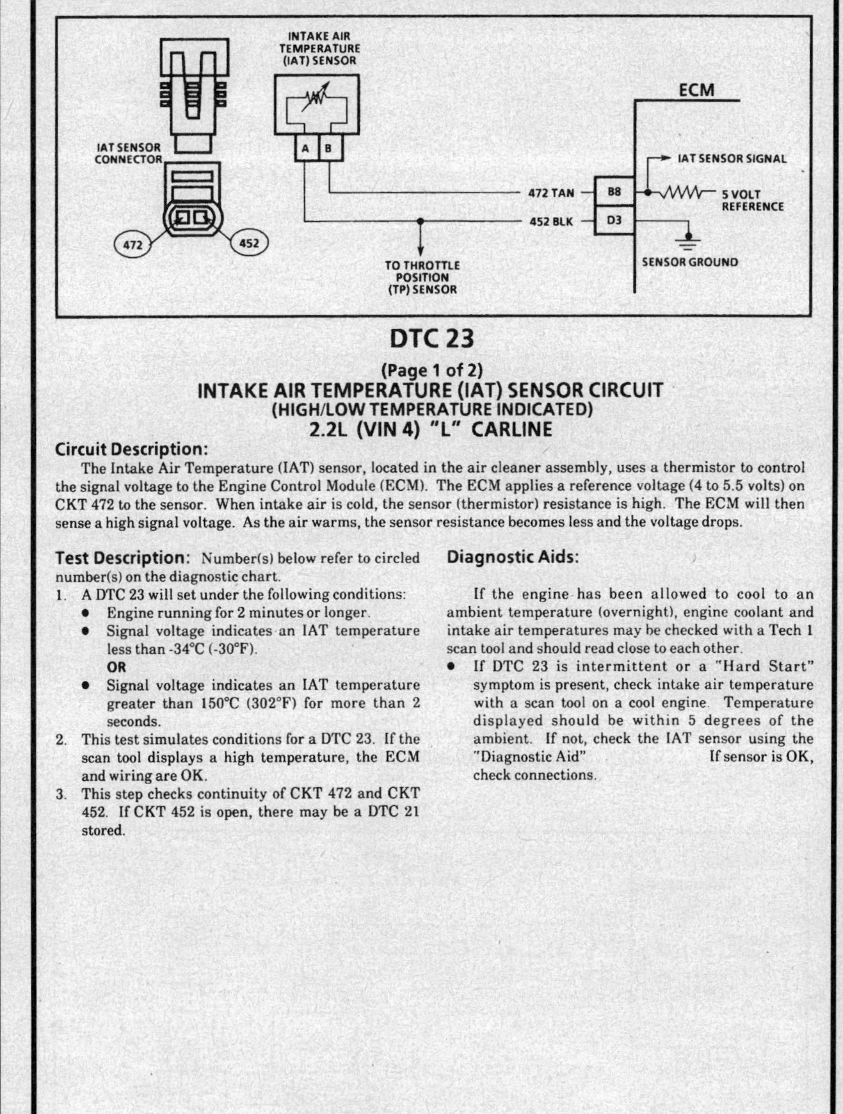

DTC 23

(Page 1 of 2)
INTAKE AIR TEMPERATURE (IAT) SENSOR CIRCUIT
(HIGH/LOW TEMPERATURE INDICATED)
2.2L (VIN 4) "L" CARLINE

Circuit Description:

The Intake Air Temperature (IAT) sensor, located in the air cleaner assembly, uses a thermistor to control the signal voltage to the Engine Control Module (ECM). The ECM applies a reference voltage (4 to 5.5 volts) on CKT 472 to the sensor. When intake air is cold, the sensor (thermistor) resistance is high. The ECM will then sense a high signal voltage. As the air warms, the sensor resistance becomes less and the voltage drops.

Test Description: Number(s) below refer to circled number(s) on the diagnostic chart.

1. A DTC 23 will set under the following conditions:
   - Engine running for 2 minutes or longer.
   - Signal voltage indicates an IAT temperature less than -34°C (-30°F).
   OR
   - Signal voltage indicates an IAT temperature greater than 150°C (302°F) for more than 2 seconds.
2. This test simulates conditions for a DTC 23. If the scan tool displays a high temperature, the ECM and wiring are OK.
3. This step checks continuity of CKT 472 and CKT 452. If CKT 452 is open, there may be a DTC 21 stored.

Diagnostic Aids:

If the engine has been allowed to cool to an ambient temperature (overnight), engine coolant and intake air temperatures may be checked with a Tech 1 scan tool and should read close to each other.
- If DTC 23 is intermittent or a "Hard Start" symptom is present, check intake air temperature with a scan tool on a cool engine. Temperature displayed should be within 5 degrees of the ambient. If not, check the IAT sensor using the "Diagnostic Aid"   If sensor is OK, check connections.

DTC 23

(Page 1 of 2)
INTAKE AIR TEMPERATURE (IAT) SENSOR CIRCUIT
(HIGH/LOW TEMPERATURE INDICATED)
2.2L (VIN 4) "L" CARLINE

(1) • DOES SCAN TOOL DISPLAY IAT -34°C (-30°F) OR COLDER?

YES

NO → REFER TO DTC 23 (PAGE 2 OF 2)

(2) • DISCONNECT IAT SENSOR CONNECTOR.
• JUMPER HARNESS TERMINALS TOGETHER.
• SCAN TOOL SHOULD DISPLAY TEMPERATURE OVER 150°C (302°F). DOES IT?

YES → FAULTY CONNECTION OR SENSOR.

NO

(3) • JUMPER CKT 472 TO GROUND.
• SCAN TOOL SHOULD DISPLAY TEMPERATURE OVER 150°C (302°F). DOES IT?

YES → OPEN SENSOR GROUND CIRCUIT, OR FAULTY CONNECTION OR FAULTY ECM.*

NO → OPEN CKT 472, OR FAULTY CONNECTION OR FAULTY ECM.*

DIAGNOSTIC AID

IAT SENSOR		
TEMPERATURE VS. RESISTANCE VALUES (APPROXIMATE)		
°C	°F	OHMS
100	210	185
70	160	450
38	100	1,800
20	70	3,400
4	40	7,500
-7	20	13,500
-18	0	25,000
-40	-40	100,700

* IF ECM IS FAULTY AND MUST BE REPLACED, THE NEW ECM MUST BE PROGRAMMED. REFER TO ECM REPLACEMENT AND PROGRAMMING PROCEDURES

"AFTER REPAIRS," REFER TO DTC CRITERIA   AND CONFIRM DTC DOES NOT RESET.

INTAKE AIR TEMPERATURE (IAT) SENSOR
ECM
IAT SENSOR CONNECTOR
A B
472 TAN
452 BLK
88
D3
IAT SENSOR SIGNAL
5 VOLT REFERENCE
SENSOR GROUND
472
452
TO THROTTLE POSITION (TP) SENSOR

DTC 23

(Page 2 of 2)
INTAKE AIR TEMPERATURE (IAT) SENSOR CIRCUIT
(HIGH/LOW TEMPERATURE INDICATED)
2.2L (VIN 4) "L" CARLINE

Circuit Description:

The Intake Air Temperature (IAT) sensor, located in the rear air duct assembly, uses a thermistor to control the signal voltage to the Engine Control Module (ECM). The ECM applies a reference voltage (4 to 5.5 volts) on CKT 472 to the sensor. When intake air is cold, the sensor (thermistor) resistance is high. Therefore, the ECM will sense a high signal voltage. As the air warms, the sensor resistance becomes less and the voltage drops.

Test Description: Number(s) below refer to circled number(s) on the diagnostic chart.

1. This step determines if DTC 23 is the result of a hard failure or an intermittent condition.
2. If the ECM recognizes the open circuit (high voltage) and displays a low temperature, the ECM and wiring are OK.

Diagnostic Aids:

If the engine has been allowed to cool to an ambient temperature (overnight), engine coolant and intake air temperatures may be checked with a scan tool and should read close to each other.

• If DTC 23 is intermittent or a "Hard Start" symptom is present, check intake air temperature with a scan tool on a cool engine. Temperature displayed should be within 5 degrees of the ambient. If not, check the IAT sensor using the "Diagnostic Aid"   If sensor is OK, check connections.

2.2L (VIN 4) ENGINE — DIAGNOSTIC TROUBLE CODE CHART — CORSICA AND BERETTA

DTC 23
(Page 2 of 2)
INTAKE AIR TEMPERATURE (IAT) SENSOR CIRCUIT
(HIGH/LOW TEMPERATURE INDICATED)
2.2L (VIN 4) "L" CARLINE

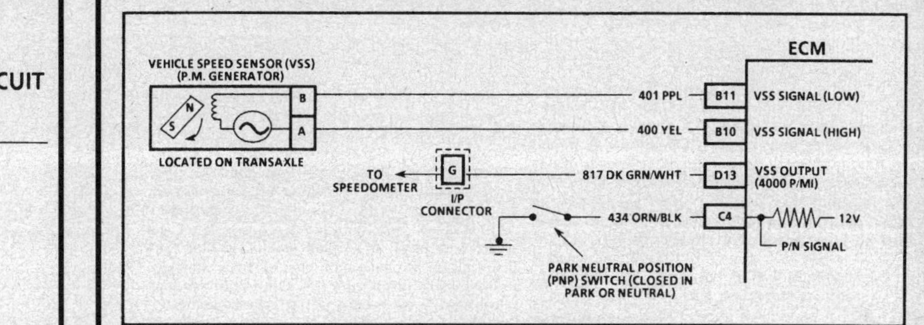

DIAGNOSTIC AID

IAT SENSOR		
TEMPERATURE VS. RESISTANCE VALUES (APPROXIMATE)		
°C	°F	OHMS
100	210	185
70	160	450
38	100	1,800
20	70	3,400
4	40	7,500
-7	20	13,500
-18	0	25,000
-40	-40	100,700

* IF ECM IS FAULTY AND MUST BE REPLACED, THE NEW ECM MUST BE PROGRAMMED. REFER TO ECM REPLACEMENT AND PROGRAMMING PROCEDURES

"AFTER REPAIRS," REFER TO DTC CRITERIA

AND CONFIRM DTC DOES NOT RESET.

2.2L (VIN 4) ENGINE — DIAGNOSTIC TROUBLE CODE CHART — CORSICA AND BERETTA

DTC 24
VEHICLE SPEED SENSOR (VSS) CIRCUIT
2.2L (VIN 4) "L" CARLINE

Circuit Description:

Vehicle speed information is provided to the Engine Control Module (ECM) by the Vehicle Speed Sensor (VSS), which is a Permanent Magnet (PM) generator, and it is mounted in the transaxle. The PM generator produces a pulsing voltage, whenever vehicle speed is over about 3 mph. The AC voltage level and the number of pulses increases with vehicle speed. The ECM then converts the pulsing voltage to mph, which is used for calculations, and the mph can be displayed with a scan tool.

The function of VSS buffer used in past model years has been incorporated into the ECM. The ECM then supplies the necessary signal to the instrument panel (4000 pulses per mile) for operating the speedometer and the odometer.

Test Description: Number(s) below refer to circled number(s) on the diagnostic chart.

1. DTC 24 will set if vehicle speed equals 0 mph when:
   - Engine speed is between about 1400 and 3600 RPM.
   - Low load condition (low MAP voltage), high manifold vacuum).
   - Transaxle not in park or neutral.
   - All above conditions are met for 4 seconds.
   These conditions are met during a road load deceleration.
   Disregard a DTC 24 that sets when the drive wheels are not turning. This can be caused by a faulty PNP switch circuit.
   The PM generator only produces a signal if the drive wheels are turning greater than 3 mph.
2. At this point, the ECM is not sending vehicle speed data to the scan tool. If the scan tool is connected and functioning properly, the ECM is at fault.*

Diagnostic Aids:

Scan tool should indicate a vehicle speed whenever the drive wheels are turning greater than 3 mph.

A problem in CKT 817 will not affect the VSS input or the readings on a scan tool.

Check CKTs 400 and 401 for proper connections to be sure they are clean and tight and the harness is routed correctly.

- A faulty or misadjusted Park/Neutral Position (PNP) switch can result in a false DTC 24. Use a scan tool and check for the proper signal while in a drive range. Refer to CHART C-1A for the PNP switch check.

2.2L (VIN 4) ENGINE — DIAGNOSTIC TROUBLE CODE CHART — CORSICA AND BERETTA

DTC 24
VEHICLE SPEED SENSOR (VSS) CIRCUIT
2.2L (VIN 4) "L" CARLINE

DISREGARD DTC 24 IF SET WHILE DRIVE WHEELS ARE NOT TURNING.

(1) • RAISE DRIVE WHEELS.

NOTICE: DO NOT PERFORM THIS TEST WITHOUT SUPPORTING THE LOWER CONTROL ARMS SO THAT THE DRIVE AXLES ARE IN A NORMAL HORIZONTAL POSITION. RUNNING THE VEHICLE IN GEAR WITH THE WHEELS HANGING DOWN AT FULL TRAVEL MAY DAMAGE THE DRIVE AXLES.

• WITH ENGINE IDLING IN GEAR, SCAN TOOL SHOULD DISPLAY VEHICLE SPEED ABOVE 0 MPH. DOES IT?

NO → DOES SPEEDOMETER WORK PROPERLY?

YES → DTC 24 IS INTERMITTENT. IF NO ADDITIONAL DTC(S) WERE STORED, REFER TO "DIAGNOSTIC AIDS"

NO →
• IGNITION "OFF."
• DISCONNECT VSS HARNESS CONNECTOR AT TRANSAXLE.
• CONNECT SIGNAL GENERATOR TESTER J 33431-8 OR EQUIVALENT TO VSS HARNESS CONNECTOR.
• IGNITION "ON," TOOL "ON" AND SET TO GENERATE A VSS SIGNAL.
• SCAN TOOL SHOULD DISPLAY VEHICLE SPEED ABOVE 0 MPH. DOES IT?

YES → (2) REPLACE ECM. *

NO →
CKT 400
OR
401 OPEN, SHORTED TO GROUND, SHORTED TOGETHER, FAULTY CONNECTIONS
OR
FAULTY ECM. *

YES → REPLACE VEHICLE SPEED SENSOR.

* IF ECM IS FAULTY AND MUST BE REPLACED, THE NEW ECM MUST BE PROGRAMMED. REFER TO ECM REPLACEMENT AND PROGRAMMING PROCEDURES

"AFTER REPAIRS," REFER TO DTC CRITERIA   AND CONFIRM DTC DOES NOT RESET.

2.2L (VIN 4) ENGINE — DIAGNOSTIC TROUBLE CODE CHART — CORSICA AND BERETTA

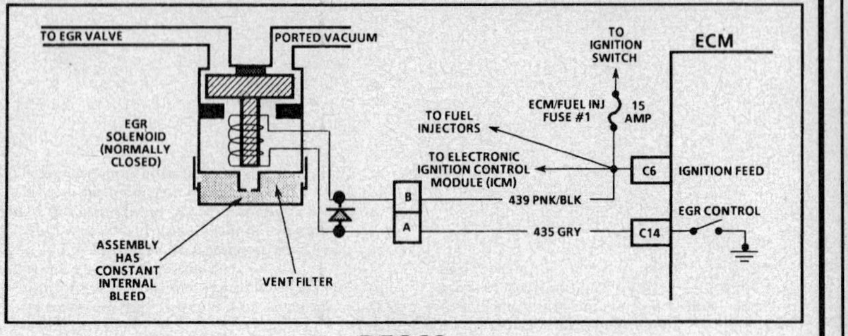

DTC 32
(Page 1 of 2)
EXHAUST GAS RECIRCULATION (EGR) SYSTEM FAILURE
2.2L (VIN 4) "L" CARLINE

Circuit Description:

The Exhaust Gas Recirculation (EGR) system is controlled by the ECM. The ECM controls the vacuum being supplied to the valve by energizing and de-energizing a solenoid.

The ECM uses information from various engine sensors to determine when EGR is necessary. Once the ECM has requested EGR by grounding the solenoid circuit, the ECM will monitor engine operating conditions to determine if exhaust gas flow has entered the intake manifold. When the ECM tests for EGR operation and no change in engine operating conditions is indicated, a DTC 32 will set.

The difference in long term fuel trim values between the idle (closed throttle) fuel trim cell and cell 2 is used to monitor EGR system performance. When the difference between the two long term fuel trim values is greater than about 15 and the long term fuel trim value in cell 2 is greater than 135 for 80 seconds or more, DTC 32 is set. The system operates in fuel trim cell 2 during a cruise condition at approximately 55 mph.

Test Description: Number(s) below refer to circled number(s) on the diagnostic chart.

1. Intake Passage: Shut "OFF" engine and remove the EGR valve from the manifold. Plug the exhaust side hole with a suitable stopper. Leaving the intake side hole open, attempt to start the engine. If the engine runs at a high idle (up to 3000 RPM is possible) or starts and stalls, the EGR intake passage is not restricted. If the engine starts and idles normally, the EGR intake passage is restricted.
   Exhaust Passage: With EGR valve still removed, plug the intake side hole with a suitable stopper. With the exhaust side hole open, check for the presence of exhaust gas. If no exhaust gas is present, the EGR exhaust side passage is restricted.

2. By grounding the diagnostic "test" terminal, the EGR solenoid should be energized and allow vacuum to be applied to the gage. The vacuum at the gage may or may not slowly bleed off. It is important that the gage is able to read the amount of vacuum being applied.

3. When the diagnostic "test" terminal is ungrounded, the vacuum gage should bleed off completely through a vent in the solenoid. The vacuum pump gage may or may not bleed off but this does not indicate a problem.

4. This test will determine if the electrical control part of the system is at fault or if the connector or solenoid is at fault.

5. At this point, it has been determined that the EGR solenoid, the ECM and the vacuum supply are OK.

Diagnostic Aids:

Vacuum lines should be thoroughly checked for proper routing. Refer to "Vehicle Emission Control Information" label.

The DTC 32 chart is a functional check of the EGR system. If the EGR system works properly but a DTC 32 has been set, check other items that could result in high block learn values during a cruise condition at approximately 55 mph. Low fuel pressure or lean fuel injector(s) may set a DTC 32. It may be necessary to monitor fuel pressure while driving the vehicle at various road speeds and/or loads. Refer to "Fuel System Diagnosis," CHART A-7.

2.2L (VIN 4) ENGINE — DIAGNOSTIC TROUBLE CODE CHART — CORSICA AND BERETTA

DTC 32
(Page 1 of 2)
EXHAUST GAS RECIRCULATION (EGR) SYSTEM FAILURE
2.2L (VIN 4) "L" CARLINE

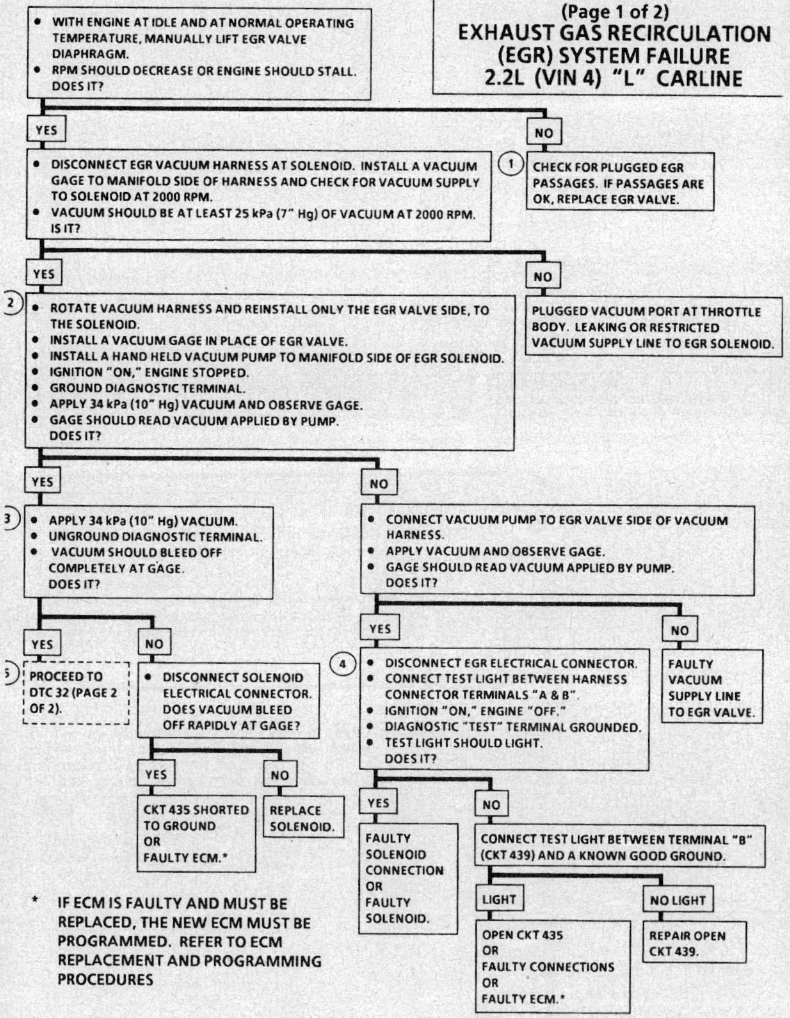

• WITH ENGINE AT IDLE AND AT NORMAL OPERATING TEMPERATURE, MANUALLY LIFT EGR VALVE DIAPHRAGM.
• RPM SHOULD DECREASE OR ENGINE SHOULD STALL. DOES IT?
YES
NO
1
CHECK FOR PLUGGED EGR PASSAGES. IF PASSAGES ARE OK, REPLACE EGR VALVE.
• DISCONNECT EGR VACUUM HARNESS AT SOLENOID. INSTALL A VACUUM GAGE TO MANIFOLD SIDE OF HARNESS AND CHECK FOR VACUUM SUPPLY TO SOLENOID AT 2000 RPM.
• VACUUM SHOULD BE AT LEAST 25 kPa (7" Hg) OF VACUUM AT 2000 RPM. IS IT?
YES
NO
PLUGGED VACUUM PORT AT THROTTLE BODY. LEAKING OR RESTRICTED VACUUM SUPPLY LINE TO EGR SOLENOID.
2
• ROTATE VACUUM HARNESS AND REINSTALL ONLY THE EGR VALVE SIDE, TO THE SOLENOID.
• INSTALL A VACUUM GAGE IN PLACE OF EGR VALVE.
• INSTALL A HAND HELD VACUUM PUMP TO MANIFOLD SIDE OF EGR SOLENOID.
• IGNITION "ON," ENGINE STOPPED.
• GROUND DIAGNOSTIC TERMINAL.
• APPLY 34 kPa (10" Hg) VACUUM AND OBSERVE GAGE.
• GAGE SHOULD READ VACUUM APPLIED BY PUMP. DOES IT?
YES
NO
3
• APPLY 34 kPa (10" Hg) VACUUM.
• UNGROUND DIAGNOSTIC TERMINAL.
• VACUUM SHOULD BLEED OFF COMPLETELY AT GAGE. DOES IT?
• CONNECT VACUUM PUMP TO EGR VALVE SIDE OF VACUUM HARNESS.
• APPLY VACUUM AND OBSERVE GAGE.
• GAGE SHOULD READ VACUUM APPLIED BY PUMP. DOES IT?
YES
NO
5
PROCEED TO DTC 32 (PAGE 2 OF 2).
• DISCONNECT SOLENOID ELECTRICAL CONNECTOR. DOES VACUUM BLEED OFF RAPIDLY AT GAGE?
YES
NO
CKT 435 SHORTED TO GROUND OR FAULTY ECM.*
REPLACE SOLENOID.
YES
NO
4
• DISCONNECT EGR ELECTRICAL CONNECTOR.
• CONNECT TEST LIGHT BETWEEN HARNESS CONNECTOR TERMINALS "A & B".
• IGNITION "ON," ENGINE "OFF."
• DIAGNOSTIC "TEST" TERMINAL GROUNDED.
• TEST LIGHT SHOULD LIGHT. DOES IT?
FAULTY VACUUM SUPPLY LINE TO EGR VALVE.
YES
NO
FAULTY SOLENOID CONNECTION OR FAULTY SOLENOID.
CONNECT TEST LIGHT BETWEEN TERMINAL "B" (CKT 439) AND A KNOWN GOOD GROUND.
LIGHT
NO LIGHT
OPEN CKT 435 OR FAULTY CONNECTIONS OR FAULTY ECM.*
REPAIR OPEN CKT 439.
* IF ECM IS FAULTY AND MUST BE REPLACED, THE NEW ECM MUST BE PROGRAMMED. REFER TO ECM REPLACEMENT AND PROGRAMMING PROCEDURES

2.2L (VIN 4) ENGINE — DIAGNOSTIC TROUBLE CODE CHART — CORSICA AND BERETTA

TO EGR VALVE
PORTED VACUUM
EGR SOLENOID (NORMALLY CLOSED)
ASSEMBLY HAS CONSTANT INTERNAL BLEED
VENT FILTER
TO IGNITION SWITCH
ECM
ECM/FUEL INJ FUSE #1
15 AMP
TO FUEL INJECTORS
TO ELECTRONIC IGNITION CONTROL MODULE (ICM)
B
A
439 PNK/BLK
435 GRY
C6
C14
IGNITION FEED
EGR CONTROL

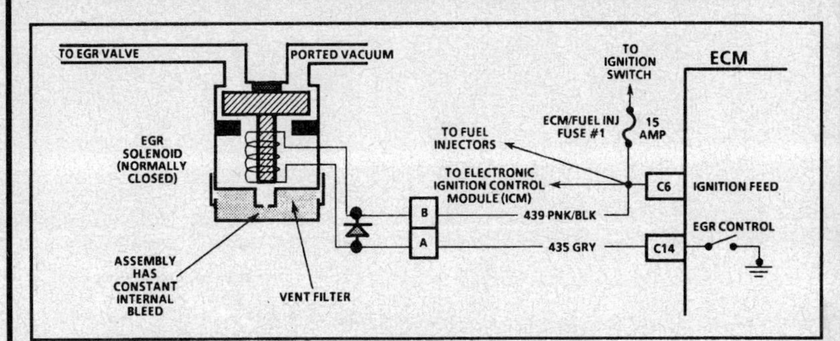

DTC 32
(Page 2 of 2)
EXHAUST GAS RECIRCULATION (EGR) SYSTEM FAILURE
2.2L (VIN 4) "L" CARLINE

Circuit Description:

The Exhaust Gas Recirculation (EGR) system is controlled by the ECM. The ECM controls the vacuum being supplied to the valve by energizing and de-energizing a solenoid.

The ECM uses information from various engine sensors to determine when EGR is necessary. Once the ECM has requested EGR by grounding the solenoid circuit, the ECM will monitor engine operating conditions to determine if exhaust gas flow has entered the intake manifold. When the ECM tests for EGR operation and no change in engine operating conditions is indicated, a DTC 32 will set.

Test Description: Number(s) below refer to circled number(s) on the diagnostic chart.

1. The remaining tests check the ability of the EGR valve to interact with the exhaust system. This system uses a negative backpressure EGR valve which should hold vacuum with engine "OFF." **Be sure shop exhaust hose is not connected during Steps 1 and 2.**
2. When engine is started, exhaust backpressure at the base of the EGR valve should open the valve's internal bleed and vent the applied vacuum allowing the valve to seat. **Because the shop exhaust hose is not installed at this time, do not allow the engine to run longer than 15 seconds.**

Diagnostic Aids:

Low fuel pressure or lean fuel injectors may cause a DTC 32 to set. Use CHART A-7. If fuel pressure is normal, perform the injector balance test, CHART C2-A It may be necessary to monitor fuel pressur ewhile driving the vehicle at various road speeds and/or loads.

2.2L (VIN 4) ENGINE — DIAGNOSTIC TROUBLE CODE CHART — CORSICA AND BERETTA

DTC 32
(Page 2 of 2)
EXHAUST GAS RECIRCULATION (EGR) SYSTEM FAILURE
2.2L (VIN 4) "L" CARLINE

CONTINUED FROM DTC 32 CHART (1 OF 2)

(1)
- IGNITION "OFF," SHOP EXHAUST HOSE NOT INSTALLED TO TAILPIPE.
- CONNECT A VACUUM PUMP TO EGR VALVE.
- OBSERVE EGR DIAPHRAGM WHILE APPLYING VACUUM.
- DIAPHRAGM SHOULD MOVE FREELY AND HOLD VACUUM FOR AT LEAST 20 SECONDS. DOES IT?

YES →

NO → REPLACE EGR VALVE.

(2)
- APPLY 34 kPa (10" HG) VACUUM TO EGR VALVE.
- BRIEFLY START ENGINE AND IMMEDIATELY OBSERVE GAGE ON VACUUM PUMP.
- ENGINE "OFF."
- EGR VALVE DIAPHRAGM SHOULD HAVE MOVED TO SEATED POSITION AND VACUUM SHOULD HAVE DROPPED FROM PUMP GAGE WHILE STARTING ENGINE. DID IT?

NO →
- REMOVE EGR VALVE.
- CHECK FOR PLUGGED OR RESTRICTED EXHAUST PASSAGES.

YES →
EGR CIRCUIT IS OPERATING PROPERLY. REFER TO DIAGNOSTIC AIDS

PASSAGES OK → REPLACE EGR VALVE.

PASSAGES NOT OK →
- CLEAN PASSAGES.
- RECHECK EGR VALVE.

2.2L (VIN 4) ENGINE — DIAGNOSTIC TROUBLE CODE CHART — CORSICA AND BERETTA

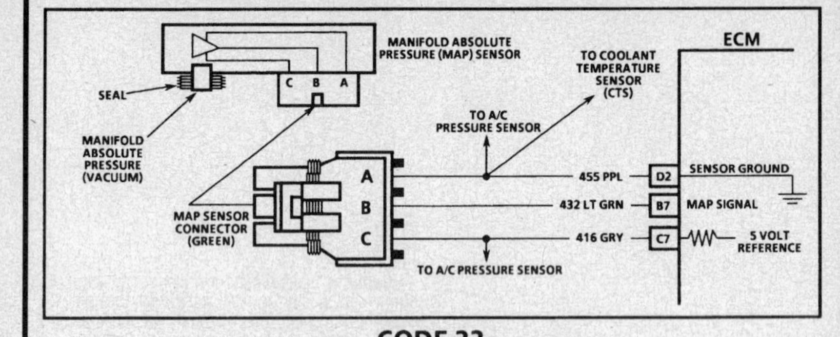

CODE 33
(Page 1 of 2)
MANIFOLD ABSOLUTE PRESSURE (MAP) SENSOR CIRCUIT
(SIGNAL VOLTAGE HIGH/LOW - LOW/HIGH VACUUM)
2.2L (VIN 4) "L" CARLINE (PORT)

Circuit Description:

The Manifold Absolute Pressure (MAP) sensor responds to changes in manifold pressure (vacuum). The ECM receives this information as a signal voltage that will vary from about 1 to 1.5 volts at closed throttle (idle), to about 4.5 volts at Wide Open Throttle (WOT) (low vacuum).

If the MAP sensor fails, the Electronic Control Module (ECM) will substitute a fixed MAP value based on engine rpm and use the Throttle Position Sensor (TPS) to control fuel delivery.

Test Description: Number(s) below refer to circled number(s) on the diagnostic chart.

1. A Code 33 will set under the following conditions:
   - The engine is running.
   - MAP signal indicates greater than 3.7 volts (80 kPa).
   - Throttle position is less than 5%.
   - These conditions exist for a time period longer than 5 seconds.

   OR
   - MAP signal indicates less than .3 volt (15 kPa).
   - Engine speed is less than 1200 rpm's and TPS position is greater than 15%.
2. If the ECM recognizes the change, the ECM and CKTs 416 and 432 are OK. If CKT 455 is open, there may also be other codes stored.

Diagnostic Aids:

With the ignition "ON" and the engine stopped, the manifold pressure is equal to atmospheric pressure and the signal voltage will be high. This information is used by the ECM as an indication of vehicle altitude and is referred to as BARO. Comparison of this BARO reading with a known good vehicle with the same sensor is a good way to check accuracy of a "suspect" sensor. Reading should be within ± .4 volt.

If Code 33 is intermittent, refer to "Symptoms,"

2.2L (VIN 4) ENGINE — DIAGNOSTIC TROUBLE CODE CHART — 1992 CORSICA AND BERETTA

CODE 33
(Page 1 of 2)
MANIFOLD ABSOLUTE PRESSURE (MAP) SENSOR CIRCUIT
(SIGNAL VOLTAGE HIGH/LOW - LOW/HIGH VACUUM)
2.2L (VIN 4) "L" CARLINE (PORT)

(1)
- IF ENGINE IDLE IS ROUGH, UNSTABLE, OR INCORRECT, CORRECT CONDITION BEFORE USING CHART.
- ENGINE IDLING.
- DOES "SCAN" TOOL DISPLAY A MAP VOLTAGE OF 3.7 VOLTS OR GREATER?

YES → (2)
- DISCONNECT MAP SENSOR ELECTRICAL CONNECTOR.
- ENGINE IDLING.
- "SCAN" TOOL SHOULD READ A VOLTAGE OF .3 VOLT OR LESS. DOES IT?

NO → REFER TO CODE 33 (PAGE 2 OF 2).

YES →
- PROBE SENSOR GROUND CIRCUIT WITH A TEST LIGHT TO BATTERY VOLTAGE.
- TEST LIGHT SHOULD LIGHT. DOES IT?

NO → CKT 432 SHORTED TO VOLTAGE, SHORTED TO CKT 416 OR FAULTY ECM.

YES →
PLUGGED SENSOR INLET PORT OR LEAKING SENSOR SEAL OR FAULTY MAP SENSOR.

NO → OPEN SENSOR GROUND CIRCUIT.

"AFTER REPAIRS," REFER TO CODE CRITERIA AND CONFIRM CODE DOES NOT RESET.

2.2L (VIN 4) ENGINE — DIAGNOSTIC TROUBLE CODE CHART — 1992 CORSICA AND BERETTA

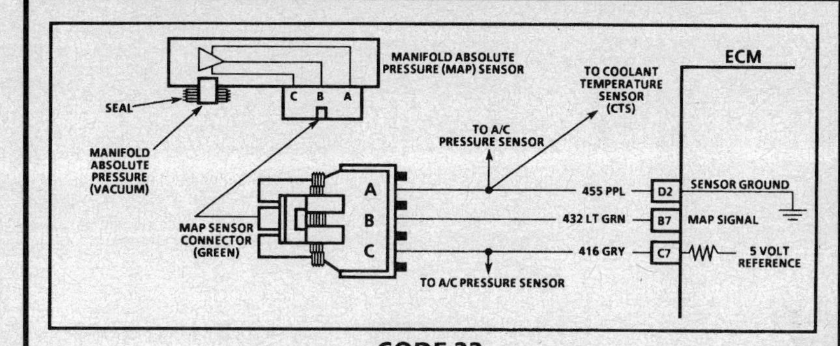

CODE 33
(Page 2 of 2)
MANIFOLD ABSOLUTE PRESSURE (MAP) SENSOR CIRCUIT
(SIGNAL VOLTAGE HIGH/LOW - LOW/HIGH VACUUM)
2.2L (VIN 4) "L" CARLINE (PORT)

Circuit Description:

The Manifold Absolute Pressure (MAP) sensor responds to changes in manifold pressure (vacuum). The ECM receives this information as a signal voltage that will vary from about 1 to 1.5 volts at closed throttle (idle), to about 4.5 volts at Wide Open Throttle (WOT) (low vacuum).

If the MAP sensor fails, the Electronic Control Module (ECM) will substitute a fixed MAP value based on engine rpm and use the Throttle Position Sensor (TPS) to control fuel delivery.

Test Description: Number(s) below refer to circled number(s) on the diagnostic chart.

1. Jumpering harness terminals "B" to "C" (5 volts to signal circuit), will determine if the sensor is at fault, or if there is a problem with the ECM or wiring.

   The "Scan" tool may not display 5 volts. The important thing is that the ECM recognizes the voltage as more than 4 volts, indicating that the ECM CKTs 432, and 416 are OK.

2. This step determines if CKT 416 or CKT 432 is faulty. The "Scan" tool will not display battery voltage, but should indicate over 4 volts.

Diagnostic Aids:

With the ignition "ON" and the engine stopped, the manifold pressure is equal to atmospheric pressure and the signal voltage will be high. This information is used by the ECM as an indication of vehicle altitude and is referred to as BARO. Comparison of this BARO reading with a known good vehicle with the same sensor is a good way to check accuracy of a "suspect" sensor. Reading should be within ± .4 volt.

If Code 33 is intermittent, refer to "Symptoms,"

2.2L (VIN 4) ENGINE — DIAGNOSTIC TROUBLE CODE CHART — 1992 CORSICA AND BERETTA

CODE 33
(Page 2 of 2)
MANIFOLD ABSOLUTE PRESSURE (MAP) SENSOR CIRCUIT
(SIGNAL VOLTAGE HIGH/LOW - LOW/HIGH VACUUM)
2.2L (VIN 4) "L" CARLINE (PORT)

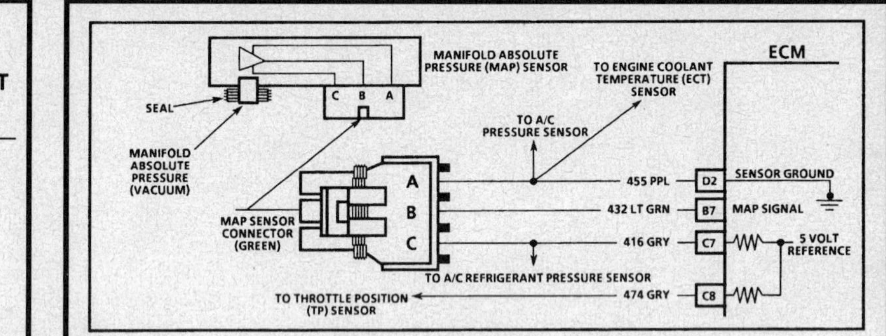

2.2L (VIN 4) ENGINE — DIAGNOSTIC TROUBLE CODE CHART — 1993–94 CORSICA AND BERETTA

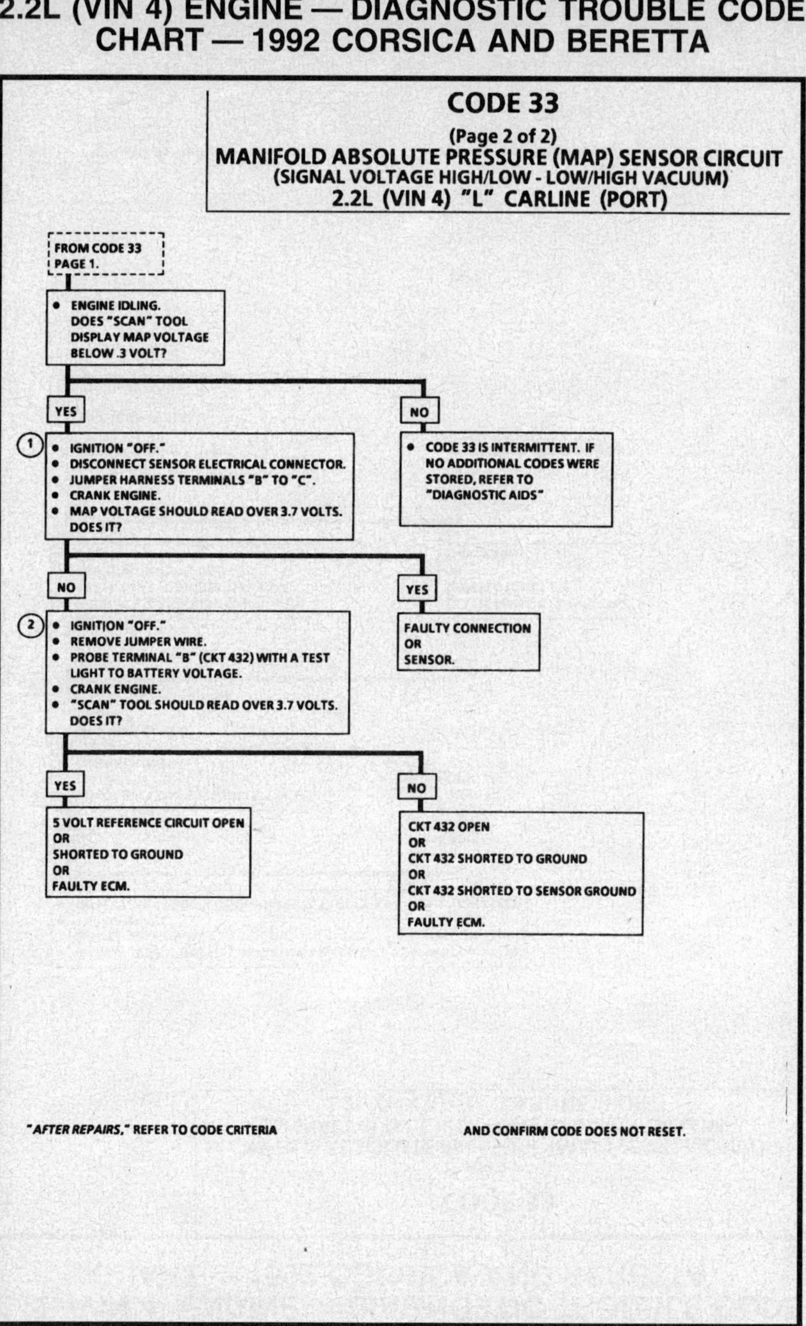

DTC 33
(Page 1 of 2)
MANIFOLD ABSOLUTE PRESSURE (MAP) SENSOR CIRCUIT
(SIGNAL VOLTAGE HIGH/LOW - LOW/HIGH VACUUM)
2.2L (VIN 4) "L" CARLINE

Circuit Description:

The Manifold Absolute Pressure (MAP) sensor responds to changes in manifold pressure (vacuum). The ECM receives this information as a signal voltage that will vary from about 1 to 1.5 volts at closed throttle (idle), to about 4.5 volts at Wide Open Throttle (WOT) (low vacuum).

If the MAP sensor fails, the Engine Control Module (ECM) will substitute a fixed MAP value based on engine RPM.

Test Description: Number(s) below refer to circled number(s) on the diagnostic chart.

1. A DTC 33 will set under the following conditions:
   - MAP signal indicates greater than 3.7 volts (80 kPa).
   - Throttle position is less than 5%.
   - These conditions exist for a time period longer than 5 seconds.
     OR
   - MAP signal indicates less than .3 volt (15 kPa).
2. If the ECM recognizes the change, the ECM and CKTs 416 and 432 are OK. If CKT 455 is open, there may also be other DTC(s) stored.

Diagnostic Aids:

With the ignition "ON" and the engine stopped, the manifold pressure is equal to atmospheric pressure and the signal voltage will be high. This information is used by the ECM as an indication of vehicle altitude and is referred to as BARO. Comparison of this BARO reading with a known good vehicle with the same sensor is a good way to check accuracy of a "suspect" sensor. Reading should be within ± .4 volt.

If DTC 33 is intermittent, refer to "Symptoms,"

2.2L (VIN 4) ENGINE — DIAGNOSTIC TROUBLE CODE CHART — 1993–94 CORSICA AND BERETTA

DTC 33

(Page 1 of 2)
MANIFOLD ABSOLUTE PRESSURE (MAP) SENSOR CIRCUIT
(SIGNAL VOLTAGE HIGH/LOW - LOW/HIGH VACUUM)
2.2L (VIN 4) "L" CARLINE

1. • IF ENGINE IDLE IS ROUGH, UNSTABLE, OR INCORRECT, CORRECT CONDITION BEFORE USING CHART.
   • ENGINE IDLING.
   • DOES SCAN TOOL DISPLAY A MAP VOLTAGE OF 3.7 VOLTS OR GREATER?

   YES →
   NO → REFER TO DTC 33 (PAGE 2 OF 2).

2. • DISCONNECT MAP SENSOR ELECTRICAL CONNECTOR.
   • ENGINE IDLING.
   • SCAN TOOL SHOULD DISPLAY A VOLTAGE OF .3 VOLT OR LESS.
     DOES IT?

   YES →
   NO → CKT 432 SHORTED TO VOLTAGE, SHORTED TO CKT 416 OR FAULTY ECM.*

   • PROBE SENSOR GROUND CIRCUIT WITH A TEST LIGHT TO BATTERY VOLTAGE.
     TEST LIGHT SHOULD LIGHT.
     DOES IT?

   YES →
   NO → OPEN SENSOR GROUND CIRCUIT.

   PLUGGED SENSOR INLET PORT OR LEAKING SENSOR SEAL OR FAULTY MAP SENSOR.

* IF ECM IS FAULTY AND MUST BE REPLACED, THE NEW ECM MUST BE PROGRAMMED. REFER TO ECM REPLACEMENT AND PROGRAMMING PROCEDURES

"AFTER REPAIRS," REFER TO DTC CRITERIA   AND CONFIRM DTC DOES NOT RESET.

2.2L (VIN 4) ENGINE — DIAGNOSTIC TROUBLE CODE CHART — 1993–94 CORSICA AND BERETTA

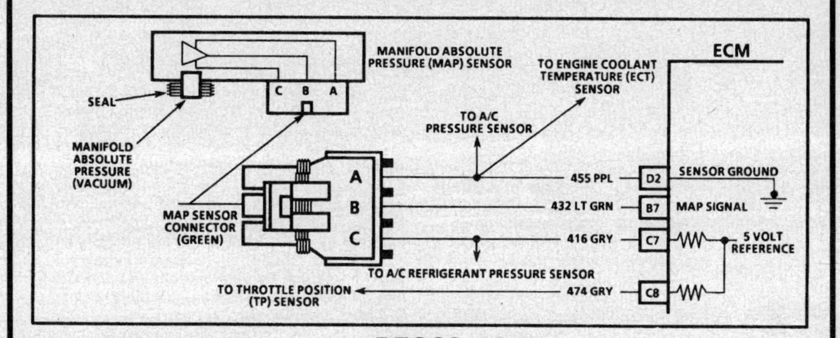

DTC 33

(Page 2 of 2)
MANIFOLD ABSOLUTE PRESSURE (MAP) SENSOR CIRCUIT
(SIGNAL VOLTAGE HIGH/LOW - LOW/HIGH VACUUM)
2.2L (VIN 4) "L" CARLINE

Circuit Description:

The Manifold Absolute Pressure (MAP) sensor responds to changes in manifold pressure (vacuum). The ECM receives this information as a signal voltage that will vary from about 1 to 1.5 volts at closed throttle (idle), to about 4.5 volts at Wide Open Throttle (WOT) (low vacuum).

If the MAP sensor fails, the Engine Control Module (ECM) will substitute a fixed MAP value based on engine RPM and use the Throttle Position (TP) sensor to control fuel delivery.

Test Description: Number(s) below refer to circled number(s) on the diagnostic chart.

1. Jumpering harness terminals "B" to "C" (5 volts to signal circuit), will determine if the sensor is at fault, or if there is a problem with the ECM or wiring.

   The scan tool may not display 5 volts. The important thing is that the ECM recognizes the voltage as more than 4 volts, indicating that the ECM CKTs 432, and 416 are OK.
2. This step determines if CKT 474 or CKT 432 is faulty. The scan tool will not display battery voltage, but should indicate over 4 volts.
3. The 5 volt reference Throttle Position (TP) sensor connects to the same 5 volt power supply in the ECM. If the 5 volt reference circuit to the TP sensor has a short to ground, the MAP sensor and A/C refrigerant pressure sensor 5 volt reference circuit will also have a short to ground because they connect to the same power supply in the ECM. A short to ground on either 5 volt reference circuit does not damage the ECM 5 volt power supply. When the short is repaired, the ECM 5 volt supply will return to normal.

Diagnostic Aids:

With the ignition "ON" and the engine stopped, the manifold pressure is equal to atmospheric pressure and the signal voltage will be high. This information is used by the ECM as an indication of vehicle altitude and is referred to as BARO. Comparison of this BARO reading with a known good vehicle with the same sensor is a good way to check accuracy of a "suspect" sensor. Reading should be within ± .4 volt.

If DTC 33 is intermittent, refer to "Symptoms,"

DTC 33

(Page 2 of 2)
MANIFOLD ABSOLUTE PRESSURE (MAP) SENSOR CIRCUIT
(SIGNAL VOLTAGE HIGH/LOW - LOW/HIGH VACUUM)
2.2L (VIN 4) "L" CARLINE

FROM DTC 33 PAGE 1.

- ENGINE IDLING.
  DOES SCAN TOOL
  DISPLAY MAP VOLTAGE
  BELOW .3 VOLT?

YES

(1)
- IGNITION "OFF."
- DISCONNECT SENSOR ELECTRICAL CONNECTOR.
- JUMPER HARNESS TERMINALS "B" TO "C".
- CRANK ENGINE.
- MAP VOLTAGE SHOULD READ OVER 3.7 VOLTS. DOES IT?

NO

- DTC 33 IS INTERMITTENT. IF NO ADDITIONAL DTC(S) WERE STORED, REFER TO "DIAGNOSTIC AIDS"

NO

(2)
- IGNITION "OFF."
- REMOVE JUMPER WIRE.
- PROBE TERMINAL "B" (CKT 432) WITH A TEST LIGHT TO BATTERY VOLTAGE.
- CRANK ENGINE.
- SCAN TOOL SHOULD READ OVER 3.7 VOLTS. DOES IT?

YES

FAULTY CONNECTION OR SENSOR.

YES

(3)
5 VOLT REFERENCE CIRCUIT OPEN
OR
SHORTED TO GROUND
OR
FAULTY ECM.*

NO

CKT 432 OPEN
OR
CKT 432 SHORTED TO GROUND
OR
CKT 432 SHORTED TO SENSOR GROUND
OR
FAULTY ECM.*

* IF ECM IS FAULTY AND MUST BE REPLACED, THE NEW ECM MUST BE PROGRAMMED. REFER TO ECM REPLACEMENT AND PROGRAMMING PROCEDURES

"AFTER REPAIRS," REFER TO DTC CRITERIA

AND CONFIRM DTC DOES NOT RESET.

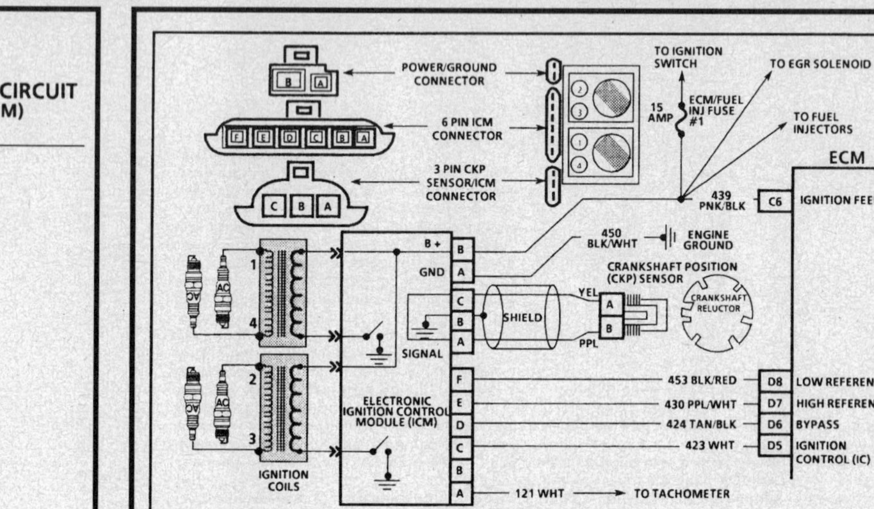

DTC 42

IGNITION CONTROL (IC) CIRCUIT
2.2L (VIN 4) "L" CARLINE

Circuit Description:

The electronic Ignition Control Module (ICM) sends a reference signal to the Engine Control Module (ECM) when the engine is cranking. While the engine speed is under 400 RPM, the ICM controls the ignition timing. When the system is running on the ICM (no voltage on the bypass line), the ICM grounds the Ignition Control (IC) signal. The ECM expects to sense no voltage on the Ignition Control (IC) line during this condition. If it senses a voltage, it sets DTC 42 and will not enter the ignition control mode.

When the engine speed exceeds 400 RPM, the ECM applies 5 volts to the bypass line to switch the timing to ECM control (ignition control). If the bypass line is open or grounded, once the RPM for ignition control is reached, the ICM will not switch to Ignition Control (IC) mode. This results in low Ignition Control (IC) voltage and the setting of DTC 42. If the IC line is grounded, the ICM will switch to Ignition Control (IC), but because the line is grounded, there will be no IC signal. A DTC 42 will be set.

Test Description: Number(s) below refer to circled number(s) on the diagnostic chart.

1. DTC 42 means the ECM has sensed an open or short to ground in the IC or bypass circuits. This test confirms DTC 42 and that the fault causing the DTC 42 is present.
2. Checks for a normal IC ground path through the ICM. An IC CKT 423, shorted to ground, will also read less than 500 ohms, but this will be checked later.
3. As the test light voltage contacts CKT 424, the ICM should switch, causing the ohmmeter to "overrange" if the meter is in the 1000 to 2000 ohms position. Selecting the 10,000 to 20,000 ohms position will indicate a reading above 5000 ohms.

The important thing is that the module "switched."

4. The module did not switch and this step checks for:
   - IC CKT 423 shorted to ground.
   - Bypass CKT 424 open.
   - Faulty ICM connection or ICM.
5. Confirms that DTC 42 is a faulty ECM and not an intermittent in CKTs 423 or 424.

Diagnostic Aids:

If DTC 42 is intermittent, refer to "Symptoms,"

2.2L (VIN 4) ENGINE — DIAGNOSTIC TROUBLE CODE CHART — CORSICA AND BERETTA

DTC 42
IGNITION CONTROL (IC) CIRCUIT
2.2L (VIN 4) "L" CARLINE

1. • CLEAR DTC(S).
   • IDLE ENGINE FOR 1 MINUTE OR UNTIL DTC 42 SETS.
   DOES DTC 42 SET?

YES →

NO → DTC 42 INTERMITTENT. REFER TO "DIAGNOSTIC AIDS"

2. • IGNITION "OFF."
   • DISCONNECT ECM CONNECTORS.
   • IGNITION "ON."
   • SET OHMMETER SELECTOR SWITCH IN THE 1000 TO 2000 OHMS RANGE.
   • GROUND THE BLACK (-) OHMMETER LEAD.
   • PROBE ECM HARNESS IGNITION CONTROL CKT 423 USING THE RED (+) OHMMETER LEAD. THE OHMMETER SHOULD READ LESS THAN 500 OHMS.
   DOES IT?

YES →

NO → OPEN CKT 423, FAULTY CONNECTION, OR FAULTY IGNITION CONTROL MODULE (ICM).

• PROBE ECM HARNESS CONNECTOR CKT 424 WITH A TEST LIGHT TO BATTERY VOLTAGE AND OBSERVE LIGHT.

LIGHT "OFF" →

LIGHT "ON" →

• DISCONNECT IGNITION CONTROL MODULE 6-PIN CONNECTOR.

LIGHT "ON" → CKT 424 SHORTED TO GROUND.

LIGHT "OFF" → REPLACE IGNITION CONTROL MODULE.

3. • WITH OHMMETER STILL CONNECTED TO ECM HARNESS CKT 423 AND GROUND AS DESCRIBED IN STEP #2, AGAIN PROBE ECM HARNESS CKT 424 WITH THE TEST LIGHT CONNECTED TO BATTERY VOLTAGE. AS TEST LIGHT CONTACTS CKT 424, RESISTANCE SHOULD SWITCH FROM HUNDREDS TO THOUSANDS OHMS.
   DOES IT?

NO →

YES →

4. • DISCONNECT IGNITION CONTROL MODULE 6-PIN CONNECTOR. NOTE OHMMETER THAT IS STILL CONNECTED TO CKT 423 AND GROUND. RESISTANCE SHOULD HAVE BECOME VERY HIGH (OPEN CIRCUIT).
   DOES IT?

YES → CKT 424 OPEN, FAULTY CONNECTIONS, OR FAULTY IGNITION CONTROL MODULE.

NO → CKT 423 SHORTED TO GROUND.

5. • RECONNECT ECM AND IDLE ENGINE FOR ONE MINUTE OR UNTIL DTC 42 SETS. DOES DTC 42 SET?

YES → REPLACE ECM. *

NO → DTC 42 INTERMITTENT. REFER TO "DIAGNOSTIC AIDS"

* IF ECM IS FAULTY AND MUST BE REPLACED, THE NEW ECM MUST BE PROGRAMMED. REFER TO ECM REPLACEMENT AND PROGRAMMING PROCEDURES

"AFTER REPAIRS," REFER TO DTC CRITERIA AND CONFIRM DTC DOES NOT RESET.

2.2L (VIN 4) ENGINE — DIAGNOSTIC TROUBLE CODE CHART — CORSICA AND BERETTA

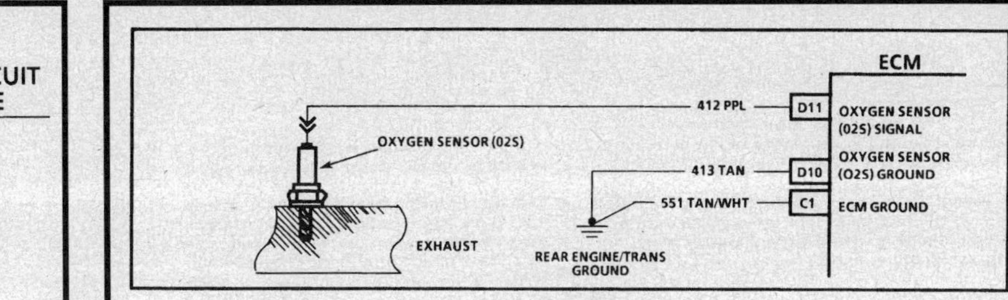

DTC 44
OXYGEN SENSOR (O2S) CIRCUIT
(LEAN EXHAUST INDICATED)
2.2L (VIN 4) "L" CARLINE

Circuit Description:

The Engine Control Module (ECM) supplies a voltage of about .45 volt between terminals "D10" and "D11". (If measured with a 10 megohm digital voltmeter, this may read as low as .32 volt.)

When the O2S reaches operating temperature, it varies this voltage from about .1 volt (exhaust is lean) to about .9 volt (exhaust is rich).

The sensor is like an open circuit and produces no voltage when it is below 316°C (600°F). An open sensor circuit, or cold sensor, causes "Open Loop" operation.

Test Description: Number(s) below refer to circled number(s) on the diagnostic chart.

1. DTC 44 is set when the O2S signal voltage on CKT 412 remains below .28 volt (280 mV) for 25 seconds or more.

Diagnostic Aids:

Using the scan tool, observe the long term fuel trim value at different engine speeds. If the conditions for DTC 44 exists, the long term fuel trim values will be around 160 or higher.

Check the following possible causes:

- Oxygen Sensor Wire - Sensor pigtail may be mispositioned and contacting the exhaust manifold. Check for ground in wire between connector and sensor.

- Fuel Contamination - Water, even in small amounts, near the in-tank fuel pump inlet can be delivered to the injectors. The water causes a lean exhaust and can set a DTC 44.

- Fuel Pressure - System will be lean if fuel pressure is too low. It may be necessary to monitor fuel pressure while driving the vehicle at various road speeds and/or loads to confirm. Refer to "Fuel System Diagnosis," CHART A-7.

- Exhaust Leaks - If there is an exhaust leak, the engine can cause outside air to be pulled into the exhaust and past the sensor. Vacuum or crankcase leaks can cause a lean condition.

- If DTC 44 is intermittent, refer to "Symptoms."

- A cracked or otherwise damaged O2S may set an intermittent DTC 44.

DTC 44

OXYGEN SENSOR (O2S) CIRCUIT
(LEAN EXHAUST INDICATED)
2.2L (VIN 4) "L" CARLINE

① • RUN WARM ENGINE (75°C/167°F TO 95°C/203°F) AT 1200 RPM.
   DOES SCAN TOOL INDICATE OXYGEN SENSOR VOLTAGE FIXED BELOW .28 VOLT (280 mV)?

YES

• DISCONNECT OXYGEN SENSOR ELECTRICAL CONNECTOR.
• WITH ENGINE IDLING, SCAN TOOL SHOULD DISPLAY
  OXYGEN SENSOR VOLTAGE BETWEEN .28 VOLT AND .7
  VOLT (280 mV AND 700 MV).
  DOES IT?

NO

DTC 44 IS INTERMITTENT. IF NO ADDITIONAL
DTC(S) WERE STORED, REFER TO "DIAGNOSTIC
AIDS"

YES

REFER TO "DIAGNOSTIC
AIDS"

NO

CKT 412 SHORTED TO GROUND
OR
FAULTY ECM.*

* IF ECM IS FAULTY AND MUST BE REPLACED, THE NEW ECM MUST BE PROGRAMMED.
  REFER TO ECM REPLACEMENT AND PROGRAMMING PROCEDURES

"*AFTER REPAIRS,*" REFER TO DTC CRITERIA AND CONFIRM DTC DOES NOT RESET.

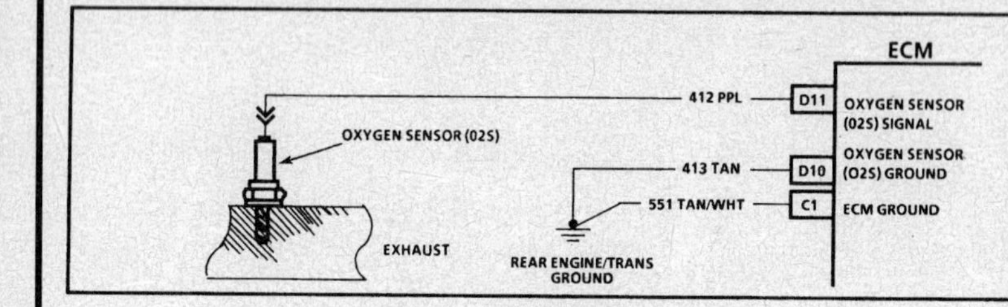

DTC 45

OXYGEN SENSOR (O2S) CIRCUIT
(RICH EXHAUST INDICATED)
2.2L (VIN 4) "L" CARLINE

Circuit Description:

The Engine Control Module (ECM) supplies a voltage of about .45 volt between terminals "D10" and "D11". (If measured with a 10 megohm digital voltmeter, this may read as low as .32 volt.)

When the Oxygen Sensor (O2S) reaches operating temperature, it varies this voltage from about .1 volt (exhaust is lean) to about .9 volt (exhaust is rich).

The sensor is like an open circuit and produces no voltage when it is below 316°C (600°F). An open sensor circuit, or cold sensor, causes "Open Loop" operation.

Test Description: Number(s) below refer to circled number(s) on the diagnostic chart.
1. DTC 45 is set when the O2S signal voltage on CKT 412 remains above .7 volt (700 mV) under the following conditions:
   • Engine run time after start is 2 minutes or more.
   • System is operating in "Closed Loop."
   • Throttle angle is greater than 5%.
   • Above conditions exist for 51 seconds or more.

Diagnostic Aids:

Using the scan tool, observe the long term fuel trim value at different engine speeds. If the conditions for DTC 45 exist, the long term fuel trim values will be around 90 or lower.

Check the following possible causes:
• Fuel Pressure - System will go rich, if pressure is too high. The ECM can compensate for some increase. However, if it gets too high, a DTC 45 will be set. Refer to "Fuel System Diagnosis" CHART A-7.
• Leaking Fuel Injector(s) - Refer to CHART A-7.
• An open ground CKT 453 - May result in induced electrical "noise." The ECM interprets this "noise" as reference pulses. The additional pulses result in a higher than actual engine speed signal. The ECM then delivers too much fuel causing the system to go rich. The engine tachometer will also show higher than actual engine speed, which can help in diagnosing this problem.

• Canister Purge - Check for fuel saturation. If evaporative emission canister is full of fuel, check EVAP canister control valve and hoses. Refer to "Canister Purge," in "Evaporative Emission (EVAP) Control System

• MAP Sensor - An output that causes the ECM to sense a higher than normal manifold pressure (low vacuum) can cause the system to go rich. Disconnecting the Manifold Absolute Pressure (MAP) sensor will allow the ECM to set a fixed value for the MAP sensor. Substitute a different MAP sensor if the rich condition is gone, while the sensor is disconnected.
• TP Sensor - An intermittent Throttle Position (TP) sensor output will cause the system to operate richly due to a false indication of the engine accelerating.
• O2S Contamination - Inspect Oxygen Sensor (O2S) for silicone contamination from fuel, or use of improper RTV sealant. The sensor may have a white, powdery coating and result in a high but false signal voltage (rich exhaust indication). The ECM will then reduce the amount of fuel delivered to the engine causing a severe surge driveability problem.
• EGR Valve - Exhaust Gas Recirculation (EGR) sticking open at idle is usually accompanied by a rough idle and/or stall condition.
• If DTC 45 is intermittent, refer to "Symptoms,"

• Engine Oil Contamination - Fuel fouled engine oil could cause the O2S to sense a rich air/fuel mixture and set a DTC 45.

2.2L (VIN 4) ENGINE — DIAGNOSTIC TROUBLE CODE CHART — CORSICA AND BERETTA

DTC 45

OXYGEN SENSOR (O2S) CIRCUIT
(RICH EXHAUST INDICATED)
2.2L (VIN 4) "L" CARLINE

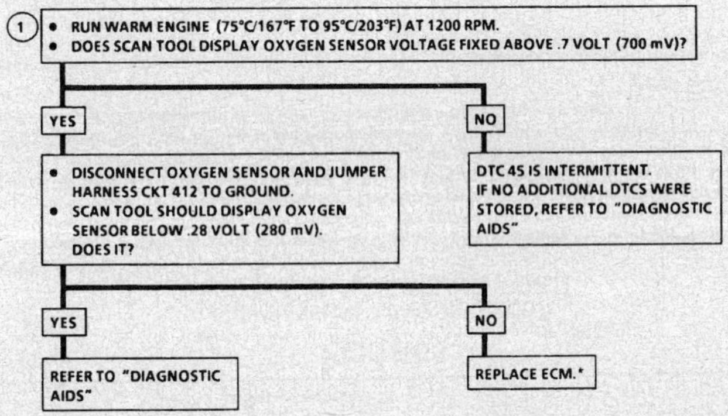

① • RUN WARM ENGINE (75°C/167°F TO 95°C/203°F) AT 1200 RPM.
• DOES SCAN TOOL DISPLAY OXYGEN SENSOR VOLTAGE FIXED ABOVE .7 VOLT (700 mV)?

YES
• DISCONNECT OXYGEN SENSOR AND JUMPER HARNESS CKT 412 TO GROUND.
• SCAN TOOL SHOULD DISPLAY OXYGEN SENSOR BELOW .28 VOLT (280 mV). DOES IT?

NO
DTC 45 IS INTERMITTENT. IF NO ADDITIONAL DTCS WERE STORED, REFER TO "DIAGNOSTIC AIDS"

YES
REFER TO "DIAGNOSTIC AIDS"

NO
REPLACE ECM.*

*IF ECM IS FAULTY AND MUST BE REPLACED, THE NEW ECM MUST BE PROGRAMMED. REFER TO ECM REPLACEMENT AND PROGRAMMING PROCEDURES

"AFTER REPAIRS," REFER TO DTC CRITERIA   AND CONFIRM DTC DOES NOT RESET.

2.2L (VIN 4) ENGINE — DIAGNOSTIC TROUBLE CODE CHART — CORSICA AND BERETTA

DTC 51

EEPROM OR ECM FAILURE
2.2L (VIN 4) "L" CARLINE

DIAGNOSTIC TROUBLE CODE (DTC) 51

ECM FAILURE
(ECM FAILED OR EEPROM FAILURE)

CHECK THAT ALL ECM CONNECTIONS ARE GOOD.
IF OK, CLEAR MEMORY AND RECHECK ECM.
IF DTC 51 REAPPEARS, REPLACE ECM AND PROGRAM NEW ECM.

*   IF ECM IS FAULTY AND MUST BE REPLACED, THE NEW ECM MUST BE PROGRAMMED. REFER TO ECM REPLACEMENT AND PROGRAMMING PROCEDURES

2.2L (VIN 4) ENGINE — DIAGNOSTIC TROUBLE CODE CHART — 1992 CORSICA AND BERETTA

2.2L (VIN 4) ENGINE — DIAGNOSTIC TROUBLE CODE CHART — 1992 CORSICA AND BERETTA

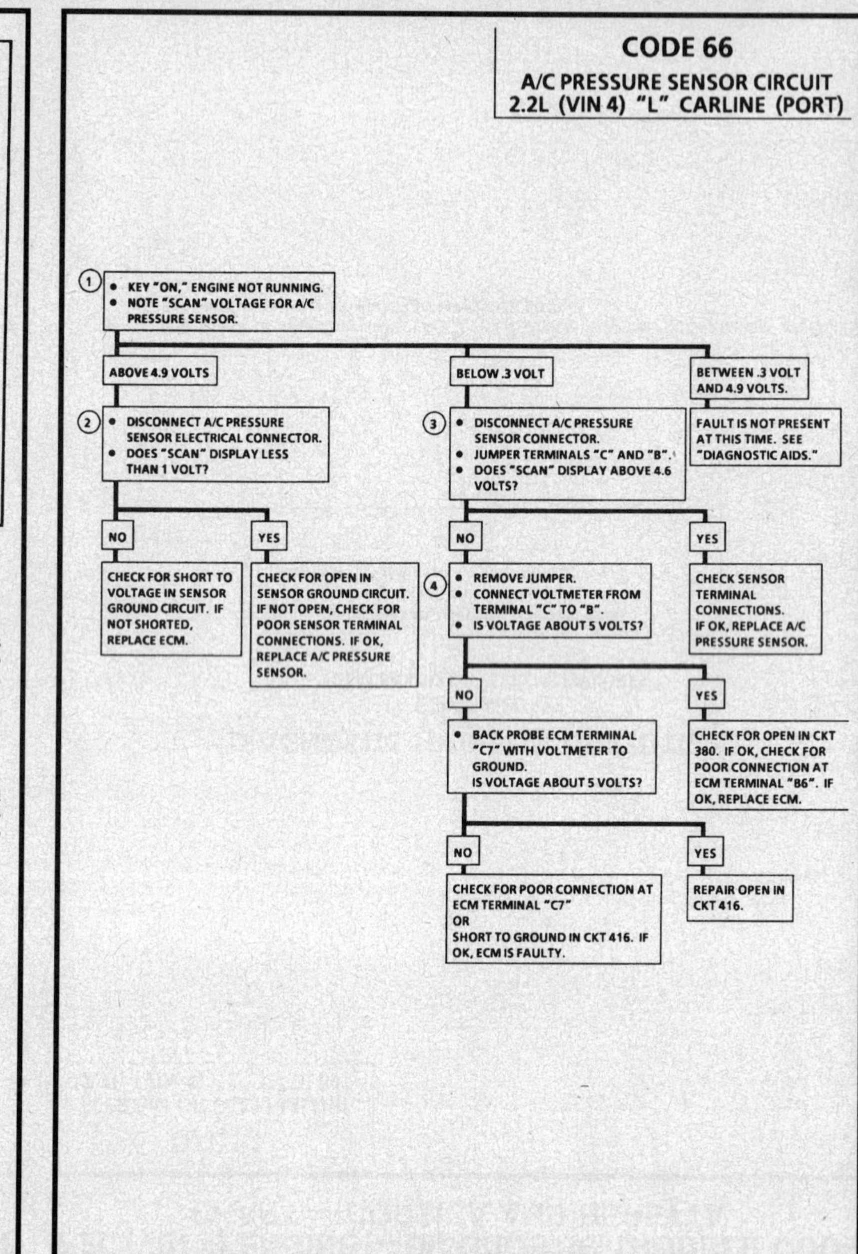

CODE 66
A/C PRESSURE SENSOR CIRCUIT
2.2L (VIN 4) "L" CARLINE (PORT)

Circuit Description:

The A/C pressure sensor responds to changes in A/C refrigerant system high side pressure. This input indicates how much load the A/C compressor is putting on the engine and is one of the factors used by the ECM to determine IAC valve position for idle speed control. The circuit consists of a 5 volt reference and a ground, both provided by the ECM, and a signal line to the ECM. The signal is a voltage which is proportional to the pressure. The sensor's range of operation is 0 to 454 psi. At 0 psi, the signal will be about .1 volt, varying up to about 4.9 volts at 454 psi or above. Code 66 sets if the voltage is above 4.9 volts (454 psi) or below .3 volt (7.5 psi) for 5 seconds or more. The A/C compressor is disabled by the ECM if fault is currently present.

Test Description: Number(s) below refer to circled number(s) on the diagnostic chart.

1. This step checks the voltage signal being received by the ECM from the A/C pressure sensor. The normal operating range is between .1 volt and 4.9 volts.

2. Checks to see if the high voltage signal is from a shorted sensor or a short to voltage in the circuit. Normally, disconnecting the sensor would make a normal circuit go to near zero volt.

3. Checks to see if low voltage signal is from the sensor or the circuit. Jumpering the sensor signal CKT 380 to 5 volts, checks the circuit, connections, and ECM.

4. This step checks to see if the low voltage signal was due to an open in the sensor circuit or the 5 volt reference circuit since the prior step eliminated the pressure sensor.

Diagnostic Aids:

Code 66 sets when signal voltage falls outside the normal possible range of the sensor and is not due to a refrigerant system problem. If problem is intermittent, check for opens or shorts in harness or poor connections.

2.2L (VIN 4) ENGINE — DIAGNOSTIC TROUBLE CODE CHART — 1993–94 CORSICA AND BERETTA

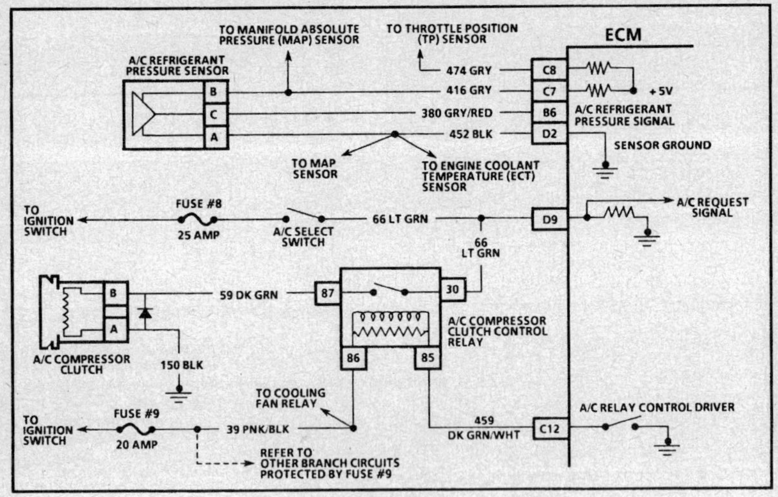

DTC 66
A/C REFRIGERANT PRESSURE SENSOR CIRCUIT
2.2L (VIN 4) "L" CARLINE

Circuit Description:

The A/C refrigerant pressure sensor responds to changes in A/C refrigerant system high side pressure. This input indicates how much load the A/C compressor is putting on the engine and is one of the factors used by the ECM to determine IAC valve position for idle speed control. The circuit consists of a 5 volt reference and a ground, both provided by the ECM, and a signal line to the ECM. The signal is a voltage which is proportional to the pressure. The sensor's range of operation is 0 to 454 psi. At 0 psi, the signal will be about .1 volt, varying up to about 4.9 volts at 454 psi or above. DTC 66 sets if the voltage is above 4.9 volts (454 psi) or below .3 volt (7.5 psi) for 5 seconds or more. The A/C compressor clutch is disabled by the ECM if fault is present.

Test Description: Number(s) below refer to circled number(s) on the diagnostic chart.

1. This step checks the voltage signal being received by the ECM from the A/C refrigerant pressure sensor.
2. Checks to see if the high voltage signal is from a shorted sensor or a short to voltage in the circuit. Normally, disconnecting the sensor would make a normal circuit go to near zero volt.
3. Checks to see if low voltage signal is from the sensor or the circuit. Jumpering the sensor signal CKT 380 to 5 volts, checks the circuit, connections, and ECM.
4. This step checks to see if the low voltage signal was due to an open in the sensor circuit or the 5 volt reference circuit since the prior step eliminated the pressure sensor.
5. The 5 volt reference for the Throttle Position (TP) sensor connects to the same 5 volt power supply in the ECM. If the 5 volt reference circuit to TP sensor has a short to ground, the MAP sensor and A/C refrigerant pressure sensor 5 volt reference circuit will also have a short to ground because they connect to the same power supply in the ECM. A short to ground on either 5 volt reference circuit does not damage the ECM 5 volt power supply. When the short is repaired, the ECM 5 volt supply will return to normal.

Diagnostic Aids:

At temperatures of -20 degrees fahrenheit and below, the A/C pressure can drop below the .3 volt (7.5 psi) threshold. If this happens the DTC 66 will set and disable the A/C system. Clear DTC 66 and check A/C pressure in a heated garage, if pressure is between .3 volt (7.5 psi) and 4.9 volts (454 psi), the A/C pressure transducer is functioning correctly.

DTC 66 sets when the signal voltage falls outside the normal possible range of the sensor. If signal voltage is greater than .1 volt, but less than .3 volt, there could be a low refrigerant pressure problem with the A/C system.

NOTICE: TO INSURE PROPER DIAGNOSIS OF THE DTC 66 CIRCUIT, THE A/C REFRIGERANT PRESSURE SHOULD BE WITHIN LIMITS

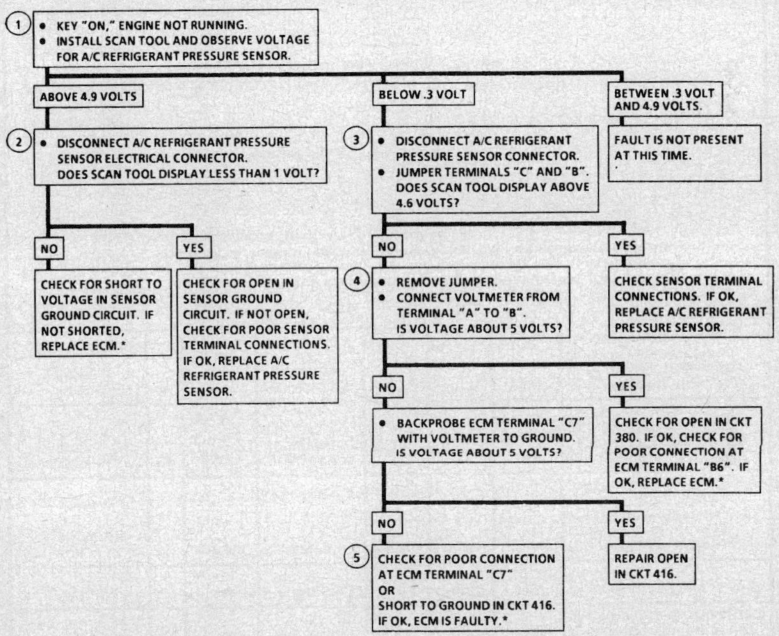

(1)
- KEY "ON," ENGINE NOT RUNNING.
- INSTALL SCAN TOOL AND OBSERVE VOLTAGE FOR A/C REFRIGERANT PRESSURE SENSOR.

- ABOVE 4.9 VOLTS
- BELOW .3 VOLT
- BETWEEN .3 VOLT AND 4.9 VOLTS.

(2)
- DISCONNECT A/C REFRIGERANT PRESSURE SENSOR ELECTRICAL CONNECTOR.
- DOES SCAN TOOL DISPLAY LESS THAN 1 VOLT?

(3)
- DISCONNECT A/C REFRIGERANT PRESSURE SENSOR CONNECTOR.
- JUMPER TERMINALS "C" AND "B". DOES SCAN TOOL DISPLAY ABOVE 4.6 VOLTS?

FAULT IS NOT PRESENT AT THIS TIME.

(from step 2) NO / YES

- NO: CHECK FOR SHORT TO VOLTAGE IN SENSOR GROUND CIRCUIT. IF NOT SHORTED, REPLACE ECM.*
- YES: CHECK FOR OPEN IN SENSOR GROUND CIRCUIT. IF NOT OPEN, CHECK FOR POOR SENSOR TERMINAL CONNECTIONS. IF OK, REPLACE A/C REFRIGERANT PRESSURE SENSOR.

(from step 3) NO / YES

- NO **(4)**:
  - REMOVE JUMPER.
  - CONNECT VOLTMETER FROM TERMINAL "A" TO "B". IS VOLTAGE ABOUT 5 VOLTS?
- YES: CHECK SENSOR TERMINAL CONNECTIONS. IF OK, REPLACE A/C REFRIGERANT PRESSURE SENSOR.

(from step 4) NO / YES

- NO:
  - BACKPROBE ECM TERMINAL "C7" WITH VOLTMETER TO GROUND. IS VOLTAGE ABOUT 5 VOLTS?
- YES: CHECK FOR OPEN IN CKT 380. IF OK, CHECK FOR POOR CONNECTION AT ECM TERMINAL "B6". IF OK, REPLACE ECM.*

NO / YES

- NO **(5)**: CHECK FOR POOR CONNECTION AT ECM TERMINAL "C7" OR SHORT TO GROUND IN CKT 416. IF OK, ECM IS FAULTY.*
- YES: REPAIR OPEN IN CKT 416.

---

* IF ECM IS FAULTY AND MUST BE REPLACED, THE NEW ECM MUST BE PROGRAMMED. REFER TO ECM REPLACEMENT AND PROGRAMMING PROCEDURES

DTC 66
A/C REFRIGERANT PRESSURE SENSOR CIRCUIT
2.2L (VIN 4) "L" CARLINE

2.2L (VIN 4) ENGINE — ECM SYMPTOM CHART — 1992 CORSICA AND BERETTA

	PINK A-B 24 PIN ECM CONNECTOR				NORMAL VOLTAGES			
ECM PIN/FUNCTION		CKT #	WIRE COLOR	COMPONENT CONNECTOR CAVITY	KEY "ON"	ENG RUN	CODES AFFECT	POSSIBLE SYMPTOMS FROM FAULTY CIRCUIT
A4	PEAK AND HOLD JUMPER	887	TAN	ECM-"A12"	(8)	(8)		(3) CRANKS, BUT WON'T RUN.
A5	INJECTOR DRIVER	467	DK BLU	INJECTOR JUMPER HARNESS CONNECTOR (GRY) "B"	B +	B +		(3) CRANKS, BUT WON'T RUN. (4) POSSIBLE NO START (BLOWN 10 AMP ECM FUSE) OR INJECTORS ENERGIZED AT ALL TIMES WITH KEY "ON"
A7	FUEL PUMP RELAY CONTROL	465	DK GRN/WHT	FUEL PUMP RELAY - "D"	(6)	B +		(3) LONG CRANKING TIME BEFORE ENGINE STARTS
A8	IAC "A" HIGH	441	LT BLU/WHT	IAC VALVE "D"	NOT USABLE	NOT USABLE		(5) INCORRECT, UNSTABLE IDLE
A9	IAC "A" LOW	442	LT BLU/BLK	IAC VALVE "C"	NOT USABLE	NOT USABLE		(5) INCORRECT, UNSTABLE IDLE
A10	IAC "B" LOW	444	LT GRN/BLK	IAC VALVE "A"	NOT USABLE	NOT USABLE		(5) INCORRECT, UNSTABLE IDLE
A11	IAC "B" HIGH	443	LT GRN/WHT	IAC VALVE "B"	NOT USABLE	NOT USABLE		(5) INCORRECT, UNSTABLE IDLE
A12	PEAK AND HOLD JUMPER	887	TAN	ECM "A4"	(8)	(8)		(3) CRANKS, BUT WON'T RUN.

NOTICE: Voltage may vary due to low battery charge or other reasons, but should be very close. All voltages shown in the "ENG RUN" column are typical with engine at idle, closed throttle, normal operating temperature, park or neutral, system in "closed loop" all accessories "OFF", brake not applied.

(1) Increases with vehicle speed (Measure on "AC volts" scale).
(2) Varies
(3) Open circuit.
(4) Grounded circuit.
(5) Open or grounded circuit.
(6) Measures B + for 2 seconds after KEY "ON", then measures "0" volts.
(7) Varies with temperature.
(8) Less than .5 volt (500 mV.)

2.2L (VIN 4) ENGINE — ECM SYMPTOM CHART — 1992 CORSICA AND BERETTA

	PINK A-B 24 PIN ECM CONNECTOR				NORMAL VOLTAGES			
ECM PIN/FUNCTION		CKT #	WIRE COLOR	COMPONENT CONNECTOR CAVITY	KEY "ON"	ENG RUN	CODES AFFECT	POSSIBLE SYMPTOMS FROM FAULTY CIRCUIT
B4	CTS SIGNAL	410	YEL	CTS "D"	(7)	(7)	14(5)	(5) INCORRECT IDLE, COOLING FAN RUNS AT ALL TIMES
B5	TPS SIGNAL	417	DK BLU	TPS "C"	6V (9)	6V (9)	21(5)	(5) POOR PERFORMANCE, HESITATION
B6	A/C PRESSURE SENSOR SIGNAL	380	GRY/RED	A/C PRESSURE SENSOR "C"	(2)	(2)	66(5)	(5) NO A/C COOLING, A/C CLUTCH INOPERATIVE, REFER TO CODE 66
B7	MAP SIGNAL	432	LT GRN	MAP SENSOR "B"	4.75V	(2)	33(5)	(5) POOR PERFORMANCE, ROUGH IDLE, SURGE REFER TO CHART C-1D
B8	IAT SIGNAL	472	TAN	IAT SENSOR "B"	(7)	(7)	23(5)	(5) POOR PERFORMANCE, HARD TO START
B9	ALDL DIAGNOSTIC TEST TERMINAL	451	WHT/BLK	ALDL "B"	5.0V	5.0V		(3) NO ALDL DATA, WON'T FLASH CODES. (4) KEY "ON" - CODES DISPLAYED ENG RUN - FIELD SERVICE MODE "ON".
B10	VSS SIGNAL (HIGH)	400	YEL	VEHICLE SPEED SENSOR (VSS) "A"	**	**	24(5)	(5) NO VSS SIGNAL, INACCURATE OR INOPERATIVE SPEEDOMETER, INOPERATIVE CRUISE CONTROL, TCC INOPERATIVE.
B11	VSS SIGNAL (LOW)	401	PPL	VEHICLE SPEED SENSOR (VSS) "B"	**	**	24(5)	(5) NO VSS SIGNAL, INACCURATE OR INOPERATIVE SPEEDOMETER, INOPERATIVE CRUISE CONTROL, TCC INOPERATIVE.

NOTICE: Voltages may vary due to low battery charge or other reasons, but should be very close. All voltages shown in the "ENG. RUN" column are typical with engine at idle, closed throttle, normal operating temperature, park or neutral, system in "closed loop" all accessories "OFF", brake not applied.

(1) Increases with vehicle speed (Measure on "AC volts" scale).
(2) Varies
(3) Open circuit.
(4) Grounded circuit.
(5) Open or grounded circuit.
(6) Measures B + for 2 seconds after KEY "ON", then measures "0" volt.
(7) Varies with temperature
(8) Less than .5 volt (500 mV.)
(9) Within the range of .33-1.33 volts

2.2L (VIN 4) ENGINE — ECM SYMPTOM CHART — 1992 CORSICA AND BERETTA

PINK C-D 32 PIN ECM CONNECTOR

ECM PIN/FUNCTION	CKT #	WIRE COLOR	COMPONENT CONNECTOR CAVITY	NORMAL VOLTAGES KEY "ON"	NORMAL VOLTAGES ENG RUN	CODES AFFECT	POSSIBLE SYMPTOMS FROM FAULTY CIRCUIT
C1 ECM GROUND	551	TAN/WHT	ENGINE GROUND	(8)	(8)		(3) NO EFFECT UNLESS "D1" IS ALSO OPEN. IF BOTH ARE OPEN, CRANKS, BUT WON'T RUN
C4 PARK/NEUTRAL SWITCH	434	ORN/BLK	NEUTRAL START BACK-UP SWITCH - "A"	(8)	(8)		(5) INCORRECT IDLE
C6 IGNITION FEED	439	PNK/BLK	"DIS" MODULE 2 PIN CONNECTOR - "B", FUEL INJECTOR JUMPER HARNESS CONNECTOR - "A", EGR SOLENOID - "B"	B +	B +		(5) CRANKS, BUT WON'T RUN. (4) CRANKS, BUT WON'T RUN, BLOWN 15 AMP FUSE #1
C7 MAP & A/C PRESSURE SENSORS 5V REF	416	GRY	MAP - "C", AND A/C PRESSURE SENSOR- "B"	5.0V	5.0V	33(5) 66(5)	(5) DEGRADED PERFORMANCE NO A/C COOLING
C8 TPS 5 VOLT REFERENCE	474	GRY	TPS - "A"	5.0V	5.0V	21(5)	(3) DEGRADED PERFORMANCE
C9 KEEP ALIVE MEMORY (B +)	340	ORN	FUEL PUMP RELAY - "A", AND OIL PRESS SWITCH - "C"	B +	B +		(3) CRANKS, BUT WON'T RUN. (4) CRANKS, BUT WON'T RUN, BLOWN UNDERHOOD 20 AMP FUSE.
C11 TCC (A/T) OR SHIFT LIGHT (M/T) CONTROL	422 / 456	TAN/BLK	TCC SOLENOID - "D", AND ALDL - "F"	A/T - (8) M/T - B +	A/T - (8) M/T - B +		(5) TCC OR SHIFT LIGHT INOPERATIVE, POOR FUEL ECONOMY (4) TCC ENGAGES AT VERY LOW ROAD SPEED - (A/T) OR SHIFT LIGHT "ON" AT ALL TIMES - (M/T)
C12 A/C RELAY CONTROL	459	DK GRN/WHT	A/C COMPRESSOR RELAY - "F"	B +	B +		(3) NO A/C COOLING (4) NO A/C COOLING, ENGINE OVERHEATS. BLOWN 20 AMP FUSE #9.
C13 COOLING FAN RELAY CONTROL	335	DK GRN/WHT	COOLING FAN RELAY - "F"	B +	B +		(3) ENGINE OVERHEATS. (4) COOLING FAN RUNS AT ALL TEMPERATURES.
C14 EGR SOLENOID CONTROL	435	GRY	EGR SOLENOID - "A"	B +	B +	32(5)	(3) EGR VALVE DOESN'T OPEN DETONATION, SPARK KNOCK. (4) DEGRADED PERFORMANCE IMPORTANT: EGR SOLENOID IS DISABLED IF CODE 14, 21, 23, 32 OR 33 IS SET.

NOTICE: Voltages may vary due to low battery charge or other reasons, but should be very close. All voltages shown in the "ENG RUN" column are typical with engine at idle, closed throttle, normal operating temperature, park or neutral, system in "closed loop" all accessories "OFF", brake not applied.

(2) Varies.
(3) Open circuit.
(4) Grounded circuit.
(5) Open or grounded circuit.
(7) Varies with temperature.
(8) Less than .5 volt (500 mV.)

2.2L (VIN 4) ENGINE — ECM SYMPTOM CHART — 1992 CORSICA AND BERETTA

PINK C-D 24 PIN ECM CONNECTOR

ECM PIN/FUNCTION	CKT #	WIRE COLOR	COMPONENT CONNECTOR CAVITY	NORMAL VOLTAGES KEY "ON"	NORMAL VOLTAGES ENG RUN	CODES AFFECT	POSSIBLE SYMPTOMS FROM FAULTY CIRCUIT
D1 ECM GROUND	450	BLK/WHT	ENGINE GROUND	(8)	(8)		(3) NO EFFECT UNLESS "C1" IS ALSO OPEN. IF BOTH ARE OPEN, CRANKS, BUT WON'T RUN.
D2 SENSOR GROUND	455	PPL	MAP - "A", - CTS - "C", AND A/C PRESSURE SENSOR - "A"	(8)	(8)	14(3) 33(3) 66(3)	(3) HARD START OR CRANKS BUT WON'T RUN.
D3 SENSOR GROUND	452	BLK	TPS - "B", AND IAT - "A"	(8)	(8)	21(3) 23(3)	(3) HARD START OR CRANKS BUT WON'T RUN.
D5 EST	423	WHT	"DIS" MODULE 6 PIN CONN. "C"	(8)	2.4V	42(5)	(5) STUMBLES, UNSTABLE IDLE, DEGRADED PERFORMANCE
D6 BYPASS	424	TAN/BLK	"DIS" MODULE 6 PIN CONN. "D"	(8)	4.8V	42(5)	(5) DEGRADED PERFORMANCE
D7 IGNITION REFERENCE HIGH	430	PPL/WHT	"DIS" MODULE 6 PIN CONN. "E"	(8)	3.2V		(5) CRANKS BUT WON'T RUN.
D8 IGNITION REFERENCE LOW	453	BLK/RED	"DIS" MODULE 6 PIN CONN. "F"	(8)	(8)		(3) REDUCED PERFORMANCE
D9 A/C REQUEST	66	LT GRN	A/C RELAY CONNECTOR "A"	(8) "OFF" B + "ON"	(8) "OFF" B + "ON"		(3) NO A/C COOLING A/C CLUTCH WON'T ENGAGE.
D10 OXYGEN SENSOR REFERENCE LOW	413	TAN	ENGINE GROUND	(8)	(8)	13(3)	(3) OPEN LOOP OPERATION REDUCED PERFORMANCE
D11 OXYGEN SENSOR SIGNAL	412	PPL	OXYGEN SENSOR	.01-.55V	1.9V (10)	13 44 45	(5) OPEN LOOP OPERATION, STRONG EXHAUST ODOR, REDUCED PERFORMANCE.
D12 SERIAL DATA	461	ORN	ALDL CONNECTOR "M"	4.7V	4.7V		(5) NO DATA
D13 VEHICLE SPEED SENSOR OUTPUT	389	DK GRN	I/P CONNECTOR "A", AND I/P CLUSTER CONNECTOR "L"				(5) INCORRECT SPEEDOMETER READING, CRUISE CONTROL INOPERATIVE
D14 "SERVICE ENGINE SOON" LIGHT CONTROL	419	BRN/WHT	I/P CONNECTOR "B", AND I/P CLUSTER CONNECTOR "L"	(8)	B +		(3) LIGHT NOT "ON" FOR "BULB CHECK," WON'T FLASH CODES. (4) LIGHT "ON" ALL THE TIME, DOES NOT FLASH, REFER TO "DIAGNOSTIC CIRCUIT CHECK."

NOTICE: Voltages may vary due to low battery charge or other reasons, but should be very close. All voltages shown in the "ENG RUN" column are typical with engine at idle, closed throttle, normal operating temperature, park or neutral, system in "closed loop" all accessories "OFF", brake not applied.

(3) Open circuit.
(4) Grounded circuit.
(5) Open or grounded circuit.
(8) Less than .5 volt (500 mV.)

(10) Actively varies with indicated range.

2.2L (VIN 4) ENGINE — ECM SYMPTOM CHART — 1993–94 CORSICA AND BERETTA

ECM Connector and Driveability Symptoms Identification

This ECM voltage chart is for use with a J 39200 to further aid in diagnosis. These voltages were derived from a known good vehicle. The voltages you get may vary due to low battery charge or other reasons, but they should be very close.

THE FOLLOWING CONDITIONS MUST BE MET BEFORE TESTING:
- Engine at operating temperature • Closed Loop • Engine idling (for "Engine Run" column)
- Test terminal <u>not</u> grounded • Scan tool <u>not</u> installed • Brake <u>not</u> applied.

ECM PIN/FUNCTION	CKT #	WIRE COLOR	COMPONENT CONNECTOR CAVITY	NORMAL VOLTAGES KEY "ON"	NORMAL VOLTAGES ENG RUN	DTC(s) AFFECTED	POSSIBLE SYMPTOMS FROM FAULTY CIRCUIT
A4 PEAK AND HOLD JUMPER	887	TAN	ECM-"A12"	(8)	(8)		(3) CRANKS, BUT WON'T RUN.
A5 INJECTOR DRIVER	467	DK BLU	INJECTOR JUMPER HARNESS CONNECTOR (GRY) "B"	B+	B+ (13)		(3) CRANKS, BUT WON'T RUN. (4) POSSIBLE NO START (OPEN 15 AMP FUSE #1) OR INJECTORS ENERGIZED AT ALL TIMES WITH KEY "ON."
A7 FUEL PUMP RELAY CONTROL	465	DK GRN/WHT	FUEL PUMP RELAY - "86"	(6)	B+		(3) LONG CRANKING TIME BEFORE ENGINE STARTS.
A8 IAC "A" HIGH	441	LT BLU/WHT	IAC VALVE "D"	NOT USABLE	NOT USABLE		(5) INCORRECT, UNSTABLE IDLE.
A9 IAC "A" LOW	442	LT BLU/BLK	IAC VALVE "C"	NOT USABLE	NOT USABLE		(5) INCORRECT, UNSTABLE IDLE.
A10 IAC "B" LOW	444	LT GRN/BLK	IAC VALVE "A"	NOT USABLE	NOT USABLE		(5) INCORRECT, UNSTABLE IDLE.
A11 IAC "B" HIGH	443	LT GRN/WHT	IAC VALVE "B"	NOT USABLE	NOT USABLE		(5) INCORRECT, UNSTABLE IDLE.
A12 PEAK AND HOLD JUMPER	887	TAN	ECM "A4"	(8)	(8)		(3) CRANKS, BUT WON'T RUN.

(2) VARIES.
(3) OPEN CIRCUIT.
(4) GROUNDED CIRCUIT.
(5) OPEN OR GROUNDED CIRCUIT.
(7) VARIES WITH TEMPERATURE.
(9) WITHIN THE RANGE OF .33-1.33 VOLTS.

24 PIN A-B CONNECTOR

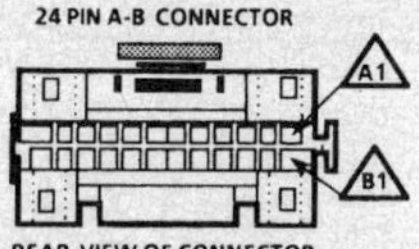

A1
B1

REAR VIEW OF CONNECTOR (PINK)

2.2L (VIN 4) ENGINE — ECM SYMPTOM CHART — 1993–94 CORSICA AND BERETTA

ECM Connector and Driveability Symptoms Identification

This ECM voltage chart is for use with a J 39200 to further aid in diagnosis. These voltages were derived from a known good vehicle. The voltages you get may vary due to low battery charge or other reasons, but they should be very close.

THE FOLLOWING CONDITIONS MUST BE MET BEFORE TESTING:
- Engine at operating temperature • Closed Loop • Engine idling (for "Engine Run" column)
- Test terminal <u>not</u> grounded • Scan tool <u>not</u> installed • Brake <u>not</u> applied.

ECM PIN/FUNCTION	CKT #	WIRE COLOR	COMPONENT CONNECTOR CAVITY	NORMAL VOLTAGES KEY "ON"	NORMAL VOLTAGES ENG RUN	DTC(S) AFFECT	POSSIBLE SYMPTOMS FROM FAULTY CIRCUIT
B4 ECT SIGNAL	410	YEL	ECT SENSOR "D"	(7)	(7)	14 (5)	(5) INCORRECT IDLE, COOLING FAN RUNS AT ALL TIMES.
B5 TP SIGNAL	417	DK BLU	TP SENSOR "C"	.6V (9)	.6V (9)	21 (5)	(5) POOR PERFORMANCE, HESITATION.
B6 A/C REFRIGERANT PRESSURE SENSOR SIGNAL	380	GRY/RED	A/C REFRIGERANT PRESSURE SENSOR "C"	(2)	(2)	66 (5)	(5) NO A/C COOLING, A/C COMPRESSOR CLUTCH INOPERATIVE, REFER TO <u>DTC 66</u>.
B7 MAP SIGNAL	432	LT GRN	MAP SENSOR "B"	4.75V	(2)	33 (5)	(5) POOR PERFORMANCE, ROUGH IDLE, SURGE REFER TO CHART C-1D.
B8 IAT SIGNAL	472	TAN	IAT SENSOR "B"	(7)	(7)	23 (5)	(5) POOR PERFORMANCE, HARD TO START.
B9 DATA LINK TEST TERMINAL	451	WHT/BLK	DATA LINK CONNECTOR "B"	5.0V	5.0V		(3) NO DATA, WON'T FLASH DTC(S). (4) KEY "ON" - DTC(S) DISPLAYED ENG RUN - FIELD SERVICE MODE "ON."
B10 VSS SIGNAL (HIGH)	400	YEL	VEHICLE SPEED SENSOR (VSS) "A"			24 (5)	(5) NO VSS SIGNAL, INACCURATE OR INOPERATIVE SPEEDOMETER, INOPERATIVE CRUISE CONTROL, TCC INOPERATIVE.
B11 VSS SIGNAL (LOW)	401	PPL	VEHICLE SPEED SENSOR (VSS) "B"			24 (5)	(5) NO VSS SIGNAL, INACCURATE OR INOPERATIVE SPEEDOMETER, INOPERATIVE CRUISE CONTROL, TCC INOPERATIVE.

(2) VARIES.
(3) OPEN CIRCUIT.
(4) GROUNDED CIRCUIT.
(5) OPEN OR GROUNDED CIRCUIT.
(7) VARIES WITH TEMPERATURE.
(9) WITHIN THE RANGE OF .33-1.33 VOLTS.

24 PIN A-B CONNECTOR

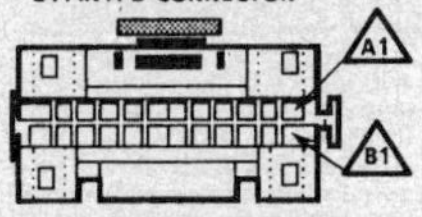

A1
B1

REAR VIEW OF CONNECTOR (PINK)

2.2L (VIN 4) ENGINE — ECM SYMPTOM CHART — 1993–94 CORSICA AND BERETTA

ECM Connector and Driveability Symptoms Identification

This ECM voltage chart is for use with a J 39200 to further aid in diagnosis. These voltages were derived from a known good vehicle. The voltages you get may vary due to low battery charge or other reasons, but they should be very close.

THE FOLLOWING CONDITIONS MUST BE MET BEFORE TESTING:
- Engine at operating temperature • Closed Loop • Engine idling (for "Engine Run" column)
- Test terminal not grounded • Scan tool not installed • Brake not applied.

ECM PIN/FUNCTION	CKT #	WIRE COLOR	COMPONENT CONNECTOR CAVITY	NORMAL VOLTAGES KEY "ON"	NORMAL VOLTAGES ENG RUN	DTC(s) AFFECTED	POSSIBLE SYMPTOMS FROM FAULTY CIRCUIT
C1 ECM GROUND	551	TAN/WHT	ENGINE GROUND	(8) (15)	(8) (15)		(3) NO EFFECT UNLESS "D1" IS ALSO OPEN. IF BOTH ARE OPEN, CRANKS, BUT WON'T RUN.
C4 PARK/NEUTRAL POSITION (PNP) SWITCH	434	ORN/BLK	PARK/NEUTRAL POSITION (PNP) SWITCH "A"	(8)	(8)		(5) INCORRECT IDLE.
C6 IGNITION FEED	439	PNK/BLK	ICM 2 PIN CONN "B", FUEL INJ JUMPER HARNESS CONN "A", EGR SOLENOID "B"	B + (15)	B + (15)		(5) CRANKS, BUT WON'T RUN. (4) CRANKS, BUT WON'T RUN, OPEN 15 AMP ECM FUSE.
C7 MAP AND A/C REFRIGERANT PRESSURE SENSORS 5V REF	416	GRY	MAP - "C" AND A/C REFRIGERANT PRESSURE SENSOR "B"	5.0V	5.0V	33(5) 66(5)	(5) DEGRADED PERFORMANCE. NO A/C COOLING.
C8 TP SENSOR 5 VOLT REFERENCE	474	GRY	TP SENSOR "A"	5.0V	5.0V	21(5)	(3) DEGRADED PERFORMANCE.
C9 KEEP ALIVE MEMORY (B +)	340	ORN	FUEL PUMP RELAY "87" AND OIL PRESS SWITCH "C" AND 20 AMP UNDERHOOD FUSE	B + (15)	B + (15)		(3) CRANKS, BUT WON'T RUN. (4) CRANKS, BUT WON'T RUN, OPEN UNDERHOOD 20 AMP FUSE.
C11 TCC (A/T) OR SHIFT LIGHT (M/T) CONTROL	422 456	TAN/BLK	TCC SOLENOID "D" AND DATA LINK CONNECTOR "F"	A/T - (8) M/T - B +	A/T - (8) M/T - B +		(5) TCC OR SHIFT LIGHT INOPERATIVE, POOR FUEL ECONOMY (4) TCC ENGAGES AT VERY LOW ROAD SPEED - (A/T) OR SHIFT LIGHT "ON" AT ALL TIMES - (M/T).
C12 A/C COMPRESSOR CLUTCH RELAY CONTROL	459	DK GRN/WHT	A/C COMPRESSOR CLUTCH CONTROL RELAY "85"	B +	B +		(3) NO A/C COOLING. (4) NO A/C COOLING, ENGINE OVERHEATS. OPEN 10 AMP ENG/A/C FUSE.
C13 COOLING FAN RELAY CONTROL	335	DK GRN/WHT	COOLING FAN CONTROL RELAY "85"	B +	B +		(3) ENGINE OVERHEATS. (4) COOLING FAN RUNS AT ALL TEMPERATURES.
C14 EGR SOLENOID CONTROL	435	GRY	EGR CONTROL VALVE RELAY SOLENOID "A"	B +	B +	32(5)	(3) EGR VALVE DOESN'T OPEN, HIGH NOx DETONATION, SPARK KNOCK. (4) DEGRADED PERFORMANCE. IMPORTANT: EGR SOLENOID IS DISABLED IF DTC 14, 21, 23, 32 OR 33 IS SET.

(3) OPEN CIRCUIT.
(4) GROUNDED CIRCUIT.
(5) OPEN OR GROUNDED CIRCUIT.
(8) LESS THAN .5 VOLT (500 mV).
(15) THIS TYPE OF CIRCUIT SHOULD ALSO BE CHECKED USING J 34142-B TEST LIGHT.

32 PIN C-D CONNECTOR

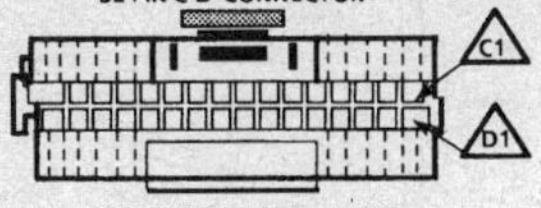

REAR VIEW OF CONNECTOR (PINK)

2.2L (VIN 4) ENGINE — ECM SYMPTOM CHART — 1993–94 CORSICA AND BERETTA

ECM Connector and Driveability Symptoms Identification

This ECM voltage chart is for use with a J 39200 to further aid in diagnosis. These voltages were derived from a known good vehicle. The voltages you get may vary due to low battery charge or other reasons, but they should be very close.

THE FOLLOWING CONDITIONS MUST BE MET BEFORE TESTING:
- Engine at operating temperature • Closed Loop • Engine idling (for "Engine Run" column)
- Test terminal not grounded • Scan tool not installed • Brake not applied.

ECM PIN/FUNCTION	CKT #	WIRE COLOR	COMPONENT CONNECTOR CAVITY	NORMAL VOLTAGES KEY "ON"	NORMAL VOLTAGES ENG RUN	DTC(S) AFFECT	POSSIBLE SYMPTOMS FROM FAULTY CIRCUIT
D1 ECM GROUND	450	BLK/WHT	ENGINE GROUND	(8) (15)	(8) (15)		(3) NO EFFECT UNLESS "C1" IS ALSO OPEN. IF BOTH ARE OPEN, CRANKS, BUT WON'T RUN.
D2 SENSOR GROUND	455	PPL	MAP SENSOR "A", ECT SENSOR "C" AND A/C REFRIGERANT PRESSURE SENSOR "A"	(8)	(8)	14 (3) 33 (3) 66 (3)	(3) HARD START OR CRANKS, BUT WON'T RUN.
D3 SENSOR GROUND	452	BLK	TP SENSOR "B" AND IAT SENSOR "A"	(8)	(8)	21 (3) 23 (3)	(3) HARD START OR CRANKS, BUT WON'T RUN.
D5 IGNITION CONTROL (IC)	423	WHT	IGNITION CONTROL MODULE 6 PIN CONN. "C"	(8)	2.4V	42 (5)	(5) STUMBLES, UNSTABLE IDLE, DEGRADED PERFORMANCE.
D6 BYPASS	424	TAN/BLK	IGNITION CONTROL MODULE 6 PIN CONN. "D"	(8)	4.8V	42 (5)	(5) DEGRADED PERFORMANCE.
D7 IGNITION REFERENCE HIGH	430	PPL/WHT	IGNITION CONTROL MODULE 6 PIN CONN. "E"	(8)	3.2V (14)		(5) CRANKS, BUT WON'T RUN.
D8 IGNITION REFERENCE LOW	453	BLK/RED	IGNITION CONTROL MODULE 6 PIN CONN. "F"	(8)	(8)		(3) REDUCED PERFORMANCE
D9 A/C REQUEST	66	LT GRN	8 PIN I/P CONN. "A" AND A/C COMP CLUTCH CONTROL RELAY "30"	(8) "OFF" B + "ON"	(8) "OFF" B + "ON"		(3) NO A/C COOLING A/C COMPRESSOR CLUTCH WON'T ENGAGE. (4) SAME AS OPEN - FUSE #8 OPEN.
D10 OXYGEN SENSOR (O2S) REFERENCE LOW	413	TAN	ENGINE GROUND	(8)	(8)	13 (3)	(3) "OPEN LOOP" OPERATION REDUCED PERFORMANCE.
D11 OXYGEN SENSOR (O2S) SIGNAL	412	PPL	OXYGEN SENSOR	.01-.55V	1.9V (10)	13 44 45	(5) "OPEN LOOP" OPERATION, STRONG EXHAUST ODOR, REDUCED PERFORMANCE.
D12 SERIAL DATA	461	ORN	DATA LINK CONNECTOR "M"	4.7V	4.7V		(5) NO DATA.
D13 VEHICLE SPEED SENSOR (VSS) OUTPUT	817	DK GRN/WHT	8 PIN I/P CONN "G" (SEE SECTION 8A FOR ADDITIONAL INFO				(5) INCORRECT SPEEDOMETER READING, CRUISE CONTROL INOPERATIVE.
D14 MIL (SERVICE ENGINE SOON) LAMP CONTROL	419	BRN/WHT	8 PIN I/P CONN "B" (SEE SECTION 8A FOR ADDITIONAL INFO	.01-1.0V	B +		(3) LAMP NOT "ON" FOR "BULB CHECK," WON'T FLASH DTC(S). (4) LAMP "ON" ALL THE TIME, DOES NOT FLASH, REFER TO "OBD SYSTEM CHECK."

(3) OPEN CIRCUIT.
(4) GROUNDED CIRCUIT.
(8) LESS THAN .5 VOLT (500 mV).
(10) ACTIVELY VARIES WITH INDICATED RANGE.
(14) J 39200 RED LEAD TO REF HIGH ("D7" AT ECM) AND BLACK LEAD TO ECM GROUND ("D1" AT ECM); SHOULD MEASURE ABOUT 25 - 30 Hz IN DC VOLTAGE RANGE.
(15) THIS TYPE OF CIRCUIT SHOULD ALSO BE CHECKED USING J 34142-B TEST LIGHT.

32 PIN C-D CONNECTOR

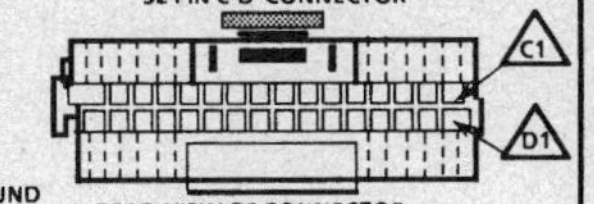

REAR VIEW OF CONNECTOR (PINK)

2.2L (VIN 4) ENGINE — COMPONENT DIAGNOSTIC CHART — CORSICA AND BERETTA

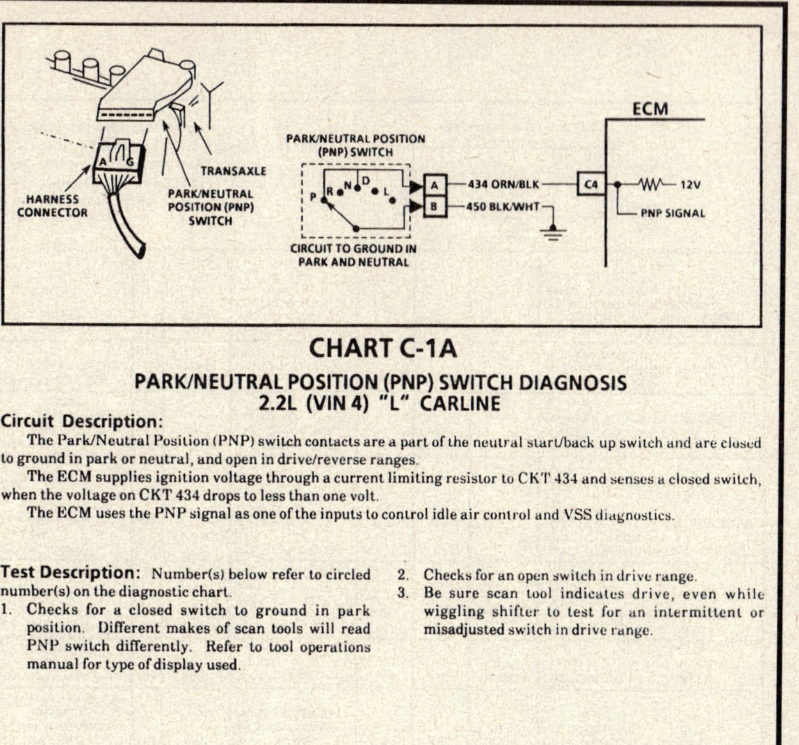

CHART C-1A
PARK/NEUTRAL POSITION (PNP) SWITCH DIAGNOSIS
2.2L (VIN 4) "L" CARLINE

Circuit Description:

The Park/Neutral Position (PNP) switch contacts are a part of the neutral start/back up switch and are closed to ground in park or neutral, and open in drive/reverse ranges.

The ECM supplies ignition voltage through a current limiting resistor to CKT 434 and senses a closed switch, when the voltage on CKT 434 drops to less than one volt.

The ECM uses the PNP signal as one of the inputs to control idle air control and VSS diagnostics.

Test Description: Number(s) below refer to circled number(s) on the diagnostic chart.

1. Checks for a closed switch to ground in park position. Different makes of scan tools will read PNP switch differently. Refer to tool operations manual for type of display used.

2. Checks for an open switch in drive range.
3. Be sure scan tool indicates drive, even while wiggling shifter to test for an intermittent or misadjusted switch in drive range.

2.2L (VIN 4) ENGINE — COMPONENT DIAGNOSTIC CHART — CORSICA AND BERETTA

CHART C-1A
PARK/NEUTRAL POSITION (PNP) SWITCH DIAGNOSIS
2.2L (VIN 4) "L" CARLINE

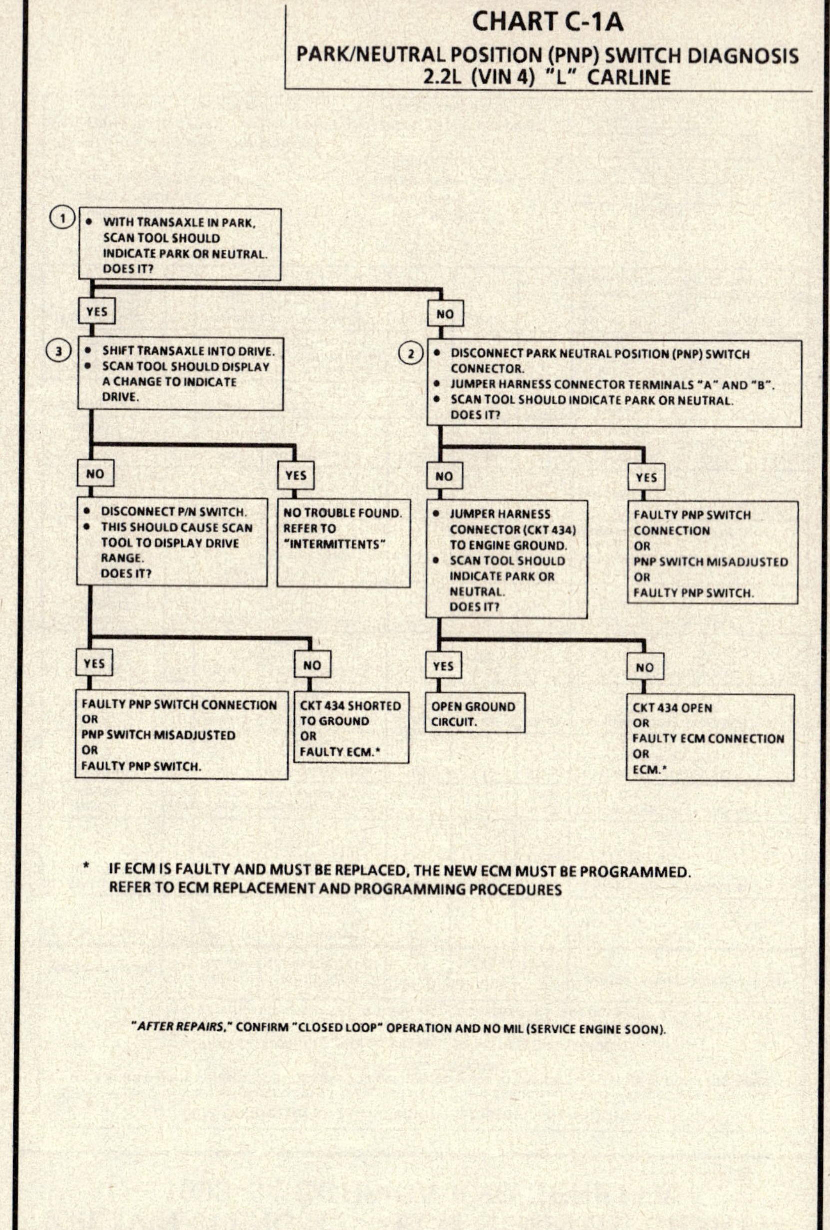

* IF ECM IS FAULTY AND MUST BE REPLACED, THE NEW ECM MUST BE PROGRAMMED. REFER TO ECM REPLACEMENT AND PROGRAMMING PROCEDURES

"AFTER REPAIRS," CONFIRM "CLOSED LOOP" OPERATION AND NO MIL (SERVICE ENGINE SOON).

2.2L (VIN 4) ENGINE — COMPONENT DIAGNOSTIC CHART — CORSICA AND BERETTA

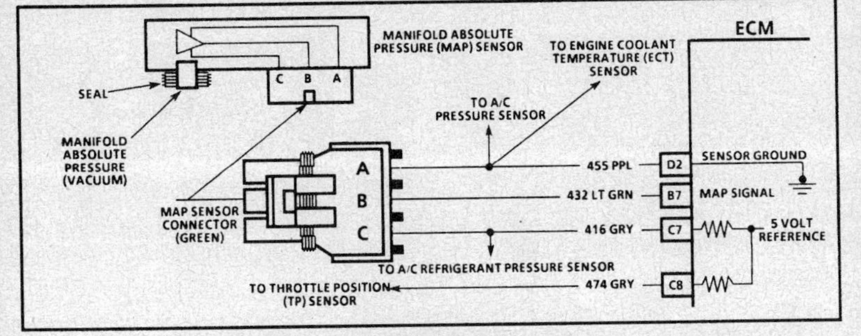

CHART C-1D
MANIFOLD ABSOLUTE PRESSURE (MAP) SENSOR OUTPUT CHECK
2.2L (VIN 4) "L" CARLINE

Circuit Description:

The Manifold Absolute Pressure (MAP) sensor measures the changes in the intake manifold pressure which result from engine load (intake manifold vacuum) and RPM changes; and converts these into a voltage output. The ECM sends a 5 volts reference voltage to the MAP sensor. As the manifold pressure changes, the output voltage of the sensor also changes. By monitoring the sensor output voltage, the ECM calculates the manifold pressure. A lower pressure (low voltage) output voltage will be about 1 to 2 volts at idle. While higher pressure (high voltage) output voltage will be about 4 to 4.8 at Wide Open Throttle (WOT). The MAP sensor is also used, under certain conditions, to measure barometric pressure, allowing the ECM to make adjustments for altitude changes. The ECM uses the MAP sensor to control fuel delivery and ignition timing.

Test Description: Number(s) below refer to circled number(s) on the diagnostic chart.

Important
- Be sure to use the same Diagnostic Test Equipment for all measurements.

1. When comparing scan readings to a known good vehicle, it is important to compare vehicles that use a MAP sensor having the same color insert and the same "Hot Stamped" number. See Figures 1 and 2

2. Applying 34 kPa (10" Hg) vacuum to the MAP sensor should result in voltage readings of 1.5 to 2.1 volts less than the voltage in Step 1. Upon applying vacuum to the sensor, the change in voltage should be instantaneous. A slow voltage change indicates a faulty sensor.

3. Check vacuum seal to sensor for leaking or restriction.

NOTICE: Make sure electrical connector remains securely fastened.

4. Remove sensor from the intake plenum and twist sensor **(by hand only)** to check for intermittent connection. Output changes greater than .10 volt indicate a faulty sensor or connection. If OK, replace sensor.

2.2L (VIN 4) ENGINE — COMPONENT DIAGNOSTIC CHART — CORSICA AND BERETTA

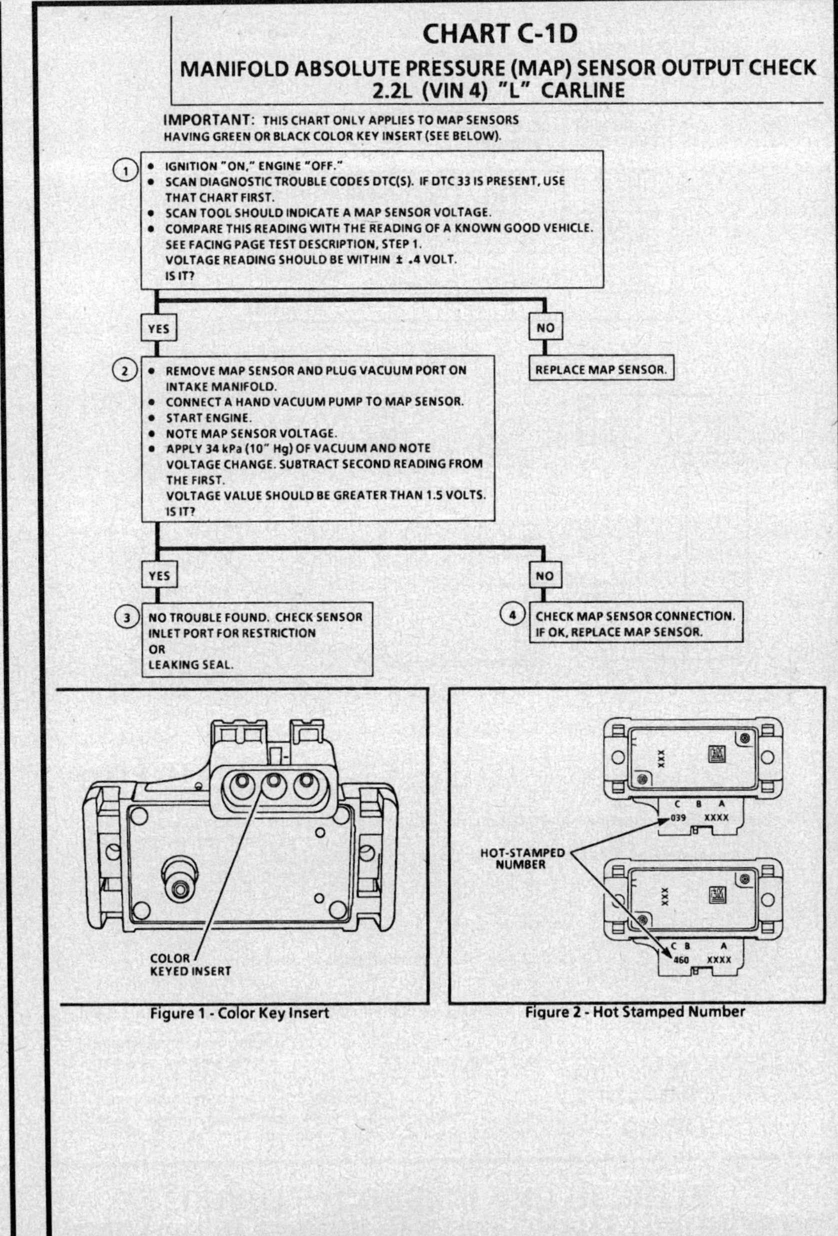

CHART C-1D
MANIFOLD ABSOLUTE PRESSURE (MAP) SENSOR OUTPUT CHECK
2.2L (VIN 4) "L" CARLINE

IMPORTANT: THIS CHART ONLY APPLIES TO MAP SENSORS HAVING GREEN OR BLACK COLOR KEY INSERT (SEE BELOW).

(1)
- IGNITION "ON," ENGINE "OFF."
- SCAN DIAGNOSTIC TROUBLE CODES DTC(S). IF DTC 33 IS PRESENT, USE THAT CHART FIRST.
- SCAN TOOL SHOULD INDICATE A MAP SENSOR VOLTAGE.
- COMPARE THIS READING WITH THE READING OF A KNOWN GOOD VEHICLE. SEE FACING PAGE TEST DESCRIPTION, STEP 1. VOLTAGE READING SHOULD BE WITHIN ± .4 VOLT. IS IT?

YES → (2)

NO → REPLACE MAP SENSOR.

(2)
- REMOVE MAP SENSOR AND PLUG VACUUM PORT ON INTAKE MANIFOLD.
- CONNECT A HAND VACUUM PUMP TO MAP SENSOR.
- START ENGINE.
- NOTE MAP SENSOR VOLTAGE.
- APPLY 34 kPa (10" Hg) OF VACUUM AND NOTE VOLTAGE CHANGE. SUBTRACT SECOND READING FROM THE FIRST. VOLTAGE VALUE SHOULD BE GREATER THAN 1.5 VOLTS. IS IT?

YES → (3)

NO → (4)

(3) NO TROUBLE FOUND. CHECK SENSOR INLET PORT FOR RESTRICTION OR LEAKING SEAL.

(4) CHECK MAP SENSOR CONNECTION. IF OK, REPLACE MAP SENSOR.

Figure 1 - Color Key Insert

Figure 2 - Hot Stamped Number

2.2L (VIN 4) ENGINE — COMPONENT DIAGNOSTIC CHART — CORSICA AND BERETTA

CHART C-2A

INJECTOR BALANCE TEST
2.2L (VIN 4) "L" CARLINE (PORT)

The injector balance tester is a tool used to turn the injector "ON" for a precise amount of time, thus spraying a measured amount of fuel into the manifold. This causes a drop in fuel rail pressure that we can record and compare between each injector. All injectors should have the same amount of pressure drop ($\pm$10 kPa). Any injector with a pressure drop that is 10 kPa (or more) greater or less than the average drop of the other injectors should be considered faulty and replaced.

STEP 1

Engine "cool down" period (10 minutes) is necessary to avoid irregular readings due to "Hot Soak" fuel boiling. Relieve fuel pressure in the fuel rail using the "fuel pressure relief procedure" described previously in this section. With ignition "OFF" connect fuel gage J 29658-D or equivalent. (Refer to CHART A-7 Page 3 of 3 for connecting fuel pressure gage).

Disconnect electrical connectors at all injectors except #2. Then, connect injector adaptor harness J 34730-200 to the injector jumper harness connector. Now, connect tester J 34730-3 to adaptor harness J 34730-200. Ignition must be "OFF" at least 10 seconds to complete ECM shutdown cycle. Fuel pump should run about 2 seconds after ignition is turned "ON." At this point, insert clear tubing attached to vent valve into a suitable container and bleed air from gauge and hose to insure accurate gauge operation. Repeat this step until all air is bled from gauge.

STEP 2

Turn ignition "OFF" for 10 seconds and then "ON" again to get fuel pressure to its maximum. Record this initial pressure reading. Energize tester one time and note pressure drop at its lowest point. (Disregard any slight pressure increase after drop hits low point.) By subtracting this second pressure reading from the initial pressure, we have the actual amount of injector pressure drop.

STEP 3

Disconnect injector adaptor harness J 34730-200 and reconnect injector jumper harness connector to the engine harness connector. Now connect J 34730-200 directly to each of the remaining injectors. Repeat Step 2 on each injector and compare the amount of drop. Usually, good injectors will have virtually the same drop. Retest any injector that has a pressure difference of 10 kPa, either more or less than the average of the other injectors on the engine. Replace any injector that also fails the retest. If the pressure drop of all injectors is within 10 kPa of this average, the injectors appear to be flowing properly.

NOTICE: The entire test should not be repeated more than once without running the engine to prevent flooding. (This includes any retest on faulty injectors).

2.2L (VIN 4) ENGINE — COMPONENT DIAGNOSTIC CHART — CORSICA AND BERETTA

NOTICE: The entire test should <u>NOT</u> be repeated more than once without running the engine to prevent flooding. (This includes any retest on faulty injectors.)

CHART C-2A

INJECTOR BALANCE TEST
2.2L (VIN 4) "L" CARLINE

The fuel pressure test in Chart A-7, should be completed prior to this test.

Step 1. If engine is at operating temperature, allow a 10 minute "cool down" period then connect fuel pressure gage and injector tester.
1. Ignition "OFF."
2. Connect fuel pressure gage and injector tester. Refer to Step 1
3. Ignition "ON."
4. Bleed off air in gage. Repeat until all air is bled from gauge.

Step 2. Run test:
1. Ignition "OFF" for 10 seconds.
2. Ignition "ON". Record gage pressure. (Pressure must hold steady, if not see the Fuel System diagnosis, Chart A-7
3. Turn injector on, by depressing button on injector tester, and note pressure at the instant the gauge needle stops.

Step 3. Refer to Step 3

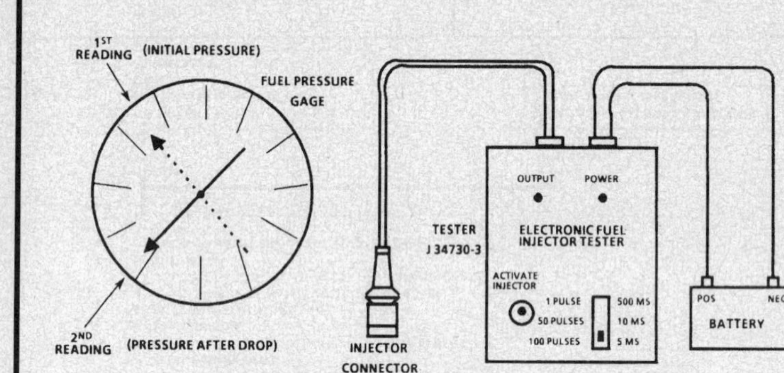

EXAMPLE

CYLINDER	1	2	3
1ST READING	293 kPa (43 psi)	293 kPa (43 psi)	293 kPa (43 psi)
2ND READING	131 kPa (19 psi)	115 kPa (17 psi)	145 kPa (21 psi)
AMOUNT OF DROP	162 kPa (24 psi)	178 kPa (26 psi)	148 kPa (21 psi)
	OK	FAULTY, RICH (TOO MUCH FUEL DROP)	FAULTY, LEAN (TOO LITTLE FUEL DROP)

2.2L (VIN 4) ENGINE — COMPONENT DIAGNOSTIC CHART — CORSICA AND BERETTA

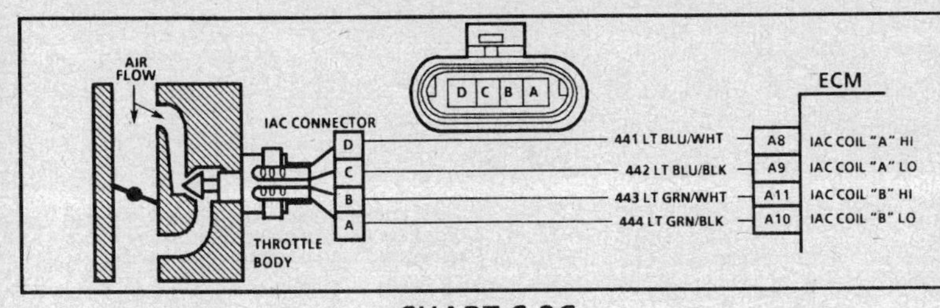

CHART C-2C
IDLE AIR CONTROL (IAC) SYSTEM CHECK
2.2L (VIN 4) "L" CARLINE

Circuit Description:

The ECM controls engine idle speed with the IAC valve. To increase idle speed, the ECM retracts the IAC valve pintle away from its seat, allowing more air to bypass the throttle bore. To decrease idle speed, it extends the IAC valve pintle towards its seat, reducing bypass air flow. A Tech 1 scan tool will read the ECM commands to the IAC valve in counts. Higher the counts indicate more air bypass (higher idle). The lower the counts indicate less air is allowed to bypass (lower idle).

Test Description: Number(s) below refer to circled numbers on the diagnostic chart.

1. The Tech 1 RPM control mode is used to extend and retract the IAC valve. The valve should move smoothly within the specified range. If the idle speed commanded (IAC extended) too low (below 700 RPM), the engine may stall. This may be normal and would not indicate a problem. Retracting the IAC beyond its controlled range (above 1500 RPM) will cause a delay before the RPMs start dropping. This too is normal.
2. This test uses the Tech 1 to command the IAC controlled idle speed. The ECM issues commands to obtain commanded idle speed. The node lights each should flash red and green to indicate a good circuit as the ECM issues commands. While the sequence of color is not important if either light is "OFF" or does not flash red and green, check the circuits for faults, beginning with poor terminal contacts.

Diagnostic Aids:

A slow, unstable, or fast idle may be caused by a non-IAC system problem that cannot be overcome by the IAC valve. Out of control range IAC scan tool counts will be above 50 if idle is too low, and zero counts if idle is too high. The following checks should be made to repair a non-IAC system problem:

- Vacuum Leak (High Idle) - If idle is too high, stop the engine. Fully extend (low) IAC with tester. Start engine. If idle speed is above 800 RPM, locate and correct vacuum leak including crankcase ventilation system. Also check for binding of throttle blade or linkage.

- Crankcase Ventilation Valve - If a high idle condition exists (800 to 1000 RPM), check for vacuum leaks and proper crankcase ventilation valve operation. All throttle bodies are preset at the factory and do not need adjustment. A missing crankcase ventilation valve or grommet or stuck valve can cause this condition.
- System Too Lean (High Air/Fuel Ratio) - The idle speed may be too high or too low. Engine speed may vary up and down and disconnecting the IAC valve does not help. DTC 44 may be set. Scan oxygen voltage will be less than 300 mV (.3 volt). Check for low regulated fuel pressure, water in the fuel or a restricted injector.
- System Too Rich (Low Air/Fuel Ratio) - The idle speed will be too low. Scan tool IAC counts will usually be above 50. System is obviously rich and may exhibit black smoke in exhaust. Scan tool oxygen voltage will be fixed above 800 mV (.8 volt).
  Check for high fuel pressure, leaking or sticking injector. Silicone contaminated oxygen sensors scan voltage will be slow to respond.
- Throttle Body - Remove IAC valve and inspect bore for foreign material.
- IAC Valve Electrical Connections - IAC valve connections should be carefully checked for proper contact.

- If intermittent poor driveability or idle symptoms are resolved by disconnecting the IAC, carefully recheck connections, valve terminal resistance, or replace IAC.

2.2L (VIN 4) ENGINE — COMPONENT DIAGNOSTIC CHART — CORSICA AND BERETTA

CHART C-2C
IDLE AIR CONTROL (IAC) SYSTEM CHECK
2.2L (VIN 4) "L" CARLINE

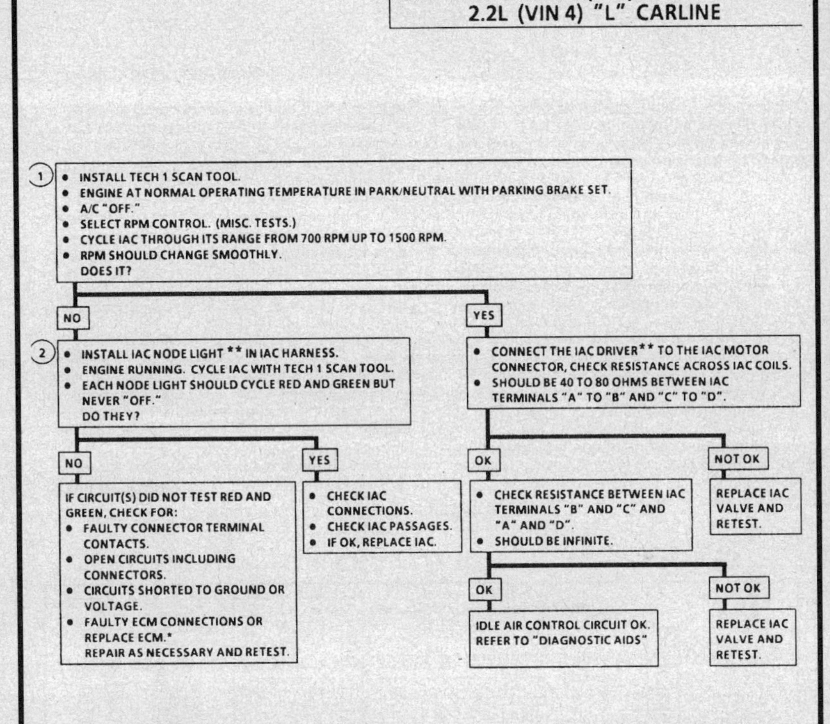

** IAC DRIVER AND NODE LIGHT REQUIRED KIT
222-L FROM: CONCEPT TECHNOLOGY, INC.
J 37027 FROM: KENT-MOORE, INC.

CLEAR DIAGNOSTIC TROUBLE CODES, CONFIRM "CLOSED LOOP" OPERATION, NO MALFUNCTION INDICATOR LAMP (MIL) "SERVICE ENGINE SOON," PERFORM IAC RESET PROCEDURE (REFER TO SECTION "6E3-C2" FOR IAC RESET PROCEDURE AND VERIFY CONTROLLED IDLE SPEED IS CORRECT.

* IF ECM IS FAULTY AND MUST BE REPLACED, THE NEW ECM MUST BE PROGRAMMED. REFER TO ECM REPLACEMENT AND PROGRAMMING PROCEDURES

2.2L (VIN 4) ENGINE — COMPONENT DIAGNOSTIC CHART — CORSICA AND BERETTA

CHART C-3

EVAPORATIVE EMISSION CANISTER CONTROL VALVE CHECK
(NON-ECM CONTROLLED)
2.2L (VIN 4) "L" CARLINE

- DISCONNECT THE VACUUM HOSE AT THE EVAP CANISTER CONTROL VALVE LABELED "CONTROL VAC."
- START ENGINE AND IDLE.
- MANIFOLD VACUUM SHOULD BE PRESENT AT THE DISCONNECTED HOSE.
  IS IT?

NO → REPAIR PLUGGED OR LEAKING MANIFOLD VACUUM SOURCE.

YES ↓

- REMOVE PURGE VALVE.
- APPLY A CLEAN SHORT HOSE TO THE TUBE LABELED "PURGE VAC" OF THE VALVE AND BLOW INTO THE HOSE. NO AIR SHOULD PASS THROUGH THE VALVE.
  DOES VALVE HOLD?

NO → REPLACE VALVE.

YES ↓

- INSTALL A HAND VACUUM PUMP TO THE TUBE OF THE VALVE LABELED "CONTROL VAC."
- APPLY 51 kPa (15" Hg) VACUUM TO THE DIAPHRAGM. DIAPHRAGM SHOULD HOLD VACUUM FOR AT LEAST 20 SECONDS.
  DOES IT?

NO → REPLACE VALVE.

YES ↓

- WITH THE VACUUM STILL APPLIED TO THE "CONTROL VAC" TUBE, AGAIN TRY TO BLOW INTO A CLEAN HOSE ON THE TUBE LABELED "PURGE VAC" OF THE VALVE. AIR SHOULD PASS THROUGH FREELY.
  DOES IT?

NO → REPLACE VALVE.

YES ↓

VALVE OK.

CHECK ALL EMISSION CONTROL LINES, HOSES AND FITTINGS FOR BLOCKAGE, LEAKAGE OR DETERIORATION AND REPAIR AS NECESSARY.

2.2L (VIN 4) ENGINE — COMPONENT DIAGNOSTIC CHART — CORSICA AND BERETTA

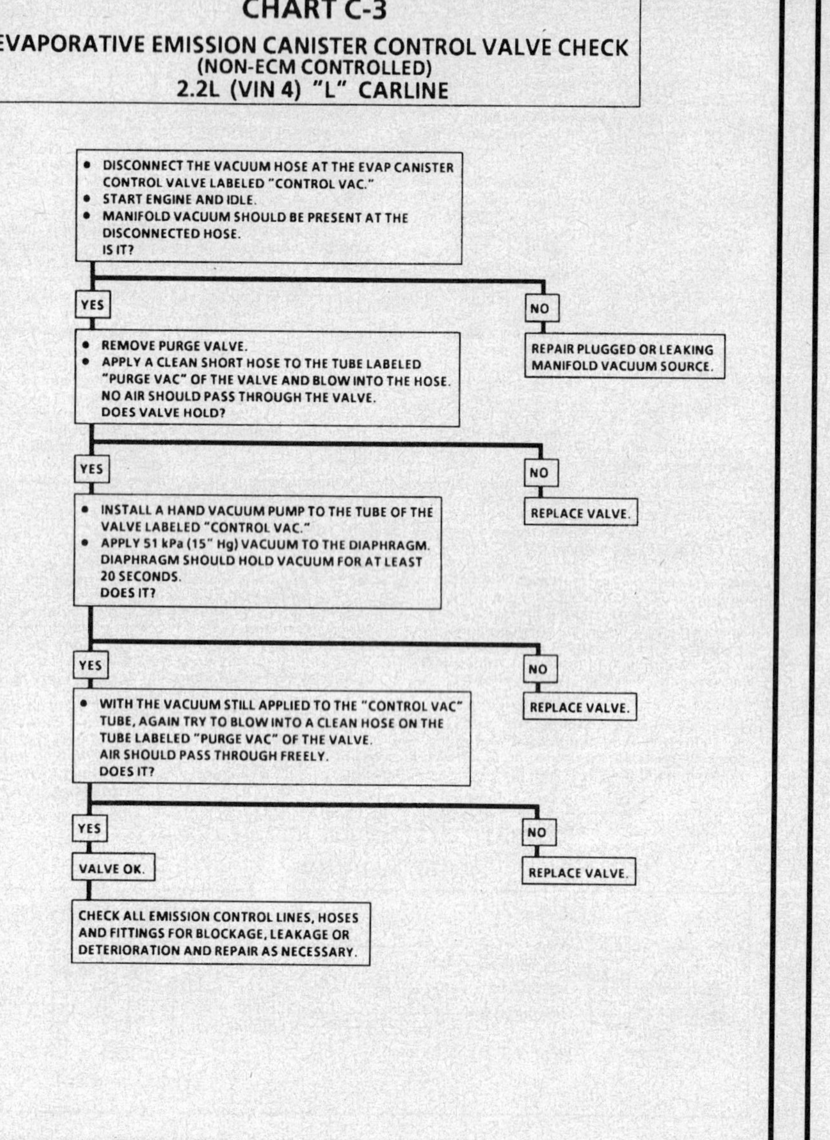

CHART C-4D-1

MISFIRE AT IDLE
2.2L (VIN 4) "L" CARLINE

Circuit Description:

The Electronic Ignition (EI) system uses a waste spark method of distribution. In this type of system, the electronic Ignition Control Module (ICM) triggers the number 1/4 coil pair resulting in both number 1 and number 4 spark plugs firing at the same time. Number 1 cylinder is on the compression stroke at the same time number 4 is on the exhaust stroke, resulting in a lower energy requirement to fire number 4 spark plug. This leaves the remainder of the high voltage to be used to fire number 1 spark plug. The Crankshaft Position (CKP) sensor is remotely mounted beside the ignition coil assembly and protrudes through the block to within approximately .050" of the crankshaft reluctor. Since the reluctor is a machined portion of the crankshaft and the CKP sensor is mounted in a fixed position on the block, timing adjustments are not possible or necessary.

Test Description: Number(s) below refer to circled number(s) on the diagnostic chart.

1. If the "Misfire" complaint exists <u>under load only</u>, CHART C-4D-2 must be used. Engine RPM should drop approximately equally on all plug leads.
2. A spark tester, such as a ST-125, must be used because it is essential to verify adequate available secondary voltage at the spark plug (25,000 volts). If the spark jumps the tester gap after grounding the companion plug cable, it indicates excessive resistance in the plug which was bypassed. A faulty or poor connection at that plug could also result in a misfire condition. Also, check for carbon deposits inside the spark plug boot.
3. If carbon tracking is evident, replace coil and be sure plug cables relating to that coil are clean and tight. Excessive wire resistance or faulty connections could have caused the coil to be damaged.
4. By checking the secondary resistance, a coil with an open secondary may be located.
5. If the no spark condition follows the suspected coil, that coil is faulty. Otherwise, the electronic Ignition Control Module (ICM) is the cause of no spark. This test could also be performed by substituting a known good coil for the one causing the no spark condition.

2.2L (VIN 4) ENGINE — COMPONENT DIAGNOSTIC CHART — CORSICA AND BERETTA

CHART C-4D-1
MISFIRE AT IDLE
2.2L (VIN 4) "L" CARLINE

★CAUTION: WHEN HANDLING SECONDARY SPARK PLUG LEADS WITH ENGINE RUNNING, J 21350 OR EQUIVALENT INSULATED PLIERS MUST BE USED AND CARE EXERCISED TO PREVENT A POSSIBLE ELECTRICAL SHOCK.

1.
- IF ENGINE MISFIRES UNDER LOAD ONLY. REFER TO CHART C-4D-2.
- ENGINE IDLING AT NORMAL OPERATING TEMPERATURE, DISCONNECT IAC.
- MOMENTARILY DISCONNECT (ONE AT A TIME) EACH SPARK PLUG LEAD, USING J 21350 OR EQUIVALENT INSULATED PLIERS, WHILE OBSERVING ENGINE RPM FOR A DROP. [DISCONNECTED LEAD SHOULD BE GROUNDED TO AVOID OVER-STRESSING IN THE IGNITION COIL(S).]
- SEE CAUTION★.
- EACH DISCONNECTED PLUG LEAD SHOULD HAVE RESULTED IN A FAIRLY EQUAL RPM DROP.

NO →

YES → REFER TO "ROUGH, UNSTABLE OR INCORRECT IDLE OR STALLING"

2.
- WITH IGNITION "OFF," INSTALL SPARK TESTER (ST-125) J 26792 ON PLUG LEAD(S) WHOSE REMOVAL DID NOT RESULT IN RPM DROP (AT PLUG END AND CLIP TO GROUND).
- CONNECT PLUG END OF COMPANION CYLINDER SPARK PLUG CABLE TO GROUND AT ENGINE TO AVOID OVER-STRESSING THE IGNITION COILS.
- DISCONNECT INJECTOR JUMPER HARNESS CONNECTOR TO DISABLE FUEL INJECTORS.
- OBSERVE TESTER WHILE CRANKING ENGINE.
- SPARK SHOULD JUMP TESTER GAP WHILE CRANKING ENGINE. DOES IT?

NO →

YES → CHECK FOR:
- FAULTY, WORN OR CRACKED SPARK PLUG(S).
- PLUG FOULING OR OVERHEATING DUE TO ENGINE MECHANICAL OR FUEL SYSTEM FAULT.
- PERFORM INJECTOR BALANCE TEST, CHART C-2A.
- REFER TO "CUTS OUT, MISSES"

- CHECK THE RESISTANCE OF EACH SPARK PLUG CABLE WHICH DID NOT FIRE THE SPARK TESTER.
- CABLE RESISTANCE SHOULD BE LESS THAN 30K OHMS EACH AND CABLES SHOULD NOT BE GROUNDED. ARE CABLES OK?

YES →

NO → REPLACE FAULTY CABLE(S) AND RETEST.

3.
- REMOVE THE IGNITION COIL FROM IGNITION CONTROL MODULE WHICH DID NOT FIRE SPARK TESTER.
- INSPECT COIL, SPARK PLUG CABLES AND CABLE BOOTS. THEY SHOULD BE FREE OF CARBON TRACKING. ARE THEY?

YES →

NO → REPLACE FAULTY COMPONENT.

4.
- CHECK SECONDARY COIL RESISTANCE. SHOULD BE 5-8K OHMS RESISTANCE. IS IT?

YES →

NO → REPLACE IGNITION COIL. ALSO, CHECK FOR FAULTY SPARK PLUG CABLE CONNECTION(S) AND CABLE BOOT(S) FOR CARBON TRACKING.

5.
- INSTALL A KNOWN GOOD IGNITION COIL.
- SPARK SHOULD JUMP TESTER GAP AT PROBLEM CYLINDER WITH ENGINE IDLING. DID IT?

YES → ORIGINAL IGNITION COIL IS FAULTY.

NO → REPLACE ELECTRONIC IGNITION CONTROL MODULE.

2.2L (VIN 4) ENGINE — COMPONENT DIAGNOSTIC CHART — CORSICA AND BERETTA

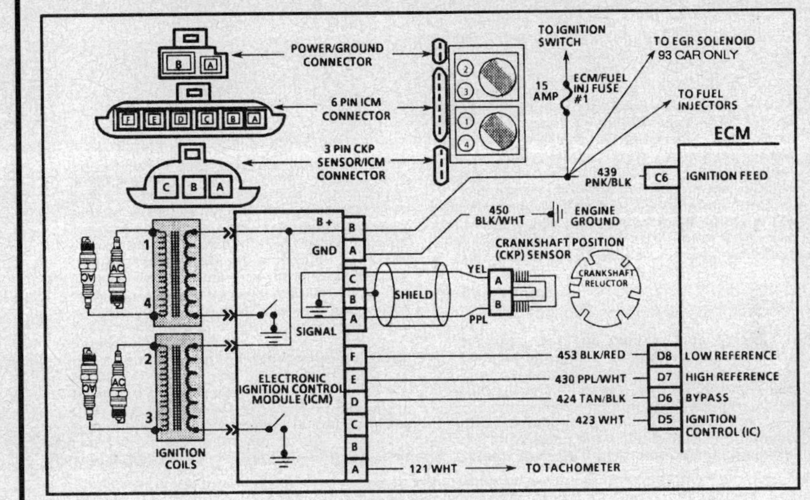

CHART C-4D-2
MISFIRE UNDER LOAD
2.2L (VIN 4) "L" CARLINE

Circuit Description:

The Electronic Ignition (EI) system uses a waste spark method of distribution. In this type of system, the electronic Ignition Control Module (ICM) triggers the number 1/4 coil pair resulting in both number 1 and number 4 spark plugs firing at the same time. Number 1 cylinder is on the compression stroke at the same time number 4 is on the exhaust stroke, resulting in a lower energy requirement to fire number 4 spark plug. This leaves the remainder of the high voltage to be used to fire number 1 spark plug. The Crankshaft Position (CKP) sensor is remotely mounted beside the ignition coil assembly and protrudes through the block to within approximately .050″ of the crankshaft reluctor. Since the reluctor is a machined portion of the crankshaft and the CKP sensor is mounted in a fixed position on the block, timing adjustments are not possible or necessary.

Test Description: Number(s) below refer to circled number(s) on the diagnostic chart.

If the "Misfire" complaint exists at idle only, CHART C-4D-1 must be used.

1. Simulates an engine load. Cranking the engine with the ST-125 installed on any cylinder should produce a crisp, blue spark. A cylinder that produces a spark significantly weaker than the others should be considered as having no spark.

2.2L (VIN 4) ENGINE — COMPONENT DIAGNOSTIC CHART — CORSICA AND BERETTA

CHART C-4D-2
MISFIRE UNDER LOAD
2.2L (VIN 4) "L" CARLINE

① • IF ENGINE MISFIRES AT IDLE, REFER TO CHART C-4D-1.
• IGNITION "OFF," INSTALL SPARK TESTER (ST-125) J 26792 TO SPARK PLUG CABLE AT PLUG END AND CLIP TO GROUND.
• CONNECT PLUG END OF COMPANION CYLINDER PLUG CABLE TO ENGINE GROUND TO AVOID OVER-STRESSING THE IGNITION COILS.
• DISCONNECT INJECTOR JUMPER HARNESS CONNECTOR TO DISABLE FUEL INJECTORS.
• CRANK ENGINE, SPARK SHOULD JUMP TESTER GAP WHILE CRANKING.
• REPEAT ABOVE STEPS FOR EACH CYLINDER, RECORDING ANY CYLINDER(S) THAT DID NOT FIRE. RECONNECT EACH SPARK PLUG CABLE TO PLUG BEFORE MOVING ON TO NEXT CYLINDER.
• SPARK SHOULD JUMP TESTER GAP AT ALL PLUG CABLES WHILE CRANKING.

OK
• CHECK FOR FAULTY, CRACKED, FOULED OR WORN SPARK PLUG(S).
• PERFORM INJECTOR BALANCE TEST, CHART C-2A.
• REFER TO "CUTS OUT, MISSES" IN "SYMPTOMS."

CYLINDER PAIR (1-4, OR 2-3) NOT FIRING
• CHECK FOR AN OPEN SPARK PLUG CABLE FOR CYLINDER PAIR THAT DID NOT FIRE (CABLES SHOULD MEASURE LESS THAN 30K Ω). IF OK, SWAP PROBLEM COIL WITH ONE FOR CYLINDER PAIR THAT DOES FIRE AND RETEST. IF PROBLEM FOLLOWS COIL, REPLACE COIL. IF NOT, REPLACE ELECTRONIC IGNITION CONTROL MODULE (ICM).

SINGLE CYLINDER NOT FIRING
• CHECK FOR GROUNDED SPARK PLUG CABLE(S) FOR CYLINDER(S) THAT DID NOT FIRE AND REPLACE AS NECESSARY. IF OK, SWAP PROBLEM COIL WITH ONE THAT OPERATES PROPERLY. IF PROBLEM FOLLOWS COIL, REPLACE IT. IF NOT, REPLACE SPARK PLUG CABLE FOR CYLINDER(S) THAT DID NOT FIRE.

★CAUTION: When handling secondary spark plug leads with engine running, insulated pliers must be used and care exercised to prevent a possible electrical shock.

"AFTER REPAIRS," CONFIRM "CLOSED LOOP" OPERATION AND NO MIL (SERVICE ENGINE SOON).

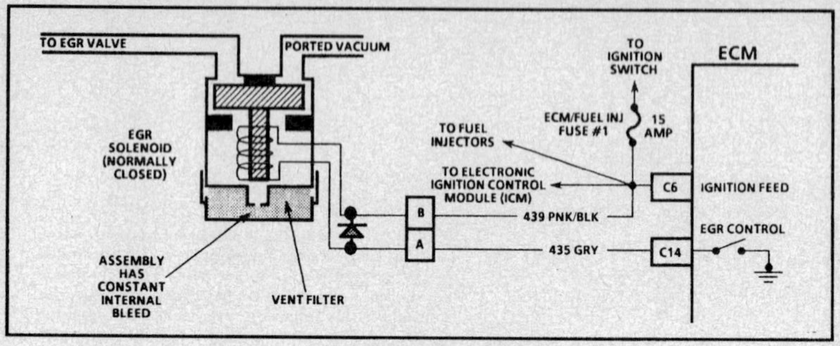

2.2L (VIN 4) ENGINE — COMPONENT DIAGNOSTIC CHART — CORSICA AND BERETTA

CHART C-7
EXHAUST GAS RECIRCULATION (EGR) FLOW CHECK
2.2L (VIN 4) "L" CARLINE

Circuit Description:

A properly operating EGR system will directly affect the air/fuel mixture requirements of the engine. Since the exhaust gas introduced into the air/fuel mixture cannot be used in combustion (contains very little oxygen), less fuel is required to maintain a correct air/fuel ratio. If the EGR system were to fail in a closed position, the exhaust gas would be replaced with air, and the air/fuel mixture would be leaner. The ECM would compensate for the lean condition by adding fuel, resulting in higher long term fuel trim values.

The fuel control on this engine is conducted within four fuel trim cells. Since EGR is not used at idle, the idle cell would not be affected by EGR system operation. The other fuel trim cells are affected by EGR operation, and, when the EGR system is operating properly, the long term fuel trim values in all cells should be close to the same. If the EGR system becomes inoperative, the long term fuel trim values in the open throttle cells would change to compensate for the resulting lean or rich mixtures, but the long term fuel trim value in the closed throttle cell would not change.

The difference in long term fuel trim values between the idle (closed throttle) cell and cell 2 is used to monitor EGR system performance. When the difference between the two long term fuel trim values is greater than 10, and the long term fuel trim value in cell 2 is greater than 135, Diagnostic Trouble Code (DTC) 32 is set. The system operates in fuel trim cell 2 during a cruise condition at approximately 55 mph.

Test Description: Number(s) below refer to circled number(s) on the diagnostic chart.

1. DTC(s) should be diagnosed using appropriate chart before preparing a functional check. If Diagnostic Trouble Codes 14, 21, 23, 32, or 33 are set, use those charts first.
Be sure shop exhaust hose is not connected during Steps 2 and 4.
2. The tail pipe ventilation hose must be removed for this test. The ventilation hose can sometimes cause a good EGR valve to fail this test.
3. Intake Passage: Shut "OFF" engine and remove the EGR valve from the manifold. Plug the exhaust side hole with a suitable stopper. Leaving the intake side hole open, attempt to start the engine. If the engine runs at a high idle (up to 3000 RPM is possible) or starts and stalls, the EGR intake passage is not restricted. If the engine starts and idles normally, the EGR intake passage is restricted.
4. Because the shop exhaust hose is not installed at this point, don't allow the engine to run longer than 15 seconds.
Exhaust Passage: With EGR valve still removed, plug the intake side hole with a suitable stopper. With the exhaust side hole open, check for the presence of exhaust gas. If no exhaust gas is present, the EGR exhaust side passage is restricted.

Diagnostic Aids:

This chart is a functional check of the EGR system. If the EGR system works properly, check other items that result in high long term fuel trim values in fuel trim cell 2, but not in the closed throttle cell.

Incorrect fuel pressure or lean/rich fuel injectors can also cause incorrect long term fuel trim values. Refer to "Fuel Pressure Test" in CHART A-7. It may be necessary to monitor fuel pressure while driving the vehicle at various road speeds and/or loads. If fuel pressure checks out OK, proceed to "Fuel Injector Balance Test" in CHART C-2A.

2.2L (VIN 4) ENGINE — COMPONENT DIAGNOSTIC CHART — CORSICA AND BERETTA

CHART C-7
EXHAUST GAS RECIRCULATION (EGR) FLOW CHECK
2.2L (VIN 4) "L" CARLINE

(1) SCAN DTC(S).
IF DTC(S) ARE PRESENT, REFER TO THOSE CHARTS FIRST.

- WITH ENGINE AT IDLE, MANUALLY LIFT THE EGR VALVE DIAPHRAGM, RPM SHOULD DECREASE OR ENGINE STALL. DOES IT?

YES

(2)
- IGNITION "OFF."
- DISCONNECT VACUUM LINE FROM EGR VALVE.
- DISCONNECT SHOP EXHAUST HOSE.
- APPLY 34 kPa (10 Hg.) OF VACUUM TO THE EGR VALVE AND OBSERVE EGR VALVE DIAPHRAGM, SHOULD MOVE. DOES IT?

NO

(3) CHECK EGR VALVE, GASKETS, AND ALL PASSAGES FOR DAMAGE, LEAKAGE, OR PLUGGING. IF OK, REPLACE EGR VALVE.

YES

(4)
- RECONNECT VACUUM LINE TO EGR VALVE.
- START ENGINE AND IDLE.
- EXHAUST HOSE DISCONNECTED.
- OBSERVE EGR VALVE DIAPHRAGM WHILE SNAPPING THE THROTTLE FROM IDLE TO WIDE OPEN THROTTLE AND BACK TO IDLE.
  THE EGR VALVE SHOULD OPEN ON ENGINE DECELERATION. DOES IT?

NO

REPLACE EGR VALVE.

YES

EGR SYSTEM IS OK, NO PROBLEM FOUND. SEE "DIAGNOSTIC AIDS"

NO

USE DTC 32 CHART.

2.2L (VIN 4) ENGINE — COMPONENT DIAGNOSTIC CHART — CORSICA AND BERETTA

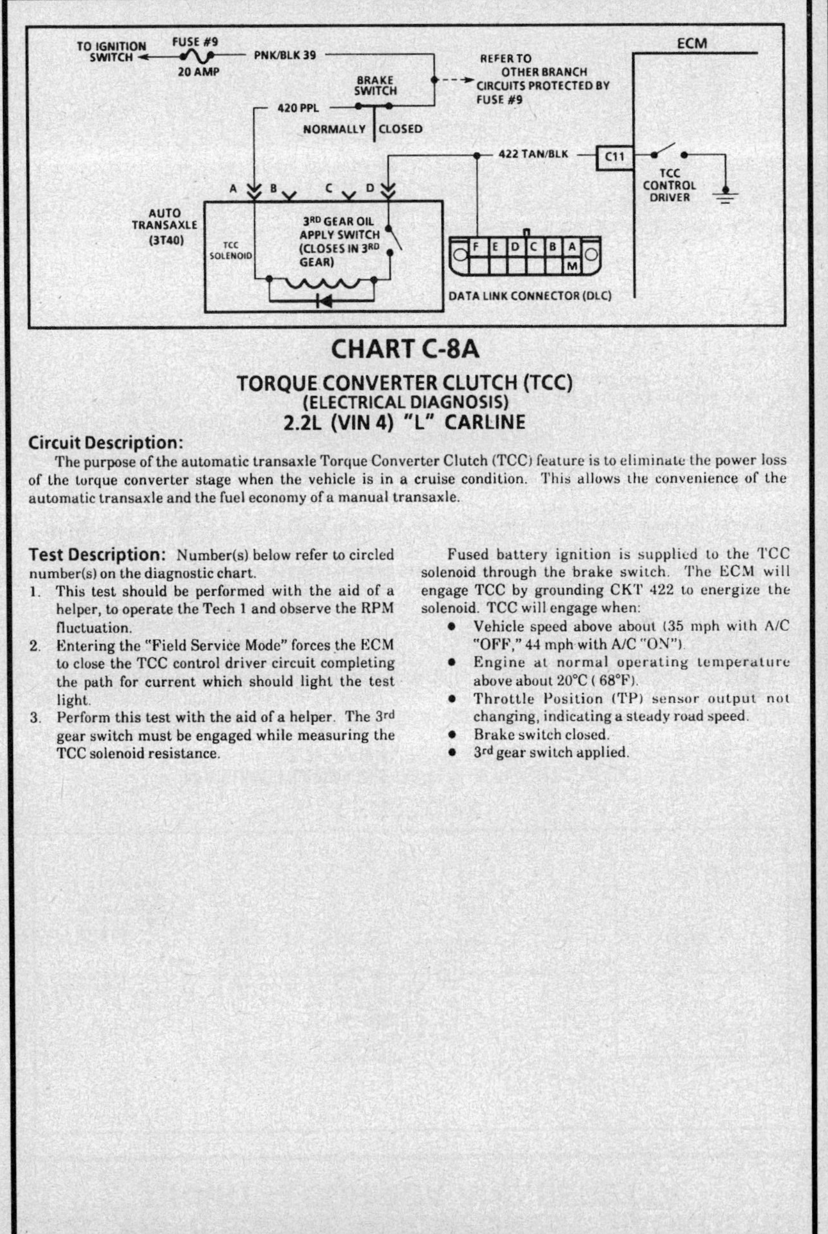

CHART C-8A
TORQUE CONVERTER CLUTCH (TCC)
(ELECTRICAL DIAGNOSIS)
2.2L (VIN 4) "L" CARLINE

Circuit Description:

The purpose of the automatic transaxle Torque Converter Clutch (TCC) feature is to eliminate the power loss of the torque converter stage when the vehicle is in a cruise condition. This allows the convenience of the automatic transaxle and the fuel economy of a manual transaxle.

Test Description: Number(s) below refer to circled number(s) on the diagnostic chart.

1. This test should be performed with the aid of a helper, to operate the Tech 1 and observe the RPM fluctuation.
2. Entering the "Field Service Mode" forces the ECM to close the TCC control driver circuit completing the path for current which should light the test light.
3. Perform this test with the aid of a helper. The 3rd gear switch must be engaged while measuring the TCC solenoid resistance.

Fused battery ignition is supplied to the TCC solenoid through the brake switch. The ECM will engage TCC by grounding CKT 422 to energize the solenoid. TCC will engage when:

- Vehicle speed above about (35 mph with A/C "OFF," 44 mph with A/C "ON").
- Engine at normal operating temperature above about 20°C (68°F).
- Throttle Position (TP) sensor output not changing, indicating a steady road speed.
- Brake switch closed.
- 3rd gear switch applied.

2.2L (VIN 4) ENGINE — COMPONENT DIAGNOSTIC CHART — CORSICA AND BERETTA

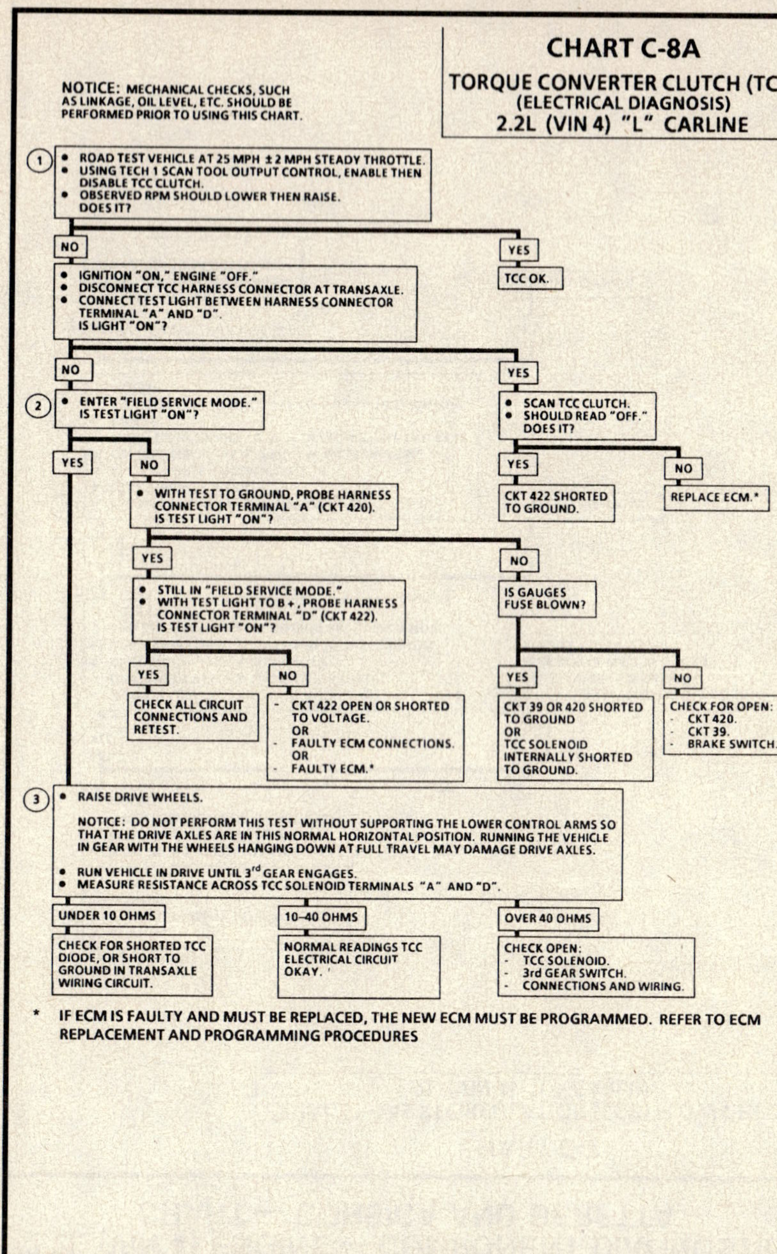

2.2L (VIN 4) ENGINE — COMPONENT DIAGNOSTIC CHART — CORSICA AND BERETTA

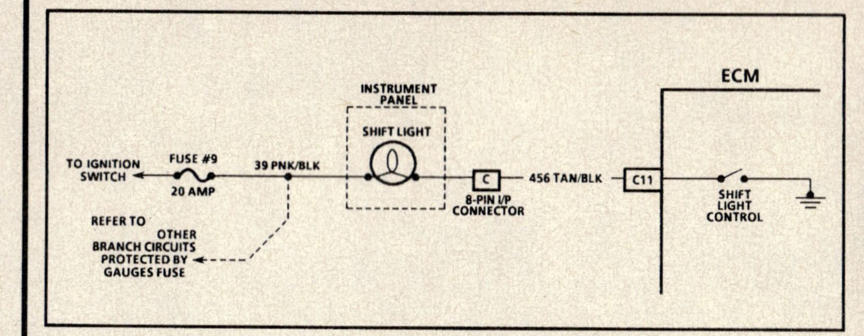

CHART C-8B
MANUAL TRANSAXLE (M/T) SHIFT LIGHT CHECK
2.2L (VIN 4) "L" CARLINE

Circuit Description:

The shift light indicates the best transaxle shift point for maximum fuel economy. The light is controlled by the ECM and is turned "ON" by grounding CKT 456.

The ECM uses information from the following inputs to control the shift light:
- Engine Coolant Temperature (ECT) sensor.
- Throttle Position (TP) sensor.
- Vehicle Speed Sensor (VSS).
- RPM.

The ECM uses the measured RPM and the vehicle speed to calculate what gear the vehicle is in. It's this calculation that determines when the shift light should be turned "ON."

Test Description: Number(s) below refer to circled number(s) on the diagnostic chart.

1. This should not turn "ON" the shift light. If the light is "ON," there is a short to ground in CKT 456 wiring or a fault in the ECM.
2. When the diagnostic terminal is grounded, the ECM should ground CKT 456 and the shift light should come "ON."
3. This checks the shift light circuit up to the ECM connector. If the shift light illuminates, then the ECM connector is faulty or the ECM does not have the ability to ground the circuit.

Diagnostic Aids:

Be sure to consider branch circuits and splices to other components if a fuse is blown.

2.2L (VIN 4) ENGINE — COMPONENT DIAGNOSTIC CHART — CORSICA AND BERETTA

CHART C-8B

**MANUAL TRANSAXLE (M/T) SHIFT LIGHT CHECK
2.2L (VIN 4) "L" CARLINE**

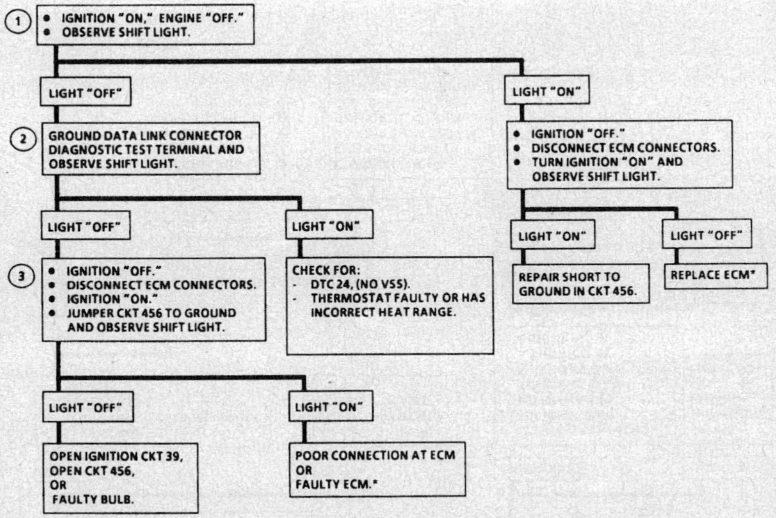

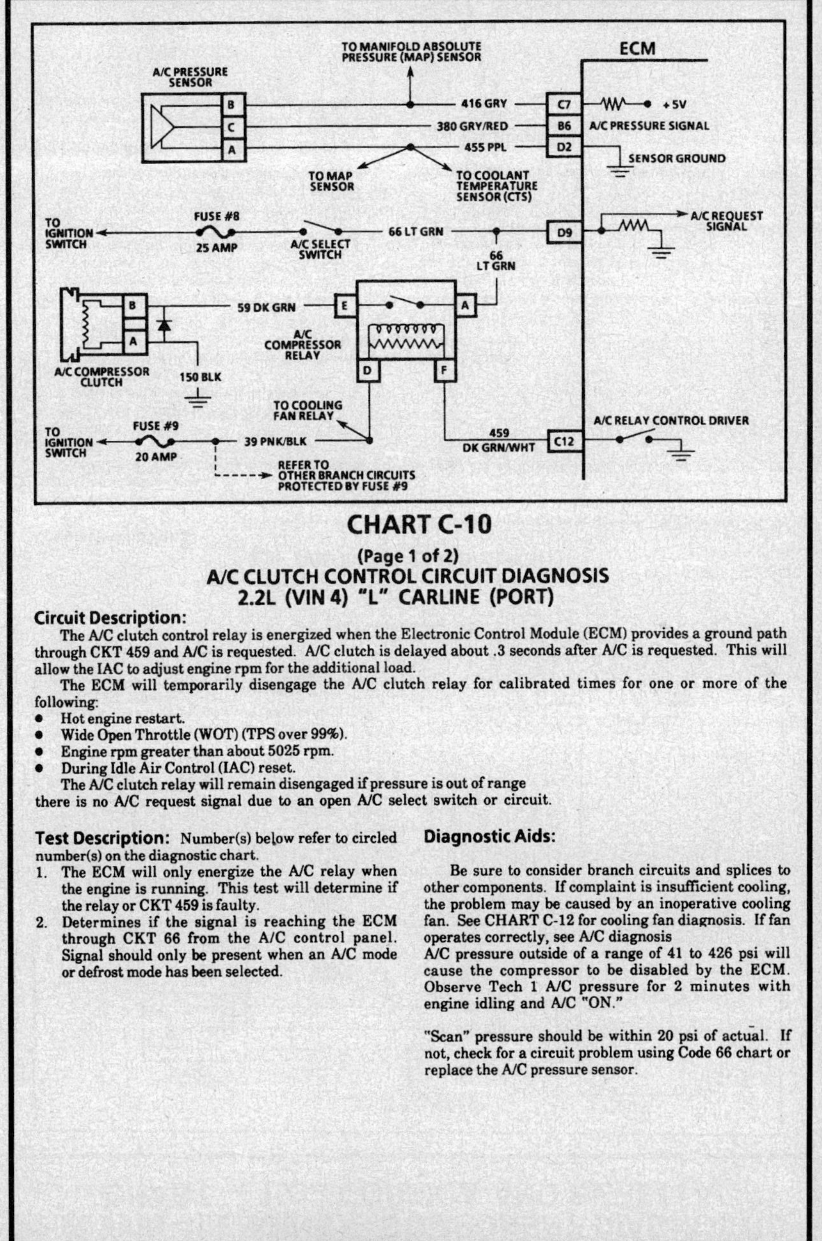

* IF ECM IS FAULTY AND MUST BE REPLACED, THE NEW ECM MUST BE PROGRAMMED. REFER TO ECM REPLACEMENT AND PROGRAMMING PROCEDURES

2.2L (VIN 4) ENGINE — COMPONENT DIAGNOSTIC CHART — 1992 CORSICA AND BERETTA

CHART C-10

(Page 1 of 2)
**A/C CLUTCH CONTROL CIRCUIT DIAGNOSIS
2.2L (VIN 4) "L" CARLINE (PORT)**

Circuit Description:

The A/C clutch control relay is energized when the Electronic Control Module (ECM) provides a ground path through CKT 459 and A/C is requested. A/C clutch is delayed about .3 seconds after A/C is requested. This will allow the IAC to adjust engine rpm for the additional load.

The ECM will temporarily disengage the A/C clutch relay for calibrated times for one or more of the following:

- Hot engine restart.
- Wide Open Throttle (WOT) (TPS over 99%).
- Engine rpm greater than about 5025 rpm.
- During Idle Air Control (IAC) reset.

The A/C clutch relay will remain disengaged if pressure is out of range there is no A/C request signal due to an open A/C select switch or circuit.

Test Description: Number(s) below refer to circled number(s) on the diagnostic chart.

1. The ECM will only energize the A/C relay when the engine is running. This test will determine if the relay or CKT 459 is faulty.
2. Determines if the signal is reaching the ECM through CKT 66 from the A/C control panel. Signal should only be present when an A/C mode or defrost mode has been selected.

Diagnostic Aids:

Be sure to consider branch circuits and splices to other components. If complaint is insufficient cooling, the problem may be caused by an inoperative cooling fan. See CHART C-12 for cooling fan diagnosis. If fan operates correctly, see A/C diagnosis

A/C pressure outside of a range of 41 to 426 psi will cause the compressor to be disabled by the ECM. Observe Tech 1 A/C pressure for 2 minutes with engine idling and A/C "ON."

"Scan" pressure should be within 20 psi of actual. If not, check for a circuit problem using Code 66 chart or replace the A/C pressure sensor.

2.2L (VIN 4) ENGINE — COMPONENT DIAGNOSTIC CHART — 1992 CORSICA AND BERETTA

CHART C-10
(Page 1 of 2)
A/C CLUTCH CONTROL CIRCUIT DIAGNOSIS
2.2L (VIN 4) "L" CARLINE (PORT)

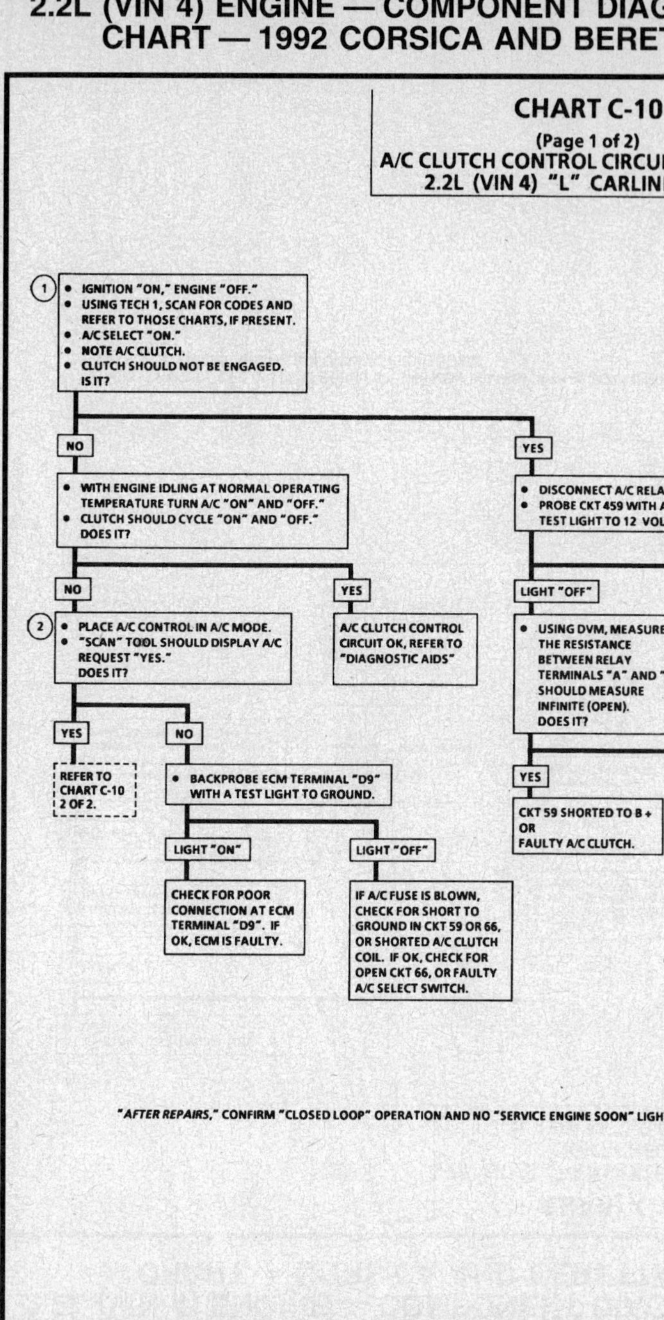

2.2L (VIN 4) ENGINE — COMPONENT DIAGNOSTIC CHART — 1992 CORSICA AND BERETTA

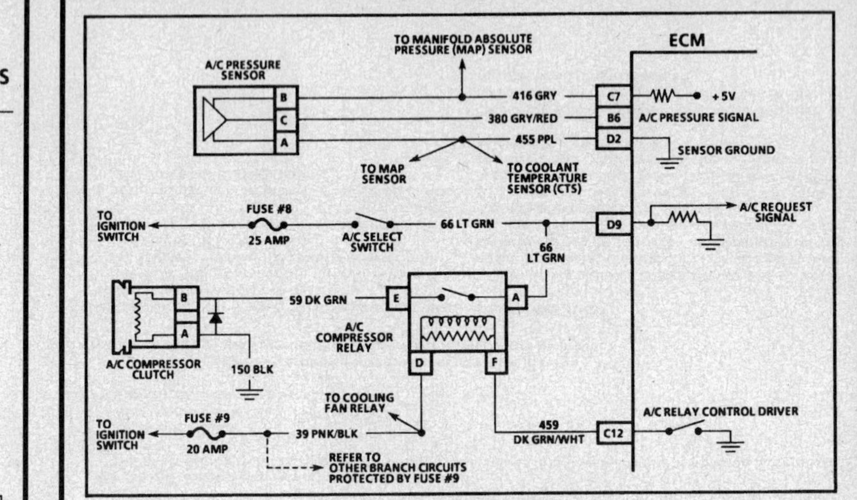

CHART C-10
(Page 2 of 2)
A/C CLUTCH CONTROL CIRCUIT DIAGNOSIS
2.2L (VIN 4) "L" CARLINE (PORT)

Circuit Description:

The A/C clutch control relay is energized when the ECM provides a ground path through CKT 459 and A/C is requested. A/C clutch is delayed about .3 seconds after A/C is requested. This will allow the IAC to adjust engine rpm for the additional load.

The ECM will temporarily disengage the A/C clutch relay for calibrated times for one or more of the following:

- Hot engine restart.
- Wide Open Throttle (WOT) (TPS over 99%).
- Engine rpm greater than about 5025 rpm.
- During IAC reset.

The A/C clutch relay will remain disengaged if pressure is out of range there is no A/C request signal due to an open A/C select switch or circuit.

Test Description: Number(s) below refer to circled number(s) on the diagnostic chart.

1. Determines if the pressure transducer is out of range causing the compressor clutch to be disengaged.
2. With the engine stopped and field service mode activated, the ECM should be grounding CKT 459, which should cause the test light to be "ON."

Diagnostic Aids:

If complaint is insufficient cooling, the problem may be caused by an inoperative cooling fan. See CHART C-12 for cooling fan diagnosis. If fan operates correctly, see A/C diagnosis A/C pressure outside of a range of 41 to 426 psi will cause the compressor to be disabled by the ECM. Observe Tech 1 A/C pressure for 2 minutes with engine idling and A/C "ON."

Tech 1 pressure should be within 20 psi of actual. If not, check for a circuit problem using Code 66 chart or replace A/C pressure sensor.

2.2L (VIN 4) ENGINE — COMPONENT DIAGNOSTIC CHART — 1992 CORSICA AND BERETTA

CHART C-10

(2 of 2)
A/C CLUTCH CONTROL CIRCUIT DIAGNOSIS
2.2L (VIN 4) "L" CARLINE (PORT)

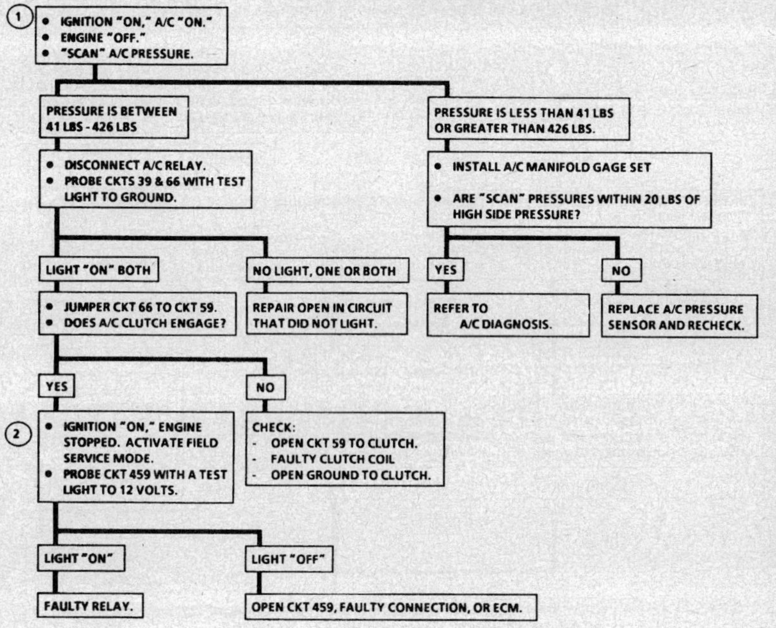

"AFTER REPAIRS," CONFIRM "CLOSED LOOP" OPERATION AND NO "SERVICE ENGINE SOON" LIGHT.

2.2L (VIN 4) ENGINE — COMPONENT DIAGNOSTIC CHART — 1993–94 CORSICA AND BERETTA

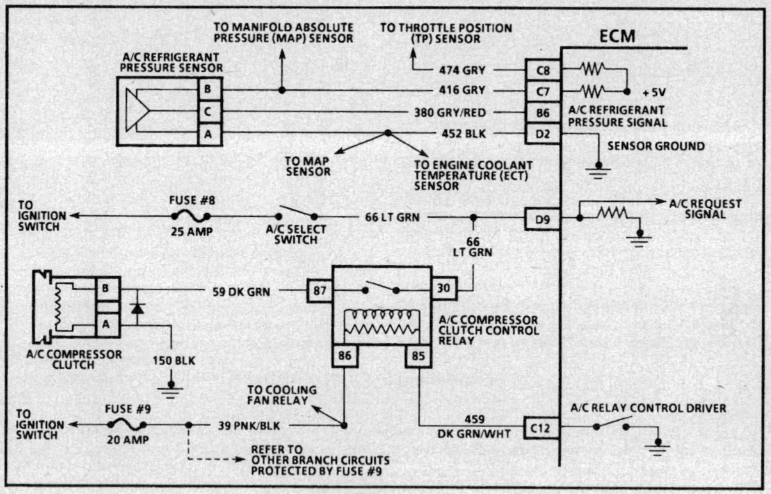

CHART C-10

(Page 1 of 2)
A/C COMPRESSOR CLUTCH CONTROL CIRCUIT DIAGNOSIS
2.2L (VIN 4) "L" CARLINE

Circuit Description:

The A/C compressor clutch control relay is energized when the Engine Control Module (ECM) provides a ground path through CKT 459 and A/C is requested. A/C compressor clutch is delayed about .3 second after A/C is requested. This will allow the IAC to adjust engine RPM for the additional load.

The ECM will disable the A/C compressor clutch for any of the following reasons:

- Engine not running.
- Diagnostic Trouble Code (DTC) 66 is set, and fault is current.
- Throttle position is about 99 percent or greater.
- Engine coolant temperature is about 122°C (251°F) or greater.
- During an IAC motor reset.
- Engine RPM is greater than about 5025 RPM.
- A/C refrigerant pressure is about 426 psi or greater.
- A/C refrigerant pressure is about 41 psi or less.

If the ECM temporarily disabled the A/C compressor clutch, the clutch will be renabled as follows:

- Throttle position drops below about 89 percent.
- Coolant temperature drops below about 119°C (246°F).
- Engine RPM drops below about 4500 RPM.
- A/C refrigerant pressure drops below about 201 psi.
- A/C refrigerant pressure rises above about 51 psi.
- IAC motor reset is complete.

Test Description: Number(s) below refer to circled number(s) on the diagnostic chart.

1. The ECM will only energize the A/C relay when the engine is running. This test will determine if the relay or CKT 459 is faulty.
2. Determines if the signal is reaching the ECM through CKT 66 from the A/C control panel. Signal should only be present when an A/C mode or defrost mode has been selected.

Diagnostic Aids:

Be sure to consider branch circuits and splices to other components. If complaint is insufficient cooling, the problem may be caused by an inoperative cooling fan. Refer to CHART C-12 for cooling fan diagnosis. If fan operates correctly, see "A/C Diagnosis"

A/C refrigerant pressure outside of a range of 41 to 426 psi will cause the compressor clutch to be disabled by the ECM. Observe Tech 1 A/C refrigerant pressure for 2 minutes with engine idling and A/C "ON."

Scanned refrigerant pressure should be within 20 psi of actual. If not, check for a circuit problem using DTC 66 chart or replace the A/C refrigerant pressure sensor.

2.2L (VIN 4) ENGINE — COMPONENT DIAGNOSTIC CHART — 1993–94 CORSICA AND BERETTA

2.2L (VIN 4) ENGINE — COMPONENT DIAGNOSTIC CHART — 1993–94 CORSICA AND BERETTA

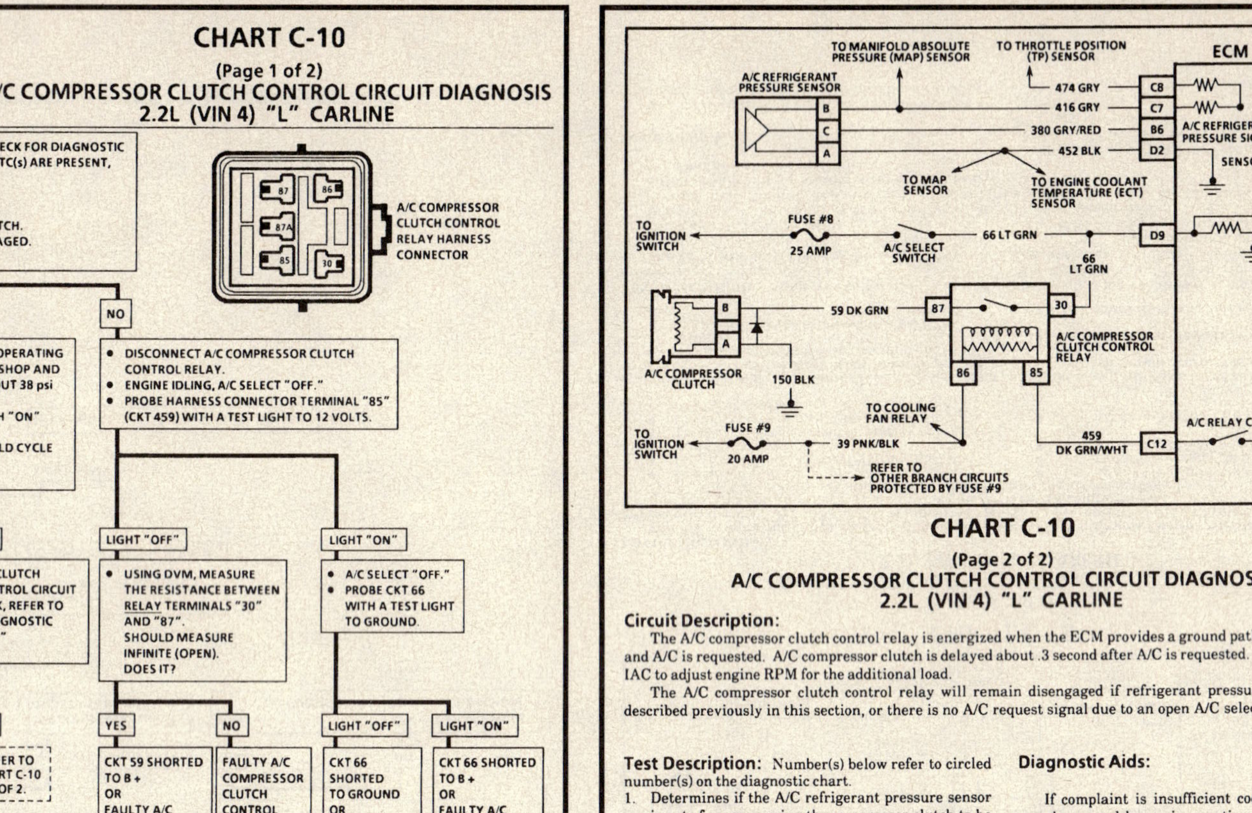

CHART C-10
(Page 2 of 2)
A/C COMPRESSOR CLUTCH CONTROL CIRCUIT DIAGNOSIS
2.2L (VIN 4) "L" CARLINE

Circuit Description:

The A/C compressor clutch control relay is energized when the ECM provides a ground path through CKT 366 and A/C is requested. A/C compressor clutch is delayed about .3 second after A/C is requested. This will allow the IAC to adjust engine RPM for the additional load.

The A/C compressor clutch control relay will remain disengaged if refrigerant pressure is out of range described previously in this section, or there is no A/C request signal due to an open A/C select switch or circuit.

Test Description: Number(s) below refer to circled number(s) on the diagnostic chart.
1. Determines if the A/C refrigerant pressure sensor is out of range causing the compressor clutch to be disengaged.
2. With the engine stopped and "Field Service Mode" activated, the ECM should be grounding CKT 459, which should cause the test light to be "ON."

Diagnostic Aids:

If complaint is insufficient cooling, the problem may be caused by an inoperative engine cooling fan. Refer to CHART C-12 for cooling fan diagnosis. If fan operates correctly, refer to "A/C Diagnosis"

A/C refrigerant pressure outside of a range of 41 to 426 psi will cause the compressor to be disabled by the ECM. With a scan tool, observe A/C refrigerant pressure for 2 minutes with engine idling and A/C "ON."

Scanned refrigerant pressure should be within 20 psi of actual. If not, check for a circuit problem using DTC 66 chart or replace A/C refrigerant pressure sensor.

2.2L (VIN 4) ENGINE — COMPONENT DIAGNOSTIC CHART — 1992 CORSICA AND BERETTA

CHART C-12
ECM CONTROLLED COOLING FAN
2.2L (VIN 4) "L" CARLINE (PORT)

Circuit Description:

Battery voltage to operate the cooling fan motor is supplied to relay by CKT 2. Ignition voltage to energize the relay is supplied to relay by CKT 39. When the ECM grounds CKT 335, the relay is energized and the cooling fan is turned "ON." When the engine is running, the ECM will turn the cooling fan "ON" if:

- A/C is "ON."
- Coolant temperature greater than 106°C (223°F).
- Code 14, coolant sensor failure.

Diagnostic Aids:

If the owner complained of an overheating problem, it must be determined if the complaint was due to an actual boil over, or the hot light, or temperature gage indicated overheating.

If the gage or light indicates overheating, but no boil over is detected, the gage circuit should be checked. The gage accuracy can also be checked by comparing the coolant sensor reading using a "Scan" tool and comparing its reading with the gage reading.

If the engine is actually overheating and the gage indicates overheating, but the cooling fan is not coming "ON," the coolant sensor has probably shifted out of calibration and should be replaced.

2.2L (VIN 4) ENGINE — COMPONENT DIAGNOSTIC CHART — 1993–94 CORSICA AND BERETTA

CHART C-10
(Page 2 of 2)
A/C COMPRESSOR CLUTCH CONTROL CIRCUIT DIAGNOSIS
2.2L (VIN 4) "L" CARLINE

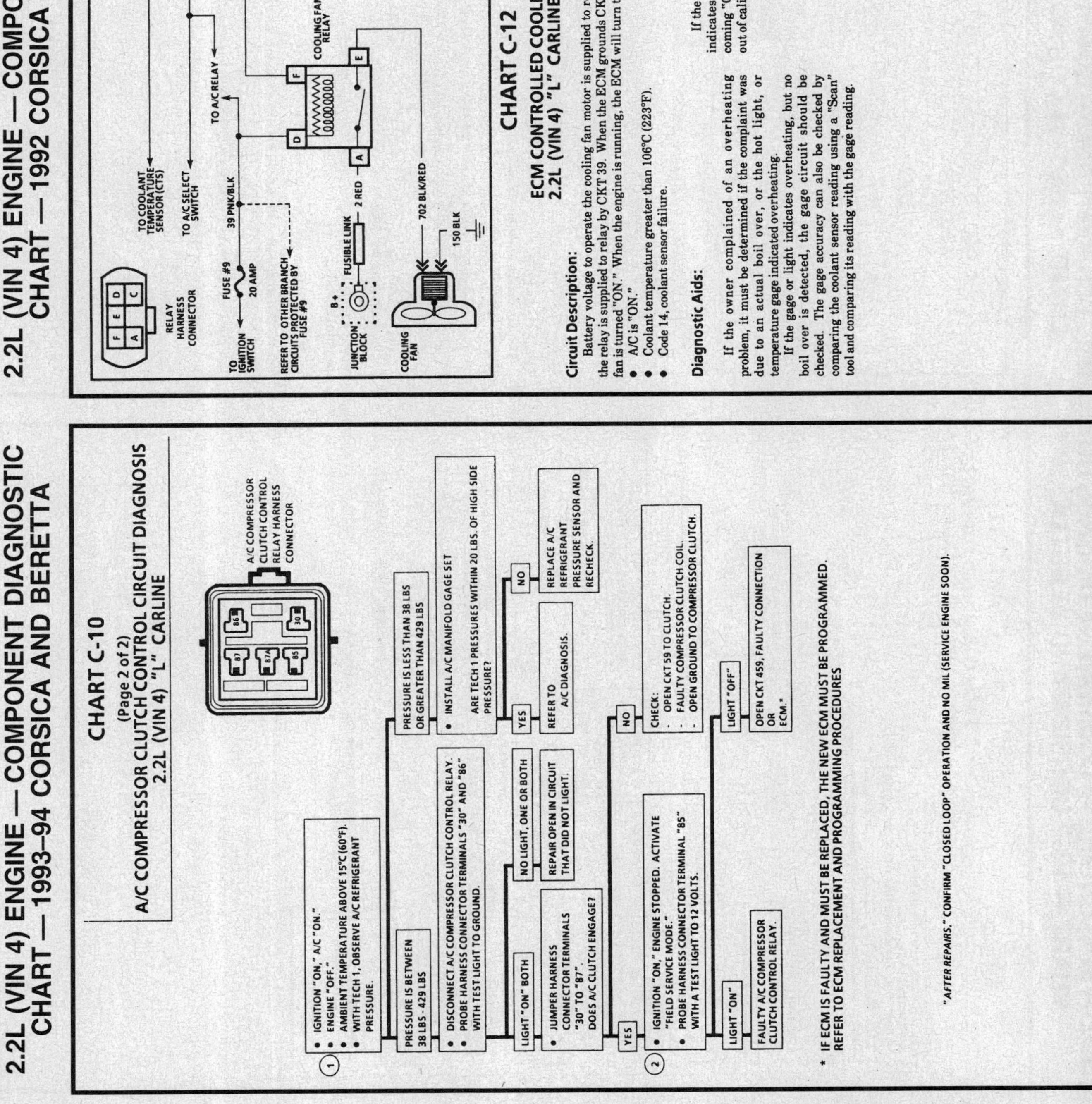

2.2L (VIN 4) ENGINE — COMPONENT DIAGNOSTIC CHART — 1992 CORSICA AND BERETTA

CHART C-12
ECM CONTROLLED COOLING FAN
2.2L (VIN 4) "L" CARLINE (PORT)

IF CODE 14 IS PRESENT, USE THAT CHART FIRST.
- IGNITION "ON," ENGINE "OFF."
- A/C "OFF," COOLANT TEMPERATURE BELOW 106°C (223°F).
IS FAN "ON"?

NO →

- USING TECH 1 "SCAN" TOOL ENTER "FIELD SERVICE MODE." IS FAN "ON"?

NO →

- DISCONNECT FAN CONTROL RELAY.
- PROBE HARNESS CONNECTOR TERMINALS "D" & "A" WITH A TEST LIGHT TO GROUND. DOES LIGHT TURN "ON" FOR BOTH TERMINALS?

YES →
- "FIELD SERVICE MODE" STILL ENGAGED."
- PROBE HARNESS CONNECTOR TERMINAL "F" WITH A TEST LIGHT TO BATTERY VOLTAGE. DOES LIGHT COME "ON"?

YES →
- JUMPER HARNESS CONNECTOR TERMINALS "A" & "E". DOES FAN RUN?

YES → REPLACE COOLING FAN RELAY.

NO → OPEN CKT 702, FAULTY FAN MOTOR, OR FAULTY GROUND CKT 150.

NO → CHECK CKT 335 FOR OPEN OR SHORT TO VOLTAGE AND ECM CONNECTIONS. IF OK, REPLACE ECM.

YES → NON-A/C → NO TROUBLE FOUND.

YES → REPAIR OPEN IN CIRCUIT THAT DID NOT LIGHT.

WITH A/C →
- EXIT "FIELD SERVICE MODE."
- ENGINE IDLING.
- A/C "ON."
IS FAN "ON"?

YES → NO TROUBLE FOUND.

NO → BACKPROBE ECM CONNECTOR TERMINAL "D9" WITH A TEST LIGHT TO GROUND.

LIGHT "OFF" → SEE A/C CHART C-10.

LIGHT "ON" → FAULTY ECM CONNECTION OR ECM.

YES →
- DISCONNECT COOLING FAN RELAY.
- JUMPER HARNESS CONNECTOR TERMINALS "D" & "F" WITH A TEST LIGHT.

LIGHT "ON" → CHECK FOR SHORT TO GROUND IN CKT 335. IF NOT GROUNDED, REPLACE ECM.

LIGHT "OFF" → CHECK CKT 702 FOR SHORT TO VOLTAGE. IF OK, REPLACE COOLING FAN RELAY.

2.2L (VIN 4) ENGINE — COMPONENT DIAGNOSTIC CHART — 1993–94 CORSICA AND BERETTA

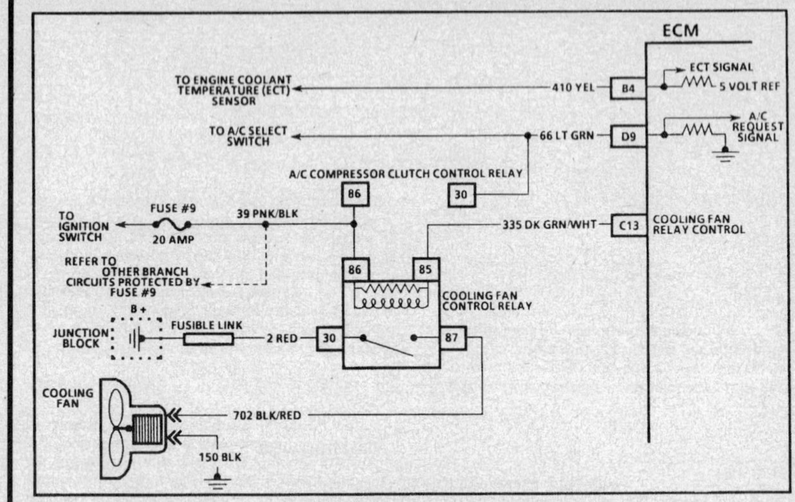

CHART C-12
COOLING FAN CONTROL DIAGNOSIS
2.2L (VIN 4) "L" CARLINE

Circuit Description:

Battery voltage to operate the cooling fan motor is supplied to relay by CKT 2. Ignition voltage to energize the cooling fan control is supplied by CKT 39. When the ECM grounds CKT 335, the relay is energized and the cooling fan is turned "ON."

When the engine is running, the ECM will turn the engine cooling fan "ON" for any of the following reasons:
- Diagnostic Trouble Code (DTC) 14 is set.
- Coolant temperature is 106°C (223°F) or greater.
- A/C refrigerant pressure is 189 psi or greater with A/C selected and a Diagnostic Trouble Code (DTC) 66 is not set.

Once the cooling fan is enabled, the ECM will turn the fan "OFF" when:
- Engine is turned "OFF."
- Coolant temperature drops to 99°C (203°F) or less.
- A/C refrigerant pressure drops to 38 psi or less with A/C selected.

Diagnostic Aids:

If the owner complained of an overheating problem, it must be determined if the complaint was due to an actual boil over, or the hot light, or temperature gage indicated overheating.

If the gage or light indicates overheating, but no boil over is detected, the gage circuit should be checked. The gage accuracy can also be checked by checking the engine coolant temperature reading using a scan tool and comparing its display with the instrument cluster gage.

If the engine is actually overheating and the gage indicates overheating, but the cooling fan is not coming "ON," the ECT sensor has probably shifted out of calibration and should be replaced.

2.2L (VIN 4) ENGINE — COMPONENT DIAGNOSTIC CHART — 1993–94 CORSICA AND BERETTA

CHART C-12
COOLING FAN CONTROL DIAGNOSIS
2.2L (VIN 4) "L" CARLINE

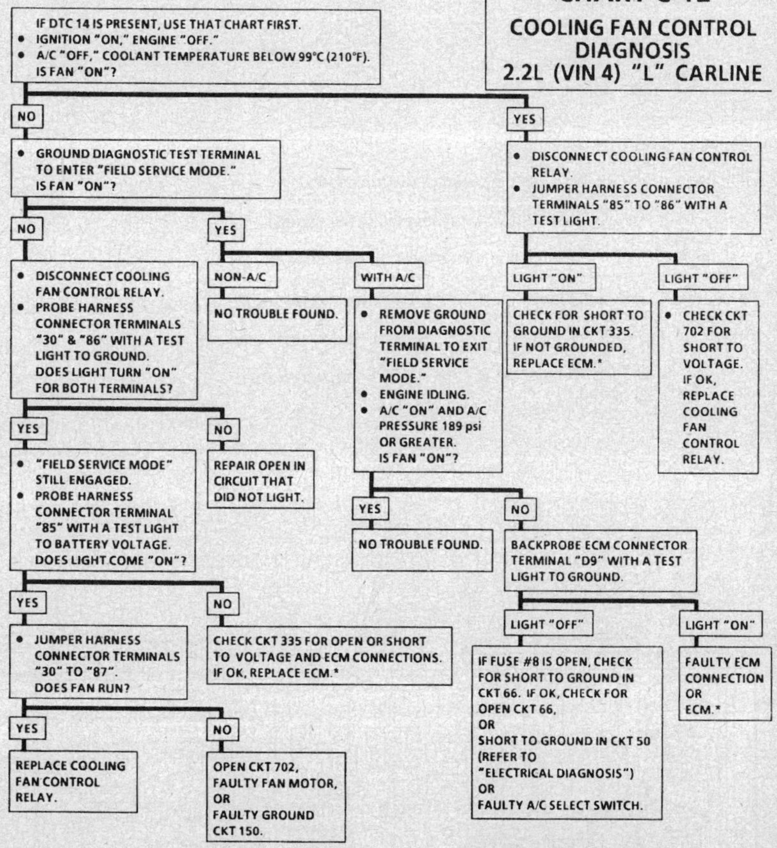
IF DTC 14 IS PRESENT, USE THAT CHART FIRST.
• IGNITION "ON," ENGINE "OFF."
• A/C "OFF," COOLANT TEMPERATURE BELOW 99°C (210°F).
IS FAN "ON"?
NO
• GROUND DIAGNOSTIC TEST TERMINAL TO ENTER "FIELD SERVICE MODE." IS FAN "ON"?
NO
• DISCONNECT COOLING FAN CONTROL RELAY.
• PROBE HARNESS CONNECTOR TERMINALS "30" & "86" WITH A TEST LIGHT TO GROUND. DOES LIGHT TURN "ON" FOR BOTH TERMINALS?
YES
• "FIELD SERVICE MODE" STILL ENGAGED.
• PROBE HARNESS CONNECTOR TERMINAL "85" WITH A TEST LIGHT TO BATTERY VOLTAGE. DOES LIGHT COME "ON"?
YES
• JUMPER HARNESS CONNECTOR TERMINALS "30" TO "87". DOES FAN RUN?
YES
REPLACE COOLING FAN CONTROL RELAY.
NO
OPEN CKT 702, FAULTY FAN MOTOR, OR FAULTY GROUND CKT 150.
NO
CHECK CKT 335 FOR OPEN OR SHORT TO VOLTAGE AND ECM CONNECTIONS. IF OK, REPLACE ECM.*
NO
REPAIR OPEN IN CIRCUIT THAT DID NOT LIGHT.
YES
NON-A/C
NO TROUBLE FOUND.
WITH A/C
• REMOVE GROUND FROM DIAGNOSTIC TERMINAL TO EXIT "FIELD SERVICE MODE."
• ENGINE IDLING.
• A/C "ON" AND A/C PRESSURE 189 psi OR GREATER. IS FAN "ON"?
YES
NO TROUBLE FOUND.
NO
BACKPROBE ECM CONNECTOR TERMINAL "D9" WITH A TEST LIGHT TO GROUND.
LIGHT "OFF"
IF FUSE #8 IS OPEN, CHECK FOR SHORT TO GROUND IN CKT 66. IF OK, CHECK FOR OPEN CKT 66, OR SHORT TO GROUND IN CKT 50 (REFER TO "ELECTRICAL DIAGNOSIS") OR FAULTY A/C SELECT SWITCH.
LIGHT "ON"
FAULTY ECM CONNECTION OR ECM.*
YES
• DISCONNECT COOLING FAN CONTROL RELAY.
• JUMPER HARNESS CONNECTOR TERMINALS "85" TO "86" WITH A TEST LIGHT.
LIGHT "ON"
CHECK FOR SHORT TO GROUND IN CKT 335. IF NOT GROUNDED, REPLACE ECM.*
LIGHT "OFF"
• CHECK CKT 702 FOR SHORT TO VOLTAGE. IF OK, REPLACE COOLING FAN CONTROL RELAY.

* IF ECM IS FAULTY AND MUST BE REPLACED, THE NEW ECM MUST BE PROGRAMMED. REFER TO ECM REPLACEMENT AND PROGRAMMING PROCEDURES

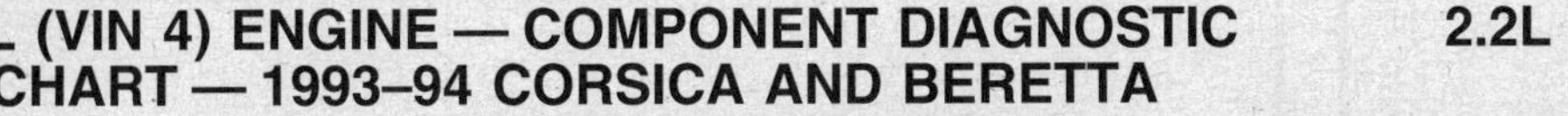

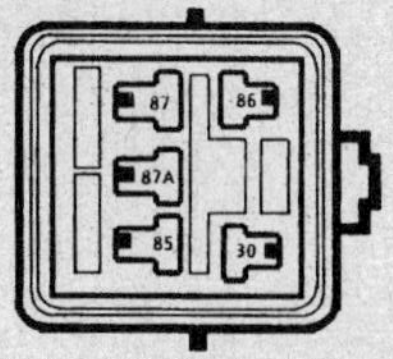
87
86
87A
85
30

COOLING FAN CONTROL RELAY CONNECTOR

2.2L (VIN 4) ENGINE — ENGINE COMPONENT LOCATION CHART — LUMINA

"W" CARLINE RPO: LN2 VIN CODE: 4 2.2L L4 MFI

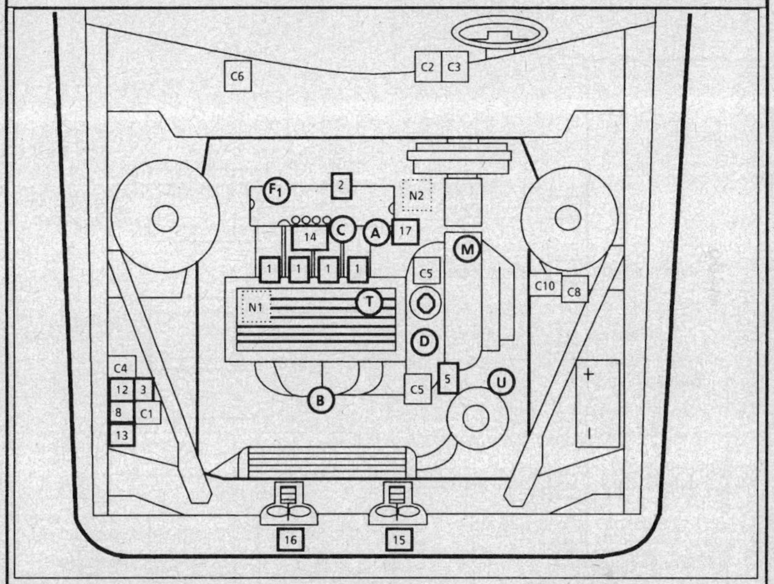
C6
C2 C3
F1
2
N2
14
C
A
17
M
1 1 1 1
C5
N1
T
C10
C8
D
C4
12 3
8 C1
13
B
C5
5
U
16
15

COMPUTER HARNESS
- C1 Engine Control Module (ECM) (Beneath Passenger Side Underhood Electrical Center)
- C2 Data Link Connector (DLC)
- C3 Malfunction Indicator Lamp (MIL) "Service Engine Soon"
- C4 Fuel/ECM Power Fuse**
- C5 ECM Harness Grounds
- C6 Fuse Panel
- C8 Fuel Pump Test Connector
- C10 Remote Battery Feed Stud

NOT ECM CONNECTED
- N1 Crankcase Vent Valve (PCV)
- N2 Evaporative Emission Canister Control Valve

NOTICE: Evaporative Emission Canister (not shown) located at rear of vehicle next to fuel filler neck.

CONTROLLED DEVICES
- 1 Fuel Injector
- 2 Idle Air Control (IAC) Valve
- 3 Fuel Pump Relay**
- 5 Torque Converter Clutch (TCC) Solenoid
- 8 Cooling Fan #1 Control (FC) Relay**
- 12 Cooling Fan #2 Control (FC) Relay
- 13 A/C Compressor Clutch Control Relay**
- 14 Ignition Coil Assembly
- 15 Primary Cooling Fan #1
- 16 Secondary Cooling Fan #2 (A/C equipped vehicles only)
- 17 EGR Control Valve Relay Solenoid (Beneath MAP Sensor)

Exhaust Gas Recirculation (EGR) Valve

INFORMATION SENSORS
- A Manifold Absolute Pressure (MAP) Sensor
- B Oxygen Sensor (O2S)
- C Throttle Position (TP) Sensor
- D Engine Coolant Temperature (ECT) Sensor
- F_1 Vehicle Speed Sensor (VSS)
- M Park/Neutral Position (PNP) Switch
- T Intake Air Temperature (IAT) Sensor (On Rear Air Duct Assembly)
- U A/C Refrigerant Pressure Sensor

** Passenger Side Underhood Electrical Center

MULTIPORT FUEL INJECTION (MFI) SYSTEMS
EXCEPT LIGHT TRUCKS, VANS, GEO AND SATURN

2.2L (VIN 4) ENGINE —ECM WIRING SCHEMATIC— LUMINA

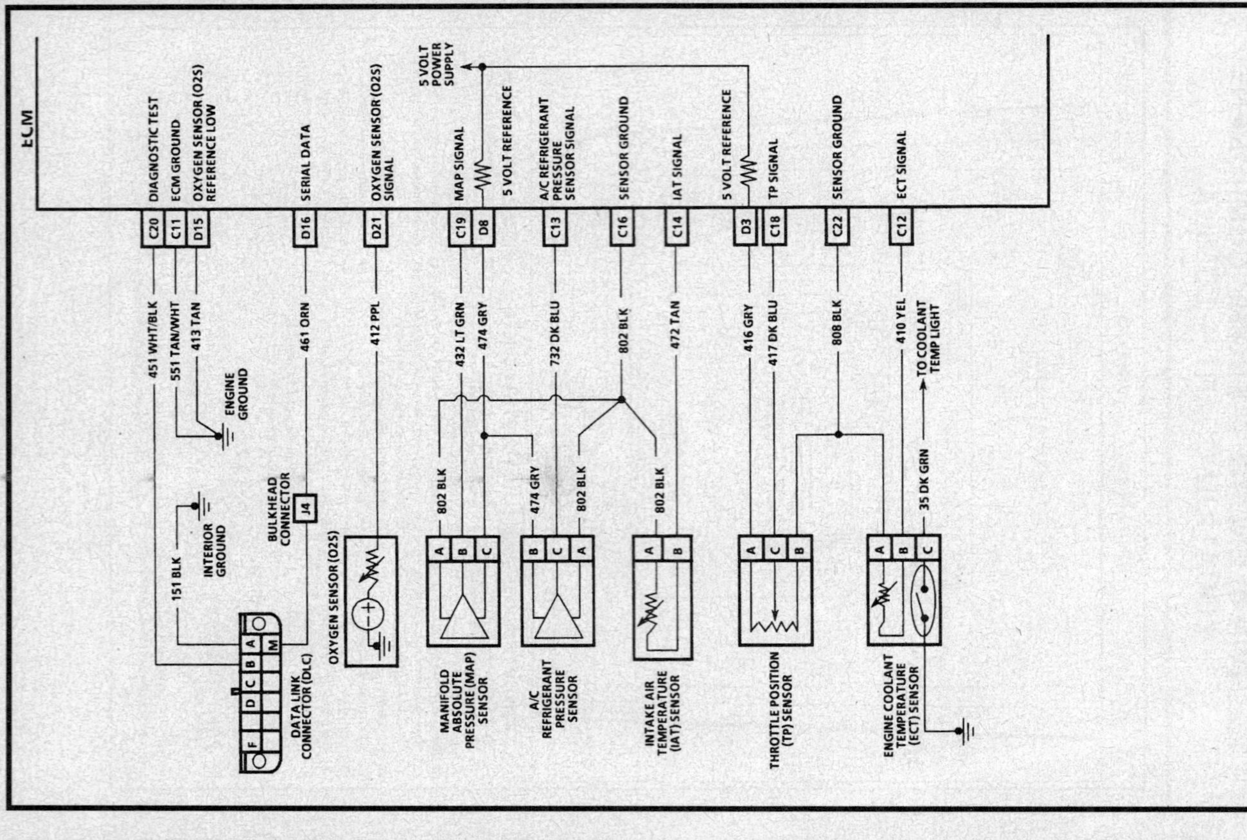

2.2L (VIN 4) ENGINE —ECM WIRING SCHEMATIC— LUMINA

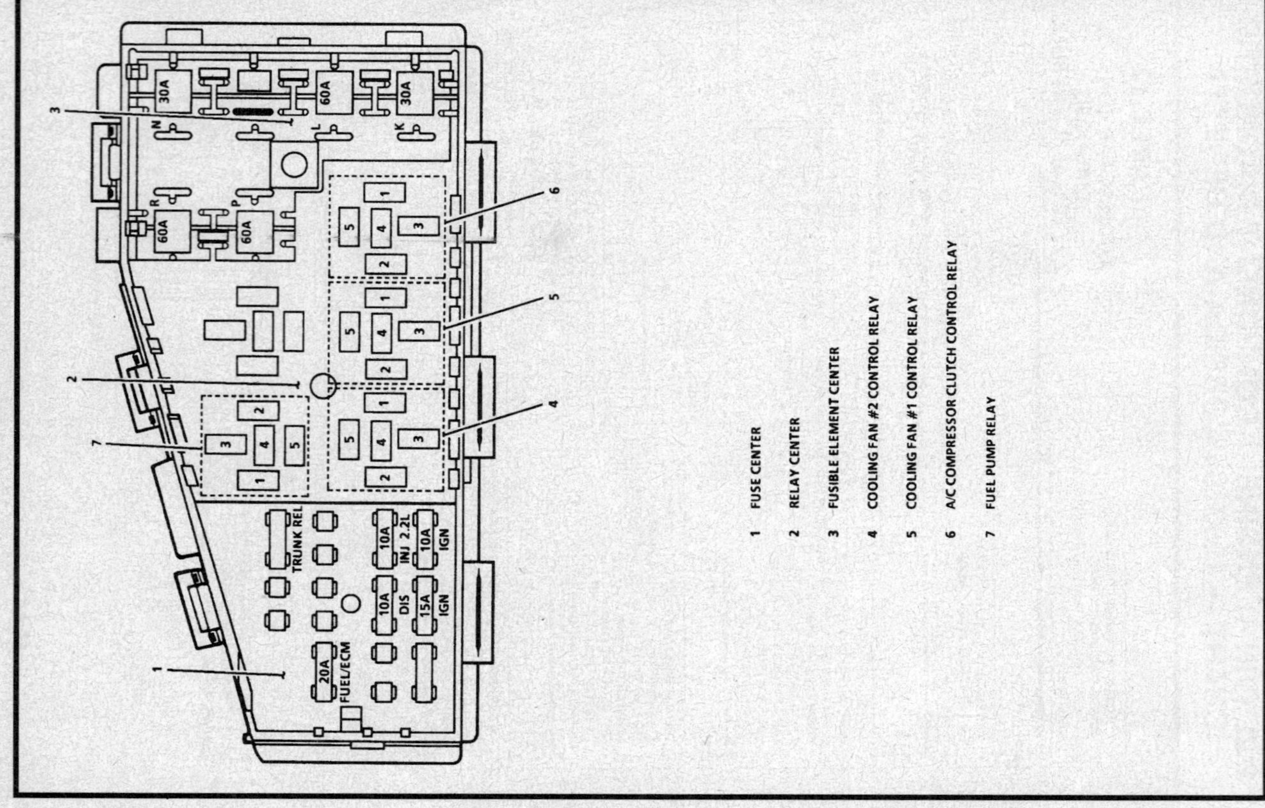

2.2L (VIN 4) ENGINE — ECM WIRING SCHEMATIC — LUMINA

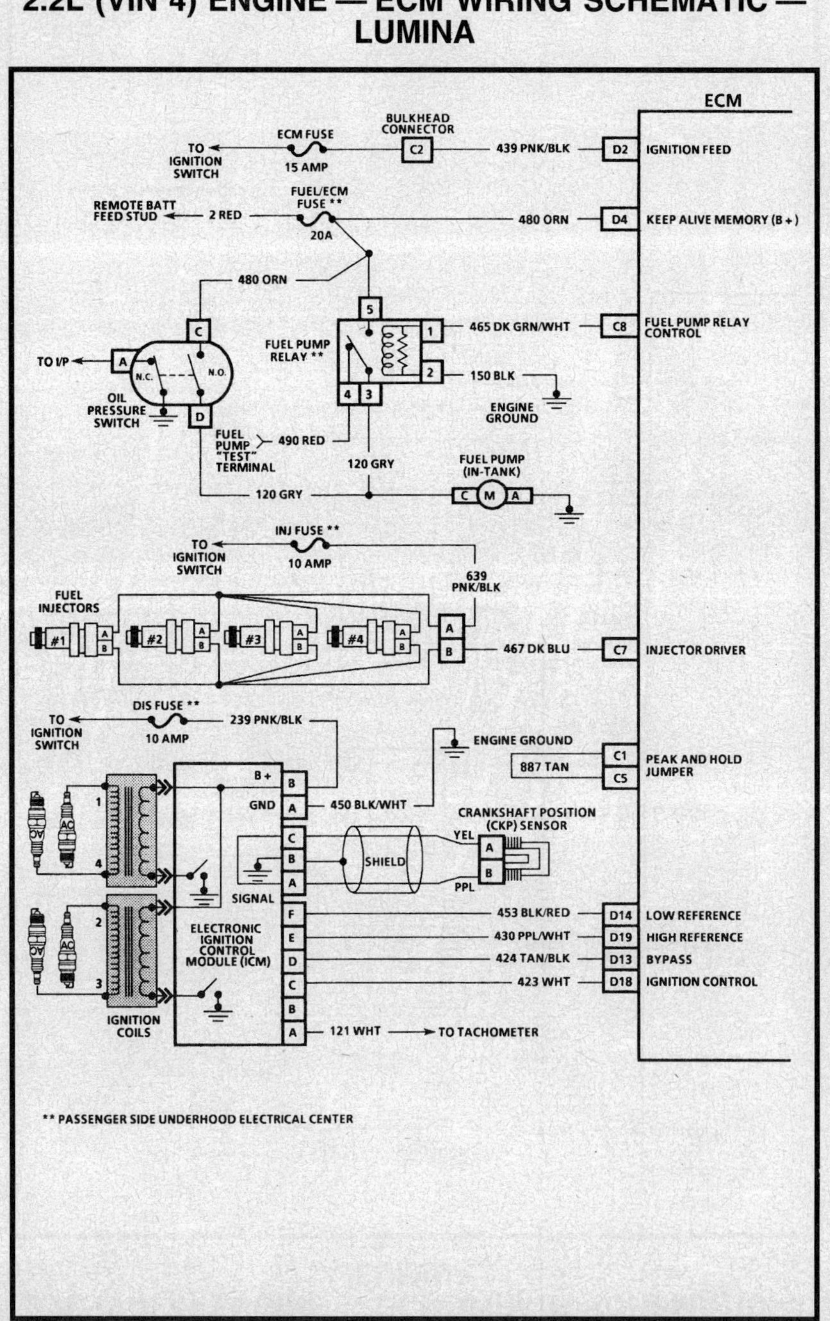

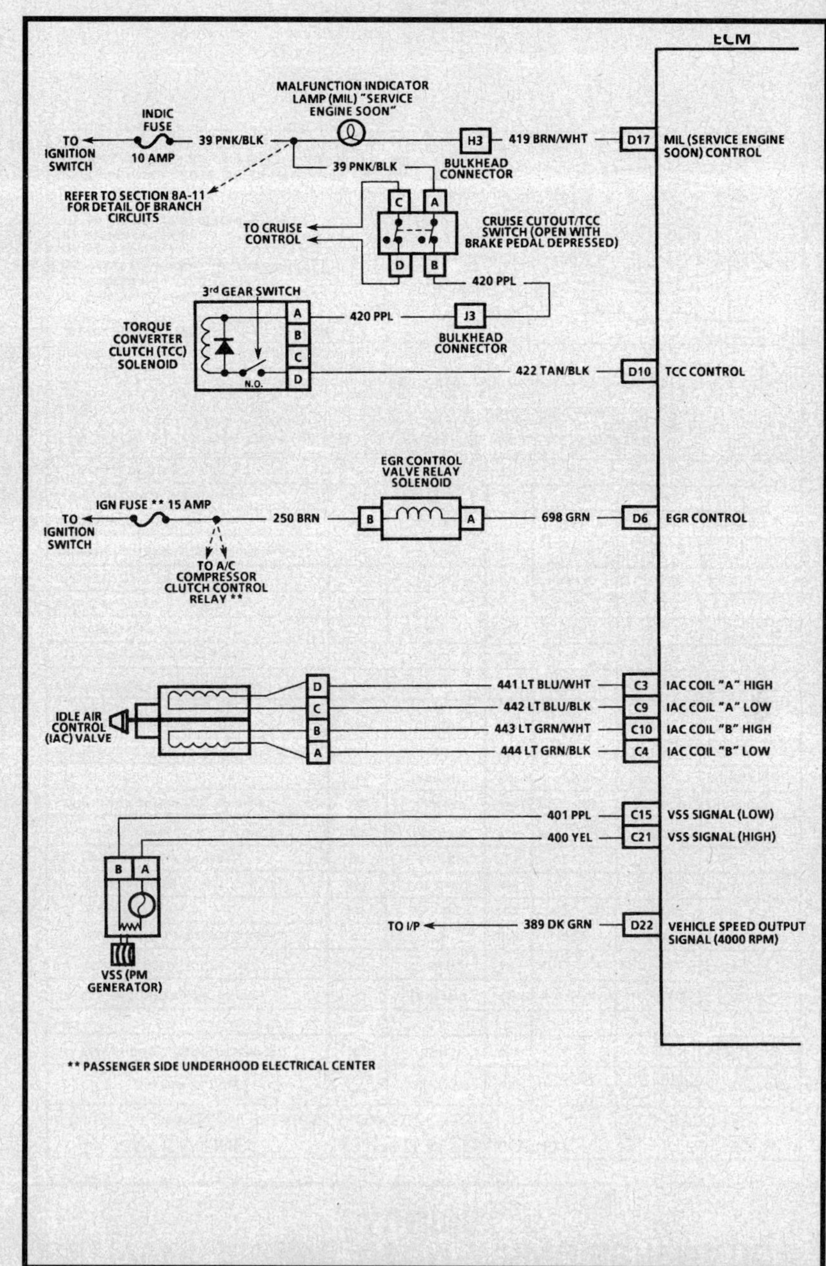

2.2L (VIN 4) ENGINE — ECM WIRING SCHEMATIC — LUMINA

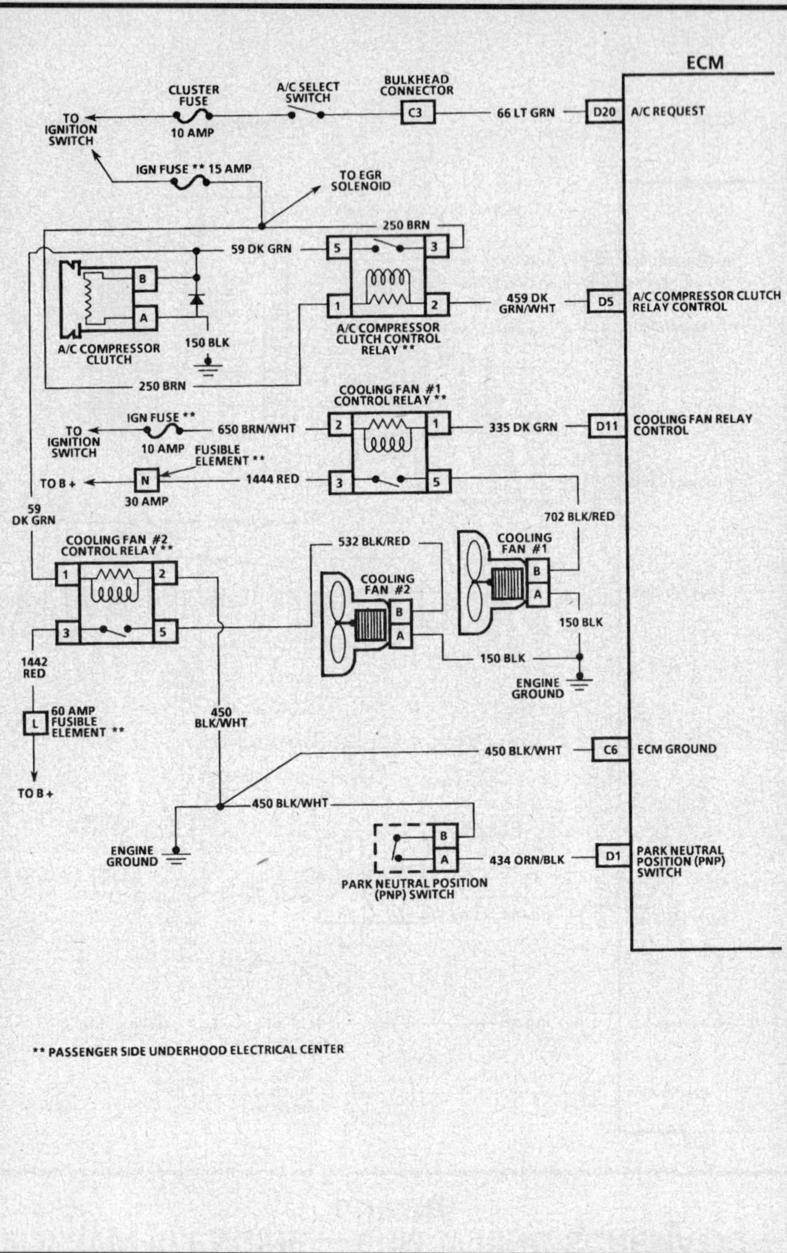

1993 "W" CARLINE	22 PIN ECM CONNECTOR				2.2L (VIN 4)	
* USE T-100 YELLOW BREAKOUT BOX (BOB)					RPO: LN2	
ECM PIN/FUNCTION	BOB PIN #	WIRE COLOR	CKT #	KEY "ON"	ENG "RUN"	REFERENCE
C1 PEAK AND HOLD (INJECTOR JUMPER)	304	TAN	887	0 (6)	0 (6)	CHART A-3
C2	301					
C3 IDLE AIR CONTROL COIL "A" HIGH	312	LT BLU/WHT	441	NOT USABLE	NOT USABLE	CHART C-2C
C4 IDLE AIR CONTROL COIL "B" LOW	309	LT GRN/BLK	444	NOT USABLE	NOT USABLE	CHART C-2C
C5 PEAK AND HOLD (INJECTOR JUMPER)	319	TAN	887	0 (6)	0 (6)	CHART A-3
C6 ECM GROUND	317	BLK/WHT	450	0 (6) (14)	0 (6) (14)	CHART A-3
C7 INJECTOR DRIVER	303	DK BLU	467	B+	B + (13)	CHART A-3
C8 FUEL PUMP RELAY DRIVER	302	DK GRN/WHT	465	0 (7)	B+	CHART A-5
C9 IDLE AIR CONTROL COIL "A" LOW	311	LT BLU/BLK	442	NOT USABLE	NOT USABLE	CHART C-2C
C10 IDLE AIR CONTROL COIL "B" HIGH	310	LT GRN/WHT	443	NOT USABLE	NOT USABLE	CHART C-2C
C11 ECM GROUND	318	TAN/WHT	551	0 (6)	0 (6)	FIGURE A-3
C12 ENGINE COOLANT TEMPERATURE SIGNAL	308	YEL	410	(1)	(1)	DTC 14
C13 A/C REFRIGERANT PRESSURE SENSOR SIGNAL	305	DK BLU	732	VARIES	VARIES	DTC 66
C14 INTAKE AIR TEMPERATURE SIGNAL	316	TAN	472	(1)	(1)	DTC 23
C15 VEHICLE SPEED SIGNAL (LOW)	313	PPL	401	(4)	(4)	DTC 24
C16 SENSOR GROUND	322	BLK	802	0 (6)	0 (6)	DTC 23, 33, 66
C17	320					
C18 THROTTLE POSITION SIGNAL	307	DK BLU	417	.33-1.33	.33-1.33	DTC 21
C19 MANIFOLD ABSOLUTE PRESSURE SIGNAL	306	LT GRN	432	(3)	(2)	DTC 33
C20 DIAGNOSTIC TEST	315	WHT/BLK	451	5.0V	5.0V	CHART A-2
C21 VEHICLE SPEED SIGNAL (HIGH)	314	YEL	400	(4) (8)	(4) (8)	DTC 24
C22 SENSOR GROUND	321	BLK	808	0 (6)	0 (6)	DTC 14, 21

***NOTICE:** IF TOOL IS NOT AVAILABLE, DO NOT "BACKPROBE" OR PIERCE INSULATION! This Chart may be used in conjunction with the T-100 Yellow Breakout Box (48921) to obtain voltage present for each circuit listed. Install the BOB between the ECM harness connectors and the ECM, then probe the pin listed under "BOB PIN#." Voltage may vary due to low battery charge or other reasons, but should be very close. All voltages shown in the ENG "RUN" column are typical with engine at idle, closed throttle, normal operating temperature, park or neutral, system in "Closed Loop," all accessories "OFF," brake not applied and scan tool not installed.

DVM NEGATIVE (BLACK) LEAD MUST BE CONNECTED TO A KNOWN GOOD GROUND.

(1) VARIES WITH TEMPERATURE.
(2) VARIES WITH MANIFOLD PRESSURE.
(3) VARIES WITH BAROMETRIC PRESSURE.
(4) VARIES DEPENDING ON POSITION OF DRIVE WHEELS.
(6) LESS THAN 0.5 VOLT.
(7) B + FOR 2 SECONDS AFTER KEY "ON."

(13) ALTERNATE TEST: SET J 39200 DVM TO DC VOLTAGE FREQUENCY SCALE. CONNECT RED LEAD TO IGNITION FEED (BOB PIN #401) AND BLACK LEAD TO INJECTOR DRIVER (BOB PIN #303); SHOULD MEASURE ABOUT 9-15 Hz.
(14) THIS TYPE OF CIRCUIT SHOULD ALSO BE CHECKED USING J 34142-B TEST LIGHT.

GREEN ECM CONNECTOR C

2.2L (VIN 4) ENGINE — ECM CONNECTOR END VIEW — LUMINA

1993 "W" CARLINE 22 PIN ECM CONNECTOR 2.2L (VIN 4) RPO: LN2						
* USE T-100 YELLOW BREAKOUT BOX (BOB)						
ECM PIN/FUNCTION	BOB PIN #	WIRE COLOR	CKT #	VOLTAGE KEY "ON"	VOLTAGE ENG "RUN"	REFERENCE
D1 PARK/NEUTRAL POSITION (PNP) SWITCH INPUT	404	ORN/BLK	434	0 (6)	0 (6)	CHART C-1A
D2 IGNITION FEED	401	PNK/BLK	439	B+	B+	ON-BOARD DIAGNOSTIC SYSTEM CHECK
D3 5 VOLT REFERENCE (TP SENSOR)	412	GRY	416	5.0V	5.0V	DTC 21
D4 KEEP ALIVE MEMORY (B+)	409	ORN	480	B+	B+	ON-BOARD DIAGNOSTIC SYSTEM CHECK
D5 A/C COMPRESSOR CLUTCH CONTROL	419	DK GRN/WHT	459	B+	B+	CHART C-10
D6 EXHAUST GAS RECIRCULATION CONTROL	417	GRN	698	B+	B+	CHART C-7
D7	403					
D8 5 VOLT REFERENCE	402	GRY	474	5.0V	5.0V	DTC 33, 66
D9	411					
D10 TORQUE CONVERTER CLUTCH CONTROL	410	TAN/BLK	422	0 (6)	0 (6)	CHART C-8A
D11 COOLING FAN CONTROL	418	DK GRN	335	B+	B+	CHART C-12
D12	408					
D13 IGNITION BYPASS	405	TAN/BLK	424	0 (6)	4.8V	
D14 IGNITION LOW REFERENCE	416	BLK/RED	453	0 (6)	0 (6)	
D15 OXYGEN SENSOR LOW REFERENCE	413	TAN	413	0 (6)	0 (6)	DTC 13, 44, 45
D16 SERIAL DATA	422	ORN	461	4.7V	4.7V	ON-BOARD DIAGNOSTIC SYSTEM CHECK
D17 MALFUNCTION INDICATOR LAMP (MIL) "SERVICE ENGINE SOON"	420	BRN/WHT	419	0-1.0V	B+	ON-BOARD DIAGNOSTIC SYSTEM CHECK
D18 IGNITION CONTROL	407	WHT	423	0 (6)	2.4V	
D19 IGNITION HIGH REFERENCE	406	PPL/WHT	430	0 (6)	3.2V (15)	
D20 A/C REQUEST INPUT	415	LT GRN	66	0 (6)	0 (6)	CHART C-10
D21 OXYGEN SENSOR SIGNAL	414	PPL	412	.01-.55V	.1-.9V (5)	DTC 13, 44, 45
D22 VEHICLE SPEED OUTPUT SIGNAL	421	DK GRN	389	(4)	(4)	

***NOTICE:** IF TOOL IS NOT AVAILABLE, DO NOT "BACKPROBE" OR PIERCE INSULATION! This Chart may be used in conjunction with the T-100 Yellow Breakout Box (48921) to obtain voltage present for each circuit listed. Install the BOB between the ECM harness connectors and the ECM, then probe the pin listed under "BOB PIN#." Voltage may vary due to low battery charge or other reasons, but should be very close. All voltages shown in the ENG "RUN" column are typical with engine at idle, closed throttle, normal operating temperature, park or neutral, system in "Closed Loop," all accessories "OFF," brake not applied and scan tool not installed.

DVM NEGATIVE (BLACK) LEAD MUST BE CONNECTED TO A KNOWN GOOD GROUND.

(5) VARIES ACTIVELY WITHIN INDICATED RANGE.
(6) LESS THAN 0.5 VOLT.

(15) ALTERNATE TEST: SET J 39200 DVM TO DC VOLTAGE FREQUENCY SCALE. CONNECT RED LEAD TO IGNITION HIGH REFERENCE (BOB PIN #406) AND BLACK LEAD TO ECM GROUND (BOB PIN #317); SHOULD MEASURE ABOUT 25-30 Hz.

BROWN ECM CONNECTOR D

2.2L (VIN 4) ENGINE — ECM CONNECTOR END VIEW — LUMINA

ON-BOARD DIAGNOSTIC (OBD) SYSTEM CHECK
2.2L (VIN 4) "W" CARLINE

Circuit Description:

The OBD system check is an organized approach to identifying a problem created by an electronic engine control system malfunction. It must be the starting point for any driveability complaint diagnosis, because it directs the service technician to the next logical step in diagnosing the complaint. Understanding the chart and using it correctly will reduce diagnostic time and prevent the unnecessary replacement of good parts.

Test Description: Number(s) below refer to circled number(s) on the diagnostic chart.

1. This step is a check for the proper operation of the Malfunction Indicator Lamp (MIL) "Service Engine Soon." The MIL should be "ON" steady.
2. No MIL at this point indicates that there is a problem with the MIL circuit or the ECM control of that circuit.
3. This test checks the ability of the ECM to control the MIL. With the diagnostic terminal grounded, the MIL should flash a DTC 12 three times, followed by any trouble code stored in memory. Depending upon the type of ECM, an EEPROM error may result in the inability to flash DTC 12.
4. Most of the 6E procedures use a Tech 1 to aid diagnosis, therefore, serial data must be available. If an EEPROM error is present, the ECM may have been able to flash DTC 12/51, but not enable serial data.
5. Although the ECM is powered up, a "Cranks But Will Not Run" symptom could exist because of an ECM or system problem.
6. This step will isolate if the customer complaint is a MIL or a driveability problem with no MIL. Refer to "ECM Diagnostic Trouble Codes (DTCs)" for valid DTCs. An invalid DTC may be the result of a faulty scan tool, EEPROM or ECM.
7. Comparison of actual control system data with the typical values is a quick check to determine if any parameter is not within limits. Keep in mind that a base engine problem (i.e. advanced cam timing) may substantially alter sensor values.
8. Installation of a scan tool will provide a good ground path for the ECM and may hide a driveability complaint due to poor ECM grounds.
9. If the actual data is not within the typical values established, the charts "Component Systems," will provide a functional check of the suspect component or system.

2.2L (VIN 4) ENGINE — ON-BOARD DIAGNOSTIC SYSTEM CHART — LUMINA

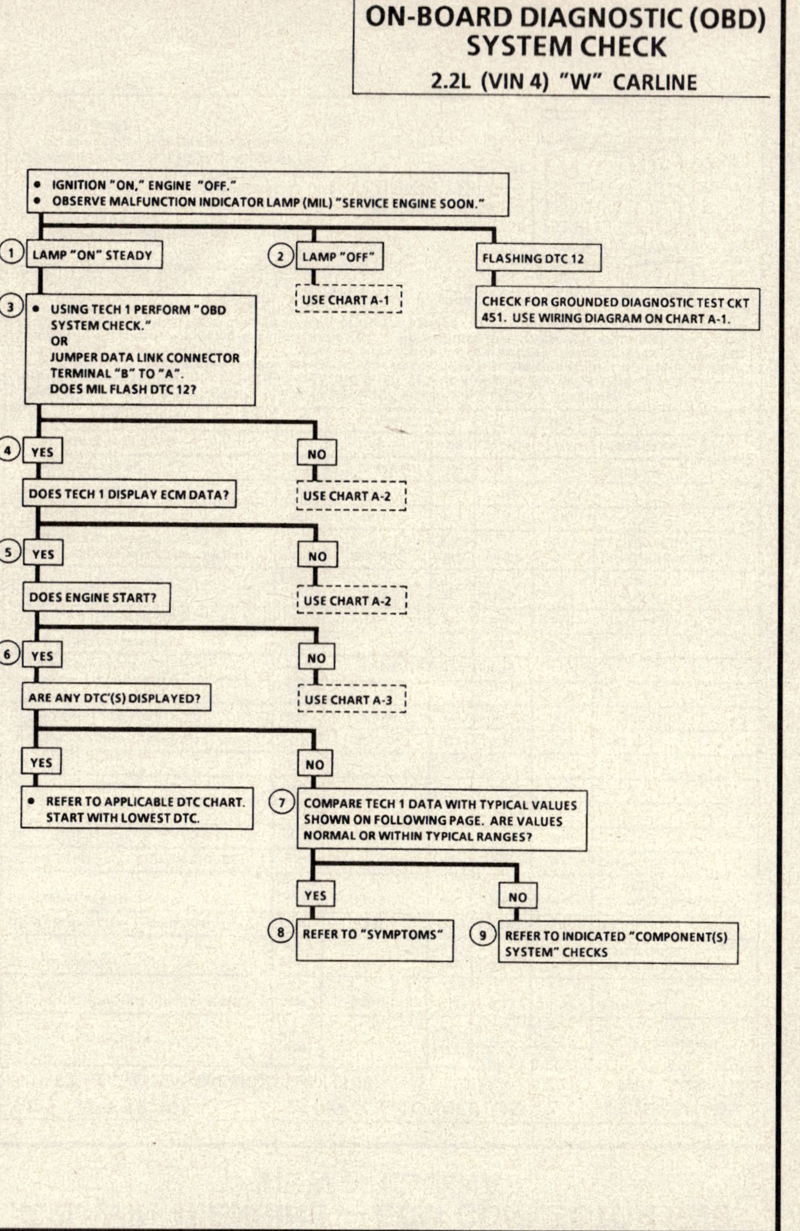

If after completing the "On-Board Diagnostic (OBD) System Check" and finding the Tech 1 diagnostics functioning properly and no diagnostic trouble codes displayed, the "Typical Tech 1 Data Values" may be used for comparison with values obtained on the vehicle being diagnosed. The "Typical Tech 1 Data Values," are an average of display values recorded from normally operating vehicles and are intended to represent what a normally functioning system would display.

A SCAN TOOL THAT DISPLAYS FAULTY DATA SHOULD NOT BE USED, AND THE PROBLEM SHOULD BE REPORTED TO THE MANUFACTURER. THE USE OF A FAULTY SCAN TOOL CAN RESULT IN MISDIAGNOSIS AND UNNECESSARY PARTS REPLACEMENT.

Only the parameters listed are used for diagnosis. If a scan tool reads other parameters, the values are not recommended by General Motors for use in diagnosis. For more description on the values and use of the Tech 1 to diagnosis ECM inputs, refer to the applicable "Component Systems,"

TYPICAL TECH 1 DATA VALUES
2.2L (VIN 4) "W" CARLINE

Idle / Upper Radiator Hose Hot / Closed Throttle / Park or Neutral / Closed Loop / Acc. OFF/ Brake Not Applied

Tech 1 Parameter	Units Displayed	Typical Data Value
Engine Speed	RPM	700 ± 75 RPM
Desired Idle	RPM	ECM commanded (varies with temp.)
Engine Cool Temp.	°C/ °F	85°-105°C (185°-220°F)
Intake Air Temp.	°C/ °F	Varies with air temp.
MAP	kPa/Volts	29-48 kPa/1-2 volts (varies with manifold and barometric pressures)
BARO	kPa/Volts	Varies with barometric pressure
Throt Position	Volts	.6 (within the range of .33-1.33)
Throttle Angle	0-100%	0%
Oxygen Sensor	Millivolts	100-999 (actively varying)
Inj. Pulse Width	Milliseconds	.8-3.0
Spark Advance	Degrees	18° (varies with rpm and engine load)
Short Term (ST) Fuel Trim	Counts	Varies
Long Term (LT) Fuel Trim	Counts	90-160
Idle Air Control	Counts	1-50
Park/Neutral	P-N/R-D-L	P-N
Fuel Trim Cell	Cell Number	0-3 (depends on air flow and RPM)
Fuel Trim Enable	Yes/No	YES/NO (depends on short term fuel trim)
Loop Status	Open/Closed	"Closed Loop"
EGR	"ON"/"OFF"	"OFF"
MPH/Kph	0-255	0
TCC/Shift Light	"ON"/"OFF"	"OFF"
Fan Relay	"ON"/"OFF"	"OFF" ("ON" when coolant temp. reaches 108°C or when A/C head pressure reaches 189 psi with A/C or defrost selected)
A/C Request	Yes/No	NO
A/C Clutch	"ON"/"OFF"	"OFF"
A/C Pressure	psi/Volts	Varies with temp.
System Voltage	Volts	13.5-14.5
Calibration ID	0-9999	Internal ID only
Time From Start	Hrs/Min	Varies

2.2L (VIN 4) ENGINE — ON-BOARD DIAGNOSTIC SYSTEM CHART — LUMINA

2.2L (VIN 4) ENGINE — SYSTEM DIAGNOSTIC CHARTS — LUMINA

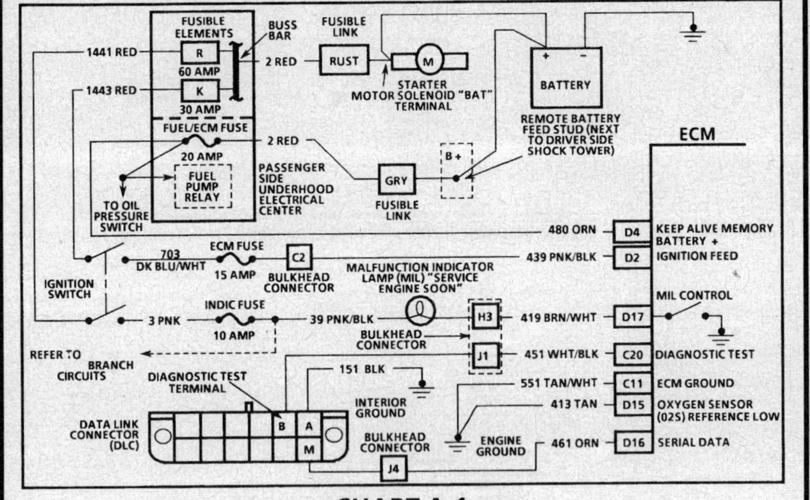

CHART A-1

NO MALFUNCTION INDICATOR LAMP (MIL) "SERVICE ENGINE SOON"
2.2L (VIN 4) "W" CARLINE

Circuit Description:

There should always be a steady MIL (Service Engine Soon) when the ignition is "ON" and engine "OFF." Battery voltage is supplied directly to the lamp. The Engine Control Module (ECM) will control the lamp and turn it "ON" by providing a ground path through CKT 419 to the ECM.

Test Description: Number(s) below refer to circled number(s) on the diagnostic chart.

1. This check determines if there is a fault in the MIL circuit or in the ECM.
2. If CKTs 480 and 439 have voltage and the engine cranks but won't run, the ECM grounds or the ECM is faulty.

   Using a test light connect to Battery +, probe each of the system ground circuits to be sure a good ground is present. See "ECM Connector End View" in front of this section for ECM pin locations of ground circuits.

Diagnostic Aids:

If engine runs OK, check the following:
- Faulty lamp.
- CKT 419 open.
- "INDIC" fuse open. This will result in no oil, or generator lights, seat belt reminder, etc.

   Refer to "Fuse Block Details, for other branch circuits that could cause a blown "INDIC" fuse.

   If engine cranks, but will not run, use CHART A-3.

CHART A-2

NO DATA LINK CONNECTOR (DLC) DATA OR WON'T FLASH DTC 12
MALFUNCTION INDICATOR LAMP (MIL) "SERVICE ENGINE SOON" "ON" STEADY
2.2L (VIN 4) "W" CARLINE

Circuit Description:

There should always be a steady MIL (Service Engine Soon) when the ignition is "ON" and the engine is "OFF." Battery voltage is supplied directly to the lamp. The Engine Control Module (ECM) will control the lamp and turn it "ON" by providing a ground path through CKT 419 to the ECM.

With the "Diagnostic Test Terminal" grounded, the lamp should flash a Diagnostic Trouble Code (DTC) 12, followed by any diagnostic trouble code(s) stored in memory. A steady lamp suggests a short to ground in the lamp control CKT 419, or an open in diagnostic CKT 451.

Test Description: Number(s) below refer to circled number(s) on the diagnostic chart.

1. If there is a problem with the ECM that causes a scan tool to not read data from the ECM, then the ECM should not flash a DTC 12. If DTC 12 does flash, be sure that the scan tool is working properly on another vehicle. If the scan is functioning properly and CKT 461 is OK, the EEPROM or ECM may be at fault for the "No Data Link Connector Data" symptom.

2. If the lamp turns "OFF" when the ECM connector is disconnected, then CKT 419 is not shorted to ground.

3. This step will check for an open diagnostic CKT 451 or faulty ground CKT 151.

4. At this point, the MIL (Service Engine Soon) wiring is OK. The problem is a faulty ECM. If DTC 12 does not flash, the ECM should be replaced and the new ECM must be programmed.

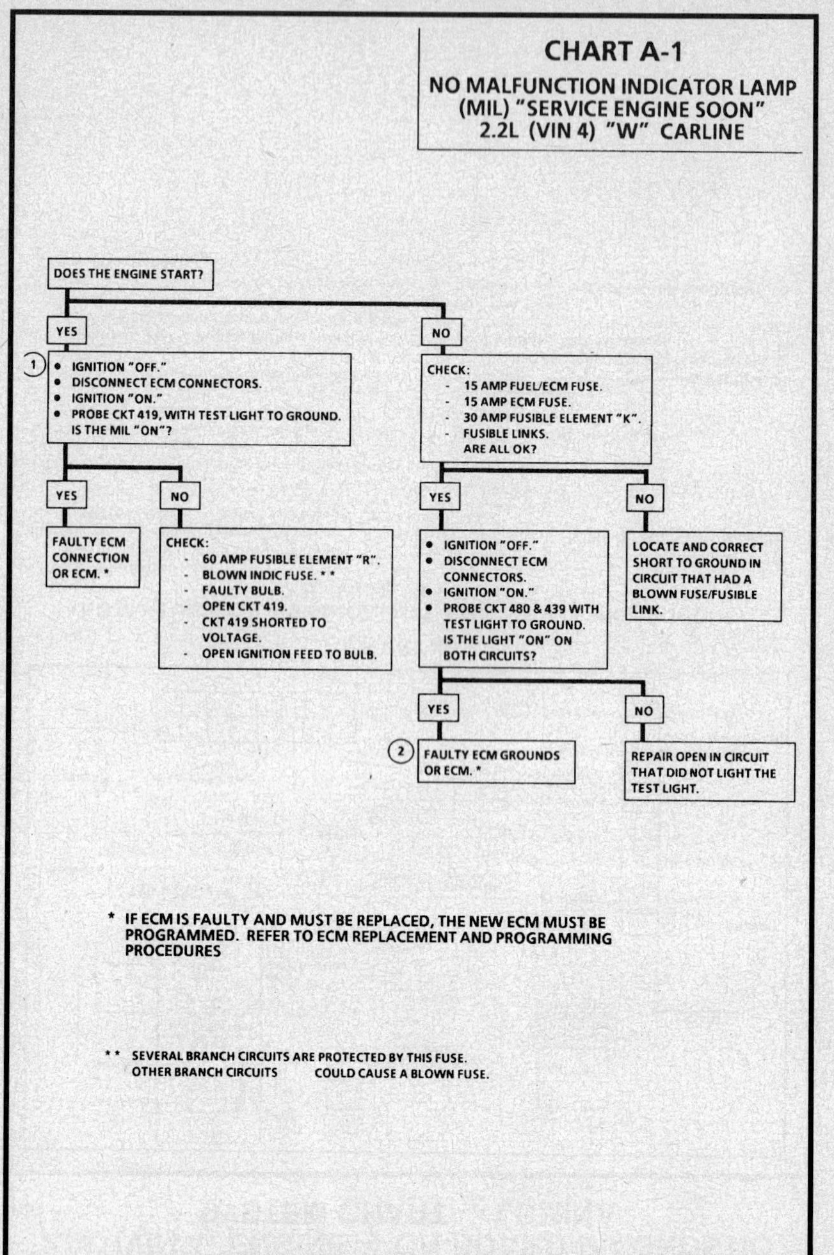
CHART A-1
NO MALFUNCTION INDICATOR LAMP (MIL) "SERVICE ENGINE SOON" 2.2L (VIN 4) "W" CARLINE
DOES THE ENGINE START?
YES
NO
1 IGNITION "OFF." DISCONNECT ECM CONNECTORS. IGNITION "ON." PROBE CKT 419, WITH TEST LIGHT TO GROUND. IS THE MIL "ON"?
CHECK:
- 15 AMP FUEL/ECM FUSE.
- 15 AMP ECM FUSE.
- 30 AMP FUSIBLE ELEMENT "K".
- FUSIBLE LINKS.
ARE ALL OK?
YES
NO
FAULTY ECM CONNECTION OR ECM. *
CHECK:
- 60 AMP FUSIBLE ELEMENT "R".
- BLOWN INDIC FUSE. * *
- FAULTY BULB.
- OPEN CKT 419.
- CKT 419 SHORTED TO VOLTAGE.
- OPEN IGNITION FEED TO BULB.
IGNITION "OFF." DISCONNECT ECM CONNECTORS. IGNITION "ON." PROBE CKT 480 & 439 WITH TEST LIGHT TO GROUND. IS THE LIGHT "ON" ON BOTH CIRCUITS?
LOCATE AND CORRECT SHORT TO GROUND IN CIRCUIT THAT HAD A BLOWN FUSE/FUSIBLE LINK.
YES
NO
2 FAULTY ECM GROUNDS OR ECM. *
REPAIR OPEN IN CIRCUIT THAT DID NOT LIGHT THE TEST LIGHT.
* IF ECM IS FAULTY AND MUST BE REPLACED, THE NEW ECM MUST BE PROGRAMMED. REFER TO ECM REPLACEMENT AND PROGRAMMING PROCEDURES
* * SEVERAL BRANCH CIRCUITS ARE PROTECTED BY THIS FUSE. OTHER BRANCH CIRCUITS COULD CAUSE A BLOWN FUSE.

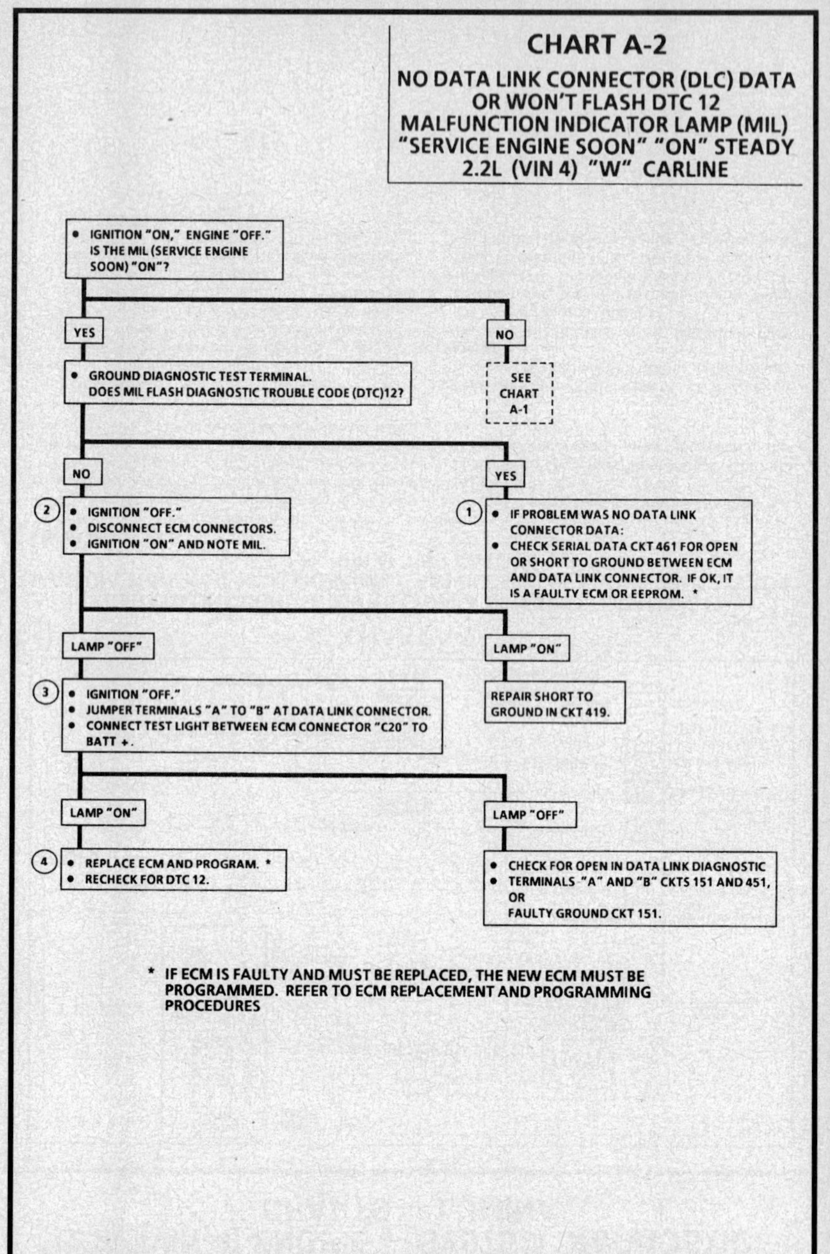
CHART A-2
NO DATA LINK CONNECTOR (DLC) DATA OR WON'T FLASH DTC 12 MALFUNCTION INDICATOR LAMP (MIL) "SERVICE ENGINE SOON" "ON" STEADY 2.2L (VIN 4) "W" CARLINE
IGNITION "ON," ENGINE "OFF." IS THE MIL (SERVICE ENGINE SOON) "ON"?
YES
NO
GROUND DIAGNOSTIC TEST TERMINAL. DOES MIL FLASH DIAGNOSTIC TROUBLE CODE (DTC)12?
SEE CHART A-1
NO
YES
2 IGNITION "OFF." DISCONNECT ECM CONNECTORS. IGNITION "ON" AND NOTE MIL.
1 IF PROBLEM WAS NO DATA LINK CONNECTOR DATA: CHECK SERIAL DATA CKT 461 FOR OPEN OR SHORT TO GROUND BETWEEN ECM AND DATA LINK CONNECTOR. IF OK, IT IS A FAULTY ECM OR EEPROM. *
LAMP "OFF"
LAMP "ON"
REPAIR SHORT TO GROUND IN CKT 419.
3 IGNITION "OFF." JUMPER TERMINALS "A" TO "B" AT DATA LINK CONNECTOR. CONNECT TEST LIGHT BETWEEN ECM CONNECTOR "C20" TO BATT +.
LAMP "ON"
LAMP "OFF"
4 REPLACE ECM AND PROGRAM. * RECHECK FOR DTC 12.
CHECK FOR OPEN IN DATA LINK DIAGNOSTIC TERMINALS "A" AND "B" CKTS 151 AND 451, OR FAULTY GROUND CKT 151.
* IF ECM IS FAULTY AND MUST BE REPLACED, THE NEW ECM MUST BE PROGRAMMED. REFER TO ECM REPLACEMENT AND PROGRAMMING PROCEDURES

2.2L (VIN 4) ENGINE — SYSTEM DIAGNOSTIC CHARTS — LUMINA

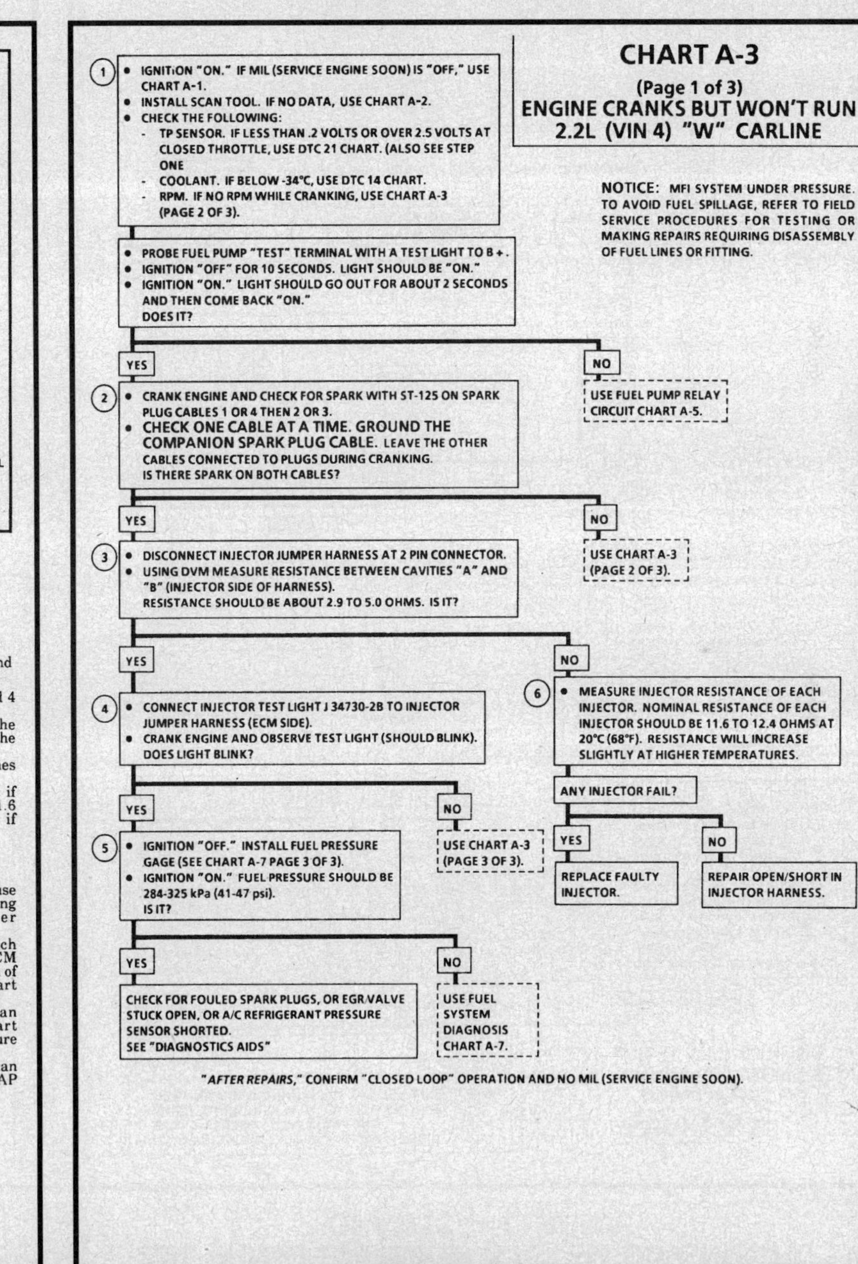

CHART A-3
(Page 1 of 3)
ENGINE CRANKS BUT WON'T RUN
2.2L (VIN 4) "W" CARLINE

Circuit Description:
Before using this chart, battery condition, engine cranking speed, and fuel quantity should be checked and verified as being OK.

Test Description: Number(s) below refer to circled number(s) on the diagnostic chart.
1. • A Malfunction Indicator Lamp (MIL) "Service Engine Soon" "ON" is a basic test to determine if there is battery and ignition voltage at the ECM.
   • No DLC data may be the result of an ECM problem, and CHART A-2 will diagnose an ECM problem.
   • If Throttle Position (TP) sensor is less than .2 volts, the TP sensor 5 volt reference circuit could be shorted to ground. If TP sensor is over 2.5 volts, the ECM could be in the "Clear Flood Mode" which may cause the engine to not start.
   • Compare coolant temperature with intake air temperature when engine is cold. If coolant temperature reading is 10 degrees greater or less than intake air temperature on a cold engine, check resistance of the Engine Coolant Temperature (ECT) sensor circuit or sensor. Compare resistance value to the "Diagnostic Aid" chart found on DTC 14 chart.
   • The scan tool should display RPM during cranking.
2. Because the Electronic Ignition (EI) uses two spark plugs and cables to complete the circuit of each coil, the opposite spark plug cable should be connected to a good ground to prevent overloading the coil.

3. This test is performed with injectors 1, 2, 3, and 4 in parallel.
4. The test light should flash, indicating that the ECM is controlling the injectors. How bright the light flashes is not important.
5. Ignition may have to be cycled "ON" several times to obtain maximum fuel pressure.
6. Damage to the ECM injector driver may occur if any injector resistance measures less than 11.6 ohms (internal injector short to CKT 639), or if CKT 467 shorts to CKT 639.

Diagnostic Aids:
• Water or contamination in fuel system may cause a no start condition during very cold or freezing weather. The engine may start after approximately 5 minutes in a heated shop.
• An EGR valve sticking open can cause a rich air/fuel ratio during cranking. Unless the ECM enters "Clear Flood Mode" at the first indication of a flooding condition, it may result in a no start condition.
• An A/C refrigerant pressure sensor with an internal short to ground can cause a no start condition. Disconnect the A/C refrigerant pressure sensor. If vehicle starts, replace faulty sensor.
• A MAP sensor stuck between .5 and 2.5 volts can cause a no start condition. Disconnect the MAP sensor. If vehicle starts, replace faulty sensor.

2.2L (VIN 4) ENGINE — SYSTEM DIAGNOSTIC CHARTS — LUMINA

CHART A-3
(Page 1 of 3)
ENGINE CRANKS BUT WON'T RUN
2.2L (VIN 4) "W" CARLINE

NOTICE: MFI SYSTEM UNDER PRESSURE. TO AVOID FUEL SPILLAGE, REFER TO FIELD SERVICE PROCEDURES FOR TESTING OR MAKING REPAIRS REQUIRING DISASSEMBLY OF FUEL LINES OR FITTING.

① • IGNITION "ON." IF MIL (SERVICE ENGINE SOON) IS "OFF," USE CHART A-1.
   • INSTALL SCAN TOOL. IF NO DATA, USE CHART A-2.
   • CHECK THE FOLLOWING:
     - TP SENSOR. IF LESS THAN .2 VOLTS OR OVER 2.5 VOLTS AT CLOSED THROTTLE, USE DTC 21 CHART. (ALSO SEE STEP ONE
     - COOLANT. IF BELOW -34°C, USE DTC 14 CHART.
     - RPM. IF NO RPM WHILE CRANKING, USE CHART A-3 (PAGE 2 OF 3).

• PROBE FUEL PUMP "TEST" TERMINAL WITH A TEST LIGHT TO B+.
• IGNITION "OFF" FOR 10 SECONDS. LIGHT SHOULD BE "ON."
• IGNITION "ON." LIGHT SHOULD GO OUT FOR ABOUT 2 SECONDS AND THEN COME BACK "ON."
DOES IT?

→ YES ↓ | NO → USE FUEL PUMP RELAY CIRCUIT CHART A-5.

② • CRANK ENGINE AND CHECK FOR SPARK WITH ST-125 ON SPARK PLUG CABLES 1 OR 4 THEN 2 OR 3.
   • CHECK ONE CABLE AT A TIME. GROUND THE COMPANION SPARK PLUG CABLE. LEAVE THE OTHER CABLES CONNECTED TO PLUGS DURING CRANKING. IS THERE SPARK ON BOTH CABLES?

→ YES ↓ | NO → USE CHART A-3 (PAGE 2 OF 3).

③ • DISCONNECT INJECTOR JUMPER HARNESS AT 2 PIN CONNECTOR.
   • USING DVM MEASURE RESISTANCE BETWEEN CAVITIES "A" AND "B" (INJECTOR SIDE OF HARNESS). RESISTANCE SHOULD BE ABOUT 2.9 TO 5.0 OHMS. IS IT?

→ YES ↓ | NO → ⑥ • MEASURE INJECTOR RESISTANCE OF EACH INJECTOR. NOMINAL RESISTANCE OF EACH INJECTOR SHOULD BE 11.6 TO 12.4 OHMS AT 20°C (68°F). RESISTANCE WILL INCREASE SLIGHTLY AT HIGHER TEMPERATURES.
   ANY INJECTOR FAIL?
   YES → REPLACE FAULTY INJECTOR. | NO → REPAIR OPEN/SHORT IN INJECTOR HARNESS.

④ • CONNECT INJECTOR TEST LIGHT J 34730-2B TO INJECTOR JUMPER HARNESS (ECM SIDE).
   • CRANK ENGINE AND OBSERVE TEST LIGHT (SHOULD BLINK). DOES LIGHT BLINK?

→ YES ↓ | NO → USE CHART A-3 (PAGE 3 OF 3).

⑤ • IGNITION "OFF." INSTALL FUEL PRESSURE GAGE (SEE CHART A-7 PAGE 3 OF 3).
   • IGNITION "ON." FUEL PRESSURE SHOULD BE 284-325 kPa (41-47 psi). IS IT?

→ YES ↓ | NO ↓

YES → CHECK FOR FOULED SPARK PLUGS, OR EGR VALVE STUCK OPEN, OR A/C REFRIGERANT PRESSURE SENSOR SHORTED. SEE "DIAGNOSTICS AIDS"

NO → USE FUEL SYSTEM DIAGNOSIS CHART A-7.

"AFTER REPAIRS," CONFIRM "CLOSED LOOP" OPERATION AND NO MIL (SERVICE ENGINE SOON).

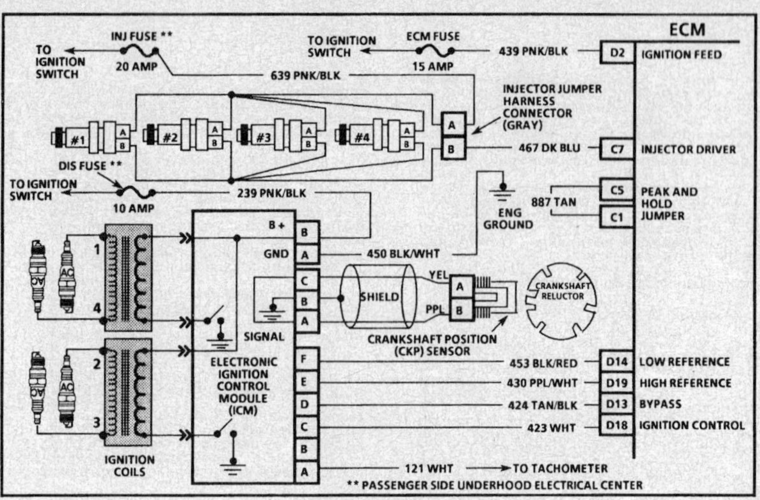

CHART A-3

(Page 2 of 3)
**ENGINE CRANKS BUT WON'T RUN
2.2L (VIN 4) "W" CARLINE**

Circuit Description

A magnetic Crankshaft Position (CKP) sensor is used to determine engine crankshaft position, much the same way as the pick-up coil did in Distributor Ignition (DI) type systems. The sensor is mounted in the block, near a slotted wheel on the crankshaft. The rotation of the wheel creates a flux change in the sensor, which produces a voltage signal. The electronic Ignition Control Module (ICM) module processes this signal and creates the reference pulses needed by the Engine Control Module (ECM) to trigger the correct coil at the correct time.

If the scan tool did not indicate cranking RPM, and there is no spark present at the plugs, the problem lies in the electronic Ignition Control Module (ICM), CKP sensor or the power and ground supplies to the ICM.

Test Description: Number(s) below refer to circled number(s) on the diagnostic chart.

1. The electronic Ignition Control Module (ICM) system uses two plugs and cables to complete the circuit of each coil. The companion spark plug cable in the circuit must be connected to a good ground to create a spark on the ST-125 and to avoid overloading the ignition coils.

2. This test will determine if the 12 volt supply and a good ground is available at the ICM.

3. This test will determine if the ignition module is not generating the reference pulse, or if the wiring or ECM are at fault. By touching and removing a test light to 12 volts on CKT 430, a reference pulse should be generated. If RPM is indicated, the ECM and wiring are OK.

4. This test will determine if the ignition module is not triggering the problem coil, or if the tested coil is at fault. This test could also be performed by substituting a known good coil. The secondary coil winding can be checked with a Digital Voltmeter (DVM). There should be 5,000 to 10,000 ohms across the coil towers. There should not be any continuity from either coil tower to ground.

5. Checks for continuity of the crankshaft position sensor and connections.

6. Normal crankshaft position sensor voltage output range is .8 to 1.4 volts (800 to 1400 mV) with a fully charged battery and engine at room temperature. Minimum output voltage (slow cranking, low battery) can be as low as .3 volt (300 mV).

CHART A-3

(Page 2 of 3)
**ENGINE CRANKS BUT WON'T RUN
2.2L (VIN 4) "W" CARLINE**

① • FOLLOW THESE STEPS TO CHECK FOR SPARK WITH ST-125 ON SPARK PLUG CABLES 1 OR 4, THEN 2 OR 3.
• CHECK ONE SPARK PLUG CABLE AT A TIME. GROUND COMPANION SPARK PLUG CABLE, LEAVING THE OTHER CABLES CONNECTED TO THE SPARK PLUGS.
• CRANK ENGINE.
IS THERE SPARK?

NO SPARK.

② • IGNITION "OFF."
• DISCONNECT 2-PIN CONNECTOR AT IGNITION CONTROL MODULE.
• IGNITION "ON."
• PROBE IGNITION CKT 239 WITH A TEST LIGHT TO GROUND.
IS THE TEST LIGHT "ON"?

YES → • PROBE GROUND CKT 450 WITH A TEST LIGHT TO BATTERY VOLTAGE. IS TEST LIGHT "ON"?

NO → OPEN CKT 239.

YES →

⑤ • IGNITION "OFF."
• DISCONNECT CRANKSHAFT POSITION SENSOR HARNESS AT IGNITION CONTROL MODULE.
• MEASURE SENSOR HARNESS TERMINALS "A" AND "C" WITH OHMMETER. DOES IT READ BETWEEN 800 AND - 1200 OHMS?

NO → OPEN CKT 450.

YES →

⑥ • SET VOLTMETER ON AC VOLTS POSITION.
• CRANK ENGINE AND OBSERVE VOLTAGE READING. IS READING GREATER THAN .3 VOLT (300 mV)?

YES → REPLACE IGNITION CONTROL MODULE.

NO → 800 OHMS OR LESS.
CRANKSHAFT POSITION SENSOR LEADS SHORTED TOGETHER OR FAULTY SENSOR.

NO → FAULTY CONNECTION OR FAULTY CRANKSHAFT POSITION SENSOR.

1200 OHMS OR GREATER.
OPEN CRANKSHAFT POSITION SENSOR CIRCUIT, FAULTY CONNECTION, OR FAULTY SENSOR.

SPARK ON BOTH CABLES.

③ • IGNITION "OFF."
• DISCONNECT 6-PIN CONNECTOR AT IGNITION CONTROL MODULE.
• IGNITION "ON."
• REPEATEDLY TOUCH CKT 430 WITH A TEST LIGHT TO B + WHILE OBSERVING CRANK RPM ON SCAN TOOL.
IS RPM INDICATED?

YES → REPLACE IGNITION CONTROL MODULE.

NO → OPEN OR SHORT TO GROUND ON CKT 430, OR FAULTY ECM. *

SPARK ON ONE CABLE.

• NOTE WHICH CABLE HAS NO SPARK.
• IGNITION "OFF."
• CHECK RESISTANCE OF SUSPECT SPARK PLUG CABLE. SHOULD BE LESS THAN 30,000 OHMS AND NOT SHORTED TO GROUND. IF CABLES ARE OK,
• REMOVE IGNITION COILS. IS THERE EVIDENCE OF CARBON TRACKING ARCING OR A FAULTY CONNECTION BETWEEN COIL(S) AND MODULE?

NO →

④ • SWITCH COIL POSITIONS ON MODULE AND RE-INSTALL.
• CHECK FOR SPARK WITH ST-125 ON PREVIOUSLY NOTED CABLE WITH NO SPARK.
IS SPARK PRESENT NOW?

YES → REPLACE AFFECTED COIL.

NO → REPLACE IGNITION CONTROL MODULE.

YES → REPLACE COIL ORIGINALLY CONNECTED TO THIS SPARK PLUG CABLE.

---

* IF ECM IS FAULTY AND MUST BE REPLACED, THE NEW ECM MUST BE PROGRAMMED. REFER TO ECM REPLACEMENT AND PROGRAMMING PROCEDURES

"AFTER REPAIRS," CONFIRM "CLOSED LOOP" OPERATION AND NO MIL (SERVICE ENGINE SOON).

2.2L (VIN 4) ENGINE — SYSTEM DIAGNOSTIC CHARTS — LUMINA

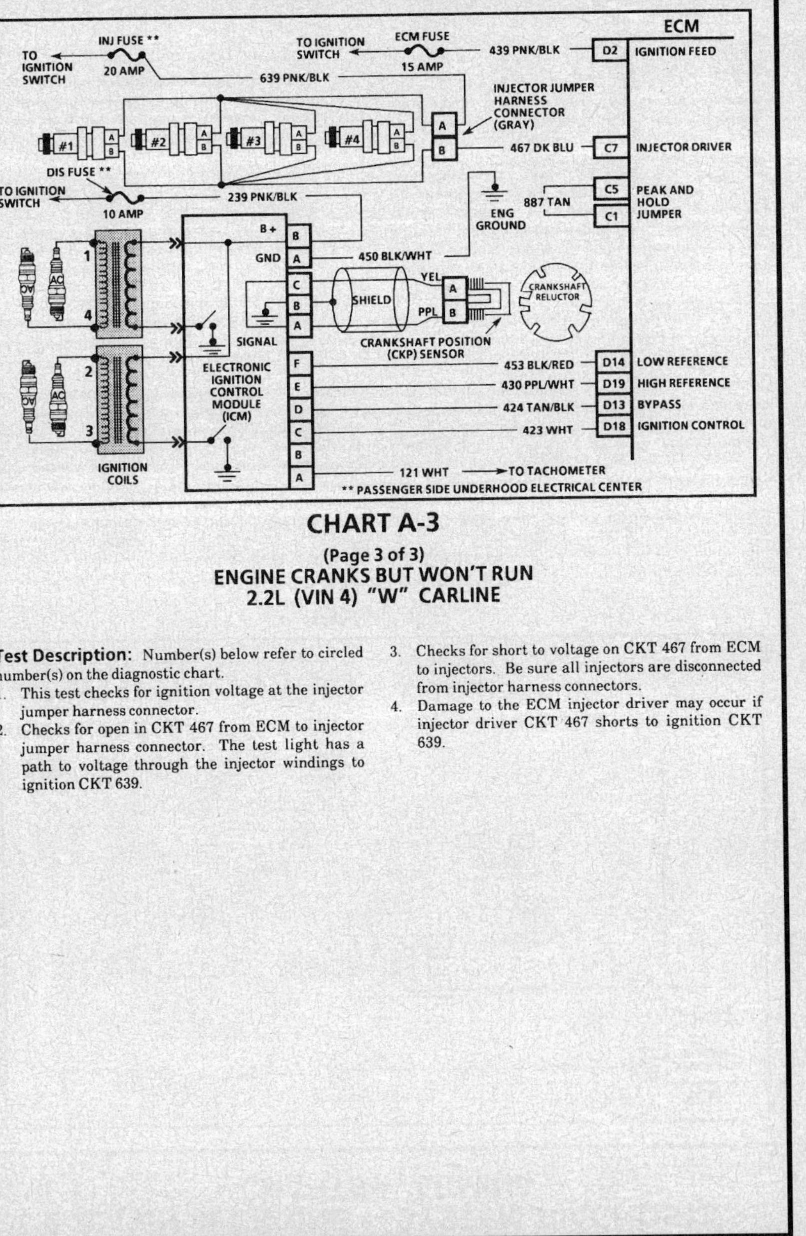

CHART A-3
(Page 3 of 3)
ENGINE CRANKS BUT WON'T RUN
2.2L (VIN 4) "W" CARLINE

Test Description: Number(s) below refer to circled number(s) on the diagnostic chart.

1. This test checks for ignition voltage at the injector jumper harness connector.
2. Checks for open in CKT 467 from ECM to injector jumper harness connector. The test light has a path to voltage through the injector windings to ignition CKT 639.
3. Checks for short to voltage on CKT 467 from ECM to injectors. Be sure all injectors are disconnected from injector harness connectors.
4. Damage to the ECM injector driver may occur if injector driver CKT 467 shorts to ignition CKT 639.

2.2L (VIN 4) ENGINE — SYSTEM DIAGNOSTIC CHARTS — LUMINA

CHART A-3
(Page 3 of 3)
ENGINE CRANKS BUT WON'T RUN
2.2L (VIN 4) "W" CARLINE

2.2L (VIN 4) ENGINE — SYSTEM DIAGNOSTIC CHARTS — LUMINA

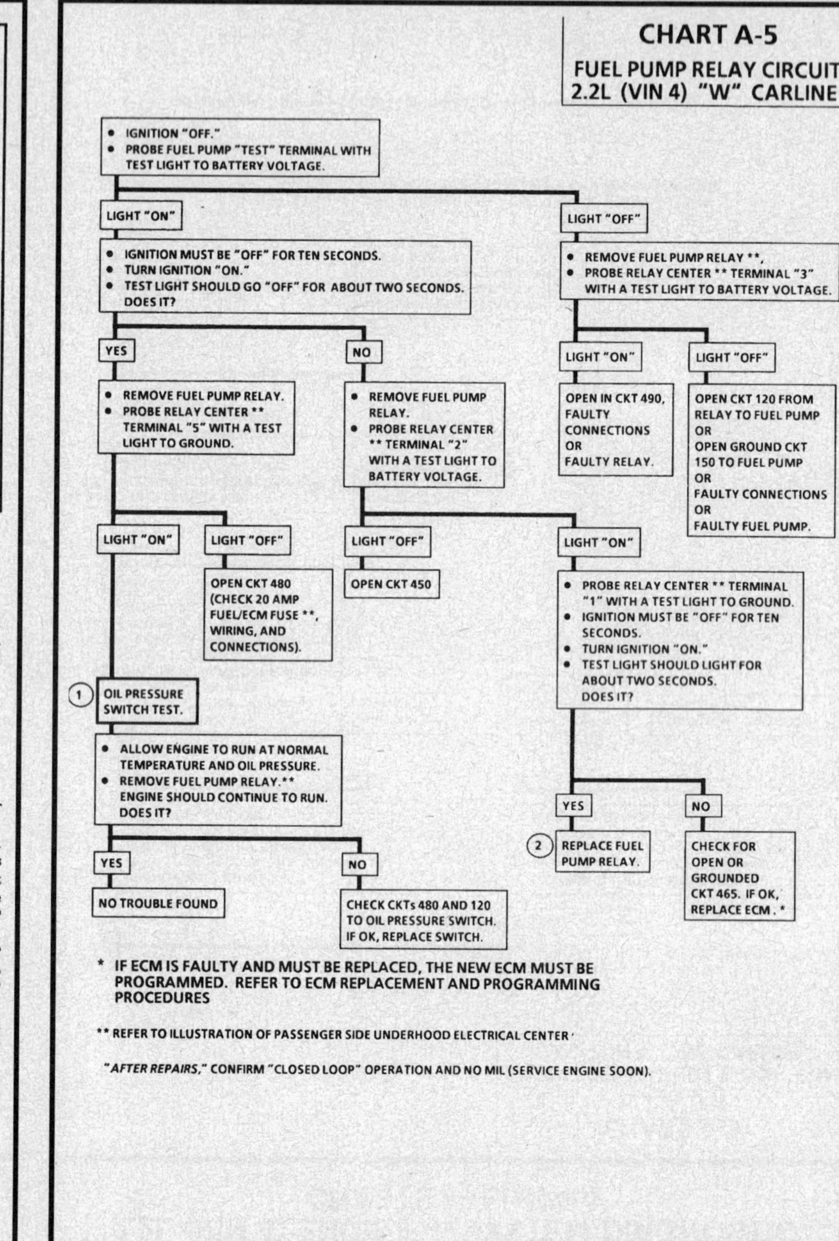

CHART A-5
FUEL PUMP RELAY CIRCUIT
2.2L (VIN 4) "W" CARLINE

Circuit Description:

When the ignition switch is turned "ON," the Engine Control Module (ECM) will activate the fuel pump relay with a 12 volt supply and run the in-tank fuel pump. The fuel pump will operate as long as the engine is cranking or running and the ECM is receiving ignition reference pulses. If there are no ignition reference pulses, the ECM will no longer supply the fuel pump relay signal within 2 seconds after key "ON."

Should the fuel pump relay or the 12 volt relay drive from the ECM fail, the fuel pump will receive supply current through the oil pressure switch back-up circuit.

The fuel pump test terminal is located in the driver's side of the engine compartment. When the engine is stopped, the pump can be turned "ON" by applying battery voltage to the test terminal.

Test Description: Number(s) below refer to circled number(s) on the diagnostic chart.
1. At this point, the fuel pump relay is operating correctly. The back-up circuit through the oil pressure switch is now tested.
2. After the fuel pump relay is replaced, continue with "Oil Pressure Switch Test."

Diagnostic Aids:

An inoperative fuel pump relay can result in long cranking times. The extended crank period is caused by the time necessary for oil pressure to reach the pressure required to close the oil pressure switch and supply the necessary current for the fuel pump.

If the fuel pump relay circuit checks out OK, refer to "Fuel System Diagnosis" CHART A-7.

Excess fuel may also cause long cranking times. This would usually be accompanied by a start that is not as fast as normal (once fired, the engine does not build up speed as fast) and a puff of black smoke at the tailpipe. An improperly connected or faulty EVAP canister control valve can cause this problem. Disconnect the "EVAP purge hose" from the "EVAP canister control valve" to diagnose. Refer to "Evaporative Emission (EVAP) Control System,"

One or more leaking fuel injectors may also extend the cranking time. Perform the "Injector Balance Test" in CHART C2-A

*   IF ECM IS FAULTY AND MUST BE REPLACED, THE NEW ECM MUST BE PROGRAMMED. REFER TO ECM REPLACEMENT AND PROGRAMMING PROCEDURES

** REFER TO ILLUSTRATION OF PASSENGER SIDE UNDERHOOD ELECTRICAL CENTER

"AFTER REPAIRS," CONFIRM "CLOSED LOOP" OPERATION AND NO MIL (SERVICE ENGINE SOON).

2.2L (VIN 4) ENGINE — SYSTEM DIAGNOSTIC CHARTS — LUMINA

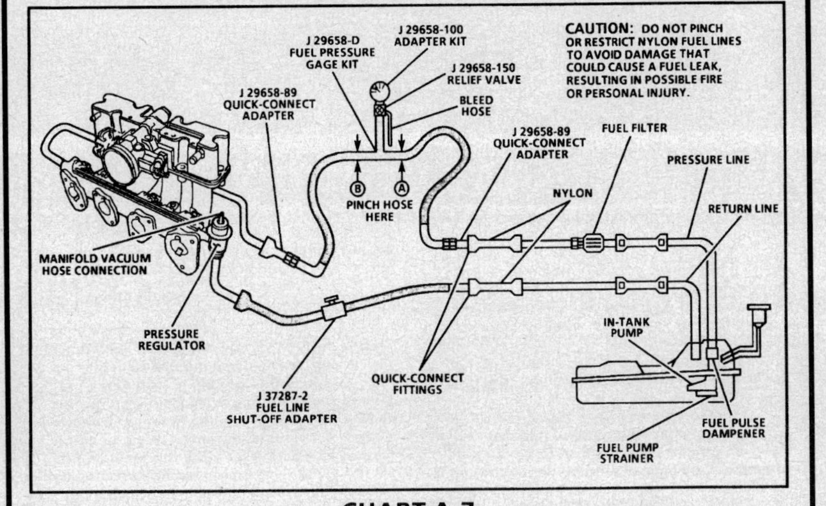

CHART A-7
(Page 1 of 3)
FUEL SYSTEM DIAGNOSIS
2.2L (VIN 4) "W" CARLINE

Circuit Description:

When the ignition switch is turned "ON," the Engine Control Module (ECM) will turn "ON" the in-tank fuel pump. It will remain "ON" as long as the engine is cranking or running, and the ECM is receiving reference pulses. If there are no reference pulses, the ECM will shut "OFF" the fuel pump within 2 seconds after ignition "ON" or engine stops.

An electric fuel pump, attached to the fuel sender assembly (inside the fuel tank), pumps fuel through an in-line filter to the fuel passage within the lower manifold assembly. The pump is designed to provide fuel at a pressure above the regulated pressure needed by the injectors. A pressure regulator, attached to the lower manifold assembly, keeps fuel available to the injectors at a regulated pressure. Unused fuel is returned to the fuel tank by a separate line.

Test Description: Number(s) below refer to circled number(s) on the diagnostic chart.

1. Install fuel pressure gage as shown in illustration. See "Fuel Pressure Relief Procedure" and for "Servicing Quick-Connect Fittings." Ignition "ON" pump pressure should be 284-325 kPa to (41-47 psi). This pressure is controlled by spring pressure within the regulator assembly.

2. When the engine is idling, the manifold pressure is low (high vacuum) and is applied to the fuel regulator diaphragm. This will offset the spring and result in a lower fuel pressure.

This idle pressure will vary somewhat depending on barometric pressure, however, the pressure idling should be less, indicating pressure regulator control.

3. Pressure that continues to fall is caused by one of the following:
   - In-tank fuel pump check valve not holding.
   - Partially disconnected fuel pulse dampener.
   - Fuel pressure regulator valve leaking.
   - Injector(s) sticking open.

4. An injector sticking open can best be determined by checking for a fouled or saturated spark plug(s).

2.2L (VIN 4) ENGINE — SYSTEM DIAGNOSTIC CHARTS — LUMINA

CHART A-7
(Page 1 of 3)
FUEL SYSTEM DIAGNOSIS
2.2L (VIN 4) "W" CARLINE

FROM CHART A-3

(1) • INSTALL FUEL PRESSURE GAGE AS SHOWN (SEE CHART A-7, PAGE 3 OF 3 FOR FUEL PRESSURE RELIEF PROCEDURE AND FOR SERVICING QUICK-CONNECT FITTINGS).
- IGNITION "OFF" FOR 10 SECONDS. A/C "OFF."
- IGNITION "ON." FUEL PUMP WILL RUN FOR ABOUT 2 SECONDS. THE IGNITION MAY HAVE TO BE CYCLED "ON" MORE THAN ONCE TO OBTAIN MAXIMUM PRESSURE.
- NOTE FUEL PRESSURE, WITH PUMP RUNNING, PRESSURE SHOULD BE 284-325 kPa (41-47 psi). WHEN PUMP STOPS, PRESSURE MAY VARY SLIGHTLY THEN SHOULD HOLD STEADY. IS PRESSURE CORRECT AND DOES IT HOLD?

FROM CHART A-5

YES →
IF FUEL PRESSURE IS WITHIN NORMAL RANGE, BUT IS SUSPECTED OF DROPPING OFF DURING ACCELERATION, CRUISE OR HARD CORNERING SEE CHART A-7 (2 OF 3).

(2) • START ENGINE AND ALLOW IT TO IDLE AT NORMAL OPERATING TEMPERATURE.
- FUEL PRESSURE NOTED IN STEP 1 SHOULD DROP 21-69 kPa (3-10 psi). DOES IT?

YES → NO TROUBLE FOUND.

NO → • DISCONNECT VACUUM HOSE FROM PRESSURE REGULATOR ASSEMBLY.
- WITH ENGINE IDLING, APPLY 12-14 INCHES OF VACUUM TO PRESSURE REGULATOR. FUEL PRESSURE NOTED IN STEP 1 SHOULD DROP 21-69 kPa (3-10 psi). DOES IT?

YES → LOCATE AND REPAIR LOSS OF VACUUM TO PRESSURE REGULATOR.

NO → REPLACE PRESSURE REGULATOR.

NO →
(3) FUEL PRESSURE WITHIN SPEC., BUT DOES NOT HOLD.

- IGNITION "OFF."
- USING A 10 AMP FUSED JUMPER WIRE, APPLY B + TO FUEL PUMP "TEST" CONNECTOR FOR 2 SECONDS.
- BLOCK GAGE INLET HOSE BY PINCHING AT "A" |

PRESSURE SHOULD HOLD. DOES IT?

NO → • IGNITION "OFF."
- INSTALL J 37287-2 FUEL LINE SHUT-OFF ADAPTER IN FUEL RETURN LINE, (SEE PAGE 3 OF 3 FOR FUEL PRESSURE RELIEF PROCEDURE AND FOR SERVICING QUICK-CONNECT FITTINGS).
- USING A 10 AMP FUSED JUMPER WIRE, APPLY B + TO FUEL PUMP "TEST" CONNECTOR FOR 2 SECONDS.
- CLOSE SHUT-OFF VALVE.
- PRESSURE SHOULD HOLD. DOES IT?

NO →
(4) LOCATE AND CORRECT LEAKING INJECTOR(s).

YES → REPLACE PRESSURE REGULATOR.

YES → CHECK FOR:
- LEAKING FUEL PUMP FEED HOSE.
- FAULTY FUEL PUMP.

FUEL PRESSURE OUT OF SPEC.
SEE CHART A-7 (2 OF 3)

NO PRESSURE, FUEL PUMP ELECTRICAL CIRCUIT OK.
CHECK FOR:
- PLUGGED IN-LINE FILTER.
- PLUGGED FUEL PUMP STRAINER.
- RESTRICTED FUEL PRESSURE LINE.
- LEAKING FUEL PUMP FEED HOSE.

IF OK → REPLACE FUEL PUMP.

"AFTER REPAIRS," CONFIRM "CLOSED LOOP" OPERATION AND NO MIL (SERVICE ENGINE SOON).

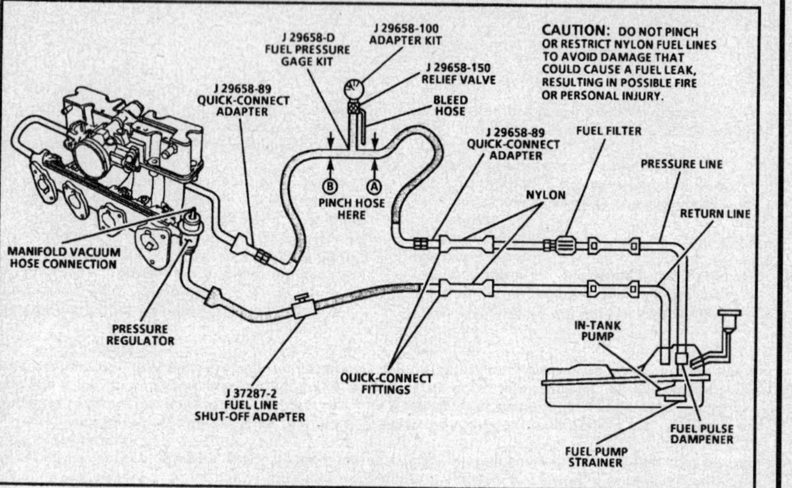

CHART A-7
(Page 2 of 3)
FUEL SYSTEM DIAGNOSIS
2.2L (VIN 4) "W" CARLINE

Test Description: Number(s) below refer to circled number(s) on the diagnostic chart.

5. Fuel pressure that drops off during acceleration, cruise or hard cornering may cause a lean condition and result in a loss of power, surging or misfire. This condition can be diagnosed using a Tech 1 scan tool. If the fuel system is very lean, the oxygen sensor will stop toggling and output voltage will drop below 500 mV. Also, injector pulse width will increase.

> **Important**
> - Make sure system is not operating at "fuel cut-off" which may cause false readings on the scan tool.

6. Fuel pressure below 284 kPa (42 psi) may cause a lean condition and may set a DTC 44 or a DTC 32. Driveability conditions can include hard starting cold, hesitation, poor driveability, lack of power, surging or misfire.

7. Restricting fuel flow at the fuel pressure gage (at "B") causes fuel pressure to build above regulated pressure. With battery voltage applied the pump "test" terminal pressure should rise above 325 kPa (47 psi) as the gage outlet hose is pinched.

NOTICE: Do not allow pressure to exceed 414 kPa (60 psi) as damage to the regulator may result.

8. Fuel pressure above 325 kPa (47 psi) may cause a rich condition and may set a DTC 45. Driveability conditions can include hard starting (followed by black smoke) and a strong sulphur smell in the exhaust.

9. This test determines if the high fuel pressure is due to a restricted fuel return line or a faulty fuel pressure regulator.

CHART A-7
(Page 2 of 3)
FUEL SYSTEM DIAGNOSIS
2.2L (VIN 4) "W" CARLINE

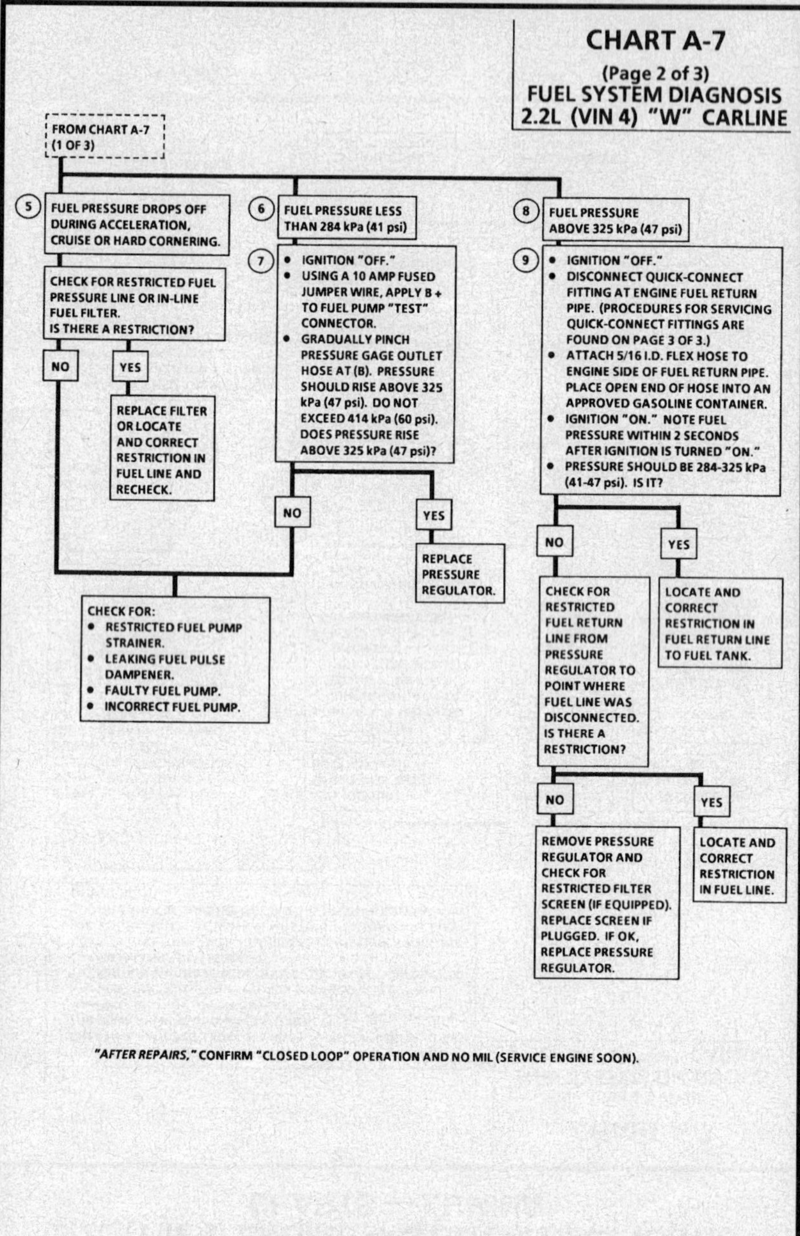

"AFTER REPAIRS," CONFIRM "CLOSED LOOP" OPERATION AND NO MIL (SERVICE ENGINE SOON).

2.2L (VIN 4) ENGINE — SYSTEM DIAGNOSTIC CHARTS — LUMINA

CHART A-7
(Page 3 of 3)
FUEL SYSTEM DIAGNOSIS
2.2L (VIN 4) "W" CARLINE

FUEL SYSTEM PRESSURE RELIEF PROCEDURE

MFI Engines Without Fuel Pressure Connection

(Must Be Performed Before Disconnecting Fuel Line Fittings)

CAUTION:
- To reduce the risk of fire and personal injury, it is necessary to relieve fuel system pressure before disconnecting fuel line fittings.
- After relieving system pressure, a small amount of fuel may be released when disconnecting fuel line fittings. In order to reduce the chance of personal injury, cover fuel line fittings with a shop towel before disconnecting, to catch any fuel that may leak out. Place the towel in an approved container when disconnect is completed.

1. Loosen fuel filler cap to relieve tank pressure.
2. Raise vehicle.
3. Disconnect fuel pump electrical connector.
4. Lower vehicle.
5. Start engine and run until fuel supply remaining in fuel pipes is consumed. Engage starter for 3.0 seconds to assure relief of any remaining pressure.
6. Raise vehicle.
7. Connect fuel pump electrical connector.
8. Lower vehicle.
9. Disconnect negative battery cable to avoid possible fuel discharge if an accidental attempt is made to start the engine.
10. Fuel line fittings are now safe for servicing.
11. Perform service required.
    - If performing fuel pressure check with a gage equipped with a bleed hose, fuel pressure can be relieved through the gage following test. Place bleed hose into approved gasoline container and open valve to bleed system pressure.
12. Tighten fuel filler cap.
13. Connect negative battery cable.
14. Cycle ignition "ON" and "OFF" twice, waiting ten seconds between cycles, then check for fuel leaks.

2.2L (VIN 4) ENGINE — SYSTEM DIAGNOSTIC CHARTS — LUMINA

CHART A-7
(Page 3 of 3)
FUEL SYSTEM DIAGNOSIS
2.2L (VIN 4) "W" CARLINE

SERVICING QUICK-CONNECT FITTINGS

Important
- In order to install fuel system diagnostic equipment on vehicles equipped with plastic quick-connect fittings, fuel line separator tools must be used to disconnect the fittings. Use of the separator tools will cause the plastic retainer to remain inside the female connector allowing diagnostic equipment to be connected.

Tools required:
    J 37088-A tool set, fuel line quick-connect separator;
    J 39504 tool set, fuel line quick-connect separator (restricted access).

Remove or Disconnect

1. Grasp both sides of fitting. Twist female connector 1/4 turn in each direction to loosen any dirt within fitting.

    CAUTION: Safety glasses must be worn when using compressed air, as flying dirt particles may cause eye injury.

2. Using compressed air, blow dirt out of fitting.
3. Choose correct tool from J 37088-A or J 39504 tool set for size of fitting. Insert tool into female connector, then push/pull inward to release locking tabs.
4. Pull connection apart.

Clean and Inspect

NOTICE: If it is necessary to remove rust or burrs from fuel pipe, use emery cloth in a radial motion with the pipe end to prevent damage to O-ring sealing surface.

- Using a clean shop towel, wipe off male pipe end.
- Inspect both ends of fitting for dirt and burrs. Clean or replace components/assemblies as required.

Install or Connect

CAUTION: To Reduce the Risk of Fire and Personal Injury:
- Before connecting fitting, always apply a few drops of clean engine oil to the male pipe end of engine fuel pipe, pressure gage adapter or fuel line shut-off adapter. This will ensure proper reconnection and prevent a possible fuel leak. (During normal operation, the O-rings located in the female connector will swell and may prevent proper reconnection if not lubricated.)

1. Apply a few drops of clean engine oil to the male pipe end of engine fuel pipe, pressure gage adapter or fuel line shut-off adapter.
2. Push both sides of fitting together to cause the retaining tabs/fingers to snap into place.
3. Once installed, pull on both sides of fitting to make sure connection is secure.

"AFTER REPAIRS," CONFIRM "CLOSED LOOP" OPERATION AND NO MIL (SERVICE ENGINE SOON).

DTC 13
OXYGEN SENSOR (O2S) CIRCUIT
(OPEN CIRCUIT)
2.2L (VIN 4) "W" CARLINE

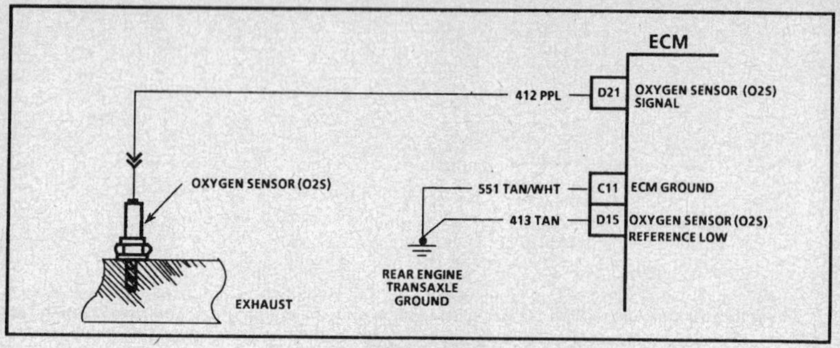

Circuit Description:

The Engine Control Module (ECM) supplies a voltage of about .45 volt between terminals "C11" and "D21". (If measured with a 10 megohm digital voltmeter, this may read as low as .32 volt.)

When the Oxygen Sensor (O2S) reaches operating temperature, it varies this voltage from about .1 volt (exhaust is lean) to about .9 volt (exhaust is rich).

The sensor is like an open circuit and produces no voltage when it is below 316°C (600°F). An open sensor circuit, or cold sensor, causes "Open Loop" operation.

Test Description: Number(s) below refer to circled number(s) on the diagnostic chart.

1. DTC 13 will set under the following conditions:
   - Engine at normal operating temperature.
   - At least 2 minutes has elapsed since engine start-up.
   - O2S signal voltage is steady between .35 and .55 volt.
   - Throttle angle is above 5%.
   - All above conditions are met for 40.3 seconds or more.

   If the conditions for a DTC 13 exist, the system will not operate in "Closed Loop."

2. This test determines if the Oxygen Sensor (O2S) is the problem or if the ECM and wiring are at fault.

3. In doing this test, use only a 10 megohm digital voltmeter. This test checks the continuity of CKT 412 and CKT 413. If CKT 413 is open, the ECM voltage on CKT 412 will be over .6 volt (600 mV).

Diagnostic Aids:

Normal Tech 1 scan tool Oxygen Sensor (O2S) voltage varies between 100 mV to 999 mV (.1 and 1.0 volt) while in "Closed Loop." DTC 13 sets in about 40 seconds if sensor signal voltage remains between .35 and .55 volt.

Verify a clean, tight ground connection for CKT 413. Open CKT 412 or CKT 413 will result in a DTC 13. If DTC 13 is intermittent, refer to "Symptoms,"

DTC 13
OXYGEN SENSOR (O2S) CIRCUIT
(OPEN CIRCUIT)
2.2L (VIN 4) "W" CARLINE

* IF ECM IS FAULTY AND MUST BE REPLACED, THE NEW ECM MUST BE PROGRAMMED. REFER TO ECM REPLACEMENT AND PROGRAMMING PROCEDURES

"AFTER REPAIRS," REFER TO DTC CRITERIA AND CONFIRM DTC DOES NOT RESET.

2.2L (VIN 4) ENGINE — DIAGNOSTIC TROUBLE CODE CHART — LUMINA

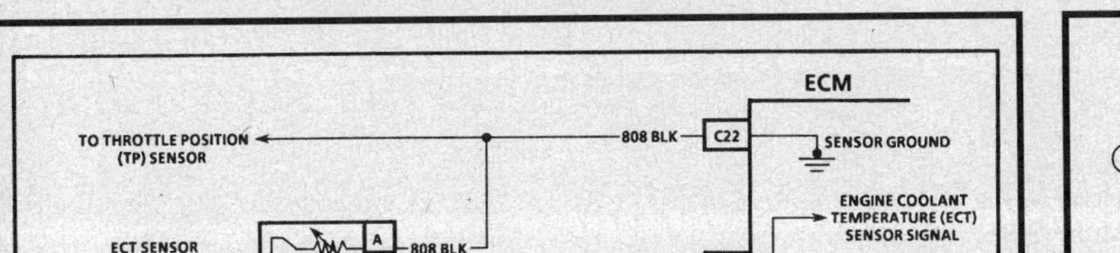

DTC 14

(Page 1 of 2)
ENGINE COOLANT TEMPERATURE (ECT) SENSOR CIRCUIT
(HIGH/LOW TEMPERATURE INDICATED)
2.2L (VIN 4) "W" CARLINE

Circuit Description:

The Engine Coolant Temperature (ECT) sensor utilizes a thermistor to control the signal voltage to the Engine Control Module (ECM). The ECM applies a reference voltage on CKT 410 to the sensor. When the engine is cold, the sensor (thermistor) resistance is high. The ECM will then sense a high signal voltage.

As the engine warms up, the sensor resistance decreases and the voltage drops. At normal engine operating temperature, the voltage will measure about 1.5 to 2.0 volts at ECM terminal "C12".

Coolant temperature is one of the inputs used to control the following:

- Cooling fan.
- Fuel delivery.
- Ignition Control (IC).
- Idle Air Control (IAC).
- Torque Converter Clutch (TCC).

A separate thermistor within the ECT sensor provides a signal to the coolant temperature gage located in the instrument panel.

Test Description: Number(s) below refer to circled number(s) on the diagnostic chart.

1. DTC 14 will set if:
   - The engine has been running for 2 minutes or more.
   - Signal voltage indicates a coolant temperature below -34°C (-30°F).
   OR
   - Signal voltage indicates a coolant temperature above 143°C (289°F) for 3 seconds or more.
2. If the ECM recognizes the grounded circuit (low voltage) and displays a high temperature, the ECM and wiring are OK.
3. This test will determine if there is a wiring problem or a faulty ECM. If CKT 808 is open, there may also be a DTC 21 stored.

Diagnostic Aids:

The Tech 1 scan tool reads engine temperature in degrees Celsius.

After the engine is started, the temperature should rise steadily to about 90°C (194°F), then stabilize when the thermostat opens.

If the engine has been allowed to cool to an ambient temperature (overnight), coolant temperature and Intake Air Temperature (IAT) may be checked with a scan tool and should read close to each other.

When a DTC 14 is set, the ECM will turn "ON" the engine cooling fan.

If DTC 14 is intermittent or a "Hard Start" symptom is present, check engine coolant temperature with a scan tool on a cool engine. Temperature displayed should be within 5 degrees of the ambient. If not, check the ECT sensor using the "Diagnostic Aid" If sensor is OK, check connections.

2.2L (VIN 4) ENGINE — DIAGNOSTIC TROUBLE CODE CHART — LUMINA

DTC 14

(Page 1 of 2)
ENGINE COOLANT TEMPERATURE (ECT) SENSOR CIRCUIT
(HIGH/LOW TEMPERATURE INDICATED)
2.2L (VIN 4) "W" CARLINE

(1) DOES SCAN TOOL DISPLAY COOLANT TEMPERATURE OF -34°C (-30°F) OR LESS?

- YES
- NO

(2)
- DISCONNECT SENSOR CONNECTOR.
- JUMPER HARNESS CKT 410 TO SENSOR GROUND CIRCUIT.
- SCAN TOOL SHOULD DISPLAY 143°C (289°F) OR HIGHER. DOES IT?

REFER TO DTC 14 (PAGE 2 OF 2)

- NO
- YES

(3)
- JUMPER CKT 410 TO A KNOWN GOOD GROUND.
- SCAN TOOL SHOULD DISPLAY 143°C (289°F) OR HIGHER. DOES IT?

FAULTY SENSOR CONNECTOR OR FAULTY SENSOR.

- YES
- NO

OPEN SENSOR GROUND CIRCUIT, FAULTY CONNECTION OR FAULTY ECM.*

OPEN CKT 410, FAULTY CONNECTION OR FAULTY ECM.*

* IF ECM IS FAULTY AND MUST BE REPLACED, THE NEW ECM MUST BE PROGRAMMED. REFER TO ECM REPLACEMENT AND PROGRAMMING PROCEDURES

DIAGNOSTIC AID		
ECT SENSOR		
TEMPERATURE VS. RESISTANCE VALUES (APPROXIMATE)		
°C	°F	OHMS
100	212	177
90	194	241
80	176	332
70	158	467
60	140	667
50	122	973
45	113	1188
40	104	1459
35	95	1802
30	86	2238
25	77	2796
20	68	3520
15	59	4450
10	50	5670
5	41	7280
0	32	9420
-5	23	12300
-10	14	16180
-15	5	21450
-20	-4	28680
-30	-22	52700
-40	-40	100700

"AFTER REPAIRS," REFER TO DTC CRITERIA   AND CONFIRM DTC DOES NOT RESET.

2.2L (VIN 4) ENGINE — DIAGNOSTIC TROUBLE CODE CHART — LUMINA

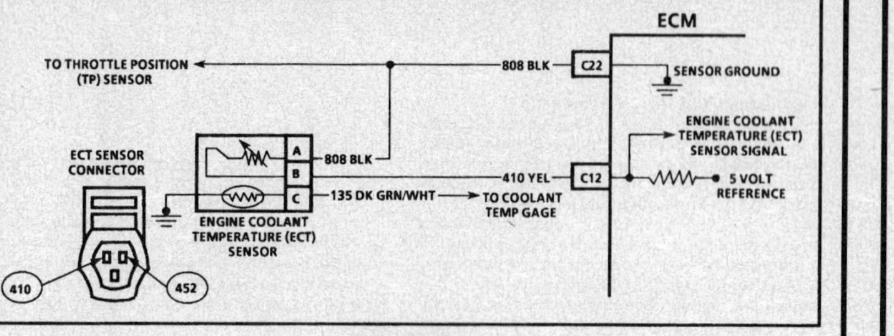

DTC 14
(Page 2 of 2)
ENGINE COOLANT TEMPERATURE (ECT) SENSOR CIRCUIT
(HIGH/LOW TEMPERATURE INDICATED)
2.2L (VIN 4) "W" CARLINE

Circuit Description:

The Engine Coolant Temperature (ECT) sensor utilizes a thermistor to control the signal voltage to the Engine Control Module (ECM). The ECM applies a reference voltage on CKT 410 to the sensor. When the engine is cold, the sensor (thermistor) resistance is high. The ECM will then sense a high signal voltage.

As the engine warms up, the sensor resistance decreases and the voltage drops. At normal engine operating temperature, the voltage will measure about 1.5 to 2.0 volts at ECM terminal "C12".

Coolant temperature is one of the inputs used to control the following:

- Fuel delivery.
- Ignition Control (IC).
- Cooling fan.
- Torque Converter Clutch (TCC).
- Idle Air Control (IAC).

Test Description: Number(s) below refer to circled number(s) on the diagnostic chart.

1. Checks to see if a diagnostic trouble code was set as a result of hard failure or intermittent condition.
2. If the ECM recognizes the open circuit (high voltage), and displays a low temperature, the ECM and wiring are OK.

Diagnostic Aids:

The Tech 1 scan tool reads engine temperature in degrees Celsius.

After the engine is started, the temperature should rise steadily to about 90°C (194°F), then stabilize when the thermostat opens.

If the engine has been allowed to cool to an ambient temperature (overnight), coolant temperature and Intake Air Temperature (IAT) may be checked with a scan tool and should read close to each other.

When a DTC 14 is set, the ECM will turn "ON" the engine cooling fan.

If DTC 14 is intermittent or a "Hard Start" symptom is present, check engine coolant temperature with a scan tool on a cool engine. Temperature displayed should be within 5 degrees of the ambient. If not, check the ECT sensor using the "Diagnostic Aid"

If sensor is OK, check connections.

2.2L (VIN 4) ENGINE — DIAGNOSTIC TROUBLE CODE CHART — LUMINA

DTC 14
(Page 2 of 2)
ENGINE COOLANT TEMPERATURE (ECT) SENSOR CIRCUIT
(HIGH/LOW TEMPERATURE INDICATED)
2.2L (VIN 4) "W" CARLINE

```
FROM DTC 14
PAGE 1 OF 2
```

(1) DOES SCAN TOOL DISPLAY COOLANT TEMPERATURE OF 143°C (289°F) OR HIGHER?

- YES → (2) DISCONNECT SENSOR CONNECTOR. SCAN TOOL SHOULD DISPLAY TEMPERATURE BELOW -34°C (-30°F) DOES IT?
  - YES → REPLACE ENGINE COOLANT TEMPERATURE (ECT) SENSOR.
  - NO → CKT 410 SHORTED TO GROUND, OR CKT 410 SHORTED TO SENSOR GROUND CIRCUIT OR FAULTY ECM.*
- NO → • DTC 14 IS INTERMITTENT. IF NO ADDITIONAL DTCS WERE STORED, REFER TO "DIAGNOSTIC AIDS"

DIAGNOSTIC AID

ECT SENSOR		
TEMPERATURE VS. RESISTANCE VALUES		
(APPROXIMATE)		
°C	°F	OHMS
100	212	177
90	194	241
80	176	332
70	158	467
60	140	667
50	122	973
45	113	1188
40	104	1459
35	95	1802
30	86	2238
25	77	2796
20	68	3520
15	59	4450
10	50	5670
5	41	7280
0	32	9420
-5	23	12300
-10	14	16180
-15	5	21450
-20	-4	28680
-30	-22	52700
-40	-40	100700

* IF ECM IS FAULTY AND MUST BE REPLACED, THE NEW ECM MUST BE PROGRAMMED. REFER TO ECM REPLACEMENT AND PROGRAMMING PROCEDURES

"AFTER REPAIRS," REFER TO DTC CRITERIA   AND CONFIRM DTC DOES NOT RESET.

2.2L (VIN 4) ENGINE — DIAGNOSTIC TROUBLE CODE CHART — LUMINA

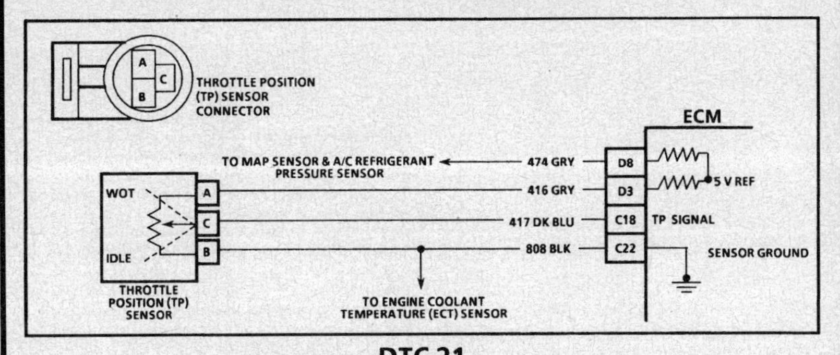

DTC 21

(Page 1 of 2)
THROTTLE POSITION (TP) SENSOR CIRCUIT
(SIGNAL VOLTAGE HIGH/LOW)
2.2L (VIN 4) "W" CARLINE

Circuit Description:

The Throttle Position (TP) sensor provides a voltage signal that changes relative to the throttle valve. Signal voltage will vary from .33 to 1.33 volts at idle to about 4.5 volts at Wide Open Throttle (WOT).

The TP sensor signal is one of the most important inputs used by the Engine Control Module (ECM) for fuel control and for many of the ECM controlled outputs.

Test Description: Number(s) below refer to circled number(s) on the diagnostic chart.
1. A DTC 21 will set under the following conditions:
   - TP sensor reading below 1.9 volts or above 4.9 volts for about 64 seconds or more.
   OR
   - TP sensor reading above 3.9 volts, but below 4.9 volts.
   - MAP reading below 65 kPa.
   - Engine speed less than 1,750 RPM.
   - All of the above conditions present for about 5 seconds.

   The TP sensor has an auto zeroing feature. If the voltage reading is within the range of about .33 to 1.33 volts, the ECM will use that value as closed throttle. If the voltage reading is outside of the auto zero range at closed throttle, check for a binding throttle cable or damaged linkage. If OK, continue with diagnosis.

2. If the ECM recognizes the change of state, the ECM and CKTs 416 and 417 are OK.
3. This step isolates a faulty sensor, ECM, or an open CKT 808. If CKT 808 is open, DTC 14 may also be stored.

Diagnostic Aids:

A scan tool displays throttle position in volts. Closed throttle voltage should be .33 to 1.33 volts. TP sensor voltage should increase at a steady rate as throttle is moved to WOT.

If DTC 21 is intermittent, refer to "Symptoms,"

2.2L (VIN 4) ENGINE — DIAGNOSTIC TROUBLE CODE CHART — LUMINA

DTC 21

(Page 1 of 2)
THROTTLE POSITION (TP) SENSOR CIRCUIT
(SIGNAL VOLTAGE HIGH/LOW)
2.2L (VIN 4) "W" CARLINE

1. • THROTTLE CLOSED.
   DOES SCAN TOOL DISPLAY TP SENSOR OVER 3.9 VOLTS?

   YES → 2
   NO → REFER TO DTC 21 (PAGE 2 OF 2)

2. • DISCONNECT SENSOR CONNECTOR.
   SCAN TOOL SHOULD DISPLAY TP SENSOR BELOW .2 VOLT (200mV).
   DOES IT?

   YES → 3
   NO → TP SENSOR SIGNAL CIRCUIT SHORTED TO VOLTAGE OR FAULTY ECM.*

3. • PROBE SENSOR GROUND CIRCUIT WITH A TEST LIGHT CONNECTED TO BATTERY VOLTAGE.

   LIGHT "ON" → FAULTY CONNECTION OR THROTTLE POSITION SENSOR.

   LIGHT "OFF" → OPEN SENSOR GROUND CIRCUIT OR FAULTY ECM.*

* IF ECM IS FAULTY AND MUST BE REPLACED, THE NEW ECM MUST BE PROGRAMMED. REFER TO ECM REPLACEMENT AND PROGRAMMING PROCEDURES

"AFTER REPAIRS," REFER TO DTC CRITERIA   AND CONFIRM DTC DOES NOT RESET.

2.2L (VIN 4) ENGINE — DIAGNOSTIC TROUBLE CODE CHART — LUMINA

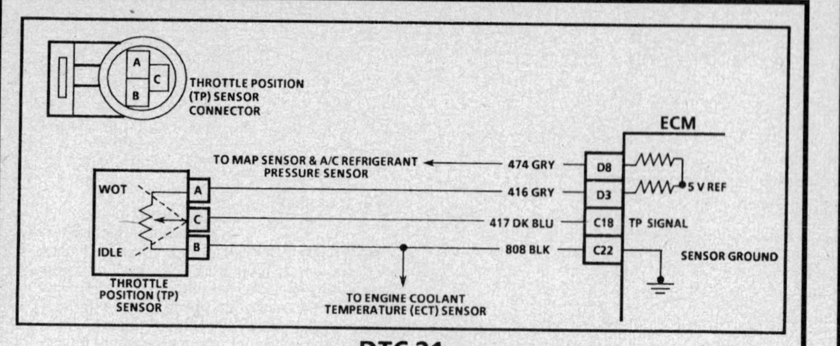

DTC 21
(Page 2 of 2)
THROTTLE POSITION (TP) SENSOR CIRCUIT
(SIGNAL VOLTAGE HIGH/LOW)
2.2L (VIN 4) "W" CARLINE

Circuit Description:

The Throttle Position (TP) sensor provides a voltage signal that changes relative to the throttle valve. Signal voltage will vary from .33 to 1.33 volts at idle to about 4.5 volts at Wide Open Throttle (WOT).

The TP sensor signal is one of the most important inputs used by the Engine Control Module (ECM) for fuel control and for many of the ECM controlled outputs.

Test Description: Number(s) below refer to circled number(s) on the diagnostic chart.
1. This step checks to see if DTC 21 is the result of a hard failure or an intermittent condition.
2. This step simulates conditions for a DTC 21. If the scan tool displays over 4 volts, the ECM and wiring are OK.
3. This simulates a high signal voltage to check for an open or short in CKT 417. The Tech 1 scan tool will not read battery voltage, but the ECM should recognize the signal on CKT 417.
4. Refer to "Fuel Metering System,"
   for TP sensor replacement procedures.
5. The 5 volt reference for the MAP sensor and A/C refrigerant pressure sensor connects to the same 5 volt power supply in the ECM. If the 5 volt reference circuit to the MAP sensor and A/C refrigerant pressure sensor has a short to ground, the Throttle Position (TP) sensor 5 volt reference circuit will also have a short to ground because they connect to the same power supply in the ECM. A short to ground on either 5 volt reference circuit does not damage the ECM 5 volt power supply. When the short is repaired, the ECM 5 volt supply will return to normal.

Diagnostic Aids:

A scan tool displays throttle position in volts. Closed throttle voltage should be .33 to 1.33 volts. TP sensor voltage should increase at a steady rate as throttle is moved to WOT.

If DTC 21 is intermittent, refer to "Symptoms,"

2.2L (VIN 4) ENGINE — DIAGNOSTIC TROUBLE CODE CHART — LUMINA

DTC 21
(Page 2 of 2)
THROTTLE POSITION (TP) SENSOR CIRCUIT
(SIGNAL VOLTAGE HIGH/LOW)
2.2L (VIN 4) "W" CARLINE

FROM DTC 21
(PAGE 1 OF 2).

(1) • THROTTLE CLOSED.
  DOES SCAN TOOL DISPLAY LESS THAN .2 VOLTS?

YES

(2) • DISCONNECT TP SENSOR SENSOR CONNECTOR.
  • JUMPER 5 VOLT REFERENCE CIRCUIT AND CKT 417 TOGETHER.
  SCAN SHOULD DISPLAY THROTTLE POSITION OVER 4.0 V (4000 mV).
  DOES IT?

NO

DTC 21 IS INTERMITTENT. IF NO ADDITIONAL DTCS WERE STORED, REFER TO "DIAGNOSTIC AIDS"

NO

(3) • PROBE CKT 417 WITH A TEST LIGHT CONNECTED TO BATTERY VOLTAGE.
  SCAN TOOL SHOULD DISPLAY THROTTLE POSITION OVER 4.0V (4000 mV).
  DOES IT?

YES

(4) • RECONNECT TP SENSOR CONNECTOR.
  • CHECK FOR FAULTY OR INTERMITTENT CONNECTION. IF OK, TP SENSOR IS FAULTY.

YES

(5) 5 VOLT REFERENCE CIRCUIT OPEN
OR
SHORTED TO GROUND
OR
FAULTY CONNECTION
OR
FAULTY ECM.*

NO

CKT 417 OPEN
OR
SHORTED TO GROUND
OR
SHORTED TO SENSOR GROUND CIRCUIT
OR
FAULTY ECM CONNECTION
OR
FAULTY ECM.*

* IF ECM IS FAULTY AND MUST BE REPLACED, THE NEW ECM MUST BE PROGRAMMED. REFER TO ECM REPLACEMENT AND PROGRAMMING PROCEDURES

"AFTER REPAIRS," REFER TO DTC CRITERIA   AND CONFIRM DTC DOES NOT RESET.

2.2L (VIN 4) ENGINE — DIAGNOSTIC TROUBLE CODE CHART — LUMINA

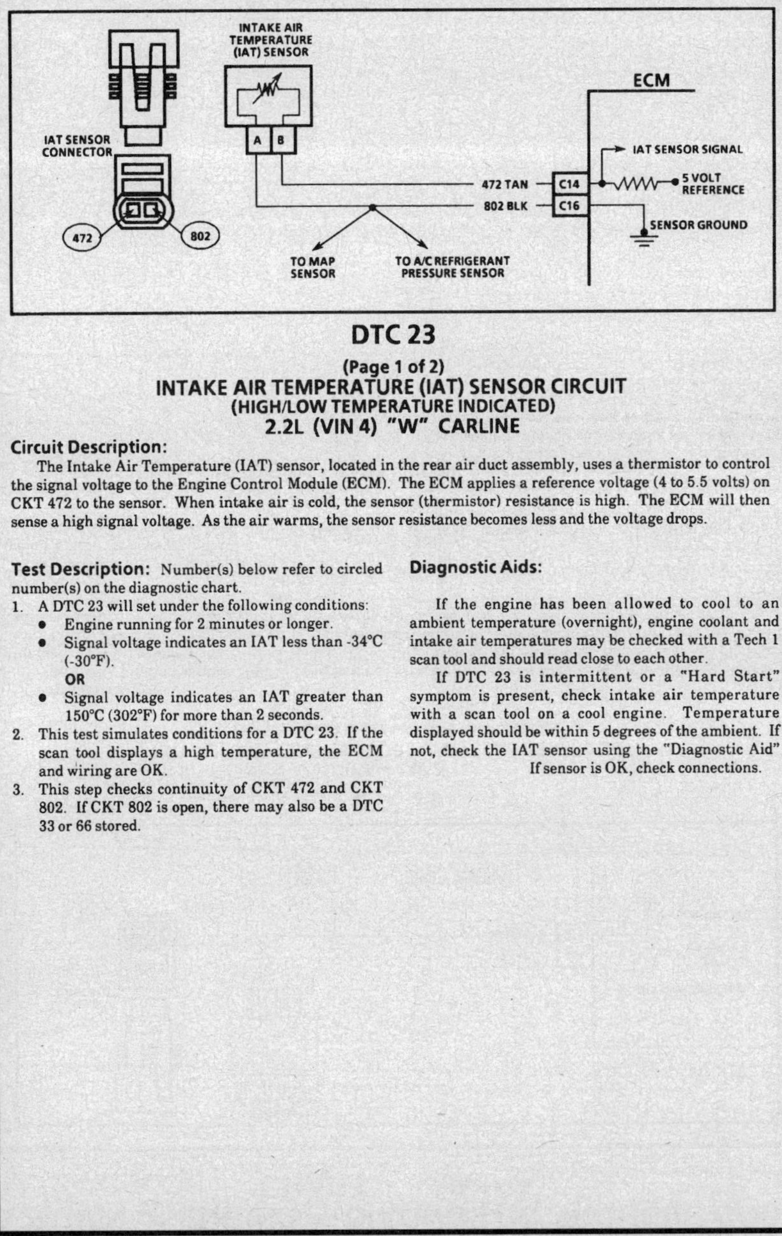

DTC 23
(Page 1 of 2)
INTAKE AIR TEMPERATURE (IAT) SENSOR CIRCUIT
(HIGH/LOW TEMPERATURE INDICATED)
2.2L (VIN 4) "W" CARLINE

Circuit Description:

The Intake Air Temperature (IAT) sensor, located in the rear air duct assembly, uses a thermistor to control the signal voltage to the Engine Control Module (ECM). The ECM applies a reference voltage (4 to 5.5 volts) on CKT 472 to the sensor. When intake air is cold, the sensor (thermistor) resistance is high. The ECM will then sense a high signal voltage. As the air warms, the sensor resistance becomes less and the voltage drops.

Test Description: Number(s) below refer to circled number(s) on the diagnostic chart.
1. A DTC 23 will set under the following conditions:
   - Engine running for 2 minutes or longer.
   - Signal voltage indicates an IAT less than -34°C (-30°F).

   OR
   - Signal voltage indicates an IAT greater than 150°C (302°F) for more than 2 seconds.
2. This test simulates conditions for a DTC 23. If the scan tool displays a high temperature, the ECM and wiring are OK.
3. This step checks continuity of CKT 472 and CKT 802. If CKT 802 is open, there may also be a DTC 33 or 66 stored.

Diagnostic Aids:

If the engine has been allowed to cool to an ambient temperature (overnight), engine coolant and intake air temperatures may be checked with a Tech 1 scan tool and should read close to each other.

If DTC 23 is intermittent or a "Hard Start" symptom is present, check intake air temperature with a scan tool on a cool engine. Temperature displayed should be within 5 degrees of the ambient. If not, check the IAT sensor using the "Diagnostic Aid" If sensor is OK, check connections.

2.2L (VIN 4) ENGINE — DIAGNOSTIC TROUBLE CODE CHART — LUMINA

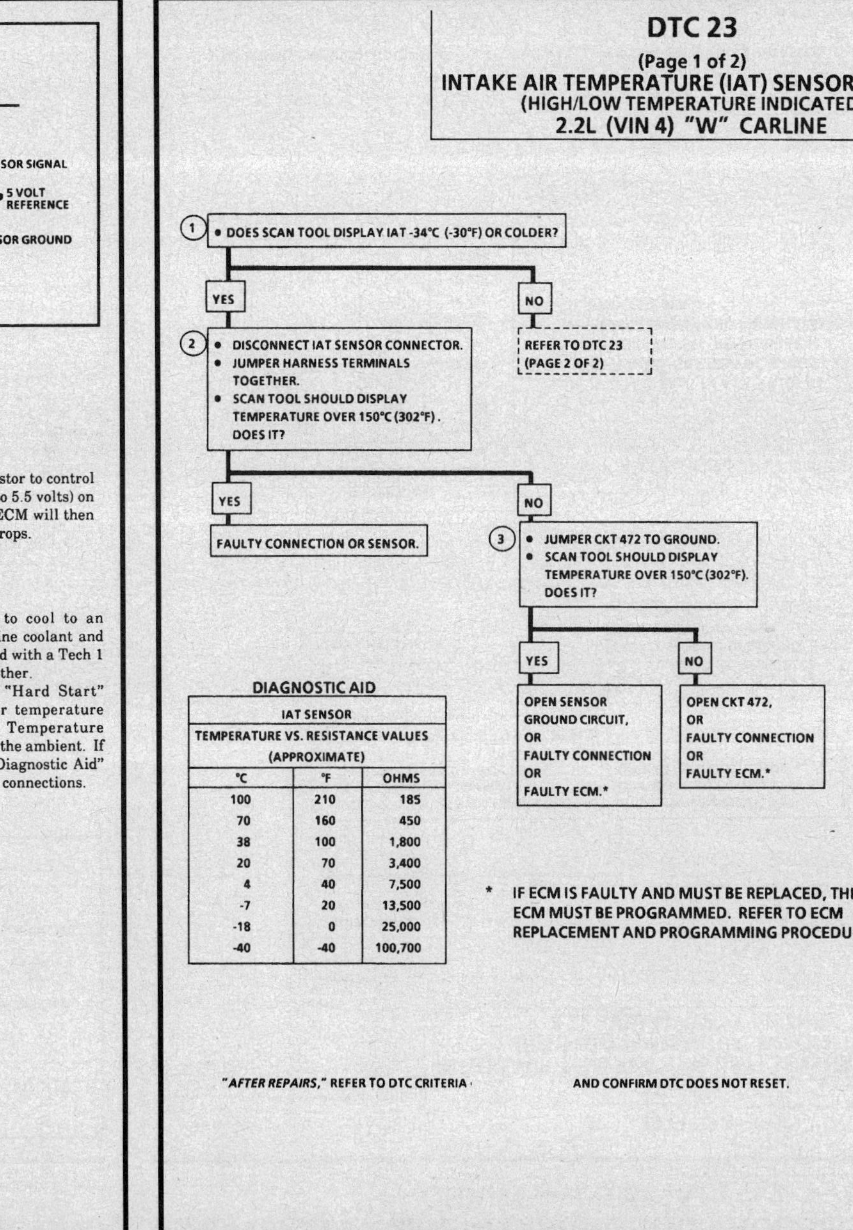

DIAGNOSTIC AID

IAT SENSOR TEMPERATURE VS. RESISTANCE VALUES (APPROXIMATE)		
°C	°F	OHMS
100	210	185
70	160	450
38	100	1,800
20	70	3,400
4	40	7,500
-7	20	13,500
-18	0	25,000
-40	-40	100,700

2.2L (VIN 4) ENGINE — DIAGNOSTIC TROUBLE CODE CHART — LUMINA

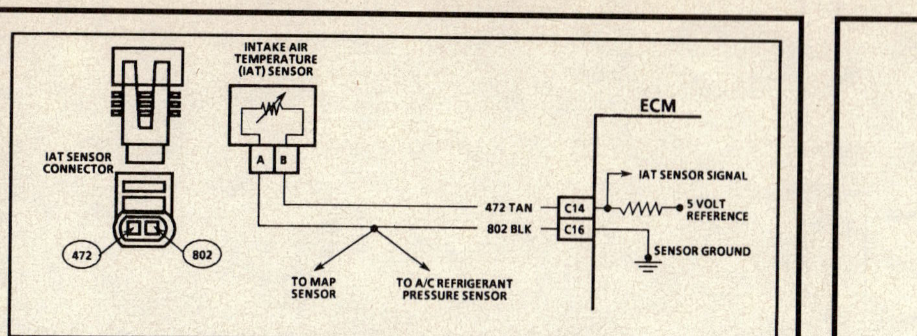

DTC 23
(Page 2 of 2)
INTAKE AIR TEMPERATURE (IAT) SENSOR CIRCUIT
(HIGH/LOW TEMPERATURE INDICATED)
2.2L (VIN 4) "W" CARLINE

Circuit Description:

The Intake Air Temperature (IAT) sensor, located in the air cleaner assembly, uses a thermistor to control the signal voltage to the Engine Control Module (ECM). The ECM applies a reference voltage (4 to 5.5 volts) on CKT 472 to the sensor. When intake air is cold, the sensor (thermistor) resistance is high. Therefore, the ECM will sense a high signal voltage. As the air warms, the sensor resistance becomes less and the voltage drops.

Test Description: Number(s) below refer to circled number(s) on the diagnostic chart.
1. This step determines if DTC 23 is the result of a hard failure or an intermittent condition.
2. If the ECM recognizes the open circuit (high voltage) and displays a low temperature, the ECM and wiring are OK.

Diagnostic Aids:

If the engine has been allowed to cool to an ambient temperature (overnight), engine coolant and intake air temperatures may be checked with a scan tool and should read close to each other.
- If DTC 23 is intermittent or a "Hard Start" symptom is present, check intake air temperature with a scan tool on a cool engine. Temperature displayed should be within 5 degrees of the ambient. If not, check the IAT sensor using the "Diagnostic Aid"   If sensor is OK, check connections.

2.2L (VIN 4) ENGINE — DIAGNOSTIC TROUBLE CODE CHART — LUMINA

DTC 23
(Page 2 of 2)
INTAKE AIR TEMPERATURE (IAT) SENSOR CIRCUIT
(HIGH/LOW TEMPERATURE INDICATED)
2.2L (VIN 4) "W" CARLINE

FROM DTC 23 PAGE 1 OF 2

(1) DOES SCAN TOOL DISPLAY IAT OF 150°C (302°F) OR HOTTER?

YES → (2) • DISCONNECT IAT SENSOR CONNECTOR. SCAN TOOL SHOULD DISPLAY TEMPERATURE BELOW -34°C (-30°F). DOES IT?

NO → DTC 23 IS INTERMITTENT. IF NO ADDITIONAL DTC(S) WERE STORED, REFER TO "DIAGNOSTIC AIDS"

YES → REPLACE IAT SENSOR.

NO → CKT 472 SHORTED TO GROUND, OR TO SENSOR GROUND, OR ECM IS FAULTY.*

DIAGNOSTIC AID		
IAT SENSOR		
TEMPERATURE VS. RESISTANCE VALUES		
(APPROXIMATE)		
°C	°F	OHMS
100	210	185
70	160	450
38	100	1,800
20	70	3,400
4	40	7,500
-7	20	13,500
-18	0	25,000
-40	-40	100,700

*   IF ECM IS FAULTY AND MUST BE REPLACED, THE NEW ECM MUST BE PROGRAMMED. REFER TO ECM REPLACEMENT AND PROGRAMMING PROCEDURES

"AFTER REPAIRS," REFER TO DTC CRITERIA   AND CONFIRM DTC DOES NOT RESET.

2.2L (VIN 4) ENGINE — DIAGNOSTIC TROUBLE CODE CHART — LUMINA

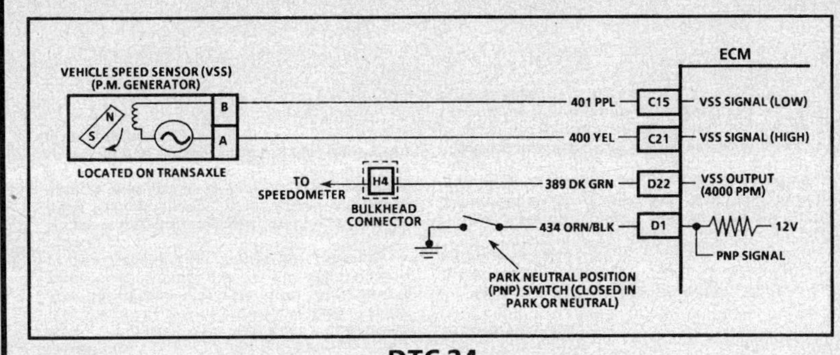

DTC 24
VEHICLE SPEED SENSOR (VSS) CIRCUIT
2.2L (VIN 4) "W" CARLINE

Circuit Description:

Vehicle speed information is provided to the Engine Control Module (ECM) by the Vehicle Speed Sensor (VSS), which is a Permanent Magnet (PM) generator, and it is mounted in the transaxle. The PM generator produces a pulsing voltage, whenever vehicle speed is over about 3 mph. The AC voltage level and the number of pulses increases with vehicle speed. The ECM then converts the pulsing voltage to mph, which is used for calculations, and the mph can be displayed with a scan tool.

The function of VSS buffer used in past model years has been incorporated into the ECM. The ECM then supplies the necessary signal to the instrument panel (4000 pulses per mile) for operating the speedometer and the odometer.

Test Description: Number(s) below refer to circled number(s) on the diagnostic chart.

1. DTC 24 will set if vehicle speed equals 0 mph when:
   - Engine speed is between about 1400 and 3600 RPM.
   - Low load condition (low MAP voltage), high manifold vacuum).
   - Transaxle not in park or neutral.
   - All above conditions are met for 4 seconds.

   These conditions are met during a road load deceleration.

   Disregard a DTC 24 that sets when the drive wheels are not turning. This can be caused by a faulty Park/Neutral Position (PNP) switch circuit. The PM generator only produces a signal if the drive wheels are turning greater than 3 mph.

2. At this point, the ECM is not sending vehicle speed data to the scan tool. If the scan tool is connected and functioning properly, the ECM is at fault.

Diagnostic Aids:

Scan tool should indicate a vehicle speed whenever the drive wheels are turning greater than 3 mph.

A problem in CKT 389 will not affect the VSS input or the readings on a scan tool.

Check CKTs 400 and 401 for proper connections to be sure they are clean and tight and the harness is routed correctly.

- A faulty or misadjusted Park/Neutral Position (PNP) switch can result in a false DTC 24. Use a scan tool and check for the proper signal while in a drive range. Refer to CHART C-1A for the PNP switch check.

2.2L (VIN 4) ENGINE — DIAGNOSTIC TROUBLE CODE CHART — LUMINA

DTC 24
VEHICLE SPEED SENSOR (VSS) CIRCUIT
2.2L (VIN 4) "W" CARLINE

DISREGARD DTC 24 IF SET WHILE DRIVE WHEELS ARE NOT TURNING.

(1)
- RAISE DRIVE WHEELS.

  NOTICE: DO NOT PERFORM THIS TEST WITHOUT SUPPORTING THE LOWER CONTROL ARMS SO THAT THE DRIVE AXLES ARE IN A NORMAL HORIZONTAL POSITION. RUNNING THE VEHICLE IN GEAR WITH THE WHEELS HANGING DOWN AT FULL TRAVEL MAY DAMAGE THE DRIVE AXLES.

- WITH ENGINE IDLING IN GEAR, SCAN TOOL SHOULD DISPLAY VEHICLE SPEED ABOVE 0 MPH. DOES IT?

NO →

DOES SPEEDOMETER WORK PROPERLY?

YES → DTC 24 IS INTERMITTENT. IF NO ADDITIONAL DTC(S) WERE STORED, REFER TO "DIAGNOSTIC AIDS"

NO →

- IGNITION "OFF."
- DISCONNECT VSS HARNESS CONNECTOR AT TRANSAXLE.
- CONNECT SIGNAL GENERATOR TESTER J 33431-B OR EQUIVALENT TO VSS HARNESS CONNECTOR.
- IGNITION "ON," TOOL "ON" AND SET TO GENERATE A VSS SIGNAL.
- SCAN TOOL SHOULD DISPLAY VEHICLE SPEED ABOVE 0 MPH. DOES IT?

YES → (2) REPLACE ECM.*

NO →

CKT 400 OR 401 OPEN, SHORTED TO GROUND, SHORTED TOGETHER, FAULTY CONNECTIONS OR FAULTY ECM.*

YES → REPLACE VEHICLE SPEED SENSOR.

---

* IF ECM IS FAULTY AND MUST BE REPLACED, THE NEW ECM MUST BE PROGRAMMED. REFER TO ECM REPLACEMENT AND PROGRAMMING PROCEDURES

"AFTER REPAIRS," REFER TO DTC CRITERIA   AND CONFIRM DTC DOES NOT RESET.

2.2L (VIN 4) ENGINE — DIAGNOSTIC TROUBLE CODE CHART — LUMINA

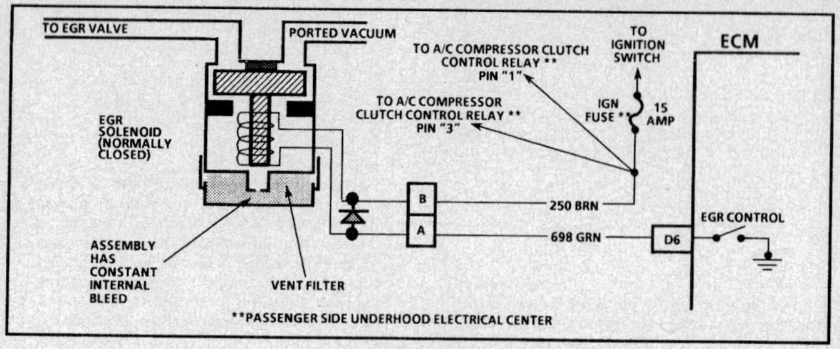

DTC 32
(Page 1 of 2)
EXHAUST GAS RECIRCULATION (EGR) SYSTEM FAILURE
2.2L (VIN 4) "W" CARLINE

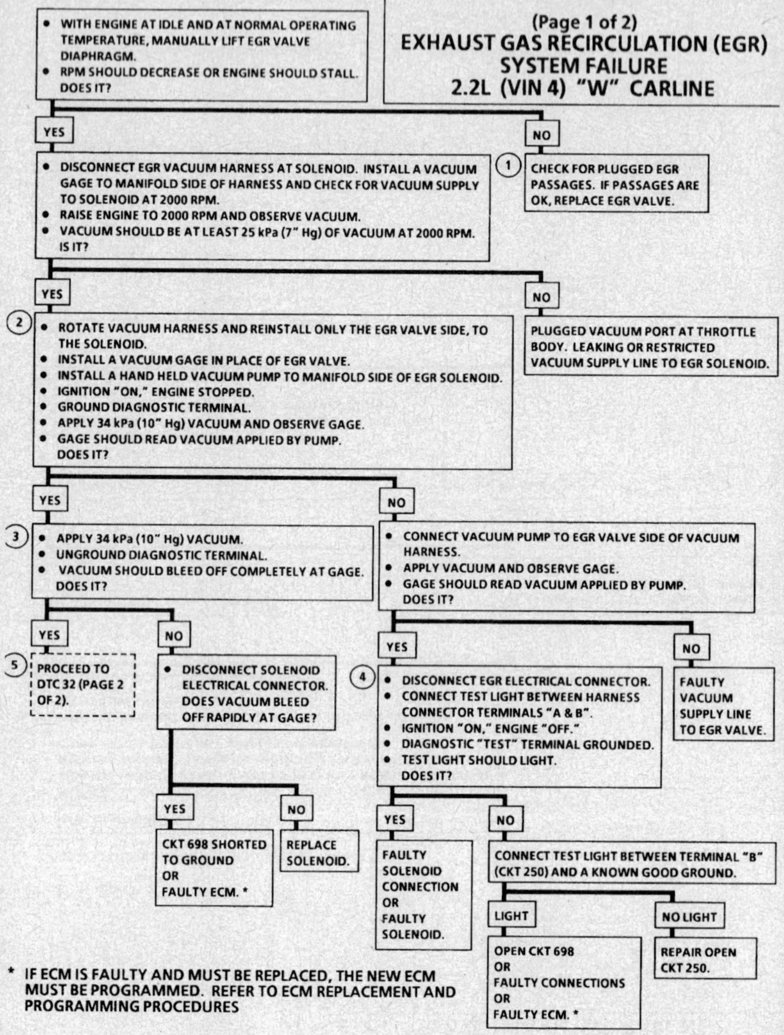

Circuit Description:

The Exhaust Gas Recirculation (EGR) system is controlled by the ECM. The ECM controls the vacuum being supplied to the valve by energizing and de-energizing a solenoid.

The ECM uses information from various engine sensors to determine when EGR is necessary. Once the ECM has requested EGR by grounding the solenoid circuit, the ECM will monitor engine operating conditions to determine if exhaust gas flow has entered the intake manifold. When the ECM tests for EGR operation and no change in engine operating conditions is indicated, a DTC 32 will set.

The difference in long term fuel trim values between the idle (closed throttle) cell and cell 2 is used to monitor EGR system performance. When the difference between the two long term fuel trim values is greater than about 15, and the long term fuel trim value in cell 2 is greater than 135 for 80 seconds or more, DTC 32 is set. The system operates in long term fuel trim cell 2 during a cruise condition at approximately 55 mph.

Test Description: Number(s) below refer to circled number(s) on the diagnostic chart.

1. **Intake Passage:** Shut "OFF" engine and remove the EGR valve from the manifold. Plug the exhaust side hole with a suitable stopper. Leaving the intake side hole open, attempt to start the engine. If the engine runs at a high idle (up to 3000 RPM is possible) or starts and stalls, the EGR intake passage is not restricted. If the engine starts and idles normally, the EGR intake passage is restricted.
Exhaust Passage: With EGR valve still removed, plug the intake side hole with a suitable stopper. With the exhaust side hole open, check for the presence of exhaust gas. If no exhaust gas is present, the EGR exhaust side passage is restricted.

2. By grounding the diagnostic "test" terminal, the EGR solenoid should be energized and allow vacuum to be applied to the gage. The vacuum at the gage may or may not <u>slowly</u> bleed off. It is important that the gage is able to read the amount of vacuum being applied.

3. When the diagnostic "test" terminal is ungrounded, the vacuum gage should bleed off completely through a vent in the solenoid. The vacuum pump gage may or may not bleed off but this does not indicate a problem.

4. This test will determine if the electrical control part of the system is at fault or if the connector or solenoid is at fault.

5. At this point, it has been determined that the EGR solenoid, the ECM and the vacuum supply are OK.

Diagnostic Aids:

Vacuum lines should be thoroughly checked for proper routing. Refer to "Vehicle Emission Control Information" label.

The DTC 32 chart is a functional check of the EGR system. If the EGR system works properly but a DTC 32 has been set, check other items that could result in high block learn values during a cruise condition at approximately 55 mph. Low fuel pressure or lean fuel injector(s) may set a DTC 32. It may be necessary to monitor fuel pressure while driving the vehicle at various road speeds and/or loads. Refer to "Fuel System Diagnosis," CHART A-7.

2.2L (VIN 4) ENGINE — DIAGNOSTIC TROUBLE CODE CHART — LUMINA

2.2L (VIN 4) ENGINE — DIAGNOSTIC TROUBLE CODE CHART — LUMINA

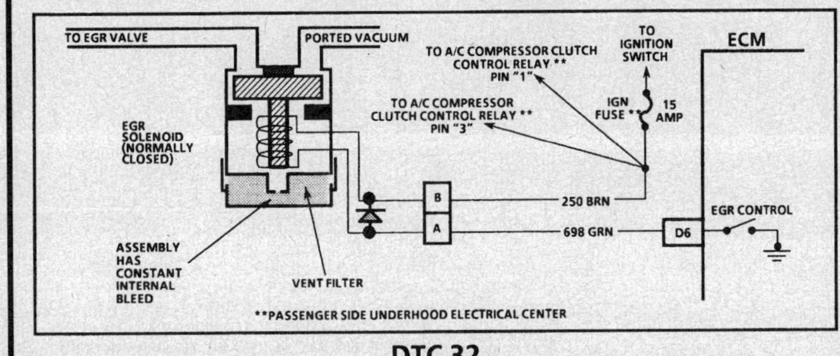

DTC 32
(Page 2 of 2)
**EXHAUST GAS RECIRCULATION (EGR) SYSTEM FAILURE
2.2L (VIN 4) "W" CARLINE**

Circuit Description:

The Exhaust Gas Recirculation (EGR) system is controlled by the ECM. The ECM controls the vacuum being supplied to the valve by energizing and de-energizing a solenoid.

The ECM uses information from various engine sensors to determine when EGR is necessary. Once the ECM has requested EGR by grounding the solenoid circuit, the ECM will monitor engine operating conditions to determine if exhaust gas flow has entered the intake manifold. When the ECM tests for EGR operation and no change in engine operating conditions is indicated, a DTC 32 will set.

Test Description: Number(s) below refer to circled number(s) on the diagnostic chart.

1. The remaining tests check the ability of the EGR valve to interact with the exhaust system. This system uses a negative backpressure EGR valve which should hold vacuum with engine "OFF." **Be sure shop exhaust hose is not connected during Steps 1 and 2.**
2. When engine is started, exhaust backpressure at the base of the EGR valve should open the valve's internal bleed and vent the applied vacuum allowing the valve to seat. **Because the shop exhaust hose is not installed at this time, do not allow the engine to run longer than 15 seconds.**

Diagnostic Aids:

Low fuel pressure or lean fuel injectors may cause DTC 32 to set. Use CHART A-7. It may be necessary to monitor fuel pressure while driving the vehicle at various road speeds and/or loads. If fuel pressure is normal, perform the injector balance test, CHART C2-A

2.2L (VIN 4) ENGINE — DIAGNOSTIC TROUBLE CODE CHART — LUMINA

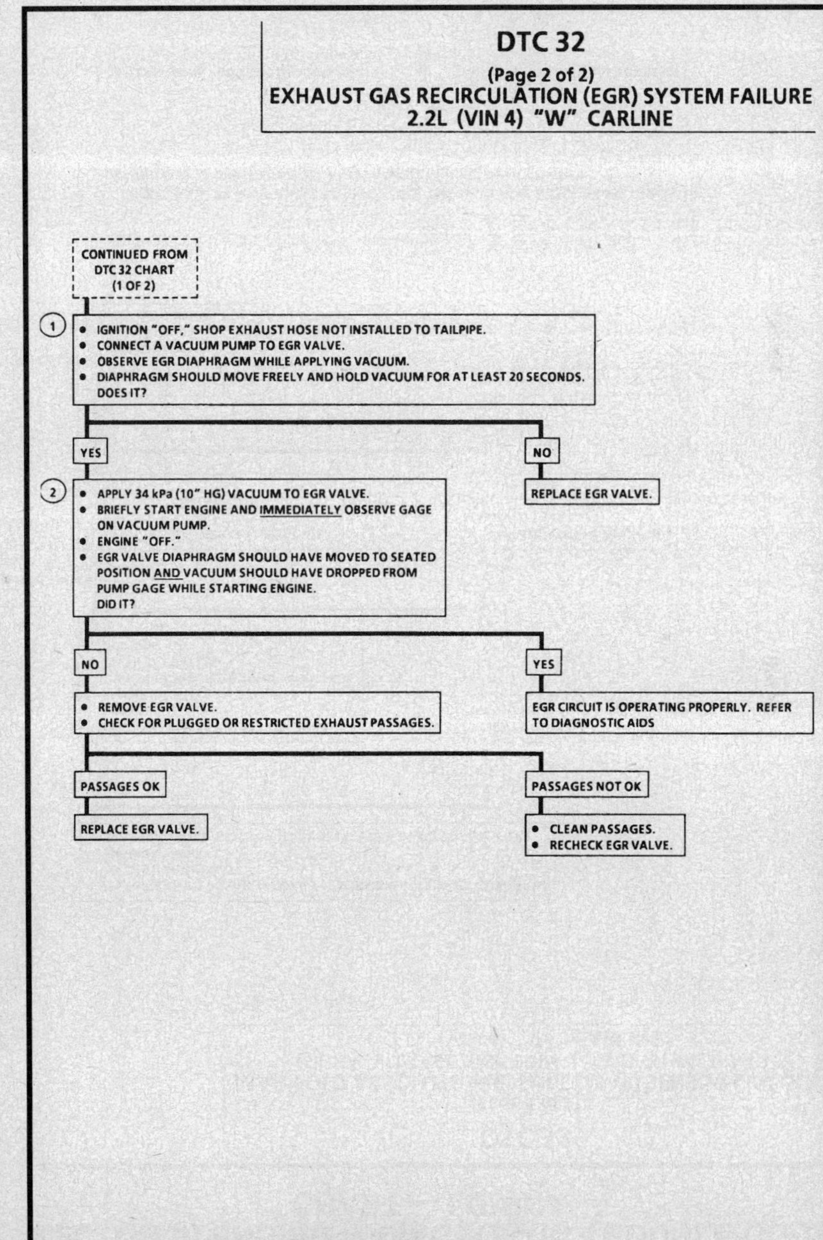

2.2L (VIN 4) ENGINE — DIAGNOSTIC TROUBLE CODE CHART — LUMINA

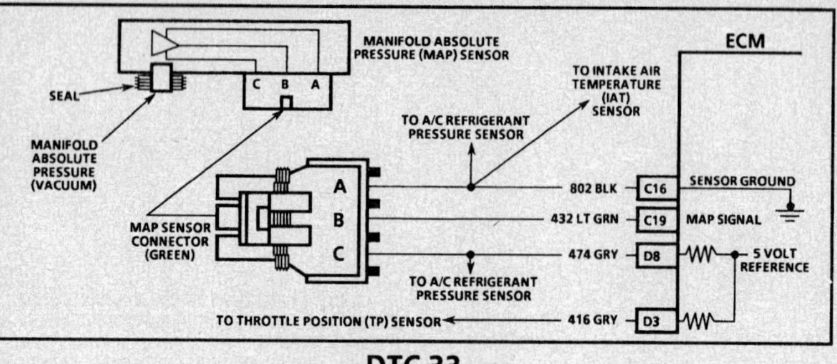

DTC 33
(Page 1 of 2)
MANIFOLD ABSOLUTE PRESSURE (MAP) SENSOR CIRCUIT
(SIGNAL VOLTAGE HIGH/LOW - LOW/HIGH VACUUM)
2.2L (VIN 4) "W" CARLINE

Circuit Description:

The Manifold Absolute Pressure (MAP) sensor responds to changes in manifold pressure (vacuum). The ECM receives this information as a signal voltage that will vary from about 1 to 1.5 volts at closed throttle (idle), to about 4.5 volts at Wide Open Throttle (WOT) (low vacuum).

If the MAP sensor fails, the Engine Control Module (ECM) will substitute a fixed MAP value based on engine RPM.

Test Description: Number(s) below refer to circled number(s) on the diagnostic chart.

1. A DTC 33 will set under the following conditions:
   - MAP signal indicates greater than 3.7 volts (80 kPa).
   - Throttle position is less than 5%.
   - These conditions exist for a time period longer than 5 seconds.
   OR
   - MAP signal indicates less than .3 volt (15 kPa).
2. If the ECM recognizes the change, the ECM and CKTs 474 and 432 are OK. If CKT 802 is open, there may also be other DTCs stored.

Diagnostic Aids:

With the ignition "ON" and the engine stopped, the manifold pressure is equal to atmospheric pressure and the signal voltage will be high. This information is used by the ECM as an indication of vehicle altitude and is referred to as BARO. Comparison of this BARO reading with a known good vehicle with the same sensor is a good way to check accuracy of a "suspect" sensor. Reading should be within ± .4 volt.

If DTC 33 is intermittent, refer to "Symptoms,"

2.2L (VIN 4) ENGINE — DIAGNOSTIC TROUBLE CODE CHART — LUMINA

DTC 33
(Page 1 of 2)
MANIFOLD ABSOLUTE PRESSURE (MAP) SENSOR CIRCUIT
(SIGNAL VOLTAGE HIGH/LOW - LOW/HIGH VACUUM)
2.2L (VIN 4) "W" CARLINE

(1)
- IF ENGINE IDLE IS ROUGH, UNSTABLE, OR INCORRECT, CORRECT CONDITION BEFORE USING CHART.
- ENGINE IDLING.
DOES SCAN TOOL DISPLAY A MAP VOLTAGE OF 3.7 VOLTS OR GREATER?

YES	NO

(2)
- DISCONNECT MAP SENSOR ELECTRICAL CONNECTOR.
- ENGINE IDLING.
SCAN TOOL SHOULD DISPLAY A VOLTAGE OF .3 VOLT OR LESS.
DOES IT?

NO → REFER TO DTC 33 (PAGE 2 OF 2).

YES	NO

YES:
- PROBE SENSOR GROUND CIRCUIT WITH A TEST LIGHT TO BATTERY VOLTAGE. TEST LIGHT SHOULD LIGHT. DOES IT?

NO:
CKT 432 SHORTED TO VOLTAGE, SHORTED TO CKT 474
OR
FAULTY ECM.*

YES	NO

YES:
PLUGGED SENSOR INLET PORT
OR
LEAKING SENSOR SEAL
OR
FAULTY MAP SENSOR.

NO:
OPEN SENSOR GROUND CIRCUIT.

* IF ECM IS FAULTY AND MUST BE REPLACED, THE NEW ECM MUST BE PROGRAMMED. REFER TO ECM REPLACEMENT AND PROGRAMMING PROCEDURES

"AFTER REPAIRS," REFER TO DTC CRITERIA AND CONFIRM DTC DOES NOT RESET.

2.2L (VIN 4) ENGINE — DIAGNOSTIC TROUBLE CODE CHART — LUMINA

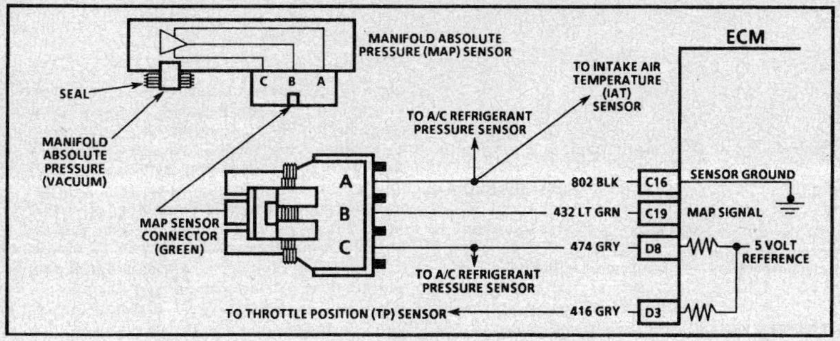

DTC 33
(Page 2 of 2)
MANIFOLD ABSOLUTE PRESSURE (MAP) SENSOR CIRCUIT
(SIGNAL VOLTAGE HIGH/LOW - LOW/HIGH VACUUM)
2.2L (VIN 4) "W" CARLINE

Circuit Description:

The Manifold Absolute Pressure (MAP) sensor responds to changes in manifold pressure (vacuum). The ECM receives this information as a signal voltage that will vary from about 1 to 1.5 volts at closed throttle (idle), to about 4.5 volts at Wide Open Throttle (WOT) (low vacuum).

If the MAP sensor fails, the Engine Control Module (ECM) will substitute a fixed MAP value based on engine RPM and use the Throttle Position (TP) sensor to control fuel delivery.

Test Description: Number(s) below refer to circled number(s) on the diagnostic chart.

1. Jumpering harness terminals "B" to "C" (5 volts to signal circuit), will determine if the sensor is at fault, or if there is a problem with the ECM or wiring.

   The scan tool may not display 5 volts. The important thing is that the ECM recognizes the voltage as more than 4 volts, indicating that the ECM CKTs 432, and 474 are OK.
2. This step determines if CKT 474 or CKT 432 is faulty. The scan tool will not display battery voltage, but should indicate over 4 volts.
3. The 5 volt reference for the Throttle Position (TP) sensor connects to the same 5 volt power supply in the ECM. If the 5 volt reference circuit to the TP sensor has a short to ground, the MAP sensor and A/C refrigerant pressure sensor 5 volt reference circuit will also have a short to ground because they connect to the same power supply in the ECM. A short to ground on either 5 volt reference circuit does not damage the ECM 5 volt power supply. When the short is repaired, the ECM 5 volt supply will return to normal.

Diagnostic Aids:

With the ignition "ON" and the engine stopped, the manifold pressure is equal to atmospheric pressure and the signal voltage will be high. This information is used by the ECM as an indication of vehicle altitude and is referred to as BARO. Comparison of this BARO reading with a known good vehicle with the same sensor is a good way to check accuracy of a "suspect" sensor. Reading should be within ± .4 volt.

If DTC 33 is intermittent, refer to "Symptoms,"

2.2L (VIN 4) ENGINE — DIAGNOSTIC TROUBLE CODE CHART — LUMINA

DTC 33
(Page 2 of 2)
MANIFOLD ABSOLUTE PRESSURE (MAP) SENSOR CIRCUIT
(SIGNAL VOLTAGE HIGH/LOW - LOW/HIGH VACUUM)
2.2L (VIN 4) "W" CARLINE

* IF ECM IS FAULTY AND MUST BE REPLACED, THE NEW ECM MUST BE PROGRAMMED. REFER TO ECM REPLACEMENT AND PROGRAMMING PROCEDURES

"AFTER REPAIRS," REFER TO DTC CRITERIA   AND CONFIRM DTC DOES NOT RESET.

2.2L (VIN 4) ENGINE — DIAGNOSTIC TROUBLE CODE CHART — LUMINA

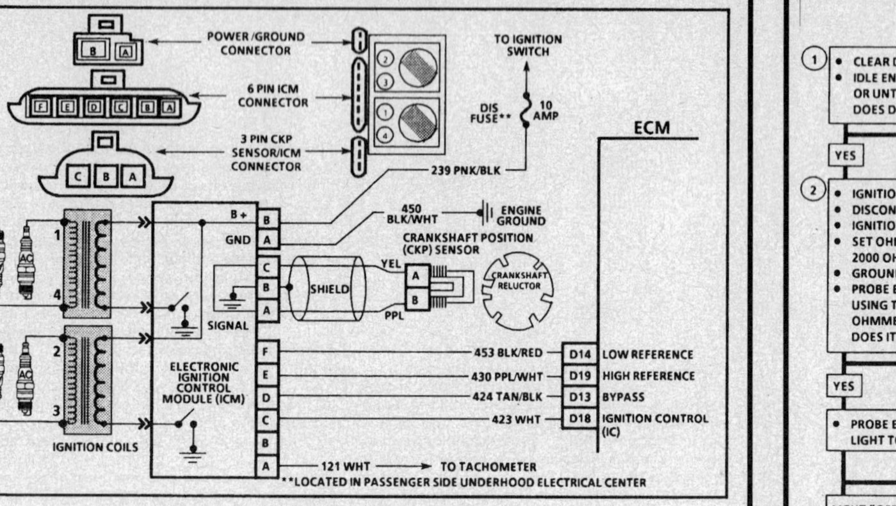

DTC 42
IGNITION CONTROL (IC) CIRCUIT
2.2L (VIN 4) "W" CARLINE

Circuit Description:

The Electronic Ignition (EI) system module sends a reference signal to the Engine Control Module (ECM) when the engine is cranking. While the engine speed is under 400 RPM, the EI module controls the ignition timing. When the system is running on the ignition module (no voltage on the bypass line), the ignition module grounds the ignition control signal. The ECM expects to sense no voltage on the Ignition Control (IC) line during this condition. If it senses a voltage, it sets DTC 42 and will not enter the ignition control mode.

When the engine speed exceeds 400 RPM, the ECM applies 5 volts to the bypass line to switch the timing to ECM control ignition control. If the bypass line is open or grounded, once the RPM for ignition control is reached, the ignition module will not switch to ignition control mode. This results in low ignition control voltage and the setting of DTC 42. If the ignition control line is grounded, the ignition module will switch to ignition control, but because the line is grounded, there will be no ignition control signal. A DTC 42 will be set.

Test Description: Number(s) below refer to circled number(s) on the diagnostic chart.

1. DTC 42 means the ECM has sensed an open or short to ground in the IC or bypass circuits. This test confirms DTC 42 and that the fault causing the code is present.
2. Checks for a normal IC ground path through the ignition module. An IC CKT 423, shorted to ground, will also read less than 500 ohms, but this will be checked later.
3. As the test light voltage contacts CKT 424, the module should switch, causing the ohmmeter to "overrange" if the meter is in the 1000 to 2000 ohms position. Selecting the 10,000 to 20,000 ohms position will indicate a reading above 5000 ohms.

The important thing is that the module "switched."
4. The module did not switch and this step checks for:
   - IC CKT 423 shorted to ground.
   - Bypass CKT 424 open.
   - Faulty ignition module connection or module.
5. Confirms that DTC 42 is a faulty ECM and not an intermittent in CKTs 423 or 424.

Diagnostic Aids:

If DTC 42 is intermittent, refer to "Symptoms,"

2.2L (VIN 4) ENGINE — DIAGNOSTIC TROUBLE CODE CHART — LUMINA

DTC 42
IGNITION CONTROL (IC) CIRCUIT
2.2L (VIN 4) "W" CARLINE

1.
- CLEAR DTC(S).
- IDLE ENGINE FOR 1 MINUTE OR UNTIL DTC 42 SETS. DOES DTC 42 SET?

YES →

2.
- IGNITION "OFF."
- DISCONNECT ECM CONNECTORS.
- IGNITION "ON."
- SET OHMMETER SELECTOR SWITCH IN THE 1000 TO 2000 OHMS RANGE.
- GROUND THE BLACK (-) OHMMETER LEAD.
- PROBE ECM HARNESS IGNITION CONTROL CKT 423 USING THE RED (+) OHMMETER LEAD. THE OHMMETER SHOULD READ LESS THAN 500 OHMS. DOES IT?

NO → DTC 42 INTERMITTENT. REFER TO "DIAGNOSTIC AIDS"

YES →
- PROBE ECM HARNESS CONNECTOR CKT 424 WITH A TEST LIGHT TO BATTERY VOLTAGE AND OBSERVE LIGHT.

NO → OPEN CKT 423, FAULTY CONNECTION, OR FAULTY IGNITION CONTROL MODULE (ICM).

LIGHT "OFF" →

3.
- WITH OHMMETER STILL CONNECTED TO ECM HARNESS CKT 423 AND GROUND AS DESCRIBED IN STEP #2, AGAIN PROBE ECM HARNESS CKT 424 WITH THE TEST LIGHT CONNECTED TO BATTERY VOLTAGE. AS TEST LIGHT CONTACTS CKT 424, RESISTANCE SHOULD SWITCH FROM HUNDREDS TO THOUSANDS OHMS. DOES IT?

LIGHT "ON" →
- DISCONNECT IGNITION CONTROL MODULE 6-PIN CONNECTOR.

LIGHT "ON" → CKT 424 SHORTED TO GROUND.

LIGHT "OFF" → REPLACE IGNITION CONTROL MODULE.

NO →

4.
- DISCONNECT IGNITION CONTROL MODULE 6-PIN CONNECTOR. NOTE OHMMETER THAT IS STILL CONNECTED TO CKT 423 AND GROUND. RESISTANCE SHOULD HAVE BECOME VERY HIGH (OPEN CIRCUIT). DOES IT?

YES → CKT 424 OPEN, FAULTY CONNECTIONS, OR FAULTY IGNITION CONTROL MODULE.

NO → CKT 423 SHORTED TO GROUND.

YES →

5.
- RECONNECT ECM AND IDLE ENGINE FOR ONE MINUTE OR UNTIL DTC 42 SETS. DOES DTC 42 SET?

YES → REPLACE ECM.*

NO → DTC 42 INTERMITTENT. REFER TO "DIAGNOSTIC AIDS"

*   IF ECM IS FAULTY AND MUST BE REPLACED, THE NEW ECM MUST BE PROGRAMMED. REFER TO ECM REPLACEMENT AND PROGRAMMING PROCEDURES

"AFTER REPAIRS," REFER TO DTC CRITERIA   AND CONFIRM DTC DOES NOT RESET.

2.2L (VIN 4) ENGINE — DIAGNOSTIC TROUBLE CODE CHART — LUMINA

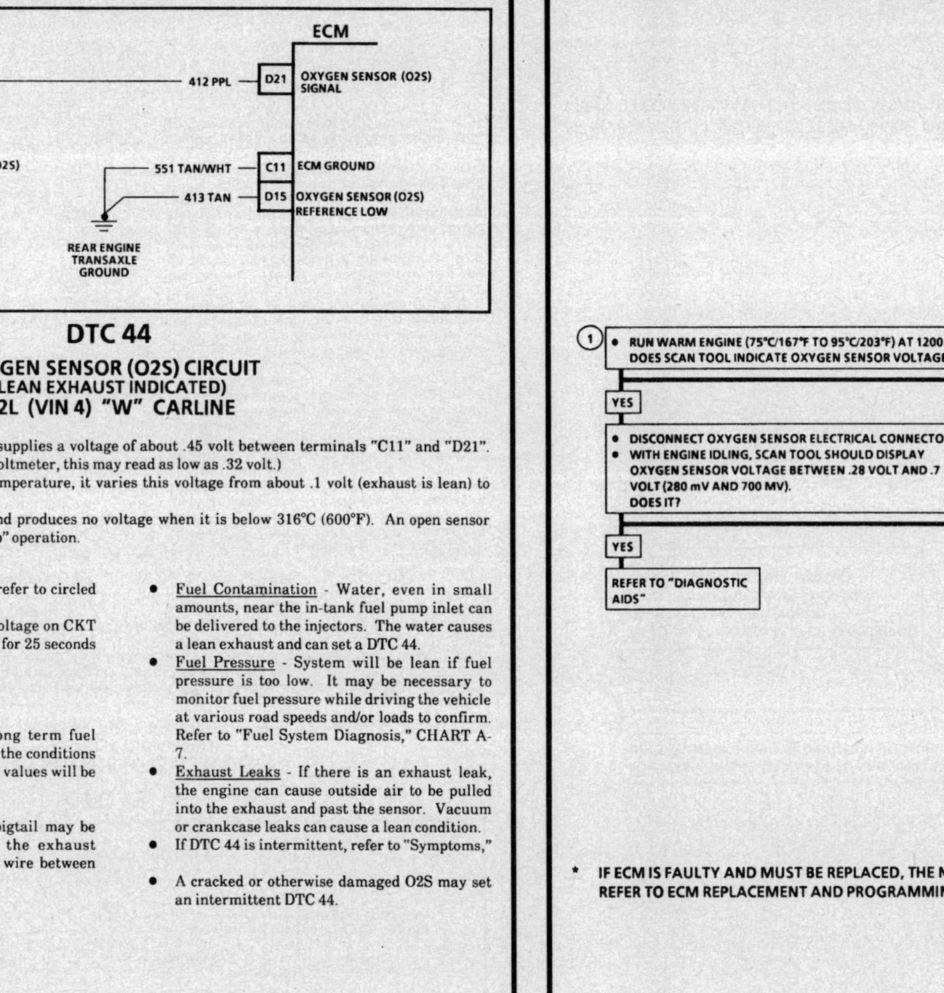

DTC 44
OXYGEN SENSOR (O2S) CIRCUIT
(LEAN EXHAUST INDICATED)
2.2L (VIN 4) "W" CARLINE

Circuit Description:

The Engine Control Module (ECM) supplies a voltage of about .45 volt between terminals "C11" and "D21". (If measured with a 10 megohm digital voltmeter, this may read as low as .32 volt.)

When the O2S reaches operating temperature, it varies this voltage from about .1 volt (exhaust is lean) to about .9 volt (exhaust is rich).

The sensor is like an open circuit and produces no voltage when it is below 316°C (600°F). An open sensor circuit, or cold sensor, causes "Open Loop" operation.

Test Description: Number(s) below refer to circled number(s) on the diagnostic chart.
1. DTC 44 is set when the O2S signal voltage on CKT 412 remains below .28 volt (200 mV) for 25 seconds or more.

Diagnostic Aids:

Using the scan tool, observe the long term fuel trim value at different engine speeds. If the conditions for DTC 44 exist, the long term fuel trim values will be around 160 or higher.
Check the following possible causes:
- Oxygen Sensor Wire - Sensor pigtail may be mispositioned and contacting the exhaust manifold. Check for ground in wire between connector and sensor.
- Fuel Contamination - Water, even in small amounts, near the in-tank fuel pump inlet can be delivered to the injectors. The water causes a lean exhaust and can set a DTC 44.
- Fuel Pressure - System will be lean if fuel pressure is too low. It may be necessary to monitor fuel pressure while driving the vehicle at various road speeds and/or loads to confirm. Refer to "Fuel System Diagnosis," CHART A-7.
- Exhaust Leaks - If there is an exhaust leak, the engine can cause outside air to be pulled into the exhaust and past the sensor. Vacuum or crankcase leaks can cause a lean condition.
- If DTC 44 is intermittent, refer to "Symptoms,"
- A cracked or otherwise damaged O2S may set an intermittent DTC 44.

2.2L (VIN 4) ENGINE — DIAGNOSTIC TROUBLE CODE CHART — LUMINA

DTC 44
OXYGEN SENSOR (O2S) CIRCUIT
(LEAN EXHAUST INDICATED)
2.2L (VIN 4) "W" CARLINE

(1)
- RUN WARM ENGINE (75°C/167°F TO 95°C/203°F) AT 1200 RPM.
DOES SCAN TOOL INDICATE OXYGEN SENSOR VOLTAGE FIXED BELOW .28 VOLT (280 mV)?

YES
- DISCONNECT OXYGEN SENSOR ELECTRICAL CONNECTOR.
- WITH ENGINE IDLING, SCAN TOOL SHOULD DISPLAY OXYGEN SENSOR VOLTAGE BETWEEN .28 VOLT AND .7 VOLT (280 mV AND 700 mV).
DOES IT?

NO
DTC 44 IS INTERMITTENT. IF NO ADDITIONAL DTC(S) WERE STORED, REFER TO "DIAGNOSTIC AIDS"

YES
REFER TO "DIAGNOSTIC AIDS"

NO
CKT 412 SHORTED TO GROUND
OR
FAULTY ECM.*

* IF ECM IS FAULTY AND MUST BE REPLACED, THE NEW ECM MUST BE PROGRAMMED.
REFER TO ECM REPLACEMENT AND PROGRAMMING PROCEDURES

"AFTER REPAIRS," REFER TO DTC CRITERIA   AND CONFIRM DTC DOES NOT RESET.

2.2L (VIN 4) ENGINE — DIAGNOSTIC TROUBLE CODE CHART — LUMINA

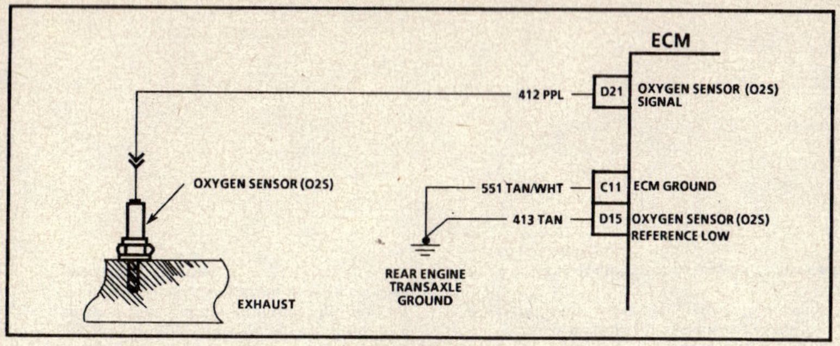

ECM

412 PPL — D21 — OXYGEN SENSOR (O2S) SIGNAL

OXYGEN SENSOR (O2S)

551 TAN/WHT — C11 — ECM GROUND

413 TAN — D15 — OXYGEN SENSOR (O2S) REFERENCE LOW

REAR ENGINE TRANSAXLE GROUND

EXHAUST

DTC 45
OXYGEN SENSOR (O2S) CIRCUIT
(RICH EXHAUST INDICATED)
2.2L (VIN 4) "W" CARLINE

Circuit Description:

The Engine Control Module (ECM) supplies a voltage of about .45 volt between terminals "C11" and "D21". (If measured with a 10 megohm digital voltmeter, this may read as low as .32 volt.)

When the Oxygen Sensor (O2S) reaches operating temperature, it varies this voltage from about .1 volt (exhaust is lean) to about .9 volt (exhaust is rich).

The sensor is like an open circuit and produces no voltage when it is below 316°C (600°F). An open sensor circuit, or cold sensor, causes "Open Loop" operation.

Test Description: Number(s) below refer to circled number(s) on the diagnostic chart.
1. DTC 45 is set when the O2S signal voltage on CKT 412 remains above .7 volt (700 mV) under the following conditions:
   - Engine run time after start is 2 minutes or more.
   - System is operating in "Closed Loop."
   - Throttle angle is greater than 5%.
   - Above conditions exist for 51 seconds or more.

Diagnostic Aids:

Using the scan tool, observe the long term fuel trim value at different engine speeds. If the conditions for DTC 45 exist, the long term fuel trim values will be around 90 or lower.

Check the following possible causes:
- <u>Fuel Pressure</u> - System will go rich, if pressure is too high. The ECM can compensate for some increase. However, if it gets too high, a DTC 45 will be set. Refer to "Fuel System Diagnosis" CHART A-7.
- <u>Leaking Injector</u> - Refer to CHART A-7.
- <u>An open ground CKT 453</u> - May result in induced electrical "noise." The ECM interprets this "noise" as reference pulses. The additional pulses result in a higher than actual engine speed signal. The ECM then delivers too much fuel causing the system to go rich.
- <u>Canister Purge</u> - Check for fuel saturation. If full of fuel, check canister control and hoses.

- <u>MAP Sensor</u> - An output that causes the ECM to sense a higher than normal manifold pressure (low vacuum) can cause the system to go rich. Disconnecting the Manifold Absolute Pressure (MAP) sensor will allow the ECM to set a fixed value for the MAP sensor. Substitute a different MAP sensor if the rich condition is gone, while the sensor is disconnected.
- <u>TP Sensor</u> - An intermittent Throttle Position (TP) sensor output will cause the system to operate richly due to a false indication of the engine accelerating.
- <u>O2S Contamination</u> - Inspect Oxygen Sensor (O2S) for silicone contamination from fuel, or use of improper RTV sealant. The sensor may have a white, powdery coating and result in a high but false signal voltage (rich exhaust indication). The ECM will then reduce the amount of fuel delivered to the engine causing a severe surge driveability problem.
- <u>EGR Valve</u> - Exhaust Gas Recirculation (EGR) sticking open at idle is usually accompanied by a rough idle and/or stall condition.
- If DTC 45 is intermittent, refer to "Symptoms,"
- <u>Engine Oil Contamination</u> - Fuel fouled engine oil could cause the O2S to sense a rich air/fuel mixture and set a DTC 45.

2.2L (VIN 4) ENGINE — DIAGNOSTIC TROUBLE CODE CHART — LUMINA

DTC 45
OXYGEN SENSOR (O2S) CIRCUIT
(RICH EXHAUST INDICATED)
2.2L (VIN 4) "W" CARLINE

(1)
- RUN WARM ENGINE (75°C/167°F TO 95°C/203°F) AT 1200 RPM.
- DOES SCAN TOOL DISPLAY OXYGEN SENSOR VOLTAGE FIXED ABOVE .7 VOLT (700 mV)?

YES
- DISCONNECT OXYGEN SENSOR AND JUMPER HARNESS CKT 412 TO GROUND.
- SCAN TOOL SHOULD DISPLAY OXYGEN SENSOR BELOW .28 VOLT (280 mV). DOES IT?

NO
DTC 45 IS INTERMITTENT. IF NO ADDITIONAL DTCS WERE STORED, REFER TO "DIAGNOSTIC AIDS"

YES
REFER TO "DIAGNOSTIC AIDS"

NO
REPLACE ECM.*

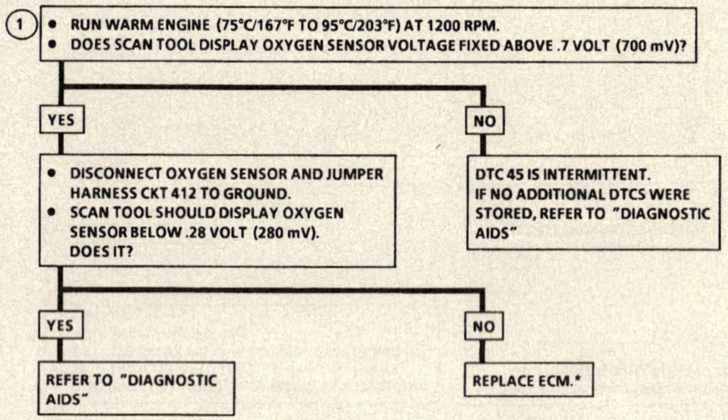

***IF ECM IS FAULTY AND MUST BE REPLACED, THE NEW ECM MUST BE PROGRAMMED. REFER TO ECM REPLACEMENT AND PROGRAMMING PROCEDURES**

"AFTER REPAIRS," REFER TO DTC CRITERIA   AND CONFIRM DTC DOES NOT RESET.

2.2L (VIN 4) ENGINE — DIAGNOSTIC TROUBLE CODE CHART — LUMINA

DTC 51

ECM FAILURE
(ECM FAILED OR EEPROM FAILURE)
2.2L (VIN 4) "W" CARLINE

DIAGNOSTIC TROUBLE CODE (DTC) 51

ECM FAILURE
(ECM FAILED OR EEPROM FAILURE)

CHECK THAT ALL ECM CONNECTIONS ARE GOOD.
IF OK, CLEAR MEMORY AND RECHECK ECM.
IF DTC 51 REAPPEARS, REPLACE ECM AND PROGRAM NEW ECM.

*   IF ECM IS FAULTY AND MUST BE REPLACED, THE NEW ECM MUST BE PROGRAMMED. REFER TO ECM REPLACEMENT AND PROGRAMMING PROCEDURES.

2.2L (VIN 4) ENGINE — DIAGNOSTIC TROUBLE CODE CHART — LUMINA

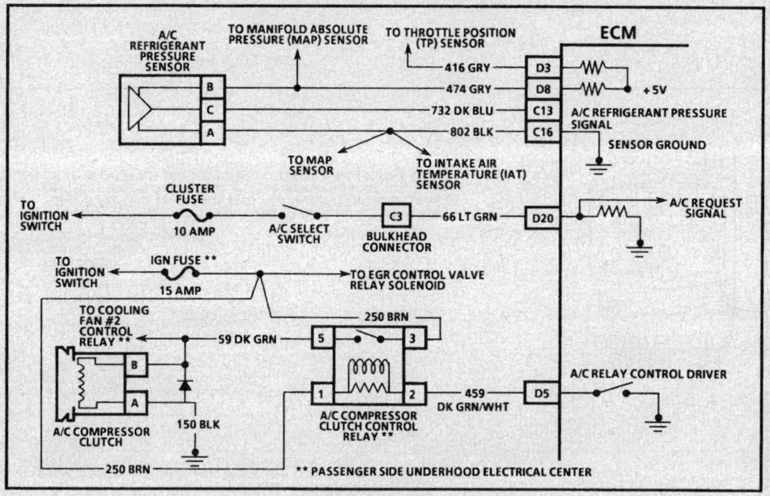

DTC 66

A/C REFRIGERANT PRESSURE SENSOR CIRCUIT
2.2L (VIN 4) "W" CARLINE

Circuit Description:

The A/C refrigerant pressure sensor responds to changes in A/C refrigerant system high side pressure. This input indicates how much load the A/C compressor is putting on the engine and is one of the factors used by the ECM to determine IAC valve position for idle speed control. The circuit consists of a 5 volt reference and a ground, both provided by the ECM, and a signal line to the ECM. The signal is a voltage which is proportional to the pressure. The sensor's range of operation is 0 to 454 psi. At 0 psi, the signal will be about .1 volt, varying up to about 4.9 volts at 454 psi or above. DTC 66 sets if the voltage is above 4.9 volts (454 psi) or below .3 volt (7.5 psi) for 5 seconds or more. The A/C compressor clutch is disabled by the ECM if fault is present.

Test Description: Numbers(s) below refer to circled number(s) on the diagnostic chart.

1. This step checks the voltage signal being received by the ECM from the A/C pressure sensor.
2. Checks to see if the high voltage signal is from a shorted sensor or a short to voltage in the circuit. Normally, disconnecting the sensor would make a normal circuit go to near zero volts.
3. Checks to see if low voltage signal is from the sensor or the circuit. Jumpering the sensor signal CKT 380 to 5 volts, checks the circuit, connections, and ECM.
4. This step checks to see if the low voltage signal was due to an open in the sensor circuit of the 5 volt reference circuit since the prior step eliminated the pressure sensor.
5. The 5 volt reference for the Throttle Position (TP) sensor connects to the same 5 volt power supply in the ECM. If the 5 volt reference circuit to the TP sensor has a short to ground, the MAP sensor and A/C refrigerant pressure sensor 5 volt reference circuit will also have a short to ground because they connect to the same power supply in the ECM.

A short to ground on either 5 volt reference circuit does not damage the ECM 5 volt power supply. When the short is repaired, the ECM 5 volt supply will return to normal.

Diagnostic Aids:

At temperatures of -20 degrees fahrenheit and below, the A/C pressure can drop below the .3 volt (7.5 psi) threshold. If this happens, the DTC 66 will set and disable the A/C system. Clear DTC 66 and check A/C pressure in a heated garage. If pressure is between .3 volt (7.5 psi) and 4.9 volts (454 psi), the A/C refrigerant pressure sensor is functioning correctly.

DTC 66 sets when the signal voltage falls outside the normal possible range of the sensor. If signal voltage is greater than .1 volt, but less than .3 volt, there could be a low refrigerant pressure problem with the A/C system.

2.2L (VIN 4) ENGINE — DIAGNOSTIC TROUBLE CODE CHART — LUMINA

DTC 66
A/C REFRIGERANT PRESSURE SENSOR CIRCUIT
2.2L (VIN 4) "W" CARLINE

IMPORTANT: TO ENSURE PROPER DIAGNOSIS OF THE DTC 66 CIRCUIT, THE A/C REFRIGERANT PRESSURE SHOULD BE WITHIN LIMITS

(1)
- KEY "ON," ENGINE NOT RUNNING.
- INSTALL SCAN TOOL AND OBSERVE VOLTAGE FOR A/C REFRIGERANT PRESSURE SENSOR.

ABOVE 4.9 VOLTS.

(2)
- DISCONNECT A/C REFRIGERANT PRESSURE SENSOR ELECTRICAL CONNECTOR. DOES SCAN TOOL DISPLAY LESS THAN 1 VOLT?

NO → CHECK FOR SHORT TO VOLTAGE IN SENSOR GROUND CIRCUIT. IF NOT SHORTED, REPLACE ECM.*

YES → CHECK FOR OPEN IN SENSOR GROUND CIRCUIT. IF NOT OPEN, CHECK FOR POOR SENSOR TERMINAL CONNECTIONS. IF OK, REPLACE A/C REFRIGERANT PRESSURE SENSOR.

BELOW .3 VOLT

(3)
- DISCONNECT A/C REFRIGERANT PRESSURE SENSOR CONNECTOR.
- JUMPER TERMINALS "C" AND "B". DOES SCAN TOOL DISPLAY ABOVE 4.6 VOLTS?

NO → (4)
- REMOVE JUMPER.
- CONNECT VOLTMETER FROM TERMINAL "A" TO "B". IS VOLTAGE ABOUT 5 VOLTS?

NO → (5) CHECK FOR OPEN/SHORT TO GROUND IN CKT 474. IF OK, REPLACE ECM.*

YES → CHECK FOR OPEN IN CKT 732. IF OK, CHECK FOR POOR CONNECTION AT ECM TERMINAL "C13". IF OK, REPLACE ECM.*

YES → CHECK SENSOR TERMINAL CONNECTIONS. IF OK, REPLACE A/C REFRIGERANT PRESSURE SENSOR.

BETWEEN .3 VOLT AND 4.9 VOLTS.

FAULT IS NOT PRESENT AT THIS TIME. SEE "DIAGNOSTIC AIDS."

* IF ECM IS FAULTY AND MUST BE REPLACED, THE NEW ECM MUST BE PROGRAMMED. REFER TO ECM REPLACEMENT AND PROGRAMMING PROCEDURES

2.2L (VIN 4) ENGINE — ECM SYMPTOM CHART — LUMINA

1993 "W" CARLINE 22 PIN ECM CONNECTOR 2.2L (VIN 4) RPO: LN2
* USE T-100 YELLOW BREAKOUT BOX (BOB)

ECM PIN	FUNCTION	BOB PIN #	WIRE COLOR	CKT #	COMPONENT CONNECTOR CAVITY	NORMAL VOLTAGE KEY "ON"	NORMAL VOLTAGE ENG "RUN"	DTC(S) AFFECTED	POSSIBLE SYMPTOMS FROM FAULTY CIRCUIT
C1	PEAK AND HOLD (INJECTOR JUMPER)	304	TAN	887	ECM "C5"	0(6)	0(6)		(9) CRANKS, BUT WON'T RUN.
C2		301							
C3	IDLE AIR CONTROL COIL "A" HIGH	312	LT BLU/WHT	441	IAC VALVE "D"	NOT USABLE	NOT USABLE		(11) INCORRECT, UNSTABLE IDLE.
C4	IDLE AIR CONTROL COIL "B" LOW	309	LT GRN/BLK	444	IAC VALVE "A"	NOT USABLE	NOT USABLE		(11) INCORRECT, UNSTABLE IDLE.
C5	PEAK AND HOLD (INJECTOR JUMPER)	319	TAN	887	ECM "C1"	0(6)	0(6)		(9) CRANKS, BUT WON'T RUN.
C6	ECM GROUND	317	BLK/WHT	450	PNP SWITCH "B" COOLING FAN #2 RELAY "2"	0(6) (14)	0(6) (14)		(9) NONE. IF OTHER GROUNDS ARE OPEN, CRANKS, BUT WON'T RUN.
C7	INJECTOR DRIVER	303	DK BLU	467	INJECTOR MAIN HARNESS "B"	B+	B+ (13)		(11) CRANKS, BUT WON'T RUN.
C8	FUEL PUMP RELAY DRIVER	302	DK GRN/WHT	465	FUEL PUMP RELAY "1"	0(7)	B+		(9) EXTENDED CRANK BEFORE ENGINE STARTS.
C9	IDLE AIR CONTROL COIL "A" LOW	311	LT BLU/BLK	442	IAC VALVE "C"	NOT USABLE	NOT USABLE		(11) INCORRECT, UNSTABLE IDLE.
C10	IDLE AIR CONTROL COIL "B" HIGH	310	LT GRN/WHT	443	IAC VALVE "B"	NOT USABLE	NOT USABLE		(11) INCORRECT, UNSTABLE IDLE.
C11	ECM GROUND	318	TAN/WHT	551	ENGINE GROUND	0(6) (14)	0(6) (14)		(9) NONE. IF OTHER GROUNDS ARE OPEN, CRANKS, BUT WON'T RUN.

·NOTICE: IF TOOL IS NOT AVAILABLE, DO NOT "BACKPROBE" OR PIERCE INSULATION! This Chart may be used in conjunction with the T-100 Yellow Breakout Box (48921) to obtain voltage present for each circuit listed. Install the BOB between the ECM harness connectors and the ECM, then probe the pin listed under "BOB PIN#". Voltage may vary due to low battery charge or other reasons, but should be very close. All voltages shown in the ENG "RUN" column are typical with engine at idle, closed throttle, normal operating temperature, park or neutral, system in "Closed Loop," all accessories "OFF," brake not applied and scan tool not installed.

DVM NEGATIVE (BLACK) LEAD MUST BE CONNECTED TO A KNOWN GOOD GROUND.

(1) VARIES WITH TEMPERATURE.
(2) VARIES WITH MANIFOLD PRESSURE.
(3) VARIES WITH BAROMETRIC PRESSURE.
(4) VARIES DEPENDING ON POSITION OF DRIVE WHEELS.
(6) LESS THAN 0.5 VOLT.
(7) B+ FOR 2 SECONDS AFTER KEY "ON."
(9) OPEN CIRCUIT.
(10) GROUNDED CIRCUIT.
(11) OPEN OR GROUNDED CIRCUIT.
(12) WITHIN THE RANGE OF .33-1.33 VOLTS.
(13) ALTERNATE TEST: SET J 39200 DVM TO DC VOLTAGE FREQUENCY SCALE. CONNECT RED LEAD TO IGNITION FEED (BOB PIN #401) AND BLACK LEAD TO INJECTOR DRIVER (BOB PIN #303); SHOULD MEASURE ABOUT 9-15 Hz.
(14) THIS TYPE OF CIRCUIT SHOULD ALSO BE CHECKED USING J 34142-B TEST LIGHT.

GREEN ECM CONNECTOR C

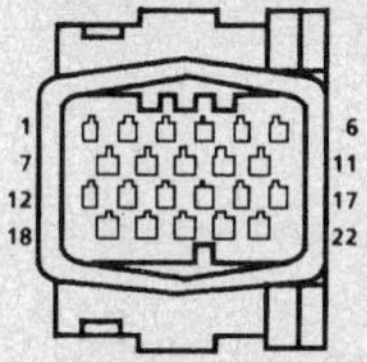

2.2L (VIN 4) ENGINE — ECM SYMPTOM CHART — LUMINA

1993 "W" CARLINE   22 PIN ECM CONNECTOR   2.2L (VIN 4)
*** USE T-100 YELLOW BREAKOUT BOX (BOB)**   RPO: LN2

ECM PIN	FUNCTION	BOB PIN #	WIRE COLOR	CKT #	COMPONENT CONNECTOR CAVITY	NORMAL VOLTAGE KEY "ON"	NORMAL VOLTAGE ENG "RUN"	DTC(S) AFFECTED	POSSIBLE SYMPTOMS FROM FAULTY CIRCUIT
C12	ENGINE COOLANT TEMPERATURE (ECT) SIGNAL	308	YEL	410	ECT SENSOR "B"	(1)	(1)	14 (11)	(11) INCORRECT IDLE, COOLING FAN #1 "ON" AT ALL TIMES.
C13	A/C REFRIGERANT PRESSURE SENSOR SIGNAL	305	DK BLU	732	A/C REFRIGERANT PRESSURE SENSOR "C"	VARIES	VARIES	66 (11)	(11) NO A/C COOLING, REFER TO DTC 66.
C14	INTAKE AIR TEMPERATURE (IAT) SIGNAL	316	TAN	472	IAT SENSOR "B"	(1)	(1)	23 (11)	(11) HARD START.
C15	VEHICLE SPEED SENSOR (VSS) SIGNAL (LOW)	313	PPL	401	VSS "B"	4	4	24 (11)	NO CRUISE CONTROL, NO SPEEDOMETER, NO TCC - POOR FUEL ECONOMY.
C16	SENSOR GROUND	322	BLK	802	MAP SENSOR "A" IAT SENSOR "A" A/C REFRIGERANT PRESSURE SENSOR "A"	0 (6)	0 (6)	23 33 66	(9) LACK OF POWER, SLUGGISH.
C17		320							
C18	THROTTLE POSITION (TP) SENSOR SIGNAL	307	DK BLU	417	TP SENSOR "C"	.6V (12)	.6V (12)	21	(11) UNSTABLE IDLE, REDUCED PERFORMANCE.
C19	MANIFOLD ABSOLUTE PRESSURE (MAP) SENSOR SIGNAL	306	LT GRN	432	MAP SENSOR "B"	(3)	(2)	33 (11)	(11) POOR PERFORMANCE, ROUGH IDLE, SURGE. REFER TO CHART C1-D.
C20	DIAGNOSTIC TEST	315	WHT/ BLK	451	DATA LINK CONNECTOR "B"	5.0V	5.0V		(9) NO DATA. (10) KEY "ON" - DTC(S) DISPLAYED. ENG "RUN" - FIELD SERVICE MODE "ON."
C21	VEHICLE SPEED SENSOR (VSS) SIGNAL (HIGH)	314	YEL	400	VSS "A"	4	4	24 (11)	NO CRUISE CONTROL, NO SPEEDOMETER, NO TCC - POOR FUEL ECONOMY.
C22	SENSOR GROUND	321	BLK	808	ECT SENSOR "A" TPS "B"	0 (6)	0 (6)		(9) INCORRECT IDLE.

•**NOTICE**: IF TOOL IS NOT AVAILABLE, DO NOT "BACKPROBE" OR PIERCE INSULATION!
This Chart may be used in conjunction with the T-100 Yellow Breakout Box (48921) to obtain voltage present for each circuit listed. Install the BOB between the ECM harness connectors and the ECM, then probe the pin listed under "BOB PIN#". Voltage may vary due to low battery charge or other reasons, but should be very close. All voltages shown in the ENG "RUN" column are typical with engine at idle, closed throttle, normal operating temperature, park or neutral, system in "Closed Loop," all accessories "OFF," brake not applied and scan tool not installed.

DVM NEGATIVE (BLACK) LEAD MUST BE CONNECTED TO A KNOWN GOOD GROUND.

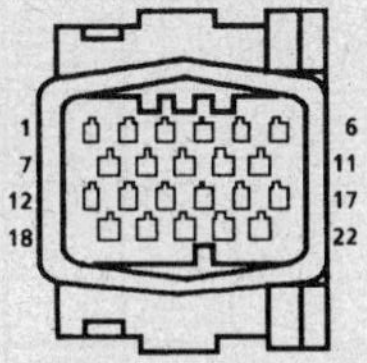

(1) VARIES WITH TEMPERATURE.
(2) VARIES WITH MANIFOLD PRESSURE.
(3) VARIES WITH BAROMETRIC PRESSURE.
(4) VARIES DEPENDING ON POSITION OF DRIVE WHEELS.
(6) LESS THAN 0.5 VOLT.
(7) B + FOR 2 SECONDS AFTER KEY "ON."

(9) OPEN CIRCUIT.
(10) GROUNDED CIRCUIT.
(11) OPEN OR GROUNDED CIRCUIT.
(12) WITHIN THE RANGE OF .33-1.33 VOLTS.

GREEN ECM CONNECTOR C

2.2L (VIN 4) ENGINE — ECM SYMPTOM CHART — LUMINA

1993 "W" CARLINE   22 PIN ECM CONNECTOR   2.2L (VIN 4)
*** USE T-100 YELLOW BREAKOUT BOX (BOB)**   RPO: LN2

ECM PIN	FUNCTION	BOB PIN #	WIRE COLOR	CKT #	COMPONENT CONNECTOR CAVITY	NORMAL VOLTAGE KEY "ON"	NORMAL VOLTAGE ENG "RUN"	DTC(S) AFFECTED	POSSIBLE SYMPTOMS FROM FAULTY CIRCUIT
D1	PARK/NEUTRAL POSITION (PNP) SWITCH INPUT	404	ORN/ BLK	434	PARK/NEUTRAL POSITION (PNP) SWITCH "A"	(6)	(6)		(11) INCORRECT IDLE.
D2	IGNITION FEED	401	PNK/ BLK	439	BULKHEAD CONNECTOR "C2"	B+ (14)	B+ (14)		(9) CRANKS, BUT WON'T RUN. (10) CRANKS, BUT WON'T RUN, BLOWN 15 AMP ECM FUSE.
D3	5 VOLT REFERENCE (TP SENSOR)	412	GRY	416	TP SENSOR "A"	5.0V	5.0V		(9) ROUGH IDLE, STALLS.
D4	KEEP ALIVE MEMORY (B+)	409	ORN	480	FUEL/ECM FUSE, FUEL PUMP RELAY "S" AND OIL PRESSURE SWITCH "C"	B+ (14)	B+ (14)		(9) CRANKS, BUT WON'T RUN. (10) CRANKS, BUT WON'T RUN, BLOWN FUEL/ECM FUSE.
D5	A/C COMPRESSOR CLUTCH CONTROL	419	DK GRN/ WHT	459	A/C COMP. CLUTCH CONTROL RELAY "2"	B+	B+		
D6	EXHAUST GAS RECIRCU- LATION (EGR) CONTROL	417	GRN	698	EGR SOLENOID "A"	B+	B+	32 (11)	(9) HIGH NOx, SPARK KNOCK. (10) HARD START, COLD HESITATION.
D7		403							
D8	5 VOLT REFERENCE (MAP & A/C REFRIGERANT PRESSURE SENSORS)	402	GRY	474	MAP SENSOR "C" A/C REFRIGERANT PRESSURE SENSOR "B"	5.0V	5.0V	33 66	(11) NO A/C COOLING
D9		411							
D10	TORQUE CONVERTER CLUTCH (TCC) CONTROL	410	TAN/ BLK	422	TCC SOLENOID "D"	(6)	(6)		(11) POOR FUEL ECONOMY (TCC INOPERATIVE). (10) LACK OF POWER AT LOW ROAD SPEEDS (TCC ENGAGES AT LOW VEHICLE SPEED).
D11	COOLING FAN CONTROL	418	DK GRN	335	COOLING FAN #1 CONTROL RELAY "1"	B+	B+		(9) OVERHEATED, MARGINAL A/C PERFORMANCE. (10) COOLING FAN #1 OPERATES WHEN KEY "ON" OR ENGINE "RUN" REGARDLESS OF COOLANT TEMPERATURE.

•**NOTICE**: IF TOOL IS NOT AVAILABLE, DO NOT "BACKPROBE" OR PIERCE INSULATION!
This Chart may be used in conjunction with the T-100 Yellow Breakout Box (48921) to obtain voltage present for each circuit listed. Install the BOB between the ECM harness connectors and the ECM, then probe the pin listed under "BOB PIN#". Voltage may vary due to low battery charge or other reasons, but should be very close. All voltages shown in the ENG "RUN" column are typical with engine at idle, closed throttle, normal operating temperature, park or neutral, system in "Closed Loop," all accessories "OFF," brake not applied and scan tool not installed.

DVM NEGATIVE (BLACK) LEAD MUST BE CONNECTED TO A KNOWN GOOD GROUND.

(6) LESS THAN 0.5 VOLT.

(9) OPEN CIRCUIT.
(10) GROUNDED CIRCUIT.
(11) OPEN OR GROUNDED CIRCUIT.
(14) THIS TYPE OF CIRCUIT SHOULD ALSO BE CHECKED USING J 34142-B TEST LIGHT.

BROWN ECM CONNECTOR D

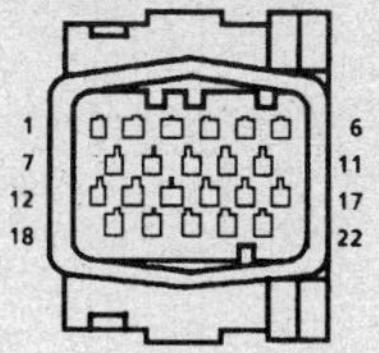

2.2L (VIN 4) ENGINE — ECM SYMPTOM CHART — LUMINA

1993 "W" CARLINE	22 PIN ECM CONNECTOR				NORMAL VOLTAGE			2.2L (VIN 4) RPO: LN2
ECM PIN	FUNCTION	BOB PIN #	WIRE COLOR	CKT #	COMPONENT CONNECTOR CAVITY	KEY "ON"	ENG "RUN"	CODES AFFECT.
* USE T-100 YELLOW BREAKOUT BOX (BOB)

ECM PIN	FUNCTION	BOB PIN #	WIRE COLOR	CKT #	COMPONENT CONNECTOR CAVITY	KEY "ON"	ENG "RUN"	CODES AFFECT.	POSSIBLE SYMPTOMS FROM FAULTY CIRCUIT
D12		408							
D13	IGNITION BYPASS	405	TAN/BLK	424	ELECTRONIC IGNITION MODULE "D"				
D14	IGNITION LOW REFERENCE	416	BLK/RED	453	ELECTRONIC IGNITION MODULE "F"	(6)	(6)		
D15	OXYGEN SENSOR LOW REFERENCE	413	TAN	413	ENGINE GROUND	(6)	(6)	13 (9)	(9) "OPEN LOOP" OPERATION, REDUCED PERFORMANCE.
D16	SERIAL DATA	422	ORN	461	DATA LINK CONNECTOR "M"	4.7V	4.7V		(11) NO DATA.
D17	MALFUNCTION INDICATOR LAMP (MIL) "SERVICE ENGINE SOON" CONTROL	420	BRN/WHT	419	BASE CLUSTER "C10" OR GAUGES CLUSTER "D16"	(6)	B+		(9) LAMP NOT "ON" FOR "BULB CHECK," WON'T FLASH DTC'S. (10) LAMP "ON" ALL THE TIME. REFER TO "DIAGNOSTIC CIRCUIT CHECK."
D18	IGNITION CONTROL (IC)	407	WHT	423	ELECTRONIC IGNITION MODULE 4 PIN CONNECTOR "C"	(6)	2.4V	42 (11)	(11) DEGRADED PERFORMANCE.
D19	IGNITION HIGH REFERENCE.	406	PPL/WHT	430	ELECTRONIC IGNITION MODULE - 4 PIN CONNECTOR "E"	(6)	3.2V		(11) CRANKS, BUT WON'T RUN.
D20	A/C REQUEST INPUT	415	LT GRN	66	HVAC CONTROL HEAD - 9 PIN BLACK CONNECTOR "C12"	(6)	(6)		(9) NO A/C COOLING.
D21	OXYGEN SENSOR SIGNAL	414	PPL	412	OXYGEN SENSOR	.01-.55V	.1-.9V (5)	13 44 45	(11) "OPEN LOOP," STRONG EXHAUST ODOR, REDUCED PERFORMANCE.
D22	VEHICLE SPEED OUTPUT SIGNAL	421	DK GRN	389	BASE CLUSTER "D13" OR GAUGES CLUSTER "D5"				(11) INCORRECT SPEEDOMETER READING, CRUISE CONTROL INOPERATIVE.

***NOTICE:** IF TOOL IS NOT AVAILABLE, DO NOT "BACKPROBE" OR PIERCE INSULATION!
This Chart may be used in conjunction with the T-100 Yellow Breakout Box (48921) to obtain voltage present for each circuit listed. Install the BOB between the ECM harness connectors and the ECM, then probe the pin listed under "BOB PIN#." Voltage may vary due to low battery charge or other reasons, but should be very close. All voltages shown in the ENG "RUN" column are typical with engine at idle, closed throttle, normal operating temperature, park or neutral, system in "Closed Loop," all accessories "OFF," brake not applied and scan tool not installed.

DVM NEGATIVE (BLACK) LEAD MUST BE CONNECTED TO A KNOWN GOOD GROUND.

BROWN ECM CONNECTOR D

(5) Varies actively within indicated range.
(6) Less than 0.5 volt.
(9) Open circuit.
(10) Grounded circuit.
(11) Open or grounded circuit.

2.2L (VIN 4) ENGINE — COMPONENT DIAGNOSTIC CHART — LUMINA

CHART C-1A
PARK/NEUTRAL POSITION (PNP) SWITCH DIAGNOSIS
2.2L (VIN 4) "W" CARLINE

Circuit Description:

The Park/Neutral Position (PNP) switch contacts are a part of the neutral start/backup switch and are closed to ground in park or neutral, and open in drive/reverse ranges.

The ECM supplies ignition voltage through a current limiting resistor to CKT 434 and senses a closed switch, when the voltage on CKT 434 drops to less than one volt.

The ECM uses the PNP signal as one of the inputs to control idle air control and VSS diagnostics.

Test Description: Number(s) below refer to circled number(s) on the diagnostic chart.

1. Checks for a closed switch to ground in park position. Different makes of scan tools will read PNP switch differently. Refer to tool operations manual for type of display used.
2. Checks for an open switch in drive range.
3. Be sure scan tool indicates drive, even while wiggling shifter to test for an intermittent or misadjusted switch in drive range.

2.2L (VIN 4) ENGINE — COMPONENT DIAGNOSTIC CHART — LUMINA

CHART C-1A

PARK/NEUTRAL POSITION (PNP) SWITCH DIAGNOSIS
2.2L (VIN 4) "W" CARLINE

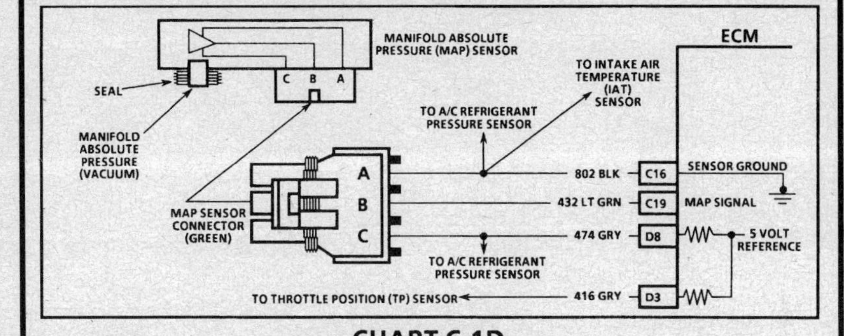

(1) • WITH TRANSAXLE IN PARK, SCAN TOOL SHOULD INDICATE PARK OR NEUTRAL. DOES IT?

YES →

(3) • SHIFT TRANSAXLE INTO DRIVE. • SCAN TOOL SHOULD DISPLAY A CHANGE TO INDICATE DRIVE.

  - **NO** → • DISCONNECT P/N SWITCH. • THIS SHOULD CAUSE SCAN TOOL TO DISPLAY DRIVE RANGE. DOES IT?
    - **YES** → FAULTY PNP SWITCH CONNECTION OR PNP SWITCH MISADJUSTED OR FAULTY PNP SWITCH.
    - **NO** → CKT 434 SHORTED TO GROUND OR FAULTY ECM.*
  - **YES** → NO TROUBLE FOUND.

NO →

(2) • DISCONNECT PARK NEUTRAL POSITION (PNP) SWITCH CONNECTOR. • JUMPER HARNESS CONNECTOR TERMINALS "A" AND "B". • SCAN TOOL SHOULD INDICATE PARK OR NEUTRAL. DOES IT?

  - **NO** → • JUMPER HARNESS CONNECTOR (CKT 434) TO ENGINE GROUND. • SCAN TOOL SHOULD INDICATE PARK OR NEUTRAL. DOES IT?
    - **YES** → OPEN GROUND CIRCUIT.
    - **NO** → CKT 434 OPEN OR FAULTY ECM CONNECTION OR ECM.*
  - **YES** → FAULTY PNP SWITCH CONNECTION OR PNP SWITCH MISADJUSTED OR FAULTY PNP SWITCH.

* IF ECM IS FAULTY AND MUST BE REPLACED, THE NEW ECM MUST BE PROGRAMMED. REFER TO ECM REPLACEMENT AND PROGRAMMING PROCEDURES

"AFTER REPAIRS," CONFIRM "CLOSED LOOP" OPERATION AND NO MIL (SERVICE ENGINE SOON).

2.2L (VIN 4) ENGINE — COMPONENT DIAGNOSTIC CHART — LUMINA

CHART C-1D

MANIFOLD ABSOLUTE PRESSURE (MAP) SENSOR OUTPUT CHECK
2.2L (VIN 4) "W" CARLINE

Circuit Description:

The Manifold Absolute Pressure (MAP) sensor measures the changes in the intake manifold pressure which result from engine load (intake manifold vacuum) and RPM changes; and converts these into a voltage output. The ECM sends a 5 volts reference voltage to the MAP sensor. As the manifold pressure changes, the output voltage of the sensor also changes. By monitoring the sensor output voltage, the ECM knows the manifold pressure. A lower pressure (low voltage) output voltage will be about 1 to 2 volts at idle. While higher pressure (high voltage) output voltage will be about 4 to 4.8 at Wide Open Throttle (WOT). The MAP sensor is also used, under certain conditions, to measure barometric pressure, allowing the ECM to make adjustments for altitude changes. The ECM uses the MAP sensor to control fuel delivery and ignition timing.

Test Description: Number(s) below refer to circled number(s) on the diagnostic chart.

Important
- Be sure to use the same diagnostic test equipment for all measurements.

1. When comparing scan readings to a known good vehicle, it is important to compare vehicles that use a MAP sensor having the same color insert and the same "Hot Stamped" number. See Figures 1 and 2

2. Applying 34 kPa (10" Hg) vacuum to the MAP sensor should result in voltage readings of 1.5 to 2.1 volts less than the voltage in Step 1. Upon applying vacuum to the sensor, the change in voltage should be instantaneous. A slow voltage change indicates a faulty sensor.

3. Check vacuum seal to sensor for leaking or restriction.

NOTICE: Make sure electrical connector remains securely fastened.

4. Remove sensor from the intake plenum and twist sensor (by hand only) to check for intermittent connection. Output changes greater than .10 volt indicate a faulty sensor or connection. If OK, replace sensor.

2.2L (VIN 4) ENGINE — COMPONENT DIAGNOSTIC CHART — LUMINA

CHART C-1D

MANIFOLD ABSOLUTE PRESSURE (MAP) SENSOR OUTPUT CHECK
2.2L (VIN 4) "W" CARLINE

IMPORTANT: THIS CHART ONLY APPLIES TO MAP SENSORS HAVING GREEN OR BLACK COLOR KEY INSERT (SEE BELOW).

1.
- IGNITION "ON," ENGINE "OFF."
- SCAN DIAGNOSTIC TROUBLE CODES DTC(S). IF DTC 33 IS PRESENT, USE THAT CHART FIRST.
- SCAN TOOL SHOULD INDICATE A MAP SENSOR VOLTAGE.
- COMPARE THIS READING WITH THE READING OF A KNOWN GOOD VEHICLE. SEE FACING PAGE TEST DESCRIPTION, STEP 1. VOLTAGE READING SHOULD BE WITHIN ± .4 VOLT. IS IT?

YES → 2 / NO → REPLACE MAP SENSOR.

2.
- REMOVE MAP SENSOR AND PLUG VACUUM PORT ON INTAKE MANIFOLD.
- CONNECT A HAND VACUUM PUMP TO MAP SENSOR.
- START ENGINE.
- NOTE MAP SENSOR VOLTAGE.
- APPLY 34 kPa (10" Hg) OF VACUUM AND NOTE VOLTAGE CHANGE. SUBTRACT SECOND READING FROM THE FIRST. VOLTAGE VALUE SHOULD BE GREATER THAN 1.5 VOLTS. IS IT?

YES → 3 / NO → 4

3. NO TROUBLE FOUND. CHECK SENSOR INLET PORT FOR RESTRICTION OR LEAKING SEAL.

4. CHECK MAP SENSOR CONNECTION. IF OK, REPLACE MAP SENSOR.

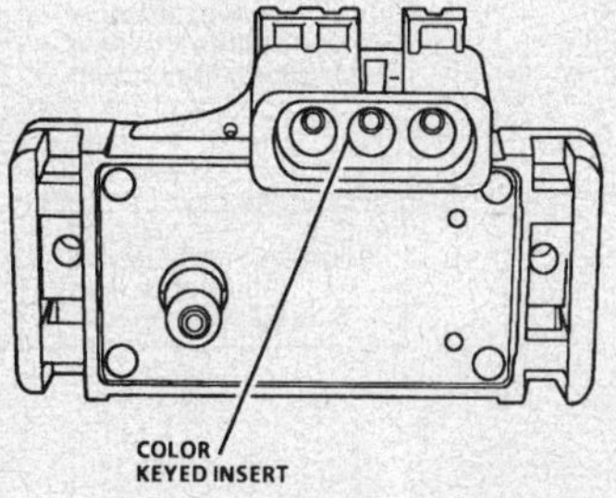

Figure 1 - Color Key Insert

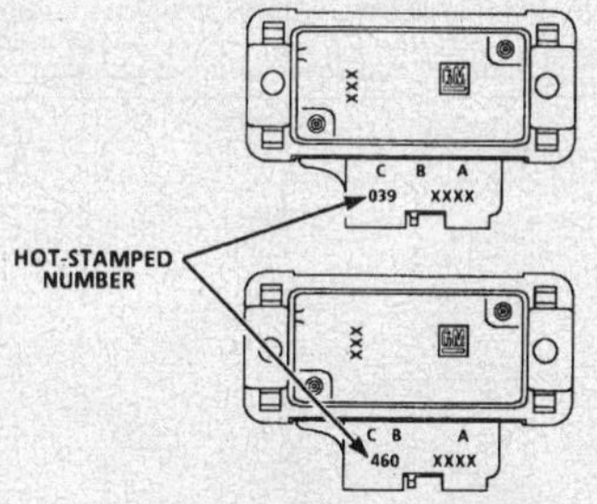

Figure 2 - Hot Stamped Number

2.2L (VIN 4) ENGINE — COMPONENT DIAGNOSTIC CHART — LUMINA

CHART C-2A

INJECTOR BALANCE TEST
2.2L (VIN 4) "W" CARLINE

The injector balance tester is a tool used to turn the injector "ON" for a precise amount of time, thus spraying a measured amount of fuel into the manifold. This causes a drop in fuel rail pressure that we can record and compare between each injector. All injectors should have the same amount of pressure drop (±10 kPa). Any injector with a pressure drop that is 10 kPa (or more) greater or less than the average drop of the other injectors should be considered faulty and replaced.

STEP 1

Engine "cool down" period (10 minutes) is necessary to avoid irregular readings due to "Hot Soak" fuel boiling. Relieve fuel pressure in the fuel rail using the "fuel pressure relief procedure" described previously in this section. With ignition "OFF" connect fuel gage J 29658-D or equivalent. (Refer to CHART A-7 Page 3 of 3 for connecting fuel pressure gage.)

Disconnect electrical connectors at all injectors <u>except</u> #2. Then, connect injector adaptor harness J 34730-200 to the injector jumper harness connector. Now, connect tester J 34730-3 to adaptor harness J 34730-200. Ignition must be "OFF" at least 10 seconds to complete ECM shutdown cycle. Fuel pump should run about 2 seconds after ignition is turned "ON." At this point, insert clear tubing attached to vent valve into a suitable container and bleed air from gauge and hose to insure accurate gauge operation. Repeat this step until all air is bled from gauge.

STEP 2

Turn ignition "OFF" for 10 seconds and then "ON" again to get fuel pressure to its maximum. Record this initial pressure reading. Energize tester one time and note pressure drop at its lowest point. (Disregard any slight pressure increase after drop hits low point.) By subtracting this second pressure reading from the initial pressure, we have the actual amount of injector pressure drop.

STEP 3

Disconnect injector adaptor harness J 34730-200 and reconnect injector jumper harness connector to the engine harness connector. Now connect J 34730-200 directly to each of the remaining injectors. Repeat Step 2 on each injector and compare the amount of drop. Usually, good injectors will have virtually the same drop. Retest any injector that has a pressure difference of 10 kPa, either more or less than the average of the other injectors on the engine. Replace any injector that also fails the retest. If the pressure drop of all injectors is within 10 kPa of this average, the injectors appear to be flowing properly.

NOTICE: The entire test should not be repeated more than once without running the engine to prevent flooding. (This includes any retest on faulty injectors.)

2.2L (VIN 4) ENGINE — COMPONENT DIAGNOSTIC CHART — LUMINA

NOTICE: The entire test should **NOT** be repeated more than once without running the engine to prevent flooding. (This includes any retest on faulty injectors.)

The fuel pressure test in Chart A-7, should be completed prior to this test.

CHART C-2A
INJECTOR BALANCE TEST
2.2L (VIN 4) "W" CARLINE

Step 1. If engine is at operating temperature, allow a 10 minute "cool down" period then connect fuel pressure gage and injector tester.
   1. Ignition "OFF."
   2. Connect fuel pressure gage and injector tester. Refer to Step 1
   3. Ignition "ON."
   4. Bleed off air in gage. Repeat until all air is bled from gauge.

Step 2. Run test:
   1. Ignition "OFF" for 10 seconds.
   2. Ignition "ON". Record gage pressure. (Pressure must hold steady, if not see the Fuel System diagnosis, Chart A-7,
   3. Turn injector on, by depressing button on injector tester, and note pressure at the instant the gauge needle stops.

Step 3. Refer to Step 3

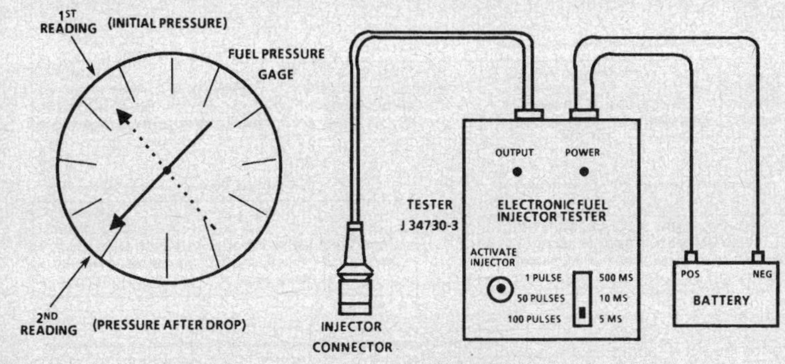

EXAMPLE

CYLINDER	1	2	3
1ST READING	293 kPa (43 psi)	293 kPa (43 psi)	293 kPa (43 psi)
2ND READING	131 kPa (19 psi)	115 kPa (17 psi)	145 kPa (21 psi)
AMOUNT OF DROP	162 kPa (24 psi)	178 kPa (26 psi)	148 kPa (21 psi)
	OK	FAULTY, RICH (TOO MUCH FUEL DROP)	FAULTY, LEAN (TOO LITTLE FUEL DROP)

2.2L (VIN 4) ENGINE — COMPONENT DIAGNOSTIC CHART — LUMINA

CHART C-2C
IDLE AIR CONTROL (IAC) SYSTEM CHECK
2.2L (VIN 4) "W" CARLINE

Circuit Description:

The ECM controls engine idle speed with the IAC valve. To increase idle speed, the ECM retracts the IAC valve pintle away from its seat, allowing more air to bypass the throttle bore. To decrease idle speed, it extends the IAC valve pintle towards its seat, reducing bypass air flow. A Tech 1 scan tool will read the ECM commands to the IAC valve in counts. Higher the counts indicate more air bypass (higher idle). The lower the counts indicate less air is allowed to bypass (lower idle).

Test Description: Number(s) below refer to circled numbers on the diagnostic chart.

1. The Tech 1 RPM control mode is used to extend and retract the IAC valve. The valve should move smoothly within the specified range. If the idle speed commanded (IAC extended) too low (below 700 RPM), the engine may stall. This may be normal and would not indicate a problem. Retracting the IAC beyond its controlled range (above 1500 RPM) will cause a delay before the RPMs start dropping. This too is normal.
2. This test uses the Tech 1 to command the IAC controlled idle speed. The ECM issues commands to obtain commanded idle speed. The node lights each should flash red and green to indicate a good circuit as the ECM issues commands. While the sequence of color is not important if either light is "OFF" or does not flash red and green, check the circuits for faults, beginning with poor terminal contacts.

Diagnostic Aids:

A slow, unstable, or fast idle may be caused by a non-IAC system problem that cannot be overcome by the IAC valve. Out of control range IAC scan tool counts will be above 50 if idle is too low, and zero counts if idle is too high. The following checks should be made to repair a non-IAC system problem:

- Vacuum Leak (High Idle) - If idle is too high, stop the engine. Fully extend (low) IAC with tester. Start engine. If idle speed is above 800 RPM, locate and correct vacuum leak including crankcase ventilation system. Also check for binding of throttle blade or linkage.
- Crankcase Ventilation Valve - If a high idle condition exists (800 to 1000 RPM), check for vacuum leaks and proper crankcase ventilation valve operation. All throttle bodies are preset at the factory and do not need adjustment. A missing crankcase ventilation valve or grommet or stuck valve can cause this condition.
- System Too Lean (High Air/Fuel Ratio) - The idle speed may be too high or too low. Engine speed may vary up and down and disconnecting the IAC valve does not help. DTC 44 may be set. Scan oxygen voltage will be less than 300 mV (.3 volt). Check for low regulated fuel pressure, water in the fuel or a restricted injector.
- System Too Rich (Low Air/Fuel Ratio) - The idle speed will be too low. Scan tool IAC counts will usually be above 50. System is obviously rich and may exhibit black smoke in exhaust. Scan tool oxygen voltage will be fixed above 800 mV (.8 volt).
  Check for high fuel pressure, leaking or sticking injector. Silicone contaminated oxygen sensors scan voltage will be slow to respond.
- Throttle Body - Remove IAC valve and inspect bore for foreign material.
- IAC Valve Electrical Connections - IAC valve connections should be carefully checked for proper contact.

- If intermittent poor driveability or idle symptoms are resolved by disconnecting the IAC, carefully recheck connections, valve terminal resistance, or replace IAC.

2.2L (VIN 4) ENGINE — COMPONENT DIAGNOSTIC CHART — LUMINA

2.2L (VIN 4) ENGINE — COMPONENT DIAGNOSTIC CHART — LUMINA

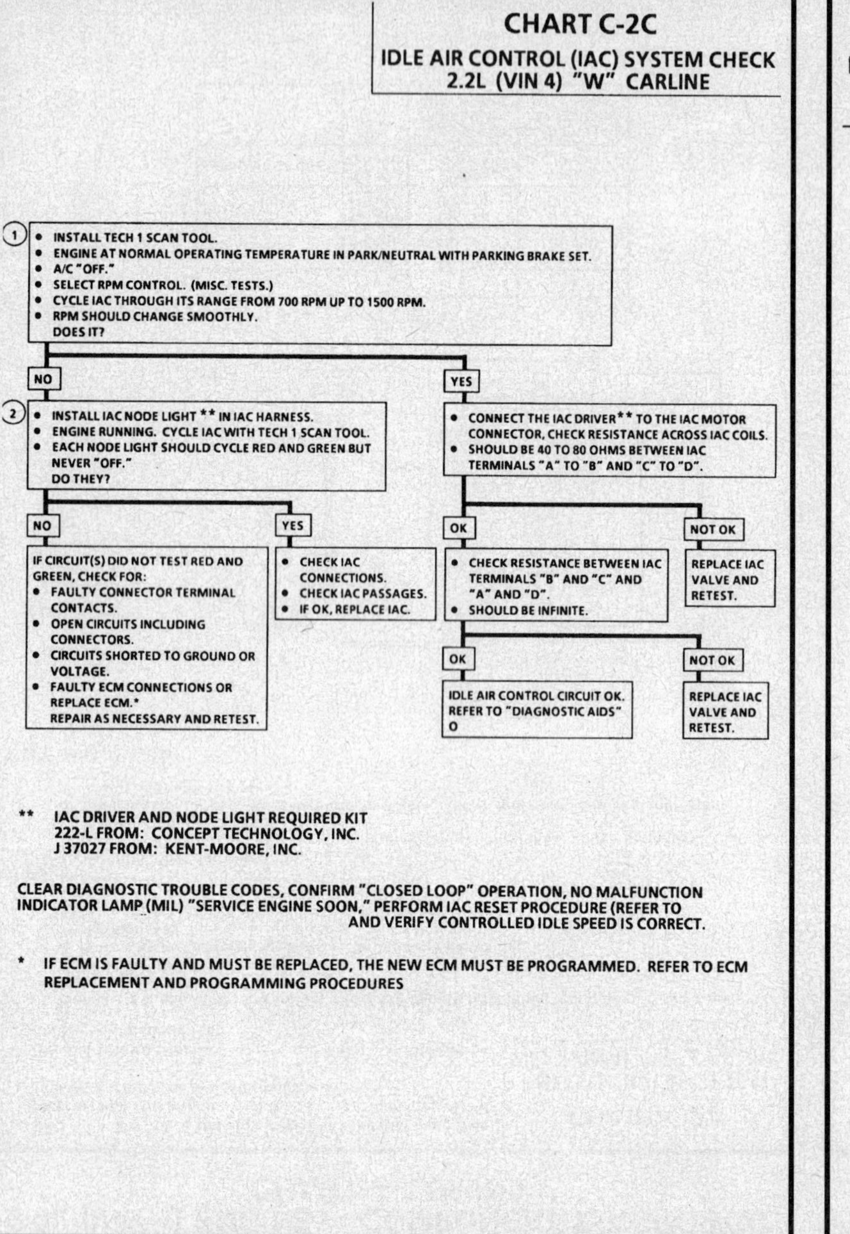

2.2L (VIN 4) ENGINE — COMPONENT DIAGNOSTIC CHART — LUMINA

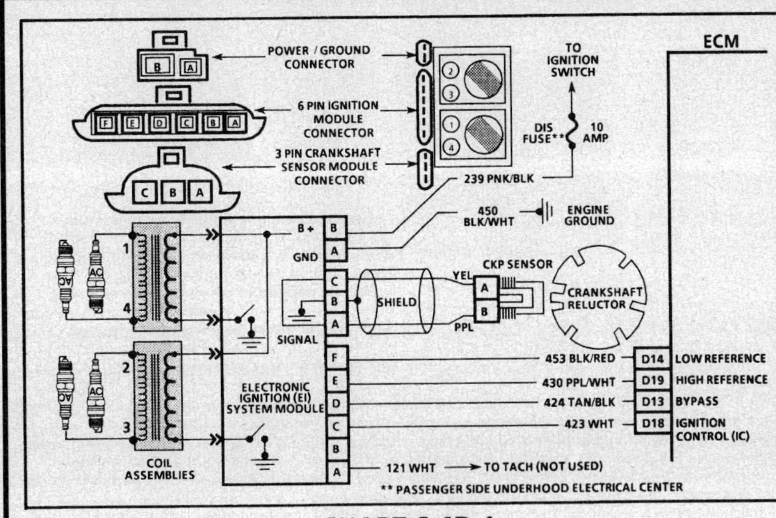

CHART C-4D-1
MISFIRE AT IDLE
2.2L (VIN 4) "W" CARLINE

Circuit Description:

The Electronic Ignition (EI) system uses a waste spark method of distribution. In this type of system, the electronic Ignition Control Module (ICM) triggers the number 1/4 coil pair resulting in both number 1 and number 4 spark plugs firing at the same time. Number 1 cylinder is on the compression stroke at the same time number 4 is on the exhaust stroke, resulting in a lower energy requirement to fire number 4 spark plug. This leaves the remainder of the high voltage to be used to fire number 1 spark plug. The Crankshaft Position (CKP) sensor is remotely mounted beside the ignition coil assembly and protrudes through the block to within approximately .050" of the crankshaft reluctor. Since the reluctor is a machined portion of the crankshaft and the CKP sensor is mounted in a fixed position on the block, timing adjustments are not possible or necessary.

Test Description: Number(s) below refer to circled number(s) on the diagnostic chart.

1. If the "Misfire" complaint exists under load only, CHART C-4D-2 must be used.
   - Keep disconnected spark plug leads away from sensors and other electronic components.
   - Move quickly through this test. Don't leave any spark plug lead disconnected for longer than 15 seconds.
   - Let the engine run normally for 30 seconds between tests to avoid an excessive buildup of fuel.
2. A spark tester, such as a ST-125, must be used because it is essential to verify adequate available secondary voltage at the spark plug (25,000 volts). If the spark jumps the tester gap after grounding the companion plug cable, it indicates excessive resistance in the plug which was bypassed. A faulty or poor connection at that plug could also result in a miss condition. Also, check for carbon deposits inside the spark plug boot.
3. If carbon tracking is evident, replace coil and be sure plug cables relating to that coil are clean and tight. Excessive wire resistance or faulty connections could have caused the coil to be damaged.
4. By checking the secondary resistance, a coil with an open secondary may be located.
5. If the no spark condition follows the suspected coil, that coil is faulty. Otherwise, the Ignition Control Module (ICM) is the cause of no spark. This test could also be performed by substituting a known good coil for the one causing the no spark condition.

2.2L (VIN 4) ENGINE — COMPONENT DIAGNOSTIC CHART — LUMINA

CHART C-4D-1
MISFIRE AT IDLE
2.2L (VIN 4) "W" CARLINE

★**CAUTION:** WHEN HANDLING SECONDARY SPARK PLUG LEADS WITH ENGINE RUNNING, J 21350 OR EQUIVALENT INSULATED PLIERS MUST BE USED AND CARE EXERCISED TO PREVENT A POSSIBLE ELECTRICAL SHOCK.

1
- IF ENGINE MISFIRES UNDER LOAD ONLY. REFER TO CHART C-4D-2.
- ENGINE IDLING AT NORMAL OPERATING TEMPERATURE, DISCONNECT IAC.
- MOMENTARILY DISCONNECT (ONE AT A TIME) EACH SPARK PLUG LEAD, USING J 21350 OR EQUIVALENT INSULATED PLIERS, WHILE OBSERVING ENGINE RPM FOR A DROP. [DISCONNECTED LEAD SHOULD BE GROUNDED TO AVOID OVER-STRESSING IN THE IGNITION COIL(S).] SEE CAUTION ★.
- EACH DISCONNECTED PLUG LEAD SHOULD HAVE RESULTED IN A FAIRLY EQUAL RPM DROP.

→ YES → REFER TO "ROUGH, UNSTABLE OR INCORRECT IDLE OR STALLING"

↓ NO

2
- WITH IGNITION "OFF," INSTALL SPARK TESTER (ST-125) J 26792 ON PLUG LEAD(S) WHOSE REMOVAL DID NOT RESULT IN RPM DROP (AT PLUG END AND CLIP TO GROUND).
- CONNECT PLUG END OF COMPANION CYLINDER SPARK PLUG CABLE TO GROUND AT ENGINE TO AVOID OVER-STRESSING THE IGNITION COILS.
- DISCONNECT INJECTOR JUMPER HARNESS CONNECTOR TO DISABLE FUEL INJECTORS.
- OBSERVE TESTER WHILE CRANKING ENGINE.
- SPARK SHOULD JUMP TESTER GAP WHILE CRANKING ENGINE. DOES IT?

↓ NO (left) / YES → (right)

Left (NO):
- CHECK THE RESISTANCE OF EACH SPARK PLUG CABLE WHICH DID NOT FIRE THE SPARK TESTER.
- CABLE RESISTANCE SHOULD BE LESS THAN 30K OHMS EACH AND CABLES SHOULD NOT BE GROUNDED. ARE CABLES OK?

Right (YES):
CHECK FOR:
- FAULTY, WORN OR CRACKED SPARK PLUG(S).
- PLUG FOULING OR OVERHEATING DUE TO ENGINE MECHANICAL OR FUEL SYSTEM FAULT.
- PERFORM INJECTOR BALANCE TEST, CHART C-2A.

YES (left) ↓ / NO (right):

3 (left, YES)
- REMOVE THE IGNITION COIL FROM IGNITION CONTROL MODULE WHICH DID NOT FIRE SPARK TESTER.
- INSPECT COIL, SPARK PLUG CABLES AND CABLE BOOTS. THEY SHOULD BE FREE OF CARBON TRACKING. ARE THEY?

Right (NO): REPLACE FAULTY CABLE(S) AND RETEST.

YES (left) ↓ / NO (right):

4 (left, YES)
- CHECK SECONDARY COIL RESISTANCE. SHOULD BE 5-8K OHMS RESISTANCE. IS IT?

Right (NO): REPLACE FAULTY COMPONENT.

YES (left) ↓ / NO (right):

5 (left, YES)
- INSTALL A KNOWN GOOD IGNITION COIL.
- SPARK SHOULD JUMP TESTER GAP AT PROBLEM CYLINDER WITH ENGINE IDLING. DID IT?

Right (NO): REPLACE IGNITION COIL. ALSO, CHECK FOR FAULTY SPARK PLUG CABLE CONNECTION(S) AND CABLE BOOT(S) FOR CARBON TRACKING.

YES (left) ↓ / NO (right):

Left (YES): ORIGINAL IGNITION COIL IS FAULTY.

Right (NO): REPLACE ELECTRONIC IGNITION CONTROL MODULE.

2.2L (VIN 4) ENGINE — COMPONENT DIAGNOSTIC CHART — LUMINA

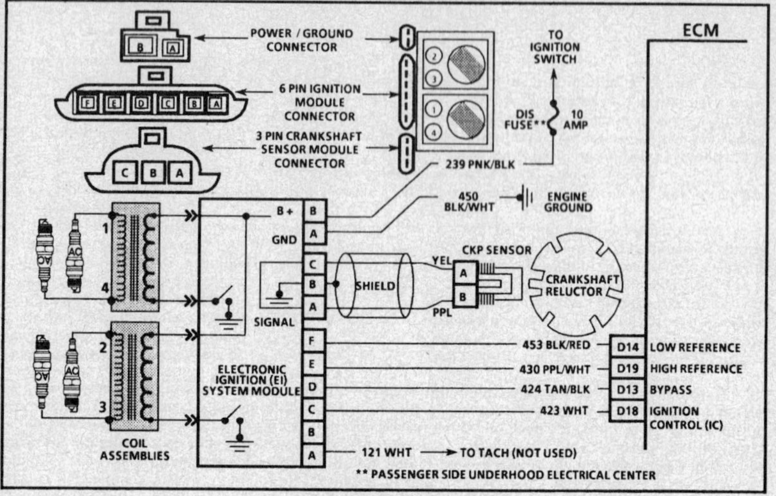

CHART C-4D-2
MISFIRE UNDER LOAD
2.2L (VIN 4) "W" CARLINE

Circuit Description:

The Electronic Ignition (EI) system uses a waste spark method of distribution. In this type of system, the ignition module triggers the number 1/4 coil pair resulting in both number 1 and number 4 spark plugs firing at the same time. Number 1 cylinder is on the compression stroke at the same time number 4 is on the exhaust stroke, resulting in a lower energy requirement to fire number 4 spark plug. This leaves the remainder of the high voltage to be used to fire number 1 spark plug. The Crankshaft Position (CKP) sensor is remotely mounted beside the module/coil assembly and protrudes through the block to within approximately .050" of the crankshaft reluctor. Since the reluctor is a machined portion of the crankshaft and the CKP sensor is mounted in a fixed position on the block, timing adjustments are not possible or necessary.

Test Description: Number(s) below refer to circled number(s) on the diagnostic chart.

If the "Misfire" complaint exists <u>at idle only</u>, CHART C-4D-1 must be used.

1. Simulates an engine load. Cranking the engine with the ST-125 installed on any cylinder should produce a crisp, blue spark. A cylinder that produces a spark significantly weaker than the others should be considered as having no spark.

2.2L (VIN 4) ENGINE — COMPONENT DIAGNOSTIC CHART — LUMINA

CHART C-4D-2
MISFIRE UNDER LOAD
2.2L (VIN 4) "W" CARLINE

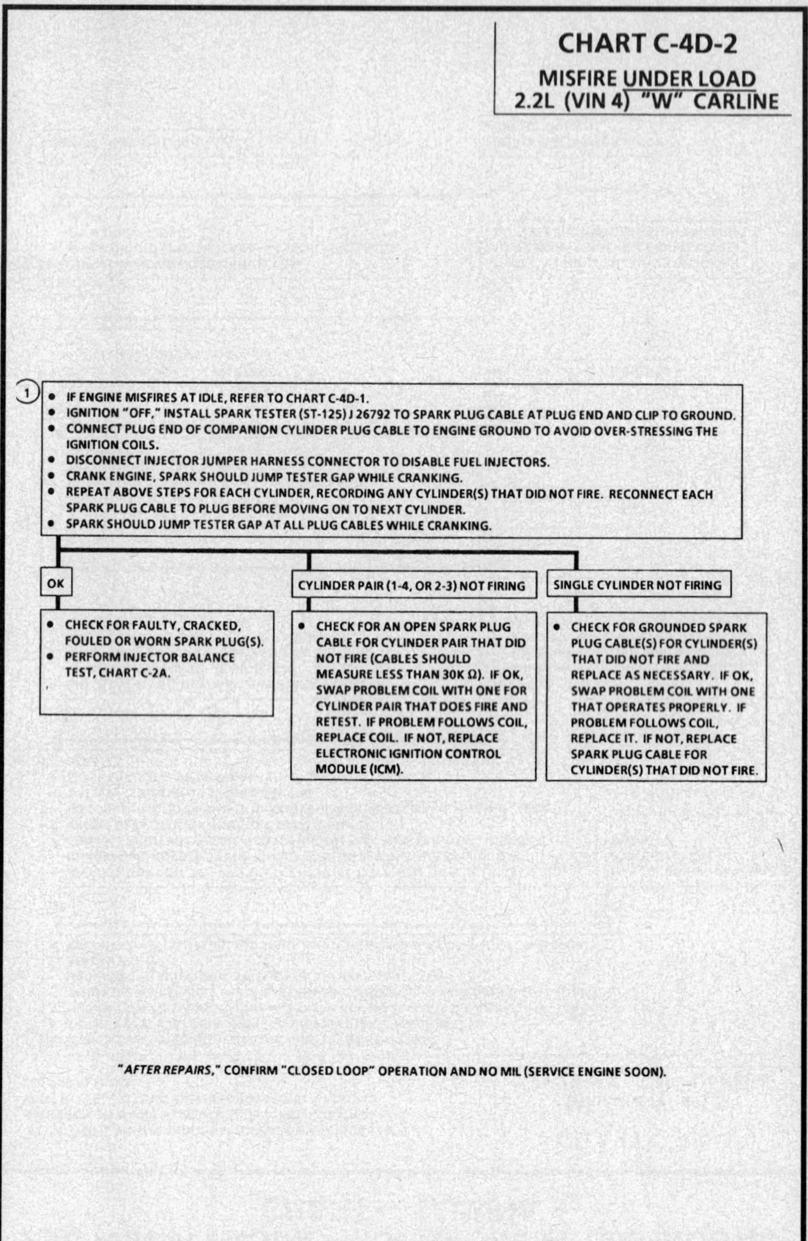

"AFTER REPAIRS," CONFIRM "CLOSED LOOP" OPERATION AND NO MIL (SERVICE ENGINE SOON).

2.2L (VIN 4) ENGINE — COMPONENT DIAGNOSTIC CHART — LUMINA

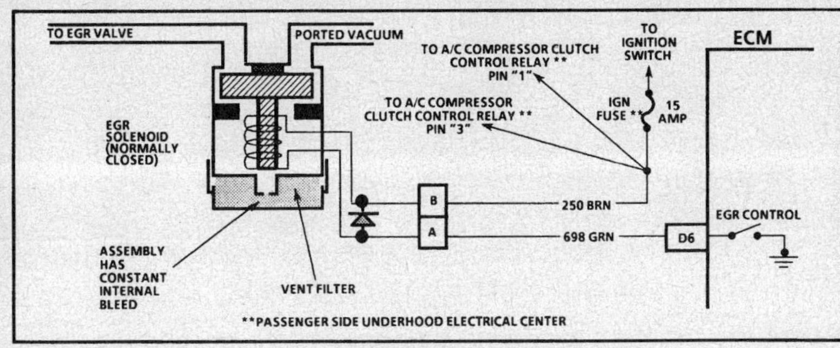

CHART C-7
EXHAUST GAS RECIRCULATION (EGR) FLOW CHECK
2.2L (VIN 4) "W" CARLINE

Circuit Description:

A properly operating EGR system will directly affect the air/fuel mixture requirements of the engine. Since the exhaust gas introduced into the air/fuel mixture cannot be used in combustion (contains very little oxygen), less fuel is required to maintain a correct air/fuel ratio. If the EGR system were to fail in a closed position, the exhaust gas would be replaced with air, and the air/fuel mixture would be leaner. The ECM would compensate for the lean condition by adding fuel, resulting in higher long term fuel trim values.

The fuel control on this engine is conducted within four fuel trim cells. Since EGR is not used at idle, the idle cell would not be affected by EGR system operation. The other fuel trim cells are affected by EGR operation, and, when the EGR system is operating properly, the long term fuel trim values in all cells should be close to the same. If the EGR system becomes inoperative, the long term fuel trim values in the open throttle cells would change to compensate for the resulting lean or rich mixtures, but the long term fuel trim value in the closed throttle cell would not change.

The difference in long term fuel trim values between the idle (closed throttle) cell and cell 2 is used to monitor EGR system performance. When the difference between the two long term fuel trim values is greater than 10, and the long term fuel trim value in cell 2 is greater than 135, Diagnostic Trouble Code (DTC) 32 is set. The system operates in fuel trim cell 2 during a cruise condition at approximately 55 mph.

Test Description: Number(s) below refer to circled number(s) on the diagnostic chart.

1. Diagnostic Trouble Code(s) [DTC(s)] should be diagnosed using appropriate chart before preparing a functional check. If DTC(s) 14, 21, 23, 32, or 33 are set, use those charts first.

   Be sure shop exhaust hose is not connected during Steps 2 and 4.
2. The tailpipe ventilation hose must be removed for this test. The ventilation hose can sometimes cause a good EGR valve to fail this test.
3. Intake Passage: Shut "OFF" engine and remove the EGR valve from the manifold. Plug the exhaust side hole with a suitable stopper. Leaving the intake side hole open, attempt to start the engine. If the engine runs at a high idle (up to 3000 RPM is possible) or starts and stalls, the EGR intake passage is not restricted. If the engine starts and idles normally, the EGR intake passage is restricted.
4. **Because the shop exhaust hose is not installed at this point, don't allow the engine to run longer than 15 seconds.**

   Exhaust Passage: With EGR valve still removed, plug the intake side hole with a suitable stopper. With the exhaust side hole open, check for the presence of exhaust gas. If no exhaust gas is present, the EGR exhaust side passage is restricted.

Diagnostic Aids:

This chart is a functional check of the EGR system. If the EGR system works properly, check other items that result in high long term fuel trim values in fuel trim cell 2, but not in the closed throttle cell.

Incorrect fuel pressure or lean/rich fuel injectors can also cause incorrect long term fuel trim values. Refer to "Fuel Pressure Test" in CHART A-7. It may be necessary to monitor fuel pressure while driving the vehicle at various road speeds and/or loads. If fuel pressure checks out OK, proceed to "Fuel Injector Balance Test" in CHART C-2A.

2.2L (VIN 4) ENGINE — COMPONENT DIAGNOSTIC CHART — LUMINA

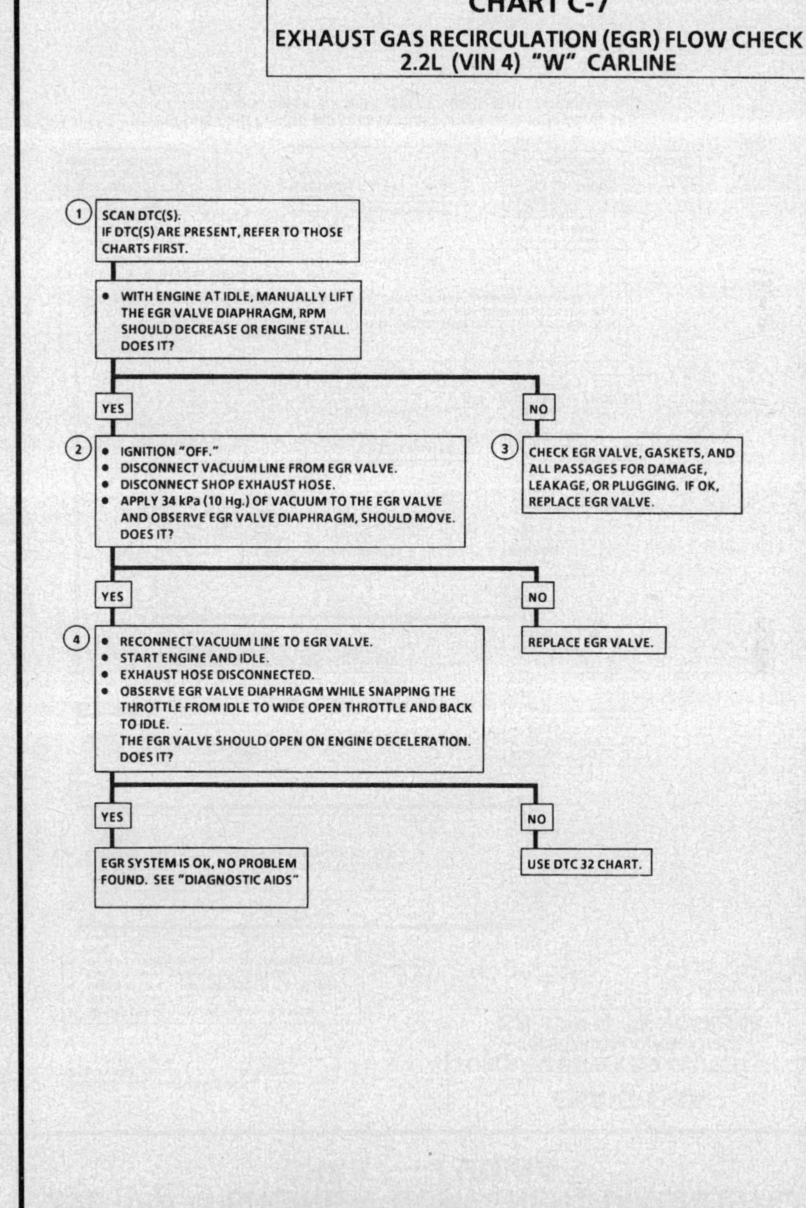

2.2L (VIN 4) ENGINE — COMPONENT DIAGNOSTIC CHART — LUMINA

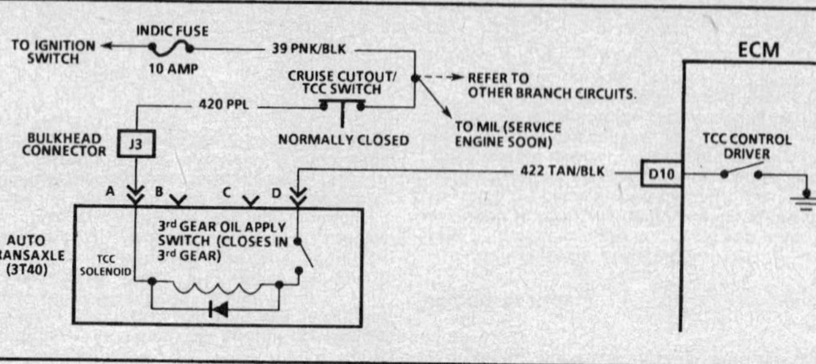

CHART C-8A
TORQUE CONVERTER CLUTCH (TCC)
(ELECTRICAL DIAGNOSIS)
2.2L (VIN 4) "W" CARLINE

Circuit Description:

The purpose of the automatic transaxle Torque Converter Clutch (TCC) feature is to eliminate the power loss of the torque converter stage when the vehicle is in a cruise condition. This allows the convenience of the automatic transaxle and the fuel economy of a manual transaxle.

Test Description: Number(s) below refer to circled number(s) on the diagnostic chart.

1. This test should be performed with the aid of a helper, to operate the Tech 1 and observe the RPM fluctuation.
2. Entering the "Field Service Mode" forces the ECM to close the TCC control driver circuit completing the path for current which should light the test light.
3. Perform this test with the aid of a helper. The 3rd gear switch must be engaged while measuring the TCC solenoid resistance.

Fused battery ignition is supplied to the TCC solenoid through the brake switch. The ECM will engage TCC by grounding CKT 422 to energize the solenoid.

TCC will engage when:

- Vehicle speed above 35 km/h (22 mph).
- Engine at normal operating temperature (above 70°C, 158°F).
- Throttle Position (TP) sensor output not changing, indicating a steady road speed.
- Brake switch closed.

2.2L (VIN 4) ENGINE — COMPONENT DIAGNOSTIC CHART — LUMINA

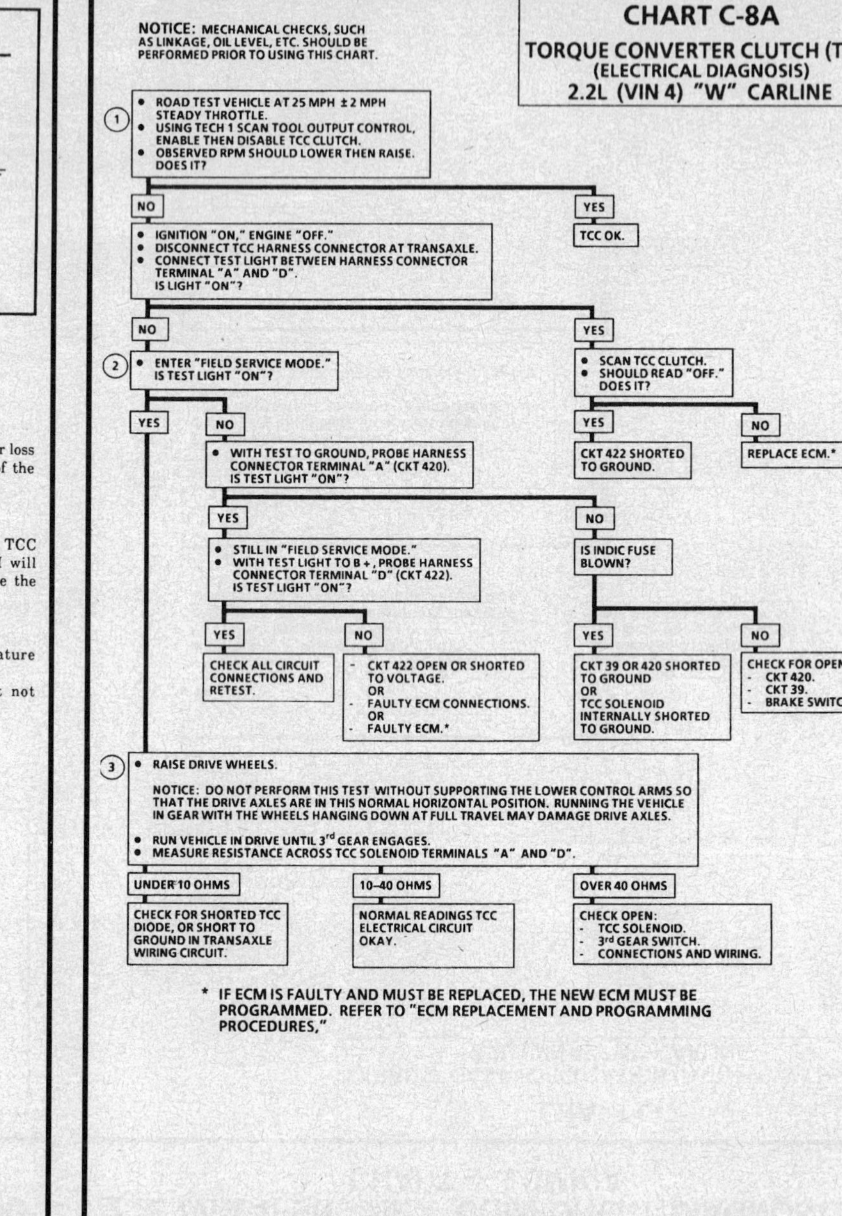

2.2L (VIN 4) ENGINE — COMPONENT DIAGNOSTIC CHART — LUMINA

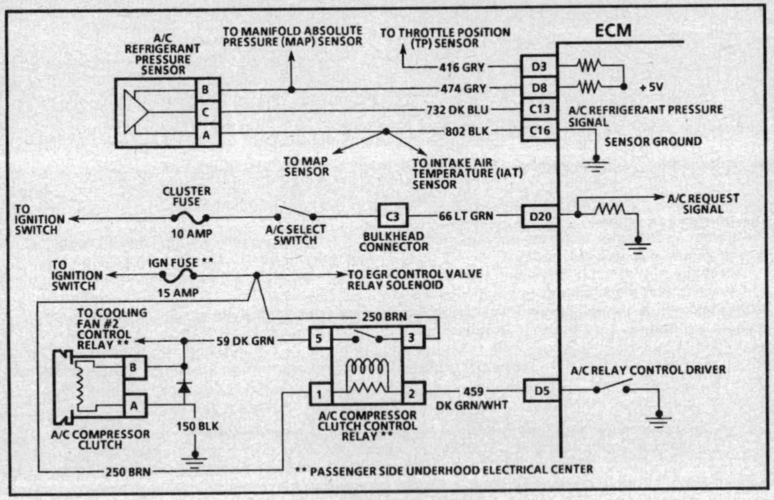

CHART C-10
(Page 1 of 2)
A/C COMPRESSOR CLUTCH CONTROL CIRCUIT DIAGNOSIS
2.2L (VIN 4) "W" CARLINE

Circuit Description:

The A/C compressor clutch control relay is energized when the Engine Control Module (ECM) provides a ground path through CKT 459 and A/C is requested. A/C compressor clutch is delayed about .3 seconds after A/C is requested. This will allow the IAC to adjust engine RPM for the additional load.

The ECM will disable the A/C compressor clutch control relay for any of the following reasons:
- Engine not running.
- Diagnostic Trouble Code (DTC) 66 is set, and fault is current.
- Throttle position is about 99 percent or greater.
- Engine coolant temperature is about 122°C (251°F) or greater.

If the ECM temporarily disabled the A/C compressor clutch, the clutch will be renabled as follows:
- Throttle position drops below about 89 percent.
- Coolant temperature drops below about 119°C (246°F).
- Engine RPM drops below about 3850 RPM.

- During an IAC motor reset.
- Engine RPM is greater than about 4224 RPM.
- A/C refrigerant pressure is about 429 psi or greater.
- A/C refrigerant pressure is about 38 psi or less.

- A/C refrigerant pressure drops below about 199 psi.
- A/C refrigerant pressure rises above about 47 psi.
- IAC motor reset is complete.

Test Description: Number(s) below refer to circled number(s) on the diagnostic chart.
1. The ECM will only energize the A/C relay when the engine is running. This test will determine if the relay or CKT 459 is faulty.
2. Determines if the signal is reaching the ECM through CKT 66 from the A/C control panel. Signal should only be present when an A/C mode or defrost mode has been selected.

Diagnostic Aids:

Be sure to consider branch circuits and splices to other components. If complaint is insufficient cooling, the problem may be caused by an inoperative cooling fan. Refer to CHART C-12 for cooling fan diagnosis. If fan operates correctly, refer to A/C diagnosis

A/C pressure outside of a range of 38 to 429 psi will cause the compressor clutch to be disabled by the ECM. Observe Tech 1 A/C refrigerant pressure for 2 minutes with engine idling and A/C "ON."

Tech 1 pressure should be within 20 psi of actual. If not, check for a circuit problem using Diagnostic Trouble Code (DTC) 66 chart or replace the A/C refrigerant pressure sensor.

2.2L (VIN 4) ENGINE — COMPONENT DIAGNOSTIC CHART — LUMINA

CHART C-10
(Page 1 of 2)
A/C COMPRESSOR CLUTCH CONTROL CIRCUIT DIAGNOSIS
2.2L (VIN 4) "W" CARLINE

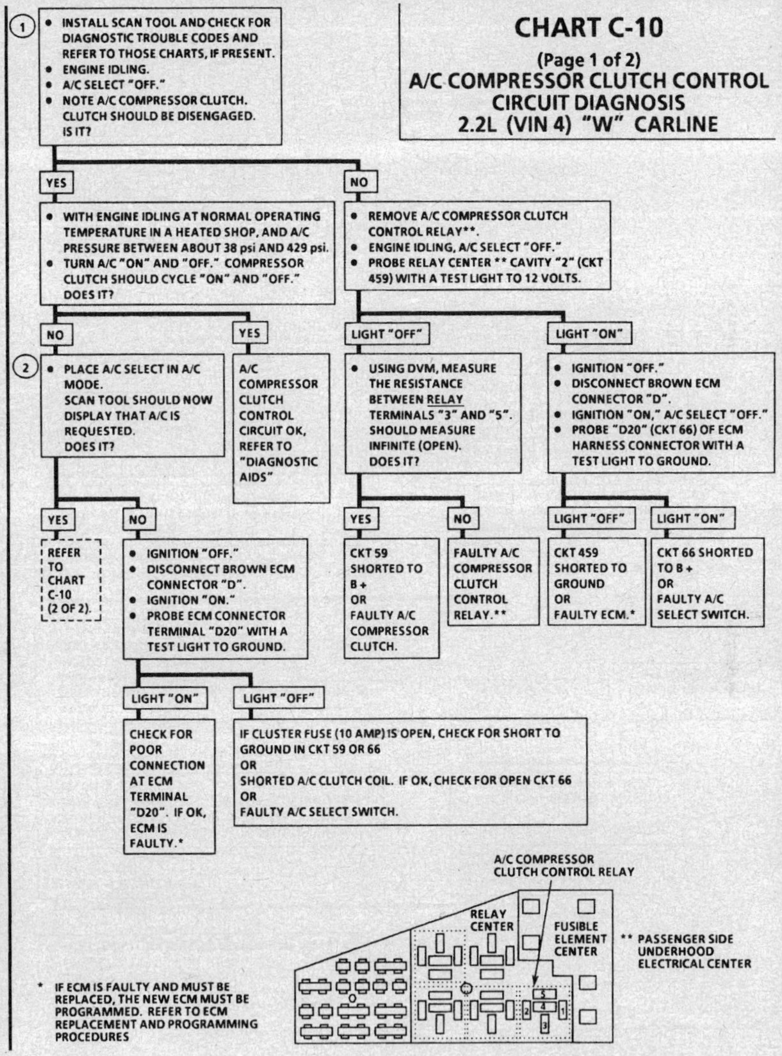

2.2L (VIN 4) ENGINE — COMPONENT DIAGNOSTIC CHART — LUMINA

A/C REFRIGERANT PRESSURE SENSOR
TO MANIFOLD ABSOLUTE PRESSURE (MAP) SENSOR
TO THROTTLE POSITION (TP) SENSOR
ECM
B
C
A
416 GRY
474 GRY
732 DK BLU
802 BLK
D3
D8
C13
C16
+5V
A/C REFRIGERANT PRESSURE SIGNAL
SENSOR GROUND
TO MAP SENSOR
TO INTAKE AIR TEMPERATURE (IAT) SENSOR
TO IGNITION SWITCH
CLUSTER FUSE
10 AMP
A/C SELECT SWITCH
BULKHEAD CONNECTOR
C3
66 LT GRN
D20
A/C REQUEST SIGNAL
TO IGNITION SWITCH
IGN FUSE **
15 AMP
TO EGR CONTROL VALVE RELAY SOLENOID
TO COOLING FAN #2 CONTROL RELAY **
250 BRN
59 DK GRN
5
3
1
2
B
A
A/C COMPRESSOR CLUTCH
150 BLK
A/C COMPRESSOR CLUTCH CONTROL RELAY **
459 DK GRN/WHT
D5
A/C RELAY CONTROL DRIVER
** PASSENGER SIDE UNDERHOOD ELECTRICAL CENTER

CHART C-10
(Page 2 of 2)
**A/C COMPRESSOR CLUTCH CONTROL CIRCUIT DIAGNOSIS
2.2L (VIN 4) "W" CARLINE**

Circuit Description:

The A/C compressor clutch control relay is energized when the ECM provides a ground path through CKT 459 when A/C is requested. A/C compressor clutch is delayed about .3 seconds after A/C is requested. This will allow the IAC to adjust engine RPM for the additional load.

The A/C compressor clutch control relay will remain disengaged if pressure is out of range or there is no A/C request signal due to an open A/C select switch or circuit.

Test Description: Number(s) below refer to circled number(s) on the diagnostic chart.

1. Determines if the A/C refrigerant pressure sensor is out of range causing the compressor clutch to be disengaged.
2. With the engine stopped and "Field Service Mode" activated, the ECM should be grounding CKT 459, which should cause the test light to be "ON."

Diagnostic Aids:

If complaint is insufficient cooling, the problem may be caused by an inoperative engine cooling fan(s). Refer to CHART C-12 for cooling fans diagnosis. If fans operate correctly, refer to A/C diagnosis

A/C pressure outside of a range of 41 to 426 psi will cause the compressor to be disabled by the ECM. Observe Tech 1 A/C pressure for 2 minutes with engine idling and A/C "ON."

Tech 1 pressure should be within 20 psi of actual. If not, check for a circuit problem using Diagnostic Trouble Code (DTC) 66 chart or replace A/C pressure sensor.

2.2L (VIN 4) ENGINE — COMPONENT DIAGNOSTIC CHART — LUMINA

CHART C-10
(Page 2 of 2)
**A/C COMPRESSOR CLUTCH CONTROL CIRCUIT DIAGNOSIS
2.2L (VIN 4) "W" CARLINE**

1
• IGNITION "ON," A/C "ON."
• ENGINE "OFF."
• WITH TECH 1, OBSERVE A/C REFRIGERANT PRESSURE.

PRESSURE IS BETWEEN 38 LBS - 429 LBS

PRESSURE IS LESS THAN 38 LBS OR GREATER THAN 429 LBS

• REMOVE A/C COMPRESSOR CLUTCH CONTROL RELAY**.
• PROBE CKT 250 AT "1" & "3" OF THE RELAY CENTER WITH TEST LIGHT TO GROUND. (SEE ILLUSTRATION BELOW.)

• INSTALL A/C MANIFOLD GAGE SET
ARE TECH 1 PRESSURES WITHIN 20 LBS OF HIGH SIDE PRESSURE?

LIGHT "ON" BOTH

NO LIGHT, ONE OR BOTH

YES

NO

• AT THE RELAY CENTER (SEE ILLUSTRATION BELOW), JUMPER CAVITY "3" (CKT 250) TO CAVITY "5" (CKT S9). DOES A/C COMPRESSOR CLUTCH ENGAGE?

REPAIR OPEN IN CIRCUIT THAT DID NOT LIGHT.

REFER TO A/C DIAGNOSIS.

REPLACE A/C REFRIGERANT PRESSURE SENSOR AND RECHECK.

YES

NO

2
• IGNITION "ON," ENGINE STOPPED. ACTIVATE FIELD SERVICE MODE.
• AT THE RELAY CENTER PROBE CAVITY "2" (CKT 459) WITH A TEST LIGHT TO 12 VOLTS.

CHECK:
- OPEN CKT 59 TO CLUTCH.
- FAULTY COMPRESSOR CLUTCH COIL.
- OPEN GROUND TO COMPRESSOR CLUTCH.

LIGHT "ON"

LIGHT "OFF"

FAULTY A/C COMPRESSOR CLUTCH CONTROL RELAY.

OPEN CKT 459, FAULTY CONNECTION, OR FAULTY ECM.*

* IF ECM IS FAULTY AND MUST BE REPLACED, THE NEW ECM MUST BE PROGRAMMED. REFER TO "ECM REPLACEMENT AND PROGRAMMING PROCEDURES,"

RELAY CENTER
FUSIBLE ELEMENT CENTER
** PASSENGER SIDE UNDERHOOD ELECTRICAL CENTER
A/C COMPRESSOR CLUTCH CONTROL RELAY

2.2L (VIN 4) ENGINE — COMPONENT DIAGNOSTIC CHART — LUMINA

CHART C-12
(Page 1 of 2)
COOLING FAN CONTROL DIAGNOSIS
2.2L (VIN 4) "W" CARLINE

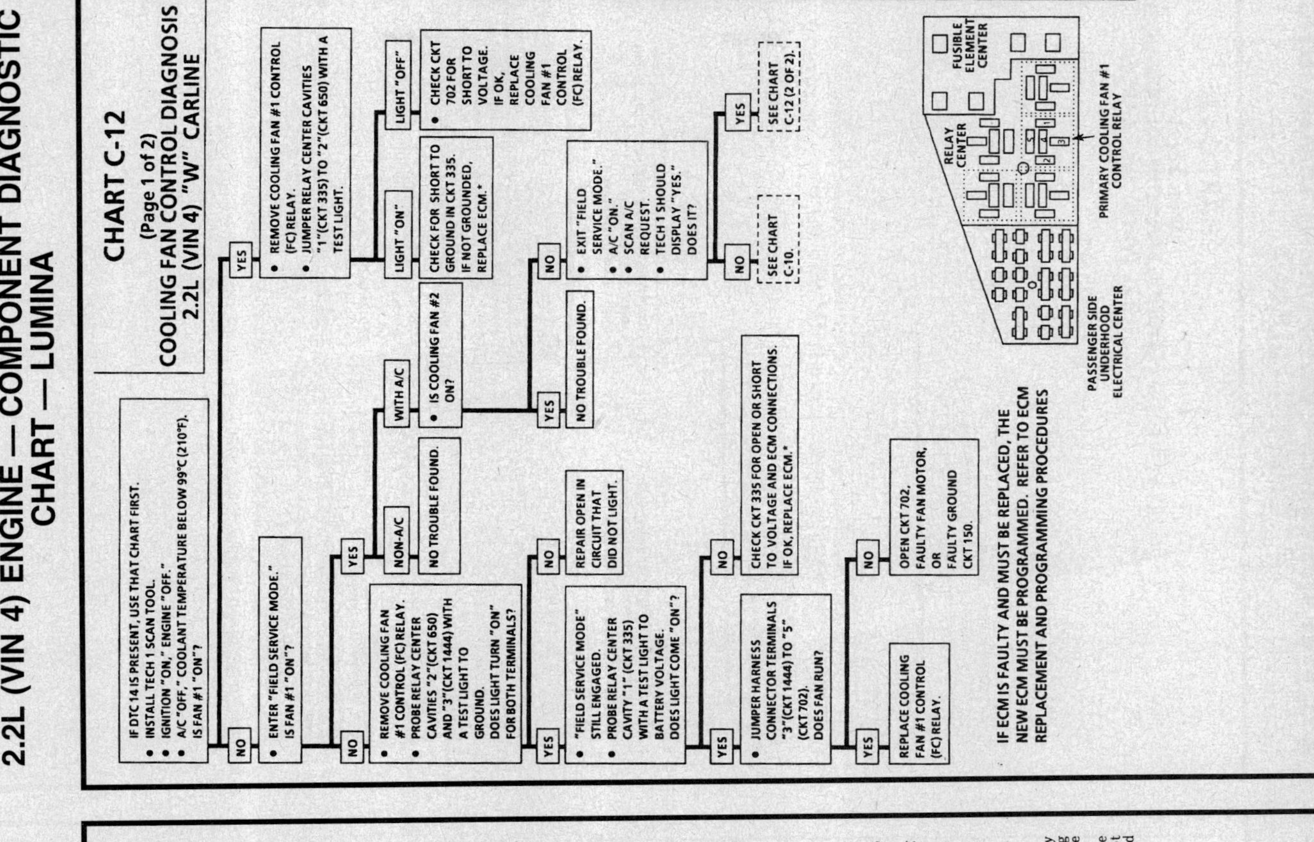

2.2L (VIN 4) ENGINE — COMPONENT DIAGNOSTIC CHART — LUMINA

CHART C-12
(Page 1 of 2)
COOLING FAN CONTROL DIAGNOSIS
2.2L (VIN 4) "W" CARLINE

Circuit Description:

Battery voltage to operate cooling fan #1 motor is supplied to cooling fan #1 control relay by CKT 1444. Ignition voltage is energized and cooling fan #1 control relay is supplied by CKT 650. When the ECM grounds CKT 335, the relay is energized and cooling fan #1 is turned "ON."

When the engine is running, the ECM will turn the engine cooling fan "ON" for any one of the following reasons:

- Diagnostic Trouble Code (DTC) 14 is set.
- Coolant temperature is 106°C (223°F) or greater.
- A/C pressure is 219 psi or greater with A/C selected and a Diagnostic Trouble Code (DTC) 66 is not set.

Once the cooling fan is enabled, the ECM will turn the fan "OFF" when:

- Engine is turned "OFF."
- Coolant temperature drops to 99°C (210°F) or less.
- A/C refrigerant pressure drops to 189 psi or less with A/C selected.

Diagnostic Aids:

If the owner comments about an overheating problem, it must be determined whether the comment was due to an actual boil over, or the hot light, or temperature gage indicated overheating.

If the gage or light indicates overheating, but no boil over is detected, the gage circuit should be checked. The gage accuracy can also be checked by comparing the engine coolant temperature reading using a scan tool and comparing its reading with the gage reading.

If the engine is actually overheating and the gage indicates overheating, but the cooling fan is not coming "ON," the coolant sensor has probably shifted out of calibration and should be replaced.

2.2L (VIN 4) ENGINE — COMPONENT DIAGNOSTIC CHART — LUMINA

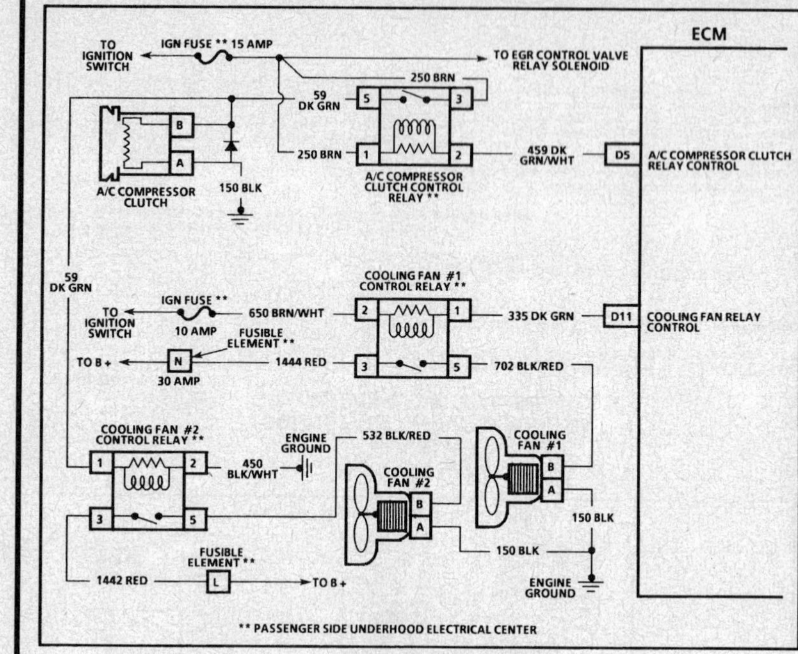

CHART C-12
(Page 2 of 2)
COOLING FAN CONTROL DIAGNOSIS
2.2L (VIN 4) "W" CARLINE

Circuit Description:

Battery voltage to operate cooling fan #2 motor is supplied to cooling fan #2 control relay by CKT 1442. Ignition voltage to energize cooling fan #2 control relay is supplied by CKT 250. CKT 59 is shared by the A/C compressor clutch and cooling fan #2 control relay. When the A/C compressor clutch is energized by selecting A/C or Defrost, the relay is energized and cooling fan #2 is turned "ON."

Diagnostic Aids:

If the owner comments about poor A/C cooling, it must be determined whether a low A/C refrigerant charge is the cause, or DTC 66 is set, or the cooling fan #2 circuit is faulty.

2.2L (VIN 4) ENGINE — COMPONENT DIAGNOSTIC CHART — LUMINA

CHART C-12
(Page 2 of 2)
COOLING FAN CONTROL DIAGNOSIS
2.2L (VIN 4) "W" CARLINE

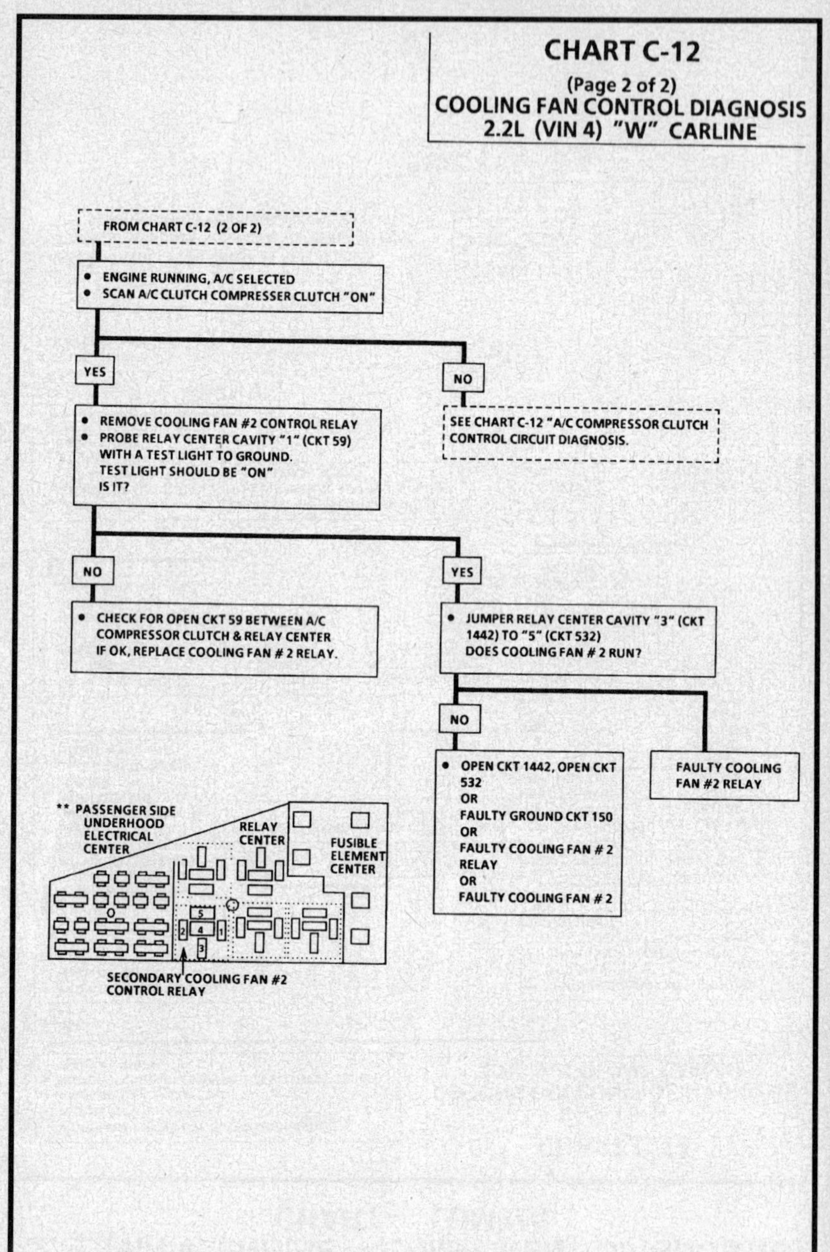

2.3L (VIN A) ENGINE — ENGINE COMPONENT LOCATION CHART — 1992 BERETTA

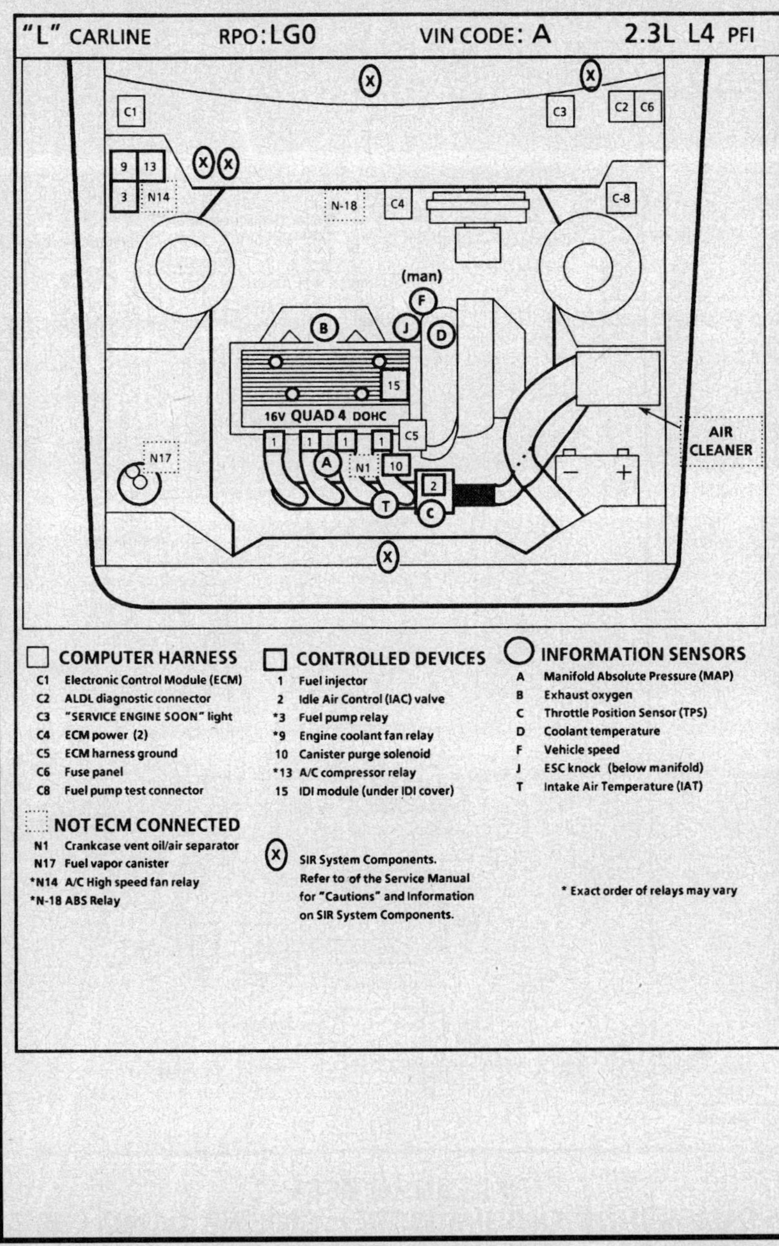

2.3L (VIN A) ENGINE — ECM WIRING SCHEMATIC — 1992 BERETTA

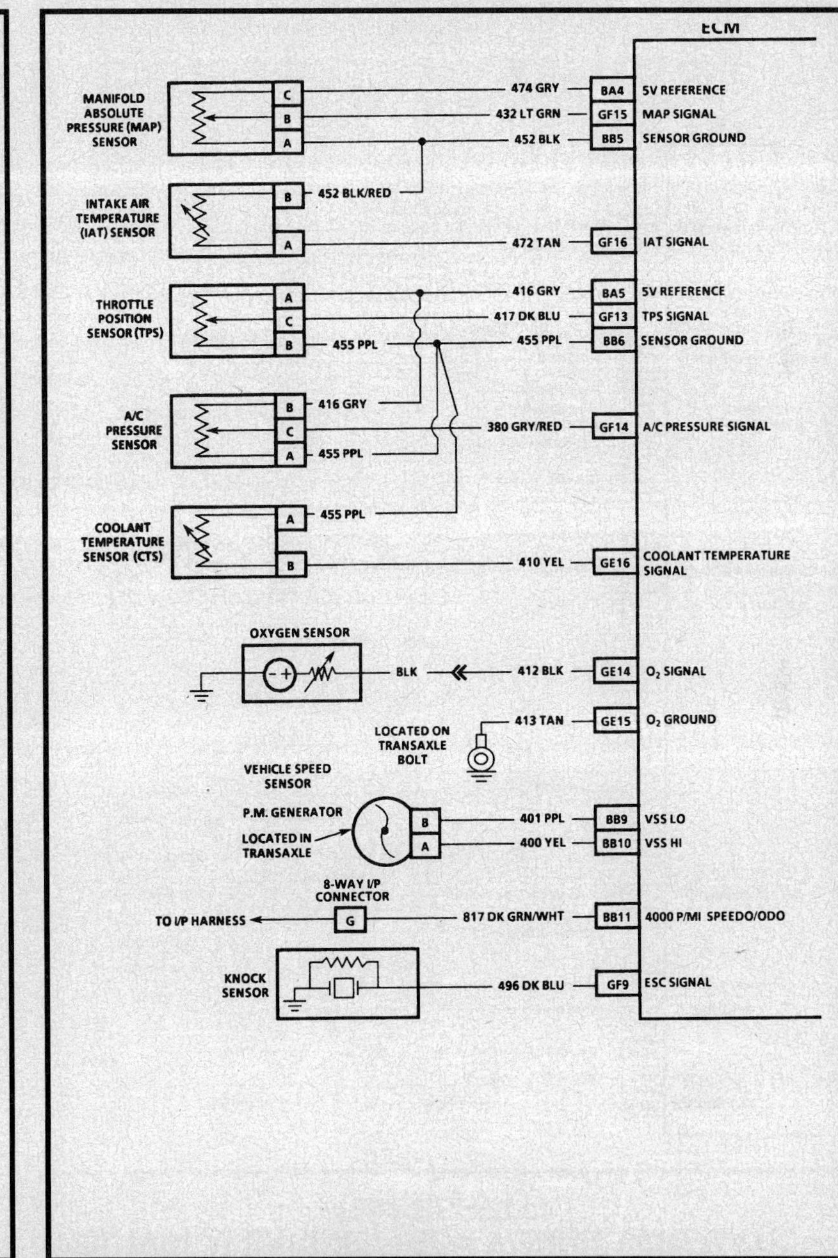

MULTIPORT FUEL INJECTION (MFI) SYSTEMS
EXCEPT LIGHT TRUCKS, VANS, GEO AND SATURN

2.3L (VIN A) ENGINE — ECM WIRING SCHEMATIC — 1992 BERETTA

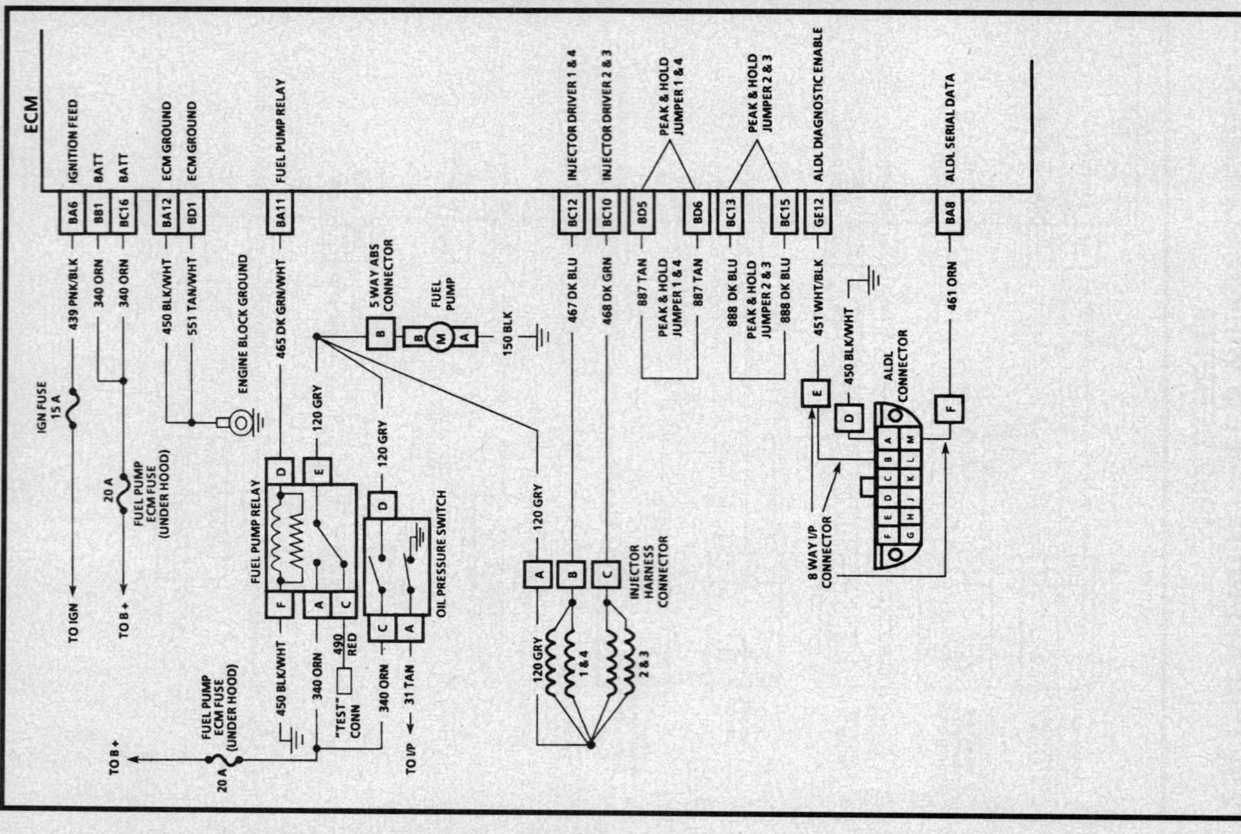

2.3L (VIN A) ENGINE — ECM WIRING SCHEMATIC — 1992 BERETTA

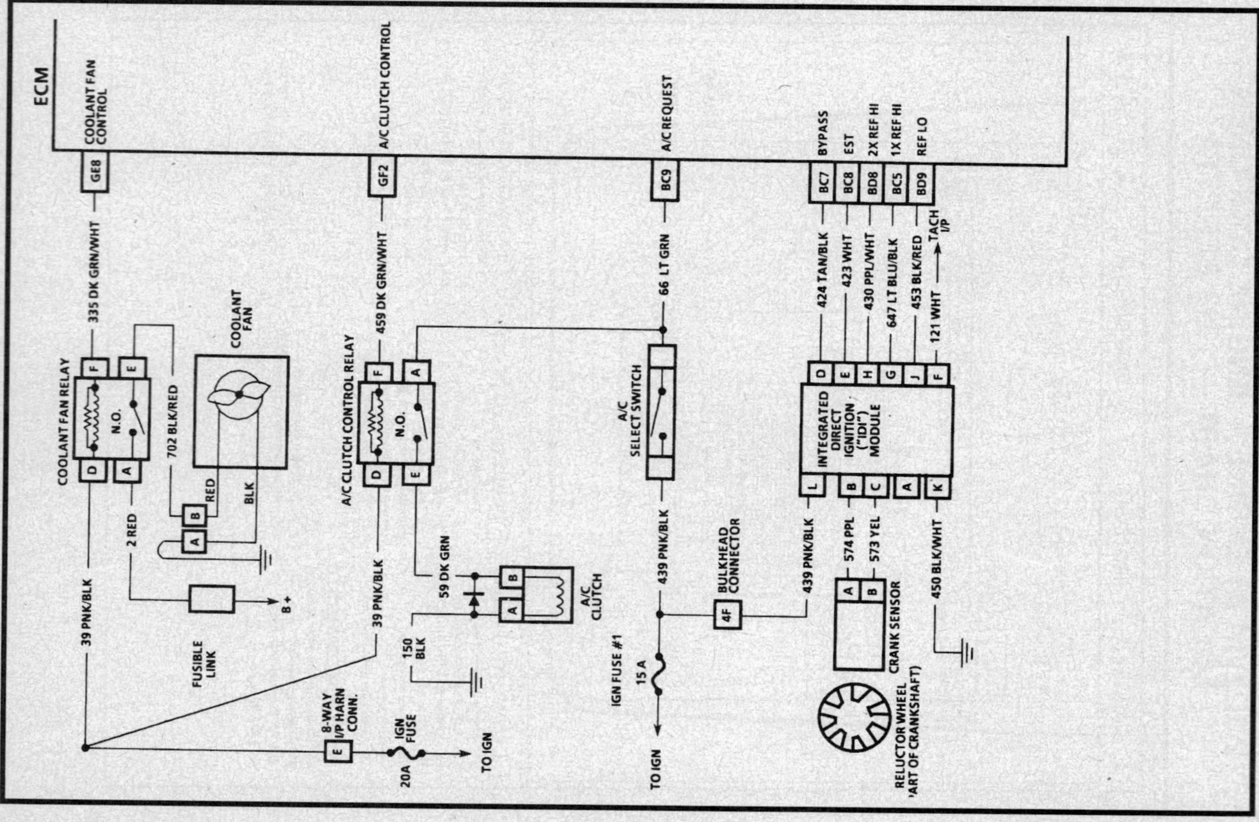

2.3L (VIN A) ENGINE — ECM WIRING SCHEMATIC — 1992 BERETTA

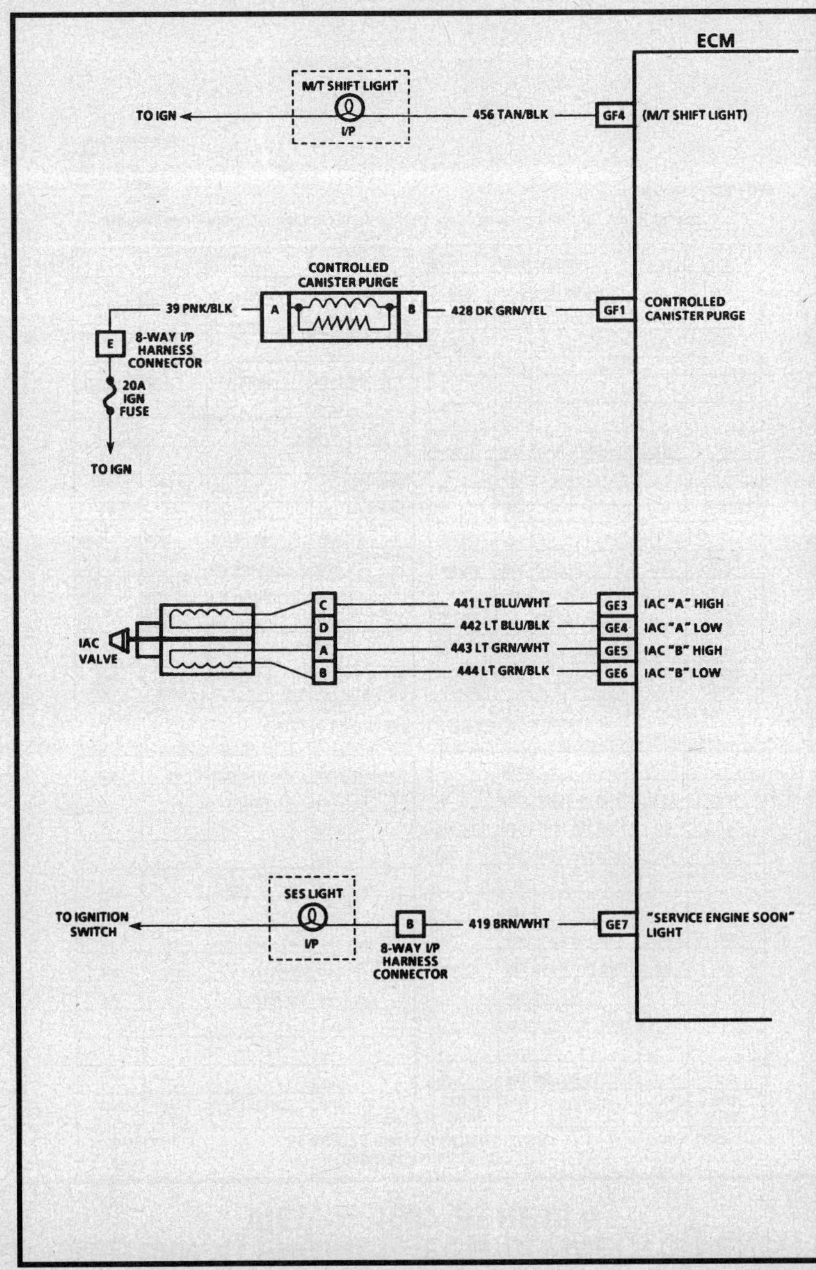

2.3L (VIN A) ENGINE — ECM CONNECTOR END VIEW — 1992 BERETTA

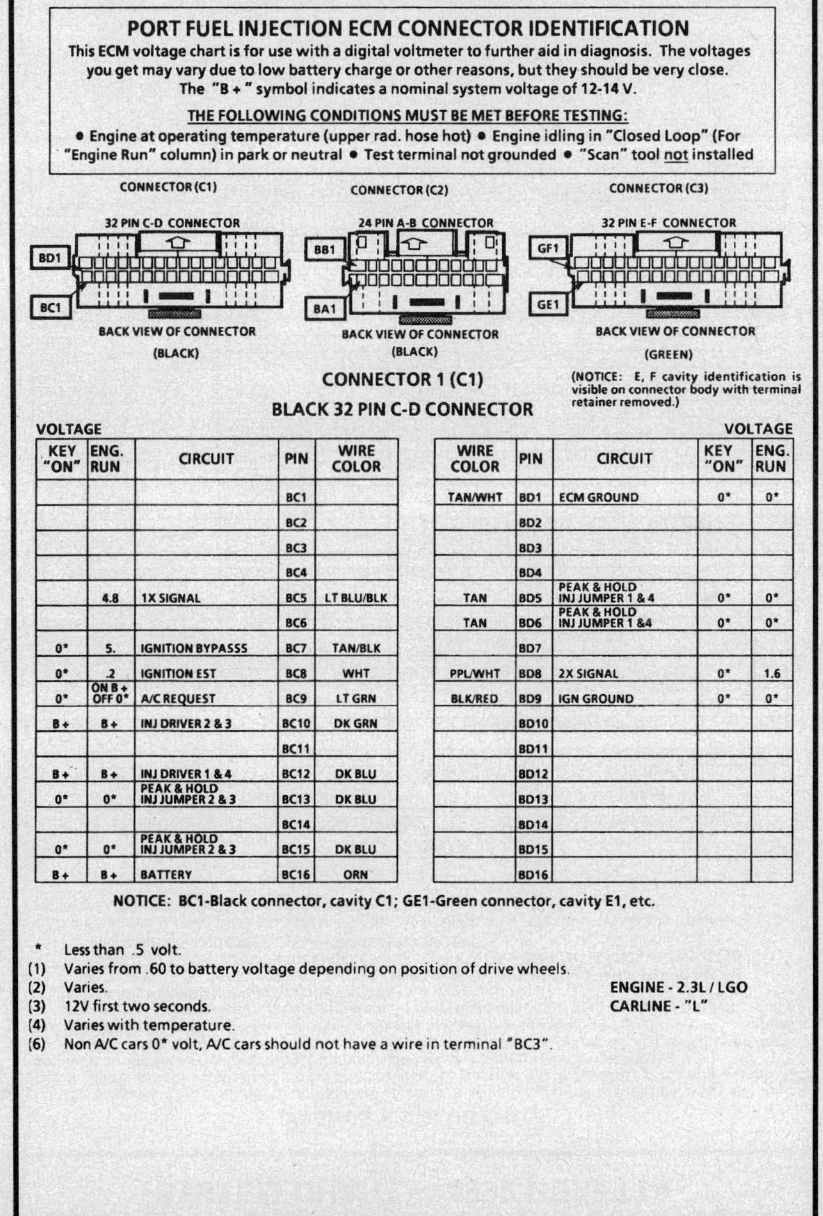

CONNECTOR 1 (C1) — BLACK 32 PIN C-D CONNECTOR

KEY "ON"	ENG. RUN	CIRCUIT	PIN	WIRE COLOR	WIRE COLOR	PIN	CIRCUIT	KEY "ON"	ENG. RUN
			BC1		TAN/WHT	BD1	ECM GROUND	0*	0*
			BC2			BD2			
			BC3			BD3			
			BC4			BD4			
	4.8	1X SIGNAL	BC5	LT BLU/BLK	TAN	BD5	PEAK & HOLD INJ JUMPER 1 & 4	0*	0*
			BC6		TAN	BD6	PEAK & HOLD INJ JUMPER 1 &4	0*	0*
0*	5.	IGNITION BYPASSS	BC7	TAN/BLK		BD7			
0*	.2	IGNITION EST	BC8	WHT	PPL/WHT	BD8	2X SIGNAL	0*	1.6
0*	ON B + OFF 0*	A/C REQUEST	BC9	LT GRN	BLK/RED	BD9	IGN GROUND	0*	0*
B +	B +	INJ DRIVER 2 & 3	BC10	DK GRN		BD10			
			BC11			BD11			
B +	B +	INJ DRIVER 1 & 4	BC12	DK BLU		BD12			
0*	0*	PEAK & HOLD INJ JUMPER 2 & 3	BC13	DK BLU		BD13			
			BC14			BD14			
0*	0*	PEAK & HOLD INJ JUMPER 2 & 3	BC15	DK BLU		BD15			
B +	B +	BATTERY	BC16	ORN		BD16			

NOTICE: BC1-Black connector, cavity C1; GE1-Green connector, cavity E1, etc.

*   Less than .5 volt.
(1)   Varies from .60 to battery voltage depending on position of drive wheels.
(2)   Varies.
(3)   12V first two seconds.
(4)   Varies with temperature.
(6)   Non A/C cars 0* volt, A/C cars should not have a wire in terminal "BC3".

ENGINE - 2.3L / LGO
CARLINE - "L"

2.3L (VIN A) ENGINE — ECM CONNECTOR END VIEW — 1992 BERETTA

CONNECTOR 2 (C-2)
BLACK 24 PIN A-B CONNECTOR

| VOLTAGE | | | | | | | | VOLTAGE | |
KEY "ON"	ENG. RUN	CIRCUIT	PIN	WIRE COLOR	WIRE COLOR	PIN	CIRCUIT	KEY "ON"	ENG. RUN
			BA1		ORN	BB1	BATTERY	B+	B+
			BA2			BB2			
			BA3			BB3			
5.0	5.0	+5V REFERENCE	BA4	GRY		BB4			
5.0	5.0	+5V REFERENCE	BA5	GRY	BLK	BB5	IAT & MAP GND	0*	0*
B+	B+	IGNITION FEED	BA6	PNK/BLK	PPL	BB6	A/C. CTS.TPS GND	0*	0*
			BA7			BB7			
		SERIAL DATA/ALDL ②	BA8	ORN		BB8			
			BA9		PPL	BB9	MAG. VSS LOW	0*	0*
			BA10		YEL	BB10	MAG. VSS HIGH	0*	0*
0* ③	B+	FUEL PUMP	BA11	DK GRN/WHT	DK GRN/WHT	BB11	4000 P/MI SPEED	4.85	5.3
0*	0*	ECM GROUND	BA12	BLK/WHT		BB12			

GREEN 32 PIN E-F CONNECTOR 3 (C-3)

| VOLTAGE | | | | | | | | VOLTAGE | |
KEY "ON"	ENG. RUN	CIRCUIT	PIN	WIRE COLOR	WIRE COLOR	PIN	CIRCUIT	KEY "ON"	ENG. RUN
			GE1		DK GRN/YEL	GF1	CANISTER PURGE	B+	.3
			GE2		DK GRN/WHT	GF2	A/C CLUTCH RELAY	B+	OFF B+ ON 0*
NOT USABLE	NOT USABLE	IAC-A-HIGH	GE3	LT BLU/WHT		GF3			
NOT USABLE	NOT USABLE	IAC-A-LOW	GE4	LT BLU/BLK	TAN/BLK	GF4	SHIFT LT	B+	OFF B+ ON 0*
NOT USABLE	NOT USABLE	IAC-B-HIGH	GE5	LT GRN/WHT		GF5			
NOT USABLE	NOT USABLE	IAC-B-LOW	GE6	LK GRN/BLK		GF6			
0*	B+	SES LIGHT	GE7	BRN/WHT		GF7			
B+	ON 0* OFF B+	CLG FAN RLY	GE8	DK GRN/WHT		GF8			
			GE9		DK BLU	GF9	KNOCK SIGNAL	2.3	2.3
			GE10			GF10			
			GE11			GF11			
5.0	5.0	ALDL/DIAG TERM	GE12	WHT/BLK		GF12			
			GE13		DK BLU	GF13	TPS SIGNAL	.54	.54
.3-.5 ②	.1-.9	O₂ SIGNAL	GE14	BLK	GRY/RED	GF14	A/C PRESS SIGNAL	1.0	1.0 ②
0*	0*	O₂ GROUND	GE15	TAN	LT GRN	GF15	MAP SIGNAL	4.7	1.4 ②
B+ ②	B+	CLNT TEMP SIGNAL	GE16	YEL	TAN	GF16	IAT SIGNAL	2.33	1.5 ②

Notice: BA1 = Black Connector, cavity A1, etc. GE1 - Green Connector, cavity E1, etc.

* Less than .5 volt.
1. Varies from .60 to battery voltage depending on position of drive wheels.
2. Varies.
3. B+ first two seconds.

ENGINE - 2.3L / LG0
CARLINE - "L" Series

2.3L (VIN A) ENGINE — ON-BOARD DIAGNOSTIC SYSTEM CHART — 1992 BERETTA

DIAGNOSTIC CIRCUIT CHECK

The Diagnostic Circuit Check is an organized approach to identifying a problem created by an electronic engine control system malfunction. It must be the starting point for any driveability complaint diagnosis, because it directs the service technician to the next logical step in diagnosing the complaint.

The Tech 1 data listed in the table may be used for comparison, after completing the diagnostic circuit check and finding the on-board diagnostics functioning properly and no trouble codes displayed. The "Typical Values" are an average of display values recorded from normally operating vehicles and are intended to represent what a normally functioning system would typically display.

A "SCAN" TOOL THAT DISPLAYS FAULTY DATA SHOULD NOT BE USED, AND THE PROBLEM SHOULD BE REPORTED TO THE MANUFACTURER. THE USE OF A FAULTY "SCAN" CAN RESULT IN MISDIAGNOSIS AND UNNECESSARY PARTS REPLACEMENT.

Only the parameters listed below are used in this manual for diagnosing. If a "Scan" reads other parameters, the values are not recommended by General Motors for use in diagnosing.

TECH 1 DATA
Idle / Upper Radiator Hose Hot / Closed Throttle / Park or Neutral / "Closed Loop" / Acc. "OFF"

"SCAN" Position	Units Displayed	Typical Data Value
Engine Speed	RPM	± 100 RPM from desired RPM (± 50 in drive)
Desired Idle	RPM	ECM idle command (varies with calibration. temp.)
Coolant Temp.	C° F°	85° – 115°C
IAT Temp.	C° F°	10° – 80°C (depends on underhood temp.)
MAP	kPa, V	1 – 3 Volts (depends on Vacuum & Baro pressure)
BARO	kPa, V	3–5 Volts (depends on altitude & Baro pressure)
Throttle Position	Volts	.400 – .900 (up to 5.0 at wide open throttle)
Throttle Angle	0 – 100%	0% (up to 100% at wide open throttle)
Oxygen Sensor	mV	1–1000 and varying
Injector Pulse Width	m Sec.	1 – 4 and varying
Spark Advance	# of Degrees	Varies
Lo Oct. Spark Ret.	# of Degrees	Varies
Fuel Integrator	Counts	Varies
Block Learn	Counts	58 – 198 (see Section "C2")
Open/Closed Loop	Open/Closed	Closed Loop (may go open with extended idle)
Block Learn Cell	Cell Number	18 to 21 at idle (depends on Air Flow, RPM,P/N&A/C)
Knock Retard	Degrees of Retard	0
Knock Signal	Yes/No	No
BYP Line Volts	LOW/HI	HI
EST Command	Yes/No	Yes
Idle Air Control	Counts (steps)	5 – 60
Park Neutral Switch	Park Neutral and RDL	P - N -- (or -R-DL manual only)
VSS	MPH/KPH	0
Torque Conv. Cl. (TCC)	On/Off	Off ("ON," with TCC commanded)
Battery Voltage	Volts	13.5 – 14.5
1X Ref. Pulse	0 – 255 Counts	0 – 255 (Useable for Code 41)
2X Ref. Pulse	0 – 255 Counts	0 – 255 (Useable for Code 16)
A/C Request	Yes/No	No (yes, with A/C requested, ie: selector "ON")
A/C Clutch	On/Off	Off ("ON," with A/C commanded on)
A/C Clutch	On/Off	Off ("ON," with A/C commanded on)
A/C Pressure	psi/Volts	0 – 450 psi (varying with high side pressure)
Cool Fan Relay	On/Off	Off ("ON," with A/C "ON" or hot eng)
Purge Duty Cycle	%	0–100%
QDM A	LOW/HI	Low
QDM B	LOW/HI	Low
2nd Gear	Yes/No (N/A L40 Only)	No (yes, when in 2nd or 3rd gear) or yes (Man. Trans. only)
3rd Gear	Yes/No (N/A L40 Only)	No (yes, when in 3rd gear) or yes (Man. Trans. only)
PROM ID	#	Production ECM/PROM ID (not useable)
Time From Start	min/sec	Varies (engine run time since start)

2.3L (VIN A) ENGINE — ON-BOARD DIAGNOSTIC SYSTEM CHART — 1992 BERETTA

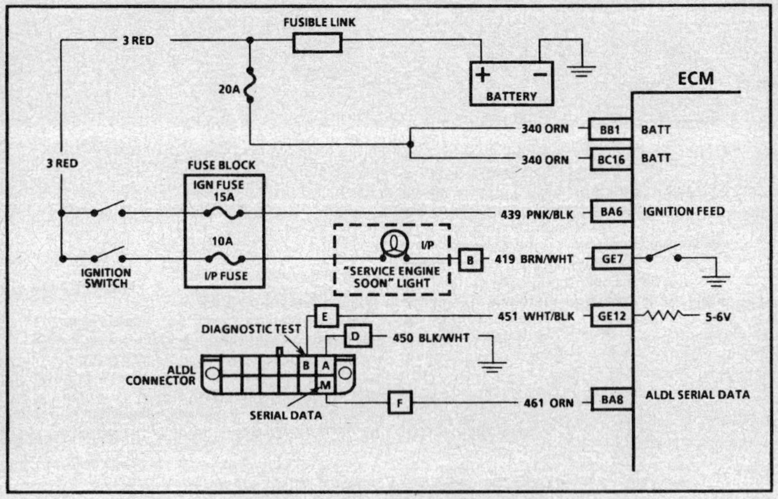

DIAGNOSTIC CIRCUIT CHECK
2.3L (VIN A) "L" CARLINE (PORT)

Circuit Description:

The diagnostic circuit check is an organized approach to identifying a problem created by an Electronic Engine Control System malfunction. It must be the starting point for any driveability complaint diagnosis, because it directs the service technician to the next logical step in diagnosing the complaint. Understanding the chart and using it correctly will reduce diagnostic time and prevent the unnecessary replacement of good parts.

Test Description: Number(s) below refer to circled number(s) on the diagnostic chart.

1. This step is a check for the proper operation of the "Service Engine Soon" light. The "SES" light should be "ON" steady.
2. No "SES" light at this point indicates that there is a problem with the "SES" light circuit or the ECM control of that circuit.
3. This test checks the ability of the ECM to control the "SES" light. With the diagnostic terminal grounded, the "SES" light should flash a Code 12 three times, followed by any trouble code stored in memory.
4. Most of the   procedures use a Tech 1 to aid diagnosis, therefore, serial data must be available. If a PROM error is present, the ECM may have been able to flash Code 12/51, but not enable serial data.
5. Although the ECM is powered up, a "Cranks But Will Not Run" symptom could exist because of an ECM or system problem.

6. This step will isolate if the customer complaint is a "SES" light or a driveability problem with no "SES" light. An invalid code may be the result of a faulty "Scan" tool.
7. Comparison of actual control system data with the typical values is a quick check to determine if any parameter is not within limits. Keep in mind that a base engine problem (i.e., advanced cam timing) may substantially alter sensor values.
8. Installation of a "Scan" tool will provide a good ground path for the ECM and may hide a driveability complaint due to poor ECM grounds.
9. If the actual data is not within the typical values established, the charts   "Component Systems," will provide a functional check of the suspect component or system.

2.3L (VIN A) ENGINE — ON-BOARD DIAGNOSTIC SYSTEM CHART — 1992 BERETTA

DIAGNOSTIC CIRCUIT CHECK
2.3L (VIN A) "L" CARLINE (PORT)

- IGNITION "ON," ENGINE "OFF."
- NOTE "SERVICE ENGINE SOON" LIGHT.

(1) STEADY LIGHT   (2) NO LIGHT → USE CHART A-1   FLASHING CODE 12 → CHECK FOR GROUNDED DIAGNOSTIC TEST CKT 451. USE WIRING DIAGRAM ON CHART A-1.

(3)
- USING TECH 1 PERFORM "DIAGNOSTIC CIRCUIT CHECK."
  OR
- JUMPER ALDL TERMINAL "B" TO "A".
- DOES "SES" LIGHT FLASH CODE 12?

(4) YES — DOES TECH 1 DISPLAY ECM DATA?   NO → USE CHART A-2

(5) YES — DOES ENGINE START?   NO → USE CHART A-2

(6) YES — ARE ANY CODES DISPLAYED?   NO → USE CHART A-3

YES →
- REFER TO APPLICABLE CODE CHART. START WITH LOWEST CODE.

NO →
(7) COMPARE TECH 1 DATA WITH TYPICAL VALUES SHOWN ON FACING PAGE. ARE VALUES NORMAL OR WITHIN TYPICAL RANGES?

YES → (8) REFER TO "SYMPTOMS"

NO → (9) REFER TO INDICATED "COMPONENT(S) SYSTEM" CHECKS

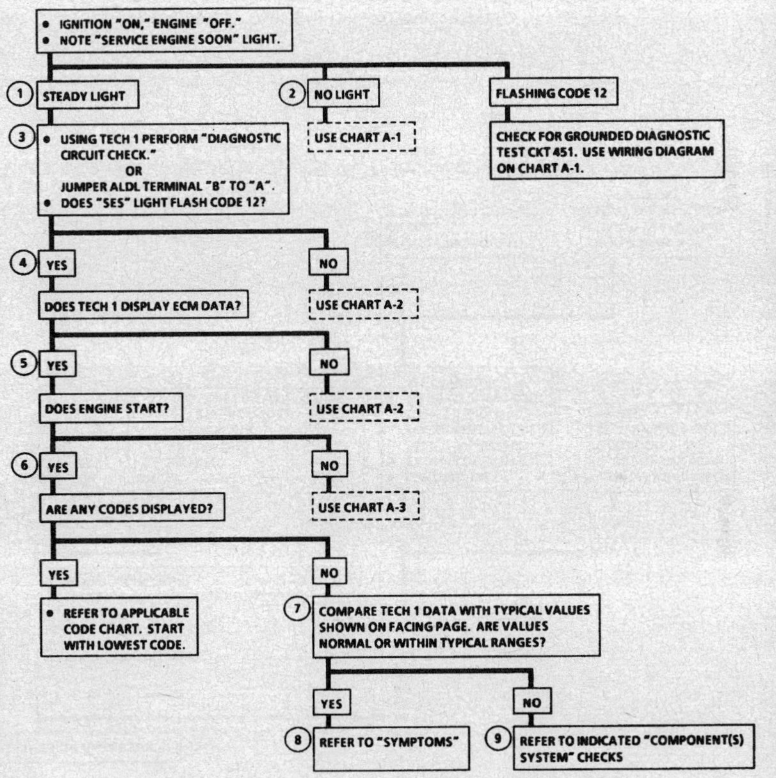

2.3L (VIN A) ENGINE — SYSTEM DIAGNOSTIC CHARTS — 1992 BERETTA

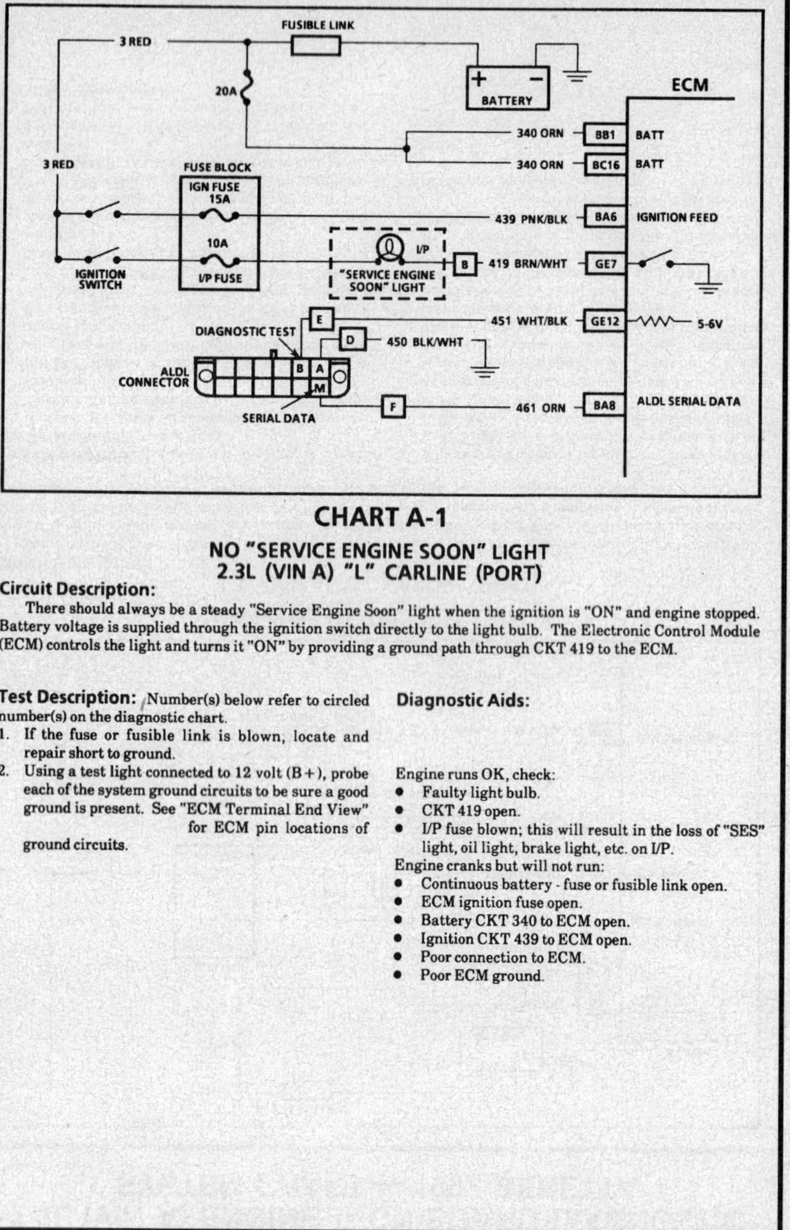

CHART A-1

NO "SERVICE ENGINE SOON" LIGHT
2.3L (VIN A) "L" CARLINE (PORT)

Circuit Description:
There should always be a steady "Service Engine Soon" light when the ignition is "ON" and engine stopped. Battery voltage is supplied through the ignition switch directly to the light bulb. The Electronic Control Module (ECM) controls the light and turns it "ON" by providing a ground path through CKT 419 to the ECM.

Test Description: Number(s) below refer to circled number(s) on the diagnostic chart.
1. If the fuse or fusible link is blown, locate and repair short to ground.
2. Using a test light connected to 12 volt (B+), probe each of the system ground circuits to be sure a good ground is present. See "ECM Terminal End View" for ECM pin locations of ground circuits.

Diagnostic Aids:

Engine runs OK, check:
- Faulty light bulb.
- CKT 419 open.
- I/P fuse blown; this will result in the loss of "SES" light, oil light, brake light, etc. on I/P.

Engine cranks but will not run:
- Continuous battery - fuse or fusible link open.
- ECM ignition fuse open.
- Battery CKT 340 to ECM open.
- Ignition CKT 439 to ECM open.
- Poor connection to ECM.
- Poor ECM ground.

2.3L (VIN A) ENGINE — SYSTEM DIAGNOSTIC CHARTS — 1992 BERETTA

CHART A-1

NO "SERVICE ENGINE SOON" LIGHT
2.3L (VIN A) "L" CARLINE (PORT)

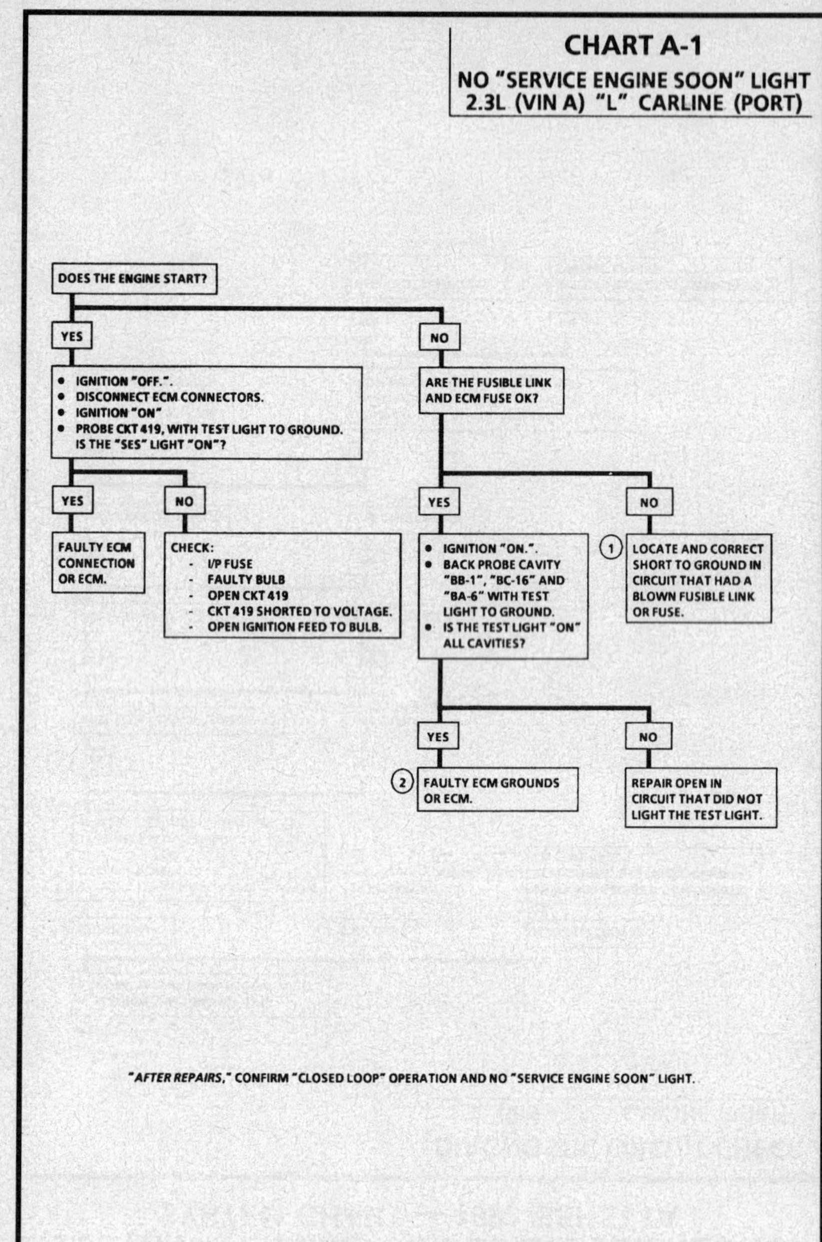

2.3L (VIN A) ENGINE — SYSTEM DIAGNOSTIC CHARTS — 1992 BERETTA

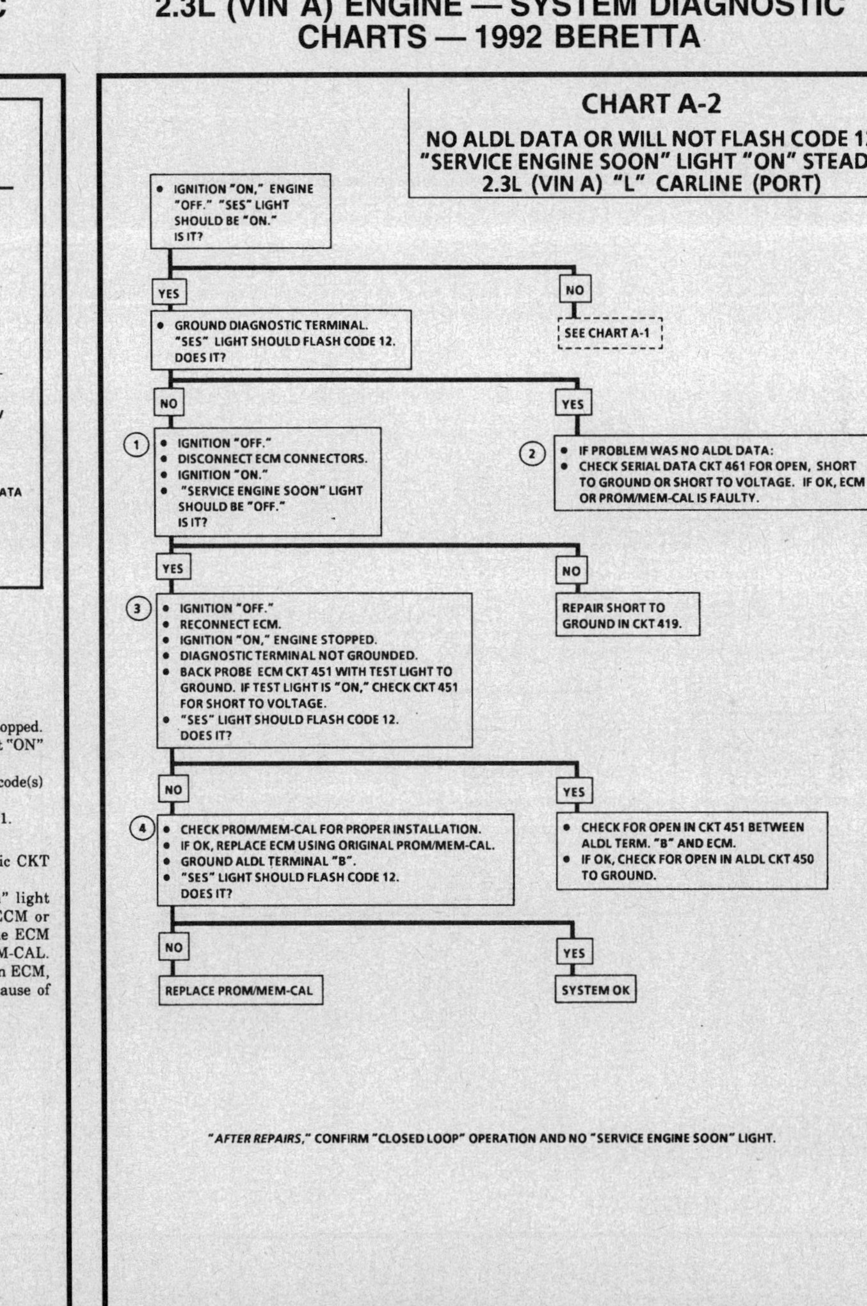

CHART A-2

**NO ALDL DATA OR WILL NOT FLASH CODE 12
"SERVICE ENGINE SOON" LIGHT "ON" STEADY
2.3L (VIN A) "L" CARLINE (PORT)**

Circuit Description:

There should always be a steady "Service Engine Soon" light when the ignition is "ON" and engine stopped. Battery ignition voltage is supplied to the light bulb. The Electronic Control Module (ECM) turns the light "ON" by grounding CKT 419 at the ECM.

With the diagnostic terminal grounded, the light should flash a Code 12, followed by any trouble code(s) stored in memory.

A steady light suggests a short to ground in the light control CKT 419, or an open in diagnostic CKT 451.

Test Description: Number(s) below refer to circled number(s) on the diagnostic chart.

1. Light "OFF" with CKT 419 disconnected from ECM indicates that ground circuit was completed through the ECM, not through external short to ground.
2. If there is a problem with the ECM that causes a Tech 1 "Scan" tool to not read serial data, the ECM should not flash a Code 12. If Code 12 is flashing, check for CKT 451 short to ground. If Code 12 does flash, be sure that the "Scan" tool is working properly on another vehicle. If the "Scan" tool is functioning properly, check CKT 461 for open or short to ground or voltage. If CKT 461 is OK, the ECM or MEM-CAL may be the fault for the "NO ALDL" symptom.
3. This step will check for an open diagnostic CKT 451.
4. At this point, the "Service Engine Soon" light wiring is OK. The problem is a faulty ECM or MEM-CAL. If Code 12 does not flash, the ECM should be replaced using the original MEM-CAL. Replace the MEM-CAL only after trying an ECM, as a defective MEM-CAL is an unlikely cause of the problem.

2.3L (VIN A) ENGINE — SYSTEM DIAGNOSTIC CHARTS — 1992 BERETTA

CHART A-2

**NO ALDL DATA OR WILL NOT FLASH CODE 12
"SERVICE ENGINE SOON" LIGHT "ON" STEADY
2.3L (VIN A) "L" CARLINE (PORT)**

2.3L (VIN A) ENGINE — SYSTEM DIAGNOSTIC CHARTS — 1992 BERETTA

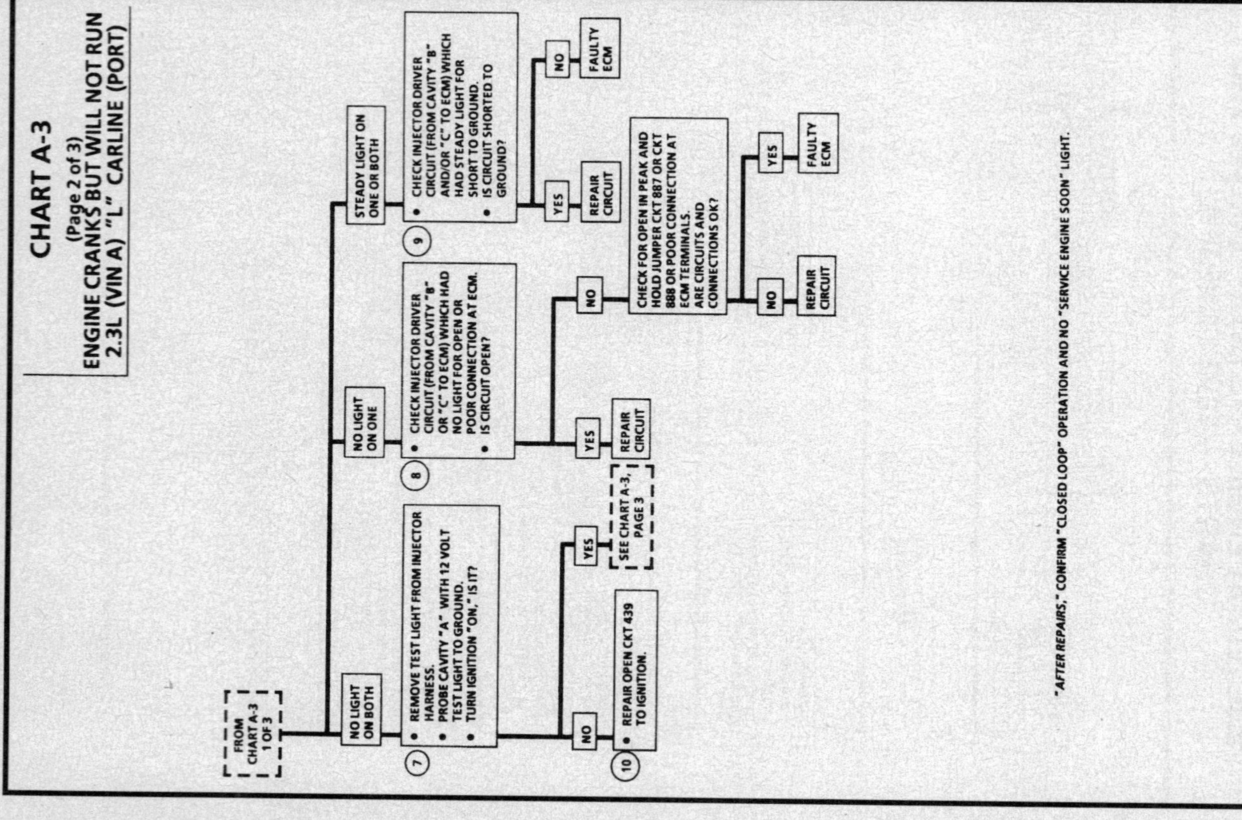

2.3L (VIN A) ENGINE — SYSTEM DIAGNOSTIC CHARTS — 1992 BERETTA

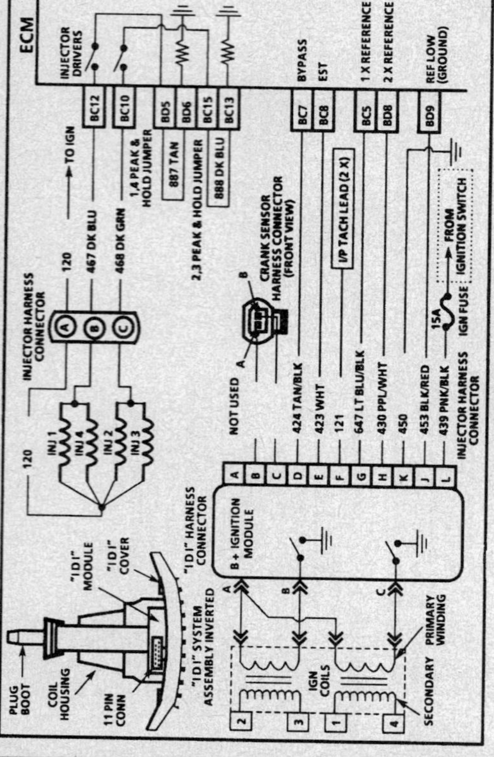

CHART A-3
(Page 1 of 3)
ENGINE CRANKS BUT WILL NOT RUN
2.3L (VIN A) "L" CARLINE (PORT)

Condition:

Engine cranks but will not run, or engine may start, but immediately stops running. Battery condition and engine cranking speed are OK and there is adequate fuel in the tank.

Circuit Description:

This engine is equipped with a distributorless ignition system called the Integrated Direct Ignition (IDI) system. The primary circuit of the IDI consists of two separate ignition coils, an IDI module and crankshaft sensor as well as the related connecting wires and the Electronic Spark Timing (EST) portion of the ECM. Each secondary circuit consists of the secondary winding of the coil, two connecting metal strips molded into the coil housing, the spark plug boot/connector assemblies and spark plugs.

Test Description: Number(s) below refer to circled number(s) on the diagnostic chart.

1. This step verifies that "SES" light operation, on-board diagnostics, cranking rpm, TPS and coolant sensor signals are normal. A blinking test light verifies that the ECM is receiving the IDI reference signal and is attempting to activate the injectors.

2. This step checks injector harness and injectors for opens or shorts. Resistance should measure half that of one injector due to parallel circuit.

3. By installing spark plug jumper leads and testing for spark on two adjacent plug leads (do not use 2 & 3 as they are on same coil), each ignition coil's ability to produce at least 25,000 volts is verified.

4. Checks to see if fuel pump and relay are operating correctly (fuel pump only "ON" 2-3 seconds) and fuel pressure is within proper range.

5. If module can make the test light blink, the fault is coil harness or connections. If not, module or its' connections are faulty.

6. This step determines whether harness or injector is cause of incorrect resistance. Nominal injector resistance is 1.9 to 2.1 ohms at 60°C (140°F). Resistance will increase slightly at higher temperatures.

Diagnostic Aids:

Check For:

- TPS binding or sticking in wide open throttle position or intermittently shorted or open.
- Water or foreign material in fuel.
- Low Compression. (Timing chain failure.)
- Verify that only resistor spark plugs are used.

2.3L (VIN A) ENGINE — SYSTEM DIAGNOSTIC CHARTS — 1992 BERETTA

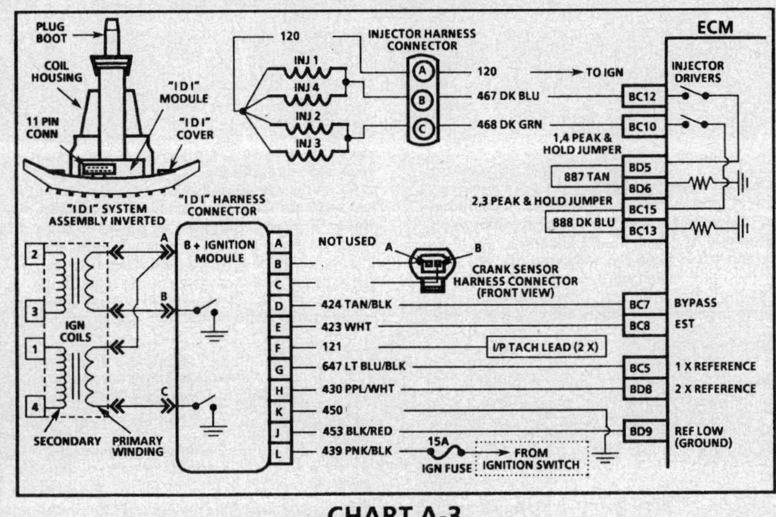

CHART A-3
(Page 3 of 3)
ENGINE CRANKS BUT WILL NOT RUN
2.3L (VIN A) "L" CARLINE (PORT)

Circuit Description:
The Integrated Direct Ignition (IDI) system uses a waste spark method of distribution. In this type of system the ignition module triggers the #1-4 coil pair resulting in both #1 and #4 spark plugs firing at the same time. #1 cylinder is on the compression stroke at the same time #4 is on the exhaust stroke, resulting in a lower energy requirement to fire # 4 spark plug. This leaves the remainder of the high voltage to be used to fire #1 spark plug. On this application, the crank sensor is mounted to, and protrudes through the block to within approximately 0.050" of the crankshaft reluctor. Since the reluctor is a machined portion of the crankshaft and the sensor is mounted in a fixed position on the block, timing adjustments are not possible or necessary.

Test Description: Number(s) below refer to circled number(s) on the diagnostic chart.

11. Battery voltage should be available at terminal "L" of the IDI 11 pin connector, and terminal "K" should be a good ground.
12. The test light to 12 volts simulates a reference signal to the ECM which will result in an injector test light blink for every other touch of the test light, if CKT 430, the ECM and the injector driver circuit are all functioning properly.
13. The crankshaft sensor should output a voltage as the crankshaft turns. If no voltage is produced, the indication is a poor sensor connection or faulty sensor.
14. The crank sensors core is a magnet, therefore, it should be magnetized and the resistance should be within a range of 500 to 900 ohms.

2.3L (VIN A) ENGINE — SYSTEM DIAGNOSTIC CHARTS — 1992 BERETTA

CHART A-3
(Page 3 of 3)
ENGINE CRANKS BUT WILL NOT RUN
2.3L (VIN A) "L" CARLINE (PORT)

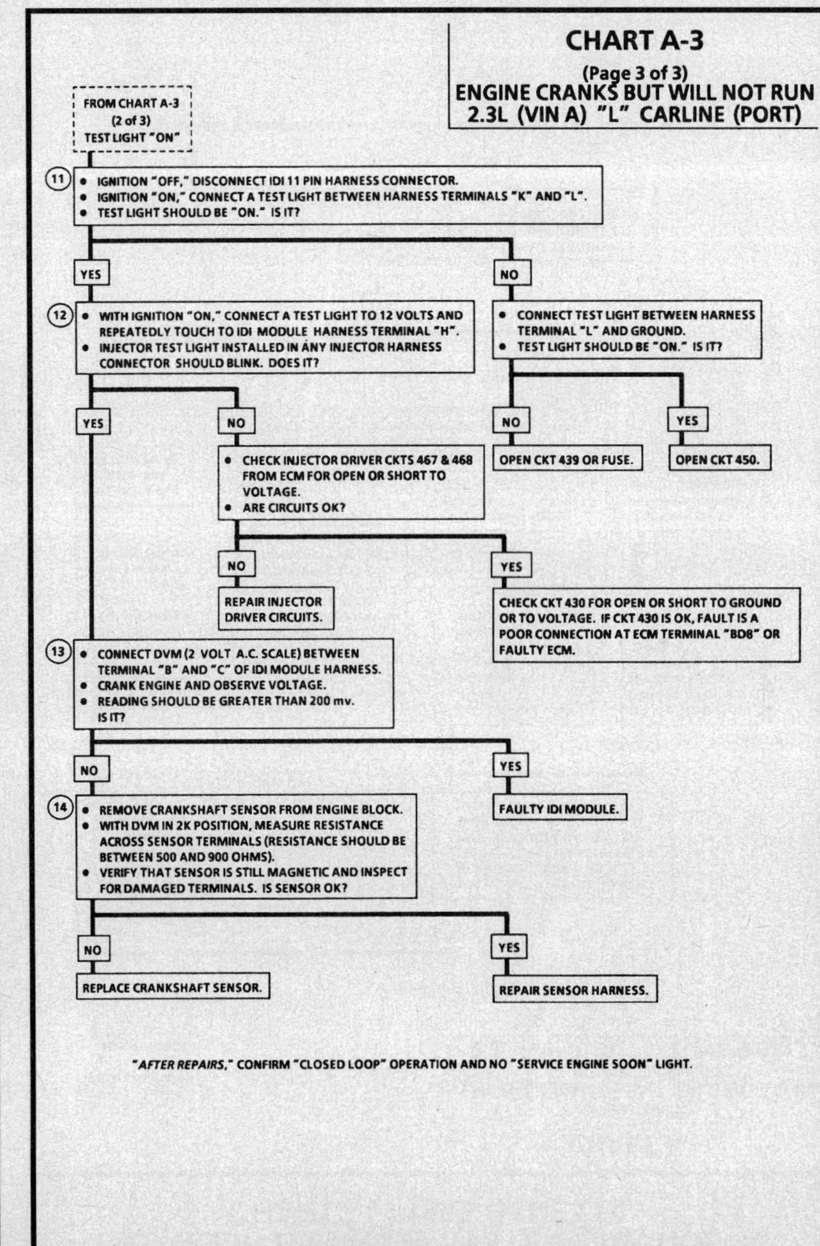

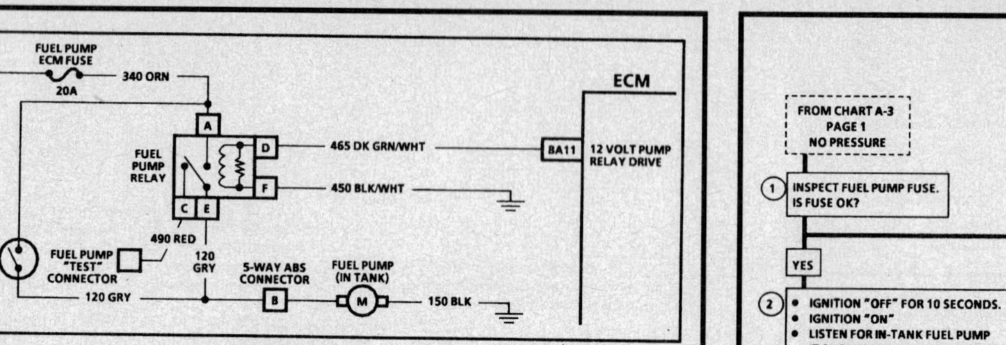

CHART A-5
(Page 1 of 2)
ENGINE CRANKS BUT WILL NOT RUN
(FUEL PUMP CIRCUIT)
2.3L (VIN A) "L" CARLINE (PORT)

Circuit Description:

When the ignition switch is turned "ON," the Electronic Control Module (ECM) turns "ON" the in-tank fuel pump. It will remain "ON" as long as the ECM is receiving ignition reference pulses from the Integrated Direct Ignition (IDI) module.

If there are no reference pulses, the ECM turns "OFF" the fuel pump about 2-3 seconds after key "ON," or about 10 seconds after reference pulses stop. If sufficient oil pressure is present to close the oil pressure switch, the fuel pump will remain "ON" during cranking without reference pulses.

The pump delivers fuel to the fuel rail and injectors, then to the pressure regulator, where the system pressure is controlled to 280 - 325 kPa (40.5 - 47 psi) with no manifold vacuum or 211 - 304 kPa (30.5 - 44 psi) at idle. Excess fuel is returned to the fuel tank.

The fuel pump "test" terminal is located in the engine compartment. When the engine is stopped, the pump can be turned "ON" by applying battery voltage to the "test" terminal.

Improper fuel system pressure will result in one or all of the following symptoms:

- Cranks but will not run.
- Code 44.
- Code 45.
- Cuts out, may feel like ignition problem.
- Poor fuel economy, loss of power.
- Hesitation.

Test Description: Number(s) below refer to circled number(s) on the diagnostic chart.

1. If the fuse is blown, a short to ground in CKTs 120, 340 or the fuel pump itself is the cause.
2. This step determines if the fuel pump circuit is being controlled by the ECM. The ECM should energize the fuel pump relay and turn the fuel pump "ON." If the engine is not cranking or running, the ECM should de-energize the relay and/or fuel pump within 2 seconds after the ignition is turned "ON."
3. Applying B+ to the pump test connector turns "ON" the fuel pump. This validates CKT 120 wiring, relay contacts and fuel pump.
4. This test will determine if a short to ground on CKT 120 caused the fuse to blow. To prevent a misdiagnosis, be sure the fuel pump is disconnected before the test.
5. Checks for a short to ground in the fuel pump relay harness CKT 340.

CHART A-5
(Page 1 of 2)
ENGINE CRANKS BUT WILL NOT RUN
(FUEL PUMP CIRCUIT)
2.3L (VIN A) "L" CARLINE (PORT)

FROM CHART A-3 PAGE 1 NO PRESSURE

(1) INSPECT FUEL PUMP FUSE. IS FUSE OK?

YES →
(2)
- IGNITION "OFF" FOR 10 SECONDS.
- IGNITION "ON"
- LISTEN FOR IN-TANK FUEL PUMP IT SHOULD RUN FOR 2 SECONDS AFTER IGNITION IS TURNED "ON." DOES IT?

YES → CHECK FOR:
- PLUGGED IN LINE FILTER.
- PLUGGED PUMP INLET FILTER.
- RESTRICTED FUEL LINE.

IF ALL CHECK OUT OK, REPLACE IN-TANK FUEL PUMP.

NO →
(3)
- IGNITION "OFF."
- USING A FUSED JUMPER WIRE, CONNECT THE FUEL PUMP TEST CONNECTOR TO B+. DOES THE PUMP RUN?

YES → SEE PAGE 2 OF THIS CHART

NO →
- DISCONNECT FUEL PUMP RELAY.
- USING A FUSED JUMPER WIRE, CONNECT CKT 120 TO B+. DOES THE FUEL PUMP RUN?

YES → FAULTY CONNECTION OR FAULTY RELAY.

NO → OPEN CKT 120, FAULTY IN-TANK PUMP, OR FAULTY FUEL PUMP GROUND CKT 450.

NO →
(4)
- DISCONNECT FUEL PUMP HARNESS AT REAR BODY CONNECTOR.
- IGNITION "OFF," PROBE FUEL PUMP PRIME CONNECTOR WITH A TEST LIGHT CONNECTED TO B+. IS THE TEST LIGHT "ON"?

NO →
(5)
- REMOVE FUEL PUMP RELAY.
- PROBE RELAY CONNECTOR TERMINAL "A" WITH A TEST LIGHT CONNECTED TO B+. IS THE LIGHT "ON"?

YES → CKT 340 SHORTED TO GROUND.

YES → CKT 120, 490 OR FUEL PUMP RELAY OR OIL PRESSURE SWITCH SHORTED TO GROUND.

NO → FUEL TANK METER ASSEMBLY HARNESS SHORTED TO GROUND OR FAULTY FUEL PUMP.

"AFTER REPAIRS," CONFIRM "CLOSED LOOP" OPERATION AND NO "SERVICE ENGINE SOON" LIGHT.

2.3L (VIN A) ENGINE — SYSTEM DIAGNOSTIC CHARTS — 1992 BERETTA

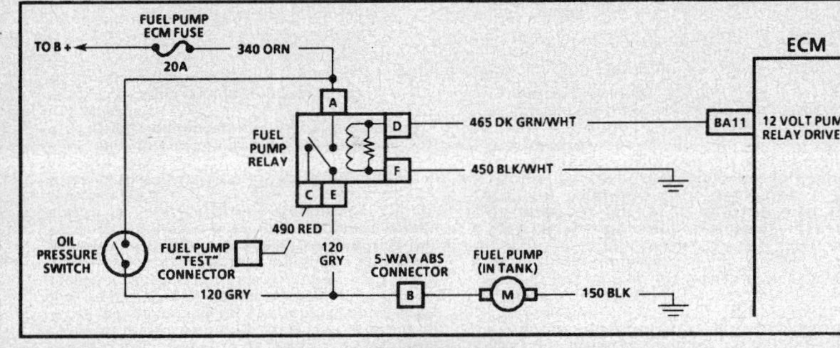

CHART A-5
(Page 2 of 2)
ENGINE CRANKS BUT WILL NOT RUN
(FUEL PUMP CIRCUIT)
2.3L (VIN A) "L" CARLINE (PORT)

Circuit Description:

When the ignition switch is turned "ON," the Electronic Control Module (ECM) turns "ON" the in-tank fuel pump. It will remain "ON" as long as the ECM is receiving ignition reference pulses from the Integrated Direct Ignition (IDI) module.

If there are no reference pulses, the ECM turns "OFF" the fuel pump about 2-3 seconds after key "ON," or about 10 seconds after reference pulses stop. If sufficient oil pressure is present to close the oil pressure switch, the fuel pump will remain "ON" during cranking without reference pulses.

The pump delivers fuel to the fuel rail and injectors, then to the pressure regulator, where the system pressure is controlled to 280 - 325 kPa (40.5 - 47 psi) with no manifold vacuum or 211 - 304 kPa (30.5 - 44 psi) at idle. Excess fuel is then returned to the fuel tank.

The fuel pump "test" terminal is located in the engine compartment. When the engine is stopped, the pump can be turned "ON" by applying battery voltage to the "test" terminal.

Improper fuel system pressure will result in one or all of the following symptoms:
- Cranks but will not run.
- Code 44.
- Code 45.
- Cuts out, may feel like ignition problem.
- Poor fuel economy, loss of power.
- Hesitation.

Test Description: Number(s) below refer to circled number(s) on the diagnostic chart.
6. Checks for open in the fuel pump relay ground CKT 150.
7. Determines if the ECM is in control of the fuel pump relay through CKT 465.
8. The fuel pump control circuit includes an engine oil pressure switch with a separate set of normally open contacts. The switch closes at about (4 lb.) 28 kPa of oil pressure and provides a second battery feed path to the fuel pump. If the relay fails, the pump will run due to the battery feed supplied by the closed oil pressure switch.

This step checks the oil pressure switch to be sure it provides battery feed to the fuel pump should the pump relay fail. A failed pump relay will result in extended engine crank time because of the time required to build enough oil pressure to close the oil pressure switch and turn "ON" the fuel pump. There may be instances when the relay has failed but the engine will not crank fast enough to build enough oil pressure to close the switch. This or a faulty oil pressure switch can result in "Engine Cranks But Will not Run."

2.3L (VIN A) ENGINE — SYSTEM DIAGNOSTIC CHARTS — 1992 BERETTA

CHART A-5
(Page 2 of 2)
ENGINE CRANKS BUT WILL NOT RUN
(FUEL PUMP CIRCUIT)
2.3L (VIN A) "L" CARLINE (PORT)

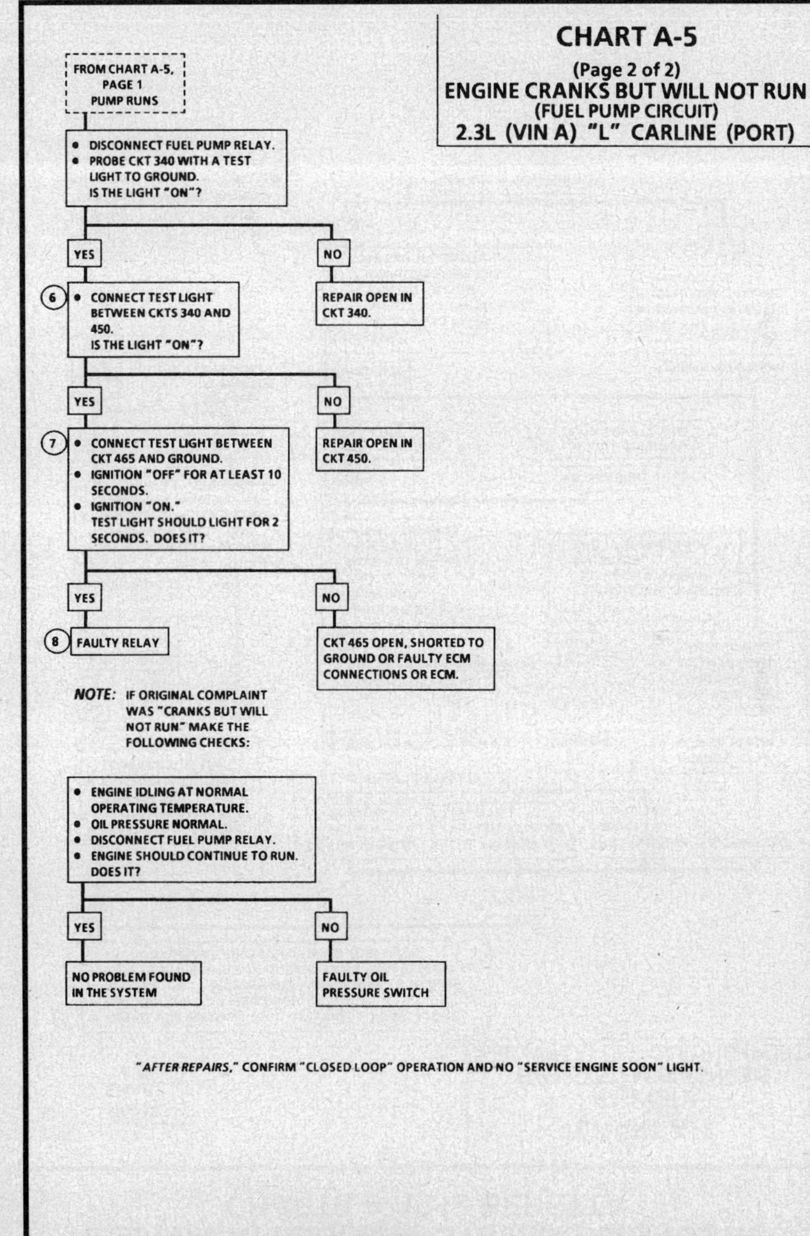

2.3L (VIN A) ENGINE — SYSTEM DIAGNOSTIC CHARTS — 1992 BERETTA

2.3L (VIN A) ENGINE — SYSTEM DIAGNOSTIC CHARTS — 1992 BERETTA

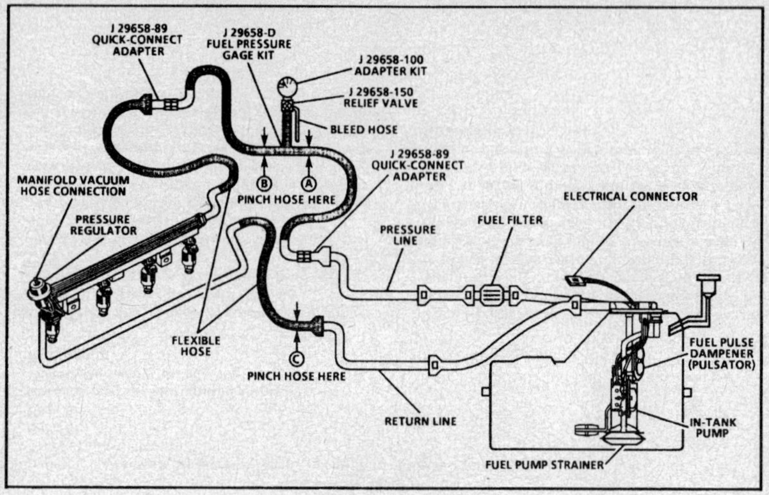

CHART A-7

(Page 1 of 3)
FUEL SYSTEM DIAGNOSIS
2.3L (VIN A) "L" CARLINE (PORT)

Circuit Description:

When the ignition switch is turned "ON," the Electronic Control Module (ECM) will turn "ON" the in-tank fuel pump. It will remain "ON" as long as the engine is cranking or running, and the ECM is receiving reference pulses. If there are no reference pulses, the ECM will shut "OFF" the fuel pump in about 2 seconds after ignition "ON" or 10 seconds after reference pulses stop.

An electric fuel pump, attached to the fuel sender assembly (inside the fuel tank) pumps fuel through an in-line filter to the fuel rail assembly. The pump is designed to provide fuel at a pressure above the regulated pressure needed by the injectors. A pressure regulator, attached to the fuel rail, keeps fuel available to the injectors at a regulated pressure. Unused fuel is returned to the fuel tank by a separate line.

Test Description: Number(s) below refer to circled number(s) on the diagnostic chart.

1. Install fuel pressure gage per instructions on Page 3 of 3. Ignition "ON," pump pressure should be 284 to 325 kPa (41-47 psi). This pressure is controlled by spring pressure within the regulator assembly

2. When the engine is idling, the manifold pressure is low (high vacuum) and is applied to the fuel regulator diaphragm. This will offset the spring and result in a lower fuel pressure. This idle pressure will vary somewhat depending on barometric pressure, however, the pressure idling should be less indicating pressure regulator control.

3. Pressure that continues to fall quickly is caused by one of the following.
   - In-tank fuel pump check valve not holding
   - Partially disconnected fuel pulse dampener (pulsator).
   - Fuel pressure regulator valve leaking.
   - Injector(s) sticking open.

4. An injector sticking open can best be determined by checking for a fouled or saturated spark plug(s). If a leaking injector can not be determined by a fouled or saturated spark plug the following procedure should be used.
   - Remove fuel rail bolts, but leave fuel lines connected.
   - Lift fuel rail out just enough to leave injector nozzles in the ports.

CAUTION: Be sure injector(s) are not allowed to spray on engine and that injector retaining clips are intact. This should be carefully followed to prevent fuel spray on engine which would cause a fire hazard.

   - Pressurize the fuel system and observe for injector(s) leaking.

CHART A-7

(Page 1 of 3)
FUEL SYSTEM DIAGNOSIS
2.3L (VIN A) "L" CARLINE (PORT)

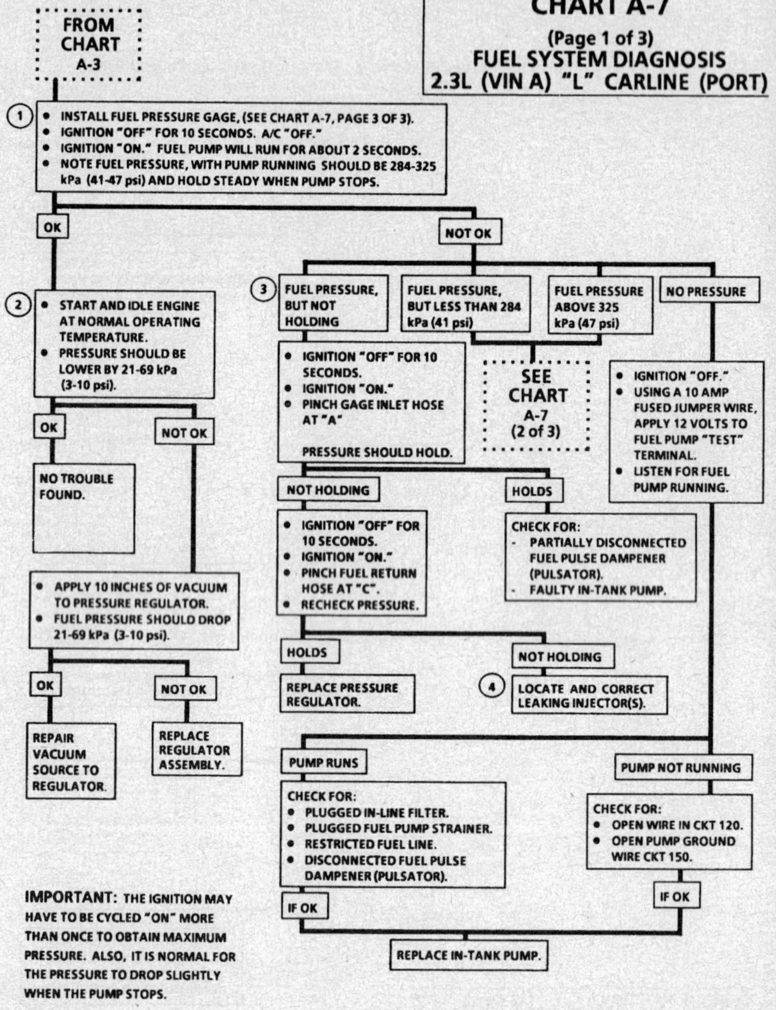

IMPORTANT: THE IGNITION MAY HAVE TO BE CYCLED "ON" MORE THAN ONCE TO OBTAIN MAXIMUM PRESSURE. ALSO, IT IS NORMAL FOR THE PRESSURE TO DROP SLIGHTLY WHEN THE PUMP STOPS.

2.3L (VIN A) ENGINE — SYSTEM DIAGNOSTIC CHARTS — 1992 BERETTA

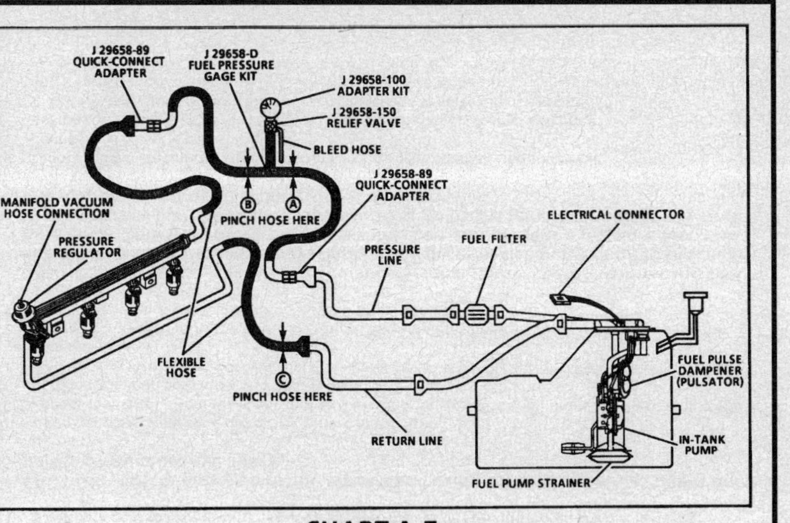

CHART A-7

(Page 2 of 3)
FUEL SYSTEM DIAGNOSIS
2.3L (VIN A) "L" CARLINE (PORT)

Test Description: Number(s) below refer to circled number(s) on the diagnostic chart.

1. Pressure below 284 kPa (41 psi) may cause a lean condition and may set a Code 44. It could also cause hard starting cold and poor driveability. Low enough pressure will cause the engine not to run at all. Restricted flow may allow the engine to run at idle, or low speeds, but may cause a surge and stall when more fuel is required, as when accelerating or driving at high speeds.

2. Restricting fuel flow at the fuel pressure gage (at "B") causes fuel pressure to build above regulated pressure. With battery voltage applied to the pump "test" terminal, pressure should rise above 325 kPa (47 psi) as the gage outlet hose is restricted.

   NOTICE: Do not allow pressure to exceed 414 kPa (60 psi), as damage to the regulator may result.

3. This test determines if the high fuel pressure is due to a restricted fuel return line or a faulty fuel pressure regulator. High fuel pressure may cause a rich condition and may set a Code 45.

2.3L (VIN A) ENGINE — SYSTEM DIAGNOSTIC CHARTS — 1992 BERETTA

CHART A-7

(Page 2 of 3)
FUEL SYSTEM DIAGNOSIS
2.3L (VIN A) "L" CARLINE (PORT)

FROM CHART A-7 (1 of 3)

(1) HAS PRESSURE BUT LESS THAN 284 kPa (41 psi)

→ CHECK FOR RESTRICTED FUEL LINES OR IN-LINE FILTER.

→ OK / NOT OK

NOT OK → REPLACE FILTER OR REPAIR FUEL LINE AND RECHECK.

(2)
- IGNITION "OFF."
- USING A 10 AMP FUSED JUMPER WIRE, APPLY 12 VOLTS TO FUEL PUMP "TEST" TERMINAL.
- GRADUALLY PINCH PRESSURE GAGE OUTLET HOSE AT (B). LOOK FOR PRESSURE ABOVE 325 kPa (47 psi). DO NOT EXCEED 414 kPa (60 psi).

ABOVE 325 kPa (47 psi) → IF LINES ARE OK, REPLACE PRESSURE REGULATOR.

PRESSURE BUT LESS THAN 284 kPa (41 psi) → CHECK FOR:
- FAULTY FUEL PUMP.
- PARTIALLY DISCONNECTED FUEL PULSE DAMPENER (PULSATOR).
- RESTRICTED FUEL PUMP STRAINER.
- INCORRECT FUEL PUMP.

FUEL PRESSURE ABOVE 325 kPa (47 psi)

(3)
- DISCONNECT ENGINE COMPARTMENT FUEL RETURN LINE QUICK-CONNECT FITTING. (PROCEDURES FOR DISCONNECTING/ CONNECTING QUICK-CONNECT FITTINGS ARE SAME AS THOSE FOR FUEL FEED LINE, SEE PAGE 3 OF 3).
- PLACE OPEN END OF FLEXIBLE HOSE INTO AN APPROVED GASOLINE CONTAINER. NOTE FUEL PRESSURE WITHIN 2 SECONDS AFTER IGNITION IS TURNED "ON."

ABOVE 325 kPa (47 psi) → CHECK FOR RESTRICTED FUEL RETURN LINE FROM FUEL PRESSURE REGULATOR TO POINT WHERE FUEL LINE WAS DISCONNECTED. → IF LINE OK, REPLACE FUEL PRESSURE REGULATOR.

284-325 kPa (41-47 psi) → LOCATE AND CORRECT RESTRICTED FUEL RETURN LINE TO FUEL TANK.

"AFTER REPAIRS," CONFIRM "CLOSED LOOP" OPERATION AND NO "SERVICE ENGINE SOON" LIGHT.

2.3L (VIN A) ENGINE — SYSTEM DIAGNOSTIC CHARTS — 1992 BERETTA

CHART A-7
(Page 3 of 3)
FUEL SYSTEM DIAGNOSIS 2.3L (VIN A) "L" CARLINE (PORT)

FUEL PRESSURE CHECK

Tools Required:
J 29658-D - Fuel Pressure Gage Kit
J 29658-150 - Fuel Pressure Gage Relief Valve
J 29658-100 - TBI Pressure Gage Modification Kit
J 29658-89 - Fuel Pressure Quick Connect Adapters
J 37088 - A - Fuel Line Quick-Connect Separators

CAUTION: To Reduce the Risk of Fire and Personal Injury:
- It is necessary to relieve fuel system pressure before connecting a fuel pressure gage.
- After relieving system pressure, a small amount of fuel may be released when disconnecting the fuel lines. Cover fuel line fittings with a shop towel before disconnecting, to catch any fuel that may leak out. Place towel in approved container when disconnect is completed.
- Do not pinch or restrict nylon fuel lines to avoid severing, which could cause a fuel leak.

NOTICE: If nylon fuel lines become kinked, and cannot be straightened, they must be replaced.

1. Loosen fuel filler cap to relieve fuel tank pressure. (Do not tighten at this time.)
2. Raise vehicle.
3. Disconnect fuel pump electrical connector.
4. Lower vehicle.
5. Start and run engine until fuel supply remaining in fuel pipes is consumed. Engage starter for three seconds to assure relief of any remaining pressure.
6. Disconnect negative battery cable.
7. Locate engine compartment fuel feed quick-connect fitting.
8. Grasp both ends of fitting, twist female end $\frac{1}{4}$ turn in each direction to loosen any dirt in fitting.

CAUTION: Safety glasses must be worn when using compressed air, as flying dirt particles may cause eye injury.

9. Using compressed air, blow dirt out of quick-connect fitting.
10. Choose correct tool from separator tool set J 37088-A for size of fitting. Insert tool into female end of connector, then push inward to release male connector.
11. If not previously installed, connect Fuel Pressure Gage Relief Valve J 29658-150, to fuel pressure gage hose assembly.
12. Connect 414 kPa (60 psi) gage from TBI Pressure Gage Modification kit J 29658-100 to hose assembly.
13. Connect gage quick-connect adapters J 29658-89 to hose assembly.

CAUTION: To Reduce the Risk of Fire and Personal Injury: Before connecting fuel line quick-connect fittings, always apply a few drops of clean engine oil to the male tube ends. This will ensure proper reconnection and prevent a possible fuel leak. (During normal operation, the O-rings located inside the female connector will swell and may prevent proper reconnection if not lubricated.)

14. Lubricate the male tube end of the fuel line and the gage adapter with engine oil.
15. Connect fuel pressure gage.
    - Push connectors together to cause the retaining tabs/fingers to snap into place.
    - Once installed, pull on both ends of each connection to make sure it is secure.

2.3L (VIN A) ENGINE — SYSTEM DIAGNOSTIC CHARTS — 1992 BERETTA

CHART A-7
(Page 3 of 3)
**FUEL SYSTEM DIAGNOSIS
2.3L (VIN A) "L" CARLINE (PORT)**

FUEL PRESSURE CHECK
-continued-

16. Connect negative battery cable.
17. Check fuel pressure.
18. Place bleed hose into an approved container and open valve to bleed system pressure.
19. Disconnect negative battery cable.
20. Disconnect fuel pressure gage.
21. Lubricate the male tube end of the fuel line, and reconnect quick-connect fitting.
    - Push connector together to cause the retaining tabs/fingers to snap into place
    - Once installed, pull on both ends of connection to make sure it is secure
22. Tighten fuel filler cap.
23. Connect negative battery cable.
24. Cycle ignition "ON" and "OFF" twice, waiting ten seconds between cycles, then check for fuel leaks.

2.3L (VIN A) ENGINE — DIAGNOSTIC TROUBLE CODE CHART — 1992 BERETTA

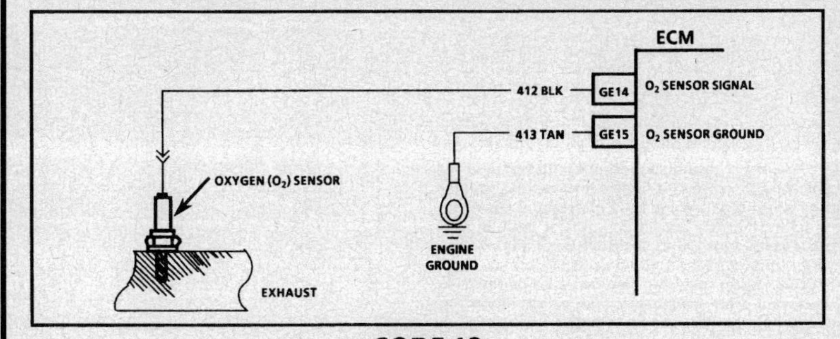

CODE 13
OXYGEN (O₂) SENSOR CIRCUIT
(OPEN CIRCUIT)
2.3L (VIN A) "L" CARLINE (PORT)

Circuit Description:

The ECM supplies a voltage of about .45 volt between terminals "GE14" and "GE15". (If measured with a 10 megohm digital voltmeter, this may read as low as .32 volt.) The O_2 sensor varies the voltage within a range of about 1 volt if the exhaust is rich, down through about .10 volt if exhaust is lean.

The sensor is like an open circuit and produces no voltage when it is below 315°C (600°F). An open sensor circuit or cold sensor causes "Open Loop" operation.

Test Description: Number(s) below refer to circled number(s) on the diagnostic chart.

1. Code 13 will set under the following conditions:
   - Engine running at least 40 seconds after start.
   - Coolant temperature at least 42.5°C (108.5°C).
   - No Code 21 or Code 22.
   - O_2 signal voltage steady between .34 and .55 volt.
   - Throttle position sensor signal above 6% for more time than TPS was below 6%. (About .3 volt above closed throttle voltage.)
   - All conditions must be met and held for at least 30 seconds.

   If the conditions for a Code 13 exist, the system will not go "Closed Loop."
2. This will determine if the sensor is at fault or the wiring or ECM is the cause of the Code 13.

3. Use only a high impedance digital volt ohmmeter for this test. This test checks the continuity of CKT 412 and CKT 413; because if CKT 413 is open, the ECM voltage on CKT 412 will be over .6 volt (600 mV).

Diagnostic Aids:

Normal Tech 1 "Scan" voltage varies between 100 mV to 999 mV (.1 volt to 1.0 volt) while in "Closed Loop." Code 13 sets in 20 seconds if voltage remains between .35 volt and .55 volt, but the system will go "Open Loop" in about 15 seconds.

2.3L (VIN A) ENGINE — DIAGNOSTIC TROUBLE CODE CHART — 1992 BERETTA

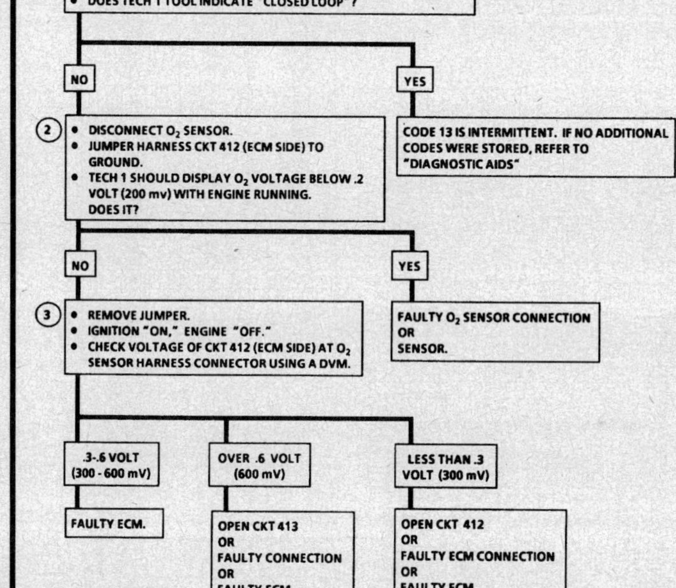

2.3L (VIN A) ENGINE — DIAGNOSTIC TROUBLE CODE CHART — 1992 BERETTA

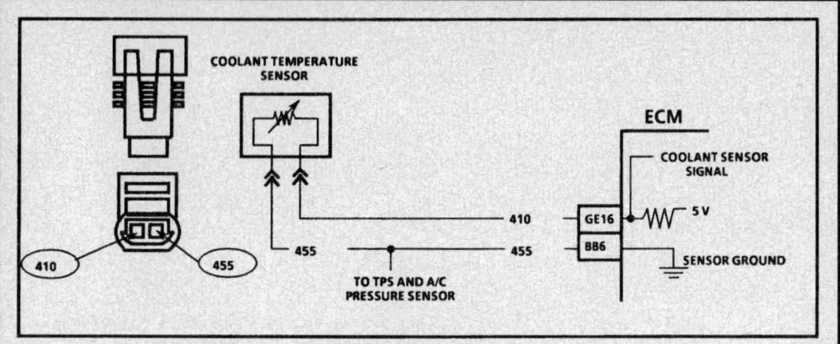

CODE 14
COOLANT TEMPERATURE SENSOR (CTS) CIRCUIT
(HIGH TEMPERATURE INDICATED)
2.3L (VIN A) "L" CARLINE (PORT)

Circuit Description:

The Coolant Temperature Sensor (CTS) uses a thermistor to control the signal voltage at the ECM. The ECM applies a voltage on CKT 410 to the sensor. When the engine is cold the sensor (thermistor) resistance is high, therefore, the ECM terminal "GE16" voltage will be high.

As the engine warms, the sensor resistance becomes less, and the voltage drops. At normal engine operating temperature, the voltage will measure about 1.5 to 2.0 volts at ECM terminal "GE16".

Coolant temperature is one of the inputs used to control:

- Fuel delivery.
- Engine Spark Timing (EST).
- Idle Air Control (IAC).
- Controlled Canister Purge (CCP).
- Cooling Fan.

Test Description: Number(s) below refer to circled number(s) on the diagnostic chart.

1. Code 14 will set if:
   - Signal voltage indicates a coolant temperature above 140°C (285°F).
   - Engine running longer than 128 seconds.
2. This test will determine if CKT 410 is shorted to ground which will cause the conditions for Code 14.

Diagnostic Aids:

Check harness routing for a potential short to ground in CKT 410.

Tech 1 "Scan" tool displays engine temperature in degrees Celsius. After engine is started, the temperature should rise steadily to about 90°C, and then stabilize when thermostat opens.

Verify that engine is not overheating and has not been subjected to conditions which could create an overheating condition (i.e., overload, trailer towing, hilly terrain, heavy stop and go traffic, etc.). The "Temperature To Resistance Value" scale at the right may be used to test the coolant sensor at various temperature levels to evaluate the possibility of a "shifted" (mis-scaled) sensor. A "shifted" sensor could result in poor driveability complaints.

2.3L (VIN A) ENGINE — DIAGNOSTIC TROUBLE CODE CHART — 1992 BERETTA

CODE 14
COOLANT TEMPERATURE SENSOR (CTS) CIRCUIT
(HIGH TEMPERATURE INDICATED)
2.3L (VIN A) "L" CARLINE (PORT)

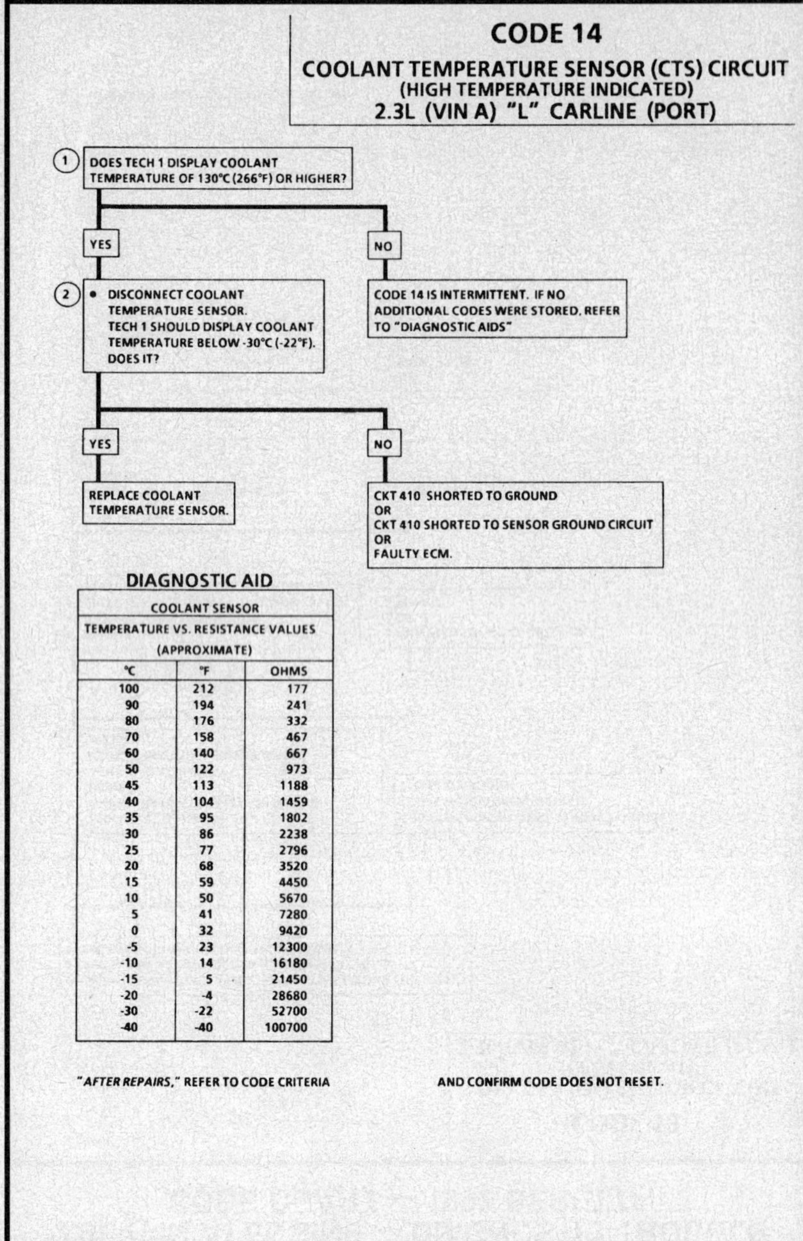

DIAGNOSTIC AID

COOLANT SENSOR		
TEMPERATURE VS. RESISTANCE VALUES		
(APPROXIMATE)		
°C	°F	OHMS
100	212	177
90	194	241
80	176	332
70	158	467
60	140	667
50	122	973
45	113	1188
40	104	1459
35	95	1802
30	86	2238
25	77	2796
20	68	3520
15	59	4450
10	50	5670
5	41	7280
0	32	9420
-5	23	12300
-10	14	16180
-15	5	21450
-20	-4	28680
-30	-22	52700
-40	-40	100700

"AFTER REPAIRS," REFER TO CODE CRITERIA   AND CONFIRM CODE DOES NOT RESET.

2.3L (VIN A) ENGINE — DIAGNOSTIC TROUBLE CODE CHART — 1992 BERETTA

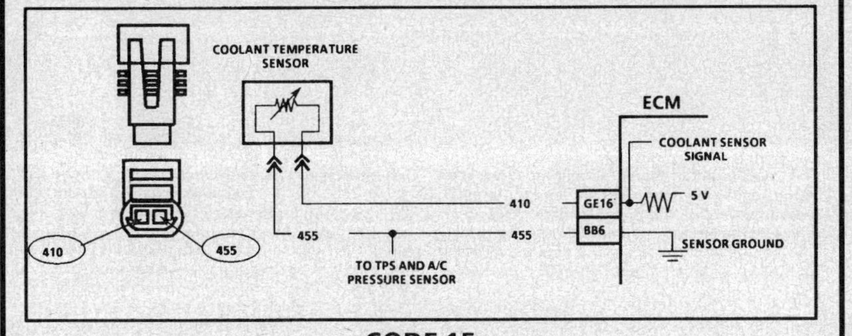

CODE 15
COOLANT TEMPERATURE SENSOR (CTS) CIRCUIT
(LOW TEMPERATURE INDICATED)
2.3L (VIN A) "L" CARLINE (PORT)

Circuit Description:

The Coolant Temperature Sensor (CTS) uses a thermistor to control the signal voltage at the ECM. The ECM applies a voltage on CKT 410 to the sensor. When the engine is cold, the sensor (thermistor) resistance is high, therefore, ECM terminal "GE16" voltage will be high.

As the engine warms, the sensor resistance becomes less, and the voltage drops. At normal engine operating temperature the voltage will measure about 1.5 to 2.0 volts at ECM terminal "GE16."
Coolant temperature is one of the inputs used to control:
- Fuel delivery.
- Engine Spark Timing (EST).
- Idle Air Control (IAC).
- Controlled Canister Purge (CCP).
- Cooling Fan.

Test Description: Number(s) below refer to circled number(s) on the diagnostic chart.
1. Code 15 will set if:
   - Signal voltage indicates a coolant temperature less than -39°C (-38°F) for 60 seconds.
2. This test simulates a Code 14. If the ECM senses the low signal voltage (high temperature) and the "Scan" reads 130°C, the ECM and wiring are OK.
3. This test will determine if CKT 410 is open. There should be 5 volts present at sensor connector if measured with a DVM.

Diagnostic Aids:

A Tech 1 "Scan" tool displays engine temperature in degrees Celsius. After the engine is started the temperature should rise steadily to about 95°C, and then stabilize when the thermostat opens. It is normal for coolant temperature to fluctuate slightly around 95°C.

A faulty connection, or an open in CKT 410 or CKT 455 can result in a Code 15.

Codes 15, 21 and 66 stored at the same time could be the result of an open CKT 455.

The "Temperature to Resistance Value" scale , may be used to test the coolant sensor at various temperature levels to evaluate the possibility of a "shifted" (mis-scaled) sensor. A "shifted" sensor could result in poor driveability complaints.

2.3L (VIN A) ENGINE — DIAGNOSTIC TROUBLE CODE CHART — 1992 BERETTA

CODE 15
COOLANT TEMPERATURE SENSOR (CTS) CIRCUIT
(LOW TEMPERATURE INDICATED)
2.3L (VIN A) "L" CARLINE (PORT)

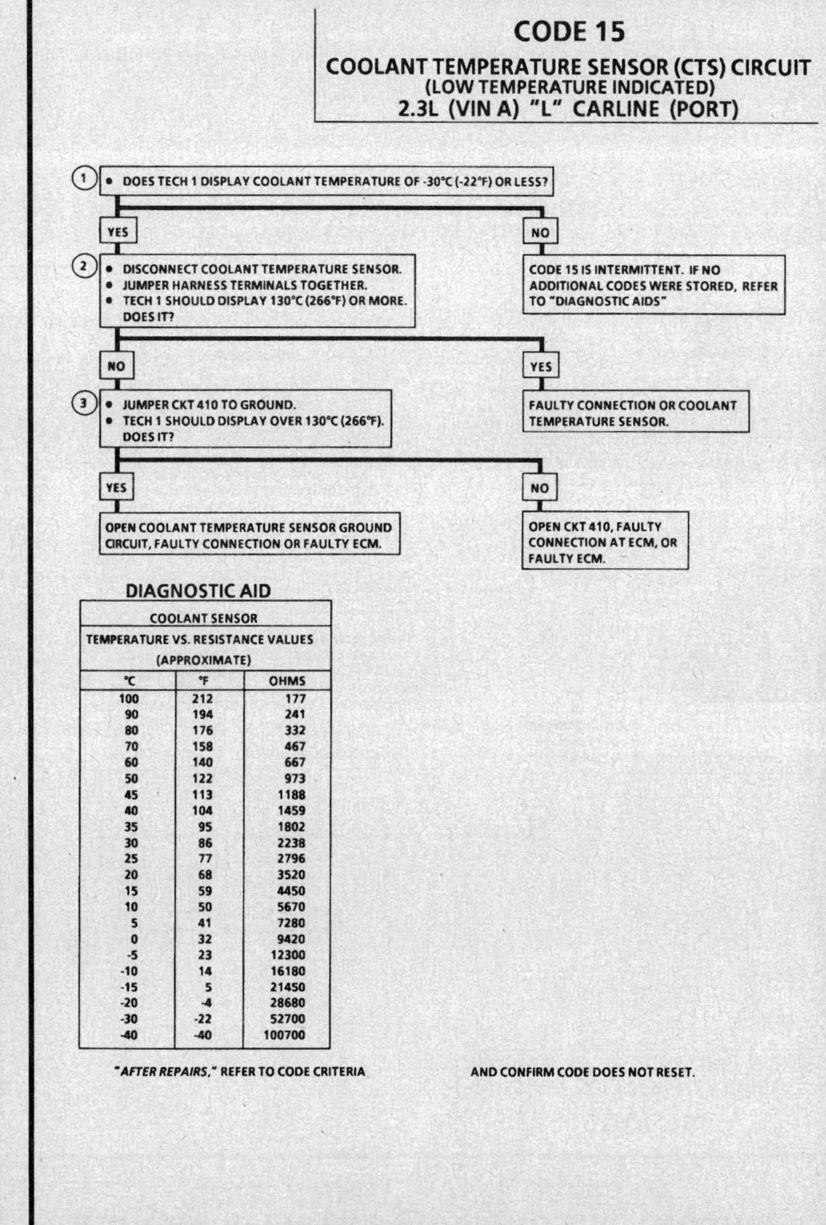

DIAGNOSTIC AID

COOLANT SENSOR		
TEMPERATURE VS. RESISTANCE VALUES		
(APPROXIMATE)		
°C	°F	OHMS
100	212	177
90	194	241
80	176	332
70	158	467
60	140	667
50	122	973
45	113	1188
40	104	1459
35	95	1802
30	86	2238
25	77	2796
20	68	3520
15	59	4450
10	50	5670
5	41	7280
0	32	9420
-5	23	12300
-10	14	16180
-15	5	21450
-20	-4	28680
-30	-22	52700
-40	-40	100700

"AFTER REPAIRS," REFER TO CODE CRITERIA   AND CONFIRM CODE DOES NOT RESET.

2.3L (VIN A) ENGINE — DIAGNOSTIC TROUBLE CODE CHART — 1992 BERETTA

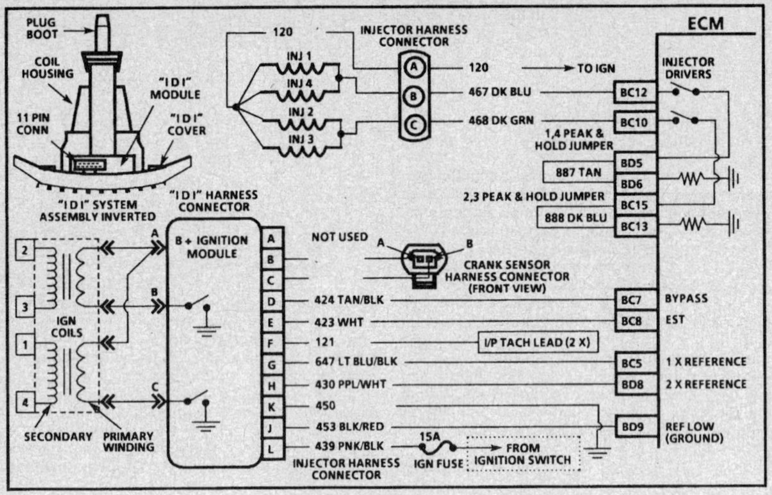

CODE 16

**MISSING 2X REFERENCE CIRCUIT
2.3L (VIN A) "L" CARLINE (PORT)**

Circuit Description:

The ignition module sends a reference signal to the ECM twice per revolution to indicate crankshaft position and rpm so that the ECM can determine when to pulse the ignition coils and control ignition timing. This signal is called the 2X reference because it occurs two times per revolution. The ignition module applies 5 volts from terminal "H" through CKT 430 to ECM terminal "BD8" and in effect, switches this circuit to ground for a very short period of time. Code 16 is set if the ECM receives 1X reference pulses with no 2X reference pulses.

Test Description: Number(s) below refer to circled number(s) on the diagnostic chart.

1. This determines if the ECM recognizes a problem. If it doesn't set Code 16 at this point, the problem is intermittent and could be due to a loose connection. See "Diagnostic Aids."
2. This step simulates the 2X signal. The ECM should recognize the drop in voltage as the test light probe is removed, if the circuit and ECM are OK. This step will give accurate results only if the chart sequence is used - ignition "OFF," ignition "ON," Tech 1 set to 2X reference and terminal "G" touched with test light probe.
3. If the ECM did not recognize the simulation of the 2X signal, CKT 430 may be open or shorted to ground or voltage. If CKT 430 is OK, the ECM is faulty.
4. Step 2 indicated that CKT 430 is OK and the ECM is capable of recognizing the simulated 2X reference pulse. This indicates either a poor connection at ignition module terminal "H" or a faulty ignition module caused the Code 16.

Diagnostic Aids:

An intermittent may be caused by a poor connection, rubbed through wire insulation, or a wire broken inside the insulation. Inspect ECM harness connector terminal "BD8" and ignition module terminal "H" for improper mating, broken locks, improperly formed or damaged terminals, poor terminal to wire connection and damaged harness.

2.3L (VIN A) ENGINE — DIAGNOSTIC TROUBLE CODE CHART — 1992 BERETTA

CODE 16

**MISSING 2X REFERENCE CIRCUIT
2.3L (VIN A) "L" CARLINE (PORT)**

(1)
- CLEAR CODES.
- IDLE ENGINE FOR 2 MINUTES OR UNTIL "SERVICE ENGINE SOON" LIGHT TURNS "ON," IF SOONER.
- DOES TECH 1 DISPLAY CODE 16?

YES → (2); NO → CODE 16 IS INTERMITTENT. IF NO ADDITIONAL CODES WERE DISPLAYED, REFER TO "DIAGNOSTIC AIDS."

(2)
- IGNITION "OFF," DISCONNECT "IDI" CONNECTOR.
- IGNITION "ON," SET "SCAN" TOOL TO DISPLAY 2X REFERENCE PULSES.
- WITH TEST LIGHT CONNECTED TO B+, TOUCH ENGINE HARNESS CONNECTOR CAVITY "H" WITH LIGHT AND REMOVE. LIGHT ON = GROUND.
- "SCAN" TOOL DISPLAY SHOULD INDICATE 2X REFERENCE PULSE INCREMENT WHEN TEST LIGHT IS REMOVED. DOES IT?

NO → (3); YES → (4)

(3)
- CHECK CKT 430 FOR OPEN OR SHORT TO GROUND OR VOLTAGE AND REPAIR AS NECESSARY. IF OK, REPLACE ECM.

(4)
- CHECK FOR POOR CONNECTION BETWEEN "IDI" CONNECTOR CAVITY "H" AND IGNITION MODULE AND REPAIR AS NECESSARY. IF OK, REPLACE IGNITION MODULE.

"AFTER REPAIRS," REFER TO CODE CRITERIA AND CONFIRM CODE DOES NOT RESET.

2.3L (VIN A) ENGINE — DIAGNOSTIC TROUBLE CODE CHART — 1992 BERETTA

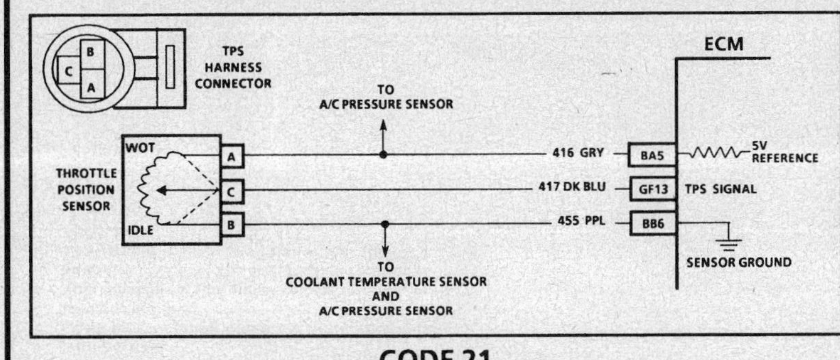

CODE 21
THROTTLE POSITION SENSOR (TPS) CIRCUIT
(SIGNAL VOLTAGE HIGH)
2.3L (VIN A) "L" CARLINE (PORT)

Circuit Description:

The Throttle Position Sensor (TPS) provides a voltage signal that changes relative to the throttle opening. Signal voltage will vary from about .5 volt at idle to about 4.9 volts at Wide Open Throttle (WOT).

The TPS signal is one of the most important inputs used by the ECM for fuel control and for most of the ECM control outputs.

Test Description: Number(s) below refer to circled number(s) on the diagnostic chart.

1. Code 21 will set if:
   - Engine is running.
   - Rpm less than 1500 rpm.
   - No Code 33 or 34.
   - Manifold Absolute Pressure (MAP) less than 65 kPa.
   - TPS signal voltage greater than approximately 4.0 volts (78%).
   - Above conditions exist for over 5 seconds.
   OR
   - TPS voltage greater than about 4.7 volts.
   With throttle closed, the TPS should read less than .900 volt. If it doesn't, replace TPS.
2. With the TPS disconnected, the TPS voltage should go low if the ECM and wiring are OK.
3. Probing CKT 455 with a test light checks the TPS ground circuit because an open or very high resistance ground circuit will cause a Code 21.

Diagnostic Aids:

A Tech 1 "Scan" tool displays throttle position in volts. It should display .400 volt to .900 volt with throttle closed and ignition "ON" or at idle. Voltage should increase at a steady rate as throttle is moved toward Wide Open Throttle (WOT).

Also Tech 1 "Scan" tools will display throttle angle %, 0% = closed throttle 100% = WOT.

An open in CKT 455 will result in a Code 21.

Codes 15, 21 and 66 stored at the same time could be the result of an open CKT 455. "Scan" TPS while depressing accelerator pedal with engine stopped and ignition "ON." Display should vary from about .5 volt (500 mV) when throttle is closed, to over 4.5 volts (4500 mV) when throttle is held wide open.

Check condition of connector and sensor terminals for moisture or corrosion, and clean or replace as necessary. If corrosion found, check condition of connector seal, repair and/or replace if necessary.

2.3L (VIN A) ENGINE — DIAGNOSTIC TROUBLE CODE CHART — 1992 BERETTA

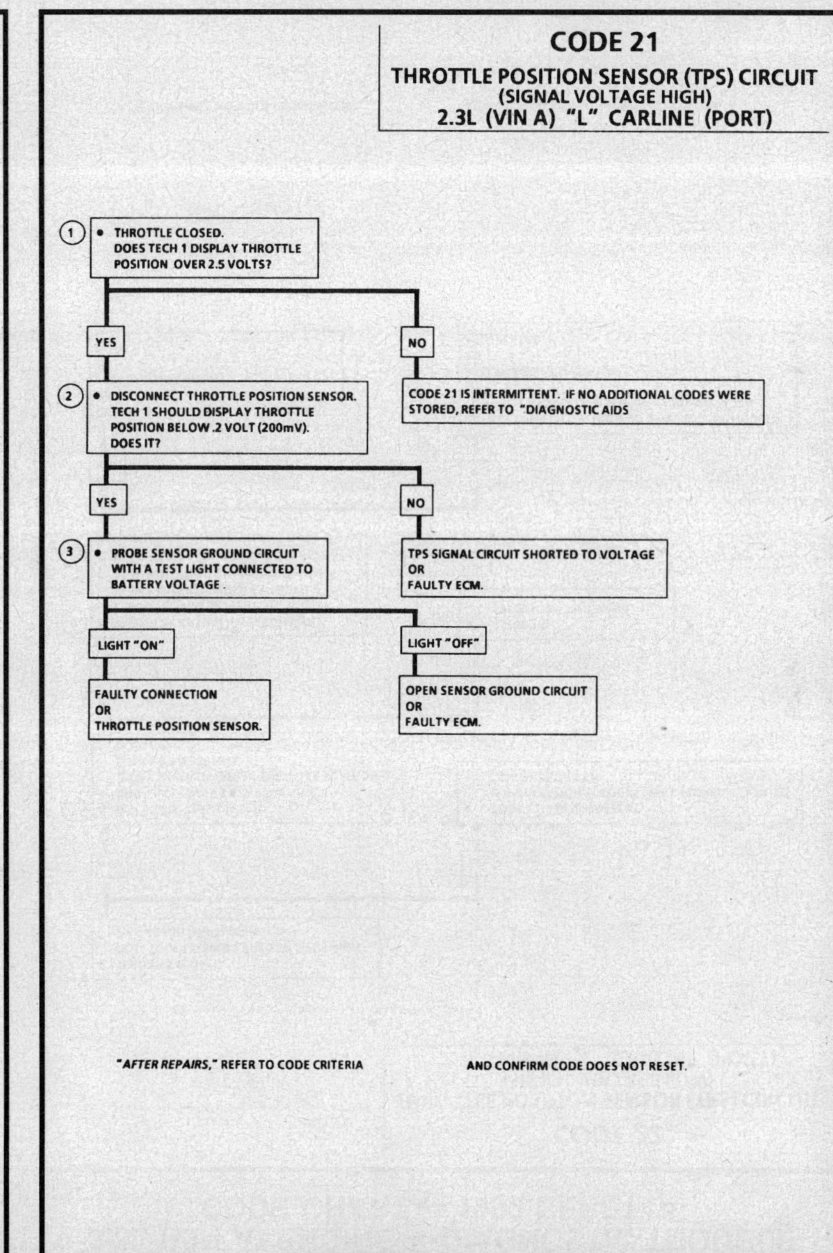

2.3L (VIN A) ENGINE — DIAGNOSTIC TROUBLE CODE CHART — 1992 BERETTA

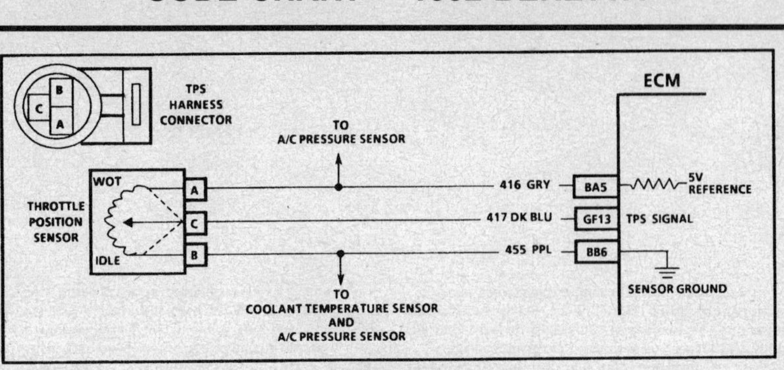

CODE 22
THROTTLE POSITION SENSOR (TPS) CIRCUIT
(SIGNAL VOLTAGE LOW)
2.3L (VIN A) "L" CARLINE (PORT)

Circuit Description:

The Throttle Position Sensor (TPS) provides a voltage signal that changes relative to the throttle opening. Signal voltage will vary from about .5 volt at idle to about 4.9 volts at Wide Open Throttle (WOT).

The TPS signal is one of the most important inputs used by the ECM for fuel control and for most of the ECM control outputs.

Test Description: Number(s) below refer to circled number(s) on the diagnostic chart.

1. Code 22 will set if:
   - Engine is running.
   - TPS signal voltage is less than about .2 volt for 5 seconds.
   The TPS has an auto zeroing feature. If the voltage reading is within the range of about .4 to .9 volt, the ECM will use that value as closed throttle. If the voltage reading is out of the auto zero range at closed throttle, check for a binding throttle cable or damaged linkage. If OK, continue with diagnosis.
2. Simulates Code 21 (high voltage). If ECM recognizes high signal voltage, the ECM and wiring are OK.
3. Check for good sensor connection. If connection is good, replace TPS.
4. This simulates a high signal voltage to check for an open in CKT 417. The Tech 1 will not read up to 12 volts, but what is important is that the ECM recognizes the signal on CKT 417.

Diagnostic Aids:

"Scan" TPS while depressing accelerator pedal with engine stopped and ignition "ON." Display should vary from about 500 mV (.5 volt) when throttle is closed, to over 4500 mV (4.5 volts) when throttle is held wide open.

Also, Tech 1 "Scan" tools will display throttle angle %: 0% = closed throttle; 100% = WOT.

If Code 22 and/or 66 are set, check CKT 416 for faulty wiring or connections.

Should check condition of connector and sensor terminals for corrosion, and clean and/or replace as necessary. If moisture or corrosion is found, check condition of connector seal and repair or replace if necessary.

2.3L (VIN A) ENGINE — DIAGNOSTIC TROUBLE CODE CHART — 1992 BERETTA

CODE 22
THROTTLE POSITION SENSOR (TPS) CIRCUIT
(SIGNAL VOLTAGE LOW)
2.3L (VIN A) "L" CARLINE (PORT)

1
- THROTTLE CLOSED.
DOES TECH 1 DISPLAY THROTTLE POSITION .2V (200 mV) OR BELOW?

YES →

NO →
- CODE 22 IS INTERMITTENT.
IF NO ADDITIONAL CODES WERE STORED, REFER TO "DIAGNOSTIC AIDS

2
- DISCONNECT TPS SENSOR.
JUMPER CKTS 416 & 417 TOGETHER.
TECH 1 SHOULD DISPLAY THROTTLE POSITION OVER 4.0 V (4000 mV).
DOES IT?

NO →

YES →

3
- REFER TO SPECIFIC INSTRUCTIONS.

4
- PROBE CKT 417 WITH A TEST LIGHT CONNECTED TO BATTERY VOLTAGE.
TECH 1 SHOULD DISPLAY THROTTLE POSITION OVER 4.0V (4000 mV).
DOES IT?

YES →
CKT 416 OPEN OR SHORTED TO GROUND
OR
FAULTY CONNECTION
OR
FAULTY ECM.

NO →
CKT 417 OPEN OR SHORTED TO GROUND, OR SHORTED TO THROTTLE POSITION SENSOR GROUND CIRCUIT
OR
FAULTY ECM CONNECTION
OR
FAULTY ECM.

"AFTER REPAIRS," REFER TO CODE CRITERIA   AND CONFIRM CODE DOES NOT RESET.

2.3L (VIN A) ENGINE — DIAGNOSTIC TROUBLE CODE CHART — 1992 BERETTA

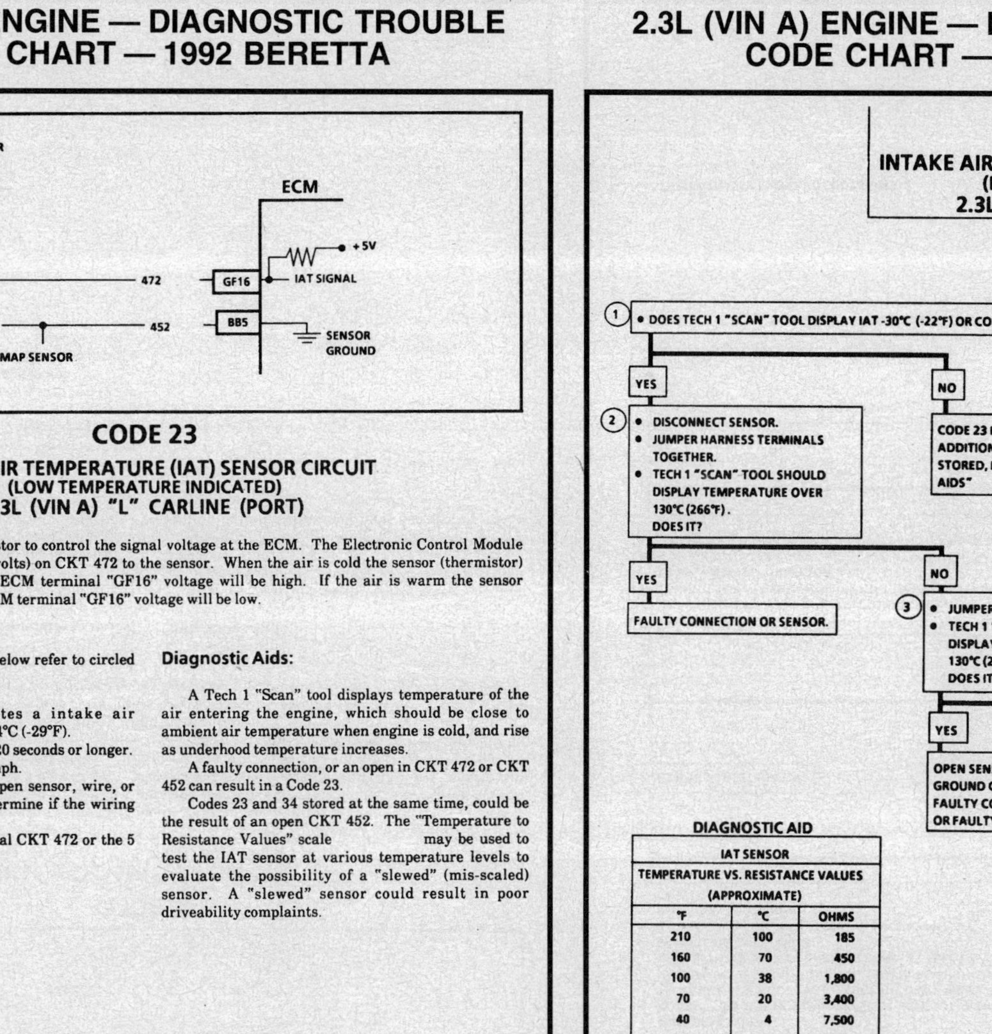

CODE 23
INTAKE AIR TEMPERATURE (IAT) SENSOR CIRCUIT
(LOW TEMPERATURE INDICATED)
2.3L (VIN A) "L" CARLINE (PORT)

Circuit Description:

The IAT sensor uses a thermistor to control the signal voltage at the ECM. The Electronic Control Module (ECM) applies a voltage (about 5 volts) on CKT 472 to the sensor. When the air is cold the sensor (thermistor) resistance is high, therefore, the ECM terminal "GF16" voltage will be high. If the air is warm the sensor resistance is low, therefore, the ECM terminal "GF16" voltage will be low.

Test Description: Number(s) below refer to circled number(s) on the diagnostic chart.

1. Code 23 will set if:
   - A signal voltage indicates a intake air temperature below about -34°C (-29°F).
   - Time since engine start is 320 seconds or longer.
   - Vehicle speed less than 15 mph.
2. A Code 23 will set due to an open sensor, wire, or connection. This test will determine if the wiring and ECM are OK.
3. This will determine if the signal CKT 472 or the 5 volts return CKT 452 is open.

Diagnostic Aids:

A Tech 1 "Scan" tool displays temperature of the air entering the engine, which should be close to ambient air temperature when engine is cold, and rise as underhood temperature increases.

A faulty connection, or an open in CKT 472 or CKT 452 can result in a Code 23.

Codes 23 and 34 stored at the same time, could be the result of an open CKT 452. The "Temperature to Resistance Values" scale   may be used to test the IAT sensor at various temperature levels to evaluate the possibility of a "slewed" (mis-scaled) sensor. A "slewed" sensor could result in poor driveability complaints.

CODE 23
INTAKE AIR TEMPERATURE (IAT) SENSOR CIRCUIT
(LOW TEMPERATURE INDICATED)
2.3L (VIN A) "L" CARLINE (PORT)

(1) • DOES TECH 1 "SCAN" TOOL DISPLAY IAT -30°C (-22°F) OR COLDER?

- **YES**
  - **(2)** • DISCONNECT SENSOR.
    • JUMPER HARNESS TERMINALS TOGETHER.
    • TECH 1 "SCAN" TOOL SHOULD DISPLAY TEMPERATURE OVER 130°C (266°F). DOES IT?
    - **YES** → FAULTY CONNECTION OR SENSOR.
    - **NO**
      - **(3)** • JUMPER CKT 472 TO GROUND.
        • TECH 1 "SCAN" TOOL SHOULD DISPLAY TEMPERATURE OVER 130°C (266°F). DOES IT?
        - **YES** → OPEN SENSOR GROUND CIRCUIT, FAULTY CONNECTION OR FAULTY ECM.
        - **NO** → OPEN CKT 472, FAULTY CONNECTION OR FAULTY ECM.
- **NO** → CODE 23 IS INTERMITTENT. IF NO ADDITIONAL CODES WERE STORED, REFER TO "DIAGNOSTIC AIDS"

DIAGNOSTIC AID		
IAT SENSOR		
TEMPERATURE VS. RESISTANCE VALUES (APPROXIMATE)		
°F	°C	OHMS
210	100	185
160	70	450
100	38	1,800
70	20	3,400
40	4	7,500
20	-7	13,500
0	-18	25,000
-40	-40	100,700

"AFTER REPAIRS," REFER TO CODE CRITERIA   AND CONFIRM CODE DOES NOT RESET.

2.3L (VIN A) ENGINE — DIAGNOSTIC TROUBLE CODE CHART — 1992 BERETTA

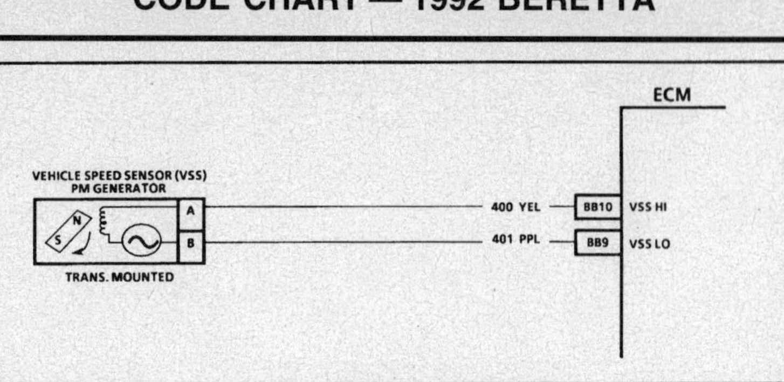

CODE 24
VEHICLE SPEED SENSOR (VSS) CIRCUIT
2.3L (VIN A) "L" CARLINE (PORT)

Circuit Description:

Vehicle speed information is provided to the ECM by the Vehicle Speed Sensor (VSS) which is a Permanent Magnet (PM) generator that is mounted in the transmission. The PM generator produces a pulsing voltage whenever vehicle speed is over about 3 mph (5 kph). The AC voltage level and the number of pulses increases with vehicle speed. The Electronic Control Module (ECM) then converts the pulsing voltage to mph which is used for calculations, and the mph can be displayed with a Tech 1 "Scan" tool. Output of the generator can also be seen by using a digital voltmeter on the AC scale while rotating the generator.

The function of VSS buffer used in past model years has been incorporated into the ECM. The ECM then supplies the necessary signal for the instrument panel for operating the speedometer, the odometer, and for the cruise control module.

Test Description: Number(s) below refer to circled number(s) on the diagnostic chart.

1. Code 24 will set if vehicle speed is less than 2 mph when:
   - Engine speed is between 1600 and 3000 rpm.
   - Throttle Position Sensor (TPS) is greater than 7% and less than 20%.
   - MAP between 40 - 60 kPa.
   - All conditions met for 20 seconds.
   - No Code 21 or 22.

   These conditions are met during a road load operation. Disregard Code 24 that sets when drive wheels are not turning.

   The PM generator only produces a signal if drive wheels are turning greater than 3 mph (5 kph).

2. Check MEM-CAL for correct application before replacing ECM.

Diagnostic Aids:

Tech 1 should indicate a vehicle speed whenever the drive wheels are turning greater than 3 mph, (5 kph).

Check CKT 400 and 401 for proper connections. Be sure they are clean and tight and the harness is routed correctly.

2.3L (VIN A) ENGINE — DIAGNOSTIC TROUBLE CODE CHART — 1992 BERETTA

CODE 24
VEHICLE SPEED SENSOR (VSS) CIRCUIT
2.3L (VIN A) "L" CARLINE (PORT)

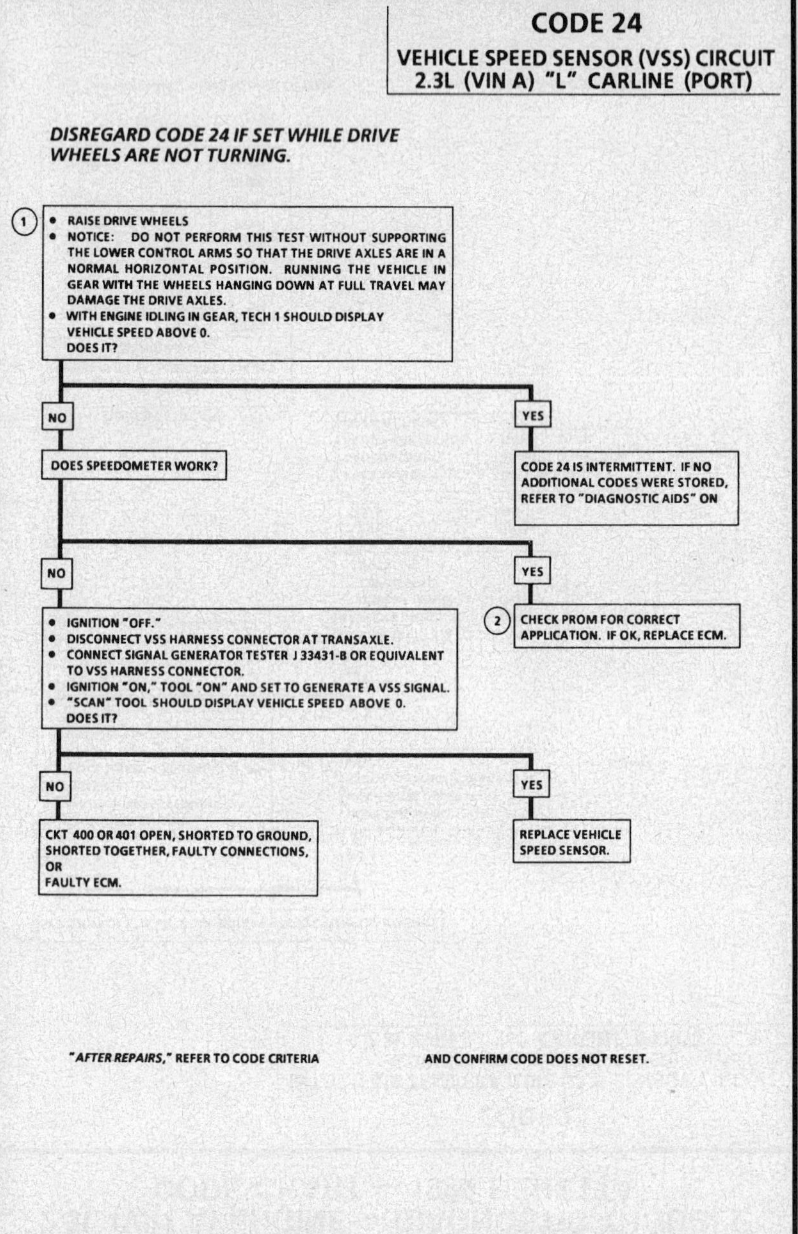

2.3L (VIN A) ENGINE — DIAGNOSTIC TROUBLE CODE CHART — 1992 BERETTA

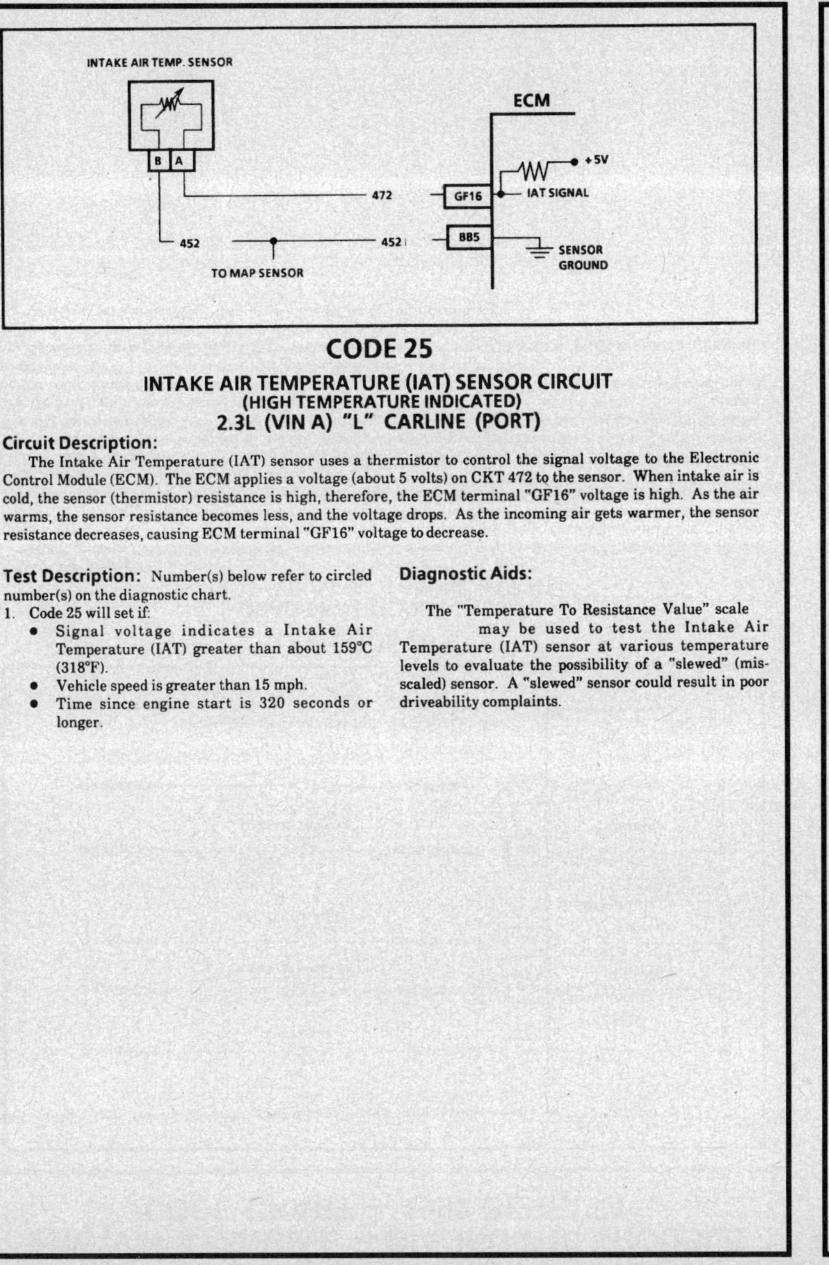

CODE 25
INTAKE AIR TEMPERATURE (IAT) SENSOR CIRCUIT
(HIGH TEMPERATURE INDICATED)
2.3L (VIN A) "L" CARLINE (PORT)

Circuit Description:

The Intake Air Temperature (IAT) sensor uses a thermistor to control the signal voltage to the Electronic Control Module (ECM). The ECM applies a voltage (about 5 volts) on CKT 472 to the sensor. When intake air is cold, the sensor (thermistor) resistance is high, therefore, the ECM terminal "GF16" voltage is high. As the air warms, the sensor resistance becomes less, and the voltage drops. As the incoming air gets warmer, the sensor resistance decreases, causing ECM terminal "GF16" voltage to decrease.

Test Description: Number(s) below refer to circled number(s) on the diagnostic chart.

1. Code 25 will set if:
   - Signal voltage indicates a Intake Air Temperature (IAT) greater than about 159°C (318°F).
   - Vehicle speed is greater than 15 mph.
   - Time since engine start is 320 seconds or longer.

Diagnostic Aids:

The "Temperature To Resistance Value" scale may be used to test the Intake Air Temperature (IAT) sensor at various temperature levels to evaluate the possibility of a "slewed" (mis-scaled) sensor. A "slewed" sensor could result in poor driveability complaints.

2.3L (VIN A) ENGINE — DIAGNOSTIC TROUBLE CODE CHART — 1992 BERETTA

CODE 25
INTAKE AIR TEMPERATURE (IAT) SENSOR CIRCUIT
(HIGH TEMPERATURE INDICATED)
2.3L (VIN A) "L" CARLINE (PORT)

(1) DOES TECH 1 "SCAN" TOOL DISPLAY IAT OF 145°C (293°F) OR HOTTER?

- **YES** → DISCONNECT SENSOR. TECH 1 "SCAN" TOOL SHOULD DISPLAY TEMPERATURE BELOW -30°C (-22°F). DOES IT?
  - **YES** → REPLACE SENSOR.
  - **NO** → CKT 472 SHORTED TO GROUND OR TO SENSOR GROUND OR ECM IS FAULTY.
- **NO** → CODE 25 IS INTERMITTENT. IF NO ADDITIONAL CODES WERE STORED, REFER TO "DIAGNOSTIC AIDS"

DIAGNOSTIC AID

INTAKE AIR TEMPERATURE SENSOR		
TEMPERATURE VS. RESISTANCE VALUES (APPROXIMATE)		
°C	°F	OHMS
100	212	177
90	194	241
80	176	332
70	158	467
60	140	667
50	122	973
45	113	1188
40	104	1459
35	95	1802
30	86	2238
25	77	2796
20	68	3520
15	59	4450
10	50	5670
5	41	7280
0	32	9420
-5	23	12300
-10	14	16180
-15	5	21450
-20	-4	28680
-30	-22	52700
-40	-40	100700

"AFTER REPAIRS," REFER TO CODE CRITERIA AND CONFIRM CODE DOES NOT RESET.

2.3L (VIN A) ENGINE — DIAGNOSTIC TROUBLE CODE CHART — 1992 BERETTA

2.3L (VIN A) ENGINE — DIAGNOSTIC TROUBLE CODE CHART — 1992 BERETTA

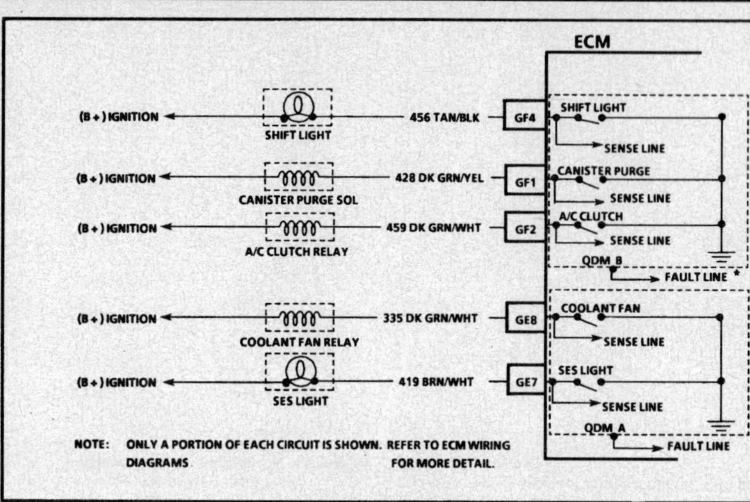

NOTE: ONLY A PORTION OF EACH CIRCUIT IS SHOWN. REFER TO ECM WIRING DIAGRAMS   FOR MORE DETAIL.

CODE 26
QUAD-DRIVER (QDM) CIRCUIT
2.3L (VIN A) "L" CARLINE (PORT)

Circuit Description:

The Electronic Control Module (ECM) controls most components with electronic switches which complete a ground circuit when turned "ON." These switches are arranged in groups of 4, called Quad-Driver Modules (QDM's) which can independently control up to 4 outputs (ECM terminals), although not all outputs are used. When an output is "ON," the terminal is grounded and its voltage normally will be low. When an output is "OFF," its terminal voltage normally will be high.

QDM's are fault protected. If a relay or solenoid coil is shorted, having very low or zero resistance, or if the control side of the circuit is shorted to voltage, it would allow too much current into the QDM. The QDM senses this and the output turns "OFF" or its internal resistance increases to limit current flow and protect the QDM. The result is high output terminal voltage when it should be low. If the circuit from B+ or the component is open, or the control side of the circuit is shorted to ground, terminal voltage will be low, even when output is commanded "OFF." Either of these conditions is considered to be a QDM fault.

Each QDM has a separate fault line to indicate the presence of a current fault to the ECM's central processor. A Tech 1 "Scan" tool displays the status of each of these fault lines as "Low" = OK, "High" = Fault.

Code 26 is set if either QDM fault line is "High" for 20 seconds or more.

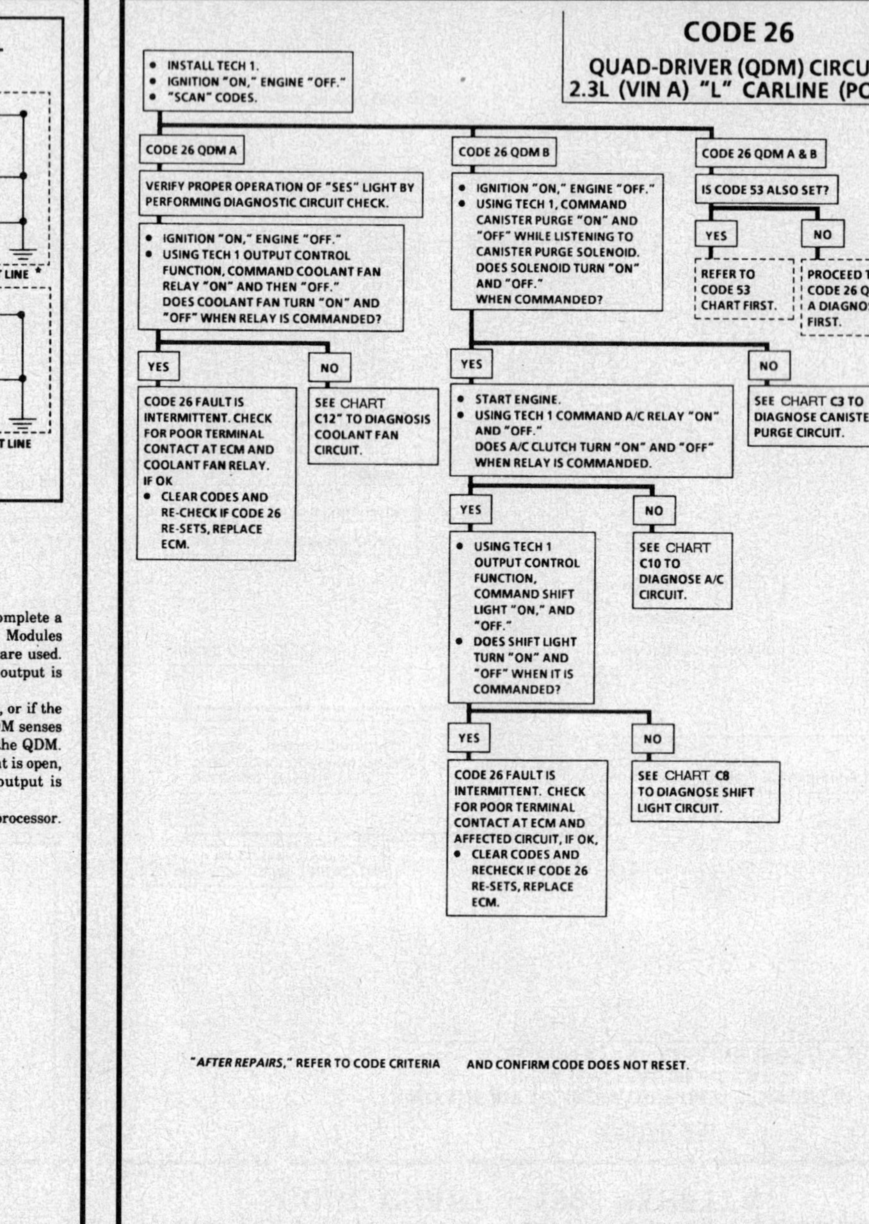

"*AFTER REPAIRS*," REFER TO CODE CRITERIA   AND CONFIRM CODE DOES NOT RESET.

2.3L (VIN A) ENGINE — DIAGNOSTIC TROUBLE CODE CHART — 1992 BERETTA

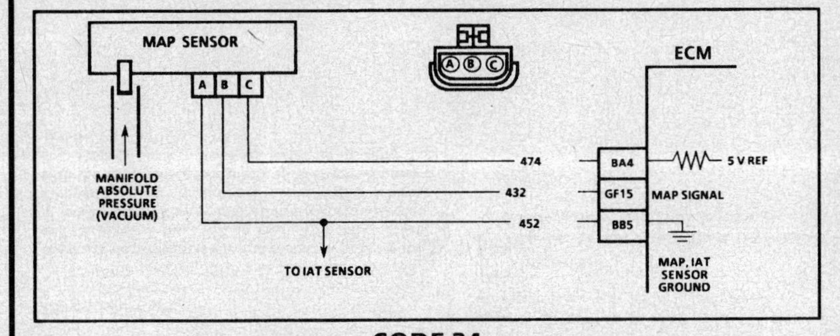

CODE 34

MANIFOLD ABSOLUTE PRESSURE (MAP) SENSOR CIRCUIT
(SIGNAL VOLTAGE LOW - HIGH VACUUM)
2.3L (VIN A) "L" CARLINE (PORT)

Circuit Description:

The Manifold Absolute Pressure (MAP) sensor responds to changes in manifold pressure (vacuum). The Electronic Control Module (ECM) receives this information as a signal voltage that will vary from about 1 to 1.5 volts at closed throttle (idle), to 4.5 - 4.8 volts at Wide Open Throttle (WOT) (low vacuum).

If the MAP sensor fails, the ECM will substitute a fixed MAP value and use the Throttle Position Sensor (TPS) to control fuel delivery.

Test Description: Number(s) below refer to circled number(s) on the diagnostic chart.

1. This step determines if Code 34 is the result of a hard failure or an intermittent condition.
   Code 34 will set when:
   - Engine running.
   - No Code 21.
   - MAP greater than 14 kPa.
   - Engine rpm less than 1200 or TPS greater than 15.2%.
   - Above conditions met for .2 second.
2. Jumpering harness terminals "B" to "C" (5 volts to signal circuit) will determine if the sensor is at fault, or if there is a problem with the ECM or wiring.
3. The Tech 1 "Scan" tool may not display 5 volts. The important thing is that the ECM recognizes the voltage as more than 4 volts, indicating that the ECM and CKT 432 are OK.

Diagnostic Aids

An intermittent open in CKT 432 or CKT 474 will result in a Code 34. With the ignition "ON" and the engine "OFF," the manifold pressure is equal to atmospheric pressure and the signal voltage will be high. This information is used by the ECM as an indication of vehicle altitude.

Comparison of this reading with a known good vehicle with the same sensor is a good way to check accuracy of a "suspect" sensor. Readings should be the same ± .4 volt. Also CHART C-1D can be used to test the MAP sensor.

- Check all connections.
- Disconnect sensor from bracket and twist sensor by hand (only) to check for intermittent connections. Output changes greater than .1 volt indicates a bad connector or connection. If OK, replace sensor.

Important
- Make sure electrical connector remains securely fastened.

- Refer to CHART C-1D, MAP sensor voltage vs. atmospheric pressure for further diagnosis.

Important
- After repairs, use Tech 1 "Block Learn Reset" function to reset block learn to 128.

2.3L (VIN A) ENGINE — DIAGNOSTIC TROUBLE CODE CHART — 1992 BERETTA

CODE 34

MANIFOLD ABSOLUTE PRESSURE (MAP) SENSOR CIRCUIT
(SIGNAL VOTAGE LOW - HIGH VACUUM)
2.3L (VIN A) "L" CARLINE (PORT)

(1)
- ENGINE IDLING.
- DOES TECH 1 DISPLAY MAP VOLTAGE BELOW .25 VOLT?

→ YES

(2)
- IGNITION "OFF."
- DISCONNECT SENSOR ELECTRICAL CONNECTOR.
- JUMPER HARNESS TERMINALS "B" TO "C".
- IGNITION "ON."
- MAP VOLTAGE SHOULD READ OVER 4.7 VOLTS. DOES IT?

→ NO

CODE 34 IS INTERMITTENT. IF NO ADDITIONAL CODES WERE STORED, REFER TO "DIAGNOSTIC AIDS"

→ NO (from 2)

(3)
- IGNITION "OFF."
- REMOVE JUMPER WIRE.
- PROBE TERMINAL "B" (CKT 432) WITH A TEST LIGHT TO BATTERY VOLTAGE.
- IGNITION "ON."
- TECH 1 SHOULD READ OVER 4 VOLTS. DOES IT?

→ YES (from 2)

FAULTY CONNECTION OR SENSOR.

→ YES (from 3)

5 VOLT REFERENCE CIRCUIT OPEN
OR
SHORTED TO GROUND
OR
FAULTY ECM.

→ NO (from 3)

CKT 432 OPEN
OR
CKT 432 SHORTED TO GROUND
OR
CKT 432 SHORTED TO SENSOR GROUND
OR
FAULTY ECM.

"AFTER REPAIRS," REFER TO CODE CRITERIA   AND CONFIRM CODE DOES NOT RESET.

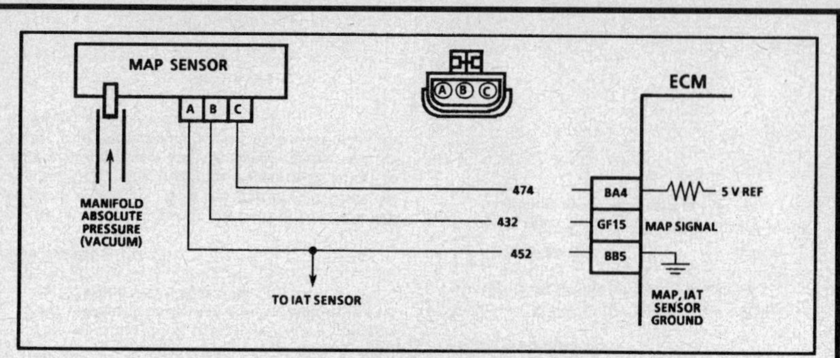

CODE 33

MANIFOLD ABSOLUTE PRESSURE (MAP) SENSOR CIRCUIT
(SIGNAL VOLTAGE HIGH - LOW VACUUM)
2.3L (VIN A) "L" CARLINE (PORT)

Circuit Description:

The Manifold Absolute Pressure (MAP) sensor responds to changes in manifold pressure (vacuum). The ECM receives this information as a signal voltage that will vary from about 1 to 1.5 volts at closed throttle (idle) to 4.5 - 4.8 volts at Wide Open Throttle (WOT) (low vacuum).

If the MAP sensor fails, the ECM will substitute a fixed MAP value and use the Throttle Position Sensor (TPS) to control fuel delivery.

Test Description: Number(s) below refer to circled number(s) on the diagnostic chart.

1. This step will determine if Code 33 is the result of a hard failure or an intermittent condition.
   A Code 33 will set under the following condition:
   - Engine running.
   - MAP signal less than 80 kPa.
   - No Codes 21 or 22.
   - TPS less than 12%.
   - VSS less than 1 mph.
   - Above conditions met for 5 seconds.
2. This step simulates conditions for a Code 34. If the ECM recognizes the change, the ECM and CKT 474 and CKT 432 are OK. If CKT 452 is open, there may also be a stored Code 23.

Diagnostic Aids:

With the ignition "ON" and the engine stopped, the manifold pressure is equal to atmospheric pressure and the signal voltage will be high. This information is used by the ECM as an indication of vehicle altitude. Comparison of this reading with a known good vehicle with the same sensor is a good way to check accuracy of a "suspect" sensor. Readings should be the same ± .4 volt.

A Code 33 will result if CKT 452 is open or if CKT 432 is shorted to voltage or to CKT 474.

- Check all connections.
- Disconnect sensor from bracket and twist sensor by hand (only) to check for intermittent connections. Output changes greater than .1 volt indicates a bad connector or connection. If OK, replace sensor.

> **Important**
> - Make sure electrical connector remains securely fastened.

- Refer to CHART C-1D, MAP sensor voltage vs. atmospheric pressure for further diagnosis.

> **Important**
> - After repairs, use Tech 1 "Block Learn Reset" function to reset block learn to 128.

CODE 33

MANIFOLD ABSOLUTE PRESSURE (MAP) SENSOR CIRCUIT
(SIGNAL VOLTAGE HIGH - LOW VACUUM)
2.3L (VIN A) "L" CARLINE (PORT)

(1)
- IF ENGINE IDLE IS ROUGH, UNSTABLE OR INCORRECT, CORRECT BEFORE USING CHART.
- ENGINE IDLING.
- DOES TECH 1 DISPLAY A MAP OF 3.75 VOLTS OR OVER?

YES →

(2)
- IGNITION "OFF."
- DISCONNECT MAP SENSOR ELECTRICAL CONNECTOR.
- IGNITION "ON."
- TECH 1 SHOULD DISPLAY A VOLTAGE OF 1 VOLT OR LESS. DOES IT?

NO → CODE 33 IS INTERMITTENT. IF NO ADDITIONAL CODES WERE STORED, REFER TO "DIAGNOSTIC AIDS"

YES →
- PROBE CKT 452 WITH A TEST LIGHT TO 12 VOLTS.
- TEST LIGHT SHOULD LIGHT. DOES IT?

NO → CKT 432 SHORTED TO VOLTAGE, SHORTED TO CKT 474, OR FAULTY ECM.

YES → PLUGGED OR LEAKING SENSOR VACUUM HOSE OR FAULTY MAP SENSOR.

NO → OPEN CKT 452.

"AFTER REPAIRS," REFER TO CODE CRITERIA AND CONFIRM CODE DOES NOT RESET.

2.3L (VIN A) ENGINE — DIAGNOSTIC TROUBLE CODE CHART — 1992 BERETTA

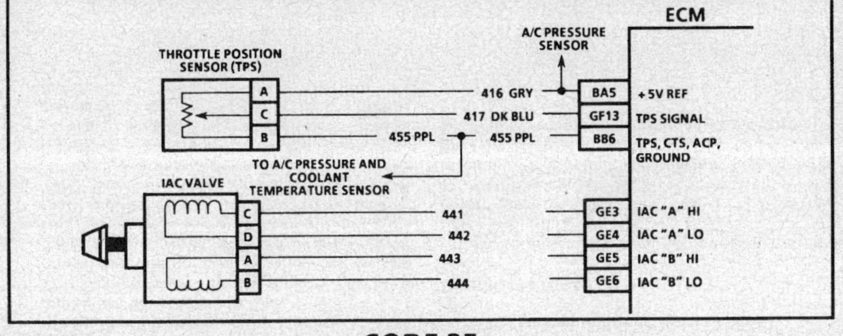

CODE 35
IDLE SPEED ERROR
2.3L (VIN A) "L" CARLINE (PORT)

Circuit Description:

The ECM controls idle speed to a calculated, "desired" rpm based on sensor inputs and actual engine rpm, determined by the time between successive 2X ignition reference pulses from the ignition module. The ECM uses 4 circuits to move an Idle Air Control (IAC) valve, which allows varying amounts of air flow into the intake manifold, controlling idle speed.

Code 35 sets when:
- Engine speed is at least 175 rpm greater or 200 rpm less than "desired."
- TPS voltage indicates throttle is open less than 1%.
- VSS indicates vehicle speed is less than 3 mph.
- All above conditions are continuously met for 5 seconds or more.
- IAC steps must be less than 10 or greater than 140.

Test Description: Number(s) below refer to circled number(s) on the diagnostic chart.

1. The IAC tester is used to extend and retract the IAC valve. Valve movement is verified by an engine speed change. If no change in engine speed occurs, the valve can be retested when removed from the throttle body.
2. This step checks the quality of the IAC movement in Step 1. Between 700 rpm and about 1500 rpm, the engine speed should change smoothly with each flash of the tester light in both extend and retract. If the IAC valve is retracted beyond the control range (about 1500 rpm), it may take many flashes in the extend position before engine speed will begin to drop. This is normal on certain engines, fully extending IAC may cause engine stall. This may be normal.

Step 1 verified proper IAC valve operation while this step checks the IAC circuits. Each lamp on the node light should flash red and green while the IAC valve is cycled. While the sequence of color is not important if either light is "OFF" or does not flash red and green, check the circuits for faults, beginning with poor terminal contacts.

Diagnostic Aids:

Check for vacuum leaks, disconnected or brittle vacuum hoses, cuts, etc. Examine manifold and throttle body gaskets for proper seal. Check for cracked intake manifold. Check open, shorts, or poor connections to IAC valve in CKTs 441, 442, 443 and 444.

An open, short, or poor connection in CKTs 441, 442, 443 or 444 will result in improper idle control and may cause Code 35.

An IAC valve which is stopped and cannot respond to the ECM, a throttle stop screw which has been tampered with, or a damaged throttle body or linkage could cause Code 35. If no problem is found and Code resets, replace ECM.

⚡ Important
- If Code 35 is currently set and speedometer is not accurately reflecting correct vehicle, refer to speed Code 24 (vehicle speed sensor circuit) diagnostics.

2.3L (VIN A) ENGINE — DIAGNOSTIC TROUBLE CODE CHART — 1992 BERETTA

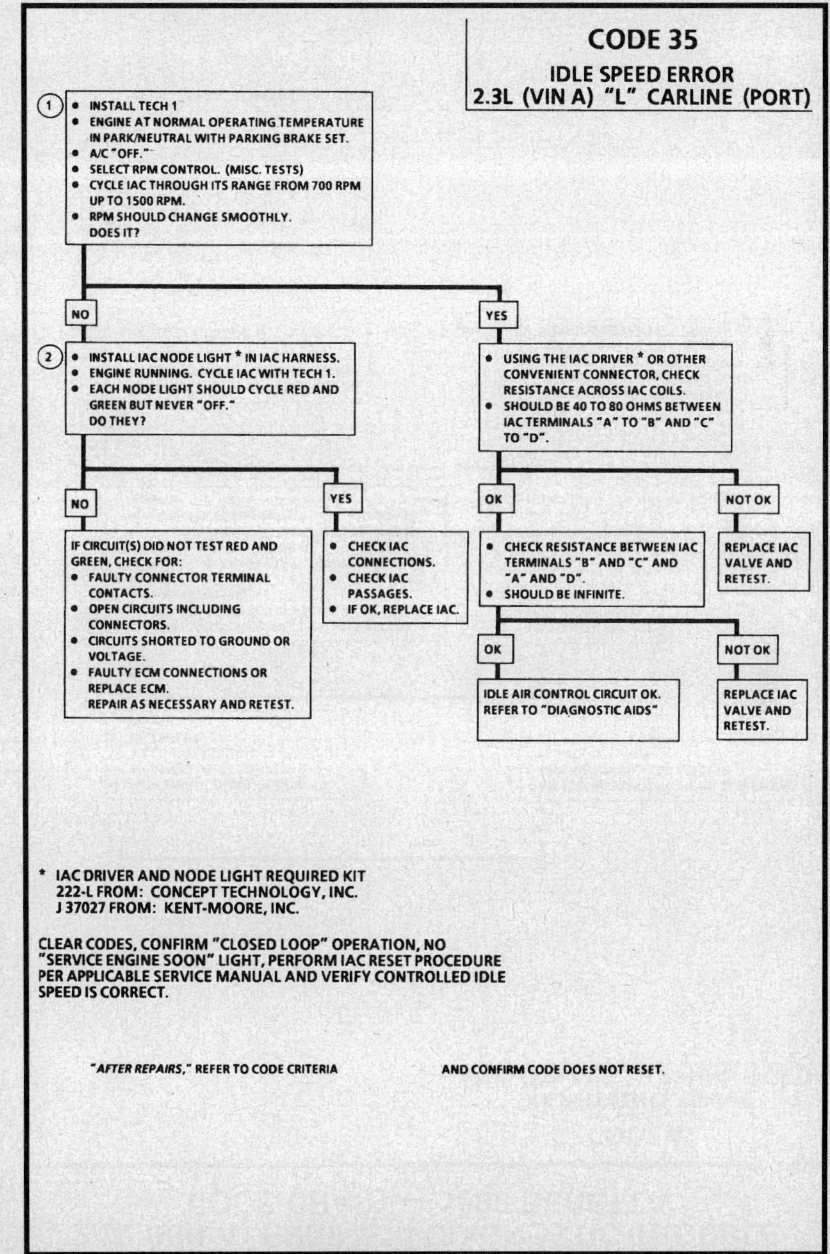

* IAC DRIVER AND NODE LIGHT REQUIRED KIT
222-L FROM: CONCEPT TECHNOLOGY, INC.
J 37027 FROM: KENT-MOORE, INC.

CLEAR CODES, CONFIRM "CLOSED LOOP" OPERATION, NO "SERVICE ENGINE SOON" LIGHT, PERFORM IAC RESET PROCEDURE PER APPLICABLE SERVICE MANUAL AND VERIFY CONTROLLED IDLE SPEED IS CORRECT.

"AFTER REPAIRS," REFER TO CODE CRITERIA   AND CONFIRM CODE DOES NOT RESET.

2.3L (VIN A) ENGINE — DIAGNOSTIC TROUBLE CODE CHART — 1992 BERETTA

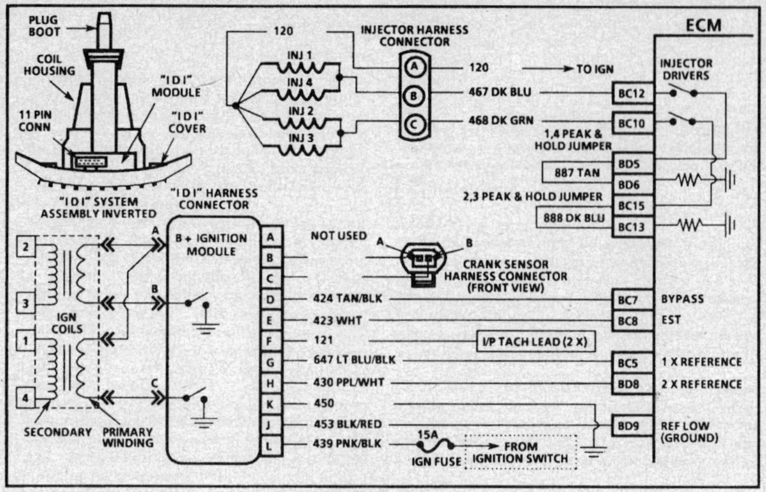

CODE 41
1X REFERENCE CIRCUIT
2.3L (VIN A) "L" CARLINE (PORT)

Circuit Description:

The ignition module sends a reference signal to the Electronic Control Module (ECM) once per revolution to indicate crankshaft position so that the ECM can determine when to pulse the injectors for cylinders 2 and 3 in the ASDF fuel control mode. This signal may be described as a synchronization signal and is called the 1X reference because it occurs one time per revolution. The ignition module applies 5 volts from terminal "G" through CKT 647 to ECM terminal "BC5" and in effect, switches this circuit to ground for a very short period of time, 125 degrees before TDC of cylinders 2 and 3. Code 41 is set if the ECM receives (8) 2X reference pulses with no 1X reference pulses. When Code 41 is present, the ECM pulses the injectors in the SSDF (simultaneous) mode.

Test Description: Number(s) below refer to circled number(s) on the diagnostic chart.

1. This determines if the ECM recognizes a problem. If it doesn't set Code 41 at this point, the problem is intermittent and could be due to a loose connection. See "Diagnostic Aids."

2. This step simulates the 1X signal. The ECM should recognize the drop in voltage as the test light probe is removed, if the circuit and ECM are OK. This step will give accurate results only if the chart sequence is used - ignition "OFF," ignition "ON," "Scan" tool set to 1X reference and terminal "G" touched with test light probe.

3. If the ECM did not recognize the simulation of the 1X signal, CKT 647 may be open or shorted to ground or voltage. If CKT 647 is OK, the ECM is faulty.

4. Step 2 indicated that CKT 647 is OK and the ECM is capable of recognizing the simulated 1X reference pulse. This indicates either a poor connection at ignition module terminal "G" or a faulty ignition module caused the Code 41.

Diagnostic Aids:

An intermittent may be caused by a poor connection, rubbed through wire insulation, or a wire broken inside the insulation. Inspect ECM harness connector terminal "BC5" and ignition module terminal "G" for improper mating, broken locks, improperly formed or damaged terminals, poor terminal to wire connection and damaged harness.

2.3L (VIN A) ENGINE — DIAGNOSTIC TROUBLE CODE CHART — 1992 BERETTA

CODE 41
1X REFERENCE CIRCUIT
2.3L (VIN A) "L" CARLINE (PORT)

2.3L (VIN A) ENGINE — DIAGNOSTIC TROUBLE CODE CHART — 1992 BERETTA

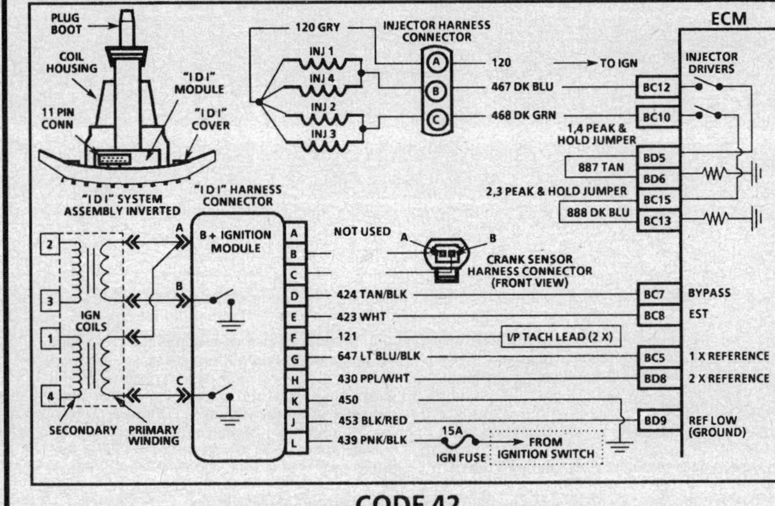

CODE 42
ELECTRONIC SPARK TIMING (EST) CIRCUIT
2.3L (VIN A) "L" CARLINE (PORT)

Circuit Description:

The ignition module sends a reference signal to the ECM when the engine is cranking or running. While the engine is under 700 rpm, the ignition module controls the ignition timing. When the engine speed exceeds 700 rpm, the ECM sends a 5 volts signal on the "Bypass" CKT 424 to switch the timing to ECM control through the EST CKT 423. Engine will remain under EST control until rpm drops below 150.

An open or ground in the EST or "Bypass" circuit will set a Code 42 and cause the engine to run on module or "Bypass" timing. This will result in poor performance and poor fuel economy.

Test Description: Number(s) below refer to circled number(s) on the diagnostic chart.

1. Checks to see if ECM recognizes a problem. If it doesn't set Code 42 at this point, it is an intermittent problem and could be due to a loose connection.
2. With the ECM disconnected, the ohmmeter should be reading less than 500 ohms, which is the normal resistance of the ignition module. A higher resistance would indicate a fault in CKT 423, a poor ignition module connection or a faulty ignition module.
3. If the test light was "ON" when connected from 12 volts to ECM harness terminal "BC7", either CKT 423 is shorted to ground or the ignition module is faulty.
4. Checks to see if ignition module switches when the bypass circuit is energized by 12 volts through the test light.

If the ignition module actually switches, the ohmmeter reading should shift to over 8000 ohms.

5. Disconnecting the ignition module should make the ohmmeter read as if it were monitoring an open circuit (infinite reading). If the ohmmeter has a reading other than infinite, CKT 423 is shorted to ground.

Diagnostic Aids:

An intermittent may be caused by a poor connection, rubbed through wire insulation, or a wire broken inside the insulation. Inspect ECM harness connectors for backed out terminals "BC7" or "BC8", improper mating, broken locks, improperly formed or damaged terminals, poor terminal to wire connection, and damaged harness.

2.3L (VIN A) ENGINE — DIAGNOSTIC TROUBLE CODE CHART — 1992 BERETTA

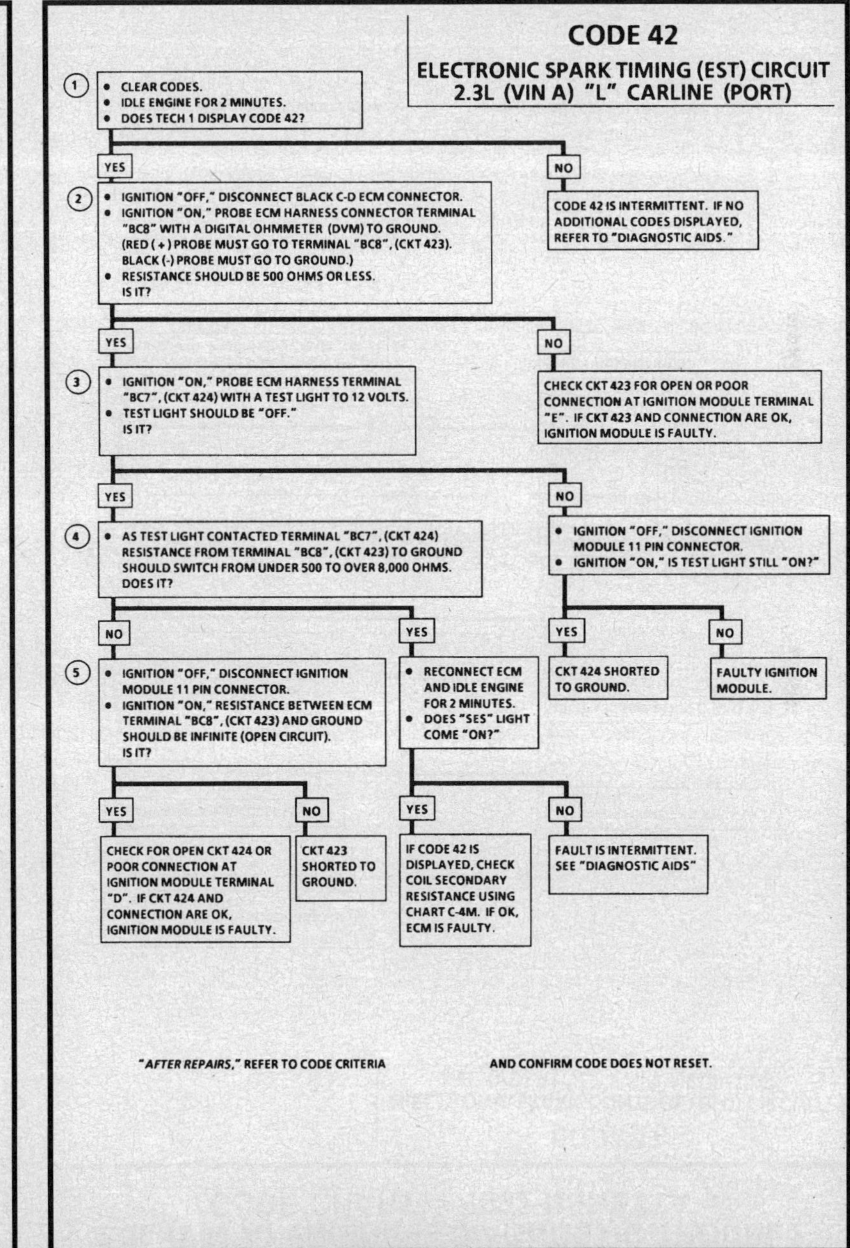

2.3L (VIN A) ENGINE — DIAGNOSTIC TROUBLE CODE CHART — 1992 BERETTA

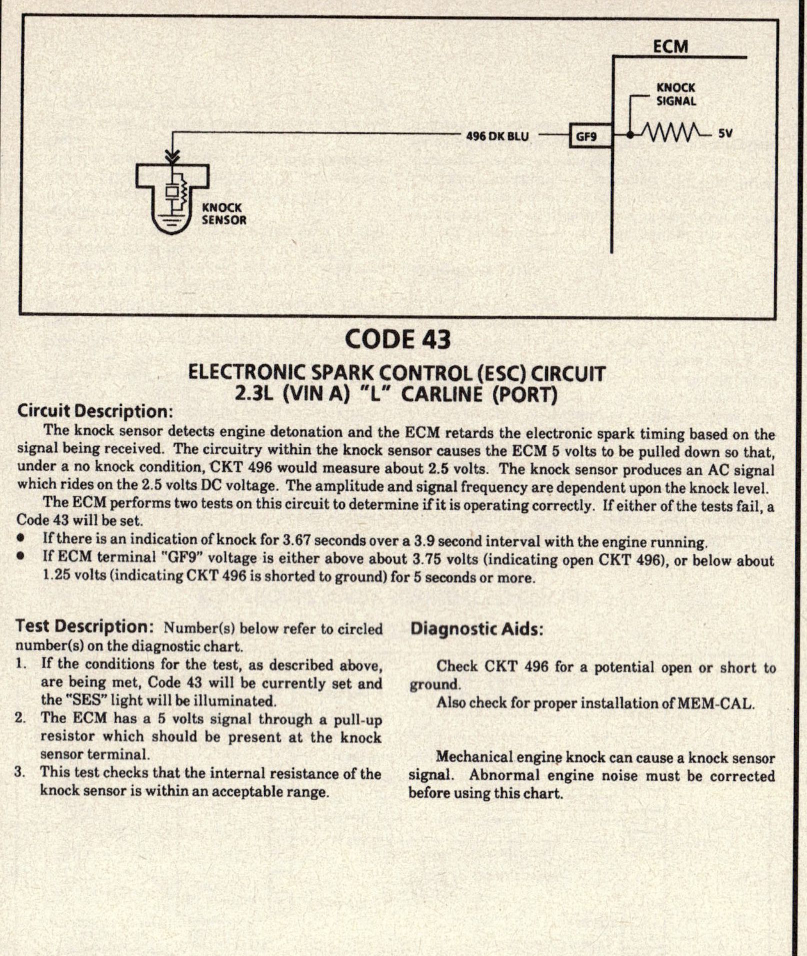

CODE 43
ELECTRONIC SPARK CONTROL (ESC) CIRCUIT
2.3L (VIN A) "L" CARLINE (PORT)

Circuit Description:

The knock sensor detects engine detonation and the ECM retards the electronic spark timing based on the signal being received. The circuitry within the knock sensor causes the ECM 5 volts to be pulled down so that, under a no knock condition, CKT 496 would measure about 2.5 volts. The knock sensor produces an AC signal which rides on the 2.5 volts DC voltage. The amplitude and signal frequency are dependent upon the knock level.

The ECM performs two tests on this circuit to determine if it is operating correctly. If either of the tests fail, a Code 43 will be set.

- If there is an indication of knock for 3.67 seconds over a 3.9 second interval with the engine running.
- If ECM terminal "GF9" voltage is either above about 3.75 volts (indicating open CKT 496), or below about 1.25 volts (indicating CKT 496 is shorted to ground) for 5 seconds or more.

Test Description: Number(s) below refer to circled number(s) on the diagnostic chart.

1. If the conditions for the test, as described above, are being met, Code 43 will be currently set and the "SES" light will be illuminated.
2. The ECM has a 5 volts signal through a pull-up resistor which should be present at the knock sensor terminal.
3. This test checks that the internal resistance of the knock sensor is within an acceptable range.

Diagnostic Aids:

Check CKT 496 for a potential open or short to ground.

Also check for proper installation of MEM-CAL.

Mechanical engine knock can cause a knock sensor signal. Abnormal engine noise must be corrected before using this chart.

2.3L (VIN A) ENGINE — DIAGNOSTIC TROUBLE CODE CHART — 1992 BERETTA

CODE 43
ELECTRONIC SPARK CONTROL (ESC) CIRCUIT
2.3L (VIN A) "L" CARLINE (PORT)

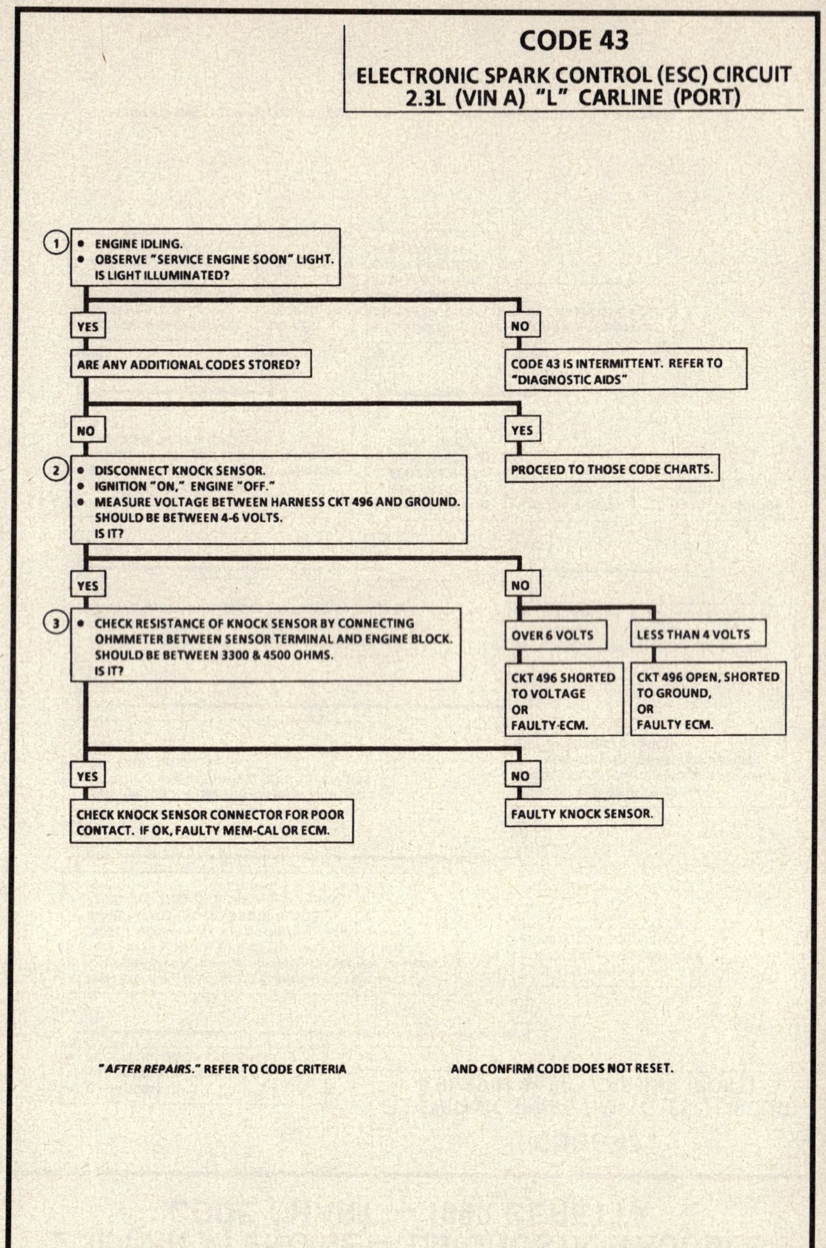

2.3L (VIN A) ENGINE — DIAGNOSTIC TROUBLE CODE CHART — 1992 BERETTA

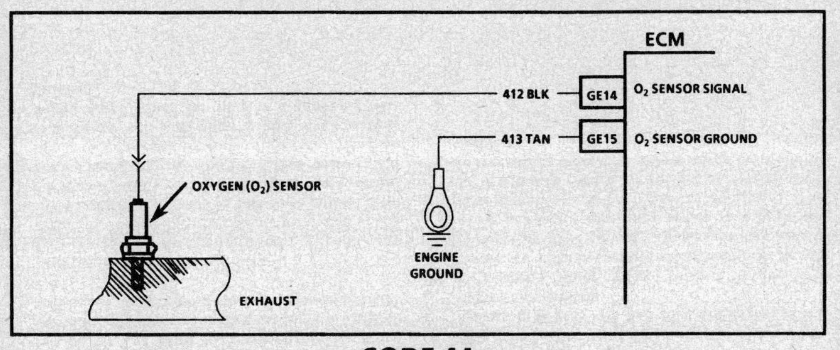

CODE 44
OXYGEN (O_2) SENSOR CIRCUIT
(LEAN EXHAUST INDICATED)
2.3L (VIN A) "L" CARLINE (PORT)

Circuit Description:

The ECM supplies a voltage of about .45 volt between terminals "GE14" and "GE15". (If measured with a 10 megohm digital voltmeter, this may read as low as .32 volt.) The O_2 sensor varies the voltage within a range of about 1 volt if the exhaust is rich, down through about .10 volt if exhaust is lean.

The sensor is like an open circuit and produces no voltage when it is below 315°C (600°F). An open sensor circuit or cold sensor causes "Open Loop" operation.

Test Description: Number(s) below refers to circled number(s) on the diagnostic chart.

1. Code 44 is set when the O_2 sensor signal voltage on CKT 412:
   - Remains below .3 volt for 90 seconds or more.
   - The system is operating in "Closed Loop."
   - No Code 33 or 34.
   - "Closed Loop" integrator active.

Diagnostic Aids:

The Code 44 or lean exhaust is most likely caused by one of the following:

- CKT 413 - If CKT 413 is open, the voltage at terminal "GE14" will be over 1 volt.
- Fuel Pressure - System will be lean if pressure is too low. It may be necessary to monitor fuel pressure while driving the car at various road speeds and/or loads to confirm.

- MAP Sensor - An output that causes the ECM to sense a lower than normal manifold pressure (high vacuum) can cause the system to go lean. Disconnecting the MAP sensor will allow the ECM to substitute a fixed (default) value for the MAP sensor. If the rich condition is gone when the sensor is disconnected, substitute a known good sensor and recheck.
- Fuel Contamination - Water, even in small amounts, near the in-tank fuel pump inlet can be delivered to the injector. The water causes a lean exhaust and can set a Code 44.
- Sensor Harness - Sensor pigtail may be mispositioned and contacting the exhaust manifold.
- Engine Misfire - A cylinder misfire will result in unburned oxygen in the exhaust, which could cause Code 44. Refer to CHART C4-M
- Cracked O_2 Sensor - A crack in the O_2 sensor ceramic or poor sensor ground could cause Code 44.
- Plugged Fuel Filter - A plugged fuel filter can cause a lean condition, and can cause a Code 44 to set.

2.3L (VIN A) ENGINE — DIAGNOSTIC TROUBLE CODE CHART — 1992 BERETTA

CODE 44
OXYGEN (O_2) SENSOR CIRCUIT
(LEAN EXHAUST INDICATED)
2.3L (VIN A) "L" CARLINE (PORT)

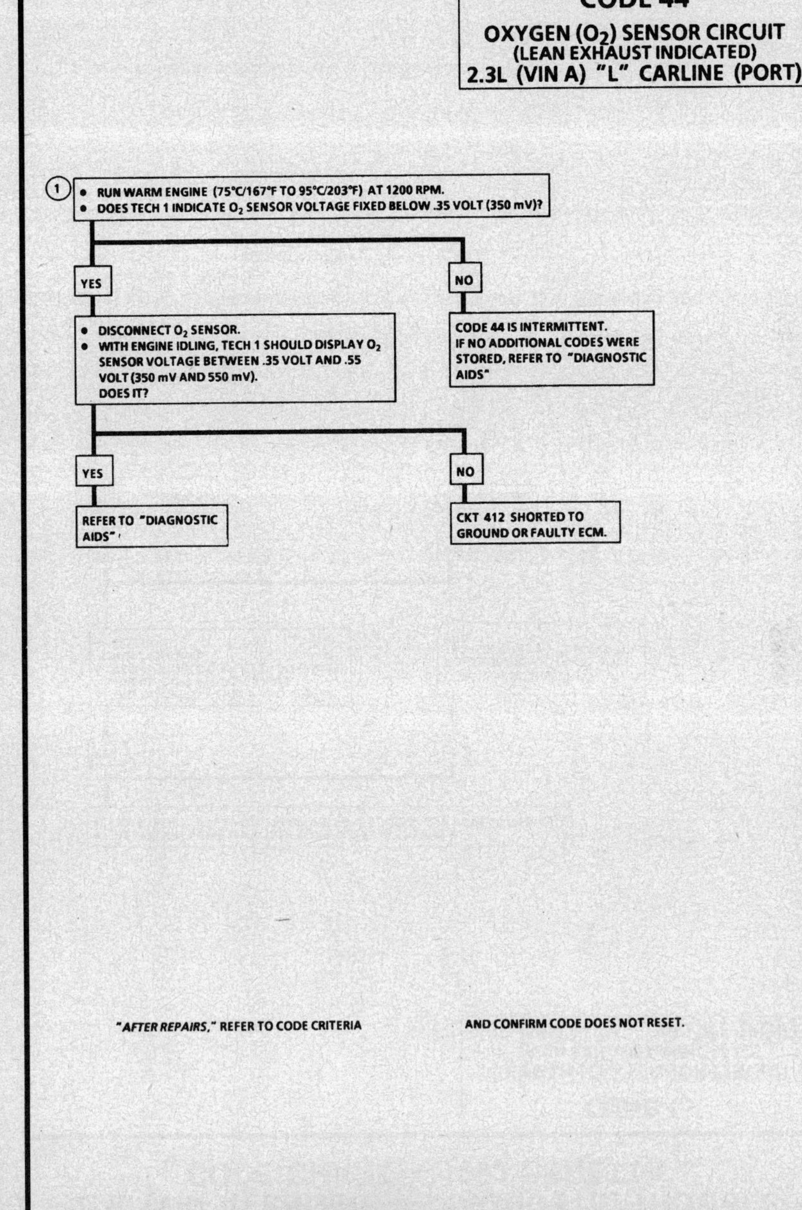

"AFTER REPAIRS," REFER TO CODE CRITERIA AND CONFIRM CODE DOES NOT RESET.

2.3L (VIN A) ENGINE — DIAGNOSTIC TROUBLE CODE CHART — 1992 BERETTA

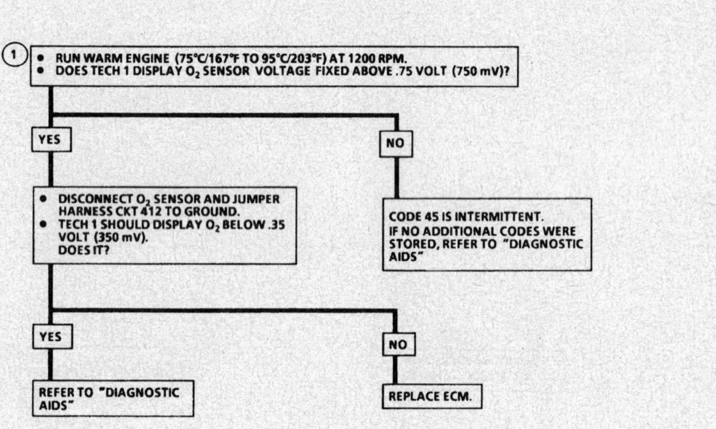

CODE 45
OXYGEN (O_2) SENSOR CIRCUIT
(RICH EXHAUST INDICATED)
2.3L (VIN A) "L" CARLINE (PORT)

Circuit Description:

The ECM supplies a voltage of about .45 volt between terminals "GE14" and "GE15". (If measured with a 10 megohm digital voltmeter, this may read as low as .32 volt.) The O_2 sensor varies the voltage within a range of about 1 volt if the exhaust is rich, down through about .10 volt if exhaust is lean.

The sensor is like an open circuit and produces no voltage when it is below 315°C (600°F). An open sensor circuit or cold sensor causes "Open Loop" operation.

Test Description: Number(s) below refer to circled number(s) on the diagnostic chart.
1. Code 45 is set when:
   - O_2 voltage is above .75 volt.
   - No Code 33 or 34.
   - Fuel system in "Closed Loop."
   - TPS above 5%.
   - Above conditions met for 30 seconds or O_2 voltage is above 1 volt for 5 seconds.

Diagnostic Aids:

The Code 45 or rich exhaust is most likely caused by one of the following:
- <u>Fuel Pressure</u> - System will go rich if pressure is too high. The ECM can compensate for some increase. However, if it gets too high, a Code 45 will be set. See "Fuel System" diagnosis CHART A-7.
- <u>Leaking Injector</u> - See CHART A-7.
- <u>HEI Shielding</u> - An open ground CKT 453 may result in EMI or induced electrical noise. The ECM looks at this noise as reference pulses. The additional pulses result in a higher than actual engine speed signal. The ECM then delivers too much fuel causing system to go rich. Engine tachometer will also show higher than actual engine speed which can help in diagnosing this problem.
- <u>Canister Purge</u> - Check for fuel saturation. If full of fuel, check canister control and hoses. See "Canister Purge," "Evaporative Emission Control System (EECS)
- <u>MAP Sensor</u> - An output that causes the ECM to sense a higher than normal manifold pressure (low vacuum) can cause the system to go rich. Disconnecting the MAP sensor will allow the ECM to set a fixed value for the MAP sensor. Substitute a different MAP sensor if the rich condition is gone while the sensor is disconnected.
- <u>Pressure Regulator</u> - Check for leaking fuel pressure regulator diaphragm by checking for the presence of liquid fuel in the vacuum line to the regulator.
- <u>TPS</u> - An intermittent TPS output will cause the system to go rich due to a false indication of the engine accelerating.
- <u>O_2 Sensor Contamination</u> - Inspect Oxygen (O_2) sensor for silicone contamination from fuel or use of improper RTV sealant. The sensor may have a white powdery coating and result in a high but false signal voltage (rich exhaust indication). The ECM will then reduce the amount of fuel delivered to the engine causing a severe surge driveability problem.

2.3L (VIN A) ENGINE — DIAGNOSTIC TROUBLE CODE CHART — 1992 BERETTA

(1)
- RUN WARM ENGINE (75°C/167°F TO 95°C/203°F) AT 1200 RPM.
- DOES TECH 1 DISPLAY O_2 SENSOR VOLTAGE FIXED ABOVE .75 VOLT (750 mV)?

YES
- DISCONNECT O_2 SENSOR AND JUMPER HARNESS CKT 412 TO GROUND.
- TECH 1 SHOULD DISPLAY O_2 BELOW .35 VOLT (350 mV). DOES IT?

NO
CODE 45 IS INTERMITTENT. IF NO ADDITIONAL CODES WERE STORED, REFER TO "DIAGNOSTIC AIDS"

YES
REFER TO "DIAGNOSTIC AIDS"

NO
REPLACE ECM.

"AFTER REPAIRS," REFER TO CODE CRITERIA   AND CONFIRM CODE DOES NOT RESET.

2.3L (VIN A) ENGINE — DIAGNOSTIC TROUBLE CODE CHART — 1992 BERETTA

CODE 51

MEM-CAL ERROR
(FAULTY OR INCORRECT MEM-CAL)
2.3L (VIN A) "L" CARLINE (PORT)

CHECK THAT ALL PINS ARE FULLY INSERTED IN THE SOCKET AND THAT MEM-CAL IS PROPERLY LATCHED. IF OK, REPLACE MEM-CAL, CLEAR MEMORY, AND RECHECK. IF CODE 51 REAPPEARS, REPLACE ECM.

NOTICE: TO PREVENT POSSIBLE ELECTROSTATIC DISCHARGE DAMAGE TO THE ECM OR MEM-CAL, DO NOT TOUCH THE COMPONENT LEADS, AND DO NOT REMOVE THE MEM CAL COVER OR THE INTEGRATED CIRCUIT FROM CARRIER.

"AFTER REPAIRS," CONFIRM "CLOSED LOOP" OPERATION AND NO "SERVICE ENGINE SOON" LIGHT.

2.3L (VIN A) ENGINE — DIAGNOSTIC TROUBLE CODE CHART — 1992 BERETTA

CODE 53

BATTERY VOLTAGE ERROR
2.3L (VIN A) "L" CARLINE (PORT)

Circuit Description:
    Code 53 will set when the ignition is "ON" and ECM terminal "BB1" and "BC16" voltages are more than 17.1 volts for about .2 seconds, or under 10 volts for more than 240 seconds.
    During the time the failure is present, all ECM outputs will be disengaged. (The setting of additional codes may result.)

Test Description: Number(s) below refer to circled number(s) on the diagnostic chart.
1. Normal battery output is between 10 - 17.1 volts.
2. Checks to see if generator is faulty under load condition. If the voltage is above 17.1 volts or under 10 volts, refer to "Alternator Diagnosis"

Note On Intermittents:

    Charging battery with a battery charger and starting engine, may set Code 53. If code sets when an accessory is operated, check for poor connections or excessive current draw.
    Also, check for poor connections at starter solenoid or fusible link junction box.

2.3L (VIN A) ENGINE — DIAGNOSTIC TROUBLE CODE CHART — 1992 BERETTA

CODE 53

BATTERY VOLTAGE ERROR
2.3L (VIN A) "L" CARLINE (PORT)

① • ENGINE RUNNING ABOVE 800 RPM.
  • NOTE BATTERY VOLTAGE ON TECH 1.

BELOW 17.1 VOLTS

ABOVE 17.1 VOLTS

② • RAISE ENGINE RPM TO 2000.
  • LOAD ELECTRICAL SYSTEM WITH HEADLAMPS AND HIGH BLOWER "ON."
  • NOTE VOLTAGE.

REMOVE GENERATOR FOR REPAIR.

BELOW 17.1 VOLTS

ABOVE 17.1 VOLTS

CHECK BATTERY VOLTAGE AT BATTERY.

REMOVE GENERATOR FOR REPAIR.

VOLTAGE ABOVE 10 VOLTS.

VOLTAGE LESS THAN 10 VOLTS.

CHECK VOLTAGE AT ECM TERMINALS "BB1", "BC16" IS VOLTAGE ABOVE 10 VOLTS?

REFER TO "CHARGING SYSTEM DIAGNOSIS,"

NO

YES

CHECK FOR HIGH RESISTANCE OR OPEN CKT 2.

REPLACE ECM

"AFTER REPAIRS," REFER TO CODE CRITERIA   AND CONFIRM CODE DOES NOT RESET.

2.3L (VIN A) ENGINE — DIAGNOSTIC TROUBLE CODE CHART — 1992 BERETTA

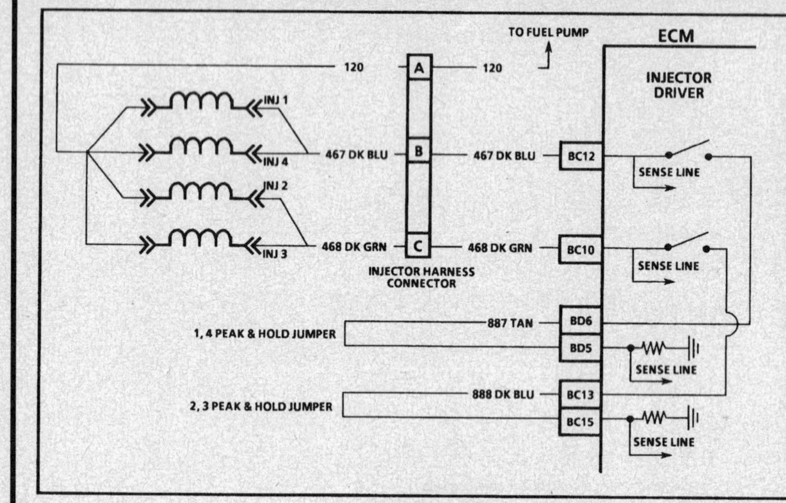

CODE 65

(Page 1 of 2)
FUEL INJECTOR CIRCUIT
(LOW CURRENT)
2.3L (VIN A) "L" CARLINE (PORT)

Circuit Description:

The ECM has two injector driver circuits, each of which controls a pair of injectors (1 and 4 or 2 and 3). The ECM monitors the current in each driver circuit by measuring voltage drop through a fixed resistor and is able to control it. The current through each driver is allowed to rise to a "peak" of 4 amps to quickly open the injectors and is then reduced to 1 amp to "hold" them open. This is called "peak and hold." If the current can't reach a 4 amp peak, Code 65 is set as noted below. This code is also set if an injector driver circuit is shorted to voltage.

Test Description: Number(s) below refer to circled number(s) on the diagnostic chart.

1. Code 65 sets when:
   • 4 amp injector driver current not reached on either circuit.
   • Battery voltage greater than 9 volts.
   • Injectors commanded "ON" longer than a calibrated pulse width.
   • Above conditions met for 10 seconds.
2. Tests ECM and harness wiring to the 3 terminal injector harness connector.
3. Tests for open or shorted injector harness or injector. A shorted harness or injector will not cause Code 65.
4. Results of Step 2 will determine which branch to follow on Page 2.
5. Checks remainder of circuit from injectors to ECM as both harnesses were confirmed OK in Steps 2 and 3.
6. Determines cause of high resistance found in Step 3. (Low resistance or a short will not cause Code 65, but should be corrected if found.)
7. Checks for grounded "peak and hold" jumpers. This fault would allow injectors to pulse but would not allow "peak and hold" operation as current would not flow through the resistor in the ECM.

Diagnostic Aids:

Open CKTs 887 or 888 or CKT 467 or 468 shorted to voltage will cause Code 65 and will also cause a misfire due to an inoperative pair of injectors. CKTs 887 and 888 shorted to ground will cause Code 65 while allowing the injectors to pulse. An intermittent problem would have to be present for at least 20 seconds to set Code 65.

2.3L (VIN A) ENGINE — DIAGNOSTIC TROUBLE CODE CHART — 1992 BERETTA

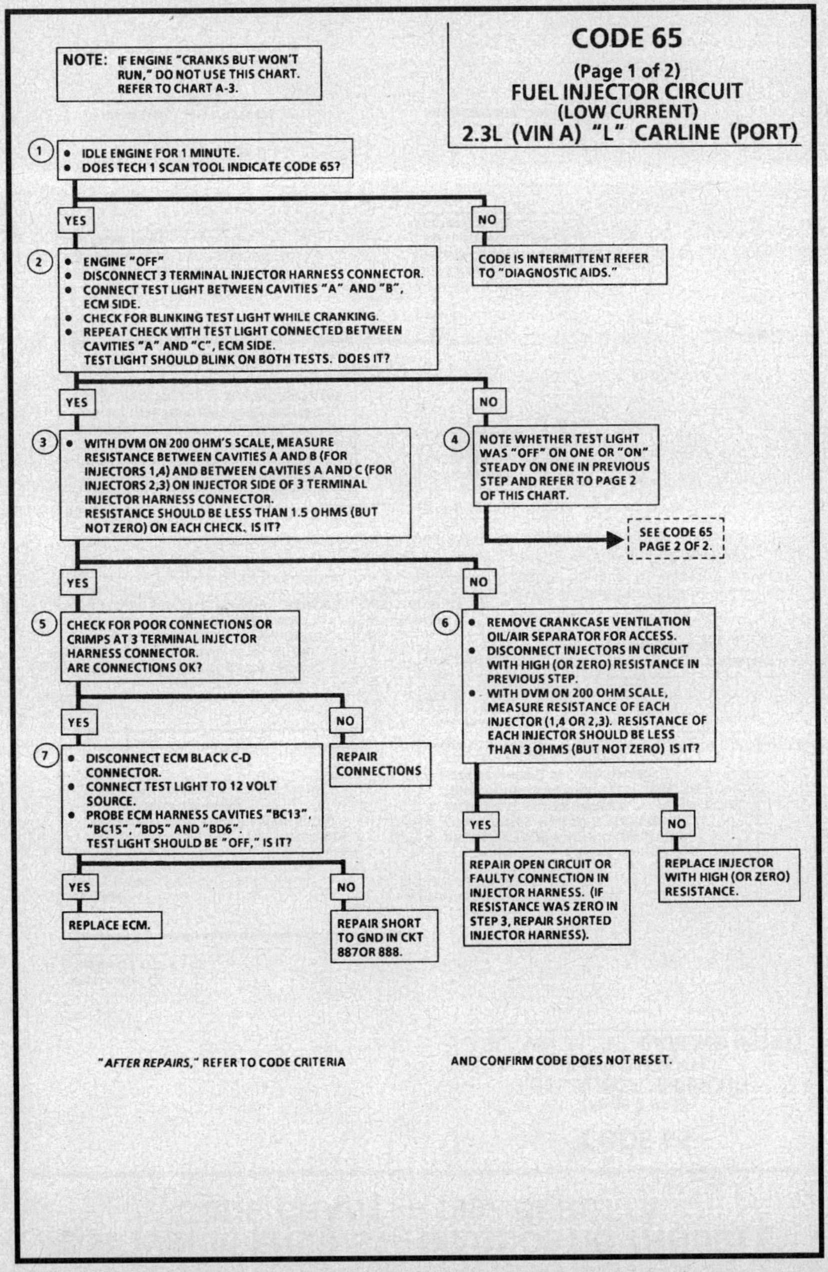

CODE 65
(Page 1 of 2)
FUEL INJECTOR CIRCUIT
(LOW CURRENT)
2.3L (VIN A) "L" CARLINE (PORT)
NOTE: IF ENGINE "CRANKS BUT WON'T RUN," DO NOT USE THIS CHART. REFER TO CHART A-3.
1 • IDLE ENGINE FOR 1 MINUTE. • DOES TECH 1 SCAN TOOL INDICATE CODE 65?
YES
NO
CODE IS INTERMITTENT REFER TO "DIAGNOSTIC AIDS."
2 • ENGINE "OFF" • DISCONNECT 3 TERMINAL INJECTOR HARNESS CONNECTOR. • CONNECT TEST LIGHT BETWEEN CAVITIES "A" AND "B", ECM SIDE. • CHECK FOR BLINKING TEST LIGHT WHILE CRANKING. • REPEAT CHECK WITH TEST LIGHT CONNECTED BETWEEN CAVITIES "A" AND "C", ECM SIDE. TEST LIGHT SHOULD BLINK ON BOTH TESTS. DOES IT?
YES
NO
4 NOTE WHETHER TEST LIGHT WAS "OFF" ON ONE OR "ON" STEADY ON ONE IN PREVIOUS STEP AND REFER TO PAGE 2 OF THIS CHART.
SEE CODE 65 PAGE 2 OF 2.
3 • WITH DVM ON 200 OHM'S SCALE, MEASURE RESISTANCE BETWEEN CAVITIES A AND B (FOR INJECTORS 1,4) AND BETWEEN CAVITIES A AND C (FOR INJECTORS 2,3) ON INJECTOR SIDE OF 3 TERMINAL INJECTOR HARNESS CONNECTOR. RESISTANCE SHOULD BE LESS THAN 1.5 OHMS (BUT NOT ZERO) ON EACH CHECK. IS IT?
YES
NO
5 CHECK FOR POOR CONNECTIONS OR CRIMPS AT 3 TERMINAL INJECTOR HARNESS CONNECTOR. ARE CONNECTIONS OK?
6 • REMOVE CRANKCASE VENTILATION OIL/AIR SEPARATOR FOR ACCESS. • DISCONNECT INJECTORS IN CIRCUIT WITH HIGH (OR ZERO) RESISTANCE IN PREVIOUS STEP. • WITH DVM ON 200 OHM SCALE, MEASURE RESISTANCE OF EACH INJECTOR (1,4 OR 2,3). RESISTANCE OF EACH INJECTOR SHOULD BE LESS THAN 3 OHMS (BUT NOT ZERO) IS IT?
YES
NO
REPAIR CONNECTIONS
7 • DISCONNECT ECM BLACK C-D CONNECTOR. • CONNECT TEST LIGHT TO 12 VOLT SOURCE. • PROBE ECM HARNESS CAVITIES "BC13", "BC15", "BD5" AND "BD6". TEST LIGHT SHOULD BE "OFF," IS IT?
YES
NO
REPAIR OPEN CIRCUIT OR FAULTY CONNECTION IN INJECTOR HARNESS. (IF RESISTANCE WAS ZERO IN STEP 3, REPAIR SHORTED INJECTOR HARNESS).
REPLACE INJECTOR WITH HIGH (OR ZERO) RESISTANCE.
REPLACE ECM.
REPAIR SHORT TO GND IN CKT 887OR 888.
"AFTER REPAIRS," REFER TO CODE CRITERIA AND CONFIRM CODE DOES NOT RESET.

2.3L (VIN A) ENGINE — DIAGNOSTIC TROUBLE CODE CHART — 1992 BERETTA

TO FUEL PUMP
ECM
INJECTOR DRIVER
120
A
INJ 1
INJ 4
INJ 2
INJ 3
467 DK BLU
B
BC12
SENSE LINE
468 DK GRN
C
BC10
INJECTOR HARNESS CONNECTOR
887 TAN
BD6
BD5
1, 4 PEAK & HOLD JUMPER
888 DK BLU
BC13
BC15
2, 3 PEAK & HOLD JUMPER

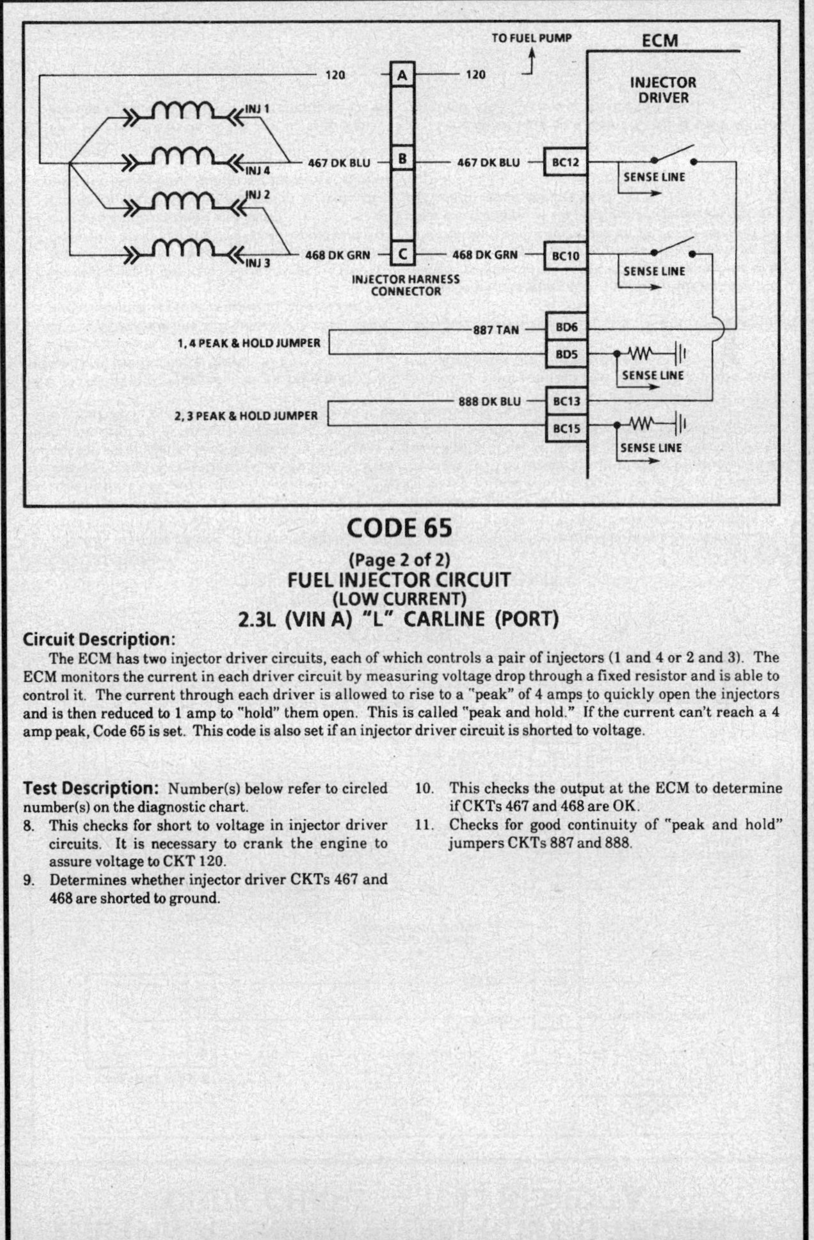

CODE 65

(Page 2 of 2)
FUEL INJECTOR CIRCUIT
(LOW CURRENT)
2.3L (VIN A) "L" CARLINE (PORT)

Circuit Description:

The ECM has two injector driver circuits, each of which controls a pair of injectors (1 and 4 or 2 and 3). The ECM monitors the current in each driver circuit by measuring voltage drop through a fixed resistor and is able to control it. The current through each driver is allowed to rise to a "peak" of 4 amps to quickly open the injectors and is then reduced to 1 amp to "hold" them open. This is called "peak and hold." If the current can't reach a 4 amp peak, Code 65 is set. This code is also set if an injector driver circuit is shorted to voltage.

Test Description: Number(s) below refer to circled number(s) on the diagnostic chart.

8. This checks for short to voltage in injector driver circuits. It is necessary to crank the engine to assure voltage to CKT 120.
9. Determines whether injector driver CKTs 467 and 468 are shorted to ground.
10. This checks the output at the ECM to determine if CKTs 467 and 468 are OK.
11. Checks for good continuity of "peak and hold" jumpers CKTs 887 and 888.

2.3L (VIN A) ENGINE — DIAGNOSTIC TROUBLE CODE CHART — 1992 BERETTA

CODE 65
(Page 2 of 2)
FUEL INJECTOR CIRCUIT
(LOW CURRENT)
2.3L (VIN A) "L" CARLINE (PORT)

FROM PAGE 1 OF THIS CHART STEP 4

TEST LIGHT "OFF" ON ONE

8 • CONNECT TEST LIGHT FROM INJECTOR DRIVER CKT WHICH DID NOT BLINK (CAVITY "B" OR "C", ECM SIDE) TO GROUND.
TEST LIGHT SHOULD REMAIN "OFF" WHILE CRANKING.
DOES IT?

TEST LIGHT "ON" STEADY ON ONE

9 • DISCONNECT ECM BLACK C-D CONNECTOR.
• CONNECT TEST LIGHT TO 12 VOLT SOURCE.
• PROBE ECM HARNESS TERMINAL "BC10" OR "BC12" FOR INJECTOR CIRCUIT WITH TEST LIGHT "ON" STEADY IN STEP 2, PREVIOUS PAGE.
TEST LIGHT SHOULD BE "OFF."
IS IT?

YES → 10 • CONNECT TEST LIGHT TO 12 VOLT SOURCE.
• BACKPROBE ECM TERMINAL "BC10" OR "BC12" FOR INJECTOR CIRCUIT WHICH DID NOT BLINK IN STEP 2, PREVIOUS PAGE.
• TEST LIGHT SHOULD BLINK WHILE CRANKING.
DOES IT?

NO → REPAIR SHORT TO VOLTAGE IN CKT 467 OR 468.

YES → REPLACE ECM

NO → REPAIR SHORT TO GROUND IN CKT 467 OR 468

NO → 11 • DISCONNECT ECM BLACK C-D CONNECTOR.
• WITH DVM ON 200 OHM SCALE, MEASURE RESISTANCE OF PEAK AND HOLD JUMPERS IN HARNESS (BC13 TO BC15 OR BD5 TO BD6) FOR CIRCUIT WITH TEST LIGHT "OFF" IN STEP 8.
RESISTANCE SHOULD BE ALMOST ZERO (LESS THAN 1 OHM)
IS IT?

YES → REPAIR OPEN CKT 467 OR CKT 468

YES → CHECK FOR POOR CONNECTION AT ECM TERMINALS "BC13" AND "BC15" OR "BD5" AND "BD6". IF OK, REPLACE ECM.

NO → REPAIR OPEN CIRCUIT OR FAULTY CONNECTION IN PEAK AND HOLD JUMPER CKT 887 OR 888.

"AFTER REPAIRS," REFER TO CODE CRITERIA   AND CONFIRM CODE DOES NOT RESET.

2.3L (VIN A) ENGINE — DIAGNOSTIC TROUBLE CODE CHART — 1992 BERETTA

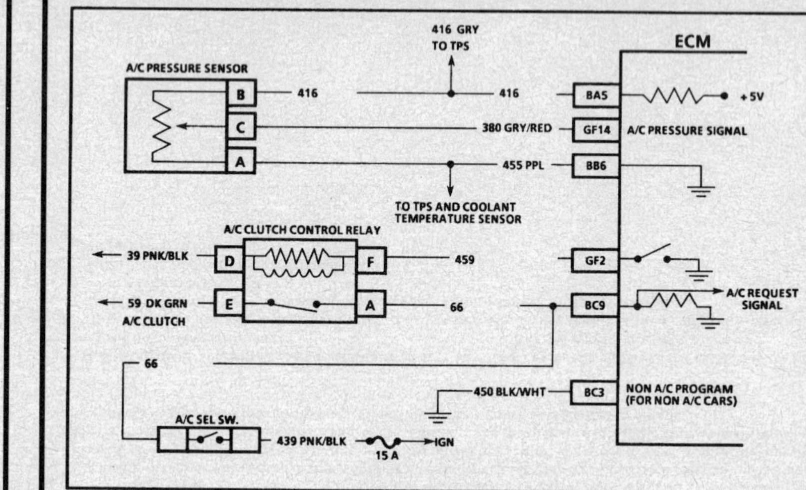

CODE 66

A/C PRESSURE SENSOR CIRCUIT
2.3L (VIN A) "L" CARLINE (PORT)

Circuit Description:

The A/C pressure sensor responds to changes in A/C refrigerant system high side pressure. This input indicates how much load the A/C compressor is putting on the engine and is one of the factors used by the ECM to determine IAC valve position for idle speed control. The circuit consists of a 5 volts reference and a ground, both provided by the ECM, and a signal line to the ECM. The signal is a voltage which is proportional to the pressure. The sensor's range of operation is 0 to 450 psi. At 0 psi, the signal will be about .1 volt, varying up to about 4.9 volts at 450 psi or above. Code 66 sets if the voltage is above 4.9 volts or below .28 volt 15 seconds or more. Code 66 will also set if A/C is not requested and voltage is greater than 3.9 volts. The A/C compressor is disabled by the ECM if Code 66 is present, or if pressure is above or below calibrated values

Test Description: Number(s) below refer to circled number(s) on the diagnostic chart.

1. This step checks the voltage signal being received by the ECM from the A/C pressure sensor. The normal operating range is between .28 volt and 4.9 volts.
2. Checks to see if the high voltage signal is from a shorted sensor or a short to voltage in the circuit. Normally, disconnecting the sensor would make a normal circuit go to near zero volt.
3. Checks to see if low voltage signal is from the sensor or the circuit. Jumpering the sensor signal CKT 380 to 5 volts, checks the circuit, connections, and ECM.
4. This step checks to see if the low voltage signal was due to an open in the sensor circuit or the 5 volts reference circuit since the prior step eliminated the pressure sensor.

Diagnostic Aids:

Code 66 sets when signal voltage falls outside the normal possible range of the sensor and is not due to a refrigerant system problem. If problem is intermittent, check for opens or shorts in harness or poor connections. If OK, replace A/C pressure sensor. If Code 66 resets, replace ECM.

Non-A/C Program

Code 66 will set on a Non-A/C car if CKT 450 to terminal "BC3" is open or shorted to B+.

2.3L (VIN A) ENGINE — DIAGNOSTIC TROUBLE CODE CHART — 1992 BERETTA

CODE 66
A/C PRESSURE SENSOR CIRCUIT
2.3L (VIN A) "L" CARLINE (PORT)

NOTE: IF CAR IS NOT EQUIPPED WITH AIR CONDITIONING, DO NOT USE THIS CHART, SEE "DIAGNOSTIC AIDS".

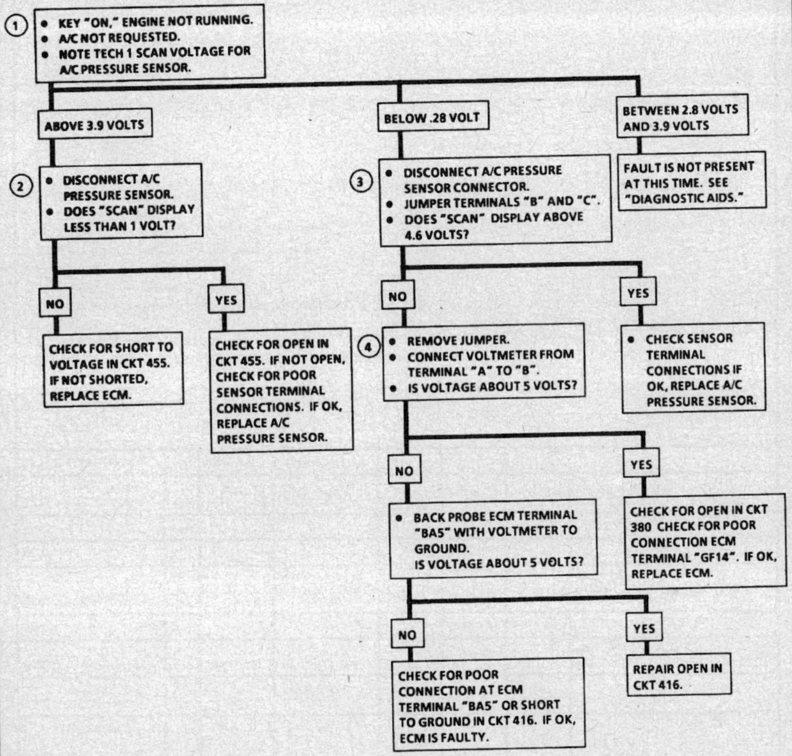

(1)
- KEY "ON," ENGINE NOT RUNNING.
- A/C NOT REQUESTED.
- NOTE TECH 1 SCAN VOLTAGE FOR A/C PRESSURE SENSOR.

ABOVE 3.9 VOLTS

BELOW .28 VOLT

BETWEEN 2.8 VOLTS AND 3.9 VOLTS

(2)
- DISCONNECT A/C PRESSURE SENSOR.
- DOES "SCAN" DISPLAY LESS THAN 1 VOLT?

(3)
- DISCONNECT A/C PRESSURE SENSOR CONNECTOR.
- JUMPER TERMINALS "B" AND "C".
- DOES "SCAN" DISPLAY ABOVE 4.6 VOLTS?

FAULT IS NOT PRESENT AT THIS TIME. SEE "DIAGNOSTIC AIDS."

NO

YES

NO

YES

CHECK FOR SHORT TO VOLTAGE IN CKT 455. IF NOT SHORTED, REPLACE ECM.

CHECK FOR OPEN IN CKT 455. IF NOT OPEN, CHECK FOR POOR SENSOR TERMINAL CONNECTIONS. IF OK, REPLACE A/C PRESSURE SENSOR.

(4)
- REMOVE JUMPER.
- CONNECT VOLTMETER FROM TERMINAL "A" TO "B".
- IS VOLTAGE ABOUT 5 VOLTS?

- CHECK SENSOR TERMINAL CONNECTIONS IF OK, REPLACE A/C PRESSURE SENSOR.

NO

YES

- BACK PROBE ECM TERMINAL "BA5" WITH VOLTMETER TO GROUND. IS VOLTAGE ABOUT 5 VOLTS?

CHECK FOR OPEN IN CKT 380 CHECK FOR POOR CONNECTION ECM TERMINAL "GF14". IF OK, REPLACE ECM.

NO

YES

CHECK FOR POOR CONNECTION AT ECM TERMINAL "BA5" OR SHORT TO GROUND IN CKT 416. IF OK, ECM IS FAULTY.

REPAIR OPEN IN CKT 416.

"AFTER REPAIRS," REFER TO CODE CRITERIA   AND CONFIRM CODE DOES NOT RESET.

2.3L (VIN A) ENGINE — EXHAUST SYSTEM DIAGNOSTIC CHART — 1992 BERETTA

CHART B-1
RESTRICTED EXHAUST SYSTEM CHECK
2.3L (VIN A)

Proper diagnosis for a restricted exhaust system is essential before any components are replaced. The following procedure may be used for diagnosis.

CHECK AT O$_2$ SENSOR:

1. Carefully remove O$_2$ sensor.
2. Install Borroughs exhaust backpressure tester (BT 8515 or BT 8603) or equivalent in place of O$_2$ sensor (see illustration).
3. After completing test described below, be sure to coat threads of O$_2$ sensor with anti-seize compound P/N 5613695 or equivalent prior to re-installation.

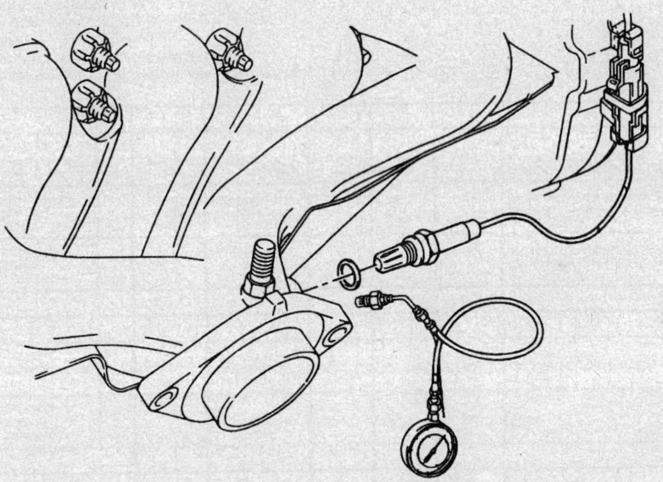

DIAGNOSIS:

1. With the engine idling at normal operating temperature, transaxle in park or neutral, observe the exhaust system backpressure reading on the gauge. Reading should not exceed 3.4 kPa (.5 psi).
2. Increase engine speed to 3000 rpm and observe gauge. Reading should not exceed 5 kPa (.75 psi).
3. If the backpressure at either speed exceeds specification, a restricted exhaust system is indicated.
4. Inspect the entire exhaust system for a collapsed pipe, heat distress, or possible internal muffler failure.
5. If there are no obvious reasons for the excessive backpressure, the catalytic converter is suspected to be restricted and should be replaced using current recommended procedures.

2.3L (VIN A) ENGINE — ECM SYMPTOM CHART — 1992 BERETTA

ECM CONNECTOR "A"				NORMAL VOLTAGES			
PIN FUNCTION	CKT #	WIRE COLOR	COMPONENT CONNECTOR CAVITY	KEY "ON"	ENG RUN	CODES AFFECT	POSSIBLE SYMPTOMS FROM FAULTY CIRCUIT
BA1							
BA2							
BA3							
BA4 MAP SENSOR 5 VOLT REFERENCE	474	GRY	MAP SENSOR "C"	5V	5V	34 (10)	LACK OF POWER ROUGH IDLE SURGE
BA5 TPS 5 VOLT REFERENCE	416	GRY	TPS "A" A/C SENSOR "B"	5V	5V	22 (10) 66 (10)	HIGH IDLE
BA6 IGNITION FEED	439	PNK/BLK	IGN FUSE	B+	B+		NO SES LIGHT, ENGINE CRANKS BUT WILL NOT START (8)
BA7							
BA8 SERIAL DATA	461	ORN	ALDL CONNECTOR TERMINAL "M"				NO SERIAL DATA "SCAN" TOOL WILL NOT READ DATA (10)
BA9							
BA10							
BA11 FUEL PUMP RELAY DRIVE	465	DK GRN/ WHT	FUEL PUMP RELAY "D"	B + (4)	B + (4)		LONG CRANKING TIME BEFORE ENGINE STARTS (8)
BA12 ECM GROUND	450	BLK/WHT	ENGINE BLOCK	0*	0*		

(1) VARIES FROM .60 TO BATTERY VOLTAGE, DEPENDING ON POSITION OF DRIVE WHEELS.
(2) BATTERY VOLTAGE FOR FIRST TWO SECONDS
(3) VARIES
(4) BATTERY VOLTAGE WHEN FUEL PUMP IS RUNNING
(5) VARIES WITH TEMPERATURE
(6) READS BATTERY VOLTAGE IN GEAR
(7) BATTERY VOLTAGE WHEN ENGINE IS CRANKING
(8) OPEN CIRCUIT
(9) GROUNDED CIRCUIT
(10) OPEN/GROUNDED CIRCUIT
(11) LESS THAN 1 VOLT
* LESS THAN .5 VOLT (500 MV)

2.3L (VIN A) ENGINE — ECM SYMPTOM CHART — 1992 BERETTA

ECM CONNECTOR "B"				NORMAL VOLTAGES			
PIN FUNCTION	CKT #	WIRE COLOR	COMPONENT CONNECTOR CAVITY	KEY "ON"	ENG RUN	CODES AFFECT	POSSIBLE SYMPTOMS FROM FAULTY CIRCUIT
BB1 BATTERY FEED	340	ORN	FUEL PUMP RELAY TERMINAL "A" AND OIL PRESSURE TERM "C" ECM	B+	B+		CRANKS BUT WILL NOT START
BB2							
BB3							
BB4							
BB5 MAP AND IAT SENSOR GROUND	452	BLK	IAT TERM "B" MAP TERM "A"	0*	0*	23 (8) 33 (8)	STALLING AT IDLE AND RUNS ROUGH
BB6 TPS, CTS AND A/C SENSOR GROUND	455	PPL	TPS TERM "B" A/C TERM "A" CTS TERM "A"	0*	0*	15 (8) 21 (8) 66 (8)	ROUGH IDLE LACK OF PERFORMANCE EXHAUST ODOR
BB7							
BB8							
BB9 VEHICLE SPEED SENSOR (VSS) SIGNAL LOW	401	PPL	VEHICLE SPEED SENSOR (VSS) TERM "B"	0*	0*	24 (10) 35 (10)	NO VSS SIGNAL INOPERATIVE SPEEDOMETER INOPERATIVE CRUISE CONTROL
BB10 VEHICLE SPEED SENSOR (VSS) SIGNAL HIGH	400	YEL	VEHICLE SPEED SENSOR (VSS) TERM "A"	0*	0*	24 (10) 35 (10)	NO VSS SIGNAL INOPERATIVE SPEEDOMETER INOPERATIVE CRUISE CONTROL
BB11 ECM TO INSTRUMENT CLUSTER VSS	817	DK GRN/WHT	I/P CLUSTER TERM "G"	(1) 4.85V	(1) 5.3V		INOPERATIVE SPEEDOMETER
BB12							

(1) VARIES FROM .60 TO BATTERY VOLTAGE, DEPENDING ON POSITION OF DRIVE WHEELS.
(2) BATTERY VOLTAGE FOR FIRST TWO SECONDS
(3) VARIES
(4) BATTERY VOLTAGE WHEN FUEL PUMP IS RUNNING
(5) VARIES WITH TEMPERATURE
(6) READS BATTERY VOLTAGE IN GEAR
(7) BATTERY VOLTAGE WHEN ENGINE IS CRANKING
(8) OPEN CIRCUIT
(9) GROUNDED CIRCUIT
(10) OPEN/GROUNDED CIRCUIT
(11) LESS THAN 1 VOLT
* LESS THAN .5 VOLT (500 MV)

2.3L (VIN A) ENGINE — ECM SYMPTOM CHART — 1992 BERETTA

ECM CONNECTOR "C"

PIN FUNCTION		CKT #	WIRE COLOR	COMPONENT CONNECTOR CAVITY	NORMAL VOLTAGES		CODES AFFECT	POSSIBLE SYMPTOMS FROM FAULTY CIRCUIT
					KEY "ON"	ENG RUN		
BC1								
BC2								
BC3								
BC4								
BC5	1X REF HI	647	LT BLU/BLK	IGNITION MODULE TERM G		4.8V	41 (10)	LACK OF PERFORMANCE
BC6								
BC7	IGNITION BYPASS	424	TAN/BLK	IGNITION MODULE TERM D	0*	5V	42 (10)	LACK OF POWER HUNTING IDLE, STALLING
BC8	ELECTRONIC SPARK TIMING (EST)	423	WHT	IGNITION MODULE TERM E	0*	2 V	42 (10)	LACK OF POWER STALLS, SURGES
BC9	A/C REQUEST SIGNAL	66	LT GRN	A/C SELECT SWITCH	0*	ON B + OFF 0*		INOPERATIVE A/C INCORRECT IDLE
BC10	INJECTOR DRIVERS 2 & 3	468	DK GRN	INJECTOR CONNECTOR C	B +	B +	65 (10)	ROUGH IDLE, HARD TO START LACK OF PERFORMANCE
BC11								
BC12	INJECTOR DRIVERS 1 & 4	467	DK BLU	INJECTOR CONNECTOR B	B +	B +	65 (10)	ROUGH IDLE, HARD TO START LACK OF PERFORMANCE
BC13	PEAK & HOLD INJ. JUMPER 2 & 3	888	DK BLU	ECM CONNECTOR BC 15	0*	0*	65 (10)	ROUGH IDLE, HARD TO START LACK OF PERFORMANCE
BC14								
BC15	PEAK & HOLD INJ. JUMPER 2 & 3	888	DK BLU	ECM CONNECTOR BC 13	0*	0*	65 (10)	ROUGH IDLE, HARD TO START LACK OF PERFORMANCE
BC16	BATTERY FEED	340	ORN	FUEL PUMP ECM FUSE 20A	B +	B +		CRANKS BUT WILL NOT START

(1) VARIES FROM .60 TO BATTERY VOLTAGE, DEPENDING ON POSITION OF DRIVE WHEELS.
(2) BATTERY VOLTAGE FOR FIRST TWO SECONDS
(3) VARIES
(4) BATTERY VOLTAGE WHEN FUEL PUMP IS RUNNING
(5) VARIES WITH TEMPERATURE
(6) READS BATTERY VOLTAGE IN GEAR
(7) BATTERY VOLTAGE WHEN ENGINE IS CRANKING
(8) OPEN CIRCUIT
(9) GROUNDED CIRCUIT
(10) OPEN/GROUNDED CIRCUIT
(11) LESS THAN 1 VOLT
* LESS THAN .5 VOLT (500 MV)

2.3L (VIN A) ENGINE — ECM SYMPTOM CHART — 1992 BERETTA

ECM CONNECTOR "D"

PIN FUNCTION		CKT #	WIRE COLOR	COMPONENT CONNECTOR CAVITY	NORMAL VOLTAGES		CODES AFFECT	POSSIBLE SYMPTOMS FROM FAULTY CIRCUIT
					KEY "ON"	ENG RUN		
BD1	ECM GROUND	551	TAN/WHT	ENGINE BLOCK	0*	0*		
BD2								
BD3								
BD4								
BD5	PEAK & HOLD INJ. JUMPER 1 & 4	887	TAN	ECM TERMINAL BD6	0*	0*	65 (10)	ROUGH IDLE, HARD TO START, LACK OF PERFORMANCE
BD6	PEAK & HOLD INJ. JUMPER 1 & 4	887	TAN	ECM TERMINAL BD5	0*	0*	65 (10)	ROUGH IDLE, HARD TO START, LACK OF PERFORMANCE
BD7								
BD8	2X REF HI	430	PPL/WHT	IGNITION MODULE TERMINAL "H"	0*	1.6V	16 (10)	
BD9	REFERENCE GROUND	453	BLK/RED	IGNITION MODULE TERMINAL "J"	0*	0*		
BD10								
BD11								
BD12								
BD13								
BD14								

(1) VARIES FROM .60 TO BATTERY VOLTAGE, DEPENDING ON POSITION OF DRIVE WHEELS.
(2) BATTERY VOLTAGE FOR FIRST TWO SECONDS
(3) VARIES
(4) BATTERY VOLTAGE WHEN FUEL PUMP IS RUNNING
(5) VARIES WITH TEMPERATURE
(6) READS BATTERY VOLTAGE IN GEAR
(7) BATTERY VOLTAGE WHEN ENGINE IS CRANKING
(8) OPEN CIRCUIT
(9) GROUNDED CIRCUIT
(10) OPEN/GROUNDED CIRCUIT
(11) LESS THAN 1 VOLT
* LESS THAN .5 VOLT (500 MV)

2.3L (VIN A) ENGINE — ECM SYMPTOM CHART — 1992 BERETTA

	PIN FUNCTION	CKT #	WIRE COLOR	COMPONENT CONNECTOR CAVITY	NORMAL VOLTAGES KEY "ON"	NORMAL VOLTAGES ENG RUN	CODES AFFECT	POSSIBLE SYMPTOMS FROM FAULTY CIRCUIT
	ECM CONNECTOR "E"							
GE1								
GE2								
GE3	IAC COIL "A" HIGH	441	LT BLU/ WHT	IAC VALVE "C"	0 OR B +	0 OR B +	35 (10)	INCORRECT IDLE SURGES
GE4	IAC COIL "A" LOW	442	LT BLU/ BLK	IAC VALVE "D"	0 OR B +	0 OR B +	35 (10)	INCORRECT IDLE SURGES
GE5	IAC COIL "B" HIGH	443	LT GRN/ WHT	IAC VALVE "A"	0 OR B +	0 OR B +	35 (10)	INCORRECT IDLE SURGES
GE6	IAC COIL "B" LOW	444	LT GRN/ BLK	IAC VALVE "B"	0 OR B +	0 OR B +	35 (10)	INCORRECT IDLE SURGES
GE7	SERVICE ENGINE SOON LIGHT	419	BRN/WHT	I/P HARNESS CONNECTOR "B"	0*	B +		NO SES LIGHT (8) SES LIGHT ON CONSTANTLY (9)
GE8	PRIMARY COOLING FAN (FAN 1) CONTROL	335	DK GRN/WHT	PRIMARY FAN (FAN 1) RELAY TERM "F"	ON 0* OFF B +	ON 0* OFF B +		INOPERATIVE FAN 1 (8) FAN 1 RUNS ALL THE TIME (9)
GE9								
GE10								
GE11								
GE12	DIAGNOSTIC ENABLE TERMINAL	451	WHT/BLK	ALDL CONNECTOR "B"	5V (2)	5V (2)		SES LIGHT FLASHES ALL THE TIME (9) NO FIELD SERVICE MODE (8)
GE13								
GE14	OXYGEN (O2) SENSOR SIGNAL	412	BLK	OXYGEN (O2) SENSOR	.35V-.55V	(3) .1V - .9V	13 (10) 44 (10)	EXHAUST ODOR POOR PERFORMANCE
GE15	OXYGEN (O2) SENSOR GROUND	413	TAN	ON TRANSAXLE BOLT	0*	0*	13 (8)	EXHAUST ODOR POOR PERFORMANCE
GE16	COOLANT TEMP SENSOR	410	YEL	CTS "B"	1.8V (5)	1.8V (5)	14 (9) 15 (8)	LACK OF PERFORMANCE

(1)   VARIES FROM .60 TO BATTERY VOLTAGE, DEPENDING ON POSITION OF DRIVE WHEELS.
(2)   BATTERY VOLTAGE FOR FIRST TWO SECONDS
(3)   VARIES
(4)   BATTERY VOLTAGE WHEN FUEL PUMP IS RUNNING
(5)   VARIES WITH TEMPERATURE
(6)   READS BATTERY VOLTAGE IN GEAR
(7)   BATTERY VOLTAGE WHEN ENGINE IS CRANKING
(8)   OPEN CIRCUIT
(9)   GROUNDED CIRCUIT
(10) OPEN/GROUNDED CIRCUIT
(11) LESS THAN 1 VOLT
*   LESS THAN .5 VOLT (500 MV)

2.3L (VIN A) ENGINE — COMPONENT DIAGNOSTIC CHART — 1992 BERETTA

	PIN FUNCTION	CKT #	WIRE COLOR	COMPONENT CONNECTOR CAVITY	NORMAL VOLTAGES KEY "ON"	NORMAL VOLTAGES ENG RUN	CODES AFFECT	POSSIBLE SYMPTOMS FROM FAULTY CIRCUIT
	ECM CONNECTOR "F"							
GF1	CANISTER PURGE	428	DK GRY/YEL	CANISTER PURGE TERM. "B"	B +	.3V	26 (10)	
GF2	A/C CLUTCH CONTROL	459	DK GRN/ WHT	A/C RELAY TERM. "F"	B +	OFF B + ON 0*	26 (10)	A/C CLUTCH INOPERATIVE (8)
GF3								
GF4	SHIFT LIGHT	456	TAN/BLK	M/T SHIFT LIGHT IN I/P	B +	OFF B + ON 0*		SHIFT LIGHT INOPERATIVE ((8) SHIFT LIGHT ON (9)
GF5								
GF6								
GF7								
GF8								
GF9	ESC KNOCK SENSOR SIGNAL	496	DK BLU	KNOCK SENSOR	2.3V	2.3V	43 (10)	
GF10								
GF11								
GF12								
GF13	TPS SIGNAL	417	DK BLU	TPS "C"	.54V	.54V	22 (10)	LACK OF PERFORMANCE
GF14	A/C PRESS SIGNAL	380	GRY/RED	A/C SENSOR TERM. "C"	1.0V (3)	1.0V (3)	66 (10)	A/C CLUTCH INOPERATIVE
GF15	MAP SIGNAL	432	LT GRN	MAP SENSOR "B"	4.7V	1.4V	34 (10)	LACK OF PERFORMANCE ROUGH IDLE SURGE
GF16	IAT SIGNAL	472	TAN	IAT SENSOR TERM. "A"	2.33V	1.5V	23 (8) 25 (9)	

(1)   VARIES FROM .60 TO BATTERY VOLTAGE, DEPENDING ON POSITION OF DRIVE WHEELS.
(2)   BATTERY VOLTAGE FOR FIRST TWO SECONDS
(3)   VARIES
(4)   BATTERY VOLTAGE WHEN FUEL PUMP IS RUNNING
(5)   VARIES WITH TEMPERATURE
(6)   READS BATTERY VOLTAGE IN GEAR
(7)   BATTERY VOLTAGE WHEN ENGINE IS CRANKING
(8)   OPEN CIRCUIT
(9)   GROUNDED CIRCUIT
(10) OPEN/GROUNDED CIRCUIT
(11) LESS THAN 1 VOLT
*   LESS THAN .5 VOLT (500 MV)

2.3L (VIN A) ENGINE — COMPONENT DIAGNOSTIC CHART — 1992 BERETTA

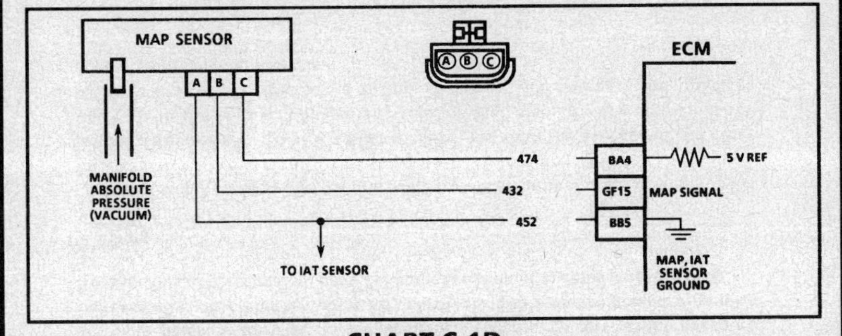

CHART C-1D
MANIFOLD ABSOLUTE PRESSURE (MAP) OUTPUT CHECK
2.3L (VIN A) "L" CARLINE (PORT)

Circuit Description:

The Manifold Absolute Pressure (MAP) sensor measures the changes in the intake manifold pressure which result from engine load (intake manifold vacuum) and rpm changes, and converts these into a voltage output. The ECM sends a 5 volt reference voltage to the MAP sensor. As the manifold pressure changed, the output voltage of the sensor also changes. By monitoring the sensor output voltage, the ECM knows the manifold pressure. A lower pressure (low voltage) output voltage will be about 1-2 volts at idle. While higher pressure (high voltage) output voltage will be about 4 - 4.8 at Wide Open Throttle (WOT). The MAP sensor is also used, under certain conditions, to measure barometric pressure, allowing the ECM to make adjustments for different altitudes. The ECM uses the MAP sensor to control fuel delivery and ignition thing.

Test Description: Number(s) below refer to circled number(s) on the diagnostic chart.

Important
- Be sure to use the same Diagnostic Test Equipment for all measurements.

1. When comparing "Scan" readings to a known good vehicle, it is important to compare vehicles that use a MAP sensor having the same color insert or having the same "Hot Stamped" number. See figures
2. Applying 34 kPa (10" Hg) vacuum to the MAP sensor should cause the voltage to change. Subtract second reading from the first. Voltage value should be greater than 1.5 volts. Upon applying vacuum to the sensor, the change in voltage should be instantaneous. A slow voltage change indicates a faulty sensor.

3. Check vacuum hose to sensor for leaking or restriction. Be sure that no other vacuum devices are connected to the MAP hose.

Important
- Make sure electrical connector remains securely fastened.

4. Disconnect sensor from bracket and twist sensor by hand (only) to check for intermittent connection. Output changes greater than .1 volt indicate a bad connector or connection. If OK, replace sensor.

2.3L (VIN A) ENGINE — COMPONENT DIAGNOSTIC CHART — 1992 BERETTA

CHART C-1D
MANIFOLD ABSOLUTE PRESSURE (MAP) OUTPUT CHECK
2.3L (VIN A) "L" CARLINE (PORT)

NOTICE: THIS CHART ONLY APPLIES TO MAP SENSORS HAVING GREEN OR BLACK COLOR KEY INSERT (SEE BELOW).

1.
- IGNITION "ON," ENGINE "OFF."
- TECH 1 "SCAN" TOOL SHOULD INDICATE A MAP SENSOR VOLTAGE.
- COMPARE THIS READING WITH THE READING OF A KNOWN GOOD VEHICLE. SEE FACING PAGE TEST DESCRIPTION, STEP 1. VOLTAGE READING SHOULD BE WITHIN ± .4 VOLT. IS IT?

YES → 2.
- DISCONNECT AND PLUG VACUUM SOURCE TO MAP SENSOR.
- CONNECT A HAND VACUUM PUMP TO MAP SENSOR.
- START ENGINE.
- NOTE MAP SENSOR VOLTAGE.
- APPLY 34 kPa (10" Hg) OF VACUUM AND NOTE VOLTAGE CHANGE. SUBTRACT SECOND READING FROM THE FIRST. VOLTAGE VALUE SHOULD BE GREATER THAN 1.5 VOLTS. IS IT?

NO → REPLACE MAP SENSOR.

YES → 3. NO TROUBLE FOUND. CHECK MAP SENSOR VACUUM SOURCE FOR LEAKAGE OR RESTRICTION. BE SURE THIS SOURCE SUPPLIES VACUUM TO MAP SENSOR ONLY.

NO → 4. CHECK MAP SENSOR CONNECTION. IF OK, REPLACE MAP SENSOR.

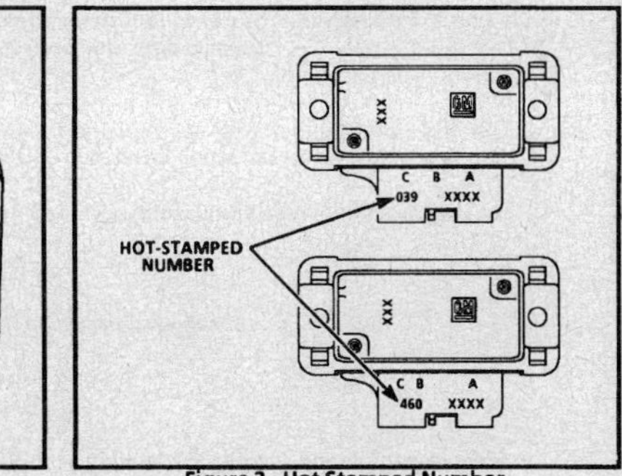

"AFTER REPAIRS," CONFIRM "CLOSED LOOP" OPERATION AND NO "SERVICE ENGINE SOON" LIGHT.

CHART C-2A

INJECTOR BALANCE TEST
2.3L (VIN A) "L" CARLINE (PORT)

The injector balance tester is a tool used to turn the injector on for a precise amount of time, thus spraying a measured amount of fuel into the manifold. This causes a drop in fuel rail pressure that we can record and compare between each injector. All injectors should have the same amount of pressure drop (± 10 kPa). Any injector with a pressure drop that is 10 kPa (or more) greater or less than the average drop of the other injectors should be considered faulty and replaced.

STEP 1

Engine "cool down" period (10 minutes) is necessary to avoid irregular readings due to "Hot Soak" fuel boiling. Relieve fuel pressure in the fuel rail using the "fuel pressure relief procedure" described previously in this section. With ignition "OFF" connect fuel gauge J 347301 or equivalent to fuel pressure tap.

Disconnect harness connectors at all injectors, and connect injector tester J 34730-3, or equivalent, to one injector. On Turbo equipped engines, use adaptor harness furnished with injector tester to energize injectors that are not accessible. Follow manufacturers instructions for use of adaptor harness. Ignition must be "OFF" at least 10 seconds to complete ECM shutdown cycle. Fuel pump should run about 2 seconds after ignition is turned "ON." At this point, insert clear tubing attached to vent valve into a suitable container and bleed air from gauge and hose to insure accurate gauge operation. Repeat this step until all air is bled from gauge.

STEP 2

Turn ignition "OFF" for 10 seconds and then "ON" again to get fuel pressure to its maximum. Record this initial pressure reading. Energize tester one time and note pressure drop at its lowest point. (Disregard any slight pressure increase after drop hits low point.) By subtracting this second pressure reading from the initial pressure, we have the actual amount of injector pressure drop.

STEP 3

Repeat step 2 on each injector and compare the amount of drop. Usually, good injectors will have virtually the same drop. Retest any injector that has a pressure difference of 10 kPa, either more or less than the average of the other injectors on the engine. Replace any injector that also fails the retest. If the pressure drop of all injectors is within 10 kPa of this average, the injectors appear to be flowing properly.

NOTE: *The entire test should not be repeated more than once without running the engine to prevent flooding. (This includes any retest on faulty injectors).*

NOTE: If injectors are suspected of being dirty, they should be cleaned using an approved tool and procedure prior to performing this test. The fuel pressure test in CHART A-7, should be completed prior to this test.

CHART C-2A

INJECTOR BALANCE TEST
2.3L (VIN A) "L" CARLINE (PORT)

Step 1. If engine is at operating temperature, allow a 10 minute "cool down" period then connect fuel pressure gauge and injector tester.
1. Ignition "OFF".
2. Connect fuel pressure gauge and injector tester.
3. Ignition "ON".
4. Bleed off air in gauge. Repeat until all air is bled from gauge.

Step 2. Run test:
1. Ignition "OFF" for 10 seconds.
2. Ignition "ON". Record gauge pressure. (Pressure must hold steady, if not see the Fuel System Diagnosis, CHART A-7
3. Turn injector "ON", by depressing button on injector tester, and note pressure at the instant the gauge needle stops.

Step 3.
1. Repeat step 2 on all injectors and record pressure drop on each.
Retest injectors that appear faulty (any injectors that have a 10 kPa difference, either more or less, in pressure from the average).

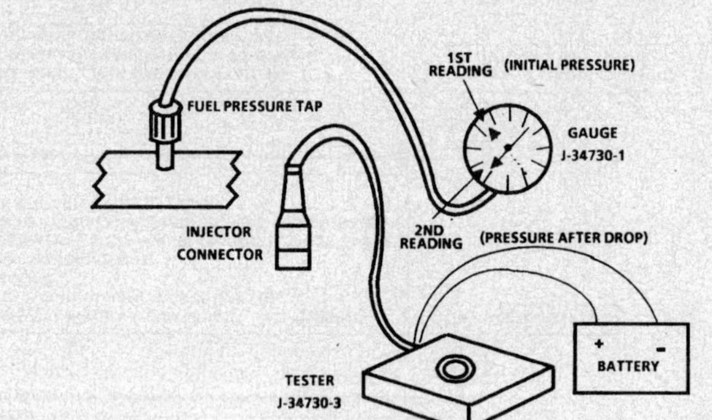

— EXAMPLE —

CYLINDER	1	2	3	4
1ST READING	225	225	225	225
2ND READING	100	115	100	85
AMOUNT OF DROP	125	110	125	140
	OK	FAULTY LEAN (TOO LITTLE) (FUEL DROP)	OK	FAULTY RICH (TOO MUCH) (FUEL DROP)

2.3L (VIN A) ENGINE — COMPONENT DIAGNOSTIC CHART — 1992 BERETTA

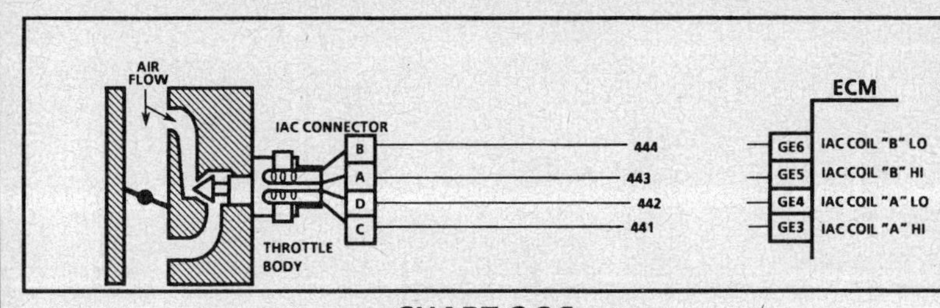

CHART C-2C
IDLE AIR CONTROL (IAC) VALVE CHECK
2.3L (VIN A) "L" CARLINE (PORT)

Circuit Description:

The ECM controls idle rpm with the IAC valve. To increase idle rpm, the ECM moves the IAC valve out, allowing more air to bypass the throttle plate. To decrease rpm, it moves the IAC valve in, reducing air flow by-passing the throttle plate. A Tech 1 "Scan" tool will read the ECM commands to the IAC valve in counts. The higher the counts, the more air allowed (higher idle). The lower the counts, the less air allowed (lower idle).

Test Description: Number(s) below refer to circled numbers on the diagnostic chart.

1. The IAC tester is used to extend and retract the IAC valve. Valve movement is verified by an engine speed change. If no change in engine speed occurs, the valve can be retested when removed from the throttle body.
   This step checks the quality of the IAC movement in Step 1. Between 700 rpm and about 1500 rpm, the engine speed should change smoothly with each flash of the tester light in both extend and retract. If the IAC valve is retracted beyond the control range (about 1500 rpm), it may take many flashes in the extend position before engine speed will begin to drop. This is normal on certain engines, fully extending IAC may cause engine stall. This may be normal.
2. Step 1 verified proper IAC valve operation while this step checks IAC circuits. Each lamp on the node light should flash red and green while the IAC valve is cycled. While the sequence of color is not important if either light is "OFF" or does not flash red and green, check the circuits for faults, beginning with poor terminal contacts.

Diagnostic Aids:

A slow, unstable, or fast idle may be caused by a non-IAC system problem that cannot be overcome by the IAC valve. Out of control range IAC Tech 1 counts will be above 60 if idle is too low, and zero counts if idle is too high. The following checks should be made to repair a non-IAC system problem.

- **Vacuum Leak (High Idle)** - If idle is too high, stop the engine. Fully extend (low) IAC with tester. Start engine. If Idle speed is above 800 rpm, locate and correct vacuum leak including CV system. Also check for binding of throttle blade or linkage.
- **System too lean (High Air/Fuel Ratio)** - Idle speed may be too high or too low. Engine speed may vary up and down and disconnecting IAC does not help. Code 44 may be set. "Scan" O_2 voltage will be less than 300 mV (.3 volt). Check for low regulated fuel pressure, water in the fuel or a restricted injector.
- **System too rich (Low Air/Fuel Ratio)** - The idle speed will be too low. Tech 1 "Scan" tool IAC counts will usually be above 80. System is obviously rich and may exhibit black smoke exhaust.
   "Scan" tool O_2 voltage will be fixed above 800 mV (.8 volt). Check for high fuel pressure, leaking or sticking injector. Silicone contaminated O_2 sensor will "Scan" an O_2 voltage slow to respond.
- **Throttle Body** - Remove IAC and inspect bore for foreign material.

- If intermittent, poor driveability, or idle symptoms are resolved by disconnecting the IAC, carefully recheck connections, valve terminal resistance, or replace IAC.

2.3L (VIN A) ENGINE — COMPONENT DIAGNOSTIC CHART — 1992 BERETTA

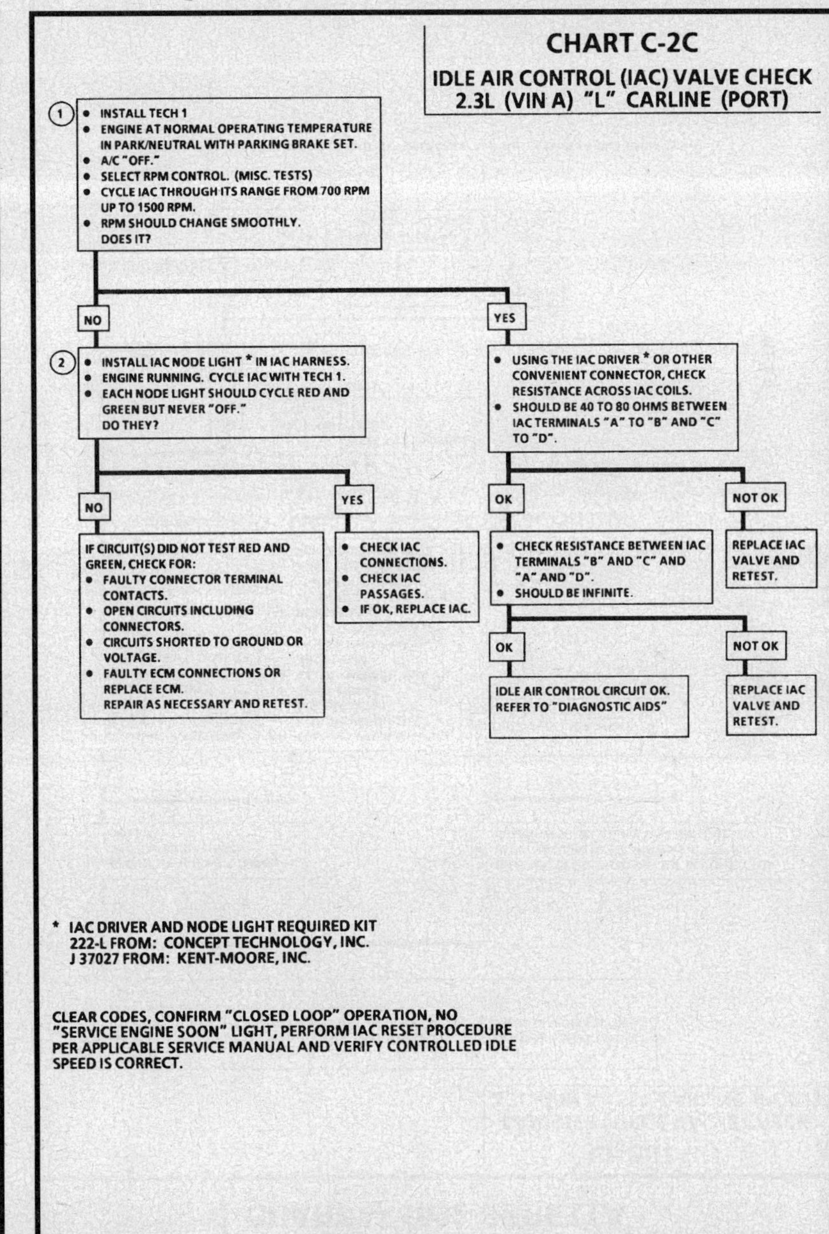

* IAC DRIVER AND NODE LIGHT REQUIRED KIT
222-L FROM: CONCEPT TECHNOLOGY, INC.
J 37027 FROM: KENT-MOORE, INC.

CLEAR CODES, CONFIRM "CLOSED LOOP" OPERATION, NO "SERVICE ENGINE SOON" LIGHT, PERFORM IAC RESET PROCEDURE PER APPLICABLE SERVICE MANUAL AND VERIFY CONTROLLED IDLE SPEED IS CORRECT.

2.3L (VIN A) ENGINE — COMPONENT DIAGNOSTIC CHART — 1992 BERETTA

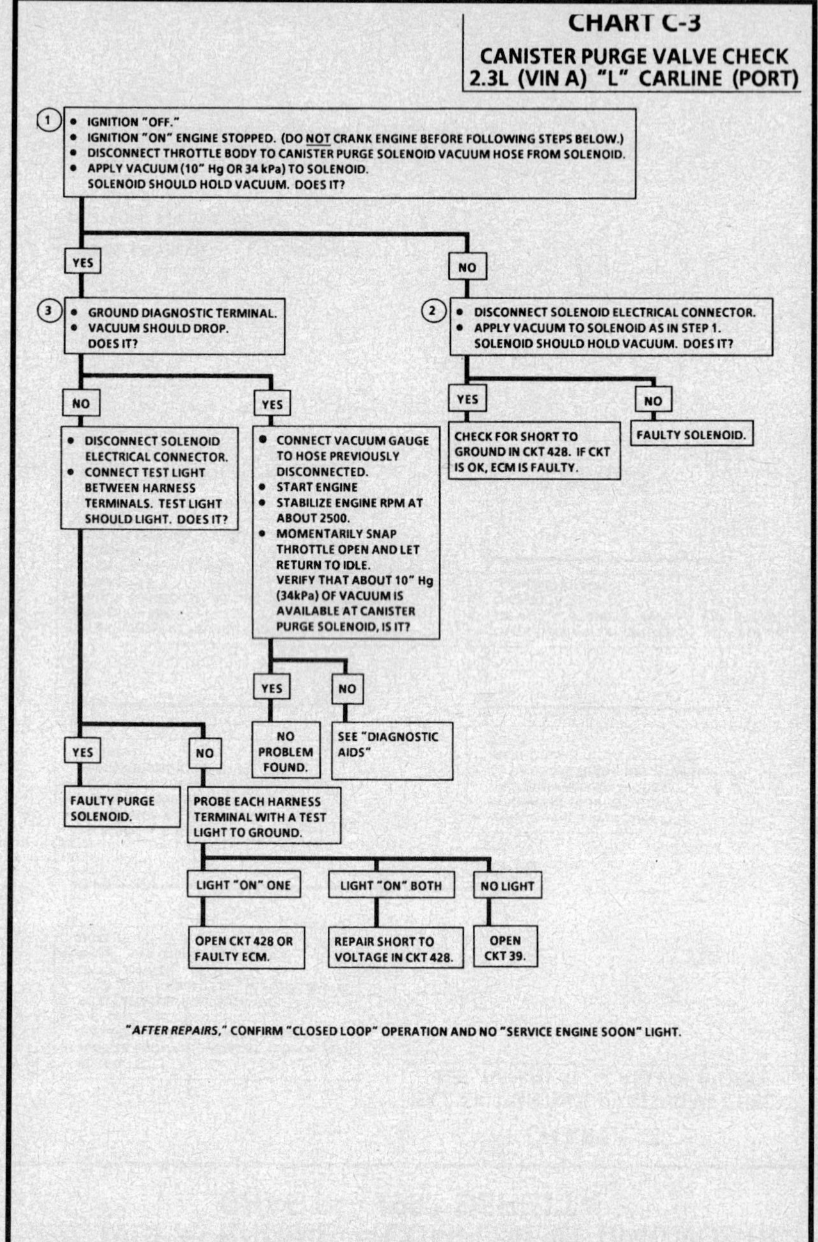

CHART C-3
CANISTER PURGE VALVE CHECK
2.3L (VIN A) "L" CARLINE (PORT)

Circuit Description:

Canister purge is controlled by a solenoid that allows manifold and/or ported vacuum to purge the canister when energized. The Electronic Control Module (ECM) supplies a ground to energize the solenoid (purge "ON"). The purge solenoid control by the ECM is pulse width modulated (turned "ON" and "OFF" several times a second). The duty cycle (pulse width) is determined by "Closed Loop" feed back from the O_2 sensor. The duty cycle is calculated by the ECM and the output commanded when the following conditions have been met:

- Engine run time after start more than 65 seconds.
- Coolant temperature above 56°C.

Also, if the diagnostic "test" terminal is grounded with the engine stopped, the purge solenoid is energized (purge "ON").

Test Description: Number(s) below refer to circled number(s) on the diagnostic chart.
1. Checks to see if the solenoid is opened or closed. The solenoid is normally de-energized in this step, so it should be closed.
2. Checks to determine if solenoid was open due to electrical circuit problem or defective solenoid.
3. Completes functional check by grounding "test" terminal. This should normally energize the solenoid opening the valve which should allow the vacuum to drop (purge "ON").

Diagnostic Aids:

Make a visual check of vacuum hose(s). Check throttle body for possible cracked, broken, or plugged vacuum block. Check engine for possible mechanical problem.

2.3L (VIN A) ENGINE — COMPONENT DIAGNOSTIC CHART — 1992 BERETTA

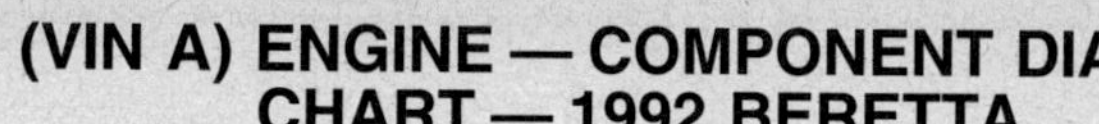

CHART C-3
CANISTER PURGE VALVE CHECK
2.3L (VIN A) "L" CARLINE (PORT)

2.3L (VIN A) ENGINE — COMPONENT DIAGNOSTIC CHART — 1992 BERETTA

2.3L (VIN A) ENGINE — COMPONENT DIAGNOSTIC CHART — 1992 BERETTA

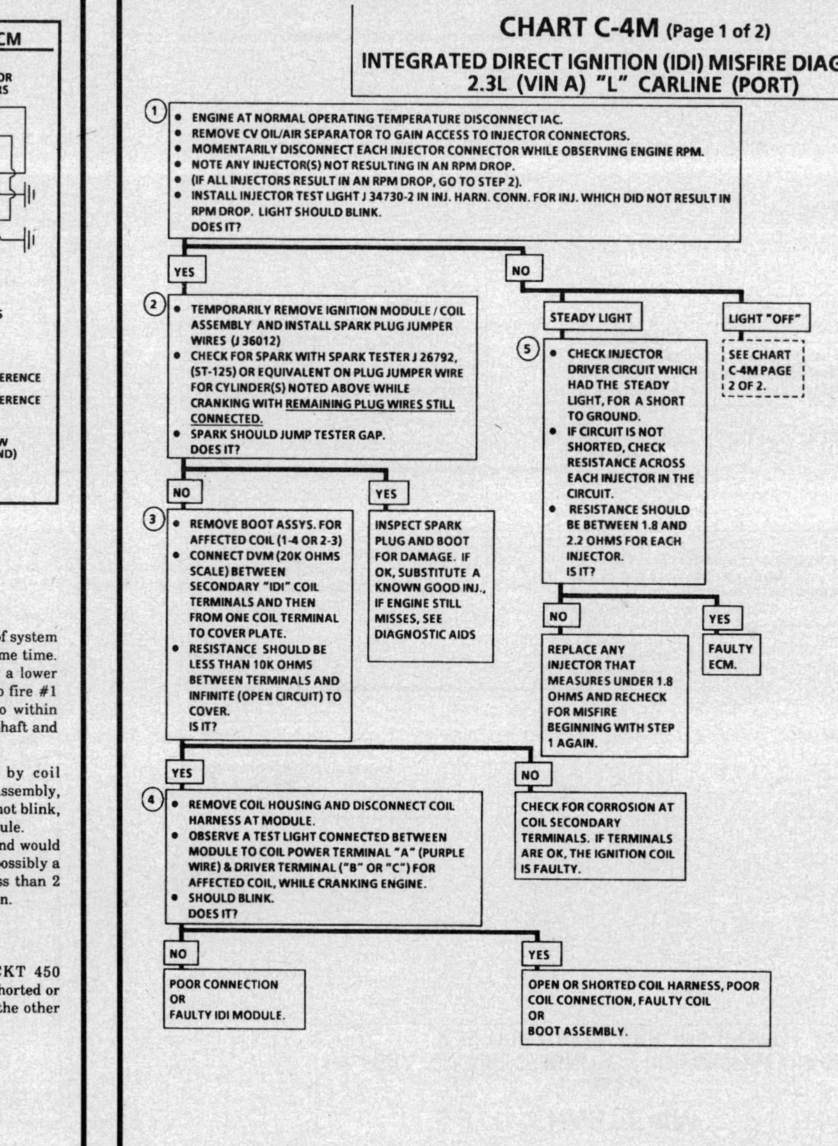

CHART C-4M

(Page 1 of 2)
INTEGRATED DIRECT IGNITION (IDI) MISFIRE DIAGNOSIS
2.3L (VIN A) "L" CARLINE (PORT)

Circuit Description:

The Integrated Direct Ignition (IDI) system uses a waste spark method of distribution. In this type of system the ignition module triggers the #1-4 coil pair resulting in both #1 and #4 spark plugs firing at the same time. #1 cylinder is on the compression stroke at the same time #4 is on the exhaust stroke, resulting in a lower energy requirement to fire # 4 spark plug. This leaves the remainder of the high voltage to be used to fire #1 spark plug. On this application, the crank sensor is mounted to, and protrudes through the block to within approximately 0.050" of the crankshaft reluctor. Since the reluctor is a machined portion of the crankshaft and the sensor is mounted in a fixed position on the block, timing adjustments are not possible or necessary.

Test Description: Number(s) below refer to circled number(s) on the diagnostic chart.

1. This checks for equal relative power output between the cylinders. Any injector which when disconnected did not result in an rpm drop approximately equal to the others, is located on the misfiring cylinder.
2. If a plug boot is burned, the other plug on that coil may still fire at idle. This step tests the system's ability to produce at least 25,000 volts at each spark plug.
3. No spark, on one coil, may be caused by an open secondary circuit. Therefore, the coil's secondary resistance should be checked. Resistance readings above 20,000 ohms, but not infinite, will probably not cause a no start but may cause an engine miss under certain conditions.
4. If the no spark condition is caused by coil connections, a coil or a secondary boot assembly, the test light will blink. If the light does not blink, the fault is module connections or the module.
5. An injector driver circuit shorted to ground would result in the test light "ON" steady, and possibly a flooded condition. A shorted injector (less than 2 ohms) could cause incorrect ECM operation.

Diagnostic Aid:

Verify IDI connector terminal "K", CKT 450 resistance to ground is less than .5 ohm. A shorted or low resistance injector may cause a miss in the other injector in that pair (1 & 4 or 2 & 3).

CHART C-4M (Page 1 of 2)

INTEGRATED DIRECT IGNITION (IDI) MISFIRE DIAGNOSIS
2.3L (VIN A) "L" CARLINE (PORT)

2.3L (VIN A) ENGINE — COMPONENT DIAGNOSTIC CHART — 1992 BERETTA

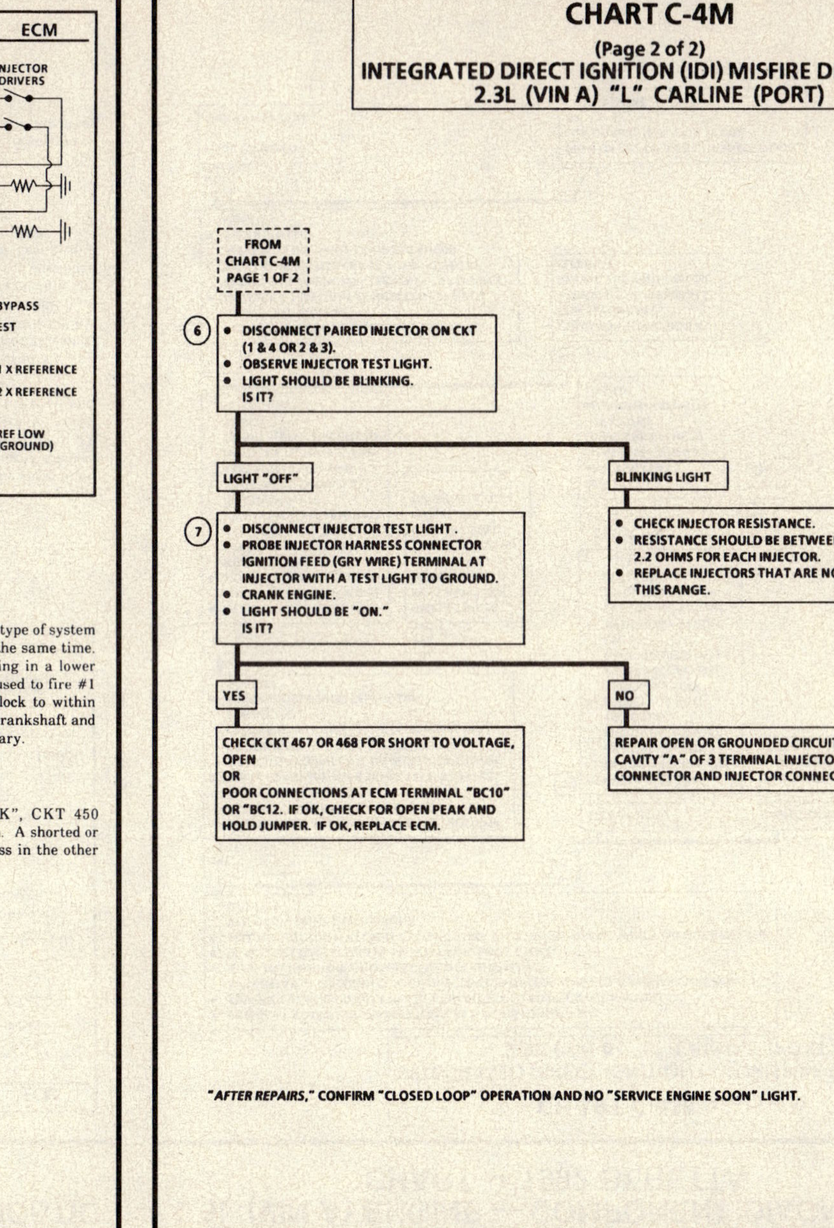

CHART C-4M

(Page 2 of 2)

INTEGRATED DIRECT IGNITION (IDI) MISFIRE DIAGNOSIS
2.3L (VIN A) "L" CARLINE (PORT)

Circuit Description:

The Integrated Direct Ignition (IDI) system uses a waste spark method of distribution. In this type of system the ignition module triggers the #1-4 coil pair resulting in both #1 and #4 spark plugs firing at the same time. #1 cylinder is on the compression stroke at the same time #4 is on the exhaust stroke, resulting in a lower energy requirement to fire # 4 spark plug. This leaves the remainder of the high voltage to be used to fire #1 spark plug. On this application, the crank sensor is mounted to, and protrudes through the block to within approximately 0.050" of the crankshaft reluctor. Since the reluctor is a machined portion of the crankshaft and the sensor is mounted in a fixed position on the block, timing adjustments are not possible or necessary.

Test Description: Number(s) below refer to circled number(s) on the diagnostic chart.

6. Checks for grounded or low resistance injector. This condition will not able the paired injector to properly deliver fuel.
7. Checks for ignition voltage feed to injector and open injector driver circuit.

Diagnostic Aid:

Verify IDI connector terminal "K", CKT 450 resistance to ground is less than .5 ohm. A shorted or low resistance injector may cause a miss in the other injector in that pair (1 & 4 or 2 & 3).

2.3L (VIN A) ENGINE — COMPONENT DIAGNOSTIC CHART — 1992 BERETTA

CHART C-4M

(Page 2 of 2)

INTEGRATED DIRECT IGNITION (IDI) MISFIRE DIAGNOSIS
2.3L (VIN A) "L" CARLINE (PORT)

FROM CHART C-4M PAGE 1 OF 2

6
- DISCONNECT PAIRED INJECTOR ON CKT (1 & 4 OR 2 & 3).
- OBSERVE INJECTOR TEST LIGHT.
- LIGHT SHOULD BE BLINKING. IS IT?

LIGHT "OFF"

7
- DISCONNECT INJECTOR TEST LIGHT.
- PROBE INJECTOR HARNESS CONNECTOR IGNITION FEED (GRY WIRE) TERMINAL AT INJECTOR WITH A TEST LIGHT TO GROUND.
- CRANK ENGINE.
- LIGHT SHOULD BE "ON." IS IT?

BLINKING LIGHT
- CHECK INJECTOR RESISTANCE.
- RESISTANCE SHOULD BE BETWEEN 1.8 AND 2.2 OHMS FOR EACH INJECTOR.
- REPLACE INJECTORS THAT ARE NOT WITHIN THIS RANGE.

YES

CHECK CKT 467 OR 468 FOR SHORT TO VOLTAGE, OPEN
OR
POOR CONNECTIONS AT ECM TERMINAL "BC10" OR "BC12. IF OK, CHECK FOR OPEN PEAK AND HOLD JUMPER. IF OK, REPLACE ECM.

NO

REPAIR OPEN OR GROUNDED CIRCUIT BETWEEN CAVITY "A" OF 3 TERMINAL INJECTOR HARNESS CONNECTOR AND INJECTOR CONNECTOR.

"AFTER REPAIRS," CONFIRM "CLOSED LOOP" OPERATION AND NO "SERVICE ENGINE SOON" LIGHT.

2.3L (VIN A) ENGINE — COMPONENT DIAGNOSTIC CHART — 1992 BERETTA

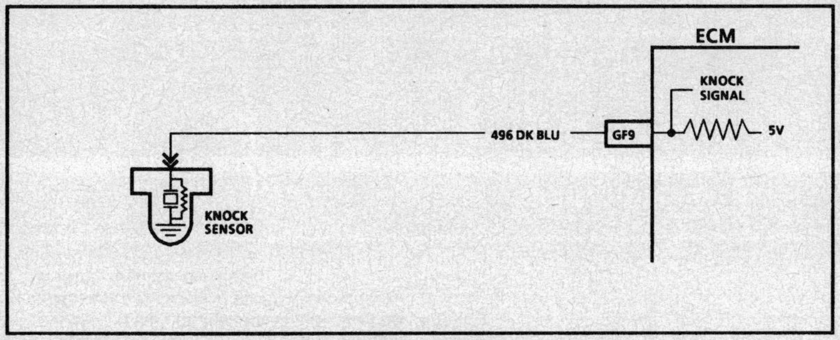

CHART C-5
ELECTRONIC SPARK CONTROL (ESC) SYSTEM CHECK
2.3L (VIN A) "L" CARLINE (PORT)

Circuit Description:

The knock sensor is used to detect engine detonation and the Electronic Control Module (ECM) will retard the Electronic Spark Timing (EST) based on the signal being received. The circuitry within the knock sensor causes the ECM's 5 volts to be pulled down so that CKT 496 would measure about 2.5 volts. The knock sensor produces an AC signal which rides on the 2.5 volts DC voltage. The amplitude and frequency are dependent upon the knock level.

The MEM-CAL used with this engine contains the functions which were part of remotely mounted ESC modules used on other GM vehicles. The ESC portion of the MEM-CAL then sends a signal to other parts of the ECM which adjusts the spark timing to retard the spark and reduce the detonation.

Test Description: Number(s) below refer to circled number(s) on the diagnostic chart.

1. With engine idling, there should not be a knock signal present at the ECM because detonation is not likely under a no load condition.
2. Tapping on the engine lift hook bracket should simulate a knock signal to determine if the sensor is capable of detecting detonation. If no knock is detected, try tapping on engine block closer to sensor before replacing sensor.
3. If the engine has an internal problem which is creating a knock, the knock sensor may be responding to the internal failure.

4. This test determines if the knock sensor is faulty or if the ESC portion of the MEM-CAL is faulty. If it is determined that the MEM-CAL is faulty, be sure that is is properly installed and latched into place. If not properly installed, repair and retest.

Diagnostic Aids:

While observing knock signal on the Tech 1, there should be an indication that knock is present when detonation can be heard. Detonation is most likely to occur under high engine load conditions.

2.3L (VIN A) ENGINE — COMPONENT DIAGNOSTIC CHART — 1992 BERETTA

CHART C-5
ELECTRONIC SPARK CONTROL (ESC) SYSTEM CHECK
2.3L (VIN A) "L" CARLINE (PORT)

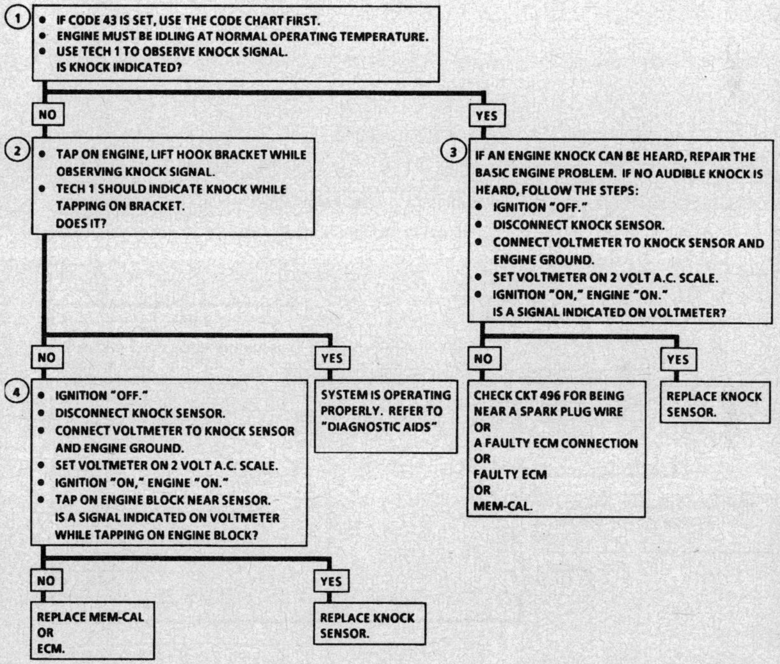

2.3L (VIN A) ENGINE — COMPONENT DIAGNOSTIC CHART — 1992 BERETTA

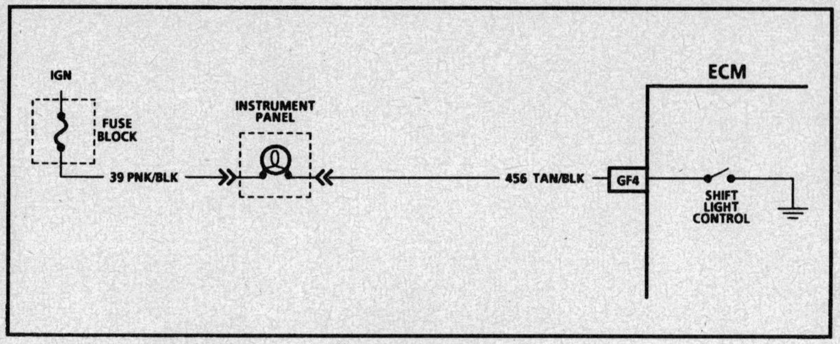

CHART C-8C
MANUAL TRANSAXLE (M/T) SHIFT LIGHT CHECK
2.3L (VIN A) "L" CARLINE (PORT)

Circuit Description:

The shift light indicates the best transaxle shift point for maximum fuel economy. The light is controlled by the Electronic Control Module (ECM) and is turned "ON" by grounding CKT 456.

The ECM uses inputs from various sensors to calculate when the shift light should be turned "ON" as follows:
- Coolant temperature must be above -10°C (14°F).
- ECM can determine the transaxle has been in a gear for at least 1.2 seconds, by comparison of vehicle speed (from VSS) with engine rpm.
- Throttle Position Sensor (TPS) is between minimum and maximum calibrated values for each gear.
- Rpm is above a calibrated value for each gear (maximum 6500 rpm).

The light will be turned "ON" after a calibrated delay time which is dependent on last gear change or downshift, and will remain on for a minimum of 10 seconds.

Test Description: Numbers below refer to circled number(s) on the diagnostic chart.

1. This should not turn "ON" the shift light. If the light is "ON," there is a short to ground in CKT 456 wiring or a fault in the ECM.
2. When the diagnostic terminal is grounded, the ECM should ground CKT 456 and the shift light should come "ON."
3. This checks the shift light circuit up to the ECM connector. If the shift light illuminates, then the ECM connector is faulty or the ECM does not have the ability to ground the circuit.

Diagnostic Aids:

Check for Code 24 (no VSS). Faulty thermostat or incorrect heat range. Incorrect or faulty MEM-CAL.

A faulty or improper VSS may result in a shift light that does not operate.

2.3L (VIN A) ENGINE — COMPONENT DIAGNOSTIC CHART — 1992 BERETTA

CHART C-8C
MANUAL TRANSAXLE (M/T) SHIFT LIGHT CHECK
2.3L (VIN A) "L" CARLINE (PORT)

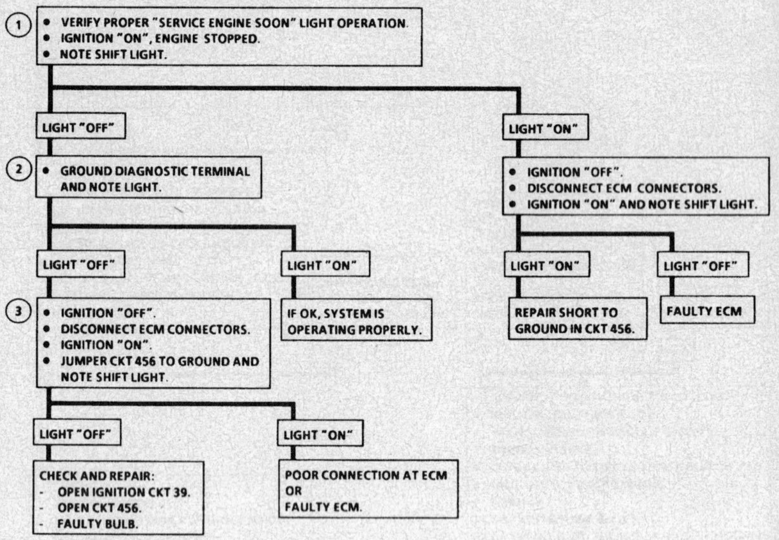

2.3L (VIN A) ENGINE — COMPONENT DIAGNOSTIC CHART — 1992 BERETTA

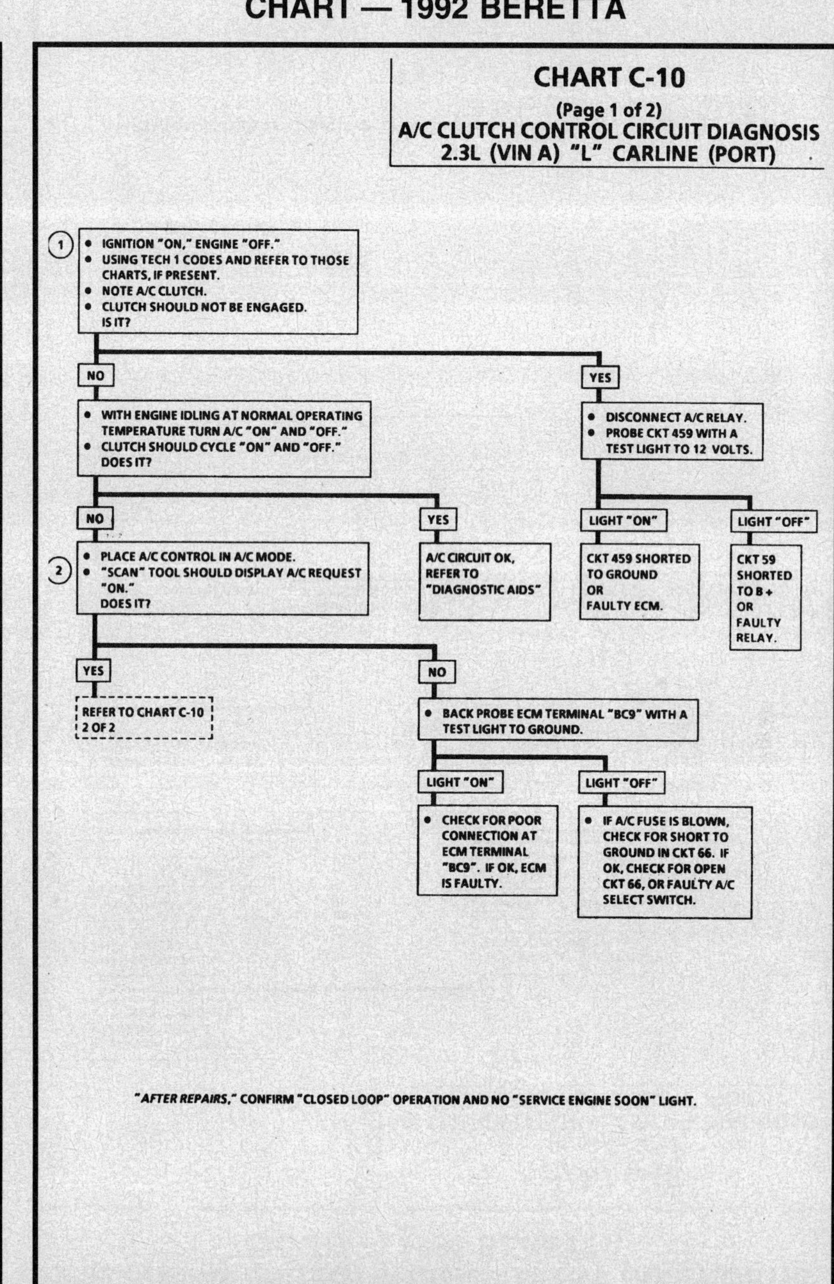

CHART C-10
(Page 1 of 2)
**A/C CLUTCH CONTROL CIRCUIT DIAGNOSIS
2.3L (VIN A) "L" CARLINE (PORT)**

Circuit Description:

The A/C clutch control relay is energized when the Electronic Control Module (ECM) provides a ground path through CKT 459 and A/C is requested. A/C clutch is delayed about .3 second after A/C is requested. This will allow the IAC to adjust engine rpm for the additional load.

The ECM will temporarily disengage the A/C clutch relay for calibrated times for one or more of the following:
- Hot engine restart.
- Wide Open Throttle (WOT) (TPS over 90%).
- Engine rpm greater than about 6000 rpm.
- During Idle Air Control (IAC) reset.

The A/C clutch relay will remain disengaged when a Code 66 is present, if pressure is out of range ' . or there is no A/C request signal due to an open A/C select switch or circuit.

Test Description: Number(s) below refer to circled number(s) on the diagnostic chart.
1. The ECM will only energize the A/C relay when the engine is running. This test will determine if the relay or CKT 459 is faulty.
2. Determines if the signal is reaching the ECM through CKT 66 from the A/C control panel. Signal should only be present when an A/C mode or defrost mode has been selected.

2.3L (VIN A) ENGINE — COMPONENT DIAGNOSTIC CHART — 1992 BERETTA

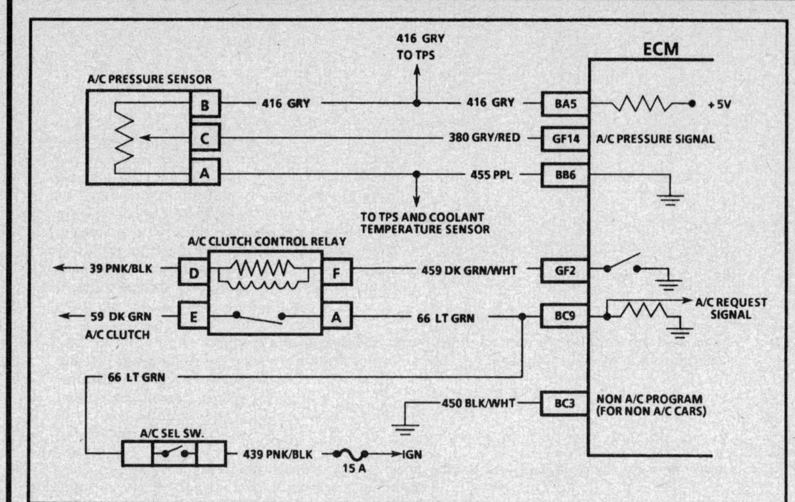

CHART C-10
(Page 2 of 2)
**A/C CLUTCH CONTROL CIRCUIT DIAGNOSIS
2.3L (VIN A) "L" CARLINE (PORT)**

Circuit Description:

The A/C clutch control relay is energized when the ECM provides a ground path through CKT 459 and A/C is requested. A/C clutch is delayed about .3 second after A/C is requested. This will allow the IAC to adjust engine rpm for the additional load.

The ECM will temporarily disengage the A/C clutch relay for calibrated times for one or more of the following:
- Hot engine restart.
- Wide open throttle (TPS over 90%).
- Engine rpm greater than about 6000 rpm.
- During IAC reset.

The A/C clutch relay will remain disengaged when a Code 66 is present, if pressure is out of range or there is no A/C request signal due to an open A/C select switch or circuit.

Test Description: Number(s) below refer to circled number(s) on the diagnostic chart.
1. Determines if the pressure transducer is out of range causing the compressor clutch to be disengaged.
2. With the engine stopped and field service mode activated, the ECM should be grounding CKT 459, which should cause the test light to be "ON."

Diagnostic Aids:

If complaint is insufficient cooling, the problem may be caused by an inoperative cooling fan. See CHART C-12 for cooling fan diagnosis. If fan operates correctly, see "A/C Diagnosis" A/C pressure outside of a range of 43 to 428 psi will cause the compressor to be disabled by the ECM. Observe Tech 1 A/C pressure for 2 minutes with engine idling and A/C "ON."

"Scan" pressure should be within 20 psi of actual. If not, check for a circuit problem using Code 66 chart or replace sensor.

2.3L (VIN A) ENGINE — COMPONENT DIAGNOSTIC CHART — 1992 BERETTA

CHART C-10
(Page 2 of 2)
**A/C CLUTCH CONTROL CIRCUIT DIAGNOSIS
2.3L (VIN A) "L" CARLINE (PORT)**

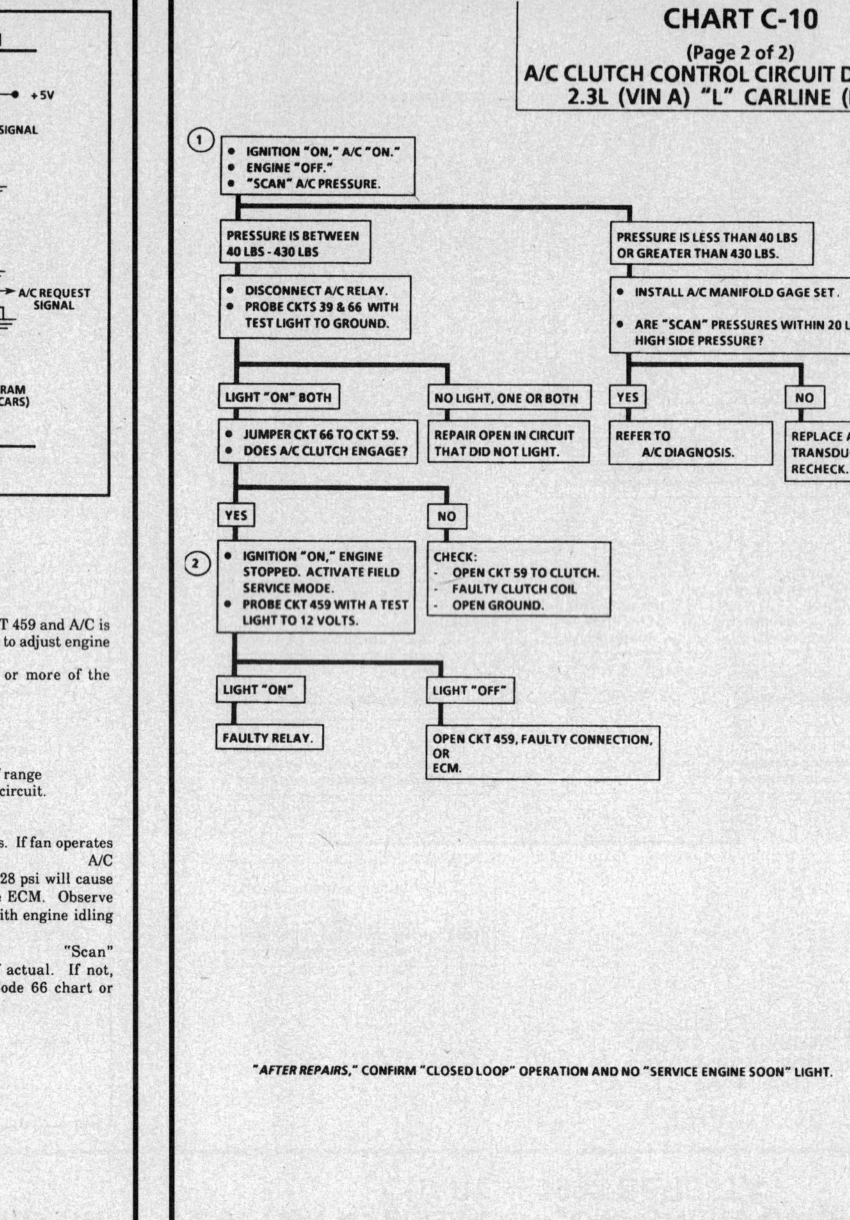

"*AFTER REPAIRS*," CONFIRM "CLOSED LOOP" OPERATION AND NO "SERVICE ENGINE SOON" LIGHT.

2.3L (VIN A) ENGINE — COMPONENT DIAGNOSTIC CHART — 1992 BERETTA

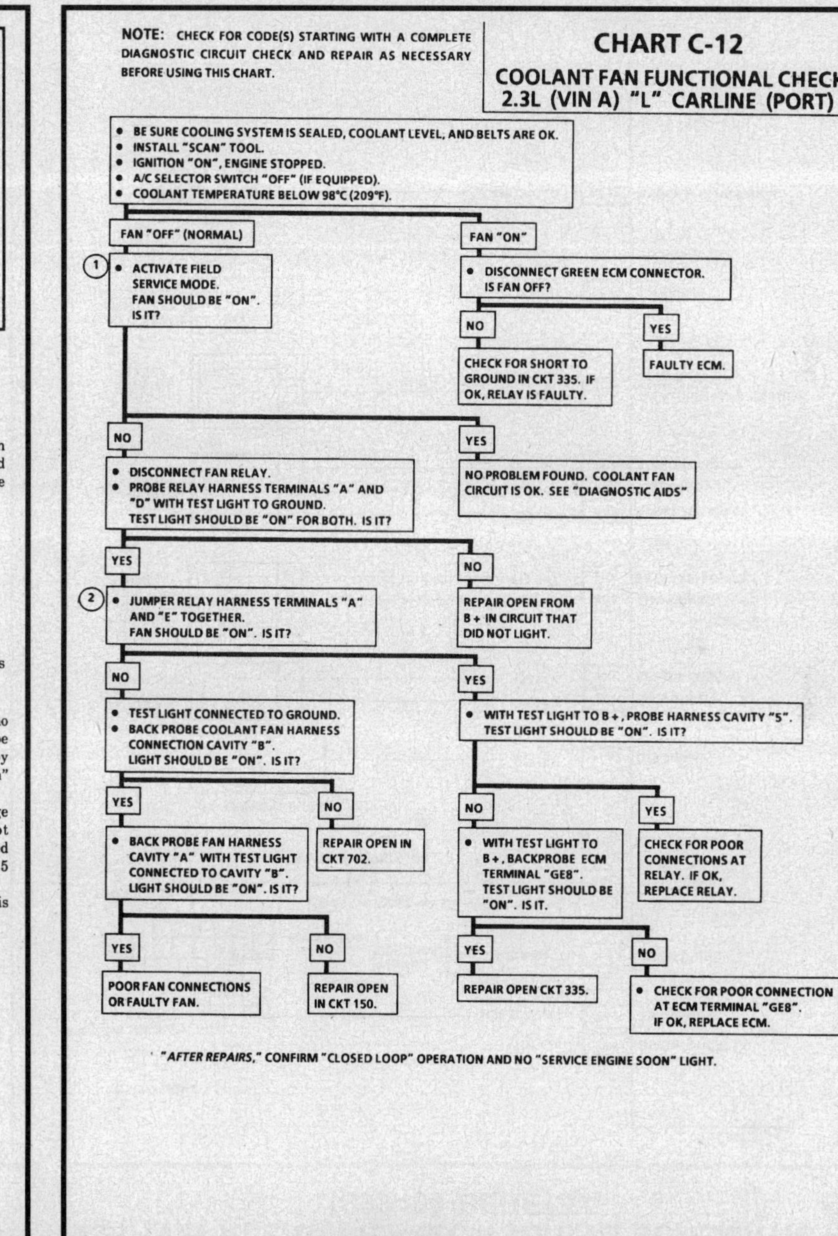

Circuit Description:

The electric coolant fan is controlled by the Electronic Control Module (ECM) through the fan relay based on inputs from the coolant and manifold air temperature sensors, the A/C control switch, A/C pressure sensor and the Vehicle Speed Sensor (VSS). The ECM controls the coolant fan by grounding CKT 335 which turns "ON" the fan relay.

The fan relay will be commanded "ON" when:
- Coolant temperature 103°C - 106°C or more.
- A/C clutch requested.
- Vehicle speed is less than 35 mph.

The fan relay will be commanded "ON" regardless of vehicle speed when:
- Code 14 or 15 is set.
- Coolant temperature 115°C - 118°C or more.
- A/C pressure is high.
- The coolant fan may be commanded "ON" when the engine is not running under fan "Run-On" conditions

Test Description:
Number(s) below refer to circled number(s) on the diagnostic chart.
1. With the field service mode activated, the coolant fan control driver should close, which should energize the fan control relay.
2. Test to see if fault is in wiring to the fan or the fan/fan connection.

Diagnostic Aids:

If the owner complained of an overheating problem, it must be determined if the complaint was due to an actual boil over, or the "hot light," or temperature gage indicated overheating.

If the gage, or light, indicates overheating, but no boil over is detected, the gage or light circuit should be checked. The gage accuracy can also be checked by comparing the coolant sensor reading using a "Scan" tool with the gage reading.

If the engine is actually overheating, and the gage indicates overheating, but the coolant fan is not coming "ON," the coolant sensor has probably shifted out of calibration and should be replaced. See Code 15 chart for a temperature to resistance chart.

If the engine is overheating, and the coolant fan is "ON," the cooling system should be checked.

2.3L (VIN A) ENGINE — COMPONENT DIAGNOSTIC CHART — 1992 BERETTA

NOTE: CHECK FOR CODE(S) STARTING WITH A COMPLETE DIAGNOSTIC CIRCUIT CHECK AND REPAIR AS NECESSARY BEFORE USING THIS CHART.

CHART C-12
COOLANT FAN FUNCTIONAL CHECK
2.3L (VIN A) "L" CARLINE (PORT)

- BE SURE COOLING SYSTEM IS SEALED, COOLANT LEVEL, AND BELTS ARE OK.
- INSTALL "SCAN" TOOL.
- IGNITION "ON", ENGINE STOPPED.
- A/C SELECTOR SWITCH "OFF" (IF EQUIPPED).
- COOLANT TEMPERATURE BELOW 98°C (209°F).

FAN "OFF" (NORMAL)

(1) ACTIVATE FIELD SERVICE MODE. FAN SHOULD BE "ON". IS IT?

FAN "ON"
- DISCONNECT GREEN ECM CONNECTOR. IS FAN OFF?

NO: CHECK FOR SHORT TO GROUND IN CKT 335. IF OK, RELAY IS FAULTY.
YES: FAULTY ECM.

NO:
- DISCONNECT FAN RELAY.
- PROBE RELAY HARNESS TERMINALS "A" AND "D" WITH TEST LIGHT TO GROUND. TEST LIGHT SHOULD BE "ON" FOR BOTH. IS IT?

YES: NO PROBLEM FOUND. COOLANT FAN CIRCUIT IS OK. SEE "DIAGNOSTIC AIDS"

YES:
(2) JUMPER RELAY HARNESS TERMINALS "A" AND "E" TOGETHER. FAN SHOULD BE "ON". IS IT?

NO: REPAIR OPEN FROM B + IN CIRCUIT THAT DID NOT LIGHT.

NO:
- TEST LIGHT CONNECTED TO GROUND.
- BACK PROBE COOLANT FAN HARNESS CONNECTION CAVITY "B". LIGHT SHOULD BE "ON". IS IT?

YES:
- WITH TEST LIGHT TO B +, PROBE HARNESS CAVITY "5". TEST LIGHT SHOULD BE "ON". IS IT?

YES:
- BACK PROBE FAN HARNESS CAVITY "A" WITH TEST LIGHT CONNECTED TO CAVITY "B". LIGHT SHOULD BE "ON". IS IT?

NO: REPAIR OPEN IN CKT 702.

NO:
- WITH TEST LIGHT TO B +, BACKPROBE ECM TERMINAL "GE8". TEST LIGHT SHOULD BE "ON". IS IT.

YES: CHECK FOR POOR CONNECTIONS AT RELAY. IF OK, REPLACE RELAY.

YES: POOR FAN CONNECTIONS OR FAULTY FAN.

NO: REPAIR OPEN IN CKT 150.

YES: REPAIR OPEN CKT 335.

NO:
- CHECK FOR POOR CONNECTION AT ECM TERMINAL "GE8". IF OK, REPLACE ECM.

"AFTER REPAIRS," CONFIRM "CLOSED LOOP" OPERATION AND NO "SERVICE ENGINE SOON" LIGHT.

MULTIPORT FUEL INJECTION (MFI) SYSTEMS
EXCEPT LIGHT TRUCKS, VANS, GEO AND SATURN

2.3L (VIN A) ENGINE — ECM WIRING SCHEMATIC — 1993–94 BERETTA

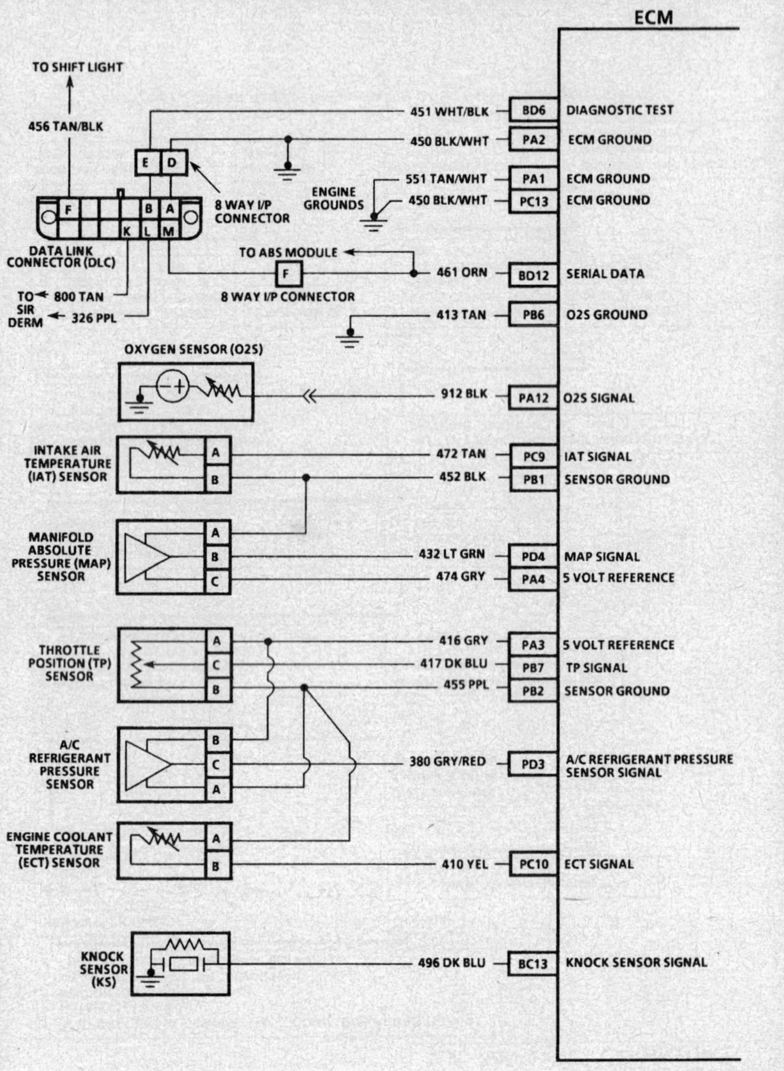

2.3L (VIN A) ENGINE — ENGINE COMPONENT LOCATION CHART — 1993–94 BERETTA

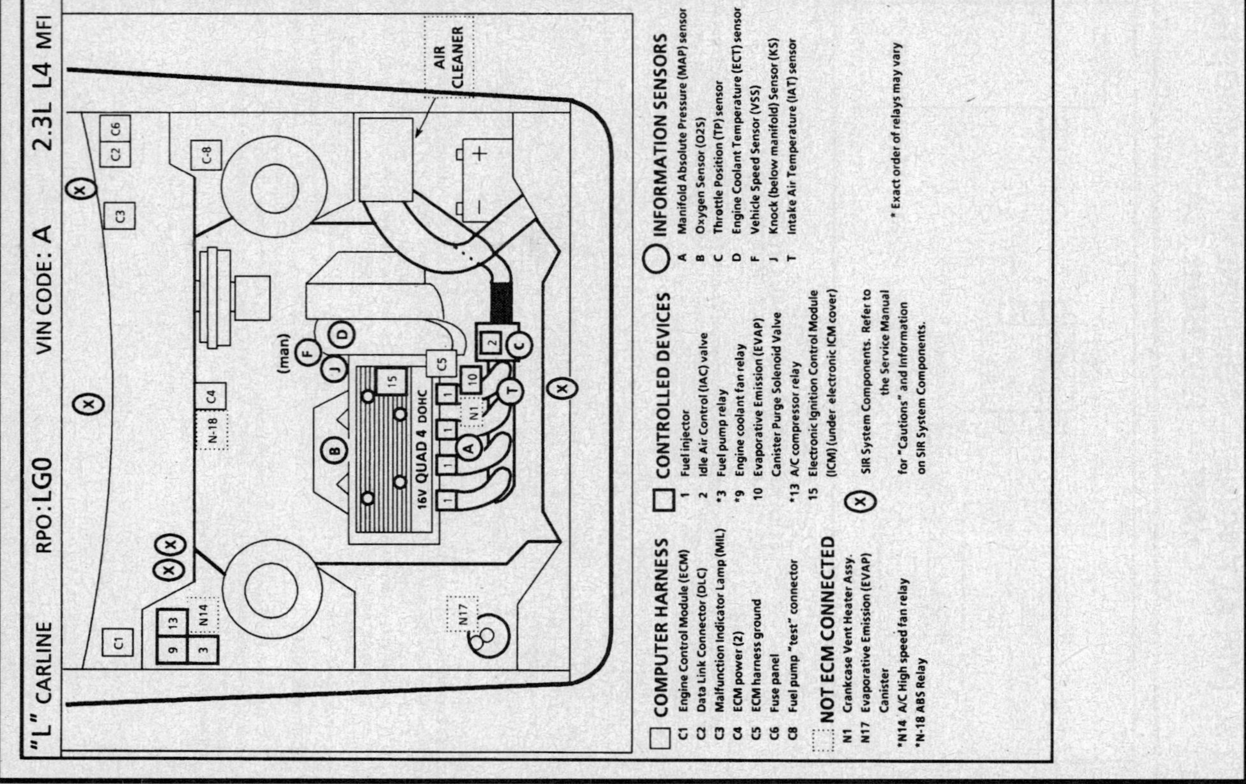

2.3L (VIN A) ENGINE — ECM WIRING SCHEMATIC — 1993–94 BERETTA

2.3L (VIN A) ENGINE — ECM WIRING SCHEMATIC — 1993–94 BERETTA

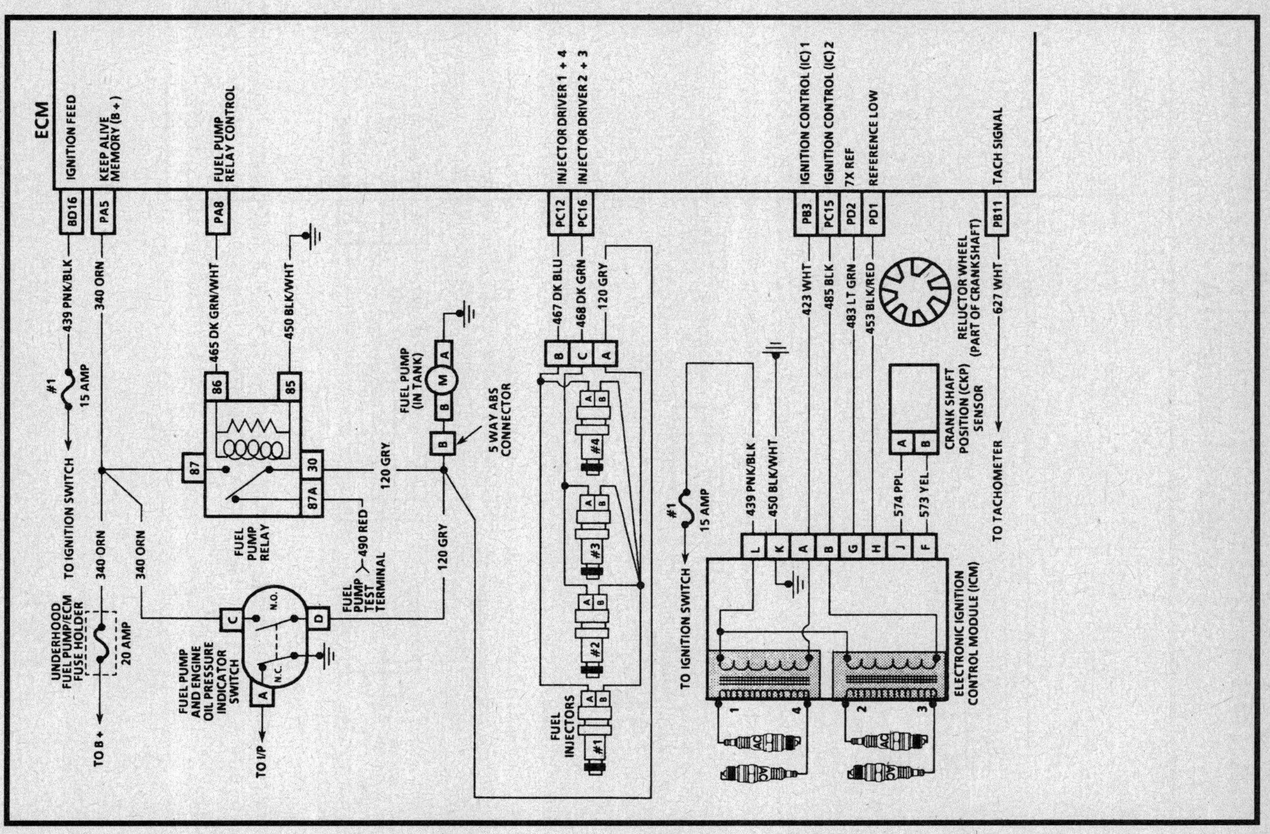

2.3L (VIN A) ENGINE — ECM WIRING SCHEMATIC — 1993–94 BERETTA

2.3L (VIN A) ENGINE — ECM CONNECTOR END VIEW — 1993–94 BERETTA

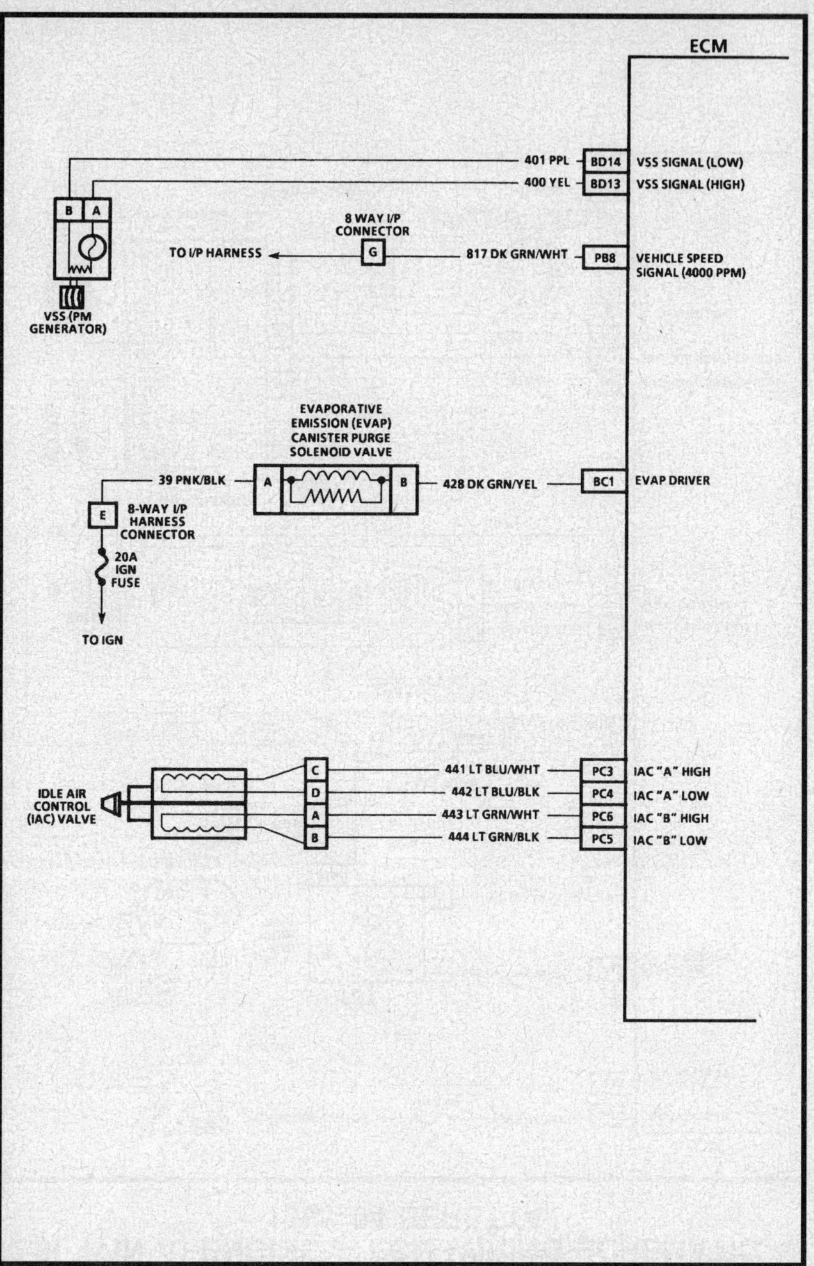

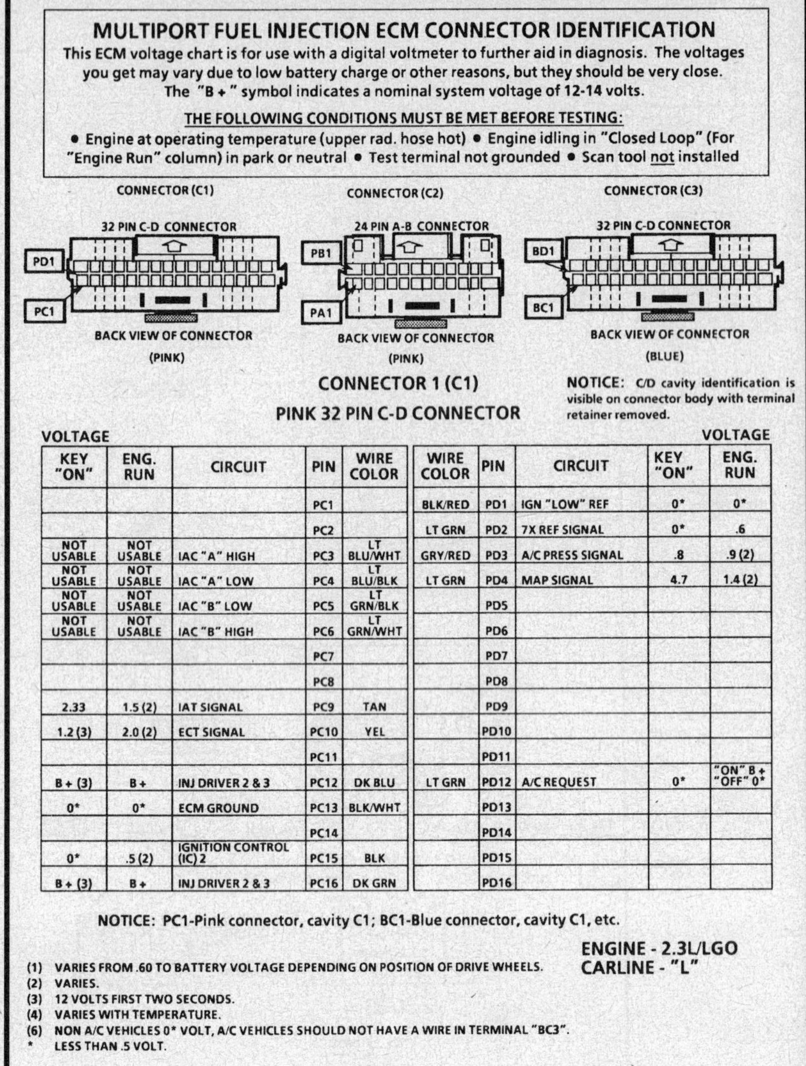

MULTIPORT FUEL INJECTION ECM CONNECTOR IDENTIFICATION

This ECM voltage chart is for use with a digital voltmeter to further aid in diagnosis. The voltages you get may vary due to low battery charge or other reasons, but they should be very close. The "B +" symbol indicates a nominal system voltage of 12-14 volts.

THE FOLLOWING CONDITIONS MUST BE MET BEFORE TESTING:

- Engine at operating temperature (upper rad. hose hot)
- Engine idling in "Closed Loop" (For "Engine Run" column) in park or neutral
- Test terminal not grounded
- Scan tool not installed

CONNECTOR 1 (C1)

PINK 32 PIN C-D CONNECTOR

NOTICE: C/D cavity identification is visible on connector body with terminal retainer removed.

KEY "ON"	ENG. RUN	CIRCUIT	PIN	WIRE COLOR	WIRE COLOR	PIN	CIRCUIT	KEY "ON"	ENG. RUN
			PC1		BLK/RED	PD1	IGN "LOW" REF	0*	0*
			PC2		LT GRN	PD2	7X REF SIGNAL	0*	.6
NOT USABLE	NOT USABLE	IAC "A" HIGH	PC3	LT BLU/WHT	GRY/RED	PD3	A/C PRESS SIGNAL	.8	.9 (2)
NOT USABLE	NOT USABLE	IAC "A" LOW	PC4	LT BLU/BLK	LT GRN	PD4	MAP SIGNAL	4.7	1.4 (2)
NOT USABLE	NOT USABLE	IAC "B" LOW	PC5	LT GRN/BLK		PD5			
NOT USABLE	NOT USABLE	IAC "B" HIGH	PC6	LT GRN/WHT		PD6			
			PC7			PD7			
			PC8			PD8			
2.33	1.5 (2)	IAT SIGNAL	PC9	TAN		PD9			
1.2 (3)	2.0 (2)	ECT SIGNAL	PC10	YEL		PD10			
			PC11			PD11			
B + (3)	B +	INJ DRIVER 2 & 3	PC12	DK BLU	LT GRN	PD12	A/C REQUEST	0*	"ON" B + "OFF" 0*
0*	0*	ECM GROUND	PC13	BLK/WHT		PD13			
			PC14			PD14			
0*	.5 (2)	IGNITION CONTROL (IC) 2	PC15	BLK		PD15			
B + (3)	B +	INJ DRIVER 2 & 3	PC16	DK GRN		PD16			

NOTICE: PC1-Pink connector, cavity C1; BC1-Blue connector, cavity C1, etc.

ENGINE - 2.3L/LGO
CARLINE - "L"

(1) VARIES FROM .60 TO BATTERY VOLTAGE DEPENDING ON POSITION OF DRIVE WHEELS.
(2) VARIES.
(3) 12 VOLTS FIRST TWO SECONDS.
(4) VARIES WITH TEMPERATURE.
(6) NON A/C VEHICLES 0* VOLT, A/C VEHICLES SHOULD NOT HAVE A WIRE IN TERMINAL "BC3".
* LESS THAN .5 VOLT.

2.3L (VIN A) ENGINE — ECM CONNECTOR END VIEW — 1993–94 BERETTA

CONNECTOR 2 (C-2)
PINK 24 PIN A-B CONNECTOR

| VOLTAGE | | | | | | | | | VOLTAGE | |
KEY "ON"	ENG. RUN	CIRCUIT	PIN	WIRE COLOR	WIRE COLOR	PIN	CIRCUIT	KEY "ON"	ENG. RUN
0*	0*	ECM GROUND	PA1	TAN/WHT	BLK	PB1	IAT AND MAP GROUND	0*	0*
0*	0*	ECM GROUND	PA2	BLK/WHT	PPL	PB2	A/C, ECT, TP GROUND	0*	0*
5.0	5.0	+5V REFERENCE	PA3	GRY	WHT	PB3	IGNITION CONTROL (IC) (1)	0*	.5 (2)
5.0	5.0	+5V REFERENCE	PA4	GRY		PB4			
B+	B+	BATTERY	PA5	ORN		PB5			
			PA6		TAN	PB6	O2S GROUND	0*	0*
			PA7		DK BLU	PB7	TP SIGNAL	.49	.5 (2)
B+ (3)	B+	FUEL PUMP	PA8	DK GRN/WHT	DK GRN/WHT	PB8	4000 P/MI SPEED	8.3 (2)	9.4
			PA9			PB9			
			PA10			PB10			
			PA11		WHT	PB11	TACH	4.9	1.0
.35 (2)	.1-.9 (2)	O2S SIGNAL	PA12	BLK		PB12			

BLUE 32 PIN C-D CONNECTOR 3 (C-3)

| VOLTAGE | | | | | | | | | VOLTAGE | |
KEY "ON"	ENG. RUN	CIRCUIT	PIN	WIRE COLOR	WIRE COLOR	PIN	CIRCUIT	KEY "ON"	ENG. RUN
B+	.3	EVAP DRIVER	BC1	DK GRN/YEL		BD1			
B+	"OFF" B+ "ON" 0*	SHIFT LIGHT	BC2	TAN/BLK		BD2			
			BC3			BD3			
			BC4			BD4			
			BC5			BD5			
			BC6		WHT/BLK	BD6	DLC DIAG TERMINAL	B+	13.6
			BC7			BD7			
0*	B+	MIL	BC8	BRN/WHT		BD8			
B+	"ON" 0* "OFF" B+	CLG FAN RELAY	BC9	DK GRN/WHT		BD9			
			BC10			BD10			
0*	OFF" B+ "ON" 0*	A/C CLUTCH RELAY	BC11	DK GRN/WHT		BD11			
			BC12		ORN	BD12	DLC SERIAL DATA	4.2	4.5
2.3	2.3	KNOCK SIGNAL	BC13	DK BLU	YEL	BD13	VSS "HIGH"	0*	0*
			BC14		PPL	BD14	VSS "LOW"	0*	0*
			BC15			BD15			
			BC16		PNK/BLK	BD16	IGNITION FEED	B+	B+

NOTICE: PA1 = Pink Connector, cavity A1, etc. BL1 - Blue Connector, cavity C1, etc.

(1) VARIES FROM .60 TO BATTERY VOLTAGE DEPENDING ON POSITION OF DRIVE WHEELS.
(2) VARIES.
(3) B+ FIRST TWO SECONDS.
* LESS THAN .5 VOLT.

ENGINE - 2.3L / LG0
CARLINE - "L" Series

2.3L (VIN A) ENGINE — ON-BOARD DIAGNOSTIC SYSTEM CHART — 1993–94 BERETTA

ON-BOARD DIAGNOSTIC (OBD) SYSTEM CHECK

The On-Board Diagnostic (OBD) system check is an organized approach to identifying a problem created by an electronic engine control system malfunction. It must be the starting point for any driveability complaint diagnosis, because it directs the service technician to the next logical step in diagnosing the complaint.

The Tech 1 data listed in the table may be used for comparison, after completing the OBD system check and finding the on-board diagnostics functioning properly and no diagnostic trouble codes displayed. The "Typical Values" are an average of display values recorded from normally operating vehicles and are intended to represent what a normally functioning system would typically display.

A SCAN TOOL THAT DISPLAYS FAULTY DATA SHOULD NOT BE USED, AND THE PROBLEM SHOULD BE REPORTED TO THE MANUFACTURER. THE USE OF A FAULTY SCAN CAN RESULT IN MISDIAGNOSIS AND UNNECESSARY PARTS REPLACEMENT.

Only the parameters listed below are used in this manual for diagnosing. If a scan reads other parameters, the values are not recommended by General Motors for use in diagnosing.

TECH 1 DATA
Idle / Upper Radiator Hose Hot / Closed Throttle / Park or Neutral / "Closed Loop" / Acc. "OFF"

Scan Position	Units Displayed	Typical Data Value
Engine Speed	RPM	± 100 RPM from desired RPM (± 50 in drive)
Desired Idle	RPM	ECM idle command (varies with calibration. temp.)
Eng. Coolant Temp.	C° F°	85°–115°C
Intake Air Temp.	C° F°	10°–80°C (depends on underhood temp.)
MAP	kPa, V	1–3 volts (depends on vacuum & baro pressure)
BARO	kPa, V	3–5 volts (depends on altitude & baro pressure)
Throt Position	Volts	200–.900 (up to 5.0 at wide open throttle)
Throttle Angle	0–100%	0% (up to 100% at wide open throttle)
O2S	mV	1–1000 and varying
Injector Pulse Width	m Sec.	1–4 and varying
Spark Advance	# of Degrees	Varies
Lo Oct. Knock Ret.	# of Degrees	Varies
S.T. Fuel Trim	Counts	Varies
L.T. Fuel Trim	Counts	58–198 (see Section "C2")
Loop Status	Open/Closed	"Closed Loop" (may go open with extended idle)
Fuel Trim Cell	Cell Number	18-21 at idle (depends on air flow, RPM, P/N & A/C)
Knock Retard	Degrees of Retard	0
Knock Signal	Yes/No	No
KS Activity	0-255 Counts	16-22 (usable for DTC 43)
7x Resync Counter	0-255 Counts	0 (usable for DTC 19)
Ign. Control 7x Ref.	0-255 Counts	0-255 (usable for DTC 19)
Ign Control 7x Syn.	"ON/OFF"	"ON" (usable for DTC 19 and for misfire diagnosis)
Idle Air Control	Counts (steps)	5–60
Park/Neutral Position	PN and RDL	P - N -- (or -R-DL manual only)
VSS	Mph and km/h	0
TCC/Shift Light	"ON/OFF"	"OFF" ("ON," with TCC commanded)
System Voltage	Volts	13.5 – 14.5
2nd Gear	Yes/No (N/A L40 only)	No (yes in 2nd or 3rd gear) or yes (man. trans. only)
3rd Gear	Yes/No (N/A L40 only)	No (yes in 3rd gear) or yes (man. trans. only)
A/C Request	Yes/No	No (yes, with A/C requested, ie: selector "ON")
A/C Clutch	"ON/OFF"	"OFF" ("ON," with A/C commanded "ON")
A/C Clutch	"ON/OFF"	"OFF" ("ON," with A/C commanded "ON")
A/C Ref. Pressure	psi/Volts	0–450 psi (varying with high side refrigerant pressure)
Cool Fan Relay	"ON/OFF"	"OFF" ("ON," with A/C "ON" or hot eng)
Fuel EVAP Purge	%	0–100%
QDM A	Low/Hi	Low
QDM B	Low/Hi	Low
QDM C	Low/Hi	Low
QDM D	Low/Hi	Low
Calibration ID	#	Production ECM/Calibration ID (not usable)
Time From Start	Min/Sec	Varies (engine run time since start)

2.3L (VIN A) ENGINE — ON-BOARD DIAGNOSTIC SYSTEM CHART — 1993–94 BERETTA

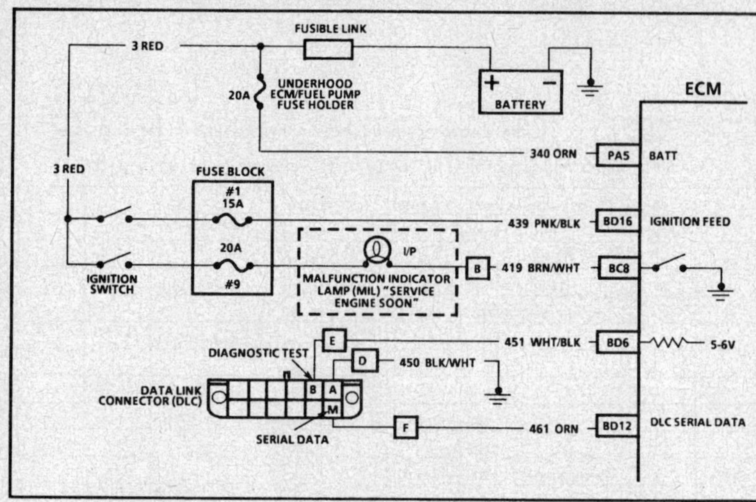

ON-BOARD DIAGNOSTIC (OBD) SYSTEM CHECK
2.3L (VIN A) "L" CARLINE

Circuit Description:

The On-Board Diagnostic (OBD) system check is an organized approach to identifying a problem created by an electronic engine control system malfunction. It must be the starting point for any driveability complaint diagnosis, because it directs the service technician to the next logical step in diagnosing the complaint. Understanding the chart and using it correctly will reduce diagnostic time and prevent the unnecessary replacement of good parts.

Test Description: Number(s) below refer to circled number(s) on the diagnostic chart.

1. This step is a check for the proper operation of the Malfunction Indicator Lamp (MIL). The MIL should be "ON" steady.
2. No MIL at this point indicates that there is a problem with the MIL circuit or the Engine Control Module (ECM) control of that circuit.
3. This test checks the ability of the ECM to control the MIL. With the diagnostic terminal grounded, the MIL should flash a DTC 12 three times, followed by any DTC stored in memory.
4. Most of the procedures use a Tech 1 scan tool to aid diagnosis, therefore, serial data must be available. If a PROM error is present, the ECM may have been able to flash DTC 12/51, but not enable serial data.
5. Although the ECM is powered up, a "Cranks But Will Not Run" symptom could exist because of an ECM or system problem.

6. This step will isolate if the customer complaint is a MIL or a driveability problem with no MIL. An invalid code may be the result of a faulty scan tool.
7. Comparison of actual control system data with the typical values is a quick check to determine if any parameter is not within limits. Keep in mind that a base engine problem (i.e., advanced cam timing) may substantially alter sensor values.
8. Installation of a Tech 1 scan tool will provide a good ground path for the ECM and may hide a driveability complaint due to poor ECM grounds.
9. If the actual data is not within the typical values established, the charts in "Component Systems," will provide a functional check of the suspect component or system.

2.3L (VIN A) ENGINE — ON-BOARD DIAGNOSTIC SYSTEM CHART — 1993–94 BERETTA

ON-BOARD DIAGNOSTIC (OBD) SYSTEM CHECK
2.3L (VIN A) "L" CARLINE

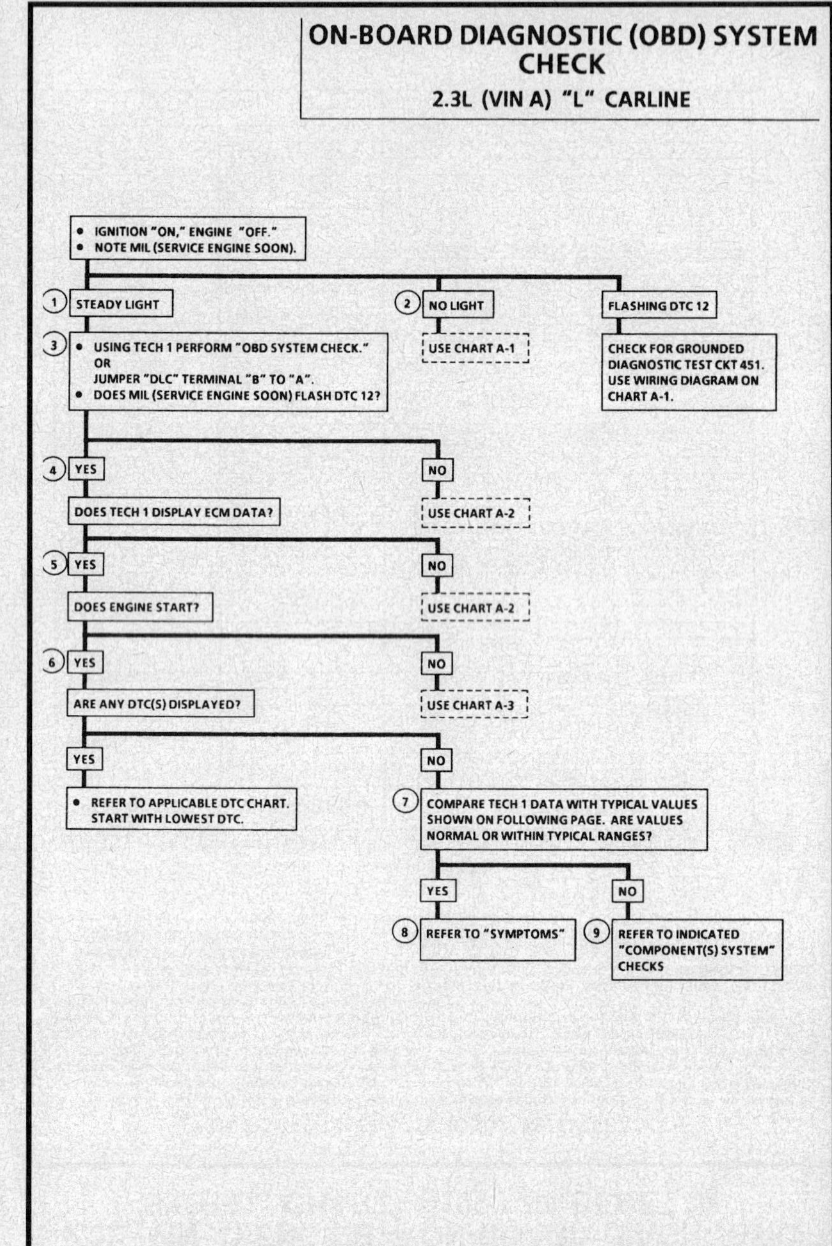

2.3L (VIN A) ENGINE — SYSTEM DIAGNOSTIC CHARTS — 1993–94 BERETTA

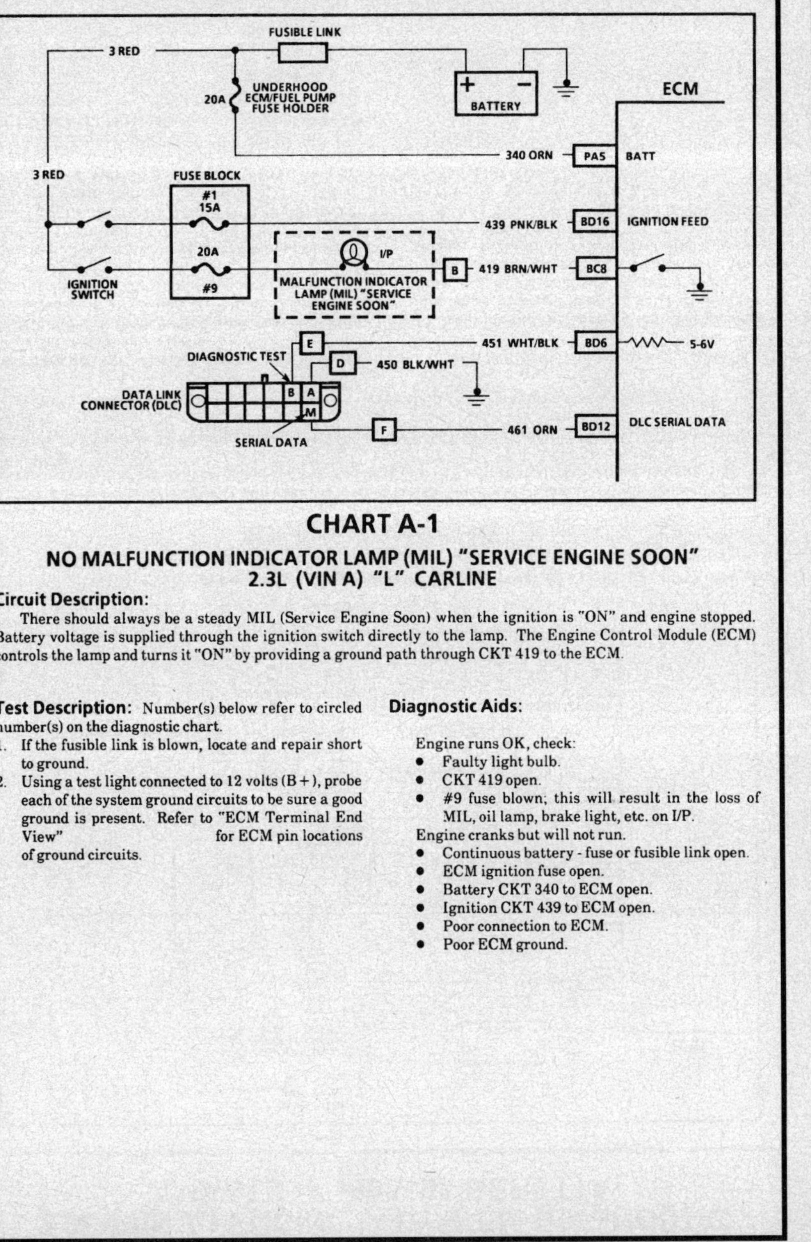

CHART A-1
NO MALFUNCTION INDICATOR LAMP (MIL) "SERVICE ENGINE SOON"
2.3L (VIN A) "L" CARLINE

Circuit Description:

There should always be a steady MIL (Service Engine Soon) when the ignition is "ON" and engine stopped. Battery voltage is supplied through the ignition switch directly to the lamp. The Engine Control Module (ECM) controls the lamp and turns it "ON" by providing a ground path through CKT 419 to the ECM.

Test Description: Number(s) below refer to circled number(s) on the diagnostic chart.
1. If the fusible link is blown, locate and repair short to ground.
2. Using a test light connected to 12 volts (B+), probe each of the system ground circuits to be sure a good ground is present. Refer to "ECM Terminal End View" for ECM pin locations of ground circuits.

Diagnostic Aids:

Engine runs OK, check:
- Faulty light bulb.
- CKT 419 open.
- #9 fuse blown; this will result in the loss of MIL, oil lamp, brake light, etc. on I/P.

Engine cranks but will not run.
- Continuous battery - fuse or fusible link open.
- ECM ignition fuse open.
- Battery CKT 340 to ECM open.
- Ignition CKT 439 to ECM open.
- Poor connection to ECM.
- Poor ECM ground.

2.3L (VIN A) ENGINE — SYSTEM DIAGNOSTIC CHARTS — 1993–94 BERETTA

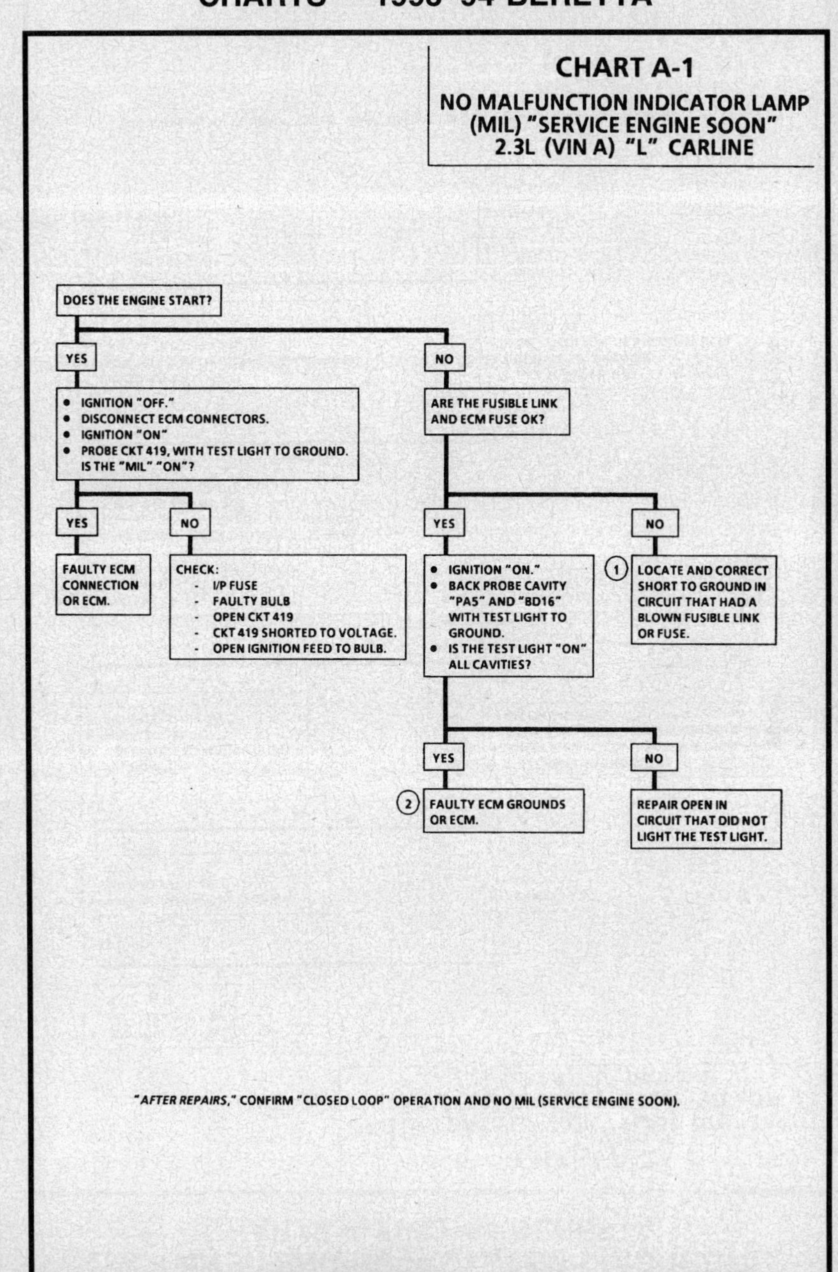

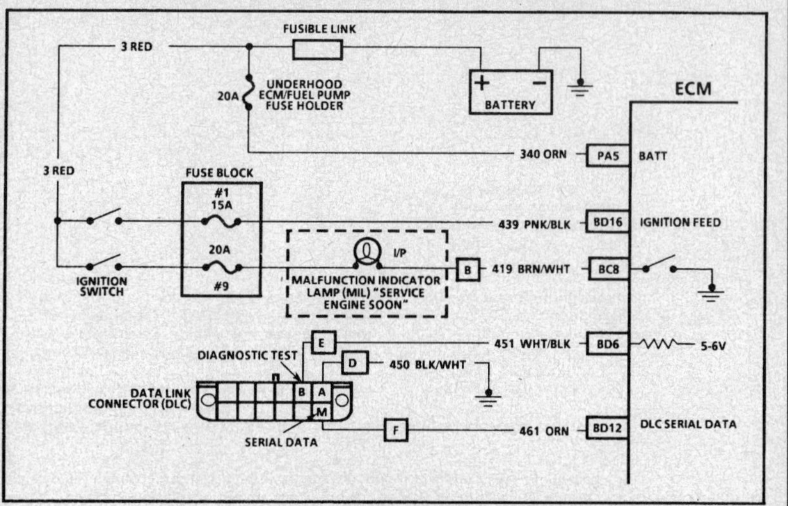

CHART A-2
**NO DLC DATA OR WON'T FLASH DTC 12
MIL (SERVICE ENGINE SOON) "ON" STEADY
2.3L (VIN A) "L" CARLINE**

Circuit Description:

There should always be a steady MIL when the ignition is "ON" and engine stopped. Battery ignition voltage is supplied to the light bulb. The Engine Control Module (ECM) will turn the MIL "ON" by grounding CKT 419 at the ECM.

With the diagnostic terminal grounded, the light should flash a DTC 12, followed by any DTC(s) stored in memory.

A steady MIL suggests a short to ground in the light control CKT 419, or an open in diagnostic CKT 451.

Test Description: Number(s) below refer to circled number(s) on the diagnostic chart.

1. MIL "OFF" with CKT 419 disconnected from ECM indicates that ground circuit was completed through the ECM, not through external short to ground.
2. If there is a problem with the ECM that causes a Tech 1 scan tool to not read serial data, the ECM should not flash a DTC 12. If DTC 12 is flashing, check for CKT 451 short to ground. If DTC 12 does flash, be sure that the scan tool is working properly on another vehicle. If the scan tool is functioning properly and CKT 461 is OK, the PROM or ECM may be at fault for the "NO DLC" symptom.
3. This step will check for an open diagnostic CKT 451.
4. At this point, the MIL wiring is OK. The problem is a faulty ECM or PROM. If DTC 12 does not flash, the ECM should be replaced using the original PROM. Replace the PROM only after trying an ECM, as a defective PROM is an unlikely cause of the problem.

CHART A-2
**NO DLC DATA OR WON'T FLASH DTC 12 MIL
(SERVICE ENGINE SOON) "ON" STEADY
2.3L (VIN A) "L" CARLINE**

- IGNITION "ON," ENGINE "OFF." MIL SHOULD BE "ON." IS IT?

YES → NO

NO → SEE CHART A-1.

- GROUND DIAGNOSTIC TERMINAL. MIL SHOULD FLASH DTC 12. DOES IT?

NO → YES

(1)
- IGNITION "OFF."
- DISCONNECT ECM CONNECTORS.
- IGNITION "ON."
- MIL SHOULD BE "OFF." IS IT?

(2)
- IF PROBLEM WAS NO DLC DATA:
- CHECK SERIAL DATA CKT 461 FOR OPEN, SHORT TO GROUND OR SHORT TO VOLTAGE. IF OK, ECM OR PROM IS FAULTY.

YES → NO

(3)
- IGNITION "OFF."
- RECONNECT ECM.
- IGNITION "ON," ENGINE STOPPED.
- DIAGNOSTIC TERMINAL NOT GROUNDED.
- BACK PROBE ECM CKT 451 WITH TEST LIGHT TO GROUND. IF TEST LIGHT IS "ON," CHECK CKT 451 FOR SHORT TO VOLTAGE.
- MIL SHOULD FLASH DTC 12. DOES IT?

NO → REPAIR SHORT TO GROUND IN CKT 419.

(4)
- CHECK PROM FOR PROPER INSTALLATION.
- IF OK, REPLACE ECM USING ORIGINAL PROM.
- GROUND DLC TERMINAL "B".
- MIL SHOULD FLASH DTC 12. DOES IT?

YES
- CHECK FOR OPEN IN CKT 451 BETWEEN DLC TERM. "B" AND ECM.
- IF OK, CHECK FOR OPEN IN DLC CKT 450 TO GROUND.

NO → REPLACE PROM.

YES → SYSTEM OK.

"AFTER REPAIRS," CONFIRM "CLOSED LOOP" OPERATION AND NO MIL (SERVICE ENGINE SOON).

2.3L (VIN A) ENGINE — SYSTEM DIAGNOSTIC CHARTS — 1993–94 BERETTA

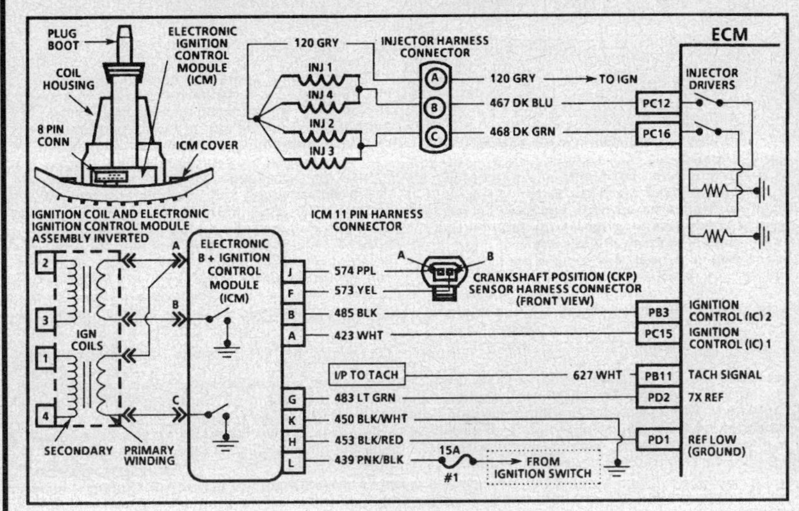

CHART A-3

(Page 1 of 4)
ENGINE CRANKS BUT WILL NOT RUN
2.3L (VIN A) "L" CARLINE

Condition:

Engine cranks but will not run, or engine may start, but immediately stops running. Battery condition and engine cranking speed are OK and there is adequate fuel in the tank.

Circuit Description:

This engine is equipped with a distributorless ignition system called the Electronic Ignition (EI) system. The primary circuit of the EI system consists of two separate ignition coils, an electronic Ignition Control Module (ICM) and Crankshaft Position (CKP) sensor as well as the related connecting wires and the Ignition Control (IC) portion of the Engine Control Module (ECM). Each secondary circuit consists of the secondary winding of the coil, two connecting metal strips molded into the coil housing, the spark plug boot/connector assemblies and spark plugs.

Test Description: Number(s) below refer to circled number(s) on the diagnostic chart.

1. This step verifies that MIL operation, on-board diagnostics, cranking RPM, TP sensor, MAP sensor and Engine Coolant Temperature (ECT) sensor signals are normal. A blinking injector test light verifies that the ECM is receiving the 7X reference signal and is attempting to activate the injectors.
2. This step determines whether harness or injector is cause of incorrect resistance. Nominal injector resistance is 1.9 to 2.1 ohms at 60°C (140°F). Resistance will increase slightly at higher temperatures.
3. By installing spark plug jumper leads and testing for spark on all 4 plug leads each ignition coil's ability to produce at least 25,000 volts is verified.

4. Checks to see if fuel pump and relay are operating correctly (fuel pump only "ON" 2-3 seconds) and fuel pressure is within proper range.
5. If module can make the test light blink, the fault is coil harness or connections. If not, module or it's connections are faulty.

Diagnostic Aids:

Check for:
- TP sensor binding or sticking in Wide Open Throttle (WOT) position or intermittently shorted or open.
- Water or foreign material in fuel.
- Low compression. (Timing chain failure.)
- Verify that only <u>resistor</u> spark plugs are used.

2.3L (VIN A) ENGINE — SYSTEM DIAGNOSTIC CHARTS — 1993–94 BERETTA

① CHECK ALL ECM GROUNDS AND MAKE SURE THEY ARE CLEAN AND TIGHT.
- IF DTC 19 IS PRESENT, REFER TO THE CHART BEFORE PROCEEDING.
- FUEL PUMP/INJ FUSE OK.
- FUEL QUANTITY OK.
- IGNITION "ON," THROTTLE CLOSED.
- MALFUNCTION INDICATOR LAMP (MIL) ("SERVICE ENGINE SOON") SHOULD BE "ON." (IF NOT SEE CHART A-1.)
- IF TECH 1 SCAN TOOL INDICATES NO DATA LINK CONNECTION (DLC), SEE CHART A-2.
- TP SENSOR SHOULD READ LESS THAN 2.5V. (IF NOT SEE DIAGNOSTIC TROUBLE CODE (DTC) 21 CHART.)
- ENGINE COOLANT TEMP. SHOULD BE CLOSE TO AMBIENT TEMPERATURE (IF NOT SEE DTC 14 OR 15 CHART).
- MAP SENSOR READINGS SHOULD CHANGE WHILE CRANKING ENGINE.
- SCAN 7X REFERENCE PULSES OR CRANK RPM WHILE CRANKING ENGINE. (IF "NO" 7X REFERENCE PULSES OR RPM IS "0", BEGIN AT STEP #10 ON PAGE 3.)
- REMOVE CRANKCASE OIL/AIR SEPARATOR FOR ACCESS TO INJECTOR CONNECTORS.
- DISCONNECT ALL INJECTORS AND INSTALL AN INJECTOR TEST LIGHT ON INJECTOR CONNECTOR #1 OR 4.
- CRANK ENGINE AND NOTE LIGHT.
- REPEAT INJECTOR CIRCUIT TEST WITH INJECTOR CONNECTOR #2 OR 3. LIGHT SHOULD BLINK ON BOTH TESTS. DOES IT?

CHART A-3
(Page 1 of 4)
ENGINE CRANKS BUT WILL NOT RUN
2.3L (VIN A) "L" CARLINE

NOTICE: FUEL SYSTEM IS UNDER PRESSURE. TO AVOID FUEL SPILLAGE, REFER TO FIELD SERVICE PROCEDURES FOR TESTING OR REPAIRS REQUIRING DISASSEMBLY OF FUEL LINES OR FITTINGS.

① → YES

② WITH DVM ON 200 OHM SCALE, MEASURE RESISTANCE OF EACH INJECTOR. RESISTANCE SHOULD BE ABOUT 1.9 TO 2.1 OHMS*. IS IT?

① → NO

NOTE WHETHER THERE WAS "NO LIGHT ON BOTH," "NO LIGHT ON ONE," OR "STEADY LIGHT ON ONE OR BOTH" AND REFER TO CHART A-3, PAGE 2.

SEE CHART A-3, PAGE 2.

② → YES

③ REMOVE IGNITION COIL AND ELECTRONIC IGNITION CONTROL MODULE ASSEMBLY AND INSTALL SPARK PLUG JUMPER WIRES (J 36012).
- CHECK FOR SPARK WITH SPARK TESTER J 26792, (ST-125) <u>ON ALL 4 PLUG WIRES</u> (ONE AT A TIME).
- REMOVE SPARK PLUG BOOT FROM COMPANION CYLINDER OF THE IGNITION COIL HOUSING AND INSTALL A JUMPER WIRE FROM THE SPARK PLUG BOOT CONNECTOR OF THE IGNITION COIL HOUSING TO GROUND (1-4 AND 2-3).
- CRANK ENGINE WITH <u>REMAINING PLUG WIRES STILL CONNECTED.</u>
- SPARK SHOULD JUMP TESTER GAP ON ALL WIRES. DOES IT?

② → NO

REPLACE INJECTOR WITH INCORRECT RESISTANCE AND CONFIRM NO OPEN OR SHORT IN INJECTOR HARNESS.

③ → YES

④ INSTALL FUEL PRESS. GAGE (SEE CHART A-7 FOR INSTRUCTIONS) AND NOTE PRESSURE AFTER IGN "ON" FOR 2 SECONDS.
- PRESSURE SHOULD BE 284-325 kPa (41-47 PSI). IS IT?

③ → NO

⑤ IGN. "OFF." DISCONNECT PLUG JUMPER WIRES AND REMOVE IGNITION COIL HOUSING.
- DISCONNECT IGNITION COIL HARN. CONN. AT MOD. AND INSTALL A TEST LIGHT BETWEEN ICM TERM. "A" (PURPLE WIRE) AND CONTROL TERMINAL FOR IGNITION COIL(S) WHICH DID NOT SPARK.
- CRANK ENGINE AND NOTE TEST LIGHT.
- TEST LIGHT SHOULD BLINK AT BOTH TERMINALS. DOES IT?

***NOTICE:** INJECTOR RESISTANCE SPECIFICATION IS AT 60°C (140°F) AND MAY BE SLIGHTLY HIGHER IF INJECTOR IS HOTTER.

④ → YES

CHECK FOR WET SPARK PLUGS. IF PLUGS ARE OK, SEE "DIAGNOSTIC AIDS"

④ → NO

SEE CHART A-7.

⑤ → YES

FAULTY IGNITION COIL HARNESS, POOR CONNECTION OR FAULTY IGNITION COIL.

⑤ → NO

SEE CHART A-3, (4 OF 4)

"AFTER REPAIRS," CONFIRM "CLOSED LOOP" OPERATION AND NO MIL (SERVICE ENGINE SOON).

2.3L (VIN A) ENGINE — SYSTEM DIAGNOSTIC CHARTS — 1993–94 BERETTA

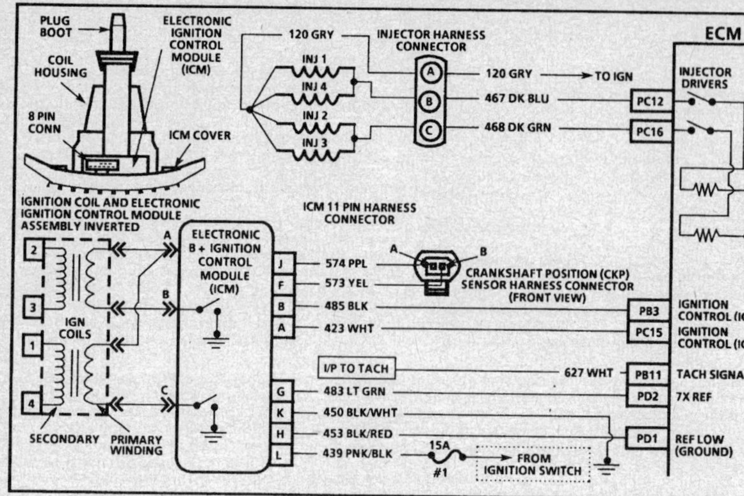

CHART A-3
(Page 2 of 4)
ENGINE CRANKS BUT WILL NOT RUN
2.3L (VIN A) "L" CARLINE

Condition:

Engine cranks but will not run, or engine may start, but immediately stops running. Battery condition and engine cranking speed are OK and there is adequate fuel in the tank.

Circuit Description:

This engine is equipped with a distributorless ignition system called the Electronic Ignition (EI) system. The primary circuit of the EI system consists of two separate ignition coils, an electronic Ignition Control Module (ICM) and Crankshaft Position (CKP) sensor as well as the related connecting wires and the Ignition Control (IC) portion of the Engine Control Module (ECM). Each secondary circuit consists of the secondary winding of the coil, two connecting metal strips molded into the coil housing, the spark plug boot/connector assemblies and spark plugs.

Test Description: Number(s) below refer to circled number(s) on the diagnostic chart.

6. Battery voltage should be available at CKT 120 whenever the fuel pump power feed circuit is switched "ON." The ECM should switch the fuel pump "ON" for 2-3 seconds after ignition is turned "ON" (and when ECM is receiving ignition reference pulses, as while cranking or running). The ignition must be turned "OFF" for at least 10 seconds to assure that the ECM powers down and will then switch the fuel pump back "ON" for 2-3 seconds when ignition is turned back "ON."

7. Light "ON" one circuit only indicates power is available at cavity "A", but grounded circuit is not being completed on the other circuit. This could be due to open circuit or ECM not switching the injector driver circuit to ground.

8. Steady light indicates ground circuit is always completed and is not being switched. This could be due to short to ground in circuit, or faulty ECM injector driver.

9. The fuel pump should be switched "ON" by the ECM for 2-3 seconds after ignition is first turned "ON." It is necessary to turn the ignition "OFF" for at least 10 seconds to assure that the ECM powers down and will then switch the fuel pump back "ON." If the fuel pump operates, but power is not available at injector harness, circuit must be open. If the fuel pump does not operate, CHART A-5 should be used to diagnose the cause.

2.3L (VIN A) ENGINE — SYSTEM DIAGNOSTIC CHARTS — 1993–94 BERETTA

CHART A-3
(Page 2 of 4)
ENGINE CRANKS BUT WILL NOT RUN
2.3L (VIN A) "L" CARLINE

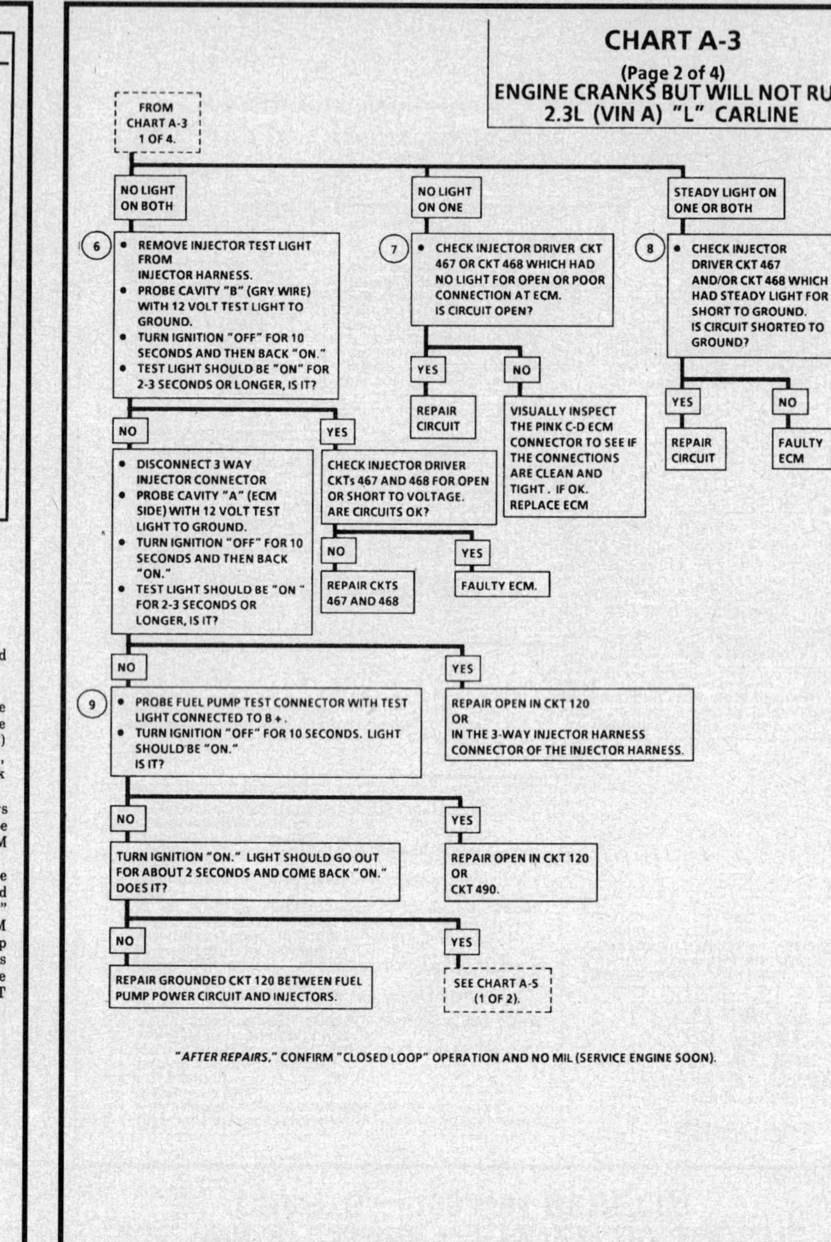

"AFTER REPAIRS," CONFIRM "CLOSED LOOP" OPERATION AND NO MIL (SERVICE ENGINE SOON).

2.3L (VIN A) ENGINE — SYSTEM DIAGNOSTIC CHARTS — 1993–94 BERETTA

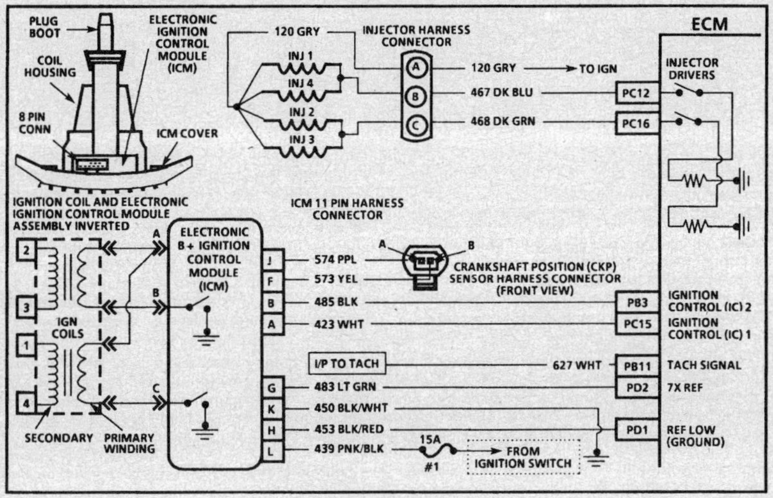

CHART A-3
(Page 3 of 4)
**ENGINE CRANKS BUT WILL NOT RUN
2.3L (VIN A) "L" CARLINE**

Circuit Description:

The Electronic Ignition (EI) system uses a waste spark method of distribution. In this type of system the electronic Ignition Control Module (ICM) triggers the #1 - 4 coil pair resulting in both #1 and #4 spark plugs firing at the same time. #1 cylinder is on the compression stroke at the same time #4 is on the exhaust stroke, resulting in a lower energy requirement to fire #4 spark plug. This leaves the remainder of the high voltage to be used to fire #1 spark plug. On this application, the Crankshaft Position (CKP) sensor is mounted to, and protrudes through the block to within approximately 0.050" of the crankshaft reluctor. Since the reluctor is a machined portion of the crankshaft and the CKP sensor is mounted in a fixed position on the block, timing adjustments are not possible or necessary.

Test Description: Number(s) below refer to circled number(s) on the diagnostic chart.

10. Battery voltage should be available at terminal "L" of the electronic Ignition Control Module (ICM) 11 pin connector, and terminal "K" should be a good ground.

11. The CKP sensor should output a voltage as the crankshaft turns. If no voltage is produced, the indication is a poor sensor connection or faulty sensor.

12. The CKP sensors core is a magnet, therefore, it should be magnetized and the resistance should be within a range of 500 to 900 ohms.

13. The test light to 12 volts simulates a reference signal to the Engine Control Module (ECM) which will result in an injector test light blink for every other touch of the test light, if CKT 483, the ECM and the injector driver circuit are all functioning properly.

2.3L (VIN A) ENGINE — SYSTEM DIAGNOSTIC CHARTS — 1993–94 BERETTA

CHART A-3
(Page 3 of 4)
**ENGINE CRANKS BUT WILL NOT RUN
2.3L (VIN A) "L" CARLINE**

FROM CHART A-3
(1 of 4)
STEP 1

(10)
- IGNITION "OFF," DISCONNECT ELECTRONIC IGNITION CONTROL MODULE (ICM) 11 PIN HARNESS CONNECTOR.
- IGNITION "ON," CONNECT A TEST LIGHT BETWEEN HARNESS TERMINALS "K" AND "L".
- TEST LIGHT SHOULD BE "ON." IS IT?

YES

(11)
- CONNECT DVOM (2 VOLT A.C. SCALE) BETWEEN TERMINAL "F" AND "J" OF ELECTRONIC ICM HARNESS.
- CRANK ENGINE AND OBSERVE VOLTAGE.
- READING SHOULD BE GREATER THAN 200 mV. IS IT?

NO

(12)
- REMOVE CRANKSHAFT POSITION (CKP) SENSOR FROM ENGINE BLOCK.
- WITH DVOM IN 2K POSITION, MEASURE RESISTANCE ACROSS SENSOR TERMINALS (RESISTANCE SHOULD BE BETWEEN 500 AND 900 OHMS).
- VERIFY THAT SENSOR IS STILL MAGNETIC AND INSPECT FOR DAMAGED TERMINALS. IS SENSOR OK?

NO

REPLACE CKP SENSOR.

NO (from box 10)

- CONNECT TEST LIGHT BETWEEN HARNESS TERMINAL "L" AND GROUND.
- TEST LIGHT SHOULD BE "ON." IS IT?

NO → OPEN CKT 439 OR FUSE.

YES → OPEN CKT 450.

YES (from box 11)

(13)
- IGNITION "ON", SET TECH 1 SCAN TOOL TO DISPLAY 7X REFERENCE PULSES.
- WITH TEST LIGHT CONNECTED TO B+, "TOUCH ENGINE HARNESS CONNECTOR CAVITY "G" WITH LIGHT AND REMOVE.
- SCAN TOOL DISPLAY SHOULD INDICATE 7X REFERENCE PULSE INCREMENT WHEN TEST LIGHT IS REMOVED. DOES IT?

NO → CHECK CKT 483 FOR OPEN OR SHORT TO GROUND OR VOLTAGE AND REPAIRS AS NECESSARY. IF OK, REPLACE ECM.

YES → CHECK FOR FAULTY CONNECTIONS ICM 11 PIN HARNESS CONNECTOR. IF OK, REPLACE ICM.

"AFTER REPAIRS," CONFIRM "CLOSED LOOP" OPERATION AND NO MIL (SERVICE ENGINE SOON).

2.3L (VIN A) ENGINE — SYSTEM DIAGNOSTIC CHARTS — 1993–94 BERETTA

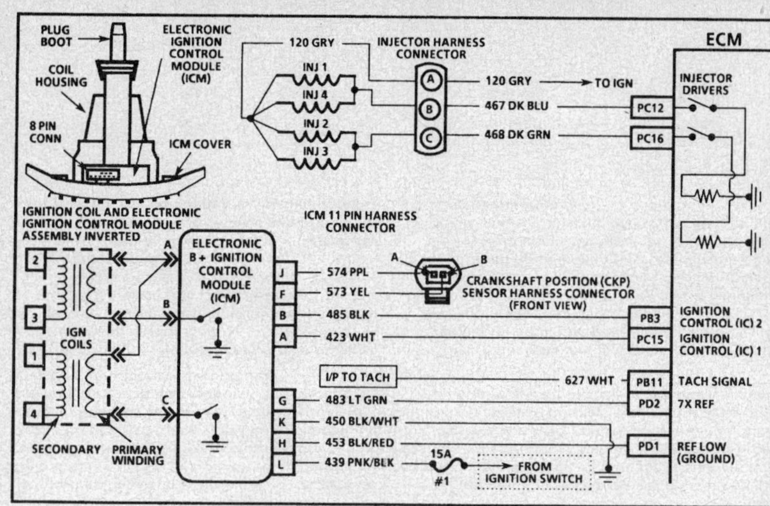

CHART A-3
(Page 4 of 4)
ENGINE CRANKS BUT WILL NOT RUN
2.3L (VIN A) "L" CARLINE

Circuit Description:
Engine cranks but will not run, or engine may start, but immediately stops running. Battery condition and engine cranking speed are OK and there is adequate fuel in the tank.

Circuit Description:
This engine is equipped with a distributorless ignition system called the Electronic Ignition (EI) system. The primary circuit of the EI system consists of two separate ignition coils, an electronic Ignition Control Module (ICM) and Crankshaft Position (CKP) sensor as well as the related connecting wires and the Ignition Control (IC) portion of the Engine Control Module (ECM). Each secondary circuit consists of the secondary winding of the coil, two connecting metal strips molded into the coil housing, the spark plug boot/connector assemblies and spark plugs.

Test Description: Number(s) below refer to circled number(s) on the diagnostic chart.

14. This step verifies if there is a frequency output coming from the ECM.
15. This step identifies a short in the circuit.
16. This step will verify if the open circuit is in the ICM or circuit.
17. If there is a short to battery voltage there will be no frequency response. This step will show if there is a short to the battery.
18. This step will determine if the circuit or the ICM is shorted to ground.
19. If CKT 423 and CKT 485 are shorted together, there will not be a frequency signal. This step determines if CKT 423 and CKT 485 are shorted together.

2.3L (VIN A) ENGINE — SYSTEM DIAGNOSTIC CHARTS — 1993–94 BERETTA

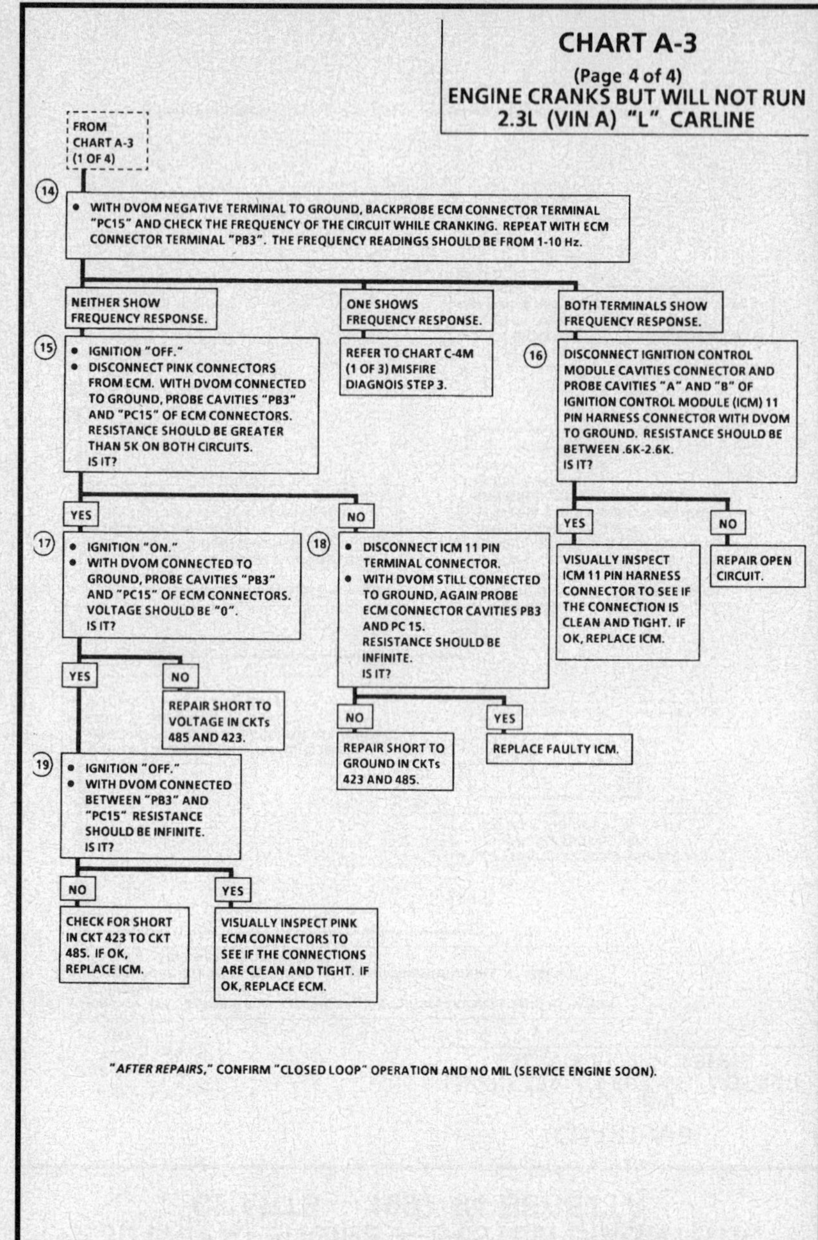

2.3L (VIN A) ENGINE — SYSTEM DIAGNOSTIC CHARTS — 1993–94 BERETTA

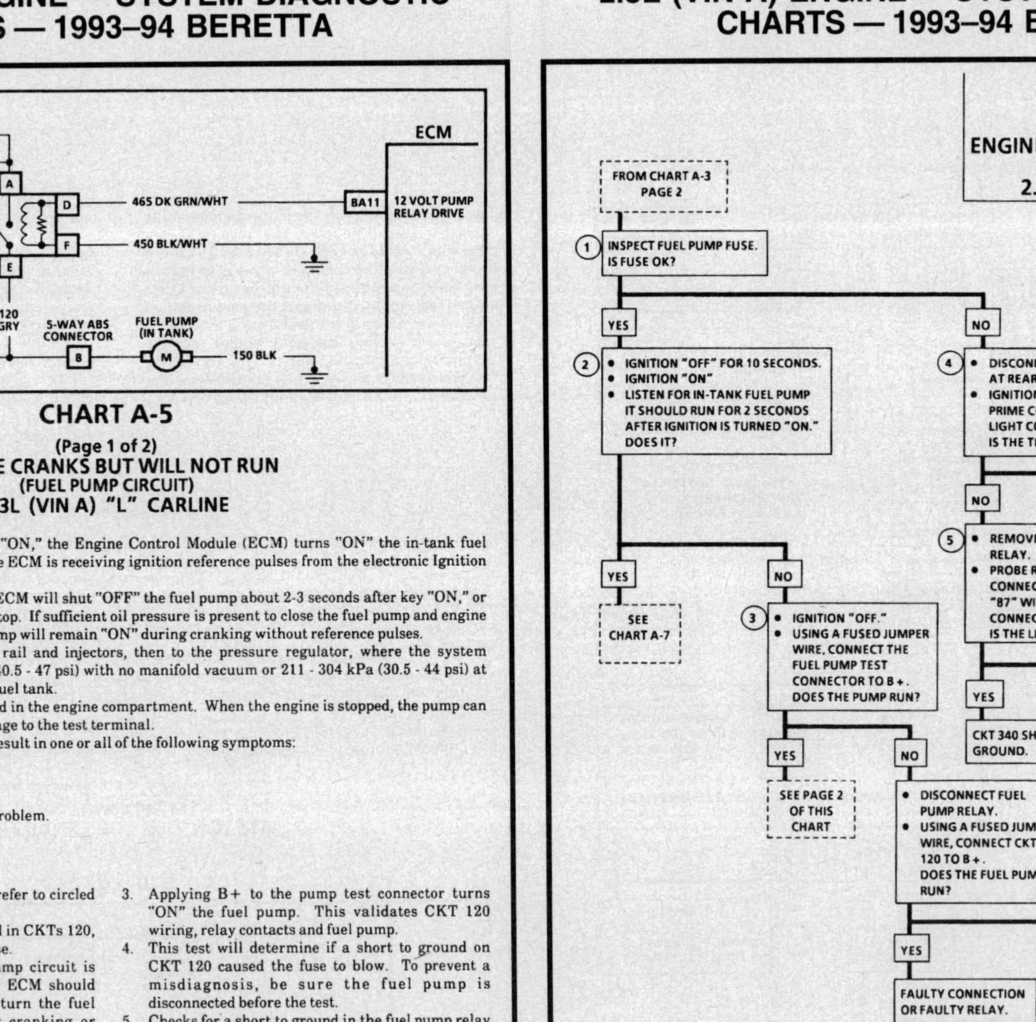

CHART A-5

(Page 1 of 2)
ENGINE CRANKS BUT WILL NOT RUN
(FUEL PUMP CIRCUIT)
2.3L (VIN A) "L" CARLINE

Circuit Description:

When the ignition switch is turned "ON," the Engine Control Module (ECM) turns "ON" the in-tank fuel pump. It will remain "ON" as long as the ECM is receiving ignition reference pulses from the electronic Ignition Control Module (ICM).

If there are no reference pulses, the ECM will shut "OFF" the fuel pump about 2-3 seconds after key "ON," or about 10 seconds after reference pulses stop. If sufficient oil pressure is present to close the fuel pump and engine oil pressure indicator switch, the fuel pump will remain "ON" during cranking without reference pulses.

The pump delivers fuel to the fuel rail and injectors, then to the pressure regulator, where the system pressure is controlled to 280 - 325 kPa (40.5 - 47 psi) with no manifold vacuum or 211 - 304 kPa (30.5 - 44 psi) at idle. Excess fuel is then returned to the fuel tank.

The fuel pump test terminal is located in the engine compartment. When the engine is stopped, the pump can be turned "ON" by applying battery voltage to the test terminal.

Improper fuel system pressure will result in one or all of the following symptoms:
- Cranks but will not run.
- DTC 44.
- DTC 45.
- Cuts out, may feel like ignition problem.
- Poor fuel economy, loss of power.
- Hesitation.

Test Description: Number(s) below refer to circled number(s) on the diagnostic chart.

1. If the fuse is blown, a short to ground in CKTs 120, 340 or the fuel pump itself is the cause.
2. This step determines if the fuel pump circuit is being controlled by the ECM. The ECM should energize the fuel pump relay and turn the fuel pump "ON." If the engine is not cranking or running, the ECM should de-energize the relay and/or fuel pump within 2 seconds after the ignition is turned "ON."
3. Applying B+ to the pump test connector turns "ON" the fuel pump. This validates CKT 120 wiring, relay contacts and fuel pump.
4. This test will determine if a short to ground on CKT 120 caused the fuse to blow. To prevent a misdiagnosis, be sure the fuel pump is disconnected before the test.
5. Checks for a short to ground in the fuel pump relay harness CKT 340.

2.3L (VIN A) ENGINE — SYSTEM DIAGNOSTIC CHARTS — 1993–94 BERETTA

CHART A-5

(Page 1 of 2)
ENGINE CRANKS BUT WILL NOT RUN
(FUEL PUMP CIRCUIT)
2.3L (VIN A) "L" CARLINE

FROM CHART A-3 PAGE 2

(1) INSPECT FUEL PUMP FUSE. IS FUSE OK?

YES →

(2)
- IGNITION "OFF" FOR 10 SECONDS.
- IGNITION "ON"
- LISTEN FOR IN-TANK FUEL PUMP IT SHOULD RUN FOR 2 SECONDS AFTER IGNITION IS TURNED "ON." DOES IT?

YES → SEE CHART A-7

NO →

(3)
- IGNITION "OFF."
- USING A FUSED JUMPER WIRE, CONNECT THE FUEL PUMP TEST CONNECTOR TO B +. DOES THE PUMP RUN?

YES → SEE PAGE 2 OF THIS CHART

NO → CKT 340 SHORTED TO GROUND.

NO →

(4)
- DISCONNECT FUEL PUMP HARNESS AT REAR BODY CONNECTOR.
- IGNITION "OFF," PROBE FUEL PUMP PRIME CONNECTOR WITH A TEST LIGHT CONNECTED TO B +. IS THE TEST LIGHT "ON"?

YES → CKT 120, 490 OR FUEL PUMP RELAY OR OIL PRESSURE SWITCH SHORTED TO GROUND.

NO →

(5)
- REMOVE FUEL PUMP RELAY.
- PROBE RELAY CONNECTOR TERMINAL "87" WITH A TEST LIGHT CONNECTED TO B +. IS THE LIGHT "ON"?

YES →
- DISCONNECT FUEL PUMP RELAY.
- USING A FUSED JUMPER WIRE, CONNECT CKT 120 TO B +. DOES THE FUEL PUMP RUN?

YES → FAULTY CONNECTION OR FAULTY RELAY.

NO → OPEN CKT 120, FAULTY IN-TANK PUMP, OR FAULTY FUEL PUMP GROUND CKT 450.

NO → FUEL TANK METER ASSEMBLY HARNESS SHORTED TO GROUND OR FAULTY FUEL PUMP.

"AFTER REPAIRS," CONFIRM "CLOSED LOOP" OPERATION AND NO MIL (SERVICE ENGINE SOON).

2.3L (VIN A) ENGINE — SYSTEM DIAGNOSTIC CHARTS — 1993–94 BERETTA

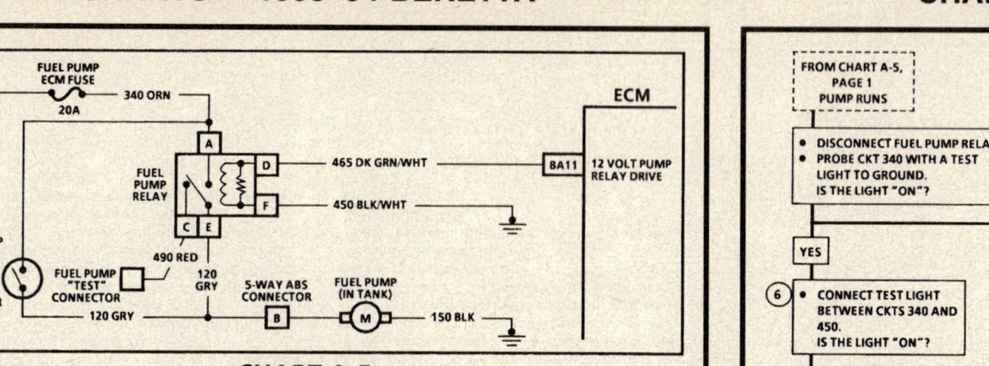

CHART A-5
(Page 2 of 2)
ENGINE CRANKS BUT WILL NOT RUN
(FUEL PUMP CIRCUIT)
2.3L (VIN A) "L" CARLINE

Circuit Description:

When the ignition switch is turned "ON," the Engine Control Module (ECM) turns "ON" the in-tank fuel pump. It will remain "ON" as long as the ECM is receiving ignition reference pulses from the electronic Ignition Control Module (ICM).

If there are no reference pulses, the ECM will shut "OFF" the fuel pump about 2-3 seconds after key "ON," or about 10 seconds after reference pulses stop. If sufficient oil pressure is present to close the fuel pump and engine oil pressure indicator switch, the fuel pump will remain "ON" during cranking without reference pulses.

The pump delivers fuel to the fuel rail and injectors, then to the pressure regulator, where the system pressure is controlled to 280 - 325 kPa (40.5 - 47 psi) with no manifold vacuum or 211 - 304 kPa (30.5 - 44 psi) at idle. Excess fuel is then returned to the fuel tank.

The fuel pump test terminal is located in the engine compartment. When the engine is stopped, the pump can be turned "ON" by applying battery voltage to the "test" terminal.

Improper fuel system pressure will result in one or all of the following symptoms:
- Cranks but will not run.
- DTC 44.
- DTC 45.
- Cuts out, may feel like ignition problem.
- Poor fuel economy, loss of power.
- Hesitation.

Test Description: Number(s) below refer to circled number(s) on the diagnostic chart.
6. Checks for open in the fuel pump relay ground CKT 450.
7. Determines if the ECM is in control of the fuel pump relay through CKT 465.
8. The fuel pump control circuit includes an engine oil pressure switch with a separate set of normally open contacts. The switch closes at about (4 lb.) 28 kPa of oil pressure and provides a second battery feed path to the fuel pump. If the relay fails, the pump will run due to the battery feed supplied by the closed oil pressure switch.

This step checks the oil pressure switch to be sure it provides battery feed to the fuel pump should the pump relay fail. A failed pump relay will result in extended engine crank time because of the time required to build enough oil pressure to close the oil pressure switch and turn "ON" the fuel pump. There may be instances when the relay has failed but the engine will not crank fast enough to build enough oil pressure to close the switch. This or a faulty fuel pump and engine oil pressure indicator switch can result in "Engine Cranks But Will not Run."

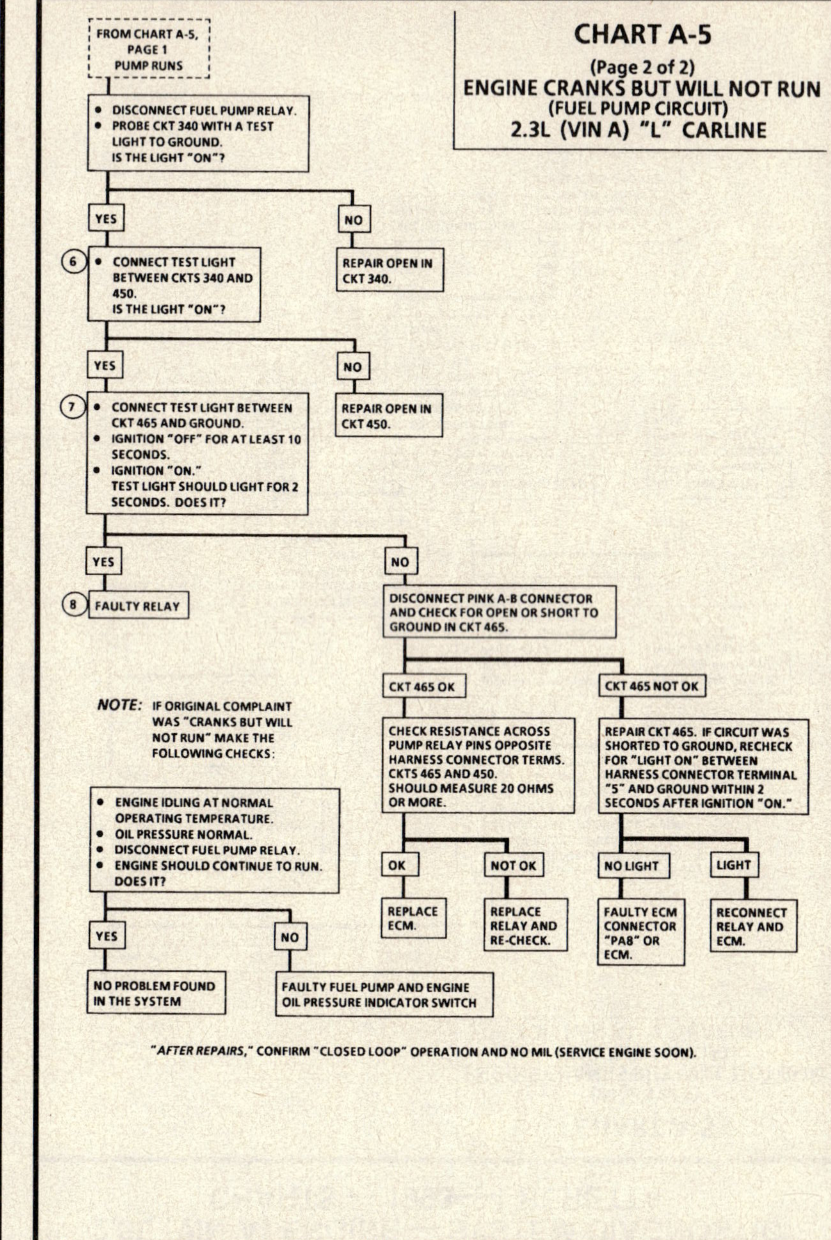

2.3L (VIN A) ENGINE — SYSTEM DIAGNOSTIC CHARTS — 1993–94 BERETTA

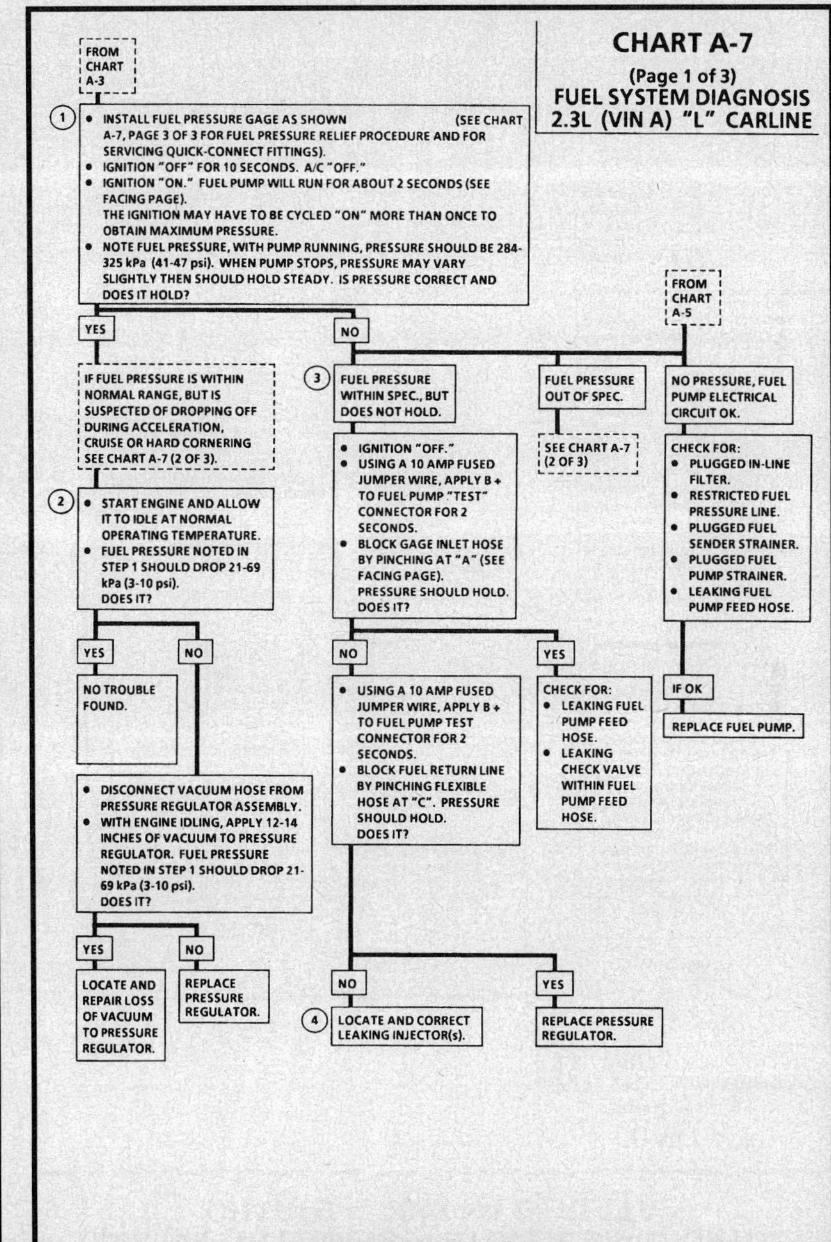

CHART A-7 (Page 1 of 3)

FUEL SYSTEM DIAGNOSIS
2.3L (VIN A) "L" CARLINE

Circuit Description:
When the ignition switch is turned "ON," the Engine Control Module (ECM) will turn "ON" the in-tank fuel pump. It will remain "ON" as long as the engine is cranking or running, and the ECM is receiving reference pulses. If there are no reference pulses, the ECM will shut "OFF" the fuel pump in about 2 seconds after ignition "ON" or 10 seconds after reference pulses stop.

An electric fuel pump, attached to the fuel sender assembly (inside the fuel tank) pumps fuel through an in-line filter to the fuel rail assembly. The pump is designed to provide fuel at a pressure above the regulated pressure needed by the injectors. A pressure regulator, attached to the fuel rail, keeps fuel available to the injectors at a regulated pressure. Unused fuel is returned to the fuel tank by a separate line.

Test Description: Number(s) below refer to circled number(s) on the diagnostic chart.
1. Install fuel pressure gage as shown in illustration. See Page 3 of 3 for "Fuel Pressure Relief Procedure" and for "Servicing Quick Connect Fittings." Ignition "ON," pump pressure should be 284 to 325 kPa (41-47 psi). This pressure is controlled by spring pressure within the regulator assembly.
2. When the engine is idling, the manifold pressure is low (high vacuum) and is applied to the fuel regulator diaphragm. This will offset the spring and result in a lower fuel pressure. This idle pressure will vary somewhat depending on barometric pressure, however, the pressure idling should be less indicating pressure regulator control.
3. Pressure that continues to fall quickly is caused by one of the following:
   - Leaking fuel pump feed hose.
   - Leaking check valve within fuel pump feed hose.
   - Fuel pressure regulator valve leaking.
   - Injector(s) sticking open.
4. An injector sticking open can best be determined by checking for a fouled or saturated spark plug(s). If a leaking injector can not be determined by a fouled or saturated spark plug, the following procedure should be used.
   - Remove fuel rail bolts, but leave fuel lines connected.
   - Lift fuel rail out just enough to leave injector nozzles in the ports.

CAUTION: Be sure injector(s) are not allowed to spray on engine and that injector retaining clips are intact. This should be carefully followed to prevent fuel spray on engine which would cause a fire hazard.
   - Pressurize the fuel system and observe for injector(s) leaking.

2.3L (VIN A) ENGINE — SYSTEM DIAGNOSTIC CHARTS — 1993–94 BERETTA

CHART A-7

(Page 1 of 3)
FUEL SYSTEM DIAGNOSIS
2.3L (VIN A) "L" CARLINE

FROM CHART A-3

(1)
- INSTALL FUEL PRESSURE GAGE AS SHOWN (SEE CHART A-7, PAGE 3 OF 3 FOR FUEL PRESSURE RELIEF PROCEDURE AND FOR SERVICING QUICK-CONNECT FITTINGS).
- IGNITION "OFF" FOR 10 SECONDS. A/C "OFF."
- IGNITION "ON." FUEL PUMP WILL RUN FOR ABOUT 2 SECONDS (SEE FACING PAGE). THE IGNITION MAY HAVE TO BE CYCLED "ON" MORE THAN ONCE TO OBTAIN MAXIMUM PRESSURE.
- NOTE FUEL PRESSURE, WITH PUMP RUNNING, PRESSURE SHOULD BE 284-325 kPa (41-47 psi). WHEN PUMP STOPS, PRESSURE MAY VARY SLIGHTLY THEN SHOULD HOLD STEADY. IS PRESSURE CORRECT AND DOES IT HOLD?

YES → IF FUEL PRESSURE IS WITHIN NORMAL RANGE, BUT IS SUSPECTED OF DROPPING OFF DURING ACCELERATION, CRUISE OR HARD CORNERING SEE CHART A-7 (2 OF 3).

(2)
- START ENGINE AND ALLOW IT TO IDLE AT NORMAL OPERATING TEMPERATURE.
- FUEL PRESSURE NOTED IN STEP 1 SHOULD DROP 21-69 kPa (3-10 psi). DOES IT?

YES → NO TROUBLE FOUND.

NO →
- DISCONNECT VACUUM HOSE FROM PRESSURE REGULATOR ASSEMBLY.
- WITH ENGINE IDLING, APPLY 12-14 INCHES OF VACUUM TO PRESSURE REGULATOR. FUEL PRESSURE NOTED IN STEP 1 SHOULD DROP 21-69 kPa (3-10 psi). DOES IT?

YES → LOCATE AND REPAIR LOSS OF VACUUM TO PRESSURE REGULATOR.

NO → REPLACE PRESSURE REGULATOR.

NO → (from step 1)

(3) FUEL PRESSURE WITHIN SPEC., BUT DOES NOT HOLD.
- IGNITION "OFF."
- USING A 10 AMP FUSED JUMPER WIRE, APPLY B + TO FUEL PUMP "TEST" CONNECTOR FOR 2 SECONDS.
- BLOCK GAGE INLET HOSE BY PINCHING AT "A" (SEE FACING PAGE). PRESSURE SHOULD HOLD. DOES IT?

NO →
- USING A 10 AMP FUSED JUMPER WIRE, APPLY B + TO FUEL PUMP TEST CONNECTOR FOR 2 SECONDS.
- BLOCK FUEL RETURN LINE BY PINCHING FLEXIBLE HOSE AT "C". PRESSURE SHOULD HOLD. DOES IT?

NO → (4) LOCATE AND CORRECT LEAKING INJECTOR(s).

YES → REPLACE PRESSURE REGULATOR.

YES → CHECK FOR:
- LEAKING FUEL PUMP FEED HOSE.
- LEAKING CHECK VALVE WITHIN FUEL PUMP FEED HOSE.

FUEL PRESSURE OUT OF SPEC.
SEE CHART A-7 (2 OF 3)

FROM CHART A-5

NO PRESSURE, FUEL PUMP ELECTRICAL CIRCUIT OK.

CHECK FOR:
- PLUGGED IN-LINE FILTER.
- RESTRICTED FUEL PRESSURE LINE.
- PLUGGED FUEL SENDER STRAINER.
- PLUGGED FUEL PUMP STRAINER.
- LEAKING FUEL PUMP FEED HOSE.

IF OK → REPLACE FUEL PUMP.

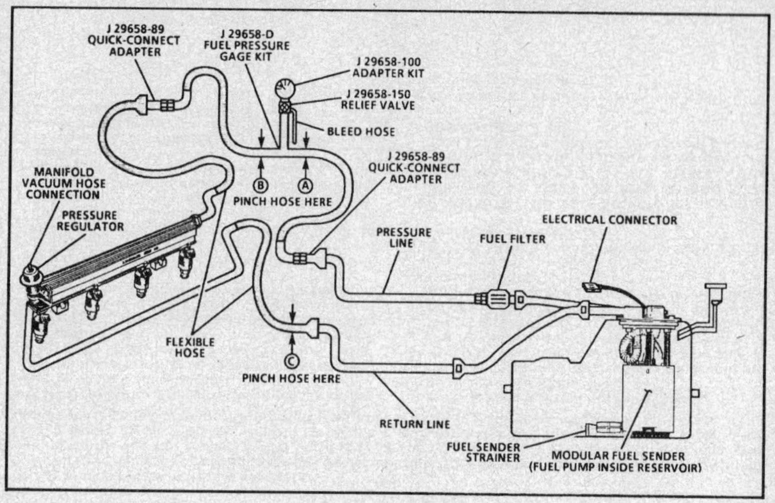

2.3L (VIN A) ENGINE — SYSTEM DIAGNOSTIC CHARTS — 1993-94 BERETTA

CHART A-7
(Page 2 of 3)
FUEL SYSTEM DIAGNOSIS
2.3L (VIN A) "L" CARLINE

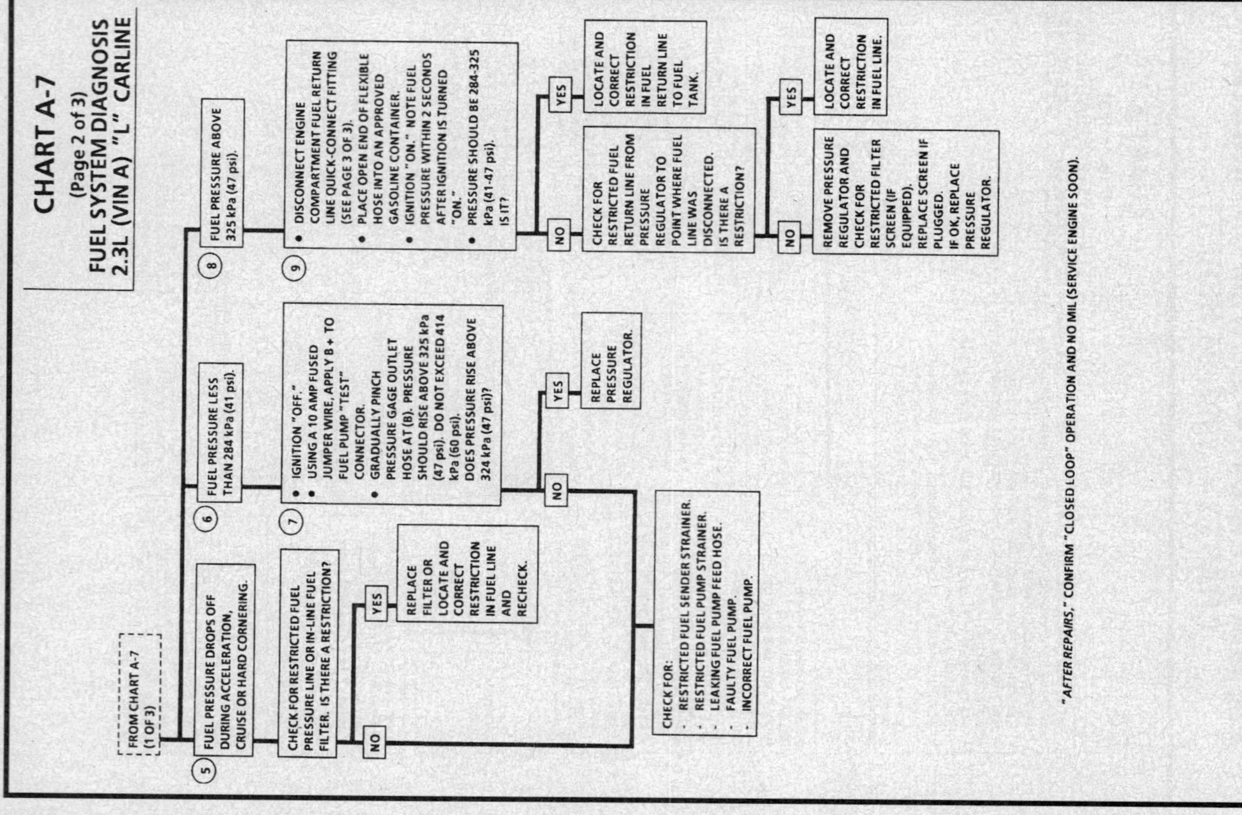

2.3L (VIN A) ENGINE — SYSTEM DIAGNOSTIC CHARTS — 1993-94 BERETTA

CHART A-7
(Page 2 of 3)
FUEL SYSTEM DIAGNOSIS
2.3L (VIN A) "L" CARLINE

Diagram labels: J 29658-89 QUICK-CONNECT ADAPTER; J 29658-D FUEL PRESSURE GAGE KIT; J 29658-100 ADAPTER KIT; J 29658-150 RELIEF VALVE; BLEED HOSE; J 29658-89 QUICK-CONNECT ADAPTER; PRESSURE LINE; ELECTRICAL CONNECTOR; FUEL FILTER; RETURN LINE; FUEL SENDER STRAINER; MODULAR FUEL SENDER (FUEL PUMP INSIDE RESERVOIR); PINCH HOSE HERE; FLEXIBLE HOSE; MANIFOLD VACUUM HOSE CONNECTION; PRESSURE REGULATOR.

Test Description: Number(s) below refer to circled number(s) on the diagnostic chart.

5. Fuel pressure that drops off during acceleration, cruise or hard cornering may cause a lean condition and result in a loss of power, surging or misfire. This condition can be diagnosed using a Tech 1 scan tool. If the fuel system is very lean, the Oxygen Sensor (O2S) will stop toggling and output voltage will drop below 500 mV. Also, injector pulse width will increase.

> **Important**
> - Make sure system is not operating at "Fuel Cut-Off" which may cause false readings on the scan tool.

6. Fuel pressure below 284 kPa (41 psi) may cause a lean condition and may set a DTC 44. Driveability conditions can include hard starting cold, hesitation, poor driveability, lack of power, surging or misfire.

7. Restricting fuel flow at the fuel pressure gage (at "B") causes fuel pressure to build above regulated pressure. With battery voltage applied the pump "test" connector, pressure should rise above 325 kPa (47 psi) as the gage outlet hose is pinched.

NOTICE: Do not allow pressure to exceed 414 kPa (60 psi) as damage to the regulator may result.

8. Fuel pressure above 325 kPa (47 psi) may cause a rich condition and may set a DTC 45. Driveability conditions can include hard starting (followed by black smoke) and a strong sulfur smell in the exhaust.

9. This test determines if the high fuel pressure is due to a restricted fuel return line or a faulty fuel pressure regulator.

2.3L (VIN A) ENGINE — SYSTEM DIAGNOSTIC CHARTS — 1993–94 BERETTA

CHART A-7
(Page 3 of 3)
FUEL SYSTEM DIAGNOSIS
2.3L (VIN A) "L" CARLINE

FUEL SYSTEM PRESSURE RELIEF PROCEDURE

PFI Engines Without Fuel Pressure Connection (Except J Car Applications)

(Must Be Performed Before Disconnecting Fuel Line Fittings)

CAUTION:
- To reduce the risk of fire and personal injury, it is necessary to relieve fuel system pressure before disconnecting fuel line fittings.
- After relieving system pressure, a small amount of fuel may be released when disconnecting fuel line fittings. In order to reduce the chance of personal injury, cover fuel line fittings with a shop towel before disconnecting, to catch any fuel that may leak out. Place the towel in an approved container when disconnect is completed.

1. Loosen fuel filler cap to relieve tank pressure.
2. Raise vehicle.
3. Disconnect fuel pump electrical connector.
4. Lower vehicle.
5. Start engine and run until fuel supply remaining in fuel pipes is consumed. Engage starter for 3.0 seconds to assure relief of any remaining pressure.
6. Raise vehicle.
7. Connect fuel pump electrical connector.
8. Lower vehicle.
9. Disconnect negative battery cable to avoid possible fuel discharge if an accidental attempt is made to start the engine.
10. Fuel line fittings are now safe for servicing.
11. Perform service required.
    - If performing fuel pressure check with a gage equipped with a bleed hose, fuel pressure can be relieved through the gage following test. Place bleed hose into approved gasoline container and open valve to bleed system pressure.
12. Tighten fuel filler cap.
13. Connect negative battery cable.
14. Cycle ignition "ON" and "OFF" twice, waiting ten seconds between cycles, then check for fuel leaks.

2.3L (VIN A) ENGINE — SYSTEM DIAGNOSTIC CHARTS — 1993–94 BERETTA

CHART A-7
(Page 3 of 3)
FUEL SYSTEM DIAGNOSIS
2.3 (VIN A) "L" CARLINE

SERVICING QUICK-CONNECT FITTINGS

Important
- In order to install fuel system diagnostic equipment on vehicles equipped with plastic quick-connect fittings, fuel line separator tools must be used to disconnect the fittings. Use of the separator tools will cause the plastic retainer to remain inside the female connector allowing diagnostic equipment to be connected.

Tools required:
J 37088-A tool set, fuel line quick-connect separator;
J 39504 tool set, fuel line quick-connect separator (restricted access).

Remove or Disconnect

1. Grasp both sides of fitting. Twist female connector 1/4 turn in each direction to loosen any dirt within fitting.

CAUTION: Safety glasses must be worn when using compressed air, as flying dirt particles may cause eye injury.

2. Using compressed air, blow dirt out of fitting.
3. Choose correct tool from J 37088-A or J 39504 tool set for size of fitting. Insert tool into female connector, then push/pull inward to release locking tabs.
4. Pull connection apart.

Clean and Inspect

NOTICE: If it is necessary to remove rust or burrs from fuel pipe, use emery cloth in a radial motion with the pipe end to prevent damage to O-ring sealing surface.

- Using a clean shop towel, wipe off male pipe end.
- Inspect both ends of fitting for dirt and burrs. Clean or replace components/assemblies as required.

Install or Connect

CAUTION: To Reduce the Risk of Fire and Personal Injury:
- Before connecting fitting, always apply a few drops of clean engine oil to the male pipe end of engine fuel pipe, pressure gage adapter or fuel line shut-off adapter. This will ensure proper reconnection and prevent a possible fuel leak. (During normal operation, the O-rings located in the female connector will swell and may prevent proper reconnection if not lubricated.)

1. Apply a few drops of clean engine oil to the male pipe end of engine fuel pipe, pressure gage adapter or fuel line shut-off adapter.
2. Push both sides of fitting together to cause the retaining tabs/fingers to snap into place.
3. Once installed, pull on both sides of fitting to make sure connection is secure.

2.3L (VIN A) ENGINE — DIAGNOSTIC TROUBLE CODE CHART — 1993–94 BERETTA

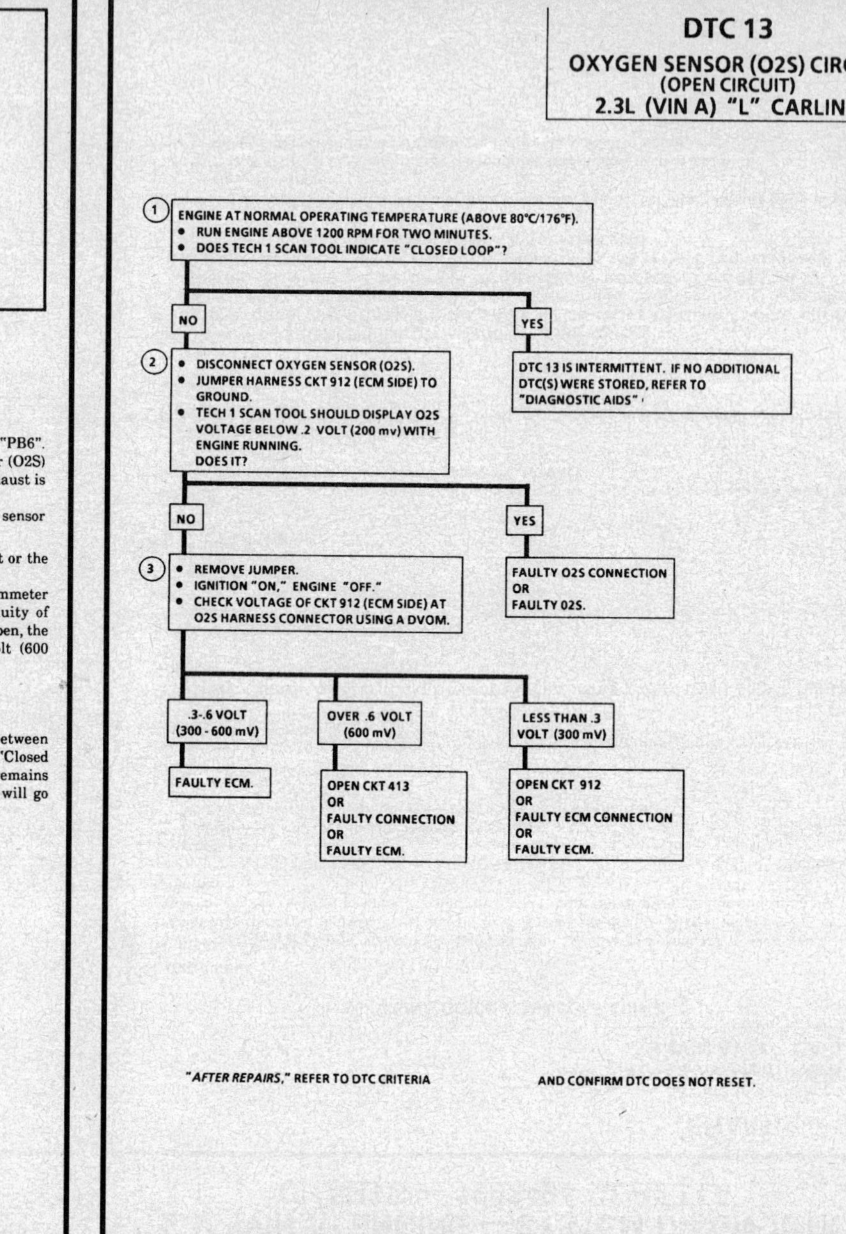

DTC 13
OXYGEN SENSOR (O2S) CIRCUIT
(OPEN CIRCUIT)
2.3L (VIN A) "L" CARLINE

Circuit Description:

The Engine Control Module (ECM) supplies a voltage of about .45 volt between terminals "PA12" and "PB6". (If measured with a 10 megohm digital voltmeter, this may read as low as .32 volt.) The Oxygen Sensor (O2S) varies the voltage within a range of about 1 volt if the exhaust is rich, down through about .10 volt if exhaust is lean.

The sensor is like an open circuit and produces no voltage when it is below 315°C (600° F). An open sensor circuit, or cold sensor, causes "Open Loop" operation.

Test Description: Number(s) below refer to circled number(s) on the diagnostic chart.

1. DTC 13 will set under the following conditions:
   - Engine running at least 40 seconds after start.
   - Coolant temperature at least 42.5°C (108°F).
   - No DTC 21 or 22.
   - O2S signal voltage steady between approximately .35 and .55 volt.
   - Throttle Position (TP) sensor signal above 6% for more time than TP sensor was below 6%. (About .3 volt above closed throttle voltage.)
   - All conditions must be met and held for at least 30 seconds.

   If the conditions for a DTC 13 exist, the system will not go "Closed Loop."

2. This will determine if the sensor is at fault or the wiring or ECM is the cause of the DTC 13.
3. Use only a high impedance digital volt ohmmeter for this test. This test checks the continuity of CKTs 912 and 413; because if CKT 413 is open, the ECM voltage on CKT 912 will be over .6 volt (600 mV).

Diagnostic Aids:

Normal Tech 1 scan tool voltage varies between 100 mV to 999 mV (.1 volt to 1.0 volt) while in "Closed Loop." DTC 13 sets in 20 seconds if voltage remains between .35 volt and .55 volt, but the system will go "Open Loop" in about 15 seconds.

2.3L (VIN A) ENGINE — DIAGNOSTIC TROUBLE CODE CHART — 1993–94 BERETTA

DTC 13
OXYGEN SENSOR (O2S) CIRCUIT
(OPEN CIRCUIT)
2.3L (VIN A) "L" CARLINE

(1)
- ENGINE AT NORMAL OPERATING TEMPERATURE (ABOVE 80°C/176°F).
- RUN ENGINE ABOVE 1200 RPM FOR TWO MINUTES.
- DOES TECH 1 SCAN TOOL INDICATE "CLOSED LOOP"?

NO →

(2)
- DISCONNECT OXYGEN SENSOR (O2S).
- JUMPER HARNESS CKT 912 (ECM SIDE) TO GROUND.
- TECH 1 SCAN TOOL SHOULD DISPLAY O2S VOLTAGE BELOW .2 VOLT (200 mv) WITH ENGINE RUNNING. DOES IT?

YES →
DTC 13 IS INTERMITTENT. IF NO ADDITIONAL DTC(S) WERE STORED, REFER TO "DIAGNOSTIC AIDS".

NO →

(3)
- REMOVE JUMPER.
- IGNITION "ON," ENGINE "OFF."
- CHECK VOLTAGE OF CKT 912 (ECM SIDE) AT O2S HARNESS CONNECTOR USING A DVOM.

YES →
FAULTY O2S CONNECTION
OR
FAULTY O2S.

.3-.6 VOLT (300 - 600 mV)
FAULTY ECM.

OVER .6 VOLT (600 mV)
OPEN CKT 413
OR
FAULTY CONNECTION
OR
FAULTY ECM.

LESS THAN .3 VOLT (300 mV)
OPEN CKT 912
OR
FAULTY ECM CONNECTION
OR
FAULTY ECM.

"AFTER REPAIRS," REFER TO DTC CRITERIA   AND CONFIRM DTC DOES NOT RESET.

2.3L (VIN A) ENGINE — DIAGNOSTIC TROUBLE CODE CHART — 1993–94 BERETTA

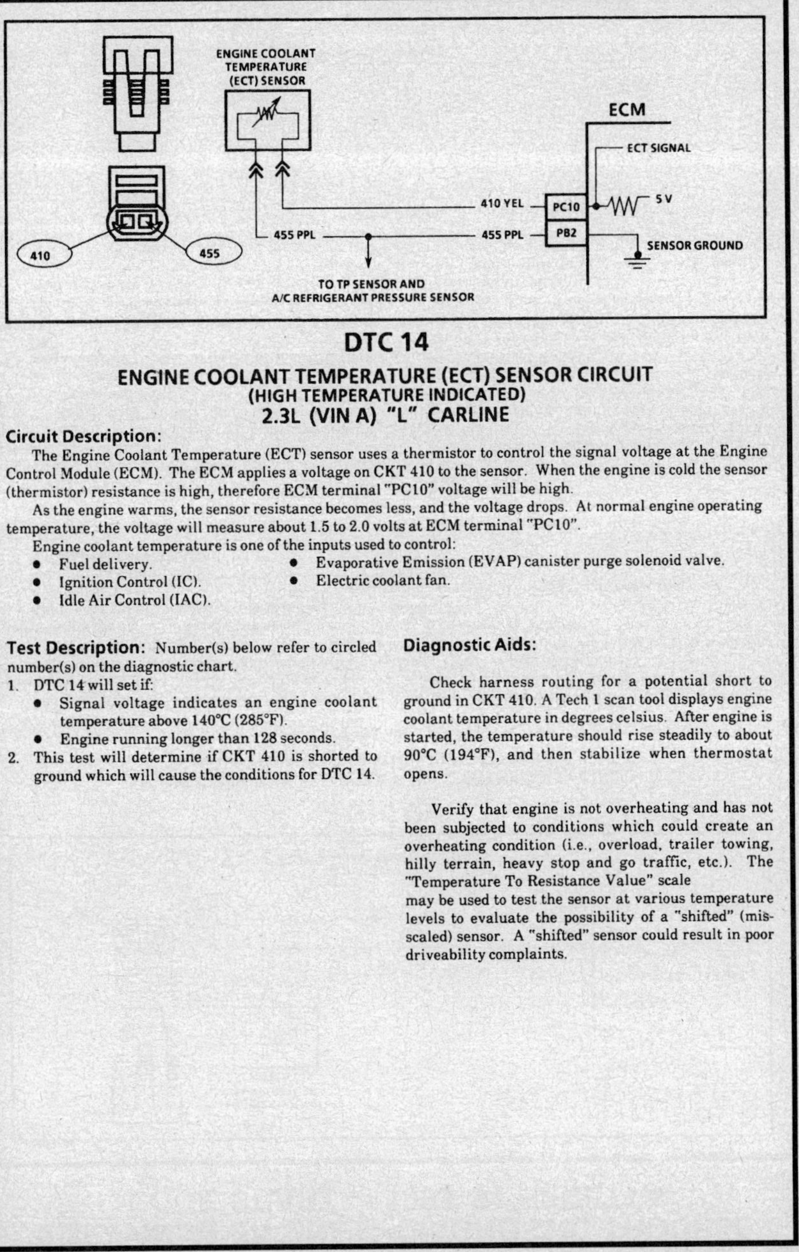

DTC 14
ENGINE COOLANT TEMPERATURE (ECT) SENSOR CIRCUIT
(HIGH TEMPERATURE INDICATED)
2.3L (VIN A) "L" CARLINE

Circuit Description:

The Engine Coolant Temperature (ECT) sensor uses a thermistor to control the signal voltage at the Engine Control Module (ECM). The ECM applies a voltage on CKT 410 to the sensor. When the engine is cold the sensor (thermistor) resistance is high, therefore ECM terminal "PC10" voltage will be high.

As the engine warms, the sensor resistance becomes less, and the voltage drops. At normal engine operating temperature, the voltage will measure about 1.5 to 2.0 volts at ECM terminal "PC10".

Engine coolant temperature is one of the inputs used to control:
- Fuel delivery.
- Ignition Control (IC).
- Idle Air Control (IAC).
- Evaporative Emission (EVAP) canister purge solenoid valve.
- Electric coolant fan.

Test Description: Number(s) below refer to circled number(s) on the diagnostic chart.
1. DTC 14 will set if:
   - Signal voltage indicates an engine coolant temperature above 140°C (285°F).
   - Engine running longer than 128 seconds.
2. This test will determine if CKT 410 is shorted to ground which will cause the conditions for DTC 14.

Diagnostic Aids:

Check harness routing for a potential short to ground in CKT 410. A Tech 1 scan tool displays engine coolant temperature in degrees celsius. After engine is started, the temperature should rise steadily to about 90°C (194°F), and then stabilize when thermostat opens.

Verify that engine is not overheating and has not been subjected to conditions which could create an overheating condition (i.e., overload, trailer towing, hilly terrain, heavy stop and go traffic, etc.). The "Temperature To Resistance Value" scale may be used to test the sensor at various temperature levels to evaluate the possibility of a "shifted" (mis-scaled) sensor. A "shifted" sensor could result in poor driveability complaints.

2.3L (VIN A) ENGINE — DIAGNOSTIC TROUBLE CODE CHART — 1993–94 BERETTA

DTC 14
ENGINE COOLANT TEMPERATURE (ECT) SENSOR CIRCUIT
(HIGH TEMPERATURE INDICATED)
2.3L (VIN A) "L" CARLINE

1. DOES SCAN TOOL DISPLAY ENGINE COOLANT TEMPERATURE OF 130°C (266°F) OR HIGHER?

- **YES**
  2. DISCONNECT ENGINE COOLANT TEMPERATURE SENSOR. SCAN TOOL SHOULD DISPLAY ENGINE COOLANT TEMPERATURE BELOW -30°C (-22°F). DOES IT?
    - **YES** → REPLACE ENGINE COOLANT TEMPERATURE SENSOR.
    - **NO** → CKT 410 SHORTED TO GROUND OR CKT 410 SHORTED TO SENSOR GROUND CIRCUIT OR FAULTY ECM.
- **NO** → DTC 14 IS INTERMITTENT. IF NO ADDITIONAL DTC(s) WERE STORED, REFER TO "DIAGNOSTIC AIDS"

DIAGNOSTIC AID

ENGINE COOLANT TEMPERATURE SENSOR
TEMPERATURE VS. RESISTANCE VALUES
(APPROXIMATE)

°C	°F	OHMS
100	212	177
90	194	241
80	176	332
70	158	467
60	140	667
50	122	973
45	113	1188
40	104	1459
35	95	1802
30	86	2238
25	77	2796
20	68	3520
15	59	4450
10	50	5670
5	41	7280
0	32	9420
-5	23	12300
-10	14	16180
-15	5	21450
-20	-4	28680
-30	-22	52700
-40	-40	100700

"AFTER REPAIRS," REFER TO DTC CRITERIA AND CONFIRM DTC DOES NOT RESET.

2.3L (VIN A) ENGINE — DIAGNOSTIC TROUBLE CODE CHART — 1993–94 BERETTA

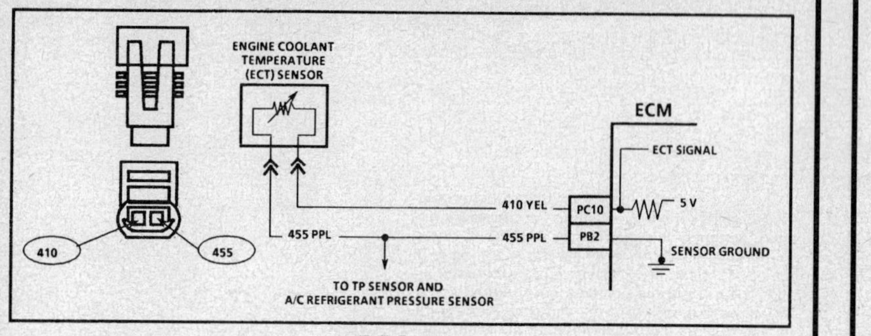

DTC 15
ENGINE COOLANT TEMPERATURE (ECT) SENSOR CIRCUIT
(LOW TEMPERATURE INDICATED)
2.3L (VIN A) "L" CARLINE

Circuit Description:

The Engine Coolant Temperature (ECT) sensor uses a thermistor to control the signal voltage at the Engine Control Module (ECM). The ECM applies a voltage on CKT 410 to the sensor. When the engine is cold, the sensor (thermistor) resistance is high, therefore, ECM terminal "PC10" voltage will be high.

As the engine warms, the sensor resistance becomes less, and the voltage drops. At normal engine operating temperature the voltage will measure about 1.5 to 2.0 volts at ECM terminal "PC10".

Engine coolant temperature is one of the inputs used to control:
- Fuel delivery.
- Ignition Control (IC).
- Idle Air Control (IAC).
- Evaporative Emission (EVAP) canister purge solenoid valve.
- Electric coolant fan.

Test Description: Number(s) below refer to circled number(s) on the diagnostic chart.

1. DTC 15 will set if:
   - Signal voltage indicates a coolant temperature less than -39°C (-38°F) for 60 seconds.
2. This test simulates a DTC 14. If the ECM senses the low signal voltage (high temperature) and the scan reads 130°C (266°F), the ECM and wiring are OK.
3. This test will determine if CKT 410 is open. There should be 5 volts present at sensor connector if measured with a DVOM.

Diagnostic Aids:

A Tech 1 scan tool displays engine temperature in degrees celsius. After the engine is started, the temperature should rise steadily to about 95°C (203°F), and then stabilize when the thermostat opens. It is normal for engine coolant temperature to fluctuate slightly around 95°C (203°F).

A faulty connection, or an open in CKT 410 or CKT 455 can result in a DTC 15.

DTCs 15, 21 and 66 stored at the same time could be the result of an open CKT 455.

The "Temperature to Resistance Value" scale may be used to test the sensor at various temperature levels to evaluate the possibility of a "shifted" (mis-scaled) sensor. A "shifted" sensor could result in poor driveability complaints.

2.3L (VIN A) ENGINE — DIAGNOSTIC TROUBLE CODE CHART — 1993–94 BERETTA

DTC 15
ENGINE COOLANT TEMPERATURE (ECT) SENSOR CIRCUIT
(LOW TEMPERATURE INDICATED)
2.3L (VIN A) "L" CARLINE

(1) • DOES TECH 1 SCAN TOOL DISPLAY ENGINE COOLANT TEMPERATURE OF -30°C (-22°F) OR LESS?

YES

(2) • DISCONNECT ENGINE COOLANT TEMPERATURE SENSOR.
• JUMPER HARNESS TERMINALS TOGETHER.
• TECH 1 SCAN TOOL SHOULD DISPLAY 130°C (266°F) OR MORE. DOES IT?

NO

DTC 15 IS INTERMITTENT. IF NO ADDITIONAL DTC(S) WERE STORED, REFER TO "DIAGNOSTIC AIDS"

NO

(3) • JUMPER CKT 410 TO GROUND.
• TECH 1 SCAN TOOL SHOULD DISPLAY OVER 130°C (266°F). DOES IT?

YES

FAULTY CONNECTION OR ENGINE COOLANT TEMPERATURE SENSOR.

YES

OPEN ENGINE COOLANT TEMPERATURE SENSOR GROUND CIRCUIT, FAULTY CONNECTION OR FAULTY ECM.

NO

OPEN CKT 410, FAULTY CONNECTION AT ECM, OR FAULTY ECM.

DIAGNOSTIC AID

ENGINE COOLANT TEMPERATURE SENSOR		
TEMPERATURE VS. RESISTANCE VALUES		
(APPROXIMATE)		
°C	°F	OHMS
100	212	177
90	194	241
80	176	332
70	158	467
60	140	667
50	122	973
45	113	1188
40	104	1459
35	95	1802
30	86	2238
25	77	2796
20	68	3520
15	59	4450
10	50	5670
5	41	7280
0	32	9420
-5	23	12300
-10	14	16180
-15	5	21450
-20	-4	28680
-30	-22	52700
-40	-40	100700

"AFTER REPAIRS," REFER TO DTC CRITERIA   AND CONFIRM DTC DOES NOT RESET.

2.3L (VIN A) ENGINE — DIAGNOSTIC TROUBLE CODE CHART — 1993–94 BERETTA

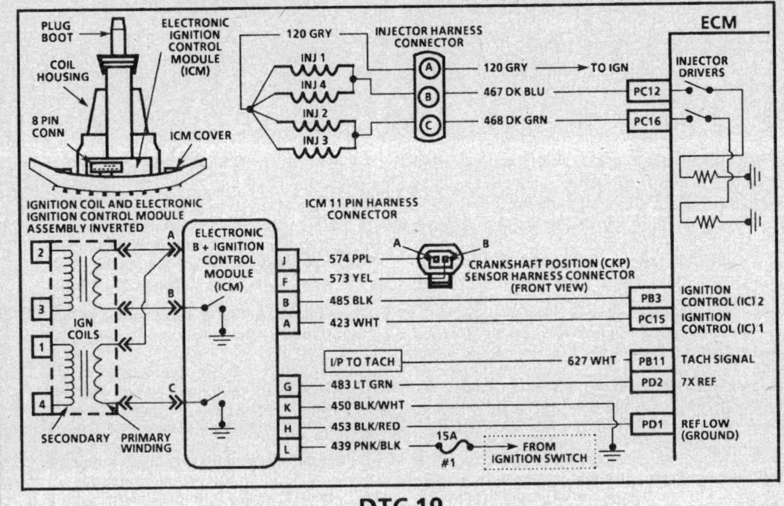

DTC 19
INTERMITTENT 7X REFERENCE CIRCUIT
2.3L (VIN A) "L" CARLINE

Circuit Description:

The electronic Ignition Control Module (ICM) sends a reference signal to the Engine Control Module (ECM) seven times per revolution to indicate crankshaft position and RPM so that the ECM can determine when to pulse the ignition coils and control ignition timing. This signal is called the 7X reference because it occurs seven times per revolution. The ICM applies 5 volts from terminal "G" through CKT 483 to ECM terminal "PD2" and in effect, switches this circuit to ground for a very short period of time. The seventh pulse is the sync pulse and is used for crankshaft position reference. DTC 19 is set if the ECM misses at least 20 resync pulses within 4 minutes and 16 seconds.

Test Description: Number(s) below refer to circled number(s) on the diagnostic chart.

1. This determines if the ECM recognizes a problem. If it doesn't set DTC 19 at this point, the problem is intermittent.
2. When a 7X resync occurs, engine stumble also occurs. If a component connection or circuitry is at fault, engine stumble may be induced by wiggling the circuit or connection. This step determines if a component connection or circuit is at fault.
3. Operating faulty non-engine related electronic components may emit Electromagnetic Interference (EMI) which may cause a 7X resync. This step determines if the non-engine related component is fault.

Diagnostic Aids:

An intermittent may be caused by a poor connection, rubbed through wire insulation, or a wire broken inside the insulation. Inspect ECM harness connector terminal "PD2" and ICM terminal "G" for improper mating, broken locks, improperly formed or damaged terminals, poor terminal to wire connection and damaged harness.

If vehicle has non-standard electrical equipment, (such as CB radios, 2-way radios, etc.) check to see if their operation may cause a 7X resync. Faulty wiring or faulty operation of the faulty non-standard electrical component may cause DTC 19.

If complaint was "Cranks But Will Not Run," DTC 19 may be set when engine start is attempted more than 20 times and ignition is not turned "OFF" between attempts.

2.3L (VIN A) ENGINE — DIAGNOSTIC TROUBLE CODE CHART — 1993–94 BERETTA

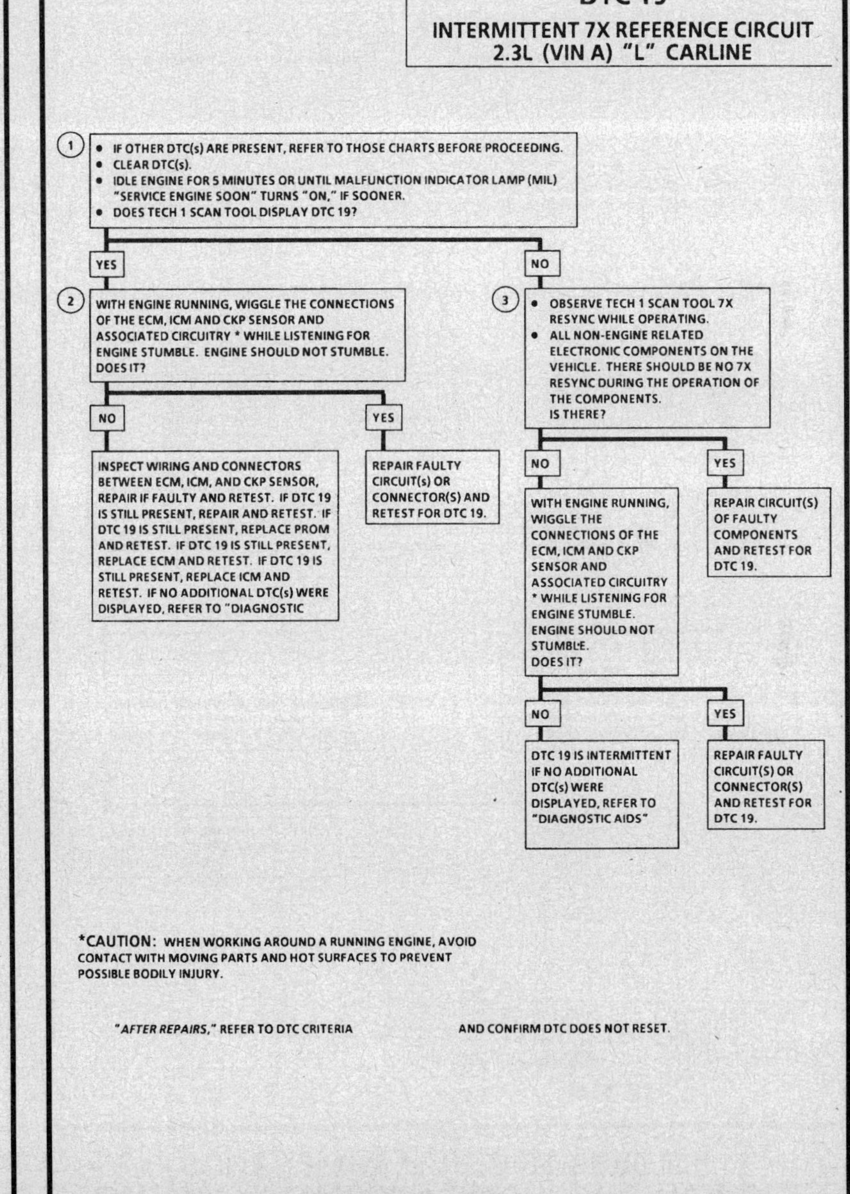

2.3L (VIN A) ENGINE — DIAGNOSTIC TROUBLE CODE CHART — 1993–94 BERETTA

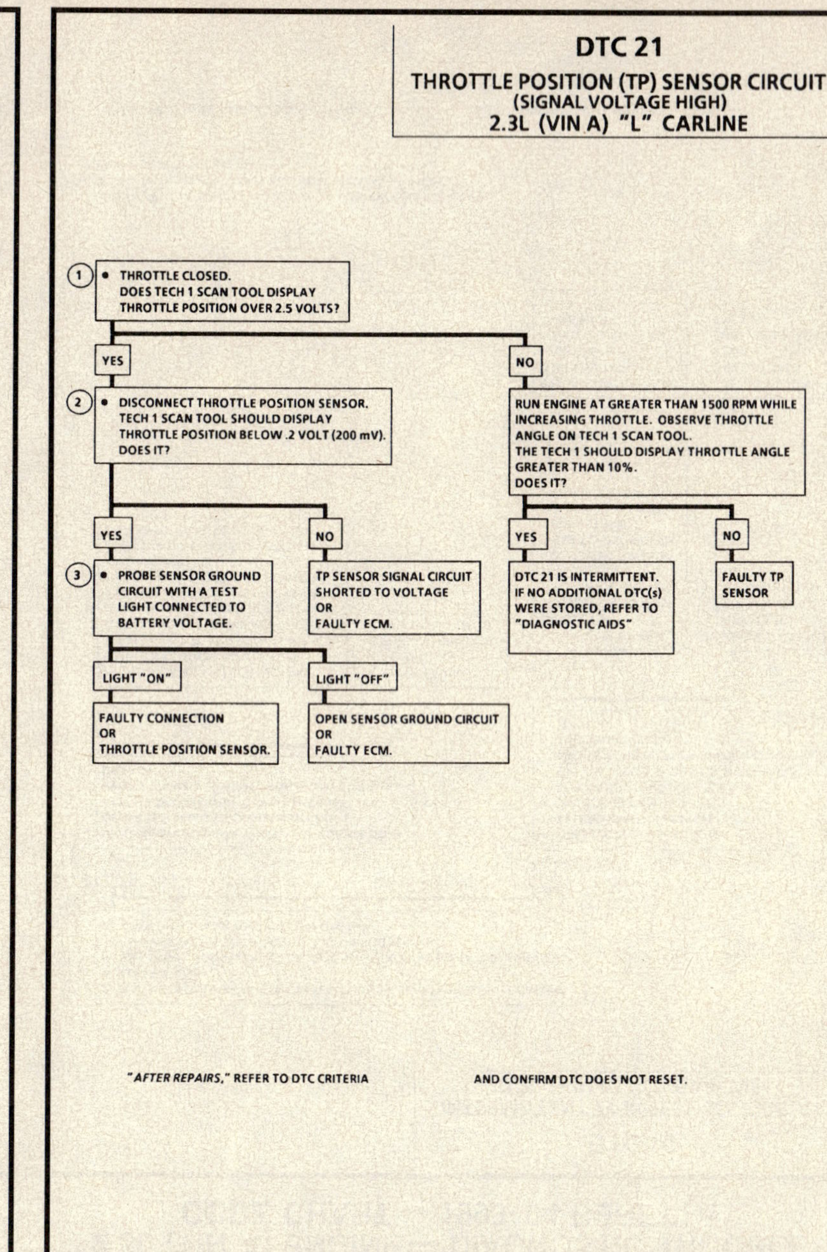

DTC 21
THROTTLE POSITION (TP) SENSOR CIRCUIT
(SIGNAL VOLTAGE HIGH)
2.3L (VIN A) "L" CARLINE

Circuit Description:

The Throttle Position (TP) sensor provides a voltage signal that changes relative to the throttle opening. Signal voltage will vary from about .6 volt at idle to about 4.7 volts at Wide Open Throttle (WOT).

The TP sensor signal is one of the most important inputs used by the Engine Control Module (ECM) for fuel control and for most of the ECM control outputs.

Test Description: Number(s) below refer to circled number(s) on the diagnostic chart.

1. DTC 21 will set if:
   - Engine is running below 1500 RPM.
   - No DTC 33 or 34.
   - MAP less than 65 kPa.
   - TP sensor signal voltage greater than approximately 3.9 volts.
   - Above conditions exist for over 5 seconds.
   OR
   - TP sensor voltage greater than about 4.7 volts.
   With throttle closed, the TP sensor should read less than .900 volt. If it doesn't, check for sticking TP sensor or throttle linkage. If OK, replace TP sensor.
2. With the TP sensor disconnected, the TP sensor voltage should go low if the ECM and wiring are OK.
3. Probing CKT 455 with a test light checks the TP sensor ground circuit because an open or very high resistance ground circuit will cause a DTC 21.

Diagnostic Aids:

A Tech 1 scan tool displays throttle position in volts. It should display .200 volt to .900 volt with throttle closed and ignition "ON" or at idle. Voltage should increase at a steady rate as throttle is moved toward Wide Open Throttle (WOT).

Also, a Tech 1 scan tool will display throttle angle %, 0% = closed throttle 100% = WOT.

An open in CKT 455 will result in a DTC 21.

DTCs 15, 21 and 66 stored at the same time could be the result of an open CKT 455. Scan TP sensor while depressing accelerator pedal with engine stopped and ignition "ON." Display should vary from about .5 volt (500 mV) when throttle is closed, to over 4500 mV (4.5 volts) when throttle is held wide open.

Check condition of connector and sensor terminals for moisture or corrosion, and clean or replace as necessary. If corrosion is found, check condition of connector seal, repair and/or replace if necessary.

2.3L (VIN A) ENGINE — DIAGNOSTIC TROUBLE CODE CHART — 1993–94 BERETTA

DTC 21
THROTTLE POSITION (TP) SENSOR CIRCUIT
(SIGNAL VOLTAGE HIGH)
2.3L (VIN A) "L" CARLINE

2.3L (VIN A) ENGINE — DIAGNOSTIC TROUBLE CODE CHART — 1993–94 BERETTA

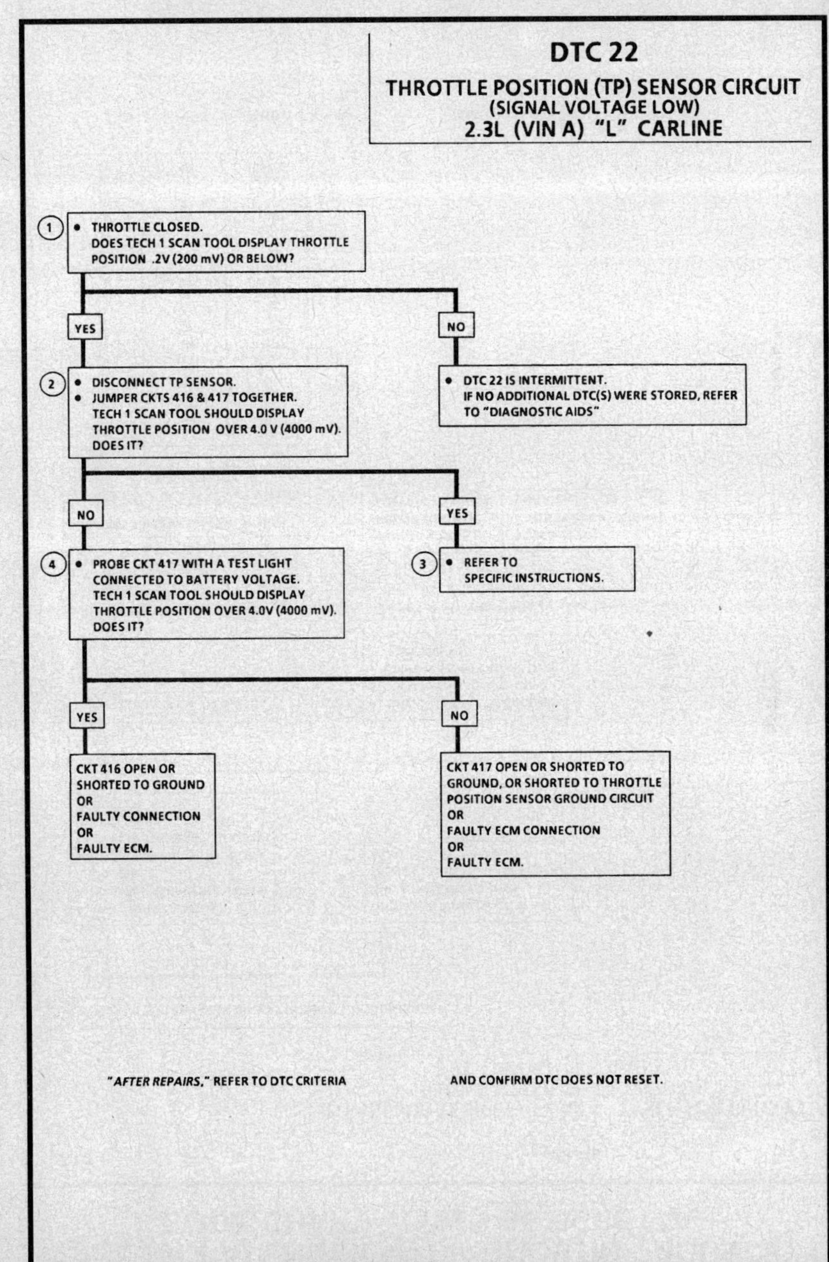

DTC 22
THROTTLE POSITION (TP) SENSOR CIRCUIT
(SIGNAL VOLTAGE LOW)
2.3L (VIN A) "L" CARLINE

Circuit Description:

The Throttle Position (TP) sensor provides a voltage signal that changes relative to the throttle opening. Signal voltage will vary from about .6 volt at idle to about 4.7 volts at Wide Open Throttle (WOT).

The TP sensor signal is one of the most important inputs used by the Engine Control Module (ECM) for fuel control and for most of the ECM control outputs.

Test Description: Number(s) below refer to circled number(s) on the diagnostic chart.
1. DTC 22 will set if:
   - Engine is running.
   - TP sensor signal voltage is less than about .16 volt.

   The TP sensor has an auto zeroing feature. If the voltage reading is within the range of about .2 to .9 volt, the ECM will use that value as closed throttle. If the voltage reading is out of the auto zero range at closed throttle, check for a binding throttle cable or damaged linkage, if OK, continue with diagnosis.
2. Simulates DTC 21: (high voltage). If the ECM recognizes the high signal voltage, then the ECM and wiring are OK.
3. Check for good sensor connection. If connection is good, replace TP sensor.
4. This simulates a high signal voltage to check for an open in CKT 417. The Tech 1 scan tool will not read up to 12 volts, but what is important is that the ECM recognizes the signal on CKT 417.

Diagnostic Aids:

Scan TP sensor while depressing accelerator pedal with engine stopped and ignition "ON." Display should vary from about 600 mV (.6 volt) when throttle is closed, to over 4500 mV (4.5 volts) when throttle is held wide open.

Also, a Tech 1 scan tool will display throttle angle %: 0% = closed throttle; 100% = WOT.

If DTC 22, 33 and/or 66 are set, check CKT 416 for faulty wiring or connections.

Should check condition of connector and sensor terminals for moisture or corrosion, and clean and/or replace as necessary. If corrosion is found, check condition of connector seal and repair or replace if necessary.

2.3L (VIN A) ENGINE — DIAGNOSTIC TROUBLE CODE CHART — 1993–94 BERETTA

DTC 22
THROTTLE POSITION (TP) SENSOR CIRCUIT
(SIGNAL VOLTAGE LOW)
2.3L (VIN A) "L" CARLINE

(1)
- THROTTLE CLOSED.
  DOES TECH 1 SCAN TOOL DISPLAY THROTTLE POSITION .2V (200 mV) OR BELOW?

YES →

(2)
- DISCONNECT TP SENSOR.
  JUMPER CKTS 416 & 417 TOGETHER.
  TECH 1 SCAN TOOL SHOULD DISPLAY THROTTLE POSITION OVER 4.0 V (4000 mV).
  DOES IT?

NO →
- DTC 22 IS INTERMITTENT.
  IF NO ADDITIONAL DTC(S) WERE STORED, REFER TO "DIAGNOSTIC AIDS"

NO (from 2) →

(4)
- PROBE CKT 417 WITH A TEST LIGHT CONNECTED TO BATTERY VOLTAGE.
  TECH 1 SCAN TOOL SHOULD DISPLAY THROTTLE POSITION OVER 4.0V (4000 mV).
  DOES IT?

YES (from 2) →

(3)
- REFER TO SPECIFIC INSTRUCTIONS.

YES (from 4) →
CKT 416 OPEN OR SHORTED TO GROUND
OR
FAULTY CONNECTION
OR
FAULTY ECM.

NO (from 4) →
CKT 417 OPEN OR SHORTED TO GROUND, OR SHORTED TO THROTTLE POSITION SENSOR GROUND CIRCUIT
OR
FAULTY ECM CONNECTION
OR
FAULTY ECM.

"AFTER REPAIRS," REFER TO DTC CRITERIA   AND CONFIRM DTC DOES NOT RESET.

2.3L (VIN A) ENGINE — DIAGNOSTIC TROUBLE CODE CHART — 1993–94 BERETTA

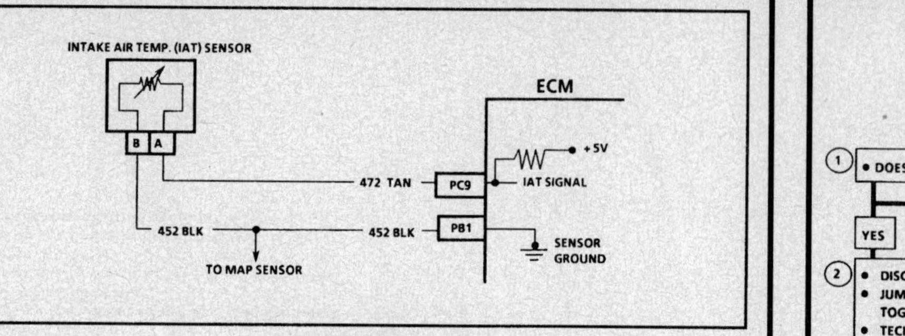

DTC 23
INTAKE AIR TEMPERATURE (IAT) SENSOR CIRCUIT
(LOW TEMPERATURE INDICATED)
2.3L (VIN A) "L" CARLINE

Circuit Description:

The Intake Air Temperature (IAT) sensor uses a thermistor to control the signal voltage at the Engine Control Module (ECM). The ECM applies a voltage (about 5 volts) on CKT 472 to the sensor. When the air is cold the sensor (thermistor) resistance is high, therefore the ECM terminal "PC9" voltage will be high. If the air is warm the sensor resistance is low, therefore the ECM terminal "PC9" voltage will be low.

Test Description: Number(s) below refer to circled number(s) on the diagnostic chart.

1. DTC 23 will set if:
   - A signal voltage indicates an intake air temperature below about -40℃ (-40°F).
   - Time since engine start is 320 seconds or longer.
   - Vehicle speed less than 15 mph.
2. A DTC 23 will set due to an open sensor, wire, or connection. This test will determine if the wiring and ECM are OK.
3. This will determine if the signal CKT 472 or the sensor ground CKT 452 is open.

Diagnostic Aids:

A Tech 1 scan tool displays temperature of the air entering the engine, which should be close to ambient air temperature when engine is cold, and rise as underhood temperature increases.

A faulty connection, or an open in CKT 472 or CKT 452 can result in a DTC 23.

DTCs 23 and 34 stored at the same time, could be the result of an open CKT 452. The "Temperature to Resistance Values" scale : may be used to test the IAT sensor at various temperature levels to evaluate the possibility of a "slewed" (mis-scaled) sensor. A "slewed" sensor could result in poor driveability complaints.

2.3L (VIN A) ENGINE — DIAGNOSTIC TROUBLE CODE CHART — 1993–94 BERETTA

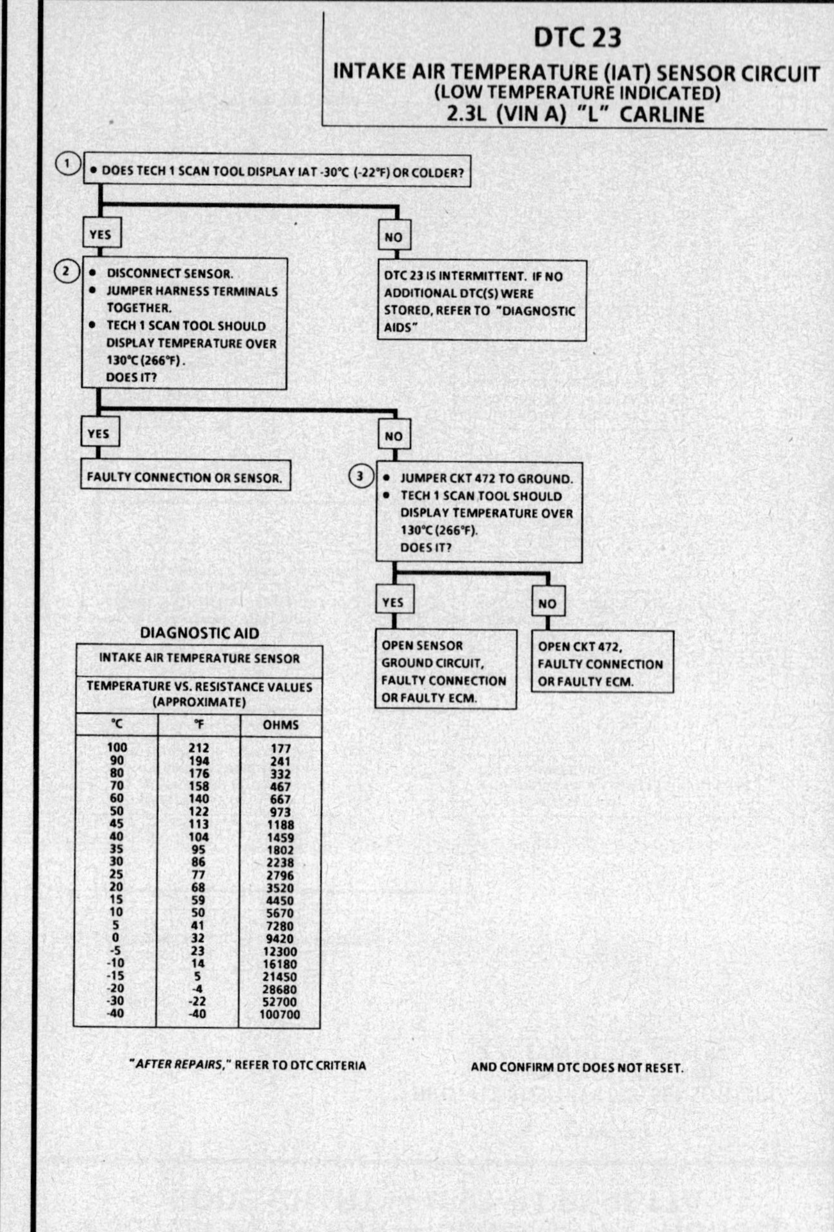

DIAGNOSTIC AID

| INTAKE AIR TEMPERATURE SENSOR | | |
| TEMPERATURE VS. RESISTANCE VALUES (APPROXIMATE) | | |
℃	°F	OHMS
100	212	177
90	194	241
80	176	332
70	158	467
60	140	667
50	122	973
45	113	1188
40	104	1459
35	95	1802
30	86	2238
25	77	2796
20	68	3520
15	59	4450
10	50	5670
5	41	7280
0	32	9420
-5	23	12300
-10	14	16180
-15	5	21450
-20	-4	28680
-30	-22	52700
-40	-40	100700

2.3L (VIN A) ENGINE — DIAGNOSTIC TROUBLE CODE CHART — 1993–94 BERETTA

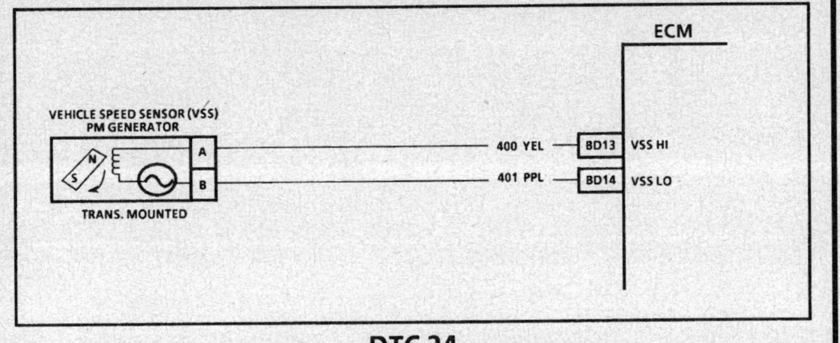

DTC 24
VEHICLE SPEED SENSOR (VSS) CIRCUIT
2.3L (VIN A) "L" CARLINE

Circuit Description:

Vehicle speed information is provided to the Engine Control Module (ECM) by the Vehicle Speed Sensor (VSS) which is a permanent magnet generator that is mounted in the transaxle. The permanent magnet generator produces a pulsing voltage whenever vehicle speed is over about 3 mph (5 km/h). The A/C voltage level and the number of pulses increases with vehicle speed. The ECM then converts the pulsing voltage to mph which is used for calculations, and the mph can be displayed with a Tech 1 scan tool. Output of the generator can also be seen by using a digital voltmeter on the AC scale while rotating the generator.

The function of VSS buffer used in past model years has been incorporated into the ECM. The ECM then supplies the necessary signal for the instrument panel for operating the speedometer, the odometer, and for the cruise control module.

Test Description: Number(s) below refer to circled number(s) on the diagnostic chart.

1. DTC 24 will set if vehicle speed is less than 2 mph when:
   - Engine speed is between 1500 and 3600 RPM.
   - TP sensor is greater than 7%.
   - All conditions met for 20 seconds.
   - No DTC 21 or 22, 33 or 34.
   - Engine vacuum between 38-58 kPa.

   The permanent magnet generator only produces a signal if drive wheels are turning greater than 3 mph (5 km/h).

2. Check PROM for correct application before replacing ECM.

Diagnostic Aids:

A Tech 1 scan tool should indicate a vehicle speed whenever the drive wheels are turning greater than 3 mph (5 km/h).

Check CKT 400 and 401 for proper connections. Be sure they are clean and tight and the harness is routed correctly.

DTC 24 can be falsely set if engine is "brake-torqued" in gear for 20 seconds.

2.3L (VIN A) ENGINE — DIAGNOSTIC TROUBLE CODE CHART — 1993–94 BERETTA

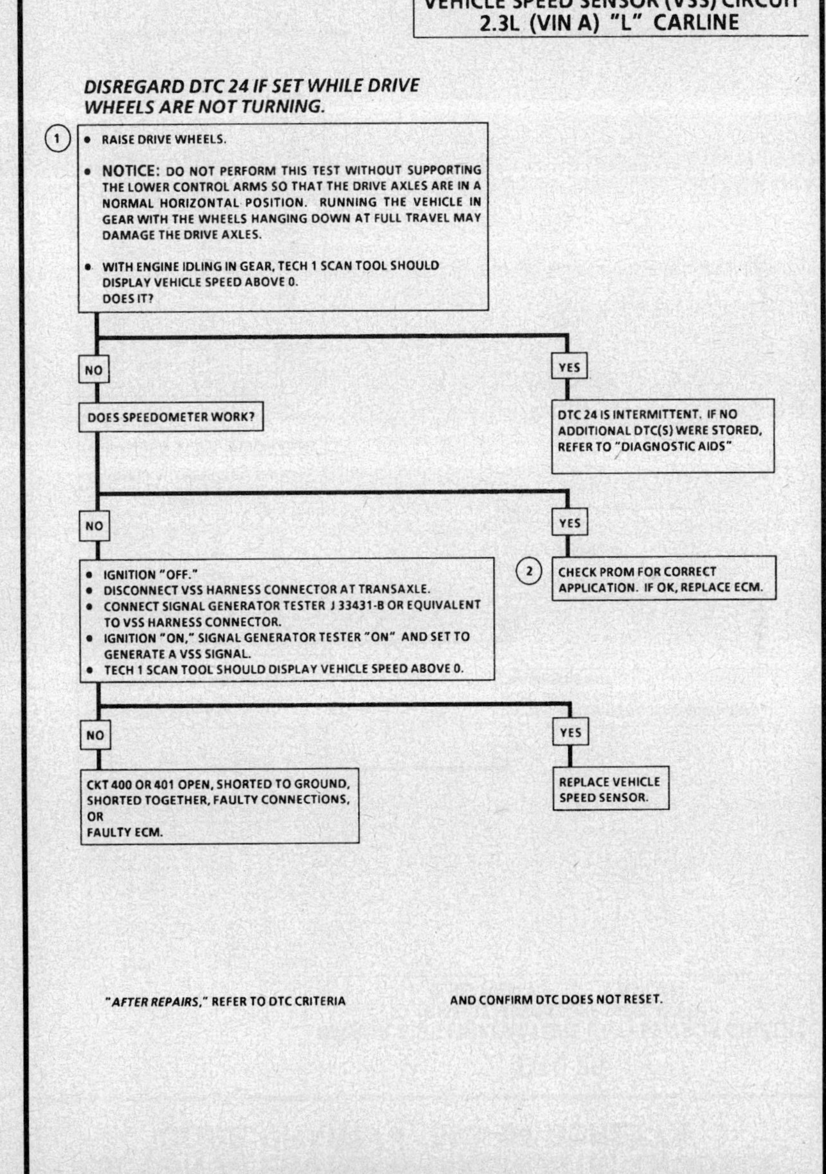

2.3L (VIN A) ENGINE — DIAGNOSTIC TROUBLE CODE CHART — 1993–94 BERETTA

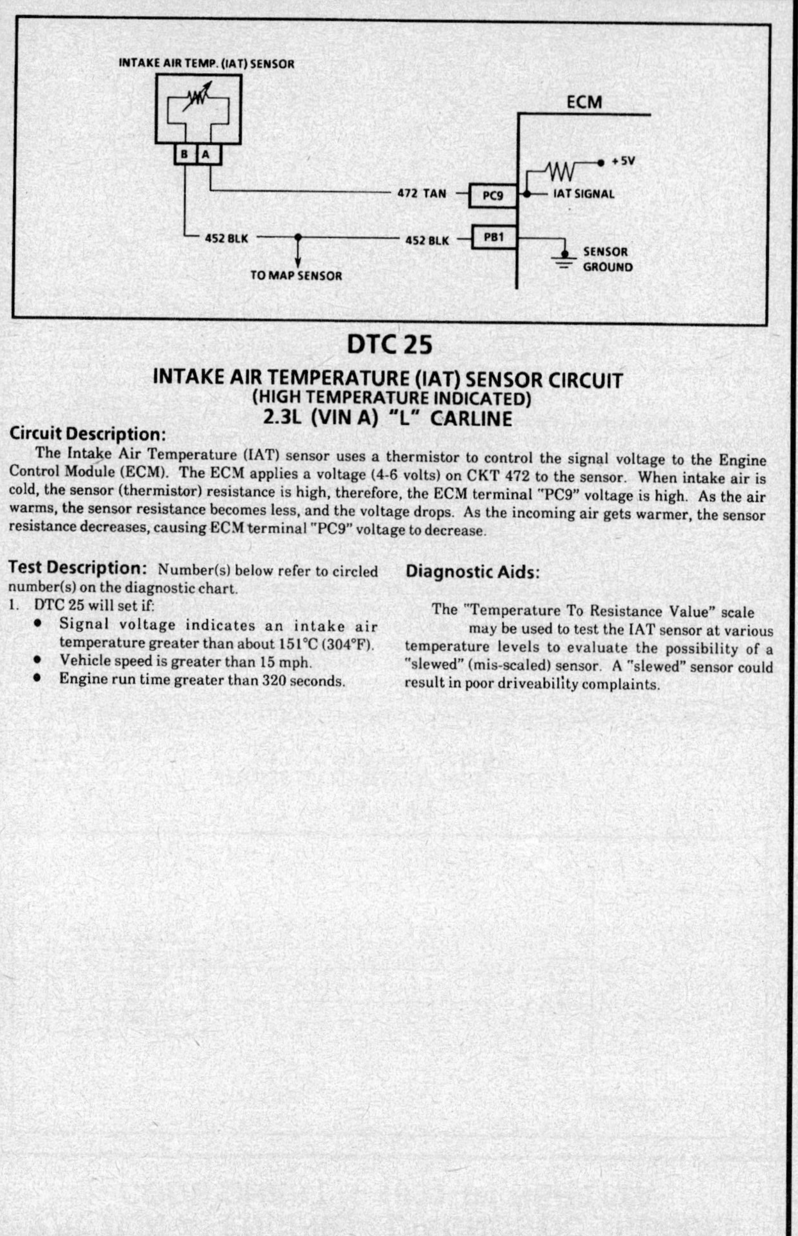

DTC 25
INTAKE AIR TEMPERATURE (IAT) SENSOR CIRCUIT
(HIGH TEMPERATURE INDICATED)
2.3L (VIN A) "L" CARLINE

Circuit Description:

The Intake Air Temperature (IAT) sensor uses a thermistor to control the signal voltage to the Engine Control Module (ECM). The ECM applies a voltage (4-6 volts) on CKT 472 to the sensor. When intake air is cold, the sensor (thermistor) resistance is high, therefore, the ECM terminal "PC9" voltage is high. As the air warms, the sensor resistance becomes less, and the voltage drops. As the incoming air gets warmer, the sensor resistance decreases, causing ECM terminal "PC9" voltage to decrease.

Test Description: Number(s) below refer to circled number(s) on the diagnostic chart.

1. DTC 25 will set if:
   - Signal voltage indicates an intake air temperature greater than about 151°C (304°F).
   - Vehicle speed is greater than 15 mph.
   - Engine run time greater than 320 seconds.

Diagnostic Aids:

The "Temperature To Resistance Value" scale may be used to test the IAT sensor at various temperature levels to evaluate the possibility of a "slewed" (mis-scaled) sensor. A "slewed" sensor could result in poor driveability complaints.

2.3L (VIN A) ENGINE — DIAGNOSTIC TROUBLE CODE CHART — 1993–94 BERETTA

DTC 25
INTAKE AIR TEMPERATURE (IAT) SENSOR CIRCUIT
(HIGH TEMPERATURE INDICATED)
2.3L (VIN A) "L" CARLINE

DIAGNOSTIC AID

INTAKE AIR TEMPERATURE SENSOR		
TEMPERATURE VS. RESISTANCE VALUES (APPROXIMATE)		
°C	°F	OHMS
100	212	177
90	194	241
80	176	332
70	158	467
60	140	667
50	122	973
45	113	1188
40	104	1459
35	95	1802
30	86	2238
25	77	2796
20	68	3520
15	59	4450
10	50	5670
5	41	7280
0	32	9420
-5	23	12300
-10	14	16180
-15	5	21450
-20	-4	28680
-30	-22	52700
-40	-40	100700

"AFTER REPAIRS," REFER TO DTC CRITERIA   AND CONFIRM DTC DOES NOT RESET.

2.3L (VIN A) ENGINE — DIAGNOSTIC TROUBLE CODE CHART — 1993–94 BERETTA

ECM

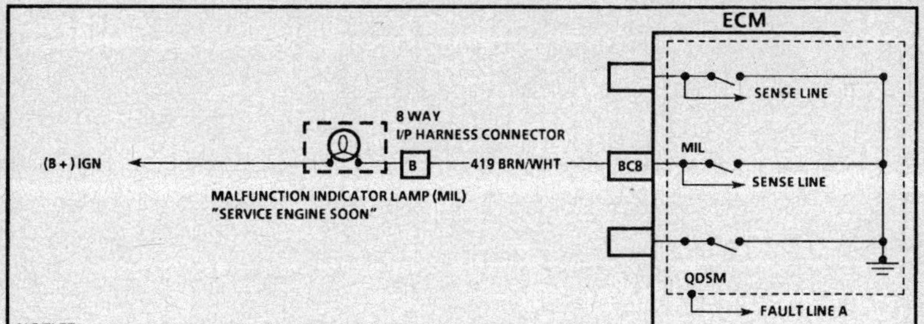

NOTICE: ONLY A PORTION OF EACH CIRCUIT IS SHOWN. REFER TO ECM WIRING DIAGRAM FOR MORE DETAIL.

DTC 26
QUAD-DRIVER MODULE (QDSM) CIRCUIT
2.3L (VIN A) "L" CARLINE

Circuit Description:

The Engine Control Module (ECM) controls most components with electronic switches which complete a ground circuit when turned "ON." These switches are arranged in groups of 4, called Quad-Driver Modules (QDMs) which can independently control up to 4 outputs (ECM terminals), although not all outputs are used. When an output is "ON," the terminal is grounded and its voltage normally will be low. When an output is "OFF," its terminal voltage normally will be high.

The QDSM is fault protected. If a relay or solenoid coil is shorted, having very low or zero resistance, or if the control side of the circuit is shorted to voltage, it would allow too much current into the QDSM. The QDSM senses this and the output turns "OFF" or its internal resistance increases to limit current flow and protect the QDSM. The result is high output terminal voltage when it should be low. If the circuit from B + or the component is open, or the control side of the circuit is shorted to ground, terminal voltage will be low, even when output is commanded "OFF." Either of these conditions is considered to be a QDSM fault.

The QDSM has a separate fault line to indicate the presence of a current fault to the ECM's central processor. A Tech 1 scan tool displays the status of each of these fault lines as "Low" = OK, "High" = Fault.

- QDM A fault line.
  - "High" for 20 seconds or more.
  - Battery voltage is greater than 10.5 volts.

2.3L (VIN A) ENGINE — DIAGNOSTIC TROUBLE CODE CHART — 1993–94 BERETTA

DTC 26
QUAD-DRIVER MODULE (QDSM) CIRCUIT
2.3L (VIN A) "L" CARLINE

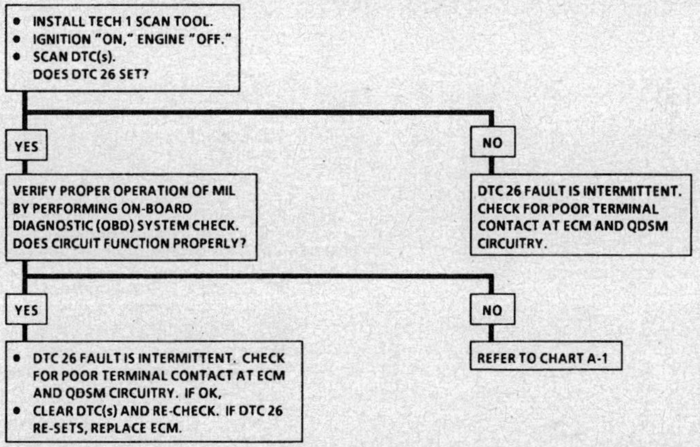

"AFTER REPAIRS," REFER TO DTC CRITERIA   AND CONFIRM DTC DOES NOT RESET.

2.3L (VIN A) ENGINE — DIAGNOSTIC TROUBLE CODE CHART — 1993–94 BERETTA

DTC 27
QUAD-DRIVER MODULE (QDM 1) CIRCUIT
2.3L (VIN A) "L" CARLINE

Circuit Description:

The Engine Control Module (ECM) controls most components with electronic switches which complete a ground circuit when turned "ON." These switches are arranged in groups of 4, called Quad-Driver Modules (QDMs) which can independently control up to 4 outputs (ECM terminals), although not all outputs are used. When an output is "ON," the terminal is grounded and its voltage normally will be low. When an output is "OFF," its terminal voltage normally will be high.

QDMs are fault protected. If a relay or solenoid coil is shorted, having very low or zero resistance, or if the control side of the circuit is shorted to voltage, it would allow too much current into the QDM. The QDM senses this and the output turns "OFF" or its internal resistance increases to limit current flow and protect the QDM. The result is high output terminal voltage when it should be low. If the circuit from B+ or the component is open, or the control side of the circuit is shorted to ground, terminal voltage will be low, even when output is commanded "OFF." Either of these conditions is considered to be a QDM fault.

QDM1 has a fault line to indicated the presence of a current fault to the ECM's central processor. A scan tool displays the status of the fault line as "Low" = OK, "High" = Fault.

- QDM B fault line.
  - High for 20 seconds or more.
  - Battery voltage is greater than 10.5 volts.

2.3L (VIN A) ENGINE — DIAGNOSTIC TROUBLE CODE CHART — 1993–94 BERETTA

DTC 27
QUAD-DRIVER MODULE (QDM 1) CIRCUIT
2.3L (VIN A) "L" CARLINE

- IGNITION "ON," ENGINE "OFF."
- USING TECH 1 SCAN TOOL, COMMAND EVAP CANISTER PURGE SOLENOID VALVE "ON" AND "OFF" WHILE LISTENING TO EVAP CANISTER PURGE SOLENOID VALVE. DOES EVAP CANISTER PURGE SOLENOID VALVE TURN "ON" AND "OFF" WHEN COMMANDED?

YES

- USING TECH 1 SCAN TOOL OUTPUT CONTROL FUNCTION, COMMAND SHIFT LIGHT "ON" AND "OFF."
- DOES SHIFT LIGHT TURN "ON" AND "OFF" WHEN IT IS COMMANDED?

NO

SEE CHART C3" TO DIAGNOSE EVAP CANISTER PURGE SOLENOID VALVE CIRCUIT.

YES

DTC 27 FAULT IS INTERMITTENT. CHECK FOR POOR TERMINAL CONTACT AT ECM AND QDM1 CIRCUITRY. IF OK,
- CLEAR DTCs AND RECHECK. IF DTC 27 RE-SETS, REPLACE ECM.

NO

SEE CHART C8 TO DIAGNOSE SHIFT LIGHT CIRCUIT.

"AFTER REPAIRS," REFER TO DTC CRITERIA AND CONFIRM DTC DOES NOT RESET.

2.3L (VIN A) ENGINE — DIAGNOSTIC TROUBLE CODE CHART — 1993–94 BERETTA

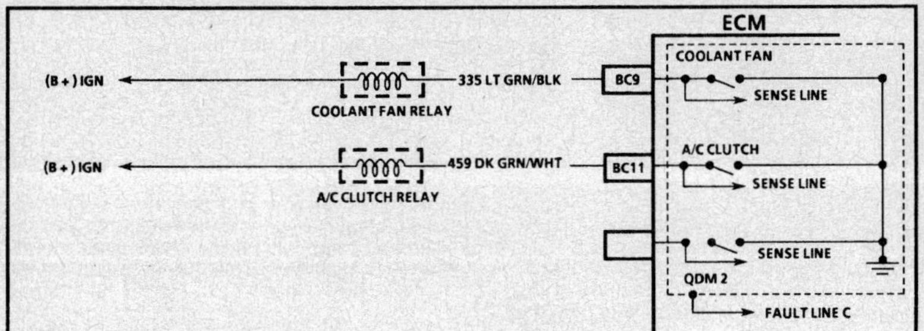

NOTICE: ONLY A PORTION OF EACH CIRCUIT IS SHOWN. REFER TO ECM WIRING DIAGRAM FOR MORE DETAIL.

DTC 28
QUAD-DRIVER MODULE (QDM 2) CIRCUIT
2.3L (VIN A) "L" CARLINE

Circuit Description:

The Engine Control Module (ECM) controls most components with electronic switches which complete a ground circuit when turned "ON." These switches are arranged in groups of 4, called Quad-Driver Modules (QDMs) which can independently control up to 4 outputs (ECM terminals), although not all outputs are used. When an output is "ON," the terminal is grounded and its voltage normally will be low. When an output is "OFF," its terminal voltage normally will be high.

QDM 2 is fault protected. If a relay or solenoid coil is shorted, having very low or zero resistance, or if the control side of the circuit is shorted to voltage, it would allow too much current into the QDM. The QDM senses this and the output turns "OFF" or its internal resistance increases to limit current flow and protect the QDM. The result is high output terminal voltage when it should be low. If the circuit from B+ or the component is open, or the control side of the circuit is shorted to ground, terminal voltage will be low, even when output is commanded "OFF." Either of these conditions is considered to be a QDM fault.

QDM 2 has a fault line to indicated the presence of a current fault to the ECM's central processor. A scan tool displays the status of the fault line as "Low" = OK, "High" = Fault.

- QDM C fault line.
  - "High" for 20 seconds or more.
  - Battery voltage is greater than 10.5 volts.

2.3L (VIN A) ENGINE — DIAGNOSTIC TROUBLE CODE CHART — 1993–94 BERETTA

DTC 28
QUAD-DRIVER MODULE (QDM 2) CIRCUIT
2.3L (VIN A) "L" CARLINE

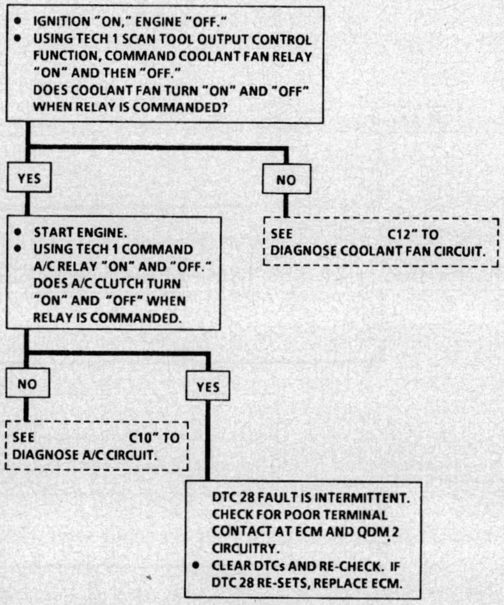

"AFTER REPAIRS," REFER TO DTC CRITERIA AND CONFIRM DTC DOES NOT RESET.

2.3L (VIN A) ENGINE — DIAGNOSTIC TROUBLE CODE CHART — 1993–94 BERETTA

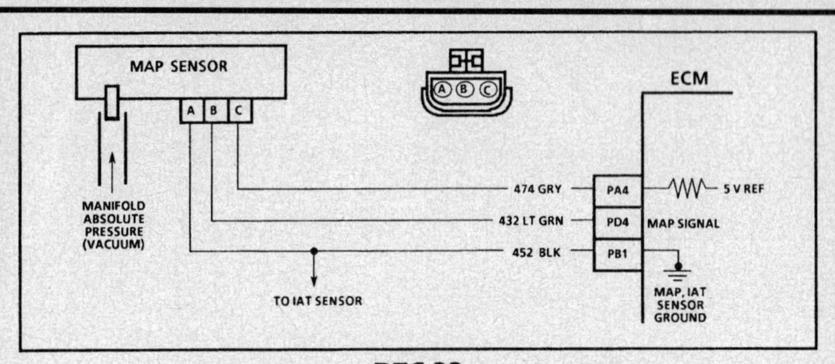

DTC 33
MANIFOLD ABSOLUTE PRESSURE (MAP) SENSOR CIRCUIT
(SIGNAL VOLTAGE HIGH - LOW VACUUM)
2.3L (VIN A) "L" CARLINE

Circuit Description:

The Manifold Absolute Pressure (MAP) sensor responds to changes in manifold pressure (vacuum). The Engine Control Module (ECM) receives this information as a signal voltage that will vary from about 1 to 1.5 volts at closed throttle (idle) to 4.5 - 4.8 volts at Wide Open Throttle (WOT) (low vacuum).

If the MAP sensor fails, the ECM will substitute a fixed MAP value and use the Throttle Position (TP) sensor to control fuel delivery.

Test Description: Number(s) below refer to circled number(s) on the diagnostic chart.
1. This step will determine if DTC 33 is the result of a hard failure or an intermittent condition. DTC 33 will set when:
   - Engine running.
   - MAP signal greater than 82 kPa.
   - TP sensor less than 12%.
   - VSS less than 1 mph.
   - DTC 21 or 22 not present.
   - Above conditions met for 5 seconds.
2. This step simulates conditions for a DTC 34. If the ECM recognizes the change, the ECM and CKT 474 and CKT 432 are OK. If CKT 452 is open, there may also be a stored DTC 23.

Diagnostic Aids:

With the ignition "ON," and the engine stopped, the manifold pressure is equal to atmospheric pressure and the signal voltage will be high. This information is used by the ECM as an indication of vehicle altitude. Comparison of this reading with a known good vehicle with the same sensor is a good way to check accuracy of a "suspect" sensor. Readings should be the same ± .4 volt.

DTC 33 will result if CKT 452 is open or if CKT 432 is shorted to voltage or to CKT 474.

- Check all connections.
- Disconnect sensor from bracket and twist sensor by hand (only) to check for intermittent connections. Output changes greater than .1 volt indicates a bad connector or connection. If OK, replace sensor.

NOTICE: Make sure electrical connector remains securely fastened.

- Refer to CHART C-1D, MAP sensor voltage vs. atmospheric pressure for further diagnosis.

NOTICE: After repairs use Tech 1 scan tool "Fuel Trim Reset" function to reset L. T. fuel trim to 128.

2.3L (VIN A) ENGINE — DIAGNOSTIC TROUBLE CODE CHART — 1993–94 BERETTA

DTC 33
MANIFOLD ABSOLUTE PRESSURE (MAP) SENSOR CIRCUIT
(SIGNAL VOLTAGE HIGH - LOW VACUUM)
2.3L (VIN A) "L" CARLINE

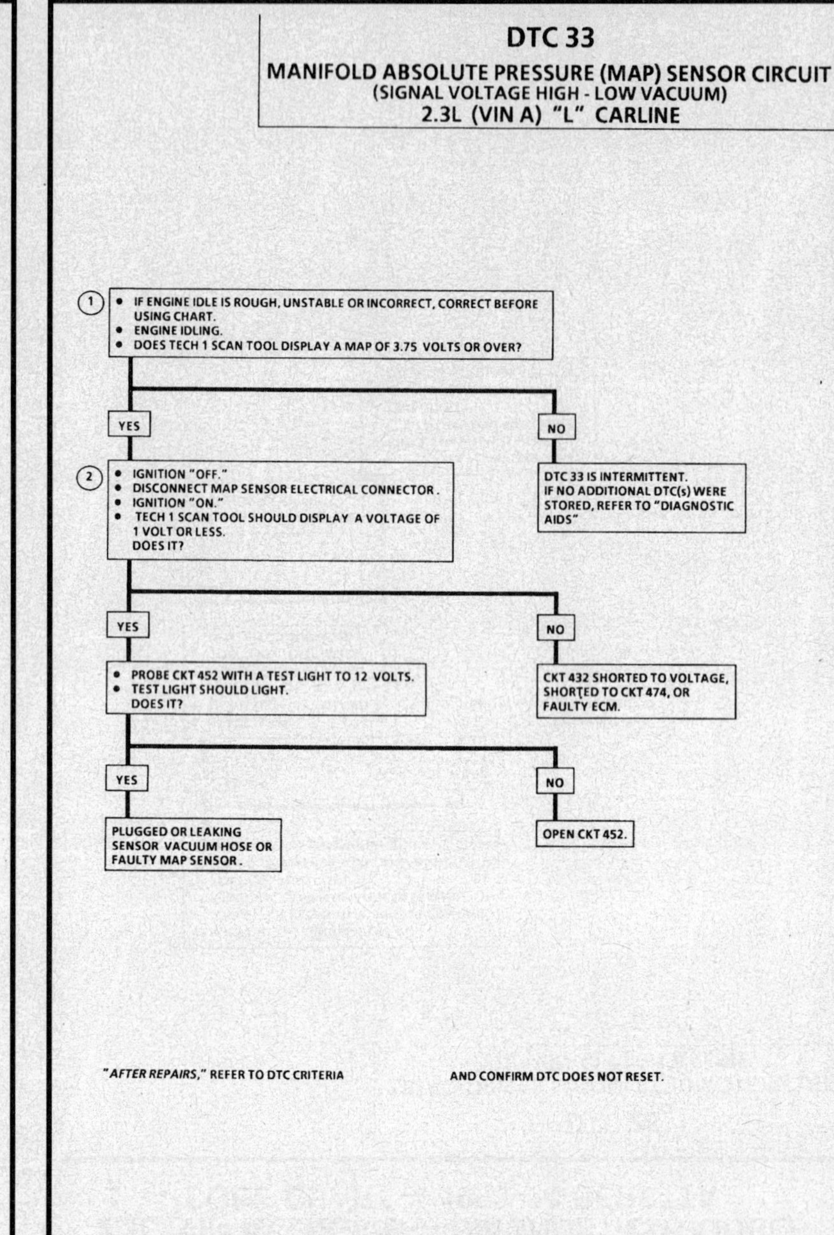

2.3L (VIN A) ENGINE — DIAGNOSTIC TROUBLE CODE CHART — 1993–94 BERETTA

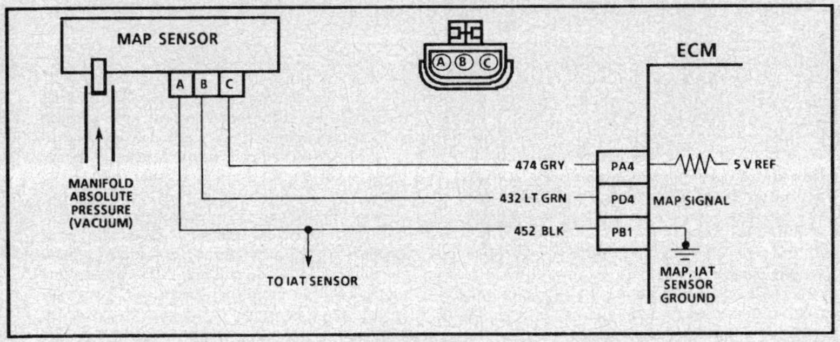

DTC 34

MANIFOLD ABSOLUTE PRESSURE (MAP) SENSOR CIRCUIT
(SIGNAL VOLTAGE LOW - HIGH VACUUM)
2.3L (VIN A) "L" CARLINE

Circuit Description:

The Manifold Absolute Pressure (MAP) sensor responds to changes in manifold pressure (vacuum). The Engine Control Module (ECM) receives this information as a signal voltage that will vary from about 1 to 1.5 volts at closed throttle (idle) to 4.5 - 4.8 volts at Wide Open Throttle (WOT) (low vacuum).

If the MAP sensor fails, the ECM will substitute a fixed MAP value and use the Throttle Position (TP) sensor to control fuel delivery.

Test Description: Number(s) below refer to circled number(s) on the diagnostic chart.

1. This step determines if DTC 34 is the result of a hard failure or an intermittent condition. DTC 34 will set when:
   - Engine running.
   - No DTC 21.
   - MAP less than 14 kPa.
   - Engine RPM less than 1200 or TP sensor greater than 15.2%.
   - Above condition met for .2 second.
2. Jumpering harness terminals "B" to "C" (5 volts to signal circuit) will determine if the sensor is at fault, or if there is a problem with the ECM or wiring.
3. The Tech 1 scan tool may not display 5 volts. The important thing is that the ECM recognizes the voltage as more than 4 volts, indicating that the ECM and CKT 432 are OK.

Diagnostic Aids:

An intermittent open in CKT 432 or CKT 474 will result in a DTC 34. With the ignition "ON," and the engine "OFF," the manifold pressure is equal to atmospheric pressure and the signal voltage will be high. This information is used by the ECM as an indication of vehicle altitude.

Comparison of this reading with a known good vehicle with the same sensor is a good way to check accuracy of a "suspect" sensor. Readings should be the same ± .4 volt. Also, CHART C-1D can be used to test the MAP sensor.

- Check all connections.
- Disconnect sensor from bracket and twist sensor by hand (only) to check for intermittent connections. Output changes greater than .1 volt indicates a bad connector or connection. If OK, replace sensor.

NOTICE: Make sure electrical connector remains securely fastened.

- Refer to CHART C-1D, MAP sensor voltage vs. atmospheric pressure for further diagnosis.

NOTICE: After repairs use Tech 1 scan tool "Fuel Trim Reset" function to reset L. T. fuel trim to 128.

2.3L (VIN A) ENGINE — DIAGNOSTIC TROUBLE CODE CHART — 1993–94 BERETTA

DTC 34

MANIFOLD ABSOLUTE PRESSURE (MAP) SENSOR CIRCUIT
(SIGNAL VOLTAGE LOW - HIGH VACUUM)
2.3L (VIN A) "L" CARLINE

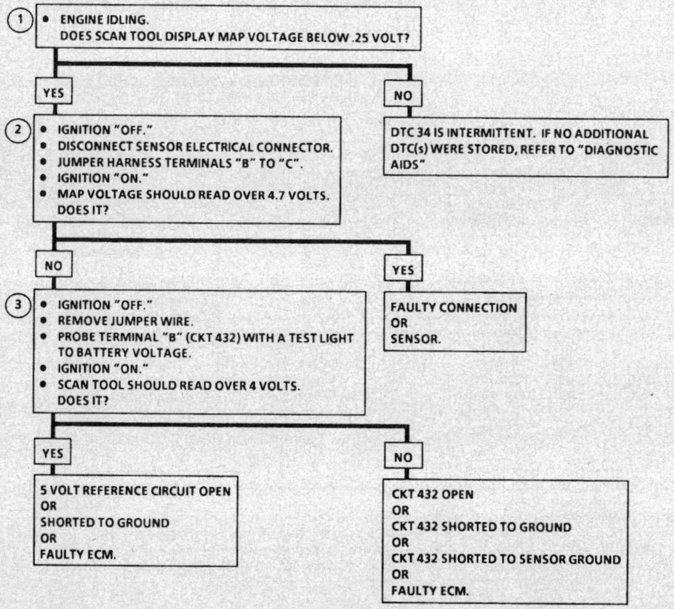

"AFTER REPAIRS," REFER TO DTC CRITERIA   AND CONFIRM DTC DOES NOT RESET.

2.3L (VIN A) ENGINE — DIAGNOSTIC TROUBLE CODE CHART — 1993–94 BERETTA

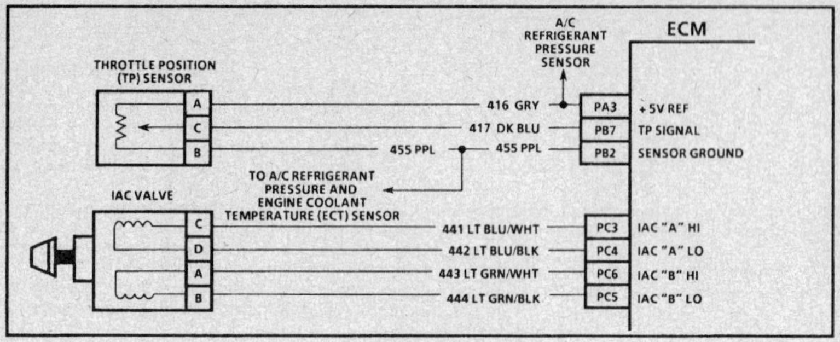

DTC 35
IDLE SPEED ERROR
2.3L (VIN A) "L" CARLINE

Circuit Description:

The Engine Control Module (ECM) controls idle speed to a calculated, "desired" RPM based on sensor inputs and actual engine RPM, determined by the time between successive 7X ignition reference pulses from the electronic Ignition Control Module (ICM). The ECM uses 4 circuits to move an Idle Air Control (IAC) valve, which allows varying amounts of air flow into the intake manifold, controlling idle speed.

DTC 35 sets when:
- Engine speed is at least 175 RPM more or 200 RPM less than "desired."
- TP sensor voltage indicates throttle is open less than 1%.
- All above conditions are continuously met for 5 seconds or more.
- IAC steps must be less than 10 or greater than 140.

Test Description: Number(s) below refer to circled number(s) on the diagnostic chart.
1. The Tech 1 scan tool is used to extend and retract the IAC valve. Valve movement is verified by an engine speed change. If no change in engine speed occurs, the valve can be retested when removed from the throttle body.
2. This step checks the quality of the IAC valve movement in Step 1. Between 900 RPM and about 1500 RPM, the engine speed should change smoothly with each flash of the tester light in both extend and retract. If the IAC valve is retracted beyond the control range (about 1500 RPM), it may take many flashes in the extend position before engine speed will begin to drop. This is normal on certain engines, fully extending IAC valve may cause engine stall. This may be normal.
Step 1 verified proper IAC valve operation while this step checks the IAC valve circuits. Each lamp on the node light should flash red and green while the IAC valve is cycled. While the sequence of color is not important if either light is "OFF" or does not flash red and green, check the circuits for faults, beginning with poor terminal contacts.

Diagnostic Aids:

Check for vacuum leaks, unconnected or brittle vacuum hoses, cuts, etc. Examine manifold and throttle body gaskets for proper seal. Check for cracked intake manifold. Check open, shorts, chafed insulation or poor connections to IAC valve in CKTs 441, 442, 443 and 444.

An open, short, or poor connection in CKTs 441, 442, 443, or 444 will result in improper idle control and may cause DTC 35.

An IAC valve which is stopped and cannot respond to the ECM, a throttle stop screw which has been tampered with, or a damaged throttle body or linkage could cause DTC 35. If no problem is found and DTC 35 resets, replace ECM.

NOTICE: If DTC 35 is currently set and speedometer is not accurately reflecting correct vehicle speed, refer to DTC 24 (vehicle speed sensor circuit) diagnostics.

* IAC DRIVER AND NODE LIGHT REQUIRED KIT 222-L FROM: CONCEPT TECHNOLOGY, INC. J 37027 FROM: KENT-MOORE, INC.

CLEAR CODES, CONFIRM "CLOSED LOOP" OPERATION, NO MIL (SERVICE ENGINE SOON) PERFORM IAC RESET PROCEDURE PER APPLICABLE SERVICE MANUAL AND VERIFY CONTROLLED IDLE SPEED IS CORRECT.

"AFTER REPAIRS," REFER TO DTC CRITERIA   AND CONFIRM DTC DOES NOT RESET.

2.3L (VIN A) ENGINE — DIAGNOSTIC TROUBLE CODE CHART — 1993–94 BERETTA

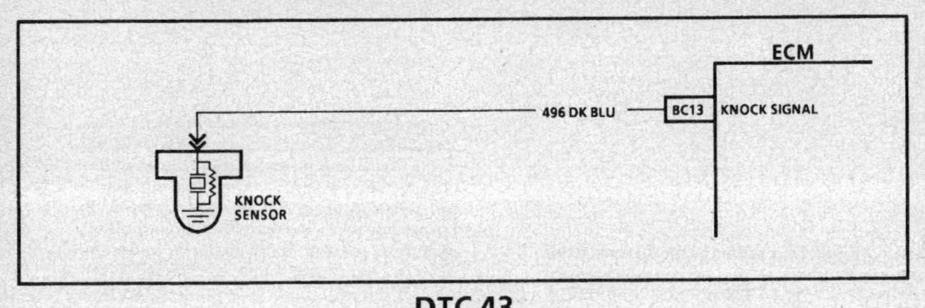

DTC 43
KNOCK SENSOR (KS) CIRCUIT
2.3L (VIN A) "L" CARLINE

Circuit Description:

The knock sensor detects engine detonation and the Engine Control Module (ECM) retards the electronic spark timing based on the signal being received. The knock sensor produces an AC signal which is used for knock detection. The amplitude and signal frequency are dependent upon the knock level.

The ECM performs two tests on this circuit to determine if it is operating correctly. If either of the tests fail, a DTC 43 will be set.

- If there is an indication of knock for more than 4.54 seconds over a 5 second interval with the engine running.
- If ECM terminal "BC13" does not indicate increasing Knock Sensor (KS) activity with engine speed.

Test Description: Number(s) below refer to circled number(s) on the diagnostic chart.

1. If the conditions for the test, as described above, are being met, DTC 43 will be currently set and the MIL (Service Engine Soon) will be illuminated.
2. The Tech 1 scan tool displays knock sensor activity in counts approximately 16-22 at idle. The counter should rise when engine speed increases or fall when engine speed decreases. This step checks knock sensor activity.
3. This step checks that the internal resistance of knock sensor is within an acceptable range.
4. If the engine has an internal problem which is creating a knock, the knock sensor may be responding to the mechanical noise.

Diagnostic Aids:

Check CKT 496 for a potential open or short to ground.

Also check for proper installation of PROM.

An "Intermittent" problem may be caused by a poor connection, rubbed through wire insulation, or a wire that is broken inside the insulation.

Any circuitry, that is suspected as causing the intermittent complaint, should be thoroughly checked for backed out terminals, improper mating, broken locks, improperly formed or damaged terminals, poor terminal to wiring connections or physical damage to the wiring harness.

Mechanical engine knock can cause a knock sensor signal. Abnormal engine noise must be corrected before using this chart.

2.3L (VIN A) ENGINE — DIAGNOSTIC TROUBLE CODE CHART — 1993–94 BERETTA

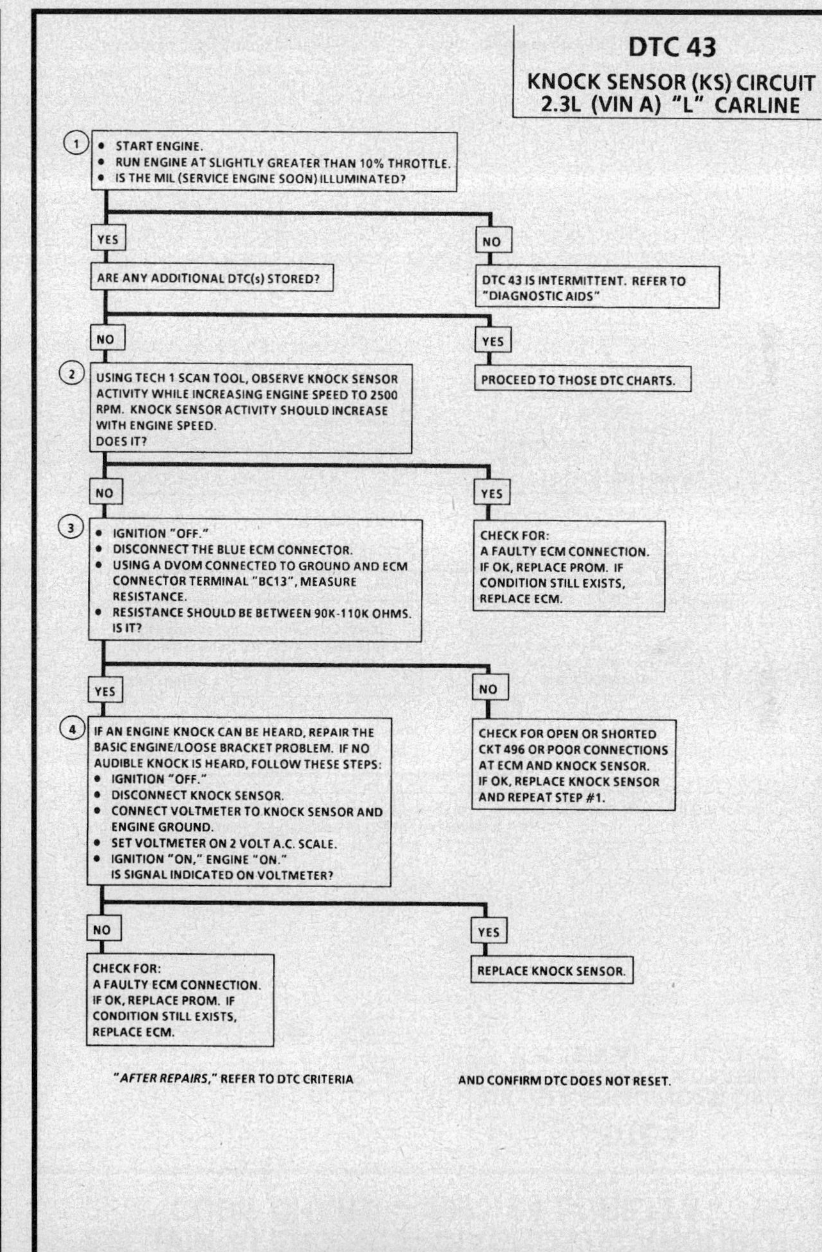

2.3L (VIN A) ENGINE — DIAGNOSTIC TROUBLE CODE CHART — 1993–94 BERETTA

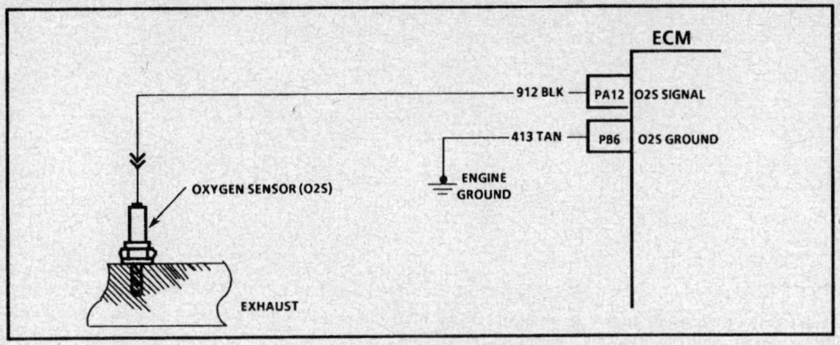

DTC 44

OXYGEN SENSOR (O2S) CIRCUIT
(LEAN EXHAUST INDICATED)
2.3L (VIN A) "L" CARLINE

Circuit Description:

The Engine Control Module (ECM) supplies a voltage of about .45 volt between terminals "PA12" and "PB6". (If measured with a 10 megohm digital voltmeter, this may read as low as .32 volt.) The Oxygen Sensor (O2S) varies the voltage within a range of about 1 volt if the exhaust is rich, down through about .10 volt if exhaust is lean.

The sensor is like an open circuit and produces no voltage when it is below 315°C (600°F). An open sensor circuit, or cold sensor causes "Open Loop" operation.

Test Description: Number(s) below refers to circled number(s) on the diagnostic chart.

1. DTC 44 is set when the O2S signal voltage on CKT 912:
   - Remains below .3 volt for 90 seconds or more.
   - The system is operating in "Closed Loop."
   - No DTC 33 or 34.
   - "Closed Loop" integrator active.
   - TP sensor above 5%.

Diagnostic Aids:

The DTC 44 or lean exhaust is most likely caused by one of the following:

- Fuel Pressure. System will be lean if pressure is too low. It may be necessary to monitor fuel pressure while driving the vehicle at various road speeds and/or loads to confirm. Refer to CHART A-7.
- MAP Sensor. An output that causes the ECM to sense a lower than normal manifold pressure (high vacuum) can cause the system to go lean. Disconnecting the MAP sensor will allow the ECM to substitute a fixed (default) value for the MAP sensor. If the rich condition is gone when the sensor is disconnected, substitute a known good sensor and recheck.
- Fuel Contamination. Water, even in small amounts, near the in-tank fuel pump inlet can be delivered to the injector. The water causes a lean exhaust and can set a DTC 44.
- Sensor Harness. Sensor pigtail may be mispositioned and contacting the exhaust manifold.
- Engine Misfire. A cylinder misfire will result in unburned oxygen in the exhaust, which could cause DTC 44. Refer to CHART C4-M
- Cracked Oxygen Sensor (O2S). A cracked O2S, or poor ground at the sensor, could cause DTC 44.
- Plugged Fuel Filter. A plugged fuel filter can cause a lean condition, and can cause a DTC 44 to set.
- Plugged Oxygen Sensor (O2S). A plugged reference port on the O2S will indicate a lower than normal voltage output from the O2S.

2.3L (VIN A) ENGINE — DIAGNOSTIC TROUBLE CODE CHART — 1993–94 BERETTA

DTC 44

OXYGEN SENSOR (O2S) CIRCUIT
(LEAN EXHAUST INDICATED)
2.3L (VIN A) "L" CARLINE

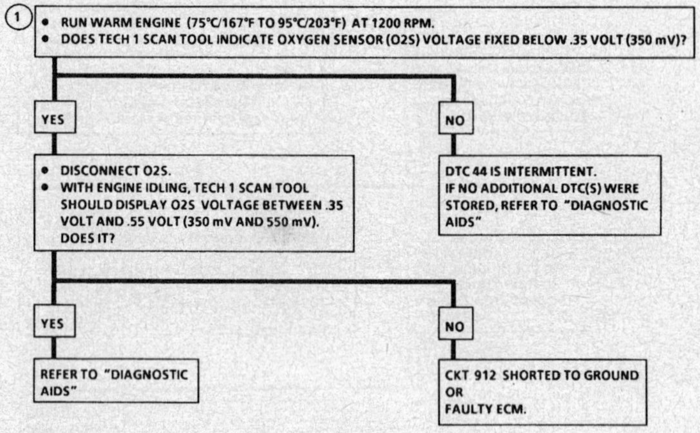

"AFTER REPAIRS," REFER TO DTC CRITERIA AND CONFIRM DTC DOES NOT RESET.

2.3L (VIN A) ENGINE — DIAGNOSTIC TROUBLE CODE CHART — 1993–94 BERETTA

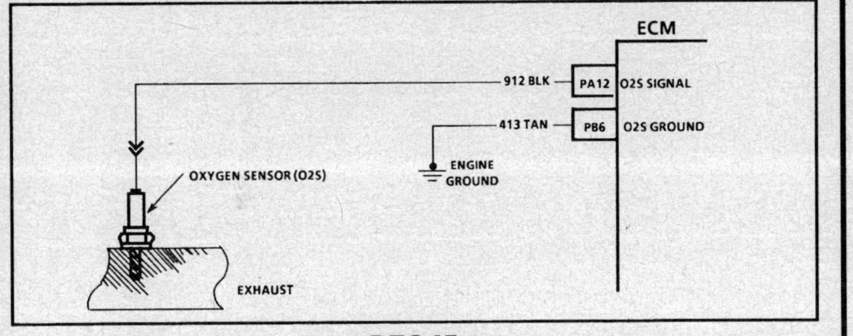

DTC 45
**OXYGEN SENSOR (O2S) CIRCUIT
(RICH EXHAUST INDICATED)
2.3L (VIN A) "L" CARLINE**

Circuit Description:

The Engine Control Module (ECM) supplies a voltage of about .45 volt between terminals "PA12" and "PB6". (If measured with a 10 megohm digital voltmeter, this may read as low as .32 volt.) The Oxygen Sensor (O2S) varies the voltage within a range of about 1 volt if the exhaust is rich, down through about .10 volt if exhaust is lean.

The sensor is like an open circuit and produces no voltage when it is below 315°C (600°F). An open sensor circuit, or cold sensor causes "Open Loop" operation.

Test Description: Number(s) below refer to circled number(s) on the diagnostic chart.
1. DTC 45 is set when:
   - O2S voltage is above .75 volt.
   - No DTC 33 or 34.
   - Fuel system in "Closed Loop."
   - TP sensor above 5%.
   - Above conditions met for 30 seconds.
   OR
   - Voltage above 1 volt for 5 seconds.

Diagnostic Aids:

The DTC 45 or rich exhaust is most likely caused by one of the following:
- **Fuel Pressure.** System will go rich if pressure is too high. The ECM can compensate for some increase. However, if it gets too high, a DTC 45 will be set. Refer to "Fuel System" diagnosis CHART A-7.
- **Leaking Injector.** Refer to CHART A-7.
- **Electronic Ignition (EI) Shielding.** An open ground CKT 453 may result in EMI or induced electrical noise. The ECM looks at this noise as Crankshaft Position (CKP) sensor pulses. The additional pulses result in a higher than actual engine speed signal.

The ECM then delivers too much fuel causing system to go rich. Engine tachometer will also show higher than actual engine speed which can help in diagnosing this problem.
- **EVAP Canister Purge.** Check for fuel saturation. If full of fuel, check canister control and hoses. See "Evaporative Emission (EVAP) Control System,"
- **MAP Sensor.** An output that causes the ECM to sense a higher than normal manifold pressure (low vacuum) can cause the system to go rich. Disconnecting the MAP sensor will allow the ECM to set a fixed value for the MAP sensor. Substitute a different MAP sensor if the rich condition is gone while the sensor is disconnected.
- **Pressure Regulator.** Check for leaking fuel pressure regulator diaphragm by checking for the presence of liquid fuel in the vacuum line to the regulator.
- **TP Sensor.** An intermittent TP sensor output will cause the system to go rich due to a false indication of the engine accelerating.
- **O2S Contamination.** Inspect O2S for silicone contamination from fuel or use of improper RTV sealant. The sensor may have a white powdery coating and result in a false signal voltage (rich exhaust indication. The ECM will then reduce the amount of fuel delivered to the engine causing a severe surge driveability problem.

2.3L (VIN A) ENGINE — DIAGNOSTIC TROUBLE CODE CHART — 1993–94 BERETTA

DTC 45
**OXYGEN SENSOR (O2S) CIRCUIT
(RICH EXHAUST INDICATED)
2.3L (VIN A) "L" CARLINE**

(1)
- RUN WARM ENGINE (75°C/167°F TO 95°C/203°F) AT 1200 RPM.
- DOES TECH 1 SCAN TOOL DISPLAY OXYGEN SENSOR (O2S) VOLTAGE FIXED ABOVE .75 VOLT (750 mV)?

YES
- DISCONNECT O2S AND JUMPER HARNESS CKT 912 TO GROUND.
- TECH 1 SCAN TOOL SHOULD DISPLAY OXYGEN SENSOR VOLTAGE BELOW .35 VOLT (350 mV). DOES IT?

NO
DTC 45 IS INTERMITTENT. IF NO ADDITIONAL DTC(S) WERE STORED, REFER TO "DIAGNOSTIC AIDS".

YES
REFER TO "DIAGNOSTIC AIDS"

NO
REPLACE ECM.

"AFTER REPAIRS," REFER TO DTC CRITERIA   AND CONFIRM DTC DOES NOT RESET.

2.3L (VIN A) ENGINE — DIAGNOSTIC TROUBLE CODE CHART — 1993–94 BERETTA

DTC 51

PROM ERROR
(FAULTY OR INCORRECT PROM)
2.3L (VIN A) "L" CARLINE

CHECK THAT ALL PINS ARE FULLY INSERTED IN THE SOCKET AND THAT PROM IS PROPERLY LATCHED. IF OK, REPLACE PROM, CLEAR MEMORY, AND RECHECK. IF DTC 51 REAPPEARS, REPLACE ECM.

NOTICE: TO PREVENT POSSIBLE ELECTROSTATIC DISCHARGE DAMAGE TO THE ECM OR PROM, DO NOT TOUCH THE COMPONENT LEADS, AND DO NOT REMOVE THE PROM COVER OR THE INTEGRATED CIRCUIT FROM CARRIER.

"AFTER REPAIRS," CONFIRM "CLOSED LOOP" OPERATION AND NO MIL (SERVICE ENGINE SOON).

2.3L (VIN A) ENGINE — DIAGNOSTIC TROUBLE CODE CHART — 1993–94 BERETTA

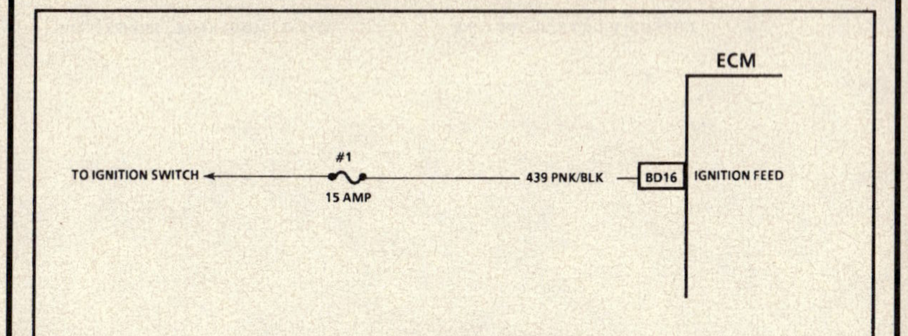

DTC 53

BATTERY VOLTAGE ERROR
2.3L (VIN A) "L" CARLINE

Circuit Description:
DTC 53 will set when the ignition is "ON" and Engine Control Module (ECM) terminal "BD16" voltage is more than 17.1 volts for about .2 second, or less than 10 volts for 240 seconds.

During the time the failure is present, all ECM outputs will be disengaged. (The setting of additional diagnostic trouble codes may result.)

Test Description: Number(s) below refer to circled number(s) on the diagnostic chart.
1. Normal battery output is between 10 - 17.0 volts.
2. Checks to see if generator is faulty under load condition.

Note On Intermittents:

Charging battery with a battery charger and starting engine, may set DTC 53. If DTC sets when an accessory is operated, check for poor connections or excessive current draw.

Also, check for poor connections at starter solenoid or fusible link junction box.

2.3L (VIN A) ENGINE — DIAGNOSTIC TROUBLE CODE CHART — 1993–94 BERETTA

DTC 53
BATTERY VOLTAGE ERROR
2.3L (VIN A) "L" CARLINE

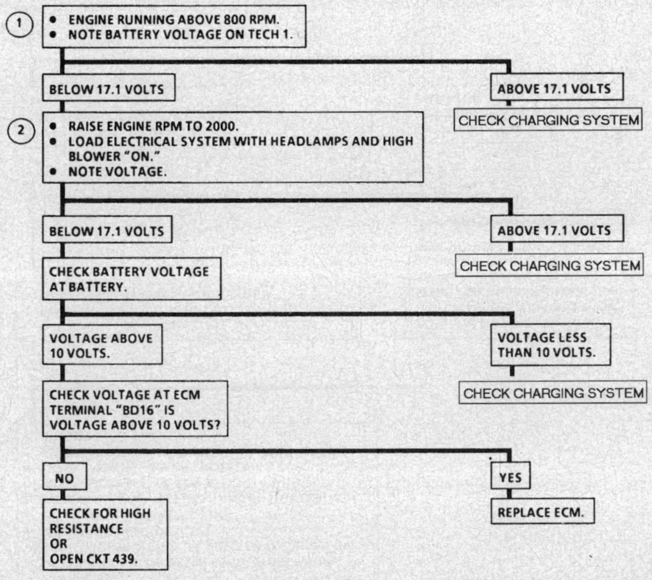

① • ENGINE RUNNING ABOVE 800 RPM.
• NOTE BATTERY VOLTAGE ON TECH 1.

BELOW 17.1 VOLTS — ABOVE 17.1 VOLTS → CHECK CHARGING SYSTEM

② • RAISE ENGINE RPM TO 2000.
• LOAD ELECTRICAL SYSTEM WITH HEADLAMPS AND HIGH BLOWER "ON."
• NOTE VOLTAGE.

BELOW 17.1 VOLTS — ABOVE 17.1 VOLTS → CHECK CHARGING SYSTEM

CHECK BATTERY VOLTAGE AT BATTERY.

VOLTAGE ABOVE 10 VOLTS. — VOLTAGE LESS THAN 10 VOLTS. → CHECK CHARGING SYSTEM

CHECK VOLTAGE AT ECM TERMINAL "BD16" IS VOLTAGE ABOVE 10 VOLTS?

NO → CHECK FOR HIGH RESISTANCE OR OPEN CKT 439.

YES → REPLACE ECM.

"AFTER REPAIRS," REFER TO DTC CRITERIA   AND CONFIRM DTC DOES NOT RESET.

2.3L (VIN A) ENGINE — DIAGNOSTIC TROUBLE CODE CHART — 1993–94 BERETTA

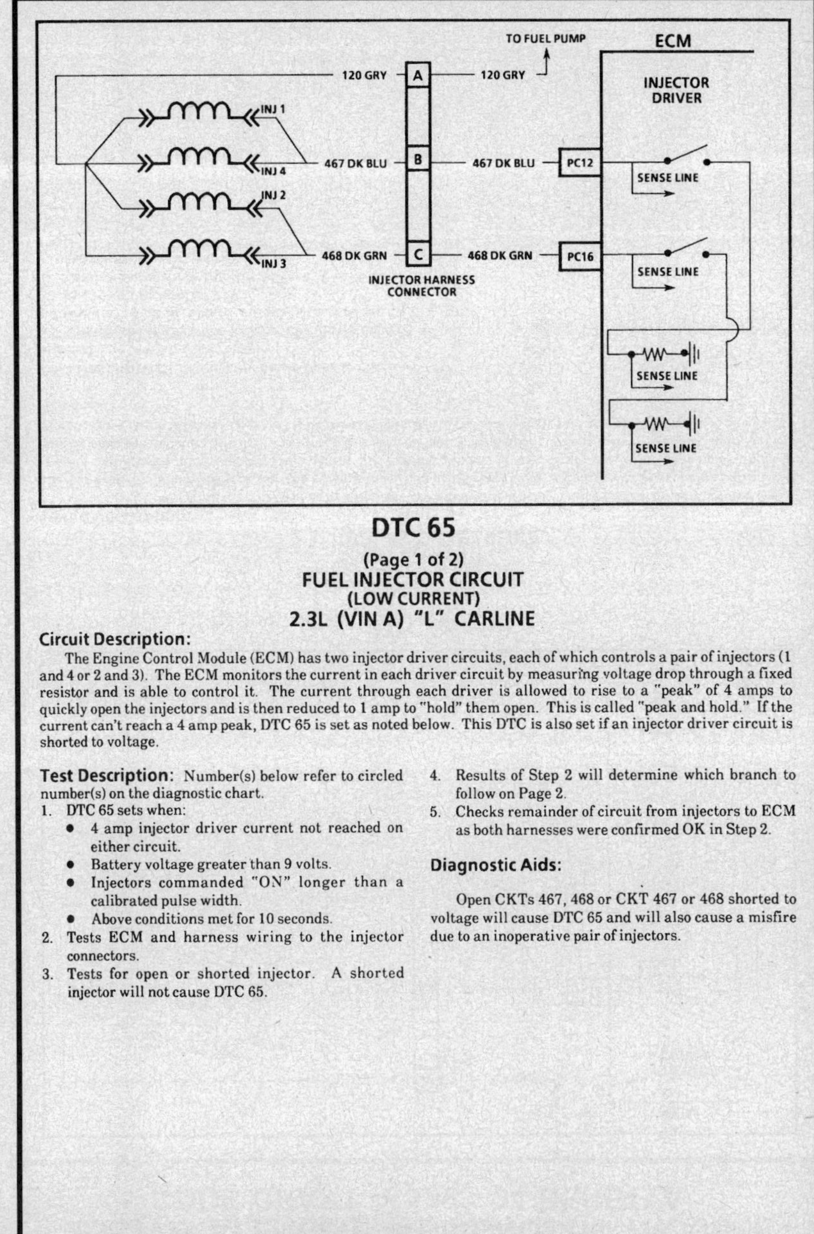

DTC 65
(Page 1 of 2)
FUEL INJECTOR CIRCUIT
(LOW CURRENT)
2.3L (VIN A) "L" CARLINE

Circuit Description:

The Engine Control Module (ECM) has two injector driver circuits, each of which controls a pair of injectors (1 and 4 or 2 and 3). The ECM monitors the current in each driver circuit by measuring voltage drop through a fixed resistor and is able to control it. The current through each driver is allowed to rise to a "peak" of 4 amps to quickly open the injectors and is then reduced to 1 amp to "hold" them open. This is called "peak and hold." If the current can't reach a 4 amp peak, DTC 65 is set as noted below. This DTC is also set if an injector driver circuit is shorted to voltage.

Test Description: Number(s) below refer to circled number(s) on the diagnostic chart.

1. DTC 65 sets when:
   • 4 amp injector driver current not reached on either circuit.
   • Battery voltage greater than 9 volts.
   • Injectors commanded "ON" longer than a calibrated pulse width.
   • Above conditions met for 10 seconds.
2. Tests ECM and harness wiring to the injector connectors.
3. Tests for open or shorted injector. A shorted injector will not cause DTC 65.

4. Results of Step 2 will determine which branch to follow on Page 2.
5. Checks remainder of circuit from injectors to ECM as both harnesses were confirmed OK in Step 2.

Diagnostic Aids:

Open CKTs 467, 468 or CKT 467 or 468 shorted to voltage will cause DTC 65 and will also cause a misfire due to an inoperative pair of injectors.

2.3L (VIN A) ENGINE — DIAGNOSTIC TROUBLE CODE CHART — 1993–94 BERETTA

2.3L (VIN A) ENGINE — DIAGNOSTIC TROUBLE CODE CHART — 1993–94 BERETTA

DTC 65
(Page 1 of 2)
FUEL INJECTOR CIRCUIT
(LOW CURRENT)
2.3L (VIN A) "L" CARLINE

NOTICE: IF ENGINE "CRANKS BUT WILL NOT RUN," DO NOT USE THIS CHART. REFER TO CHART A-3.

1
- IDLE ENGINE FOR 1 MINUTE.
- DOES TECH 1 SCAN TOOL INDICATE DTC 65?

NO → DTC 65 IS INTERMITTENT REFER TO "DIAGNOSTIC AIDS."

YES

2
- ENGINE "OFF."
- REMOVE CRANKCASE VENTILATION OIL/AIR SEPARATOR FOR ACCESS.
- DISCONNECT ALL 4 INJECTOR CONNECTORS FROM INJECTORS AND INSTALL AN INJECTOR TEST LIGHT ON INJECTOR CONNECTOR 1 OR 4.
- CRANK ENGINE AND NOTE LIGHT.
- REMOVE INJECTOR TEST LIGHT AND INSTALL LIGHT ON INJECTOR CONNECTOR 2 OR 3.
- CRANK ENGINE AGAIN AND NOTE LIGHT. INJECTOR TEST LIGHT SHOULD BLINK ON BOTH TESTS. DOES IT?

YES

3
- WITH DVOM ON 200 OHM SCALE, MEASURE THE RESISTANCE OF EACH INJECTOR. RESISTANCE SHOULD BE LESS THAN 3 OHMS (BUT NOT ZERO). IS IT?

NO →
4
NOTE WHETHER INJECTOR TEST LIGHT WAS "OFF" ON ONE OR "ON" STEADY ON ONE IN PREVIOUS STEP AND REFER TO PAGE 2 OF THIS CHART.

→ SEE DTC 65 PAGE 2 OF 2.

YES

5
CHECK FOR POOR CONNECTIONS OR CRIMPS AT INJECTOR CONNECTORS. ARE CONNECTIONS OK?

NO → REPLACE INJECTOR WITH HIGH (OR ZERO) RESISTANCE.

YES

REPLACE ECM.

NO → REPAIR CONNECTIONS.

"AFTER REPAIRS," REFER TO DTC CRITERIA   AND CONFIRM DTC DOES NOT RESET.

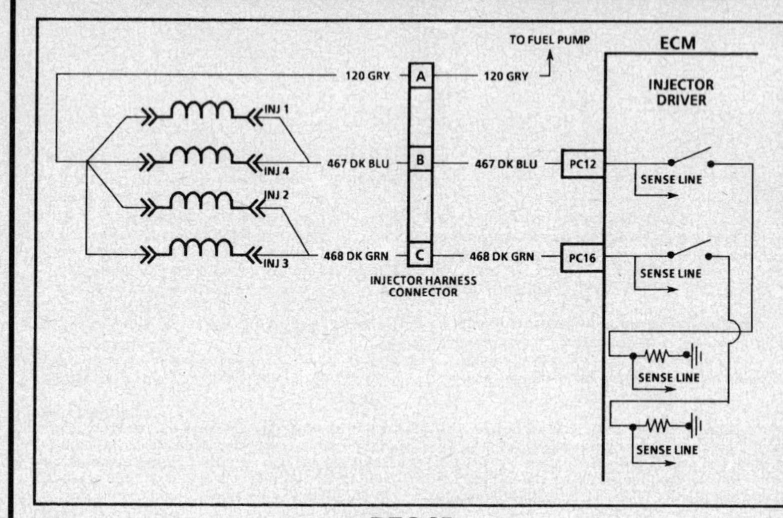

DTC 65
(Page 2 of 2)
FUEL INJECTOR CIRCUIT
(LOW CURRENT)
2.3L (VIN A) "L" CARLINE

Circuit Description:

The Engine Control Module (ECM) has two injector driver circuits, each of which controls a pair of injectors (1 and 4 or 2 and 3). The ECM monitors the current in each driver circuit by measuring voltage drop through a fixed resistor and is able to control it. The current through each driver is allowed to rise to a "peak" of 4 amps to quickly open the injectors and is then reduced to 1 amp to "hold" them open. This is called "peak and hold." If the current can't reach a 4 amp peak, DTC 65 is set. This DTC is also set if an injector driver circuit is shorted to voltage.

Test Description: Number(s) below refer to circled number(s) on the diagnostic chart.

6. This checks for short to voltage in injector driver circuits. It is necessary to crank the engine to assure voltage to CKT 120.
7. Determines whether injector driver CKTs 467 and 468 are shorted to ground.

2.3L (VIN A) ENGINE — DIAGNOSTIC TROUBLE CODE CHART — 1993–94 BERETTA

DTC 65

(Page 2 of 2)
FUEL INJECTOR CIRCUIT
(LOW CURRENT)
2.3L (VIN A) "L" CARLINE

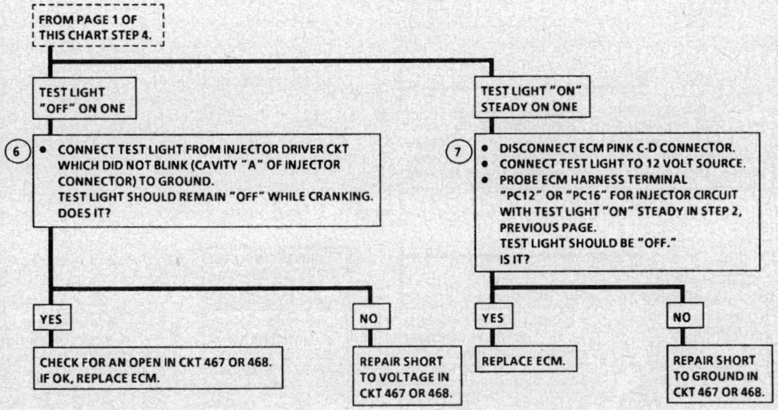

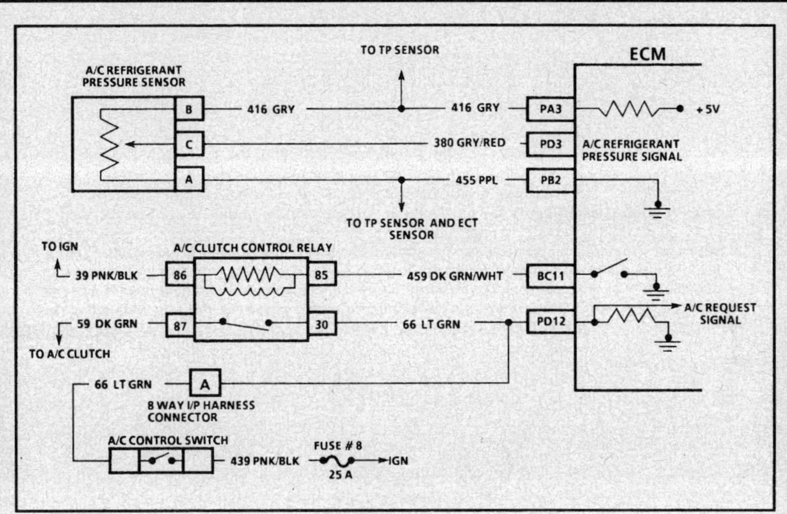

2.3L (VIN A) ENGINE — DIAGNOSTIC TROUBLE CODE CHART — 1993–94 BERETTA

DTC 66

A/C REFRIGERANT PRESSURE SENSOR CIRCUIT
2.3L (VIN A) "L" CARLINE

Circuit Description:

The A/C refrigerant pressure sensor responds to changes in A/C refrigerant system high side pressure. This input indicates how much load the A/C compressor is putting on the engine and is one of the factors used by the Engine Control Module (ECM) to determine IAC valve position for idle speed control. The circuit consists of a 5 volts reference and a ground, both provided by the ECM, and a signal line to the ECM. The signal is a voltage which is proportional to the pressure. The sensor's range of operation is 0 to 450 psi. At 0 psi, the signal will be about .1 volt, varying up to about 4.9 volts at 450 psi or above. DTC 66 sets if the voltage is above 4.9 volts or below .28 volt, 15 seconds or more or voltage is above 4 volts and A/C is not requested. The A/C compressor is disabled by the ECM if DTC 66 is present, or if refrigerant pressure is above or below calibrated values

Test Description: Number(s) below refer to circled number(s) on the diagnostic chart.

1. This step checks the voltage signal being received by the ECM from the A/C refrigerant pressure sensor. The normal operating range is between .28 volt and 4.9 volts.
2. Checks to see if the high voltage signal is from a shorted sensor or a short to voltage in the circuit. Normally, disconnecting the sensor would make a normal circuit go to near zero volt.
3. Checks to see if low voltage signal is from the sensor or the circuit. Jumpering the sensor signal CKT 380 to 5 volts, checks the circuit, connections, and ECM.
4. This step checks to see if the low voltage signal was due to an open in the sensor circuit or the 5 volts reference circuit since the prior step eliminated the refrigerant pressure sensor.

Diagnostic Aids:

DTC 66 sets when signal voltage falls outside the normal possible range of the sensor and is not due to a refrigerant system problem. If problem is intermittent, check for opens or shorts in harness or poor connections. If OK, replace A/C refrigerant pressure sensor. If DTC 66 re-sets, replace ECM.

2.3L (VIN A) ENGINE — DIAGNOSTIC TROUBLE CODE CHART — 1993–94 BERETTA

2.3L (VIN A) ENGINE — EXHAUST SYSTEM DIAGNOSTIC CHART — 1993–94 BERETTA

DTC 66
A/C REFRIGERANT PRESSURE SENSOR CIRCUIT
2.3L (VIN A) "L" CARLINE

NOTICE: IF VEHICLE IS NOT EQUIPPED WITH AIR CONDITIONING, DO NOT USE THIS CHART, SEE "DIAGNOSTIC AIDS."

1. • KEY "ON," ENGINE NOT RUNNING.
   • A/C NOT REQUESTED.
   • NOTE TECH 1 SCAN TOOL VOLTAGE FOR A/C REFRIGERANT PRESSURE SENSOR.

ABOVE 3.9 VOLTS

BELOW .28 VOLT

BETWEEN 2.8 VOLTS AND 3.9 VOLTS

2. • DISCONNECT A/C REFRIGERANT PRESSURE SENSOR.
   • DOES SCAN TOOL DISPLAY LESS THAN 1 VOLT?

3. • DISCONNECT A/C REFRIGERANT PRESSURE SENSOR CONNECTOR.
   • JUMPER TERMINALS "B" AND "C".
   • DOES SCAN TOOL DISPLAY ABOVE 4.6 VOLTS?

FAULT IS NOT PRESENT AT THIS TIME. SEE "DIAGNOSTIC AIDS."

NO → CHECK FOR SHORT TO VOLTAGE IN CKT 455. IF NOT SHORTED, REPLACE ECM.

YES → CHECK FOR OPEN IN CKT 455. IF NOT OPEN, CHECK FOR POOR SENSOR TERMINAL CONNECTIONS. IF OK, REPLACE A/C REFRIGERANT PRESSURE SENSOR.

NO →
4. • REMOVE JUMPER.
   • CONNECT VOLTMETER FROM TERMINAL "A" TO "B".
   • IS VOLTAGE ABOUT 5 VOLTS?

YES → • CHECK SENSOR TERMINAL CONNECTIONS IF OK, REPLACE A/C REFRIGERANT PRESSURE SENSOR.

NO → • BACK PROBE ECM TERMINAL "PA3" WITH VOLTMETER TO GROUND. IS VOLTAGE ABOUT 5 VOLTS?

YES → CHECK FOR OPEN IN CKT 380 CHECK FOR POOR CONNECTION ECM TERMINAL "PD3". IF OK, REPLACE ECM.

NO → CHECK FOR POOR CONNECTION AT ECM TERMINAL "PA3" OR SHORT TO GROUND IN CKT 416. IF OK, ECM IS FAULTY.

YES → REPAIR OPEN IN CKT 416.

"AFTER REPAIRS," REFER TO DTC CRITERIA   AND CONFIRM DTC DOES NOT RESET.

CHART B-1
RESTRICTED EXHAUST SYSTEM CHECK
2.3L (VIN A) "L" CARLINE

Proper diagnosis for a restricted exhaust system is essential before any components are replaced. The following procedure may be used for diagnosis.

CHECK AT OXYGEN SENSOR (O2S):

1. Carefully remove O2S.
2. Install Borroughs exhaust backpressure tester (BT-8515 or BT-8603) in place of O2S (see illustration).
3. After completing test described below, be sure to coat threads of O2S with anti-seize compound P/N 5613695 prior to re-installation.

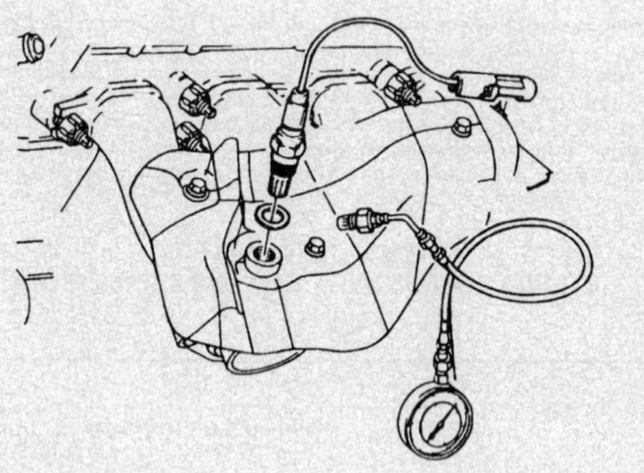

DIAGNOSIS:

1. With the engine idling at normal operating temperature, transaxle in park or neutral, observe the exhaust system backpressure reading on the gauge. Reading should not exceed 3.4 kPa (.5 psi).
2. Increase engine speed to 3000 RPM and observe gauge. Reading should not exceed 5 kPa (.75 psi).
3. If the backpressure at either speed exceeds specification, a restricted exhaust system is indicated.
4. Inspect the entire exhaust system for a collapsed pipe, heat distress, or possible internal muffler failure.
5. If there are no obvious reasons for the excessive backpressure, the catalytic converter is suspected to be restricted and should be replaced using current recommended procedures.

2.3L (VIN A) ENGINE — ECM SYMPTOM CHART — 1993–94 BERETTA

PIN FUNCTION	CKT #	WIRE COLOR	COMPONENT CONNECTOR CAVITY	NORMAL VOLTAGES KEY "ON"	NORMAL VOLTAGES ENG RUN	DTC(S) AFFECTED	POSSIBLE SYMPTOMS FROM FAULTY CIRCUIT
ECM CONNECTOR "PC"							
PC1							
PC2							
PC3 IAC "A" HIGH	441	LT BLU/WHT	IAC TERM "C"	0 OR B+	0 OR B+	35 (10)	INCORRECT IDLE SURGES
PC4 IAC "A" LOW	442	LT BLU/BLK	IAC TERM "D"	0 OR B+	0 OR B+	35 (10)	INCORRECT IDLE SURGES
PC5 IAC "B" LOW	444	LT GRN/BLK	IAC TERM "B"	0 OR B+	0 OR B+	35 (10)	INCORRECT IDLE SURGES
PC6 IAC "B" HIGH	443	LT GRN/WHT	IAC TERM "A"	0 OR B+	0 OR B+	35 (10)	INCORRECT IDLE SURGE
PC7							
PC8							
PC9 IAT SIGNAL	472	TAN	IAT TERM "A"	3.6	1.5 (3)	23 (8) 25 (9)	
PC10 ECT SIGNAL	410	YEL	ECT TERM "B"	1.2 (5)	2.0 (5)	14 (9) 15 (8)	LACK OF PERFORMANCE
PC11							
PC12 INJECTOR DRIVERS 1 & 4	467	DK BLU	INJECTOR CONNECTOR "B"	B+ (2)	B+	65 (10)	ROUGH IDLE, HARD TO START LACK OF PERFORMANCE
PC13 ECM GROUND	450	BLK/WHT	ENGINE GROUND	0*	0*		
PC14							
PC15 IGNITION CONTROL (IC) 2	485	BLK	ELECTRONIC IGNITION CONTROL MODULE (ICM) TERM "B"	0*	.5 (3)		ROUGH IDLE, HARD TO START LACK OF PERFORMANCE
PC16 INJECTOR DRIVERS 2 & 3	468	DK GRN	INJECTOR CONNECTOR "C"	B+ (2)	B+	65 (10)	ROUGH IDLE, HARD TO START LACK OF PERFORMANCE

(1) VARIES FROM .60 TO BATTERY VOLTAGE, DEPENDING ON POSITION OF DRIVE WHEELS.
(2) BATTERY VOLTAGE FOR FIRST TWO SECONDS.
(3) VARIES.
(4) BATTERY VOLTAGE WHEN FUEL PUMP IS RUNNING.
(5) VARIES WITH TEMPERATURE.
(6) READS BATTERY VOLTAGE IN GEAR.
(7) BATTERY VOLTAGE WHEN ENGINE IS CRANKING.
(8) OPEN CIRCUIT.
(9) GROUNDED CIRCUIT.
(10) OPEN/GROUNDED CIRCUIT.
(11) LESS THAN 1 VOLT.
* LESS THAN .5 VOLT (500 mV).

2.3L (VIN A) ENGINE — ECM SYMPTOM CHART — 1993–94 BERETTA

PIN FUNCTION	CKT #	WIRE COLOR	COMPONENT CONNECTOR CAVITY	NORMAL VOLTAGES KEY "ON"	NORMAL VOLTAGES ENG RUN	DTC(S) AFFECTED	POSSIBLE SYMPTOMS FROM FAULTY CIRCUIT
ECM CONNECTOR "PD"							
PD1 IGN "LOW" REF	453	BLK/RED	ELECTRONIC IGNITION CONTROL MODULE (ICM) TERM "H"	0*	0*		
PD2 7X REF	483	LT GRN	ELECTRONIC IGNITION CONTROL MODULE (ICM) TERM "G"	0*	.6 (3)	19	VEHICLE STALLS
PD3 A/C REFRIGERANT PRESSURE SIGNAL	380	GRY/RED	A/C REFRIGERANT PRESSURE SENSOR TERM "C"	.8 (3)	.9 (3)	66 (10)	A/C CLUTCH INOPERATIVE
PD4 MAP SIGNAL	432	LT GRN	MAP SENSOR "B"	3.0	1.3 (3)	34 (10) 33 (10)	LACK OF PERFORMANCE, ROUGH IDLE, SURGE
PD5							
PD6							
PD7							
PD8							
PD9							
PD10							
PD11							
PD12 A/C REQUEST	66	LT GRN	8-WAY I/P HARNESS TERM "A"	0* B+	0* B+		INOPERATIVE A/C INCORRECT IDLE
PD13							
PD14							
PD15							
PD16							

(1) VARIES FROM .60 TO BATTERY VOLTAGE, DEPENDING ON POSITION OF DRIVE WHEELS.
(2) BATTERY VOLTAGE FOR FIRST TWO SECONDS.
(3) VARIES.
(4) BATTERY VOLTAGE WHEN FUEL PUMP IS RUNNING.
(5) VARIES WITH TEMPERATURE.
(6) READS BATTERY VOLTAGE IN GEAR.
(7) BATTERY VOLTAGE WHEN ENGINE IS CRANKING.
(8) OPEN CIRCUIT.
(9) GROUNDED CIRCUIT.
(10) OPEN/GROUNDED CIRCUIT.
(11) LESS THAN 1 VOLT.
* LESS THAN .5 VOLT (500 mV).

2.3L (VIN A) ENGINE — ECM SYMPTOM CHART — 1993–94 BERETTA

ECM CONNECTOR "PA"

PIN FUNCTION	CKT #	WIRE COLOR	COMPONENT CONNECTOR CAVITY	NORMAL VOLTAGES		DTC(S) AFFECTED	POSSIBLE SYMPTOMS FROM FAULTY CIRCUIT
				KEY "ON"	ENG RUN		
PA1 ENGINE GROUND	551	TAN/WHT	ENGINE	0*	0*		
PA2 ECM GROUND	450	BLK/WHT	ENGINE	0*	0*		
PA3 TP, A/C REFRIGERANT PRESSURE 5V REF	416	GRY	TP SENSOR TERM "A" A/C REFRIGERANT PRESS TERM "B"	5	5	22 (10) 66 (10)	HIGH IDLE
PA4 5V MAP REF	474	GRY	MAP SENSOR TERM "C"	5	5	34 (10)	LACK OF POWER ROUGH IDLE SURGE
PA5 BATTERY FEED	340	ORN	FUEL PUMP RELAY TERMINAL "87" AND OIL PRESSURE TERM "C"	B+	B+		CRANKS BUT WILL NOT START
PA6							
PA7							
PA8 FUEL PUMP RELAY	465	DK GRY/WHT	FUEL PUMP RELAY TERM "86"	B+ (4)	B+ (4)		LONG CRANKING TIME BEFORE ENGINE STARTS
PA9							
PA10							
PA11							
PA12 O2S SIGNAL	912	BLK	O2S	.35-.55	.1-.9 (3)	13 (8)	EXHAUST ODOR POOR PERFORMANCE

(1) VARIES FROM .60 TO BATTERY VOLTAGE, DEPENDING ON POSITION OF DRIVE WHEELS.
(2) BATTERY VOLTAGE FOR FIRST TWO SECONDS.
(3) VARIES.
(4) BATTERY VOLTAGE WHEN FUEL PUMP IS RUNNING.
(5) VARIES WITH TEMPERATURE.
(6) READS BATTERY VOLTAGE IN GEAR.
(7) BATTERY VOLTAGE WHEN ENGINE IS CRANKING.
(8) OPEN CIRCUIT.
(9) GROUNDED CIRCUIT.
(10) OPEN/GROUNDED CIRCUIT.
(11) LESS THAN 1 VOLT.
* LESS THAN .5 VOLT (500 mV).

2.3L (VIN A) ENGINE — ECM SYMPTOM CHART — 1993–94 BERETTA

ECM CONNECTOR "PB"

PIN FUNCTION	CKT #	WIRE COLOR	COMPONENT CONNECTOR CAVITY	NORMAL VOLTAGES		DTC(S) AFFECTED	POSSIBLE SYMPTOMS FROM FAULTY CIRCUIT
				KEY "ON"	ENG RUN		
PB1 MAP AND IAT SENSOR GROUND	452	BLK	MAP TERM "A" IAT SENSOR TERM "B"	0*	0*	23 (8) 33 (8)	STALLING AT IDLE AND RUNS ROUGH
PB2 TP, ECT, AND A/C SENSOR GROUND	455	PPL	TP TERM "B" A/C REFRIGERANT PRESS SENSOR TERM "A" ECT SENSOR TERM "A"	0*	0*	15 (8) 21 (8) 56 (8)	ROUGH IDLE LACK OF PERFORMANCE EXHAUST ODOR
PB3 IGNITION CONTROL (IC) 1	423	WHT	ELECTRONIC IGNITION CONTROL MODULE (ICM) TERM "A"	0*	.5 (3)		LACK OF POWER, STALLS, SURGES
PB4							
PB5							
PB6 O2S GROUND	413	TAN	ON TRANSAXLE BOLT	0*	0*	13 (8)	EXHAUST ODOR POOR PERFORMANCE
PB7 TP SIGNAL	417	DK BLU	TP TERM "C"	.49	.49	22 (10)	LACK OF PERFORMANCE
PB8 4000 P/MI SPEED	817	DK GRY/WHT	I/P TERM "G"	8.3 (1)	9.4 (1)		INOPERATIVE SPEEDOMETER ODOMETER
PB9							
PB10							
PB11 TACH	627	WHT	TO TACH	4.9	1.0		NO TACH
PB12							

(1) VARIES FROM .60 TO BATTERY VOLTAGE, DEPENDING ON POSITION OF DRIVE WHEELS.
(2) BATTERY VOLTAGE FOR FIRST TWO SECONDS.
(3) VARIES.
(4) BATTERY VOLTAGE WHEN FUEL PUMP IS RUNNING.
(5) VARIES WITH TEMPERATURE.
(6) READS BATTERY VOLTAGE IN GEAR.
(7) BATTERY VOLTAGE WHEN ENGINE IS CRANKING.
(8) OPEN CIRCUIT.
(9) GROUNDED CIRCUIT.
(10) OPEN/GROUNDED CIRCUIT.
(11) LESS THAN 1 VOLT.
* LESS THAN .5 VOLT (500 mV).

2.3L (VIN A) ENGINE — ECM SYMPTOM CHART — 1993–94 BERETTA

ECM CONNECTOR "BC"

PIN FUNCTION	CKT #	WIRE COLOR	COMPONENT CONNECTOR CAVITY	NORMAL VOLTAGES KEY "ON"	NORMAL VOLTAGES ENG RUN	DTC(S) AFFECTED	POSSIBLE SYMPTOMS FROM FAULTY CIRCUIT
BC1 EVAP DRIVER	428	DK GRN/YEL	EVAP "B"	B+	3	27(10)	
BC2 SHIFT LIGHT	456	TAN/BLK	M/T SHIFT LIGHT IN I/P	B+	"OFF" B+ ON 0*	27(10)	SHIFT LIGHT INOPERATIVE (8) SHIFT LIGHT ON (9)
BC3							
BC4							
BC5							
BC6							
BC7							
BC8 MIL	419	BRN/WHT	8-WAY I/P HARNESS CONNECTOR "B"	0*	B+	26(10)	NO MIL (8) MIL ON CONSTANTLY (9)
BC9 COOLANT FAN CONTROL	335	DK GRN/WHT	COOLING FAN RELAY TERM "85"	B+	"ON" (0) "OFF" B+	28	INOPERATIVE FAN (8) FAN RUNS CONSTANTLY (9)
BC10							
BC11 A/C CLUTCH CONTROL	459	DK GRN/WHT	A/C CLUTCH CONTROL TERM "85"	B+	"OFF" B+ "ON" 0*	28(10)	A/C CLUTCH INOPERATIVE (8) A/C CLUTCH STAYS "ON" (9)
BC12							
BC13 KNOCK SIGNAL	496	DK BLU	KNOCK SENSOR	2.3	2.3	43(10)	
BC14							
BC15							
BC16							

(1) VARIES FROM .60 TO BATTERY VOLTAGE, DEPENDING ON POSITION OF DRIVE WHEELS.
(2) BATTERY VOLTAGE FOR FIRST TWO SECONDS.
(3) VARIES.
(4) BATTERY VOLTAGE WHEN FUEL PUMP IS RUNNING.
(5) VARIES WITH TEMPERATURE.
(6) READS BATTERY VOLTAGE IN GEAR.
(7) BATTERY VOLTAGE WHEN ENGINE IS CRANKING.
(8) OPEN CIRCUIT.
(9) GROUNDED CIRCUIT.
(10) OPEN/GROUNDED CIRCUIT.
(11) LESS THAN 1 VOLT.
* LESS THAN .5 VOLT (500 mV).

2.3L (VIN A) ENGINE — ECM SYMPTOM CHART — 1993–94 BERETTA

ECM CONNECTOR "BD"

PIN FUNCTION	CKT #	WIRE COLOR	COMPONENT CONNECTOR CAVITY	NORMAL VOLTAGES KEY "ON"	NORMAL VOLTAGES ENG RUN	DTC(S) AFFECTED	POSSIBLE SYMPTOMS FROM FAULTY CIRCUIT
BD1							
BD2							
BD3							
BD4							
BD5							
BD6 DATA LINK CONNECTOR (DLC) DIAGNOSTIC ENABLE	451	WHT/BLK	8-WAY I/P HARNESS CONNECTOR TERM "E"	12.0	13.6		MIL FLASHES ALL THE TIME (9) NO FIELD SERVICE MODE (8)
BD7							
BD8							
BD9							
BD10							
BD11							
BD12 DLC SERIAL DATA	461	ORN	8-WAY I/P HARNESS CONNECTOR TERM "F"	4.2 (3)	4.5 (3)		NO SERIAL DATA SCAN TOOL WILL NOT READ DATA (10)
BD13 VSS "HI"	400	YEL	VSS TERM "A"	0* (1)	0* (1)	24 (10) 35 (10)	NO VSS SIGNAL INOPERATIVE SPEEDOMETER INOPERATIVE CRUISE CONTROL
BD14 VSS "LO"	401	PPL	VSS TERM "B"	0*	0*	24 (10) 35 (10)	NO VSS SIGNAL INOPERATIVE SPEEDOMETER INOPERATIVE CRUISE CONTROL
BD15							
BD16 IGNITION FEED	439	PNK/BLK	IGN FUSE	B+	B+	53	NO MIL, ENGINE CRANKS BUT WILL NOT START (8)

(1) VARIES FROM .60 TO BATTERY VOLTAGE, DEPENDING ON POSITION OF DRIVE WHEELS.
(2) BATTERY VOLTAGE FOR FIRST TWO SECONDS.
(3) VARIES.
(4) BATTERY VOLTAGE WHEN FUEL PUMP IS RUNNING.
(5) VARIES WITH TEMPERATURE.
(6) READS BATTERY VOLTAGE IN GEAR.
(7) BATTERY VOLTAGE WHEN ENGINE IS CRANKING.
(8) OPEN CIRCUIT.
(9) GROUNDED CIRCUIT.
(10) OPEN/GROUNDED CIRCUIT.
(11) LESS THAN 1 VOLT.
* LESS THAN .5 VOLT (500 mV).

2.3L (VIN A) ENGINE — COMPONENT DIAGNOSTIC CHART — 1993–94 BERETTA

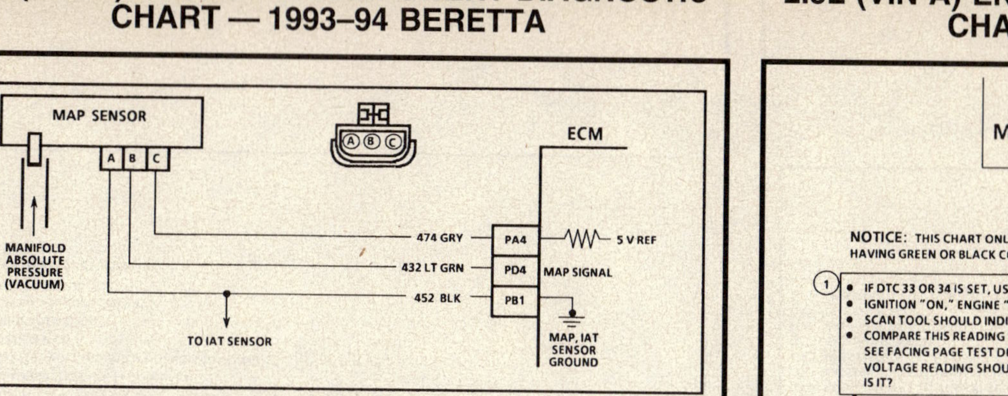

CHART C-1D
MANIFOLD ABSOLUTE PRESSURE (MAP) OUTPUT CHECK
2.3L (VIN A) "L" CARLINE

Circuit Description:

The Manifold Absolute Pressure (MAP) sensor measures the changes in the intake manifold pressure which result from engine load (intake manifold vacuum) and RPM changes; and converts these into a voltage output. The Engine Control Module (ECM) sends a 5 volt reference voltage to the MAP sensor. As the manifold pressure changed, the output voltage of the sensor also changes. By monitoring the sensor output voltage, the ECM knows the manifold pressure. A lower pressure (low voltage) output voltage will be about 1 - 2 volts at idle. While higher pressure (high voltage) output voltage will be about 4 - 4.8 at Wide Open Throttle (WOT). The MAP sensor is also used, under certain conditions, to measure barometric pressure, allowing the ECM to make adjustments for different altitudes. The ECM uses the MAP sensor to control fuel delivery and ignition timing.

Test Description: Number(s) below refer to circled number(s) on the diagnostic chart.

⚠ Important

- Be sure to use the same diagnostic test equipment for all measurements.

1. When comparing Tech 1 scan tool readings to a known good vehicle, it is important to compare vehicles that use a MAP sensor having the same color insert or having the same "Hot Stamped" number. Refer to figures.

2. Applying 34 kPa (10" Hg) vacuum to the MAP sensor should cause the voltage to be 1.5 to 2.1 volts less than the voltage at Step 1. Upon applying vacuum to the sensor, the change in voltage should be instantaneous. A slow voltage change indicates a faulty sensor.

3. Check vacuum hose to sensor for leaking or restriction. Be sure that no other vacuum devices are connected to the MAP hose.

NOTICE: Make sure electrical connector remains securely fastened.

4. Disconnect sensor from bracket and twist sensor by hand (only) to check for intermittent connection. Output changes greater than .10 volt indicate a bad sensor.

2.3L (VIN A) ENGINE — COMPONENT DIAGNOSTIC CHART — 1993–94 BERETTA

CHART C-1D
MANIFOLD ABSOLUTE PRESSURE (MAP) OUTPUT CHECK
2.3L (VIN A) "L" CARLINE

NOTICE: THIS CHART ONLY APPLIES TO MAP SENSORS HAVING GREEN OR BLACK COLOR KEY INSERT (SEE BELOW).

1.
- IF DTC 33 OR 34 IS SET, USE THOSE CHARTS FIRST.
- IGNITION "ON," ENGINE "OFF."
- SCAN TOOL SHOULD INDICATE A MAP SENSOR VOLTAGE.
- COMPARE THIS READING WITH THE READING OF A KNOWN GOOD VEHICLE. SEE FACING PAGE TEST DESCRIPTION, STEP 1. VOLTAGE READING SHOULD BE WITHIN ± .4 VOLT. IS IT?

YES →

2.
- DISCONNECT AND PLUG VACUUM SOURCE TO MAP SENSOR.
- CONNECT A HAND VACUUM PUMP TO MAP SENSOR.
- START ENGINE.
- NOTE MAP SENSOR VOLTAGE.
- APPLY 34 kPa (10" Hg) OF VACUUM AND NOTE VOLTAGE CHANGE. SUBTRACT SECOND READING FROM THE FIRST. VOLTAGE VALUE SHOULD BE GREATER THAN 1.5 VOLTS. IS IT?

YES →

3. NO TROUBLE FOUND. CHECK MAP SENSOR VACUUM SOURCE FOR LEAKAGE OR RESTRICTION. BE SURE THIS SOURCE SUPPLIES VACUUM TO MAP SENSOR ONLY.

NO → REPLACE MAP SENSOR.

NO →

4. CHECK MAP SENSOR CONNECTION. IF OK, REPLACE MAP SENSOR.

Figure 1 - Color Key Insert

Figure 2 - Hot Stamped Number

2.3L (VIN A) ENGINE — COMPONENT DIAGNOSTIC CHART — 1993–94 BERETTA

CHART C-2A

INJECTOR BALANCE TEST
2.3L (VIN A) "L" CARLINE

The injector balance tester is a tool used to turn the injector on for a precise amount of time, thus spraying a measured amount of fuel into the manifold. This causes a drop in fuel rail pressure that we can record and compare between each injector. All injectors should have the same amount of pressure drop (± 10 kPa). Any injector with a pressure drop that is 10 kPa (or more) greater or less than the average drop of the other injectors should be considered faulty and replaced.

STEP 1

Engine "cool down" period (10 minutes) is necessary to avoid irregular readings due to "Hot Soak" fuel boiling. Relieve fuel pressure in the fuel rail using the "fuel pressure relief procedure" described previously in this section. With ignition "OFF" connect fuel gauge J 347301 or equivalent to fuel pressure tap.

Disconnect harness connectors at all injectors, and connect injector tester J 34730-3A, to one injector. On Turbo equipped engines, use adaptor harness furnished with injector tester to energize injectors that are not accessible. Follow manufacturers instructions for use of adaptor harness. Ignition must be "OFF" at least 10 seconds to complete ECM shutdown cycle. Fuel pump should run about 2 seconds after ignition is turned "ON." At this point, insert clear tubing attached to vent valve into a suitable container and bleed air from gauge and hose to insure accurate gauge operation. Repeat this step until all air is bled from gauge.

STEP 2

Turn ignition "OFF" for 10 seconds and then "ON" again to get fuel pressure to its maximum. Record this initial pressure reading. Energize tester one time and note pressure drop at its lowest point. (Disregard any slight pressure increase after drop hits low point.) By subtracting this second pressure reading from the initial pressure, we have the actual amount of injector pressure drop.

STEP 3

Repeat step 2 on each injector and compare the amount of drop. Usually, good injectors will have virtually the same drop. Retest any injector that has a pressure difference of 10 kPa, either more or less than the average of the other injectors on the engine. Replace any injector that also fails the retest. If the pressure drop of all injectors is within 10 kPa of this average, the injectors appear to be flowing properly.

NOTE: *The entire test should <u>not</u> be repeated more than once without running the engine to prevent flooding. (This includes any retest on faulty injectors).*

2.3L (VIN A) ENGINE — COMPONENT DIAGNOSTIC CHART — 1993–94 BERETTA

NOTICE: The entire test should <u>NOT</u> be repeated more than once without running the engine to prevent flooding. (This includes any retest on faulty injectors.)

The fuel pressure test in Chart A-7, should be completed prior to this test.

CHART C-2A

INJECTOR BALANCE TEST
2.3L (VIN A) "L" CARLINE

Step 1. If engine is at operating temperature, allow a 10 minute "cool down" period then connect fuel pressure gage and injector tester.
1. Ignition "OFF."
2. Connect fuel pressure gage and injector tester.
3. Ignition "ON."
4. Bleed off air in gage. Repeat until all air is bled from gauge.

Step 2. Run test:
1. Ignition "OFF" for 10 seconds.
2. Ignition "ON." Record gage pressure. (Pressure must hold steady, if not, see "Fuel System Diagnosis," CHART A-7.
3. Turn injector on, by depressing button on injector tester, and note pressure at the instant the gauge needle stops.

Step 3.
1. Repeat Step 2 on all injectors and record pressure drop on each. Retest injectors that appear faulty (Any injectors that have a 10 kPa (1.5 psi) difference, either more or less, in pressure from the average).

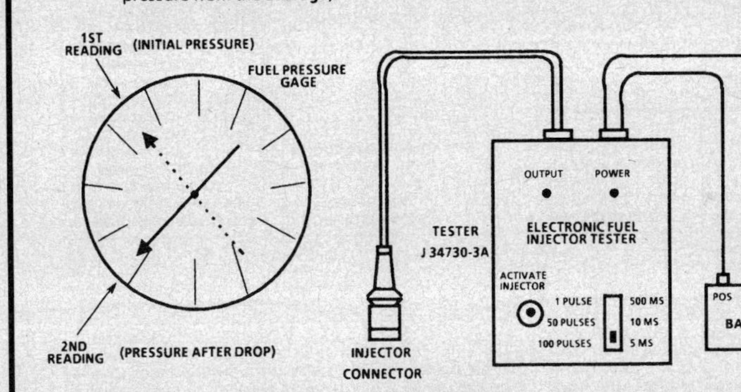

EXAMPLE

CYLINDER	1	2	3
1ST READING	293 kPa (43 psi)	293 kPa (43 psi)	293 kPa (43 psi)
2ND READING	131 kPa (19 psi)	115 kPa (17 psi)	145 kPa (21 psi)
AMOUNT OF DROP	162 kPa (24 psi)	178 kPa (26 psi)	148 kPa (21 psi)
	OK	FAULTY, RICH (TOO MUCH FUEL DROP)	FAULTY, LEAN (TOO LITTLE FUEL DROP)

2.3L (VIN A) ENGINE — COMPONENT DIAGNOSTIC CHART — 1993–94 BERETTA

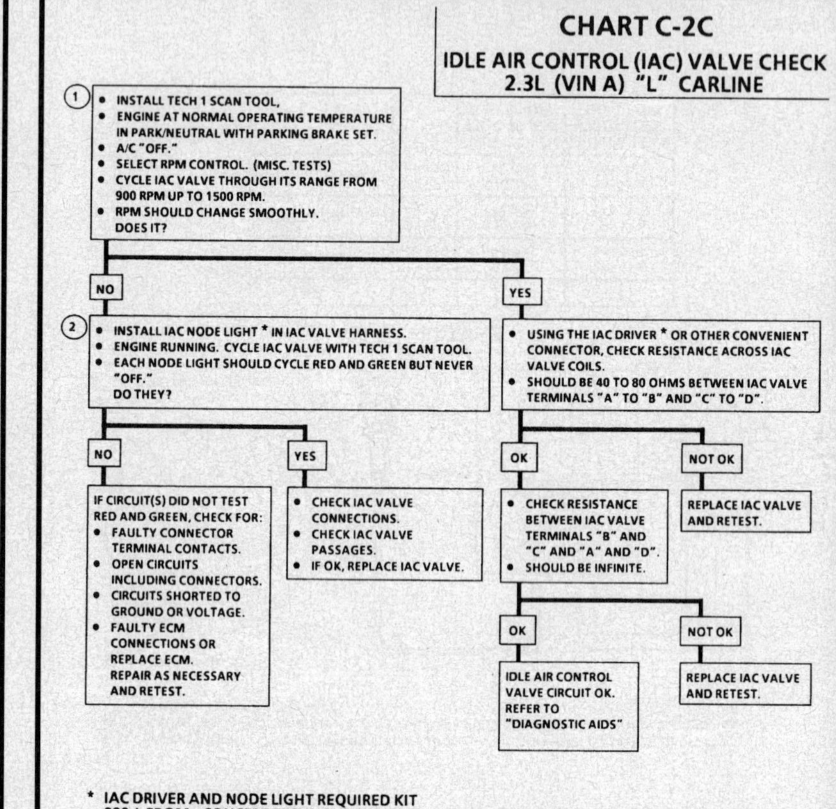

CHART C-2C
IDLE AIR CONTROL (IAC) VALVE CHECK
2.3L (VIN A) "L" CARLINE

Circuit Description:

The Engine Control Module (ECM) controls idle RPM with the IAC valve. To increase idle RPM, the ECM moves the IAC valve out, allowing more air to bypass the throttle plate. To decrease RPM, it moves the IAC valve in, reducing air flow by- passing the throttle plate. A Tech 1 will read the ECM commands to the IAC valve in counts. The higher the counts, the more air allowed (higher idle). The lower the counts, the less air allowed (lower idle).

Test Description: Number(s) below refer to circled numbers on the diagnostic chart.

1. The Tech 1 scan tool is used to extend and retract the IAC valve. Valve movement is verified by an engine speed change. If no change in engine speed occurs, the valve can be retested when removed from the throttle body.
2. This step checks the quality of the IAC valve movement in Step 1. Between 900 RPM and about 1500 RPM, the engine speed should change smoothly with each flash of the tester light in both extend and retract. If the IAC valve is retracted beyond the control range (about 1500 RPM), it may take many flashes in the extend position before engine speed will begin to drop. This is normal on certain engines, fully extending IAC valve may cause engine stall. This may be normal. Step 1 verified proper IAC valve operation while this step checks IAC valve circuits. Each lamp on the node light should flash red and green while the IAC valve is cycled. While the sequence of color is not important if either light is "OFF" or does not flash red and green, check the circuits for faults, beginning with poor terminal contacts.

Diagnostic Aids:

A slow, unstable, or fast idle may be caused by a non-IAC valve system problem that cannot be overcome by the IAC valve. Out of control range IAC valve scan tool counts will be above 60 if idle is too low, and zero counts if idle is too high. The following checks should be made to repair a non-IAC valve system problem.

- Vacuum Leak (High Idle). - If idle is too high, stop the engine. Fully extend (low) IAC valve with tester. Start engine. If idle speed is above 800 RPM, locate and correct vacuum leak including crankcase ventilation system. Also, check for binding of throttle blade or linkage.
- System too Lean (High Air/Fuel Ratio). - Idle speed may be too high or too low. Engine speed may vary up and down and disconnecting IAC does not help. DTC 44 may be set. Tech 1 scan tool Oxygen Sensor (O2S) voltage will be less than 300 mV (.3 volt.) Check for low regulated fuel pressure, water in the fuel or a restricted injector.
- System too Rich (Low Air/Fuel Ratio). - The idle speed will be too low. Tech 1 scan tool IAC valve counts will usually be above 80. System is obviously rich and may exhibit black smoke exhaust.
Tech 1 scan tool O2S voltage will be fixed above 800 mV (.8 volt). Check for high fuel pressure, leaking or sticking injector. Silicone contaminated O2S sensor will scan an O2S voltage slow to respond.
- Throttle Body. - Remove IAC valve and inspect bore for foreign material.
- If intermittent poor driveability or idle symptoms are resolved by disconnecting the IAC valve, carefully recheck connections, valve terminal resistance, or replace IAC valve.

2.3L (VIN A) ENGINE — COMPONENT DIAGNOSTIC CHART — 1993–94 BERETTA

CHART C-2C
IDLE AIR CONTROL (IAC) VALVE CHECK
2.3L (VIN A) "L" CARLINE

(1)
- INSTALL TECH 1 SCAN TOOL,
- ENGINE AT NORMAL OPERATING TEMPERATURE IN PARK/NEUTRAL WITH PARKING BRAKE SET.
- A/C "OFF."
- SELECT RPM CONTROL. (MISC. TESTS)
- CYCLE IAC VALVE THROUGH ITS RANGE FROM 900 RPM UP TO 1500 RPM.
- RPM SHOULD CHANGE SMOOTHLY. DOES IT?

NO →

(2)
- INSTALL IAC NODE LIGHT * IN IAC VALVE HARNESS.
- ENGINE RUNNING. CYCLE IAC VALVE WITH TECH 1 SCAN TOOL.
- EACH NODE LIGHT SHOULD CYCLE RED AND GREEN BUT NEVER "OFF." DO THEY?

NO:
IF CIRCUIT(S) DID NOT TEST RED AND GREEN, CHECK FOR:
- FAULTY CONNECTOR TERMINAL CONTACTS.
- OPEN CIRCUITS INCLUDING CONNECTORS.
- CIRCUITS SHORTED TO GROUND OR VOLTAGE.
- FAULTY ECM CONNECTIONS OR REPLACE ECM. REPAIR AS NECESSARY AND RETEST.

YES:
- CHECK IAC VALVE CONNECTIONS.
- CHECK IAC VALVE PASSAGES.
- IF OK, REPLACE IAC VALVE.

YES →
- USING THE IAC DRIVER * OR OTHER CONVENIENT CONNECTOR, CHECK RESISTANCE ACROSS IAC VALVE COILS.
- SHOULD BE 40 TO 80 OHMS BETWEEN IAC VALVE TERMINALS "A" TO "B" AND "C" TO "D".

OK:
- CHECK RESISTANCE BETWEEN IAC VALVE TERMINALS "B" AND "C" AND "A" AND "D". SHOULD BE INFINITE.

  OK: IDLE AIR CONTROL VALVE CIRCUIT OK. REFER TO "DIAGNOSTIC AIDS"

  NOT OK: REPLACE IAC VALVE AND RETEST.

NOT OK: REPLACE IAC VALVE AND RETEST.

* IAC DRIVER AND NODE LIGHT REQUIRED KIT
222-L FROM: CONCEPT TECHNOLOGY, INC.
J 37027 FROM: KENT-MOORE, INC.

CLEAR CODES, CONFIRM "CLOSED LOOP" OPERATION, NO MIL (SERVICE ENGINE SOON) PERFORM IAC RESET PROCEDURE PER APPLICABLE SERVICE MANUAL AND VERIFY CONTROLLED IDLE SPEED IS CORRECT.

2.3L (VIN A) ENGINE — COMPONENT DIAGNOSTIC CHART — 1993–94 BERETTA

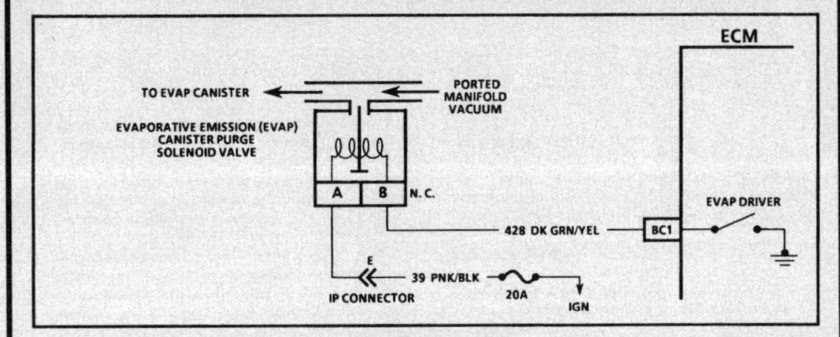

CHART C-3

EVAPORATIVE EMISSION (EVAP) CANISTER PURGE SOLENOID VALVE CHECK
2.3L (VIN A) "L" CARLINE

Circuit Description:

Canister purge is controlled by a solenoid valve that allows manifold and/or ported vacuum to purge the canister when energized. The Engine Control Module (ECM) supplies a ground to energize the valve (purge "ON"). The purge solenoid valve control by the ECM is pulse width modulated (turned "ON" and "OFF" several times a second). The duty cycle (pulse width) is determined by "Closed Loop" feed back from the Oxygen Sensor (O2S). The duty cycle is calculated by the ECM and the output commanded when the following conditions have been met:

- Engine run time after start more than 65 seconds.
- Engine coolant temperature above 56°C (133°F).

Also, if the diagnostic "test" terminal is grounded with the engine stopped, the EVAP canister purge solenoid valve is energized (purge "ON").

Test Description: Number(s) below refer to circled number(s) on the diagnostic chart.

1. Checks to see if the solenoid valve is opened or closed. The solenoid valve is normally de-energized in this step, so it should be closed.
2. Checks to determine if solenoid valve was open due to electrical circuit problem or defective solenoid.
3. Completes functional check by grounding test terminal. This should normally energize the solenoid valve opening the valve which should allow the vacuum to drop (purge "ON").

Diagnostic Aids:

Make a visual check of vacuum hose(s). Check throttle body for possible cracked, broken, or plugged vacuum block. Check engine for possible mechanical problem.

2.3L (VIN A) ENGINE — COMPONENT DIAGNOSTIC CHART — 1993–94 BERETTA

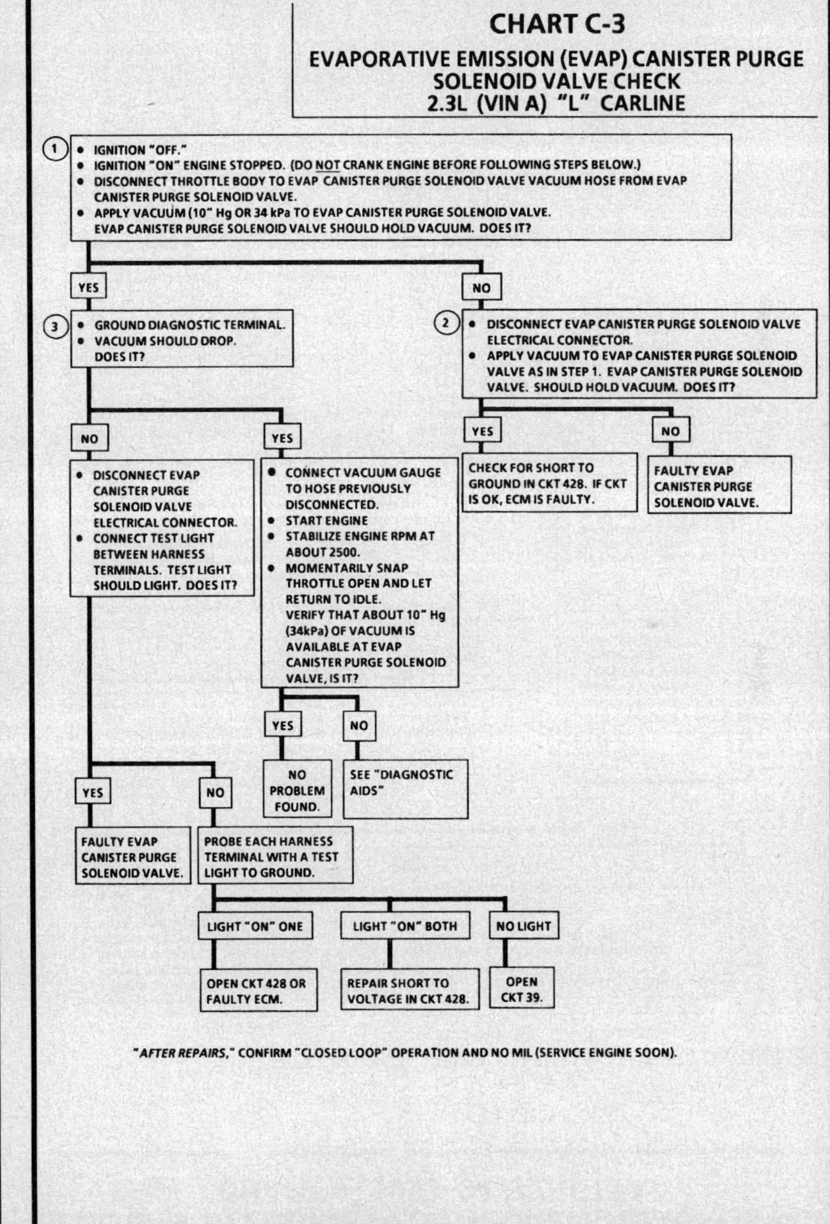

2.3L (VIN A) ENGINE — COMPONENT DIAGNOSTIC CHART — 1993–94 BERETTA

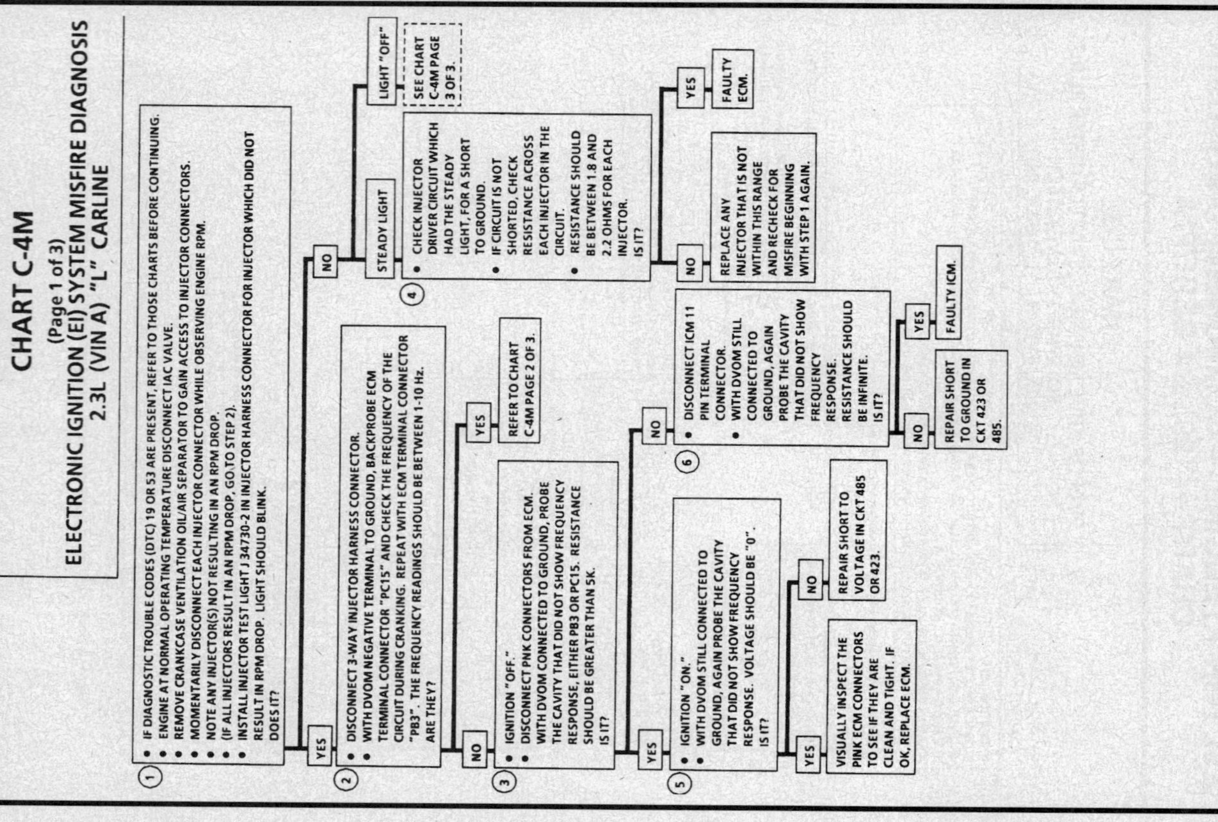

2.3L (VIN A) ENGINE — COMPONENT DIAGNOSTIC CHART — 1993–94 BERETTA

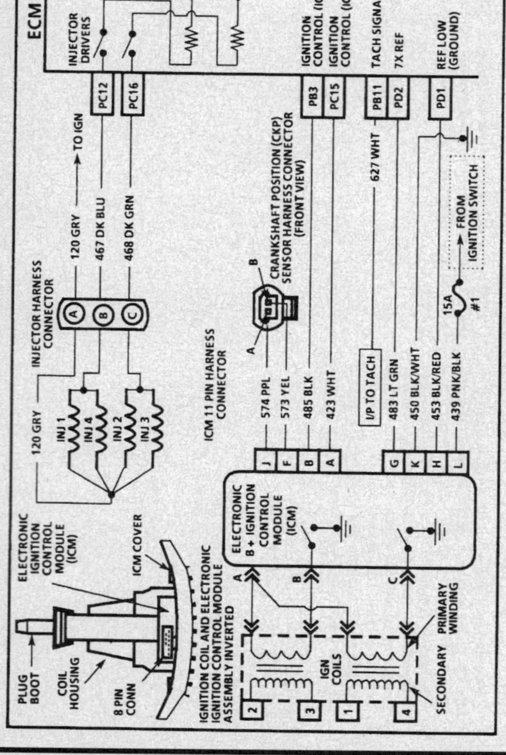

2.3L (VIN A) ENGINE — COMPONENT DIAGNOSTIC CHART — 1993–94 BERETTA

CHART C-4M
(Page 1 of 3)
ELECTRONIC IGNITION (EI) SYSTEM MISFIRE DIAGNOSIS
2.3L (VIN A) "L" CARLINE

Circuit Description:

The Electronic Ignition (EI) system uses a waste spark method of distribution. In this type of system the electronic Ignition Control Module (ICM) triggers the #1-4 coil pair resulting in both #1 and #4 spark plugs firing at the same time. #1 cylinder is on the compression stroke at the same time #4 is on the exhaust stroke, resulting in a lower energy requirement to fire #4 spark plug. This leaves the remainder of the high voltage to be used to fire #1 spark plug. On this application, the Crankshaft Position (CKP) sensor is mounted to, and protrudes through the block to within approximately 0.050" of the crankshaft reluctor. Since the reluctor is a machined portion of the crankshaft and the sensor is mounted in a fixed position on the block, timing adjustments are not possible or necessary.

Test Description: Number(s) below refer to circled number(s) on the diagnostic chart.

1. This checks for equal relative power output between the cylinders. Any injector which when disconnected did not result in an RPM drop approximately equal to the others, is located on the misfiring cylinder.

2. This step checks for Ignition Control (IC) frequency output from the ECM.

3. This step checks if the circuit is shorted to ground.

4. An injector driver circuit shorted to ground would result in the test light "ON" steady, and possibly a flooded condition. A shorted injector (less than 1.8 ohms) could cause incorrect ECM operation.

5. This step will determine if there is a short to voltage on the circuit that did not show frequency response.

6. This step will determine if the short to ground is in the circuit or the ICM.

Diagnostic Aid:

Verify ICM connector terminal "K", CKT 450 resistance to ground is less than .5 ohm. A shorted or low resistance injector may cause a miss in the other injector in that pair (1 & 4 or 2 & 3). A poor connection or open in CKT 453 may cause a miss.

2.3L (VIN A) ENGINE — COMPONENT DIAGNOSTIC CHART — 1993–94 BERETTA

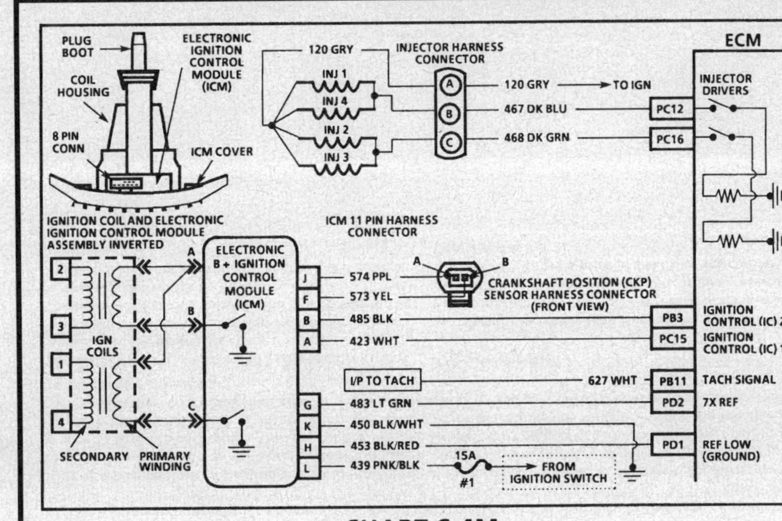

CHART C-4M
(Page 2 of 3)
ELECTRONIC IGNITION (EI) SYSTEM MISFIRE DIAGNOSIS
2.3L (VIN A) "L" CARLINE

Circuit Description:

The Electronic Ignition (EI) system uses a waste spark method of distribution. In this type of system the electronic Ignition Control Module (ICM) triggers the #1-4 coil pair resulting in both #1 and #4 spark plugs firing at the same time. #1 cylinder is on the compression stroke at the same time #4 is on the exhaust stroke, resulting in a lower energy requirement to fire #4 spark plug. This leaves the remainder of the high voltage to be used to fire #1 spark plug. On this application, the Crankshaft Position (CKP) sensor is mounted to, and protrudes through the block to within approximately 0.050" of the crankshaft reluctor. Since the reluctor is a machined portion of the crankshaft and the sensor is mounted in a fixed position on the block, timing adjustments are not possible or necessary.

Test Description: Number(s) below refer to circled number(s) on the diagnostic chart.

7. This step checks for an open in CKTs 485 or 423.
8. If a plug boot is burned, the other plug on that ignition coil may still fire at idle. This step tests the system's ability to produce at least 25,000 volts at each spark plug.
9. No spark, on one coil, may be caused by an open secondary circuit. Therefore, the coil's secondary resistance should be checked. Resistance readings above 20,000 ohms, but not infinite, will probably not cause a no start but may cause an engine miss under certain conditions.
10. If the no spark condition is caused by coil connections, a coil or a secondary boot assembly, the test light will blink. If the light does not blink, the fault is module connections or the module.

Diagnostic Aid:

Verify electronic Ignition Control Module (ICM) connector terminal "K", CKT 450 resistance to ground is less than .5 ohm. A shorted or low resistance injector may cause a miss in the other injector in that pair (1 & 4 or 2 & 3).

2.3L (VIN A) ENGINE — COMPONENT DIAGNOSTIC CHART — 1993–94 BERETTA

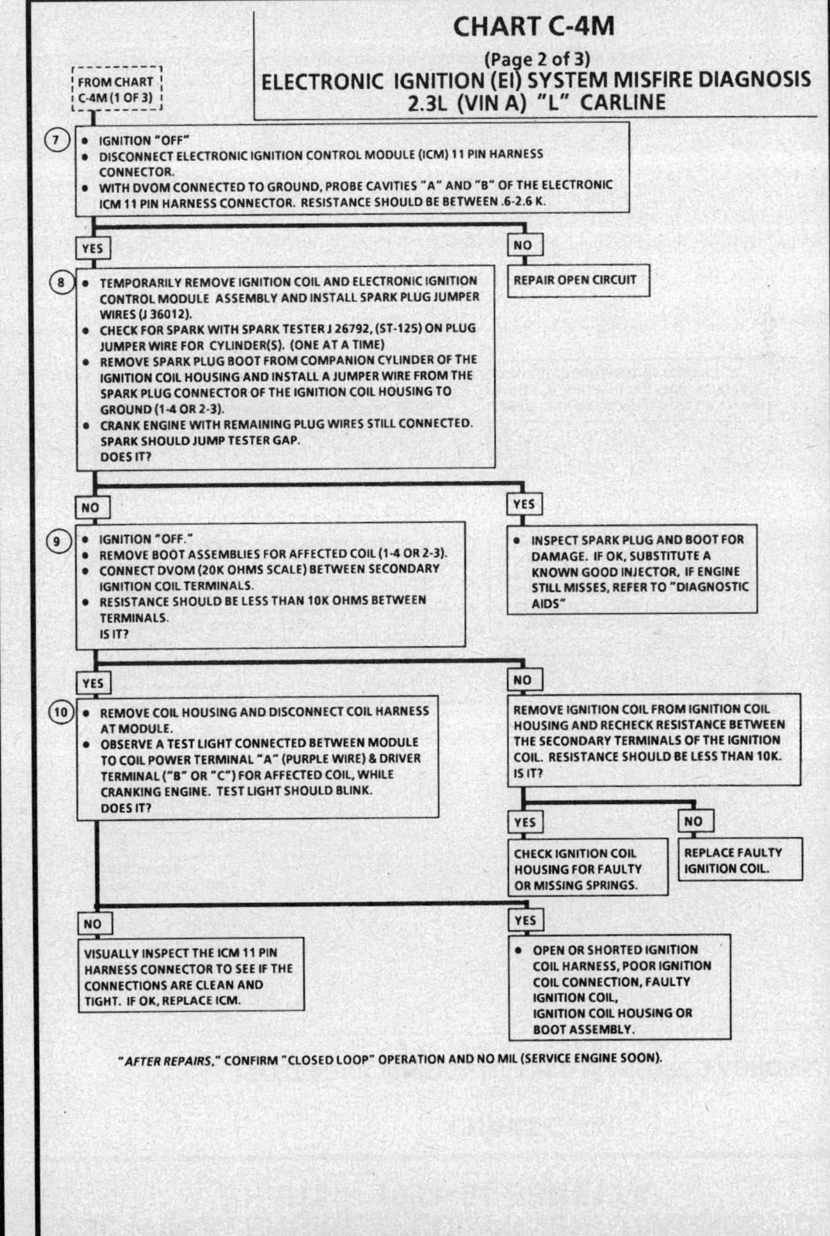

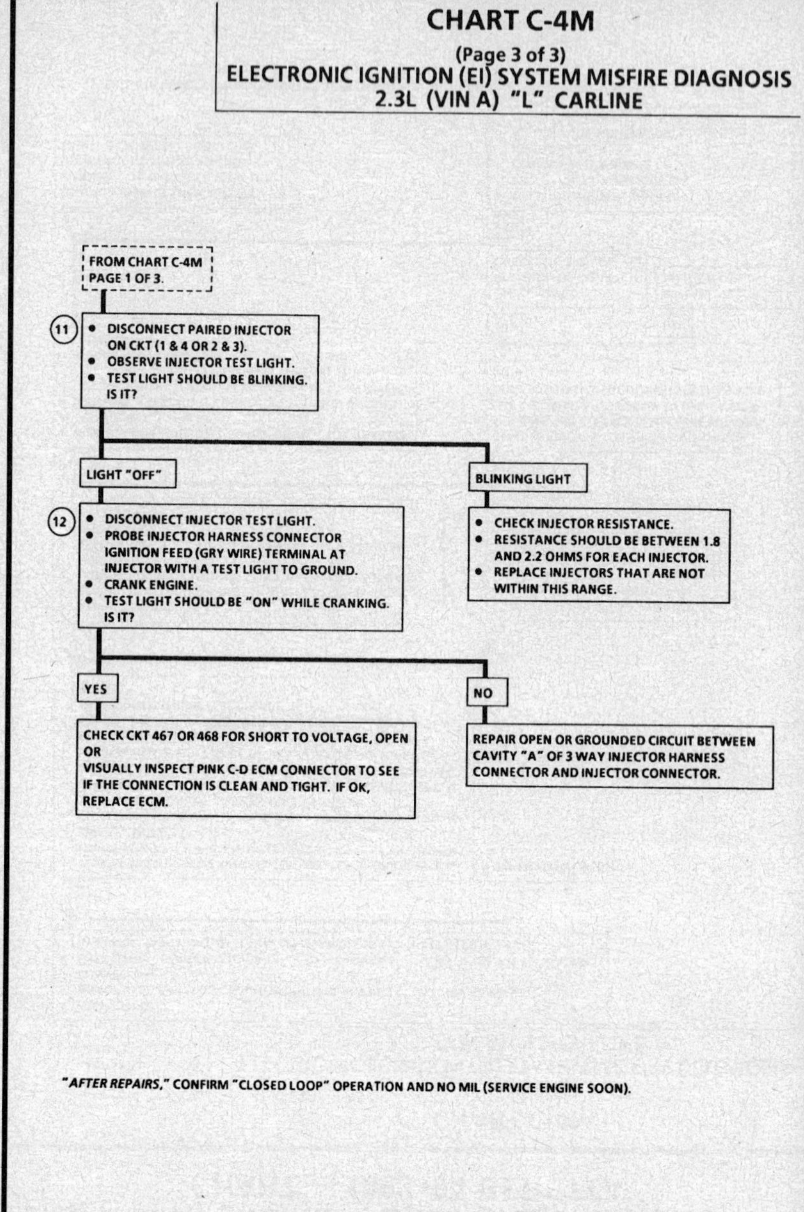

CHART C-4M
(Page 3 of 3)
ELECTRONIC IGNITION (EI) SYSTEM MISFIRE DIAGNOSIS
2.3L (VIN A) "L" CARLINE

Circuit Description:

The Electronic Ignition (EI) system uses a waste spark method of distribution. In this type of system the electronic Ignition Control Module (ICM) triggers the #1-4 coil pair resulting in both #1 and #4 spark plugs firing at the same time. #1 cylinder is on the compression stroke at the same time #4 is on the exhaust stroke, resulting in a lower energy requirement to fire #4 spark plug. This leaves the remainder of the high voltage to be used to fire #1 spark plug. On this application, the Crankshaft Position (CKP) sensor is mounted to, and protrudes through the block to within approximately 0.050" of the crankshaft reluctor. Since the reluctor is a machined portion of the crankshaft and the sensor is mounted in a fixed position on the block, timing adjustments are not possible or necessary.

Test Description: Number(s) below refer to circled number(s) on the diagnostic chart.
11. Checks for a grounded or low resistance injector. This condition will not able the paired injector to properly deliver fuel.
12. Checks for voltage available at the injector driver circuit.

Diagnostic Aid:

Verify electronic Ignition Control Module (ICM) connector terminal "K", CKT 450 resistance to ground is less than .5 ohm. A shorted or low resistance injector may cause a miss in the other injector in that pair (1 & 4 or 2 & 3).

CHART C-4M
(Page 3 of 3)
ELECTRONIC IGNITION (EI) SYSTEM MISFIRE DIAGNOSIS
2.3L (VIN A) "L" CARLINE

2.3L (VIN A) ENGINE — COMPONENT DIAGNOSTIC CHART — 1993–94 BERETTA

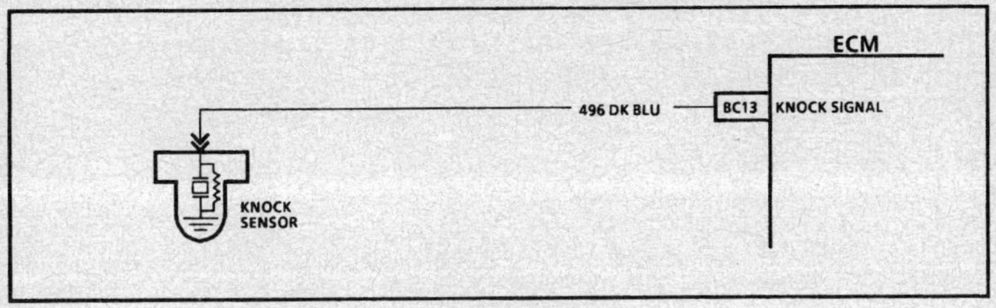

ECM

496 DK BLU — BC13 — KNOCK SIGNAL

KNOCK SENSOR

CHART C-5

KNOCK SENSOR (KS) SYSTEM CHECK
2.3L (VIN A) "L" CARLINE

Circuit Description:

The knock sensor is used to detect engine detonation and the ECM will retard the electronic spark timing based on the signal being received. The knock sensor produces an AC signal. The amplitude and frequency are dependent upon the knock level.

The PROM used with this engine contains the functions which were part of remotely mounted Knock Sensor (KS) modules used on other GM vehicles. The KS portion of the PROM then sends a signal to other parts of the Engine Control Module (ECM) which adjusts the spark timing to retard the spark and reduce the detonation.

Test Description: Number(s) below refer to circled number(s) on the diagnostic chart.

1. The Tech 1 scan tool displays knock sensor activity in counts approximately 16-22 at idle. The counts should rise when engine speed increases or fall when engine speed decreases. This step checks knock sensor activity.
2. This step checks that the internal resistance of knock sensor is within an acceptable range.
3. If the engine has an internal problem which is creating a knock, the knock sensor may be responding to the mechanical noise.

Diagnostic Aids:

While observing knock signal on the Tech 1 scan tool, there should be an indication that knock is present when detonation can be heard. Detonation is most likely to occur under high engine load conditions.

2.3L (VIN A) ENGINE — COMPONENT DIAGNOSTIC CHART — 1993–94 BERETTA

CHART C-5

KNOCK SENSOR (KS) SYSTEM CHECK
2.3L (VIN A) "L" CARLINE

(1) USE TECH 1 SCAN TOOL TO OBSERVE KS ACTIVITY WHILE RAISING AND LOWERING ENGINE SPEED. KNOCK SENSOR ACTIVITY SHOULD CHANGE WITH ENGINE SPEED. DOES IT?

NO — (2)
- IGNITION "OFF."
- DISCONNECT THE BLUE ECM CONNECTOR.
- USING A DVOM CONNECTED TO GROUND AND ECM CONNECTOR TERMINAL "BC13", MEASURE RESISTANCE.
- RESISTANCE SHOULD BE BETWEEN 90K-110K OHMS. IS IT?

YES — SYSTEM IS OPERATING PROPERLY. REFER TO "DIAGNOSTIC AIDS"

YES — (3) IF AN ENGINE KNOCK CAN BE HEARD, REPAIR THE BASIC ENGINE/LOOSE BRACKET PROBLEM. IF NO AUDIBLE KNOCK IS HEARD, FOLLOW THESE STEPS:
- IGNITION "OFF."
- DISCONNECT KNOCK SENSOR.
- CONNECT VOLTMETER TO KNOCK SENSOR AND ENGINE GROUND.
- SET VOLTMETER ON 2 VOLTS AC SCALE.
- IGNITION "ON," ENGINE "ON.
- IS SIGNAL INDICATED ON VOLTMETER?

NO — CHECK FOR OPEN OR SHORTED CKT 496 OR POOR CONNECTIONS AT ECM AND KNOCK SENSOR. IF OK, REPLACE KNOCK SENSOR AND REPEAT STEP #1.

NO — CHECK FOR: A FAULTY ECM CONNECTION. IF OK, REPLACE PROM. IF CONDITION STILL EXISTS, REPLACE ECM.

YES — REPLACE KNOCK SENSOR.

"AFTER REPAIRS," CONFIRM "CLOSED LOOP" OPERATION AND NO MIL (SERVICE ENGINE SOON).

CHART C-8C
MANUAL TRANSAXLE (M/T) SHIFT LIGHT CHECK
2.3L (VIN A) "L" CARLINE

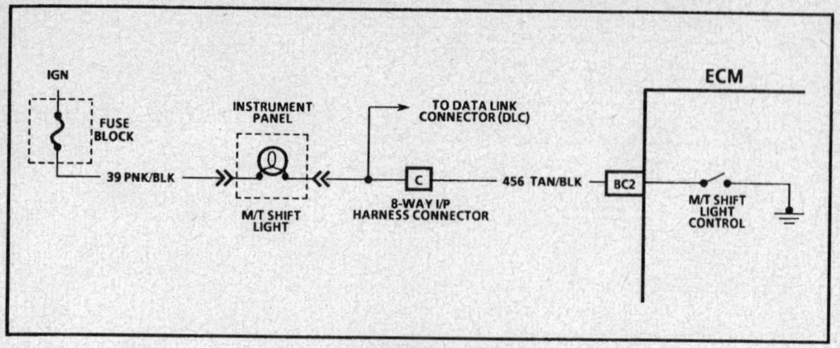

Circuit Description:

The shift light indicates the best transaxle shift point for maximum fuel economy. The light is controlled by the Engine Control Module (ECM) and is turned "ON" by grounding CKT 456.

The ECM uses inputs from various sensors to calculate when the shift light should be turned "ON" as follows:

- Coolant temperature must be above -10°C (14°F).
- ECM can determine the transaxle has been in a gear for at least 1.2 seconds, by comparison of vehicle speed (from VSS) with engine RPM.
- Throttle Position (TP) sensor is between minimum and maximum calibrated values for each gear.
- RPM is above a calibrated value for each gear (maximum 6500 RPM).

The light will be turned "ON" after a calibrated delay time which is dependent on last gear change or downshift, and will remain on for a minimum of 10 seconds.

Test Description: Numbers below refer to circled number(s) on the diagnostic chart.

1. This should not turn "ON" the shift light. If the light is "ON," there is a short to ground in CKT 456 wiring or a fault in the ECM.
2. When the diagnostic terminal is grounded, the ECM should ground CKT 456 and the shift light should come "ON."
3. This checks the shift light circuit up to the ECM connector. If the shift light illuminates, then the ECM connector is faulty or the ECM does not have the ability to ground the circuit.

Diagnostic Aids:

Check for DTC 24 (no VSS). Faulty thermostat or incorrect heat range. Incorrect or faulty PROM.

A faulty or improper VSS may result in a shift light that does not operate.

CHART C-8C
MANUAL TRANSAXLE (M/T) SHIFT LIGHT CHECK
2.3L (VIN A) "L" CARLINE

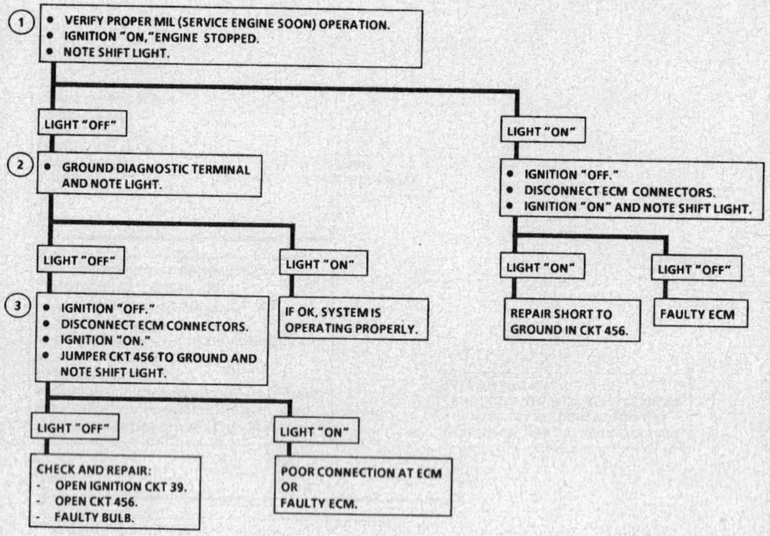

2.3L (VIN A) ENGINE — COMPONENT DIAGNOSTIC CHART — 1993–94 BERETTA

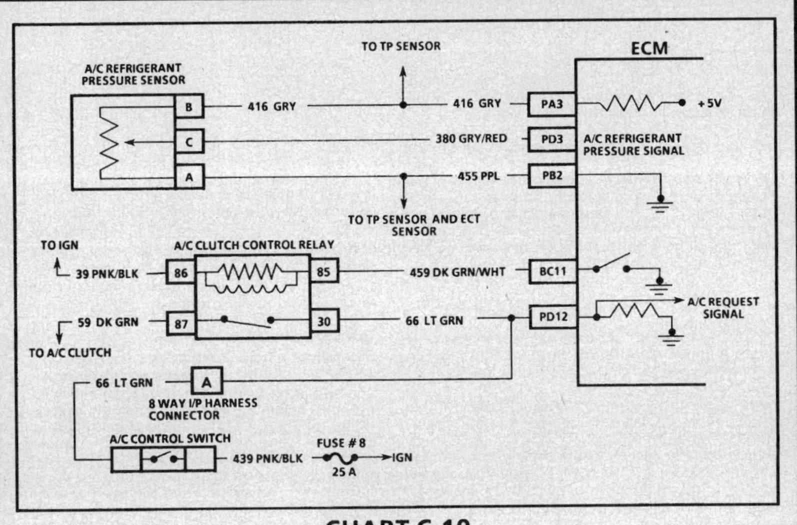

CHART C-10
(Page 1 of 2)
A/C CLUTCH CONTROL CIRCUIT DIAGNOSIS
2.3L (VIN A) "L" CARLINE

Circuit Description:
The A/C clutch control relay is energized when the Engine Control Module (ECM) provides a ground path through CKT 459 and A/C is requested. A/C clutch is delayed about .3 second after A/C is requested. This will allow the IAC valve to adjust engine RPM for the additional load.

The ECM will temporarily disengage the A/C clutch relay for calibrated times for one or more of the following:
- Hot engine restart.
- Wide Open Throttle (WOT) (TP sensor over 90%).
- Engine RPM greater than about 6000 RPM.
- During Idle Air Control (IAC) valve reset.

The A/C clutch relay will remain disengaged when DTC 66 is present, if AC refrigerant pressure is out of range or there is no A/C request signal due to an open A/C select switch or circuit.

Test Description: Number(s) below refer to circled number(s) on the diagnostic chart.

1. The ECM will only energize the A/C relay when the engine is running. This test will determine if the relay or CKT 459 is faulty.

2. Determines if the signal is reaching the ECM through CKT 66 from the A/C control panel. Signal should only be present when an A/C mode or defrost mode has been selected.

2.3L (VIN A) ENGINE — COMPONENT DIAGNOSTIC CHART — 1993–94 BERETTA

CHART C-10
(Page 1 of 2)
A/C CLUTCH CONTROL CIRCUIT DIAGNOSIS
2.3L (VIN A) "L" CARLINE

(1)
- IGNITION "ON," ENGINE "OFF."
- USING TECH 1 SCAN TOOL, SCAN FOR DIAGNOSTIC TROUBLE CODES (DTC) AND USE THOSE CHARTS, IF PRESENT.
- NOTE A/C CLUTCH.
- CLUTCH SHOULD NOT BE ENGAGED.
  IS IT?

NO
- WITH ENGINE IDLING AT NORMAL OPERATING TEMPERATURE TURN A/C "ON" AND "OFF."
- CLUTCH SHOULD CYCLE "ON" AND "OFF."
  DOES IT?

YES
- DISCONNECT A/C RELAY.
- PROBE CKT 459 WITH A TEST LIGHT TO 12 VOLTS.

NO / (2)
- PLACE A/C CONTROL IN A/C MODE.
- SCAN TOOL SHOULD DISPLAY A/C REQUEST "ON."
  DOES IT?

YES
A/C CIRCUIT OK, REFER TO "DIAGNOSTIC AIDS"

LIGHT "ON"
CKT 459 SHORTED TO GROUND OR FAULTY ECM.

LIGHT "OFF"
CKT 59 SHORTED TO B + OR FAULTY RELAY.

YES
REFER TO CHART C-10 (2 OF 2).

NO
- BACKPROBE ECM TERMINAL "BC11" WITH A TEST LIGHT TO GROUND.

LIGHT "ON"
- CHECK FOR POOR CONNECTION AT ECM TERMINAL "BC11". IF OK, ECM IS FAULTY.

LIGHT "OFF"
- IF A/C FUSE IS BLOWN, CHECK FOR SHORT TO GROUND IN CKT 66. IF OK, CHECK FOR OPEN CKT 66, OR FAULTY A/C SELECT SWITCH.

"AFTER REPAIRS," CONFIRM "CLOSED LOOP" OPERATION AND NO MIL (SERVICE ENGINE SOON).

2.3L (VIN A) ENGINE — COMPONENT DIAGNOSTIC CHART — 1993–94 BERETTA

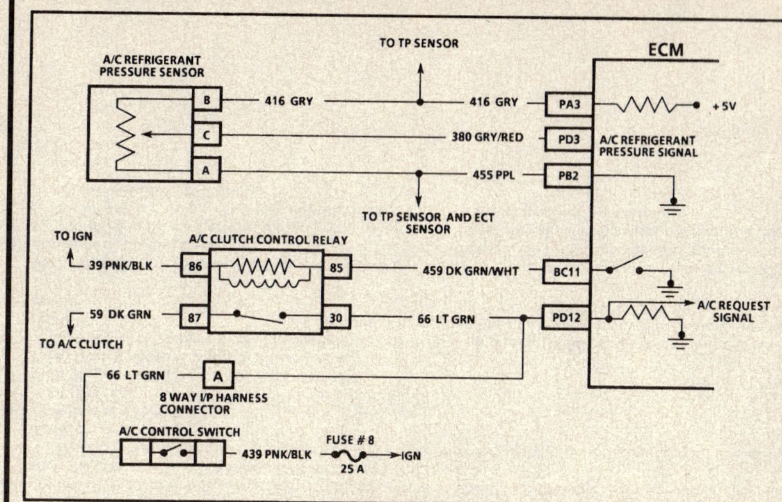

CHART C-10
(Page 2 of 2)
A/C CLUTCH CONTROL CIRCUIT DIAGNOSIS
2.3L (VIN A) "L" CARLINE

Circuit Description:

The A/C clutch control relay is energized when the Engine Control Module (ECM) provides a ground path through CKT 459 and A/C is requested. A/C clutch is delayed about .3 second after A/C is requested. This will allow the Idle Air Control (IAC) valve to adjust engine RPM for the additional.

The ECM will temporarily disengage the A/C clutch relay for calibrated times for one or more of the following:

- Hot engine restart.
- Wide open throttle (sensor over 90%).
- Engine RPM greater than about 6000 RPM.
- During IAC valve reset.

The A/C clutch relay will remain disengaged when a DTC 66 is present, if AC refrigerant pressure is out of range or there is no A/C request signal due to an open A/C select switch or circuit.

Test Description: Number(s) below refer to circled number(s) on the diagnostic chart.

1. Determines if the A/C refrigerant pressure sensor is out of range causing the compressor clutch to be disengaged.
2. With the engine stopped and field service mode activated, the ECM should be grounding CKT 459, which should cause the test light to be "ON."

Diagnostic Aids:

If complaint is insufficient cooling, the problem may be caused by an inoperative cooling fan. See CHART C-12 for cooling fan diagnosis. If fan operates correctly, see A/C diagnosis A/C refrigerant pressure outside of a range of 43 to 428 psi will cause the compressor to be disabled by the ECM. Observe Tech 1 scan tool A/C refrigerant pressure for 2 minutes with engine idling and A/C "ON."

Scan pressure should be within 20 psi of actual. If not, check for a circuit problem using DTC 66 chart or replace sensor.

2.3L (VIN A) ENGINE — COMPONENT DIAGNOSTIC CHART — 1993–94 BERETTA

CHART C-10
(Page 2 of 2)
A/C CLUTCH CONTROL CIRCUIT DIAGNOSIS
2.3L (VIN A) "L" CARLINE

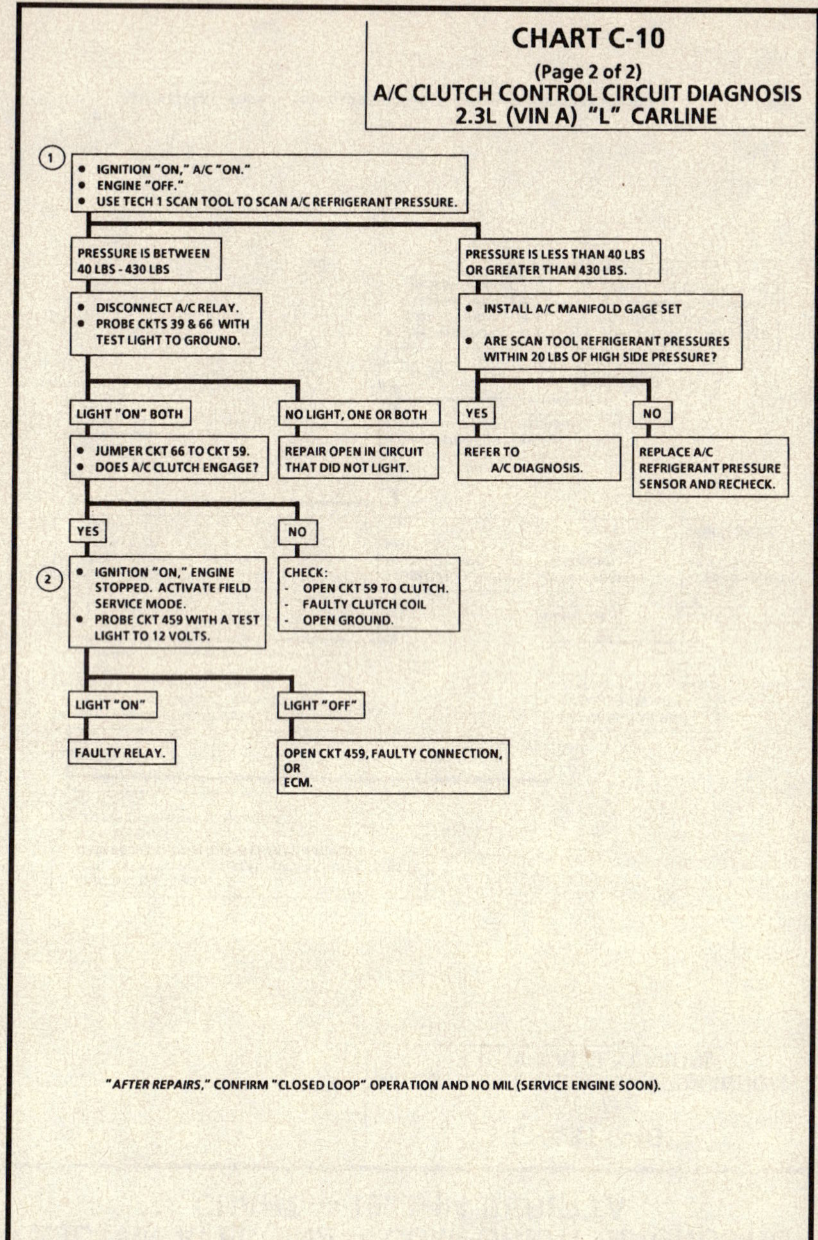

2.3L (VIN A) ENGINE — COMPONENT DIAGNOSTIC CHART — 1993–94 BERETTA

CHART C-12
COOLANT FAN FUNCTIONAL CHECK
2.3L (VIN A) "L" CARLINE

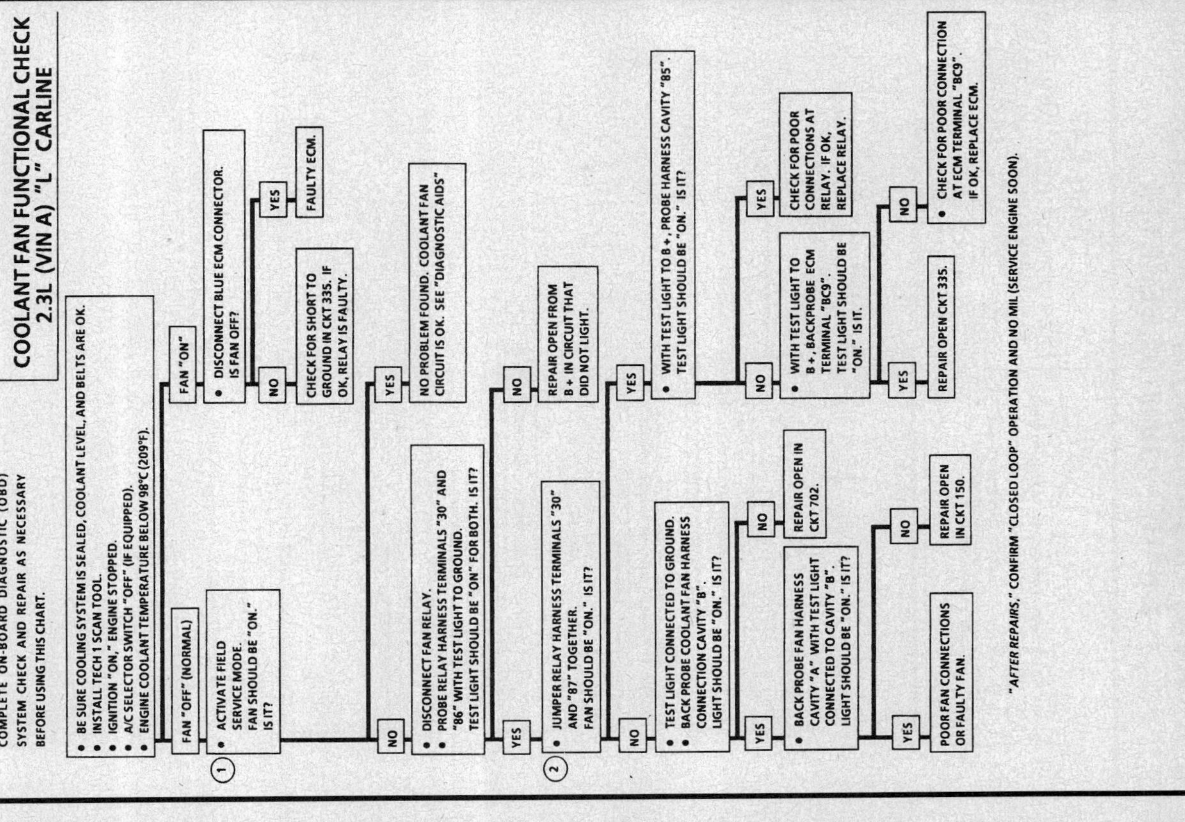

2.3L (VIN A) ENGINE — COMPONENT DIAGNOSTIC CHART — 1993–94 BERETTA

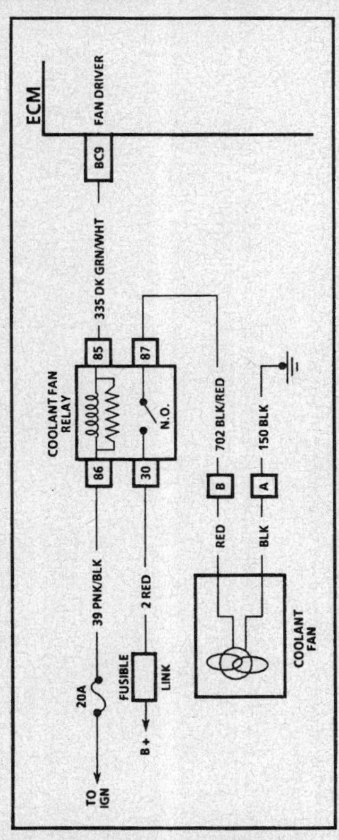

CHART C-12
COOLANT FAN FUNCTIONAL CHECK
2.3L (VIN A) "L" CARLINE

Circuit Description:

The electric coolant fan is controlled by the Engine Control Module (ECM) through the fan relay based on inputs from the engine coolant and intake air temperature sensors, the A/C control switch, A/C refrigerant pressure sensor and the vehicle speed sensor. The ECM controls the coolant fan by grounding CKT 335 which turns "ON" the fan relay.

The fan relay will be commanded "ON" when:
- Engine coolant temperature 103°C (217°F) - 106°C (223°F) or more.
- A/C clutch requested.
- Vehicle speed is less than 35 mph.

The fan relay will be commanded "ON" regardless of vehicle speed when:
- DTC 14 or 15 are set.
- Engine coolant temperature 115°C (239°F) - 118°C (224°F) or more.
- A/C refrigerant pressure is high.

The coolant fan may be commanded "ON" when the engine is not running under fan "Run-ON" conditions.

Test Description: Number(s) below refer to circled number(s) on the diagnostic chart.

1. With the field service mode activated, the coolant fan control driver should close, which should energize the fan control relay.
2. Test to see if fault is in wiring to the fan or the fan/relay connection.

If the gage, or light, indicates overheating, but no boil over is detected, the gage or light circuit should be checked. The gage accuracy can also be checked by comparing the coolant sensor reading using a scan tool with the gage reading.

If the engine is actually overheating, and the gage indicates overheating, but the coolant fan is not coming "ON," the Engine Coolant Temperature (ECT) sensor has probably shifted out of calibration and should be replaced. Refer to DTC 15 chart for a temperature to resistance chart.

If the engine is overheating, and the coolant fan is "ON," the cooling system should be checked.

Diagnostic Aids:

If the owner complained of an overheating problem, it must be determined if the complaint was due to an actual boil over, or the "Temp" light, or temperature gage indicated overheating.

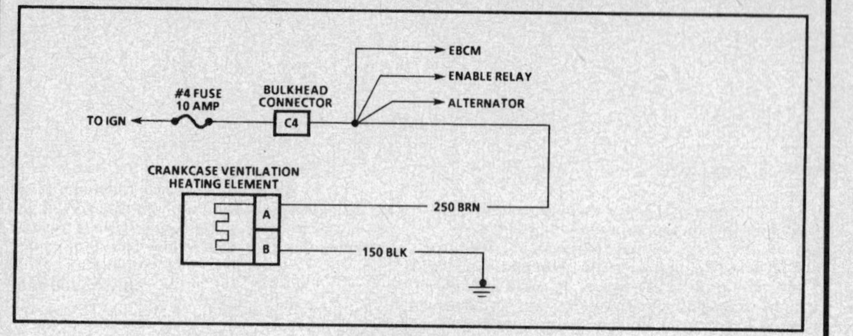

CHART C-13
CRANKCASE VENTILATION HEATER ASSEMBLY
FUNCTIONAL CHECK
2.3L (VIN A) "L" CARLINE

Circuit Description:

The crankcase ventilation heater assembly is used to prevent icing in the crankcase ventilation system. The heating element inside the vent hose is used to prevent icing from the oil/air separator to the intake manifold.

The heating element consists of two parallel wires that extend the length of the vent hose. One wire supplies the current when the ignition is turned "ON" while the second wire provides a constant path to ground.

Although the wires are not physically attached, the material between the two wires is conductive. Current is passed through the material between the wires. As the material is heated, electrical resistance increases and current flow decreases. When the material is cooled, electrical resistance decreases, current flow and heat output increases.

By this means, the temperature of the heating element is controlled to maintain approximately 115°F.

Test Description: Number(s) below refer to circled number(s) on the diagnostic chart.
1. This step verifies voltage is present to the crankcase ventilation heater assembly system.
2. This step insures a good path to ground in CKT 150.
3. This step identifies a visual failure.
4. This step identifies an open, or out of range crankcase ventilation heater element.
5. This confirms that the crankcase ventilation heating element is operational and is functioning properly.

Diagnostic Aids:

Check to insure that both battery and associated vehicle electrical connections are clean and tight. The electrical connection in the electrical system could lead to voltage spikes in the crankcase ventilation heater assembly system, causing damage to the crankcase ventilation heating element.

CHART C-13
CRANKCASE VENTILATION HEATER
ASSEMBLY FUNCTIONAL CHECK
2.3L (VIN A) "L" CARLINE

NOTICE: CONFIRM BOTH BATTERY AND VEHICLE ELECTRICAL CONNECTIONS ARE CLEAN & TIGHT BEFORE PROCEEDING.

- (1) • IGNITION "OFF."
  - DISCONNECT 2 WAY CRANKCASE VENTILATION HEATER ASSEMBLY CONNECTOR.
  - IGNITION "ON."
  - USING A TEST LIGHT CONNECTED TO GROUND, PROBE TERMINAL "A" OF THE CRANKCASE VENTILATION HEATER ASSEMBLY CONNECTOR.
  - TEST LIGHT SHOULD LIGHT. DOES IT?

- NO → INSPECT #4 10 AMP FUSE. IS FUSE BLOWN?
  - YES → (goes to step 3)
  - NO → REPAIR OPEN IN CKT 250 AND REPEAT STEP #1.

- YES → (2) • CONNECT TEST LIGHT BETWEEN TERMINALS "A" & "B" OF THE 2 WAY CRANKCASE VENTILATION HEATER ASSEMBLY CONNECTOR. TEST LIGHT SHOULD LIGHT. DOES IT?
  - YES → (goes to step 3)
  - NO → REPAIR OPEN IN CKT 150 AND REPEAT STEP #1.

- (3) • IGNITION "OFF."
  - REMOVE CRANKCASE VENTILATION HEATER ASSEMBLY AND INSPECT HOSE ASSEMBLY FOR A VISUAL FAILURE.

- NO VISUAL FAILURE PRESENT → (goes to step 4)
- VISUAL FAILURE PRESENT → REPLACE CRANKCASE VENTILATION HEATER ASSEMBLY AND PROCEED TO STEP #5.

- (4) • USING A DVOM CONNECTED BETWEEN TERMINALS "A" & "B" OF THE CRANKCASE VENTILATION HEATER ASSEMBLY, MEASURE RESISTANCE AT ROOM TEMPERATURE. RESISTANCE SHOULD BE BETWEEN 2 AND 6 OHMS. IS IT?

- YES → (goes to step 5)
- NO → REPLACE CRANKCASE VENTILATION HEATER ASSEMBLY.

- (5) • WITH CRANKCASE VENTILATION HEATER ASSEMBLY REMOVED, RECONNECT CRANKCASE VENTILATION HEATER ASSEMBLY ELECTRICAL CONNECTOR.
  - IGNITION "ON" FOR 1 MINUTE.
  - HEATING ELEMENT SHOULD BE WARM TO THE TOUCH. IS IT?

- YES → LUBRICATE HOSE ENDS WITH MILD SOAP AND REINSTALL CRANKCASE VENTILATION HEATER ASSEMBLY ON VEHICLE.
- NO → INSPECT ELECTRICAL CONNECTOR FOR PROPER CONNECTION. IF OK, REPLACE CRANKCASE VENTILATION HEATER AND REPEAT STEP #5.

2.3L (VIN D & 3) ENGINE — ENGINE COMPONENT LOCATION CHART — 1992 SKYLARK

"N" CARLINE   RPO:L40   VIN CODE: 3   2.3L L4 PFI

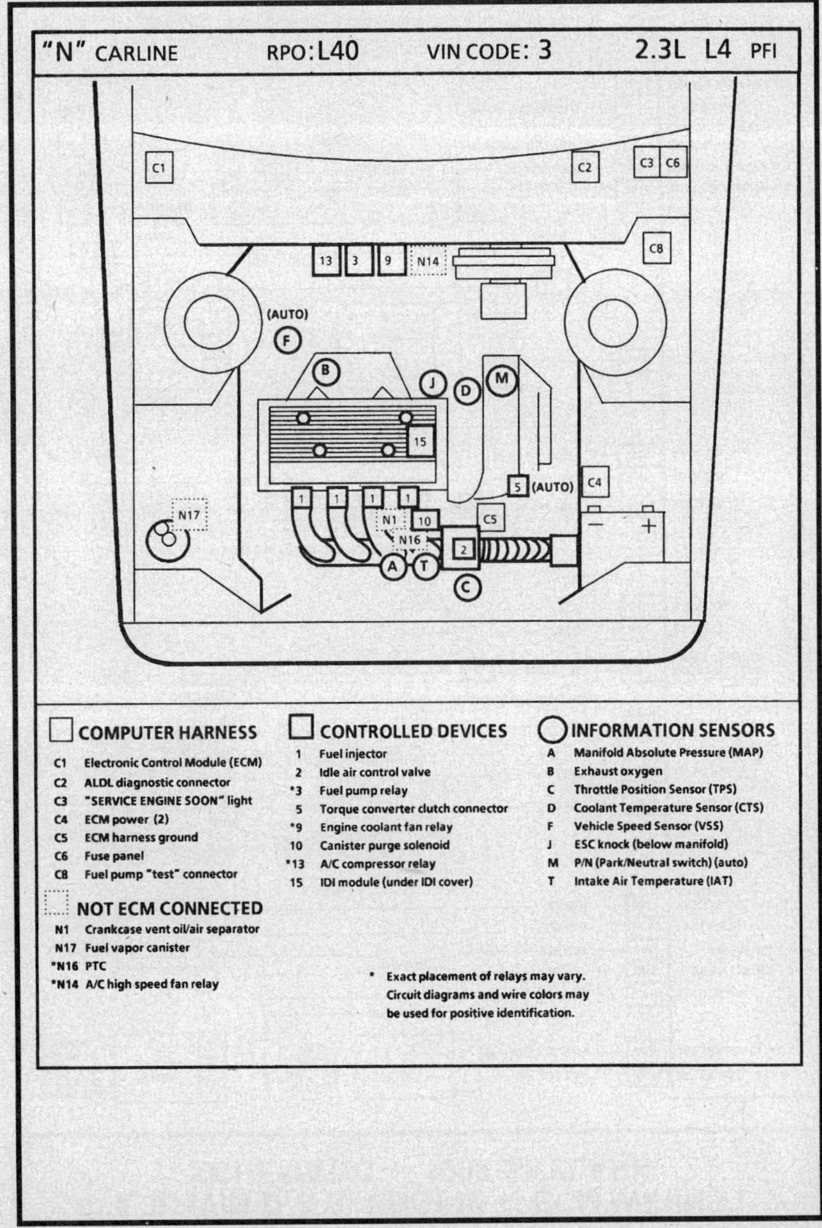

COMPUTER HARNESS
C1 Electronic Control Module (ECM)
C2 ALDL diagnostic connector
C3 "SERVICE ENGINE SOON" light
C4 ECM power (2)
C5 ECM harness ground
C6 Fuse panel
C8 Fuel pump "test" connector

NOT ECM CONNECTED
N1 Crankcase vent oil/air separator
N17 Fuel vapor canister
*N16 PTC
*N14 A/C high speed fan relay

CONTROLLED DEVICES
1 Fuel injector
2 Idle air control valve
*3 Fuel pump relay
5 Torque converter clutch connector
*9 Engine coolant fan relay
10 Canister purge solenoid
*13 A/C compressor relay
15 IDI module (under IDI cover)

INFORMATION SENSORS
A Manifold Absolute Pressure (MAP)
B Exhaust oxygen
C Throttle Position Sensor (TPS)
D Coolant Temperature Sensor (CTS)
F Vehicle Speed Sensor (VSS)
J ESC knock (below manifold)
M P/N (Park/Neutral switch) (auto)
T Intake Air Temperature (IAT)

* Exact placement of relays may vary. Circuit diagrams and wire colors may be used for positive identification.

2.3L (VIN D, A & 3) ENGINE — ENGINE COMPONENT LOCATION CHART — 1992 ACHIEVA AND GRAND AM

'N' CARLINE   RPO:LD2/LG0   VIN CODE: D, A & 3   2.3L L4 PFI

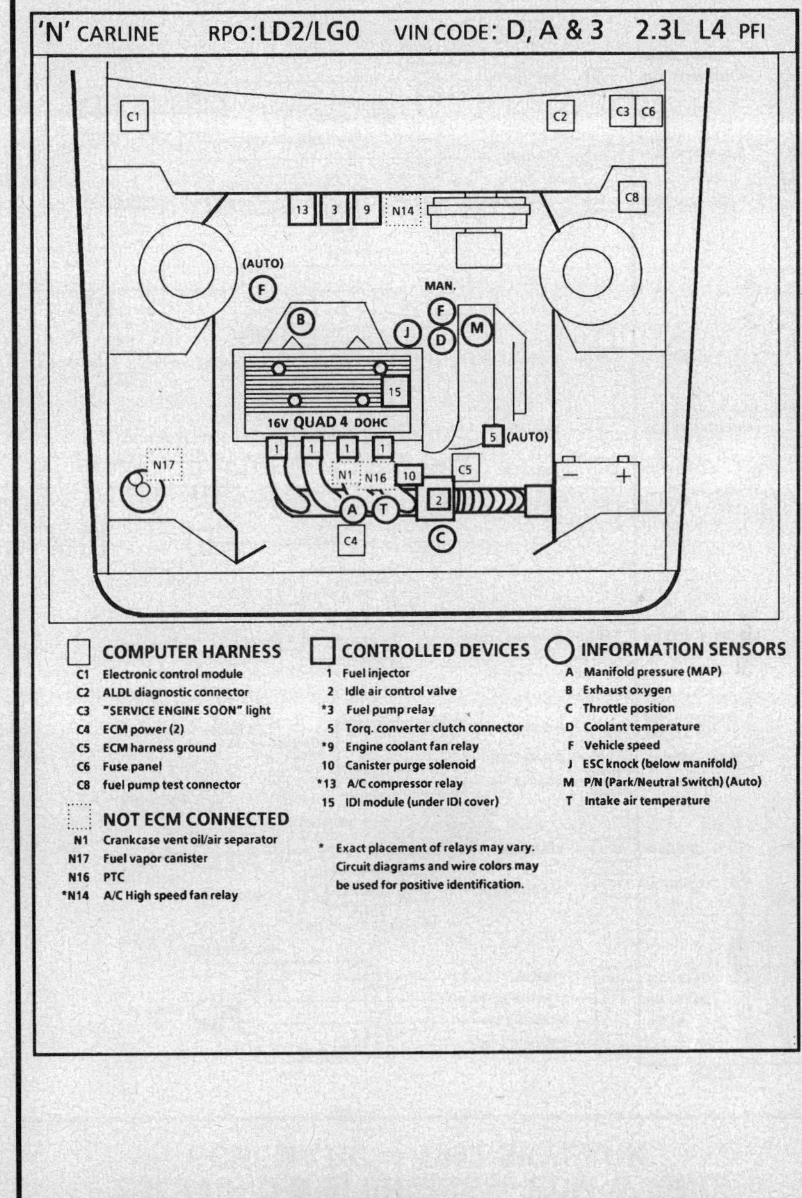

COMPUTER HARNESS
C1 Electronic control module
C2 ALDL diagnostic connector
C3 "SERVICE ENGINE SOON" light
C4 ECM power (2)
C5 ECM harness ground
C6 Fuse panel
C8 fuel pump test connector

NOT ECM CONNECTED
N1 Crankcase vent oil/air separator
N17 Fuel vapor canister
N16 PTC
*N14 A/C High speed fan relay

CONTROLLED DEVICES
1 Fuel injector
2 Idle air control valve
*3 Fuel pump relay
5 Torq. converter clutch connector
*9 Engine coolant fan relay
10 Canister purge solenoid
*13 A/C compressor relay
15 IDI module (under IDI cover)

INFORMATION SENSORS
A Manifold pressure (MAP)
B Exhaust oxygen
C Throttle position
D Coolant temperature
F Vehicle speed
J ESC knock (below manifold)
M P/N (Park/Neutral Switch) (Auto)
T Intake air temperature

* Exact placement of relays may vary. Circuit diagrams and wire colors may be used for positive identification.

2.3L (VIN D & 3) ENGINE — ECM WIRING SCHEMATIC — 1992 SKYLARK

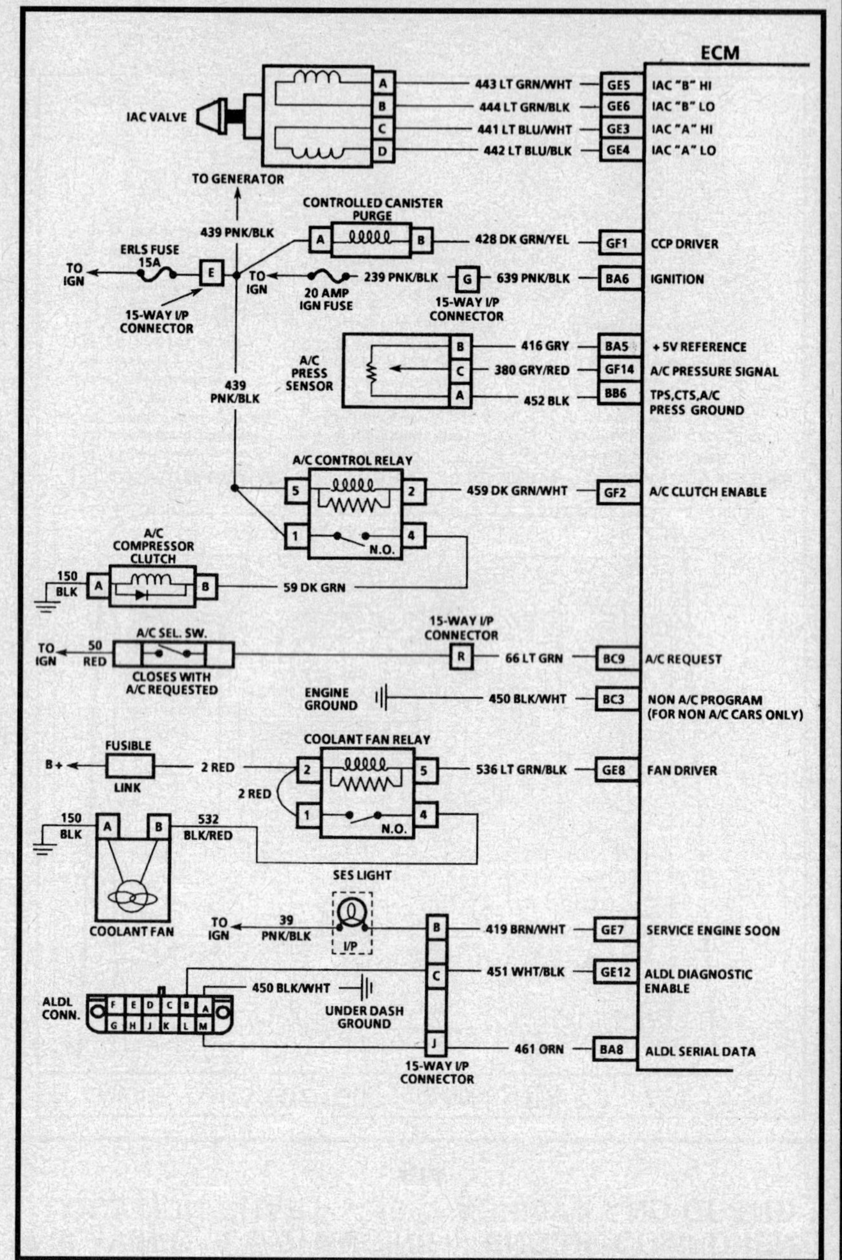

2.3L (VIN D & 3) ENGINE — ECM WIRING SCHEMATIC — 1992 SKYLARK

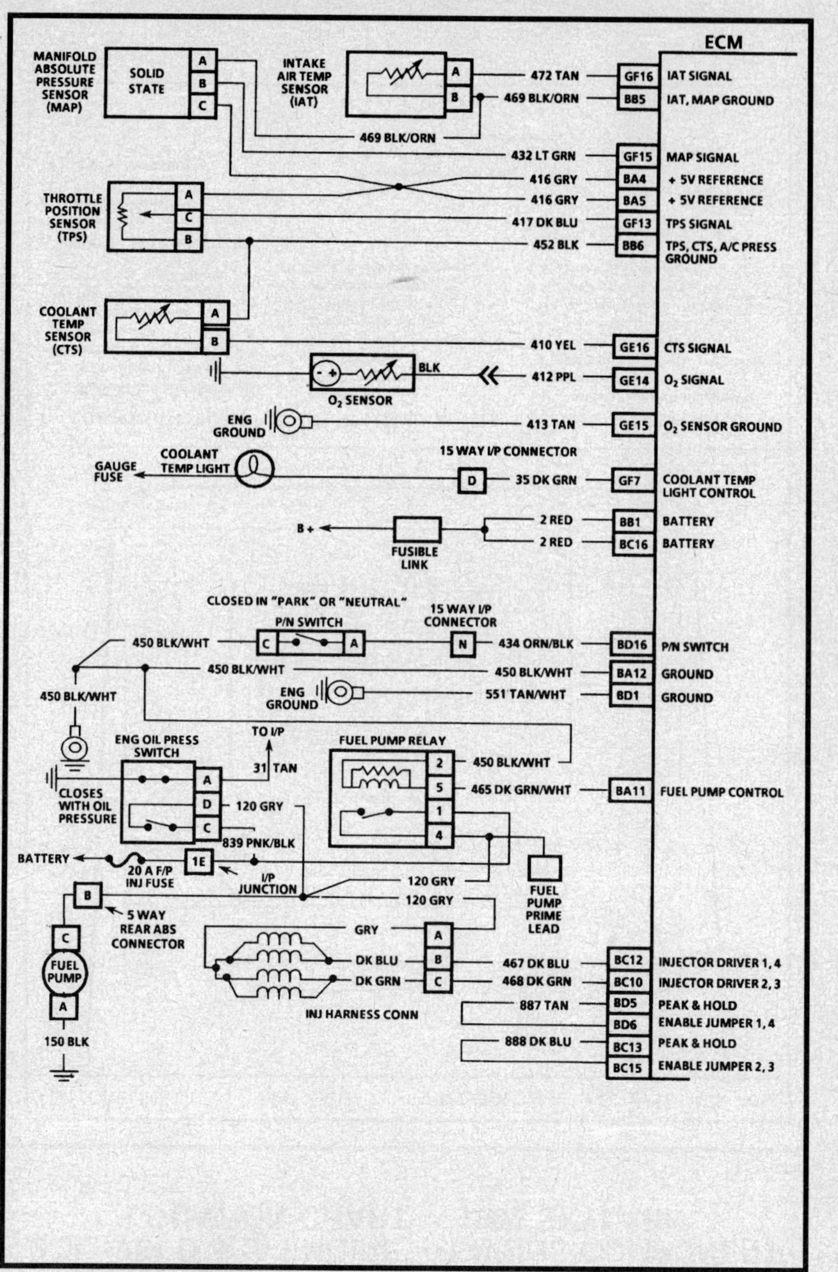

2.3L (VIN D, A & 3) ENGINE — ECM WIRING SCHEMATIC — 1992 ACHIEVA AND GRAND AM

2.3L (VIN D & 3) ENGINE — ECM WIRING SCHEMATIC — 1992 SKYLARK

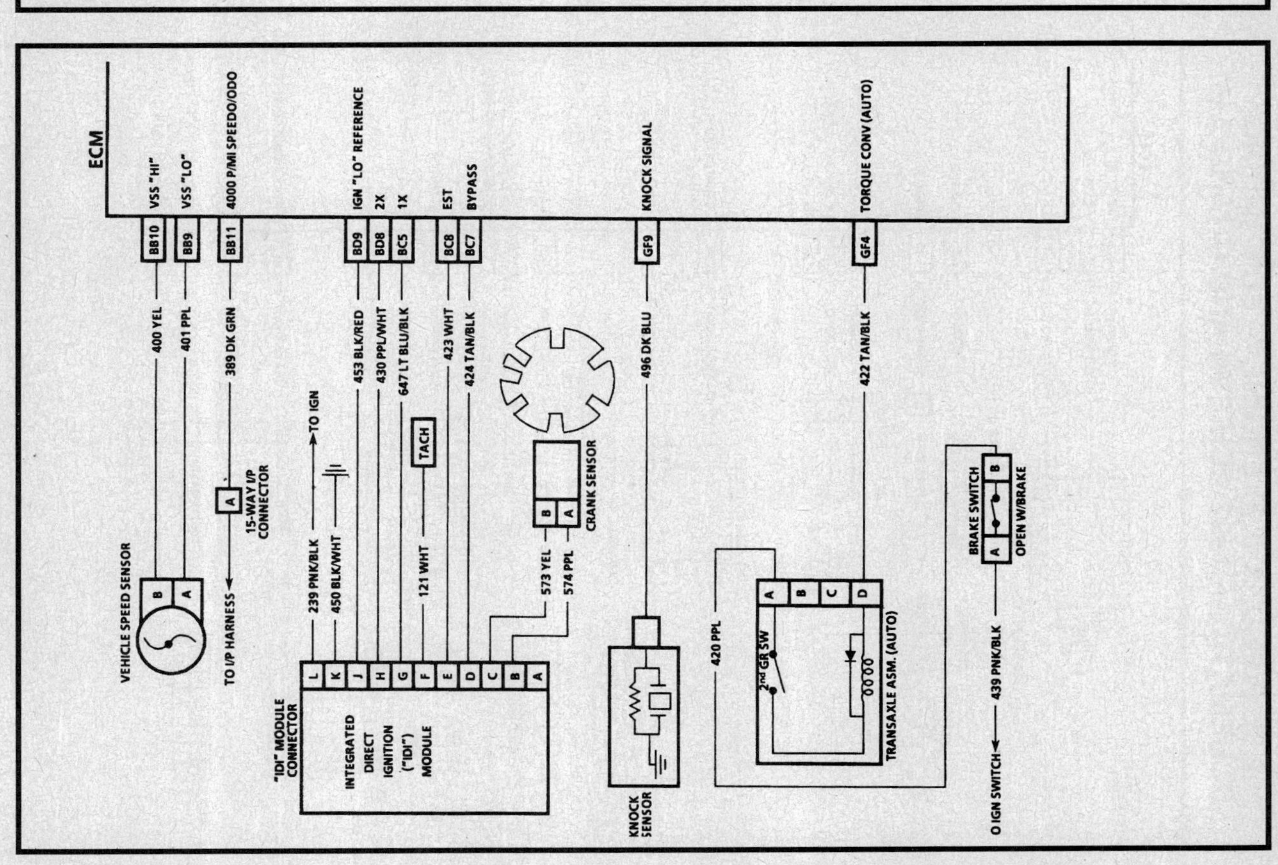

2.3L (VIN D, A & 3) ENGINE — ECM WIRING SCHEMATIC — 1992 ACHIEVA AND GRAND AM

2.3L (VIN D, A & 3) ENGINE — ECM WIRING SCHEMATIC — 1992 ACHIEVA AND GRAND AM

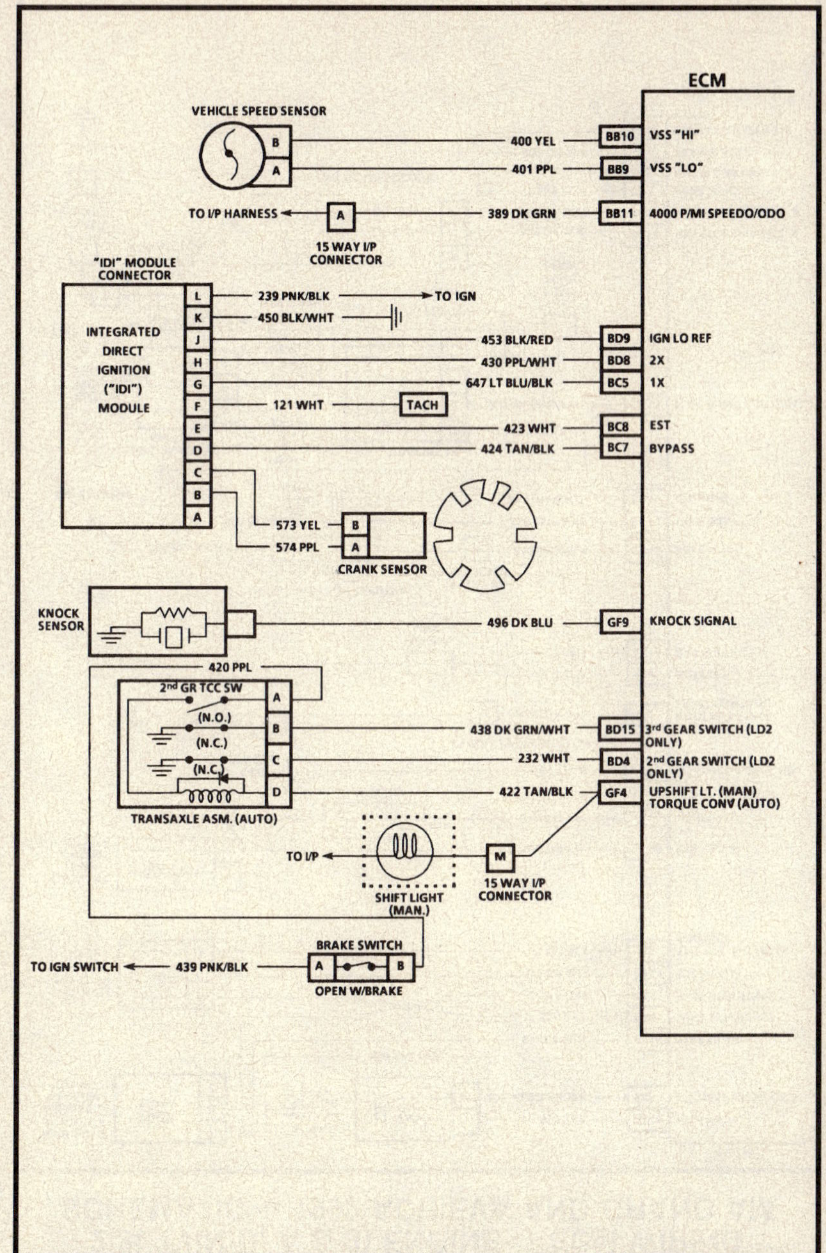

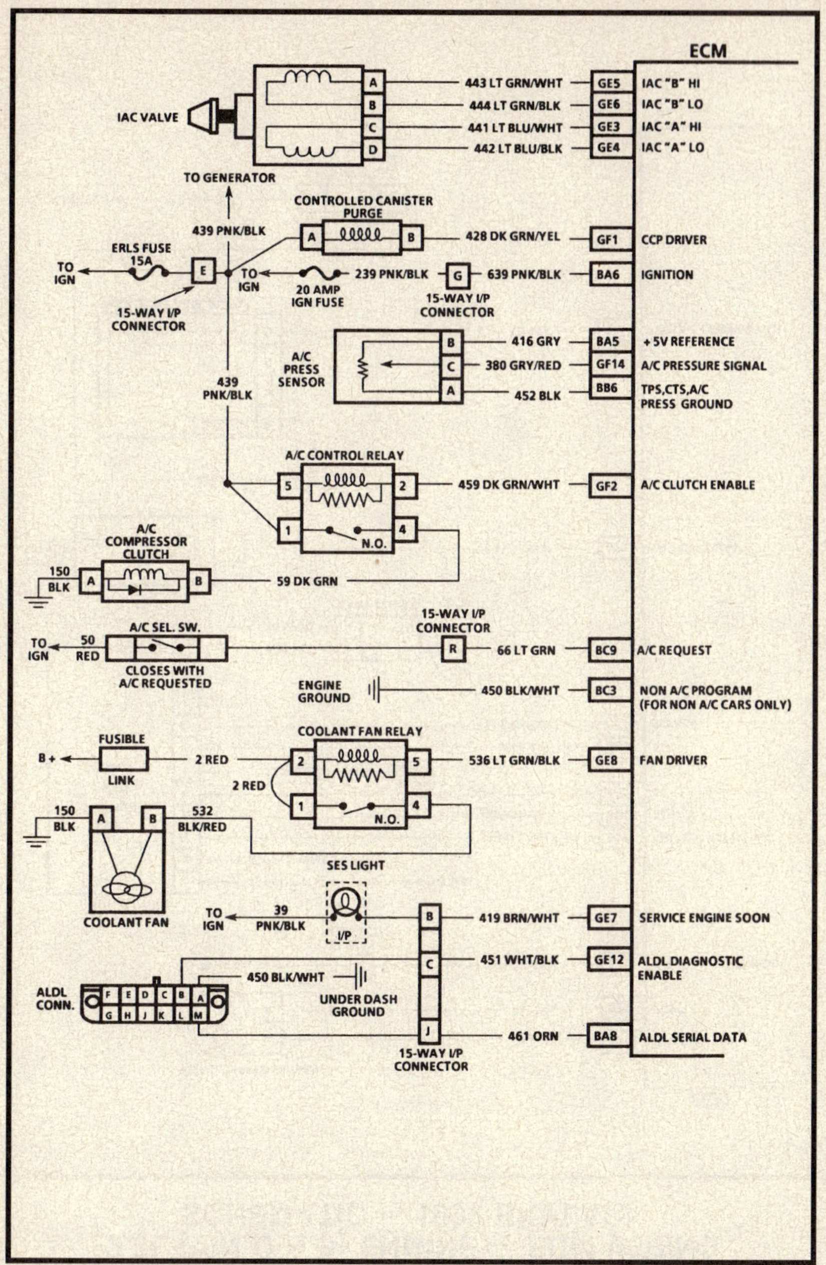

2.3L (VIN D & 3) ENGINE — ECM CONNECTOR END VIEW — 1992 SKYLARK

PORT FUEL INJECTION ECM CONNECTOR IDENTIFICATION

This ECM voltage chart is for use with a digital voltmeter to further aid in diagnosis. The voltages you get may vary due to low battery charge or other reasons, but they should be very close. The "B +" symbol indicates a nominal system voltage of 12-14 V.

THE FOLLOWING CONDITIONS MUST BE MET BEFORE TESTING:
- Engine at operating temperature (upper rad. hose hot) ● Engine idling in "Closed Loop" (For "Engine Run" column) in park or neutral ● Test terminal not grounded ● Tech 1 not installed

CONNECTOR (C1) — 32 PIN C-D CONNECTOR — BD1, BC1 — BACK VIEW OF CONNECTOR (BLACK)

CONNECTOR (C2) — 24 PIN A-B CONNECTOR — BB1, BA1 — BACK VIEW OF CONNECTOR (BLACK)

CONNECTOR (C3) — 32 PIN E-F CONNECTOR — GF1, GE1 — BACK VIEW OF CONNECTOR (GREEN)

(NOTICE: E, F CAVITY IDENTIFICATION IS VISIBLE ON CONNECTOR BODY WITH TERMINAL RETAINERS REMOVED.)

CONNECTOR 1 (C1) — BLACK 32 PIN C-D CONNECTOR

| VOLTAGE | | | | | | VOLTAGE | | | |
KEY "ON"	ENG. RUN	CIRCUIT	PIN	WIRE COLOR	WIRE COLOR	PIN	CIRCUIT	KEY "ON"	ENG. RUN
			BC1		TAN/WHT	BD1	ECM GROUND	0★	0★
			BC2			BD2			
B + (5)	B+	NON A/C PROG.	BC3	BLK/WHT		BD3			
			BC4						
		1X SIGNAL	BC5	BLK/LT GRN	LT GRN/BLK	BD5	PEAK & HOLD INJ JUMPER 1 & 4	0★	0★
			BC6		LT GRN/BLK	BD6	PEAK & HOLD INJ JUMPER 1 & 4	0★	0★
0★	5.	IGNITION BYPASS	BC7	TAN/BLK		BD7			
0★	.2	IGNITION EST	BC8	WHT	PPL/WHT	BD8	2X SIGNAL	0★	2.9
0★	ON B+ OFF 0★	A/C REQUEST	BC9	LT GRN	BLK/RED	BD9	IGN GROUND	0★	0★
B+	B+	INJ DRIVER 2 & 3	BC10	LT GRN		BD10			
			BC11			BD11			
B+	B+	INJ DRIVER 1 & 4	BC12	LT BLU		BD12			
0★	0★	PEAK & HOLD INJ JUMPER 2 & 3	BC13	ORN		BD13			
			BC14			BD14			
0★	0★	PEAK & HOLD INJ JUMPER 2 & 3	BC15	ORN					
B+	B+	BATTERY	BC16	RED	ORN/BLK	BD16	P/N SWITCH A/T	ON 0★ OFF B+	ON 0★ OFF B+

NOTICE: BC1-Black connector, cavity C1; GE1-Green connector, cavity E1, etc.

★ Less than .5 volt.
1. Varies from .60 to battery voltage depending on position of drive wheels.
2. Varies.
3. 12 Volts first two seconds.
4. Varies with temperature.
5. Non A/C cars 0★ volt. A/C cars should not have a wire in terminal "BC3".

ENGINE - 2.3L / LD2
CARLINE - "N" Series

2.3L (VIN D & 3) ENGINE — ECM CONNECTOR END VIEW — 1992 SKYLARK

CONNECTOR 2 (C-2) BLACK 24 PIN A-B CONNECTOR

| VOLTAGE | | | | | | VOLTAGE | | | |
ENG. RUN	KEY "ON"	CIRCUIT	PIN	WIRE COLOR	WIRE COLOR	PIN	CIRCUIT	KEY "ON"	ENG. RUN
			BA1		RED	BB1	BATTERY	B+	B+
			BA2			BB2			
			BA3			BB3			
5.0	5.0	+5V REFERENCE	BA4	GRY		BB4			
5.0	5.0	+5V REFERENCE	BA5	GRY	BLK/ORN	BB5	IAT & MAP GND	0★	0★
B+	B+	IGNITION FEED	BA6	PNK/BLK	BLK	BB6	A/C,CTS,TPS GND	0★	0★
			BA7			BB7			
(2)		SERIAL DATA/ALDL	BA8	ORN		BB8			
			BA9		PPL	BB9	MAG. VSS LOW	0★	0★
			BA10		YEL	BB10	MAG. VSS HIGH	0★	0★
0★ (3)	B+	FUEL PUMP	BA11	GRN/WHT	DK GRN	BB11	4000 P/MI SPEED	10.0	11.4
0★	0★	ECM GROUND	BA12	BLK/WHT		BB12			

GREEN 32 PIN E-F CONNECTOR 3 (C-3)

| VOLTAGE | | | | | | VOLTAGE | | | |
ENG. RUN	KEY "ON"	CIRCUIT	PIN	WIRE COLOR	WIRE COLOR	PIN	CIRCUIT	KEY "ON"	ENG. RUN
			GE1		GRN/YEL	GF1	CANISTER PURGE	B+	.3
			GE2		DK GRN/WHT	GF2	A/C CLUTCH RELAY	B+	OFF B+ ON 0★
NOT	USEABLE	IAC - A - HIGH	GE3	LT BLU/WHT		GF3			
NOT	USEABLE	IAC - A - LOW	GE4	LT BLU/BLK	TAN/BLK	GF4	TCC	0	0
NOT	USEABLE	IAC - B - HIGH	GE5	LT GRN/WHT		GF5			
NOT	USEABLE	IAC - B - LOW	GE6	LT GRN/BLK		GF6			
0★	B+	SES LIGHT	GE7	BRN/WHT	DK GRN	GF7	COOLANT TEMP LIGHT		
B+	ON 0★ OFF B+	CLG FAN RLY	GE8	LT GRN/BLK		GF8			
			GE9		DK BLU	GF9	KNOCK SIGNAL	2.3	2.3
			GE10			GF10			
			GE11			GF11			
5.0	5.0	ALDL/DIAG TERM	GE12	WHT/BLK		GF12			
			GE13		DK BLU	GF13	TPS SIGNAL	.54	.54
.3 - .5	.1 - .9	O$_2$ SIGNAL	GE14	PPL	GRY/RED	GF14	A/C PRESS SIGNAL		(2)
0★ (2)	0★ (2)	O$_2$ GROUND	GE15	TAN	GRN	GF15	MAP SIGNAL	4.7	1.4 (2)
1.8 (2)	1.8 (2)	CLNT TEMP SIGNAL	GE16	YEL	TAN	GF16	IAT SIGNAL	3.6	1.5 (2)

Notice: BA1 = Black Connector, cavity A1, etc. GE1 - Green Connector, cavity E1, etc.

★ Less than .5 volt.
1. Varies from .60 to battery voltage depending on position of drive wheels.
2. Varies.
3. B+ first two seconds.

ENGINE - 2.3L / LD2
CARLINE - "N" SERIES

2.3L (VIN D, A & 3) ENGINE — ECM CONNECTOR END VIEW — 1992 ACHIEVA AND GRAND AM

CONNECTOR 2 (C-2) — BLACK 24 PIN A-B CONNECTOR

VOLTAGE KEY ON	ENG. RUN	CIRCUIT	PIN	WIRE COLOR
			BA1	
			BA2	
			BA3	
5.0	5.0	+5V REFERENCE	BA4	GRY
5.0	5.0	+5V REFERENCE	BA5	GRY
B+	B+	IGNITION FEED	BA6	PNK/BLK
			BA7	
		SERIAL DATA/ALDL	BA8	ORN (2)
			BA9	
			BA10	
0*	B+	FUEL PUMP	BA11	DK GRN/WHT (3)
0*	0*	ECM GROUND	BA12	BLK/WHT

WIRE COLOR	PIN	CIRCUIT	KEY ON	ENG. RUN VOLTAGE
RED	BB1	BATTERY	B+	B+
	BB2			
	BB3			
	BB4			
BLK/ORN	BB5	IAT & MAP GND	0*	0*
BLK	BB6	A/C,CTS,TPS GND	0*	0*
	BB7			
	BB8			
PPL	BB9	MAG. VSS LOW	0*	0*
YEL	BB10	MAG. VSS HIGH	0*	0*
DK GRN	BB11	4000 P/MI SPEED	10.0	11.4
	BB12			

GREEN 32 PIN E-F CONNECTOR 3 (C-3)

VOLTAGE KEY ON	ENG. RUN	CIRCUIT	PIN	WIRE COLOR
			GE1	
			GE2	
NOT USEABLE		IAC - A - HIGH	GE3	LT BLU/WHT
NOT USEABLE		IAC - A - LOW	GE4	LT BLU/BLK
NOT USEABLE		IAC - B - HIGH	GE5	LT GRN/WHT
NOT USEABLE		IAC - B - LOW	GE6	LT GRN/BLK
0*	B+	SES LIGHT	GE7	LT BRN/WHT
B+	OFF B+ ON 0*	CLG FAN RLY	GE8	LT GRN/BLK
			GE9	
			GE10	
			GE11	
5.0	5.0	ALDL/DIAG TERM	GE12	WHT/BLK
			GE13	
.3 - .5	.1 - .9	O₂ SIGNAL	GE14	PPL (2)
0*	0*	O₂ GROUND	GE15	TAN
1.8	1.8	CLNT TEMP SIGNAL	GE16	YEL (2)

WIRE COLOR	PIN	CIRCUIT	KEY ON	ENG. RUN VOLTAGE
YEL	GF1	CANISTER PURGE	B+	.3
DK GRN/WHT	GF2	A/C CLUTCH RELAY	B+	OFF B+ ON 0*
	GF3			
TAN/BLK	GF4	TCC / SHIFT LT	12 / 0	12 / 0
	GF5			
	GF6			
DK GRN	GF7	COOLANT TEMP LIGHT		
	GF8			
DK BLU	GF9	KNOCK SIGNAL	2.3	2.3
	GF10			
	GF11			
	GF12			
DK BLU	GF13	TPS SIGNAL	.54	.54
GRY/RED	GF14	A/C PRESS SIGNAL			(2)
LT GRN	GF15	MAP SIGNAL	4.7	1.4	(2)
TAN	GF16	IAT SIGNAL	3.6	1.5	(2)

* Less than .5 volt.
1. Varies from .60 to battery voltage depending on position of drive wheels.
2. Varies.
3. B + first two seconds.

NOTICE: BA1 = Black Connector, cavity A1, etc.   GE1 - Green Connector, cavity E1, etc.

ENGINE - 2.3L / LD2/LG0/L40
CARLINE - "N" Series

2.3L (VIN D, A & 3) ENGINE — ECM CONNECTOR END VIEW — 1992 ACHIEVA AND GRAND AM

PORT FUEL INJECTION ECM CONNECTOR IDENTIFICATION

This ECM voltage chart is for use with a digital voltmeter to further aid in diagnosis. The voltages you get may vary due to low battery charge or other reasons, but they should be very close. The "B +" symbol indicates a nominal system voltage of 12-14 V.

THE FOLLOWING CONDITIONS MUST BE MET BEFORE TESTING:
- Engine at operating temperature (upper rad. hose hot) • Engine idling in "Closed Loop" (For "Engine Run" column) in park or neutral • Test terminal not grounded • Tech 1 not installed

CONNECTOR (C1) 32 PIN C-D CONNECTOR — BACK VIEW OF CONNECTOR (BLACK) — BD1, BC1

CONNECTOR (C2) 24 PIN A-B CONNECTOR — BACK VIEW OF CONNECTOR (BLACK) — BB1, BA1

CONNECTOR (C3) 32 PIN E-F CONNECTOR — BACK VIEW OF CONNECTOR (GREEN) — GF1, GE1

(NOTICE: E, F CAVITY IDENTIFICATION IS VISIBLE ON CONNECTOR BODY WITH TERMINAL RETAINERS REMOVED.)

CONNECTOR 1 (C1) — BLACK 32 PIN C-D CONNECTOR

VOLTAGE KEY "ON"	ENG. RUN	CIRCUIT	PIN	WIRE COLOR
			BC1	
			BC2	
(5) B+	B+	NON A/C PROG.	BC3	BLK/WHT
			BC4	
		1X SIGNAL	BC5	LT BLU/BLK
			BC6	
0*	.5	IGNITION BYPASS	BC7	TAN/BLK
0*	.2	IGNITION EST	BC8	WHT
0*	ON B+ OFF 0*	A/C REQUEST	BC9	LT GRN
B+	B+	INJ DRIVER 2 & 3	BC10	DK GRN
			BC11	
B+	B+	INJ DRIVER 1 & 4	BC12	DK BLU
0*	0*	PEAK & HOLD INJ JUMPER 2 & 3	BC13	DK BLU
			BC14	
0*	0*	PEAK & HOLD INJ JUMPER 2 & 3	BC15	DK BLU
B+	B+	BATTERY	BC16	RED

WIRE COLOR	PIN	CIRCUIT	KEY "ON"	ENG. RUN VOLTAGE
TAN/WHT	BD1	ECM GROUND	0*	0*
	BD2			
	BD3			
WHT	BD4	2nd GEAR SW A/T	ON B+ OFF 0*	ON B+ OFF 0*
TAN	BD5	PEAK & HOLD INJ JUMPER 1 & 4	0*	0*
TAN	BD6	PEAK & HOLD INJ JUMPER 1 & 4	0*	0*
	BD7			
PPL/WHT	BD8	2X SIGNAL	0*	2.9
BLK/RED	BD9	IGN GROUND	0*	0*
	BD10			
	BD11			
	BD12			
	BD13			
	BD14			
DK GRN/WHT	BD15	3rd GEAR SW A/T	ON 0* OFF B+ (6)	ON 0* OFF B+
ORN/BLK	BD16	P/N SWITCH A/T	ON 0* OFF B+	ON 0* OFF B+

NOTICE: BC1=Black connector, cavity C1; GE1=Green connector, cavity E1, etc.

* Less than .5 volt.
(1) Varies from .60 to battery voltage depending on position of drive wheels.
(2) Varies.
(3) 12Volt first two seconds.
(4) Varies with temperature.
(5) Non A/C cars 0* volt. A/C cars should not have a wire in term. "BC3".
(6) LD2 Only.

ENGINE - 2.3L / LD2/LG0/L40
CARLINE - "N" Series

2.3L (VIN D, A & 3) ENGINE — ON-BOARD DIAGNOSTIC SYSTEM CHART — 1992 ACHIEVA AND GRAND AM

DIAGNOSTIC CIRCUIT CHECK

The Diagnostic Circuit Check is an organized approach to identifying a problem created by an electronic engine control system malfunction. It must be the starting point for any driveability complaint diagnosis, because it directs the service technician to the next logical step in diagnosing the complaint.

The Tech 1 data listed in the table may be used for comparison, after completing the diagnostic circuit check and finding the on-board diagnostics functioning properly and no trouble codes displayed. The "Typical Values" are an average of display values recorded from normally operating vehicles and are intended to represent what a normally functioning system would typically display.

A "SCAN" TOOL THAT DISPLAYS FAULTY DATA SHOULD NOT BE USED, AND THE PROBLEM SHOULD BE REPORTED TO THE MANUFACTURER. THE USE OF A FAULTY "SCAN" CAN RESULT IN MISDIAGNOSIS AND UNNECESSARY PARTS REPLACEMENT.

Only the parameters listed below are used in this manual for diagnosing. If a "Scan" reads other parameters, the values are not recommended by General Motors for use in diagnosing.

TECH 1 DATA

Idle / Upper Radiator Hose Hot / Closed Throttle / Park or Neutral / "Closed Loop" / Acc. "OFF"

"SCAN" Position	Units Displayed	Typical Data Value
Engine Speed	RPM	± 100 RPM from desired RPM (± 50 in drive)
Desired Idle	RPM	ECM idle command (varies with calibration. temp.)
Coolant Temp.	C° F°	85° – 115°C
IAT Temp.	C° F°	10° – 80°C (depends on underhood temp.)
MAP	kPa, V	1 – 3 Volts (depends on Vacuum & Baro pressure)
BARO	kPa, V	3–5 Volts (depends on altitude & Baro pressure)
Throttle Position	Volts	.400 – .900 (up to 5.0 at wide open throttle)
Throttle Angle	0 – 100%	0% (up to 100% at wide open throttle)
Oxygen Sensor	mV	1–1000 and varying
Injector Pulse Width	m Sec.	1 – 4 and varying
Spark Advance	# of Degrees	Varies
Lo Oct. Spark Ret.	# of Degrees	Varies
Fuel Integrator	Counts	Varies
Block Learn	Counts	58 – 198 (see Section "C2")
Open/Closed Loop	Open/Closed	Closed Loop (may go open with extended idle)
Block Learn Cell	Cell Number	18 to 21 at idle (depends on Air Flow, RPM,P/N&A/C)
Knock Retard	Degrees of Retard	0
Knock Signal	Yes/No	No
BYP Line Volts	LOW/HI	HI
EST Command	Yes/No	Yes
Idle Air Control	Counts (steps)	5 – 60
Park Neutral Switch	Park Neutral and RDL	P – N – (or -R-DL manual only)
VSS	MPH/KPH	0
Torque Conv. Cl. (TCC)	On/Off	Off ("ON," with TCC commanded)
Battery Voltage	Volts	13.5 – 14.5
1X Ref. Pulse	0 – 255 Counts	0 – 255 (Useable for Code 41)
2X Ref. Pulse	0 – 255 Counts	0 – 255 (Useable for Code 16)
A/C Request	Yes/No	No (yes, with A/C requested, ie: selector "ON")
A/C Clutch	On/Off	Off ("ON," with A/C commanded on)
A/C Clutch	On/Off	Off ("ON," with A/C commanded on)
A/C Pressure	psi/Volts	0 – 450 psi (varying with high side pressure)
Cool Fan Relay	On/Off	Off ("ON," with A/C "ON" or hot eng)
Purge Duty Cycle	%	0–100%
QDM A	LOW/HI	Low
QDM B	LOW/HI	Low
2nd Gear	Yes/No (N/A L40 Only)	No (yes, when in 2nd or 3rd gear) or yes (Man. Trans. only)
3rd Gear	Yes/No (N/A L40 Only)	No (yes, when in 3rd gear) or yes (Man. Trans. only)
PROM ID	#	Production ECM/PROM ID (not useable)
Time From Start	min/sec	Varies (engine run time since start)

2.3L (VIN D, A & 3) ENGINE — ON-BOARD DIAGNOSTIC SYSTEM CHART — 1992 ACHIEVA, GRAND AM AND SKYLARK

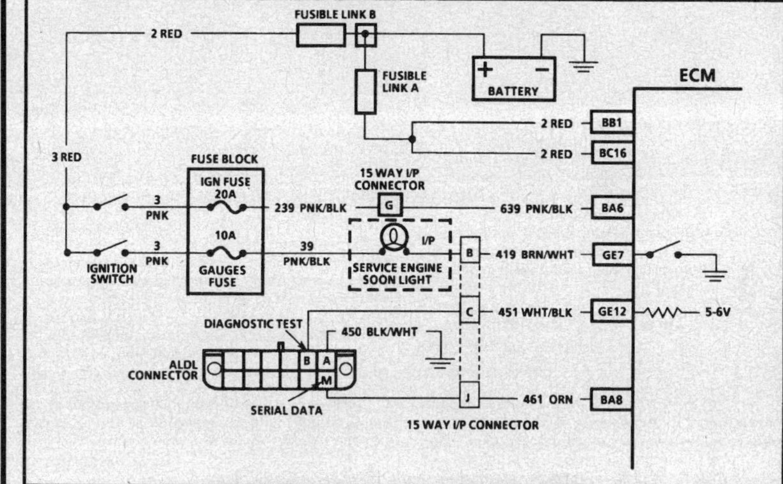

DIAGNOSTIC CIRCUIT CHECK
2.3L (VIN D, A & 3) "N" CARLINE (PORT)

Circuit Description:

The diagnostic circuit check is an organized approach to identifying a problem created by an Electronic Engine Control System malfunction. It must be the starting point for any driveability complaint diagnosis, because it directs the service technician to the next logical step in diagnosing the complaint. Understanding the chart and using it correctly will reduce diagnostic time and prevent the unnecessary replacement of good parts.

Test Description: Number(s) below refer to circled number(s) on the diagnostic chart.

1. This step is a check for the proper operation of the "Service Engine Soon" light. The "SES" light should be "ON" steady.
2. No "SES" light at this point indicates that there is a problem with the "SES" light circuit or the ECM control of that circuit.
3. This test checks the ability of the ECM to control the "SES" light. With the diagnostic terminal grounded, the "SES" light should flash a Code 12 three times, followed by any trouble code stored in memory.
4. Most of the   procedures use a Tech 1 to aid diagnosis, therefore, serial data must be available. If a PROM error is present, the ECM may have been able to flash Code 12/51, but not enable serial data.
5. Although the ECM is powered up, a "Cranks But Will Not Run" symptom could exist because of an ECM or system problem.
6. This step will isolate if the customer complaint is a "SES" light or a driveability problem with no "SES" light. An invalid code may be the result of a faulty "Scan" tool.
7. Comparison of actual control system data with the typical values is a quick check to determine if any parameter is not within limits. Keep in mind that a base engine problem (i.e., advanced cam timing) may substantially alter sensor values.
8. Installation of a "Scan" tool will provide a good ground path for the ECM and may hide a driveability complaint due to poor ECM grounds.
9. If the actual data is not within the typical values established, the charts   "Component Systems," will provide a functional check of the suspect component or system.

2.3L (VIN D, A & 3) ENGINE — ON-BOARD DIAGNOSTIC SYSTEM CHART — 1992 ACHIEVA, GRAND AM AND SKYLARK

DIAGNOSTIC CIRCUIT CHECK
2.3L (VIN D, A & 3) "N" CARLINE

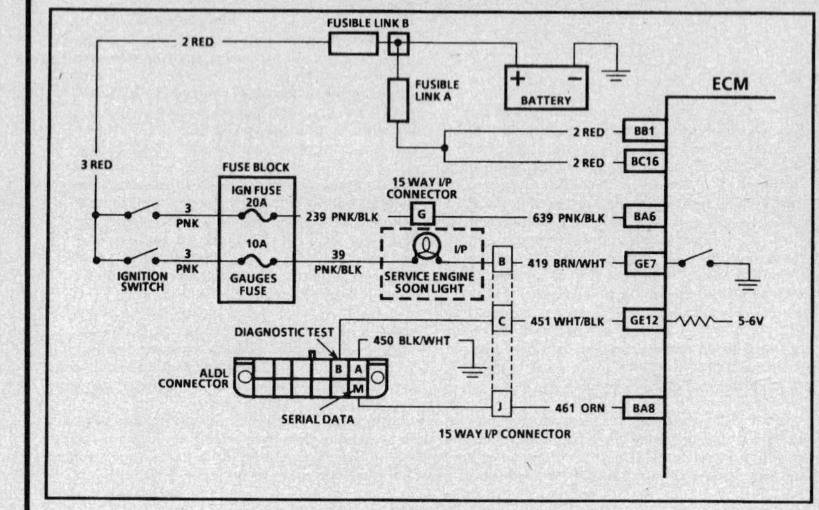

2.3L (VIN D, A & 3) ENGINE — SYSTEM DIAGNOSTIC CHARTS — 1992 ACHIEVA, GRAND AM AND SKYLARK

CHART A-1
NO "SERVICE ENGINE SOON" LIGHT
2.3L (VIN D, A & 3) "N" CARLINE (PORT)

Circuit Description:

There should always be a steady "Service Engine Soon" light when the ignition is "ON" and engine stopped. Battery voltage is supplied through the ignition switch directly to the light bulb. The Electronic Control Module (ECM) controls the light and turns it "ON" by providing a ground path through CKT 419 to the ECM.

Test Description: Number(s) below refer to circled number(s) on the diagnostic chart.

1. If the fusible link is blown, locate and repair short to ground.
2. Using a test light connected to 12 volts (B+) probe each of the system ground circuits to be sure a good ground is present. See ECM Terminal End View for ECM pin locations of ground circuits.

Diagnostic Aids:

Engine runs OK, check:
- Faulty light bulb.
- CKT 419 open.
- I/P fuse blown; this will result in the loss of "SES" light, oil light, brake light, etc. on I/P.

Engine cranks but will not run.
- Continuous battery - fuse or fusible link open.
- ECM ignition fuse open.
- Battery CKT 2 to ECM open.
- Ignition CKT 639 to ECM open.
- Poor connection to ECM.
- Poor ECM ground.

2.3L (VIN D, A & 3) ENGINE — SYSTEM DIAGNOSTIC CHARTS — 1992 ACHIEVA, GRAND AM AND SKYLARK

CHART A-1
NO "SERVICE ENGINE SOON" LIGHT
2.3L (VIN D, A & 3) "N" CARLINE (PORT)

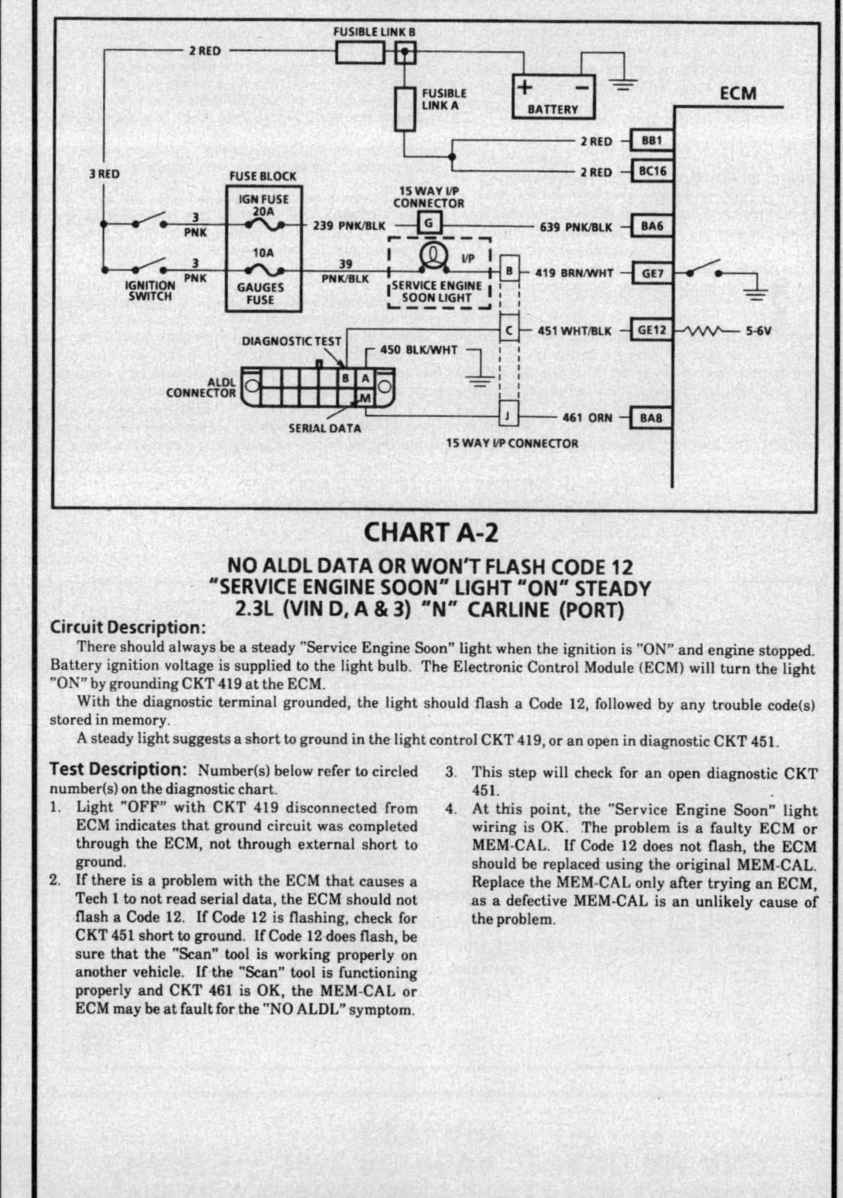

Diagnostic flow (Chart A-1):

DOES THE ENGINE START?

- **YES** →
  - IGNITION "OFF".
  - DISCONNECT ECM CONNECTORS.
  - IGNITION "ON"
  - PROBE CKT 419, WITH TEST LIGHT TO GROUND. IS THE "SES" LIGHT "ON"?
    - **YES** → FAULTY ECM CONNECTION OR ECM.
    - **NO** → CHECK:
      - I/P FUSE
      - FAULTY BULB
      - OPEN CKT 419
      - CKT 419 SHORTED TO VOLTAGE.
      - OPEN IGNITION FEED TO BULB.

- **NO** → ARE THE FUSIBLE LINK AND ECM FUSE OK?
  - **YES** →
    - IGNITION "ON".
    - BACK PROBE CAVITY "BB-1", "BC-16" AND "BA-6" WITH TEST LIGHT TO GROUND.
    - IS THE TEST LIGHT "ON" ALL CAVITIES?
      - **YES** → (2) FAULTY ECM GROUNDS OR ECM.
      - **NO** → REPAIR OPEN IN CIRCUIT THAT DID NOT LIGHT THE TEST LIGHT.
  - **NO** → (1) LOCATE AND CORRECT SHORT TO GROUND IN CIRCUIT THAT HAD A BLOWN FUSIBLE LINK OR FUSE.

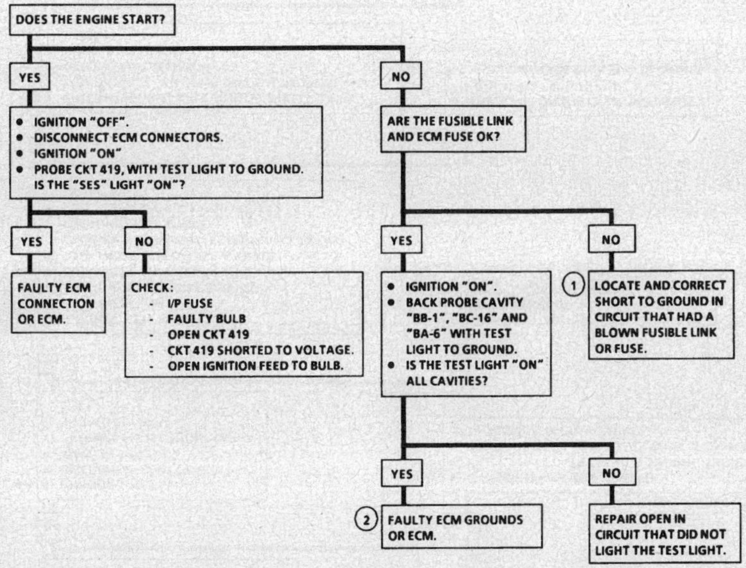

"AFTER REPAIRS," CONFIRM "CLOSED LOOP" OPERATION AND NO "SERVICE ENGINE SOON" LIGHT.

2.3L (VIN D, A & 3) ENGINE — SYSTEM DIAGNOSTIC CHARTS — 1992 ACHIEVA, GRAND AM AND SKYLARK

CHART A-2
NO ALDL DATA OR WON'T FLASH CODE 12
"SERVICE ENGINE SOON" LIGHT "ON" STEADY
2.3L (VIN D, A & 3) "N" CARLINE (PORT)

Circuit Description:

There should always be a steady "Service Engine Soon" light when the ignition is "ON" and engine stopped. Battery ignition voltage is supplied to the light bulb. The Electronic Control Module (ECM) will turn the light "ON" by grounding CKT 419 at the ECM.

With the diagnostic terminal grounded, the light should flash a Code 12, followed by any trouble code(s) stored in memory.

A steady light suggests a short to ground in the light control CKT 419, or an open in diagnostic CKT 451.

Test Description: Number(s) below refer to circled number(s) on the diagnostic chart.

1. Light "OFF" with CKT 419 disconnected from ECM indicates that ground circuit was completed through the ECM, not through external short to ground.
2. If there is a problem with the ECM that causes a Tech 1 to not read serial data, the ECM should not flash a Code 12. If Code 12 is flashing, check for CKT 451 short to ground. If Code 12 does flash, be sure that the "Scan" tool is working properly on another vehicle. If the "Scan" tool is functioning properly and CKT 461 is OK, the MEM-CAL or ECM may be at fault for the "NO ALDL" symptom.
3. This step will check for an open diagnostic CKT 451.
4. At this point, the "Service Engine Soon" light wiring is OK. The problem is a faulty ECM or MEM-CAL. If Code 12 does not flash, the ECM should be replaced using the original MEM-CAL. Replace the MEM-CAL only after trying an ECM, as a defective MEM-CAL is an unlikely cause of the problem.

2.3L (VIN D, A & 3) ENGINE — SYSTEM DIAGNOSTIC CHARTS — 1992 ACHIEVA, GRAND AM AND SKYLARK

CHART A-2

**NO ALDL DATA OR WON'T FLASH CODE 12
"SERVICE ENGINE SOON" LIGHT "ON" STEADY
2.3L (VIN D, A & 3) "N" CARLINE (PORT)**

- IGNITION "ON," ENGINE "OFF." "SES" LIGHT SHOULD BE "ON." IS IT?

YES

- GROUND DIAGNOSTIC TERMINAL. "SES" LIGHT SHOULD FLASH CODE 12. DOES IT?

NO → SEE CHART A-1

(1)
- IGNITION "OFF."
- DISCONNECT ECM CONNECTORS.
- IGNITION "ON."
- "SERVICE ENGINE SOON" LIGHT SHOULD BE "OFF." IS IT?

YES
(2)
- IF PROBLEM WAS NO ALDL DATA:
- CHECK SERIAL DATA CKT 461 FOR OPEN, SHORT TO GROUND OR SHORT TO VOLTAGE. IF OK, ECM OR PROM/MEM-CAL IS FAULTY.

(3)
- IGNITION "OFF."
- RECONNECT ECM.
- IGNITION "ON," ENGINE STOPPED.
- DIAGNOSTIC TERMINAL NOT GROUNDED.
- BACK PROBE ECM CKT 451 WITH TEST LIGHT TO GROUND. IF TEST LIGHT IS "ON," CHECK CKT 451 FOR SHORT TO VOLTAGE.
- "SES" LIGHT SHOULD FLASH CODE 12. DOES IT?

NO → REPAIR SHORT TO GROUND IN CKT 419.

(4)
- CHECK PROM/MEM-CAL FOR PROPER INSTALLATION.
- IF OK, REPLACE ECM USING ORIGINAL PROM/MEM-CAL.
- GROUND ALDL TERMINAL "B".
- "SES" LIGHT SHOULD FLASH CODE 12. DOES IT?

YES
- CHECK FOR OPEN IN CKT 451 BETWEEN ALDL TERM. "B" AND ECM.
- IF OK, CHECK FOR OPEN IN ALDL CKT 450 TO GROUND.

NO → REPLACE PROM/MEM-CAL

YES → SYSTEM OK

"AFTER REPAIRS," CONFIRM "CLOSED LOOP" OPERATION AND NO "SERVICE ENGINE SOON" LIGHT.

2.3L (VIN D, A & 3) ENGINE — SYSTEM DIAGNOSTIC CHARTS — 1992 ACHIEVA, GRAND AM AND SKYLARK

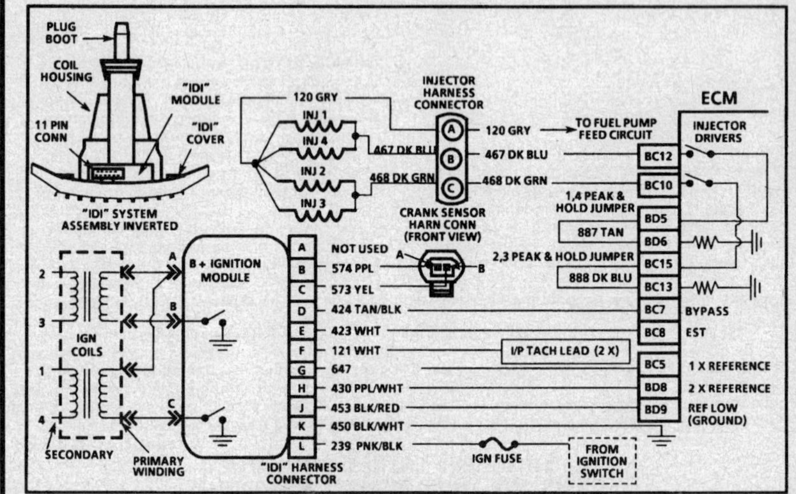

CHART A-3

**(Page 1 of 3)
ENGINE CRANKS BUT WILL NOT RUN
2.3L (VIN D, A & 3) "N" CARLINE (PORT)**

Condition:

Engine cranks but will not run, or engine may start, but immediately stops running. Battery condition and engine cranking speed are OK and there is adequate fuel in the tank.

Circuit Description:

This engine is equipped with a distributorless ignition system called the Integrated Direct Ignition (IDI) system. The primary circuit of the IDI consists of two separate ignition coils, an IDI (ignition) module and crankshaft sensor as well as the related connecting wires and the Electronic Spark Timing (EST) portion of the ECM. Each secondary circuit consists of the secondary winding of the coil, two connecting metal strips molded into the coil housing, the spark plug boot/connector assemblies and spark plugs.

Test Description: Number(s) below refer to circled number(s) on the diagnostic chart.

1. This step verifies that "SES" light operation, on-board diagnostics, cranking rpm, TPS and coolant sensor signals are normal. A blinking test light verifies that the ECM is receiving the IDI reference signal and is attempting to activate the injectors.
2. This step checks injector harness and injectors for opens or shorts. Resistance should measure half that of one injector due to parallel circuit.
3. By installing spark plug jumper leads and testing for spark on two adjacent plug leads (do not use 2 & 3 as they are on same coil), each ignition coil's ability to produce at least 25,000 volts is verified.
4. Checks to see if fuel pump and relay are operating correctly (fuel pump only "ON" 2-3 seconds) and fuel pressure is within proper range.
5. If module can make the test light blink, the fault is coil harness or connections. If not, module or it's connections are faulty.
6. This step determines whether harness or injector is cause of incorrect resistance. Nominal injector resistance is 1.9 to 2.1 ohms at 60°C (140°F). Resistance will increase slightly at higher temperatures.

Diagnostic Aids:

Check For:
- TPS binding or sticking in Wide Open Throttle (WOT) position or intermittently shorted or open.
- Water or foreign material in fuel.
- Low Compression (Timing chain failure). Verify that only resistor spark plugs are used.

2.3L (VIN D, A & 3) ENGINE — SYSTEM DIAGNOSTIC CHARTS — 1992 ACHIEVA, GRAND AM AND SKYLARK

NOTICE: FUEL SYSTEM IS UNDER PRESSURE. TO AVOID FUEL SPILLAGE, REFER TO FIELD SERVICE PROCEDURES FOR TESTING OR REPAIRS REQUIRING DISASSEMBLY OF FUEL LINES OR FITTINGS.

CHART A-3

(Page 1 of 3)
ENGINE CRANKS BUT WILL NOT RUN
2.3L (VIN D, A & 3)
"N" CARLINE (PORT)

(1)
- FUEL PUMP/INJ FUSE OK.
- FUEL QUANTITY OK.
- IGNITION "ON," THROTTLE CLOSED.
- "SERVICE ENGINE SOON LIGHT" SHOULD BE "ON." (IF NOT SEE CHART A-1).
- IF "SCAN" TOOL INDICATES "NO ALDL," SEE CHART A-2.
- TPS SHOULD "SCAN" LESS THAN 2.5V. (IF NOT SEE CODE 21 CHART)
- COOLANT TEMP. SHOULD "SCAN" BETWEEN -30°C & 130°C. (IF NOT SEE CODE 14 OR 15 CHART.
- "SCAN" 2X REFERENCE PULSES OR CRANK RPM WHILE CRANKING ENGINE. (IF "NO" 2X REFERENCE PULSES OR RPM IS "0", BEGIN AT STEP #11 ON PAGE 3).
- DISCONNECT 3 TERM. INJ HARNESS CONNECTOR AND CONNECT TEST LIGHT BETWEEN CAVITIES "A" AND "B" ON ECM SIDE OF HARNESS.
- CRANK ENGINE AND OBSERVE TEST LIGHT. (SHOULD BLINK.)
- PERFORM TEST AGAIN WITH TEST LIGHT BETWEEN CAVITIES "A" AND "C".
- LIGHT SHOULD BLINK ON BOTH TESTS. DOES IT?

YES

(2)
- WITH DVM ON 200 OHM SCALE, MEASURE RESISTANCE BETWEEN CAVITIES "A" AND "B" ON INJECTOR SIDE OF 3 TERMINAL INJECTOR HARNESS CONNECTOR.
- PERFORM MEASUREMENT AGAIN BETWEEN CAVITIES "A" AND "C".
- RESISTANCE MEASUREMENT SHOULD BE BETWEEN ABOUT .9 TO 1.1 OHM* ON EACH TEST. IS IT?

YES

(3)
- TEMPORARILY REMOVE "IDI" ASSEMBLY AND INSTALL SPARK PLUG JUMPER WIRES (J 36012).
- REMOVE TEST LIGHT FROM INJ HARNESS.
- CHECK FOR SPARK WITH SPARK TESTER J 26792, (ST-125) OR EQUIV. ON 2 ADJACENT PLUG WIRES (1&2 OR 3&4 NOT 2 & 3) WHILE CRANKING WITH REMAINING PLUG WIRES STILL CONNECTED.
- SPARK SHOULD JUMP TESTER GAP ON BOTH WIRES. DOES IT?

YES

(4)
- INSTALL FUEL PRESS. GAGE AND NOTE PRESS. AFTER IGN "ON" FOR 2 SECONDS.
- PRESSURE SHOULD BE 280-325 kPa (40.5-47 PSI). IS IT?

YES → CHECK FOR WET SPARK PLUGS. IF PLUGS ARE OK, SEE "DIAGNOSTIC AIDS"

NO → SEE CHART A-7

(5)
- IGN. "OFF." DISCONNECT PLUG JUMPER WIRES AND REMOVE COIL HOUSING.
- DISCONNECT COIL HARN. CONN. AT MOD. AND INSTALL A TEST LIGHT BETWEEN MODULE TERM. "A" (PURPLE WIRE) AND CONTROL TERMINAL FOR COIL(S) WHICH DID NOT SPARK.
- CRANK ENGINE AND NOTE TEST LIGHT.
- TEST LIGHT SHOULD BLINK AT BOTH TERMINALS. DOES IT?

YES → FAULTY HARNESS, POOR CONNECTION OR FAULTY COIL.

NO → FAULTY "IDI" MODULE CONNECTIONS OR MODULE.

NO (from step 2) → NOTE WHETHER THERE WAS "NO LIGHT ON BOTH," "NO LIGHT ON ONE," OR "STEADY LIGHT ON ONE OR BOTH" AND REFER TO CHART A-3, PAGE 2.

SEE CHART A-3, PAGE 2

(6)
- REMOVE CV OIL/AIR SEPARATOR FOR ACCESS TO INJECTOR CONNECTORS
- DISCONNECT INJECTORS ON CIRCUIT(S) WITH INCORRECT RESISTANCE (CAVITY "B" - INJECTORS #1 AND #4, CAVITY "C" - INJECTORS #2 AND #3).
- WITH DVM ON 200 OHM SCALE, MEASURE RESISTANCE OF EACH INJECTOR.
- RESISTANCE SHOULD BE ABOUT 1.9 TO 2.1 OHMS* IS IT?

YES → REPAIR OPEN, SHORT OR POOR CONNECTION IN INJECTOR HARNESS.

NO → REPLACE INJECTOR WITH INCORRECT RESISTANCE AND CONFIRM NO OPEN OR SHORT IN INJECTOR HARNESS.

***NOTE:** INJECTOR RESISTANCE SPECIFICATION IS AT 60°C (140°F) AND MAY BE SLIGHTLY HIGHER IF INJECTOR IS HOTTER.

"AFTER REPAIRS," CONFIRM "CLOSED LOOP" OPERATION AND NO "SERVICE ENGINE SOON" LIGHT.

2.3L (VIN D, A & 3) ENGINE — SYSTEM DIAGNOSTIC CHARTS — 1992 ACHIEVA, GRAND AM AND SKYLARK

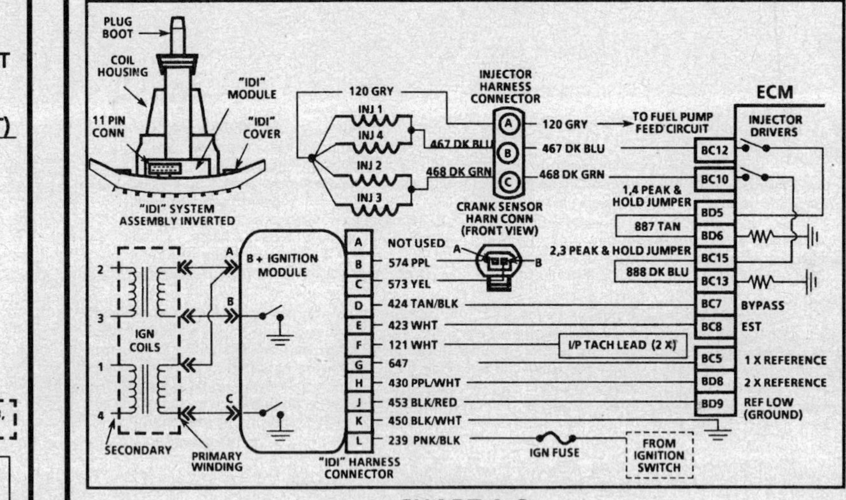

CHART A-3

(Page 2 of 3)
ENGINE CRANKS BUT WILL NOT RUN
2.3L (VIN D, A & 3) "N" CARLINE (PORT)

Condition:
Engine cranks but will not run, or engine may start, but immediately stops running. Battery condition and engine cranking speed are OK and there is adequate fuel in the tank.

Circuit Description:
This engine is equipped with a distributorless ignition system called the Integrated Direct Ignition (IDI) system. The primary circuit of the IDI consists of two separate ignition coils, an IDI ignition module and crankshaft sensor as well as the related connecting wires and the Electronic Spark Timing (EST) portion of the ECM. Each secondary circuit consists of the secondary winding of the coil, two connecting metal strips molded into the coil housing, the spark plug boot/connector assemblies and spark plugs.

Test Description: Number(s) below refer to circled number(s) on the diagnostic chart.

7. Battery voltage should be available at cavity "A" whenever the fuel pump power feed circuit is switched "ON." The ECM should switch the fuel pump "ON" for 2-3 seconds after ignition is turned "ON" (and when ECM is receiving ignition reference pulses, as while cranking or running). The ignition must be turned "OFF" for at least 10 seconds to assure that the ECM powers down and will then switch the fuel pump back "ON" for 2-3 seconds when ignition is turned back "ON."

8. Light "ON" one circuit only indicates power is available at cavity "A", but grounded circuit is not being completed on the other circuit. This could be due to open circuit or ECM not switching the injector driver circuit to ground.

9. Steady light indicates ground circuit is always completed and is not being switched. This could be due to short to ground in circuit, or faulty ECM injector driver.

10. The fuel pump should be switched "ON" by the ECM for 2-3 seconds after ignition is first turned "ON." It is necessary to turn the ignition "OFF" for at least 10 seconds to assure that the ECM powers down and will then switch the fuel pump back "ON." If the fuel pump operates, but power is not available at injector harness, circuit must be open. If the fuel pump does not operate, CHART A-5 should be used to diagnose the cause.

MULTIPORT FUEL INJECTION (MFI) SYSTEMS
EXCEPT LIGHT TRUCKS, VANS, GEO AND SATURN

2.3L (VIN D, A & 3) ENGINE — SYSTEM DIAGNOSTIC CHARTS — 1992 ACHIEVA, GRAND AM AND SKYLARK

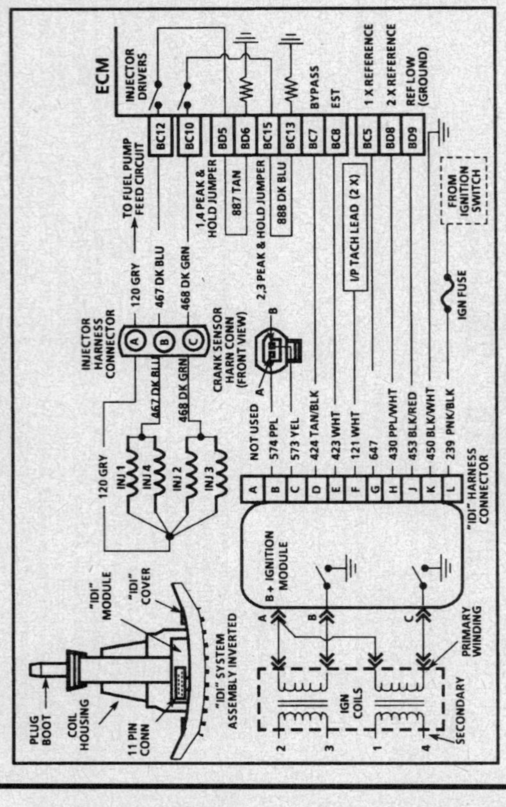

CHART A-3
(Page 3 of 3)
ENGINE CRANKS BUT WILL NOT RUN
2.3L (VIN D, A & 3) "N" CARLINE (PORT)

Circuit Description:

The Integrated Direct Ignition (IDI) system uses a waste spark method of distribution. In this type of system, the ignition module triggers the #1-4 coil pair resulting in both #1 and #4 spark plugs firing at the same time. #1 cylinder is on the compression stroke at the same time #4 is on the exhaust stroke, resulting in a lower energy requirement to fire #4 spark plug. This leaves the remainder of the high voltage to be used to fire #1 spark plug. On this application, the crank sensor is mounted to, and protrudes through the block to within approximately 0.050" of the crankshaft reluctor. Since the reluctor is a machined portion of the crankshaft and the sensor is mounted in a fixed position on the block, timing adjustments are not possible or necessary.

Test Description: Number(s) below refer to circled number(s) on the diagnostic chart.

11. Battery voltage should be available at terminal "L" of the IDI 11 pin connector, and terminal "K" should be a good ground.

12. The test light to 12 volts simulates a reference signal to the ECM which will result in an injector test light blink for every other touch of the test light, if CKT 430, the ECM and the injector driver circuit are all functioning properly.

13. The crankshaft sensor should output a voltage as the crankshaft turns. If no voltage is produced, the indication is a poor sensor connection or faulty sensor.

14. The crank sensors core is a magnet, therefore, it should be magnetized and the resistance should be within a range of 500 to 900 ohms.

2.3L (VIN D, A & 3) ENGINE — SYSTEM DIAGNOSTIC CHARTS — 1992 ACHIEVA, GRAND AM AND SKYLARK

CHART A-3
(Page 2 of 3)
ENGINE CRANKS BUT WILL NOT RUN
2.3L (VIN D, A & 3) "N" CARLINE (PORT)

"AFTER REPAIRS," CONFIRM "CLOSED LOOP" OPERATION AND NO "SERVICE ENGINE SOON" LIGHT.

2.3L (VIN D, A & 3) ENGINE — SYSTEM DIAGNOSTIC CHARTS — 1992 ACHIEVA, GRAND AM AND SKYLARK

CHART A-3
(Page 3 of 3)
ENGINE CRANKS BUT WILL NOT RUN
2.3L (VIN D, A & 3) "N" CARLINE (PORT)

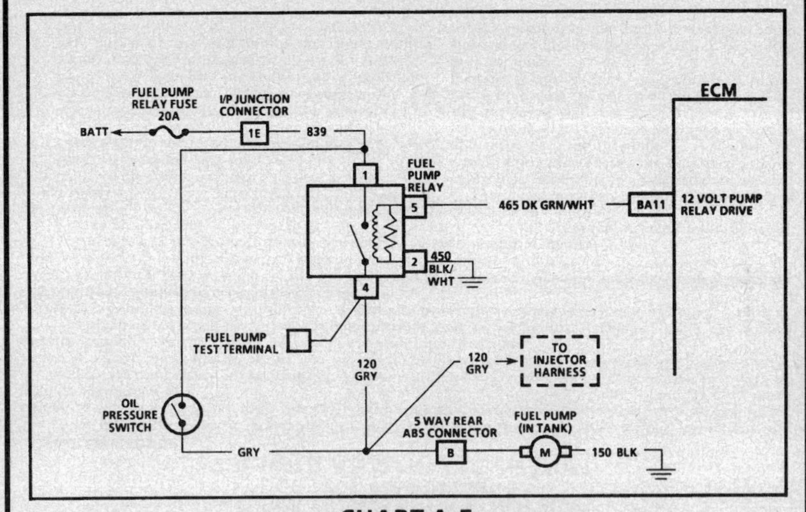

2.3L (VIN D, A & 3) ENGINE — SYSTEM DIAGNOSTIC CHARTS — 1992 ACHIEVA, GRAND AM AND SKYLARK

CHART A-5
(Page 1 of 2)
ENGINE CRANKS BUT WILL NOT RUN
(FUEL PUMP CIRCUIT)
2.3L (VIN D, A & 3) "N" CARLINE (PORT)

Circuit Description:

When the ignition switch is turned "ON," the Electronic Control Module (ECM) turns "ON" the in-tank fuel pump. It will remain "ON" as long as the ECM is receiving ignition reference pulses from the Integrated Direct Ignition (IDI) module.

If there are no reference pulses, the ECM will shut "OFF" the fuel pump about 2-3 seconds after key "ON," or about 10 seconds after reference pulses stop. If sufficient oil pressure is present to close the oil pressure switch, the fuel pump will remain "ON" during cranking without reference pulses.

The pump delivers fuel to the fuel rail and injectors, then to the pressure regulator, where the system pressure is controlled to 280 - 325 kPa (40.5 - 47 psi) with no manifold vacuum or 211 - 304 kPa (30.5 - 44 psi) at idle. Excess fuel is then returned to the fuel tank.

The fuel pump "test" terminal is located in the engine compartment. When the engine is stopped, the pump can be turned "ON" by applying battery voltage to the "test" terminal.

Improper fuel system pressure will result in one or all of the following symptoms:
- Cranks but won't run.
- Code 44.
- Code 45.
- Cuts out, may feel like ignition problem.
- Poor fuel economy, loss of power.
- Hesitation.

Test Description: Number(s) below refer to circled number(s) on the diagnostic chart.

1. Determines if the pump circuit is ECM controlled. The ECM will turn "ON" the pump relay. Engine is not cranking or running so the ECM will turn "OFF" the relay within 2 seconds after ignition is turned "ON."

2. If the fuse is blown, this test will confirm a short to ground on CKT 120. To prevent misdiagnosis, be sure fuel pump is disconnected before test.

3. Turns "ON" the fuel pump if CKT 120 wiring is OK. If the pump runs, it is a basic fuel delivery problem which the following steps will locate.

4. Check for battery voltage at the pump relay.

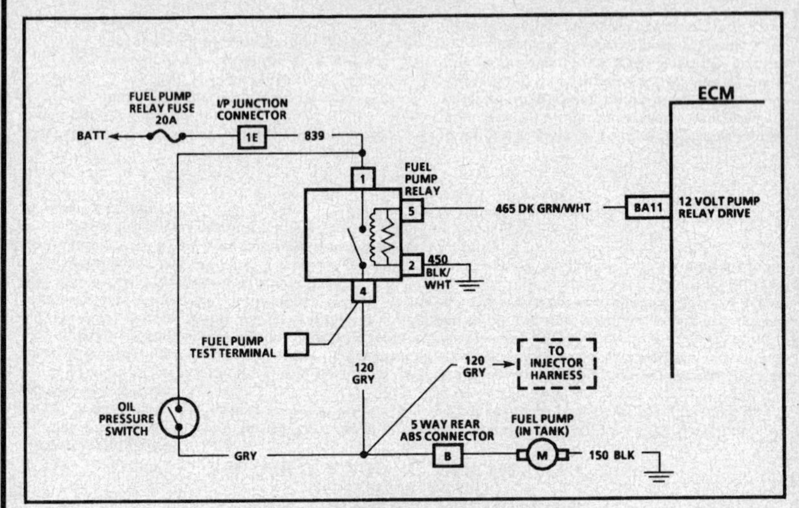

CHART A-5

(Page 2 of 2)
ENGINE CRANKS BUT WILL NOT RUN
(FUEL PUMP CIRCUIT)
2.3L (VIN D, A & 3) "N" CARLINE (PORT)

Circuit Description:

When the ignition switch is turned "ON," the Electronic Control Module (ECM) turns "ON" the in-tank fuel pump. It will remain "ON" as long as the ECM is receiving ignition reference pulses from the Integrated Direct Ignition (IDI) module.

If there are no reference pulses, the ECM will shut "OFF" the fuel pump about 2-3 seconds after key "ON," or about 10 seconds after reference pulses stop. If sufficient oil pressure is present to close the oil pressure switch, the fuel pump will remain "ON" during cranking without reference pulses.

The pump delivers fuel to the fuel rail and injectors, then to the pressure regulator, where the system pressure is controlled to 280 - 325 kPa (40.5 - 47 psi) with no manifold vacuum or 211 - 304 kPa (30.5 - 44 psi) at idle. Excess fuel is then returned to the fuel tank.

The fuel pump "test" terminal is located in the engine compartment. When the engine is stopped, the pump can be turned "ON" by applying battery voltage to the "test" terminal.

Improper fuel system pressure will result in one or all of the following symptoms:

- Cranks but won't run.
- Code 44.
- Code 45.
- Cuts out, may feel like ignition problem.
- Poor fuel economy, loss of power.
- Hesitation.

Test Description: Number(s) below refer to circled number(s) on the diagnostic chart.

5. Check relay ground CKT 450.
6. Check for ECM control of relay through CKT 465.
7. The fuel pump voltage control circuit includes an engine oil pressure switch with a separate set of normally open contacts. The switch closes at about 28 kPa (4 psi) of oil pressure and provides a second battery feed path to the fuel pump. If the relay fails, the pump will continue to run using the battery feed supplied by the closed oil pressure switch.

A failed pump relay will result in extended engine crank time, because of the time required to build enough oil pressure to close the oil pressure switch and turn "ON" the fuel pump. There may be instances when the relay has failed but the engine will not crank fast enough to build enough oil pressure to close the switch. This or a faulty oil pressure switch can result in "Engine Cranks But Will Not Run."

8. Check the oil pressure switch to be sure it provides battery feed to the fuel pump should the pump relay fail.

2.3L (VIN D, A & 3) ENGINE — SYSTEM DIAGNOSTIC CHARTS — 1992 ACHIEVA, GRAND AM AND SKYLARK

CHART A-5
(Page 2 of 2)
ENGINE CRANKS BUT WILL NOT RUN
(FUEL PUMP CIRCUIT)
2.3L (VIN D, A & 3) "N" CARLINE (PORT)

FROM CHART A-5 (1 of 2)

(5) CONNECT A TEST LIGHT BETWEEN HARNESS CONN CKT 839 AND 450.

LIGHT "ON" → (6)
- IGNITION "OFF" FOR TEN SECONDS.
- CONNECT TEST LIGHT BETWEEN CKT 465 AND GROUND.
- NOTE TEST LIGHT WITHIN 2 SECONDS AFTER IGNITION "ON."

LIGHT "OFF" → REPAIR OPEN GROUND CKT 450.

LIGHT "ON" → REPLACE RELAY

(7) NOTE: IF ORIGINAL COMPLAINT WAS "CRANKS BUT WILL NOT RUN" MAKE THE FOLLOWING CHECKS:

(8)
- ENGINE IDLING AT NORMAL OPERATING TEMPERATURE.
- OIL PRESSURE NORMAL.
- DISCONNECT FUEL PUMP RELAY.
- ENGINE SHOULD CONTINUE TO RUN. DOES IT?

YES → NO PROBLEM FOUND IN FUEL SYSTEM.

NO → SEE OIL PRESSURE SWITCH DIAGNOSIS

LIGHT "OFF" → DISCONNECT BLACK A-B CONNECTOR AND CHECK FOR OPEN OR SHORT TO GROUND IN CKT 465.

CKT 465 OK → CHECK RESISTANCE ACROSS PUMP RELAY PINS OPPOSITE HARNESS CONNECTOR TERMS. CKTS 465 AND 450. SHOULD MEASURE 20 OHMS OR MORE.

OK → REPLACE ECM

NOT OK → REPLACE RELAY AND RE-CHECK

CKT 465 NOT OK → REPAIR CKT 465. IF CIRCUIT WAS SHORTED TO GROUND, RECHECK FOR "LIGHT ON" BETWEEN HARNESS CONNECTOR TERMINAL "5" AND GROUND WITHIN 2 SECONDS AFTER IGNITION "ON."

NO LIGHT → FAULTY ECM CONNECTOR "BA11" OR ECM.

LIGHT → RECONNECT RELAY AND ECM.

"AFTER REPAIRS," CONFIRM "CLOSED LOOP" OPERATION AND NO "SERVICE ENGINE SOON" LIGHT.

2.3L (VIN D, A & 3) ENGINE — SYSTEM DIAGNOSTIC CHARTS — 1992 ACHIEVA, GRAND AM AND SKYLARK

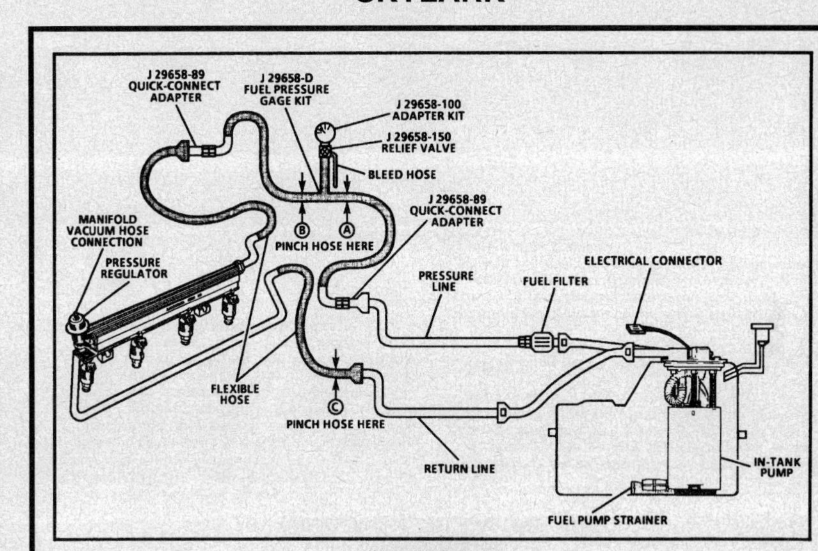

CHART A-7
(Page 1 of 3)
FUEL SYSTEM DIAGNOSIS
2.3L (VIN D, A & 3) "N" CARLINE (PORT)

Circuit Description:
When the ignition switch is turned "ON," the Electronic Control Module (ECM) will turn "ON" the in-tank fuel pump. It will remain "ON" as long as the engine is cranking or running, and the ECM is receiving reference pulses. If there are no reference pulses, the ECM will shut "OFF" the fuel pump in about 2 seconds after ignition "ON" or 10 seconds after reference pulses stop.

An electric fuel pump, part of the modular fuel sender and located inside the fuel tank, pumps fuel through an in-line filter to the fuel rail assembly. The pump is designed to provide fuel at a pressure above the regulated pressure needed by the injectors. A pressure regulator, attached to the fuel rail, keeps fuel available to the injectors at a regulated pressure. Unused fuel is returned to the fuel tank by a separate line.

Test Description: Number(s) below refer to circled number(s) on the diagnostic chart.

1. Install fuel pressure gage per instructions on Page 3 of 3. Ignition "ON" pump pressure should be 284-325 kPa (41-47 psi). This pressure is controlled by spring pressure within the regulator assembly.
2. When the engine is idling, the manifold pressure is low (high vacuum) and is applied to the fuel regulator diaphragm. This will offset the spring and result in a lower fuel pressure. This idle pressure will vary somewhat depending on barometric pressure, however, the pressure idling should be less indicating pressure regulator control.
3. Pressure that continues to fall quickly is caused by one of the following:
   - In-tank fuel pump check valve not holding.
   - Fuel pressure regulator valve leaking.
   - Injector(s) sticking open.
4. An injector sticking open can best be determined by checking for a fouled or saturated spark plug(s). If a leaking injector can not be determined by a fouled or saturated spark plug the following procedure should be used.
   - Remove fuel rail bolts, but leave fuel lines connected.

CAUTION: Be sure injector(s) are not allowed to spray on engine and that injector retaining clips are intact. This should be carefully followed to prevent fuel spray on engine which would cause a fire hazard.

- Pressurize the fuel system and observe for injector(s) leaking.

CHART A-7

(Page 1 of 3)
FUEL SYSTEM DIAGNOSIS
2.3L (VIN D, A & 3) "N" CARLINE (PORT)

(1)
- INSTALL FUEL PRESSURE GAGE (SEE CHART A-7, PAGE 3 OF 3)
- IGNITION "OFF" FOR 10 SECONDS. A/C "OFF."
- IGNITION "ON." FUEL PUMP WILL RUN FOR ABOUT 2 SECONDS.
- NOTE FUEL PRESSURE, WITH PUMP RUNNING SHOULD BE 284-325 kPa (41-47 psi) AND HOLD STEADY WHEN PUMP STOPS.*

FROM CHART A-3

OK → NOT OK

(2)
- START AND IDLE ENGINE AT NORMAL OPERATING TEMPERATURE.
- PRESSURE SHOULD BE LOWER BY 21-69 kPa (3-10 psi).

(3)
FUEL PRESSURE WITHIN SPECIFICATION, BUT NOT HOLDING.

FUEL PRESSURE LESS THAN 284 kPa (41 psi).

FUEL PRESSURE ABOVE 325 kPa (47 psi).

NO PRESSURE

OK → NOT OK

NO TROUBLE FOUND.

- APPLY 10 INCHES OF VACUUM TO PRESSURE REGULATOR.
- FUEL PRESSURE SHOULD DROP 21-69 kPa (3-10 psi).

- IGNITION "OFF" FOR 10 SECONDS.
- IGNITION "ON."
- PINCH GAGE INLET HOSE AT "A" PRESSURE SHOULD HOLD.

SEE CHART A-7 (2 of 2)

SEE CHART A-5 (1 of 2)

NOT HOLDING

HOLDS

- IGNITION "OFF" FOR 10 SECONDS.
- IGNITION "ON."
- PINCH FUEL RETURN HOSE AT "C".
- RECHECK PRESSURE.

CHECK FOR:
- FAULTY IN-TANK PUMP.

HOLDS

NOT HOLDING

REPLACE PRESSURE REGULATOR.

(4) LOCATE AND CORRECT LEAKING INJECTOR(S).

NOT OK → OK

REPLACE PRESSURE REGULATOR.

REPAIR VACUUM SOURCE TO REGULATOR.

* IMPORTANT: THE IGNITION MAY HAVE TO BE CYCLED "ON" MORE THAN ONCE TO OBTAIN MAXIMUM PRESSURE. ALSO, IT IS NORMAL FOR THE PRESSURE TO DROP SLIGHTLY WHEN THE PUMP STOPS.

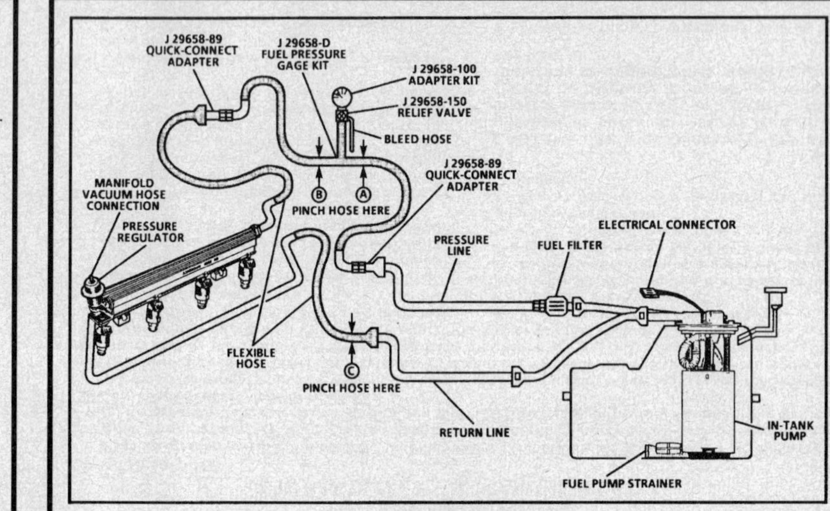

CHART A-7

(Page 2 of 3)
FUEL SYSTEM DIAGNOSIS
2.3L (VIN D, A & 3) "N" CARLINE (PORT)

Test Description: Number(s) below refer to circled number(s) on the diagnostic chart.

1. Pressure below 284 kPa (41 psi) may cause a lean condition and may set a Code 44. It could also cause hard starting cold and poor driveability. Low enough pressure will cause the engine not to run at all. Restricted flow may allow the engine to run at idle, or low speeds, but may cause a surge and stall when more fuel is required, as when accelerating or driving at high speeds.

2. Restricting fuel flow at the fuel pressure gage (at B) causes fuel pressure to build above regulated pressure. With battery voltage applied to the pump "test" terminal, pressure should rise above 325 kPa (47 psi) as the gage outlet hose is restricted.

NOTICE: Do not allow pressure to exceed 414 kPa (60 psi), as damage to the regulator may result.

3. This test determines if the high fuel pressure is due to a restricted fuel return line or a faulty fuel pressure regulator. High fuel pressure may cause a rich condition and may set a Code 45 or cause driveability problems.

2.3L (VIN D, A & 3) ENGINE — SYSTEM DIAGNOSTIC CHARTS — 1992 ACHIEVA, GRAND AM AND SKYLARK

CHART A-7
(Page 2 of 3)
FUEL SYSTEM DIAGNOSIS
2.3L (VIN D, A & 3) "N" CARLINE (PORT)

```
FROM CHART A-7 (1 of 2)

(1) HAS FUEL PRESSURE BUT LESS THAN 284 kPa (41 psi).
    • CHECK FOR RESTRICTED FUEL LINES OR IN-LINE FILTER.
        OK → 
        NOT OK → REPLACE FILTER OR REPAIR FUEL LINE AND RECHECK.

(2) • IGNITION "OFF."
    • USING A 10 AMP FUSED JUMPER WIRE, APPLY BATTERY VOLTAGE TO FUEL PUMP "TEST" CONNECTOR.
    • GRADUALLY PINCH PRESSURE GAGE OUTLET HOSE AT (B). LOOK FOR PRESSURE ABOVE 325 kPa (47 psi). DO NOT EXCEED 414 kPa (60 psi).

        ABOVE 325 kPa (47 psi). → IF LINES ARE OK, REPLACE PRESSURE REGULATOR.
        PRESSURE BUT LESS THAN 284 kPa (41 psi). → CHECK FOR;
            - FAULTY FUEL PUMP.
            - RESTRICTED FUEL PUMP STRAINER.
            - INCORRECT MODULAR FUEL SENDER ASSEMBLY.

FUEL PRESSURE ABOVE 325 kPa (47 psi).

(3) • DISCONNECT ENGINE COMPARTMENT FUEL RETURN LINE QUICK-CONNECT FITTING. (PROCEDURES FOR DISCONNECTING/CONNECTING QUICK-CONNECT FITTINGS ARE SAME AS THOSE FOR FUEL FEED LINE, SEE PAGE 3 OF 3).
    • PLACE OPEN END IF FLEXIBLE HOSE INTO AN APPROVED GASOLINE CONTAINER.
    • NOTE FUEL PRESSURE WITHIN 2 SECONDS AFTER IGNITION IS TURNED "ON."

        ABOVE 325 kPa (47 psi) → CHECK FOR RESTRICTED FUEL RETURN LINE FROM FUEL PRESSURE REGULATOR TO POINT WHERE FUEL LINE WAS DISCONNECTED.
            → IF LINE IS OK, REPLACE PRESSURE REGULATOR.
        284-325 kPa (41-47 psi). → LOCATE AND CORRECT RESTRICTED FUEL RETURN LINE TO FUEL TANK.
```

"AFTER REPAIRS," CONFIRM "CLOSED LOOP" OPERATION AND NO "SERVICE ENGINE SOON" LIGHT.

2.3L (VIN D, A & 3) ENGINE — SYSTEM DIAGNOSTIC CHARTS — 1992 ACHIEVA, GRAND AM AND SKYLARK

CHART A-7
(Page 3 of 3)
FUEL SYSTEM DIAGNOSIS 2.3L (VIN D, A & 3) "N" CARLINE (PORT)

FUEL PRESSURE CHECK

Tools Required: J 29658-D - Fuel Pressure Gage Kit
J 29658-150 - Fuel Pressure Gage Relief Valve
J 29658-100 - TBI Pressure Gage Modification Kit
J 29658-89 - Fuel Pressure Quick Connect Adapters
J 37088 - A - Fuel Line Quick-Connect Separators

CAUTION: To Reduce the Risk of Fire and Personal Injury:
- It is necessary to relieve fuel system pressure before connecting a fuel pressure gage.
- After relieving system pressure, a small amount of fuel may be released when disconnecting the fuel lines. Cover fuel line fittings with a shop towel before disconnecting, to catch any fuel that may leak out. Place towel in approved container when disconnect is completed.
- Do not pinch or restrict nylon fuel lines to avoid severing, which could cause a fuel leak.

NOTICE: If nylon fuel lines become kinked, and cannot be straightened, they must be replaced.

1. Loosen fuel filler cap to relieve fuel tank pressure. (Do not tighten at this time.)
2. Raise vehicle.
3. Disconnect fuel pump electrical connector.
4. Lower vehicle.
5. Start and run engine until fuel supply remaining in fuel pipes is consumed. Engage starter for three seconds to assure relief of any remaining pressure.
6. Disconnect negative battery cable.
7. Locate engine compartment fuel feed quick-connect fitting.
8. Grasp both ends of fitting, twist female end $\frac{1}{4}$ turn in each direction to loosen any dirt in fitting.

CAUTION: Safety glasses must be worn when using compressed air, as flying dirt particles may cause eye injury.

9. Using compressed air, blow dirt out of quick-connect fitting.
10. Choose correct tool from separator tool set J 37088-A for size of fitting. Insert tool into female end of connector, then push inward to release male connector.
11. If not previously installed, connect Fuel Pressure Gage Relief Valve J 29658-150, to fuel pressure gage hose assembly.
12. Connect 414 kPa (60 psi) gage from TBI Pressure Gage Modification kit J 29658-100 to hose assembly.
13. Connect gage quick-connect adapters J 29658-89 to hose assembly.

CAUTION: To Reduce the Risk of Fire and Personal Injury: Before connecting fuel line quick-connect fittings, always apply a few drops of clean engine oil to the male tube ends. This will ensure proper reconnection and prevent a possible fuel leak. (During normal operation, the O-rings located inside the female connector will swell and may prevent proper reconnection if not lubricated.)

14. Lubricate the male tube end of the fuel line and the gage adapter with engine oil.
15. Connect fuel pressure gage.
    - Push connectors together to cause the retaining tabs/fingers to snap into place.
    - Once installed, pull on both ends of each connection to make sure it is secure.

2.3L (VIN D, A & 3) ENGINE — SYSTEM DIAGNOSTIC CHARTS — 1992 ACHIEVA, GRAND AM AND SKYLARK

CHART A-7
(Page 3 of 3)
FUEL SYSTEM DIAGNOSIS
2.3L (VIN D, A & 3) "N" CARLINE (PORT)

FUEL PRESSURE CHECK
-continued

16. Connect negative battery cable.
17. Check fuel pressure.
18. Place bleed hose into an approved container and open valve to bleed system pressure.
19. Disconnect negative battery cable.
20. Disconnect fuel pressure gage.
21. Lubricate the male tube end of the fuel line, and reconnect quick-connect fitting.
    - Push connector together to cause the retaining tabs/fingers to snap into place.
    - Once installed, pull on both ends of connection to make sure it is secure.
22. Tighten fuel filler cap.
23. Connect negative battery cable.
24. Cycle ignition "ON" and "OFF" twice, waiting ten seconds between cycles, then check for fuel leaks.

2.3L (VIN D, A & 3) ENGINE — DIAGNOSTIC TROUBLE CODE CHART — 1992 ACHIEVA, GRAND AM AND SKYLARK

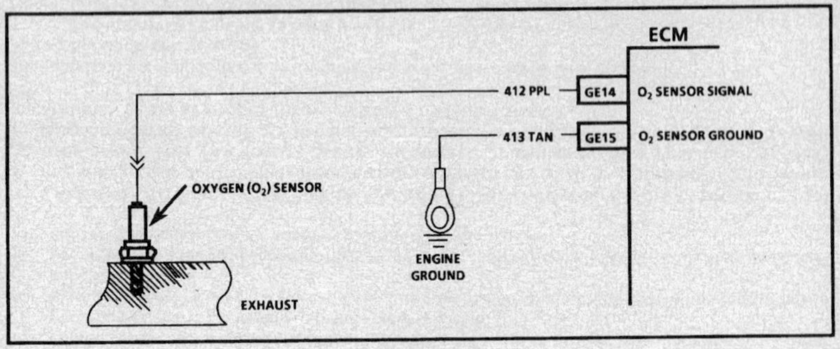

CODE 13
OXYGEN (O_2) SENSOR CIRCUIT
(OPEN CIRCUIT)
2.3L (VIN D, A & 3) "N" CARLINE (PORT)

Circuit Description:

The ECM supplies a voltage of about .45 volt between terminals "GE14" and "GE15". (If measured with a 10 megohm digital voltmeter, this may read as low as .32 volt). The O_2 sensor varies the voltage within a range of about 1 volt if the exhaust is rich, down through about .10 volt if exhaust is lean.

The sensor is like an open circuit and produces no voltage when it is below 315°C (600°F). An open sensor circuit or cold sensor causes "Open Loop" operation.

Test Description: Number(s) below refer to circled number(s) on the diagnostic chart.

1. Code 13 will set under the following conditions:
   - Engine running at least 40 seconds after start.
   - Coolant temperature at least 42.5°C (108°F).
   - No Code 21 or 22.
   - O_2 signal voltage steady between .34 and .55 volts.
   - Throttle Position Sensor (TPS) signal above 6% for more time than TPS was below 6%. (About .3 volt above closed throttle voltage)
   - All conditions must be met and held for at least 30 seconds.
     If the conditions for a Code 13 exist, the system will not go "Closed Loop."

2. This will determine if the sensor is at fault or the wiring or ECM is the cause of the Code 13.

3. Use only a high impedance digital volt ohmmeter for this test. This test checks the continuity of CKT 412 and CKT 413; because if CKT 413 is open, the ECM voltage on CKT 412 will be over .6 volt (600 mV).

Diagnostic Aids:

Normal "Scan" voltage varies between 100 mV to 999 mV (.1 volt to 1.0 volt) while in "Closed Loop." Code 13 sets in 20 seconds if voltage remains between .35 volt and .55 volt, but the system will go "Open Loop" in about 15 seconds.

2.3L (VIN D, A & 3) ENGINE — DIAGNOSTIC TROUBLE CODE CHART — 1992 ACHIEVA, GRAND AM AND SKYLARK

CODE 13
OXYGEN (O₂) SENSOR CIRCUIT
(OPEN CIRCUIT)
2.3L (VIN D, A & 3) "N" CARLINE (PORT)

① ENGINE AT NORMAL OPERATING TEMPERATURE (ABOVE 80°C/176°F).
- RUN ENGINE ABOVE 1200 RPM FOR TWO MINUTES.
- DOES TECH 1 TOOL INDICATE "CLOSED LOOP"?

NO →
② • DISCONNECT O₂ SENSOR.
- JUMPER HARNESS CKT 412 (ECM SIDE) TO GROUND.
- TECH 1 SHOULD DISPLAY O₂ VOLTAGE BELOW .2 VOLT (200 mv) WITH ENGINE RUNNING. DOES IT?

YES →
CODE 13 IS INTERMITTENT. IF NO ADDITIONAL CODES WERE STORED, REFER TO "DIAGNOSTIC AIDS"

NO →
③ • REMOVE JUMPER.
- IGNITION "ON," ENGINE "OFF."
- CHECK VOLTAGE OF CKT 412 (ECM SIDE) AT O₂ SENSOR HARNESS CONNECTOR USING A DVM.

YES →
FAULTY O₂ SENSOR CONNECTION OR SENSOR.

.3-.6 VOLT (300 - 600 mV)
→ FAULTY ECM.

OVER .6 VOLT (600 mV)
→ OPEN CKT 413 OR FAULTY CONNECTION OR FAULTY ECM.

LESS THAN .3 VOLT (300 mV)
→ OPEN CKT 412 OR FAULTY ECM CONNECTION OR FAULTY ECM.

"AFTER REPAIRS," REFER TO CODE CRITERIA   AND CONFIRM CODE DOES NOT RESET.

2.3L (VIN D, A & 3) ENGINE — DIAGNOSTIC TROUBLE CODE CHART — 1992 ACHIEVA, GRAND AM AND SKYLARK

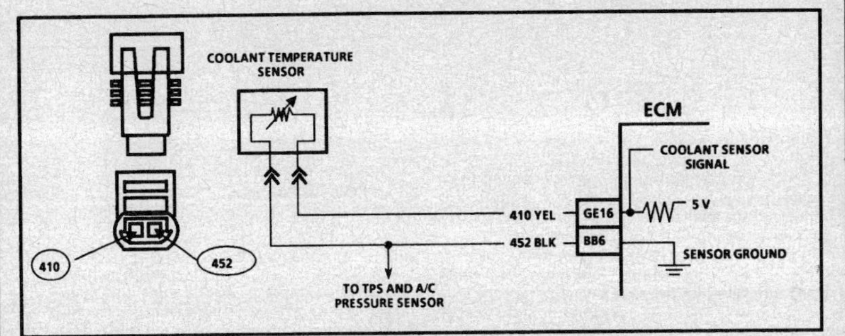

CODE 14
COOLANT TEMPERATURE SENSOR (CTS) CIRCUIT
(HIGH TEMPERATURE INDICATED)
2.3L (VIN D, A & 3) "N" CARLINE (PORT)

Circuit Description:

The Coolant Temperature Sensor (CTS) uses a thermistor to control the signal voltage at the ECM. The ECM applies a voltage on CKT 410 to the sensor. When the engine is cold the sensor (thermistor) resistance is high, therefore ECM terminal "GE16" voltage will be high.

As the engine warms, the sensor resistance becomes less, and the voltage drops. At normal engine operating temperature, the voltage will measure about 1.5 to 2.0 volts at ECM terminal "GE16".

Coolant temperature is one of the inputs used to control:
- Fuel delivery.
- Engine Spark Timing (EST).
- Idle Air Control (IAC).
- Torque Convertor Clutch (TCC).
- Controlled Canister Purge (CCP).
- Cooling Fan.

Test Description: Number(s) below refer to circled number(s) on the diagnostic chart.
1. Code 14 will set if:
   - Signal voltage indicates a coolant temperature above 140°C (285°F).
   - Engine running longer than 128 seconds.
2. This test will determine if CKT 410 is shorted to ground which will cause the conditions for Code 14.

Diagnostic Aids:

Check harness routing for a potential short to ground in CKT 410. A Tech 1 displays engine temperature in degrees celsius. After engine is started, the temperature should rise steadily to about 90°C, and then stabilize when thermostat opens.

Verify that engine is not overheating and has not been subjected to conditions which could create an overheating condition (i.e. overload, trailer towing, hilly terrain, heavy stop and go traffic, etc.). The "Temperature To Resistance Value" scale may be used to test the coolant sensor at various temperature levels to evaluate the possibility of a "shifted" (mis-scaled) sensor. A "shifted" sensor could result in poor driveability complaints.

2.3L (VIN D, A & 3) ENGINE — DIAGNOSTIC TROUBLE CODE CHART — 1992 ACHIEVA, GRAND AM AND SKYLARK

CODE 14

COOLANT TEMPERATURE SENSOR (CTS) CIRCUIT
(HIGH TEMPERATURE INDICATED)
2.3L (VIN D, A & 3) "N" CARLINE (PORT)

① DOES TECH 1 DISPLAY COOLANT TEMPERATURE OF 130°C (266°F) OR HIGHER?

YES / NO

② • DISCONNECT COOLANT TEMPERATURE SENSOR. TECH 1 SHOULD DISPLAY COOLANT TEMPERATURE BELOW -30°C (-22°F). DOES IT?

CODE 14 IS INTERMITTENT. IF NO ADDITIONAL CODES WERE STORED, REFER TO "DIAGNOSTIC AIDS"

YES / NO

REPLACE COOLANT TEMPERATURE SENSOR.

CKT 410 SHORTED TO GROUND
OR
CKT 410 SHORTED TO SENSOR GROUND CIRCUIT
OR
FAULTY ECM.

DIAGNOSTIC AID

COOLANT SENSOR		
TEMPERATURE VS. RESISTANCE VALUES (APPROXIMATE)		
°C	°F	OHMS
100	212	177
90	194	241
80	176	332
70	158	467
60	140	667
50	122	973
45	113	1188
40	104	1459
35	95	1802
30	86	2238
25	77	2796
20	68	3520
15	59	4450
10	50	5670
5	41	7280
0	32	9420
-5	23	12300
-10	14	16180
-15	5	21450
-20	-4	28680
-30	-22	52700
-40	-40	100700

"AFTER REPAIRS," REFER TO CODE CRITERIA   AND CONFIRM CODE DOES NOT RESET.

2.3L (VIN D, A & 3) ENGINE — DIAGNOSTIC TROUBLE CODE CHART — 1992 ACHIEVA, GRAND AM AND SKYLARK

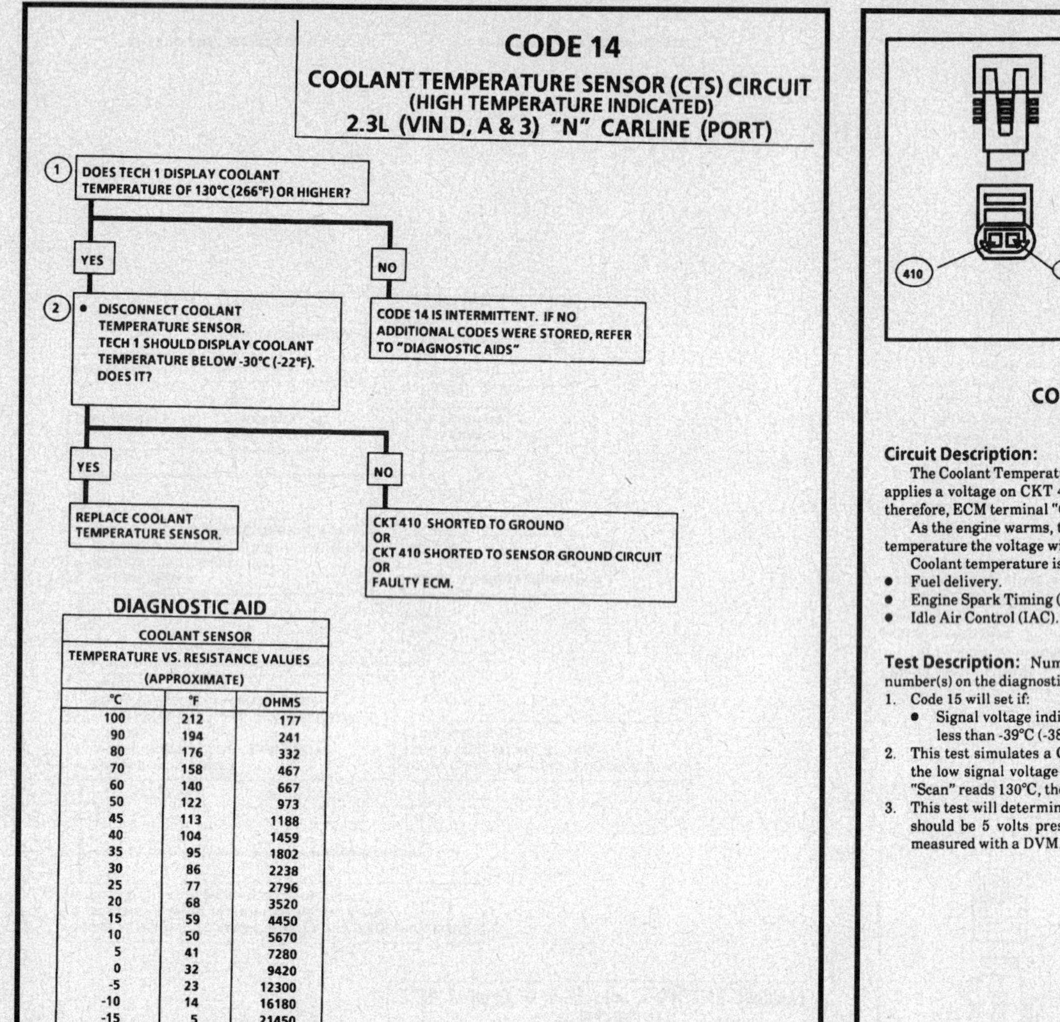

CODE 15

COOLANT TEMPERATURE SENSOR (CTS) CIRCUIT
(LOW TEMPERATURE INDICATED)
2.3L (VIN D, A & 3) "N" CARLINE (PORT)

Circuit Description:

The Coolant Temperature Sensor (CTS) uses a thermistor to control the signal voltage at the ECM. The ECM applies a voltage on CKT 410 to the sensor. When the engine is cold, the sensor (thermistor) resistance is high, therefore, ECM terminal "GE16" voltage will be high.

As the engine warms, the sensor resistance becomes less, and the voltage drops. At normal engine operating temperature the voltage will measure about 1.5 to 2.0 volts at ECM terminal "GE16".

Coolant temperature is one of the inputs used to control:

- Fuel delivery.
- Engine Spark Timing (EST).
- Idle Air Control (IAC).
- Torque Convertor Clutch (TCC).
- Controlled Canister Purge (CCP).
- Cooling Fan.

Test Description: Number(s) below refer to circled number(s) on the diagnostic chart.

1. Code 15 will set if:
   - Signal voltage indicates a coolant temperature less than -39°C (-38°F) for 60 seconds.
2. This test simulates a Code 14. If the ECM senses the low signal voltage (high temperature) and the "Scan" reads 130°C, the ECM and wiring are OK.
3. This test will determine if CKT 410 is open. There should be 5 volts present at sensor connector if measured with a DVM.

Diagnostic Aids:

A Tech 1 displays engine temperature in degrees celsius. After the engine is started the temperature should rise steadily to about 95°C, and then stabilize when the thermostat opens. It is normal for coolant temperature to fluctuate slightly around 95°C.

A faulty connection, or an open in CKT 410 or CKT 452 can result in a Code 15.

Codes 15, 21 and 66 stored at the same time could be the result of an open CKT 452.

The "Temperature to Resistance Value" scale may be used to test the coolant sensor at various temperature levels to evaluate the possibility of a "shifted" (mis-scaled) sensor. A "shifted" sensor could result in poor driveability complaints.

2.3L (VIN D, A & 3) ENGINE — DIAGNOSTIC TROUBLE CODE CHART — 1992 ACHIEVA, GRAND AM AND SKYLARK

CODE 15
COOLANT TEMPERATURE SENSOR (CTS) CIRCUIT
(LOW TEMPERATURE INDICATED)
2.3L (VIN D, A & 3) "N" CARLINE (PORT)

1. • DOES TECH 1 DISPLAY COOLANT TEMPERATURE OF -30°C (-22°F) OR LESS?

YES

2. • DISCONNECT COOLANT TEMPERATURE SENSOR.
   • JUMPER HARNESS TERMINALS TOGETHER.
   • TECH 1 SHOULD DISPLAY 130°C (266°F) OR MORE. DOES IT?

NO

CODE 15 IS INTERMITTENT. IF NO ADDITIONAL CODES WERE STORED, REFER TO "DIAGNOSTIC AIDS"

NO

3. • JUMPER CKT 410 TO GROUND.
   • TECH 1 SHOULD DISPLAY OVER 130°C (266 °F). DOES IT?

YES

FAULTY CONNECTION OR COOLANT TEMPERATURE SENSOR.

YES

OPEN COOLANT TEMPERATURE SENSOR GROUND CIRCUIT, FAULTY CONNECTION OR FAULTY ECM.

NO

OPEN CKT 410, FAULTY CONNECTION AT ECM, OR FAULTY ECM.

DIAGNOSTIC AID

COOLANT SENSOR		
TEMPERATURE VS. RESISTANCE VALUES (APPROXIMATE)		
°C	°F	OHMS
100	212	177
90	194	241
80	176	332
70	158	467
60	140	667
50	122	973
45	113	1188
40	104	1459
35	95	1802
30	86	2238
25	77	2796
20	68	3520
15	59	4450
10	50	5670
5	41	7280
0	32	9420
-5	23	12300
-10	14	16180
-15	5	21450
-20	-4	28680
-30	-22	52700
-40	-40	100700

"AFTER REPAIRS," REFER TO CODE CRITERIA

AND CONFIRM CODE DOES NOT RESET.

2.3L (VIN D, A & 3) ENGINE — DIAGNOSTIC TROUBLE CODE CHART — 1992 ACHIEVA, GRAND AM AND SKYLARK

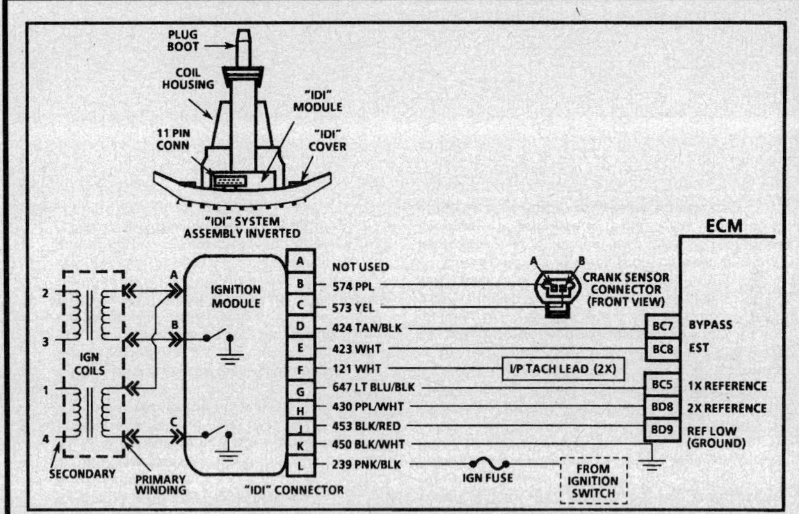

CODE 16
MISSING 2X REFERENCE CIRCUIT
2.3L (VIN D, A & 3) "N" CARLINE (PORT)

Circuit Description:

The ignition module sends a reference signal to the ECM twice per revolution to indicate crankshaft position and rpm so that the ECM can determine when to pulse the ignition coils and control ignition timing. This signal is called the 2X reference because it occurs two times per revolution. The ignition module applies 5 volts from terminal "H" through CKT 430 to ECM terminal "BD8" and in effect, switches this circuit to ground for a very short period of time. Code 16 is set if the ECM receives 1X reference pulses with no 2X reference pulses.

Test Description: Number(s) below refer to circled number(s) on the diagnostic chart.

1. This determines if the ECM recognizes a problem. If it doesn't set Code 16 at this point, the problem is intermittent and could be due to a loose connection. See "Diagnostic Aids."

2. This step simulates the 2X signal. The ECM should recognize the drop in voltage as the test light probe is removed, if the circuit and ECM are OK. This step will give accurate results only if the chart sequence is used - ignition "OFF," ignition "ON," Tech 1 set to 2X reference and terminal "G" touched with test light probe.

3. If the ECM did not recognize the simulation of the 2X signal, CKT 430 may be open or shorted to ground or voltage. If CKT 430 is OK, the ECM is faulty.

4. Step 2 indicated that CKT 430 is OK and the ECM is capable of recognizing the simulated 2X reference pulse. This indicates either a poor connection at ignition module terminal "H" or a faulty ignition module caused the Code 16.

Diagnostic Aids:

An intermittent may be caused by a poor connection, rubbed through wire insulation, or a wire broken inside the insulation. Inspect ECM harness connector terminal "BD8" and ignition module terminal "H" for improper mating, broken locks, improperly formed or damaged terminals, poor terminal to wire connection and damaged harness.

2.3L (VIN D, A & 3) ENGINE — DIAGNOSTIC TROUBLE CODE CHART — 1992 SKYLARK

CODE 16
MISSING 2X REFERENCE CIRCUIT
2.3L (VIN D, A & 3) "N" CARLINE (PORT)

1. • CLEAR CODES.
   • IDLE ENGINE FOR 2 MINUTES OR UNTIL "SERVICE ENGINE SOON" LIGHT TURNS "ON," IF SOONER.
   • DOES TECH 1 DISPLAY CODE 16?

YES

2. • IGNITION "OFF," DISCONNECT "IDI" CONNECTOR.
   • IGNITION "ON," SET "SCAN" TOOL TO DISPLAY 2X REFERENCE PULSES.
   • WITH TEST LIGHT CONNECTED TO B +, TOUCH ENGINE HARNESS CONNECTOR CAVITY "H" WITH LIGHT AND REMOVE. LIGHT ON = GROUND.
   • "SCAN" TOOL DISPLAY SHOULD INDICATE 2X REFERENCE PULSE INCREMENT WHEN TEST LIGHT IS REMOVED. DOES IT?

NO

3. • CHECK CKT 430 FOR OPEN OR SHORT TO GROUND OR VOLTAGE AND REPAIR AS NECESSARY. IF OK, REPLACE ECM.

NO

CODE 16 IS INTERMITTENT. IF NO ADDITIONAL CODES WERE DISPLAYED, REFER TO "DIAGNOSTIC AIDS."

YES

4. • CHECK FOR POOR CONNECTION BETWEEN "IDI" CONNECTOR CAVITY "H" AND IGNITION MODULE AND REPAIR AS NECESSARY. IF OK, REPLACE IGNITION MODULE.

"AFTER REPAIRS," REFER TO CODE CRITERIA AND CONFIRM CODE DOES NOT RESET.

2.3L (VIN D, A & 3) ENGINE — DIAGNOSTIC TROUBLE CODE CHART — 1992 ACHIEVA AND GRAND AM

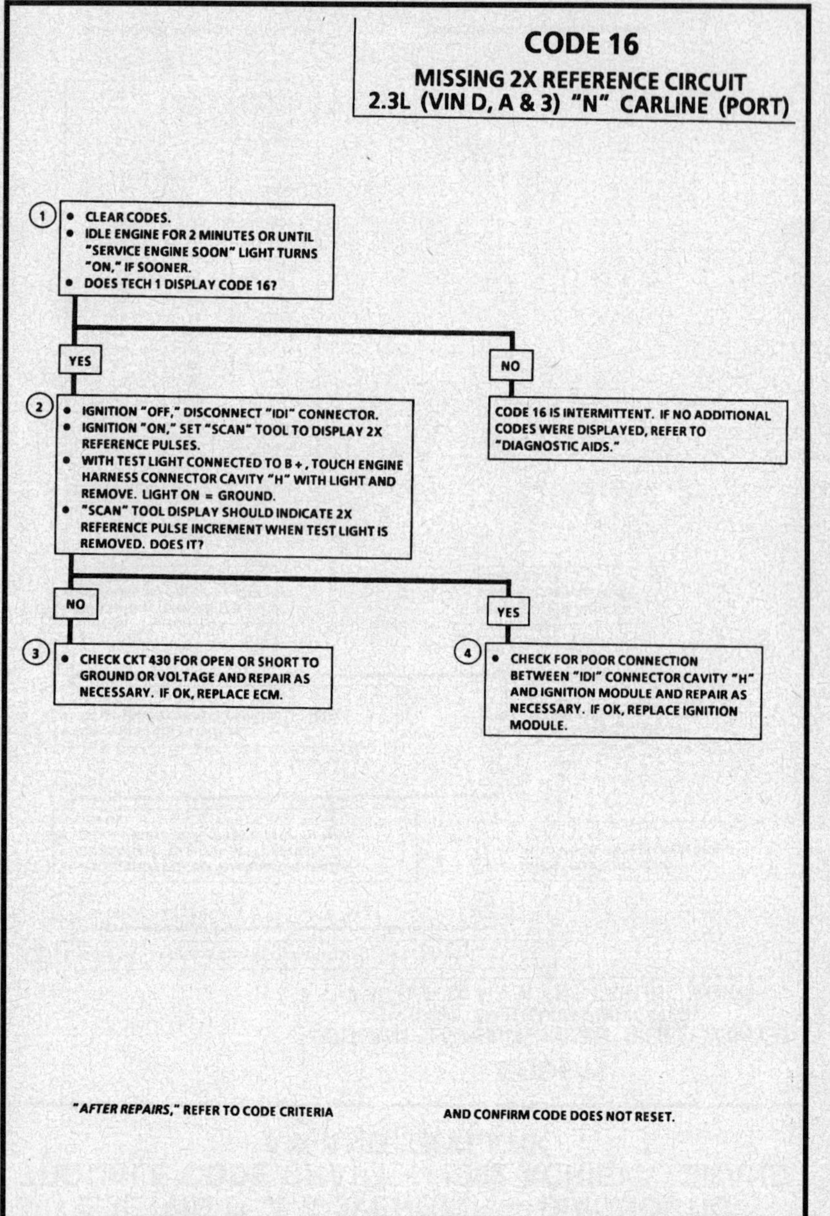

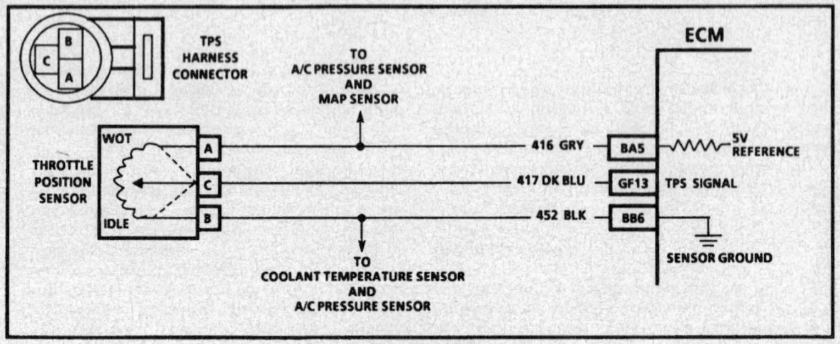

CODE 21
THROTTLE POSITION SENSOR (TPS) CIRCUIT
(SIGNAL VOLTAGE HIGH)
2.3L (VIN D, A & 3) "N" CARLINE (PORT)

Circuit Description:

The Throttle Position Sensor (TPS) provides a voltage signal that changes relative to the throttle opening. Signal voltage will vary from about .5 volt at idle to about 4.9 volts at Wide Open Throttle (WOT).

The TPS signal is one of the most important inputs used by the ECM for fuel control and for most of the ECM control outputs.

Test Description: Number(s) below refer to circled number(s) on the diagnostic chart.

1. Code 21 will set if:
   • Engine is running below 1500 rpm.
   • No Code 33 or 34.
   • MAP less than 65 kPa.
   • TPS signal voltage greater than approximately 4.0 volts (78%).
   • Above conditions exist for over 5 seconds.
   OR
   • TPS voltage greater than about 4.7 volts.
   With throttle closed the TPS should read less than .900 volt. If it doesn't, check for sticking TPS or throttle linkage. If OK, replace TPS.
2. With the TPS disconnected, the TPS voltage should go low if the ECM and wiring are OK.
3. Probing CKT 452 with a test light checks the TPS ground circuit because an open or very high resistance ground circuit will cause a Code 21.

Diagnostic Aids:

A Tech 1 displays throttle position in volts. It should display .400 volt to .900 volt with throttle closed and ignition "ON" or at idle. Voltage should increase at a steady rate as throttle is moved toward Wide Open Throttle (WOT).

Also some "Scan" tools will display throttle angle %, 0% = closed throttle 100% = WOT.

An open in CKT 452 will result in a Code 21.

Codes 15, 21 and 66 stored at the same time could be the result of an open CKT 452. "Scan" TPS while depressing accelerator pedal with engine stopped and ignition "ON." Display should vary from about .5 volt (500 mV) when throttle is closed, to over 4500 mV (4.5 volts) when throttle is held wide open.

Check condition of connector and sensor terminals for moisture or corrosion, and clean or replace as necessary. If corrosion is found, check condition of connector seal, repair and/or replace if necessary.

2.3L (VIN D, A & 3) ENGINE — DIAGNOSTIC TROUBLE CODE CHART — 1992 ACHIEVA, GRAND AM AND SKYLARK

CODE 21
THROTTLE POSITION SENSOR (TPS) CIRCUIT
(SIGNAL VOLTAGE HIGH)
2.3L (VIN D, A & 3) "N" CARLINE (PORT)

① • THROTTLE CLOSED.
  DOES TECH 1 DISPLAY THROTTLE POSITION OVER 2.5 VOLTS?

YES

NO → CODE 21 IS INTERMITTENT. IF NO ADDITIONAL CODES WERE STORED, REFER TO "DIAGNOSTIC AIDS"

② • DISCONNECT THROTTLE POSITION SENSOR. TECH 1 SHOULD DISPLAY THROTTLE POSITION BELOW .2 VOLT (200mV). DOES IT?

YES

NO → TPS SIGNAL CIRCUIT SHORTED TO VOLTAGE OR FAULTY ECM.

③ • PROBE SENSOR GROUND CIRCUIT WITH A TEST LIGHT CONNECTED TO BATTERY VOLTAGE.

LIGHT "ON" → FAULTY CONNECTION OR THROTTLE POSITION SENSOR.

LIGHT "OFF" → OPEN SENSOR GROUND CIRCUIT OR FAULTY ECM.

"AFTER REPAIRS," REFER TO CODE CRITERIA   AND CONFIRM CODE DOES NOT RESET.

2.3L (VIN D, A & 3) ENGINE — DIAGNOSTIC TROUBLE CODE CHART — 1992 ACHIEVA, GRAND AM AND SKYLARK

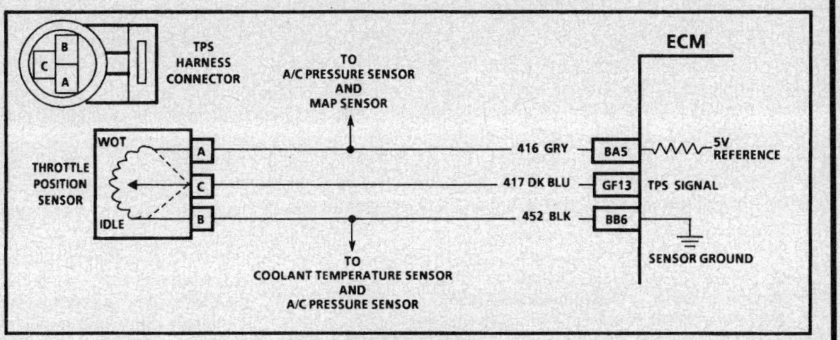

CODE 22
THROTTLE POSITION SENSOR (TPS) CIRCUIT
(SIGNAL VOLTAGE LOW)
2.3L (VIN D, A & 3) "N" CARLINE (PORT)

Circuit Description:

The Throttle Position Sensor (TPS) provides a voltage signal that changes relative to the throttle opening. Signal voltage will vary from about .5 volt at idle to about 4.9 volts at Wide Open Throttle (WOT).

The TPS signal is one of the most important inputs used by the ECM for fuel control and for most of the ECM control outputs.

Test Description: Number(s) below refer to circled number(s) on the diagnostic chart.

1. Code 22 will set if:
   - Engine is running.
   - TPS signal voltage is less than about .20 volt.
   - Above conditions present for 5 seconds.

   The TPS has an auto zeroing feature. If the voltage reading is within the range of about .4 to .9 volt, the ECM will use that value as closed throttle. If the voltage reading is out of the auto zero range at closed throttle, check for a binding throttle cable or damaged linkage, if OK, continue with diagnosis.
2. Simulates Code 21: (high voltage). If the ECM recognizes the high signal voltage then the ECM and wiring are OK.
3. Check for good sensor connection. If connection is good, replace TPS.
4. This simulates a high signal voltage to check for an open in CKT 417. The Tech 1 will not read up to 12 volts, but what is important is that the ECM recognizes the signal on CKT 417.

Diagnostic Aids:

"Scan" TPS while depressing accelerator pedal with engine stopped and ignition "ON." Display should vary from about 500 mV (.5 volt) when throttle is closed, to over 4500 mV (4.5 volts) when throttle is held wide open.

Also, some "Scan" tools will display throttle angle %: 0% = closed throttle; 100% = WOT.

If Code 22, 33 and/or 66 are set, check CKT 416 for faulty wiring or connections.

Should check condition of connector and sensor terminals for moisture or corrosion, and clean and/or replace as necessary. If corrosion is found, check condition of connector seal and repair or replace if necessary.

2.3L (VIN D, A & 3) ENGINE — DIAGNOSTIC TROUBLE CODE CHART — 1992 ACHIEVA, GRAND AM AND SKYLARK

CODE 22

THROTTLE POSITION SENSOR (TPS) CIRCUIT
(SIGNAL VOLTAGE LOW)
2.3L (VIN D, A & 3) "N" CARLINE (PORT)

1. • THROTTLE CLOSED.
   DOES TECH 1 DISPLAY THROTTLE POSITION
   .2V (200 mV) OR BELOW?

   YES →

2. • DISCONNECT TPS SENSOR.
   • JUMPER CKTS 416 & 417 TOGETHER.
   TECH 1 SHOULD DISPLAY THROTTLE POSITION
   OVER 4.0 V (4000 mV).
   DOES IT?

   NO →

4. • PROBE CKT 417 WITH A TEST LIGHT
   CONNECTED TO BATTERY VOLTAGE.
   TECH 1 SHOULD DISPLAY THROTTLE
   POSITION OVER 4.0V (4000 mV).
   DOES IT?

   YES →

   CKT 416 OPEN OR SHORTED TO GROUND
   OR
   FAULTY CONNECTION
   OR
   FAULTY ECM.

   NO →

   CODE 22 IS INTERMITTENT.
   IF NO ADDITIONAL CODES WERE STORED, REFER TO
   "DIAGNOSTIC AIDS"

   YES →

3. REFER TO
   TEST DESCRIPTION 3

   NO →

   CKT 417 OPEN OR SHORTED TO GROUND, OR SHORTED
   TO THROTTLE POSITION SENSOR GROUND CIRCUIT
   OR
   FAULTY ECM CONNECTION
   OR
   FAULTY ECM.

"AFTER REPAIRS," REFER TO CODE CRITIERIA AND CONFIRM CODE DOES NOT RESET.

2.3L (VIN D, A & 3) ENGINE — DIAGNOSTIC TROUBLE CODE CHART — 1992 ACHIEVA, GRAND AM AND SKYLARK

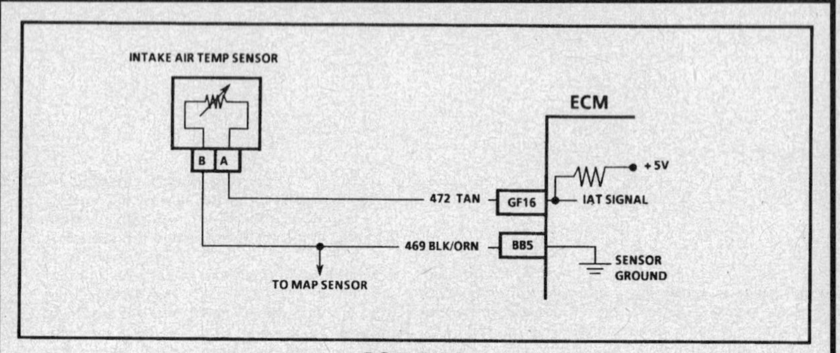

CODE 23

INTAKE AIR TEMPERATURE (IAT) SENSOR CIRCUIT
(LOW TEMPERATURE INDICATED)
2.3L (VIN D, A & 3) "N" CARLINE (PORT)

Circuit Description:

The Intake Air Temperature (IAT) sensor uses a thermistor to control the signal voltage at the ECM. The ECM applies a voltage (about 5 volts) on CKT 472 to the sensor. When the air is cold the sensor (thermistor) resistance is high, therefore the ECM terminal "GF16" voltage will be high. If the air is warm the sensor resistance is low, therefore the ECM terminal "GF16" voltage will be low.

Test Description: Number(s) below refer to circled number(s) on the diagnostic chart.

1. Code 23 will set if:
   • A signal voltage indicates a intake air temperature below about -34°C (-29°F).
   • Time since engine start is 320 seconds or longer.
   • Vehicle speed less than 15 mph.
2. A Code 23 will set due to an open sensor, wire, or connection. This test will determine if the wiring and ECM are OK.
3. This will determine if the signal CKT 472 or the sensor ground CKT 469 is open.

Diagnostic Aids:

A Tech 1 displays temperature of the air entering the engine, which should be close to ambient air temperature when engine is cold, and rise as underhood temperature increases.

A faulty connection, or an open in CKT 472 or CKT 469 can result in a Code 23.

Codes 23 and 34 stored at the same time, could be the result of an open CKT 469. The "Temperature to Resistance Values" scale at the right may be used to test the IAT sensor at various temperature levels to evaluate the possibility of a "slewed" (mis-scaled) sensor. A "slewed" sensor could result in poor driveability complaints.

2.3L (VIN D, A & 3) ENGINE — DIAGNOSTIC TROUBLE CODE CHART — 1992 ACHIEVA, GRAND AM AND SKYLARK

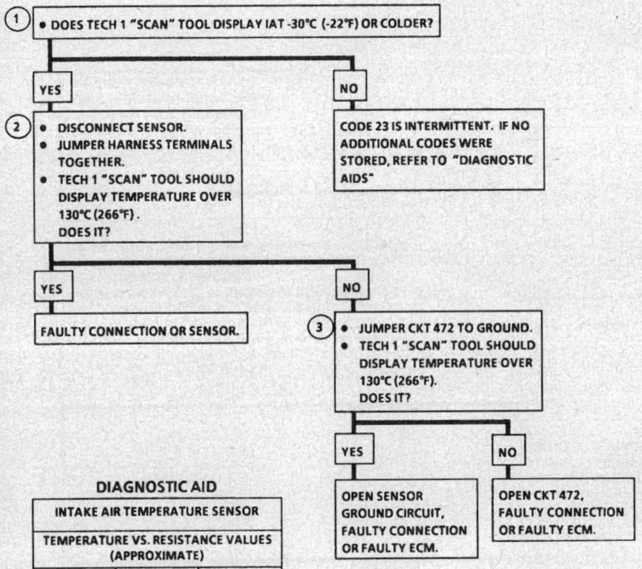

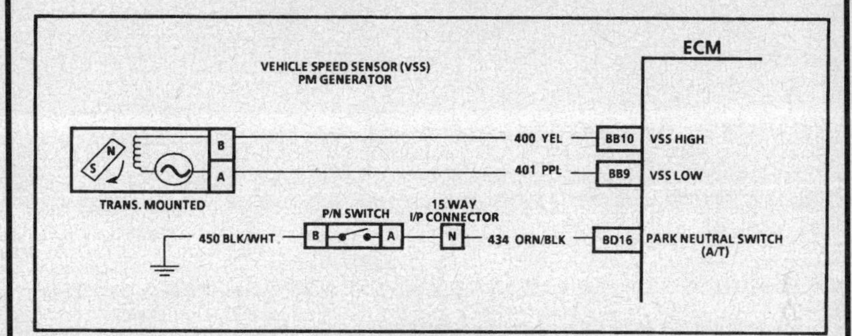

DIAGNOSTIC AID

INTAKE AIR TEMPERATURE SENSOR		
TEMPERATURE VS. RESISTANCE VALUES (APPROXIMATE)		
°C	°F	OHMS
100	212	177
90	194	241
80	176	332
70	158	467
60	140	667
50	122	973
45	113	1188
40	104	1459
35	95	1802
30	86	2238
25	77	2796
20	68	3520
15	59	4450
10	50	5670
5	41	7280
0	32	9420
-5	23	12300
-10	14	16180
-15	5	21450
-20	-4	28680
-30	-22	52700
-40	-40	100700

2.3L (VIN D, A & 3) ENGINE — DIAGNOSTIC TROUBLE CODE CHART — 1992 ACHIEVA, GRAND AM AND SKYLARK

CODE 24
VEHICLE SPEED SENSOR (VSS) CIRCUIT
2.3L (VIN 3) "N" CARLINE (PORT)

Circuit Description:

Vehicle speed information is provided to the ECM by the Vehicle Speed Sensor (VSS) which is a Permanent Magnet (PM) generator that is mounted in the transaxle. The PM generator produces a pulsing voltage whenever vehicle speed is over about 3 mph (5 kph). The A/C voltage level and the number of pulses increases with vehicle speed. The ECM then converts the pulsing voltage to mph which is used for calculations, and the mph can be displayed with a Tech 1. Output of the generator can also be seen by using a digital voltmeter on the AC scale while rotating the generator.

The function of VSS buffer used in past model years has been incorporated into the ECM. The ECM then supplies the necessary signal for the instrument panel for operating the speedometer, the odometer, and for the cruise control module.

Test Description: Number(s) below refer to circled number(s) on the diagnostic chart.
1. Code 24 will set if vehicle speed is less than 2 mph when:
   - Engine speed is between 1500 and 3600 rpm.
   - TPS is less than 2%.
   - Not in park or neutral.
   - All conditions met for 5 seconds.
   - No Code 21 or 22, 33 or 34.
   - MAP between 12-30 kPa.

   These conditions are met during a road load operation. Disregard Code 24 that sets when drive wheels are not turning.

   The PM generator only produces a signal if drive wheels are turning greater than 3 mph (5 kph).
2. Check MEM-CAL for correct application before replacing ECM.

Diagnostic Aids:

A Tech 1 should indicate a vehicle speed whenever the drive wheels are turning greater than 3 mph (5 kph).

A problem in CKT 434 will not affect the VSS input or the readings on a "Scan."

Check CKT 400 and 401 for proper connections. Be sure they are clean and tight and the harness is routed correctly.

(A/T) A faulty or misadjusted park/neutral switch can result in a false Code 24. Use a "Scan" and check for proper signal while in drive (3T40). Refer to CHART C-1A for P/N switch diagnosis check.

Code 24 can be falsely set if engine is "brake-torqued" in gear for 20 seconds.

2.3L (VIN D, A & 3) ENGINE — DIAGNOSTIC TROUBLE CODE CHART — 1992 ACHIEVA, GRAND AM AND SKYLARK

CODE 24
VEHICLE SPEED SENSOR (VSS) CIRCUIT
2.3L (VIN D, A & 3) "N" CARLINE (PORT)

DISREGARD CODE 24 IF SET WHILE DRIVE WHEELS ARE NOT TURNING.

(1)
- RAISE DRIVE WHEELS
- "NOTICE": *DO NOT PERFORM THIS TEST WITHOUT SUPPORTING THE LOWER CONTROL ARMS SO THAT THE DRIVE AXLES ARE IN A NORMAL HORIZONTAL POSITION. RUNNING THE VEHICLE IN GEAR WITH THE WHEELS HANGING DOWN AT FULL TRAVEL MAY DAMAGE THE DRIVE AXLES.*
- WITH ENGINE IDLING IN GEAR, TECH 1 SHOULD DISPLAY VEHICLE SPEED ABOVE 0.
  DOES IT?

NO → DOES SPEEDOMETER WORK?

YES → CODE 24 IS INTERMITTENT. IF NO ADDITIONAL CODES WERE STORED, REFER TO "DIAGNOSTIC AIDS" F

NO
- IGNITION "OFF."
- DISCONNECT VSS HARNESS CONNECTOR AT TRANSAXLE.
- CONNECT SIGNAL GENERATOR TESTER J 33431-B OR EQUIVALENT TO VSS HARNESS CONNECTOR.
- IGNITION "ON," TOOL "ON" AND SET TO GENERATE A VSS SIGNAL.
- "SCAN" TOOL SHOULD DISPLAY VEHICLE SPEED ABOVE 0.
  DOES IT?

YES → (2) CHECK PROM FOR CORRECT APPLICATION. IF OK, REPLACE ECM.

NO → CKT 400 OR 401 OPEN, SHORTED TO GROUND, SHORTED TOGETHER, FAULTY CONNECTIONS, OR FAULTY ECM.

YES → REPLACE VEHICLE SPEED SENSOR.

"AFTER REPAIRS," REFER TO CODE CRITERIA   AND CONFIRM CODE DOES NOT RESET.

2.3L (VIN D, A & 3) ENGINE — DIAGNOSTIC TROUBLE CODE CHART — 1992 ACHIEVA, GRAND AM AND SKYLARK

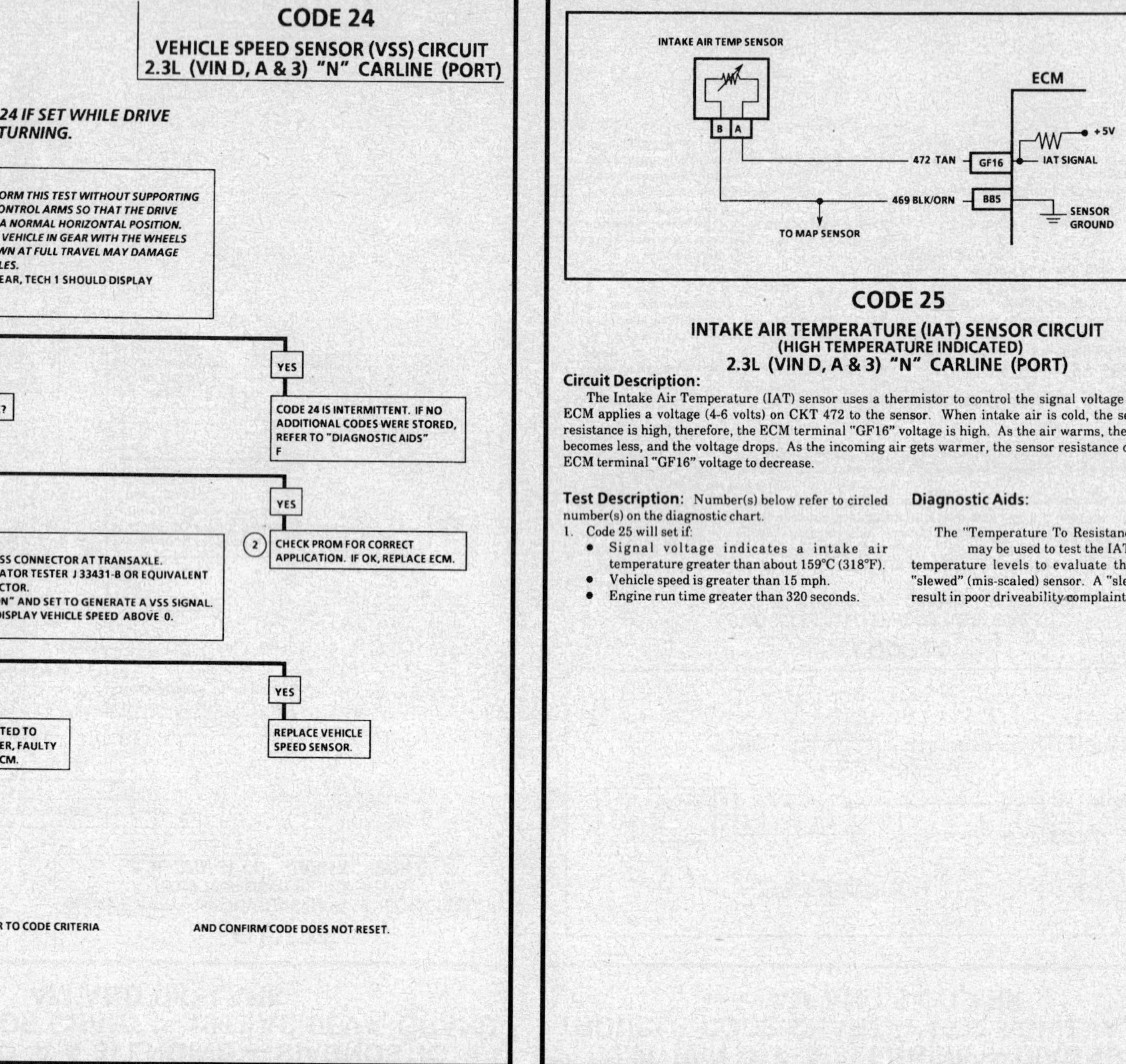

CODE 25
INTAKE AIR TEMPERATURE (IAT) SENSOR CIRCUIT
(HIGH TEMPERATURE INDICATED)
2.3L (VIN D, A & 3) "N" CARLINE (PORT)

Circuit Description:

The Intake Air Temperature (IAT) sensor uses a thermistor to control the signal voltage to the ECM. The ECM applies a voltage (4-6 volts) on CKT 472 to the sensor. When intake air is cold, the sensor (thermistor) resistance is high, therefore, the ECM terminal "GF16" voltage is high. As the air warms, the sensor resistance becomes less, and the voltage drops. As the incoming air gets warmer, the sensor resistance decreases, causing ECM terminal "GF16" voltage to decrease.

Test Description: Number(s) below refer to circled number(s) on the diagnostic chart.
1. Code 25 will set if:
   - Signal voltage indicates a intake air temperature greater than about 159°C (318°F).
   - Vehicle speed is greater than 15 mph.
   - Engine run time greater than 320 seconds.

Diagnostic Aids:

The "Temperature To Resistance Value" scale may be used to test the IAT sensor at various temperature levels to evaluate the possibility of a "slewed" (mis-scaled) sensor. A "slewed" sensor could result in poor driveability complaints.

2.3L (VIN D, A & 3) ENGINE — DIAGNOSTIC TROUBLE CODE CHART — 1992 ACHIEVA, GRAND AM AND SKYLARK

CODE 25
INTAKE AIR TEMPERATURE (IAT) SENSOR CIRCUIT
(HIGH TEMPERATURE INDICATED)
2.3L (VIN 3) "N" CARLINE (PORT)

① DOES TECH 1 "SCAN" TOOL DISPLAY IAT OF 145°C (293°F) OR HOTTER?

YES
- DISCONNECT SENSOR. TECH 1 "SCAN" TOOL SHOULD DISPLAY TEMPERATURE BELOW -30°C (-22°F). DOES IT?

YES
REPLACE SENSOR.

NO
CODE 25 IS INTERMITTENT. IF NO ADDITIONAL CODES WERE STORED, REFER TO "DIAGNOSTIC AIDS"

NO
CKT 472 SHORTED TO GROUND
OR
TO SENSOR GROUND
OR
ECM IS FAULTY.

DIAGNOSTIC AID

INTAKE AIR TEMPERATURE SENSOR
TEMPERATURE VS. RESISTANCE VALUES (APPROXIMATE)

°C	°F	OHMS
100	212	177
90	194	241
80	176	332
70	158	467
60	140	667
50	122	973
45	113	1188
40	104	1459
35	95	1802
30	86	2238
25	77	2796
20	68	3520
15	59	4450
10	50	5670
5	41	7280
0	32	9420
-5	23	12300
-10	14	16180
-15	5	21450
-20	-4	28680
-30	-22	52700
-40	-40	100700

"AFTER REPAIRS," REFER TO CODE CRITERIA   AND CONFIRM CODE DOES NOT RESET.

2.3L (VIN D, A & 3) ENGINE — DIAGNOSTIC TROUBLE CODE CHART — 1992 SKYLARK

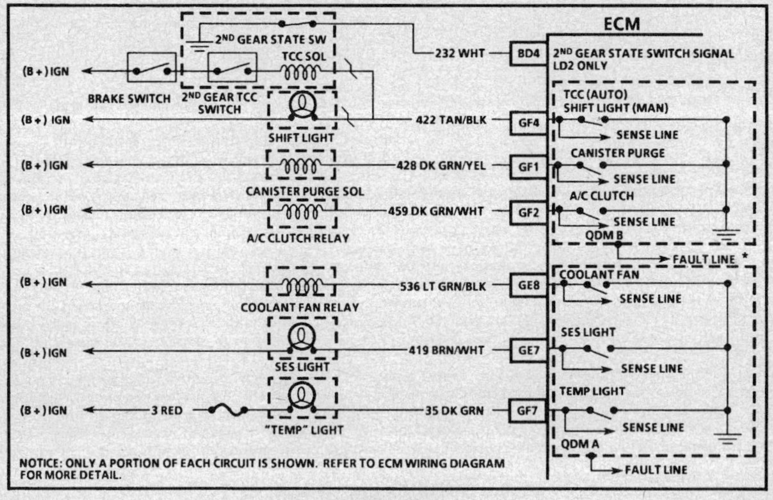

CODE 26
QUAD-DRIVER (QDM) CIRCUIT
2.3L (VIN D, A & 3) "N" CARLINE (PORT)

Circuit Description:

The ECM controls most components with electronic switches which complete a ground circuit when turned "ON." These switches are arranged in groups of 4, called Quad-Driver Modules (QDM's) which can independently control up to 4 outputs (ECM terminals), although not all outputs are used. When an output is "ON," the terminal is grounded and its voltage normally will be low. When an output is "OFF," its terminal voltage normally will be high, except for the TCC, as noted below, which depends on the brake and 2nd gear TCC switches.

QDM's are fault protected. If a relay or solenoid coil is shorted, having very low or zero resistance, or if the control side of the circuit is shorted to voltage, it would allow too much current into the QDM. The QDM senses this and the output turns "OFF" or its internal resistance increases to limit current flow and protect the QDM. The result is high output terminal voltage when it should be low. If the circuit from B + or the component is open, or the control side of the circuit is shorted to ground, terminal voltage will be low, even when output is commanded "OFF." Either of these conditions is considered to be a QDM fault. *See **NOTE** below!

Each QDM has a separate fault line to indicate the presence of a current fault to the ECM's central processor. A "Scan" tool displays the status of each of these fault lines as "Low" = OK, "High" = Fault*. Because of the brake and 2nd gear switches in the TCC circuit, Code 26 is set under different conditions for QDM A and QDM B as follows:
- QDM A fault line = "High" for 20 seconds or more.
- QDM B fault line = "High" for 20 seconds or more and
  - Brake switch signal indicates brake switch is closed and 2nd gear state switch indicates transaxle is in 2nd or 3rd gear.
  OR
  - TCC is commanded "ON."

2.3L (VIN D, A & 3) ENGINE — DIAGNOSTIC TROUBLE CODE CHART — 1992 SKYLARK

CODE 26
QUAD-DRIVER (QDM) CIRCUIT
2.3L   "N" CARLINE (PORT)

- INSTALL TECH 1.
- IGNITION "ON," ENGINE "OFF."
- "SCAN" CODES.

CODE 26 QDM A

VERIFY PROPER OPERATION OF "SES" LIGHT BY PERFORMING DIAGNOSTIC CIRCUIT CHECK.

IGNITION "ON," ENGINE "OFF." NOTE TEMPERATURE LIGHT.

"ON" / "OFF"

START ENGINE. IS TEMPERATURE LIGHT "ON"?

NO / YES

SEE CHART C16 TO DIAGNOSIS TEMPERATURE LIGHT CIRCUIT.

- IGNITION "ON," ENGINE "OFF."
- USING TECH 1 OUTPUT CONTROL FUNCTION, COMMAND COOLANT FAN RELAY "ON" AND THEN "OFF."
- DOES COOLANT FAN TURN "ON" AND "OFF" WHEN RELAY IS COMMANDED?

YES / NO

CODE 26 FAULT IS INTERMITTENT. CHECK FOR POOR TERMINAL CONTACT AT ECM AND QDM A CIRCUITRY. IF OK
- CLEAR CODES AND RE-CHECK, IF CODE 26 RE-SETS, REPLACE ECM.

SEE CHART C12 TO DIAGNOSIS COOLANT FAN CIRCUIT.

CODE 26 QDM B

- IGNITION "ON," ENGINE "OFF."
- USING TECH 1, COMMAND CANISTER PURGE "ON" AND "OFF" WHILE LISTENING TO CANISTER PURGE SOLENOID. DOES SOLENOID TURN "ON" AND "OFF." WHEN COMMANDED?

YES / NO

IS VEHICLE EQUIPPED WITH AIR CONDITIONING?

YES / NO

SEE CHART C3 TO DIAGNOSE CANISTER PURGE CIRCUIT.

- START ENGINE.
- USING TECH 1 COMMAND A/C RELAY "ON" AND "OFF." DOES A/C CLUTCH TURN "ON" AND "OFF" WHEN RELAY IS COMMANDED?

NO / YES

WITH TEST LIGHT CONNECTED TO B+, BACKPROBE ECM TERMINAL "BC3" LIGHT SHOULD BE "ON," IS IT?

YES / NO

SEE CHART C10 TO DIAGNOSE A/C CIRCUIT.

REPAIR OPEN GROUND CKT 450.

- IGNITION "OFF."
- RAISE DRIVE WHEELS.
- NOTICE: DO NOT PERFORM THIS TEST WITHOUT SUPPORTING THE LOWER CONTROL ARMS SO THAT THE DRIVE AXLES ARE IN A NORMAL HORIZONTAL POSITION. RUNNING THE VEHICLE IN GEAR WITH THE WHEELS HANGING DOWN AT FULL TRAVEL MAY DAMAGE THE DRIVE AXLES.
- USING TECH 1, OBSERVE TCC AND RPM.
- RUN VEHICLE IN DRIVE AT APPROXIMATELY 35 MPH.
- WHEN TCC ENGAGES, RPM SHOULD DROP WITH STEADY THROTTLE, DOES IT?

NO / YES

SEE CHART C8 TO DIAGNOSE TCC CIRCUIT.

CODE 26 FAULT IS INTERMITTENT. CHECK FOR POOR TERMINAL CONTACT AT ECM AND QDM B CIRCUITRY, IF OK,
- CLEAR CODES AND RECHECK, IF CODE 26 RE-SETS, REPLACE ECM.

CODE 26 QDM A & B

IS CODE 53 ALSO SET?

YES / NO

REFER TO CODE 53 CHART FIRST.

PROCEED TO CODE 26 QDM A DIAGNOSTICS FIRST.

"AFTER REPAIRS," REFER TO CODE CRITERIA AND CONFIRM CODE DOES NOT RESET.

2.3L (VIN D, A & 3) ENGINE — DIAGNOSTIC TROUBLE CODE CHART — 1992 ACHIEVA AND GRAND AM

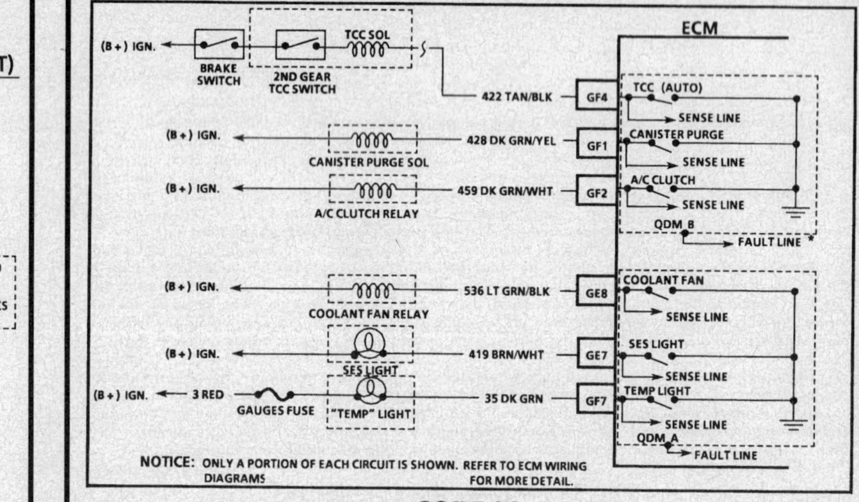

CODE 26
QUAD-DRIVER (QDM) CIRCUIT
2.3L (VIN 3) "N" CARLINE (PORT)

Circuit Description:

The ECM controls most components with electronic switches which complete a ground circuit when turned "ON." These switches are arranged in groups of 4, called Quad-Driver Modules (QDM's) which can independently control up to 4 outputs (ECM terminals), although not all outputs are used. When an output is "ON," the terminal is grounded and its voltage normally will be low. When an output is "OFF," its terminal voltage normally will be high, except for the TCC, as noted below, which depends on the brake and 2nd gear TCC switches.

QDM's are fault protected. If a relay or solenoid coil is shorted, having very low or zero resistance, or if the control side of the circuit is shorted to voltage, it would allow too much current into the QDM. The QDM senses this and the output turns "OFF" or its internal resistance increases to limit current flow and protect the QDM. The result is high output terminal voltage when it should be low. If the circuit from B+ or the component is open, or the control side of the circuit is shorted to ground, terminal voltage will be low, even when output is commanded "OFF." Either of these conditions is considered to be a QDM fault. *SEE NOTICE BELOW!*

Each QDM has a separate fault line to indicate the presence of a current fault to the ECM's central processor. A Tech 1 displays the status of each of these fault lines as "Low" = OK, "High" = Fault*. Because of the brake switch in the TCC circuit, Code 26 is set under different conditions for QDM A and QDM B as follows:

- QDM A fault line = "High" for 20 seconds or more.
- QDM B fault line = "High" for 20 seconds or more <u>and</u>
  - Brake switch signal indicates brake switch is closed.
    OR
  - TCC is commanded "ON."

2.3L (VIN D, A & 3) ENGINE — DIAGNOSTIC TROUBLE CODE CHART — 1992 ACHIEVA, GRAND AM AND SKYLARK

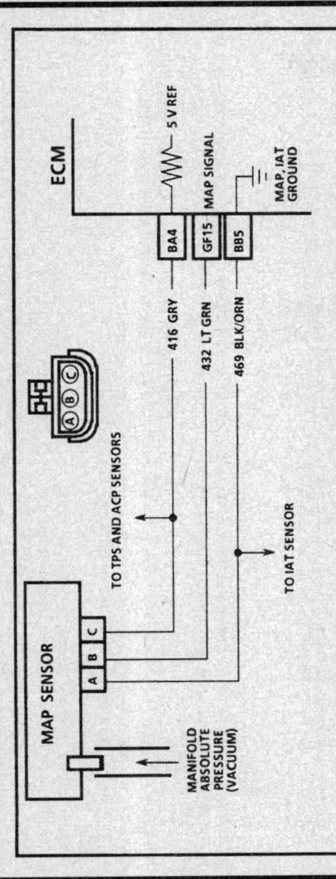

CODE 33

MANIFOLD ABSOLUTE PRESSURE (MAP) SENSOR CIRCUIT
(SIGNAL VOLTAGE HIGH - LOW VACUUM)
2.3L (VIN D, A & 3) "N" CARLINE (PORT)

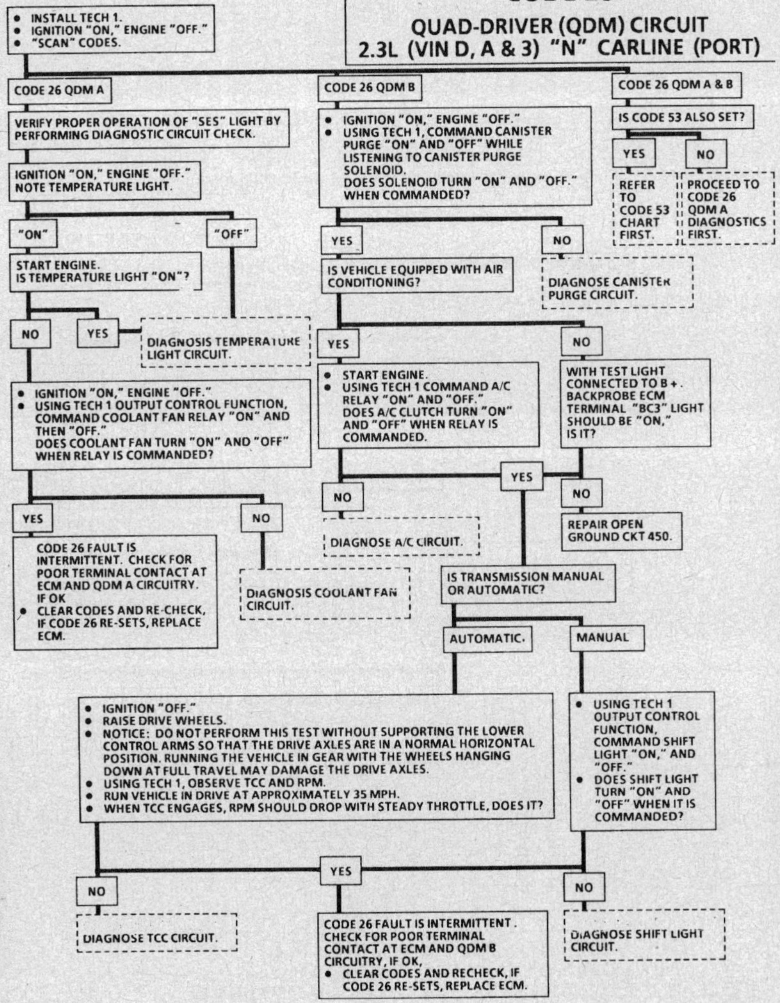

Circuit Description:

The Manifold Absolute Pressure (MAP) sensor responds to changes in manifold pressure (vacuum). The ECM receives this information as a signal voltage that will vary from about 1 to 1.5 volts at closed throttle (idle) to 4.5 - 4.8 volts at Wide Open Throttle (WOT) (low vacuum).

If the MAP sensor fails, the ECM will substitute a fixed MAP value and use the Throttle Position Sensor (TPS) to control fuel delivery.

Test Description: Number(s) below refer to circled number(s) on the diagnostic chart.

1. This step will determine if Code 33 is the result of a hard failure or an intermittent condition. Code 33 will set when:
   - MAP greater than 80 kPa.
   - TPS less than 12%.
   - VSS less than 1 mph.
   - Code 21 or 22 not present.
   - Above conditions met for 5 seconds.
2. This step simulates conditions for a Code 34. If the ECM recognizes the change, the ECM and CKT 416 and CKT 432 are OK. If CKT 469 is open, there may also be a stored Code 23.

A Code 33 will result if CKT 469 is open or if CKT 432 is shorted to voltage or to CKT 416.

If Code 33 is intermittent, refer to "Symptoms."

- Check all connections.
- Disconnect sensor from bracket and twist sensor by hand (only) to check for intermittent connections. Output changes greater than .1 volt indicates a bad connector or connection. if OK, replace sensor.

NOTICE: Make sure electrical connector remains securely fastened.

- Refer to CHART C-1D, MAP sensor voltage vs. atmospheric pressure for further diagnosis.

NOTICE: After repairs use Tech 1 "Block Learn Reset" function to reset block learn to 128.

Diagnostic Aids:

With the ignition "ON" and the engine stopped, the manifold pressure is equal to atmospheric pressure and the signal voltage will be high. This information is used by the ECM as an indication of vehicle altitude. Comparison of this reading with a known good vehicle with the same sensor is a good way to check accuracy of a "suspect" sensor. Readings should be the same ± .4 volt.

2.3L (VIN D, A & 3) ENGINE — DIAGNOSTIC TROUBLE CODE CHART — 1992 ACHIEVA AND GRAND AM

CODE 26

QUAD-DRIVER (QDM) CIRCUIT
2.3L (VIN D, A & 3) "N" CARLINE (PORT)

2.3L (VIN D, A & 3) ENGINE — DIAGNOSTIC TROUBLE CODE CHART — 1992 ACHIEVA, GRAND AM AND SKYLARK

CODE 33

MANIFOLD ABSOLUTE PRESSURE (MAP) SENSOR CIRCUIT
(SIGNAL VOLTAGE HIGH - LOW VACUUM)
2.3L (VIN D, A & 3) "N" CARLINE (PORT)

(1)
- IF ENGINE IDLE IS ROUGH, UNSTABLE, OR INCORRECT, CORRECT CONDITION BEFORE USING CHART.
- ENGINE IDLING.
- DOES TECH 1 DISPLAY A MAP VOLTAGE OF 4.0 VOLTS OR OVER?

YES / NO

(2)
- IGNITION "OFF."
- DISCONNECT MAP SENSOR ELECTRICAL CONNECTOR.
- IGNITION "ON."
- TECH 1 SHOULD READ A VOLTAGE OF 1 VOLT OR LESS.
  DOES IT?

NO → CODE 33 IS INTERMITTENT. IF NO ADDITIONAL CODES WERE STORED, REFER TO "DIAGNOSTIC AIDS"

YES / NO

- PROBE SENSOR GROUND CIRCUIT WITH A TEST LIGHT TO BATTERY VOLTAGE.
- TEST LIGHT SHOULD LIGHT.
  DOES IT?

NO → MAP SIGNAL CIRCUIT SHORTED TO VOLTAGE, SHORTED TO 5 VOLT REFERENCE CIRCUIT OR FAULTY ECM.

YES / NO

PLUGGED OR LEAKING SENSOR VACUUM HOSE OR FAULTY MAP SENSOR.

NO → OPEN SENSOR GROUND CIRCUIT.

"AFTER REPAIRS," REFER TO CODE CRITERIA   AND CONFIRM CODE DOES NOT RESET.

2.3L (VIN D, A & 3) ENGINE — DIAGNOSTIC TROUBLE CODE CHART — 1992 ACHIEVA, GRAND AM AND SKYLARK

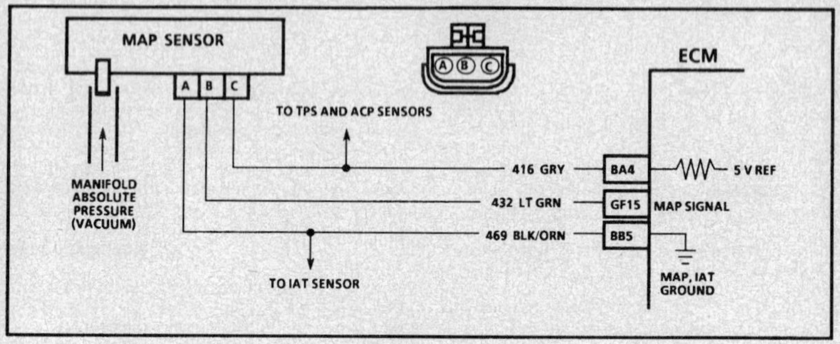

CODE 34

MANIFOLD ABSOLUTE PRESSURE (MAP) SENSOR CIRCUIT
(SIGNAL VOLTAGE LOW - HIGH VACUUM)
2.3L (VIN D, A & 3) "N" CARLINE (PORT)

Circuit Description:

The Manifold Absolute Pressure (MAP) sensor responds to changes in manifold pressure (vacuum). The ECM receives this information as a signal voltage that will vary from about 1 to 1.5 volts at closed throttle (idle) to 4.5 - 4.8 volts at Wide Open Throttle (WOT) (low vacuum).

If the MAP sensor fails, the ECM will substitute a fixed MAP value and use the Throttle Position Sensor (TPS) to control fuel delivery.

Test Description: Number(s) below refer to circled number(s) on the diagnostic chart.

1. This step determines if Code 34 is the result of a hard failure or an intermittent condition. Code 34 will set when:
   - Engine running.
   - No Code 21.
   - MAP less than 14 kPa.
   - Engine rpm less than 1200 or TPS greater than 15.2%.
   - Above conditions met for .2 seconds.
2. Jumpering harness terminals "B" to "C" (5 volts to signal circuit) will determine if the sensor is at fault, or if there is a problem with the ECM or wiring.
3. The Tech 1 may not display 5 volts. The important thing is that the ECM recognizes the voltage as more than 4 volts, indicating that the ECM and CKT 432 are OK.

Diagnostic Aids:

An intermittent open in CKT 432 or CKT 416 will result in a Code 34. With the ignition "ON" and the engine "OFF," the manifold pressure is equal to atmospheric pressure and the signal voltage will be high. This information is used by the ECM as an indication of vehicle altitude.

Comparison of this reading with a known good vehicle with the same sensor is a good way to check accuracy of a "suspect" sensor. Readings should be the same ± .4 volt. Also CHART C-1D can be used to test the MAP sensor.

- Check all connections
- Disconnect sensor from bracket and twist sensor by hand (only) to check for intermittent connections. Output changes greater than .1 volt indicates a bad connector or connection. If OK, replace sensor.

NOTICE: Make sure electrical connector remains securely fastened.

- Refer to CHART C-1D, MAP sensor voltage vs. atmospheric pressure for further diagnosis.

NOTICE: After repairs use Tech 1 "Block Learn Reset" function to reset block learn to 128.

2.3L (VIN D, A & 3) ENGINE — DIAGNOSTIC TROUBLE CODE CHART — 1992 ACHIEVA, GRAND AM AND SKYLARK

CODE 34

MANIFOLD ABSOLUTE PRESSURE (MAP) SENSOR CIRCUIT
(SIGNAL VOLTAGE LOW - HIGH VACUUM)
2.3L (VIN D, A & 3) "N" CARLINE (PORT)

① ENGINE IDLING.
• DOES TECH 1 DISPLAY MAP VOLTAGE BELOW .25 VOLT?

YES → ② IGNITION "OFF."
• DISCONNECT SENSOR ELECTRICAL CONNECTOR.
• JUMPER HARNESS TERMINALS "B" TO "C".
• IGNITION "ON."
• MAP VOLTAGE SHOULD READ OVER 4.7 VOLTS. DOES IT?

NO → CODE 34 IS INTERMITTENT. IF NO ADDITIONAL CODES WERE STORED, REFER TO "DIAGNOSTIC AIDS"

NO → ③ IGNITION "OFF."
• REMOVE JUMPER WIRE.
• PROBE TERMINAL "B" (CKT 432) WITH A TEST LIGHT TO BATTERY VOLTAGE.
• IGNITION "ON."
• TECH 1 SHOULD READ OVER 4 VOLTS. DOES IT?

YES → FAULTY CONNECTION OR SENSOR.

YES → 5 VOLT REFERENCE CIRCUIT OPEN
OR
SHORTED TO GROUND
OR
FAULTY ECM.

NO → CKT 432 OPEN
OR
CKT 432 SHORTED TO GROUND
OR
CKT 432 SHORTED TO SENSOR GROUND
OR
FAULTY ECM.

"AFTER REPAIRS," REFER TO CODE CRITERIA AND CONFIRM CODE DOES NOT RESET.

2.3L (VIN D, A & 3) ENGINE — DIAGNOSTIC TROUBLE CODE CHART — 1992 ACHIEVA, GRAND AM AND SKYLARK

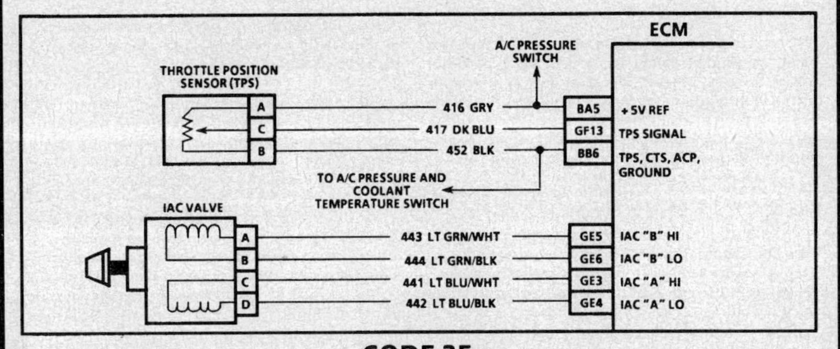

CODE 35

IDLE SPEED ERROR
2.3L (VIN D, A & 3) "N" CARLINE (PORT)

Circuit Description:
The ECM controls idle speed to a calculated, "desired" rpm based on sensor inputs and actual engine rpm, determined by the time between successive 2X ignition reference pulses from the ignition module. The ECM uses 4 circuits to move an Idle Air Control (IAC) valve, which allows varying amounts of air flow into the intake manifold, controlling idle speed.

Code 35 sets when:
• Engine speed is at least 175 rpm more or 200 rpm less than "desired."
• TPS voltage indicates throttle is open less than 1%.
• VSS indicates vehicle speed is less than 3 mph.
• All above conditions are continuously met for 5 seconds or more.
• IAC steps must be less than 10, or greater than 140.

Test Description: Number(s) below refer to circled number(s) on the diagnostic chart.
1. The Tech 1 is used to extend and retract the IAC valve. Valve movement is verified by an engine speed change. If no change in engine speed occurs, the valve can be retested when removed from the throttle body.
2. This step checks the quality of the IAC movement in Step 1. Between 900 rpm and about 1500 rpm, the engine speed should change smoothly with each flash of the tester light in both extend and retract. If the IAC valve is retracted beyond the control range (about 1500 rpm), it may take many flashes in the extend position before engine speed will begin to drop. This is normal on certain engines, fully extending IAC may cause engine stall. This may be normal.
Step 1 verifies proper IAC valve operation while this step checks the IAC circuits. Each lamp on the node light should flash red and green while the IAC valve is cycled. While the sequence of color is not important if either light is "OFF" or does not flash red and green, check the circuits for faults, beginning with poor terminal contacts.

Diagnostic Aids:
Check for vacuum leaks, unconnected or brittle vacuum hoses, cuts, etc. Examine manifold and throttle body gaskets for proper seal. Check for cracked intake manifold. Check open, shorts, or poor connections to IAC valve in CKTs 441, 442, 443 and 444.

An open, short, or poor connection in CKTs 441, 442, 443, or 444 will result in improper idle control and may cause Code 35.

An IAC valve which is stopped and cannot respond to the ECM, a throttle stop screw which has been tampered with, or a damaged throttle body or linkage could cause Code 35. If no problem is found and Code resets, replace ECM.

NOTICE: If Code 35 is currently set and speedometer is not accurately reflecting correct vehicle speed, refer to Code 24 (vehicle speed sensor circuit) diagnostics.

2.3L (VIN D, A & 3) ENGINE — DIAGNOSTIC TROUBLE CODE CHART — 1992 ACHIEVA, GRAND AM AND SKYLARK

2.3L (VIN D, A & 3) ENGINE — DIAGNOSTIC TROUBLE CODE CHART — 1992 ACHIEVA, GRAND AM AND SKYLARK

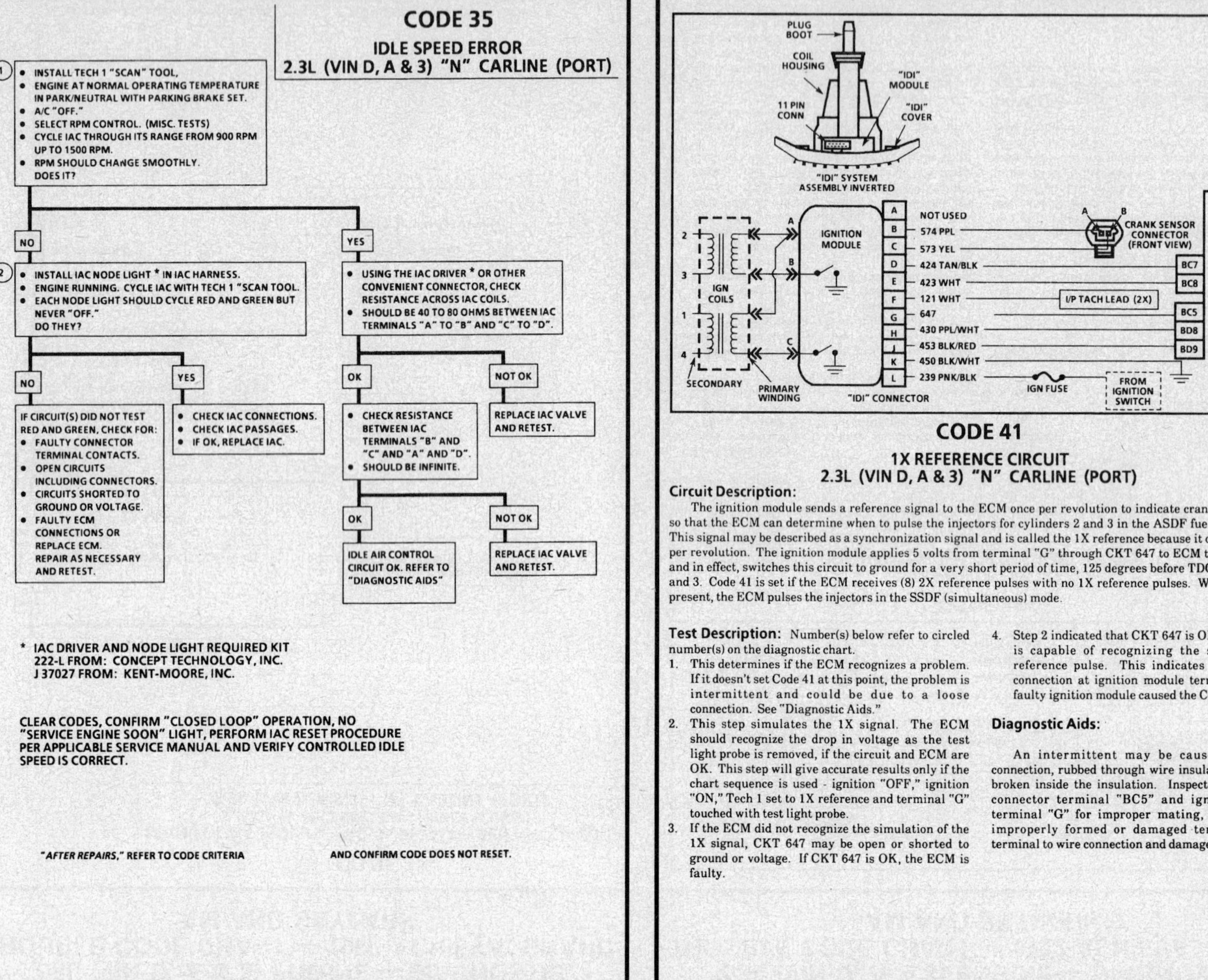

CODE 41

1X REFERENCE CIRCUIT
2.3L (VIN D, A & 3) "N" CARLINE (PORT)

Circuit Description:

The ignition module sends a reference signal to the ECM once per revolution to indicate crankshaft position so that the ECM can determine when to pulse the injectors for cylinders 2 and 3 in the ASDF fuel control mode. This signal may be described as a synchronization signal and is called the 1X reference because it occurs one time per revolution. The ignition module applies 5 volts from terminal "G" through CKT 647 to ECM terminal "BC5" and in effect, switches this circuit to ground for a very short period of time, 125 degrees before TDC of cylinders 2 and 3. Code 41 is set if the ECM receives (8) 2X reference pulses with no 1X reference pulses. When Code 41 is present, the ECM pulses the injectors in the SSDF (simultaneous) mode.

Test Description: Number(s) below refer to circled number(s) on the diagnostic chart.

1. This determines if the ECM recognizes a problem. If it doesn't set Code 41 at this point, the problem is intermittent and could be due to a loose connection. See "Diagnostic Aids."

2. This step simulates the 1X signal. The ECM should recognize the drop in voltage as the test light probe is removed, if the circuit and ECM are OK. This step will give accurate results only if the chart sequence is used - ignition "OFF," ignition "ON," Tech 1 set to 1X reference and terminal "G" touched with test light probe.

3. If the ECM did not recognize the simulation of the 1X signal, CKT 647 may be open or shorted to ground or voltage. If CKT 647 is OK, the ECM is faulty.

4. Step 2 indicated that CKT 647 is OK and the ECM is capable of recognizing the simulated 1X reference pulse. This indicates either a poor connection at ignition module terminal "G" or a faulty ignition module caused the Code 41.

Diagnostic Aids:

An intermittent may be caused by a poor connection, rubbed through wire insulation, or a wire broken inside the insulation. Inspect ECM harness connector terminal "BC5" and ignition module terminal "G" for improper mating, broken locks, improperly formed or damaged terminals, poor terminal to wire connection and damaged harness.

2.3L (VIN D, A & 3) ENGINE — DIAGNOSTIC TROUBLE CODE CHART — 1992 ACHIEVA, GRAND AM AND SKYLARK

CODE 41

1X REFERENCE CIRCUIT
2.3L (VIN D, A & 3) "N" CARLINE (PORT)

① • CLEAR CODES.
  • IDLE ENGINE FOR 2 MINUTES OR UNTIL "SERVICE ENGINE SOON" LIGHT TURNS "ON," IF SOONER.
  • DOES TECH 1 DISPLAY CODE 41?

YES

• WITH ENGINE IDLING, OBSERVE 1 X REFERENCE PULSES ON TECH 1.
• 1 X COUNTER SHOULD CONTINUOUSLY INCREMENT
  DOES IT?

NO

CODE 41 IS INTERMITTENT. IF NO ADDITIONAL CODES WERE DISPLAYED, REFER TO "DIAGNOSTIC AIDS."

② • IGNITION "OFF," DISCONNECT "IDI" CONNECTOR.
  • IGNITION "ON," SET "SCAN" TOOL TO DISPLAY 1X REFERENCE PULSES.
  • WITH TEST LIGHT CONNECTED TO B +, TOUCH ENGINE HARNESS CONNECTOR CAVITY "G" WITH LIGHT AND REMOVE. LIGHT ON = GROUND.
  • "SCAN" TOOL DISPLAY SHOULD INDICATE 1X REFERENCE PULSE INCREMENT WHEN TEST LIGHT IS REMOVED.
    DOES IT?

YES

FAULTY ECM, REPLACE ECM AND RECHECK.

NO

③ • CHECK CKT 647 FOR OPEN OR SHORT TO GROUND OR VOLTAGE AND REPAIR AS NECESSARY. IF OK, REPLACE ECM.

YES

④ • CHECK FOR POOR CONNECTION BETWEEN "IDI" CONNECTOR CAVITY "G" AND IGNITION MODULE AND REPAIR AS NECESSARY. IF OK, REPLACE IGNITION MODULE.

"AFTER REPAIRS," REFER TO CODE CRITERIA   AND CONFIRM CODE DOES NOT RESET.

2.3L (VIN D, A & 3) ENGINE — DIAGNOSTIC TROUBLE CODE CHART — 1992 ACHIEVA, GRAND AM AND SKYLARK

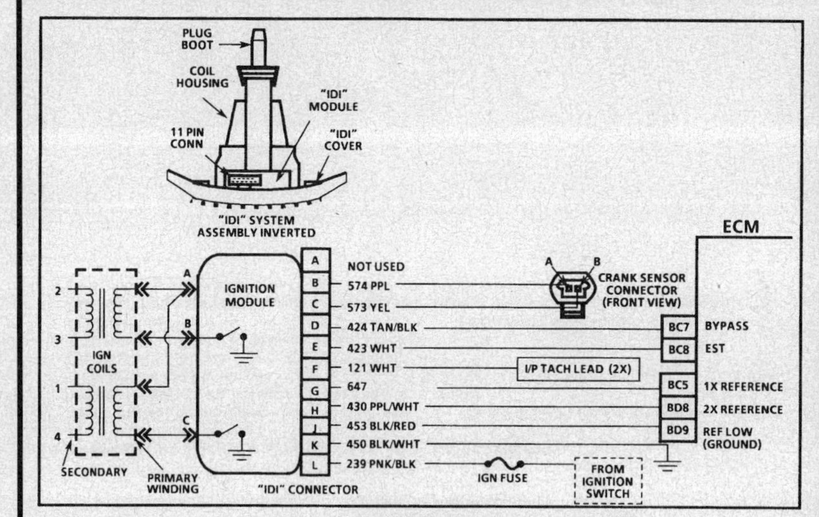

CODE 42

ELECTRONIC SPARK TIMING (EST) CIRCUIT
2.3L (VIN D, A & 3) "N" CARLINE (PORT)

Circuit Description:

The ignition module sends a reference signal to the ECM when the engine is cranking or running. While the engine is under 700 rpm, the ignition module controls the ignition timing. When the engine speed exceeds 700 rpm, the ECM sends a 5 volts signal on the "bypass" CKT 424 to switch the timing to ECM control through the EST CKT 423. Engine will remain under EST control until rpm drops below 150.

An open or ground in the EST or "bypass" circuit will set a Code 42 and cause the engine to run on module or "bypass" timing. This will result in poor performance and poor fuel economy.

Test Description: Number(s) below refer to circled number(s) on the diagnostic chart.

1. Checks to see if ECM recognizes a problem. If it doesn't set Code 42 at this point, it is an intermittent problem and could be due to a loose connection.

2. With the ECM disconnected, the ohmmeter should be reading less than 500 ohms, which is the normal resistance of the ignition module. A higher resistance would indicate a fault in CKT 423, a poor ignition module connection or a faulty ignition module.

3. If the test light was "ON" when connected from 12 volts to ECM harness terminal "BC7", either CKT 423 is shorted to ground or the ignition module is faulty.

4. Checks to see if ignition module switches when the bypass circuit is energized by 12 volts through the test light.

If the ignition module actually switches, the ohmmeter reading should shift to over 8000 ohms.

5. Disconnecting the ignition module should make the ohmmeter read as if it were monitoring an open circuit (infinite reading). If the ohmmeter has a reading other than infinite, CKT 423 is shorted to ground.

Diagnostic Aids:

An intermittent may be caused by a poor connection, rubbed through wire insulation, or a wire broken inside the insulation. Inspect ECM harness connectors for backed out terminals "BC7" or "BC8", improper mating, broken locks, improperly formed or damaged terminals, poor terminal to wire connection, and damaged harness.

2.3L (VIN D, A & 3) ENGINE — DIAGNOSTIC TROUBLE CODE CHART — 1992 ACHIEVA, GRAND AM AND SKYLARK

CODE 42
ELECTRONIC SPARK TIMING (EST) CIRCUIT
2.3L (VIN D, A & 3) "N" CARLINE (PORT)

(1)
- CLEAR CODES.
- IDLE ENGINE FOR 2 MINUTES.
- DOES TECH 1 DISPLAY CODE 42?

YES

(2)
- IGNITION "OFF," DISCONNECT BLACK C-D ECM CONNECTOR.
- IGNITION "ON," PROBE ECM HARNESS CONNECTOR TERMINAL "BC8" WITH A DIGITAL OHMMETER (DVM) TO GROUND. (RED (+) PROBE MUST GO TO TERMINAL "BC8", (CKT 423). BLACK (-) PROBE MUST GO TO GROUND.)
- RESISTANCE SHOULD BE 500 OHMS OR LESS. IS IT?

NO → CODE 42 IS INTERMITTENT. IF NO ADDITIONAL CODES DISPLAYED, REFER TO "DIAGNOSTIC AIDS."

YES

(3)
- IGNITION "ON," PROBE ECM HARNESS TERMINAL "BC7", (CKT 424) WITH A TEST LIGHT TO 12 VOLTS.
- TEST LIGHT SHOULD BE "OFF." IS IT?

NO → CHECK CKT 423 FOR OPEN OR POOR CONNECTION AT IGNITION MODULE TERMINAL "E". IF CKT 423 AND CONNECTION ARE OK, IGNITION MODULE IS FAULTY.

YES

(4)
- AS TEST LIGHT CONTACTED TERMINAL "BC7", (CKT 424) RESISTANCE FROM TERMINAL "BC8", (CKT 423) TO GROUND SHOULD SWITCH FROM UNDER 500 TO OVER 8,000 OHMS. DOES IT?

NO
- IGNITION "OFF," DISCONNECT IGNITION MODULE 11 PIN CONNECTOR.
- IGNITION "ON," IS TEST LIGHT STILL "ON?"

    YES → CKT 424 SHORTED TO GROUND.

    NO → FAULTY IGNITION MODULE.

YES

(5)
- IGNITION "OFF," DISCONNECT IGNITION MODULE 11 PIN CONNECTOR.
- IGNITION "ON," RESISTANCE BETWEEN ECM TERMINAL "BC8", (CKT 423) AND GROUND SHOULD BE INFINITE (OPEN CIRCUIT). IS IT?

YES → CHECK FOR OPEN CKT 424 OR POOR CONNECTION AT IGNITION MODULE TERMINAL "D". IF CKT 424 AND CONNECTION ARE OK, IGNITION MODULE IS FAULTY.

NO → CKT 423 SHORTED TO GROUND.

(from RECONNECT step)
- RECONNECT ECM AND IDLE ENGINE FOR 2 MINUTES.
- DOES "SES" LIGHT COME "ON?"

YES → IF CODE 42 IS DISPLAYED, CHECK COIL SECONDARY RESISTANCE USING CHART C-4M. IF OK, ECM IS FAULTY.

NO → FAULT IS INTERMITTENT. SEE "DIAGNOSTIC AIDS"

"AFTER REPAIRS," REFER TO CODE CRITERIA AND CONFIRM CODE DOES NOT RESET.

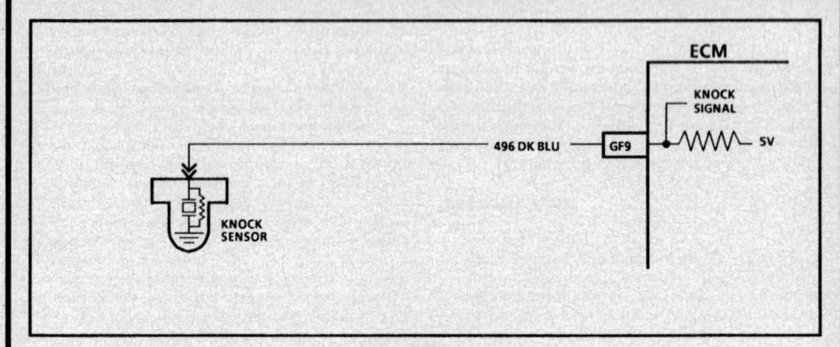

CODE 43
ELECTRONIC SPARK CONTROL (ESC) CIRCUIT
2.3L (VIN D, A & 3) "N" CARLINE (PORT)

Circuit Description:

The knock sensor detects engine detonation and the ECM retards the electronic spark timing based on the signal being received. The circuitry within the knock sensor causes the ECM 5 volts to be pulled down so that, under a no knock condition, CKT 496 would measure about 2.5 volts. The knock sensor produces an AC signal which rides on the 2.5 volts DC voltage. The amplitude and signal frequency are dependent upon the knock level.

The ECM performs two tests on this circuit to determine if it is operating correctly. If either of the tests fail, a Code 43 will be set.

- If there is an indication of knock for 3.67 seconds over a 3.9 second interval with the engine running.
- If ECM terminal "GF9" voltage is either above about 3.75 volts (indicating open CKT 496), or below about 1.25 volts (indicating CKT 496 is shorted to ground) for 5 seconds or more.

Test Description: Number(s) below refer to circled number(s) on the diagnostic chart.

1. If the conditions for the test as described above are being met, Code 43 will be currently set and the "SES" light will be illuminated.
2. The ECM has a 5 volts signal through a pull-up resistor which should be present at the knock sensor terminal.
3. This step checks that the internal resistance of knock sensor is within an acceptable range.

Diagnostic Aids:

Check CKT 496 for a potential open or short to ground.

Also check for proper installation of MEM-CAL.

Mechanical engine knock can cause a knock sensor signal. Abnormal engine noise must be corrected before using this chart.

2.3L (VIN D, A & 3) ENGINE — DIAGNOSTIC TROUBLE CODE CHART — 1992 ACHIEVA, GRAND AM AND SKYLARK

CODE 43
ELECTRONIC SPARK CONTROL (ESC) CIRCUIT
2.3L (VIN D, A & 3) "N" CARLINE (PORT)

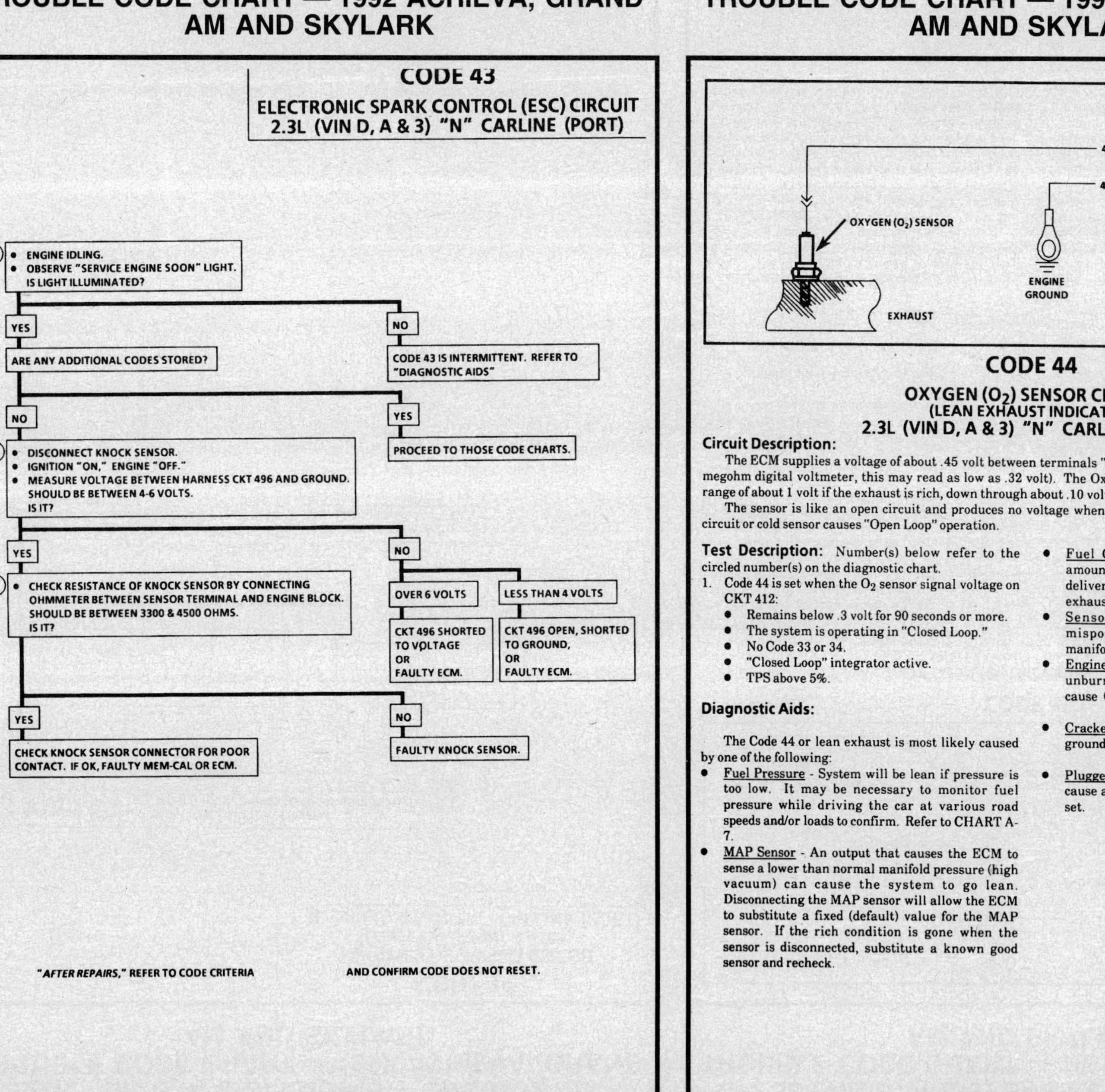

2.3L (VIN D, A & 3) ENGINE — DIAGNOSTIC TROUBLE CODE CHART — 1992 ACHIEVA, GRAND AM AND SKYLARK

CODE 44
OXYGEN (O$_2$) SENSOR CIRCUIT
(LEAN EXHAUST INDICATED)
2.3L (VIN D, A & 3) "N" CARLINE (PORT)

Circuit Description:

The ECM supplies a voltage of about .45 volt between terminals "GE14" and "GE15". (If measured with a 10 megohm digital voltmeter, this may read as low as .32 volt). The Oxygen O$_2$ sensor varies the voltage within a range of about 1 volt if the exhaust is rich, down through about .10 volt if exhaust is lean.

The sensor is like an open circuit and produces no voltage when it is below 315°C (600°F). An open sensor circuit or cold sensor causes "Open Loop" operation.

Test Description: Number(s) below refer to the circled number(s) on the diagnostic chart.
1. Code 44 is set when the O$_2$ sensor signal voltage on CKT 412:
   ● Remains below .3 volt for 90 seconds or more.
   ● The system is operating in "Closed Loop."
   ● No Code 33 or 34.
   ● "Closed Loop" integrator active.
   ● TPS above 5%.

Diagnostic Aids:

The Code 44 or lean exhaust is most likely caused by one of the following:
● _Fuel Pressure_ - System will be lean if pressure is too low. It may be necessary to monitor fuel pressure while driving the car at various road speeds and/or loads to confirm. Refer to CHART A-7.
● _MAP Sensor_ - An output that causes the ECM to sense a lower than normal manifold pressure (high vacuum) can cause the system to go lean. Disconnecting the MAP sensor will allow the ECM to substitute a fixed (default) value for the MAP sensor. If the rich condition is gone when the sensor is disconnected, substitute a known good sensor and recheck.

● _Fuel Contamination_ - Water, even in small amounts, near the in-tank fuel pump inlet can be delivered to the injector. The water causes a lean exhaust and can set a Code 44.
● _Sensor Harness_ - Sensor pigtail may be mispositioned and contacting the exhaust manifold.
● _Engine Misfire_ - A cylinder misfire will result in unburned oxygen in the exhaust, which could cause Code 44. Refer to CHART C4-M

● _Cracked O$_2$ Sensor_ - A Cracked O$_2$ Sensor or poor ground at the sensor, could cause Code 44.

● _Plugged Fuel Filter_ - A plugged fuel filter can cause a lean condition, and can cause a Code 44 to set.

2.3L (VIN D, A & 3) ENGINE — DIAGNOSTIC TROUBLE CODE CHART — 1992 ACHIEVA, GRAND AM AND SKYLARK

CODE 44

OXYGEN (O₂) SENSOR CIRCUIT
(LEAN EXHAUST INDICATED)
2.3L (VIN D, A & 3) "N" CARLINE (PORT)

①
- RUN WARM ENGINE (75°C/167°F TO 95°C/203°F) AT 1200 RPM.
- DOES TECH 1 INDICATE O₂ SENSOR VOLTAGE FIXED BELOW .35 VOLT (350 mV)?

YES
- DISCONNECT O₂ SENSOR.
- WITH ENGINE IDLING, TECH 1 SHOULD DISPLAY O₂ SENSOR VOLTAGE BETWEEN .35 VOLT AND .55 VOLT (350 mV AND 550 mV). DOES IT?

NO
- CODE 44 IS INTERMITTENT. IF NO ADDITIONAL CODES WERE STORED, REFER TO "DIAGNOSTIC AIDS"

YES
- REFER TO "DIAGNOSTIC AIDS"

NO
- CKT 412 SHORTED TO GROUND OR FAULTY ECM.

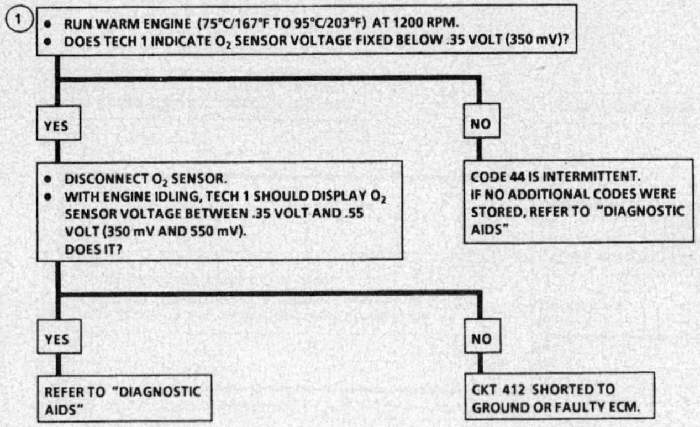

"AFTER REPAIRS," REFER TO CODE CRITERIA   AND CONFIRM CODE DOES NOT RESET.

2.3L (VIN D, A & 3) ENGINE — DIAGNOSTIC TROUBLE CODE CHART — 1992 ACHIEVA, GRAND AM AND SKYLARK

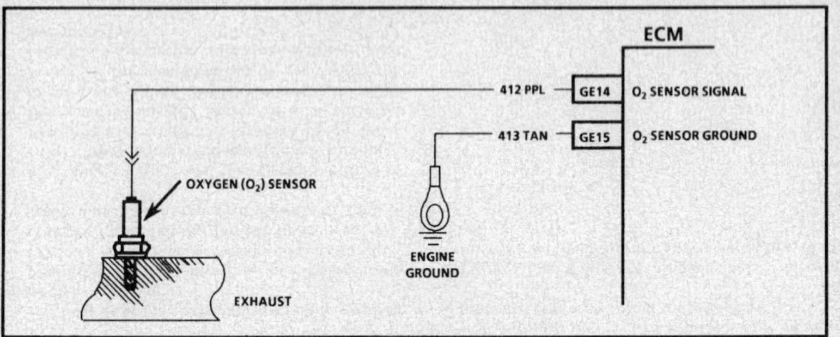

CODE 45

OXYGEN (O₂) SENSOR CIRCUIT
(RICH EXHAUST INDICATED)
2.3L (VIN D, A & 3) "N" CARLINE (PORT)

Circuit Description:

The ECM supplies a voltage of about .45 volt between terminals "GE14" and "GE15". (If measured with a 10 megohm digital voltmeter, this may read as low as .32 volt.) The Oxygen O₂ sensor varies the voltage within a range of about 1 volt if the exhaust is rich, down through about .10 volt if exhaust is lean.

The sensor is like an open circuit and produces no voltage when it is below 315°C (600°F). An open sensor circuit or cold sensor causes "Open Loop" operation.

Test Description: Number(s) below refer to the circled number(s) on the diagnostic chart.
1. Code 45 is set when:
   - O₂ voltage is above .75 volt.
   - No Code 33 or 34.
   - Fuel system in "Closed Loop."
   - TPS above 5%
   - Above conditions met for 30 seconds.
   OR
   - Voltage above 1 volt for 5 seconds.

Diagnostic Aids:

The Code 45 or rich exhaust is most likely caused by one of the following:
- **Fuel Pressure** - System will go rich if pressure is too high. The ECM can compensate for some increase. However, if it gets too high, a Code 45 will be set. See "Fuel System" diagnosis CHART A-7.
- **Leaking Injector** - See CHART A-7.
- **HEI Shielding** - An open ground CKT 453 may result in EMI or induced electrical noise. The ECM looks at this noise as crank sensor pulses. The additional pulses result in a higher than actual engine speed signal. The ECM then delivers too much fuel causing system to go rich. Engine tachometer will also show higher than actual engine speed which can help in diagnosing this problem.
- **Canister Purge** - Check for fuel saturation. If full of fuel, check canister control and hoses. See "Canister Purge," in "Evaporative Emission Control System (EECS),"
- **MAP Sensor** - An output that causes the ECM to sense a higher than normal manifold pressure (low vacuum) can cause the system to go rich. Disconnecting the MAP sensor will allow the ECM to set a fixed value for the MAP sensor. Substitute a different MAP sensor if the rich condition is gone while the sensor is disconnected.
- **Pressure Regulator** - Check for leaking fuel pressure regulator diaphragm by checking for the presence of liquid fuel in the vacuum line to the regulator.
- **TPS** - An intermittent TPS output will cause the system to go rich due to a false indication of the engine accelerating.
- **O₂ Sensor Contamination** - Inspect oxygen sensor for silicone contamination from fuel or use of improper RTV sealant. The sensor may have a white powdery coating and result in a high but false signal voltage (rich exhaust indication). The ECM will then reduce the amount of fuel delivered to the engine causing a severe surge driveability problem.

2.3L (VIN D, A & 3) ENGINE — DIAGNOSTIC TROUBLE CODE CHART — 1992 ACHIEVA, GRAND AM AND SKYLARK

CODE 45

OXYGEN (O_2) SENSOR CIRCUIT
(RICH EXHAUST INDICATED)
2.3L (VIN D, A & 3) "N" CARLINE (PORT)

① • RUN WARM ENGINE (75°C/167°F TO 95°C/203°F) AT 1200 RPM.
• DOES TECH 1 DISPLAY O_2 SENSOR VOLTAGE FIXED ABOVE .75 VOLT (750 mV)?

YES

• DISCONNECT O_2 SENSOR AND JUMPER HARNESS CKT 412 TO GROUND.
• TECH 1 SHOULD DISPLAY O_2 BELOW .35 VOLT (350 mV).
DOES IT?

NO

CODE 45 IS INTERMITTENT.
IF NO ADDITIONAL CODES WERE STORED, REFER TO "DIAGNOSTIC AIDS"

YES

REFER TO "DIAGNOSTIC AIDS"

NO

REPLACE ECM.

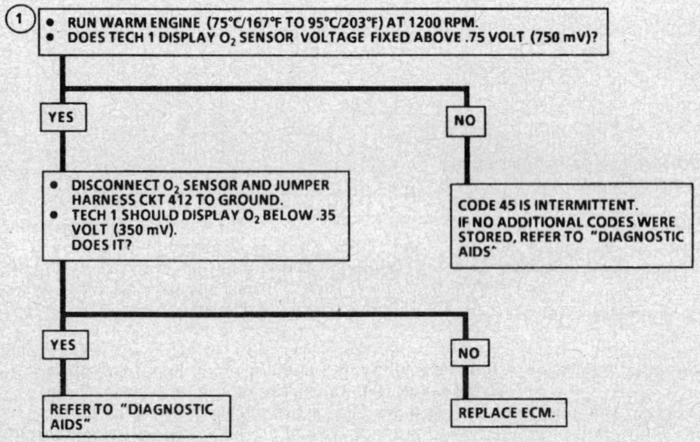

"AFTER REPAIRS," REFER TO CODE CRITERIA   AND CONFIRM CODE DOES NOT RESET.

2.3L (VIN D, A & 3) ENGINE — DIAGNOSTIC TROUBLE CODE CHART — 1992 ACHIEVA, GRAND AM AND SKYLARK

CODE 51

MEM-CAL ERROR
(FAULTY OR INCORRECT MEM-CAL)
2.3L (VIN D, A & 3) "N" CARLINE (PORT)

CHECK THAT ALL PINS ARE FULLY INSERTED IN THE SOCKET AND THAT MEM-CAL IS PROPERLY LATCHED.
IF OK, REPLACE MEM-CAL, CLEAR MEMORY, AND RECHECK. IF CODE 51 REAPPEARS, REPLACE ECM.

NOTICE: TO PREVENT POSSIBLE ELECTROSTATIC DISCHARGE DAMAGE TO THE ECM OR MEM-CAL, DO NOT TOUCH THE COMPONENT LEADS, AND DO NOT REMOVE THE MEM CAL COVER OR THE INTEGRATED CIRCUIT FROM CARRIER.

"AFTER REPAIRS," CONFIRM "CLOSED LOOP" OPERATION AND NO "SERVICE ENGINE SOON" LIGHT.

2.3L (VIN D, A & 3) ENGINE — DIAGNOSTIC TROUBLE CODE CHART — 1992 ACHIEVA, GRAND AM AND SKYLARK

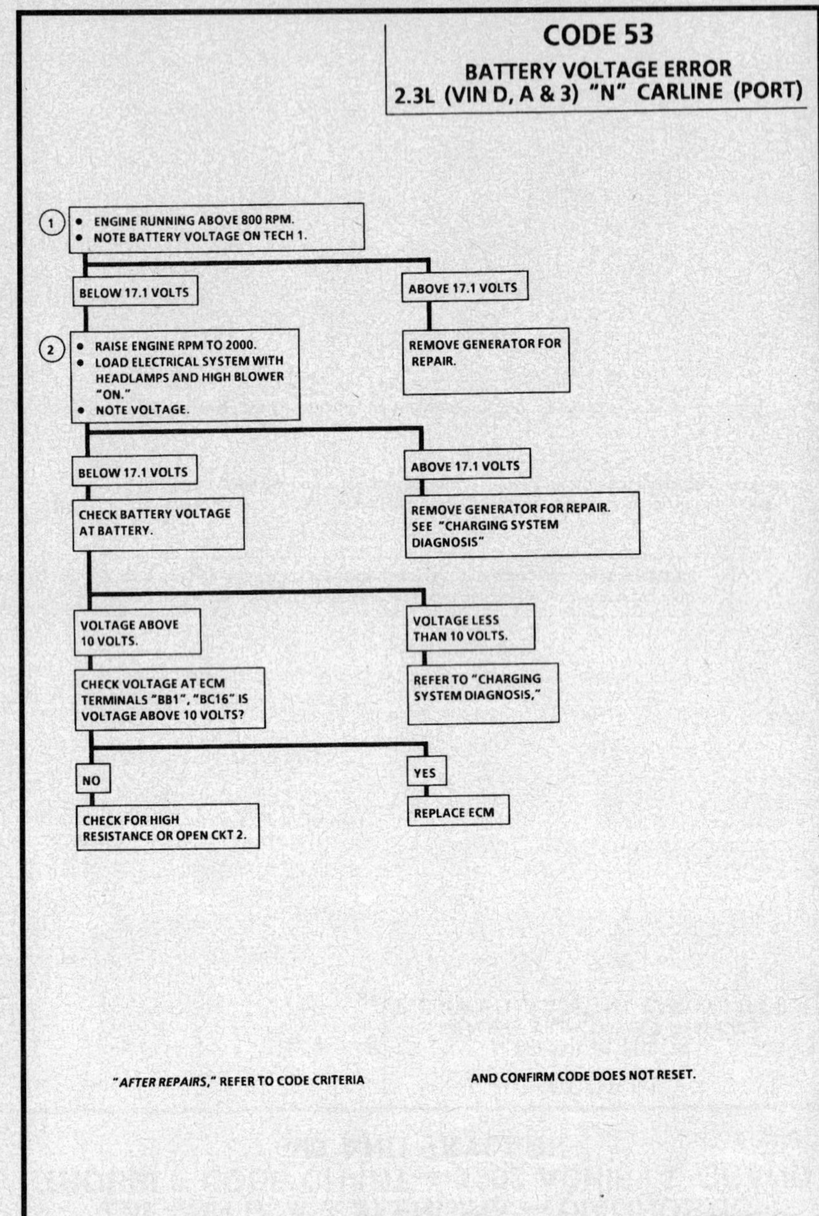

CODE 53
BATTERY VOLTAGE ERROR
2.3L (VIN D, A & 3) "N" CARLINE (PORT)

Circuit Description:

Code 53 will set when the ignition is "ON" and ECM terminal "BB1" and "BC16" voltages are more than 17.1 volts for about .2 seconds, or less than 10 volts for 240 seconds.

During the time the failure is present, all ECM outputs will be disengaged. (The setting of additional codes may result).

Test Description: Number(s) below refer to circled number(s) on the diagnostic chart.
1. Normal battery output is between 10 - 17.0 volts.
2. Checks to see if generator is faulty under load condition.

Note On Intermittents:

Charging battery with a battery charger and starting engine, may set Code 53. If code sets when an accessory is operated, check for poor connections or excessive current draw.

Also, check for poor connections at starter solenoid or fusible link junction box.

2.3L (VIN D, A & 3) ENGINE — DIAGNOSTIC TROUBLE CODE CHART — 1992 ACHIEVA, GRAND AM AND SKYLARK

CODE 53
BATTERY VOLTAGE ERROR
2.3L (VIN D, A & 3) "N" CARLINE (PORT)

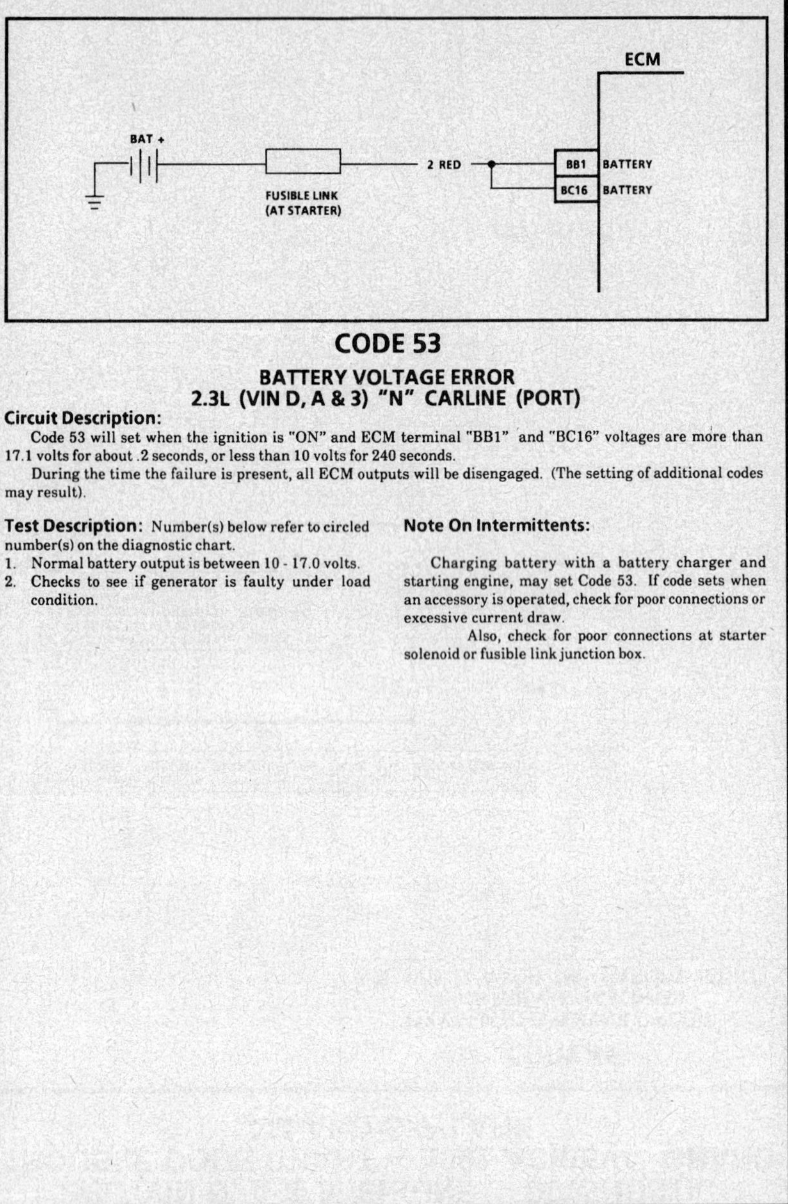

2.3L (VIN D, A & 3) ENGINE — DIAGNOSTIC TROUBLE CODE CHART — 1992 ACHIEVA, GRAND AM AND SKYLARK

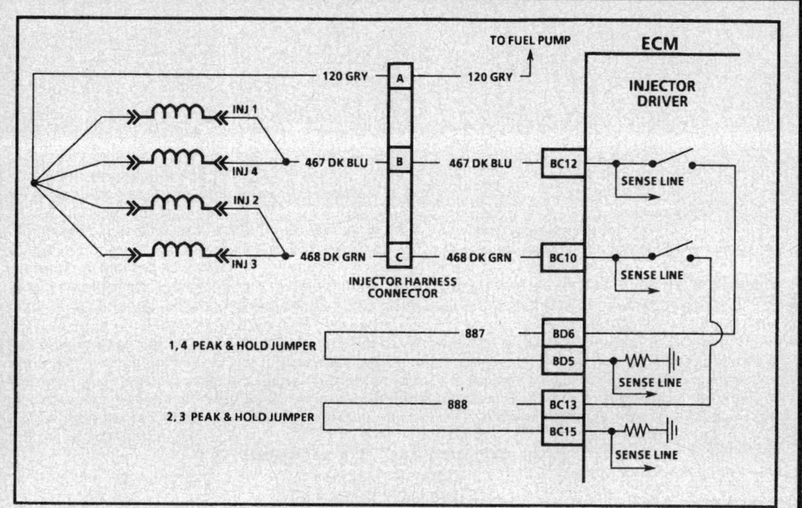

CODE 65 (Page 1 of 2)
FUEL INJECTOR CIRCUIT
(LOW CURRENT)
2.3L (VIN D, A & 3) "N" CARLINE (PORT)

Circuit Description:

The ECM has two injector driver circuits, each of which controls a pair of injectors (1 and 4 or 2 and 3). The ECM monitors the current in each driver circuit by measuring voltage drop through a fixed resistor and is able to control it. The current through each driver is allowed to rise to a "peak" of 4 amps to quickly open the injectors and is then reduced to 1 amp to "hold" them open. This is called "peak and hold." If the current can't reach a 4 amp peak, Code 65 is set as noted below. This code is also set if an injector driver circuit is shorted to voltage.

Test Description: Number(s) below refer to circled number(s) on the diagnostic chart.

1. Code 65 sets when:
   - 4 amp injector driver current not reached on either circuit.
   - Battery voltage greater than 9 volts.
   - Injectors commanded "ON" longer than a calibrated pulse width.
   - Above conditions met for 10 seconds.
2. Tests ECM and harness wiring to the 3 terminal injector harness connector.
3. Tests for open or shorted injector harness or injector. A shorted harness or injector will not cause Code 65.
4. Results of Step 2 will determine which branch to follow on Page 2.
5. Checks remainder of circuit from injectors to ECM as both harnesses were confirmed OK in Steps 2 and 3.

6. Determines cause of high resistance found in Step 3. (Low resistance or a short will not cause Code 65, but should be corrected if found.)
7. Checks for grounded "peak and hold" jumpers. This fault would allow injectors to pulse but would not allow "peak and hold" operation as current would not flow through the resistor in the ECM.

Diagnostic Aids:

Open CKTs 467, 468, 887 or 888 or CKT 467 or 468 shorted to voltage will cause Code 65 and will also cause a misfire due to an inoperative pair of injectors. CKTs 887 and 888 shorted to ground will cause Code 65 while allowing the injectors to pulse. An intermittent problem would have to be present for at least 20 seconds to set Code 65.

2.3L (VIN D, A & 3) ENGINE — DIAGNOSTIC TROUBLE CODE CHART — 1992 ACHIEVA, GRAND AM AND SKYLARK

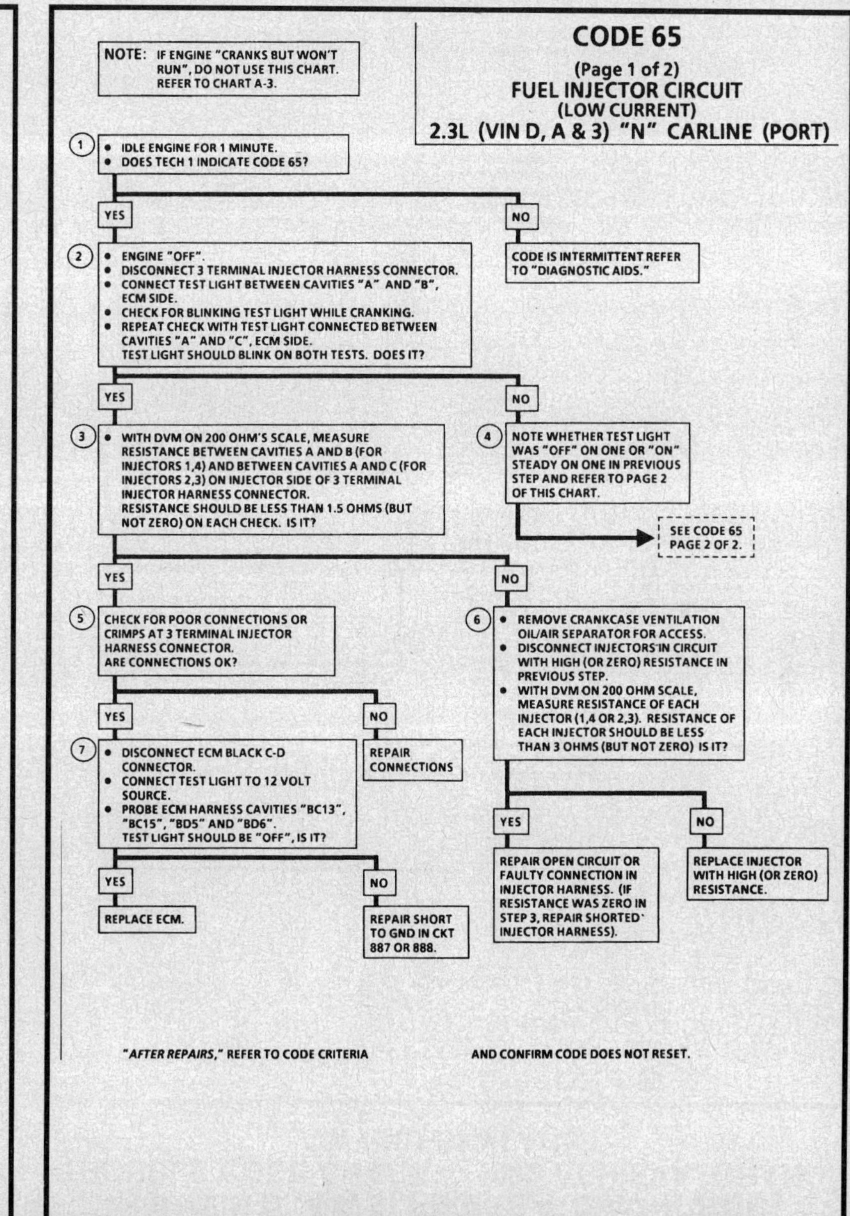

2.3L (VIN D, A & 3) ENGINE — DIAGNOSTIC TROUBLE CODE CHART — 1992 ACHIEVA, GRAND AM AND SKYLARK

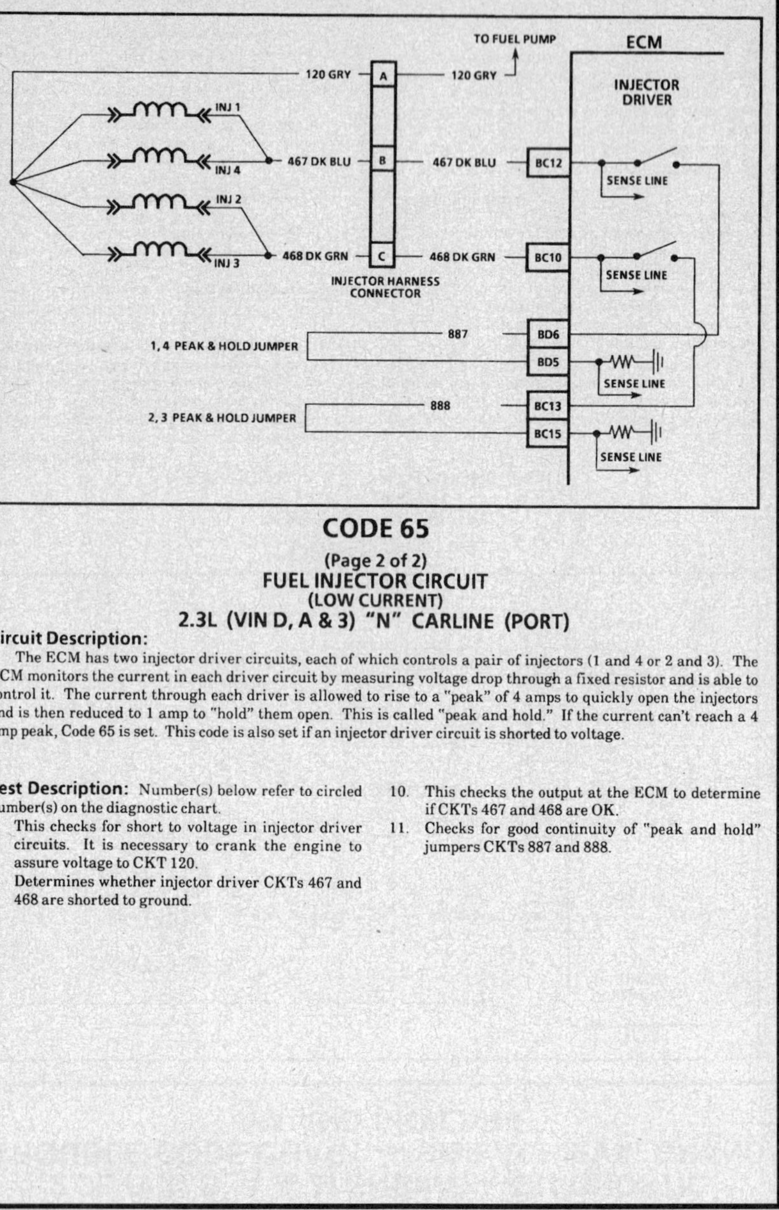

CODE 65
(Page 2 of 2)
FUEL INJECTOR CIRCUIT
(LOW CURRENT)
2.3L (VIN D, A & 3) "N" CARLINE (PORT)

Circuit Description:

The ECM has two injector driver circuits, each of which controls a pair of injectors (1 and 4 or 2 and 3). The ECM monitors the current in each driver circuit by measuring voltage drop through a fixed resistor and is able to control it. The current through each driver is allowed to rise to a "peak" of 4 amps to quickly open the injectors and is then reduced to 1 amp to "hold" them open. This is called "peak and hold." If the current can't reach a 4 amp peak, Code 65 is set. This code is also set if an injector driver circuit is shorted to voltage.

Test Description: Number(s) below refer to circled number(s) on the diagnostic chart.

8. This checks for short to voltage in injector driver circuits. It is necessary to crank the engine to assure voltage to CKT 120.
9. Determines whether injector driver CKTs 467 and 468 are shorted to ground.
10. This checks the output at the ECM to determine if CKTs 467 and 468 are OK.
11. Checks for good continuity of "peak and hold" jumpers CKTs 887 and 888.

2.3L (VIN D, A & 3) ENGINE — DIAGNOSTIC TROUBLE CODE CHART — 1992 ACHIEVA, GRAND AM AND SKYLARK

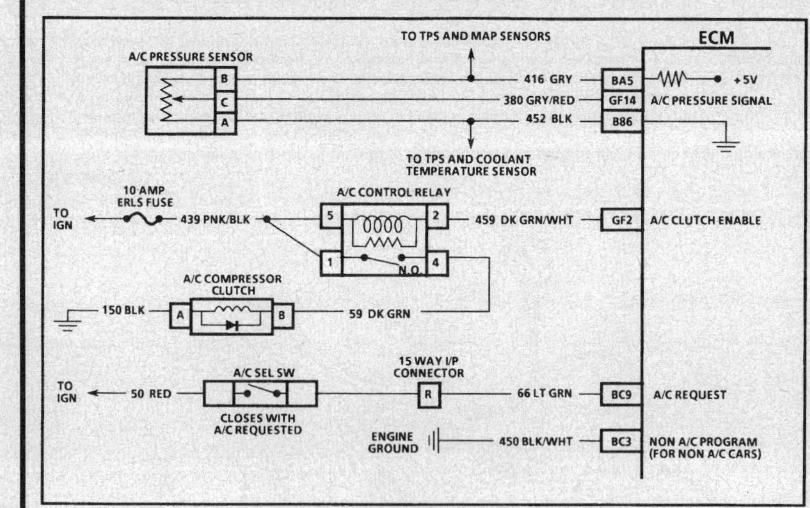

CODE 66

A/C PRESSURE SENSOR CIRCUIT
2.3L (VIN D, A & 3) "N" CARLINE (PORT)

Circuit Description:

The A/C pressure sensor responds to changes in A/C refrigerant system high side pressure. This input indicates how much load the A/C compressor is putting on the engine and is one of the factors used by the ECM to determine IAC valve position for idle speed control. The circuit consists of a 5 volt reference and a ground, both provided by the ECM, and a signal line to the ECM. The signal is a voltage which is proportional to the pressure. The sensor's range of operation is 0 to 450 psi. At 0 psi, the signal will be about .1 volt, varying up to about 4.9 volts at 450 psi or above. Code 66 sets if the voltage is above 4.9 volts or below .28 volt 15 seconds or more, or voltage is above 4 volts and A/C is not requested. The A/C compressor is disabled by the ECM if Code 66 is present, or if pressure is above or below calibrated values

Test Description: Number(s) below refer to circled number(s) on the diagnostic chart.

1. This step checks the voltage signal being received by the ECM from the A/C pressure sensor. The normal operating range is between .28 volt and 4.9 volts.
2. Checks to see if the high voltage signal is from a shorted sensor or a short to voltage in the circuit. Normally, disconnecting the sensor would make a normal circuit go to near zero volt.
3. Checks to see if low voltage signal is from the sensor or the circuit. Jumpering the sensor signal CKT 380 to 5 volts, checks the circuit, connections and ECM.
4. This step checks to see if the low voltage signal was due to an open in the sensor circuit or the 5 volt reference circuit since the prior step eliminated the pressure sensor.

Diagnostic Aids:

Code 66 sets when signal voltage falls outside the normal possible range of the sensor and is not due to a refrigerant system problem. If problem is intermittent, check for opens or shorts in harness or poor connections. If OK, replace A/C pressure sensor. If Code 66 re-sets, replace ECM.

Non-A/C Program

Code 66 will set on a Non-A/C car if CKT 450 to terminal "BC3" is open or shorted to B+.

2.3L (VIN D, A & 3) ENGINE — DIAGNOSTIC TROUBLE CODE CHART — 1992 ACHIEVA, GRAND AM AND SKYLARK

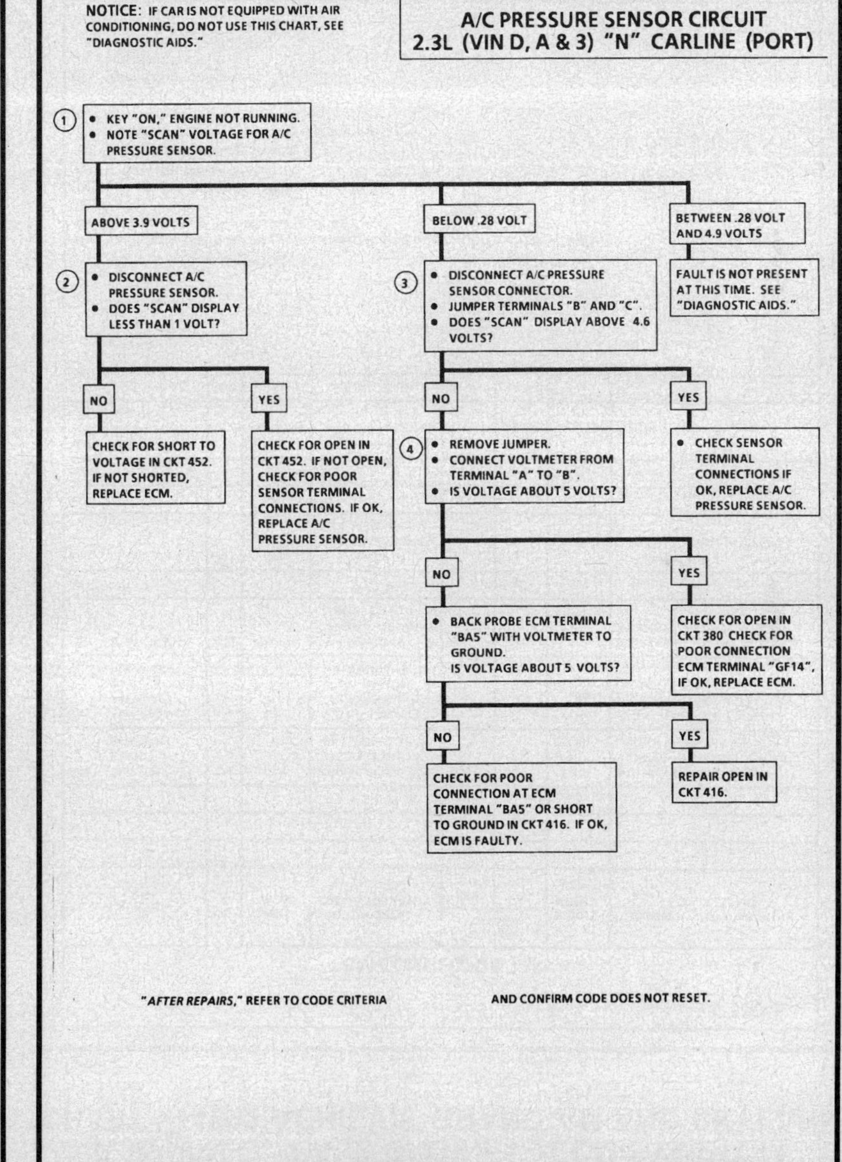

2.3L (VIN D, A & 3) ENGINE — EXHAUST SYSTEM DIAGNOSTIC CHART — 1992 ACHIEVA, GRAND AM AND SKYLARK

CHART B-1

RESTRICTED EXHAUST SYSTEM CHECK

2.3L VIN A

Proper diagnosis for a restricted exhaust system is essential before any components are replaced. The following procedure may be used for diagnosis.

CHECK AT O_2 SENSOR:

1. Carefully remove O_2 sensor.
2. Install Borroughs exhaust backpressure tester (BT 8515 or BT 8603) or equivalent in place of O_2 sensor (see illustration).
3. After completing test described below, be sure to coat threads of O_2 sensor with anti-seize compound P/N 5613695 or equivalent prior to re-installation.

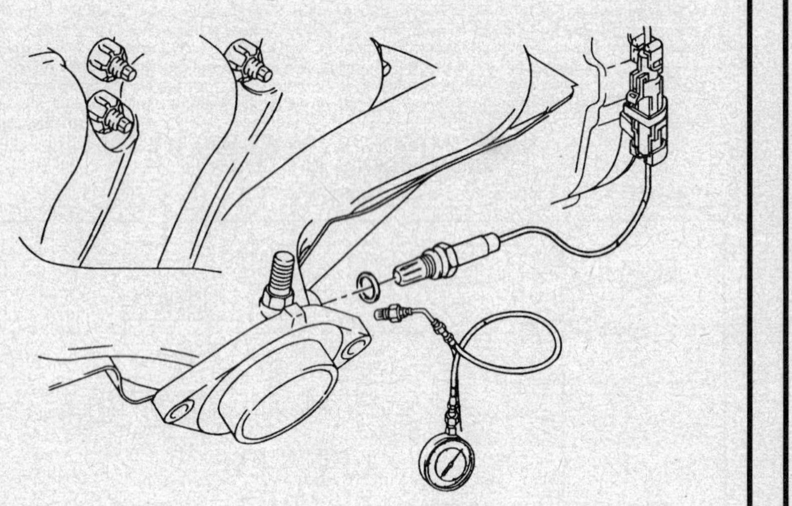

DIAGNOSIS:

1. With the engine idling at normal operating temperature, transaxle in park or neutral, observe the exhaust system backpressure reading on the gauge. Reading should not exceed 3.4 kPa (.5 psi).
2. Increase engine speed to 3000 rpm and observe gauge. Reading should not exceed 5 kPa (.75 psi).
3. If the backpressure at either speed exceeds specification, a restricted exhaust system is indicated.
4. Inspect the entire exhaust system for a collapsed pipe, heat distress, or possible internal muffler failure.
5. If there are no obvious reasons for the excessive backpressure, the catalytic converter is suspected to be restricted and should be replaced using current recommended procedures.

2.3L (VIN D, A & 3) ENGINE — ECM SYMPTOM CHART — 1992 ACHIEVA, GRAND AM AND SKYLARK

PIN FUNCTION		CKT #	WIRE COLOR	COMPONENT CONNECTOR CAVITY	NORMAL VOLTAGES KEY "ON"	NORMAL VOLTAGES ENG RUN	CODES AFFECT	POSSIBLE SYMPTOMS FROM FAULTY CIRCUIT
BA1								
BA2								
BA3								
BA4	MAP, TPS 5 VOLT REFERENCE	416	GRY	MAP TERMINAL "C" TPS TERMINAL "A" BA4 SPLICE TO BA5	5V	5V	22 (10) 34 (10) 66 (10)	WITH BA4 AND BA5 OPEN OR SHORTED TO GROUND. STALLS, LACK OF PERFORMANCE
BA5	MAP, TPS A/C PRESSURE SWITCH 5 VOLT REFERENCE	416	GRY	MAP TERMINAL "C" TPS TERMINAL "A" A/C PRESSURE TERMINAL "B"	5V	5V	22 (10) 34 (10) 66 (10)	WITH BA4 AND BA5 OPEN OR SHORTED TO GROUND. STALLS, LACK OF PERFORMANCE.
BA6	IGNITION FEED	639	PNK/BLK	15 WAY I/P TERMINAL "G" AND ECM 10 A FUSE	B +	B +		NO SES LIGHT, ENGINE CRANKS BUT WILL NOT START, NO DATA. (8)
BA7								
BA8	SERIAL DATA	461	ORN	15 WAY I/P TERMINAL "J" AND ALDL TERMINAL "M"	2.5 (3)	2.5 (3)		NO SERIAL DATA "SCAN" TOOL WILL NOT READ DATA (10)
BA9								
BA10								
BA11	FUEL PUMP RELAY CONTROL	465	DK GRN/WHT	FUEL PUMP RELAY 5	0	B + (4)		LONG CRANKING TIME BEFORE ENGINE STARTS (8)
BA12	ECM GROUND	450	BLK/WHT	ENGINE BLOCK	0	0		

(1) Varies from .60 to battery voltage, depending on position of drive wheels.
(2) Battery voltage for first two seconds
(3) Varies
(4) Battery voltage when fuel pump is running
(5) Varies with temperature
(6) Reads battery voltage in gear
(7) Battery voltage when engine is cranking
(8) Open circuit
(9) Grounded circuit
(10) Open/Grounded circuit
(11) Less than 1 volt
* Less than .5 volt (500 mV)

2.3L (VIN D, A & 3) ENGINE — ECM SYMPTOM CHART — 1992 ACHIEVA, GRAND AM AND SKYLARK

ECM CONNECTOR "B"

PIN FUNCTION	CKT #	WIRE COLOR	COMPONENT CONNECTOR CAVITY	NORMAL VOLTAGES KEY "ON"	NORMAL VOLTAGES ENG RUN	CODES AFFECT	POSSIBLE SYMPTOMS FROM FAULTY CIRCUIT
BB1 BATTERY FEED	2	RED	FUSIBLE LINK AND B+	B+	B+		CRANKS BUT WILL NOT START
BB2							
BB3							
BB4							
BB5 IAT, MAP GROUND	469	BLK/ORN	IAT TERMINAL "B" MAP TERMINAL "A"	0	0	23 (8) 33	STALLS AT IDLE. POOR PERFORMANCE.
BB6 TPS, CTS, A/C PRESS GROUND	452	BLK	TPS TERMINAL "B" CTS TERMINAL "A" A/C TERMINAL "A"	0	0	15 21 (8) 66	HESITATION ON ACCELERATION. LACK OF PERFORMANCE, EXHAUST ODOR
BB7							
BB8							
BB9 VEHICLE SPEED SENSOR (VSS) SIGNAL LOW	401	PPL	VEHICLE SPEED SENSOR (VSS) TERMINAL "A"	0	0	24 (10) 35 (10)	NO VSS SIGNAL INOPERATIVE SPEEDOMETER INOPERATIVE CRUISE CONTROL NO TCC
BB10 VEHICLE SPEED SENSOR (VSS) SIGNAL HIGH	400	YEL	VEHICLE SPEED SENSOR (VSS) TERMINAL "B"	(1) 0	(1) 0	24 (10) 35 (10)	NO VSS SIGNAL INOPERATIVE SPEEDOMETER INOPERATIVE CRUISE CONTROL NO TCC
BB11 ECM TO INSTRUMENT CLUSTER VSS	389	DK GRN	15 WAY I/P CLUSTER TERMINAL "A"	(1) 10.0	(1) 11.4		INOPERATIVE SPEEDOMETER/ODO
BB12							

(1) Varies from .60 to battery voltage, depending on position of drive wheels.
(2) Battery voltage for first two seconds
(3) Varies
(4) Battery voltage when fuel pump is running
(5) Varies with temperature
(6) Reads battery voltage in gear
(7) Battery voltage when engine is cranking
(8) Open circuit
(9) Grounded circuit
(10) Open/Grounded circuit
(11) Less than 1 volt
* Less than .5 volt (500 mV)

2.3L (VIN D, A & 3) ENGINE — ECM SYMPTOM CHART — 1992 ACHIEVA, GRAND AM AND SKYLARK

ECM CONNECTOR "C"

PIN FUNCTION	CKT #	WIRE COLOR	COMPONENT CONNECTOR CAVITY	NORMAL VOLTAGES KEY "ON"	NORMAL VOLTAGES ENG RUN	CODES AFFECT	POSSIBLE SYMPTOMS FROM FAULTY CIRCUIT
BC1							
BC2							
BC3 NON A/C PROGRAM	450	BLK/WHT	GROUND	0	0	26 (B)	
BC4							
BC5 1X SIGNAL	647	LT BLU/ BLK	IGNITION MODULE TERMINAL "G"		4.8 V	41 (10)	POOR PERFORMANCE
BC6							
BC7 IGNITION BYPASS	424	TAN/BLK	IGNITION MODULE TERMINAL "D"	0 *	5V	42 (10)	LACK OF POWER HUNTING IDLE, STALLING
BC8 ELECTRONIC SPARK TIMING (EST)	423	WHT	IGNITION MODULE TERMINAL "E"	0 *	0.2V	42 (10)	LACK OF POWER STALLS, SURGES
BC9 A/C REQUEST SIGNAL	66	LT GRN	A/C SELECT SWITCH 15 WAY I/P CONNECTOR "R"	0 * B+	0 * B+		INOPERATIVE A/C INCORRECT IDLE
BC10 INJECTOR DRIVER 2,3	468	DK GRN	INJ HARNESS CONNECTOR TERMINAL "C"	B+	B+	65 (10)	ROUGH IDLE, LACK OF PERFORMANCE, HARD TO START
BC11							
BC12 INJ DRIVER 1,4	467	DK BLU	INJ HARNESS CONNECTOR TERMINAL "B"	B+	B+	65 (10)	ROUGH IDLE, LACK OF PERFORMANCE, HARD TO START.
BC13 PEAK & HOLD INJ JUMPER 2,3	888	DK BLU	ECM TERMINAL "BC15"	0	0	65 (10)	ROUGH IDLE, LACK OF PERFORMANCE, HARD TO START.
BC14							
BC15 PEAK & HOLD INJ JUMPER 2,3	888	DK BLU	ECM TERMINAL "BC13"	0	0	65 (10)	ROUGH IDLE, LACK OF PERFORMANCE, HARD TO START
BC16 BATTERY FEED	2	RED	FUSIBLE LINK AND B+	B+	B+		CRANKS BUT WILL NOT START

(1) Varies from .60 to battery voltage, depending on position of drive wheels.
(2) Battery voltage for first two seconds
(3) Varies
(4) Battery voltage when fuel pump is running
(5) Varies with temperature
(6) Reads battery voltage in gear
(7) Battery voltage when engine is cranking
(8) Open circuit
(9) Grounded circuit
(10) Open/Grounded circuit
(11) Less than 1 volt
* Less than .5 volt (500 mV)

2.3L (VIN D, A & 3) ENGINE — ECM SYMPTOM CHART — 1992 ACHIEVA, GRAND AM AND SKYLARK

ECM CONNECTOR "D"

PIN FUNCTION		CKT #	WIRE COLOR	COMPONENT CONNECTOR CAVITY	NORMAL VOLTAGES KEY "ON"	NORMAL VOLTAGES ENG RUN	CODES AFFECT	POSSIBLE SYMPTOMS FROM FAULTY CIRCUIT
BD1	ECM GROUND	551	TAN/WHT	ENGINE BLOCK	0	0		
BD2								
BD3								
BD4	2ND GEAR SWITCH A/T	232	WHT	TRANSAXLE ASM (AUTO) TERMINAL "C"	ON B+	ON B+(6) OFF 0		IMPROPER TCC OPERATION
BD5	PEAK AND HOLD INJ JUMPER 1 & 4	887	TAN	ECM TERMINAL "BD6"	0	0	65 (10)	ROUGH IDLE, LACK OF PERFORMANCE, HARD TO START.
BD6	PEAK AND HOLD INJ JUMPER 1 & 4	887	TAN	ECM TERMINAL "BD5"	0	0	65 (10)	ROUGH IDLE, LACK OF PERFORMANCE, HARD TO START.
BD7								
BD8	REFERENCE 2X SIGNAL	430	PPL/WHT	IGNITION MODULE TERMINAL "H"	0	2.9V	16 (10)	
BD9	REFERENCE IGNITION LOW	453	BLK/RED	IGNITION MODULE TERMINAL "J"	0*	0*		
BD10								
BD11								
BD12								
BD13								
BD14								
BD15	3RD GEAR SWITCH A/T	438	DK GRN/WHT	TRANSAXLE ASM (AUTO) TERMINAL "B"	ON 0	ON 0 OFF B+(6)		IMPROPER TCC OPERATION
BD16	PARK/NEUTRAL (P/N) SWITCH SIGNAL	434	ORN/BLK	PARK/NEUTRAL (P/N) SWITCH "A"	ON 0 OFF B+	ON 0 OFF B+		INCORRECT IDLE

(1) Varies from .60 to battery voltage, depending on position of drive wheels.
(2) Battery voltage for first two seconds
(3) Varies
(4) Battery voltage when fuel pump is running
(5) Varies with temperature
(6) Reads battery voltage in gear
(7) Battery voltage when engine is cranking
(8) Open circuit
(9) Grounded circuit
(10) Open/Grounded circuit
(11) Less than 1 volt
* Less than .5 volt (500 mV)

ECM CONNECTOR "E"

PIN FUNCTION		CKT #	WIRE COLOR	COMPONENT CONNECTOR CAVITY	NORMAL VOLTAGES KEY "ON"	NORMAL VOLTAGES ENG RUN	CODES AFFECT	POSSIBLE SYMPTOMS FROM FAULTY CIRCUIT
GE1								
GE2								
GE3	IAC COIL "A" HIGH	441	LT BLU/WHT	IAC VALVE "C"	0 OR B+	0 OR B+	35 (10)	INCORRECT IDLE SURGE
GE4	IAC COIL "A" LOW	442	LT BLU/BLK	IAC VALVE "D"	0 OR B+	0 OR B+	35 (10)	INCORRECT IDLE SURGE
GE5	IAC COIL "B" HIGH	443	LT GRN/WHT	IAC VALVE "A"	0 OR B+	0 OR B+	35 (10)	INCORRECT IDLE SURGE
GE6	IAC COIL "B" LOW	444	LT GRN/BLK	IAC VALVE "B"	0 OR B+	0 OR B+	35 (10)	INCORRECT IDLE SURGE
GE7	SERVICE ENGINE SOON LIGHT	419	BRN/WHT	I/P PRINTED CIRCUIT CONNECTOR "B"	0*	B+		NO SES LIGHT (8) SES LIGHT ON CONSTANTLY (9)
GE8	PRIMARY COOLING FAN (FAN 1) CONTROL RELAY	536	LT GRN/BLK	COOLANT FAN RELAY TERMINAL "5"	B+	ON 0 OFF B+		INOPERATIVE FAN 1 (8) FAN 1 RUNS ALL THE TIME (9)
GE9								
GE10								
GE11								
GE12	DIAGNOSTIC ENABLE TERMINAL	451	WHT/BLK	ALDL CONNECTOR "B"	5V	5V		SES LIGHT FLASHES ALL THE TIME (9) NO FIELD SERVICE MODE (8)
GE13								
GE14	OXYGEN (O_2) SENSOR SIGNAL	412	PPL	OXYGEN (O_2) SENSOR	.35 - .55	(3) .1 - .9	13 (8)	EXHAUST ODOR POOR PERFORMANCE
GE15	OXYGEN (O_2) SENSOR GROUND	413	TAN	ENGINE GROUND	0	0	13 (8)	EXHAUST ODOR POOR PERFORMANCE
GE16	COOLANT TEMP SENSOR SIGNAL	410	YEL	CTS "B"	1.8V (5)	1.8V (5)	14 (9) 15 (8)	LACK OF PERFORMANCE

(1) Varies from .60 to battery voltage, depending on position of drive wheels.
(2) Battery voltage for first two seconds
(3) Varies
(4) Battery voltage when fuel pump is running
(5) Varies with temperature
(6) Reads battery voltage in gear
(7) Battery voltage when engine is cranking
(8) Open circuit
(9) Grounded circuit
(10) Open/Grounded circuit
(11) Less than 1 volt
* Less than .5 volt (500 mV)

2.3L (VIN D, A & 3) ENGINE — ECM SYMPTOM CHART — 1992 ACHIEVA, GRAND AM AND SKYLARK

				ECM CONNECTOR "F"				
PIN FUNCTION		CKT #	WIRE COLOR	COMPONENT CONNECTOR CAVITY	NORMAL VOLTAGES		CODES AFFECT	POSSIBLE SYMPTOMS FROM FAULTY CIRCUIT
					KEY "ON"	ENG RUN		
GF1	CANISTER PURGE	428	DK GRN/YEL	CANISTER PURGE TERMINAL "B"	B+	.3	26 (10)	
GF2	A/C CLUTCH ENABLE	459	DK GRN/WHT	A/C CONTROL RELAY TERMINAL "2"	B+	OFF B+ ON 0	26 (10)	A/C CLUTCH INOPERATIVE (8) A/C CLUTCH STAYS ON (9)
GF3								
GF4	UPSHIFT LIGHT (MAN)	456	TAN/BLK	I/P CONNECTOR TERMINAL "M"	B+	B + OR 0	26 (10)	SHIFT LIGHT INOPERATIVE (8) SHIFT LIGHT "ON" (9)
GF4	TORQUE CONVERTER (AUTO)	422	TAN/BLK	TRANSAXLE CONNECTOR TERMINAL "D"	B+	OFF B+ ON 0	26 (10)	TORQUE CONVERTER INOPERATIVE (8) TORQUE CONVERTER "ON" (9)
GF5								
GF6								
GF7	COOLANT TEMP LIGHT	35	DK GRN	15 WAY I/P CONNECTOR TERMINAL "D"	0*	B+	26 (10)	TEMP LIGHT OFF (8) TEMP LIGHT STAYS "ON" (9)
GF8								
GF9	ESC KNOCK SENSOR SIGNAL	496	DK BLU	KNOCK SENSOR	2.3V	2.3V	43 (10)	
GF10								
GF11								
GF12								
GF13	TPS SIGNAL	417	DK BLU	TPS "C"	.54V (3)	.54V (3)	22 (10)	LACK OF PERFORMANCE
GF14	A/C PRESSURE SIGNAL	380	GRY/RED	A/C PRESSURE SENSOR TERMINAL "C"	1.0V (3)	1.0V (3)	66 (10)	A/C CLUTCH INOPERATIVE
GF15	MAP SIGNAL	432	LT GRN	MAP SENSOR "B"	(3) 4.7V	(3) 1.4V	34 (10) 33 (10)	LACK OF PERFORMANCE ROUGH IDLE, SURGE
GF16	IAT SIGNAL	472	TAN	IAT SENSOR "A"	(3) 3.6V	(3) 1.5V	23 (8) 25 (9)	

(1) Varies from .60 to battery voltage, depending on position of drive wheels
(2) Battery voltage for first two seconds
(3) Varies
(4) Battery voltage when fuel pump is running
(5) Varies with temperature
(6) Reads battery voltage in gear
(7) Battery voltage when engine is cranking
(8) Open circuit
(9) Grounded circuit
(10) Open/grounded circuit
(11) Less than 1 volt
* Less than .5 volt (500 mV)

2.3L (VIN D, A & 3) ENGINE — COMPONENT DIAGNOSTIC CHART — 1992 ACHIEVA, GRAND AM AND SKYLARK

HARNESS CONNECTOR
A B C D E F G
FRONT VIEW
P/N SWITCH CONNECTOR
NEUTRAL START AND BACK-UP SWITCH TRANSAXLE MOUNTED
PARK/NEUTRAL SWITCH
R N D L
P
CIRCUIT TO GROUND IN PARK AND NEUTRAL
15 WAY I/P CONNECTOR
A N
C
ORN/BLK 434
BLK/WHT 450
BD16
12V
ECM
P/N SIGNAL

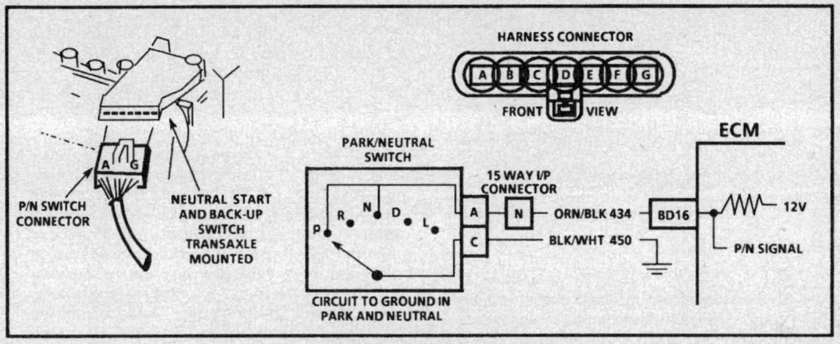

CHART C-1A
PARK/NEUTRAL (P/N) SWITCH DIAGNOSIS
(AUTO TRANSAXLE ONLY)
2.3L (VIN D & 3) "N" CARLINE (PORT)

Circuit Description:

The Park/Neutral (P/N) switch contacts are a part of the neutral start switch and are closed to ground in park or neutral, and open in drive ranges and reverse.

The Electronic Control Module (ECM) supplies ignition voltage through a current limiting resistor to CKT 434 and senses a closed switch when the voltage on CKT 434 drops to less than 1 volt.

The ECM uses the P/N signal as one of the inputs to control:

Idle Air Control (IAC).

Code 24 VSS diagnostics.

If CKT 434 indicates drive (open) a dip in the idle may exist when the gear selector is moved into drive range.

Test Description: Number(s) below refer to circled number(s) on the diagnostic chart.

1. Checks for a closed switch to ground in park position. Different makes of "Scan" tools will display P/N differently. Refer to "Tool Operator's" manual for type of display used for a specific tool.

2. Checks for an open switch in drive range.

3. Be sure "Scan" indicates drive, even while wiggling shifter, to test for an intermittent or misadjusted switch in drive range.

2.3L (VIN D, A & 3) ENGINE — COMPONENT DIAGNOSTIC CHART — 1992 ACHIEVA, GRAND AM AND SKYLARK

CHART C-1A
PARK/NEUTRAL (P/N) SWITCH DIAGNOSIS
(AUTOMATIC TRANSAXLE ONLY)
2.3L (VIN D & 3) "N" CARLINE (PORT)

① • WITH TRANSMISSION IN PARK, TECH 1 "SCAN" TOOL SHOULD INDICATE PARK OR NEUTRAL. DOES IT?

YES → ③ • SHIFT TRANSMISSION INTO DRIVE. • TECH 1 "SCAN" TOOL SHOULD DISPLAY A CHANGE TO INDICATE DRIVE. DOES IT?

NO → ② • DISCONNECT PARK/NEUTRAL SWITCH CONNECTOR. • JUMPER HARNESS CONNECTOR TERMINALS "A" AND "C". • TECH 1 "SCAN" TOOL SHOULD INDICATE PARK OR NEUTRAL. DOES IT?

(from ③) NO → • DISCONNECT P/N SWITCH. • THIS SHOULD CAUSE TECH 1 "SCAN" TOOL TO DISPLAY DRIVE RANGE. DOES IT?

(from ③) YES → NO TROUBLE FOUND.

(from ②) NO → • JUMPER HARNESS CONNECTOR (CKT 434) TO ENGINE GROUND. • TECH 1 "SCAN" TOOL SHOULD INDICATE PARK OR NEUTRAL. DOES IT?

(from ②) YES → FAULTY P/N SWITCH CONNECTION OR P/N SWITCH MISADJUSTED OR FAULTY P/N SWITCH.

YES → FAULTY P/N SWITCH CONNECTION OR P/N SWITCH MISADJUSTED OR FAULTY P/N SWITCH.

NO → CKT 434 SHORTED TO GROUND OR FAULTY ECM.

YES → OPEN GROUND CIRCUIT.

NO → CKT 434 OPEN OR FAULTY ECM CONNECTION OR ECM.

"AFTER REPAIRS," CONFIRM "CLOSED LOOP" OPERATION AND NO "SERVICE ENGINE SOON" LIGHT.

2.3L (VIN D, A & 3) ENGINE — COMPONENT DIAGNOSTIC CHART — 1992 ACHIEVA, GRAND AM AND SKYLARK

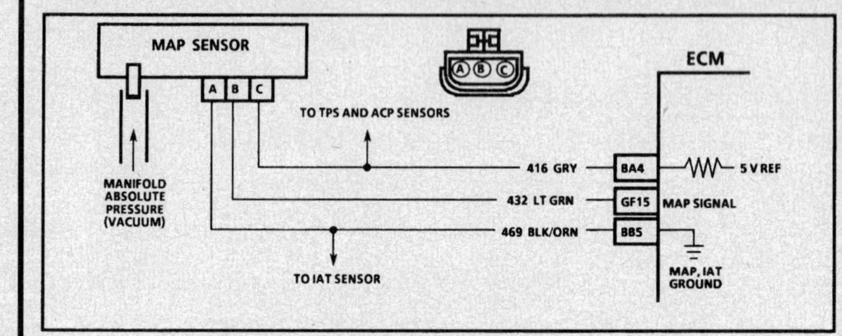

CHART C-1D
MANIFOLD ABSOLUTE PRESSURE (MAP) OUTPUT CHECK
2.3L (VIN D, A & 3) "N" CARLINE (PORT)

Circuit Description:

The Manifold Absolute Pressure (MAP) sensor measures the changes in the intake manifold pressure which results from engine load (intake manifold vacuum) and rpm changes; and converts these into a voltage output. The ECM sends a 5 volt reference voltage to the MAP sensor. As the manifold pressure changed, the output voltage of the sensor also changes. By monitoring the sensor output voltage, the ECM knows the manifold pressure. A lower pressure (low voltage) output voltage will be about 1 - 2 volts at idle. While higher pressure (high voltage) output voltage will be about 4 - 4.8 at Wide Open Throttle (WOT). The MAP sensor is also used, under certain conditions, to measure barometric pressure, allowing the ECM to make adjustments for different altitudes. The ECM uses the map sensor to control fuel delivery and ignition timing.

Test Description: Number(s) below refer to circled number(s) on the diagnostic chart.

⚠ Important

• Be sure to use the same Diagnostic Test Equipment for all measurements.

1. When comparing Tech 1 readings to a known good vehicle, it is important to compare vehicles that use a MAP sensor having the same color insert or having the same "Hot Stamped" number. See figures on facing page.

2. Applying 34 kPa (10" Hg) vacuum to the MAP sensor should cause the voltage to change. Subtract second reading from the first. Voltage value should be greater than 1.5 volts. Upon applying vacuum to the sensor, the change in voltage should be instantaneous. A slow voltage change indicates a faulty sensor.

3. Check vacuum hose to sensor for leaking or restriction. Be sure no other vacuum devices are connected to the MAP hose.

NOTICE: Make sure electrical connector remains securely fastened.

4. Disconnect sensor from bracket and twist sensor by hand (only) to check for intermittent connection. Output changes greater than .1 volt indicate a bad connector or connection. If OK, replace sensor.

2.3L (VIN D, A & 3) ENGINE — COMPONENT DIAGNOSTIC CHART — 1992 ACHIEVA, GRAND AM AND SKYLARK

CHART C–1D

MANIFOLD ABSOLUTE PRESSURE (MAP) OUTPUT CHECK
2.3L (VIN D, A & 3) "N" CARLINE (PORT)

NOTE: THIS CHART ONLY APPLIES TO MAP SENSORS HAVING GREEN OR BLACK COLOR KEY INSERT (SEE BELOW).

1.
- IGNITION "ON," ENGINE "OFF."
- TECH 1 SHOULD INDICATE A MAP SENSOR VOLTAGE.
- COMPARE THIS READING WITH THE READING OF A KNOWN GOOD VEHICLE. SEE FACING PAGE TEST DESCRIPTION, STEP 1.
- VOLTAGE READING SHOULD BE WITHIN, ± .4 VOLT.
  IS IT?

YES → 2. / NO → REPLACE SENSOR.

2.
- DISCONNECT AND PLUG VACUUM SOURCE TO MAP SENSOR.
- CONNECT A HAND VACUUM PUMP TO MAP SENSOR.
- START ENGINE.
- NOTE MAP SENSOR VOLTAGE.
- APPLY 34 kPa (10" Hg) OF VACUUM AND NOTE VOLTAGE CHANGE. SUBTRACT SECOND READING FROM THE FIRST. VOLTAGE VALUE SHOULD BE GREATER THAN 1.5 VOLTS.
  IS IT?

YES → 3. / NO → 4. CHECK SENSOR CONNECTION. IF OK, REPLACE SENSOR.

3. NO TROUBLE FOUND. CHECK SENSOR VACUUM SOURCE FOR LEAKAGE OR RESTRICTION. BE SURE THIS SOURCE SUPPLIES VACUUM TO MAP SENSOR ONLY.

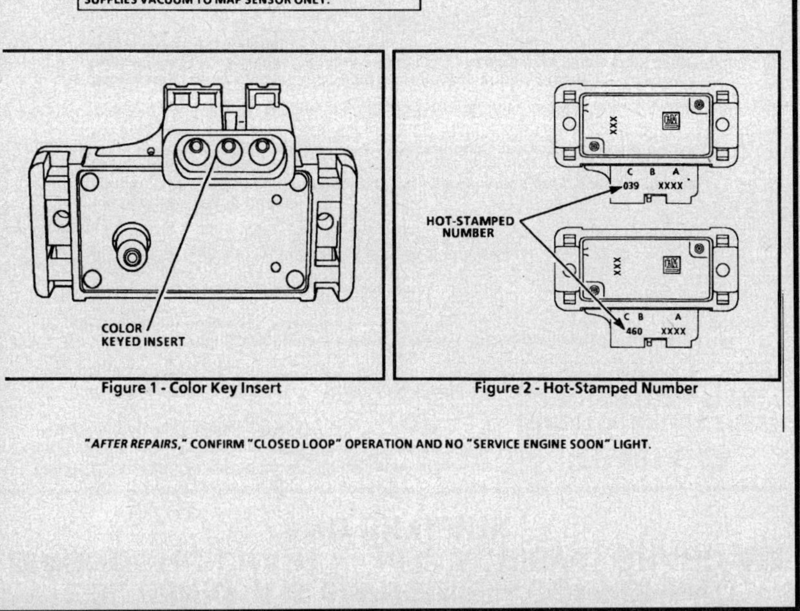

"AFTER REPAIRS," CONFIRM "CLOSED LOOP" OPERATION AND NO "SERVICE ENGINE SOON" LIGHT.

2.3L (VIN D, A & 3) ENGINE — COMPONENT DIAGNOSTIC CHART — 1992 ACHIEVA, GRAND AM AND SKYLARK

CHART C-2A

INJECTOR BALANCE TEST
2.3L (VIN D, A & 3) "N" CARLINE (PORT)

The injector balance tester is a tool used to turn the injector on for a precise amount of time, thus spraying a measured amount of fuel into the manifold. This causes a drop in fuel rail pressure that we can record and compare between each injector. All injectors should have the same amount of pressure drop (± 10 kpa). Any injector with a pressure drop that is 10 kpa (or more) greater or less than the average drop of the other injectors should be considered faulty and replaced.

STEP 1

Engine "cool down" period (10 minutes) is necessary to avoid irregular readings due to "Hot Soak" fuel boiling. With ignition "OFF" connect fuel gauge. Wrap a shop towel around fitting while connecting gage to avoid fuel spillage.

Disconnect harness connectors at all injectors, and connect injector tester J-34730-3, or equivalent, to one injector. Ignition must be "OFF" at least 10 seconds to complete ECM shutdown cycle. Fuel pump should run about 2 seconds after ignition is turned "ON". At this point, insert clear tubing attached to vent valve into a suitable container and bleed air from gauge and hose to insure accurate gauge operation. Repeat this step until all air is bled from gauge.

STEP 2

Turn ignition "OFF" for 10 seconds and then "ON" again to get fuel pressure to its maximum. Record this initial pressure reading. Energize tester one time and note pressure drop at its lowest point (Disregard any slight pressure increase after drop hits low point.). By subtracting this second pressure reading from the initial pressure, we have the actual amount of injector pressure drop.

STEP 3

Repeat Step 2 on each injector and compare the amount of drop. Usually, good injectors will have virtually the same drop. Retest any injector that has a pressure difference of 10kPa, either more or less than the average of the other injectors on the engine. Replace any injector that also fails the retest. If the pressure drop of all injectors is within 10 kPa of this average, the injectors appear to be flowing properly.

NOTE: The entire test should not be repeated more than once without running the engine to prevent flooding. (This includes any retest on faulty injectors).

2.3L (VIN D, A & 3) ENGINE — COMPONENT DIAGNOSTIC CHART — 1992 ACHIEVA, GRAND AM AND SKYLARK

2.3L (VIN D, A & 3) ENGINE — COMPONENT DIAGNOSTIC CHART — 1992 ACHIEVA, GRAND AM AND SKYLARK

CHART C-2A
INJECTOR BALANCE TEST
2.3L (VIN D, A & 3) "N" CARLINE (PORT)

NOTICE: The entire test should <u>NOT</u> be repeated more than once without running the engine to prevent flooding. (This includes any retest on faulty injectors.)

The fuel pressure test in Chart A-7, should be completed prior to this test.

Step 1. If engine is at operating temperature, allow a 10 minute "cool down" period then connect fuel pressure gauge and injector tester.
1. Ignition "OFF."
2. Connect fuel pressure gauge and injector tester.
3. Ignition "ON."
4. Bleed off air in gauge. Repeat until all air is bled from gauge.

Step 2. Run test:
1. Ignition "OFF" for 10 seconds.
2. Ignition "ON". Record gauge pressure. (Pressure must hold steady, if not see the Fuel System diagnosis, Chart A-7
3. Turn injector on, by depressing button on injector tester, and note pressure at the instant the gauge needle stops.

Step 3.
1. Repeat step 2 on all injectors and record pressure drop on each. Retest injectors that appear faulty (Any injectors that have a 10 kPa (1.5 psi) difference, either more or less, in pressure from the average).

1ST READING (INITIAL PRESSURE)
FUEL PRESSURE GAGE
2ND READING (PRESSURE AFTER DROP)
INJECTOR CONNECTOR
TESTER J 34730-3
ELECTRONIC FUEL INJECTOR TESTER
OUTPUT POWER
ACTIVATE INJECTOR
1 PULSE 500 MS
50 PULSES 50 MS
100 PULSES 5 MS
POS NEG
BATTERY

EXAMPLE

CYLINDER	1	2	3
1ST READING	293 kPa (43 psi)	293 kPa (43 psi)	293 kPa (43 psi)
2ND READING	131 kPa (19 psi)	115 kPa (17 psi)	145 kPa (21 psi)
AMOUNT OF DROP	162 kPa (24 psi)	178 kPa (26 psi)	148 kPa (21 psi)
	OK	FAULTY, RICH (TOO MUCH FUEL DROP)	FAULTY, LEAN (TOO LITTLE FUEL DROP)

AIR FLOW
IAC CONNECTOR
A
B
C
D
THROTTLE BODY
ECM
443 LT GRN/WHT — GE5 — IAC COIL "B" HI
444 LT GRN/BLK — GE6 — IAC COIL "B" LO
441 LT BLU/WHT — GE3 — IAC COIL "A" HI
442 LT BLU/BLK — GE4 — IAC COIL "A" LO

CHART C-2C
IDLE AIR CONTROL (IAC) VALVE CHECK
2.3L (VIN D, A & 3) "N" CARLINE (PORT)

Circuit Description:

The ECM controls idle rpm with the IAC valve. To increase idle rpm, the ECM moves the IAC valve out, allowing more air to bypass the throttle plate. To decrease rpm, it moves the IAC valve in, reducing air flow bypassing the throttle plate. A Tech 1 will read the ECM commands to the IAC valve in counts. The higher the counts, the more air allowed (higher idle). The lower the counts, the less air allowed (lower idle).

Test Description: Number(s) below refer to circled number(s) on the diagnostic chart.
1. The Tech 1 is used to extend and retract the IAC valve. Valve movement is verified by an engine speed change. If no change in engine speed occurs, the valve can be retested when removed from the throttle body.
2. This step checks the quality of the IAC movement in Step 1. Between 900 rpm and about 1500 rpm, the engine speed should change smoothly with each flash of the tester light in both extend and retract. If the IAC valve is retracted beyond the control range (about 1500 rpm), it may take many flashes in the extend position before engine speed will begin to drop. This is normal on certain engines, fully extending IAC may cause engine stall. This may be normal. Step 1 verified proper IAC valve operation while this step checks IAC circuits. Each lamp on the node light should flash red and green while the IAC valve is cycled. While the sequence of color is not important if either light is "OFF" or does not flash red and green, check the circuits for faults, beginning with poor terminal contacts.

Diagnostic Aids:

A slow, unstable, or fast idle may be caused by a non-IAC system problem that cannot be overcome by the IAC valve. Out of control range IAC "Scan" tool counts will be above 60 if idle is too low, and zero counts if idle is too high. The following checks should be made to repair a non-IAC system problem.

- Vacuum Leak (High Idle) - If idle is too high, stop the engine. Fully extend (low) IAC with tester. Start engine. If Idle speed is above 800 rpm, locate and correct vacuum leak including PCV system. Also check for binding of throttle blade or linkage.
- System too lean (High Air/Fuel Ratio) - Idle speed may be too high or too low. Engine speed may vary up and down and disconnecting IAC does not help. Code 44 may be set. "Scan" O_2 voltage will be less than 300 mV (.3 volt.) Check for low regulated fuel pressure, water in the fuel or a restricted injector.
- System too rich (Low Air/Fuel Ratio) - The idle speed will be too low. "Scan" tool IAC counts will usually be above 80. System is obviously rich and may exhibit black smoke exhaust. "Scan" tool O_2 voltage will be fixed above 800 mV (.8 volt). Check for high fuel pressure, leaking or sticking injector. Silicone contaminated O_2 sensor will "Scan" an O_2 voltage slow to respond.
- Throttle Body - Remove IAC and inspect bore for foreign material.

- If intermittent poor driveability or idle symptoms are resolved by disconnecting the IAC, carefully recheck connections, valve terminal resistance, or replace IAC.

2.3L (VIN D, A & 3) ENGINE — COMPONENT DIAGNOSTIC CHART — 1992 ACHIEVA, GRAND AM AND SKYLARK

CHART C-2C

IDLE AIR CONTROL (IAC) VALVE CHECK
2.3L (VIN D, A & 3) "N" CARLINE (PORT)

1
• INSTALL TECH 1 "SCAN" TOOL.
• ENGINE AT NORMAL OPERATING TEMPERATURE IN PARK/NEUTRAL WITH PARKING BRAKE SET.
• A/C "OFF."
• SELECT RPM CONTROL. (MISC. TESTS)
• CYCLE IAC THROUGH ITS RANGE FROM 900 RPM UP TO 1500 RPM.
• RPM SHOULD CHANGE SMOOTHLY.
DOES IT?
NO
YES
2
• INSTALL IAC NODE LIGHT * IN IAC HARNESS.
• ENGINE RUNNING. CYCLE IAC WITH TECH 1 "SCAN TOOL.
• EACH NODE LIGHT SHOULD CYCLE RED AND GREEN BUT NEVER "OFF."
DO THEY?
• USING THE IAC DRIVER * OR OTHER CONVENIENT CONNECTOR, CHECK RESISTANCE ACROSS IAC COILS.
• SHOULD BE 40 TO 80 OHMS BETWEEN IAC TERMINALS "A" TO "B" AND "C" TO "D".
NO
YES
OK
NOT OK
IF CIRCUIT(S) DID NOT TEST RED AND GREEN, CHECK FOR:
• FAULTY CONNECTOR TERMINAL CONTACTS.
• OPEN CIRCUITS INCLUDING CONNECTORS.
• CIRCUITS SHORTED TO GROUND OR VOLTAGE.
• FAULTY ECM CONNECTIONS OR REPLACE ECM.
REPAIR AS NECESSARY AND RETEST.
• CHECK IAC CONNECTIONS.
• CHECK IAC PASSAGES.
• IF OK, REPLACE IAC.
• CHECK RESISTANCE BETWEEN IAC TERMINALS "B" AND "C" AND "A" AND "D".
• SHOULD BE INFINITE.
REPLACE IAC VALVE AND RETEST.
OK
NOT OK
IDLE AIR CONTROL CIRCUIT OK. REFER TO "DIAGNOSTIC AIDS"
REPLACE IAC VALVE AND RETEST.

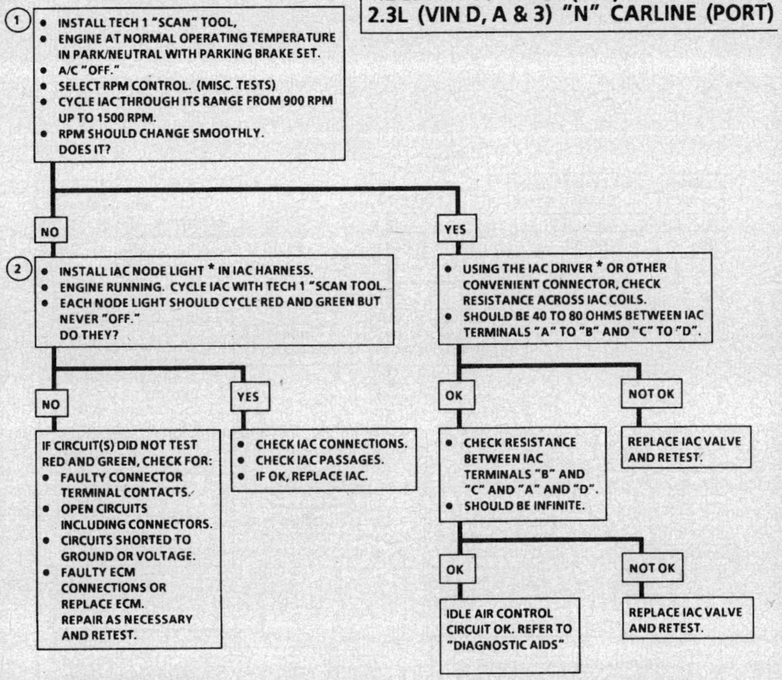

* IAC DRIVER AND NODE LIGHT REQUIRED KIT
222-L FROM: CONCEPT TECHNOLOGY, INC.
J 37027 FROM: KENT-MOORE, INC.

CLEAR CODES, CONFIRM "CLOSED LOOP" OPERATION, NO "SERVICE ENGINE SOON" LIGHT, PERFORM IAC RESET PROCEDURE PER APPLICABLE SERVICE MANUAL AND VERIFY CONTROLLED IDLE SPEED IS CORRECT.

"AFTER REPAIRS," CONFIRM "CLOSED LOOP" OPERATION AND NO "SERVICE ENGINE SOON" LIGHT.

2.3L (VIN D, A & 3) ENGINE — COMPONENT DIAGNOSTIC CHART — 1992 ACHIEVA, GRAND AM AND SKYLARK

ECM
TO CANISTER
PORTED MANIFOLD VACUUM
CANISTER PURGE SOLENOID
A
B
N. C.
CANISTER PURGE SOLENOID CONTROL DRIVER
428 DK GRN/YEL
GF1
ERLS FUSE
439 PNK/BLK
10A

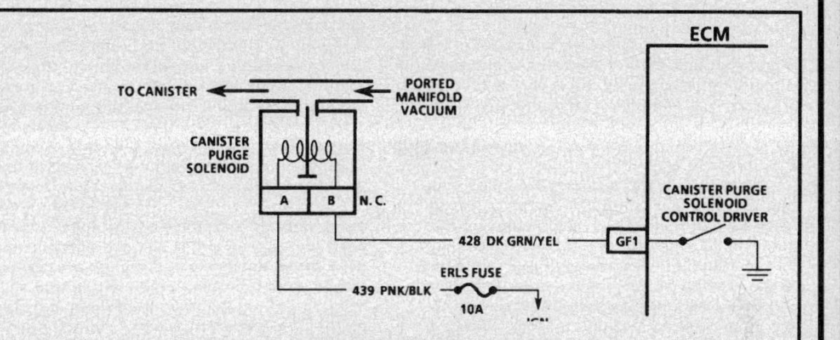

CHART C-3

CANISTER PURGE VALVE CHECK
2.3L (VIN D, A & 3) "N" CARLINE (PORT)

Circuit Description:

Canister purge is controlled by a solenoid that allows manifold and/or ported vacuum to purge the canister when energized. The Electronic Control Module (ECM) supplies a ground to energize the solenoid (purge "ON"). The purge solenoid control by the ECM is pulse width modulated (turned "ON" and "OFF" several times a second). The duty cycle (pulse width) is determined by "Closed Loop" feed back from the O_2 sensor. The duty cycle is calculated by the ECM and the output commanded when the following conditions have been met:

- Engine run time after start more than 65 seconds.
- Coolant temperature above 56°C.

Also, if the diagnostic "test" terminal is grounded with the engine stopped, the purge solenoid is energized (purge "ON").

Test Description: Number(s) below refer to circled number(s) on the diagnostic chart.

1. Checks to see if the solenoid is opened or closed. The solenoid is normally de-energized in this step, so it should be closed.
2. Checks to determine if solenoid was open due to electrical circuit problem or defective solenoid.
3. Completes functional check by grounding "test" terminal. This should normally energize the solenoid opening the valve which should allow the vacuum to drop (purge "ON").

Diagnostic Aids:

Make a visual check of vacuum hose(s). Check throttle body for possible cracked, broken, or plugged vacuum block. Check engine for possible mechanical problem.

2.3L (VIN D, A & 3) ENGINE — COMPONENT DIAGNOSTIC CHART — 1992 ACHIEVA, GRAND AM AND SKYLARK

CHART C-3

CANISTER PURGE VALVE CHECK
2.3L (VIN D, A & 3) "N" CARLINE (PORT)

"AFTER REPAIRS," CONFIRM "CLOSED LOOP" OPERATION AND NO "SERVICE ENGINE SOON" LIGHT.

2.3L (VIN D, A & 3) ENGINE — COMPONENT DIAGNOSTIC CHART — 1992 ACHIEVA, GRAND AM AND SKYLARK

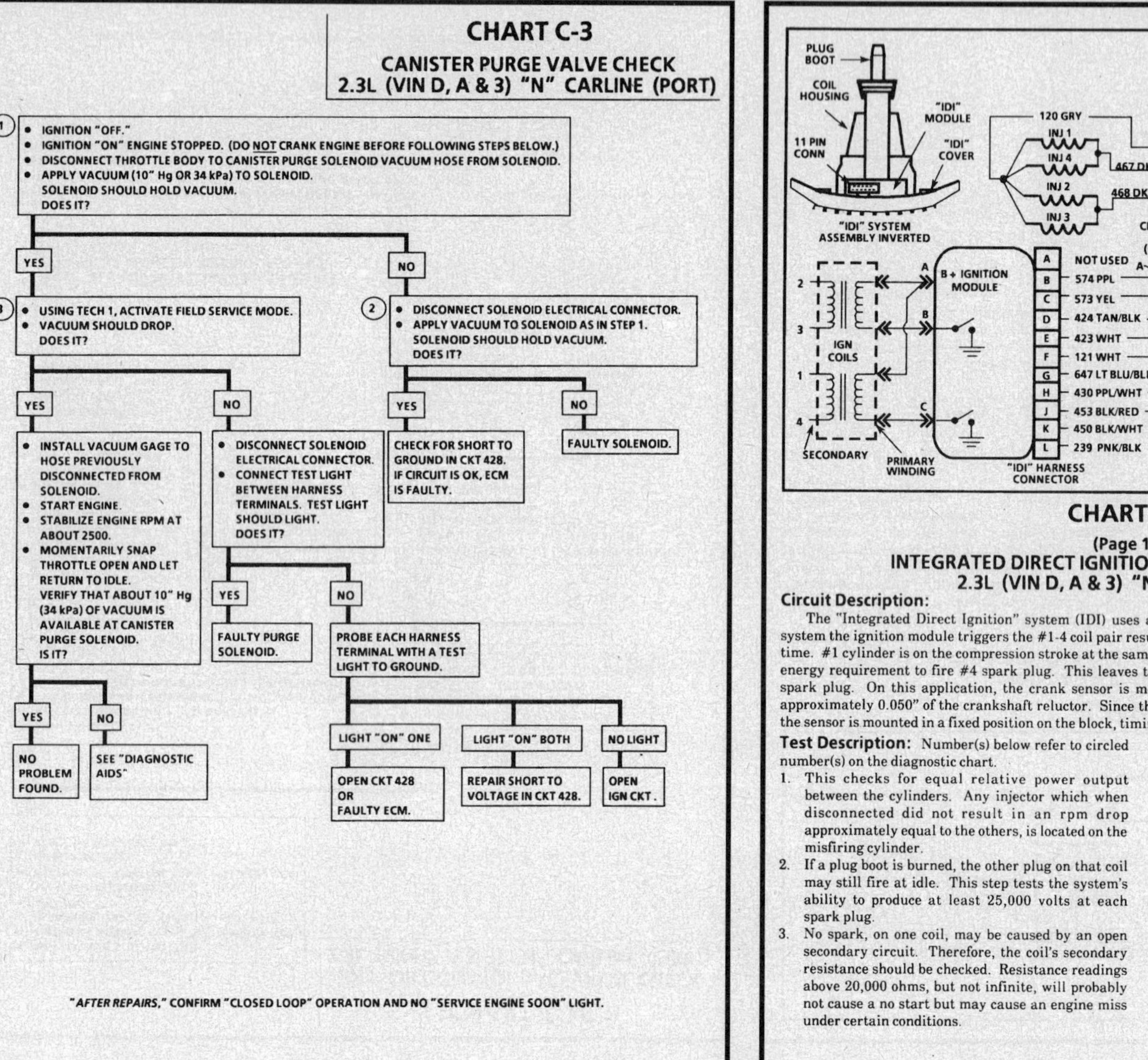

CHART C-4M

(Page 1 of 2)
INTEGRATED DIRECT IGNITION (IDI) MISFIRE DIAGNOSIS
2.3L (VIN D, A & 3) "N" CARLINE (PORT)

Circuit Description:

The "Integrated Direct Ignition" system (IDI) uses a waste spark method of distribution. In this type of system the ignition module triggers the #1-4 coil pair resulting in both #1 and #4 spark plugs firing at the same time. #1 cylinder is on the compression stroke at the same time #4 is on the exhaust stroke, resulting in a lower energy requirement to fire #4 spark plug. This leaves the remainder of the high voltage to be used to fire #1 spark plug. On this application, the crank sensor is mounted to, and protrudes through the block to within approximately 0.050" of the crankshaft reluctor. Since the reluctor is a machined portion of the crankshaft and the sensor is mounted in a fixed position on the block, timing adjustments are not possible or necessary.

Test Description: Number(s) below refer to circled number(s) on the diagnostic chart.

1. This checks for equal relative power output between the cylinders. Any injector which when disconnected did not result in an rpm drop approximately equal to the others, is located on the misfiring cylinder.
2. If a plug boot is burned, the other plug on that coil may still fire at idle. This step tests the system's ability to produce at least 25,000 volts at each spark plug.
3. No spark, on one coil, may be caused by an open secondary circuit. Therefore, the coil's secondary resistance should be checked. Resistance readings above 20,000 ohms, but not infinite, will probably not cause a no start but may cause an engine miss under certain conditions.
4. If the no spark condition is caused by coil connections, a coil or a secondary boot assembly, the test light will blink. If the light does not blink, the fault is module connections or the module.
5. An injector driver circuit shorted to ground would result in the test light "ON" steady, and possibly a flooded condition. A shorted injector (less than ohms) could cause incorrect ECM operation.

Diagnostic Aid:

Verify IDI connector terminal "K", CKT 450 resistance to ground is less than .5 ohm. A shorted or low resistance injector may cause a miss in the other injector in that pair (1 & 4 or 2 & 3).

2.3L (VIN D, A & 3) ENGINE — COMPONENT DIAGNOSTIC CHART — 1992 ACHIEVA, GRAND AM AND SKYLARK

CHART C-4M (Page 1 of 2)
INTEGRATED DIRECT IGNITION (IDI) MISFIRE DIAGNOSIS
2.3L (VIN D, A & 3) "N" CARLINE (PORT)

1.
- ENGINE AT NORMAL OPERATING TEMPERATURE DISCONNECT IAC.
- REMOVE CV OIL/AIR SEPARATOR TO GAIN ACCESS TO INJECTOR CONNECTORS.
- MOMENTARILY DISCONNECT EACH INJECTOR CONNECTOR WHILE OBSERVING ENGINE RPM.
- NOTE ANY INJECTOR(S) NOT RESULTING IN AN RPM DROP.
- (IF ALL INJECTORS RESULT IN AN RPM DROP, GO TO STEP 2).
- INSTALL INJECTOR TEST LIGHT J 34730-2 IN INJ. HARN. CONN. FOR INJ. WHICH DID NOT RESULT IN RPM DROP. LIGHT SHOULD BLINK.
  DOES IT?

YES →

2.
- TEMPORARILY REMOVE IGNITION MODULE / COIL ASSEMBLY AND INSTALL SPARK PLUG JUMPER WIRES (J 36012)
- CHECK FOR SPARK WITH SPARK TESTER J 26792, (ST-125) OR EQUIVALENT ON PLUG JUMPER WIRE FOR CYLINDER(S) NOTED ABOVE WHILE CRANKING WITH REMAINING PLUG WIRES STILL CONNECTED.
- SPARK SHOULD JUMP TESTER GAP.
  DOES IT?

NO →

3.
- REMOVE BOOT ASSYS. FOR AFFECTED COIL (1-4 OR 2-3)
- CONNECT DVM (20K OHMS SCALE) BETWEEN SECONDARY "IDI" COIL TERMINALS AND THEN FROM ONE COIL TERMINAL TO COVER PLATE.
- RESISTANCE SHOULD BE LESS THAN 10K OHMS BETWEEN TERMINALS AND INFINITE (OPEN CIRCUIT) TO COVER.
  IS IT?

YES →

4.
- REMOVE COIL HOUSING AND DISCONNECT COIL HARNESS AT MODULE.
- OBSERVE A TEST LIGHT CONNECTED BETWEEN MODULE TO COIL POWER TERMINAL "A" (PURPLE WIRE) & DRIVER TERMINAL ("B" OR "C") FOR AFFECTED COIL, WHILE CRANKING ENGINE.
- SHOULD BLINK.
  DOES IT?

NO →
POOR CONNECTION OR FAULTY IDI MODULE.

YES →
OPEN OR SHORTED COIL HARNESS, POOR COIL CONNECTION, FAULTY COIL OR BOOT ASSEMBLY.

YES (step 2) →
INSPECT SPARK PLUG AND BOOT FOR DAMAGE. IF OK, SUBSTITUTE A KNOWN GOOD INJ., IF ENGINE STILL MISSES, SEE DIAGNOSTIC AIDS

NO (step 1) →

STEADY LIGHT

LIGHT "OFF" → SEE CHART C-4M PAGE 2 OF 2.

5.
- CHECK INJECTOR DRIVER CIRCUIT WHICH HAD THE STEADY LIGHT, FOR A SHORT TO GROUND.
- IF CIRCUIT IS NOT SHORTED, CHECK RESISTANCE ACROSS EACH INJECTOR IN THE CIRCUIT.
- RESISTANCE SHOULD BE BETWEEN 1.8 AND 2.2 OHMS FOR EACH INJECTOR. IS IT?

NO →
REPLACE ANY INJECTOR THAT MEASURES UNDER 1.8 OHMS AND RECHECK FOR MISFIRE BEGINNING WITH STEP 1 AGAIN.

YES →
FAULTY ECM.

NO →
CHECK FOR CORROSION AT COIL SECONDARY TERMINALS. IF TERMINALS ARE OK, THE IGNITION COIL IS FAULTY.

"AFTER REPAIRS," CONFIRM "CLOSED LOOP" OPERATION AND NO "SERVICE ENGINE SOON" LIGHT.

2.3L (VIN D, A & 3) ENGINE — COMPONENT DIAGNOSTIC CHART — 1992 ACHIEVA, GRAND AM AND SKYLARK

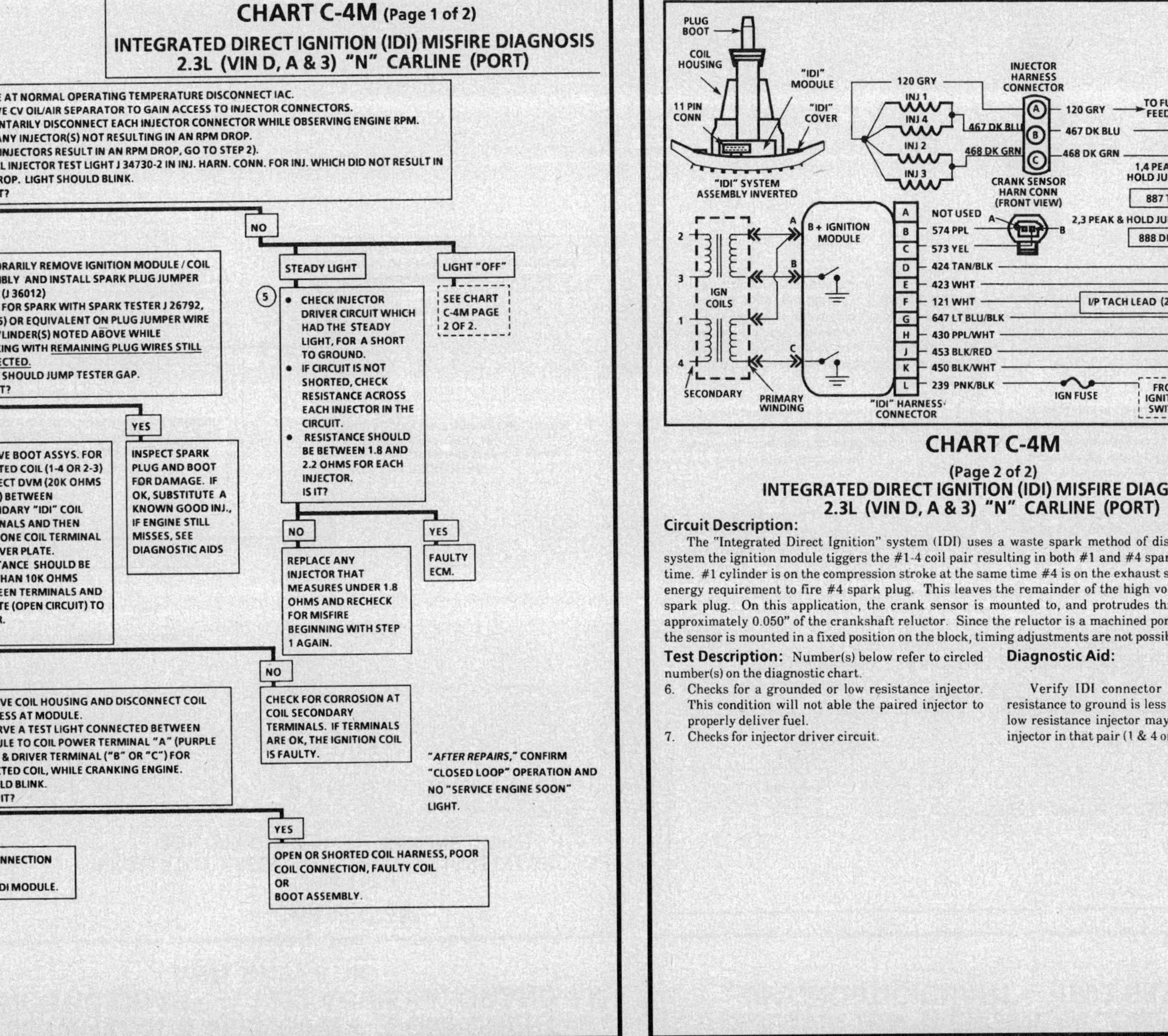

CHART C-4M
(Page 2 of 2)
INTEGRATED DIRECT IGNITION (IDI) MISFIRE DIAGNOSIS
2.3L (VIN D, A & 3) "N" CARLINE (PORT)

Circuit Description:
The "Integrated Direct Ignition" system (IDI) uses a waste spark method of distribution. In this type of system the ignition module tiggers the #1-4 coil pair resulting in both #1 and #4 spark plugs firing at the same time. #1 cylinder is on the compression stroke at the same time #4 is on the exhaust stroke, resulting in a lower energy requirement to fire #4 spark plug. This leaves the remainder of the high voltage to be used to fire #1 spark plug. On this application, the crank sensor is mounted to, and protrudes through the block to within approximately 0.050" of the crankshaft reluctor. Since the reluctor is a machined portion of the crankshaft and the sensor is mounted in a fixed position on the block, timing adjustments are not possible or necessary.

Test Description: Number(s) below refer to circled number(s) on the diagnostic chart.
6. Checks for a grounded or low resistance injector. This condition will not able the paired injector to properly deliver fuel.
7. Checks for injector driver circuit.

Diagnostic Aid:
Verify IDI connector terminal "K", CKT 450 resistance to ground is less than .5 ohm. A shorted or low resistance injector may cause a miss in the other injector in that pair (1 & 4 or 2 & 3).

2.3L (VIN D, A & 3) ENGINE — COMPONENT DIAGNOSTIC CHART — 1992 ACHIEVA, GRAND AM AND SKYLARK

CHART C-4M

(Page 2 of 2)
INTEGRATED DIRECT IGNITION (IDI) MISFIRE DIAGNOSIS
2.3L (VIN D, A & 3) "N" CARLINE (PORT)

```
FROM
CHART C-4M
PAGE 1 OF 2

6  • DISCONNECT PAIRED INJECTOR ON CKT
     (1 & 4 OR 2 & 3).
   • OBSERVE INJECTOR TEST LIGHT.
   • LIGHT SHOULD BE BLINKING.
     IS IT?

LIGHT "OFF"                    BLINKING LIGHT

7  • DISCONNECT INJECTOR TEST LIGHT.      • CHECK INJECTOR RESISTANCE.
   • PROBE INJECTOR HARNESS CONNECTOR     • RESISTANCE SHOULD BE BETWEEN 1.8 AND
     IGNITION FEED (GRY WIRE) TERMINAL AT    2.2 OHMS FOR EACH INJECTOR.
     INJECTOR WITH A TEST LIGHT TO GROUND. • REPLACE INJECTORS THAT ARE NOT WITHIN
   • CRANK ENGINE.                           THIS RANGE.
   • LIGHT SHOULD BE "ON."
     IS IT?

YES                            NO

CHECK CKT 467 OR 468 FOR SHORT TO VOLTAGE,   REPAIR OPEN OR GROUNDED CIRCUIT BETWEEN
OPEN                                         CAVITY "A" OF 3 TERMINAL INJECTOR HARNESS
OR                                           CONNECTOR AND INJECTOR CONNECTOR.
POOR CONNECTIONS AT ECM TERMINAL "BC10"
OR "BC12. IF OK, CHECK FOR OPEN PEAK AND
HOLD JUMPER. IF OK, REPLACE ECM.
```

"AFTER REPAIRS." CONFIRM "CLOSED LOOP" OPERATION AND NO "SERVICE ENGINE SOON" LIGHT.

2.3L (VIN D & 3) ENGINE — COMPONENT DIAGNOSTIC CHART — 1992 SKYLARK

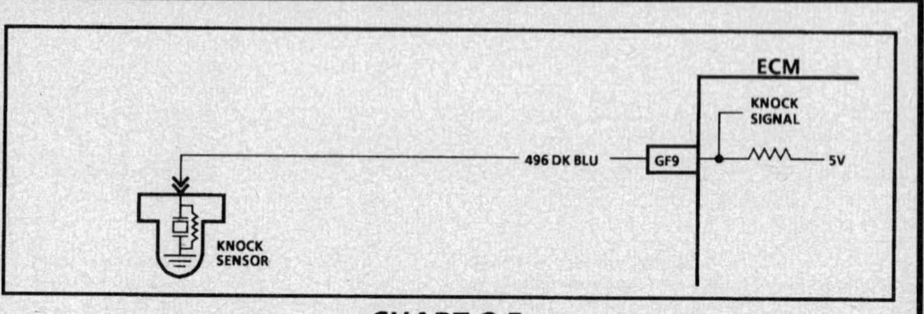

CHART C-5

ELECTRONIC SPARK CONTROL (ESC) SYSTEM CHECK
2.3L (VIN 3) "N" CARLINE (PORT)

Circuit Description:

The knock sensor is used to detect engine detonation and the ECM will retard the electronic spark timing based on the signal being received. The circuitry within the knock sensor causes the ECM's 5 volts to be pulled down so that CKT 496 would measure about 2.5 volts. The knock sensor produces an AC signal which rides on the 2.5 volts DC voltage. The amplitude and frequency are dependent upon the knock level.

The MEM-CAL used with this engine contains the functions which were part of remotely mounted ESC modules used on other GM vehicles. The ESC portion of the MEM-CAL then sends a signal to other parts of the ECM which adjusts the spark timing to retard the spark and reduce the detonation.

Test Description: Number(s) below refer to circled number(s) on the diagnostic chart.

1. With engine idling, there should not be a knock signal present at the ECM because detonation is not likely under a no load condition.
2. Tapping on the engine lift hook bracket should simulate a knock signal to determine if the sensor is capable of detecting detonation. If no knock is detected, try tapping on engine block closer to sensor before replacing sensor.
3. If the engine has an internal problem which is creating a knock, the knock sensor may be responding to the internal failure.
4. This test determines if the knock sensor is faulty or if the ESC portion of the MEM-CAL is faulty. If it is determined that the MEM-CAL is faulty, be sure that it is properly installed and latched into place. If not properly installed, repair and retest.

Diagnostic Aids:

While observing knock signal on the Tech 1 there should be an indication that knock is present when detonation can be heard. Detonation is most likely to occur under high engine load conditions.

2.3L (VIN D & 3) ENGINE — COMPONENT DIAGNOSTIC CHART — 1992 SKYLARK

CHART C-5
ELECTRONIC SPARK CONTROL (ESC) SYSTEM CHECK
2.3L   "N" CARLINE (PORT)

① • IF CODE 43 IS SET, USE THE CODE CHART FIRST.
• ENGINE MUST BE IDLING AT NORMAL OPERATING TEMPERATURE.
• USE TECH 1 TO OBSERVE KNOCK SIGNAL.
IS KNOCK INDICATED?

NO →

② • TAP ON ENGINE, LIFT HOOK BRACKET WHILE OBSERVING KNOCK SIGNAL.
• TECH 1 SHOULD INDICATE KNOCK WHILE TAPPING ON BRACKET.
DOES IT?

YES →

③ IF AN ENGINE KNOCK CAN BE HEARD, REPAIR THE BASIC ENGINE PROBLEM. IF NO AUDIBLE KNOCK IS HEARD, FOLLOW THE STEPS:
• IGNITION "OFF."
• DISCONNECT KNOCK SENSOR.
• CONNECT VOLTMETER TO KNOCK SENSOR AND ENGINE GROUND.
• SET VOLTMETER ON 2 VOLT A.C. SCALE.
• IGNITION "ON," ENGINE "ON."
IS A SIGNAL INDICATED ON VOLTMETER?

NO →

④ • IGNITION "OFF."
• DISCONNECT KNOCK SENSOR.
• CONNECT VOLTMETER TO KNOCK SENSOR AND ENGINE GROUND.
• SET VOLTMETER ON 2 VOLT A.C. SCALE.
• IGNITION "ON," ENGINE "ON."
• TAP ON ENGINE BLOCK NEAR SENSOR.
IS A SIGNAL INDICATED ON VOLTMETER WHILE TAPPING ON ENGINE BLOCK?

YES → SYSTEM IS OPERATING PROPERLY. REFER TO "DIAGNOSTIC AIDS"

NO → CHECK CKT 496 FOR BEING NEAR A SPARK PLUG WIRE
OR
A FAULTY ECM CONNECTION
OR
FAULTY ECM
OR
MEM-CAL.

YES → REPLACE KNOCK SENSOR.

NO → REPLACE MEM-CAL OR ECM.

YES → REPLACE KNOCK SENSOR.

"AFTER REPAIRS," CONFIRM "CLOSED LOOP" OPERATION AND NO "SERVICE ENGINE SOON" LIGHT.

2.3L (VIN D, A & 3) ENGINE — COMPONENT DIAGNOSTIC CHART — 1992 ACHIEVA AND GRAND AM

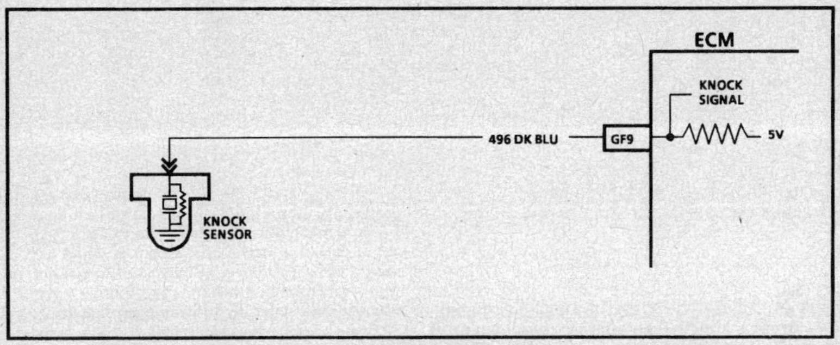

CHART C-5
ELECTRONIC SPARK CONTROL (ESC) SYSTEM CHECK
2.3L (VIN D, A & 3) "N" CARLINE (PORT)

Circuit Description:
The knock sensor is used to detect engine detonation and the ECM will retard the electronic spark timing based on the signal being received. The circuitry within the knock sensor causes the ECM's 5 volts to be pulled down so that CKT 496 would measure about 2.5 volts. The knock sensor produces an AC signal which rides on the 2.5 volts DC voltage. The amplitude and frequency are dependent upon the knock level.

The MEM-CAL used with this engine contains the functions which were part of remotely mounted ESC modules used on other GM vehicles. The ESC portion of the MEM-CAL then sends a signal to other parts of the ECM which adjusts the spark timing to retard the spark and reduce the detonation.

Test Description: Number(s) below refer to circled number(s) on the diagnostic chart.
1. With engine idling, there should not be a knock signal present at the ECM because detonation is not likely under a no load condition.
2. Tapping on the engine lift hook bracket should simulate a knock signal to determine if the sensor is capable of detecting detonation. If no knock is detected, try tapping on engine block closer to sensor before replacing sensor.
3. If the engine has an internal problem which is creating a knock, the knock sensor may be responding to the internal failure.

4. This test determines if the knock sensor is faulty or if the ESC portion of the MEM-CAL is faulty. If it is determined that the MEM-CAL is faulty, be sure that is is properly installed and latched into place. If not properly installed, repair and retest.

Diagnostic Aids:

While observing knock signal on the Tech 1, there should be an indication that knock is present when detonation can be heard. Detonation is most likely to occur under high engine load conditions.

2.3L (VIN D, A & 3) ENGINE — COMPONENT DIAGNOSTIC CHART — 1992 ACHIEVA AND GRAND AM

CHART C-5

ELECTRONIC SPARK CONTROL (ESC) SYSTEM CHECK 2.3L (VIN D, A & 3) "N" CARLINE (PORT)

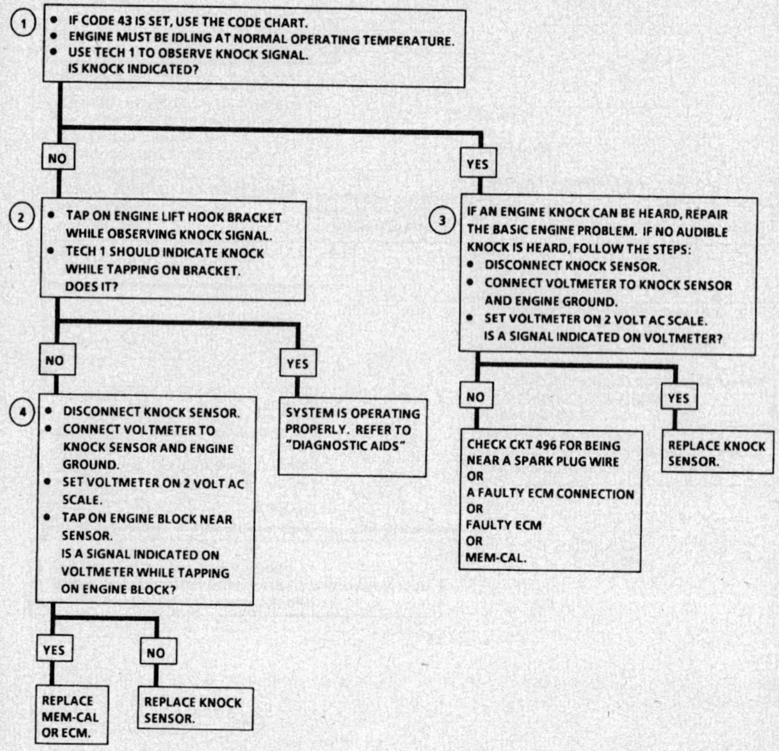
1
• IF CODE 43 IS SET, USE THE CODE CHART.
• ENGINE MUST BE IDLING AT NORMAL OPERATING TEMPERATURE.
• USE TECH 1 TO OBSERVE KNOCK SIGNAL.
IS KNOCK INDICATED?
NO
YES
2
• TAP ON ENGINE LIFT HOOK BRACKET WHILE OBSERVING KNOCK SIGNAL.
• TECH 1 SHOULD INDICATE KNOCK WHILE TAPPING ON BRACKET.
DOES IT?
3
IF AN ENGINE KNOCK CAN BE HEARD, REPAIR THE BASIC ENGINE PROBLEM. IF NO AUDIBLE KNOCK IS HEARD, FOLLOW THE STEPS:
• DISCONNECT KNOCK SENSOR.
• CONNECT VOLTMETER TO KNOCK SENSOR AND ENGINE GROUND.
• SET VOLTMETER ON 2 VOLT AC SCALE.
IS A SIGNAL INDICATED ON VOLTMETER?
NO
YES
SYSTEM IS OPERATING PROPERLY. REFER TO "DIAGNOSTIC AIDS"
NO
YES
CHECK CKT 496 FOR BEING NEAR A SPARK PLUG WIRE OR A FAULTY ECM CONNECTION OR FAULTY ECM OR MEM-CAL.
REPLACE KNOCK SENSOR.
4
• DISCONNECT KNOCK SENSOR.
• CONNECT VOLTMETER TO KNOCK SENSOR AND ENGINE GROUND.
• SET VOLTMETER ON 2 VOLT AC SCALE.
• TAP ON ENGINE BLOCK NEAR SENSOR.
IS A SIGNAL INDICATED ON VOLTMETER WHILE TAPPING ON ENGINE BLOCK?
YES
NO
REPLACE MEM-CAL OR ECM.
REPLACE KNOCK SENSOR.

"AFTER REPAIRS," CONFIRM "CLOSED LOOP" OPERATION AND NO "SERVICE ENGINE SOON" LIGHT.

2.3L (VIN D, A & 3) ENGINE — COMPONENT DIAGNOSTIC CHART — 1992 ACHIEVA, GRAND AM AND SKYLARK

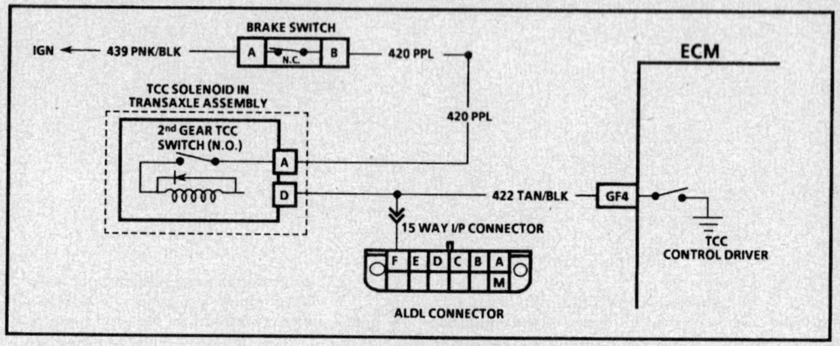
BRAKE SWITCH
IGN
439 PNK/BLK
A
N.C.
B
420 PPL
420 PPL
ECM
TCC SOLENOID IN TRANSAXLE ASSEMBLY
2nd GEAR TCC SWITCH (N.O.)
A
D
422 TAN/BLK
GF4
15 WAY I/P CONNECTOR
F E D C B A
M
TCC CONTROL DRIVER
ALDL CONNECTOR

CHART C-8A

3T40 TORQUE CONVERTER CLUTCH (TCC) (ELECTRICAL DIAGNOSIS) 2.3L (VIN D & 3) "N" CARLINE (PORT)

Circuit Description:

The purpose of the Torque Converter Clutch (TCC) feature is to eliminate the power loss of the transaxle converter stage when the vehicle is in a cruise condition. This allows the convenience of the automatic transaxle and the fuel economy of a manual transaxle.

Fused battery ignition is supplied to the TCC solenoid through the brake switch, and transaxle second gear apply switch. The ECM will engage TCC by grounding CKT 422 to energize the solenoid.

TCC will engage when:

- Vehicle speed above a calibrated value (about 34 mph) (55 km/h).
- Throttle position sensor output not changing, indicating a steady road speed.
- Transaxle second gear switch closed.
- Brake switch closed.

Test Description: Number(s) below refer to circled number(s) on the diagnostic chart.

1. Light "OFF" confirms transaxle second gear apply switch is open.
2. By 25 mph, the transaxle second gear TCC switch should close. Test light will come "ON" and confirm battery supply and closed brake switch.
3. Grounding the diagnostic terminal with ignition "ON," engine "OFF," should energize the TCC solenoid by grounding CKT 422. This test checks the ability of the ECM to supply a ground to the TCC solenoid. The test light connected from 12 volts to ALDL terminal "F" will turn "ON" as CKT 422 is grounded.

Diagnostic Aids:

A Tech 1 only indicates when the ECM has turned "ON" the TCC driver and this does not confirm that the TCC has engaged. To determine if TCC is functioning properly, engine rpm should decrease when the "Scan" indicates the TCC driver has turned "ON."

2.3L (VIN D, A & 3) ENGINE — COMPONENT DIAGNOSTIC CHART — 1992 ACHIEVA, GRAND AM AND SKYLARK

CHART C-8A
3T40 TORQUE CONVERTER CLUTCH (TCC)
(ELECTRICAL DIAGNOSIS)
2.3L (VIN D & 3) "N" CARLINE (PORT)

USING A TECH 1 CHECK THE FOLLOWING AND CORRECT IF NECESSARY
- TPS - BE SURE TPS SIGNAL IS NOT ERRATIC.
- VSS - SHOULD INDICATE VSS WITH WHEELS TURNING.
- CODES - IF 24 IS PRESENT, SEE CODE 24 CHART.

(1)
- PERFORM MECHANICAL CHECKS, SUCH AS LINKAGE, OIL LEVEL, ETC., BEFORE USING THIS CHART. VERIFY THAT TRANS IS STARTING OUT IN FIRST GEAR AND CORRECT AS NECESSARY.
- CONNECT TEST LIGHT FROM TCC TEST POINT, ALDL TERMINAL "F" TO GROUND.
- RAISE DRIVE WHEELS.
- START AND IDLE ENGINE WITH TRANS IN DRIVE. DO NOT DEPRESS BRAKE PEDAL.
- *"NOTICE:" DO NOT PERFORM THIS TEST WITHOUT SUPPORTING THE LOWER CONTROL ARMS SO THAT THE DRIVE AXLES ARE IN A NORMAL HORIZONTAL POSITION. RUNNING THE VEHICLE IN GEAR WITH THE WHEELS HANGING DOWN AT FULL TRAVEL MAY DAMAGE THE DRIVE AXLES.*
- NOTE LIGHT.

LIGHT "OFF"

(2)
- VEHICLE IN DRIVE.
- INCREASE SPEED SLOWLY UNTIL TRANS. SHIFTS INTO 2ND GEAR TO CLOSE 2ND GEAR TCC SWITCH.
- NOTE TEST LIGHT.

LIGHT "ON"

CKT 422 SHORTED TO VOLTAGE
OR
FAULTY TRANSAXLE SECOND GEAR SWITCH.

LIGHT "ON"

TEST LIGHT SHOULD GO OUT AS BRAKE PEDAL IS DEPRESSED. DOES IT?

LIGHT "OFF"

- CHECK FOR BLOWN FUSE. IF OK, DISCONNECT CONNECTOR AT TRANS.
- CONNECT TEST LIGHT FROM HARNESS CONNECTOR "A" TO "D".
- IGNITION "ON", ENGINE STOPPED.

YES

(3)
- IGNITION "ON", ENGINE STOPPED.
- INSTEAD OF GROUND, CONNECT TEST LIGHT TO 12 VOLTS AND PROBE ALDL TERMINAL "F".
- GROUND DIAGNOSTIC TERMINAL AND NOTE LIGHT.

NO

FAULTY BRAKE SWITCH OR ADJUSTMENT.

LIGHT "OFF"

- CONNECT A TEST LIGHT FROM TERMINAL "A" TO GROUND.

LIGHT "ON"

- CHECK FOR SHORT TO GROUND IN CKT 422. IF NOT GROUNDED, REPLACE ECM.

LIGHT "ON"

- GROUND TCC TEST POINT AND AGAIN CONNECT TEST LIGHT BETWEEN HARNESS CONNECTOR TERMINALS "A" AND "D".

LIGHT "OFF"

BRAKE SWITCH MISADJUSTED
OR
OPEN CKT 420
OR
FAULTY BRAKE SWITCH.

LIGHT "ON"

CHECK FOR CORRECT MEM-CAL. IF OK, TCC ELECTRICAL CONTROL IS OK. REFER TO "DIAGNOSTIC AIDS"

LIGHT "OFF"

CHECK FOR OPEN CKT 422 FROM ALDL TO ECM CONNECTOR TERMINAL. IF CKT 422 IS OK, THE ECM IS FAULTY.

LIGHT "ON"

FAULTY TRANS. TCC CONNECTION
OR
TCC SOLENOID
OR
SECOND GEAR TCC SWITCH.

LIGHT "OFF"

REPAIR OPEN IN WIRE FROM TRANS. TO ALDL TEST POINT TERMINAL "F".

"AFTER REPAIRS," CONFIRM "CLOSED LOOP" OPERATION AND NO "SERVICE ENGINE SOON" LIGHT.

2.3L (VIN D, A & 3) ENGINE — COMPONENT DIAGNOSTIC CHART — 1992 ACHIEVA AND GRAND AM

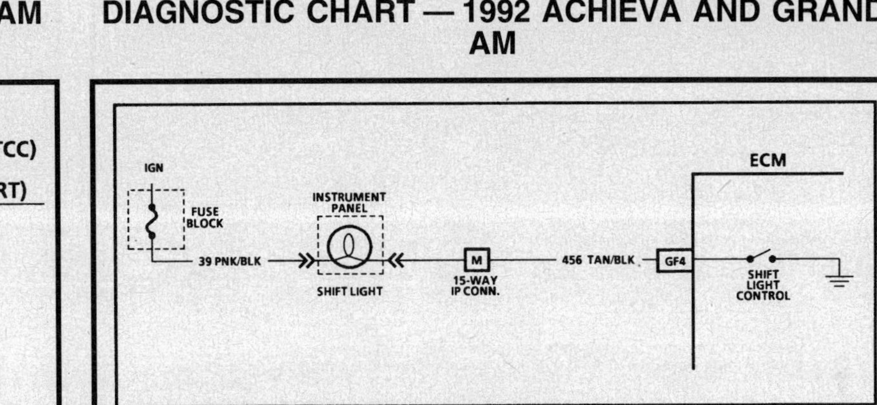

CHART C-8C
MANUAL TRANSAXLE (M/T) SHIFT LIGHT CHECK
2.3L (VIN A) "N" CARLINE (PORT)

Circuit Description:

The shift light indicates the best transaxle shift point for maximum fuel economy. The light is controlled by the ECM and is turned "ON" by grounding CKT 456.

The ECM uses inputs from various sensors to calculate when the shift light should be turned "ON" as follows:
- Coolant temperature must be above -10°C (14°F).
- ECM can determine the transaxle has been in a gear for at least 1.2 seconds, by comparison of vehicle speed (from VSS) with engine rpm.
- TPS is between minimum and maximum calibrated values for each gear.
- RPM is above a calibrated value for each gear (maximum 6500 rpm).

The light will be turned "ON" after a calibrated delay time which is dependent on last gear change or downshift, and will remain on for a maximum of 10 seconds.

Test Description: Number(s) below refer to circled number(s) on the diagnostic chart.

1. This should not turn "ON" the shift light. If the light is "ON," there is a short to ground in CKT 456 wiring or a fault in the ECM.
2. When the diagnostic terminal is grounded, the ECM should ground CKT 456 and the shift light should come "ON."
3. This checks the shift light circuit up to the ECM connector. If the shift light illuminates, then the ECM connector is faulty or the ECM does not have the ability to ground the circuit.

Diagnostic Aids:

Check for Code 24 (no VSS). Faulty thermostat or incorrect heat range. Incorrect or faulty MEM-CAL. A faulty or incorrect VSS may result in a shift light that does not operate.

2.3L (VIN D, A & 3) ENGINE — COMPONENT DIAGNOSTIC CHART — 1992 ACHIEVA AND GRAND AM

CHART C-8C
**MANUAL TRANSAXLE (M/T) SHIFT LIGHT CHECK
2.3L (VIN A) "N" CARLINE (PORT)**

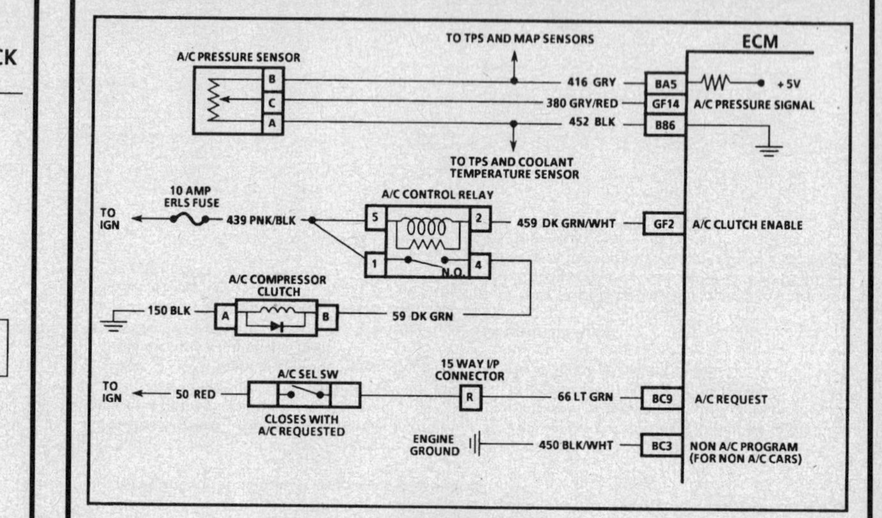

"AFTER REPAIRS," CONFIRM "CLOSED LOOP" OPERATION AND NO "SERVICE ENGINE SOON" LIGHT.

2.3L (VIN D, A & 3) ENGINE — COMPONENT DIAGNOSTIC CHART — 1992 ACHIEVA, GRAND AM AND SKYLARK

CHART C-10
(Page 1 of 2)
**A/C CLUTCH CONTROL CIRCUIT DIAGNOSIS
2.3L (VIN D, A & 3) "N" CARLINE (PORT)**

Circuit Description:

The A/C clutch control relay is energized when the ECM provides a ground path through CKT 459 and A/C is requested. A/C clutch is delayed about .3 seconds after A/C is requested. This will allow the IAC to adjust engine rpm for the additional load.

The ECM will temporarily disengage the A/C clutch relay for calibrated times for one or more of the following:

- Hot engine restart.
- Wide Open Throttle (WOT) (TPS over 90%).
- Engine rpm greater than about 6000 rpm.
- During IAC reset.

The A/C clutch relay will remain disengaged when a Code 66 is present, if pressure is out of range or there is no A/C request signal due to an open A/C select switch or circuit.

Test Description: Number(s) below refer to circled number(s) on the diagnostic chart.

1. The ECM will only energize the A/C relay when the engine is running. This test will determine if the relay or CKT 459 is faulty.

2. Determines if the signal is reaching the ECM through CKT 66 from the A/C control panel. Signal should only be present when an A/C mode or defrost mode has been selected.

2.3L (VIN D, A & 3) ENGINE — COMPONENT DIAGNOSTIC CHART — 1992 ACHIEVA, GRAND AM AND SKYLARK

CHART C-10
(Page 1 of 2)
A/C CLUTCH CONTROL CIRCUIT DIAGNOSIS
2.3L (VIN D, A & 3) "N" CARLINE (PORT)

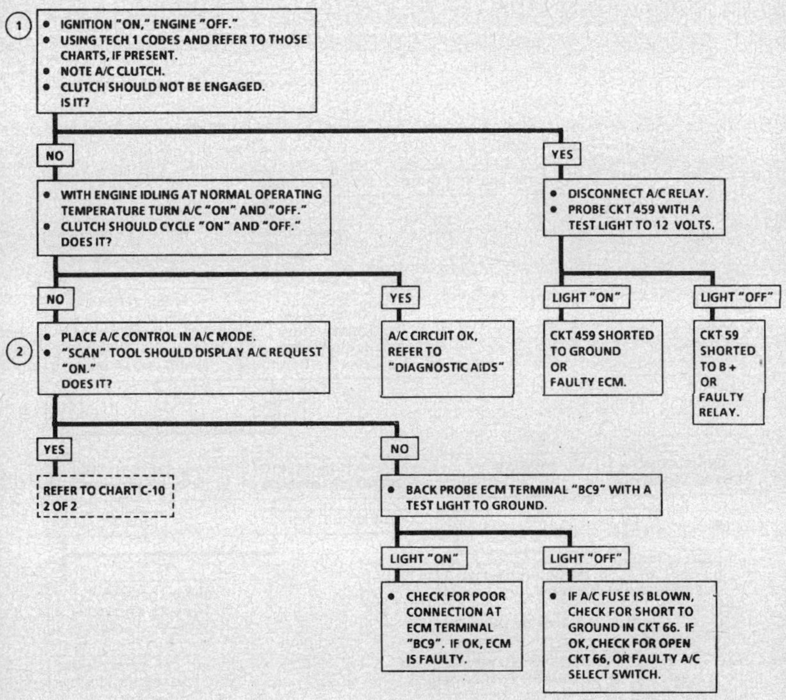

"*AFTER REPAIRS,*" CONFIRM "CLOSED LOOP" OPERATION AND NO "SERVICE ENGINE SOON" LIGHT.

2.3L (VIN D, A & 3) ENGINE — COMPONENT DIAGNOSTIC CHART — 1992 ACHIEVA, GRAND AM AND SKYLARK

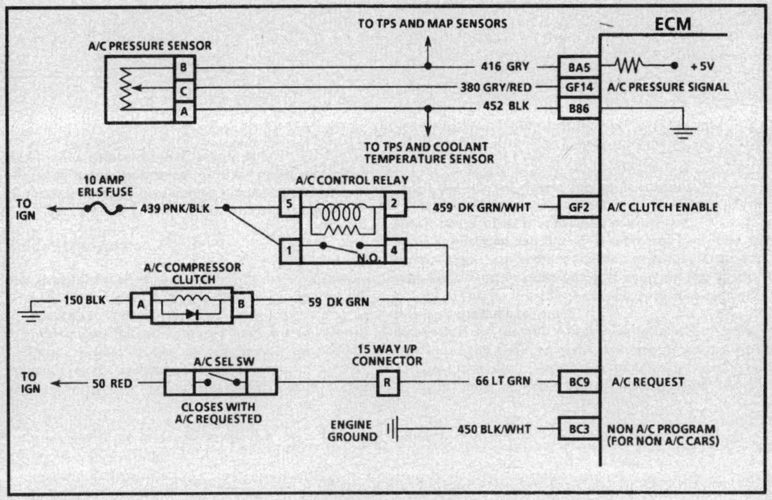

CHART C-10
(Page 2 of 2)
A/C CLUTCH CONTROL CIRCUIT DIAGNOSIS
2.3L (VIN D, A & 3) "N" CARLINE (PORT)

Circuit Description:

The A/C clutch control relay is energized when the ECM provides a ground path through CKT 459 and A/C is requested. A/C clutch is delayed about .3 seconds after A/C is requested. This will allow the IAC to adjust engine rpm for the additional load.

The ECM will temporarily disengage the A/C clutch relay for calibrated times for one or more of the following:

- Hot engine restart.
- Wide Open Throttle (WOT) (TPS over 90%).
- Engine rpm greater than about 6000 rpm.
- During IAC reset.

The A/C clutch relay will remain disengaged when a Code 66 is present, if pressure is out of range or there is no A/C request signal due to an open A/C select switch or circuit.

Test Description: Number(s) below refer to circled number(s) on the diagnostic chart.

1. Determines if the pressure transducer is out of range causing the compressor clutch to be disengaged.
2. With the engine stopped and field service mode activated, the ECM should be grounding CKT 459, which should cause the test light to be "ON."

Diagnostic Aids:

If complaint is insufficient cooling, the problem may be caused by an inoperative cooling fan.

See CHART C-12 for cooling fan diagnosis. If fan operates correctly, see A/C diagnosis

A/C pressure outside of a range of 43 to 428 psi will cause the compressor to be disabled by the ECM. Observe "Scan" A/C pressure for 2 minutes with engine idling and A/C "ON."

"Scan" pressure should be within 20 psi of actual. If not, check for a circuit problem using Code 66 chart or replace sensor.

2.3L (VIN D, A & 3) ENGINE — COMPONENT DIAGNOSTIC CHART — 1992 ACHIEVA, GRAND AM AND SKYLARK

CHART C-10
(Page 2 of 2)
A/C CLUTCH CONTROL CIRCUIT DIAGNOSIS
2.3L (VIN D, A & 3) "N" CARLINE (PORT)

① • IGNITION "ON," A/C "ON."
• ENGINE "OFF."
• "SCAN" A/C PRESSURE.

PRESSURE IS BETWEEN 40 LBS - 430 LBS
• DISCONNECT A/C RELAY.
• PROBE CKTS 439 WITH TEST LIGHT TO GROUND.

PRESSURE IS LESS THAN 40 LBS OR GREATER THAN 430 LBS.
• INSTALL A/C MANIFOLD GAGE SET
• ARE "SCAN" PRESSURES WITHIN 20 LBS OF HIGH SIDE PRESSURE?

LIGHT "ON" BOTH
• JUMPER CKT 439 TO CKT 59.
• DOES A/C CLUTCH ENGAGE?

NO LIGHT, ONE OR BOTH
REPAIR OPEN IN CIRCUIT THAT DID NOT LIGHT.

YES
REFER TO A/C DIAGNOSIS.

NO
REPLACE A/C PRESSURE TRANSDUCER AND RECHECK.

YES
② • IGNITION "ON," ENGINE STOPPED. ACTIVATE FIELD SERVICE MODE.
• PROBE CKT 459 WITH A TEST LIGHT TO 12 VOLTS.

NO
CHECK:
- OPEN CKT 59 TO CLUTCH.
- FAULTY CLUTCH COIL.
- OPEN GROUND.

LIGHT "ON"
FAULTY RELAY.

LIGHT "OFF"
OPEN CKT 459, FAULTY CONNECTION, OR ECM.

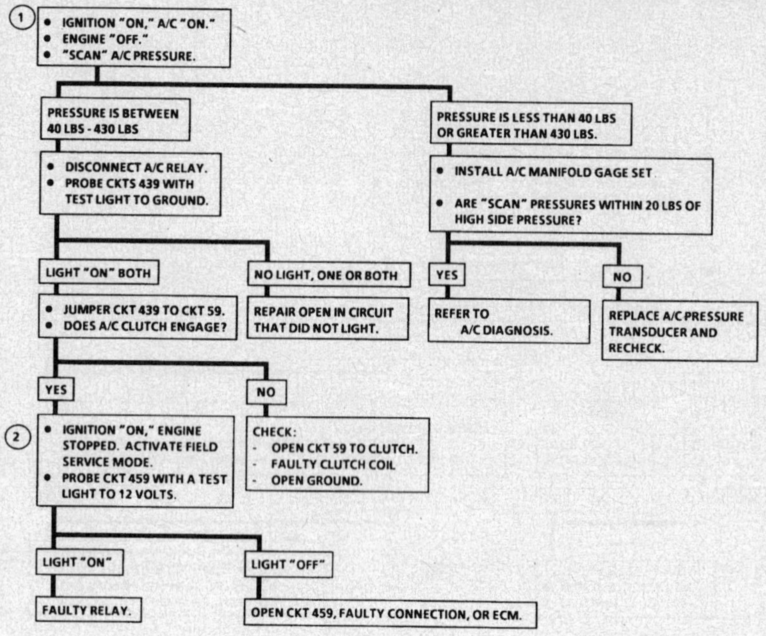

"AFTER REPAIRS," CONFIRM "CLOSED LOOP" OPERATION AND NO "SERVICE ENGINE SOON" LIGHT.

2.3L (VIN D, A & 3) ENGINE — COMPONENT DIAGNOSTIC CHART — 1992 ACHIEVA, GRAND AM AND SKYLARK

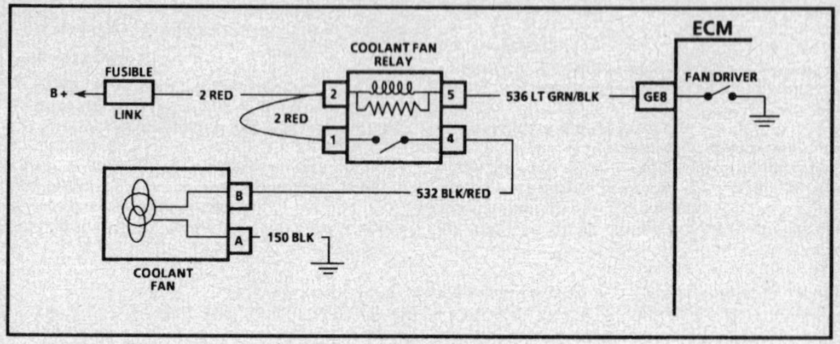

CHART C-12
COOLANT FAN FUNCTIONAL CHECK
2.3L (VIN D, A & 3) "N" CARLINE (PORT)

Circuit Description:

The electric coolant fan is controlled by the ECM through the fan relay based on inputs from the coolant and manifold air temperature sensors, the A/C control switch, A/C pressure sensor and the vehicle speed sensor. The ECM controls the coolant fan by grounding CKT 536 which turns "ON" the fan relay.

The fan relay will be commanded "ON" when:
• Coolant temperature 103°C - 106°C or more.
• A/C clutch requested.
• Vehicle speed is less than 35 mph.

The fan relay will be commanded "ON" regardless of vehicle speed when:
• Code 14 or 15 are set.
• Coolant temperature 115°C - 118°C or more.
• A/C pressure is high.

The coolant fan may be commanded "ON" when the engine is not running under fan "Run-ON" conditions

Test Description: Number(s) below refer to circled number(s) on the diagnostic chart.

1. With the field service mode activated, the coolant fan control driver should close, which should energize the fan control relay.
2. Test to see if fault is in wiring to the fan or the fan/fan connection.

Diagnostic Aids:

If the owner complained of an overheating problem, it must be determined if the complaint was due to an actual boil over, or the "temp light," or temperature gage indicated overheating.

If the gage, or light, indicates overheating, but no boil over is detected, the gage or light circuit should be checked. The gage accuracy can also be checked by comparing the coolant sensor reading using a "Scan" tool with the gage reading.

If the engine is actually overheating, and the gage indicates overheating, but the coolant fan is not coming "ON," the coolant sensor has probably shifted out of calibration and should be replaced. See Code 15 chart for a temperature to resistance chart.

If the engine is overheating, and the coolant fan is "ON," the cooling system should be checked.

2.3L (VIN D, A & 3) ENGINE — COMPONENT DIAGNOSTIC CHART — 1992 ACHIEVA, GRAND AM AND SKYLARK

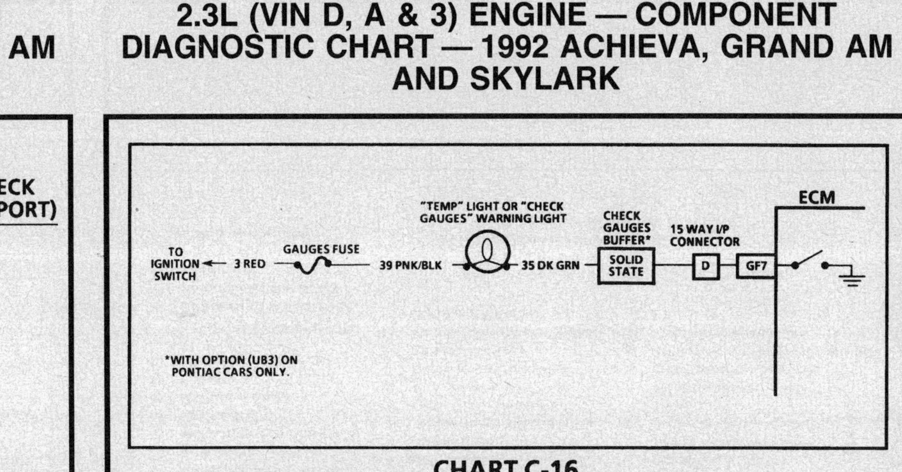

2.3L (VIN D, A & 3) ENGINE — COMPONENT DIAGNOSTIC CHART — 1992 ACHIEVA, GRAND AM AND SKYLARK

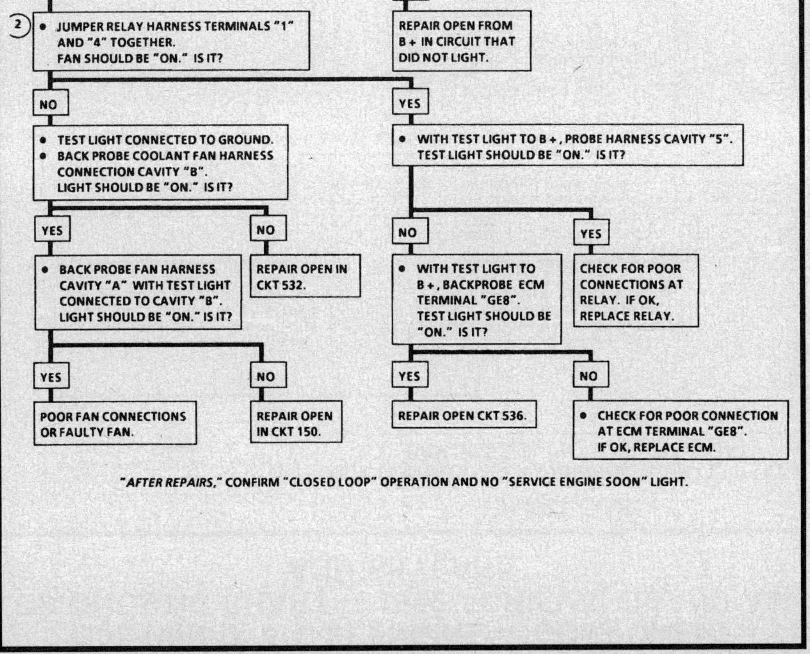

CHART C-16
ECM CONTROLLED "WARNING" LIGHT CIRCUIT
2.3L (VIN D, A & 3) "N" CARLINE (PORT)

Circuit Description:

The "Temp" light or optional "Check Gauges" light, is connected to battery voltage through the ignition switch. The ECM energizes the bulb by supplying a path to ground through "Quad-Driver" module #2.

Test Description: Number(s) below refer to circled number(s) on the diagnostic chart.

1. With the ignition "ON" and engine "OFF," the ECM should be grounding CKT 35.
2. While the engine is running, the "Temp" light or optional "Check Gauges" light should be turned "ON" by the ECM only when the coolant temperature is above 123°C (253°F). The "Temp" light or "Check Gauges" light should be turned "OFF" when the engine is running and the coolant temperature goes lower than 120°C (249°F).

Diagnostic Aids:

The coolant temperature sensor, in rare cases, may fail to indicate the correct engine coolant temperature without setting a malfunction code. This could result in turning "ON" the "Temp" light or optional "Check Gauges" light without having an overheating condition. It could also result in overheating without the "Temp/Check Gauges" light being turned "ON."

2.3L (VIN D, A & 3) ENGINE — COMPONENT DIAGNOSTIC CHART — 1992 ACHIEVA, GRAND AM AND SKYLARK

2.3L (VIN D & 3) ENGINE — ENGINE COMPONENT LOCATION CHART — 1993–94 SKYLARK

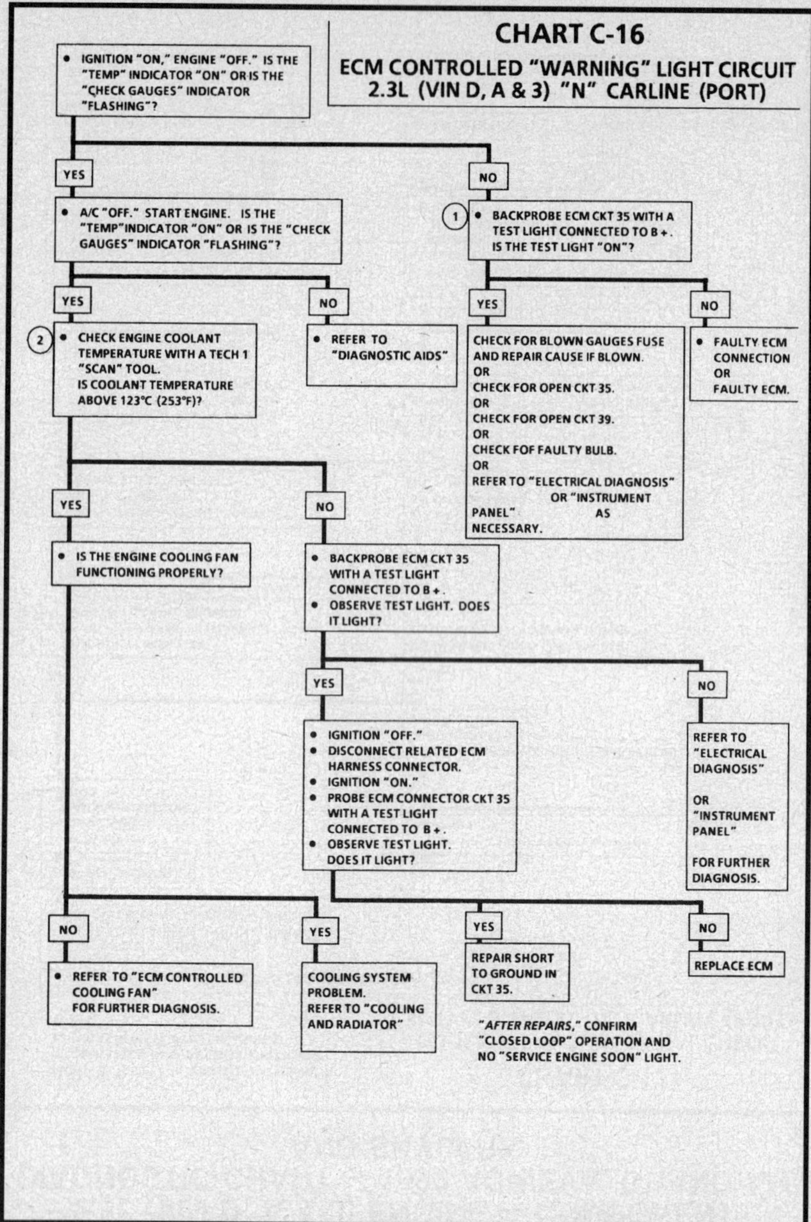

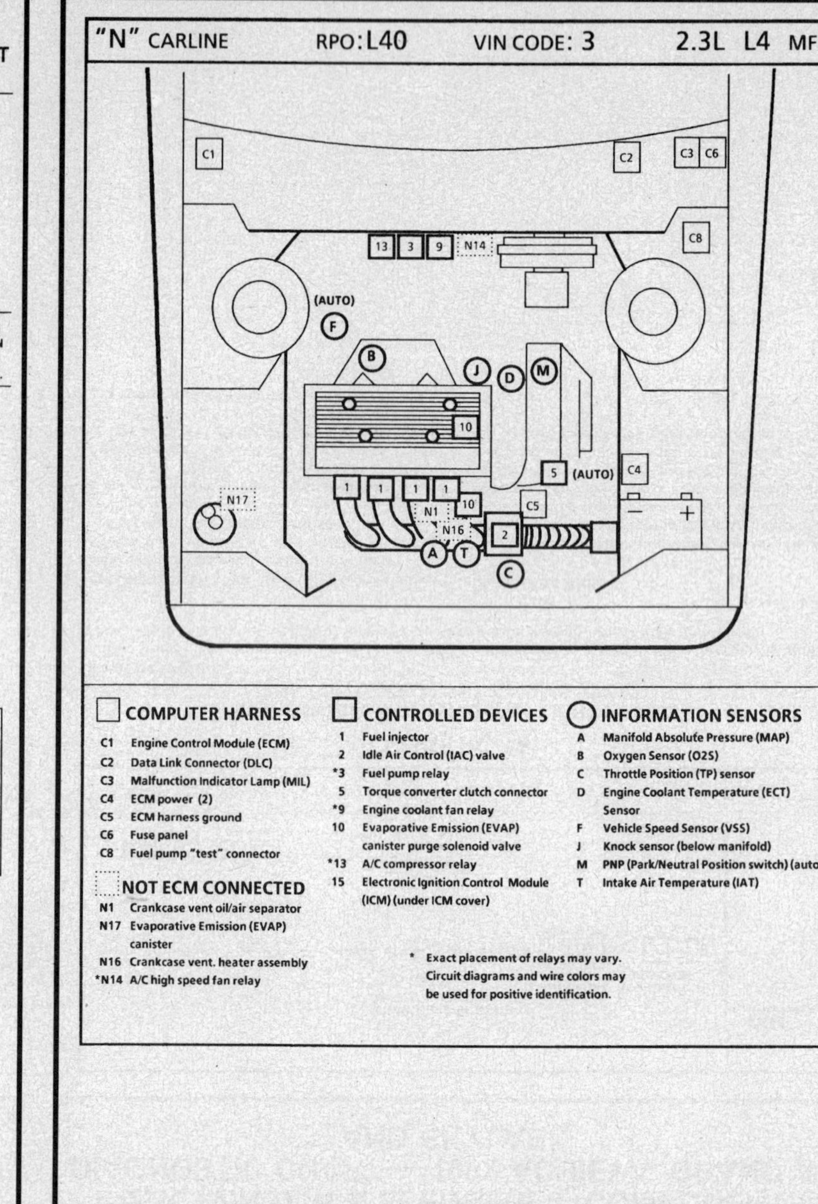

2.3L (VIN D & 3) ENGINE — ECM WIRING SCHEMATIC — 1993–94 SKYLARK

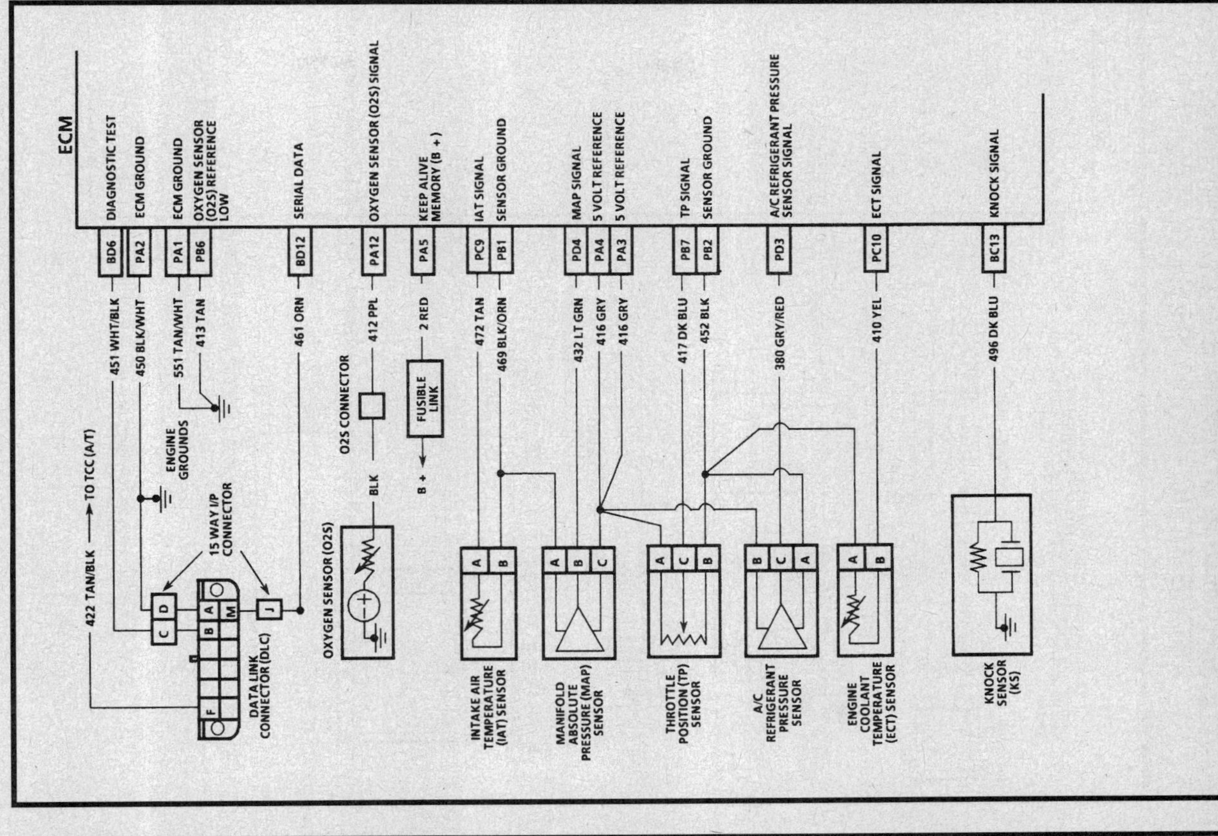

2.3L (VIN D, A & 3) ENGINE — ENGINE COMPONENT LOCATION CHART — 1993–94 ACHIEVA AND GRAND AM

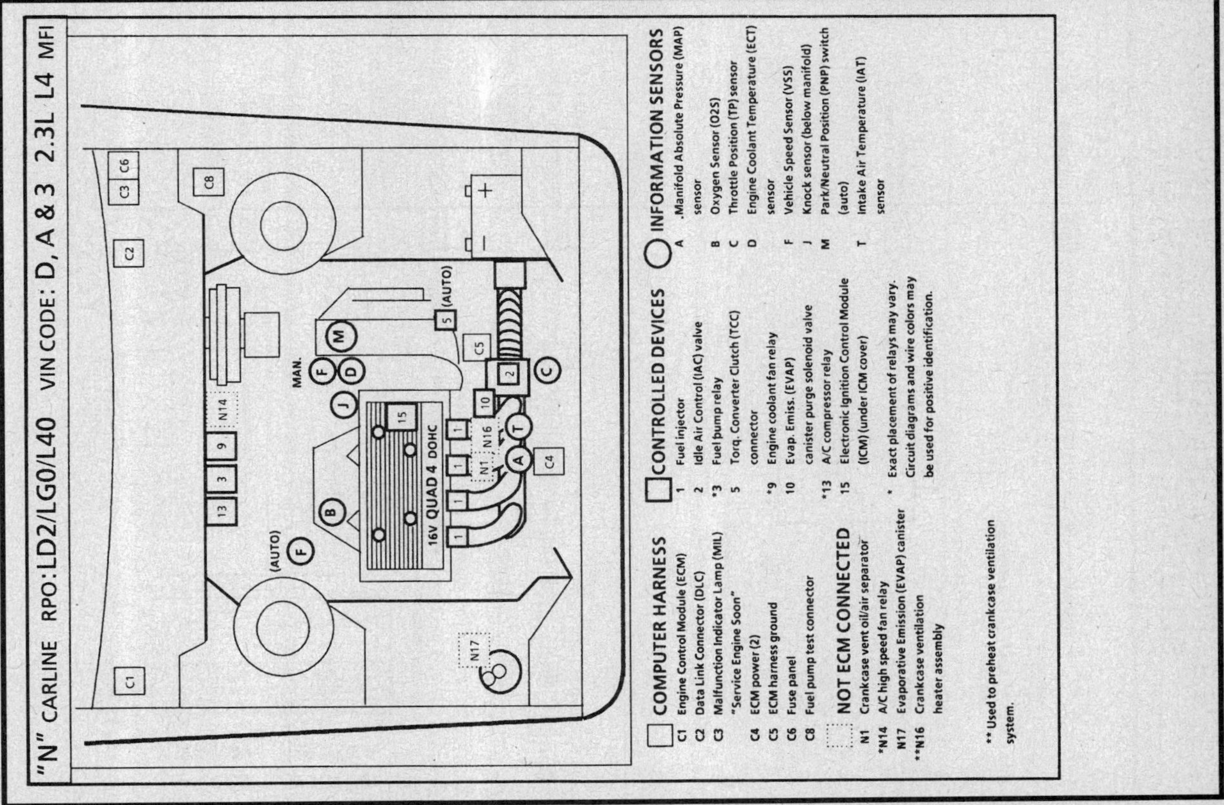

MULTIPORT FUEL INJECTION (MFI) SYSTEMS
EXCEPT LIGHT TRUCKS, VANS, GEO AND SATURN

2.3L (VIN D & 3) ENGINE — ECM WIRING SCHEMATIC — 1993–94 SKYLARK

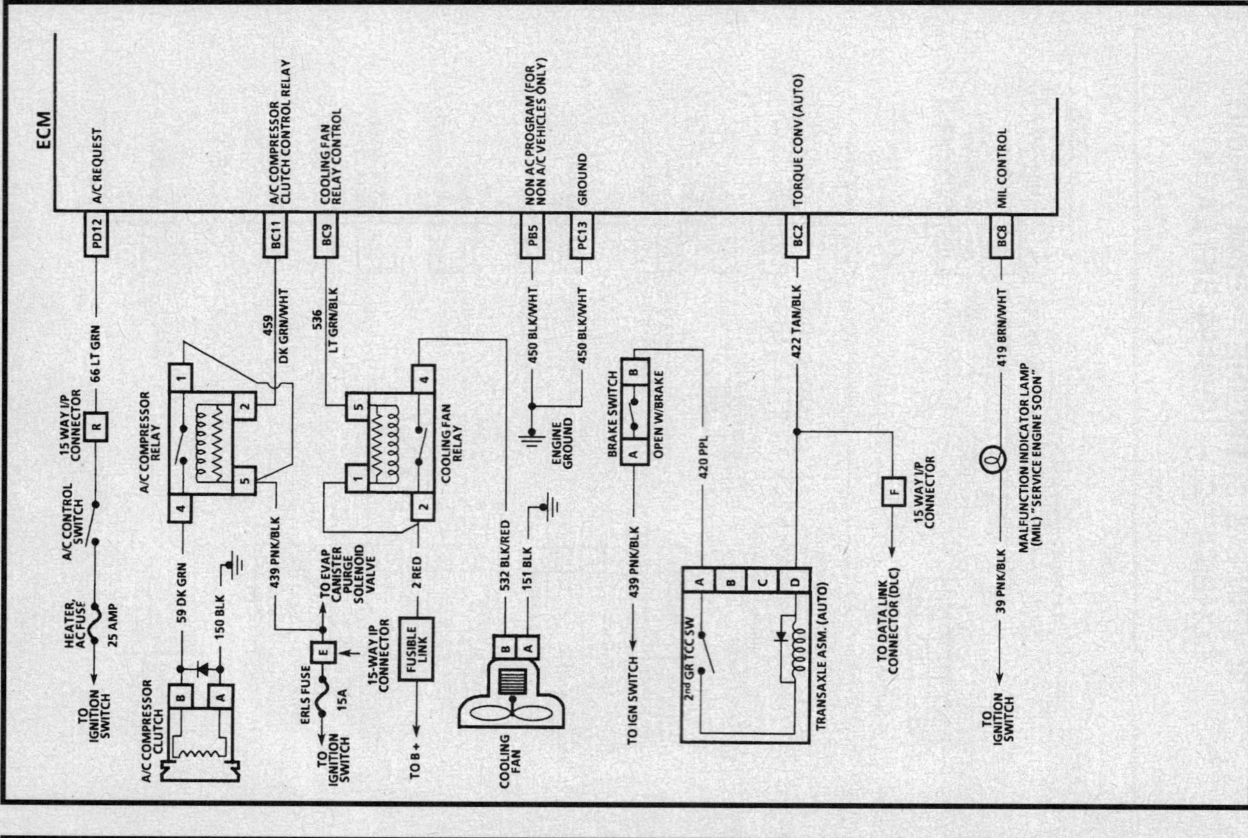

2.3L (VIN D & 3) ENGINE — ECM WIRING SCHEMATIC — 1993–94 SKYLARK

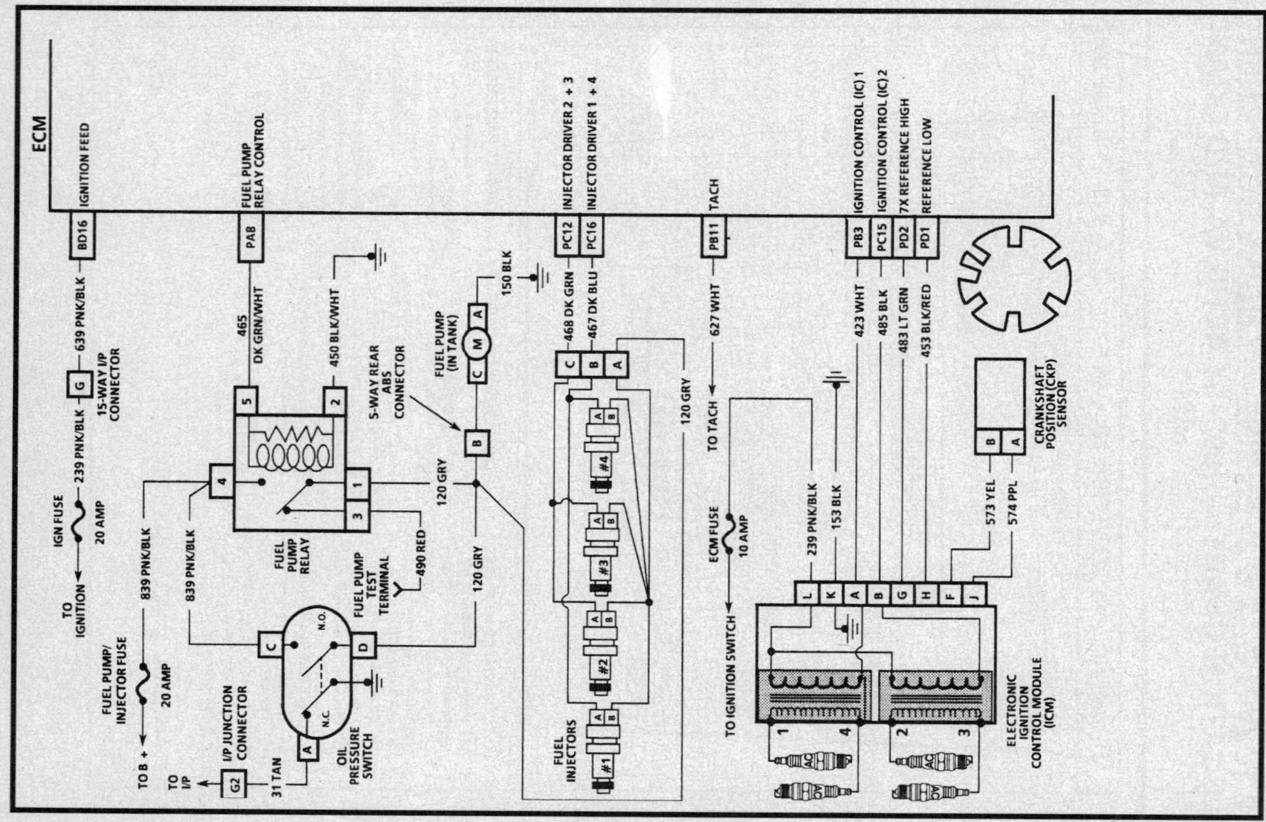

2.3L (VIN D, A & 3) ENGINE — ECM WIRING SCHEMATIC — 1993–94 ACHIEVA AND GRAND AM

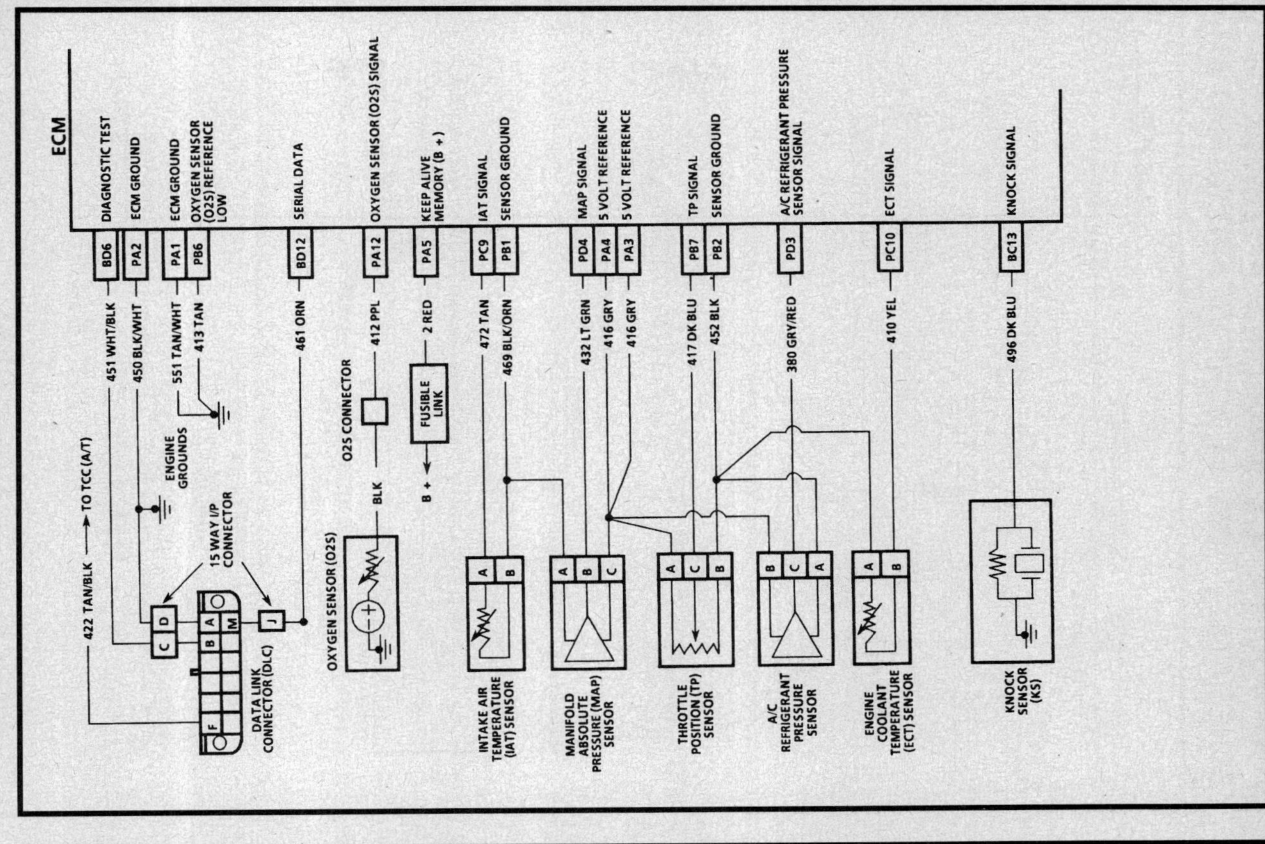

2.3L (VIN D & 3) ENGINE — ECM WIRING SCHEMATIC — 1993–94 SKYLARK

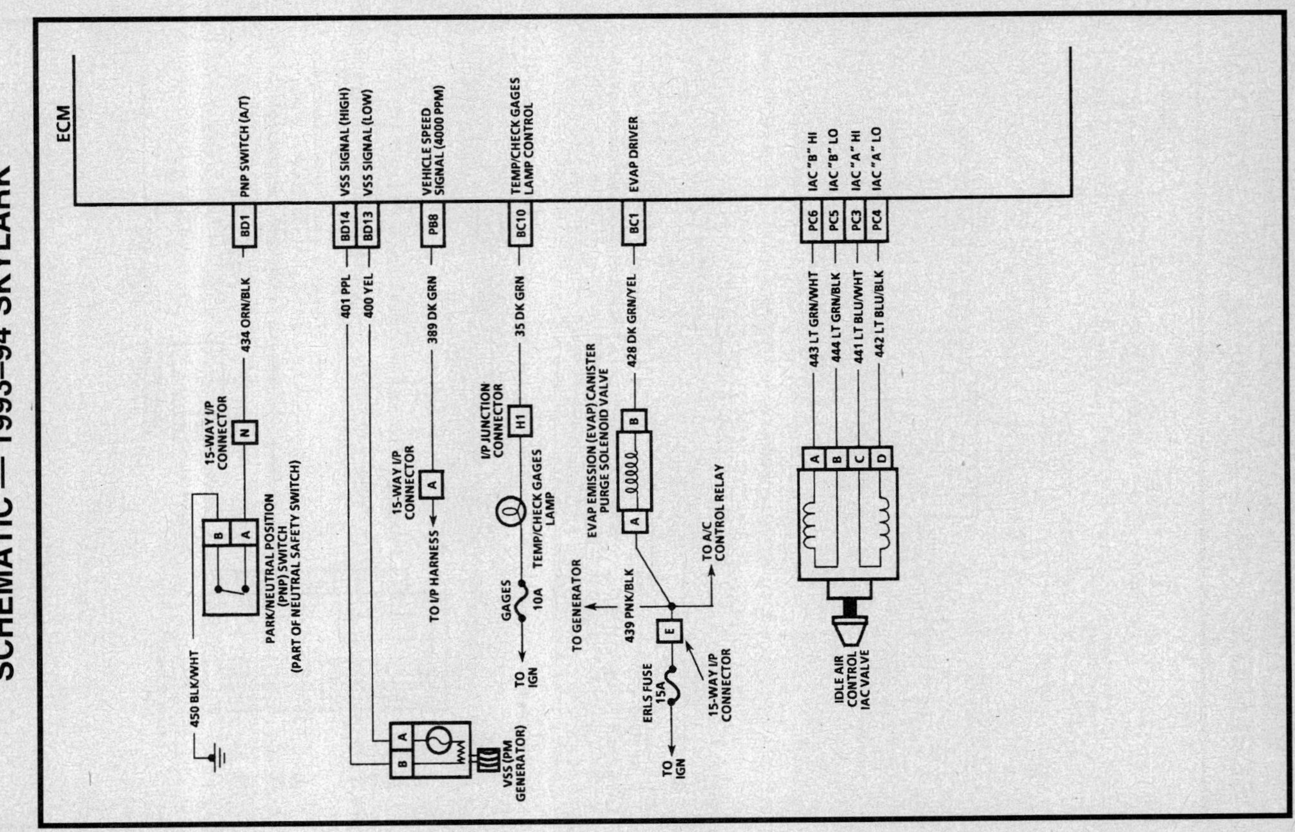

2.3L (VIN D, A & 3) ENGINE — ECM WIRING SCHEMATIC — 1993–94 ACHIEVA AND GRAND AM

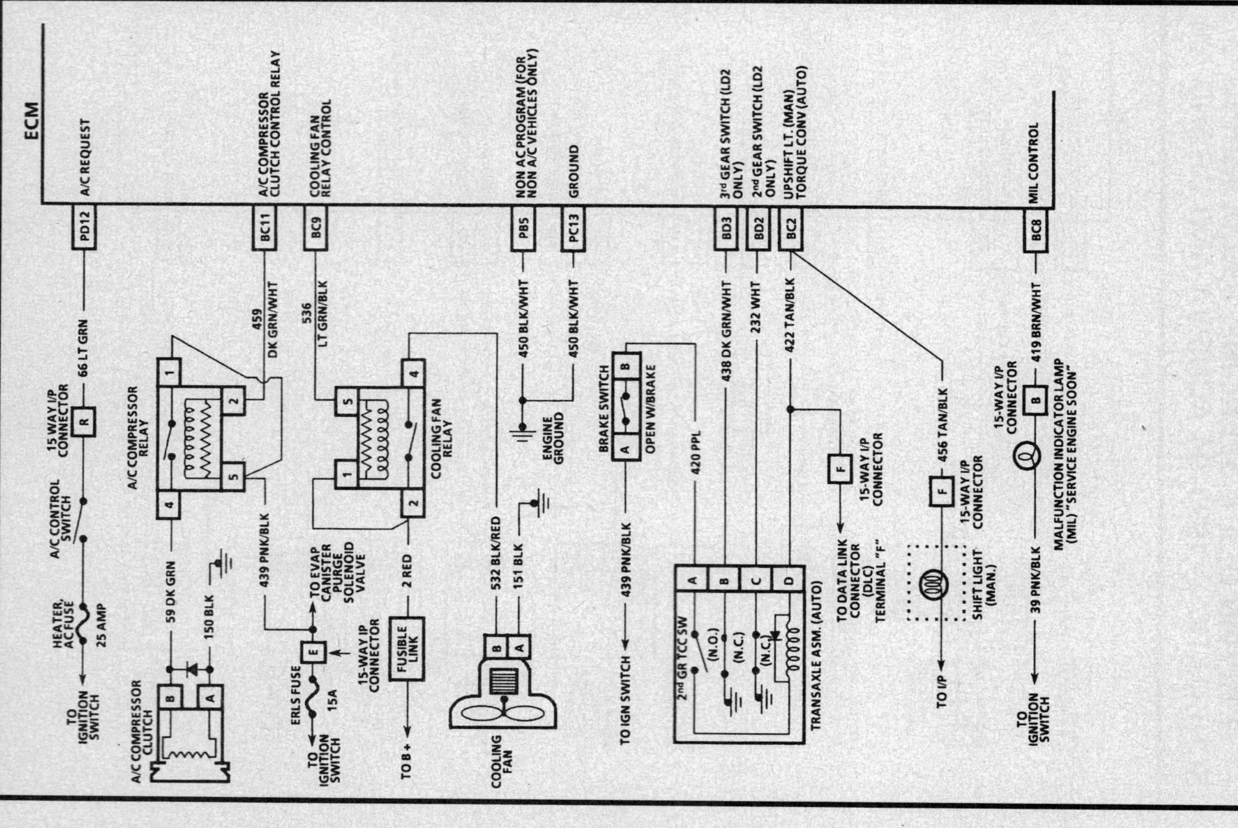

2.3L (VIN D, A & 3) ENGINE — ECM WIRING SCHEMATIC — 1993–94 ACHIEVA AND GRAND AM

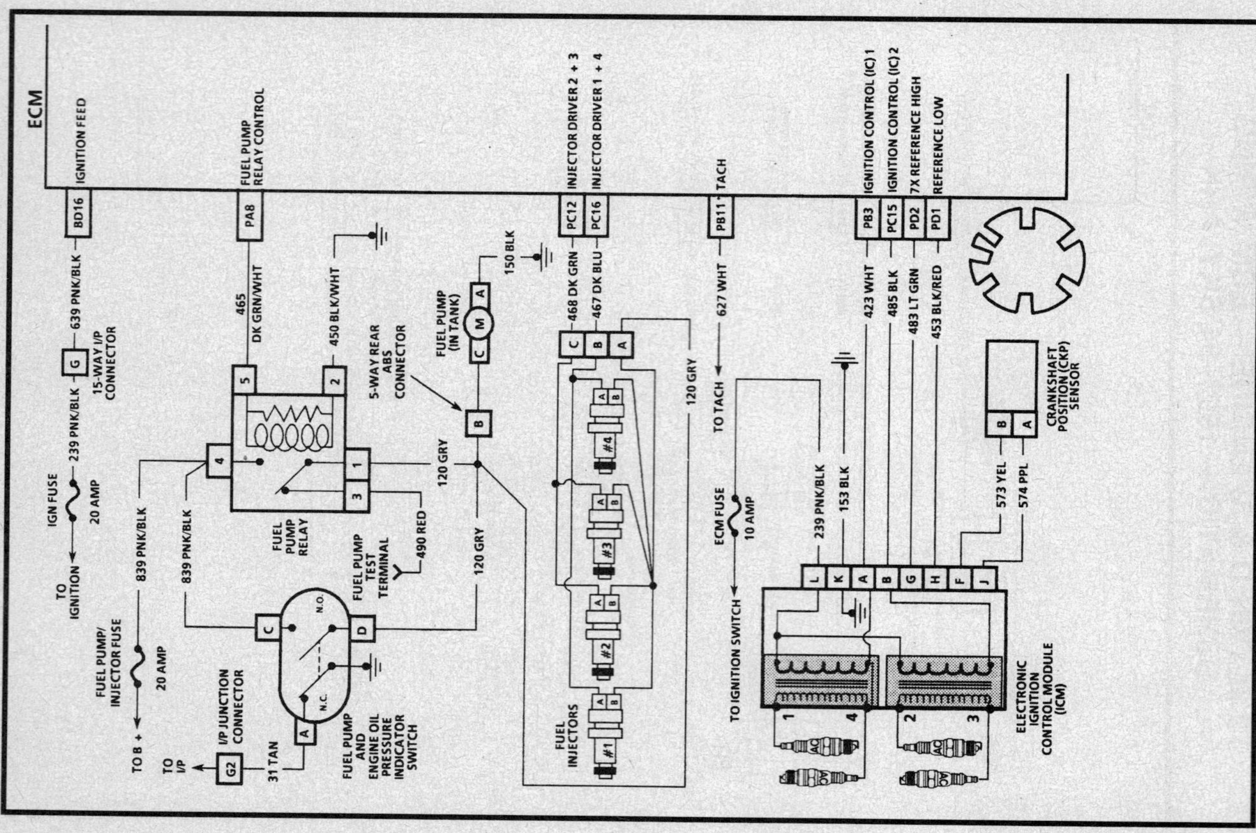

2.3L (VIN D, A & 3) ENGINE — ECM WIRING SCHEMATIC — 1993–94 ACHIEVA AND GRAND AM

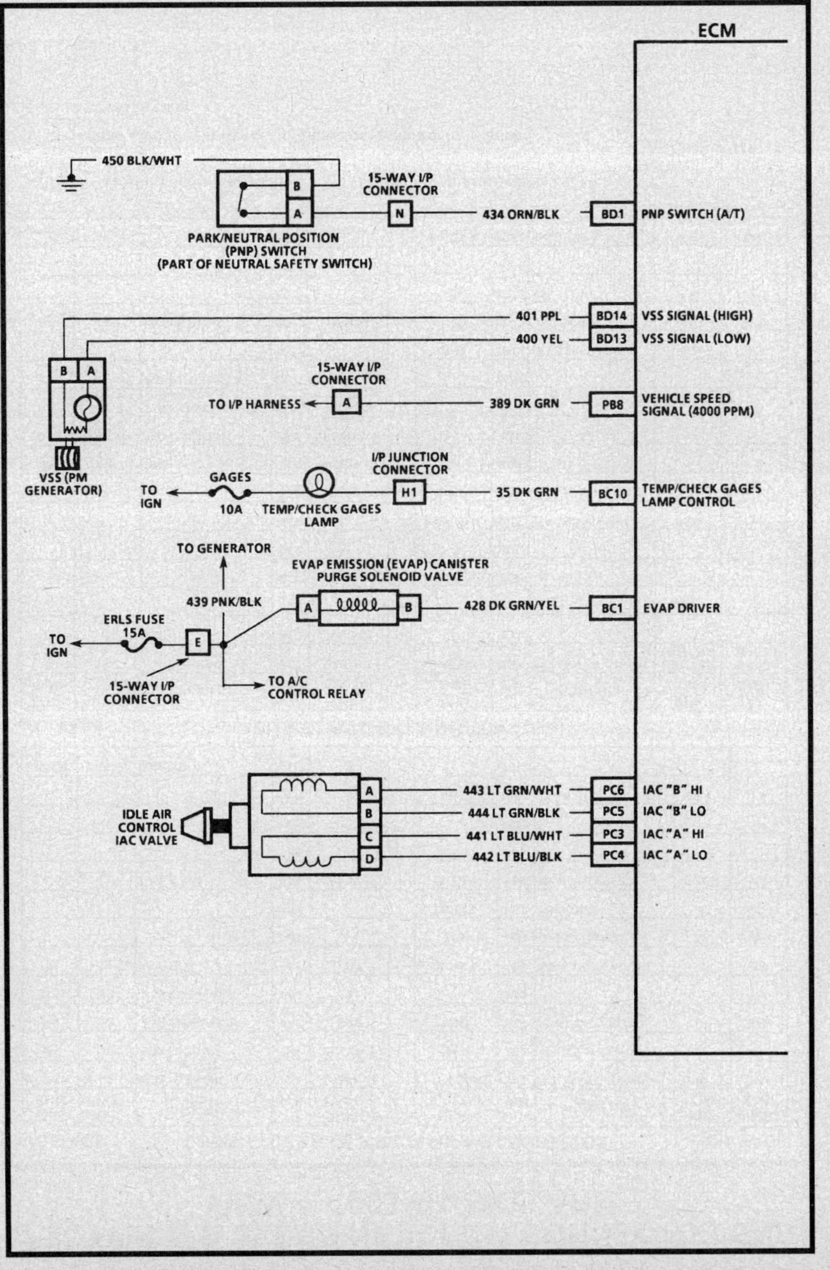

2.3L (VIN D & 3) ENGINE — ECM CONNECTOR END VIEW — 1993–94 SKYLARK

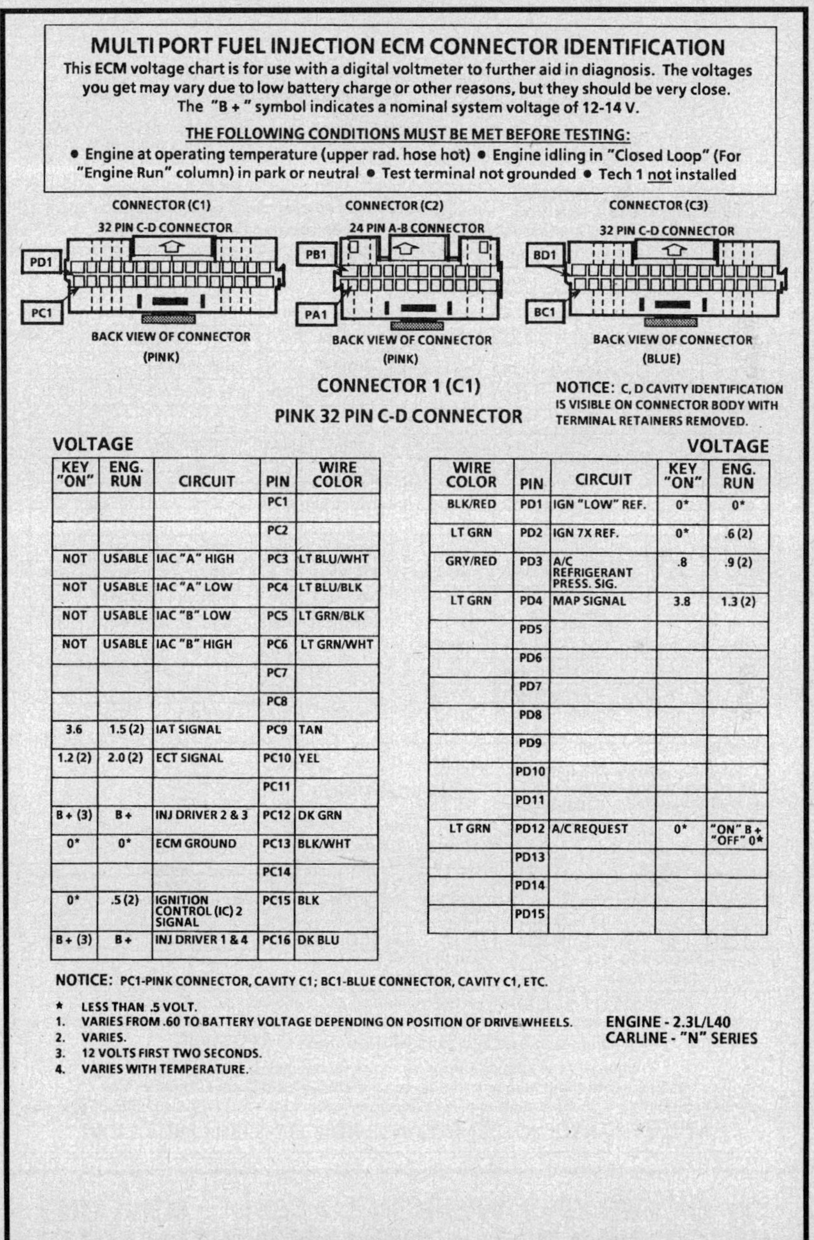

MULTIPORT FUEL INJECTION ECM CONNECTOR IDENTIFICATION

This ECM voltage chart is for use with a digital voltmeter to further aid in diagnosis. The voltages you get may vary due to low battery charge or other reasons, but they should be very close. The "B +" symbol indicates a nominal system voltage of 12-14 V.

THE FOLLOWING CONDITIONS MUST BE MET BEFORE TESTING:

• Engine at operating temperature (upper rad. hose hot) • Engine idling in "Closed Loop" (For "Engine Run" column) in park or neutral • Test terminal not grounded • Tech 1 not installed

NOTICE: C, D CAVITY IDENTIFICATION IS VISIBLE ON CONNECTOR BODY WITH TERMINAL RETAINERS REMOVED.

CONNECTOR 1 (C1)
PINK 32 PIN C-D CONNECTOR

VOLTAGE

KEY "ON"	ENG. RUN	CIRCUIT	PIN	WIRE COLOR
			PC1	
			PC2	
NOT	USABLE	IAC "A" HIGH	PC3	LT BLU/WHT
NOT	USABLE	IAC "A" LOW	PC4	LT BLU/BLK
NOT	USABLE	IAC "B" LOW	PC5	LT GRN/BLK
NOT	USABLE	IAC "B" HIGH	PC6	LT GRN/WHT
			PC7	
			PC8	
3.6	1.5 (2)	IAT SIGNAL	PC9	TAN
1.2 (2)	2.0 (2)	ECT SIGNAL	PC10	YEL
			PC11	
B + (3)	B +	INJ DRIVER 2 & 3	PC12	DK GRN
0*	0*	ECM GROUND	PC13	BLK/WHT
			PC14	
0*	.5 (2)	IGNITION CONTROL (IC) 2 SIGNAL	PC15	BLK
B + (3)	B +	INJ DRIVER 1 & 4	PC16	DK BLU

VOLTAGE

WIRE COLOR	PIN	CIRCUIT	KEY "ON"	ENG. RUN
BLK/RED	PD1	IGN "LOW" REF.	0*	0*
LT GRN	PD2	IGN 7X REF.	0*	.6 (2)
GRY/RED	PD3	A/C REFRIGERANT PRESS. SIG.	.8	.9 (2)
LT GRN	PD4	MAP SIGNAL	3.8	1.3 (2)
	PD5			
	PD6			
	PD7			
	PD8			
	PD9			
	PD10			
	PD11			
LT GRN	PD12	A/C REQUEST	0*	"ON" B+ "OFF" 0*
	PD13			
	PD14			
	PD15			

NOTICE: PC1-PINK CONNECTOR, CAVITY C1; BC1-BLUE CONNECTOR, CAVITY C1, ETC.

*   LESS THAN .5 VOLT.
1.   VARIES FROM .60 TO BATTERY VOLTAGE DEPENDING ON POSITION OF DRIVE WHEELS.
2.   VARIES.
3.   12 VOLTS FIRST TWO SECONDS.
4.   VARIES WITH TEMPERATURE.

ENGINE - 2.3L/L40
CARLINE - "N" SERIES

2.3L (VIN D & 3) ENGINE — ECM CONNECTOR END VIEW — 1993–94 SKYLARK

CONNECTOR 2 (C-2) PINK 24 PIN A-B CONNECTOR

VOLTAGE / VOLTAGE

ENG. RUN	KEY "ON"	CIRCUIT	PIN	WIRE COLOR	WIRE COLOR	PIN	CIRCUIT	KEY "ON"	ENG. RUN
0*	0*	ECM TO ENGINE GND	PA1	TAN/WHT	BLK/ORN	PB1	IAT, MAP GROUND	0*	0*
0*	0*	ECM GROUND	PA2	BLK/WHT	BLK	PB2	TP, ECT, A/C GND	0*	0*
5.0	5.0	+5V REFERENCE	PA3	GRY	WHT	PB3	IGNITION CONTROL (IC)1 SIGNAL	0*	.5 (2)
5.0	5.0	+5V REFERENCE	PA4	GRY		PB4			
B+	B+	BATTERY	PA5	RED	BLK/WHT	PB5	NON A/C PROGRAM	0*	0*
			PA6		TAN	PB6	O2S GROUND	0*	0*
			PA7		DK BLU	PB7	TP SIGNAL	.49	.5 (2)
B+ (3)	B+	FUEL PUMP	PA8	GRN/WHT	DK GRN	PB8	4000 P/MI SPEED	8.3 (2)	9.4
			PA9			PB9			
			PA10			PB10			
			PA11		WHT	PB11	TACH	4.9	1.0
.35 (2)	.1-.9 (2)	O2S SIGNAL	PA12	PPL		PB12			

BLUE 32 PIN C-D CONNECTOR 3 (C-3)

VOLTAGE / VOLTAGE

ENG. RUN	KEY "ON"	CIRCUIT	PIN	WIRE COLOR	WIRE COLOR	PIN	CIRCUIT	KEY "ON"	ENG. RUN
.3	12.0	EVAP SIGNAL	BC1	GRN/YEL	ORN/BLK	BD1	PNP SWITCH SIGNAL	"ON" 0* "OFF" B+	"ON" 0* "OFF" B+
0	0	TCC	BC2	TAN/BLK		BD2			
			BC3			BD3			
			BC4			BD4			
			BC5			BD5			
			BC6		WHT/BLK	BD6	DIAGNOSTIC DLC	12.0	13.6
			BC7			BD7			
0*	B+	MIL	BC8	BRN/WHT		BD8			
B+	"ON" 0* "OFF" B+	CLG FAN RELAY	BC9	LT GRN/BLK		BD9			
B+	"ON" 0* "OFF" B+	TEMP/CHECK GAGES LAMP	BC10	DK GRN		BD10			
B+	"ON" 0* "OFF" B+	A/C CLUTCH RELAY	BC11	DK GRN/WHT		BD11			
			BC12		ORN	BD12	DLC SERIAL DATA	4.2	4.5
4.9	4.8	KNOCK SIGNAL	BC13	DK BLU	YEL	BD13	VSS "HIGH"	0*	0*
			BC14		PPL	BD14	VSS "LOW"	0*	0*
			BC15			BD15			
			BC16		PNK/BLK	BD16	IGNITION	B+	B+

NOTICE: PA1 = Pink Connector, cavity A1, etc. BC1-Blue Connector, cavity C1, etc.

ENGINE - 2.3L/L40
CARLINE - "N"

* LESS THAN .5 VOLT.
1. VARIES FROM .60 TO BATTERY VOLTAGE DEPENDING ON POSITION OF DRIVE WHEELS.
2. VARIES.
3. B+ FIRST TWO SECONDS.

2.3L (VIN D, A & 3) ENGINE — ECM CONNECTOR END VIEW — 1993–94 ACHIEVA AND GRAND AM

MULTIPORT FUEL INJECTION ECM CONNECTOR IDENTIFICATION

This ECM voltage chart is for use with a digital voltmeter to further aid in diagnosis. The voltages you get may vary due to low battery charge or other reasons, but they should be very close. The "B +" symbol indicates a nominal system voltage of 12-14 volts.

THE FOLLOWING CONDITIONS MUST BE MET BEFORE TESTING:
- Engine at operating temperature (upper rad. hose hot) • Engine idling in "Closed Loop" (For "Engine Run" column) in park or neutral • Test terminal not grounded • Tech 1 not installed

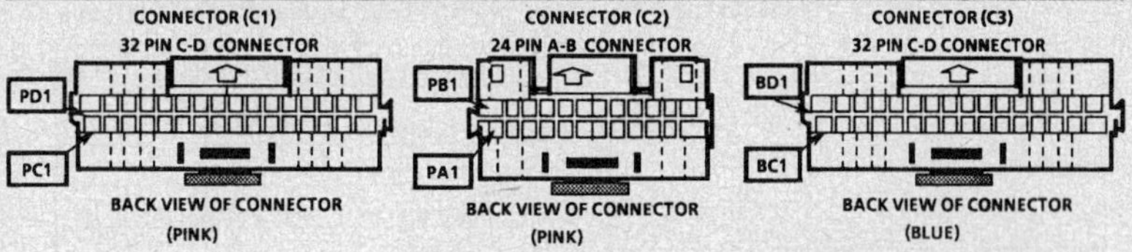

NOTICE: C, D CAVITY IDENTIFICATION IS VISIBLE ON CONNECTOR BODY WITH TERMINAL RETAINERS REMOVED.

CONNECTOR 1 (C1) — PINK 32 PIN C-D CONNECTOR

VOLTAGE / VOLTAGE

KEY "ON"	ENG. RUN	CIRCUIT	PIN	WIRE COLOR	WIRE COLOR	PIN	CIRCUIT	KEY "ON"	ENG. RUN
			PC1		BLK/RED	PD1	IGN REF LO	0*	0*
			PC2		LT GRN	PD2	IGN 7X REF	0*	.6 (2)
NOT	USABLE	IAC "A" HIGH	PC3	LT BLU/WHT	GRY/RED	PD3	A/C REFRIGERANT PRESSURE SIGNAL	.8	.9 (2)
NOT	USABLE	IAC "A" LOW	PC4	LT BLU/BLK					
NOT	USABLE	IAC "B" LOW	PC5	LT GRN/BLK	LT GRN	PD4	MAP SIGNAL	3.8	1.3 (2)
NOT	USABLE	IAC "B" HIGH	PC6	LT GRN/WHT		PD5			
			PC7			PD6			
			PC8			PD7			
3.6	1.5 (2)	IAT SIGNAL	PC9	TAN		PD8			
1.2 (2)	2.0 (2)	ECT SIGNAL	PC10	YEL		PD9			
			PC11			PD10			
B+ (3)	B+	INJ DRIVER 2 & 3	PC12	DK GRN		PD11			
0*	0*	ECM GROUND	PC13	BLK/WHT	LT GRN	PD12	A/C REQUEST	0*	"ON" B+ "OFF" 0*
			PC14			PD13			
0*	.5 (2)	IGNITION CONTROL (IC) 2 SIGNAL	PC15	BLK		PD14			
						PD15			
B+ (3)	B+	INJ DRIVER 1 & 4	PC16	DK BLU		PD16			

NOTICE: PC1 - PINK CONNECTOR, CAVITY C1; BC1 - BLUE CONNECTOR, CAVITY C1, ETC.

(1) VARIES FROM .60 TO BATTERY VOLTAGE DEPENDING ON POSITION OF DRIVE WHEELS.
(2) VARIES.
(3) B+ FIRST TWO SECONDS.
(4) VARIES WITH TEMPERATURE.
(5) NON A/C VEHICLES 0* VOLT, A/C VEHICLES SHOULD NOT HAVE A WIRE IN TERMINAL "BC3".
(6) LD2 ONLY.
* ALL VOLTAGES SHOWN "0" SHOULD READ LESS THAN .5 VOLT.

ENGINE - 2.3L/LD2/LGO/L40
CARLINE - "N"

2.3L (VIN D, A & 3) ENGINE — ECM CONNECTOR END VIEW — 1993–94 ACHIEVA AND GRAND AM

CONNECTOR 2 (C-2) — PINK 24 PIN A-B CONNECTOR

VOLTAGE KEY "ON"	VOLTAGE ENG. RUN	CIRCUIT	PIN	WIRE COLOR	WIRE COLOR	PIN	CIRCUIT	VOLTAGE KEY "ON"	VOLTAGE ENG. RUN
0*	0*	ECM TO ENG GND	PA1	TAN/WHT	BLK/ORN	PB1	IAT, MAP GROUND	0*	0*
0*	0*	ECM GROUND	PA2	BLK/WHT	BLK	PB2	TP, ECT, AC GND	0*	0*
5.0	5.0	5V REFERENCE	PA3	GRY	WHT	PB3	IGNITION CONTROL (IC)1 SIGNAL	0*	.5 (2)
5.0	5.0	5V REFERENCE	PA4	GRY		PB4			
B+	B+	BATTERY	PA5	RED	BLK/WHT	PB5	NON AC PROGRAM	0*	0*
			PA6		TAN	PB6	O2S GROUND	0*	0*
			PA7		DK BLU	PB7	TP SIGNAL	.49	.5(2)
B+(3)	B+	FUEL PUMP	PA8	GRN/WHT	DK GRN	PB8	4000 P/ML SPEED	8.3 (2)	9.4
			PA9			PB9			
			PA10			PB10			
			PA11		WHT	PB11	TACH	4.9	1.0
.32(2)	.1-.9 (2)	O2S SIGNAL	PA12	PPL		PB12			

BLUE 32 PIN C-D CONNECTOR 3 (C-3)

VOLTAGE KEY "ON"	VOLTAGE ENG. RUN	CIRCUIT	PIN	WIRE COLOR	WIRE COLOR	PIN	CIRCUIT	VOLTAGE KEY "ON"	VOLTAGE ENG. RUN
.3	12.0	EVAP SIGNAL	BC1	GRN/YEL	ORN/BLK	BD1	PNP SWITCH SIGNAL	ON 0* OFF B+	ON 0* OFF B+
0	0	TCC/SHIFT LT	BC2	TAN/BLK	WHT	BD2	2nd GEAR SWITCH	ON 0* OFF B+	ON 0* OFF B+ (6)
			BC3		DK GRN/WHT	BD3	3rd GEAR SWITCH	ON 0* OFF B+	ON 0* OFF B+ (6)
			BC4			BD4			
			BC5			BD5			
			BC6		WHT/BLK	BD6	DIAGNOSTIC DLC	12.0	13.6
			BC7			BD7			
0*	B+	"MIL"	BC8	BRN/WHT		BD8			
B+	ON 0* OFF B+	COOLANT FAN RELAY	BC9	LT GRN/BLK		BD9			
B+	ON 0* OFF B+	TEMP/CHECK GAUGES LAMP	BC10	DK GRN		BD10			
B+	ON 0* OFF B+	A/C CLUTCH RELAY	BC11	DK GRN/WHT		BD11			
			BC12		ORN	BD12	DLC SERIAL DATA	4.2	4.5
4.9	4.8	KNOCK SIGNAL	BC13	DK BLU	YEL	BD13	VSS "HI"	0*	0*
			BC14		PPL	BD14	VSS "LO"	0*	0*
			BC15			BD15			
			BC16		PNK/BLK	BD16	IGNITION	B+	B+

NOTICE: PA1 = Pink Connector, cavity A1, etc.   BC1 - Blue Connector, cavity C1, etc.

* ALL VOLTAGES SHOWN "0" SHOULD READ LESS THAN .5 VOLT.
1. VARIES FROM .60 TO BATTERY VOLTAGE DEPENDING ON POSITION OF DRIVE WHEELS.
2. VARIES.
3. B+ FIRST TWO SECONDS.
4. VARIES WITH TEMPERATURE.
5. NON A/C VEHICLES 0* VOLT, A/C VEHICLES SHOULD NOT HAVE A WIRE IN TERMINAL "BC3".
6. LD2 ONLY.

ENGINE - 2.3L / LD2/LG0/L40
CARLINE - "N"

2.3L (VIN D, A & 3) ENGINE — ON-BOARD DIAGNOSTIC SYSTEM CHART — 1993–94 ACHIEVA, GRAND AM AND SKYLARK

ON-BOARD DIAGNOSTIC (OBD) SYSTEM CHECK

The On-Board Diagnostic (OBD) system check is an organized approach to identifying a problem created by an electronic engine control system malfunction. It must be the starting point for any driveability complaint diagnosis, because it directs the service technician to the next logical step in diagnosing the complaint.

The Tech 1 data listed in the table may be used for comparison, after completing the OBD system check and finding the on-board diagnostics functioning properly and no diagnostic trouble codes displayed. The "Typical Values" are an average of display values recorded from normally operating vehicles and are intended to represent what a normally functioning system would typically display.

A SCAN TOOL THAT DISPLAYS FAULTY DATA SHOULD NOT BE USED, AND THE PROBLEM SHOULD BE REPORTED TO THE MANUFACTURER. THE USE OF A FAULTY SCAN CAN RESULT IN MISDIAGNOSIS AND UNNECESSARY PARTS REPLACEMENT.

Only the parameters listed below are used in this manual for diagnosing. If a scan reads other parameters, the values are not recommended by General Motors for use in diagnosing.

If all values are within the range illustrated, refer to "Symptoms,"

TECH 1 DATA
Idle / Upper Radiator Hose Hot / Closed Throttle / Park or Neutral / "Closed Loop" / Acc. "OFF"

Scan Position	Units Displayed	Typical Data Value
Engine Speed	RPM	± 100 RPM from desired RPM (± 50 in drive)
Desired Idle	RPM	ECM idle command (varies with calibration. temp.)
Eng. Coolant Temp.	C° F°	85°–115°C
Intake Air Temp.	C° F°	10°–80°C (depends on underhood temp.)
MAP	kPa, V	1–3 volts (depends on vacuum & baro pressure)
BARO	kPa, V	3–5 volts (depends on altitude & baro pressure)
Throt Position	Volts	.200–.900 (up to 5.0 at wide open throttle)
Throttle Angle	0–100%	0% (up to 100% at wide open throttle)
O2S	mV	1–1000 and varying
Injector Pulse Width	m Sec.	1–4 and varying
Spark Advance	# of Degrees	Varies
Lo Oct. Knock Ret.	# of Degrees	Varies
S.T. Fuel Trim	Counts	Varies
L.T. Fuel Trim	Counts	58–198 (see Section "C2")
Loop Status	Open/Closed	"Closed Loop" (may go open with extended idle)
Fuel Trim Cell	Cell Number	18-21 at idle (depends on air flow, RPM, P/N & A/C)
Knock Retard	Degrees of Retard	0
Knock Signal	Yes/No	No
KS Activity	0-255 Counts	0-255 (usable for DTC 43)
7x Resync Counter	0-255 Counts	0 (usable for DTC 19)
Ign. Control 7x Ref.	0-255 Counts	0-255 (usable for DTC 19)
Ign Control 7x Syn.	"ON/OFF"	"ON" (usable for DTC 19 and for misfire diagnosis)
Idle Air Control	Counts (steps)	5–60
Park/Neutral Position	PN and RDL	P - N -- (or -R-DL manual only)
VSS	Mph and km/h	0
TCC/Shift Light	"ON/OFF"	"OFF" ("ON," with TCC commanded)
System Voltage	Volts	13.5 – 14.5
2nd Gear	Yes/No (N/A L40 only)	No (yes in 2nd or 3rd gear) or yes (man. trans. only)
3rd Gear	Yes/No (N/A L40 only)	No (yes in 3rd gear) or yes (man. trans. only)
A/C Request	Yes/No	No (yes, with A/C requested, ie: selector "ON")
A/C Clutch	"ON/OFF"	"OFF" ("ON," with A/C commanded "ON")
A/C Clutch	"ON/OFF"	"OFF" ("ON," with A/C commanded "ON")
A/C Ref. Pressure	psi/Volts	0–450 psi (varying with high side refrigerant pressure)
Cool Fan Relay	"ON/OFF"	"OFF" ("ON," with A/C "ON" or hot eng)
Fuel EVAP Purge	%	0–100%
QDM A	Low/Hi	Low
QDM B	Low/Hi	Low
QDM C	Low/Hi	Low
QDM D	Low/Hi	Low
Calibration ID	#	Production ECM/Calibration ID (not usable)
Time From Start	Min/Sec	Varies (engine run time since start)

2.3L (VIN D, A & 3) ENGINE — ON-BOARD DIAGNOSTIC SYSTEM CHART — 1993–94 ACHIEVA, GRAND AM AND SKYLARK

2.3L (VIN D, A & 3) ENGINE — ON-BOARD DIAGNOSTIC SYSTEM CHART — 1993–94 ACHIEVA, GRAND AM AND SKYLARK

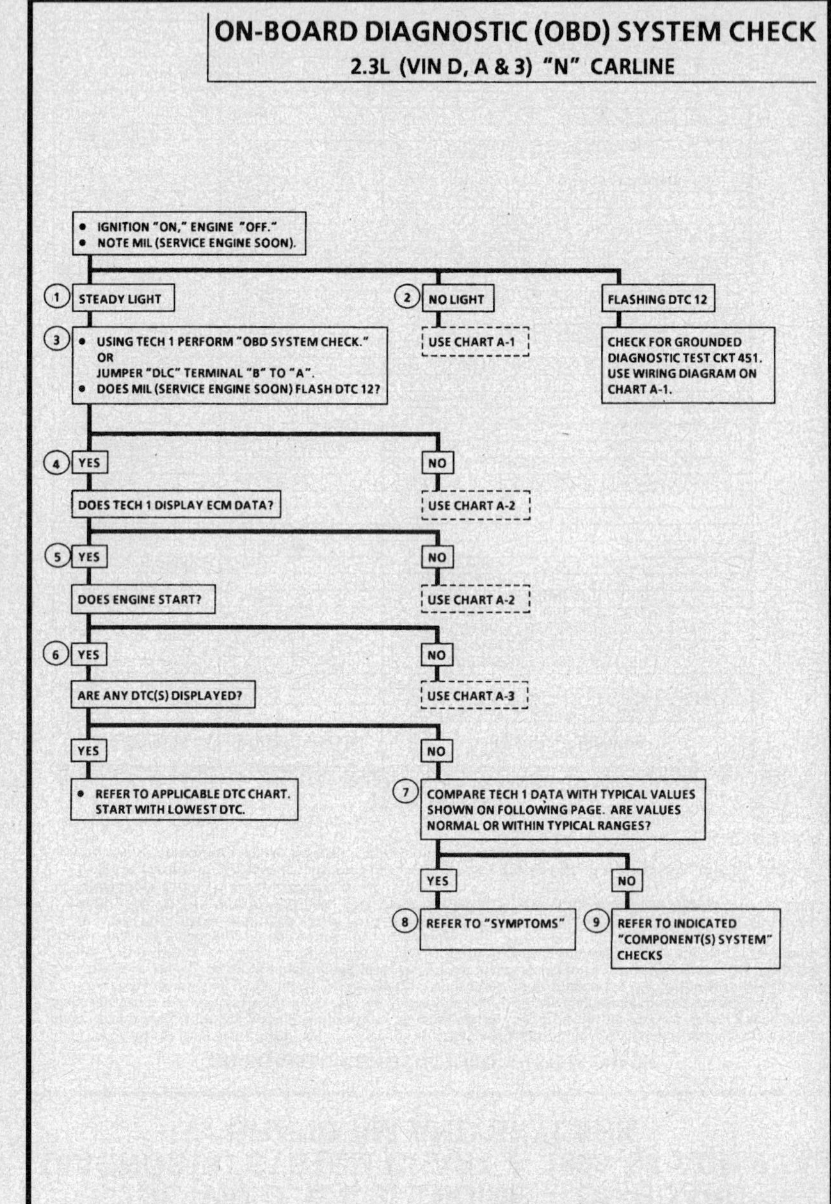

ON-BOARD DIAGNOSTIC (OBD) SYSTEM CHECK
2.3L (VIN D, A & 3) "N" CARLINE

Circuit Description:

The On-Board Diagnostic (OBD) system check is an organized approach to identifying a problem created by an electronic engine control system malfunction. It must be the starting point for any driveability complaint diagnosis, because it directs the service technician to the next logical step in diagnosing the complaint. Understanding the chart and using it correctly will reduce diagnostic time and prevent the unnecessary replacement of good parts.

Test Description: Number(s) below refer to circled number(s) on the diagnostic chart.

1. This step is a check for the proper operation of the Malfunction Indicator Lamp (MIL). The MIL should be "ON" steady.
2. No MIL at this point indicates that there is a problem with the MIL circuit or the Engine Control Module (ECM) control of that circuit.
3. This test checks the ability of the ECM to control the MIL. With the diagnostic terminal grounded, the MIL should flash a DTC 12 three times, followed by any DTC stored in memory.
4. Most of the procedures use a Tech 1 scan tool to aid diagnosis, therefore, serial data must be available. If a PROM error is present, the ECM may have been able to flash DTC 12/51, but not enable serial data.
5. Although the ECM is powered up, a "Cranks But Will Not Run" symptom could exist because of an ECM or system problem.

6. This step will isolate if the customer complaint is a MIL or a driveability problem with no MIL. An invalid code may be the result of a faulty scan tool.
7. Comparison of actual control system data with the typical values is a quick check to determine if any parameter is not within limits. Keep in mind that a base engine problem (i.e., advanced cam timing) may substantially alter sensor values.
8. Installation of a Tech 1 scan tool will provide a good ground path for the ECM and may hide a driveability complaint due to poor ECM grounds.
9. If the actual data is not within the typical values established, the charts "Component Systems," will provide a functional check of the suspect component or system.

ON-BOARD DIAGNOSTIC (OBD) SYSTEM CHECK
2.3L (VIN D, A & 3) "N" CARLINE

2.3L (VIN D, A & 3) ENGINE — SYSTEM DIAGNOSTIC CHARTS — 1993–94 ACHIEVA, GRAND AM AND SKYLARK

CHART A-1
NO MALFUNCTION INDICATOR LAMP (MIL) "SERVICE ENGINE SOON"
2.3L (VIN D, A & 3) "N" CARLINE

2.3L (VIN D, A & 3) ENGINE — SYSTEM DIAGNOSTIC CHARTS — 1993–94 ACHIEVA, GRAND AM AND SKYLARK

CHART A-1
NO MALFUNCTION INDICATOR LAMP (MIL) "SERVICE ENGINE SOON"
2.3L (VIN D, A & 3) "N" CARLINE

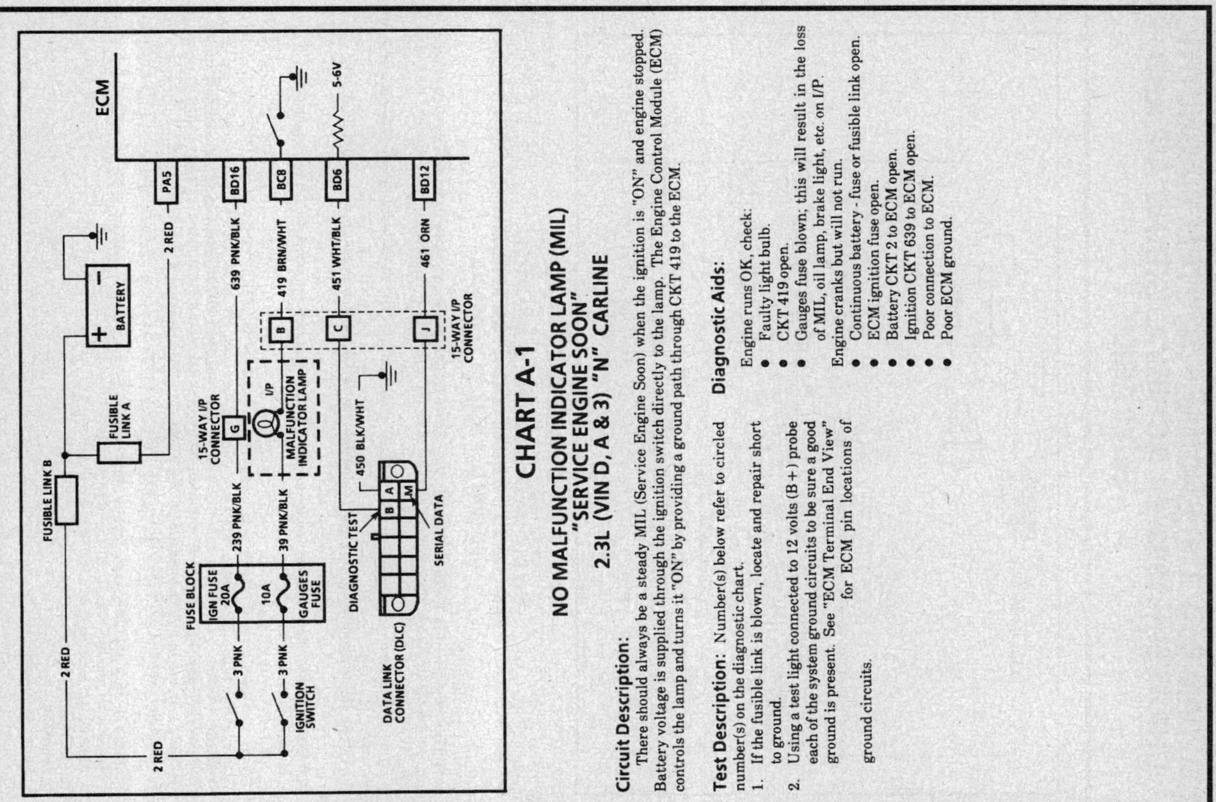

Circuit Description:

There should always be a steady MIL (Service Engine Soon) when the ignition is "ON" and engine stopped. Battery voltage is supplied through the ignition switch directly to the lamp. The Engine Control Module (ECM) controls the lamp and turns it "ON" by providing a ground path through CKT 419 to the ECM.

Test Description: Number(s) below refer to circled number(s) on the diagnostic chart.

1. If the fusible link is blown, locate and repair short to ground.
2. Using a test light connected to 12 volts (B+) probe each of the system ground circuits to be sure a good ground is present. See "ECM Terminal End View" for ECM pin locations of ground circuits.

Diagnostic Aids:

Engine runs OK, check:
- Faulty light bulb.
- CKT 419 open.
- Gauges fuse blown; this will result in the loss of MIL, oil lamp, brake light, etc. on I/P.

Engine cranks but will not run.
- Continuous battery - fuse or fusible link open.
- ECM ignition fuse open.
- Battery CKT 2 to ECM open.
- Ignition CKT 639 to ECM open.
- Poor connection to ECM.
- Poor ECM ground.

2.3L (VIN D, A & 3) ENGINE — SYSTEM DIAGNOSTIC CHARTS — 1993–94 ACHIEVA, GRAND AM AND SKYLARK

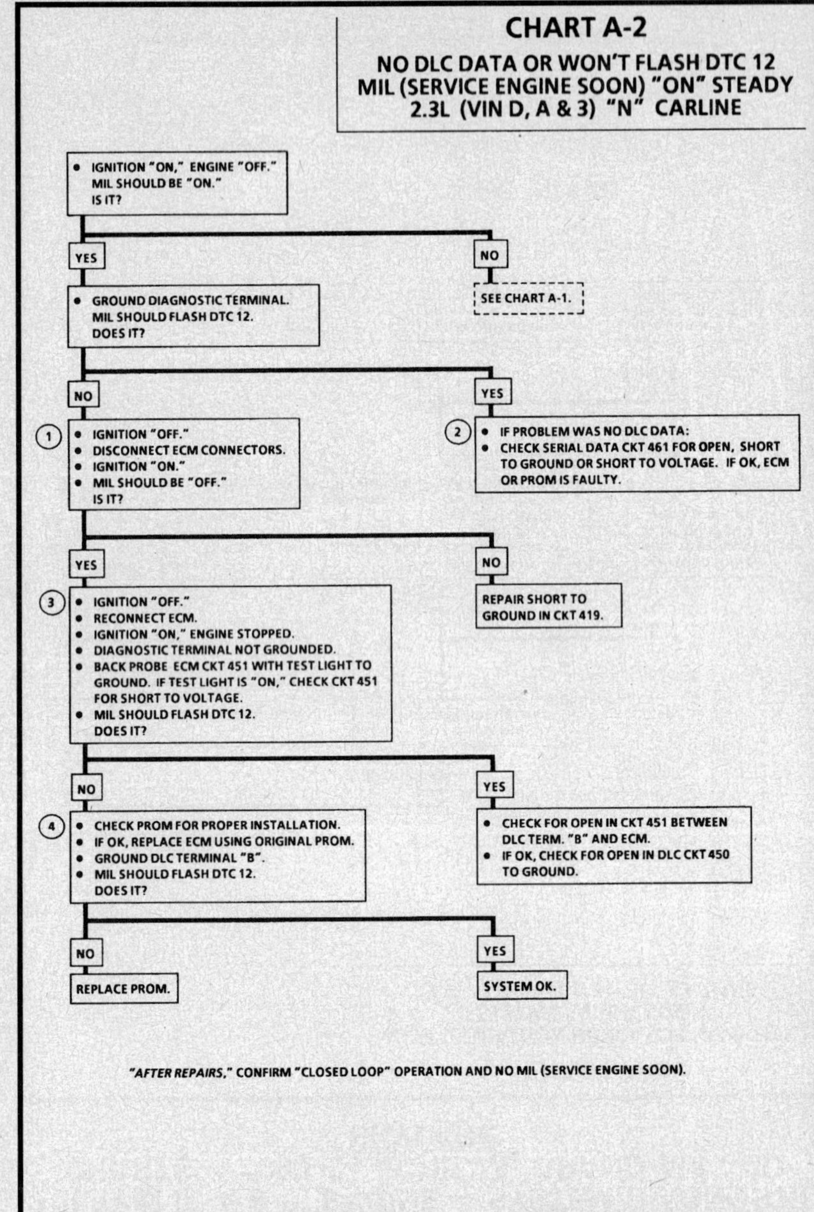

CHART A-2

NO DLC DATA OR WON'T FLASH DTC 12
MIL (SERVICE ENGINE SOON) "ON" STEADY
2.3L (VIN D, A & 3) "N" CARLINE

Circuit Description:

There should always be a steady MIL when the ignition is "ON" and engine stopped. Battery ignition voltage is supplied to the light bulb. The Engine Control Module (ECM) will turn the MIL "ON" by grounding CKT 419 at the ECM.

With the diagnostic terminal grounded, the light should flash a DTC 12, followed by any DTCs stored in memory.

A steady MIL suggests a short to ground in the light control CKT 419, or an open in diagnostic CKT 451.

Test Description: Number(s) below refer to circled number(s) on the diagnostic chart.

1. MIL "OFF" with CKT 419 disconnected from ECM indicates that ground circuit was completed through the ECM, not through external short to ground.
2. If there is a problem with the ECM that causes a Tech 1 scan tool to not read serial data, the ECM should not flash a DTC 12. If DTC 12 is flashing, check for CKT 451 short to ground. If DTC 12 does flash, be sure that the scan tool is working properly on another vehicle. If the scan tool is functioning properly and CKT 461 is OK, the PROM or ECM may be at fault for the "NO DLC" symptom.
3. This step will check for an open diagnostic CKT 451.
4. At this point, the MIL wiring is OK. The problem is a faulty ECM or PROM. If DTC 12 does not flash, the ECM should be replaced using the original PROM. Replace the PROM only after trying an ECM, as a defective PROM is an unlikely cause of the problem.

2.3L (VIN D, A & 3) ENGINE — SYSTEM DIAGNOSTIC CHARTS — 1993–94 ACHIEVA, GRAND AM AND SKYLARK

CHART A-2

NO DLC DATA OR WON'T FLASH DTC 12
MIL (SERVICE ENGINE SOON) "ON" STEADY
2.3L (VIN D, A & 3) "N" CARLINE

- IGNITION "ON," ENGINE "OFF."
  MIL SHOULD BE "ON."
  IS IT?

YES → GROUND DIAGNOSTIC TERMINAL. MIL SHOULD FLASH DTC 12. DOES IT?

NO → SEE CHART A-1.

(1)
- IGNITION "OFF."
- DISCONNECT ECM CONNECTORS.
- IGNITION "ON."
- MIL SHOULD BE "OFF."
  IS IT?

(2)
- IF PROBLEM WAS NO DLC DATA:
- CHECK SERIAL DATA CKT 461 FOR OPEN, SHORT TO GROUND OR SHORT TO VOLTAGE. IF OK, ECM OR PROM IS FAULTY.

(3)
- IGNITION "OFF."
- RECONNECT ECM.
- IGNITION "ON," ENGINE STOPPED.
- DIAGNOSTIC TERMINAL NOT GROUNDED.
- BACK PROBE ECM CKT 451 WITH TEST LIGHT TO GROUND. IF TEST LIGHT IS "ON," CHECK CKT 451 FOR SHORT TO VOLTAGE.
- MIL SHOULD FLASH DTC 12.
  DOES IT?

NO → REPAIR SHORT TO GROUND IN CKT 419.

(4)
- CHECK PROM FOR PROPER INSTALLATION.
- IF OK, REPLACE ECM USING ORIGINAL PROM.
- GROUND DLC TERMINAL "B."
- MIL SHOULD FLASH DTC 12.
  DOES IT?

- CHECK FOR OPEN IN CKT 451 BETWEEN DLC TERM. "B" AND ECM.
- IF OK, CHECK FOR OPEN IN DLC CKT 450 TO GROUND.

NO → REPLACE PROM.

YES → SYSTEM OK.

"AFTER REPAIRS," CONFIRM "CLOSED LOOP" OPERATION AND NO MIL (SERVICE ENGINE SOON).

2.3L (VIN D, A & 3) ENGINE — SYSTEM DIAGNOSTIC CHARTS — 1993–94 ACHIEVA, GRAND AM AND SKYLARK

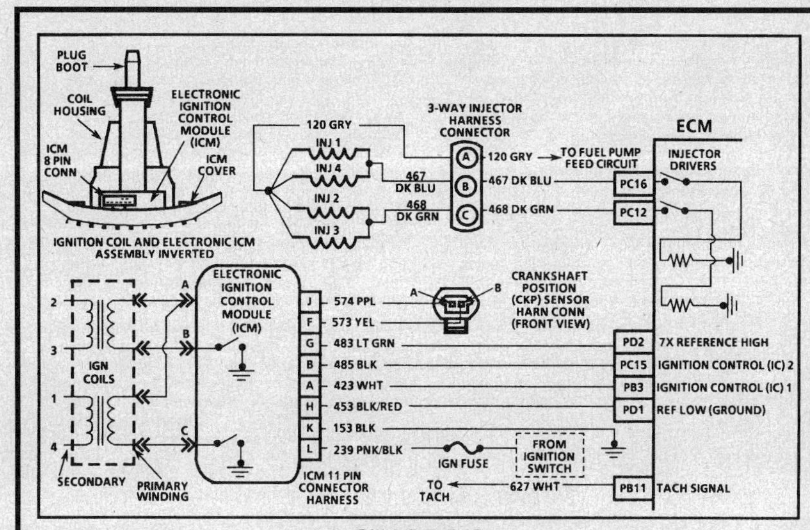

CHART A-3
(Page 1 of 4)
ENGINE CRANKS BUT WILL NOT RUN
2.3L (VIN D, A & 3) "N" CARLINE

Condition:

Engine cranks but will not run, or engine may start, but immediately stops running. Battery condition and engine cranking speed are OK and there is adequate fuel in the tank.

Circuit Description:

This engine is equipped with a distributorless ignition system called the Electronic Ignition (EI) system. The primary circuit of the EI system consists of two separate ignition coils, an Electronic Ignition Control Module (ICM) and Crankshaft Position (CKP) sensor as well as the related connecting wires and the Ignition Control (IC) portion of the ECM. Each secondary circuit consists of the secondary winding of the coil, two connecting metal strips molded into the coil housing, the spark plug boot/connector assemblies and spark plugs.

Test Description: Number(s) below refer to circled number(s) on the diagnostic chart.
1. This step verifies that MIL operation, on-board diagnostics, cranking RPM, TP sensor, MAP sensor and engine coolant sensor signals are normal. A blinking injector test light verifies that the ECM is receiving the 7X reference signal and is attempting to activate the injectors.
2. This step determines whether harness or injector is cause of incorrect resistance. Nominal injector resistance is 1.9 to 2.1 ohms at 60°C (140°F). Resistance will increase slightly at higher temperatures.
3. By installing spark plug jumper leads and testing for spark on all 4 plug leads, each ignition coil's ability to produce at least 25,000 volts is verified.

4. Checks to see if fuel pump and relay are operating correctly (fuel pump only "ON" 2-3 seconds) and fuel pressure is within proper range.
5. If module can make the test light blink, the fault is coil harness or connections. If not, module or it's connections are faulty.

Diagnostic Aids:

Check for:
- TP sensor binding or sticking in Wide Open Throttle (WOT) position or intermittently shorted or open.
- Water or foreign material in fuel.
- Low compression (Timing chain failure).
- Verify that only <u>resistor</u> spark plugs are used.

2.3L (VIN D, A & 3) ENGINE — SYSTEM DIAGNOSTIC CHARTS — 1993–94 ACHIEVA, GRAND AM AND SKYLARK

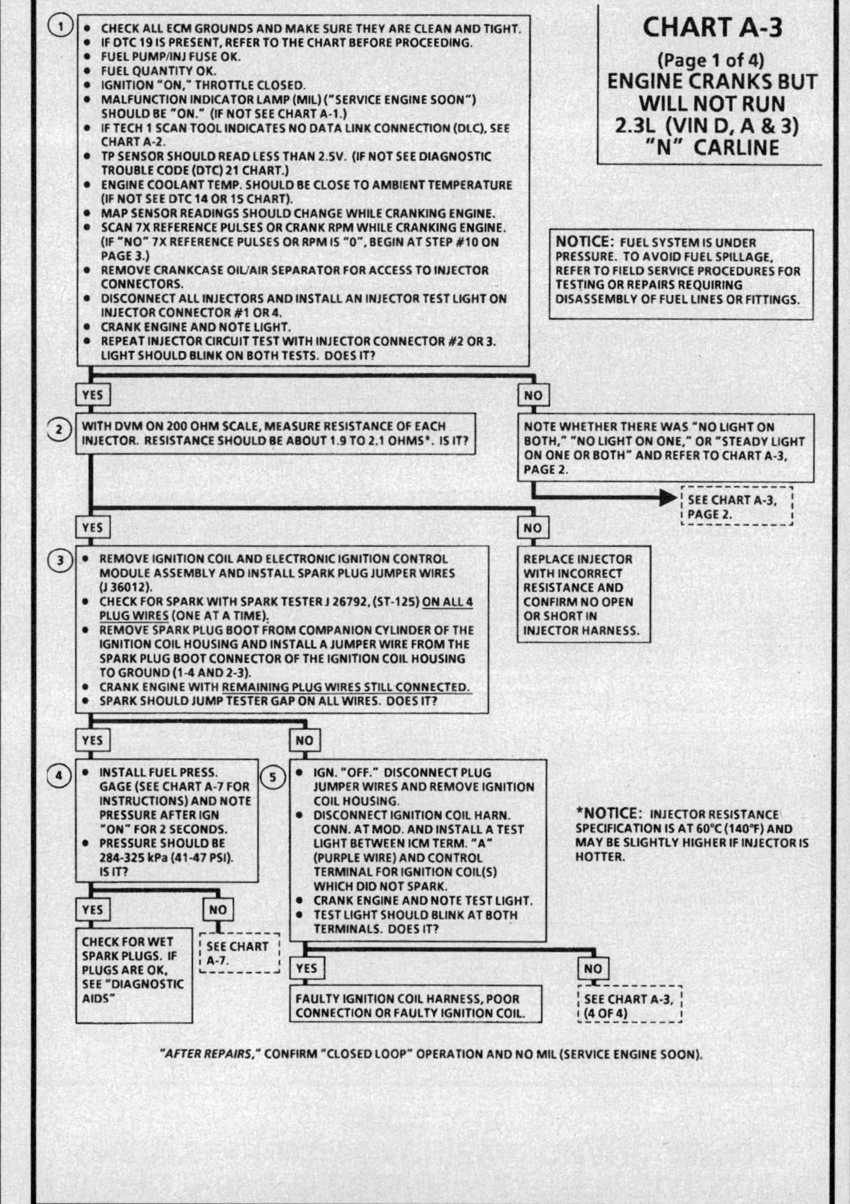

2.3L (VIN D, A & 3) ENGINE — SYSTEM DIAGNOSTIC CHARTS — 1993–94 ACHIEVA, GRAND AM AND SKYLARK

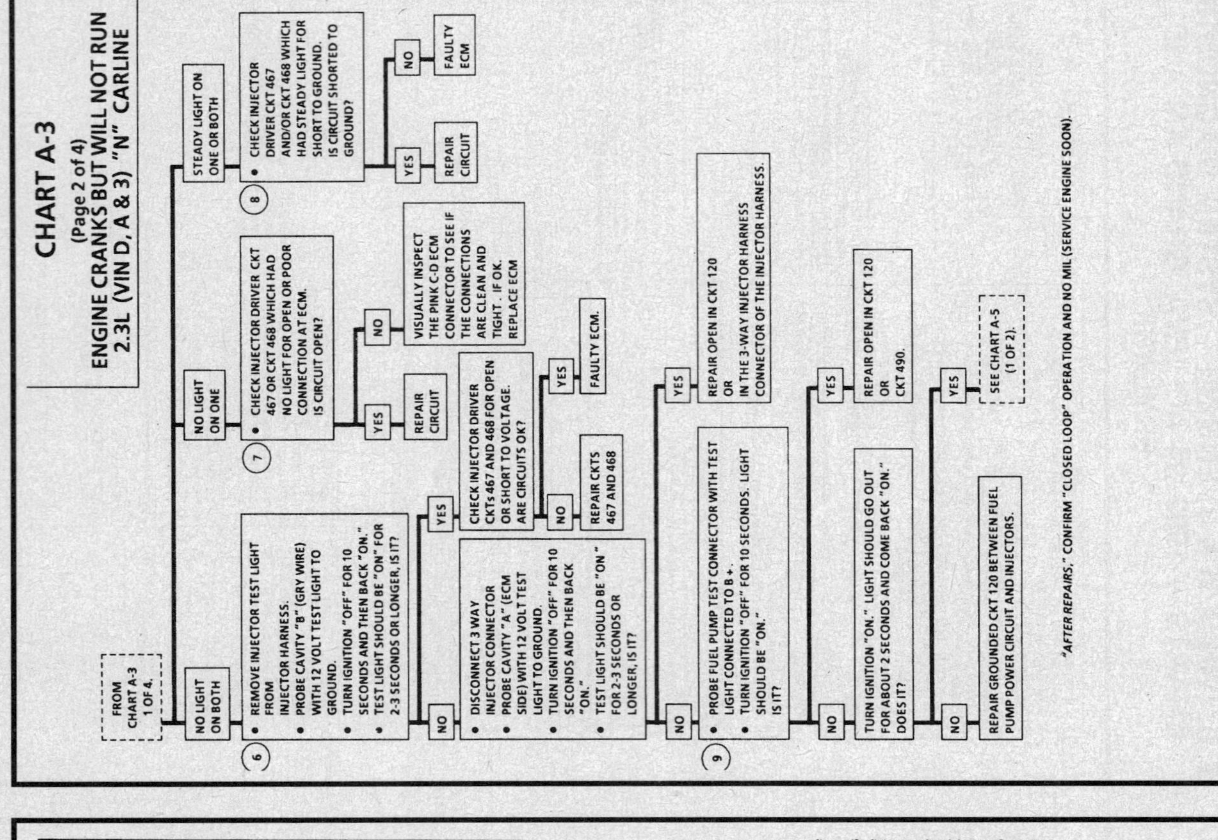

CHART A-3
(Page 2 of 4)
ENGINE CRANKS BUT WILL NOT RUN
2.3L (VIN D, A & 3) "N" CARLINE

2.3L (VIN D, A & 3) ENGINE — SYSTEM DIAGNOSTIC CHARTS — 1993–94 ACHIEVA, GRAND AM AND SKYLARK

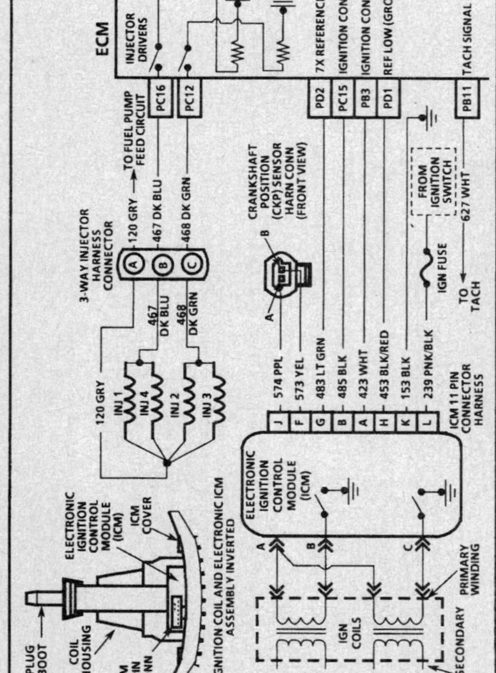

CHART A-3
(Page 2 of 4)
ENGINE CRANKS BUT WILL NOT RUN
2.3L (VIN D, A & 3) "N" CARLINE

Condition:

Engine cranks but will not run, or engine may start, but immediately stops running. Battery condition and engine cranking speed are OK and there is adequate fuel in the tank.

Circuit Description:

This engine is equipped with a distributorless ignition system called the Electronic Ignition (EI) system. The primary circuit of the EI system consists of two separate ignition coils, an Electronic Ignition Control Module, (ICM) and Crankshaft Position (CKP) sensor as well as the related connecting wires and the Ignition Control (IC) portion of the ECM. Each secondary circuit consists of the secondary winding of the coil, two connecting metal strips molded into the coil housing, the spark plug boot/connector assemblies and spark plugs.

Test Description: Number(s) below refer to circled number(s) on the diagnostic chart.

6. Battery voltage should be available at CKT 120 whenever the fuel pump power feed circuit is switched "ON." The ECM should switch the fuel pump "ON" for 2-3 seconds after ignition is turned "ON" (and when ECM is receiving ignition reference pulses, as while cranking or running). The ignition must be turned "OFF" for at least 10 seconds to assure that the ECM powers down and will then switch the fuel pump back "ON" for 2-3 seconds when ignition is turned back "ON."

7. Light "ON" one circuit only indicates power is available at cavity "A", but grounded circuit is not being completed on the other circuit. This could be due to open circuit or ECM not switching the injector driver circuit to ground.

8. Steady light indicates ground circuit is always completed and is not being switched. This could be due to short to ground in circuit, or faulty ECM injector driver.

9. The fuel pump should be switched "ON" by the ECM for 2-3 seconds after ignition is first turned "ON." It is necessary to turn the ignition "OFF" for at least 10 seconds to assure that the ECM powers down and will then switch the fuel pump back "ON." If the fuel pump operates, but power is not available at injector harness, circuit must be open. If the fuel pump does not operate, CHART A-5 should be used to diagnose the cause.

2.3L (VIN D, A & 3) ENGINE — SYSTEM DIAGNOSTIC CHARTS — 1993–94 ACHIEVA, GRAND AM AND SKYLARK

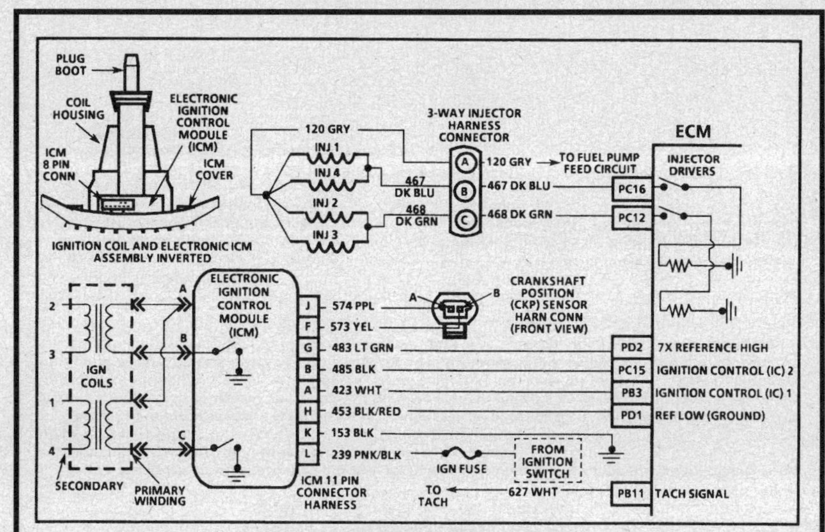

CHART A-3
(Page 3 of 4)
ENGINE CRANKS BUT WILL NOT RUN
2.3L (VIN D, A & 3) "N" CARLINE

Circuit Description:
The Electronic Ignition (EI) system uses a waste spark method of distribution. In this type of system, the Electronic Ignition Control Module (ICM) triggers the #1-4 coil pair resulting in both #1 and #4 spark plugs firing at the same time. #1 cylinder is on the compression stroke at the same time #4 is on the exhaust stroke, resulting in a lower energy requirement to fire #4 spark plug. This leaves the remainder of the high voltage to be used to fire #1 spark plug. On this application, the Crankshaft Position (CKP) sensor is mounted to, and protrudes through the block to within approximately 0.050" of the crankshaft reluctor. Since the reluctor is a machined portion of the crankshaft and the (CKP) sensor is mounted in a fixed position on the block, timing adjustments are not possible or necessary.

Test Description: Number(s) below refer to circled number(s) on the diagnostic chart.
10. Battery voltage should be available at terminal "L" of the electronic ICM 11 pin connector, and terminal "K" should be a good ground.
11. The CKP sensor should output a voltage as the crankshaft turns. If no voltage is produced, the indication is a poor sensor connection or faulty sensor.
12. The CKP sensors core is a magnet, therefore, it should be magnetized and the resistance should be within a range of 500 to 900 ohms.
13. The test light to 12 volts simulates a reference signal to the ECM which will result in an injector test light blink for every other touch of the test light, if CKT 483, the ECM and the injector driver circuit are all functioning properly.

2.3L (VIN D, A & 3) ENGINE — SYSTEM DIAGNOSTIC CHARTS — 1993–94 ACHIEVA, GRAND AM AND SKYLARK

CHART A-3
(Page 3 of 4)
ENGINE CRANKS BUT WILL NOT RUN
2.3L (VIN D, A & 3) "N" CARLINE

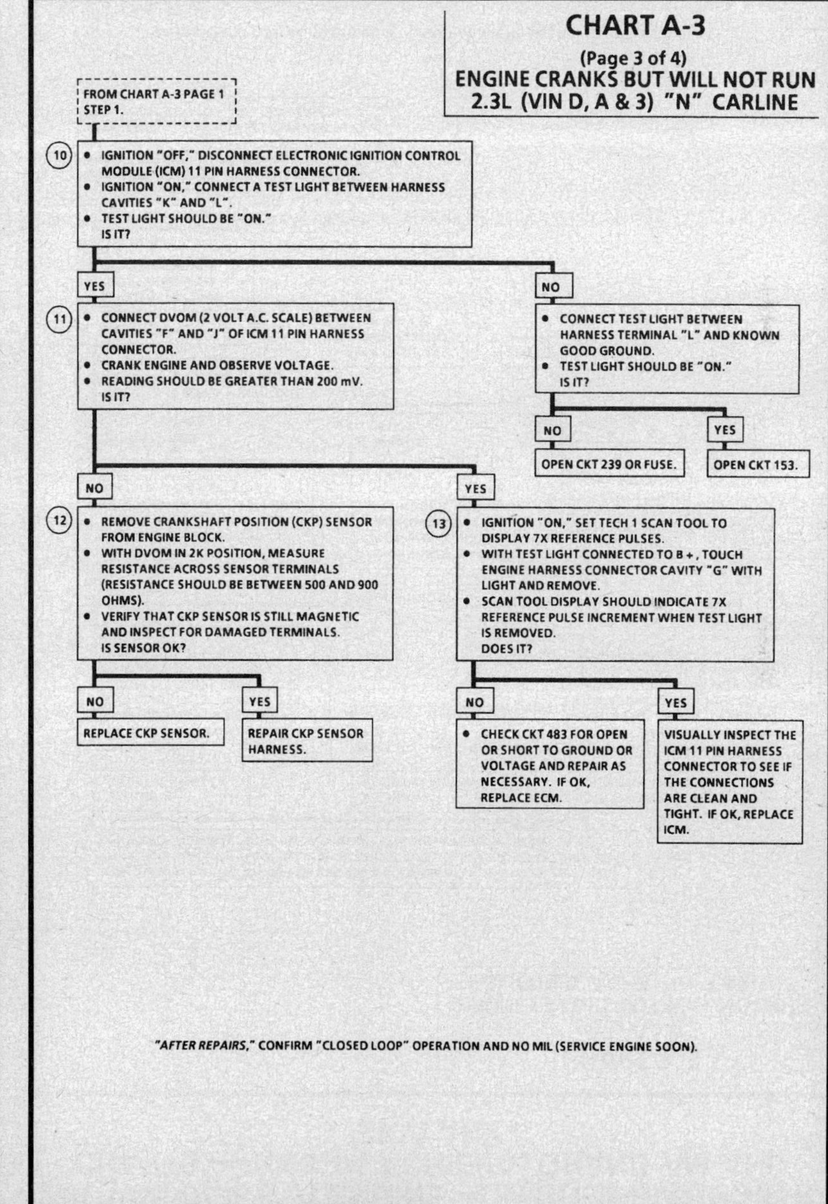

2.3L (VIN D, A & 3) ENGINE — SYSTEM DIAGNOSTIC CHARTS — 1993–94 ACHIEVA, GRAND AM AND SKYLARK

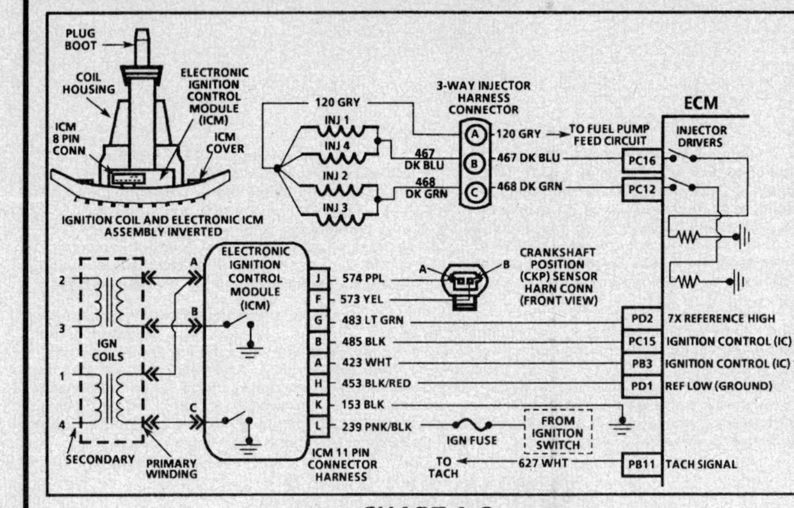

CHART A-3
(Page 4 of 4)
ENGINE CRANKS BUT WILL NOT RUN
2.3L (VIN D, A & 3) "N" CARLINE

Condition:
Engine cranks but will not run, or engine may start, but immediately stops running. Battery condition and engine cranking speed are OK and there is adequate fuel in the tank.

Circuit Description:
This engine is equipped with a distributorless ignition system called the Electronic Ignition (EI) system. The primary circuit of the EI system consists of two separate ignition coils, an Ignition Control Module (ICM) and Crankshaft Position (CKP) sensor as well as the related connecting wires and the Ignition Control (IC) portion of the Engine Control Module (ECM). Each secondary circuit consists of the secondary winding of the coil, two connecting metal strips molded into the coil housing, the spark plug boot/connector assemblies and spark plugs.

Test Description: Number(s) below refer to circled number(s) on the diagnostic chart.

14. This step verifies if there is a frequency output coming from the ECM.
15. This step identifies a short in the circuit.
16. This step will verify if the open circuit is in the ICM or circuit.
17. If there is a short to battery voltage there will be no frequency response. This step will show if there is a short to the battery.
18. This step will determine if the circuit or the ICM is shorted to ground.
19. If CKT 423 and CKT 485 are shorted together, there will not be a frequency signal. This step determines if CKT 423 and CKT 485 are shorted together.

2.3L (VIN D, A & 3) ENGINE — SYSTEM DIAGNOSTIC CHARTS — 1993–94 ACHIEVA, GRAND AM AND SKYLARK

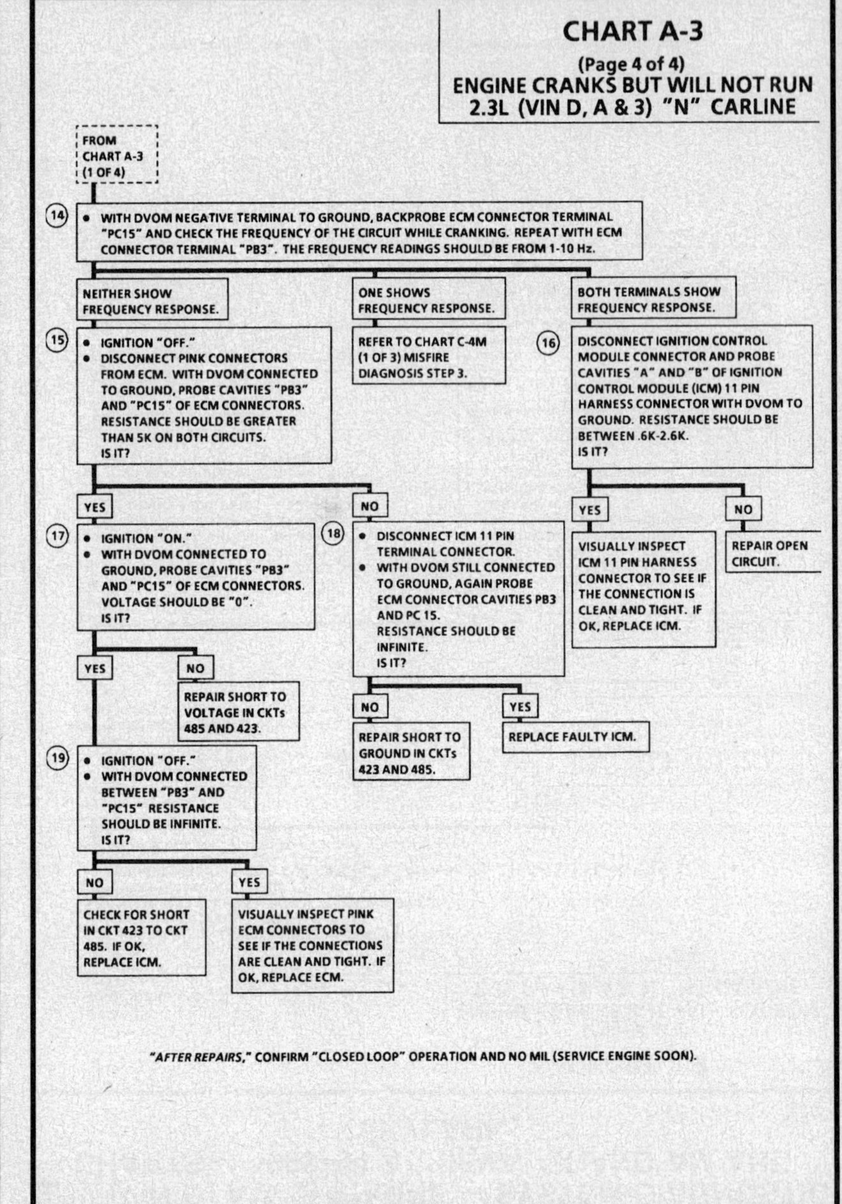

2.3L (VIN D, A & 3) ENGINE — SYSTEM DIAGNOSTIC CHARTS — 1993–94 ACHIEVA, GRAND AM AND SKYLARK

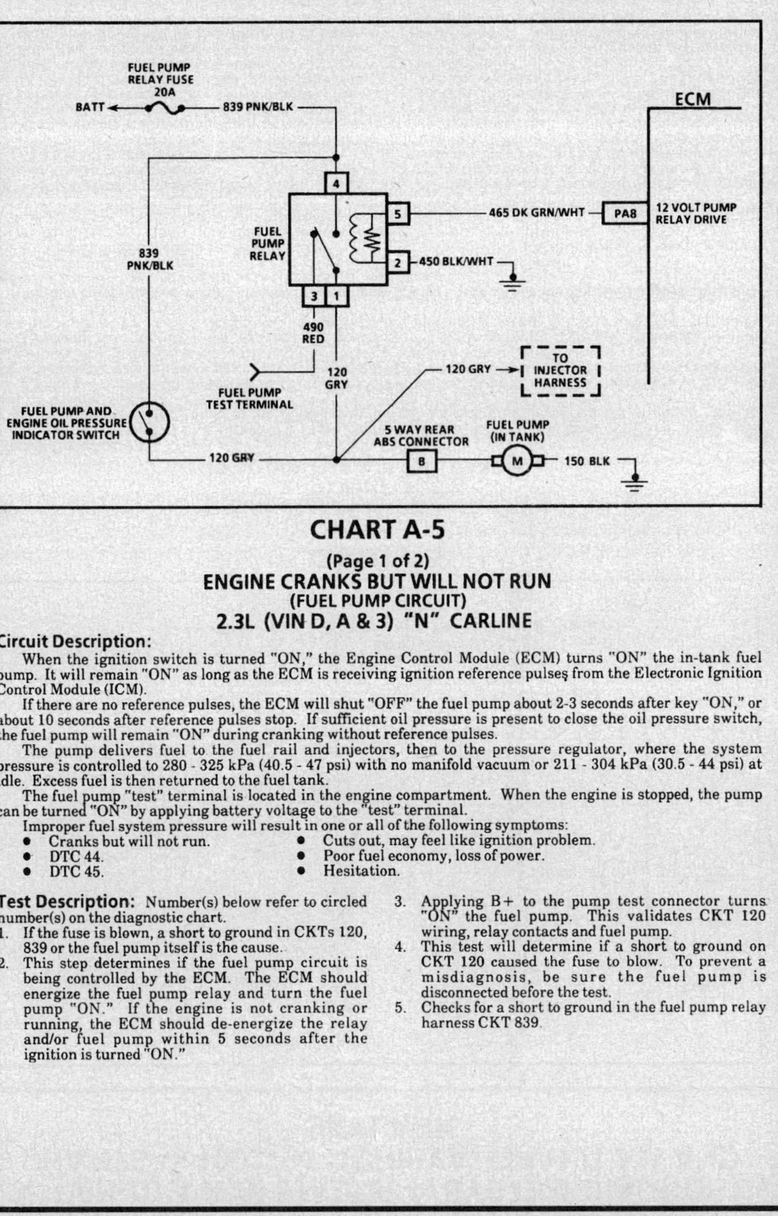

CHART A-5
(Page 1 of 2)
ENGINE CRANKS BUT WILL NOT RUN
(FUEL PUMP CIRCUIT)
2.3L (VIN D, A & 3) "N" CARLINE

Circuit Description:

When the ignition switch is turned "ON," the Engine Control Module (ECM) turns "ON" the in-tank fuel pump. It will remain "ON" as long as the ECM is receiving ignition reference pulses from the Electronic Ignition Control Module (ICM).

If there are no reference pulses, the ECM will shut "OFF" the fuel pump about 2-3 seconds after key "ON," or about 10 seconds after reference pulses stop. If sufficient oil pressure is present to close the oil pressure switch, the fuel pump will remain "ON" during cranking without reference pulses.

The pump delivers fuel to the fuel rail and injectors, then to the pressure regulator, where the system pressure is controlled to 280 - 325 kPa (40.5 - 47 psi) with no manifold vacuum or 211 - 304 kPa (30.5 - 44 psi) at idle. Excess fuel is then returned to the fuel tank.

The fuel pump "test" terminal is located in the engine compartment. When the engine is stopped, the pump can be turned "ON" by applying battery voltage to the "test" terminal.

Improper fuel system pressure will result in one or all of the following symptoms:
- Cranks but will not run.
- DTC 44.
- DTC 45.
- Cuts out, may feel like ignition problem.
- Poor fuel economy, loss of power.
- Hesitation.

Test Description: Number(s) below refer to circled number(s) on the diagnostic chart.
1. If the fuse is blown, a short to ground in CKTs 120, 839 or the fuel pump itself is the cause.
2. This step determines if the fuel pump circuit is being controlled by the ECM. The ECM should energize the fuel pump relay and turn the fuel pump "ON." If the engine is not cranking or running, the ECM should de-energize the relay and/or fuel pump within 5 seconds after the ignition is turned "ON."
3. Applying B+ to the pump test connector turns "ON" the fuel pump. This validates CKT 120 wiring, relay contacts and fuel pump.
4. This test will determine if a short to ground on CKT 120 caused the fuse to blow. To prevent a misdiagnosis, be sure the fuel pump is disconnected before the test.
5. Checks for a short to ground in the fuel pump relay harness CKT 839.

2.3L (VIN D, A & 3) ENGINE — SYSTEM DIAGNOSTIC CHARTS — 1993–94 ACHIEVA, GRAND AM AND SKYLARK

CHART A-5
(Page 1 of 2)
ENGINE CRANKS BUT WILL NOT RUN
(FUEL PUMP CIRCUIT)
2.3L (VIN D, A & 3) "N" CARLINE

FROM CHART A-3 PAGE 2

(1) INSPECT FUEL PUMP FUSE. IS FUSE OK?

YES → (2)
- IGNITION "OFF" FOR 10 SECONDS.
- IGNITION "ON"
- LISTEN FOR IN-TANK FUEL PUMP IT SHOULD RUN FOR 2 SECONDS AFTER IGNITION IS TURNED "ON." DOES IT?

NO → (4)
- DISCONNECT FUEL PUMP HARNESS AT REAR BODY CONNECTOR.
- IGNITION "OFF," PROBE FUEL PUMP PRIME CONNECTOR WITH A TEST LIGHT CONNECTED TO B+. IS THE TEST LIGHT "ON"?

YES → SEE CHART A-7

NO → (3)
- IGNITION "OFF."
- USING A FUSED JUMPER WIRE, CONNECT THE FUEL PUMP TEST CONNECTOR TO B+. DOES THE PUMP RUN?

YES → SEE PAGE 2 OF THIS CHART

NO → CKT 839 SHORTED TO GROUND.

NO (from 4) → (5)
- REMOVE FUEL PUMP RELAY.
- PROBE RELAY CONNECTOR TERMINAL "1" WITH A TEST LIGHT CONNECTED TO B+. IS THE LIGHT "ON"?

YES (from 4) → CKT 120, 490 OR FUEL PUMP RELAY OR OIL PRESSURE SWITCH SHORTED TO GROUND.

YES (from 5) →
- DISCONNECT FUEL PUMP RELAY.
- USING A FUSED JUMPER WIRE, CONNECT CKT 120 TO B+. DOES THE FUEL PUMP RUN?

NO (from 5) → FUEL TANK METER ASSEMBLY HARNESS SHORTED TO GROUND OR FAULTY FUEL PUMP.

YES → FAULTY CONNECTION OR FAULTY RELAY.

NO → OPEN CKT 120, FAULTY IN-TANK PUMP, OR FAULTY FUEL PUMP GROUND CKT 450.

"AFTER REPAIRS," CONFIRM "CLOSED LOOP" OPERATION AND NO MIL (SERVICE ENGINE SOON).

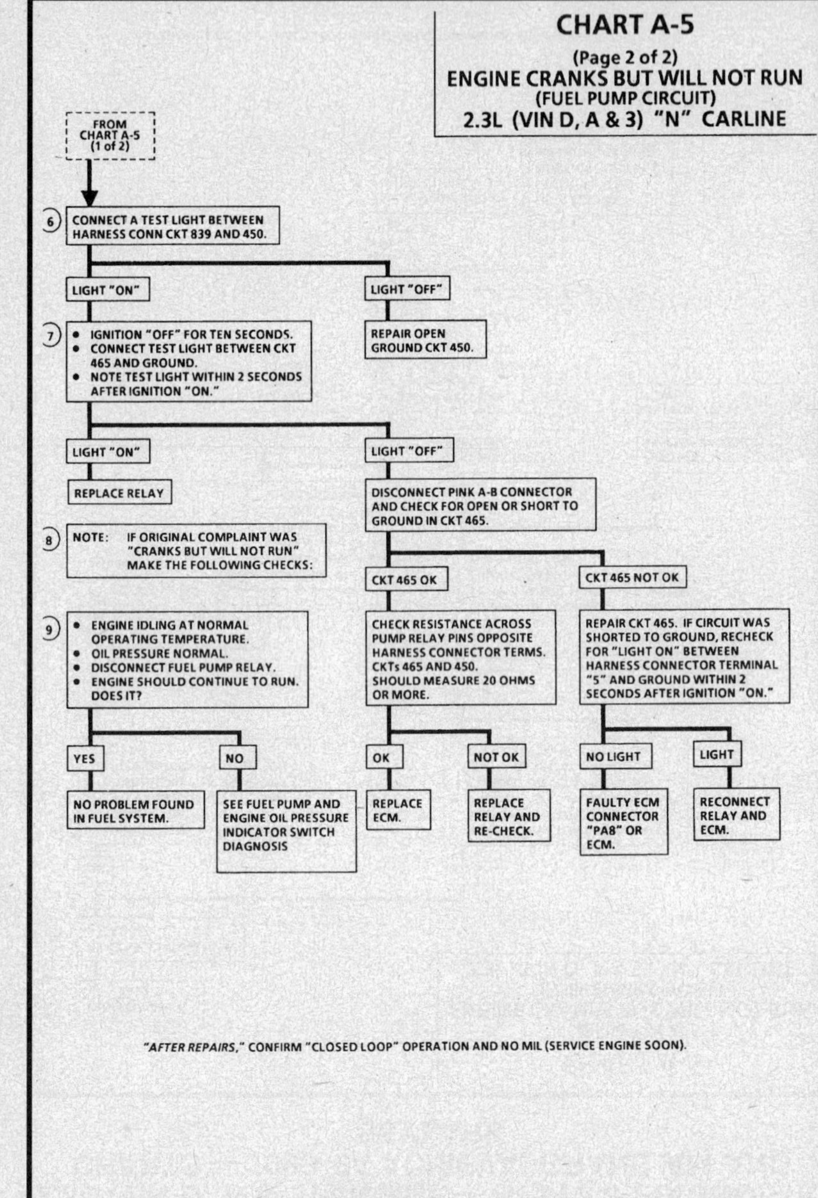

CHART A-5

(Page 2 of 2)
ENGINE CRANKS BUT WILL NOT RUN
(FUEL PUMP CIRCUIT)
2.3L (VIN D, A & 3) "N" CARLINE

Circuit Description:

When the ignition switch is turned "ON," the Engine Control Module (ECM) turns "ON" the in-tank fuel pump. It will remain "ON" as long as the ECM is receiving ignition reference pulses from the Electronic Ignition Control Module (ICM).

If there are no reference pulses, the ECM will shut "OFF" the fuel pump about 2-3 seconds after key "ON," or about 10 seconds after reference pulses stop. If sufficient oil pressure is present to close the fuel pump switch and oil pressure switch, the fuel pump will remain "ON" during cranking without reference pulses.

The pump delivers fuel to the fuel rail and injectors, then to the pressure regulator, where the system pressure is controlled to 280 - 325 kPa (41 - 47 psi) with no manifold vacuum or 211 - 304 kPa (30.5 - 44 psi) at idle. Excess fuel is then returned to the fuel tank.

The fuel pump "test" terminal is located in the engine compartment. When the engine is stopped, the pump can be turned "ON" by applying battery voltage to the "test" terminal.

Improper fuel system pressure will result in one or all of the following symptoms:
- Cranks but will not run.
- DTC 44.
- DTC 45.
- Cuts out, may feel like ignition problem.
- Poor fuel economy, loss of power.
- Hesitation.

Test Description: Number(s) below refer to circled number(s) on the diagnostic chart.
6. Check relay ground CKT 450.
7. Check for ECM control of relay through CKT 465.
8. The fuel pump voltage control circuit includes an engine oil pressure switch with a separate set of normally open contacts. The switch closes at about 28 kPa (4 psi) of oil pressure and provides a second battery feed path to the fuel pump. If the relay fails, the pump will continue to run using the battery feed supplied by the closed oil pressure switch.

A failed pump relay will result in extended engine crank time, because of the time required to build enough oil pressure to close the oil pressure switch and turn "ON" the fuel pump. There may be instances when the relay has failed but the engine will not crank fast enough to build enough oil pressure to close the switch. This or a faulty oil pressure switch can result in "Engine Cranks But Will Not Run."
9. Check the fuel pump switch and oil pressure switch to be sure it provides battery feed to the fuel pump should the pump relay fail.

2.3L (VIN D, A & 3) ENGINE — SYSTEM DIAGNOSTIC CHARTS — 1993–94 ACHIEVA, GRAND AM AND SKYLARK

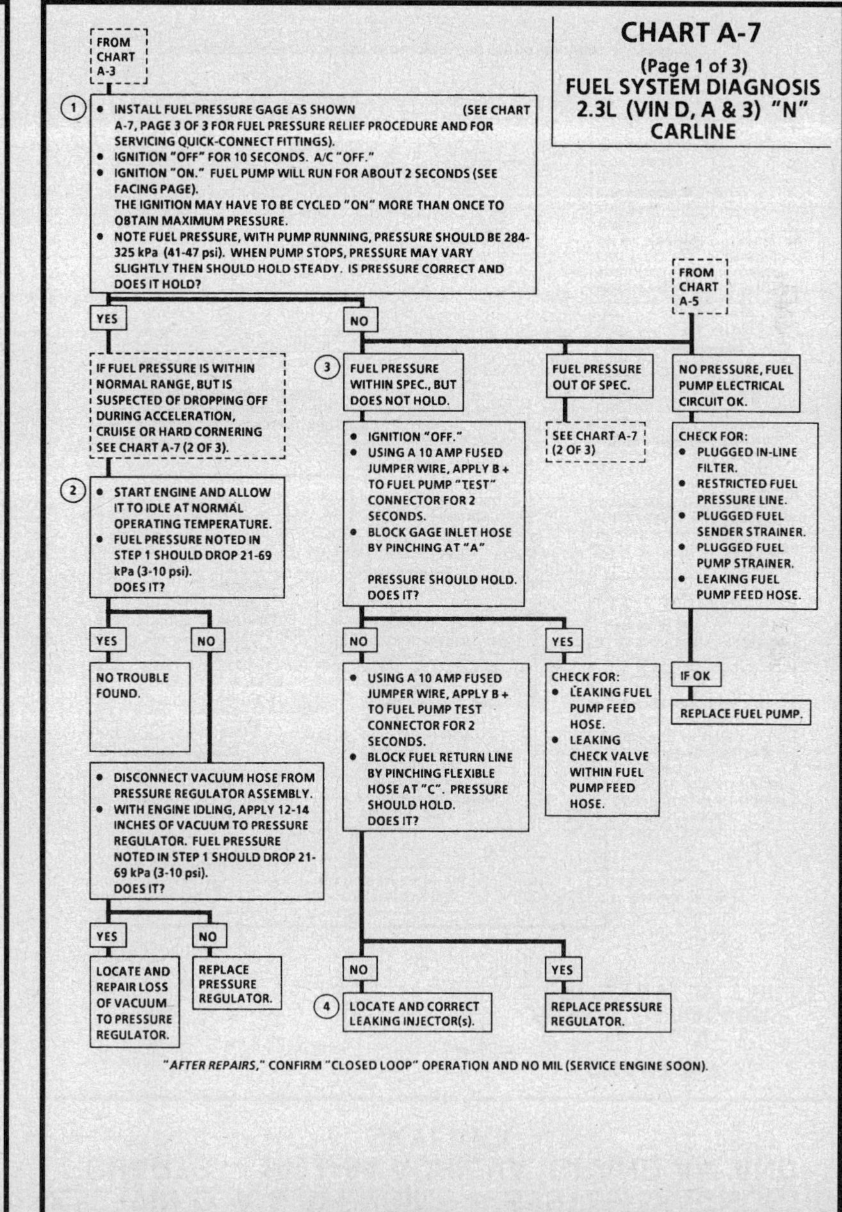

CHART A-7
(Page 1 of 3)
FUEL SYSTEM DIAGNOSIS
2.3L (VIN D, A & 3) "N" CARLINE

Circuit Description:

When the ignition switch is turned "ON," the Engine Control Module (ECM) will turn "ON" the in-tank fuel pump. It will remain "ON" as long as the engine is cranking or running, and the ECM is receiving reference pulses. If there are no reference pulses, the ECM will shut "OFF" the fuel pump in about 2 seconds after ignition "ON" or 10 seconds after reference pulses stop.

An electric fuel pump, part of the modular fuel sender and located inside the fuel tank, pumps fuel through an in-line filter to the fuel rail assembly. The pump is designed to provide fuel at a pressure above the regulated pressure needed by the injectors. A pressure regulator, attached to the fuel rail, keeps fuel available to the injectors at a regulated pressure. Unused fuel is returned to the fuel tank by a separate line.

Test Description: Number(s) below refer to circled number(s) on the diagnostic chart.

1. Install fuel pressure gage as shown in illustration. See Page 3 of 3 for "Fuel Pressure Relief Procedure" and for "Servicing Quick-Connect Fittings." Ignition "ON" pump pressure should be 284-325 kPa (41-47 psi). This pressure is controlled by spring pressure within the regulator assembly.

2. When the engine is idling, the manifold pressure is low (high vacuum) and is applied to the fuel regulator diaphragm. This will offset the spring and result in a lower fuel pressure. This idle pressure will vary somewhat depending on barometric pressure, however, the pressure idling should be less indicating pressure regulator control.

3. Pressure that continues to fall quickly is caused by one of the following:
   - Leaking fuel pump feed hose.
   - Leaking check valve within fuel pump feed hose.
   - Fuel pressure regulator valve leaking.
   - Injector(s) sticking open.

4. An injector sticking open can best be determined by checking for a fouled or saturated spark plug(s). If a leaking injector can not be determined by a fouled or saturated spark plug the following procedure should be used.
   - Remove fuel rail bolts, but leave fuel lines connected.

CAUTION: Be sure injector(s) are not allowed to spray on engine and that injector retaining clips are intact. This should be carefully followed to prevent fuel spray on engine which would cause a fire hazard.

   - Pressurize the fuel system and observe for injector(s) leaking.

2.3L (VIN D, A & 3) ENGINE — SYSTEM DIAGNOSTIC CHARTS — 1993–94 ACHIEVA, GRAND AM AND SKYLARK

CHART A-7
(Page 1 of 3)
FUEL SYSTEM DIAGNOSIS
2.3L (VIN D, A & 3) "N" CARLINE

FROM CHART A-3

① • INSTALL FUEL PRESSURE GAGE AS SHOWN (SEE CHART A-7, PAGE 3 OF 3 FOR FUEL PRESSURE RELIEF PROCEDURE AND FOR SERVICING QUICK-CONNECT FITTINGS).
• IGNITION "OFF" FOR 10 SECONDS. A/C "OFF."
• IGNITION "ON." FUEL PUMP WILL RUN FOR ABOUT 2 SECONDS (SEE FACING PAGE).
THE IGNITION MAY HAVE TO BE CYCLED "ON" MORE THAN ONCE TO OBTAIN MAXIMUM PRESSURE.
• NOTE FUEL PRESSURE, WITH PUMP RUNNING, PRESSURE SHOULD BE 284-325 kPa (41-47 psi). WHEN PUMP STOPS, PRESSURE MAY VARY SLIGHTLY THEN SHOULD HOLD STEADY. IS PRESSURE CORRECT AND DOES IT HOLD?

YES → IF FUEL PRESSURE IS WITHIN NORMAL RANGE, BUT IS SUSPECTED OF DROPPING OFF DURING ACCELERATION, CRUISE OR HARD CORNERING SEE CHART A-7 (2 OF 3).

② • START ENGINE AND ALLOW IT TO IDLE AT NORMAL OPERATING TEMPERATURE.
• FUEL PRESSURE NOTED IN STEP 1 SHOULD DROP 21-69 kPa (3-10 psi). DOES IT?

YES → NO TROUBLE FOUND.

NO → • DISCONNECT VACUUM HOSE FROM PRESSURE REGULATOR ASSEMBLY.
• WITH ENGINE IDLING, APPLY 12-14 INCHES OF VACUUM TO PRESSURE REGULATOR. FUEL PRESSURE NOTED IN STEP 1 SHOULD DROP 21-69 kPa (3-10 psi). DOES IT?

YES → LOCATE AND REPAIR LOSS OF VACUUM TO PRESSURE REGULATOR.

NO → REPLACE PRESSURE REGULATOR.

NO (from ①) →

③ FUEL PRESSURE WITHIN SPEC., BUT DOES NOT HOLD.
• IGNITION "OFF."
• USING A 10 AMP FUSED JUMPER WIRE, APPLY B + TO FUEL PUMP "TEST" CONNECTOR FOR 2 SECONDS.
• BLOCK GAGE INLET HOSE BY PINCHING AT "A"
PRESSURE SHOULD HOLD. DOES IT?

NO → • USING A 10 AMP FUSED JUMPER WIRE, APPLY B + TO FUEL PUMP TEST CONNECTOR FOR 2 SECONDS.
• BLOCK FUEL RETURN LINE BY PINCHING FLEXIBLE HOSE AT "C". PRESSURE SHOULD HOLD. DOES IT?

NO → ④ LOCATE AND CORRECT LEAKING INJECTOR(S).

YES → REPLACE PRESSURE REGULATOR.

YES (from ③ block) → CHECK FOR:
• LEAKING FUEL PUMP FEED HOSE.
• LEAKING CHECK VALVE WITHIN FUEL PUMP FEED HOSE.

FUEL PRESSURE OUT OF SPEC. → SEE CHART A-7 (2 OF 3)

FROM CHART A-5

NO PRESSURE, FUEL PUMP ELECTRICAL CIRCUIT OK.
CHECK FOR:
• PLUGGED IN-LINE FILTER.
• RESTRICTED FUEL PRESSURE LINE.
• PLUGGED FUEL SENDER STRAINER.
• PLUGGED FUEL PUMP STRAINER.
• LEAKING FUEL PUMP FEED HOSE.

IF OK → REPLACE FUEL PUMP.

"AFTER REPAIRS," CONFIRM "CLOSED LOOP" OPERATION AND NO MIL (SERVICE ENGINE SOON).

2.3L (VIN D, A & 3) ENGINE — SYSTEM DIAGNOSTIC CHARTS — 1993–94 ACHIEVA, GRAND AM AND SKYLARK

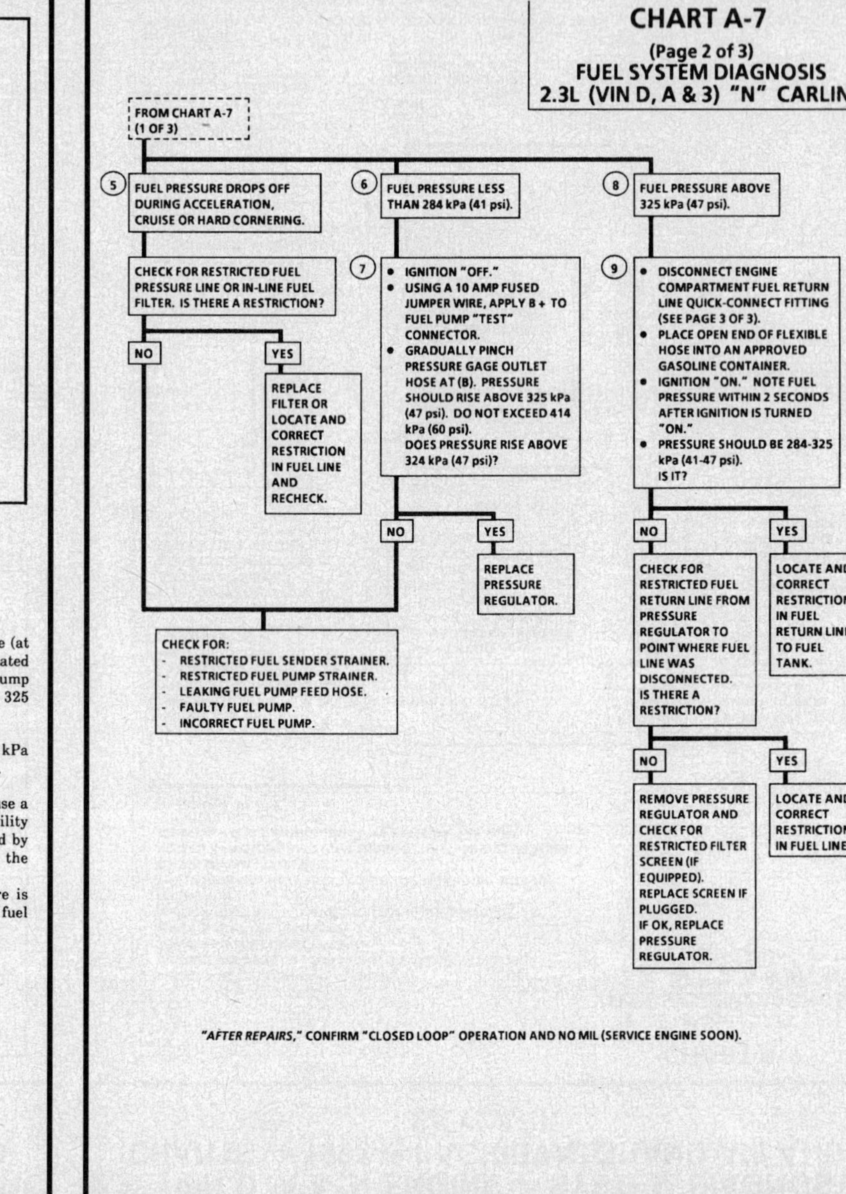

CHART A-7
(Page 2 of 3)
FUEL SYSTEM DIAGNOSIS
2.3L (VIN D, A & 3) "N" CARLINE

Test Description: Number(s) below refer to circled number(s) on the diagnostic chart.

5. Fuel pressure that drops off during acceleration, cruise or hard concerning may cause a lean condition and result in a loss of power, surging or misfire. This condition can be diagnosed using a Tech 1 scan tool. If the fuel system is very lean, the Oxygen Sensor (O2S) will stop toggling and output voltage will drop below 500 mV. Also, injector pulse width will increase.

Important
- Make sure system is not operating at "Fuel Cut-Off" which may cause false reading on the scan tool.

6. Fuel pressure below 284 kPa (41 psi) may cause a lean condition and may set a DTC 44. Driveability conditions can include hard starting cold, hesitation, poor driveability, lack of power, surging or misfire.

7. Restricting fuel flow at the fuel pressure gage (at "B") causes fuel pressure to build above regulated pressure. With battery voltage applied the pump "test" connector, pressure should rise above 325 kPa (47 psi) as the gage outlet hose is pinched.

NOTICE: Do not allow pressure to exceed 414 kPa (60 psi), as damage to the regulator may result.

8. Fuel pressure above 325 kPa (47 psi) may cause a rich condition and may set a DTC 45. Driveability conditions can include hard starting (followed by black smoke) and a strong sulfur smell in the exhaust.

9. This test determines if the high fuel pressure is due to a restricted fuel return line or a faulty fuel pressure regulator.

CHART A-7
(Page 2 of 3)
FUEL SYSTEM DIAGNOSIS
2.3L (VIN D, A & 3) "N" CARLINE

FROM CHART A-7 (1 OF 3)

(5) FUEL PRESSURE DROPS OFF DURING ACCELERATION, CRUISE OR HARD CORNERING.

CHECK FOR RESTRICTED FUEL PRESSURE LINE OR IN-LINE FUEL FILTER. IS THERE A RESTRICTION?
- NO → CHECK FOR:
  - RESTRICTED FUEL SENDER STRAINER.
  - RESTRICTED FUEL PUMP STRAINER.
  - LEAKING FUEL PUMP FEED HOSE.
  - FAULTY FUEL PUMP.
  - INCORRECT FUEL PUMP.
- YES → REPLACE FILTER OR LOCATE AND CORRECT RESTRICTION IN FUEL LINE AND RECHECK.

(6) FUEL PRESSURE LESS THAN 284 kPa (41 psi).

(7)
- IGNITION "OFF."
- USING A 10 AMP FUSED JUMPER WIRE, APPLY B + TO FUEL PUMP "TEST" CONNECTOR.
- GRADUALLY PINCH PRESSURE GAGE OUTLET HOSE AT (B). PRESSURE SHOULD RISE ABOVE 325 kPa (47 psi). DO NOT EXCEED 414 kPa (60 psi). DOES PRESSURE RISE ABOVE 324 kPa (47 psi)?
- NO → REPLACE PRESSURE REGULATOR.
- YES

(8) FUEL PRESSURE ABOVE 325 kPa (47 psi).

(9)
- DISCONNECT ENGINE COMPARTMENT FUEL RETURN LINE QUICK-CONNECT FITTING (SEE PAGE 3 OF 3).
- PLACE OPEN END OF FLEXIBLE HOSE INTO AN APPROVED GASOLINE CONTAINER.
- IGNITION "ON." NOTE FUEL PRESSURE WITHIN 2 SECONDS AFTER IGNITION IS TURNED "ON."
- PRESSURE SHOULD BE 284-325 kPa (41-47 psi). IS IT?
- NO → CHECK FOR RESTRICTED FUEL RETURN LINE FROM PRESSURE REGULATOR TO POINT WHERE FUEL LINE WAS DISCONNECTED. IS THERE A RESTRICTION?
  - NO → REMOVE PRESSURE REGULATOR AND CHECK FOR RESTRICTED FILTER SCREEN (IF EQUIPPED). REPLACE SCREEN IF PLUGGED. IF OK, REPLACE PRESSURE REGULATOR.
  - YES → LOCATE AND CORRECT RESTRICTION IN FUEL LINE.
- YES → LOCATE AND CORRECT RESTRICTION IN FUEL RETURN LINE TO FUEL TANK.

"AFTER REPAIRS," CONFIRM "CLOSED LOOP" OPERATION AND NO MIL (SERVICE ENGINE SOON).

2.3L (VIN D, A & 3) ENGINE — SYSTEM DIAGNOSTIC CHARTS — 1993–94 ACHIEVA, GRAND AM AND SKYLARK

CHART A-7
(Page 3 of 3)
FUEL SYSTEM DIAGNOSIS
2.3L (VIN D, A & 3) "N" CARLINE

FUEL SYSTEM PRESSURE RELIEF PROCEDURE

PFI Engines Without Fuel Pressure Connection (Except J Car Applications)

(Must Be Performed Before Disconnecting Fuel Line Fittings)

CAUTION:
- **To reduce the risk of fire and personal injury, it is necessary to relieve fuel system pressure before disconnecting fuel line fittings.**
- **After relieving system pressure, a small amount of fuel may be released when disconnecting fuel line fittings. In order to reduce the chance of personal injury, cover fuel line fittings with a shop towel before disconnecting, to catch any fuel that may leak out. Place the towel in an approved container when disconnect is completed.**

1. Loosen fuel filler cap to relieve tank pressure.
2. Raise vehicle.
3. Disconnect fuel pump electrical connector.
4. Lower vehicle.
5. Start engine and run until fuel supply remaining in fuel pipes is consumed. Engage starter for 3.0 seconds to assure relief of any remaining pressure.
6. Raise vehicle.
7. Connect fuel pump electrical connector.
8. Lower vehicle.
9. Disconnect negative battery cable to avoid possible fuel discharge if an accidental attempt is made to start the engine.
10. Fuel line fittings are now safe for servicing.
11. Perform service required.
    - If performing fuel pressure check with a gage equipped with a bleed hose, fuel pressure can be relieved through the gage following test. Place bleed hose into approved gasoline container and open valve to bleed system pressure.
12. Tighten fuel filler cap.
13. Connect negative battery cable.
14. Cycle ignition "ON" and "OFF" twice, waiting ten seconds between cycles, then check for fuel leaks.

2.3L (VIN D, A & 3) ENGINE — SYSTEM DIAGNOSTIC CHARTS — 1993–94 ACHIEVA, GRAND AM AND SKYLARK

CHART A-7
(Page 3 of 3)
FUEL SYSTEM DIAGNOSIS
2.3L (VIN D, A & 3) "N" CARLINE

SERVICING QUICK-CONNECT FITTINGS

Important
- In order to install fuel system diagnostic equipment on vehicles equipped with plastic quick-connect fittings, fuel line separator tools must be used to disconnect the fittings. Use of the separator tools will cause the plastic retainer to remain inside the female connector allowing diagnostic equipment to be connected.

    Tools required:
    J 37088-A tool set, fuel line quick-connect separator;
    J 39504 tool set, fuel line quick-connect separator (restricted access).

Remove or Disconnect
1. Grasp both sides of fitting. Twist female connector 1/4 turn in each direction to loosen any dirt within fitting.

    CAUTION: Safety glasses must be worn when using compressed air, as flying dirt particles may cause eye injury.

2. Using compressed air, blow dirt out of fitting.
3. Choose correct tool from J 37088-A or J 39504 tool set for size of fitting. Insert tool into female connector, then push/pull inward to release locking tabs.
4. Pull connection apart.

Clean and Inspect

NOTICE: If it is necessary to remove rust or burrs from fuel pipe, use emery cloth in a radial motion with the pipe end to prevent damage to O-ring sealing surface.

- Using a clean shop towel, wipe off male pipe end.
- Inspect both ends of fitting for dirt and burrs. Clean or replace components/assemblies as required.

Install or Connect

CAUTION: To Reduce the Risk of Fire and Personal Injury:
- **Before connecting fitting, always apply a few drops of clean engine oil to the male pipe end of engine fuel pipe, pressure gage adapter or fuel line shut-off adapter. This will ensure proper reconnection and prevent a possible fuel leak. (During normal operation, the O-rings located in the female connector will swell and may prevent proper reconnection if not lubricated.)**

1. Apply a few drops of clean engine oil to the male pipe end of engine fuel pipe, pressure gage adapter or fuel line shut-off adapter.
2. Push both sides of fitting together to cause the retaining tabs/fingers to snap into place.
3. Once installed, pull on both sides of fitting to make sure connection is secure.

2.3L (VIN D, A & 3) ENGINE — DIAGNOSTIC TROUBLE CODE CHART — 1993–94 ACHIEVA, GRAND AM AND SKYLARK

DTC 13
OXYGEN SENSOR (O2S) CIRCUIT
(OPEN CIRCUIT)
2.3L (VIN D, A & 3) "N" CARLINE

Circuit Description:

The Engine Control Module (ECM) supplies a voltage of about .45 volt between terminals "PA12" and "PB6". (If measured with a 10 megohm digital voltmeter, this may read as low as .32 volt). The Oxygen Sensor (O2S) varies the voltage within a range of about 1 volt if the exhaust is rich, down through about .10 volt if exhaust is lean.

The sensor is like an open circuit and produces no voltage when it is below 315°C (600°F). An open sensor circuit or cold sensor causes "Open Loop" operation.

Test Description: Number(s) below refer to circled number(s) on the diagnostic chart.

1. DTC 13 will set under the following conditions:
   - Engine running at least 40 seconds after start.
   - Coolant temperature at least 42.5°C (108°F).
   - No DTC 21 or 22.
   - O2S signal voltage steady between approximately .35 and .55 volts.
   - Throttle Position (TP) sensor signal above 6% for more time than TP sensor was below 6%. (About .3 volt above closed throttle voltage.)
   - All conditions must be met and held for at least 30 seconds.
   If the conditions for a DTC 13 exist, the system will not go "Closed Loop."

2. This will determine if the sensor is at fault or the wiring or ECM is the cause of the DTC 13.
3. Use only a high impedance digital volt ohmmeter for this test. This test checks the continuity of CKT 412 and CKT 413; because if CKT 413 is open, the ECM voltage on CKT 412 will be over .6 volt (600 mV).

Diagnostic Aids:

Normal Tech 1 scan tool voltage varies between 100 mV to 999 mV (.1 volt to 1.0 volt) while in "Closed Loop." DTC 13 sets in 20 seconds if voltage remains between .35 volt and .55 volt, but the system will go "Open Loop" in about 15 seconds.

2.3L (VIN D, A & 3) ENGINE — DIAGNOSTIC TROUBLE CODE CHART — 1993–94 ACHIEVA, GRAND AM AND SKYLARK

DTC 13
OXYGEN SENSOR (O2S) CIRCUIT
(OPEN CIRCUIT)
2.3L (VIN D, A & 3) "N" CARLINE

2.3L (VIN D, A & 3) ENGINE — DIAGNOSTIC TROUBLE CODE CHART — 1993–94 ACHIEVA, GRAND AM AND SKYLARK

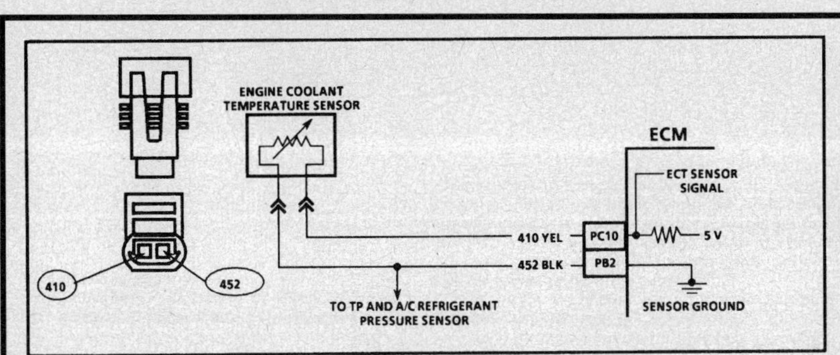

DTC 14
ENGINE COOLANT TEMPERATURE (ECT) SENSOR CIRCUIT
(HIGH TEMPERATURE INDICATED)
2.3L (VIN D, A & 3) "N" CARLINE

Circuit Description:
The Engine Coolant Temperature (ECT) sensor uses a thermistor to control the signal voltage at the Engine Control Module (ECM). The ECM applies a voltage on CKT 410 to the sensor. When the engine is cold the sensor (thermistor) resistance is high, therefore ECM terminal "PC10" voltage will be high.

As the engine warms, the sensor resistance becomes less, and the voltage drops. At normal engine operating temperature, the voltage will measure about 1.5 to 2.0 volts at ECM terminal "PC10".

Engine coolant temperature is one of the inputs used to control:
- Fuel delivery.
- Ignition Control (IC).
- Idle Air Control (IAC).
- Torque Convertor Clutch (TCC).
- Evaporative Emission (EVAP) canister purge solenoid valve.
- Electric coolant fan.

Test Description: Number(s) below refer to circled number(s) on the diagnostic chart.
1. DTC 14 will set if:
   - Signal voltage indicates an engine coolant temperature above about 140°C (285°F).
   - Engine running longer than 128 seconds.
2. This test will determine if CKT 410 is shorted to ground which will cause the conditions for DTC 14.

Diagnostic Aids:

Check harness routing for a potential short to ground in CKT 410. A Tech 1 scan tool displays engine coolant temperature in degrees celsius. After engine is started, the temperature should rise steadily to about 90°C, and then stabilize when thermostat opens.

Verify that engine is not overheating and has not been subjected to conditions which could create an overheating condition (i.e. overload, trailer towing, hilly terrain, heavy stop and go traffic, etc.). The "Temperature To Resistance Value" scale may be used to test the sensor at various temperature levels to evaluate the possibility of a "shifted" (mis-scaled) sensor. A "shifted" sensor could result in poor driveability complaints.

2.3L (VIN D, A & 3) ENGINE — DIAGNOSTIC TROUBLE CODE CHART — 1993–94 ACHIEVA, GRAND AM AND SKYLARK

DTC 14
ENGINE COOLANT TEMPERATURE (ECT) SENSOR CIRCUIT
(HIGH TEMPERATURE INDICATED)
2.3L (VIN D, A & 3) "N" CARLINE

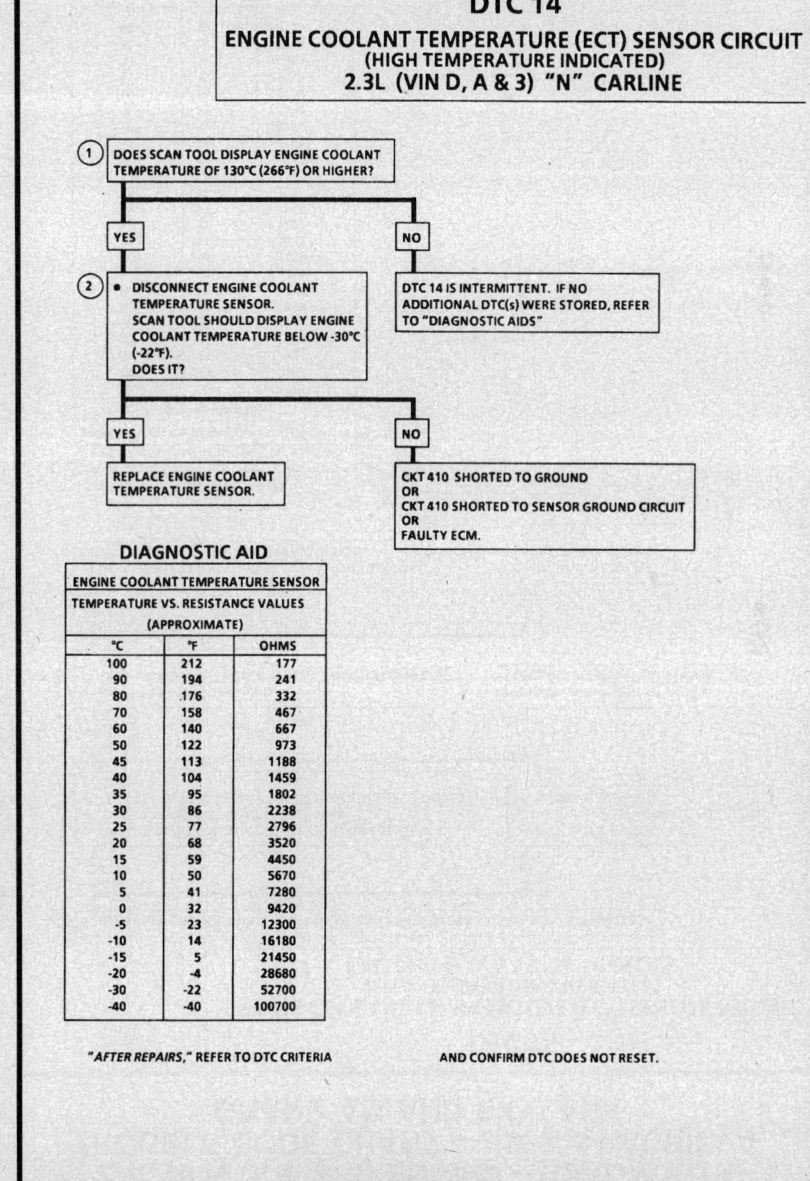

DIAGNOSTIC AID

ENGINE COOLANT TEMPERATURE SENSOR		
TEMPERATURE VS. RESISTANCE VALUES		
(APPROXIMATE)		
°C	°F	OHMS
100	212	177
90	194	241
80	176	332
70	158	467
60	140	667
50	122	973
45	113	1188
40	104	1459
35	95	1802
30	86	2238
25	77	2796
20	68	3520
15	59	4450
10	50	5670
5	41	7280
0	32	9420
-5	23	12300
-10	14	16180
-15	5	21450
-20	-4	28680
-30	-22	52700
-40	-40	100700

"AFTER REPAIRS," REFER TO DTC CRITERIA   AND CONFIRM DTC DOES NOT RESET.

2.3L (VIN D, A & 3) ENGINE — DIAGNOSTIC TROUBLE CODE CHART — 1993–94 ACHIEVA, GRAND AM AND SKYLARK

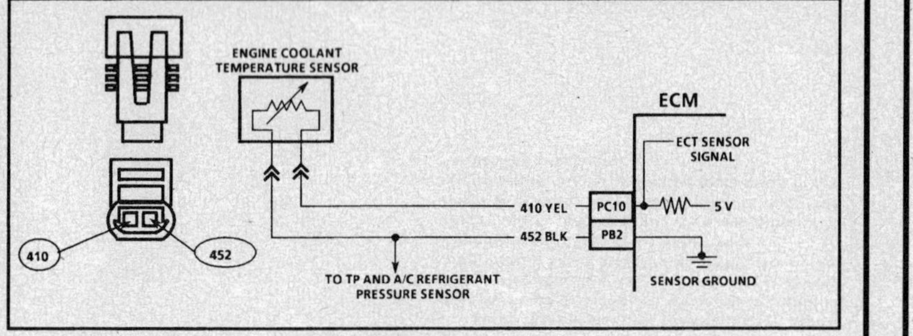

DTC 15
ENGINE COOLANT TEMPERATURE (ECT) SENSOR CIRCUIT
(LOW TEMPERATURE INDICATED)
2.3L (VIN D, A & 3) "N" CARLINE

Circuit Description:

The Engine Coolant Temperature (ECT) sensor uses a thermistor to control the signal voltage at the Engine Control Module (ECM). The ECM applies a voltage on CKT 410 to the sensor. When the engine is cold, the sensor (thermistor) resistance is high, therefore, ECM terminal "PC10" voltage will be high.

As the engine warms, the sensor resistance becomes less, and the voltage drops. At normal engine operating temperature the voltage will measure about 1.5 to 2.0 volts at ECM terminal "PC10".

Engine coolant temperature is one of the inputs used to control:

- Fuel delivery.
- Ignition Control (IC).
- Idle Air Control (IAC).
- Torque Convertor Clutch (TCC).
- Evaporative Emission (EVAP) canister purge solenoid valve.
- Electric coolant fan.

Test Description: Number(s) below refer to circled number(s) on the diagnostic chart.

1. DTC 15 will set if:
   - Signal voltage indicates an engine coolant temperature less than -39°C (-38°F) for 60 seconds.
2. This test simulates a DTC 14. If the ECM senses the low signal voltage (high temperature) and the scan reads 130°C, the ECM and wiring are OK.
3. This test will determine if CKT 410 is open. There should be 5 volts present at sensor connector if measured with a DVOM.

Diagnostic Aids:

A Tech 1 scan tool displays engine coolant temperature in degrees celsius. After the engine is started the temperature should rise steadily to about 95°C, and then stabilize when the thermostat opens. It is normal for engine coolant temperature to fluctuate slightly around 95°C.

A faulty connection, or an open in CKT 410 or CKT 452 can result in a DTC 15.

DTCs 15, 21 and 66 stored at the same time could be the result of an open CKT 452.

The "Temperature to Resistance Value" scale may be used to test the sensor at various temperature levels to evaluate the possibility of a "shifted" (mis-scaled) sensor. A "shifted" sensor could result in poor driveability complaints.

2.3L (VIN D, A & 3) ENGINE — DIAGNOSTIC TROUBLE CODE CHART — 1993–94 ACHIEVA, GRAND AM AND SKYLARK

DTC 15
ENGINE COOLANT TEMPERATURE (ECT) SENSOR CIRCUIT
(LOW TEMPERATURE INDICATED)
2.3L (VIN D, A & 3) "N" CARLINE

1. • DOES TECH 1 SCAN TOOL DISPLAY ENGINE COOLANT TEMPERATURE OF -30°C (-22°F) OR LESS?

   YES →

   2. • DISCONNECT ENGINE COOLANT TEMPERATURE SENSOR.
      • JUMPER HARNESS TERMINALS TOGETHER.
      • TECH 1 SCAN TOOL SHOULD DISPLAY 130°C (266°F) OR MORE. DOES IT?

   NO → DTC 15 IS INTERMITTENT. IF NO ADDITIONAL DTC(S) WERE STORED, REFER TO "DIAGNOSTIC AIDS"

   (from 2) NO →

   3. • JUMPER CKT 410 TO GROUND.
      • TECH 1 SCAN TOOL SHOULD DISPLAY OVER 130°C (266°F). DOES IT?

   (from 2) YES → FAULTY CONNECTION OR ENGINE COOLANT TEMPERATURE SENSOR.

   (from 3) YES → OPEN ENGINE COOLANT TEMPERATURE SENSOR GROUND CIRCUIT, FAULTY CONNECTION OR FAULTY ECM.

   (from 3) NO → OPEN CKT 410, FAULTY CONNECTION AT ECM, OR FAULTY ECM.

DIAGNOSTIC AID

ENGINE COOLANT TEMPERATURE SENSOR		
TEMPERATURE VS. RESISTANCE VALUES (APPROXIMATE)		
°C	°F	OHMS
100	212	177
90	194	241
80	176	332
70	158	467
60	140	667
50	122	973
45	113	1188
40	104	1459
35	95	1802
30	86	2238
25	77	2796
20	68	3520
15	59	4450
10	50	5670
5	41	7280
0	32	9420
-5	23	12300
-10	14	16180
-15	5	21450
-20	-4	28680
-30	-22	52700
-40	-40	100700

"AFTER REPAIRS," REFER TO DTC CRITERIA   AND CONFIRM DTC DOES NOT RESET.

2.3L (VIN D, A & 3) ENGINE — DIAGNOSTIC TROUBLE CODE CHART — 1993–94 ACHIEVA, GRAND AM AND SKYLARK

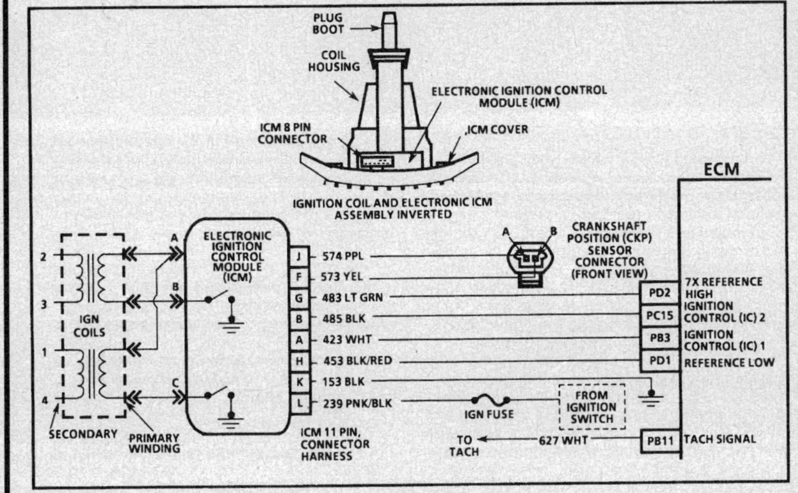

DTC 19
INTERMITTENT 7X REFERENCE CIRCUIT
2.3L (VIN D, A & 3) "N" CARLINE

Circuit Description:

The Electronic Ignition Control Module (ICM) sends a reference signal to the Engine Control Module (ECM) seven times per revolution to indicate crankshaft position and RPM so that the ECM can determine when to pulse the ignition coils and control ignition timing. This signal is called the 7X reference because it occurs seven times per revolution. The ICM applies 5 volts from terminal "G" through CKT 483 to ECM terminal "PD2" and in effect, switches this circuit to ground for a very short period of time. The seventh pulse is the sync pulse and is used for crankshaft position reference. DTC 19 is set if the ECM misses at least 20 resync pulses within 4 minutes and 16 seconds.

Test Description: Number(s) below refer to circled number(s) on the diagnostic chart.

1. This determines if the ECM recognizes a problem. If it doesn't set DTC 19 at this point, the problem is intermittent.
2. When a 7X resync occurs, engine stumble also occurs. If a component connection or circuitry is at fault, engine stumble may be induced by wiggling the circuit or connection. This step determines if a component connection or circuit is at fault.
3. Operating faulty non-engine related electronic components may emit Electromagnetic Interference (EMI) which may cause a 7X resync. This step determines if the non-engine related component is at fault.

Diagnostic Aids:

An intermittent may be caused by a poor connection, rubbed through wire insulation, or a wire broken inside the insulation. Inspect ECM harness connector terminal "PD2" and ICM 11 pin connector terminal "G" for improper mating, broken locks, improperly formed or damaged terminals, poor terminal to wire connection and damaged harness.

If vehicle has non-standard electrical equipment, (such as CB radios, 2-way radios, etc.) check to see if their operation may cause a 7X resync. Faulty wiring or faulty non-standard electrical component operation may cause DTC 19.

2.3L (VIN D, A & 3) ENGINE — DIAGNOSTIC TROUBLE CODE CHART — 1993–94 ACHIEVA, GRAND AM AND SKYLARK

DTC 19
INTERMITTENT 7X REFERENCE CIRCUIT
2.3L (VIN D, A & 3) "N" CARLINE

1.
   - IF OTHER DTC(s) ARE PRESENT, REFER TO THOSE CHARTS BEFORE PROCEEDING.
   - CLEAR DTC(s).
   - IDLE ENGINE FOR 5 MINUTES OR UNTIL MALFUNCTION INDICATOR LAMP (MIL) "SERVICE ENGINE SOON" TURNS "ON," IF SOONER.
   - DOES TECH 1 SCAN TOOL DISPLAY DTC 19?

YES →

2. WITH ENGINE RUNNING, WIGGLE THE CONNECTIONS OF THE ECM, ICM AND CKP SENSOR AND ASSOCIATED CIRCUITRY * WHILE LISTENING FOR ENGINE STUMBLE. ENGINE SHOULD NOT STUMBLE. DOES IT?

- **NO** → INSPECT WIRING AND CONNECTORS BETWEEN ECM, ICM, AND CKP SENSOR. IF DTC 19 IS STILL PRESENT, REPAIR AND RETEST. IF DTC 19 IS STILL PRESENT, REPLACE PROM AND RETEST. IF DTC 19 IS STILL PRESENT, REPLACE ECM AND RETEST. IF DTC 19 IS STILL PRESENT, REPLACE ICM AND RETEST. IF NO ADDITIONAL DTC(s) WERE DISPLAYED, REFER TO "DIAGNOSTIC AIDS."
- **YES** → REPAIR FAULTY CIRCUIT(s) OR CONNECTOR(s) AND RETEST FOR DTC 19.

NO →

3.
   - OBSERVE TECH 1 SCAN TOOL 7X RESYNC WHILE OPERATING.
   - ALL NON-ENGINE RELATED ELECTRONIC COMPONENTS ON THE VEHICLE. THERE SHOULD BE NO 7X RESYNC DURING THE OPERATION OF THE COMPONENTS. IS THERE?

- **NO** → WITH ENGINE RUNNING, WIGGLE THE CONNECTIONS OF THE ECM, ICM AND CKP SENSOR AND ASSOCIATED CIRCUITRY * WHILE LISTENING FOR ENGINE STUMBLE. ENGINE SHOULD NOT STUMBLE. DOES IT?
  - **NO** → DTC 19 IS INTERMITTENT IF NO ADDITIONAL DTC(s) WERE DISPLAYED, REFER TO "DIAGNOSTIC AIDS"
  - **YES** → REPAIR FAULTY CIRCUIT(s) OR CONNECTOR(s) AND RETEST FOR DTC 19.
- **YES** → REPAIR CIRCUIT(s) OF FAULTY COMPONENTS AND RETEST FOR DTC 19.

***CAUTION:** WHEN WORKING AROUND A RUNNING ENGINE, AVOID CONTACT WITH MOVING PARTS AND HOT SURFACES TO PREVENT POSSIBLE BODILY INJURY.

"AFTER REPAIRS," REFER TO DTC CRITERIA AND CONFIRM DTC DOES NOT RESET.

2.3L (VIN D, A & 3) ENGINE — DIAGNOSTIC TROUBLE CODE CHART — 1993–94 ACHIEVA, GRAND AM AND SKYLARK

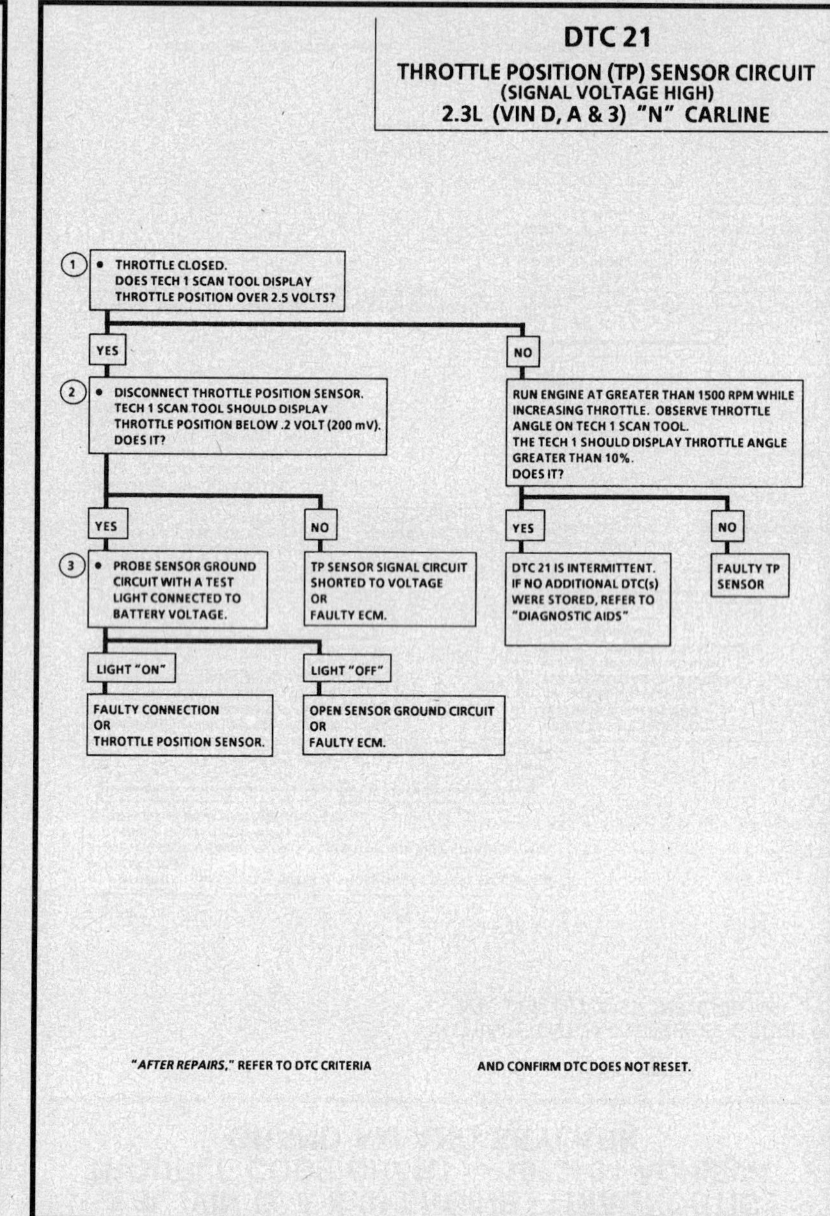

DTC 21
THROTTLE POSITION (TP) SENSOR CIRCUIT
(SIGNAL VOLTAGE HIGH)
2.3L (VIN D, A & 3) "N" CARLINE

Circuit Description:

The Throttle Position (TP) sensor provides a voltage signal that changes relative to the throttle opening. Signal voltage will vary from approximately .6 volt at idle to about 4.7 volts at Wide Open Throttle (WOT).

The TP sensor signal is one of the most important inputs used by the Engine Control Module (ECM) for fuel control and for most of the ECM control outputs.

Test Description: Number(s) below refer to circled number(s) on the diagnostic chart.

1 DTC 21 will set if:
- Engine is running below 1500 RPM.
- No DTC 33 or 34.
- MAP less than 65 kPa.
- TP sensor signal voltage greater than approximately 3.9 volts.
- Above conditions exist for over 5 seconds.
  OR
- TP sensor voltage greater than about 4.7 volts. With throttle closed the TP sensor should read less than .900 volt. If it doesn't, check for sticking TP sensor for throttle linkage. If OK, replace TP sensor.

2 With the TP sensor disconnected, the TP sensor voltage should go low if the ECM and wiring are OK.

3 Probing CKT 452 with a test light checks the TP sensor ground circuit because an open or very high resistance ground circuit will cause a DTC 21.

Diagnostic Aids:

A Tech 1 scan tool displays throttle position in volts. It should display .200 volt to .900 volt with throttle closed and ignition "ON" or at idle. Voltage should increase at a steady rate as throttle is moved toward Wide Open Throttle (WOT).

Also a Tech 1 scan tool will display throttle angle %, 0% = closed throttle 100% = WOT.

An open in CKT 452 will result in a DTC 21.

DTCs 15, 21 and 66 stored at the same time could be the result of an open CKT 452. Scan TP sensor while depressing accelerator pedal with engine stopped and ignition "ON." Display should vary from about .6 volt (600 mV) when throttle is closed, to over 4500 mV (4.5 volts) when throttle is held wide open.

Check condition of connector and sensor terminals for moisture or corrosion, and clean or replace as necessary. If corrosion is found, check condition of connector seal, repair and/or replace if necessary.

2.3L (VIN D, A & 3) ENGINE — DIAGNOSTIC TROUBLE CODE CHART — 1993–94 ACHIEVA, GRAND AM AND SKYLARK

DTC 21
THROTTLE POSITION (TP) SENSOR CIRCUIT
(SIGNAL VOLTAGE HIGH)
2.3L (VIN D, A & 3) "N" CARLINE

1 ● THROTTLE CLOSED. DOES TECH 1 SCAN TOOL DISPLAY THROTTLE POSITION OVER 2.5 VOLTS?

YES →

2 ● DISCONNECT THROTTLE POSITION SENSOR. TECH 1 SCAN TOOL SHOULD DISPLAY THROTTLE POSITION BELOW .2 VOLT (200 mV). DOES IT?

NO → RUN ENGINE AT GREATER THAN 1500 RPM WHILE INCREASING THROTTLE. OBSERVE THROTTLE ANGLE ON TECH 1 SCAN TOOL. THE TECH 1 SHOULD DISPLAY THROTTLE ANGLE GREATER THAN 10%. DOES IT?

YES →

3 ● PROBE SENSOR GROUND CIRCUIT WITH A TEST LIGHT CONNECTED TO BATTERY VOLTAGE.

NO → TP SENSOR SIGNAL CIRCUIT SHORTED TO VOLTAGE OR FAULTY ECM.

YES → DTC 21 IS INTERMITTENT. IF NO ADDITIONAL DTC(s) WERE STORED, REFER TO "DIAGNOSTIC AIDS"

NO → FAULTY TP SENSOR

LIGHT "ON" → FAULTY CONNECTION OR THROTTLE POSITION SENSOR.

LIGHT "OFF" → OPEN SENSOR GROUND CIRCUIT OR FAULTY ECM.

"AFTER REPAIRS," REFER TO DTC CRITERIA   AND CONFIRM DTC DOES NOT RESET.

2.3L (VIN D, A & 3) ENGINE — DIAGNOSTIC TROUBLE CODE CHART — 1993–94 ACHIEVA, GRAND AM AND SKYLARK

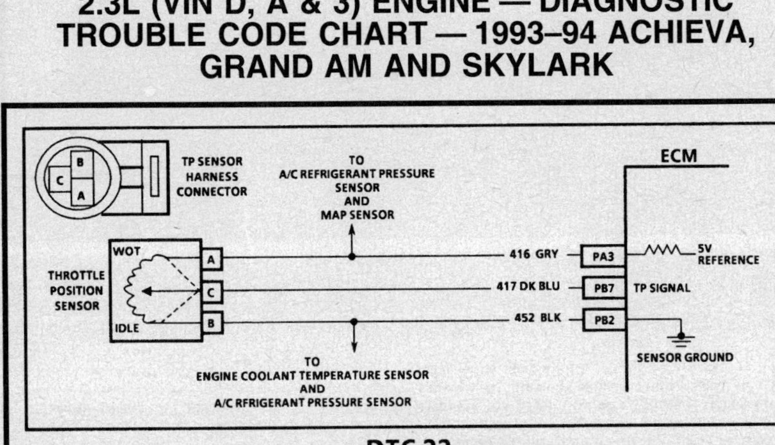

DTC 22
THROTTLE POSITION (TP) SENSOR CIRCUIT
(SIGNAL VOLTAGE LOW)
2.3L (VIN D, A & 3) "N" CARLINE

Circuit Description:

The Throttle Position (TP) sensor provides a voltage signal that changes relative to the throttle opening. Signal voltage will vary from about .6 volt at idle to about 4.7 volts at Wide Open Throttle (WOT).

The TP sensor signal is one of the most important inputs used by the Engine Control Module (ECM) for fuel control and for most of the ECM control outputs.

Test Description: Number(s) below refer to circled number(s) on the diagnostic chart.

1. DTC 22 will set if:
   - Engine is running.
   - TP sensor signal voltage is less than about .16 volt.
   - Above conditions present for 5 seconds.

   The TP sensor has an auto zeroing feature. If the voltage reading is within the range of about .2 to .9 volt, the ECM will use that value as closed throttle. If the voltage reading is out of the auto zero range at closed throttle, check for a binding throttle cable or damaged linkage, if OK, continue with diagnosis.
2. Simulates DTC 21: (high voltage). If the ECM recognizes the high signal voltage then the ECM and wiring are OK.
3. Check for good sensor connection. If connection is good, replace TP sensor.
4. This simulates a high signal voltage to check for an open in CKT 417. The Tech 1 scan tool will not read up to 12 volts, but what is important is that the ECM recognizes the signal on CKT 417.

Diagnostic Aids:

Scan TP sensor while depressing accelerator pedal with engine stopped and ignition "ON." Display should vary from about 600 mV (.6 volt) when throttle is closed, to over 4500 mV (4.5 volts) when throttle is held wide open.

Also, a Tech 1 scan tool will display throttle angle %: 0% = closed throttle; 100% = WOT.

If DTC 22, 33 and/or 66 are set, check CKT 416 for faulty wiring or connections.

Should check condition of connector and sensor terminals for moisture or corrosion, and clean and/or replace as necessary. If corrosion is found, check condition of connector seal and repair or replace if necessary.

2.3L (VIN D, A & 3) ENGINE — DIAGNOSTIC TROUBLE CODE CHART — 1993–94 ACHIEVA, GRAND AM AND SKYLARK

DTC 22
THROTTLE POSITION (TP) SENSOR CIRCUIT
(SIGNAL VOLTAGE LOW)
2.3L (VIN D, A & 3) "N" CARLINE

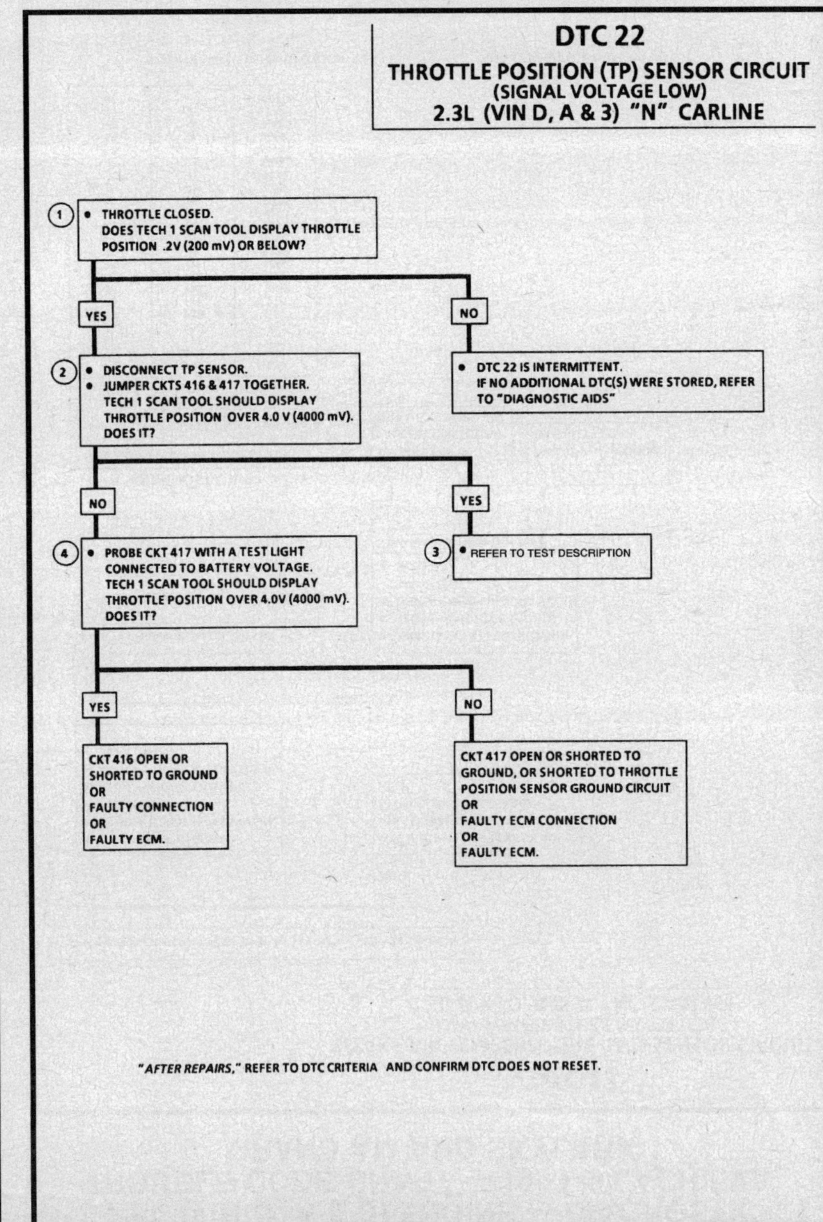

2.3L (VIN D, A & 3) ENGINE — DIAGNOSTIC TROUBLE CODE CHART — 1993–94 ACHIEVA, GRAND AM AND SKYLARK

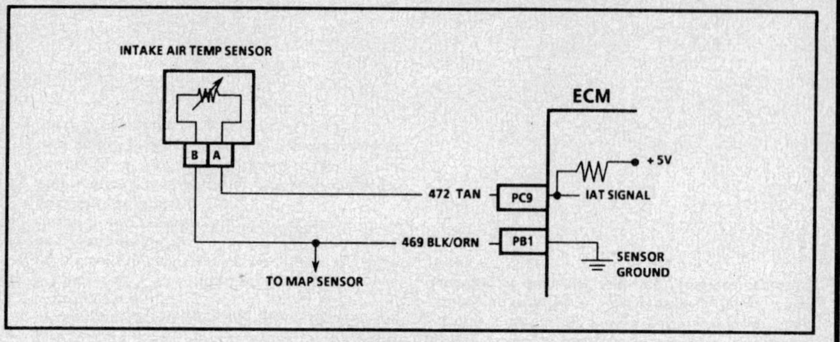

DTC 23
INTAKE AIR TEMPERATURE (IAT) SENSOR CIRCUIT
(LOW TEMPERATURE INDICATED)
2.3L (VIN D, A & 3) "N" CARLINE

Circuit Description:

The Intake Air Temperature (IAT) sensor uses a thermistor to control the signal voltage at the Engine Control Module (ECM). The ECM applies a voltage (about 5 volts) on CKT 472 to the sensor. When the air is cold the sensor (thermistor) resistance is high, therefore the ECM terminal "PC9" voltage will be high. If the air is warm the sensor resistance is low, therefore the ECM terminal "PC9" voltage will be low.

Test Description: Number(s) below refer to circled number(s) on the diagnostic chart.

1. DTC 23 will set if:
   - A signal voltage indicates a intake air temperature below about -40°C (-40°F).
   - Time since engine start is 320 seconds or longer.
   - Vehicle speed less than 15 mph.
2. A DTC 23 will set due to an open sensor, wire, or connection. This test will determine if the wiring and ECM are OK.
3. This will determine if the signal CKT 472 or the sensor ground CKT 469 is open.

Diagnostic Aids:

A Tech 1 scan tool displays temperature of the air entering the engine, which should be close to ambient air temperature when engine is cold, and rise as underhood temperature increases.

A faulty connection, or an open in CKT 472 or CKT 469 can result in a DTC 23.

DTCs 23 and 34 stored at the same time, could be the result of an open CKT 469. The "Temperature to Resistance Values" scale may be used to test the IAT sensor at various temperature levels to evaluate the possibility of a "slewed" (mis-scaled) sensor. A "slewed" sensor could result in poor driveability complaints.

2.3L (VIN D, A & 3) ENGINE — DIAGNOSTIC TROUBLE CODE CHART — 1993–94 ACHIEVA, GRAND AM AND SKYLARK

DTC 23
INTAKE AIR TEMPERATURE (IAT) SENSOR CIRCUIT
(LOW TEMPERATURE INDICATED)
2.3L (VIN D, A & 3) "N" CARLINE

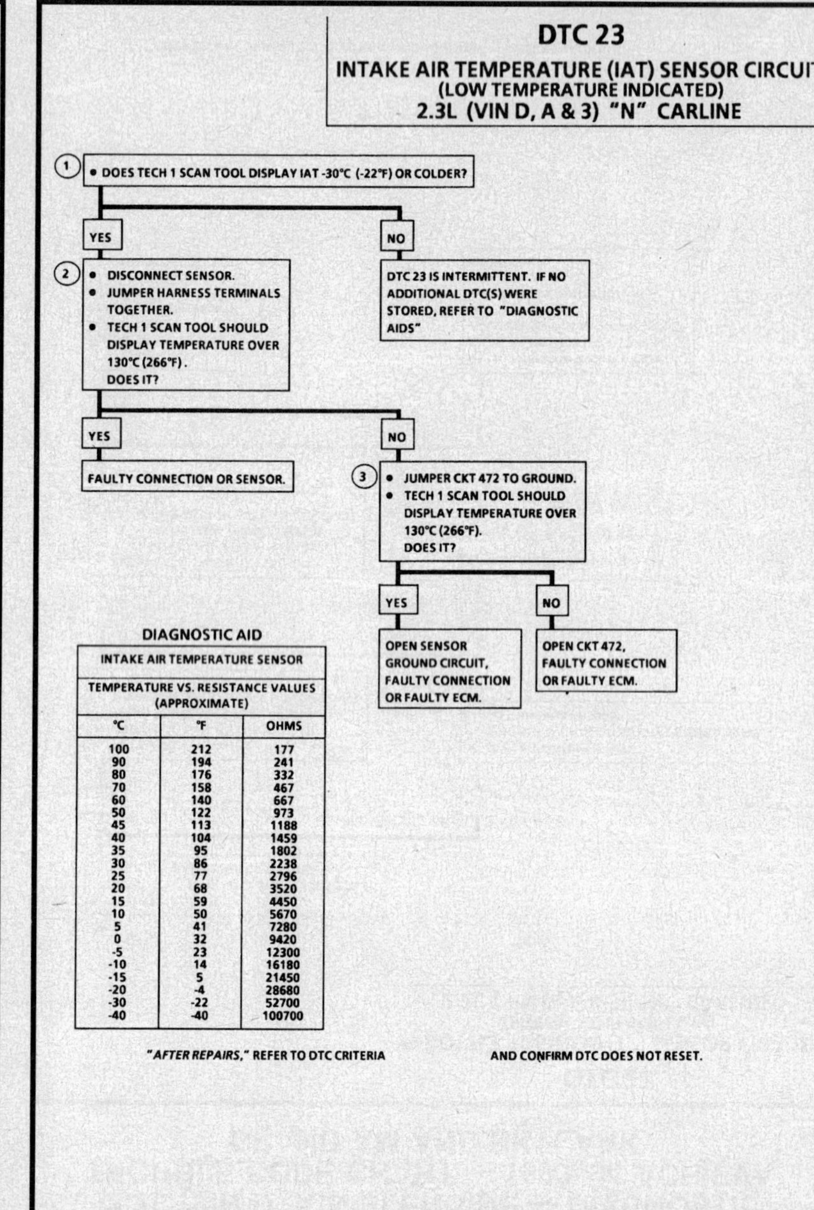

DIAGNOSTIC AID		
INTAKE AIR TEMPERATURE SENSOR		
TEMPERATURE VS. RESISTANCE VALUES (APPROXIMATE)		
°C	°F	OHMS
100	212	177
90	194	241
80	176	332
70	158	467
60	140	667
50	122	973
45	113	1188
40	104	1459
35	95	1802
30	86	2238
25	77	2796
20	68	3520
15	59	4450
10	50	5670
5	41	7280
0	32	9420
-5	23	12300
-10	14	16180
-15	5	21450
-20	-4	28680
-30	-22	52700
-40	-40	100700

"AFTER REPAIRS," REFER TO DTC CRITERIA   AND CONFIRM DTC DOES NOT RESET.

2.3L (VIN D, A & 3) ENGINE — DIAGNOSTIC TROUBLE CODE CHART — 1993–94 ACHIEVA, GRAND AM AND SKYLARK

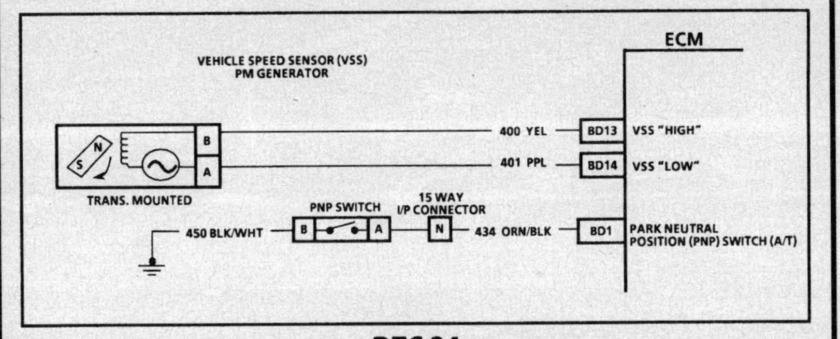

DTC 24
VEHICLE SPEED SENSOR (VSS) CIRCUIT
2.3L (VIN D, A & 3) "N" CARLINE

Circuit Description:

Vehicle speed information is provided to the Engine Control Module (ECM) by the Vehicle Speed Sensor (VSS) which is a permanent magnet generator that is mounted in the transaxle. The permanent magnet generator produces a pulsing voltage whenever vehicle speed is over about 3 mph (5 km/h). The AC voltage level and the number of pulses increases with vehicle speed. The ECM then converts the pulsing voltage to mph which is used for calculations, and the mph can be displayed with a Tech 1 scan tool. Output of the generator can also be seen by using a digital voltmeter on the AC scale while rotating the generator.

The function of VSS buffer used in past model years has been incorporated into the ECM. The ECM then supplies the necessary signal for the instrument panel for operating the speedometer, the odometer, and for the cruise control module.

Test Description: Number(s) below refer to circled number(s) on the diagnostic chart.

1. DTC 24 will set if vehicle speed is less than 2 mph when:
   - Engine speed is between 1500 and 3600 RPM.
   - TP sensor is less than 2% (Auto) or greater than 7% (Man).
   - Not in park or neutral.
   - All conditions met for 20 seconds.
   - No DTC 21, 22, 33 or 34.
   - Engine vacuum between 68 - 80 kPa (Auto), or 38-58 kPa (Man).

   These conditions are met during a road load operation.

   The permanent magnet generator only produces a signal if drive wheels are turning greater than 3 mph (5 km/h).

2. Check PROM for correct application before replacing ECM.

Diagnostic Aids:

A Tech 1 scan tool should indicate a vehicle speed whenever the drive wheels are turning greater than 3 mph (5 km/h).

A problem in CKT 434 will not affect the VSS input or the readings on a scan.

Check CKT 400 and CKT 401 for proper connections. Be sure they are clean and tight and the harness is routed correctly.

(A/T) A faulty or misadjusted Park/Neutral Position (PNP) switch can result in a false DTC 24. Use a Tech 1 scan tool and check for proper signal while in drive (3T40). Refer to CHART C-1A for Park/Neutral Position (PNP) switch diagnosis check.

DTC 24 can be falsely set if engine is "brake-torqued" in gear for 20 seconds.

2.3L (VIN D, A & 3) ENGINE — DIAGNOSTIC TROUBLE CODE CHART — 1993–94 ACHIEVA, GRAND AM AND SKYLARK

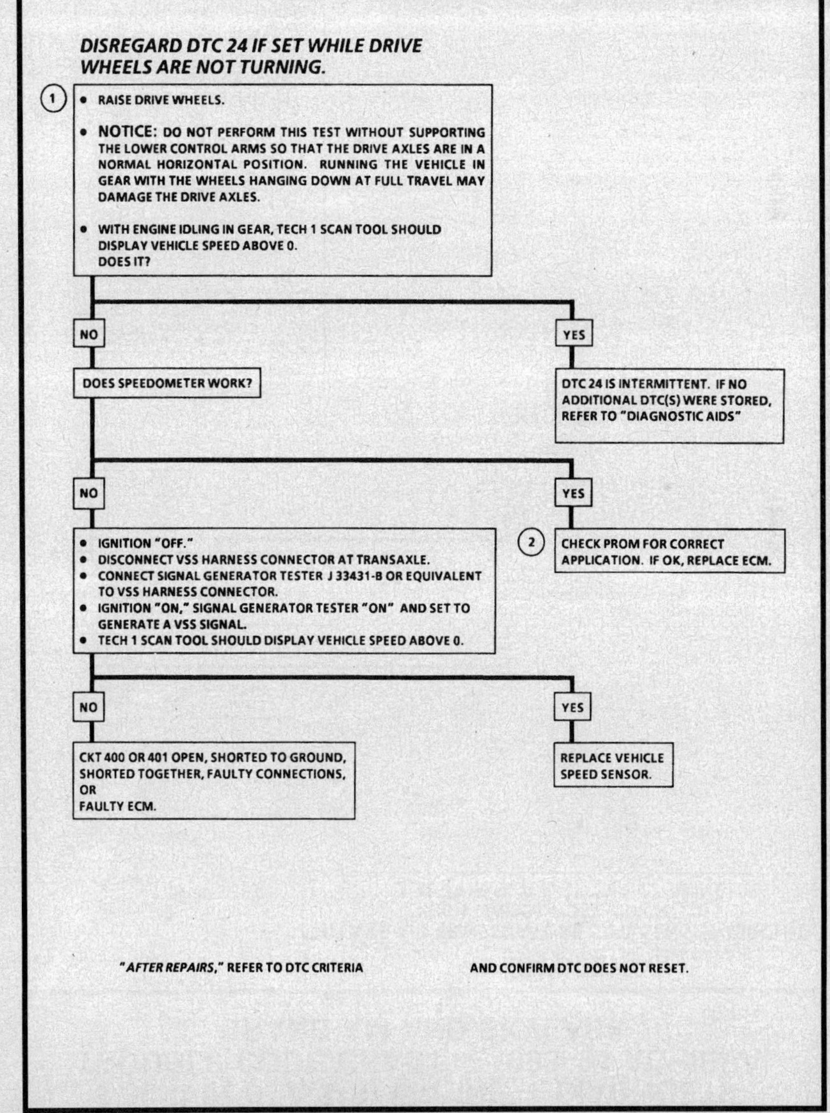

2.3L (VIN D, A & 3) ENGINE — DIAGNOSTIC TROUBLE CODE CHART — 1993–94 ACHIEVA, GRAND AM AND SKYLARK

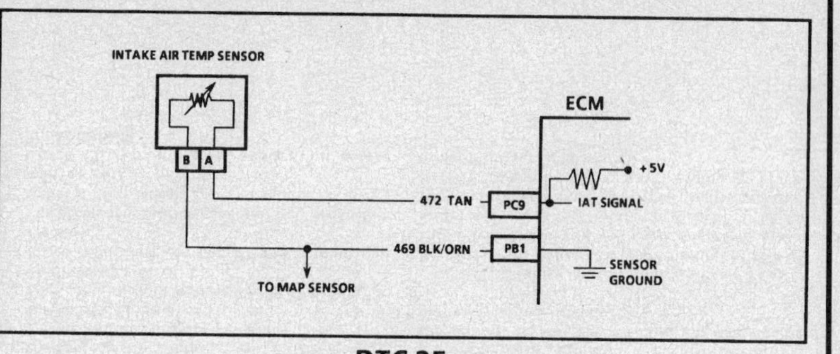

DTC 25
INTAKE AIR TEMPERATURE (IAT) SENSOR CIRCUIT
(HIGH TEMPERATURE INDICATED)
2.3L (VIN D, A & 3) "N" CARLINE

Circuit Description:

The Intake Air Temperature (IAT) sensor uses a thermistor to control the signal voltage to the Engine Control Module (ECM). The ECM applies a voltage (4-6 volts) on CKT 472 to the sensor. When intake air is cold, the sensor (thermistor) resistance is high, therefore, the ECM terminal "PC9" voltage is high. As the air warms, the sensor resistance becomes less, and the voltage drops. As the incoming air gets warmer, the sensor resistance decreases, causing ECM terminal "PC9" voltage to decrease.

Test Description: Number(s) below refer to circled number(s) on the diagnostic chart.

1. DTC 25 will set if:
   - Signal voltage indicates a intake air temperature greater than about 151°C (304°F).
   - Vehicle speed is greater than 15 mph.
   - Engine run time greater than 320 seconds.

Diagnostic Aids:

The "Temperature To Resistance Value" scale may be used to test the IAT sensor at various temperature levels to evaluate the possibility of a "slewed" (mis-scaled) sensor. A "slewed" sensor could result in poor driveability complaints.

2.3L (VIN D, A & 3) ENGINE — DIAGNOSTIC TROUBLE CODE CHART — 1993–94 ACHIEVA, GRAND AM AND SKYLARK

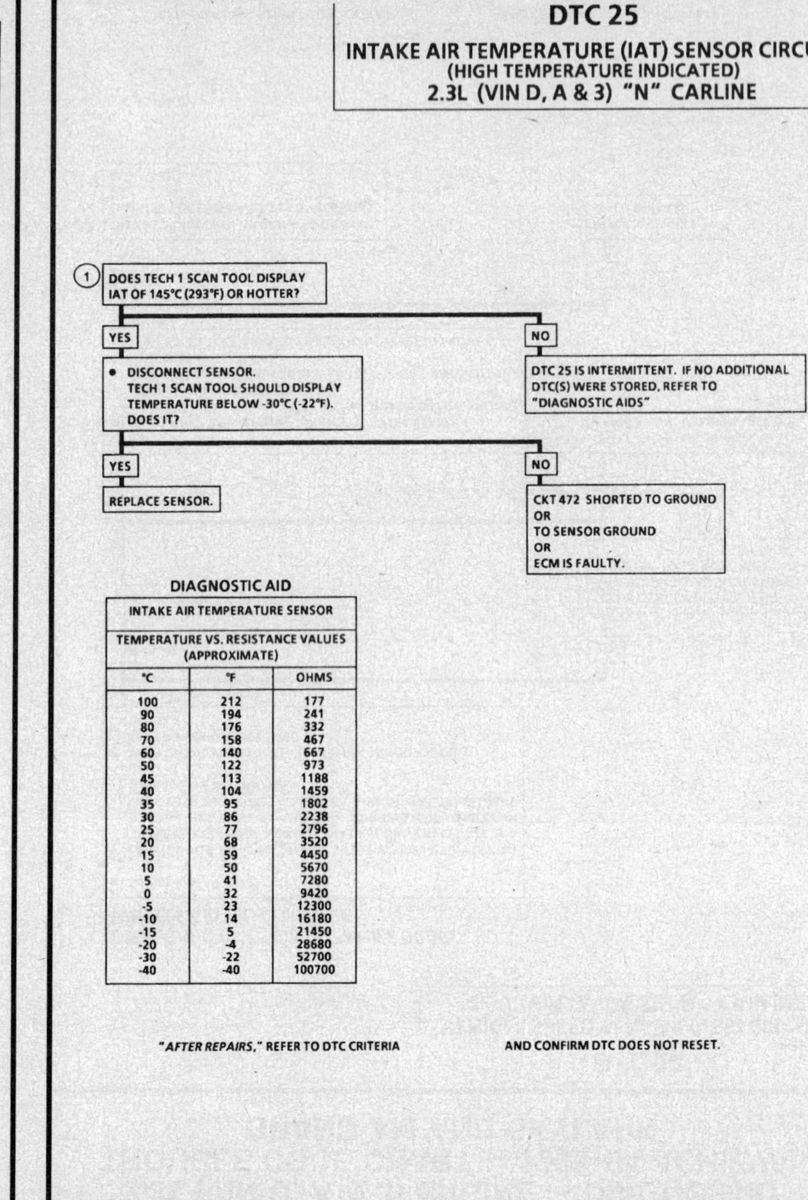

DIAGNOSTIC AID

INTAKE AIR TEMPERATURE SENSOR		
TEMPERATURE VS. RESISTANCE VALUES (APPROXIMATE)		
°C	°F	OHMS
100	212	177
90	194	241
80	176	332
70	158	467
60	140	667
50	122	973
45	113	1188
40	104	1459
35	95	1802
30	86	2238
25	77	2796
20	68	3520
15	59	4450
10	50	5670
5	41	7280
0	32	9420
-5	23	12300
-10	14	16180
-15	5	21450
-20	-4	28680
-30	-22	52700
-40	-40	100700

"AFTER REPAIRS," REFER TO DTC CRITERIA   AND CONFIRM DTC DOES NOT RESET.

2.3L (VIN D, A & 3) ENGINE — DIAGNOSTIC TROUBLE CODE CHART — 1993–94 ACHIEVA, GRAND AM AND SKYLARK

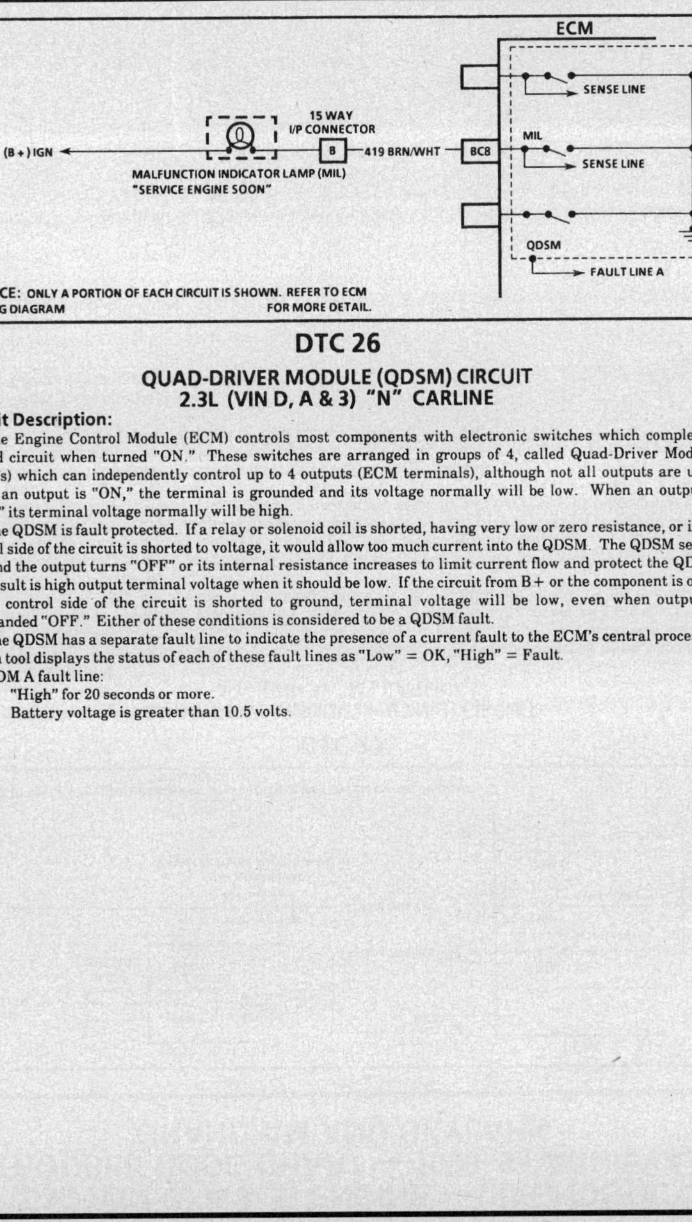

DTC 26

**QUAD-DRIVER MODULE (QDSM) CIRCUIT
2.3L (VIN D, A & 3) "N" CARLINE**

Circuit Description:

The Engine Control Module (ECM) controls most components with electronic switches which complete a ground circuit when turned "ON." These switches are arranged in groups of 4, called Quad-Driver Modules (QDM's) which can independently control up to 4 outputs (ECM terminals), although not all outputs are used. When an output is "ON," the terminal is grounded and its voltage normally will be low. When an output is "OFF," its terminal voltage normally will be high.

The QDSM is fault protected. If a relay or solenoid coil is shorted, having very low or zero resistance, or if the control side of the circuit is shorted to voltage, it would allow too much current into the QDSM. The QDSM senses this and the output turns "OFF" or its internal resistance increases to limit current flow and protect the QDSM. The result is high output terminal voltage when it should be low. If the circuit from B+ or the component is open, or the control side of the circuit is shorted to ground, terminal voltage will be low, even when output is commanded "OFF." Either of these conditions is considered to be a QDSM fault.

The QDSM has a separate fault line to indicate the presence of a current fault to the ECM's central processor. A scan tool displays the status of each of these fault lines as "Low" = OK, "High" = Fault.

- QDM A fault line:
  - "High" for 20 seconds or more.
  - Battery voltage is greater than 10.5 volts.

2.3L (VIN D, A & 3) ENGINE — DIAGNOSTIC TROUBLE CODE CHART — 1993–94 ACHIEVA, GRAND AM AND SKYLARK

DTC 26

**QUAD-DRIVER MODULE (QDSM) CIRCUIT
2.3L (VIN D, A & 3) "N" CARLINE**

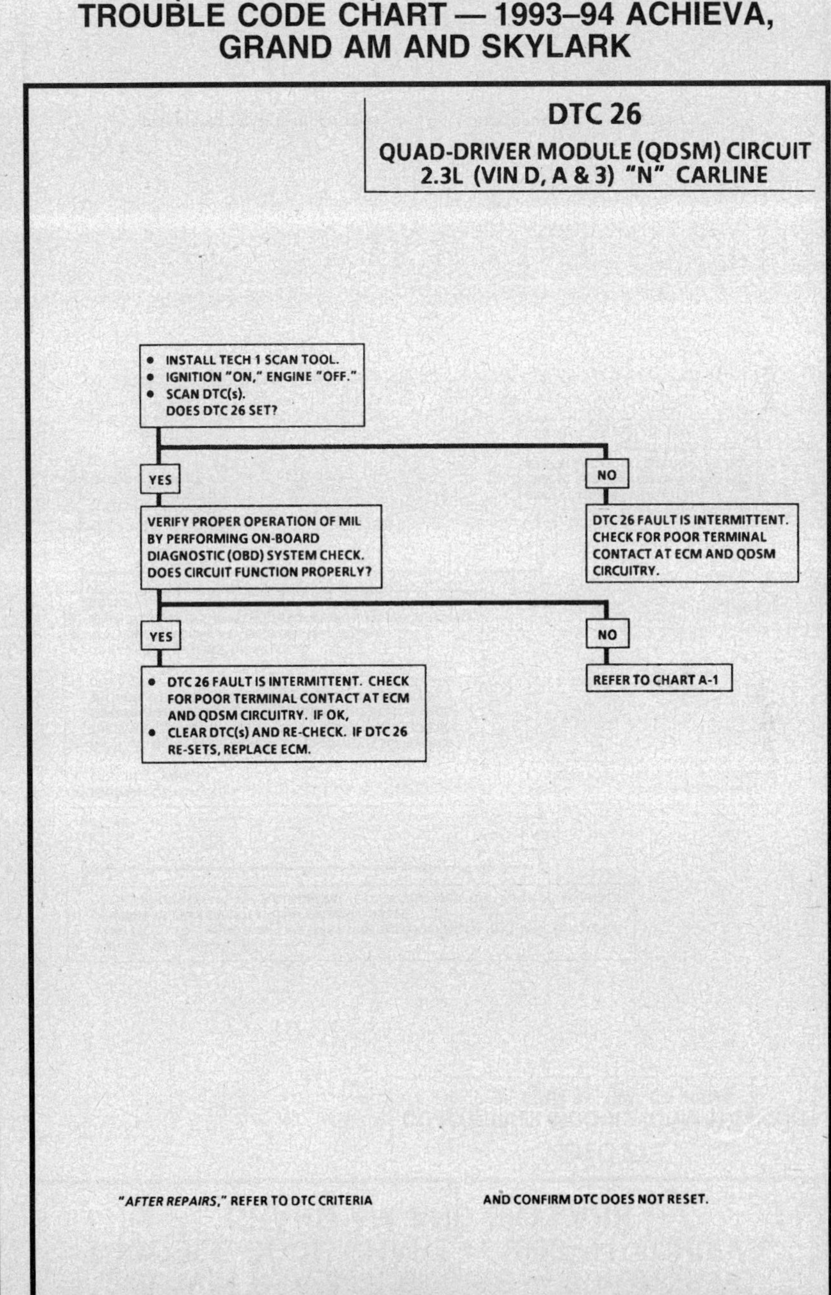

"AFTER REPAIRS," REFER TO DTC CRITERIA   AND CONFIRM DTC DOES NOT RESET.

2.3L (VIN D, A & 3) ENGINE — DIAGNOSTIC TROUBLE CODE CHART — 1993–94 ACHIEVA, GRAND AM AND SKYLARK

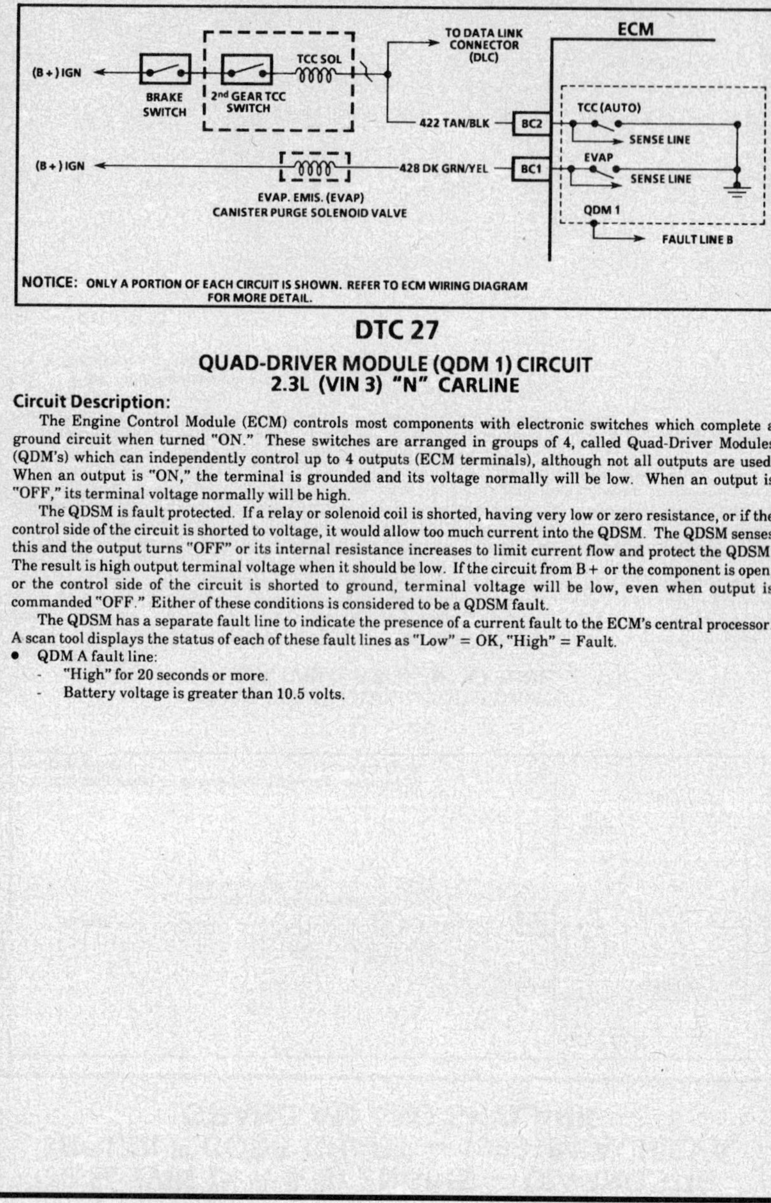

DTC 27
QUAD-DRIVER MODULE (QDM 1) CIRCUIT
2.3L (VIN 3) "N" CARLINE

Circuit Description:

The Engine Control Module (ECM) controls most components with electronic switches which complete a ground circuit when turned "ON." These switches are arranged in groups of 4, called Quad-Driver Modules (QDM's) which can independently control up to 4 outputs (ECM terminals), although not all outputs are used. When an output is "ON," the terminal is grounded and its voltage normally will be low. When an output is "OFF," its terminal voltage normally will be high.

The QDSM is fault protected. If a relay or solenoid coil is shorted, having very low or zero resistance, or if the control side of the circuit is shorted to voltage, it would allow too much current into the QDSM. The QDSM senses this and the output turns "OFF" or its internal resistance increases to limit current flow and protect the QDSM. The result is high output terminal voltage when it should be low. If the circuit from B+ or the component is open, or the control side of the circuit is shorted to ground, terminal voltage will be low, even when output is commanded "OFF." Either of these conditions is considered to be a QDSM fault.

The QDSM has a separate fault line to indicate the presence of a current fault to the ECM's central processor. A scan tool displays the status of each of these fault lines as "Low" = OK, "High" = Fault.

- QDM A fault line:
  - "High" for 20 seconds or more.
  - Battery voltage is greater than 10.5 volts.

2.3L (VIN D, A & 3) ENGINE — DIAGNOSTIC TROUBLE CODE CHART — 1993–94 ACHIEVA, GRAND AM AND SKYLARK

DTC 27
QUAD DRIVER MODULE (QDM 1) CIRCUIT
2.3L (VIN 3) "N" CARLINE

- IGNITION "ON," ENGINE "OFF."
- USING TECH 1, COMMAND EVAP CANISTER PURGE SOLENOID VALVE "ON" AND "OFF" WHILE LISTENING TO EVAP CANISTER PURGE SOLENOID VALVE.
  DOES EVAP CANISTER PURGE SOLENOID VALVE TURN "ON" AND "OFF" WHEN COMMANDED?

YES

- IGNITION "OFF."
- RAISE DRIVE WHEELS.
- NOTICE: DO NOT PERFORM THIS TEST WITHOUT SUPPORTING THE LOWER CONTROL ARMS SO THAT THE DRIVE AXLES ARE IN A NORMAL HORIZONTAL POSITION. RUNNING THE VEHICLE IN GEAR WITH THE WHEELS HANGING DOWN AT FULL TRAVEL MAY DAMAGE THE DRIVE AXLES.
- USING TECH 1, OBSERVE TCC AND RPM.
- RUN VEHICLE IN DRIVE AT APPROXIMATELY 35 MPH.
- WHEN TCC ENGAGES, RPM SHOULD DROP WITH STEADY THROTTLE, DOES IT?

NO

NO

SEE: C8" TO DIAGNOSE TCC CIRCUIT.

SEE C3 TO DIAGNOSE EVAP CANISTER PURGE SOLENOID VALVE CIRCUIT.

YES

DTC 27 FAULT IS INTERMITTENT. CHECK FOR POOR TERMINAL CONTACT AT ECM AND QDM 1 CIRCUITRY. IF OK,
- CLEAR DTCs AND RECHECK. IF DTC 27 RE-SETS, REPLACE ECM.

"AFTER REPAIRS," REFER TO DTC CRITERIA   AND CONFIRM DTC DOES NOT RESET.

2.3L (VIN D, A & 3) ENGINE — DIAGNOSTIC TROUBLE CODE CHART — 1993–94 ACHIEVA, GRAND AM AND SKYLARK

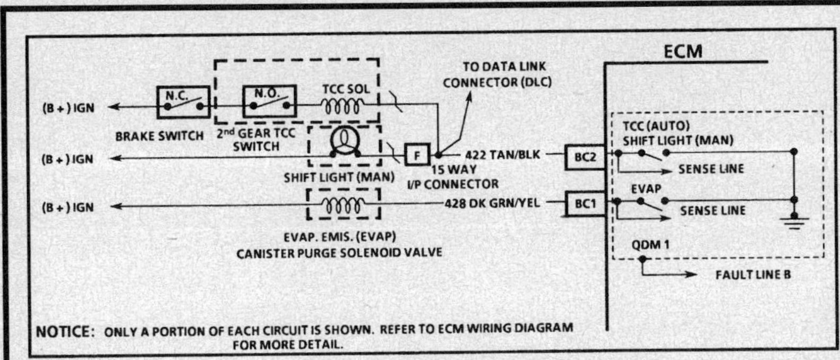

DTC 27
QUAD-DRIVER MODULE (QDM 1) CIRCUIT
2.3L (VIN D, A & 3) "N" CARLINE

Circuit Description:

The ECM controls most components with electronic switches which complete a ground circuit when turned "ON." These switches are arranged in groups of 4, called Quad-Driver Modules (QDM's) which can independently control up to 4 outputs (ECM terminals), although not all outputs are used. When an output is "ON," the terminal is grounded and its voltage normally will be low. When an output is "OFF," its terminal voltage normally will be high, except for the TCC, as noted below, which depends on the brake and 2nd gear TCC switches.

QDM's are fault protected. If a relay or solenoid coil is shorted, having very low or zero resistance, or if the control side of the circuit is shorted to voltage, it would allow too much current into the QDM. The QDM senses this and the output turns "OFF" or its internal resistance increases to limit current flow and protect the QDM. The result is high output terminal voltage when it should be low. If the circuit from B + or the component is open, or the control side of the circuit is shorted to ground, terminal voltage will be low, even when output is commanded "OFF." Either of these conditions is considered to be a QDM fault.

QDM 1 has a fault line to indicate the presence of a current fault to the ECM's central processor. A scan tool displays the status of the fault line as "Low" = OK, "High" = Fault. Because of the brake and 2nd gear switches in the TCC circuit, DTC 27 is set under different conditions as follows:

- QDM B fault line:
  - "High" for 20 seconds or more.
  - TCC is commanded "ON."
  - Battery voltage is greater than 10.5 volts.

2.3L (VIN D, A & 3) ENGINE — DIAGNOSTIC TROUBLE CODE CHART — 1993–94 ACHIEVA, GRAND AM AND SKYLARK

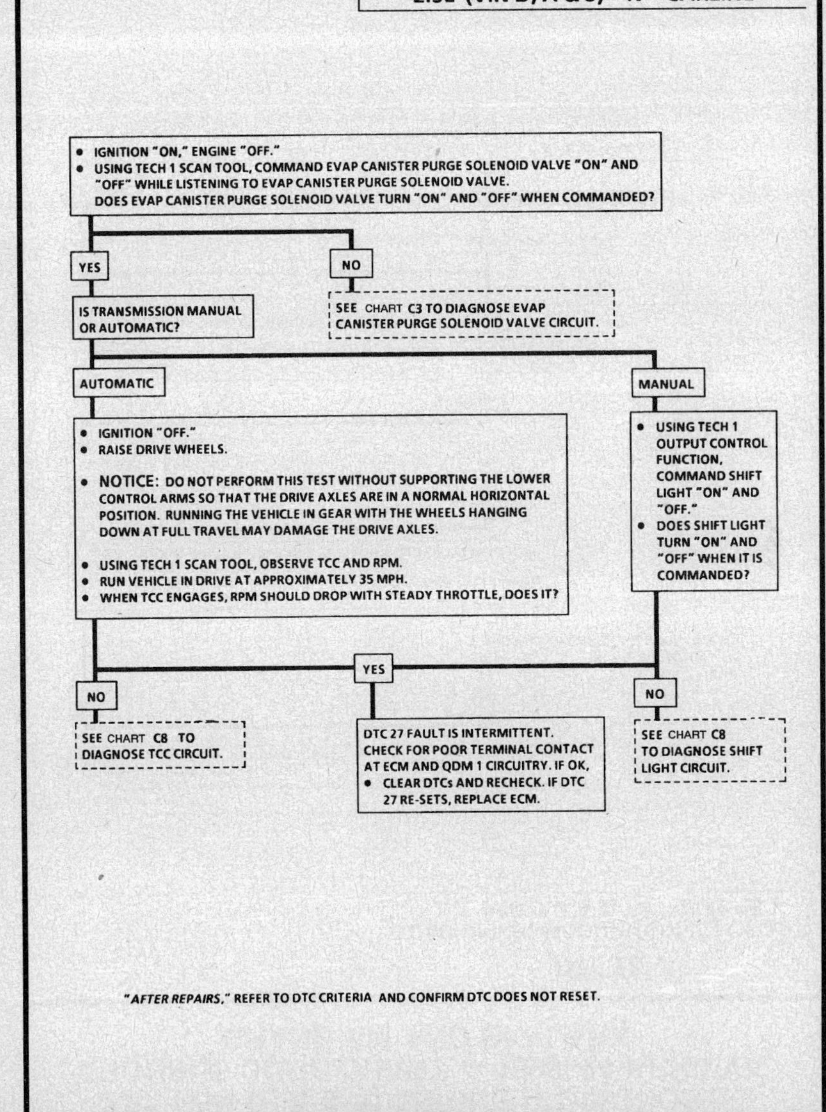

2.3L (VIN D, A & 3) ENGINE — DIAGNOSTIC TROUBLE CODE CHART — 1993–94 ACHIEVA, GRAND AM AND SKYLARK

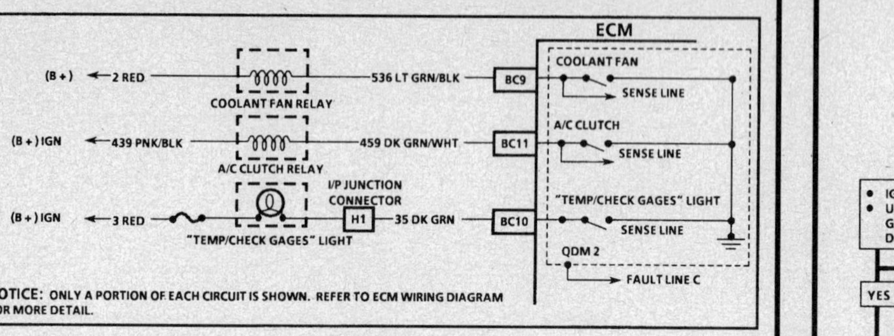

NOTICE: ONLY A PORTION OF EACH CIRCUIT IS SHOWN. REFER TO ECM WIRING DIAGRAM FOR MORE DETAIL.

DTC 28
QUAD-DRIVER MODULE (QDM 2) CIRCUIT
2.3L (VIN D, A & 3) "N" CARLINE

Circuit Description:

The Engine Control Module (ECM) controls most components with electronic switches which complete a ground circuit when turned "ON." These switches are arranged in groups of 4, called Quad-Driver Modules (QDM's) which can independently control up to 4 outputs (ECM terminals), although not all outputs are used. When an output is "ON," the terminal is grounded and its voltage normally will be low. When an output is "OFF," its terminal voltage normally will be high, except for the TCC, as noted below, which depends on the brake and 2nd gear TCC switches.

QDM's are fault protected. If a relay or solenoid coil is shorted, having very low or zero resistance, or if the control side of the circuit is shorted to voltage, it would allow too much current into the QDM. The QDM senses this and the output turns "OFF" or its internal resistance increases to limit current flow and protect the QDM. The result is high output terminal voltage when it should be low. If the circuit from B+ or the component is open, or the control side of the circuit is shorted to ground, terminal voltage will be low, even when output is commanded "OFF." Either of these conditions is considered to be a QDM fault.

QDM 2 has a fault line to indicate the presence of a current fault to the ECM's central processor. A scan tool displays the status of the fault line as "Low" = OK, "High" = Fault.

- QDM C fault line:
  - "High" for 20 seconds or more.
  - Battery voltage is greater than 10.5 volts.

2.3L (VIN D, A & 3) ENGINE — DIAGNOSTIC TROUBLE CODE CHART — 1993–94 ACHIEVA, GRAND AM AND SKYLARK

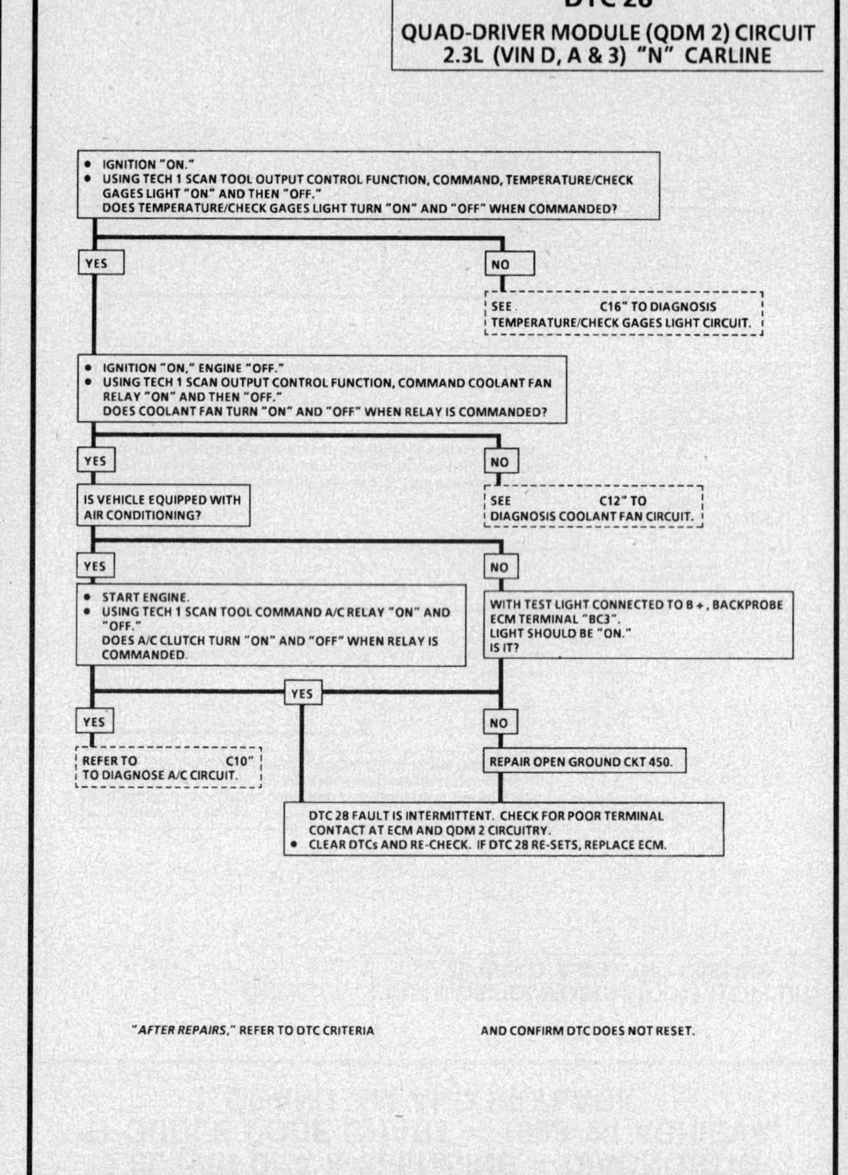

2.3L (VIN D, A & 3) ENGINE — DIAGNOSTIC TROUBLE CODE CHART — 1993–94 ACHIEVA, GRAND AM AND SKYLARK

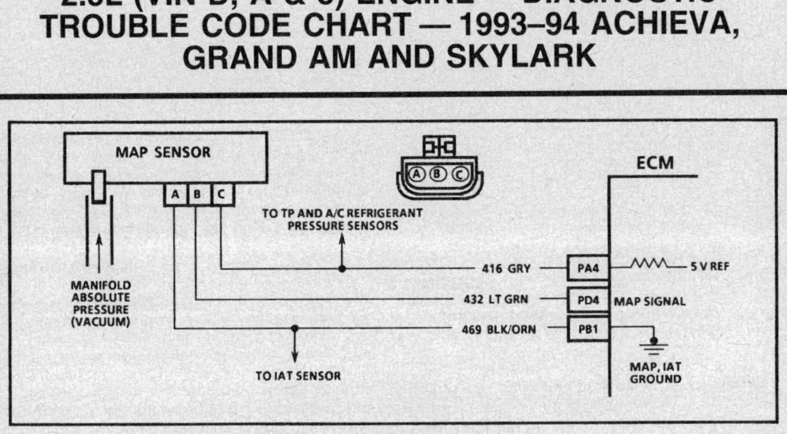

DTC 33

MANIFOLD ABSOLUTE PRESSURE (MAP) SENSOR CIRCUIT
(SIGNAL VOLTAGE HIGH - LOW VACUUM)
2.3L (VIN D, A & 3) "N" CARLINE

Circuit Description:

The Manifold Absolute Pressure (MAP) sensor responds to changes in manifold pressure (vacuum). The Engine Control Module (ECM) receives this information as a signal voltage that will vary from about 1 to 1.5 volts at closed throttle (idle) to 4.5 - 4.8 volts at Wide Open Throttle (WOT) (low vacuum).

If the MAP sensor fails, the ECM will substitute a fixed MAP value and use the Throttle Position (TP) sensor to control fuel delivery.

Test Description: Number(s) below refer to circled number(s) on the diagnostic chart.
1. This step will determine if <u>DTC 33</u> is the result of a hard failure or an intermittent condition.
   DTC 33 will set when:
   - Engine running.
   - MAP greater than 82 kPa.
   - TP sensor less than 12%.
   - VSS less than 1 mph.
   - DTC 21 or 22 not present.
   - Above conditions met for 5 seconds.
2. This step simulates conditions for a <u>DTC 34</u>. If the ECM recognizes the change, the ECM and CKT 416 and CKT 432 are OK. If CKT 469 is open, there may also be a stored DTC 23.

Diagnostic Aids:

With the ignition "ON" and the engine stopped, the manifold pressure is equal to atmospheric pressure and the signal voltage will be high. This information is used by the ECM as an indication of vehicle altitude. Comparison of this reading with a known good vehicle with the same sensor is a good way to check accuracy of a "suspect" sensor. Readings should be the same ± .4 volt.

A DTC 33 will result if CKT 469 is open or if CKT 432 is shorted to voltage or to CKT 416.

- Check all connections.
- Disconnect sensor from bracket and twist sensor by hand (only) to check for intermittent connections. Output changes greater than .1 volt indicates a bad connector or connection. If OK, replace sensor.

NOTICE: Make sure electrical connector remains securely fastened.

- Refer to CHART C-1D, MAP sensor voltage vs. atmospheric pressure for further diagnosis.

NOTICE: After repairs use Tech 1 scan tool "Fuel Trim Reset" function to reset L. T. fuel trim to 128.

2.3L (VIN D, A & 3) ENGINE — DIAGNOSTIC TROUBLE CODE CHART — 1993–94 ACHIEVA, GRAND AM AND SKYLARK

DTC 33

MANIFOLD ABSOLUTE PRESSURE (MAP) SENSOR CIRCUIT
(SIGNAL VOTAGE HIGH - LOW VACUUM)
2.3L (VIN D, A & 3) "N" CARLINE

(1)
- IF ENGINE IDLE IS ROUGH, UNSTABLE, OR INCORRECT, CORRECT CONDITION BEFORE USING CHART.
- ENGINE IDLING.
- DOES TECH 1 SCAN TOOL DISPLAY A MAP VOLTAGE OF 4.0 VOLTS OR OVER?

YES

(2)
- IGNITION "OFF."
- DISCONNECT MAP SENSOR ELECTRICAL CONNECTOR.
- IGNITION "ON."
- TECH 1 SCAN TOOL SHOULD READ A VOLTAGE OF 1 VOLT OR LESS.
  DOES IT?

NO
DTC 33 IS INTERMITTENT.
IF NO ADDITIONAL DTC(S) WERE STORED, REFER TO "DIAGNOSTIC AIDS"

YES
- PROBE SENSOR GROUND CIRCUIT WITH A TEST LIGHT TO BATTERY VOLTAGE.
- TEST LIGHT SHOULD LIGHT.
  DOES IT?

NO
MAP SIGNAL CIRCUIT SHORTED TO VOLTAGE, SHORTED TO 5 VOLT REFERENCE CIRCUIT
OR
FAULTY ECM.

YES
PLUGGED
OR
LEAKING SENSOR VACUUM HOSE
OR
FAULTY MAP SENSOR.

NO
OPEN SENSOR GROUND CIRCUIT.

"AFTER REPAIRS," REFER TO DTC CRITERIA   AND CONFIRM DTC DOES NOT RESET.

2.3L (VIN D, A & 3) ENGINE — DIAGNOSTIC TROUBLE CODE CHART — 1993–94 ACHIEVA, GRAND AM AND SKYLARK

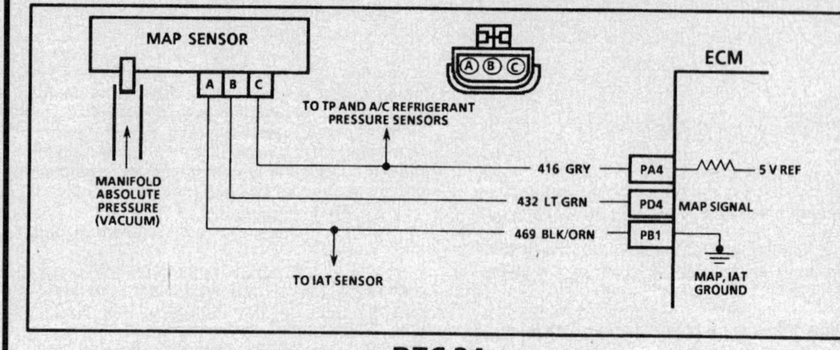

DTC 34
MANIFOLD ABSOLUTE PRESSURE (MAP) SENSOR CIRCUIT
(SIGNAL VOLTAGE LOW - HIGH VACUUM)
2.3L (VIN D, A & 3) "N" CARLINE

Circuit Description:

The Manifold Absolute Pressure (MAP) sensor responds to changes in manifold pressure (vacuum). The Engine Control Module (ECM) receives this information as a signal voltage that will vary from about 1 to 1.5 volts at closed throttle (idle) to 4.5 - 4.8 volts at Wide Open Throttle (WOT) (low vacuum).

If the MAP sensor fails, the ECM will substitute a fixed MAP value and use the Throttle Position (TP) sensor to control fuel delivery.

Test Description: Number(s) below refer to circled number(s) on the diagnostic chart.
1. This step determines if DTC 34 is the result of a hard failure or an intermittent condition.
   DTC 34 will set when:
   - Engine running.
   - No DTC 21.
   - MAP less than 14 kPa.
   - Engine RPM less than 1200 or TP sensor greater than 15.2%.
   - Above conditions met for .2 seconds.
2. Jumpering harness terminals "B" to "C" (5 volts to signal circuit) will determine if the sensor is at fault, or if there is a problem with the ECM or wiring.
3. The Tech 1 scan tool may not display 5 volts. The important thing is that the ECM recognizes the voltage as more than 4 volts, indicating that the ECM and CKT 432 are OK.

Diagnostic Aids:

An intermittent open in CKT 432 or CKT 416 will result in a DTC 34. With the ignition "ON" and the engine "OFF," the manifold pressure is equal to atmospheric pressure and the signal voltage will be high. This information is used by the ECM as an indication of vehicle altitude.

Comparison of this reading with a known good vehicle with the same sensor is a good way to check accuracy of a "suspect" sensor. Readings should be the same ± .4 volt. Also CHART C-1D can be used to test the MAP sensor. Refer to "Intermittents" in

- Check all connections.
- Disconnect sensor from bracket and twist sensor by hand (only) to check for intermittent connections. Output changes greater than .1 volt indicates a bad connector or connection. If OK, replace sensor.

NOTICE: Make sure electrical connector remains securely fastened.

- Refer to CHART C-1D, MAP sensor voltage vs. atmospheric pressure for further diagnosis.

NOTICE: After repairs use Tech 1 scan tool "Fuel Trim Reset" function to reset L. T. fuel trim to 128.

DTC 34
MANIFOLD ABSOLUTE PRESSURE (MAP) SENSOR CIRCUIT
(SIGNAL VOLTAGE LOW - HIGH VACUUM)
2.3L (VIN D, A & 3) "N" CARLINE

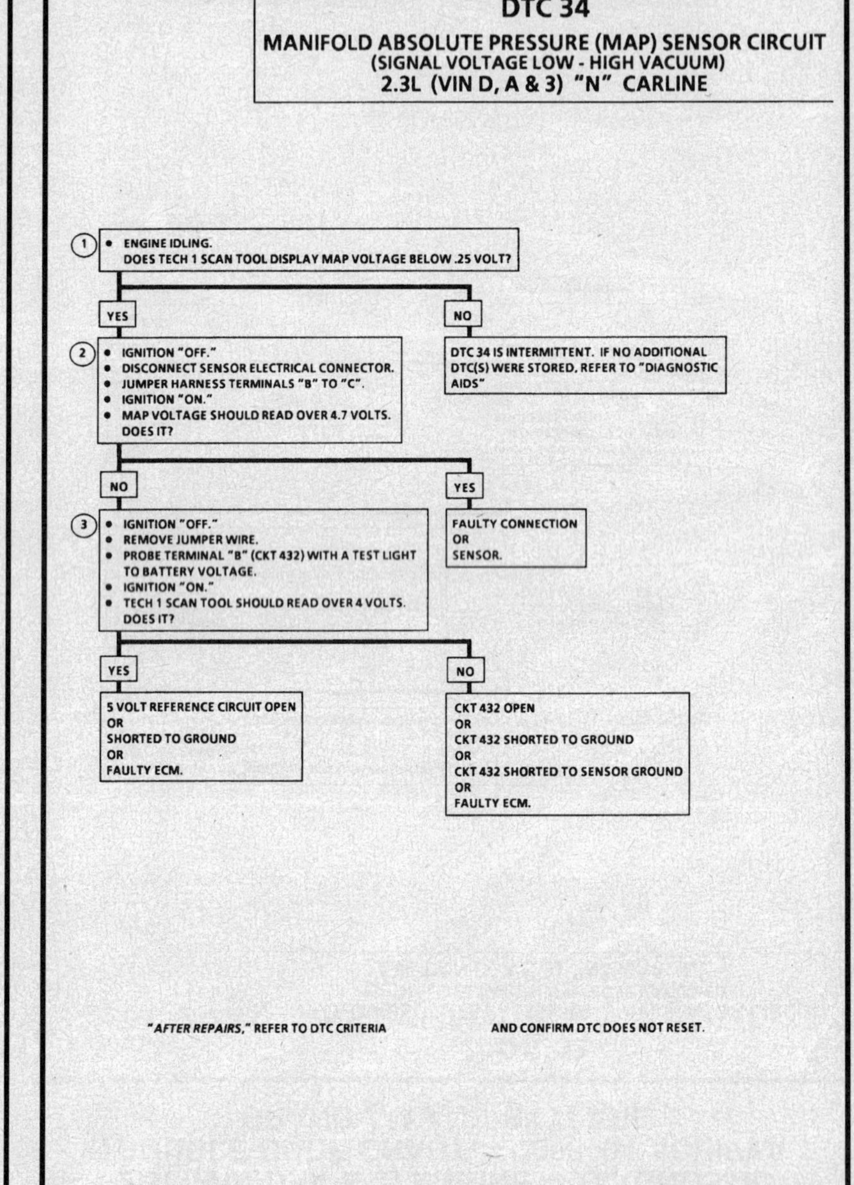

2.3L (VIN D, A & 3) ENGINE — DIAGNOSTIC TROUBLE CODE CHART — 1993–94 ACHIEVA, GRAND AM AND SKYLARK

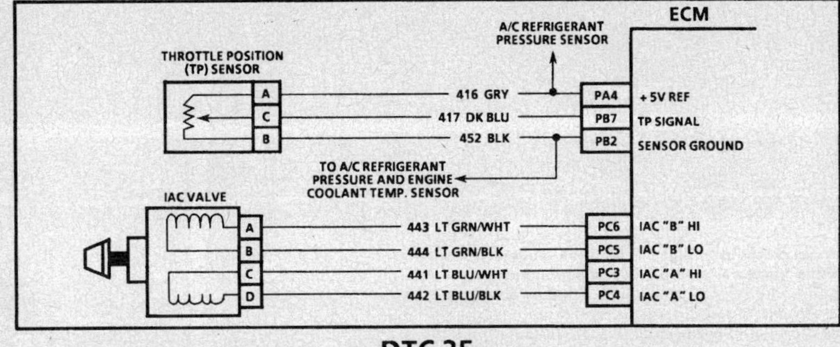

DTC 35
IDLE SPEED ERROR
2.3L (VIN D, A & 3) "N" CARLINE

Circuit Description:

The Engine Control Module (ECM) controls idle speed to a calculated, "desired" RPM based on sensor inputs and actual engine RPM, determined by the time between successive 7X reference pulses from the Electronic Ignition Control Module (ICM). The ECM uses 4 circuits to move an Idle Air Control (IAC) valve, which allows varying amounts of air flow into the intake manifold, controlling idle speed.

DTC 35 sets when:
- Engine speed is at least 175 RPM more or 200 RPM less than "desired."
- TP sensor voltage indicates throttle is open less than 1%.
- All above conditions are continuously met for 5 seconds or more.
- IAC steps must be less than 10, or greater than 140.

Test Description: Number(s) below refer to circled number(s) on the diagnostic chart.
1. The Tech 1 scan tool is used to extend and retract the IAC valve. Valve movement is verified by an engine speed change. If no change in engine speed occurs, the valve can be retested when removed from the throttle body.
2. This step checks the quality of the IAC valve movement in Step 1. Between 900 RPM and about 1500 RPM, the engine speed should change smoothly with each flash of the tester light in both extend and retract. If the IAC valve is retracted beyond the control range (about 1500 RPM), it may take many flashes in the extend position before engine speed will begin to drop. This is normal on certain engines, fully extending IAC valve may cause engine stall. This may be normal.

Step 1 verifies proper IAC valve operation while this step checks the IAC valve circuits. Each lamp on the node light should flash red and green while the IAC valve is cycled. While the sequence of color is not important if either light is "OFF" or does not flash red and green, check the circuits for faults, beginning with poor terminal contacts.

Diagnostic Aids:

Check for vacuum leaks, unconnected or brittle vacuum hoses, cuts, etc. Examine manifold and throttle body gaskets for proper seal. Check for cracked intake manifold. Check open, shorts, chafed insulation or poor connections to IAC valve in CKTs 441, 442, 443 and 444.

An open, short, or poor connection in CKTs 441, 442, 443, or 444 will result in improper idle control and may cause DTC 35.

An IAC valve which is stopped and cannot respond to the ECM, a throttle stop screw which has been tampered with, or a damaged throttle body or linkage could cause DTC 35. If no problem is found and DTC 35 resets, replace ECM.

NOTICE: If DTC 35 is currently set and speedometer is not accurately reflecting correct vehicle speed, refer to DTC 24 (vehicle speed sensor circuit) diagnostics.

2.3L (VIN D, A & 3) ENGINE — DIAGNOSTIC TROUBLE CODE CHART — 1993–94 ACHIEVA, GRAND AM AND SKYLARK

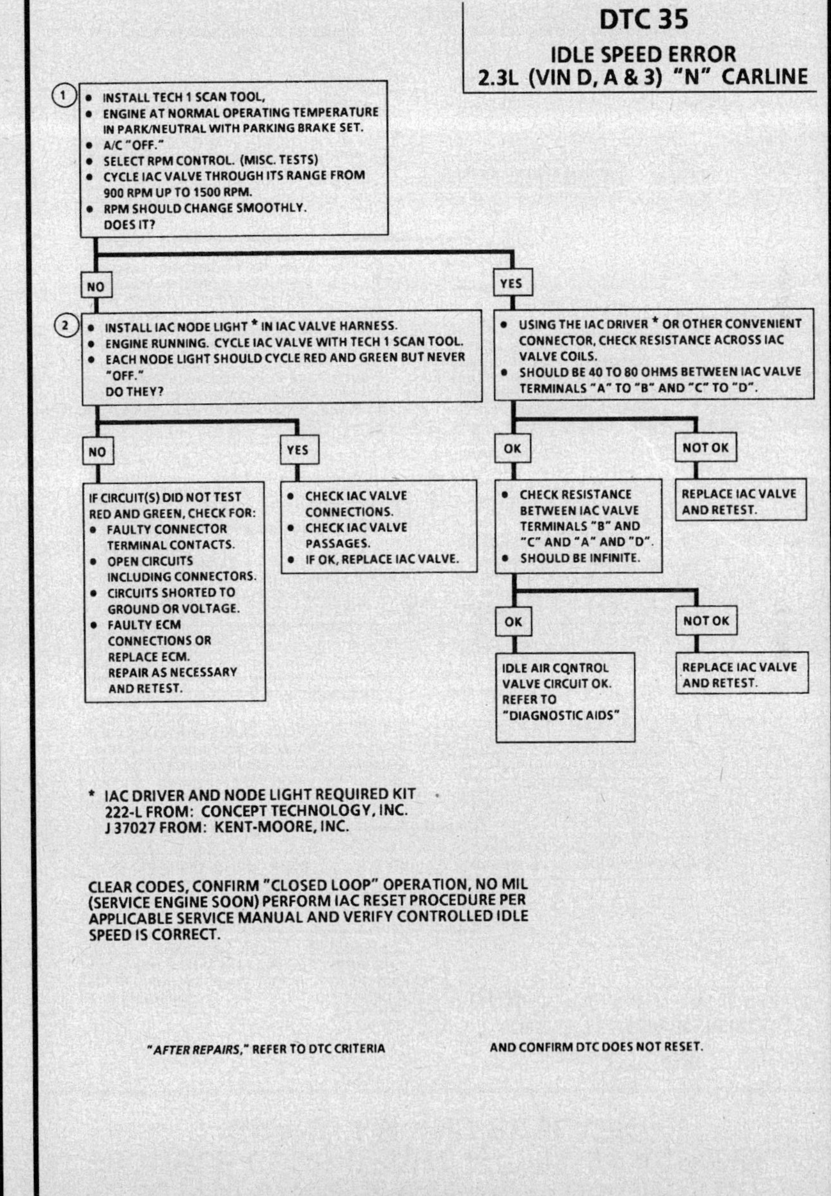

2.3L (VIN D, A & 3) ENGINE — DIAGNOSTIC TROUBLE CODE CHART — 1993–94 ACHIEVA, GRAND AM AND SKYLARK

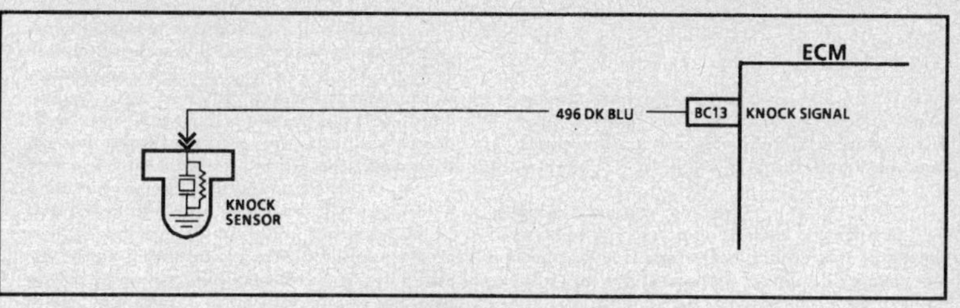

DTC 43
KNOCK SENSOR (KS) CIRCUIT
2.3L (VIN D, A & 3) "N" CARLINE

Circuit Description:

The knock sensor detects engine detonation and the Engine Control Module (ECM) retards the electronic spark timing based on the signal being received. The knock sensor produces an AC signal which is used for knock detection. The amplitude and signal frequency are dependent upon the knock level.

The ECM performs two tests on this circuit to determine if it is operating correctly. If either of the tests fail, a DTC 43 will be set.

- If there is an indication of knock for more than 4.54 seconds over a 5 second interval with the engine running.
- If ECM terminal "BC13" does not indicate increasing Knock Sensor (KS) activity with engine speed.

Test Description: Number(s) below refer to circled number(s) on the diagnostic chart.

1. If the conditions for the test as described above are being met, DTC 43 will be currently set and the MIL (Service Engine Soon) will be illuminated.
2. The Tech 1 scan tool displays knock sensor activity in counts. The counts should rise when engine speed increases or fall when engine speed decreases. This step checks knock sensor activity
3. This step checks that the internal resistance of knock sensor is within an acceptable range.
4. If the engine has an internal problem which is creating a knock, the knock sensor may be responding to the mechanical noise.

Diagnostic Aids:

Check CKT 496 for a potential open or short to ground.

Also check for proper installation of the PROM.

An "Intermittent" problem may be caused by a poor connection, rubbed through wire insulation, or a wire that is broken inside the insulation.

Any circuitry, that is suspected as causing the intermittent complaint, should be thoroughly checked for backed out terminals, improper mating, broken locks, improperly formed or damaged terminals, poor terminal to wiring connections or physical damage to the wiring harness.

Mechanical engine knock can cause a knock sensor signal. Abnormal engine noise must be corrected before using this chart.

2.3L (VIN D, A & 3) ENGINE — DIAGNOSTIC TROUBLE CODE CHART — 1993–94 ACHIEVA, GRAND AM AND SKYLARK

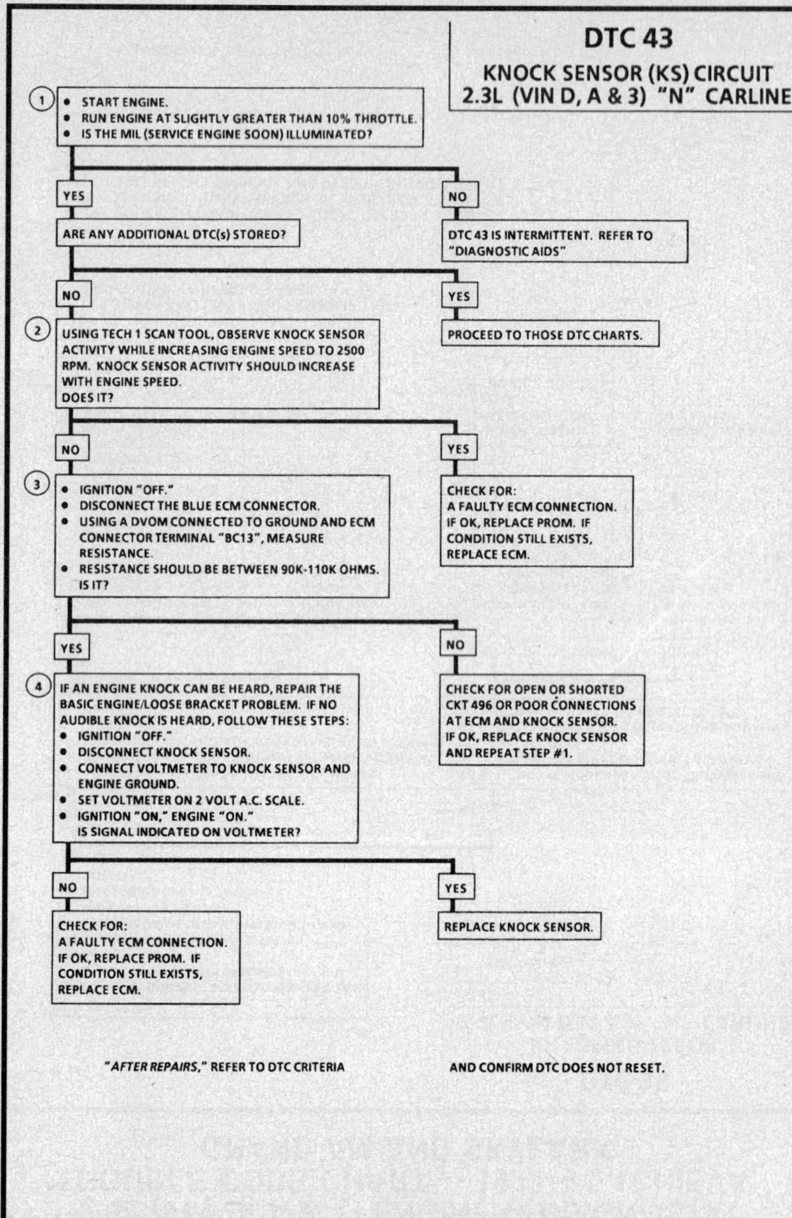

2.3L (VIN D, A & 3) ENGINE — DIAGNOSTIC TROUBLE CODE CHART — 1993–94 ACHIEVA, GRAND AM AND SKYLARK

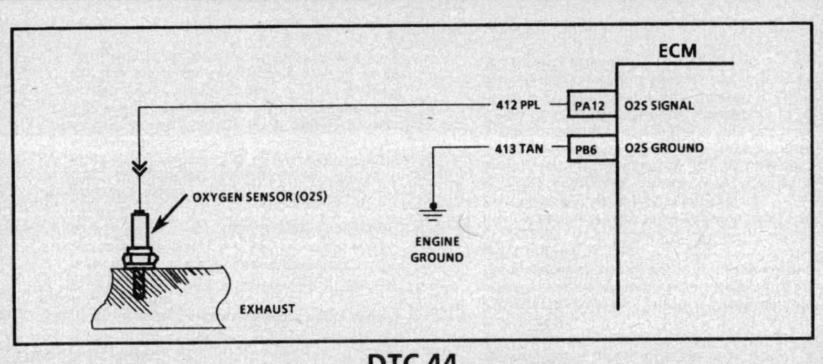

DTC 44
OXYGEN SENSOR (O2S) CIRCUIT
(LEAN EXHAUST INDICATED)
2.3L (VIN D, A & 3) "N" CARLINE

Circuit Description:

The Engine Control Module (ECM) supplies a voltage of about .45 volt between terminals "PA12" and "PB6". (If measured with a 10 megohm digital voltmeter, this may read as low as .32 volt). The Oxygen Sensor (O2S) varies the voltage within a range of about 1 volt if the exhaust is rich, down through about .10 volt if exhaust is lean.

The sensor is like an open circuit and produces no voltage when it is below 315°C (600°F). An open sensor circuit or cold sensor causes "Open Loop" operation.

Test Description: Number(s) below refer to the circled number(s) on the diagnostic chart.

1. DTC 44 is set when the O2S signal voltage on CKT 412:
   - Remains below .3 volt for 90 seconds or more.
   - The system is operating in "Closed Loop."
   - No DTC 33 or 34.
   - "Closed Loop" integrator active.
   - TP sensor above 5%.

Diagnostic Aids:

The DTC 44 or lean exhaust is most likely caused by one of the following:

- Fuel Pressure - System will be lean if pressure is too low. It may be necessary to monitor fuel pressure while driving the vehicle at various road speeds and/or loads to confirm. Refer to CHART A-7.
- MAP Sensor - An output that causes the ECM to sense a lower than normal manifold pressure (high vacuum) can cause the system to go lean. Disconnecting the MAP sensor will allow the ECM to substitute a fixed (default) value for the MAP sensor. If the rich condition is gone when the sensor is disconnected, substitute a known good sensor and recheck.
- Fuel Contamination - Water, even in small amounts, near the in-tank fuel pump inlet can be delivered to the injector. The water causes a lean exhaust and can set a DTC 44.
- Sensor Harness - Sensor pigtail may be mispositioned and contacting the exhaust manifold.
- Engine Misfire - A cylinder misfire will result in unburned oxygen in the exhaust, which could cause DTC 44. Refer to CHART C4-M
- Cracked Oxygen Sensor (O2S) - A cracked O2S or poor ground at the sensor, could cause DTC 44.
- Plugged Fuel Filter - A plugged fuel filter can cause a lean condition, and can cause a DTC 44 to set.
- Plugged Oxygen Sensor (O2S) - A plugged reference port on the Oxygen Sensor (O2S) will indicate a lower than normal voltage output from the O2S.

2.3L (VIN D, A & 3) ENGINE — DIAGNOSTIC TROUBLE CODE CHART — 1993–94 ACHIEVA, GRAND AM AND SKYLARK

DTC 44
OXYGEN SENSOR (O2S) CIRCUIT
(LEAN EXHAUST INDICATED)
2.3L (VIN D, A & 3) "N" CARLINE

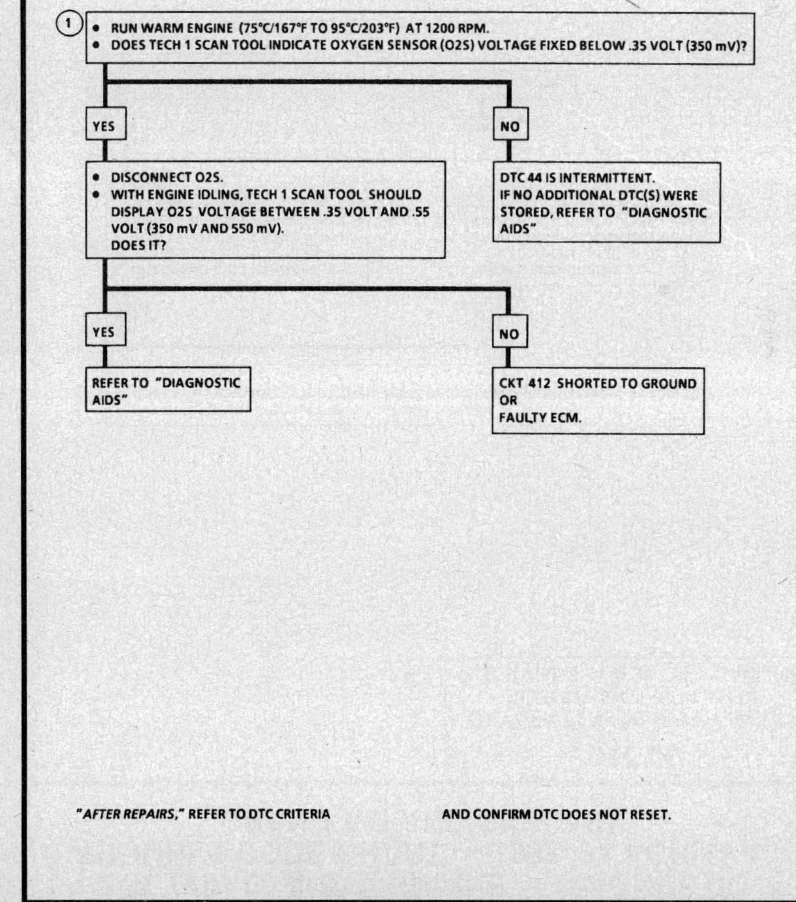

"AFTER REPAIRS," REFER TO DTC CRITERIA   AND CONFIRM DTC DOES NOT RESET.

2.3L (VIN D, A & 3) ENGINE — DIAGNOSTIC TROUBLE CODE CHART — 1993–94 ACHIEVA, GRAND AM AND SKYLARK

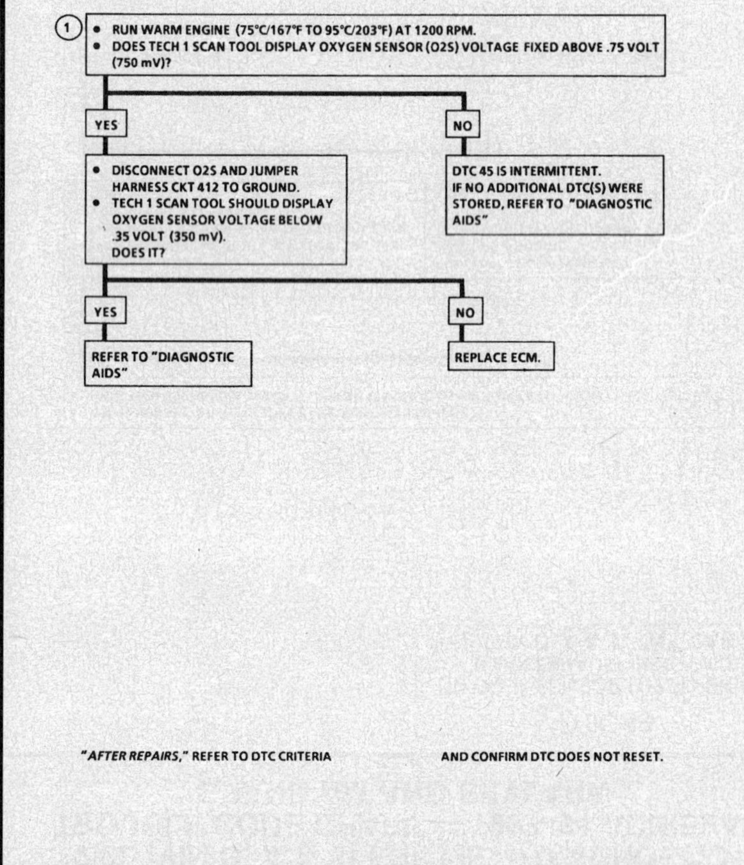

DTC 45
OXYGEN SENSOR (O2S) CIRCUIT
(RICH EXHAUST INDICATED)
2.3L (VIN D, A & 3) "N" CARLINE

Circuit Description:

The Engine Control Module (ECM) supplies a voltage of about .45 volt between terminals "PA12" and "PB6". (If measured with a 10 megohm digital voltmeter, this may read as low as .32 volt.) The Oxygen Sensor (O2S) varies the voltage within a range of about 1 volt if the exhaust is rich, down through about .10 volt if exhaust is lean.

The sensor is like an open circuit and produces no voltage when it is below 315°C (600°F). An open sensor circuit or cold sensor causes "Open Loop" operation.

Test Description: Number(s) below refer to the circled number(s) on the diagnostic chart.
1. DTC 45 is set when:
   - O2S voltage is above .75 volt.
   - No DTC 33 or 34.
   - Fuel system in "Closed Loop."
   - TP sensor above 5%.
   - Above conditions met for 30 seconds.
   OR
   - Voltage above 1 volt for 5 seconds.

Diagnostic Aids:

The DTC 45 or rich exhaust is most likely caused by one of the following:
- Fuel Pressure - System will go rich if pressure is too high. The ECM can compensate for some increase. However, if it gets too high, a DTC 45 will be set. See "Fuel System Diagnosis," CHART A-7.
- Leaking Injector - See CHART A-7.
- Electronic Ignition (EI) Shielding - An open ground CKT 453 may result in EMI or induced electrical noise. The ECM looks at this noise as Crankshaft Position (CKP) sensor pulses. The additional pulses result in a higher than actual engine speed signal. The ECM then delivers too much fuel causing system to go rich. Engine tachometer will also show higher than actual engine speed which can help in diagnosing this problem.
- EVAP Canister Purge - Check for fuel saturation. If full of fuel, check canister control and hoses. See "Evaporative Emission (EVAP) Control System"
- MAP Sensor - An output that causes the ECM to sense a higher than normal manifold pressure (low vacuum) can cause the system to go rich. Disconnecting the MAP sensor will allow the ECM to set a fixed value for the MAP sensor. Substitute a different MAP sensor if the rich condition is gone while the sensor is disconnected.
- Pressure Regulator - Check for leaking fuel pressure regulator diaphragm by checking for the presence of liquid fuel in the vacuum line to the regulator.
- TP Sensor - An intermittent TP sensor output will cause the system to go rich due to a false indication of the engine accelerating.
- O2S Contamination - Inspect O2S for silicone contamination from fuel or use of improper RTV sealant. The sensor may have a white powdery coating and result in a high but false signal voltage (rich exhaust indication). The ECM will then reduce the amount of fuel delivered to the engine causing a severe surge driveability problem.

2.3L (VIN D, A & 3) ENGINE — DIAGNOSTIC TROUBLE CODE CHART — 1993–94 ACHIEVA, GRAND AM AND SKYLARK

DTC 45
OXYGEN SENSOR (O2S) CIRCUIT
(RICH EXHAUST INDICATED)
2.3L (VIN D, A & 3) "N" CARLINE

(1)
- RUN WARM ENGINE (75°C/167°F TO 95°C/203°F) AT 1200 RPM.
- DOES TECH 1 SCAN TOOL DISPLAY OXYGEN SENSOR (O2S) VOLTAGE FIXED ABOVE .75 VOLT (750 mV)?

YES
- DISCONNECT O2S AND JUMPER HARNESS CKT 412 TO GROUND.
- TECH 1 SCAN TOOL SHOULD DISPLAY OXYGEN SENSOR VOLTAGE BELOW .35 VOLT (350 mV). DOES IT?

NO
DTC 45 IS INTERMITTENT. IF NO ADDITIONAL DTC(S) WERE STORED, REFER TO "DIAGNOSTIC AIDS"

YES
REFER TO "DIAGNOSTIC AIDS"

NO
REPLACE ECM.

"AFTER REPAIRS," REFER TO DTC CRITERIA   AND CONFIRM DTC DOES NOT RESET.

2.3L (VIN D, A & 3) ENGmdash;DIAGNOSTIC TROUBLE CODE CHART — 1993–94 ACHIEVA, GRAND AM AND SKYLARK

2.3L (VIN D, A & 3) ENGINE — DIAGNOSTIC TROUBLE CODE CHART — 1993–94 ACHIEVA, GRAND AM AND SKYLARK

DTC 51

PROM ERROR
(FAULTY OR INCORRECT PROM)
2.3L (VIN D, A & 3) "N" CARLINE

CHECK THAT ALL PINS ARE FULLY INSERTED IN THE SOCKET AND THAT PROM IS PROPERLY LATCHED. IF OK, REPLACE PROM, CLEAR MEMORY, AND RECHECK. IF DTC 51 REAPPEARS, REPLACE ECM.

NOTICE: TO PREVENT POSSIBLE ELECTROSTATIC DISCHARGE DAMAGE TO THE ECM OR PROM, DO NOT TOUCH THE COMPONENT LEADS, AND DO NOT REMOVE THE PROM COVER OR THE INTEGRATED CIRCUIT FROM CARRIER.

CLEAR DIAGNOSTIC TROUBLE CODES (DTC) AND CONFIRM "CLOSED LOOP" OPERATION AND NO MIL (SERVICE ENGINE SOON).

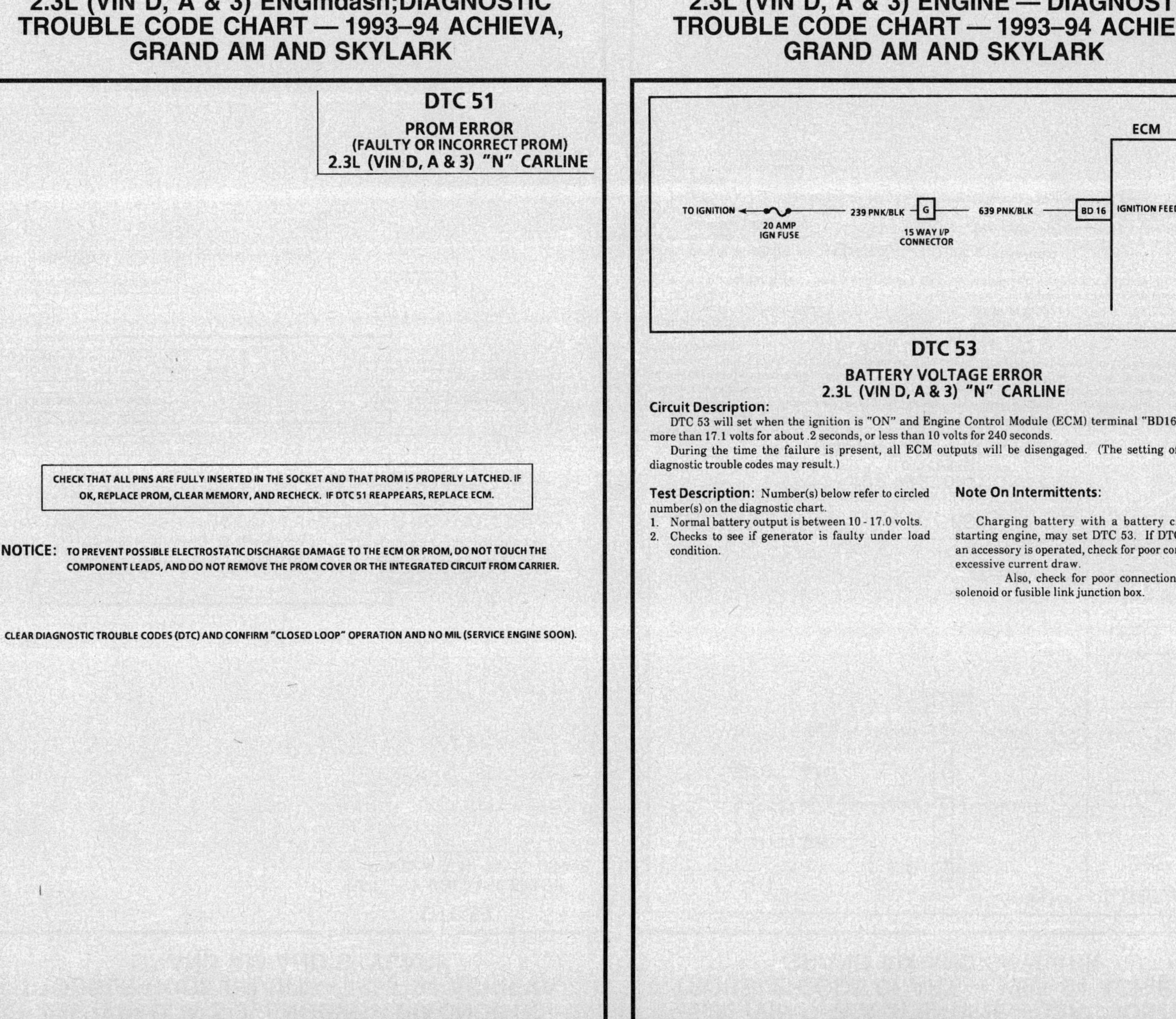

DTC 53

BATTERY VOLTAGE ERROR
2.3L (VIN D, A & 3) "N" CARLINE

Circuit Description:

DTC 53 will set when the ignition is "ON" and Engine Control Module (ECM) terminal "BD16" voltage is more than 17.1 volts for about .2 seconds, or less than 10 volts for 240 seconds.

During the time the failure is present, all ECM outputs will be disengaged. (The setting of additional diagnostic trouble codes may result.)

Test Description: Number(s) below refer to circled number(s) on the diagnostic chart.
1. Normal battery output is between 10 - 17.0 volts.
2. Checks to see if generator is faulty under load condition.

Note On Intermittents:

Charging battery with a battery charger and starting engine, may set DTC 53. If DTC sets when an accessory is operated, check for poor connections or excessive current draw.

Also, check for poor connections at starter solenoid or fusible link junction box.

2.3L (VIN D, A & 3) ENGINE — DIAGNOSTIC TROUBLE CODE CHART — 1993–94 ACHIEVA, GRAND AM AND SKYLARK

2.3L (VIN D, A & 3) ENGINE — DIAGNOSTIC TROUBLE CODE CHART — 1993–94 ACHIEVA, GRAND AM AND SKYLARK

DTC 53
BATTERY VOLTAGE ERROR
2.3L (VIN D, A & 3) "N" CARLINE

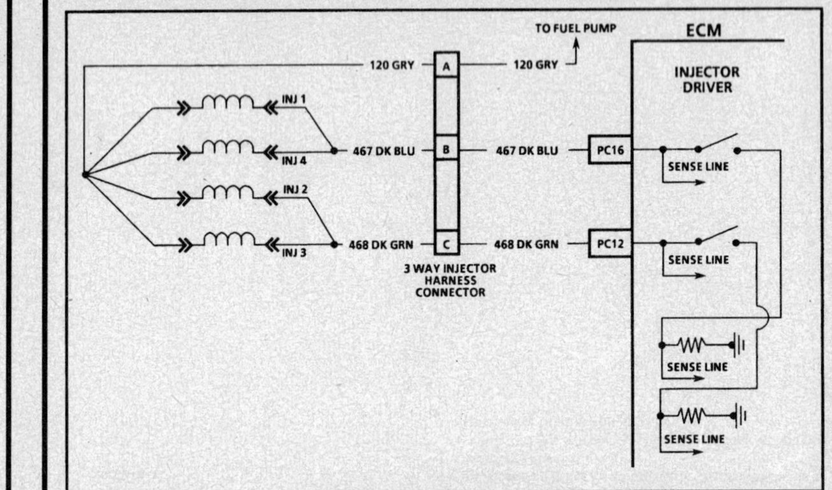

DTC 53 flow chart:

1. • ENGINE RUNNING ABOVE 800 RPM.
   • NOTE BATTERY VOLTAGE ON TECH 1 SCAN TOOL.

- BELOW 17.1 VOLTS
- ABOVE 17.1 VOLTS → CHECK CHARGING SYSTEM

2. • RAISE ENGINE RPM TO 2000.
   • LOAD ELECTRICAL SYSTEM WITH HEADLAMPS AND HIGH BLOWER "ON."
   • NOTE VOLTAGE.

- BELOW 17.1 VOLTS
- ABOVE 17.1 VOLTS → CHECK CHARGING SYSTEM

- CHECK BATTERY VOLTAGE AT BATTERY.

- VOLTAGE ABOVE 10 VOLTS
- VOLTAGE LESS THAN 10 VOLTS → CHECK CHARGING SYSTEM

- CHECK VOLTAGE AT ECM TERMINAL "BD16" IS VOLTAGE ABOVE 10 VOLTS?

- NO → CHECK FOR HIGH RESISTANCE OR OPEN CKT 639.
- YES → REPLACE ECM.

DTC 65
(Page 1 of 2)
FUEL INJECTOR CIRCUIT
(LOW CURRENT)
2.3L (VIN D, A & 3) "N" CARLINE

Circuit Description:

The Engine Control Module (ECM) has two injector driver circuits, each of which controls a pair of injectors (1 and 4 or 2 and 3). The ECM monitors the current in each driver circuit by measuring voltage drop through a fixed resistor and is able to control it. The current through each driver is allowed to rise to a "peak" of 4 amps to quickly open the injectors and is then reduced to 1 amp to "hold" them open. This is called "peak and hold." If the current can't reach a 4 amp peak, DTC 65 is set as noted below. This DTC is also set if an injector driver circuit is shorted to voltage.

Test Description: Number(s) below refer to circled number(s) on the diagnostic chart.

1. DTC 65 sets when:
   - 4 amp injector driver current not reached on either circuit.
   - Battery voltage greater than 9 volts.
   - Injectors commanded "ON" longer than a calibrated pulse width.
   - Above conditions met for 10 seconds.
2. Tests ECM and harness wiring to the injector connectors.
3. Tests for open or shorted injector. A shorted injector will not cause DTC 65.

4. Results of Step 2 will determine which branch to follow on Page 2.
5. Checks remainder of circuit from injectors to ECM as both harnesses were confirmed OK in Step 2.

Diagnostic Aids:

Open CKTs 467, 468 or CKT 467 or 468 shorted to voltage will cause DTC 65 and will also cause a misfire due to an inoperative pair of injectors.

"AFTER REPAIRS," REFER TO DTC CRITERIA AND CONFIRM DTC DOES NOT RESET.

2.3L (VIN D, A & 3) ENGINE — DIAGNOSTIC TROUBLE CODE CHART — 1993–94 ACHIEVA, GRAND AM AND SKYLARK

DTC 65

(Page 1 of 2)
FUEL INJECTOR CIRCUIT
(LOW CURRENT)
2.3L (VIN D, A & 3) "N" CARLINE

NOTICE: IF ENGINE "CRANKS BUT WILL NOT RUN," DO NOT USE THIS CHART. REFER TO CHART A-3.

1.
- IDLE ENGINE FOR 1 MINUTE.
- DOES TECH 1 SCAN TOOL INDICATE DTC 65?

YES

2.
- ENGINE "OFF."
- REMOVE CRANKCASE VENTILATION OIL/AIR SEPARATOR FOR ACCESS.
- DISCONNECT ALL 4 INJECTOR CONNECTORS FROM INJECTORS AND INSTALL AN INJECTOR TEST LIGHT ON INJECTOR CONNECTOR 1 OR 4.
- CRANK ENGINE AND NOTE LIGHT.
- REMOVE INJECTOR TEST LIGHT AND INSTALL LIGHT ON INJECTOR CONNECTOR 2 OR 3.
- CRANK ENGINE AGAIN AND NOTE LIGHT. INJECTOR TEST LIGHT SHOULD BLINK ON BOTH TESTS. DOES IT?

NO — DTC 65 IS INTERMITTENT REFER TO "DIAGNOSTIC AIDS."

YES

3.
- WITH DVOM ON 200 OHM SCALE, MEASURE THE RESISTANCE OF EACH INJECTOR. RESISTANCE SHOULD BE LESS THAN 3 OHMS (BUT NOT ZERO). IS IT?

NO

4.
- NOTE WHETHER INJECTOR TEST LIGHT WAS "OFF" ON ONE OR "ON" STEADY ON ONE IN PREVIOUS STEP AND REFER TO PAGE 2 OF THIS CHART.

SEE DTC 65 PAGE 2 OF 2.

YES

5.
- CHECK FOR POOR CONNECTIONS OR CRIMPS AT INJECTOR CONNECTORS. ARE CONNECTIONS OK?

NO — REPLACE INJECTOR WITH HIGH (OR ZERO) RESISTANCE.

YES — REPLACE ECM.

NO — REPAIR CONNECTIONS.

"AFTER REPAIRS," REFER TO DTC CRITERIA AND CONFIRM DTC DOES NOT RESET.

2.3L (VIN D, A & 3) ENGINE — DIAGNOSTIC TROUBLE CODE CHART — 1993–94 ACHIEVA, GRAND AM AND SKYLARK

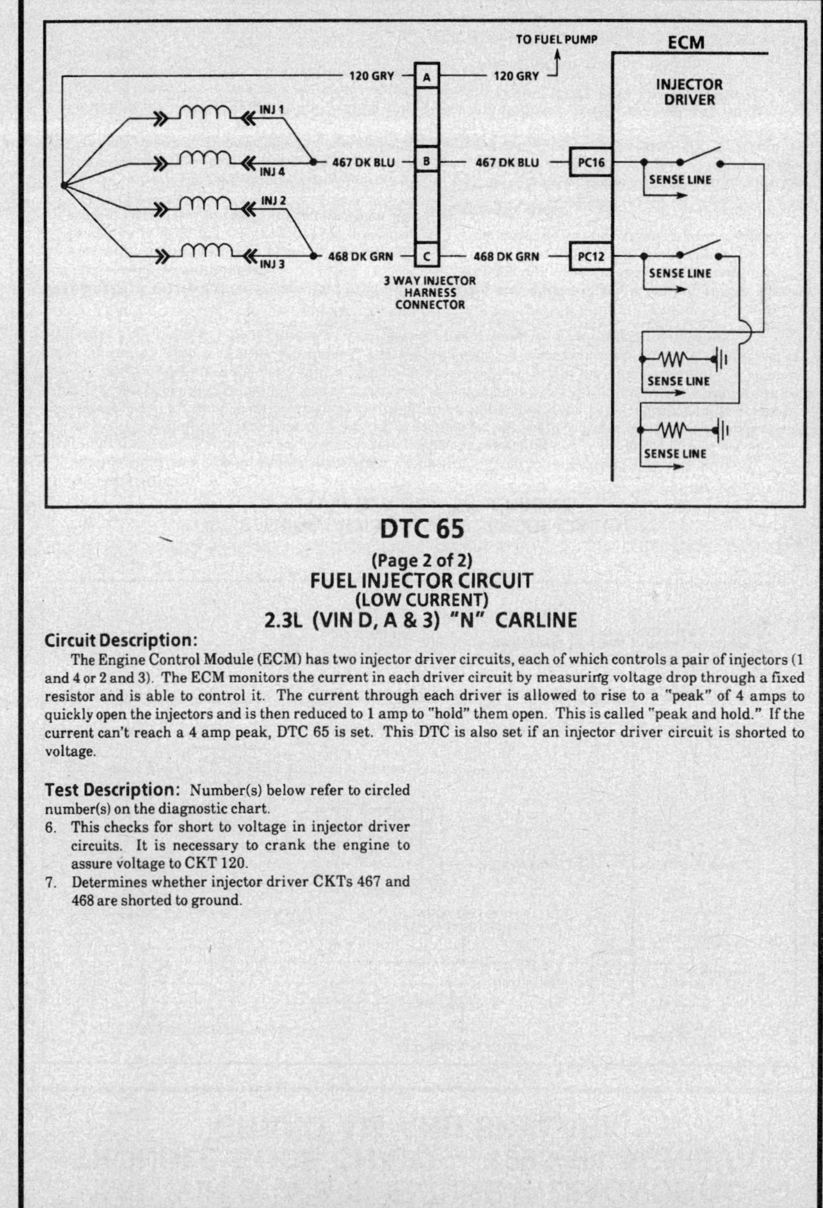

DTC 65

(Page 2 of 2)
FUEL INJECTOR CIRCUIT
(LOW CURRENT)
2.3L (VIN D, A & 3) "N" CARLINE

Circuit Description:
The Engine Control Module (ECM) has two injector driver circuits, each of which controls a pair of injectors (1 and 4 or 2 and 3). The ECM monitors the current in each driver circuit by measuring voltage drop through a fixed resistor and is able to control it. The current through each driver is allowed to rise to a "peak" of 4 amps to quickly open the injectors and is then reduced to 1 amp to "hold" them open. This is called "peak and hold." If the current can't reach a 4 amp peak, DTC 65 is set. This DTC is also set if an injector driver circuit is shorted to voltage.

Test Description: Number(s) below refer to circled number(s) on the diagnostic chart.
6. This checks for short to voltage in injector driver circuits. It is necessary to crank the engine to assure voltage to CKT 120.
7. Determines whether injector driver CKTs 467 and 468 are shorted to ground.

2.3L (VIN D, A & 3) ENGINE — DIAGNOSTIC TROUBLE CODE CHART — 1993–94 ACHIEVA, GRAND AM AND SKYLARK

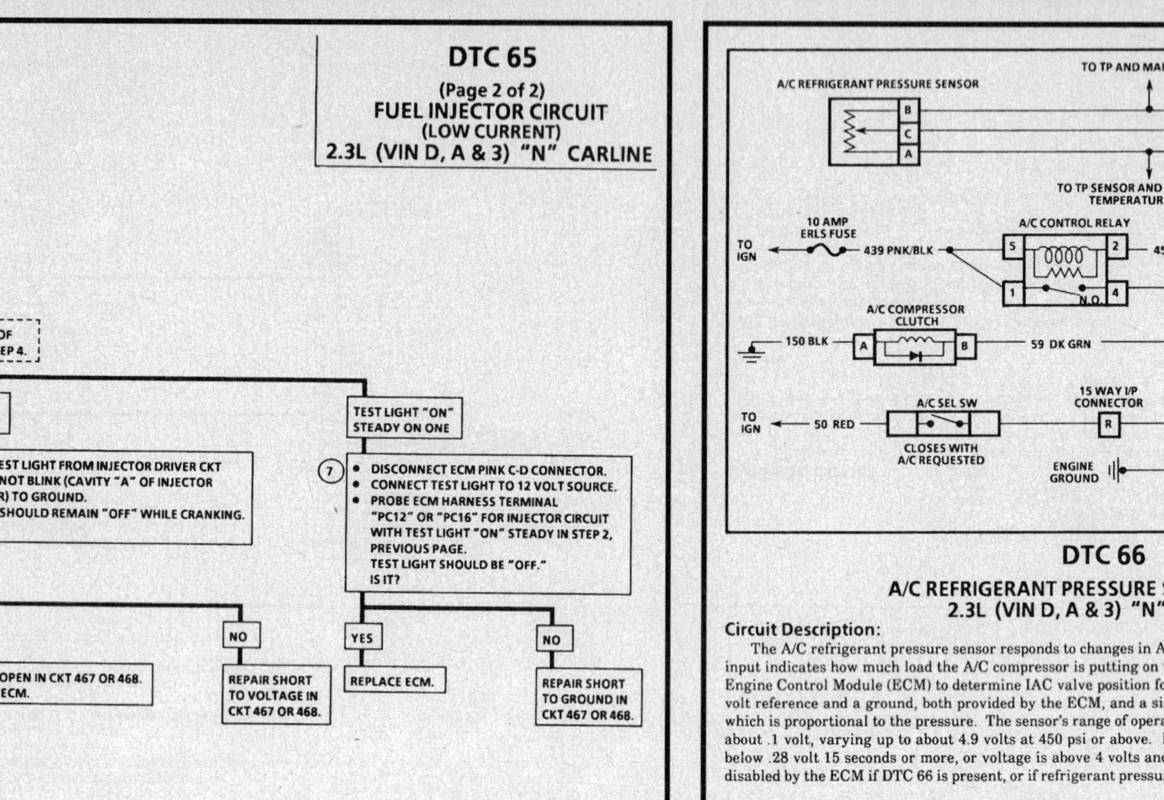

2.3L (VIN D, A & 3) ENGINE — DIAGNOSTIC TROUBLE CODE CHART — 1993–94 ACHIEVA, GRAND AM AND SKYLARK

DTC 66
A/C REFRIGERANT PRESSURE SENSOR CIRCUIT
2.3L (VIN D, A & 3) "N" CARLINE

Circuit Description:

The A/C refrigerant pressure sensor responds to changes in A/C refrigerant system high side pressure. This input indicates how much load the A/C compressor is putting on the engine and is one of the factors used by the Engine Control Module (ECM) to determine IAC valve position for idle speed control. The circuit consists of a 5 volt reference and a ground, both provided by the ECM, and a signal line to the ECM. The signal is a voltage which is proportional to the pressure. The sensor's range of operation is 0 to 450 psi. At 0 psi, the signal will be about .1 volt, varying up to about 4.9 volts at 450 psi or above. DTC 66 sets if the voltage is above 4.9 volts or below .28 volt 15 seconds or more, or voltage is above 4 volts and A/C is not requested. The A/C compressor is disabled by the ECM if DTC 66 is present, or if refrigerant pressure is above or below calibrated values

Test Description: Number(s) below refer to circled number(s) on the diagnostic chart.

1. This step checks the voltage signal being received by the ECM from the A/C refrigerant pressure sensor. The normal operating range is between .28 volt and 4.9 volts.
2. Checks to see if the high voltage signal is from a shorted sensor or a short to voltage in the circuit. Normally, disconnecting the sensor would make a normal circuit go to near zero volt.
3. Checks to see if low voltage signal is from the sensor or the circuit. Jumpering the sensor signal CKT 380 to 5 volts, checks the circuit, connections and ECM.

4. This step checks to see if the low voltage signal was due to an open in the sensor circuit or the 5 volt reference circuit since the prior step eliminated the A/C refrigerant pressure sensor.

Diagnostic Aids:

DTC 66 sets when signal voltage falls outside the normal possible range of the sensor and is not due to a refrigerant system problem. If problem is intermittent, check for opens or shorts in harness or poor connections. If OK, replace A/C refrigerant pressure sensor. If DTC 66 re-sets, replace ECM.

Non-A/C Program

DTC 66 will set on a Non-A/C vehicle if CKT 450 to terminal "PB5" is open or shorted to B+.

2.3L (VIN D, A & 3) ENGINE — DIAGNOSTIC TROUBLE CODE CHART — 1993–94 ACHIEVA, GRAND AM AND SKYLARK

NOTICE: IF VEHICLE IS NOT EQUIPPED WITH AIR CONDITIONING, DO NOT USE THIS CHART, SEE "DIAGNOSTIC AIDS."

DTC 66
A/C REFRIGERANT PRESSURE SENSOR CIRCUIT
2.3L (VIN D, A & 3) "N" CARLINE

(1)
- KEY "ON," ENGINE NOT RUNNING.
- NOTE TECH 1 SCAN TOOL VOLTAGE FOR A/C REFRIGERANT PRESSURE.

ABOVE 3.9 VOLTS

(2)
- DISCONNECT A/C REFRIGERANT PRESSURE SENSOR.
- DOES SCAN TOOL DISPLAY LESS THAN 1 VOLT?

NO → CHECK FOR SHORT TO VOLTAGE IN CKT 452. IF NOT SHORTED, REPLACE ECM.

YES → CHECK FOR OPEN IN CKT 452. IF NOT OPEN, CHECK FOR POOR SENSOR TERMINAL CONNECTIONS. IF OK, REPLACE A/C REFRIGERANT PRESSURE SENSOR.

BELOW .28 VOLT

(3)
- DISCONNECT A/C REFRIGERANT PRESSURE SENSOR CONNECTOR.
- JUMPER TERMINALS "B" AND "C".
- DOES SCAN TOOL DISPLAY ABOVE 4.6 VOLTS?

BETWEEN .28 VOLT AND 4.9 VOLTS

FAULT IS NOT PRESENT AT THIS TIME. SEE "DIAGNOSTIC AIDS."

NO
(4)
- REMOVE JUMPER.
- CONNECT VOLTMETER FROM TERMINAL "A" TO "B".
- IS VOLTAGE ABOUT 5 VOLTS?

YES → CHECK SENSOR TERMINAL CONNECTIONS IF OK, REPLACE A/C REFRIGERANT PRESSURE SENSOR.

NO
- BACK PROBE ECM TERMINAL "PA4" WITH VOLTMETER TO GROUND. IS VOLTAGE ABOUT 5 VOLTS?

YES → CHECK FOR OPEN IN CKT 380 CHECK FOR POOR CONNECTION ECM TERMINAL "PD3", IF OK, REPLACE ECM.

NO → CHECK FOR POOR CONNECTION AT ECM TERMINAL "PA4" OR SHORT TO GROUND IN CKT 416. IF OK, ECM IS FAULTY.

YES → REPAIR OPEN IN CKT 416.

"AFTER REPAIRS," REFER TO DTC CRITERIA   AND CONFIRM DTC DOES NOT RESET.

2.3L (VIN D, A & 3) ENGINE — EXHAUST SYSTEM DIAGNOSTIC CHART — 1993–94 ACHIEVA, GRAND AM AND SKYLARK

CHART B-1
RESTRICTED EXHAUST SYSTEM CHECK
2.3L (VIN D, A & 3)

Proper diagnosis for a restricted exhaust system is essential before any components are replaced. The following procedure may be used for diagnosis.

CHECK AT OXYGEN SENSOR (O2S):

1. Carefully remove O2S.
2. Install Borroughs exhaust backpressure tester (BT-8515 or BT-8603) in place of O2S (see illustration).
3. After completing test described below, be sure to coat threads of O2S with anti-seize compound P/N 5613695 prior to re-installation.

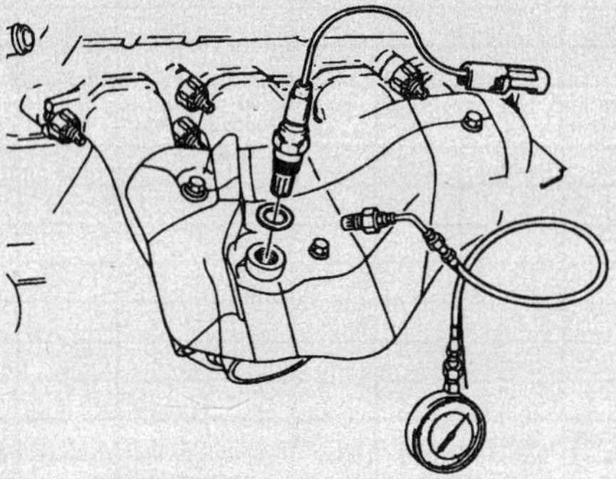

DIAGNOSIS:

1. With the engine idling at normal operating temperature, transaxle in park or neutral, observe the exhaust system backpressure reading on the gauge. Reading should not exceed 3.4 kPa (.5 psi).
2. Increase engine speed to 3000 RPM and observe gauge. Reading should not exceed 5 kPa (.75 psi).
3. If the backpressure at either speed exceeds specification, a restricted exhaust system is indicated.
4. Inspect the entire exhaust system for a collapsed pipe, heat distress, or possible internal muffler failure.
5. If there are no obvious reasons for the excessive backpressure, the catalytic converter is suspected to be restricted and should be replaced using current recommended procedures.

2.3L (VIN D, A & 3) ENGINE — ECM SYMPTOM CHART — 1993–94 ACHIEVA, GRAND AM AND SKYLARK

ECM CONNECTOR "PC"

PIN FUNCTION	CKT #	WIRE COLOR	COMPONENT CONNECTOR CAVITY	NORMAL VOLTAGES		DTC(s) AFFECTED	POSSIBLE SYMPTOMS FROM FAULTY CIRCUIT
				KEY "ON"	ENG RUN		
PC1							
PC2							
PC3 IAC "A" HI	441	LT BLU/WHT	IAC VALVE "C"	0 OR B+	0 OR B+	35 (10)	INCORRECT IDLE SURGE
PC4 IAC "A" LOW	442	LT BLU/BLK	IAC VALVE "D"	0 OR B+	0 OR B+	35 (10)	INCORRECT IDLE SURGE
PC5 IAC "B" LOW	444	LT GRN/BLK	IAC VALVE "B"	0 OR B+	0 OR B+	35 (10)	INCORRECT IDLE SURGE
PC6 IAC "B" HI	443	LT GRN/WHT	IAC VALVE "A"	0 OR B+	0 OR B+	35 (10)	INCORRECT IDLE SURGE
PC7							
PC8							
PC9 IAT SIGNAL	472	TAN	IAT SENSOR "A"	3.6 (3)	1.5 (3)	23 (8) 25 (9)	
PC10 ECT SIGNAL	410	YEL	ECT "B"	1.2 (5)	2.0 (5)	14 (9) 15 (8)	LACK OF PERFORMANCE
PC11							
PC12 INJ DRIVER 2 & 3	468	DK GRN	INJECTOR HARNESS TERM. "C"	B+	B+	65 (10)	ROUGH IDLE, LACK OF PERFORMANCE, HARD TO START.
PC13 ECM GROUND	450	BLK/WHT	ENGINE BLOCK	0*	0*		
PC14							
PC15 IGNITION CONTROL (IC) 2	485	BLK	ELECTRONIC IGNITION CONTROL MODULE (ICM) CONN TERM. "B"	0*	.5		LACK OF POWER, STALLS, SURGES
PC16 INJ DRIVER 1 & 4	467	DK BLU	INJECTOR HARNESS TERM. "B"	B+	B+	65 (10)	ROUGH IDLE, LACK OF PERFORMANCE, HARD TO START.

(1) VARIES FROM .60 TO BATTERY VOLTAGE, DEPENDING ON POSITION OF DRIVE WHEELS.
(2) BATTERY VOLTAGE FOR FIRST TWO SECONDS.
(3) VARIES.
(4) BATTERY VOLTAGE WHEN FUEL PUMP IS RUNNING.
(5) VARIES WITH TEMPERATURE.
(6) READS BATTERY VOLTAGE IN GEAR.
(7) BATTERY VOLTAGE WHEN ENGINE IS CRANKING.
(8) OPEN CIRCUIT.
(9) GROUNDED CIRCUIT.
(10) OPEN/GROUNDED CIRCUIT.
(11) LESS THAN 1 VOLT.
* LESS THAN .5 VOLT (500 MV).

2.3L (VIN D, A & 3) ENGINE — ECM SYMPTOM CHART — 1993–94 ACHIEVA, GRAND AM AND SKYLARK

ECM CONNECTOR "PD"

PIN FUNCTION	CKT #	WIRE COLOR	COMPONENT CONNECTOR CAVITY	NORMAL VOLTAGES		DTC(s) AFFECTED	POSSIBLE SYMPTOMS FROM FAULTY CIRCUIT
				KEY "ON"	ENG RUN		
PD1 IGN REF LOW	453	BLK/RED	ELECTRONIC IGNITION CONTROL MODULE (ICM) TERM. "H"	0*	0*		
PD2 7X REF	483	LT GRN	ELECTRONIC IGNITION CONTROL MODULE (ICM) TERM. "G"	0	6	19	VEHICLE STALLS, LOSE RPM.
PD3 A/C REFRIGERANT PRESSURE SIGNAL	380	GRY/RED	A/C REFRIGERANT PRESS. SENSOR TERM. "C"	.8 (3)	.9 (3)	66 (10)	A/C CLUTCH INOPERATIVE
PD4 MAP SIGNAL	432	LT GRN	MAP SENSOR "B"	3.8 (3)	1.3 (3)	34 (10) 33 (10)	LACK OF PERFORMANCE, ROUGH IDLE, SURGE
PD5							
PD6							
PD7							
PD8							
PD9							
PD10							
PD11							
PD12 A/C REQUEST	66	LT GRN	15 WAY I/P CONN. "R" A/C SEL. SW.	0* B+	0* B+		INOPERATIVE A/C INCORRECT IDLE
PD13							
PD14							
PD15							
PD16							

(1) VARIES FROM .60 TO BATTERY VOLTAGE, DEPENDING ON POSITION OF DRIVE WHEELS.
(2) BATTERY VOLTAGE FOR FIRST TWO SECONDS.
(3) VARIES.
(4) BATTERY VOLTAGE WHEN FUEL PUMP IS RUNNING.
(5) VARIES WITH TEMPERATURE.
(6) READS BATTERY VOLTAGE IN GEAR.
(7) BATTERY VOLTAGE WHEN ENGINE IS CRANKING.
(8) OPEN CIRCUIT.
(9) GROUNDED CIRCUIT.
(10) OPEN/GROUNDED CIRCUIT.
(11) LESS THAN 1 VOLT.
* LESS THAN .5 VOLT (500 mV).

2.3L (VIN D, A & 3) ENGINE — ECM SYMPTOM CHART — 1993–94 ACHIEVA, GRAND AM AND SKYLARK

ECM CONNECTOR "PA"

PIN FUNCTION		CKT #	WIRE COLOR	COMPONENT CONNECTOR CAVITY	NORMAL VOLTAGES		DTC(s) AFFECTED	POSSIBLE SYMPTOMS FROM FAULTY CIRCUIT
					KEY "ON"	ENG RUN		
PA1	ECM GROUND	551	TAN/WHT	ENGINE BLOCK	0*	0*		
PA2	ECM GROUND	450	BLK/WHT	ENGINE GROUND	0*	0*		
PA3	MAP, TP + 5V REFERENCE	416	GRY	MAP TERM. "C" TP TERMINAL "A"	5	5	22 34 (10) 66	WITH PA3 AND PA4 OPEN OR SHORTED TO GROUND STALLS, LACK OF PERFORMANCE.
PA4	+ 5V REFERENCE, MAP, TP, A/C REFRIGERANT PRESSURE SENSOR	416	GRY	MAP TERM. "C" TP TERM. "A" A/C TERMINAL "B"	5	5	22 34 (10) 66	WITH PA3 AND PA4 OPEN OR SHORTED TO GROUND STALLS.
PA5	BATTERY	2	RED	FUSIBLE LINK AND B +	B +	B +		CRANKS BUT WILL NOT START.
PA6								
PA7								
PA8	FUEL PUMP	465	DK GRN/WHT	FUEL PUMP RELAY "5"	B + (2)	B + (4)		LONG CRANKING TIME BEFORE ENGINE STARTS (8).
PA9								
PA10								
PA11								
PA12	O2S SIGNAL	412	PPL	OXYGEN SENSOR (O2S)	.35 (3)	.1-.9 (3)	13 (8)	EXHAUST ODOR, POOR PERFORMANCE.

(1) VARIES FROM .60 TO BATTERY VOLTAGE, DEPENDING ON POSITION OF DRIVE WHEELS.
(2) BATTERY VOLTAGE FOR FIRST TWO SECONDS.
(3) VARIES.
(4) BATTERY VOLTAGE WHEN FUEL PUMP IS RUNNING.
(5) VARIES WITH TEMPERATURE.
(6) READS BATTERY VOLTAGE IN GEAR.
(7) BATTERY VOLTAGE WHEN ENGINE IS CRANKING.
(8) OPEN CIRCUIT.
(9) GROUNDED CIRCUIT.
(10) OPEN/GROUNDED CIRCUIT.
(11) LESS THAN 1 VOLT.
* LESS THAN .5 VOLT (500 mV).

2.3L (VIN D, A & 3) ENGINE — ECM SYMPTOM CHART — 1993–94 ACHIEVA, GRAND AM AND SKYLARK

ECM CONNECTOR "PB"

PIN FUNCTION		CKT #	WIRE COLOR	COMPONENT CONNECTOR CAVITY	NORMAL VOLTAGES		DTC(s) AFFECTED	POSSIBLE SYMPTOMS FROM FAULTY CIRCUIT
					KEY "ON"	ENG RUN		
PB1	IAT, MAP GND	469	BLK/ORN	IAT TERM. "B" MAP TERM. "A"	0*	0*	25 (8) 33 (8)	LACK OF PERFORMANCE HESITATION ON ACCELERATION
PB2	TP, ECT, A/C REFRIGERANT PRESSURE SENSOR GROUND	452	BLK	TP TERM. "B" ECT TERM. "A" A/C TERM. "A"	0	0	15 21 (8) 66	HESITATION ON ACCEL., LACK OF PERFORMANCE, EXHAUST ODOR.
PB3	IGNITION CONTROL (IC) 1	423	WHT	ELECTRONIC IGNITION CONTROL MODULE (ICM) TERM. "A"	0*	.5 (3)		LACK OF POWER, STALLS, SURGES
PB4								
PB5	NON A/C PROGRAM	450		ENGINE GROUND	0*	0*	66 (8)	
PB6	O2S GND	413	TAN	ENGINE GND	0*	0*	13 (8)	EXHAUST ODOR, POOR PERFORMANCE
PB7	TP SIGNAL	417	DK BLU	TP TERM. "C"	.49	.49	22 (10)	LACK OF PERFORMANCE
PB8	4000 P/MI SPEED	389	DK GRN	15 WAY I/P CONNECTOR TERM. "A"	(1) 8.3	(1) 9.4		INOPERATIVE SPEEDOMETER, ODOMETER
PB9								
PB10								
PB11	TACH	627	WHT	TO TACH	4.9	1.0		NO TACH
PB12								

(1) VARIES FROM .60 TO BATTERY VOLTAGE, DEPENDING ON POSITION OF DRIVE WHEELS.
(2) BATTERY VOLTAGE FOR FIRST TWO SECONDS.
(3) VARIES.
(4) BATTERY VOLTAGE WHEN FUEL PUMP IS RUNNING.
(5) VARIES WITH TEMPERATURE.
(6) READS BATTERY VOLTAGE IN GEAR.
(7) BATTERY VOLTAGE WHEN ENGINE IS CRANKING.
(8) OPEN CIRCUIT.
(9) GROUNDED CIRCUIT.
(10) OPEN/GROUNDED CIRCUIT.
(11) LESS THAN 1 VOLT.
* LESS THAN .5 VOLT (500 mV).

MULTIPORT FUEL INJECTION (MFI) SYSTEMS
EXCEPT LIGHT TRUCKS, VANS, GEO AND SATURN

2.3L (VIN D, A & 3) ENGINE — ECM SYMPTOM CHART — 1993–94 ACHIEVA, GRAND AM AND SKYLARK

ECM CONNECTOR "BD"

PIN	PIN FUNCTION	CKT #	WIRE COLOR	COMPONENT CONNECTOR CAVITY	NORMAL VOLTAGES KEY "ON"	NORMAL VOLTAGES ENG RUN	DTC(S) AFFECTED	POSSIBLE SYMPTOMS FROM FAULTY CIRCUIT
BD1	PNP SW. SIG.	434	ORN/BLK	PNP SWITCH TERM. "A"	"ON" 0 "OFF" B+	"ON" 0 "OFF" B+		INCORRECT IDLE
BD2	2nd GEAR SWITCH (LD2 ONLY)	232	WHT	TRANSAXLE CONNECTOR TERMINAL "C"	0*	"ON" B+ "OFF" 0*		
BD3	3rd GEAR SWITCH (LD2 ONLY)	438	DK GRN/WHT	TRANSAXLE CONNECTOR TERMINAL "B"	0*	"ON" B+ "OFF" 0*		
BD4								
BD5								
BD6	DATA LINK CONNECTOR (DLC) DIAGNOSTIC ENABLE	451	WHT/BLK	DLC CONN. "B"	12.0	13.6		MIL FLASHES (9) NO FIELD SERVICE MODE (8)
BD7								
BD8								
BD9								
BD10								
BD11								
BD12	DLC SERIAL DATA	461	ORN	15 WAY I/P CONN TERM. "J"	4.2 (3)	4.5 (3)		NO DATA, TECH 1 SCAN TOOL WON'T READ DATA (10)
BD13	VSS "HI"	400	YEL	VSS TERM. "A"	0* (1)	0* (1)	24 (10) 35 (10)	NO VSS SIG., INOPERATIVE SPEEDOMETER, INOPERATIVE CRUISE, NO TCC.
BD14	VSS "LO"	401	PPL	VSS TERM. "B"	0*	0*	24 (10) 35 (10)	NO VSS SIG., INOPERATIVE SPEEDOMETER, INOPERATIVE CRUISE, NO TCC.
BD15								
BD16	IGNITION	639	PNK/BLK	15 WAY I/P TERM. "G"	B+	B+	53	NO MIL, ENGINE CRANKS BUT WILL NOT START, NO DATA (8)

(1) VARIES FROM .60 TO BATTERY VOLTAGE, DEPENDING ON POSITION OF DRIVE WHEELS.
(2) BATTERY VOLTAGE FOR FIRST TWO SECONDS.
(3) VARIES.
(4) BATTERY VOLTAGE WHEN FUEL PUMP IS RUNNING.
(5) VARIES WITH TEMPERATURE.
(6) READS BATTERY VOLTAGE IN GEAR.
(7) BATTERY VOLTAGE WHEN ENGINE IS CRANKING.
(8) OPEN CIRCUIT.
(9) GROUNDED CIRCUIT.
(10) OPEN/GROUNDED CIRCUIT.
(11) LESS THAN 1 VOLT.
* LESS THAN .5 VOLT (500 MV).

2.3L (VIN D, A & 3) ENGINE — ECM SYMPTOM CHART — 1993–94 ACHIEVA, GRAND AM AND SKYLARK

ECM CONNECTOR "BC"

PIN	PIN FUNCTION	CKT #	WIRE COLOR	COMPONENT CONNECTOR CAVITY	NORMAL VOLTAGES KEY "ON"	NORMAL VOLTAGES ENG RUN	DTC(s) AFFECTED	POSSIBLE SYMPTOMS FROM FAULTY CIRCUIT
BC1	(EVAP) CAN. PURGE SOLENOID VALVE	428	DK GRN/YEL	(EVAP) DRIVER TERM. "B"	B+	.3	27 (10)	
BC2	TCC/SHIFT LIGHT	422/456	TAN/BLK	TRANSAXLE CONN TERM. "D"	B+	"OFF" B+ "ON" 0	27 (10)	TCC INOPERATIVE (8) TCC OPERATIVE (9)
BC3								
BC4								
BC5								
BC6								
BC7								
BC8	MIL	419	BRN/WHT	15 WAY I/P CONN TERM. "B"	0*	B+	26 (10)	NO MIL (8) MIL "ON" CONSTANTLY (9)
BC9	CLG FAN RELAY	536	LT GRN/BLK	COOLING FAN RELAY TERM. "5"	B+	"ON" 0 "OFF" B+	28 (10)	INOPERATIVE FAN (8) FAN RUNS CONSISTENTLY (9)
BC10	COOLANT TEMP./CHECK GAUGES LIGHT	35	DK GRN	15 WAY JUNCTION CONN "H1"	0	"ON" 0 "OFF" B+	28 (10)	TEMP LIGHT "OFF" (8) TEMP LIGHT STAYS "ON" (9)
BC11	A/C CLUTCH RELAY	459	DK GRN/WHT	A/C CONTROL RELAY TERM. "2"	B+	"ON" 0 "OFF" B+	28 (10)	A/C CLUTCH INOPERATIVE (8) A/C CLUTCH STAYS "ON" (9)
BC12								
BC13	KNOCK SIGNAL	496	DK BLU	KNOCK SENSOR	2.3	2.3	43 (10)	
BC14								
BC15								
BC16								

(1) VARIES FROM .60 TO BATTERY VOLTAGE, DEPENDING ON POSITION OF DRIVE WHEELS.
(2) BATTERY VOLTAGE FOR FIRST TWO SECONDS.
(3) VARIES.
(4) BATTERY VOLTAGE WHEN FUEL PUMP IS RUNNING.
(5) VARIES WITH TEMPERATURE.
(6) READS BATTERY VOLTAGE IN GEAR.
(7) BATTERY VOLTAGE WHEN ENGINE IS CRANKING.
(8) OPEN CIRCUIT.
(9) GROUNDED CIRCUIT.
(10) OPEN/GROUNDED CIRCUIT.
(11) LESS THAN 1 VOLT.
* LESS THAN .5 VOLT (500 MV).

2.3L (VIN D, A & 3) ENGINE — COMPONENT DIAGNOSTIC CHART — 1993–94 ACHIEVA, GRAND AM AND SKYLARK

CHART C-1A
PARK/NEUTRAL POSITION (PNP) SWITCH DIAGNOSIS
(AUTO TRANSAXLE ONLY)
2.3L (VIN D, A & 3) "N" CARLINE

① WITH TRANSAXLE IN PARK, TECH 1 SCAN TOOL SHOULD INDICATE PARK OR NEUTRAL. DOES IT?

- **YES → ③ SHIFT TRANSAXLE INTO DRIVE. TECH 1 SCAN TOOL SHOULD DISPLAY A CHANGE TO INDICATE DRIVE. DOES IT?**
  - **YES → NO TROUBLE FOUND.**
  - **NO → DISCONNECT PNP SWITCH. THIS SHOULD CAUSE SCAN TOOL TO DISPLAY DRIVE RANGE. DOES IT?**
    - **YES → FAULTY PNP SWITCH CONNECTION OR PNP SWITCH MISADJUSTED OR FAULTY PNP SWITCH.**
    - **NO → CKT 434 SHORTED TO GROUND OR FAULTY ECM.**

- **NO → ② DISCONNECT PARK NEUTRAL POSITION (PNP) SWITCH CONNECTOR. JUMPER HARNESS CONNECTOR TERMINALS "A" AND "C". SCAN TOOL SHOULD INDICATE PARK OR NEUTRAL. DOES IT?**
  - **YES → FAULTY PNP SWITCH CONNECTION OR PNP SWITCH MISADJUSTED OR FAULTY PNP SWITCH.**
  - **NO → JUMPER HARNESS CONNECTOR (CKT 434) TO ENGINE GROUND. TECH 1 SCAN TOOL SHOULD INDICATE PARK OR NEUTRAL. DOES IT?**
    - **YES → OPEN GROUND CIRCUIT.**
    - **NO → CKT 434 OPEN OR FAULTY ECM CONNECTION OR ECM.**

"AFTER REPAIRS," CONFIRM "CLOSED LOOP" OPERATION AND NO MIL (SERVICE ENGINE SOON).

2.3L (VIN D, A & 3) ENGINE — COMPONENT DIAGNOSTIC CHART — 1993–94 ACHIEVA, GRAND AM AND SKYLARK

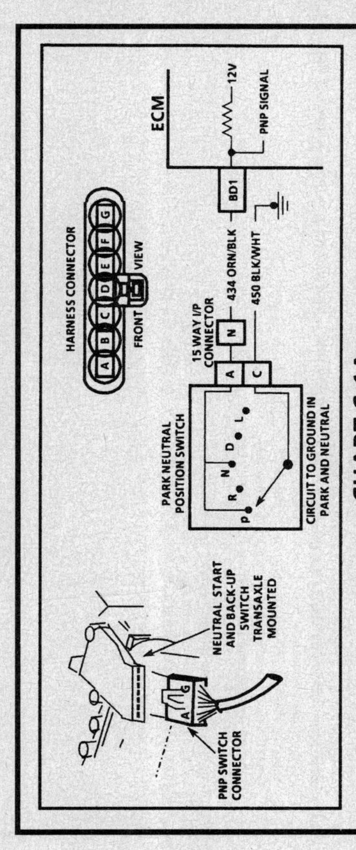

CHART C-1A
PARK/NEUTRAL POSITION (PNP) SWITCH DIAGNOSIS
(AUTO TRANSAXLE ONLY)
2.3L (VIN D, A & 3) "N" CARLINE

Circuit Description:

The Park/Neutral Position (PNP) switch contacts are a part of the neutral start switch and are closed to ground in park or neutral, and open in drive ranges and reverse.

The Engine Control Module (ECM) supplies ignition voltage through a current limiting resistor to CKT 434 and senses a closed switch when the voltage on CKT 434 drops to less than 1 volt.

The ECM uses the PNP signal as one of the inputs to control:
- *Idle Air Control (IAC) valve.*
- DTC 24 VSS diagnostics.

If CKT 434 indicates drive (open) a dip in the idle may exist when the gear selector is moved into drive range.

Test Description: Number(s) below refer to circled number(s) on the diagnostic chart.

1. Checks for a closed switch to ground in park position. Different makes of scan tools will display PNP differently. Refer to "Tool Operator's" manual for type of display used for a specific tool.

2. Checks for an open switch in drive range.

3. Be sure Tech 1 scan tool indicates drive, even while wiggling shifter, to test for an intermittent or misadjusted switch in drive range.

2.3L (VIN D, A & 3) ENGINE — COMPONENT DIAGNOSTIC CHART — 1993–94 ACHIEVA, GRAND AM AND SKYLARK

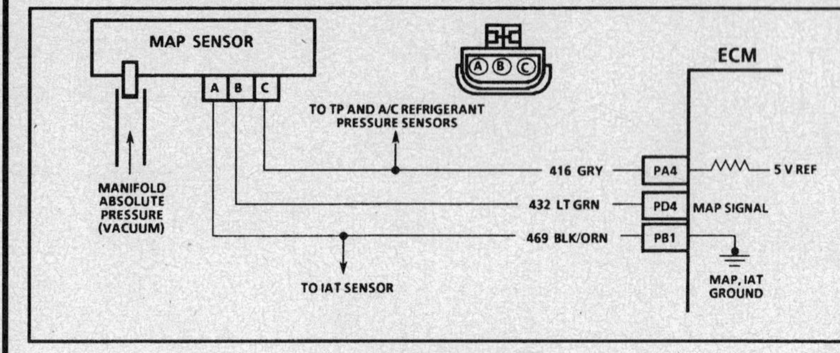

CHART C-1D
MANIFOLD ABSOLUTE PRESSURE (MAP) OUTPUT CHECK
2.3L (VIN D, A & 3) "N" CARLINE

Circuit Description:

The Manifold Absolute Pressure (MAP) sensor measures the changes in the intake manifold pressure which results from engine load (intake manifold vacuum) and RPM changes; and converts these into a voltage output. The Engine Control Module (ECM) sends a 5 volt reference voltage to the MAP sensor. As the manifold pressure changed, the output voltage of the sensor also changes. By monitoring the sensor output voltage, the ECM knows the manifold pressure. A lower pressure (low voltage) output voltage will be about 1 - 2 volts at idle. While higher pressure (high voltage) output voltage will be about 4 - 4.8 at Wide Open Throttle (WOT). The MAP sensor is also used, under certain conditions, to measure barometric pressure, allowing the ECM to make adjustments for different altitudes. The ECM uses the map sensor to control fuel delivery and ignition timing.

Test Description: Number(s) below refer to circled number(s) on the diagnostic chart.

Important
- Be sure to use the same Diagnostic Test Equipment for all measurements.

1. When comparing Tech 1 scan tool readings to a known good vehicle, it is important to compare vehicles that use a MAP sensor having the same color insert or having the same "Hot Stamped" number. See figures
2. Applying 34 kPa (10" Hg) vacuum to the MAP sensor should cause the voltage to change. Subtract second reading from the first. Voltage value should be greater than 1.5 volts. Upon applying vacuum to the sensor, the change in voltage should be instantaneous. A slow voltage change indicates a faulty sensor.

3. Check vacuum hose to sensor for leaking or restriction. Be sure no other vacuum devices are connected to the MAP hose.

NOTICE: Make sure electrical connector remains securely fastened.

4. Disconnect sensor from bracket and twist sensor by hand (only) to check for intermittent connection. Output changes greater than .1 volt indicate a bad connector or connection. If OK, replace sensor.

2.3L (VIN D, A & 3) ENGINE — COMPONENT DIAGNOSTIC CHART — 1993–94 ACHIEVA, GRAND AM AND SKYLARK

CHART C-1D
MANIFOLD ABSOLUTE PRESSURE (MAP) OUTPUT CHECK
2.3L (VIN D, A & 3) "N" CARLINE

NOTICE: THIS CHART ONLY APPLIES TO MAP SENSORS HAVING GREEN OR BLACK COLOR KEY INSERT (SEE BELOW).

①
- IF DTC 33 OR 34 IS SET, USE THOSE CHARTS FIRST.
- IGNITION "ON," ENGINE "OFF."
- SCAN TOOL SHOULD INDICATE A MAP SENSOR VOLTAGE.
- COMPARE THIS READING WITH THE READING OF A KNOWN GOOD VEHICLE. SEE FACING PAGE TEST DESCRIPTION, STEP 1. VOLTAGE READING SHOULD BE WITHIN ± .4 VOLT. IS IT?

YES → ②
- DISCONNECT AND PLUG VACUUM SOURCE TO MAP SENSOR.
- CONNECT A HAND VACUUM PUMP TO MAP SENSOR.
- START ENGINE.
- NOTE MAP SENSOR VOLTAGE.
- APPLY 34 kPa (10" Hg) OF VACUUM AND NOTE VOLTAGE CHANGE. SUBTRACT SECOND READING FROM THE FIRST. VOLTAGE VALUE SHOULD BE GREATER THAN 1.5 VOLTS. IS IT?

NO → REPLACE MAP SENSOR.

YES → ③ NO TROUBLE FOUND. CHECK MAP SENSOR VACUUM SOURCE FOR LEAKAGE OR RESTRICTION. BE SURE THIS SOURCE SUPPLIES VACUUM TO MAP SENSOR ONLY.

NO → ④ CHECK MAP SENSOR CONNECTION. IF OK, REPLACE MAP SENSOR.

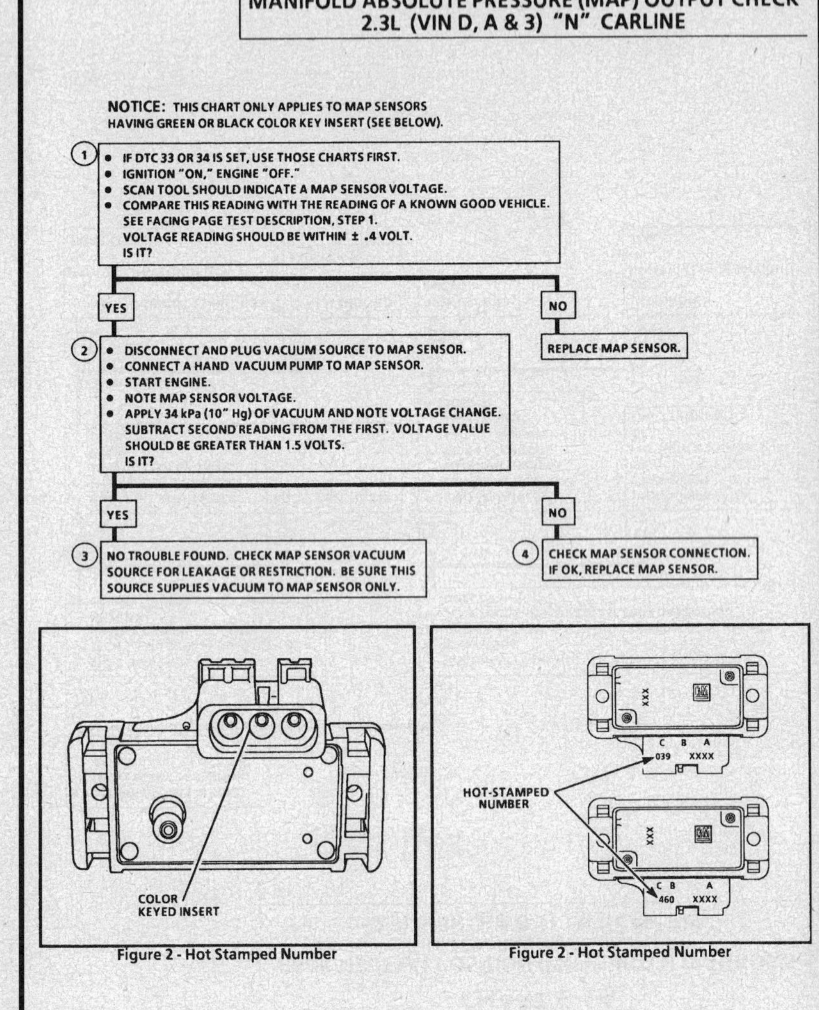

Figure 2 - Hot Stamped Number

Figure 2 - Hot Stamped Number

2.3L (VIN D, A & 3) ENGINE — COMPONENT DIAGNOSTIC CHART — 1993–94 ACHIEVA, GRAND AM AND SKYLARK

CHART C-2A

INJECTOR BALANCE TEST
2.3L (VIN D, A & 3) "N" CARLINE

The injector balance tester is a tool used to turn the injector on for a precise amount of time, thus spraying a measured amount of fuel into the manifold. This causes a drop in fuel rail pressure that we can record and compare between each injector. All injectors should have the same amount of pressure drop (± 10 kpa). Any injector with a pressure drop that is 10 kpa (or more) greater or less than the average drop of the other injectors should be considered faulty and replaced.

STEP 1

Engine "cool down" period (10 minutes) is necessary to avoid irregular readings due to "Hot Soak" fuel boiling. With ignition "OFF" connect fuel gauge. Wrap a shop towel around fitting while connecting gage to avoid fuel spillage.

Disconnect harness connectors at all injectors, and connect injector tester J-34730-3A to one injector. Ignition must be "OFF" at least 10 seconds to complete ECM shutdown cycle. Fuel pump should run about 2 seconds after ignition is turned "ON". At this point, insert clear tubing attached to vent valve into a suitable container and bleed air from gauge and hose to insure accurate gauge operation. Repeat this step until all air is bled from gauge.

STEP 2

Turn ignition "OFF" for 10 seconds and then "ON" again to get fuel pressure to its maximum. Record this initial pressure reading. Energize tester one time and note pressure drop at its lowest point (Disregard any slight pressure increase after drop hits low point.). By subtracting this second pressure reading from the initial pressure, we have the actual amount of injector pressure drop.

STEP 3

Repeat Step 2 on each injector and compare the amount of drop. Usually, good injectors will have virtually the same drop. Retest any injector that has a pressure difference of 10 kPa, either more or less than the average of the other injectors on the engine. Replace any injector that also fails the retest. If the pressure drop of all injectors is within 10 kPa of this average, the injectors appear to be flowing properly.

NOTE: *The entire test should <u>not</u> be repeated more than once without running the engine to prevent flooding. (This includes any retest on faulty injectors).*

2.3L (VIN D, A & 3) ENGINE — COMPONENT DIAGNOSTIC CHART — 1993–94 ACHIEVA, GRAND AM AND SKYLARK

NOTICE: The entire test should <u>NOT</u> be repeated more than once without running the engine to prevent flooding. (This includes any retest on faulty injectors.)

The fuel pressure test in Chart A-7, should be completed prior to this test.

CHART C-2A

INJECTOR BALANCE TEST
2.3L (VIN D, A & 3) "N" CARLINE

Step 1. If engine is at operating temperature, allow a 10 minute "cool down" period then connect fuel pressure gage and injector tester.
1. Ignition "OFF."
2. Connect fuel pressure gage and injector tester.
3. Ignition "ON."
4. Bleed off air in gage. Repeat until all air is bled from gauge.

Step 2. Run test:
1. Ignition "OFF" for 10 seconds.
2. Ignition "ON." Record gage pressure. (Pressure must hold steady, if not, see "Fuel System Diagnosis," CHART A-7
3. Turn injector on, by depressing button on injector tester, and note pressure at the instant the gauge needle stops.

Step 3.
1. Repeat Step 2 on all injectors and record pressure drop on each. Retest injectors that appear faulty (Any injectors that have a 10 kPa (1.5 psi) difference, either more or less, in pressure from the average).

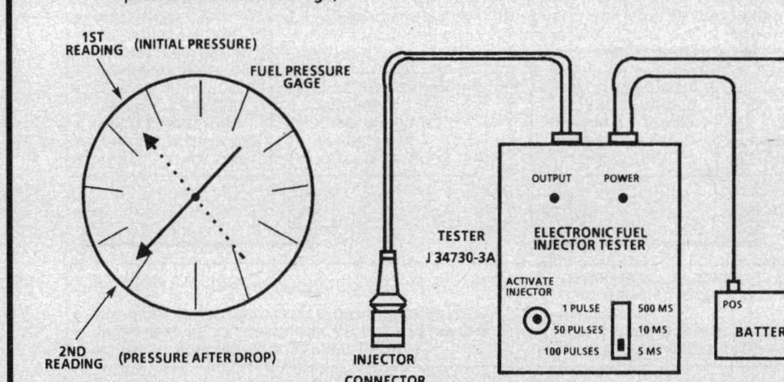

EXAMPLE

CYLINDER	1	2	3
1ST READING	293 kPa (43 psi)	293 kPa (43 psi)	293 kPa (43 psi)
2ND READING	131 kPa (19 psi)	115 kPa (17 psi)	145 kPa (21 psi)
AMOUNT OF DROP	162 kPa (24 psi)	178 kPa (26 psi)	148 kPa (21 psi)
	OK	FAULTY, RICH (TOO MUCH FUEL DROP)	FAULTY, LEAN (TOO LITTLE FUEL DROP)

2.3L (VIN D, A & 3) ENGINE — COMPONENT DIAGNOSTIC CHART — 1993–94 ACHIEVA, GRAND AM AND SKYLARK

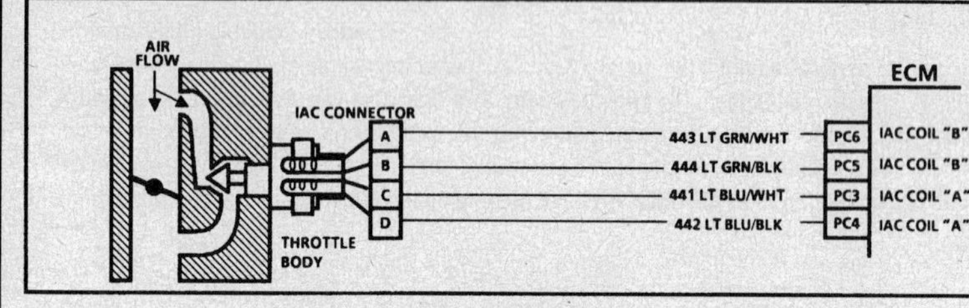

CHART C-2C
IDLE AIR CONTROL (IAC) VALVE CHECK
2.3L (VIN D, A & 3) "N" CARLINE

Circuit Description:

The Engine Control Module (ECM) controls idle RPM with the IAC valve. To increase idle RPM, the ECM moves the IAC valve out, allowing more air to bypass the throttle plate. To decrease RPM, it moves the IAC valve in, reducing air flow by-passing the throttle plate. A Tech 1 scan tool will read the ECM commands to the IAC valve in counts. The higher the counts, the more air allowed (higher idle). The lower the counts, the less air allowed (lower idle).

Test Description: Number(s) below refer to circled number(s) on the diagnostic chart.

1. The Tech 1 scan tool is used to extend and retract the IAC valve valve. Valve movement is verified by an engine speed change. If no change in engine speed occurs, the valve can be retested when removed from the throttle body.
2. This step checks the quality of the IAC valve movement in Step 1. Between 900 RPM and about 1500 RPM, the engine speed should change smoothly with each flash of the tester light in both extend and retract. If the IAC valve is retracted beyond the control range (about 1500 RPM), it may take many flashes in the extend position before engine speed will begin to drop. This is normal on certain engines, fully extending IAC valve may cause engine stall. This may be normal. Step 1 verified proper IAC valve operation while this step checks IAC valve circuits. Each lamp on the node light should flash red and green while the IAC valve is cycled. While the sequence of color is not important if either light is "OFF" or does not flash red and green, check the circuits for faults, beginning with poor terminal contacts.

Diagnostic Aids:

A slow, unstable, or fast idle may be caused by a non-IAC valve system problem that cannot be overcome by the IAC valve. Out of control range IAC scan tool counts will be above 60 if idle is too low, and zero counts if idle is too high. The following checks should be made to repair a non-IAC valve system problem.

- **Vacuum Leak (High Idle)** - If idle is too high, stop the engine. Fully extend (low) IAC valve with tester. Start engine. If Idle speed is above 800 RPM, locate and correct vacuum leak including crankcase ventilation system. Also check for binding of throttle blade or linkage.
- **System too lean (High Air/Fuel Ratio)** - Idle speed may be too high or too low. Engine speed may vary up and down and disconnecting IAC does not help. DTC 44 may be set. Tech 1 scan tool Oxygen Sensor (O2S) voltage will be less than 300 mV (.3 volt). Check for low regulated fuel pressure, water in the fuel or a restricted injector.
- **System too rich (Low Air/Fuel Ratio)** - The idle speed will be too low. Tech 1 scan tool IAC valve counts will usually be above 80. System is obviously rich and may exhibit black smoke exhaust.
Scan tool O2S voltage will be fixed above 800 mV (.8 volt). Check for high fuel pressure, leaking or sticking injector. Silicone contaminated O2S will scan an O2S voltage slow to respond.
- **Throttle Body** - Remove IAC valve and inspect bore for foreign material.

- If intermittent poor driveability or idle symptoms are resolved by disconnecting the IAC valve, carefully recheck connections, valve terminal resistance, or replace IAC valve.

2.3L (VIN D, A & 3) ENGINE — COMPONENT DIAGNOSTIC CHART — 1993–94 ACHIEVA, GRAND AM AND SKYLARK

CHART C-2C
IDLE AIR CONTROL (IAC) VALVE CHECK
2.3L (VIN D, A & 3) "N" CARLINE

(1)
- INSTALL TECH 1 SCAN TOOL.
- ENGINE AT NORMAL OPERATING TEMPERATURE IN PARK/NEUTRAL WITH PARKING BRAKE SET.
- A/C "OFF."
- SELECT RPM CONTROL. (MISC. TESTS)
- CYCLE IAC VALVE THROUGH ITS RANGE FROM 900 RPM UP TO 1500 RPM.
- RPM SHOULD CHANGE SMOOTHLY. DOES IT?

NO →

(2)
- INSTALL IAC NODE LIGHT * IN IAC VALVE HARNESS.
- ENGINE RUNNING. CYCLE IAC VALVE WITH TECH 1 SCAN TOOL.
- EACH NODE LIGHT SHOULD CYCLE RED AND GREEN BUT NEVER "OFF." DO THEY?

NO →

IF CIRCUIT(S) DID NOT TEST RED AND GREEN, CHECK FOR:
- FAULTY CONNECTOR TERMINAL CONTACTS.
- OPEN CIRCUITS INCLUDING CONNECTORS.
- CIRCUITS SHORTED TO GROUND OR VOLTAGE.
- FAULTY ECM CONNECTIONS OR REPLACE ECM. REPAIR AS NECESSARY AND RETEST.

YES →
- CHECK IAC VALVE CONNECTIONS.
- CHECK IAC VALVE PASSAGES.
- IF OK, REPLACE IAC VALVE.

YES →
- USING THE IAC DRIVER * OR OTHER CONVENIENT CONNECTOR, CHECK RESISTANCE ACROSS IAC VALVE COILS.
- SHOULD BE 40 TO 80 OHMS BETWEEN IAC VALVE TERMINALS "A" TO "B" AND "C" TO "D".

OK →
- CHECK RESISTANCE BETWEEN IAC VALVE TERMINALS "B" AND "C" AND "A" AND "D".
- SHOULD BE INFINITE.

OK →
IDLE AIR CONTROL VALVE CIRCUIT OK. REFER TO "DIAGNOSTIC AIDS"

NOT OK →
REPLACE IAC VALVE AND RETEST.

NOT OK →
REPLACE IAC VALVE AND RETEST.

* IAC DRIVER AND NODE LIGHT REQUIRED KIT 222-L FROM: CONCEPT TECHNOLOGY, INC. J 37027 FROM: KENT-MOORE, INC.

CLEAR CODES, CONFIRM "CLOSED LOOP" OPERATION, NO MIL (SERVICE ENGINE SOON) PERFORM IAC RESET PROCEDURE PER APPLICABLE SERVICE MANUAL AND VERIFY CONTROLLED IDLE SPEED IS CORRECT.

"AFTER REPAIRS," CONFIRM "CLOSED LOOP" OPERATION AND NO MIL (SERVICE ENGINE SOON).

2.3L (VIN D, A & 3) ENGINE — COMPONENT DIAGNOSTIC CHART — 1993–94 ACHIEVA, GRAND AM AND SKYLARK

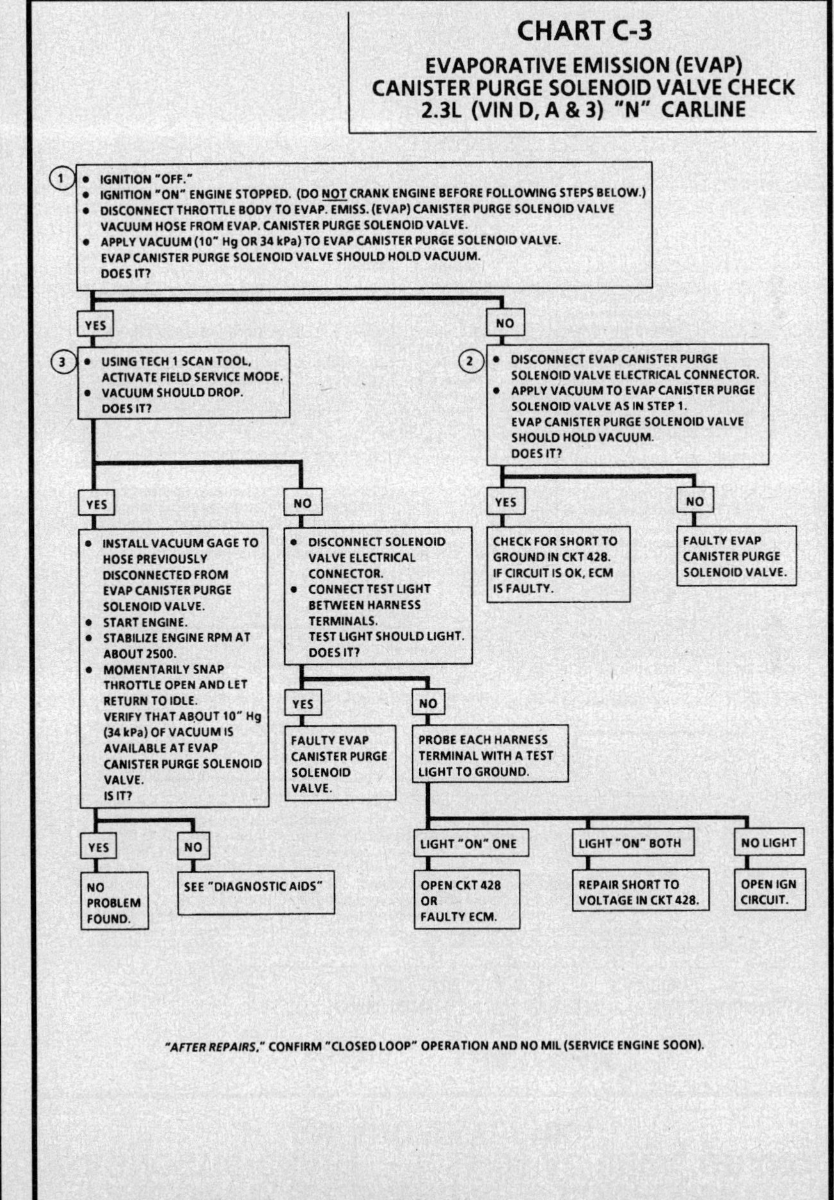

CHART C-3
EVAPORATIVE EMISSION (EVAP)
CANISTER PURGE SOLENOID VALVE CHECK
2.3L (VIN D, A & 3) "N" CARLINE

Circuit Description:

Canister purge is controlled by a solenoid valve that allows manifold and/or ported vacuum to purge the canister when energized. The Engine Control Module (ECM) supplies a ground to energize the solenoid valve (purge "ON"). The purge solenoid control by the ECM is pulse width modulated (turned "ON" and "OFF" several times a second). The duty cycle (pulse width) is determined by "Closed Loop" feed back from the Oxygen Sensor (O2S). The duty cycle is calculated by the ECM and the output commanded when the following conditions have been met:

- Engine run time after start more than 65 seconds.
- Engine coolant temperature above 56°C.

Also, if the diagnostic "test" terminal is grounded with the engine stopped, the EVAP canister purge solenoid valve is energized (purge "ON").

Test Description: Number(s) below refer to circled number(s) on the diagnostic chart.

1. Checks to see if the solenoid valve is opened or closed. The solenoid valve is normally de-energized in this step, so it should be closed.
2. Checks to determine if solenoid valve was open due to electrical circuit problem or defective solenoid.
3. Completes functional check by grounding "test" terminal. This should normally energize the solenoid valve opening the valve which should allow the vacuum to drop (purge "ON").

Diagnostic Aids:

Make a visual check of vacuum hose(s). Check throttle body for possible cracked, broken, or plugged vacuum block. Check engine for possible mechanical problem.

2.3L (VIN D, A & 3) ENGINE — COMPONENT DIAGNOSTIC CHART — 1993–94 ACHIEVA, GRAND AM AND SKYLARK

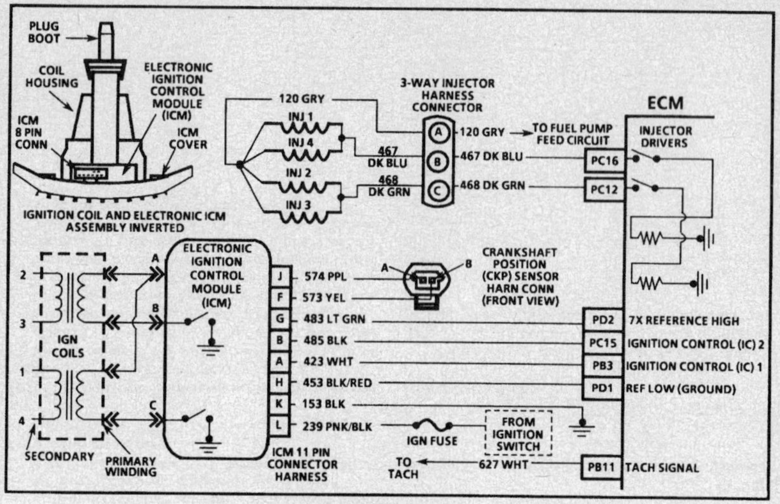

CHART C-4M
(Page 1 of 3)
ELECTRONIC IGNITION (EI) SYSTEM MISFIRE DIAGNOSIS
2.3L (VIN D, A & 3) "N" CARLINE

Circuit Description:

The Electronic Ignition (EI) system uses a waste spark method of distribution. In this type of system the Electronic Ignition Control Module (ICM) triggers the #1-4 coil pair resulting in both #1 and #4 spark plugs firing at the same time. #1 cylinder is on the compression stroke at the same time #4 is on the exhaust stroke, resulting in a lower energy requirement to fire #4 spark plug. This leaves the remainder of the high voltage to be used to fire #1 spark plug. On this application, the Crankshaft Position (CKP) sensor is mounted to, and protrudes through the block to within approximately 0.050" of the crankshaft reluctor. Since the reluctor is a machined portion of the crankshaft and the sensor is mounted in a fixed position on the block, timing adjustments are not possible or necessary.

Test Description: Number(s) below refer to circled number(s) on the diagnostic chart.

1. This checks for equal relative power output between the cylinders. Any injector which when disconnected did not result in an RPM drop approximately equal to the others, is located on the misfiring cylinder.
2. This step checks for Ignition Control (IC) frequency output from the ECM.
3. This step checks if the circuit is shorted to ground.
4. An injector driver circuit shorted to ground would result in the test light "ON" steady, and possibly a flooded condition. A shorted injector (less than 1.8 ohms) could cause incorrect ECM operation.
5. This step will determine if there is a short to voltage on the circuit that did not show frequency response.
6. This step will determine if the short to ground is in the circuit or the ICM.

Diagnostic Aid:

Verify ICM 11 pin harness connector terminal "K", CKT 153 resistance to ground is less than .5 ohm. A shorted or low resistance injector may cause a miss in the other injector in that pair (1 & 4 or 2 & 3). A poor connection or open in CKT 453 may cause a miss.

CHART C-4M
(Page 1 of 3)
ELECTRONIC IGNITION (EI) SYSTEM MISFIRE DIAGNOSIS
2.3L (VIN D, A & 3) "N" CARLINE

(1)
- IF DIAGNOSTIC TROUBLE CODES (DTC) 19 OR 53 ARE PRESENT, REFER TO THOSE CHARTS BEFORE CONTINUING.
- ENGINE AT NORMAL OPERATING TEMPERATURE DISCONNECT IAC VALVE.
- REMOVE CRANKCASE VENTILATION OIL/AIR SEPARATOR TO GAIN ACCESS TO INJECTOR CONNECTORS.
- MOMENTARILY DISCONNECT EACH INJECTOR CONNECTOR WHILE OBSERVING ENGINE RPM.
- NOTE ANY INJECTOR(S) NOT RESULTING IN AN RPM DROP.
- (IF ALL INJECTORS RESULT IN AN RPM DROP, GO TO STEP 2).
- INSTALL INJECTOR TEST LIGHT J 34730-2 IN INJECTOR HARNESS CONNECTOR FOR INJECTOR WHICH DID NOT RESULT IN RPM DROP. LIGHT SHOULD BLINK.
- DOES IT?

YES → (2)
- DISCONNECT 3-WAY INJECTOR HARNESS CONNECTOR.
- WITH DVOM NEGATIVE TERMINAL TO GROUND, BACKPROBE ECM TERMINAL CONNECTOR "PC15" AND CHECK THE FREQUENCY OF THE CIRCUIT DURING CRANKING. REPEAT WITH ECM TERMINAL CONNECTOR "PB3". THE FREQUENCY READINGS SHOULD BE BETWEEN 1-10 Hz. ARE THEY?

NO → (3)
- IGNITION "OFF."
- DISCONNECT PNK CONNECTORS FROM ECM. WITH DVOM CONNECTED TO GROUND, PROBE THE CAVITY THAT DID NOT SHOW FREQUENCY RESPONSE, EITHER PB3 OR PC15. RESISTANCE SHOULD BE GREATER THAN 5K. IS IT?

YES → REFER TO CHART C-4M PAGE 2 OF 3.

YES → (5)
- IGNITION "ON."
- WITH DVOM STILL CONNECTED TO GROUND, AGAIN PROBE THE CAVITY THAT DID NOT SHOW FREQUENCY RESPONSE. VOLTAGE SHOULD BE "0". IS IT?

NO → (6)
- DISCONNECT ICM 11 PIN TERMINAL CONNECTOR.
- WITH DVOM STILL CONNECTED TO GROUND, AGAIN PROBE THE CAVITY THAT DID NOT SHOW FREQUENCY RESPONSE. RESISTANCE SHOULD BE INFINITE. IS IT?

YES → VISUALLY INSPECT THE PINK ECM CONNECTORS TO SEE IF THEY ARE CLEAN AND TIGHT. IF OK, REPLACE ECM.

NO → REPAIR SHORT TO VOLTAGE IN CKT 485 OR 423.

NO → REPAIR SHORT TO GROUND IN CKT 423 OR 485.

YES → FAULTY ICM.

NO → STEADY LIGHT → (4)
- CHECK INJECTOR DRIVER CIRCUIT WHICH HAD THE STEADY LIGHT, FOR A SHORT TO GROUND.
- IF CIRCUIT IS NOT SHORTED, CHECK RESISTANCE ACROSS EACH INJECTOR IN THE CIRCUIT.
- RESISTANCE SHOULD BE BETWEEN 1.8 AND 2.2 OHMS FOR EACH INJECTOR. IS IT?

LIGHT "OFF" → SEE CHART C-4M PAGE 3 OF 3.

NO → REPLACE ANY INJECTOR THAT IS NOT WITHIN THIS RANGE AND RECHECK FOR MISFIRE BEGINNING WITH STEP 1 AGAIN.

YES → FAULTY ECM.

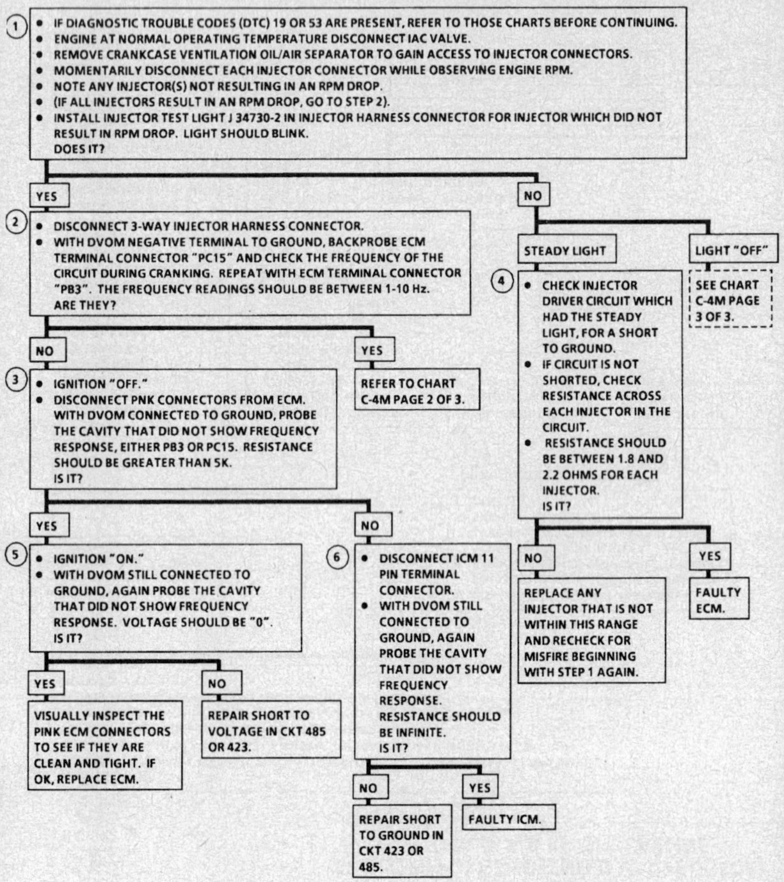

2.3L (VIN D, A & 3) ENGINE — COMPONENT DIAGNOSTIC CHART — 1993–94 ACHIEVA, GRAND AM AND SKYLARK

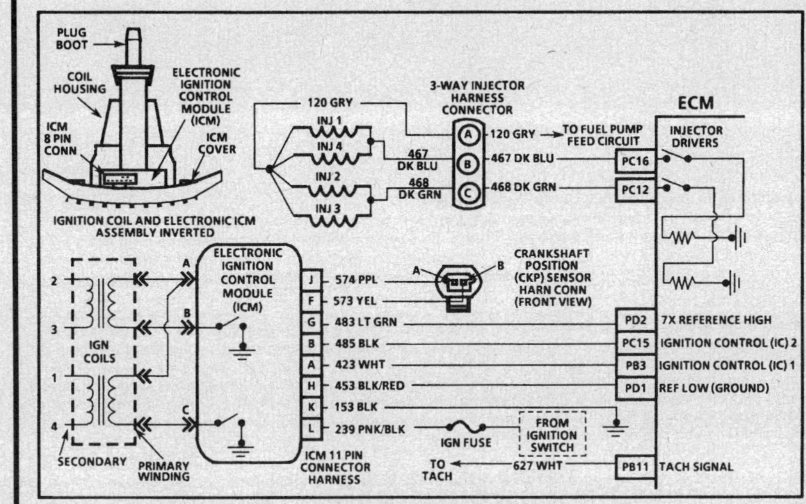

CHART C-4M
(Page 2 of 3)
ELECTRONIC IGNITION (EI) SYSTEM MISFIRE DIAGNOSIS
2.3L (VIN D, A & 3) "N" CARLINE

Circuit Description:

The Electronic Ignition (EI) system uses a waste spark method of distribution. In this type of system the Electronic Ignition Control Module (ICM) triggers the #1-4 coil pair resulting in both #1 and #4 spark plugs firing at the same time. #1 cylinder is on the compression stroke at the same time #4 is on the exhaust stroke, resulting in a lower energy requirement to fire #4 spark plug. This leaves the remainder of the high voltage to be used to fire #1 spark plug. On this application, the Crankshaft Position (CKP) sensor is mounted to, and protrudes through the block to within approximately 0.050" of the crankshaft reluctor. Since the reluctor is a machined portion of the crankshaft and the sensor is mounted in a fixed position on the block, timing adjustments are not possible or necessary.

Test Description: Number(s) below refer to circled number(s) on the diagnostic chart.

7. This step checks for an open in CKTs 485 or 423.
8. If a plug boot is burned, the other plug on that ignition coil may still fire at idle. This step tests the system's ability to produce at least 25,000 volts at each spark plug.
9. No spark, on one coil, may be caused by an open secondary circuit. Therefore, the coil's secondary resistance should be checked. Resistance readings above 20,000 ohms, but not infinite, will probably not cause a no start but may cause an engine miss under certain conditions.
10. If the no spark condition is caused by coil connections, a coil or a secondary boot assembly, the test light will blink. If the light does not blink, the fault is module connections or the module.

Diagnostic Aid:

Verify ICM 11 pin harness connector terminal "K", CKT 153 resistance to ground is less than .5 ohm. A shorted or low resistance injector may cause a miss in the other injector in that pair (1 & 4 or 2 & 3).

2.3L (VIN D, A & 3) ENGINE — COMPONENT DIAGNOSTIC CHART — 1993–94 ACHIEVA, GRAND AM AND SKYLARK

CHART C-4M
(Page 2 of 3)
ELECTRONIC IGNITION (EI) SYSTEM MISFIRE DIAGNOSIS
2.3L (VIN D, A & 3) "N" CARLINE

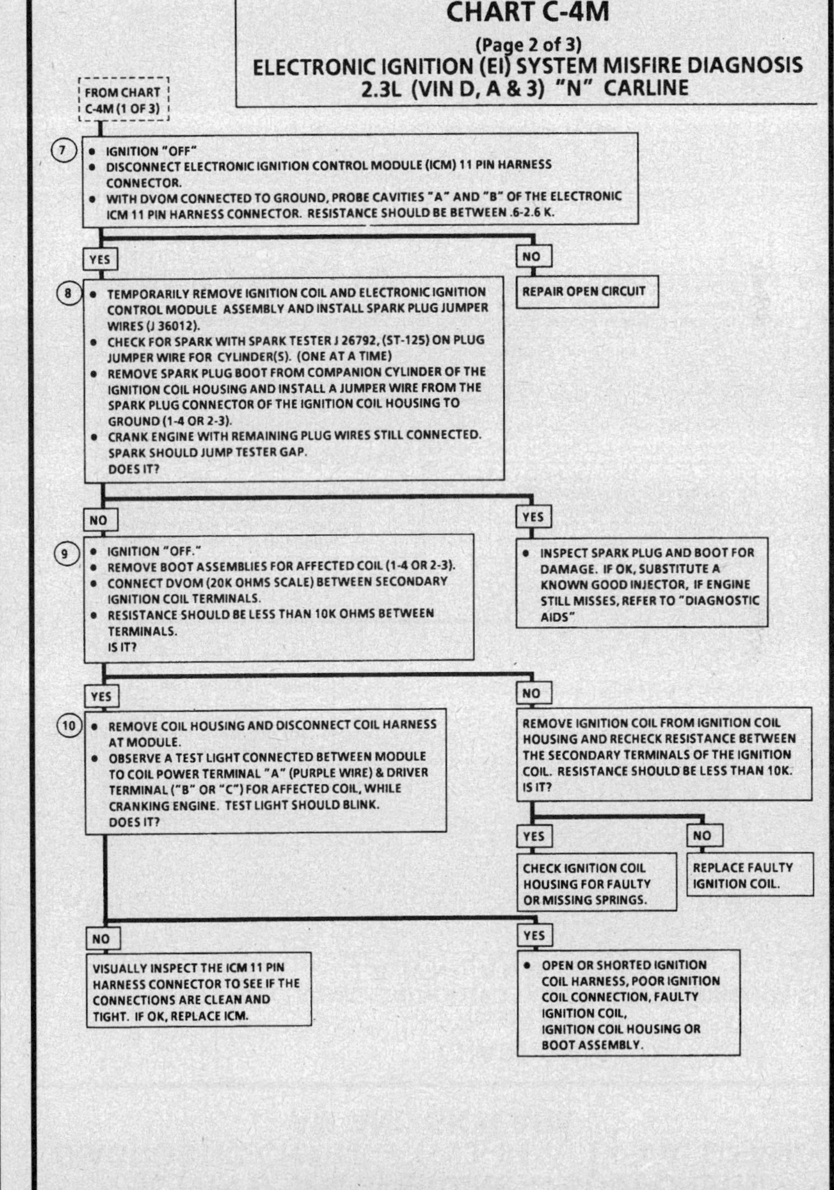

2.3L (VIN D, A & 3) ENGINE — COMPONENT DIAGNOSTIC CHART — 1993–94 ACHIEVA, GRAND AM AND SKYLARK

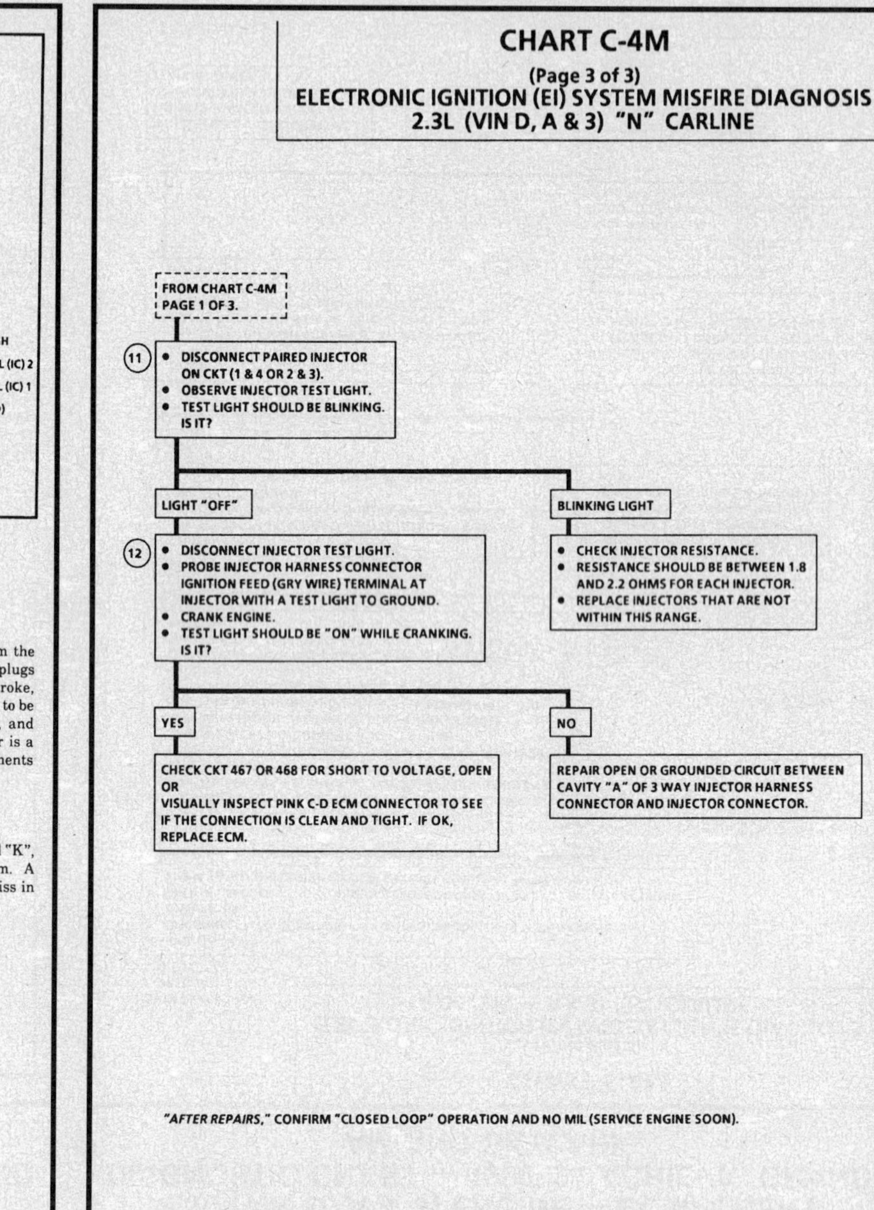

CHART C-4M
(Page 3 of 3)
ELECTRONIC IGNITION (EI) SYSTEM MISFIRE DIAGNOSIS
2.3L (VIN D, A & 3) "N" CARLINE

Circuit Description:

The Electronic Ignition (EI) system uses a waste spark method of distribution. In this type of system the Electronic Ignition Control Module (ICM) triggers the #1-4 coil pair resulting in both #1 and #4 spark plugs firing at the same time. #1 cylinder is on the compression stroke at the same time #4 is on the exhaust stroke, resulting in a lower energy requirement to fire #4 spark plug. This leaves the remainder of the high voltage to be used to fire #1 spark plug. On this application, the Crankshaft Position (CKP) sensor is mounted to, and protrudes through the block to within approximately 0.050" of the crankshaft reluctor. Since the reluctor is a machined portion of the crankshaft and the sensor is mounted in a fixed position on the block, timing adjustments are not possible or necessary.

Test Description: Number(s) below refer to circled number(s) on the diagnostic chart.

11. Checks for a grounded or low resistance injector. This condition will not able the paired injector to properly deliver fuel.
12. Checks for voltage available at the injector driver circuit.

Diagnostic Aid:

Verify ICM 11 pin harness connector terminal "K", CKT 153 resistance to ground is less than .5 ohm. A shorted or low resistance injector may cause a miss in the other injector in that pair (1 & 4 or 2 & 3).

2.3L (VIN D, A & 3) ENGINE — COMPONENT DIAGNOSTIC CHART — 1993–94 ACHIEVA, GRAND AM AND SKYLARK

CHART C-4M
(Page 3 of 3)
ELECTRONIC IGNITION (EI) SYSTEM MISFIRE DIAGNOSIS
2.3L (VIN D, A & 3) "N" CARLINE

FROM CHART C-4M PAGE 1 OF 3.

11
- DISCONNECT PAIRED INJECTOR ON CKT (1 & 4 OR 2 & 3).
- OBSERVE INJECTOR TEST LIGHT.
- TEST LIGHT SHOULD BE BLINKING. IS IT?

LIGHT "OFF"

BLINKING LIGHT
- CHECK INJECTOR RESISTANCE.
- RESISTANCE SHOULD BE BETWEEN 1.8 AND 2.2 OHMS FOR EACH INJECTOR.
- REPLACE INJECTORS THAT ARE NOT WITHIN THIS RANGE.

12
- DISCONNECT INJECTOR TEST LIGHT.
- PROBE INJECTOR HARNESS CONNECTOR IGNITION FEED (GRY WIRE) TERMINAL AT INJECTOR WITH A TEST LIGHT TO GROUND.
- CRANK ENGINE.
- TEST LIGHT SHOULD BE "ON" WHILE CRANKING. IS IT?

YES

NO

CHECK CKT 467 OR 468 FOR SHORT TO VOLTAGE, OPEN OR VISUALLY INSPECT PINK C-D ECM CONNECTOR TO SEE IF THE CONNECTION IS CLEAN AND TIGHT. IF OK, REPLACE ECM.

REPAIR OPEN OR GROUNDED CIRCUIT BETWEEN CAVITY "A" OF 3 WAY INJECTOR HARNESS CONNECTOR AND INJECTOR CONNECTOR.

"AFTER REPAIRS," CONFIRM "CLOSED LOOP" OPERATION AND NO MIL (SERVICE ENGINE SOON).

2.3L (VIN D, A & 3) ENGINE — COMPONENT DIAGNOSTIC CHART — 1993–94 ACHIEVA, GRAND AM AND SKYLARK

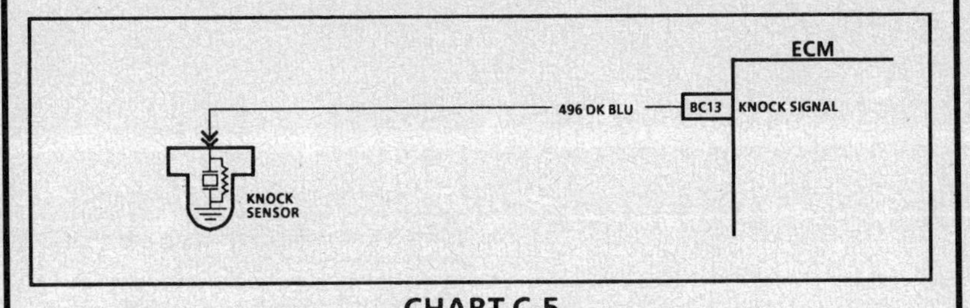

CHART C-5
KNOCK SENSOR (KS) SYSTEM CHECK
2.3L (VIN D, A & 3) "N" CARLINE

Circuit Description:

The knock sensor is used to detect engine detonation and the Engine Control Module (ECM) will retard the electronic spark timing based on the signal being received. The knock sensor produces an AC signal. The amplitude and frequency are dependent upon the knock level.

The PROM used with this engine contains the functions which were part of remotely mounted Knock Sensor (KS) modules used on other GM vehicles. The KS portion of the PROM then sends a signal to other parts of the ECM which adjusts the spark timing to retard the spark and reduce the detonation.

Test Description: Number(s) below refer to circled number(s) on the diagnostic chart.

1. The Tech 1 scan tool displays knock sensor activity in counts. The counts should rise when engine speed increases or fall when engine speed decreases. This step checks knock sensor activity.
2. This step checks that the internal resistance of knock sensor is within an acceptable range.

3. If the engine has an internal problem which is creating a knock, the knock sensor may be responding to the mechanical noise.

Diagnostic Aids:

While observing knock signal on the Tech 1 scan tool, there should be an indication that knock is present when detonation can be heard. Detonation is most likely to occur under high engine load conditions.

2.3L (VIN D, A & 3) ENGINE — COMPONENT DIAGNOSTIC CHART — 1993–94 ACHIEVA, GRAND AM AND SKYLARK

CHART C-5
KNOCK SENSOR (KS) SYSTEM CHECK
2.3L (VIN D, A & 3) "N" CARLINE

2.3L (VIN D, A & 3) ENGINE — COMPONENT DIAGNOSTIC CHART — 1993–94 ACHIEVA, GRAND AM AND SKYLARK

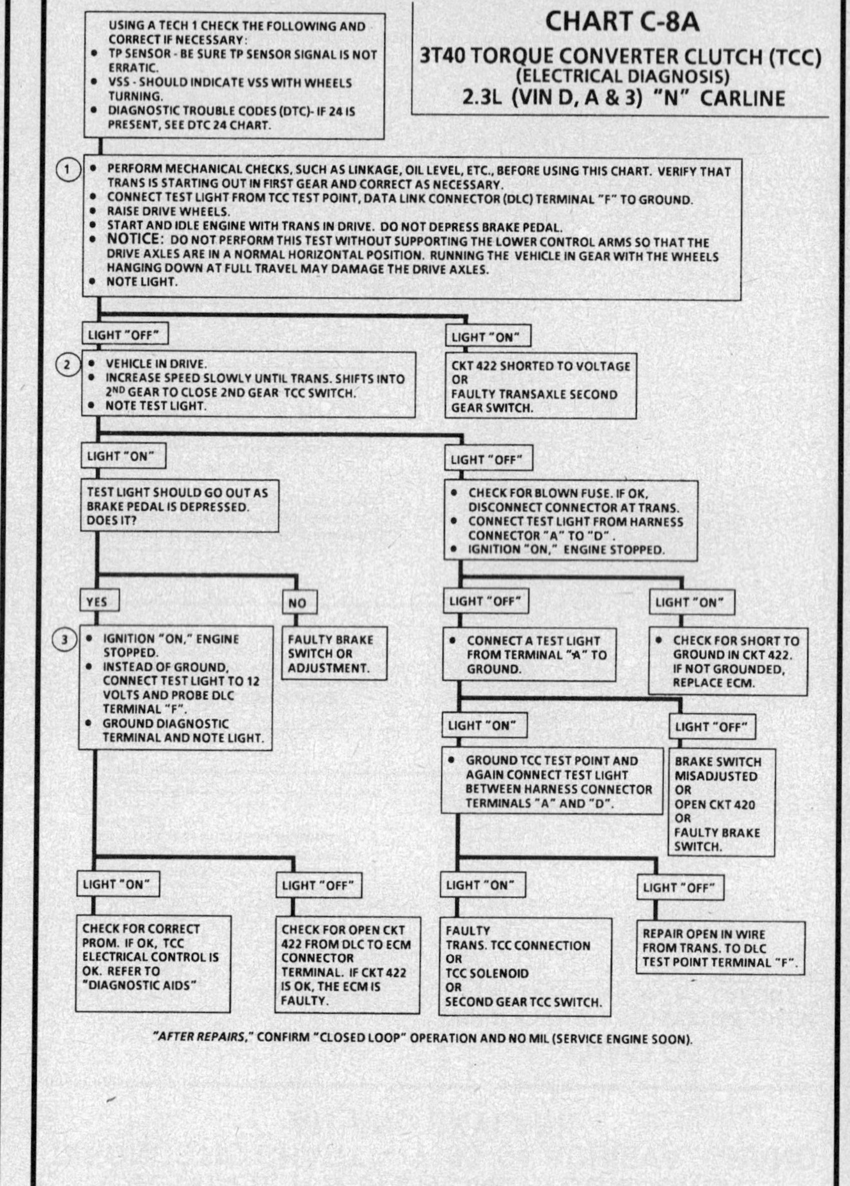

CHART C-8A
3T40 TORQUE CONVERTER CLUTCH (TCC)
(ELECTRICAL DIAGNOSIS)
2.3L (VIN D, A & 3) "N" CARLINE

Circuit Description:

The purpose of the Torque Converter Clutch (TCC) feature is to eliminate the power loss of the transaxle converter stage when the vehicle is in a cruise condition. This allows the convenience of the automatic transaxle and the fuel economy of a manual transaxle.

Fused battery ignition is supplied to the TCC solenoid through the brake switch, and transaxle 2nd gear apply switch. The Engine Control Module (ECM) will engage TCC by grounding CKT 422 to energize the solenoid.

TCC will engage when:
- Vehicle speed above a calibrated value (about 34 mph) (55 km/h).
- Throttle position sensor output not changing, indicating a steady road speed.
- Transaxle 2nd gear switch closed.
- Brake switch closed.

Test Description: Number(s) below refer to circled number(s) on the diagnostic chart.
1. Light "OFF" confirms transaxle 2nd gear apply switch is open.
2. By 25 mph, the transaxle 2nd gear TCC switch should close. Test light will come "ON" and confirm battery supply and closed brake switch.
3. Grounding the diagnostic terminal with ignition "ON," engine "OFF," should energize the TCC solenoid by grounding CKT 422. This test checks the ability of the ECM to supply a ground to the TCC solenoid. The test light connected from 12 volts to the Data Link Connector (DLC) terminal "F" will turn "ON" as CKT 422 is grounded.

Diagnostic Aids:

A Tech 1 scan tool only indicates when the ECM has turned "ON" the TCC driver and this does not confirm that the TCC has engaged. To determine if TCC is functioning properly, engine RPM should decrease when the scan indicates the TCC driver has turned "ON."

2.3L (VIN D, A & 3) ENGINE — COMPONENT DIAGNOSTIC CHART — 1993–94 ACHIEVA, GRAND AM AND SKYLARK

CHART C-8A
3T40 TORQUE CONVERTER CLUTCH (TCC)
(ELECTRICAL DIAGNOSIS)
2.3L (VIN D, A & 3) "N" CARLINE

"AFTER REPAIRS," CONFIRM "CLOSED LOOP" OPERATION AND NO MIL (SERVICE ENGINE SOON).

2.3L (VIN D, A & 3) ENGINE — COMPONENT DIAGNOSTIC CHART — 1993–94 ACHIEVA AND GRAND AM

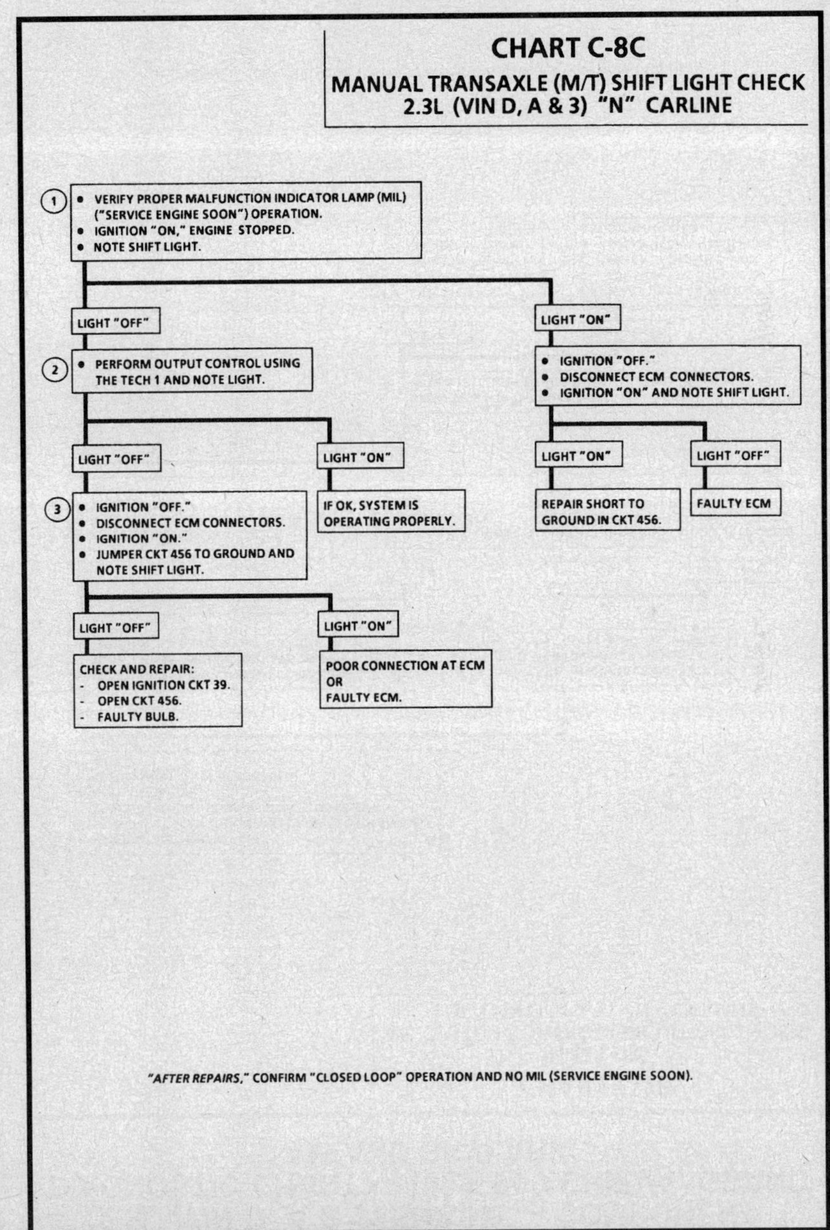

CHART C-8C
MANUAL TRANSAXLE (M/T) SHIFT LIGHT CHECK
2.3L (VIN D, A & 3) "N" CARLINE

Circuit Description:

The shift light indicates the best transaxle shift point for maximum fuel economy. The light is controlled by the Engine Control Module (ECM) and is turned "ON" by grounding CKT 456.

The Engine Control Module (ECM) uses inputs from various sensors to calculate when the shift light should be turned "ON" as follows:

- Engine coolant temperature must be above -10°C (14°F).
- ECM can determine the transaxle has been in a gear for at least 1.2 seconds, by comparison of vehicle speed (from VSS) with engine RPM.
- TP sensor is between minimum and maximum calibrated values for each gear.
- RPM is above a calibrated value for each gear (maximum 6500 RPM).

The light will be turned "ON" after a calibrated delay time which is dependent on last gear change or downshift, and will remain "ON" for a maximum of 10 seconds.

Test Description: Number(s) below refer to circled number(s) on the diagnostic chart.

1. This should not turn "ON" the shift light. If the light is "ON," there is a short to ground in CKT 456 wiring or a fault in the ECM.
2. When the diagnostic terminal is grounded, the ECM should ground CKT 456 and the shift light should come "ON."
3. This checks the shift light circuit up to the ECM connector. If the shift light illuminates, then the ECM connector is faulty or the ECM does not have the ability to ground the circuit.

Diagnostic Aids:

Check for DTC 24 (no VSS). Faulty thermostat or incorrect heat range. Incorrect or faulty PROM. A faulty or incorrect VSS may result in a shift light that does not operate.

2.3L (VIN D, A & 3) ENGINE — COMPONENT DIAGNOSTIC CHART — 1993–94 ACHIEVA, GRAND AM AND SKYLARK

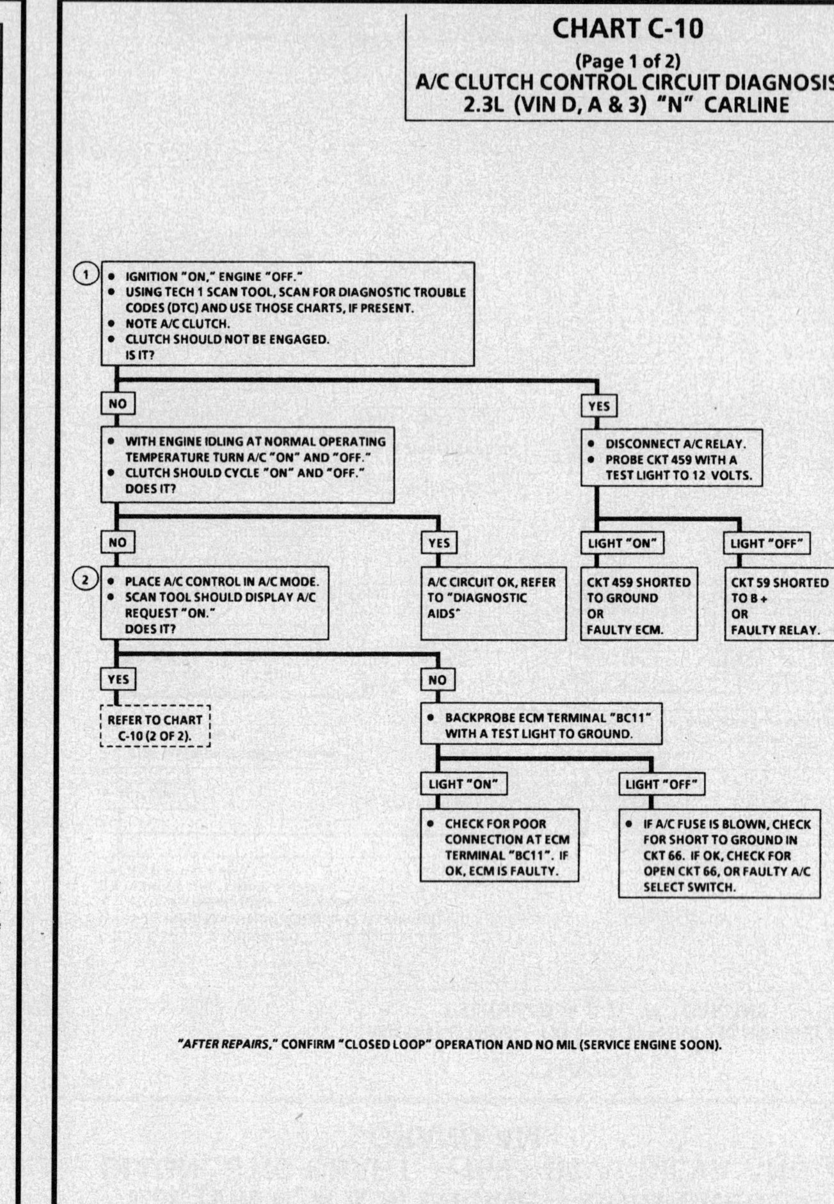

CHART C-10
(Page 1 of 2)
A/C CLUTCH CONTROL CIRCUIT DIAGNOSIS
2.3L (VIN D, A & 3) "N" CARLINE

Circuit Description:

The A/C clutch control relay is energized when the Engine Control Module (ECM) provides a ground path through CKT 459 and A/C is requested. A/C clutch is delayed about .3 seconds after A/C is requested. This will allow the IAC valve to adjust engine RPM for the additional load.

The ECM will temporarily disengage the A/C clutch relay for calibrated times for one or more of the following:
- Hot engine restart.
- Wide Open Throttle (WOT) (TP sensor over 90%).
- Engine RPM greater than about 6000 RPM.
- During IAC valve reset.

The A/C clutch relay will remain disengaged when DTC 66 is present, if A/C refrigerant pressure is out of range described previously in this section, or there is no A/C request signal due to an open A/C select switch or circuit.

Test Description: Number(s) below refer to circled number(s) on the diagnostic chart.

1. The ECM will only energize the A/C relay when the engine is running. This test will determine if the relay or CKT 459 is faulty.

2. Determines if the signal is reaching the ECM through CKT 66 from the A/C control panel. Signal should only be present when an A/C mode or defrost mode has been selected.

2.3L (VIN D, A & 3) ENGINE — COMPONENT DIAGNOSTIC CHART — 1993–94 ACHIEVA, GRAND AM AND SKYLARK

CHART C-10
(Page 1 of 2)
A/C CLUTCH CONTROL CIRCUIT DIAGNOSIS
2.3L (VIN D, A & 3) "N" CARLINE

(1)
- IGNITION "ON," ENGINE "OFF."
- USING TECH 1 SCAN TOOL, SCAN FOR DIAGNOSTIC TROUBLE CODES (DTC) AND USE THOSE CHARTS, IF PRESENT.
- NOTE A/C CLUTCH.
- CLUTCH SHOULD NOT BE ENGAGED. IS IT?

NO →
- WITH ENGINE IDLING AT NORMAL OPERATING TEMPERATURE TURN A/C "ON" AND "OFF."
- CLUTCH SHOULD CYCLE "ON" AND "OFF." DOES IT?

YES →
- DISCONNECT A/C RELAY.
- PROBE CKT 459 WITH A TEST LIGHT TO 12 VOLTS.

From "DISCONNECT A/C RELAY":
- LIGHT "ON" → CKT 459 SHORTED TO GROUND OR FAULTY ECM.
- LIGHT "OFF" → CKT 59 SHORTED TO B+ OR FAULTY RELAY.

From "WITH ENGINE IDLING... CLUTCH SHOULD CYCLE":
- **NO (2)**
  - PLACE A/C CONTROL IN A/C MODE.
  - SCAN TOOL SHOULD DISPLAY A/C REQUEST "ON." DOES IT?
- **YES** → A/C CIRCUIT OK, REFER TO "DIAGNOSTIC AIDS"

From "PLACE A/C CONTROL IN A/C MODE":
- **YES** → REFER TO CHART C-10 (2 OF 2).
- **NO** → BACKPROBE ECM TERMINAL "BC11" WITH A TEST LIGHT TO GROUND.

From "BACKPROBE ECM TERMINAL":
- LIGHT "ON" → CHECK FOR POOR CONNECTION AT ECM TERMINAL "BC11". IF OK, ECM IS FAULTY.
- LIGHT "OFF" → IF A/C FUSE IS BLOWN, CHECK FOR SHORT TO GROUND IN CKT 66. IF OK, CHECK FOR OPEN CKT 66, OR FAULTY A/C SELECT SWITCH.

"AFTER REPAIRS," CONFIRM "CLOSED LOOP" OPERATION AND NO MIL (SERVICE ENGINE SOON).

2.3L (VIN D, A & 3) ENGINE — COMPONENT DIAGNOSTIC CHART — 1993–94 ACHIEVA, GRAND AM AND SKYLARK

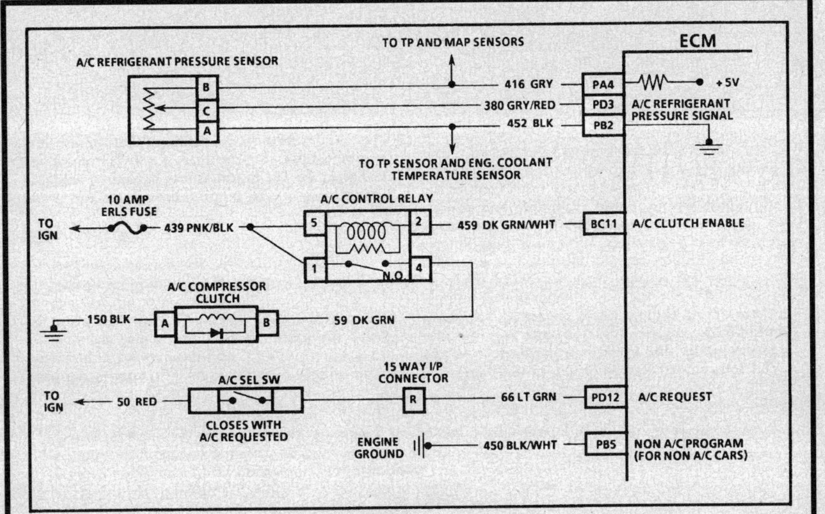

CHART C-10
(Page 2 of 2)
A/C CLUTCH CONTROL CIRCUIT DIAGNOSIS
2.3L (VIN D, A & 3) "N" CARLINE

Circuit Description:
The A/C clutch control relay is energized when the Engine Control Module (ECM) provides a ground path through CKT 459 and A/C is requested. A/C clutch is delayed about .3 seconds after A/C is requested. This will allow the IAC valve to adjust engine RPM for the additional load.

The ECM will temporarily disengage the A/C clutch relay for calibrated times for one or more of the following:
- Hot engine restart.
- Wide Open Throttle (WOT) (TP sensor over 90%).
- Engine RPM greater than about 6000 RPM.
- During IAC valve reset.

The A/C clutch relay will remain disengaged when DTC 66 is present, if A/C refrigerant pressure is out of range as described previously in this section, or there is no A/C request signal due to an open A/C select switch or circuit.

Test Description: Number(s) below refer to circled number(s) on the diagnostic chart.
1. Determines if the A/C refrigerant pressure sensor is out of range causing the compressor clutch to be disengaged.
2. With the engine stopped and field service mode activated, the ECM should be grounding CKT 459, which should cause the test light to be "ON."

Diagnostic Aids:
If complaint is insufficient cooling, the problem may be caused by an inoperative cooling fan.

See CHART C-12 for cooling fan diagnosis. If fan operates correctly, see A/C diagnosis
A/C refrigerant pressure outside of a range of 43 to 428 psi will cause the compressor to be disabled by the ECM. Observe Tech 1 scan tool A/C refrigerant pressure for 2 minutes with engine idling and A/C "ON."

Scan pressure should be within 20 psi of actual. If not, check for a circuit problem using DTC 66 chart or replace sensor.

2.3L (VIN D, A & 3) ENGINE — COMPONENT DIAGNOSTIC CHART — 1993–94 ACHIEVA, GRAND AM AND SKYLARK

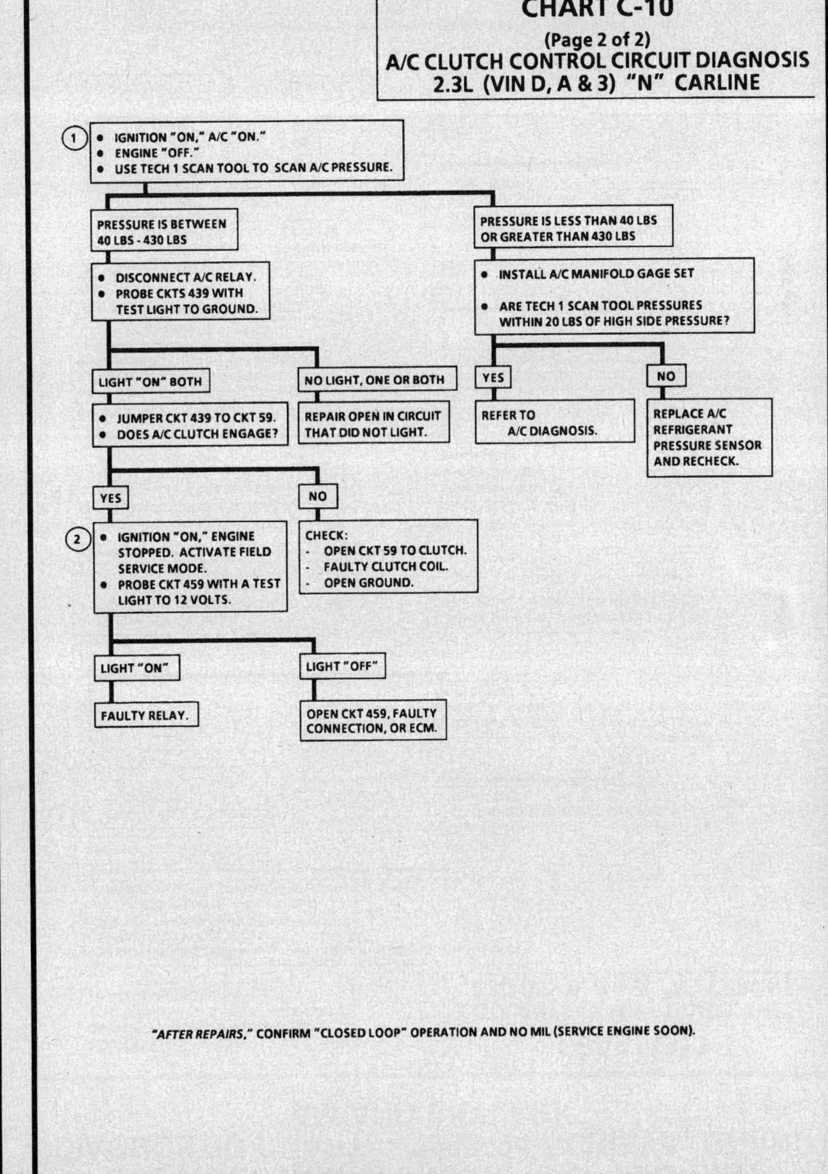

"AFTER REPAIRS," CONFIRM "CLOSED LOOP" OPERATION AND NO MIL (SERVICE ENGINE SOON).

2.3L (VIN D, A & 3) ENGINE — COMPONENT DIAGNOSTIC CHART — 1993–94 ACHIEVA, GRAND AM AND SKYLARK

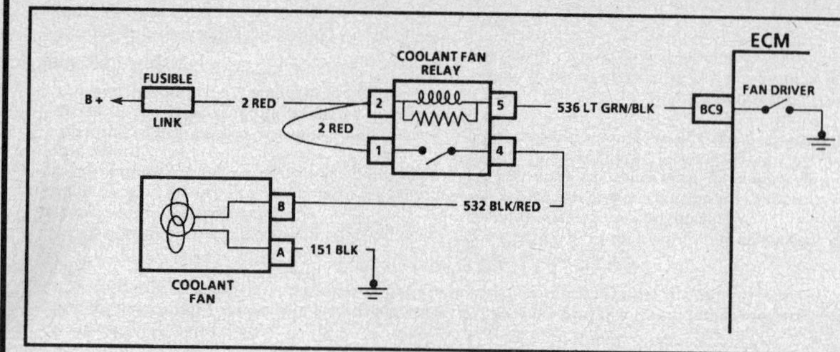

CHART C-12
COOLANT FAN FUNCTIONAL CHECK
2.3L (VIN D, A & 3) "N" CARLINE

Circuit Description:

The electric coolant fan is controlled by the Engine Control Module (ECM) through the fan relay based on inputs from the engine coolant and intake air temperature sensors, the A/C control switch, A/C refrigerant pressure sensor and the vehicle speed sensor. The ECM controls the coolant fan by grounding CKT 536 which turns "ON" the fan relay.

The fan relay will be commanded "ON" when:
- Engine coolant temperature 103°C - 106°C or more.
- A/C clutch requested.
- Vehicle speed is less than 35 mph.

The fan relay will be commanded "ON" regardless of vehicle speed when:
- DTC 14 or 15 are set.
- Engine coolant temperature 115°C - 118°C or more.
- A/C refrigerant pressure is high.

The coolant fan may be commanded "ON" when the engine is not running under fan "Run-ON" conditions

Test Description: Number(s) below refer to circled number(s) on the diagnostic chart.
1. With the field service mode activated, the coolant fan control driver should close, which should energize the fan control relay.
2. Test to see if fault is in wiring to the fan or the fan/fan connection.

Diagnostic Aids:

If the owner complained of an overheating problem, it must be determined if the complaint was due to an actual boil over, or the "temp light," or temperature gage indicated overheating.

If the gage, or light, indicates overheating, but no boil over is detected, the gage or light circuit should be checked. The gage accuracy can also be checked by comparing the coolant sensor reading using a scan tool with the gage reading.

If the engine is actually overheating, and the gage indicates overheating, but the coolant fan is not coming "ON," the Engine Coolant Temperature (ECT) sensor has probably shifted out of calibration and should be replaced. See DTC 15 chart for a temperature to resistance chart.

If the engine is overheating, and the coolant fan is "ON," the cooling system should be checked.

2.3L (VIN D, A & 3) ENGINE — COMPONENT DIAGNOSTIC CHART — 1993–94 ACHIEVA, GRAND AM AND SKYLARK

CHART C-12
COOLANT FAN FUNCTIONAL CHECK
2.3L (VIN D, A & 3) "N" CARLINE

NOTICE: CHECK FOR DTC(S) STARTING WITH A COMPLETE ON-BOARD DIAGNOSTIC (OBD) SYSTEM CHECK AND REPAIR AS NECESSARY BEFORE USING THIS CHART.

- BE SURE COOLING SYSTEM IS SEALED, COOLANT LEVEL, AND BELTS ARE OK.
- INSTALL TECH 1 SCAN TOOL.
- IGNITION "ON," ENGINE STOPPED.
- A/C SELECTOR SWITCH "OFF" (IF EQUIPPED).
- ENGINE COOLANT TEMPERATURE BELOW 98°C (209°F).

FAN "OFF" (NORMAL)

(1) • ACTIVATE FIELD SERVICE MODE FAN SHOULD BE "ON." IS IT?

FAN "ON"
- DISCONNECT BLUE ECM CONNECTOR. IS FAN OFF?
  - NO → CHECK FOR SHORT TO GROUND IN CKT 536. IF OK, RELAY IS FAULTY.
  - YES → FAULTY ECM.

NO →
- DISCONNECT FAN RELAY.
- PROBE RELAY HARNESS TERMINALS "1" AND "2" WITH TEST LIGHT TO GROUND. TEST LIGHT SHOULD BE "ON" FOR BOTH. IS IT?

YES → NO PROBLEM FOUND. COOLANT FAN CIRCUIT IS OK. SEE "DIAGNOSTIC AIDS"

YES →
(2) • JUMPER RELAY HARNESS TERMINALS "1" AND "4" TOGETHER. FAN SHOULD BE "ON." IS IT?

NO → REPAIR OPEN FROM B + IN CIRCUIT THAT DID NOT LIGHT.

NO →
- TEST LIGHT CONNECTED TO GROUND.
- BACK PROBE COOLANT FAN HARNESS CONNECTION CAVITY "B". LIGHT SHOULD BE "ON." IS IT?

YES →
- WITH TEST LIGHT TO B +, PROBE HARNESS CAVITY "5". TEST LIGHT SHOULD BE "ON." IS IT?

YES →
- BACK PROBE FAN HARNESS CAVITY "A" WITH TEST LIGHT CONNECTED TO CAVITY "B". LIGHT SHOULD BE "ON." IS IT?

NO → REPAIR OPEN IN CKT 532.

NO →
- WITH TEST LIGHT TO B +, BACKPROBE ECM TERMINAL "BC9". TEST LIGHT SHOULD BE "ON." IS IT?

YES → CHECK FOR POOR CONNECTIONS AT RELAY. IF OK, REPLACE RELAY.

YES → POOR FAN CONNECTIONS OR FAULTY FAN.

NO → REPAIR OPEN IN CKT 150.

YES → REPAIR OPEN CKT 536.

NO → • CHECK FOR POOR CONNECTION AT ECM TERMINAL "BC9". IF OK, REPLACE ECM.

"AFTER REPAIRS," CONFIRM "CLOSED LOOP" OPERATION AND NO MIL (SERVICE ENGINE SOON).

2.3L (VIN D, A & 3) ENGINE — COMPONENT DIAGNOSTIC CHART — 1993–94 ACHIEVA, GRAND AM AND SKYLARK

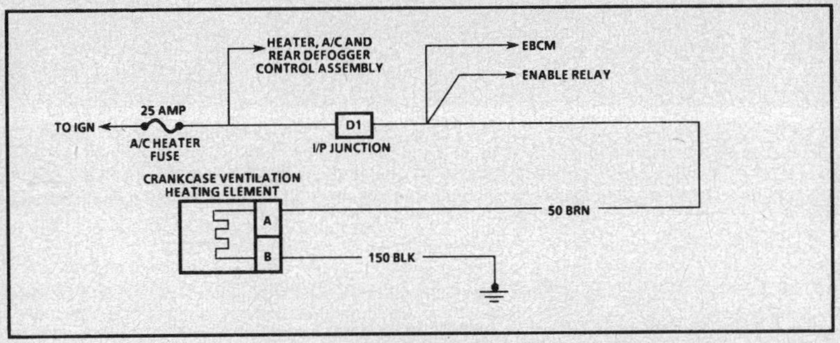

CHART C-13

CRANKCASE VENTILATION HEATER ASSEMBLY FUNCTIONAL CHECK
2.3L (VIN D, A & 3) "N" CARLINE

Circuit Description:

The crankcase ventilation heater assembly is used to prevent icing in the crankcase ventilation system. The heating element inside the vent hose is used to prevent icing from the oil/air separator to the intake manifold.

The heating element consists of two parallel wires that extend the length of the vent hose. One wire supplies the current when the ignition is turned "ON" while the second wire provides a constant path to ground.

Although the wires are not physically attached, the material between the two wires is conductive. Current is passed through the material between the wires. As the material is heated, electrical resistance increases and current flow decreases. When the material is cooled, electrical resistance decreases, current flow and heat output increases.

By this means, the temperature of the heating element is controlled to maintain approximately 115°F.

Test Description: Number(s) below refer to circled number(s) on the diagnostic chart.

1. This step verifies voltage is present to the crankcase ventilation heater assembly system.
2. This step insures a good path to ground in CKT 150.
3. This step identifies a visual failure.
4. This step identifies an open, or out of range crankcase ventilation heater element.
5. This confirms that the crankcase ventilation heating element is operational and is functioning properly.

Diagnostic Aids:

Check to insure that both battery and associated vehicle electrical connections are clean & tight.

2.3L (VIN D, A & 3) ENGINE — COMPONENT DIAGNOSTIC CHART — 1993–94 ACHIEVA, GRAND AM AND SKYLARK

CHART C-13

CRANKCASE VENTILATION HEATER ASSEMBLY FUNCTIONAL CHECK
2.3L (VIN D, A & 3) "N" CARLINE

NOTICE: CONFIRM BOTH BATTERY AND VEHICLE ELECTRICAL CONNECTIONS ARE CLEAN & TIGHT BEFORE PROCEEDING.

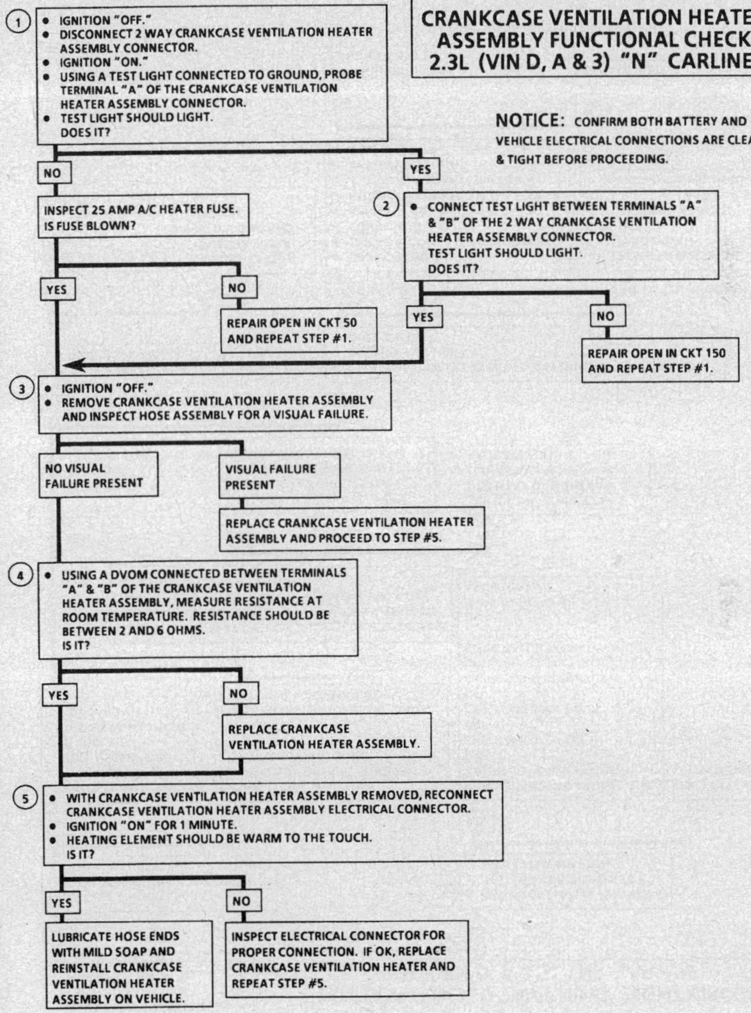

2.3L (VIN D, A & 3) ENGINE — COMPONENT DIAGNOSTIC CHART — 1993–94 ACHIEVA, GRAND AM AND SKYLARK

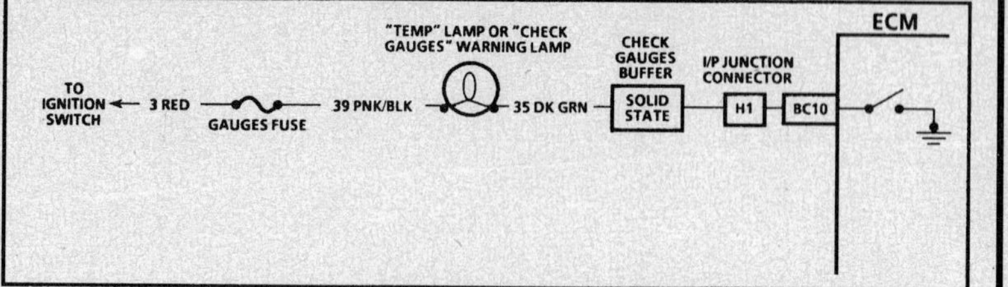

CHART C-16
ECM CONTROLLED "WARNING" LAMP CIRCUIT
2.3L (VIN D, A & 3) "N" CARLINE

Circuit Description:

The "Temp" lamp or optional "Check Gauges" lamp, is connected to battery voltage through the ignition switch. The ECM energizes the bulb by supplying a path to ground through "Quad-Driver" module #2 (QDM 2).

Test Description: Number(s) below refer to circled number(s) on the diagnostic chart.
1. With the ignition "ON" and engine "OFF," the ECM should be grounding CKT 35.
2. While the engine is running, the "Temp" lamp or optional "Check Gauges" lamp should be turned "ON" by the ECM only when the engine coolant temperature is above 123°C (253°F). The "Temp" lamp or "Check Gauges" lamp should be turned "OFF" when the engine is running and the engine coolant temperature goes lower than 120°C (249°F).

Diagnostic Aids:

The engine coolant temperature sensor, in rare cases, may fail to indicate the correct engine coolant temperature without setting a DTC. This could result in turning "ON" the "Temp" lamp or optional "Check Gauges" lamp without having an overheating condition. It could also result in overheating without the "Temp/Check Gauges" lamp being turned "ON."

An intermittent problem may be caused by a poor or corroded connection, rubbed through wire connection, a wire that is broken inside the insulation, or a corroded wire.

Any circuitry, that is suspected as causing the intermittent complaint, should be thoroughly checked for backed out terminals, improper mating, broken locks, improperly formed or damaged terminals, poor terminal to wiring connections, corroded terminals and/or wiring, or physical damage to the wiring harness.

2.3L (VIN D, A & 3) ENGINE — COMPONENT DIAGNOSTIC CHART — 1993–94 ACHIEVA, GRAND AM AND SKYLARK

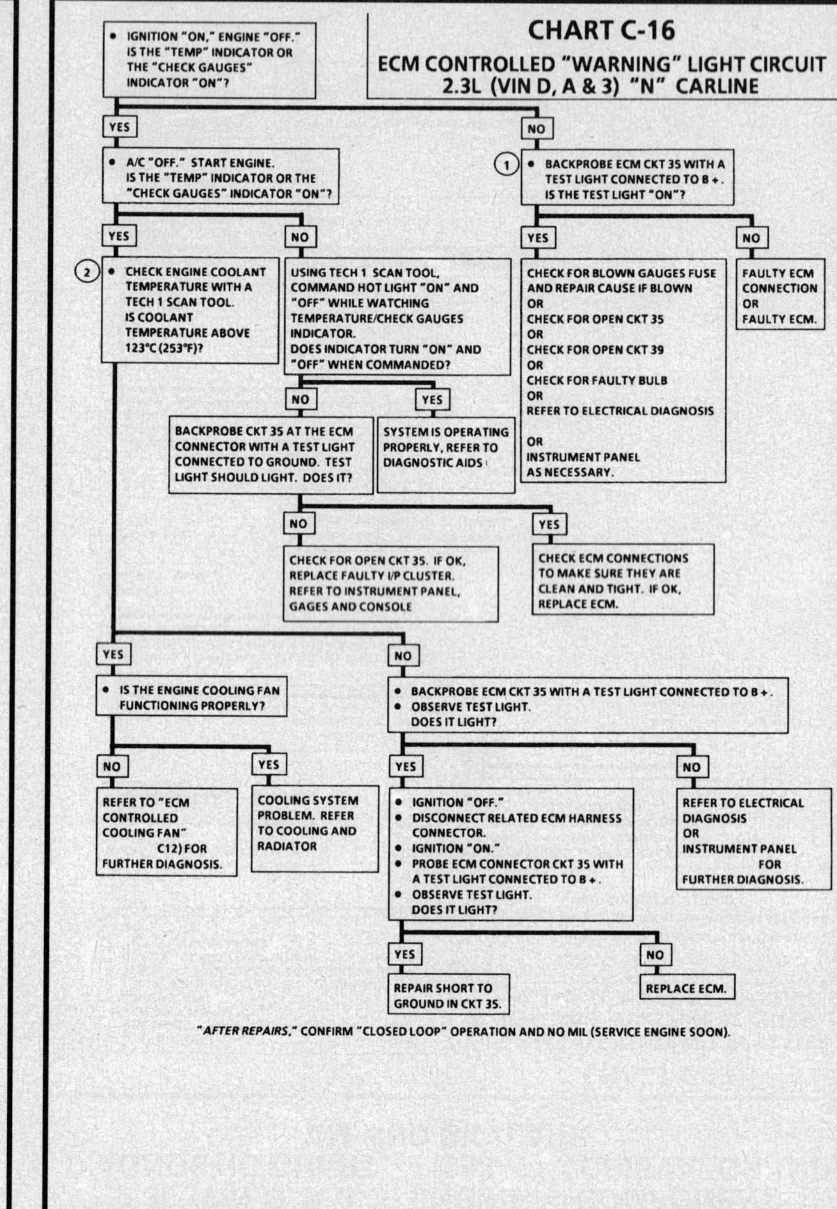

3.1L (VIN T) ENGINE — ENGINE COMPONENT LOCATION CHART — CAMARO AND FIREBIRD

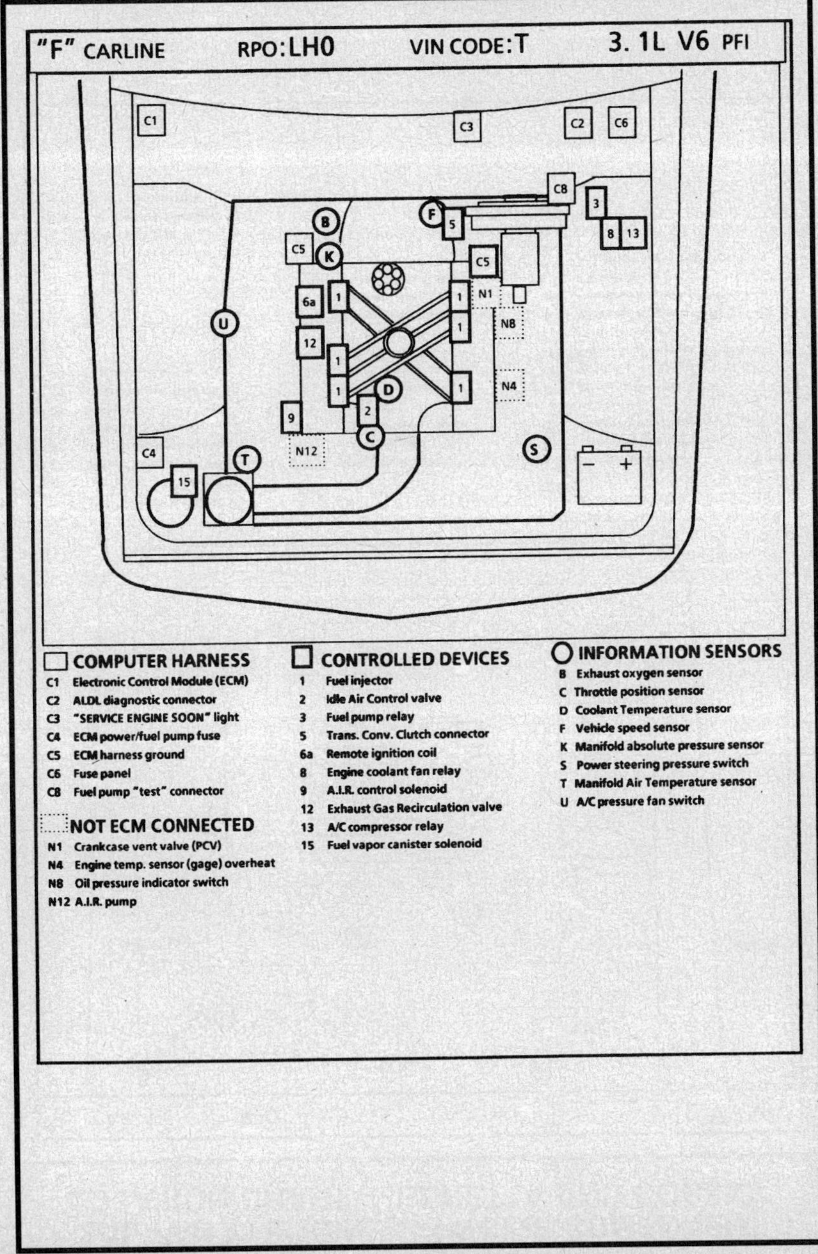

COMPUTER HARNESS

C1 Electronic Control Module (ECM)
C2 ALDL diagnostic connector
C3 "SERVICE ENGINE SOON" light
C4 ECM power/fuel pump fuse
C5 ECM harness ground
C6 Fuse panel
C8 Fuel pump "test" connector

NOT ECM CONNECTED

N1 Crankcase vent valve (PCV)
N4 Engine temp. sensor (gage) overheat
N8 Oil pressure indicator switch
N12 A.I.R. pump

CONTROLLED DEVICES

1 Fuel injector
2 Idle Air Control valve
3 Fuel pump relay
5 Trans. Conv. Clutch connector
6a Remote ignition coil
8 Engine coolant fan relay
9 A.I.R. control solenoid
12 Exhaust Gas Recirculation valve
13 A/C compressor relay
15 Fuel vapor canister solenoid

INFORMATION SENSORS

B Exhaust oxygen sensor
C Throttle position sensor
D Coolant Temperature sensor
F Vehicle speed sensor
K Manifold absolute pressure sensor
S Power steering pressure switch
T Manifold Air Temperature sensor
U A/C pressure fan switch

3.1L (VIN T) ENGINE — ENGINE COMPONENT LOCATION CHART — CAVALIER AND SUNBIRD

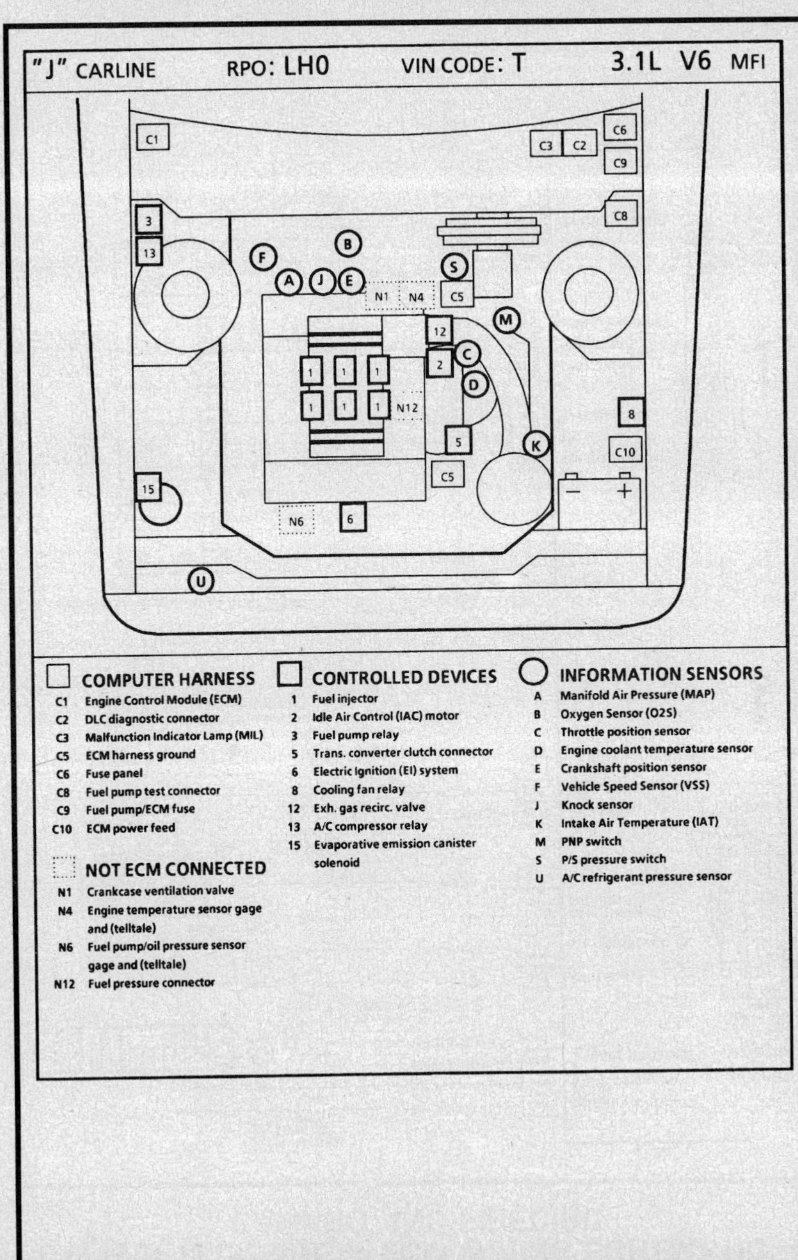

COMPUTER HARNESS

C1 Engine Control Module (ECM)
C2 DLC diagnostic connector
C3 Malfunction Indicator Lamp (MIL)
C5 ECM harness ground
C6 Fuse panel
C8 Fuel pump test connector
C9 Fuel pump/ECM fuse
C10 ECM power feed

NOT ECM CONNECTED

N1 Crankcase ventilation valve
N4 Engine temperature sensor gage and (telltale)
N6 Fuel pump/oil pressure sensor gage and (telltale)
N12 Fuel pressure connector

CONTROLLED DEVICES

1 Fuel injector
2 Idle Air Control (IAC) motor
3 Fuel pump relay
5 Trans. converter clutch connector
6 Electric Ignition (EI) system
8 Cooling fan relay
12 Exh. gas recirc. valve
13 A/C compressor relay
15 Evaporative emission canister solenoid

INFORMATION SENSORS

A Manifold Air Pressure (MAP)
B Oxygen Sensor (O2S)
C Throttle position sensor
D Engine coolant temperature sensor
E Crankshaft position sensor
F Vehicle Speed Sensor (VSS)
J Knock sensor
K Intake Air Temperature (IAT)
M PNP switch
S P/S pressure switch
U A/C refrigerant pressure sensor

3.1L (VIN T) ENGINE — ENGINE COMPONENT LOCATION CHART — BERETTA AND CORSICA

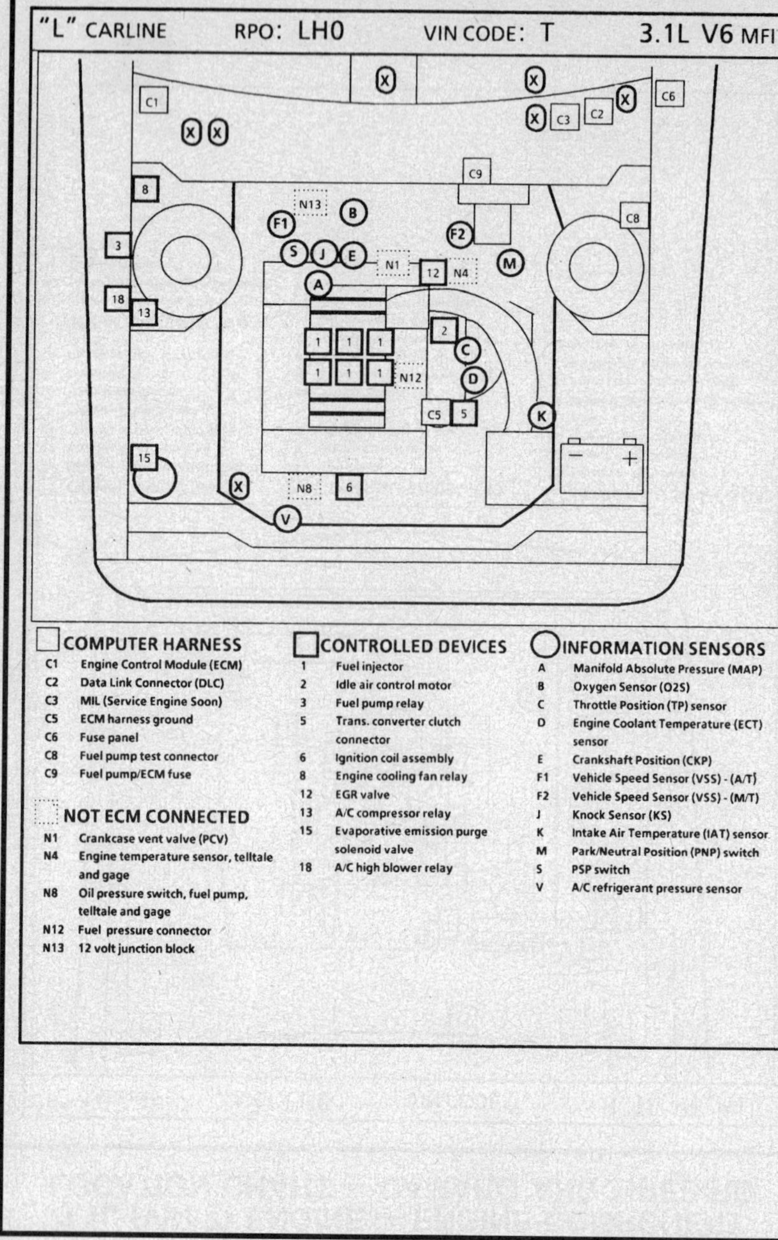

3.1L (VIN T) ENGINE — ECM WIRING SCHEMATIC — CAMARO AND FIREBIRD

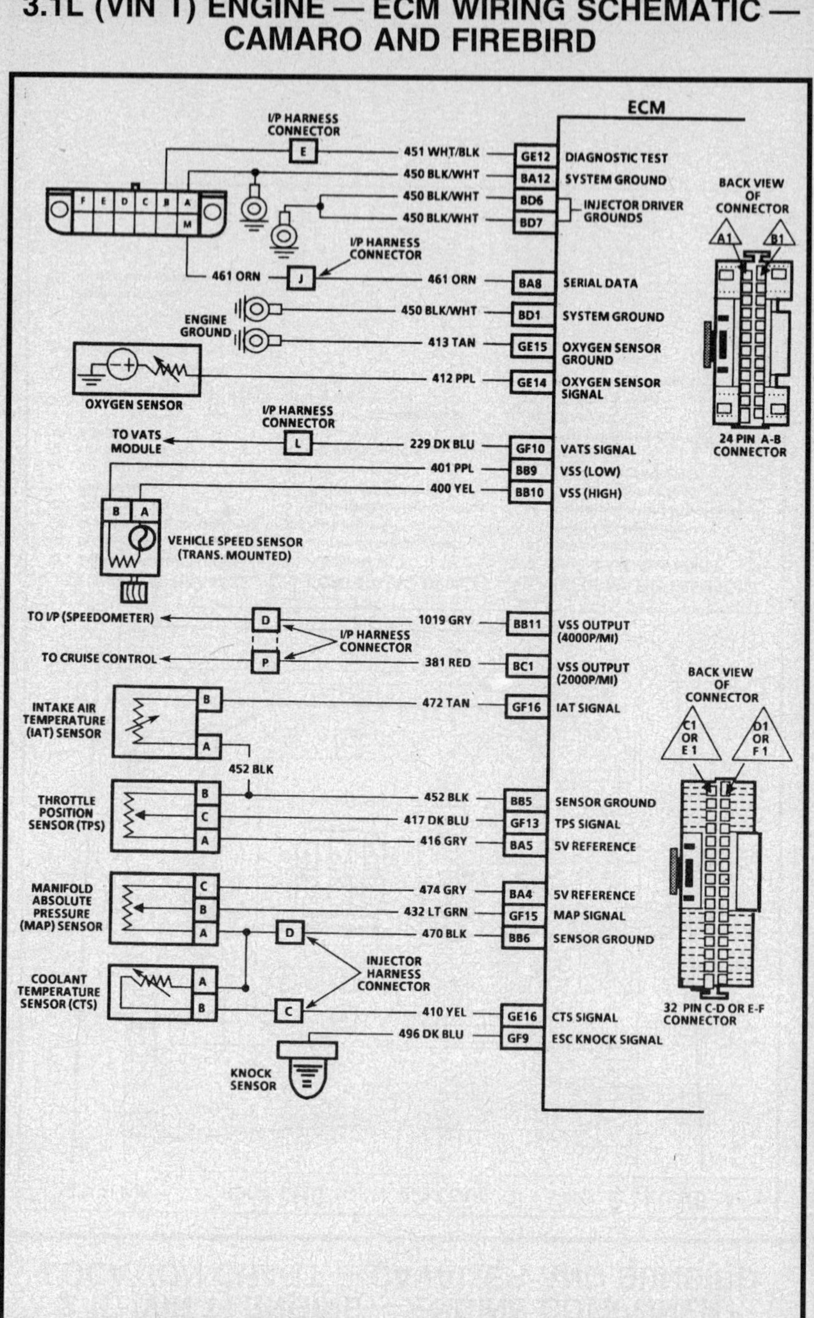

3.1L (VIN T) ENGINE — ECM WIRING SCHEMATIC — CAMARO AND FIREBIRD

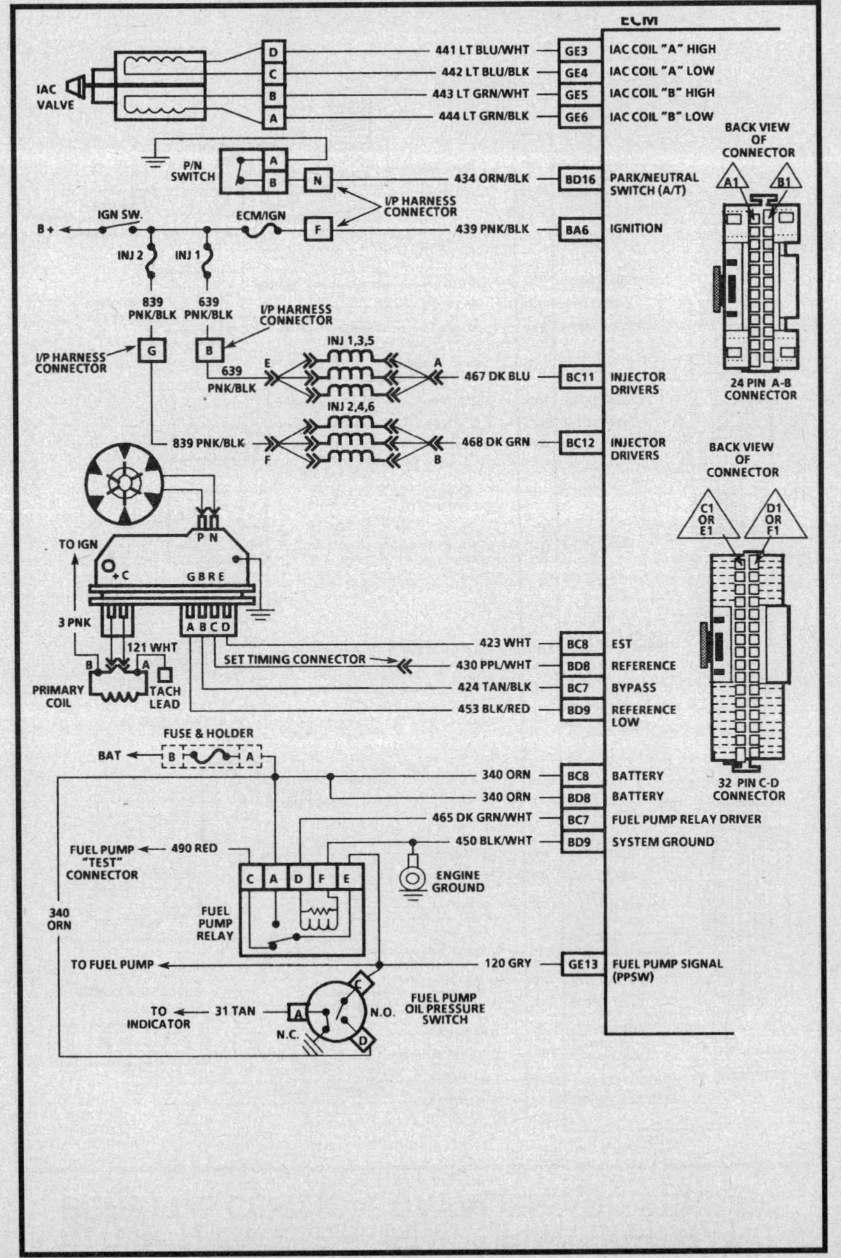

3.1L (VIN T) ENGINE — ECM WIRING SCHEMATIC — CAMARO AND FIREBIRD

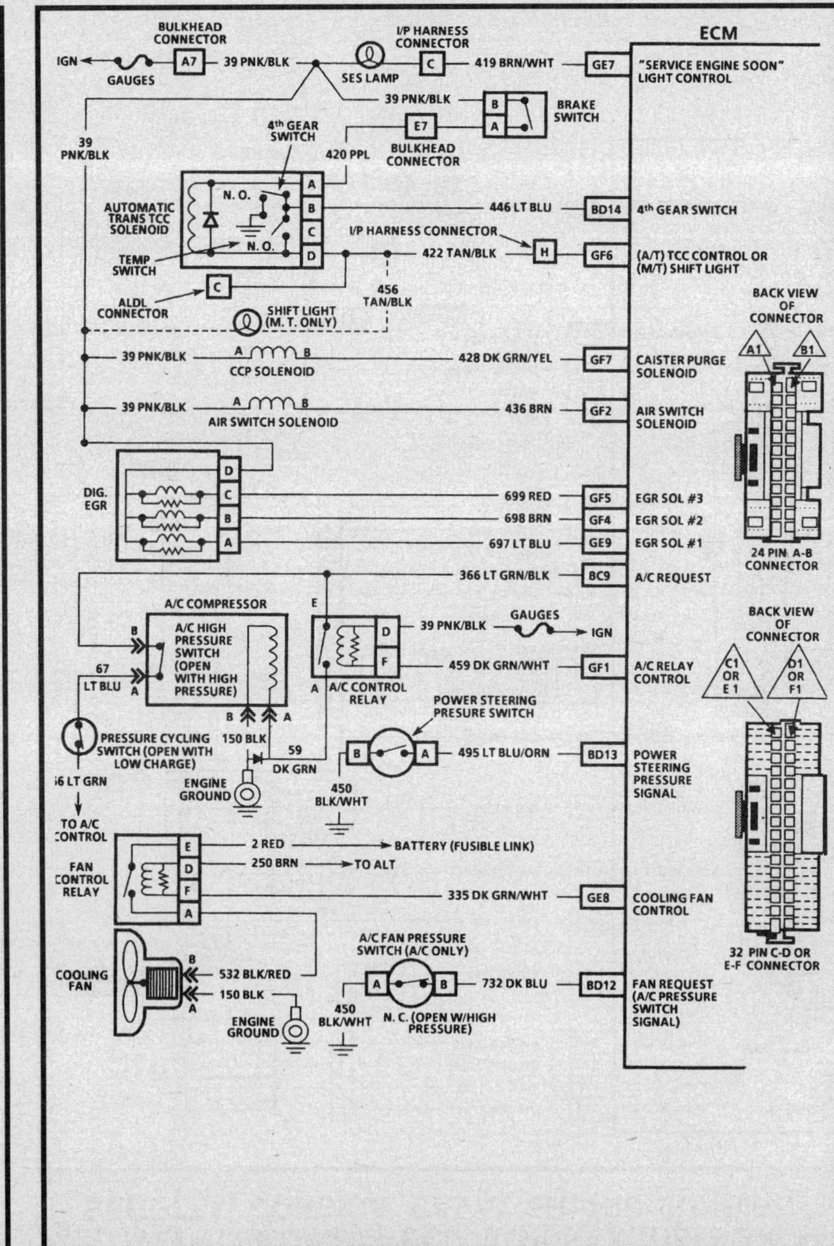

MULTIPORT FUEL INJECTION (MFI) SYSTEMS
EXCEPT LIGHT TRUCKS, VANS, GEO AND SATURN

3.1L (VIN T) ENGINE — ECM WIRING SCHEMATIC — BERETTA, CORSICA, CAVALIER AND SUNBIRD

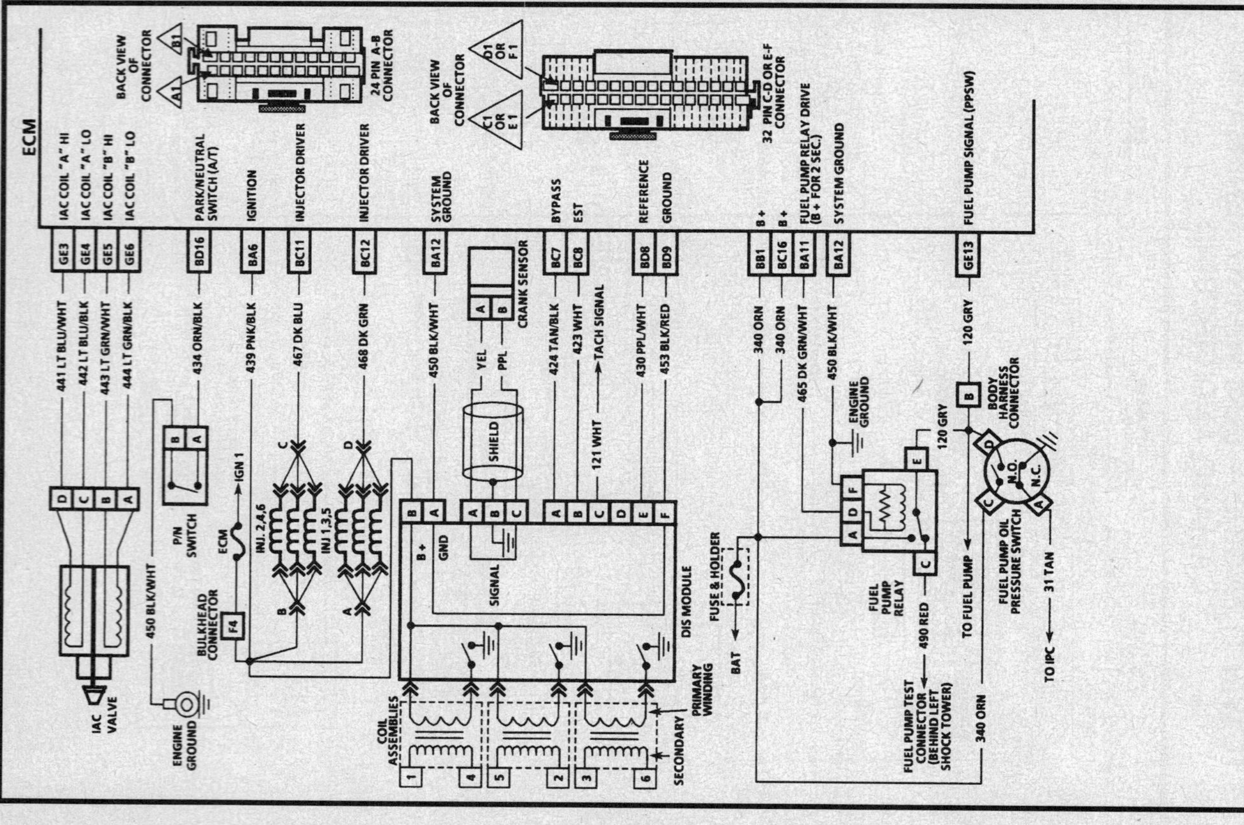

3.1L (VIN T) ENGINE — ECM WIRING SCHEMATIC — BERETTA, CORSICA, CAVALIER AND SUNBIRD

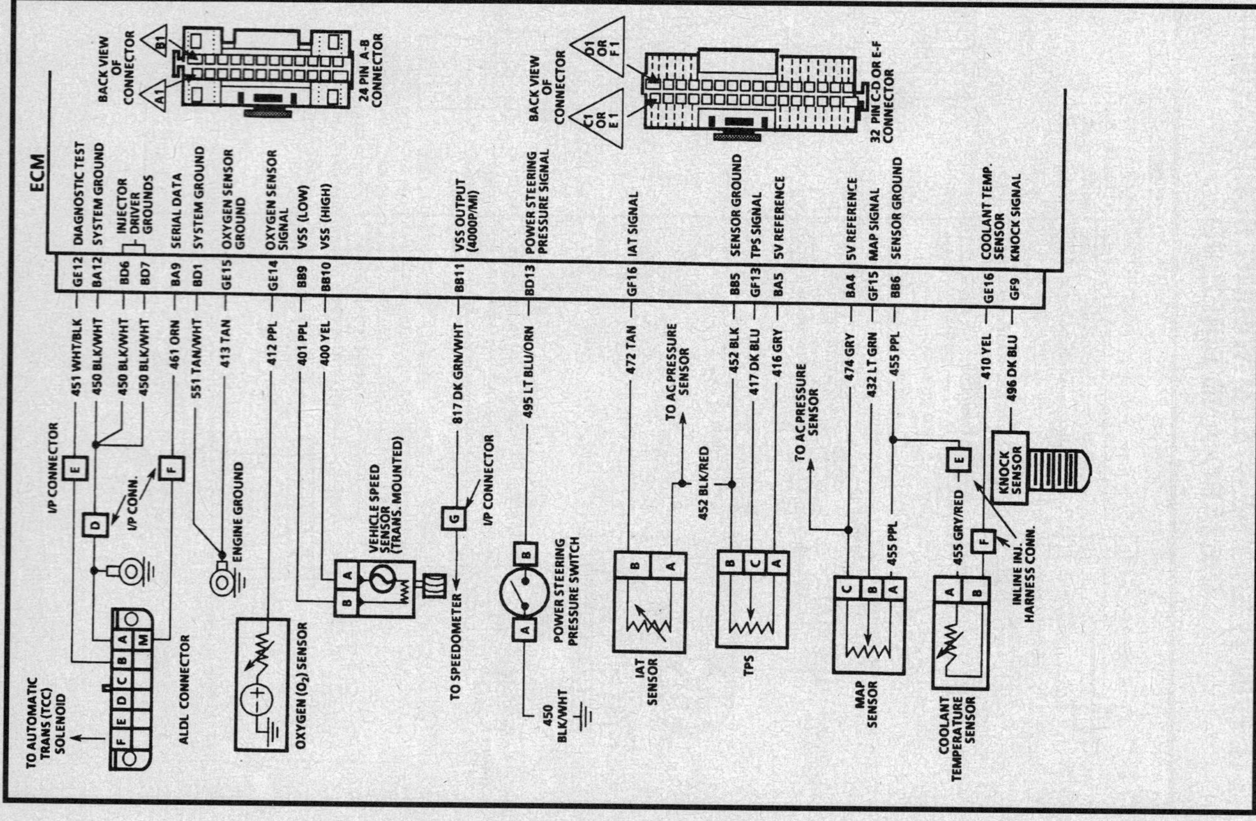

3.1L (VIN T) ENGINE — ECM WIRING SCHEMATIC — BERETTA, CORSICA, CAVALIER AND SUNBIRD

3.1L (VIN T) ENGINE — ECM CONNECTOR END VIEW — CAMARO AND FIREBIRD

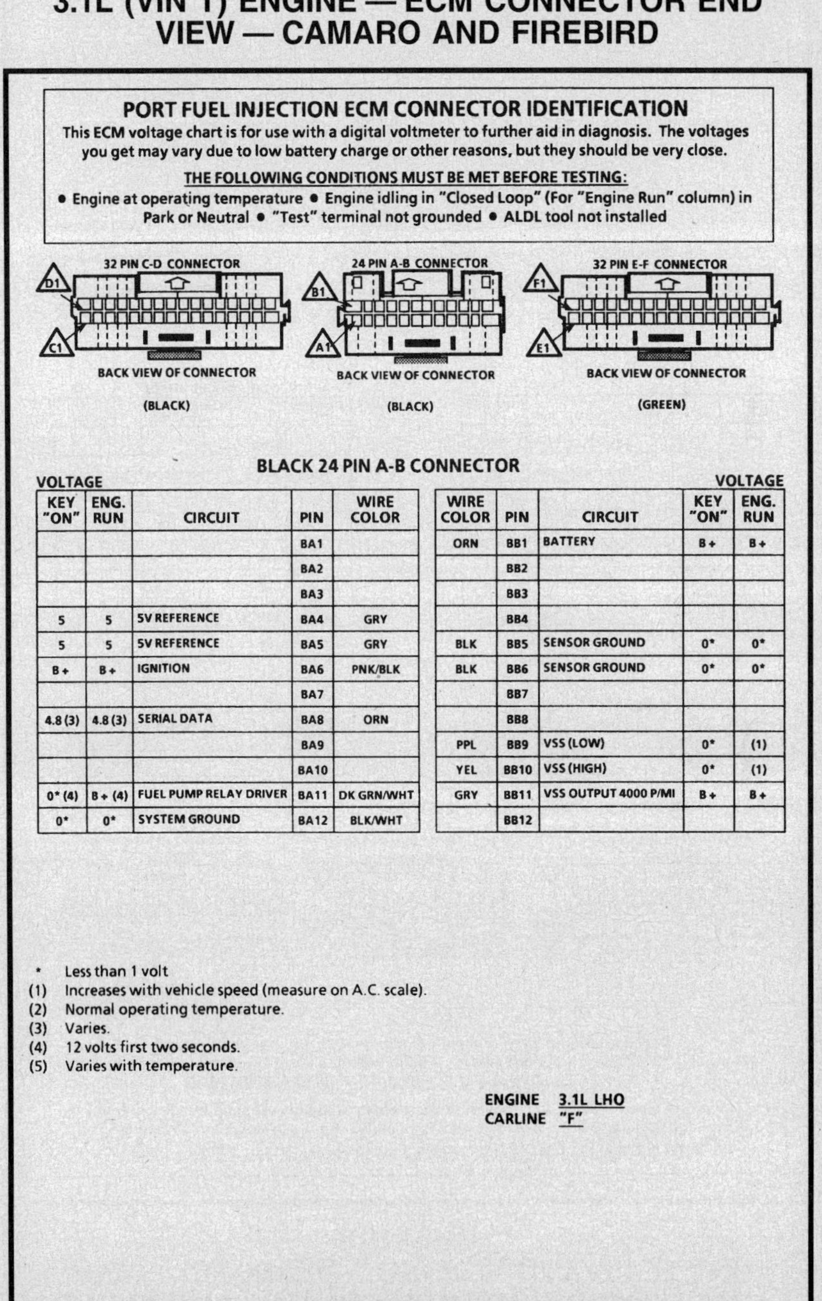

BLACK 24 PIN A-B CONNECTOR

| VOLTAGE | | | | |
KEY "ON"	ENG. RUN	CIRCUIT	PIN	WIRE COLOR
			BA1	
			BA2	
			BA3	
5	5	5V REFERENCE	BA4	GRY
5	5	5V REFERENCE	BA5	GRY
B+	B+	IGNITION	BA6	PNK/BLK
			BA7	
4.8 (3)	4.8 (3)	SERIAL DATA	BA8	ORN
			BA9	
			BA10	
0* (4)	B+ (4)	FUEL PUMP RELAY DRIVER	BA11	DK GRN/WHT
0*	0*	SYSTEM GROUND	BA12	BLK/WHT

WIRE COLOR	PIN	CIRCUIT	KEY "ON"	ENG. RUN
ORN	BB1	BATTERY	B+	B+
	BB2			
	BB3			
	BB4			
BLK	BB5	SENSOR GROUND	0*	0*
BLK	BB6	SENSOR GROUND	0*	0*
	BB7			
	BB8			
PPL	BB9	VSS (LOW)	0*	(1)
YEL	BB10	VSS (HIGH)	0*	(1)
GRY	BB11	VSS OUTPUT 4000 P/MI	B+	B+
	BB12			

*   Less than 1 volt
(1) Increases with vehicle speed (measure on A.C. scale).
(2) Normal operating temperature.
(3) Varies.
(4) 12 volts first two seconds.
(5) Varies with temperature.

ENGINE   3.1L LHO
CARLINE "F"

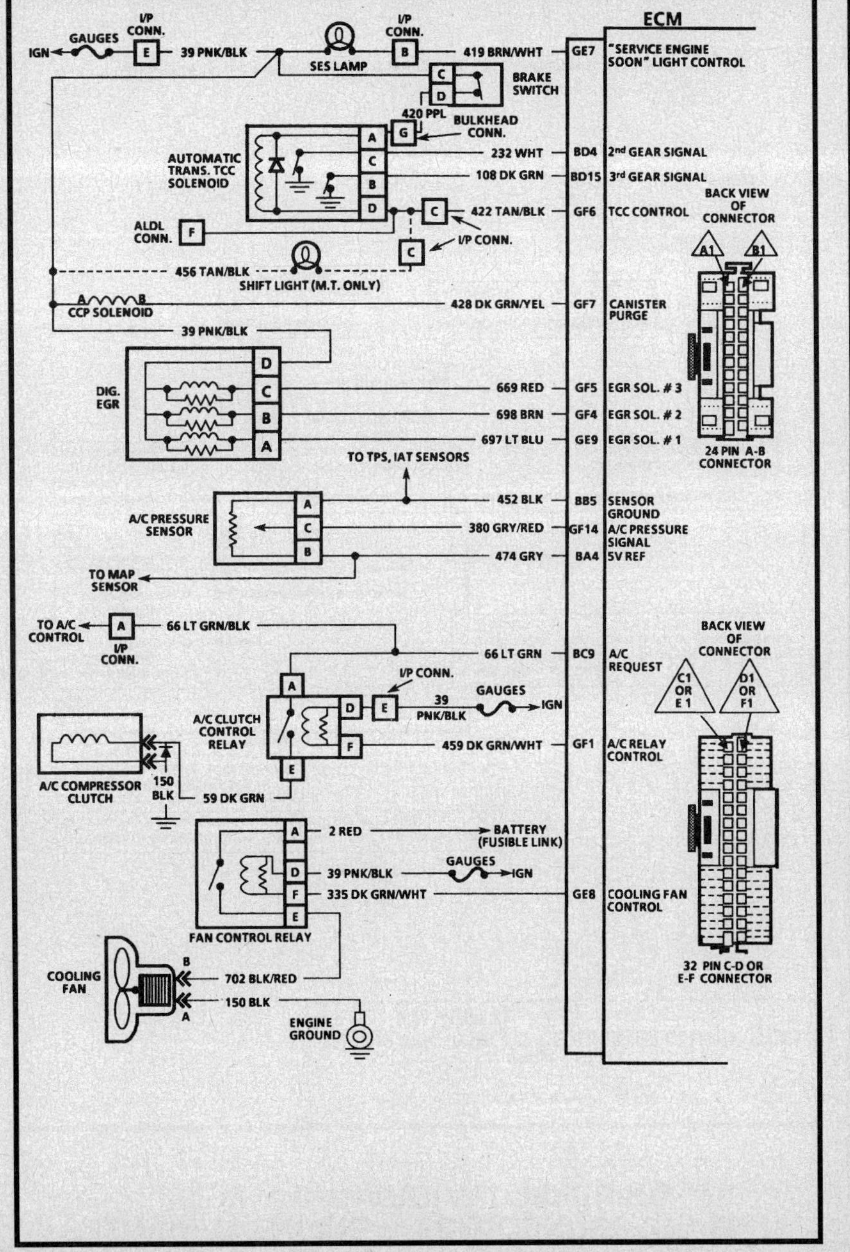

3.1L (VIN T) ENGINE — ECM CONNECTOR END VIEW — CAMARO AND FIREBIRD

CHART C-8
(Page 2 of 2)
AUTOMATIC TORQUE CONVERTER CLUTCH (TCC)
3.1L (VIN T)

(1) CHECKS MADE ON THIS PAGE WILL NOT PREVENT THE TCC FROM WORKING, BUT WILL AFFECT ENGAGEMENT OR DISENGAGEMENT POINTS.

- IGNITION "ON," ENGINE "OFF." DOES TECH 1 "SCAN" TOOL INDICATE TRANSMISSION IS IN 4th GEAR?

NO → (2)
- RAISE DRIVE WHEELS.
- START ENGINE.
- SHIFT VEHICLE INTO OVERDRIVE.
- INCREASE SPEED SLOWLY UNTIL TRANSMISSION SHIFTS INTO 4th GEAR. DOES TECH 1 "SCAN" TOOL INDICATE TRANSMISSION IS IN 4th GEAR?

  YES → 4th GEAR SWITCH OK. REFER TO "DIAGNOSTIC AIDS" ON FACING PAGE.

  NO → (4)
  - DISCONNECT TCC ELECTRICAL CONNECTOR.
  - JUMPER HARNESS TERMINAL "B" (CKT 446) TO GROUND. DOES TECH 1 "SCAN" TOOL INDICATE TRANSMISSION IS IN 4th GEAR?

    YES → FAULTY CONNECTION OR FAULTY 4th GEAR SWITCH IN TRANSMISSION.

    NO → OPEN CKT 446, FAULTY CONNECTION OR FAULTY ECM.

YES → (3)
- IGNITION "ON," ENGINE "OFF."
- DISCONNECT TRANSMISSION ELECTRICAL CONNECTOR. DOES TECH 1 "SCAN" TOOL INDICATE TRANSMISSION IS IN 4th GEAR?

  YES → CKT 446 SHORTED TO GROUND OR FAULTY ECM.

  NO → FAULTY 4th GEAR SWITCH OR CKT 446 GROUNDED INTERNALLY IN TRANSMISSION.

"*AFTER REPAIRS*," CONFIRM "CLOSED LOOP" OPERATION AND NO "SERVICE ENGINE SOON" LIGHT.

3.1L (VIN T) ENGINE — ECM CONNECTOR END VIEW — BERETTA, CORSICA, CAVALIER AND SUNBIRD

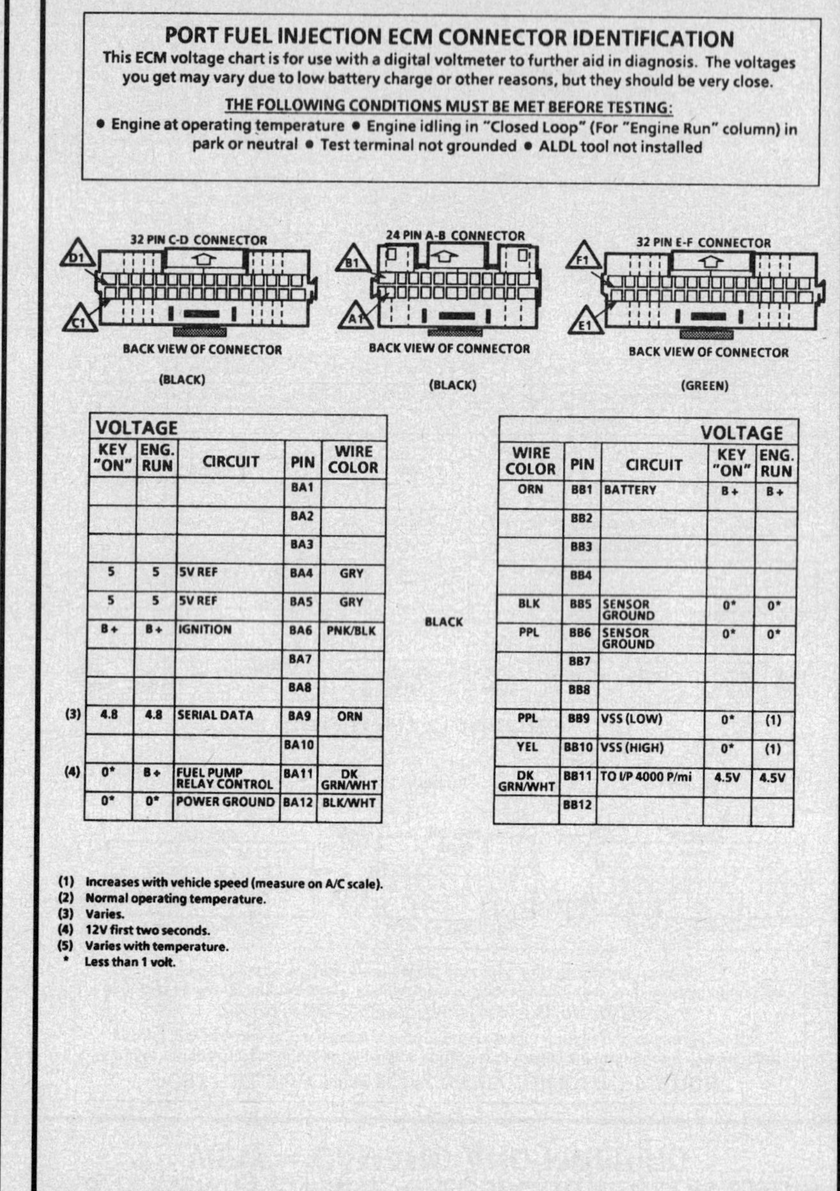

PORT FUEL INJECTION ECM CONNECTOR IDENTIFICATION

This ECM voltage chart is for use with a digital voltmeter to further aid in diagnosis. The voltages you get may vary due to low battery charge or other reasons, but they should be very close.

THE FOLLOWING CONDITIONS MUST BE MET BEFORE TESTING:

- Engine at operating temperature • Engine idling in "Closed Loop" (For "Engine Run" column) in park or neutral • Test terminal not grounded • ALDL tool not installed

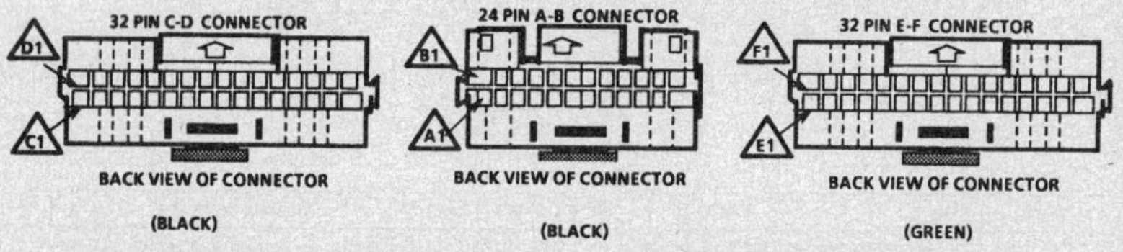

	VOLTAGE KEY "ON"	VOLTAGE ENG. RUN	CIRCUIT	PIN	WIRE COLOR
				BA1	
				BA2	
				BA3	
	5	5	5V REF	BA4	GRY
	5	5	5V REF	BA5	GRY
	B+	B+	IGNITION	BA6	PNK/BLK
				BA7	
				BA8	
(3)	4.8	4.8	SERIAL DATA	BA9	ORN
				BA10	
(4)	0*	B+	FUEL PUMP RELAY CONTROL	BA11	DK GRN/WHT
	0*	0*	POWER GROUND	BA12	BLK/WHT

BLACK

WIRE COLOR	PIN	CIRCUIT	VOLTAGE KEY "ON"	VOLTAGE ENG. RUN
ORN	BB1	BATTERY	B+	B+
	BB2			
	BB3			
	BB4			
BLK	BB5	SENSOR GROUND	0*	0*
PPL	BB6	SENSOR GROUND	0*	0*
	BB7			
	BB8			
PPL	BB9	VSS (LOW)	0*	(1)
YEL	BB10	VSS (HIGH)	0*	(1)
DK GRN/WHT	BB11	TO I/P 4000 P/mi	4.5V	4.5V
	BB12			

(1) Increases with vehicle speed (measure on A/C scale).
(2) Normal operating temperature.
(3) Varies.
(4) 12V first two seconds.
(5) Varies with temperature.
* Less than 1 volt.

3.1L (VIN T) ENGINE — ECM CONNECTOR END VIEW — BERETTA, CORSICA, CAVALIER AND SUNBIRD

BLACK

VOLTAGE				
KEY "ON"	ENG. RUN	CIRCUIT	PIN	WIRE COLOR
			BC1	
			BC2	
			BC3	
			BC4	
			BC5	
			BC6	
0*	4.7	BYPASS	BC7	TAN/BLK
0*	1.3	EST	BC8	WHT
B+ / 0*	B+ / 0*	WITH A/C "ON" A/C REQUEST	BC9	LT GRN
			BC10	
B+	B+	INJECTOR 2, 4, 6	BC11	DK BLU
B+	B+	INJECTOR 1, 3, 5	BC12	DK GRN
			BC13	
			BC14	
			BC15	
B+	B+	BATTERY	BC16	ORN

WIRE COLOR	PIN	CIRCUIT	VOLTAGE	
			KEY "ON"	ENG. RUN
TAN/WHT	BD1	POWER GROUND	0*	0*
	BD2			
	BD3			
WHT	BD4	2nd GEAR SW	0*	0*
	BD5			
BLK/WHT	BD6	INJ DRIVE LOW	0*	0*
BLK/WHT	BD7	INJ DRIVE LOW	0*	0*
PPL/WHT	BD8	REFERENCE	0*	2.3
BLK/RED	BD9	REFERENCE LOW	0*	0*
	BD10			
	BD11			
	BD12			
LT BLU/ORN	BD13	PSPS	B+	B+
	BD14			
DK GRN	BD15	3nd GEAR SW	0*	0*
ORN/BLK	BD16	P/N SWITCH	0*	0*

GREEN

VOLTAGE				
KEY "ON"	ENG. RUN	CIRCUIT	PIN	WIRE COLOR
			GE1	
			GE2	
NOT	USE-ABLE	IAC "A" HI	GE3	LT BLU/WHT
NOT	USE-ABLE	IAC "A" LO	GE4	LT BLU/BLK
NOT	USE-ABLE	IAC "B" HI	GE5	LT GRN/WHT
NOT	USE-ABLE	IAC "B" LO	GE6	LT GRN/BLK
0*	B+	"SERVICE ENGINE SOON" LIGHT	GE7	BRN/WHT
B+	B+	FAN RELAY CONTROL	GE8	DK GRN/WHT
B+	B+	EGR SOL. #1	GE9	LT BLU
			GE10	
			GE11	
5	5	DIAG. TERMINAL	GE12	WHT/BLK
(4)	B+	FUEL PUMP SIGNAL	GE13	GRY
.35 / .55	(3)	O₂ SIGNAL	GE14	PPL
0*	0*	O₂ GROUND	GE15	TAN
(5)	(5)	COOLANT TEMP.	GE16	YEL

WIRE COLOR	PIN	CIRCUIT	VOLTAGE	
			KEY "ON"	ENG. RUN
DK GRN/WHT	GF1	A/C RELAY CONTROL	B+	B+
	GF2			
	GF3			
BRN	GF4	EGR SOL. #2	B+	B+
RED	GF5	EGR SOL. #3	B+	B+
TAN/BLK	GF6	TCC CONTROL A/T SHIFT LIGHT M/T	0* / B+	0* / B+
DK GRN/YEL	GF7	PURGE CONTROL	0*	.25 (3)
	GF8			
DK BLU	GF9	ESC SIGNAL	2.5	2.5
	GF10			
	GF11			
	GF12			
DK BLU	GF13	TPS SIGNAL	.65	.6
GRY/RED	GF14	A/C PRESSURE SIGNAL	(3)	(3)
LT GRN	GF15	MAP SIGNAL	4.57	1.7 (3)
TAN	GF16	IAT SIGNAL	3.1	3.2 (5)

(1) Increases with vehicle speed (measure on A/C scale).
(2) Normal operating temperature.
(3) Varies.
(4) 12 volts first two seconds.
(5) Varies with temperature.
* Less than 1 volt.

3.1L (VIN T) ENGINE — ON-BOARD DIAGNOSTIC SYSTEM CHART — CAMARO, FIREBIRD, BERETTA, CORSICA, CAVALIER AND SUNBIRD

DIAGNOSTIC CIRCUIT CHECK
3.1L (VIN T) "F" CARLINE (PORT)

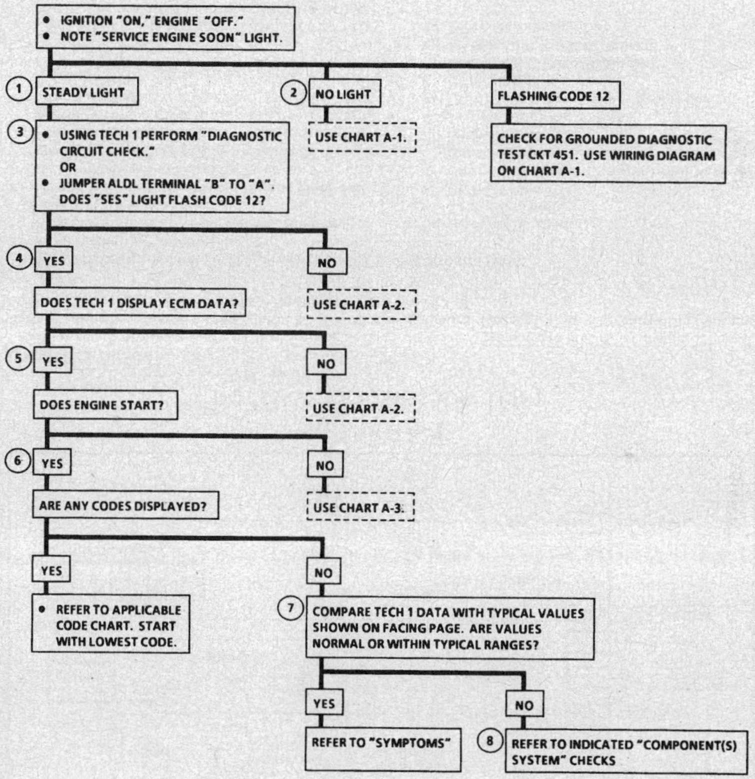

- IGNITION "ON," ENGINE "OFF."
- NOTE "SERVICE ENGINE SOON" LIGHT.

(1) STEADY LIGHT | (2) NO LIGHT | FLASHING CODE 12

(3)
- USING TECH 1 PERFORM "DIAGNOSTIC CIRCUIT CHECK." OR
- JUMPER ALDL TERMINAL "B" TO "A". DOES "SES" LIGHT FLASH CODE 12?

USE CHART A-1.

CHECK FOR GROUNDED DIAGNOSTIC TEST CKT 451. USE WIRING DIAGRAM ON CHART A-1.

(4) YES — DOES TECH 1 DISPLAY ECM DATA? | NO — USE CHART A-2.

(5) YES — DOES ENGINE START? | NO — USE CHART A-2.

(6) YES — ARE ANY CODES DISPLAYED? | NO — USE CHART A-3.

YES
- REFER TO APPLICABLE CODE CHART. START WITH LOWEST CODE.

NO

(7) COMPARE TECH 1 DATA WITH TYPICAL VALUES SHOWN ON FACING PAGE. ARE VALUES NORMAL OR WITHIN TYPICAL RANGES?

YES — REFER TO "SYMPTOMS" | NO — (8) REFER TO INDICATED "COMPONENT(S) SYSTEM" CHECKS

3.1L (VIN T) ENGINE — ON-BOARD DIAGNOSTIC SYSTEM CHART — CAMARO, FIREBIRD, BERETTA, CORSICA, CAVALIER AND SUNBIRD

3.1L (VIN T) ENGINE — SYSTEM DIAGNOSTIC CHARTS — CAMARO, FIREBIRD, BERETTA, CORSICA, CAVALIER AND SUNBIRD

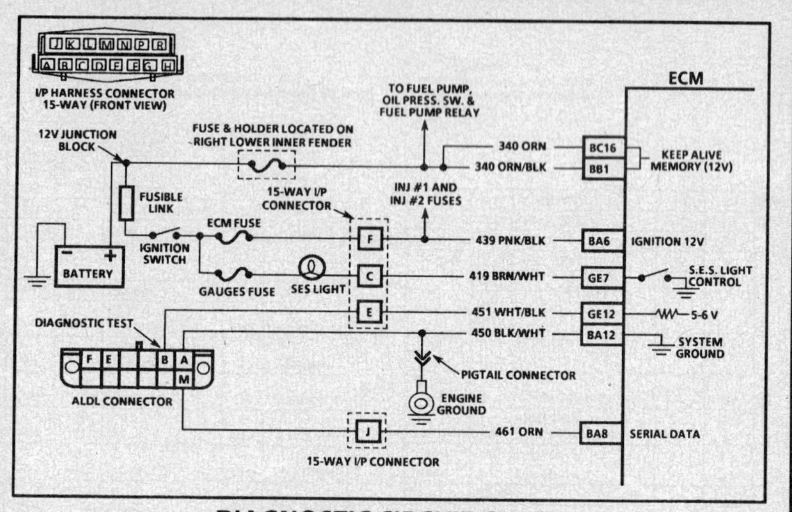

DIAGNOSTIC CIRCUIT CHECK
3.1L (VIN T)

Circuit Description:

The diagnostic circuit check is an organized approach to identifying a problem created by an Electronic Engine Control System malfunction. It must be the starting point for any driveability complaint diagnosis, because it directs the service technician to the next logical step in diagnosing the complaint. Understanding the chart and using it correctly will reduce diagnostic time and prevent the unnecessary replacement of good parts.

Test Description: Number(s) below refer to circled number(s) on the diagnostic chart.

1. This step is a check for the proper operation of "Service Engine Soon" light. The "SES" light should be "ON" steady.
2. No "SES" light at this point indicates that there is a problem with the "SES" light circuit or the ECM control of that circuit.
3. This test checks the ability of the ECM to control the "SES" light. With the diagnostic terminal grounded, the "SES" light should flash a Code 12 three times, followed by any trouble code stored in memory. A MEM-CAL error may result in the inability to flash Code 12.
4. Most of the diagnostic charts use a Tech 1 to aid diagnosis, therefore, serial data must be available. If a MEM-CAL error is present, the ECM may have been able to flash Code 12 or 51, but not transmit serial data.
5. Although the ECM is powered up, a "Cranks But Will Not Run" symptom could exist because of an ECM or system problem.
6. This step will isolate if the customer complaint is a "SES" light or a driveability problem with no "SES" light. Refer to ECM Diagnostic Codes chart in this section for a list of valid codes. An invalid code may be the result of a faulty "Scan" tool, MEM-CAL or ECM.
7. Comparison of actual control system data with the typical values is a quick check to determine if any parameter is not within limits. Keep in mind that a base engine problem (i.e. advanced cam timing) may substantially alter sensor values.
8. If the actual data is not within the typical values established, the charts in Section "C" will provide a functional check of the suspect component of system.

CHART A-1
NO "SERVICE ENGINE SOON" LIGHT
3.1L (VIN T)

Circuit Description:

There should always be a steady "Service Engine Soon" light when the ignition is "ON" and engine stopped. Battery is supplied directly to the light bulb. The Electronic Control Module (ECM) will control the light and turn it "ON" by providing a ground path through CKT 419 to the ECM.

Test Description: Number(s) below refer to circled number(s) on the diagnostic chart.

1. If the fuse in holder is blown, refer to facing page of Code 54 for complete circuit.
2. Using a test light connected to 12 volts probe each of the system ground circuits to be sure a good ground is present. See ECM terminal end view in front of this section for ECM pin locations of ground circuits. If the ECM has to be replaced, refer to "On-Vehicle Service" in "Electronic Control Module (ECM) and Sensors,"

CAUTION: Damage can occur to controller circuit board by pressing down hard on the ends of MEM-CAL, to secure the locking tabs.

Diagnostic Aids:

Engine runs OK, check:
- Faulty light bulb.
- CKT 419 open.
- Gauges fuse blown. This will result in no stop lights, oil or generator lights, seat belt reminder, etc.

Engine cranks but will not run:
- Continuous battery - fuse or fusible link open.
- ECM ignition fuse open.
- Battery CKT 340 to ECM open.
- Ignition CKT 439 to ECM open.
- Poor connection to ECM.

3.1L (VIN T) ENGINE — SYSTEM DIAGNOSTIC CHARTS — CAMARO, FIREBIRD, BERETTA, CORSICA, CAVALIER AND SUNBIRD

CHART A-1
NO "SERVICE ENGINE SOON" LIGHT
3.1L (VIN T)

DOES THE ENGINE START?

YES
- IGNITION "OFF."
- DISCONNECT ECM CONNECTORS.
- IGNITION "ON."
- PROBE CKT 419, WITH TEST LIGHT TO GROUND.
  IS THE "SES" LIGHT "ON"?

YES
FAULTY ECM CONNECTION OR ECM.

NO
CHECK:
- GAUGE FUSE.
- FAULTY BULB.
- OPEN CKT 419.
- CKT 419 SHORTED TO VOLTAGE.
- OPEN IGNITION FEED TO BULB.

NO
IS THE 20A CONTINUOUS BATTERY FUSE (IN HOLDER) AND ECM FUSE OK?

YES
- IGNITION "ON."
- PROBE CKT 340 & 439 WITH TEST LIGHT TO GROUND.
  IS THE LIGHT "ON" ON BOTH CIRCUITS?

YES
(2) FAULTY ECM GROUNDS OR ECM.

NO
REPAIR OPEN IN CIRCUIT THAT DID NOT LIGHT THE TEST LIGHT.

NO
(1) LOCATE AND CORRECT SHORT TO GROUND IN CIRCUIT THAT HAD AN OPEN FUSE.

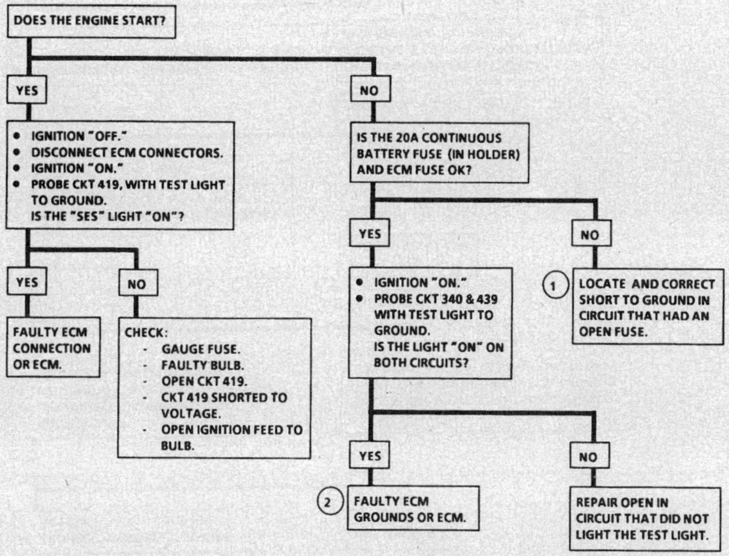

"AFTER REPAIRS," CONFIRM "CLOSED LOOP" OPERATION AND NO "SERVICE ENGINE SOON" LIGHT.

3.1L (VIN T) ENGINE — SYSTEM DIAGNOSTIC CHARTS — CAMARO, FIREBIRD, BERETTA, CORSICA, CAVALIER AND SUNBIRD

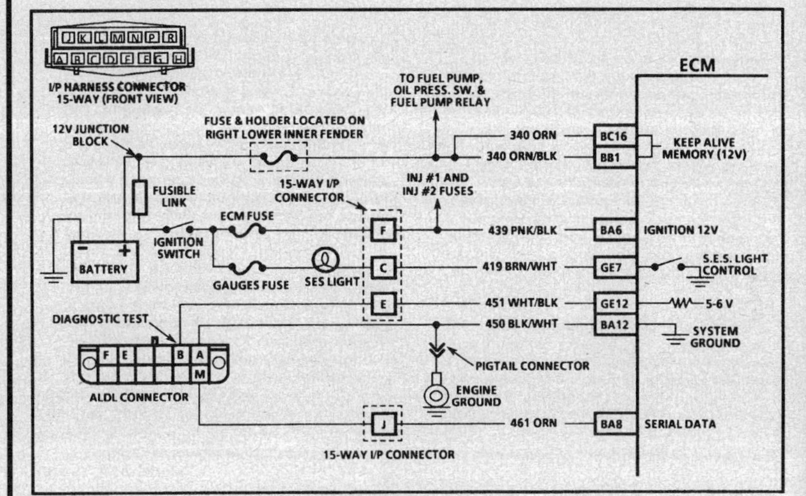

CHART A-2
NO ALDL DATA OR WILL NOT FLASH CODE 12
"SERVICE ENGINE SOON" LIGHT "ON" STEADY
3.1L (VIN T)

Circuit Description:

There should always be a steady "Service Engine Soon" light when the ignition is "ON" and engine stopped. Battery ignition voltage is supplied to the light bulb. The Electronic Control Module (ECM) will turn the light "ON" by grounding CKT 419 at the ECM.

With the diagnostic terminal grounded, the light should flash a Code 12, followed by any trouble code(s) stored in memory.

A steady light suggests a short to ground in the light control CKT 419, or an open in diagnostic CKT 451.

Test Description: Number(s) below refer to circled number(s) on the diagnostic chart.

1. If there is a problem with the ECM that causes a Tech 1 "Scan" tool to not read serial data, the ECM should not flash a Code 12. If Code 12 is flashing, check for CKT 451 short to ground. If Code 12 does flash, be sure that the Tech 1 "Scan" tool is working properly on another vehicle. If the Tech 1 "Scan" tool is functioning properly and CKT 461 is OK, the MEM-CAL or ECM may be at fault for the no ALDL symptom.

2. If the light goes "OFF" when the ECM connector is disconnected, CKT 419 is not shorted to ground.

3. This step will check for an open diagnostic CKT 451.

4. At this point, the "Service Engine Soon" light wiring is OK. The problem is a faulty ECM or MEM-CAL. If Code 12 does not flash, the ECM should be replaced using the original MEM-CAL. Replace the MEM-CAL only after trying an ECM, as a defective MEM-CAL is an unlikely cause of the problem. If the ECM has to be replaced, refer to "On-Vehicle Service" in "Electronic Control Module (ECM) and Sensors,"

CAUTION: Damage can occur to controller circuit board by pressing down hard on the ends of MEM-CAL, to secure the locking tabs.

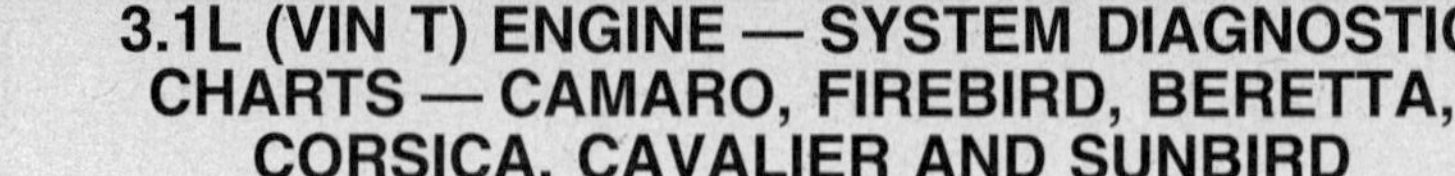

3.1L (VIN T) ENGINE — SYSTEM DIAGNOSTIC CHARTS — CAMARO, FIREBIRD, BERETTA, CORSICA, CAVALIER AND SUNBIRD

CHART A-2

NO ALDL DATA OR WILL NOT FLASH CODE 12
"SERVICE ENGINE SOON" LIGHT "ON" STEADY
3.1L (VIN T)

- IGNITION "ON," ENGINE "OFF." IS THE "SES" LIGHT "ON"?

YES →
- GROUND DIAGNOSTIC TERMINAL. DOES LIGHT FLASH CODE 12?

NO → USE CHART A-1

(2)
- IGNITION "OFF"
- DISCONNECT ECM CONNECTORS.
- IGNITION "ON" AND NOTE "SERVICE ENGINE SOON" LIGHT.

YES →
(1) IF PROBLEM WAS NO ALDL DATA: CHECK SERIAL DATA CKT 461 FOR OPEN OR SHORT TO GROUND BETWEEN ECM AND ALDL CONNECTOR. IF OK, IT IS A FAULTY ECM OR MEM-CAL.

(3) LIGHT "OFF"
- IGNITION "OFF"
- RECONNECT ECM
- IGNITION "ON," ENGINE "OFF," DIAGNOSTIC TERMINAL NOT GROUNDED. BACK PROBE ECM CKT 451 WITH TEST LIGHT TO GROUND. LEAVE CONNECTED AND WATCH "SES" LIGHT.

LIGHT "ON" → REPAIR SHORT TO GROUND IN CKT 419.

NO CODE 12 →
(4)
- CHECK MEM-CAL FOR PROPER INSTALLATION.
- IF OK, REPLACE ECM USING ORIGINAL MEM-CAL
- RECHECK FOR CODE 12

CODE 12 → CHECK FOR OPEN CKT 451 TO ECM. IF OK, CHECK FOR OPEN CIRCUIT BETWEEN ALDL TERMINAL "A" AND ECM.

NO CODE 12 → REPLACE MEM-CAL

CODE 12 → SYSTEM OK

"AFTER REPAIRS," CONFIRM "CLOSED LOOP" OPERATION AND NO "SERVICE ENGINE SOON" LIGHT.

3.1L (VIN T) ENGINE — SYSTEM DIAGNOSTIC CHARTS — CAMARO, FIREBIRD, BERETTA, CORSICA, CAVALIER AND SUNBIRD

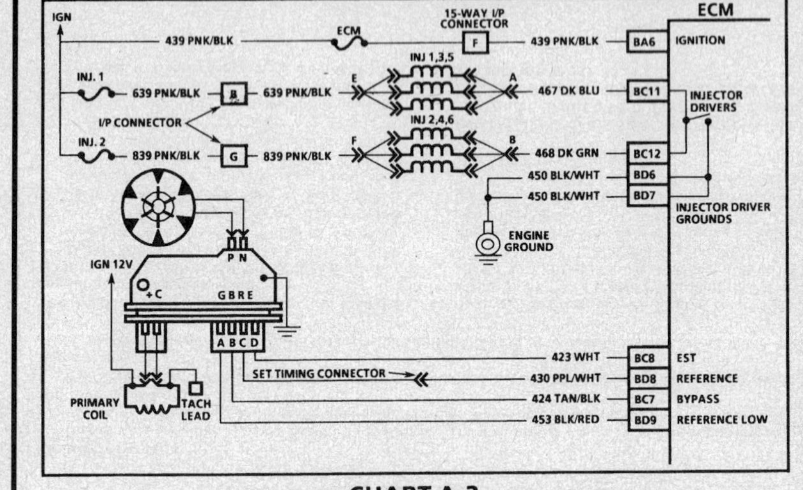

CHART A-3

(Page 1 of 2)
ENGINE CRANKS BUT WILL NOT RUN
3.1L (VIN T)

Circuit Description:

This chart assumes that battery condition and engine cranking speed are OK, and there is adequate fuel in the tank.

Test Description: Number(s) below refer to circled number(s) on the diagnostic chart.

1. This chart assumes that the battery condition and engine cranking speed are OK, and there is adequate fuel in the tank. If the engine starts but immediately stalls, see "Symptoms," Section "6E3-B" (Hard Start). A "Service Engine Soon" light "ON" is a basic check for ignition and battery supply to the Electronic Control Module (ECM).
2. No spark indicates a basic HEI problem.
3. This test will determine if the ECM is receiving the reference signal and controlling the injectors. This test could also be performed at the 4-way injector connector by using a test light between terminals "B" & "F" and "A" & "E".
   If the injector harness test light "blinks" when cranking, then ECM control should be considered OK. How bright the injector harness test light "blinks" is not important. However, the test light should be a J 34730-2A or equivalent.
4. Use fuel pressure gage J 34730-1. Wrap a shop towel around the fuel pressure tap to absorb any small amount of fuel leakage that may occur when installing the gage.
5. This test will determine if the injectors are open or shorted. Each injector should have a resistance of 12.2 ± .4 ohm at 20°C (68°F). There should be no more than .8 ohm difference between injectors.

Diagnostic Aids:

- An EGR valve sticking open can cause a low air/fuel ratio during cranking. Unless engine enters "Clear Flood" at the first indication of a flooding condition, it can result in a no start.
- Check for fouled plugs:
  - If the TPS is sticking or binding in the Wide Open Throttle (WOT) position, the ECM will be in the "Clear Flood" mode.
  - Also check that injectors on both sides of engine will cause a test light to "blink." Checking of two injectors on each bank in this manner will locate a shorted injector.
  - If above are all OK, refer to "Symptoms"

3.1L (VIN T) ENGINE — SYSTEM DIAGNOSTIC CHARTS — CAMARO, FIREBIRD, BERETTA, CORSICA, CAVALIER AND SUNBIRD

CHART A-3
(Page 1 of 2)
ENGINE CRANKS BUT WILL NOT RUN
3.1L (VIN T)

NOTICE: FUEL SYSTEM UNDER PRESSURE. TO AVOID FUEL SPILLAGE, REFER TO FIELD SERVICE PROCEDURES FOR TESTING OR MAKING REPAIRS REQUIRING DISASSEMBLY OF FUEL LINES OR FITTINGS.

1.
- FUEL QUANTITY OK.
- IGNITION "ON" - IF "SES" LIGHT IS "OFF," SEE CHART A-1.
- INSTALL "SCAN" TOOL - IF "NO ALDL," SEE CHART A-2.
- CHECK THE FOLLOWING:
  - TPS - IF OVER 2.5V AT CLOSED THROTTLE, SEE CODE 21.
  - IF CODE 54 IS SET, SEE CODE 54 CHART.

2.
- CONNECT ST-125 (SPARK CHECKER) J 26792 OR EQUIVALENT.
- CHECK FOR SPARK WHILE CRANKING.
- CHECK AT LEAST TWO WIRES.

3.
SPARK
- CHECK EACH BANK OF INJECTORS AS FOLLOWS:
  - DISCONNECT ONE INJECTOR.
  - CONNECT TEST LIGHT J-34730-2 OR EQUIVALENT TO INJECTOR HARNESS CONNECTOR TERMINALS.
  - CHECK FOR BLINKING LIGHT WHILE CRANKING.
  - RECONNECT INJECTOR.
  - REPEAT TEST ON ANOTHER INJECTOR ON THE SAME BANK.
  - LIGHT SHOULD BLINK ON BOTH.

NO SPARK
NOTICE : DO NOT ALLOW INJECTOR, TERMINALS TO SHORT OR TOUCH TOGETHER WHILE CRANKING OR ECM MAY BE DAMAGED.

CHECK FOR BATTERY VOLTAGE TO IGNITION SYSTEM. IF OK, THERE IS A BASIC HEI PROBLEM. REFER TO APPROPRIATE CHART C-4.

BLINKING LIGHT

4.
- INSTALL FUEL PRESSURE GAGE AND NOTE PRESSURE AFTER IGNITION "ON" AND FUEL PUMP STOPS.
  RUNNING SHOULD BE 284-325 kPa (41-47 psi).

OK → SEE "DIAGNOSTIC AIDS" ON FACING PAGE FOR ADDITIONAL ITEMS TO CHECK.
IF ALL CHECK OK, THERE IS NO TROUBLE FOUND. REVIEW "SYMPTOMS

NOT OK → SEE FUEL SYSTEM DIAGNOSTIC CHART A-7

STEADY LIGHT
- CHECK INJECTOR DRIVER CIRCUIT WITH STEADY LIGHT FOR SHORT TO GROUND.
- IF CIRCUIT IS NOT SHORTED, CHECK RESISTANCE AT INJECTOR HARNESS CONNECTOR. MEASURE FROM TERMINAL "A" ACROSS "E" AND "B" ACROSS "F". RESISTANCE SHOULD BE 3.9 TO 4.2 OHMS.

OK → FAULTY ECM.

NOT OK →
5.
- CHECK RESISTANCE ACROSS EACH INJECTOR..
- SHOULD BE 12.2 ± .4 OHMS.
- THERE SHOULD BE NO MORE THAN .8 OHMS DIFFERENCE BETWEEN INJECTORS. ANY INJECTOR THAT DOES NOT HAVE THE CORRECT RESISTANCE VALUES SHOULD BE REPLACED.

NO LIGHT
SEE CHART A-3 (PAGE 2 OF 2).

"AFTER REPAIRS," CONFIRM "CLOSED LOOP" OPERATION AND NO "SERVICE ENGINE SOON" LIGHT.

3.1L (VIN T) ENGINE — SYSTEM DIAGNOSTIC CHARTS — CAMARO, FIREBIRD, BERETTA, CORSICA, CAVALIER AND SUNBIRD

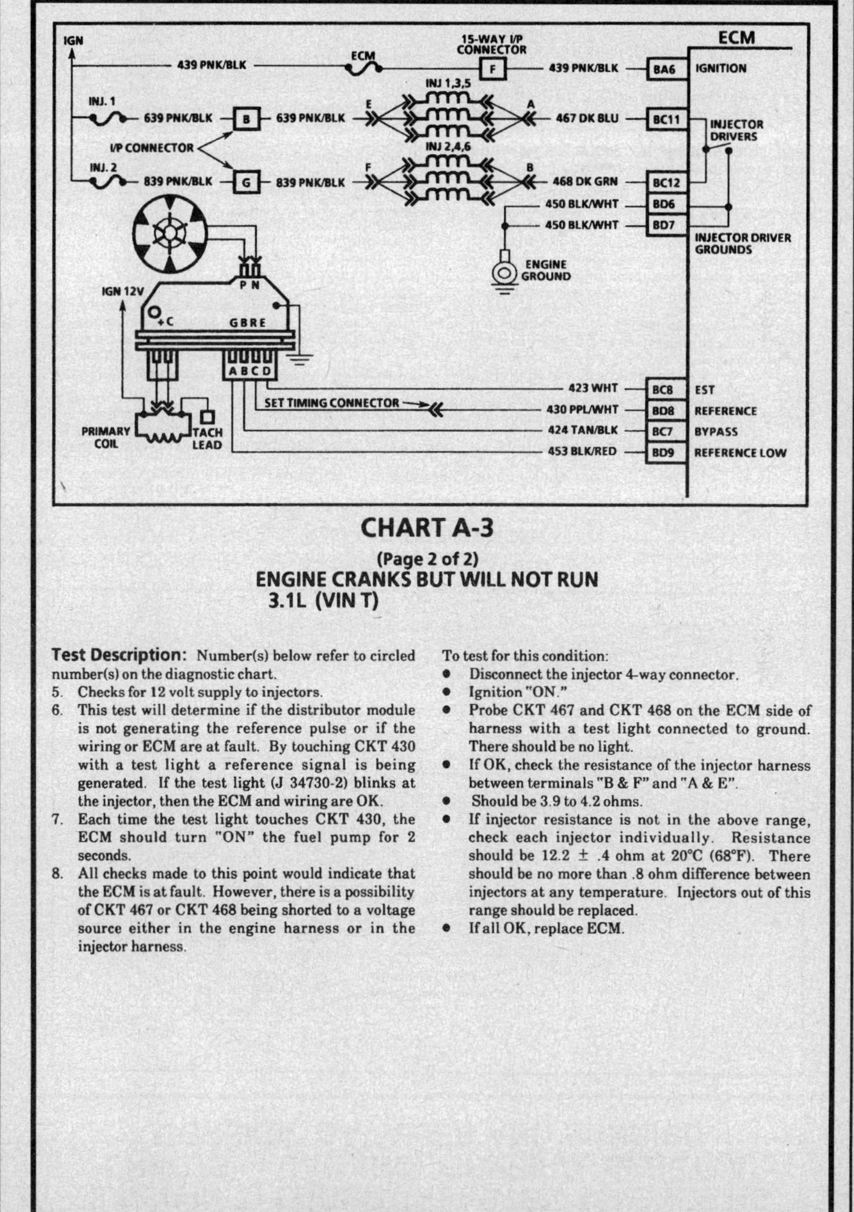

CHART A-3
(Page 2 of 2)
ENGINE CRANKS BUT WILL NOT RUN
3.1L (VIN T)

Test Description: Number(s) below refer to circled number(s) on the diagnostic chart.

5. Checks for 12 volt supply to injectors.
6. This test will determine if the distributor module is not generating the reference pulse or if the wiring or ECM are at fault. By touching CKT 430 with a test light a reference signal is being generated. If the test light (J 34730-2) blinks at the injector, then the ECM and wiring are OK.
7. Each time the test light touches CKT 430, the ECM should turn "ON" the fuel pump for 2 seconds.
8. All checks made to this point would indicate that the ECM is at fault. However, there is a possibility of CKT 467 or CKT 468 being shorted to a voltage source either in the engine harness or in the injector harness.

To test for this condition:
- Disconnect the injector 4-way connector.
- Ignition "ON."
- Probe CKT 467 and CKT 468 on the ECM side of harness with a test light connected to ground. There should be no light.
- If OK, check the resistance of the injector harness between terminals "B & F" and "A & E". Should be 3.9 to 4.2 ohms.
- If injector resistance is not in the above range, check each injector individually. Resistance should be 12.2 ± .4 ohm at 20°C (68°F). There should be no more than .8 ohm difference between injectors at any temperature. Injectors out of this range should be replaced.
- If all OK, replace ECM.

3.1L (VIN T) ENGINE — SYSTEM DIAGNOSTIC CHARTS — CAMARO, FIREBIRD, BERETTA, CORSICA, CAVALIER AND SUNBIRD

CHART A-3

(Page 2 of 2)
ENGINE CRANKS BUT WILL NOT RUN
3.1L (VIN T)

NOTICE: EFI SYSTEM UNDER PRESSURE. TO AVOID FUEL SPILLAGE, REFER TO FIELD SERVICE PROCEDURES FOR TESTING OR MAKING REPAIRS REQUIRING DISASSEMBLY OF FUEL LINES OR FITTINGS

CONTINUED FROM
CHART A-3
(PAGE 1 OF 2)

NO LIGHT

5
- DISCONNECT INJECTOR HARNESS CONNECTOR.
- IGNITION "ON."
- PROBE INJECTOR HARNESS TERMINALS "E" AND "F" WITH A TEST LIGHT TO GROUND. LIGHT SHOULD BE "ON" AT BOTH TERMINALS. IS IT?

LIGHT "ON" BOTH

6
- RECONNECT INJECTOR HARNESS CONNECTOR.
- DISCONNECT ANY INJECTOR.
- CONNECT TOOL J 34730-2 OR EQUIVALENT TEST LIGHT TO INJECTOR HARNESS CONNECTOR TERMINALS.
- DISCONNECT DISTRIBUTOR 4-WAY CONNECTOR.
- MOMENTARILY TOUCH HARNESS CONNECTOR TERMINAL CKT 430 WITH TEST LIGHT TO 12 VOLTS. USE A TEST LIGHT ONLY. TOUCH TERMINAL ONLY MOMENTARILY AND NOTE INJECTOR TEST LIGHT. SHOULD "BLINK" EACH TIME THE TEST LIGHT IS REMOVED FROM CKT 430.

LIGHT "ON" ONE
- REPAIR OPEN OR SHORT TO INJECTOR FEED CIRCUIT THAT DID NOT LIGHT.

NO LIGHT
REPAIR OPEN IN INJECTOR FEED CIRCUITS.

NO BLINKING LIGHT AT INJECTOR

7
REPEAT TEST AND OBSERVE FOR FUEL PUMP RUNNING FOR 2 SECONDS
OR
FUEL PUMP RELAY CLICK.

INJECTOR LIGHT "BLINKS"
FAULTY IGNITION MODULE OR CONNECTION.

OK
- RECONNECT INJECTOR(S).
- IGNITION "OFF."
- DISCONNECT ECM.
- IGNITION "ON."
- PROBE HARNESS TERMINALS "BC11" AND "BC12" WITH A TEST LIGHT TO GROUND.

NOT OK
- OPEN OR GROUNDED CKT 430.
- FAULTY CONNECTION AT "BD8" OR FAULTY ECM.

LIGHT

8
SEE FACING PAGE
TEST DESCRIPTION 8.

NO LIGHT
OPEN CKT 467 OR CKT 468.

"AFTER REPAIRS," CONFIRM "CLOSED LOOP" OPERATION AND NO "SERVICE ENGINE SOON" LIGHT.

3.1L (VIN T) ENGINE — SYSTEM DIAGNOSTIC CHARTS — CAMARO, FIREBIRD, BERETTA, CORSICA, CAVALIER AND SUNBIRD

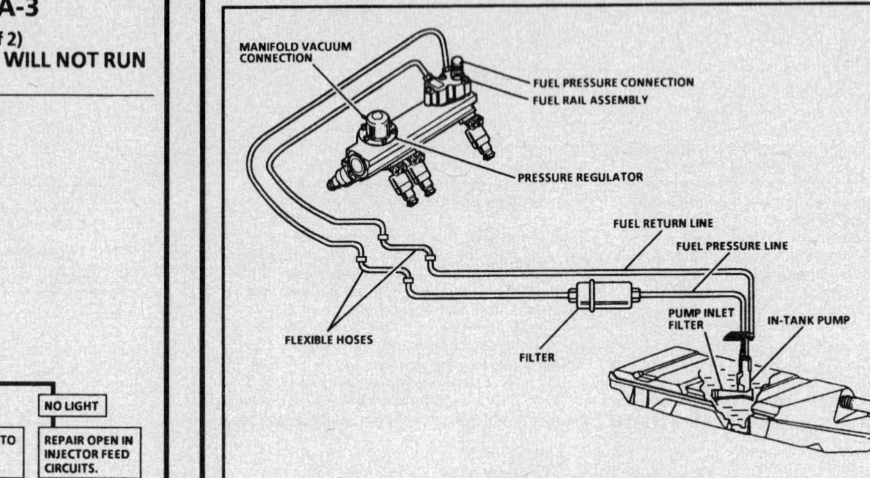

CHART A-7 (Page 1 of 2)

FUEL SYSTEM DIAGNOSIS
3.1L (VIN T)

Circuit Description:

When the ignition switch is turned "ON," the Electronic Control Module (ECM) will turn "ON" the in-tank fuel pump. It will remain "ON" as long as the engine is cranking or running, and the ECM is receiving reference pulses. If there are no reference pulses, the ECM will shut "OFF" the fuel pump within 2 seconds after ignition "ON" or engine stops.

The pump will deliver fuel to the fuel rail and injectors, then to the pressure regulator, where the system pressure is controlled to about 234 to 325 kPa (34 to 47 psi). Excess fuel is then returned to the fuel tank.

Test Description: Number(s) below refer to circled number(s) on the diagnostic chart.

1. Wrap a shop towel around the fuel pressure connector to absorb any small amount of fuel leakage that may occur when installing the gage. Ignition "ON," pump pressure should be 280-325 kPa (40.5-47 psi). This pressure is controlled by spring pressure within the regulator assembly.

2. When the engine is idling, the manifold pressure is low (high vacuum) and is applied to the fuel regulator diaphragm. This will offset the spring and result in a lower fuel pressure. This idle pressure will vary somewhat depending on barometric pressure, however, the pressure idling should be less indication pressure regulator control.

3. Pressure that continues to fall is caused by one of the following:
   - In-tank fuel pump check valve not holding.
   - Partially disconnected fuel pulse dampener (pulsator).
   - Fuel pressure regulator valve leaking.

- Injectors(s) sticking open.

4. An injector sticking open can best be determined by checking for a fouled or saturated spark plug(s). If a leaking injector cannot be determined by a fouled or saturated spark plug, the following procedure should be used.
   - Remove plenum, and remove fuel rail bolts. Follow procedures in the Fuel Metering Section of this manual, but leave fuel lines connected.
   - Lift fuel rail out just enough to leave injector nozzles in the ports.

CAUTION: Be sure injectors(s) are not allowed to spray on engine and that injector retaining clips are intact. This should be carefully followed to prevent fuel spray on engine which would cause a fire hazard.

- Pressurize the fuel system and observe injector nozzles.

3.1L (VIN T) ENGINE — SYSTEM DIAGNOSTIC CHARTS — CAMARO, FIREBIRD, BERETTA, CORSICA, CAVALIER AND SUNBIRD

CHART A-7
(Page 1 of 2)
FUEL SYSTEM DIAGNOSIS
3.1L (VIN T)

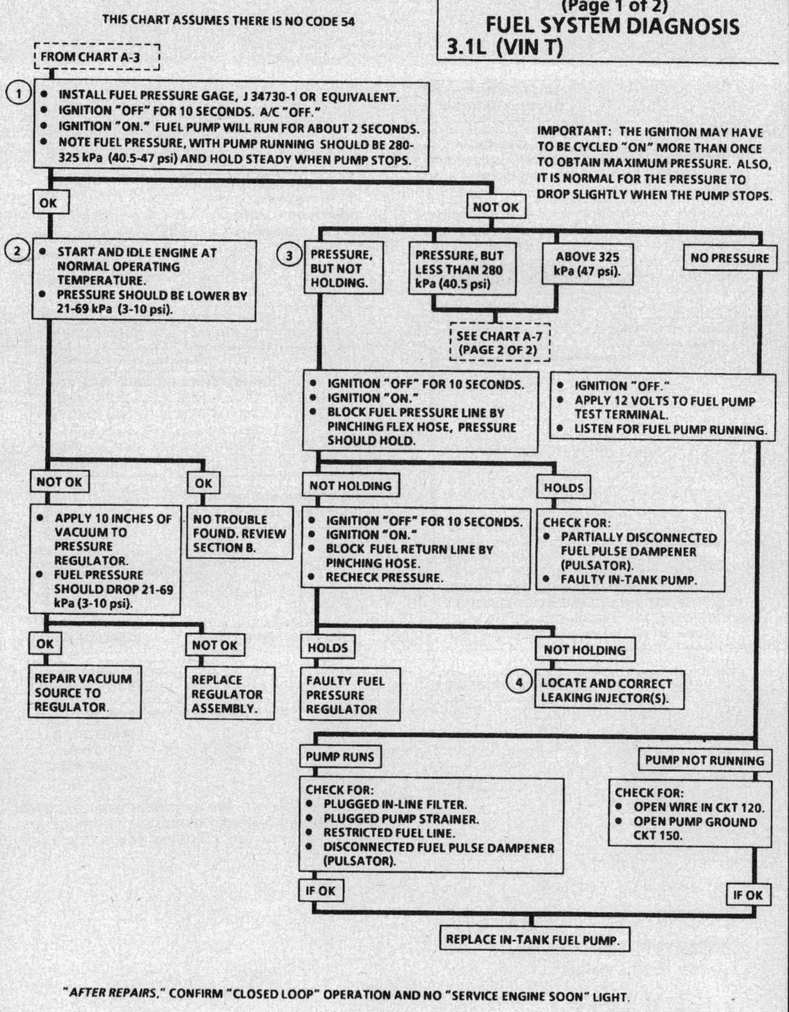

3.1L (VIN T) ENGINE — SYSTEM DIAGNOSTIC CHARTS — CAMARO, FIREBIRD, BERETTA, CORSICA, CAVALIER AND SUNBIRD

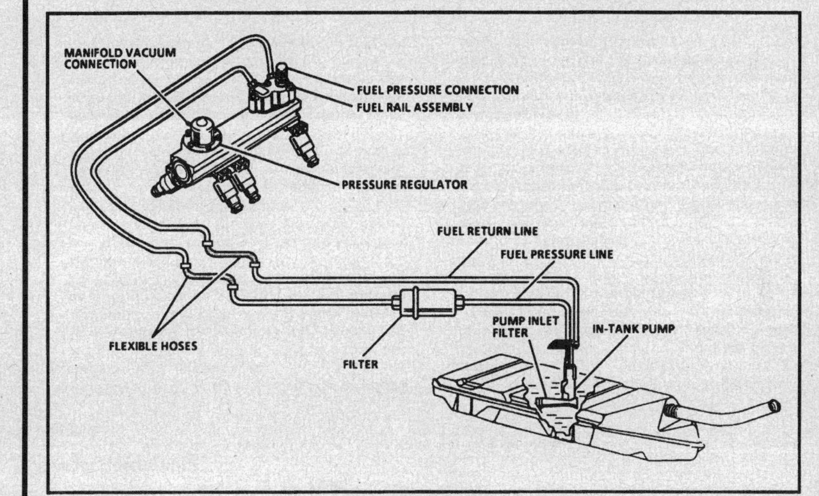

CHART A-7
(Page 2 of 2)
FUEL SYSTEM DIAGNOSIS
3.1L (VIN T)

Test Description: Number(s) below refer to circled number(s) on the diagnostic chart.

1. Pressure but less than 280 kPa (40.5 psi) falls into two areas:
   - Regulated pressure but less than 280 kPa (40.5 psi). Amount of fuel to injectors OK but pressure is too low. System will be lean running and may set Code 44. Also, hard starting cold and overall poor performance.
   - Restricted flow causing pressure drop - Normally, a vehicle with a fuel pressure of less than 165 kPa (24 psi) at idle will not be driveable. However, if the pressure drop occurs only while driving, the engine will normally surge then stop running as pressure begins to drop rapidly. This is most likely caused be a restricted fuel line or plugged in-line filter.

2. Restricting the fuel return line allows fuel pressure to build above regulated pressure. With battery voltage applied to the pump "test" terminal, pressure should rise above 325 kPa (47 psi) as the fuel return hose is gradually pinched.

   NOTICE: Do Not allow pressure to exceed 414 kPa (60 psi), as damage to the regulator may result.

3. This test determines if the high fuel pressure is due to a restricted fuel return line or a pressure regulator problem.

3.1L (VIN T) ENGINE — SYSTEM DIAGNOSTIC CHARTS — CAMARO, FIREBIRD, BERETTA, CORSICA, CAVALIER AND SUNBIRD

CHART A-7
(Page 2 of 2)
FUEL SYSTEM DIAGNOSIS
3.1L (VIN T)

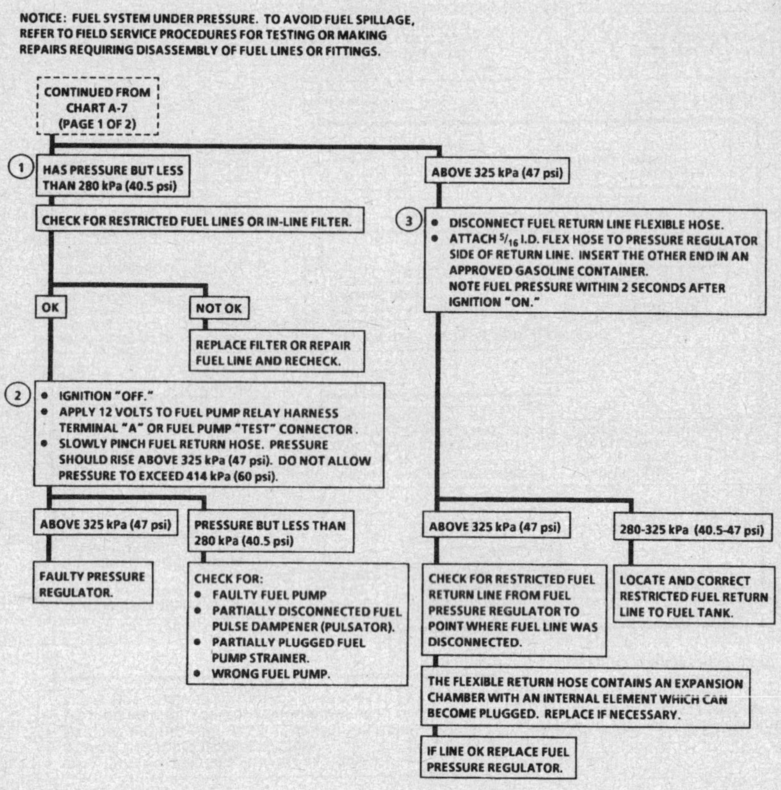

NOTICE: FUEL SYSTEM UNDER PRESSURE. TO AVOID FUEL SPILLAGE, REFER TO FIELD SERVICE PROCEDURES FOR TESTING OR MAKING REPAIRS REQUIRING DISASSEMBLY OF FUEL LINES OR FITTINGS.

CONTINUED FROM CHART A-7 (PAGE 1 OF 2)

(1) HAS PRESSURE BUT LESS THAN 280 kPa (40.5 psi)

CHECK FOR RESTRICTED FUEL LINES OR IN-LINE FILTER.

OK

NOT OK

REPLACE FILTER OR REPAIR FUEL LINE AND RECHECK.

(2)
- IGNITION "OFF."
- APPLY 12 VOLTS TO FUEL PUMP RELAY HARNESS TERMINAL "A" OR FUEL PUMP "TEST" CONNECTOR.
- SLOWLY PINCH FUEL RETURN HOSE. PRESSURE SHOULD RISE ABOVE 325 kPa (47 psi). DO NOT ALLOW PRESSURE TO EXCEED 414 kPa (60 psi).

ABOVE 325 kPa (47 psi)

FAULTY PRESSURE REGULATOR.

PRESSURE BUT LESS THAN 280 kPa (40.5 psi)

CHECK FOR:
- FAULTY FUEL PUMP
- PARTIALLY DISCONNECTED FUEL PULSE DAMPENER (PULSATOR).
- PARTIALLY PLUGGED FUEL PUMP STRAINER.
- WRONG FUEL PUMP.

ABOVE 325 kPa (47 psi)

(3)
- DISCONNECT FUEL RETURN LINE FLEXIBLE HOSE.
- ATTACH 5/16 I.D. FLEX HOSE TO PRESSURE REGULATOR SIDE OF RETURN LINE. INSERT THE OTHER END IN AN APPROVED GASOLINE CONTAINER. NOTE FUEL PRESSURE WITHIN 2 SECONDS AFTER IGNITION "ON."

ABOVE 325 kPa (47 psi)

CHECK FOR RESTRICTED FUEL RETURN LINE FROM FUEL PRESSURE REGULATOR TO POINT WHERE FUEL LINE WAS DISCONNECTED.

THE FLEXIBLE RETURN HOSE CONTAINS AN EXPANSION CHAMBER WITH AN INTERNAL ELEMENT WHICH CAN BECOME PLUGGED. REPLACE IF NECESSARY.

IF LINE OK REPLACE FUEL PRESSURE REGULATOR.

280-325 kPa (40.5-47 psi)

LOCATE AND CORRECT RESTRICTED FUEL RETURN LINE TO FUEL TANK.

"AFTER REPAIRS," CONFIRM "CLOSED LOOP" OPERATION AND NO "SERVICE ENGINE SOON" LIGHT.

3.1L (VIN T) ENGINE — SYSTEM DIAGNOSTIC CHARTS — CAMARO, FIREBIRD, BERETTA, CORSICA, CAVALIER AND SUNBIRD

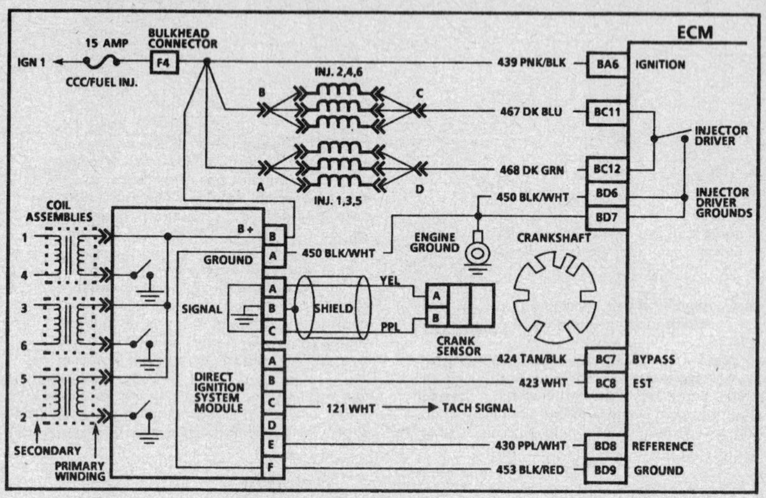

CHART A-3
(Page 1 of 3)
ENGINE CRANKS BUT WILL NOT RUN
3.1L (VIN T)

Circuit Description:

This chart assumes that battery condition and engine cranking speed are OK, and there is adequate fuel in the tank.

Test Description: Number(s) below refer to circled number(s) on the diagnostic chart.

1. A "Service Engine Soon" light "ON" is a basic test to determine if there is a 12 volt supply and ignition 12 volts to ECM. No ALDL may be due to an ECM problem and CHART A-2 will diagnose the ECM. If TPS is over 62%, the engine may be in the clear flood mode which will cause starting problems. The engine will not start without reference pulses and, therefore, the "Scan" should read rpm (reference) during crank.

2. For the first two seconds with ignition "ON," or whenever reference pulses are being received, PPSW should indicate fuel pump circuit voltage (8 to 12 volts).

3. Because the direct ignition system uses two plugs and wires to complete the circuit of each coil, the opposite spark should be left connected. If rpm was indicated during crank, the ignition module is receiving a crank signal, but no spark at this test indicates the ignition module is not triggering the coils.

4. The test light should "blink" indicating the ECM is controlling the injectors OK. How bright the light "blinks" is not important. However, the test light should be a J 34730-3 or equivalent.

5. Use fuel pressure gage J 34730-1 or equivalent. Wrap a shop towel around the fuel pressure tap to absorb any small amount of fuel leakage that may occur when installing the gage.

6. This test will determine if the ignition module is not generating the reference pulse or if the wiring or ECM are at fault. By touching and removing a test light to 12 volts on CKT 430, a reference pulse should be generated. If rpm is indicated, the ECM and wiring are OK.

7. This test will determine if the ignition module is not triggering the problem coil or if the tested coil is at fault. This test could also be performed by using another known good coil.

3.1L (VIN T) ENGINE — SYSTEM DIAGNOSTIC CHARTS — CAMARO, FIREBIRD, BERETTA, CORSICA, CAVALIER AND SUNBIRD

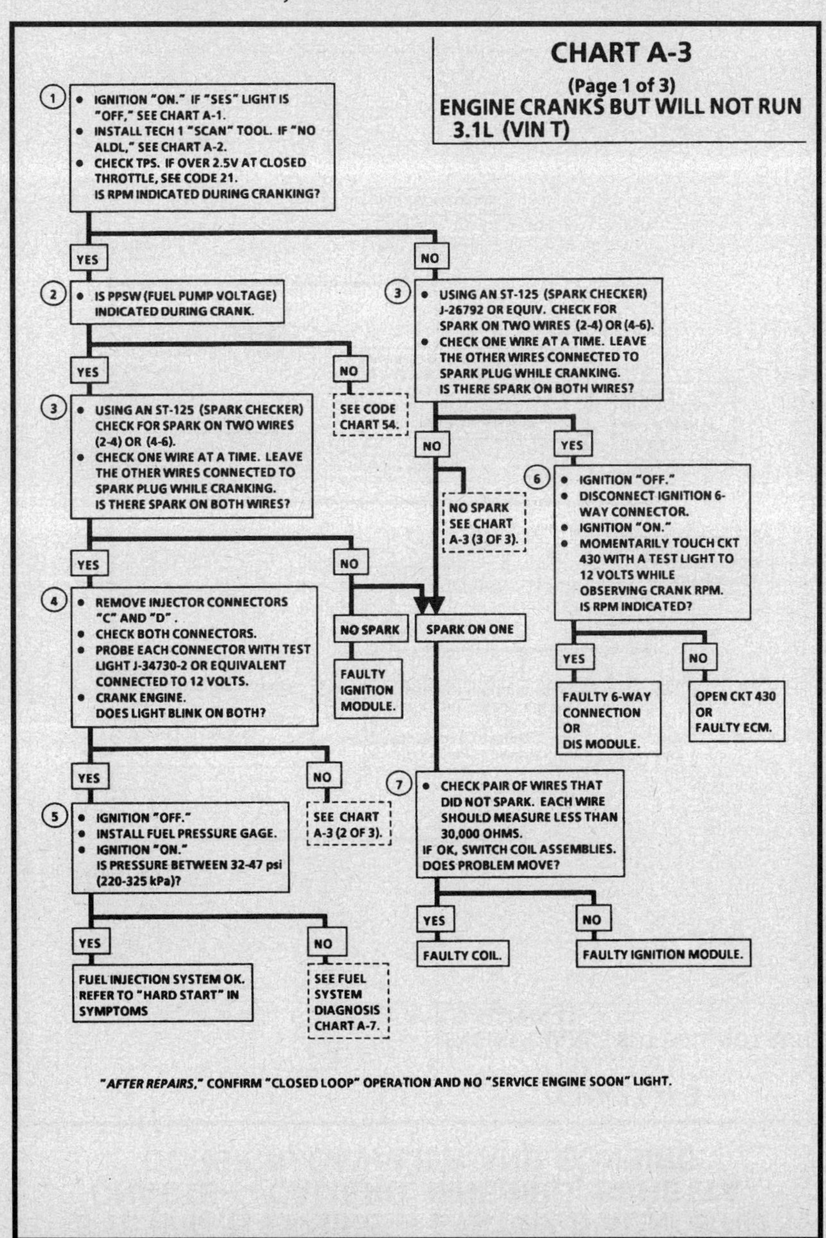

3.1L (VIN T) ENGINE — SYSTEM DIAGNOSTIC CHARTS — CAMARO, FIREBIRD, BERETTA, CORSICA, CAVALIER AND SUNBIRD

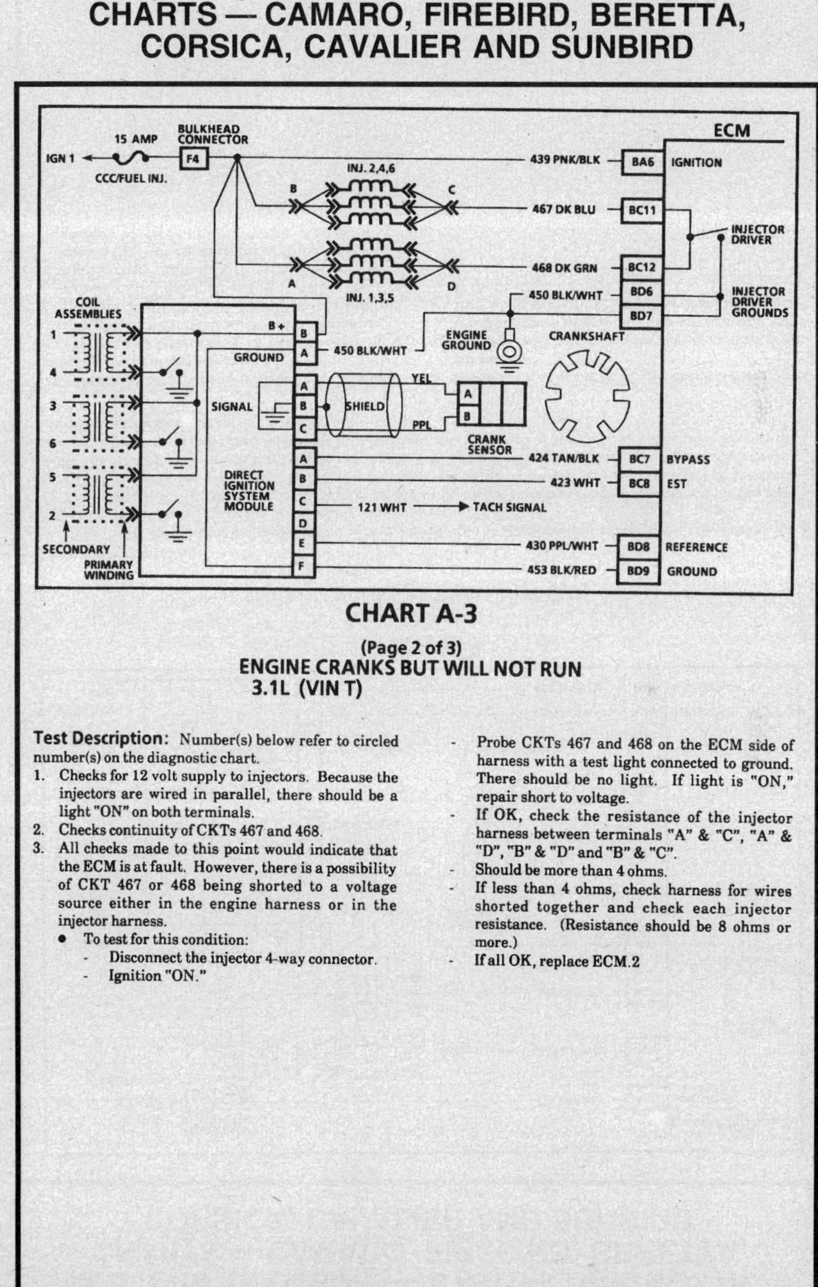

Test Description: Number(s) below refer to circled number(s) on the diagnostic chart.

1. Checks for 12 volt supply to injectors. Because the injectors are wired in parallel, there should be a light "ON" on both terminals.
2. Checks continuity of CKTs 467 and 468.
3. All checks made to this point would indicate that the ECM is at fault. However, there is a possibility of CKT 467 or 468 being shorted to a voltage source either in the engine harness or in the injector harness.
   - To test for this condition:
     - Disconnect the injector 4-way connector.
     - Ignition "ON."

- Probe CKTs 467 and 468 on the ECM side of harness with a test light connected to ground. There should be no light. If light is "ON," repair short to voltage.
- If OK, check the resistance of the injector harness between terminals "A" & "C", "A" & "D", "B" & "D" and "B" & "C".
- Should be more than 4 ohms.
- If less than 4 ohms, check harness for wires shorted together and check each injector resistance. (Resistance should be 8 ohms or more.)
- If all OK, replace ECM.2

3.1L (VIN T) ENGINE — SYSTEM DIAGNOSTIC CHARTS — CAMARO, FIREBIRD, BERETTA, CORSICA, CAVALIER AND SUNBIRD

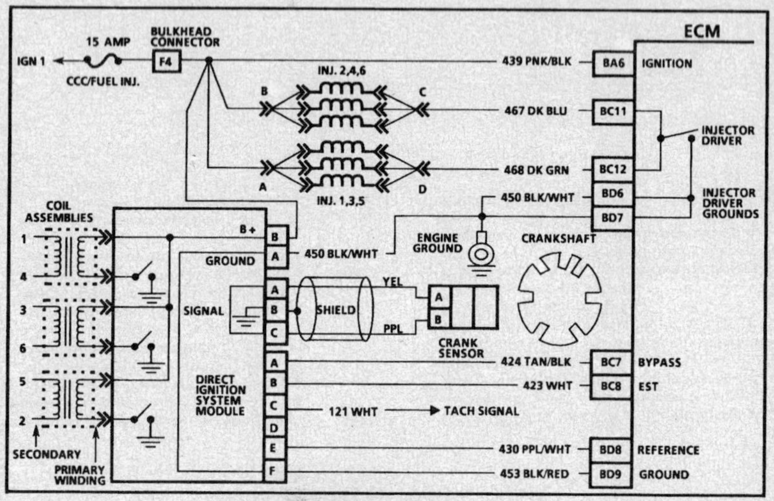

CHART A-3
(Page 2 of 3)
ENGINE CRANKS BUT WILL NOT RUN
3.1L (VIN T)

FROM A-3 (1 OF 3)

NO BLINKING LIGHT AT INJECTOR(S).

NO LIGHT

(1)
- IGNITION "ON"
- PROBE INJECTOR HARNESS TERMINALS WITH A TEST LIGHT TO GROUND.
- LIGHT SHOULD BE "ON" AT BOTH TERMINALS.

STEADY LIGHT

- CHECK INJECTOR DRIVER CKT WITH STEADY LIGHT FOR SHORT TO GROUND.
- IF CKT IS NOT SHORTED, CHECK RESISTANCE ACROSS EACH INJECTOR IN THE CIRCUIT.
- RESISTANCE SHOULD BE GREATER THAN 8 OHMS.

OK → FAULTY ECM.

NOT OK → REPLACE ANY INJECTOR THAT MEASURES UNDER 8 OHMS.

LIGHT "ON" BOTH

(2)
- RECONNECT INJECTOR(S)
- IGNITION "OFF".
- DISCONNECT ECM
- IGNITION "ON".
- PROBE TERMINALS BC11 AND BC12 WITH A TEST LIGHT TO GROUND.

LIGHT "ON" ONE → DUE TO INJECTORS WIRED IN PARALLEL, THERE SHOULD BE A LIGHT ON BOTH TERMINALS. IF NOT, THE PROBLEM IS IN THE HARNESS TO THE TESTED INJECTOR.

LIGHT "OFF" → REPAIR OPEN IN INJECTOR FEED CIRCUIT.

LIGHT "ON"

LIGHT "OFF" → OPEN CKT 467 OR 468.

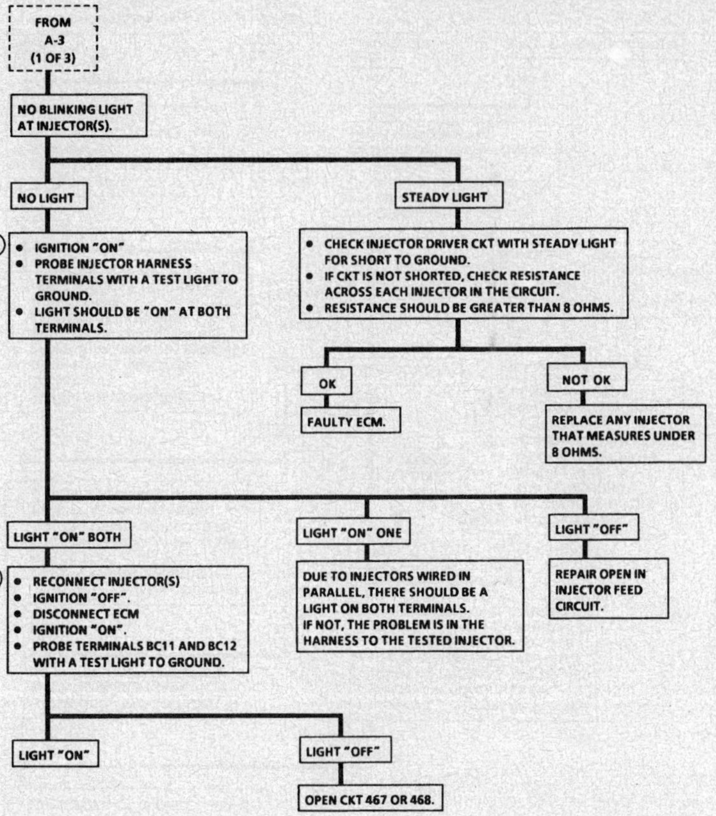

"AFTER REPAIRS," CONFIRM "CLOSED LOOP" OPERATION AND NO "SERVICE ENGINE SOON" LIGHT.

CHART A-3
(Page 3 of 3)
ENGINE CRANKS BUT WILL NOT RUN
3.1L (VIN T)

Circuit Description:

If the "Scan" tool did not indicate a cranking rpm, and there is no spark present at the plugs, the problem lies in the direct ignition system or the power and ground supplies to the module.

The magnetic crank sensor is used to determine engine crankshaft position much the same way as the pick-up coil did in distributor type systems. The sensor is mounted in the block near a seven slot wheel on the crank shaft. The rotation of the wheel creates a flux change in the sensor which produces a voltage signal. The ignition module then processes this signal and creates the reference pulses needed by the ECM and the signal triggers the correct coil at the correct time.

Test Description: Number(s) below refer to circled number(s) on the diagnostic chart.

1. This test will determine if the 12 volt supply and a good ground is available at the ignition module.
2. Tests for continuity of CKT 439 to the ignition module. If test light does not light but the "Service Engine Soon" light is "ON" with ignition repair open in CKT 439 between DIS ignition module and splice.
3. Checks for continuity of the crank sensor and connections.
4. Voltage will vary in this test depending on cranking speed of engine. The voltage will vary from about 500 mV at very slow cranking speeds, to about 100 mV at high speeds.

3.1L (VIN T) ENGINE — SYSTEM DIAGNOSTIC CHARTS — CAMARO, FIREBIRD, BERETTA, CORSICA, CAVALIER AND SUNBIRD

CHART A-3
(Page 3 of 3)
ENGINE CRANKS BUT WILL NOT RUN
3.1L (VIN T)

FROM CHART A-3 (1 OF 3)

NO CRANKING RPM INDICATED AND NO SPARK.

① • DISCONNECT DIS MODULE 2 PIN CONNECTOR
• IGNITION "ON".
• CONNECT TEST LIGHT BETWEEN HARNESS TERMINALS.

LIGHT "OFF"

LIGHT "ON"

② • CONNECT TEST LIGHT FROM HARNESS TERMINAL "B" TO GROUND.

③ • DISCONNECT CRANK SENSOR 3-WAY CONNECTOR FROM DIS MODULE.
• WITH OHMMETER IN 2K OHMS POSITION, PROBE HARNESS TERMINALS "A&C" (SHOULD READ BETWEEN 900-1200 OHMS).

LIGHT "ON"

LIGHT "OFF"

OK

NOT OK

REPAIR OPEN GROUND CKT 450.

REPAIR OPEN IGNITION FEED CKT 439.

④ • SET VOLTMETER ON 2 VOLT AC POSITION.
• CRANK ENGINE AND OBSERVE VOLTAGE READING. READING SHOULD BE GREATER THAN .1 VOLTS (100 mV). IS IT?

LESS THAN 900 OHMS

GREATER THAN 1200 OHMS

CRANK SENSOR LEADS SHORTED TOGETHER OR FAULTY CRANK SENSOR.

OPEN CRANK SENSOR CKT FAULTY CONNECTION OR FAULTY CRANK SENSOR.

YES

NO

REPLACE DIS MODULE

FAULTY CONNECTION OR FAULTY CRANK SENSOR.

"AFTER REPAIRS," CONFIRM "CLOSED LOOP" OPERATION AND NO "SERVICE ENGINE SOON" LIGHT.

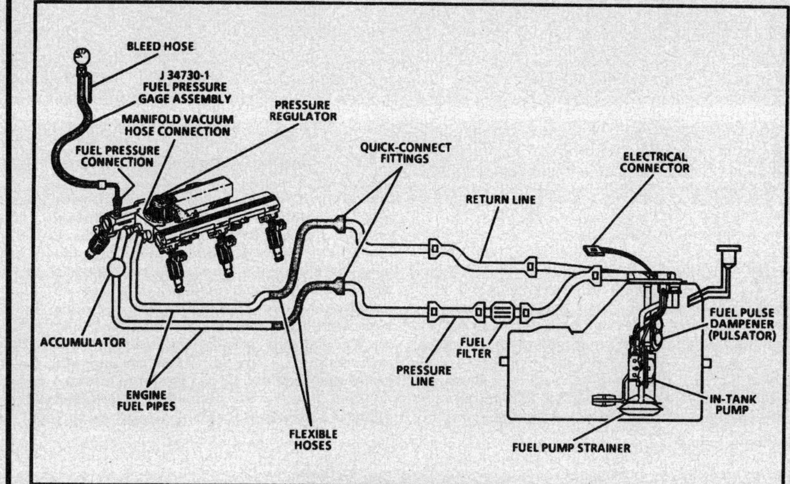

CHART A-7
(Page 1 of 3)
FUEL SYSTEM DIAGNOSIS
3.1L (VIN T)

Circuit Description:

When the ignition switch is turned "ON," the Electronic Control Module (ECM) will turn "ON" the in-tank fuel pump. It will remain "ON" as long as the engine is cranking or running, and the ECM is receiving reference pulses. If there are no reference pulses, the ECM will shut "OFF" the fuel pump within 2 seconds after ignition "ON" or engine stops.

An electric fuel pump, attached to the fuel sender assembly (inside the fuel tank), pumps fuel through and in-line filter to the fuel rail assembly. The pump is designed to provide fuel at a pressure above the regulated pressure needed by the injector. A pressure regulator attached to the fuel rail, keep fuel available to the injectors at a regulated pressure. Unused fuel is returned to the fuel tank by a separate line.

Test Description: Number(s) below refer to circled number(s) on the diagnostic chart.

1. Wrap a shop towel around the fuel pressure connector to absorb any small amount of fuel leakage that may occur when installing the gage. Ignition "ON," engine "OFF," pump pressure should be 284-325 kPa (41-47 psi). This pressure is controlled by spring pressure within the regulator assembly.

2. When the engine is idling, the manifold pressure is low (high vacuum) and is applied to the fuel regulator diaphragm. This will offset the spring and result in a lower fuel pressure. This idle pressure will vary somewhat depending on barometric pressure, however, the pressure idling should be less indicating pressure regulator control.

3. Pressure that continues to fall is caused by one of the following:
• In-tank fuel pump check valve not holding.
• Partially disconnected fuel pulse dampener (pulsator).

• Fuel pressure regulator valve leaking.
• Injector(s) sticking open.

4. An injector striking open can best be determined by checking for a fouled or saturated spark plug(s). If a leaking injector can not be determined by a fouled or saturated spark plug, the following procedure should be used.
• Remove plenum, and remove fuel rail bolts.

• Lift fuel rail out just enough to leave injector nozzles in the ports.

CAUTION: Be sure injector(s) are not allowed to spray on engine and that injector retaining clips are intact. This should be carefully followed to prevent fuel spray on engine which would cause a fire hazard.

• Pressurize the fuel system and observe for injector(s) leaking.

3.1L (VIN T) ENGINE — SYSTEM DIAGNOSTIC CHARTS — CAMARO, FIREBIRD, BERETTA, CORSICA, CAVALIER AND SUNBIRD

CAUTION: TO REDUCE THE RISK OF FIRE AND PERSONAL INJURY, WRAP A SHOP TOWEL AROUND THE FUEL PRESSURE CONNECTION TO ABSORB ANY FUEL LEAKAGE THAT MAY OCCUR WHEN INSTALLING THE PRESSURE GAGE. PLACE TOWEL IN APPROVED CONTAINER.

CHART A-7
(Page 1 of 3)
FUEL SYSTEM DIAGNOSIS
3.1L (VIN T)

FROM CHART A-3

(1)
- THIS CHART ASSUMES THERE IS NO CODE 54.
- INSTALL FUEL PRESSURE GAGE, J 34730-1 OR EQUIVALENT.
- IGNITION "OFF" FOR 10 SECONDS. A/C "OFF."
- IGNITION "ON." FUEL PUMP WILL RUN FOR ABOUT 2 SECONDS. NOTE FUEL PRESSURE, WITH PUMP RUNNING SHOULD BE 284-325 kPa (41-47 psi) AND HOLD STEADY WHEN PUMP STOPS.

OK → NOT OK

(2)
- START AND IDLE ENGINE AT NORMAL OPERATING TEMPERATURE. PRESSURE SHOULD BE LOWER BY 21-69 kPa (3-10 psi).

(3)
- FUEL PRESSURE, BUT NOT HOLDING
- FUEL PRESSURE, BUT LESS THAN 284 kPa (41 psi)
- FUEL PRESSURE ABOVE 325 kPa (47 psi)
- NO PRESSURE

SEE CHART A-7 (PAGE 2 of 3)

- IGNITION "OFF" FOR 10 SECONDS.
- IGNITION "ON."
- BLOCK FUEL PRESSURE LINE BY PINCHING FLEX HOSE, PRESSURE SHOULD HOLD.

- IGNITION "OFF."
- USING A 10 AMP FUSED JUMPER WIRE, APPLY 12 VOLTS TO FUEL PUMP "TEST" TERMINAL.
- LISTEN FOR FUEL PUMP RUNNING.

NOT OK → OK

- APPLY 10 INCHES OF VACUUM TO PRESSURE REGULATOR. FUEL PRESSURE SHOULD DROP 21-69 kPa (3-10 psi).

NO TROUBLE FOUND.

NOT HOLDING → HOLDS

- IGNITION "OFF" FOR 10 SECONDS.
- IGNITION "ON."
- BLOCK FUEL RETURN LINE BY PINCHING HOSE.
- RECHECK PRESSURE.

CHECK FOR:
- PARTIALLY DISCONNECTED FUEL PULSE DAMPENER (PULSATOR).
- FAULTY IN-TANK PUMP.

OK → NOT OK

REPAIR VACUUM SOURCE TO REGULATOR.

REPLACE REGULATOR ASSEMBLY.

HOLDS → NOT HOLDING

REPLACE REGULATOR ASSEMBLY.

(4) LOCATE AND CORRECT LEAKING INJECTOR(S).

PUMP RUNS → PUMP NOT RUNNING

CHECK FOR:
- PLUGGED IN-LINE FILTER.
- PLUGGED FUEL PUMP STRAINER.
- RESTRICTED FUEL LINE.
- DISCONNECTED FUEL PULSE DAMPENER (PULSATOR).

CHECK FOR:
- OPEN WIRE IN CKT 120.
- OPEN PUMP GROUND WIRE CKT 150.

IF OK → IF OK

REPLACE IN-TANK PUMP.

IMPORTANT: THE IGNITION MAY HAVE TO BE CYCLED "ON" MORE THAN ONCE TO OBTAIN MAXIMUM PRESSURE. ALSO, IT IS NORMAL FOR THE PRESSURE TO DROP SLIGHTLY WHEN THE PUMP STOPS.

"AFTER REPAIRS," CONFIRM "CLOSED LOOP" OPERATION AND NO "SERVICE ENGINE SOON" LIGHT.

3.1L (VIN T) ENGINE — SYSTEM DIAGNOSTIC CHARTS — CAMARO, FIREBIRD, BERETTA, CORSICA, CAVALIER AND SUNBIRD

BLEED HOSE
J 34730-1 FUEL PRESSURE GAGE ASSEMBLY
PRESSURE REGULATOR
MANIFOLD VACUUM HOSE CONNECTION
FUEL PRESSURE CONNECTION
QUICK-CONNECT FITTINGS
RETURN LINE
ELECTRICAL CONNECTOR
FUEL PULSE DAMPENER (PULSATOR)
ACCUMULATOR
FUEL FILTER
PRESSURE LINE
IN-TANK PUMP
ENGINE FUEL PIPES
FLEXIBLE HOSES
FUEL PUMP STRAINER

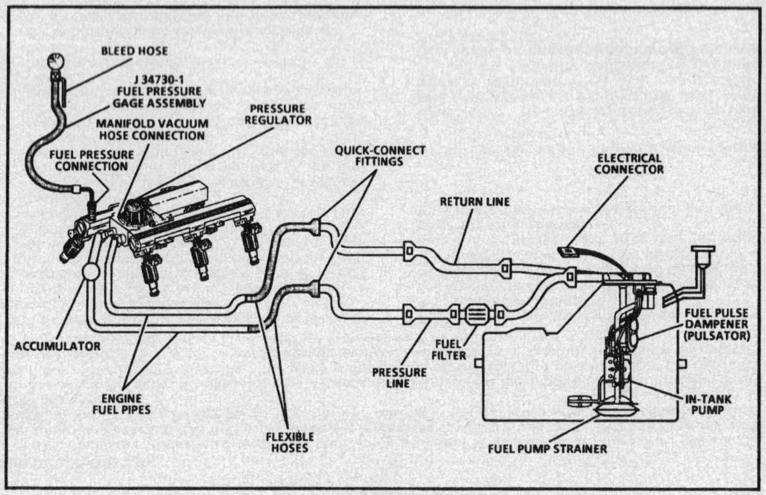

CHART A-7
(Page 2 of 3)
FUEL SYSTEM DIAGNOSIS
3.1L (VIN T)

Test Description: Number(s) below refer to circled number(s) on the diagnostic chart.

1. Pressure below 284 kPa (41 psi) may cause a lean condition and may set a Code 44. It could also cause hard starting cold and poor driveability. Low enough pressure will cause the engine not to run at all. Restricted flow may allow the engine to run at idle, or low speeds, but may cause a surge and stall when more fuel is required, as when accelerating or driving at high speeds.

2. Restricting the fuel return line allows fuel pressure to build above regulated pressure. With battery voltage applied to the pump "test" terminal, pressure should rise above 325 kPa (47 psi) as the fuel return hose is restricted.

NOTICE: Do not allow pressure to exceed 414 kPa (60 psi), as damage to the regulator may result.

3. This test determines if the high fuel pressure is due to a restricted fuel return line or a faulty pressure regulator. High fuel pressure may cause a rich condition and may set a Code 45.

3.1L (VIN T) ENGINE — SYSTEM DIAGNOSTIC CHARTS — CAMARO, FIREBIRD, BERETTA, CORSICA, CAVALIER AND SUNBIRD

CHART A-7
(Page 2 of 3)
FUEL SYSTEM DIAGNOSIS
3.1L (VIN T)

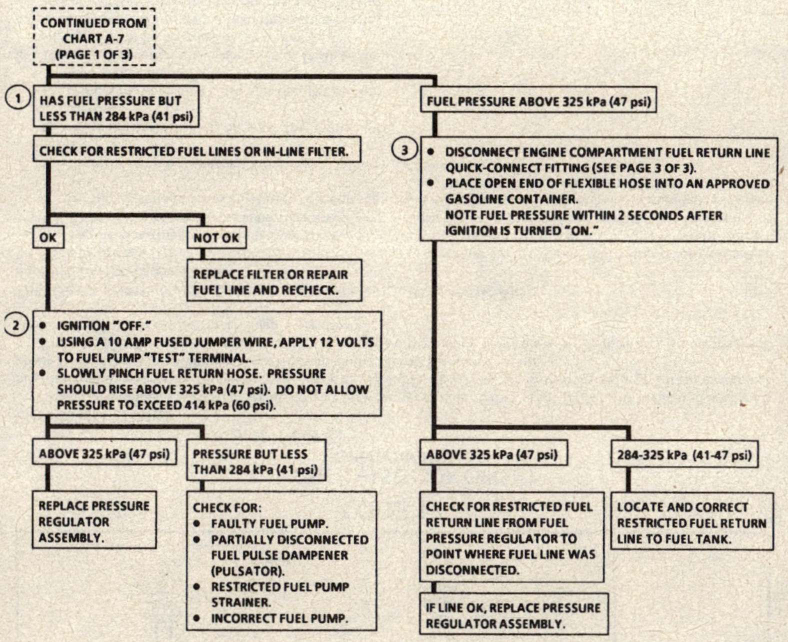

CONTINUED FROM CHART A-7 (PAGE 1 OF 3)

(1) HAS FUEL PRESSURE BUT LESS THAN 284 kPa (41 psi)

CHECK FOR RESTRICTED FUEL LINES OR IN-LINE FILTER.

OK

NOT OK

REPLACE FILTER OR REPAIR FUEL LINE AND RECHECK.

(2)
- IGNITION "OFF."
- USING A 10 AMP FUSED JUMPER WIRE, APPLY 12 VOLTS TO FUEL PUMP "TEST" TERMINAL.
- SLOWLY PINCH FUEL RETURN HOSE. PRESSURE SHOULD RISE ABOVE 325 kPa (47 psi). DO NOT ALLOW PRESSURE TO EXCEED 414 kPa (60 psi).

ABOVE 325 kPa (47 psi)

REPLACE PRESSURE REGULATOR ASSEMBLY.

PRESSURE BUT LESS THAN 284 kPa (41 psi)

CHECK FOR:
- FAULTY FUEL PUMP.
- PARTIALLY DISCONNECTED FUEL PULSE DAMPENER (PULSATOR).
- RESTRICTED FUEL PUMP STRAINER.
- INCORRECT FUEL PUMP.

FUEL PRESSURE ABOVE 325 kPa (47 psi)

(3)
- DISCONNECT ENGINE COMPARTMENT FUEL RETURN LINE QUICK-CONNECT FITTING (SEE PAGE 3 OF 3).
- PLACE OPEN END OF FLEXIBLE HOSE INTO AN APPROVED GASOLINE CONTAINER.
  NOTE FUEL PRESSURE WITHIN 2 SECONDS AFTER IGNITION IS TURNED "ON."

ABOVE 325 kPa (47 psi)

CHECK FOR RESTRICTED FUEL RETURN LINE FROM FUEL PRESSURE REGULATOR TO POINT WHERE FUEL LINE WAS DISCONNECTED.

IF LINE OK, REPLACE PRESSURE REGULATOR ASSEMBLY.

284-325 kPa (41-47 psi)

LOCATE AND CORRECT RESTRICTED FUEL RETURN LINE TO FUEL TANK.

"AFTER REPAIRS," CONFIRM "CLOSED LOOP" OPERATION AND NO "SERVICE ENGINE SOON" LIGHT.

3.1L (VIN T) ENGINE — SYSTEM DIAGNOSTIC CHARTS — CAMARO, FIREBIRD, BERETTA, CORSICA, CAVALIER AND SUNBIRD

CHART A-7
(Page 3 of 3)
FUEL SYSTEM PRESSURE TEST
3.1L (VIN T)

FUEL PRESSURE CHECK

(Instructions for Disconnecting and Reconnecting Return Line Quick-Connect Fitting)

Tools Required: J 37088-A - Fuel Line Quick-Connect Separators

1. Locate engine compartment fuel return line quick-connect fitting.
2. Grasp both ends of fitting, twist female end ¼ turn in each direction to loosen any dirt in fitting.

CAUTION: Safety glasses must be worn when using compressed air, as flying dirt particles may cause eye injury.

3. Using compressed air, blow dirt out of quick-connect fitting.
4. Using J 37088-2 from separator tool set J 37088-A, insert tool into female end of connector, push inward to release retaining tabs/fingers and pull connection apart.

CAUTION: To Reduce the Risk of Fire and Personal Injury: Before reconnecting the fuel return line quick-connect fitting, always apply a few drops of clean engine oil to the male tube end. This will ensure proper reconnection and prevent a possible fuel leak. (During normal operation, the O-rings located inside the female connector will swell and may prevent proper reconnection if not lubricated.)

5. After performing check, lubricate the male tube end of the fuel line with engine oil and reconnect quick-connect fitting.
   - Push connectors together to cause the retaining tabs/fingers to snap into place.
   - Once installed, pull on both ends of connection to make sure it is secure.
6. Cycle ignition "ON" and "OFF" twice, waiting ten seconds between cycles, then check for fuel leaks.

3.1L (VIN T) ENGINE — DIAGNOSTIC TROUBLE CODE CHART — CAMARO, FIREBIRD, BERETTA, CORSICA, CAVALIER AND SUNBIRD

ECM

412 PPL — GE14 — O₂ SENSOR SIGNAL

413 TAN — GE15 — O₂ SENSOR GROUND

OXYGEN (O₂) SENSOR

EXHAUST

ENGINE GROUND

CODE 13
OXYGEN (O₂) SENSOR CIRCUIT
(OPEN CIRCUIT)
3.1L (VIN T)

Circuit Description:

The ECM supplies a voltage of about .45 volt between terminals "GE14" and "GE15". (If measured with a 10 megohm digital voltmeter, this may read as low as .32 volt.) The O₂ sensor varies the voltage within a range of about 1 volt if the exhaust is rich, down through about .10 volt if exhaust is lean.

The sensor is like an open circuit and produces no voltage when it is below 315°C (600°F). An open sensor circuit or cold sensor causes "Open Loop" operation.

Test Description: Number(s) below refer to circled number(s) on the diagnostic chart.

1. Code 13 will set:
   - Engine at normal operating temperature.
   - At least 2 minutes engine time after start.
   - O₂ signal voltage steady between .35 and .55 volt.
   - Throttle Position Sensor (TPS) signal above 4%.
   - All conditions must be met for about 60 seconds.
     If the conditions for a Code 13 exist, the system will not go "Closed Loop."
2. This will determine if the sensor is at fault or the wiring or ECM is the cause of the Code 13.
3. In doing this test use only a high impedance digital volt ohmmeter. This test checks the continuity of CKT 412 and CKT 413 because if CKT 413 is open the ECM voltage on CKT 412 will be over .6 volt (600 mV).

Diagnostic Aids:

Normal Tech 1 "Scan" tool voltage varies between 100 mV to 999 mV (.1 and 1.0 volt) while in "Closed Loop." Code 13 sets in one minute if voltage remains between .35 and .55 volt, but the system will go "Open Loop" in about 15 seconds. Refer to "Intermittents" in "Symptoms,".

3.1L (VIN T) ENGINE — DIAGNOSTIC TROUBLE CODE CHART — CAMARO, FIREBIRD, BERETTA, CORSICA, CAVALIER AND SUNBIRD

CODE 13
OXYGEN (O₂) SENSOR CIRCUIT
(OPEN CIRCUIT)
3.1L (VIN T)

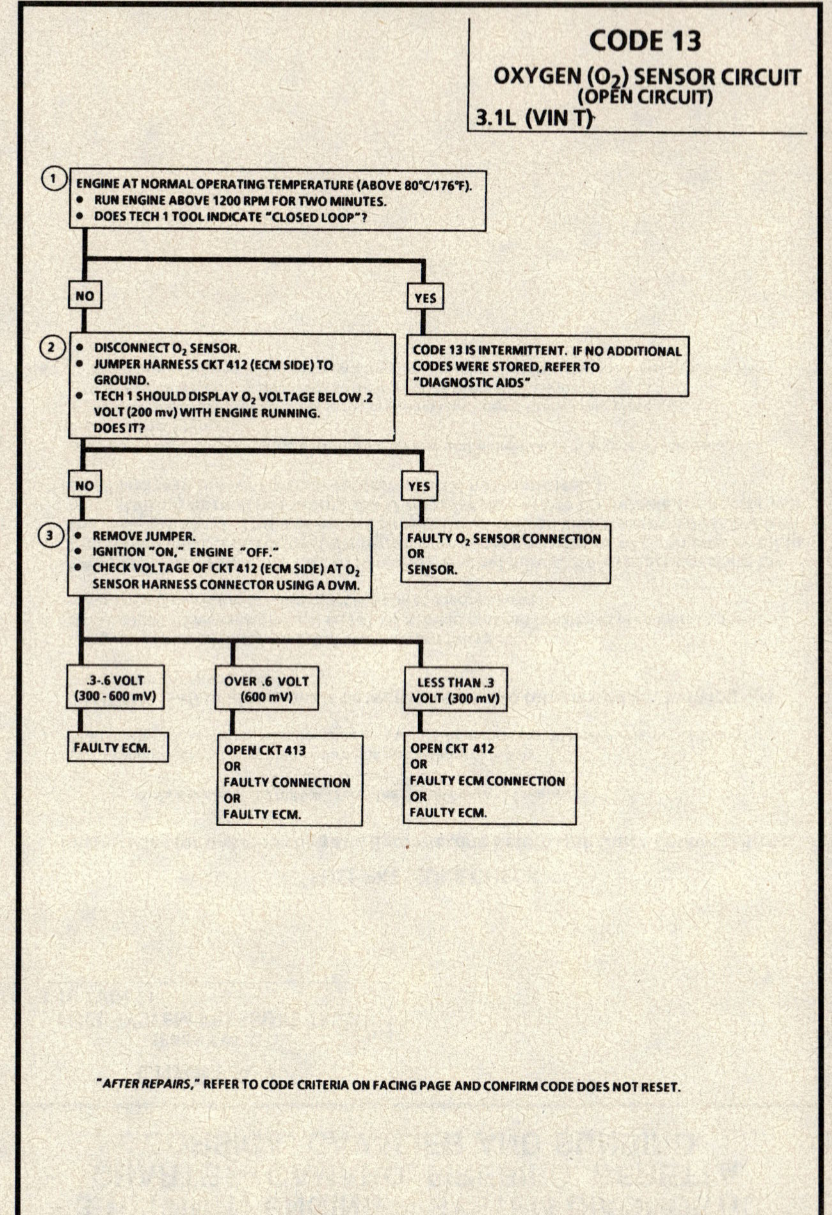

3.1L (VIN T) ENGINE — DIAGNOSTIC TROUBLE CODE CHART — CAMARO, FIREBIRD, BERETTA, CORSICA, CAVALIER AND SUNBIRD

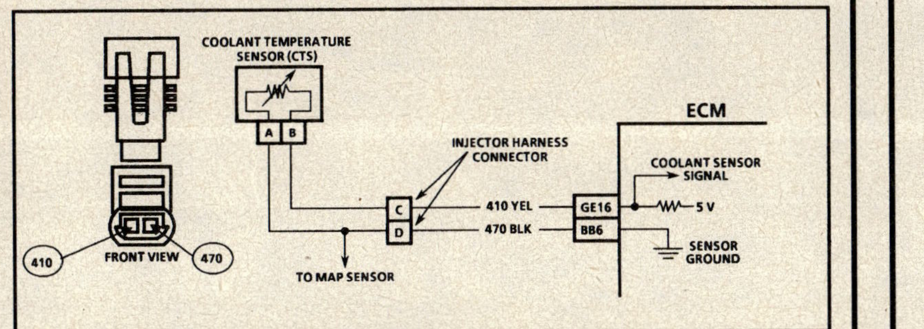
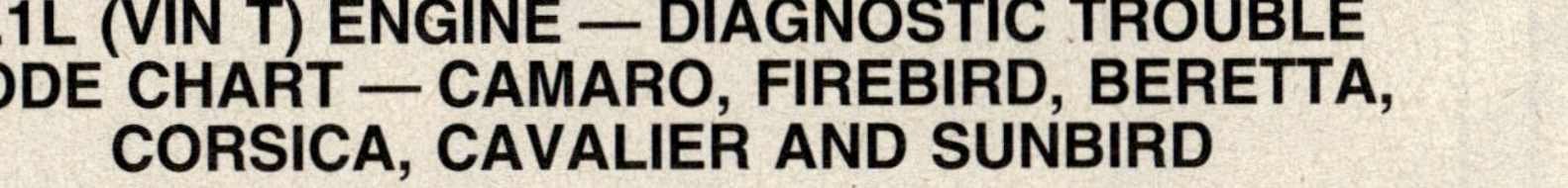

CODE 14
COOLANT TEMPERATURE SENSOR (CTS) CIRCUIT
(HIGH TEMPERATURE INDICATED)
3.1L (VIN T)

Circuit Description:

The Coolant Temperature Sensor (CTS) uses a thermistor to control the signal voltage to the ECM. The ECM applies a voltage on CKT 410 to the sensor. When the engine is cold, the sensor (thermistor) resistance is high, therefore, the ECM will see high signal voltage.

As the engine warms, the sensor resistance becomes less, and the voltage drops. At normal engine operating temperature (85°C to 95°C), the voltage will measure about 1.5 to 2.0 volts.

Test Description: Number(s) below refer to circled number(s) on the diagnostic chart.

1. Code 14 will set if:
   - Signal voltage indicates a coolant temperature above 135°C (275°F) for 20 seconds.
2. This test will determine if CKT 410 is shorted to ground which will cause the conditions for Code 14.

Diagnostic Aids:

Check harness routing for a potential short to ground in CKT 410. Tech 1 "Scan" tool displays engine temperature in degrees centigrade. After engine is started, the temperature should rise steadily to about 90°C, then stabilize when thermostat opens. Refer to "Intermittents" in "Symptoms."

3.1L (VIN T) ENGINE — DIAGNOSTIC TROUBLE CODE CHART — CAMARO, FIREBIRD, BERETTA, CORSICA, CAVALIER AND SUNBIRD

CODE 14
COOLANT TEMPERATURE SENSOR (CTS) CIRCUIT
(HIGH TEMPERATURE INDICATED)
3.1L (VIN T)

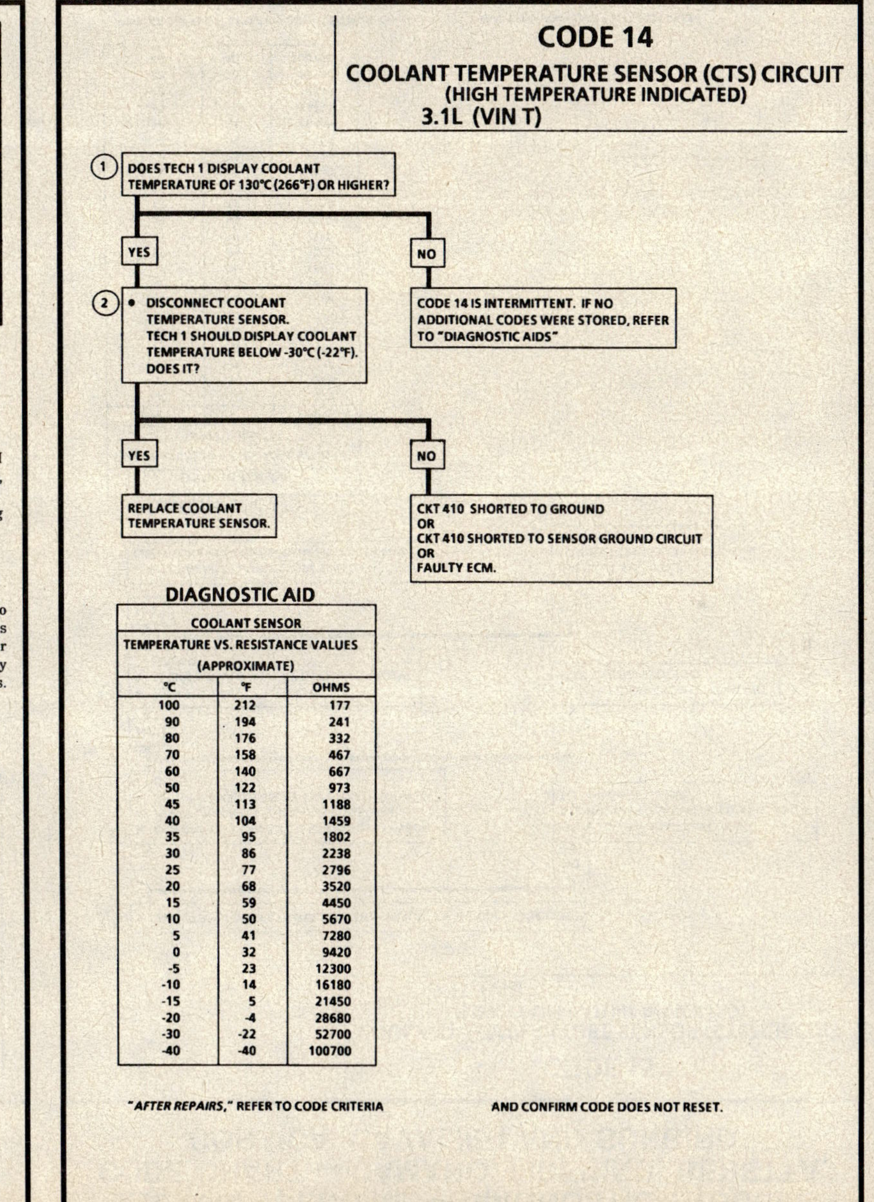

DIAGNOSTIC AID

COOLANT SENSOR		
TEMPERATURE VS. RESISTANCE VALUES (APPROXIMATE)		
°C	°F	OHMS
100	212	177
90	194	241
80	176	332
70	158	467
60	140	667
50	122	973
45	113	1188
40	104	1459
35	95	1802
30	86	2238
25	77	2796
20	68	3520
15	59	4450
10	50	5670
5	41	7280
0	32	9420
-5	23	12300
-10	14	16180
-15	5	21450
-20	-4	28680
-30	-22	52700
-40	-40	100700

"AFTER REPAIRS," REFER TO CODE CRITERIA   AND CONFIRM CODE DOES NOT RESET.

3.1L (VIN T) ENGINE — DIAGNOSTIC TROUBLE CODE CHART — CAMARO, FIREBIRD, BERETTA, CORSICA, CAVALIER AND SUNBIRD

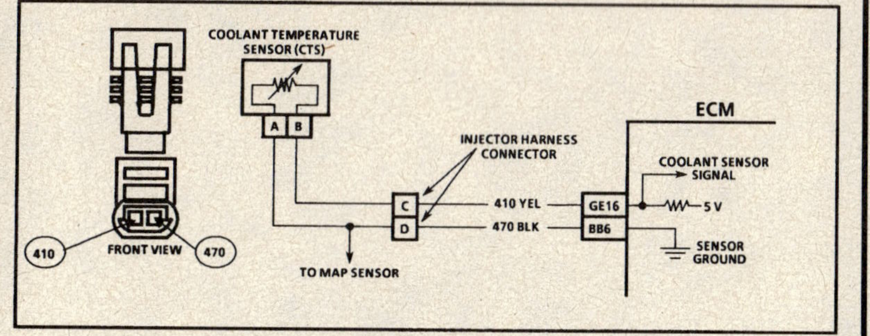

CODE 15
COOLANT TEMPERATURE SENSOR (CTS) CIRCUIT
(LOW TEMPERATURE INDICATED)
3.1L (VIN T)

Circuit Description:

The Coolant Temperature Sensor (CTS) uses a thermistor to control the signal voltage to the ECM. The ECM applies a voltage on CKT 410 to the sensor. When the engine is cold, the sensor (thermistor) resistance is high, therefore, the ECM will see high signal voltage.

As the engine warms, the sensor resistance becomes less and the voltage drops. At normal engine operating temperature (85°C to 95°C), the voltage will measure about 1.5 to 2.0 volts at the ECM.

Test Description: Number(s) below refer to circled number(s) on the diagnostic chart.

1. Code 15 will set if:
   - Signal voltage indicates a coolant temperature less than -39°C (-38°F) for 20 seconds.
2. This test simulates a Code 14. If the ECM recognizes the low signal voltage (high temperature), and the "Scan" reads 130°C, the ECM and wiring are OK.
3. This test will determine if CKT 410 is open. There should be 5 volts present at sensor connector, if measured with a DVOM.

Diagnostic Aids:

A Tech 1 "Scan" tool reads engine temperature in degrees centigrade. After engine is started, the temperature should rise steadily to about 90°C then stabilize when thermostat opens.

A faulty connection, or an open in CKT 410 or CKT 470 will result in a Code 15.

If Code 23 or Code 33 is also set, check CKT 470 for faulty wiring or connections. Check terminals at sensor for good contact. Refer to "Intermittents" in "Symptoms,"

3.1L (VIN T) ENGINE — DIAGNOSTIC TROUBLE CODE CHART — CAMARO, FIREBIRD, BERETTA, CORSICA, CAVALIER AND SUNBIRD

CODE 15
COOLANT TEMPERATURE SENSOR (CTS) CIRCUIT
(LOW TEMPERATURE INDICATED)
3.1L (VIN T)

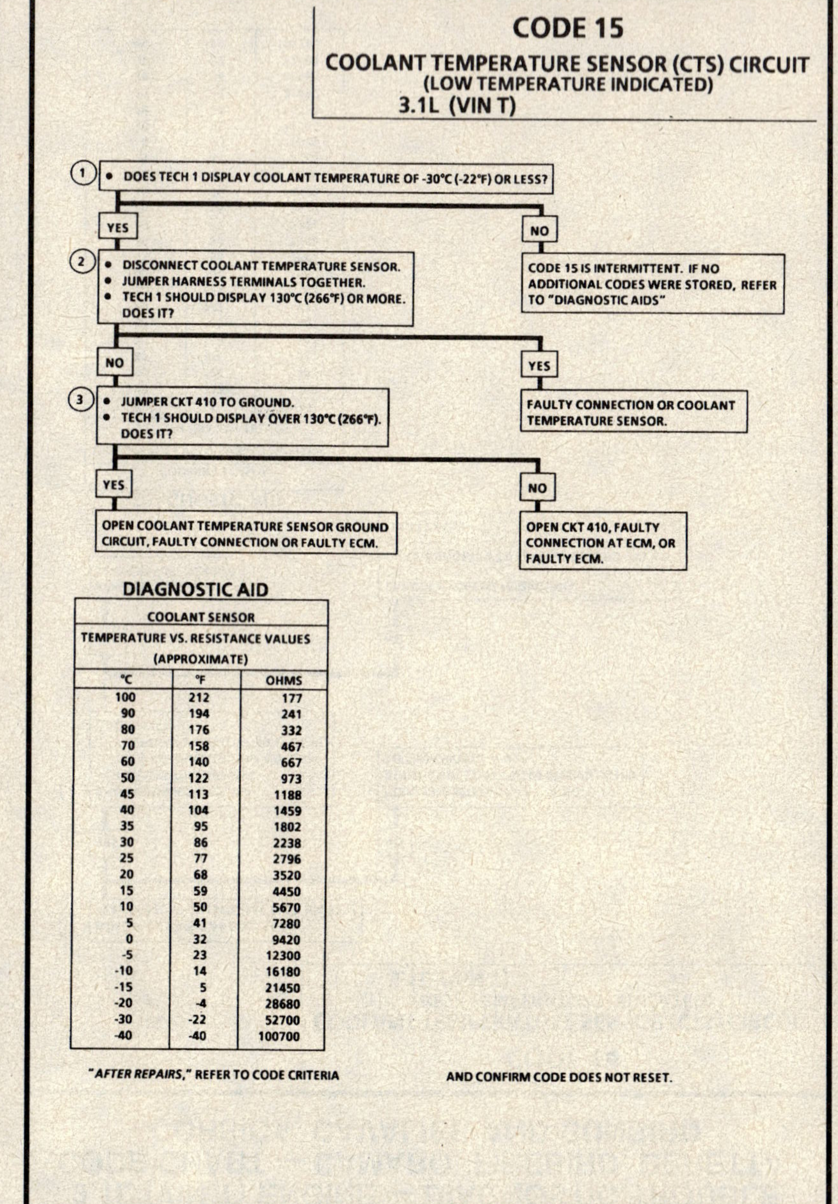

DIAGNOSTIC AID

COOLANT SENSOR		
TEMPERATURE VS. RESISTANCE VALUES (APPROXIMATE)		
°C	°F	OHMS
100	212	177
90	194	241
80	176	332
70	158	467
60	140	667
50	122	973
45	113	1188
40	104	1459
35	95	1802
30	86	2238
25	77	2796
20	68	3520
15	59	4450
10	50	5670
5	41	7280
0	32	9420
-5	23	12300
-10	14	16180
-15	5	21450
-20	-4	28680
-30	-22	52700
-40	-40	100700

"AFTER REPAIRS," REFER TO CODE CRITERIA

AND CONFIRM CODE DOES NOT RESET.

3.1L (VIN T) ENGINE — DIAGNOSTIC TROUBLE CODE CHART — CAMARO, FIREBIRD, BERETTA, CORSICA, CAVALIER AND SUNBIRD

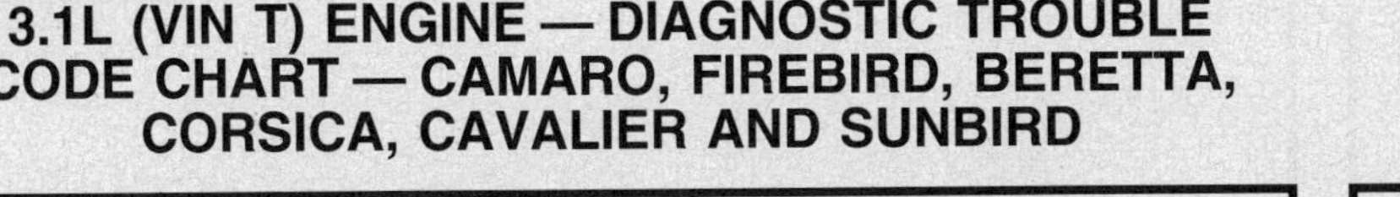

CODE 21
THROTTLE POSITION SENSOR (TPS) CIRCUIT
(SIGNAL VOLTAGE HIGH)
3.1L (VIN T)

Circuit Description:

The Throttle Position Sensor (TPS) provides a voltage signal that changes relative to the throttle blade. Signal voltage will vary from about .29 volt at idle to about 4.8 volts at Wide Open Throttle (WOT).

The TPS signal is one of the most important inputs used by the ECM for fuel control and for most of the ECM control outputs.

Test Description: Number(s) below refer to circled number(s) on the diagnostic chart.

1. Code 21 will set if:
   - Engine is running.
   - TPS signal voltage is greater than 3.8 volts.
   - All conditions met for 10 seconds.
   - Air flow is less than 17 gm/sec.
   OR
   - TPS check: The TPS has an auto zeroing feature. If the voltage reading is within the range of 0.29 to 0.98 volt, the ECM will use that value as closed throttle. If the voltage reading is out of the auto zero range on an existing or replacement TPS, make sure the cruise control and throttle cables are not being held open.

2. With the TPS sensor disconnected, the TPS voltage should go low if the ECM and wiring are OK.
3. Probing CKT 452 with a test light checks the 5 volts return circuit, because a faulty 5 volts return will cause a Code 21.

Diagnostic Aids:

Tech 1 "Scan" tool reads throttle position in volts. Voltage should increase at a steady rate as throttle is moved toward WOT.

Tech 1 "Scan" tool will read throttle angle 0% = closed throttle, 100% = WOT.

An open in CKT 452 will result in a Code 21. Refer to "Intermittents" in "Symptoms,"

3.1L (VIN T) ENGINE — DIAGNOSTIC TROUBLE CODE CHART — CAMARO, FIREBIRD, BERETTA, CORSICA, CAVALIER AND SUNBIRD

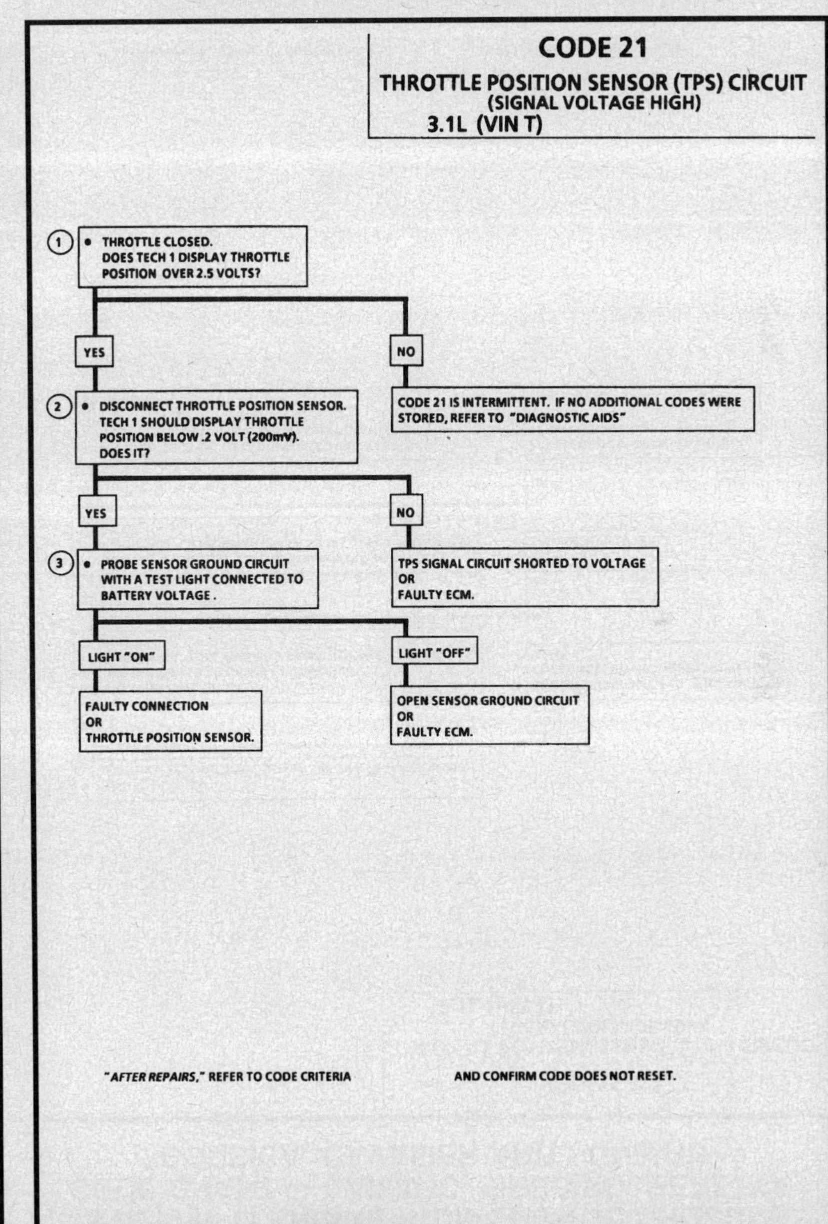

3.1L (VIN T) ENGINE — DIAGNOSTIC TROUBLE CODE CHART — CAMARO, FIREBIRD, BERETTA, CORSICA, CAVALIER AND SUNBIRD

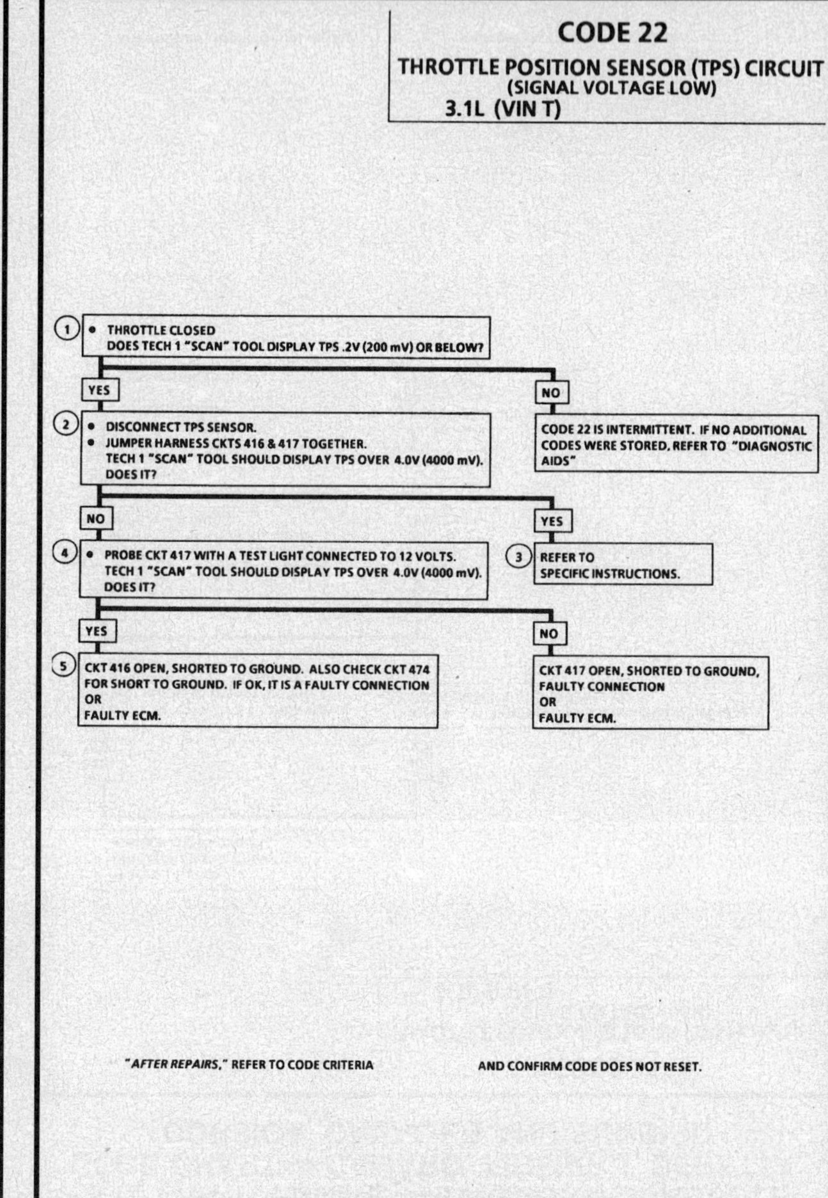

CODE 22
THROTTLE POSITION SENSOR (TPS) CIRCUIT
(SIGNAL VOLTAGE LOW)
3.1L (VIN T)

Circuit Description:

The Throttle Position Sensor (TPS) provides a voltage signal that changes relative to the throttle blade. Signal voltage will vary from about .29 volt at idle to about 4.8 volts at Wide Open Throttle (WOT).

The TPS signal is one of the most important inputs used by the ECM for fuel control and for most of the ECM control outputs.

Test Description: Number(s) below refer to circled number(s) on the diagnostic chart.

1. Code 22 will set if:
   - Engine running.
   - TPS signal voltage is less than about .2 volt.
2. Simulates Code 21: (high voltage) if the ECM recognizes the high signal voltage, the ECM and wiring are OK.
3. TPS check: The TPS has an auto zeroing feature. If the voltage reading is within the range of 0.29 to 0.98 volt, the ECM will use that value as closed throttle. If the voltage reading is out of the auto zero range on an existing or replacement TPS, make sure the cruise control and throttle cables are not being held open.
4. This simulates a high signal voltage to check for an open in CKT 417.

5. CKT 416 and CKT 474 share a common sensor ground buffered reference signal. If either of these circuits is shorted to ground, Code 22 will set. To determine if the MAP sensor is causing the Code 22 problem, disconnect it to see if Code 22 resets. Be sure TPS is connected and clear codes before testing.

Diagnostic Aids:

Tech 1 "Scan" tool reads throttle position in volts. Voltage should increase at a steady rate as throttle is moved toward WOT.

Tech 1 "Scan" tool will read: throttle angle 0% = closed throttle, 100% = WOT.

An open or short to ground in CKT 416 or CKT 417 will result in a Code 22. Also, a short to ground in CKT 474 will result in a Code 22.

Refer to "Intermittents" in "Symptoms,"

3.1L (VIN T) ENGINE — DIAGNOSTIC TROUBLE CODE CHART — CAMARO, FIREBIRD, BERETTA, CORSICA, CAVALIER AND SUNBIRD

CODE 22
THROTTLE POSITION SENSOR (TPS) CIRCUIT
(SIGNAL VOLTAGE LOW)
3.1L (VIN T)

1. • THROTTLE CLOSED
   DOES TECH 1 "SCAN" TOOL DISPLAY TPS .2V (200 mV) OR BELOW?

   YES →
   NO → CODE 22 IS INTERMITTENT. IF NO ADDITIONAL CODES WERE STORED, REFER TO "DIAGNOSTIC AIDS"

2. • DISCONNECT TPS SENSOR.
   • JUMPER HARNESS CKTS 416 & 417 TOGETHER.
   TECH 1 "SCAN" TOOL SHOULD DISPLAY TPS OVER 4.0V (4000 mV).
   DOES IT?

   NO →
   YES →

4. • PROBE CKT 417 WITH A TEST LIGHT CONNECTED TO 12 VOLTS.
   TECH 1 "SCAN" TOOL SHOULD DISPLAY TPS OVER 4.0V (4000 mV).
   DOES IT?

3. REFER TO SPECIFIC INSTRUCTIONS.

   YES →
   NO →

5. CKT 416 OPEN, SHORTED TO GROUND. ALSO CHECK CKT 474 FOR SHORT TO GROUND. IF OK, IT IS A FAULTY CONNECTION OR FAULTY ECM.

   CKT 417 OPEN, SHORTED TO GROUND, FAULTY CONNECTION OR FAULTY ECM.

"AFTER REPAIRS," REFER TO CODE CRITERIA   AND CONFIRM CODE DOES NOT RESET.

3.1L (VIN T) ENGINE — DIAGNOSTIC TROUBLE CODE CHART — CAMARO, FIREBIRD, BERETTA, CORSICA, CAVALIER AND SUNBIRD

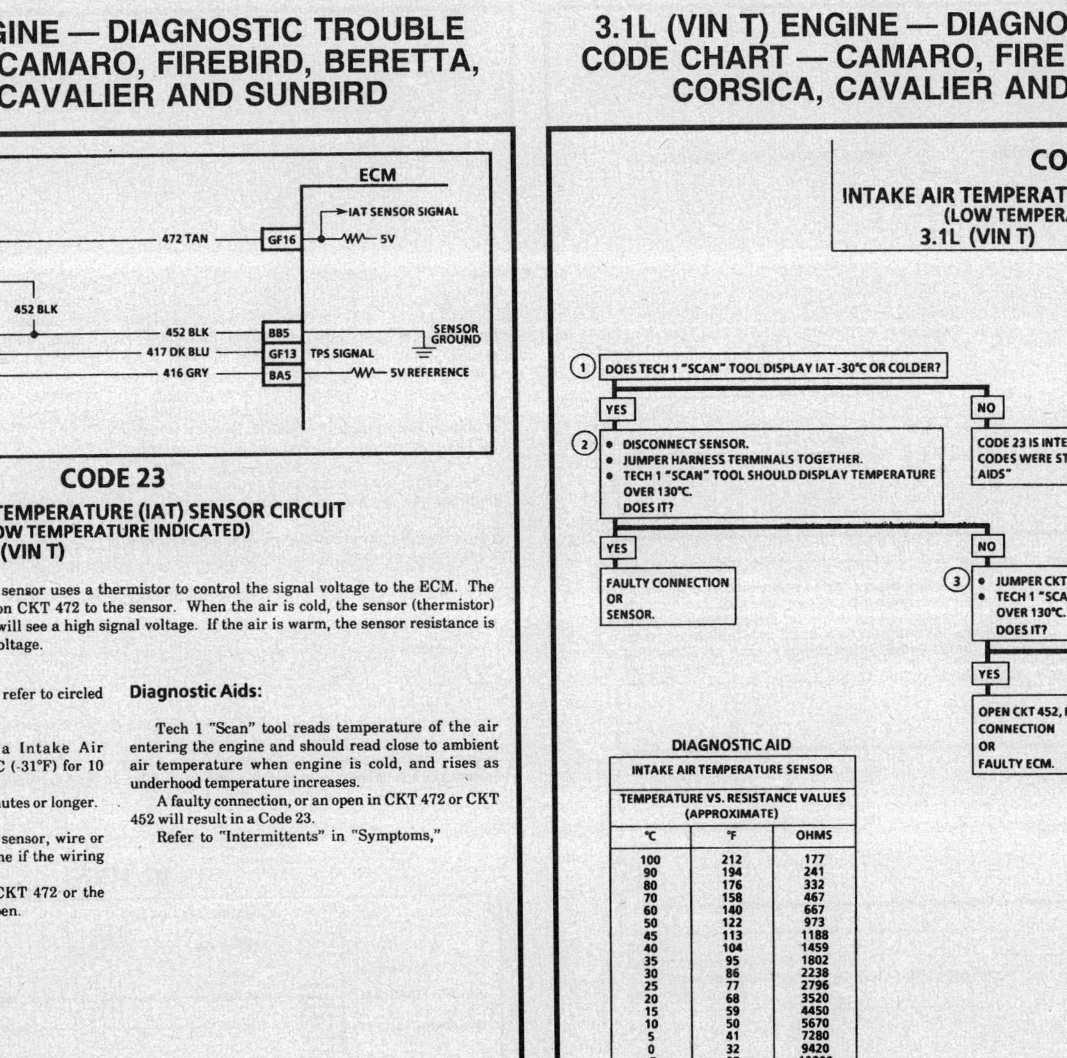

CODE 23
INTAKE AIR TEMPERATURE (IAT) SENSOR CIRCUIT
(LOW TEMPERATURE INDICATED)
3.1L (VIN T)

Circuit Description:

The Intake Air Temperature (IAT) sensor uses a thermistor to control the signal voltage to the ECM. The ECM applies a voltage (about 5 volts) on CKT 472 to the sensor. When the air is cold, the sensor (thermistor) resistance is high, therefore, the ECM will see a high signal voltage. If the air is warm, the sensor resistance is low, therefore, the ECM will see a low voltage.

Test Description: Number(s) below refer to circled number(s) on the diagnostic chart.

1. Code 23 will set if:
   - A signal voltage indicates a Intake Air Temperature (IAT) below -35°C (-31°F) for 10 seconds.
   - Time since engine start is 4 minutes or longer.
   - No VSS.
2. A Code 23 will set, due to an open sensor, wire or connection. This test will determine if the wiring and ECM are OK.
3. This will determine if the signal CKT 472 or the sensor ground return CKT 452 is open.

Diagnostic Aids:

Tech 1 "Scan" tool reads temperature of the air entering the engine and should read close to ambient air temperature when engine is cold, and rises as underhood temperature increases.

A faulty connection, or an open in CKT 472 or CKT 452 will result in a Code 23.

Refer to "Intermittents" in "Symptoms,"

3.1L (VIN T) ENGINE — DIAGNOSTIC TROUBLE CODE CHART — CAMARO, FIREBIRD, BERETTA, CORSICA, CAVALIER AND SUNBIRD

CODE 23
INTAKE AIR TEMPERATURE (IAT) SENSOR CIRCUIT
(LOW TEMPERATURE INDICATED)
3.1L (VIN T)

1. DOES TECH 1 "SCAN" TOOL DISPLAY IAT -30°C OR COLDER?

YES →

2.
   - DISCONNECT SENSOR.
   - JUMPER HARNESS TERMINALS TOGETHER.
   - TECH 1 "SCAN" TOOL SHOULD DISPLAY TEMPERATURE OVER 130°C. DOES IT?

NO → CODE 23 IS INTERMITTENT. IF NO ADDITIONAL CODES WERE STORED, REFER TO "DIAGNOSTIC AIDS"

YES → FAULTY CONNECTION OR SENSOR.

NO →

3.
   - JUMPER CKT 472 TO GROUND.
   - TECH 1 "SCAN" TOOL SHOULD DISPLAY TEMPERATURE OVER 130°C. DOES IT?

YES → OPEN CKT 452, FAULTY CONNECTION OR FAULTY ECM.

NO → OPEN CKT 472, FAULTY CONNECTION OR FAULTY ECM.

DIAGNOSTIC AID

INTAKE AIR TEMPERATURE SENSOR		
TEMPERATURE VS. RESISTANCE VALUES (APPROXIMATE)		
°C	°F	OHMS
100	212	177
90	194	241
80	176	332
70	158	467
60	140	667
50	122	973
45	113	1188
40	104	1459
35	95	1802
30	86	2238
25	77	2796
20	68	3520
15	59	4450
10	50	5670
5	41	7280
0	32	9420
-5	23	12300
-10	14	16180
-15	5	21450
-20	-4	28680
-30	-22	52700
-40	-40	100700

"AFTER REPAIRS," REFER TO CODE CRITERIA   AND CONFIRM CODE DOES NOT RESET.

3.1L (VIN T) ENGINE — DIAGNOSTIC TROUBLE CODE CHART — CAMARO, FIREBIRD, BERETTA, CORSICA, CAVALIER AND SUNBIRD

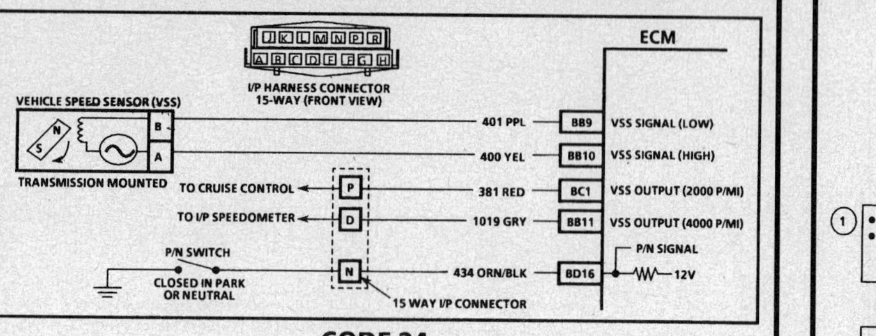

CODE 24
VEHICLE SPEED SENSOR (VSS) CIRCUIT
3.1L (VIN T)

Circuit Description:

Vehicle speed information is provided to the ECM by the Vehicle Speed Sensor (VSS), which is a Permanent Magnet (PM) generator and it is mounted in the transmission. The PM generator produces a pulsing voltage, whenever vehicle speed is over 3 pulse per mile. The A/C voltage level and the number of pulses increases with vehicle speed. The ECM then converts the pulsing voltage to mph which is used for calculations, and the mph can be displayed with a Tech 1 "Scan" tool.

The function of VSS buffer used in past model years has been incorporated into the ECM. The ECM then supplies the necessary signal for the instrument panel (4000 pulses per mile) for operating the speedometer and the odometer. If the vehicle is equipped with cruise control, the ECM also provides a signal (2000 pulses per mile) to the cruise control module.

Test Description: Number(s) below refer to circled number(s) on the diagnostic chart.

1. Code 24 will set if vehicle speed equals 0 mph when:
   - Coolant temperature is greater than 85°C.
   - Engine speed is between 2200 and 4400 rpm.
   - TPS is less than 2%.
   - Low load condition (low air flow).
   - Not in park or neutral.
   - All conditions met for 3 seconds.

   These conditions are met during a road load deceleration. Disregard Code 24 that sets when drive wheels are not turning.
   - The PM generator only produces a signal if drive wheels are turning greater than 3 mph.

2. CKTs 400, 401 and 381 are OK if the speedometer works properly. Code 24 is being caused by a faulty ECM, faulty MEM-CAL or an incorrect MEM-CAL.

Diagnostic Aids:

Tech 1 "Scan" tool should indicate a vehicle speed whenever the drive wheels are turning greater than 3 mph.

A problem in CKT 437 or CKT 381 will not affect the VSS input or the readings on a Tech 1 "Scan" tool.

Check CKT 400 and CKT 401 for proper connections to be sure they're clean and tight and the harness is routed correctly. Refer to "Intermittents" in "Symptoms," Section

(A/T) A faulty or misadjusted Park/Neutral (P/N) switch can result in a false Code 24. Use a Tech 1 "Scan" tool and check for proper signal while in drive or overdrive. Refer to CHART C-1A for P/N switch diagnosis check.

3.1L (VIN T) ENGINE — DIAGNOSTIC TROUBLE CODE CHART — CAMARO, FIREBIRD, BERETTA, CORSICA, CAVALIER AND SUNBIRD

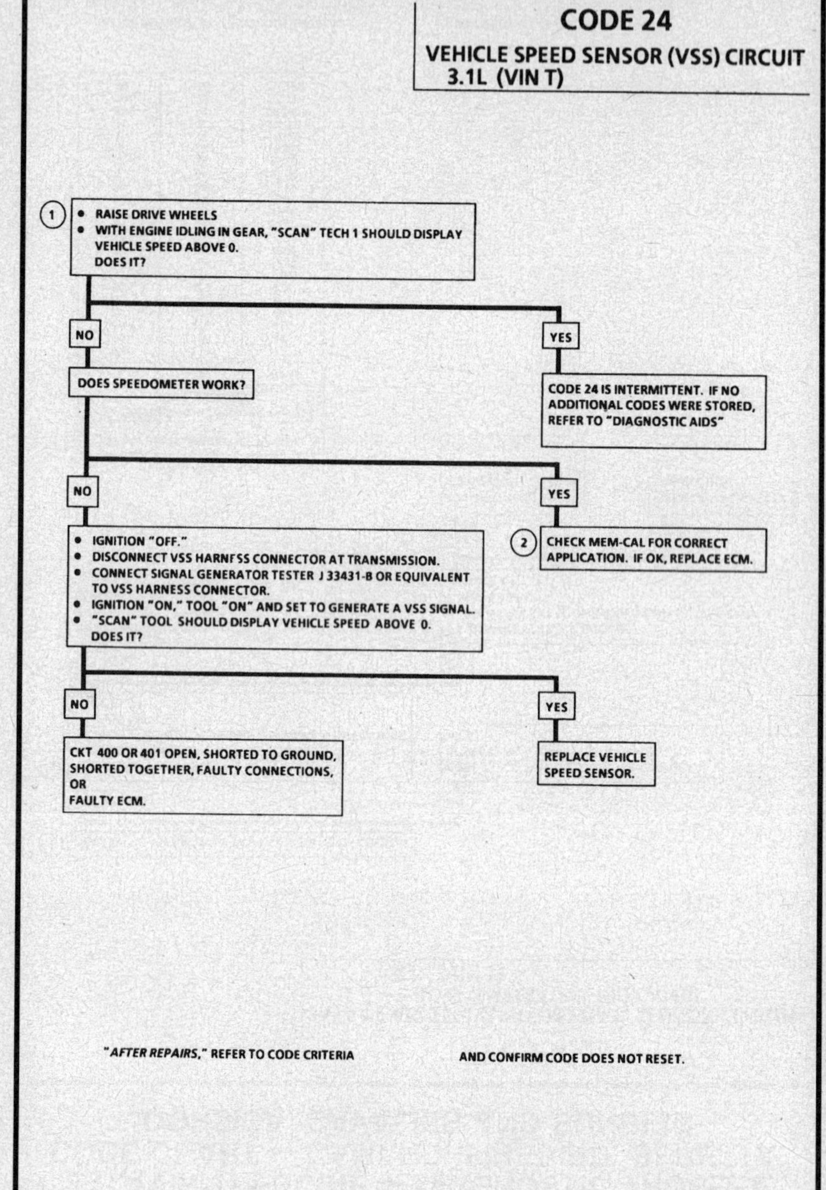

3.1L (VIN T) ENGINE — DIAGNOSTIC TROUBLE CODE CHART — CAMARO, FIREBIRD, BERETTA, CORSICA, CAVALIER AND SUNBIRD

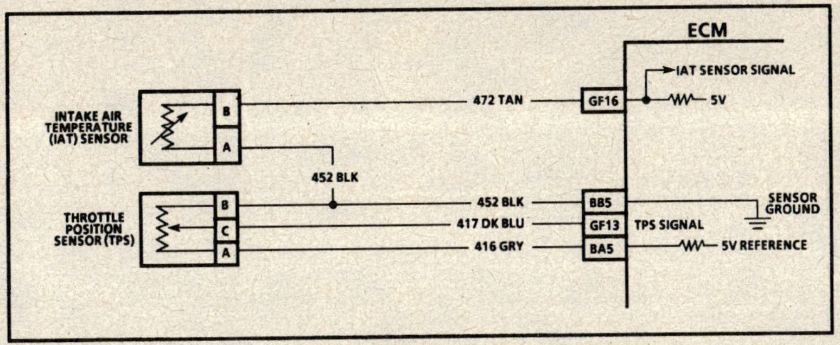

CODE 25
INTAKE AIR TEMPERATURE (IAT) SENSOR CIRCUIT
(HIGH TEMPERATURE INDICATED)
3.1L (VIN T)

Circuit Description:

The Intake Air Temperature (IAT) sensor uses a thermistor to control the signal voltage to the ECM. The ECM applies a voltage (about 5 volts) on CKT 472 to the sensor. When intake air is cold, the sensor (thermistor) resistance is high, therefore, the ECM will see a high signal voltage. As the air warms, the sensor resistance becomes less, and the voltage drops.

Test Description: Number(s) below refers to circled number(s) on the diagnostic chart.

1. Code 25 will set if:
   - IAT signal voltage is less than or equal to .04 volts.
   - Signal voltage indicates an Intake Air Temperature (IAT) greater than 145°C (293°F) for .2 second.
   - Time since engine start is 4 minutes or longer.
   - A vehicle speed is present.

Diagnostic Aids:

Tech 1 "Scan" tool reads temperature of the air entering the engine and should read close to ambient air temperature when engine is cold and rises as underhood temperature increases.

A short to ground in CKT 472 will result in a Code 25.

Refer to "Intermittents" in "Symptoms," Section

3.1L (VIN T) ENGINE — DIAGNOSTIC TROUBLE CODE CHART — CAMARO, FIREBIRD, BERETTA, CORSICA, CAVALIER AND SUNBIRD

CODE 25
INTAKE AIR TEMPERATURE (IAT) SENSOR CIRCUIT
(HIGH TEMPERATURE INDICATED)
3.1L (VIN T)

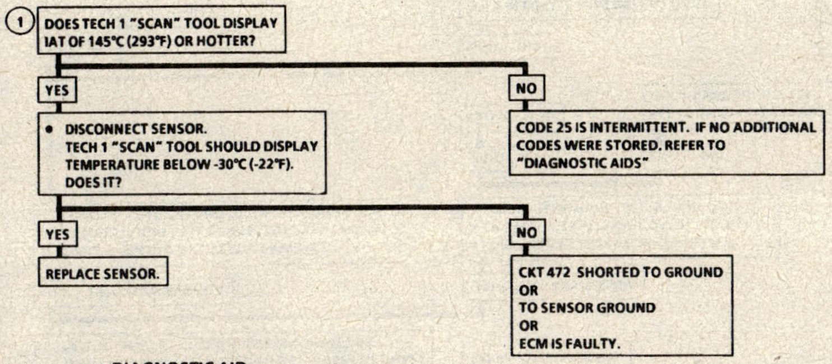

DIAGNOSTIC AID

INTAKE AIR TEMPERATURE SENSOR		
TEMPERATURE VS. RESISTANCE VALUES (APPROXIMATE)		
°C	°F	OHMS
100	212	177
90	194	241
80	176	332
70	158	467
60	140	667
50	122	973
45	113	1188
40	104	1459
35	95	1802
30	86	2238
25	77	2796
20	68	3520
15	59	4450
10	50	5670
5	41	7280
0	32	9420
-5	23	12300
-10	14	16180
-15	5	21450
-20	-4	28680
-30	-22	52700
-40	-40	100700

"AFTER REPAIRS," REFER TO CODE CRITERIA   AND CONFIRM CODE DOES NOT RESET.

3.1L (VIN T) ENGINE — DIAGNOSTIC TROUBLE CODE CHART — CAMARO, FIREBIRD, BERETTA, CORSICA, CAVALIER AND SUNBIRD

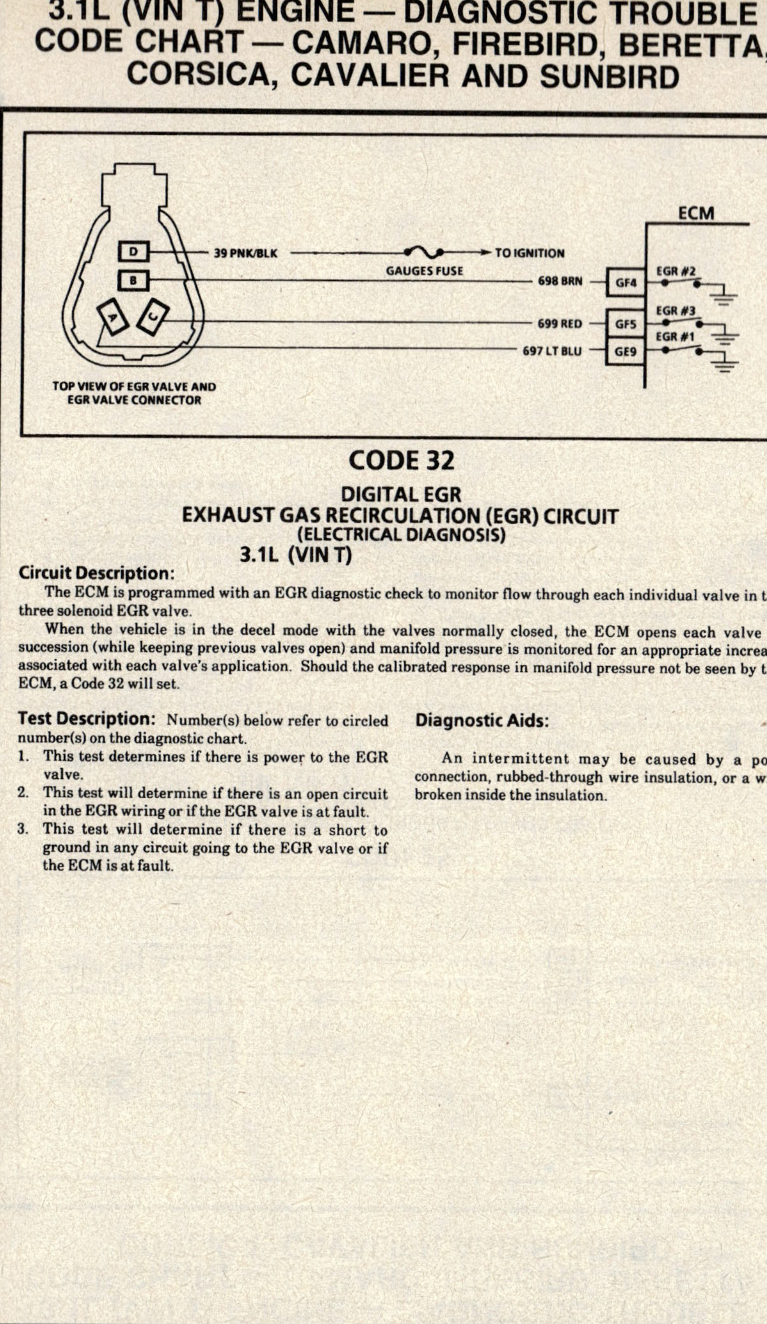

CODE 32

DIGITAL EGR
EXHAUST GAS RECIRCULATION (EGR) CIRCUIT
(ELECTRICAL DIAGNOSIS)
3.1L (VIN T)

Circuit Description:

The ECM is programmed with an EGR diagnostic check to monitor flow through each individual valve in the three solenoid EGR valve.

When the vehicle is in the decel mode with the valves normally closed, the ECM opens each valve in succession (while keeping previous valves open) and manifold pressure is monitored for an appropriate increase associated with each valve's application. Should the calibrated response in manifold pressure not be seen by the ECM, a Code 32 will set.

Test Description: Number(s) below refer to circled number(s) on the diagnostic chart.

1. This test determines if there is power to the EGR valve.
2. This test will determine if there is an open circuit in the EGR wiring or if the EGR valve is at fault.
3. This test will determine if there is a short to ground in any circuit going to the EGR valve or if the ECM is at fault.

Diagnostic Aids:

An intermittent may be caused by a poor connection, rubbed-through wire insulation, or a wire broken inside the insulation.

3.1L (VIN T) ENGINE — DIAGNOSTIC TROUBLE CODE CHART — CAMARO, FIREBIRD, BERETTA, CORSICA, CAVALIER AND SUNBIRD

3.1L (VIN T) ENGINE — DIAGNOSTIC TROUBLE CODE CHART — CAMARO, FIREBIRD, BERETTA, CORSICA, CAVALIER AND SUNBIRD

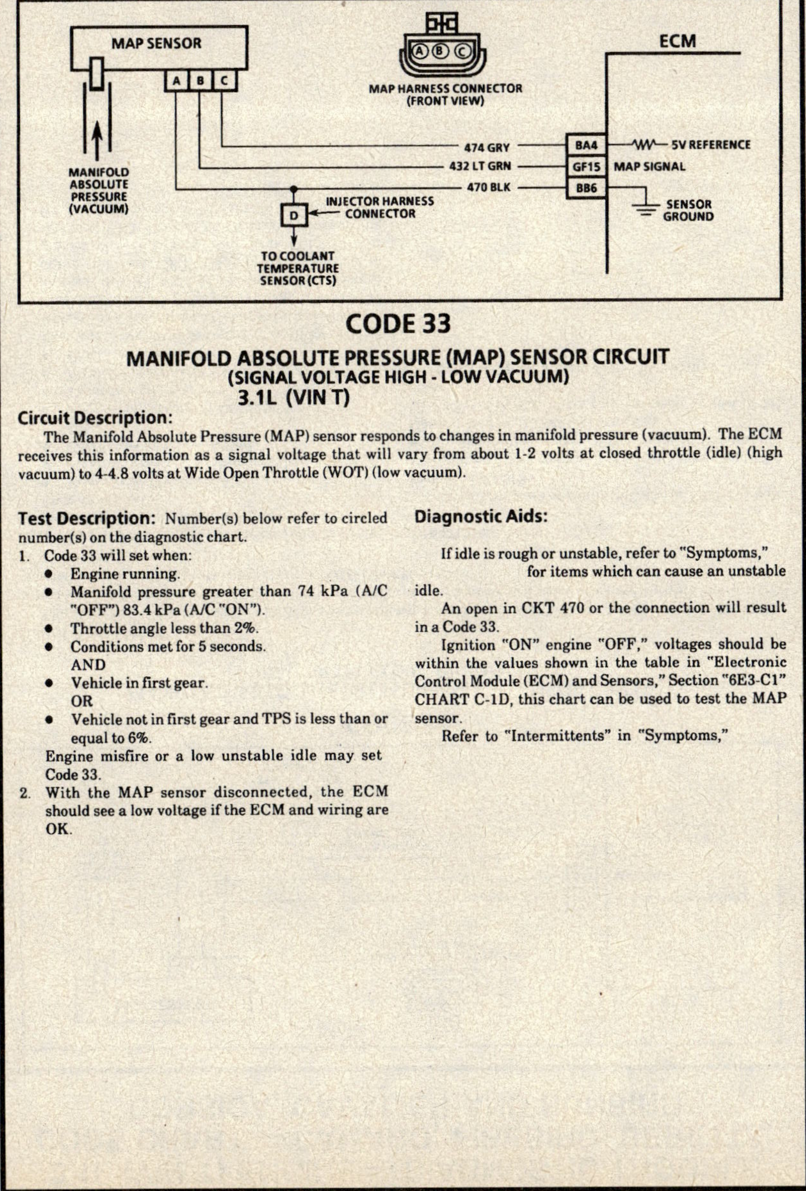

CODE 33
MANIFOLD ABSOLUTE PRESSURE (MAP) SENSOR CIRCUIT
(SIGNAL VOLTAGE HIGH - LOW VACUUM)
3.1L (VIN T)

Circuit Description:
The Manifold Absolute Pressure (MAP) sensor responds to changes in manifold pressure (vacuum). The ECM receives this information as a signal voltage that will vary from about 1-2 volts at closed throttle (idle) (high vacuum) to 4-4.8 volts at Wide Open Throttle (WOT) (low vacuum).

Test Description: Number(s) below refer to circled number(s) on the diagnostic chart.
1. Code 33 will set when:
   - Engine running.
   - Manifold pressure greater than 74 kPa (A/C "OFF") 83.4 kPa (A/C "ON").
   - Throttle angle less than 2%.
   - Conditions met for 5 seconds.
     AND
   - Vehicle in first gear.
     OR
   - Vehicle not in first gear and TPS is less than or equal to 6%.

   Engine misfire or a low unstable idle may set Code 33.
2. With the MAP sensor disconnected, the ECM should see a low voltage if the ECM and wiring are OK.

Diagnostic Aids:
If idle is rough or unstable, refer to "Symptoms," for items which can cause an unstable idle.

An open in CKT 470 or the connection will result in a Code 33.

Ignition "ON" engine "OFF," voltages should be within the values shown in the table in "Electronic Control Module (ECM) and Sensors," Section "6E3-C1" CHART C-1D, this chart can be used to test the MAP sensor.

Refer to "Intermittents" in "Symptoms,"

3.1L (VIN T) ENGINE — DIAGNOSTIC TROUBLE CODE CHART — CAMARO, FIREBIRD, BERETTA, CORSICA, CAVALIER AND SUNBIRD

CODE 33
MANIFOLD ABSOLUTE PRESSURE (MAP) SENSOR CIRCUIT
(SIGNAL VOLTAGE HIGH - LOW VACUUM)
3.1L (VIN T)

1.
- IF ENGINE IDLE IS ROUGH, UNSTABLE OR INCORRECT, CORRECT BEFORE USING CHART. SEE "SYMPTOMS" IN SECTION "B".
- ENGINE IDLING.
  DOES TECH 1 "SCAN" TOOL DISPLAY A MAP OF 3.75 VOLTS OR OVER?

YES → 2.
- IGNITION "OFF."
- DISCONNECT MAP SENSOR ELECTRICAL CONNECTOR.
- IGNITION "ON."
  TECH 1 "SCAN" TOOL SHOULD READ A VOLTAGE OF 1 VOLT OR LESS.
  DOES IT?

NO → CODE 33 IS INTERMITTENT. IF NO ADDITIONAL CODES WERE STORED, REFER TO "DIAGNOSTIC AIDS"

YES →
- PROBE CKT 470 WITH A TEST LIGHT TO 12 VOLTS. TEST LIGHT SHOULD LIGHT.
  DOES IT?

NO → CKT 432 SHORTED TO VOLTAGE, SHORTED TO CKT 474 OR FAULTY ECM.

YES → PLUGGED OR LEAKING MAP SENSOR VACUUM HOSE OR FAULTY MAP SENSOR.

NO → OPEN CKT 470.

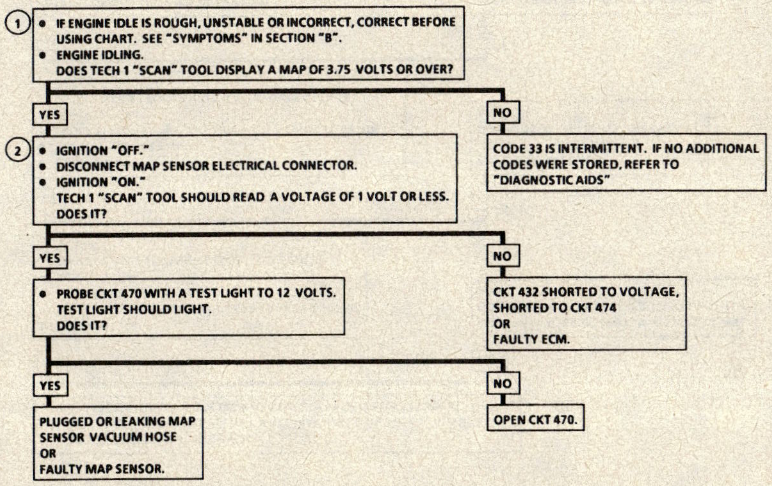

"AFTER REPAIRS," REFER TO CODE CRITERIA   AND CONFIRM CODE DOES NOT RESET.

3.1L (VIN T) ENGINE — DIAGNOSTIC TROUBLE CODE CHART — CAMARO, FIREBIRD, BERETTA, CORSICA, CAVALIER AND SUNBIRD

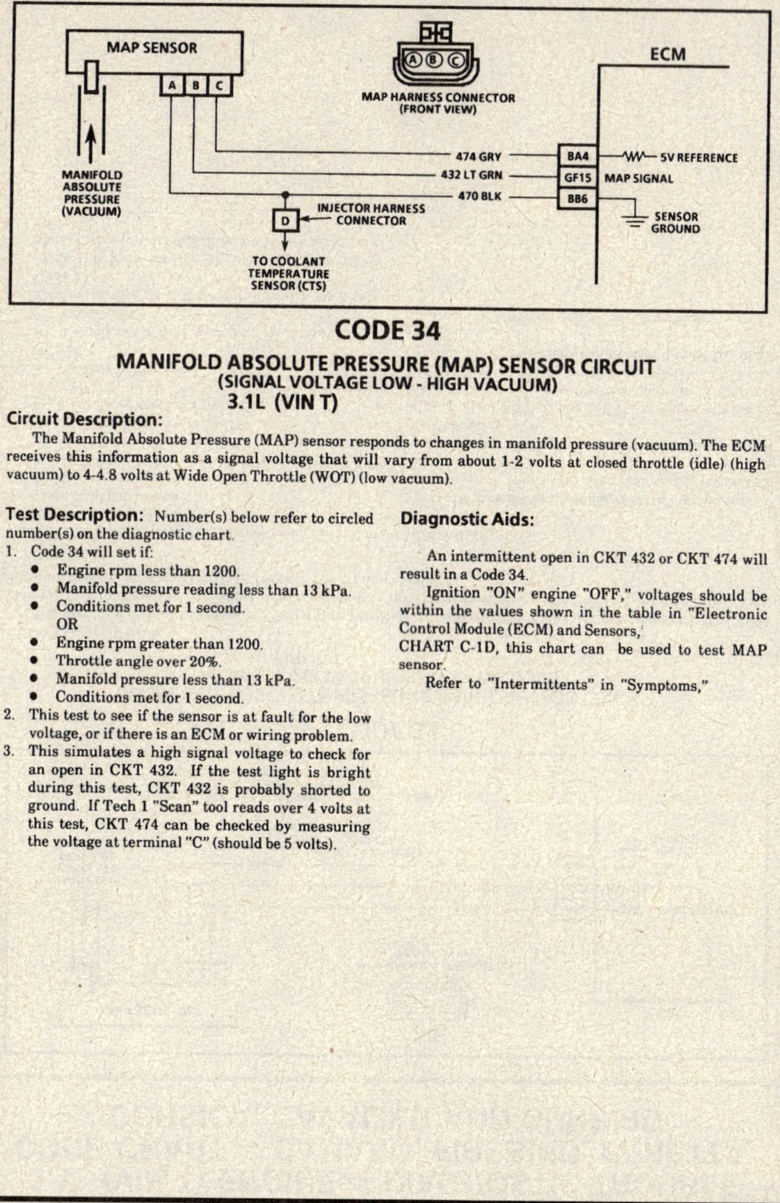

CODE 34
MANIFOLD ABSOLUTE PRESSURE (MAP) SENSOR CIRCUIT
(SIGNAL VOLTAGE LOW - HIGH VACUUM)
3.1L (VIN T)

Circuit Description:

The Manifold Absolute Pressure (MAP) sensor responds to changes in manifold pressure (vacuum). The ECM receives this information as a signal voltage that will vary from about 1-2 volts at closed throttle (idle) (high vacuum) to 4-4.8 volts at Wide Open Throttle (WOT) (low vacuum).

Test Description: Number(s) below refer to circled number(s) on the diagnostic chart.

1. Code 34 will set if:
   - Engine rpm less than 1200.
   - Manifold pressure reading less than 13 kPa.
   - Conditions met for 1 second.
   OR
   - Engine rpm greater than 1200.
   - Throttle angle over 20%.
   - Manifold pressure less than 13 kPa.
   - Conditions met for 1 second.
2. This test to see if the sensor is at fault for the low voltage, or if there is an ECM or wiring problem.
3. This simulates a high signal voltage to check for an open in CKT 432. If the test light is bright during this test, CKT 432 is probably shorted to ground. If Tech 1 "Scan" tool reads over 4 volts at this test, CKT 474 can be checked by measuring the voltage at terminal "C" (should be 5 volts).

Diagnostic Aids:

An intermittent open in CKT 432 or CKT 474 will result in a Code 34.

Ignition "ON" engine "OFF," voltages should be within the values shown in the table in "Electronic Control Module (ECM) and Sensors," CHART C-1D, this chart can be used to test MAP sensor.

Refer to "Intermittents" in "Symptoms,"

3.1L (VIN T) ENGINE — DIAGNOSTIC TROUBLE CODE CHART — CAMARO, FIREBIRD, BERETTA, CORSICA, CAVALIER AND SUNBIRD

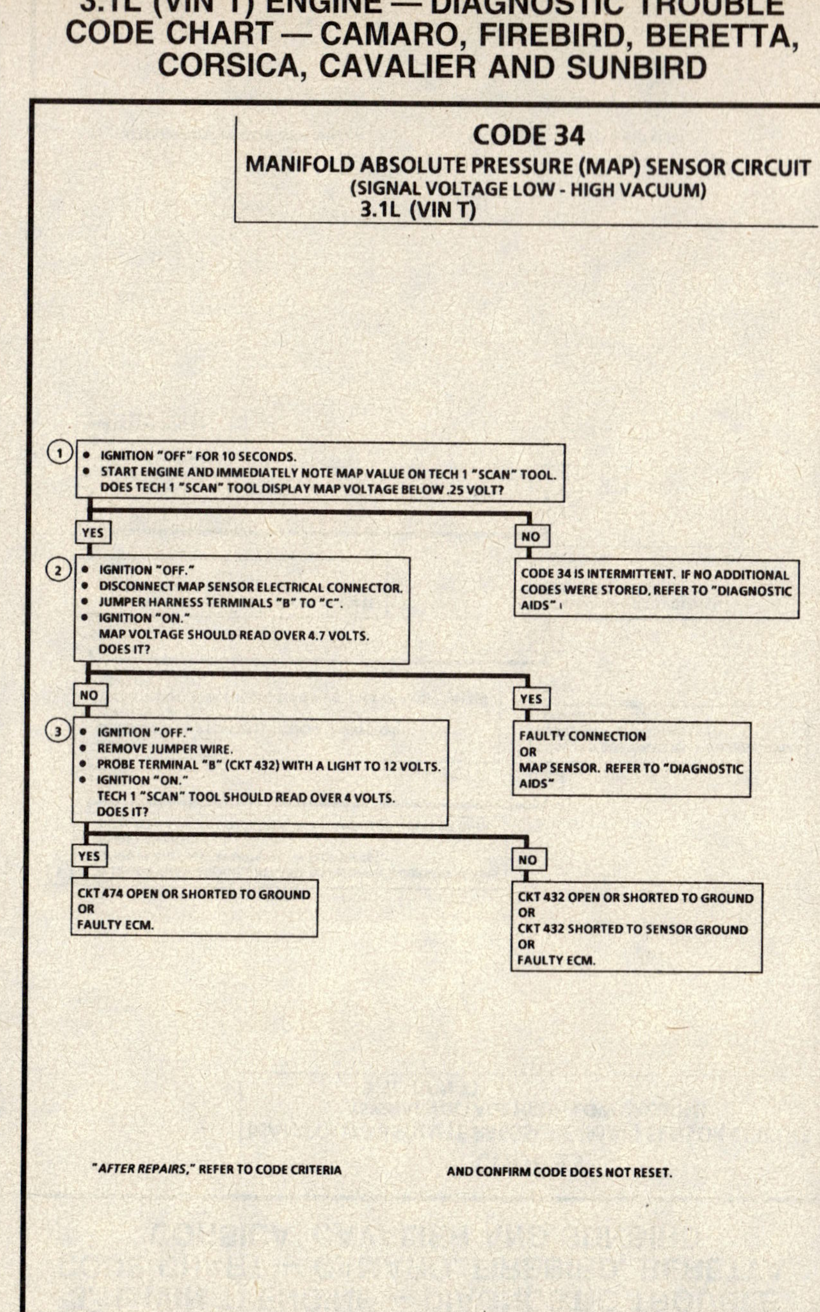

3.1L (VIN T) ENGINE — DIAGNOSTIC TROUBLE CODE CHART — CAMARO, FIREBIRD, BERETTA, CORSICA, CAVALIER AND SUNBIRD

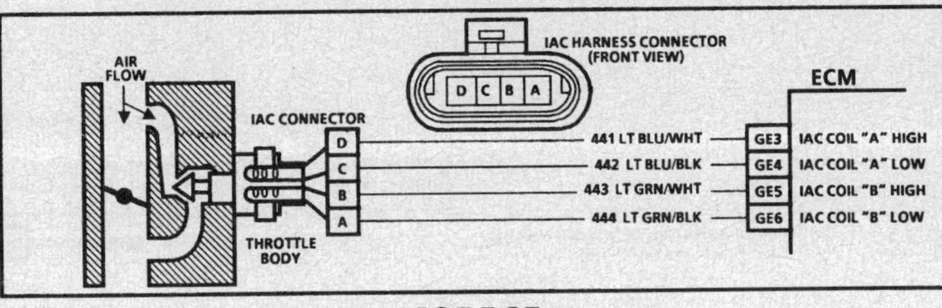

CODE 35
IDLE AIR CONTROL (IAC) CIRCUIT SYSTEM CHECK
3.1L (VIN T)

Circuit Description:

Code 35 will set when the closed throttle engine speed is 125 rpm above or below the desired (commanded) idle speed for 45 seconds. Review the general description of the IAC operation in "Fuel Metering System'" Section "6E3-C2".

Test Description: Number(s) below refer to circled number(s) on the diagnostic chart.

1. The Tech 1 rpm control mode is used to extend and retract the IAC valve. The valve should move smoothly within the specified range. If the idle speed is commanded (IAC extended) too low (below 700 rpm), the engine may stall. This may be normal and would not indicate a problem. Retracting the IAC beyond its controlled range (above 1500 rpm) will cause a delay before the rpm's start dropping. This too is normal.
2. This test uses the Tech 1 to command the IAC controlled idle speed. The ECM issues commands to obtain commended idle speed. The node lights each should flash red and green to indicate a good circuit as the ECM issues commands. While the sequence of color is not important if either light is "OFF" or does not flash red _and_ green, check the circuits for faults, beginning with poor terminal contacts.

Diagnostic Aids:

A slow unstable, or fast idle may be caused by a non-IAC system problem that cannot be overcome by the IAC valve. Out of control range IAC "Scan" tool counts will be above 60 if idle is too low, and zero counts if idle is too high. The following checks should be made to repair a non-IAC system problem:

- _Vacuum Leak (High Idle)_ - If idle is too high, stop the engine. Fully extend (low) IAC with Tech 1. Start engine. If idle speed is above 800 rpm, locate and correct vacuum leak including PCV system. Also check for binding of throttle blade or linkage.
- _System too lean (High Air/Fuel Ratio)_ - The idle speed may be too high or too low. Engine speed may vary up and down and disconnecting the IAC valve does not help. Code 44 may be set. "Scan" O_2 voltage will be less than 300 mV (.3 volt). Check for low regulated fuel pressure, water in the fuel or a restricted injector.
- _System too rich (low air/fuel ratio)_ - Idle speed too low. "Scan" tool IAC counts will usually be above 80. System is obviously rich and may exhibit black smoke in exhaust.
  "Scan" O_2 voltage will be fixed above 800 mV (.8 volt).
  Check for high fuel pressure, leaking or sticking injector. Silicone contaminated O_2 sensors "Scan" voltage will be slow to respond.
- _Throttle body_ - Remove IAC valve and inspect bore for foreign material.
- _IAC Valve Electrical Connections_ - IAC valve connections should be carefully checked for proper contact.
- _PCV Valve_ - An incorrect or faulty PCV valve may result in an incorrect idle speed.
- Refer to "Rough, Unstable, Incorrect Idle or Stalling" in "Symptoms," Section '
- If intermittent poor driveability or idle symptoms are resolved by disconnecting the IAC, carefully recheck connections, valve terminal resistance, or replace IAC.

3.1L (VIN T) ENGINE — DIAGNOSTIC TROUBLE CODE CHART — CAMARO, FIREBIRD, BERETTA, CORSICA, CAVALIER AND SUNBIRD

CODE 35
IDLE AIR CONTROL (IAC) CIRCUIT SYSTEM CHECK·
3.1L (VIN T)

(1)
- INSTALL TECH 1
- ENGINE AT NORMAL OPERATING TEMPERATURE IN PARK/NEUTRAL WITH PARKING BRAKE SET.
- A/C "OFF."
- SELECT RPM CONTROL. (MISC. TESTS)
- CYCLE IAC THROUGH ITS RANGE FROM 700 RPM UP TO 1500 RPM.
- RPM SHOULD CHANGE SMOOTHLY. DOES IT?

NO →

(2)
- INSTALL IAC NODE LIGHT * IN IAC HARNESS.
- ENGINE RUNNING. CYCLE IAC WITH TECH 1.
- EACH NODE LIGHT SHOULD CYCLE RED AND GREEN BUT NEVER "OFF." DO THEY?

NO →

IF CIRCUIT(S) DID NOT TEST RED AND GREEN, CHECK FOR:
- FAULTY CONNECTOR TERMINAL CONTACTS.
- OPEN CIRCUITS INCLUDING CONNECTORS.
- CIRCUITS SHORTED TO GROUND OR VOLTAGE.
- FAULTY ECM CONNECTIONS OR REPLACE ECM. REPAIR AS NECESSARY AND RETEST.

YES →

- CHECK IAC CONNECTIONS.
- CHECK IAC PASSAGES.
- IF OK, REPLACE IAC.

YES →

- USING THE IAC DRIVER * OR OTHER CONVENIENT CONNECTOR, CHECK RESISTANCE ACROSS IAC COILS.
- SHOULD BE 40 TO 80 OHMS BETWEEN IAC TERMINALS "A" TO "B" AND "C" TO "D".

OK →

- CHECK RESISTANCE BETWEEN IAC TERMINALS "B" AND "C" AND "A" AND "D".
- SHOULD BE INFINITE.

OK →

IDLE AIR CONTROL CIRCUIT OK. REFER TO "DIAGNOSTIC AIDS"

NOT OK →

REPLACE IAC VALVE AND RETEST.

NOT OK →

REPLACE IAC VALVE AND RETEST.

* IAC DRIVER AND NODE LIGHT REQUIRED KIT
222-L FROM: CONCEPT TECHNOLOGY, INC.
J 37027 FROM: KENT-MOORE, INC.

CLEAR CODES, CONFIRM "CLOSED LOOP" OPERATION, NO "SERVICE ENGINE SOON" LIGHT, PERFORM IAC RESET PROCEDURE PER APPLICABLE SERVICE MANUAL AND VERIFY CONTROLLED IDLE SPEED IS CORRECT.

"AFTER REPAIRS," REFER TO CODE CRITERIA   AND CONFIRM CODE DOES NOT RESET.

3.1L (VIN T) ENGINE — DIAGNOSTIC TROUBLE CODE CHART — CAMARO, FIREBIRD, BERETTA, CORSICA, CAVALIER AND SUNBIRD

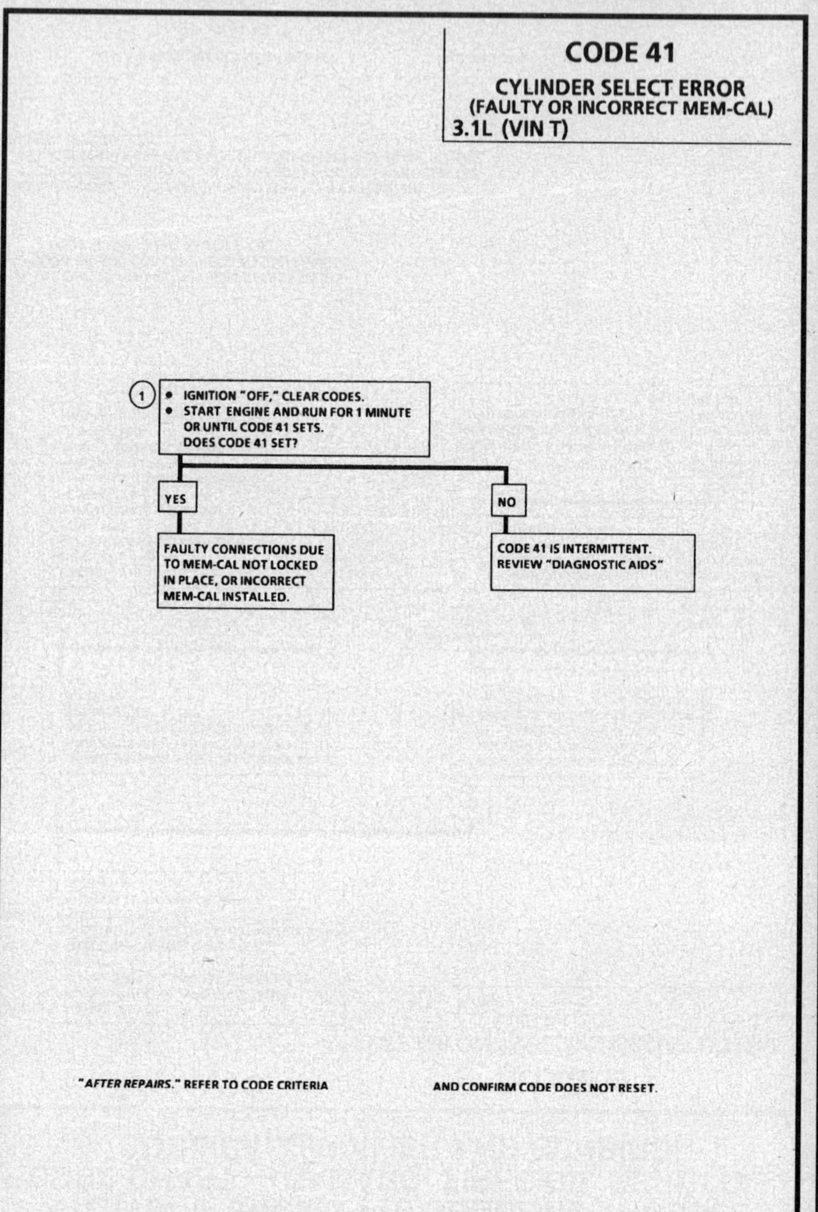

CODE 41
CYLINDER SELECT ERROR
(FAULTY OR INCORRECT MEM-CAL)
3.1L (VIN T)

Test Description: Number(s) below refer to circled number(s) on the diagnostic chart.

1. The ECM used for this engine can also be used for other engines, and the difference is in the MEM-CAL. If a Code 41 sets, the incorrect MEM-CAL has been installed, or it is faulty and must be replaced.

Diagnostic Aids:

Check MEM-CAL to be sure locking tabs are secure.

Important
- Before checking MEM-CAL to be sure locking tabs are secure, refer to "Electronic Control Module (ECM) and Sensor," Section "C-1," "On-Vehicle Service," "ECM or MEM-CAL replacement," and follow installation instruction.

CAUTION: Damage can occur to controller circuit board by pressing down hard on the ends of MEM-CAL, to secure the locking tabs.

Also, check the pins on both the MEM-CAL and ECM to assure they are making proper contact. Check the MEM-CAL part number to assure it is the correct part. If the MEM-CAL is faulty it must be replaced. It is also possible that the ECM is faulty, however, it should not be replaced until all of the above have been checked.

For additional information, refer to "Intermittents" in "Symptoms," Section

3.1L (VIN T) ENGINE — DIAGNOSTIC TROUBLE CODE CHART — CAMARO, FIREBIRD, BERETTA, CORSICA, CAVALIER AND SUNBIRD

CODE 41
CYLINDER SELECT ERROR
(FAULTY OR INCORRECT MEM-CAL)
3.1L (VIN T)

① • IGNITION "OFF," CLEAR CODES.
   • START ENGINE AND RUN FOR 1 MINUTE OR UNTIL CODE 41 SETS. DOES CODE 41 SET?

YES
FAULTY CONNECTIONS DUE TO MEM-CAL NOT LOCKED IN PLACE, OR INCORRECT MEM-CAL INSTALLED.

NO
CODE 41 IS INTERMITTENT. REVIEW "DIAGNOSTIC AIDS"

"AFTER REPAIRS." REFER TO CODE CRITERIA   AND CONFIRM CODE DOES NOT RESET.

3.1L (VIN T) ENGINE — DIAGNOSTIC TROUBLE CODE CHART — CAMARO, FIREBIRD, BERETTA, CORSICA, CAVALIER AND SUNBIRD

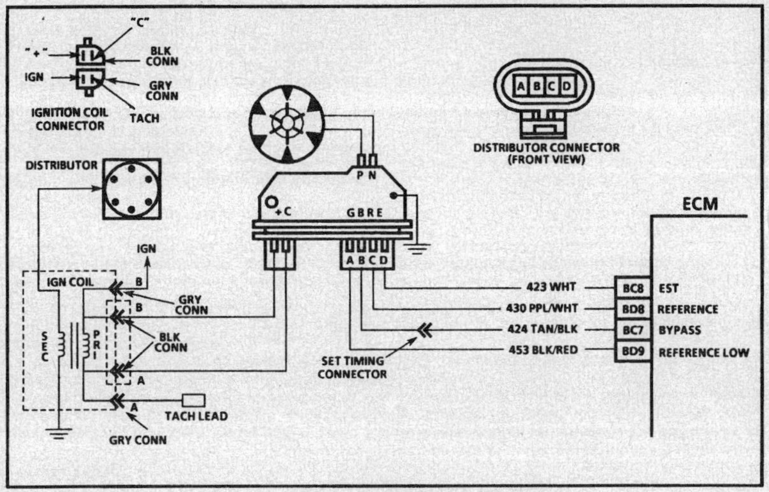

CODE 42
ELECTRONIC SPARK TIMING (EST) CIRCUIT
3.1L (VIN T) "F" CARLINE (PORT)

Circuit Description:

When the system is running on the ignition module, that is, no voltage on the bypass line, the ignition module grounds the EST signal. The ECM expects to see no voltage on the EST line during this condition. If it sees a voltage, it sets Code 42 and will not go into the EST mode.

When the rpm for EST is reached (about 400 rpm), and bypass voltage applied, the EST should no longer be grounded in the ignition module so the EST voltage should be varying.

If the bypass line is open or grounded, the ignition module will not switch to EST mode so the EST voltage will be low and Code 42 will be set.

If the EST line is grounded, the ignition module will switch to EST, but because the line is grounded there will be no EST signal. A Code 42 will be set.

Test Description: Number(s) below refer to circled number(s) on the diagnostic chart.

1. Code 42 means the ECM has seen an open or short to ground in the EST or bypass circuits. This test confirms Code 42 and that the fault causing the code is present.
2. Checks for a normal EST ground path through the ignition module. An EST CKT 423 shorted to ground will also read less than 500 ohms; however, this will be checked later.
3. As the test light voltage touches CKT 424, the module should switch causing the ohmmeter to "overrange" if the meter is in the 1000-2000 ohms position. Selecting the 10-20,000 ohms position will indicate above 5000 ohms. The important thing is that the module "switched."
4. The module did not switch and this step checks for:
   - EST CKT 423 shorted to ground.
   - Bypass CKT 424 open.
   - Faulty ignition module connection or module.
5. Confirms that Code 42 is a faulty ECM and not an intermittent in CKTs 423 or 424.

Diagnostic Aids:

The Tech 1 "Scan" tool does not have any ability to help diagnose a Code 42 problem.

A MEM-CAL not fully seated in the ECM can result in a Code 42.

Refer to "Intermittents" in "Symptoms," Section

3.1L (VIN T) ENGINE — DIAGNOSTIC TROUBLE CODE CHART — CAMARO, FIREBIRD, BERETTA, CORSICA, CAVALIER AND SUNBIRD

CODE 42
ELECTRONIC SPARK TIMING (EST) CIRCUIT
3.1L (VIN T) "F" CARLINE (PORT)

3.1L (VIN T) ENGINE — DIAGNOSTIC TROUBLE CODE CHART — CAMARO, FIREBIRD, BERETTA, CORSICA, CAVALIER AND SUNBIRD

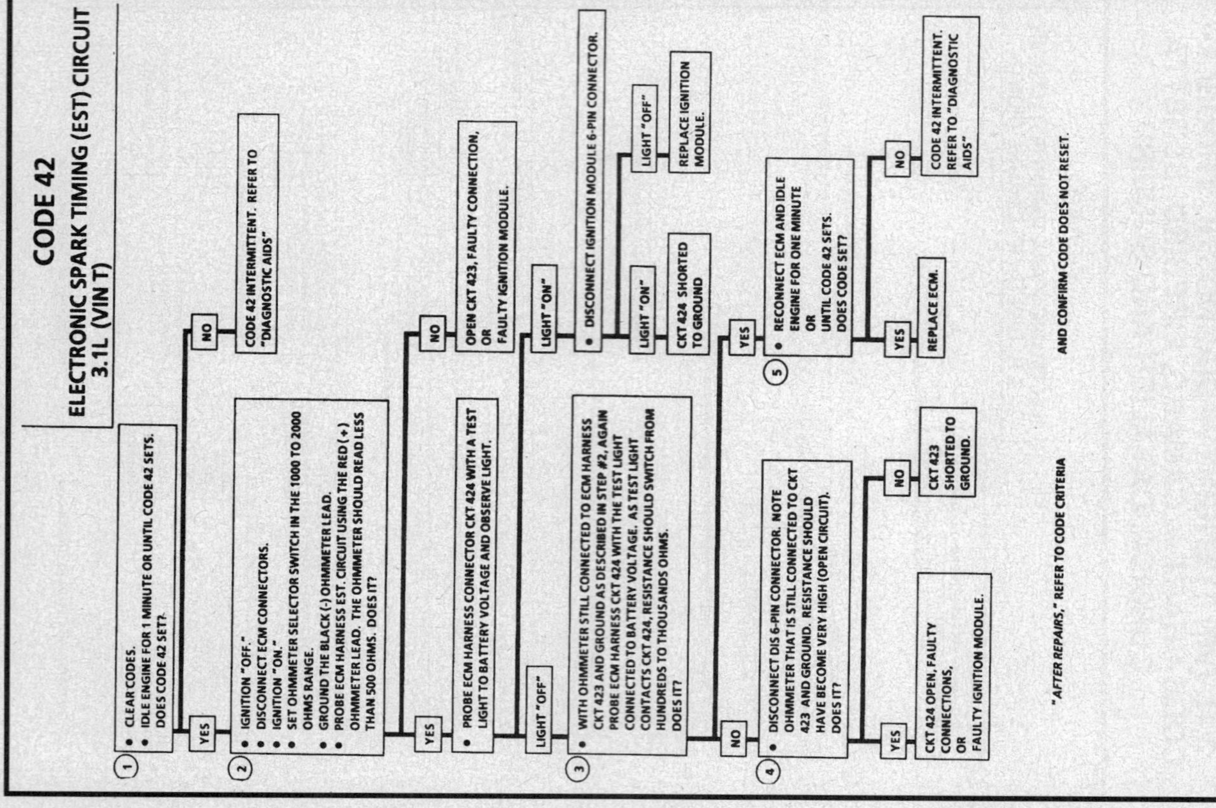

3.1L (VIN T) ENGINE — DIAGNOSTIC TROUBLE CODE CHART — CAMARO, FIREBIRD, BERETTA, CORSICA, CAVALIER AND SUNBIRD

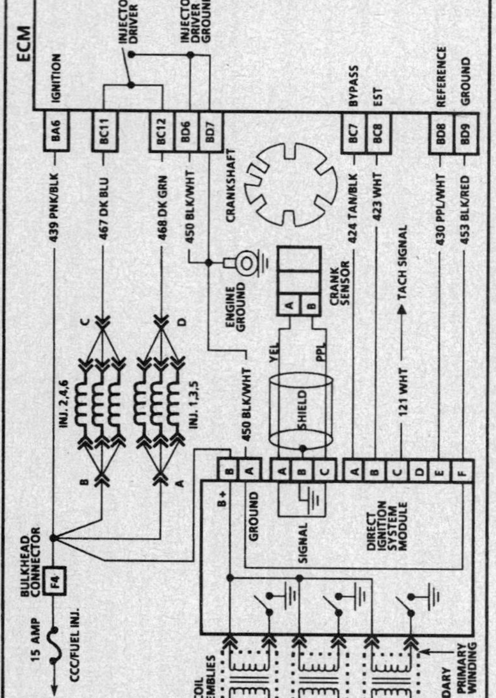

CODE 42
ELECTRONIC SPARK TIMING (EST) CIRCUIT
3.1L (VIN T) "L" CARLINE (PORT)

Circuit Description:

When the system is running on the ignition module, that is no voltage on the bypass line, the ignition module grounds the Electronic Spark Timing (EST) signal. The ECM expects to see no voltage on the EST line during this condition. If it sees a voltage, it sets Code 42 and will not go into the EST mode.

When the rpm for EST is reached (about 600 rpm) and bypass voltage applied, the ignition module will no longer be grounded in the ignition module so the EST voltage should be varying.

If the bypass line is open or grounded, the ignition module will not switch to EST mode so the EST voltage will be low and Code 42 will be set.

If the EST line is grounded, the ignition module will switch to EST, but because the line is grounded there will be no EST signal. A Code 42 will be set.

Test Description: Number(s) below refer to circled number(s) on the diagnostic chart.

1. Code 42 means the ECM has seen an open or short to ground in the EST or bypass circuits. This test confirms Code 42 and that the fault causing the code is present.
2. Checks for a normal EST ground path through the ignition module. An EST CKT 423 shorted to ground will also read less than 500 ohms, however, this will be checked later.
3. As the test light voltage touches CKT 424, the module should switch causing the ohmmeter to "overrange" if the meter is in the 1000-2000 ohms position. Selecting the 10-20,000 ohms position will indicate above 5000 ohms. The important thing is that the module "switched."

4. The module did not switch and this step checks for
   • EST CKT 423 shorted to ground.
   • Bypass CKT 424 open.
   • Faulty ignition module connection or module
5. Confirms that Code 42 is a faulty ECM and not an intermittent in CKT 423 or 424.

Diagnostic Aids:

The Tech 1 "Scan" tool does not have any ability to help diagnose a Code 42 problem.

A MEM-CAL not fully seated in the ECM can result in a Code 42.

Refer to "Intermittents" in "Symptoms," Section

3.1L (VIN T) ENGINE — DIAGNOSTIC TROUBLE CODE CHART — CAMARO, FIREBIRD, BERETTA, CORSICA, CAVALIER AND SUNBIRD

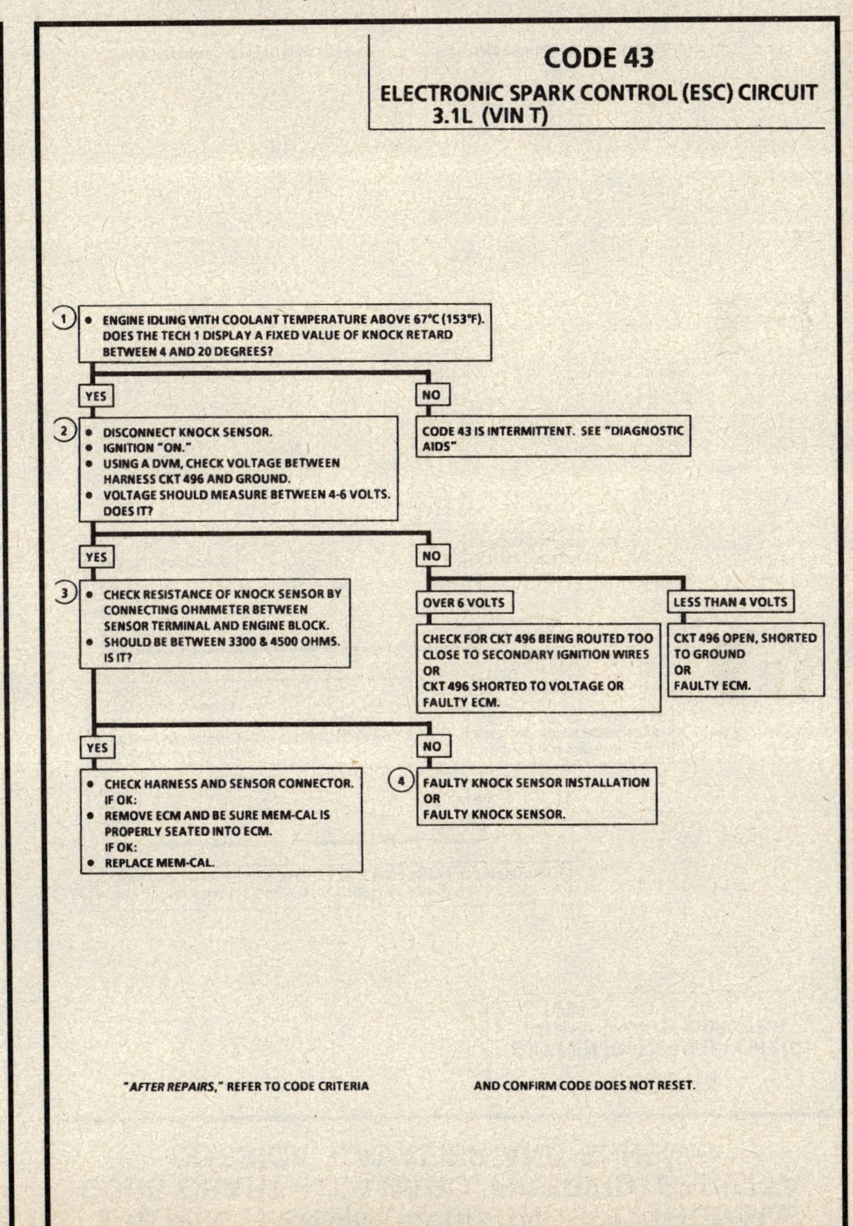

CODE 43
ELECTRONIC SPARK CONTROL (ESC) CIRCUIT
3.1L (VIN T)

Circuit Description:

The knock sensor is used to detect engine detonation and the ECM will retard the electronic spark timing based on the signal being received. The circuitry within the knock sensor causes the ECM 5 volts to be pulled down so that, under a no knock condition, CKT 496 would measure about 2.5 volts. The knock sensor produces an A.C. signal which rides on the 2.5 volts DC voltage. The amplitude and signal frequency is dependent upon the knock level.

If CKT 496 becomes open or shorted to ground, the voltage will either go above 4.8 volts or below .62 volt. If either of these conditions are met for about 10 seconds, a Code 43 will be stored.

Test Description: Number(s) below refer to circled number(s) on the diagnostic chart.

1. This step determines if conditions for Code 43 still exist (voltage on CKT 496 above 4.8 volts or below .62 volt). The system is designed to retard the spark 15° if either condition exists.
2. The ECM has a 5 volt pullup resistor, which applies 5 volts to CKT 496. The 5 volt signal should be present at the knock sensor terminal during these test conditions.
3. This step determines if the knock sensor resistance is 3300 to 4500 ohms the sensor is OK.
4. Knock sensor case is the ground. Do Not use Teflon tape on threads and the threads are free of corrosion.

Diagnostic Aids:

If CKT 496 is not open or shorted to ground and the voltage reading is below 4 volts, the most likely cause is an open circuit in the ECM. It is possible that a faulty MEM-CAL could be drawing the 5 volt signal down, and it should be replaced, if a replacement ECM did not correct the problem.

Refer to "Intermittents" in "Symptoms," Section

3.1L (VIN T) ENGINE — DIAGNOSTIC TROUBLE CODE CHART — CAMARO, FIREBIRD, BERETTA, CORSICA, CAVALIER AND SUNBIRD

CODE 43
ELECTRONIC SPARK CONTROL (ESC) CIRCUIT
3.1L (VIN T)

1. ● ENGINE IDLING WITH COOLANT TEMPERATURE ABOVE 67°C (153°F). DOES THE TECH 1 DISPLAY A FIXED VALUE OF KNOCK RETARD BETWEEN 4 AND 20 DEGREES?

YES

2. ● DISCONNECT KNOCK SENSOR.
● IGNITION "ON."
● USING A DVM, CHECK VOLTAGE BETWEEN HARNESS CKT 496 AND GROUND.
● VOLTAGE SHOULD MEASURE BETWEEN 4-6 VOLTS. DOES IT?

NO → CODE 43 IS INTERMITTENT. SEE "DIAGNOSTIC AIDS"

YES

3. ● CHECK RESISTANCE OF KNOCK SENSOR BY CONNECTING OHMMETER BETWEEN SENSOR TERMINAL AND ENGINE BLOCK.
● SHOULD BE BETWEEN 3300 & 4500 OHMS. IS IT?

NO

OVER 6 VOLTS → CHECK FOR CKT 496 BEING ROUTED TOO CLOSE TO SECONDARY IGNITION WIRES OR CKT 496 SHORTED TO VOLTAGE OR FAULTY ECM.

LESS THAN 4 VOLTS → CKT 496 OPEN, SHORTED TO GROUND OR FAULTY ECM.

YES

● CHECK HARNESS AND SENSOR CONNECTOR. IF OK:
● REMOVE ECM AND BE SURE MEM-CAL IS PROPERLY SEATED INTO ECM. IF OK:
● REPLACE MEM-CAL.

NO

4. FAULTY KNOCK SENSOR INSTALLATION OR FAULTY KNOCK SENSOR.

"AFTER REPAIRS," REFER TO CODE CRITERIA   AND CONFIRM CODE DOES NOT RESET.

3.1L (VIN T) ENGINE — DIAGNOSTIC TROUBLE CODE CHART — CAMARO, FIREBIRD, BERETTA, CORSICA, CAVALIER AND SUNBIRD

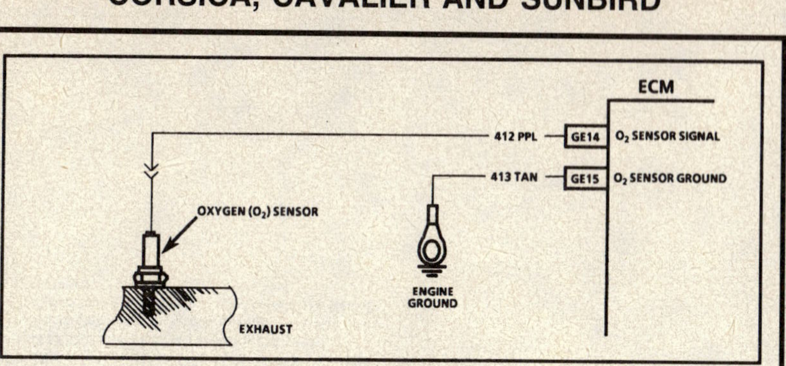

CODE 44
OXYGEN (O₂) SENSOR CIRCUIT
(LEAN EXHAUST INDICATED)
3.1L (VIN T)

Circuit Description:

The ECM supplies a voltage of about .45 volt between terminals "GE14" and "GE15". (If measured with a 10 megohm digital voltmeter, this may read as low as .32 volt.) The O_2 sensor varies the voltage within a range of about 1 volt if the exhaust is rich, down through about .10 volt if exhaust is lean.

The sensor is like an open circuit and produces no voltage when it is below about 315°C (600°F). An open sensor circuit or cold sensor causes "Open Loop" operation.

Test Description: Number(s) below refers to circled number(s) on the diagnostic chart.

1. Code 44 is set when the O_2 sensor signal voltage on CKT 412.
   - Remains below .2 volt for 60 seconds or more.
   - And the system is operating "Closed Loop."

Diagnostic Aids:

Using the Tech 1 "Scan" tool, observe the block learn values at different rpm and air flow conditions. The Tech 1 "Scan" tool also displays the block cells, so the block learn values can be checked in each of the cells to determine when the Code 44 may have been set. If the conditions for Code 44 exist, the block learn values will be around 150.

- **O₂ Sensor Wire** - Sensor pigtail may be mispositioned and contacting the exhaust manifold.
- Check for intermittent ground in wire between connector and sensor.

- **Lean Injector(s)** - Perform injector balance test CHART C-2A.
- **Fuel Contamination** - Water, even in small amounts, near the in-tank fuel pump inlet can be delivered to the injectors. The water causes a lean exhaust and can set a Code 44.
- **Fuel Pressure** - System will be lean if pressure is too low. It may be necessary to monitor fuel pressure while driving the car at various road speeds and/or loads to confirm. See "Fuel System Diagnosis" CHART A-7.
- **Exhaust Leaks** - If there is an exhaust leak, the engine can cause outside air to be pulled into the exhaust and past the sensor. Vacuum or crankcase leaks can cause a lean condition.
- **A.I.R. System** - Be sure air is not being directed to the exhaust ports while in "Closed Loop." If the block learn value goes down while squeezing air hose to exhaust ports, refer to CHART C-6.
- If the above are OK, it is a faulty oxygen sensor.

3.1L (VIN T) ENGINE — DIAGNOSTIC TROUBLE CODE CHART — CAMARO, FIREBIRD, BERETTA, CORSICA, CAVALIER AND SUNBIRD

CODE 44
OXYGEN (O₂) SENSOR CIRCUIT
(LEAN EXHAUST INDICATED)
3.1L (VIN T)

1.
- RUN WARM ENGINE (75°C/167°F TO 95°C/203°F) AT 1200 RPM.
- DOES TECH 1 INDICATE O₂ SENSOR VOLTAGE FIXED BELOW .35 VOLT (350 mV)?

YES
- DISCONNECT O₂ SENSOR.
- WITH ENGINE IDLING, TECH 1 SHOULD DISPLAY O₂ SENSOR VOLTAGE BETWEEN .35 VOLT AND .55 VOLT (350 mV AND 550 mV). DOES IT?

NO
- CODE 44 IS INTERMITTENT. IF NO ADDITIONAL CODES WERE STORED, REFER TO "DIAGNOSTIC AIDS"

YES → REFER TO "DIAGNOSTIC AIDS"

NO → CKT 412 SHORTED TO GROUND OR FAULTY ECM.

"AFTER REPAIRS," REFER TO CODE CRITERIA AND CONFIRM CODE DOES NOT RESET.

3.1L (VIN T) ENGINE — DIAGNOSTIC TROUBLE CODE CHART — CAMARO, FIREBIRD, BERETTA, CORSICA, CAVALIER AND SUNBIRD

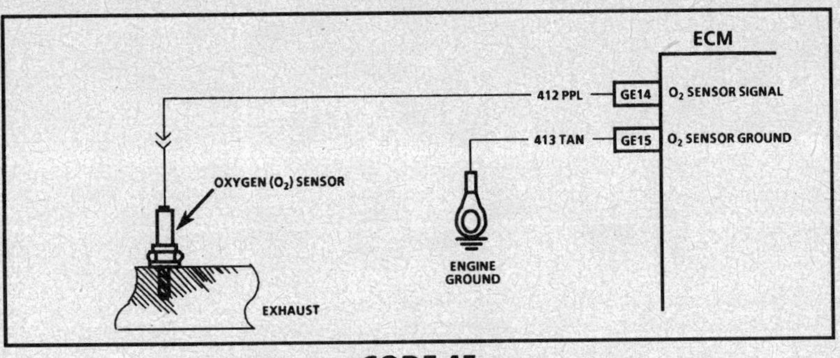

CODE 45
OXYGEN (O₂) SENSOR CIRCUIT
(RICH EXHAUST INDICATED)
3.1L (VIN T)

Circuit Description:

The ECM supplies a voltage of about .45 volt between terminals "GE14" and "GE15". (If measured with a 10 megohm digital voltmeter, this may read as low as .32 volt). The O_2 sensor varies the voltage within a range of about 1 volt if the exhaust is rich, down through about .10 volt if exhaust is lean.

The sensor is like an open circuit and produces no voltage when it is below about 315°C (600°F). An open sensor circuit or cold sensor causes "Open Loop" operation.

Test Description: Number(s) below refers to circled number(s) on the diagnostic chart.

1. Code 45 is set when the O_2 sensor signal voltage or CKT 412.
   - Remains above .7 volt for 50 seconds, and in "Closed Loop."
   - Engine time after start is 1 minute or more.
   - Throttle angle between 3% and 45%.

Diagnostic Aids:

Using the Tech 1 "Scan" tool, observe the block learn values at different rpm and air flow conditions. The Tech 1 "Scan" tool also displays the block cells, so the block learn values can be checked in each of the cells to determine when the Code 45 may have been set. If the conditions for Code 45 exist, the block learn values will be around 115.

- Check for short to voltage on CKT 412.
- Fuel Pressure - System will go rich if pressure is too high. The ECM can compensate for some increase. However, if it gets too high, a Code 45 may be set. See "Fuel System Diagnosis" CHART A-7.
- Rich Injector - Perform injector balance test CHART C-2A.
- Leaking Injector - See CHART A-7.
- Check for fuel contaminated oil.
- HEI Shielding - An open ground CKT 453 (ignition system reference low) may result in EMI, or induced electrical "noise." The ECM looks at this "noise" as reference pulses. The additional pulses result in a higher than actual engine speed signal. The ECM then delivers too much fuel, causing system to go rich. Engine tachometer will also show higher than actual engine speed, which can help in diagnosing this problem.
- Canister Purge - Check for fuel saturation. If full of fuel, check canister control and hoses. See "Canister Purge," in "Evaporative Emission Control System (EECS)" Section
- Check for leaking fuel pressure regulator diaphragm by checking vacuum line to regulator for fuel.
- TPS - An intermittent TPS output will cause the system to go rich, due to a false indication of the engine accelerating.
- EGR - An EGR staying open (especially at idle) will cause the O_2 sensor to indicate a rich exhaust, and this could result in a Code 45.

3.1L (VIN T) ENGINE — DIAGNOSTIC TROUBLE CODE CHART — CAMARO, FIREBIRD, BERETTA, CORSICA, CAVALIER AND SUNBIRD

CODE 45
OXYGEN (O₂) SENSOR CIRCUIT
(RICH EXHAUST INDICATED)
3.1L (VIN T)

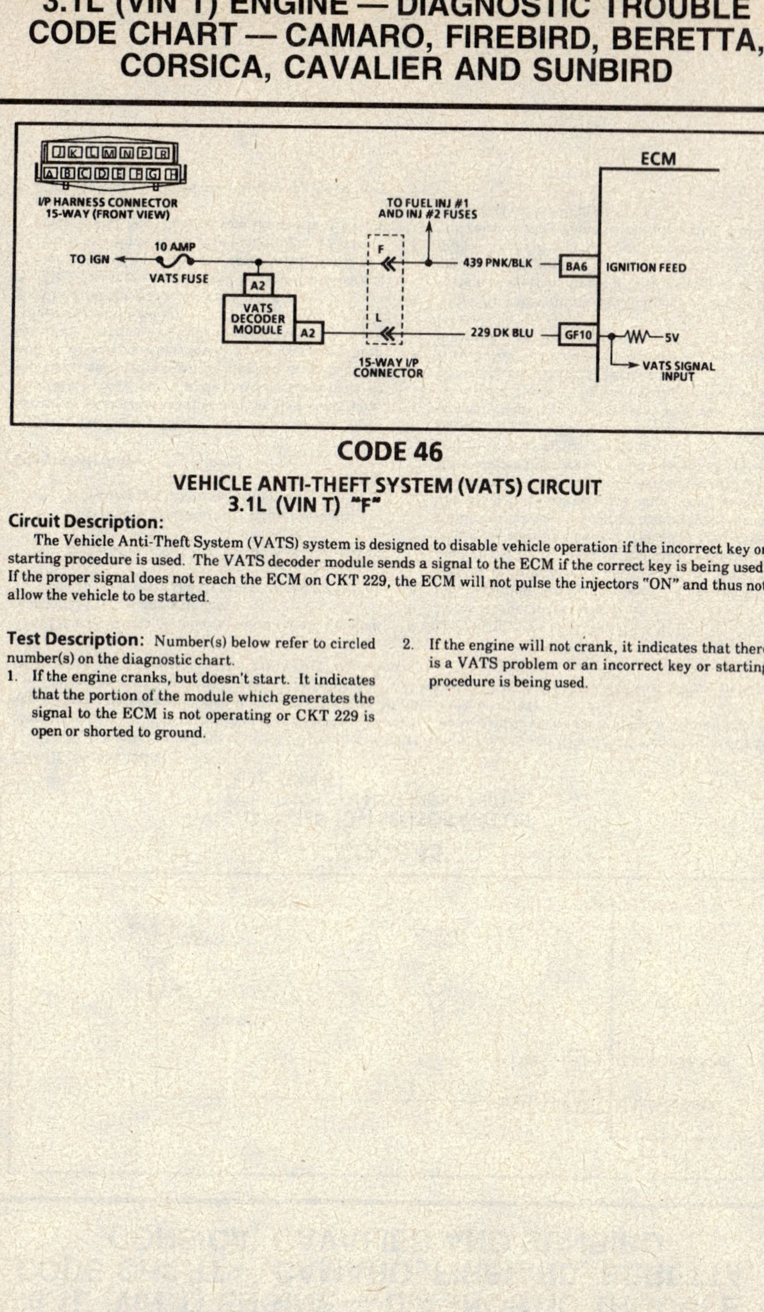

CODE 46
VEHICLE ANTI-THEFT SYSTEM (VATS) CIRCUIT
3.1L (VIN T) "F"

Circuit Description:

The Vehicle Anti-Theft System (VATS) system is designed to disable vehicle operation if the incorrect key or starting procedure is used. The VATS decoder module sends a signal to the ECM if the correct key is being used. If the proper signal does not reach the ECM on CKT 229, the ECM will not pulse the injectors "ON" and thus not allow the vehicle to be started.

Test Description: Number(s) below refer to circled number(s) on the diagnostic chart.

1. If the engine cranks, but doesn't start. It indicates that the portion of the module which generates the signal to the ECM is not operating or CKT 229 is open or shorted to ground.

2. If the engine will not crank, it indicates that there is a VATS problem or an incorrect key or starting procedure is being used.

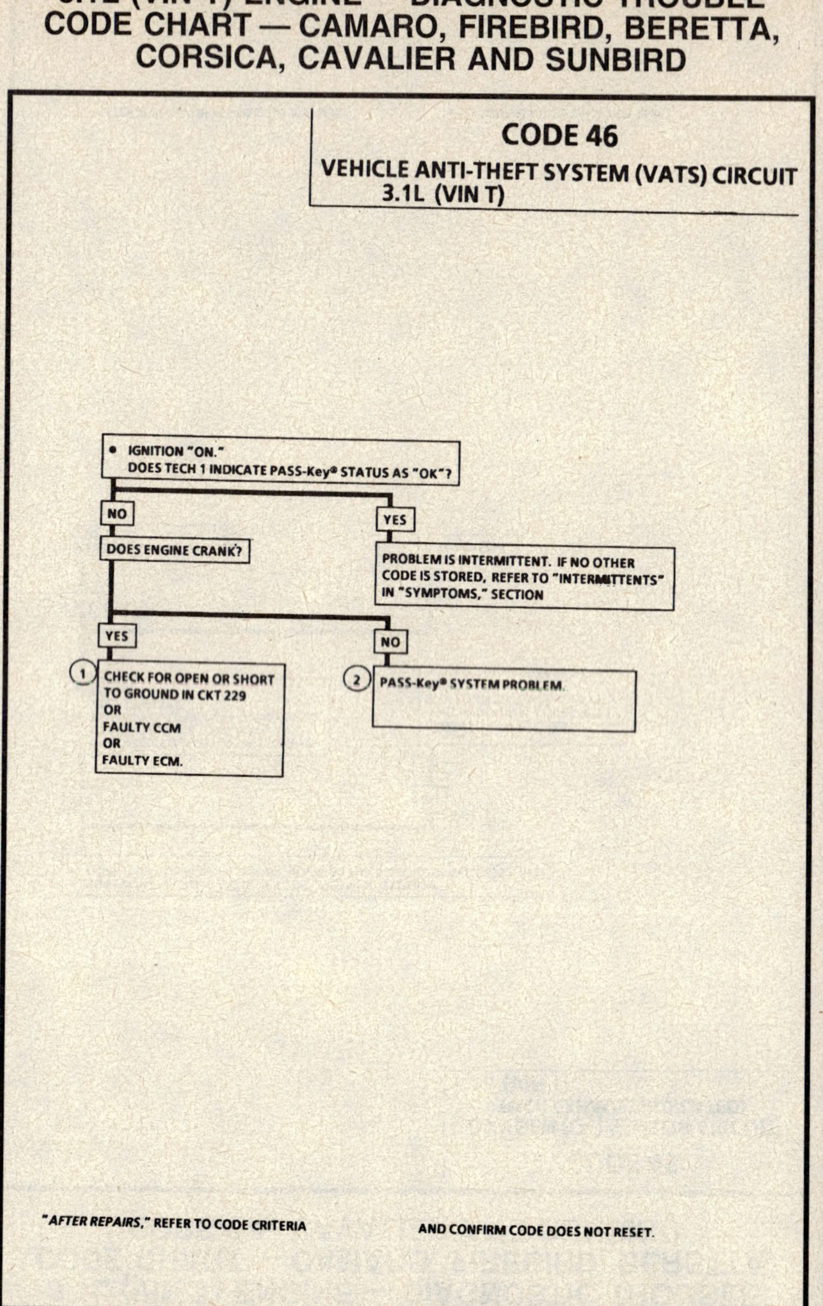

3.1L (VIN T) ENGINE — DIAGNOSTIC TROUBLE CODE CHART — CAMARO, FIREBIRD, BERETTA, CORSICA, CAVALIER AND SUNBIRD

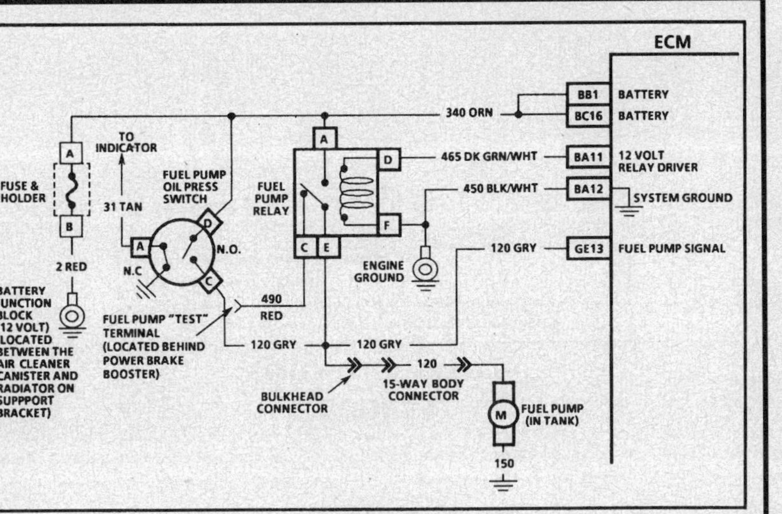

CODE 54
FUEL PUMP CIRCUIT
(LOW VOLTAGE)
3.1L (VIN T)

Circuit Description:

The status of the fuel pump CKT 120 is monitored by the ECM at terminal "GE13" and is used to compensate fuel delivery based on system voltage. This signal is also used to store a trouble code if the fuel pump relay is defective or fuel pump voltage is lost while the engine is running. There should be about 12 volts on CKT 120 for 2 seconds after the ignition is turned or any time references pulses are being received by the ECM.

Code 54 will set if the voltage at terminal "GE13" is less than 4 volts for .4 second since the last reference pulse was received. This code is designed to detect a faulty relay, causing extended crank time, and the code will help the diagnosis of an engine that "Cranks But Will Not Run."

If a fault is detected during start-up, the "Service Engine Soon" light will stay "ON" until the ignition is cycled "OFF." However, if the voltage is detected below 4 volts with the engine running, the light will only remain "ON" while the condition exists.

3.1L (VIN T) ENGINE — DIAGNOSTIC TROUBLE CODE CHART — CAMARO, FIREBIRD, BERETTA, CORSICA, CAVALIER AND SUNBIRD

CODE 54
FUEL PUMP CIRCUIT
(LOW VOLTAGE)
3.1L (VIN T)

- IGNITION "OFF" FOR 10 SECONDS.
- IGNITION "ON".
- LISTEN FOR IN-TANK FUEL PUMP.
- PUMP SHOULD RUN FOR 2 SECONDS AFTER IGNITION "ON". DOES IT?

NO →
- IGNITION "OFF".
- USING A FUSED JUMPER WIRE, CONNECT FUEL PUMP TEST CONNECTOR TO 12 VOLTS.
- DOES PUMP RUN?

YES →
- IGNITION "OFF".
- DISCONNECT FUEL PUMP RELAY.
- PROBE CKT 340 WITH A TEST LIGHT TO GROUND.

NO →
- DISCONNECT FUEL PUMP RELAY.
- USING THE FUSED JUMPER WIRE, CONNECT CKT 120 TO 12 VOLTS. DOES PUMP RUN?

LIGHT "ON" → CONNECT TEST LIGHT BETWEEN CKTS 340 & 450.

LIGHT "OFF" → REPAIR OPEN IN CKT 340.

YES → FAULTY RELAY.

NO → OPEN CKT 120, FAULTY IN-TANK PUMP OR FAULTY PUMP GROUND.

LIGHT "ON" →
- CONNECT TEST LIGHT BETWEEN HARNESS CKT 465 AND GROUND.
- IGNITION "OFF" FOR 10 SECONDS.
- NOTE TEST LIGHT WITHIN 2 SECONDS AFTER IGNITION "ON".

LIGHT "ON" → FAULTY RELAY.

YES →
- CLEAR CODES.
- START AND RUN ENGINE FOR 30 SECONDS OR UNTIL CODE 54 SETS. DOES CODE SET?

YES →
- AT THE ECM, BACK PROBE CKT 120 WITH A TEST LIGHT TO GROUND.
- IGNITION "OFF" FOR 10 SECONDS.
- NOTE LIGHT WITHIN 2 SECONDS AFTER IGNITION "ON".

NO → CODE 54 IS INTERMITTENT. REFER TO "INTERMITTENTS"

LIGHT "ON" → FAULTY CONNECTION AT ECM OR FAULTY ECM.

LIGHT "OFF" → OPEN CKT 120 TO ECM.

LIGHT "OFF" → REPAIR OPEN CKT 450.

LIGHT "OFF" → CKT 465 OPEN, SHORTED TO GROUND, OR FAULTY ECM.

NOTE: IF ORIGINAL COMPLAINT WAS "CRANKS BUT WILL NOT RUN" MAKE THE FOLLOWING ADDITIONAL CHECKS:
- ENGINE IDLING AT NORMAL OPERATING TEMPERATURE.
- OIL PRESSURE NORMAL.
- DISCONNECT FUEL PUMP RELAY.
- ENGINE SHOULD CONTINUE TO RUN. DOES IT?

YES →
- RECONNECT FUEL PUMP RELAY.
- IGNITION "OFF"
- PROBE FUEL PUMP TEST TERMINAL WITH A TEST LIGHT TO GROUND.

NO → FAULTY OIL PRESSURE SWITCH.

LIGHT "OFF" → FUEL PUMP CIRCUIT OK

LIGHT "ON" → FAULTY OIL PRESSURE SWITCH

3.1L (VIN T) ENGINE — DIAGNOSTIC TROUBLE CODE CHART — CAMARO, FIREBIRD, BERETTA, CORSICA, CAVALIER AND SUNBIRD

CODE 51
CODE 53

3.1L (VIN T)

CODE 51

MEM-CAL ERROR
(FAULTY OR INCORRECT MEM-CAL)

CHECK THAT ALL PINS ARE FULLY INSERTED IN THE SOCKET AND THAT MEM-CAL IS PROPERLY LATCHED. IF OK, REPLACE MEM/CAL, CLEAR MEMORY, AND RECHECK. IF CODE 51 REAPPEARS, REPLACE ECM.

CODE 53

SYSTEM OVER VOLTAGE

THIS CODE INDICATES THERE IS A BASIC GENERATOR PROBLEM.
- CODE 53 WILL SET, IF VOLTAGE AT ECM IGNITION INPUT PIN IS GREATER THAN 16.9 VOLTS FOR 10 SECONDS.
- CHECK AND REPAIR CHARGING SYSTEM.

3.1L (VIN T) ENGINE — DIAGNOSTIC TROUBLE CODE CHART — CAMARO, FIREBIRD, BERETTA, CORSICA, CAVALIER AND SUNBIRD

CODE 61

DEGRADED OXYGEN SENSOR
3.1L (VIN T)

IF A CODE 61 IS STORED IN MEMORY THE ECM HAS DETERMINED THE OXYGEN SENSOR IS CONTAMINATED OR DEGRADED, BECAUSE THE VOLTAGE CHANGE TIME IS SLOW OR SLUGGISH.

THE ECM PERFORMS THE OXYGEN SENSOR RESPONSE TIME TEST WHEN:

COOLANT TEMPERATURE IS GREATER THAN 85°C.

IAT TEMPERATURE IS GREATER THAN 10°C.

IN CLOSED LOOP.

IN DECEL FUEL CUT-OFF MODE.

IF A CODE 61 IS STORED THE OXYGEN SENSOR SHOULD BE REPLACED. A CONTAMINATED SENSOR CAN BE CAUSED BY FUEL ADDITIVES, SUCH AS SILICON, OR BY USE OF NON-GM APPROVED LUBRICANTS OR SEALANTS. SILICON CONTAMINATION IS USUALLY INDICATED BY A WHITE POWDERY SUBSTANCE ON THE SENSOR FINS.

3.1L (VIN T) ENGINE — RESTRICTED EXHAUST SYSTEM CHART — CAMARO, FIREBIRD, BERETTA, CORSICA, CAVALIER AND SUNBIRD

CHART B-1

RESTRICTED EXHAUST SYSTEM CHECK

ALL ENGINES

Proper diagnosis for a restricted exhaust system is essential before any components are replaced. Either of the following procedures may be used for diagnosis, depending upon engine or tool used:

CHECK AT A. I. R. PIPE:

1. Remove the rubber hose at the exhaust manifold A.I.R. pipe check valve. Remove check valve.
2. Connect a fuel pump pressure gauge to a hose and nipple from a Propane Enrichment Device (J 26911) (see illustration).
3. Insert the nipple into the exhaust manifold A.I.R. pipe.

OR CHECK AT O₂ SENSOR:

1. Carefully remove Oxygen (O$_2$) sensor.
2. Install Borroughs exhaust backpressure tester (BT 8515 or BT 8603) or equivalent in place of O$_2$ sensor (see illustration).
3. After completing test described below, be sure to coat threads of O$_2$ sensor with anti-seize compound P/N 5613695 or equivalent prior to re-installation.

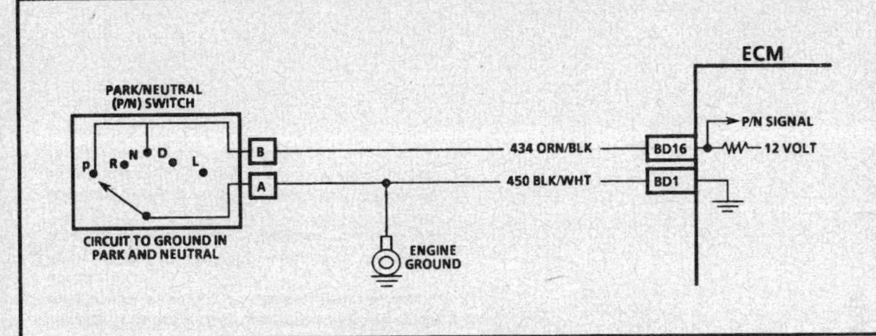

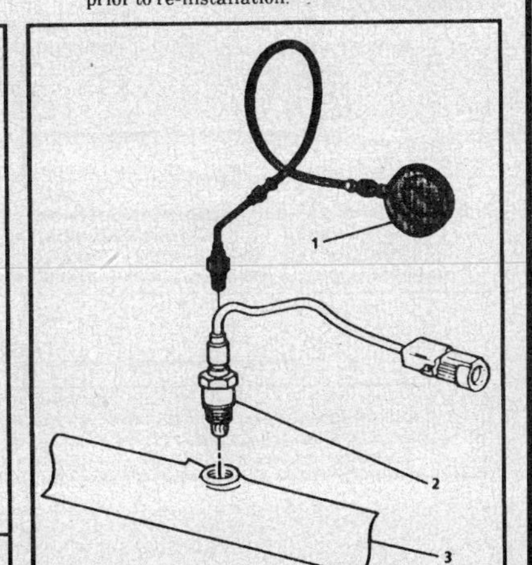

DIAGNOSIS:

1. With the engine idling at normal operating temperature, observe the exhaust system backpressure reading on the gage. Reading should not exceed 8.6 kPa (1.25 psi).
2. Increase engine speed to 2000 rpm and observe gage. Reading should not exceed 20.7 kPa (3 psi).
3. If the backpressure at either speed exceeds specification, a restricted exhaust system is indicated.
4. Inspect the entire exhaust system for a collapsed pipe, heat distress, or possible internal muffler failure.
5. If there are no obvious reasons for the excessive backpressure, the catalytic converter is suspected to be restricted and should be replaced using current recommended procedures.

3.1L (VIN T) ENGINE — COMPONENT DIAGNOSTIC CHART — CAMARO, FIREBIRD, BERETTA, CORSICA, CAVALIER AND SUNBIRD

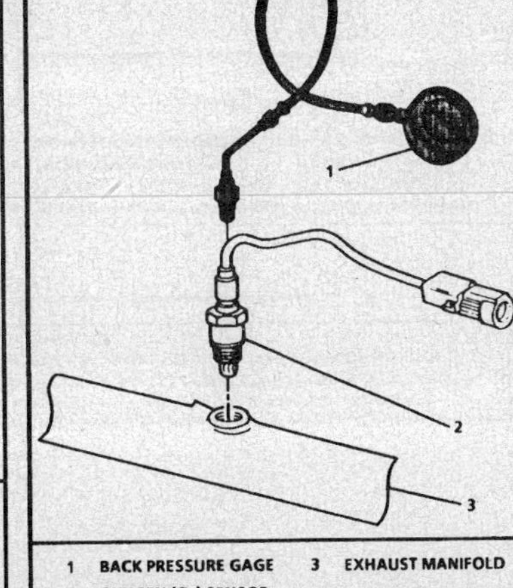

CHART C-1A

PARK/NEUTRAL (P/N) SWITCH DIAGNOSIS
3.1L (VIN T)

Circuit Description:

The Park/Neutral (P/N) switch contacts are a part of the neutral start switch and are closed to ground in park or neutral, and open in drive ranges.

The ECM supplies ignition voltage through a pull-up resistor to CKT 434 and senses a closed switch when the voltage on CKT 434 drops to less than one volt.

The ECM uses the P/N signal as one of the inputs to control:

- Idle Air Control.
- VSS diagnostics.
- EGR.
  If CKT 434 indicates P/N (grounded) while in drive range, the EGR would be inoperative, resulting in possible detonation.
  If CKT 434 indicates drive (open) a dip in the idle may exist when the gear selector is moved into drive range.

Test Description: Number(s) below refer to circled number(s) on the diagnostic chart.

1. Checks for a closed switch to ground in park position. Different makes of "Scan" tools will read P/N differently. Refer to tool operator's manual for type of display used for a specific tool.
2. Checks for an open switch in drive range.
3. Be sure Tech 1 "Scan" tool indicates drive, even while wiggling shifter, to test for an intermittent or misadjusted switch in drive or overdrive range.

CHART C-1A

PARK/NEUTRAL (P/N) SWITCH DIAGNOSIS
3.1L (VIN T)

(1)
- WITH TRANSMISSION IN PARK, TECH 1 "SCAN" TOOL SHOULD INDICATE PARK OR NEUTRAL. DOES IT?

YES →

(3)
- SHIFT TRANSMISSION INTO DRIVE.
- "SCAN" TOOL SHOULD DISPLAY A CHANGE TO INDICATE DRIVE. DOES IT?

NO →

- DISCONNECT P/N SWITCH.
- THIS SHOULD CAUSE "SCAN" TOOL TO DISPLAY DRIVE RANGE. DOES IT?

YES →

FAULTY P/N SWITCH CONNECTION OR P/N SWITCH MISADJUSTED OR FAULTY P/N SWITCH.

YES →

NO TROUBLE FOUND. REFER TO "INTERMITTENTS" IN "SYMPTOM,"

NO →

CKT 434 SHORTED TO GROUND OR FAULTY ECM.

NO →

(2)
- DISCONNECT PARK/NEUTRAL SWITCH CONNECTOR.
- JUMPER HARNESS CONNECTOR TERMINALS "A" AND "B".
- "SCAN" TOOL SHOULD INDICATE PARK OR NEUTRAL. DOES IT?

NO →

- JUMPER HARNESS CONNECTOR (CKT 434) TO ENGINE GROUND.
- "SCAN" TOOL SHOULD INDICATE PARK OR NEUTRAL. DOES IT?

YES →

OPEN GROUND CIRCUIT.

YES →

FAULTY P/N SWITCH CONNECTION OR P/N SWITCH MISADJUSTED OR FAULTY P/N SWITCH.

NO →

CKT 434 OPEN OR FAULTY ECM CONNECTION OR ECM.

"AFTER REPAIRS," CONFIRM "CLOSED LOOP" OPERATION AND NO "SERVICE ENGINE SOON" LIGHT.

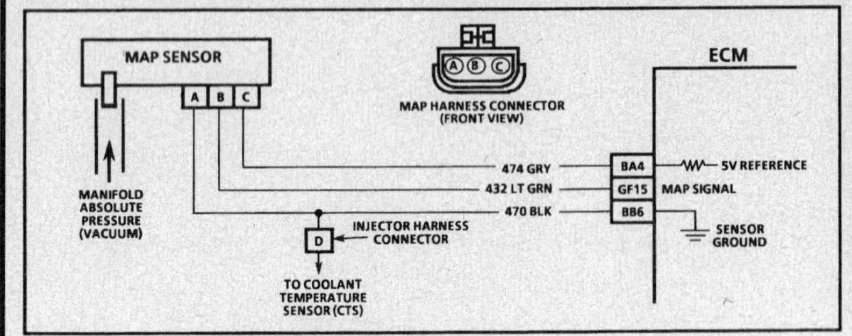

CHART C-1D

MANIFOLD ABSOLUTE PRESSURE (MAP) OUTPUT CHECK
3.1L (VIN T)

Circuit Description:

The Manifold Absolute Pressure (MAP) sensor measures the changes in the intake manifold pressure which result from engine load (intake manifold vacuum) and rpm changes: and converts theses into a voltage output. The ECM sends a 5 volt reference voltage to the MAP sensor. As the manifold pressure changed, the output voltage of the sensor also changes. By monitoring the sensor output voltage, the ECM knows the manifold pressure. A lower pressure (low voltage will be about 1 - 2 volts at idle. While higher pressure (high voltage) output voltage will be about 4 - 4.8 at Wide Open Throttle (WOT). The MAP sensor is also used, under certain conditions, to measure barometric pressure, allowing the ECM to make adjustments for different altitudes. The ECM used the MAP sensor to control fuel delivery and ignition timing.

Test Description: Number(s) below refer to circled number(s) on the diagnostic chart.

Important
- Be sure to use the same Diagnostic Test Equipment for all measurements.

1. When comparing Tech 1 "Scan" tool readings to a known good vehicle, it is important to compare vehicles that use a MAP sensor having the same color insert or having the same "Hot Stamped" number.
2. Applying 34 kPa (10" Hg) vacuum to the MAP sensor should cause the voltage to be 1.5 to 2.1 volts less than the voltage at Step 1. Upon applying vacuum to the sensor, the change in voltage should be instantaneous. A slow voltage change indicates a faulty sensor.

3. Check vacuum hose to sensor for leaking or restriction. Be sure that no other vacuum devices are connected to the MAP hose.

NOTICE: Make sure electrical connector remains securely fastened.

4. Disconnect sensor from bracket and twist sensor by hand (only) to check for intermittent connection. Output changes greater than .1 volt indicates a bad sensor.

3.1L (VIN T) ENGINE — COMPONENT DIAGNOSTIC CHART — CAMARO, FIREBIRD, BERETTA, CORSICA, CAVALIER AND SUNBIRD

CHART C-1D
MANIFOLD ABSOLUTE PRESSURE (MAP) OUTPUT CHECK
3.1L (VIN T)

NOTICE: THIS CHART ONLY APPLIES TO MAP SENSORS HAVING GREEN OR BLACK COLOR KEY INSERT (SEE BELOW).

1. • IGNITION "ON," ENGINE "OFF."
   • TECH 1 "SCAN" TOOL SHOULD INDICATE A MAP SENSOR VOLTAGE.
   • COMPARE THIS READING WITH THE READING OF A KNOWN GOOD VEHICLE. SEE FACING PAGE TEST DESCRIPTION, STEP 1.
   VOLTAGE READING SHOULD BE WITHIN ± .4 VOLT.
   IS IT?

   YES →
   NO → REPLACE MAP SENSOR.

2. • DISCONNECT AND PLUG VACUUM SOURCE TO MAP SENSOR.
   • CONNECT A HAND VACUUM PUMP TO MAP SENSOR.
   • START ENGINE.
   • NOTE MAP SENSOR VOLTAGE.
   • APPLY 34 kPa (10" Hg) OF VACUUM AND NOTE VOLTAGE CHANGE. SUBTRACT SECOND READING FROM THE FIRST. VOLTAGE VALUE SHOULD BE GREATER THAN 1.5 VOLTS.
   IS IT?

   YES →
   NO → CHECK MAP SENSOR CONNECTION. IF OK, REPLACE MAP SENSOR.

3. NO TROUBLE FOUND. CHECK MAP SENSOR VACUUM SOURCE FOR LEAKAGE OR RESTRICTION. BE SURE THIS SOURCE SUPPLIES VACUUM TO MAP SENSOR ONLY.

4. CHECK MAP SENSOR CONNECTION. IF OK, REPLACE MAP SENSOR.

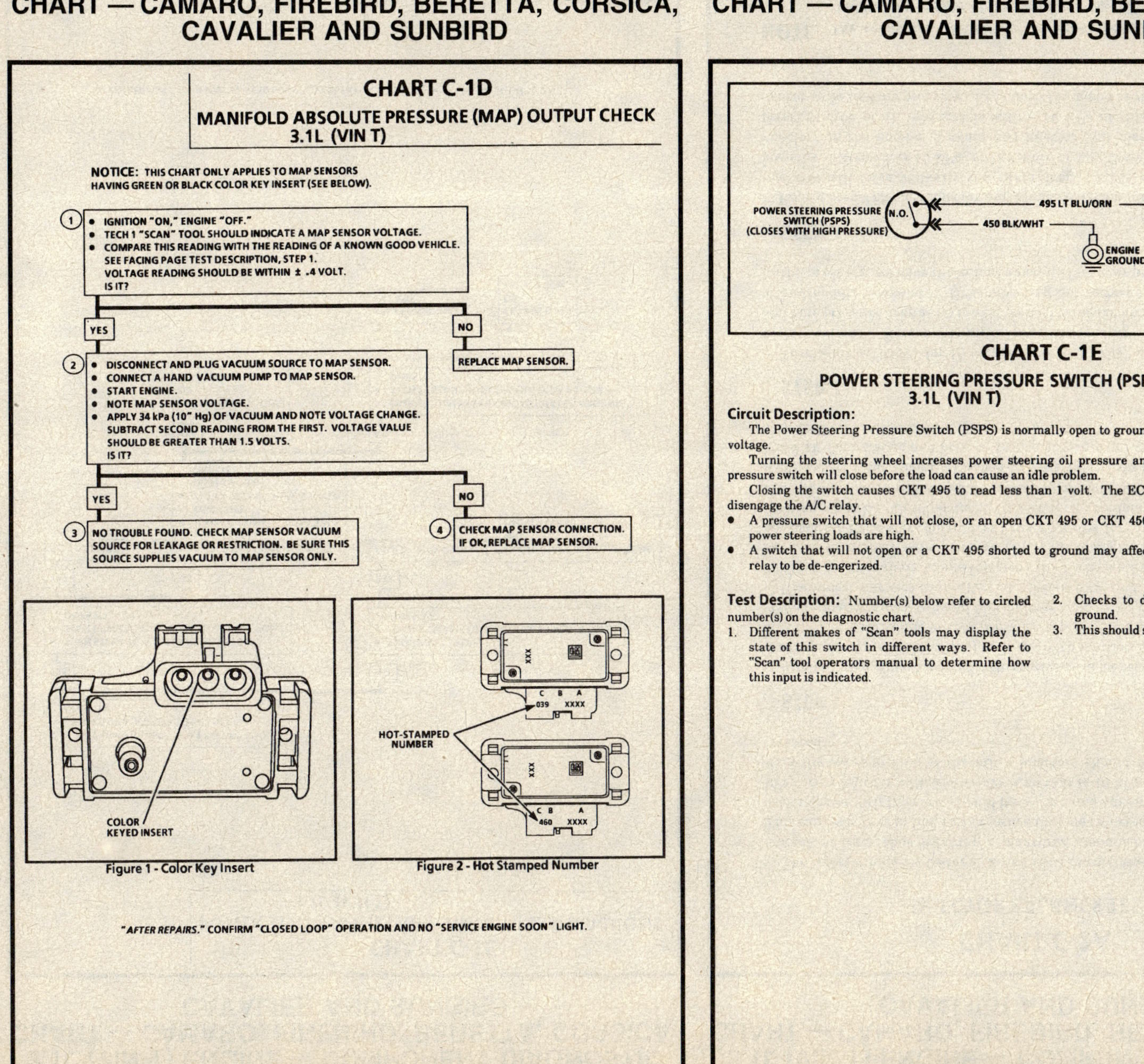

Figure 1 - Color Key Insert

Figure 2 - Hot Stamped Number

"AFTER REPAIRS," CONFIRM "CLOSED LOOP" OPERATION AND NO "SERVICE ENGINE SOON" LIGHT.

3.1L (VIN T) ENGINE — COMPONENT DIAGNOSTIC CHART — CAMARO, FIREBIRD, BERETTA, CORSICA, CAVALIER AND SUNBIRD

CHART C-1E
POWER STEERING PRESSURE SWITCH (PSPS) DIAGNOSIS
3.1L (VIN T)

Circuit Description:

The Power Steering Pressure Switch (PSPS) is normally open to ground, and CKT 495 will near the battery voltage.

Turning the steering wheel increases power steering oil pressure and its load on an idling engine. The pressure switch will close before the load can cause an idle problem.

Closing the switch causes CKT 495 to read less than 1 volt. The ECM will increase the idle air rate and disengage the A/C relay.

• A pressure switch that will not close, or an open CKT 495 or CKT 450, may cause the engine to stop when power steering loads are high.

• A switch that will not open or a CKT 495 shorted to ground may affect idle quality and will cause the A/C relay to be de-engerized.

Test Description: Number(s) below refer to circled number(s) on the diagnostic chart.

1. Different makes of "Scan" tools may display the state of this switch in different ways. Refer to "Scan" tool operators manual to determine how this input is indicated.

2. Checks to determine if CKT 495 is shorted to ground.

3. This should simulate a closed switch.

3.1L (VIN T) ENGINE — COMPONENT DIAGNOSTIC CHART — CAMARO, FIREBIRD, BERETTA, CORSICA, CAVALIER AND SUNBIRD

CHART C-1E
POWER STEERING PRESSURE SWITCH (PSPS) DIAGNOSIS
3.1L (VIN T)

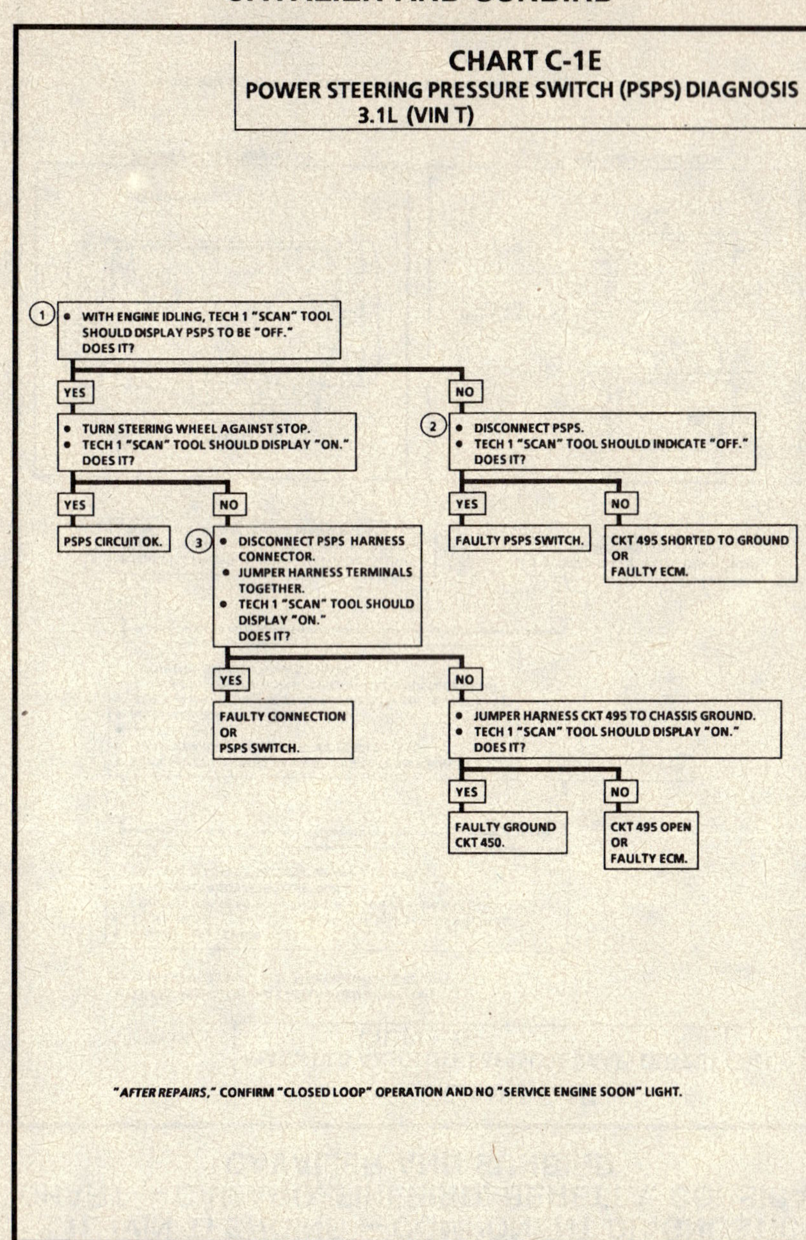

3.1L (VIN T) ENGINE — COMPONENT DIAGNOSTIC CHART — CAMARO, FIREBIRD, BERETTA, CORSICA, CAVALIER AND SUNBIRD

CHART C-2A
INJECTOR BALANCE TEST

The injector balance tester is a tool used to turn the injector on for a precise amount of time, thus spraying a measured amount of fuel into the manifold. This causes a drop in fuel rail pressure that we can record and compare between each injector. All injectors should have the same amount of pressure drop ($\pm$ 10 kPa). Any injector with a pressure drop that is 10 kPa (or more) greater or less than the average drop of the other injectors should be considered faulty and replaced.

STEP 1

Engine "cool down" period (10 minutes) is necessary to avoid irregular readings due to "Hot Soak" fuel boiling. With ignition "OFF" connect fuel gauge J 347301 or equivalent to fuel pressure tap. Wrap a shop towel around fitting while connecting gage to avoid fuel spillage.

Disconnect harness connectors at all injectors, and connect injector tester J 34730-3, or equivalent, to one injector. On Turbo equipped engines, use adaptor harness furnished with injector tester to energize injectors that are not accessible. Follow manufacturers instructions for use of adaptor harness. Ignition must be "OFF" at least 10 seconds to complete ECM shutdown cycle. Fuel pump should run about 2 seconds after ignition is turned "ON". At this point, insert clear tubing attached to vent valve into a suitable container and bleed air from gauge and hose to insure accurate gauge operation. Repeat this step until all air is bled from gauge.

STEP 2

Turn ignition "OFF" for 10 seconds and then "ON" again to get fuel pressure to its maximum. Record this initial pressure reading. Energize tester one time and note pressure drop at its lowest point (Disregard any slight pressure increase after drop hits low point.). By subtracting this second pressure reading from the initial pressure, we have the actual amount of injector pressure drop.

STEP 3

Repeat step 2 on each injector and compare the amount of drop. Usually, good injectors will have virtually the same drop. Retest any injector that has a pressure difference of 10 kPa, either more or less than the average of the other injectors on the engine. Replace any injector that also fails the retest. If the pressure drop of all injectors is within 10 kPa of this average, the injectors appear to be flowing properly. Reconnect them and review "Symptoms"

NOTE: *The entire test should <u>not</u> be repeated more than once without running the engine to prevent flooding. (This includes any retest on faulty injectors).*

3.1L (VIN T) ENGINE — COMPONENT DIAGNOSTIC CHART — CAMARO, FIREBIRD, BERETTA, CORSICA, CAVALIER AND SUNBIRD

NOTE: The fuel pressure test in Section "A", CHART A-7, should be completed prior to this test.

CHART C-2A

INJECTOR BALANCE TEST
3.1L (VIN T)

Step 1. If engine is at operating temperature, allow a 10 minute "cool down" period then connect fuel pressure gauge and injector tester.
1. Ignition "OFF".
2. Connect fuel pressure gauge and injector tester.
3. Ignition "ON".
4. Bleed off air in gauge. Repeat until all air is bled from gauge.

Step 2. Run test:
1. Ignition "OFF" for 10 seconds.
2. Ignition "ON". Record gauge pressure. (Pressure must hold steady, if not see the Fuel System diagnosis, Chart A-7.
3. Turn injector on, by depressing button on injector tester, and note pressure at the instant the gauge needle stops.

Step 3.
1. Repeat step 2 on all injectors and record pressure drop on each.
Retest injectors that appear faulty (Any injectors that have a 10 kPa difference, either more or less, in pressure from the average). If no problem is found, review "Symptoms"

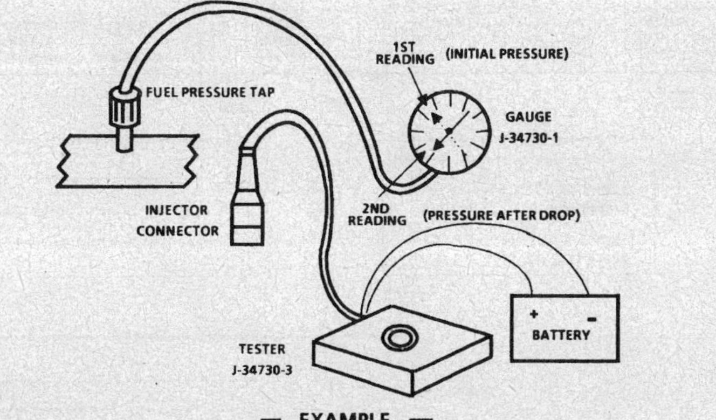

— EXAMPLE —

CYLINDER	1	2	3	4	5	6
1ST READING	225	225	225	225	225	225
2ND READING	100	100	100	90	100	115
AMOUNT OF DROP	125	125	125	135	125	110
	OK	OK	OK	FAULTY, RICH (TOO MUCH) (FUEL DROP)	OK	FAULTY, LEAN (TOO LITTLE) (FUEL DROP)

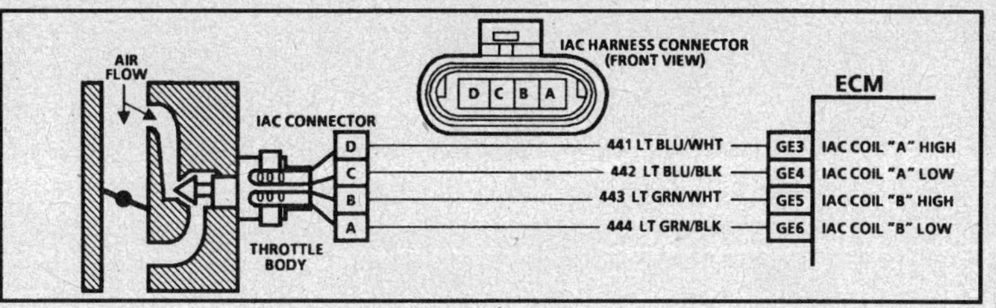

CHART C-2B

IDLE AIR CONTROL (IAC) SYSTEM CHECK
3.1L (VIN T)

Circuit Description:

The ECM controls engine idle speed with the IAC valve. To increase idle speed, the ECM retracts the IAC valve pintle away from its seat, allowing more air to bypass the throttle bore. To decrease idle speed, it extends the IAC valve pintle towards its seat, reducing bypass air flow. A "Scan" tool will read the ECM commands to the IAC valve in counts. Higher the counts indicate more air bypass (higher idle). The lower the counts indicate less air is allowed to bypass (lower idle).

Test Description: Number(s) below refer to circled number(s) on the diagnostic chart.
1. The Tech 1 rpm control mode is used to extend and retract the IAC valve. The valve should move smoothly within the specified range. If the idle speed is commanded (IAC extended) too low (below 700 rpm), the engine may stall. This may be normal and would not indicate a problem. Retracting the IAC beyond its controlled range (above 1500 rpm) will cause a delay before the rpm's start dropping. This too is normal.
2. This test uses the Tech 1 to command the IAC controlled idle speed. The ECM issues commands to obtain commended idle speed. The node lights each should flash red and green to indicate a good circuit as the ECM issues commands. While the sequence of color is not important if either light is "OFF" or does not flash red _and_ green, check the circuits for faults, beginning with poor terminal contacts.

Diagnostic Aids:

A slow, unstable, or fast idle may be caused by a non-IAC system problem that cannot be overcome by the IAC valve. Out of control range IAC "Scan" tool counts will be above 60 if idle is too low, and zero counts if idle is too high. The following checks should be made to repair a non-IAC system problem:
- Vacuum Leak (High Idle) - If idle is too high, stop the engine. Fully extend (low) IAC with tester. Start engine. If idle speed is above 800 rpm, locate and correct vacuum leak including PCV system. Also check for binding of throttle blade or linkage.
- System too lean (High Air/Fuel Ratio) - The idle speed may be too high or too low. Engine speed may vary up and down and disconnecting the IAC valve does not help. Code 44 may be set. "Scan" O$_2$ voltage will be less than 300 mV (.3 volt). Check for low regulated fuel pressure, water in the fuel or a restricted injector.
- System too rich (Low Air/Fuel Ratio) - The idle speed will be too low. "Scan" tool IAC counts will usually be above 80. System is obviously rich and may exhibit black smoke in exhaust. "Scan" tool O$_2$ voltage will be fixed above 800 mV (.8 volt). Check for high fuel pressure, leaking or sticking injector. Silicone contaminated O$_2$ sensors "Scan" voltage will be slow to respond.
- Throttle Body - Remove IAC valve and inspect bore for foreign material.
- IAC Valve Electrical Connections - IAC valve connections should be carefully checked for proper contact.
- PCV Valve - An incorrect or faulty PCV valve may result in an incorrect idle speed.
- Refer to "Rough, Unstable, Incorrect Idle or Stalling" in "Symptoms,"
- If intermittent poor driveability or idle symptoms are resolved by disconnecting the IAC, carefully recheck connections, valve terminal resistance, or replace IAC.

3.1L (VIN T) ENGINE — COMPONENT DIAGNOSTIC CHART — CAMARO, FIREBIRD, BERETTA, CORSICA, CAVALIER AND SUNBIRD

CHART C-2B
IDLE AIR CONTROL (IAC) SYSTEM CHECK
3.1L (VIN T)

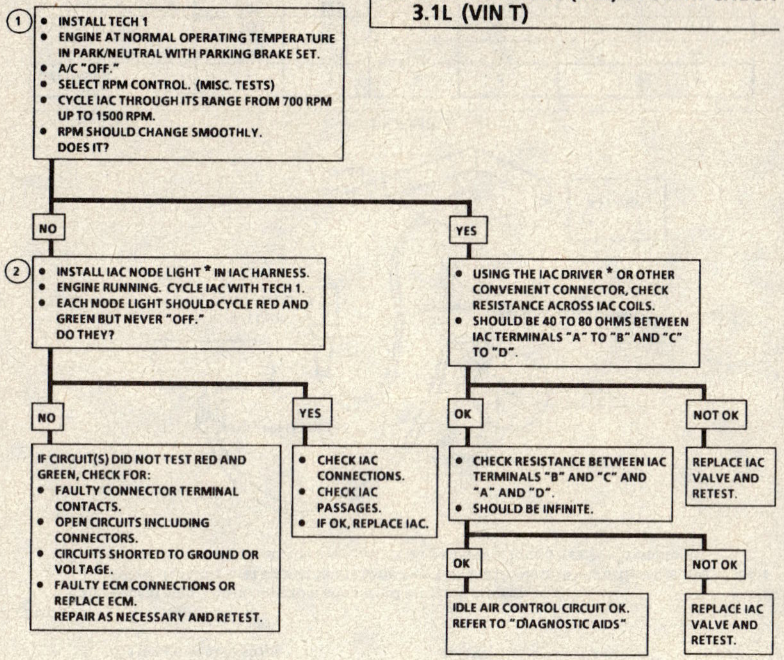

1.
- INSTALL TECH 1
- ENGINE AT NORMAL OPERATING TEMPERATURE IN PARK/NEUTRAL WITH PARKING BRAKE SET.
- A/C "OFF."
- SELECT RPM CONTROL. (MISC. TESTS)
- CYCLE IAC THROUGH ITS RANGE FROM 700 RPM UP TO 1500 RPM.
- RPM SHOULD CHANGE SMOOTHLY. DOES IT?

NO →

2.
- INSTALL IAC NODE LIGHT * IN IAC HARNESS.
- ENGINE RUNNING. CYCLE IAC WITH TECH 1.
- EACH NODE LIGHT SHOULD CYCLE RED AND GREEN BUT NEVER "OFF." DO THEY?

YES →
- USING THE IAC DRIVER * OR OTHER CONVENIENT CONNECTOR, CHECK RESISTANCE ACROSS IAC COILS.
- SHOULD BE 40 TO 80 OHMS BETWEEN IAC TERMINALS "A" TO "B" AND "C" TO "D".

NO
IF CIRCUIT(S) DID NOT TEST RED AND GREEN, CHECK FOR:
- FAULTY CONNECTOR TERMINAL CONTACTS.
- OPEN CIRCUITS INCLUDING CONNECTORS.
- CIRCUITS SHORTED TO GROUND OR VOLTAGE.
- FAULTY ECM CONNECTIONS OR REPLACE ECM. REPAIR AS NECESSARY AND RETEST.

YES
- CHECK IAC CONNECTIONS.
- CHECK IAC PASSAGES.
- IF OK, REPLACE IAC.

OK
- CHECK RESISTANCE BETWEEN IAC TERMINALS "B" AND "C" AND "A" AND "D".
- SHOULD BE INFINITE.

NOT OK
REPLACE IAC VALVE AND RETEST.

OK
IDLE AIR CONTROL CIRCUIT OK. REFER TO "DIAGNOSTIC AIDS"

NOT OK
REPLACE IAC VALVE AND RETEST.

* IAC DRIVER AND NODE LIGHT REQUIRED KIT 222-L FROM: CONCEPT TECHNOLOGY, INC. J 37027 FROM: KENT-MOORE, INC.

CLEAR CODES, CONFIRM "CLOSED LOOP" OPERATION, NO "SERVICE ENGINE SOON" LIGHT, PERFORM IAC RESET PROCEDURE PER APPLICABLE SERVICE MANUAL AND VERIFY CONTROLLED IDLE SPEED IS CORRECT.

"AFTER REPAIRS," CONFIRM "CLOSED LOOP" OPERATION AND NO "SERVICE ENGINE SOON" LIGHT.

3.1L (VIN T) ENGINE — COMPONENT DIAGNOSTIC CHART — CAMARO, FIREBIRD, BERETTA, CORSICA, CAVALIER AND SUNBIRD

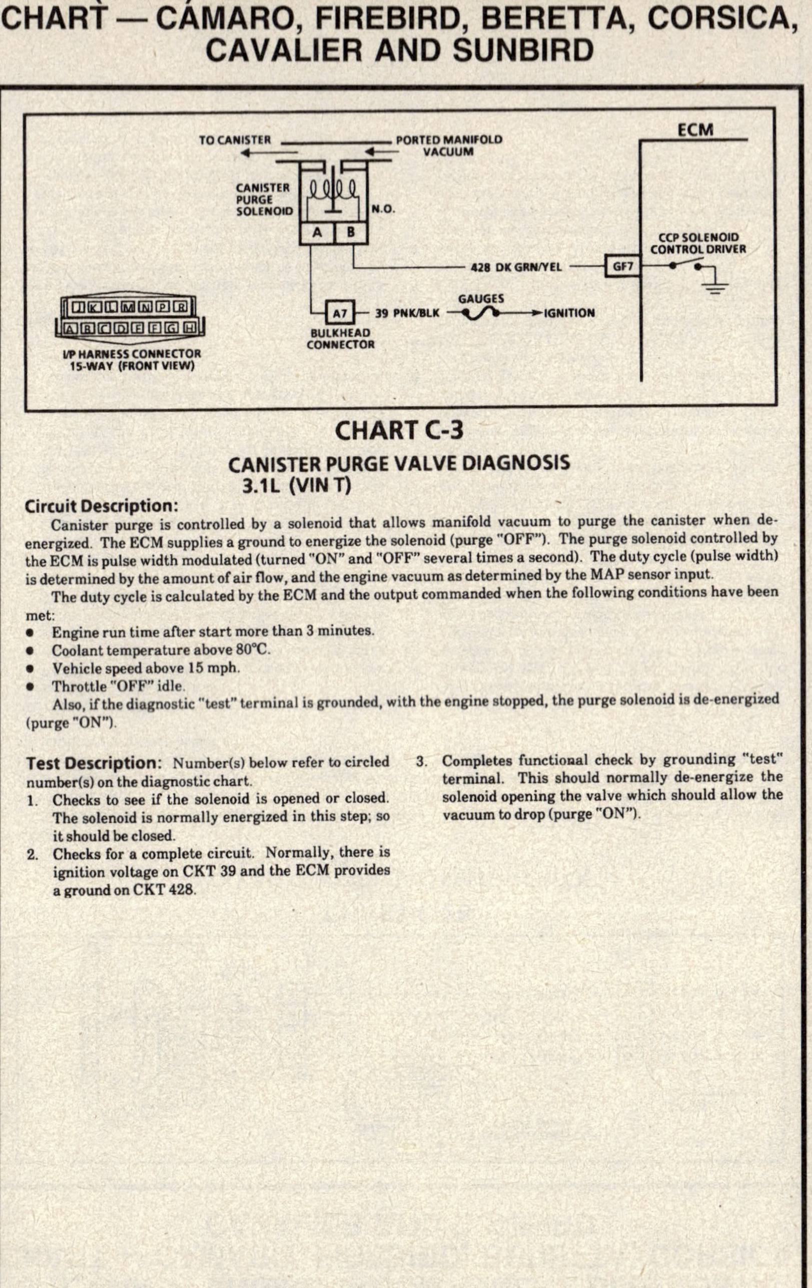

CHART C-3
CANISTER PURGE VALVE DIAGNOSIS
3.1L (VIN T)

Circuit Description:

Canister purge is controlled by a solenoid that allows manifold vacuum to purge the canister when de-energized. The ECM supplies a ground to energize the solenoid (purge "OFF"). The purge solenoid controlled by the ECM is pulse width modulated (turned "ON" and "OFF" several times a second). The duty cycle (pulse width) is determined by the amount of air flow, and the engine vacuum as determined by the MAP sensor input.

The duty cycle is calculated by the ECM and the output commanded when the following conditions have been met:

- Engine run time after start more than 3 minutes.
- Coolant temperature above 80°C.
- Vehicle speed above 15 mph.
- Throttle "OFF" idle.

Also, if the diagnostic "test" terminal is grounded, with the engine stopped, the purge solenoid is de-energized (purge "ON").

Test Description: Number(s) below refer to circled number(s) on the diagnostic chart.

1. Checks to see if the solenoid is opened or closed. The solenoid is normally energized in this step; so it should be closed.
2. Checks for a complete circuit. Normally, there is ignition voltage on CKT 39 and the ECM provides a ground on CKT 428.

3. Completes functional check by grounding "test" terminal. This should normally de-energize the solenoid opening the valve which should allow the vacuum to drop (purge "ON").

3.1L (VIN T) ENGINE — COMPONENT DIAGNOSTIC CHART — CAMARO, FIREBIRD, BERETTA, CORSICA, CAVALIER AND SUNBIRD

CHART C-3

CANISTER PURGE VALVE DIAGNOSIS 3.1L (VIN T)

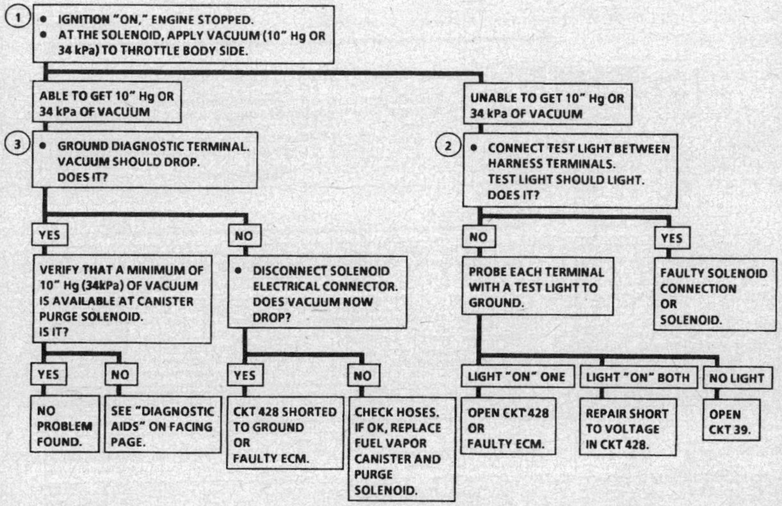

1
• IGNITION "ON," ENGINE STOPPED.
• AT THE SOLENOID, APPLY VACUUM (10" Hg OR 34 kPa) TO THROTTLE BODY SIDE.
ABLE TO GET 10" Hg OR 34 kPa OF VACUUM
UNABLE TO GET 10" Hg OR 34 kPa OF VACUUM
3
• GROUND DIAGNOSTIC TERMINAL. VACUUM SHOULD DROP. DOES IT?
2
• CONNECT TEST LIGHT BETWEEN HARNESS TERMINALS. TEST LIGHT SHOULD LIGHT. DOES IT?
YES
NO
NO
YES
VERIFY THAT A MINIMUM OF 10" Hg (34kPa) OF VACUUM IS AVAILABLE AT CANISTER PURGE SOLENOID. IS IT?
• DISCONNECT SOLENOID ELECTRICAL CONNECTOR. DOES VACUUM NOW DROP?
PROBE EACH TERMINAL WITH A TEST LIGHT TO GROUND.
FAULTY SOLENOID CONNECTION OR SOLENOID.
YES
NO
YES
NO
LIGHT "ON" ONE
LIGHT "ON" BOTH
NO LIGHT
NO PROBLEM FOUND.
SEE "DIAGNOSTIC AIDS" ON FACING PAGE.
CKT 428 SHORTED TO GROUND OR FAULTY ECM.
CHECK HOSES. IF OK, REPLACE FUEL VAPOR CANISTER AND PURGE SOLENOID.
OPEN CKT 428 OR FAULTY ECM.
REPAIR SHORT TO VOLTAGE IN CKT 428.
OPEN CKT 39.

"AFTER REPAIRS," CONFIRM "CLOSED LOOP" OPERATION AND NO "SERVICE ENGINE SOON" LIGHT.

3.1L (VIN T) ENGINE — COMPONENT DIAGNOSTIC CHART — CAMARO, FIREBIRD, BERETTA, CORSICA, CAVALIER AND SUNBIRD

"C"
"+"
IGN
BLK CONN
GRY CONN
TACH
IGNITION COIL CONNECTOR
DISTRIBUTOR
P N
+C
G B R E
A B C D
DISTRIBUTOR CONNECTOR (FRONT VIEW)
ECM
IGN
IGN COIL
S E C
P R I
B
GRY CONN
BLK CONN
A
TACH LEAD
GRY CONN
SET TIMING CONNECTOR
423 WHT
430 PPL/WHT
424 TAN/BLK
453 BLK/RED
BC8
BD8
BC7
BD9
EST
REFERENCE
BYPASS
REFERENCE LOW

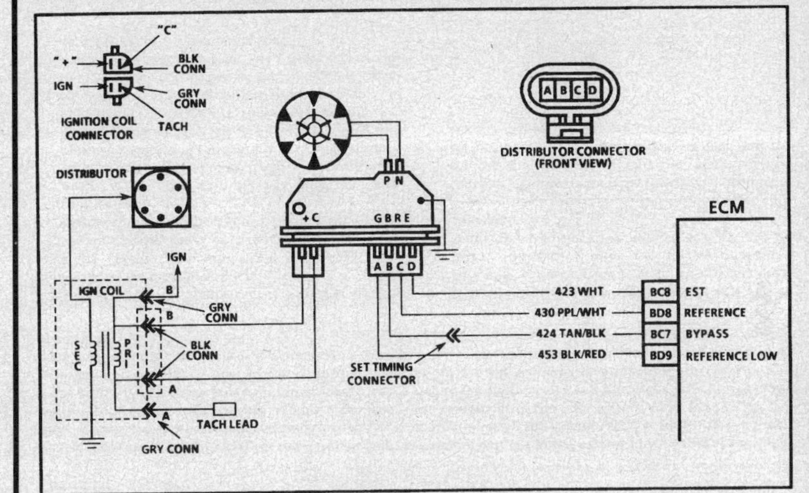

CHART C-4A

IGNITION SYSTEM CHECK
(REMOTE COIL/SEALED MODULE CONNECTOR DISTRIBUTOR)
3.1L (VIN T) "F" CARLINE (PORT)

Test Description: Number(s) below refer to circled number(s) on the diagnostic chart.

1. Two wires are checked, to ensure that an open is not present in a spark plug wire.

1A. If spark occurs with EST connector disconnected, pick-up coil output is too low for EST operation.

2. A spark indicates the problem must be the distributor cap or rotor.

3. Normally, there should be battery voltage at the "C" and "+" terminals. Low voltage would indicate an open or a high resistance circuit from the distributor to the coil or ignition switch. If "C" terminal voltage was low, but "+" terminal voltage is 10 volts or more, circuit from "C" terminal to ignition coil or ignition coil primary winding is open.

4. Checks for a shorted module or grounded circuit from the ignition coil to the module. The distributor module should be turned "OFF," so normal voltage should be about 12 volts.
   If the module is turned "ON," the voltage would be low, but above 1 volt. This could cause the ignition coil to fail from excessive heat. With an open ignition coil primary winding, a small amount of voltage will leak through the module from the "Batt" to the "tach" terminal.

5. Applying a voltage (1.5 to 8 volts) to module terminal "P" should turn the module "ON" and the "tach" terminal voltage should drop to about 7-9 volts. This test will determine whether the module or coil is faulty or if the pick-up coil is not generating the proper signal to turn the module "ON." This test can be performed by using a DC battery with a rating of 1.5 to 8 volts. The use of the test light is mainly to allow the "P" terminal to be probed more easily.
   Some digital multi-meters can also be used to trigger the module by selecting ohms, usually the diode position. In this position, the meter may have a voltage across its terminals which can be used to trigger the module. The voltage in the ohm's position can be checked by using a second meter or by checking the manufacturer's specification of the tool being used.

6. This should turn "OFF" the module and cause a spark. If no spark occurs, the fault is most likely in the ignition coil because most module problems would have been found before this point in the procedure. A module tester could determine which is at fault.

3.1L (VIN T) ENGINE — COMPONENT DIAGNOSTIC CHART — CAMARO, FIREBIRD, BERETTA, CORSICA, CAVALIER AND SUNBIRD

3.1L (VIN T) ENGINE — COMPONENT DIAGNOSTIC CHART — BERETTA, CORSICA, CAVALIER AND SUNBIRD

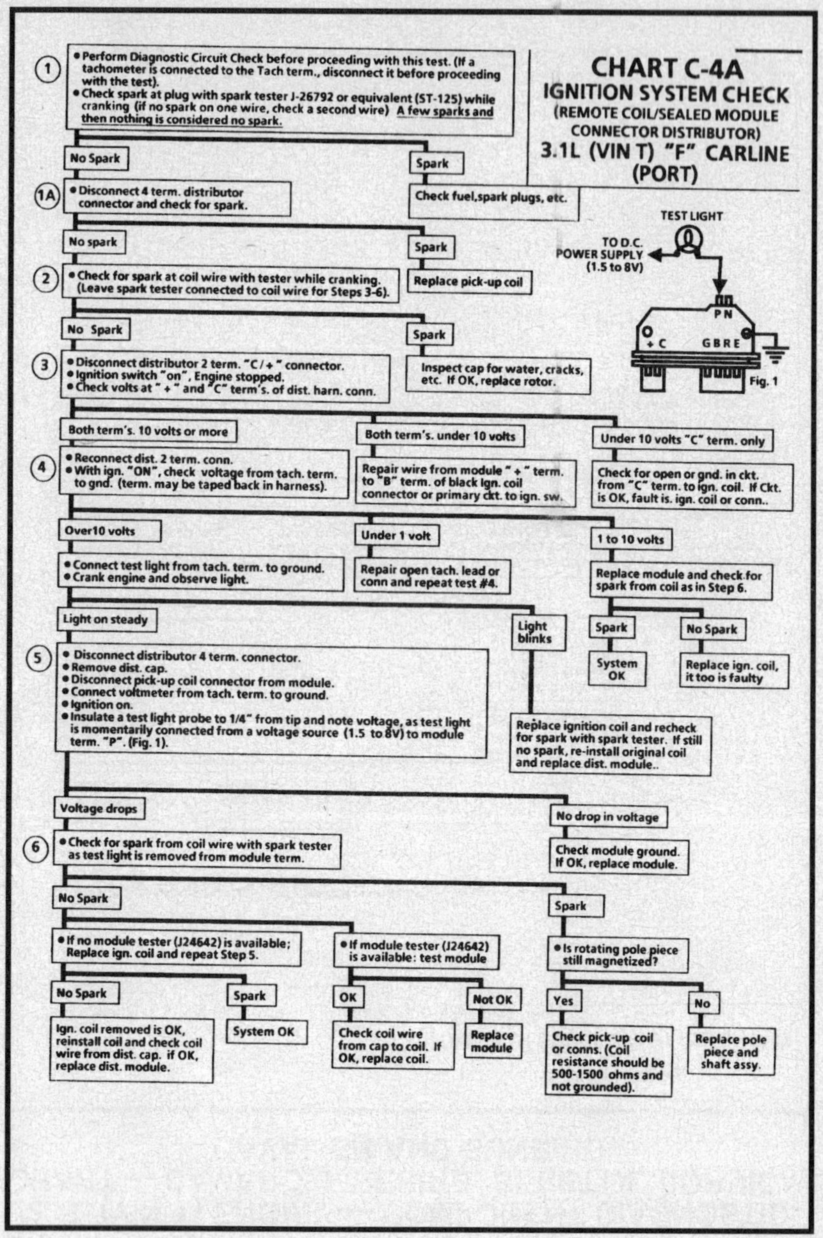

CHART C-4A
IGNITION SYSTEM CHECK
(REMOTE COIL/SEALED MODULE CONNECTOR DISTRIBUTOR)
3.1L (VIN T) "F" CARLINE
(PORT)

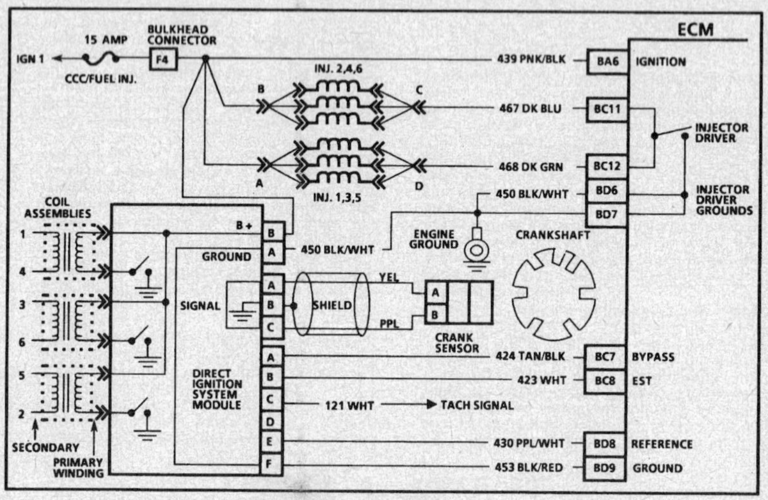

CHART C-4D-1
DIS MISFIRE AT IDLE
3.1L (VIN T)

Circuit Description:

The Direct Ignition System (DIS) uses a waste spark method of distribution. For example, in this type of system the ignition module triggers the #1/4 coil pair resulting in both #1 and #4 spark plugs firing at the same time. #1 cylinder is on the compression stroke at the same time #4 is on the exhaust stroke, resulting in a lower energy requirement to fire #4 spark plug. This leaves the remainder of the high voltage to be used to fire #1 spark plug. On this application, the crank sensor is mounted to the engine block and protrudes through the block to within approximately .050" of the crankshaft reluctor. Since the reluctor is a machined portion of the crankshaft and the crank sensor is mounted in a fixed position on the block, timing adjustments are not possible or necessary.

Test Description: Number(s) below refer to circled number(s) on the diagnostic chart.

1. If the "Misfire" complaint exists under load only, the diagnostic chart on page 2 must be used. Engine rpm should drop approximately equally on all plug leads.
2. A spark test such as a ST-125 must be used because it is essential to verify adequate available secondary voltage at the spark plug (25,000 volts).
3. If the spark jumps the test gap after grounding the opposite plug wire, it indicates excessive resistance in the plug which was bypassed. A faulty or poor connection at that plug could also result in the miss condition. Also, check for carbon deposits inside the spark plug boot.
4. If carbon tracking is evident, replace coil and be sure plug wires relating to that coil are clean and tight. Excessive wire resistance or faulty connections could have caused the coil to be damaged.
5. If the no spark condition follows the suspected coil, that coil is faulty, otherwise, the ignition module is the cause of no spark. This test could also be performed by substituting a known good coil for the one causing the no spark condition.

3.1L (VIN T) ENGINE — COMPONENT DIAGNOSTIC CHART — BERETTA, CORSICA, CAVALIER AND SUNBIRD

CHART C-4D-1
DIS MISFIRE AT IDLE
3.1L (VIN T)

1.
- IF ENGINE MISFIRES UNDER LOAD ONLY, SEE CHART C-4D-2.
- ENGINE IDLING AT NORMAL OPERATING TEMPERATURE, DISCONNECT IAC.
- MOMENTARILY DISCONNECT EACH SPARK PLUG LEAD, USING INSULATED PLIERS, WHILE OBSERVING ENGINE RPM. SEE CAUTION ★.
- ALL PLUG LEAD(S) SHOULD RESULT IN AN RPM DROP.
DID THEY?

NO →

YES → SEE "ROUGH, UNSTABLE OR INCORRECT IDLE OR STALLING" IN "SYMPTOMS" SECTION

2.
- WITH IGNITION "OFF," INSTALL SPARK TESTER (ST-125) J 26792 OR EQUIVALENT ON PLUG LEAD(S) WHOSE REMOVAL DID NOT RESULT IN RPM DROP.
- SPARK SHOULD JUMP TESTER GAP WHILE CRANKING ENGINE. DOES IT?

NO →

YES → CHECK FOR:
- FAULTY, WORN OR CRACKED SPARK PLUG(S).
- PLUG FOULING DUE TO ENGINE MECHANICAL FAULT. IF SPARK PLUGS CHECK OUT OK, SEE "CUTS OUT, MISSES" IN "SYMPTOMS" SECTION

3.
- WITH IGNITION "OFF," GROUND THE OPPOSITE PLUG LEAD OF THE AFFECTED COIL AT SPARK PLUG.
- SPARK SHOULD JUMP TESTER GAP WHILE CRANKING ENGINE. DOES IT?

NO →

YES → REPLACE THE SPARK PLUG FOR THE LEAD WHICH WAS JUMPERED TO GROUND. IF MISFIRE IS STILL PRESENT, START MISFIRE TEST AGAIN AT STEP #1.

- CHECK THE RESISTANCE OF EACH PLUG CABLE OF THE COIL WHICH DID NOT FIRE THE SPARK TESTER.
- CABLE RESISTANCE SHOULD BE LESS THAN 30,000 OHMS EACH AND CABLES SHOULD NOT BE GROUNDED. ARE CABLES OK?

YES →

NO → REPLACE FAULTY CABLE(S).

4.
- REMOVE COIL RETAINING NUTS AND REMOVE COILS.
- COILS SHOULD BE FREE OF CARBON TRACKING. ARE THEY?

YES →

NO → REPLACE IGNITION COIL. ALSO CHECK FOR FAULTY PLUG CABLE CONNECTION(S) AND CABLE NIPPLE(S) FOR CARBON TRACKING.

5.
- SWITCH A NORMALLY OPERATING COIL WITH THE COIL FROM PROBLEM CYLINDER.
- SPARK SHOULD JUMP TESTER GAP AT PROBLEM CYLINDER WHILE CRANKING ENGINE. DID IT?

YES → ORIGINAL IGNITION COIL IS FAULTY.

NO → REPLACE DIS MODULE.

★**CAUTION:** When handling secondary spark plug leads with engine running, insulated pliers must be used and care exercised to prevent a possible electrical shock.

"AFTER REPAIRS," CONFIRM "CLOSED LOOP" OPERATION AND NO "SERVICE ENGINE SOON" LIGHT.

3.1L (VIN T) ENGINE — COMPONENT DIAGNOSTIC CHART — BERETTA, CORSICA, CAVALIER AND SUNBIRD

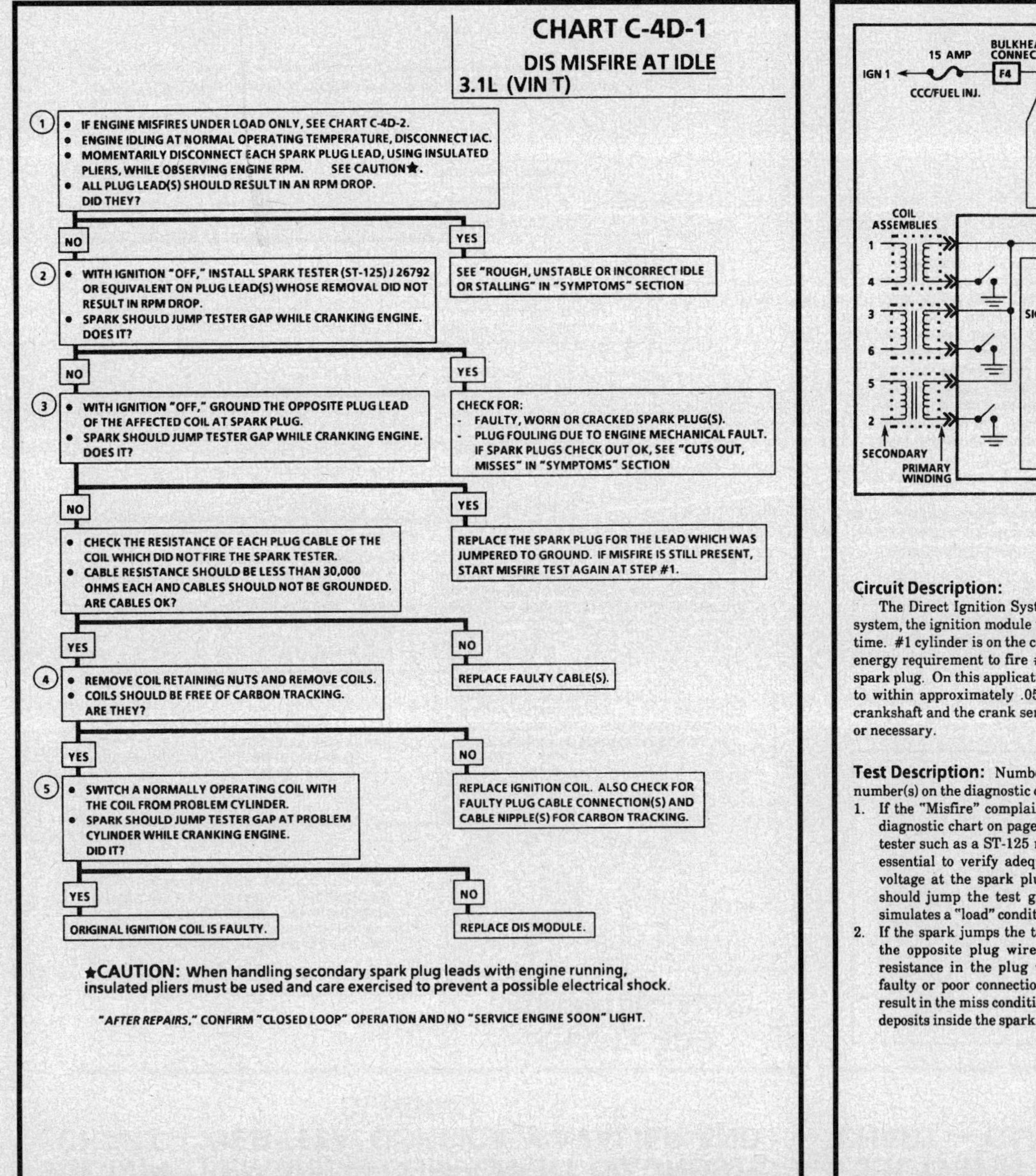

CHART C-4D-2
DIS MISFIRE UNDER LOAD
3.1L (VIN T)

Circuit Description:

The Direct Ignition System (DIS) uses a waste spark method of distribution. For example, in this type of system, the ignition module triggers the #1/4 coil pair resulting in both #1 and #4 spark plugs firing at the same time. #1 cylinder is on the compression stroke at the same time #4 is on the exhaust stroke, resulting in a lower energy requirement to fire #4 spark plug. This leaves the remainder of the high voltage to be used to fire #1 spark plug. On this application, the crank sensor is mounted to the engine block and protrudes through the block to within approximately .050" of the crankshaft reluctor. Since the reluctor is a machined portion of the crankshaft and the crank sensor is mounted in a fixed position on the block, timing adjustments are not possible or necessary.

Test Description: Number(s) below refer to circled number(s) on the diagnostic chart.

1. If the "Misfire" complaint exists at idle only, the diagnostic chart on page 1 must be used. A spark tester such as a ST-125 must be used because it is essential to verify adequate available secondary voltage at the spark plug (25,000 volts). Spark should jump the test gap on all 4 leads. This simulates a "load" condition.

2. If the spark jumps the tester gap after grounding the opposite plug wire, it indicates excessive resistance in the plug which was bypassed. A faulty or poor connection at that plug could also result in the miss condition. Also, check for carbon deposits inside the spark plug boot.

3. If carbon tracing is evident replace coil and be sure plug wires relating to that coil are clean and tight. Excessive wire resistance or faulty connections could have caused the coil to be damaged.

4. If the no spark condition follows the suspected coil, that coil is faulty, otherwise, the ignition module is the cause of no spark. This test could also be performed by substituting a known good coil for the one causing the no spark condition.

3.1L (VIN T) ENGINE — COMPONENT DIAGNOSTIC CHART — BERETTA, CORSICA, CAVALIER AND SUNBIRD

3.1L (VIN T) ENGINE — COMPONENT DIAGNOSTIC CHART — CAMARO, FIREBIRD, BERETTA, CORSICA, CAVALIER AND SUNBIRD

CHART C-4D-2

DIS MISFIRE UNDER LOAD
3.1L (VIN T)

(1)
- IF ENGINE MISFIRES AT IDLE ONLY, SEE CHART C-4D-1.
- IGNITION "OFF."
- DISCONNECT ONE SPARK PLUG LEAD AT A TIME AND, INSTALL SPARK TESTER (ST-125) J 26792 OR EQUIVALENT.
- OBSERVE SPARK TESTER WITH ENGINE IDLING. REPEAT THIS TEST FOR ALL PLUG LEADS.   SEE CAUTION★
- SPARK SHOULD JUMP TESTER GAP ON ALL LEADS WITH ENGINE IDLING.
  DID IT?

NO → (2)
YES →

(2)
- WITH IGNITION "OFF," GROUND THE OPPOSITE PLUG LEAD OF THE AFFECTED COIL AT SPARK PLUG.
- SPARK SHOULD JUMP TESTER GAP WHILE CRANKING ENGINE.
  DOES IT?

CHECK FOR:
- FAULTY, WORN OR CRACKED SPARK PLUG(S).
- PLUG FOULING DUE TO ENGINE MECHANICAL FAULT. IF SPARK PLUGS CHECK OUT OK, SEE "CUTS OUT, MISSES" IN SYMPTOMS SECTION

NO →
YES →

- CHECK THE RESISTANCE OF EACH PLUG CABLE OF THE COIL WHICH DID NOT FIRE THE SPARK TESTER.
- CABLE RESISTANCE SHOULD BE LESS THAN 30,000 OHMS EACH AND CABLES SHOULD NOT BE GROUNDED. ARE CABLES OK?

REPLACE THE SPARK PLUG FOR THE LEAD WHICH WAS JUMPERED TO GROUND. IF MISFIRE IS STILL PRESENT, START MISFIRE TEST AGAIN AT STEP #1.

YES → (3)
NO →

(3)
- REMOVE COIL RETAINING NUTS AND REMOVE COILS.
- COILS SHOULD BE FREE OF CARBON TRACKING. ARE THEY?

REPLACE FAULTY CABLE(S).

YES → (4)
NO →

(4)
- SWITCH A NORMALLY OPERATING COIL WITH THE COIL FROM PROBLEM CYLINDER.
- SPARK SHOULD JUMP TESTER GAP WITH ENGINE IDLING.
  DID IT?

REPLACE IGNITION COIL. ALSO CHECK FOR FAULTY PLUG CABLE CONNECTIONS AND CABLE NIPPLES FOR CARBON TRACKING.

YES →
NO →

ORIGINAL IGNITION COIL IS FAULTY.

REPLACE DIS MODULE.

★CAUTION: When handling secondary spark plug leads with engine running, insulated pliers must be used and care exercised to prevent a possible electrical shock.

"AFTER REPAIRS," CONFIRM "CLOSED LOOP" OPERATION AND NO "SERVICE ENGINE SOON" LIGHT.

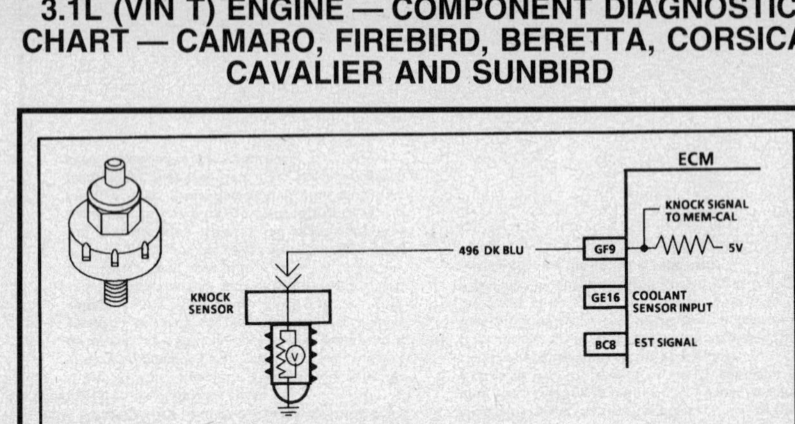

CHART C-5

ELECTRONIC SPARK CONTROL (ESC) SYSTEM CHECK
3.1L (VIN T)

Circuit Description:

The knock sensor is used to detect engine detonation and the ECM will retard the electronic spark timing based on the signal being received. The circuitry within the knock sensor causes the ECM's 5 volts to be pulled down so that under a no knock condition, CKT 496 would measure about 2.5 volts. The knock sensor produces an A.C. signal which rides on the 2.5 volt DC voltage. The amplitude and frequency are dependent upon the knock level.

The MEM-CAL used with this engine, contains the functions which were part of remotely mounted ESC modules used on other GM vehicles. The ESC portion of the MEM-CAL then sends a signal to other parts of the ECM which adjusts the spark timing to retard the spark and reduce the detonation.

Test Description: Number(s) below refer to circled number(s) on the diagnostic chart.

1. With engine idling, there should not be a knock signal present at the ECM because detonation is not likely under a no load condition.
2. Tapping on the engine lift hood bracket should simulate a knock signal to determine if the sensor is capable of detecting detonation. If no knock is detected, try tapping on engine block closer to sensor before replacing sensor.
3. If the engine has an internal problem which is creating a knock, the knock sensor may be responding to the internal failure.
4. This test determines if the knock sensor is faulty or if the ESC portion of the MEM-CAL is faulty. If it is determined that the MEM-CAL is faulty, be sure that it is properly installed and latched into place. If not properly installed, repair and retest.

Diagnostic Aids:

While observing knock signal on the Tech 1 "Scan," tool there should be an indication that knock is present when detonation can be heard. Detonation is most likely to occur under high engine load conditions.

3.1L (VIN T) ENGINE — COMPONENT DIAGNOSTIC CHART — CAMARO, FIREBIRD, BERETTA, CORSICA, CAVALIER AND SUNBIRD

CHART C-5
ELECTRONIC SPARK CONTROL (ESC) SYSTEM CHECK
3.1L (VIN T)

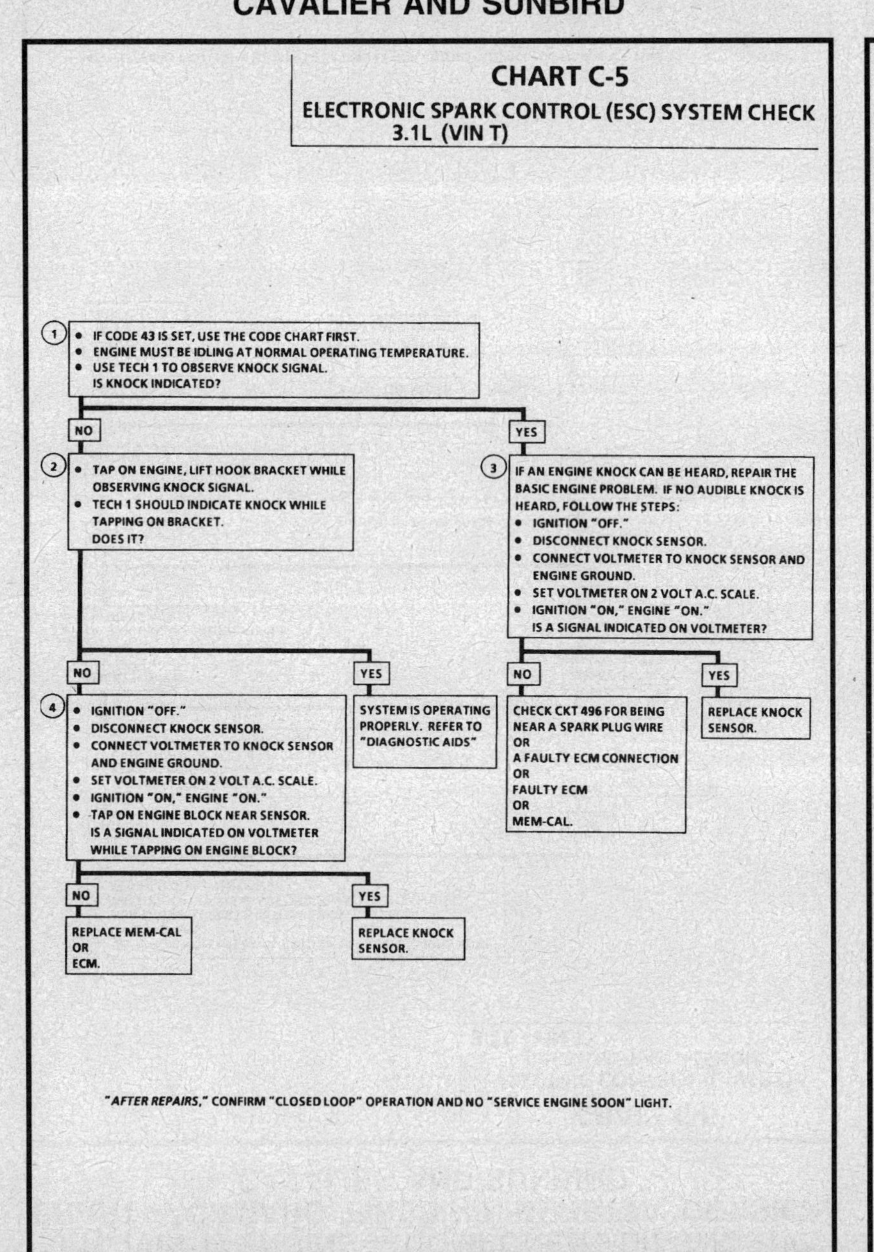

"AFTER REPAIRS," CONFIRM "CLOSED LOOP" OPERATION AND NO "SERVICE ENGINE SOON" LIGHT.

3.1L (VIN T) ENGINE — COMPONENT DIAGNOSTIC CHART — CAMARO, FIREBIRD, BERETTA, CORSICA, CAVALIER AND SUNBIRD

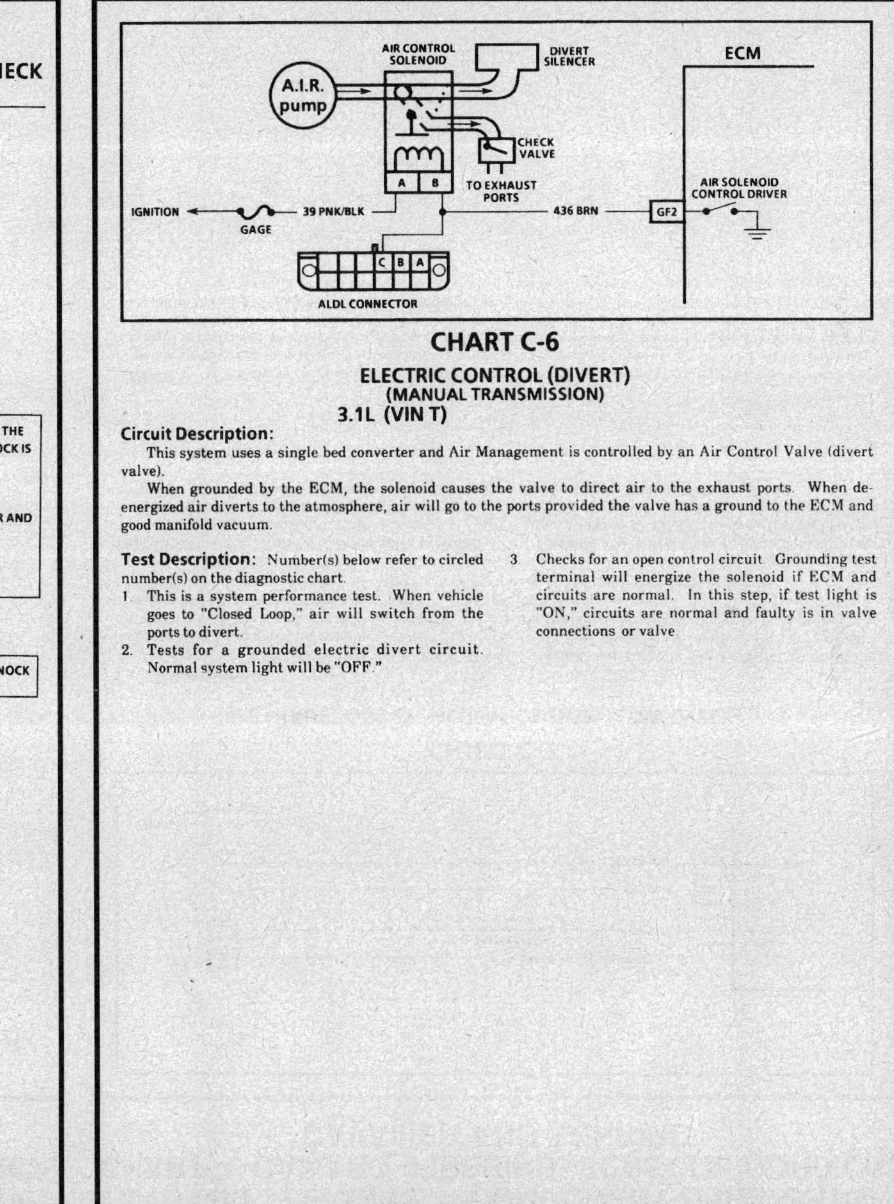

CHART C-6
ELECTRIC CONTROL (DIVERT)
(MANUAL TRANSMISSION)
3.1L (VIN T)

Circuit Description:
This system uses a single bed converter and Air Management is controlled by an Air Control Valve (divert valve).

When grounded by the ECM, the solenoid causes the valve to direct air to the exhaust ports. When de-energized air diverts to the atmosphere, air will go to the ports provided the valve has a ground to the ECM and good manifold vacuum.

Test Description: Number(s) below refer to circled number(s) on the diagnostic chart.

1. This is a system performance test. When vehicle goes to "Closed Loop," air will switch from the ports to divert.
2. Tests for a grounded electric divert circuit. Normal system light will be "OFF."
3. Checks for an open control circuit. Grounding test terminal will energize the solenoid if ECM and circuits are normal. In this step, if test light is "ON," circuits are normal and faulty is in valve connections or valve.

3.1L (VIN T) ENGINE — COMPONENT DIAGNOSTIC CHART — CAMARO, FIREBIRD, BERETTA, CORSICA, CAVALIER AND SUNBIRD

CHART C-6

ELECTRIC CONTROL (DIVERT)
(MANUAL TRANSMISISON)
3.1L (VIN T)

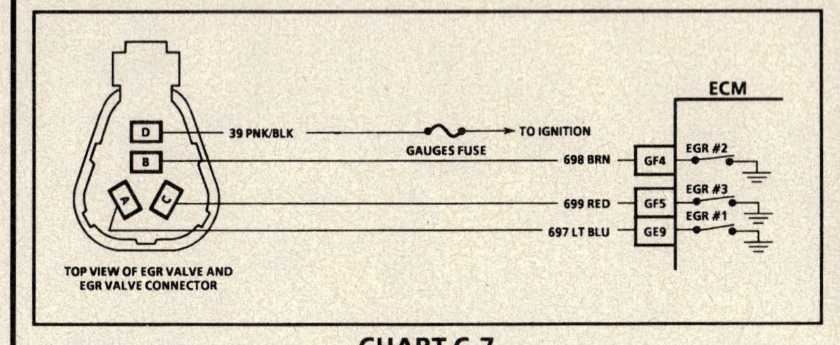

"AFTER REPAIRS," CONFIRM "CLOSED LOOP" OPERATION AND NO "SERVICE ENGINE SOON" LIGHT.

3.1L (VIN T) ENGINE — COMPONENT DIAGNOSTIC CHART — CAMARO, FIREBIRD, BERETTA, CORSICA, CAVALIER AND SUNBIRD

CHART C-7

EXHAUST GAS RECIRCULATION (EGR) FLOW CHECK
3.1L (VIN T)

Circuit Description:

The digital Exhaust Gas Recirculation (EGR) valve is designed to accurately supply EGR to an engine independent of intake manifold vacuum. The valve controls EGR flow from the exhaust to the intake manifold through three orifices which increment in size to produce seven combinations. When a solenoid is energized, the armature with attached shaft and swivel pintle is lifted opening the orifice.

The flow accuracy is dependent on metering orifice size only, which results in improved control.

Test Description: Number(s) below refer to circled number(s) on diagnostic chart.

1. Codes should be diagnosed using appropriate chart before preparing a functional check.

2. This step activates each solenoid individually. As you energize #1 or #2 solenoid, the engine rpm should drop. #3 solenoid has the large port and may stall the engine when energized.

NOTICE: If the digital EGR valve shows signs of excessive heat, a melted condition. Check the exhaust system for blockage (possibly a plugged converter) using the procedure found on CHART B-1. If the exhaust system is restricted repair the cause, one of which might be an injector which is open due to one of the following: a. stuck, b. grounded driver circuit, c. possibly defective ECM. If this condition is found, the oil should be checked for possible fuel contamination.

3.1L (VIN T) ENGINE — COMPONENT DIAGNOSTIC CHART — CAMARO, FIREBIRD, BERETTA, CORSICA, CAVALIER AND SUNBIRD

CHART C-7
EXHAUST GAS RECIRCULATION (EGR) FLOW CHECK
3.1L (VIN T)

① • IGNITION "ON," ENGINE "OFF."
• TECH 1 "SCAN" TOOL TROUBLE CODES.
• IF CODES ARE PRESENT, REFER TO THOSE CHARTS FIRST.

② • USING TECH 1 "SCAN" TOOL, SELECT EGR CONTROL (MISCELLANEOUS).
• START ENGINE AND ALLOW IDLE TO STABILIZE.
• ENERGIZE EGR SOL #1, RPM SHOULD DROP SLIGHTLY.*
• ENERGIZE EGR SOL #2, ENGINE SHOULD HAVE A ROUGH IDLE.*
• ENERGIZE EGR SOL #3, ENGINE SHOULD IDLE ROUGH OR STALL.*
DID RPM DROP AND ENGINE IDLE ROUGH ON ALL THREE SOLENOIDS?

YES
EGR SYSTEM IS OK.
NO PROBLEM FOUND.

NO
CHECK EGR VALVE, EGR TRANSFER TUBE, ADAPTOR, GASKETS, FITTINGS AND ALL PASSAGES FOR DAMAGE, LEAKAGE OR PLUGGING. IF OK, REPLACE EGR VALVE.

* THESE STEPS MUST BE DONE VERY QUICKLY, AS THE ECM WILL ADJUST THE IDLE AIR CONTROL VALVE TO CORRECT IDLE SPEED.

"AFTER REPAIRS," CONFIRM "CLOSED LOOP" OPERATION AND NO "SERVICE ENGINE SOON" LIGHT.

3.1L (VIN T) ENGINE — COMPONENT DIAGNOSTIC CHART — CAMARO, FIREBIRD, BERETTA, CORSICA, CAVALIER AND SUNBIRD

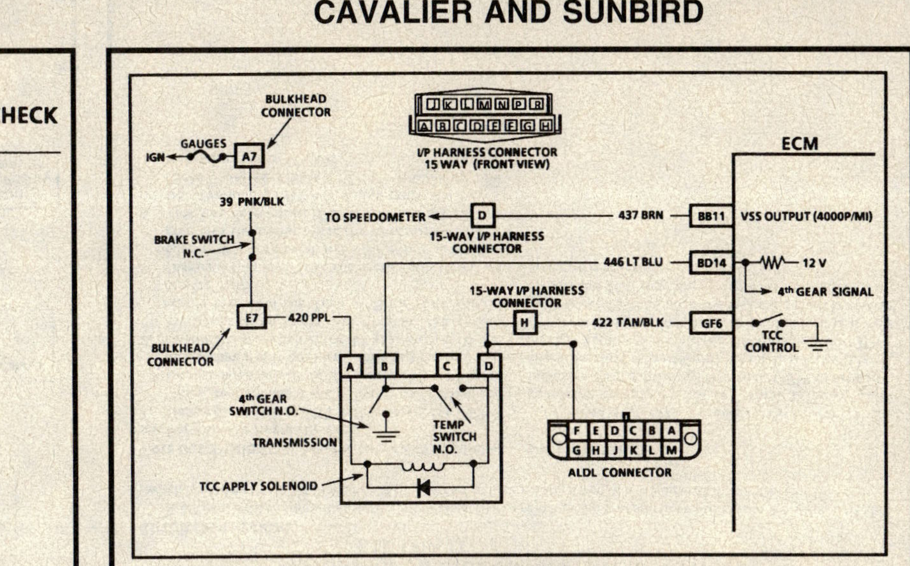

CHART C-8
(Page 1 of 2)
AUTOMATIC TORQUE CONVERTER CLUTCH (TCC)
3.1L (VIN T)

Circuit Description:

The purpose of the automatic transmission Torque Converter Clutch (TCC) feature is to eliminate the power loss of the torque converter stage when the vehicle is in a cruise condition. This allows the convenience of the automatic transmission and the fuel economy of a manual transmission. The heart of the system is a solenoid located inside the automatic transmission which is controlled by the ECM.

When the solenoid coil is activated ("ON"), the Torque Converter Clutch (TCC) is applied which results in straight through mechanical coupling from the engine to transmission. When the transmission solenoid is deactivated, the Torque Converter Clutch (TCC) is released which allows the torque converter to operate in the conventional manner (fluidic coupling between engine and transmission).

The ECM turns "ON" the TCC when coolant temperature is above 65°C (149°F), TPS not changing, and vehicle speed above a specified value.

Test Description: Number(s) below refer to circled number(s) on the diagnostic chart.
1. When a test light is connected from ALDL terminal "F" to ground, a test light "ON" indicates battery voltage is OK and the TCC solenoid is disengaged.
2. When the diagnostic terminal is grounded, the ECM should energize the TCC solenoid and the test light should go out.

Diagnostic Aids:

A Tech 1 "Scan" tool only indicates when the ECM has turned "ON" the TCC driver (grounded CKT 422), but this does not confirm that the TCC has engaged. To determine if TCC is functioning properly, engine rpm should decrease when the Tech 1 "Scan" tool indicates the TCC driver has turned "ON." To determine if the 4th gear switch is functioning properly, perform the checks in CHART C-8 (Page 2 of 2). The switches will not prevent TCC from functioning but will affect TCC lock and unlock points.

3.1L (VIN T) ENGINE — COMPONENT DIAGNOSTIC CHART — CAMARO, FIREBIRD, BERETTA, CORSICA, CAVALIER AND SUNBIRD

CHART C-8
(Page 1 of 2)
AUTOMATIC TORQUE CONVERTER CLUTCH (TCC)
3.1L (VIN T)

(1) USING A TECH 1 "SCAN" TOOL, CHECK THE FOLLOWING AND CORRECT IF NECESSARY.
- COOLANT TEMPERATURE SHOULD BE ABOVE 65°C.
- TPS - BE SURE TPS SIGNAL IS NOT ERRATIC.
- VSS - BE SURE TECH 1 "SCAN" TOOL DISPLAYS VSS WITH DRIVE WHEELS TURNING. IF CODE 24 IS PRESENT, SEE CODE CHART 24.

- MECHANICAL CHECKS, SUCH AS LINKAGE, OIL LEVEL, ETC., SHOULD BE PERFORMED PRIOR TO USING THIS CHART.
- IGNITION "ON."
- CONNECT TEST LIGHT TO ALDL CONNECTOR TERMINAL "F" AND GROUND. BULB SHOULD "LIGHT." DOES IT?

YES →

- DEPRESS BRAKE PEDAL. LIGHT SHOULD GO OUT. DOES IT?

YES →

(2)
- IGNITION "ON," ENGINE "OFF."
- RELEASE BRAKE PEDAL.
- GROUND DIAGNOSTIC TERMINAL. LIGHT SHOULD GO OUT. DOES IT?

YES →

TCC CIRCUIT OK. BE SURE VEHICLE IS EQUIPPED WITH THE CORRECT PROM OR MEM-CAL. TO CHECK 4th GEAR SWITCH, SEE CHART C-8 (2 OF 2).

NO →

OPEN CKT 422 OR FAULTY ECM.

NO → (from brake pedal step)

BRAKE SWITCH OUT OF ADJUSTMENT OR FAULTY OR CKT 422 SHORTED TO VOLTAGE.

NO → (from bulb should light step)

- DISCONNECT TCC ELECTRICAL CONNECTOR.
- CONNECT TEST LIGHT BETWEEN TERMINALS "A" & "D". BULB SHOULD <u>NOT</u> LIGHT. DOES IT?

NO →

- CONNECT TEST LIGHT FROM TERMINAL "A" TO GROUND. BULB SHOULD LIGHT. DOES IT?

YES →

- GROUND ALDL TERMINAL "F".
- WITH TEST LIGHT CONNECTED BETWEEN TRANSMISSION CONNECTOR TERMINALS "A" AND "D". THE BULB SHOULD LIGHT. DOES IT?

YES →

FAULTY TCC CONNECTION OR TCC SOLENOID.

NO →

REPAIR OPEN CIRCUIT BETWEEN TRANSMISSION AND ALDL TERMINAL "F".

NO → (from connect test light from terminal "A" to ground step)

OPEN CKT 39, TCC BRAKE SWITCH CIRCUIT OR ADJUST SWITCH.

YES → (from disconnect TCC step)

CKT 422 SHORTED TO GROUND OR FAULTY ECM.

"AFTER REPAIRS," CONFIRM "CLOSED LOOP" OPERATION AND NO "SERVICE ENGINE SOON" LIGHT.

3.1L (VIN T) ENGINE — COMPONENT DIAGNOSTIC CHART — CAMARO, FIREBIRD, BERETTA, CORSICA, CAVALIER AND SUNBIRD

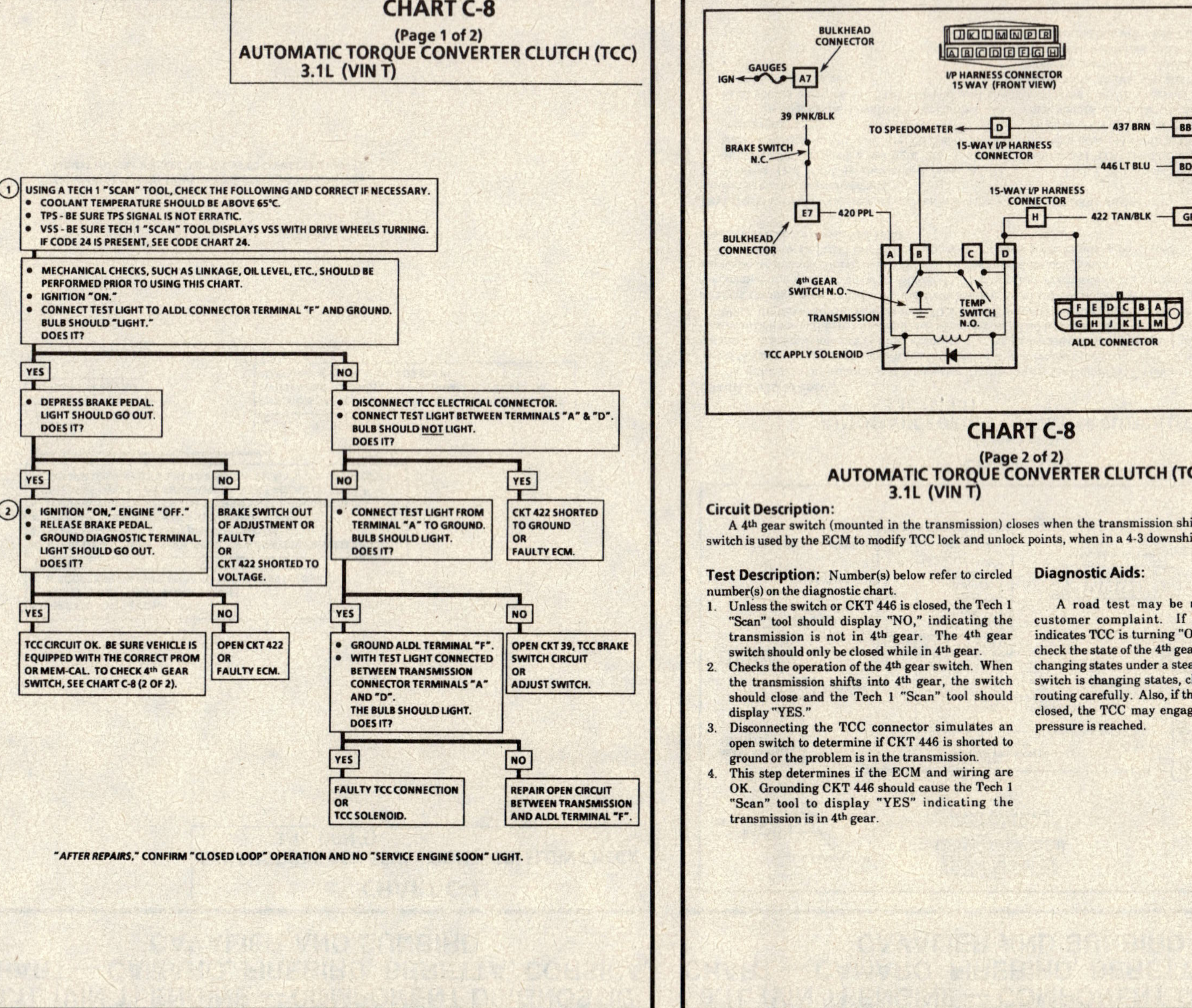

CHART C-8
(Page 2 of 2)
AUTOMATIC TORQUE CONVERTER CLUTCH (TCC)
3.1L (VIN T)

Circuit Description:

A 4th gear switch (mounted in the transmission) closes when the transmission shifts into 4th gear, and this switch is used by the ECM to modify TCC lock and unlock points, when in a 4-3 downshift maneuver.

Test Description: Number(s) below refer to circled number(s) on the diagnostic chart.

1. Unless the switch or CKT 446 is closed, the Tech 1 "Scan" tool should display "NO," indicating the transmission is not in 4th gear. The 4th gear switch should only be closed while in 4th gear.
2. Checks the operation of the 4th gear switch. When the transmission shifts into 4th gear, the switch should close and the Tech 1 "Scan" tool should display "YES."
3. Disconnecting the TCC connector simulates an open switch to determine if CKT 446 is shorted to ground or the problem is in the transmission.
4. This step determines if the ECM and wiring are OK. Grounding CKT 446 should cause the Tech 1 "Scan" tool to display "YES" indicating the transmission is in 4th gear.

Diagnostic Aids:

A road test may be necessary to verify the customer complaint. If the Tech 1 "Scan" tool indicates TCC is turning "ON" and "OFF" erratically, check the state of the 4th gear switch to be sure it is not changing states under a steady throttle position. If the switch is changing states, check connections and wire routing carefully. Also, if the 4th gear switch is always closed, the TCC may engage as soon as sufficient oil pressure is reached.

3.1L (VIN T) ENGINE — COMPONENT DIAGNOSTIC CHART — CAMARO, FIREBIRD, BERETTA, CORSICA, CAVALIER AND SUNBIRD

CHART C-8
(Page 2 of 2)
AUTOMATIC TORQUE CONVERTER CLUTCH (TCC)
3.1L (VIN T)

"AFTER REPAIRS," CONFIRM "CLOSED LOOP" OPERATION AND NO "SERVICE ENGINE SOON" LIGHT.

3.1L (VIN T) ENGINE — COMPONENT DIAGNOSTIC CHART — CAMARO, FIREBIRD, BERETTA, CORSICA, CAVALIER AND SUNBIRD

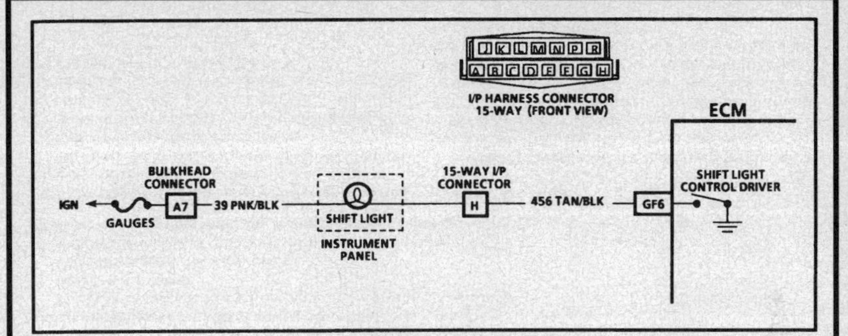

CHART C-8A
M/T SHIFT LIGHT CHECK
3.1L (VIN T) "F" CARLINE (PORT)

Circuit Description:
The shift light indicates the best transmission shift point for maximum fuel economy. The light is controlled by the ECM and is turned "ON" by grounding CKT 456. The ECM uses information from the following inputs to control the shift light:
• Coolant temperature.
• TPS.
• VSS.
• rpm.
The ECM uses the measured rpm and the vehicle speed to calculate what gear the vehicle is in. It's this calculation that determines when the shift light should be turned "ON."

Test Description: Number(s) below refer to circled number(s) on the diagnostic chart.
1. This should not turn "ON" the shift light. If the light is "ON," there is a short to ground in CKT 456 wiring or a fault in the ECM.
2. When the diagnostic terminal is grounded, the ECM should ground CKT 456 and the shift light should come "ON."
3. This checks the shift light circuit up to the ECM connector. If the shift light illuminates, then the ECM connector is faulty or the ECM does not have the ability to ground the circuit.

3.1L (VIN T) ENGINE — COMPONENT DIAGNOSTIC CHART — CAMARO, FIREBIRD, BERETTA, CORSICA, CAVALIER AND SUNBIRD

CHART C-8A

M/T SHIFT LIGHT CHECK
3.1L (VIN T)

(1)
• IGNITION "ON", ENGINE "OFF".
• OBSERVE SHIFT LIGHT.

LIGHT "OFF"

(2) GROUND ALDL DIAGNOSTIC TERMINAL AND OBSERVE SHIFT LIGHT.

LIGHT "ON"
• IGNITION "OFF"
• DISCONNECT ECM CONNECTORS.
• TURN IGNITION "ON" AND OBSERVE SHIFT LIGHT.

LIGHT "ON"
REPAIR SHORT TO GROUND IN CKT 456.

LIGHT "OFF"
REPLACE ECM.

LIGHT "OFF"

(3)
• IGNITION "OFF".
• DISCONNECT ECM CONNECTORS.
• IGNITION "ON".
• JUMPER CKT 456 TO GROUND AND OBSERVE SHIFT LIGHT.

LIGHT "ON"
CHECK FOR:
- CODE 24, (NO VSS).
- THERMOSTAT FAULTY OR HAS INCORRECT HEAT RANGE. IF OK, REVIEW SYMPTOMS

LIGHT "OFF"
OPEN IGNITION CKT 39, OPEN CKT 456, OR FAULTY BULB.

LIGHT "ON"
POOR CONNECTION AT ECM OR FAULTY ECM.

"AFTER REPAIRS," CONFIRM "CLOSED LOOP" OPERATION AND NO "SERVICE ENGINE SOON" LIGHT.

3.1L (VIN T) ENGINE — COMPONENT DIAGNOSTIC CHART — BERETTA, CORSICA, CAVALIER AND SUNBIRD

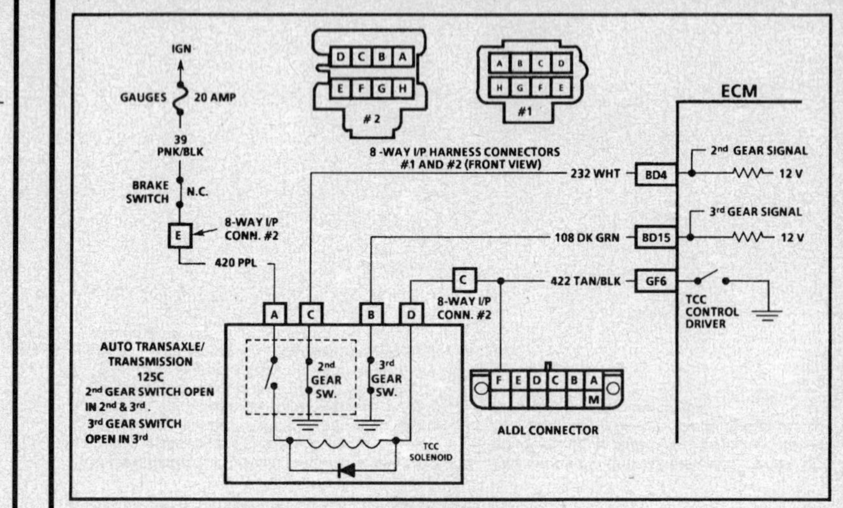

CHART C-8

(Page 1 of 2)
3T40 TORQUE CONVERTER CLUTCH (TCC)
(ELECTRICAL DIAGNOSIS)
3.1L (VIN T)

Circuit Description:

The purpose of the automatic transmission Torque Converter Clutch (TCC) feature is to eliminate the power loss of the torque converter when the vehicle is in a cruise condition. This allows the convenience of the automatic transmission and the fuel economy of a manual transmission. The heart of the system is a solenoid, located inside the automatic transmission which is controlled by the ECM.

When the solenoid coil is activated ("ON"), the torque converter clutch is applied which results in straight through mechanical coupling from the engine to transmission. When the transmission solenoid is deactivated, the torque converter clutch is released, which allows the torque converter to operate in the conventional manner (fluidic coupling between engine and transmission).

The TCC will engage on a warm engine under given road load in 2nd and 3rd gears. TCC will engage when:

• Brake switch closed.
• Coolant temperature is above 65°C.
• Vehicle speed above a calibrated value (about 28 mph 45km/h).
• Throttle position not changing, indicating a steady speed.

Test Description: Number(s) below refer to circled number(s) on the diagnostic chart.

1. This test checks the continuity of the TCC circuit from the fuse to the ALDL connector.
2. When the brake pedal is released, the light should come back "ON" and then go "OFF" when the diagnostic terminal is grounded. This tests CKT 422 and the TCC driver in the ECM.

Diagnostic Aids:

A Tech 1 "Scan" tool only indicates when the ECM has turned "ON" the TCC driver, and this does not confirm that the TCC has engaged. To determine if TCC is functioning properly, engine rpm should decrease when the Tech 1 "Scan" tool indicates the TCC driver has turned "ON." A thermostat, that is stuck open or opens at too low a temperature, may result in an inoperative TCC.

3.1L (VIN T) ENGINE — COMPONENT DIAGNOSTIC CHART — BERETTA, CORSICA, CAVALIER AND SUNBIRD

CHART C-8
(Page 1 of 2)
3T40 TORQUE CONVERTER CLUTCH (TCC)
(ELECTRICAL DIAGNOSIS)
3.1L (VIN T)

- USING A TECH 1 "SCAN" TOOL CHECK THE FOLLOWING AND CORRECT IF NECESSARY.
- COOLANT TEMPERATURE SHOULD BE ABOVE 65°C.
- TPS - BE SURE TPS SIGNAL IS NOT ERRATIC.
- VSS - BE SURE TECH 1 "SCAN" TOOL DISPLAYS VSS WITH DRIVE WHEELS TURNING.

(1)
- MECHANICAL CHECKS, SUCH AS LINKAGE, OIL LEVEL, ETC. SHOULD BE PERFORMED PRIOR TO USING THIS CHART.
- IGNITION "ON."
- CONNECT TEST LIGHT TO ALDL CONNECTOR TERMINAL "F" AND GROUND.
BULB SHOULD "LIGHT."
DOES IT?

YES
- DEPRESS BRAKE PEDAL.
LIGHT SHOULD GO OUT.
DOES IT?

NO
- DISCONNECT TCC ELECTRICAL CONNECTOR.
- CONNECT TEST LIGHT BETWEEN TERM. "A" AND "D".
BULB SHOULD NOT "LIGHT."
DOES IT?

YES
(2)
- IGNITION "ON," ENGINE "OFF."
- RELEASE BRAKE PEDAL.
- GROUND DIAG. TERMINAL.
LIGHT SHOULD GO OUT.
DOES IT?

NO
BRAKE SWITCH OUT OF ADJUSTMENT
OR
FAULTY
OR
CKT 422 SHORTED TO VOLTAGE.

NO
- CONNECT TEST LIGHT FROM TERMINAL "A" TO GROUND.
BULB SHOULD "LIGHT."
DOES IT?

YES
CKT 422 SHORTED TO GROUND
OR
FAULTY ECM.

YES
TCC CIRCUIT OK. REFER TO "DIAGNOSTIC AIDS" ON FACING PAGE. BE SURE VEHICLE IS EQUIPPED WITH THE CORRECT MEM-CAL. TO CHECK GEAR SWITCHES REFER TO APPROPRIATE TRANSAXLE GEAR SWITCH CHART.

NO
OPEN CKT 422
OR
FAULTY ECM.

YES
- GROUND ALDL TERM. "F".
- WITH TEST LIGHT CONN. BETWEEN TRANS. CONNECTOR TERMINALS "A" AND "D". THE BULB SHOULD "LIGHT."
DOES IT?

NO
TCC BRAKE SWITCH NOT ADJUSTED PROPERLY
OR
OPEN IN CKT'S 39 OR 420.

YES
FAULTY TCC CONNECTION
OR
TCC SOLENOID.

NO
REPAIR OPEN CIRCUIT BETWEEN TRANS. AND ALDL TERMINAL "F".

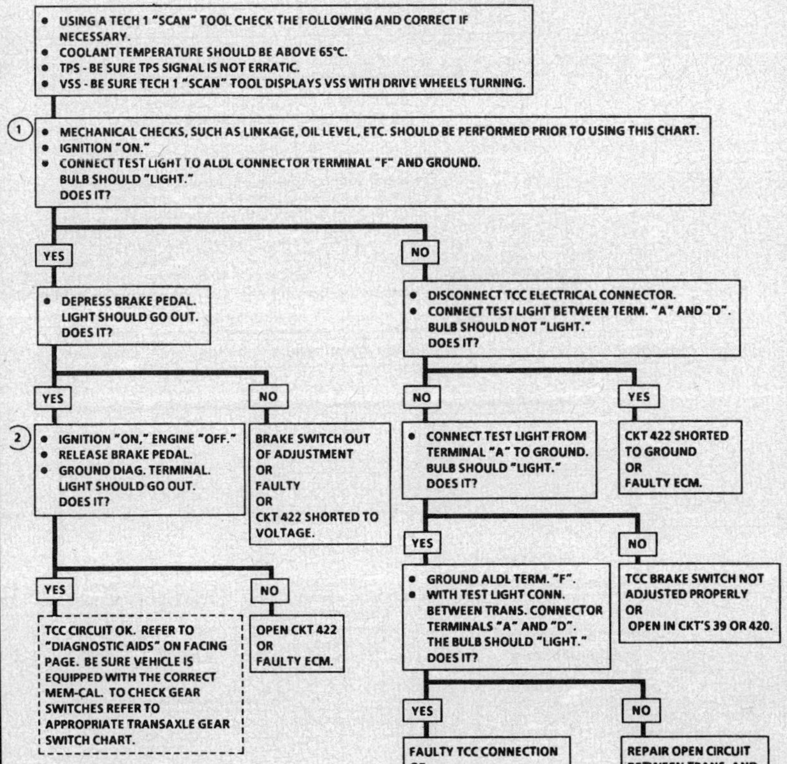

"AFTER REPAIRS," CONFIRM "CLOSED LOOP" OPERATION AND NO "SERVICE ENGINE SOON" LIGHT.

3.1L (VIN T) ENGINE — COMPONENT DIAGNOSTIC CHART — BERETTA, CORSICA, CAVALIER AND SUNBIRD

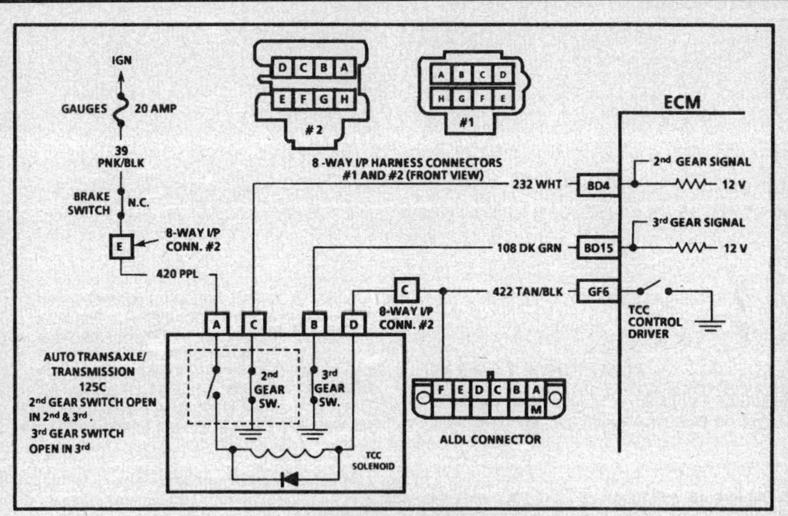

CHART C-8
(Page 2 of 2)
3T40 TORQUE CONVERTER CLUTCH (TCC)
(ELECTRICAL DIAGNOSIS)
3.1L (VIN T)

Circuit Description:
The 2nd gear switch in this vehicle will be open in 2nd and 3rd gear. The ECM uses this signal to disengage the TCC when going into a downshift.
The 3rd gear switch will be open in 3rd gear.

Test Description: Number(s) below refer to circled number(s) on the diagnostic chart.
1. Some "Scan" tools display the state of these switches in different ways. Be familiar with the type of tool being used. Since both switches should be in the closed state during this test, the tool should read the same for either the 2nd or 3rd gear switch.
2. Determines whether the switch or signal circuit is open. The circuit can be checked for an open by measuring the voltage (with a voltmeter) at the TCC connector (should be about 12 volts).
3. Because the switch(s) should be grounded in this step, disconnecting the TCC connector should cause the Tech 1 "Scan" tool switch state to change.
4. The switch state should change when the vehicle shifts into 2nd gear.

Diagnostic Aids:

If vehicle is road tested because of a TCC related problem, be sure the switch states do not change while in 3rd gear because the TCC will disengage. If switches change state, carefully check wire routing and connections.

3.1L (VIN T) ENGINE — COMPONENT DIAGNOSTIC CHART — BERETTA, CORSICA, CAVALIER AND SUNBIRD

CHART C-8

(Page 2 of 2)
3T40 TORQUE CONVERTER CLUTCH (TCC)
(ELECTRICAL DIAGNOSIS)
3.1L (VIN T)

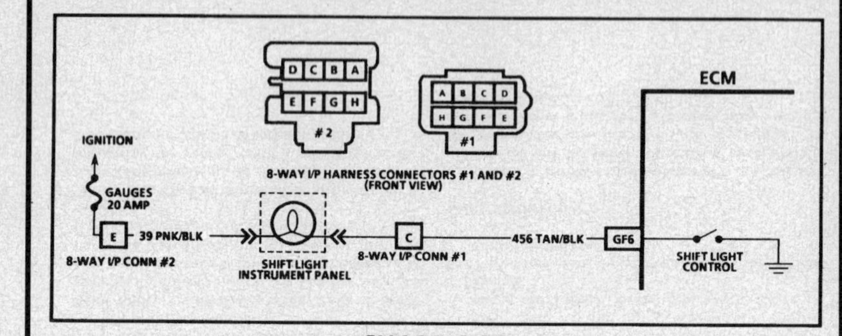

3.1L (VIN T) ENGINE — COMPONENT DIAGNOSTIC CHART — BERETTA, CORSICA, CAVALIER AND SUNBIRD

CHART C-8C

MANUAL TRANSMISSION (M/T) SHIFT LIGHT CHECK
3.1L (VIN T)

Circuit Description:

The shift light indicates the best transmission shift point for maximum fuel economy. The light is controlled by the ECM and is turned "ON" by grounding CKT 456. The ECM uses information from the following inputs to control the shift light:

- Coolant temperature must be above 16°C (61°F).
- TPS above 4%.
- VSS.
- RPM above about 1900.
- Calculate Air Flow - The ECM uses rpm, air flow VSS to calculate what gear the vehicle is in.

It's this calculation that determines when the shift light should be turned "ON." The shift light will only stay "ON" 5 seconds after the conditions were met to turn it "ON."

Test Description: Number(s) below refer to circled number(s) on the diagnostic chart.

1. This should not turn "ON" the shift light. If the light is "ON," there is a short to ground in CKT 456 wiring or a fault in the ECM.
2. When the diagnostic terminal is grounded, the ECM should ground CKT 456 and the shift light should come "ON."
3. This checks the shift light circuit up to the ECM connector. If the shift light illuminates, then the ECM connector is faulty or the ECM does not have the ability to ground the circuit.

3.1L (VIN T) ENGINE — COMPONENT DIAGNOSTIC CHART — BERETTA, CORSICA, CAVALIER AND SUNBIRD

CHART C-8C

MANUAL TRANSMISSION (M/T) SHIFT LIGHT CHECK
3.1L (VIN T)

1. • IGNITION "ON," ENGINE STOPPED
   • NOTE SHIFT LIGHT

LIGHT "OFF"

2. • GROUND DIAGNOSTIC TERMINAL AND NOTE LIGHT

LIGHT "OFF"

3. • IGNITION "OFF"
   • DISCONNECT ECM CONNECTORS
   • IGNITION "ON"
   • JUMPER CKT 456 TO GROUND AND NOTE SHIFT LIGHT

LIGHT "OFF"
- CHECK AND REPAIR
  - OPEN IGNITION CKT 39
  - OPEN CKT 456
  - FAULTY BULB

LIGHT "ON"
- POOR CONNECTION AT ECM OR FAULTY ECM

CHECK FOR
• CODE 24 (NO VSS)
• REFER TO "DIAGNOSTIC AIDS" ON FACING PAGE

LIGHT "ON"
• IGNITION "OFF"
• DISCONNECT ECM CONNECTORS
• IGNITION "ON" AND NOTE SHIFT LIGHT

LIGHT "ON"
REPAIR SHORT TO GROUND IN CKT 456

LIGHT "OFF"
FAULTY ECM

"AFTER REPAIRS," CONFIRM "CLOSED LOOP" OPERATION AND NO "SERVICE ENGINE SOON" LIGHT.

3.1L (VIN T) ENGINE — COMPONENT DIAGNOSTIC CHART — CAMARO, FIREBIRD, BERETTA, CORSICA, CAVALIER AND SUNBIRD

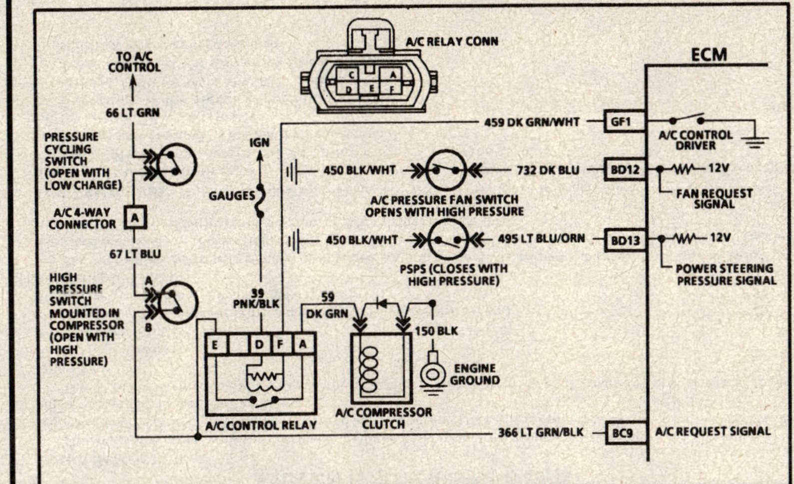

CHART C-10

A/C CLUTCH CONTROL CIRCUIT DIAGNOSIS
3.1L (VIN T) "F" CARLINE (PORT)

Circuit Description:

The A/C clutch control relay is ECM controlled to delay A/C clutch engagement about .4 second after A/C is turned "ON." This allows the IAC to adjust engine rpm, before the A/C clutch engages. The ECM, also, causes the relay to disengage the A/C clutch during WOT, when high power steering pressure is present, or if engine is overheating. The A/C clutch control relay is energized, when the ECM provides a ground path for CKT 459. The low pressure switch will open if A/C pressure is less than 40 psi (276 kPa). The high pressure switch will open, if A/C pressure exceeds about 440 psi (3034 kPa). The A/C pressure fan switch opens when A/C pressure exceeds about 200 psi (1380 kPa).

Test Description: Number(s) below refer to circled number(s) on the diagnostic chart.

1. The ECM will only energize the A/C relay, when the engine is running. This test will determine if the relay or CKT 459 is faulty.
2. In order for the clutch to properly be engaged, the pressure cycling switch must be closed to provide 12 volts to the relay, and the high pressure switch must be closed, so the A/C request (12 volts) will be present at the ECM.
3. Determines if the signal is reaching the ECM on CKT 366 from the A/C control panel. Signal should only be present when the A/C mode or defrost mode has been selected.
4. A short to ground in any part of the A/C request circuit, CKT 67 to the relay, CKT 59 to the A/C clutch, or the A/C clutch, could be the cause of the blown fuse.
5. If the ECM is seeing a high power steering pressure signal, the A/C clutch will be disengaged by the ECM.
6. With the engine idling and A/C "ON," the ECM should be grounding CKT 459, which should cause the test light to be "ON."

Diagnostic Aids:

If complaint was insufficient cooling, the problem may be caused by a inoperative cooling fan, or A/C pressure fan switch. The engine cooling fan should turn "ON," when A/C pressure exceeds a value to open the switch, which causes the ECM to energize the cooling fan relay. See CHART C-12, for cooling fan diagnosis.

3.1L (VIN T) ENGINE — COMPONENT DIAGNOSTIC CHART — CAMARO, FIREBIRD, BERETTA, CORSICA, CAVALIER AND SUNBIRD

CHART C-10

A/C CLUTCH CONTROL CIRCUIT DIAGNOSIS
3.1L (VIN T) "F" CARLINE (PORT)

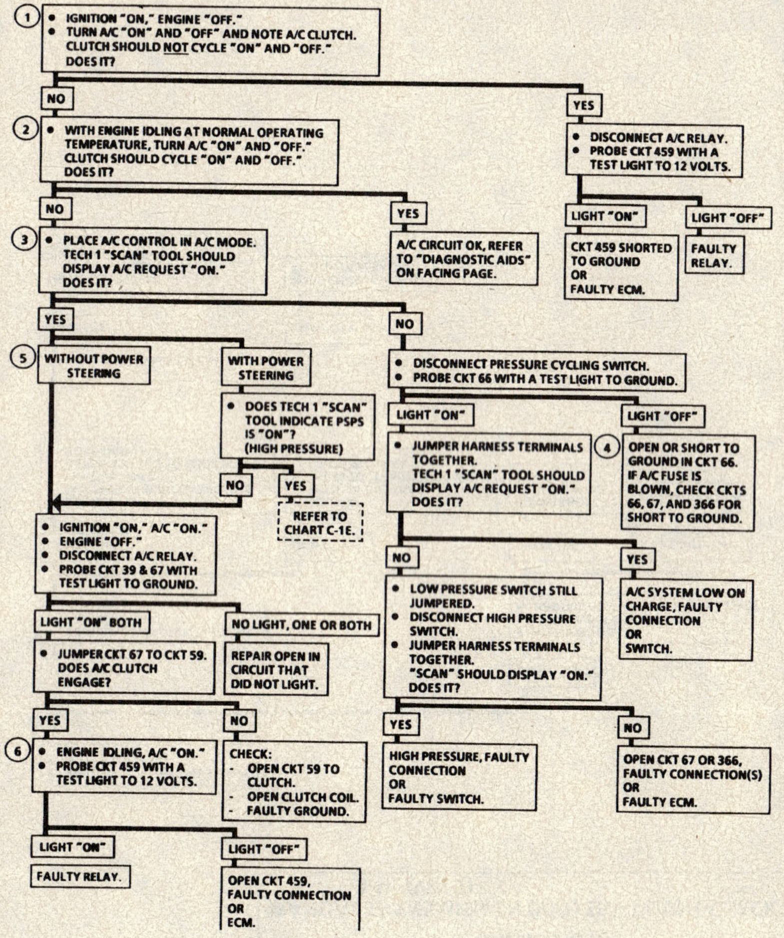

3.1L (VIN T) ENGINE — COMPONENT DIAGNOSTIC CHART — BERETTA, CORSICA, CAVALIER AND SUNBIRD

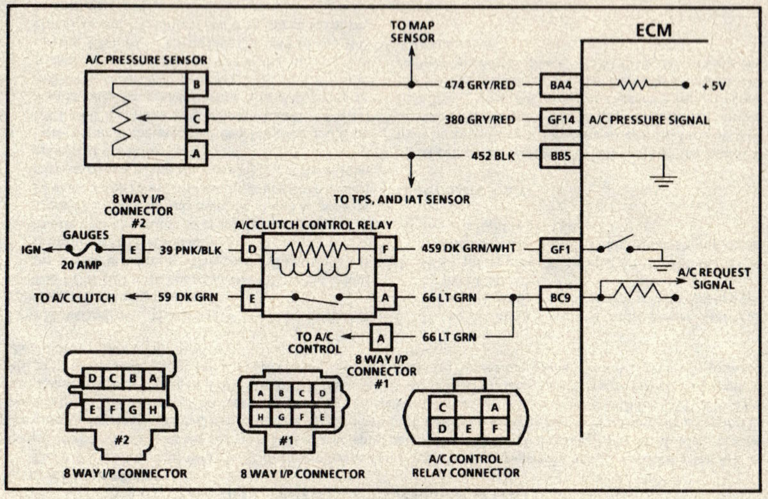

CHART C-10

(Page 1 of 2)
A/C CLUTCH CONTROL CIRCUIT DIAGNOSIS
3.1L (VIN T) "L" CARLINE (PORT)

Circuit Description:

The A/C clutch control relay is energized when the ECM provides a ground path through CKT 459 and A/C is requested via the switch. A/C clutch is delayed about .3 second after A/C is requested. This will allow the IAC to adjust engine rpm for the additional load.

The ECM will temporarily disengage the A/C clutch relay for calibrated times for one or more of the following:

- Hot engine restart.
- Wide open throttle (TPS over 90%).
- Power steering pressure high (open power steering pressure switch).
- Engine rpm greater than about 6000 rpm.
- During IAC reset.

The A/C clutch relay will remain disengaged when a Code 66 is present, if pressure is out of range as described previously in this section, or there is no A/C request signal due to an open A/C select switch or circuit. Refer to SECTION 1B for more information on A/C refrigerant systems.

Test Description: Number(s) below refer to circled number(s) on the diagnostic chart.

1. The ECM will only energize the A/C relay when the engine is running. This test will determine if the relay or CKT 459 is faulty.

2. Determines if the signal is reaching the ECM through CKT 66 from the A/C control panel. Signal should only be present when an A/C mode or defrost mode has been selected.

3. If the ECM is receiving a high power steering pressure signal, the A/C clutch will be disengaged by the ECM.

3.1L (VIN T) ENGINE — COMPONENT DIAGNOSTIC CHART — BERETTA, CORSICA, CAVALIER AND SUNBIRD

CHART C-10
(Page 1 of 2)
A/C CLUTCH CONTROL CIRCUIT DIAGNOSIS
3.1L (VIN T)

1
- IGNITION "ON," ENGINE "OFF."
- "SCAN" CODES AND REFER TO THOSE CHARTS, IF PRESENT.
- NOTE A/C CLUTCH.
CLUTCH SHOULD NOT BE ENGAGED.
IS IT?

NO →
- WITH ENGINE IDLING AT NORMAL OPERATING TEMPERATURE TURN A/C "ON" AND "OFF."
- CLUTCH SHOULD CYCLE "ON" AND "OFF."
DOES IT?

YES →
- DISCONNECT A/C RELAY.
- PROBE CKT 459 WITH A TEST LIGHT TO 12 VOLTS.

NO →
2
- PLACE A/C CONTROL IN A/C MODE.
- TECH 1 "SCAN" TOOL SHOULD DISPLAY A/C REQUEST "ON."
DOES IT?

YES → A/C CIRCUIT OK, REFER TO "DIAGNOSTIC AIDS"

LIGHT "ON" → CKT 459 SHORTED TO GROUND OR FAULTY ECM.

LIGHT "OFF" → CKT 59 SHORTED TO B + OR FAULTY RELAY.

YES →
- KEY "ON" ENGINE "OFF."
- ALLOW A/C PRESSURES TO EQUALIZE.
- "SCAN" A/C PRESSURE.
IS PRESSURE OVER 150 psi?

NO →
- BACK PROBE ECM TERMINAL "BC9" WITH A TEST LIGHT TO GROUND.

YES → OPEN CKT 452 OR CONNECTION OR A/C PRESSURE SENSOR

NO →
3
- WITH ENGINE IDLING, DOES TECH 1 "SCAN" TOOL INDICATE PSPS IS "ON"? (HIGH PRESSURE)

LIGHT "ON" → CHECK FOR POOR CONNECTION AT ECM TERMINAL "BC9". IF OK, ECM IS FAULTY.

LIGHT "OFF" → IF A/C FUSE IS BLOWN, CHECK FOR SHORT TO GROUND IN CKT 650, OR SHORTED A/C CLUTCH COIL. IF OK, CHECK FOR OPEN CKT 650, OR FAULTY A/C SELECT

YES → REFER TO CHART C-1E.

NO → REFER TO CHART C-10 2 OF 2.

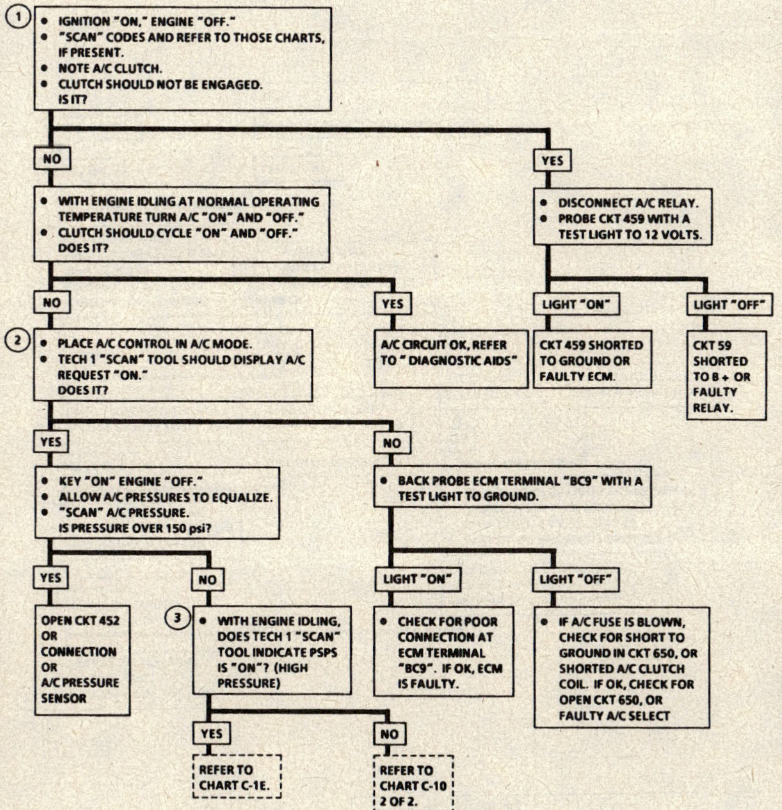

"AFTER REPAIRS," CONFIRM "CLOSED LOOP" OPERATION AND NO "SERVICE ENGINE SOON" LIGHT.

3.1L (VIN T) ENGINE — COMPONENT DIAGNOSTIC CHART — BERETTA, CORSICA, CAVALIER AND SUNBIRD

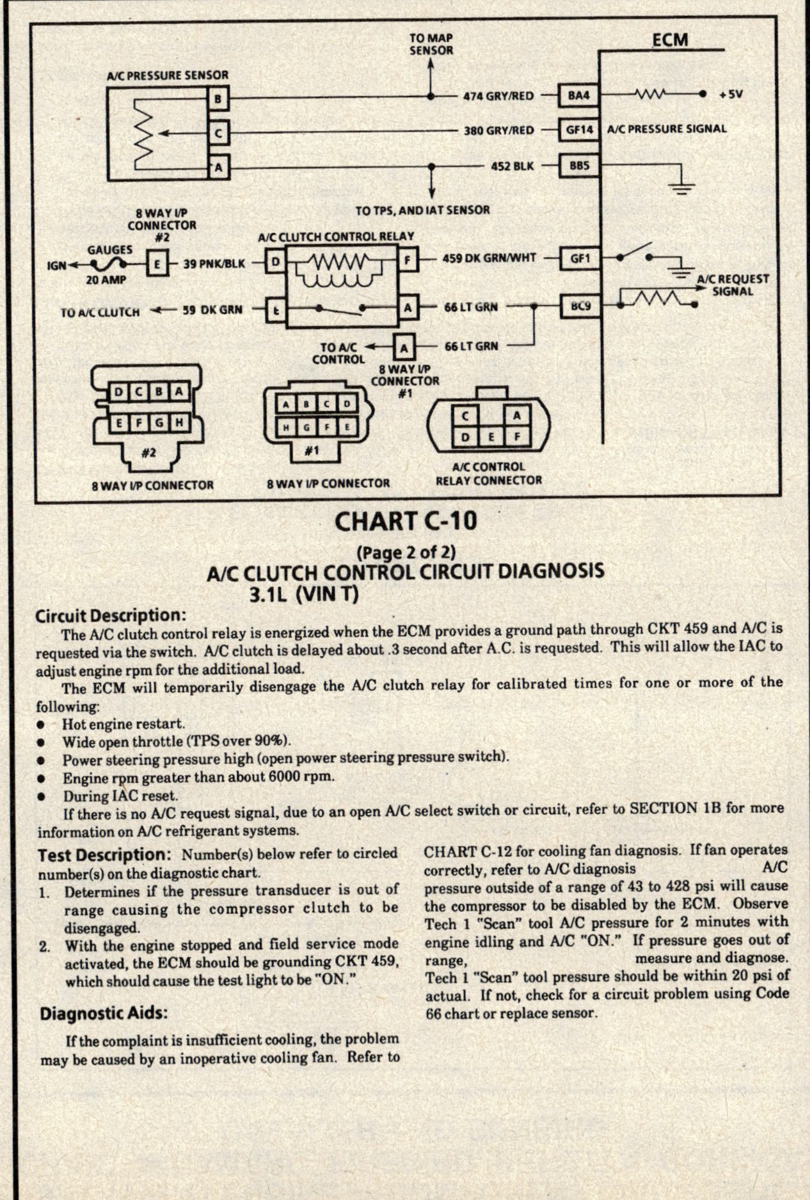

CHART C-10
(Page 2 of 2)
A/C CLUTCH CONTROL CIRCUIT DIAGNOSIS
3.1L (VIN T)

Circuit Description:

The A/C clutch control relay is energized when the ECM provides a ground path through CKT 459 and A/C is requested via the switch. A/C clutch is delayed about .3 second after A.C. is requested. This will allow the IAC to adjust engine rpm for the additional load.

The ECM will temporarily disengage the A/C clutch relay for calibrated times for one or more of the following:
- Hot engine restart.
- Wide open throttle (TPS over 90%).
- Power steering pressure high (open power steering pressure switch).
- Engine rpm greater than about 6000 rpm.
- During IAC reset.

If there is no A/C request signal, due to an open A/C select switch or circuit, refer to SECTION 1B for more information on A/C refrigerant systems.

Test Description: Number(s) below refer to circled number(s) on the diagnostic chart.

1. Determines if the pressure transducer is out of range causing the compressor clutch to be disengaged.
2. With the engine stopped and field service mode activated, the ECM should be grounding CKT 459, which should cause the test light to be "ON."

Diagnostic Aids:

If the complaint is insufficient cooling, the problem may be caused by an inoperative cooling fan. Refer to CHART C-12 for cooling fan diagnosis. If fan operates correctly, refer to A/C diagnosis A/C pressure outside of a range of 43 to 428 psi will cause the compressor to be disabled by the ECM. Observe Tech 1 "Scan" tool A/C pressure for 2 minutes with engine idling and A/C "ON." If pressure goes out of range, measure and diagnose. Tech 1 "Scan" tool pressure should be within 20 psi of actual. If not, check for a circuit problem using Code 66 chart or replace sensor.

3.1L (VIN T) ENGINE — COMPONENT DIAGNOSTIC CHART — BERETTA, CORSICA, CAVALIER AND SUNBIRD

CHART C-10
(Page 2 of 2)
A/C CLUTCH CONTROL CIRCUIT DIAGNOSIS
3.1L (VIN T)

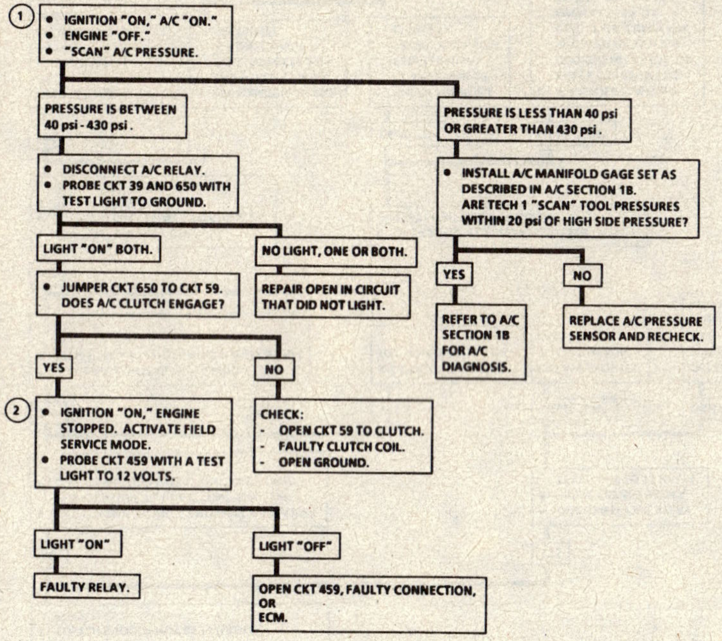

3.1L (VIN T) ENGINE — COMPONENT DIAGNOSTIC CHART — CAMARO, FIREBIRD, BERETTA, CORSICA, CAVALIER AND SUNBIRD

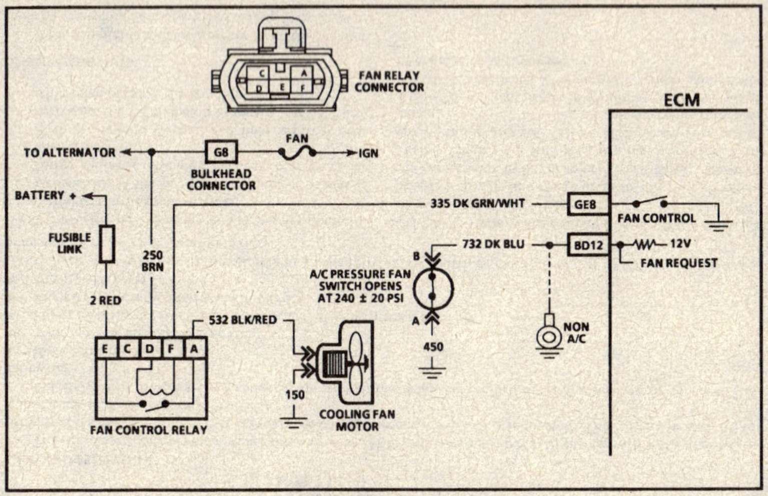

CHART C-12
(Page 1 of 2)
ELECTRIC COOLING FAN CONTROL CIRCUIT
3.1L (VIN T) "F" CARLINE (PORT)

Circuit Description:

The electric cooling fan is controlled by the ECM, based on inputs from the Coolant Temperature Sensor (CTS), the A/C fan control switch, and vehicle speed. The ECM controls the fan by grounding CKT 335, which energizes the fan control relay. Battery voltage is then supplied to the fan motor.

The ECM grounds CKT 335, when coolant temperature is over 109°C (228°F), or when A/C has been requested, and the fan control switch opens with high A/C pressure, about 240 psi (1655 kPa). Once the ECM turns the relay "ON," it will keep it "ON" for a minimum of 30 seconds, or until vehicle speed exceeds 70 mph.

Also, if Code 14 or Code 15 sets, or the ECM is in fuel back up mode, the fan will run at all times. On a vehicle not equipped with A/C, CKT 732 is jumpered to ground so that the fan does not run at all times.

Test Description: Number(s) below refer to circled number(s) on the diagnostic chart.
1. With the diagnostic terminal grounded, the cooling fan control driver will close, which should energize the fan control relay.
2. If the A/C fan control switch or circuit is open, the fan would run whenever the engine is running.
3. With A/C clutch engaged, the A/C fan control switch should open, when A/C high pressure exceeds about 200 psi (1380 kPa). This signal should cause the ECM to energize the fan control relay.

Diagnostic Aids:

If the owner complained of an overheating problem, it must be determined if the complaint was due to an actual boil over, or the hot light, or temperature gage indicated over heating.

If the gage, or light, indicates overheating, but no boil over is detected, the gage circuit should be checked. The gage accuracy can, also, be checked by comparing the coolant sensor reading using a Tech 1 "Scan" tool and comparing its reading with the gage reading.

If the engine is actually overheating, and the gage indicates overheating, but the cooling fan is not coming "ON," the coolant sensor has probably shifted out of calibration and should be replaced.

If the engine is overheating, and the cooling fan is "ON," the cooling system should be checked.

3.1L (VIN T) ENGINE — COMPONENT DIAGNOSTIC CHART — CAMARO, FIREBIRD, BERETTA, CORSICA, CAVALIER AND SUNBIRD

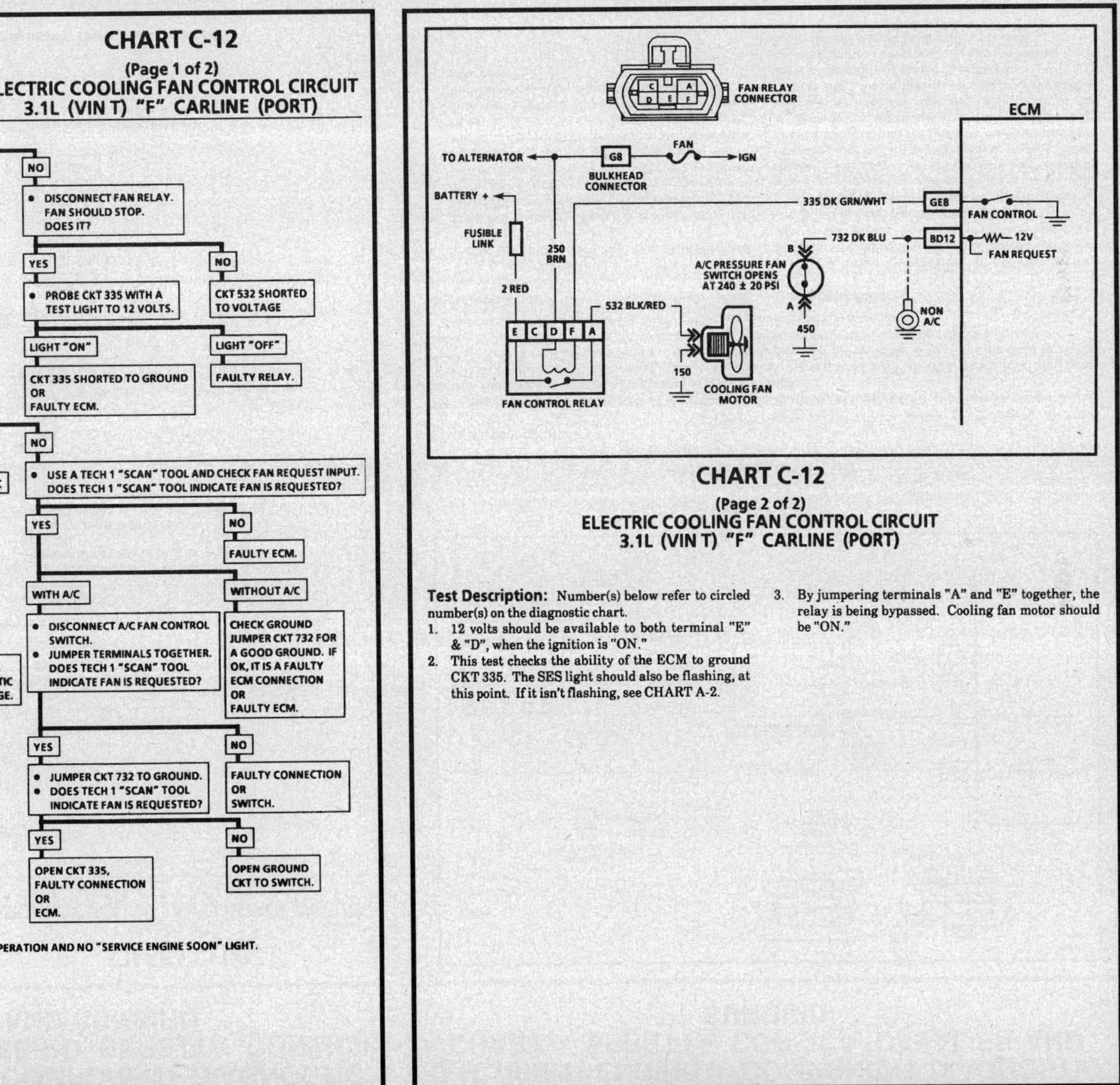

3.1L (VIN T) ENGINE — COMPONENT DIAGNOSTIC CHART — CAMARO, FIREBIRD, BERETTA, CORSICA, CAVALIER AND SUNBIRD

Test Description: Number(s) below refer to circled number(s) on the diagnostic chart.

1. 12 volts should be available to both terminal "E" & "D," when the ignition is "ON."
2. This test checks the ability of the ECM to ground CKT 335. The SES light should also be flashing, at this point. If it isn't flashing, see CHART A-2.
3. By jumpering terminals "A" and "E" together, the relay is being bypassed. Cooling fan motor should be "ON."

3.1L (VIN T) ENGINE — COMPONENT DIAGNOSTIC CHART — CAMARO, FIREBIRD, BERETTA, CORSICA, CAVALIER AND SUNBIRD

CHART C-12
(Page 2 of 2)
ELECTRIC COOLING FAN CONTROL CIRCUIT
3.1L (VIN T) "F" CARLINE (PORT)

"AFTER REPAIRS," CONFIRM "CLOSED LOOP" OPERATION AND NO "SERVICE ENGINE SOON" LIGHT.

3.1L (VIN T) ENGINE — COMPONENT DIAGNOSTIC CHART — BERETTA, CORSICA, CAVALIER AND SUNBIRD

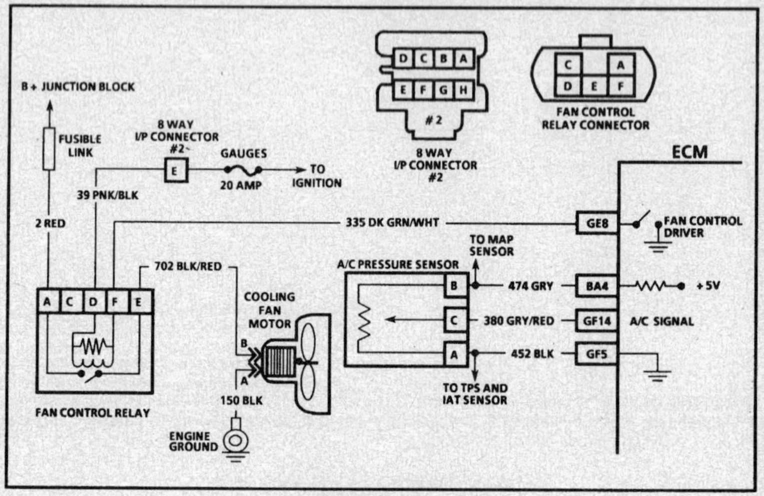

CHART C-12
(Page 1 of 2)
FAN CONTROL CIRCUIT
3.1L (VIN T)

Circuit Description:

The electric cooling fan is controlled by the ECM, based on inputs from the coolant temperature sensor, the A/C pressure sensor and vehicle speed. The ECM controls the fan by grounding CKT 335 which energizes the fan control relay. Battery voltage is then supplied to the fan motor.

The ECM grounds CKT 335 when coolant temperature is over about 109°C (228°F), or when A/C has been requested and the A/C pressure sensor indicates high A/C pressure, 200 psi (1380 kPa). Once the ECM turns the relay "ON," it will keep it "ON" for a minimum of 25 seconds, or until vehicle speed exceeds 70 mph.

Also, if Code 14 or 15 sets or the ECM is in back up, the fan will run at all times.

Test Description: Number(s) below refer to circled number(s) on the diagnostic chart.

1. With the diagnostic terminal grounded, the cooling fan control driver will close, which should energize the fan control relay.
2. If the A/C fan control switch or circuit is open, the fan would run whenever A/C is requested.
3. With A/C clutch engaged, the A/C fan control switch should open when A/C high pressure exceeds about 200 psi (1380 kPa). This signal should cause the ECM to energize the fan control relay.
4. Disconnecting the A/C pressure sensor will cause a Code 66 to set. After finishing this step, be sure to clear codes.

Diagnostic Aids:

If the owner complained of an overheating problem, it must be determined if the complaint was due to an actual boil over, or the hot light (temperature gage) indicated overheating.

If the gage (light) indicates overheating, but no boil over is detected, the gage circuit should be checked. The gage accuracy can also be checked by comparing the coolant sensor reading using a "Scan" tool and comparing its reading with the gage reading.

If the engine is actually overheating, and the gage indicates overheating, but the cooling fan is not coming "ON," the coolant sensor has probably shifted out of calibration and should be replaced.

If the engine is overheating, and the cooling fan is "ON," the cooling system should be checked.

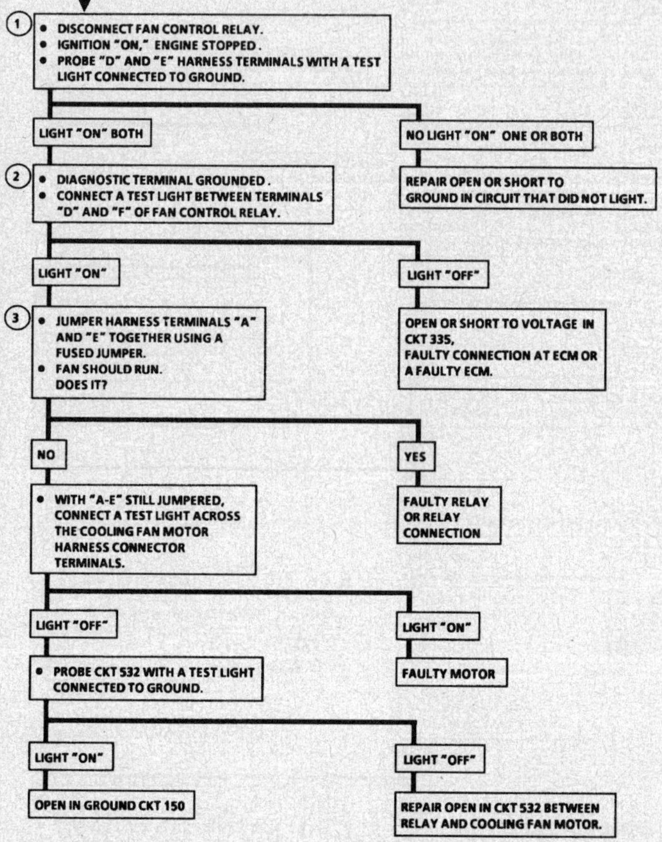

3.1L (VIN T) ENGINE — COMPONENT DIAGNOSTIC CHART — BERETTA, CORSICA, CAVALIER AND SUNBIRD

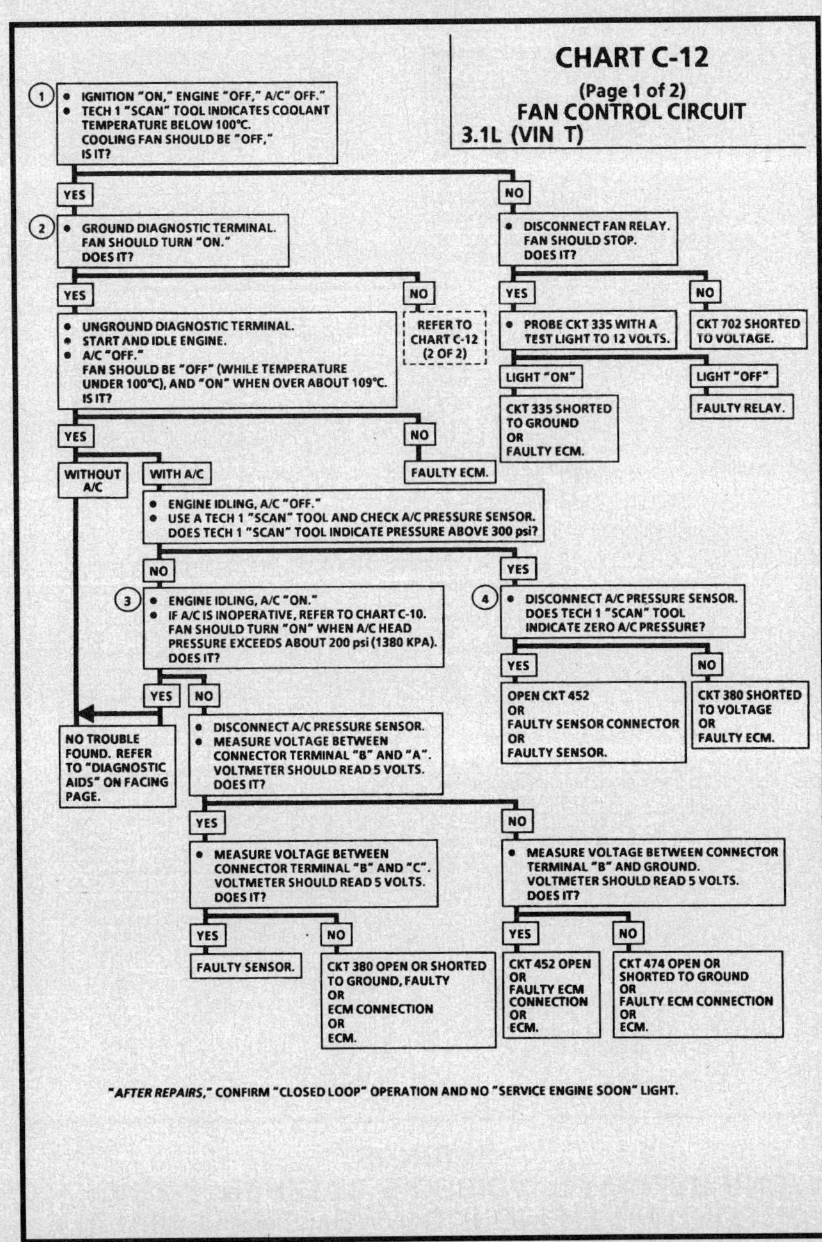

3.1L (VIN T) ENGINE — COMPONENT DIAGNOSTIC CHART — BERETTA, CORSICA, CAVALIER AND SUNBIRD

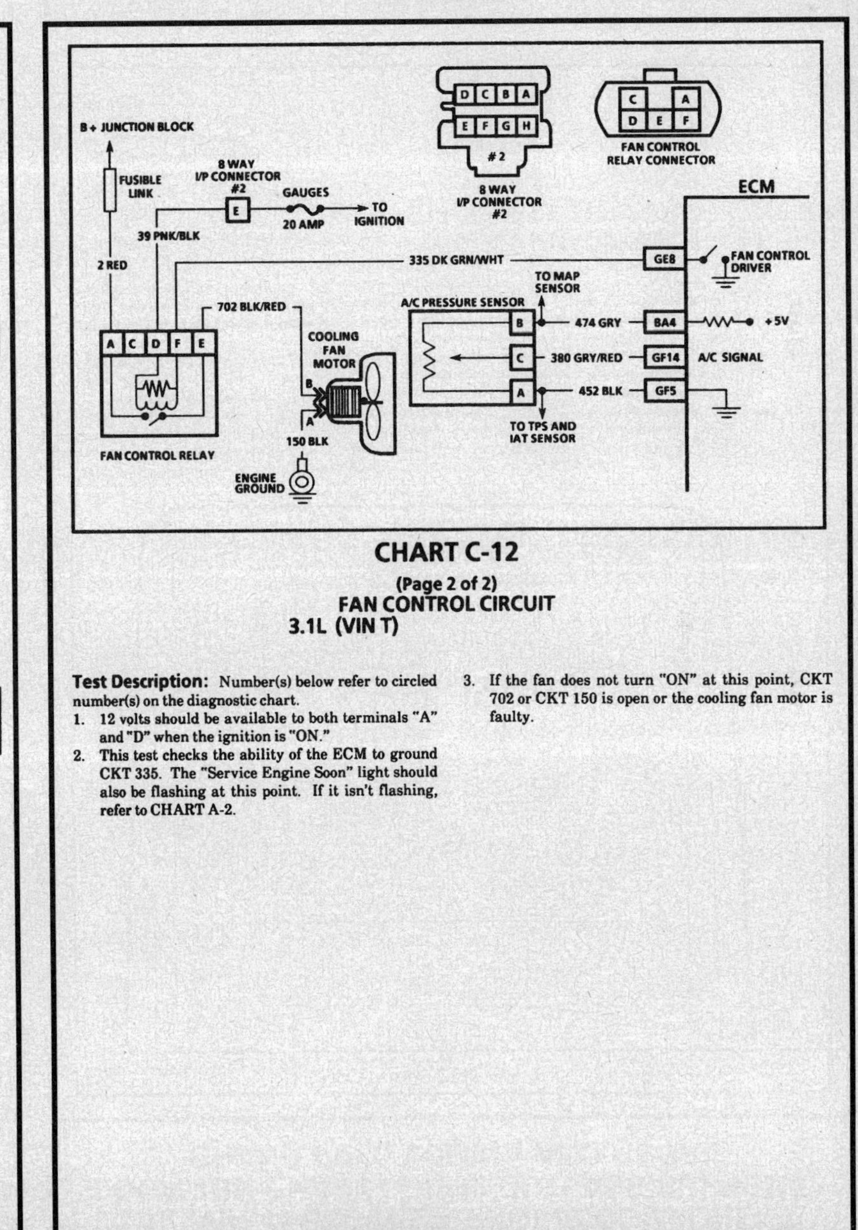

Test Description: Number(s) below refer to circled number(s) on the diagnostic chart.

1. 12 volts should be available to both terminals "A" and "D" when the ignition is "ON."
2. This test checks the ability of the ECM to ground CKT 335. The "Service Engine Soon" light should also be flashing at this point. If it isn't flashing, refer to CHART A-2.
3. If the fan does not turn "ON" at this point, CKT 702 or CKT 150 is open or the cooling fan motor is faulty.

3.1L (VIN T) ENGINE — COMPONENT DIAGNOSTIC CHART — BERETTA, CORSICA, CAVALIER AND SUNBIRD

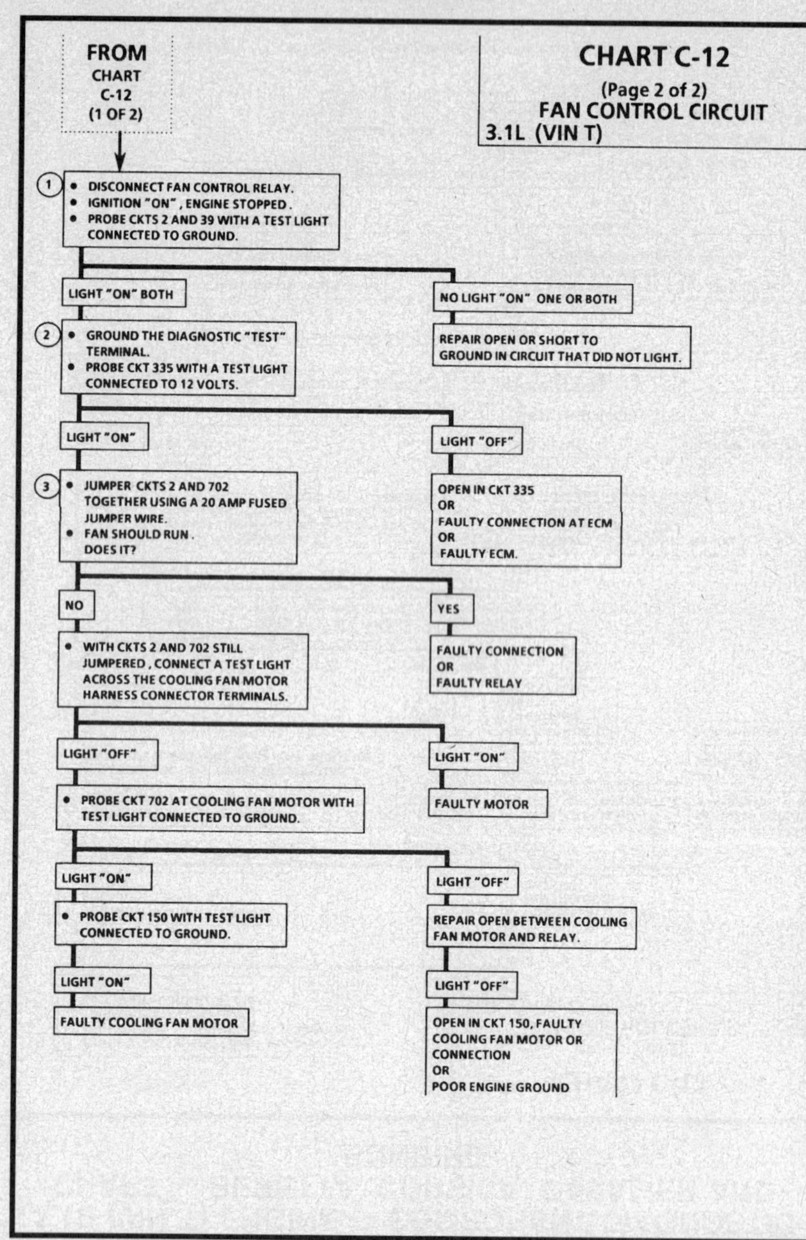

3.1L (VIN T) ENGINE — ENGINE COMPONENT LOCATION CHART — 1992 CUTLASS SUPREME, GRAND PRIX, LUMINA AND REGAL

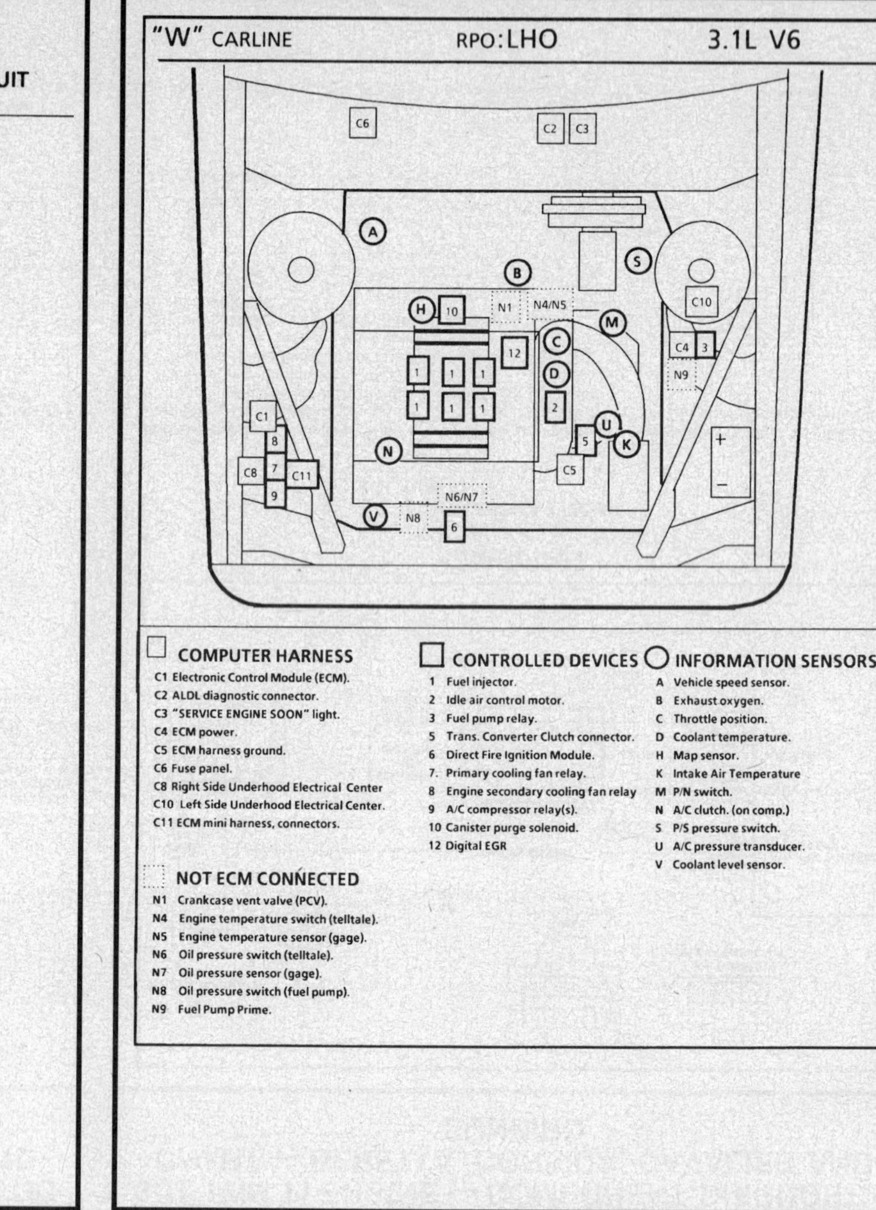

3.4L (VIN) ENGINE — ENGINE COMPONENT LOCATION CHART — 1992 CUTLASS SUPREME, GRAND PRIX, LUMINA AND REGAL

"W" CARLINE   RPO:LQI   3.4L V6

COMPUTER HARNESS

- C1 Electronic Control Module (ECM).
- C2 ALDL diagnostic connector.
- C3 "SERVICE ENGINE SOON" light.
- C4 ECM power.
- C5 ECM harness ground.
- C6 Fuse panel.
- C8 Right Side Underhood Electrical Center
- C10 Left Side Underhood Electrical Center.
- C11 ECM mini harness, connectors.

CONTROLLED DEVICES

- 1 Fuel injector.
- 2 Idle air control motor.
- 3 Fuel pump relay.
- 5 Trans. Converter Clutch connector.
- 6 Direct Fire Ignition Module.
- 7. Primary cooling fan relay.
- 8 Engine secondary cooling fan relay
- 9 A/C compressor relay(s).
- 10 Canister purge solenoid.
- 12 Digital EGR

INFORMATION SENSORS

- A Vehicle speed sensor.
- B Exhaust oxygen.
- C Throttle position.
- D Coolant temperature.
- H Map sensor.
- K Intake Air Temperature
- M P/N switch.
- N A/C clutch. (on comp.)
- S P/S pressure switch.
- U A/C pressure transducer.
- V Coolant level sensor.

NOT ECM CONNECTED

- N1 Crankcase vent valve (PCV).
- N4 Engine temperature switch (telltale).
- N5 Engine temperature sensor (gage).
- N6 Oil pressure switch (telltale).
- N7 Oil pressure sensor (gage).
- N8 Oil pressure switch (fuel pump).
- N9 Fuel Pump Prime.

3.1L (VIN T) ENGINE — ENGINE COMPONENT LOCATION CHART — 1993-94 CUTLASS SUPREME, GRAND PRIX, LUMINA AND REGAL

"W" CARLINE   RPO:LH0   3.1L V6 MFI

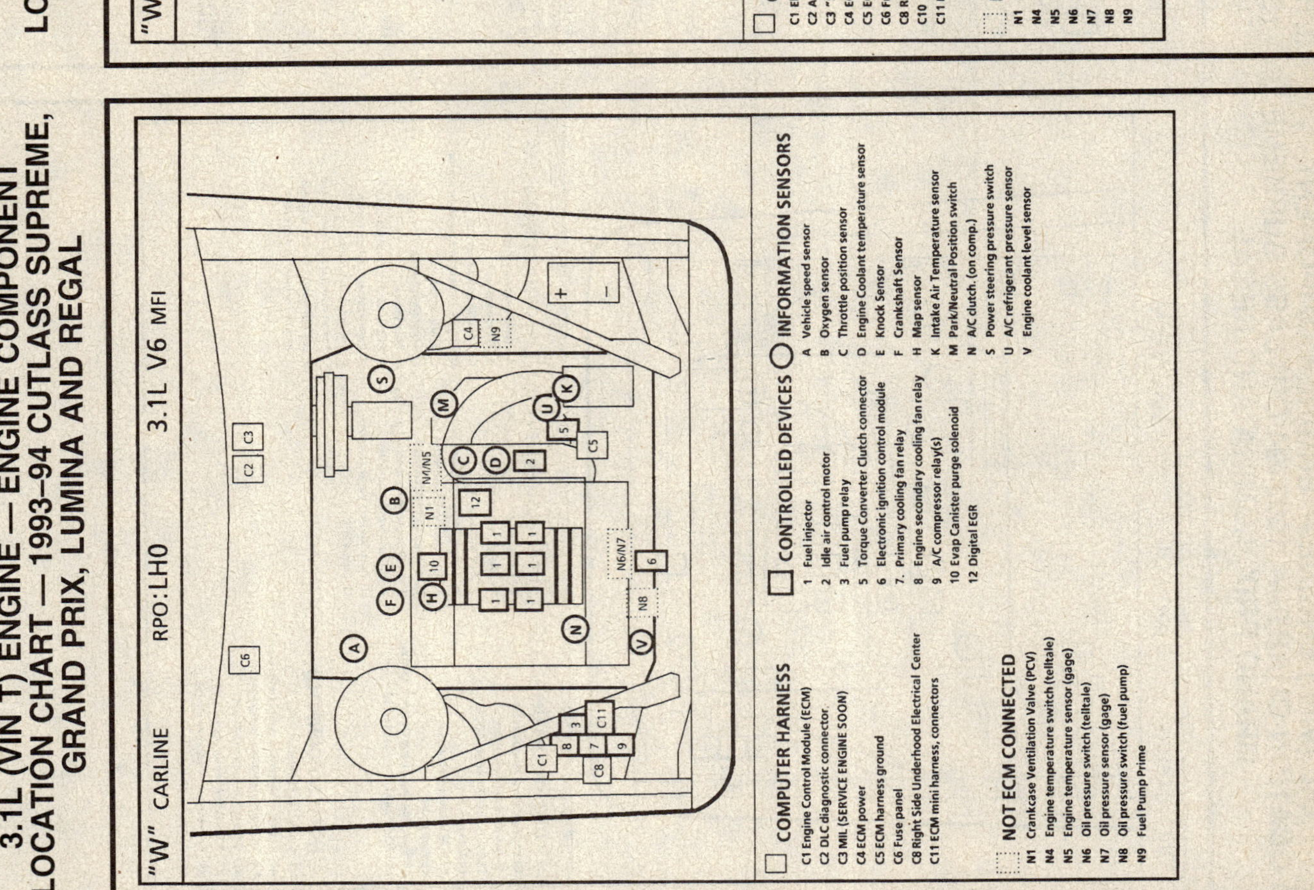

COMPUTER HARNESS

- C1 Engine Control Module (ECM)
- C2 DLC diagnostic connector.
- C3 MIL (SERVICE ENGINE SOON)
- C4 ECM power
- C5 ECM harness ground
- C6 Fuse panel
- C8 Right Side Underhood Electrical Center
- C11 ECM mini harness, connectors

CONTROLLED DEVICES

- 1 Fuel injector
- 2 Idle air control motor
- 3 Fuel pump relay
- 5 Torque Converter Clutch connector
- 6 Electronic ignition control module
- 7. Primary cooling fan relay
- 8 Engine secondary cooling fan relay
- 9 A/C compressor relay(s)
- 10 Evap Canister purge solenoid
- 12 Digital EGR

INFORMATION SENSORS

- A Vehicle speed sensor
- B Oxygen sensor
- C Throttle position sensor
- D Engine Coolant temperature sensor
- E Knock Sensor
- F Crankshaft Sensor
- H Map sensor
- K Intake Air Temperature sensor
- M Park/Neutral Position switch
- N A/C clutch. (on comp.)
- S Power steering pressure switch
- U A/C refrigerant pressure sensor
- V Engine coolant level sensor

NOT ECM CONNECTED

- N1 Crankcase Ventilation Valve (PCV)
- N4 Engine temperature switch (telltale)
- N5 Engine temperature sensor (gage)
- N6 Oil pressure switch (telltale)
- N7 Oil pressure sensor (gage)
- N8 Oil pressure switch (fuel pump)
- N9 Fuel Pump Prime

3.1L (VIN T) ENGINE — ECM WIRING SCHEMATIC — 1992 CUTLASS SUPREME, GRAND PRIX, LUMINA AND REGAL

ECM

Pin	Signal	Wire
B3	DIAGNOSTIC TEST	451 WHT/BLK
B5	SERIAL DATA INPUT	461 ORN
B7	A/T TCC CONTROL	422 TAN/BLK
C20	2ND GEAR SIGNAL (440-T4 ONLY)	232 WHT
D6	3RD GEAR SIGNAL (440-T4)	108 DK GRN
D22	4TH GEAR SIGNAL (440-T4)	446 LT BLU
D11	PARK/NEUTRAL SWITCH (A/T)	434 ORN/BLK
C2	VSS (INPUT)	401 PPL
C8	VSS (INPUT)	400 YEL
B1	SERVICE ENGINE SOON LIGHT CONTROL	419 BRN/WHT
B7	SHIFT LIGHT CONTROL (M/T)	456 TAN/BLK
B8	BUFFERED SPEED OUTPUT	389 DK GRN
A18	AIR DIVERT CONTROL (M/T)	429 BLK/PNK
A10	CANISTER PURGE CONTROL	428 DK GRN/YEL
C13	EGR SOLENOID # 3	699 RED
A19	EGR SOLENOID # 2	698 BRN
A4	EGR SOLENOID # 1	697 LT BLU
C5	SENSOR GROUND	802 BLK
C22	MANIFOLD ABSOLUTE PRESSURE SENSOR SIGNAL	432 LT GRN
C7	+5V REFERENCE	474 GRY
C10	SENSOR GROUND	808 BLK
C16	COOLANT TEMPERATURE SIGNAL	410 YEL

3.4L (VIN) ENGINE — ENGINE COMPONENT LOCATION CHART — 1993–94 CUTLASS SUPREME, GRAND PRIX, LUMINA AND REGAL

"W" CARLINE RPO: LQ1 3.4L V6 MFI

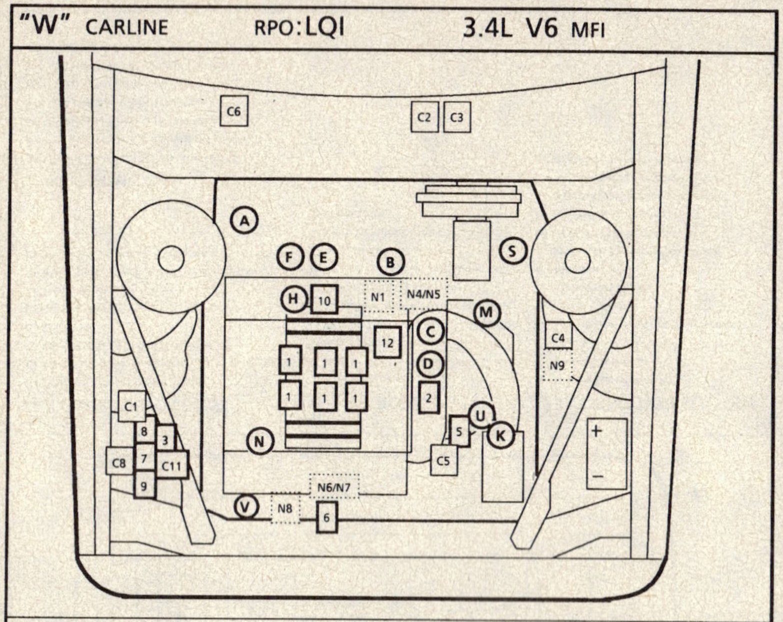

COMPUTER HARNESS
- C1 Engine Control Module (ECM)
- C2 DLC diagnostic connector
- C3 MIL (SERVICE ENGINE SOON)
- C4 ECM power
- C5 ECM harness ground
- C6 Fuse panel
- C8 Right Side Underhood Electrical Center
- C11 ECM mini harness, connectors

NOT ECM CONNECTED
- N1 Crankcase Ventilation Valve (PCV)
- N4 Engine temperature switch (telltale)
- N5 Engine temperature sensor (gage)
- N6 Oil pressure switch (telltale)
- N7 Oil pressure sensor (gage)
- N8 Oil pressure switch (fuel pump)
- N9 Fuel Pump Prime

CONTROLLED DEVICES
- 1 Fuel injector
- 2 Idle air control motor
- 3 Fuel pump relay
- 5 Torque Converter Clutch connector
- 6 Electronic ignition control module
- 7 Primary cooling fan relay
- 8 Engine secondary cooling fan relay
- 9 A/C compressor relay(s)
- 10 Evap Canister purge solenoid
- 12 Digital EGR

INFORMATION SENSORS
- A Vehicle speed sensor
- B Oxygen sensor
- C Throttle position sensor
- D Engine Coolant temperature sensor
- E Knock Sensor
- F Crankshaft Sensor
- H Map sensor
- K Intake Air Temperature sensor
- M Park/Neutral Position switch
- S A/C clutch. (on comp.)
- U A/C refrigerant pressure sensor
- V Engine coolant level sensor

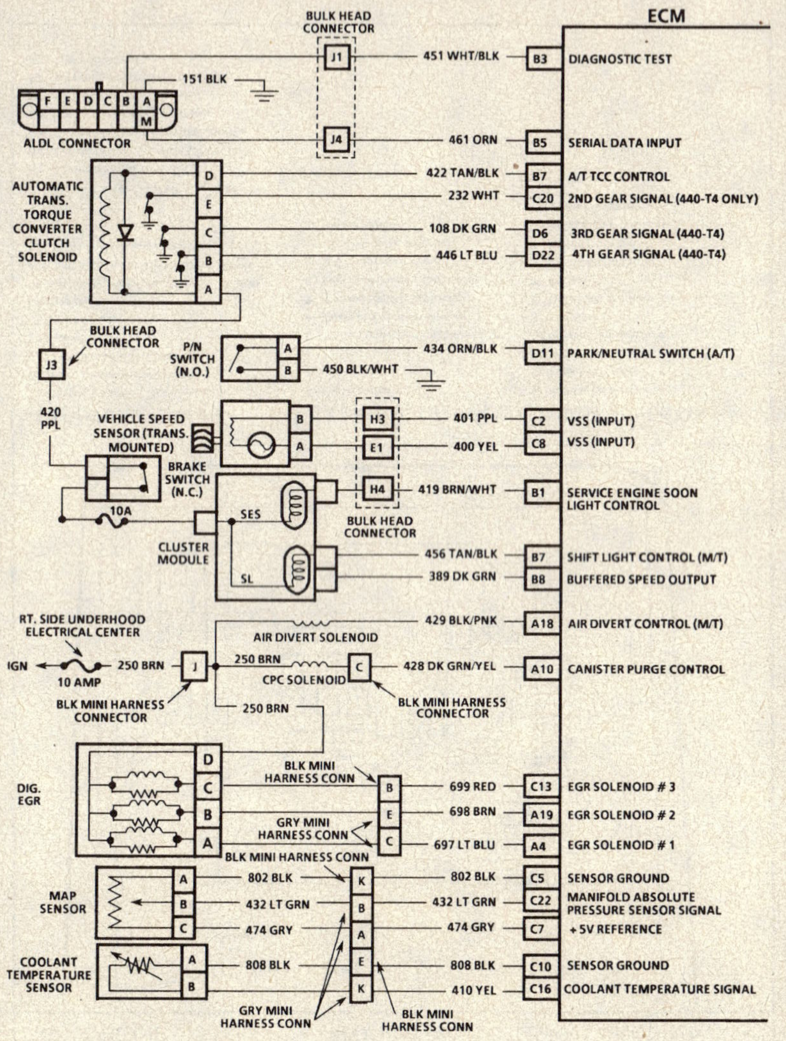

3.1L (VIN T) ENGINE — ECM WIRING SCHEMATIC — 1992 CUTLASS SUPREME, GRAND PRIX, LUMINA AND REGAL

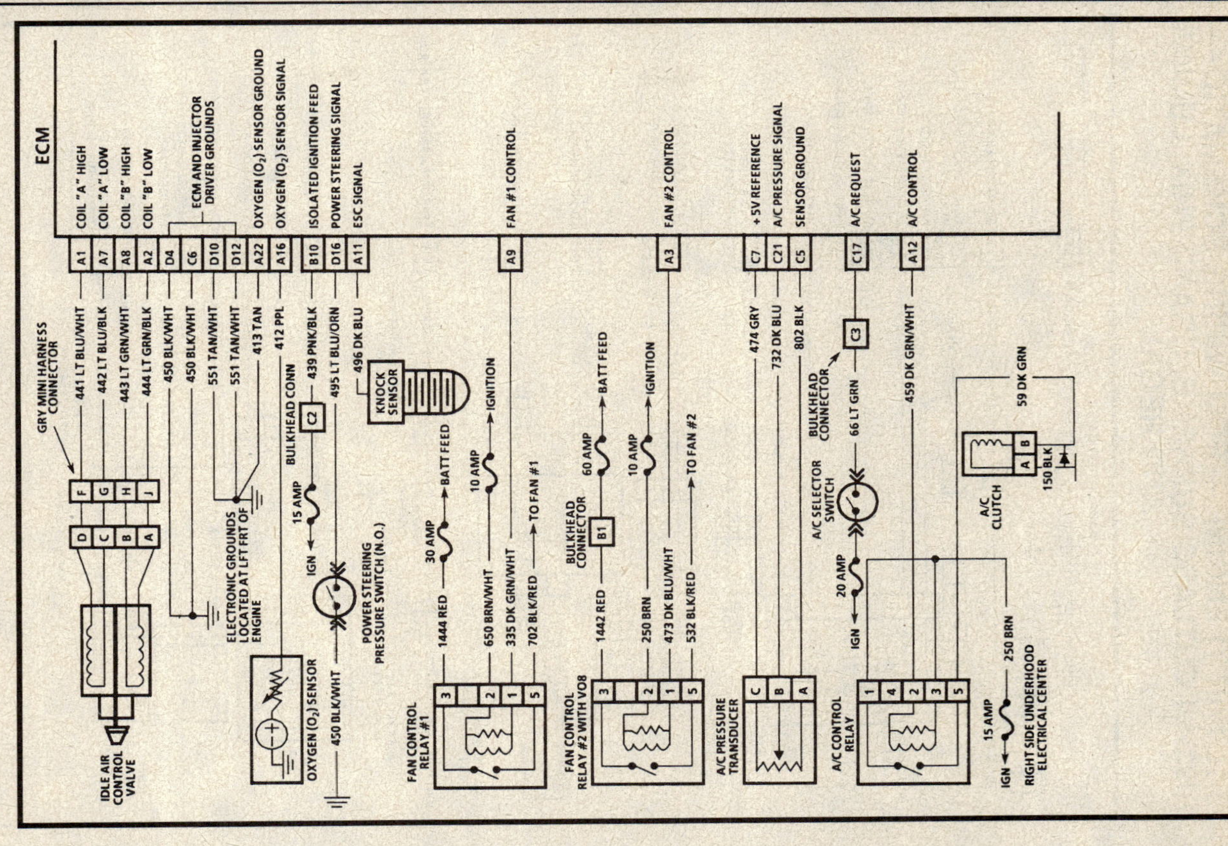

3.1L (VIN T) ENGINE — ECM WIRING SCHEMATIC — 1992 CUTLASS SUPREME, GRAND PRIX, LUMINA AND REGAL

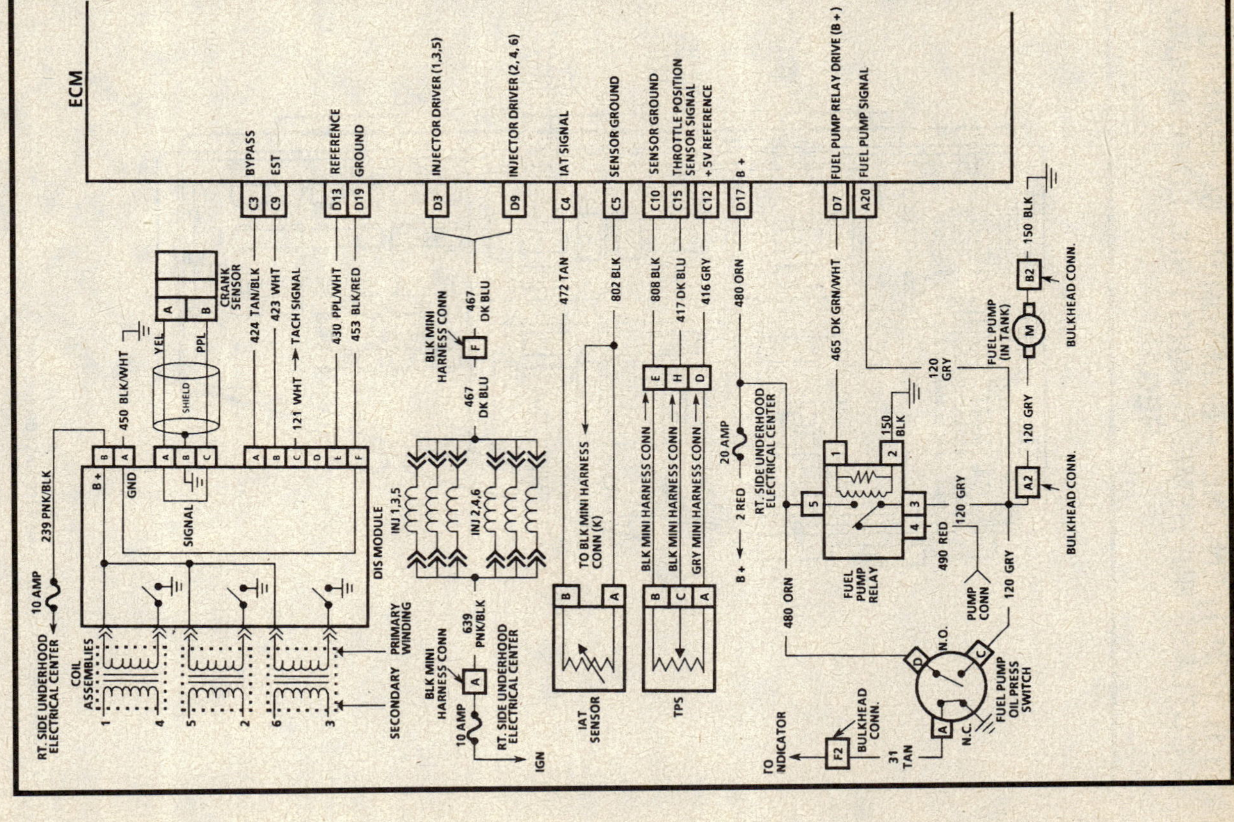

MULTIPORT FUEL INJECTION (MFI) SYSTEMS
EXCEPT LIGHT TRUCKS, VANS, GEO AND SATURN

3.1L (VIN T) ENGINE — ECM WIRING SCHEMATIC — 1993–94 CUTLASS SUPREME, GRAND PRIX, LUMINA AND REGAL

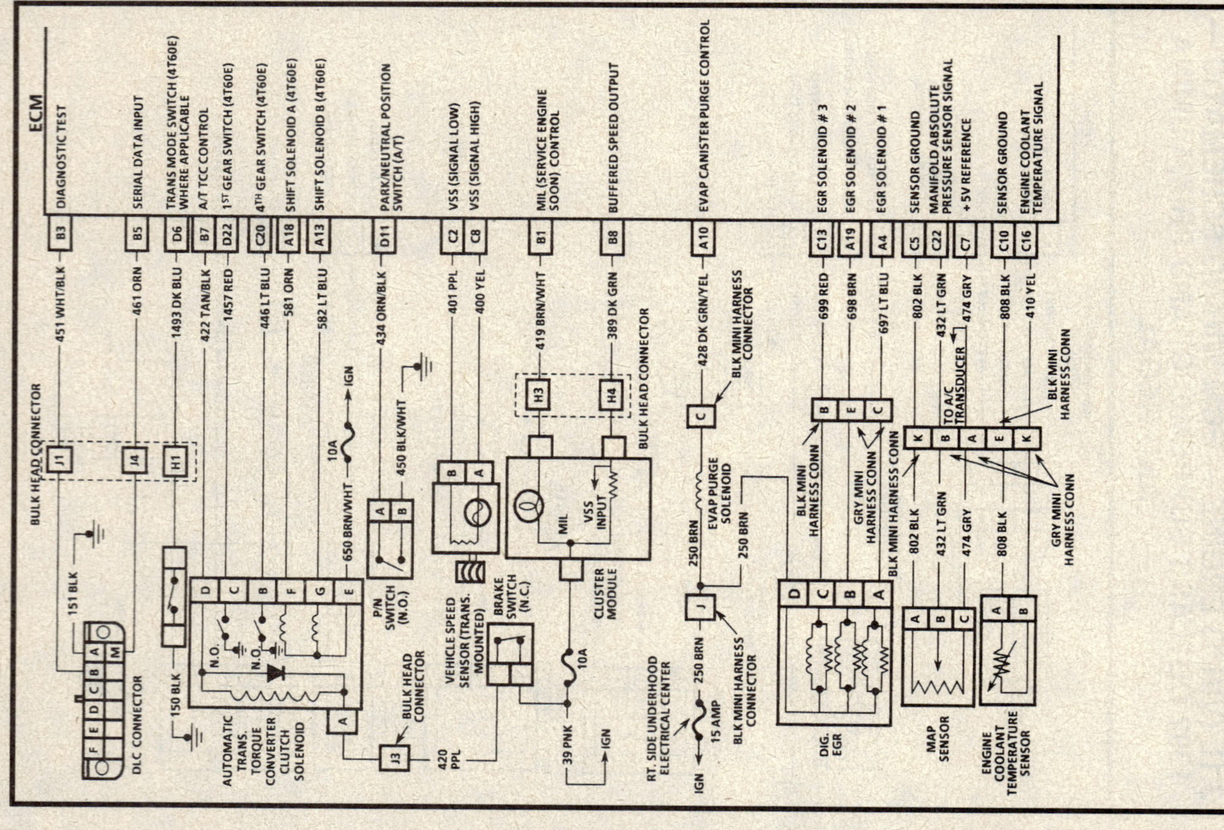

3.1L (VIN T) ENGINE — ECM WIRING SCHEMATIC — CUTLASS SUPREME, GRAND PRIX, LUMINA AND REGAL

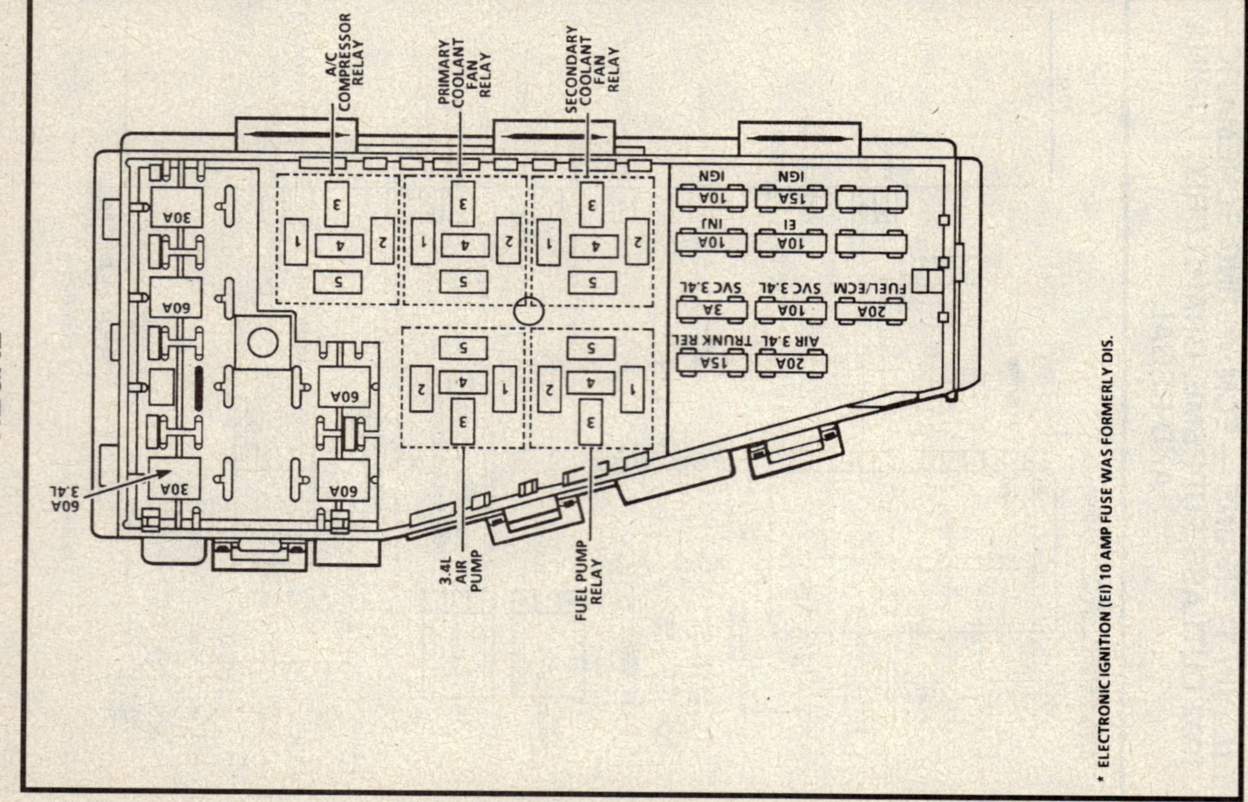

* ELECTRONIC IGNITION (EI) 10 AMP FUSE WAS FORMERLY DIS.

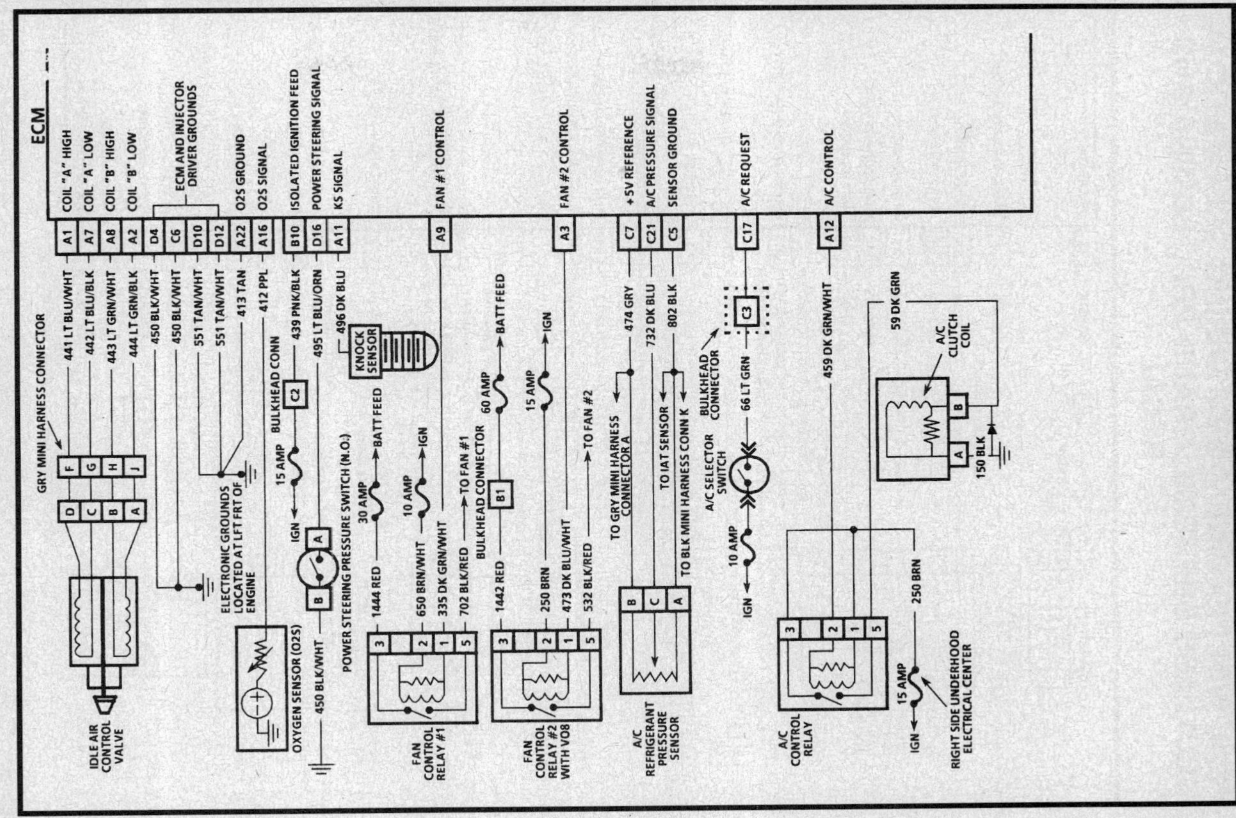

3.1L (VIN T) ENGINE — ECM WIRING SCHEMATIC — 1993–94 CUTLASS SUPREME, GRAND PRIX, LUMINA AND REGAL

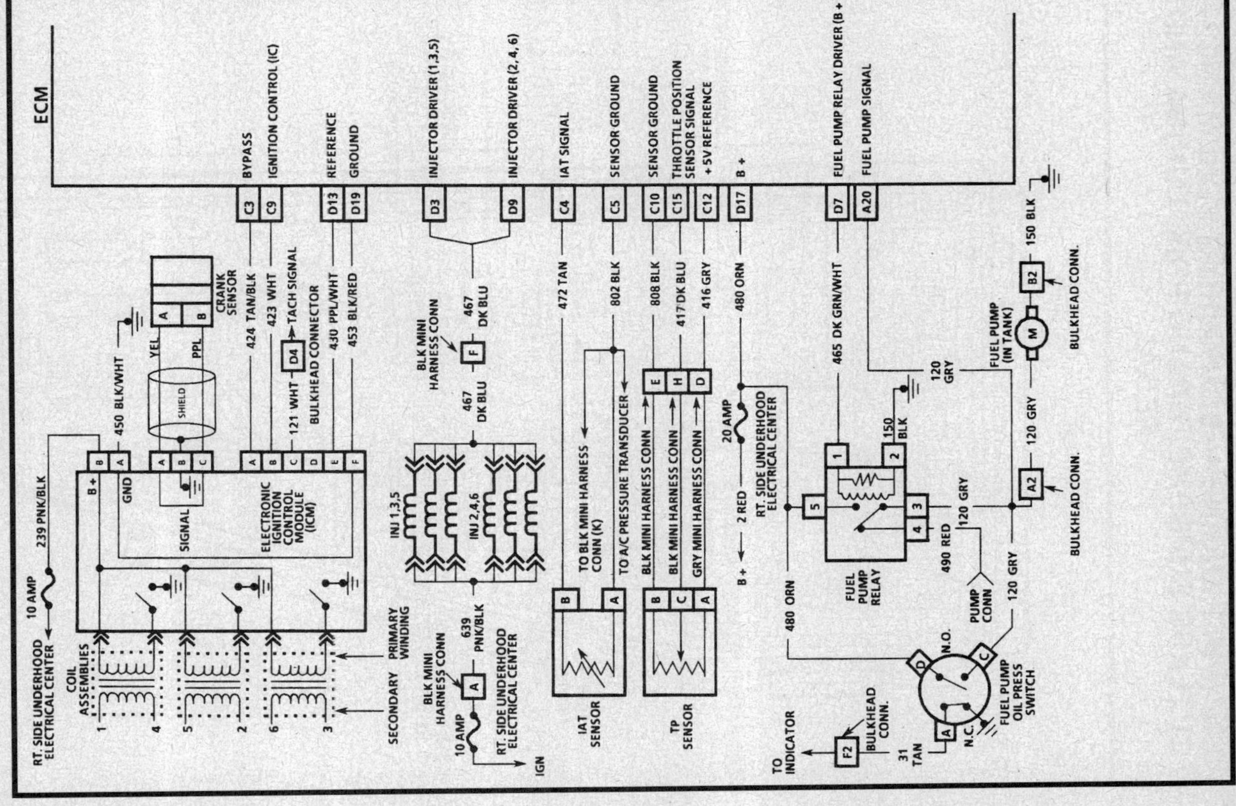

3.1L (VIN T) ENGINE — ECM WIRING SCHEMATIC — 1993–94 CUTLASS SUPREME, GRAND PRIX, LUMINA AND REGAL

MULTIPORT FUEL INJECTION (MFI) SYSTEMS
EXCEPT LIGHT TRUCKS, VANS, GEO AND SATURN

3.4L (VIN) ENGINE — ECM WIRING SCHEMATIC — CUTLASS SUPREME, GRAND PRIX, LUMINA AND REGAL

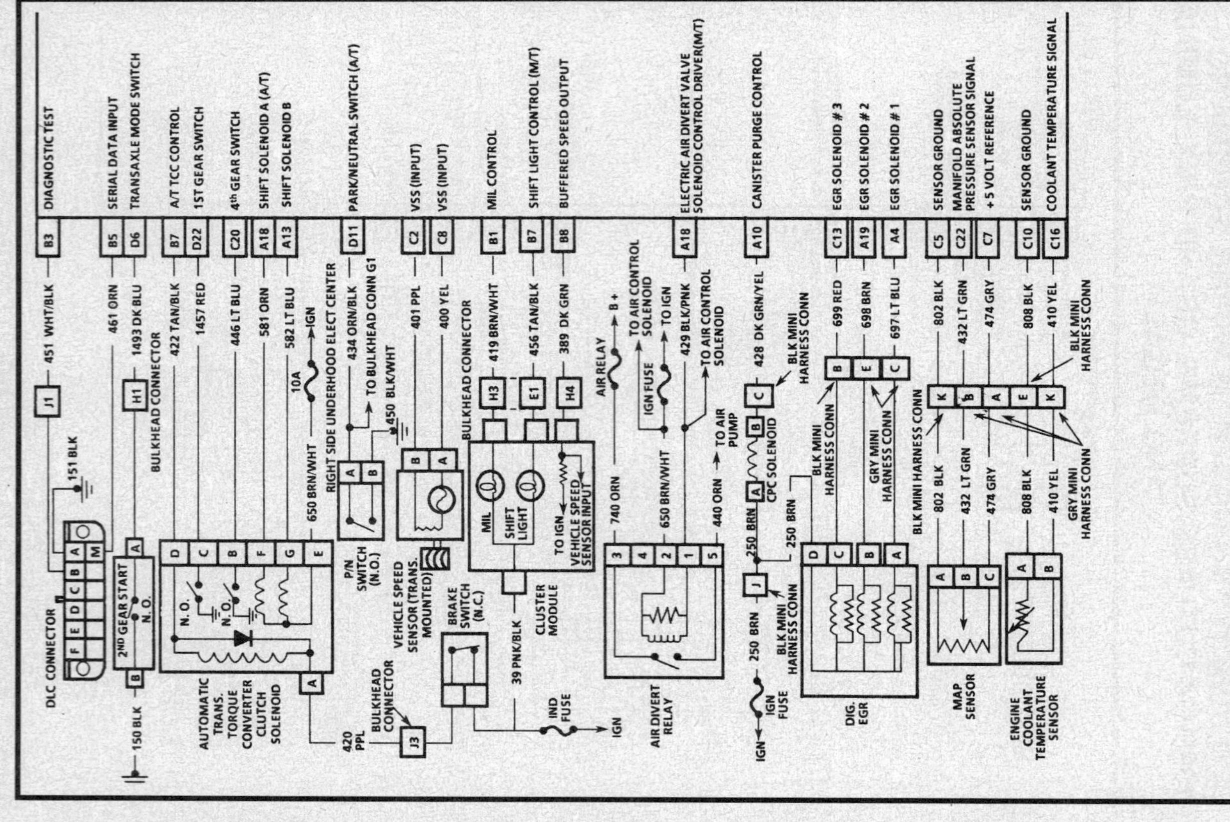

3.4L (VIN) ENGINE — ECM WIRING SCHEMATIC — CUTLASS SUPREME, GRAND PRIX, LUMINA AND REGAL

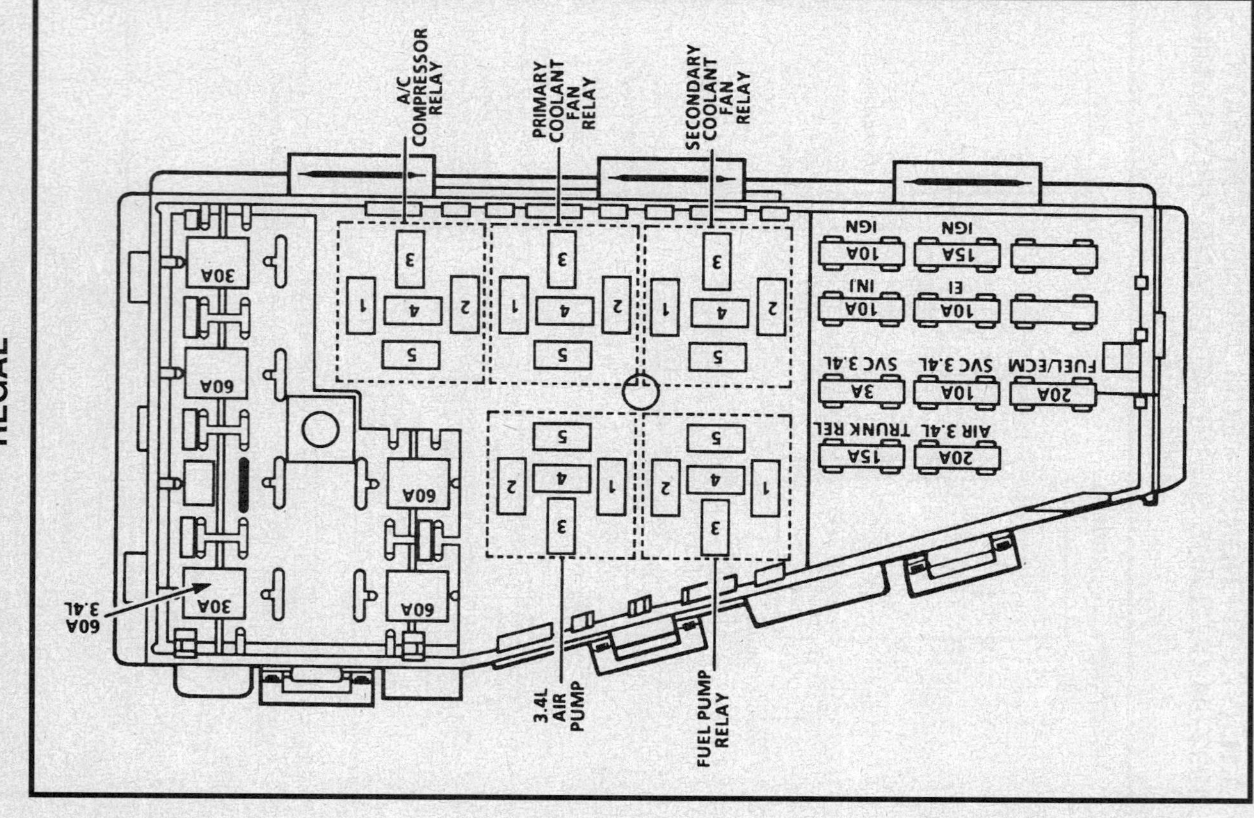

3.4L (VIN) ENGINE — ECM WIRING SCHEMATIC — CUTLASS SUPREME, GRAND PRIX, LUMINA AND REGAL

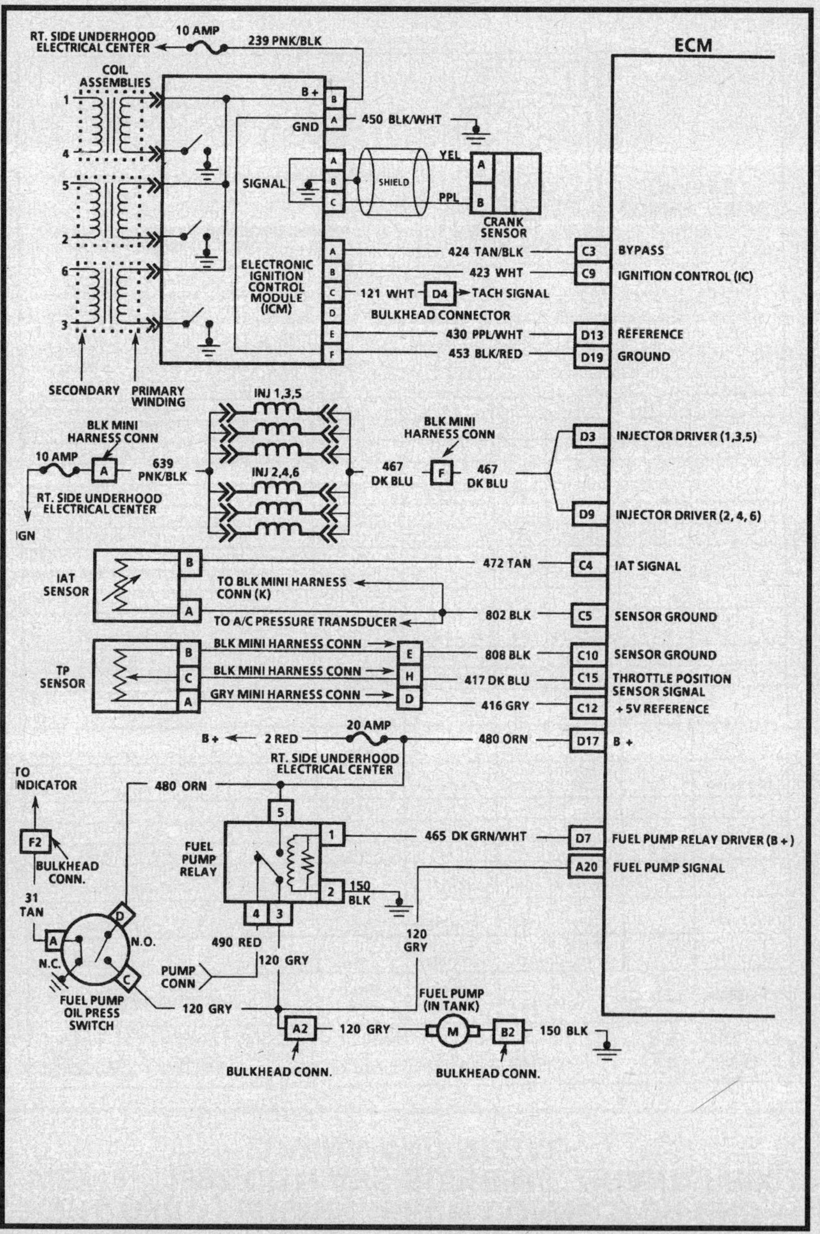

3.4L (VIN) ENGINE — ECM WIRING SCHEMATIC — CUTLASS SUPREME, GRAND PRIX, LUMINA AND REGAL

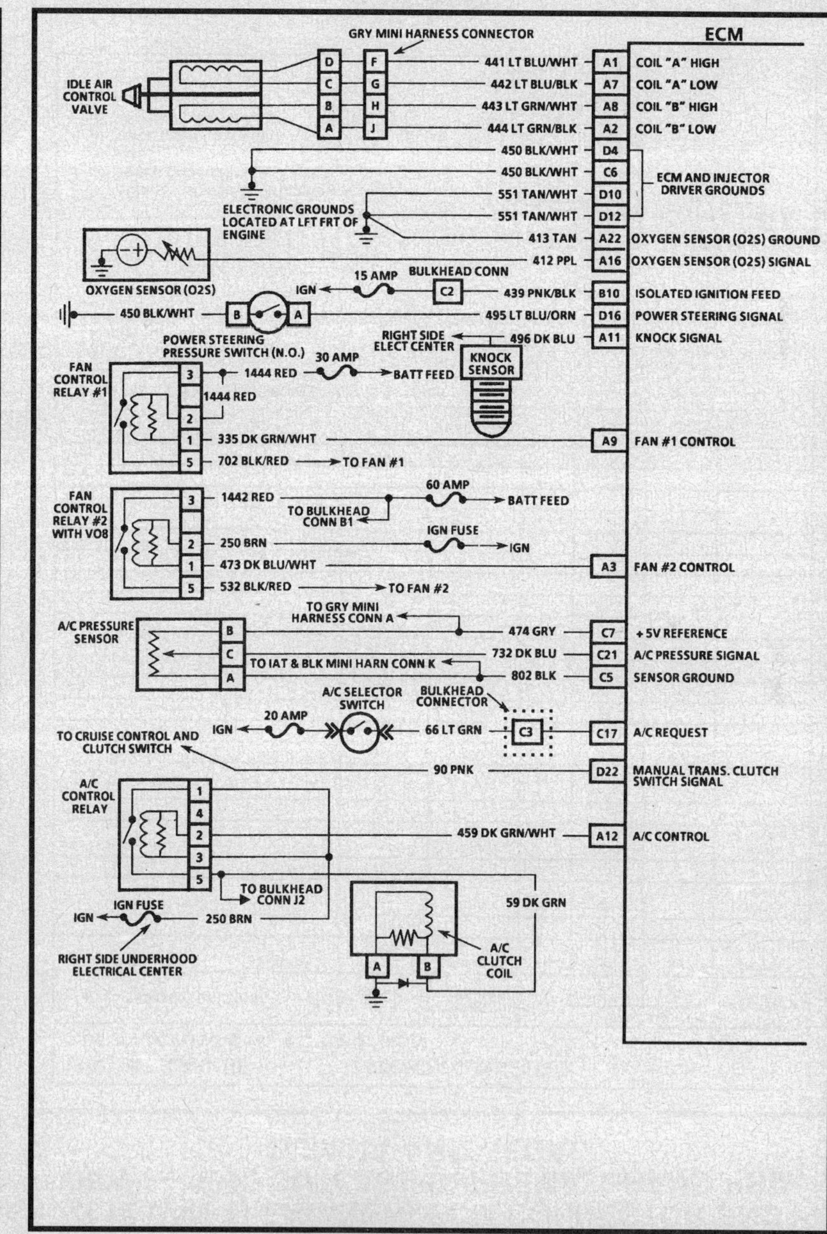

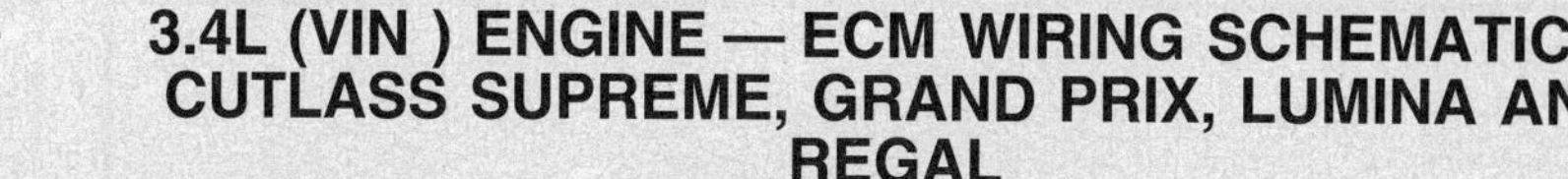

3.1L (VIN T) ENGINE — ECM CONNECTOR END VIEW — 1992 CUTLASS SUPREME, GRAND PRIX, LUMINA AND REGAL

1992 "W" CARLINE — 22 PIN ECM CONNECTOR — 3.1L (VIN T) RPO: LHO

* USE T-100 YELLOW BREAKOUT BOX (BOB)

ECM PIN/FUNCTION	BOB PIN #	WIRE COLOR	CKT #	VOLTAGE KEY "ON"	VOLTAGE ENG "RUN"	REFERENCE
A1 IAC "A" HIGH	104	LT BLU/WHT	441	NOT USEABLE		CHART C-2B
A2 IAC "B" LOW	101	LT GRN/BLK	444	NOT USEABLE		CHART C-2B
A3 FAN #2 CONTROL	112	DK BLU/WHT	473	FAN "OFF" B+	FAN "ON" 0*	
A4 EGR SOLENOID #1	109	LT BLU	697	B+	B+	
A5	119					
A6	117					
A7 IAC "A" LOW	103	LT BLU/BLK	442	NOT USEABLE		CHART C-2B
A8 IAC "B" HIGH	102	LT GRN/WHT	443	NOT USEABLE		CHART C-2B
A9 FAN #1 CONTROL	111	DK GRN/WHT	335	FAN "OFF" B+	FAN "ON" 0*	
A10 CANISTER PURGE	110	DK GRN/YEL	428	0*	0*	
A11 ESC SIGNAL	118	DK BLU	496	2.5	2.5	
A12 A/C RELAY CONTROL	108	DK GRN/WHT	459	A/C "OFF" B+	A/C "ON" 0*	
A13	105					
A14	116					
A15	113					
A16 O_2 SIGNAL	122	PPL	412	(3)	(3)	
A17	120					
A18 AIR SOLENOID (M/T)	107	BLK/PNK	429	B+	*	
A19 EGR SOLENOID #2	106	BRN	698	B+	B+	
A20 FUEL PUMP SIGNAL	115	GRY	120	(4)	B+	
A21	114					
A22 SENSOR GROUND	121	TAN	413	0*	0*	

***NOTICE: DO NOT BACKPROBE ECM CONNECTORS!**
This Chart may be used in conjunction with the T-100 Yellow Breakout Box (48921) to obtain voltage present for each circuit listed. Install the BOB between the ECM harness connectors and the ECM, then probe the pin listed under "BOB PIN#". Voltage may vary due to low battery charge or other reasons, but should be very close. All voltages shown in the ENG "RUN" column are typical with engine at idle, closed throttle, normal operating temperature, park or neutral, system in "Closed Loop," all accessories "OFF," and "Scan" tool not installed.

DVM NEGATIVE (BLACK) LEAD MUST BE CONNECTED TO A KNOWN GOOD GROUND.

(1) INCREASES WITH VEHICLE SPEED (MEASURE ON A/C SCALE).
(2) NORMAL OPERATING TEMPERATURE.
(3) VARIES.
(4) 12 VOLTS FIRST TWO SECONDS.
(5) VARIES WITH TEMPERATURE.
* LESS THAN 1 VOLT.

ECM CONNECTOR A ORANGE

3.1L (VIN T) ENGINE — ECM CONNECTOR END VIEW — 1992 CUTLASS SUPREME, GRAND PRIX, LUMINA AND REGAL

1992 "W" CARLINE — 22 PIN ECM CONNECTOR — 3.1L (VIN T) RPO: LHO

* USE T-100 YELLOW BREAKOUT BOX (BOB)

ECM PIN/FUNCTION	BOB PIN #	WIRE COLOR	CKT #	VOLTAGE KEY "ON"	VOLTAGE ENG "RUN"	REFERENCE
B1 "SES" LIGHT	204	BRN/WHT	419	0*	B+	CHARTS A-1 AND A-2
B2	201					
B3 DIAGNOSTIC / TEST	212	WHT/BLK	451	(5)	(5)	CHARTS A-1 AND A-2
B4	209					
B5 SERIAL DATA	219	ORN		4.8 (3)	4.8 (3)	CHARTS A-1 AND A-2
B6	217					
B7 TCC (A/T) SHIFT LIGHT (M/T)	203	TAN/BLK		A/T 0* M/T B+	A/T 0* M/T B+	
B8 BUFFERED SPEED OUT	202	DK GRN		B+	B+	
B9	211					
B10 ISOLATED IGNITION FEED	210	PNK/BLK		B+	B+	
B11	218					
B12	208					
B13	205					
B14	216					
B15	213					
B16	222					
B17	220					
B18	207					
B19	206					
B20	215					
B21	214					
B22	221					

***NOTICE: DO NOT BACKPROBE ECM CONNECTORS!**
This Chart may be used in conjunction with the T-100 Yellow Breakout Box (48921) to obtain voltage present for each circuit listed. Install the BOB between the ECM harness connectors and the ECM, then probe the pin listed under "BOB PIN#". Voltage may vary due to low battery charge or other reasons, but should be very close. All voltages shown in the ENG "RUN" column are typical with engine at idle, closed throttle, normal operating temperature, park or neutral, system in "Closed Loop," all accessories "OFF," and "Scan" tool not installed.

DVM NEGATIVE (BLACK) LEAD MUST BE CONNECTED TO A KNOWN GOOD GROUND.

(1) INCREASES WITH VEHICLE SPEED (MEASURE ON A/C SCALE).
(2) NORMAL OPERATING TEMPERATURE.
(3) VARIES.
(4) 12 VOLTS FIRST TWO SECONDS.
(5) VARIES WITH TEMPERATURE.
* LESS THAN 1 VOLT.

ECM CONNECTOR B WHITE

3.1L (VIN T) ENGINE — ECM CONNECTOR END VIEW — 1992 CUTLASS SUPREME, GRAND PRIX, LUMINA AND REGAL

1992 "W" CARLINE **22 PIN ECM CONNECTOR** **3.1L (VIN T) RPO: LHO**
* USE T-100 YELLOW BREAKOUT BOX (BOB)

ECM PIN/FUNCTION	BOB PIN #	WIRE COLOR	CKT #	VOLTAGE KEY "ON"	VOLTAGE ENG "RUN"
C1	304				
C2 MAG. VSS SIGNAL LOW	301	PPL	401	0*	(1)
C3 DIS BYPASS	312	TAN/BLK	424	0*	5
C4 IAT SENSOR SIGNAL	309	TAN	472	(5)	(5)
C5 SENSOR GROUND	319	BLK	802	0*	0*
C6 GROUND	317	BLK/WHT	450	0*	0*
C7 + 5 VOLT REFERENCE (MAP)	303	GRY	474	5	5
C8 MAG. VSS SIGNAL HIGH	302	YEL	400	0*	(1)
C9 EST CONTROL	311	WHT	423	0*	1.3 (3)
C10 SENSOR GROUND	310	BLK	808	0*	0*
C11	318				
C12 + 5 VOLT REFERENCE (TPS)	308	GRY	416	5	5
C13 EGR SOLENOID #3	305	RED	699	B+	B+
C14	316				
C15 TPS SIGNAL	313	DK BLU	417	.88	.88
C16 COOLANT TEMPERATURE SIGNAL	322	YEL	410	(5)	(5)
C17 A/C REQUEST	320	LT GRN	66	A/C REQUEST 0*	A/C "ON" B+
C18	307				
C19	306				
C20 2nd GEAR SIGNAL	315	WHT	232	0*	0*
C21 A/C PRESSURE SIGNAL	314	DK BLU	732	A/C OFF 0* A/C ON 12V	A/C OFF 0* A/C ON 12V
C22 MAP SIGNAL	321	LT GRN	432	4.75	(3)

*NOTICE: DO NOT BACKPROBE ECM CONNECTORS!
This Chart may be used in conjunction with the T-100 Yellow Breakout Box (48921) to obtain voltage present for each circuit listed. Install the BOB between the ECM harness connectors and the ECM, then probe the pin listed under "BOB PIN#". Voltage may vary due to low battery charge or other reasons, but should be very close. All voltages shown in the ENG "RUN" column are typical with engine at idle, closed throttle, normal operating temperature, park or neutral, system in "Closed Loop," all accessories "OFF," and "Scan" tool not installed.

DVM NEGATIVE (BLACK) LEAD MUST BE CONNECTED TO A KNOWN GOOD GROUND.

(1) INCREASES WITH VEHICLE SPEED (MEASURE ON A/C SCALE).
(2) NORMAL OPERATING TEMPERATURE.
(3) VARIES.
(4) 12 VOLTS FIRST TWO SECONDS.
(5) VARIES WITH TEMPERATURE.
* LESS THAN 1 VOLT.

ECM CONNECTOR C GREEN

3.1L (VIN T) ENGINE — ECM CONNECTOR END VIEW — 1992 CUTLASS SUPREME, GRAND PRIX, LUMINA AND REGAL

1992 "W" CARLINE **22 PIN ECM CONNECTOR** **3.1L (VIN T) RPO: LHO**
* USE T-100 YELLOW BREAKOUT BOX (BOB)

ECM PIN/FUNCTION	BOB PIN #	WIRE COLOR	CKT #	VOLTAGE KEY "ON"	VOLTAGE ENG "RUN"
D1	404				
D2	401				
D3 INJECTOR DRIVER (1,3,5)	412	DK BLU	467	B+	B+
D4 GROUND	409	BLK/WHT	450	0*	0*
D5	419				
D6 3rd GEAR SIGNAL	417	DK GRN	108	0*	0*
D7 FUEL PUMP RELAY DRIVER	403	DK GRN/WHT	465	0* (4)	B+
D8	402				
D9 INJECTOR DRIVER (2,4,6)	411	DK BLU	467	B+	B+
D10 GROUND	410	TAN/WHT	551	0*	0*
D11 P/N SWITCH (A/T)	418	ORN/BLK	434	0*	0*
D12 GROUND	408	TAN/WHT	551	0*	0*
D13 DIS REFERENCE HIGH	405	PPL/WHT	430	0*	2.3 (3)
D14	416				
D15	413				
D16 P/S PRESSURE SIGNAL	422	LT BLU/ORN	495	B+	B+
D17 BATTERY FEED	420	ORN	480	B+	B+
D18	407				
D19 DIS REFERENCE LOW	406	BLK/RED	453	0*	0*
D20	415				
D21	414				
D22 4th GEAR SIGNAL	421	LT BLU	446	0*	0*

*NOTICE: DO NOT BACKPROBE ECM CONNECTORS!
This Chart may be used in conjunction with the T-100 Yellow Breakout Box (48921) to obtain voltage present for each circuit listed. Install the BOB between the ECM harness connectors and the ECM, then probe the pin listed under "BOB PIN#". Voltage may vary due to low battery charge or other reasons, but should be very close. All voltages shown in the ENG "RUN" column are typical with engine at idle, closed throttle, normal operating temperature, park or neutral, system in "Closed Loop," all accessories "OFF," and "Scan" tool not installed.

DVM NEGATIVE (BLACK) LEAD MUST BE CONNECTED TO A KNOWN GOOD GROUND.

(1) INCREASES WITH VEHICLE SPEED (MEASURE ON A/C SCALE).
(2) NORMAL OPERATING TEMPERATURE.
(3) VARIES.
(4) 12 VOLTS FIRST TWO SECONDS.
(5) VARIES WITH TEMPERATURE.
* LESS THAN 1 VOLT.

ECM CONNECTOR D BLUE

3.1L (VIN T) ENGINE — ECM CONNECTOR END VIEW — 1993–94 CUTLASS SUPREME, GRAND PRIX, LUMINA AND REGAL

"W" CARLINE   3.1L (VIN T)   RPO: LHO
* USE T-100 YELLOW BREAKOUT BOX (BOB)

22 PIN ECM CONNECTOR

ECM PIN/FUNCTION	BOB PIN #	WIRE COLOR	CKT #	VOLTAGE KEY "ON"	VOLTAGE ENG "RUN"	REFERENCE
B1 MIL LIGHT	204	BRN/WHT	419	0*	B+	CHARTS A-1 AND A-2
B2	201					
B3 DIAGNOSTIC/TEST	212	WHT/BLK	451	(5)	(5)	CHARTS A-1 AND A-2
B4	209					
B5 SERIAL DATA	219	ORN	461	4.8 (3)	4.8 (3)	CHARTS A-1 AND A-2
B6	217					
B7 TCC (A/T)	203	TAN/BLK	422	A/T 0*	A/T 0*	
B8 BUFFERED SPEED OUT	202	DK GRN	389	B+	B+	
B9	211					
B10 ISOLATED IGNITION FEED	210	PNK/BLK	439	B+	B+	
B11	218					
B12	208					
B13	205					
B14	216					
B15	213					
B16	222					
B17	220					
B18	207					
B19	206					
B20	215					
B21	214					
B22	221					

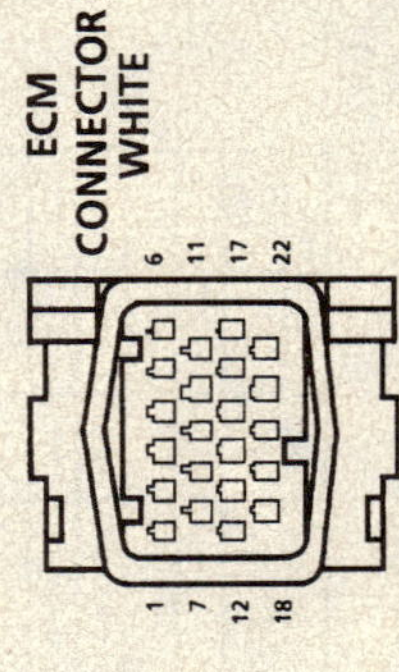

*NOTICE: DO NOT BACKPROBE ECM CONNECTORS!
This Chart may be used in conjunction with the T-100 Yellow Breakout Box (48921) to obtain voltage present for each circuit listed. Install the BOB between the ECM harness connectors and the ECM, then probe the pin listed under "BOB PIN#". Voltage may vary due to low battery charge or other reasons, but should be very close. All voltages shown in the ENG "RUN" column are typical with engine at idle, closed throttle, normal operating temperature, park or neutral, system in "Closed Loop," all accessories "OFF," and Scan tool not installed.

DVM NEGATIVE (BLACK) LEAD MUST BE CONNECTED TO A KNOWN GOOD GROUND.

(1) INCREASES WITH VEHICLE SPEED (MEASURE ON A/C SCALE).
(2) NORMAL OPERATING TEMPERATURE.
(3) VARIES.
(4) 12 VOLTS FIRST TWO SECONDS.
(5) VARIES WITH TEMPERATURE.
* LESS THAN 1 VOLT.

3.1L (VIN T) ENGINE — ECM CONNECTOR END VIEW — 1993–94 CUTLASS SUPREME, GRAND PRIX, LUMINA AND REGAL

"W" CARLINE   3.1L (VIN T)   RPO: LHO
* USE T-100 YELLOW BREAKOUT BOX (BOB)

22 PIN ECM CONNECTOR

ECM PIN/FUNCTION	BOB PIN #	WIRE COLOR	CKT #	VOLTAGE KEY "ON"	VOLTAGE ENG "RUN"	REFERENCE
A1 IAC "A" HIGH	104	LT BLU/WHT	441	NOT USEABLE	NOT USEABLE	CHART C-2B
A2 IAC "B" LOW	101	LT GRN/BLK	444	NOT USEABLE	NOT USEABLE	CHART C-2B
A3 FAN #2 CONTROL	112	DK BLU/WHT	473	FAN "OFF" B+	FAN "ON" 0*	
A4 EGR SOLENOID #1	109	LT BLU	697	B+	B+	
A5	119					
A6	117					
A7 IAC "A" LOW	103	LT BLU/BLK	442	NOT USEABLE	NOT USEABLE	CHART C-2B
A8 IAC "B" HIGH	102	LT GRN/WHT	443	NOT USEABLE	NOT USEABLE	CHART C-2B
A9 FAN #1 CONTROL	111	DK GRN/WHT	335	FAN "OFF" B+	FAN "ON" 0*	
A10 EVAP CANISTER PURGE	110	DK GRN/YEL	428	0*	0*	
A11 KNOCK SENSOR SIGNAL	118	DK BLU	496	2.5	2.5	
A12 A/C RELAY CONTROL	108	DK GRN/WHT	459	A/C "OFF" B+	A/C "ON" 0*	
A13 SHIFT SOLENOID B (4T60E)	105	LT BLU	582	0*	0*	
A14	116					
A15	113					
A16 O2S SIGNAL	122	PPL	412	(3)	(3)	
A17	120					
A18 SHIFT SOLENOID #2 (4T60E)	107	ORN	581	0*	0*	
A19 EGR SOLENOID #2	106	BRN	698	B+	B+	
A20 FUEL PUMP SIGNAL	115	GRY	120	(4)	B+	
A21	114					
A22 SENSOR GROUND	121	TAN	413	0*	0*	

*NOTICE: DO NOT BACKPROBE ECM CONNECTORS!
This Chart may be used in conjunction with the T-100 Yellow Breakout Box (48921) to obtain voltage present for each circuit listed. Install the BOB between the ECM harness connectors and the ECM, then probe the pin listed under "BOB PIN#". Voltage may vary due to low battery charge or other reasons, but should be very close. All voltages shown in the ENG "RUN" column are typical with engine at idle, closed throttle, normal operating temperature, park or neutral, system in "Closed Loop," all accessories "OFF," and Scan tool not installed.

DVM NEGATIVE (BLACK) LEAD MUST BE CONNECTED TO A KNOWN GOOD GROUND.

(1) INCREASES WITH VEHICLE SPEED (MEASURE ON A/C SCALE).
(2) NORMAL OPERATING TEMPERATURE.
(3) VARIES.
(4) 12 VOLTS FIRST TWO SECONDS.
(5) VARIES WITH TEMPERATURE.
* LESS THAN 1 VOLT.

3.1L (VIN T) ENGINE — ECM CONNECTOR END VIEW — 1993–94 CUTLASS SUPREME, GRAND PRIX, LUMINA AND REGAL

"W" CARLINE — 22 PIN ECM CONNECTOR — 3.1L (VIN T) RPO: LHO
* USE T-100 YELLOW BREAKOUT BOX (BOB)

ECM PIN/FUNCTION		BOB PIN #	WIRE COLOR	CKT #	VOLTAGE KEY "ON"	VOLTAGE ENG "RUN"
D1		404				
D2		401				
D3	INJECTOR DRIVER (1,3,5)	412	DK BLU	467	B+	B+
D4	GROUND	409	BLK/WHT	450	0*	0*
D5		419				
D6	TRANSAXLE MODE SWITCH (4T60E)	417	DK BLU	1493	B+	B+
D6	3rd GEAR SIGNAL (3T40)	417	DK GRN	108	0*	0*
D7	FUEL PUMP RELAY DRIVER	403	DK GRN/WHT	465	0* (4)	B+
D8		402				
D9	INJECTOR DRIVER (2,4,6)	411	DK BLU	467	B+	B+
D10	GROUND	410	TAN/WHT	551	0*	0*
D11	PNP SWITCH (A/T)	418	ORN/BLK	434	0*	0*
D12	GROUND	408	TAN/WHT	551	0*	0*
D13	EI REFERENCE HIGH	405	PPL/WHT	430	0*	2.3 (3)
D14		416				
D15		413				
D16	P/S PRESSURE SIGNAL	422	LT BLU/ORN	495	B+	B+
D17	BATTERY FEED	420	ORN	480	B+	B+
D18		407				
D19	EI REFERENCE LOW	406	BLK/RED	453	0*	0*
D20		415				
D21		414				
D22	1st GEAR SIGNAL (4T60E)	421	RED	1457	0*	0*

***NOTICE:** DO NOT BACKPROBE ECM CONNECTORS! This Chart may be used in conjunction with the T-100 Yellow Breakout Box (48921) to obtain voltage present for each circuit listed. Install the BOB between the ECM harness connectors and the ECM, then probe the pin listed under "BOB PIN#". Voltage may vary due to low battery charge or other reasons, but should be very close. All voltages shown in the ENG "RUN" column are typical with engine at idle, closed throttle, normal operating temperature, park or neutral, system in "Closed Loop," all accessories "OFF," and Scan tool not installed.

DVM NEGATIVE (BLACK) LEAD MUST BE CONNECTED TO A KNOWN GOOD GROUND.

(1) INCREASES WITH VEHICLE SPEED (MEASURE ON A/C SCALE).
(2) NORMAL OPERATING TEMPERATURE.
(3) VARIES.
(4) 12 VOLTS FIRST TWO SECONDS.
(5) VARIES WITH TEMPERATURE.
* LESS THAN 1 VOLT.

3.1L (VIN T) ENGINE — ECM CONNECTOR END VIEW — 1993–94 CUTLASS SUPREME, GRAND PRIX, LUMINA AND REGAL

"W" CARLINE — 22 PIN ECM CONNECTOR — 3.1L (VIN T) RPO: LHO
* USE T-100 YELLOW BREAKOUT BOX (BOB)

ECM PIN/FUNCTION		BOB PIN #	WIRE COLOR	CKT #	VOLTAGE KEY "ON"	VOLTAGE ENG "RUN"
C1		304				
C2	MAG. VSS SIGNAL LOW	301	PPL	401	0*	(1)
C3	EI BYPASS	312	TAN/BLK	424	0*	5
C4	IAT SENSOR SIGNAL	309	TAN	472	(5)	(5)
C5	SENSOR GROUND	319	BLK	802	0*	0*
C6	GROUND	317	BLK/WHT	450	0*	0*
C7	+ 5 VOLT REFERENCE (MAP)	303	GRY	474	5	5
C8	MAG. VSS SIGNAL HIGH	302	YEL	400	0*	(1)
C9	IGNITION CONTROL (IC)	311	WHT	423	0*	1.3 (3)
C10	SENSOR GROUND	310	BLK	808	0*	0*
C11		318				
C12	+ 5 VOLT REFERENCE (TP SENSOR)	308	GRY	416	5	5
C13	EGR SOLENOID #3	305	RED	699	B+	B+
C14		316				
C15	TP SENSOR SIGNAL	313	DK BLU	417	.88	.88
C16	ENGINE COOLANT TEMPERATURE SIGNAL	322	YEL	410	(5)	(5)
C17	A/C REQUEST	320	LT GRN	66	A/C REQUEST 0*	A/C "ON" B+
C18		307				
C19		306				
C20	4th GEAR SIGNAL (4T60E)	315	LT BLU	446	0*	0*
C20	2nd GEAR SIGNAL (3T40)	315	WHT	232	0*	0*
C21	A/C PRESSURE SIGNAL	314	DK BLU	732	A/C OFF 0* A/C ON 12V	A/C OFF 0* A/C ON 12V
C22	MAP SIGNAL	321	LT GRN	432	4.75	(3)

***NOTICE:** DO NOT BACKPROBE ECM CONNECTORS! This Chart may be used in conjunction with the T-100 Yellow Breakout Box (48921) to obtain voltage present for each circuit listed. Install the BOB between the ECM harness connectors and the ECM, then probe the pin listed under "BOB PIN#". Voltage may vary due to low battery charge or other reasons, but should be very close. All voltages shown in the ENG "RUN" column are typical with engine at idle, closed throttle, normal operating temperature, park or neutral, system in "Closed Loop," all accessories "OFF," and Scan tool not installed.

DVM NEGATIVE (BLACK) LEAD MUST BE CONNECTED TO A KNOWN GOOD GROUND.

(1) INCREASES WITH VEHICLE SPEED (MEASURE ON A/C SCALE).
(2) NORMAL OPERATING TEMPERATURE.
(3) VARIES.
(4) 12 VOLTS FIRST TWO SECONDS.
(5) VARIES WITH TEMPERATURE.
* LESS THAN 1 VOLT.

3.4L (VIN X) ENGINE — ECM CONNECTOR END VIEW — CUTLASS SUPREME, GRAND PRIX, LUMINA AND REGAL

"W" CARLINE — 22 PIN ECM CONNECTOR — 3.4L (VIN X) RPO: LQ1
* USE T-100 YELLOW BREAKOUT BOX (BOB)

ECM PIN/FUNCTION	BOB PIN #	WIRE COLOR	CKT #	VOLTAGE KEY "ON"	VOLTAGE ENG "RUN"	REFERENCE
A1 IAC "A" HIGH	104	LT BLU/WHT	441	NOT USEABLE		CHART C-2B
A2 IAC "B" LOW	101	LT GRN/BLK	444	NOT USEABLE		CHART C-2B
A3 FAN #2 CONTROL	112	DK BLU/WHT	473	FAN "OFF" B+	FAN "ON" 0*	
A4 EGR SOLENOID #1	109	LT BLU	697	B+	B+	
A5	119					
A6	117					
A7 IAC "A" LOW	103	LT BLU/BLK	442	NOT USEABLE		CHART C-2B
A8 IAC "B" HIGH	102	LT GRN/WHT	443	NOT USEABLE		CHART C-2B
A9 FAN #1 CONTROL	111	DK GRN/WHT	335	FAN "OFF" B+	FAN "ON" 0*	
A10 EVAP CANISTER PURGE	110	DK GRN/YEL	428	0*	0*	
A11 KS SIGNAL	118	DK BLU	496	2.5	2.5	
A12 A/C RELAY CONTROL	108	DK GRN/WHT	459	A/C "OFF" B+	A/C "ON" 0*	
A13 SHIFT SOLENOID "B"	105	LT BLU	582	0*	0*	
A14	116					
A15	113					
A16 O2S SIGNAL	122	PPL	412	(3)	(3)	
A17	120					
A18 SHIFT SOLENOID "A"	107	ORN	581	0*	0*	
A18 AIR PUMP RELAY (M/T 3.4L)	107	BLK/PNK	429	B+	*	
A19 EGR SOLENOID #2	106	BRN	698	B+	B+	
A20 FUEL PUMP SIGNAL	115	GRY	120	(4)	B+	
A21	114					
A22 SENSOR GROUND	121	TAN	413	0*	0*	

*NOTICE: DO NOT BACKPROBE ECM CONNECTORS!
This Chart may be used in conjunction with the T-100 Yellow Breakout Box (48921) to obtain voltage present for each circuit listed. Install the BOB between the ECM harness connectors and the ECM, then probe the pin listed under "BOB PIN#". Voltage may vary due to low battery charge or other reasons, but should be very close. All voltages shown in the ENG "RUN" column are typical with engine at idle, closed throttle, normal operating temperature, park or neutral, system in "Closed Loop," all accessories "OFF," and "Scan" tool not installed.

DVM NEGATIVE (BLACK) LEAD MUST BE CONNECTED TO A KNOWN GOOD GROUND.

ECM CONNECTOR A ORANGE

(1) INCREASES WITH VEHICLE SPEED (MEASURE ON A/C SCALE).
(2) NORMAL OPERATING TEMPERATURE.
(3) VARIES.
(4) 12 VOLTS FIRST TWO SECONDS.
(5) VARIES WITH TEMPERATURE.
* LESS THAN 1 VOLT.

3.4L (VIN X) ENGINE — ECM CONNECTOR END VIEW — CUTLASS SUPREME, GRAND PRIX, LUMINA AND REGAL

"W" CARLINE — 22 PIN ECM CONNECTOR — 3.4L (VIN X) RPO: LQ1
* USE T-100 YELLOW BREAKOUT BOX (BOB)

ECM PIN/FUNCTION	BOB PIN #	WIRE COLOR	CKT #	VOLTAGE KEY "ON"	VOLTAGE ENG "RUN"	REFERENCE
B1 MIL LIGHT	204	BRN/WHT	419	0*	B+	CHARTS A-1 AND A-2
B2	201					
B3 DIAGNOSTIC / TEST	212	WHT/BLK	451	(5)	(5)	CHARTS A-1 AND A-2
B4	209					
B5 SERIAL DATA	219	ORN	461	4.8 (3)	4.8 (3)	CHARTS A-1 AND A-2
B6	217					
B7 TCC (A/T) SHIFT LIGHT (M/T)	203	TAN/BLK	422	A/T 0*	A/T 0*	
B7 SHIFT LIGHT (M/T)	203	TAN/BLK	422	M/T B+	M/T B+	
B8 BUFFERED SPEED OUT	202	DK GRN	389	B+	B+	
B9	211					
B10 ISOLATED IGNITION FEED	210	PNK/BLK	439	B+	B+	
B11	218					
B12	208					
B13	205					
B14	216					
B15	213					
B16	222					
B17	220					
B18	207					
B19	206					
B20	215					
B21	214					
B22	221					

*NOTICE: DO NOT BACKPROBE ECM CONNECTORS!
This Chart may be used in conjunction with the T-100 Yellow Breakout Box (48921) to obtain voltage present for each circuit listed. Install the BOB between the ECM harness connectors and the ECM, then probe the pin listed under "BOB PIN#". Voltage may vary due to low battery charge or other reasons, but should be very close. All voltages shown in the ENG "RUN" column are typical with engine at idle, closed throttle, normal operating temperature, park or neutral, system in "Closed Loop," all accessories "OFF," and "Scan" tool not installed.

DVM NEGATIVE (BLACK) LEAD MUST BE CONNECTED TO A KNOWN GOOD GROUND.

ECM CONNECTOR B WHITE

(1) INCREASES WITH VEHICLE SPEED (MEASURE ON A/C SCALE).
(2) NORMAL OPERATING TEMPERATURE.
(3) VARIES.
(4) 12 VOLTS FIRST TWO SECONDS.
(5) VARIES WITH TEMPERATURE.
* LESS THAN 1 VOLT.

3.4L (VIN X) ENGINE — ECM CONNECTOR END VIEW — CUTLASS SUPREME, GRAND PRIX, LUMINA AND REGAL

"W" CARLINE — 22 PIN ECM CONNECTOR — 3.4L (VIN X) RPO: LQ1 * USE T-100 YELLOW BREAKOUT BOX (BOB)					
ECM PIN/FUNCTION	BOB PIN #	WIRE COLOR	CKT #	VOLTAGE KEY "ON"	VOLTAGE ENG "RUN"
C1	304				
C2 MAG. VSS SIGNAL LOW	301	PPL	401	0*	(1)
C3 EI BYPASS	312	TAN/BLK	424	0*	5
C4 IAT SENSOR SIGNAL	309	TAN	472	(5)	(5)
C5 SENSOR GROUND	319	BLK	802	0*	0*
C6 GROUND	317	BLK/WHT	450	0*	0*
C7 +5 VOLT REFERENCE (MAP)	303	GRY	474	5	5
C8 MAG. VSS SIGNAL HIGH	302	YEL	400	0*	(1)
C9 IGNITION CONTROL	311	WHT	423	0*	1.3 (3)
C10 SENSOR GROUND	310	BLK	808	0*	0*
C11	318				
C12 +5 VOLT REFERENCE (TP SENSOR)	308	GRY	416	5	5
C13 EGR SOLENOID #3	305	RED	699	B+	B+
C14	316				
C15 TP SENSOR SIGNAL	313	DK BLU	417	.88	.88
C16 ENGINE COOLANT TEMPERATURE SIGNAL	322	YEL	410	(5)	(5)
C17 A/C REQUEST	320	LT GRN	66	A/C REQUEST 0*	A/C "ON" B+
C18	307				
C19	306				
C20 4th GEAR SIGNAL	315	LT BLU	446	0*	0*
C21 A/C PRESSURE SIGNAL	314	DK BLU	732	A/C OFF 0* A/C ON 12V	A/C OFF 0* A/C ON 12V
C22 MAP SIGNAL	321	LT GRN	432	4.75	(3)

***NOTICE:** DO NOT BACKPROBE ECM CONNECTORS!
This Chart may be used in conjunction with the T-100 Yellow Breakout Box (48921) to obtain voltage present for each circuit listed. Install the BOB between the ECM harness connectors and the ECM, then probe the pin listed under "BOB PIN#". Voltage may vary due to low battery charge or other reasons, but should be very close. All voltages shown in the ENG "RUN" column are typical with engine at idle, closed throttle, normal operating temperature, park or neutral, system in "Closed Loop," all accessories "OFF," and scan tool not installed.

DVM NEGATIVE (BLACK) LEAD MUST BE CONNECTED TO A KNOWN GOOD GROUND.

(1) INCREASES WITH VEHICLE SPEED (MEASURE ON A/C SCALE).
(2) NORMAL OPERATING TEMPERATURE.
(3) VARIES.
(4) 12 VOLTS FIRST TWO SECONDS.
(5) VARIES WITH TEMPERATURE.
* LESS THAN 1 VOLT.

ECM CONNECTOR C GREEN

3.4L (VIN X) ENGINE — ECM CONNECTOR END VIEW — CUTLASS SUPREME, GRAND PRIX, LUMINA AND REGAL

"W" CARLINE — 22 PIN ECM CONNECTOR — 3.4L (VIN X) RPO: LQ1 * USE T-100 YELLOW BREAKOUT BOX (BOB)					
ECM PIN/FUNCTION	BOB PIN #	WIRE COLOR	CKT #	VOLTAGE KEY "ON"	VOLTAGE ENG "RUN"
D1	404				
D2	401				
D3 INJECTOR DRIVER (1,3,5)	412	DK BLU	467	B+	B+
D4 GROUND	409	BLK/WHT	450	0*	0*
D5	419				
D6 TRANSAXLE MODE SWITCH	417	DK BLU	1493	B+	B+
D7 FUEL PUMP RELAY DRIVER	403	DK GRN/WHT	465	0* (4)	B+
D8	402				
D9 INJECTOR DRIVER (2,4,6)	411	DK BLU	467	B+	B+
D10 GROUND	410	TAN/WHT	551	0*	0*
D11 PNP SWITCH (A/T)	418	ORN/BLK	434	0*	0*
D12 GROUND	408	TAN/WHT	551	0*	0*
D13 EI REFERENCE HIGH	405	PPL/WHT	430	0*	2.3 (3)
D14	416				
D15	413				
D16 P/S PRESSURE SIGNAL	422	LT BLU/ORN	495	B+	B+
D17 BATTERY FEED	420	ORN	480	B+	B+
D18	407				
D19 EI REFERENCE LOW	406	BLK/RED	453	0*	0*
D20	415				
D21 NOT USED	414	WHT	85		
D22 MANUAL TRANSMISSION CLUTCH SWITCH	421	PNK	90	0*	DEPRESSED B+
D22 1st GEAR SIGNAL	421	RED	1457	0*	0*

***NOTICE:** DO NOT BACKPROBE ECM CONNECTORS!
This Chart may be used in conjunction with the T-100 Yellow Breakout Box (48921) to obtain voltage present for each circuit listed. Install the BOB between the ECM harness connectors and the ECM, then probe the pin listed under "BOB PIN#". Voltage may vary due to low battery charge or other reasons, but should be very close. All voltages shown in the ENG "RUN" column are typical with engine at idle, closed throttle, normal operating temperature, park or neutral, system in "Closed Loop," all accessories "OFF," and scan tool not installed.

DVM NEGATIVE (BLACK) LEAD MUST BE CONNECTED TO A KNOWN GOOD GROUND.

(1) INCREASES WITH VEHICLE SPEED (MEASURE ON A/C SCALE).
(2) NORMAL OPERATING TEMPERATURE.
(3) VARIES.
(4) 12 VOLTS FIRST TWO SECONDS.
(5) VARIES WITH TEMPERATURE.
* LESS THAN 1 VOLT.

ECM CONNECTOR D BLUE

3.1L (VIN T) AND 3.4L (VIN X) ENGINES — ON-BOARD DIAGNOSTIC SYSTEM CHART — CUTLASS SUPREME, GRAND PRIX, LUMINA AND REGAL

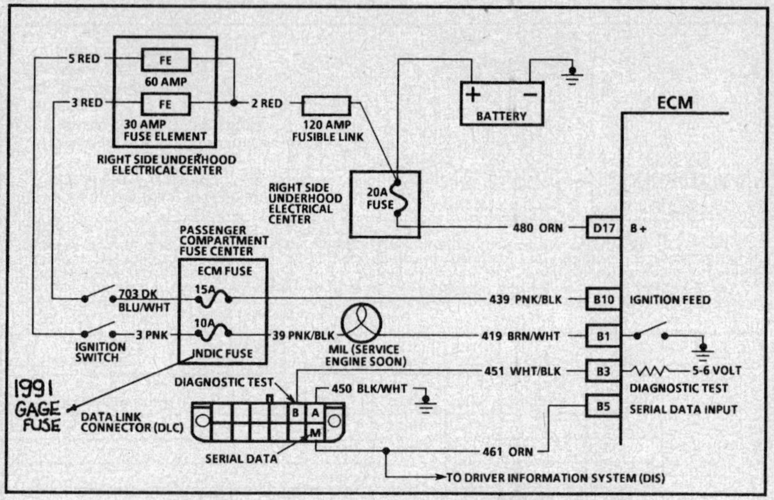

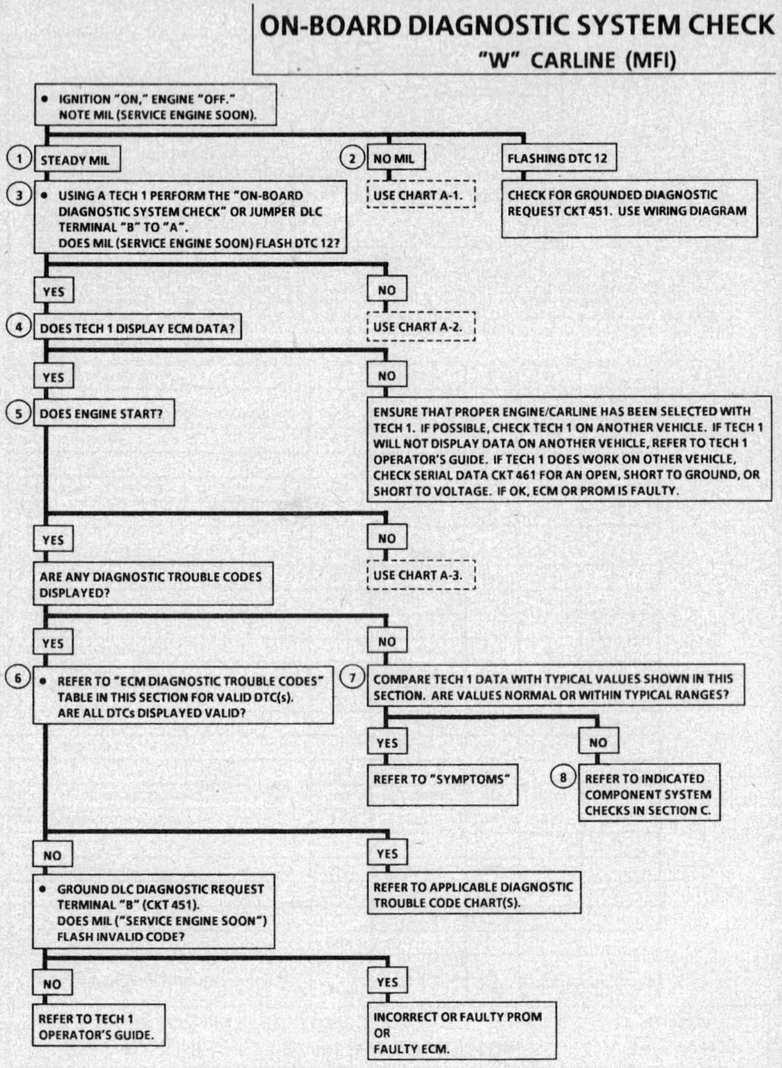

ON-BOARD DIAGNOSTIC SYSTEM CHECK
"W" CARLINE (MFI)

Circuit Description:

The on-board diagnostic system check is an organized approach to identifying a problem created by an electronic engine control system malfunction. It must be the starting point for any driveability complaint diagnosis, because it directs the service technician to the next logical step in diagnosing the complaint. Understanding the chart and using it correctly will reduce diagnostic time and prevent the unnecessary replacement of good parts.

Test Description: Number(s) below refer to circled number(s) on the diagnostic chart.

1. This step is a check for the proper operation of the Malfunction Indicator Lamp (MIL) "Service Engine Soon." The MIL should be "ON" steady.
2. No MIL, at this point indicates that there is a problem with the MIL circuit or the ECM control of that circuit.
3. This test checks the ability of the ECM to control the MIL. With the diagnostic terminal grounded, the MIL should flash a DTC 12 three times, followed by any diagnostic trouble code stored in memory. A PROM error may result in the inability to flash DTC 12.
4. Most of the diagnostic charts use a Tech 1 to aid diagnosis, therefore, serial data must be available. If a PROM error is present, the ECM may have been able to flash Diagnostic Trouble Code (DTC) 12 or 51, but not transmit serial data.
5. Although the ECM is powered up, a "Cranks But Will Not Run" symptom could exist because of an ECM or system problem.
6. This step will isolate if the customer complaint is a MIL or a driveability problem with no MIL. Refer to ECM diagnostic trouble codes chart in this section for a list of valid DTC(s). An invalid DTC may be the result of a faulty scan tool, PROM or ECM.
7. Comparison of actual control system data with the typical valves is a quick check to determine if any parameter is not within limits. Keep in mind that a base engine problem (i.e. advanced cam timing) may substantially alter sensor values.
8. If the actual data is not within the typical values established, the charts in "Component Systems," will provide a functional check of the suspect component of system.

ON-BOARD DIAGNOSTIC SYSTEM CHECK
"W" CARLINE (MFI)

3.1L (VIN T) AND 3.4L (VIN X) ENGINES — ECM DIAGNOSTIC TROUBLE CODES — CUTLASS SUPREME, GRAND PRIX, LUMINA AND REGAL

DTC	DESCRIPTION	ILLUMINATE MIL
13	Oxygen Sensor - open circuit	YES
14	Engine Coolant Temperatuare Sensor - high temperature indicated	YES
15	Engine Coolant Temperature Sensor - low temperature indicated	YES
21	Throttle Position Sensor - voltage high	YES
22	Throttle Position Sensor - voltage low	YES
23	Intake Air Temperature Sensor - low temperature indicated	YES
24	Vehicle Speed Sensor Circuit	YES
25	Intake Air Temperature Sensor - high temperature indicated	YES
32	Exhaust gas recirculation circuit	YES
33	MAP Sensor - high voltage, low vacuum indicated	YES
34	MAP Sensor - high voltage, low vacuum indicated	YES
35	Idle Air Control (IAC) Circuit	YES
41	Cylinder Select Error	YES
42	Knock Sensor (KS) Circuit	YES
43	Ignition Control (IC) Circuit	YES
44	Oxygen Sensor - lean exhaust indicated	YES
45	Oxygen Sensor - rich exhaust indicated	YES
53	System Over Voltage	YES
54	Fuel Pump Circuit - low voltage	YES
51	PROM Error (Faulty or Incorrect PROM)	YES
55	ECM Error	YES
61	Degraded Oxygen Sensor (O2S)	YES
62	Transaxle Gear Switch Signal Circuit	YES
66	A/C Refrigerant Pressure Sensor Circuit	NO

If a DTC not listed above appears on Tech 1, ground DLC diagnostic terminal "B" and observe flashed DTC(s).
If DTC does not reappear, refer to Tech 1 Operator's Guide.
If DTC does reappear, check for incorrect or faulty PROM.

3.1L (VIN T) AND 3.4L (VIN X) ENGINES — SYSTEM DIAGNOSTIC CHARTS — CUTLASS SUPREME, GRAND PRIX, LUMINA AND REGAL

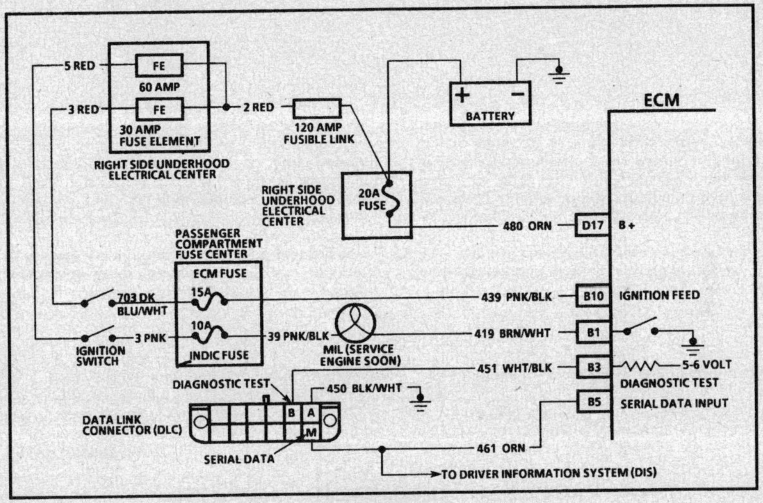

CHART A-1

NO MIL (SERVICE ENGINE SOON)
"W" CARLINE (MFI)

Circuit Description:
There should always be a steady Malfunction Indicator Lamp (MIL) "Service Engine Soon" when the ignition is "ON" and engine stopped. Battery voltage is supplied directly to the light bulb. The Engine Control Module (ECM) will control the light and turn it "ON" by providing a ground path through CKT 419 to the ECM.

Test Description: Number(s) below refer to circled number(s) on the diagnostic chart.
1. If the fuse in holder is blown refer to facing page of DTC 54 for complete circuit.
2. Using a test light connected to 12 volts probe each of the system ground circuits to be sure a good ground is present. Refer to ECM terminal end view in front of this section for ECM pin locations of ground circuits.

Diagnostic Aids:

Engine runs OK, check:
- Faulty light bulb.
- CKT 419 open.
- INDIC fuse blown. This will result in no stop lights, oil or generator lights, seat belt reminder, etc.

Engine cranks but will not run.
- Continuous battery - fuse or fusible link open.
- ECM ignition fuse open.
- Battery CKT 480 to ECM open.
- Ignition CKT 439 to ECM open.
- Poor connection to ECM.

3.1L (VIN T) AND 3.4L (VIN X) ENGINES — SYSTEM DIAGNOSTIC CHARTS — CUTLASS SUPREME, GRAND PRIX, LUMINA AND REGAL

CHART A-1

NO MIL (SERVICE ENGINE SOON)
"W" CARLINE (MFI)

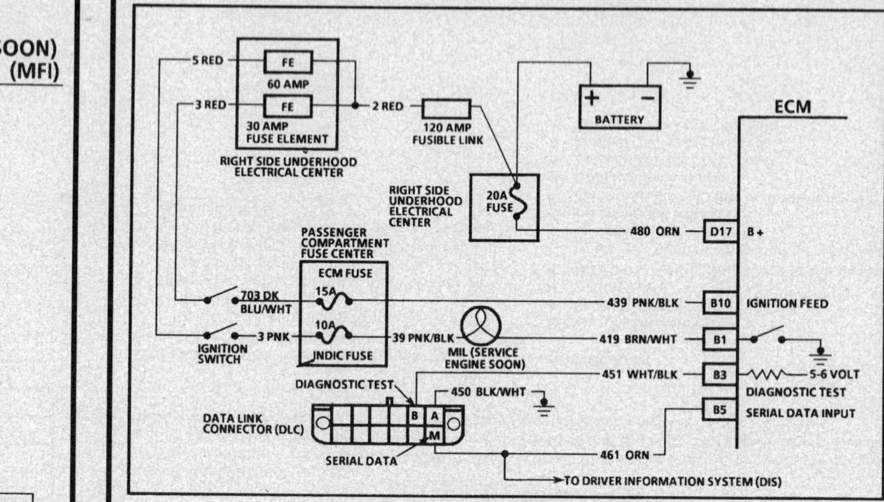

"AFTER REPAIRS," CONFIRM "CLOSED LOOP" OPERATION AND NO MIL (SERVICE ENGINE SOON).

3.1L (VIN T) AND 3.4L (VIN X) ENGINES — SYSTEM DIAGNOSTIC CHARTS — CUTLASS SUPREME, GRAND PRIX, LUMINA AND REGAL

CHART A-2

NO DLC DATA OR WON'T FLASH DTC 12
MIL (SERVICE ENGINE SOON) "ON" STEADY
"W" CARLINE (MFI)

Circuit Description:

There should always be a steady Malfunction Indicator Lamp (MIL) "Service Engine Soon" when the ignition is "ON" and engine stopped. Battery ignition voltage is supplied to the light bulb. The Engine Control Module (ECM) will turn the light "ON" by grounding CKT 419 at the ECM.

With the diagnostic terminal grounded, the light should flash a Diagnostic Trouble Code (DTC) 12, followed by any DTC(s) stored in memory.

A steady light suggests a short to ground in the light control CKT 419 or an open in diagnostic CKT 451.

Test Description: Number(s) below refer to circled number(s) on the diagnostic chart.

1. If there is a problem with the ECM that causes a Tech 1 scan tool not to read serial data, the ECM should not flash a DTC 12. If DTC 12 is flashing check for CKT 451 short to ground. If DTC 12 does flash be sure that the Tech 1 scan tool is working properly on another vehicle. If the Tech 1 scan is functioning properly and CKT 461 is OK, the PROM or ECM may be at fault for the no DLC symptom.

2. If the light goes "OFF" when the ECM connector is disconnected, CKT 419 is not shorted to ground.

3. This step will check for an open diagnostic CKT 451.

4. At this point the MIL wiring is OK. The problem is a faulty ECM or PROM. If DTC 12 does not flash, the ECM should be replaced using the original PROM. Replace the PROM only after trying an ECM, as a defective PROM is an unlikely cause of the problem.

3.1L (VIN T) AND 3.4L (VIN X) ENGINES — SYSTEM DIAGNOSTIC CHARTS — CUTLASS SUPREME, GRAND PRIX, LUMINA AND REGAL

CHART A-2

NO DLC DATA OR WON'T FLASH DTC 12 MIL (SERVICE ENGINE SOON) "ON" STEADY "W" CARLINE (MFI)

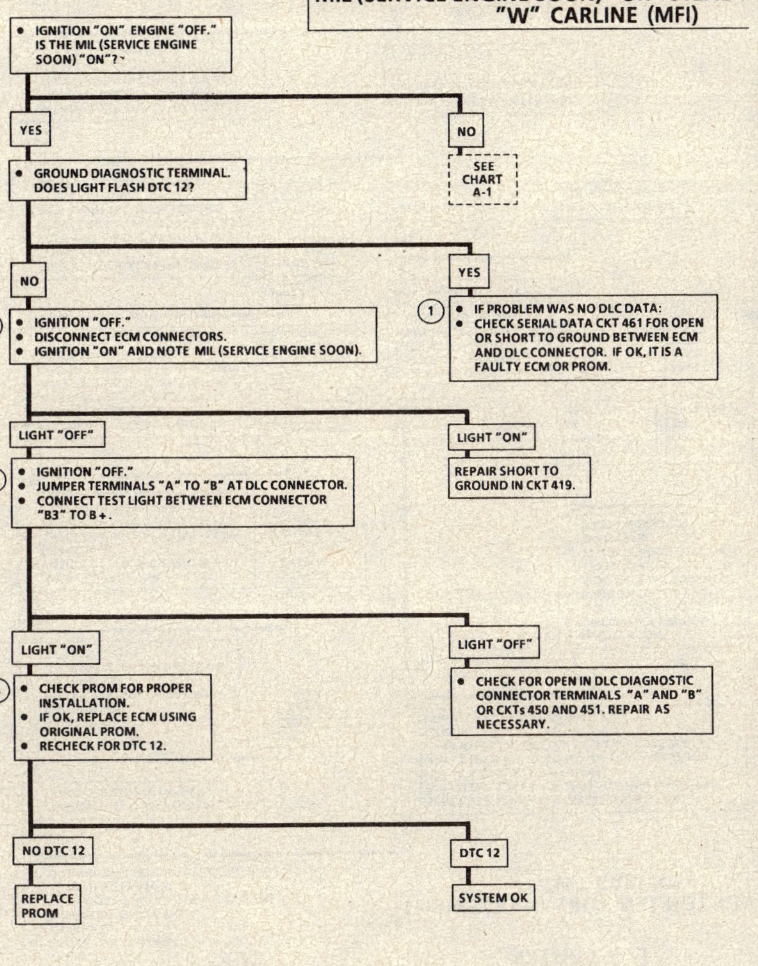

- IGNITION "ON" ENGINE "OFF." IS THE MIL (SERVICE ENGINE SOON) "ON"?

YES

- GROUND DIAGNOSTIC TERMINAL. DOES LIGHT FLASH DTC 12?

NO → **SEE CHART A-1**

NO

(2)
- IGNITION "OFF."
- DISCONNECT ECM CONNECTORS.
- IGNITION "ON" AND NOTE MIL (SERVICE ENGINE SOON).

YES

(1)
- IF PROBLEM WAS NO DLC DATA:
- CHECK SERIAL DATA CKT 461 FOR OPEN OR SHORT TO GROUND BETWEEN ECM AND DLC CONNECTOR. IF OK, IT IS A FAULTY ECM OR PROM.

LIGHT "OFF"

(3)
- IGNITION "OFF."
- JUMPER TERMINALS "A" TO "B" AT DLC CONNECTOR.
- CONNECT TEST LIGHT BETWEEN ECM CONNECTOR "B3" TO B+.

LIGHT "ON"

- REPAIR SHORT TO GROUND IN CKT 419.

LIGHT "ON"

(4)
- CHECK PROM FOR PROPER INSTALLATION.
- IF OK, REPLACE ECM USING ORIGINAL PROM.
- RECHECK FOR DTC 12.

LIGHT "OFF"

- CHECK FOR OPEN IN DLC DIAGNOSTIC CONNECTOR TERMINALS "A" AND "B" OR CKTs 450 AND 451. REPAIR AS NECESSARY.

NO DTC 12

REPLACE PROM

DTC 12

SYSTEM OK

"AFTER REPAIRS," CONFIRM "CLOSED LOOP" OPERATION AND NO MIL (SERVICE ENGINE SOON).

3.1L (VIN T) AND 3.4L (VIN X) ENGINES — SYSTEM DIAGNOSTIC CHARTS — CUTLASS SUPREME, GRAND PRIX, LUMINA AND REGAL

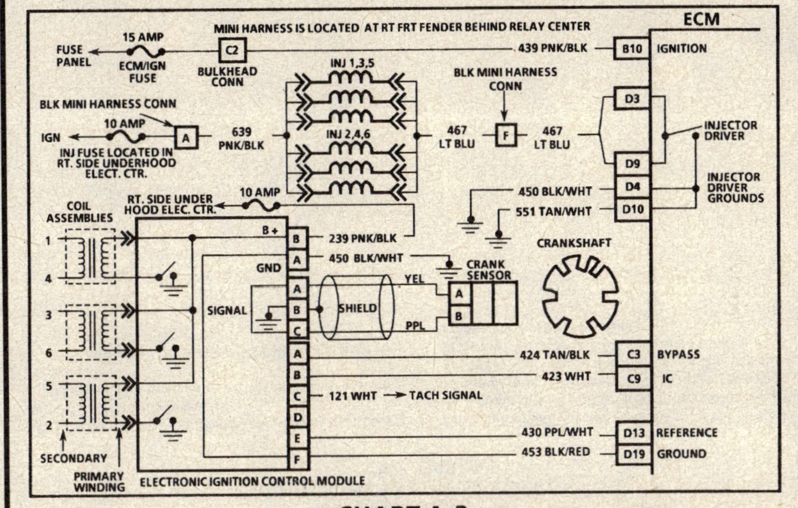

CHART A-3

(Page 1 of 3)
ENGINE CRANKS BUT WILL NOT RUN
"W" CARLINE

Circuit Description:

This chart assumes that battery condition and engine cranking speed are OK, and there is adequate fuel in the tank.

Test Description: Number(s) below refer to circled number(s) on the diagnostic chart.

1. A MIL (Service Engine Soon) "ON" is a basic test to determine if there is a 12 volts supply and ignition 12 volts to ECM. No DLC may be due to an ECM problem and CHART A-2 will diagnose the ECM. If TP sensor is over 2.5 volts, the engine may be in the clear flood mode which will cause starting problems. The engine will not start without reference pulses and therefore the Tech 1 scan tool should read RPM (reference) during crank.

2. For the first two seconds with ignition "ON," or whenever reference pulses are being received, PPSW should indicate fuel pump circuit voltage (8 to 12 volts).

3. Because the electronic ignition system uses two plugs and wires to complete the circuit of each coil, the opposite spark should be left connected. If RPM was indicated during crank, the ignition control module is receiving a crank signal, but no spark at this test indicates the ignition control module is not triggering the coils.

4. This test will determine if there is B+ at the injectors. The injectors are powered by a 10 amp fuse located in the right side underhood electrical center.

5. This test will check the integrity of the ECM injector driver.

6. This test will determine if CKT 467 is open between the mini harness connector and the injectors.

7. This test will determine if the ignition control module is not triggering the problem coil or if the tested coil is at fault. This test could also be performed by using another known good coil.

8. This test will determine if the ignition control module is not generating the reference pulse or if the wiring or ECM are at fault. By touching and removing a test light to 12 volts on CKT 430, a reference pulse should be generated. If RPM is indicated, the ECM and wiring are OK.

3.1L (VIN T) AND 3.4L (VIN X) ENGINES — SYSTEM DIAGNOSTIC CHARTS — CUTLASS SUPREME, GRAND PRIX, LUMINA AND REGAL

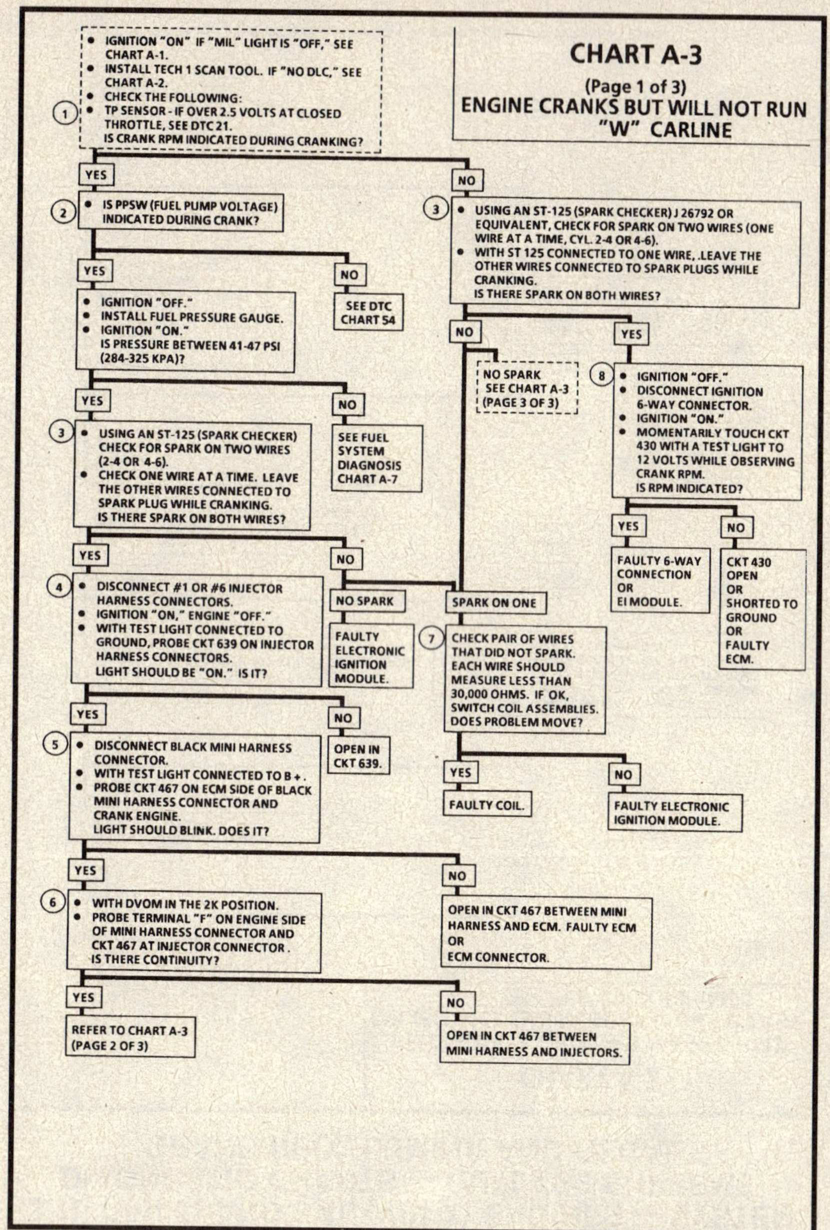

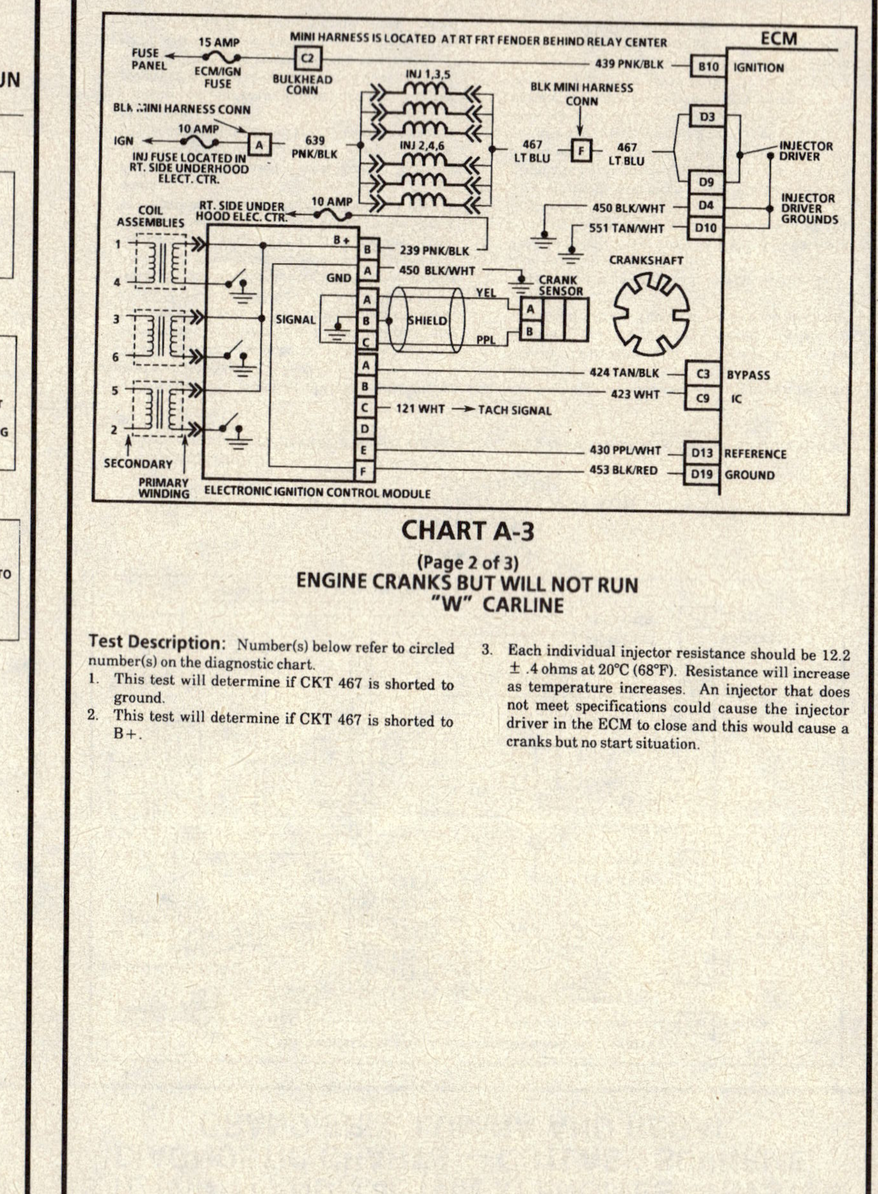

Test Description: Number(s) below refer to circled number(s) on the diagnostic chart.

1. This test will determine if CKT 467 is shorted to ground.
2. This test will determine if CKT 467 is shorted to B+.
3. Each individual injector resistance should be 12.2 ± .4 ohms at 20°C (68°F). Resistance will increase as temperature increases. An injector that does not meet specifications could cause the injector driver in the ECM to close and this would cause a cranks but no start situation.

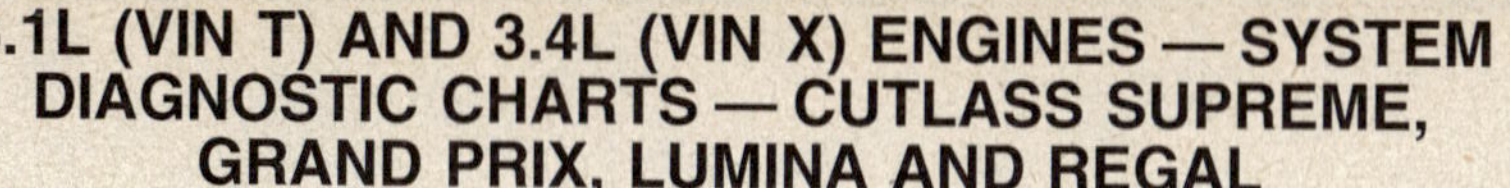

3.1L (VIN T) AND 3.4L (VIN X) ENGINES — SYSTEM DIAGNOSTIC CHARTS — CUTLASS SUPREME, GRAND PRIX, LUMINA AND REGAL

CHART A-3
(Page 2 of 3)
ENGINE CRANKS BUT WILL NOT RUN
"W" CARLINE

"AFTER REPAIRS," CONFIRM "CLOSED LOOP" OPERATION AND NO MIL (SERVICE ENGINE SOON).

3.1L (VIN T) AND 3.4L (VIN X) ENGINES — SYSTEM DIAGNOSTIC CHARTS — CUTLASS SUPREME, GRAND PRIX, LUMINA AND REGAL

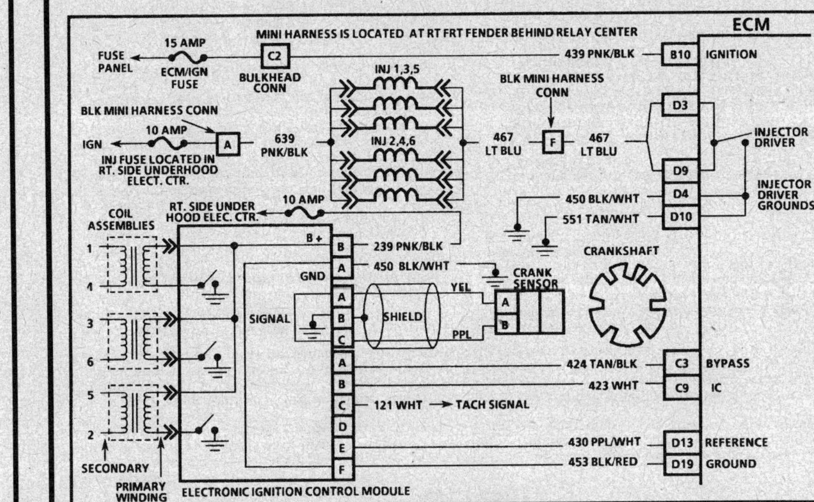

CHART A-3
(Page 3 of 3)
ENGINE CRANKS BUT WILL NOT RUN
"W" CARLINE

Circuit Description:

If the Tech 1 scan tool did not indicate a cranking RPM and there is no spark present at the plugs, the problem lies in the electronic ignition system or the power and ground supplies to the module.

The magnetic crank sensor is used to determine engine crankshaft position much the same way as the pick-up coil did in distributor type systems. The sensor is mounted in the block near a seven slot wheel on the crank shaft. The rotation of the wheel creates a flux change in the sensor which produces a voltage signal. The electronic ignition module then processes this signal and creates the reference pulses needed by the ECM and the signal triggers the correct coil at the correct time.

Test Description: Number(s) below refer to circled number(s) on the diagnostic chart.

1. This test will determine if the 12 volts supply and a good ground is available at the electronic ignition module.
2. Tests integrity of CKT 239 to the electronic ignition module. If test light does not light but the MIL is "ON" with ignition "ON," repair open in CKT 239.
3. Checks for continuity of the crank sensor and connections.
4. Voltage will vary in this test depending on cranking speed of engine.

3.1L (VIN T) AND 3.4L (VIN X) ENGINES — SYSTEM DIAGNOSTIC CHARTS — CUTLASS SUPREME, GRAND PRIX, LUMINA AND REGAL

CHART A-3
(Page 3 of 3)
ENGINE CRANKS BUT WILL NOT RUN
"W" CARLINE

CONTINUED FROM CHART A-3 (1 OF 3)

NO CRANKING RPM INDICATED AND NO SPARK.

① • DISCONNECT IGNITION CONTROL MODULE 2 PIN CONNECTOR.
  • IGNITION "ON."
  • CONNECT TEST LIGHT BETWEEN HARNESS TERMINALS.

LIGHT "OFF"

LIGHT "ON"

② • CONNECT TEST LIGHT FROM HARNESS TERMINAL "B" TO GROUND.

③ • DISCONNECT CRANK SENSOR 3-WAY CONNECTOR FROM IGNITION CONTROL MODULE.
  • WITH OHMMETER IN 2K OHMS POSITION, PROBE HARNESS TERMINALS "A" & "C" (SHOULD READ BETWEEN 900-1200 OHMS).

LIGHT "ON"

LIGHT "OFF"

OK

NOT OK

REPAIR OPEN GROUND CKT 450.

REPAIR OPEN IGNITION FEED CKT 239.

④ • SET VOLTMETER ON 2 VOLT AC POSITION.
  • CRANK ENGINE AND OBSERVE VOLTAGE READING. READING SHOULD BE GREATER THAN .1 VOLT (100 mV). IS IT?

LESS THAN 900 OHMS

GREATER THAN 1200 OHMS

CRANK SENSOR LEADS SHORTED TOGETHER OR FAULTY CRANK SENSOR.

OPEN CRANK SENSOR CIRCUIT, FAULTY CONNECTION OR FAULTY CRANK SENSOR.

YES

NO

REPLACE IGNITION CONTROL MODULE.

FAULTY CONNECTION OR FAULTY CRANK SENSOR.

"AFTER REPAIRS," CONFIRM "CLOSED LOOP" OPERATION AND NO MIL (SERVICE ENGINE SOON).

3.1L (VIN T) AND 3.4L (VIN X) ENGINES — SYSTEM DIAGNOSTIC CHARTS — CUTLASS SUPREME, GRAND PRIX, LUMINA AND REGAL

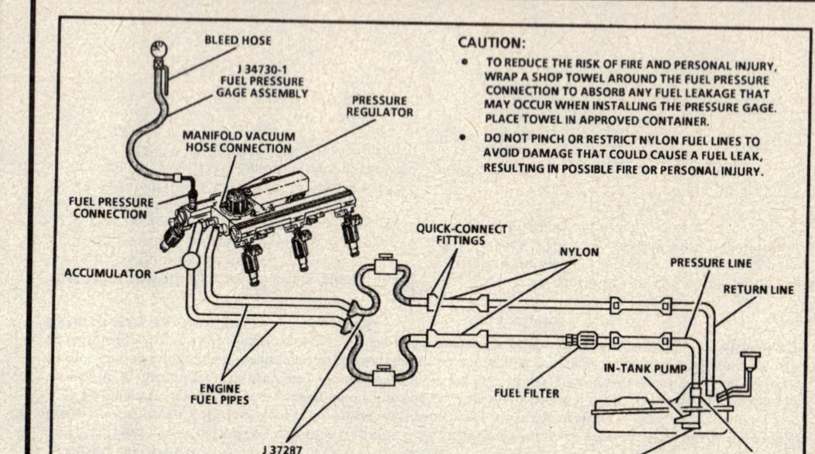

CHART A-7 (Page 1 of 3)

FUEL SYSTEM DIAGNOSIS
"W" CARLINE

Circuit Description:

When the ignition switch is turned "ON," the Engine Control Module (ECM) will turn "ON" the in-tank fuel pump. It will remain "ON" as long as the engine is cranking or running, and the ECM is receiving reference pulses. If there are no reference pulses, the ECM will shut "OFF" the fuel pump within 2 seconds after ignition "ON" or engine stops.

An electric fuel pump, attached to the fuel sender assembly (inside the fuel tank), pumps fuel through an in-line filter to the fuel rail assembly. The pump is designed to provide fuel at a pressure above the regulated pressure needed by the injector. A pressure regulator attached to the fuel rail, keeps fuel available to the injectors at a regulated pressure. Unused fuel is returned to the fuel tank by a separate line.

Test Description: Number(s) below refer to circled number(s) on the diagnostic chart.

1. Wrap a shop towel around the fuel pressure connector to absorb any small amount of fuel leakage that may occur when installing the gage. Ignition "ON," engine "OFF," pump pressure should be 284-325 kPa (41-47 psi). This pressure is controlled by spring pressure within the regulator assembly.
2. When the engine is idling, the manifold pressure is low (high vacuum) and is applied to the fuel regulator diaphragm. This will offset the spring and result in a lower fuel pressure. This idle pressure will vary somewhat depending on barometric pressure, however, the pressure idling should be less indicating pressure regulator control.
3. Pressure that continues to fall is caused by one of the following:
   • Fuel pump check valve not holding.
   • Leaking fuel pulse dampener.
   • Fuel pressure regulator valve leaking.
   • Injector(s) sticking open.
4. An injector sticking open can best be determined by checking for a fouled or saturated spark plug(s). If a leaking injector can not be determined by a fouled or saturated spark plug the following procedure should be used:
   • Remove plenum, and remove fuel rail bolts.

   • Lift fuel rail out just enough to leave injector nozzles in the ports.

   CAUTION: Be sure injector(s) are not allowed to spray on engine and that injector retaining clips are intact. This should be carefully followed to prevent fuel spray on engine which would cause a fire hazard.

   • Pressurize the fuel system and observe for injector(s) leaking

3.1L (VIN T) AND 3.4L (VIN X) ENGINES — SYSTEM DIAGNOSTIC CHARTS — CUTLASS SUPREME, GRAND PRIX, LUMINA AND REGAL

CHART A-7
(Page 1 of 3)
FUEL SYSTEM DIAGNOSIS
"W" CARLINE

(1)
- THIS CHART ASSUMES THERE IS NO CODE 54.
- INSTALL FUEL PRESSURE GAGE AS SHOWN ON FACING PAGE.
- IGNITION "OFF" FOR 10 SECONDS. A/C "OFF."
- IGNITION "ON." FUEL PUMP WILL RUN FOR ABOUT 2 SECONDS. THE IGNITION MAY HAVE TO BE CYCLED "ON" MORE THAN ONCE TO OBTAIN MAXIMUM PRESSURE.
- NOTE FUEL PRESSURE, WITH PUMP RUNNING, PRESSURE SHOULD BE 284-325 kPa (41-47 psi). WHEN PUMP STOPS, PRESSURE MAY VARY SLIGHTLY THEN SHOULD HOLD STEADY. IS PRESSURE CORRECT AND DOES IT HOLD?

YES →
- IF FUEL PRESSURE IS WITHIN NORMAL RANGE, BUT IS SUSPECTED OF DROPPING OFF DURING ACCELERATION, CRUISE OR HARD CORNERING, SEE CHART A-7 (2 OF 3).

(2)
- START AND IDLE ENGINE AT NORMAL OPERATING TEMPERATURE.
- FUEL PRESSURE NOTED IN STEP 1 SHOULD DROP BY 21-69 kPa (3-10 psi). DOES IT?

YES → NO TROUBLE FOUND, REVIEW "SYMPTOMS," SECTION "B".

NO →
- DISCONNECT VACUUM HOSE FROM PRESSURE REGULATOR ASSEMBLY.
- WITH ENGINE IDLING, APPLY 12-14 INCHES OF VACUUM TO PRESSURE REGULATOR. FUEL PRESSURE NOTED IN STEP 1 SHOULD DROP 21-69 kPa (3-10 psi). DOES IT?

YES → LOCATE AND REPAIR LOSS OF VACUUM TO PRESSURE REGULATOR.

NO → REPLACE PRESSURE REGULATOR.

FROM CHART A-3 → **NO**

(3)
- FUEL PRESSURE WITHIN SPEC., BUT DOES NOT HOLD.

FUEL PRESSURE OUT OF SPEC. → SEE CHART A-7 (2 OF 3)

NO PRESSURE, FUEL PUMP ELECTRICAL CIRCUIT OK. → CHECK FOR:
- PLUGGED IN-LINE FILTER.
- PLUGGED FUEL PUMP STRAINER.
- RESTRICTED FUEL PRESSURE LINE.
- LEAKING FUEL PULSE DAMPENER.

IF OK → REPLACE FUEL PUMP.

(for FUEL PRESSURE WITHIN SPEC., BUT DOES NOT HOLD)
- INSTALL J 37287 FUEL LINE SHUTOFF ADAPTORS, REFER TO PAGE 3 OF 3 FOR FUEL PRESSURE RELIEF PROCEDURE AND FOR SERVICING QUICK-CONNECT FITTINGS.
- USING A 10 AMP FUSED JUMPER WIRE, APPLY B+ TO FUEL PUMP TEST CONNECTOR FOR 2 SECONDS.
- BLOCK FUEL PRESSURE LINE BY CLOSING VALVE. PRESSURE SHOULD HOLD. DOES IT?

NO →
- OPEN SHUTOFF VALVE ON FUEL PRESSURE LINE.
- CLOSE SHUTOFF VALVE ON FUEL RETURN LINE.
- USING A 10 AMP FUSED JUMPER WIRE, APPLY B+ TO FUEL PUMP TEST CONNECTOR FOR 2 SECONDS. PRESSURE SHOULD HOLD. DOES IT?

YES → CHECK FOR:
- LEAKING FUEL PULSE DAMPENER.
- FAULTY FUEL PUMP.

NO → (4) LOCATE AND CORRECT LEAKING INJECTOR(S).

YES → REPLACE PRESSURE REGULATOR.

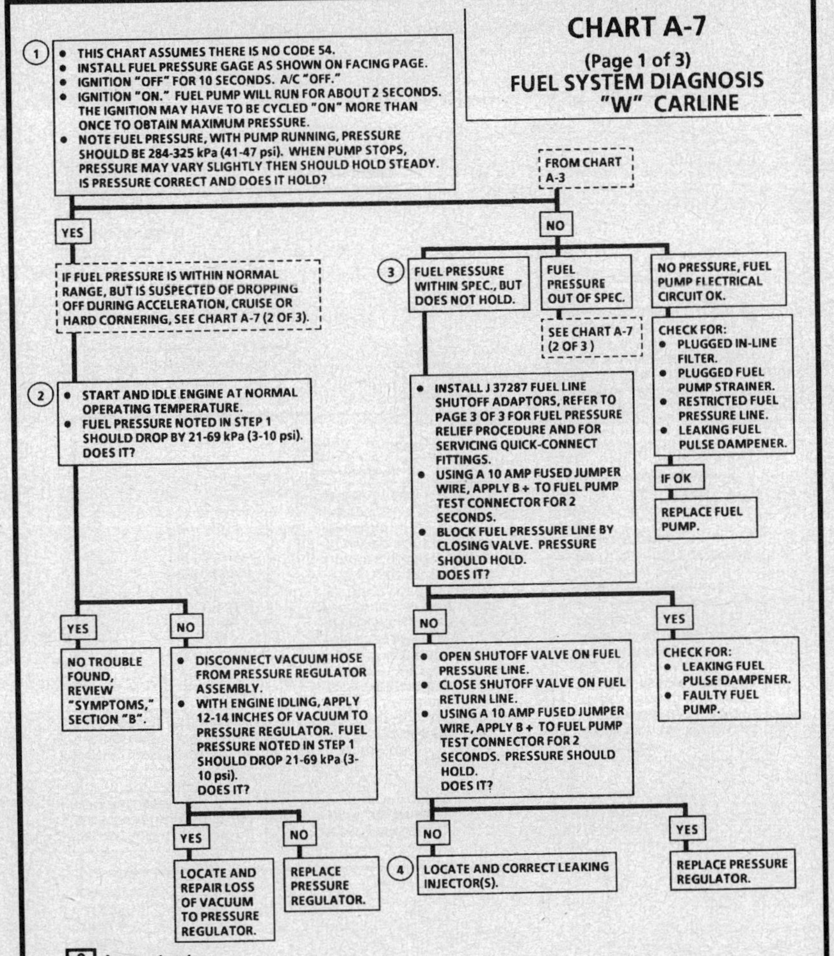

Important
- (3.1L NON SFI AND 3.4L) IF NEGATIVE BATTERY TERMINAL IS DISCONNECTED DURING DIAGNOSIS, SEE "FUEL METERING SYSTEM" FOR "IDLE LEARN PROCEDURE."

"AFTER REPAIRS," CONFIRM "CLOSED LOOP" OPERATION AND NO MIL (SERVICE ENGINE SOON).

3.1L (VIN T) AND 3.4L (VIN X) ENGINES — SYSTEM DIAGNOSTIC CHARTS — CUTLASS SUPREME, GRAND PRIX, LUMINA AND REGAL

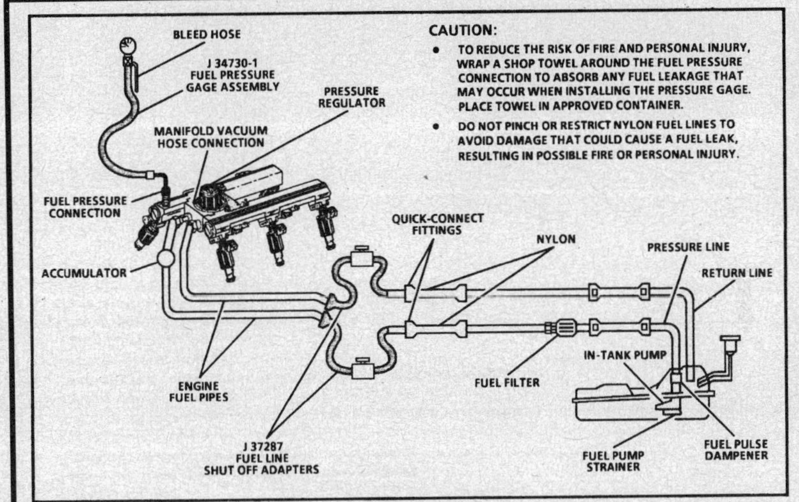

CHART A-7
(Page 2 of 3)
FUEL SYSTEM DIAGNOSIS
"W" CARLINE

Test Description: Number(s) below refer to circled number(s) on the diagnostic chart.

5. Fuel pressure that drops off during acceleration, cruise or hard cornering may cause a lean condition and result in a loss of power, surging or misfire. This condition can be diagnosed using a Tech 1 scan tool. If the fuel system is very lean, the O2S will stop toggling and output voltage will drop below 500 mV. Also, injector pulse width will increase.

Important
- Make sure system is not operating at "Fuel-Cut-Off" which may cause false readings on the scan tool.

6. Fuel pressure below 284 kPa (41 psi) may cause a lean condition and may set a DTC 44 or a DTC 32. Driveability conditions can include hard starting cold, hesitation, poor driveability, lack of power, surging or misfire.

7. Restricting the fuel return line causes fuel pressure to build above regulated pressure. With battery voltage applied to the pump "test" connector, pressure should rise above 325 kPa (47 psi) as the valve in the return line is partially closed.

NOTICE: Do not allow pressure to exceed 414 kPa (60 psi) as damage to the regulator may result.

8. Fuel pressure above 325 kPa (47 psi) may cause a rich condition and may set a DTC 45. Driveability conditions can include hard starting (followed by black smoke) and a strong sulphur smell in the exhaust.

9. This test determines if the high fuel pressure is due to a restricted fuel return line or a faulty fuel pressure regulator.

3.1L (VIN T) AND 3.4L (VIN X) ENGINES — SYSTEM DIAGNOSTIC CHARTS — CUTLASS SUPREME, GRAND PRIX, LUMINA AND REGAL

CHART A-7
(Page 2 of 3)
FUEL SYSTEM DIAGNOSIS
"W" CARLINE

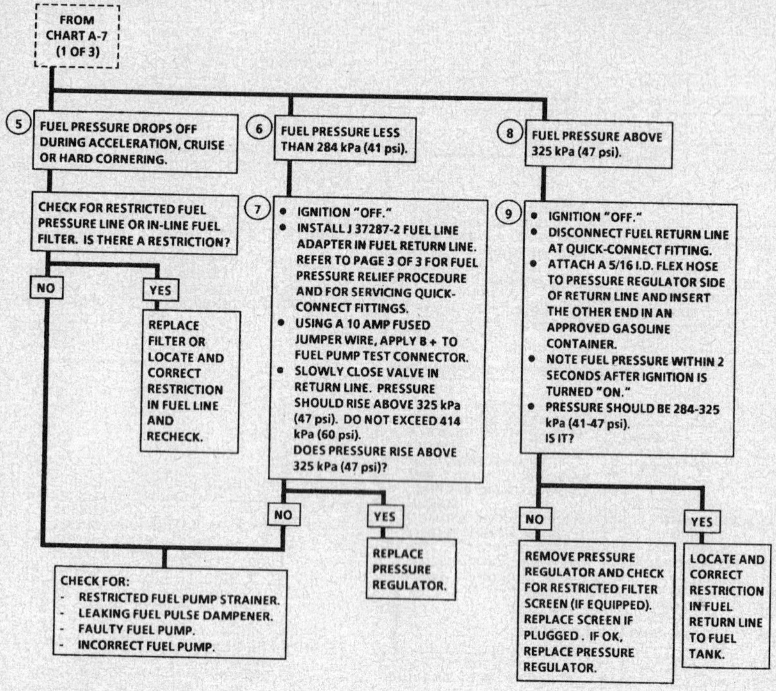

FROM CHART A-7 (1 OF 3)

5. FUEL PRESSURE DROPS OFF DURING ACCELERATION, CRUISE OR HARD CORNERING.

CHECK FOR RESTRICTED FUEL PRESSURE LINE OR IN-LINE FUEL FILTER. IS THERE A RESTRICTION?

NO

YES

REPLACE FILTER OR LOCATE AND CORRECT RESTRICTION IN FUEL LINE AND RECHECK.

CHECK FOR:
- RESTRICTED FUEL PUMP STRAINER.
- LEAKING FUEL PULSE DAMPENER.
- FAULTY FUEL PUMP.
- INCORRECT FUEL PUMP.

6. FUEL PRESSURE LESS THAN 284 kPa (41 psi).

7.
- IGNITION "OFF."
- INSTALL J 37287-2 FUEL LINE ADAPTER IN FUEL RETURN LINE. REFER TO PAGE 3 OF 3 FOR FUEL PRESSURE RELIEF PROCEDURE AND FOR SERVICING QUICK-CONNECT FITTINGS.
- USING A 10 AMP FUSED JUMPER WIRE, APPLY B + TO FUEL PUMP TEST CONNECTOR.
- SLOWLY CLOSE VALVE IN RETURN LINE. PRESSURE SHOULD RISE ABOVE 325 kPa (47 psi). DO NOT EXCEED 414 kPa (60 psi). DOES PRESSURE RISE ABOVE 325 kPa (47 psi)?

NO

YES

REPLACE PRESSURE REGULATOR.

8. FUEL PRESSURE ABOVE 325 kPa (47 psi).

9.
- IGNITION "OFF."
- DISCONNECT FUEL RETURN LINE AT QUICK-CONNECT FITTING. ATTACH A 5/16 I.D. FLEX HOSE TO PRESSURE REGULATOR SIDE OF RETURN LINE AND INSERT THE OTHER END IN AN APPROVED GASOLINE CONTAINER.
- NOTE FUEL PRESSURE WITHIN 2 SECONDS AFTER IGNITION IS TURNED "ON."
- PRESSURE SHOULD BE 284-325 kPa (41-47 psi). IS IT?

NO

YES

REMOVE PRESSURE REGULATOR AND CHECK FOR RESTRICTED FILTER SCREEN (IF EQUIPPED). REPLACE SCREEN IF PLUGGED . IF OK, REPLACE PRESSURE REGULATOR.

LOCATE AND CORRECT RESTRICTION IN FUEL RETURN LINE TO FUEL TANK.

⚑ **Important**
(3.1L NON SFI AND 3.4L)
IF NEGATIVE BATTERY TERMINAL IS DISCONNECTED DURING DIAGNOSIS, REFER TO   "FUEL METERING SYSTEM" FOR "IDLE LEARN PROCEDURE."

"AFTER REPAIRS," CONFIRM "CLOSED LOOP" OPERATION AND NO MIL (SERVICE ENGINE SOON).

3.1L (VIN T) AND 3.4L (VIN X) ENGINES — SYSTEM DIAGNOSTIC CHARTS — CUTLASS SUPREME, GRAND PRIX, LUMINA AND REGAL

CHART A-7
(Page 3 of 3)
FUEL SYSTEM PRESSURE TEST
"W" CARLINE

FUEL SYSTEM PRESSURE RELIEF PROCEDURE

PFI Engines With Fuel Pressure Connection

(Must Be Performed Before Disconnecting Fuel Line Fittings)

CAUTION:
- To reduce the risk of fire and personal injury, it is necessary to relieve fuel system pressure before disconnecting fuel line fittings.
- After relieving system pressure, a small amount of fuel may be released when disconnecting fuel line fittings. In order to reduce the chance of personal injury, cover fuel line fittings with a shop towel before disconnecting, to catch any fuel that may leak out. Place the towel in an approved container when disconnect is completed.

Tool Required: J 34730-1 Fuel Pressure Gage

1. Disconnect negative battery cable to avoid possible fuel discharge if an accidental attempt is made to start the engine.
2. Loosen fuel filler cap to relieve tank vapor pressure.
3. Connect gage J 34730-1 to fuel pressure connection. Wrap a shop towel around fitting while connecting gage to avoid spillage.
4. Install bleed hose into an approved container and open valve to bleed system pressure. Fuel line fittings are now safe for servicing.
5. Drain any fuel remaining in gage into an approved gasoline container.
6. Perform service required.
7. Tighten fuel filler cap.
8. Connect negative battery cable.
9. Cycle ignition "ON" and "OFF" twice, waiting ten seconds between cycles, then check for fuel leaks.

"AFTER REPAIRS," CONFIRM "CLOSED LOOP" OPERATION AND NO MIL (SERVICE ENGINE SOON).

3.1L (VIN T) AND 3.4L (VIN X) ENGINES — DIAGNOSTIC TROUBLE CODE CHART — CUTLASS SUPREME, GRAND PRIX, LUMINA AND REGAL

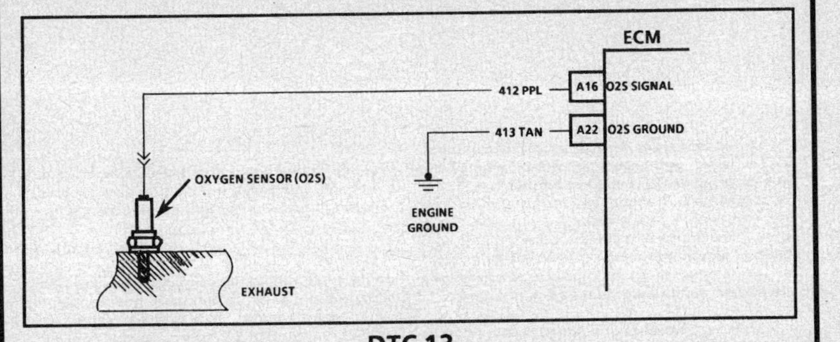

DTC 13

OXYGEN SENSOR (O2S) CIRCUIT
(OPEN CIRCUIT)
"W" CARLINE (MFI)

Circuit Description:

The ECM supplies a voltage of about .55 volt between terminals "A16" and "A22". (If measured with a 10 megohm digital voltmeter, this may read as low as .35 volt.) The Oxygen Sensor (O2S) varies the voltage within a range of about 1 volt. If the exhaust is rich, the O2S will display near 1 volt and when the exhaust is lean the O2S will display close to .1 volt.

The sensor is like an open circuit and produces no voltage when it is below 315°C (600°F). An open O2S circuit or cold O2S causes "Open Loop" operation.

Test Description: Number(s) below refer to circled number(s) on the diagnostic chart.

1. DTC 13 will set under the following conditions:
   - Engine running at least 2 minutes after start.
   - Engine coolant temperature at least 50°C (122°F).
   - No DTC 21 or 22.
   - O2S signal voltage steady between .35 and .55 volt.
   - Throttle position above 4% for more time than TP sensor was below 4%. (About .3 volt above closed throttle voltage.)
   - All conditions must be met and held for at least 25 seconds.
   If the conditions for a DTC 13 exist, the system will not go "Closed Loop."

2. This will determine if the sensor is at fault or the wiring or ECM is the cause of the DTC 13.

3. In doing this test use only a high impedance digital volt ohmmeter. This test checks the continuity of CKT 412 and CKT 413 because if CKT 413 is open the ECM voltage on CKT 412 will be over .6 volt (600 mV).

Diagnostic Aids:

Normal Tech 1 scan voltage varies between 100 mV to 999 mV (.1 and 1.0 volt) while in "Closed Loop." DTC 13 sets if voltage remains between .35 and .55 volt, but the system will go "Open Loop" in about 15 seconds.

3.1L (VIN T) AND 3.4L (VIN X) ENGINES — DIAGNOSTIC TROUBLE CODE CHART — CUTLASS SUPREME, GRAND PRIX, LUMINA AND REGAL

DTC 13

OXYGEN SENSOR (O2S) CIRCUIT
(OPEN CIRCUIT)
"W" CARLINE (MFI)

1. - ENGINE AT NORMAL OPERATING TEMPERATURE (ABOVE 80°C/176°F).
   - RUN ENGINE ABOVE 1200 RPM FOR TWO MINUTES.
   - DOES SCAN TOOL INDICATE "CLOSED LOOP"?

   NO → 2.
   YES → DTC 13 IS INTERMITTENT. IF NO ADDITIONAL DTC(s) WERE STORED, REFER TO "DIAGNOSTIC AIDS"

2. - DISCONNECT OXYGEN SENSOR (O2S).
   - JUMPER HARNESS CKT 412 (ECM SIDE) TO GROUND.
   - SCAN TOOL SHOULD DISPLAY O2S VOLTAGE BELOW .2 VOLT (200 mv) WITH ENGINE RUNNING. DOES IT?

   NO → 3.
   YES → FAULTY O2S CONNECTION OR FAULTY O2S.

3. - REMOVE JUMPER.
   - IGNITION "ON," ENGINE "OFF."
   - CHECK VOLTAGE ON CKT 412 (ECM SIDE) AT O2S HARNESS CONNECTOR USING J 39200.

   .3-.6 VOLT (300 - 600 mV) → FAULTY ECM.

   OVER .6 VOLT (600 mV) → OPEN CKT 413 OR FAULTY CONNECTION OR FAULTY ECM.

   LESS THAN .3 VOLT (300 mV) → OPEN CKT 412 OR FAULTY ECM CONNECTION OR FAULTY ECM.

"AFTER REPAIRS," REFER TO DTC CRITERIA   AND CONFIRM DTC DOES NOT RESET.

3.1L (VIN T) AND 3.4L (VIN X) ENGINES — DIAGNOSTIC TROUBLE CODE CHART — CUTLASS SUPREME, GRAND PRIX, LUMINA AND REGAL

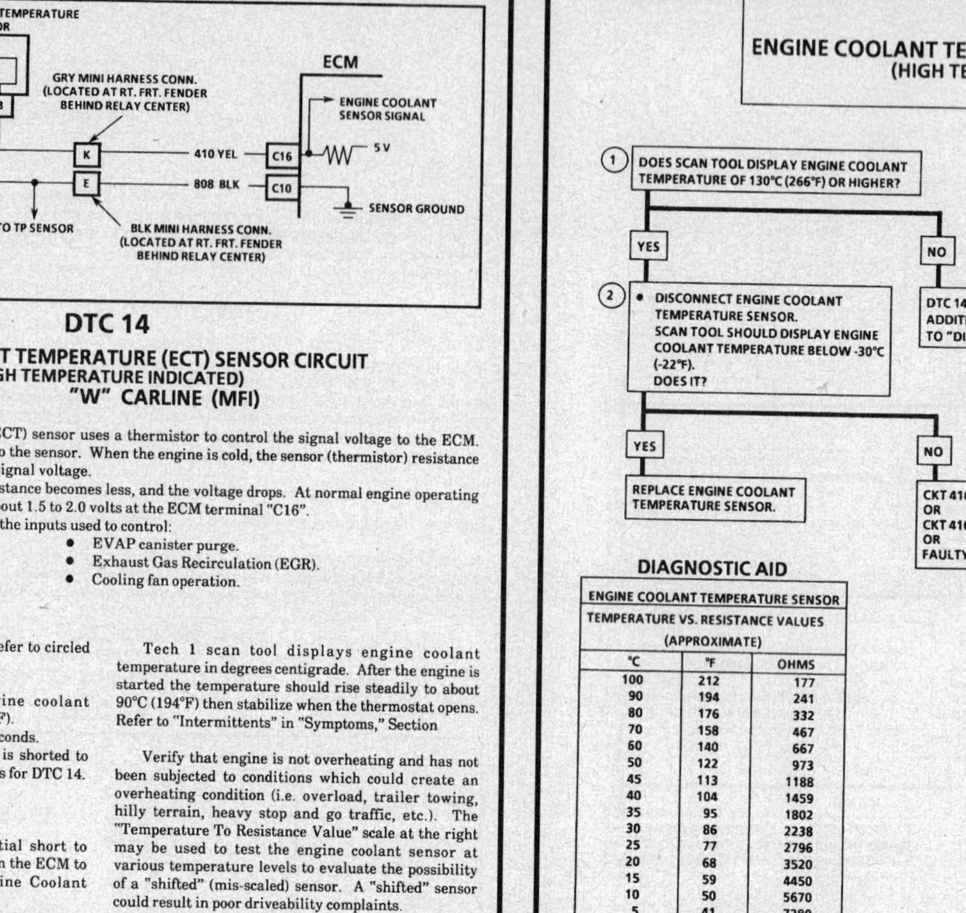

DTC 14
ENGINE COOLANT TEMPERATURE (ECT) SENSOR CIRCUIT
(HIGH TEMPERATURE INDICATED)
"W" CARLINE (MFI)

Circuit Description:

The Engine Coolant Temperature (ECT) sensor uses a thermistor to control the signal voltage to the ECM. The ECM applies a voltage on CKT 410 to the sensor. When the engine is cold, the sensor (thermistor) resistance is high, therefore the ECM will see high signal voltage.

As the engine warms, the sensor resistance becomes less, and the voltage drops. At normal engine operating temperature, the voltage will measure about 1.5 to 2.0 volts at the ECM terminal "C16".

Engine coolant temperature is one of the inputs used to control:
- Fuel delivery.
- Ignition Control (IC).
- Idle Air Control (IAC).
- Torque Converter Clutch (TCC).
- EVAP canister purge.
- Exhaust Gas Recirculation (EGR).
- Cooling fan operation.

Test Description: Number(s) below refer to circled number(s) on the diagnostic chart.
1. DTC 14 will set if:
   - Signal voltage indicates engine coolant temperature is above 135°C (270°F).
   - Engine running longer than 20 seconds.
2. This test will determine if CKT 410 is shorted to ground which will cause the conditions for DTC 14.

Diagnostic Aids:

Check harness routing for a potential short to ground in CKT 410. Circuit is routed from the ECM to a mini harness, and then to the Engine Coolant Temperature (ECT) sensor.

Tech 1 scan tool displays engine coolant temperature in degrees centigrade. After the engine is started the temperature should rise steadily to about 90°C (194°F) then stabilize when the thermostat opens. Refer to "Intermittents" in "Symptoms," Section

Verify that engine is not overheating and has not been subjected to conditions which could create an overheating condition (i.e. overload, trailer towing, hilly terrain, heavy stop and go traffic, etc.). The "Temperature To Resistance Value" scale at the right may be used to test the engine coolant sensor at various temperature levels to evaluate the possibility of a "shifted" (mis-scaled) sensor. A "shifted" sensor could result in poor driveability complaints.

3.1L (VIN T) AND 3.4L (VIN X) ENGINES — DIAGNOSTIC TROUBLE CODE CHART — CUTLASS SUPREME, GRAND PRIX, LUMINA AND REGAL

DTC 14
ENGINE COOLANT TEMPERATURE (ECT) SENSOR CIRCUIT
(HIGH TEMPERATURE INDICATED)
"W" CARLINE (MFI)

1. DOES SCAN TOOL DISPLAY ENGINE COOLANT TEMPERATURE OF 130°C (266°F) OR HIGHER?

- **YES**
- **NO**

2. • DISCONNECT ENGINE COOLANT TEMPERATURE SENSOR. SCAN TOOL SHOULD DISPLAY ENGINE COOLANT TEMPERATURE BELOW -30°C (-22°F). DOES IT?

DTC 14 IS INTERMITTENT. IF NO ADDITIONAL DTC(s) WERE STORED, REFER TO "DIAGNOSTIC AIDS"

- **YES**
- **NO**

REPLACE ENGINE COOLANT TEMPERATURE SENSOR.

CKT 410 SHORTED TO GROUND
OR
CKT 410 SHORTED TO SENSOR GROUND CIRCUIT
OR
FAULTY ECM.

DIAGNOSTIC AID

ENGINE COOLANT TEMPERATURE SENSOR

TEMPERATURE VS. RESISTANCE VALUES

(APPROXIMATE)

°C	°F	OHMS
100	212	177
90	194	241
80	176	332
70	158	467
60	140	667
50	122	973
45	113	1188
40	104	1459
35	95	1802
30	86	2238
25	77	2796
20	68	3520
15	59	4450
10	50	5670
5	41	7280
0	32	9420
-5	23	12300
-10	14	16180
-15	5	21450
-20	-4	28680
-30	-22	52700
-40	-40	100700

"AFTER REPAIRS," REFER TO DTC CRITERIA   AND CONFIRM DTC DOES NOT RESET.

3.1L (VIN T) AND 3.4L (VIN X) ENGINES — DIAGNOSTIC TROUBLE CODE CHART — CUTLASS SUPREME, GRAND PRIX, LUMINA AND REGAL

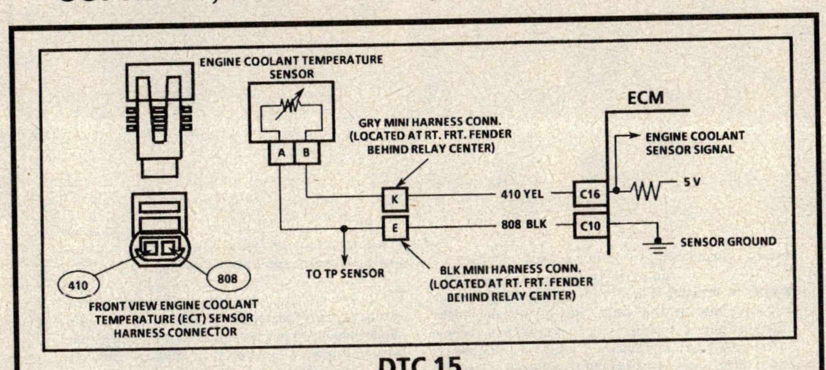

DTC 15
ENGINE COOLANT TEMPERATURE (ECT) SENSOR CIRCUIT
(LOW TEMPERATURE INDICATED)
"W" CARLINE (MFI)

Circuit Description:

The Engine Coolant Temperature (ECT) sensor uses a thermistor to control the signal voltage to the ECM. The ECM applies a voltage on CKT 410 to the sensor. When the engine is cold the sensor (thermistor) resistance is high, therefore the ECM will see high signal voltage.

As the engine warms, the sensor resistance becomes less, and the voltage drops. At normal engine operating temperature the voltage will measure about 1.5 to 2.0 volts at the ECM terminal "C16".

Coolant temperature is one of the inputs used to control:
- Fuel delivery.
- Ignition Control (IC).
- Idle Air Control (IAC).
- Torque Converter Clutch (TCC).
- EVAP canister purge.
- Exhaust Gas Recirculation (EGR).
- Cooling fan operation.

Test Description: Number(s) below refer to circled number(s) on the diagnostic chart.
1. DTC 15 will set if:
   - Signal voltage indicates engine coolant temperature is less than -38.5°C (-37.30°F).
2. This test simulates a DTC 14. If the ECM recognizes the low signal voltage (high temperature), and the Tech 1 scan tool reads 130°C (266°F), the ECM and wiring are OK.
3. This test will determine if CKT 410 is open. There should be 5 volts present at sensor connector if measured with a DVM.

A faulty connection, or an open in CKT 410 or 808 will result in a DTC 15.

DTCs 15 and 21 stored at the same time could be the result of an open CKT 808 which would also turn the temperature warning indicator "ON." The "Temperature to Resistance Value" scale at the right may be used to test the engine coolant temperature sensor at various temperature levels to evaluate the possibility of a "shifted" (mis-scaled) sensor. A "shifted" sensor could result in poor driveability complaints.

Refer to "Intermittents" in "Symptoms," Section

Diagnostic Aids:

A Tech 1 scan tool reads engine temperature in degrees centigrade. After engine is started the temperature should rise steadily to about 95°C (203°F) then stabilize when thermostat opens. CKT 410 is routed from the ECM to a mini harness, and then to the Engine Coolant Temperature (ECT) sensor.

3.1L (VIN T) AND 3.4L (VIN X) ENGINES — DIAGNOSTIC TROUBLE CODE CHART — CUTLASS SUPREME, GRAND PRIX, LUMINA AND REGAL

DTC 15
ENGINE COOLANT TEMPERATURE (ECT) SENSOR CIRCUIT
(LOW TEMPERATURE INDICATED)
"W" CARLINE (MFI)

1.
- DOES TECH 1 SCAN TOOL DISPLAY ENGINE COOLANT TEMPERATURE OF -30°C (-22°F) OR LESS?

YES → 2.
- DISCONNECT ENGINE COOLANT TEMPERATURE SENSOR.
- JUMPER HARNESS TERMINALS TOGETHER.
- TECH 1 SCAN TOOL SHOULD DISPLAY 130°C (266°F) OR MORE. DOES IT?

NO → DTC 15 IS INTERMITTENT. IF NO ADDITIONAL DTC(S) WERE STORED, REFER TO "DIAGNOSTIC AIDS"

2. NO → 3.
- JUMPER CKT 410 TO GROUND.
- TECH 1 SCAN TOOL SHOULD DISPLAY OVER 130°C (266°F). DOES IT?

2. YES → FAULTY CONNECTION OR ENGINE COOLANT TEMPERATURE SENSOR.

3. YES → OPEN ENGINE COOLANT TEMPERATURE SENSOR GROUND CIRCUIT, FAULTY CONNECTION OR FAULTY ECM.

3. NO → OPEN CKT 410, FAULTY CONNECTION AT ECM, OR FAULTY ECM.

DIAGNOSTIC AID

ENGINE COOLANT TEMPERATURE SENSOR TEMPERATURE VS. RESISTANCE VALUES (APPROXIMATE)		
°C	°F	OHMS
100	212	177
90	194	241
80	176	332
70	158	467
60	140	667
50	122	973
45	113	1188
40	104	1459
35	95	1802
30	86	2238
25	77	2796
20	68	3520
15	59	4450
10	50	5670
5	41	7280
0	32	9420
-5	23	12300
-10	14	16180
-15	5	21450
-20	-4	28680
-30	-22	52700
-40	-40	100700

"AFTER REPAIRS," REFER TO DTC CRITERIA   AND CONFIRM DTC DOES NOT RESET.

3.1L (VIN T) AND 3.4L (VIN X) ENGINES — DIAGNOSTIC TROUBLE CODE CHART — CUTLASS SUPREME, GRAND PRIX, LUMINA AND REGAL

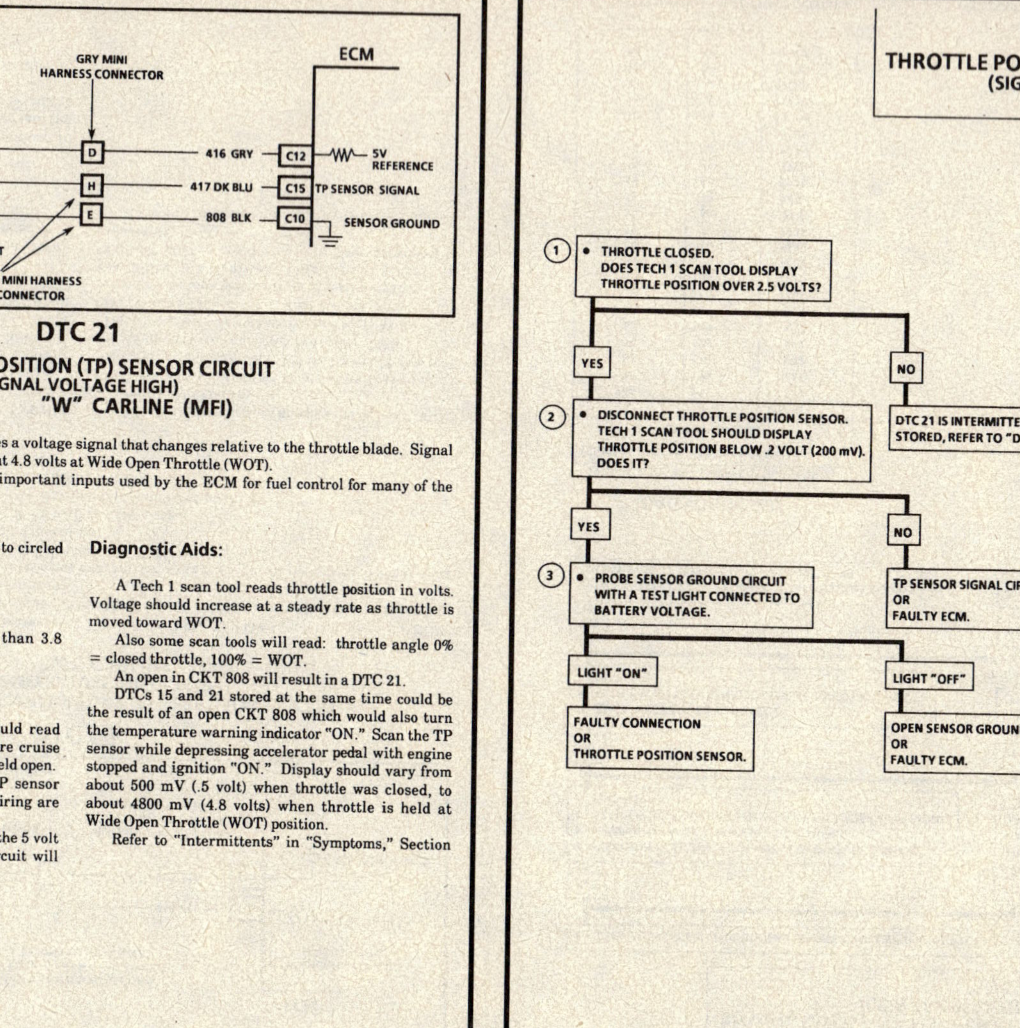

DTC 21
THROTTLE POSITION (TP) SENSOR CIRCUIT
(SIGNAL VOLTAGE HIGH)
"W" CARLINE (MFI)

Circuit Description:

The Throttle Position (TP) sensor provides a voltage signal that changes relative to the throttle blade. Signal voltage will vary from about .5 at idle to about 4.8 volts at Wide Open Throttle (WOT).

The TP sensor signal is one of the most important inputs used by the ECM for fuel control for many of the ECM control outputs.

Test Description: Number(s) below refer to circled number(s) on the diagnostic chart.

1. DTC 21 will set if:
   - No DTC 33 or DTC 34.
   - Engine is running.
   - TP sensor signal voltage is greater than 3.8 volts.
   - Air flow is less than 17 gm/sec.
   - All conditions met for 10 seconds.

   OR

   With throttle closed, the TP sensor should read less than .98 volt. If it doesn't, make sure cruise control and throttle cables are not being held open.
2. With the TP sensor disconnected, the TP sensor voltage should go low, if the ECM and wiring are OK.
3. Probing CKT 808 with a test light checks the 5 volt return circuit. Faulty sensor ground circuit will cause a DTC 21.

Diagnostic Aids:

A Tech 1 scan tool reads throttle position in volts. Voltage should increase at a steady rate as throttle is moved toward WOT.

Also some scan tools will read: throttle angle 0% = closed throttle, 100% = WOT.

An open in CKT 808 will result in a DTC 21.

DTCs 15 and 21 stored at the same time could be the result of an open CKT 808 which would also turn the temperature warning indicator "ON." Scan the TP sensor while depressing accelerator pedal with engine stopped and ignition "ON." Display should vary from about 500 mV (.5 volt) when throttle was closed, to about 4800 mV (4.8 volts) when throttle is held at Wide Open Throttle (WOT) position.

Refer to "Intermittents" in "Symptoms," Section

3.1L (VIN T) AND 3.4L (VIN X) ENGINES — DIAGNOSTIC TROUBLE CODE CHART — CUTLASS SUPREME, GRAND PRIX, LUMINA AND REGAL

DTC 21
THROTTLE POSITION (TP) SENSOR CIRCUIT
(SIGNAL VOLTAGE HIGH)
"W" CARLINE (MFI)

(1) • THROTTLE CLOSED.
DOES TECH 1 SCAN TOOL DISPLAY THROTTLE POSITION OVER 2.5 VOLTS?

YES → (2) • DISCONNECT THROTTLE POSITION SENSOR. TECH 1 SCAN TOOL SHOULD DISPLAY THROTTLE POSITION BELOW .2 VOLT (200 mV). DOES IT?

NO → DTC 21 IS INTERMITTENT. IF NO ADDITIONAL DTC(s) WERE STORED, REFER TO "DIAGNOSTIC AIDS"

YES → (3) • PROBE SENSOR GROUND CIRCUIT WITH A TEST LIGHT CONNECTED TO BATTERY VOLTAGE.

NO → TP SENSOR SIGNAL CIRCUIT SHORTED TO VOLTAGE OR FAULTY ECM.

LIGHT "ON" → FAULTY CONNECTION OR THROTTLE POSITION SENSOR.

LIGHT "OFF" → OPEN SENSOR GROUND CIRCUIT OR FAULTY ECM.

"AFTER REPAIRS," REFER TO DTC CRITERIA   AND CONFIRM DTC DOES NOT RESET.

3.1L (VIN T) AND 3.4L (VIN X) ENGINES — DIAGNOSTIC TROUBLE CODE CHART — CUTLASS SUPREME, GRAND PRIX, LUMINA AND REGAL

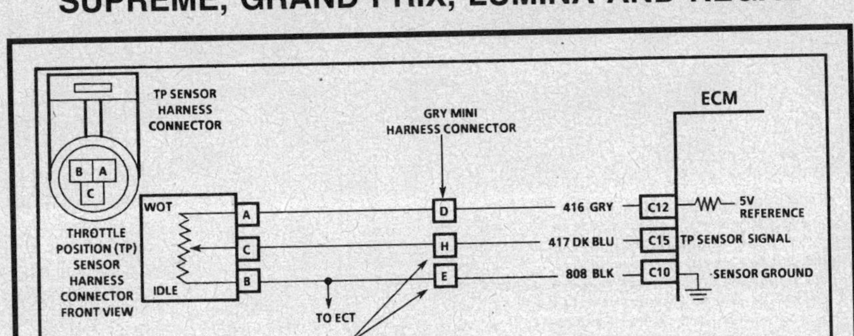

DTC 22
THROTTLE POSITION (TP) SENSOR CIRCUIT
(SIGNAL VOLTAGE LOW)
"W" CARLINE (MFI)

Circuit Description:

The Throttle Position (TP) sensor provides a voltage signal that changes relative to the throttle blade. Signal voltage will vary from about .5 at idle to about 4.8 volts at Wide Open Throttle (WOT).

The TP sensor signal is one of the most important inputs used by the ECM for fuel control and for many of the ECM control outputs.

Test Description: Number(s) below refer to circled number(s) on the diagnostic chart.

1. DTC 22 will set if:
   - Engine running.
   - TP sensor signal voltage is less than about .25 volt for 10 seconds.
2. Simulates DTC 21: (high voltage) If the ECM recognizes the high signal voltage, the ECM and wiring are OK.
3. TP sensor check: The TP sensor has an auto zeroing feature. If the voltage reading is within the range of 0.29 to 0.98 volt, the ECM will use that value as closed throttle. If the voltage reading is out of the auto zero range on an existing or replacement TP sensor; check cruise control and throttle cables for being held open.
4. This simulates a high signal voltage to check for an open in CKT 417.
5. CKT 416 and CKT 474 share a common 5 volts buffered reference signal. If either of these circuits is shorted to ground, DTC 22 will set. To determine if the MAP sensor is causing the DTC 22 problem, disconnect it to see if DTC 22 resets. Be sure TP sensor is connected and clear DTC(s) before testing.

Diagnostic Aids:

A Tech 1 scan tool reads throttle position in volts. Voltage should increase at a steady rate as throttle is moved toward WOT.

Also some scan tools will read: throttle angle 0% = closed throttle 100% = WOT.

An open or short to ground in CKTs 416 or 417 will result in a DTC 22.

CKTs 416 and 417 are routed through a mini harness. CKT 416 is connected to terminal "D" at the gray connector, CKT 417 is connected to terminal "H" at the black connector.

Scan TP sensor while depressing accelerator pedal with engine stopped and ignition "ON." Display should vary from about 500 mV (.5 volt) when throttle was closed, to about 4800 mV (4.8 volts) when throttle is held at Wide Open Throttle (WOT) position.

Also some scan tools will read throttle angle.
0% = closed throttle.
100% = open throttle.

If DTC 22 is set, check CKT 416 for faulty wiring or connections.

Refer to "Intermittents" in "Symptoms," Section

3.1L (VIN T) AND 3.4L (VIN X) ENGINES — DIAGNOSTIC TROUBLE CODE CHART — CUTLASS SUPREME, GRAND PRIX, LUMINA AND REGAL

DTC 22
THROTTLE POSITION (TP) SENSOR CIRCUIT
(SIGNAL VOLTAGE LOW)
"W" CARLINE (MFI)

1. THROTTLE CLOSED
   DOES TECH 1 SCAN TOOL DISPLAY TP SENSOR .2V (200 mV) OR BELOW?

   - **NO →** DTC 22 IS INTERMITTENT. IF NO ADDITIONAL DTCs WERE STORED, REFER TO "DIAGNOSTIC AIDS"

   - **YES ↓**

2. - DISCONNECT TP SENSOR.
   - JUMPER HARNESS CKTs 416 & 417 TOGETHER.
   TECH 1 SCAN TOOL SHOULD DISPLAY TP SENSOR OVER 4.0V (4000 mV).
   DOES IT?

   - **YES →** 3 REFER TO FACING PAGE FOR SPECIFIC INSTRUCTIONS.

   - **NO ↓**

4. - PROBE CKT 417 WITH A TEST LIGHT CONNECTED TO 12 VOLTS.
   TECH 1 SCAN TOOL SHOULD DISPLAY TP SENSOR OVER 4.0V (4000 mV).
   DOES IT?

   - **YES →** 5 CKT 416 OPEN, SHORTED TO GROUND. ALSO CHECK CKT 474 FOR SHORT TO GROUND. IF OK, IT IS A FAULTY CONNECTION OR FAULTY ECM.

   - **NO →** CKT 417 OPEN, SHORTED TO GROUND, FAULTY CONNECTION OR FAULTY ECM.

"AFTER REPAIRS," REFER TO DTC CRITERIA   AND CONFIRM DTC DOES NOT RESET.

3.1L (VIN T) AND 3.4L (VIN X) ENGINES — DIAGNOSTIC TROUBLE CODE CHART — CUTLASS SUPREME, GRAND PRIX, LUMINA AND REGAL

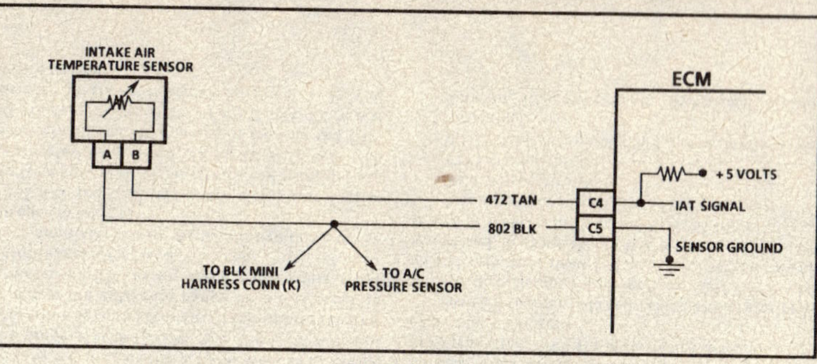

DTC 23
INTAKE AIR TEMPERATURE (IAT) SENSOR CIRCUIT
(LOW TEMPERATURE INDICATED)
"W" CARLINE (MFI)

Circuit Description:

The Intake Air Temperature (IAT) sensor uses a thermistor to control the signal voltage to the ECM. The ECM applies a voltage (about 5 volts) on CKT 472 to the sensor. When the air is cold the sensor (thermistor) resistance is high, therefore, the ECM will see a high signal voltage. If the air is warm the sensor resistance is low, therefore, the ECM will see a low voltage.

The IAT sensor is located in the air cleaner.

Test Description: Number(s) below refer to circled number(s) on the diagnostic chart.

1. DTC 23 will set if:
   - A signal voltage indicates an Intake Air Temperature (IAT) below -35°C (-31°F).
   - Time since engine start is 4 minutes or longer.
   - Vehicle speed less than 1 mph.
   - Start-up engine coolant temperature is less than or equal to -35.5°C (31.9°F).
   - All conditions met for 10 seconds.
2. A DTC 23 will set due to an open sensor, wire, or connection. This test will determine if the wiring and ECM are OK.
3. This will determine if the signal CKT 472 or the 5 volts return CKT 802 is open.

Diagnostic Aids:

A Tech 1 scan tool reads the temperature of the air entering the engine and should read close to ambient air temperature when the engine is cold and rises as underhood temperature increases.

A faulty connection or an open in CKT 472 or CKT 802 will result in a DTC 23.

DTCs 23 and 34 stored at the same time could be the result of an open CKT 802, which would also turn the temperature warning indicator "ON." CKT 802 is routed through a mini harness. A faulty connection could result in intermittent failures. The "Temperature to Resistance Values" scale at the right may be used to test the IAT sensor at various temperature levels to evaluate the possibility of a "shifted" (mis-scaled) sensor. A "slewed" sensor could result in poor driveability complaints.

Refer to "Intermittents" in "Symptoms," Section

3.1L (VIN T) AND 3.4L (VIN X) ENGINES — DIAGNOSTIC TROUBLE CODE CHART — CUTLASS SUPREME, GRAND PRIX, LUMINA AND REGAL

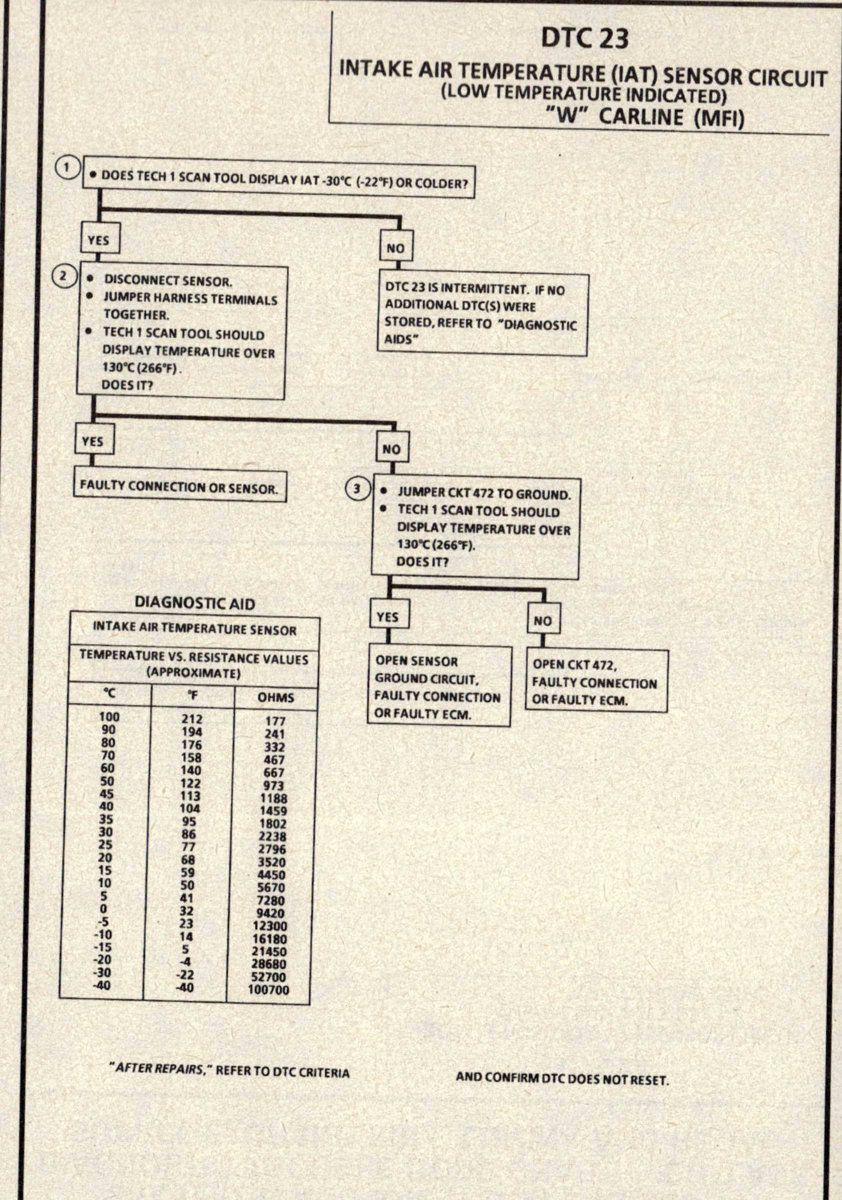

DIAGNOSTIC AID		
INTAKE AIR TEMPERATURE SENSOR		
TEMPERATURE VS. RESISTANCE VALUES (APPROXIMATE)		
°C	°F	OHMS
100	212	177
90	194	241
80	176	332
70	158	467
60	140	667
50	122	973
45	113	1188
40	104	1459
35	95	1802
30	86	2238
25	77	2796
20	68	3520
15	59	4450
10	50	5670
5	41	7280
0	32	9420
-5	23	12300
-10	14	16180
-15	5	21450
-20	-4	28680
-30	-22	52700
-40	-40	100700

"AFTER REPAIRS," REFER TO DTC CRITERIA   AND CONFIRM DTC DOES NOT RESET.

3.1L (VIN T) AND 3.4L (VIN X) ENGINES — DIAGNOSTIC TROUBLE CODE CHART — CUTLASS SUPREME, GRAND PRIX, LUMINA AND REGAL

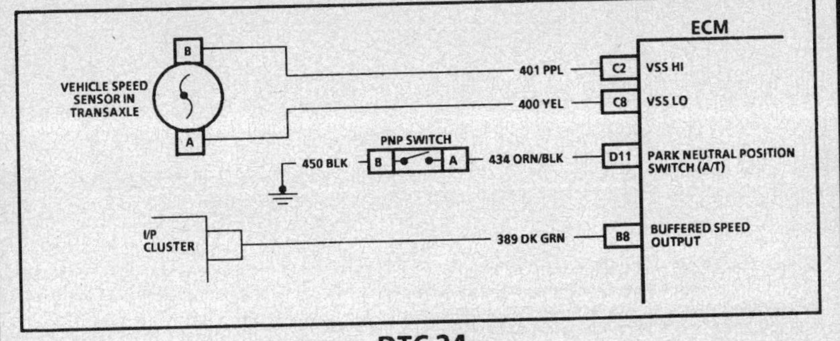

DTC 24
VEHICLE SPEED SENSOR (VSS) CIRCUIT "W" CARLINE (MFI)

Circuit Description:

Vehicle speed information is provided to the ECM by the vehicle speed sensor which uses a Permanent Magnet (PM) generator and is mounted in the transaxle. The PM generator produces a pulsing voltage whenever vehicle speed is over about 3 mph. The AC voltage level and the number of pulses increases with vehicle speed. The ECM then converts the pulsing voltage to mph which is used for calculations and the mph can be displayed with a scan tool. Output of the generator can also be seen by using a digital voltmeter on the AC scale while rotating the generator.

The function of VSS buffer used in past model years has been incorporated into the ECM. The ECM then supplies the necessary signal to the instrument panel (4000 pulses per mile) for operating the speedometer and the odometer. If the vehicle is equipped with cruise control, the ECM also provides a signal (2000 pulses per mile) to the cruise control module.

Test Description: Number(s) below refer to circled number(s) on the diagnostic chart.

1. DTC 24 will set if vehicle speed equals 0 mph when:
   - VSS indicates less than 3 mph.
   - MAP is less than 30 kPa.
   - Engine speed is between 2200 and 4400 RPM.
   - TP sensor is less than 2%.
   - Not in park or neutral.
   - No DTC 21, 22, 33 or 34.
   - All conditions met for 3 seconds.

   These conditions are met during a road load deceleration. Disregard DTC 24 that sets when drive wheels are not turning.
   - The PM generator only produces a signal if drive wheels are turning greater than 3 mph.
2. If CKTs 400, 401 and 389 are OK, and if the speedometer works properly, DTC 24 is being caused by a faulty ECM, faulty PROM or an incorrect PROM.

Diagnostic Aids:

Tech 1 scan tool should indicate a vehicle speed whenever the drive wheels are turning greater than 3 mph.

A problem in CKT 389 will not affect the VSS input or the readings on a Tech 1 scan tool.

Check CKT 400 and CKT 401 for proper connections to be sure there clean and tight and the harness is routed correctly. Refer to "Intermittents" in "Symptoms," Section "6E3-B".

(A/T) A faulty or misadjusted Park/Neutral Position (PNP) switch can result in a false DTC 24. Use a Tech 1 scan tool and check for proper signal while in overdrive (4T60E-T4). Refer to CHART C-1A for PNP switch diagnosis check.

3.1L (VIN T) AND 3.4L (VIN X) ENGINES — DIAGNOSTIC TROUBLE CODE CHART — CUTLASS SUPREME, GRAND PRIX, LUMINA AND REGAL

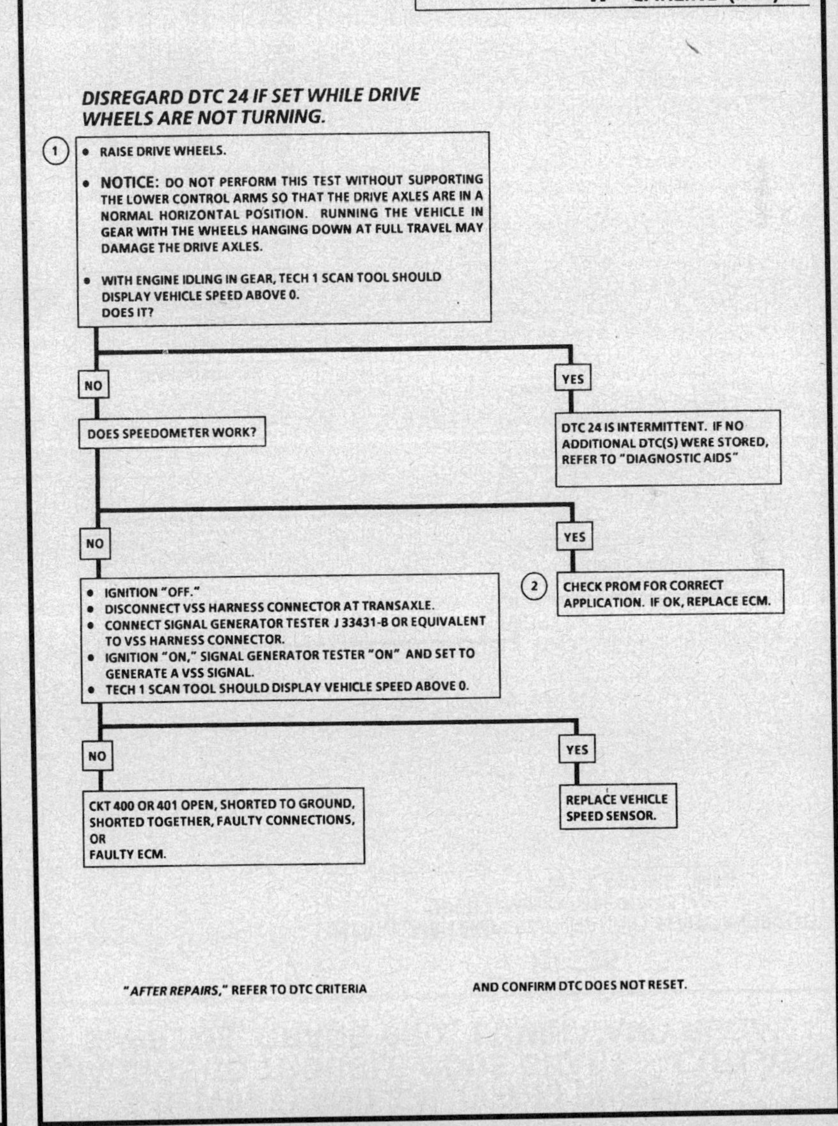

3.1L (VIN T) AND 3.4L (VIN X) ENGINES — DIAGNOSTIC TROUBLE CODE CHART — CUTLASS SUPREME, GRAND PRIX, LUMINA AND REGAL

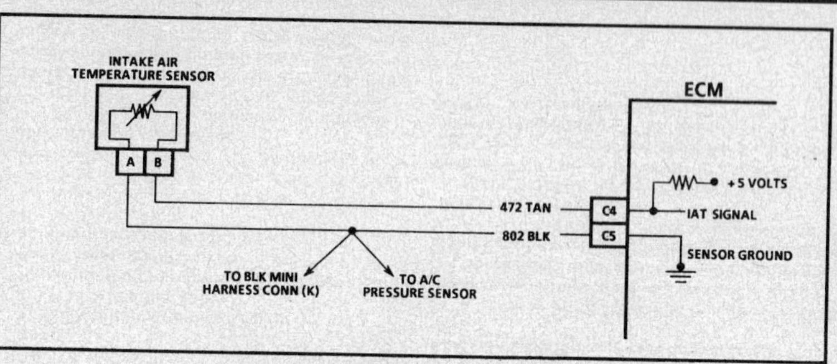

DTC 25
INTAKE AIR TEMPERATURE (IAT) SENSOR CIRCUIT
(HIGH TEMPERATURE INDICATED)
"W" CARLINE (MFI)

Circuit Description:

The Intake Air Temperature (IAT) sensor uses a thermistor to control the signal voltage to the ECM. The ECM applies a voltage (about 5 volts) on CKT 472 to the sensor. When the air is cold, the sensor (thermistor) resistance is high, therefore the ECM will see a high signal voltage. If the air is warm the sensor resistance is low, therefore the ECM will see a low voltage.

The IAT sensor is located in the air cleaner.

Test Description: Number(s) below refer to circled number(s) on the diagnostic chart.

1. DTC 25 will set if:
   - A signal voltage indicates an Intake Air Temperature (IAT) greater than 135°C (293°F) for .2 second.
   - A vehicle speed over 1 mph is present.

   Due to the conditions necessary to set a DTC 25 the MIL (Service Engine Soon) will remain "ON" while the signal is low and vehicle speed is present.

Diagnostic Aids:

A Tech 1 scan tool reads temperature of the air entering the engine and should read close to ambient air temperature when engine is cold and rises as underhood temperature increases.

A short to ground in CKT 472 will result in a DTC 25.

The "Temperature to Resistance Values" scale at the right may be used to test the IAT sensor at various temperature levels to evaluate the possibility of a "shifted" (mis-scaled) sensor. A "slewed" sensor could result in poor driveability complaints.

Refer to "Intermittents" in "Symptoms," Section

3.1L (VIN T) AND 3.4L (VIN X) ENGINES — DIAGNOSTIC TROUBLE CODE CHART — CUTLASS SUPREME, GRAND PRIX, LUMINA AND REGAL

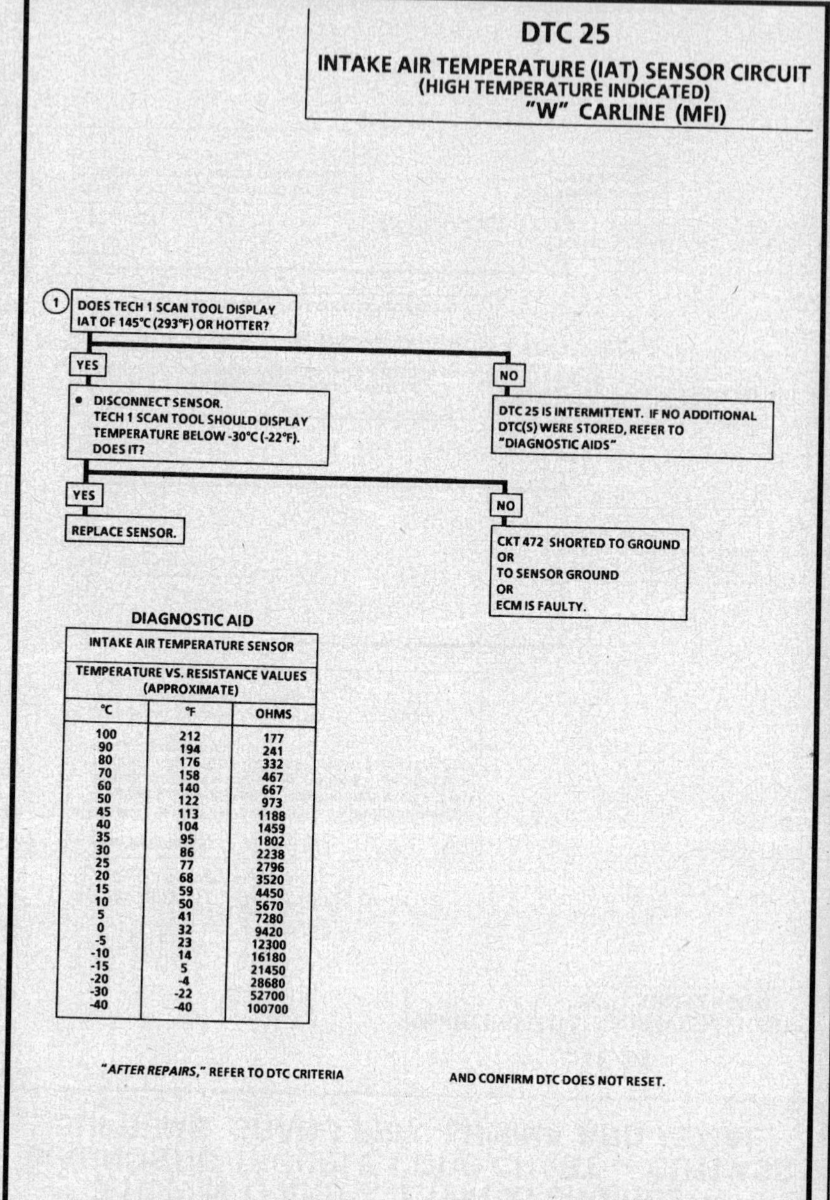

DIAGNOSTIC AID

INTAKE AIR TEMPERATURE SENSOR

TEMPERATURE VS. RESISTANCE VALUES (APPROXIMATE)		
°C	°F	OHMS
100	212	177
90	194	241
80	176	332
70	158	467
60	140	667
50	122	973
45	113	1188
40	104	1459
35	95	1802
30	86	2238
25	77	2796
20	68	3520
15	59	4450
10	50	5670
5	41	7280
0	32	9420
-5	23	12300
-10	14	16180
-15	5	21450
-20	-4	28680
-30	-22	52700
-40	-40	100700

"AFTER REPAIRS," REFER TO DTC CRITERIA   AND CONFIRM DTC DOES NOT RESET.

3.1L (VIN T) AND 3.4L (VIN X) ENGINES — DIAGNOSTIC TROUBLE CODE CHART — CUTLASS SUPREME, GRAND PRIX, LUMINA AND REGAL

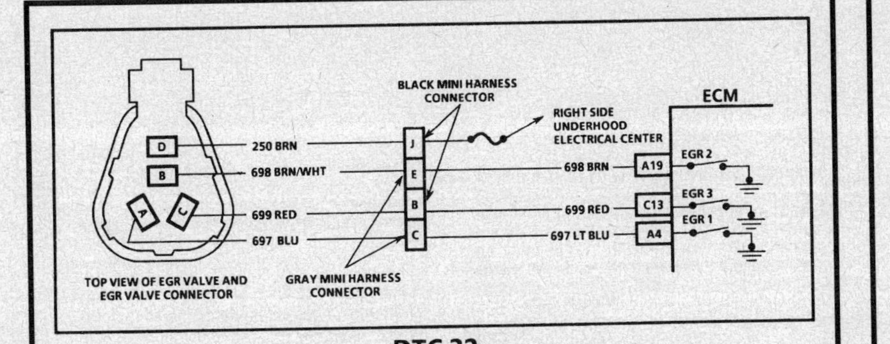

DTC 32
DIGITAL EGR
EXHAUST GAS RECIRCULATION (EGR) CIRCUIT
(ELECTRICAL DIAGNOSIS)
"W" CARLINE

Circuit Description:

The ECM is programmed with an EGR diagnostic check to monitor flow through each individual valve in the three solenoid EGR valves.

When the vehicle is in the decel mode with the valves normally closed, the ECM opens each valve in succession (while keeping previous valves open) and manifold pressure is monitored for an appropriate increase associated with each valve's application. Should the calibrated response in manifold pressure not be seen by the ECM, a DTC 32 will set.

Test Description: Number(s) below refer to circled number(s) on the diagnostic chart.

1. This test determines if there is power to the EGR valve.
2. This test will determine if there is an open circuit in the EGR wiring or if the EGR valve is at fault.
3. This test will determine if there is a short to ground in any circuit going to the EGR valve or if the ECM is at fault.

Diagnostic Aids:

An intermittent may be caused by a poor connection, rubbed-through wire insulation, or a wire broken inside the insulation.

DTC 32
DIGITAL EGR
EXHAUST GAS RECIRCULATION (EGR) CIRCUIT
(ELECTRICAL DIAGNOSIS)
"W" CARLINE

- KEY "ON", ENGINE "OFF."
- IS DTC 32 PRESENT?

YES

(1)
- DISCONNECT EGR VALVE ELECTRICAL CONNECTOR.
- KEY "ON" ENGINE "OFF."
- WITH TEST LIGHT CONNECTED TO GROUND, PROBE TERMINAL "D" OF EGR HARNESS CONNECTOR.

NO

- DTC 32 INTERMITTENT. SEE "DIAGNOSTIC AIDS"

LIGHT "ON"
- WITH TEST LIGHT CONNECTED TO B +, PROBE TERMINALS "A", "B" AND "C" OF EGR HARNESS CONNECTOR.

LIGHT "OFF"
- REPAIR OPEN IN CKT 250.

LIGHT "OFF"

(2)
- GROUND DIAGNOSTIC TEST TERMINALS.
- WITH TEST LIGHT STILL CONNECTED TO B +, PROBE TERMINALS "A", "B" AND "C" OF EGR CONNECTOR.

LIGHT "ON"

(3)
- DISCONNECT YELLOW AND GREEN ECM CONNECTOR.
- WITH TEST LIGHT STILL CONNECTED TO B +, PROBE TERMINALS "A", "B" AND "C".

LIGHT "OFF"
- FAULTY ECM

LIGHT "ON"
- REPAIR SHORT TO GROUND IN CIRCUIT WITH LIGHT "ON.".

LIGHT "ON"
- FAULTY EGR CONNECTION OR EGR VALVE.

LIGHT "OFF"
- OPEN IN CIRCUIT THAT DID NOT LIGHT
OR
FAULTY CONNECTOR AT MINI HARNESS OR ECM CONNECTOR
OR
FAULTY ECM.

"AFTER REPAIRS," REFER TO DTC CRITERIA   AND CONFIRM DTC DOES NOT RESET.

3.1L (VIN T) AND 3.4L (VIN X) ENGINES — DIAGNOSTIC TROUBLE CODE CHART — CUTLASS SUPREME, GRAND PRIX, LUMINA AND REGAL

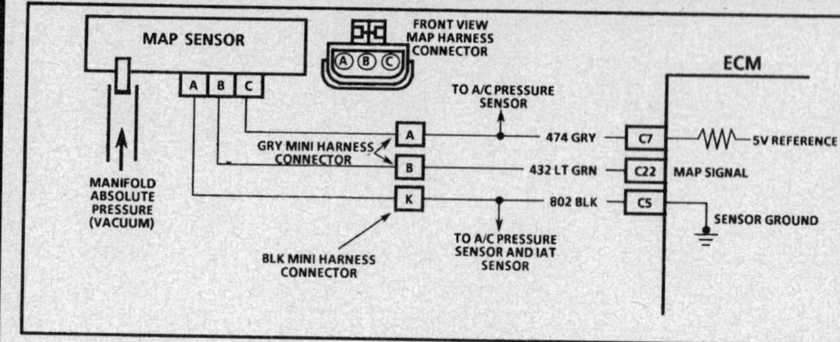

DTC 33

MANIFOLD ABSOLUTE PRESSURE (MAP) SENSOR CIRCUIT
(SIGNAL VOLTAGE HIGH - LOW VACUUM)
"W" CARLINE

Circuit Description:

The Manifold Absolute Pressure (MAP) sensor responds to changes in manifold pressure (vacuum). The ECM receives this information as a signal voltage that will vary from about 1/1.2 volts at idle (high vacuum) to 4/4.8 volts at Wide Open Throttle (WOT) (low vacuum). If the MAP sensor fails, the ECM will substitute a fixed MAP valve and use the TP sensor to control fuel delivery.

Test Description: Number(s) below refer to circled number(s) on the diagnostic chart.

1. DTC 33 will set when:
   - No DTC 21 or DTC 22.
   - Engine running.
   - Manifold pressure greater than 74 kPa (A/C "OFF") 83.4 kPa (A/C "ON").
   - Throttle angle less than 2%.
   - Conditions met for 4.8 seconds.
   Engine misfire or a low unstable idle may set DTC 33.
2. With the MAP sensor disconnected, the ECM should see a low voltage if the ECM and wiring are OK.

Diagnostic Aids:

If idle is rough or unstable, refer to "Symptoms," Section   for items which can cause an unstable idle.

An open in CKT 802 or the connection will result in a DTC 33.

With the ignition "ON" and the engine stopped, the manifold pressure is equal to atmospheric pressure and the signal voltage will be high. This information is used by the ECM as an indication of vehicle altitude. Comparison of this reading with a known good vehicle with the same sensor is a good way to check accuracy of a "suspect" sensor. Readings should be the same ± .4 volt.
- Check all connections.

Important
- Make sure electrical connector remains securely fastened.

- Disconnect sensor from bracket and twist sensor by hand (only) to check for intermittent connections. Output changes greater than .1 volt indicates a bad connector or connections. If OK, replace sensor.
- Refer to CHART C-1D, MAP sensor voltage output check for further diagnosis.
  Refer to "Intermittents" in "Symptoms," Section

3.1L (VIN T) AND 3.4L (VIN X) ENGINES — DIAGNOSTIC TROUBLE CODE CHART — CUTLASS SUPREME, GRAND PRIX, LUMINA AND REGAL

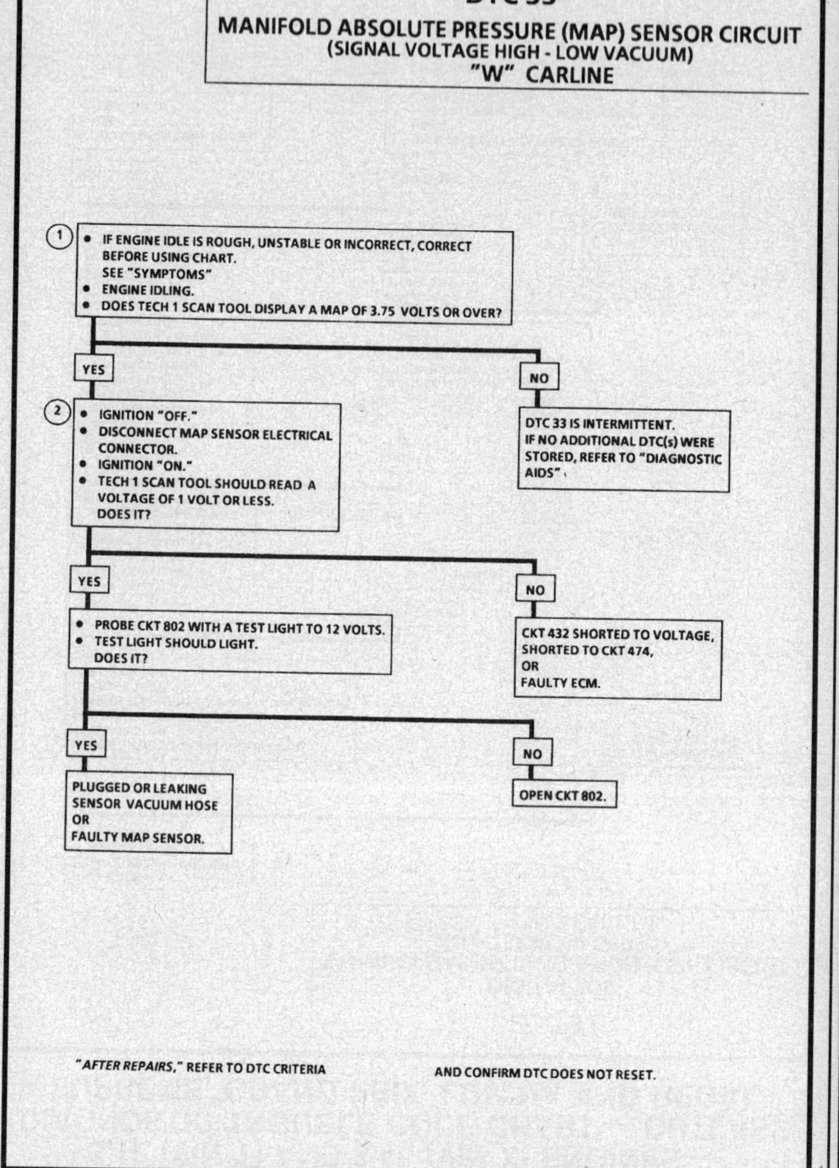

3.1L (VIN T) AND 3.4L (VIN X) ENGINES — DIAGNOSTIC TROUBLE CODE CHART — CUTLASS SUPREME, GRAND PRIX, LUMINA AND REGAL

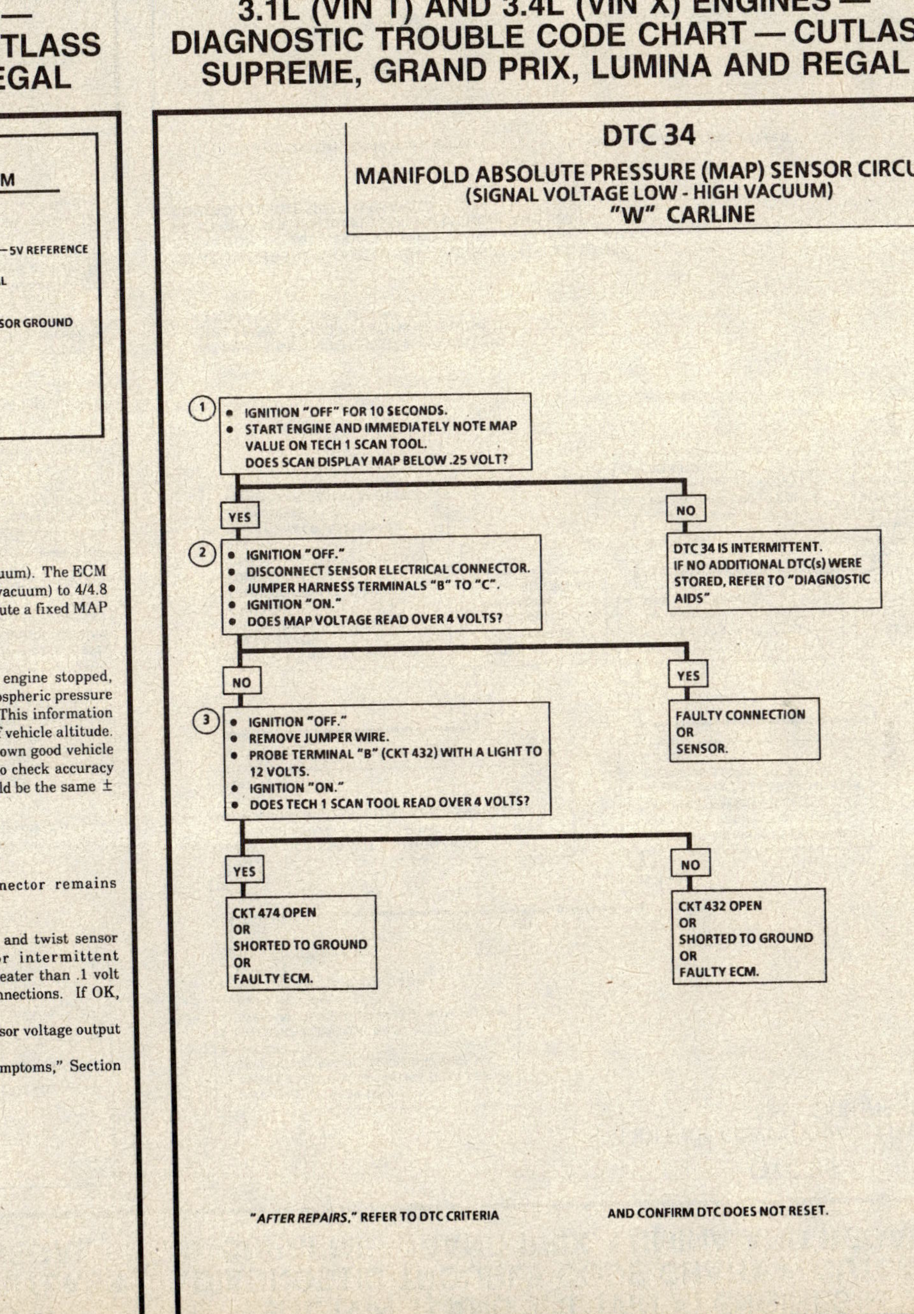

DTC 34
MANIFOLD ABSOLUTE PRESSURE (MAP) SENSOR CIRCUIT
(SIGNAL VOLTAGE LOW - HIGH VACUUM)
"W" CARLINE

Circuit Description:

The Manifold Absolute Pressure (MAP) sensor responds to changes in manifold pressure (vacuum). The ECM receives this information as a signal voltage that will vary from about 1/1.2 volts at idle (high vacuum) to 4/4.8 volts at Wide Open Throttle (WOT) (low vacuum). If the MAP sensor fails, the ECM will substitute a fixed MAP valve and use the TP sensor to control fuel delivery.

Test Description: Number(s) below refer to circled number(s) on the diagnostic chart.
1. DTC 34 will set if:
   - Engine RPM less than 600.
   - Manifold pressure reading less than 13 kPa.
   - Conditions met for .22 second.
   OR
   - Engine RPM greater than 600.
   - Throttle angle over 20%.
   - Manifold pressure less than 13 kPa.
   - Conditions met for .22 second.
2. This test is to see if the sensor is at fault for the low voltage or if there is an ECM or wiring problem.
3. This simulates a high signal voltage to check for an open in CKT 432. If the test light is bright during this test, CKT 432 is probably shorted to ground. If Tech 1 scan tool reads over 4 volts at this test, CKT 474 can be checked by measuring the voltage at terminal "C" (should be 5 volts).

Diagnostic Aids:

An intermittent open in CKT 432 or CKT 474 will result in a DTC 34.

With the ignition "ON" and the engine stopped, the manifold pressure is equal to atmospheric pressure and the signal voltage will be high. This information is used by the ECM as an indication of vehicle altitude. Comparison of this reading with a known good vehicle with the same sensor is a good way to check accuracy of a "suspect" sensor. Readings should be the same ± .4 volt.
- Check all connections.

> **Important**
> - Make sure electrical connector remains securely fastened.

- Disconnect sensor from bracket and twist sensor by hand (only) to check for intermittent connections. Output changes greater than .1 volt indicates a bad connector or connections. If OK, replace sensor.
- Refer to CHART C-1D, MAP sensor voltage output check for further diagnosis.
Refer to "Intermittents" in "Symptoms," Section "

3.1L (VIN T) AND 3.4L (VIN X) ENGINES — DIAGNOSTIC TROUBLE CODE CHART — CUTLASS SUPREME, GRAND PRIX, LUMINA AND REGAL

DTC 34
MANIFOLD ABSOLUTE PRESSURE (MAP) SENSOR CIRCUIT
(SIGNAL VOLTAGE LOW - HIGH VACUUM)
"W" CARLINE

(1)
- IGNITION "OFF" FOR 10 SECONDS.
- START ENGINE AND IMMEDIATELY NOTE MAP VALUE ON TECH 1 SCAN TOOL.
DOES SCAN DISPLAY MAP BELOW .25 VOLT?

YES →

(2)
- IGNITION "OFF."
- DISCONNECT SENSOR ELECTRICAL CONNECTOR.
- JUMPER HARNESS TERMINALS "B" TO "C".
- IGNITION "ON."
- DOES MAP VOLTAGE READ OVER 4 VOLTS?

NO → DTC 34 IS INTERMITTENT. IF NO ADDITIONAL DTC(s) WERE STORED, REFER TO "DIAGNOSTIC AIDS"

NO (from 2) →

(3)
- IGNITION "OFF."
- REMOVE JUMPER WIRE.
- PROBE TERMINAL "B" (CKT 432) WITH A LIGHT TO 12 VOLTS.
- IGNITION "ON."
- DOES TECH 1 SCAN TOOL READ OVER 4 VOLTS?

YES (from 2) → FAULTY CONNECTION OR SENSOR.

YES (from 3) → CKT 474 OPEN OR SHORTED TO GROUND OR FAULTY ECM.

NO (from 3) → CKT 432 OPEN OR SHORTED TO GROUND OR FAULTY ECM.

"AFTER REPAIRS," REFER TO DTC CRITERIA AND CONFIRM DTC DOES NOT RESET.

3.1L (VIN T) AND 3.4L (VIN X) ENGINES — DIAGNOSTIC TROUBLE CODE CHART — CUTLASS SUPREME, GRAND PRIX, LUMINA AND REGAL

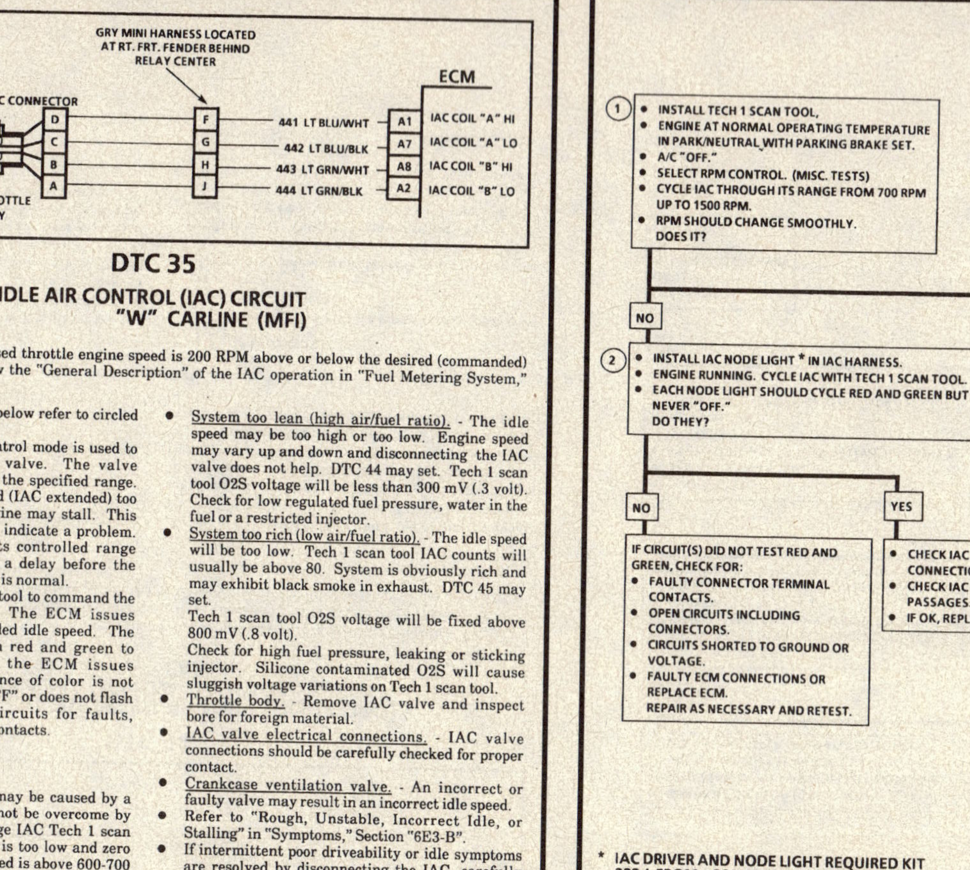

DTC 35
IDLE AIR CONTROL (IAC) CIRCUIT "W" CARLINE (MFI)

Circuit Description:

DTC 35 will set when the closed throttle engine speed is 200 RPM above or below the desired (commanded) idle speed for 50 seconds. Review the "General Description" of the IAC operation in "Fuel Metering System," Section "6E3-C2".

Test Description: Number(s) below refer to circled number(s) on the diagnostic chart.

1. The Tech 1 scan tool RPM control mode is used to extend and retract the IAC valve. The valve should move smoothly within the specified range. If the idle speed is commanded (IAC extended) too low (below 700 RPM), the engine may stall. This may be normal and would not indicate a problem. Retracting the IAC beyond its controlled range (above 1500 RPM) will cause a delay before the RPMs start dropping. This too is normal.
2. This test uses the Tech 1 scan tool to command the IAC controlled idle speed. The ECM issues commands to obtain commanded idle speed. The node lights each should flash red and green to indicate a good circuit as the ECM issues commands. While the sequence of color is not important if either light is "OFF" or does not flash red and green, check the circuits for faults, beginning with poor terminal contacts.

Diagnostic Aids:

A slow, unstable, or fast idle may be caused by a non-IAC system problem that cannot be overcome by the IAC valve. Out of control range IAC Tech 1 scan tool counts will be above 60 if idle is too low and zero counts if idle is too high. If idle speed is above 600-700 RPM in drive with an A/T, locate and correct vacuum leak. If RPM is below spec., check for foreign material around throttle plates. Refer to "Fuel Metering System," Section "6E3-C2". The following checks should be made to repair a non-IAC system problem.

- Vacuum leak (high idle). - If idle is too high, stop the engine. Fully extend (low) IAC with tester. Start engine. If idle speed is above 800 RPM, locate and correct vacuum leak including crankcase ventilation system. Also check for binding of throttle blade or linkage.

- System too lean (high air/fuel ratio). - The idle speed may be too high or too low. Engine speed may vary up and down and disconnecting the IAC valve does not help. DTC 44 may set. Tech 1 scan tool O2S voltage will be less than 300 mV (.3 volt). Check for low regulated fuel pressure, water in the fuel or a restricted injector.
- System too rich (low air/fuel ratio). - The idle speed will be too low. Tech 1 scan tool IAC counts will usually be above 80. System is obviously rich and may exhibit black smoke in exhaust. DTC 45 may set.

  Tech 1 scan tool O2S voltage will be fixed above 800 mV (.8 volt).

  Check for high fuel pressure, leaking or sticking injector. Silicone contaminated O2S will cause sluggish voltage variations on Tech 1 scan tool.
- Throttle body. - Remove IAC valve and inspect bore for foreign material.
- IAC valve electrical connections. - IAC valve connections should be carefully checked for proper contact.
- Crankcase ventilation valve. - An incorrect or faulty valve may result in an incorrect idle speed. Refer to "Rough, Unstable, Incorrect Idle, or Stalling" in "Symptoms," Section "6E3-B".
- If intermittent poor driveability or idle symptoms are resolved by disconnecting the IAC, carefully recheck connections, valve terminal resistance, or replace IAC.
- A/C compressor or relay failure. - Refer to CHART C-10 if the A/C control relay drive circuit is shorted to ground or if the relay is faulty, an idle problem may exist.
- If above are all OK, refer to "Rough, Unstable, Incorrect Idle, or Stalling" in "Symptoms," Section

3.1L (VIN T) AND 3.4L (VIN X) ENGINES — DIAGNOSTIC TROUBLE CODE CHART — CUTLASS SUPREME, GRAND PRIX, LUMINA AND REGAL

DTC 35
IDLE AIR CONTROL (IAC) CIRCUIT "W" CARLINE (MFI)

* IAC DRIVER AND NODE LIGHT REQUIRED KIT
222-L FROM: CONCEPT TECHNOLOGY, INC.
J 37027 FROM: KENT-MOORE, INC.

CLEAR DIAGNOSTIC TROUBLE CODES, CONFIRM "CLOSED LOOP" OPERATION, NO MIL (SERVICE ENGINE SOON). PERFORM IAC RESET PROCEDURE PER APPLICABLE SERVICE MANUAL AND VERIFY CONTROLLED IDLE SPEED IS CORRECT.

"AFTER REPAIRS," REFER TO DTC CRITERIA   AND CONFIRM DTC DOES NOT RESET.

3.4L (VIN X) ENGINE — DIAGNOSTIC TROUBLE CODE CHART — CUTLASS SUPREME, GRAND PRIX, LUMINA AND REGAL

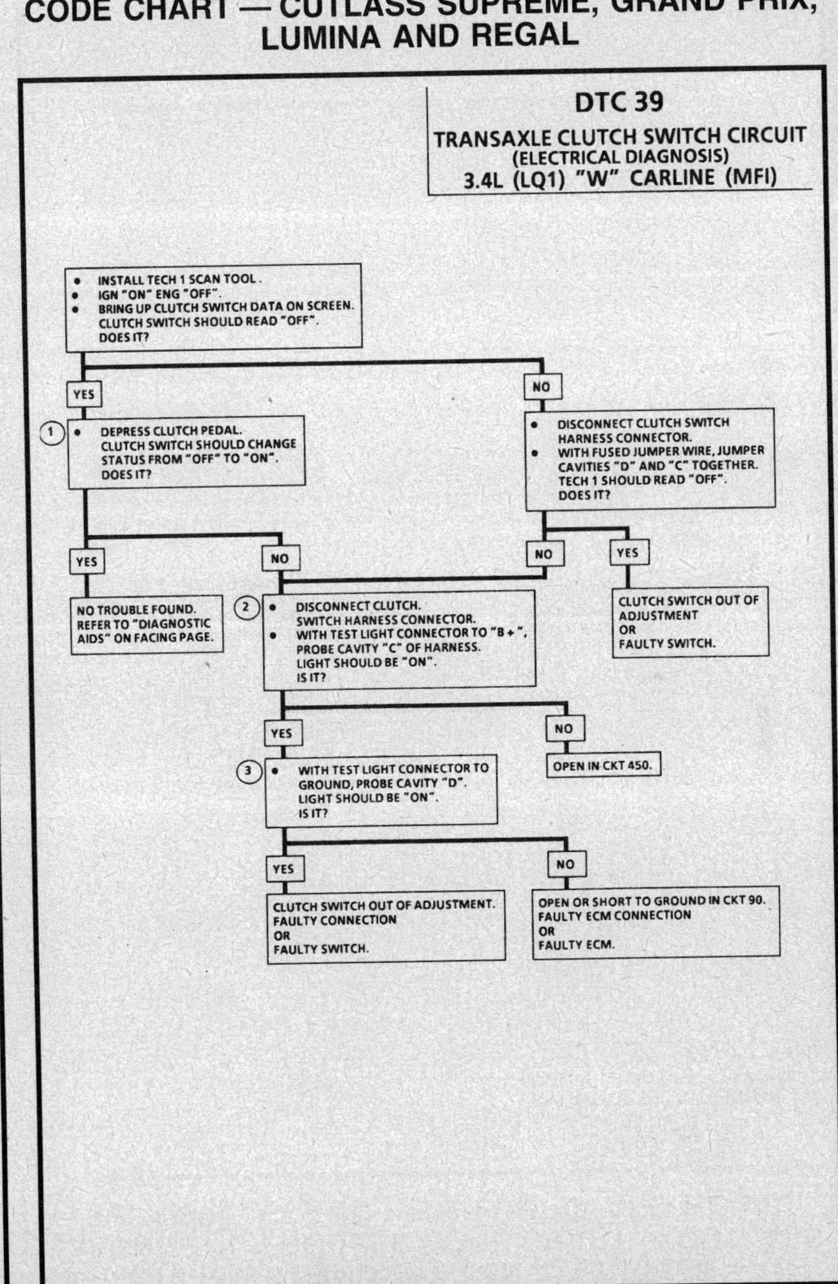

**DTC 39
TRANSAXLE CLUTCH SWITCH CIRCUIT
(ELECTRICAL DIAGNOSIS)
3.4L (LQ1) "W" CARLINE (MFI)**

Circuit Description:

The clutch switch is normally closed when the clutch pedal is at rest. When the clutch pedal is depressed (clutch disengaged), the switch will change from "OFF" to "ON." The ECM runs a diagnostic check that looks for a change of state in the switch logic. If there is a failure in the clutch switch circuit, a DTC 39 will be stored in memory and the MIL (Service Engine Soon) will be illuminated.

DTC 39 will set:
• Engine at normal operating temperature [above 90°C (194°F)].
• Vehicle speed greater than or equal to 40 mph, the ECM must see VSS go from 0 to 40 mph and back to 0 mph for the clutch diagnostic to run.
• The clutch switch must fail four consecutive times per ignition cycle.

Test Description: Number(s) below refer to circled number(s) on the diagnostic chart.
1. When the clutch pedal is depressed, the switch should go from closed to open and the status on the Tech 1 scan tool should change from "OFF" to "ON."
2. This test will check for a poor ground to the clutch switch.
3. This test will determine if the clutch switch is being supplied with "B+" or if CKT 90 is shorted to ground.

Diagnostic Aids:

If DTC 39 was stored in memory and no problem was found, make sure clutch switch fastener is secure and is not sticking or binding. The clutch and the cruise control share the same switch.

3.1L (VIN T) AND 3.4L (VIN X) ENGINES — DIAGNOSTIC TROUBLE CODE CHART — CUTLASS SUPREME, GRAND PRIX, LUMINA AND REGAL

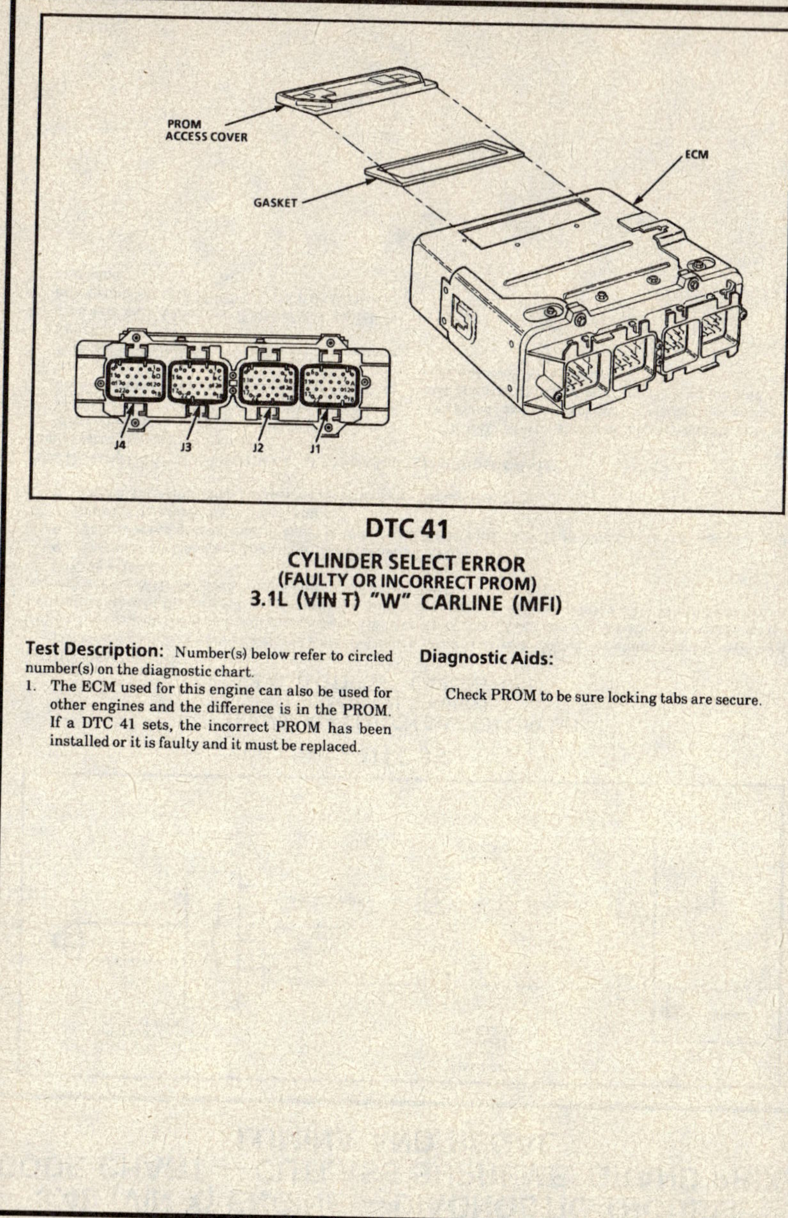

DTC 41
CYLINDER SELECT ERROR
(FAULTY OR INCORRECT PROM)
3.1L (VIN T) "W" CARLINE (MFI)

Test Description: Number(s) below refer to circled number(s) on the diagnostic chart.

1. The ECM used for this engine can also be used for other engines and the difference is in the PROM. If a DTC 41 sets, the incorrect PROM has been installed or it is faulty and it must be replaced.

Diagnostic Aids:

Check PROM to be sure locking tabs are secure.

3.1L (VIN T) AND 3.4L (VIN X) ENGINES — DIAGNOSTIC TROUBLE CODE CHART — CUTLASS SUPREME, GRAND PRIX, LUMINA AND REGAL

DTC 41
CYLINDER SELECT ERROR
(FAULTY OR INCORRECT PROM)
"W" CARLINE (MFI)

(1)
- IGNITION "OFF," CLEAR DTC(S).
- START ENGINE AND RUN FOR 1 MINUTE OR UNTIL DTC 41 SETS.
  DOES DTC 41 SET?

YES

FAULTY CONNECTIONS DUE TO PROM NOT LOCKED IN PLACE, OR INCORRECT PROM INSTALLED.

NO

DTC 41 IS INTERMITTENT. REVIEW "DIAGNOSTIC AIDS"

"AFTER REPAIRS," REFER TO DTC CRITERIA AND CONFIRM DTC DOES NOT RESET.

3.1L (VIN T) AND 3.4L (VIN X) ENGINES — DIAGNOSTIC TROUBLE CODE CHART — CUTLASS SUPREME, GRAND PRIX, LUMINA AND REGAL

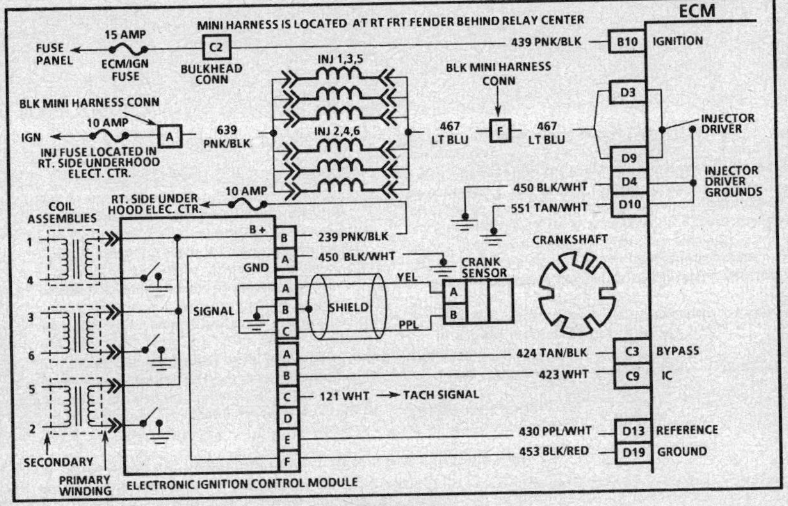

DTC 42
IGNITION CONTROL (IC) CIRCUIT "W" CARLINE (MFI)

Circuit Description:

When the system is running on the electronic ignition control module, that is, no voltage on the bypass line, the ignition control module grounds the Ignition Control (IC) signal. The ECM expects to see no voltage on the IC line during this condition. If it sees a voltage, it sets DTC 42 and will not go into the IC timing mode.

When the RPM for Ignition Control (IC) is reached (about 400 RPM), and bypass voltage applied, the IC circuit should on longer be grounded in the ignition control module so the IC voltage should be varying.

If the bypass line is open or grounded, the ignition control module will not switch to IC timing mode so the IC voltage will be low and DTC 42 will be set.

If the IC line is grounded, the ignition control module will switch to IC, but because the line is grounded there will be no IC signal. A DTC 42 will be set.

Test Description: Number(s) below refer to circled number(s) on the diagnostic chart.

1. DTC 42 means the ECM has seen an open or short to ground in the IC or bypass circuits. This test confirms DTC 42 and that the fault causing the DTC is present.
2. Checks for a normal IC ground path through the ignition control module. An IC CKT 423 shorted to ground will also read less than 500 ohms; however, this will be checked later.
3. As the test light voltage touches CKT 424, the module should switch causing the ohmmeter to "overrange" if the meter is in the 1000-2000 ohms position. Selecting the 10-20,000 ohms position will indicate above 5000 ohms. The important thing is that the module "switched."
4. The module did not switch and this step checks for:
   - IC CKT 423 shorted to ground.
   - Bypass CKT 424 open.
   - Faulty electronic ignition control module connection or module.
5. Confirms that DTC 42 is a faulty ECM and not an intermittent in CKTs 423 or 424.

Diagnostic Aids:

The Tech 1 scan tool does not have any ability to help diagnose a DTC 42 problem.

A PROM not fully seated in the ECM can result in a DTC 42.

Refer to "Intermittents" in "Symptoms," Section

3.1L (VIN T) AND 3.4L (VIN X) ENGINES — DIAGNOSTIC TROUBLE CODE CHART — CUTLASS SUPREME, GRAND PRIX, LUMINA AND REGAL

DTC 42
IGNITION CONTROL (IC) CIRCUIT "W" CARLINE (MFI)

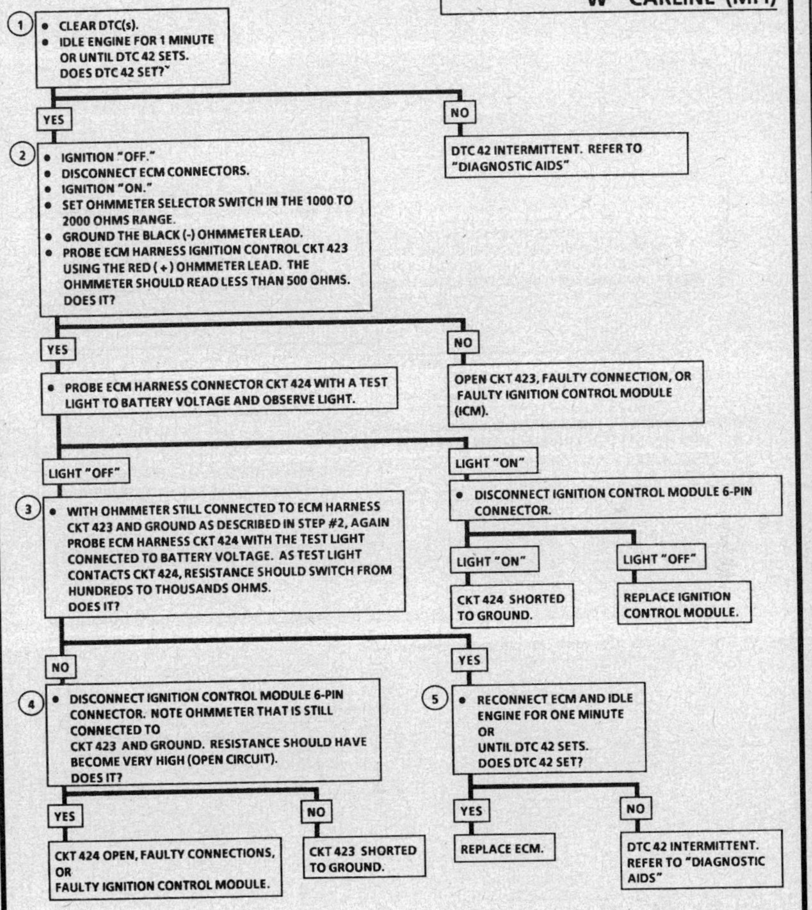

"AFTER REPAIRS," REFER TO DTC CRITERIA AND CONFIRM DTC DOES NOT RESET.

3.1L (VIN T) AND 3.4L (VIN X) ENGINES — DIAGNOSTIC TROUBLE CODE CHART — CUTLASS SUPREME, GRAND PRIX, LUMINA AND REGAL

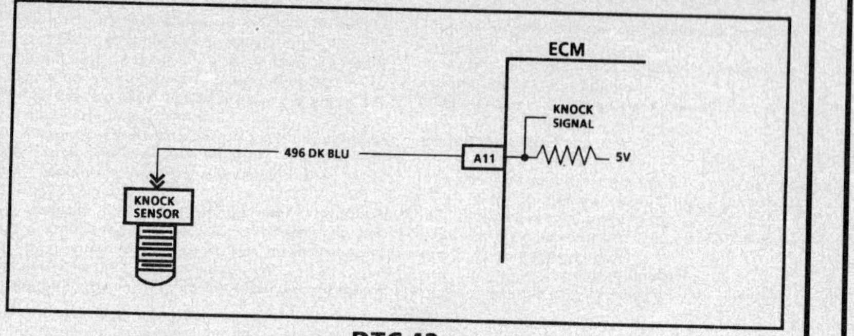

DTC 43

KNOCK SENSOR (KS) CIRCUIT "W" CARLINE (MFI)

Circuit Description:

The knock sensor is used to detect engine detonation and the ECM will retard the Ignition Control (IC) timing based on the signal being received. The circuitry within the knock sensor causes the ECM 5 volts to be pulled down so that under a no knock condition, CKT 496 would measure about 2.5 volts. The knock sensor produces an AC signal which rides on the 2.5 volts DC voltage. The amplitude and signal frequency is dependent upon the knock level.

If CKT 496 becomes open or shorted to ground, the voltage will either go above 4.8 volts or below .64 volts. If either of these conditions are met for about 10 seconds a DTC 43 will be stored.

The test is performed when:

- Engine coolant temperature is over 90°C (194°F).
- IAT temperature is over 0°C (32°F).
- High engine load based on air flow and RPM between 3400 and 4400.

Test Description: Number(s) below refer to circled number(s) on the diagnostic chart.

1. If the conditions for DTC 43, as described above, are being met the Tech 1 scan tool will always indicate "Yes" when the knock signal position is selected. If an audible knock is heard from the engine, repair the internal engine problem, as normally no knock should be detected at idle.
2. If tapping on the engine lift hook does not produce a knock signal, try tapping engine closer to sensor.
3. The ECM has a 5 volts pull-up resistor which should be present at the knock sensor terminal.

Diagnostic Aids:

Check CKT 496 for a potential open or short to ground.

Also check for proper installation of PROM. Refer to "Intermittents" in "Symptoms," Section "6E3-B".

If the customer's complaint is the MIL (Service Engine Soon) comes "ON" when in acceleration, the B portion of the code is failing. There is a possibility that the electronic ignition system was in bypass mode when the 43 test was run. An intermittent open in the IC circuit will put the EI control module in bypass which will not allow the spark to be advanced so the 43 test would fail. If ECM also had a DTC 42 stored, then the IC circuit is likely the cause of the DTC 43.

3.1L (VIN T) AND 3.4L (VIN X) ENGINES — DIAGNOSTIC TROUBLE CODE CHART — CUTLASS SUPREME, GRAND PRIX, LUMINA AND REGAL

DTC 43

KNOCK SENSOR (KS) CIRCUIT "W" CARLINE (MFI)

(1)
- INSTALL TECH 1.
- ENGINE IDLING, ENGINE COOLANT TEMP ABOVE 67°C. DOES TECH 1 INDICATE A FIXED VALUE OF KNOCK RETARD BETWEEN 4 AND 20 DEGREES?

YES →

(2)
- DISCONNECT KNOCK SENSOR.
- IGNITION "ON."
- USING A DVM, MEASURE VOLTAGE BETWEEN HARNESS CKT 496 AND GROUND. VOLTAGE SHOULD BE 4-6 VOLTS. IS IT?

NO → DTC 43 IS INTERMITTENT. SEE "DIAGNOSTIC AIDS"

YES →

(3)
- MEASURE RESISTANCE OF KNOCK SENSOR BY CONNECTING OHMMETER BETWEEN SENSOR TERMINAL AND ENGINE BLOCK. SHOULD BE BETWEEN 3.3KΩ & 4.5KΩ. IS IT?

NO →

OVER 6 VOLTS

CHECK FOR CKT 496 BEING ROUTED TOO CLOSE TO SECONDARY IGNITION WIRES
OR
CKT 496 SHORTED TO VOLTAGE
OR
FAULTY ECM.

LESS THAN 4 VOLTS

CKT 496 OPEN, SHORTED TO GROUND
OR
FAULTY ECM.

YES →
- CHECK HARNESS AND SENSOR CONNECTOR. IF OK:
- REMOVE ECM AND BE SURE PROM IS PROPERLY SEATED INTO ECM. IF OK:
- REPLACE PROM.

NO → REMOVE KNOCK SENSOR AND ENSURE THAT THREADS ON SENSOR AND BLOCK ARE CLEAN AND FREE OF CORROSION, TEFLON TAPE, ANTI-SEIZE COMPOUNDS, OR SEALANTS. IF OK, REPLACE FAULTY KNOCK SENSOR.

"AFTER REPAIRS," REFER TO DTC CRITERIA   AND CONFIRM DTC DOES NOT RESET.

3.1L (VIN T) AND 3.4L (VIN X) ENGINES — DIAGNOSTIC TROUBLE CODE CHART — CUTLASS SUPREME, GRAND PRIX, LUMINA AND REGAL

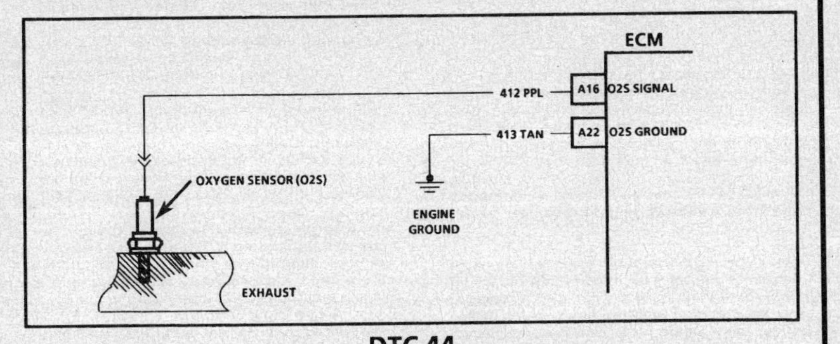

DTC 44
OXYGEN SENSOR (O2S) CIRCUIT
(LEAN EXHAUST INDICATED)
"W" CARLINE (MFI)

Circuit Description:

The ECM supplies a voltage of about .55 volt between terminals "A16" and "A22". (If measured with a 10 megohm digital voltmeter, this may read as low as .35 volt.) The Oxygen Sensor (O2S) varies the voltage within a range of about 1 volt. If the exhaust is rich, the O2S signal will display near 1 volt and when the exhaust is lean the O2S signal will display close to .1 volt.

The sensor is like an open circuit and produces no voltage when it is below about 315°C (600°F). An open sensor circuit or cold sensor causes "Open Loop" operation.

Test Description: Number(s) below refer to circled number(s) on the diagnostic chart.
1. DTC 44 will set if:
   - No DTC 33 or DTC 34.
   - Voltage on CKT 412 remains below .2 volt for 60 seconds or more.
   - The system is operating in "Closed Loop."

Diagnostic Aids:

Using the Tech 1 scan tool, observe the long term fuel trim at different RPM and air flow conditions. The Tech 1 scan tool also displays the block cells, so the long term fuel trim values can be checked in each of the cells to determine when the DTC 44 may have been set. If the conditions for DTC 44 exists the long term fuel trim values will be around 150.
- O2S wire. Sensor pigtail may be mispositioned and contacting the exhaust manifold.
- Check for intermittent ground in wire between connector and sensor.

- Lean injector(s). Perform injector balance test CHART C-2A.
- Fuel contamination. Water, even in small amounts, near the in-tank fuel pump inlet can be delivered to the injectors. The water causes a lean O2S signal and can set a DTC 44.
- Fuel pressure. System will be lean if pressure is too low. It may be necessary to monitor fuel pressure while driving the vehicle at various road speeds and/or loads to confirm. Refer to "Fuel System Diagnosis," CHART A-7.
- Exhaust leaks. If there is an exhaust leak, the engine can cause outside air to be pulled into the exhaust and past the sensor. Vacuum or crankcase leaks can cause a lean condition.
- If the above are OK, it is a faulty Oxygen Sensor (O2S).

3.1L (VIN T) AND 3.4L (VIN X) ENGINES — DIAGNOSTIC TROUBLE CODE CHART — CUTLASS SUPREME, GRAND PRIX, LUMINA AND REGAL

DTC 44
OXYGEN SENSOR (O2S) CIRCUIT
(LEAN EXHAUST INDICATED)
"W" CARLINE (MFI)

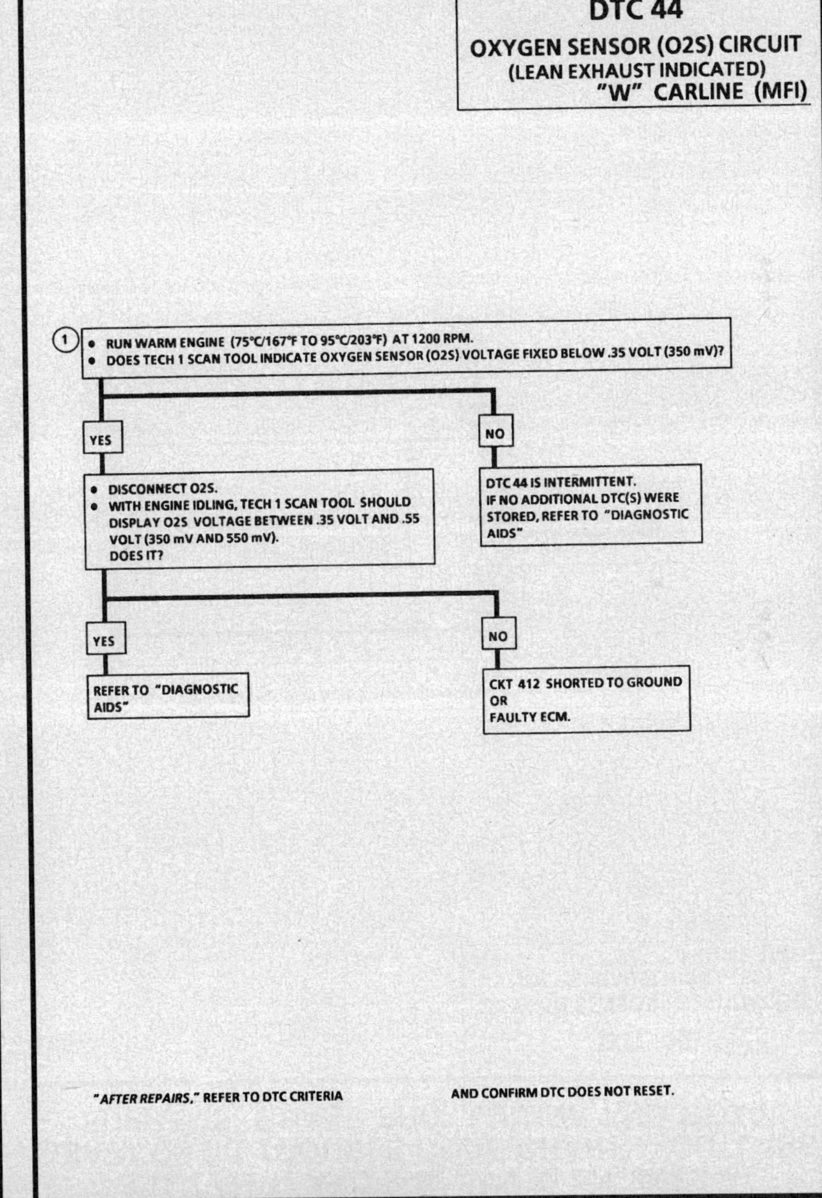

"AFTER REPAIRS," REFER TO DTC CRITERIA   AND CONFIRM DTC DOES NOT RESET.

3.1L (VIN T) AND 3.4L (VIN X) ENGINES — DIAGNOSTIC TROUBLE CODE CHART — CUTLASS SUPREME, GRAND PRIX, LUMINA AND REGAL

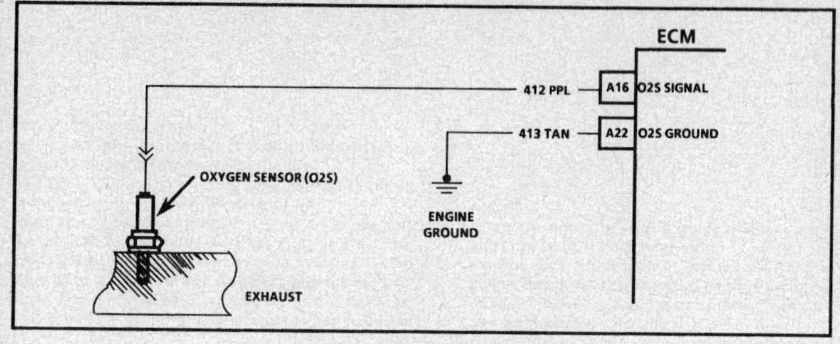

DTC 45
OXYGEN SENSOR (O2S) CIRCUIT
(RICH EXHAUST INDICATED)
"W" CARLINE (MFI)

Circuit Description:

The ECM supplies a voltage of about .55 volt between terminals "A16" and "A22". (If measured with a 10 megohm digital voltmeter, this may read as low as .35 volt.) The Oxygen Sensor (O2S) varies the voltage within a range of approximately 1 volt. If the exhaust is rich, the O2S will display near 1 volt and when the exhaust is lean, the O2S signal will display close to 100 mV.

The sensor is like an open circuit and produces no voltage when it is below about 315°C (600°F). An open O2S circuit or cold O2S causes "Open Loop" operation.

Test Description: Number(s) below refer to circled number(s) on the diagnostic chart.

1. DTC 45 will set if:
   - Voltage on CKT 412 remains above .7 volt for 50 seconds.
   - Engine time after start is 1 minute or more.
   - Throttle angle between 3% and 45%.
   - Operation is in "Closed Loop."

Diagnostic Aids:

Using the Tech 1 scan tool, observe the long term fuel trim values at different RPM and air flow conditions. The Tech 1 scan tool also displays the fuel trim cells, so the long term fuel trim values can be checked in each of the cells to determine when the DTC 45 may have been set. If the conditions for DTC 45 exists, the long term fuel trim values will be around 115.

- Fuel pressure. System will go rich if pressure is too high. The ECM can compensate for some increase. However, if it gets too high, a DTC 45 may be set. Refer to "Fuel System Diagnosis," CHART A-7.
- Rich injector. Perform "Injector Balance Test," CHART C-2A.
- Leaking injector. Refer to CHART A-7.
- Check for fuel contaminated oil.
- O2S contamination. Inspect Oxygen Sensor (O2S) for silicone contamination from fuel, or use of improper RTV sealant. The sensor may have a

white, powdery coating and result in a high but false signal voltage (rich O2S signal indication). The ECM will then reduce the amount of fuel delivered to the engine, causing a severe surge driveability problem.

- Electronic ignition. An open ground CKT 453 may result in EMI, or induced electrical "noise." The ECM looks at this "noise" as reference pulses. The additional pulses result in a higher than actual engine speed signal. The ECM then delivers too much fuel, causing system to go rich. Engine tachometer will also show higher than actual engine speed, which can help in diagnosing this problem.
- EVAP canister purge. Check for fuel saturation. If full of fuel, check canister control and hoses.

- Check for leaking fuel pressure regulator diaphragm by checking vacuum line to regulator for fuel.
- TP sensor. An intermittent TP sensor output will cause the system to go rich, due to a false indication of the engine accelerating.
- EGR. An EGR staying open (especially at idle) will cause the Oxygen Sensor (O2S) to indicate a rich O2S signal and this could result in a DTC 45.

DTC 45
OXYGEN SENSOR (O2S) CIRCUIT
(RICH EXHAUST INDICATED)
"W" CARLINE (MFI)

(1)
- RUN WARM ENGINE (75°C/167°F TO 95°C/203°F) AT 1200 RPM.
- DOES TECH 1 SCAN TOOL DISPLAY O2S VOLTAGE FIXED ABOVE .75 VOLT (750 mV)?

YES
- DISCONNECT O2S AND JUMPER HARNESS CKT 412 TO GROUND.
- TECH 1 SCAN TOOL SHOULD DISPLAY O2S VOLTAGE BELOW .35 VOLT (350 mV). DOES IT?

NO
DTC 45 IS INTERMITTENT. IF NO ADDITIONAL DTC(S) WERE STORED, REFER TO "DIAGNOSTIC AIDS"

YES
REFER TO "DIAGNOSTIC AIDS"

NO
REPLACE ECM.

"AFTER REPAIRS," REFER TO DTC CRITERIA   AND CONFIRM DTC DOES NOT RESET.

3.1L (VIN T) AND 3.4L (VIN X) ENGINES — DIAGNOSTIC TROUBLE CODE CHART — CUTLASS SUPREME, GRAND PRIX, LUMINA AND REGAL

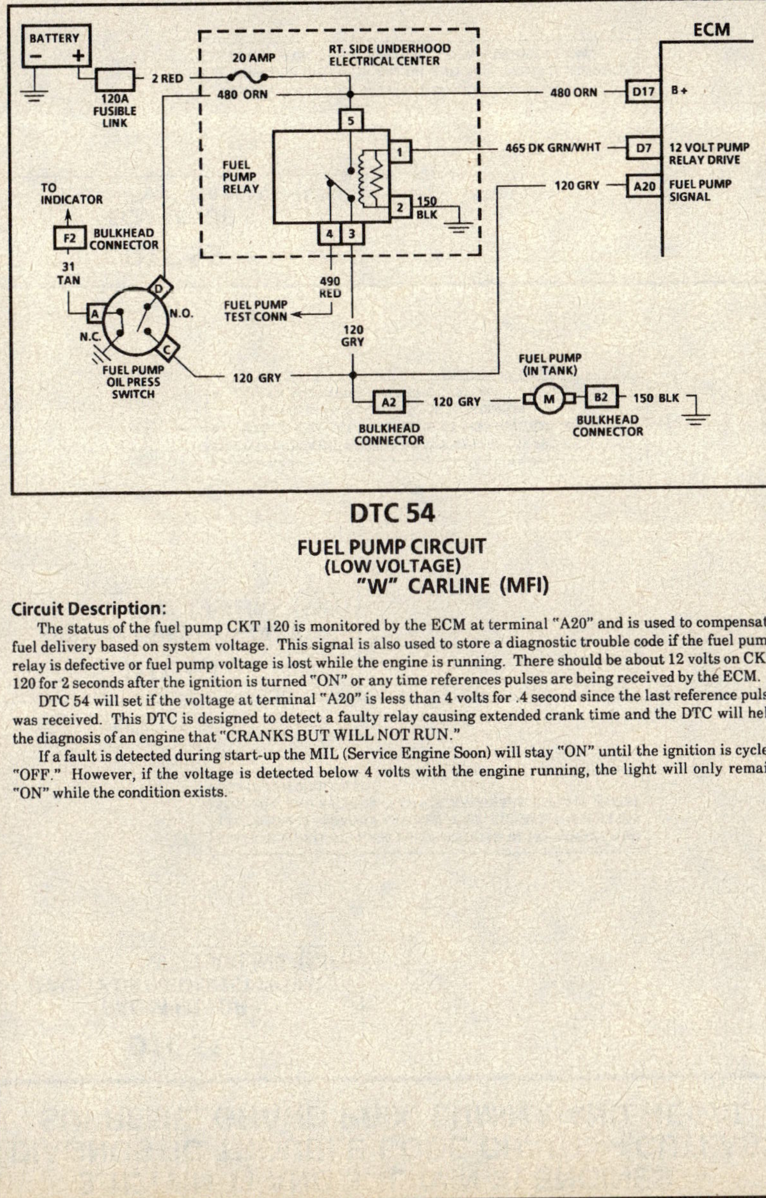

DTC 54

FUEL PUMP CIRCUIT
(LOW VOLTAGE)
"W" CARLINE (MFI)

Circuit Description:

The status of the fuel pump CKT 120 is monitored by the ECM at terminal "A20" and is used to compensate fuel delivery based on system voltage. This signal is also used to store a diagnostic trouble code if the fuel pump relay is defective or fuel pump voltage is lost while the engine is running. There should be about 12 volts on CKT 120 for 2 seconds after the ignition is turned "ON" or any time references pulses are being received by the ECM.

DTC 54 will set if the voltage at terminal "A20" is less than 4 volts for .4 second since the last reference pulse was received. This DTC is designed to detect a faulty relay causing extended crank time and the DTC will help the diagnosis of an engine that "CRANKS BUT WILL NOT RUN."

If a fault is detected during start-up the MIL (Service Engine Soon) will stay "ON" until the ignition is cycled "OFF." However, if the voltage is detected below 4 volts with the engine running, the light will only remain "ON" while the condition exists.

3.1L (VIN T) AND 3.4L (VIN X) ENGINES — DIAGNOSTIC TROUBLE CODE CHART — CUTLASS SUPREME, GRAND PRIX, LUMINA AND REGAL

DTC 54

FUEL PUMP CIRCUIT
(LOW VOLTAGE)
"W" CARLINE (MFI)

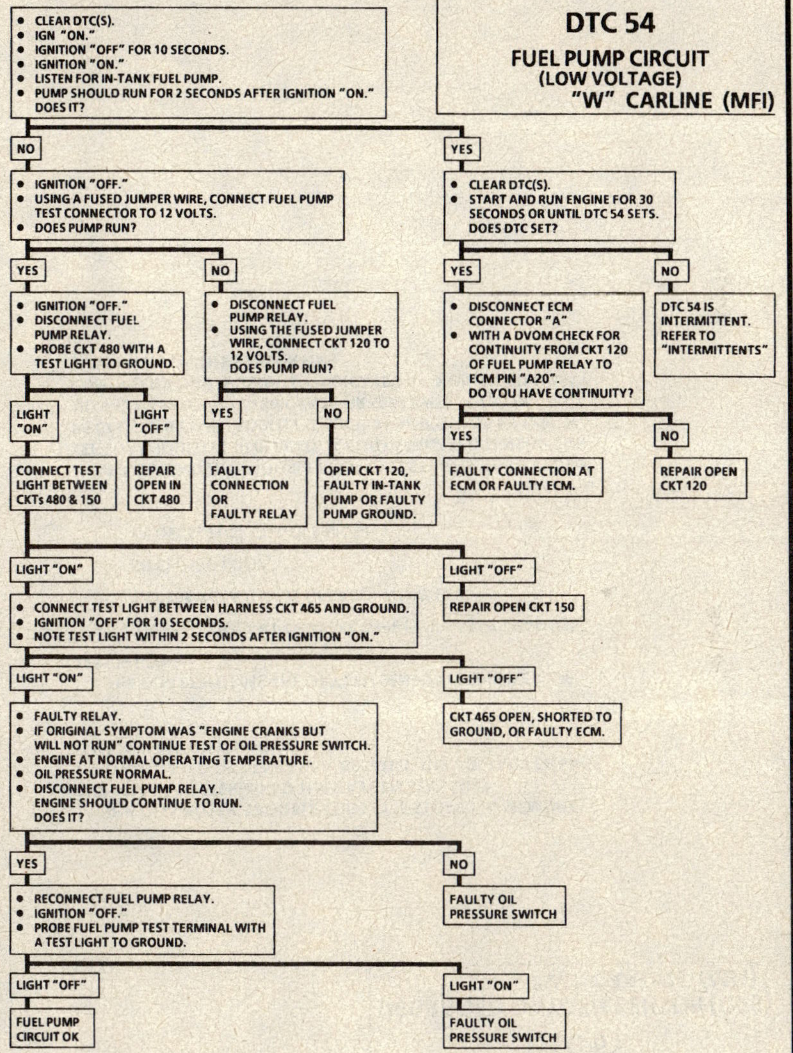

3.1L (VIN T) AND 3.4L (VIN X) ENGINES — DIAGNOSTIC TROUBLE CODE CHART — CUTLASS SUPREME, GRAND PRIX, LUMINA AND REGAL

DTC 51
PROM ERROR
(FAULTY OR INCORRECT PROM)
"W" CARLINE (MFI)

CHECK THAT ALL PINS ARE FULLY INSERTED IN THE SOCKET AND THAT PROM IS PROPERLY LATCHED. IF OK, REPLACE, PROM, CLEAR MEMORY, AND RECHECK. IF DTC 51 REAPPEARS, REPLACE ENGINE CONTROL MODULE (ECM).

DTC 53
SYSTEM OVER VOLTAGE
"W" CARLINE (MFI)

THIS DTC INDICATES THERE IS A BASIC GENERATOR PROBLEM.
- DTC 53 WILL SET IF VOLTAGE AT ECM IGNITION INPUT PIN IS GREATER THAN 16.9 VOLTS FOR 10 SECONDS.
- CHECK AND REPAIR CHARGING SYSTEM.

DTC 55
ECM ERROR
"W" CARLINE (MFI)

BE SURE ECM GROUNDS ARE OK AND THAT PROM IS PROPERLY LATCHED. IF OK, REPLACE ENGINE CONTROL MODULE (ECM).

"AFTER REPAIRS," REFER TO DTC CRITERIA   AND CONFIRM DTC DOES NOT RESET.

3.1L (VIN T) AND 3.4L (VIN X) ENGINES — DIAGNOSTIC TROUBLE CODE CHART — CUTLASS SUPREME, GRAND PRIX, LUMINA AND REGAL

DTC 61
DEGRADED OXYGEN SENSOR (O2S)
"W" CARLINE (MFI)

IF A DIAGNOSTIC TROUBLE CODE 61 IS STORED IN MEMORY THE ECM HAS DETERMINED THE OXYGEN SENSOR IS CONTAMINATED OR DEGRADED, BECAUSE THE VOLTAGE CHANGE TIME IS SLOW OR SLUGGISH.

THE ECM PERFORMS THE OXYGEN SENSOR RESPONSE TIME TEST WHEN:

ENGINE COOLANT TEMPERATURE IS GREATER THAN 85°C.

IAT TEMPERATURE IS GREATER THAN 10°C.

IN CLOSED LOOP.

IN DECEL FUEL CUT-OFF MODE.

IF A DIAGNOSTIC TROUBLE CODE 61 IS STORED, THE OXYGEN SENSOR SHOULD BE REPLACED. A CONTAMINATED SENSOR CAN BE CAUSED BY FUEL ADDITIVES SUCH AS SILICON, OR BY USE OF NON-GM APPROVED LUBRICANTS OR SEALANTS. SILICON CONTAMINATION IS USUALLY INDICATED BY A WHITE POWDERY SUBSTANCE ON THE SENSOR FINS.

"AFTER REPAIRS," REFER TO DTC CRITERIA   AND CONFIRM DTC DOES NOT RESET.

3.1L (VIN T) AND 3.4L (VIN X) ENGINES — DIAGNOSTIC TROUBLE CODE CHART — CUTLASS SUPREME, GRAND PRIX, LUMINA AND REGAL

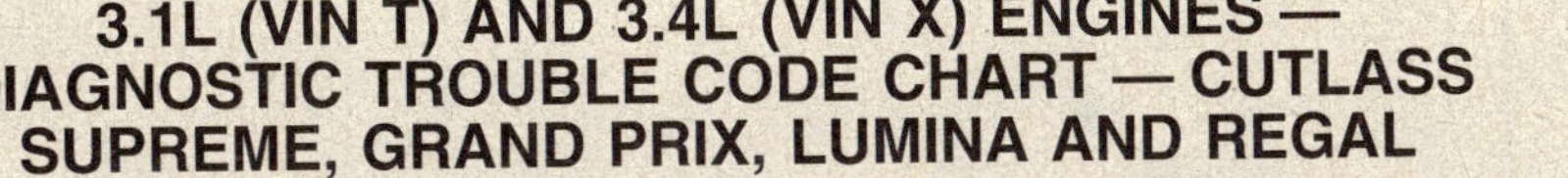

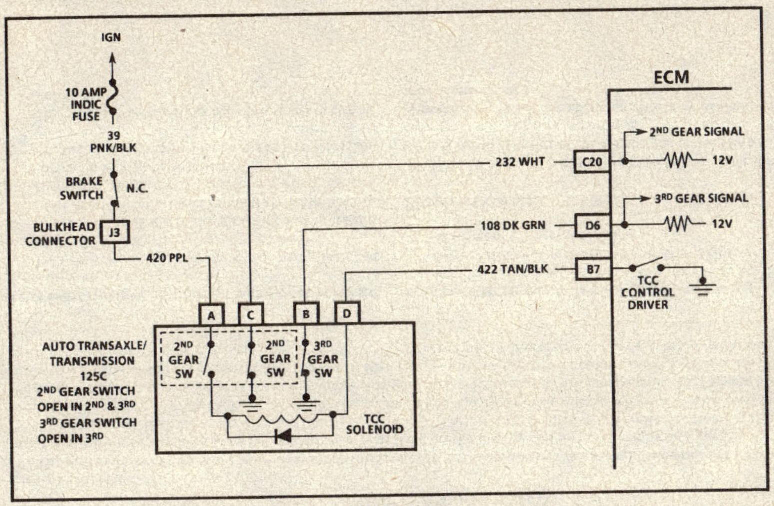

DTC 62
TRANSAXLE GEAR SWITCH SIGNAL CIRCUIT
(ELECTRICAL DIAGNOSIS)
"W" CARLINE (MFI)

Circuit Description:

The 2nd gear signal switch in this vehicle should be open in 2nd and 3rd gear. The ECM uses this 2nd gear signal to disengage the TCC when going into a downshift.

The 3rd gear switch should be open in 3rd gear.

    1st gear = 2nd gear switch open
              2nd gear signal switch closed
              3rd gear signal switch closed
    2nd gear = 2nd gear switch closed
              2nd gear signal switch open
              3rd gear signal switch closed
    3rd gear = 2nd gear switch closed
              2nd gear signal switch open
              3rd gear signal switch open

Test Description: Number(s) below refer to circled number(s) on the diagnostic chart.

1. Some Tech 1 scan tools display the state of these switches in different ways. Be familiar with the type of tool being used. Since both switches should be in the closed state during this test, the tool should read the same for either the 2nd or 3rd gear switch.
2. Determine whether the switch or signal circuit is open. This circuit can also be checked for an open by measuring the voltage (with a voltmeter) at the TCC connector (should be about 12 volts).
3. Because the switch(s) should be grounded in this step, disconnecting the TCC connector should cause the scan switch state to change.
4. The switch state should change when the vehicle shifts into 2nd gear.

Diagnostic Aids:

If vehicle is road tested because of a TCC related problem, be sure the switch states do not change while in 3rd gear because the TCC will disengage. If switches change state, carefully check wire routing and connections.

3.1L (VIN T) AND 3.4L (VIN X) ENGINES — DIAGNOSTIC TROUBLE CODE CHART — CUTLASS SUPREME, GRAND PRIX, LUMINA AND REGAL

DTC 62
TRANSAXLE GEAR SWITCH SIGNAL CIRCUIT
(ELECTRICAL DIAGNOSIS)
"W" CARLINE (MFI)

CHECKS MADE ON THIS PAGE WILL NOT PREVENT THE TCC FROM WORKING, BUT WILL AFFECT ENGAGEMENT OR DISENGAGEMENT POINTS.

(1)
- IGNITION "ON," ENGINE "OFF."
- DOES TECH 1 SCAN TOOL INDICATE TRANS IS IN 2ND GEAR?

NO → DOES TECH 1 SCAN TOOL INDICATE VEHICLE IS IN 3RD GEAR?

YES → (2)
- DISCONNECT ELECTRICAL CONNECTOR.
- JUMPER HARNESS TERMINAL "B" (CKT 581) TO GROUND.
- DOES TECH 1 SCAN TOOL INDICATE TRANS IS IN 2ND GEAR?

(3)
- DISCONNECT TCC ELECTRICAL CONNECTOR.
- DOES TECH 1 SCAN TOOL INDICATE TRANS IS IN 2ND GEAR?

NO (under 3) / **YES** → (2)
- DISCONNECT TCC ELECTRICAL CONNECTOR.
- JUMPER HARNESS TERMINAL "C" (CKT 108) TO GROUND.
- DOES TECH 1 SCAN TOOL INDICATE TRANS IS IN 3RD GEAR?

NO (under 2 left) → FAULTY CONNECTION OR 2ND GEAR SWITCH.

YES (under 2 right) → OPEN CKT 232, FAULTY CONNECTION OR ECM.

NO (under 2 middle) → FAULTY CONNECTION OR 3RD GEAR SWITCH.

YES (under 2 middle) → OPEN CKT 108, FAULTY CONNECTION, OR ECM.

YES → DOES TECH 1 SCAN TOOL INDICATE TRANS IS IN 3RD GEAR?

NO → CKT 232 SHORTED TO GROUND OR FAULTY ECM.

YES → (4)
- RECONNECT TCC ELECTRICAL CONNECTOR.
- RAISE DRIVE WHEELS.

NO → CKT 108 SHORTED TO GROUND OR FAULTY ECM.

NOTICE: DO NOT PERFORM THIS TEST WITHOUT SUPPORTING THE LOWER CONTROL ARMS SO THAT THE DRIVE AXLES ARE IN A NORMAL HORIZONTAL POSITION. RUNNING THE VEHICLE IN GEAR WITH THE WHEELS HANGING DOWN AT FULL TRAVEL MAY DAMAGE THE DRIVE AXLES.

- START AND IDLE ENGINE IN OVERDRIVE.
- INCREASE SPEED SLOWLY UNTIL TRANS SHIFTS INTO 2ND GEAR.
- DOES TECH 1 SCAN TOOL INDICATE TRANS IS IN 2ND GEAR?

YES →
- INCREASE SPEED UNTIL TRANS SHIFTS INTO 3RD GEAR.
- DOES TECH 1 SCAN TOOL INDICATE TRANS IS IN 3RD GEAR?

NO → FAULTY 2ND GEAR SWITCH.

YES → TRANS SWITCHES OK. CHECK WIRE ROUTING AND CONNECTIONS. REFER TO "DIAGNOSTIC AIDS"

NO → FAULTY 3RD GEAR SWITCH.

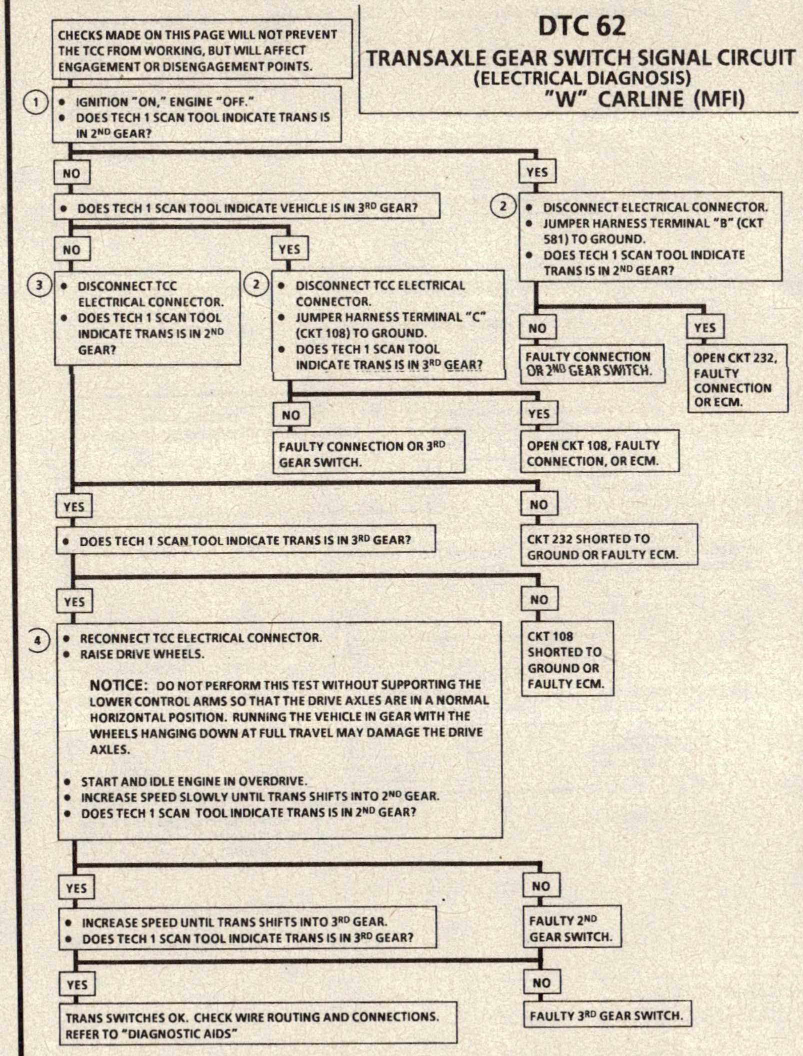

"AFTER REPAIRS," REFER TO DTC CRITERIA   AND CONFIRM DTC DOES NOT RESET.

3.1L (VIN T) AND 3.4L (VIN X) ENGINES — DIAGNOSTIC TROUBLE CODE CHART — CUTLASS SUPREME, GRAND PRIX, LUMINA AND REGAL

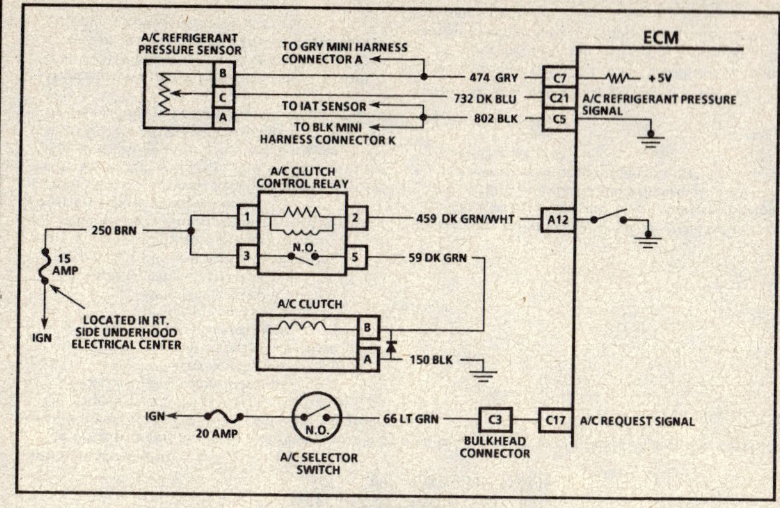

DTC 66
A/C PRESSURE SENSOR CIRCUIT "W" CARLINE (MFI)

Circuit Description:

The A/C refrigerant pressure sensor responds to changes in A/C refrigerant system high side pressure. This input indicates how much load the A/C compressor is putting on the engine and is one of the factors used by the ECM to determine IAC valve position for idle speed control. The circuit consists of a 5 volts reference and a ground, both provided by the ECM, and a signal line to the ECM. The signal is a voltage which is proportional to the pressure. The sensor's range of operation is 0 to 450 psi. At 0 psi, the signal will be about .1 volt, varying up to about 4.9 volts at 450 psi or above. DTC 66 sets if the voltage is above 4.9 volts or below .1 volt for 5 seconds or more. The A/C compressor is disabled by the ECM if DTC 66 is present or if pressure is above or below calibrated values

Test Description: Number(s) below refer to circled number(s) on the diagnostic chart.

1. This step checks the voltage signal being received by the ECM from the A/C pressure sensor.
2. Checks to see if the high voltage signal is from a shorted sensor or a short to voltage in the circuit. Normally, disconnecting the sensor would make a normal circuit go to near zero volt.
3. Checks to see if low voltage signal is from the sensor or the circuit. Jumpering the sensor signal CKT 732 to 5 volts checks the circuit, connections, and ECM.
4. This step checks to see if the low voltage signal was due to an open in the sensor circuit or the 5 volts reference circuit since the prior step eliminated the pressure sensor.

Diagnostic Aids:

DTC 66 sets when signal voltage falls outside the normal possible range of the sensor and is not due to a refrigerant system problem. If problem is intermittent, check for opens or shorts in harness or poor connections.

3.1L (VIN T) AND 3.4L (VIN X) ENGINES — DIAGNOSTIC TROUBLE CODE CHART — CUTLASS SUPREME, GRAND PRIX, LUMINA AND REGAL

DTC 66
A/C PRESSURE SENSOR CIRCUIT "W" CARLINE (MFI)

1. • KEY "ON," ENGINE NOT RUNNING.
   • NOTE TECH 1 SCAN TOOL VOLTAGE FOR A/C REFRIGERANT PRESSURE SENSOR.

- **BELOW .1 VOLT**
- **ABOVE 1.8 VOLTS**
- **BETWEEN .1 VOLT AND 1.8 VOLTS** → FAULT IS NOT PRESENT AT THIS TIME. REFER TO "DIAGNOSTIC AIDS."

BELOW .1 VOLT:

3. • DISCONNECT A/C REFRIGERANT PRESSURE SENSOR CONNECTOR.
   • JUMPER TERMINALS "B" AND "C". DOES TECH 1 SCAN TOOL DISPLAY ABOVE 4.6 VOLTS?

ABOVE 1.8 VOLTS:

2. • DISCONNECT A/C REFRIGERANT PRESSURE SENSOR. DOES TECH 1 SCAN TOOL DISPLAY LESS THAN 1 VOLT?

- NO → CHECK FOR SHORT TO VOLTAGE IN CKT 802. IF NOT SHORTED, REPLACE ECM.
- YES → CHECK FOR OPEN IN CKT 802. IF NOT OPEN, CHECK FOR POOR SENSOR TERMINAL CONNECTIONS. IF OK, REPLACE A/C REFRIGERANT PRESSURE SENSOR.

(from step 3)

- NO → 4. • REMOVE JUMPER. • CONNECT VOLTMETER FROM TERMINAL "A" TO "B". IS VOLTAGE ABOUT 5 VOLTS?
- YES → • CHECK SENSOR TERMINAL CONNECTIONS, IF OK, REPLACE A/C REFRIGERANT PRESSURE SENSOR.

(from step 4)

- NO → • BACK PROBE ECM TERMINAL "C7" WITH VOLTMETER TO GROUND. IS VOLTAGE ABOUT 5 VOLTS?
- YES → CHECK FOR OPEN IN CKT 732. CHECK FOR POOR CONNECTION AT ECM TERMINAL "C21". IF OK, REPLACE ECM.

- NO → CHECK FOR POOR CONNECTION AT ECM TERMINAL "C7" OR SHORT TO GROUND IN CKT 474. IF OK, ECM IS FAULTY.
- YES → REPAIR OPEN IN CKT 474.

"AFTER REPAIRS," REFER TO DTC CRITERIA AND CONFIRM DTC DOES NOT RESET.

3.1L (VIN T) AND 3.4L (VIN X) ENGINES — RESTRICTED EXHAUST SYSTEM DIAGNOSTIC CHART — CUTLASS SUPREME, GRAND PRIX, LUMINA AND REGAL

CHART B-1
RESTRICTED EXHAUST SYSTEM CHECK

Proper diagnosis for a restricted exhaust system is essential before any components are replaced. Either of the following procedures may be used for diagnosis, depending upon engine or tool used:

CHECK AT AIR PIPE:

1. Remove the rubber hose at the exhaust manifold AIR pipe check valve. Remove check valve.
2. Connect exhaust backpressure kit (with air adapter) J 35314-A.
3. Insert the adapter into the exhaust manifold AIR pipe.

OR

CHECK AT OXYGEN SENSOR (O2S):

1. Carefully remove oxygen sensor.
2. Install exhaust backpressure gage J 35314-A in place of oxygen sensor (refer to illustration).
3. After completing test described below, coat threads of oxygen sensor with anti-seize compound P/N 5613695 or equivalent prior to re-installation.

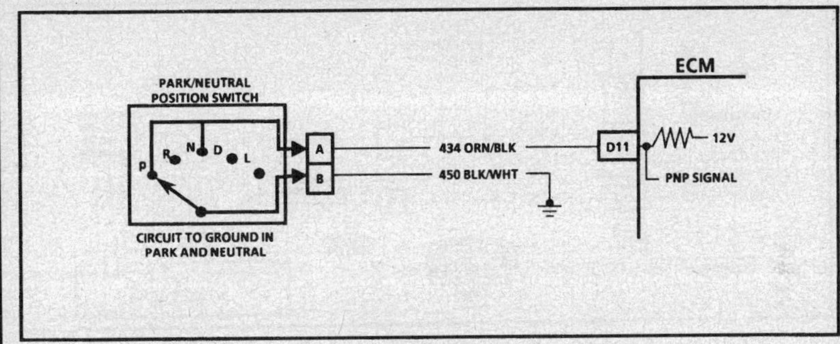

DIAGNOSIS:

1. With the engine idling at normal operating temperature, observe the exhaust system backpressure reading on the gage. Reading should not exceed 8.6 kPa (1.25 psi).
2. Increase engine speed to 2000 RPM and observe gage. Reading should not exceed 20.7 kPa (3 psi).
3. If the backpressure at either speed exceeds specification, a restricted exhaust system is indicated.
4. Inspect the entire exhaust system for a collapsed pipe, heat distress, or possible internal muffler failure.
5. If there are no obvious reasons for the excessive backpressure, the catalytic converter is suspected to be restricted and should be replaced using current recommended procedures.

3.1L (VIN T) AND 3.4L (VIN X) ENGINES — COMPONENT DIAGNOSTIC CHART — CUTLASS SUPREME, GRAND PRIX, LUMINA AND REGAL

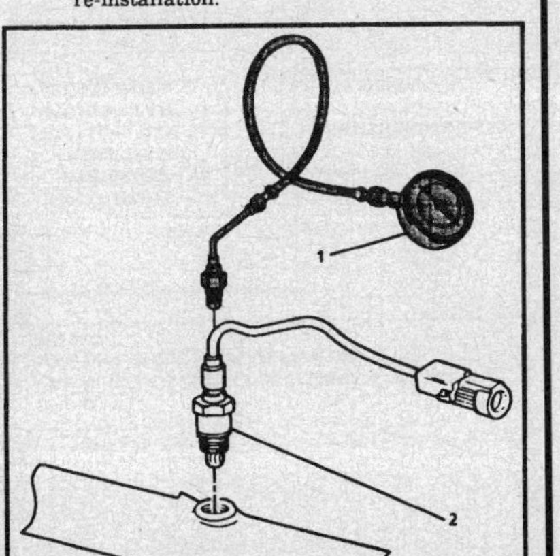

CHART C-1A
PARK/NEUTRAL POSITION (PNP) SWITCH DIAGNOSIS
"W" CARLINE (MFI)

Circuit Description:

The Park/Neutral Position (PNP) switch contacts are a part of the neutral start switch and are closed to ground in park or neutral, and open in drive ranges.

The ECM supplies ignition voltage through a current limiting resistor to CKT 434 and senses a closed switch when the voltage on CKT 434 drops to less than 1 volt.

The ECM uses the PNP signal as one of the inputs to control:

- Idle Air Control (IAC).
- Vehicle Speed Sensor (VSS) diagnostics.
- EGR.
- Torque Converter Clutch (TCC).

If CKT 434 indicates PNP switch is grounded while in a drive range, the EGR would be inoperative, resulting in possible detonation.

If CKT 434 indicates drive (open), when the vehicle is actually in park or neutral, a dip in the idle may exist when the gear selector is moved into a drive range.

Test Description: Number(s) below refer to circled number(s) on the diagnostic chart.

1. Checks for a closed switch to ground in park position. Different makes of scan tools will read PNP differently. Refer to tool operator's manual for type of display used for a specific tool.
2. Checks for an open switch in drive range.
3. Be sure Tech 1 scan tool indicates drive even while wiggling the shifter, to test for an intermittent or misadjusted switch in drive or overdrive range.

3.1L (VIN T) AND 3.4L (VIN X) ENGINES — COMPONENT DIAGNOSTIC CHART — CUTLASS SUPREME, GRAND PRIX, LUMINA AND REGAL

CHART C-1A
PARK/NEUTRAL POSITION (PNP) SWITCH DIAGNOSIS "W" CARLINE (MFI)

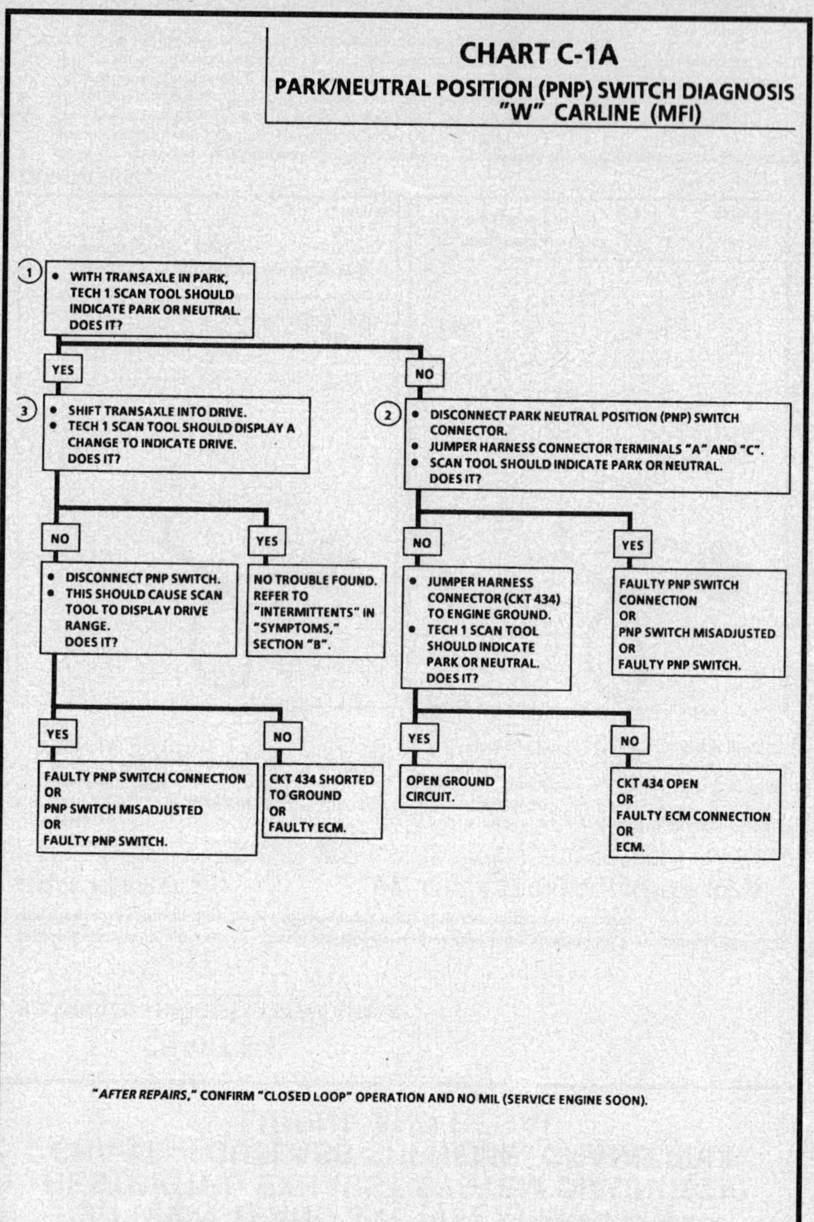

"AFTER REPAIRS," CONFIRM "CLOSED LOOP" OPERATION AND NO MIL (SERVICE ENGINE SOON).

3.1L (VIN T) AND 3.4L (VIN X) ENGINES — COMPONENT DIAGNOSTIC CHART — CUTLASS SUPREME, GRAND PRIX, LUMINA AND REGAL

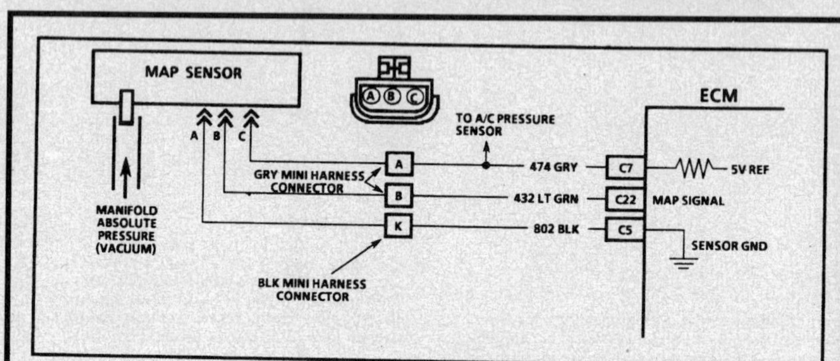

CHART C-1D
MANIFOLD ABSOLUTE PRESSURE (MAP) VOLTAGE OUTPUT CHECK "W" CARLINE

Circuit Description:

The Manifold Absolute Pressure (MAP) sensor measures manifold pressure (vacuum) and sends that signal to the ECM. The MAP sensor is mainly used for fuel calculation when the ECM is running in the throttle body backup mode. The MAP sensor is also used to determine the barometric pressure and to help calculate fuel delivery.

Test Description: Number(s) below refer to circled number(s) on the diagnostic chart.

1. Checks MAP sensor output voltage to the ECM. This voltage, without engine running, represents a barometer reading to the ECM.
2. Applying 34 kPa (10 inch Hg) vacuum to the MAP sensor should cause the voltage to be 1.2 volts less than the voltage at Step 1. Upon applying vacuum to the sensor, the change in voltage should be instantaneous. A slow voltage change indicates a faulty sensor.

The engine must be running in this step or the "scanner" will not indicate a change in voltage. It is normal for the "Service Engine Soon" light to come "ON" and for the system to set a Code 33 during this step. Make sure the code is cleared when this test is completed.

3. Check vacuum hose to sensor for leaking or restriction. Be sure no other vacuum devices are connected to the MAP hose.

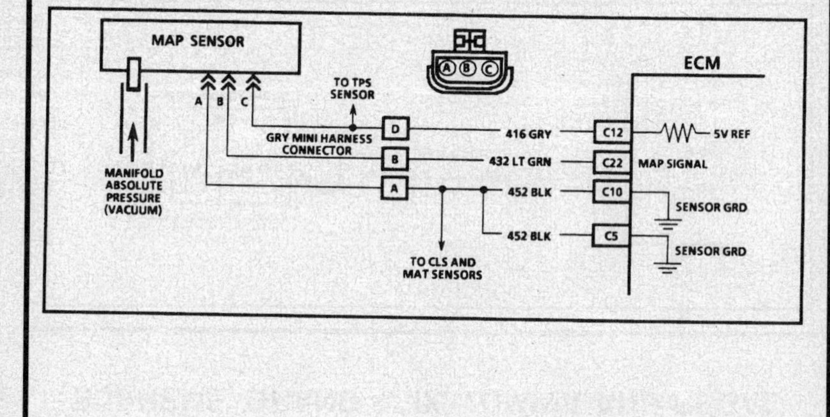

3.1L (VIN T) AND 3.4L (VIN X) ENGINES — COMPONENT DIAGNOSTIC CHART — CUTLASS SUPREME, GRAND PRIX, LUMINA AND REGAL

CHART C-1D
MANIFOLD ABSOLUTE PRESSURE (MAP) OUTPUT CHECK "W" CARLINE

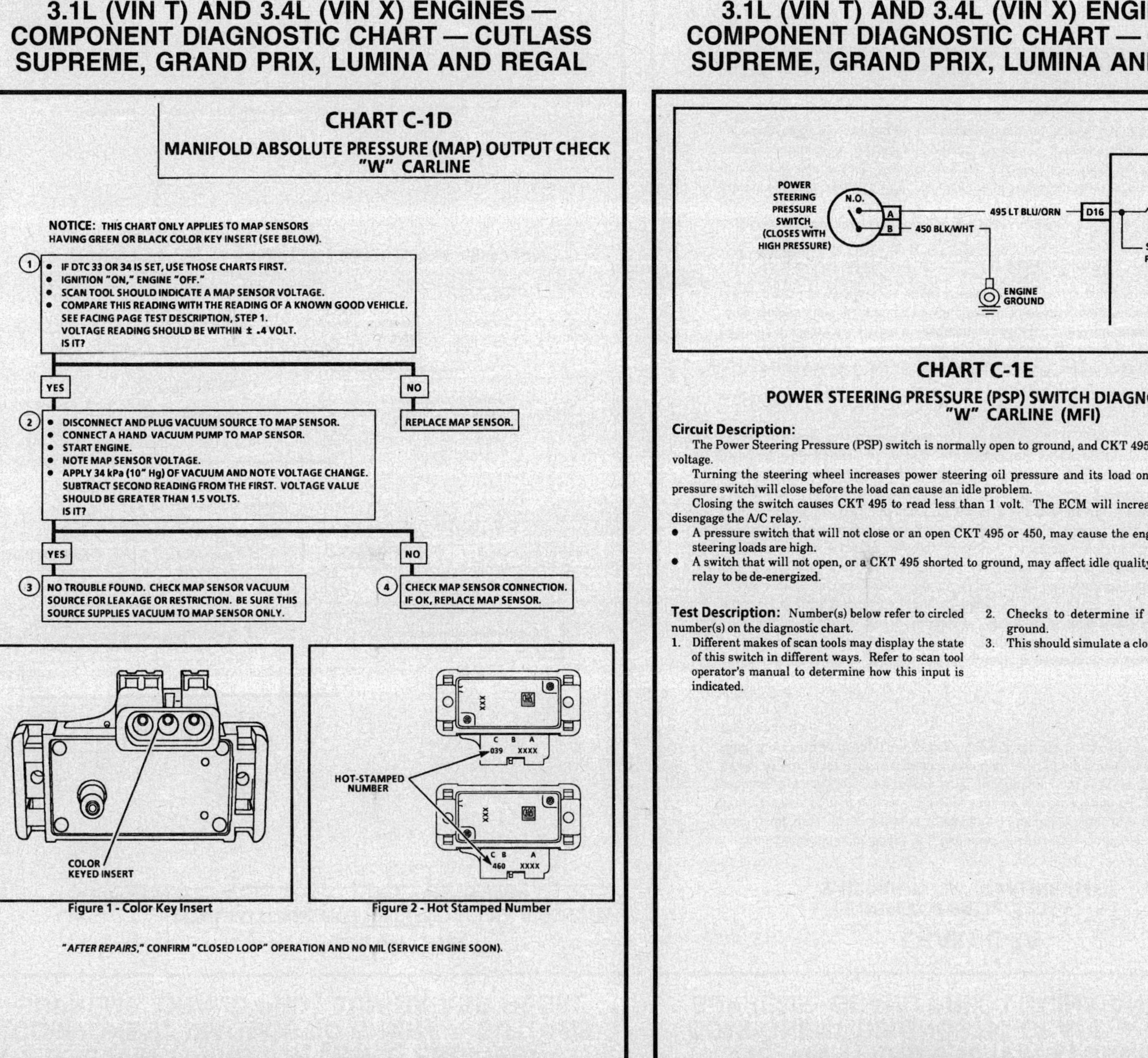

3.1L (VIN T) AND 3.4L (VIN X) ENGINES — COMPONENT DIAGNOSTIC CHART — CUTLASS SUPREME, GRAND PRIX, LUMINA AND REGAL

CHART C-1E
POWER STEERING PRESSURE (PSP) SWITCH DIAGNOSIS "W" CARLINE (MFI)

Circuit Description:

The Power Steering Pressure (PSP) switch is normally open to ground, and CKT 495 will be near the battery voltage.

Turning the steering wheel increases power steering oil pressure and its load on an idling engine. The pressure switch will close before the load can cause an idle problem.

Closing the switch causes CKT 495 to read less than 1 volt. The ECM will increase the idle air rate and disengage the A/C relay.

• A pressure switch that will not close or an open CKT 495 or 450, may cause the engine to stall when power steering loads are high.
• A switch that will not open, or a CKT 495 shorted to ground, may affect idle quality and will cause the A/C relay to be de-energized.

Test Description: Number(s) below refer to circled number(s) on the diagnostic chart.

1. Different makes of scan tools may display the state of this switch in different ways. Refer to scan tool operator's manual to determine how this input is indicated.

2. Checks to determine if CKT 495 is shorted to ground.

3. This should simulate a closed switch.

3.1L (VIN T) AND 3.4L (VIN X) ENGINES — COMPONENT DIAGNOSTIC CHART — CUTLASS SUPREME, GRAND PRIX, LUMINA AND REGAL

CHART C-1E
POWER STEERING PRESSURE (PSP) SWITCH DIAGNOSIS "W" CARLINE (MFI)

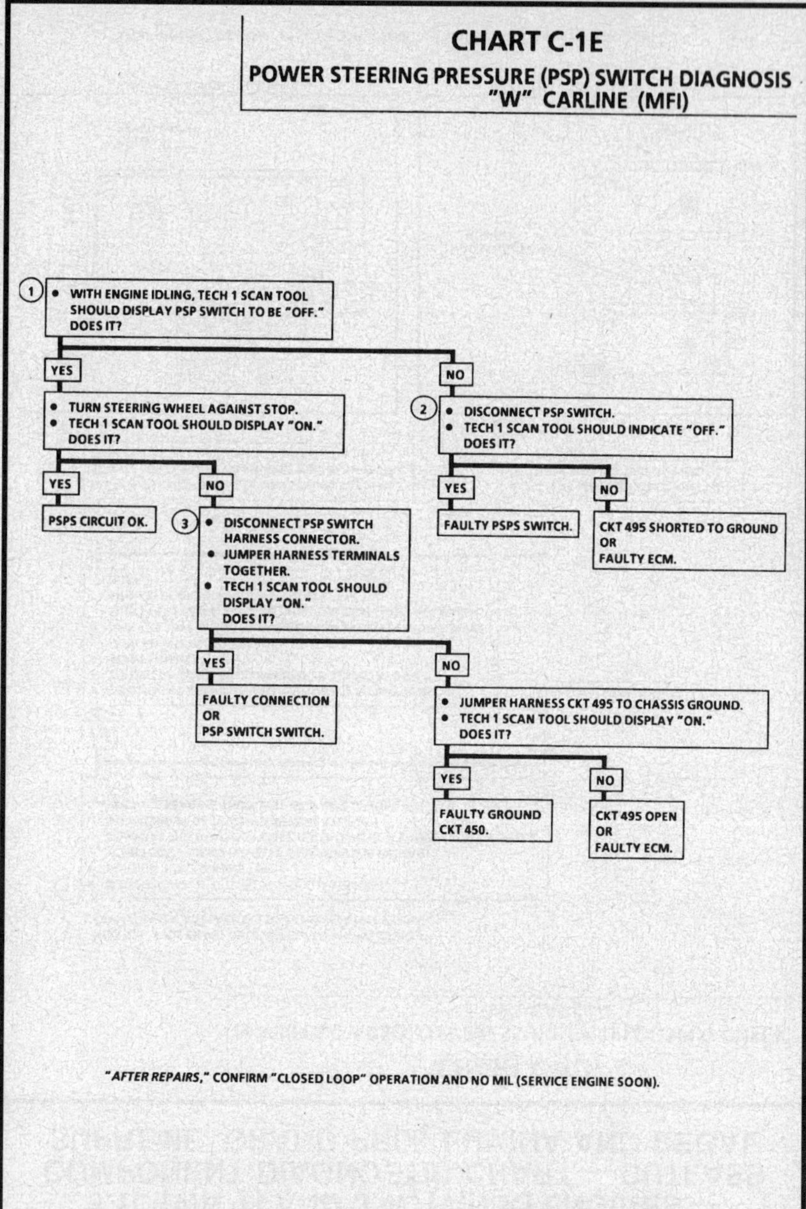

3.1L (VIN T) AND 3.4L (VIN X) ENGINES — COMPONENT DIAGNOSTIC CHART — CUTLASS SUPREME, GRAND PRIX, LUMINA AND REGAL

CHART C-2A
INJECTOR BALANCE TEST
3.1L (VIN T) "W" CARLINE (MFI)

The injector balance tester is a tool used to turn the injector on for a precise amount of time, thus spraying a measured amount of fuel into the manifold. This causes a drop in fuel rail pressure that we can record and compare between each injector. All injectors should have the same amount of pressure drop (± 10 kPa). Any injector with a pressure drop that is 10 kPa (or more) greater or less than the average drop of the other injectors should be considered faulty and replaced.

STEP 1

Engine "cool down" period (10 minutes) is necessary to avoid irregular readings due to "Hot Soak" fuel boiling. With ignition "OFF" connect fuel gauge J 347301 or equivalent to fuel pressure tap. Wrap a shop towel around fitting while connecting gage to avoid fuel spillage.

Disconnect harness connectors at all injectors, and connect injector tester J 34730-3, or equivalent, to one injector. On Turbo equipped engines, use adaptor harness furnished with injector tester to energize injectors that are not accessible. Follow manufacturers instructions for use of adaptor harness. Ignition must be "OFF" at least 10 seconds to complete ECM shutdown cycle. Fuel pump should run about 2 seconds after ignition is turned "ON". At this point, insert clear tubing attached to vent valve into a suitable container and bleed air from gauge and hose to insure accurate gauge operation. Repeat this step until all air is bled from gauge.

STEP 2

Turn ignition "OFF" for 10 seconds and then "ON" again to get fuel pressure to its maximum. Record this initial pressure reading. Energize tester one time and note pressure drop at its lowest point (Disregard any slight pressure increase after drop hits low point.). By subtracting this second pressure reading from the initial pressure, we have the actual amount of injector pressure drop.

STEP 3

Repeat step 2 on each injector and compare the amount of drop. Usually, good injectors will have virtually the same drop. Retest any injector that has a pressure difference of 10 kPa, either more or less than the average of the other injectors on the engine. Replace any injector that also fails the retest. If the pressure drop of all injectors is within 10 kPa of this average, the injectors appear to be flowing properly. Reconnect them and review "Symptoms", Section

NOTE: *The entire test should <u>not</u> be repeated more than once without running the engine to prevent flooding. (This includes any retest on faulty injectors).*

3.1L (VIN T) AND 3.4L (VIN X) ENGINES — COMPONENT DIAGNOSTIC CHART — CUTLASS SUPREME, GRAND PRIX, LUMINA AND REGAL

CHART C-2A
INJECTOR BALANCE TEST
"W" CARLINE
(MFI)

NOTICE: The entire test should <u>NOT</u> be repeated more than once without running the engine to prevent flooding. (This includes any retest on faulty injectors.)

The fuel pressure test in Section "A" Chart A-7, should be completed prior to this test.

Step 1. If engine is at operating temperature, allow a 10 minute "cool down" period then connect fuel pressure gage and injector tester.
1. Ignition "OFF."
2. Connect fuel pressure gage and injector tester.
3. Ignition "ON."
4. Bleed off air in gage. Repeat until all air is bled from gauge.

Step 2. Run test:
1. Ignition "OFF" for 10 seconds.
2. Ignition "ON." Record gage pressure. (Pressure must hold steady, if not, see "Fuel System Diagnosis," CHART A-7, in Section "A").
3. Turn injector on, by depressing button on injector tester, and note pressure at the instant the gauge needle stops.

Step 3.
1. Repeat Step 2 on all injectors and record pressure drop on each. Retest injectors that appear faulty (Any injectors that have a 10 kPa (1.5 psi) difference, either more or less, in pressure from the average). If no problem is found, review "Symptoms" Section

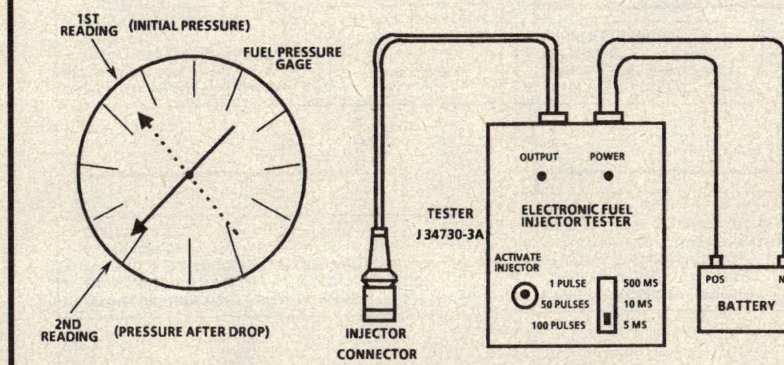

EXAMPLE

CYLINDER	1	2	3
1ST READING	293 kPa (43 psi)	293 kPa (43 psi)	293 kPa (43 psi)
2ND READING	131 kPa (19 psi)	115 kPa (17 psi)	145 kPa (21 psi)
AMOUNT OF DROP	162 kPa (24 psi)	178 kPa (26 psi)	148 kPa (21 psi)
	OK	FAULTY, RICH (TOO MUCH FUEL DROP)	FAULTY, LEAN (TOO LITTLE FUEL DROP)

3.1L (VIN T) AND 3.4L (VIN X) ENGINES — COMPONENT DIAGNOSTIC CHART — CUTLASS SUPREME, GRAND PRIX, LUMINA AND REGAL

CHART C-2B
IDLE AIR CONTROL (IAC) CIRCUIT
"W" CARLINE (MFI)

Circuit Description:

The ECM controls engine idle speed with the IAC valve. To increase idle speed, the ECM retracts the IAC valve pintle away from its seat, allowing more air to bypass the throttle bore. To decrease idle speed, it extends the IAC valve pintle towards its seat, reducing bypass air flow. A Tech 1 scan tool will read the ECM commands to the IAC valve in counts. Higher counts indicate more air bypass (higher idle). The lower the counts indicates less air is allowed to bypass (lower idle).

Test Description: Number(s) below refer to circled number(s) on the diagnostic chart.

1. The Tech 1 RPM control mode is used to extend and retract the IAC valve. The valve should move smoothly within the specified range. If the idle speed is commanded (IAC extended) too low (below 700 RPM), the engine may stall. This may be normal and would not indicate a problem. Retracting the IAC beyond its controlled range (above 1500 RPM) will cause a delay before the RPM's start dropping. This too is normal.

2. This test uses the Tech 1 to command the IAC controlled idle speed. The ECM issues commands to obtain commanded idle speed. The node lights each should flash red and green to indicate a good circuit as the ECM issues commands. While the sequence of color is not important if either light is "OFF" or does not flash red <u>and</u> green, check the circuits for faults, beginning with poor terminal contacts.

Diagnostic Aids:

A slow, unstable, or fast idle may be caused by a non-IAC system problem that cannot be overcome by the IAC valve. Out of control range IAC Tech 1 scan tool counts will be above 60 if idle is too low, and zero counts if idle is too high. The following checks should be made to repair a non-IAC system problem:
- <u>Vacuum Leak (High Idle)</u>.
  If idle is too high, stop the engine. Fully extend (low) IAC with tester. Start engine. If idle speed is above 800 RPM, locate and correct vacuum leak including PCV system. Also check for binding of throttle blade or linkage.

- <u>System too lean (High Air/Fuel Ratio)</u>.
  The idle speed may be too high or too low. Engine speed may vary up and down and disconnecting the IAC valve does not help. DTC 44 may be set. Tech 1 scan tool O2 voltage will be less than 300 mV (.3 volt). Check for low regulated fuel pressure, water in the fuel or a restricted injector.
- <u>System too rich (Low Air/Fuel Ratio)</u>.
  The idle speed will be too low. Tech 1 scan tool IAC counts will usually be above 80. System is obviously rich and may exhibit black smoke in exhaust.
  Tech 1 scan tool O2 voltage will be fixed above 800 mV (.8 volt).
  Check for high fuel pressure, leaking or sticking injector. Silicone contaminated O2 sensors Tech 1 scan tool voltage will be slow to respond.
- <u>Throttle body</u>.
  Remove IAC valve and inspect bore for foreign material.
- <u>IAC Valve Electrical Connections</u>.
  IAC valve connections should be carefully checked for proper contact.
- <u>PCV Valve</u>.
  An incorrect or faulty PCV valve may result in an incorrect idle speed.
- Refer to "Rough, Unstable, Incorrect Idle or Stalling" in "Symptoms," Section
- If intermittent poor driveability or idle symptoms are resolved by disconnecting the IAC, carefully recheck connections, valve terminal resistance, or replace IAC.

3.1L (VIN T) AND 3.4L (VIN X) ENGINES — COMPONENT DIAGNOSTIC CHART — CUTLASS SUPREME, GRAND PRIX, LUMINA AND REGAL

CHART C-2B
IDLE AIR CONTROL (IAC) CIRCUIT "W" CARLINE (MFI)

1.
- INSTALL TECH 1 SCAN TOOL.
- ENGINE AT NORMAL OPERATING TEMPERATURE IN PARK/NEUTRAL WITH PARKING BRAKE SET.
- A/C "OFF."
- SELECT RPM CONTROL. (MISC. TESTS)
- CYCLE IAC THROUGH ITS RANGE FROM 700 RPM UP TO 1500 RPM.
- RPM SHOULD CHANGE SMOOTHLY. DOES IT?

NO →

2.
- INSTALL IAC NODE LIGHT * IN IAC HARNESS.
- ENGINE RUNNING. CYCLE IAC WITH TECH 1 SCAN TOOL.
- EACH NODE LIGHT SHOULD CYCLE RED AND GREEN BUT NEVER "OFF." DO THEY?

NO →

IF CIRCUIT(S) DID NOT TEST RED AND GREEN, CHECK FOR:
- FAULTY CONNECTOR TERMINAL CONTACTS.
- OPEN CIRCUITS INCLUDING CONNECTORS.
- CIRCUITS SHORTED TO GROUND OR VOLTAGE.
- FAULTY ECM CONNECTIONS OR REPLACE ECM.
REPAIR AS NECESSARY AND RETEST.

YES →

- CHECK IAC CONNECTIONS.
- CHECK IAC PASSAGES.
- IF OK, REPLACE IAC.

YES →

- USING THE IAC DRIVER * OR OTHER CONVENIENT CONNECTOR, CHECK RESISTANCE ACROSS IAC COILS.
- SHOULD BE 40 TO 80 OHMS BETWEEN IAC TERMINALS "A" TO "B" AND "C" TO "D".

OK →

- CHECK RESISTANCE BETWEEN IAC TERMINALS "B" AND "C" AND "A" AND "D".
- SHOULD BE INFINITE.

OK →

IDLE AIR CONTROL CIRCUIT OK. REFER TO "DIAGNOSTIC AIDS"

NOT OK →

REPLACE IAC VALVE AND RETEST.

NOT OK →

REPLACE IAC VALVE AND RETEST.

* IAC DRIVER AND NODE LIGHT REQUIRED KIT
222-L FROM: CONCEPT TECHNOLOGY, INC.
J 37027 FROM: KENT-MOORE, INC.

CLEAR DIAGNOSTIC TROUBLE CODES, CONFIRM "CLOSED LOOP" OPERATION, NO MIL (SERVICE ENGINE SOON). PERFORM IAC RESET PROCEDURE PER APPLICABLE SERVICE MANUAL AND VERIFY CONTROLLED IDLE SPEED IS CORRECT.

"AFTER REPAIRS," REFER TO DTC CRITERIA   AND CONFIRM DTC DOES NOT RESET.

3.1L (VIN T) AND 3.4L (VIN X) ENGINES — COMPONENT DIAGNOSTIC CHART — CUTLASS SUPREME, GRAND PRIX, LUMINA AND REGAL

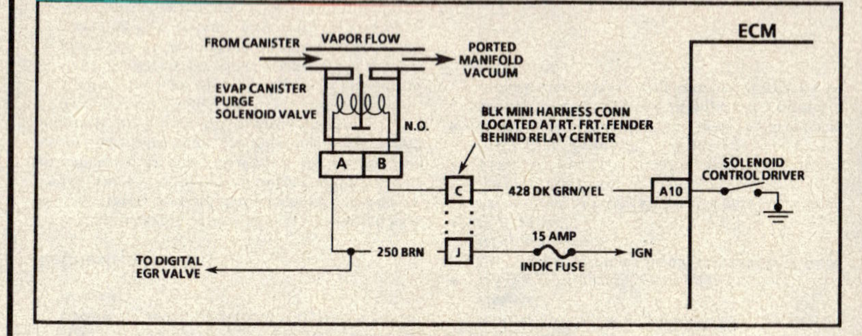

CHART C-3
EVAP CANISTER PURGE SOLENOID VALVE DIAGNOSIS "W" CARLINE (MFI)

Circuit Description:

EVAP canister purge is controlled by a solenoid that allows manifold vacuum to purge the canister when it is de-energized. The ECM supplies a ground to energize the solenoid (purge "OFF"). When the ECM is not commanding purge, the EVAP purge solenoid, controlled by the ECM, is pulse width modulated (turned "ON" and "OFF" several times a second). The duty cycle (pulse width) is determined by the amount of air flow and engine vacuum as determined by the MAP sensor input. The duty cycle is calculated by the ECM and the output commanded when the following conditions have been met:
- Engine run time after start more than 3 minutes.
- Engine coolant temperature above 80°C (176°F).
- Vehicle speed above 15 mph.
- Throttle at idle (about 3%).
Also, if the diagnostic "test" terminal is grounded with the engine stopped, the EVAP purge solenoid is de-energized (purge "ON").

Test Description: Number(s) below refer to circled number(s) on the diagnostic chart.

1. Checks to see if the solenoid is opened or closed. The solenoid is normally energized in this step; so it should be closed.
2. Checks for a complete circuit. Normally there is ignition voltage on CKT 250 and the ECM provides a ground on CKT 428.
3. Completes functional check by grounding "test" terminal. This should normally de-energize the solenoid opening the valve which should allow the vacuum to drop (purge "ON").

Diagnostic Aids:

Normal operation of the EVAP canister purge solenoid is described as follows:

With Ignition "ON," engine "OFF," diagnostic "test" terminal ungrounded, the EVAP canister purge solenoid will be energized.

With ignition "ON," engine "OFF," diagnostic "test" terminal grounded, the EVAP canister purge solenoid valve will be de-energized.

An inoperative EVAP canister purge system can cause a rich Oxygen Sensor (O2S) signal condition DTC 45. If the solenoid valve should become stuck open or the carbon canister become saturated, a continuous rich fuel condition may result.

3.1L (VIN T) AND 3.4L (VIN X) ENGINES — COMPONENT DIAGNOSTIC CHART — CUTLASS SUPREME, GRAND PRIX, LUMINA AND REGAL

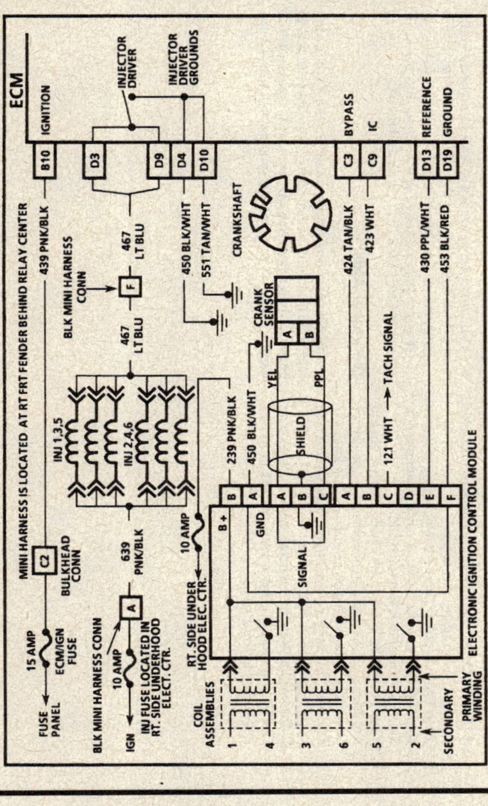

CHART C-4F
"EI" MISFIRE "W" CARLINE (MFI)

Circuit Description:

The Electronic Ignition (EI) system uses a waste spark method of distribution. In this type of system, the electronic ignition control module triggers the #1/4 coil pair resulting in both #1 and #4 spark plugs firing at the same time. #1 cylinder is on the compression stroke at the same time #4 is on the exhaust stroke, resulting in a lower energy requirement to fire #4 spark plug. This leaves the remainder of the high voltage to be used to fire #1 spark plug. On this application, the crank sensor is mounted to the engine block and protrudes through the block to within approximately .050" of the crankshaft reluctor. Since the reluctor is a machined portion of the crankshaft and the crankshaft position sensor is mounted in a fixed position on the block, timing adjustments are not possible or necessary.

Test Description: Number(s) below refer to circled number(s) on the diagnostic chart.

1. Checks for voltage output of ignition system. The spark tester must be used, as this tool requires 25,000 volts to trigger. This checks for a potential weak coil.

2. If the spark tester fires on all wires, the ignition system, with the exception of the spark plugs, may be considered in good working order. If the spark plugs show no evidence of wear, damage or fouling, an engine mechanical fault should be suspected. Refer to "Cuts Out, Misses," in "Symptoms."

3. If the spark jumps the tester gap after grounding the opposite plug wire, it indicates excessive resistance in the plug which was bypassed.

4. If carbon tracking is evident replace coil and be sure plug wires relating to that coil are clean and tight. Excessive wire resistance or faulty connections could have caused the coil to be damaged.

5. If the no spark condition follows the suspected coil, that coil is faulty. Otherwise, the electronic ignition control module is the cause of no spark. This test could also be performed by substituting a known good coil for the one causing the no spark condition.

A faulty or poor connection at that plug could also result in the miss condition. Also, check for carbon deposits inside the spark plug boot.

3.1L (VIN T) AND 3.4L (VIN X) ENGINES — COMPONENT DIAGNOSTIC CHART — CUTLASS SUPREME, GRAND PRIX, LUMINA AND REGAL

CHART C-3
EVAP CANISTER PURGE SOLENOID VALVE DIAGNOSIS "W" CARLINE (MFI)

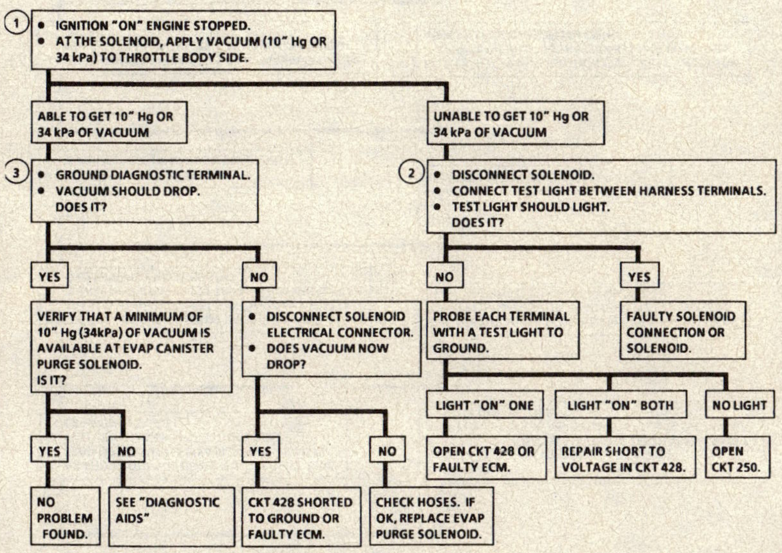

1. • IGNITION "ON" ENGINE STOPPED.
   • AT THE SOLENOID, APPLY VACUUM (10" Hg OR 34 kPa) TO THROTTLE BODY SIDE.

- ABLE TO GET 10" Hg OR 34 kPa OF VACUUM

- UNABLE TO GET 10" Hg OR 34 kPa OF VACUUM →
  2. • DISCONNECT SOLENOID.
     • CONNECT TEST LIGHT BETWEEN HARNESS TERMINALS.
     • TEST LIGHT SHOULD LIGHT. DOES IT?
     - YES → FAULTY SOLENOID CONNECTION OR SOLENOID.
     - NO → PROBE EACH TERMINAL WITH A TEST LIGHT TO GROUND.
       - LIGHT "ON" BOTH → REPAIR SHORT TO VOLTAGE IN CKT 428.
       - LIGHT "ON" ONE → OPEN CKT 428 OR FAULTY ECM.
       - NO LIGHT → OPEN CKT 250.

3. • GROUND DIAGNOSTIC TERMINAL.
   • VACUUM SHOULD DROP. DOES IT?
   - YES → VERIFY THAT A MINIMUM OF 10" Hg (34kPa) OF VACUUM IS AVAILABLE AT EVAP CANISTER PURGE SOLENOID. IS IT?
     - YES → NO PROBLEM FOUND.
     - NO → SEE "DIAGNOSTIC AIDS"
   - NO → DISCONNECT SOLENOID ELECTRICAL CONNECTOR. DOES VACUUM NOW DROP?
     - YES → CKT 428 SHORTED TO GROUND OR FAULTY ECM.
     - NO → CHECK HOSES. IF OK, REPLACE EVAP PURGE SOLENOID.

"AFTER REPAIRS," CONFIRM "CLOSED LOOP" OPERATION AND NO MIL (SERVICE ENGINE SOON).

3.1L (VIN T) AND 3.4L (VIN X) ENGINES — COMPONENT DIAGNOSTIC CHART — CUTLASS SUPREME, GRAND PRIX, LUMINA AND REGAL

CHART C-4F

"EI" MISFIRE
"W" CARLINE (MFI)

1
- IGNITION "OFF."
- INSTALL SPARK TESTER J-26792 (ST-125) OR EQUIV. TO ONE SPARK PLUG WIRE.
- IDLE ENGINE AND CHECK FOR SPARK AT TESTER.
- REPEAT TEST ON ALL PLUG WIRES.
- DOES TESTER SHOW SPARK ON ALL WIRES?

NO → **2** CHECK FOR:
- FAULTY, WORN, OR DAMAGED SPARK PLUG(S).
- PLUG FOULING, DUE TO ENGINE MECHANICAL FAULT.
- IF PLUGS ARE OK, REFER TO SYMPTOMS, SECTION B; CUTS OUT, MISSES.

YES

1 (NO branch):
- IGNITION "OFF," GROUND THE OPPOSITE PLUG LEAD OF THE AFFECTED COIL AT SPARK PLUG. WILL SPARK JUMP TESTER GAP WHILE CRANKING ENGINE?

2 (YES branch): 3 REPLACE THE SPARK PLUG FOR THE LEAD WHICH WAS JUMPERED TO GROUND. IF MISFIRE IS STILL PRESENT, START MISFIRE TEST AGAIN AT STEP # 1.

NO →
- CHECK THE RESISTANCE OF EACH PLUG WIRE OF THE COIL WHICH DID NOT FIRE THE SPARK TESTER. IS WIRE RESISTANCE LESS THAN 30,000 OHMS EACH AND WIRE NOT GROUNDED?

YES

NO → REPLACE FAULTY WIRE(S).

YES
- REMOVE AFFECTED COIL(S). IS COIL(S) FREE OF CARBON TRACKING?

YES

NO → **4** FAULTY ELECTRONIC IGNITION COIL. ALSO CHECK FOR FAULTY PLUG WIRE CONNECTION(S) AND PLUG WIRE NIPPLE(S) FOR CARBON TRACKING.

5
- SWITCH POSITION OF COILS AT PROBLEM CYLINDER. WILL SPARK JUMP TESTER GAP WHILE CRANKING ENGINE?

YES → FAULTY ELECTRONIC IGNITION COIL.

NO → FAULTY ELECTRONIC IGNITION CONTROL MODULE.

"AFTER REPAIRS," CONFIRM "CLOSED LOOP" OPERATION AND NO MIL (SERVICE ENGINE SOON).

3.1L (VIN T) AND 3.4L (VIN X) ENGINES — COMPONENT DIAGNOSTIC CHART — CUTLASS SUPREME, GRAND PRIX, LUMINA AND REGAL

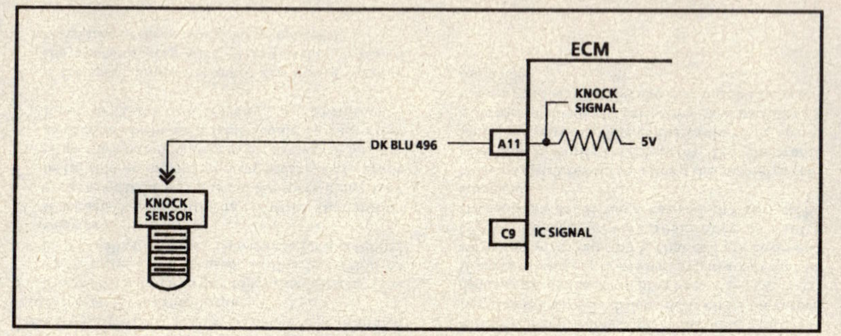

CHART C-5

KNOCK SENSOR (KS) SYSTEM CHECK
"W" CARLINE (MFI)

Circuit Description:

The Knock Sensor (KS) is used to detect engine detonation and the ECM will retard the ignition control timing based on the signal being received. The circuitry within the knock sensor causes the ECM's 5 volts to be pulled down so that under a no knock condition CKT 496 would measure about 2.5 volts. The knock sensor produces an AC signal which rides on the 2.5 volts DC voltage. The amplitude and frequency are dependent upon the knock level.

The PROM used with this engine contains the functions which were part of remotely mounted KS modules used on other GM vehicles. The KS portion of the PROM then sends a signal to other parts of the ECM which adjusts the spark timing to retard the spark and reduce the detonation.

Test Description: Number(s) below refer to circled number(s) on the diagnostic chart.

1. With engine idling, there should not be a knock signal present at the ECM, because detonation is not likely under a no load condition.
2. Tapping on the engine lift hood bracket should simulate a knock signal to determine if the sensor is capable of detecting detonation. If no knock is detected, try tapping on engine block closer to sensor before replacing sensor.
3. If the engine has an internal problem which is creating a knock, the knock sensor may be responding to the internal failure.
4. This test determines if the knock sensor is faulty or if the KS portion of the PROM is faulty. If it is determined that the PROM is faulty, be sure that it is properly installed and latched into place. If not properly installed, repair and retest.

Diagnostic Aids:

While observing knock signal on the Tech 1 scan tool, there should be an indication that knock is present when detonation can be heard. Detonation is most likely to occur under high engine load conditions.

3.1L (VIN T) AND 3.4L (VIN X) ENGINES — COMPONENT DIAGNOSTIC CHART — CUTLASS SUPREME, GRAND PRIX, LUMINA AND REGAL

CHART C-5

KNOCK SENSOR (KS) SYSTEM CHECK "W" CARLINE (MFI)

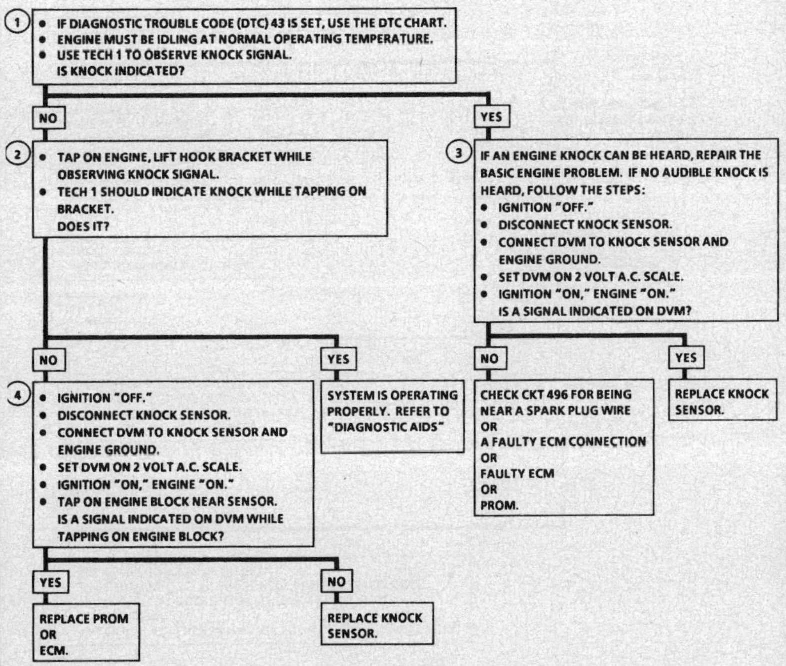

1
• IF DIAGNOSTIC TROUBLE CODE (DTC) 43 IS SET, USE THE DTC CHART.
• ENGINE MUST BE IDLING AT NORMAL OPERATING TEMPERATURE.
• USE TECH 1 TO OBSERVE KNOCK SIGNAL.
IS KNOCK INDICATED?
NO
YES
2
• TAP ON ENGINE, LIFT HOOK BRACKET WHILE OBSERVING KNOCK SIGNAL.
• TECH 1 SHOULD INDICATE KNOCK WHILE TAPPING ON BRACKET.
DOES IT?
3
IF AN ENGINE KNOCK CAN BE HEARD, REPAIR THE BASIC ENGINE PROBLEM. IF NO AUDIBLE KNOCK IS HEARD, FOLLOW THE STEPS:
• IGNITION "OFF."
• DISCONNECT KNOCK SENSOR.
• CONNECT DVM TO KNOCK SENSOR AND ENGINE GROUND.
• SET DVM ON 2 VOLT A.C. SCALE.
• IGNITION "ON," ENGINE "ON."
IS A SIGNAL INDICATED ON DVM?
NO
YES
NO
YES
4
• IGNITION "OFF."
• DISCONNECT KNOCK SENSOR.
• CONNECT DVM TO KNOCK SENSOR AND ENGINE GROUND.
• SET DVM ON 2 VOLT A.C. SCALE.
• IGNITION "ON," ENGINE "ON."
• TAP ON ENGINE BLOCK NEAR SENSOR.
IS A SIGNAL INDICATED ON DVM WHILE TAPPING ON ENGINE BLOCK?
SYSTEM IS OPERATING PROPERLY. REFER TO "DIAGNOSTIC AIDS"
CHECK CKT 496 FOR BEING NEAR A SPARK PLUG WIRE OR A FAULTY ECM CONNECTION OR FAULTY ECM OR PROM.
REPLACE KNOCK SENSOR.
YES
NO
REPLACE PROM OR ECM.
REPLACE KNOCK SENSOR.

"AFTER REPAIRS," CONFIRM "CLOSED LOOP" OPERATION AND NO MIL (SERVICE ENGINE SOON).

3.1L (VIN T) AND 3.4L (VIN X) ENGINES — COMPONENT DIAGNOSTIC CHART — CUTLASS SUPREME, GRAND PRIX, LUMINA AND REGAL

DIVERT SILENCER
AIR CONTROL SOLENOID
A.I.R. pump
CHECK VALVE
TO EXHAUST PORTS
ECM
A
B
IGNITION
GAGE
250 BRN
BLK/PNK 429
A18
AIR SOLENOID CONTROL DRIVER

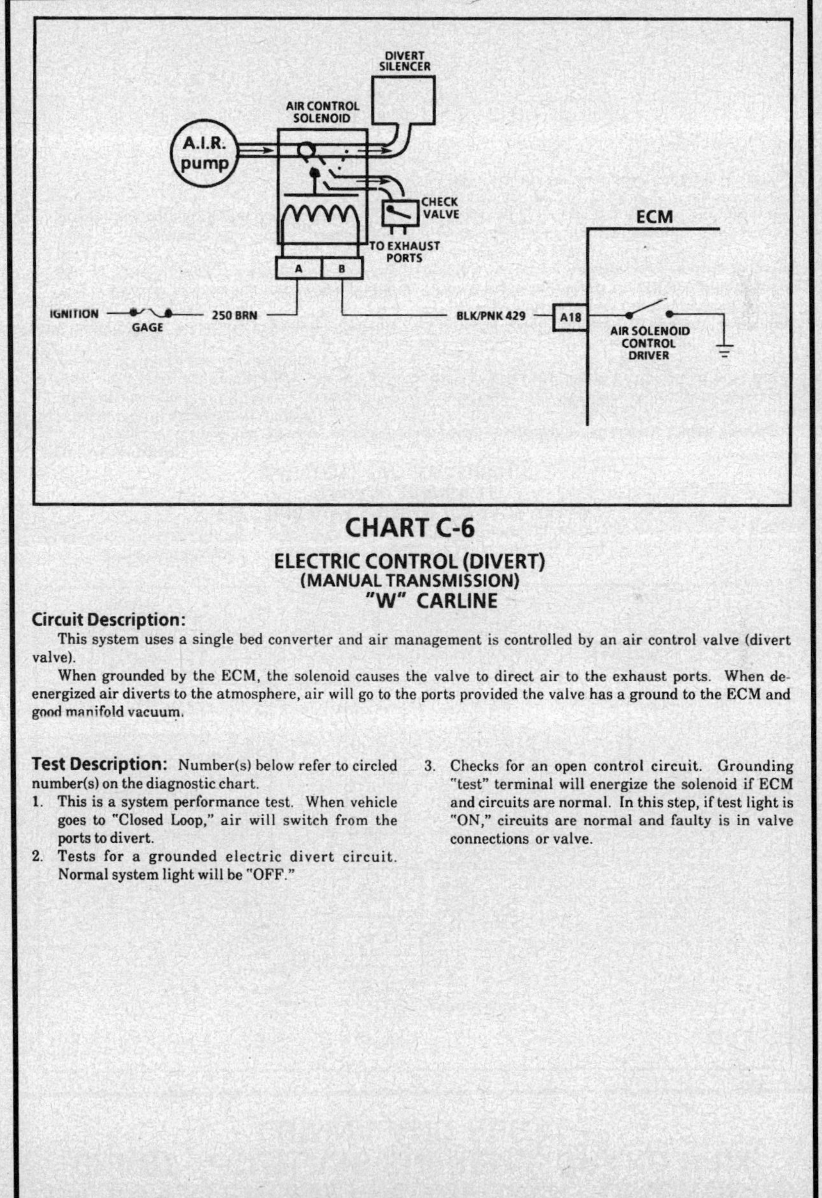

CHART C-6

ELECTRIC CONTROL (DIVERT) (MANUAL TRANSMISSION) "W" CARLINE

Circuit Description:

This system uses a single bed converter and air management is controlled by an air control valve (divert valve).

When grounded by the ECM, the solenoid causes the valve to direct air to the exhaust ports. When de-energized air diverts to the atmosphere, air will go to the ports provided the valve has a ground to the ECM and good manifold vacuum.

Test Description: Number(s) below refer to circled number(s) on the diagnostic chart.

1. This is a system performance test. When vehicle goes to "Closed Loop," air will switch from the ports to divert.
2. Tests for a grounded electric divert circuit. Normal system light will be "OFF."
3. Checks for an open control circuit. Grounding "test" terminal will energize the solenoid if ECM and circuits are normal. In this step, if test light is "ON," circuits are normal and faulty is in valve connections or valve.

3.1L (VIN T) AND 3.4L (VIN X) ENGINES — COMPONENT DIAGNOSTIC CHART — CUTLASS SUPREME, GRAND PRIX, LUMINA AND REGAL

3.4L (VIN X) ENGINE — COMPONENT DIAGNOSTIC CHART — CUTLASS SUPREME, GRAND PRIX, LUMINA AND REGAL

CHART C-6
ELECTRIC CONTROL (DIVERT)
(MANUAL TRANSMISSION)
"W" CARLINE

(1)
- CHECK FOR AT LEAST 34 KPA (10") OF VACUUM AT VALVE WITH ENGINE IDLING.
- RUN ENGINE AT PART THROTTLE (UNDER 2000 RPM).
- AIR SHOULD GO TO EXHAUST PORTS UNTIL SYSTEM GOES "CLOSED LOOP", THEN DIVERT.

NOT OK

OK → NO TROUBLE FOUND

(2)
- IGN "ON", ENGINE STOPPED.
- REMOVE CONNECTOR FROM DIVERT VALVE AND CONNECT A TEST LIGHT BETWEEN CONNECTOR TERMINALS.

LIGHT "OFF"

LIGHT "ON" → CKT 429 SHORTED TO GROUND OR FAULTY ECM.

(3)
- GROUND DIAGNOSTIC TERM.
- NOTE LIGHT.

LIGHT "OFF"

LIGHT "ON" → FAULTY DIVERT VALVE CONNECTIONS OR VALVE.

PROBE EACH TERMINAL WITH A TEST LIGHT TO GROUND.

LIGHT "ON" ONE → OPEN CKT 429 OR FAULTY ECM.

LIGHT "ON" BOTH → REPAIR SHORT TO VOLTAGE IN CKT 429.

NO LIGHT → OPEN CKT 250

"AFTER REPAIRS," CONFIRM "CLOSED LOOP" OPERATION AND NO "SERVICE ENGINE SOON" LIGHT.

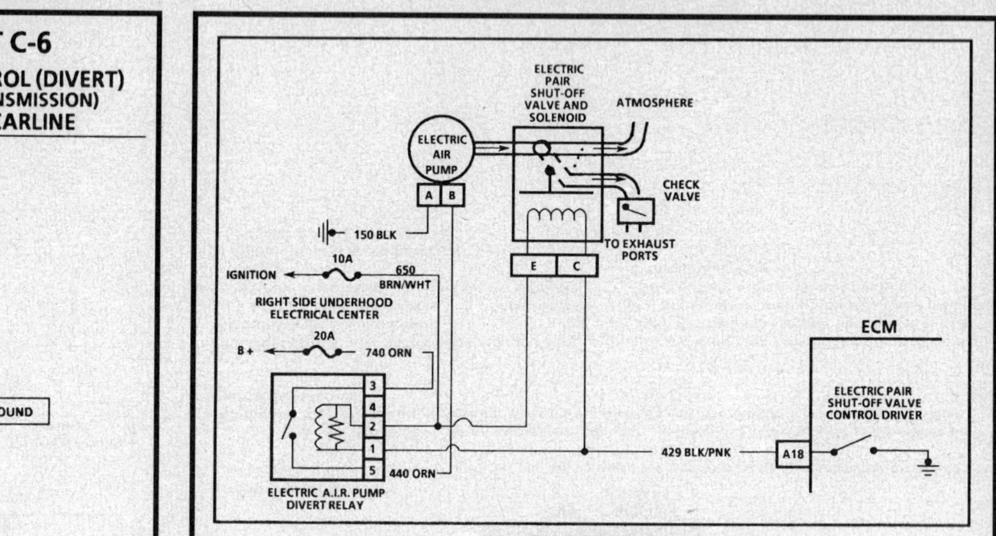

CHART C-6
(Page 1 of 2)
ELECTRIC AIR CONTROL (BYPASS VALVE)
(MANUAL TRANSAXLE)
3.4L (LQ1) "W" CARLINE (MFI)

Circuit Description:

This system uses a single bed converter and air management is controlled by an electric Pulsed Secondary Air Injection (PAIR) system shut-off valve.

When grounded by the ECM, the electric shut-off valve causes the air to direct to the exhaust ports. When de-energized, air diverts to the atmosphere. Air will go to the ports provided the electric PAIR shut-off valve has a ground to the ECM and good manifold vacuum.

Test Description: Number(s) below refer to circled number(s) on the diagnostic chart.

1. This is a system performance test. When vehicle goes to "Closed Loop," air will switch from the ports to divert.

2. Tests for a grounded electric PAIR shut-off valve circuit. In a normal system the test light will be "ON."

3. Checks for an open control circuit. Grounding "test" terminal will energize the solenoid if ECM and circuits are normal. In this step, if test light is "ON," circuits are normal and fault is in valve connections or valve.

4. Checks for an open control circuit. "Field Service Mode" will energize the valve, if the ECM and circuits are normal. In this step, if the test light is "ON," circuits are normal and fault is in the valve.

3.4L (VIN X) ENGINE — COMPONENT DIAGNOSTIC CHART — CUTLASS SUPREME, GRAND PRIX, LUMINA AND REGAL

CHART C-6
(Page 1 of 2)
ELECTRIC AIR CONTROL (BYPASS VALVE) (MANUAL TRANSAXLE)
3.4L (LQ1) "W" CARLINE (MFI)

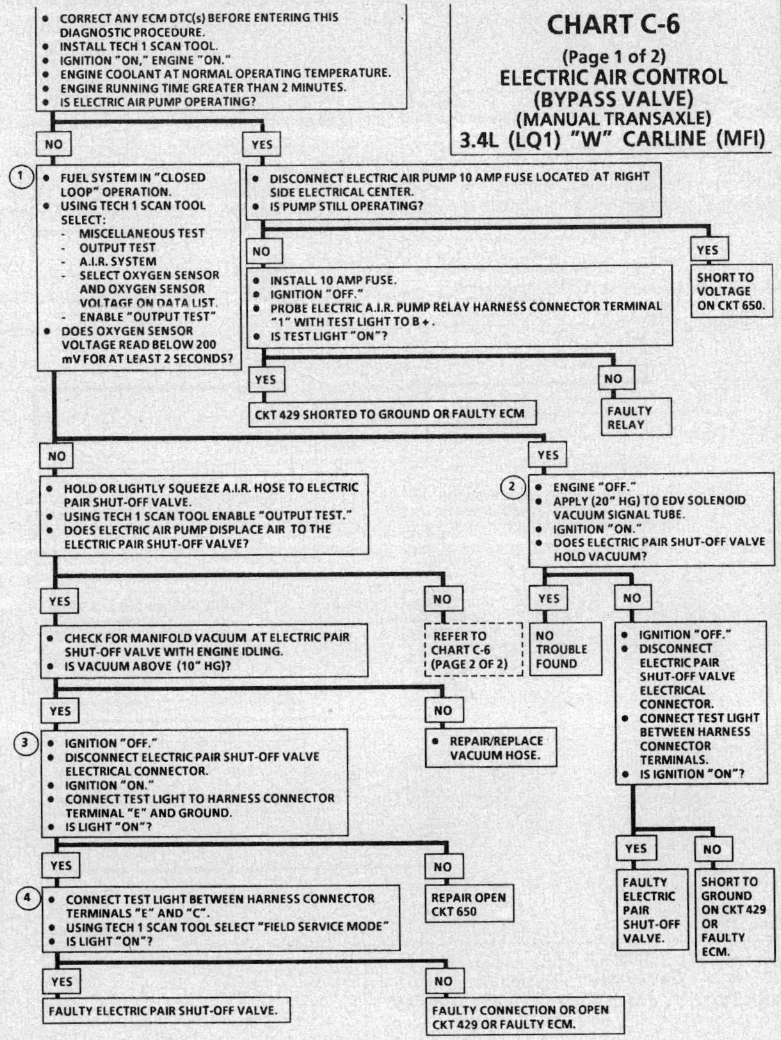

• CORRECT ANY ECM DTC(s) BEFORE ENTERING THIS DIAGNOSTIC PROCEDURE.
• INSTALL TECH 1 SCAN TOOL.
• IGNITION "ON," ENGINE "ON."
• ENGINE COOLANT AT NORMAL OPERATING TEMPERATURE.
• ENGINE RUNNING TIME GREATER THAN 2 MINUTES.
• IS ELECTRIC AIR PUMP OPERATING?
NO
YES
1
• FUEL SYSTEM IN "CLOSED LOOP" OPERATION.
• USING TECH 1 SCAN TOOL SELECT:
- MISCELLANEOUS TEST
- OUTPUT TEST
- A.I.R. SYSTEM
- SELECT OXYGEN SENSOR AND OXYGEN SENSOR VOLTAGE ON DATA LIST.
- ENABLE "OUTPUT TEST"
• DOES OXYGEN SENSOR VOLTAGE READ BELOW 200 mV FOR AT LEAST 2 SECONDS?
• DISCONNECT ELECTRIC AIR PUMP 10 AMP FUSE LOCATED AT RIGHT SIDE ELECTRICAL CENTER.
• IS PUMP STILL OPERATING?
NO
YES
SHORT TO VOLTAGE ON CKT 650.
• INSTALL 10 AMP FUSE.
• IGNITION "OFF."
• PROBE ELECTRIC A.I.R. PUMP RELAY HARNESS CONNECTOR TERMINAL "1" WITH TEST LIGHT TO B+.
• IS TEST LIGHT "ON"?
YES
NO
CKT 429 SHORTED TO GROUND OR FAULTY ECM
FAULTY RELAY
NO
YES
• HOLD OR LIGHTLY SQUEEZE A.I.R. HOSE TO ELECTRIC PAIR SHUT-OFF VALVE.
• USING TECH 1 SCAN TOOL ENABLE "OUTPUT TEST."
• DOES ELECTRIC AIR PUMP DISPLACE AIR TO THE ELECTRIC PAIR SHUT-OFF VALVE?
2
• ENGINE "OFF."
• APPLY (20" HG) TO EDV SOLENOID VACUUM SIGNAL TUBE.
• IGNITION "ON."
• DOES ELECTRIC PAIR SHUT-OFF VALVE HOLD VACUUM?
YES
NO
YES
NO
REFER TO CHART C-6 (PAGE 2 OF 2)
NO TROUBLE FOUND
• IGNITION "OFF."
• DISCONNECT ELECTRIC PAIR SHUT-OFF VALVE ELECTRICAL CONNECTOR.
• CONNECT TEST LIGHT BETWEEN HARNESS CONNECTOR TERMINALS.
• IS IGNITION "ON"?
• CHECK FOR MANIFOLD VACUUM AT ELECTRIC PAIR SHUT-OFF VALVE WITH ENGINE IDLING.
• IS VACUUM ABOVE (10" HG)?
YES
NO
REPAIR/REPLACE VACUUM HOSE.
3
• IGNITION "OFF."
• DISCONNECT ELECTRIC PAIR SHUT-OFF VALVE ELECTRICAL CONNECTOR.
• IGNITION "ON."
• CONNECT TEST LIGHT TO HARNESS CONNECTOR TERMINAL "E" AND GROUND.
• IS LIGHT "ON"?
YES
NO
REPAIR OPEN CKT 650
YES
NO
FAULTY ELECTRIC PAIR SHUT-OFF VALVE.
SHORT TO GROUND ON CKT 429 OR FAULTY ECM.
4
• CONNECT TEST LIGHT BETWEEN HARNESS CONNECTOR TERMINALS "E" AND "C".
• USING TECH 1 SCAN TOOL SELECT "FIELD SERVICE MODE"
• IS LIGHT "ON"?
YES
NO
FAULTY ELECTRIC PAIR SHUT-OFF VALVE.
FAULTY CONNECTION OR OPEN CKT 429 OR FAULTY ECM.

3.4L (VIN X) ENGINE — COMPONENT DIAGNOSTIC CHART — CUTLASS SUPREME, GRAND PRIX, LUMINA AND REGAL

ELECTRIC PAIR SHUT-OFF VALVE AND SOLENOID
ATMOSPHERE
ELECTRIC AIR PUMP
A
B
CHECK VALVE
TO EXHAUST PORTS
150 BLK
10A
650 BRN/WHT
IGNITION
E
C
RIGHT SIDE UNDERHOOD ELECTRICAL CENTER
20A
B+
740 ORN
ECM
3
4
2
1
5
440 ORN
ELECTRIC PAIR SHUT-OFF VALVE CONTROL DRIVER
429 BLK/PNK
A18
ELECTRIC A.I.R. PUMP DIVERT RELAY

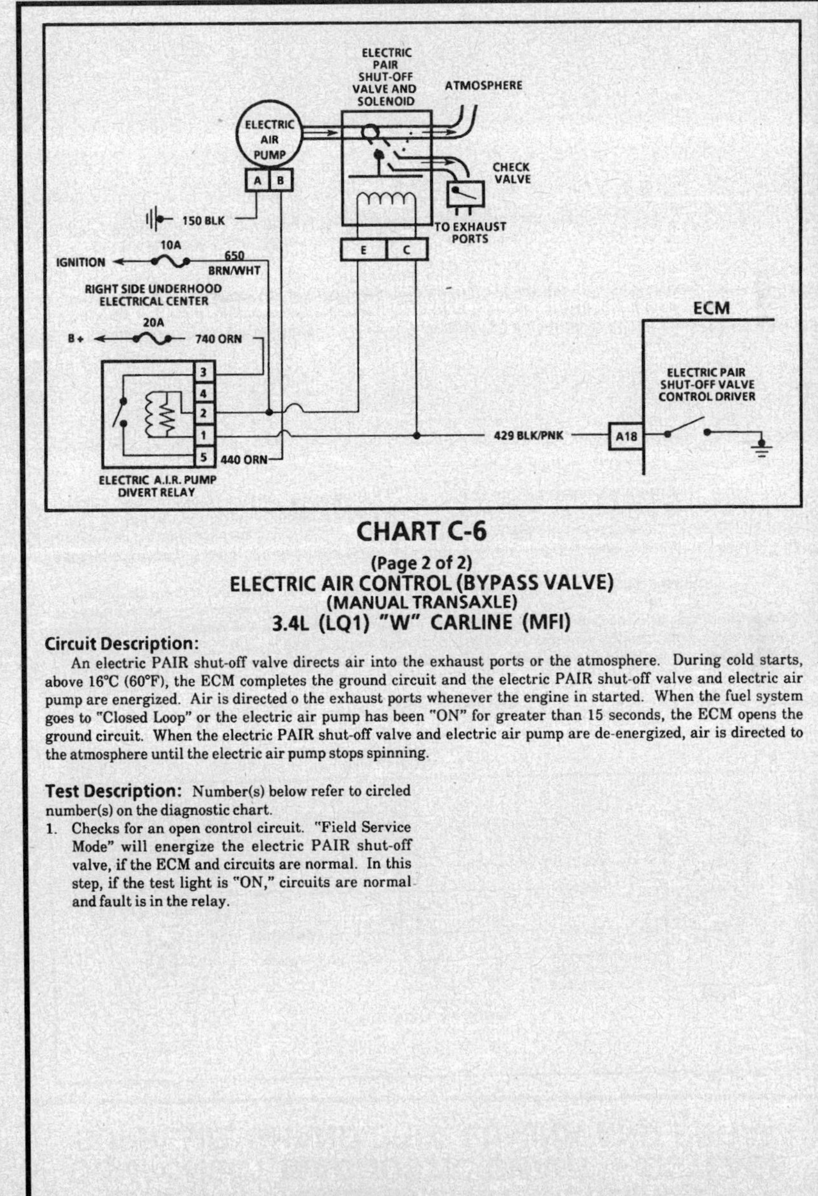

CHART C-6
(Page 2 of 2)
ELECTRIC AIR CONTROL (BYPASS VALVE)
(MANUAL TRANSAXLE)
3.4L (LQ1) "W" CARLINE (MFI)

Circuit Description:

An electric PAIR shut-off valve directs air into the exhaust ports or the atmosphere. During cold starts, above 16°C (60°F), the ECM completes the ground circuit and the electric PAIR shut-off valve and electric air pump are energized. Air is directed o the exhaust ports whenever the engine in started. When the fuel system goes to "Closed Loop" or the electric air pump has been "ON" for greater than 15 seconds, the ECM opens the ground circuit. When the electric PAIR shut-off valve and electric air pump are de-energized, air is directed to the atmosphere until the electric air pump stops spinning.

Test Description: Number(s) below refer to circled number(s) on the diagnostic chart.

1. Checks for an open control circuit. "Field Service Mode" will energize the electric PAIR shut-off valve, if the ECM and circuits are normal. In this step, if the test light is "ON," circuits are normal and fault is in the relay.

3.4L (VIN X) ENGINE — COMPONENT DIAGNOSTIC CHART — CUTLASS SUPREME, GRAND PRIX, LUMINA AND REGAL

3.1L (VIN T) AND 3.4L (VIN X) ENGINES — COMPONENT DIAGNOSTIC CHART — CUTLASS SUPREME, GRAND PRIX, LUMINA AND REGAL

CHART C-6

(Page 2 of 2)
ELECTRIC AIR CONTROL (BYPASS VALVE)
(MANUAL TRANSAXLE)
3.4L (LQ1) "W" CARLINE (MFI)

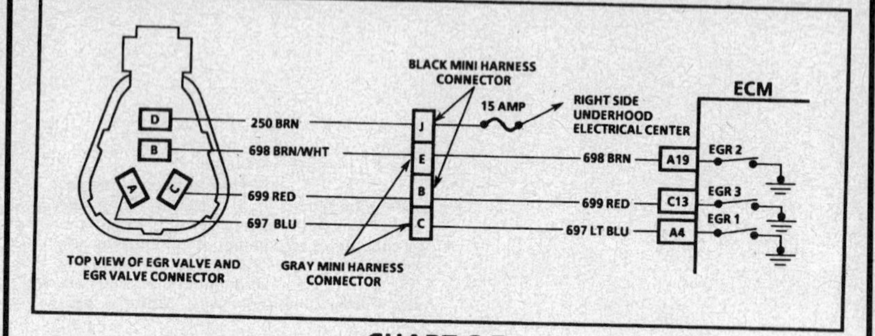

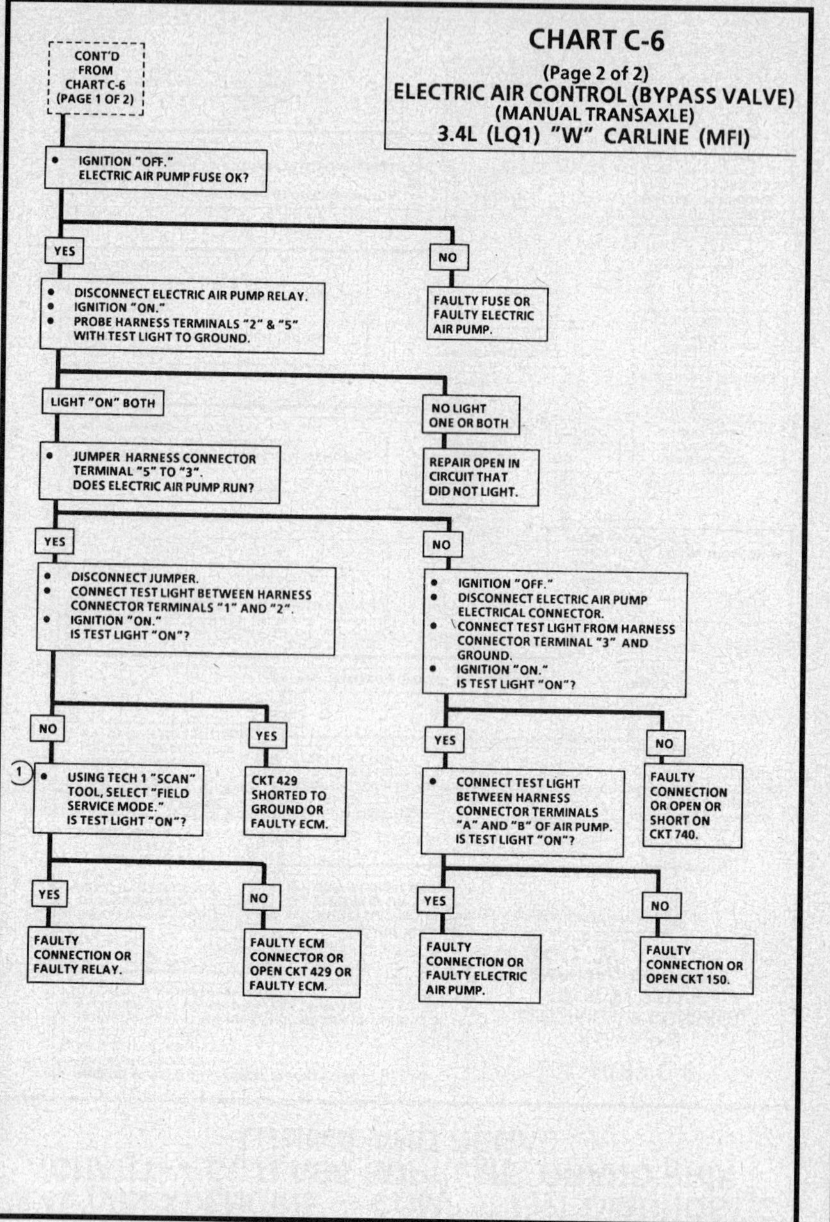

CHART C-7

EXHAUST GAS RECIRCULATION (EGR) FLOW CHECK
"W" CARLINE

Circuit Description:

The digital (EGR) valve is designed to accurately supply EGR to an engine independent of intake manifold vacuum. The valve controls EGR flow from the exhaust to the intake manifold through three orifices which increment in size to produce seven combinations. When a solenoid is energized, the armature with attached shaft and swivel pintle is lifted, opening the orifice.

The flow accuracy is dependent on metering orifice size only, which results in improved control.

Test Description: Number(s) below refer to circled number(s) on the diagnostic chart.

1. DTC(s) should be diagnosed using appropriate charts before preparing a functional check.

2. This step activates each solenoid individually. As you energize #1 or #2 solenoid, the engine RPM should drop. #3 solenoid has the largest port and may stall the engine when energized.

Important

• If the digital EGR valve shows signs of excessive heat (a melted condition), check the exhaust system for blockage (possibly a plugged converter) using the procedure found on CHART B-1. If the exhaust system is restricted, repair the cause. One possibility is an injector which is open due to one of the following:

A. Stuck.
B. Grounded driver circuit.
C. Possibly defective ECM.

If this condition is found, the oil should be checked for possible fuel contamination.

3.1L (VIN T) AND 3.4L (VIN X) ENGINES — COMPONENT DIAGNOSTIC CHART — CUTLASS SUPREME, GRAND PRIX, LUMINA AND REGAL

CHART C-7

EXHAUST GAS RECIRCULATION (EGR) FLOW CHECK
"W" CARLINE

(1)
- IGNITION "ON," ENGINE "OFF."
- CHECK TECH 1 SCAN TOOL FOR DIAGNOSTIC TROUBLE CODES.
- IF DIAGNOSTIC TROUBLE CODES ARE PRESENT, REFER TO THOSE CHARTS FIRST.

(2)
- USING TECH 1 SCAN TOOL, SELECT EGR CONTROL (MISCELLANEOUS).
- START ENGINE AND ALLOW IDLE TO STABILIZE.
- ENERGIZE EGR SOL #1, RPM SHOULD DROP SLIGHTLY.*
- ENERGIZE EGR SOL #2, ENGINE SHOULD HAVE A ROUGH IDLE.*
- ENERGIZE EGR SOL #3, ENGINE SHOULD IDLE ROUGH OR STALL.*
- DID RPM DROP AND ENGINE IDLE ROUGH ON ALL THREE SOLENOIDS?

YES → EGR SYSTEM IS OK. NO PROBLEM FOUND.

NO → CHECK EGR VALVE, EGR TRANSFER TUBE, ADAPTOR, GASKETS, FITTINGS AND ALL PASSAGES FOR DAMAGE, LEAKAGE OR PLUGGING. IF OK, REPLACE EGR VALVE.

* THESE STEPS MUST BE DONE VERY QUICKLY, AS THE ECM WILL ADJUST THE IDLE AIR CONTROL VALVE TO CORRECT IDLE SPEED.

"AFTER REPAIRS," CONFIRM "CLOSED LOOP" OPERATION AND NO MIL (SERVICE ENGINE SOON).

3.1L (VIN T) AND 3.4L (VIN X) ENGINES — COMPONENT DIAGNOSTIC CHART — CUTLASS SUPREME, GRAND PRIX, LUMINA AND REGAL

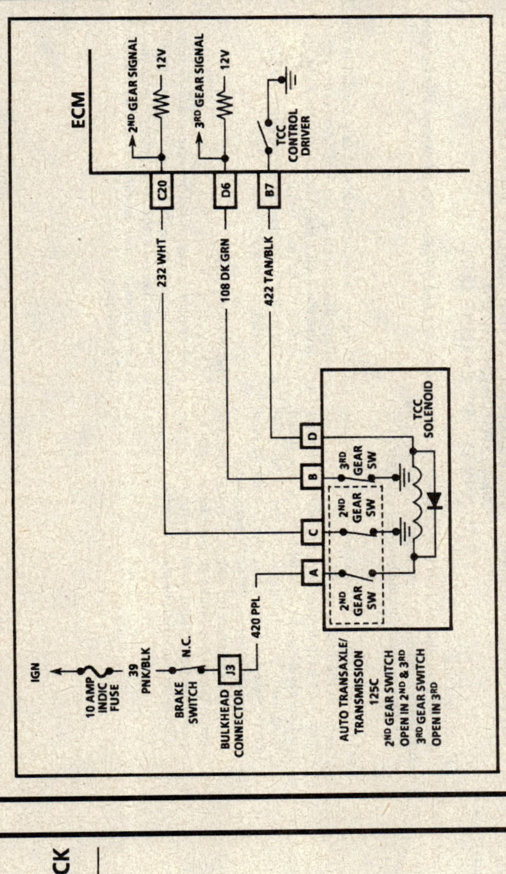

CHART C-8A (Page 1 of 2)

3T40 TORQUE CONVERTER CLUTCH (TCC) WITH TORQUE MANAGEMENT
(ELECTRICAL DIAGNOSIS)
3.1L (VIN T) "W" CARLINE (MFI)

Circuit Description:

The purpose of the automatic transmission torque converter clutch feature is to eliminate the power loss of the torque converter when the vehicle is in a cruise condition. This allows the convenience of the automatic transmission and the fuel economy of a manual transmission. The heart of the system is a solenoid located inside the automatic transmission which is controlled by the ECM.

When the solenoid coil is activated ("ON"), the Torque Converter Clutch (TCC) is applied which results in straight through mechanical coupling from the engine to transmission. When the transmission solenoid is deactivated, the Torque Converter Clutch (TCC) is released, which allows the torque converter to operate in the conventional manner (fluidic coupling between engine and transmission).

The TCC will engage on a warm engine under given road load in 2nd and 3rd gears.

TCC will engage when:
- Brake switch closed.
- Engine coolant temperature is above 65°C (149°F).
- Engine "Closed Loop" condition.
- Vehicle speed above a calibrated value (about 28 mph 45 km/h).
- Throttle position not changing, indicating a steady speed.

Test Description: Number(s) below refer to circled number(s) on the diagnostic chart.

1. This test checks the functional operation of the TCC circuit.
2. This test checks the TCC control driver in the ECM.
3. This test will confirm that there is B+ to terminal "A".
4. This test confirms that the ECM has the ability to turn the TCC "ON."

Diagnostic Aids:

A Tech 1 scan tool only indicates when the ECM has turned "ON" the TCC driver, and this does not confirm that the TCC has engaged. To determine if TCC is functioning properly, engine RPM should decrease when the Tech 1 scan tool indicates the TCC driver has turned "ON." A thermostat that is stuck open or opens at too low a temperature may result in an inoperative TCC.

3.1L (VIN T) AND 3.4L (VIN X) ENGINES — COMPONENT DIAGNOSTIC CHART — CUTLASS SUPREME, GRAND PRIX, LUMINA AND REGAL

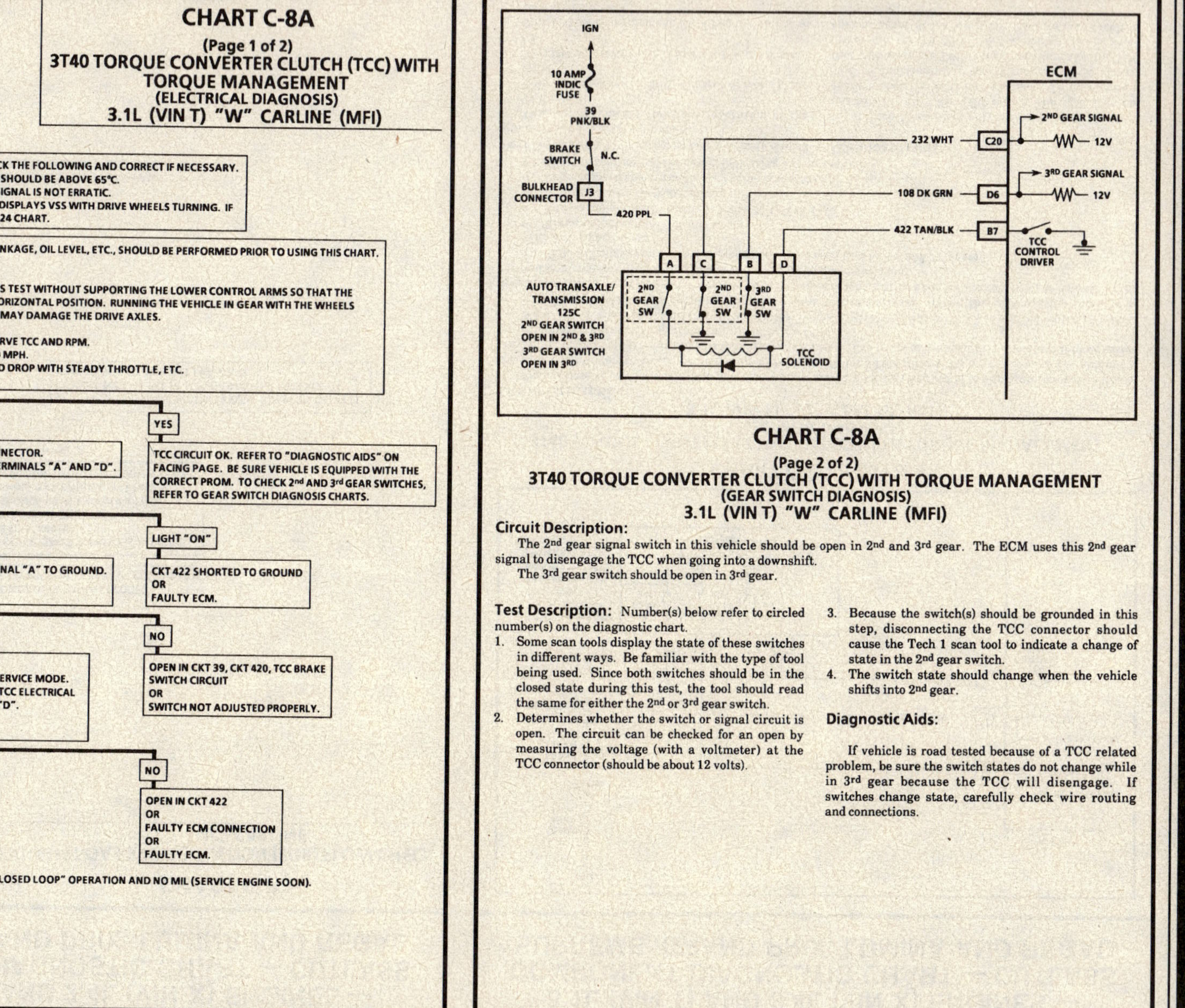

Circuit Description:

The 2nd gear signal switch in this vehicle should be open in 2nd and 3rd gear. The ECM uses this 2nd gear signal to disengage the TCC when going into a downshift.

The 3rd gear switch should be open in 3rd gear.

Test Description: Number(s) below refer to circled number(s) on the diagnostic chart.

1. Some scan tools display the state of these switches in different ways. Be familiar with the type of tool being used. Since both switches should be in the closed state during this test, the tool should read the same for either the 2nd or 3rd gear switch.

2. Determines whether the switch or signal circuit is open. The circuit can be checked for an open by measuring the voltage (with a voltmeter) at the TCC connector (should be about 12 volts).

3. Because the switch(s) should be grounded in this step, disconnecting the TCC connector should cause the Tech 1 scan tool to indicate a change of state in the 2nd gear switch.

4. The switch state should change when the vehicle shifts into 2nd gear.

Diagnostic Aids:

If vehicle is road tested because of a TCC related problem, be sure the switch states do not change while in 3rd gear because the TCC will disengage. If switches change state, carefully check wire routing and connections.

3.1L (VIN T) AND 3.4L (VIN X) ENGINES — COMPONENT DIAGNOSTIC CHART — CUTLASS SUPREME, GRAND PRIX, LUMINA AND REGAL

CHART C-8A
(Page 2 of 2)
3T40 TORQUE CONVERTER CLUTCH (TCC) WITH TORQUE MANAGEMENT
(GEAR SWITCH DIAGNOSIS)
3.1L (VIN T) "W" CARLINE (MFI)

CHECKS MADE ON THIS PAGE WILL NOT PREVENT THE TCC FROM WORKING, BUT WILL AFFECT ENGAGEMENT OR DISENGAGEMENT POINTS.

(1)
- IGNITION "ON," ENGINE "OFF."
- DOES TECH 1 SCAN TOOL INDICATE TRANS. IS IN 2ND GEAR?

NO →
- DOES TECH 1 SCAN TOOL INDICATE VEHICLE IS IN 3RD GEAR?

YES → (2)
- DISCONNECT TCC ELECTRICAL CONNECTOR.
- JUMPER HARNESS TERMINAL "C" (CKT 232) TO GROUND.
- DOES TECH 1 SCAN TOOL INDICATE TRANS. IS IN 2ND GEAR?

NO → FAULTY CONNECTION OR 2ND GEAR SWITCH.

YES → OPEN CKT 232, FAULTY CONNECTION OR ECM.

NO → (3)
- DISCONNECT TCC ELECTRICAL CONNECTOR.
- DOES TECH 1 SCAN TOOL INDICATE TRANS. IS IN 2ND GEAR?

YES → (2)
- DISCONNECT TCC ELECTRICAL CONNECTOR.
- JUMPER HARNESS TERMINAL "B" (CKT 108) TO GROUND.
- DOES TECH 1 SCAN TOOL INDICATE TRANS. IS IN 3RD GEAR?

NO → FAULTY CONNECTION OR 3RD GEAR SWITCH.

YES → OPEN CKT 108, FAULTY CONNECTION, OR ECM.

- DOES TECH 1 SCAN TOOL INDICATE TRANS. IS IN 3RD GEAR?

NO → CKT 232 SHORTED TO GROUND OR FAULTY ECM.

YES → (4)
- RECONNECT TCC ELECTRICAL CONNECTOR.
- RAISE DRIVE WHEELS.

NOTICE: *DO NOT PERFORM THIS TEST WITHOUT SUPPORTING THE LOWER CONTROL ARMS SO THAT THE DRIVE AXLES ARE IN A NORMAL HORIZONTAL POSITION. RUNNING THE VEHICLE IN GEAR WITH THE WHEELS HANGING DOWN AT FULL TRAVEL MAY DAMAGE THE DRIVE AXLES.*

- START AND IDLE ENGINE IN OVERDRIVE.
- INCREASE SPEED SLOWLY UNTIL TRANS. SHIFTS INTO 2ND GEAR.
- DOES TECH 1 SCAN TOOL INDICATE TRANS. IS IN 2ND GEAR?

NO → CKT 108 SHORTED TO GROUND OR FAULTY ECM.

YES →
- INCREASE SPEED UNTIL TRANS. SHIFTS INTO 3RD GEAR.
- DOES TECH 1 SCAN TOOL INDICATE TRANS. IS IN 3RD GEAR?

NO → FAULTY 2ND GEAR SWITCH.

YES → TRANS. SWITCHES OK. CHECK WIRE ROUTING AND CONNECTIONS. REFER TO "DIAGNOSTIC AIDS" ON FACING PAGE.

NO → FAULTY 3RD GEAR SWITCH.

"AFTER REPAIRS," CONFIRM "CLOSED LOOP" OPERATION AND NO MIL (SERVICE ENGINE SOON).

3.1L (VIN T) AND 3.4L (VIN X) ENGINES — COMPONENT DIAGNOSTIC CHART — 1993–94 CUTLASS SUPREME, GRAND PRIX, LUMINA AND REGAL

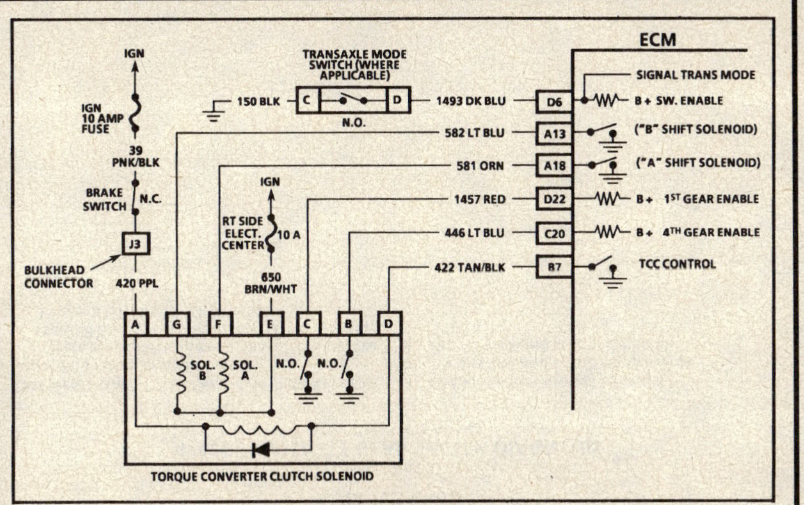

CHART C-8B
(Page 1 of 3)
4T60E TORQUE CONVERTER CLUTCH (TCC) WITH TORQUE MANAGEMENT
(ELECTRICAL DIAGNOSIS)
"W" CARLINE

Circuit Description:

The purpose of the automatic transaxle torque converter clutch feature is to eliminate the power loss of the torque converter when the vehicle is in a cruise condition. This allows the convenience of the automatic transaxle and the fuel economy of a manual transaxle. The heart of the system is a solenoid located inside the automatic transaxle which is controlled by the ECM.

When the solenoid coil is activated ("ON"), the Torque Converter Clutch (TCC) is applied which results in straight through mechanical coupling from the engine to transaxle. When the TCC solenoid is deactivated, the torque converter clutch is released, which allows the torque converter to operate in the conventional manner (fluid coupling between engine and transaxle).

The TCC will engage on a warm engine under given road load in 2nd, 3rd and 4th gears. TCC will engage when:

- Engine warmed up and in "Closed Loop" condition.
- Vehicle speed above a calibrated value (about 45 mph/72km/h).
- Throttle position not changing, indicating a steady road speed.
- Brake switch closed.

Test Description: Number(s) below refer to circled number(s) on the diagnostic chart.

1. This test checks the functional operation of the TCC circuit.
2. This test determines if there is B+ at terminal "A". If there is no B+, check 10 amp (IGN) fuse.
3. This test checks the TCC control driver in the ECM.
4. This test confirms that the ECM has the ability to turn the TCC "ON."

Diagnostic Aids:

A Tech 1 scan tool only indicates when the ECM has turned "ON" the TCC driver, and this does not confirm that the TCC has engaged. To determine if TCC is functioning properly, engine RPM should decrease when the Tech 1 scan tool indicates the TCC driver has turned "ON."

3.1L (VIN T) AND 3.4L (VIN X) ENGINES — COMPONENT DIAGNOSTIC CHART — 1993–94 CUTLASS SUPREME, GRAND PRIX, LUMINA AND REGAL

3.1L (VIN T) AND 3.4L (VIN X) ENGINES — COMPONENT DIAGNOSTIC CHART — 1993–94 CUTLASS SUPREME, GRAND PRIX, LUMINA AND REGAL

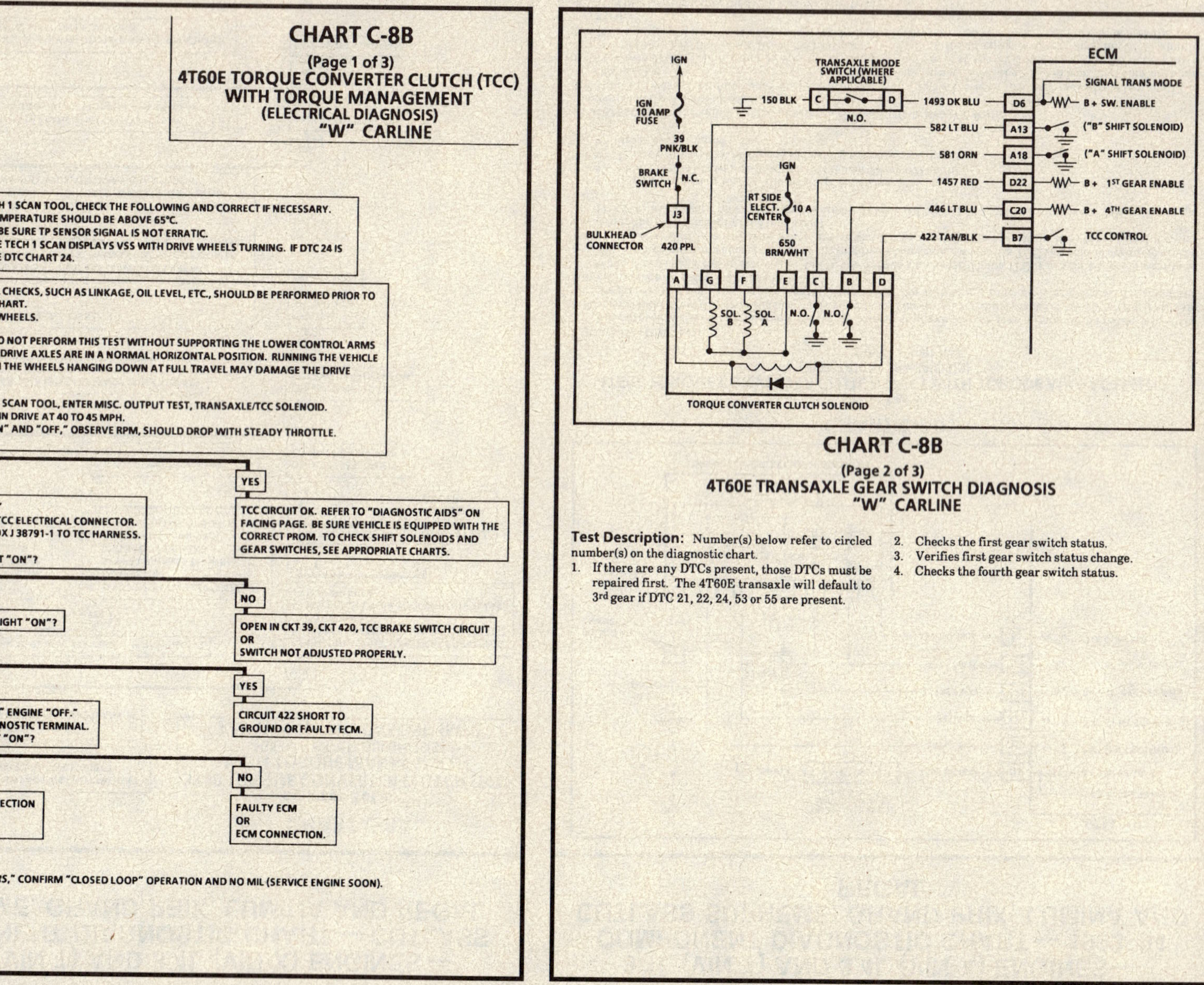

3.1L (VIN T) AND 3.4L (VIN X) ENGINES — COMPONENT DIAGNOSTIC CHART — 1993–94 CUTLASS SUPREME, GRAND PRIX, LUMINA AND REGAL

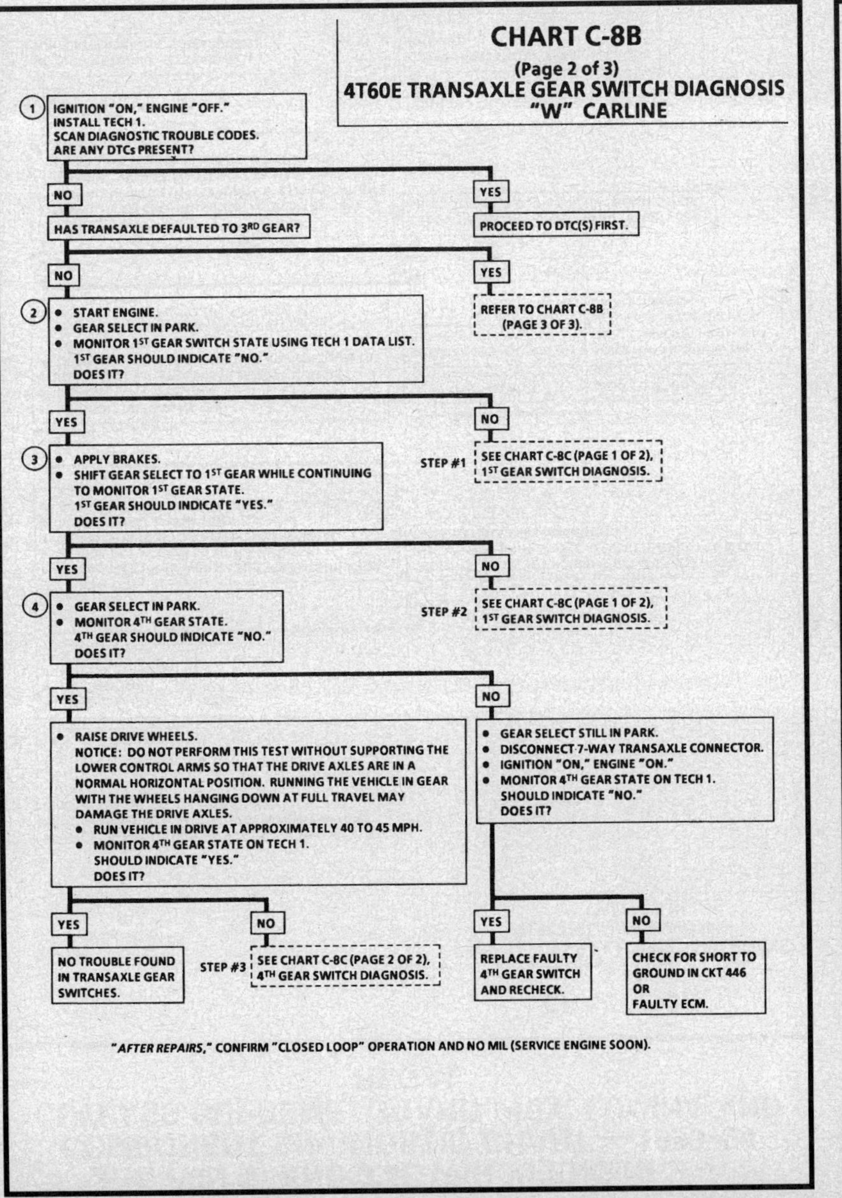

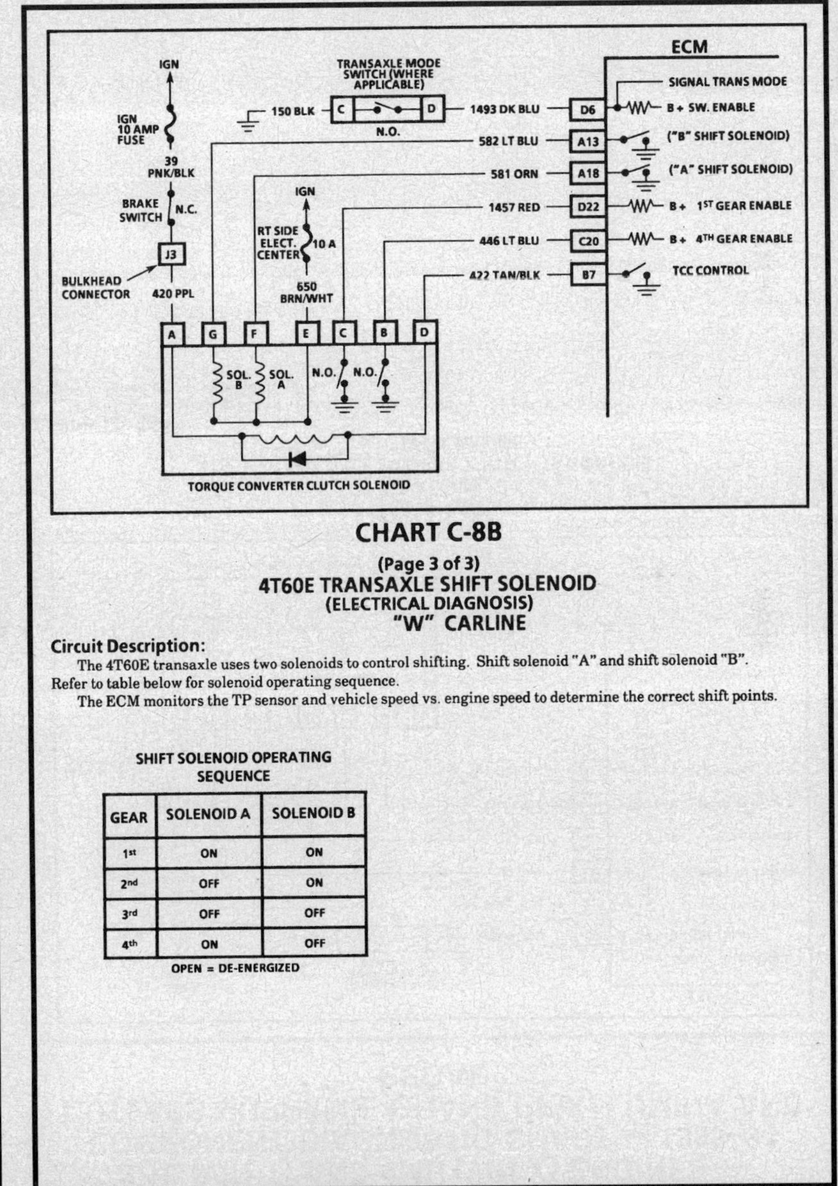

CHART C-8B
(Page 3 of 3)
4T60E TRANSAXLE SHIFT SOLENOID
(ELECTRICAL DIAGNOSIS)
"W" CARLINE

Circuit Description:

The 4T60E transaxle uses two solenoids to control shifting. Shift solenoid "A" and shift solenoid "B". Refer to table below for solenoid operating sequence.

The ECM monitors the TP sensor and vehicle speed vs. engine speed to determine the correct shift points.

SHIFT SOLENOID OPERATING SEQUENCE

GEAR	SOLENOID A	SOLENOID B
1st	ON	ON
2nd	OFF	ON
3rd	OFF	OFF
4th	ON	OFF

OPEN = DE-ENERGIZED

3.1L (VIN T) AND 3.4L (VIN X) ENGINES — COMPONENT DIAGNOSTIC CHART — 1993–94 CUTLASS SUPREME, GRAND PRIX, LUMINA AND REGAL

CHART C-8B

(Page 3 of 3)
4T60E TRANSAXLE SHIFT SOLENOID
(ELECTRICAL DIAGNOSIS)
"W" CARLINE

NOTICE: IF ANY DIAGNOSTIC TROUBLE CODES ARE STORED IN THE ECM MEMORY, THEY MUST BE REPAIRED BEFORE DIAGNOSING THE TRANSAXLE CIRCUITRY.

CONTINUED FROM CHART C-8B (PAGE 2 OF 3).

- SHIFT GEAR SELECT TO PARK.
- IGNITION "ON."
- INSPECT 10 AMP IGN FUSE.
  IS FUSE OK?

YES

- DISCONNECT THE TRANSAXLE ELECTRICAL CONNECTOR.
- CONNECT T BOX J 38791-1 TO TCC HARNESS.
- GEAR SELECTOR ROTARY SWITCH TO "NORMAL."
- IGNITION "ON," ENGINE "ON."
- SHIFT SOLENOID "A" & "B" LIGHTS SHOULD BE "ON" ON BOTH.

NO — CHECK FOR SHORT TO GROUND IN CKT 650 AND REPAIR. REPLACE FUSE AND RECHECK. IF FUSE CONTINUES TO BLOW, REFER TO ELECTRICAL DIAGNOSIS

YES

- IGNITION "ON," ENGINE "OFF."
- INSTALL TECH 1 SCAN TOOL.
- UNDER MISCELLANEOUS TEST, SELECT TRANSAXLE OUTPUT AND CYCLE SOLENOID "A" AND SOLENOID "B" "OFF" AND "ON."
  DO THE "A" AND "B" SOLENOID LIGHTS FLASH "ON" & "OFF" AS THEY ARE CYCLED USING TECH 1?

NO

LIGHT "OFF" ON BOTH — REPAIR OPEN IN CKT 650 TO HARNESS TERMINAL "E".

LIGHT "OFF" ON ONE — CHECK FOR AN OPEN IN THE CIRCUIT WHICH DID NOT LIGHT, 581 OR 582. IF THE CIRCUIT IS OK, REPLACE ECM.

YES

- IGNITION "OFF."
- WITH A DVOM CHECK THE RESISTANCE OF THE SHIFT SOLENOIDS BETWEEN TEST POINTS FOR "IGNITION" AND "SHIFT A." REPEAT RESISTANCE CHECK FOR "SHIFT B." RESISTANCE SHOULD BE 20-40 OHMS FOR EACH SOLENOID. ARE THEY?

NO — CHECK THE CIRCUITS WHICH DID NOT FLASH FOR A SHORT TO GROUND. IF OK, REPLACE FAULTY ECM.

YES

NO TROUBLE FOUND. SEE TRANSAXLE FOR FURTHER TRANSAXLE DIAGNOSIS.

NO — FAULTY TRANSAXLE INTERNAL WIRING OR SOLENOID.

"AFTER REPAIRS," CONFIRM "CLOSED LOOP" OPERATION AND NO MIL (SERVICE ENGINE SOON).

3.1L (VIN T) AND 3.4L (VIN X) ENGINES — COMPONENT DIAGNOSTIC CHART — 1993–94 CUTLASS SUPREME, GRAND PRIX, LUMINA AND REGAL

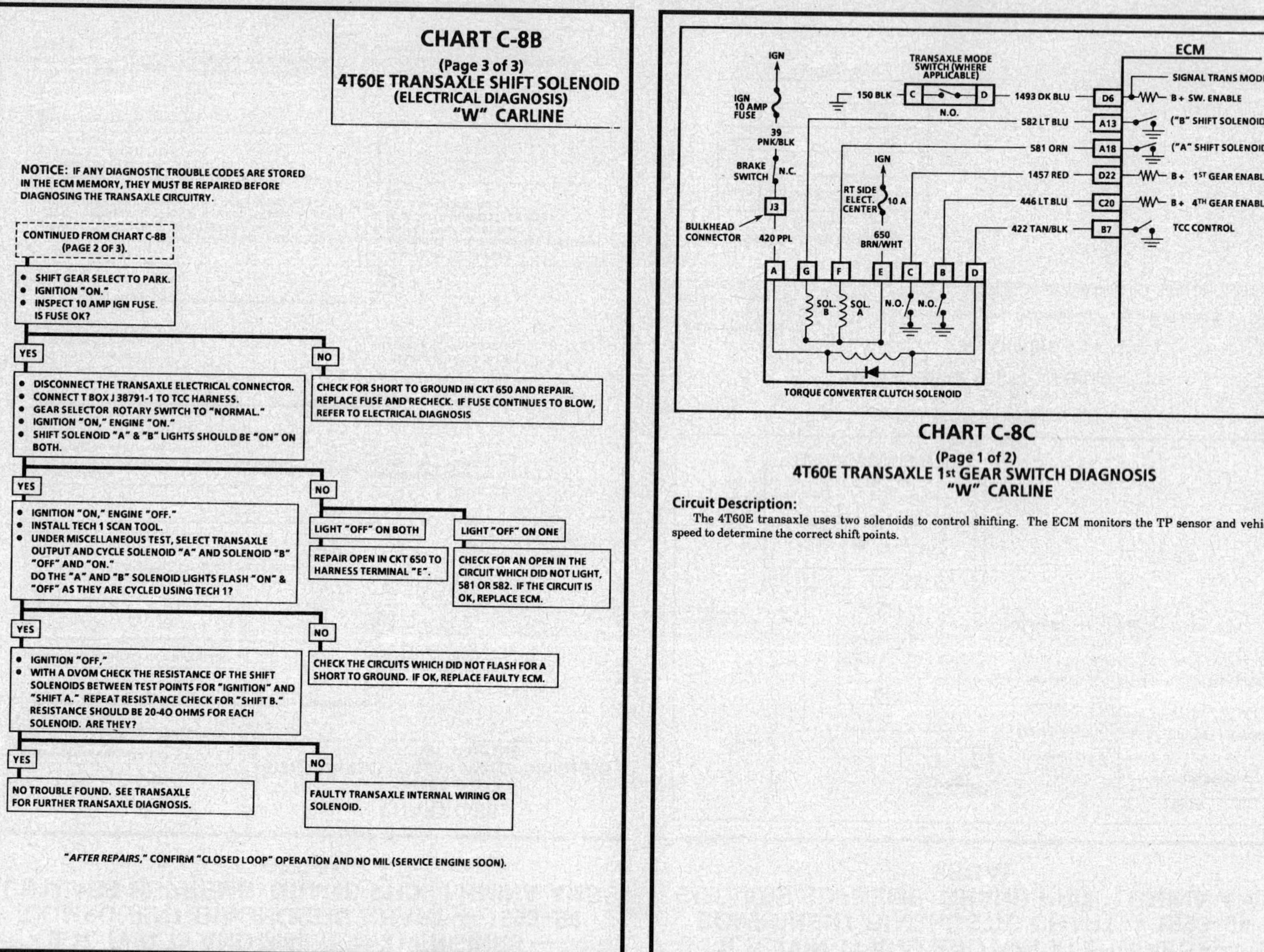

CHART C-8C

(Page 1 of 2)
4T60E TRANSAXLE 1st GEAR SWITCH DIAGNOSIS
"W" CARLINE

Circuit Description:

The 4T60E transaxle uses two solenoids to control shifting. The ECM monitors the TP sensor and vehicle speed to determine the correct shift points.

3.1L (VIN T) AND 3.4L (VIN X) ENGINES — COMPONENT DIAGNOSTIC CHART — 1993–94 CUTLASS SUPREME, GRAND PRIX, LUMINA AND REGAL

CHART C-8C

(Page 1 of 2)
4T60E TRANSAXLE 1st GEAR SWITCH DIAGNOSIS
"W" CARLINE

CONTINUED FROM STEP #1 OF CHART C-8B (PAGE 2 OF 3).

- GEAR SELECT IN PARK.
- DISCONNECT 7-WAY TRANSAXLE CONNECTOR.
- IGNITION "ON," ENGINE "ON."
- MONITOR 1ST GEAR STATE ON TECH 1.
  1ST GEAR SHOULD INDICATE "NO."
  DOES IT?

YES → REPLACE FAULTY 1ST GEAR SWITCH AND RECHECK.

NO → CHECK FOR SHORT TO GROUND IN CKT 1457 AND REPAIR. IF CKT 1457 IS NOT SHORTED, CHECK CONNECTOR OR FAULTY ECM.

CONTINUED FROM STEP #2 OF CHART C-8B (PAGE 2 OF 3).

- IGNITION "ON," ENGINE "OFF."
- DISCONNECT 7-WAY TRANSAXLE CONNECTOR.
- CONNECT DVOM (ON D.C. VOLT SCALE) BETWEEN TERMINAL "C" AND GROUND.
  SHOULD HAVE ABOVE 10 VOLTS.
  IS IT?

NO →
- DISCONNECT ECM CONNECTOR "D".
- WITH DVOM (ON D.C. VOLT SCALE), PROBE TERMINAL "D22" AT ECM TO GROUND.
  SHOULD HAVE ABOVE 10 VOLTS.
  IS IT?

YES → OPEN IN CKT 1457 OR FAULTY ECM CONNECTION.

NO → FAULTY ECM.

YES → FAULTY POOR CONNECTION, REPAIR AND RECHECK OR FAULTY 1ST GEAR SWITCH.

"AFTER REPAIRS," CONFIRM "CLOSED LOOP" OPERATION AND NO MIL (SERVICE ENGINE SOON).

3.1L (VIN T) AND 3.4L (VIN X) ENGINES — COMPONENT DIAGNOSTIC CHART — 1993–94 CUTLASS SUPREME, GRAND PRIX, LUMINA AND REGAL

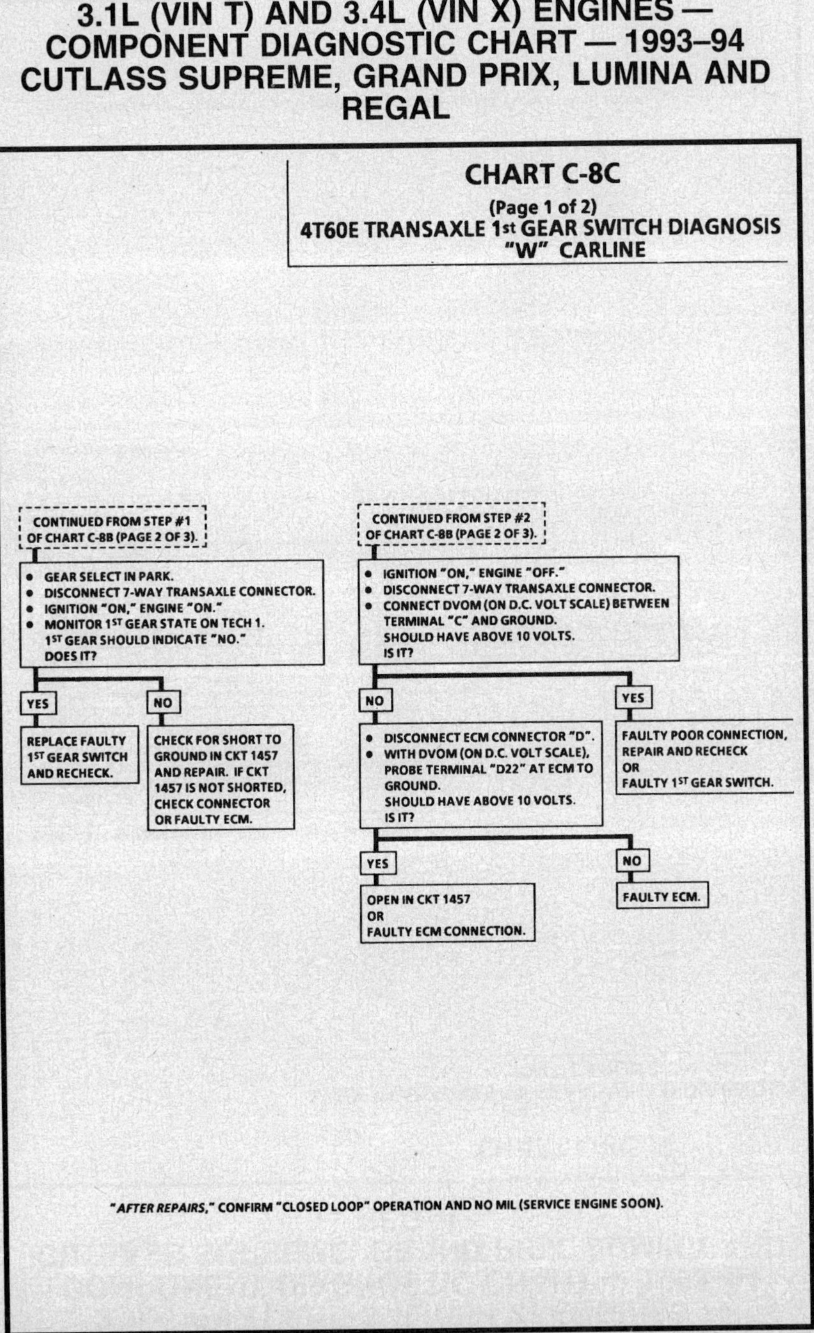

CHART C-8C

(Page 2 of 2)
4T60E TRANSAXLE 4th GEAR SWITCH DIAGNOSIS
"W" CARLINE

Circuit Description:

The 4T60E transaxle uses two solenoids to control shifting. The ECM monitors the TP sensor and vehicle speed to determine the correct shift points.

3.1L (VIN T) AND 3.4L (VIN X) ENGINES — COMPONENT DIAGNOSTIC CHART — 1993–94 CUTLASS SUPREME, GRAND PRIX, LUMINA AND REGAL

CHART C-8C
(Page 2 of 2)
4T60E TRANSAXLE 4th GEAR SWITCH DIAGNOSIS
"W" CARLINE

CONTINUED FROM STEP #3 OF CHART C-8B (PAGE 2 OF 3).

- ENGINE "OFF," IGNITION "ON."
- DISCONNECT 7-WAY TRANSAXLE CONNECTOR.
- CONNECT DVOM (ON D.C. VOLT SCALE) BETWEEN HARNESS CONNECTOR TERMINAL "B" AND GROUND. SHOULD HAVE ABOVE 10 VOLTS. IS IT?

NO →
- DISCONNECT ECM "C" CONNECTOR.
- WITH DVOM ON D.C. VOLT SCALE, PROBE TERMINAL "C20" AT ECM TO GROUND. SHOULD HAVE ABOVE 10 VOLTS. IS IT?

YES → FAULTY 4TH GEAR SWITCH CONNECTION OR FAULTY 4TH GEAR SWITCH.

YES → OPEN IN CKT 446 OR FAULTY ECM CONNECTION.

NO → FAULTY ECM.

"AFTER REPAIRS," CONFIRM "CLOSED LOOP" OPERATION AND NO MIL (SERVICE ENGINE SOON).

3.1L (VIN T) AND 3.4L (VIN X) ENGINES — COMPONENT DIAGNOSTIC CHART — 1993–94 CUTLASS SUPREME, GRAND PRIX, LUMINA AND REGAL

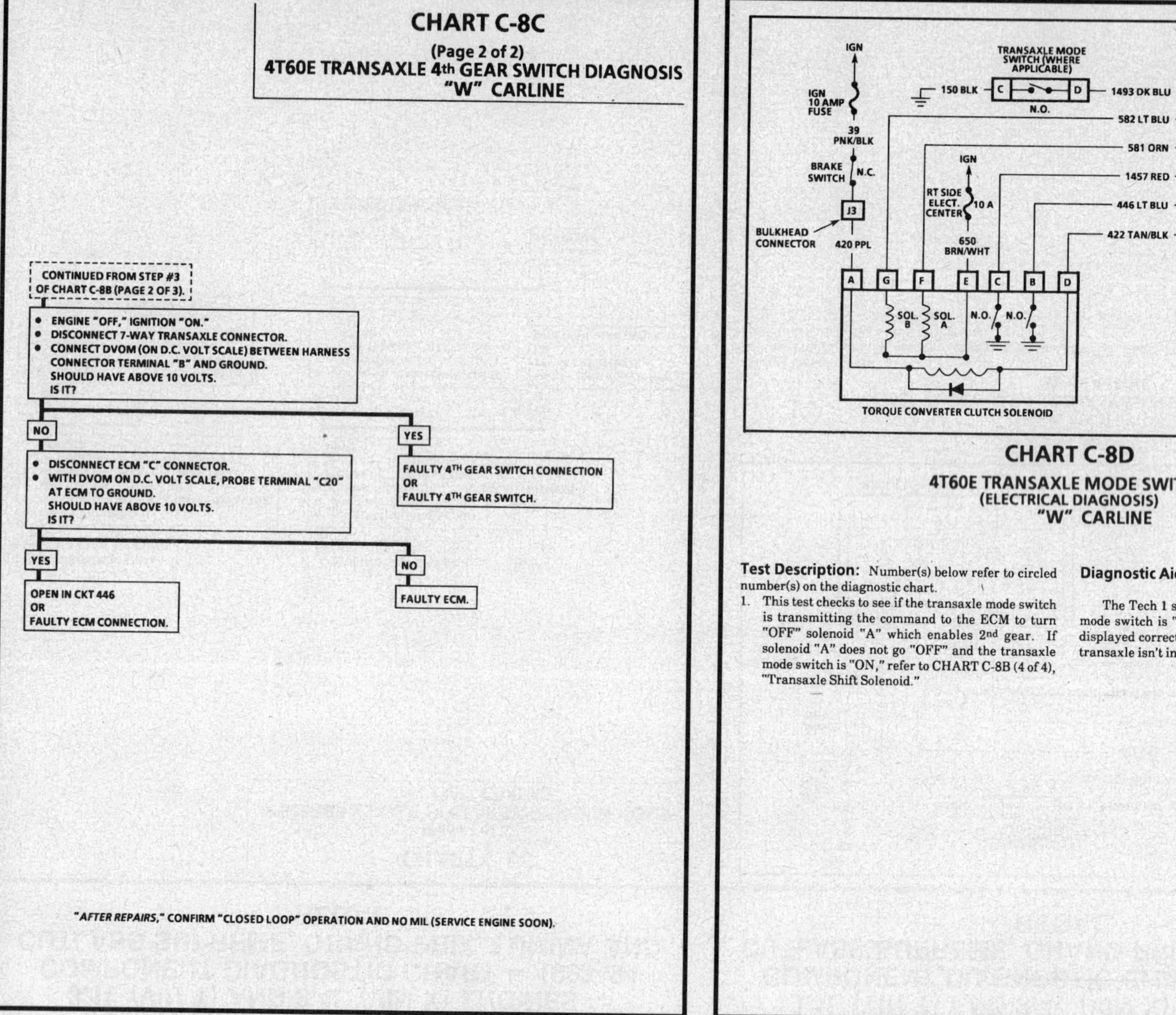

CHART C-8D
4T60E TRANSAXLE MODE SWITCH
(ELECTRICAL DIAGNOSIS)
"W" CARLINE

Test Description: Number(s) below refer to circled number(s) on the diagnostic chart.

1. This test checks to see if the transaxle mode switch is transmitting the command to the ECM to turn "OFF" solenoid "A" which enables 2nd gear. If solenoid "A" does not go "OFF" and the transaxle mode switch is "ON," refer to CHART C-8B (4 of 4), "Transaxle Shift Solenoid."

Diagnostic Aids:

The Tech 1 scan tool only displays if the transaxle mode switch is "ON" or "OFF." If the information is displayed correctly and solenoid "A" is "OFF" and the transaxle isn't in 2nd gear, refer to "Transaxle Repair,"

3.1L (VIN T) AND 3.4L (VIN X) ENGINES — COMPONENT DIAGNOSTIC CHART — 1993–94 CUTLASS SUPREME, GRAND PRIX, LUMINA AND REGAL

CHART C-8D
4T60E TRANSAXLE MODE SWITCH
(ELECTRICAL DIAGNOSIS)
"W" CARLINE

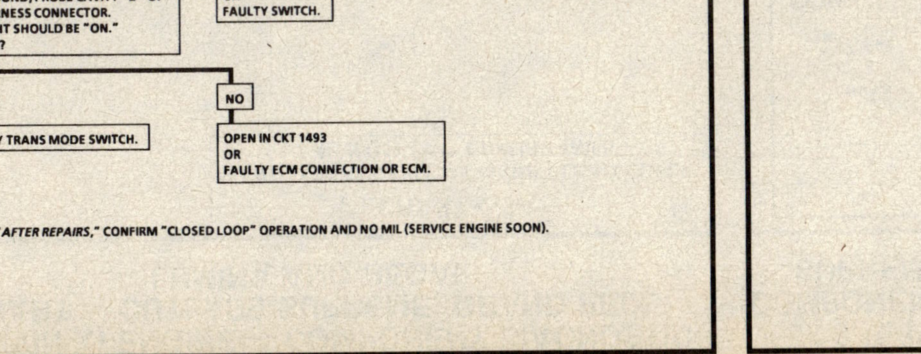

"AFTER REPAIRS," CONFIRM "CLOSED LOOP" OPERATION AND NO MIL (SERVICE ENGINE SOON).

3.4L (VIN X) ENGINE — COMPONENT DIAGNOSTIC CHART — CUTLASS SUPREME, GRAND PRIX, LUMINA AND REGAL

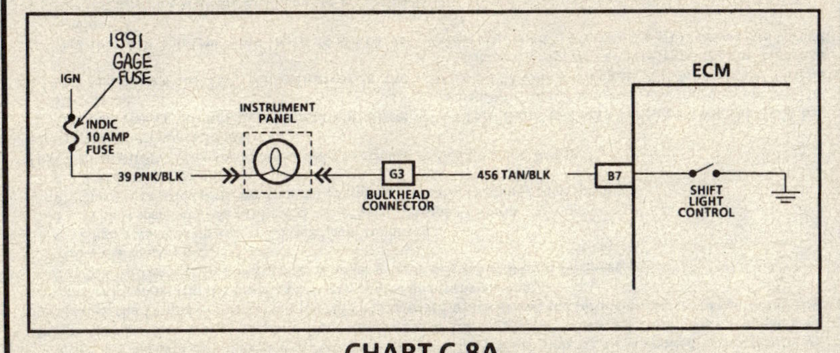

CHART C-8A
MANUAL TRANSAXLE (M/T) SHIFT LIGHT CHECK
3.4L (LQ1) "W" CARLINE (MFI)

Circuit Description:

The shift light indicates the best transaxle shift point for maximum fuel economy. The light is controlled by the ECM and is turned "ON" by grounding CKT 456. The ECM uses information from the following inputs to control the shift light:

- Engine coolant temperature must be above 16°C (61°F).
- Throttle Position (TP) sensor above 4%.
- Vehicle Speed Sensor (VSS).
- RPM above about 1900.
- Airflow - The ECM uses RPM, airflow VSS to calculate what gear the vehicle is in.

It's this calculation that determines when the shift light should be turned "ON." The shift light will only stay "ON" 5 seconds after the conditions were met to turn it "ON."

Test Description: Number(s) below refer to circled number(s) on the diagnostic chart.

1. This should not turn "ON" the shift light. If the light is "ON," there is a short to ground in CKT 456 wiring or a fault in the ECM.
2. When the diagnostic terminal is grounded, the ECM should ground CKT 456 and the shift light should come "ON."
3. This checks the shift light circuit up to the ECM connector. If the shift light illuminates, then the ECM connector is faulty or the ECM does not have the ability to ground the circuit.

3.1L (VIN T) AND 3.4L (VIN X) ENGINES — COMPONENT DIAGNOSTIC CHART — 1992 CUTLASS SUPREME, GRAND PRIX, LUMINA AND REGAL

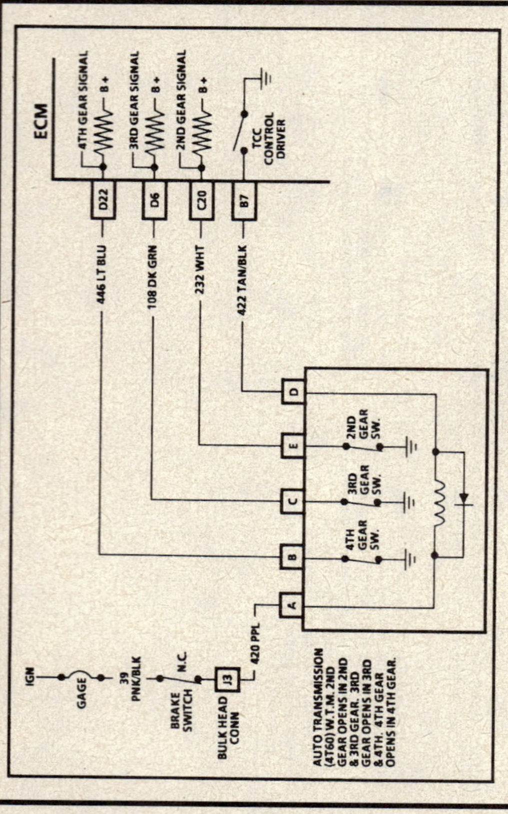

CHART C-8B
(Page 1 of 3)

4T60 TORQUE CONVERTER CLUTCH (TCC) WITH TORQUE MANAGEMENT
(ELECTRICAL DIAGNOSIS)
"W" CARLINE

Circuit Description:

The purpose of the automatic transmission torque converter clutch feature is to eliminate the power loss of the torque converter stage when the vehicle is in a cruise condition. This allows the convenience of the automatic transmission and the fuel economy of a manual transmission. The heart of the system is a solenoid located inside the automatic transmission which is controlled by the ECM.

When the solenoid coil is activated "ON," the Torque Converter Clutch (TCC) is applied which results in straight through mechanical coupling from the engine to transmission. When the transmission solenoid is deactivated, the Torque Converter Clutch (TCC) is released which allows the torque converter to operate in the conventional manner (fluidic coupling between engine and transmission).

The TCC will engage on a warm engine under given road load in 3rd and 4th gears. TCC will engage when:
- Engine warmed up and in "Closed Loop" condition.
- Vehicle speed above a calibrated value (about 28 mph 45 km/h).
- Throttle position sensor output not changing, indicating a steady road speed.
- Brake switch closed.

Test Description: Number(s) below refer to circled number(s) on the diagnostic chart.

1. This test checks the functional operation of the TCC circuit.
2. This test checks the TCC control driver in the ECM.
3. This test will confirm that there is BATT to terminal "A".
4. This test confirms that the ECM has the ability to turn the TCC "ON."

Diagnostic Aids:

The "Scan" tool only indicates when the ECM has turned "ON" the TCC driver, and this does not confirm that the TCC has engaged. To determine if TCC is functioning properly, engine rpm should decrease when the "Scan" indicates the TCC driver has turned "ON."

3.4L (VIN X) ENGINE — COMPONENT DIAGNOSTIC CHART — CUTLASS SUPREME, GRAND PRIX, LUMINA AND REGAL

CHART C-8A

MANUAL TRANSAXLE (M/T) SHIFT LIGHT CHECK
3.4L (LQ1) "W" CARLINE (MFI)

3.1L (VIN T) AND 3.4L (VIN X) ENGINES — COMPONENT DIAGNOSTIC CHART — 1992 CUTLASS SUPREME, GRAND PRIX, LUMINA AND REGAL

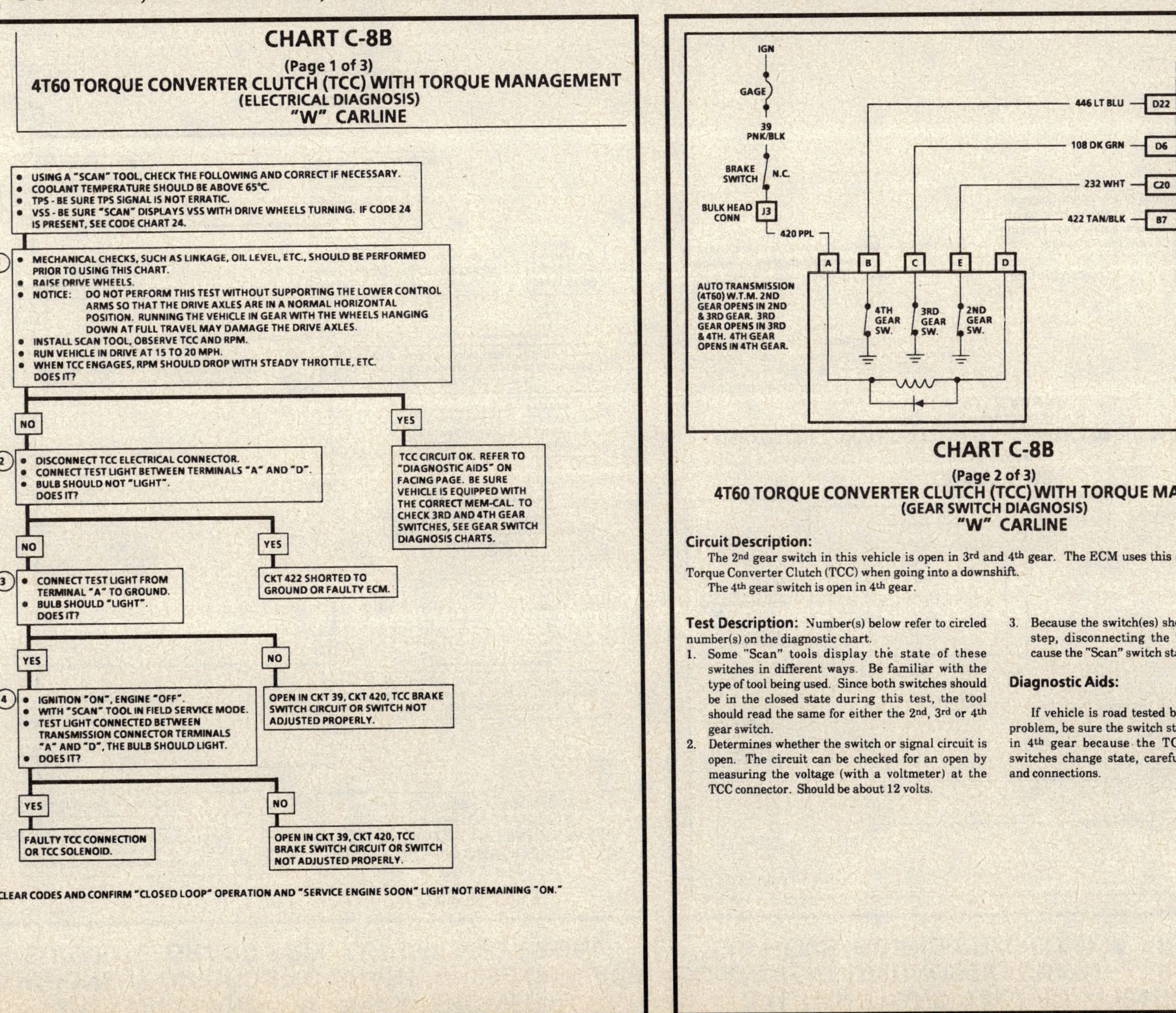

3.1L (VIN T) AND 3.4L (VIN X) ENGINES — COMPONENT DIAGNOSTIC CHART — 1992 CUTLASS SUPREME, GRAND PRIX, LUMINA AND REGAL

CHART C-8B
(Page 2 of 3)
4T60 TORQUE CONVERTER CLUTCH (TCC) WITH TORQUE MANAGEMENT
(GEAR SWITCH DIAGNOSIS)
"W" CARLINE

Circuit Description:
The 2nd gear switch in this vehicle is open in 3rd and 4th gear. The ECM uses this signal to disengage the Torque Converter Clutch (TCC) when going into a downshift.
The 4th gear switch is open in 4th gear.

Test Description: Number(s) below refer to circled number(s) on the diagnostic chart.
1. Some "Scan" tools display the state of these switches in different ways. Be familiar with the type of tool being used. Since both switches should be in the closed state during this test, the tool should read the same for either the 2nd, 3rd or 4th gear switch.
2. Determines whether the switch or signal circuit is open. The circuit can be checked for an open by measuring the voltage (with a voltmeter) at the TCC connector. Should be about 12 volts.
3. Because the switch(es) should be grounded in this step, disconnecting the TCC connector should cause the "Scan" switch state to change.

Diagnostic Aids:

If vehicle is road tested because of a TCC related problem, be sure the switch states do not change while in 4th gear because the TCC will disengage. If switches change state, carefully check wire routing and connections.

3.1L (VIN T) AND 3.4L (VIN X) ENGINES — COMPONENT DIAGNOSTIC CHART — 1992 CUTLASS SUPREME, GRAND PRIX, LUMINA AND REGAL

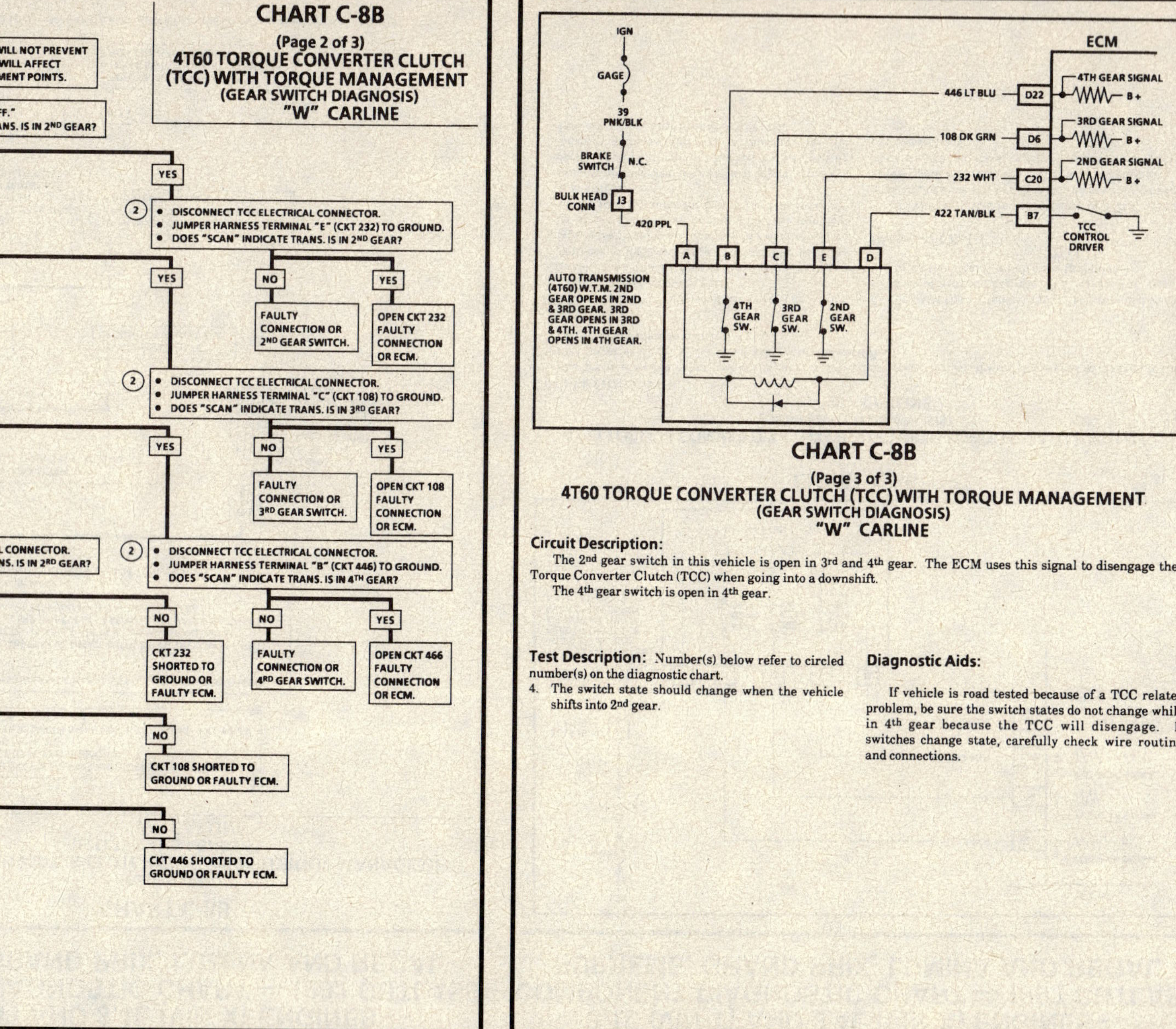

Circuit Description:
The 2nd gear switch in this vehicle is open in 3rd and 4th gear. The ECM uses this signal to disengage the Torque Converter Clutch (TCC) when going into a downshift.
The 4th gear switch is open in 4th gear.

Test Description: Number(s) below refer to circled number(s) on the diagnostic chart.
4. The switch state should change when the vehicle shifts into 2nd gear.

Diagnostic Aids:
If vehicle is road tested because of a TCC related problem, be sure the switch states do not change while in 4th gear because the TCC will disengage. If switches change state, carefully check wire routing and connections.

3.1L (VIN T) AND 3.4L (VIN X) ENGINES — COMPONENT DIAGNOSTIC CHART — 1992 CUTLASS SUPREME, GRAND PRIX, LUMINA AND REGAL

CHART C-8B

(Page 3 of 3)
4T60 TORQUE CONVERTER CLUTCH (TCC)
WITH TORQUE MANAGEMENT
(GEAR SWITCH DIAGNOSIS)
"W" CARLINE

FROM CHART C-8B (2 OF 3)

WITH TCC ELECTRICAL CONNECTOR DISCONNECTED "SCAN" INDICATES TRANS. IS IN 2ND, 3RD, AND 4TH GEAR.

④
- RECONNECT TRANS.
- RAISE DRIVE WHEELS.
- START AND IDLE ENGINE IN OVERDRIVE.
- *NOTICE:* DO NOT PERFORM THIS TEST WITHOUT SUPPORTING THE LOWER CONTROL ARMS SO THAT THE DRIVE AXLES ARE IN A NORMAL HORIZONTAL POSITION. RUNNING THE VEHICLE IN GEAR WITH THE WHEELS HANGING DOWN AT FULL TRAVEL MAY DAMAGE THE DRIVE AXLES.
- INCREASE SPEED SLOWLY UNTIL TRANS. SHIFTS INTO GEAR.
- DOES "SCAN" INDICATE TRANS. IS IN 2ND GEAR?

YES
- INCREASE SPEED UNTIL TRANS. SHIFTS INTO 3RD GEAR.
- DOES "SCAN" INDICATE TRANS. IS IN 3RD GEAR?

NO → FAULTY 2ND GEAR SWITCH.

YES
- INCREASE SPEED UNTIL TRANS. SHIFTS INTO 4TH GEAR.
- DOES "SCAN" INDICATE TRANS. IS IN 4TH GEAR?

NO → FAULTY 3RD GEAR SWITCH.

YES
TRANS. SWITCHES OK. CHECK WIRE ROUTING AND CONNECTIONS. REFER TO "DIAGNOSTIC AIDS"

NO → FAULTY 4TH GEAR SWITCH.

3.1L (VIN T) AND 3.4L (VIN X) ENGINES — COMPONENT DIAGNOSTIC CHART — CUTLASS SUPREME, GRAND PRIX, LUMINA AND REGAL

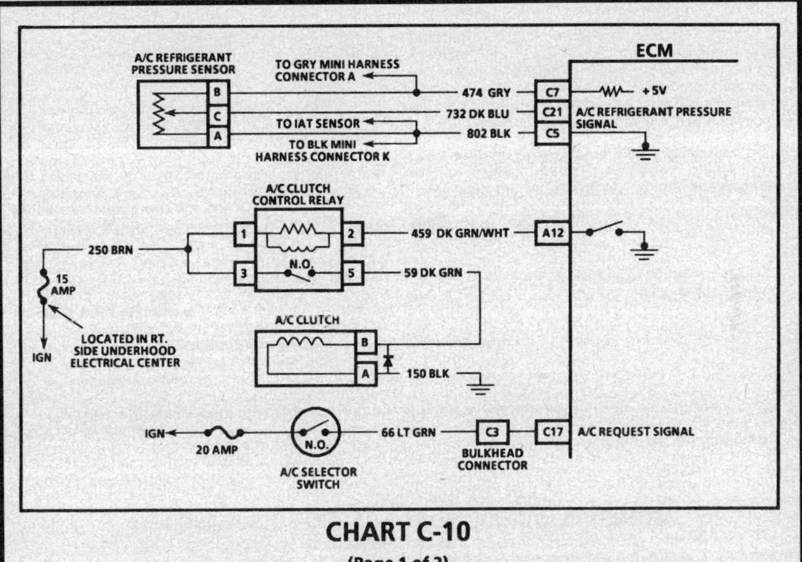

CHART C-10

(Page 1 of 2)
A/C COMPRESSOR CLUTCH CONTROL CIRCUIT DIAGNOSIS
"W" CARLINE

Circuit Description:

The A/C clutch control relay is ECM controlled to delay A/C clutch engagement about .4 second after A/C is turned "ON." This allows the IAC to adjust engine RPM before the A/C clutch engages. The ECM also causes the relay to disengage the A/C clutch during WOT when high power steering pressure is present or if engine is overheating. The A/C clutch control relay is energized when the ECM provides a ground path for CKT 459. The A/C refrigerant pressure sensor is used to determine high and low pressure in the system and also turns "ON" the coolant fan when it is needed.

Test Description: Number(s) below refer to circled number(s) on the diagnostic chart.

1. The ECM will only energize the A/C relay, when the engine is running. This test will determine if the relay or CKT 459 is faulty.
2. Determines if the signal is reaching the ECM through CKT 66 from the A/C control panel. Signal should only be present when an A/C mode or defrost mode has been selected.
3. If the ECM is seeing a high power steering pressure signal, the A/C clutch will be disengaged by the ECM.

Diagnostic Aids:

If complaint was insufficient cooling, the problem may be caused by an inoperative cooling fan. Refer to CHART C-12 for cooling fan diagnosis.

A/C refrigerant pressure outside of a range of 43 to 428 psi will cause the compressor to be disabled by the ECM. Observe Tech 1 scan tool A/C pressure for 2 minutes with engine idling and A/C "ON." If pressure goes out of range, to measure and diagnose. Tech 1 scan tool pressure should be within 20 psi of actual. If not, check for a circuit problem using DTC 66 chart or replace sensor.

3.1L (VIN T) AND 3.4L (VIN X) ENGINES — COMPONENT DIAGNOSTIC CHART — CUTLASS SUPREME, GRAND PRIX, LUMINA AND REGAL

CHART C-10

(Page 1 of 2)

A/C COMPRESSOR CLUTCH CONTROL CIRCUIT DIAGNOSIS "W" CARLINE

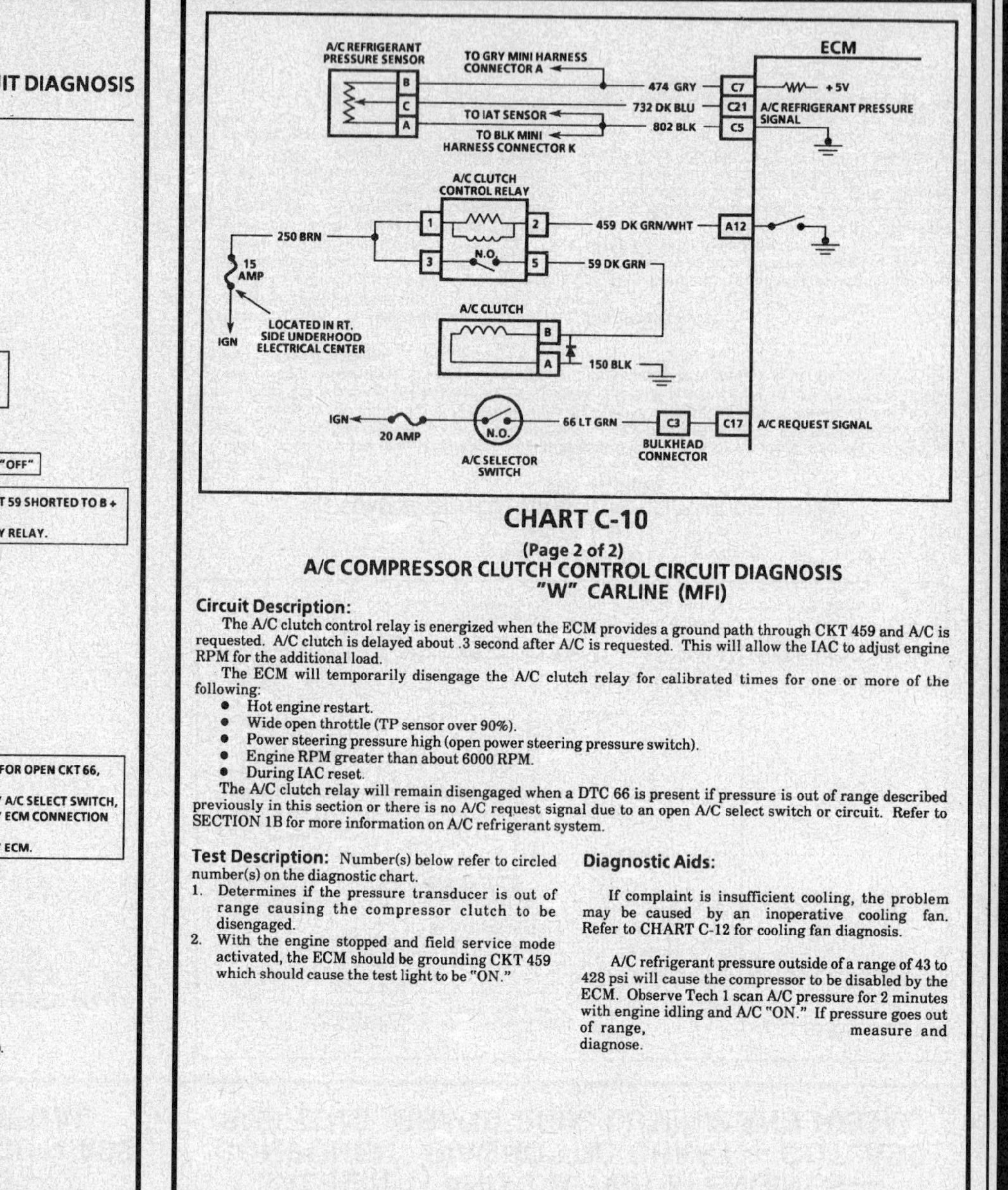

Flow chart (Page 1 of 2):

(1)
- IGNITION "ON," ENGINE "OFF."
- SCAN DTC(s) AND REFER TO THOSE CHARTS, IF PRESENT.
- NOTE A/C CLUTCH.
- CLUTCH SHOULD NOT BE ENGAGED.
IS IT?

NO →
- WITH ENGINE IDLING AT NORMAL OPERATING TEMPERATURE TURN A/C "ON" AND "OFF."
- CLUTCH SHOULD CYCLE "ON" AND "OFF."
DOES IT?

YES →
- DISCONNECT A/C RELAY.
- PROBE CKT 459 WITH A TEST LIGHT TO 12 VOLTS.

From "WITH ENGINE IDLING..." — NO:

(2)
- PLACE A/C CONTROL IN A/C MODE.
- TECH 1 SCAN TOOL SHOULD DISPLAY A/C REQUEST "ON."
DOES IT?

YES: A/C CIRCUIT OK, REFER TO "DIAGNOSTIC AIDS"

From "DISCONNECT A/C RELAY..." — LIGHT "ON": CKT 459 SHORTED TO GROUND OR FAULTY ECM.

LIGHT "OFF": CIRCUIT 59 SHORTED TO B+ OR FAULTY RELAY.

From (2) — YES:
- KEY "ON" ENGINE "OFF."
- ALLOW A/C PRESSURES TO EQUALIZE.
- SCAN A/C PRESSURE.
IS PRESSURE OVER 150 psi?

From CKT 459 — NO: IS A/C FUSE BLOWN?

YES: OPEN CKT 802, FAULTY SENSOR CONNECTION OR FAULTY SENSOR.

NO →
(3)
- WITH ENGINE IDLING DOES TECH 1 SCAN TOOL INDICATE PSP SWITCH IS "ON"? (HIGH PRESSURE)

IS A/C FUSE BLOWN? — YES: CHECK FOR SHORT TO GROUND IN CKT 66.

NO: CHECK FOR OPEN CKT 66, OR FAULTY A/C SELECT SWITCH, FAULTY ECM CONNECTION OR FAULTY ECM.

From (3) — YES: REFER TO CHART C-1E.

NO: REFER TO CHART C-10 (2 OF 2)

"AFTER REPAIRS," CONFIRM "CLOSED LOOP" OPERATION AND NO MIL (SERVICE ENGINE SOON).

Circuit Description:

The A/C clutch control relay is energized when the ECM provides a ground path through CKT 459 and A/C is requested. A/C clutch is delayed about .3 second after A/C is requested. This will allow the IAC to adjust engine RPM for the additional load.

The ECM will temporarily disengage the A/C clutch relay for calibrated times for one or more of the following:

- Hot engine restart.
- Wide open throttle (TP sensor over 90%).
- Power steering pressure high (open power steering pressure switch).
- Engine RPM greater than about 6000 RPM.
- During IAC reset.

The A/C clutch relay will remain disengaged when a DTC 66 is present if pressure is out of range described previously in this section or there is no A/C request signal due to an open A/C select switch or circuit. Refer to SECTION 1B for more information on A/C refrigerant system.

Test Description: Number(s) below refer to circled number(s) on the diagnostic chart.

1. Determines if the pressure transducer is out of range causing the compressor clutch to be disengaged.
2. With the engine stopped and field service mode activated, the ECM should be grounding CKT 459 which should cause the test light to be "ON."

Diagnostic Aids:

If complaint is insufficient cooling, the problem may be caused by an inoperative cooling fan. Refer to CHART C-12 for cooling fan diagnosis.

A/C refrigerant pressure outside of a range of 43 to 428 psi will cause the compressor to be disabled by the ECM. Observe Tech 1 scan A/C pressure for 2 minutes with engine idling and A/C "ON." If pressure goes out of range, measure and diagnose.

3.1L (VIN T) AND 3.4L (VIN X) ENGINES — COMPONENT DIAGNOSTIC CHART — CUTLASS SUPREME, GRAND PRIX, LUMINA AND REGAL

CHART C-10
(Page 2 of 2)
A/C COMPRESSOR CLUTCH CONTROL CIRCUIT DIAGNOSIS "W" CARLINE (MFI)

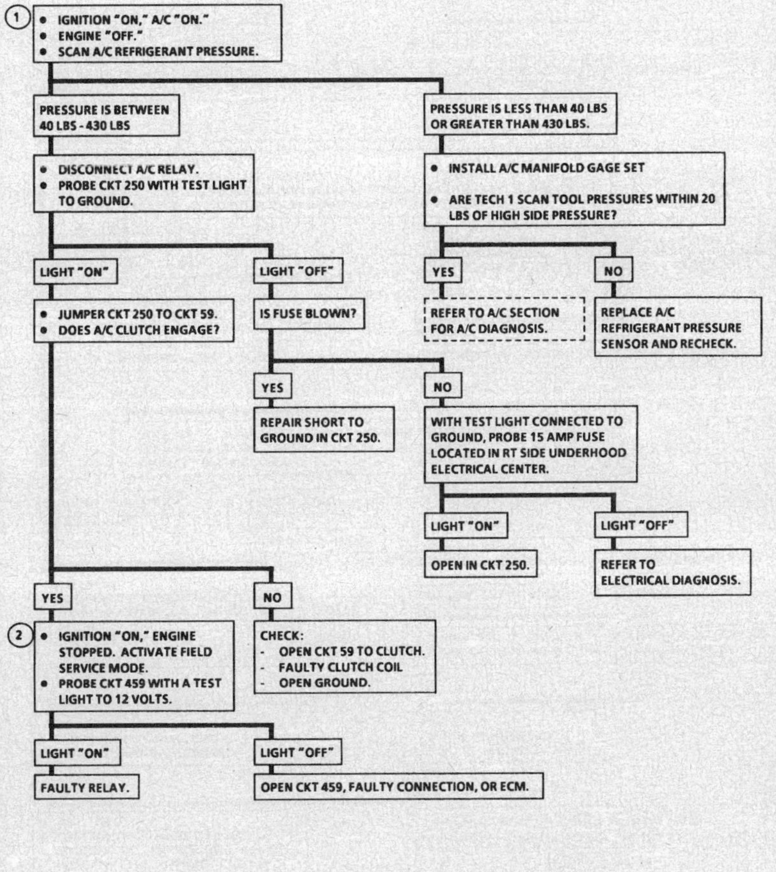

① • IGNITION "ON," A/C "ON."
• ENGINE "OFF."
• SCAN A/C REFRIGERANT PRESSURE.

PRESSURE IS BETWEEN 40 LBS - 430 LBS

PRESSURE IS LESS THAN 40 LBS OR GREATER THAN 430 LBS.

• DISCONNECT A/C RELAY.
• PROBE CKT 250 WITH TEST LIGHT TO GROUND.

• INSTALL A/C MANIFOLD GAGE SET
• ARE TECH 1 SCAN TOOL PRESSURES WITHIN 20 LBS OF HIGH SIDE PRESSURE?

LIGHT "ON"

LIGHT "OFF"

YES

NO

• JUMPER CKT 250 TO CKT 59.
• DOES A/C CLUTCH ENGAGE?

IS FUSE BLOWN?

REFER TO A/C SECTION FOR A/C DIAGNOSIS.

REPLACE A/C REFRIGERANT PRESSURE SENSOR AND RECHECK.

YES

NO

REPAIR SHORT TO GROUND IN CKT 250.

WITH TEST LIGHT CONNECTED TO GROUND, PROBE 15 AMP FUSE LOCATED IN RT SIDE UNDERHOOD ELECTRICAL CENTER.

LIGHT "ON"

LIGHT "OFF"

OPEN IN CKT 250.

REFER TO ELECTRICAL DIAGNOSIS.

YES

NO

② • IGNITION "ON," ENGINE STOPPED. ACTIVATE FIELD SERVICE MODE.
• PROBE CKT 459 WITH A TEST LIGHT TO 12 VOLTS.

CHECK:
- OPEN CKT 59 TO CLUTCH.
- FAULTY CLUTCH COIL
- OPEN GROUND.

LIGHT "ON"

LIGHT "OFF"

FAULTY RELAY.

OPEN CKT 459, FAULTY CONNECTION, OR ECM.

"AFTER REPAIRS," CONFIRM "CLOSED LOOP" OPERATION AND NO MIL (SERVICE ENGINE SOON).

3.1L (VIN T) AND 3.4L (VIN X) ENGINES — COMPONENT DIAGNOSTIC CHART — CUTLASS SUPREME, GRAND PRIX, LUMINA AND REGAL

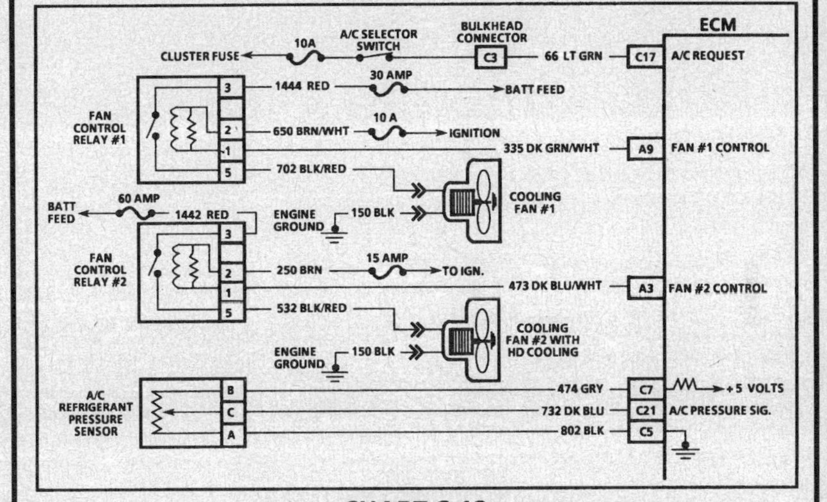

CHART C-12
(Page 1 of 2)
COOLANT FAN CONTROL DIAGNOSIS "W" CARLINE

Circuit Description:

The primary and secondary electric cooling fan(s) are controlled by the ECM, based on inputs from the Engine Coolant Temperature (ECT) sensor, the A/C control switches, vehicle speed and state of the A/C pressure sensor. The ECM controls the fan(s) by grounding CKT 335 and/or CKT 473, which energizes the fan control relay. Battery voltage is then supplied to the fan motor.

The ECM grounds CKT 335 and/or CKT 473 when engine coolant temperature is over about 106°C (223°F), or when A/C has been requested and the A/C pressure is about 1380 kPa (200 psi). Once the ECM turns the relay "ON," it will keep it "ON" for a minimum of 30 seconds, or until vehicle speed exceeds 70 mph (40 mph for secondary fan).

Also, if DTC 14 or 15 sets or the ECM is in back up, the primary fan will run at all time.

Test Description: Number(s) below refer to circled number(s) on the diagnostic chart.

1. With the diagnostic terminal grounded, the cooling fan control driver(s) will close, which should energize the fan control relay(s).
2. If the A/C pressure is above 300 psi (2069 kPa) or circuit is open, the fan would run whenever A/C is requested.
3. With A/C clutch engaged and the A/C pressure sensor is functioning properly, the fan should come "ON" when pressure exceeds about 1380 kPa (200 psi). This signal should cause the ECM to energize the fan control relay(s).
4. This will determine if the A/C pressure sensor is faulty or if the ECM or circuitry is faulty.

Diagnostic Aids:

If the owner complained of an overheating problem, it must be determined if the complaint was due to an actual boilover, a hot light or temp. gage indicated over heating.

If the gage or light indicates overheating, but no boilover is detected, the gage circuit should be checked. The gage accuracy can also be checked by comparing the coolant sensor reading using a Tech 1 scan tool and comparing its reading with the gage reading. If the engine is actually overheating and the gage indicates overheating but the cooling fan is not coming "ON," the coolant sensor has probably shifted out of calibration and should be replaced.

If the engine is overheating, and the cooling fan is "ON," the cooling system should be checked.

3.1L (VIN T) AND 3.4L (VIN X) ENGINES — COMPONENT DIAGNOSTIC CHART — CUTLASS SUPREME, GRAND PRIX, LUMINA AND REGAL

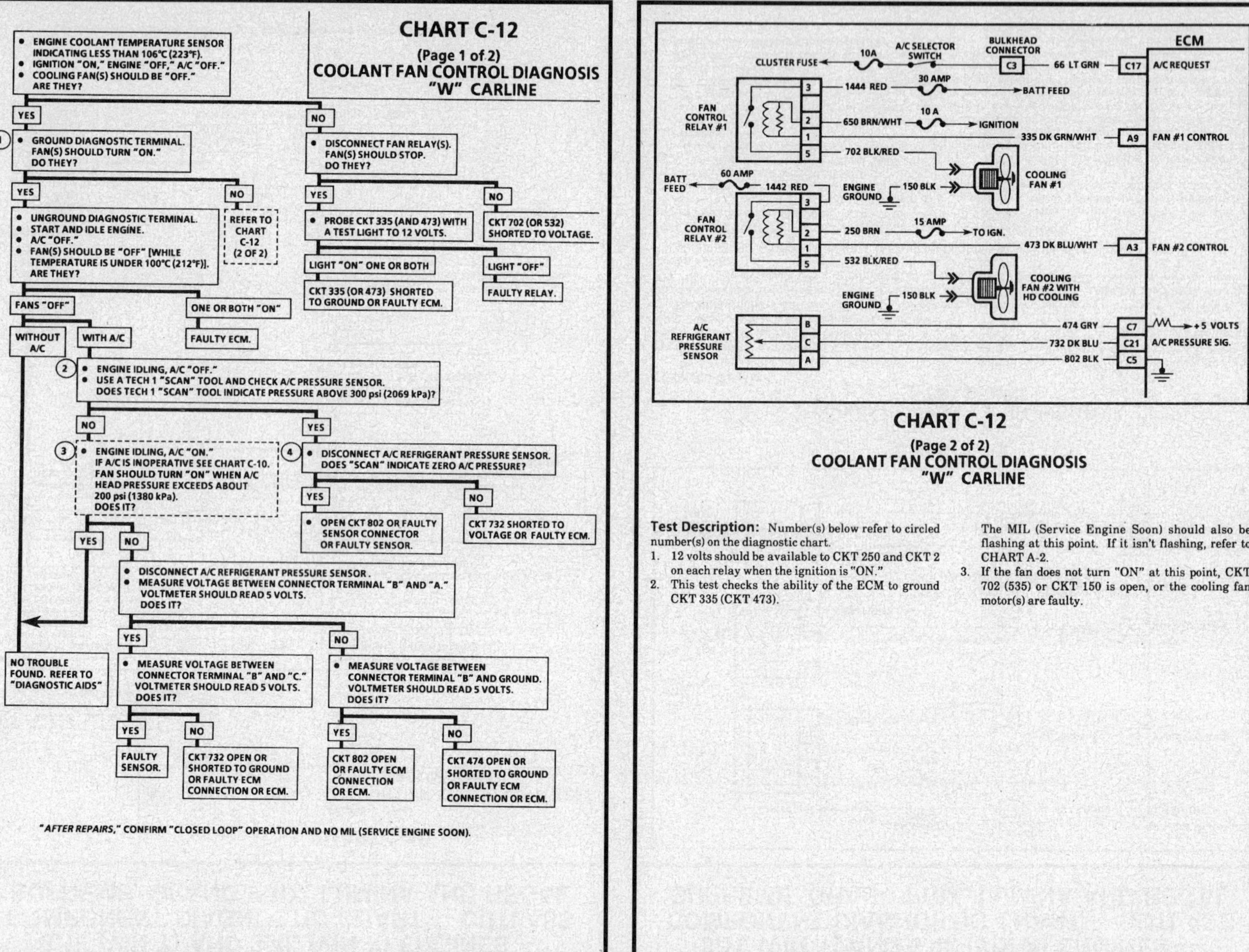

3.1L (VIN T) AND 3.4L (VIN X) ENGINES — COMPONENT DIAGNOSTIC CHART — CUTLASS SUPREME, GRAND PRIX, LUMINA AND REGAL

Test Description: Number(s) below refer to circled number(s) on the diagnostic chart.

1. 12 volts should be available to CKT 250 and CKT 2 on each relay when the ignition is "ON."
2. This test checks the ability of the ECM to ground CKT 335 (CKT 473).

The MIL (Service Engine Soon) should also be flashing at this point. If it isn't flashing, refer to CHART A-2.
3. If the fan does not turn "ON" at this point, CKT 702 (535) or CKT 150 is open, or the cooling fan motor(s) are faulty.

3.1L (VIN T) AND 3.4L (VIN X) ENGINES — COMPONENT DIAGNOSTIC CHART — CUTLASS SUPREME, GRAND PRIX, LUMINA AND REGAL

FROM CHART C-12 (1 OF 2)

CHART C-12
(Page 2 of 2)
COOLANT FAN CONTROL DIAGNOSIS
"W" CARLINE

1. • **DISCONNECT FAN CONTROL RELAY(S). *
   • IGNITION "ON," ENGINE STOPPED.
   • PROBE CAVITIES "2" AND "3" OF BOTH FAN RELAYS WITH A TEST LIGHT CONNECTED TO GROUND.

LIGHT "ON" BOTH (FOR EITHER RELAY)

LIGHT "OFF" ONE OR BOTH (FOR EITHER RELAY)

2. • DIAGNOSTIC TERMINAL GROUNDED.
   • PROBE CKTS 335 (AND 473) WITH A TEST LIGHT CONNECTED TO 12 VOLTS.

REPAIR OPEN OR SHORT TO GROUND IN CIRCUIT THAT DID NOT LIGHT.

LIGHT "ON" (BOTH)

LIGHT "OFF" BOTH

3. • JUMPER CAVITIES "5" AND "3" TOGETHER USING A FUSED JUMPER (ON EACH RELAY).
   • FAN(S) SHOULD RUN. DO THEY?

OPEN OR SHORT TO VOLTAGE IN CKT 335, (473) OR FAULTY CONNECTION AT ECM OR A FAULTY ECM.

NO

YES (ONE OR BOTH)

FAULTY RELAY OR RELAY CONNECTION.

• WITH CAVITIES STILL JUMPERED, AS ABOVE, CONNECT A TEST LIGHT ACROSS THE COOLING FAN MOTOR HARNESS CONNECTOR "A" AND "B" TERMINALS (ON EACH MOTOR).

LIGHT "OFF" (ONE OR BOTH)

LIGHT "ON" ONE OR BOTH

FAULTY MOTOR

• PROBE TERMINAL "B" ON EACH FAN MOTOR HARNESS CONNECTOR WITH A TEST LIGHT CONNECTED TO GROUND.

LIGHT "ON" (ONE OR BOTH MOTORS)

LIGHT "OFF"

OPEN IN GROUND CKT 150.

REPAIR OPEN IN CIRCUITS BETWEEN RELAY AND COOLING FAN MOTORS.

** SOME VEHICLES ARE EQUIPPED WITH TWO COOLING FANS. THE INFORMATION IN PARENTHESIS ( ) APPLIES ONLY TO THE SECOND FAN, IF EQUIPPED.

3.1L (VIN T) ENGINE — ENGINE COMPONENT LOCATION CHART — LUMINA WITH VFV

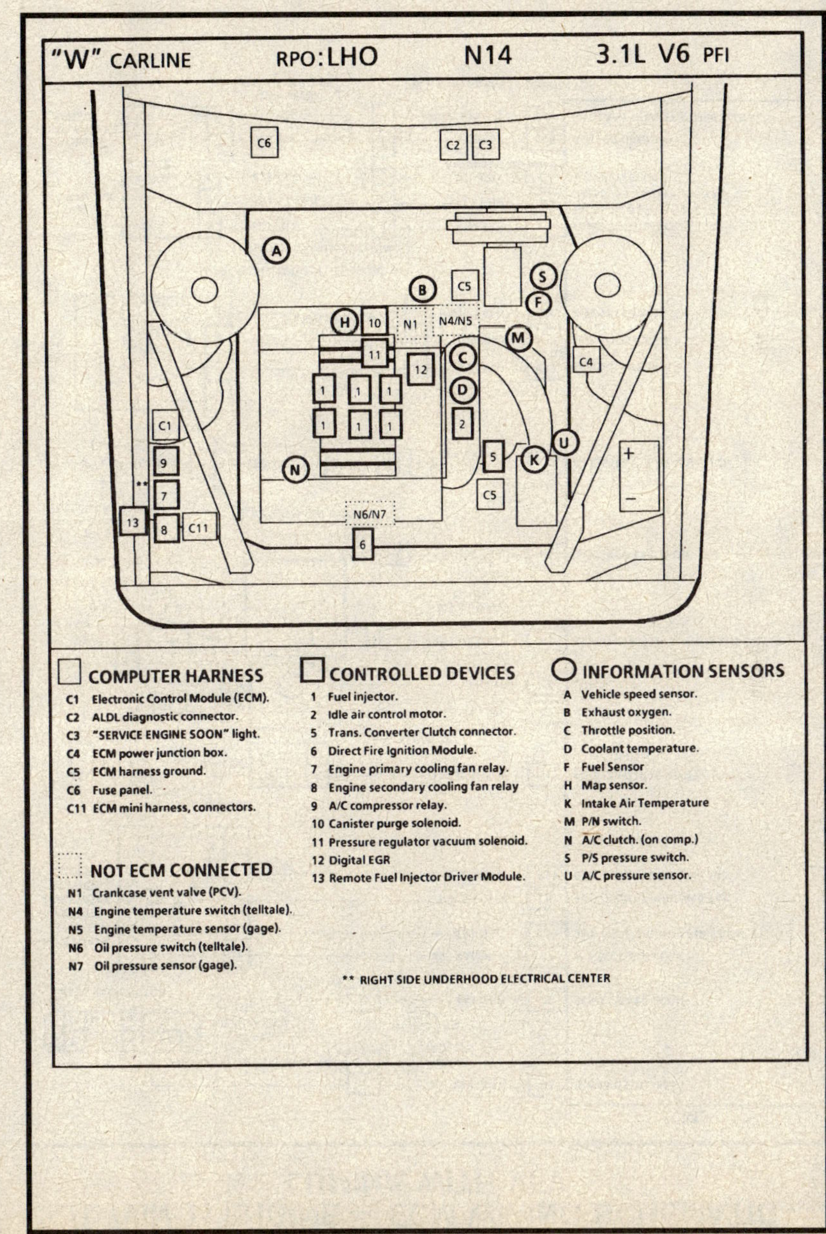

COMPUTER HARNESS
C1 Electronic Control Module (ECM).
C2 ALDL diagnostic connector.
C3 "SERVICE ENGINE SOON" light.
C4 ECM power junction box.
C5 ECM harness ground.
C6 Fuse panel.
C11 ECM mini harness, connectors.

NOT ECM CONNECTED
N1 Crankcase vent valve (PCV).
N4 Engine temperature switch (telltale).
N5 Engine temperature sensor (gage).
N6 Oil pressure switch (telltale).
N7 Oil pressure sensor (gage).

CONTROLLED DEVICES
1   Fuel injector.
2   Idle air control motor.
5   Trans. Converter Clutch connector.
6   Direct Fire Ignition Module.
7   Engine primary cooling fan relay.
8   Engine secondary cooling fan relay.
9   A/C compressor relay.
10 Canister purge solenoid.
11 Pressure regulator vacuum solenoid.
12 Digital EGR
13 Remote Fuel Injector Driver Module.

INFORMATION SENSORS
A   Vehicle speed sensor.
B   Exhaust oxygen.
C   Throttle position.
D   Coolant temperature.
F   Fuel Sensor
H   Map sensor.
K   Intake Air Temperature
M   P/N switch.
N   A/C clutch. (on comp.)
S   P/S pressure switch.
U   A/C pressure sensor.

** RIGHT SIDE UNDERHOOD ELECTRICAL CENTER

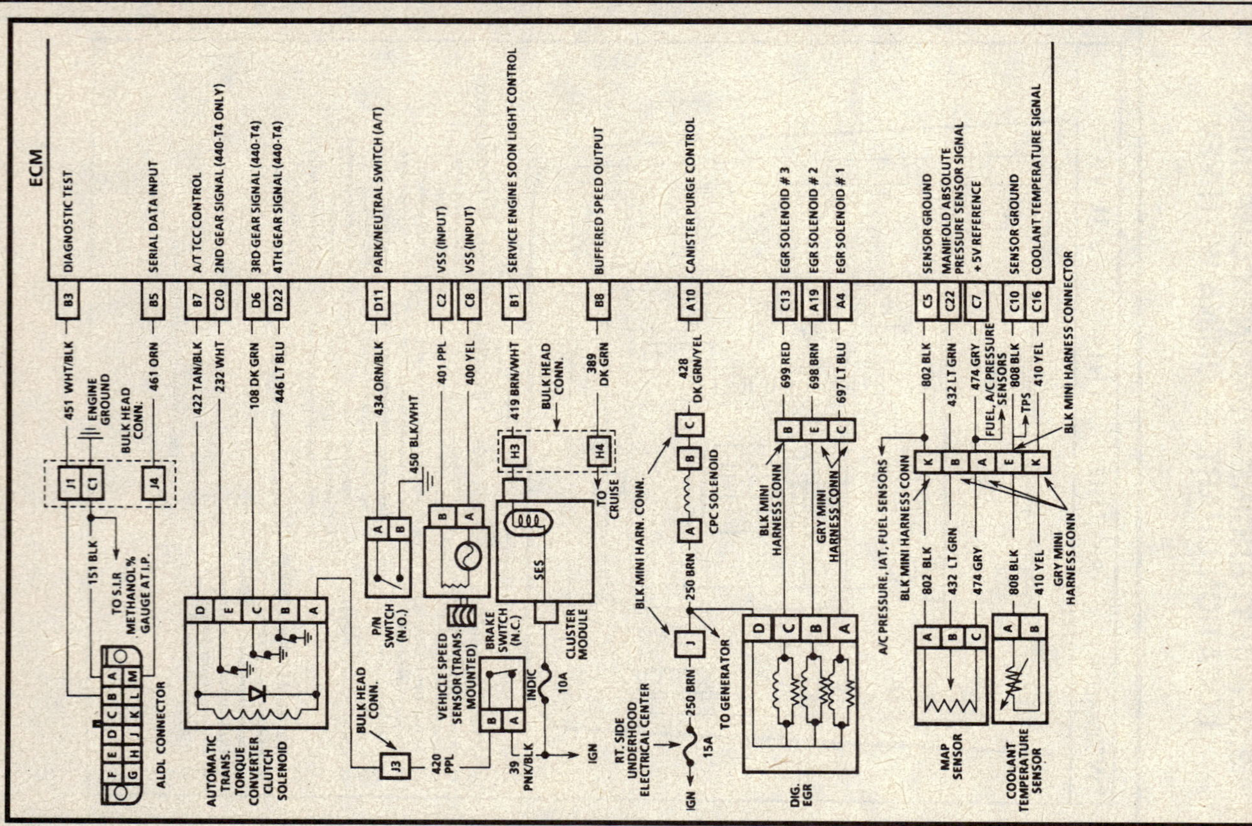

3.1L (VIN T) ENGINE —ECM WIRING SCHEMATIC— LUMINA WITH VFV
ECM
DIAGNOSTIC TEST
SERIAL DATA INPUT
A/T TCC CONTROL
2ND GEAR SIGNAL (440-T4 ONLY)
3RD GEAR SIGNAL (440-T4)
4TH GEAR SIGNAL (440-T4)
PARK/NEUTRAL SWITCH (A/T)
VSS (INPUT)
VSS (INPUT)
SERVICE ENGINE SOON LIGHT CONTROL
BUFFERED SPEED OUTPUT
CANISTER PURGE CONTROL
EGR SOLENOID # 3
EGR SOLENOID # 2
EGR SOLENOID # 1
SENSOR GROUND
MANIFOLD ABSOLUTE PRESSURE SENSOR SIGNAL
+ 5V REFERENCE
SENSOR GROUND
COOLANT TEMPERATURE SIGNAL
B3
B5
B7
C20
D6
D22
D11
C2
C8
B1
B8
A10
C13
A19
A4
C5
C22
C7
C10
C16
451 WHT/BLK
461 ORN
422 TAN/BLK
232 WHT
108 DK GRN
446 LT BLU
434 ORN/BLK
401 PPL
400 YEL
419 BRN/WHT
389 DK GRN
428 DK GRN/YEL
699 RED
698 BRN
697 LT BLU
802 BLK
432 LT GRN
474 GRY
808 BLK
410 YEL
ENGINE GROUND
BULK HEAD CONN.
151 BLK
TO S.I.R METHANOL% GAUGE AT I.P.
450 BLK/WHT
H3
H4
BULK HEAD CONN.
TO CRUISE
BLK MINI HARN. CONN.
CPC SOLENOID
BLK MINI HARNESS CONN
GRY MINI HARNESS CONN.
FUEL A/C PRESSURE SENSORS
TPS
BLK MINI HARNESS CONNECTOR
ALDL CONNECTOR
AUTOMATIC TRANS. TORQUE CONVERTER CLUTCH SOLENOID
P/N SWITCH (N.O.)
VEHICLE SPEED SENSOR (TRANS. MOUNTED)
BRAKE SWITCH (N.C.)
SES
CLUSTER MODULE
BULK HEAD CONN.
420 PPL
J3
39 PNK/BLK
INDIC
10A
IGN
250 BRN
250 BRN
TO GENERATOR
RT. SIDE UNDERHOOD ELECTRICAL CENTER
IGN 15A
DIG. EGR
A/C PRESSURE, IAT, FUEL SENSORS
BLK MINI HARNESS CONN
802 BLK
432 LT GRN
474 GRY
808 BLK
410 YEL
GRY MINI HARNESS CONN
MAP SENSOR
COOLANT TEMPERATURE SENSOR

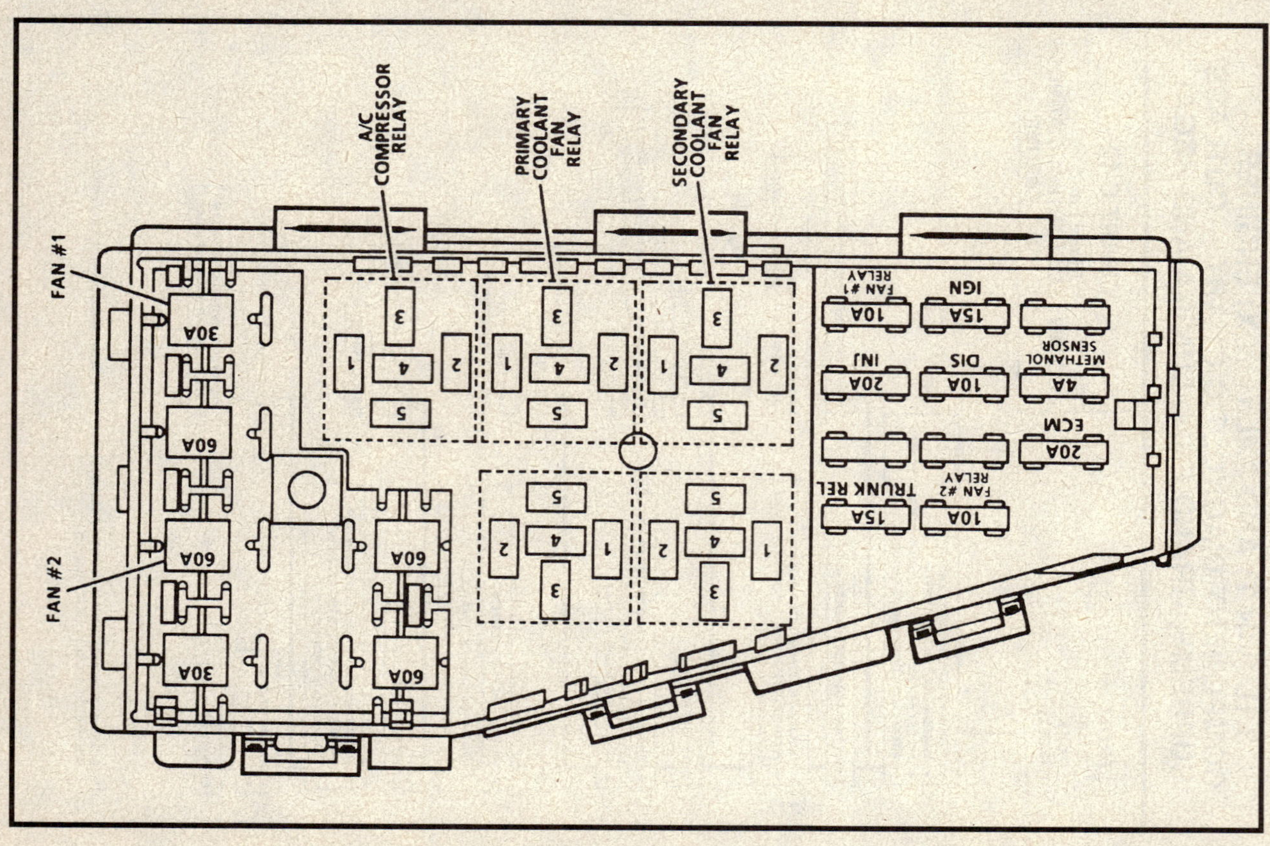

3.1L (VIN T) ENGINE —RIGHT SIDE UNDERHOOD ELECTRICAL CENTER —LUMINA WITH VFV
A/C COMPRESSOR RELAY
PRIMARY COOLANT FAN RELAY
SECONDARY COOLANT FAN RELAY
FAN #1
FAN #2
30A
60A
60A
60A
30A
60A
FAN #1 RELAY
IGN
10A
15A
INJ
DIS
METHANOL SENSOR
20A
10A
4A
FAN #2 RELAY
TRUNK REL
ECM
15A
10A
20A

3.1L (VIN T) ENGINE—ECM WIRING SCHEMATIC—LUMINA WITH VFV

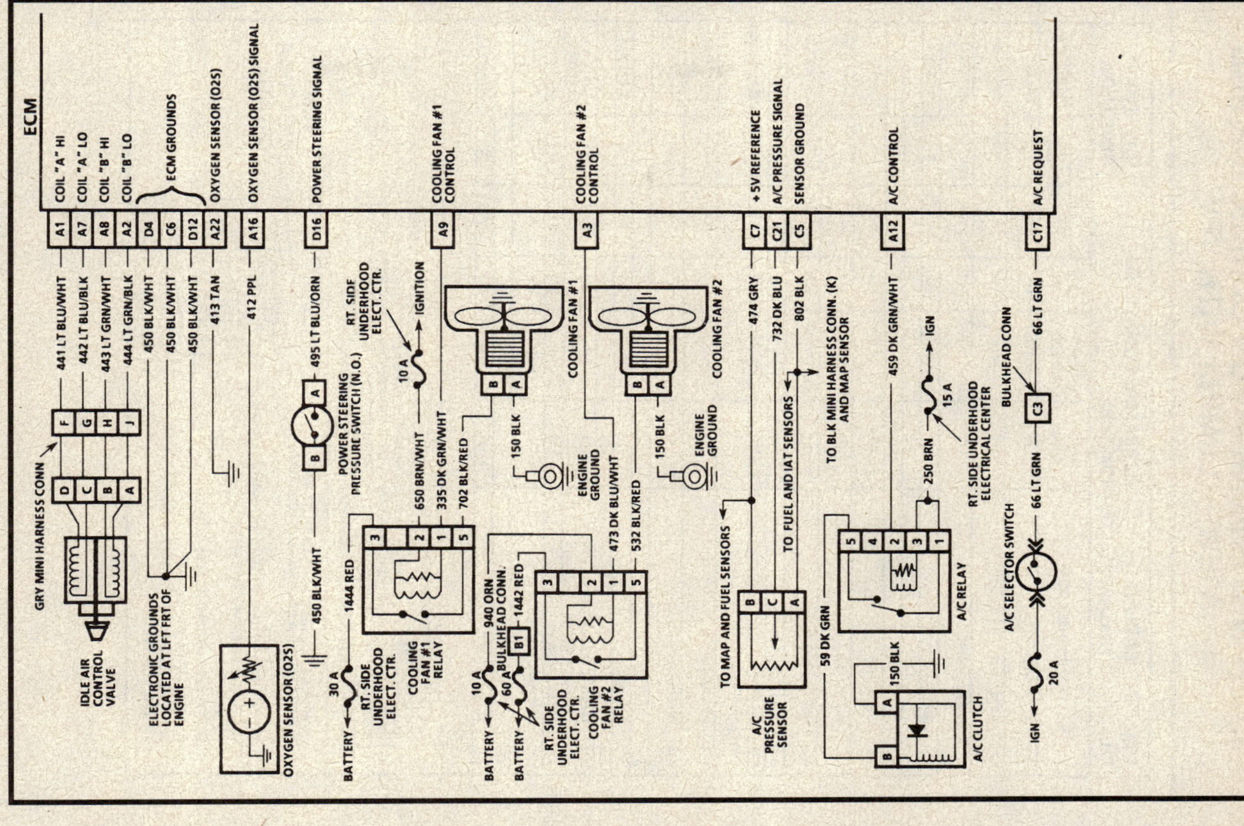

ECM
COIL "A" HI
COIL "A" LO
COIL "B" HI
COIL "B" LO
ECM GROUNDS
OXYGEN SENSOR (O2S)
OXYGEN SENSOR (O2S) SIGNAL
POWER STEERING SIGNAL
COOLING FAN #1 CONTROL
COOLING FAN #2 CONTROL
+5V REFERENCE
A/C PRESSURE SIGNAL
SENSOR GROUND
A/C CONTROL
A/C REQUEST
A1 A7 A8 A2 D4 C6 D12 A22 A16 D16 A9 A3 C7 C21 C5 A12 C17
441 LT BLU/WHT
442 LT BLU/BLK
443 LT GRN/WHT
444 LT GRN/BLK
450 BLK/WHT
450 BLK/WHT
450 BLK/WHT
413 TAN
412 PPL
495 LT BLU/ORN
474 GRY
732 DK BLU
802 BLK
459 DK GRN/WHT
66 LT GRN
IGNITION
RT. SIDE UNDERHOOD ELECT. CTR.
10 A
COOLING FAN #1
COOLING FAN #2
TO FUEL AND IAT SENSORS
TO BLK MINI HARNESS CONN. (K) AND MAP SENSOR
IGN
15 A
RT. SIDE UNDERHOOD ELECTRICAL CENTER
BULKHEAD CONN
66 LT GRN
C3
250 BRN
A/C SELECTOR SWITCH
GRY MINI HARNESS CONN
IDLE AIR CONTROL VALVE
ELECTRONIC GROUNDS LOCATED AT LFT FRT OF ENGINE
450 BLK/WHT
POWER STEERING PRESSURE SWITCH (N.O.)
650 BRN/WHT
335 DK GRN/WHT
702 BLK/RED
473 DK BLU/WHT
532 BLK/RED
150 BLK
150 BLK
ENGINE GROUND
ENGINE GROUND
OXYGEN SENSOR (O2S)
BATTERY
30 A
RT. SIDE UNDERHOOD ELECT. CTR.
1444 RED
940 ORN
BULKHEAD CONN.
B1
1442 RED
COOLING FAN #1 RELAY
COOLING FAN #2 RELAY
BATTERY
BATTERY
10 A
60 A
RT. SIDE UNDERHOOD ELECT. CTR.
TO MAP AND FUEL SENSORS
A/C PRESSURE SENSOR
59 DK GRN
A/C RELAY
150 BLK
A/C CLUTCH
IGN
20 A

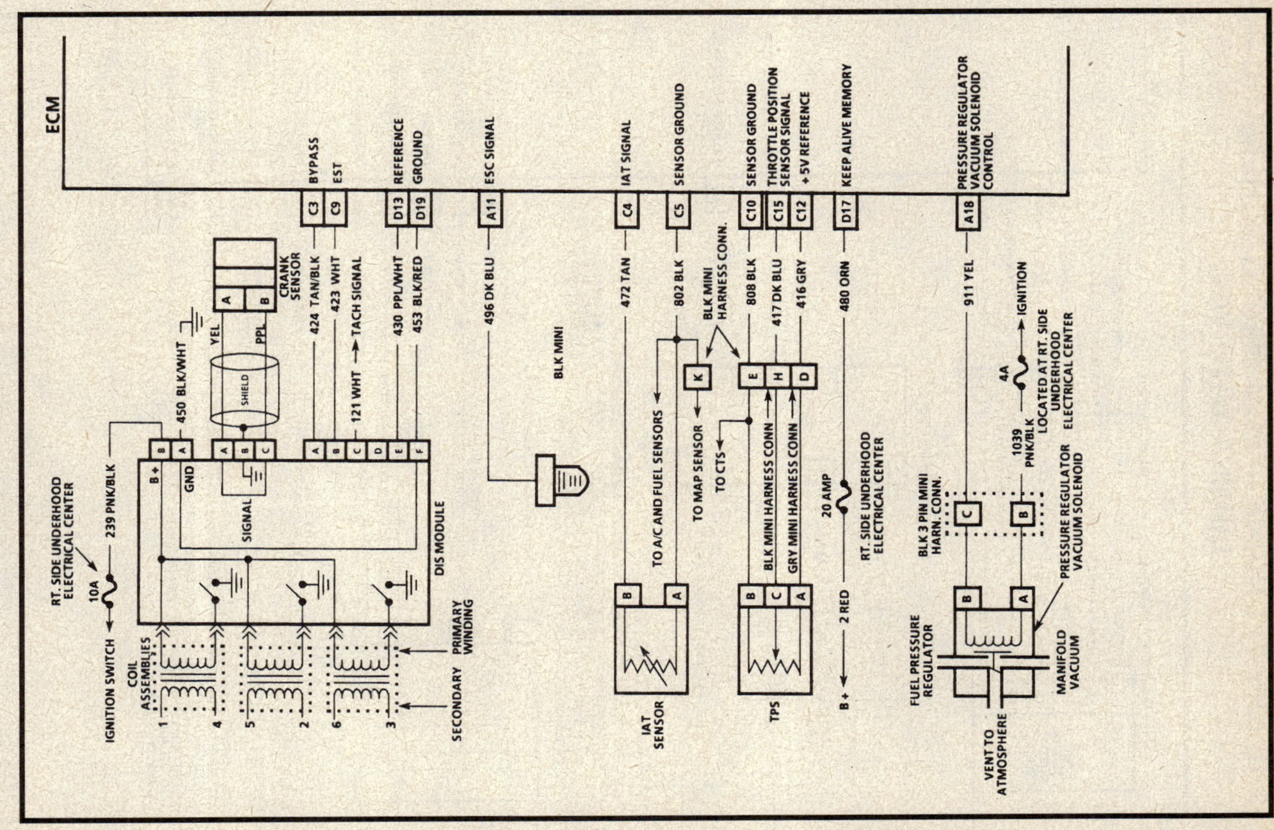

3.1L (VIN T) ENGINE—ECM WIRING SCHEMATIC—LUMINA WITH VFV
ECM
BYPASS
EST
REFERENCE
GROUND
ESC SIGNAL
IAT SIGNAL
SENSOR GROUND
SENSOR GROUND
THROTTLE POSITION SENSOR SIGNAL
+5V REFERENCE
KEEP ALIVE MEMORY
PRESSURE REGULATOR VACUUM SOLENOID CONTROL
C3 C9 D13 D19 A11 C4 C5 C10 C15 C12 D17 A18
424 TAN/BLK
423 WHT
430 PPL/WHT
453 BLK/RED
496 DK BLU
472 TAN
802 BLK
808 BLK
417 DK BLU
416 GRY
480 ORN
911 YEL
IGNITION
4A
CRANK SENSOR
YEL
450 BLK/WHT
SHIELD
SIGNAL
121 WHT
TACH SIGNAL
BLK MINI
BLK MINI HARNESS CONN.
TO A/C AND FUEL SENSORS
TO MAP SENSOR
TO CTS
BLK MINI HARNESS CONN
GRY MINI HARNESS CONN
20 AMP
RT. SIDE UNDERHOOD ELECTRICAL CENTER
BLK 3 PIN MINI HARN. CONN.
1039 PNK/BLK
LOCATED AT RT. SIDE UNDERHOOD ELECTRICAL CENTER
RT. SIDE UNDERHOOD ELECTRICAL CENTER
IGNITION SWITCH
10A
239 PNK/BLK
COIL ASSEMBLIES
B+
GND
DIS MODULE
SECONDARY
PRIMARY WINDING
IAT SENSOR
TPS
B+
2 RED
FUEL PRESSURE REGULATOR
VENT TO ATMOSPHERE
PRESSURE REGULATOR VACUUM SOLENOID
MANIFOLD VACUUM

3.1L (VIN T) ENGINE — ECM WIRING SCHEMATIC — LUMINA WITH VFV

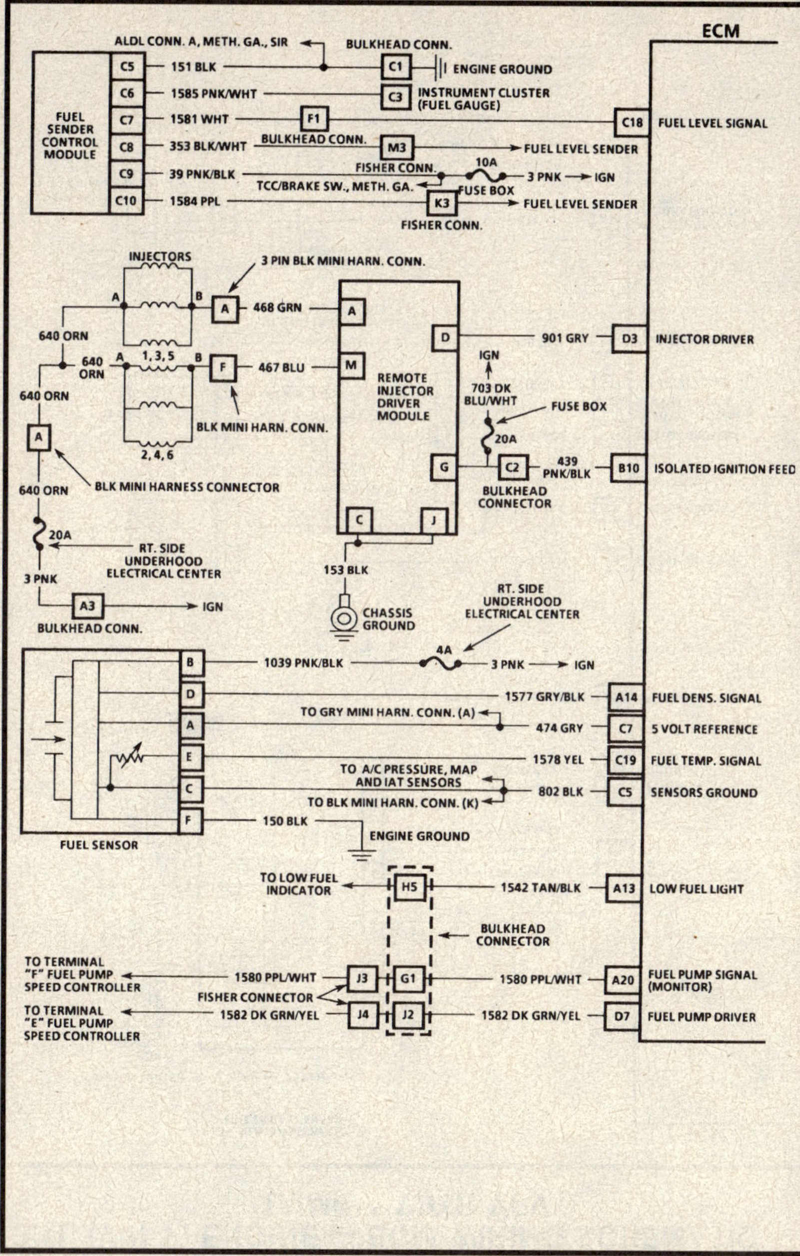

3.1L (VIN T) ENGINE — ECM CONNECTOR END VIEW — LUMINA WITH VFV

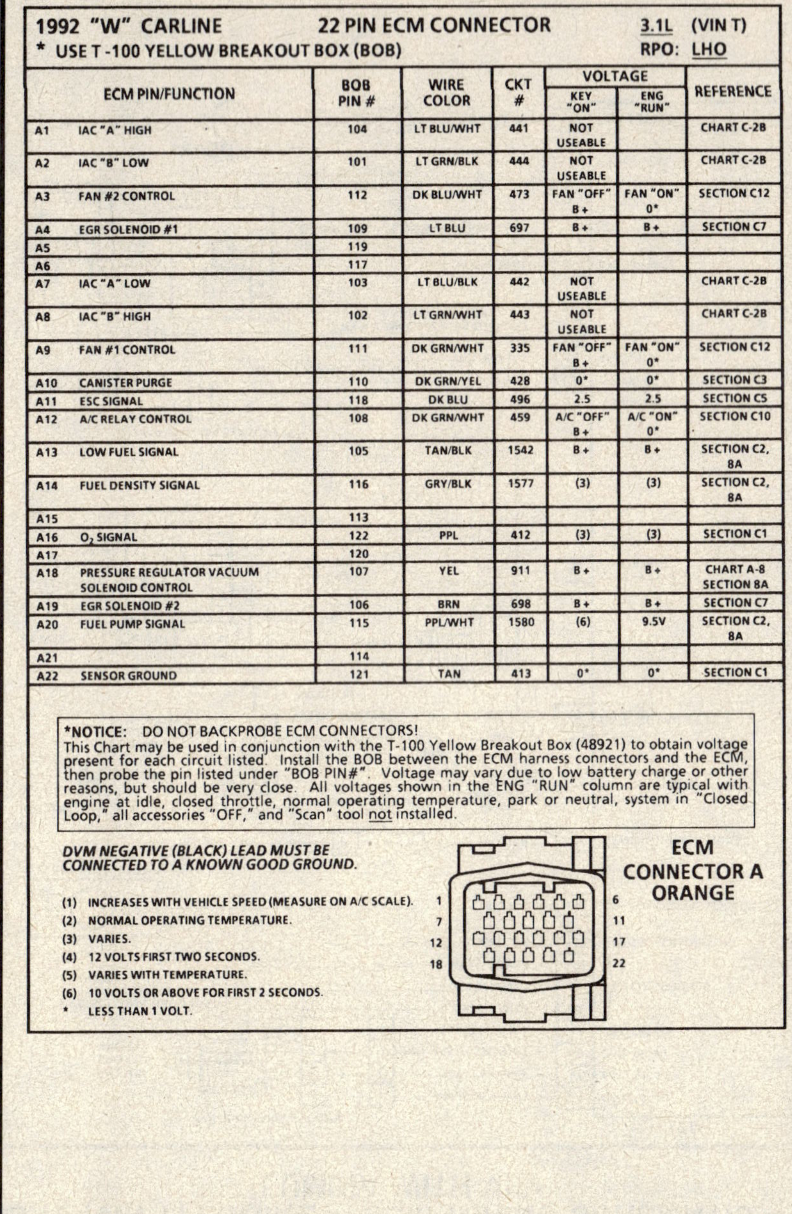

1992 "W" CARLINE 22 PIN ECM CONNECTOR 3.1L (VIN T) * USE T-100 YELLOW BREAKOUT BOX (BOB) RPO: LHO						
ECM PIN/FUNCTION	BOB PIN #	WIRE COLOR	CKT #	KEY "ON"	ENG "RUN"	REFERENCE
A1 IAC "A" HIGH	104	LT BLU/WHT	441	NOT USEABLE		CHART C-2B
A2 IAC "B" LOW	101	LT GRN/BLK	444	NOT USEABLE		CHART C-2B
A3 FAN #2 CONTROL	112	DK BLU/WHT	473	FAN "OFF" B+	FAN "ON" 0*	SECTION C12
A4 EGR SOLENOID #1	109	LT BLU	697	B+	B+	SECTION C7
A5	119					
A6	117					
A7 IAC "A" LOW	103	LT BLU/BLK	442	NOT USEABLE		CHART C-2B
A8 IAC "B" HIGH	102	LT GRN/WHT	443	NOT USEABLE		CHART C-2B
A9 FAN #1 CONTROL	111	DK GRN/WHT	335	FAN "OFF" B+	FAN "ON" 0*	SECTION C12
A10 CANISTER PURGE	110	DK GRN/YEL	428	0*	0*	SECTION C3
A11 ESC SIGNAL	118	DK BLU	496	2.5	2.5	SECTION C5
A12 A/C RELAY CONTROL	108	DK GRN/WHT	459	A/C "OFF" B+	A/C "ON" 0*	SECTION C10
A13 LOW FUEL SIGNAL	105	TAN/BLK	1542	B+	B+	SECTION C2, 8A
A14 FUEL DENSITY SIGNAL	116	GRY/BLK	1577	(3)	(3)	SECTION C2, 8A
A15	113					
A16 O₂ SIGNAL	122	PPL	412	(3)	(3)	SECTION C1
A17	120					
A18 PRESSURE REGULATOR VACUUM SOLENOID CONTROL	107	YEL	911	B+	B+	CHART A-8 SECTION 8A
A19 EGR SOLENOID #2	106	BRN	698	B+	B+	SECTION C7
A20 FUEL PUMP SIGNAL	115	PPL/WHT	1580	(6)	9.5V	SECTION C2, 8A
A21	114					
A22 SENSOR GROUND	121	TAN	413	0*	0*	SECTION C1

***NOTICE: DO NOT BACKPROBE ECM CONNECTORS!**
This Chart may be used in conjunction with the T-100 Yellow Breakout Box (48921) to obtain voltage present for each circuit listed. Install the BOB between the ECM harness connectors and the ECM, then probe the pin listed under "BOB PIN#". Voltage may vary due to low battery charge or other reasons, but should be very close. All voltages shown in the ENG "RUN" column are typical with engine at idle, closed throttle, normal operating temperature, park or neutral, system in "Closed Loop," all accessories "OFF," and "Scan" tool not installed.

DVM NEGATIVE (BLACK) LEAD MUST BE CONNECTED TO A KNOWN GOOD GROUND.

(1) INCREASES WITH VEHICLE SPEED (MEASURE ON A/C SCALE).
(2) NORMAL OPERATING TEMPERATURE.
(3) VARIES.
(4) 12 VOLTS FIRST TWO SECONDS.
(5) VARIES WITH TEMPERATURE.
(6) 10 VOLTS OR ABOVE FOR FIRST 2 SECONDS.
* LESS THAN 1 VOLT.

3.1L (VIN T) ENGINE — ECM CONNECTOR END VIEW — LUMINA WITH VFV

1992 "W" CARLINE 22 PIN ECM CONNECTOR						3.1L (VIN T) RPO: LHO
* USE T-100 YELLOW BREAKOUT BOX (BOB)				VOLTAGE		
ECM PIN/FUNCTION	BOB PIN #	WIRE COLOR	CKT #	KEY "ON"	ENG "RUN"	REFERENCE
B1 "SES" LIGHT	204	BRN/WHT	419	0*	B+	CHARTS A-1 AND A-2
B2	201					
B3 DIAGNOSTIC/TEST	212	WHT/BLK	451	(5)	(5)	CHARTS A-1 AND A-2
B4	209					
B5 SERIAL DATA	219	ORN	461	4.8 (3)	4.8 (3)	CHARTS A-1 AND A-2
B6	217					
B7 TCC	203	TAN/BLK	422	0*	0*	
B8 BUFFERED SPEED OUT	202	DK GRN	389	B+	B+	
B9	211					
B10 ISOLATED IGNITION FEED	210	PNK/BLK	439	B+	B+	
B11	218					
B12	208					
B13	205					
B14	216					
B15	213					
B16	222					
B17	220					
B18	207					
B19	206					
B20	215					
B21	214					
B22	221					

*NOTICE: DO NOT BACKPROBE ECM CONNECTORS!
This Chart may be used in conjunction with the T-100 Yellow Breakout Box (48921) to obtain voltage present for each circuit listed. Install the BOB between the ECM harness connectors and the ECM, then probe the pin listed under "BOB PIN#". Voltage may vary due to low battery charge or other reasons, but should be very close. All voltages shown in the ENG "RUN" column are typical with engine at idle, closed throttle, normal operating temperature, park or neutral, system in "Closed Loop," all accessories "OFF," and "Scan" tool not installed.

DVM NEGATIVE (BLACK) LEAD MUST BE CONNECTED TO A KNOWN GOOD GROUND.

(1) INCREASES WITH VEHICLE SPEED (MEASURE ON A/C SCALE).
(2) NORMAL OPERATING TEMPERATURE.
(3) VARIES.
(4) 12 VOLTS FIRST TWO SECONDS.
(5) VARIES WITH TEMPERATURE.
(6) 10 VOLTS OR ABOVE FOR FIRST 2 SECONDS.
* LESS THAN 1 VOLT.

ECM CONNECTOR B WHITE

3.1L (VIN T) ENGINE — ECM CONNECTOR END VIEW — LUMINA WITH VFV

1992 "W" CARLINE 22 PIN ECM CONNECTOR						3.1L (VIN T) RPO: LHO
* USE T-100 YELLOW BREAKOUT BOX (BOB)				VOLTAGE		
ECM PIN/FUNCTION	BOB PIN #	WIRE COLOR	CKT #	KEY "ON"	ENG "RUN"	REFERENCE
C1	304					
C2 MAG. VSS SIGNAL LOW	301	PPL	401	0*	(1)	
C3 DIS BYPASS	312	TAN/BLK	424	0*	5	
C4 IAT SENSOR SIGNAL	309	TAN	472	(5)	(5)	
C5 SENSOR GROUND	319	BLK	802	0*	0*	
C6 GROUND	317	BLK/WHT	450	0*	0*	
C7 +5 VOLT REFERENCE (MAP) A/C PRESSURE	303	GRY	474	5	5	CHART C-1D
C8 MAG. VSS SIGNAL HIGH	302	YEL	400	0*	(1)	
C9 EST CONTROL	311	WHT	423	0*	1.3 (3)	
C10 SENSOR GROUND	310	BLK	808	0*	0*	
C11	318					
C12 +5 VOLT REFERENCE (TPS)	308	GRY	416	5	5	
C13 EGR SOLENOID #3	305	RED	699	B+	B+	
C14	316					
C15 TPS SIGNAL	313	DK BLU	417	.88	.88	
C16 COOLANT TEMPERATURE SIGNAL	322	YEL	410	(5)	(5)	
C17 A/C REQUEST	320	LT GRN	66	A/C REQUEST 0*	A/C "ON" B+	
C18 FUEL LEVEL SIGNAL	307	WHT	1581	(3)	(3)	
C19 FUEL TEMPERATURE SIGNAL	306	YEL	1578	(3)	(3)	
C20 2nd GEAR SIGNAL (440-T4)	315	WHT	232	0*	0*	
C21 A/C PRESSURE SIGNAL	314	DK BLU	732	A/C OFF 0* A/C ON 12V	A/C OFF 0* A/C ON 12V	
C22 MAP SIGNAL	321	LT GRN	432	4.75	(3)	CHART C-1D

*NOTICE: DO NOT BACKPROBE ECM CONNECTORS!
This Chart may be used in conjunction with the T-100 Yellow Breakout Box (48921) to obtain voltage present for each circuit listed. Install the BOB between the ECM harness connectors and the ECM, then probe the pin listed under "BOB PIN#". Voltage may vary due to low battery charge or other reasons, but should be very close. All voltages shown in the ENG "RUN" column are typical with engine at idle, closed throttle, normal operating temperature, park or neutral, system in "Closed Loop," all accessories "OFF," and "Scan" tool not installed.

DVM NEGATIVE (BLACK) LEAD MUST BE CONNECTED TO A KNOWN GOOD GROUND.

(1) INCREASES WITH VEHICLE SPEED (MEASURE ON A/C SCALE).
(2) NORMAL OPERATING TEMPERATURE.
(3) VARIES.
(4) 12 VOLTS FIRST TWO SECONDS.
(5) VARIES WITH TEMPERATURE.
(6) 10 VOLTS OR ABOVE FOR FIRST 2 SECONDS.
* LESS THAN 1 VOLT.

ECM CONNECTOR C GREEN

3.1L (VIN T) ENGINE — ECM CONNECTOR END VIEW — LUMINA WITH VFV

1992 "W" CARLINE	22 PIN ECM CONNECTOR			3.1L (VIN T) RPO: LHO		
* USE T-100 YELLOW BREAKOUT BOX (BOB)						
ECM PIN/FUNCTION	BOB PIN #	WIRE COLOR	CKT #	KEY "ON"	ENG "RUN"	REFERENCE
D1	404					
D2	401					
D3 INJECTOR DRIVER	412	GRY	901	B+	B+	
D4 GROUND	409	BLK/WHT	450	0*	0*	
D5	419					
D6 3rd GEAR SIGNAL (440-T4)	417	DK GRN	108	0*	0*	
D7 FUEL PUMP RELAY DRIVER	403	DK GRN/YEL	1582	0* (4)	B+	
D8	402					
D9	411					
D10	410					
D11 P/N SWITCH (A/T)	418	ORN/BLK	434	0*	0*	CHART C-1A
D12 GROUND	408	BLK/WHT	450	0*	0*	
D13 DIS REFERENCE HIGH	405	PPL/WHT	430	0*	2.3 (3)	
D14	416					
D15	413					
D16 P/S PRESSURE SIGNAL	422	LT BLU/ORN	495	B+	B+	CHART C-1E
D17 BATTERY FEED	420	ORN	480	B+	B+	
D18	407					
D19 DIS REFERENCE LOW	406	BLK/RED	453	0*	0*	
D20	415					
D21	414					
D22 4th GEAR SIGNAL (440-T4)	421	LT BLU	446	0*	0*	

*NOTICE: DO NOT BACKPROBE ECM CONNECTORS!
This Chart may be used in conjunction with the T-100 Yellow Breakout Box (48921) to obtain voltage present for each circuit listed. Install the BOB between the ECM harness connectors and the ECM, then probe the pin listed under "BOB PIN#". Voltage may vary due to low battery charge or other reasons, but should be very close. All voltages shown in the ENG "RUN" column are typical with engine at idle, closed throttle, normal operating temperature, park or neutral, system in "Closed Loop," all accessories "OFF," and "Scan" tool not installed.

DVM NEGATIVE (BLACK) LEAD MUST BE CONNECTED TO A KNOWN GOOD GROUND.

(1) INCREASES WITH VEHICLE SPEED (MEASURE ON A/C SCALE).
(2) NORMAL OPERATING TEMPERATURE.
(3) VARIES.
(4) 12 VOLTS FIRST TWO SECONDS.
(5) VARIES WITH TEMPERATURE.
(6) 10 VOLTS OR ABOVE FOR FIRST 2 SECONDS.
*   LESS THAN 1 VOLT.

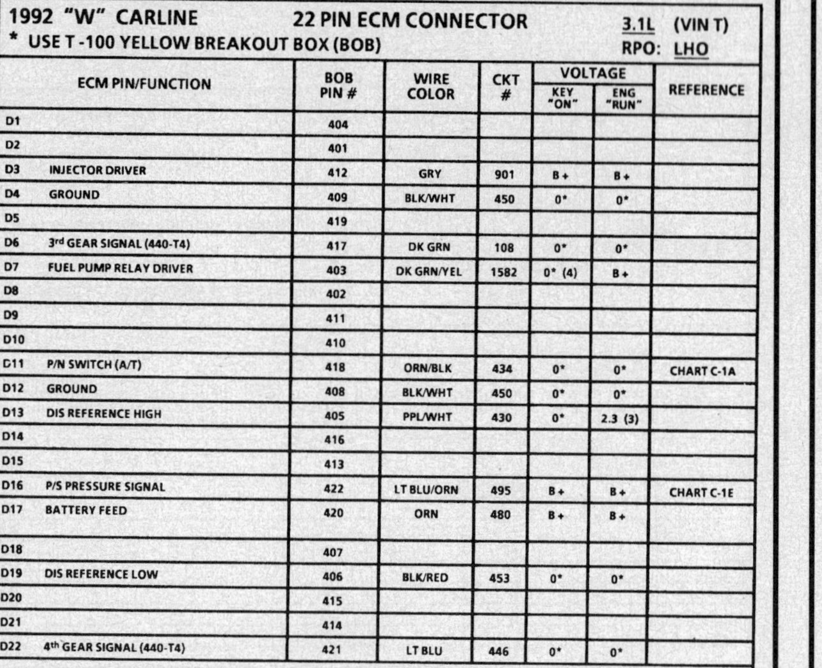

3.1L (VIN T) ENGINE — ON-BOARD DIAGNOSTIC SYSTEM CHART — LUMINA WITH VFV

DIAGNOSTIC CIRCUIT CHECK
3.1L (VIN T) "W" CARLINE (PORT)

Circuit Description:

The diagnostic circuit check is an organized approach to identifying a problem created by an Electronic Engine Control System malfunction. It must be the starting point for any driveability complaint diagnosis, because it directs the service technician to the next logical step in diagnosing the complaint. Understanding the chart and using it correctly will reduce diagnostic time and prevent the unnecessary replacement of good parts.

Test Description: Number(s) refer to circled number(s) on the diagnostic chart.

1. This step is a check for the proper operation of the "Service Engine Soon" light. The "SES" light should be "ON" steady.
2. No "SES" light at this point indicates that there is a problem with the "SES" light circuit or the ECM control of that circuit.
3. This test checks the ability of the ECM to control the "SES" light. With the diagnostic terminal grounded, the "SES" light should flash a Code 12 three times, followed by any trouble code stored in memory. A MEM-CAL error may result in the inability to flash Code 12.
4. Most of the diagnostic charts use a Tech 1 to aid diagnosis, therefore, serial data must be available. If a MEM-CAL error is present, the ECM may have been able to flash Code 12 or 51, but not transmit serial data.
5. Although the ECM is powered up, a "Cranks But Will Not Run" symptom could exist because of an ECM or system problem.
6. This step will isolate if the customer complaint is a "SES" light or a driveability problem with no "SES" light. Refer to ECM Diagnostic Codes chart in this section for a list of valid codes. An invalid code may be the result of a faulty "Scan" tool, MEM-CAL or ECM.
7. Comparison of actual control system data with the typical valves is a quick check to determine if any parameter is not within limits. Keep in mind that a base engine problem (i.e. advanced cam timing) may substantially alter sensor values.
8. If the actual data is not within the typical values established, the charts in "Component Systems," Section   will provide a functional check of the suspect component of system.

3.1L (VIN T) ENGINE — ON-BOARD DIAGNOSTIC SYSTEM CHART — LUMINA WITH VFV

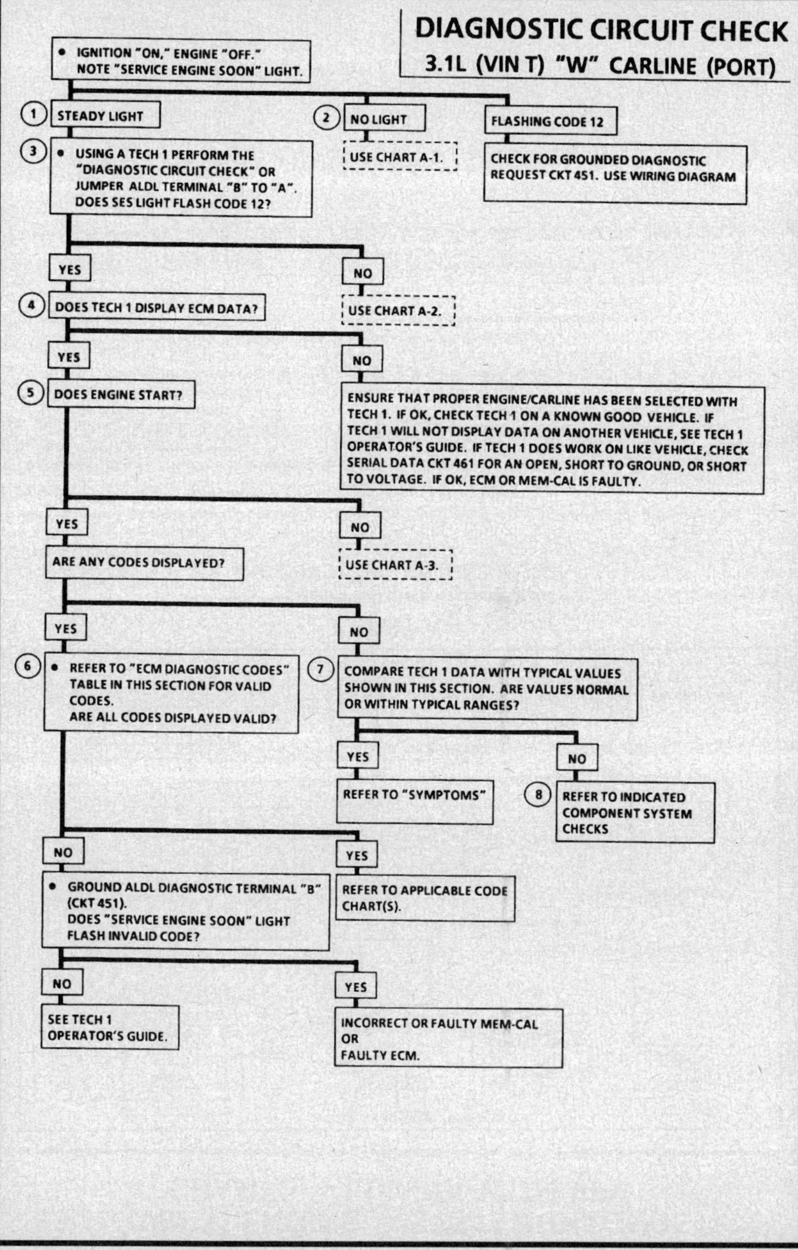

3.1L (VIN T) ENGINE — ECM DIAGNOSTIC TROUBLE CODES — LUMINA WITH VFV

CODE	DESCRIPTION	ILLUMINATE SES
13	Oxygen Sensor - open circuit	YES
14	Coolant Temperature Sensor - high temperature indicated	YES
15	Coolant Temperature Sensor - low temperature indicated	YES
21	Throttle Position Sensor - signal voltage high	YES
22	Throttle Position Sensor - signal voltage low	YES
23	Intake Air Temperature Sensor - low temperature indicated	YES
24	Vehicle Speed Sensor Circuit	YES
25	Intake Air Temperature Sensor - high temperature indicated	YES
32	Digital EGR - Exhaust Gas Recirculation (EGR) circuit (Electrical Diagnosis)	YES
33	MAP Sensor - high voltage, low vacuum indicated	YES
34	MAP Sensor - low voltage, high vacuum indicated	YES
35	Idle Air Control (IAC) Circuit	YES
41	Cylinder Select Error (Faulty or incorrect MEM-CAL)	YES
42	Electronic Spark Timing (EST) Circuit	YES
43	Electronic Spark Control (ESC) Circuit	YES
44	Oxygen Sensor - lean exhaust indicated	YES
45	Oxygen Sensor - rich exhaust indicated	YES
51	MEM-CAL Error (Faulty or incorrect MEM-CAL)	YES
53	System Over Voltage	YES
54	Fuel Pump Circuit - low voltage	YES
55	Fuel Pump Circuit - high voltage	YES
56	Fuel Sensor Failure - low signal voltage	YES
57	Fuel Sensor Failure - high signal voltage	YES
58	Degraded Fuel Sensor	YES
61	Degraded Oxygen (O_2) Sensor	YES
64	Fuel Temperature Sensor High - low voltage	YES
65	Fuel Temperature Sensor Low - high voltage	YES
66	A/C Pressure Sensor Circuit	NO

If a code not listed above appears on Tech 1, ground ALDL diagnostic terminal "B" and observe flashed codes. If code does not reappear, refer to Tech 1 Operator's Guide.
If code does reappear, check for incorrect or faulty MEM-CAL.

3.1L (VIN T) ENGINE — SYSTEM DIAGNOSTIC CHARTS — LUMINA WITH VFV

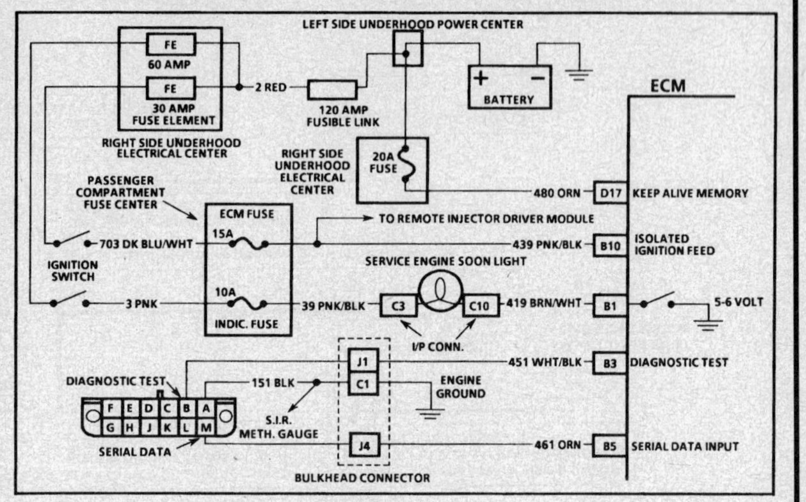

CHART A-1

**NO "SERVICE ENGINE SOON" LIGHT
3.1L (VIN T) "W" CARLINE (PORT)**

Circuit Description:

There should always be a steady "Service Engine Soon" light when the ignition is "ON" and engine stopped. Battery voltage is supplied directly to the light bulb. The Electronic Control Module (ECM) will control the light and turn it "ON" by providing a ground path through CKT 419 to the ECM.

Test Description: Number(s) below refer to circled number(s) on the diagnostic chart.
1. If the fuse in holder is blown refer to facing page of Code 54 for complete circuit.
2. Using a test light connected to 12 volts probe each of the system ground circuits to be sure a good ground is present. See ECM terminal end view in front of this section for ECM pin locations of ground circuits.

Diagnostic Aids:

Engine runs OK, check:
- Faulty light bulb.
- CKT 419 open.
- Indic. fuse blown.

Engine cranks but will not run.
- Fusible link open.
- ECM ignition fuse open.
- Battery CKT 480 to ECM open.
- Ignition CKT 439 to ECM open.
- Poor connection to ECM.

3.1L (VIN T) ENGINE — SYSTEM DIAGNOSTIC CHARTS — LUMINA WITH VFV

CHART A-1

**NO "SERVICE ENGINE SOON" LIGHT
3.1L (VIN T) "W" CARLINE (PORT)**

DOES THE ENGINE START?

YES
- IGNITION "OFF."
- DISCONNECT ECM CONNECTORS.
- IGNITION "ON."
- PROBE CKT 419, WITH TEST LIGHT TO GROUND. IS THE "SES" LIGHT "ON"?

  YES
  FAULTY ECM CONNECTION OR ECM.

  NO
  CHECK:
  - INDIC. FUSE
  - FAULTY BULB
  - OPEN CKT 419
  - CKT 419 SHORTED TO VOLTAGE
  - OPEN IGNITION FEED TO BULB.

NO
IS THE "20A" FUSE LOCATED IN THE RIGHT SIDE UNDERHOOD ELECTRICAL CENTER, 120 AMP FUSIBLE LINK AND ECM FUSE OK?

  YES
  - IGNITION "OFF."
  - DISCONNECT ECM CONNECTORS.
  - IGNITION "ON."
  - PROBE CKT 480 & 439 WITH TEST LIGHT TO GROUND. IS THE LIGHT "ON" ON BOTH CIRCUITS?

    YES
    (2) FAULTY ECM GROUNDS OR ECM.

    NO
    REPAIR OPEN IN CIRCUIT THAT DID NOT LIGHT THE TEST LIGHT.

  NO
  (1) LOCATE AND CORRECT SHORT TO GROUND IN CIRCUIT THAT HAD A BLOWN FUSE.

"AFTER REPAIRS," CONFIRM "CLOSED LOOP" OPERATION AND NO "SERVICE ENGINE SOON" LIGHT.

3.1L (VIN T) ENGINE — SYSTEM DIAGNOSTIC CHARTS — LUMINA WITH VFV

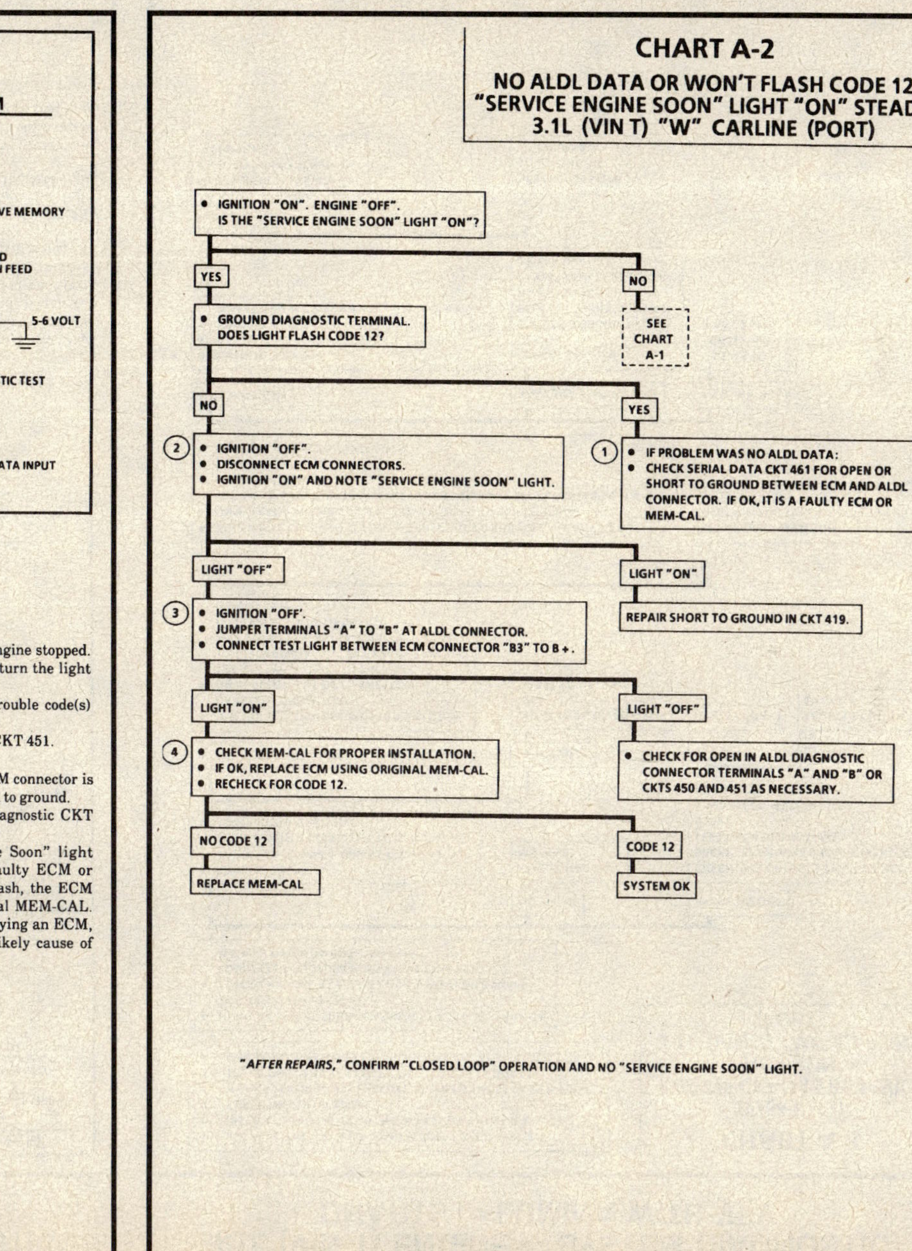

CHART A-2
**NO ALDL DATA OR WON'T FLASH CODE 12
"SERVICE ENGINE SOON" LIGHT "ON" STEADY
3.1L (VIN T) "W" CARLINE (PORT)**

Circuit Description:

There should always be a steady "Service Engine Soon" light when the ignition is "ON" and engine stopped. Battery ignition voltage is supplied to the light bulb. The Electronic Control Module (ECM) will turn the light "ON" by grounding CKT 419 at the ECM.

With the diagnostic terminal grounded, the light should flash a Code 12, followed by any trouble code(s) stored in memory.

A steady light suggests a short to ground in the light control CKT 419, or an open in diagnostic CKT 451.

Test Description: Number(s) below refer to circled number(s) on the diagnostic chart.

1. If there is a problem with the ECM that causes a "Scan" tool not to read serial data, the ECM should not flash a Code 12. If Code 12 is flashing check for CKT 451 short to ground. If Code 12 does flash be sure that the "Scan" tool is working properly on another vehicle. If the "Scan" is functioning properly and CKT 461 is OK the MEM-CAL or ECM may be at fault for the no ALDL symptom.

2. If the light goes "OFF" when the ECM connector is disconnected, CKT 419 is not shorted to ground.

3. This step will check for an open diagnostic CKT 451.

4. At this point the "Service Engine Soon" light wiring is OK. The problem is a faulty ECM or MEM-CAL. If Code 12 does not flash, the ECM should be replaced using the original MEM-CAL. Replace the MEM-CAL only after trying an ECM, as a defective MEM-CAL is an unlikely cause of the problem.

3.1L (VIN T) ENGINE — SYSTEM DIAGNOSTIC CHARTS — LUMINA WITH VFV

CHART A-2
**NO ALDL DATA OR WON'T FLASH CODE 12
"SERVICE ENGINE SOON" LIGHT "ON" STEADY
3.1L (VIN T) "W" CARLINE (PORT)**

3.1L (VIN T) ENGINE — SYSTEM DIAGNOSTIC CHARTS — LUMINA WITH VFV

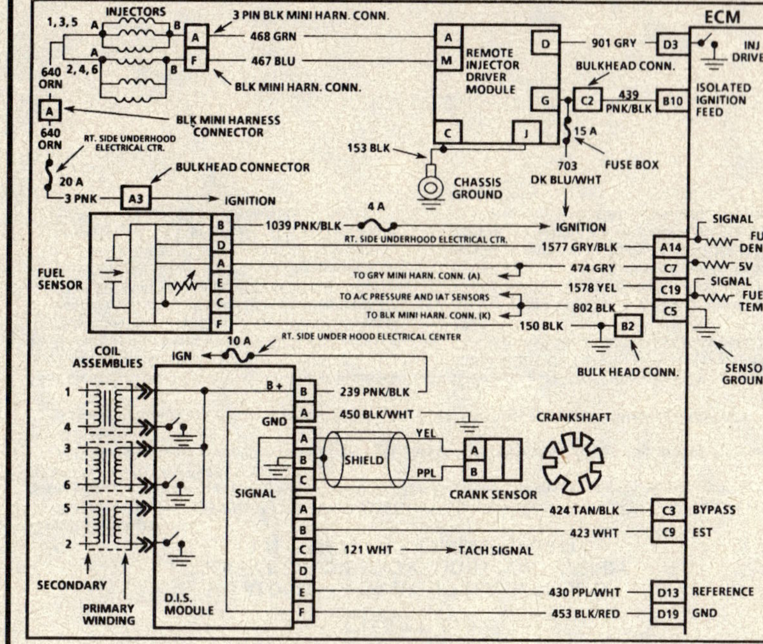

CHART A-3
(Page 1 of 3)
ENGINE CRANKS BUT WILL NOT RUN
3.1L (VIN T) "W" CARLINE (PORT)

Circuit Description:

This chart assumes that battery condition and engine cranking speed are OK, and there is adequate fuel in the tank.

Test Description: Number(s) below refer to circled number(s) on the diagnostic chart.

1. A "Service Engine Soon" light "ON" is a basic test to determine if there is a 12 volts supply and ignition 12 volts to ECM. No ALDL may be due to an ECM problem and CHART A-2 will diagnose the ECM. If TPS is over 2.5 volts, the engine may be in the clear flood mode which will cause starting problems. Code 54 or 56 can cause a "Cranks But Won't Start." The engine will not start without reference pulses and therefore the "Scan" should read rpm (reference) during crank.

2. Because the direct ignition system uses two plugs and wires to complete the circuit of each coil, the opposite spark should be left connected. If rpm was indicated during crank, the ignition module is receiving a crank signal, but no spark at this test indicates the ignition module is not triggering the coils.

3. The test light should blink, indicating the ECM is in control of the injectors. How bright the light blinks is not important. However, the test light should be a J 34730-3 or equivalent.

4. Use fuel pressure gage J 34730-FF or equivalent. Wrap a shop towel around the fuel pressure tap to absorb any small amount of fuel leakage that may occur when installing the gage.

3.1L (VIN T) ENGINE — SYSTEM DIAGNOSTIC CHARTS — LUMINA WITH VFV

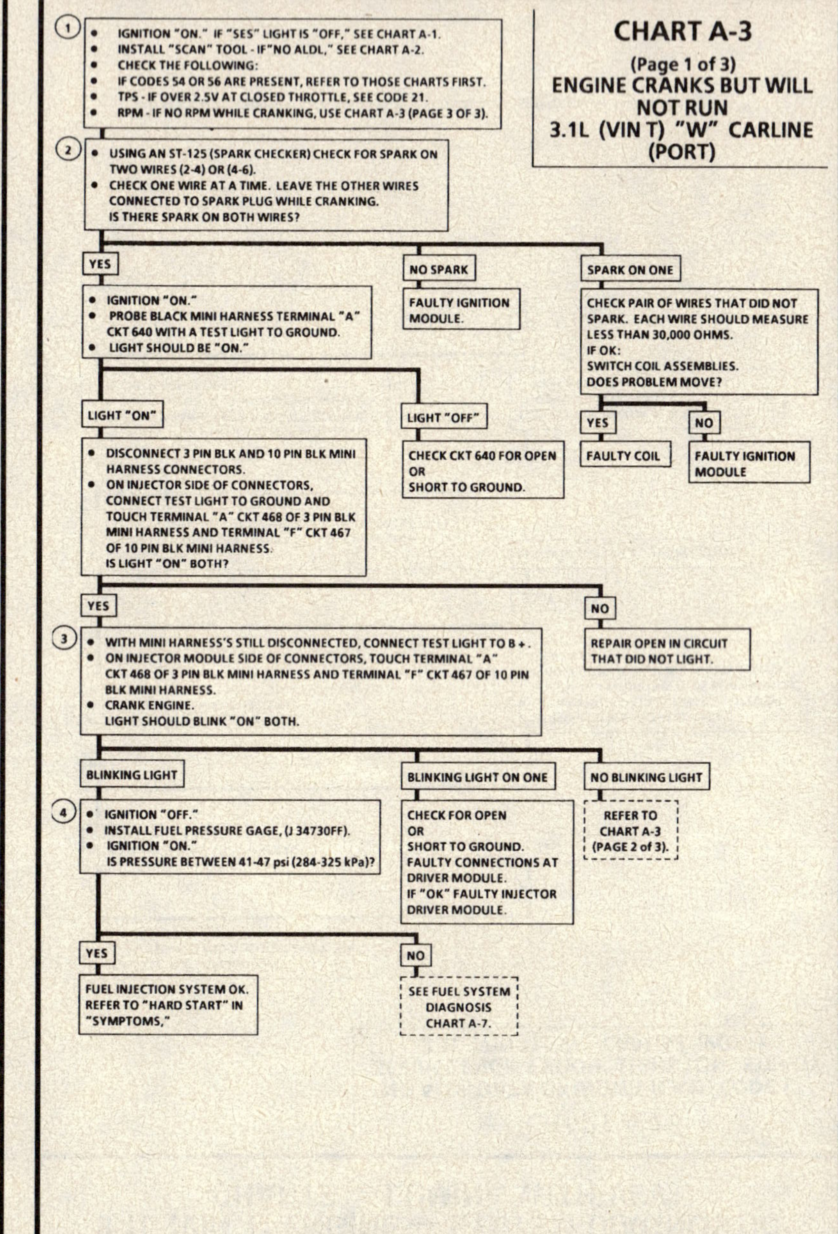

CHART A-3
(Page 1 of 3)
ENGINE CRANKS BUT WILL NOT RUN
3.1L (VIN T) "W" CARLINE (PORT)

3.1L (VIN T) ENGINE — SYSTEM DIAGNOSTIC CHARTS — LUMINA WITH VFV

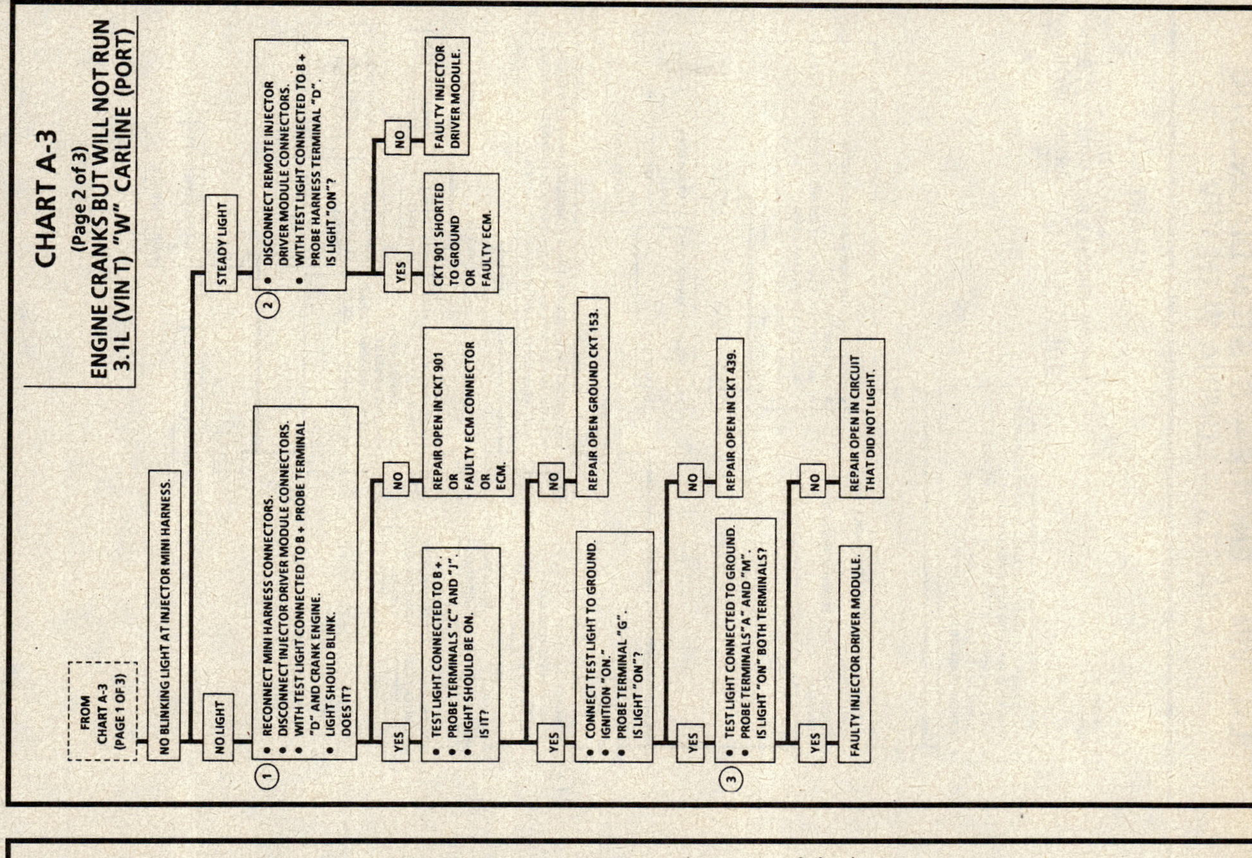

3.1L (VIN T) ENGINE — SYSTEM DIAGNOSTIC CHARTS — LUMINA WITH VFV

Test Description: Number(s) below refer to circled number(s) on the diagnostic chart.

1. This test will determine if the ECM and CKT 901 are working properly.

2. This test will determine if CKT 901 is shorted to ground or if the ECM driver or injector driver is OK.

3. This test will determine if the remote injector driver module is functioning properly.

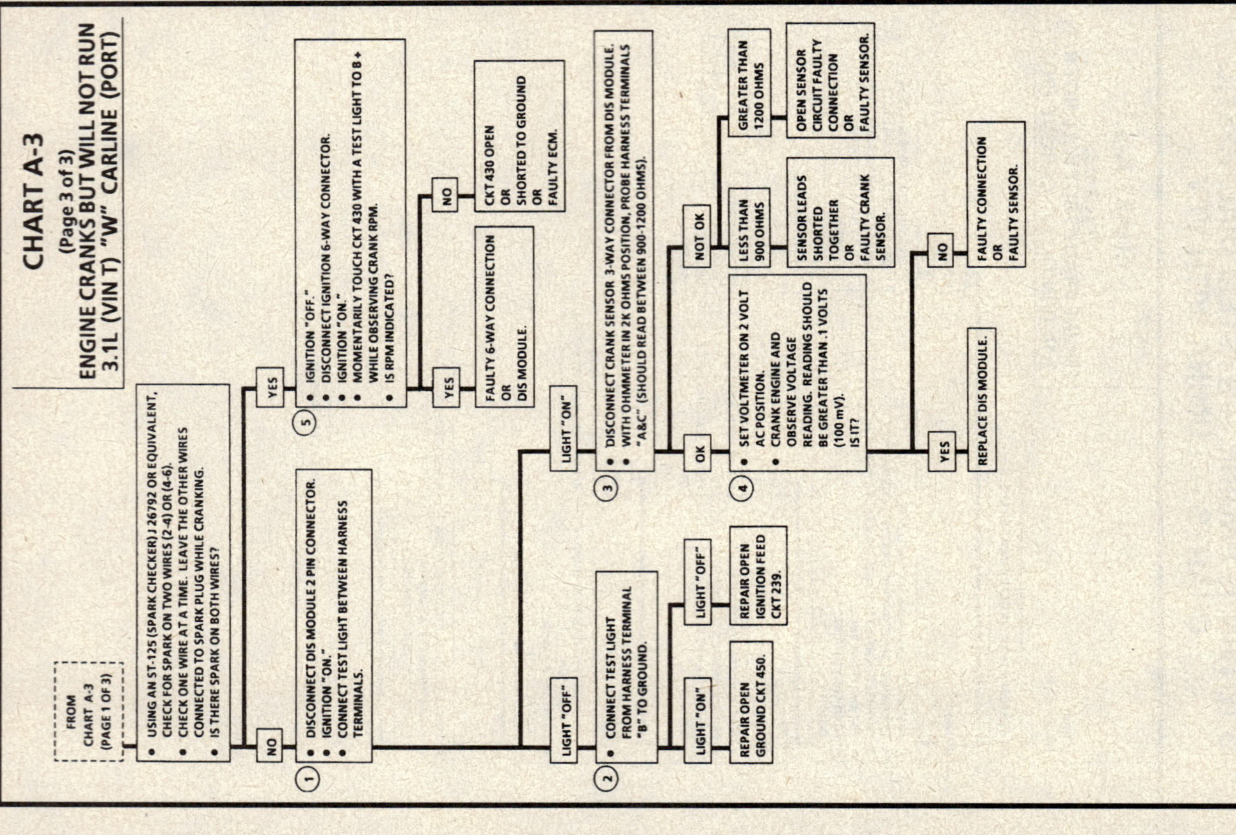

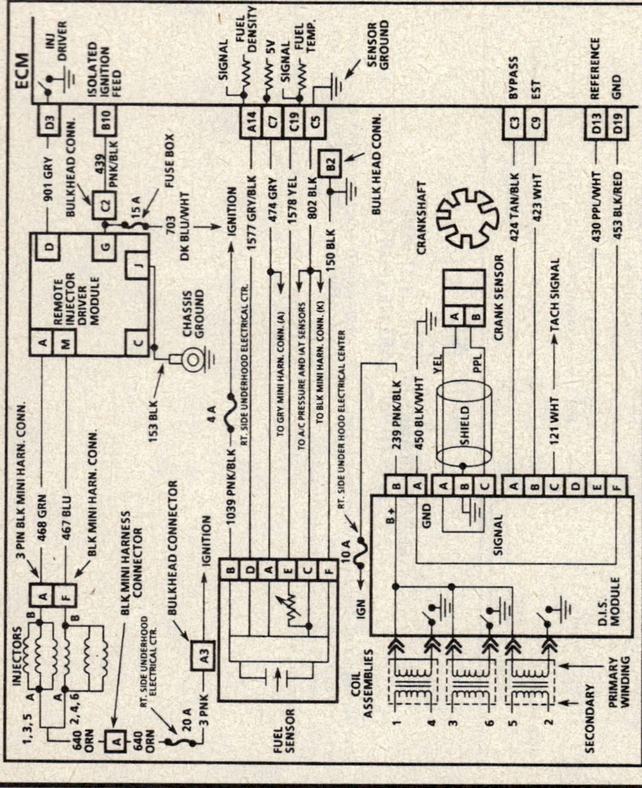

Circuit Description:
If the "Scan" tool did not indicate a cranking rpm and there is no spark present at the plugs, the problem lies in the direct ignition system or the power and ground supplies to the module.

The magnetic crank sensor is used to determine engine crankshaft position much the same way as the pick-up coil did in distributor type systems. The sensor is mounted in the block near a seven slot wheel on the crank shaft. The rotation of the wheel creates a flux change in the sensor which produces a voltage signal. The ignition module then processes this signal and creates the reference pulses needed by the ECM and the signal triggers the correct coil at the correct time.

Test Description: Number(s) below refer to circled number(s) on the diagnostic chart.
1. This test will determine if the 12 volts supply and a good ground is available at the ignition module.
2. Tests for continuity of CKT 239 to the ignition module. If test light does not light but the "SES" light is "ON" with ignition "ON," repair open in CKT 239 between DIS ignition module and splice.
3. Checks for continuity of the crank sensor and connections.
4. Voltage will vary in this test depending on cranking speed of engine.
5. This test will determine if this ignition module is not generating the reference pulse or if the wiring or ECM is at fault. By touching and removing a test light to B+ on CKT 430, a reference pulse should be generated. If rpm is indicated, the ECM and wiring are OK.

3.1L (VIN T) ENGINE — SYSTEM DIAGNOSTIC CHARTS — LUMINA WITH VFV

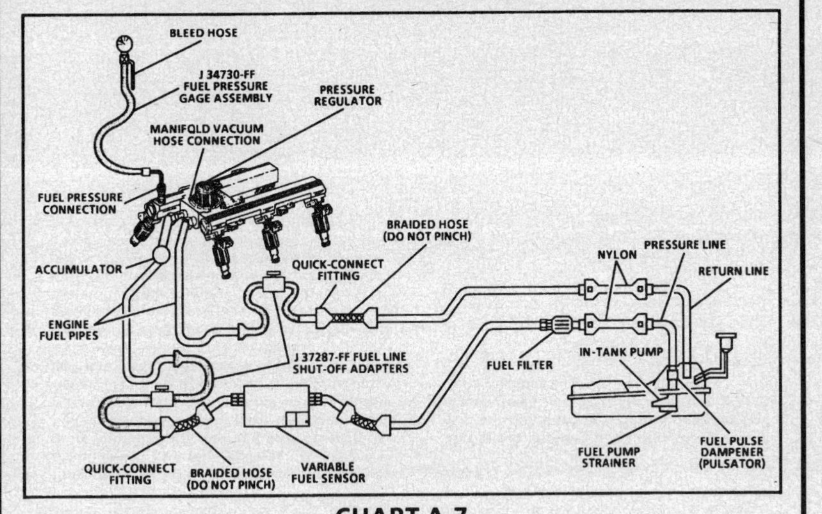

CHART A-7
(Page 1 of 3)
1992 - FUEL SYSTEM DIAGNOSIS
3.1L (VIN T) "W" CARLINE (PORT)

Circuit Description:
When the ignition switch is turned "ON," the Electronic Control Module (ECM) will turn "ON" the in-tank fuel pump. It will remain "ON" as long as the engine is cranking or running, and the ECM is receiving reference pulses. If there are no reference pulses, the ECM will shut "OFF" the fuel pump within 2 seconds after ignition "ON" or engine stops.

An electric fuel pump, attached to the fuel sender assembly (inside the fuel tank) pump fuel through and in-line fuel filter to the fuel rail assembly. The pump is designed to provide fuel at a pressure above the regulated pressure needed by the injectors. A pressure regulator attached to the fuel rail, keep fuel available to the injectors at a regulated pressure. Unused fuel is returned to the fuel tank by a separate line.

Test Description: Number(s) below refer to circled number(s) on the diagnostic chart.

1. Wrap a shop towel around the fuel pressure connector to absorb any small amount of fuel leakage that may occur when installing the gage. Ignition "ON," engine "OFF," pump pressure should be 333-376 kPa (48-55 psi). This pressure is controlled by spring pressure within the regulator assembly.
2. This system contains a pressure regulator vacuum solenoid that may be energized after hot engine restart preventing vacuum from entering the pressure regulator. Engine must be running for 2 minutes at normal operating temperature to assure solenoid is de-energized (allowing vacuum to enter regulator which should result in lower fuel pressure).
3. Pressure that continues to fall is caused by one of the following:
   - In-tank fuel pump check valve not holding.
   - Partially disconnected fuel pulse dampener.
   - Fuel pressure regulator valve leaking.
   - Injector(s) sticking open.

4. An injector sticking open can best be determined by checking for a fouled or saturated spark plug(s). If a leaking injector can not be determined by a fouled or saturated spark plug the following procedure should be used.
   - Remove plenum, and remove fuel rail bolts.

   - Lift fuel rail out just enough to leave injector nozzles in the ports.

   CAUTION: Be sure injector(s) are not allowed to spray on engine and that injector retaining clips are intact. This should be carefully followed to prevent fuel spray on engine which would cause a fire hazard.

   - Pressurize the fuel system and observe for injector(s) leaking.

3.1L (VIN T) ENGINE — SYSTEM DIAGNOSTIC CHARTS — LUMINA WITH VFV

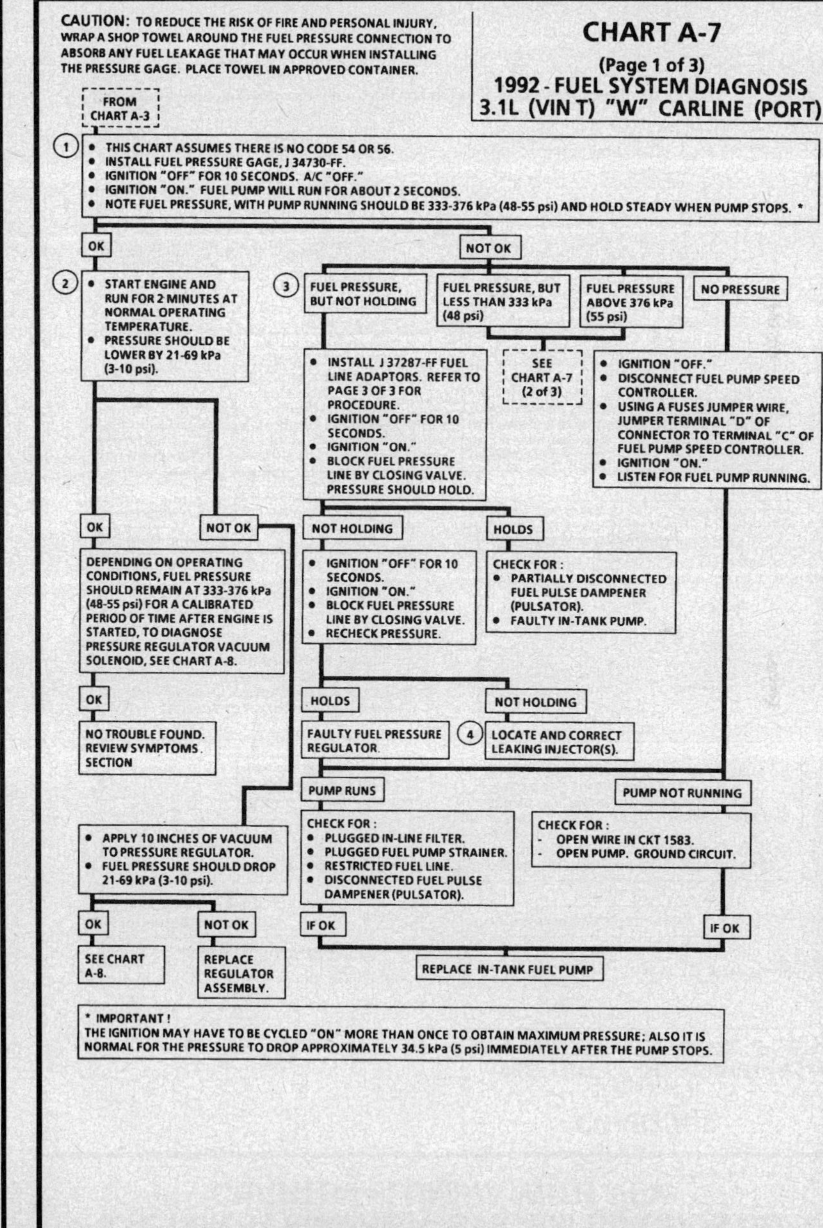

3.1L (VIN T) ENGINE — SYSTEM DIAGNOSTIC CHARTS — LUMINA WITH VFV

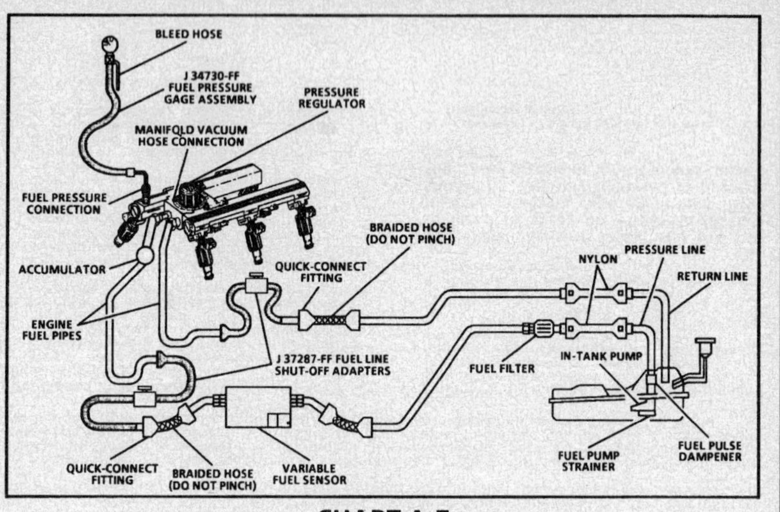

CHART A-7

(Page 2 of 3)
1992 - FUEL SYSTEM DIAGNOSIS
3.1L (VIN T) "W" CARLINE (PORT)

Test Description: Number(s) below refer to circled number(s) on the diagnostic chart.

5. Pressure below 333 kPa (48 psi) may cause a lean condition and may set a Code 44. It could also cause hard starting cold and poor driveability. Low enough pressure will cause the engine not to run at all. Restricted flow may allow the engine to run at idle, or low speeds, but may cause a surge and stall when more fuel is required, as when accelerating or driving at high speeds.

6. Restricting the fuel return line allows fuel pressure to build above regulated pressure. With battery voltage applied to the pump "test" terminal, pressure should rise above 376 kPa (55 psi) as the valve in the return line is partially closed.

NOTICE: Do not allow pressure to exceed 414 kPa (60 psi), as damage to the regulator may result.

7. This test determines if the high fuel pressure is due to a restricted fuel return line or a faulty pressure regulator. High fuel pressure may cause a rich condition and may set a Code 45.

3.1L (VIN T) ENGINE — SYSTEM DIAGNOSTIC CHARTS — LUMINA WITH VFV

CHART A-7

(Page 2 of 3)
1992 - FUEL SYSTEM DIAGNOSIS
3.1L (VIN T) "W" CARLINE (PORT)

FROM CHART A-7 (1 OF 3)

5. HAS FUEL PRESSURE BUT LESS THAN 333 kPa (48 psi).

CHECK FOR RESTRICTED FUEL LINES OR IN-LINE FILTER.

OK

NOT OK

REPLACE FILTER OR REPAIR FUEL LINE AND RECHECK.

6.
- IGNITION "OFF."
- INSTALL SHUT-OFF VALVE (FROM J 37287-FF) IN FUEL RETURN LINE. REFER TO PAGE 3 OF 3 FOR PROCEDURE.
- DISCONNECT FUEL PUMP SPEED CONTROLLER.
- USING A FUSED JUMPER WIRE, JUMPER TERMINAL "D" OF CONNECTOR TO TERMINAL "C" OF FUEL PUMP SPEED CONTROLLER.
- IGNITION "ON."
- SLOWLY CLOSE VALVE IN RETURN LINE. LOOK FOR PRESSURE ABOVE 376kPa (55 psi). DO NOT EXCEED 414 kPa (60 psi).

ABOVE 376 kPa (55 psi).

PRESSURE BUT LESS THAN 333 kPa (48 psi).

FAULTY PRESSURE REGULATOR.

CHECK FOR:
- FAULTY FUEL PUMP.
- PARTIALLY DISCONNECTED FUEL PULSE DAMPENER (PULSATOR).
- RESTRICTED FUEL PUMP STRAINER.
- INCORRECT FUEL PUMP.

FUEL PRESSURE ABOVE 376 kPa (55 psi).

7.
- DISCONNECT QUICK-CONNECT FITTING AT ENGINE FUEL RETURN PIPE. FOLLOW APPROPRIATE STEPS ON PAGE 3 OF 3.
- ATTACH 5/16 I.D. FLEX HOSE TO ENGINE SIDE OF FUEL RETURN PIPE. INSERT THE OTHER END INTO AN APPROVED GASOLINE CONTAINER.
- NOTE FUEL PRESSURE WITHIN 2 SECONDS AFTER IGNITION "ON."

ABOVE 376 kPa (55 psi).

333-376 kPa (48-55 psi).

CHECK FOR RESTRICTED FUEL RETURN LINE FROM FUEL PRESSURE REGULATOR TO POINT WHERE FUEL LINE WAS DISCONNECTED.

IF LINE OK, REPLACE FUEL PRESSURE REGULATOR.

LOCATE AND CORRECT RESTRICTED FUEL RETURN LINE TO FUEL TANK.

CLEAR CODES AND CONFIRM "CLOSED LOOP" OPERATION AND NO "SERVICE ENGINE SOON" LIGHT.

3.1L (VIN T) ENGINE — SYSTEM DIAGNOSTIC CHARTS — LUMINA WITH VFV

CHART A-7

(Page 2 of 3)
1992 - FUEL SYSTEM DIAGNOSIS
3.1L (VIN T) "W" CARLINE (PORT)

-continued-

There are two versions of the variable fuel vehicle, one is calibrated to operate on methanol fuels, the other is calibrated for ethanol fuels. Before attempting any diagnosis or service, know which fuel you are working with.

CAUTION:
- METHANOL, like all fuels, is poisonous and should be handled with care. It can cause headaches, blindness, and even death. Methanol can enter your body through your skin much faster than gasoline will. Drinking only a mouthful of methanol can cause death.
- Avoid spills. If methanol fuel gets on you, see "FIRST AID FOR METHANOL EXPOSURE" on following page.
- Always wear methanol-resistant gloves if you expect to get fuel on your hands.
- Always wear goggles or a face shield.
- Never try to siphon methanol by mouth.
- Never use it for cleaning.
- Never let anyone drink it.
- Always work in a well ventilated area.
- Do not breathe the vapor from methanol.
  - Breathing methanol vapors over a long period of time can also cause blindness and death.
- Partial combustion of methanol yields formaldehyde, which is produced in somewhat greater quantities with M85 than with gasoline. As of yet, there are no EPA regulations on formaldehyde emissions, but the California Air Resources Board has established formaldehyde emission limits which the Chevrolet Luminas meet. Much more investigation needs to be done into the effects of formaldehyde, which is suspected to be cancer-causing and is emitted from many industrial chemical processes. Exposure to formaldehyde can cause burning of the eyes, nose and throat, and even death. Formaldehyde gives off sharp odors you should notice before it becomes harmful.

CAUTION:
- ETHANOL fuel, like all fuels, is hazardous and should be handled with care. It can cause temporary loss of memory and concentration or even death. Drinking even small amounts of ethanol fuel can cause death.
- Avoid spills. If ethanol fuel gets on you, wash your skin right away. If ethanol fuel spills on your clothing, change clothes and wash your skin immediately.
- Always wear ethanol-resistant gloves if you expect to get fuel on your hands.
- Always wear goggles or a face shield.
- Never try to siphon ethanol by mouth.
- Never use it for cleaning.
- Never let anyone drink it.
- Always work in a well ventilated area.
- Do not breathe the vapor from ethanol.
  - Breathing ethanol vapors over a long period of time can also cause serious illness and death.
- A by-product of ethanol fuel combustion is acetaldehyde, which is part of the engine exhaust. Exposure to acetaldehyde can cause burning of the eyes, nose, and throat. Acetaldehyde gives off sharp odors you should notice before it becomes harmful.

"AFTER REPAIRS," CONFIRM "CLOSED LOOP" OPERATION AND NO "SERVICE ENGINE SOON" LIGHT.

3.1L (VIN T) ENGINE — SYSTEM DIAGNOSTIC CHARTS — LUMINA WITH VFV

CHART A-7

(Page 2 of 3)
1992 - FUEL SYSTEM DIAGNOSIS
3.1L (VIN T) "W" CARLINE (PORT)

FIRST AID FOR METHANOL EXPOSURE

- Move to an area with fresh air if fumes become prominent.

- Wash exposed skin with ample quantities of soap and water.
  - Allow contaminated clothing to air dry. Machine wash any clothing that has been contaminated with methanol fuel.

- If methanol comes in contact with eyes, flush with water for at least fifteen minutes, see a doctor for treatment.

- If ingestion of methanol fuel has occurred:
  - Have the person lay down and keep warm.
  - DO NOT induce vomiting.
  - If the person is unconscious, do not attempt to make him/her swallow anything.
  - Contact emergency personal immediately and advise of the methanol exposure.

- Methanol poisoning from M85 is treatable if prompt medical attention is sought.

CHART A-7

(Page 3 of 3)
1992 - FUEL SYSTEM DIAGNOSIS
3.1L (VIN T) "W" CARLINE (PORT)

FUEL PRESSURE CHECK

(Instructions for Installing Fuel Line Shut-Off Adapters)

Tools Required: J 34730-FF - Fuel Pressure Gage

J 37287-FF - Inlet/Return Fuel Line Shut-Off Adapters

J 37088-A - Fuel Line Quick-Connect Separators

CAUTION: To Reduce the Risk of Fire and Personal Injury:
- It is necessary to relieve fuel system pressure before disconnecting fuel lines.
- After relieving system pressure, a small amount of fuel may be released when disconnecting the fuel lines. Cover fuel line fittings with a shop towel before disconnecting, to catch any fuel that may leak out. Place towel in approved container when disconnect is completed.

1. Disconnect negative battery cable.
2. Loosen fuel filler cap to relieve fuel tank pressure. (Do not tighten at this time.)
3. Connect gage J 34730-FF to fuel pressure connection. Wrap a shop towel around fitting while connecting gage to avoid spillage.
4. Place bleed hose into an approved container and open valve to bleed system pressure.
5. Locate quick-connect fittings at engine fuel feed and return pipes.
6. Clean all fuel pipe and hose connections and surrounding areas before disconnecting to avoid possible contamination of the fuel system.
7. Pull back quick-connect dust covers to gain access to quick-connect fittings.
8. Choose correct tool from separator tool set J 37088-A for size of fitting. Insert tool into female end of connector, push inward and pull connection apart.
   - Repeat for other fitting.

CAUTION: To Reduce the Risk of Fire and Personal Injury: Before connecting fuel line quick-connect fittings, always apply a few drops of clean engine oil to the male tube ends. This will ensure proper reconnection and prevent a possible fuel leak. (During normal operation, the O-rings located inside the female connector will swell and may prevent proper reconnection if not lubricated.)

9. Lubricate the male tube end of the engine fuel pipes and the shut-off adapters J 37287-FF with engine oil.
10. Connect adapters to fuel feed and return lines.
    - Push connectors together to cause the retaining tabs/fingers to snap into place.
    - Once installed, pull on both ends of each connection to make sure it is secure.

"AFTER REPAIRS," CONFIRM "CLOSED LOOP" OPERATION AND NO "SERVICE ENGINE SOON" LIGHT.

CHART A-7

(Page 3 of 3)
1992 - FUEL SYSTEM DIAGNOSIS
3.1L (VIN T) "W" CARLINE (PORT)

FUEL PRESSURE CHECK
-continued-

11. Connect negative battery cable.
12. Check fuel pressure.
13. Disconnect negative battery cable.
14. Place gage bleed hose into an approved container and open valve to bleed system pressure.
15. Disconnect fuel pressure gage. Drain any fuel remaining in gage into an approved container.
16. Disconnect fuel shut-off adapters from fuel feed and return lines.
17. Lubricate the male tube end of the fuel pipes, and reconnect quick-connect fittings.
    - Push each connector together to cause the retaining tabs/fingers to snap into place.
    - Once installed, pull on both ends of connection to make sure it is secure.
    - If equipped, slide quick-connect dust covers in place over quick-connect fittings.
18. Tighten fuel filler cap.
19. Connect negative battery cable.
20. Cycle ignition "ON" and "OFF" twice (Do Not start engine), waiting ten seconds between cycles, then check for fuel leaks.

Important
- Before starting engine, follow the "Idle Learn Procedure" as follows:

21. Install Tech 1 scan tool.
22. Ignition "ON," engine "OFF."
23. Select "IAC SYSTEM" then "IDLE LEARN" in the "MISC TEST" mode.
24. Proceed with idle learn as directed.

This procedure allows the ECM memory to be updated with the correct IAC valve pintle position for the vehicle and provides for a stable idle speed.

3.1L (VIN T) ENGINE — SYSTEM DIAGNOSTIC CHARTS — LUMINA WITH VFV

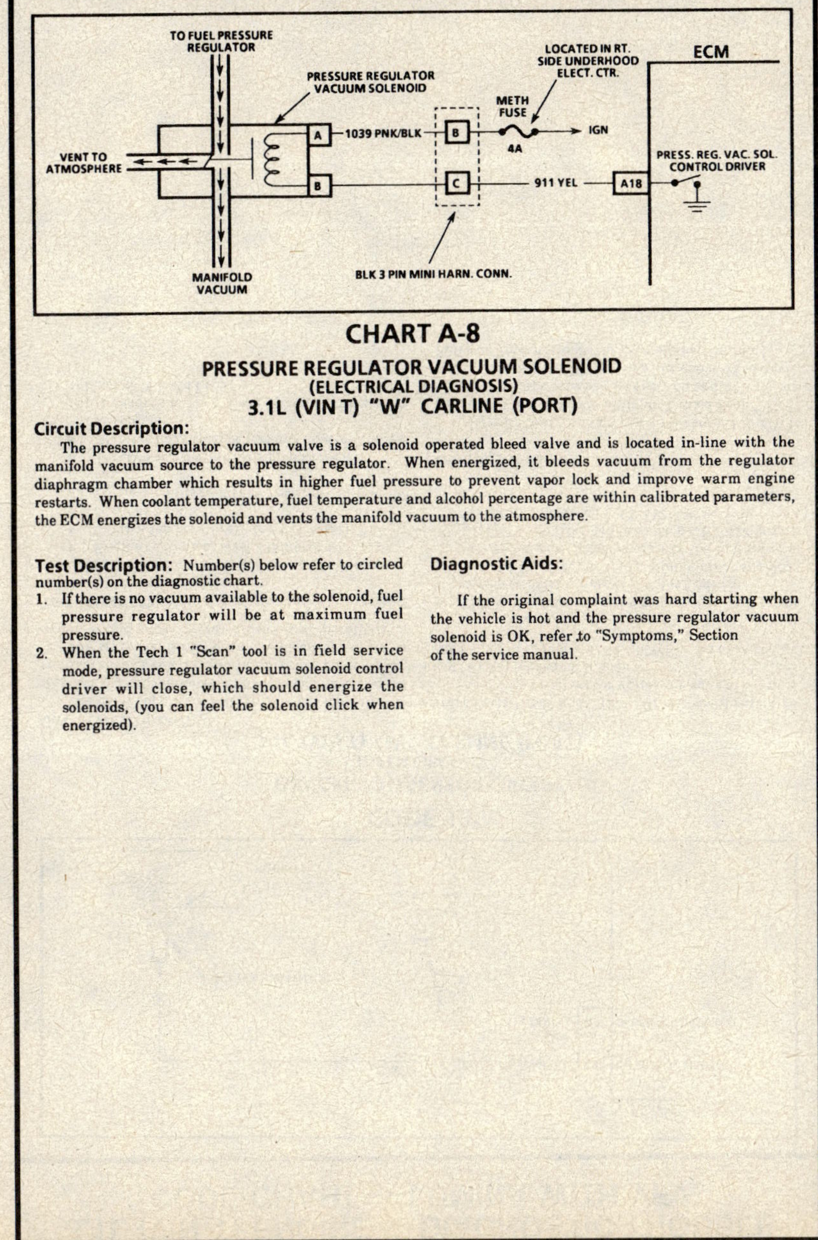

CHART A-8

PRESSURE REGULATOR VACUUM SOLENOID
(ELECTRICAL DIAGNOSIS)
3.1L (VIN T) "W" CARLINE (PORT)

Circuit Description:

The pressure regulator vacuum valve is a solenoid operated bleed valve and is located in-line with the manifold vacuum source to the pressure regulator. When energized, it bleeds vacuum from the regulator diaphragm chamber which results in higher fuel pressure to prevent vapor lock and improve warm engine restarts. When coolant temperature, fuel temperature and alcohol percentage are within calibrated parameters, the ECM energizes the solenoid and vents the manifold vacuum to the atmosphere.

Test Description: Number(s) below refer to circled number(s) on the diagnostic chart.
1. If there is no vacuum available to the solenoid, fuel pressure regulator will be at maximum fuel pressure.
2. When the Tech 1 "Scan" tool is in field service mode, pressure regulator vacuum solenoid control driver will close, which should energize the solenoids, (you can feel the solenoid click when energized).

Diagnostic Aids:

If the original complaint was hard starting when the vehicle is hot and the pressure regulator vacuum solenoid is OK, refer to "Symptoms," Section of the service manual.

3.1L (VIN T) ENGINE — SYSTEM DIAGNOSTIC CHARTS — LUMINA WITH VFV

CHART A-8

PRESSURE REGULATOR VACUUM SOLENOID
(ELECTRICAL DIAGNOSIS)
3.1L (VIN T) "W" CARLINE (PORT)

(1)
- DISCONNECT MANIFOLD VACUUM SOURCE HOSE FROM PRESSURE REGULATOR VACUUM SOLENOID.
- CONNECT VACUUM GAGE TO VACUUM SOURCE HOSE.
- START AND IDLE ENGINE AT NORMAL OPERATING TEMPERATURE. THERE SHOULD BE ABOVE 10" Hg VACUUM. IS THERE?

YES
- RECONNECT VACUUM HOSE TO SOLENOID.
- DISCONNECT (AT SOLENOID) VACUUM HOSE TO FUEL PRESSURE REGULATOR.
- CONNECT VACUUM GAGE TO OPEN PORT ON SOLENOID. WITH ENGINE RUNNING THERE SHOULD BE ABOVE 10" Hg VACUUM. IS THERE?

NO
RESTRICTED VACUUM HOSE OR INTERNAL ENGINE VACUUM LEAK.

YES
(2)
- RECONNECT VACUUM HOSE TO FUEL PRESSURE REGULATOR.
- INSTALL TECH 1 SCAN TOOL. ENTER FIELD SERVICE MODE.
- IGNITION "ON," ENGINE "OFF."
- COMMAND FIELD SERVICE MODE WHILE TOUCHING PRESSURE REGULATOR VACUUM SOLENOID. SOLENOID SHOULD ENERGIZE. DOES IT?

NO
- DISCONNECT SOLENOID ELECTRICAL CONNECTOR. DOES GAGE SHOW VACUUM?

YES
CKT 911 SHORTED TO GROUND OR FAULTY ECM.

NO
FAULTY SOLENOID.

YES
NO TROUBLE FOUND. REFER TO "DIAGNOSTIC AIDS" ON FACING PAGE.

NO
- WITH TEST LIGHT CONNECTED TO GROUND.
- IGNITION "ON," ENGINE "OFF."
- PROBE TERMINAL "A" OR "B" OF SOLENOID RELAY ELECTRICAL CONNECTOR. IS LIGHT "ON"?

YES
- WITH TEST LIGHT CONNECTED TO B+, PROBE TERMINAL "B" OF SOLENOID ELECTRICAL CONNECTOR.
- ENTER FIELD SERVICE MODE ON TECH 1 SCAN TOOL. IS LIGHT ON"?

NO
OPEN IN CKT 1039, IF FUSE IS BLOWN, CHECK FOR SHORT TO GROUND IN CKT 1039.

YES
FAULTY SOLENOID ELECTRICAL CONNECTOR OR SOLENOID.

NO
OPEN CKT 911, FAULTY ECM CONNECTION OR ECM.

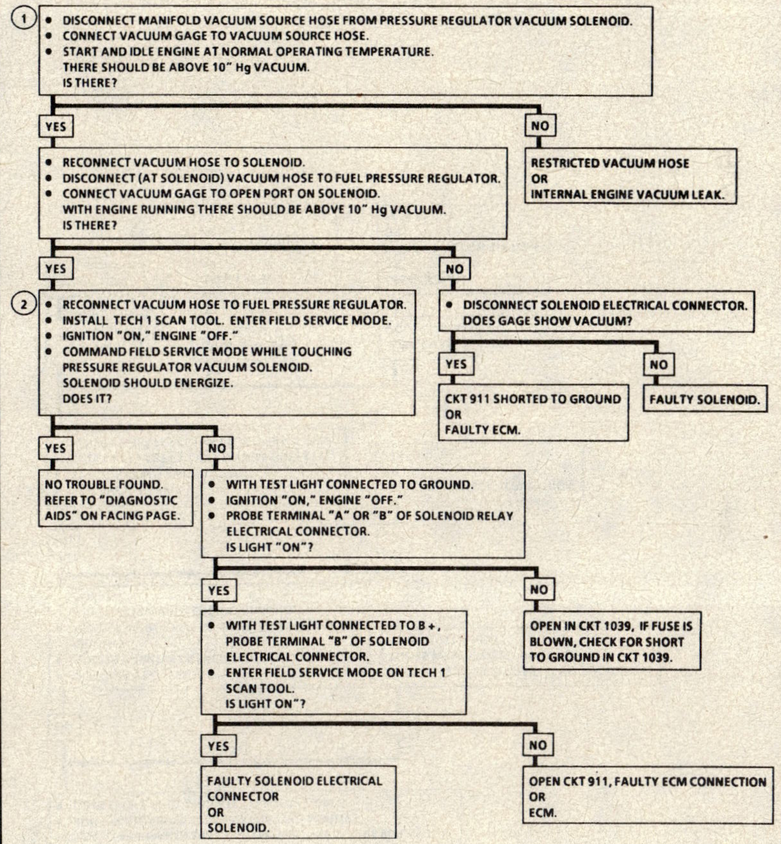

3.1L (VIN T) ENGINE — DIAGNOSTIC TROUBLE CODE CHART — LUMINA WITH VFV

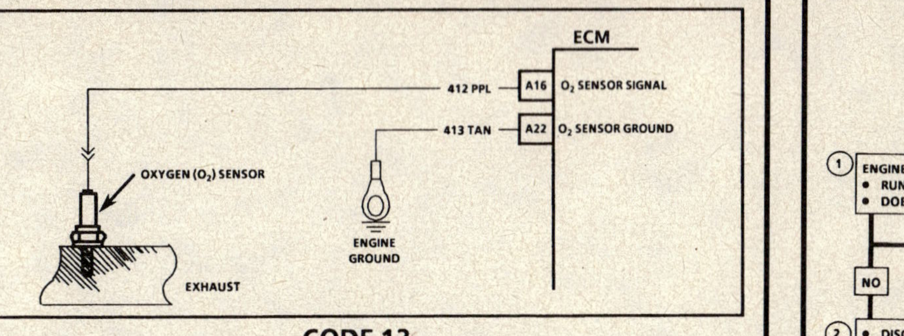

CODE 13
OXYGEN (O₂) SENSOR CIRCUIT
(OPEN CIRCUIT)
3.1L (VIN T) "W" CARLINE (PORT)

Circuit Description:

The ECM supplies a voltage of about .55 volt between terminals "A16" and "A22". (If measured with a 10 megohm digital voltmeter, this may read as low as .35 volt.) The Oxygen (O₂) sensor varies the voltage within a range of about 1 volt if the exhaust is rich, down through about .10 volt if exhaust is lean.

The sensor is like an open circuit and produces no voltage when it is below 315°C (600°F). An open sensor circuit or cold sensor causes "Open Loop" operation.

Test Description: Number(s) below refer to circled number(s) on the diagnostic chart.

1. Code 13 will set under the following conditions:
   - Engine running at least 2 minutes after start.
   - Coolant temperature at least 50°C (122°F).
   - No Code 21 or 22.
   - O₂ signal voltage steady between .35 and .55 volt.
   - Throttle position sensor signal above 4%.
   - All conditions must be met and held for at least 25 seconds.

   If the conditions for a Code 13 exist, the system will not go "Closed Loop."

2. This will determine if the sensor is at fault or the wiring or ECM is the cause of the Code 13.

3. In doing this test use only a high impedance digital volt ohmmeter. This test checks the continuity of CKT 412 and CKT 413 because if CKT 413 is open the ECM voltage on CKT 412 will be over .6 volt (600 mV).

Diagnostic Aids:

Normal "Scan" voltage varies between 100 mV to 999 mV (.1 and 1.0 volt) while in "Closed Loop." Code 13 sets if voltage remains between .35 and .55 volt, but the system will go "Open Loop" in about 15 seconds. Refer to "Intermittents" in "Symptoms," Section

---

3.1L (VIN T) ENGINE — DIAGNOSTIC TROUBLE CODE CHART — LUMINA WITH VFV

CODE 13
OXYGEN (O₂) SENSOR CIRCUIT
(OPEN CIRCUIT)
3.1L (VIN T) "W" CARLINE (PORT)

(1)
- ENGINE AT NORMAL OPERATING TEMPERATURE (ABOVE 80°C/176°F).
- RUN ENGINE ABOVE 1200 RPM FOR TWO MINUTES.
- DOES TECH 1 TOOL INDICATE "CLOSED LOOP"?

NO

YES → CODE 13 IS INTERMITTENT. IF NO ADDITIONAL CODES WERE STORED, REFER TO "DIAGNOSTIC AIDS"

(2)
- DISCONNECT O₂ SENSOR.
- JUMPER HARNESS CKT 412 (ECM SIDE) TO GROUND.
- TECH 1 SHOULD DISPLAY O₂ VOLTAGE BELOW .2 VOLT (200 mv) WITH ENGINE RUNNING. DOES IT?

NO

YES → FAULTY O₂ SENSOR CONNECTION OR SENSOR.

(3)
- REMOVE JUMPER.
- IGNITION "ON," ENGINE "OFF."
- CHECK VOLTAGE OF CKT 412 (ECM SIDE) AT O₂ SENSOR HARNESS CONNECTOR USING A DVM.

.3-.6 VOLT (300 - 600 mV) → FAULTY ECM.

OVER .6 VOLT (600 mV) → OPEN CKT 413 OR FAULTY CONNECTION OR FAULTY ECM.

LESS THAN .3 VOLT (300 mV) → OPEN CKT 412 OR FAULTY ECM CONNECTION OR FAULTY ECM.

"AFTER REPAIRS," REFER TO CODE CRITERIA   AND CONFIRM CODE DOES NOT RESET.

3.1L (VIN T) ENGINE — DIAGNOSTIC TROUBLE CODE CHART — LUMINA WITH VFV

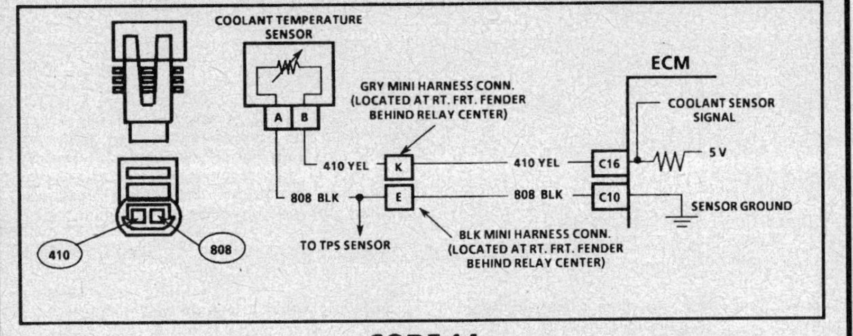

CODE 14
COOLANT TEMPERATURE SENSOR (CTS) CIRCUIT
(HIGH TEMPERATURE INDICATED)
3.1L (VIN T) "W" CARLINE (PORT)

Circuit Description:

The Coolant Temperature Sensor (CTS) uses a thermistor to control the signal voltage to the ECM. The ECM applies a voltage on CKT 410 to the sensor. When the engine is cold, the sensor (thermistor) resistance is high, therefore the ECM will see high signal voltage.

As the engine warms, the sensor resistance becomes less, and the voltage drops. At normal engine operating temperature, the voltage will measure about 1.5 to 2.0 volts at the ECM terminal "C16".

Coolant temperature is one of the inputs used to control:

- Fuel delivery.
- Electronic Spark Timing (EST).
- Idle Air Control (IAC).
- Torque Converter Clutch (TCC).
- Controlled Canister Purge (CCP).
- Exhaust Gas Recirculation (EGR).
- Cooling fan.

Test Description: Number(s) below refer to circled number(s) on the diagnostic chart.

1. Code 14 will set if:
   - Signal voltage indicates a coolant temperature above 135°C (270°F).
   - Engine running longer than 20 seconds.
2. This test will determine if CKT 410 is shorted to ground which will cause the conditions for Code 14.

Diagnostic Aids:

Check harness routing for a potential short to ground in CKT 410. Circuit is routed from the ECM to a mini harness, and then to the Coolant Temperature Sensor (CTS).

The "Scan" tool displays engine temperature in degrees centigrade. After engine is started, the temperature should rise steadily to about 90°C (194°F) then stabilize when thermostat opens. Refer to "Intermittents" in "Symptoms," Section "6E3-B".

Verify that engine is not overheating and has not been subjected to conditions which could create an overheating condition (i.e. overload, trailer towing, hilly terrain, heavy stop and go traffic, etc.). The "Temperature To Resistance Value" scale at the right may be used to test the coolant sensor at various temperature levels to evaluate the possibility of a "shifted" (mis-scaled) sensor. A "shifted" sensor could result in poor driveability complaints.

3.1L (VIN T) ENGINE — DIAGNOSTIC TROUBLE CODE CHART — LUMINA WITH VFV

CODE 14
COOLANT TEMPERATURE SENSOR (CTS) CIRCUIT
(HIGH TEMPERATURE INDICATED)
3.1L (VIN T) "W" CARLINE (PORT)

1. DOES TECH 1 DISPLAY COOLANT TEMPERATURE OF 130°C (266°F) OR HIGHER?

- YES
- NO

2. DISCONNECT COOLANT TEMPERATURE SENSOR. TECH 1 SHOULD DISPLAY COOLANT TEMPERATURE BELOW -30°C (-22°F). DOES IT?

- CODE 14 IS INTERMITTENT. IF NO ADDITIONAL CODES WERE STORED, REFER TO "DIAGNOSTIC AIDS"

- YES → REPLACE COOLANT TEMPERATURE SENSOR.

- NO → CKT 410 SHORTED TO GROUND OR CKT 410 SHORTED TO SENSOR GROUND CIRCUIT OR FAULTY ECM.

DIAGNOSTIC AID

COOLANT SENSOR		
TEMPERATURE VS. RESISTANCE VALUES		
(APPROXIMATE)		
°C	°F	OHMS
100	212	177
90	194	241
80	176	332
70	158	467
60	140	667
50	122	973
45	113	1188
40	104	1459
35	95	1802
30	86	2238
25	77	2796
20	68	3520
15	59	4450
10	50	5670
5	41	7280
0	32	9420
-5	23	12300
-10	14	16180
-15	5	21450
-20	-4	28680
-30	-22	52700
-40	-40	100700

"AFTER REPAIRS," REFER TO CODE CRITERIA   AND CONFIRM CODE DOES NOT RESET.

3.1L (VIN T) ENGINE — DIAGNOSTIC TROUBLE CODE CHART — LUMINA WITH VFV

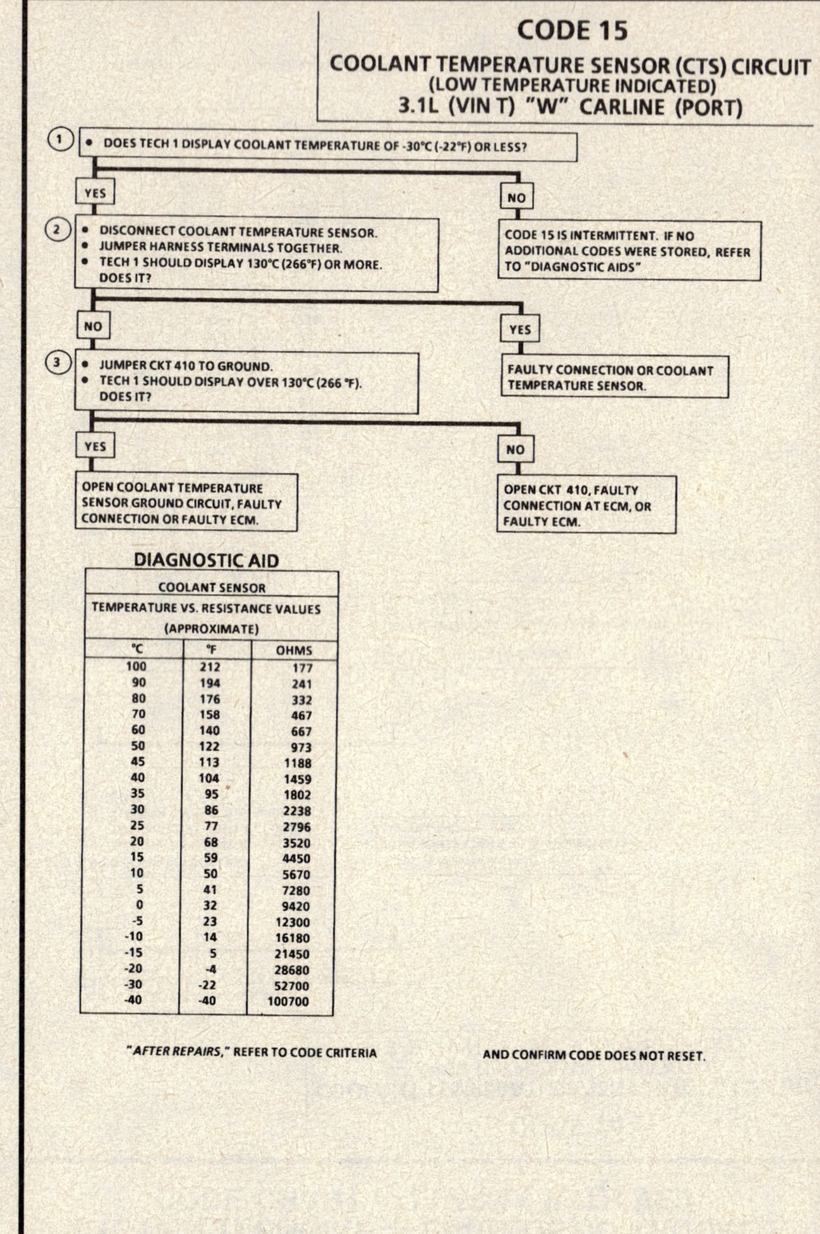

CODE 15

COOLANT TEMPERATURE SENSOR (CTS) CIRCUIT
(LOW TEMPERATURE INDICATED)
3.1L (VIN T) "W" CARLINE (PORT)

Circuit Description:

The Coolant Temperature Sensor (CTS) uses a thermistor to control the signal voltage to the ECM. The ECM applies a voltage on CKT 410 to the sensor. When the engine is cold the sensor (thermistor) resistance is high, therefore the ECM will see high signal voltage.

As the engine warms, the sensor resistance becomes less, and the voltage drops. At normal engine operating temperature the voltage will measure about 1.5 to 2.0 volts at the ECM terminal "C16".

Coolant temperature is one of the inputs used to control:
- Fuel delivery.
- Engine Spark Timing (EST).
- Idle Air Control (IAC).
- Torque Converter Clutch (TCC).
- Controlled Canister Purge (CCP).
- Electronic Gas Recirculation (EGR).
- Cooling fan.

Test Description: Number(s) below refer to circled number(s) on the diagnostic chart.
1. Code 15 will set if:
   - Signal voltage indicates a coolant temperature less than -38.5°C (-37.30°F) for one second.
2. This test simulates a Code 14. If the ECM recognizes the low signal voltage, (high temperature) and the "Scan" reads 130°C (266°F), the ECM and wiring are OK.
3. This test will determine if CKT 410 is open. There should be 5 volts present at sensor connector if measured with a DVM.

Diagnostic Aids:

A "Scan" tool reads engine temperature in degrees centigrade. After engine is started the temperature should rise steadily to about 95°C (203°F) then stabilize when thermostat opens. CKT 410 is routed from the ECM to a mini harness, and then to the Coolant Temperature Sensor (CTS).

A faulty connection, or an open in CKT 410 or 808 will result in a Code 15.

Codes 15 and 21 stored at the same time could be the result of an open CKT 808 which would also turn the temperature warning indicator "ON." The "Temperature to Resistance Value" scale at the right may be used to test the coolant sensor at various temperature levels to evaluate the possibility of a "shifted" (mis-scaled) sensor. A "shifted" sensor could result in poor driveability complaints.

Refer to "Intermittents" in "Symptoms," Section

3.1L (VIN T) ENGINE — DIAGNOSTIC TROUBLE CODE CHART — LUMINA WITH VFV

CODE 15

COOLANT TEMPERATURE SENSOR (CTS) CIRCUIT
(LOW TEMPERATURE INDICATED)
3.1L (VIN T) "W" CARLINE (PORT)

① **DOES TECH 1 DISPLAY COOLANT TEMPERATURE OF -30°C (-22°F) OR LESS?**

- **YES** →
  ② - DISCONNECT COOLANT TEMPERATURE SENSOR.
     - JUMPER HARNESS TERMINALS TOGETHER.
     - TECH 1 SHOULD DISPLAY 130°C (266°F) OR MORE. DOES IT?
    - **NO** →
      ③ - JUMPER CKT 410 TO GROUND.
         - TECH 1 SHOULD DISPLAY OVER 130°C (266 °F). DOES IT?
        - **YES** → OPEN COOLANT TEMPERATURE SENSOR GROUND CIRCUIT, FAULTY CONNECTION OR FAULTY ECM.
        - **NO** → OPEN CKT 410, FAULTY CONNECTION AT ECM, OR FAULTY ECM.
    - **YES** → FAULTY CONNECTION OR COOLANT TEMPERATURE SENSOR.
- **NO** → CODE 15 IS INTERMITTENT. IF NO ADDITIONAL CODES WERE STORED, REFER TO "DIAGNOSTIC AIDS"

DIAGNOSTIC AID

COOLANT SENSOR		
TEMPERATURE VS. RESISTANCE VALUES (APPROXIMATE)		
°C	°F	OHMS
100	212	177
90	194	241
80	176	332
70	158	467
60	140	667
50	122	973
45	113	1188
40	104	1459
35	95	1802
30	86	2238
25	77	2796
20	68	3520
15	59	4450
10	50	5670
5	41	7280
0	32	9420
-5	23	12300
-10	14	16180
-15	5	21450
-20	-4	28680
-30	-22	52700
-40	-40	100700

"AFTER REPAIRS," REFER TO CODE CRITERIA

AND CONFIRM CODE DOES NOT RESET.

3.1L (VIN T) ENGINE — DIAGNOSTIC TROUBLE CODE CHART — LUMINA WITH VFV

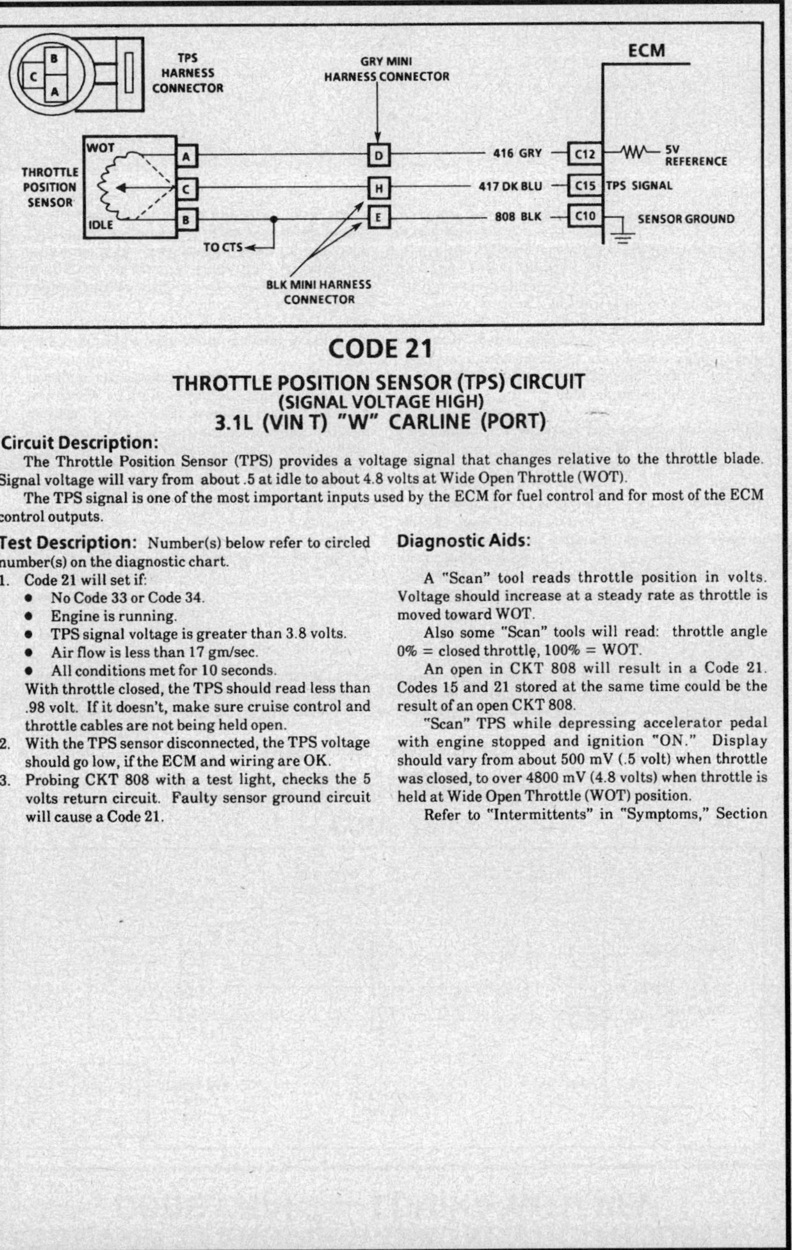

CODE 21
THROTTLE POSITION SENSOR (TPS) CIRCUIT
(SIGNAL VOLTAGE HIGH)
3.1L (VIN T) "W" CARLINE (PORT)

Circuit Description:

The Throttle Position Sensor (TPS) provides a voltage signal that changes relative to the throttle blade. Signal voltage will vary from about .5 at idle to about 4.8 volts at Wide Open Throttle (WOT).

The TPS signal is one of the most important inputs used by the ECM for fuel control and for most of the ECM control outputs.

Test Description: Number(s) below refer to circled number(s) on the diagnostic chart.

1. Code 21 will set if:
   - No Code 33 or Code 34.
   - Engine is running.
   - TPS signal voltage is greater than 3.8 volts.
   - Air flow is less than 17 gm/sec.
   - All conditions met for 10 seconds.
   With throttle closed, the TPS should read less than .98 volt. If it doesn't, make sure cruise control and throttle cables are not being held open.
2. With the TPS sensor disconnected, the TPS voltage should go low, if the ECM and wiring are OK.
3. Probing CKT 808 with a test light, checks the 5 volts return circuit. Faulty sensor ground circuit will cause a Code 21.

Diagnostic Aids:

A "Scan" tool reads throttle position in volts. Voltage should increase at a steady rate as throttle is moved toward WOT.

Also some "Scan" tools will read: throttle angle 0% = closed throttle, 100% = WOT.

An open in CKT 808 will result in a Code 21. Codes 15 and 21 stored at the same time could be the result of an open CKT 808.

"Scan" TPS while depressing accelerator pedal with engine stopped and ignition "ON." Display should vary from about 500 mV (.5 volt) when throttle was closed, to over 4800 mV (4.8 volts) when throttle is held at Wide Open Throttle (WOT) position.

Refer to "Intermittents" in "Symptoms," Section

3.1L (VIN T) ENGINE — DIAGNOSTIC TROUBLE CODE CHART — LUMINA WITH VFV

CODE 21
THROTTLE POSITION SENSOR (TPS) CIRCUIT
(SIGNAL VOLTAGE HIGH)
3.1L (VIN T) "W" CARLINE (PORT)

1. • THROTTLE CLOSED. DOES TECH 1 DISPLAY THROTTLE POSITION OVER 2.5 VOLTS?

   YES → / NO →

   NO: CODE 21 IS INTERMITTENT. IF NO ADDITIONAL CODES WERE STORED, REFER TO "DIAGNOSTIC AIDS"

2. • DISCONNECT THROTTLE POSITION SENSOR. TECH 1 SHOULD DISPLAY THROTTLE POSITION BELOW .2 VOLT (200mV). DOES IT?

   YES → / NO →

   NO: TPS SIGNAL CIRCUIT SHORTED TO VOLTAGE OR FAULTY ECM.

3. • PROBE SENSOR GROUND CIRCUIT WITH A TEST LIGHT CONNECTED TO BATTERY VOLTAGE.

   LIGHT "ON": FAULTY CONNECTION OR THROTTLE POSITION SENSOR.

   LIGHT "OFF": OPEN SENSOR GROUND CIRCUIT OR FAULTY ECM.

"AFTER REPAIRS," REFER TO CODE CRITERIA   AND CONFIRM CODE DOES NOT RESET.

3.1L (VIN T) ENGINE — DIAGNOSTIC TROUBLE CODE CHART — LUMINA WITH VFV

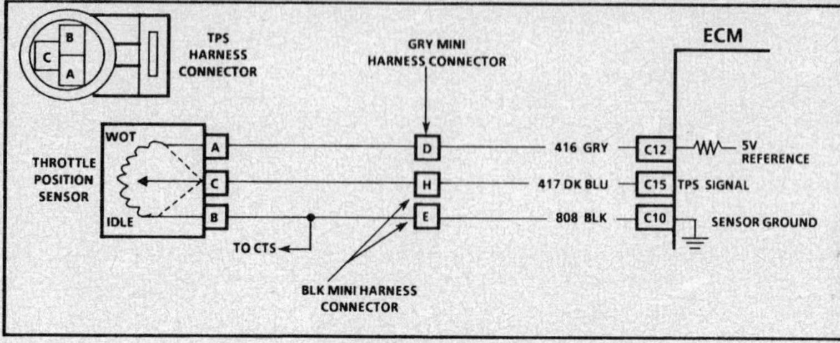

CODE 22
THROTTLE POSITION SENSOR (TPS) CIRCUIT
(SIGNAL VOLTAGE LOW)
3.1L (VIN T) "W" CARLINE (PORT)

Circuit Description:

The Throttle Position Sensor (TPS) provides a voltage signal that changes relative to the throttle blade. Signal voltage will vary from about .5 at idle to about 4.8 volts at Wide Open Throttle (WOT).

The TPS signal is one of the most important inputs used by the ECM for fuel control and for most of the ECM control outputs.

Test Description: Number(s) below refer to circled number(s) on the diagnostic chart.

1. Code 22 will set if:
   - Engine running.
   - TPS signal voltage is less than about .20 volt for 3 seconds.
2. Simulates Code 21: (High Voltage) If the ECM recognizes the high signal voltage, the ECM and wiring are OK.
3. TPS check: The TPS has an auto zeroing feature. If the voltage reading is within the range of 0.29 to 0.98 volt, the ECM will use that value as closed throttle. If the voltage reading is out of the auto zero range on an existing or replacement TPS; check for cruise control and throttle cables for being held open.
4. This simulates a high signal voltage to check for an open in CKT 417.
5. CKT 416 and CKT 474 share a common 5 volts buffered reference signal. If either of these circuits is shorted to ground, Code 22 will set. To determine if the MAP sensor is causing the Code 22 problem, disconnect it to see if Code 22 resets. Be sure TPS is connected and clear codes before testing.

Diagnostic Aids:

A "Scan" tool reads throttle position in volts. Voltage should increase at a steady rate as throttle is moved toward WOT.

Also some "Scan" tools will read throttle angle in percent. 0% = closed throttle, 100% = WOT.

An open or short to ground in CKTs 416 or 417 will result in a Code 22.

CKTs 416 and 417 are routed through a mini harness. CKT 416 is connected to terminal "D" at the gray connector, CKT 417 is connected to terminal "H" at the black connector.

"Scan" TPS while depressing accelerator pedal with engine stopped and ignition "ON." Display should vary from about 500 mV (.5 volt) when throttle was closed, to over 4800 mV (4.8 volts) when throttle is held at Wide Open Throttle (WOT) position.

Also some "Scan" tools will read throttle angle.

0% = closed throttle.

100% = open throttle.

If Code 22 is set, check CKT 416 for faulty wiring or connections.

Refer to "Intermittents" in "Symptoms," Section

3.1L (VIN T) ENGINE — DIAGNOSTIC TROUBLE CODE CHART — LUMINA WITH VFV

CODE 22
THROTTLE POSITION SENSOR (TPS) CIRCUIT
(SIGNAL VOLTAGE LOW)
3.1L (VIN T) "W" CARLINE (PORT)

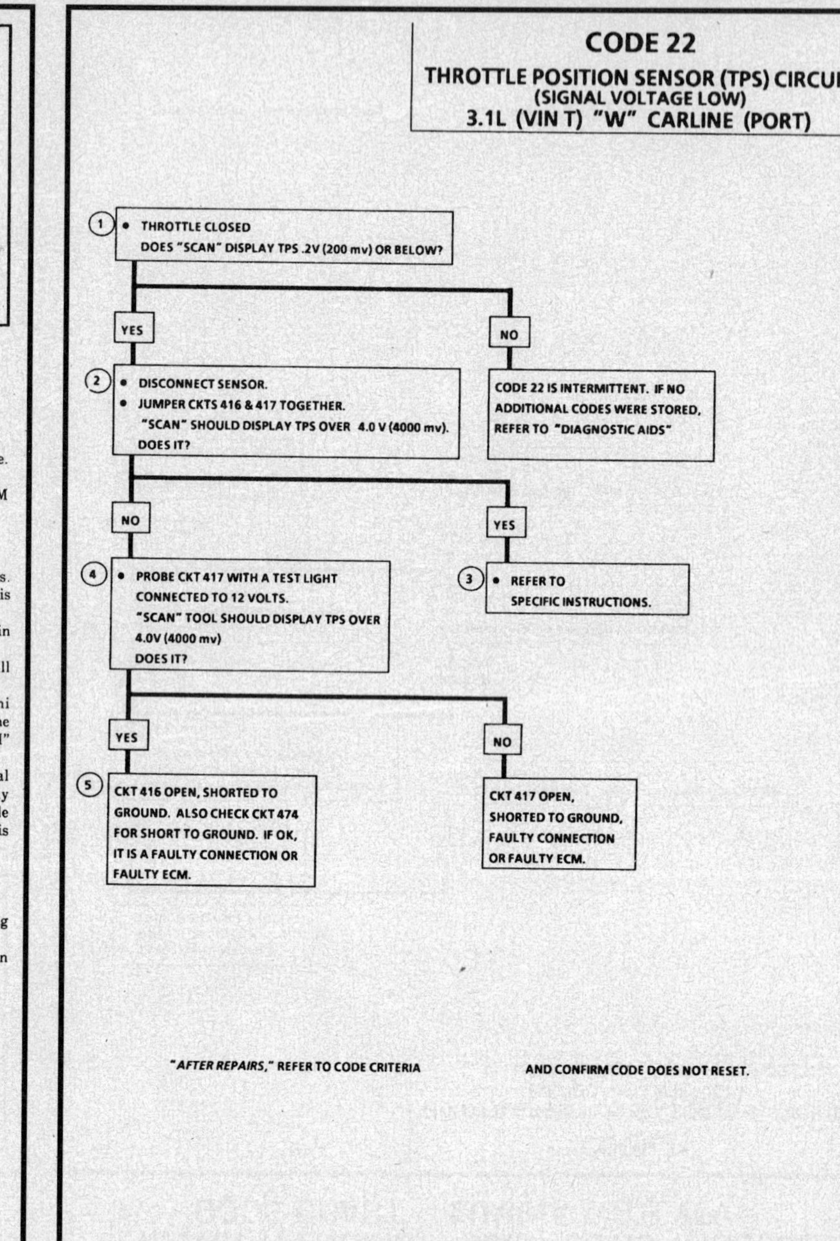

3.1L (VIN T) ENGINE — DIAGNOSTIC TROUBLE CODE CHART — LUMINA WITH VFV

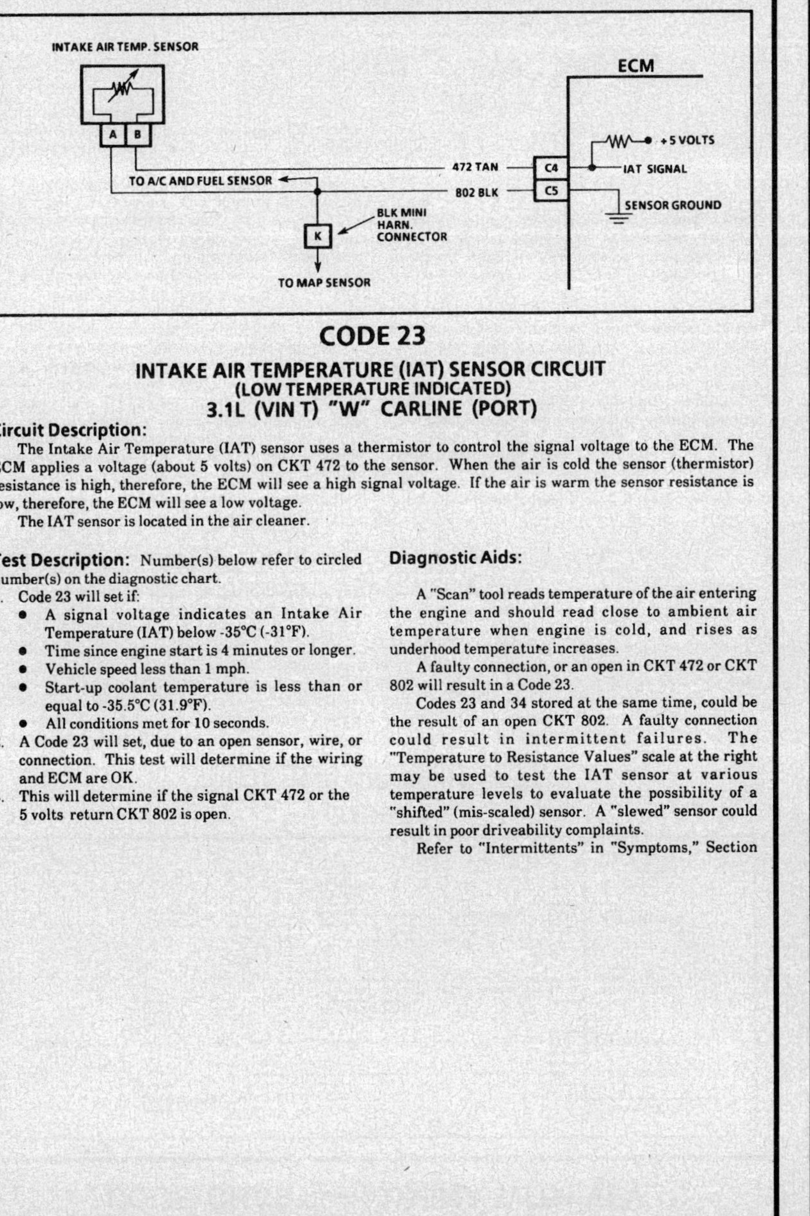

CODE 23
INTAKE AIR TEMPERATURE (IAT) SENSOR CIRCUIT
(LOW TEMPERATURE INDICATED)
3.1L (VIN T) "W" CARLINE (PORT)

Circuit Description:

The Intake Air Temperature (IAT) sensor uses a thermistor to control the signal voltage to the ECM. The ECM applies a voltage (about 5 volts) on CKT 472 to the sensor. When the air is cold the sensor (thermistor) resistance is high, therefore, the ECM will see a high signal voltage. If the air is warm the sensor resistance is low, therefore, the ECM will see a low voltage.

The IAT sensor is located in the air cleaner.

Test Description: Number(s) below refer to circled number(s) on the diagnostic chart.

1. Code 23 will set if:
   - A signal voltage indicates an Intake Air Temperature (IAT) below -35°C (-31°F).
   - Time since engine start is 4 minutes or longer.
   - Vehicle speed less than 1 mph.
   - Start-up coolant temperature is less than or equal to -35.5°C (31.9°F).
   - All conditions met for 10 seconds.
2. A Code 23 will set, due to an open sensor, wire, or connection. This test will determine if the wiring and ECM are OK.
3. This will determine if the signal CKT 472 or the 5 volts return CKT 802 is open.

Diagnostic Aids:

A "Scan" tool reads temperature of the air entering the engine and should read close to ambient air temperature when engine is cold, and rises as underhood temperature increases.

A faulty connection, or an open in CKT 472 or CKT 802 will result in a Code 23.

Codes 23 and 34 stored at the same time, could be the result of an open CKT 802. A faulty connection could result in intermittent failures. The "Temperature to Resistance Values" scale at the right may be used to test the IAT sensor at various temperature levels to evaluate the possibility of a "shifted" (mis-scaled) sensor. A "slewed" sensor could result in poor driveability complaints.

Refer to "Intermittents" in "Symptoms," Section

3.1L (VIN T) ENGINE — DIAGNOSTIC TROUBLE CODE CHART — LUMINA WITH VFV

CODE 23
INTAKE AIR TEMPERATURE (IAT) SENSOR CIRCUIT
(LOW TEMPERATURE INDICATED)
3.1L (VIN T) "W" CARLINE (PORT)

1. DOES TECH 1 "SCAN" TOOL DISPLAY IAT -30°C (-22°F) OR COLDER?

YES →
2.
- DISCONNECT SENSOR.
- JUMPER HARNESS TERMINALS TOGETHER.
- TECH 1 "SCAN" TOOL SHOULD DISPLAY TEMPERATURE OVER 130°C (266°F). DOES IT?

  - YES → FAULTY CONNECTION OR SENSOR.
  - NO →
3.
- JUMPER CKT 472 TO GROUND.
- TECH 1 "SCAN" TOOL SHOULD DISPLAY TEMPERATURE OVER 130°C (266°F). DOES IT?

  - YES → OPEN SENSOR GROUND CIRCUIT, FAULTY CONNECTION OR FAULTY ECM.
  - NO → OPEN CKT 472, FAULTY CONNECTION OR FAULTY ECM.

NO → CODE 23 IS INTERMITTENT. IF NO ADDITIONAL CODES WERE STORED, REFER TO "DIAGNOSTIC AIDS"

IAT SENSOR TEMPERATURE VS. RESISTANCE VALUES (APPROXIMATE)		
°F	°C	OHMS
210	100	185
160	70	450
100	38	1,800
70	20	3,400
40	4	7,500
20	-7	13,500
0	-18	25,000
-40	-40	100,700

DIAGNOSTIC AID

"AFTER REPAIRS," REFER TO CODE CRITERIA   AND CONFIRM CODE DOES NOT RESET.

3.1L (VIN T) ENGINE — DIAGNOSTIC TROUBLE CODE CHART — LUMINA WITH VFV

CODE 24
VEHICLE SPEED SENSOR (VSS) CIRCUIT
3.1L (VIN T) "W" CARLINE (PORT)

Circuit Description:

Vehicle speed information is provided to the ECM by the vehicle speed sensor which uses a Permanent Magnet (PM) generator and it is mounted in the transaxle. The PM generator produces a pulsing voltage whenever vehicle speed is over about 3 mph. The AC voltage level and the number of pulses increases with vehicle speed. The ECM then converts the pulsing voltage to mph which is used for calculations, and the mph can be displayed with a "Scan" tool. Output of the generator can also be seen by using a digital voltmeter on the AC scale while rotating the generator.

The function of VSS buffer used in past model years has been incorporated into the ECM. The ECM then supplies the necessary signal for the instrument panel (4000 pulses per mile) for operating the speedometer and the odometer. If the vehicle is equipped with cruise control, the ECM also provides a signal (2000 pulses per mile) to the cruise control module.

NOTE: To prevent misdiagnosis, the technician should review SECTION 8A or the Electrical Troubleshooting Manual and identify the type of vehicle speed sensor used prior to using this chart. Disregard a Code 24 set when drive wheels are not turning.

Test Description: Number(s) below refer to circled number(s) on the diagnostic chart.

1. Code 24 will set if vehicle speed equals 0 mph when:
   - VSS indicates less than 2 mph.
   - MAP is less than 30 kPa.
   - Engine speed is between 2200 and 4400 rpm.
   - TPS is less than 2%.
   - Not in park or neutral.
   - No Code 21, 22, 33 or 34.
   - All conditions met for 3 seconds.
   These conditions are met during a road load deceleration. Disregard Code 24 that sets when drive wheels are not turning.
   - The PM generator only produces a signal if drive wheels are turning greater than 3 mph.
2. If CKTs 400, 401 and 389 are OK, and if the speedometer works properly, Code 24 is being caused by a faulty ECM, faulty MEM-CAL or an incorrect MEM-CAL.

Diagnostic Aids:

"Scan" should indicate a vehicle speed whenever the drive wheels are turning greater than 3 mph.

A problem in CKT 389 will not affect the VSS input or the readings on a "Scan."

Check CKT 400 and CKT 401 for proper connections to be sure there clean and tight and the harness is routed correctly.

(A/T) A faulty or misadjusted Park/Neutral (P/N) switch can result in a false Code 24. Use a "Scan" and check for proper signal while in overdrive (440-T4). Refer to CHART C-1A for P/N switch diagnosis check.

3.1L (VIN T) ENGINE — DIAGNOSTIC TROUBLE CODE CHART — LUMINA WITH VFV

CODE 24
VEHICLE SPEED SENSOR (VSS) CIRCUIT
3.1L (VIN T) "W" CARLINE (PORT)

3.1L (VIN T) ENGINE — DIAGNOSTIC TROUBLE CODE CHART — LUMINA WITH VFV

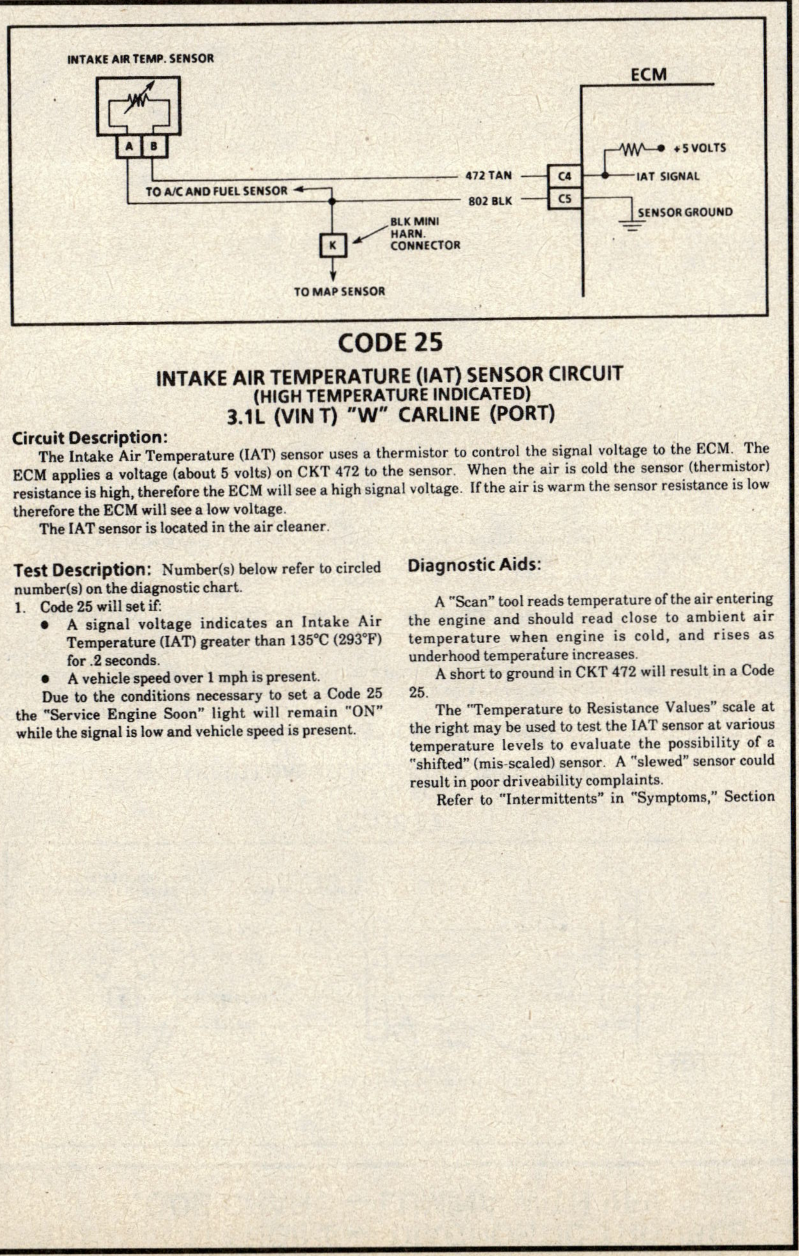

CODE 25

INTAKE AIR TEMPERATURE (IAT) SENSOR CIRCUIT
(HIGH TEMPERATURE INDICATED)
3.1L (VIN T) "W" CARLINE (PORT)

Circuit Description:

The Intake Air Temperature (IAT) sensor uses a thermistor to control the signal voltage to the ECM. The ECM applies a voltage (about 5 volts) on CKT 472 to the sensor. When the air is cold the sensor (thermistor) resistance is high, therefore the ECM will see a high signal voltage. If the air is warm the sensor resistance is low therefore the ECM will see a low voltage.

The IAT sensor is located in the air cleaner.

Test Description: Number(s) below refer to circled number(s) on the diagnostic chart.

1. Code 25 will set if:
   - A signal voltage indicates an Intake Air Temperature (IAT) greater than 135°C (293°F) for .2 seconds.
   - A vehicle speed over 1 mph is present.

   Due to the conditions necessary to set a Code 25 the "Service Engine Soon" light will remain "ON" while the signal is low and vehicle speed is present.

Diagnostic Aids:

A "Scan" tool reads temperature of the air entering the engine and should read close to ambient air temperature when engine is cold, and rises as underhood temperature increases.

A short to ground in CKT 472 will result in a Code 25.

The "Temperature to Resistance Values" scale at the right may be used to test the IAT sensor at various temperature levels to evaluate the possibility of a "shifted" (mis-scaled) sensor. A "slewed" sensor could result in poor driveability complaints.

Refer to "Intermittents" in "Symptoms," Section

3.1L (VIN T) ENGINE — DIAGNOSTIC TROUBLE CODE CHART — LUMINA WITH VFV

CODE 25

INTAKE AIR TEMPERATURE (IAT) SENSOR CIRCUIT
(HIGH TEMPERATURE INDICATED)
3.1L (VIN T) "W" CARLINE (PORT)

DIAGNOSTIC AID

IAT SENSOR		
TEMPERATURE VS. RESISTANCE VALUES		
(APPROXIMATE)		
°F	°C	OHMS
210	100	185
160	70	450
100	38	1,800
70	20	3,400
40	4	7,500
20	-7	13,500
0	-18	25,000
-40	-40	100,700

"AFTER REPAIRS," REFER TO CODE CRITERIA   AND CONFIRM CODE DOES NOT RESET.

3.1L (VIN T) ENGINE — DIAGNOSTIC TROUBLE CODE CHART — LUMINA WITH VFV

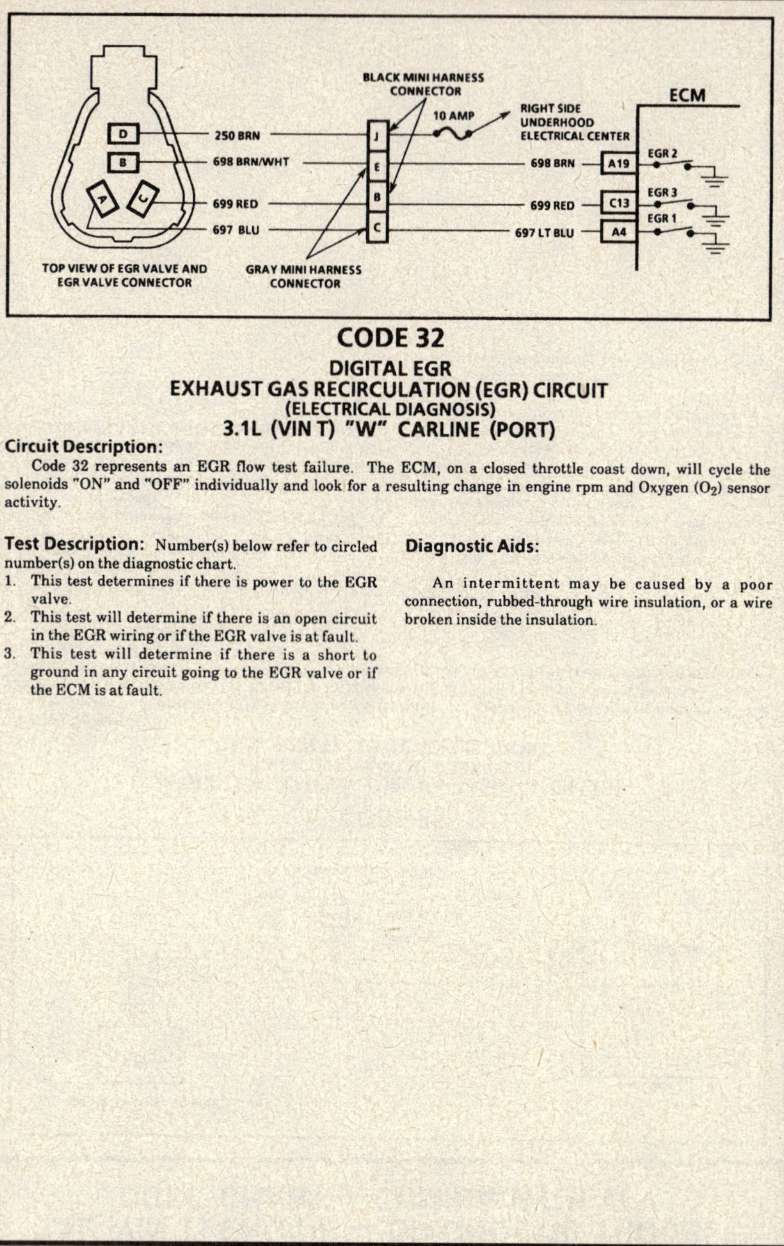

CODE 32
DIGITAL EGR
EXHAUST GAS RECIRCULATION (EGR) CIRCUIT
(ELECTRICAL DIAGNOSIS)
3.1L (VIN T) "W" CARLINE (PORT)

Circuit Description:

Code 32 represents an EGR flow test failure. The ECM, on a closed throttle coast down, will cycle the solenoids "ON" and "OFF" individually and look for a resulting change in engine rpm and Oxygen (O_2) sensor activity.

Test Description: Number(s) below refer to circled number(s) on the diagnostic chart.

1. This test determines if there is power to the EGR valve.
2. This test will determine if there is an open circuit in the EGR wiring or if the EGR valve is at fault.
3. This test will determine if there is a short to ground in any circuit going to the EGR valve or if the ECM is at fault.

Diagnostic Aids:

An intermittent may be caused by a poor connection, rubbed-through wire insulation, or a wire broken inside the insulation.

3.1L (VIN T) ENGINE — DIAGNOSTIC TROUBLE CODE CHART — LUMINA WITH VFV

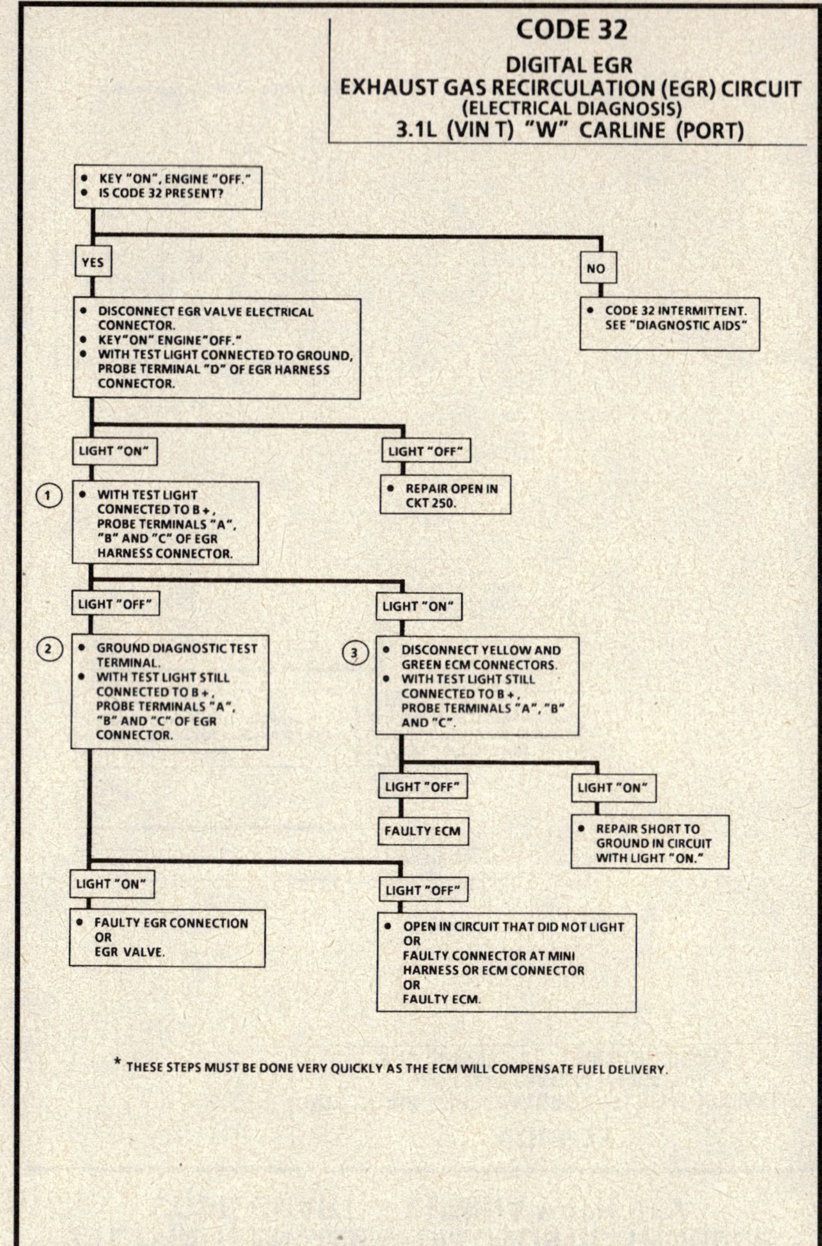

3.1L (VIN T) ENGINE — DIAGNOSTIC TROUBLE CODE CHART — LUMINA WITH VFV

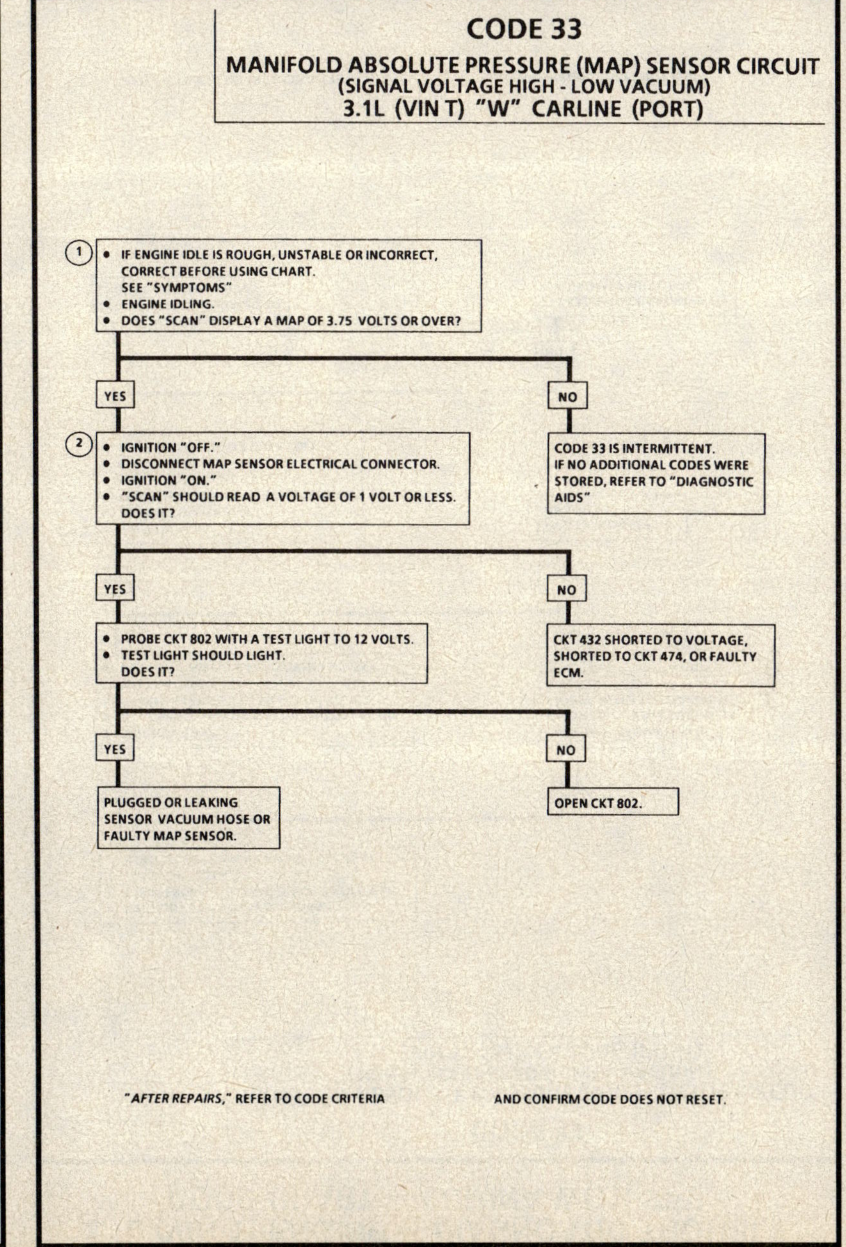

CODE 33
MANIFOLD ABSOLUTE PRESSURE (MAP) SENSOR CIRCUIT
(SIGNAL VOLTAGE HIGH - LOW VACUUM)
3.1L (VIN T) "W" CARLINE (PORT)

Circuit Description:
The Manifold Absolute Pressure (MAP) sensor responds to changes in manifold pressure (vacuum). The ECM receives this information as a signal voltage that will vary from about 1/1.5 volts at idle (high vacuum) to 4/4.5 volts at Wide Open Throttle (WOT) (low vacuum).

Test Description: Number(s) below refer to circled number(s) on the diagnostic chart.
1. Code 33 will set when:
   - No Code 21 or Code 22.
   - Engine running.
   - Manifold pressure greater than 74 kPa (A/C "OFF") 83.4 kPa (A/C "ON").
   - Throttle angle less than 2%.
   - Conditions met for 4.8 seconds.
   Engine misfire or a low unstable idle may set Code 33.
2. With the MAP sensor disconnected, the ECM should see a low voltage if the ECM and wiring are OK.

Diagnostic Aids:
If idle is rough or unstable, refer to "Symptoms," Section "6E3-B" for items which can cause an unstable idle.
An open in CKT 802 or the connection will result in a Code 33.
Refer to "Intermittents" in "Symptoms," Section

3.1L (VIN T) ENGINE — DIAGNOSTIC TROUBLE CODE CHART — LUMINA WITH VFV

CODE 33
MANIFOLD ABSOLUTE PRESSURE (MAP) SENSOR CIRCUIT
(SIGNAL VOLTAGE HIGH - LOW VACUUM)
3.1L (VIN T) "W" CARLINE (PORT)

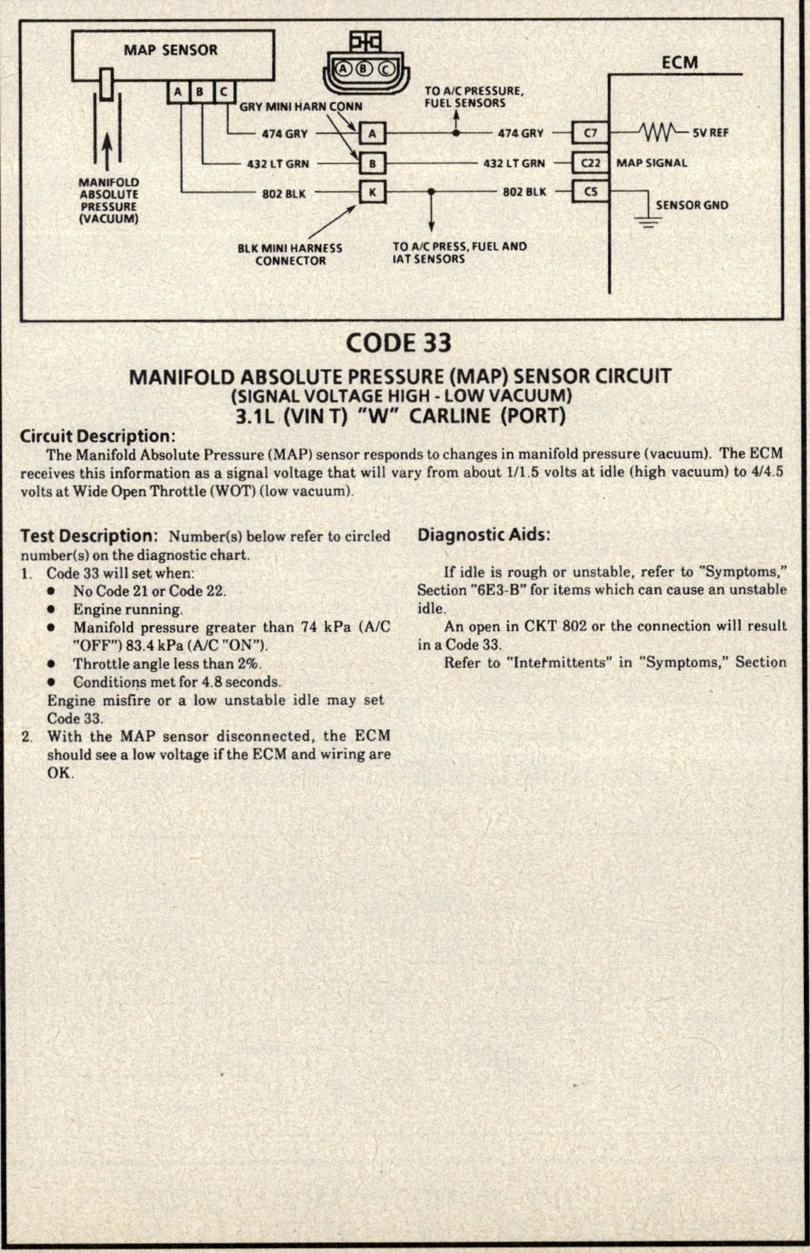

3.1L (VIN T) ENGINE — DIAGNOSTIC TROUBLE CODE CHART — LUMINA WITH VFV

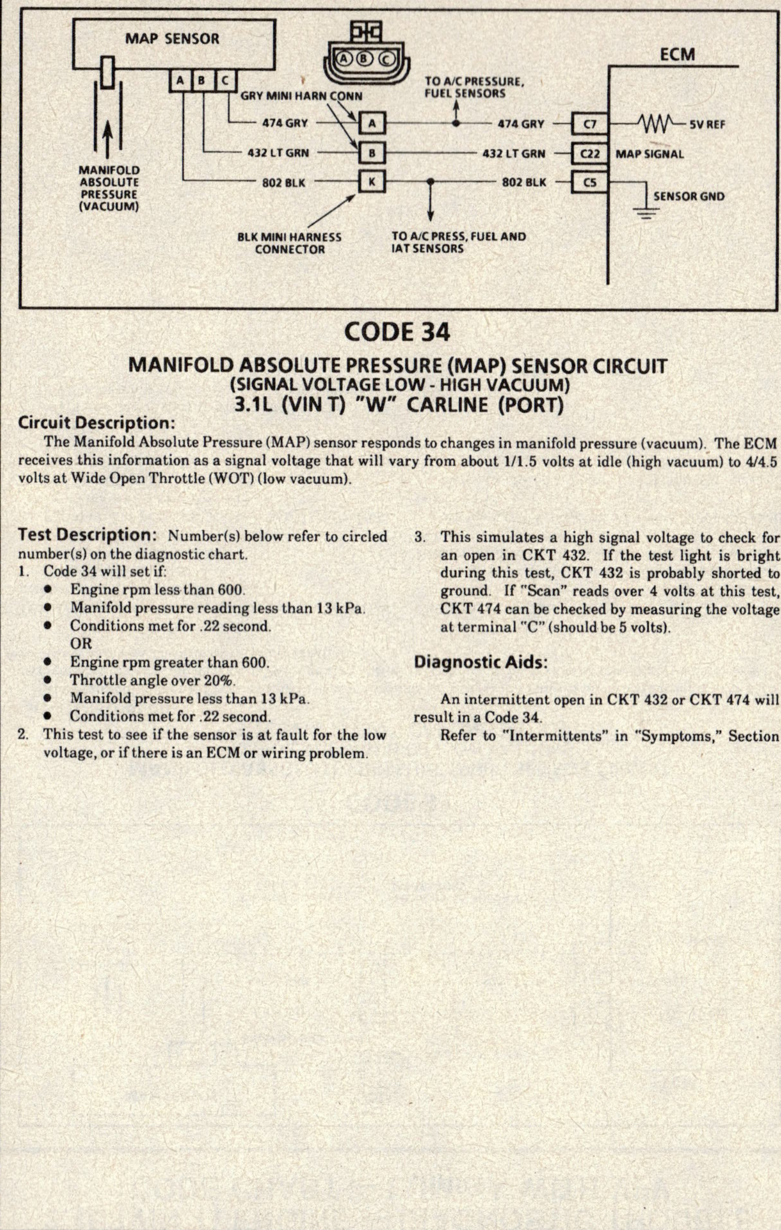

CODE 34

MANIFOLD ABSOLUTE PRESSURE (MAP) SENSOR CIRCUIT
(SIGNAL VOLTAGE LOW - HIGH VACUUM)
3.1L (VIN T) "W" CARLINE (PORT)

Circuit Description:

The Manifold Absolute Pressure (MAP) sensor responds to changes in manifold pressure (vacuum). The ECM receives this information as a signal voltage that will vary from about 1/1.5 volts at idle (high vacuum) to 4/4.5 volts at Wide Open Throttle (WOT) (low vacuum).

Test Description: Number(s) below refer to circled number(s) on the diagnostic chart.
1. Code 34 will set if:
   - Engine rpm less than 600.
   - Manifold pressure reading less than 13 kPa.
   - Conditions met for .22 second.
     OR
   - Engine rpm greater than 600.
   - Throttle angle over 20%.
   - Manifold pressure less than 13 kPa.
   - Conditions met for .22 second.
2. This test to see if the sensor is at fault for the low voltage, or if there is an ECM or wiring problem.

3. This simulates a high signal voltage to check for an open in CKT 432. If the test light is bright during this test, CKT 432 is probably shorted to ground. If "Scan" reads over 4 volts at this test, CKT 474 can be checked by measuring the voltage at terminal "C" (should be 5 volts).

Diagnostic Aids:

An intermittent open in CKT 432 or CKT 474 will result in a Code 34.

Refer to "Intermittents" in "Symptoms," Section

3.1L (VIN T) ENGINE — DIAGNOSTIC TROUBLE CODE CHART — LUMINA WITH VFV

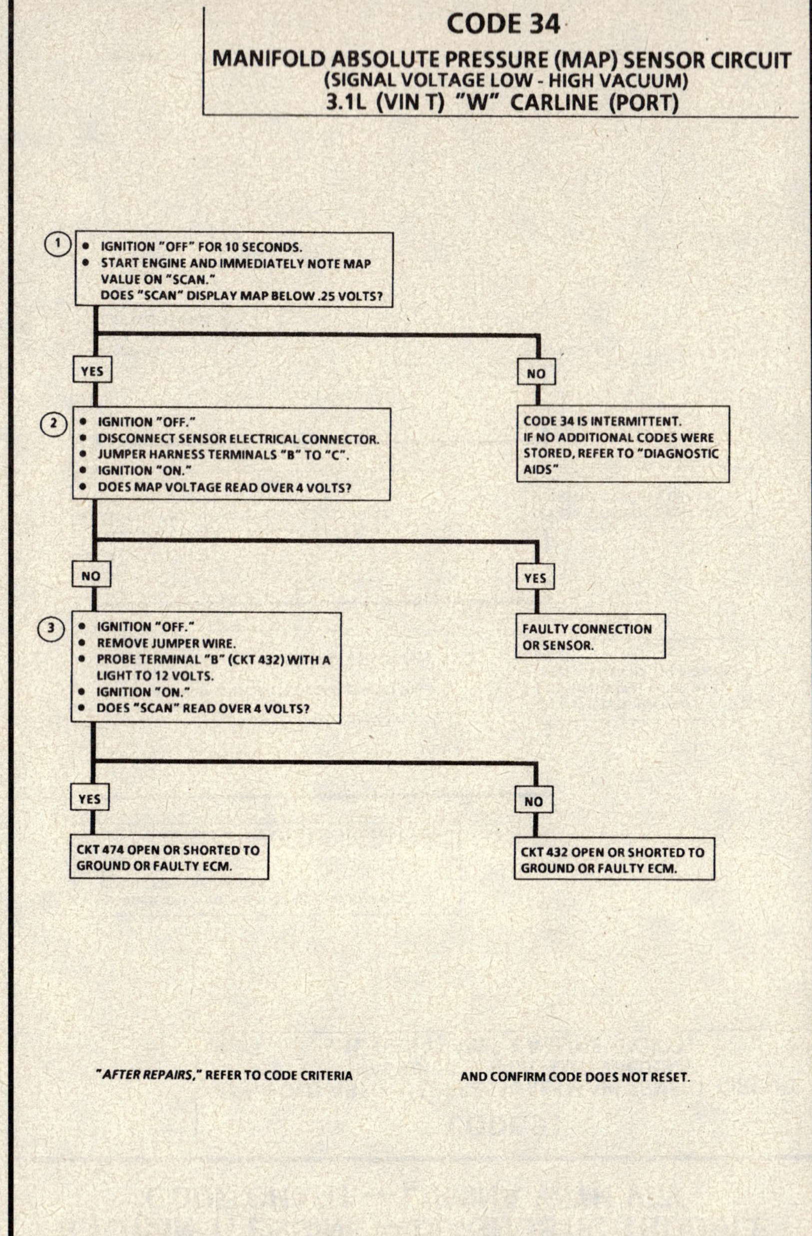

3.1L (VIN T) ENGINE — DIAGNOSTIC TROUBLE CODE CHART — LUMINA WITH VFV

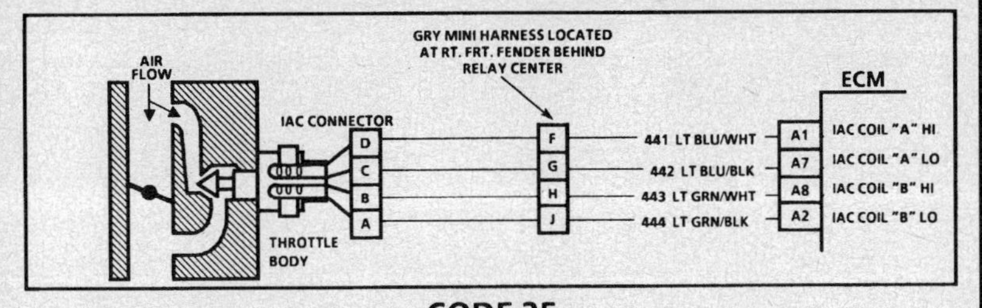

CODE 35
IDLE AIR CONTROL (IAC) CIRCUIT
3.1L (VIN T) "W" CARLINE (PORT)

Circuit Description:

Code 35 will set when the closed throttle engine speed is 200 rpm above or below the desired (commanded) idle speed for 50 seconds. Review the "General Description" of the IAC operation in "Fuel Metering System," Section "6E3-C2".

Test Description: Number(s) below refer to circled number(s) on the diagnostic chart.

1. The Tech 1 rpm control mode is used to extend and retract the IAC valve. The valve should move smoothly within the specified range. If the idle speed is commanded (IAC extended) too low (below 700 rpm), the engine may stall. This may be normal and would not indicate a problem. Retracting the IAC beyond its controlled range (above 1500 rpm) will cause a delay before the rpm's start dropping. This too is normal.
2. This test uses the Tech 1 to command the IAC controlled idle speed. The ECM issues commands to obtain commended idle speed. The node lights each should flash red and green to indicate a good circuit as the ECM issues commands. While the sequence of color is not important if either light is "OFF" or does not flash red and green, check the circuits for faults, beginning with poor terminal contacts.

Diagnostic Aids:

A slow, unstable, or fast idle may be caused by a non-IAC system problem that cannot be overcome by the IAC valve. Out of control range IAC "Scan" tool counts will be above 60 if idle is too low, and zero counts if idle is too high. If idle speed is above 600-700 rpm in drive with an A/T, locate and correct vacuum leak. If rpm is below spec., check for foreign material around throttle plates. Refer to "Fuel Metering System," Section "6E3-C2". The following checks should be made to repair a non-IAC system problem.

- Vacuum leak (high idle)
  If idle is too high, stop the engine. Fully extend (low) IAC with tester. Start engine. If idle speed is above 800 rpm, locate and correct vacuum leak including PCV system. Also check for binding of throttle blade or linkage.

- System too lean (high air/fuel ratio)
  The idle speed may be too high or too low. Engine speed may vary up and down and disconnecting the IAC valve does not help. Code 44 may set. "Scan" O_2 voltage will be less than 300 mV (.3 volt). Check for low regulated fuel pressure, water in the fuel or a restricted injector.
- System too rich (low air/fuel ratio)
  The idle speed will be too low. "Scan" tool IAC counts will usually be above 80. System is obviously rich and may exhibit black smoke in exhaust.
  "Scan" tool O_2 voltage will be fixed above 800 mV (.8 volt).
  Check for high fuel pressure, leaking or sticking injector. Silicone contaminated O_2 sensors "Scan" voltage will be slow to respond.
- Throttle body
  Remove IAC valve and inspect bore for foreign material.
- IAC valve electrical connections
  IAC valve connections should be carefully checked for proper contact.
- PCV valve
  An incorrect or faulty PCV valve may result in an incorrect idle speed.
- Refer to "Rough, Unstable, Incorrect Idle, or Stalling" in "Symptoms," Section "
- If intermittent poor driveability or idle symptoms are resolved by disconnecting the IAC, carefully recheck connections, valve terminal resistance, or replace IAC.
- A/C compressor or relay failure
  See CHART C-10 if the A/C control relay drive circuit is shorted to ground or if the relay is faulty, an idle problem may exist.
- If above are all OK, refer to "Rough, Unstable, Incorrect Idle, or Stalling" in "Symptoms," Section

---

3.1L (VIN T) ENGINE — DIAGNOSTIC TROUBLE CODE CHART — LUMINA WITH VFV

CODE 35
IDLE AIR CONTROL (IAC) CIRCUIT
3.1L (VIN T) "W" CARLINE (PORT)

(1)
- INSTALL TECH 1
- ENGINE AT NORMAL OPERATING TEMPERATURE IN PARK/NEUTRAL WITH PARKING BRAKE SET.
- A/C "OFF."
- SELECT RPM CONTROL. (MISC. TESTS)
- CYCLE IAC THROUGH ITS RANGE FROM 700 RPM UP TO 1500 RPM.
- RPM SHOULD CHANGE SMOOTHLY. DOES IT?

NO → **(2)**
- INSTALL IAC NODE LIGHT * IN IAC HARNESS.
- ENGINE RUNNING. CYCLE IAC WITH TECH 1.
- EACH NODE LIGHT SHOULD CYCLE RED AND GREEN BUT NEVER "OFF." DO THEY?

YES
- USING THE IAC DRIVER * OR OTHER CONVENIENT CONNECTOR, CHECK RESISTANCE ACROSS IAC COILS.
- SHOULD BE 40 TO 80 OHMS BETWEEN IAC TERMINALS "A" TO "B" AND "C" TO "D".

NO
IF CIRCUIT(S) DID NOT TEST RED AND GREEN, CHECK FOR:
- FAULTY CONNECTOR TERMINAL CONTACTS.
- OPEN CIRCUITS INCLUDING CONNECTORS.
- CIRCUITS SHORTED TO GROUND OR VOLTAGE.
- FAULTY ECM CONNECTIONS OR REPLACE ECM.
  REPAIR AS NECESSARY AND RETEST.

YES
- CHECK IAC CONNECTIONS.
- CHECK IAC PASSAGES.
- IF OK, REPLACE IAC.

OK
- CHECK RESISTANCE BETWEEN IAC TERMINALS "B" AND "C" AND "A" AND "D".
- SHOULD BE INFINITE.

NOT OK
REPLACE IAC VALVE AND RETEST.

OK
IDLE AIR CONTROL CIRCUIT OK. REFER TO "DIAGNOSTIC AIDS"

NOT OK
REPLACE IAC VALVE AND RETEST.

---

* IAC DRIVER AND NODE LIGHT REQUIRED KIT
222-L FROM: CONCEPT TECHNOLOGY, INC.
J 37027 FROM: KENT-MOORE, INC.

CLEAR CODES, CONFIRM "CLOSED LOOP" OPERATION, NO "SERVICE ENGINE SOON" LIGHT, PERFORM IAC RESET PROCEDURE PER APPLICABLE SERVICE MANUAL AND VERIFY CONTROLLED IDLE SPEED IS CORRECT.

"AFTER REPAIRS," REFER TO CODE CRITERIA   AND CONFIRM CODE DOES NOT RESET.

3.1L (VIN T) ENGINE — DIAGNOSTIC TROUBLE CODE CHART — LUMINA WITH VFV

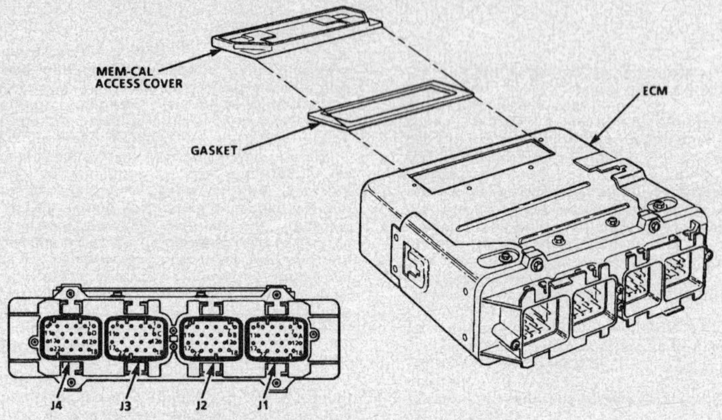

CODE 41

CYLINDER SELECT ERROR
(FAULTY OR INCORRECT MEM-CAL)
3.1L (VIN T) "W" CARLINE (PORT)

Test Description: Number(s) below refer to circled number(s) on the diagnostic chart.

1. The ECM used for this engine can also be used for other engines and the difference is in the MEM-CAL. If a Code 41 sets, the incorrect MEM-CAL has been installed or it is faulty and it must be replaced.

Diagnostic Aids:

Check MEM-CAL to be sure locking tabs are secure.

3.1L (VIN T) ENGINE — DIAGNOSTIC TROUBLE CODE CHART — LUMINA WITH VFV

CODE 41

CYLINDER SELECT ERROR
(FAULTY OR INCORRECT MEM-CAL)
3.1L (VIN T) "W" CARLINE (PORT)

(1)
- IGNITION "OFF," CLEAR CODES.
- START ENGINE AND RUN FOR 1 MINUTE OR UNTIL CODE 41 SETS.

DOES CODE 41 SET?

YES: FAULTY CONNECTIONS DUE TO MEM-CAL NOT LOCKED IN PLACE, OR INCORRECT MEM-CAL INSTALLED.

NO: CODE 41 IS INTERMITTENT. REVIEW "DIAGNOSTIC AIDS"

"AFTER REPAIRS," REFER TO CODE CRITERIA AND CONFIRM CODE DOES NOT RESET.

3.1L (VIN T) ENGINE — DIAGNOSTIC TROUBLE CODE CHART — LUMINA WITH VFV

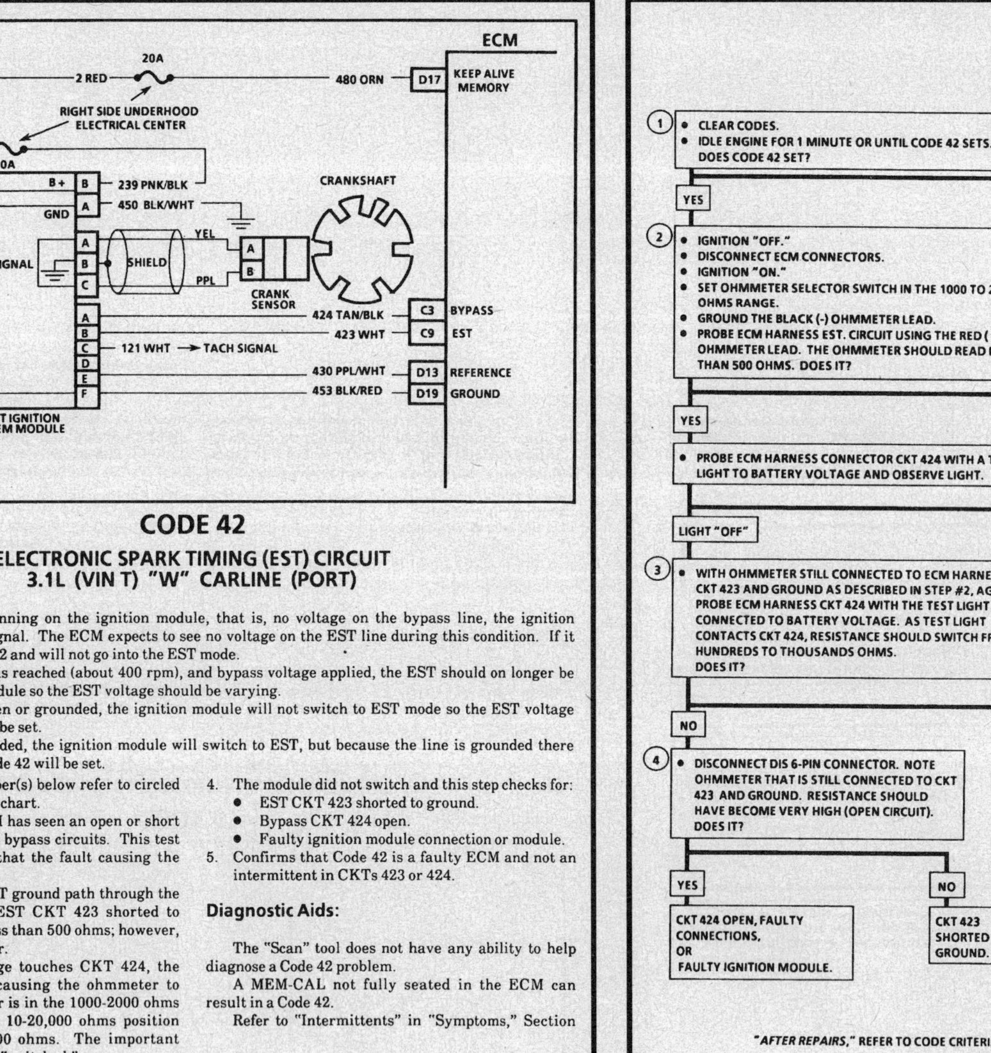

Circuit Description:

When the system is running on the ignition module, that is, no voltage on the bypass line, the ignition module grounds the EST signal. The ECM expects to see no voltage on the EST line during this condition. If it sees a voltage, it sets Code 42 and will not go into the EST mode.

When the rpm for EST is reached (about 400 rpm), and bypass voltage applied, the EST should on longer be grounded in the ignition module so the EST voltage should be varying.

If the bypass line is open or grounded, the ignition module will not switch to EST mode so the EST voltage will be low and Code 42 will be set.

If the EST line is grounded, the ignition module will switch to EST, but because the line is grounded there will be no EST signal. A Code 42 will be set.

Test Description: Number(s) below refer to circled number(s) on the diagnostic chart.

1. Code 42 means the ECM has seen an open or short to ground in the EST or bypass circuits. This test confirms Code 42 and that the fault causing the code is present.
2. Checks for a normal EST ground path through the ignition module. An EST CKT 423 shorted to ground will also read less than 500 ohms; however, this will be checked later.
3. As the test light voltage touches CKT 424, the module should switch causing the ohmmeter to "overrange" if the meter is in the 1000-2000 ohms position. Selecting the 10-20,000 ohms position will indicate above 5000 ohms. The important thing is that the module "switched."

4. The module did not switch and this step checks for:
   - EST CKT 423 shorted to ground.
   - Bypass CKT 424 open.
   - Faulty ignition module connection or module.
5. Confirms that Code 42 is a faulty ECM and not an intermittent in CKTs 423 or 424.

Diagnostic Aids:

The "Scan" tool does not have any ability to help diagnose a Code 42 problem.

A MEM-CAL not fully seated in the ECM can result in a Code 42.

Refer to "Intermittents" in "Symptoms," Section

3.1L (VIN T) ENGINE — DIAGNOSTIC TROUBLE CODE CHART — LUMINA WITH VFV

CODE 42
ELECTRONIC SPARK TIMING (EST) CIRCUIT
3.1L (VIN T) "W" CARLINE (PORT)

1. - CLEAR CODES.
   - IDLE ENGINE FOR 1 MINUTE OR UNTIL CODE 42 SETS. DOES CODE 42 SET?

YES →

2. - IGNITION "OFF."
   - DISCONNECT ECM CONNECTORS.
   - IGNITION "ON."
   - SET OHMMETER SELECTOR SWITCH IN THE 1000 TO 2000 OHMS RANGE.
   - GROUND THE BLACK (-) OHMMETER LEAD.
   - PROBE ECM HARNESS EST. CIRCUIT USING THE RED (+) OHMMETER LEAD. THE OHMMETER SHOULD READ LESS THAN 500 OHMS. DOES IT?

NO → CODE 42 INTERMITTENT. REFER TO "DIAGNOSTIC AIDS"

YES →

- PROBE ECM HARNESS CONNECTOR CKT 424 WITH A TEST LIGHT TO BATTERY VOLTAGE AND OBSERVE LIGHT.

NO → OPEN CKT 423, FAULTY CONNECTION, OR FAULTY IGNITION MODULE.

LIGHT "OFF" →

3. - WITH OHMMETER STILL CONNECTED TO ECM HARNESS CKT 423 AND GROUND AS DESCRIBED IN STEP #2, AGAIN PROBE ECM HARNESS CKT 424 WITH THE TEST LIGHT CONNECTED TO BATTERY VOLTAGE. AS TEST LIGHT CONTACTS CKT 424, RESISTANCE SHOULD SWITCH FROM HUNDREDS TO THOUSANDS OHMS. DOES IT?

LIGHT "ON" →

- DISCONNECT IGNITION MODULE 6-PIN CONNECTOR.

LIGHT "ON" → CKT 424 SHORTED TO GROUND

LIGHT "OFF" → REPLACE IGNITION MODULE.

NO →

4. - DISCONNECT DIS 6-PIN CONNECTOR. NOTE OHMMETER THAT IS STILL CONNECTED TO CKT 423 AND GROUND. RESISTANCE SHOULD HAVE BECOME VERY HIGH (OPEN CIRCUIT). DOES IT?

YES →

5. - RECONNECT ECM AND IDLE ENGINE FOR ONE MINUTE OR UNTIL CODE 42 SETS. DOES CODE SET?

YES → CKT 424 OPEN, FAULTY CONNECTIONS, OR FAULTY IGNITION MODULE.

NO → CKT 423 SHORTED TO GROUND.

YES → REPLACE ECM.

NO → CODE 42 INTERMITTENT. REFER TO "DIAGNOSTIC AIDS"

"AFTER REPAIRS," REFER TO CODE CRITERIA   AND CONFIRM CODE DOES NOT RESET.

3.1L (VIN T) ENGINE — DIAGNOSTIC TROUBLE CODE CHART — LUMINA WITH VFV

3.1L (VIN T) ENGINE — DIAGNOSTIC TROUBLE CODE CHART — LUMINA WITH VFV

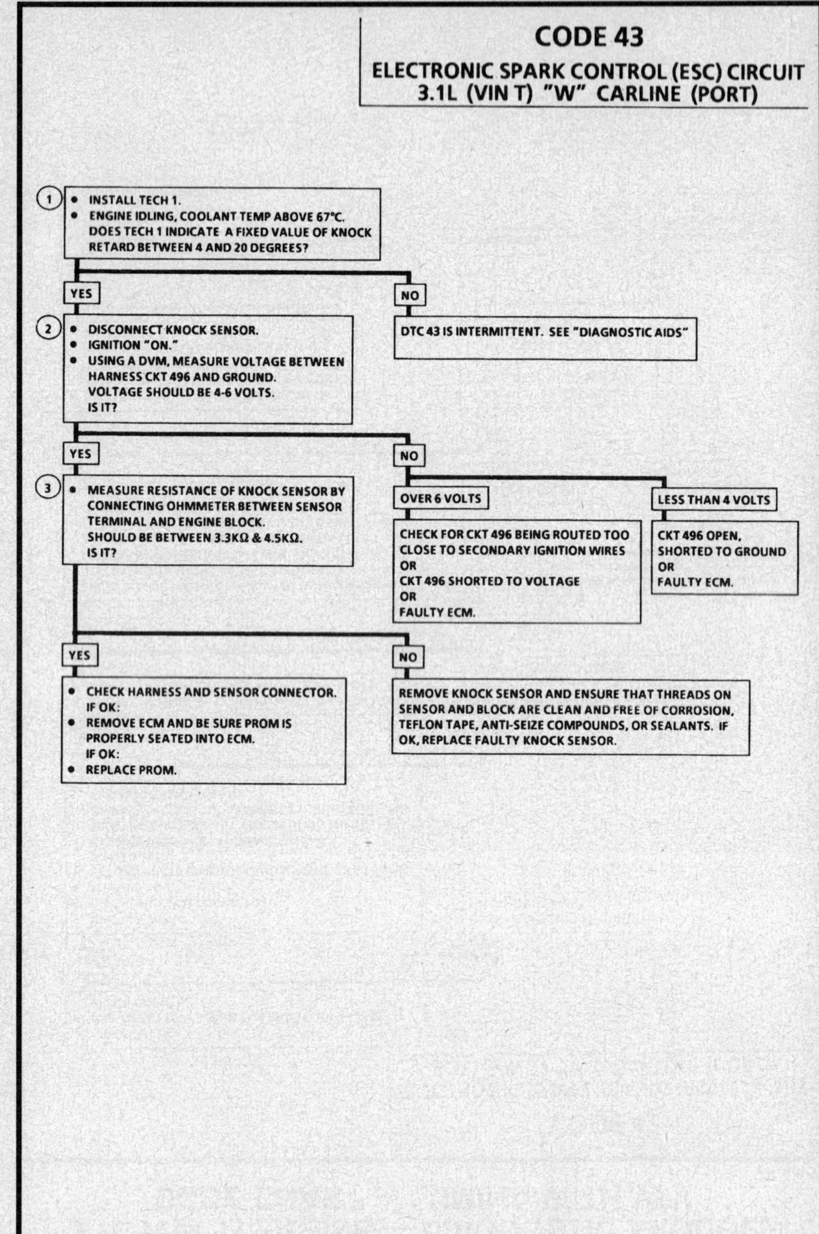

CODE 43
ELECTRONIC SPARK CONTROL (ESC) CIRCUIT
3.1L (VIN T) "W" CARLINE (PORT)

Circuit Description:

The knock sensor is used to detect engine detonation and the ECM will retard the Electronic Spark Timing (EST) based on the signal being received. The circuitry within the knock sensor causes the ECM 5 volts to be pulled down so that under a no knock condition, CKT 496 would measure about 2.5 volts. The knock sensor produces an AC signal which rides on the 2.5 volts DC voltage. The amplitude and signal frequency is dependent upon the knock level.

If CKT 496 becomes open or shorted to ground, the voltage will either go above 4.8 volts or below .64 volt. If either of these conditions are met for about 10 seconds, a Code 43 will be stored.

Test Description: Number(s) below refer to circled number(s) on the diagnostic chart.

1. This step determines if conditions for Code 43 still exist (voltage on CKT 496 above 4.6 volts or below .64 volt). the system is designed to retard the spark 15° if either condition exists.
2. The ECM has a 5 volt pull-up resistor, which applies 5 volts to CKT 496. The five volt signal should be present at the knock sensor terminal during these test conditions.
3. This step determines if the knock sensor resistance is 3900 ohms ± 15%. If the resistance is between 3300 to 4500 ohms, the sensor is OK.

Diagnostic Aids:

If CKT 496 is not open or shorted to ground and the voltage reading is below 4 volts, the most likely cause is an open circuit in the ECM. It is possible that a faulty MEM-CAL could be drawing the 5 volt signal down and it should be replaced, if a replacement ECM did not correct the problem, refer to "Intermittents" in "Symptoms," Section

3.1L (VIN T) ENGINE — DIAGNOSTIC TROUBLE CODE CHART — LUMINA WITH VFV

CODE 44
OXYGEN (O₂) SENSOR CIRCUIT
(LEAN EXHAUST INDICATED)
3.1L (VIN T) "W" CARLINE (PORT)

Circuit Description:

The ECM supplies a voltage of about .55 volt between terminals "A16" and "A22". (If measured with a 10 megohm digital voltmeter, this may read as low as .35 volt.) The Oxygen (O₂) sensor varies the voltage within a range of about 1 volt if the exhaust is rich, down through about .10 volt if exhaust is lean.

The sensor is like an open circuit and produces no voltage when it is below about 315°C (600°F). An open sensor circuit or cold sensor causes "Open Loop" operation.

Test Description: Number(s) below refer to circled number(s) on the diagnostic chart.
1. Code 44 will set if:
   - No Code 33 or Code 34.
   - Voltage on CKT 412 remains below .2 volt for 60 seconds or more.
   - The system is operating in "Closed Loop."

Diagnostic Aids:

Using the "Scan," observe the block learn values at different rpm and air flow conditions. The "Scan" also displays the block cells, so the block learn values can be checked in each of the cells to determine when the Code 44 may have been set. If the conditions for Code 44 exists the block learn values will be around 150.
- O₂ Sensor Wire. Sensor pigtail may be mispositioned and contacting the exhaust manifold.
- Check for intermittent ground in wire between connector and sensor.

- Lean Injector(s). Perform injector balance test CHART C-2A.
- Fuel Contamination. Water, even in small amounts, near the in-tank fuel pump inlet can be delivered to the injectors. The water causes a lean exhaust and can set a Code 44.
- Fuel Pressure. System will be lean if pressure is too low. It may be necessary to monitor fuel pressure while driving the vehicle at various road speeds and/or loads to confirm. See "Fuel System Diagnosis," CHART A-7.
- Exhaust Leaks. If there is an exhaust leak, the engine can cause outside air to be pulled into the exhaust and past the sensor. Vacuum or crankcase leaks can cause a lean condition.
- If the above are OK, it is a faulty Oxygen (O₂) sensor.

3.1L (VIN T) ENGINE — DIAGNOSTIC TROUBLE CODE CHART — LUMINA WITH VFV

CODE 44
OXYGEN (O₂) SENSOR CIRCUIT
(LEAN EXHAUST INDICATED)
3.1L (VIN T) "W" CARLINE (PORT)

"AFTER REPAIRS," REFER TO CODE CRITERIA   AND CONFIRM CODE DOES NOT RESET.

3.1L (VIN T) ENGINE — DIAGNOSTIC TROUBLE CODE CHART — LUMINA WITH VFV

CODE 45
OXYGEN (O₂) SENSOR CIRCUIT
(RICH EXHAUST INDICATED)
3.1L (VIN T) "W" CARLINE (PORT)

Circuit Description:

The ECM supplies a voltage of about .55 volt between terminals "A16" and "A22". (If measured with a 10 megohm digital voltmeter, this may read as low as .35 volt.) The Oxygen (O₂) sensor varies the voltage within a range of about 1 volt if the exhaust is rich, down through about .10 volt if exhaust is lean.

The sensor is like an open circuit and produces no voltage when it is below about 315°C (600°F). An open sensor circuit or cold sensor causes "Open Loop" operation.

Test Description: Number(s) below refer to circled number(s) on the diagnostic chart.

1. Code 45 will set if:
   - Voltage on CKT 412 remains above .85 volt for 50 seconds.
   - Engine time after start is 1 minute or more.
   - Throttle angle greater than 10%.
   - Operation is in "Closed Loop."

Diagnostic Aids:

Using the "Scan," observe the block learn values at different rpm and air flow conditions. The "Scan" also displays the block cells, so the block learn values can be checked in each of the cells to determine when the Code 45 may have been set. If the conditions for Code 45 exists, the block learn values will be around 115.

- **Fuel Pressure.** System will go rich if pressure is too high. The ECM can compensate for some increase. However, if it gets too high, a Code 45 may be set. See "Fuel System Diagnosis," CHART A-7.
- **Rich Injector.** Perform "Injector Balance Test," CHART C-2A.
- **Leaking Injector.** See CHART A-7.
- Check for fuel contaminated oil.
- **O₂ Sensor Contamination.** Inspect Oxygen (O₂) sensor for silicone contamination from fuel, or use of improper RTV sealant.

The sensor may have a white, powdery coating and result in a high but false signal voltage (rich exhaust indication). The ECM will then reduce the amount of fuel delivered to the engine, causing a severe surge driveability problem.

- **HEI Shielding.** An open ground CKT 453 (ignition system reflow) may result in EMI, or induced electrical "noise." The ECM looks at this "noise" as reference pulses. The additional pulses result in a higher than actual engine speed signal. The ECM then delivers too much fuel, causing system to go rich. Engine tachometer will also show higher than actual engine speed, which can help in diagnosing this problem.
- **Canister Purge.** Check for fuel saturation. If full of fuel, check canister control and hoses. See "Evaporative Emission Control System (EECS)," Section "6E3-C3".
- Check for leaking fuel pressure regulator diaphragm by checking vacuum line to regulator for fuel.
- **TPS.** An intermittent TPS output will cause the system to go rich, due to a false indication of the engine accelerating.
- **EGR.** An EGR staying open (especially at idle) will cause the Oxygen (O₂) sensor to indicate a rich exhaust, and this could result in a Code 45.

3.1L (VIN T) ENGINE — DIAGNOSTIC TROUBLE CODE CHART — LUMINA WITH VFV

CODE 45
OXYGEN (O₂) SENSOR CIRCUIT
(RICH EXHAUST INDICATED)
3.1L (VIN T) "W" CARLINE (PORT)

3.1L (VIN T) ENGINE — DIAGNOSTIC TROUBLE CODE CHART — LUMINA WITH VFV

CODE 51
**MEM-CAL ERROR
(FAULTY OR INCORRECT MEM-CAL)
"W" CARLINE (PORT)**

CHECK THAT ALL PINS ARE FULLY INSERTED IN THE SOCKET AND THAT MEM-CAL IS PROPERLY LATCHED. IF OK, REPLACE, MEM-CAL, CLEAR MEMORY, AND RECHECK. IF CODE 51 REAPPEARS, REPLACE ELECTRONIC CONTROL MODULE (ECM).

CLEAR CODES AND CONFIRM "CLOSED LOOP" OPERATION AND NO "SERVICE ENGINE SOON" LIGHT.

3.1L (VIN T) ENGINE — DIAGNOSTIC TROUBLE CODE CHART — LUMINA WITH VFV

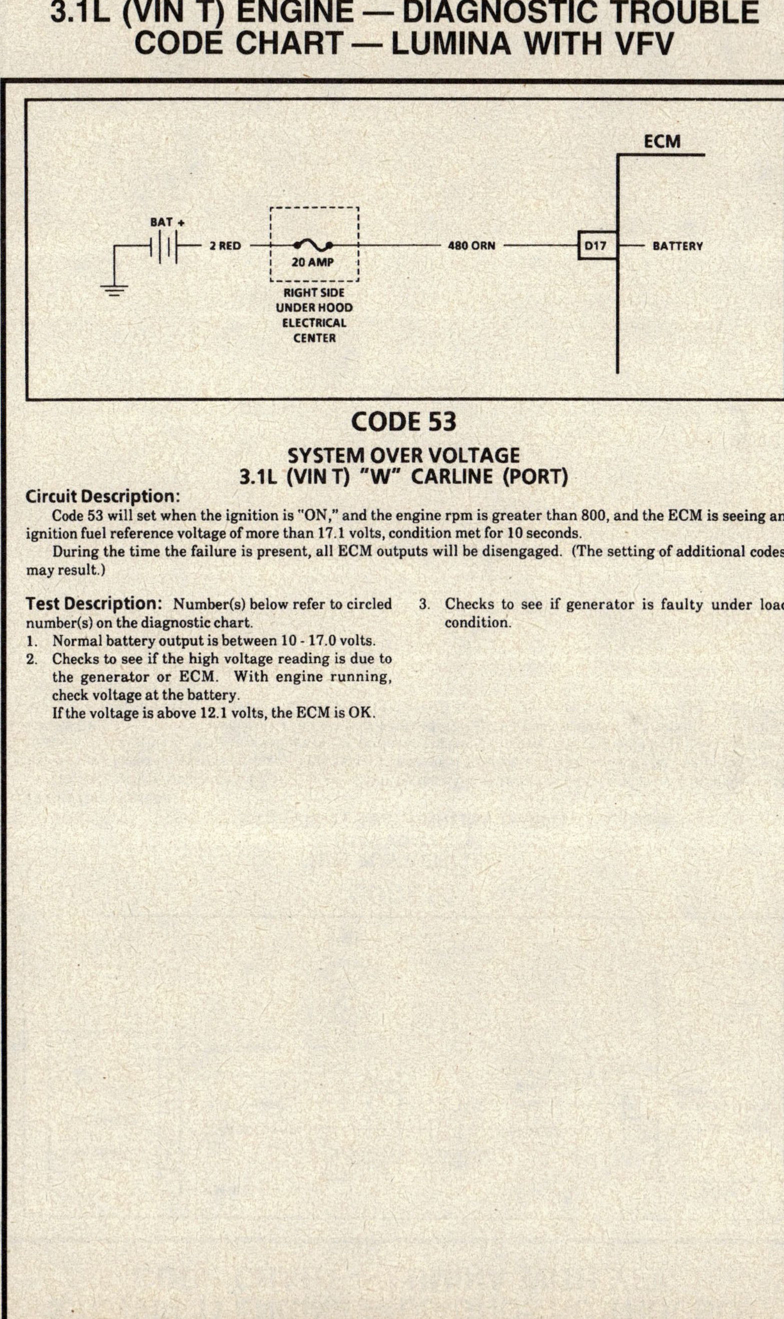

CODE 53
**SYSTEM OVER VOLTAGE
3.1L (VIN T) "W" CARLINE (PORT)**

Circuit Description:
Code 53 will set when the ignition is "ON," and the engine rpm is greater than 800, and the ECM is seeing an ignition fuel reference voltage of more than 17.1 volts, condition met for 10 seconds.
During the time the failure is present, all ECM outputs will be disengaged. (The setting of additional codes may result.)

Test Description: Number(s) below refer to circled number(s) on the diagnostic chart.
1. Normal battery output is between 10 - 17.0 volts.
2. Checks to see if the high voltage reading is due to the generator or ECM. With engine running, check voltage at the battery.
If the voltage is above 12.1 volts, the ECM is OK.
3. Checks to see if generator is faulty under load condition.

3.1L (VIN T) ENGINE — DIAGNOSTIC TROUBLE CODE CHART — LUMINA WITH VFV

CODE 53
SYSTEM OVER VOLTAGE
3.1L (VIN T) "W" CARLINE (PORT)

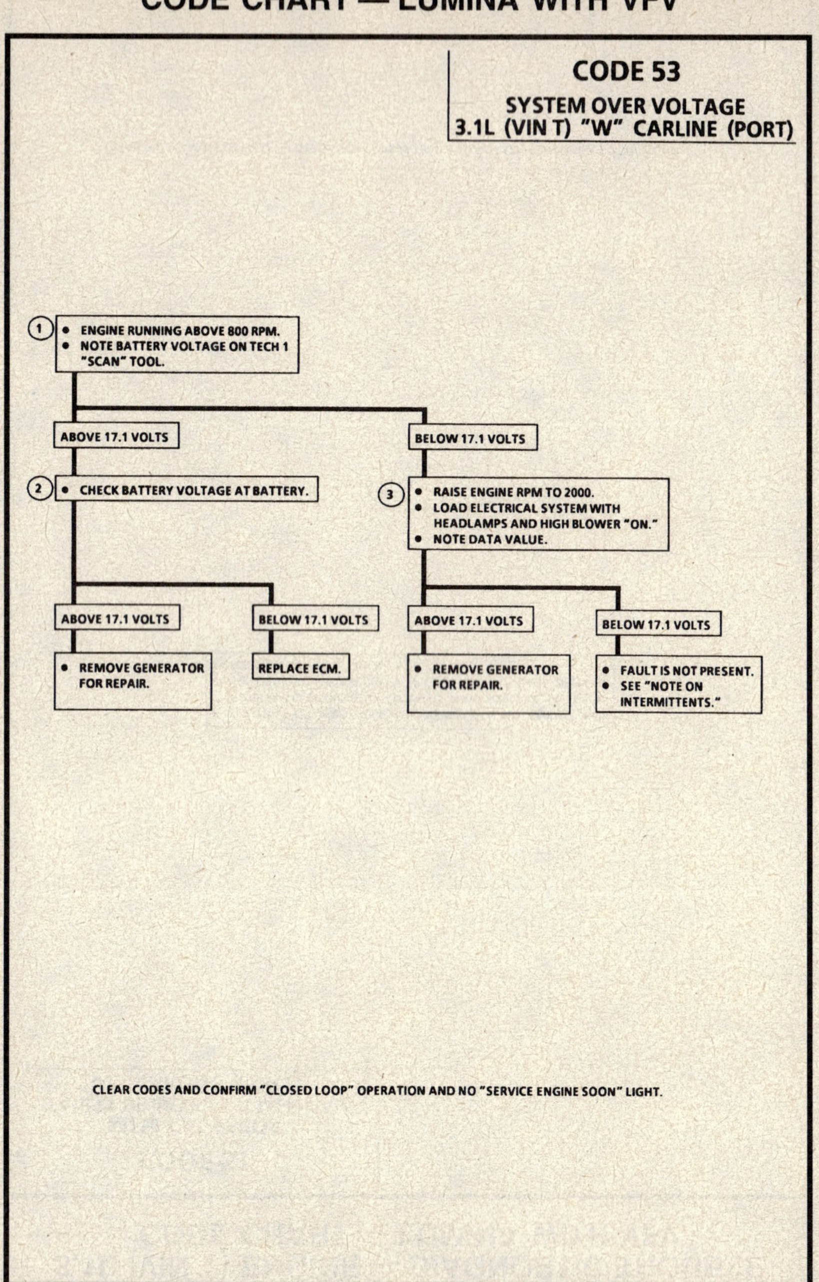

3.1L (VIN T) ENGINE — DIAGNOSTIC TROUBLE CODE CHART — LUMINA WITH VFV

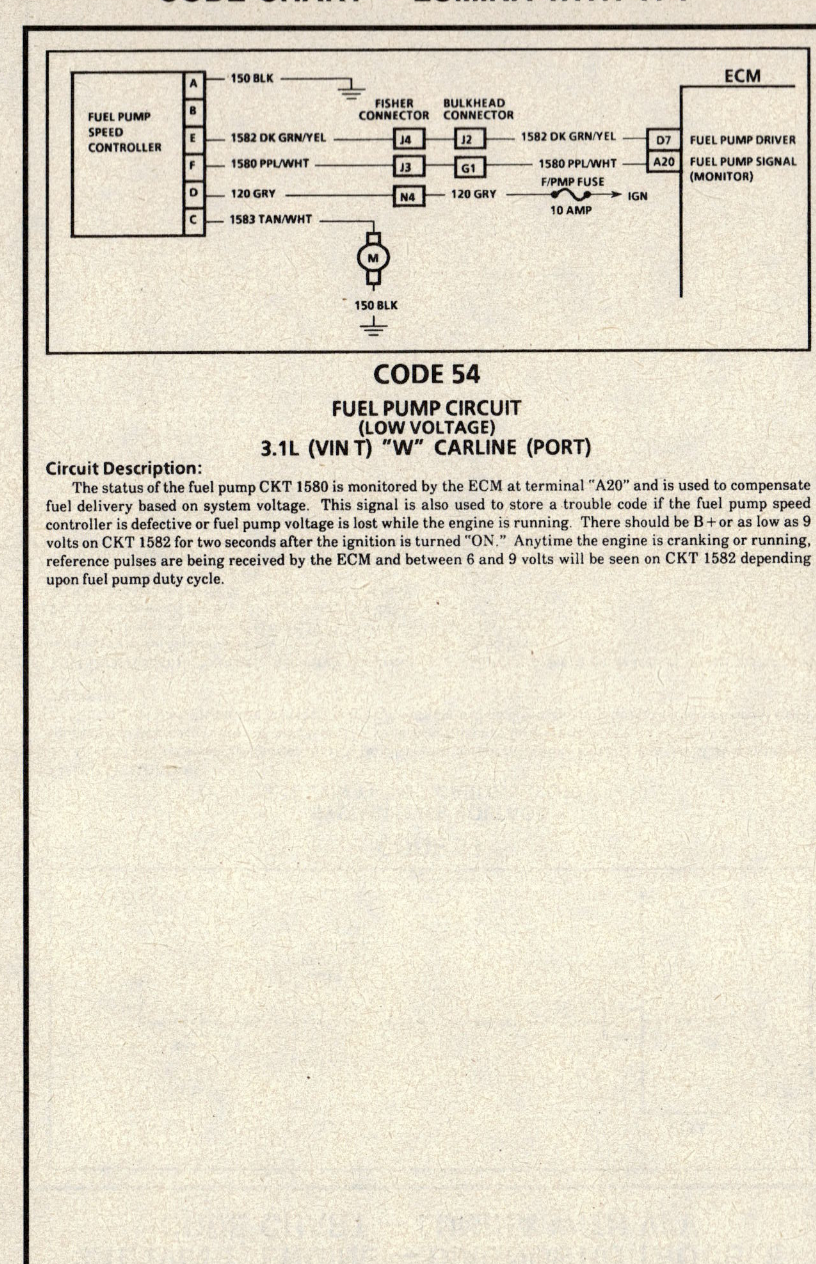

CODE 54
FUEL PUMP CIRCUIT
(LOW VOLTAGE)
3.1L (VIN T) "W" CARLINE (PORT)

Circuit Description:

The status of the fuel pump CKT 1580 is monitored by the ECM at terminal "A20" and is used to compensate fuel delivery based on system voltage. This signal is also used to store a trouble code if the fuel pump speed controller is defective or fuel pump voltage is lost while the engine is running. There should be B+ or as low as 9 volts on CKT 1582 for two seconds after the ignition is turned "ON." Anytime the engine is cranking or running, reference pulses are being received by the ECM and between 6 and 9 volts will be seen on CKT 1582 depending upon fuel pump duty cycle.

3.1L (VIN T) ENGINE — DIAGNOSTIC TROUBLE CODE CHART — LUMINA WITH VFV

CODE 54
FUEL PUMP CIRCUIT
(LOW VOLTAGE)
3.1L (VIN T) "W" CARLINE (PORT)

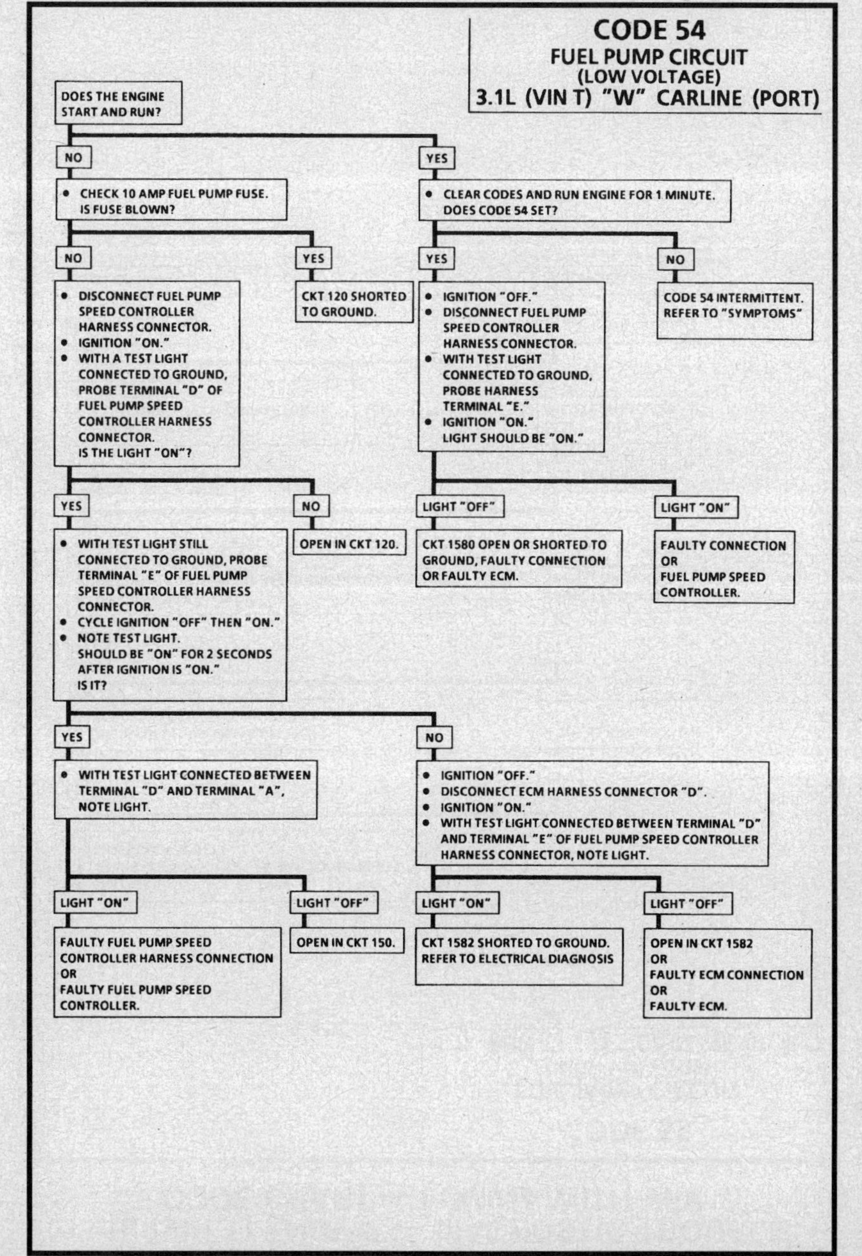

3.1L (VIN T) ENGINE — DIAGNOSTIC TROUBLE CODE CHART — LUMINA WITH VFV

CODE 55
FUEL PUMP CIRCUIT
(HIGH VOLTAGE)
3.1L (VIN T) "W" CARLINE (PORT)

Circuit Description:

The status of the fuel pump CKT 1580 is monitored by the ECM at terminal "A20" and is used to compensate fuel delivery based on system voltage. This signal is also used to store a trouble code if the fuel pump speed controller is defective or fuel pump voltage is lost while the engine is running. There should be B+ or as low as 9 volts on CKT 1582 for 2 seconds after the ignition is turned "ON," or 6 to 9 volts depending on pump duty cycle anytime the engine is cranking or running, as reference pulses are being received by the ECM.

3.1L (VIN T) ENGINE — DIAGNOSTIC TROUBLE CODE CHART — LUMINA WITH VFV

CODE 55
FUEL PUMP CIRCUIT (HIGH VOLTAGE)
3.1L (VIN T) "W" CARLINE (PORT)

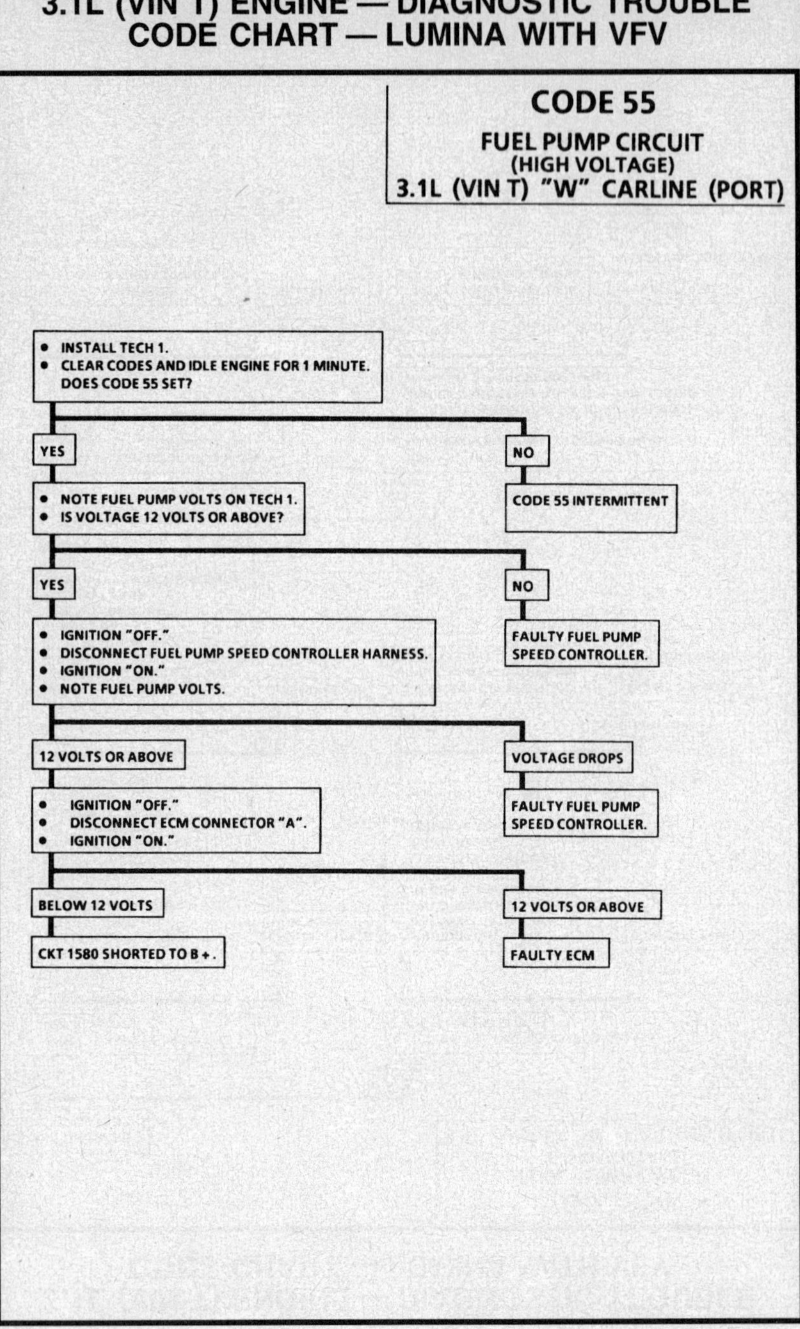

3.1L (VIN T) ENGINE — DIAGNOSTIC TROUBLE CODE CHART — LUMINA WITH VFV

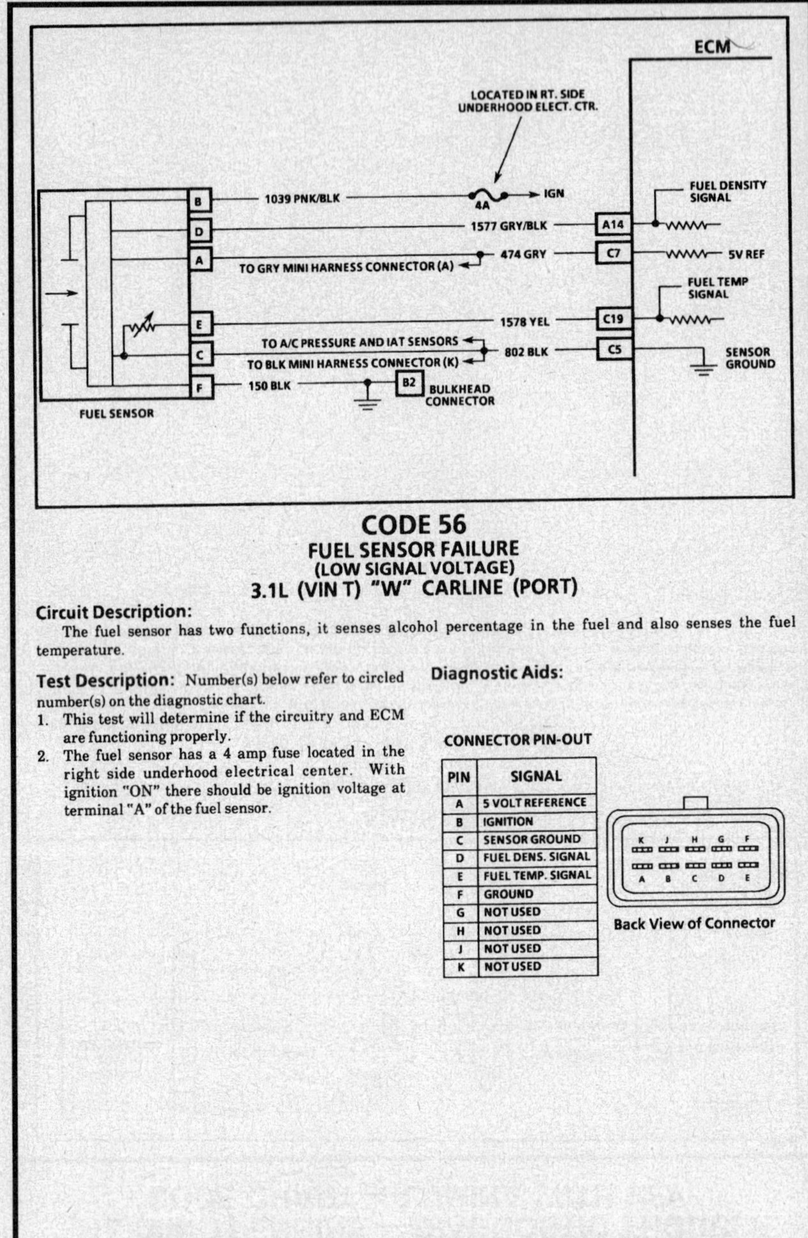

CODE 56
FUEL SENSOR FAILURE (LOW SIGNAL VOLTAGE)
3.1L (VIN T) "W" CARLINE (PORT)

Circuit Description:
The fuel sensor has two functions, it senses alcohol percentage in the fuel and also senses the fuel temperature.

Test Description: Number(s) below refer to circled number(s) on the diagnostic chart.
1. This test will determine if the circuitry and ECM are functioning properly.
2. The fuel sensor has a 4 amp fuse located in the right side underhood electrical center. With ignition "ON" there should be ignition voltage at terminal "A" of the fuel sensor.

Diagnostic Aids:

CONNECTOR PIN-OUT

PIN	SIGNAL
A	5 VOLT REFERENCE
B	IGNITION
C	SENSOR GROUND
D	FUEL DENS. SIGNAL
E	FUEL TEMP. SIGNAL
F	GROUND
G	NOT USED
H	NOT USED
J	NOT USED
K	NOT USED

3.1L (VIN T) ENGINE — DIAGNOSTIC TROUBLE CODE CHART — LUMINA WITH VFV

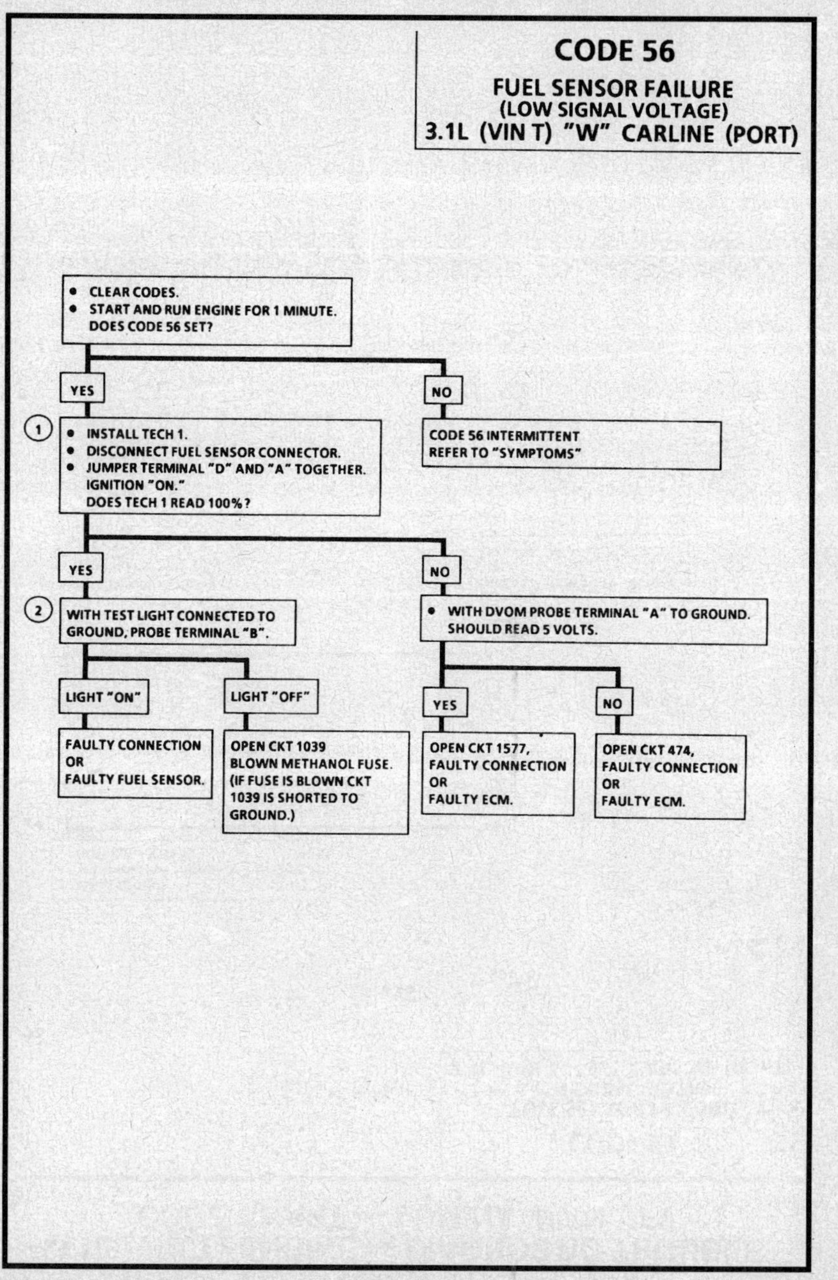

3.1L (VIN T) ENGINE — DIAGNOSTIC TROUBLE CODE CHART — LUMINA WITH VFV

CODE 57
FUEL SENSOR FAILURE
(HIGH SIGNAL VOLTAGE)
3.1L (VIN T) "W" CARLINE (PORT)

Circuit Description:

The fuel sensor has two functions, it senses alcohol percentage in the fuel and also senses the fuel temperature.

Test Description: Number(s) below refer to circled number(s) on the diagnostic chart.

1. This test will determine if the circuitry and ECM are functioning properly.

Diagnostic Aids:

CONNECTOR PIN-OUT

PIN	SIGNAL
A	5 VOLT REFERENCE
B	IGNITION
C	SENSOR GROUND
D	FUEL DENS. SIGNAL
E	FUEL TEMP. SIGNAL
F	GROUND
G	NOT USED
H	NOT USED
J	NOT USED
K	NOT USED

3.1L (VIN T) ENGINE — DIAGNOSTIC TROUBLE CODE CHART — LUMINA WITH VFV

CODE 57

FUEL SENSOR FAILURE
(HIGH SIGNAL VOLTAGE)
3.1L (VIN T) "W" CARLINE (PORT)

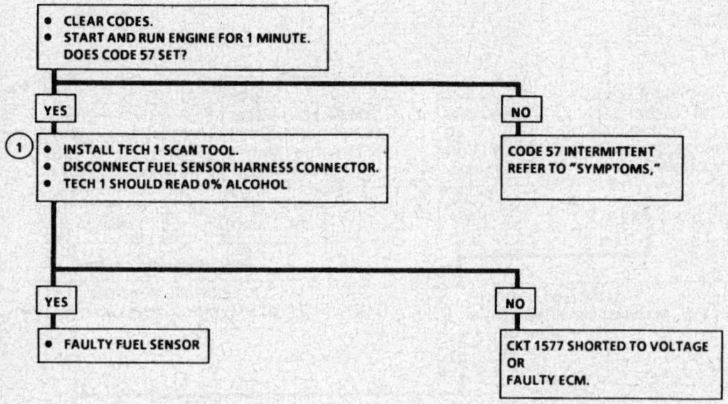

- CLEAR CODES.
- START AND RUN ENGINE FOR 1 MINUTE.
  DOES CODE 57 SET?

YES

(1)
- INSTALL TECH 1 SCAN TOOL.
- DISCONNECT FUEL SENSOR HARNESS CONNECTOR.
- TECH 1 SHOULD READ 0% ALCOHOL

NO

CODE 57 INTERMITTENT
REFER TO "SYMPTOMS,"

YES

- FAULTY FUEL SENSOR

NO

CKT 1577 SHORTED TO VOLTAGE
OR
FAULTY ECM.

3.1L (VIN T) ENGINE — DIAGNOSTIC TROUBLE CODE CHART — LUMINA WITH VFV

CODE 58

DEGRADED FUEL SENSOR
3.1L (VIN T) "W" CARLINE (PORT)

IF A CODE 58 IS STORED IN MEMORY THE ECM HAS DETERMINED THAT THE FUEL SENSOR IS CONTAMINATED OR DEGRADED. THE ECM PERFORMS A DIAGNOSTIC CHECK AFTER 45 SECONDS OF RUN TIME. THE ECM WILL CONTINUALLY PERFORM THIS DIAGNOSTIC TEST TO CHECK THE FUEL SENSORS INTEGRITY. IF THE ECM HAS STORED A CODE 58, CHECK THE FUEL FOR CONTAMINATION. IF THE FUEL HAS BEEN CONTAMINATED, THE FUEL SENSOR MUST BE REPLACED AND THE FUEL SHOULD BE DRAINED.

CLEAR CODES AND CONFIRM "CLOSED LOOP" OPERATION AND NO "SERVICE ENGINE SOON" LIGHT.

3.1L (VIN T) ENGINE — DIAGNOSTIC TROUBLE CODE CHART — LUMINA WITH VFV

CODE 61
DEGRADED OXYGEN (O_2) SENSOR
3.1L (VIN T) "W" CARLINE (PORT)

IF A CODE 61 IS STORED IN MEMORY THE ECM HAS DETERMINED THE OXYGEN SENSOR IS CONTAMINATED OR DEGRADED, BECAUSE THE VOLTAGE CHANGE TIME IS SLOW OR SLUGGISH.

THE ECM PERFORMS THE OXYGEN SENSOR RESPONSE TIME TEST WHEN:

COOLANT TEMPERATURE IS GREATER THAN 85°C.

IAT TEMPERATURE IS GREATER THAN 10°C.

IN CLOSED LOOP.

IN DECEL FUEL CUT-OFF MODE.

IF A CODE 61 IS STORED THE OXYGEN SENSOR SHOULD BE REPLACED. A CONTAMINATED SENSOR CAN BE CAUSED BY FUEL ADDITIVES, SUCH AS SILICON, OR BY USE OF NON-GM APPROVED LUBRICANTS OR SEALANTS. SILICON CONTAMINATION IS USUALLY INDICATED BY A WHITE POWDERY SUBSTANCE ON THE SENSOR FINS.

"AFTER REPAIRS," CONFIRM "CLOSED LOOP" OPERATION AND NO "SERVICE ENGINE SOON" LIGHT.

3.1L (VIN T) ENGINE — DIAGNOSTIC TROUBLE CODE CHART — LUMINA WITH VFV

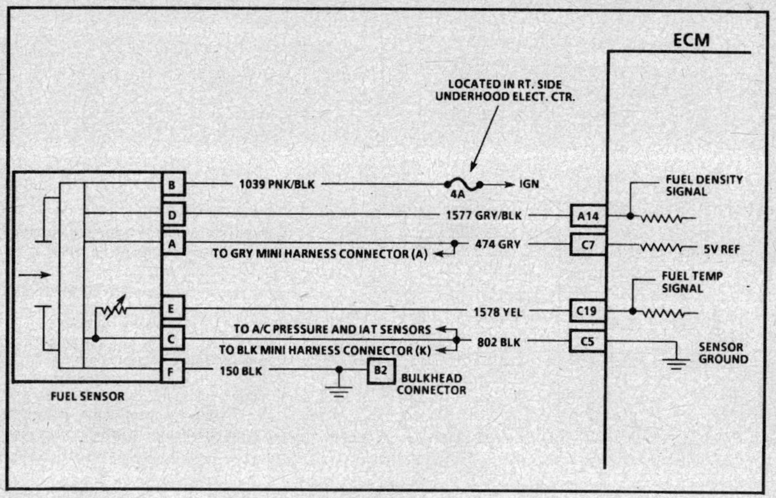

CODE 64
FUEL TEMPERATURE SENSOR HIGH
(LOW VOLTAGE)
3.1L (VIN T) "W" CARLINE (PORT)

Circuit Description:
The fuel sensor has two functions, it senses alcohol percentage in the fuel and also senses the fuel temperature.

Test Description: Number(s) below refer to circled number(s) on the diagnostic chart.
1. Tech 1 will display the actual fuel temperature on each key cycle.
2. This test will determine if the circuitry or the fuel sensor is at fault.

Diagnostic Aids:

If Code 64 is intermittent, check for faulty connections at the fuel sensor and ECM. Poor connections or intermittent shorts may cause a Code 64 or 65.

CONNECTOR PIN-OUT

PIN	SIGNAL
A	5 VOLT REFERENCE
B	IGNITION
C	SENSOR GROUND
D	FUEL DENS. SIGNAL
E	FUEL TEMP. SIGNAL
F	GROUND
G	NOT USED
H	NOT USED
J	NOT USED
K	NOT USED

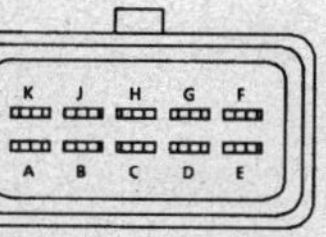

3.1L (VIN T) ENGINE — DIAGNOSTIC TROUBLE CODE CHART — LUMINA WITH VFV

CODE 64
FUEL TEMPERATURE SENSOR HIGH
(LOW VOLTAGE)
3.1L (VIN T) "W" CARLINE (PORT)

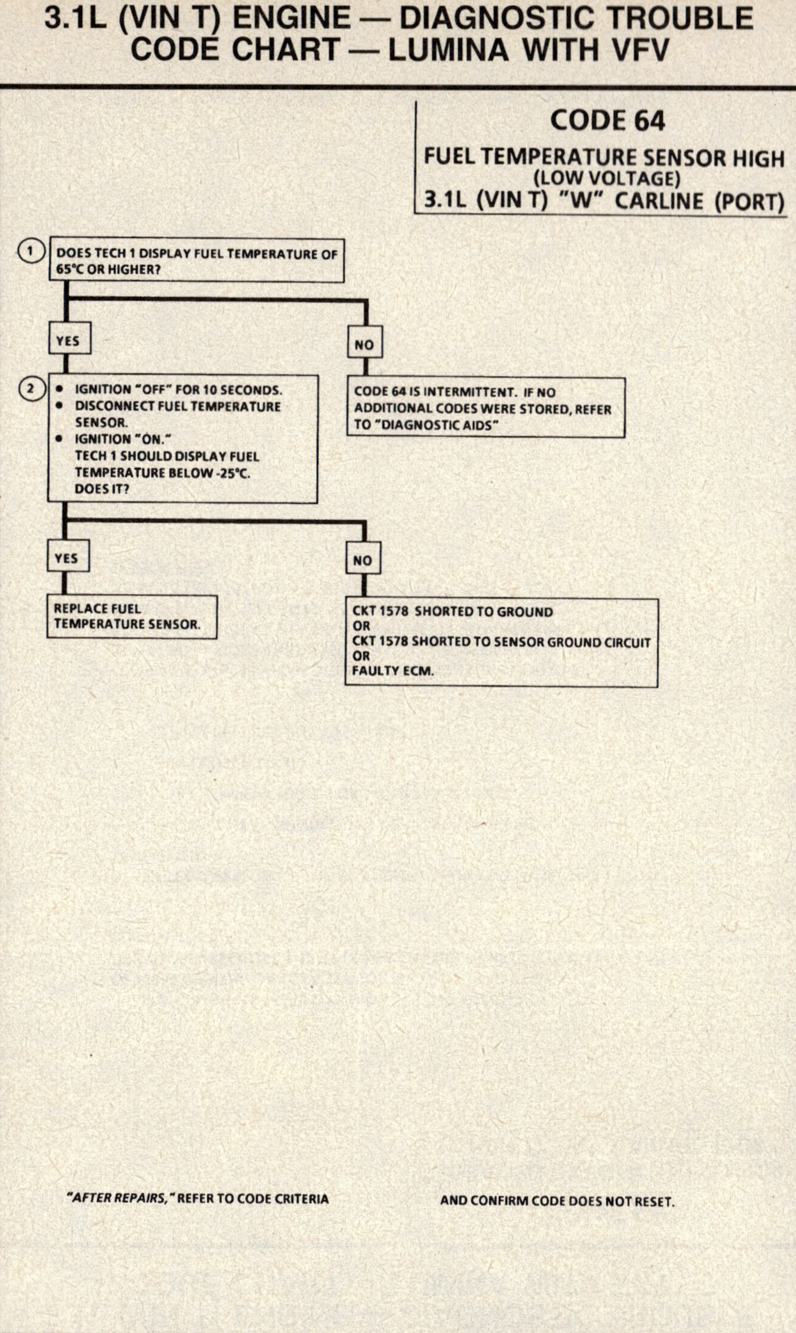

*"AFTER REPAIRS," REFER TO CODE CRITERIA   AND CONFIRM CODE DOES NOT RESET.

3.1L (VIN T) ENGINE — DIAGNOSTIC TROUBLE CODE CHART — LUMINA WITH VFV

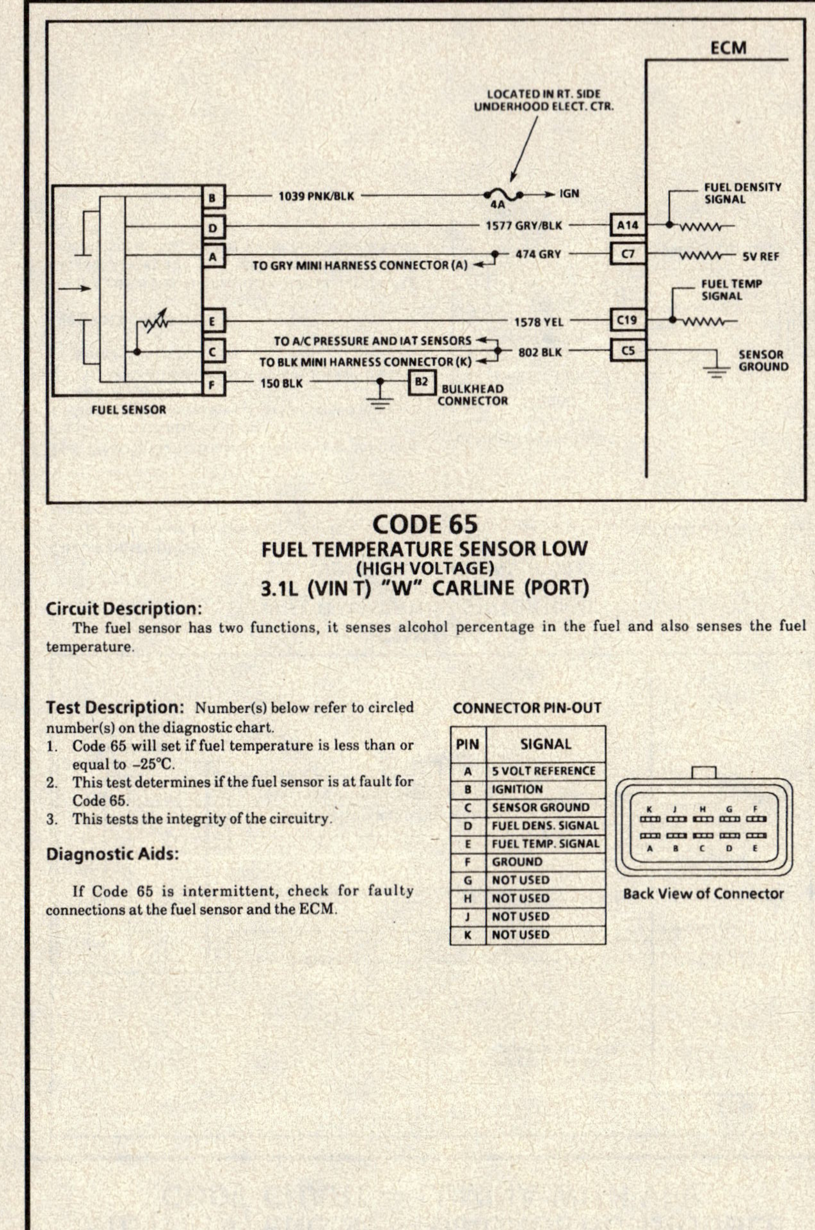

CODE 65
FUEL TEMPERATURE SENSOR LOW
(HIGH VOLTAGE)
3.1L (VIN T) "W" CARLINE (PORT)

Circuit Description:
The fuel sensor has two functions, it senses alcohol percentage in the fuel and also senses the fuel temperature.

Test Description: Number(s) below refer to circled number(s) on the diagnostic chart.
1. Code 65 will set if fuel temperature is less than or equal to −25°C.
2. This test determines if the fuel sensor is at fault for Code 65.
3. This tests the integrity of the circuitry.

Diagnostic Aids:

If Code 65 is intermittent, check for faulty connections at the fuel sensor and the ECM.

CONNECTOR PIN-OUT

PIN	SIGNAL
A	5 VOLT REFERENCE
B	IGNITION
C	SENSOR GROUND
D	FUEL DENS. SIGNAL
E	FUEL TEMP. SIGNAL
F	GROUND
G	NOT USED
H	NOT USED
J	NOT USED
K	NOT USED

3.1L (VIN T) ENGINE — DIAGNOSTIC TROUBLE CODE CHART — LUMINA WITH VFV

CODE 65
FUEL TEMPERATURE SENSOR LOW
(HIGH VOLTAGE)
3.1L (VIN T) "W" CARLINE (PORT)

1. • DOES TECH 1 DISPLAY FUEL TEMPERATURE OF -25°C OR LESS?

   YES →

2. • IGNITION "OFF" FOR 10 SECONDS.
   • DISCONNECT FUEL TEMPERATURE SENSOR.
   • JUMPER HARNESS TERMINALS "E" AND "C" TOGETHER.
   • IGNITION "ON."
   • TECH 1 SHOULD DISPLAY 65°C OR MORE.
     DOES IT?

   NO → CODE 65 IS INTERMITTENT. IF NO ADDITIONAL CODES WERE STORED, REFER TO "DIAGNOSTIC AIDS"

   NO ↓

3. • IGNITION "OFF" FOR 10 SECONDS.
   • JUMPER CKT 1578 TO GROUND.
   • IGNITION "ON."
   • TECH 1 SHOULD DISPLAY OVER 65°C.
     DOES IT?

   YES → FAULTY CONNECTOR OR FUEL TEMPERATURE SENSOR.

   YES ↓ OPEN FUEL TEMPERATURE SENSOR GROUND CIRCUIT, FAULTY CONNECTOR OR FAULTY ECM.

   NO → OPEN CKT 1578, FAULTY CONNECTOR AT ECM, OR FAULTY ECM.

3.1L (VIN T) ENGINE — DIAGNOSTIC TROUBLE CODE CHART — LUMINA WITH VFV

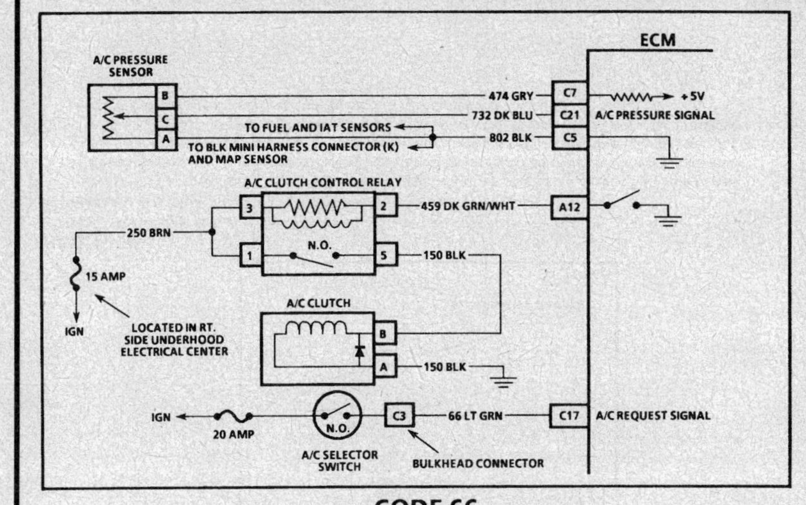

CODE 66
A/C PRESSURE SENSOR CIRCUIT
3.1L (VIN T) "W" CARLINE (PORT)

Circuit Description:

The A/C pressure sensor responds to changes in A/C refrigerant system high side pressure. This input indicates how much load the A/C compressor is putting on the engine and is one of the factors used by the ECM to determine IAC valve position for idle speed control. The circuit consists of a 5 volts reference and a ground, both provided by the ECM, and a signal line to the ECM. The signal is a voltage which is proportional to the pressure. The sensor's range of operation is 0 to 450 psi. At 0 psi, the signal will be about .1 volt, varying up to about 4.6 voltage at 450 psi or above. Code 66 sets if the voltage is above 4.6 voltage or below .1 volt 5 seconds or more. The A/C compressor is disabled by the ECM if Code 66 is present, or if pressure is above or below calibrated values described in "ECM Controlled Air Conditioning,"

Test Description: Number(s) below refer to circled number(s) on the diagnostic chart.
1. This step checks the voltage signal being received by the ECM from the A/C pressure sensor. The normal operating range is between .1 volt and 4.6 volts.
2. Checks to see if the high voltage signal is from a shorted sensor or a short to voltage in the circuit. Normally, disconnecting the sensor would make a normal circuit go to near zero volt.
3. Checks to see if low voltage signal is from the sensor or the circuit. Jumpering the sensor signal CKT 732 to 5 volts checks the circuit, connections, and ECM.

4. This step checks to see if the low voltage signal was due to an open in the sensor circuit or the 5 volts reference circuit since the prior step eliminated the pressure sensor.

Diagnostic Aids:

Code 66 sets when signal voltage falls outside the normal possible range of the sensor and may be due to a refrigerant system problem. The A/C clutch will not engage if system pressure is too low which could cause an intermittent Code 66.

If problem is intermittent, check for opens or shorts in harness or poor connections. If OK, replace A/C pressure sensor. If Code 66 re-sets, replace ECM.

3.1L (VIN T) ENGINE — DIAGNOSTIC TROUBLE CODE CHART — LUMINA WITH VFV

CODE 66
A/C PRESSURE SENSOR CIRCUIT
3.1L (VIN T) "W" CARLINE (PORT)

1. • KEY "ON," ENGINE NOT RUNNING.
   • NOTE "SCAN" VOLTAGE FOR A/C PRESSURE SENSOR.

BELOW .1 VOLT

3. • DISCONNECT A/C PRESSURE SENSOR CONNECTOR.
   • JUMPER TERMINALS "B" AND "C". DOES "SCAN" DISPLAY ABOVE 4.6 VOLTS?

ABOVE 1.8 VOLTS

2. • DISCONNECT A/C PRESSURE SENSOR. DOES "SCAN" DISPLAY LESS THAN 1 VOLT?

BETWEEN .1 VOLT AND 1.8 VOLTS

FAULT IS NOT PRESENT AT THIS TIME. SEE "DIAGNOSTIC AIDS."

NO — CHECK FOR SHORT TO VOLTAGE IN CKT 802. IF NOT SHORTED, REPLACE ECM.

YES — CHECK FOR OPEN IN CKT 802. IF NOT OPEN, CHECK FOR POOR SENSOR TERMINAL CONNECTIONS. IF OK, REPLACE A/C PRESSURE SENSOR.

NO

4. • REMOVE JUMPER.
   • CONNECT VOLTMETER FROM TERMINAL "A" TO "B". IS VOLTAGE ABOUT 5 VOLTS?

YES — • CHECK SENSOR TERMINAL CONNECTIONS, IF OK, REPLACE A/C PRESSURE SENSOR.

NO — • BACK PROBE ECM TERMINAL "C7" WITH VOLTMETER TO GROUND. IS VOLTAGE ABOUT 5 VOLTS?

YES — CHECK FOR OPEN IN CKT 732 CHECK FOR POOR CONNECTION ECM TERMINAL "C21". IF OK, REPLACE ECM.

NO — CHECK FOR POOR CONNECTION AT ECM TERMINAL "C7" OR SHORT TO GROUND IN CKT 474. IF OK, ECM IS FAULTY.

YES — REPAIR OPEN IN CKT 474.

3.1L (VIN T) ENGINE — RESTRICTED EXHAUST DIAGNOSTIC CHART — LUMINA WITH VFV

CHART B-1
RESTRICTED EXHAUST SYSTEM CHECK

Proper diagnosis for a restricted exhaust system is essential before any components are replaced. The following procedure may be used for diagnosis, depending upon engine or tool used:

CHECK AT O$_2$ SENSOR:

1. Carefully remove O$_2$ sensor.
2. Install Boroughs exhaust backpressure tester (BT 8515 or BT 8603) or equivalent in place of O$_2$ sensor (see illustration).
3. After completing test described below, be sure to coat threads of O$_2$ sensor with anti-seize compound P/N 5613695 or equivalent prior to re-installation.

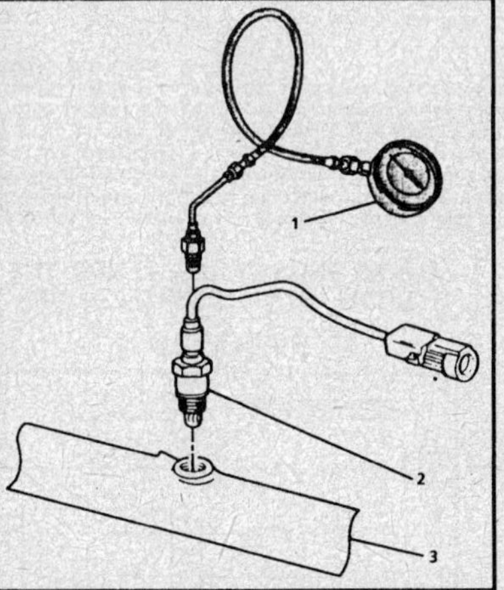

1	BACK PRESSURE GAGE	3 EXHAUST MANIFOLD
2	OXYGEN SENSOR (O2S)	7S 3338 6E

DIAGNOSIS:

1. With the engine idling at normal operating temperature, observe the exhaust system backpressure reading on the gage. Reading should not exceed 8.6 kPa (1.25 psi).
2. Increase engine speed to 2000 rpm and observe gage. Reading should not exceed 20.7 kPa (3 psi).
3. If the backpressure at either speed exceeds specification, a restricted exhaust system is indicated.
4. Inspect the entire exhaust system for a collapsed pipe, heat distress, or possible internal muffler failure.
5. If there are no obvious reasons for the excessive backpressure, the catalytic converter is suspected to be restricted and should be replaced using current recommended procedures.

3.1L (VIN T) ENGINE — COMPONENT DIAGNOSTIC CHART — LUMINA WITH VFV

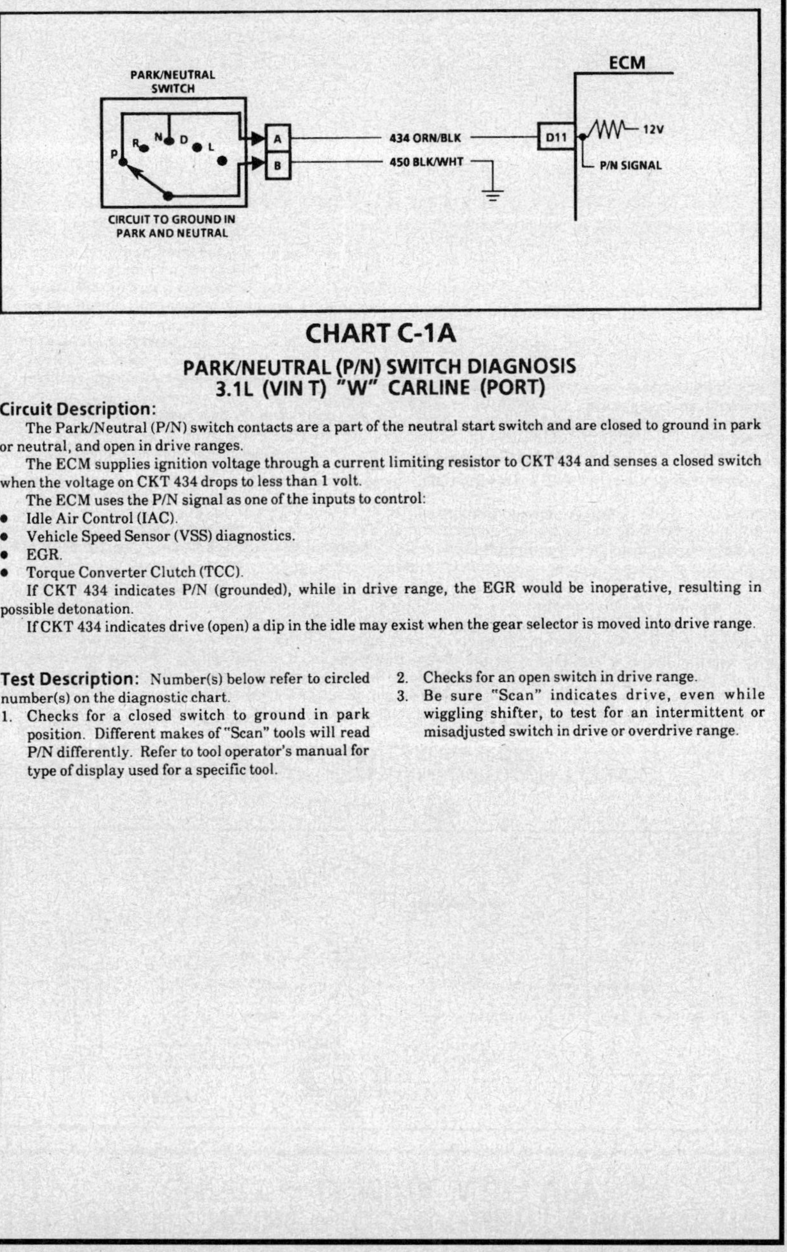

CHART C-1A
**PARK/NEUTRAL (P/N) SWITCH DIAGNOSIS
3.1L (VIN T) "W" CARLINE (PORT)**

Circuit Description:

The Park/Neutral (P/N) switch contacts are a part of the neutral start switch and are closed to ground in park or neutral, and open in drive ranges.

The ECM supplies ignition voltage through a current limiting resistor to CKT 434 and senses a closed switch when the voltage on CKT 434 drops to less than 1 volt.

The ECM uses the P/N signal as one of the inputs to control:
- Idle Air Control (IAC).
- Vehicle Speed Sensor (VSS) diagnostics.
- EGR.
- Torque Converter Clutch (TCC).

If CKT 434 indicates P/N (grounded), while in drive range, the EGR would be inoperative, resulting in possible detonation.

If CKT 434 indicates drive (open) a dip in the idle may exist when the gear selector is moved into drive range.

Test Description: Number(s) below refer to circled number(s) on the diagnostic chart.

1. Checks for a closed switch to ground in park position. Different makes of "Scan" tools will read P/N differently. Refer to tool operator's manual for type of display used for a specific tool.

2. Checks for an open switch in drive range.

3. Be sure "Scan" indicates drive, even while wiggling shifter, to test for an intermittent or misadjusted switch in drive or overdrive range.

3.1L (VIN T) ENGINE — COMPONENT DIAGNOSTIC CHART — LUMINA WITH VFV

CHART C-1A
**PARK/NEUTRAL (P/N) SWITCH DIAGNOSIS
3.1L (VIN T) "W" CARLINE (PORT)**

(1) • WITH TRANSMISSION IN PARK, TECH 1 SHOULD INDICATE PARK OR NEUTRAL. DOES IT?

— YES →

(3) • SHIFT TRANSMISSION INTO DRIVE.
• "SCAN" TOOL SHOULD DISPLAY A CHANGE TO INDICATE DRIVE. DOES IT?

— NO →

• DISCONNECT P/N SWITCH.
• THIS SHOULD CAUSE "SCAN" TOOL TO DISPLAY DRIVE RANGE. DOES IT?

— YES → FAULTY P/N SWITCH CONNECTION OR P/N SWITCH MISADJUSTED OR FAULTY P/N SWITCH.

— YES (from SHIFT TRANSMISSION) → NO TROUBLE FOUND. REFER TO "INTERMITTENTS" IN "DRIVEABILITY SYMPTOM."

— NO → CKT 434 SHORTED TO GROUND OR FAULTY ECM.

— NO (from TECH 1) →

(2) • DISCONNECT PARK/NEUTRAL SWITCH CONNECTOR.
• JUMPER HARNESS CONNECTOR TERMINALS "A" AND "B".
• "SCAN" TOOL SHOULD INDICATE PARK OR NEUTRAL. DOES IT?

— NO →

• JUMPER HARNESS CONNECTOR (CKT 434) TO ENGINE GROUND.
• "SCAN" TOOL SHOULD INDICATE PARK OR NEUTRAL. DOES IT?

— YES → OPEN GROUND CIRCUIT.

— NO → CKT 434 OPEN OR FAULTY ECM CONNECTION OR ECM.

— YES (from DISCONNECT PARK/NEUTRAL) → FAULTY P/N SWITCH CONNECTION OR P/N SWITCH MISADJUSTED OR FAULTY P/N SWITCH.

"AFTER REPAIRS," CONFIRM "CLOSED LOOP" OPERATION AND NO "SERVICE ENGINE SOON" LIGHT.

3.1L (VIN T) ENGINE — COMPONENT DIAGNOSTIC CHART — LUMINA WITH VFV

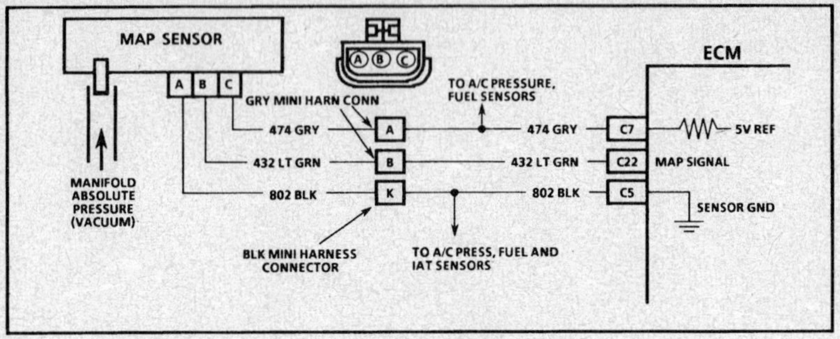

CHART C-1D

MANIFOLD ABSOLUTE PRESSURE (MAP) OUTPUT CHECK
3.1L (VIN T) "W" CARLINE (PORT)

Circuit Description:

The Manifold Absolute Pressure (MAP) sensor measures the changes in the intake manifold pressure which result from engine load (intake manifold vacuum) and rpm changes and converts these into a voltage output. The ECM sends a 5 volt reference voltage to the MAP sensor. As the manifold pressure changes, the output voltage of the sensor also changes. By monitoring the sensor output voltage, the ECM knows the manifold pressure. A lower pressure (low voltage) output voltage will be about 1–2 volts at idle. While higher pressure (high voltage) output voltage will be about 4–4.8 at Wide Open Throttle (WOT). The MAP sensor is also used, under certain conditions, to measure barometric pressure, allowing the ECM to make adjustments for different altitudes. The ECM uses the MAP sensor to control fuel delivery and ignition timing.

Test Description: Number(s) below refer to circled number(s) on the diagnostic chart.

Important

- Be sure to use the same Diagnostic Test Equipment for all measurements.

1. When comparing "Scan" readings to a known good vehicle, it is important to compare vehicles that use a MAP sensor having the same color insert or having the same "Hot Stamped" number. See figures on facing page.
2. Applying 34 kPa (10" Hg) vacuum to the MAP sensor should cause the voltage to be 1.5 to 2.1 volts less than the voltage at Step 1. Upon applying vacuum to the sensor, the change in voltage should be instantaneous. A slow voltage change indicates a faulty sensor.
3. Check vacuum hose to sensor for leaking or restriction. Be sure that no other vacuum devices are connected to the MAP hose.

NOTICE: Make sure electrical connector remains securely fastened.

4. Disconnect sensor from bracket and twist sensor by hand (only) to check for intermittent connection. Output changes greater than .1 volt indicate a bad sensor.

3.1L (VIN T) ENGINE — COMPONENT DIAGNOSTIC CHART — LUMINA WITH VFV

CHART C-1D

MANIFOLD ABSOLUTE PRESSURE (MAP) OUTPUT CHECK
3.1L (VIN T) "W" CARLINE (PORT)

NOTICE: THIS CHART ONLY APPLIES TO MAP SENSORS HAVING GREEN OR BLACK COLOR KEY INSERT (SEE BELOW).

(1)
- IGNITION "ON," ENGINE "OFF."
- TECH 1 SHOULD INDICATE A MAP SENSOR VOLTAGE.
- COMPARE THIS READING WITH THE READING OF A KNOWN GOOD VEHICLE. SEE FACING PAGE TEST DESCRIPTION, STEP 1.
- VOLTAGE READING SHOULD BE WITHIN, ± .4 VOLT.

IS IT?

YES → (2) / NO → REPLACE SENSOR.

(2)
- DISCONNECT AND PLUG VACUUM SOURCE TO MAP SENSOR.
- CONNECT A HAND VACUUM PUMP TO MAP SENSOR.
- START ENGINE.
- NOTE MAP SENSOR VOLTAGE.
- APPLY 34 kPa (10" Hg) OF VACUUM AND NOTE VOLTAGE CHANGE. SUBTRACT SECOND READING FROM THE FIRST. VOLTAGE VALUE SHOULD BE GREATER THAN 1.5 VOLTS.

IS IT?

YES → (3) / NO → (4)

(3) NO TROUBLE FOUND. CHECK SENSOR VACUUM SOURCE FOR LEAKAGE OR RESTRICTION. BE SURE THIS SOURCE SUPPLIES VACUUM TO MAP SENSOR ONLY.

(4) CHECK SENSOR CONNECTION. IF OK, REPLACE SENSOR.

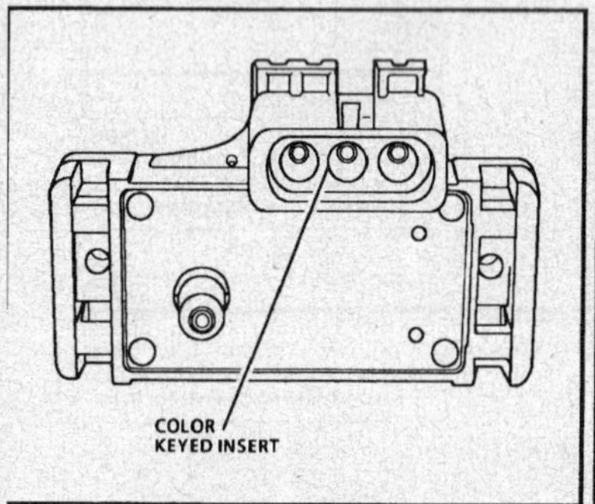

Figure 1 - Color Key Insert

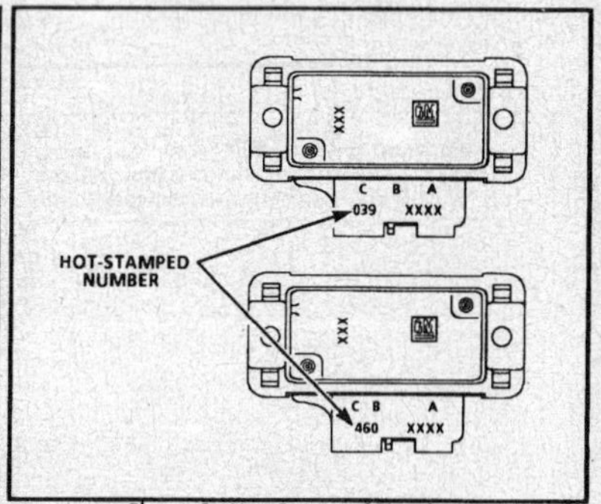

Figure 2 - Hot-Stamped Number

"AFTER REPAIRS," CONFIRM "CLOSED LOOP" OPERATION AND NO "SERVICE ENGINE SOON" LIGHT.

3.1L (VIN T) ENGINE — COMPONENT DIAGNOSTIC CHART — LUMINA WITH VFV

CHART C-1E
**POWER STEERING PRESSURE SWITCH (PSPS) DIAGNOSIS
3.1L (VIN T) "W" CARLINE (PORT)**

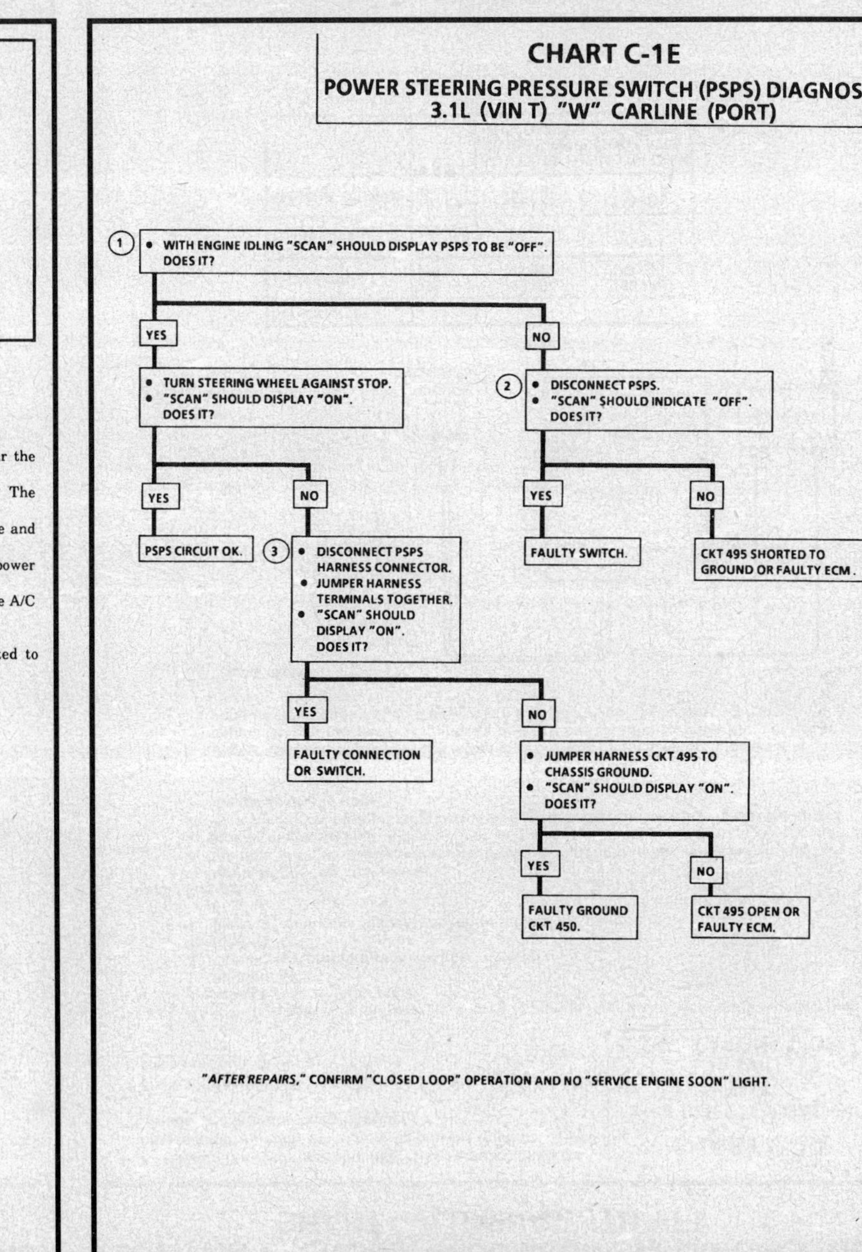

CHART C-1E
**POWER STEERING PRESSURE SWITCH (PSPS) DIAGNOSIS
3.1L (VIN T) "W" CARLINE (PORT)**

Circuit Description:

The Power Steering Pressure Switch (PSPS) is normally open to ground, and CKT 495 will be near the battery voltage.

Turning the steering wheel increases power steering oil pressure and its load on an idling engine. The pressure switch will close before the load can cause an idle problem.

Closing the switch causes CKT 495 to read less than 1 volt. The ECM will increase the idle air rate and disengage the A/C relay.

- A pressure switch that will not close, or an open CKT 495 or 450, may cause the engine to stop when power steering loads are high.
- A switch that will not open, or a CKT 495 shorted to ground, may affect idle quality and will cause the A/C relay to be de-energized.

Test Description: Number(s) below refer to circled number(s) on the diagnostic chart.

1. Different makes of "Scan" tools may display the state of this switch in different ways. Refer to "Scan" tool operator's manual to determine how this input is indicated.

2. Checks to determine if CKT 495 is shorted to ground.

3. This should simulate a closed switch.

CHART C-2A

INJECTOR BALANCE TEST

The injector balance tester is a tool used to turn the injector on for a precise amount of time, thus spraying a measured amount of fuel into the manifold. This causes a drop in fuel rail pressure that we can record and compare between each injector. All injectors should have the same amount of pressure drop ($\pm$ 10 kPa). Any injector with a pressure drop that is 10 kPa (or more) greater or less than the average drop of the other injectors should be considered faulty and replaced.

STEP 1

Engine "cool down" period (10 minutes) is necessary to avoid irregular readings due to "Hot Soak" fuel boiling. With ignition "OFF" connect fuel gauge J 347301 or equivalent to fuel pressure tap. Wrap a shop towel around fitting while connecting gage to avoid fuel spillage.

Disconnect harness connectors at all injectors, and connect injector tester J 34730-3, or equivalent, to one injector. On Turbo equipped engines, use adaptor harness furnished with injector tester to energize injectors that are not accessible. Follow manufacturers instructions for use of adaptor harness. Ignition must be "OFF" at least 10 seconds to complete ECM shutdown cycle. Fuel pump should run about 2 seconds after ignition is turned "ON". At this point, insert clear tubing attached to vent valve into a suitable container and bleed air from gauge and hose to insure accurate gauge operation. Repeat this step until all air is bled from gauge.

STEP 2

Turn ignition "OFF" for 10 seconds and then "ON" again to get fuel pressure to its maximum. Record this initial pressure reading. Energize tester one time and note pressure drop at its lowest point. (Disregard any slight pressure increase after drop hits low point.) By subtracting this second pressure reading from the initial pressure, we have the actual amount of injector pressure drop.

STEP 3

Repeat Step 2 on each injector and compare the amount of drop. Usually, good injectors will have virtually the same drop. Retest any injector that has a pressure difference of 10 kPa, either more or less than the average of the other injectors on the engine. Replace any injector that also fails the retest. If the pressure drop of all injectors is within 10 kPa of this average, the injectors appear to be flowing properly. Reconnect them and review "Symptoms," Section

NOTE: *The entire test should <u>not</u> be repeated more than once without running the engine to prevent flooding. (This includes any retest on faulty injectors.)*

NOTICE: The entire test should <u>NOT</u> be repeated more than once without running the engine to prevent flooding. (This includes any retest on faulty injectors.)

The fuel pressure test in Section "A" Chart A-7, should be completed prior to this test.

CHART C-2A

INJECTOR BALANCE TEST
3.1L (VIN T)
"W" CARLINE (PORT)

Step 1. If engine is at operating temperature, allow a 10 minute "cool down" period then connect fuel pressure gage and injector tester.
1. Ignition "OFF."
2. Connect fuel pressure gage and injector tester.
3. Ignition "ON."
4. Bleed off air in gage. Repeat until all air is bled from gauge.

Step 2. Run test:
1. Ignition "OFF" for 10 seconds.
2. Ignition "ON". Record gage pressure. (Pressure must hold steady, if not see the Fuel System diagnosis, Chart A-7, in Section "A").
3. Turn injector on, by depressing button on injector tester, and note pressure at the instant the gauge needle stops.

Step 3.
1. Repeat Step 2 on all injectors and record pressure drop on each. Retest injectors that appear faulty (Any injectors that have a 10 kPa (1.5 psi) difference, either more or less, in pressure from the average). If no problem is found, review "Symptoms" Section

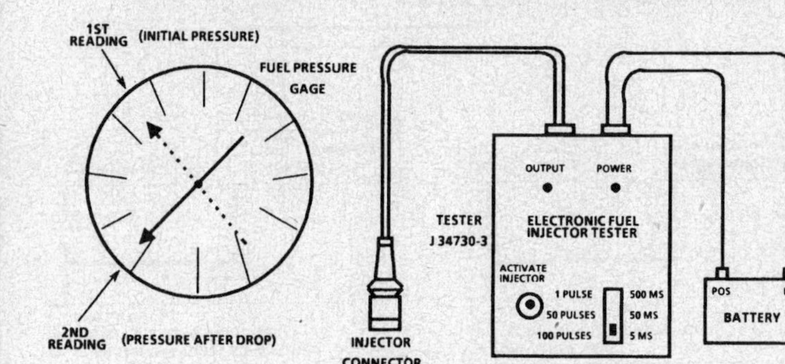

EXAMPLE

CYLINDER	1	2	3
1ST READING	293 kPa (43 psi)	293 kPa (43 psi)	293 kPa (43 psi)
2ND READING	131 kPa (19 psi)	115 kPa (17 psi)	145 kPa (21 psi)
AMOUNT OF DROP	162 kPa (24 psi)	178 kPa (26 psi)	148 kPa (21 psi)
	OK	FAULTY, RICH (TOO MUCH FUEL DROP)	FAULTY, LEAN (TOO LITTLE FUEL DROP)

3.1L (VIN T) ENGINE — COMPONENT DIAGNOSTIC CHART — LUMINA WITH VFV

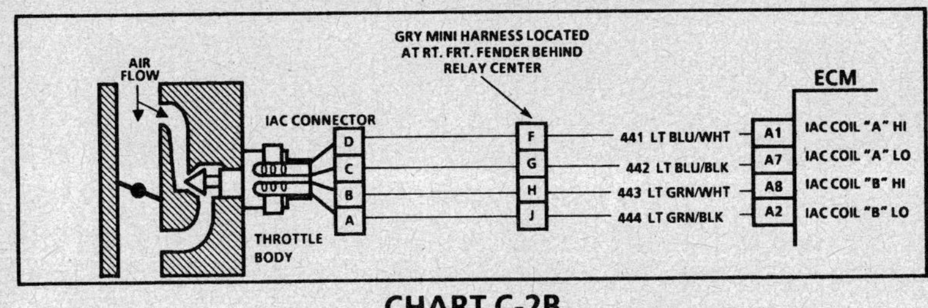

CHART C-2B
IDLE AIR CONTROL (IAC) CIRCUIT
3.1L (VIN T) "W" CARLINE (PORT)

Circuit Description:

The ECM controls engine idle speed with the IAC valve. To increase idle speed, the ECM retracts the IAC valve pintle away from its seat, allowing more air to bypass the throttle bore. To decrease idle speed, it extends the IAC valve pintle towards its seat, reducing bypass air flow. A "Scan" tool will read the ECM commands to the IAC valve in counts. Higher the counts indicate more air bypass (higher idle). The lower the counts indicate less air is allowed to bypass (lower idle).

Test Description: Number(s) below refer to circled number(s) on the diagnostic chart.

1. The Tech 1 rpm control mode is used to extend and retract the IAC valve. The valve should move smoothly within the specified range. If the idle speed is commanded (IAC extended) too low (below 700 rpm), the engine may stall. This may be normal and would not indicate a problem. Retracting the IAC beyond its controlled range (above 1500 rpm) will cause a delay before the rpm's start dropping. This too is normal.
2. This test uses the Tech 1 to command the IAC controlled idle speed. The ECM issues commands to obtain commended idle speed. The node lights each should flash red and green to indicate a good circuit as the ECM issues commands. While the sequence of color is not important if either light is "OFF" or does not flash red and green, check the circuits for faults, beginning with poor terminal contacts.

Diagnostic Aids:

A slow, unstable, or fast idle may be caused by a non-IAC system problem that cannot be overcome by the IAC system. Out of control range IAC "Scan" tool counts will be above 60 if idle is too low, and zero counts if idle is too high. The following checks should be made to repair a non-IAC system problem:

- <u>Vacuum Leak (High Idle)</u> - If idle is too high, stop the engine. Fully extend (low) IAC with tester. Start engine. If idle speed is above 800 rpm, locate and correct vacuum leak including PCV system. Also check for binding of throttle blade or linkage.
- <u>System too lean (High Air/Fuel Ratio)</u> - The idle speed may be too high or too low. Engine speed may vary up and down and disconnecting the IAC valve does not help. Code 44 may be set. "Scan" O_2 voltage will be less than 300 mV (.3 volt). Check for low regulated fuel pressure, water in the fuel or a restricted injector.
- <u>System too rich (Low Air/Fuel Ratio)</u> - The idle speed will be too low. "Scan" tool IAC counts will usually be above 80. System is obviously rich and may exhibit black smoke in exhaust. "Scan" tool O_2 voltage will be fixed above 800 mV (.8 volt). Check for high fuel pressure, leaking or sticking injector. Silicone contaminated O_2 sensors "Scan" voltage will be slow to respond.
- <u>Throttle body.</u> - Remove IAC valve and inspect bore for foreign material.
- <u>IAC Valve Electrical Connections</u> - IAC valve connections should be carefully checked for proper contact.
- <u>PCV Valve</u> - An incorrect or faulty PCV valve may result in an incorrect idle speed.
- Refer to "Rough, Unstable, Incorrect Idle or Stalling" in "Symptoms," Section "6E3-B".
- If intermittent poor driveability or idle symptoms are resolved by disconnecting the IAC, carefully recheck connections, valve terminal resistance, or replace IAC.

3.1L (VIN T) ENGINE — COMPONENT DIAGNOSTIC CHART — LUMINA WITH VFV

CHART C-2B
IDLE AIR CONTROL (IAC) CIRCUIT
3.1L (VIN T) "W" CARLINE (PORT)

* IAC DRIVER AND NODE LIGHT REQUIRED KIT
222-L FROM: CONCEPT TECHNOLOGY, INC.
J 37027 FROM: KENT-MOORE, INC.

CLEAR CODES, CONFIRM "CLOSED LOOP" OPERATION, NO "SERVICE ENGINE SOON" LIGHT, PERFORM IAC RESET PROCEDURE PER APPLICABLE SERVICE MANUAL AND VERIFY CONTROLLED IDLE SPEED IS CORRECT.

3.1L (VIN T) ENGINE — COMPONENT DIAGNOSTIC CHART — LUMINA WITH VFV

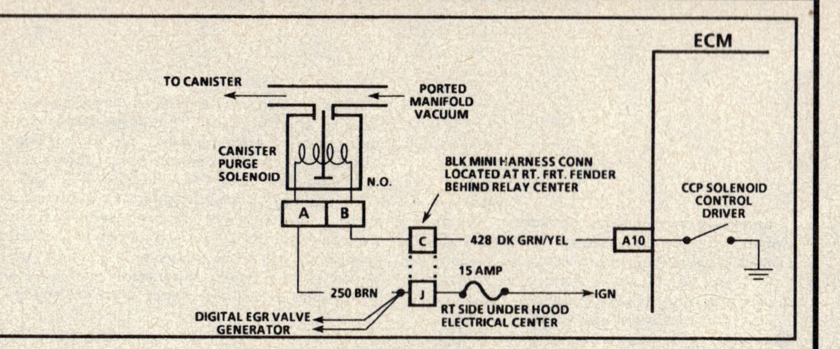

CHART C-3
CANISTER PURGE VALVE CHECK
3.1L (VIN T) "W" CARLINE (PORT)

Circuit Description:

Canister purge is controlled by a solenoid that allows manifold and/or ported vacuum to purge the canister when energized. The Electronic Control Module (ECM) supplies a ground to energize the solenoid (purge "ON"). The purge solenoid control by the ECM is pulse width modulated (turned "ON" and "OFF" several times a second). The duty cycle (pulse width) is determined by "Closed Loop" feed back from the O_2 sensor. The duty cycle is calculated by the ECM and the output commanded when the following conditions have been met:
- Engine run time after start more than 3 minutes.
- Coolant temperature above 80°C (176°F).

Also, if the diagnostic "test" terminal is grounded with the engine stopped, the purge solenoid is energized (purge "ON").

Test Description: Number(s) below refer to circled number(s) on the diagnostic chart.
1. Checks to see if the solenoid is opened or closed. The solenoid is normally de-energized in this step; so it should be closed.
2. Checks to determine if solenoid was open due to electrical circuit problem or defective solenoid.
3. Completes functional check by grounding "test" terminal. This should normally energize the solenoid opening the valve which should allow the vacuum to drop (purge "ON").

Diagnostic Aids:

Normal operation of the canister purge solenoid is described as follows:

With ignition "ON," engine "OFF," diagnostic "test" terminal ungrounded, the canister purge solenoid will be energized.

With ignition "ON," engine "OFF," diagnostic "test" terminal grounded, the canister purge solenoid will be de-energized.

An inoperative canister purge system can cause rich exhaust condition Code 45. If the solenoid valve should become stuck open of the carbon canister become saturated, a continuous rich fuel condition may result.

3.1L (VIN T) ENGINE — COMPONENT DIAGNOSTIC CHART — LUMINA WITH VFV

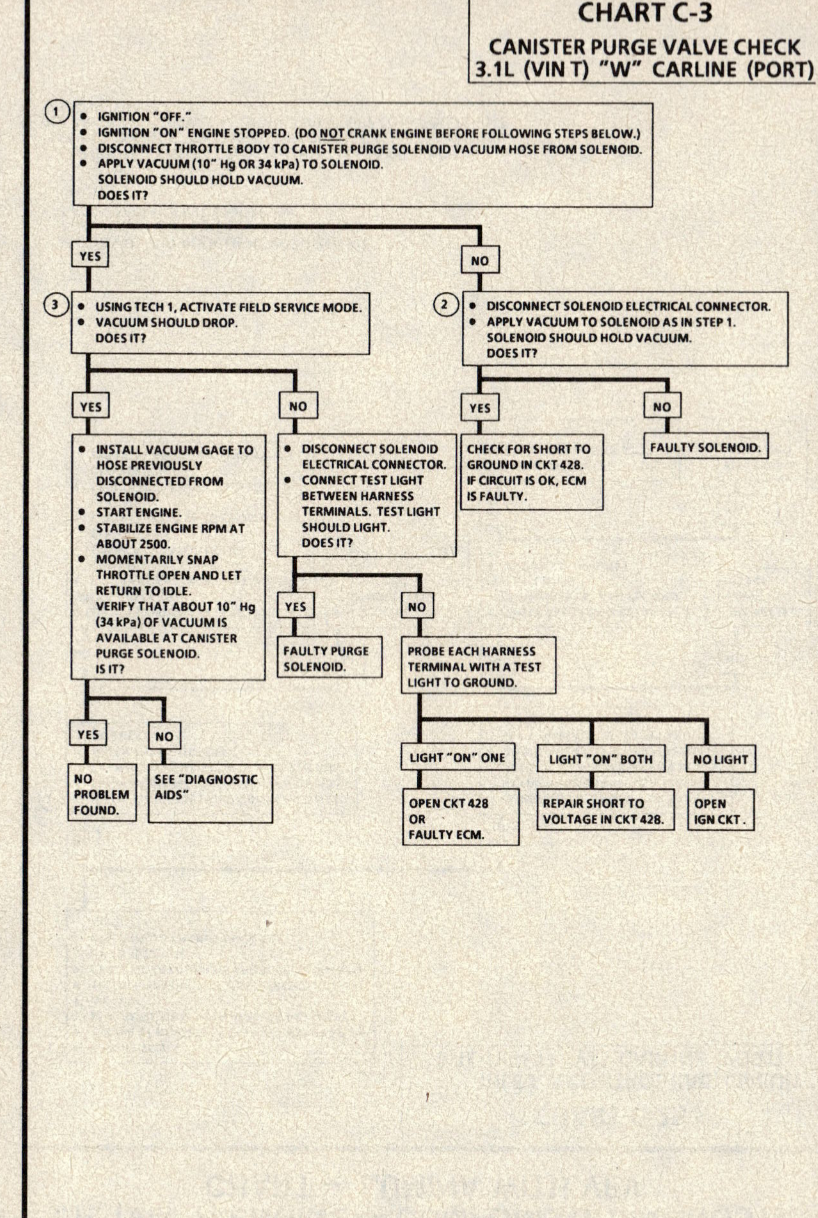

3.1L (VIN T) ENGINE — COMPONENT DIAGNOSTIC CHART — LUMINA WITH VFV

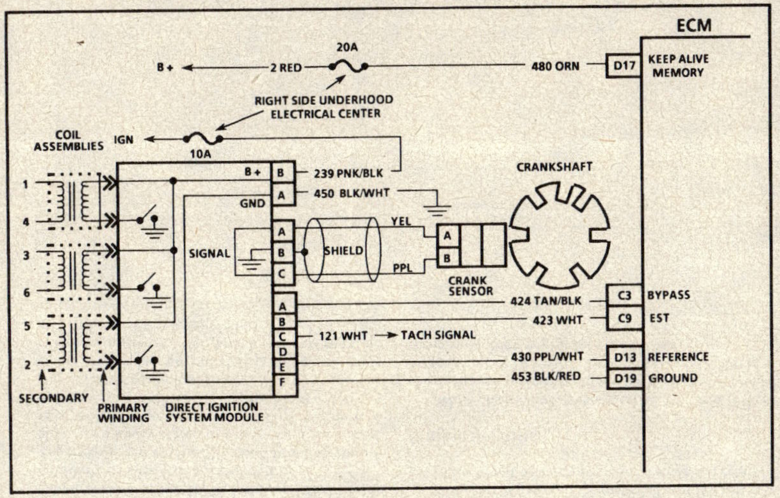

CHART C-4F
"DIS" MISFIRE
3.1L (VIN T) "W" CARLINE (PORT)

Circuit Description:
The Direct Ignition System (DIS) uses a waste spark method of distribution. In this type of system, the ignition module triggers the #1/4 coil pair resulting in both #1 and #4 spark plugs firing at the same time. #1 cylinder is on the compression stroke at the same time #4 is on the exhaust stroke, resulting in a lower energy requirement to fire #4 spark plug. This leaves the remainder of the high voltage to be used to fire #1 spark plug. On this application, the crank sensor is mounted to the engine block and protrudes through the block to within approximately .050" of the crankshaft reluctor. Since the reluctor is a machined portion of the crankshaft and the crank sensor is mounted in a fixed position on the block, timing adjustments are not possible or necessary.

Test Description: Number(s) below refer to circled number(s) on the diagnostic chart.
1. Checks for voltage output of ignition system. The spark tester must be used, as this tool requires 25,000 volts to trigger. This checks for a potential weak coil.
2. If the spark tester fires on all wires, the ignition system, with the exception of the spark plugs, may be considered in good working order. If the spark plugs show no evidence of wear, damage or fouling, an engine mechanical fault should be suspected. Refer to "Cuts Out, Misses," in "Symptoms,"
3. If the spark jumps the tester gap after grounding the opposite plug wire, it indicates excessive resistance in the plug which was bypassed.

A faulty or poor connection at that plug could also result in the miss condition. Also, check for carbon deposits inside the spark plug boot.
4. If carbon tracking is evident replace coil and be sure plug wires relating to that coil are clean and tight. Excessive wire resistance or faulty connections could have caused the coil to be damaged.
5. If the no spark condition follows the suspected coil, that coil is faulty. Otherwise, the ignition module is the cause of no spark. This test could also be performed by substituting a known good coil for the one causing the no spark condition.

3.1L (VIN T) ENGINE — COMPONENT DIAGNOSTIC CHART — LUMINA WITH VFV

CHART C-4F
"DIS" MISFIRE
3.1L (VIN T) "W" CARLINE (PORT)

1. • IGNITION "OFF".
   • INSTALL SPARK TESTER J-26792 (ST-125) OR EQUIV. TO ONE SPARK PLUG WIRE.
   • IDLE ENGINE AND CHECK FOR SPARK AT TESTER.
   • REPEAT TEST ON ALL PLUG WIRES.
   DOES TESTER SHOW SPARK ON ALL WIRES?

NO →
• IGNITION "OFF", GROUND THE OPPOSITE PLUG LEAD OF THE AFFECTED COIL AT SPARK PLUG.
• WILL SPARK JUMP TESTER GAP WHILE CRANKING ENGINE?

YES →
2. CHECK FOR:
   - FAULTY, WORN, OR DAMAGED SPARK PLUG(S).
   - PLUG FOULING, DUE TO ENGINE MECHANICAL FAULT.
   - IF PLUGS ARE OK, SEE SYMPTOMS, SECT. B; CUTS OUT, MISSES.

NO →
• CHECK THE RESISTANCE OF EACH PLUG WIRE OF THE COIL WHICH DID NOT FIRE THE SPARK TESTER.
• IS WIRE RESISTANCE LESS THAN 30,000 OHMS EACH AND WIRE NOT GROUNDED?

YES →
3. REPLACE THE SPARK PLUG FOR THE LEAD WHICH WAS JUMPERED TO GROUND. IF MISFIRE IS STILL PRESENT, START MISFIRE TEST AGAIN AT STEP # 1.

YES →
• REMOVE AFFECTED COIL(S).
• IS COIL(S) FREE OF CARBON TRACKING?

NO →
REPLACE FAULTY WIRE(S).

YES →
5. • SWITCH POSITION OF COILS AT PROBLEM CYLINDER.
   • WILL SPARK JUMP TESTER GAP WHILE CRANKING ENGINE?

NO →
4. FAULTY IGNITION COIL. ALSO CHECK FOR FAULTY PLUG WIRE CONNECTION(S) AND WIRE NIPPLE(S) FOR CARBON TRACKING.

YES →
FAULTY IGNITION COIL.

NO →
FAULTY IGNITION MODULE.

"AFTER REPAIRS," CONFIRM "CLOSED LOOP" OPERATION AND NO "SERVICE ENGINE SOON" LIGHT.

3.1L (VIN T) ENGINE — COMPONENT DIAGNOSTIC CHART — LUMINA WITH VFV

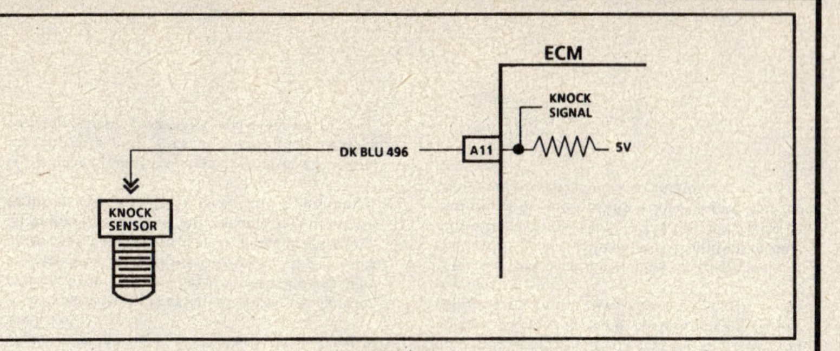

CHART C-5
ELECTRONIC SPARK CONTROL (ESC) SYSTEM CHECK
3.1L (VIN T) "W" CARLINE (PORT)

Circuit Description:

The knock sensor is used to detect engine detonation and the ECM will retard the electronic spark timing based on the signal being received. The circuitry, within the knock sensor, causes the ECM's 5 volts to be pulled down so that under a no knock condition, CKT 496 would measure about 2.5 volts. The knock sensor produces an AC signal which rides on the 2.5 volts DC voltage. The amplitude and frequency are dependent upon the knock level.

The MEM-CAL used with this engine, contains the functions which were part of remotely mounted ESC modules used on other GM vehicles. The ESC portion of the MEM-CAL, then sends a signal to other parts of the ECM which adjusts the spark timing to retard the spark and reduce the detonation.

Test Description: Number(s) below refer to circled number(s) on the diagnostic chart.
1. With engine idling, there should not be a knock signal present at the ECM, because detonation is not likely under a no load condition.
2. Tapping on the engine lift hood bracket should simulate a knock signal to determine if the sensor is capable of detecting detonation. If no knock is detected, try tapping on engine block closer to sensor before replacing sensor.
3. If the engine has an internal problem which is creating a knock, the knock sensor may be responding to the internal failure.

4. This test determines if the knock sensor is faulty or if the ESC portion of the MEM-CAL is faulty. If it is determined that the MEM-CAL is faulty, be sure that is is properly installed and latched into place. If not properly installed, repair and retest.

Diagnostic Aids:

While observing knock signal on the "Scan," there should be an indication that knock is present when detonation can be heard. Detonation is most likely to occur under high engine load conditions.

3.1L (VIN T) ENGINE — COMPONENT DIAGNOSTIC CHART — LUMINA WITH VFV

CHART C-5
ELECTRONIC SPARK CONTROL (ESC) SYSTEM CHECK
3.1L (VIN T) "W" CARLINE (PORT)

(1)
- IF CODE 43 IS SET, USE THE CODE CHART.
- ENGINE MUST BE IDLING AT NORMAL OPERATING TEMPERATURE.
- USE TECH 1 TO OBSERVE KNOCK SIGNAL.
  IS KNOCK INDICATED?

NO → (2)
- TAP ON ENGINE LIFT HOOK BRACKET WHILE OBSERVING KNOCK SIGNAL.
- TECH 1 SHOULD INDICATE KNOCK WHILE TAPPING ON BRACKET.
  DOES IT?

YES → (3)
IF AN ENGINE KNOCK CAN BE HEARD, REPAIR THE BASIC ENGINE PROBLEM. IF NO AUDIBLE KNOCK IS HEARD, FOLLOW THE STEPS:
- DISCONNECT KNOCK SENSOR.
- CONNECT VOLTMETER TO KNOCK SENSOR AND ENGINE GROUND.
- SET VOLTMETER ON 2 VOLT AC SCALE.
  IS A SIGNAL INDICATED ON VOLTMETER?

NO → (4)
- DISCONNECT KNOCK SENSOR.
- CONNECT VOLTMETER TO KNOCK SENSOR AND ENGINE GROUND.
- SET VOLTMETER ON 2 VOLT AC SCALE.
- TAP ON ENGINE BLOCK NEAR SENSOR.
  IS A SIGNAL INDICATED ON VOLTMETER WHILE TAPPING ON ENGINE BLOCK?

YES → SYSTEM IS OPERATING PROPERLY. REFER TO "DIAGNOSTIC AIDS" ON FACING PAGE.

(3) NO → CHECK CKT 496 FOR BEING NEAR A SPARK PLUG WIRE OR A FAULTY ECM CONNECTION OR FAULTY ECM OR MEM-CAL.

(3) YES → REPLACE KNOCK SENSOR.

(4) YES → REPLACE MEM-CAL OR ECM.

(4) NO → REPLACE KNOCK SENSOR.

"AFTER REPAIRS," CONFIRM "CLOSED LOOP" OPERATION AND NO "SERVICE ENGINE SOON" LIGHT.

3.1L (VIN T) ENGINE — COMPONENT DIAGNOSTIC CHART — LUMINA WITH VFV

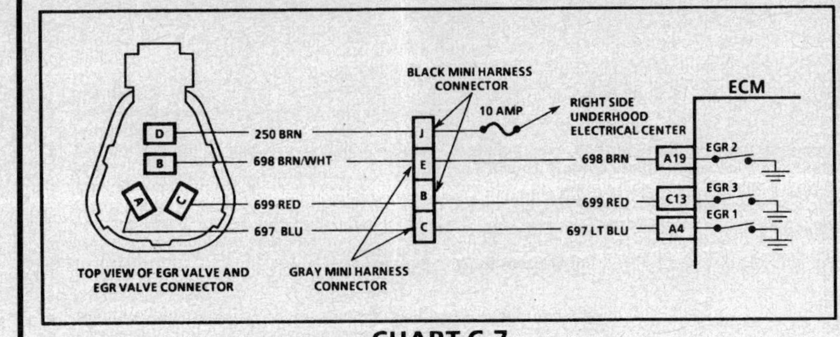

CHART C-7
EXHAUST GAS RECIRCULATION (EGR) FLOW CHECK
3.1L (VIN T) "W" CARLINE (PORT)

Circuit Description:

The digital (EGR) valve is designed to accurately supply EGR to an engine independent of intake manifold vacuum. The valve controls EGR flow from the exhaust to the intake manifold through three orifices which increment in size to produce seven combinations. When a solenoid is energized, the armature with attached shaft and swivel pintle is lifted opening the orifice.

The flow accuracy is dependent on metering orifice size only, which results in improved control.

Test Description: Number(s) below refer to circled number(s) on the diagnostic chart.

1. Codes should be diagnosed using appropriate chart before preparing a functional check.

2. This step activates each solenoid individually. As you energize #1 or #2 solenoid, the engine rpm should drop. #3 solenoid has the large port and may stall the engine when energized.

Important

- If the digital EGR valve shows signs of excessive heat, a melted condition. Check the exhaust system for blockage (possibly a plugged converter) using the procedure found on CHART B-1. If the exhaust system is restricted repair the cause, one of which might be an injector which is open due to one of the following:
  A. Stuck.
  B. Grounded driver circuit.
  C. Possibly defective ECM.
  If this condition is found, the oil should be checked for possible fuel contamination.

3.1L (VIN T) ENGINE — COMPONENT DIAGNOSTIC CHART — LUMINA WITH VFV

CHART C-7
EXHAUST GAS RECIRCULATION (EGR) FLOW CHECK
3.1L (VIN T) "W" CARLINE (PORT)

(1)
- IGNITION "ON," ENGINE "OFF."
- "SCAN" TROUBLE CODES.
- IF CODES ARE PRESENT, REFER TO THOSE CHARTS FIRST.

(2)
- USING SCAN TOOL, SELECT EGR CONTROL. (MISCELLANEOUS)
- START ENGINE AND ALLOW IDLE TO STABILIZE.
- ENERGIZE EGR SOL #1, RPM SHOULD DROP SLIGHTLY.*
- ENERGIZE EGR SOL #2, ENGINE SHOULD HAVE A ROUGH IDLE.*
- ENERGIZE EGR SOL #3, ENGINE SHOULD IDLE ROUGH OR STALL.*
- DID RPM DROP AND ENGINE IDLE ROUGH ON ALL THREE SOLENOIDS?

YES

EGR SYSTEM IS OK. NO PROBLEM FOUND.

NO

CHECK EGR VALVE, PIPE, ADAPTOR, GASKETS, FITTINGS AND ALL PASSAGES FOR DAMAGE, LEAKAGE, OR PLUGGING. IF OK, REPLACE EGR VALVE.

* THESE STEPS MUST BE DONE VERY QUICKLY, AS THE ECM WILL ADJUST THE IDLE AIR CONTROL VALVE TO CORRECT IDLE SPEED.

"AFTER REPAIRS," CONFIRM "CLOSED LOOP" OPERATION AND NO "SERVICE ENGINE SOON" LIGHT.

3.1L (VIN T) ENGINE — COMPONENT DIAGNOSTIC CHART — LUMINA WITH VFV

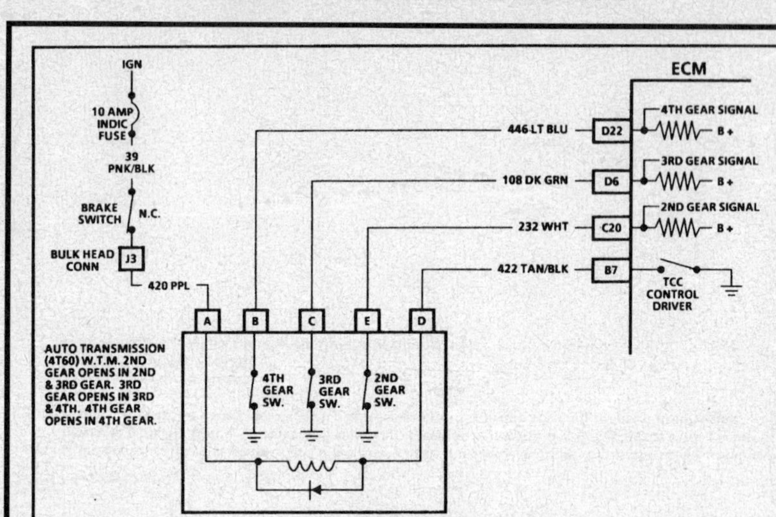

CHART C-8B

(Page 1 of 3)
4T60 TORQUE CONVERTER CLUTCH (TCC) WITH TORQUE MANAGEMENT
(ELECTRICAL DIAGNOSIS)
3.1L (VIN T) "W" CARLINE (PORT)

Circuit Description:

The purpose of the automatic transmission torque converter clutch feature·is to eliminate the power loss of the torque converter stage when the vehicle is in a cruise condition. This allows the convenience of the automatic transmission and the fuel economy of a manual transmission. The heart of the system is a solenoid located inside the automatic transmission which is controlled by the ECM.

When the solenoid coil is activated "ON," the Torque Converter Clutch (TCC) is applied which results in straight through mechanical coupling from the engine to transmission. When the transmission solenoid is deactivated, the Torque Converter Clutch (TCC) is released which allows the torque converter to operate in the conventional manner (fluidic coupling between engine and transmission).

The TCC will engage on a warm engine under given road load in 3rd and 4th gears. TCC will engage when:
- Engine warmed up and in "Closed Loop" condition.
- Vehicle speed above a calibrated value (about 28 mph 45 km/h).
- Throttle position sensor output not changing, indicating a steady road speed.
- Brake switch closed.

Test Description: Number(s) below refer to circled number(s) on the diagnostic chart.
1. This test checks the functional operation of the TCC circuit.
2. This test checks the TCC control driver in the ECM.
3. This test will confirm that there is battery voltage to terminal "A".
4. This test confirms that the ECM has the ability to turn the TCC "ON."

Diagnostic Aids:

The "Scan" tool only indicates when the ECM has turned "ON" the TCC driver, and this does not confirm that the TCC has engaged. To determine if TCC is functioning properly, engine rpm should decrease when the "Scan" indicates the TCC driver has turned "ON."

3.1L (VIN T) ENGINE — COMPONENT DIAGNOSTIC CHART — LUMINA WITH VFV

CHART C-8B

(Page 1 of 3)
4T60 TORQUE CONVERTER CLUTCH (TCC) WITH TORQUE MANAGEMENT
(ELECTRICAL DIAGNOSIS)
3.1L (VIN T) "W" CARLINE (PORT)

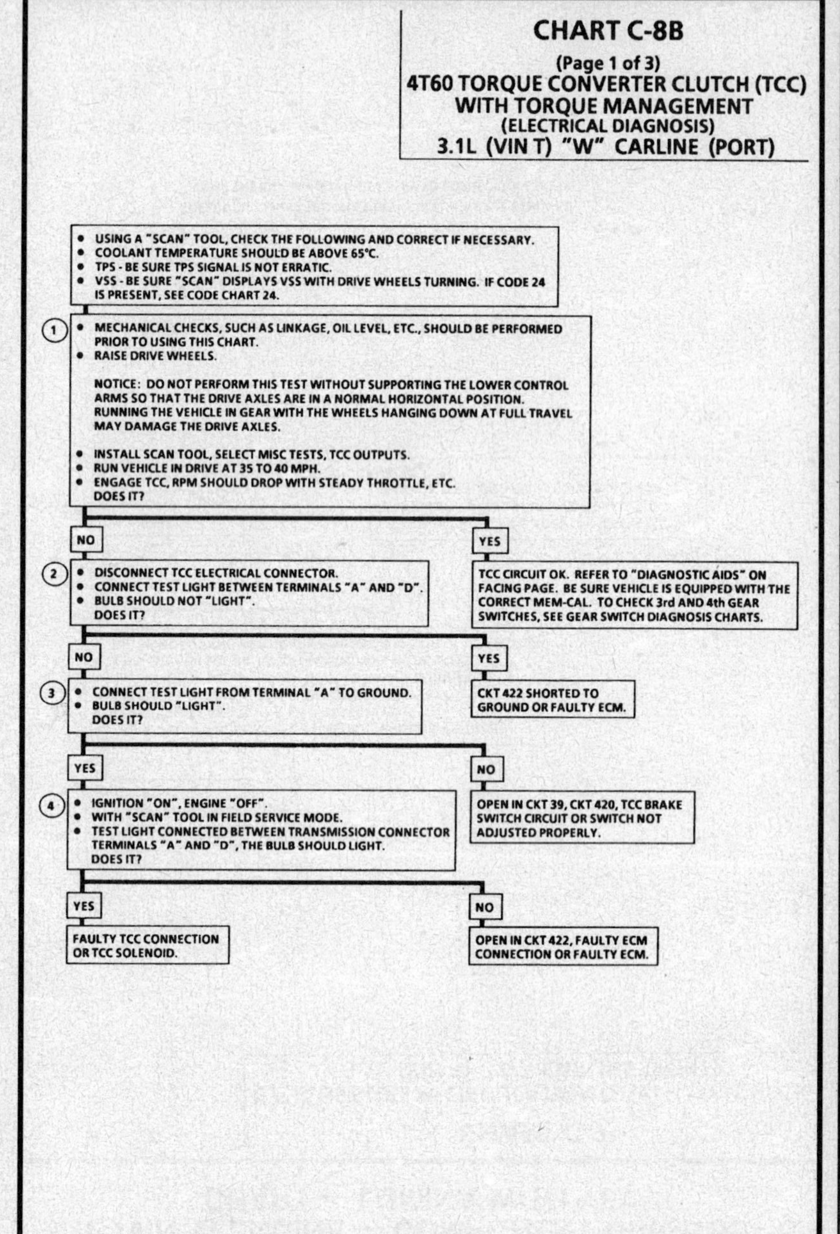

3.1L (VIN T) ENGINE — COMPONENT DIAGNOSTIC CHART — LUMINA WITH VFV

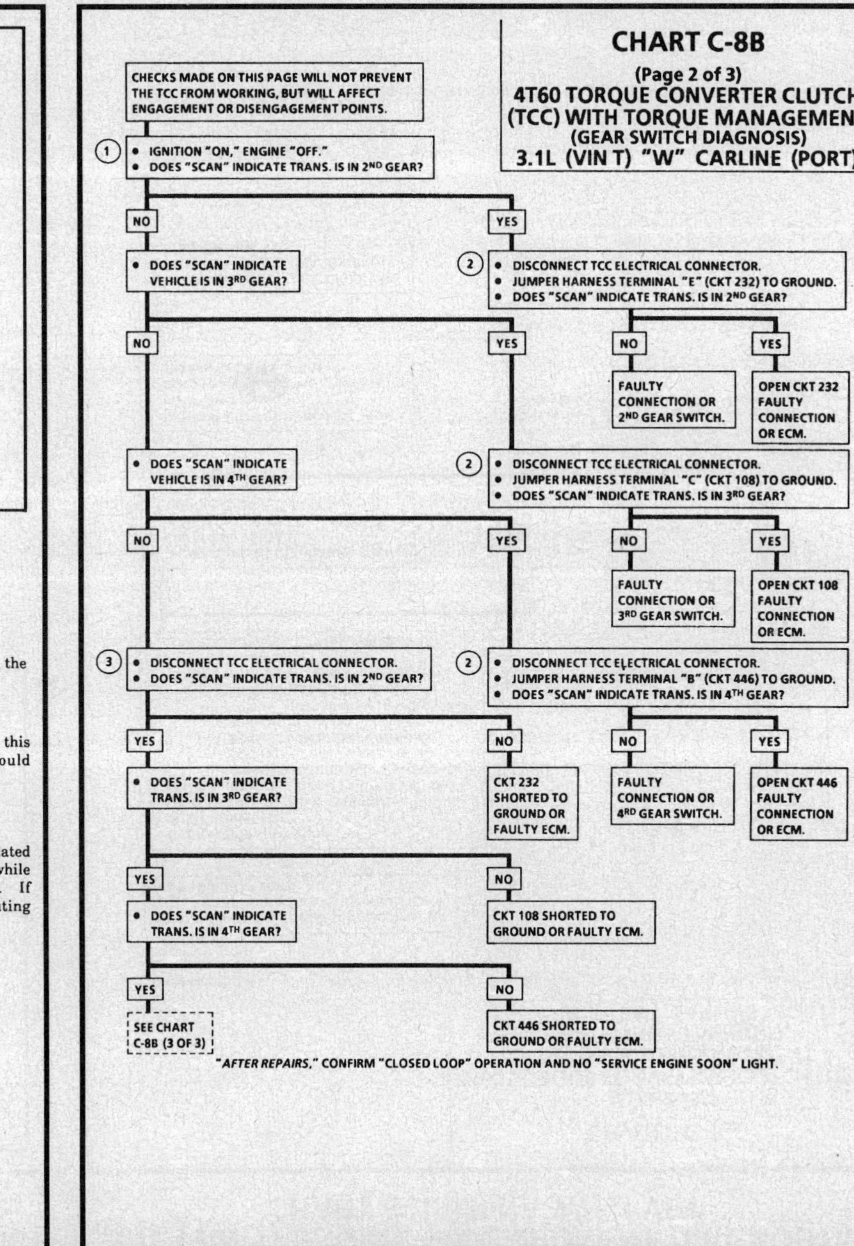

CHART C-8B
(Page 2 of 3)
4T60 TORQUE CONVERTER CLUTCH (TCC) WITH TORQUE MANAGEMENT
(GEAR SWITCH DIAGNOSIS)
3.1L (VIN T) "W" CARLINE (PORT)

Circuit Description:

The 2nd gear switch in this vehicle is open in 3rd and 4th gear. The ECM uses this signal to disengage the Torque Converter Clutch (TCC) when going into a downshift.

The fourth gear switch is open in 4th gear.

Test Description: Number(s) below refer to circled number(s) on the diagnostic chart.

1. Some "Scan" tools display the state of these switches in different ways. Be familiar with the type of tool being used. Since both switches should be in the closed state during this test, the tool should read the same for either the 2nd, 3rd or 4th gear switch.

2. Determines whether the switch or signal circuit is open. The circuit can be checked for an open by measuring the voltage (with a voltmeter) at the TCC connector. Should be about 12 volts.

3. Because the switch(es) should be grounded in this step, disconnecting the TCC connector should cause the "Scan" switch state to change.

Diagnostic Aids:

If vehicle is road tested because of a TCC related problem, be sure the switch states do not change while in 4th gear because the TCC will disengage. If switches change state, carefully check wire routing and connections.

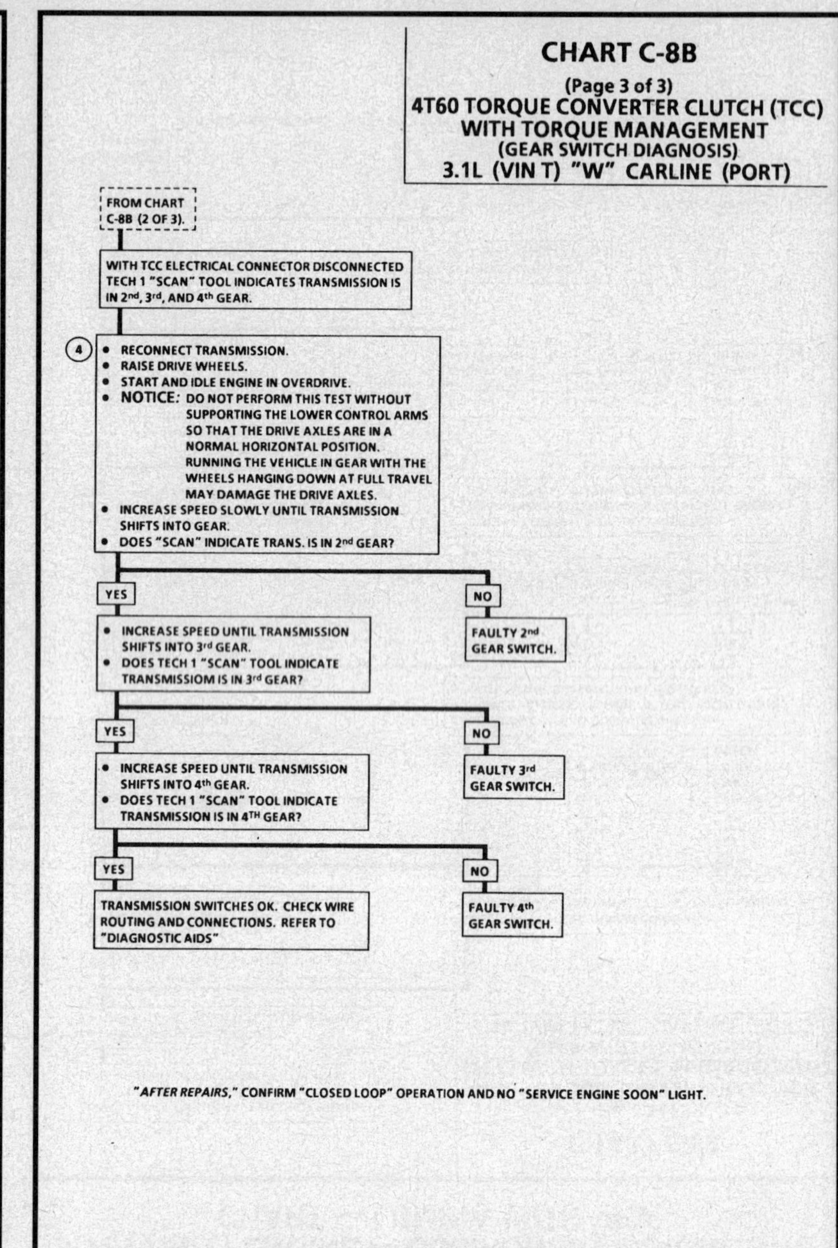

CHART C-8B

(Page 3 of 3)
4T60 TORQUE CONVERTER CLUTCH (TCC) WITH TORQUE MANAGEMENT
(GEAR SWITCH DIAGNOSIS)
3.1L (VIN T) "W" CARLINE (PORT)

Circuit Description:

The 2nd gear switch in this vehicle is open in 3rd and 4th gear. The ECM·uses this signal to disengage the Torque Converter Clutch (TCC) when going into a downshift.

The fourth gear switch is open in 4th gear.

Test Description: Number(s) below refer to circled number(s) on the diagnostic chart.

4. The switch state should change when the vehicle shifts into 2nd gear.

Diagnostic Aids:

If vehicle is road tested because of a TCC related problem, be sure the switch states do not change while in 4th gear because the TCC will disengage. If switches change state, carefully check wire routing and connections.

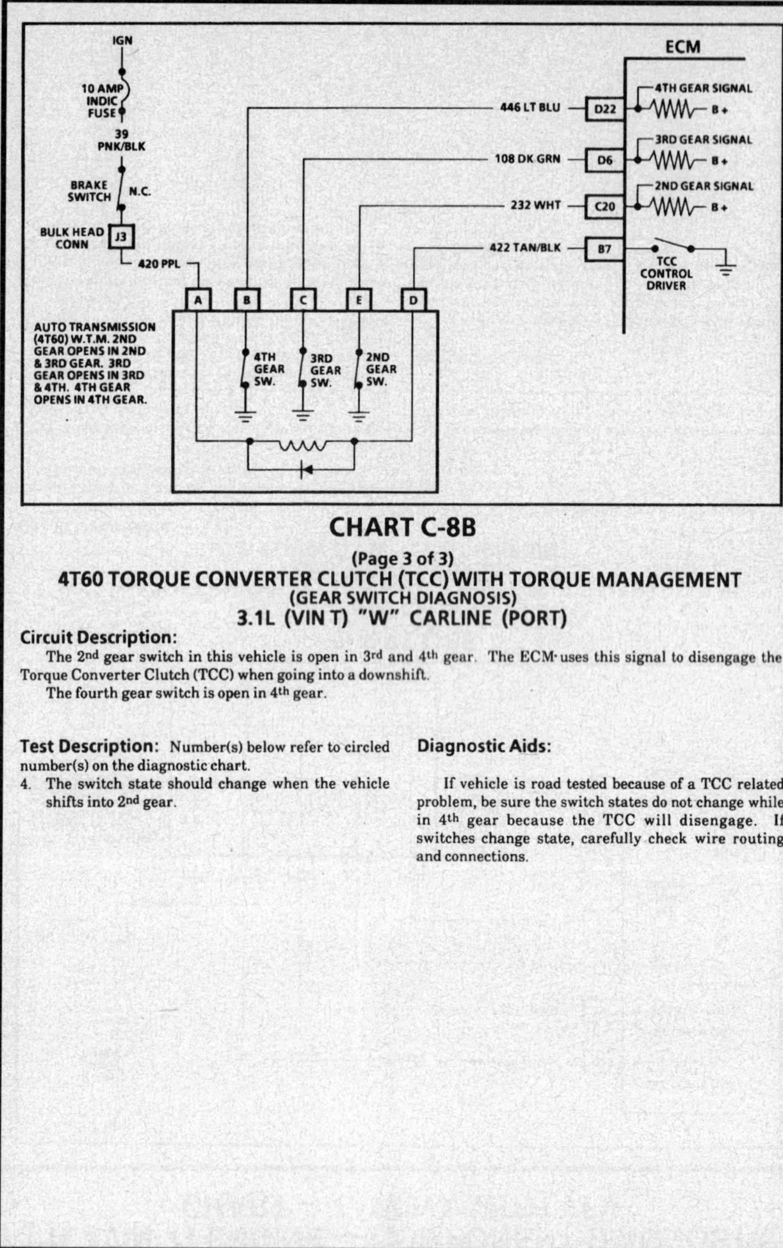

3.1L (VIN T) ENGINE — COMPONENT DIAGNOSTIC CHART — LUMINA WITH VFV

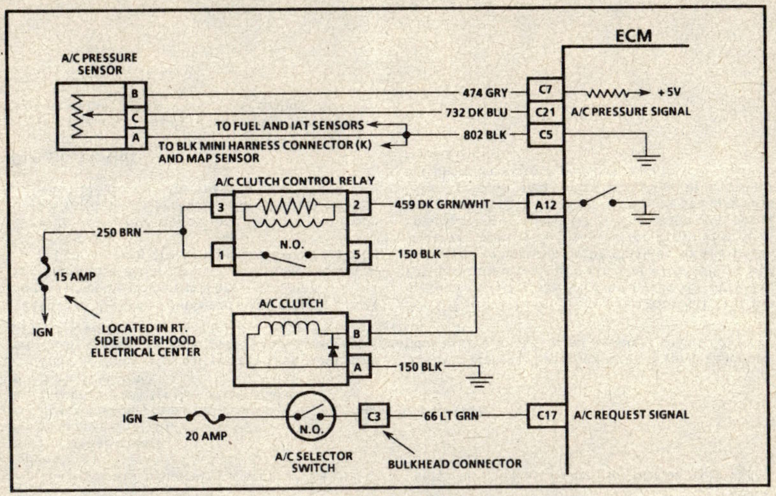

CHART C-10
(Page 1 of 2)
A/C CLUTCH CONTROL CIRCUIT DIAGNOSIS
3.1L (VIN T) "W" CARLINE (PORT)

Circuit Description:

The A/C clutch control relay is ECM controlled to delay A/C clutch engagement about .4 second after A/C is turned "ON." This allows the IAC to adjust engine rpm, before the A/C clutch engages. The ECM, also, causes the relay to disengage the A/C clutch during WOT, when high power steering pressure is present, or if engine is overheating. The A/C clutch control relay is energized, when the ECM provides a ground path for CKT 459. The A/C pressure sensor is used to determine high and low pressure in the system and also turns "ON" the coolant fan when it is needed.

Test Description: Number(s) below refer to circled number(s) on the diagnostic chart.

1. The ECM will only energize the A/C relay, when the engine is running. This test will determine if the relay, or CKT 459, is faulty.
2. Determines if the signal is reaching the ECM through CKT 66 from the A/C control panel. Signal should only be present when an A/C mode or defrost mode has been selected.
3. If the ECM is seeing a high power steering pressure signal, the A/C clutch will be disengaged by the ECM.

Diagnostic Aids:

If complaint is insufficient cooling, the problem may be caused by an inoperative cooling fan. See CHART C-12 for cooling fan diagnosis. If fan operates correctly, see A/C diagnosis in A/C pressure outside of a range of 43 to 428 psi will cause the compressor to be disabled by the ECM. Observe "Scan" A/C pressure for 2 minutes with engine idling and A/C "ON." If pressure goes out of range, refer to to measure and diagnose. "Scan" pressure should be within 20 psi of actual. If not, check for a circuit problem using Code 66 chart or replace sensor.

3.1L (VIN T) ENGINE — COMPONENT DIAGNOSTIC CHART — LUMINA WITH VFV

CHART C-10
(Page 1 of 2)
A/C CLUTCH CONTROL CIRCUIT DIAGNOSIS
3.1L (VIN T) "W" CARLINE (PORT)

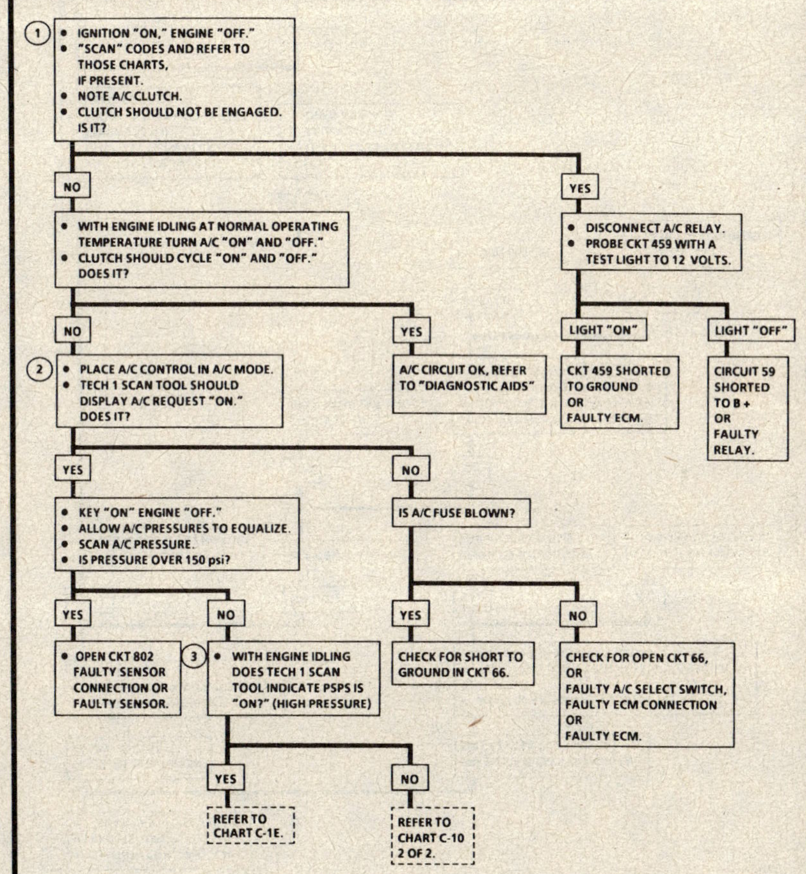

3.1L (VIN T) ENGINE — COMPONENT DIAGNOSTIC CHART — LUMINA WITH VFV

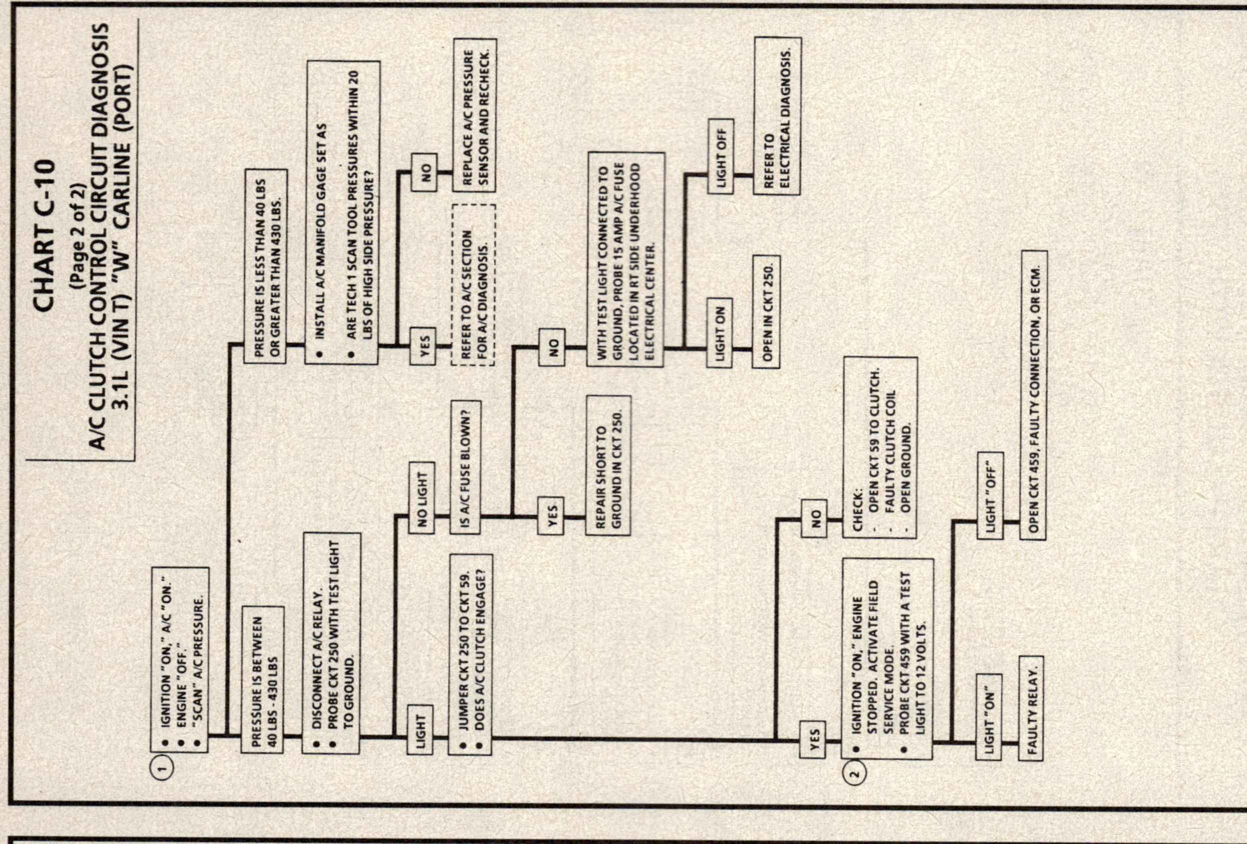

3.1L (VIN T) ENGINE — COMPONENT DIAGNOSTIC CHART — LUMINA WITH VFV

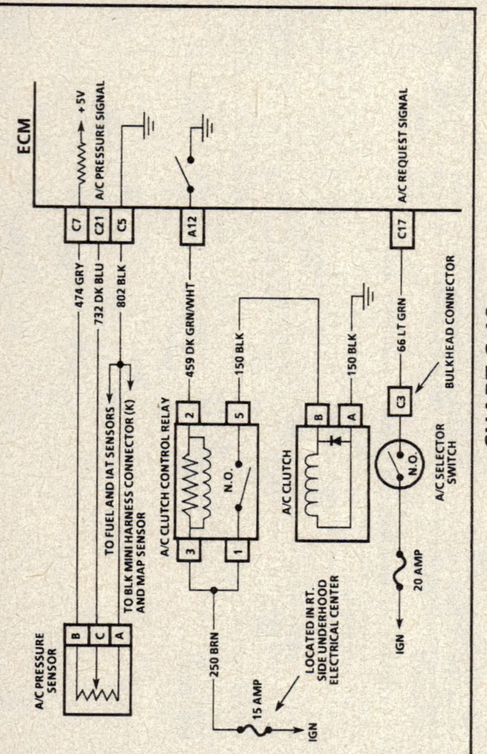

CHART C-10
(Page 2 of 2)
A/C CLUTCH CONTROL CIRCUIT DIAGNOSIS
3.1L (VIN T) "W" CARLINE (PORT)

Circuit Description:

The A/C clutch control relay is energized when the ECM provides a ground path through CKT 459 and A/C is requested. A/C clutch is delayed about 3 seconds after A/C is requested. This will allow the IAC to adjust engine rpm for the additional load.

The ECM will temporarily disengage the A/C clutch relay for calibrated times for one or more of the following:

- Hot engine restart.
- Wide open throttle (TPS over 90%).
- Power steering pressure high (open power steering pressure switch).
- Engine rpm greater than about 6000 rpm.
- During IAC reset.

The A/C clutch relay will remain disengaged when a Code 66 is present, if there is no A/C request signal due to an open A/C select switch or circuit. Refer to SECTION 1B for more information on A/C refrigerant system.

Test Description: Number(s) below refer to circled number(s) on the diagnostic chart.

1. Determines if the pressure transducer is out of range causing the compressor clutch to be disengaged.
2. With the engine stopped and field service mode activated, the ECM should be grounding CKT 459, which should cause the test light to be "ON."

See CHART C-12 for cooling fan diagnosis. If fan operates correctly, see A/C diagnosis in SECTION 1B. A/C pressure outside of a range of 43 to 428 psi will cause the compressor to be disabled by the ECM. Observe "Scan" A/C pressure for 2 minutes with engine idling and A/C "ON." If pressure goes out of range, refer to SECTION 1B to measure and diagnose. "Scan" pressure should be within 20 psi of actual. If not, check for a circuit problem using Code 66 chart or replace sensor.

Diagnostic Aids:

If complaint is insufficient cooling, the problem may be caused by an inoperative cooling fan.

3.1L (VIN T) ENGINE — COMPONENT DIAGNOSTIC CHART — LUMINA WITH VFV

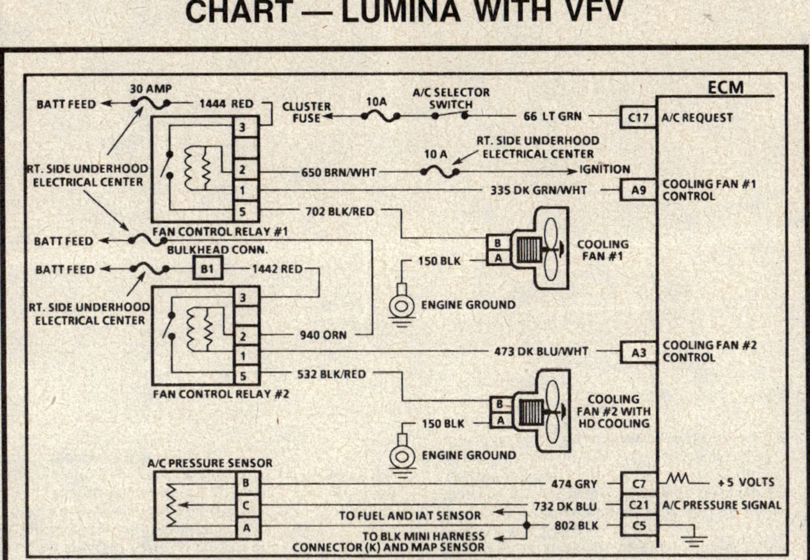

CHART C-12
(Page 1 of 2)
COOLANT FAN CONTROL CIRCUIT DIAGNOSIS
3.1L (VIN T) "W" CARLINE (PORT)

Circuit Description:

The primary and secondary electric cooling fan(s) are controlled by the ECM, based on inputs from the Coolant Temperature Sensor (CTS), the A/C control switches, vehicle speed and state of the A/C pressure sensor. The ECM controls the fan(s) by grounding CKT 335 and/or CKT 473, which energizes the fan control relay. Battery voltage is then supplied to the fan motor.

The ECM grounds CKT 335 and/or CKT 473, when coolant temperature is over about 106°C (223°F), or when A/C has been requested and the A/C pressure is about 1380 kPa (200 psi). Once the ECM turns the relay "ON," it will keep it "ON" for a minimum of 30 seconds, or until vehicle speed exceeds 70 mph (40 mph for secondary fan).

Also, if Code 14 or 15 sets, or the ECM is in throttle body back up, the primary fan will run at all times.

Test Description: Number(s) below refer to circled number(s) on the diagnostic chart.

1. With the diagnostic terminal grounded, the cooling fan control driver(s) will close, which should energize the fan control relay(s).
2. If the A/C pressure is above 300 psi (2069 kPa) or circuit is open, the fan would run whenever A/C is requested.
3. With A/C clutch engaged and the A/C pressure sensor functioning properly, the fan should come "ON" when pressure exceeds about 1380 kPa (200 psi). This signal should cause the ECM to energize the fan control relay(s).
4. This will determine if the A/C pressure sensor is faulty or if the ECM or circuitry is faulty.

Diagnostic Aids:

If the owner complained of an overheating problem, it must be determined if the complaint was due to an actual boilover, the hot light, or temp. gage indicated over heating.

If the gage, or light, indicates overheating, but no boilover is detected, the gage circuit should be checked. The gage accuracy can, also, be checked by comparing the coolant sensor reading using a "Scan" tool and comparing its reading with the gage reading.

If the engine is actually overheating, and the gage indicates overheating, but the cooling fan is not coming "ON," the coolant sensor has probably shifted out of calibration and should be replaced.

If the engine is overheating, and the cooling fan is "ON," the cooling system should be checked.

3.1L (VIN T) ENGINE — COMPONENT DIAGNOSTIC CHART — LUMINA WITH VFV

CHART C-12
(Page 1 of 2)
COOLANT FAN CONTROL CIRCUIT DIAGNOSIS
3.1L (VIN T) "W" CARLINE (PORT)

Diagnostic flow chart:

- COOLANT TEMPERATURE SENSOR INDICATING LESS THAN 106°C (223°F).
- IGNITION "ON," ENGINE "OFF," A/C "OFF."
- COOLING FAN(S) SHOULD BE "OFF."
ARE THEY?

YES → (1)
- GROUND DIAGNOSTIC TERMINAL. FAN(S) SHOULD TURN "ON." DO THEY?
  - **YES →**
    - UNGROUND DIAGNOSTIC TERMINAL.
    - START AND IDLE ENGINE.
    - A/C "OFF."
    - FAN(S) SHOULD BE "OFF" (WHILE TEMPERATURE IS UNDER 100°C [212°F]). ARE THEY?
      - **YES →** WITHOUT A/C | WITH A/C
      - **NO →** ONE OR BOTH "ON" → FAULTY ECM.
  - **NO →** REFER TO CHART C-12 (2 OF 2)

NO →
- DISCONNECT FAN RELAY(S). FAN(S) SHOULD STOP. DO THEY?
  - **YES →** PROBE CKT 335 (AND 473) WITH A TEST LIGHT TO 12 VOLTS.
    - LIGHT "ON" ONE OR BOTH → CKT 335 (OR 473) SHORTED TO GROUND OR FAULTY ECM.
    - LIGHT "OFF" → FAULTY RELAY.
  - **NO →** CKT 702 (OR 532) SHORTED TO VOLTAGE.

(2)
- ENGINE IDLING, A/C "OFF."
- USE A TECH 1 "SCAN" TOOL AND CHECK A/C PRESSURE SENSOR. DOES TECH 1 "SCAN" TOOL INDICATE PRESSURE ABOVE 300 psi (2069 kPa)?
  - **NO →** (3)
    - ENGINE IDLING, A/C "ON." IF A/C IS INOPERATIVE SEE CHART C-10. FAN SHOULD TURN "ON" WHEN A/C HEAD PRESSURE EXCEEDS ABOUT 200 psi (1380 kPa). DOES IT?
      - **YES →** NO TROUBLE FOUND. REFER TO "DIAGNOSTIC AIDS"
      - **NO →** DISCONNECT A/C PRESSURE SENSOR. MEASURE VOLTAGE BETWEEN CONNECTOR TERMINAL "B" AND "A." VOLTMETER SHOULD READ 5 VOLTS. DOES IT?
        - **YES →** MEASURE VOLTAGE BETWEEN CONNECTOR TERMINAL "B" AND "C." VOLTMETER SHOULD READ 5 VOLTS. DOES IT?
          - **YES →** FAULTY SENSOR.
          - **NO →** CKT 732 OPEN OR SHORTED TO GROUND OR FAULTY ECM CONNECTION OR ECM.
        - **NO →** MEASURE VOLTAGE BETWEEN CONNECTOR TERMINAL "B" AND GROUND. VOLTMETER SHOULD READ 5 VOLTS. DOES IT?
          - **YES →** CKT 802 OPEN OR FAULTY ECM CONNECTION OR ECM.
          - **NO →** CKT 474 OPEN OR SHORTED TO GROUND OR FAULTY ECM CONNECTION OR ECM.
  - **YES →** (4)
    - DISCONNECT A/C PRESSURE SENSOR. DOES "SCAN" INDICATE ZERO A/C PRESSURE?
      - **YES →** OPEN CKT 802 OR FAULTY SENSOR CONNECTOR OR FAULTY SENSOR.
      - **NO →** CKT 732 SHORTED TO VOLTAGE OR FAULTY ECM.

"AFTER REPAIRS," CONFIRM "CLOSED LOOP" OPERATION AND NO "SERVICE ENGINE SOON" LIGHT.

3.1L (VIN T) ENGINE — COMPONENT DIAGNOSTIC CHART — LUMINA WITH VFV

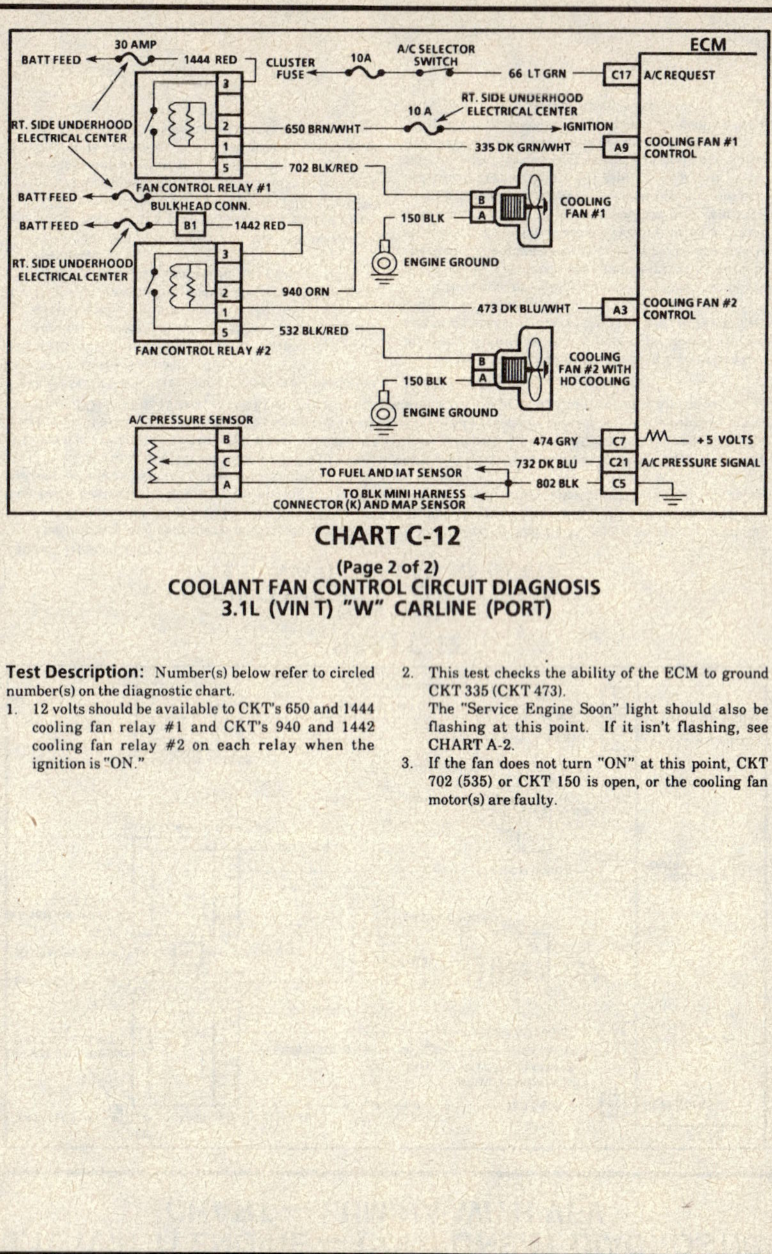

CHART C-12
(Page 2 of 2)
COOLANT FAN CONTROL CIRCUIT DIAGNOSIS
3.1L (VIN T) "W" CARLINE (PORT)

Test Description: Number(s) below refer to circled number(s) on the diagnostic chart.

1. 12 volts should be available to CKT's 650 and 1444 cooling fan relay #1 and CKT's 940 and 1442 cooling fan relay #2 on each relay when the ignition is "ON."

2. This test checks the ability of the ECM to ground CKT 335 (CKT 473).
   The "Service Engine Soon" light should also be flashing at this point. If it isn't flashing, see CHART A-2.

3. If the fan does not turn "ON" at this point, CKT 702 (535) or CKT 150 is open, or the cooling fan motor(s) are faulty.

3.1L (VIN T) ENGINE — COMPONENT DIAGNOSTIC CHART — LUMINA WITH VFV

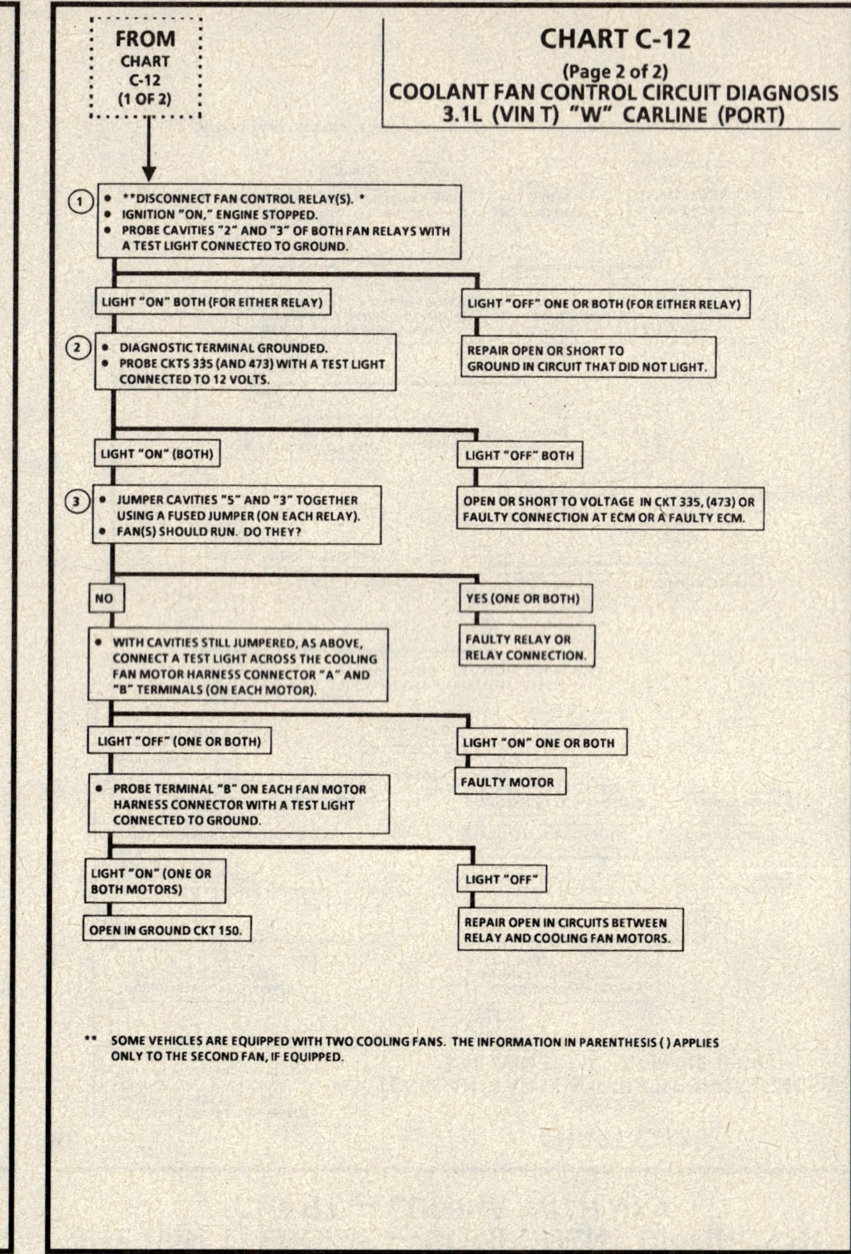

3.1L (VIN T) ENGINE — COMPONENT DIAGNOSTIC CHART — LUMINA WITH VFV

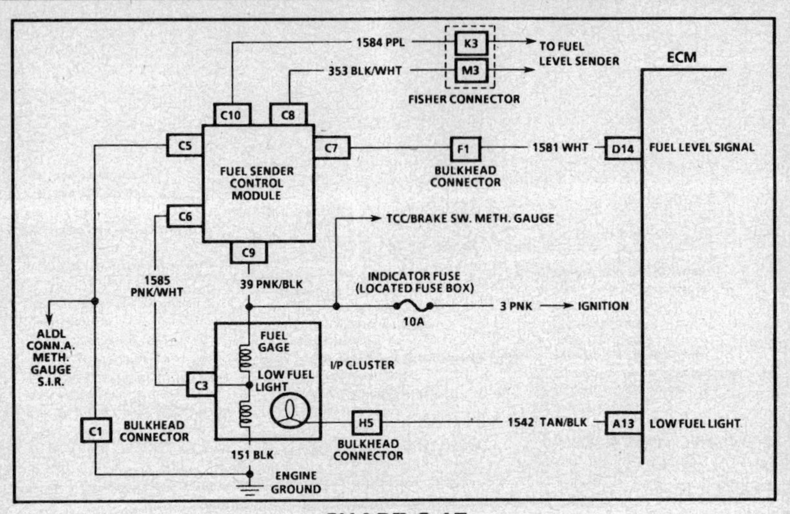

CHART C-17
LOW FUEL LIGHT INDICATOR
3.1L (VIN T) "W" CARLINE (PORT)

Circuit Description:

The Electronic Control Module (ECM) will control the light and turn it "ON" by providing a ground path through CKT 1542 to the ECM. During initial key "ON" low fuel light indicator will illuminate for approximately two seconds. A steady light suggest a short to ground in the light control CKT 1542.

Test Description: Number(s) below refer to circled number(s) on the diagnostic chart.

1. This test determines if the fuel gauge is working correctly.

2. The bulb check will determine if the low fuel light, CKT 1542, and ECM are working correctly.

3.1L (VIN T) ENGINE — COMPONENT DIAGNOSTIC CHART — LUMINA WITH VFV

CHART C-17
LOW FUEL LIGHT INDICATOR
3.1L (VIN T) "W" CARLINE (PORT)

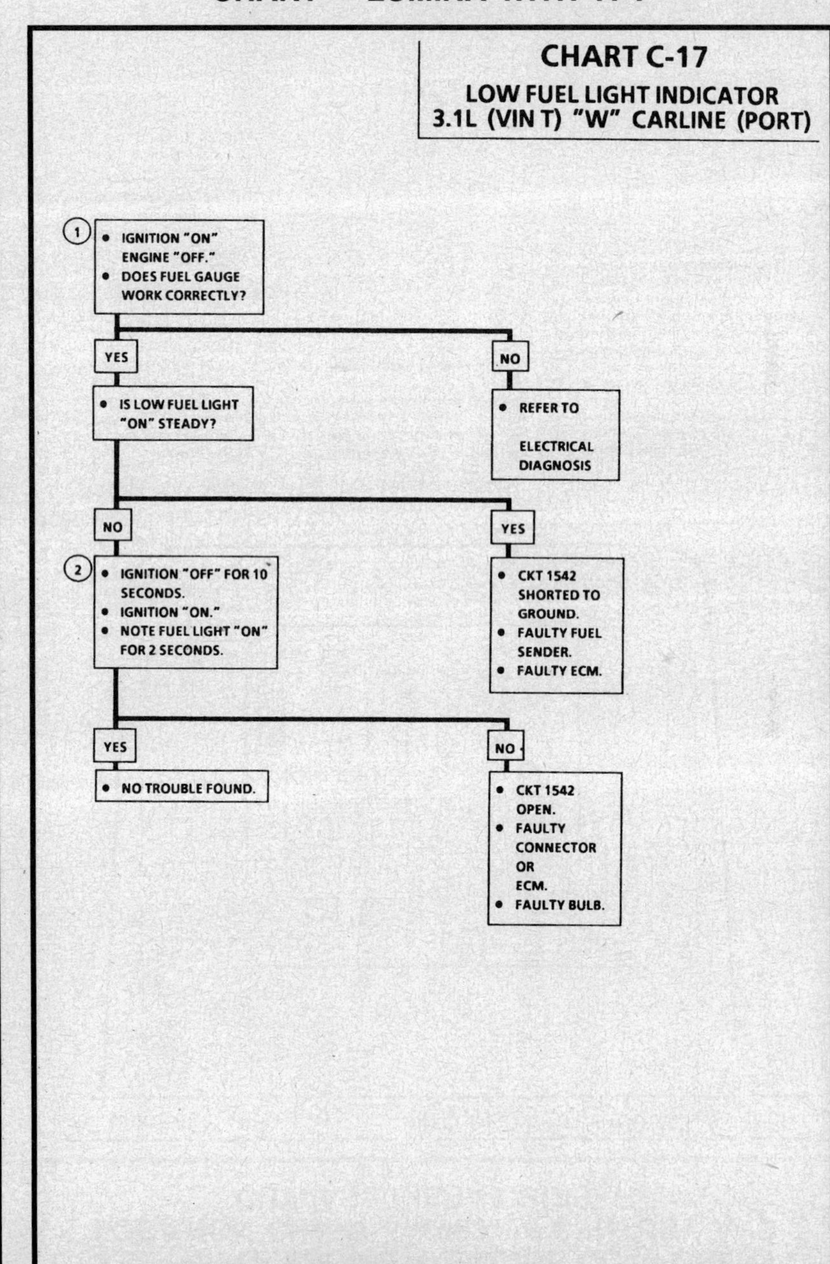

3.3L (VIN N) ENGINE — ENGINE — ENGINE COMPONENT LOCATION CHART — 1993–94 CENTURY AND CUTLASS CIERA/CRUISER

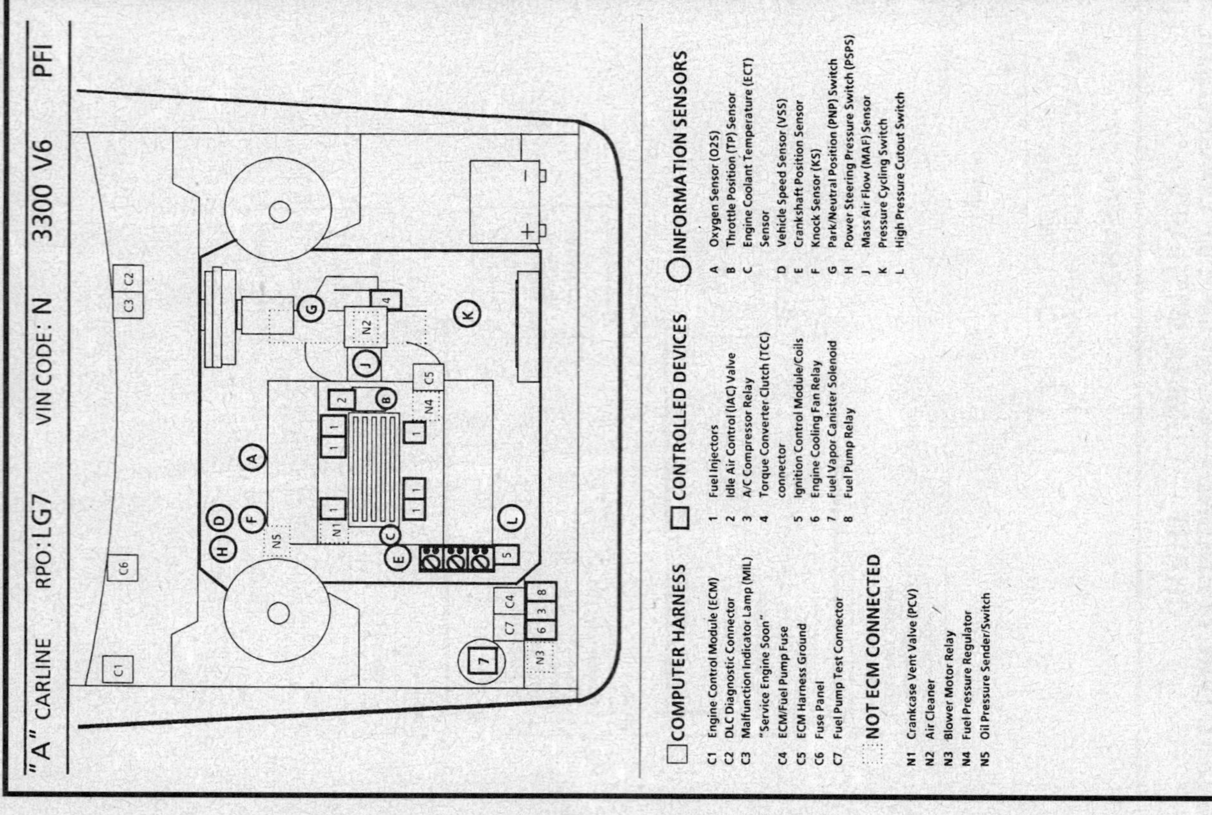

COMPUTER HARNESS

- C1 Engine Control Module (ECM)
- C2 DLC Diagnostic Connector
- C3 Malfunction Indicator Lamp (MIL) "Service Engine Soon"
- C4 ECM/Fuel Pump Fuse
- C5 ECM Harness Ground
- C6 Fuse Panel
- C7 Fuel Pump Test Connector

NOT ECM CONNECTED

- N1 Crankcase Vent Valve (PCV)
- N2 Air Cleaner
- N3 Blower Motor Relay
- N4 Fuel Pressure Regulator
- N5 Oil Pressure Sender/Switch

CONTROLLED DEVICES

- 1 Fuel Injectors
- 2 Idle Air Control (IAC) Valve
- 3 A/C Compressor Relay
- 4 Torque Converter Clutch (TCC) connector
- 5 Ignition Control Module/Coils
- 6 Engine Cooling Fan Relay
- 7 Fuel Vapor Canister Solenoid
- 8 Fuel Pump Relay

INFORMATION SENSORS

- A Oxygen Sensor (O2S)
- B Throttle Position (TP) Sensor
- C Engine Coolant Temperature (ECT) Sensor
- D Vehicle Speed Sensor (VSS)
- E Crankshaft Position Sensor
- F Knock Sensor (KS)
- G Park/Neutral Position (PNP) Switch
- H Power Steering Pressure Switch (PSPS)
- J Mass Air Flow (MAF) Sensor
- K Pressure Cycling Switch
- L High Pressure Cutout Switch

3.3L (VIN N) ENGINE — ENGINE — ENGINE COMPONENT LOCATION CHART — 1992 CENTURY AND CUTLASS CIERA/CRUISER

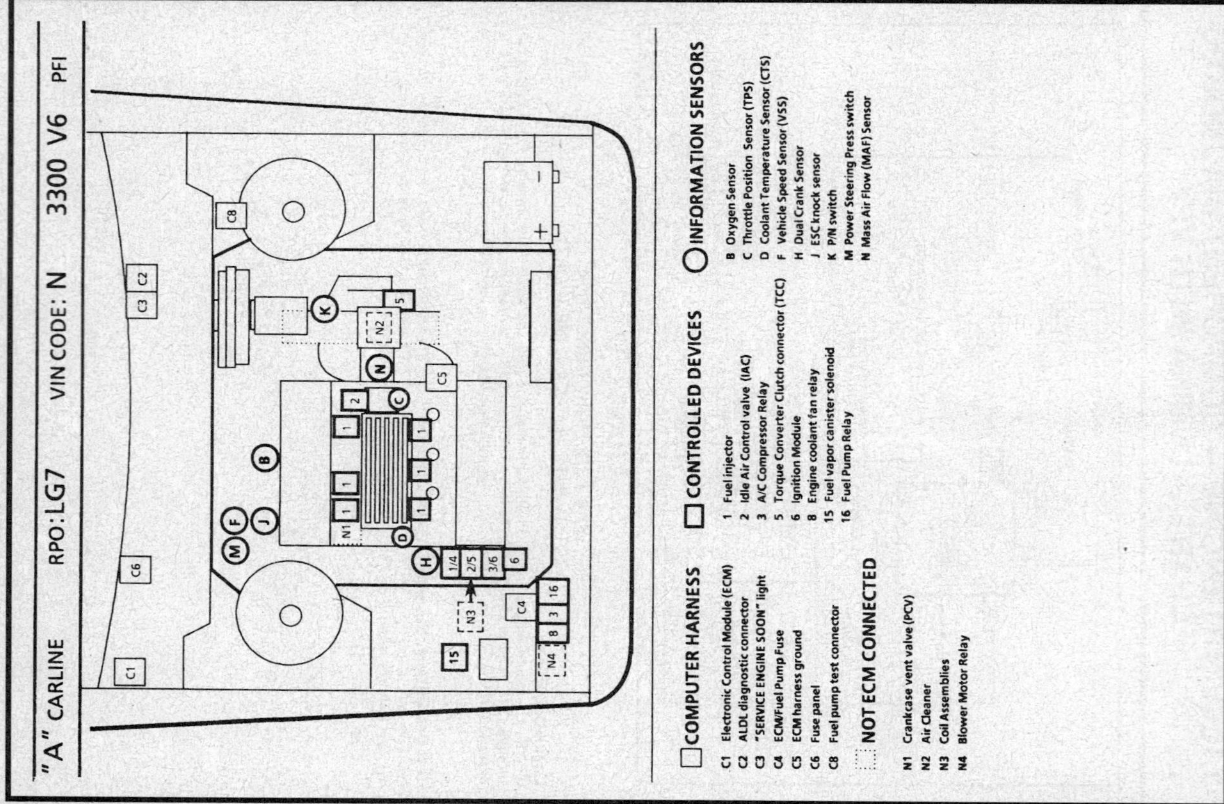

COMPUTER HARNESS

- C1 Electronic Control Module (ECM)
- C2 ALDL diagnostic connector
- C3 "SERVICE ENGINE SOON" light
- C4 ECM/Fuel Pump Fuse
- C5 ECM harness ground
- C6 Fuse panel
- C8 Fuel pump test connector

NOT ECM CONNECTED

- N1 Crankcase vent valve (PCV)
- N2 Air Cleaner
- N3 Coil Assemblies
- N4 Blower Motor Relay

CONTROLLED DEVICES

- 1 Fuel injector
- 2 Idle Air Control valve (IAC)
- 3 A/C Compressor Relay
- 5 Torque Converter Clutch connector (TCC)
- 6 Ignition Module
- 8 Engine coolant fan relay
- 15 Fuel vapor canister solenoid
- 16 Fuel Pump Relay

INFORMATION SENSORS

- B Oxygen Sensor
- C Throttle Position Sensor (TPS)
- D Coolant Temperature Sensor (CTS)
- F Vehicle Speed Sensor (VSS)
- H Dual Crank Sensor
- J ESC knock sensor
- K P/N switch
- M Power Steering Press switch
- N Mass Air Flow (MAF) Sensor

3.3L (VIN N) ENGINE — ENGINE COMPONENT LOCATION CHART — 1992 ACHIEVA, GRAND AM AND SKYLARK

'N' CARLINE   RPO: LG7   VIN CODE: N   3300 V6 PFI

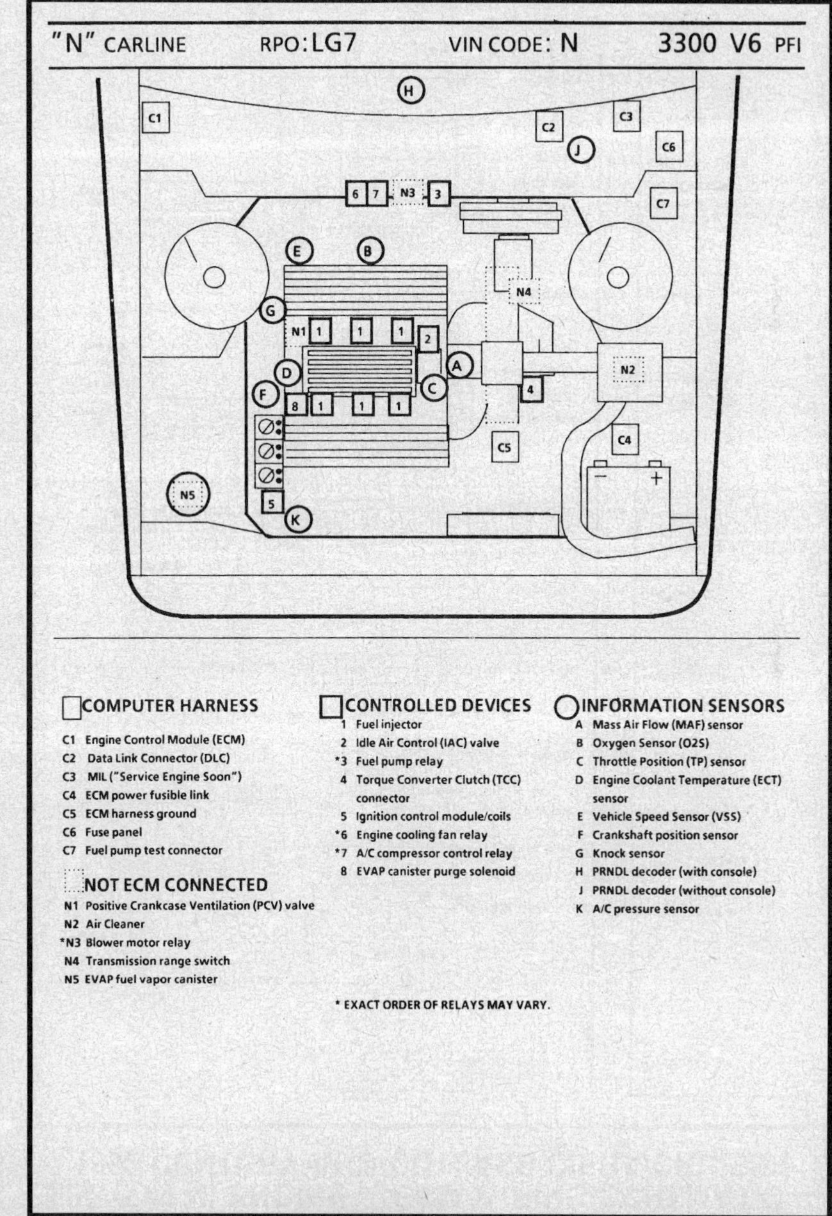

COMPUTER HARNESS

C1 Electronic Control Module (ECM)
C2 ALDL diagnostic connector
C3 "Service Engine Soon" light
C4 ECM power fusible link
C5 ECM harness ground
C6 Fuse panel
C7 Fuel pump test connector

NOT ECM CONNECTED

N1 Positive Crankcase Ventilation (PCV) valve
N2 Air Cleaner
*N3 Blower Motor Relay
N4 Neutral Safety Switch

CONTROLLED DEVICES

1 Fuel injector
2 Idle Air Control (IAC) valve
*3 Fuel pump relay
4 Torque Converter Clutch (TCC) connector
5 Ignition Module/Coils
*6 Engine coolant fan relay
*7 A/C compressor relay
8 Fuel vapor canister solenoid

INFORMATION SENSORS

A Mass Air Flow (MAF) Sensor
B Oxygen (O2) Sensor
C Throttle Position Sensor (TPS)
D Coolant Temperature Sensor (CTS)
E Vehicle Speed Sensor (VSS)
F Dual crank sensor
G ESC knock sensor
H PRNDL Decoder
J A/C Pressure Sensor

* EXACT ORDER OF RELAYS MAY VARY.

3.3L (VIN N) ENGINE — ENGINE COMPONENT LOCATION CHART — 1993–94 ACHIEVA, GRAND AM AND SKYLARK

"N" CARLINE   RPO: LG7   VIN CODE: N   3300 V6 PFI

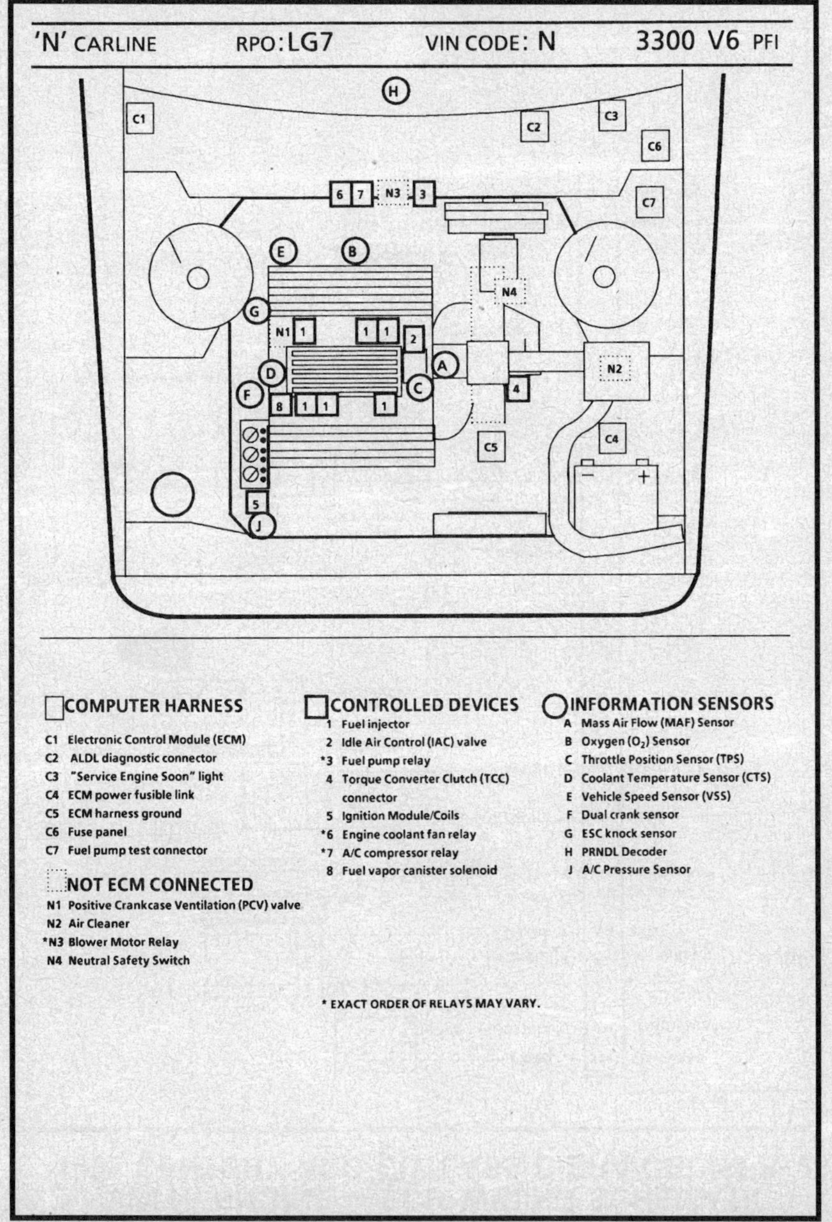

COMPUTER HARNESS

C1 Engine Control Module (ECM)
C2 Data Link Connector (DLC)
C3 MIL ("Service Engine Soon")
C4 ECM power fusible link
C5 ECM harness ground
C6 Fuse panel
C7 Fuel pump test connector

NOT ECM CONNECTED

N1 Positive Crankcase Ventilation (PCV) valve
N2 Air Cleaner
*N3 Blower motor relay
N4 Transmission range switch
N5 EVAP fuel vapor canister

CONTROLLED DEVICES

1 Fuel injector
2 Idle Air Control (IAC) valve
*3 Fuel pump relay
4 Torque Converter Clutch (TCC) connector
5 Ignition control module/coils
*6 Engine cooling fan relay
*7 A/C compressor control relay
8 EVAP canister purge solenoid

INFORMATION SENSORS

A Mass Air Flow (MAF) sensor
B Oxygen Sensor (O2S)
C Throttle Position (TP) sensor
D Engine Coolant Temperature (ECT) sensor
E Vehicle Speed Sensor (VSS)
F Crankshaft position sensor
G Knock sensor
H PRNDL decoder (with console)
J PRNDL decoder (without console)
K A/C pressure sensor

* EXACT ORDER OF RELAYS MAY VARY.

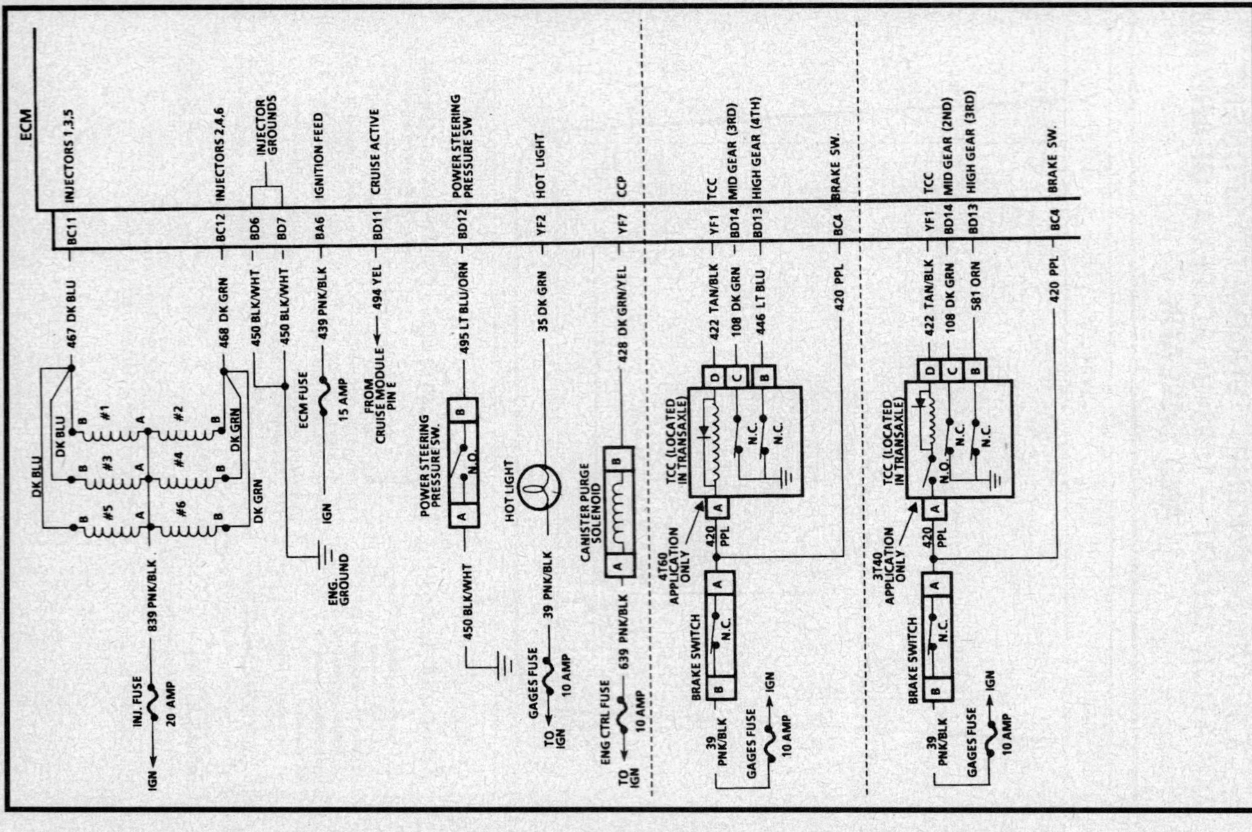
3.3L (VIN N) ENGINE — ECM WIRING SCHEMATIC — 1992 CENTURY AND CUTLASS CIERA/CRUISER

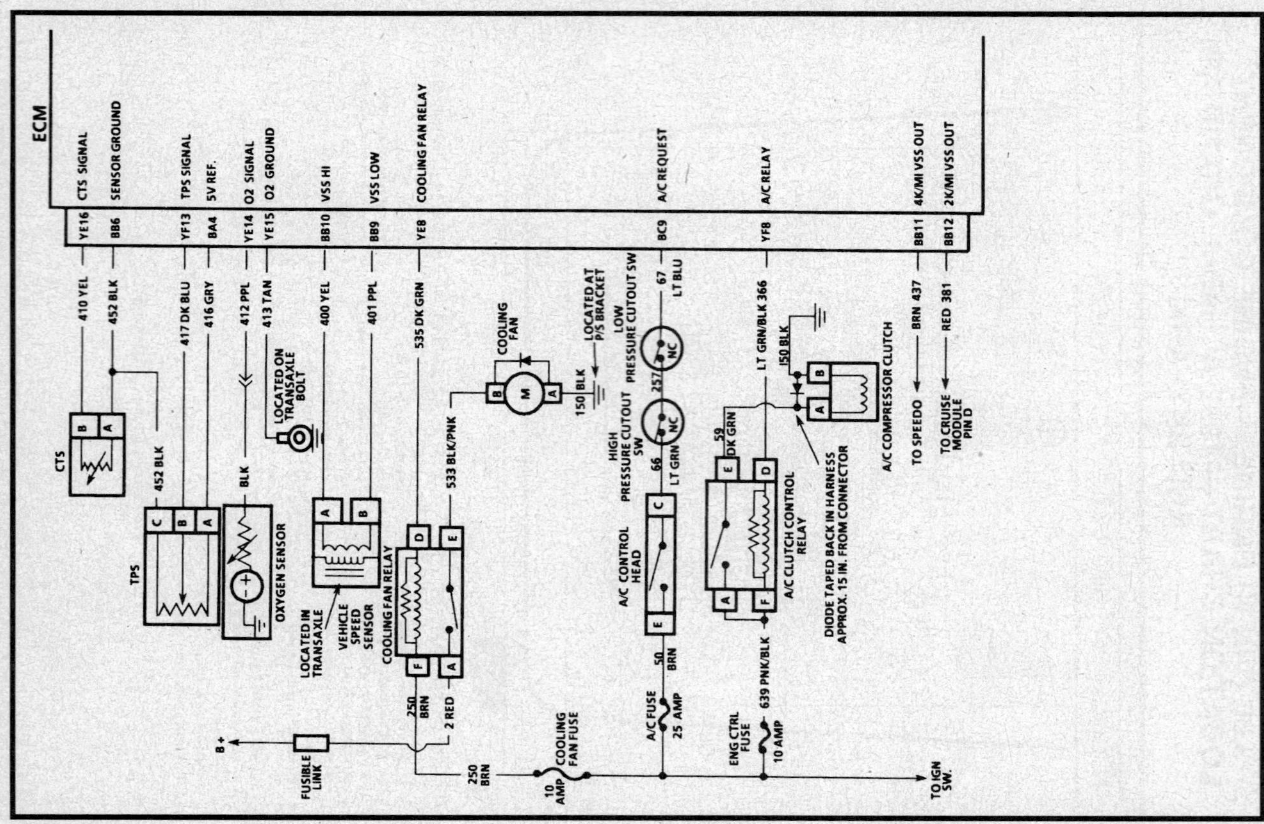
3.3L (VIN N) ENGINE — ECM WIRING SCHEMATIC — 1992 CENTURY AND CUTLASS CIERA/CRUISER

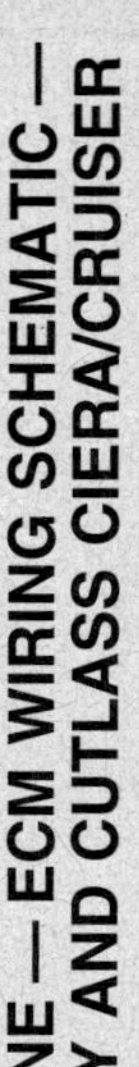

3.3L (VIN N) ENGINE — ECM WIRING SCHEMATIC — 1992 CENTURY AND CUTLASS CIERA/CRUISER

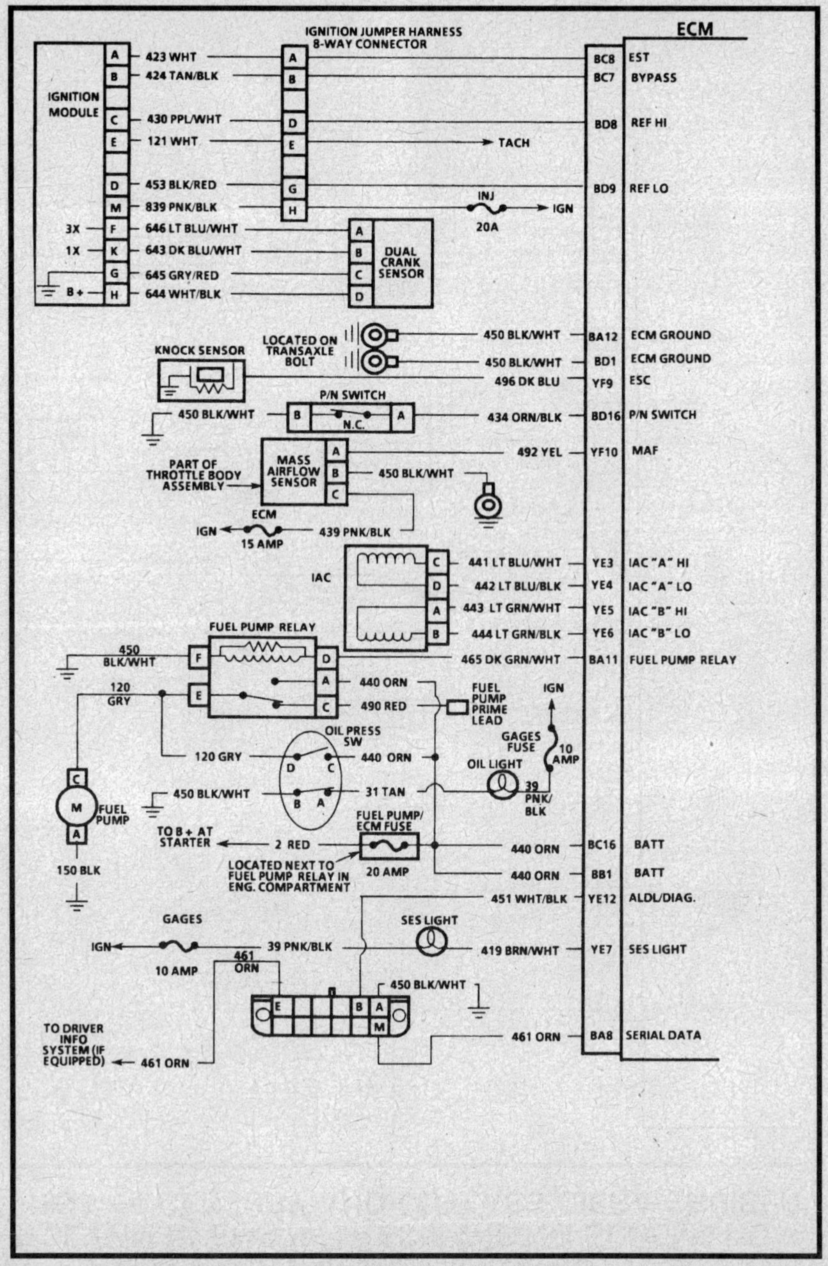

3.3L (VIN N) ENGINE — ECM WIRING SCHEMATIC — 1993–94 CENTURY AND CUTLASS CIERA/CRUISER

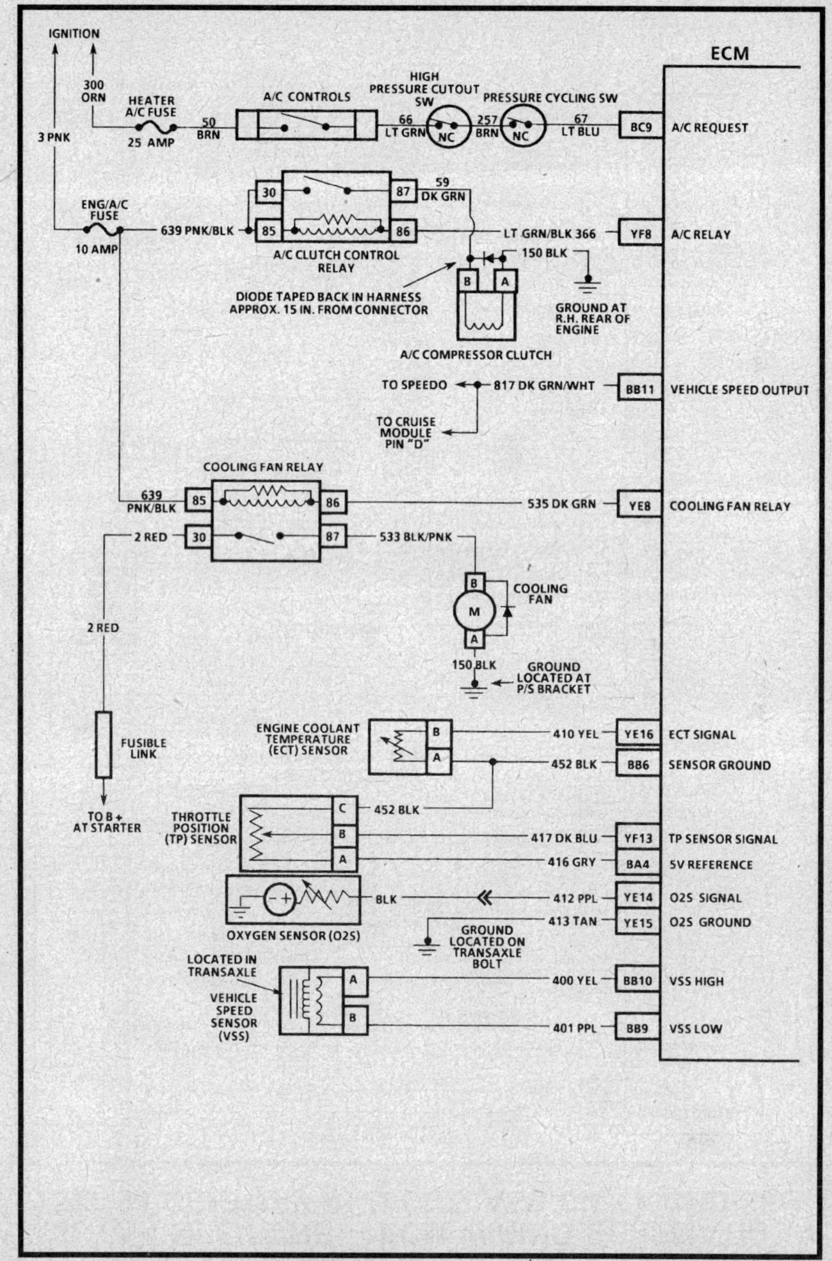

MULTIPORT FUEL INJECTION (MFI) SYSTEMS
EXCEPT LIGHT TRUCKS, VANS, GEO AND SATURN

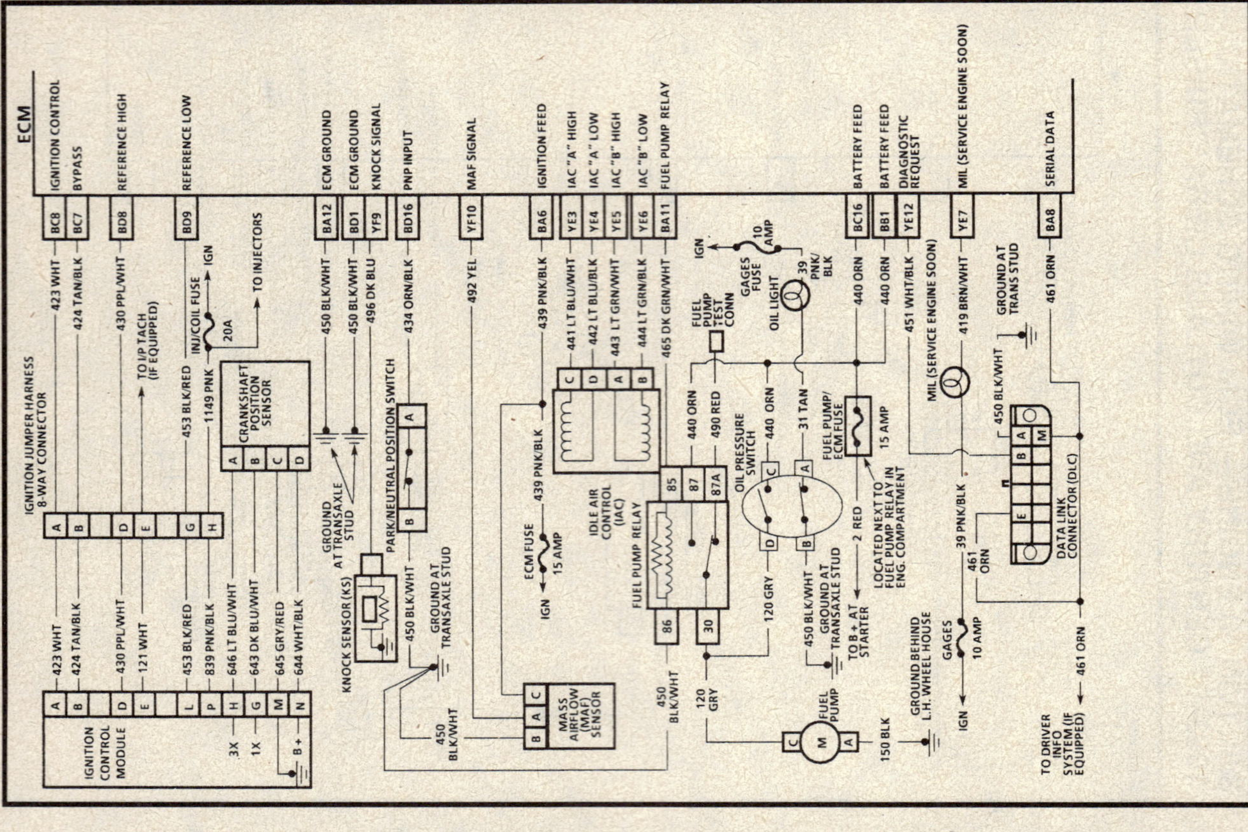

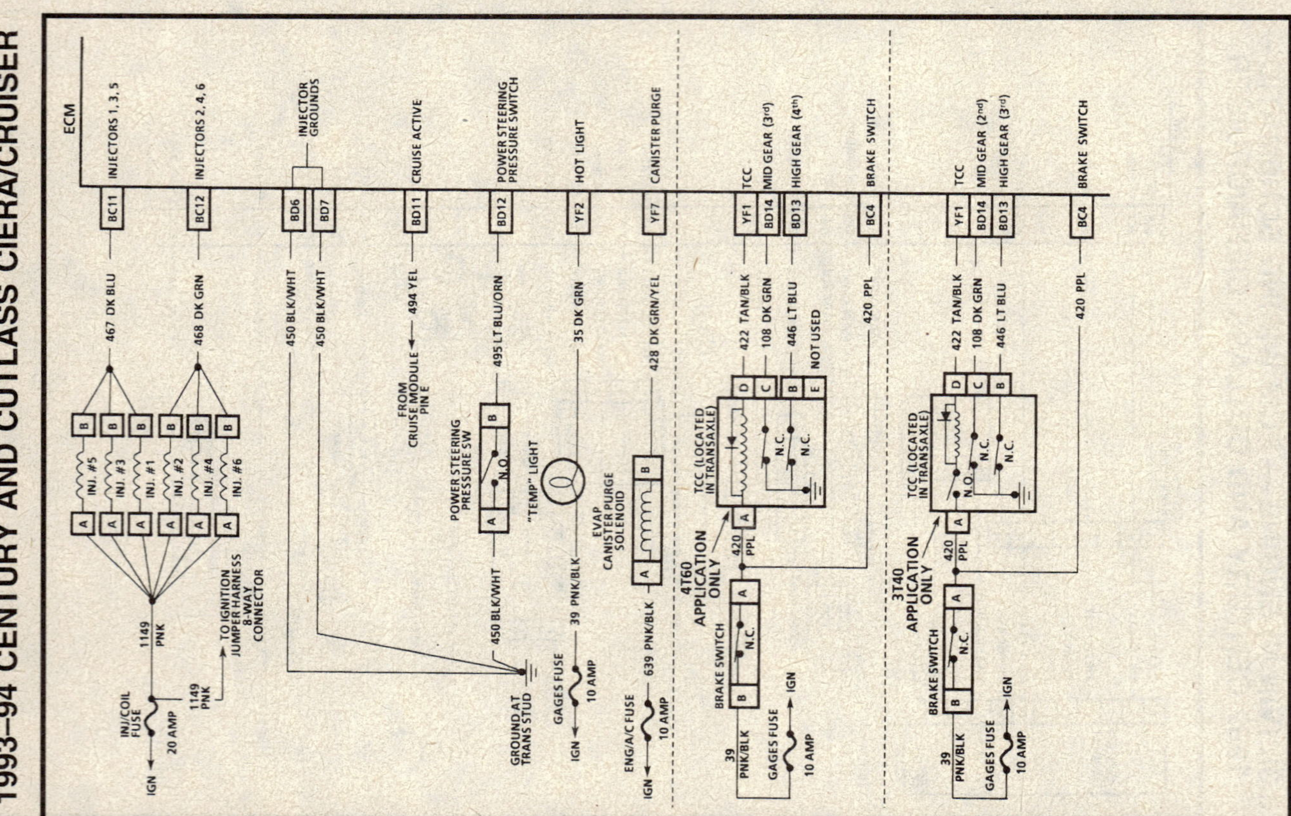

3.3L (VIN N) ENGINE — ECM WIRING SCHEMATIC — 1992 ACHIEVA, GRAND AM AND SKYLARK

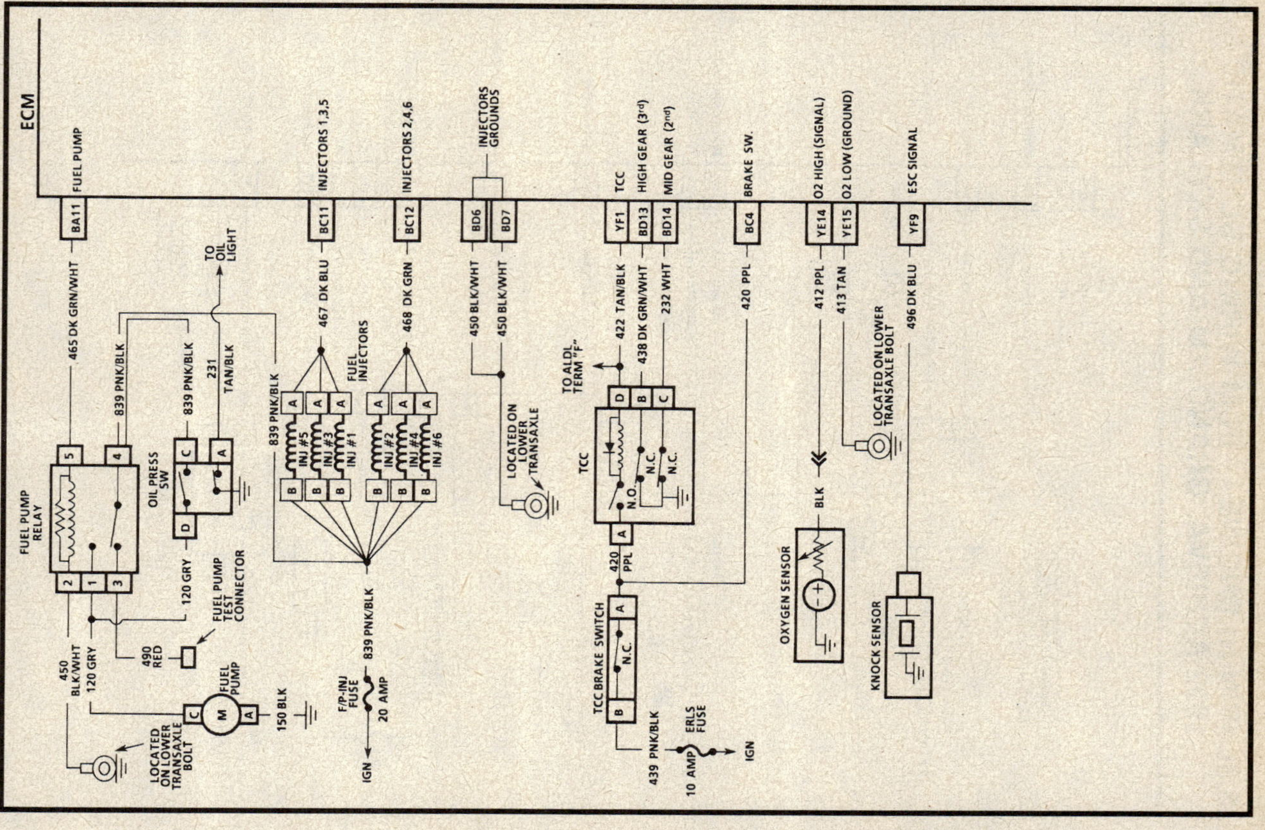

3.3L (VIN N) ENGINE — ECM WIRING SCHEMATIC — 1992 ACHIEVA, GRAND AM AND SKYLARK

MULTIPORT FUEL INJECTION (MFI) SYSTEMS
EXCEPT LIGHT TRUCKS, VANS, GEO AND SATURN

3.3L (VIN N) ENGINE — ECM WIRING SCHEMATIC — 1993–94 ACHIEVA, GRAND AM AND SKYLARK

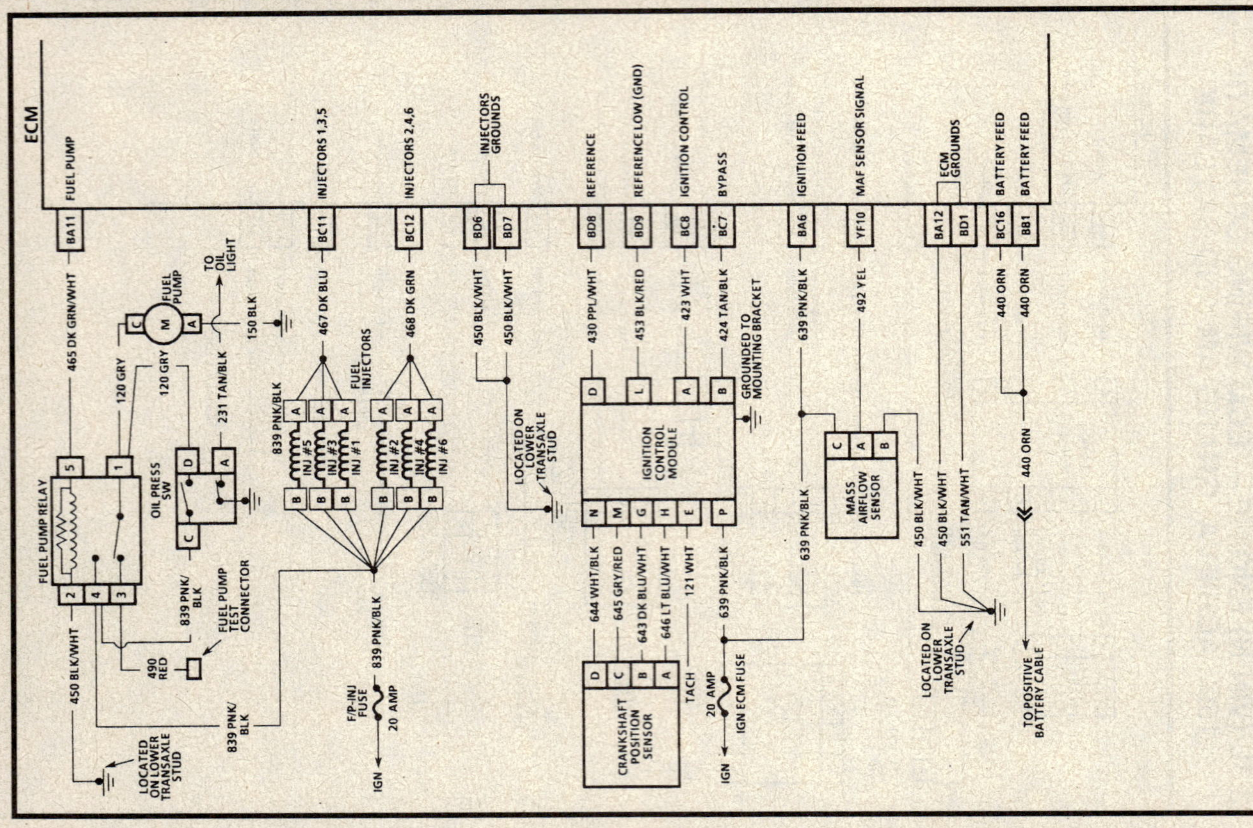

3.3L (VIN N) ENGINE — ECM WIRING SCHEMATIC — 1992 ACHIEVA, GRAND AM AND SKYLARK

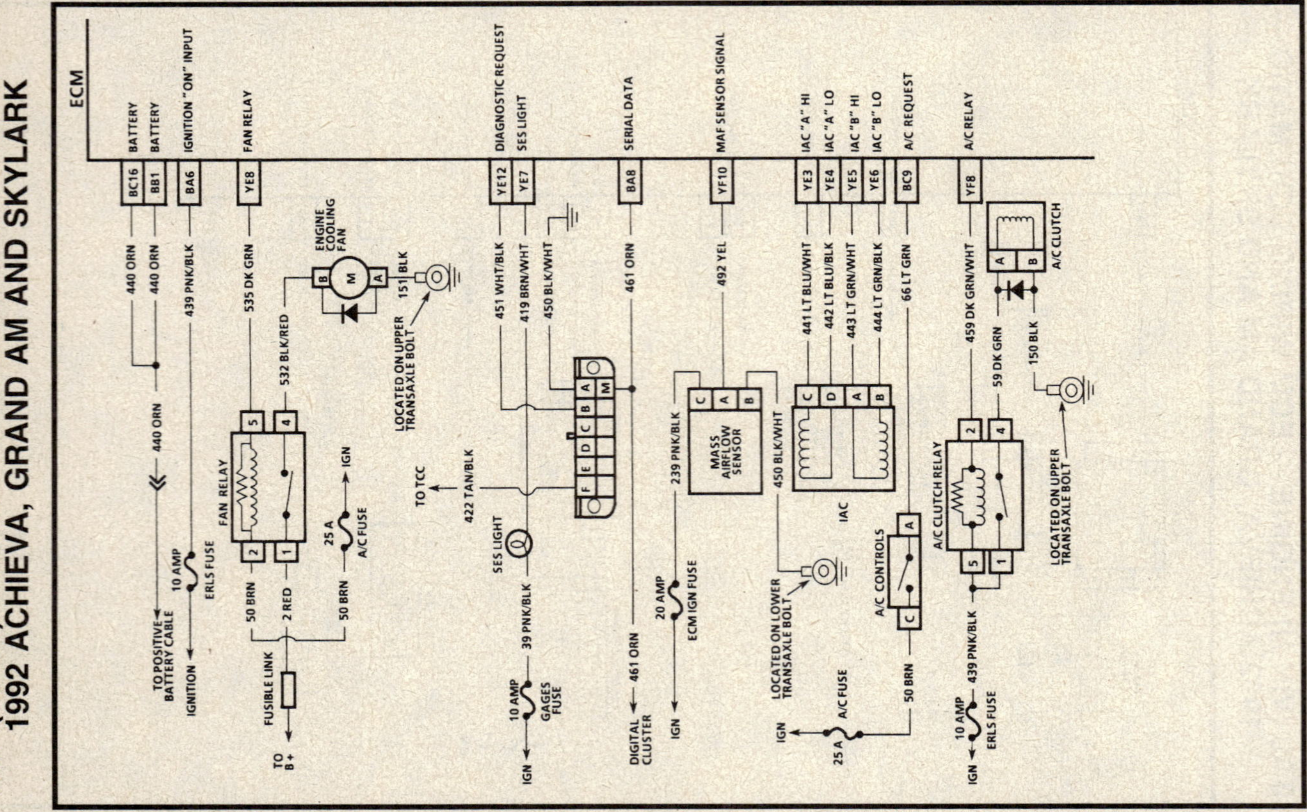

3.3L (VIN N) ENGINE — ECM WIRING SCHEMATIC — 1993–94 ACHIEVA, GRAND AM AND SKYLARK

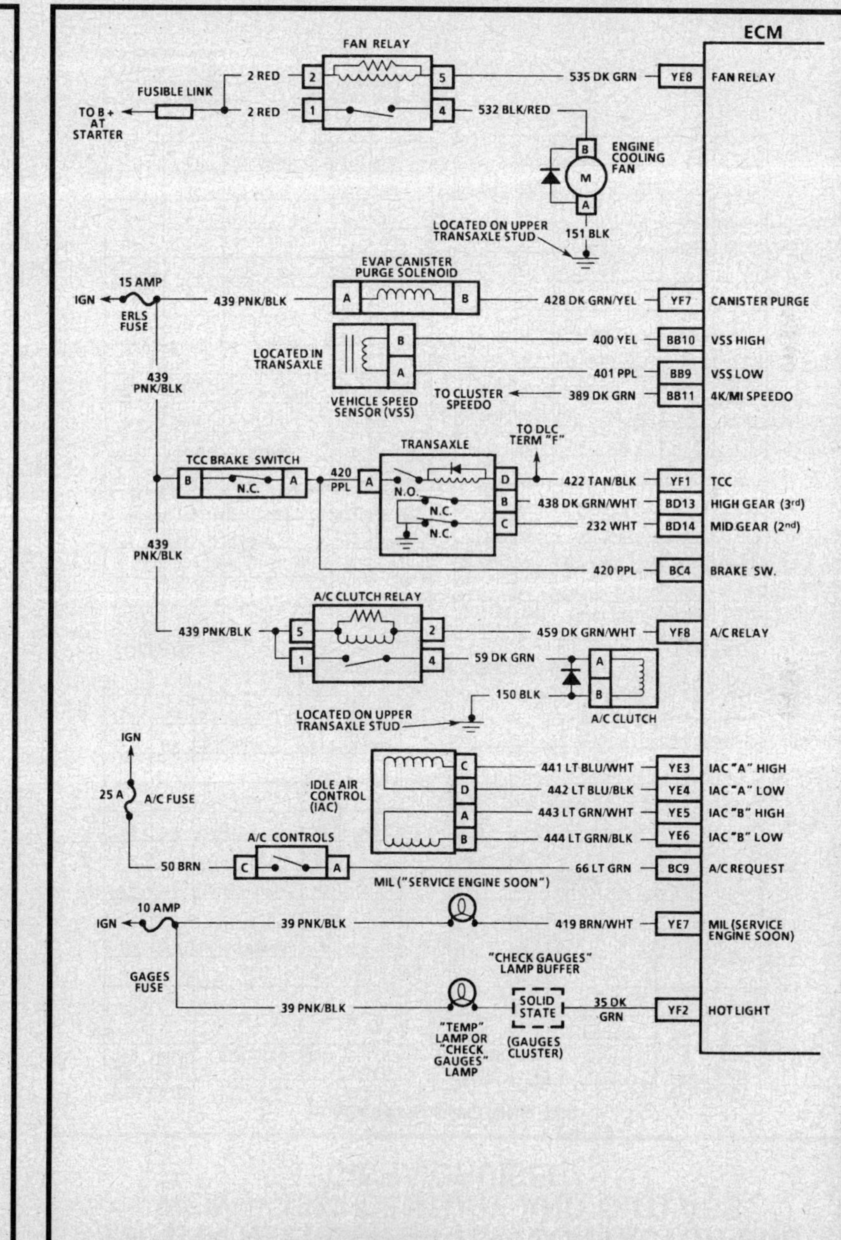

3.3L (VIN N) ENGINE — ECM WIRING SCHEMATIC — 1993–94 ACHIEVA, GRAND AM AND SKYLARK

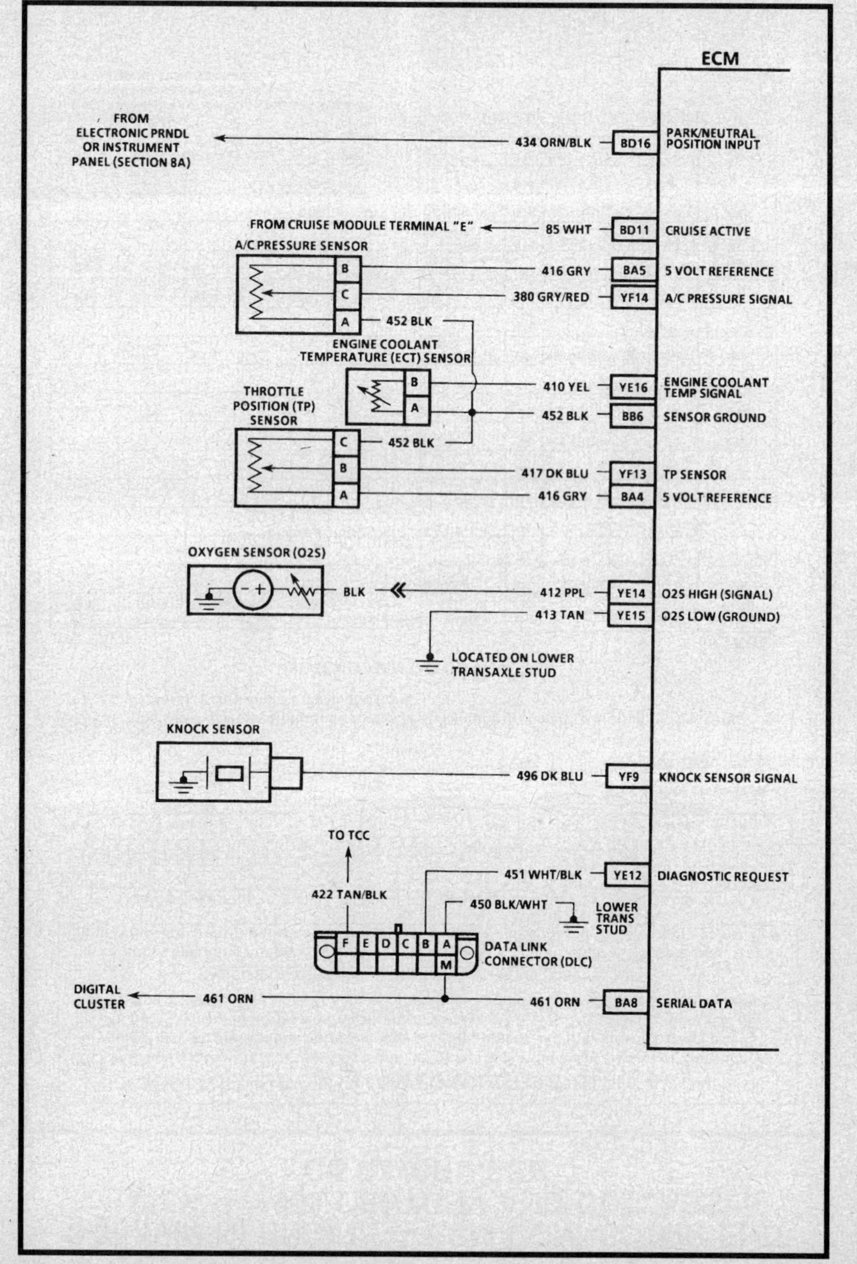

3.3L (VIN N) ENGINE — ECM CONNECTOR END VIEW — 1992 CENTURY AND CUTLASS CIERA/CRUISER

PORT FUEL INJECTION ECM CONNECTOR IDENTIFICATION

This ECM voltage chart is for use with a digital voltmeter to further aid in diagnosis. The voltages you get may vary due to low battery charge or other reasons, but they should be very close. The "B +" symbol indicates a nominal system voltage of 12-14 V. Do not probe ECM side of connector. Probing-front side (End View) of connector could cause intermittent open circuits.

THE FOLLOWING CONDITIONS MUST BE MET BEFORE TESTING:

● Engine at operating temperature (upper rad. hose hot) ● Engine idling in "Closed Loop" (For "Engine Run" column) in park or neutral ● Test terminal not grounded ● "SCAN" tool not installed

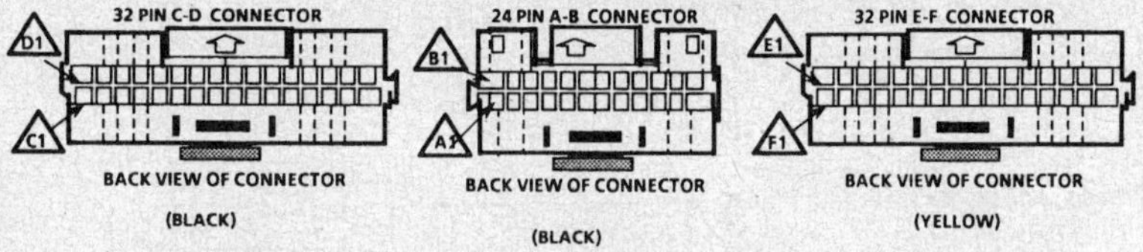

32 PIN C-D CONNECTOR — BACK VIEW OF CONNECTOR (BLACK)

24 PIN A-B CONNECTOR — BACK VIEW OF CONNECTOR (BLACK)

32 PIN E-F CONNECTOR — BACK VIEW OF CONNECTOR (YELLOW)

NOTICE: Before checking voltages, be sure ECM and engine grounds are located on the correct transaxle stud and are clean and tight.

BLACK 32 PIN C-D CONNECTOR

VOLTAGE				
KEY "ON"	ENG. RUN	CIRCUIT	PIN	WIRE COLOR
			C1	
			C2	
			C3	
OPEN 0★ CLOSE B+	OPEN 0★ CLOSE B+	BRAKE SWITCH	C4	PPL
			C5	
			C6	
0★	5V	BYPASS	C7	TAN/BLK
0★	1.5V	EST	C8	WHT
ON B+ OFF 0★	ON B+ OFF 0★	A/C REQUEST	C9	LT BLU
			C10	
B+	B+	INJ. 1, 3, 5	C11	DK BLU
B+	B+	INJ. 2, 4, 6	C12	DK GRN
			C13	
			C14	
			C15	
B+	B+	BATTERY FEED	C16	ORN

WIRE COLOR	PIN	CIRCUIT	KEY "ON"	ENG. RUN	
BLK/WHT	D1	ECM GROUND	0★	0★	
	D2				
	D3				
	D4				
	D5				
BLK/WHT	D6	INJ. GROUND	0★	0★	
BLK/WHT	D7	INJ. GROUND	0★	0★	
PPL/WHT	D8	REFERENCE HIGH	0★	3.5V	
BLK/RED	D9	REFERENCE LOW	0★	0★	
	D10				
YEL	D11	CRUISE ACTIVE	B+	B+	
LT BLU/ORN	D12	PS PS	B+	OPEN B+ CLOSED 0★	
LT BLU (4T60) ORN (3T40)	D13	HI GEAR SW.	0★	0★	②
DK GRN	D14	MID GEAR SW.	0★	0★	③
	D15				
ORN/BLK	D16	P/N SWITCH	0★	0★	①

ENGINE 3300 "A" CARLINE

★ Less than .5V (500mv)
1. 0 VOLT WHEN IN PARK/NUETRAL, B+ WHEN IN R-DL.
2. B+ WHEN HIGH GEAR IS ENGAGED.
3. B+ WHEN MID GEAR IS ENGAGED.

3.3L (VIN N) ENGINE — ECM CONNECTOR END VIEW — 1992 CENTURY AND CUTLASS CIERA/CRUISER

BLACK 24 PIN A-B CONNECTOR

VOLTAGE				
KEY "ON"	ENG. RUN	CIRCUIT	PIN	WIRE COLOR
			A1	
			A2	
			A3	
5.0	5.0	5 V REFERENCE	A4	GRY
			A5	
B+	B+	IGNITION FEED	A6	PNK/BLK
			A7	
3.5	3.5	SERIAL DATA	A8	ORN
			A9	
			A10	
★0 ③	B+	FUEL PUMP	A11	DK GRN/WHT
0★	★0	ECM GROUND	A12	BLK/WHT

WIRE COLOR	PIN	CIRCUIT	KEY "ON"	ENG. RUN
ORN	B1	BATTERY FEED	B+	B+
	B2			
	B3			
	B4			
	B5			
BLK	B6	SENSOR GROUND	0★	0★
	B7			
	B8			
PPL	B9	VSS LOW	0★	0★
YEL	B10	VSS HI	★0	★0
BRN	B11	4 K SPEEDO	0★-B+	0★-B+
RED	B12	2K CRUISE	0★-B+	0★-B+

YELLOW 32 PIN E-F CONNECTOR

VOLTAGE					
KEY "ON"	ENG. RUN	CIRCUIT	PIN	WIRE COLOR	
			E1		
			E2		
NOT	USEABLE	IAC A HI	E3	LT BLU/WHT	
NOT	USEABLE	IAC A LOW	E4	LT BLU/BLK	
NOT	USEABLE	IAC B HI	E5	LT GRN/WHT	
NOT	USEABLE	IAC B LOW	E6	LT GRN/BLK	
★0	B+	SES LIGHT	E7	BRN/WHT	
B+	ON 0★ OFF B+	COOLING FAN	E8	DK GRN	
			E9		
			E10		
			E11		
5.0	5.0	DIAG/ALDL	E12	WHT/BLK	
			E13		
0★	.1-.9	O_2 HI	E14	PPL	①
0★	0★	O_2 LOW	E15	TAN	
1.8	1.8	CTS SIGNAL	E16	YEL	②

WIRE COLOR	PIN	CIRCUIT	KEY "ON"	ENG. RUN	
TAN/BLK	F1	TCC	B+	B+ ④	
DK GRN	F2	HOT LIGHT	0★	ON 0★ OFF B+	
	F3				
	F4				
	F5				
	F6				
DK GRN/YEL	F7	CANISTER PURGE	B+	★0-B+	①
LT GRN/BLK	F8	A/C RELAY	B+	ON 0★ OFF B+	
DK BLU	F9	ESC SIGNAL	2.4	2.4	
YEL	F10	MAF SIGNAL	5.0	2.4	
	F11				
	F12				
DK BLU	F13	TPS SIGNAL	.33-.46	.33-.46	
	F14				
	F15				
	F16				

★★ , VARIES AROUND 10 VOLTS.
★ Less than .5V (500 mv).
① Varies within this range
② Varies with temperature
③ B+ for 2 sec. key up
④ 0★ With brake applied

ENGINE - 3300
A - Carline

3.3L (VIN N) ENGINE — ECM CONNECTOR END VIEW — 1993–94 CENTURY AND CUTLASS CIERA/CRUISER

PORT FUEL INJECTION ECM CONNECTOR IDENTIFICATION

This ECM voltage chart is for use with a high impedance digital multimeter (J 39200) to further aid in diagnosis. The voltages you get may vary due to low battery charge or other reasons, but they should be very close.

The "B +" symbol indicates a nominal system voltage of 12-14 V. Do not probe ECM side of connector. Probing-front side (End View) of connector could cause intermittent open circuits.

THE FOLLOWING CONDITIONS MUST BE MET BEFORE TESTING:

● Engine at operating temperature (upper rad. hose hot) ● Engine idling in "Closed Loop" (For "Engine Run" column) in park or neutral ● Test terminal not grounded ● SCAN tool not installed ● Brake not applied ● All accessories "OFF."

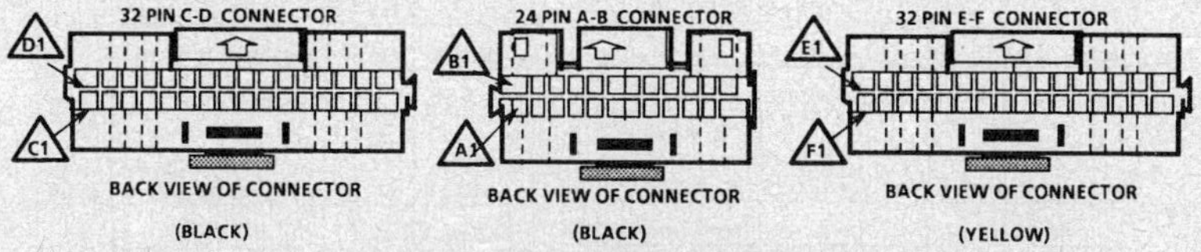

32 PIN C-D CONNECTOR — BACK VIEW OF CONNECTOR (BLACK)
24 PIN A-B CONNECTOR — BACK VIEW OF CONNECTOR (BLACK)
32 PIN E-F CONNECTOR — BACK VIEW OF CONNECTOR (YELLOW)

NOTICE: Before checking voltages, be sure ECM and engine grounds are located on the correct transaxle stud and are clean and tight.

BLACK 32 PIN C-D CONNECTOR

WIRE COLOR	PIN	FUNCTION AND CIRCUIT #	KEY "ON"	ENG. RUN
	C1			
	C2			
	C3			
PPL	C4	BRAKE SWITCH (420)	B+	B+
	C5			
	C6			
TAN/BLK	C7	BYPASS (424)	0*	5
WHT	C8	IGNITION CONTROL (423)	0*	1.5
LT BLU	C9	A/C REQUEST (67)	0*	0*
	C10			
DK BLU	C11	INJ. 1, 3, 5 (467)	B+	B+
DK GRN	C12	INJ. 2, 4, 6 (468)	B+	B+
	C13			
	C14			
	C15			
ORN	C16	BATTERY FEED (440)	B+	B+

WIRE COLOR	PIN	FUNCTION AND CIRCUIT #	KEY "ON"	ENG. RUN
BLK/WHT	D1	ECM GROUND (450)	0*	0*
	D2			
	D3			
	D4			
	D5			
BLK/WHT	D6	INJ. GROUND (450)	0*	0*
BLK/WHT	D7	INJ. GROUND (450)	0*	0*
PPL/WHT	D8	REFERENCE HIGH (430)	0*	3.5
BLK/RED	D9	REFERENCE LOW (453)	0*	0*
	D10			
YEL	D11	CRUISE ACTIVE (494)	B+	B+
LT BLU/ORN	D12	PS PS (495)	B+	0* B+
LT BLU	D13	HIGH GEAR SWITCH (446)	0*	0*
DK GRN	D14	MID GEAR SWITCH (108)	0*	0*
	D15			
ORN/BLK	D16	PNP SWITCH (434)	(1)	(1)

ENGINE 3300 "A" CARLINE

(1) 0 volts when in P-N, B + when in R-DL.
(2) B + when high gear is engaged.
(3) B + when mid gear is engaged.
* Less than .5 volt (500 mV).

3.3L (VIN N) ENGINE — ECM CONNECTOR END VIEW — 1993–94 CENTURY AND CUTLASS CIERA/CRUISER

BLACK 24 PIN A-B CONNECTOR

WIRE COLOR	PIN	FUNCTION AND CIRCUIT #	KEY "ON"	ENG. RUN
	A1			
	A2			
	A3			
GRY	A4	5 VOLT REFERENCE (416)	5	5
	A5			
PNK/BLK	A6	IGNITION FEED (439)	B+	B+
	A7			
ORN	A8	SERIAL DATA (461)	3.5	3.5
	A9			
	A10			
DK GRN/WHT	A11	FUEL PUMP (465)	0*	B+
BLK/WHT	A12	ECM GROUND (450)	0*	0*

WIRE COLOR	PIN	FUNCTION AND CIRCUIT #	KEY "ON"	ENG. RUN
ORN	B1	BATTERY FEED (440)	B+	B+
	B2			
	B3			
	B4			
	B5			
BLK	B6	SENSOR GROUND (452)	0*	0*
	B7			
	B8			
PPL	B9	VSS LOW (401)	0*	0*
YEL	B10	VSS HI (400)	0*	0*
DK GRN/WHT	B11	4 K SPEED (817)	0*-B+	0*-B+
	B12			

YELLOW 32 PIN E-F CONNECTOR

WIRE COLOR	PIN	FUNCTION AND CIRCUIT #	KEY "ON"	ENG. RUN
	E1			
	E2			
LT BLU/WHT	E3	IAC A HI (441)		
LT BLU/BLK	E4	IAC A LOW (442)		
LT GRN/WHT	E5	IAC B HI (443)		
LT GRN/BLK	E6	IAC B LOW (444)		
BRN/WHT	E7	MIL ("SERVICE ENGINE SOON") (419)	0*	B+
DK GRN	E8	COOLING FAN RELAY (535)	B+	ON 0* OFF B+
	E9			
	E10			
	E11			
WHT/BLK	E12	DIAG. REQUEST (451)	5	5
	E13			
PPL	E14	O2S HI (412)	.45	.1-.9 (1)
TAN	E15	O2S LOW (413)	0*	0*
YEL	E16	ECT SENSOR (410)	1.8 (2)	1.8 (2)

WIRE COLOR	PIN	FUNCTION AND CIRCUIT #	KEY "ON"	ENG. RUN
TAN/BLK	F1	TCC (422)	B+ (4T60) 0* (3T40)	B+ (4T60) 0* (3T40)
DK GRN	F2	HOT LIGHT (35)	0*	B+
	F3			
	F4			
	F5			
	F6			
DK GRN/YEL	F7	CANISTER PURGE (428)	B+	0*-B+ (1)
LT GRN/BLK	F8	A/C RELAY (366)	B+	B+
DK BLU	F9	KNOCK SENSOR (496)	2.4	2.4
YEL	F10	MAF (492)	5	2.4
	F11			
	F12			
DK BLU	F13	TP SENSOR (417)	.2-.74	.2-.74
	F14			
	F15			
	F16			

(1) Varies within this range.
(2) Varies with temperature.
* Less than .5 volt (500 mV).

ENGINE - 3300
A - Carline

3.3L (VIN N) ENGINE — ECM CONNECTOR END VIEW — 1992 ACHIEVA, GRAND AM AND SKYLARK

PORT FUEL INJECTION ECM CONNECTOR IDENTIFICATION

This ECM voltage chart is for use with a digital voltmeter to further aid in diagnosis. The voltages you get may vary due to low battery charge or other reasons, but they should be very close. The "B + " symbol indicates a nominal system voltage of 12-14 V. Probing front side (end view) of connector could cause intermittent open circuits.

THE FOLLOWING CONDITIONS MUST BE MET BEFORE TESTING:

• Engine at operating temperature (upper rad. hose hot) • Engine idling in "Closed Loop" (For "Engine Run" column) in park or neutral • Brake not applied • Test terminal not grounded • "SCAN" tool not installed

NOTICE: Before checking voltages be sure ECM and engine grounds are located on the correct transaxle stud and are clean and tight.

32 PIN C-D CONNECTOR — BACK VIEW OF CONNECTOR (BLACK)

24 PIN A-B CONNECTOR — BACK VIEW OF CONNECTOR (BLACK)

32 PIN E-F CONNECTOR — BACK VIEW OF CONNECTOR (YELLOW)

BLACK 32 PIN C-D CONNECTOR

VOLTAGE				
KEY "ON"	ENG. RUN	CIRCUIT	PIN	WIRE COLOR
			C1	
			C2	
			C3	
B +	B +	BRAKE	C4	PPL
			C5	
			C6	
0*	4.8	BYPASS	C7	TAN/BLK
0*	1.2-1.5	EST	C8	WHT
0* off B + on	0* off B + on	A/C REQUEST	C9	LT GRN
			C10	
B +	B +	INJ. DRIVER 1,3,5	C11	DK BLU
B +	B +	INJ. DRIVER 2,4,6	C12	DK GRN
			C13	
			C14	
			C15	
B +	B +	BATTERY	C16	ORN

(Note [1] at pin C8)

WIRE COLOR	PIN	CIRCUIT	VOLTAGE	
			KEY "ON"	ENG. RUN
TAN/WHT	D1	POWER GROUND	0*	0*
	D2			
	D3			
	D4			
	D5			
BLK/WHT	D6	INJECTOR GROUND	0*	0*
BLK/WHT	D7	INJECTOR GROUND	0*	0*
PPL/WHT	D8	REFERENCE	0*	3.6
BLK/RED	D9	REFERENCE LOW (GROUND)	0*	0*
	D10			
WHT	D11	CRUISE ACTIVE	0 - B +	0 - B +
	D12			
DK GRN/WHT	D13	3rd GEAR	0*	0*
WHT	D14	2nd GEAR	0*	0*
	D15			
ORN/BLK	D16	P/N INPUT	.5	.6

* Less than .5V (500 mV).
1. Varies within this range.

ENGINE 3300
VIN N
N CARLINE

3.3L (VIN N) ENGINE — ECM CONNECTOR END VIEW — 1992 ACHIEVA, GRAND AM AND SKYLARK

BLACK 24 PIN A-B CONNECTOR

VOLTAGE				
KEY "ON"	ENG. RUN	CIRCUIT	PIN	WIRE COLOR
			A1	
			A2	
			A3	
5	5	5 VOLTS REF	A4	GRY
5	5	5 VOLTS REF	A5	GRY
B +	B +	IGNITION "ON" INPUT	A6	PNK/BLK
			A7	
0*-5	0*-5	SERIAL DATA/ALDL	A8	ORN
			A9	
			A10	
0*	B +	FUEL PUMP	A11	DK GRN/WHT
0*	0*	ECM GROUND	A12	BLK/WHT

(Note [3] at pin A11)

WIRE COLOR	PIN	CIRCUIT	VOLTAGE	
			KEY "ON"	ENG. RUN
ORN	B1	BATTERY	B +	B +
	B2			
	B3			
	B4			
	B5			
BLK	B6	TPS, COOLANT A/C SENSOR GROUND	0*	0*
	B7			
	B8			
PPL	B9	VSS LOW	0*	0*
YEL	B10	VSS HIGH	0* · B +	0* · B +
DK GRN	B11	4K/MI SPEEDO	8	9
	B12			

YELLOW 32 PIN E-F CONNECTOR

VOLTAGE				
KEY "ON"	ENG. RUN	CIRCUIT	PIN	WIRE COLOR
			E1	
			E2	
	NOT USEABLE	IAC A HI	E3	LT BLU/WHT
	NOT USEABLE	IAC A LO	E4	LT BLU/BLK
	NOT USEABLE	IAC B HI	E5	LT GRN/WHT
	NOT USEABLE	IAC B LO	E6	LT GRN/BLK
0*	B +	SES LIGHT	E7	BRN/WHT
B +	0* ON B + OFF	FAN RELAY	E8	DK GRN
			E9	
			E10	
			E11	
5	5	DIAGNOSTIC REQUEST	E12	WHT/BLK
			E13	
.45	.1-.9	O_2 HIGH (SIGNAL)	E14	PPL
0*	0*	O_2 LOW (GROUND)	E15	TAN
1.8	1.8	CTS SIGNAL	E16	YEL

(Note [2] at pins E14 and E16)

WIRE COLOR	PIN	CIRCUIT	VOLTAGE	
			KEY "ON"	ENG. RUN
TAN/BLK	F1	TCC	0*	0*
DK GRN	F2	HOT LIGHT	0*	B +
	F3			
	F4			
	F5			
	F6			
DK GRN/YEL	F7	CANISTER PURGE	B +	0*-B +
DK GRN/WHT	F8	A/C RELAY	B +	0* ON B + OFF
DK BLU	F9	ESC SIGNAL	2.4	2.4
YEL	F10	MAF SIGNAL	5.0	2.4
	F11			
	F12			
DK BLU	F13	TPS SIGNAL	.33-.46	.33-.46
GRY/RED	F14	A/C PRESSURE SIGNAL	1	.2 - 4.5
	F15			
	F16			

(Note [1] at pin F14)

* LESS THAN .5v (500 mv).
1. VARIES WITHIN THIS RANGE.
2. VARIES WITH TEMPERATURE.
3. B + FOR 1st 2 SECONDS.

ENGINE 3300
VIN N
N CARLINE

3.3L (VIN N) ENGINE — ECM CONNECTOR END VIEW — 1993–94 ACHIEVA, GRAND AM AND SKYLARK

PORT FUEL INJECTION ECM CONNECTOR IDENTIFICATION

This ECM voltage chart is for use with a high impedance digital multimeter J 39200 to further aid in diagnosis. The voltages you get may vary due to low battery charge or other reasons, but they should be very close.

The "B +" symbol indicates a nominal system voltage of 12-14 V. Probing front side (end view) of connector could cause intermittent open circuits.

THE FOLLOWING CONDITIONS MUST BE MET BEFORE TESTING:

- Engine at operating temperature (upper rad. hose hot) • Engine idling in "Closed Loop" (For "Engine Run" column) in park or neutral • Brake not applied • Test terminal not grounded • SCAN tool not installed

NOTICE: Before checking voltages be sure ECM and engine grounds are located on the correct transaxle stud and are clean and tight.

32 PIN C-D CONNECTOR — BACK VIEW OF CONNECTOR (BLACK)

24 PIN A-B CONNECTOR — BACK VIEW OF CONNECTOR (BLACK)

32 PIN E-F CONNECTOR — BACK VIEW OF CONNECTOR (YELLOW)

BLACK 32 PIN C-D CONNECTOR

Note	KEY "ON"	ENG. RUN	CIRCUIT	PIN	WIRE COLOR	WIRE COLOR	PIN	CIRCUIT	KEY "ON"	ENG. RUN
				C1		TAN/WHT	D1	POWER GROUND	0*	0*
				C2			D2			
				C3			D3			
	B +	B +	BRAKE	C4	PPL		D4			
				C5			D5			
				C6		BLK/WHT	D6	INJECTOR GROUND	0*	0*
	0*	4.8	BYPASS	C7	TAN/BLK	BLK/WHT	D7	INJECTOR GROUND	0*	0*
(1)	0*	1.2-1.5	IGN. CONTROL	C8	WHT	PPL/WHT	D8	REFERENCE	0*	3.6
	0* off B + on	0* off B + on	A/C REQUEST	C9	LT GRN	BLK/RED	D9	REFERENCE LOW (GROUND)	0*	0*
				C10			D10			
	B +	B +	INJ. DRIVER 1,3,5	C11	DK BLU	WHT	D11	CRUISE ACTIVE	0 - B +	0 - B + -
	B +	B +	INJ. DRIVER 2,4,6	C12	DK GRN		D12			
				C13		DK GRN/WHT	D13	3rd GEAR SWITCH	0*	0*
				C14		WHT	D14	2nd GEAR SWITCH	0*	0*
				C15			D15			
	B +	B +	BATTERY FEED	C16	ORN	ORN/BLK	D16	PNP INPUT	1.0	1.0

(1) Varies within this range.
* Less than .5 volt (500 mV).

ENGINE 3300
VIN N
N CARLINE

BLACK 24 PIN A-B CONNECTOR

Note	KEY "ON"	ENG. RUN	CIRCUIT	PIN	WIRE COLOR	WIRE COLOR	PIN	CIRCUIT	KEY "ON"	ENG. RUN
				A1		ORN	B1	BATTERY FEED	B +	B +
				A2			B2			
				A3			B3			
	5	5	5 VOLTS REF	A4	GRY		B4			
	5	5	5 VOLTS REF	A5	GRY		B5			
	B +	B +	IGNITION FEED	A6	PNK/BLK	BLK	B6	THROTTLE POSITION, ECT, A/C SENSOR GROUND	0*	0*
				A7			B7			
	0*-5	0*-5	SERIAL DATA	A8	ORN		B8			
				A9		PPL	B9	VSS LOW	0*	0*
				A10		YEL	B10	VSS HIGH	0* - B +	0* - B +
(3)	0*	B +	FUEL PUMP RELAY	A11	DK GRN/WHT	DK GRN	B11	4K/MI SPEEDO	8	9
	0*	0*	ECM GROUND	A12	BLK/WHT		B12			

YELLOW 32 PIN E-F CONNECTOR

Note	KEY "ON"	ENG. RUN	CIRCUIT	PIN	WIRE COLOR	WIRE COLOR	PIN	CIRCUIT	KEY "ON"	ENG. RUN
				E1		TAN/BLK	F1	TCC	0*	0*
				E2		DK GRN	F2	HOT LIGHT	0*	B +
		NOT USEABLE	IAC A HI	E3	LT BLU/WHT		F3			
		NOT USEABLE	IAC A LO	E4	LT BLU/BLK		F4			
		NOT USEABLE	IAC B HI	E5	LT GRN/WHT		F5			
		NOT USEABLE	IAC B LO	E6	LT GRN/BLK		F6			
	0*	B +	MIL ("SERVICE ENGINE SOON")	E7	BRN/WHT	DK GRN/YEL	F7	EVAP CANISTER PURGE	B +	0*-B+
	B +	0* ON B + OFF	FAN RELAY	E8	DK GRN	DK GRN/WHT	F8	A/C RELAY	B +	0* ON B + OFF
				E9		DK BLU	F9	KNOCK SENSOR SIGNAL	2.4	2.4
				E10		YEL	F10	MAF SIGNAL	5.0	2.4
				E11			F11			
	5	5	DIAGNOSTIC REQUEST	E12	WHT/BLK		F12			
				E13		DK BLU	F13	TP SENSOR SIGNAL	.20-.74	.20-.74
(2)	.45	.1-.9	O2S HIGH (SIGNAL)	E14	PPL	GRY/RED	F14	A/C PRESSURE SIGNAL	1	.2 - 4.5
	0*	0*	O2S LOW (GROUND)	E15	TAN		F15			
(2)	1.8	1.8	ECT SIGNAL	E16	YEL		F16			

* LESS THAN .5v (500 mv).
1. VARIES WITHIN THIS RANGE.
2. VARIES WITH TEMPERATURE.
3. B + FOR FIRST 2 SECONDS.

ENGINE 3300
VIN N
N CARLINE

3.3L (VIN N) ENGINE — ON-BOARD DIAGNOSTIC SYSTEM CHART — CENTURY, CUTLASS CIERA/CRUISER, ACHIEVA, GRAND AM AND SKYLARK

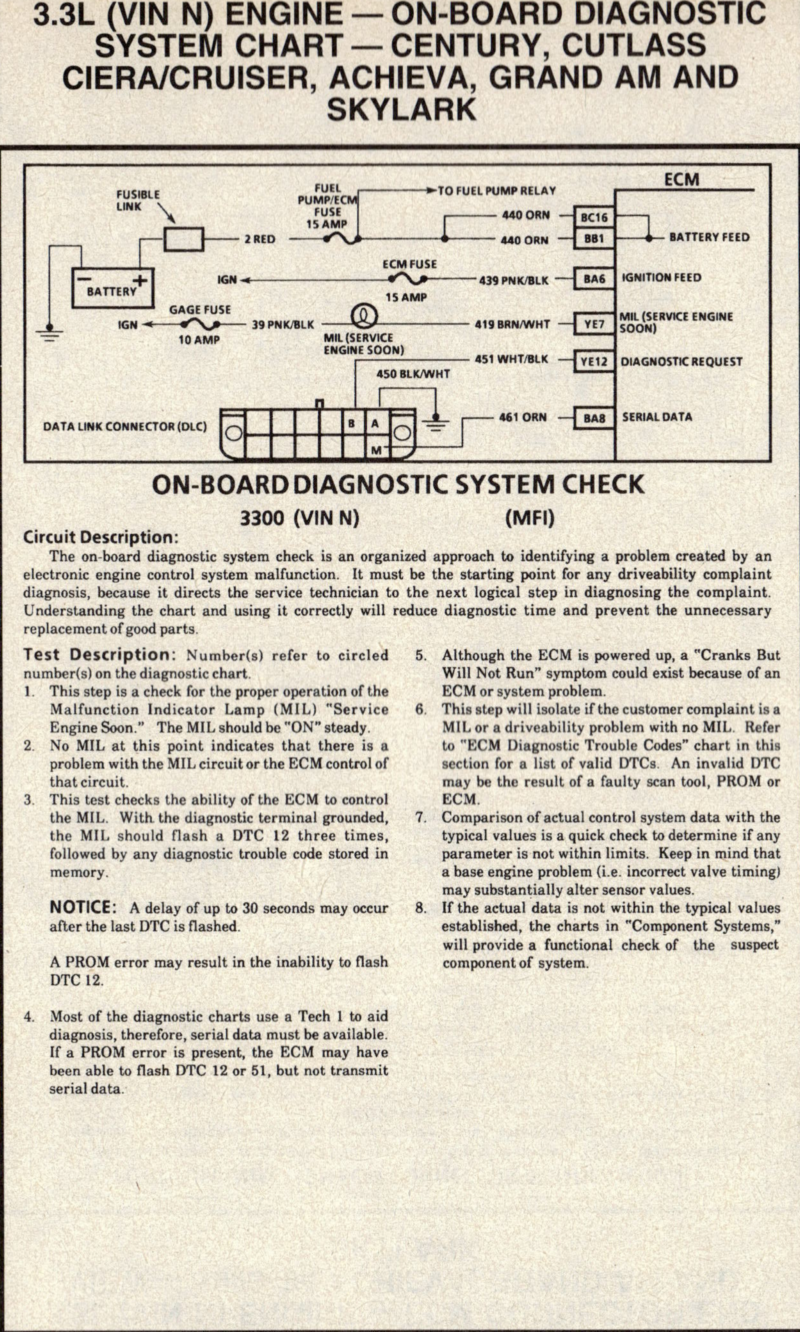

ON-BOARD DIAGNOSTIC SYSTEM CHECK
3300 (VIN N) (MFI)

Circuit Description:

The on-board diagnostic system check is an organized approach to identifying a problem created by an electronic engine control system malfunction. It must be the starting point for any driveability complaint diagnosis, because it directs the service technician to the next logical step in diagnosing the complaint. Understanding the chart and using it correctly will reduce diagnostic time and prevent the unnecessary replacement of good parts.

Test Description: Number(s) refer to circled number(s) on the diagnostic chart.

1. This step is a check for the proper operation of the Malfunction Indicator Lamp (MIL) "Service Engine Soon." The MIL should be "ON" steady.
2. No MIL at this point indicates that there is a problem with the MIL circuit or the ECM control of that circuit.
3. This test checks the ability of the ECM to control the MIL. With the diagnostic terminal grounded, the MIL should flash a DTC 12 three times, followed by any diagnostic trouble code stored in memory.

   NOTICE: A delay of up to 30 seconds may occur after the last DTC is flashed.

   A PROM error may result in the inability to flash DTC 12.

4. Most of the diagnostic charts use a Tech 1 to aid diagnosis, therefore, serial data must be available. If a PROM error is present, the ECM may have been able to flash DTC 12 or 51, but not transmit serial data.

5. Although the ECM is powered up, a "Cranks But Will Not Run" symptom could exist because of an ECM or system problem.
6. This step will isolate if the customer complaint is a MIL or a driveability problem with no MIL. Refer to "ECM Diagnostic Trouble Codes" chart in this section for a list of valid DTCs. An invalid DTC may be the result of a faulty scan tool, PROM or ECM.
7. Comparison of actual control system data with the typical values is a quick check to determine if any parameter is not within limits. Keep in mind that a base engine problem (i.e. incorrect valve timing) may substantially alter sensor values.
8. If the actual data is not within the typical values established, the charts in "Component Systems," will provide a functional check of the suspect component of system.

3.3L (VIN N) ENGINE — ON-BOARD DIAGNOSTIC SYSTEM CHART — CENTURY, CUTLASS CIERA/CRUISER, ACHIEVA, GRAND AM AND SKYLARK

ON-BOARD DIAGNOSTIC SYSTEM CHECK
3300 (MFI)

3.3L (VIN N) ENGINE — SYSTEM DIAGNOSTIC CHARTS — CENTURY, CUTLASS CIERA/CRUISER, ACHIEVA, GRAND AM AND SKYLARK

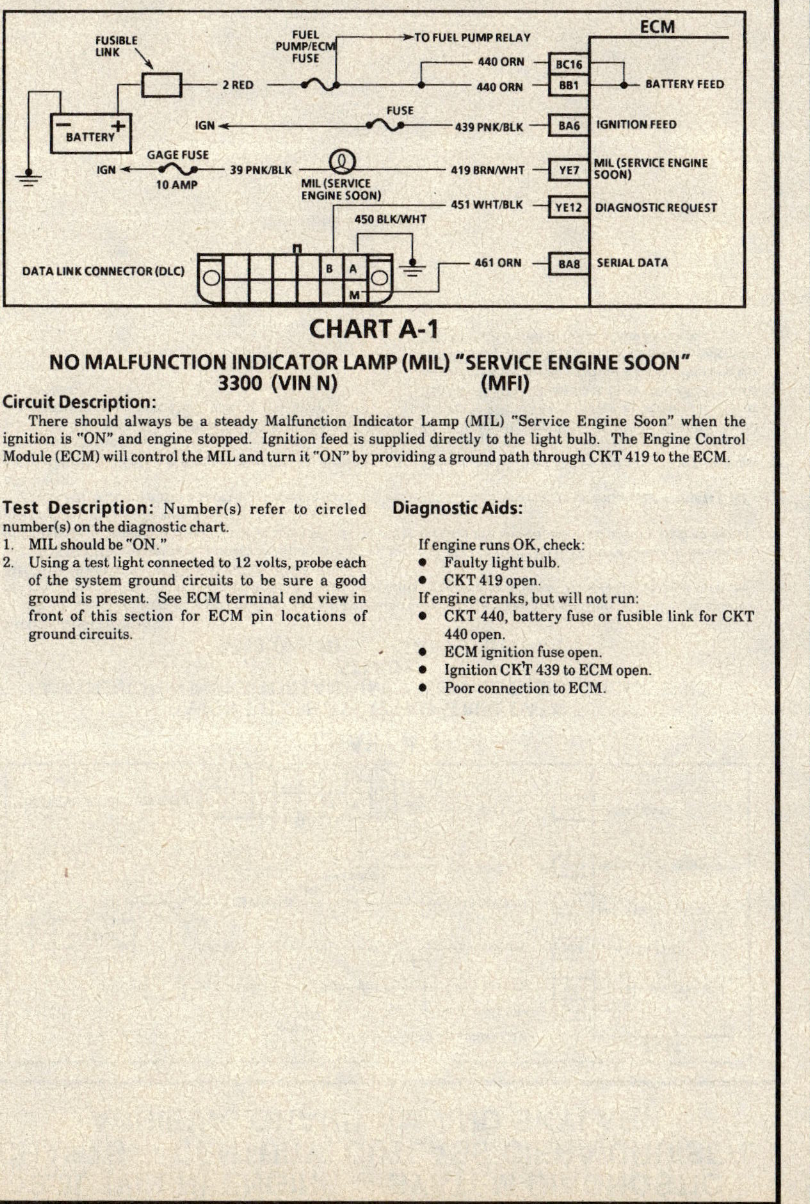

CHART A-1
NO MALFUNCTION INDICATOR LAMP (MIL) "SERVICE ENGINE SOON"
3300 (VIN N)   (MFI)

Circuit Description:

There should always be a steady Malfunction Indicator Lamp (MIL) "Service Engine Soon" when the ignition is "ON" and engine stopped. Ignition feed is supplied directly to the light bulb. The Engine Control Module (ECM) will control the MIL and turn it "ON" by providing a ground path through CKT 419 to the ECM.

Test Description: Number(s) refer to circled number(s) on the diagnostic chart.

1. MIL should be "ON."
2. Using a test light connected to 12 volts, probe each of the system ground circuits to be sure a good ground is present. See ECM terminal end view in front of this section for ECM pin locations of ground circuits.

Diagnostic Aids:

If engine runs OK, check:
- Faulty light bulb.
- CKT 419 open.

If engine cranks, but will not run:
- CKT 440, battery fuse or fusible link for CKT 440 open.
- ECM ignition fuse open.
- Ignition CKT 439 to ECM open.
- Poor connection to ECM.

3.3L (VIN N) ENGINE — SYSTEM DIAGNOSTIC CHARTS — CENTURY, CUTLASS CIERA/CRUISER, ACHIEVA, GRAND AM AND SKYLARK

CHART A-1
NO MALFUNCTION INDICATOR LAMP (MIL) "SERVICE ENGINE SOON"
3300 (VIN N)   (MFI)

DOES THE ENGINE START?

NO →
ARE THE ECM BATTERY AND IGNITION FUSES OK?

- YES →
  - IGNITION "OFF."
  - DISCONNECT ECM CONNECTORS.
  - IGNITION "ON."
  - PROBE CKTS 440 & 439 WITH TEST LIGHT TO GROUND.
  - TEST LIGHT SHOULD BE "ON" ON BOTH CIRCUITS. IS IT?
    - YES → (2) FAULTY ECM CONNECTIONS, GROUNDS, OR ECM.
    - NO → REPAIR OPEN IN CIRCUIT THAT DID NOT LIGHT THE TEST LIGHT.
- NO → LOCATE AND CORRECT SHORT TO GROUND.

YES → (1)
- IGNITION "OFF."
- DISCONNECT ECM CONNECTORS.
- IGNITION "ON."
- JUMPER CKT 419 TO GROUND.
- MALFUNCTION INDICATOR LAMP ("SERVICE ENGINE SOON") SHOULD BE "ON." IS IT?
  - YES → FAULTY ECM CONNECTION OR ECM.
  - NO → CHECK:
    - FAULTY BULB
    - OPEN CKT 419
    - CKT 419 SHORTED TO VOLTAGE
    - OPEN IGNITION FEED TO BULB.

"AFTER REPAIRS," CONFIRM "CLOSED LOOP" OPERATION AND NO MIL (SERVICE ENGINE SOON).

3.3L (VIN N) ENGINE — SYSTEM DIAGNOSTIC CHARTS — CENTURY, CUTLASS CIERA/CRUISER, ACHIEVA, GRAND AM AND SKYLARK

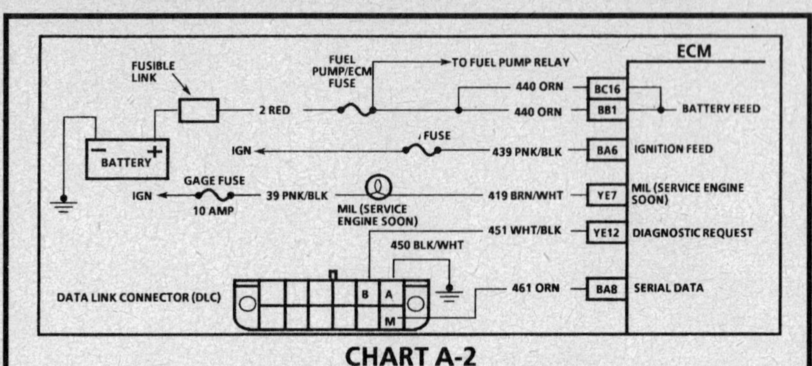

CHART A-2
WON'T FLASH DTC 12 - NO SERIAL DATA
MALFUNCTION INDICATOR LAMP (MIL) "SERVICE ENGINE SOON" "ON"
STEADY
3300 (VIN N)　　　　(MFI)

Circuit Description:

There should always be a steady Malfunction Indicator Lamp (MIL) when the ignition is "ON" and engine not running. Ignition #1 is supplied directly to the light bulb. The Engine Control Module (ECM) will turn the MIL "ON" by grounding CKT 419 at the ECM.

With the DLC diagnostic request terminal grounded, the MIL should flash DTC 12 three times, followed by any diagnostic trouble code(s) stored in memory. If no other DTC(S) are stored, DTC 12 will be flashed again after a delay of up to 30 seconds.

A steady MIL suggests a short to ground in the lamp control CKT 419 or an open in diagnostic request CKT 451.

Test Description: Number(s) refer to circled number(s) on the diagnostic chart.

1. If the MIL goes "OFF" when the ECM connector is disconnected, then CKT 419 is not shorted to ground.

2. This step will check for an open diagnostic request CKT 451.

3. At this point, the MIL (Service Engine Soon) wiring is OK. The problem is a faulty ECM or PROM. If DTC 12 does not flash, the ECM should be replaced using the original PROM. Replace the PROM only after trying an ECM, as a defective PROM is an unlikely cause of the problem.

NOTICE: WITH DIAGNOSTIC REQUEST CKT 451 GROUNDED, A DELAY MAY OCCUR AFTER DTC 12 HAS FLASHED THREE TIMES. THIS DOES NOT INDICATE A PROBLEM WITH THE MIL (SERVICE ENGINE SOON) CONTROL CIRCUITS OR THE ECM.

CHART A-2
WON'T FLASH DTC 12 - NO SERIAL DATA
MALFUNCTION INDICATOR LAMP (MIL) "SERVICE ENGINE SOON" "ON"
STEADY
3300 (VIN N)　　　　(MFI)

(1)
- IGNITION "OFF."
- DISCONNECT ECM CONNECTORS.
- IGNITION "ON."
- MIL ("SERVICE ENGINE SOON") SHOULD BE "OFF." IS IT?

YES → **(2)**

NO → REPAIR SHORT TO GROUND IN CKT 419.

(2)
- IGNITION "OFF."
- RECONNECT ECM.
- IGNITION "ON," ENGINE STOPPED.
- BACK PROBE ECM TERMINAL "YE12" (CKT 451) WITH JUMPER TO GROUND.
- MIL SHOULD FLASH DTC 12. DOES IT?

NO → **(3)**

YES →
- CHECK FOR OPEN IN CKT 451 BETWEEN DLC TERM. "B" AND ECM.
- IF OK, CHECK FOR OPEN BETWEEN DLC CKT 450 AND GROUND.

(3)
- CHECK PROM FOR PROPER INSTALLATION.
- IF OK, REPLACE ECM USING ORIGINAL PROM.
- GROUND DLC DIAGNOSTIC REQUEST TERMINAL OR ENTER FIELD SERVICE MODE WITH TECH-1.
- MIL SHOULD FLASH DTC 12. DOES IT?

NO → REPLACE PROM

YES → SYSTEM OK

"AFTER REPAIRS," CONFIRM "CLOSED LOOP" OPERATION AND NO MIL (SERVICE ENGINE SOON).

3.3L (VIN N) ENGINE — SYSTEM DIAGNOSTIC CHARTS — CENTURY, CUTLASS CIERA/CRUISER, ACHIEVA, GRAND AM AND SKYLARK

CHART A-3
(Page 1 of 2)
ENGINE CRANKS BUT WON'T RUN
3300 (VIN N) (MFI)

Circuit Description:

This chart assumes that battery condition and engine cranking speed are OK, and there is adequate fuel in the tank.

Test Description: Number(s) refer to circled number(s) on the diagnostic chart.

1. A MIL (Service Engine Soon) "ON" is a basic test to determine if there is battery and ignition voltage to the ECM. The injector test light should blink, indicating the ECM is in control of the injectors. How bright the light blinks is not important. However, the test light should be a J 34730-2A or equivalent. The Tech 1 can be used to verify that the ECM recognizes the reference signal. If it does, the Tech 1 should read RPM during crank.
2. If test light is "OFF," the 20 amp fuse could be blown or CKT 1149 open or shorted to ground.
3. Use fuel pressure gage J 34730-1 or equivalent. Wrap a shop towel around the fuel pressure tap to absorb any small amount of fuel leakage that may occur when installing the gage.
4. Because the electronic ignition system uses two plugs and wires to complete the circuit of each coil, the companion plug wire should connect to engine ground. If RPM was indicated during crank, the ignition control module is receiving a crank signal, but no spark at this test indicates the ignition control module is not triggering the coils or the coil cannot generate spark.
5. This test will determine if the ignition control module is not triggering the problem coil or if the tested coil is at fault. This test could also be performed by using another known good coil.

3.3L (VIN N) ENGINE — SYSTEM DIAGNOSTIC CHARTS — CENTURY, CUTLASS CIERA/CRUISER, ACHIEVA, GRAND AM AND SKYLARK

CAUTION: WHEN INSTALLING FUEL PRESSURE TEST GAGE, USE A SHOP TOWEL TO ABSORB ANY FUEL THAT MAY LEAK AT FUEL RAIL TEST FITTING.

CHART A-3
(Page 1 of 2)
ENGINE CRANKS BUT WON'T RUN
3300 (VIN N) (MFI)

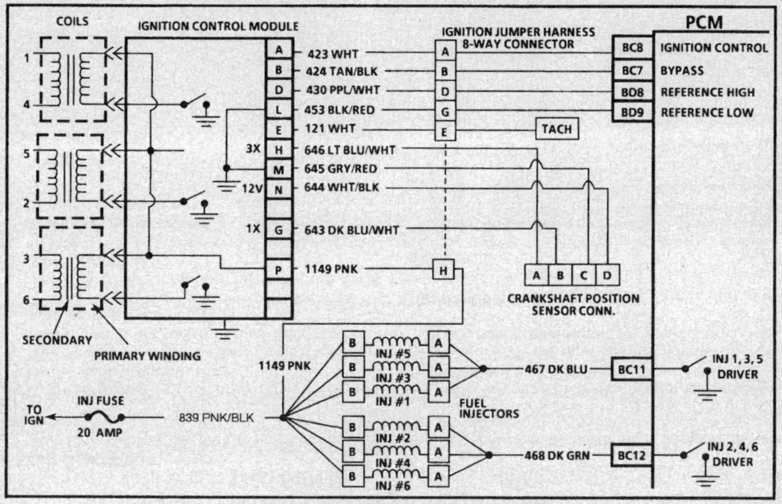

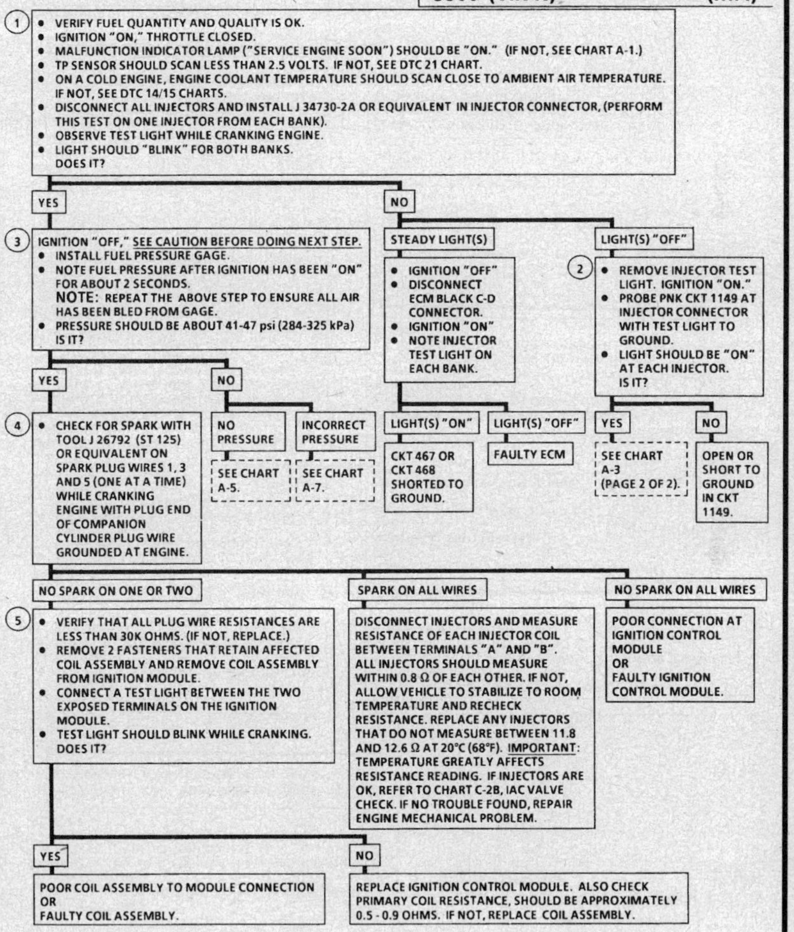

3.3L (VIN N) ENGINE — SYSTEM DIAGNOSTIC CHARTS — CENTURY, CUTLASS CIERA/CRUISER, ACHIEVA, GRAND AM AND SKYLARK

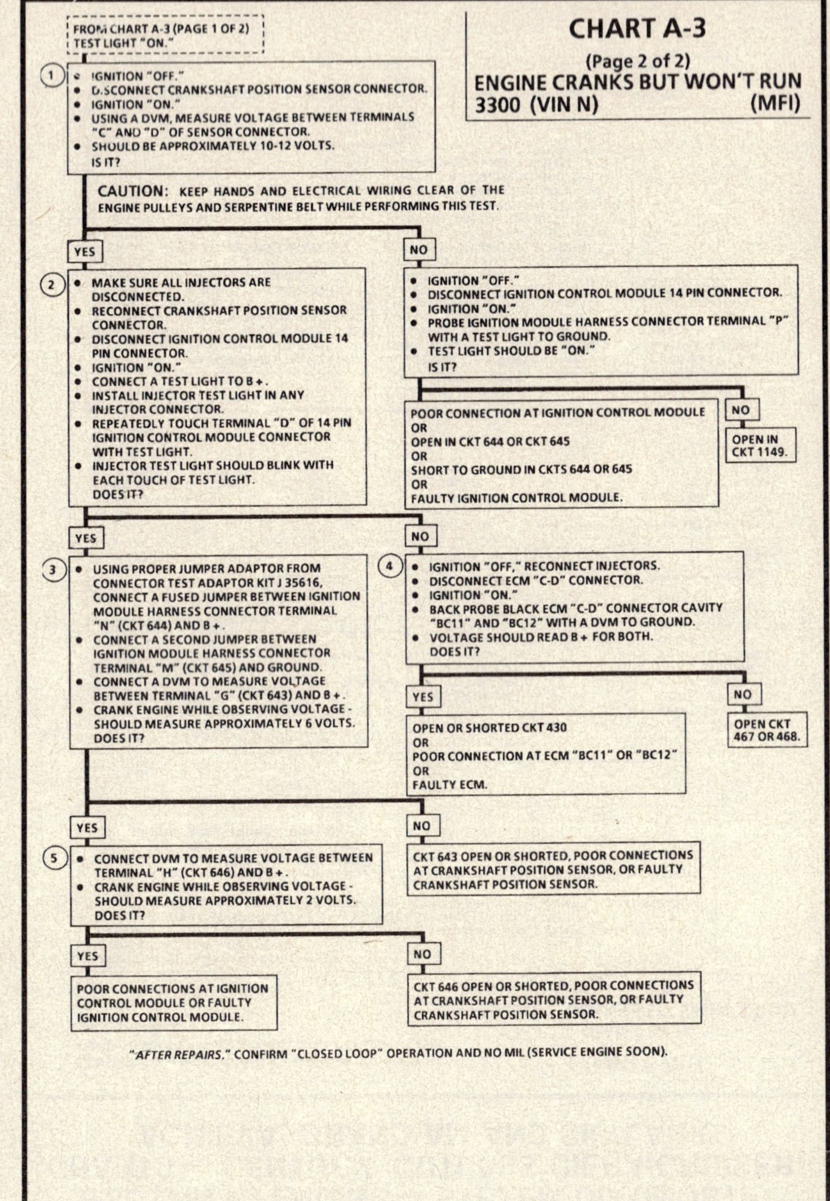

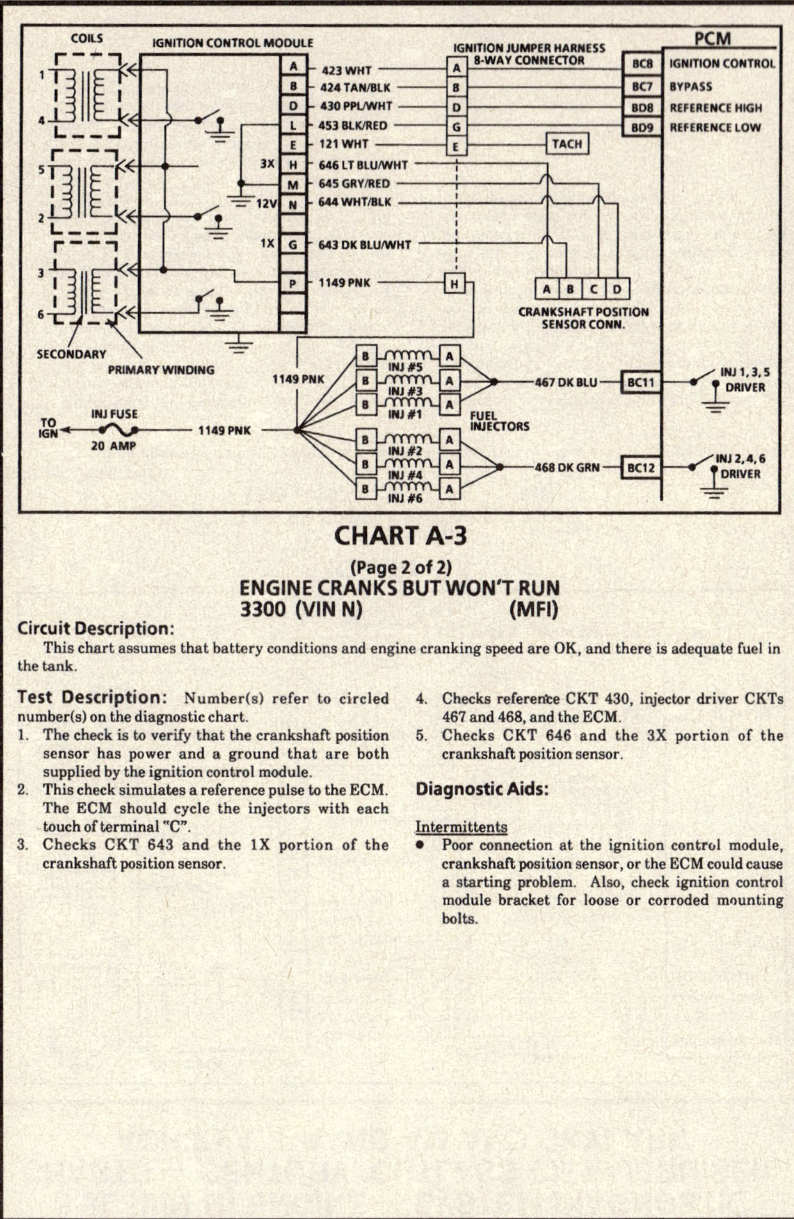

CHART A-3
(Page 2 of 2)
ENGINE CRANKS BUT WON'T RUN
3300 (VIN N) (MFI)

Circuit Description:
This chart assumes that battery conditions and engine cranking speed are OK, and there is adequate fuel in the tank.

Test Description: Number(s) refer to circled number(s) on the diagnostic chart.
1. The check is to verify that the crankshaft position sensor has power and a ground that are both supplied by the ignition control module.
2. This check simulates a reference pulse to the ECM. The ECM should cycle the injectors with each touch of terminal "C".
3. Checks CKT 643 and the 1X portion of the crankshaft position sensor.
4. Checks reference CKT 430, injector driver CKTs 467 and 468, and the ECM.
5. Checks CKT 646 and the 3X portion of the crankshaft position sensor.

Diagnostic Aids:

Intermittents
• Poor connection at the ignition control module, crankshaft position sensor, or the ECM could cause a starting problem. Also, check ignition control module bracket for loose or corroded mounting bolts.

3.3L (VIN N) ENGINE — SYSTEM DIAGNOSTIC CHARTS — CENTURY, CUTLASS CIERA/CRUISER, ACHIEVA, GRAND AM AND SKYLARK

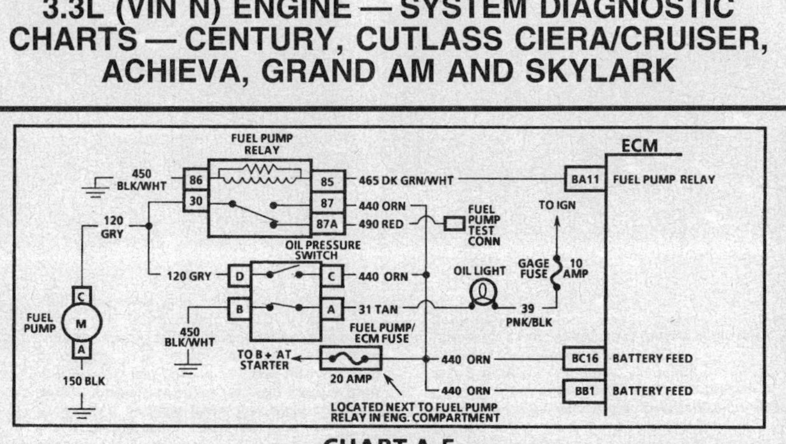

CHART A-5

**FUEL SYSTEM ELECTRICAL TEST
3300 (VIN N) "A" CARLINE (MFI)**

Circuit Description:

When the ignition switch is turned "ON," the Engine Control Module (ECM) energizes the fuel pump relay which completes the circuit to the in-tank fuel pump. It remains "ON" as long as the engine is cranking or running and the ECM is receiving reference pulses. If there are no reference pulses, the ECM de-energizes the fuel pump relay within 2 seconds after key "ON" or when the engine is stopped.

The fuel pump delivers fuel to the fuel rail and injectors, then to the pressure regulator where the system pressure is controlled. Excess fuel is bypassed back to the fuel tank. When the engine is stopped, the pump can be turned "ON" by applying fused battery voltage to the test terminal located in the engine compartment.

Improper fuel system pressure may contribute to one or all of the following symptoms:

- Cranks but won't run.
- DTC 44 or 45.
- Cuts out, may feel like ignition problem.
- Hesitation, loss of power or poor fuel economy.

Test Description: Number(s) refer to circled number(s) on the diagnostic chart.
1. Check for B + and ground at the fuel pump relay.
2. Check that ECM is supplying voltage to fuel pump relay.

3. This checks the fuel pump circuit to the tank for opens or shorts.

3.3L (VIN N) ENGINE — SYSTEM DIAGNOSTIC CHARTS — CENTURY, CUTLASS CIERA/CRUISER, ACHIEVA, GRAND AM AND SKYLARK

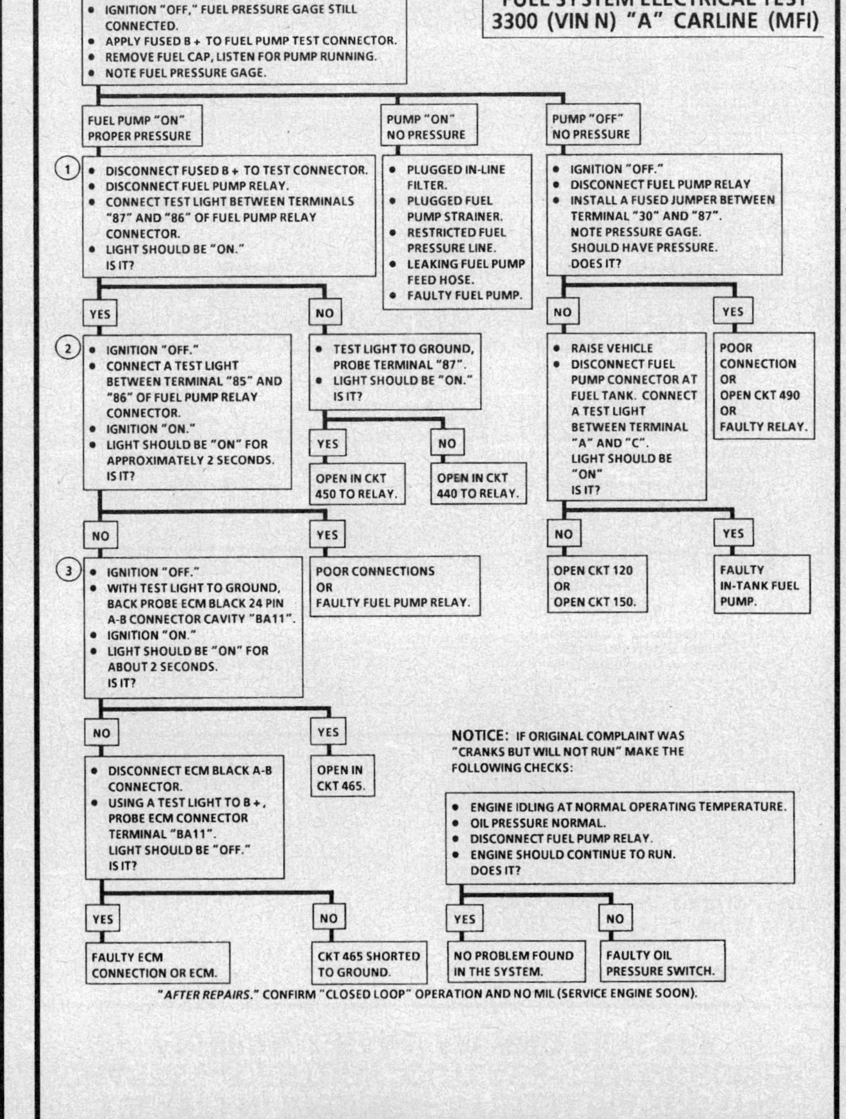

3.3L (VIN N) ENGINE — SYSTEM DIAGNOSTIC CHARTS — CENTURY, CUTLASS CIERA/CRUISER, ACHIEVA, GRAND AM AND SKYLARK

CHART A-5
(Page 1 of 2)
FUEL SYSTEM ELECTRICAL TEST
3300 (VIN N) "N" CARLINE (MFI)

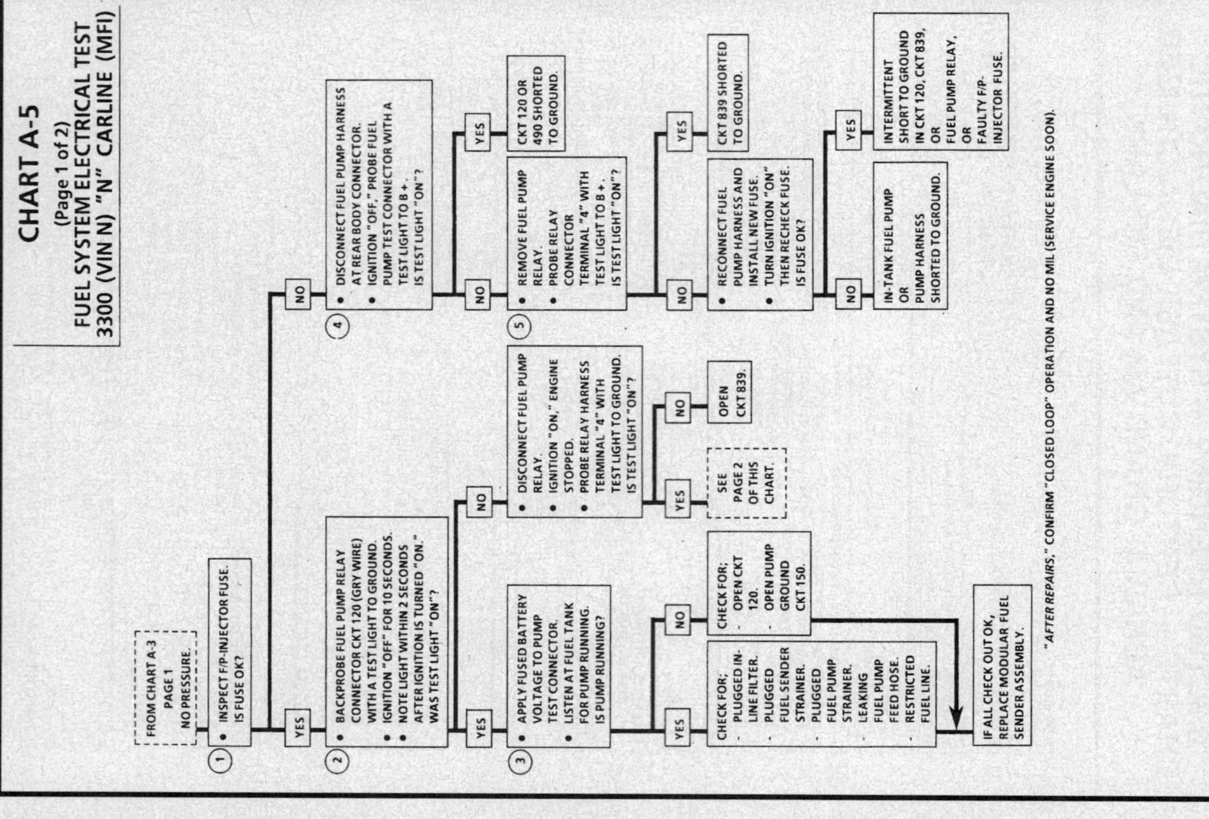

3.3L (VIN N) ENGINE — SYSTEM DIAGNOSTIC CHARTS — CENTURY, CUTLASS CIERA/CRUISER, ACHIEVA, GRAND AM AND SKYLARK

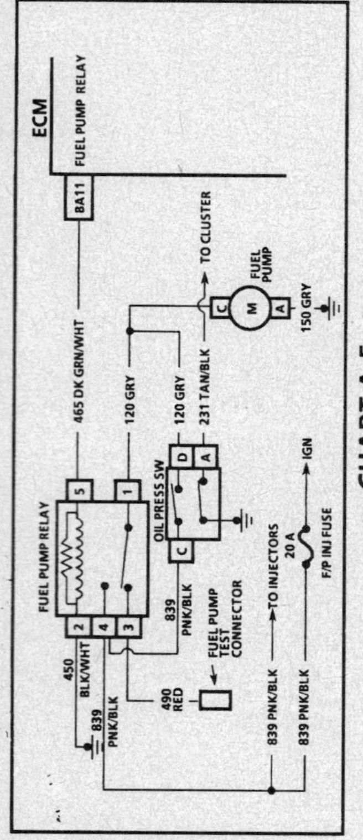

CHART A-5
(Page 1 of 2)
FUEL SYSTEM ELECTRICAL TEST
3300 (VIN N) "N" CARLINE (MFI)

Circuit Description:

When the ignition switch is turned "ON," the Engine Control Module (ECM) energizes the fuel pump relay which completes the circuit to the in-tank fuel pump. It remains "ON" as long as the engine is cranking or running and the ECM is receiving reference pulses. If there are no reference pulses, the ECM de-energizes the fuel pump relay within 2 seconds after key "ON" or when the engine is stopped.

The fuel pump delivers fuel to the fuel rail and injectors, then to the pressure regulator where the system pressure is controlled. Excess fuel is bypassed back to the fuel tank. When the engine is stopped, the pump can be turned "ON" by applying fused battery voltage to the test connector located in the engine compartment.

Improper fuel system pressure may contribute to one or all of the following symptoms:

- Cranks but won't run.
- DTC 44 or 45.
- Cuts out, may feel like ignition problem.
- Hesitation, loss of power or poor fuel economy.

Test Description: Number(s) below refer to circled number(s) on the diagnostic chart.

1. If the fuse is blown, a short to ground in CKT 120, 839, or the fuel pump itself is the cause.
2. This step determines if the fuel pump circuit is being controlled by the ECM. The ECM should energize the fuel pump relay and turn the fuel pump "ON." If the engine is not cranking or running, the ECM should de-energize the relay and/or fuel pump within 2 seconds after the ignition is turned "ON."
3. Applying B+ to the pump test connector turns "ON" the fuel pump. This verifies CKT 120 wiring. If the pump runs, it is a basic fuel delivery problem.
4. This test will determine if a short to ground on CKT 120 caused the fuse to blow. To prevent a mis-diagnosis, be sure the fuel pump is disconnected before the test.
5. Checks for a short to ground in the fuel pump relay harness CKT 839.

3.3L (VIN N) ENGINE — SYSTEM DIAGNOSTIC CHARTS — CENTURY, CUTLASS CIERA/CRUISER, ACHIEVA, GRAND AM AND SKYLARK

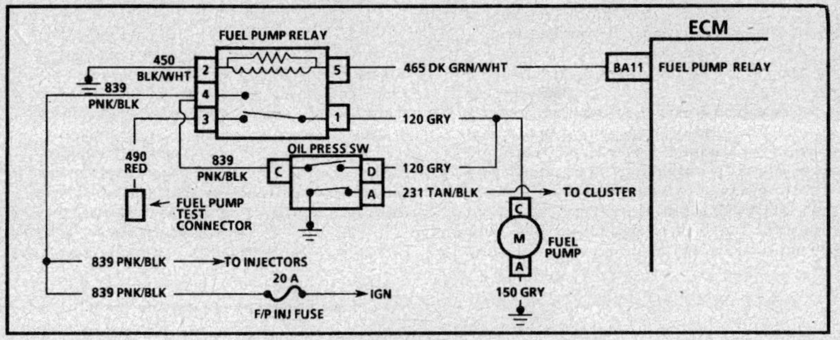

CHART A-5
(Page 2 of 2)
FUEL SYSTEM ELECTRICAL TEST
3300 (VIN N) "N" CARLINE (MFI)

Circuit Description:

When the ignition switch is turned "ON," the Engine Control Module (ECM) energizes the fuel pump relay which completes the circuit to the in-tank fuel pump. It remains "ON" as long as the engine is cranking or running and the ECM is receiving reference pulses. If there are no reference pulses, the ECM de-energizes the fuel pump relay within 2 seconds after key "ON" or when the engine is stopped.

Test Description: Number(s) below refer to circled number(s) on the diagnostic chart.

6. Checks for open in the fuel pump relay ground CKT 450.
7. Determines if the ECM is in control of the fuel pump relay through CKT 465 (terminal "5").
8. The fuel pump control circuit includes an engine oil pressure switch with a separate set of normally open contacts. The switch closes at about 28 kPa (4 psi) of oil pressure and provides a second ignition feed path to the fuel pump. If the relay fails, the pump will run using the ignition feed supplied by the closed oil pressure switch.

9. This step checks the oil pressure switch to be sure it provides ignition feed to the fuel pump should the pump relay fail. A failed pump relay will result in extended engine crank time because of the time required to build enough oil pressure to close the oil pressure switch and turn "ON" the fuel pump. There may be instances when the relay has failed but the engine will not crank fast enough to build enough oil pressure to close the switch. This or a faulty oil pressure switch can result in "Engine Cranks But Won't Run."

3.3L (VIN N) ENGINE — SYSTEM DIAGNOSTIC CHARTS — CENTURY, CUTLASS CIERA/CRUISER, ACHIEVA, GRAND AM AND SKYLARK

CHART A-5
(Page 2 of 2)
FUEL SYSTEM ELECTRICAL TEST
3300 (VIN N) "N" CARLINE (MFI)

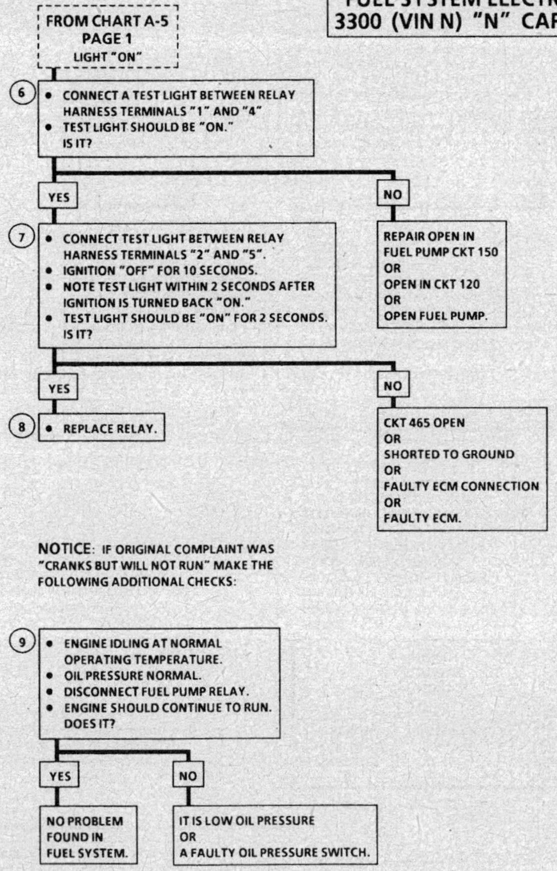

"AFTER REPAIRS," CONFIRM "CLOSED LOOP" OPERATION AND NO MIL (SERVICE ENGINE SOON).

3.3L (VIN N) ENGINE — SYSTEM DIAGNOSTIC CHARTS — CENTURY, CUTLASS CIERA/CRUISER, ACHIEVA, GRAND AM AND SKYLARK

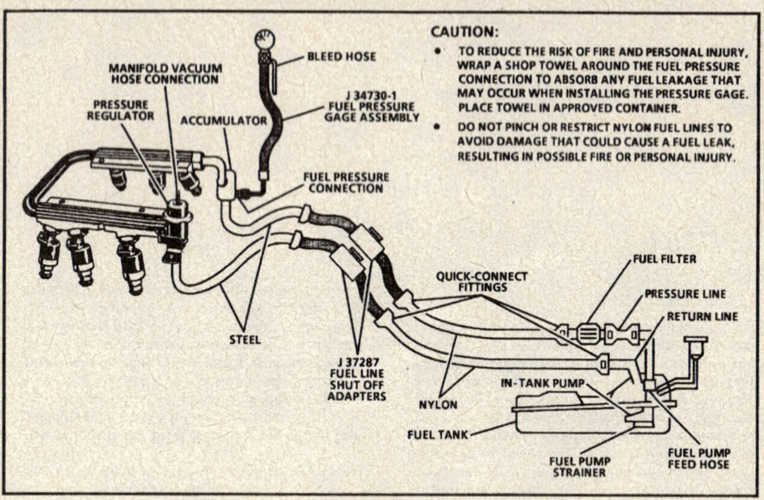

CHART A-7
(Page 1 of 3)
FUEL SYSTEM PRESSURE TEST
3300 (VIN N) "A" CARLINE (MFI)

Circuit Description:

An electric fuel pump, attached to the fuel level meter assembly (inside the fuel tank) pumps fuel through an in-line filter to the fuel rail assembly. The pump is designed to provide fuel at a pressure above the regulated pressure needed by the injectors. A pressure regulator, an integral part of the fuel rail assembly, keeps fuel available to the injectors at a regulated pressure. Unused fuel is returned to the fuel tank by a separate line. The fuel pump test terminal is located in the engine compartment. When the engine is stopped, the pump can be turned "ON" by applying battery voltage to the test connector.

Improper fuel system pressure may contribute to one or all of the following symptoms:
- Cranks but won't run.
- DTC 44 or 45.
- Cuts out, may feel like ignition problem.
- Hesitation, loss of power, or poor fuel economy.

Test Description: Number(s) refer to circled number(s) on the diagnostic chart.

1. Use pressure gage J 34730-1. Wrap a shop towel around the fuel pressure connection to absorb any small amount of fuel leakage that may occur when installing the gage.
   With ignition "ON" engine "OFF," pump pressure should be 284-325 kPa (41-47 psi). This pressure is controlled by spring tension and throttle body vacuum within the pressure regulator assembly. Pressure should not leak down after the fuel pump is shut "OFF."

2. When the engine is idling, the engine vacuum is high and is applied to the fuel regulator diaphragm. This will offset the spring and lower fuel pressure approximately 21-69 kPa (3-10 psi).

3. The application of 12-14 inches of vacuum to the pressure regulator should result in a fuel pressure drop of at least 21-69 kPa (3-10 psi).

4. Pressure that leaks down may be caused by one of the following:
   - In-tank fuel pump check valve not holding.
   - Leaking fuel pump feed hose.
   - Fuel pressure regulator valve leaking.
   - Injector sticking open.

3.3L (VIN N) ENGINE — SYSTEM DIAGNOSTIC CHARTS — CENTURY, CUTLASS CIERA/CRUISER, ACHIEVA, GRAND AM AND SKYLARK

CHART A-7
(Page 1 of 3)
FUEL SYSTEM PRESSURE TEST
3300 (VIN N) "A" CARLINE (MFI)

(1)
- INSTALL FUEL PRESSURE GAGE AS SHOWN ON FACING PAGE.
- IGNITION "OFF" FOR 10 SECONDS. A/C "OFF."
- IGNITION "ON." FUEL PUMP WILL RUN FOR ABOUT 2 SECONDS. THE IGNITION MAY HAVE TO BE CYCLED "ON" MORE THAN ONCE TO OBTAIN MAXIMUM PRESSURE.
- NOTE FUEL PRESSURE, WITH PUMP RUNNING, PRESSURE SHOULD BE 284-325 kPa (41-47 psi). WHEN PUMP STOPS, PRESSURE MAY VARY SLIGHTLY THEN SHOULD HOLD STEADY. IS PRESSURE CORRECT AND DOES IT HOLD?

YES → IF FUEL PRESSURE IS WITHIN NORMAL RANGE, BUT IS SUSPECTED OF DROPPING OFF DURING ACCELERATION, CRUISE OR HARD CORNERING, SEE PAGE 2 OF THIS CHART.

NO →
- FUEL PRESSURE WITHIN SPEC., BUT DOES NOT HOLD.
- FUEL PRESSURE OUT OF SPEC. → SEE PAGE 2 OF THIS CHART
- NO PRESSURE → SEE CHART A-5

(2)
- START AND IDLE ENGINE AT NORMAL OPERATING TEMPERATURE.
- FUEL PRESSURE NOTED IN STEP 1 SHOULD DROP BY 21-69 kPa (3-10 psi). DOES IT?

(4)
- INSTALL J 37287 FUEL LINE SHUTOFF ADAPTORS, REFER TO PAGE 3 OF 3 FOR FUEL PRESSURE RELIEF PROCEDURE AND FOR SERVICING QUICK-CONNECT FITTINGS.
- USING A 10 AMP FUSED JUMPER WIRE, APPLY B + TO FUEL PUMP TEST CONNECTOR FOR 2 SECONDS.
- BLOCK FUEL PRESSURE LINE BY CLOSING VALVE. PRESSURE SHOULD HOLD. DOES IT?

YES → NO TROUBLE FOUND, REVIEW "SYMPTOMS,"

NO (3)
- DISCONNECT VACUUM HOSE FROM PRESSURE REGULATOR ASSEMBLY.
- WITH ENGINE IDLING, APPLY 12-14 INCHES OF VACUUM TO PRESSURE REGULATOR. FUEL PRESSURE NOTED IN STEP 1 SHOULD DROP 21-69 kPa (3-10 psi). DOES IT?

NO
- OPEN SHUTOFF VALVE ON FUEL PRESSURE LINE.
- CLOSE SHUTOFF VALVE ON FUEL RETURN LINE.
- USING A 10 AMP FUSED JUMPER WIRE, APPLY B + TO FUEL PUMP TEST CONNECTOR FOR 2 SECONDS. PRESSURE SHOULD HOLD. DOES IT?

YES → CHECK FOR:
- LEAKING FUEL PUMP FEED HOSE.
- FAULTY FUEL PUMP.

YES → LOCATE AND REPAIR LOSS OF VACUUM TO PRESSURE REGULATOR.

NO → REPLACE PRESSURE REGULATOR.

NO → CHECK FOR FOULED SPARK PLUGS TO LOCATE LEAKING INJECTOR. IF LOCATION OF LEAKING INJECTOR CANNOT BE DETERMINED FROM PLUGS, REMOVE FUEL RAIL BOLTS AND LIFT RAIL SLIGHTLY, LEAVING INJECTOR NOZZLES DIRECTED INTO INTAKE PARTS. TURN IGNITION "ON" AND OBSERVE INJECTOR NOZZLES.

CAUTION: DO NOT ALLOW INJECTORS TO SPRAY ONTO ENGINE.

YES → REPLACE PRESSURE REGULATOR.

"AFTER REPAIRS." CONFIRM "CLOSED LOOP" OPERATION AND NO MIL (SERVICE ENGINE SOON).

3.3L (VIN N) ENGINE — SYSTEM DIAGNOSTIC CHARTS — CENTURY, CUTLASS CIERA/CRUISER, ACHIEVA, GRAND AM AND SKYLARK

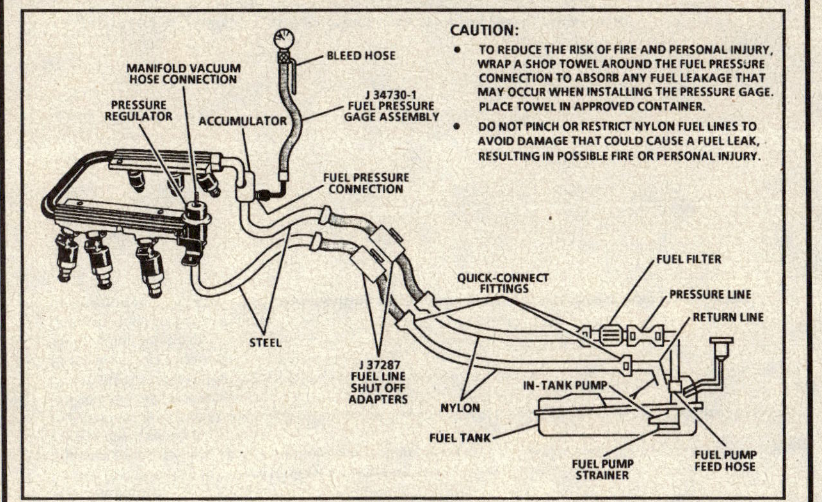

CHART A-7
(Page 2 of 3)
FUEL SYSTEM PRESSURE TEST
3300 (VIN N) "A" CARLINE (MFI)

Test Description: Number(s) refer to circled number(s) on the diagnostic chart.

5. Fuel pressure that drops off during acceleration, cruise or hard cornering may cause a lean condition and result in a loss of power, surging or misfire. This condition can be diagnosed using a Tech 1 scan tool. If the fuel system is very lean, the O2S will stop toggling and output voltage will drop below 500 mV. Also, injector pulse width will increase.

Important
- Make sure system is not operating at "Fuel Cut-Off" which may cause false readings on the scan tool.

6. Fuel pressure below 284 kPa (41 psi) may cause a lean condition and may set a DTC 44 or a DTC 32. Driveability conditions can include hard starting cold, hesitation, poor driveability, lack of power, surging or misfire.

7. Restricting the fuel return line causes fuel pressure to build above regulated pressure. With battery voltage applied to the pump "test" connector, pressure should rise above 325 kPa (47 psi) as the valve in the return line is partially closed.

NOTICE: Do not allow pressure to exceed 414 kPa (60 psi) as damage to the regulator may result.

8. Fuel pressure above 325 kPa (47 psi) may cause a rich condition and may set a DTC 45. Driveability conditions can include hard starting (followed by black smoke) and a strong sulphur smell in the exhaust.

9. This test determines if the high fuel pressure is due to a restricted fuel return line or a faulty fuel pressure regulator.

3.3L (VIN N) ENGINE — SYSTEM DIAGNOSTIC CHARTS — CENTURY, CUTLASS CIERA/CRUISER, ACHIEVA, GRAND AM AND SKYLARK

CHART A-7
(Page 2 of 3)
FUEL SYSTEM PRESSURE TEST
3300 (VIN N) "A" CARLINE (MFI)

FROM CHART A-3 PAGE 1
OR
FROM CHART A-7 PAGE 1.
PRESSURE IS SUSPECTED OF DROPPING OFF OR PRESSURE OUT OF SPECIFICATION, 284-325 kPa (41-47 psi).

5 FUEL PRESSURE DROPS OFF DURING ACCELERATION, CRUISE OR HARD CORNERING.

CHECK FOR RESTRICTED FUEL PRESSURE LINE OR IN-LINE FUEL FILTER. IS THERE A RESTRICTION?

NO → CHECK FOR:
- RESTRICTED FUEL PUMP STRAINER.
- LEAKING FUEL PUMP FEED HOSE.
- FAULTY FUEL PUMP.
- INCORRECT FUEL PUMP.

YES → REPLACE FILTER OR LOCATE AND CORRECT RESTRICTION IN FUEL LINE AND RECHECK.

6 FUEL PRESSURE LESS THAN 284 kPa (41 psi).

7
- IGNITION "OFF."
- INSTALL J 37287-2 FUEL LINE ADAPTER IN FUEL RETURN LINE. REFER TO PAGE 3 OF 3 FOR FUEL PRESSURE RELIEF PROCEDURE AND FOR SERVICING QUICK-CONNECT FITTINGS.
- USING A 10 AMP FUSED WIRE, APPLY B + TO FUEL PUMP TEST CONNECTOR.
- SLOWLY CLOSE VALVE IN RETURN LINE. PRESSURE SHOULD RISE ABOVE 325 kPa (47 psi). DO NOT EXCEED 414 kPa (60 psi). DOES PRESSURE RISE ABOVE ABOVE 325 kPa (47 psi)?

NO

YES → REPLACE PRESSURE REGULATOR.

8 FUEL PRESSURE ABOVE 325 kPa (47 psi).

9
- IGNITION "OFF."
- DISCONNECT QUICK-CONNECT FITTING AT ENGINE FUEL RETURN PIPE. (PROCEDURES FOR SERVICING QUICK-CONNECT FITTINGS ARE FOUND ON PAGE 3 OF 3.)
- ATTACH 5/16 I.D. FLEX HOSE TO ENGINE SIDE OF FUEL RETURN PIPE. PLACE OPEN END OF HOSE INTO AN APPROVED GASOLINE CONTAINER.
- IGNITION "ON." NOTE FUEL PRESSURE WITHIN 2 SECONDS AFTER IGNITION IS TURNED "ON."
- PRESSURE SHOULD BE 284-325 kPa (41-47 psi). IS IT?

NO → CHECK FOR RESTRICTED FUEL RETURN LINE FROM PRESSURE REGULATOR TO POINT WHERE FUEL LINE WAS DISCONNECTED. IS THERE A RESTRICTION?

YES → LOCATE AND CORRECT RESTRICTION IN FUEL RETURN LINE TO FUEL TANK.

NO → REMOVE PRESSURE REGULATOR AND CHECK FOR RESTRICTED FILTER SCREEN (IF EQUIPPED), REPLACE SCREEN IF PLUGGED. IF OK, REPLACE PRESSURE REGULATOR.

YES → LOCATE AND CORRECT RESTRICTION IN FUEL LINE.

"AFTER REPAIRS," CONFIRM "CLOSED LOOP" OPERATION AND NO MIL (SERVICE ENGINE SOON).

3.3L (VIN N) ENGINE — SYSTEM DIAGNOSTIC CHARTS — CENTURY, CUTLASS CIERA/CRUISER, ACHIEVA, GRAND AM AND SKYLARK

CHART A-7
(Page 3 of 3)
FUEL SYSTEM PRESSURE TEST
3300 (VIN N) "A" CARLINE (MFI)

FUEL SYSTEM PRESSURE RELIEF PROCEDURE

PFI Engines With Fuel Pressure Connection

(Must Be Performed Before Disconnecting Fuel Line Fittings)

CAUTION:
- To reduce the risk of fire and personal injury, it is necessary to relieve fuel system pressure before disconnecting fuel line fittings.
- After relieving system pressure, a small amount of fuel may be released when disconnecting fuel line fittings. In order to reduce the chance of personal injury, cover fuel line fittings with a shop towel before disconnecting, to catch any fuel that may leak out. Place the towel in an approved container when disconnect is completed.

Tool Required: J 34730-1 Fuel Pressure Gage

1. Disconnect negative battery cable to avoid possible fuel discharge if an accidental attempt is made to start the engine.
2. Loosen fuel filler cap to relieve tank vapor pressure.
3. Connect gage J 34730-1 to fuel pressure connection. Wrap a shop towel around fitting while connecting gage to avoid spillage.
4. Install bleed hose into an approved container and open valve to bleed system pressure. Fuel line fittings are now safe for servicing.
5. Drain any fuel remaining in gage into an approved gasoline container.
6. Perform service required.
7. Tighten fuel filler cap.
8. Connect negative battery cable.
9. Cycle ignition "ON" and "OFF" twice, waiting ten seconds between cycles, then check for fuel leaks.

"AFTER REPAIRS," CONFIRM "CLOSED LOOP" OPERATION AND NO MIL (SERVICE ENGINE SOON).

3.3L (VIN N) ENGINE — SYSTEM DIAGNOSTIC CHARTS — CENTURY, CUTLASS CIERA/CRUISER, ACHIEVA, GRAND AM AND SKYLARK

CHART A-7
(Page 3 of 3)
FUEL SYSTEM PRESSURE TEST
3300 (VIN N) "A" CARLINE (MFI)

SERVICING QUICK-CONNECT FITTINGS

Important
- In order to install fuel system diagnostic equipment on vehicles equipped with plastic quick-connect fittings, fuel line separator tools must be used to disconnect the fittings. Use of the separator tools will cause the plastic retainer to remain inside the female connector allowing diagnostic equipment to be connected.

Tools required:
J 37088-A tool set, fuel line quick-connect separator;
J 39504 tool set, fuel line quick-connect separator (restricted access).

Remove or Disconnect
1. Grasp both sides of fitting. Twist female connector 1/4 turn in each direction to loosen any dirt within fitting.

CAUTION: Safety glasses must be worn when using compressed air, as flying dirt particles may cause eye injury.

2. Using compressed air, blow dirt out of fitting.
3. Choose correct tool from J 37088-A or J 39504 tool set for size of fitting. Insert tool into female connector, then push/pull inward to release locking tabs.
4. Pull connection apart.

Clean and Inspect

NOTICE: If it is necessary to remove rust or burrs from fuel pipe, use emery cloth in a radial motion with the pipe end to prevent damage to O-ring sealing surface.

- Using a clean shop towel, wipe off male pipe end.
- Inspect both ends of fitting for dirt and burrs. Clean or replace components/assemblies as required.

Install or Connect

CAUTION: To Reduce the Risk of Fire and Personal Injury:
- Before connecting fitting, always apply a few drops of clean engine oil to the male pipe end of engine fuel pipe, pressure gage adapter or fuel line shut-off adapter. This will ensure proper reconnection and prevent a possible fuel leak. (During normal operation, the O-rings located in the female connector will swell and may prevent proper reconnection if not lubricated.)

1. Apply a few drops of clean engine oil to the male pipe end of engine fuel pipe, pressure gage adapter or fuel line shut-off adapter.
2. Push both sides of fitting together to cause the retaining tabs/fingers to snap into place.
3. Once installed, pull on both sides of fitting to make sure connection is secure.

"AFTER REPAIRS," CONFIRM "CLOSED LOOP" OPERATION AND NO MIL (SERVICE ENGINE SOON).

3.3L (VIN N) ENGINE — SYSTEM DIAGNOSTIC CHARTS — CENTURY, CUTLASS CIERA/CRUISER, ACHIEVA, GRAND AM AND SKYLARK

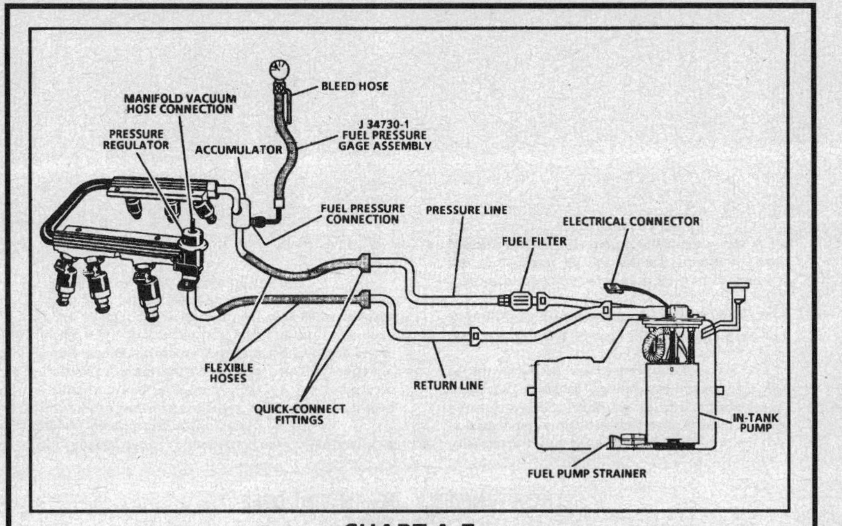

CHART A-7
(Page 1 of 3)
FUEL SYSTEM PRESSURE TEST
3300 (VIN N) "N" CARLINE (PORT)

Fuel System Description:

An electric fuel pump, part of the modular fuel sender and located inside the fuel tank, pumps fuel through an in-line filter to the fuel rail assembly. The pump is designed to provide fuel at a pressure above the regulated pressure needed by the injectors. A pressure regulator, attached to the fuel rail assembly, keeps fuel available to the injectors at a regulated pressure. Unused fuel is returned to the fuel tank by a separate line. The fuel pump test connector is located in the engine compartment. When the engine is stopped, the pump can be turned "ON" by using a fused jumper wire to apply battery voltage to the test connector.

Improper fuel system pressure may contribute to one or all of the following symptoms:

- Cranks but won't run.
- Code 44 or 45.
- Cuts out, may feel like ignition problem.
- Hesitation, loss of power or poor fuel economy.

Test Description: Number(s) below refer to circled number(s) on the diagnostic chart.

1. Use pressure gage J 34730-1. Wrap a shop towel around the fuel pressure tap to absorb any small amount of fuel leakage that may occur when installing the gage.
   With ignition "ON," engine "OFF," fuel pressure should be 284-325 kPa (41-47 psi). This pressure is controlled by spring pressure and manifold vacuum within the pressure regulator assembly. Pressure should not leak down after the fuel pump is shut "OFF."

2. When the engine is idling, the engine vacuum is high and is applied to the fuel regulator diaphragm. This will offset the spring and lower fuel pressure approximately 21-69 kPa (3-10 psi).

3. The application of 12-14 inches of vacuum to the pressure regulator should result in a fuel pressure drop of at least 21-69 kPa (3-10 psi).

4. Pressure that leaks down may be caused by one of the following:
   - In-tank fuel pump check valve not holding.
   - Fuel pressure regulator valve leaking.
   - Injector(s) sticking open.

3.3L (VIN N) ENGINE — SYSTEM DIAGNOSTIC CHARTS — CENTURY, CUTLASS CIERA/CRUISER, ACHIEVA, GRAND AM AND SKYLARK

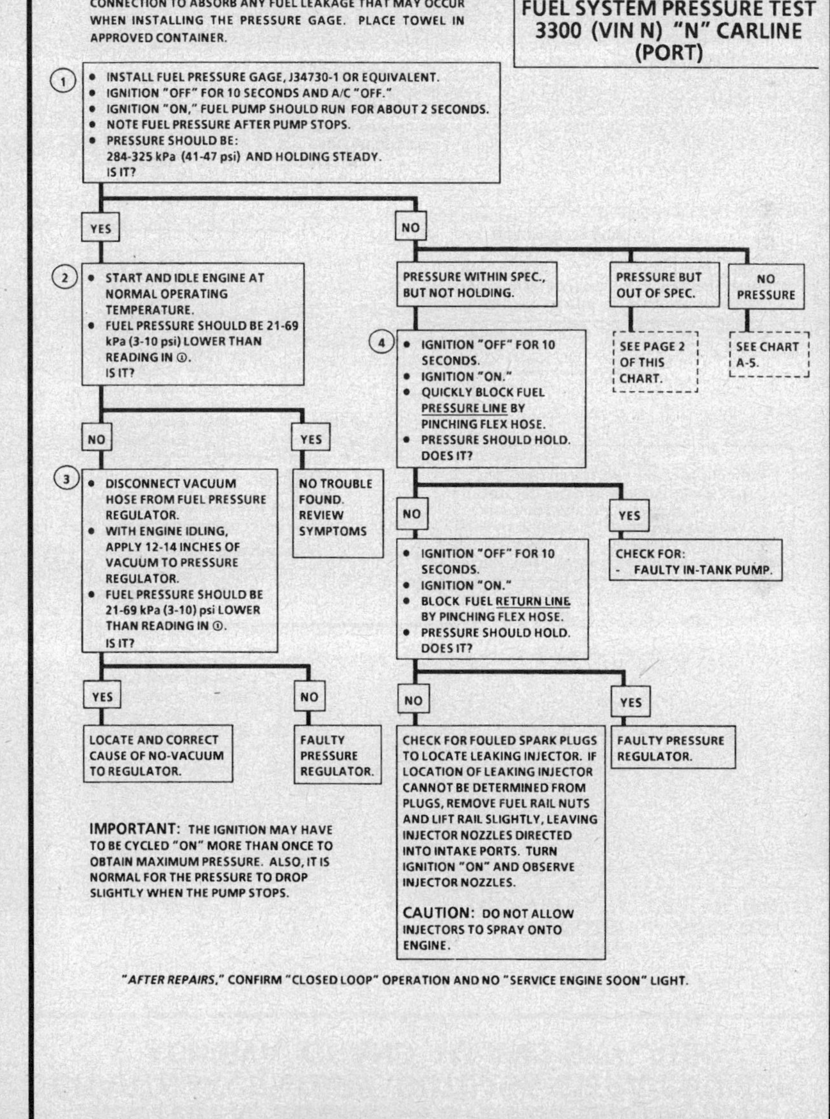

3.3L (VIN N) ENGINE — SYSTEM DIAGNOSTIC CHARTS — CENTURY, CUTLASS CIERA/CRUISER, ACHIEVA, GRAND AM AND SKYLARK

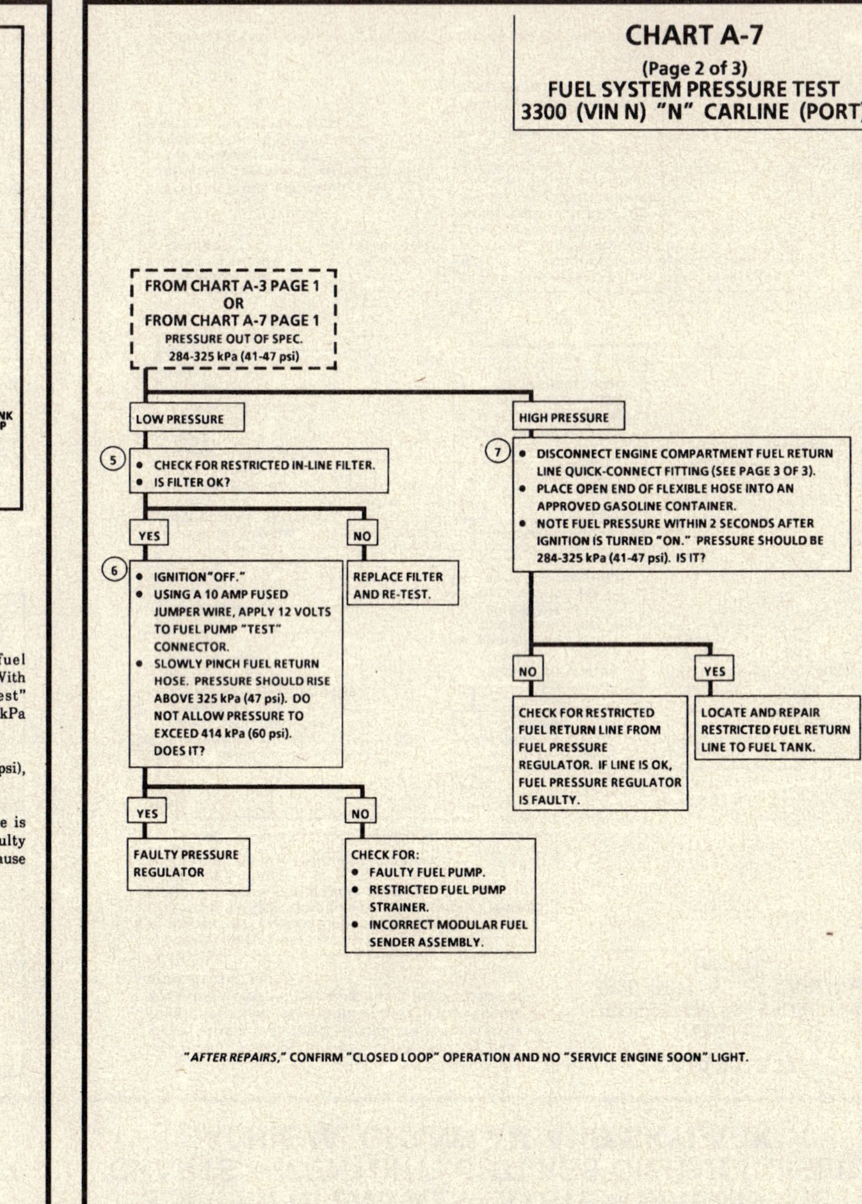

CHART A-7
(Page 2 of 3)
FUEL SYSTEM PRESSURE TEST
3300 (VIN N) "N" CARLINE (PORT)

Test Description: Number(s) below refer to circled number(s) on the diagnostic chart.

5. Pressure below 284 kPa (41 psi) may cause a lean condition and may set a Code 44. It could also cause hard starting cold and poor driveability. Low enough pressure will cause the engine not to run at all. Restricted flow may allow the engine to run at idle, or low speeds, but may cause a surge and stall when more fuel is required, as when accelerating or driving at high speeds.

6. Restricting the fuel return line allows fuel pressure to build above regulated pressure. With battery voltage applied to the pump "test" connector, pressure should rise above 325 kPa (47 psi) as the fuel return hose is restricted.

NOTICE: Do not allow to exceed 414 kPa (60 psi), as damage to the regulator may result.

7. This test determines if the high fuel pressure is due to a restricted fuel return line or a faulty pressure regulator. High fuel pressure may cause a rich condition and may set a Code 45.

3.3L (VIN N) ENGINE — SYSTEM DIAGNOSTIC CHARTS — CENTURY, CUTLASS CIERA/CRUISER, ACHIEVA, GRAND AM AND SKYLARK

CHART A-7
(Page 2 of 3)
FUEL SYSTEM PRESSURE TEST
3300 (VIN N) "N" CARLINE (PORT)

"AFTER REPAIRS," CONFIRM "CLOSED LOOP" OPERATION AND NO "SERVICE ENGINE SOON" LIGHT.

3.3L (VIN N) ENGINE — SYSTEM DIAGNOSTIC CHARTS — CENTURY, CUTLASS CIERA/CRUISER, ACHIEVA, GRAND AM AND SKYLARK

CHART A-7
(Page 3 of 3)
FUEL SYSTEM PRESSURE TEST
3300 "N" CARLINE (PORT)

FUEL PRESSURE CHECK
(Instructions for Disconnecting and Reconnecting Return Line Quick-Connect Fitting)

Tools Required: J 37088-A - Fuel Line Quick-Connect Separators

1. Locate engine compartment fuel return line quick-connect fitting.
2. Grasp both ends of fitting, twist female end ¼ turn in each direction to loosen any dirt in fitting.

CAUTION: Safety glasses must be worn when using compressed air, as flying dirt particles may cause eye injury.

3. Using compressed air, blow dirt out of quick-connect fitting.
4. Using J 37088-2 from separator tool set J 37088-A, insert tool into female end of connector, push inward to release retaining tabs/fingers and pull connection apart.

CAUTION: To Reduce the Risk of Fire and Personal Injury: Before reconnecting the fuel return line quick-connect fitting, always apply a few drops of clean engine oil to the male tube end. This will ensure proper reconnection and prevent a possible fuel leak. (During normal operation, the O-rings located inside the female connector will swell and may prevent proper reconnection if not lubricated.)

5. After performing check, lubricate the male tube end of the fuel line with engine oil and reconnect quick-connect fitting.
   - Push connectors together to cause the retaining tabs/fingers to snap into place.
   - Once installed, pull on both ends of connection to make sure it is secure.
6. Cycle ignition "ON" and "OFF" twice, waiting ten seconds between cycles, then check for fuel leaks.

3.3L (VIN N) ENGINE — DIAGNOSTIC TROUBLE CODE CHART — CENTURY, CUTLASS CIERA/CRUISER, ACHIEVA, GRAND AM AND SKYLARK

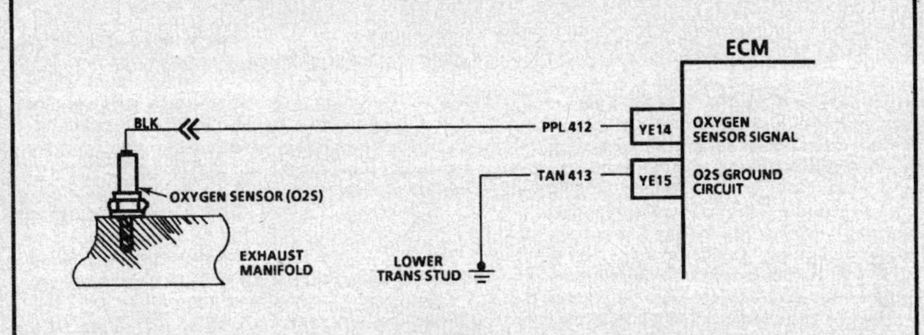

DTC 13
OXYGEN SENSOR (O2S) CIRCUIT
(OPEN CIRCUIT)
3300 (VIN N)   (MFI)

Circuit Description:

The Engine Control Module (ECM) supplies a voltage of about .45 volt between terminals "YE14" and "YE15". (If measured with a 10 megohms digital voltmeter, this may read as low as .32 volt.) The Oxygen Sensor (O2S) varies the voltage within a range of about 1 volt if the exhaust is rich, down through about .10 volt if exhaust is lean.

The sensor is like an open circuit and produces no voltage when it is below 360°C (600°F). An open oxygen sensor circuit or cold oxygen sensor causes "Open Loop" operation.

DTC 13 will set if:
- Engine coolant temperature above 50°C (122°F).
- No DTC 21 or 22.
- Engine has been running at least 40 seconds.
- O2S signal voltage is steady between .35 and .55 volt (350 mV and 550 mV).
- Throttle angle above 3.5%.
- All above conditions must be met for about 30 seconds.

Test Description: Number(s) refer to circled number(s) on the diagnostic chart.
1. If the conditions for DTC 13 exist, the system will not operate in "Closed Loop."
2. This will determine whether the sensor or the wiring is the cause of DTC 13.
3. In doing this test use only a high impedance digital multimeter (J 39200). This test checks the continuity of CKTs 412 and 413. If CKT 413 is open the ECM voltage on CKT 412 will be over .6 volt (600 mV).

Diagnostic Aids:

Normal scan voltage constantly fluctuates between 100 mV and 999 mV while in "Closed Loop." DTC 13 sets in 30 seconds if voltage remains between .35 and .55 volt, but the system will go "Open Loop" in about 15 seconds.

Refer to "Intermittents" in "Symptoms."

3.3L (VIN N) ENGINE — DIAGNOSTIC TROUBLE CODE CHART — CENTURY, CUTLASS CIERA/CRUISER, ACHIEVA, GRAND AM AND SKYLARK

DTC 13
OXYGEN SENSOR (O2S) CIRCUIT
(OPEN CIRCUIT)
3300 (VIN N)   (MFI)

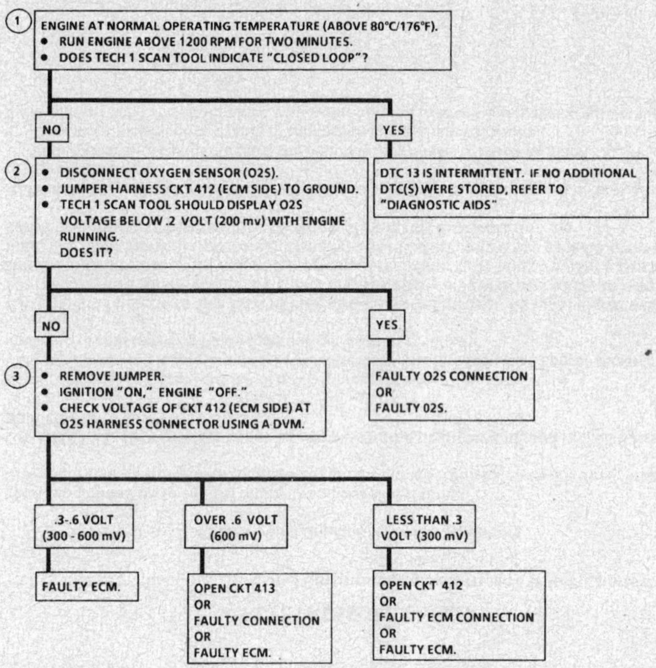

1. ENGINE AT NORMAL OPERATING TEMPERATURE (ABOVE 80°C/176°F).
   - RUN ENGINE ABOVE 1200 RPM FOR TWO MINUTES.
   - DOES TECH 1 SCAN TOOL INDICATE "CLOSED LOOP"?

NO → 2. DISCONNECT OXYGEN SENSOR (O2S).
   - JUMPER HARNESS CKT 412 (ECM SIDE) TO GROUND.
   - TECH 1 SCAN TOOL SHOULD DISPLAY O2S VOLTAGE BELOW .2 VOLT (200 mv) WITH ENGINE RUNNING. DOES IT?

YES → DTC 13 IS INTERMITTENT. IF NO ADDITIONAL DTC(S) WERE STORED, REFER TO "DIAGNOSTIC AIDS"

NO → 3. REMOVE JUMPER.
   - IGNITION "ON," ENGINE "OFF."
   - CHECK VOLTAGE OF CKT 412 (ECM SIDE) AT O2S HARNESS CONNECTOR USING A DVM.

YES → FAULTY O2S CONNECTION OR FAULTY O2S.

.3-.6 VOLT (300 - 600 mV) → FAULTY ECM.

OVER .6 VOLT (600 mV) → OPEN CKT 413 OR FAULTY CONNECTION OR FAULTY ECM.

LESS THAN .3 VOLT (300 mV) → OPEN CKT 412 OR FAULTY ECM CONNECTION OR FAULTY ECM.

"AFTER REPAIRS," REFER TO DTC CRITERIA   AND CONFIRM DTC DOES NOT RESET.

3.3L (VIN N) ENGINE — DIAGNOSTIC TROUBLE CODE CHART — CENTURY, CUTLASS CIERA/CRUISER, ACHIEVA, GRAND AM AND SKYLARK

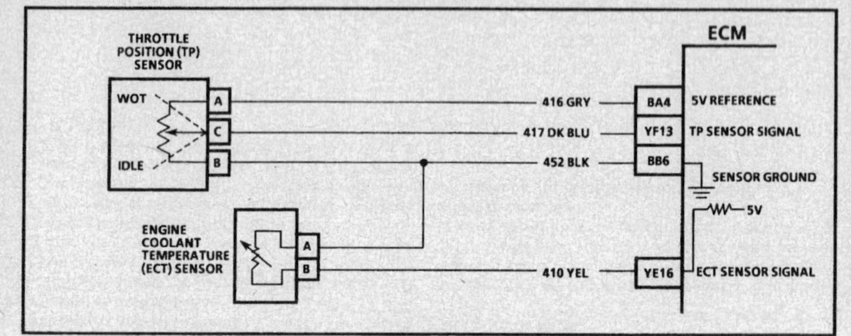

DTC 14
ENGINE COOLANT TEMPERATURE (ECT) SENSOR CIRCUIT
(HIGH TEMPERATURE INDICATED)
3300 (VIN N)   (MFI)

Circuit Description:

The Engine Coolant Temperature (ECT) sensor uses a thermistor to control the signal voltage to the ECM. The ECM applies a voltage on CKT 410 to the sensor. When the engine is cold the sensor (thermistor) resistance is high, therefore, the ECM will see high signal voltage.

As the engine warms, the sensor resistance becomes less, and the voltage drops. At normal engine operating temperature (85°C to 100°C), the ECT signal will measure about 1.5 to 2.0 volts.

DTC 14 will set if:
- Engine run time 10 seconds or more.
- ECT sensor voltage indicates a coolant temperature above 145°C (293°F) for .4 second.

Test Description: Number(s) refer to circled number(s) on the diagnostic chart.
1. This will determine if the ECT sensor is indicating a high temperature to the ECM.
2. This test will determine if CKT 410 is shorted to ground which will cause the conditions for DTC 14.

NOTICE: If DTC 14 is set, the ECM will use a fail-soft engine coolant temperature of 22°C (72°F) for fuel control. The Tech 1 will continue to display actual sensor value.

Diagnostic Aids:

The Tech 1 displays engine coolant temperature in degrees. After the engine is started, coolant temperature should rise steadily to about 90°C then stabilize when thermostat opens.

An intermittent may be caused by a poor connection, rubbed through wire insulation, or a wire broken inside the insulation.

Check For:
- Poor Connection or Damaged Harness. Inspect ECM harness connectors for rubbed through wire insulation, damaged connections, improper mating, broken locks, improperly formed or damaged terminals, poor terminal to wire connection, and damaged harness.
- Intermittent Test. If connections and harness check OK, scan coolant temperature while moving related connectors and wiring harness. If the failure is induced, the engine coolant temperature display will change. This may help to isolate the location of the malfunction.
- Shifted Sensor. The "Temperature To Resistance Value" scale may be used to test the ECT sensor at various temperature levels to evaluate the possibility of a "shifted" (mis-scaled) sensor, which may result in driveability complaints.

3.3L (VIN N) ENGINE — DIAGNOSTIC TROUBLE CODE CHART — CENTURY, CUTLASS CIERA/CRUISER, ACHIEVA, GRAND AM AND SKYLARK

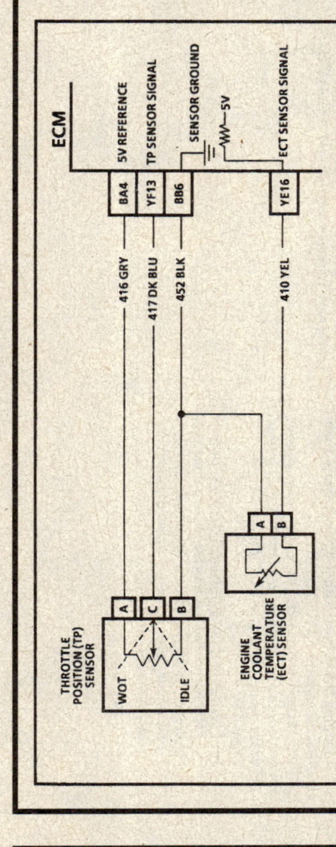

DTC 15
ENGINE COOLANT TEMPERATURE (ECT) SENSOR CIRCUIT
(LOW TEMPERATURE INDICATED)
3300 (VIN N)
(MFI)

Circuit Description:

The Engine Coolant Temperature (ECT) sensor uses a thermistor to control the signal voltage to the ECM. The ECM applies a voltage on CKT 410 to the sensor. When the engine is cold the sensor (thermistor) resistance is high, therefore, the ECM will see high signal voltage.

As the engine warms, the sensor resistance becomes less, and the voltage drops. At normal engine operating temperature (85°C to 100°C), the ECT signal will measure about 1.5 to 2.0 volts.

DTC 15 will set if:
- Engine run time 2 seconds or more.
- ECT sensor voltage indicates engine coolant temperature less than -38°C (-36°F) for at least 4 second.

Test Description: Number(s) refer to circled number(s) on the diagnostic chart.

1. This will determine if the ECT sensor is indicating a low temperature to the ECM.
2. This test simulates DTC 14. If the ECM recognizes the low signal voltage (high temperature) and the Tech 1 displays 140°C, the ECM and wiring are OK.
3. This test will determine if CKT 410 is open. There should be 5 volts present at sensor connector if measured with a DVM.

NOTICE: If DTC 15 is set, the ECM will use a fail-soft engine coolant temperature of 49°C (120°F) for fuel control. The Tech 1 will continue to display the actual sensor value.

Diagnostic Aids:

The Tech 1 reads engine coolant temperature in degrees. After the engine is started, coolant temperature should rise steadily to about 90°C then stabilize when thermostat opens.

An intermittent may be caused by a poor connection or a wire broken inside the insulation. Check For:

- **Poor Connection or Damaged Harness.** Inspect ECM harness connectors for backed out terminal "YE16" or "BB6," improper mating, broken locks, improperly formed or damaged terminals, poor terminal to wire connection and damaged harness.
- **Intermittent Test.** If connections and harness check OK, monitor engine coolant temperature while moving related connectors and wiring harness. If the failure is induced, the display will change. This may help to isolate the location of the malfunction.
- **Shifted Sensor.** The "Temperature To Resistance Value" scale may be used to test the ECT sensor at various temperature levels to evaluate the possibility of a "shifted" (mis-scaled) sensor which may result in driveability complaints.

A faulty connection, or an open in CKTs 410 or 452 will result in DTC 15.

3.3L (VIN N) ENGINE — DIAGNOSTIC TROUBLE CODE CHART — CENTURY, CUTLASS CIERA/CRUISER, ACHIEVA, GRAND AM AND SKYLARK

DTC 14
ENGINE COOLANT TEMPERATURE (ECT) SENSOR CIRCUIT
(HIGH TEMPERATURE INDICATED)
3300 (VIN N)
(MFI)

1. DOES TECH 1 SCAN TOOL DISPLAY ENGINE COOLANT TEMPERATURE OF 130°C (266°F) OR HIGHER?

   YES → 2.
   NO → DTC 14 IS INTERMITTENT. IF NO ADDITIONAL DTC(S) WERE STORED, REFER TO "DIAGNOSTIC AIDS"

2. - DISCONNECT ENGINE COOLANT TEMPERATURE SENSOR. TECH 1 SCAN TOOL SHOULD DISPLAY ENGINE COOLANT TEMPERATURE BELOW -30°C (-22°F). DOES IT?

   YES → REPLACE ENGINE COOLANT TEMPERATURE SENSOR.
   NO → CKT 410 SHORTED TO GROUND OR CKT 410 SHORTED TO SENSOR GROUND CIRCUIT OR FAULTY ECM.

DIAGNOSTIC AID

ENGINE COOLANT TEMPERATURE SENSOR
TEMPERATURE VS. RESISTANCE VALUES
(APPROXIMATE)

°C	°F	OHMS
100	212	177
90	194	241
80	176	332
70	158	467
60	140	667
50	122	973
45	113	1188
40	104	1459
35	95	1802
30	86	2238
25	77	2796
20	68	3520
15	59	4450
10	50	5670
5	41	7280
0	32	9420
-5	23	12300
-10	14	16180
-15	5	21450
-20	-4	28680
-30	-22	52700
-40	-40	100700

"AFTER REPAIRS," REFER TO DTC CRITERIA AND CONFIRM DTC DOES NOT RESET.

3.3L (VIN N) ENGINE — DIAGNOSTIC TROUBLE CODE CHART — CENTURY, CUTLASS CIERA/CRUISER, ACHIEVA, GRAND AM AND SKYLARK

DTC 15
ENGINE COOLANT TEMPERATURE (ECT) SENSOR CIRCUIT
(LOW TEMPERATURE INDICATED)
3300 (VIN N) (MFI)

1. • DOES TECH 1 SCAN TOOL DISPLAY ENGINE COOLANT TEMPERATURE OF -30°C (-22°F) OR LESS?

YES

2. • DISCONNECT ENGINE COOLANT TEMPERATURE SENSOR.
   • JUMPER HARNESS TERMINALS TOGETHER.
   • TECH 1 SCAN TOOL SHOULD DISPLAY 130°C (266°F) OR MORE.
   DOES IT?

NO

3. • JUMPER CKT 410 TO GROUND.
   • TECH 1 SCAN TOOL SHOULD DISPLAY OVER 130°C (266°F).
   DOES IT?

YES

OPEN ENGINE COOLANT TEMPERATURE SENSOR GROUND CIRCUIT, FAULTY CONNECTION OR FAULTY ECM.

NO

DTC 15 IS INTERMITTENT. IF NO ADDITIONAL DTC(S) WERE STORED. REFER TO "DIAGNOSTIC AIDS"

YES

FAULTY CONNECTION OR ENGINE COOLANT TEMPERATURE SENSOR.

NO

OPEN CKT 410, FAULTY CONNECTION AT ECM, OR FAULTY ECM.

DIAGNOSTIC AID

ENGINE COOLANT TEMPERATURE SENSOR		
TEMPERATURE VS. RESISTANCE VALUES (APPROXIMATE)		
°C	°F	OHMS
100	212	177
90	194	241
80	176	332
70	158	467
60	140	667
50	122	973
45	113	1188
40	104	1459
35	95	1802
30	86	2238
25	77	2796
20	68	3520
15	59	4450
10	50	5670
5	41	7280
0	32	9420
-5	23	12300
-10	14	16180
-15	5	21450
-20	-4	28680
-30	-22	52700
-40	-40	100700

"AFTER REPAIRS," REFER TO DTC CRITERIA   AND CONFIRM DTC DOES NOT RESET.

3.3L (VIN N) ENGINE — DIAGNOSTIC TROUBLE CODE CHART — CENTURY, CUTLASS CIERA/CRUISER, ACHIEVA, GRAND AM AND SKYLARK

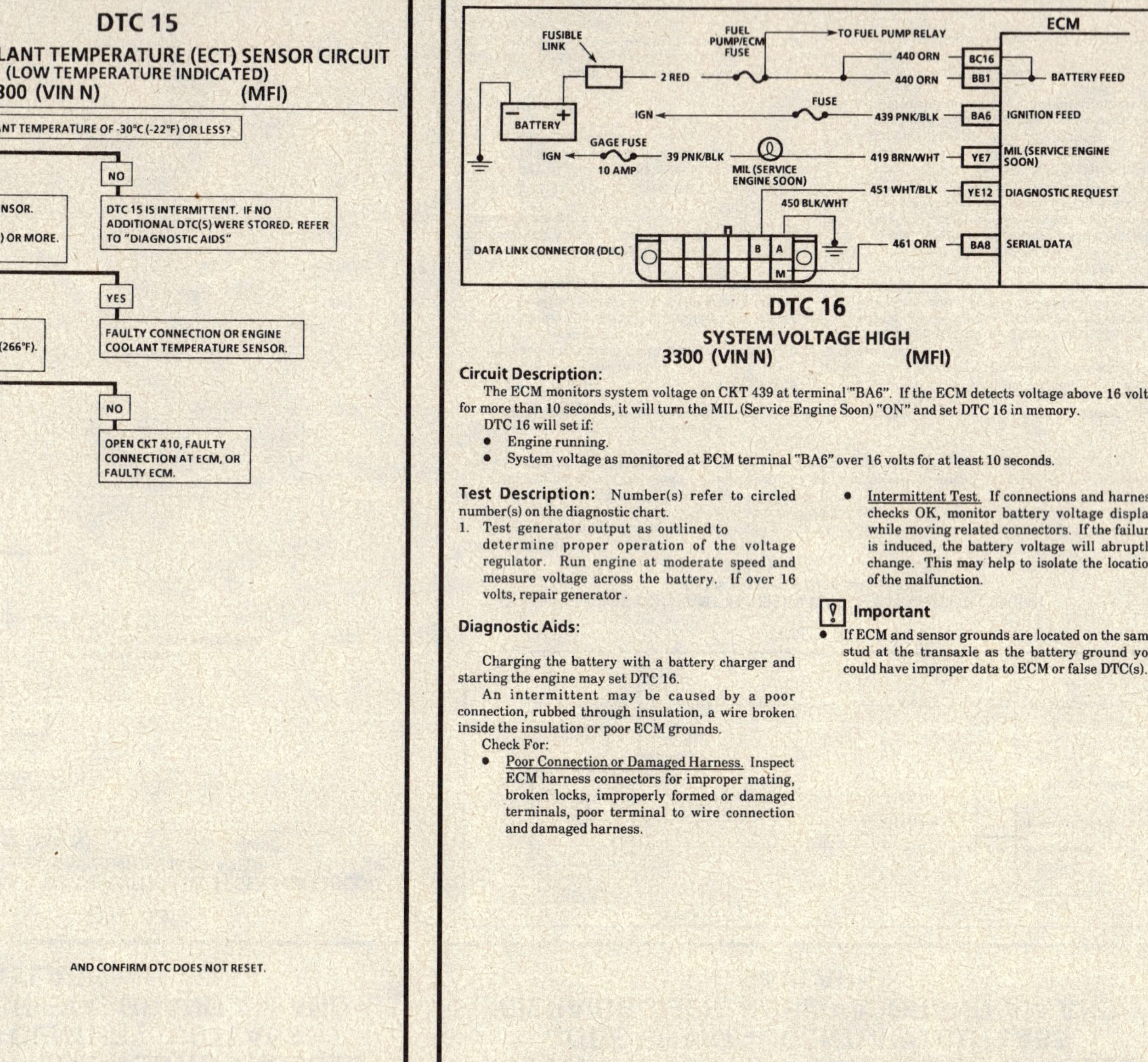

DTC 16
SYSTEM VOLTAGE HIGH
3300 (VIN N) (MFI)

Circuit Description:

The ECM monitors system voltage on CKT 439 at terminal "BA6". If the ECM detects voltage above 16 volts for more than 10 seconds, it will turn the MIL (Service Engine Soon) "ON" and set DTC 16 in memory.

DTC 16 will set if:
- Engine running.
- System voltage as monitored at ECM terminal "BA6" over 16 volts for at least 10 seconds.

Test Description: Number(s) refer to circled number(s) on the diagnostic chart.

1. Test generator output as outlined to determine proper operation of the voltage regulator. Run engine at moderate speed and measure voltage across the battery. If over 16 volts, repair generator.

Diagnostic Aids:

Charging the battery with a battery charger and starting the engine may set DTC 16.

An intermittent may be caused by a poor connection, rubbed through insulation, a wire broken inside the insulation or poor ECM grounds.

Check For:
- Poor Connection or Damaged Harness. Inspect ECM harness connectors for improper mating, broken locks, improperly formed or damaged terminals, poor terminal to wire connection and damaged harness.

- Intermittent Test. If connections and harness checks OK, monitor battery voltage display while moving related connectors. If the failure is induced, the battery voltage will abruptly change. This may help to isolate the location of the malfunction.

⚠ Important
- If ECM and sensor grounds are located on the same stud at the transaxle as the battery ground you could have improper data to ECM or false DTC(s).

3.3L (VIN N) ENGINE — DIAGNOSTIC TROUBLE CODE CHART — CENTURY, CUTLASS CIERA/CRUISER, ACHIEVA, GRAND AM AND SKYLARK

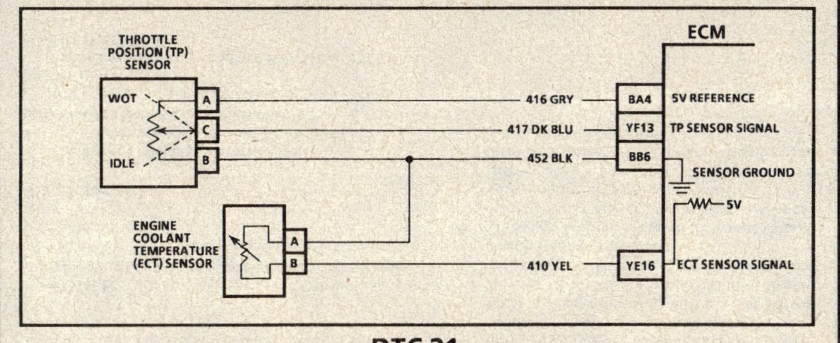

3.3L (VIN N) ENGINE — DIAGNOSTIC TROUBLE CODE CHART — CENTURY, CUTLASS CIERA/CRUISER, ACHIEVA, GRAND AM AND SKYLARK

DTC 21
THROTTLE POSITION (TP) SENSOR CIRCUIT
(SIGNAL VOLTAGE HIGH)
3300 (VIN N) (MFI)

Circuit Description:

The Throttle Position (TP) sensor provides a voltage signal that changes relative to throttle blade angle. Signal voltage will vary from about .2 - .74 volt at idle to above 4 volts at wide open throttle.

The TP sensor signal is one of the most important inputs used by the ECM for fuel control and for most of the ECM control outputs.

DTC 21 will set if:
- TP sensor signal voltage is greater than 5 volts at any time.
  OR
- Engine is running and air flow is less than 15 gm/sec.
- Throttle angle is greater than 30%.
- DTC 34 not present.
- All conditions met for 5 seconds.

Test Description: Number(s) refer to circled number(s) on the diagnostic chart.

1. With closed throttle, ignition "ON," or at idle, voltage at "YF13" should be .2-.74 volt. If not, see "Fuel Metering System."

2. When the TP sensor is disconnected, the TP sensor voltage should go low and DTC 22 will set. This test verifies that the ECM and wiring are OK.

3. Probing CKT 452 with a test light checks the sensor ground circuit. A faulty sensor ground circuit will cause DTC 21 to be set.

NOTICE: If DTC 21 is set, the ECM will use a fail-soft value for TP sensor of 3% throttle angle. The Tech 1 will continue to display actual sensor value.

Diagnostic Aids:

The Tech 1 displays throttle position in volts. With closed throttle, ignition "ON" or at idle, voltage should be .2-.74 volt. If not, see "Fuel Metering System," for replacement procedures.

An open in CKT 452 will result in DTC 21.

Check For:
- Poor Connection or Damaged Harness. Inspect ECM harness connectors for backed out terminal "BB6", improper mating, broken locks, improperly formed or damaged terminals, poor terminal to wire connection, and damaged harness.
- Intermittent Test. If connections and harness check OK, monitor TP sensor voltage while moving related connectors and wiring harness. If the failure is induced, the display will change. This may help to isolate the location of the malfunction.
- TP Sensor Scaling. Observe TP sensor voltage display while depressing accelerator pedal with engine stopped and ignition "ON." Display should vary from .2 - .74 volt when throttle is closed to over 4 volts when held at wide open throttle position.

3.3L (VIN N) ENGINE — DIAGNOSTIC TROUBLE CODE CHART — CENTURY, CUTLASS CIERA/CRUISER, ACHIEVA, GRAND AM AND SKYLARK

DTC 21
THROTTLE POSITION (TP) SENSOR CIRCUIT
(SIGNAL VOLTAGE HIGH)
3300 (VIN N)   (MFI)

1. • IGNITION "ON," INSTALL TECH 1.
   • THROTTLE CLOSED.
   • DOES TECH 1 DATA DISPLAY INDICATE THROTTLE POSITION OVER 1.5 VOLTS?

YES / NO

2. • DISCONNECT THROTTLE POSITION SENSOR. TECH 1 DATA DISPLAY SHOULD INDICATE THROTTLE POSITION BELOW .2 VOLT (200 mV). DOES IT?

• DTC 21 IS INTERMITTENT. IF NO ADDITIONAL DTC(S) WERE STORED, REFER TO "DIAGNOSTIC AIDS"

YES / NO

3. • PROBE SENSOR GROUND CIRCUIT WITH A TEST LIGHT CONNECTED TO B +.

• TP SENSOR SIGNAL CIRCUIT SHORTED TO VOLTAGE OR FAULTY ECM.

LIGHT "ON" / LIGHT "OFF"

FAULTY CONNECTION OR THROTTLE POSITION SENSOR.

OPEN SENSOR GROUND CIRCUIT OR FAULTY ECM.

"AFTER REPAIRS," REFER TO DTC CRITERIA   AND CONFIRM DTC DOES NOT RESET.

3.3L (VIN N) ENGINE — DIAGNOSTIC TROUBLE CODE CHART — CENTURY, CUTLASS CIERA/CRUISER, ACHIEVA, GRAND AM AND SKYLARK

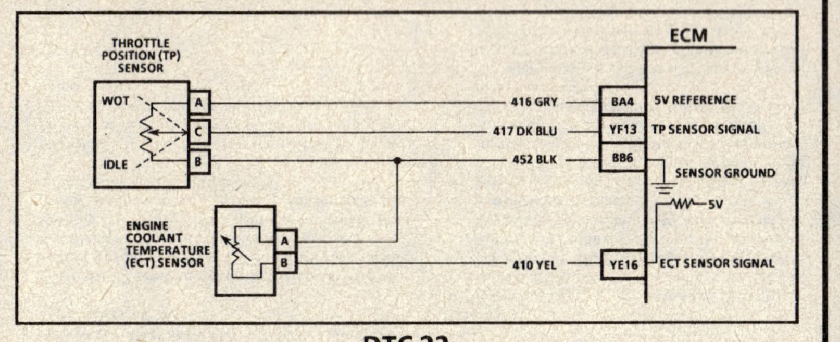

DTC 22
THROTTLE POSITION (TP) SENSOR CIRCUIT
(SIGNAL VOLTAGE LOW)
3300 (VIN N)   (MFI)

Circuit Description:

The Throttle Position (TP) sensor provides a voltage signal that changes relative to throttle blade angle. Signal voltage will vary from .2 - .74 volt at idle to above 4 volts at wide open throttle.

The TP sensor signal is one of the most important inputs used by the ECM for fuel control and for most of the ECM control outputs.

DTC 22 will set if:
- The ignition key is "ON."
- TP sensor signal voltage is less than .2 volt for 4 seconds.

Test Description: Number(s) refer to circled number(s) on the diagnostic chart.

1. With closed throttle ignition "ON," TP sensor voltage should read .2 - .74 volt.
2. Simulates DTC 21: (high voltage). If ECM recognizes the high signal voltage the ECM and wiring are OK.
3. Simulates a high signal voltage. Checks CKT 417 for an open.

NOTICE: If DTC 22 is set, the ECM will use a fail-soft value for TP sensor of about 3% throttle angle. The Tech 1 will continue to display actual sensor value.

Diagnostic Aids:

The Tech 1 displays throttle position in volts. Voltage should increase at a steady rate as throttle is moved toward Wide Open Throttle (WOT.)

The Tech 1 also displays throttle angle: 0% = closed throttle 100% = WOT.

An open or short to ground in CKTs 416 or 417 will result in DTC 22 to be set.

Check For:
- **Poor Connection or Damaged Harness.** Inspect ECM harness connectors for backed out terminal "YF13" or "BA4", improper mating, broken locks, improperly formed or damaged terminals, poor terminal to wire connection, and damaged harness.
- **Intermittent Test.** If connections and harness check OK, monitor TP sensor voltage display while moving related connectors and wiring harness. If the failure is induced, the display will change. This may help to isolate the location of the malfunction.
- **TP Sensor Scaling.** Observe TP sensor voltage display while depressing accelerator pedal with engine stopped and ignition "ON." Display should vary from .2 - .74 volt when throttle is closed to over 4 volts when throttle is held at wide open throttle position.

3.3L (VIN N) ENGINE — DIAGNOSTIC TROUBLE CODE CHART — CENTURY, CUTLASS CIERA/CRUISER, ACHIEVA, GRAND AM AND SKYLARK

DTC 22

THROTTLE POSITION (TP) SENSOR CIRCUIT
(SIGNAL VOLTAGE LOW)
3300 (VIN N)   (MFI)

1. • THROTTLE CLOSED.
   DOES TECH 1 DATA DISPLAY INDICATE THROTTLE POSITION .2 VOLT OR BELOW?

   YES →

2. • DISCONNECT TP SENSOR.
   • JUMPER CKTs 416 AND 417 TOGETHER. TECH 1 DATA DISPLAY SHOULD INDICATE THROTTLE POSITION OVER 4.0 VOLTS. DOES IT?

   NO ↓

3. • PROBE CKT 417 WITH A TEST LIGHT CONNECTED TO B +. TECH 1 DATA DISPLAY SHOULD INDICATE THROTTLE POSITION OVER 4.0 VOLTS. DOES IT?

   YES ↓

   CKT 416 OPEN OR SHORTED TO GROUND
   OR
   POOR CONNECTION
   OR
   FAULTY ECM.

   NO →
   DTC 22 IS INTERMITTENT.
   IF NO ADDITIONAL DTC(S) WERE STORED, REFER TO "DIAGNOSTIC AIDS"

   YES →
   REPLACE TP SENSOR.

   NO →
   CKT 417 OPEN OR SHORTED TO GROUND, OR SHORTED TO THROTTLE POSITION SENSOR GROUND CIRCUIT
   OR
   POOR ECM CONNECTION
   OR
   FAULTY ECM.

"AFTER REPAIRS," REFER TO DTC CRITERIA   AND CONFIRM DTC DOES NOT RESET.

3.3L (VIN N) ENGINE — DIAGNOSTIC TROUBLE CODE CHART — CENTURY, CUTLASS CIERA/CRUISER, ACHIEVA, GRAND AM AND SKYLARK

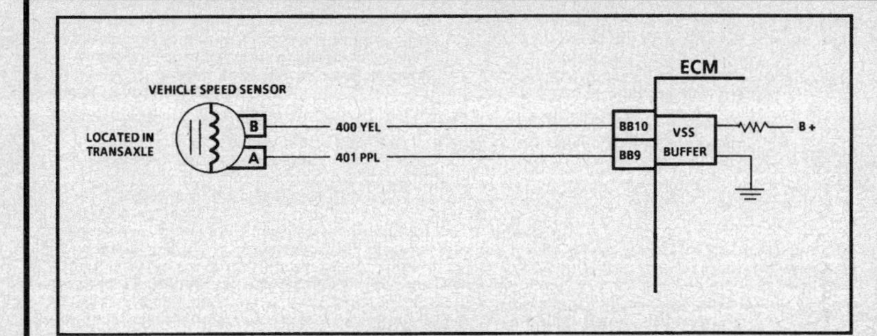

DTC 24

VEHICLE SPEED SENSOR (VSS) CIRCUIT
3300 (VIN N)   (MFI)

Circuit Description:

Vehicle speed information is provided to the ECM by the Vehicle Speed Sensor (VSS), a Permanent Magnet (PM) generator mounted in the transmission. The PM generator produces a pulsing voltage whenever vehicle speed is over about 3 mph. The AC voltage level and the number of pulses increases with vehicle speed. The ECM converts the pulsing voltage to mph, which can be monitored with a Tech 1.

The function of the VSS buffer used in past model years has been incorporated into the PROM portion of the ECM. The ECM supplies the necessary signal (4004 pulses per mile) for operating the speedometer and cruise control.

DTC 24 will set if:
- Engine speed greater than 1000 RPM.
- No DTC 31.
- No DTC 28 (3T40) or DTC 29 (4T60).
- PNP switch is open (indicating drive range).
- High gear switch is open indicating transmission is in 3rd (3T40) or 4th (4T60) gear.
- VSS signal indicates a vehicle speed less than 3 mph.
- All conditions met for 20 seconds.

Test Description: Number(s) refer to circled number(s) on the diagnostic chart.
1. The PM generator only produces a signal if drive wheels are turning greater then 3 mph.
2. Before replacing the ECM, check the PROM for correct application.

Diagnostic Aids:

The Tech 1 should indicate a vehicle speed whenever the drive wheels are turning greater than 3 mph.

Check CKTs 400 and 401 for proper connections to be sure they are clean and tight and the harness is routed correctly. Refer to "Intermittents" in "Symptoms."

3.3L (VIN N) ENGINE — DIAGNOSTIC TROUBLE CODE CHART — CENTURY, CUTLASS CIERA/CRUISER, ACHIEVA, GRAND AM AND SKYLARK

DTC 24
VEHICLE SPEED SENSOR (VSS) CIRCUIT
3300 (VIN N) **(MFI)**

① • RAISE DRIVE WHEELS.
• **NOTICE:** DO NOT PERFORM THIS TEST WITHOUT SUPPORTING THE LOWER CONTROL ARMS SO THAT THE DRIVE AXLES ARE IN A NORMAL HORIZONTAL POSITION. RUNNING THE VEHICLE IN GEAR WITH THE WHEELS HANGING DOWN AT FULL TRAVEL MAY DAMAGE THE DRIVE AXLES.
• WITH ENGINE IDLING IN GEAR, TECH 1 VEHICLE SPEED SHOULD DISPLAY MPH ABOVE 0. DOES IT?

NO
• OBSERVE SPEEDOMETER WITH WHEELS TURNING. DOES SPEEDOMETER WORK?

YES
DTC 24 IS INTERMITTENT. IF NO ADDITIONAL DTC(S) WERE STORED, REFER TO "DIAGNOSTIC AIDS" ON

NO
• SET DVM TO 20 VOLT AC SCALE.
• BACKPROBE ECM "BB10" WITH A VOLTMETER TO GROUND.
• PLACE TRANS IN DRIVE, ALLOW WHEELS TO TURN.

YES
② FAULTY ECM

• VOLTAGE BELOW .5 VOLT AC

• VOLTAGE ABOVE .5 VOLT AC

• CHECK CKT 400, 401 FOR OPEN
OR
• CKT 400 SHORTED TO GROUND.

• CHECK TERMINAL CONTACT AT ECM PINS "BB9" AND "BB10".
• IF CONTACT IS OK, REPLACE ECM.

CKT OK

POOR CONNECTION AT VEHICLE SPEED SENSOR OR FAULTY SENSOR.

OPEN OR SHORTED TO GROUND

REPAIR CIRCUIT WHICH IS OPEN OR SHORTED TO GROUND.

"AFTER REPAIRS," REFER TO DTC CRITERIA   AND CONFIRM DTC DOES NOT RESET.

3.3L (VIN N) ENGINE — DIAGNOSTIC TROUBLE CODE CHART — CENTURY, CUTLASS CIERA/CRUISER, ACHIEVA, GRAND AM AND SKYLARK

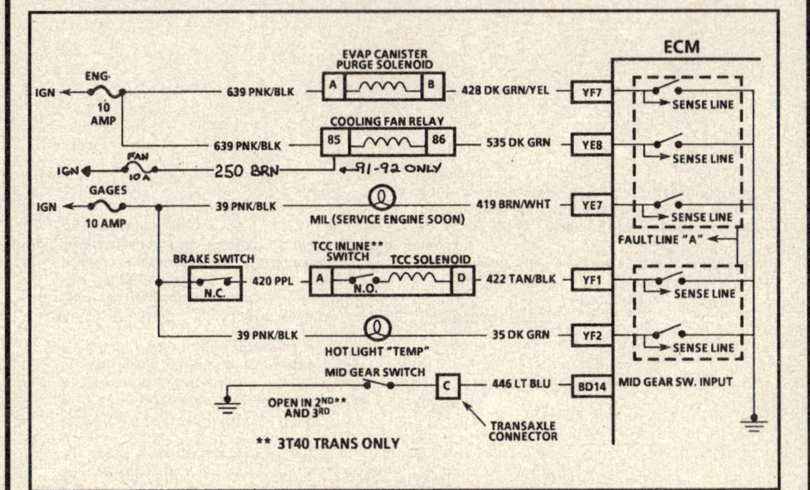

DTC 26
(Page 1 of 3)
QUAD-DRIVER (QDM) CIRCUIT
3300 (VIN N) "A" CARLINE (MFI)

Circuit Description:

Quad-Driver Modules (QDM) are used to control the components shown in the illustration above. When the ECM is commanding a component "ON," the QDM closes the switch completing the circuit to ground.

Each QDM output has a sense line. When a component is commanded "ON," the voltage potential on the sense line is low, and when the component is commanded "OFF" the voltage potential on the sense line is high.

DTC 26 will set when the ECM is commanding a component "ON" and the voltage potential on a QDM "A" sense line is high, or if the component is commanded "OFF" and the voltage potential on a QDM "A" sense line is low. Either of these conditions will cause the QDM "A" fault line to display "HIGH" and set DTC 26.

On all vehicles, the QDM "A" fault status on the Tech 1 will display "HIGH" with the brake applied or if the engine is not running. On vehicles with a 3T40 transmission, the QDM "A" fault status on the Tech 1 will read "HIGH" until the TCC inline switch is closed. These conditions will not cause DTC 26 to set even though the fault line displays "HIGH." To simulate driving in second gear and change the QDM status to low, disconnect TCC connector and connect a test light between harness terminals "A" and "D"

DTC 26 will set if:
• The system is running in "Closed Loop."
• Brake switch is closed (brake not applied).
• Transaxle in mid or high gear.
• The ECM detects an improper voltage level on a QDM "A" controlled circuit for 10 seconds.

Test Description: Number(s) refer to circled number(s) on the diagnostic chart.

1. The ECM does not know which controlled circuit caused DTC 26 so this chart will go through each of the circuits to determine which is at fault. This test checks the MIL (Service Engine Soon) driver and the MIL (Service Engine Soon) circuit.

2. QDM symptoms:
• TCC - Malfunctioning - DTC 39.
• Hot Light - "ON" all the time, "OFF" during bulb check.
• Cooling fan "ON" all the time or won't come "ON" at all, see "Electric Cooling Fan," Section "6E3-C12".
• Poor driveability due to 100% canister purge.

3.3L (VIN N) ENGINE — DIAGNOSTIC TROUBLE CODE CHART — CENTURY, CUTLASS CIERA/CRUISER, ACHIEVA, GRAND AM AND SKYLARK

DTC 26
(Page 1 of 3)
QUAD-DRIVER (QDM) CIRCUIT
3300 (VIN N) "A" CARLINE (MFI)

1. VERIFY PROPER OPERATION OF MIL (SERVICE ENGINE SOON) BY PERFORMING ONBOARD DIAGNOSTIC SYSTEM CHECK.

- KEY "ON," ENGINE "OFF."
- NOTE ENGINE COOLANT TEMPERATURE LIGHT (HOT LIGHT). LIGHT SHOULD BE "ON." IS IT?

YES →

- START ENGINE.
- NOTE ENGINE COOLANT TEMPERATURE LIGHT (HOT LIGHT). LIGHT SHOULD BE "OFF." IS IT?

NO → SEE CHART C-16 FOR HOT LIGHT DIAGNOSIS.

YES →

- INSTALL TECH 1.
- KEY "ON," ENGINE "OFF."
- NOTE DTC(s)

NO → SEE CHART C-16 FOR HOT LIGHT DIAGNOSIS.

- DTC 26 ONLY → SEE PAGE 2 OF THIS CHART APPLICABLE TO CORRECT TRANSAXLE (3T40 OR 4T60).

2. DTC 26 PLUS ADDITIONAL DTC(S) (QDM RELATED DTC(S). → GO TO THOSE DTC(S) FIRST.

"AFTER REPAIRS," REFER TO DTC CRITERIA   AND CONFIRM DTC DOES NOT RESET.

3.3L (VIN N) ENGINE — DIAGNOSTIC TROUBLE CODE CHART — CENTURY, CUTLASS CIERA/CRUISER, ACHIEVA, GRAND AM AND SKYLARK

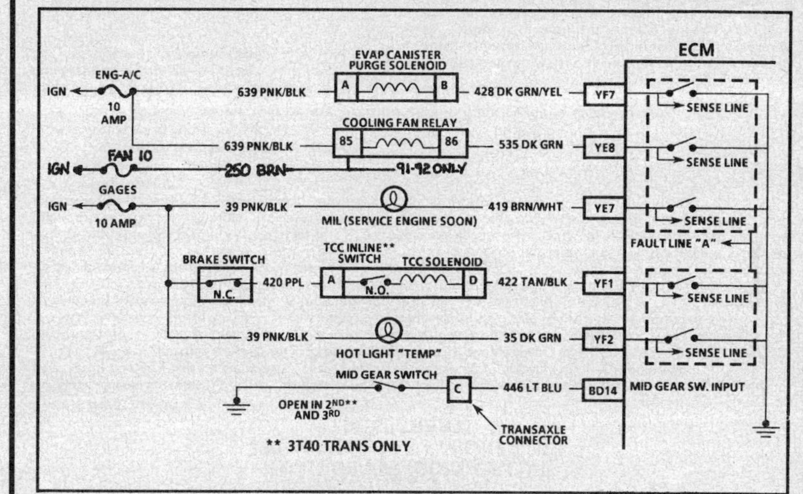

DTC 26 (Page 2 of 3)
QUAD-DRIVER (QDM) CIRCUIT
3300 (VIN N) "A" CARLINE (MFI)
(3T40 TRANS)

Circuit Description:

Quad-Driver Modules (QDM) are used to control the components shown in the illustration above. When the ECM is commanding a component "ON," the QDM closes the switch completing the circuit to ground.

Each QDM output has a sense line. When a component is commanded "ON," the voltage potential on the sense line is low, and when the component is commanded "OFF" the voltage potential on the sense line is high.

DTC 26 will set when the ECM is commanding a component "ON" and the voltage potential on a QDM "A" sense line is high, or if the component is commanded "OFF" and the voltage potential on a QDM "A" sense line is low. Either of these conditions will cause the QDM "A" fault line to display "HIGH" and set DTC 26.

On all vehicles, the QDM "A" fault status on the Tech 1 will display "HIGH" with the brake applied or if the engine is not running. On vehicles with a 3T40 transmission, the QDM "A" fault status on the Tech 1 will read "HIGH" until the TCC inline switch is closed. These conditions will not cause DTC 26 to set even though the fault line displays "HIGH." To simulate driving in second gear and change the QDM status to low, disconnect TCC connector and connect a test light between harness terminals "A" and "D".

DTC 26 will set if:
- The system is running in "Closed Loop."
- Brake switch is closed (brake not applied).
- Transaxle in mid or high gear.
- The ECM detects an improper voltage level on a QDM "A" controlled circuit for 10 seconds.

Test Description: Number(s) refer to circled number(s) on the diagnostic chart.

3. This test will determine which circuit is out of specifications.

Diagnostic Aids:

Monitor the voltage at each terminal while moving related harness connectors, including ECM harness. If the failure is induced, the voltage will change. This may help locate the intermittent. Check for bent pins at ECM and ECM connector terminals. Also, ensure that the mid gear switch and CKT 446 is not open with the vehicle operating in park/neutral or first gear. This condition can cause a false DTC 26 to be set. Refer to DTC 27 chart for diagnosis. If DTC reoccurs with no apparent connector problem, replace ECM.

3.3L (VIN N) ENGINE — DIAGNOSTIC TROUBLE CODE CHART — CENTURY, CUTLASS CIERA/CRUISER, ACHIEVA, GRAND AM AND SKYLARK

DTC 26
(Page 2 of 3)
QUAD-DRIVER (QDM) CIRCUIT 3300 (VIN N) "A" CARLINE (MFI) (3T40 TRANS)

NOTICE: IF DTC 39 IS SET ALONG WITH DTC 26, DTC 39 MUST BE DIAGNOSED FIRST.

(3)
- KEY "OFF."
- DISCONNECT YELLOW ECM CONNECTOR.
- KEY "ON," ENGINE "OFF."
- DVM SET TO 2 AMP SCALE.
- PROBE ECM HARNESS TERMINALS "YE7", "YE8", "YF2", AND "YF7".
  - RED LEAD (+) TO ECM HARNESS.
  - BLACK LEAD (-) TO GROUND.
- EACH TERMINAL SHOULD MEASURE LESS THAN .75 AMP (BUT NOT 0 AMP).

OK →
- CHECK VOLTAGE AT ECM HARNESS TERMINAL "YF1" WITH A DVM TO GROUND.
- SHOULD READ NEAR 0 VOLTS. DOES IT?

NOT OK → SEE PAGE 3

YES → CHECK CKT 422 FOR A SHORT TO GROUND AND REPAIR IF GROUNDED. IF OK, REFER TO "DIAGNOSTIC AIDS"

NO → FAULTY TCC IN-LINE SWITCH OR CKT 422 SHORTED TO VOLTAGE.

"AFTER REPAIRS," REFER TO DTC CRITERIA AND CONFIRM DTC DOES NOT RESET.

3.3L (VIN N) ENGINE — DIAGNOSTIC TROUBLE CODE CHART — CENTURY, CUTLASS CIERA/CRUISER, ACHIEVA, GRAND AM AND SKYLARK

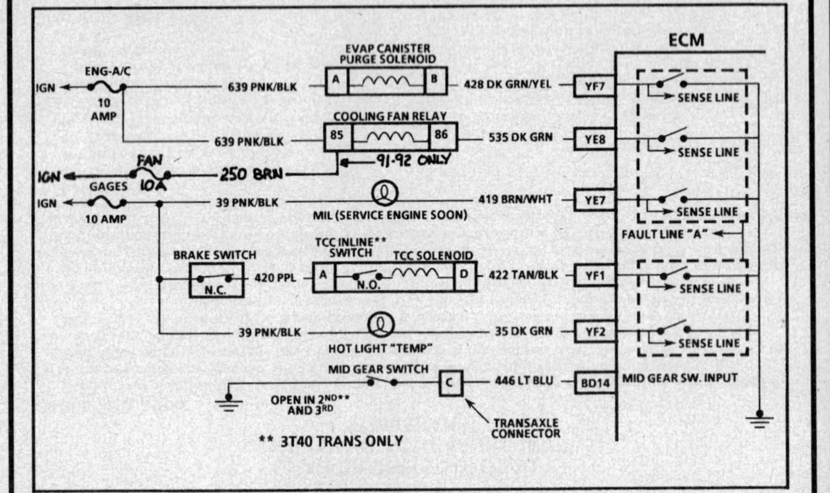

DTC 26 (Page 2 of 3)
QUAD-DRIVER (QDM) CIRCUIT 3300 (VIN N) "A" CARLINE (MFI) (4T60 TRANS)

Circuit Description:

Quad-Driver Modules (QDM) are used to control the components shown in the illustration above. When the ECM is commanding a component "ON," the QDM closes the switch completing the circuit to ground.

Each QDM output has a sense line. When a component is commanded "ON," the voltage potential on the sense line is low, and when the component is commanded "OFF" the voltage potential on the sense line is high.

DTC 26 will set when the ECM is commanding a component "ON" and the voltage potential on a QDM "A" sense line is high, or if the component is commanded "OFF" and the voltage potential on a QDM "A" sense line is low. Either of these conditions will cause the QDM "A" fault line to display "HIGH" and set DTC 26.

On all vehicles, the QDM "A" fault status on the Tech 1 will display "HIGH" with the brake applied or if the engine is not running. On vehicles with a 3T40 transmission, the QDM "A" fault status on the Tech 1 will read "HIGH" until the TCC inline switch is closed. These conditions will not cause DTC 26 to set even though the fault line displays "HIGH." To simulate driving in second gear and change the QDM status to low, disconnect TCC connector and connect a test light between harness terminals "A" and "D".

DTC 26 will set if:
- The system is running in "Closed Loop."
- Brake switch is closed (brake not applied).
- Transaxle in mid or high gear.
- The ECM detects an improper voltage level on a QDM "A" controlled circuit for 10 seconds.

Test Description: Number(s) refer to circled number(s) on the diagnostic chart.
3. This test will determine which circuit is out of specifications.

Diagnostic Aids:

Monitor the voltage at each terminal while moving related harness connectors, including ECM harness. If the failure is induced, the voltage will change. This may help locate the intermittent. Check for bent pins at ECM and ECM connector terminals. If DTC reoccurs with no apparent connector problem, replace ECM.

3.3L (VIN N) ENGINE — DIAGNOSTIC TROUBLE CODE CHART — CENTURY, CUTLASS CIERA/CRUISER, ACHIEVA, GRAND AM AND SKYLARK

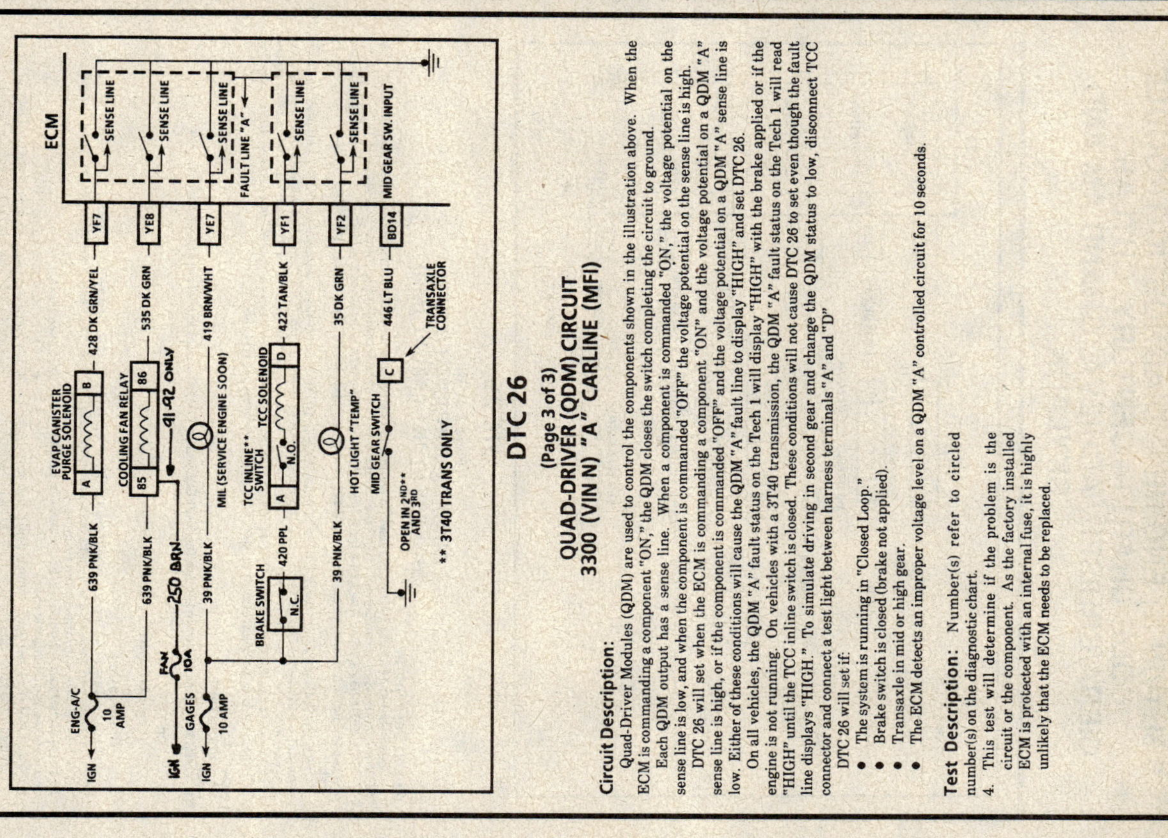

DTC 26
(Page 3 of 3)
QUAD-DRIVER (QDM) CIRCUIT
3300 (VIN N) "A" CARLINE (MFI)

Circuit Description:

Quad-Driver Modules (QDM) are used to control the components shown in the illustration above. When the ECM is commanding a component "ON," the QDM closes the switch completing the circuit to ground.

Each QDM output has a sense line. When a component is commanded "ON," the voltage potential on the sense line is low, and when the component is commanded "OFF" the voltage potential on the sense line is high.

DTC 26 will set when the ECM is commanding a component "ON" and the voltage potential on a QDM "A" sense line is high, or if the component is commanded "OFF" and the voltage potential on a QDM "A" sense line is low. Either of these conditions will cause the QDM "A" fault line to display "HIGH" and set DTC 26.

On all vehicles, the QDM "A" fault status on the Tech 1 will display "HIGH" with the brake applied or if the engine is not running. On vehicles with a 3T40 transmission, the QDM "A" fault status on the Tech 1 will read "HIGH" until the TCC inline switch is closed. These conditions will not cause DTC 26 to set even though the fault line displays "HIGH." To simulate driving in second gear and change the QDM status to low, disconnect TCC connector and connect a test light between harness terminals "A" and "D"

DTC 26 will set if:
- The system is running in "Closed Loop."
- Brake switch is closed (brake not applied).
- Transaxle in mid or high gear.
- The ECM detects an improper voltage level on a QDM "A" controlled circuit for 10 seconds.

Test Description: Number(s) refer to circled number(s) on the diagnostic chart.

4. This test will determine if the problem is the circuit or the component. As the factory installed ECM is protected with an internal fuse, it is highly unlikely that the ECM needs to be replaced.

3.3L (VIN N) ENGINE — DIAGNOSTIC TROUBLE CODE CHART — CENTURY, CUTLASS CIERA/CRUISER, ACHIEVA, GRAND AM AND SKYLARK

DTC 26
(Page 2 of 3)
QUAD-DRIVER (QDM) CIRCUIT
3300 (VIN N) "A" CARLINE (MFI) (4T60 TRANS)

(3)

- KEY "OFF."
- DISCONNECT YELLOW ECM CONNECTOR.
- KEY "ON," ENGINE "OFF."
- DVM SET TO 2 AMP SCALE.
- PROBE ECM HARNESS TERMINALS "YE7", "YE8", "YF1", "YF2", AND "YF7".
  - RED LEAD (+) TO ECM HARNESS.
  - BLACK LEAD (-) TO GROUND.
- EACH TERMINAL SHOULD MEASURE LESS THAN .75 AMP (BUT NOT 0 AMP).

OK → REFER TO "DIAGNOSTIC AIDS"

NOT OK → SEE PAGE 3

"AFTER REPAIRS," REFER TO DTC CRITERIA AND CONFIRM DTC DOES NOT RESET.

3.3L (VIN N) ENGINE — DIAGNOSTIC TROUBLE CODE CHART — CENTURY, CUTLASS CIERA/CRUISER, ACHIEVA, GRAND AM AND SKYLARK

DTC 26
(Page 3 of 3)
QUAD-DRIVER (QDM) CIRCUIT
3300 (VIN N) "A" CARLINE (MFI)

CIRCUIT ISOLATED FROM PRIOR CHARTS

(4)
- USE FACING PAGE WIRING DIAGRAM FOR SPECIFIC TERMINALS TO BE TESTED.
- RECONNECT ECM, IF APPLICABLE.
- KEY "ON," ENGINE "OFF," DLC DIAGNOSTIC REQUEST TERMINAL GROUNDED.
- REMOVE CONNECTOR FROM RELAY/SOLENOID IN AFFECTED CIRCUIT.
- PLACE TEST LIGHT ACROSS TERMINALS FOR COMPONENT IGNITION FEED AND ECM DRIVER CIRCUIT.

LIGHT SHOULD BE "ON."
IS IT?

NO
- CONNECT TEST LIGHT FROM COMPONENT IGNITION FEED CIRCUIT TO GROUND.
- NOTE LIGHT.

"ON"
- DISCONNECT ECM CONNECTOR.
- CHECK FOR OPEN IN DRIVER CIRCUIT.

NOT OK: REPAIR OPEN.

OK: ECM CONN. OR ECM.

"OFF": REPAIR OPEN IN COMPONENT IGNITION FEED CIRCUIT.

YES
- REMOVE JUMPER AT DLC.
- NOTE LIGHT.

"ON"
- DISCONNECT ECM CONNECTOR.
- NOTE LIGHT.

"ON": REPAIR GROUNDED CIRCUIT.

"OFF": REPLACE ECM

"OFF": CHECK FOR POOR CONNECTIONS. IF OK, REPLACE COMPONENT.

NOTICE: WITH THE INTRODUCTION OF GMP4 ECMS, DELCO ELECTRONICS CHANGED TO INTERNALLY PROTECTED QUAD DRIVERS. THIS INTERNAL PROTECTION CAN BE COMPARED TO THE ACTION OF A CIRCUIT BREAKER. IF TOO MUCH CURRENT FLOWS IN A CONTROLLED CIRCUIT, THIS TYPE OF QUAD DRIVER TURNS ITSELF "OFF." THIS ALLOWS THE QUAD DRIVER TO SURVIVE A SHORTED SOLENOID, RELAY, OR HARNESS. AND IT MEANS YOU DO NOT HAVE TO REPLACE THE ECM. JUST REPAIR THE FAULT IN THE OUTPUT CIRCUIT AND THE QUAD DRIVER WILL RETURN TO NORMAL OPERATION.

"*AFTER REPAIRS,*" REFER TO DTC CRITERIA AND CONFIRM DTC DOES NOT RESET.

3.3L (VIN N) ENGINE — DIAGNOSTIC TROUBLE CODE CHART — CENTURY, CUTLASS CIERA/CRUISER, ACHIEVA, GRAND AM AND SKYLARK

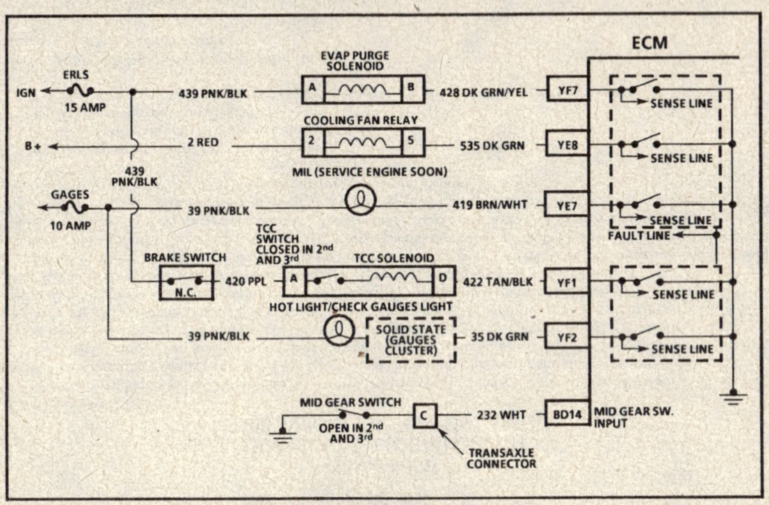

DTC 26
(Page 1 of 3)
QUAD-DRIVER (QDM) CIRCUIT
3300 (VIN N) "N" CARLINE (MFI)

Circuit Description:

Quad-Driver Modules (QDM) are used to control the components shown in the illustration above. When the ECM is commanding a component "ON," the QDM closes the switch completing the circuit to ground.

Each QDM has a sense line and a fault line. When a component is commanded "ON," the voltage potential on the sense line is low and when the component is commanded "OFF," the voltage potential on the sense line is high.

DTC 26 will set when the ECM is commanding a QDM A controlled component "ON" and the voltage potential on the sense line is high, or if the component is commanded "OFF" and the voltage potential on the sense line is low. The fault line, which can be monitored with the Tech 1, will read "HIGH" under these conditions. If the ECM detects a "HIGH" QDM A fault line when it expects a "LOW" one, it will illuminate the MIL (Service Engine Soon) and store DTC 26. QDM B will **not** set DTC 26.

Important

- DTC 26 will not set even though the fault line scans "HIGH" under the following conditions:
  - Ignition "ON"/engine not running.

  OR
  - Vehicle not operating in 2nd or 3rd gear.

  OR
  - Brake applied.

Vehicles with a 3T40 transaxle will read "HIGH" until the TCC in-line switch is closed. To check, disconnect TCC connector and connect a test light between terminals "A" and "D". This should cause the Tech 1 to display "LOW" for QDM A with the engine running.

Test Description: Number(s) below refer to circled number(s) on the diagnostic chart.

1. The ECM does not know which controlled circuit caused DTC 26 so this chart will go through each of the circuits to determine which is at fault. This test checks the MIL (Service Engine Soon) driver and the MIL (Service Engine Soon) circuit.
2. QDM symptoms:
   - TCC - malfunctioning - DTC 39.
   - "Check gauges" or hot light - "ON" all the time/"OFF" during bulb check.
   - Cooling fan "ON" all the time or won't come "ON" at all.
   - Poor driveability due to 100% canister purge.

3.3L (VIN N) ENGINE — DIAGNOSTIC TROUBLE CODE CHART — CENTURY, CUTLASS CIERA/CRUISER, ACHIEVA, GRAND AM AND SKYLARK

DTC 26
(Page 1 of 3)
QUAD-DRIVER (QDM) CIRCUIT
3300 (VIN N) "N" CARLINE (MFI)

1. VERIFY PROPER OPERATION OF MIL (SERVICE ENGINE SOON) BY PERFORMING ON-BOARD DIAGNOSTIC SYSTEM CHECK.

- KEY "ON," ENGINE "OFF."
- NOTE "CHECK GAUGES" LIGHT OR ENGINE COOLANT TEMPERATURE LIGHT (HOT LIGHT) AS APPLICABLE. LIGHT SHOULD BE "ON." IS IT?

YES
- START ENGINE.
- NOTE ENGINE COOLANT TEMPERATURE LIGHT (HOT LIGHT) OR "CHECK GAUGES" LIGHT (AS APPLICABLE). LIGHT SHOULD BE "OFF." IS IT?

NO
- SEE HOT LIGHT/"CHECK GAUGES" LIGHT DIAGNOSIS.

YES
- INSTALL TECH 1.
- KEY "ON," ENGINE "OFF."
- NOTE DTC(s).

NO
- SEE HOT LIGHT/"CHECK GAUGES" LIGHT DIAGNOSIS.

- DTC 26 ONLY

  SEE PAGE 2 OF THIS CHART.

2. DTC 26 PLUS ADDITIONAL DTC(s) [QDM RELATED DTC(s)].

  GO TO THOSE DTC(s) FIRST.

"AFTER REPAIRS," REFER TO DTC CRITERIA AND CONFIRM DTC DOES NOT RESET.

3.3L (VIN N) ENGINE — DIAGNOSTIC TROUBLE CODE CHART — CENTURY, CUTLASS CIERA/CRUISER, ACHIEVA, GRAND AM AND SKYLARK

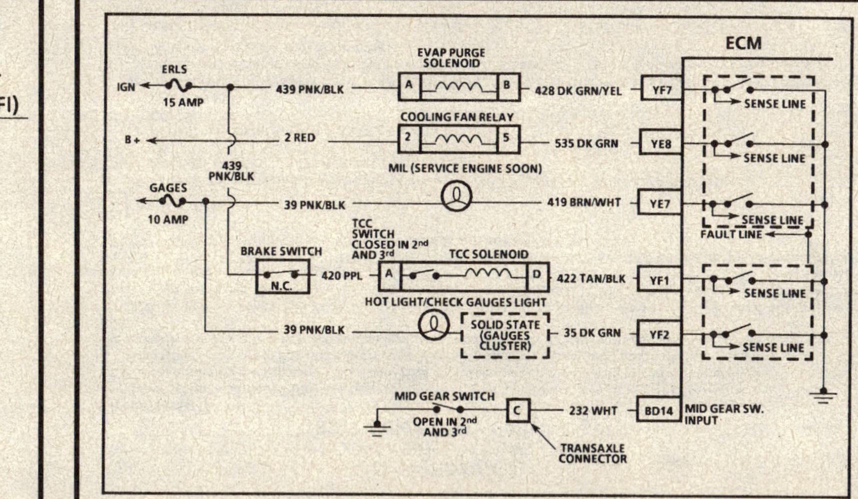

DTC 26
(Page 2 of 3)
QUAD-DRIVER (QDM) CIRCUIT
3300 (VIN N) "N" CARLINE (MFI)

Circuit Description:

Quad-Driver Modules (QDM) are used to control the components shown in the illustration above. When the ECM is commanding a component "ON," the QDM closes the switch completing the circuit to ground.

Each QDM has a sense line and a fault line. When a component is commanded "ON," the voltage potential on the sense line is low and when the component is commanded "OFF," the voltage potential on the sense line is high.

DTC 26 will set when the ECM is commanding a QDM A controlled component "ON" and the voltage potential on the sense line is high, or if the component is commanded "OFF" and the voltage potential on the sense line is low. QDM B will **not** set DTC 26.

Vehicles with a 3T40 transaxle will read "HIGH" until the TCC in-line switch is closed. To check, disconnect TCC connector and connect a test light between terminals "A" and "D". This should cause the Tech 1 to display "LOW" for QDM A with the engine running.

Test Description: Number(s) below refer to circled number(s) on the diagnostic chart.

3. This test will determine which circuit is out of specifications.

Diagnostic Aids:

Monitor the voltage at each terminal while moving related harness connectors, including ECM harness. If the failure is induced, the voltage will change. This may help locate the intermittent. Check for bent pins at ECM and ECM connector terminals. Also, ensure that the mid gear switch and CKT 232 is not open with the vehicle operating in park/neutral or 1st gear - This condition can cause a false DTC 26 to be set. Refer to DTC 27 chart for diagnosis. If DTC re-occurs with no apparent connector problem, replace ECM.

3.3L (VIN N) ENGINE — DIAGNOSTIC TROUBLE CODE CHART — CENTURY, CUTLASS CIERA/CRUISER, ACHIEVA, GRAND AM AND SKYLARK

DTC 26
(Page 2 of 3)
QUAD-DRIVER (QDM) CIRCUIT
3300 (VIN N) "N" CARLINE (MFI)

NOTICE: IF DTC 39 IS SET ALONG WITH DTC 26, DTC 39 MUST BE DIAGNOSED FIRST.

③
- KEY "OFF."
- DISCONNECT YELLOW ECM CONNECTOR.
- KEY "ON," ENGINE "OFF."
- DVM SET TO mA/A SCALE.
- PROBE ECM HARNESS TERMINALS "YE7", "YE8", AND "YF7".
  - RED LEAD (+) TO ECM HARNESS.
  - BLACK LEAD (-) TO GROUND.
- EACH TERMINAL SHOULD MEASURE LESS THAN .75 AMP (BUT NOT 0 AMP).

OK

NOT OK → SEE PAGE 3

- DVM STILL SET TO mA/A SCALE.
- PROBE ECM HARNESS TERMINAL "YF2".
  - RED LEAD (+) TO ECM HARNESS.
  - BLACK LEAD (-) TO GROUND.
- "YF2" SHOULD MEASURE LESS THAN .75A (BUT NOT 0 mA). DOES IT?

YES

- CHECK VOLTAGE AT ECM HARNESS TERMINAL "YF1" WITH A DVM TO GROUND.
- SHOULD READ NEAR 0 VOLTS. DOES IT?

NO
- IF EQUIPPED WITH "TEMP" LAMP, CHECK FOR FAULTY BULB. IF BULB IS OK, REPLACE CLUSTER.
- IF EQUIPPED WITH "CHECK GAUGES" LAMP, REPLACE CLUSTER.

YES

CHECK CKT 422 FOR A SHORT TO GROUND AND REPAIR IF GROUNDED. IF OK, REFER TO "DIAGNOSTIC AIDS"

NO

FAULTY TCC IN-LINE SWITCH OR CKT 422 SHORTED TO VOLTAGE.

"AFTER REPAIRS," REFER TO DTC CRITERIA   AND CONFIRM DTC DOES NOT RESET.

3.3L (VIN N) ENGINE — DIAGNOSTIC TROUBLE CODE CHART — CENTURY, CUTLASS CIERA/CRUISER, ACHIEVA, GRAND AM AND SKYLARK

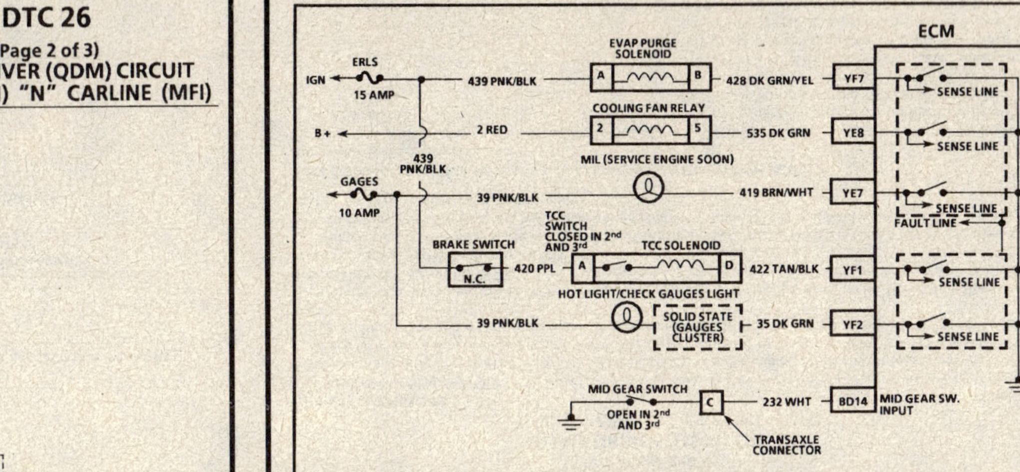

DTC 26
(Page 3 of 3)
QUAD-DRIVER (QDM) CIRCUIT
3300 (VIN N) "N" CARLINE (MFI)

Circuit Description:

Quad-Driver Modules (QDM) are used to control the components shown in the illustration above. When the ECM is commanding a component "ON," the QDM closes the switch completing the circuit to ground.

Each QDM has a sense line and a fault line. When a component is commanded "ON," the voltage potential on the sense line is low and when the component is commanded "OFF," the voltage potential on the sense line is high.

DTC 26 will set when the ECM is commanding a QDM A controlled component "ON" and the voltage potential on the sense line is high, or if the component is commanded "OFF" and the voltage potential on the sense line is low. The fault line, which can be monitored with the Tech 1, will read "HIGH" under these conditions. If the ECM detects a "HIGH" QDM A fault line when it expects a "LOW" one, it will illuminate the MIL (Service Engine Soon) and store DTC 26. QDM B will **not** set DTC 26.

Important

- DTC 26 will not set even though the fault line scans "HIGH" under the following conditions:
  - Ignition "ON"/engine not running.
  - **OR**
  - Vehicle not operating in 2nd or 3rd gear.
  - **OR**
  - Brake applied.

  Vehicles with a 3T40 transaxle will read "HIGH" until the TCC in-line switch is closed. To check, disconnect TCC connector and connect a test light between terminals "A" and "D". This should cause the Tech 1 to display "LOW" for QDM A with the engine running.

Test Description: Number(s) below refer to circled number(s) on the diagnostic chart.

4. This test will determine if the problem is the circuit or the component. As the factory installed ECM is over current protected, it is highly unlikely that the ECM needs to be replaced.

3.3L (VIN N) ENGINE — DIAGNOSTIC TROUBLE CODE CHART — CENTURY, CUTLASS CIERA/CRUISER, ACHIEVA, GRAND AM AND SKYLARK

DTC 26

(Page 3 of 3)
QUAD-DRIVER (QDM) CIRCUIT
3300 (VIN N) "N" CARLINE (MFI)

CIRCUIT ISOLATED FROM PRIOR CHARTS

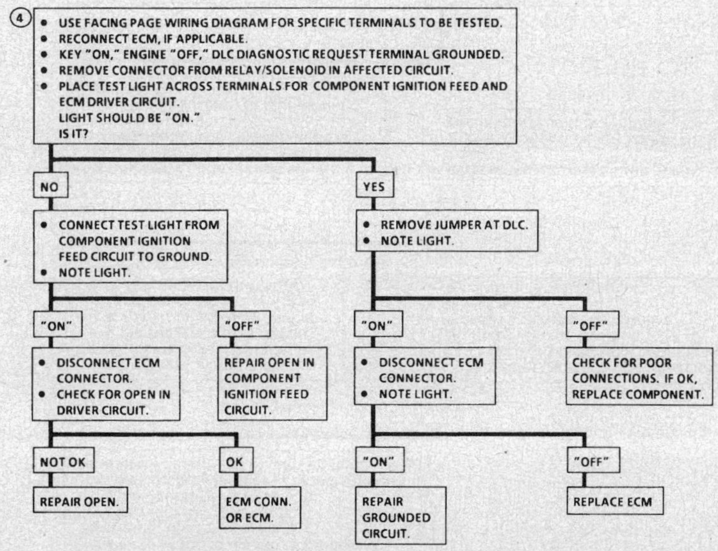

④
- USE FACING PAGE WIRING DIAGRAM FOR SPECIFIC TERMINALS TO BE TESTED.
- RECONNECT ECM, IF APPLICABLE.
- KEY "ON," ENGINE "OFF," DLC DIAGNOSTIC REQUEST TERMINAL GROUNDED.
- REMOVE CONNECTOR FROM RELAY/SOLENOID IN AFFECTED CIRCUIT.
- PLACE TEST LIGHT ACROSS TERMINALS FOR COMPONENT IGNITION FEED AND ECM DRIVER CIRCUIT.
  LIGHT SHOULD BE "ON."
  IS IT?

NO
- CONNECT TEST LIGHT FROM COMPONENT IGNITION FEED CIRCUIT TO GROUND.
- NOTE LIGHT.

"ON"
- DISCONNECT ECM CONNECTOR.
- CHECK FOR OPEN IN DRIVER CIRCUIT.

NOT OK → REPAIR OPEN.

OK → ECM CONN. OR ECM.

"OFF"
REPAIR OPEN IN COMPONENT IGNITION FEED CIRCUIT.

YES
- REMOVE JUMPER AT DLC.
- NOTE LIGHT.

"ON"
- DISCONNECT ECM CONNECTOR.
- NOTE LIGHT.

"ON" → REPAIR GROUNDED CIRCUIT.

"OFF" → REPLACE ECM

"OFF"
CHECK FOR POOR CONNECTIONS. IF OK, REPLACE COMPONENT.

NOTICE: WITH THE INTRODUCTION OF GMP4 ECMS, DELCO ELECTRONICS CHANGED TO INTERNALLY PROTECTED QUAD DRIVERS. THIS INTERNAL PROTECTION CAN BE COMPARED TO THE ACTION OF A CIRCUIT BREAKER. IF TOO MUCH CURRENT FLOWS IN A CONTROLLED CIRCUIT, THIS TYPE OF QUAD DRIVER TURNS ITSELF "OFF." THIS ALLOWS THE QUAD DRIVER TO SURVIVE A SHORTED SOLENOID, RELAY, OR HARNESS. AND IT MEANS YOU DO NOT HAVE TO REPLACE THE ECM. JUST REPAIR THE FAULT IN THE OUTPUT CIRCUIT AND THE QUAD DRIVER WILL RETURN TO NORMAL OPERATION.

"AFTER REPAIRS," REFER TO DTC CRITERIA   AND CONFIRM DTC DOES NOT RESET.

3.3L (VIN N) ENGINE — DIAGNOSTIC TROUBLE CODE CHART — CENTURY, CUTLASS CIERA/CRUISER, ACHIEVA, GRAND AM AND SKYLARK

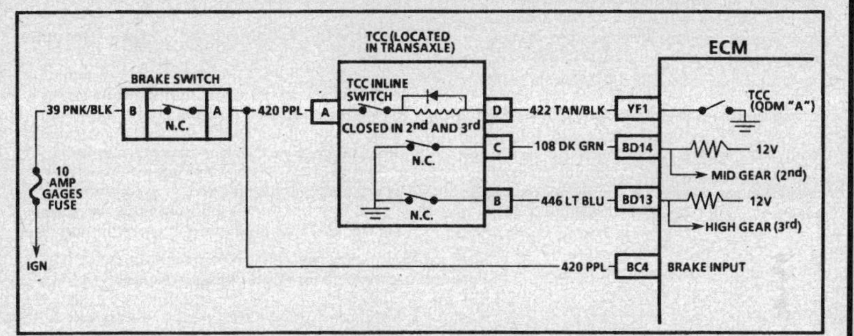

DTCs 27, 28

GEAR SWITCH CIRCUITS
3300 (VIN N) 3T40 ONLY   (MFI)

Circuit Description:

The gear switches are located inside the transaxle. They are normally closed pressure operated switches. The ECM supplies 12 volts through each selected circuit to the switch. As road speed increases, hydraulic pressure applies the specific gear clutches and the gear switch opens. At this time, the ECM monitors a high 12 volt potential, and interprets this to indicate that gear is applied. The ECM uses the gear signals to control fuel delivery and TCC.

DTC 27 will set if:
- CKT 108 is open (2nd gear) when the engine is first started.
- **OR**
- Brake not applied.
- No DTC 38.
- CKT 108 is grounded (closed switch) for 12 seconds when vehicle is operating in 3rd gear.

DTC 28 will set if:
- CKT 446 is open (3rd gear) when the engine is first started.
- **OR**
- Brake not applied.
- No DTC 38.
- Throttle angle less than 20%.
- Vehicle speed greater than 45 mph.
- CKT 446 is grounded (closed switch) for 10 seconds when vehicle is operating in 3rd gear.

DTC 27 and 28 do not illuminate the MIL (Service Engine Soon.)

Test Description: Number(s) refer to circled number(s) on the diagnostic chart.
1. Must use a DVM. A test light will not light due to the very low current being supplied by the ECM.
2. Checks to see if circuit is grounded through the switch.
3. Checks for a good, properly operating switch and checks circuit within transaxle for an improper ground.

Diagnostic Aids:

An intermittent may be caused by a poor connection, mis-routed harness, rubbed through wire insulation, or a wire broken inside the insulation.

Check for:
- Poor Connection at ECM pins. Inspect harness connectors for backed out terminals, improper mating, broken locks, improperly formed or damaged terminals, and poor terminal to wire connection.
- Mis-routed Harness. Inspect wiring harness to insure that it is not too close to high voltage wires, such as spark plug leads.
- Damaged Harness. Inspect harness for damage. If harness appears OK, observe gear switches on Tech 1 while moving related connectors and wiring harness. A change in display would indicate the intermittent fault location.

3.3L (VIN N) ENGINE — DIAGNOSTIC TROUBLE CODE CHART — CENTURY, CUTLASS CIERA/CRUISER, ACHIEVA, GRAND AM AND SKYLARK

DTCs 27, 28
GEAR SWITCH CIRCUITS
3300 (VIN N) 3T40 ONLY   (MFI)

1.
- KEY "ON," ENGINE NOT RUNNING.
- REMOVE TRANSAXLE CONNECTOR.
- CONNECT DVM TO GROUND AND PROBE TRANSAXLE HARNESS CONNECTOR TERMINALS:
  "C" FOR DTC 27 (MID GEAR SWITCH)
  "B" FOR DTC 28 (HIGH GEAR SWITCH)
- VOLTAGE SHOULD MEASURE B + FOR EACH TERMINAL. DOES IT?

YES →

2.
- CONNECT A TEST LIGHT TO THE TRANSAXLE TERMINALS (AT TRANSAXLE):
  "C" FOR DTC 27 (MID GEAR SWITCH)
  "B" FOR DTC 28 (HIGH GEAR SWITCH)
  WITH OTHER END OF TEST LIGHT TO B +.
- TEST LIGHT SHOULD LIGHT AT BOTH. DOES IT?

NO →
- DISCONNECT BLACK 32-WAY ECM CONNECTOR.
- CHECK FOR OPEN OR GROUND IN CIRCUIT WITH NEAR ZERO VOLTAGE.
- IF CIRCUIT IS OK CHECK ECM TERMINAL CONNECTIONS.
- IF CONNECTIONS ARE OK, REPLACE ECM.

YES →

3.
- SET PARKING BRAKE, BLOCK REAR WHEELS.
- RAISE DRIVE WHEELS AND IDLE VEHICLE IN DRIVE.
- AFTER VEHICLE REACHES HIGH GEAR REPEAT ABOVE TEST.
- LIGHT SHOULD BE "OFF" FOR ALL TERMINALS. IS IT?

NO →
- CHECK INTERNAL TRANSAXLE AND HARNESS CONNECTIONS.
- IF CONNECTIONS ARE OK, REPLACE SWITCH IN CIRCUIT THAT FAULTED IN PRIOR STEP.

YES →

NO TROUBLE FOUND.
SEE "DIAGNOSTIC AIDS"

NO →
- CHECK INTERNAL TRANSAXLE WIRING AND HARNESS FOR SHORTS TO GROUND.
- IF WIRING IS OK, REPLACE SWITCH IN CIRCUIT THAT FAULTED IN PRIOR STEP.

"AFTER REPAIRS," REFER TO DTC CRITERIA   AND CONFIRM DTC DOES NOT RESET.

3.3L (VIN N) ENGINE — DIAGNOSTIC TROUBLE CODE CHART — CENTURY, CUTLASS CIERA/CRUISER, ACHIEVA, GRAND AM AND SKYLARK

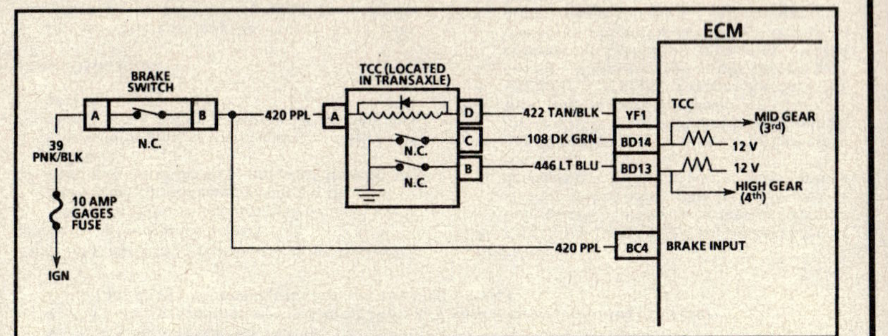

DTCs 28, 29
GEAR SWITCH CIRCUITS
3300 (VIN N) 4T60 ONLY "A" CARLINE (MFI)

Circuit Description:

The gear switches are located inside the transaxle. They are normally closed pressure operated switches. The ECM supplies 12 volts through each selected circuit to the switch. As road speed increases, hydraulic pressure applies the specific gear clutches and the gear switch opens. At this time, the ECM monitors a high 12 volt potential, and interprets this to indicate that gear is applied. The ECM uses the gear signals to control fuel delivery and TCC.

DTC 28 will set if:
- No DTC 38 or 39.
- Brake not applied.
- CKT 108 indicates ground or closed switch for 12 seconds when vehicle is in 4th gear operation TCC locked.
  OR
- CKT 108 is open (3rd gear) when the engine is first started.

DTC 29 will set if:
- No DTC 38 or 39.
- Brake not applied.
- CKT 446 indicates ground or closed switch for 10 seconds when vehicle is in 4th gear operation TCC locked.
  OR
- CKT 446 is open (4th gear) when the engine is first started.

DTC 28 and 29 do not illuminate the MIL (Service Engine Soon).

Test Description: Number(s) refer to circled number(s) on the diagnostic chart.

1. Must use a DVM. A test light will not light due to the very low current being supplied by the ECM.
2. Checks to see if circuit is grounded through the switch.
3. Checks for a good, properly operating switch and checks circuit within transaxle for an improper ground.

Diagnostic Aids:

An intermittent may be caused by a poor connection, mis-routed harness, rubbed through wire insulation, or a wire broken inside the insulation.

Check for:
- **Poor Connection at ECM pins.** Inspect harness connectors for backed out terminals, improper mating, broken locks, improperly formed or damaged terminals, and poor terminal to wire connection.
- **Mis-routed Harness.** Inspect wiring harness to insure that it is not too close to high voltage wires, such as spark plug leads.
- **Damaged Harness.** Inspect harness for damage. If harness appears OK, observe gear switches on Tech 1 while moving related connectors and wiring harness. A change in display would indicate the intermittent fault location.

3.3L (VIN N) ENGINE — DIAGNOSTIC TROUBLE CODE CHART — CENTURY, CUTLASS CIERA/CRUISER, ACHIEVA, GRAND AM AND SKYLARK

DTCs 28, 29
GEAR SWITCH CIRCUITS
3300 (VIN N) 4T60 ONLY "A" CARLINE (MFI)

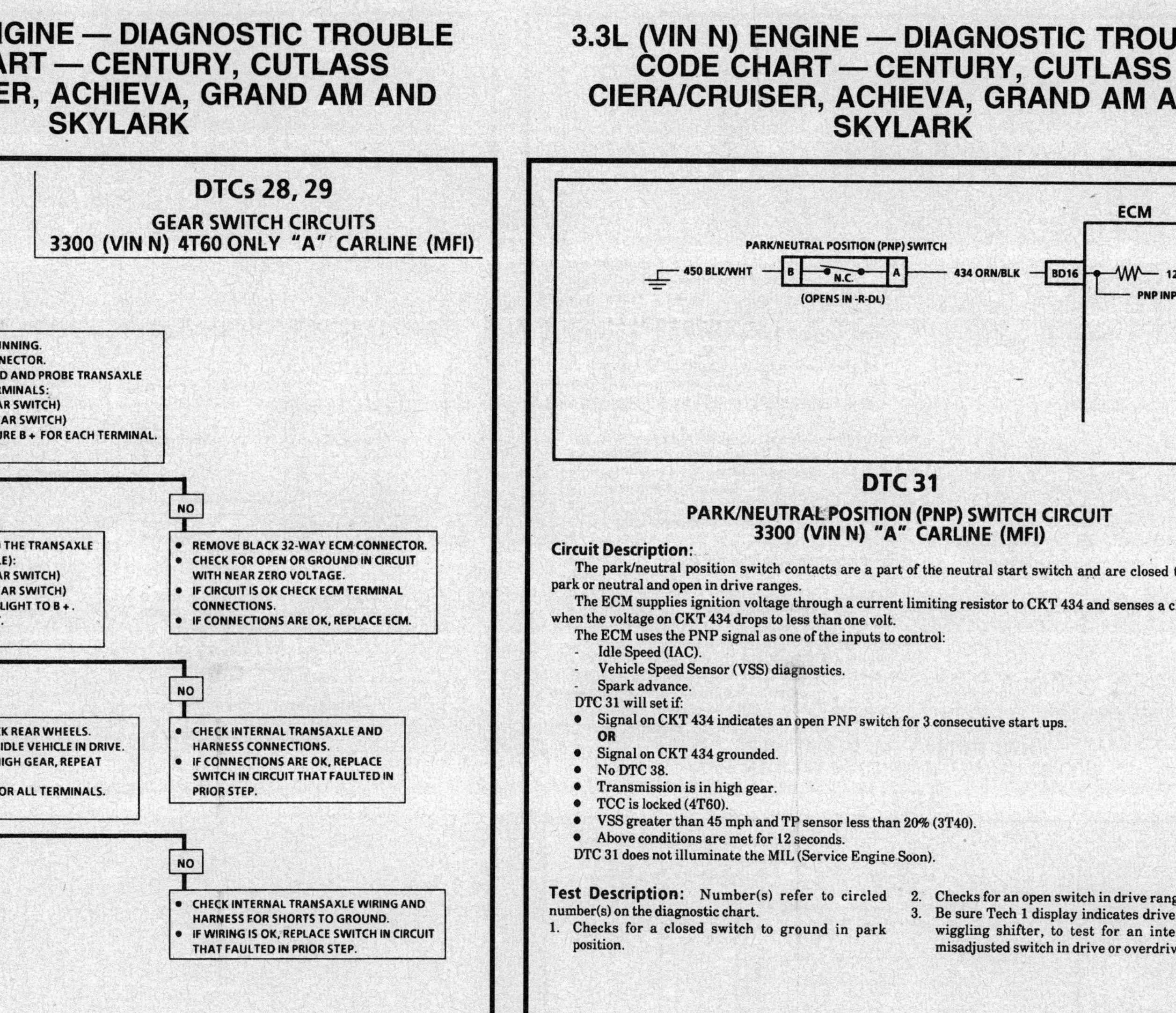

3.3L (VIN N) ENGINE — DIAGNOSTIC TROUBLE CODE CHART — CENTURY, CUTLASS CIERA/CRUISER, ACHIEVA, GRAND AM AND SKYLARK

DTC 31
PARK/NEUTRAL POSITION (PNP) SWITCH CIRCUIT
3300 (VIN N) "A" CARLINE (MFI)

Circuit Description:

The park/neutral position switch contacts are a part of the neutral start switch and are closed to ground in park or neutral and open in drive ranges.

The ECM supplies ignition voltage through a current limiting resistor to CKT 434 and senses a closed switch when the voltage on CKT 434 drops to less than one volt.

The ECM uses the PNP signal as one of the inputs to control:
- Idle Speed (IAC).
- Vehicle Speed Sensor (VSS) diagnostics.
- Spark advance.

DTC 31 will set if:
- Signal on CKT 434 indicates an open PNP switch for 3 consecutive start ups.
  OR
- Signal on CKT 434 grounded.
- No DTC 38.
- Transmission is in high gear.
- TCC is locked (4T60).
- VSS greater than 45 mph and TP sensor less than 20% (3T40).
- Above conditions are met for 12 seconds.

DTC 31 does not illuminate the MIL (Service Engine Soon).

Test Description: Number(s) refer to circled number(s) on the diagnostic chart.

1. Checks for a closed switch to ground in park position.

2. Checks for an open switch in drive range.

3. Be sure Tech 1 display indicates drive, even while wiggling shifter, to test for an intermittent or misadjusted switch in drive or overdrive range.

3.3L (VIN N) ENGINE — DIAGNOSTIC TROUBLE CODE CHART — CENTURY, CUTLASS CIERA/CRUISER, ACHIEVA, GRAND AM AND SKYLARK

DTC 31
PARK/NEUTRAL POSITION (PNP) SWITCH CIRCUIT 3300 (VIN N) "A" CARLINE (MFI)

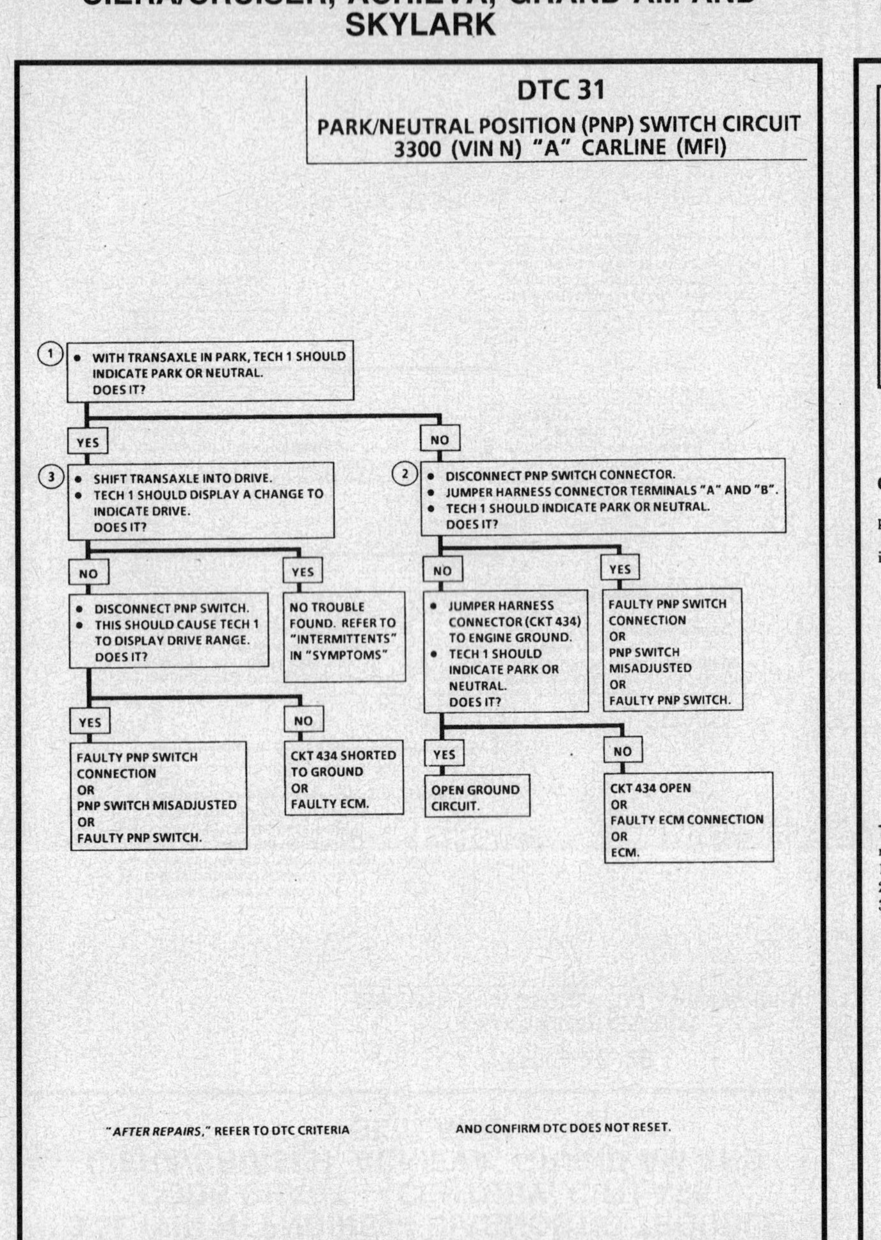

3.3L (VIN N) ENGINE — DIAGNOSTIC TROUBLE CODE CHART — CENTURY, CUTLASS CIERA/CRUISER, ACHIEVA, GRAND AM AND SKYLARK

DTC 31
PARK/NEUTRAL POSITION (PNP) INPUT CIRCUIT 3300 (VIN N) "N" CARLINE (MFI)

Circuit Description:

The Park/Neutral Position (PNP) input is grounded by the electronic transmission range (PRNDL) decoder in park or neutral and open in drive ranges.

The ECM supplies ignition voltage through a current limiting resistor to CKT 434 and senses a grounded input when the voltage on CKT 434 drops to less than one volt.

The ECM uses the PNP signal as one of the inputs to control:
- Idle speed (IAC).
- Vehicle Speed Sensor (VSS) diagnostics.
- Spark advance.

DTC 31 will set if:
- CKT 434 indicates an open for 3 consecutive starts.
  OR
- CKT 434 indicates a ground.
- No DTC 38.
- Transmission is in high gear.
- VSS greater than 45 mph and throttle angle less than 20% (.94 volt).
- Above conditions are met for 12 seconds.

DTC 31 does not illuminate the MIL (Service Engine Soon).

Test Description: Number(s) below refer to circled number(s) on the diagnostic chart.
1. Checks for a grounded PNP input in park position.
2. Checks for a faulty ECM.
3. Checks for an ungrounded PNP input in drive range. Be sure Tech 1 display indicates drive, even while wiggling shifter. If it doesn't, check for a misadjusted transmission range switch or an intermittent.

3.3L (VIN N) ENGINE — DIAGNOSTIC TROUBLE CODE CHART — CENTURY, CUTLASS CIERA/CRUISER, ACHIEVA, GRAND AM AND SKYLARK

DTC 31
PARK/NEUTRAL POSITION (PNP) INPUT CIRCUIT
3300 (VIN N) "N" CARLINE (MFI)

(1)
- IGNITION "ON."
- OBSERVE PARK/NEUTRAL POSITION DISPLAY.
- WITH TRANSAXLE/TRANSMISSION IN PARK, TECH 1 SHOULD DISPLAY "P-N".
- DOES IT?

YES → (3)
- SHIFT TRANSAXLE/TRANSMISSION INTO DRIVE.
- TECH 1 SHOULD DISPLAY "-R-DL".
- DOES IT?

NO →
- IGNITION "OFF."
- DISCONNECT ELECTRONIC PRNDL DECODER MODULE.
- IGNITION "ON."
- TECH 1 SHOULD DISPLAY "-R-DL".
- DOES IT?

YES → FAULTY ELECTRONIC PRNDL DECODER MODULE WIRING OR FAULTY DECODER MODULE -

YES →
NO TROUBLE FOUND. REFER TO "INTERMITTENTS" IN "SYMPTOMS"

NO → CKT 434 SHORTED TO GROUND OR FAULTY ECM.

NO → (2)
- BACKPROBE ECM BLACK 32-WAY CONNECTOR TERMINAL "BD16" WITH A JUMPER TO GROUND.
- TECH 1 SHOULD DISPLAY "P-N"
- DOES IT?

YES →
- IGNITION "OFF."
- DISCONNECT ELECTRONIC PRNDL DECODER MODULE CONNECTOR.
- JUMPER CKT 434 AT DECODER MODULE HARNESS CONNECTOR TO GROUND.
- IGNITION "ON."
- TECH 1 SHOULD DISPLAY "P-N".
- DOES IT?

NO → FAULTY ECM CONNECTION OR FAULTY ECM.

YES → FAULTY ELECTRONIC PRNDL DECODER MODULE WIRING OR FAULTY DECODER MODULE

NO → CKT 434 OPEN BETWEEN ECM AND DECODER MODULE.

"AFTER REPAIRS," REFER TO DTC CRITERIA AND CONFIRM DTC DOES NOT RESET.

3.3L (VIN N) ENGINE — DIAGNOSTIC TROUBLE CODE CHART — CENTURY, CUTLASS CIERA/CRUISER, ACHIEVA, GRAND AM AND SKYLARK

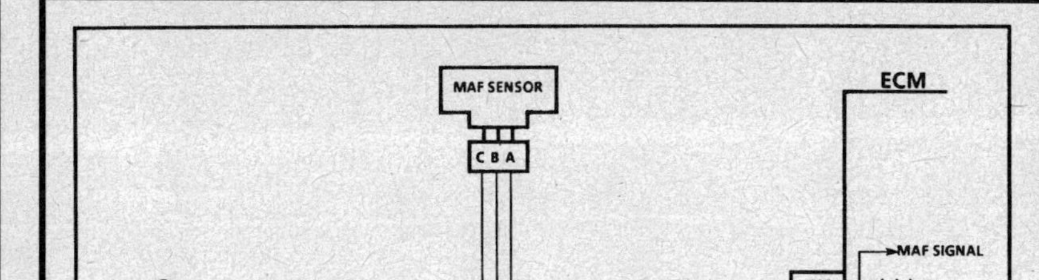

DTC 34
MASS AIR FLOW (MAF) SENSOR CIRCUIT
(GM/SEC LOW)
3300 (VIN N)   (MFI)

Circuit Description:

The Mass Air Flow (MAF) sensor measures the flow of air which passes through it in a given time. The ECM uses this information to monitor the operating condition of the engine for fuel delivery calculations. A large quantity of air movement indicates acceleration, while a small quantity indicates deceleration or idle.

The MAF sensor produces a frequency signal which cannot be easily measured. The sensor can be diagnosed using the procedures on this chart.

If the MAF sensor signal frequency is too low and Diagnostic Trouble Code (DTC) 34 is set, a substitute value for airflow is calculated based on engine RPM, TP sensor, and IAC motor position.

DTC 34 will set if the following conditions exist:
- Engine running.
- No MAF sensor signal for 4 seconds.

Test Description: Number(s) refer to circled number(s) on the diagnostic chart.
1. This step checks to see if ECM recognizes a problem.
2. A voltage reading at sensor harness connector terminal "A" of less than 4 or over 6 volts indicates a fault in CKT 492 or poor connection.
3. Verifies that both ignition voltage and a good ground circuit are available.

Diagnostic Aids:

An intermittent may be caused by a poor connection, mis-routed harness, rubbed through wire insulation, or a wire broken inside the insulation.

Check For:
- Poor connection at ECM pin "YF10" or MAF sensor. Inspect harness connectors for backed out terminals, improper mating, broken locks, improperly formed or damaged terminals, and poor terminal to wire connection.
- Mis-routed Harness. Inspect MAF sensor harness to ensure that it is not too close to high voltage wires, such as spark plug leads.
- Damaged Harness. Inspect harness for damage. If harness appears OK, observe mass air flow on Tech 1 while moving related connectors and wiring harness. A change in display would indicate the intermittent fault location.

3.3L (VIN N) ENGINE — DIAGNOSTIC TROUBLE CODE CHART — CENTURY, CUTLASS CIERA/CRUISER, ACHIEVA, GRAND AM AND SKYLARK

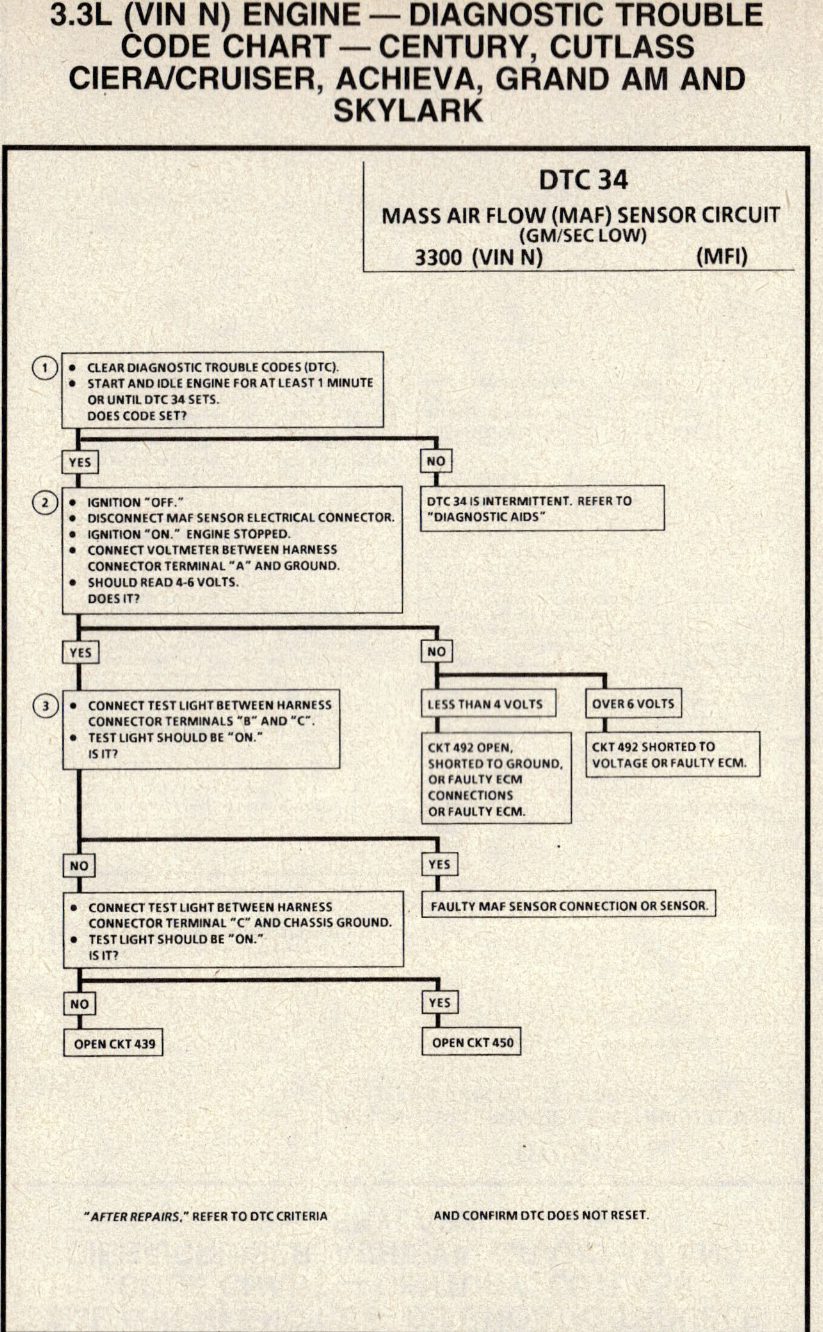

3.3L (VIN N) ENGINE — DIAGNOSTIC TROUBLE CODE CHART — CENTURY, CUTLASS CIERA/CRUISER, ACHIEVA, GRAND AM AND SKYLARK

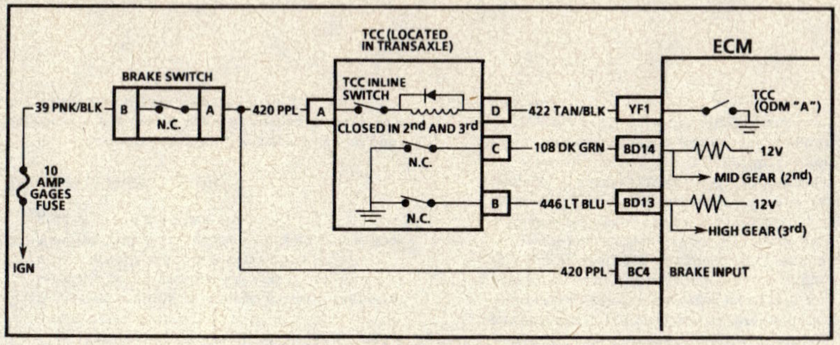

DTC 38
BRAKE SWITCH CIRCUIT
3300 (VIN N)   (MFI)

Circuit Description:

The ECM monitors the status of the brake switch circuits through ECM terminal "BC4".
DTC 38 will set if:
- No DTC 24 is present.
- Brake switch status has not changed from "released" to "applied" or "applied" to "released."
- Vehicle speed has been above 35 mph and back to 0 mph 23 times.

DTC 38 does not illuminate the MIL (Service Engine Soon).

Test Description: Number(s) refer to circled number(s) on the diagnostic chart.
1. Jumpering the brake switch will determine if the ECM and the wiring for the brake circuit are OK.
2. Verifies supply voltage to brake switch.

Diagnostic Aids:

DTC 38 accompanied by a DTC 39 would mean a problem with one or more of the following components: Fuse, CKT 639, brake switch or wire before the splice.

A single DTC 38 is the result of a wire or CKT 420 problem between the splice and the ECM, poor connection to ECM connector, or possibly the ECM.

3.3L (VIN N) ENGINE — DIAGNOSTIC TROUBLE CODE CHART — CENTURY, CUTLASS CIERA/CRUISER, ACHIEVA, GRAND AM AND SKYLARK

DTC 38
BRAKE SWITCH CIRCUIT
3300 (VIN N) (MFI)

- INSTALL TECH 1.
- IGNITION "ON."
- OBSERVE TCC BRAKE SWITCH.

"APPLIED"

(1)
- DISCONNECT BRAKE INPUT SWITCH.
- JUMPER CKTs 39 AND 420 TOGETHER.
- OBSERVE DISPLAY.

"APPLIED"

(2)
- CONNECT A TEST LIGHT TO GROUND.
- PROBE PNK/BLK WIRE CKT 39 AT BRAKE SWITCH.

LIGHT "ON"

CKT 420 OPEN, ECM CONNECTION OR FAULTY ECM.

"RELEASED"

CHECK BRAKE SWITCH ADJUSTMENT. IF OK, REPLACE BRAKE INPUT SWITCH.

LIGHT "OFF"

CHECK 10 AMP GAGES FUSE AND REPAIR SHORT TO GROUND IN CKT 39, 420, OR TCC SOLENOID IF BLOWN. IF OK, REPAIR OPEN CKT 439 TO BRAKE SWITCH.

"RELEASED"

- DEPRESS BRAKE PEDAL.
- OBSERVE DISPLAY.

"RELEASED"

- DISCONNECT BRAKE INPUT SWITCH.
- OBSERVE DISPLAY.

"APPLIED"

REPLACE BRAKE INPUT SWITCH.

"APPLIED"

NO TROUBLE FOUND, REFER TO "DIAGNOSTIC AIDS."

"RELEASED"

CKT 420 SHORTED TO VOLTAGE OR FAULTY ECM.

"AFTER REPAIRS," REFER TO DTC CRITERIA AND CONFIRM DTC DOES NOT RESET.

3.3L (VIN N) ENGINE — DIAGNOSTIC TROUBLE CODE CHART — CENTURY, CUTLASS CIERA/CRUISER, ACHIEVA, GRAND AM AND SKYLARK

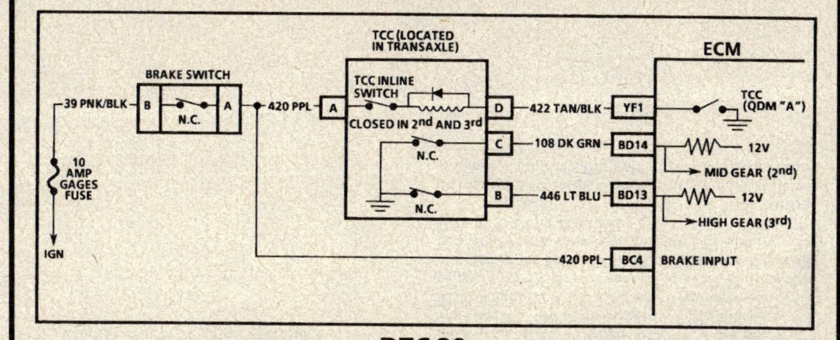

DTC 39
TORQUE CONVERTER CLUTCH (TCC) CIRCUIT
3300 (VIN N) 3T40 ONLY (MFI)

Circuit Description:

The ECM controls the Torque Converter Clutch (TCC) operation by grounding CKT 422 through a quad-driver.

DTC 39 will set if:
- No DTC 28 is present.
- Brake is not applied.
- TCC is commanded "ON" by the ECM.
- Transaxle is in 3rd gear.
- The engine speed to vehicle speed ratio indicates that the TCC is not engaged.
- All of the above for a time is greater than 15 seconds.

DTC 39 does not illuminate the MIL (Service Engine Soon).

Test Description: Number(s) refer to circled number(s) on the diagnostic chart.
1. Checks gages fuse and brake switch.
2. Checks ECM for proper operation.
3. Checks the internal switches in the transaxle.

Diagnostic Aids:

A poor connection can cause an intermittent DTC 39. Using a DVM connected to the circuit, moving the wire(s) or connector(s) may cause voltage reading to change abruptly.

DTC 39 in combination with DTC 38 indicates a problem with one or more of the following components: gages fuse, CKT 39, brake switch or CKT 420 before the splice.

DTC 39 only is the result of a wire or CKT 420 problem between the splice and the TCC solenoid, CKT 422 problem between TCC solenoid and ECM, poor connection to ECM connector, or possibly the ECM.

If the ECM, TCC solenoid, and TCC control circuitry are operating correctly but the TCC still does not engage properly, the problem could be hydraulic or mechanical

3.3L (VIN N) ENGINE — DIAGNOSTIC TROUBLE CODE CHART — CENTURY, CUTLASS CIERA/CRUISER, ACHIEVA, GRAND AM AND SKYLARK

DTC 39
TORQUE CONVERTER CLUTCH (TCC) CIRCUIT
3300 (VIN N) 3T40 ONLY (MFI)

1.
- PLACE TRANS IN PARK OR NEUTRAL.
- SET PARKING BRAKE AND BLOCK DRIVE WHEELS.
- INSTALL TECH 1.
- OBSERVE BRAKE SWITCH STATUS, SHOULD DISPLAY "RELEASED," AND SWITCH TO "APPLIED" WHEN BRAKE IS APPLIED.
- DOES THE BRAKE SWITCH STATUS DISPLAY "RELEASED" WHEN BRAKE IS NOT APPLIED?

YES → 2 / NO →

NO:
OPEN IN CKT 420
OR
OPEN IN CKT 39
OR
MISADJUSTED OR FAULTY BRAKE INPUT SWITCH
OR
GAGES FUSE.

2.
- DISCONNECT TCC CONNECTOR AT TRANSAXLE.
- CONNECT A TEST LIGHT, BETWEEN HARNESS TERMINALS "A" AND "D".
- CYCLE THE TCC SOLENOID "ON" AND "OFF" WITH THE TECH 1. TEST LIGHT SHOULD FLASH "ON" AND "OFF" AS SOLENOID IS CYCLED. DOES IT?

YES → 3 / NO →

NO:
OPEN IN CKT 420 FROM BRAKE SWITCH
OR
OPEN IN CKT 422
OR
FAULTY CONNECTION AT ECM OR FAULTY ECM.

3.
- SELECT DISPLAY DATA MODE WITH TECH 1.
- RECONNECT TCC CONNECTOR.
- DRIVE VEHICLE ABOVE 45 MPH.
- "SCAN" TCC.
- SHOULD DISPLAY "ON" AND A DROP IN RPM SHOULD OCCUR WHEN TCC ENGAGES. DOES IT?

YES → / NO →

YES:
NO PROBLEM FOUND.
SEE "DIAGNOSTIC AIDS"

NO:
POOR CONNECTION AT TCC
OR
FAULTY TRANSAXLE INTERNAL WIRING
OR
FAULTY TCC SOLENOID
OR
INLINE TCC SWITCH.

"AFTER REPAIRS," REFER TO DTC CRITERIA AND CONFIRM DTC DOES NOT RESET.

3.3L (VIN N) ENGINE — DIAGNOSTIC TROUBLE CODE CHART — CENTURY, CUTLASS CIERA/CRUISER, ACHIEVA, GRAND AM AND SKYLARK

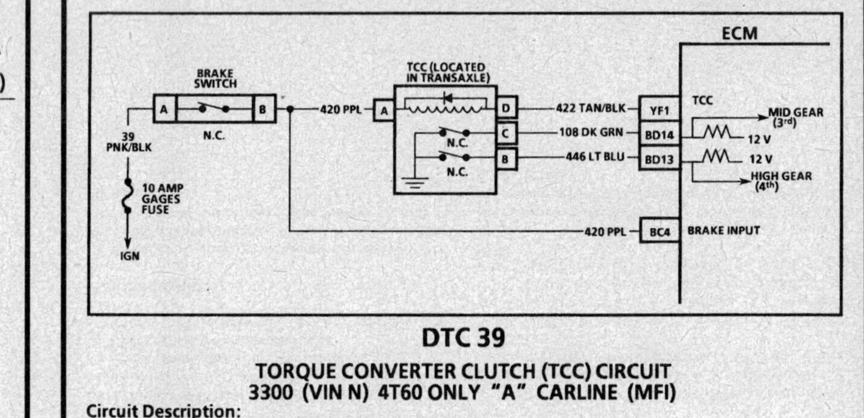

DTC 39
TORQUE CONVERTER CLUTCH (TCC) CIRCUIT
3300 (VIN N) 4T60 ONLY "A" CARLINE (MFI)

Circuit Description:
The ECM controls TCC operation by grounding CKT 422 through the quad-driver.
DTC 39 will set if:
- No DTC 28 or 29 present.
- Brake switch is closed (brake not applied).
- TCC is commanded "ON" by the ECM.
- Vehicle is in 4th gear.
- The engine speed to vehicle speed ratio indicates that the TCC is not engaged.
- All of the above for a time greater than 15 seconds.

DTC 39 does not illuminate the MIL (Service Engine Soon).

Test Description: Number(s) refer to circled number(s) on the diagnostic chart.
1. Checks fuse, brake switch and B+ circuit to the TCC solenoid.
2. Checks availability of B+ on CKT 420.
3. Checks the ECM for proper operation.

Diagnostic Aids:

DTC 39 in combination with a DTC 38 would mean a problem with one or more of the following components: Fuse, CKT 39, brake switch or wire before the splice. A single DTC 39 is the result of a wire or CKT 420 problem between the splice and the TCC solenoid, CKT 422 problem between TCC solenoid and ECM, poor connection to ECM connector, or possibly the ECM. DTC 39 could also be the result of a faulty gear switch or gear switch circuit. See CHART C-8B for diagnosis.

If the ECM, TCC solenoid, and TCC control circuitry are operating correctly but the TCC still does not engage properly, the problem could be hydraulic or mechanical

3.3L (VIN N) ENGINE — DIAGNOSTIC TROUBLE CODE CHART — CENTURY, CUTLASS CIERA/CRUISER, ACHIEVA, GRAND AM AND SKYLARK

DTC 39
TORQUE CONVERTER CLUTCH (TCC) CIRCUIT 3300 (VIN N) 4T60 ONLY "A" CARLINE (MFI)

- PLACE TRANSMISSION IN PARK OR NEUTRAL.
- SET PARKING BRAKE AND BLOCK DRIVE WHEELS.
- INSTALL TECH 1.
- IGNITION "ON," ENGINE "OFF."
- CYCLE TCC "ON" AND "OFF."
- USING A STETHOSCOPE ON THE TRANSAXLE, YOU SHOULD BE ABLE TO HEAR THE TCC SOLENOID CLICK "ON" WHEN THE " ↑ " IS PRESSED AND CLICK "OFF" WHEN " ↓ " IS PRESSED.
  DOES THE SOLENOID CLICK?

NO → (1)
YES → TCC CIRCUIT OK. SEE "DIAGNOSTIC AIDS."

(1)
- IGNITION "ON," ENGINE NOT RUNNING.
- DISCONNECT TCC CONNECTOR AT TRANSAXLE
- CONNECT A TEST LIGHT BETWEEN TERMINALS "A" AND "D" OF TCC HARNESS CONNECTOR.
- CYCLE TCC "ON" AND "OFF" USING TECH 1.
  TEST LIGHT SHOULD TURN "ON" AND "OFF."
  DOES IT?

NO → (2)
YES → REPLACE TCC SOLENOID.

(2)
- IGNITION "ON" ENGINE NOT RUNNING.
- USING TEST LIGHT CONNECTED TO GROUND, PROBE TERMINAL "A" OF TCC HARNESS CONNECTOR.
  LIGHT SHOULD BE "ON."
  IS IT?

YES → (3)
NO → OPEN IN CKT 420 OR OPEN IN CKT 39 OR MISADJUSTED OR FAULTY BRAKE INPUT SWITCH OR GAGES FUSE.

(3)
- TEST LIGHT CONNECTED TO B +.
- BACKPROBE CAVITY "YF1" OF ECM YELLOW "E-F" CONNECTOR.
- CYCLE TCC "ON" AND "OFF" USING TECH 1.
  LIGHT SHOULD TURN "ON" AND "OFF."
  DOES IT?

YES → POOR CONNECTION AT ECM OR OPEN IN CKT 422.
NO → POOR CONNECTION AT ECM OR FAULTY ECM.

"AFTER REPAIRS," REFER TO DTC CRITERIA AND CONFIRM DTC DOES NOT RESET.

3.3L (VIN N) ENGINE — DIAGNOSTIC TROUBLE CODE CHART — CENTURY, CUTLASS CIERA/CRUISER, ACHIEVA, GRAND AM AND SKYLARK

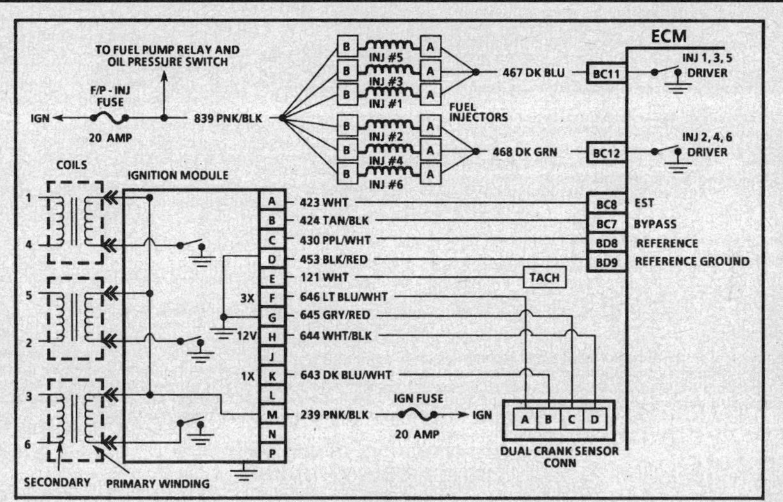

CODE 42
ELECTRONIC SPARK TIMING (EST) CIRCUIT 3300 (VIN N) "N" CARLINE (PORT)

Circuit Description:
The ignition module sends a reference signal to the ECM when the engine is cranking or running. Initially, the ignition module controls ignition timing (module mode). When the engine speed exceeds 400 rpm, the ECM applies a 5 volts to the bypass CKT 424, causing the ignition module to switch timing to ECM control. If an open or ground occurs in either CKT 424 or 423 while engine is running in EST mode, the engine will stall and set a Code 42. If CKT 424 is open or grounded while engine is cranking, the engine will start and remain in module mode timing.
To set a Code 42 one of the following conditions must be met:
- Engine speed greater than 600 rpm with no EST pulse for 200 ms (open or grounded CKT 423),
OR
- ECM commanding bypass mode (open or grounded CKT 424).

Test Description: Number(s) below refer to circled number(s) on the diagnostic chart.
1. Checks to see if ECM recognizes a problem. If it does not set Code 42, it is an intermittent problem and could be due to a loose connection.
2. With the ECM disconnected, the ohmmeter should be reading less than 200 ohms, which is the normal resistance of the EST circuit through the ignition module. A higher resistance would indicate a fault in CKT 423, a poor ignition module connection, or a faulty ignition module.
3. If test light was "ON" when connected from 12 volts to ECM harness terminal "BC7", either CKT 424 is shorted to ground or the ignition module is faulty.
4. Checks to see if ignition module switches when the bypass circuit is energized by 12 volts through the test light. If the ignition module actually switches, the ohmmeter reading should shift to over 6,000 ohms.
5. Disconnecting the ignition module should make the ohmmeter read as if it were monitoring an open circuit (infinite reading). Otherwise, CKT 423 is shorted to ground.

Diagnostic Aids:
An intermittent may be caused by a poor connection, rubbed through wire insulation, or a wire broken inside the insulation. Check For:
- Poor Connection or Damaged Harness. Inspect ECM harness connectors for backed out terminals "BC7" or "BC8", improper mating, broken locks, improperly formed or damaged terminals, poor terminal to wire connection, and damaged harness.
- Intermittent Test. If connections and harness check OK, a digital voltmeter connected from affected terminal to ground while moving related connectors and wiring harness. If the failure is induced, the voltage reading will change.

3.3L (VIN N) ENGINE — DIAGNOSTIC TROUBLE CODE CHART — CENTURY, CUTLASS CIERA/CRUISER, ACHIEVA, GRAND AM AND SKYLARK

CODE 42
ELECTRONIC SPARK TIMING (EST) CIRCUIT 3300 (VIN N) "N" CARLINE (PORT)

1.
- CLEAR CODES AND IDLE ENGINE FOR 2 MINUTES.
- DID "SERVICE ENGINE SOON" LIGHT COME "ON"?

YES →

2.
- IGNITION "ON," ENGINE STOPPED.
- "SCAN" CODES.
- IS CODE 42 DISPLAYED?

NO → CODE IS INTERMITTENT. IF NO ADDITIONAL CODES DISPLAYED, REFER TO "DIAGNOSTIC AIDS."

YES →

2.
- IGNITION "OFF," DISCONNECT BLACK 32 PIN ECM CONNECTOR.
- IGNITION "ON," PROBE ECM HARNESS CONNECTOR. TERMINAL "BC8" WITH AN OHMMETER TO GROUND RED (+) LEAD TO ECM HARNESS BLACK (-) LEAD TO GROUND.
- OHMMETER ON 200 Ω SCALE, OBSERVE RESISTANCE.
- RESISTANCE SHOULD BE LESS THAN 200 OHMS. IS IT?

NO → SEE APPLICABLE CODE CHART FOR CODE(S) DISPLAYED OTHER THAN 42.

YES →

3.
- INSTALL A TEST LIGHT BETWEEN ECM HARNESS TERMINAL "BC7" AND B +.
- TEST LIGHT SHOULD BE "OFF." IS IT?

NO → CHECK CKT 423 FOR OPEN. IF CKT 423 IS OK, IT'S POOR CONNECTION AT IGNITION MODULE TERMINAL "A" OR FAULTY IGNITION MODULE.

YES →

4.
- OHMMETER ON 20K Ω SCALE.
- NOTE RESISTANCE BETWEEN ECM HARNESS TERMINAL "BC8" AND GROUND.
- RESISTANCE READING ON BC8 SHOULD BE OVER 6,000 OHMS. IS IT?

NO (4) →

4.
- IGNITION "OFF," DISCONNECT IGNITION MODULE.
- IGNITION "ON." IS TEST LIGHT "OFF"?

YES → FAULTY IGNITION MODULE.

NO → CKT 424 SHORTED TO GROUND.

5.
- IGNITION "OFF," DISCONNECT IGNITION MODULE.
- RESISTANCE BETWEEN ECM TERMINAL "BC8" AND GROUND SHOULD BE INFINITE (OPEN CIRCUIT). IS IT?

YES → CHECK CKT 424 FOR OPEN. IF CKT 424 IS OK, IT IS A POOR CONNECTION AT IGNITION MODULE TERMINAL "B" OR FAULTY IGNITION MODULE.

NO → CKT 423 SHORTED TO GROUND.

5.
- IGNITION "OFF."
- RECONNECT ECM, START AND IDLE ENGINE FOR 2 MINUTES.
- DID "SERVICE ENGINE SOON" LIGHT COME "ON"?

YES → IF CODE 42 IS STORED, IT IS A FAULTY CONNECTION AT ECM OR A FAULTY ECM.

NO → CHECK FOR INTERMITTENT CONNECTIONS.

"AFTER REPAIRS," REFER TO CODE CRITERIA AND CONFIRM CODE DOES NOT RESET.

3.3L (VIN N) ENGINE — DIAGNOSTIC TROUBLE CODE CHART — CENTURY, CUTLASS CIERA/CRUISER, ACHIEVA, GRAND AM AND SKYLARK

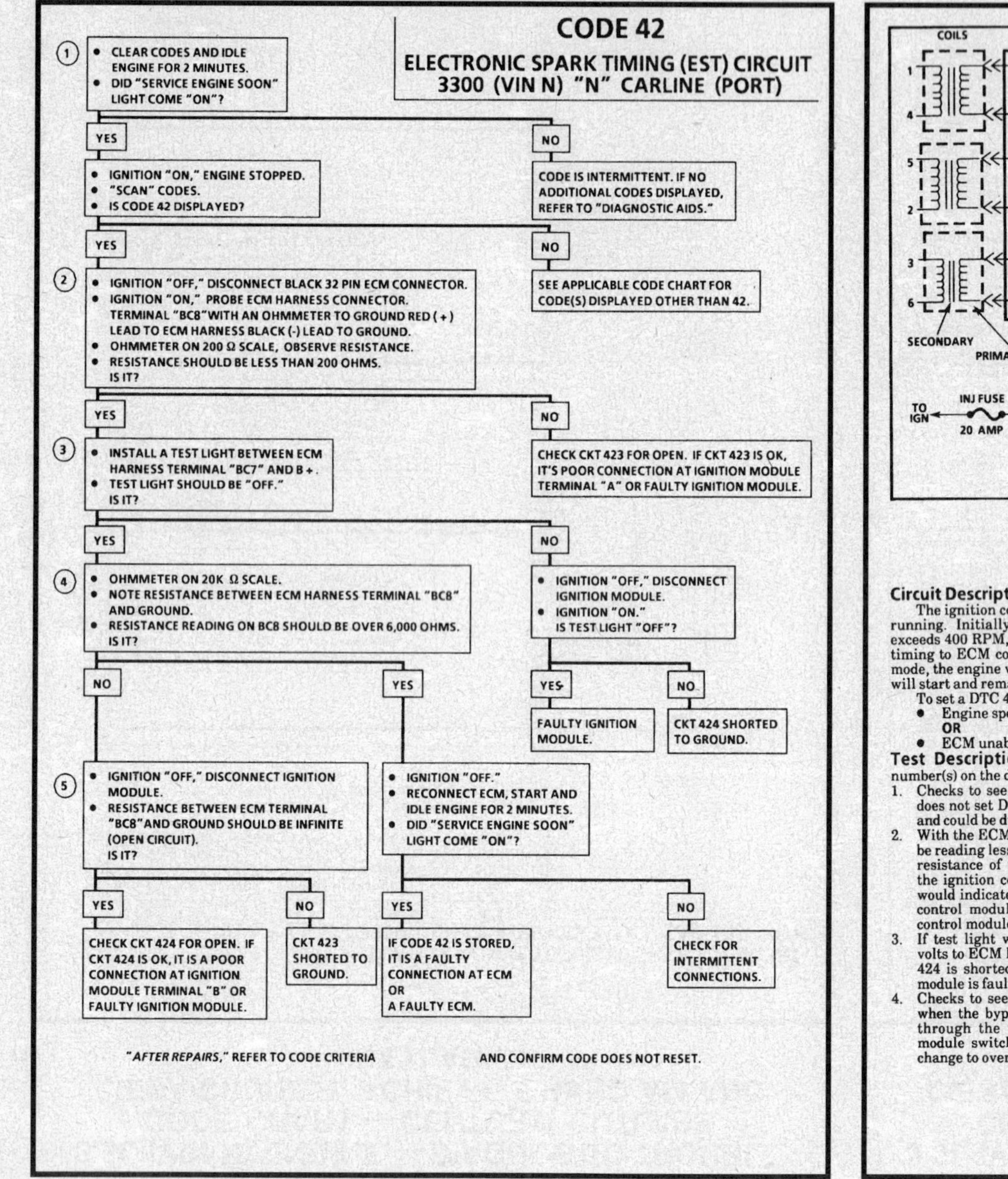

DTC 42
IGNITION CONTROL (IC) CIRCUIT 3300 (VIN N) "A" CARLINE (MFI)

Circuit Description:

The ignition control module sends a reference signal to the ECM via CKT 423 when the engine is cranking or running. Initially, the ignition control module control ignition timing (module mode). When the engine speed exceeds 400 RPM, the ECM applies 5 volts to the bypass CKT 424 causing the ignition control module to switch timing to ECM control (IC mode). If an open or ground occurs in either circuit while engine is running in IC mode, the engine will stall and set DTC 42. If CKT 424 is open or grounded while engine is cranking, the engine will start and remain in module mode timing.

To set a DTC 42 one of the following conditions must be met:
- Engine speed greater than 600 RPM with no IC pulse for .1 second (open or grounded CKT 423),
- OR
- ECM unable to command IC mode (open or grounded CKT 424).

Test Description: Number(s) refer to circled number(s) on the diagnostic chart.

1. Checks to see if ECM recognizes a problem. If it does not set DTC 42, it is an intermittent problem and could be due to a loose connection.
2. With the ECM disconnected, the ohmmeter should be reading less than 200 ohms, which is the normal resistance of the ignition control circuit through the ignition control module. A higher resistance would indicate a fault in CKT 432, a poor ignition control module connection, or a faulty ignition control module.
3. If test light was "ON" when connected from 12 volts to ECM harness terminal "BC7", either CKT 424 is shorted to ground or the ignition control module is faulty.
4. Checks to see if ignition control module switches when the bypass circuit is energized by 12 volts through the test light. If the ignition control module switches, the ohmmeter reading should change to over 6,000 ohms.
5. Disconnecting the ignition control module should make the ohmmeter read as if it were monitoring an open circuit (infinite reading). Otherwise, CKT 423 is shorted to ground.

Diagnostic Aids:

An intermittent may be caused by a poor connection, rubbed through wire insulation, or a wire broken inside the insulation.

Check for:
- Poor Connection or Damaged Harness. Inspect ECM harness connectors for backed out terminals "BC7" or "BC8", improper mating, broken locks, improperly formed or damaged terminals, poor terminal to wire connection, and damaged harness.
- Intermittent Test. If connections and harness check OK, a digital voltmeter connected from affected terminal to ground while moving related connectors and wiring harness. If the failure is induced, the voltage reading will change.

3.3L (VIN N) ENGINE — DIAGNOSTIC TROUBLE CODE CHART — CENTURY, CUTLASS CIERA/CRUISER, ACHIEVA, GRAND AM AND SKYLARK

DTC 42
IGNITION CONTROL (IC) CIRCUIT
3300 (VIN N) "A" CARLINE (MFI)

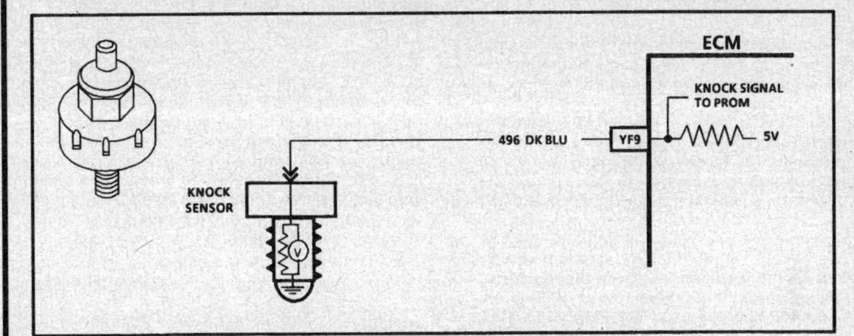

(1)
- CLEAR DTC(S) AND IDLE ENGINE FOR 2 MINUTES.
- DID MIL ("SERVICE ENGINE SOON") COME "ON"?

YES →
- IGNITION "ON," ENGINE STOPPED.
- SCAN DTC(S).
- IS DTC 42 DISPLAYED?

NO → DTC(S) IS INTERMITTENT. IF NO ADDITIONAL DTC(S) DISPLAYED, REFER TO "DIAGNOSTIC AIDS."

YES →

(2)
- IGNITION "OFF," DISCONNECT BLACK 32 PIN ECM CONNECTOR.
- IGNITION "ON," PROBE ECM HARNESS CONNECTOR. TERMINAL "BC8" WITH AN OHMMETER TO GROUND RED (+) LEAD TO ECM HARNESS BLACK (-) LEAD TO GROUND.
- OHMMETER ON 200 Ω SCALE, OBSERVE RESISTANCE.
- RESISTANCE SHOULD BE LESS THAN 200 OHMS. IS IT?

NO → SEE APPLICABLE DTC(S) CHART FOR DTC(S) DISPLAYED OTHER THAN 42.

YES →

(3)
- INSTALL A TEST LIGHT BETWEEN ECM HARNESS TERMINAL "BC7" AND B +.
- TEST LIGHT SHOULD BE "OFF." IS IT?

NO → CHECK CKT 423 FOR OPEN. IF CKT 423 IS OK, IT'S POOR CONNECTION AT IGNITION CONTROL MODULE TERMINAL "A" OR FAULTY IGNITION CONTROL MODULE.

YES →

(4)
- OHMMETER ON 20K Ω SCALE.
- NOTE RESISTANCE BETWEEN ECM HARNESS TERMINAL "BC8" AND GROUND.
- RESISTANCE READING ON BC8 SHOULD BE OVER 5,000 OHMS. IS IT?

NO →
- IGNITION "OFF," DISCONNECT IGNITION CONTROL MODULE.
- IGNITION "ON." IS TEST LIGHT "OFF"?

YES → FAULTY IGNITION CONTROL MODULE.

NO → CKT 424 SHORTED TO GROUND.

YES →

(5)
- IGNITION "OFF," DISCONNECT IGNITION CONTROL MODULE.
- RESISTANCE BETWEEN ECM TERMINAL "BC8" AND GROUND SHOULD BE INFINITE (OPEN CIRCUIT). IS IT?

YES → CHECK CKT 424 FOR OPEN. IF CKT 424 IS OK, IT IS A POOR CONNECTION AT IGNITION CONTROL MODULE TERMINAL "B" OR FAULTY IGNITION CONTROL MODULE.

NO → CKT 423 SHORTED TO GROUND.

- IGNITION "OFF."
- RECONNECT ECM, START AND IDLE ENGINE FOR 2 MINUTES.
- DID MIL ("SERVICE ENGINE SOON") COME "ON"?

YES → IF DTC 42 IS STORED, IT IS A FAULTY CONNECTION AT ECM OR A FAULTY ECM.

NO → CHECK FOR INTERMITTENT CONNECTIONS.

"AFTER REPAIRS," REFER TO DTC CRITERIA AND CONFIRM DTC DOES NOT RESET.

3.3L (VIN N) ENGINE — DIAGNOSTIC TROUBLE CODE CHART — CENTURY, CUTLASS CIERA/CRUISER, ACHIEVA, GRAND AM AND SKYLARK

DTC 43
KNOCK SENSOR (KS) CIRCUIT
3300 (VIN N)   (MFI)

Circuit Description:

The knock sensor is used to detect engine detonation and allow the ECM to retard Ignition Control (IC) timing based on the signal being received. The circuitry within the knock sensor causes the ECM supplied 5 volt signal to be pulled down, so that under a no knock condition CKT 496 should measure about 2.5 volts. The knock sensor produces an AC signal which rides on the 2.5 volts DC voltage. The amplitude and signal frequency are dependent upon the knock level.

If CKT 496 becomes open or shorted to ground, the voltage will either go above 3.5 volts or below 1.5 volts. If either of these conditions are met for 20 seconds, DTC 43 will be stored.

DTC 43 will set if:
- Ignition "ON."
- Voltage on CKT 496 above 3.4 volts or below 1.5 volts.
- Condition present for 20 seconds.

Test Description: Number(s) refer to circled number(s) on the diagnostic chart.

1. If DTC 43 is detected, "Knock Retard" on the Tech 1 will display 10 degrees.
   If an audible knock is heard from the engine, repair the internal engine problem. Normally, no knock should be detected at idle.
2. The ECM applies 5 volts to CKT 496, which should be present at the knock sensor terminal.
3. This test determines if the knock sensor is faulty or if the knock sensor portion of the PROM is faulty.

Diagnostic Aids:

Check CKT 496 for a intermittent open or short to ground. Also check for proper installation of PROM.

If the KS CKT 496 is routed too close to secondary ignition wires, it may induce a voltage and cause a false knock signal.

Refer to "Intermittents" in "Symptoms."

3.3L (VIN N) ENGINE — DIAGNOSTIC TROUBLE CODE CHART — CENTURY, CUTLASS CIERA/CRUISER, ACHIEVA, GRAND AM AND SKYLARK

DTC 43
KNOCK SENSOR (KS) CIRCUIT
3300 (VIN N) (MFI)

(1)
- INSTALL TECH 1.
- ENGINE IDLING, ENGINE COOLANT TEMP ABOVE 67°C. DOES TECH 1 INDICATE A FIXED VALUE OF KNOCK RETARD BETWEEN 4 AND 20 DEGREES?

YES → (2)
- DISCONNECT KNOCK SENSOR.
- IGNITION "ON."
- USING A DVM, MEASURE VOLTAGE BETWEEN HARNESS CKT 496 AND GROUND. VOLTAGE SHOULD BE 4-6 VOLTS. IS IT?

NO → DTC 43 IS INTERMITTENT. SEE "DIAGNOSTIC AIDS" ON FACING PAGE.

YES → (3)
- MEASURE RESISTANCE OF KNOCK SENSOR BY CONNECTING OHMMETER BETWEEN SENSOR TERMINAL AND ENGINE BLOCK. SHOULD BE BETWEEN 3.3KΩ & 4.5KΩ. IS IT?

NO →
OVER 6 VOLTS: CHECK FOR CKT 496 BEING ROUTED TOO CLOSE TO SECONDARY IGNITION WIRES OR CKT 496 SHORTED TO VOLTAGE OR FAULTY ECM.

LESS THAN 4 VOLTS: CKT 496 OPEN, SHORTED TO GROUND OR FAULTY ECM.

YES →
- CHECK HARNESS AND SENSOR CONNECTOR. IF OK:
- REMOVE ECM AND BE SURE PROM IS PROPERLY SEATED INTO ECM. IF OK:
- REPLACE PROM.

NO → REMOVE KNOCK SENSOR AND ENSURE THAT THREADS ON SENSOR AND BLOCK ARE CLEAN AND FREE OF CORROSION, TEFLON TAPE, ANTI-SEIZE COMPOUNDS, OR SEALANTS. IF OK, REPLACE FAULTY KNOCK SENSOR.

"AFTER REPAIRS," REFER TO DTC CRITERIA AND CONFIRM DTC DOES NOT RESET.

3.3L (VIN N) ENGINE — DIAGNOSTIC TROUBLE CODE CHART — CENTURY, CUTLASS CIERA/CRUISER, ACHIEVA, GRAND AM AND SKYLARK

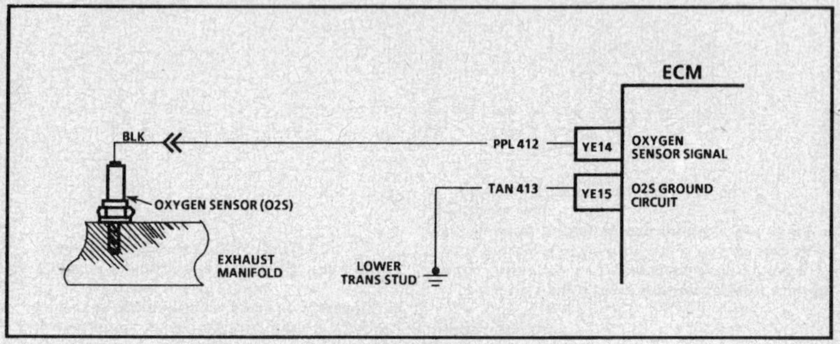

DTC 44
OXYGEN SENSOR (O2S) CIRCUIT
(LEAN EXHAUST INDICATED)
3300 (VIN N) (MFI)

Circuit Description:

The ECM supplies a voltage of about .45 volt (450 mV) between terminals "YE14" and "YE15". (If measured with a 10 megohm digital voltmeter, this may read as low as .32 volt.) The O2S varies the voltage within a range of about 1 volt (1000 mV) if the exhaust is rich, down through about .10 volt (100 mV) if the exhaust is lean.

The sensor is like an open circuit and produces no voltage when it is below about 360°C (600°F). An open sensor circuit or cold sensor causes "Open Loop" operation. DTC 44 is set when the O2S signal voltage on CKT 412:
- Remains below .3 volt (300 mV) for up to 2 minutes.
- The system is operating in "Closed Loop."

Test Description: Number(s) refer to circled number(s) on the diagnostic chart.
1. Running the engine at 1200 RPM keeps the O2S hot, so an accurate display voltage is maintained. Opening the O2S wire should result in a voltage display of between 350 and 550 mV. If the display is still fixed below 350 mV, the fault is a short to ground in CKT 412 or a faulty ECM.

Diagnostic Aids:

Use the Tech 1 to observe the long term fuel trim values at different RPM and air flow conditions. The Tech 1 also displays fuel trim cells, so that long term fuel trim values can be checked in each of the cells to determine when DTC 44 may have been set. If the conditions for DTC 44 exist, the long term fuel trim values will be around 150.
- O2S Wire. Sensor pigtail may be mispositioned and contacting the exhaust manifold.
- Check for intermittent ground in wire between connector and sensor.
- Poor ECM grounds. Check ground connections at transaxle stud and ECM connectors, making sure all connections are clean and tight.

- MAF Sensor. A Mass Air Flow (MAF) sensor output that causes the ECM to sense a lower than normal air flow will cause the system to go lean. Disconnect the MAF sensor and if the lean condition is gone, replace the MAF sensor.
- Lean Injector(s). Perform injector balance test CHART C-2A.
- Fuel Contamination. Water, even in small amounts, near the in-tank fuel pump inlet can be delivered to the injectors. The water causes a lean exhaust and can set DTC 44.
- Vacuum or crankcase leaks can cause a lean condition.
- Fuel Pressure. System will be lean if pressure is too low. It may be necessary to monitor fuel pressure while driving the vehicle at various road speeds and/or loads to confirm. See "Fuel System Diagnosis," CHART A-7.
- Exhaust Leaks. If there is an exhaust leak, the engine can cause outside air to be pulled into the exhaust and past the sensor.
- If the above are OK, it is a faulty oxygen sensor.

3.3L (VIN N) ENGINE — DIAGNOSTIC TROUBLE CODE CHART — CENTURY, CUTLASS CIERA/CRUISER, ACHIEVA, GRAND AM AND SKYLARK

DTC 44

OXYGEN SENSOR (O2S) CIRCUIT
(LEAN EXHAUST INDICATED)
3300 (VIN N) "A" CARLINE (MFI)

(1)
- RUN WARM ENGINE (75°C/167°F TO 95°C/203°F) AT 1200 RPM.
- DOES TECH 1 DATA DISPLAY INDICATE OXYGEN SENSOR (O2S) VOLTAGE FIXED BELOW .3 VOLT (300 mV)?

YES
- DISCONNECT OXYGEN SENSOR.
- WITH ENGINE IDLING, TECH 1 DATA DISPLAY SHOULD INDICATE OXYGEN SENSOR VOLTAGE BETWEEN .35 VOLT AND .55 VOLT (350 mV AND 550 mV). DOES IT?

NO
DTC 44 IS INTERMITTENT.
IF NO ADDITIONAL DTC(S) WERE STORED, REFER TO "DIAGNOSTIC AIDS"

YES
REFER TO "DIAGNOSTIC AIDS"

NO
CKT 412 SHORTED TO GROUND OR FAULTY ECM.

"AFTER REPAIRS," REFER TO DTC CRITERIA   AND CONFIRM DTC DOES NOT RESET.

3.3L (VIN N) ENGINE — DIAGNOSTIC TROUBLE CODE CHART — CENTURY, CUTLASS CIERA/CRUISER, ACHIEVA, GRAND AM AND SKYLARK

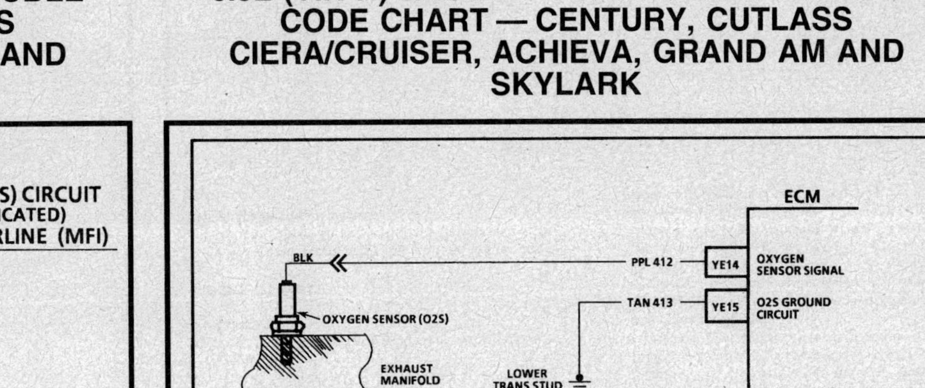

DTC 45

OXYGEN SENSOR (O2S) CIRCUIT
(RICH EXHAUST INDICATED)
3300 (VIN N)   (MFI)

Circuit Description:

The ECM supplies a voltage of about .45 volt (450 mV) between terminals "YE14" and "YE15." (If measured with a 10 megohm digital voltmeter, this may read as low as .32 volt.) The O2S varies the voltage within a range of about 1 volt (1000 mV) if the exhaust is rich, down through about .10 volt (100 mV) if exhaust is lean.

The sensor is like an open circuit and produces no voltage when it is below about 360°C (600°F). An open sensor circuit or cold sensor causes "Open Loop" operation.

DTC 45 is set if the O2S signal voltage on CKT 412:
- Remains above .75 volt for 2 minutes while in "Closed Loop."
- Throttle angle between 3.5% and 40%.
- No DTC 21 or DTC 22.

Test Description: Number(s) refer to circled number(s) on the diagnostic chart.
1. Running the engine at 1200 RPM keeps the O2S hot, so an accurate display voltage is maintained. Opening the O2S wire should result in a voltage display of between 350 and 550 mV. If the display is still fixed below 350 mV, the fault is a short to ground in CKT 412 or the ECM is faulty.

Diagnostic Aids:

Using the Tech 1 observe the long term fuel trim values at different RPM and air flow conditions. The Tech 1 also displays the fuel trim cells, so the long term fuel trim values can be checked in each of the cells to determine when the DTC 45 may have been set. If the conditions for DTC 45 exist, the long term fuel trim values will be around 115.
- Fuel Pressure. System will go rich if pressure is too high. The ECM can compensate for some increase. However, if it gets too high, DTC 45 may be set. See "Fuel System Diagnosis," CHART A-7.
- Rich Injector. Perform injector balance test CHART C-2A.
- Leaking Injector. See CHART A-7.
- Check for fuel contaminated oil.

- EVAP Canister Purge. Check for fuel saturation. If full of fuel, check canister control and hoses. See "Canister Purge" in "Evaporative Emission Control (EVAP) System," Section "6E3-C3".
- MAF Sensor. Ah output that causes the ECM to sense a higher than normal airflow can cause the system to go rich. Disconnecting the MAF sensor will allow the ECM to set a fixed value for the sensor. Substitute a different MAF sensor if the rich condition is gone while the sensor is disconnected.
- Check for leaking fuel pressure regulator diaphragm by checking vacuum line to regulator for fuel.
- TP Sensor. An intermittent TP sensor output will cause the system to go rich due to a false indication of the engine accelerating.
- False rich indication due to silicone contamination of the Oxygen Sensor (O2S). This will be indicated by DTC 45 accompanied by lean driveability conditions and a powdery white deposit on the sensor.

3.3L (VIN N) ENGINE — DIAGNOSTIC TROUBLE CODE CHART — CENTURY, CUTLASS CIERA/CRUISER, ACHIEVA, GRAND AM AND SKYLARK

DTC 45

OXYGEN SENSOR (O2S) CIRCUIT
(RICH EXHAUST INDICATED)
3300 (VIN N)   (MFI)

① • RUN WARM ENGINE (75°C/167°F TO 95°C/203°F) AT 1200 RPM.
   • DOES TECH 1 SCAN TOOL DISPLAY OXYGEN SENSOR (O2S) VOLTAGE FIXED ABOVE .75 VOLT (750 mV)?

YES
• DISCONNECT O2S AND JUMPER HARNESS CKT 412 TO GROUND.
• TECH 1 SCAN TOOL SHOULD DISPLAY OXYGEN SENSOR VOLTAGE BELOW .35 VOLT (350 mV). DOES IT?

NO
DTC 45 IS INTERMITTENT. IF NO ADDITIONAL DTC(S) WERE STORED, REFER TO "DIAGNOSTIC AIDS"

YES
REFER TO "DIAGNOSTIC AIDS"

NO
REPLACE ECM.

"AFTER REPAIRS," REFER TO DTC CRITERIA   AND CONFIRM DTC DOES NOT RESET.

3.3L (VIN N) ENGINE — DIAGNOSTIC TROUBLE CODE CHART — CENTURY, CUTLASS CIERA/CRUISER, ACHIEVA, GRAND AM AND SKYLARK

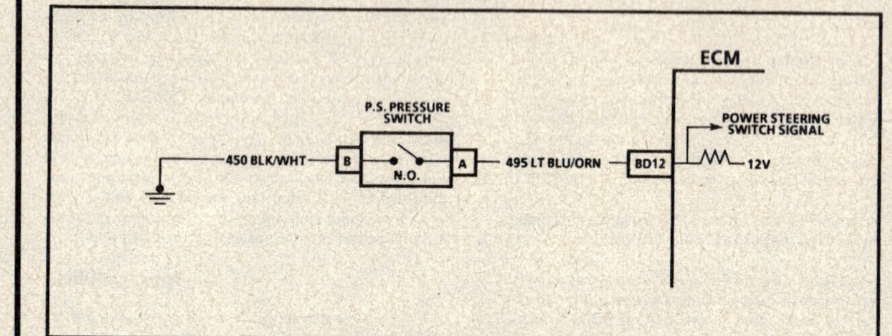

DTC 46

POWER STEERING PRESSURE SWITCH (PSPS) CIRCUIT
3300 (VIN N) "A" CARLINE (MFI)

Circuit Description:

The Power Steering Pressure Switch (PSPS) is a normally open, pressure operated switch. The ECM applies battery voltage through a current limiting resistor to CKT 495 and monitors the voltage to the switch.

Turning the steering wheel increases power steering oil pressure and its load on an idling engine. The pressure switch is designed to close CKT 495 to ground before the load can cause an idle problem.

Closing the switch causes the voltage on CKT 495 to drop to less than 1 volt. The ECM uses this as an indication to compensate for the increased engine load by increasing the idle air rate and disengaging the A/C clutch.

• A pressure switch that will not close or an open CKT 495 or 450 may cause the engine to stall when power steering loads are high.
• A switch that will not open or a CKT 495 shorted to ground may affect idle quality and will cause the A/C relay to be disabled.

To set DTC 46, the following conditions must be met:
• Closed power steering switch indicating "High Pressure" (low voltage potential).
• Vehicle speed is greater than 40 mph.
• Both conditions existing for a time greater than 15 seconds.

DTC 46 does not illuminate the MIL (Service Engine Soon).

Test Description: Number(s) refer to circled number(s) on the diagnostic chart.
1. Verifies a normally open switch.
2. Checks to determine if CKT 495 is shorted to ground.
3. This should simulate a closed switch (High Pressure).

Diagnostic Aids:

An intermittent may be caused by a poor connection, rubbed through wire insulation or a wire broken inside the insulation.

Check For:
• <u>Poor Connection or Damaged Harness.</u> Inspect ECM harness connectors for backed out terminal "BD12", improper mating, broken locks, improperly formed or damaged terminals, poor terminal to wire connection, and damaged harness.
• <u>Intermittent Test.</u> If connections and harness check OK, monitor power steering pressure switch display while moving related connectors and wiring harness. If the failure is induced, the power steering pressure switch display will abruptly change. This may help to isolate the location of the malfunction.

3.3L (VIN N) ENGINE — DIAGNOSTIC TROUBLE CODE CHART — CENTURY, CUTLASS CIERA/CRUISER, ACHIEVA, GRAND AM AND SKYLARK

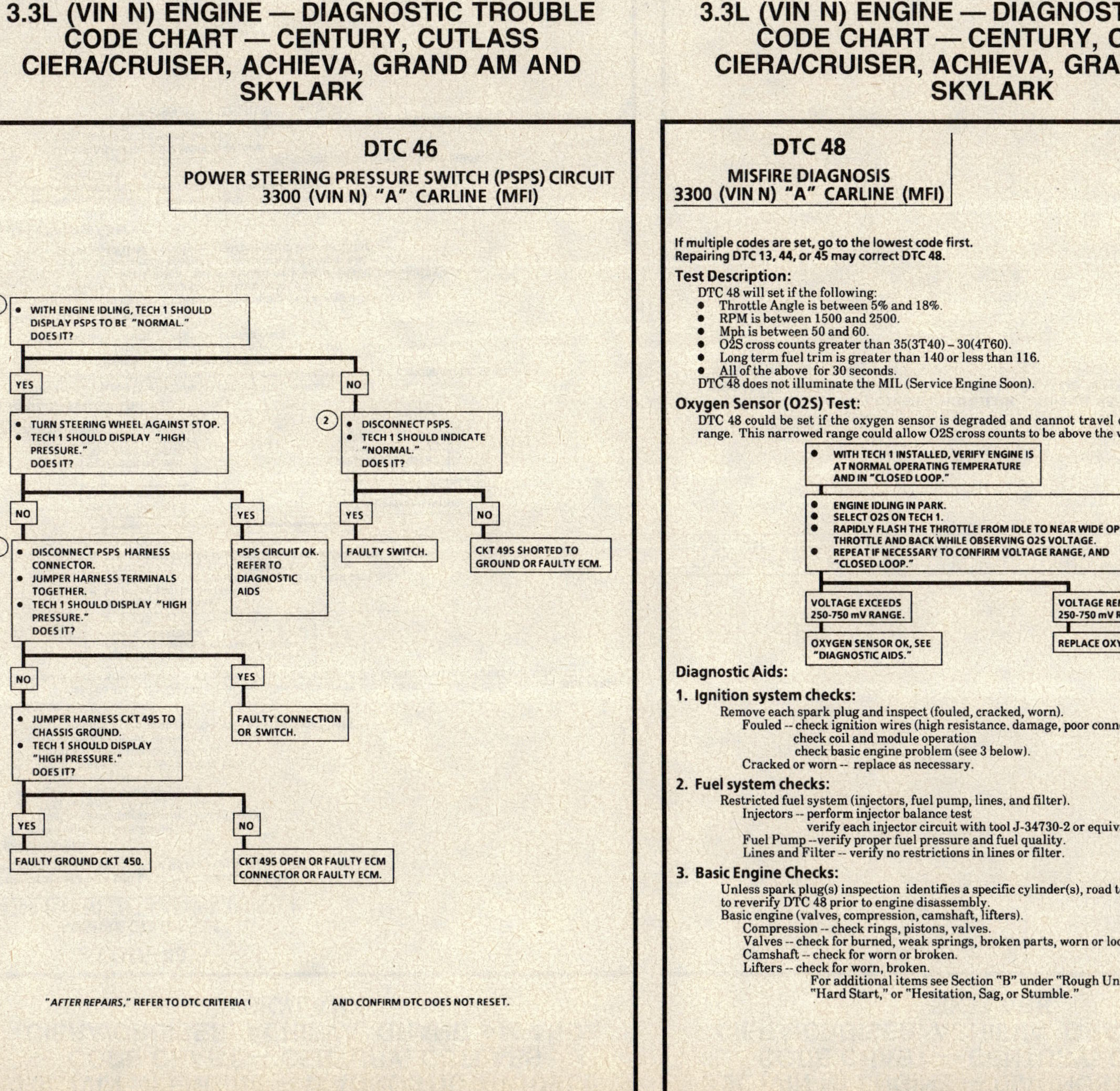

3.3L (VIN N) ENGINE — DIAGNOSTIC TROUBLE CODE CHART — CENTURY, CUTLASS CIERA/CRUISER, ACHIEVA, GRAND AM AND SKYLARK

DTC 48

MISFIRE DIAGNOSIS
3300 (VIN N) "A" CARLINE (MFI)

If multiple codes are set, go to the lowest code first. Repairing DTC 13, 44, or 45 may correct DTC 48.

Test Description:

DTC 48 will set if the following:
- Throttle Angle is between 5% and 18%.
- RPM is between 1500 and 2500.
- Mph is between 50 and 60.
- O2S cross counts greater than 35(3T40) – 30(4T60).
- Long term fuel trim is greater than 140 or less than 116.
- All of the above for 30 seconds.

DTC 48 does not illuminate the MIL (Service Engine Soon).

Oxygen Sensor (O2S) Test:

DTC 48 could be set if the oxygen sensor is degraded and cannot travel over the full rich to lean voltage range. This narrowed range could allow O2S cross counts to be above the value necessary to set the code.

Diagnostic Aids:

1. Ignition system checks:

Remove each spark plug and inspect (fouled, cracked, worn).
Fouled -- check ignition wires (high resistance. damage, poor connections, grounds).
check coil and module operation
check basic engine problem (see 3 below).
Cracked or worn -- replace as necessary.

2. Fuel system checks:

Restricted fuel system (injectors, fuel pump, lines, and filter).
Injectors -- perform injector balance test
verify each injector circuit with tool J-34730-2 or equivalent.
Fuel Pump -- verify proper fuel pressure and fuel quality.
Lines and Filter -- verify no restrictions in lines or filter.

3. Basic Engine Checks:

Unless spark plug(s) inspection identifies a specific cylinder(s), road test vehicle under test conditions to reverify DTC 48 prior to engine disassembly.
Basic engine (valves, compression, camshaft, lifters).
Compression -- check rings, pistons, valves.
Valves -- check for burned, weak springs, broken parts, worn or loose guide.
Camshaft -- check for worn or broken.
Lifters -- check for worn, broken.
For additional items see Section "B" under "Rough Unstable Idle," "Hard Start," or "Hesitation, Sag, or Stumble."

3.3L (VIN N) ENGINE — DIAGNOSTIC TROUBLE CODE CHART — CENTURY, CUTLASS CIERA/CRUISER, ACHIEVA, GRAND AM AND SKYLARK

CODE 48

MISFIRE DIAGNOSIS
3300 (VIN N) "N" CARLINE (PORT)

If multiple codes are set, go to the lowest code first.
Repairing for a Code 13, 44, or 45 may correct Code 48.

Test Description:
Code 48 will set if the following:
- Fuel System operating in "Closed Loop."
- TPS between .58 and 1.02 volts.
- Rpm between 1500 and 2500.
- Mph between 50 and 60.
- Block Learn below 116 or above 140.
- O_2 cross counts greater than 30.
- All of the above for 30 seconds.
Code 48 does not illuminate the "Service Engine Soon" light.

O_2 Sensor Test:
Code 48 could be set if the O_2 sensor is degraded and cannot travel over the full rich to lean voltage range. This narrowed range could allow O_2 cross counts to be above the value necessary to set the code.

- WITH TECH 1 INSTALLED, VERIFY ENGINE IS AT NORMAL OPERATING TEMPERATURE AND IN "CLOSED LOOP."

- ENGINE IDLING IN PARK.
- SELECT O_2 SENSOR ON TECH 1.
- RAPIDLY FLASH THE THROTTLE FROM IDLE TO NEAR WIDE OPEN THROTTLE AND BACK WHILE OBSERVING O_2 VOLTAGE.
- REPEAT IF NECESSARY TO CONFIRM VOLTAGE RANGE, AND "CLOSED LOOP."

VOLTAGE EXCEEDS 250-750 mV RANGE.

VOLTAGE REMAINS WITHIN 250-750 mV RANGE.

O_2 SENSOR OK, SEE "DIAGNOSTIC AIDS".

REPLACE O_2 SENSOR.

Diagnostic Aids:

1. **Ignition system checks:**
   Remove each spark plug and inspect (fouled, cracked, worn)
   Fouled -- check ignition wires (high resistance, damage, poor connections, grounds)
   check coil and module operation
   check basic engine problem (see 3 below)
   Cracked or worn -- replace as necessary

2. **Fuel system checks:**
   Restricted fuel system (injectors, fuel pump, lines, and filter)
   Injectors -- perform injector balance test
   verify each injector circuit with tool J-34730-2 or equivalent
   Fuel Pump --verify proper fuel pressure and fuel quality
   Lines and Filter --verify no restrictions in lines or filter

3. **Basic Engine Checks:**
   Unless spark plug(s) inspection identifies a specific cylinder(s), road test vehicle under test conditions to reverify Code 48 prior to engine disassembly.
   Basic engine (valves, compression, camshaft, lifters)
   Compression -- check rings, pistons, valves
   Valves -- check for burned, weak springs, broken parts, worn or loose guide
   Camshaft -- check for worn or broken
   Lifters--check for worn, broken
   For additional items see Section "B" under "Rough Unstable Idle", "Hard Start", or "Hesitation, Sag, or Stumble".

3.3L (VIN N) ENGINE — DIAGNOSTIC TROUBLE CODE CHART — CENTURY, CUTLASS CIERA/CRUISER, ACHIEVA, GRAND AM AND SKYLARK

DTC 51

PROM ERROR
(FAULTY OR INCORRECT PROM)
3300 (VIN N)   **(MFI)**

CHECK THAT ALL PINS ARE FULLY INSERTED IN THE SOCKET. IF OK, REPLACE PROM, CLEAR MEMORY AND RECHECK. IF DTC 51 REAPPEARS, REPLACE ECM.

NOTICE: To prevent possible Electrostatic Discharge damage to the ECM or PROM, Do Not touch the component leads.   Do Not remove integrated circuit from carrier.

3.3L (VIN N) ENGINE — DIAGNOSTIC TROUBLE CODE CHART — CENTURY, CUTLASS CIERA/CRUISER, ACHIEVA, GRAND AM AND SKYLARK

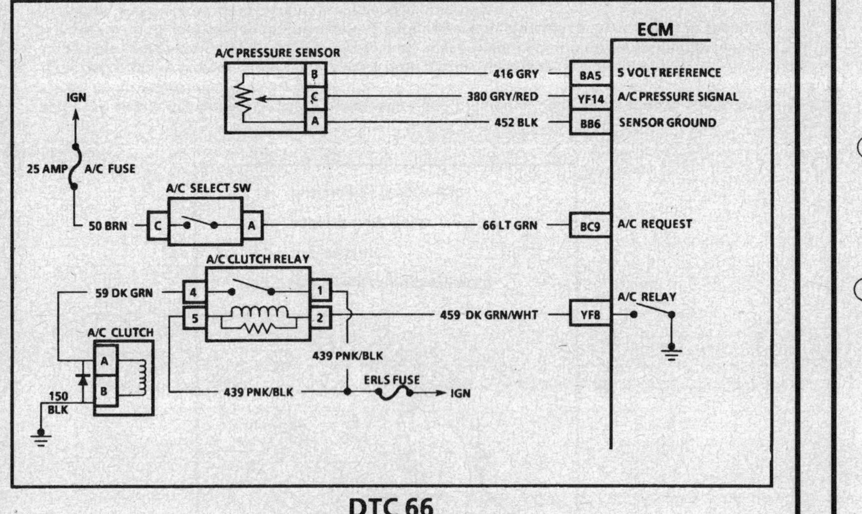

DTC 66
A/C PRESSURE SENSOR CIRCUIT
3300 (VIN N) "N" CARLINE (MFI)

Circuit Description:

The A/C pressure sensor responds to changes in A/C refrigerant system high side pressure. This input indicates how much load the A/C compressor is putting on the engine and is one of the factors used by the ECM to determine IAC valve position for idle speed control. The circuit consists of a 5 volts reference and a ground, both provided by the ECM, and a signal line to the ECM. The signal is a voltage which is proportional to the pressure. The sensor's range of operation is 0 to 468 psi. At 0 psi, the signal will be .1 volt, varying up to about 4.9 volts at 449 psi or above. The A/C compressor is disabled by the ECM if DTC 66 is present.

DTC 66 will set if:
- A/C is requested "ON."
- A/C pressure sensor signal does not indicate pressure between 0 psi (.1 volt) and 449 psi (4.9 volts) for 25 seconds.

DTC 66 does not illuminate the MIL (Service Engine Soon).

Test Description: Number(s) below refer to circled number(s) on the diagnostic chart.

1. This step checks the voltage signal being received by the ECM from the A/C pressure sensor. The normal operating range is between .1 volt and 4.9 volts.
2. Checks to see if the high voltage signal is from a shorted sensor or a short to voltage in the circuit. Normally, disconnecting the sensor would cause the signal voltage go to near zero volts.
3. Checks to see if low voltage signal is from the sensor or the circuit. Jumpering the sensor signal CKT 380 to 5 volts checks the circuit, connections, and ECM.
4. This step checks to see if the low voltage signal was due to an open in the sensor circuit or the 5 volts reference circuit since the prior step eliminated the pressure switch.

Diagnostic Aids:

DTC 66 sets when signal voltage falls outside the normal possible range of the sensor and is not due to a refrigerant system problem. If problem is intermittent, check for opens or shorts in harness or poor connections. If OK, replace A/C pressure sensor. If DTC 66 re-sets, replace ECM.

3.3L (VIN N) ENGINE — DIAGNOSTIC TROUBLE CODE CHART — CENTURY, CUTLASS CIERA/CRUISER, ACHIEVA, GRAND AM AND SKYLARK

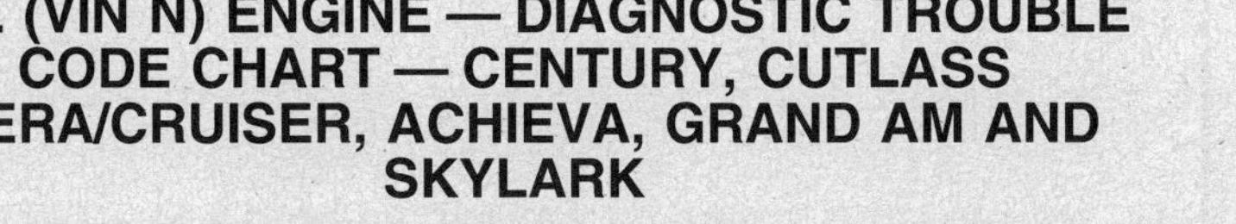

3.3L (VIN N) ENGINE — RESTRICTED EXHAUST SYSTEM DIAGNOSTIC CHART — CENTURY, CUTLASS CIERA/CRUISER, ACHIEVA, GRAND AM AND SKYLARK

3.3L (VIN N) ENGINE — COMPONENT DIAGNOSTIC CHART — CENTURY, CUTLASS CIERA/CRUISER, ACHIEVA, GRAND AM AND SKYLARK

CHART B-1
RESTRICTED EXHAUST SYSTEM CHECK

CHECK AT OXYGEN SENSOR (O2S):

1. Carefully remove oxygen sensor.
2. Install exhaust backpressure tester (BT-8515 or BT-8603) or equivalent in place of oxygen sensor (see illustration).
3. After completing test described below, be sure to coat threads of oxygen sensor with anti-seize compound P/N 5613695 or equivalent prior to re-installation.

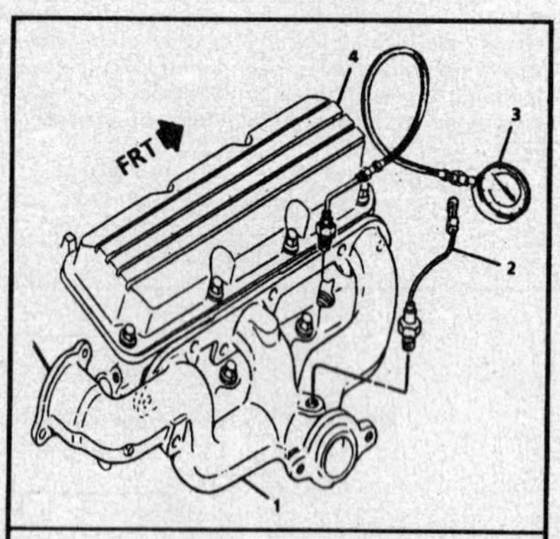

DIAGNOSIS:

1. With the engine at normal operating temperature and running at 2500 RPM, observe the exhaust system backpressure reading on the gauge.
2. If the backpressure exceeds 1 1/4 psi (8.62 kPa), a restricted exhaust system is indicated.
3. Inspect the entire exhaust system for a collapsed pipe, heat distress, or possible internal muffler failure.
4. If there are no obvious reasons for the excessive backpressure, a restricted catalytic converter should be suspected, and replaced using current recommended procedures.

CHART C-2A
INJECTOR BALANCE TEST
3300 (VIN N)   (MFI)

The injector balance tester is a tool used to turn the injector "ON" for a precise amount of time, thus spraying a measured amount of fuel into the manifold. This causes a drop in fuel rail pressure that we can record and compare between each injector. All injectors should have the same amount of pressure drop ($\pm$ 10 kPa). Any injector with a pressure drop that is 10 kPa (or more) greater than the average drop of the other injectors should be considered faulty and replaced. Any injector with a pressure drop of less than 10 kPa from the average drop of the other injectors should be replaced and retested.

STEP 1

Engine "cool down" period (10 minutes) is necessary to avoid irregular readings due to "Hot Soak" fuel boiling.

Disconnect harness connectors at all injectors, and connect injector tester J 34730-3, or equivalent, to one injector. With ignition "OFF", connect fuel gauge J 347301 or equivalent to fuel pressure tap. Wrap a shop towel around fitting while connecting gage to avoid fuel spillage. At this point, insert clear tubing attached to vent valve into a suitable container and bleed air from gauge and hose to insure accurate gauge operation, by applying B+ to fuel pump test terminal. Repeat this step until all air is bled from gauge.

STEP 2

Apply B+ to fuel pump test terminal until fuel pressure reaches its maximum, about 3 seconds. Remove B+ from test terminal and record this initial pressure reading. Energize tester one time and note pressure at its lowest point. (Disregard any slight pressure increase after drop hits low point.) By subtracting this second pressure reading from the initial pressure, we have the actual amount of injector pressure drop. See note below.

STEP 3

Repeat Step 2 on each injector and compare the amount of drop. Usually, good injectors will have virtually the same drop. Replace and retest any injector that has a pressure difference of 10 kPa, either more or less than the average of the other injectors on the engine. Replace any injector that also fails the retest. If the pressure drop of all injectors is within 10 kPa of this average, the injectors appear to be flowing properly. Reconnect them and review "Symptoms."

NOTICE: *The entire test should not be repeated more than once without running the engine to prevent flooding. (This includes any retest on faulty injectors).*

3.3L (VIN N) ENGINE — COMPONENT DIAGNOSTIC CHART — CENTURY, CUTLASS CIERA/CRUISER, ACHIEVA, GRAND AM AND SKYLARK

NOTE: The fuel pressure test in Section "A", CHART A-7, should be completed prior to this test.

CHART C-2A

INJECTOR BALANCE TEST
3300 (VIN N)   (MFI)

Step 1. If engine is at operation temperature start at step "1A," otherwise you can start at step "1B".
   A. Engine "coöl down" period (10 minutes) is necessary to avoid Irregular Readings due to "Hot Soak" Fuel Boiling.
   B. Disconnect harness connectors at all injectors and connect injector tester J 34730-3, or equivalent, to one injector.
   C. With ignition "OFF" connect fuel gauge J 347301 or equivalent to Fuel Pressure Tap.
   NOTE: Wrap a shop towel around fitting while connecting gauge to avoid fuel spray or spillage.
   D. Apply B + to fuel pump test terminal and bleed off to fuel pressure test gauge.

Step 2. Run test:
   A. Apply B + to Fuel Pump Test Terminal. Allow about 3 seconds for fuel pressure to reach its maximum.
   B. Record gauge pressure. (Pressure must hold steady, if not, see the Fuel System Diagnosis, CHART A-7, in Section "A").
   C. Turn injector "ON," by depressing button on injector tester, and note pressure at the instant the gauge needle stops.

Step 3.
   A. Repeat step 2 on all injectors and record pressure drop on each. Replace and retest injectors that appear faulty. (Any injectors that have a 10 kPa difference, either more or less, in pressure from the average). If no problem is found, review "Symptoms."

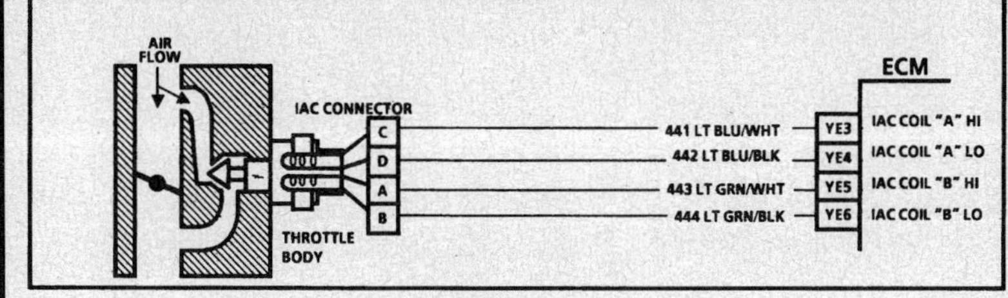

CYLINDER	1	2	3	4	5	6
HIGH READING	225	225	225	225	225	225
LOW READING	100	100	100	90	100	115
AMOUNT OF DROP	125	125	125	135	125	110
	OK	OK	OK	FAULTY, RICH (TOO MUCH) (FUEL DROP)	OK	FAULTY, LEAN (TOO LITTLE) (FUEL DROP)

3.3L (VIN N) ENGINE — COMPONENT DIAGNOSTIC CHART — CENTURY, CUTLASS CIERA/CRUISER, ACHIEVA, GRAND AM AND SKYLARK

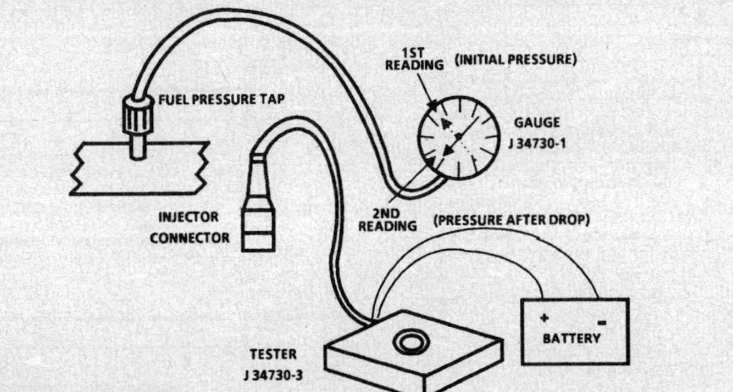

CHART C-2B

IDLE AIR CONTROL (IAC) VALVE CHECK
3300 (VIN N)   (MFI)

Circuit Description:

The ECM controls idle RPM with the Idle Air Control (IAC) valve. To increase idle RPM, the ECM retracts the IAC pintle, allowing more air to bypass the throttle plate. To decrease RPM, it extends the IAC pintle valve, reducing air flow through the IAC valve port in the throttle body. A Tech 1 will read the ECM commands to the IAC valve in counts. The higher the counts, the more air allowed (higher idle). The lower the counts, the less air allowed (lower idle).

Test Description: Number(s) below refer to circled number(s) on the diagnostic chart.
1. The Tech 1 is used to extend and retract the IAC valve. Valve movement is verified by an engine speed change.
2. This test checks all wires, connectors, and ECM.
3. This test check for shorted or open IAC valve coils.

Diagnostic Aids:

A slow, unstable idle may be caused by a system problem that cannot be overcome by the IAC. Idle air control will be above 60 counts, if idle speed is too low. If idle speed is too high, IAC counts will be "0".
- System lean (High Air/Fuel Ratio).
  Idle speed may be too high or too low. Engine speed may vary up and down, disconnecting IAC does not help. May set DTC 44.
  The Tech 1 and/or voltmeter will read an oxygen sensor output less than 300 mV (.3 volt). Check for low regulated fuel pressure or water in fuel. A lean exhaust, with an oxygen sensor output fixed above 800 mV (.8 volt indicates rich exhaust to ECM), will be a contaminated sensor, usually silicone. This may also set DTC 45.

- System rich (Low Air/Fuel Ratio).
  Idle speed too low. IAC counts usually above 80. System obviously rich and may exhibit black smoke exhaust.
  The Tech 1 and/or voltmeter will read an oxygen sensor signal fixed above 800 mV (.8 volt).
  Check:
  - High fuel pressure.
  - Injector leaking or sticking.
- Throttle Body - Remove IAC and inspect bore for foreign material or evidence of IAC pintle dragging the bore.
- Refer to "Rough, Unstable, Incorrect Idle or Stalling" in "Symptoms."

3.3L (VIN N) ENGINE — COMPONENT DIAGNOSTIC CHART — CENTURY, CUTLASS CIERA/CRUISER, ACHIEVA, GRAND AM AND SKYLARK

CHART C-2B

IDLE AIR CONTROL (IAC) VALVE CHECK
3300 (VIN N) (MFI)

CLEAR DIAGNOSTIC TROUBLE CODES, CONFIRM "CLOSED LOOP" OPERATION, NO MIL (SERVICE ENGINE SOON), PERFORM IAC RESET PROCEDURE PER APPLICABLE SERVICE MANUAL AND VERIFY CONTROLLED IDLE SPEED IS CORRECT.

3.3L (VIN N) ENGINE — COMPONENT DIAGNOSTIC CHART — CENTURY, CUTLASS CIERA/CRUISER, ACHIEVA, GRAND AM AND SKYLARK

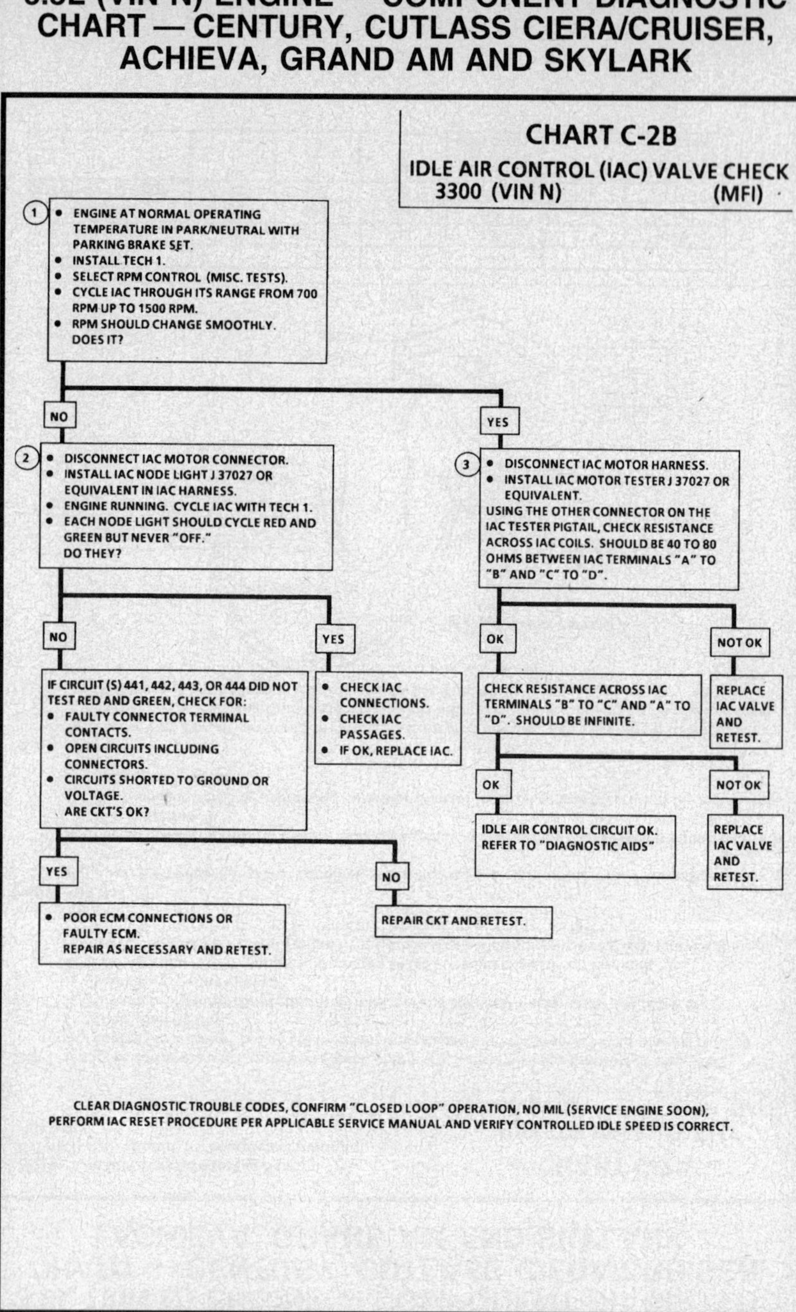

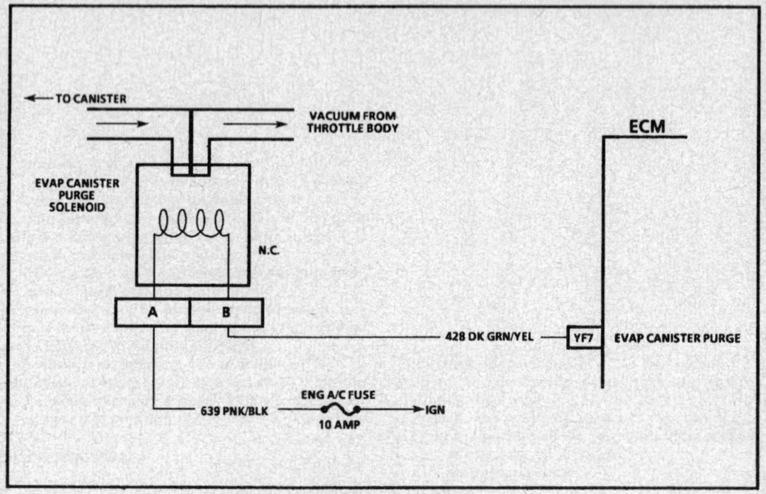

CHART C-3

EVAP CANISTER PURGE VALVE CHECK
3300 (VIN N) (MFI)

Circuit Description:

Canister purge is controlled by a solenoid that allows vacuum to purge the canister when energized. The ECM supplies a ground to energize the solenoid (purge "ON").

If the DLC diagnostic request terminal is grounded with the engine stopped or the following is met with the engine running, the purge solenoid is energized (purge "ON").

- Engine run time after start more than 45 seconds cold and/or 20 seconds when engine coolant temperature is greater than 85°C (185°F).
- Engine coolant temperature above 70°C (158°F).

Test Description: Number(s) below refer to circled number(s) on the diagnostic chart.

1. Checks the solenoid for proper operation.
2. Completes functional check. This should normally energize the solenoid and allow the vacuum to drop (purge "ON").

3.3L (VIN N) ENGINE — COMPONENT DIAGNOSTIC CHART — CENTURY, CUTLASS CIERA/CRUISER, ACHIEVA, GRAND AM AND SKYLARK

CHART C-4-1
MISFIRE AT IDLE
3300 (VIN N) "A" CARLINE (MFI)

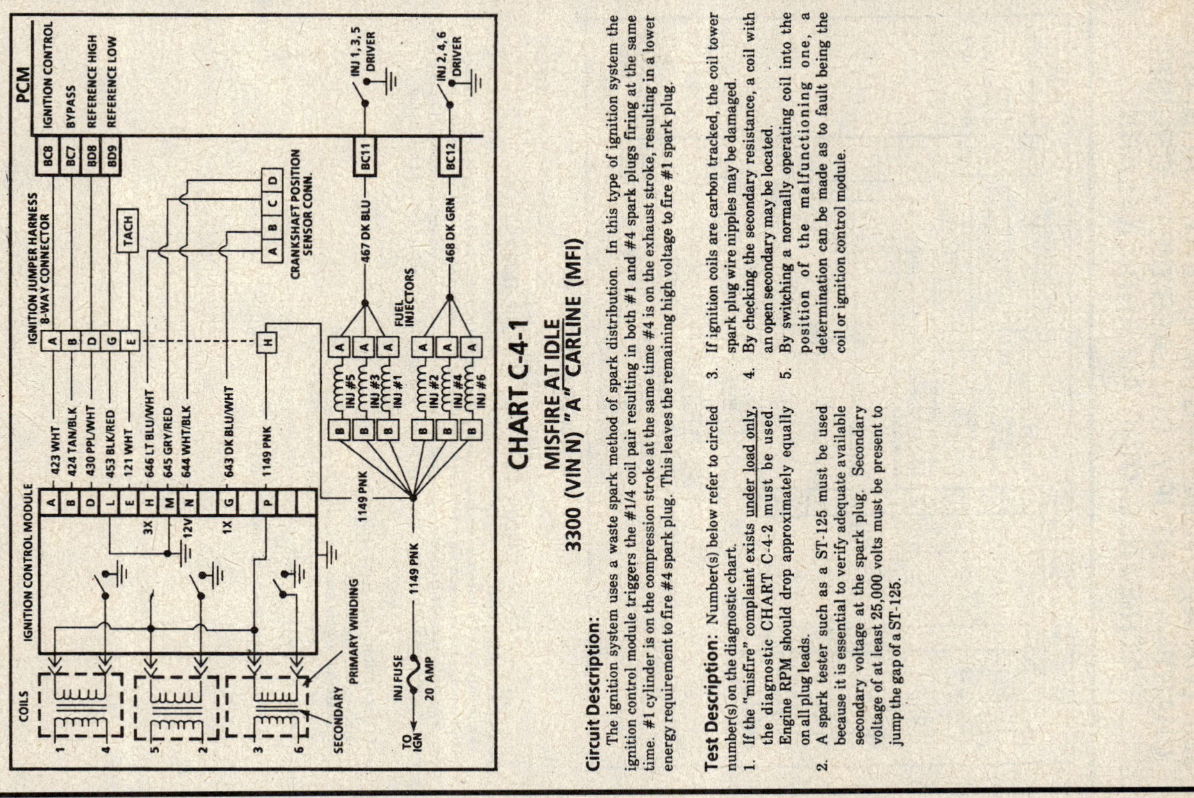

Circuit Description:

The ignition system uses a waste spark method of spark distribution. In this type of ignition system the ignition control module triggers the #1/4 coil pair resulting in both #1 and #4 spark plugs firing at the same time. #1 cylinder is on the compression stroke at the same time #4 is on the exhaust stroke, resulting in a lower energy requirement to fire the #4 spark plug. This leaves the remaining high voltage to fire #1 spark plug.

Test Description: Number(s) below refer to circled number(s) on the diagnostic chart.

1. If the "misfire" complaint exists under load only, the diagnostic CHART C-4-2 must be used. Engine RPM should drop approximately equally on all plug leads.
2. A spark tester such as a ST-125 must be used because it is essential to verify adequate available secondary voltage at the spark plug. Secondary voltage of at least 25,000 volts must be present to jump the gap of a ST-125.
3. If ignition coils are carbon tracked, the coil tower spark plug wire nipples may be damaged.
4. By checking the secondary resistance, a coil with an open secondary may be located.
5. By switching a normally operating coil into the position of the malfunctioning one, a determination can be made as to fault being the coil or ignition control module.

3.3L (VIN N) ENGINE — COMPONENT DIAGNOSTIC CHART — CENTURY, CUTLASS CIERA/CRUISER, ACHIEVA, GRAND AM AND SKYLARK

CHART C-3
EVAP CANISTER PURGE VALVE CHECK
3300 (VIN N)

(1)
- SET PARKING BRAKE AND BLOCK DRIVE WHEELS.
- CHECK VACUUM SOURCE. IF NO VACUUM, REPAIR CAUSE OF NO VACUUM. IF OK, CONTINUE WITH CHART.
- DISCONNECT CANISTER PURGE SOLENOID ELECTRICAL CONNECTOR.
- REMOVE SOLENOID VALVE FROM CANISTER AND CONNECT A VACUUM GAGE TO THE CANISTER SIDE OF THE SOLENOID VALVE.
- WITH THE ENGINE IDLING NOTE VACUUM GAGE. SHOULD READ 0 VACUUM. DOES IT?

NO → REPLACE SOLENOID.

(2)
- INSTALL TECH 1.
- RECONNECT SOLENOID ELECTRICAL CONNECTOR.
- UNDER MISC. TEST SELECT CCP CONTROL.
- CYCLE THE CCP SOLENOID "ON" AND "OFF."
- DOES THE VACUUM GO UP AND DOWN AS THE SOLENOID IS CYCLED "ON" AND "OFF"?

YES → CHECK CANISTER AND PURGE HOSE FOR DAMAGE. IF OK, NO PROBLEM FOUND.

NO
- DISCONNECT CCP SOLENOID ELECTRICAL CONNECTOR.
- KEY "ON" ENGINE "OFF."
- INSTALL A TEST LIGHT BETWEEN HARNESS CONNECTOR TERMINAL "A" AND GROUND.
- DOES THE LIGHT COME "ON"?

NO → REPAIR OPEN IN SOLENOID IGNITION FEED CIRCUIT.

YES
- CONNECT A TEST LIGHT BETWEEN HARNESS CONNECTOR TERMINALS. TEST LIGHT SHOULD BE "OFF." IS IT?

NO → CHECK CKT 428 FOR A SHORT TO GROUND. IF OK, REPLACE ECM.

YES
- WITH TEST LIGHT STILL CONNECTED BETWEEN CONNECTOR HARNESS TERMINALS, CYCLE PURGE SOLENOID "ON" WITH THE TECH 1. TEST LIGHT SHOULD BE "ON" WITH SOLENOID COMMANDED "ON." IS IT?

NO → OPEN CKT 428, POOR CONNECTION AT ECM OR FAULTY ECM.

YES → FAULTY SOLENOID.

"AFTER REPAIRS", CONFIRM "CLOSED LOOP" OPERATION AND NO MIL (SERVICE ENGINE SOON).

3.3L (VIN N) ENGINE — COMPONENT DIAGNOSTIC CHART — CENTURY, CUTLASS CIERA/CRUISER, ACHIEVA, GRAND AM AND SKYLARK

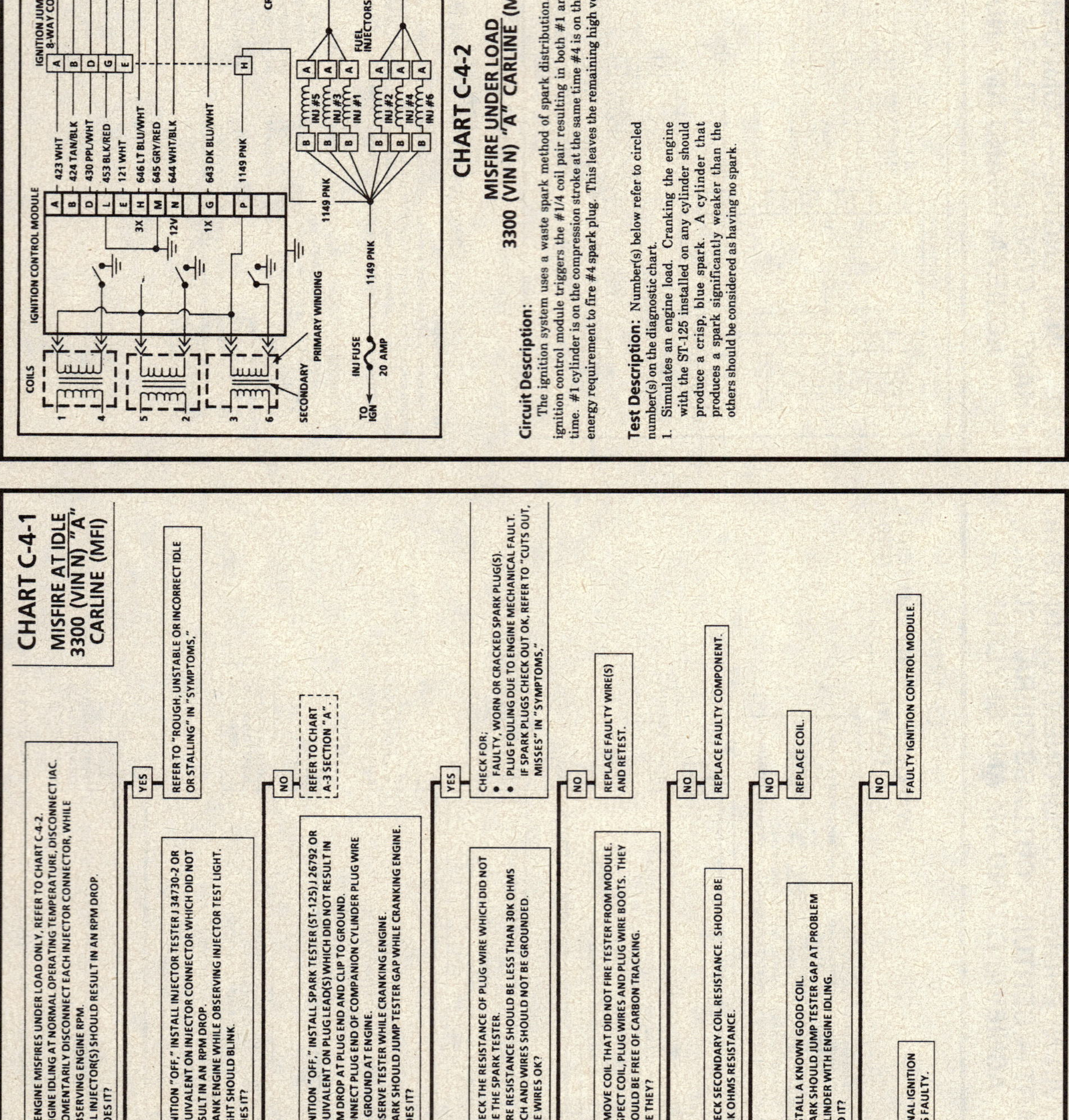

CHART C-4-2
MISFIRE UNDER LOAD
3300 (VIN N) "A" CARLINE (MFI)

Circuit Description:

The ignition system uses a waste spark method of spark distribution. In this type of ignition system the ignition control module triggers the #1/4 coil pair resulting in both #1 and #4 spark plugs firing at the same time. #1 cylinder is on the compression stroke at the same time #4 is on the exhaust stroke, resulting in a lower energy requirement to fire #4 spark plug. This leaves the remaining high voltage to fire #1 spark plug.

Test Description: Number(s) below refer to circled number(s) on the diagnostic chart.

1. Simulates an engine load. Cranking the engine with the ST-125 installed on any cylinder should produce a crisp, blue spark. A cylinder that produces a spark significantly weaker than the others should be considered as having no spark.

3.3L (VIN N) ENGINE — COMPONENT DIAGNOSTIC CHART — CENTURY, CUTLASS CIERA/CRUISER, ACHIEVA, GRAND AM AND SKYLARK

CHART C-4-1
MISFIRE AT IDLE
3300 (VIN N) "A" CARLINE (MFI)

(1)
- IF ENGINE MISFIRES UNDER LOAD ONLY, REFER TO CHART C-4-2.
- ENGINE IDLING AT NORMAL OPERATING TEMPERATURE, DISCONNECT IAC.
- MOMENTARILY DISCONNECT EACH INJECTOR CONNECTOR, WHILE OBSERVING ENGINE RPM.
- ALL INJECTOR(S) SHOULD RESULT IN AN RPM DROP.

DOES IT?

YES → REFER TO "ROUGH, UNSTABLE OR INCORRECT IDLE OR STALLING" IN "SYMPTOMS."

NO →

- IGNITION "OFF," INSTALL INJECTOR TESTER J 34730-2 OR EQUIVALENT ON INJECTOR CONNECTOR WHICH DID NOT RESULT IN AN RPM DROP.
- CRANK ENGINE WHILE OBSERVING INJECTOR TEST LIGHT. LIGHT SHOULD BLINK.

DOES IT?

YES →

NO → REFER TO CHART A-3 SECTION "A".

(2)
- IGNITION "OFF," INSTALL SPARK TESTER (ST-125) J 26792 OR EQUIVALENT ON PLUG LEAD(S) WHICH DID NOT RESULT IN RPM DROP AT PLUG END AND CLIP TO GROUND.
- CONNECT PLUG END OF COMPANION CYLINDER PLUG WIRE TO GROUND AT ENGINE.
- OBSERVE TESTER WHILE CRANKING ENGINE.
- SPARK SHOULD JUMP TESTER GAP WHILE CRANKING ENGINE.

DOES IT?

YES → CHECK FOR:
- FAULTY, WORN OR CRACKED SPARK PLUG(S).
- PLUG FOULING DUE TO ENGINE MECHANICAL FAULT.
- IF SPARK PLUGS CHECK OUT OK, REFER TO "CUTS OUT, MISSES" IN "SYMPTOMS."

NO →

- CHECK THE RESISTANCE OF PLUG WIRE WHICH DID NOT FIRE THE SPARK TESTER.
- WIRE RESISTANCE SHOULD BE LESS THAN 30K OHMS EACH AND WIRES SHOULD NOT BE GROUNDED.

ARE WIRES OK?

YES → REPLACE FAULTY WIRE(S) AND RETEST.

NO →

(3)
- REMOVE COIL THAT DID NOT FIRE TESTER FROM MODULE.
- INSPECT COIL, PLUG WIRES AND PLUG WIRE BOOTS. THEY SHOULD BE FREE OF CARBON TRACKING.

ARE THEY?

YES → REPLACE FAULTY COMPONENT.

NO →

(4)
- CHECK SECONDARY COIL RESISTANCE. SHOULD BE 5-8K OHMS RESISTANCE.

YES → REPLACE COIL.

NO →

(5)
- INSTALL A KNOWN GOOD COIL.
- SPARK SHOULD JUMP TESTER GAP AT PROBLEM CYLINDER WITH ENGINE IDLING.

DID IT?

YES → ORIGINAL IGNITION COIL IS FAULTY.

NO → FAULTY IGNITION CONTROL MODULE.

3.3L (VIN N) ENGINE — COMPONENT DIAGNOSTIC CHART — CENTURY, CUTLASS CIERA/CRUISER, ACHIEVA, GRAND AM AND SKYLARK

CHART C-4-2
MISFIRE UNDER LOAD
3300 (VIN N) "A" CARLINE (MFI)

① • IF ENGINE MISFIRES AT IDLE, REFER TO CHART C-4-1.
• IGNITION "OFF," INSTALL SPARK TESTER (ST-125) J 26792 OR EQUIVALENT TO #1 PLUG WIRE AT PLUG END AND CLIP TO GROUND.
• CONNECT PLUG END OF COMPANION CYLINDER (#4) PLUG WIRE TO ENGINE GROUND.
• REMOVE 15 AMP ECM FUSE FROM FUSE BLOCK.
• CRANK ENGINE, SPARK SHOULD JUMP TESTER GAP WHILE CRANKING.
• REPEAT ABOVE STEPS FOR EACH CYLINDER, RECORDING ANY CYLINDER(S) THAT DID NOT FIRE. RECONNECT EACH PLUG WIRE TO PLUG BEFORE MOVING ON TO NEXT CYLINDER.
• SPARK SHOULD JUMP TESTER GAP AT ALL PLUG WIRES WHILE CRANKING.

OK
• CHECK FOR FAULTY, CRACKED, FOULED OR WORN SPARK PLUG(S).
• PERFORM INJECTOR BALANCE TEST, CHART C-2A.
• REFER TO "CUTS OUT, MISSES" IN "SYMPTOMS,"

CYLINDER PAIR (1-4, 2-5, OR 3-6) NOT FIRING.
• CHECK FOR AN OPEN PLUG WIRE FOR CYLINDER PAIR THAT DID NOT FIRE (PLUG WIRES SHOULD MEASURE UNDER 30K Ω). IF OK, SWAP PROBLEM COIL WITH ONE FOR CYLINDER PAIR THAT DOES FIRE AND RETEST. IF PROBLEM FOLLOWS COIL, REPLACE COIL. IF NOT, REPLACE IGNITION CONTROL MODULE.

SINGLE CYLINDER OR UNPAIRED CYLINDERS NOT FIRING.
• CHECK FOR GROUNDED PLUG WIRE(S) FOR CYLINDER(S) THAT DID NOT FIRE AND REPLACE AS NECESSARY. IF OK, SWAP PROBLEM COIL WITH ONE THAT OPERATES PROPERLY. IF PROBLEM FOLLOWS COIL, REPLACE IT. IF NOT, REPLACE PLUG WIRE FOR CYLINDER(S) THAT DID NOT FIRE.

"AFTER REPAIRS," CONFIRM "CLOSED LOOP" OPERATION AND NO MIL (SERVICE ENGINE SOON).

3.3L (VIN N) ENGINE — COMPONENT DIAGNOSTIC CHART — CENTURY, CUTLASS CIERA/CRUISER, ACHIEVA, GRAND AM AND SKYLARK

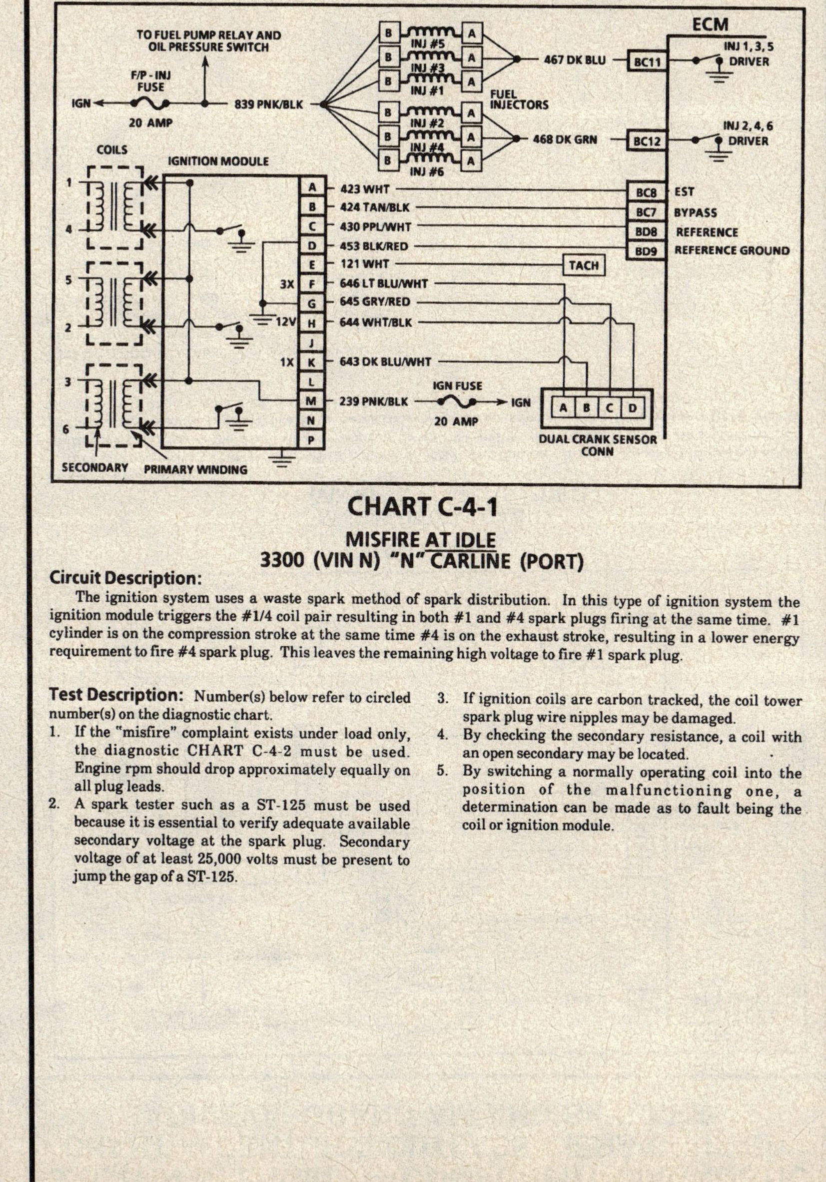

CHART C-4-1
MISFIRE AT IDLE
3300 (VIN N) "N" CARLINE (PORT)

Circuit Description:

The ignition system uses a waste spark method of spark distribution. In this type of ignition system the ignition module triggers the #1/4 coil pair resulting in both #1 and #4 spark plugs firing at the same time. #1 cylinder is on the compression stroke at the same time #4 is on the exhaust stroke, resulting in a lower energy requirement to fire #4 spark plug. This leaves the remaining high voltage to fire #1 spark plug.

Test Description: Number(s) below refer to circled number(s) on the diagnostic chart.

1. If the "misfire" complaint exists under load only, the diagnostic CHART C-4-2 must be used. Engine rpm should drop approximately equally on all plug leads.
2. A spark tester such as a ST-125 must be used because it is essential to verify adequate available secondary voltage at the spark plug. Secondary voltage of at least 25,000 volts must be present to jump the gap of a ST-125.
3. If ignition coils are carbon tracked, the coil tower spark plug wire nipples may be damaged.
4. By checking the secondary resistance, a coil with an open secondary may be located.
5. By switching a normally operating coil into the position of the malfunctioning one, a determination can be made as to fault being the coil or ignition module.

3.3L (VIN N) ENGINE — COMPONENT DIAGNOSTIC CHART — CENTURY, CUTLASS CIERA/CRUISER, ACHIEVA, GRAND AM AND SKYLARK

CHART C-4-1
MISFIRE AT IDLE
3300 (VIN N) "N" CARLINE (PORT)

(1)
- IF ENGINE MISFIRES UNDER LOAD ONLY, SEE CHART C-4-2.
- ENGINE IDLING AT NORMAL OPERATING TEMPERATURE, DISCONNECT IAC.
- MOMENTARILY DISCONNECT EACH INJECTOR CONNECTOR, WHILE OBSERVING ENGINE RPM.
- ALL INJECTOR(S) SHOULD RESULT IN AN RPM DROP. DOES IT?

YES → SEE "ROUGH, UNSTABLE OR INCORRECT IDLE OR STALLING" IN "SYMPTOMS."

NO →
- IGNITION "OFF," INSTALL INJECTOR TESTER J 34730-2 OR EQUIVALENT ON INJECTOR CONNECTOR WHICH DID NOT RESULT IN AN RPM DROP. CRANK ENGINE WHILE OBSERVING INJECTOR TEST LIGHT. LIGHT SHOULD BLINK. DOES IT?

NO → SEE CHART A-3 SECTION "A".

(2) YES →
- IGNITION "OFF," INSTALL SPARK TESTER (ST-125) J 26792 OR EQUIVALENT ON PLUG LEAD(S) WHICH DID NOT RESULT IN RPM DROP (1,3,5 AT PLUG AND 2,4,6 AT COIL).
- SPARK SHOULD JUMP TESTER GAP WHILE CRANKING ENGINE. DOES IT?

YES → CHECK FOR:
- FAULTY, WORN OR CRACKED SPARK PLUG(S)
- PLUG FOULING DUE TO ENGINE MECHANICAL FAULT.
IF SPARK PLUGS CHECK OUT OK, SEE "CUTS OUT, MISSES" IN "SYMPTOMS."

NO → CHECK THE RESISTANCE OF PLUG WIRE WHICH DID NOT FIRE THE SPARK TESTER. WIRE RESISTANCE SHOULD BE LESS THAN 30,000 OHMS EACH AND WIRES SHOULD NOT BE GROUNDED. ARE WIRES OK?

NO → REPLACE FAULTY WIRE (S) AND RETEST.

(3) YES →
- REMOVE COIL THAT DID NOT FIRE TESTER FROM MODULE.
- INSPECT COIL, PLUG WIRES AND PLUG WIRE NIPPLES, THEY SHOULD BE FREE OF CARBON TRACKING. ARE THEY?

NO → REPLACE FAULTY COMPONENT.

(4) YES →
- CHECK SECONDARY COIL RESISTANCE. SHOULD BE 5-8K OHMS RESISTANCE.

NO → REPLACE COIL.

(5) YES →
- INSTALL A KNOWN GOOD COIL.
- SPARK SHOULD JUMP TESTER GAP AT PROBLEM CYLINDER WITH ENGINE IDLING. DID IT?

YES → ORIGINAL IGNITION COIL IS FAULTY.

NO → FAULTY IGNITION MODULE.

3.3L (VIN N) ENGINE — COMPONENT DIAGNOSTIC CHART — CENTURY, CUTLASS CIERA/CRUISER, ACHIEVA, GRAND AM AND SKYLARK

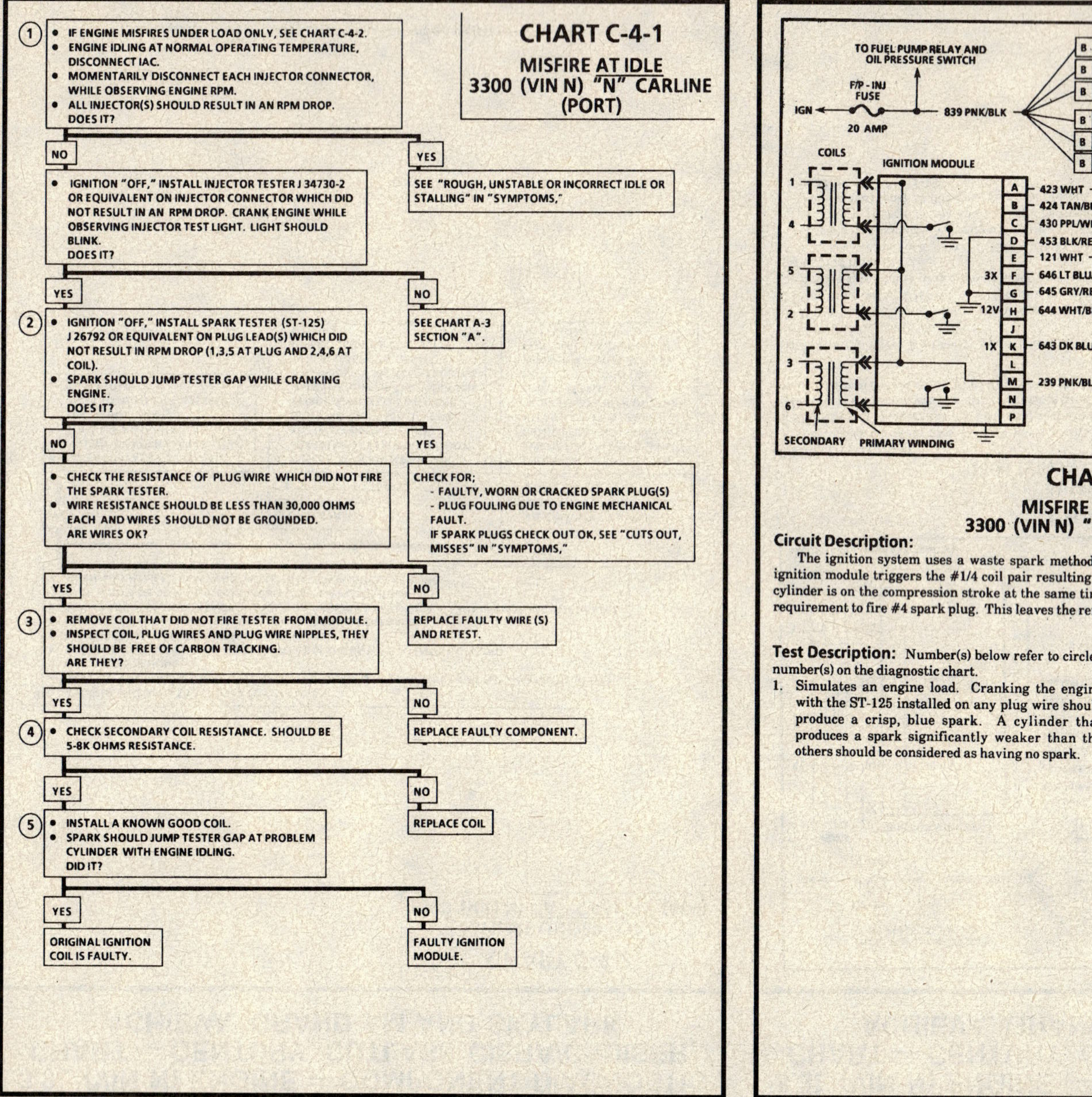

CHART C-4-2
MISFIRE UNDER LOAD
3300 (VIN N) "N" CARLINE (PORT)

Circuit Description:

The ignition system uses a waste spark method of spark distribution. In this type of ignition system the ignition module triggers the #1/4 coil pair resulting in both #1 and #4 spark plugs firing at the same time. #1 cylinder is on the compression stroke at the same time #4 is on the exhaust stroke, resulting in a lower energy requirement to fire #4 spark plug. This leaves the remaining high voltage to fire #1 spark plug.

Test Description: Number(s) below refer to circled number(s) on the diagnostic chart.
1. Simulates an engine load. Cranking the engine with the ST-125 installed on any plug wire should produce a crisp, blue spark. A cylinder that produces a spark significantly weaker than the others should be considered as having no spark.

3.3L (VIN N) ENGINE — COMPONENT DIAGNOSTIC CHART — CENTURY, CUTLASS CIERA/CRUISER, ACHIEVA, GRAND AM AND SKYLARK

CHART C-4-2
MISFIRE UNDER LOAD
3300 (VIN N) "N" CARLINE (PORT)

① • IF ENGINE MISFIRES AT IDLE, SEE CHART C-4-1.
• IGNITION "OFF," INSTALL SPARK TESTER (ST-125) J 26792 OR EQUIVALENT TO #1 PLUG WIRE AT PLUG END.
• REMOVE 10 AMP ERLS FUSE FROM FUSE BLOCK.
• CRANK ENGINE, SPARK SHOULD JUMP TESTER GAP WHILE CRANKING.
• REPEAT ABOVE STEPS FOR EACH PLUG WIRE, RECORDING ANY CYLINDER(S) THAT DID NOT FIRE. RECONNECT EACH PLUG WIRE TO PLUG BEFORE MOVING ON TO NEXT CYLINDER.
• SPARK SHOULD JUMP TESTER GAP AT ALL PLUGS WHILE CRANKING.

OK

CYLINDER PAIR (1-4, 2-5, OR 3-6) NOT FIRING.

SINGLE CYLINDER OR UNPAIRED CYLINDERS NOT FIRING.

• CHECK FOR FAULTY, CRACKED, FOULED OR WORN SPARK PLUG(S).
• PERFORM INJECTOR BALANCE TEST, CHART C-2A.
• REFER TO "CUTS OUT, MISSES" IN "SYMPTOMS,"

CHECK FOR AN OPEN PLUG WIRE FOR CYLINDER PAIR THAT DID NOT FIRE (PLUG WIRES SHOULD MEASURE UNDER 30K Ω). IF OK, SWAP PROBLEM COIL WITH ONE FOR CYLINDER PAIR THAT DOES FIRE AND RETEST. IF PROBLEM FOLLOWS COIL, REPLACE COIL. IF NOT, REPLACE IGNITION MODULE.

CHECK FOR GROUNDED PLUG WIRE(S) FOR CYLINDER(S) THAT DID NOT FIRE AND REPLACE AS NECESSARY. IF OK, SWAP PROBLEM COIL WITH ONE THAT OPERATES PROPERLY. IF PROBLEM FOLLOWS COIL, REPLACE IT. IF NOT, REPLACE PLUG WIRE FOR CYLINDER(S) THAT DID NOT FIRE.

"AFTER REPAIRS," CONFIRM "CLOSED LOOP" OPERATION AND NO "SERVICE ENGINE SOON" LIGHT.

3.3L (VIN N) ENGINE — COMPONENT DIAGNOSTIC CHART — CENTURY, CUTLASS CIERA/CRUISER, ACHIEVA, GRAND AM AND SKYLARK

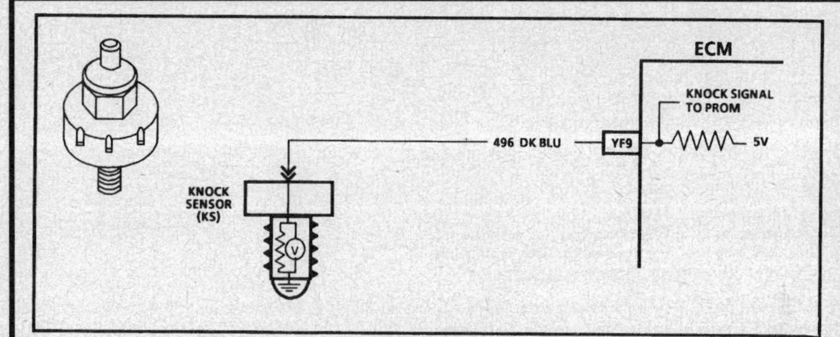

CHART C-5
KNOCK SENSOR (KS) SYSTEM CHECK
3300 (VIN N)   (MFI)

Circuit Description:

The knock sensor is used to detect engine detonation and allow the ECM to retard the ignition control timing based on the signal being received. The circuitry within the knock sensor causes the ECM's supplied 5 volt signal to be pulled down so that under a no knock condition, CKT 496 should measure about 2.5 volts. The knock sensor produces an AC signal which rides on the 2.5 volts DC voltage. The amplitude and frequency are dependent upon the knock level.

The PROM used with this engine contains the functions which were part of the remotely mounted Electronic Spark Control (ESC) modules used on past GM vehicles. The knock sensor system portion of the PROM then sends a signal to other parts of the ECM which adjusts the spark timing to retard the spark and reduce the detonation.

Test Description: Number(s) below refer to circled number(s) on the diagnostic chart.

1. With engine idling, there should not be a knock signal present at the ECM because detonation is not likely under a no load condition.
2. Tapping on the engine lift hook should simulate a knock signal to determine if the sensor is capable of detecting detonation. If no knock is detected, try tapping on engine block closer to sensor before replacing sensor.
3. If the engine has an internal problem which is creating a knock, the knock sensor may be responding to the internal failure.
4. This test determines if the knock sensor is faulty or if the knock sensor system portion of the PROM is faulty. If it is determined that the PROM is faulty, be sure that it is properly installed and latched into place. If not properly installed, repair and retest.

Diagnostic Aids:

While observing knock signal on the Tech 1 there should be an indication that knock is present when detonation can be heard. Detonation is most likely to occur under high engine load conditions.

If the knock sensor CKT 496 is routed too close to secondary ignition wires, it may induce a voltage and cause a false knock signal.

3.3L (VIN N) ENGINE — COMPONENT DIAGNOSTIC CHART — CENTURY, CUTLASS CIERA/CRUISER, ACHIEVA, GRAND AM AND SKYLARK

CHART C-5

KNOCK SENSOR (KS) SYSTEM CHECK
3300 (VIN N) **(MFI)**

1. • IF DIAGNOSTIC TROUBLE CODE (DTC) 43 IS SET, USE THE DTC CHART.
   • ENGINE MUST BE IDLING AT NORMAL OPERATING TEMPERATURE.
   • USE TECH 1 TO OBSERVE KNOCK SIGNAL.
   IS KNOCK INDICATED?

NO → 2. • TAP ON ENGINE, LIFT HOOK BRACKET WHILE OBSERVING KNOCK SIGNAL.
   • TECH 1 SHOULD INDICATE KNOCK WHILE TAPPING ON BRACKET.
   DOES IT?

YES → 3. IF AN ENGINE KNOCK CAN BE HEARD, REPAIR THE BASIC ENGINE PROBLEM. IF NO AUDIBLE KNOCK IS HEARD, FOLLOW THE STEPS:
   • IGNITION "OFF."
   • DISCONNECT KNOCK SENSOR.
   • CONNECT VOLTMETER TO KNOCK SENSOR AND ENGINE GROUND.
   • SET VOLTMETER ON 2 VOLT A.C. SCALE.
   • IGNITION "ON," ENGINE "ON."
   IS A SIGNAL INDICATED ON VOLTMETER?

NO → 4. • IGNITION "OFF."
   • DISCONNECT KNOCK SENSOR.
   • CONNECT VOLTMETER TO KNOCK SENSOR AND ENGINE GROUND.
   • SET VOLTMETER ON 2 VOLT A.C. SCALE.
   • IGNITION "ON," ENGINE "ON."
   • TAP ON ENGINE BLOCK NEAR SENSOR.
   IS A SIGNAL INDICATED ON VOLTMETER WHILE TAPPING ON ENGINE BLOCK?

YES → SYSTEM IS OPERATING PROPERLY. REFER TO "DIAGNOSTIC AIDS" ON FACING PAGE.

NO → CHECK CKT 496 FOR BEING NEAR A SPARK PLUG WIRE
OR
A FAULTY ECM CONNECTION
OR
FAULTY ECM
OR
PROM.

YES → REPLACE KNOCK SENSOR.

NO → REPLACE PROM OR ECM.

YES → REPLACE KNOCK SENSOR.

"AFTER REPAIRS," CONFIRM "CLOSED LOOP" OPERATION AND NO MIL (SERVICE ENGINE SOON).

3.3L (VIN N) ENGINE — COMPONENT DIAGNOSTIC CHART — CENTURY, CUTLASS CIERA/CRUISER, ACHIEVA, GRAND AM AND SKYLARK

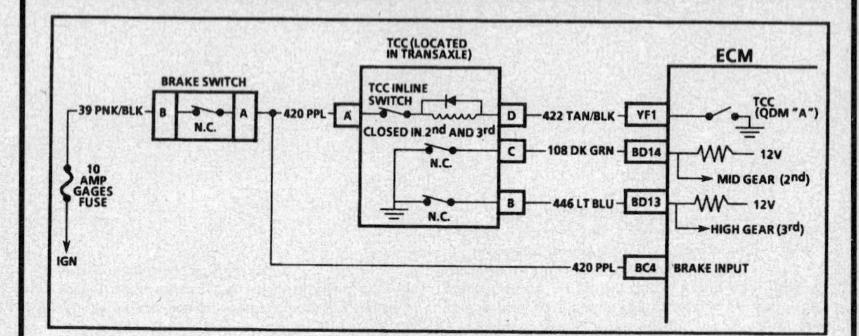

CHART C-8A

TORQUE CONVERTER CLUTCH (TCC)
3300 (VIN N) **(MFI)**
3T40 TRANSAXLE

Circuit Description:

The purpose of the Torque Converter Clutch (TCC) is to eliminate the power loss of the transmission converter stage when the vehicle is in a cruise condition. This allows the convenience of the automatic transmission and the fuel economy of a manual transaxle.

Fused ignition feed is supplied to the TCC solenoid through the brake switch and transmission TCC in-line switch. The ECM will engage TCC by grounding CKT 422 to energize the solenoid.

TCC will engage when:
- Engine warmed up.
- Vehicle speed above a calibrated value (about 33 mph 51 km/h).
- Transmission 3rd gear switch open.
- Throttle position sensor output not changing, indicating a steady road speed.
- Transmission TCC in-line switch closed.
- Brake switch closed.

Test Description: Number(s) below refer to circled number(s) on the diagnostic chart.
1. Checks gages fuse and brake switch.
2. Checks ECM for proper operation.
3. Checks the internal switches in the transaxle.

Diagnostic Aids:

The Tech 1 only indicates when the ECM has turned "ON" the TCC driver and this does not confirm that the TCC has engaged. To determine if TCC is functioning properly, engine RPM should decrease when the Tech 1 indicates the TCC driver has turned "ON."

If the ECM, TCC solenoid, and control circuitry are functioning correctly and the TCC still does not apply properly, the problem may be hydraulic or mechanical.

3.3L (VIN N) ENGINE — COMPONENT DIAGNOSTIC CHART — CENTURY, CUTLASS CIERA/CRUISER, ACHIEVA, GRAND AM AND SKYLARK

CHART C-8A

TORQUE CONVERTER CLUTCH (TCC)
3300 (VIN N)   (MFI)
3T40 TRANSAXLE

①
- PLACE TRANS IN PARK OR NEUTRAL.
- SET PARKING BRAKE AND BLOCK DRIVE WHEELS.
- INSTALL TECH 1.
- OBSERVE BRAKE SWITCH STATUS, SHOULD DISPLAY "RELEASED," AND SWITCH TO "APPLIED" WHEN BRAKE IS APPLIED.
  DOES THE BRAKE SWITCH STATUS DISPLAY "RELEASED" WHEN BRAKE IS NOT APPLIED?

YES / NO

②
- DISCONNECT TCC CONNECTOR AT TRANSAXLE.
- CONNECT A TEST LIGHT BETWEEN HARNESS TERMINALS "A" AND "D".
- CYCLE THE TCC SOLENOID "ON" AND "OFF" WITH THE TECH 1. TEST LIGHT SHOULD FLASH "ON" AND "OFF" AS SOLENOID IS CYCLED. DOES IT?

NO (from ①):
OPEN IN CKT 420
OR
OPEN IN CKT 39
OR
MISADJUSTED OR FAULTY BRAKE INPUT SWITCH
OR
GAGES FUSE.

YES / NO

③
- SELECT DISPLAY DATA MODE WITH TECH 1.
- RECONNECT TCC CONNECTOR.
- DRIVE VEHICLE ABOVE 45 MPH.
- "SCAN" TCC.
- SHOULD DISPLAY "ON" AND A DROP IN RPM SHOULD OCCUR WHEN TCC ENGAGES. DOES IT?

NO (from ②):
OPEN IN CKT 420 FROM BRAKE SWITCH
OR
OPEN IN CKT 422
OR
FAULTY CONNECTION AT ECM OR FAULTY ECM.

YES:
NO PROBLEM FOUND. SEE "DIAGNOSTIC AIDS"

NO (from ③):
POOR CONNECTION AT TCC
OR
FAULTY TRANSAXLE INTERNAL WIRING
OR
FAULTY TCC SOLENOID
OR
INLINE TCC SWITCH.

"AFTER REPAIRS," CONFIRM "CLOSED LOOP" OPERATION AND NO MIL (SERVICE ENGINE SOON).

3.3L (VIN N) ENGINE — COMPONENT DIAGNOSTIC CHART — CENTURY, CUTLASS CIERA/CRUISER, ACHIEVA, GRAND AM AND SKYLARK

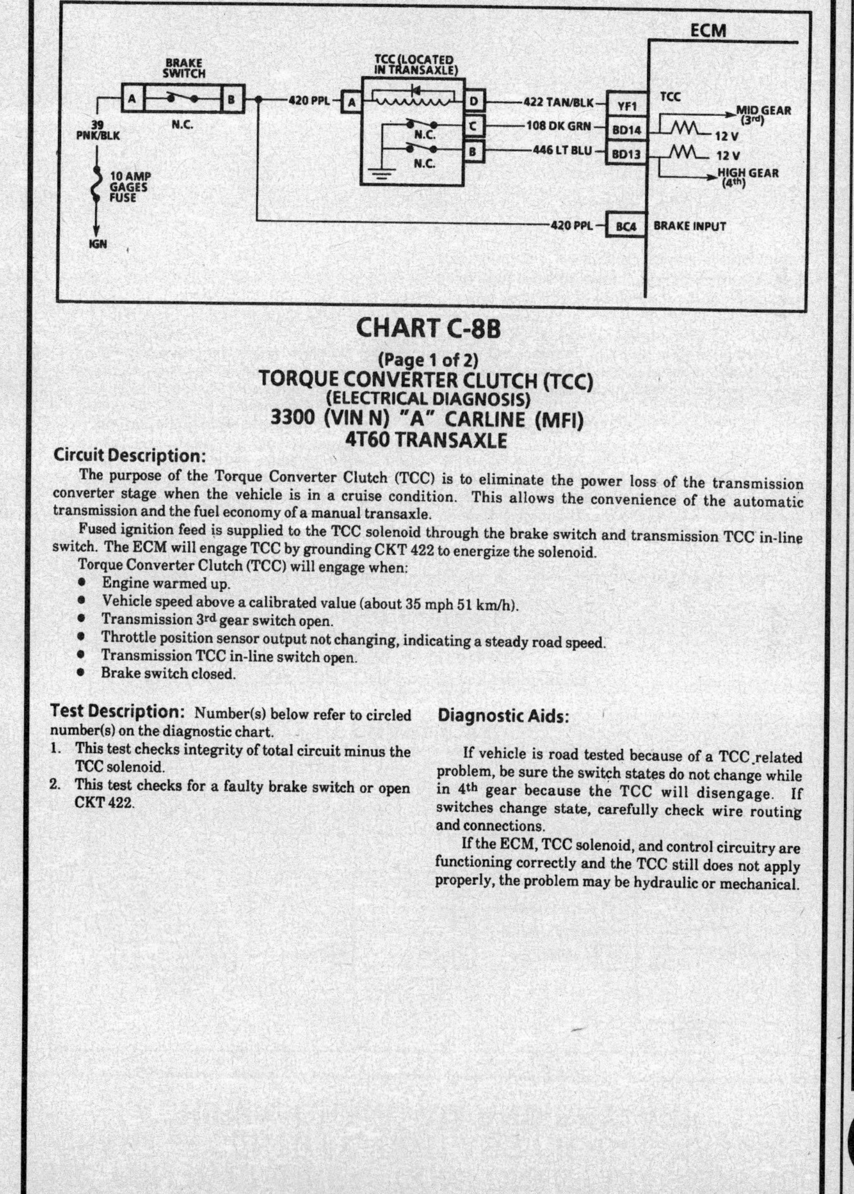

CHART C-8B

(Page 1 of 2)
TORQUE CONVERTER CLUTCH (TCC)
(ELECTRICAL DIAGNOSIS)
3300 (VIN N) "A" CARLINE (MFI)
4T60 TRANSAXLE

Circuit Description:

The purpose of the Torque Converter Clutch (TCC) is to eliminate the power loss of the transmission converter stage when the vehicle is in a cruise condition. This allows the convenience of the automatic transmission and the fuel economy of a manual transaxle.

Fused ignition feed is supplied to the TCC solenoid through the brake switch and transmission TCC in-line switch. The ECM will engage TCC by grounding CKT 422 to energize the solenoid.

Torque Converter Clutch (TCC) will engage when:
- Engine warmed up.
- Vehicle speed above a calibrated value (about 35 mph 51 km/h).
- Transmission 3rd gear switch open.
- Throttle position sensor output not changing, indicating a steady road speed.
- Transmission TCC in-line switch open.
- Brake switch closed.

Test Description: Number(s) below refer to circled number(s) on the diagnostic chart.
1. This test checks integrity of total circuit minus the TCC solenoid.
2. This test checks for a faulty brake switch or open CKT 422.

Diagnostic Aids:

If vehicle is road tested because of a TCC related problem, be sure the switch states do not change while in 4th gear because the TCC will disengage. If switches change state, carefully check wire routing and connections.

If the ECM, TCC solenoid, and control circuitry are functioning correctly and the TCC still does not apply properly, the problem may be hydraulic or mechanical.

3.3L (VIN N) ENGINE — COMPONENT DIAGNOSTIC CHART — CENTURY, CUTLASS CIERA/CRUISER, ACHIEVA, GRAND AM AND SKYLARK

CHART C-8B

(Page 1 of 2)
TORQUE CONVERTER CLUTCH (TCC)
(ELECTRICAL DIAGNOSIS)
3300 (VIN N) "A" CARLINE (MFI)
4T60 TRANSAXLE

- PLACE TRANSMISSION IN PARK OR NEUTRAL.
- SET PARKING BRAKE AND BLOCK DRIVE WHEELS.
- INSTALL TECH I.
- IGNITION "ON". ENGINE "OFF".
- CYCLE TCC SOLENOID "ON" AND "OFF".
- USING A STETHOSCOPE ON THE TRANSAXLE, YOU SHOULD BE ABLE TO HEAR THE TCC SOLENOID CLICK "ON" WHEN THE " ↑ " IS PRESSED AND CLICK "OFF" WHEN " ↓ " IS PRESSED.

NO →
- IGNITION "ON". ENGINE "OFF".
- CHECK TCC-GAGE FUSE.
- IS FUSE OK?

YES →
- TCC CIRUIT OK. SEE FACING PAGE "DIAGNOSTIC AIDS".
- TO CHECK GEAR SWITCHES, SEE PAGE TWO OF THIS CHART.

YES →
(1)
- IGNITION "ON". ENGINE "OFF".
- DISCONNECT TCC SOLENOID CONNECTOR AT TRANSAXLE.
- CONNECT A TEST LIGHT BETWEEN HARNESS CONNECTOR TERMINALS "A" and "D".
- CYCLE THE "ON" (↑) AND "OFF" (↓) BUTTONS ON TECH 1.
- LIGHT SHOULD FLASH "ON" AND "OFF" AS BUTTONS ARE CYCLED. DOES IT?

NO → REPLACE FUSE AND RETEST.

NO →
(2)
- CONNECT A TEST LIGHT BETWEEN CONNECTOR TERMINAL "A" AND GROUND.
- DOES THE LIGHT COME "ON"?

YES → FAULTY TCC SOLENOID.

NO →
- DISCONNECT BRAKE SWITCH.
- CONNECT A TEST LIGHT BETWEEN CKT 39 AND GROUND. IS THE LIGHT "ON"?

YES → CHECK FOR OPEN IN CKT 422. IF OK, THEN IT IS A FAULTY ECM CONNECTION OR ECM.

YES →
- JUMPER BRAKE SWITCH CONNECTOR TERMINALS TOGETHER.
- DOES TEST LIGHT COME "ON" AT TCC SOLENOID CONNECTOR TERMINAL "A"?

NO → REPAIR OPEN CKT 39.

YES → BRAKE SWITCH MISADJUSTED OR FAULTY.

NO → REPAIR OPEN CKT 420.

"AFTER REPAIRS," CONFIRM "CLOSED LOOP" OPERATION AND NO MIL (SERVICE ENGINE SOON).

3.3L (VIN N) ENGINE — COMPONENT DIAGNOSTIC CHART — CENTURY, CUTLASS CIERA/CRUISER, ACHIEVA, GRAND AM AND SKYLARK

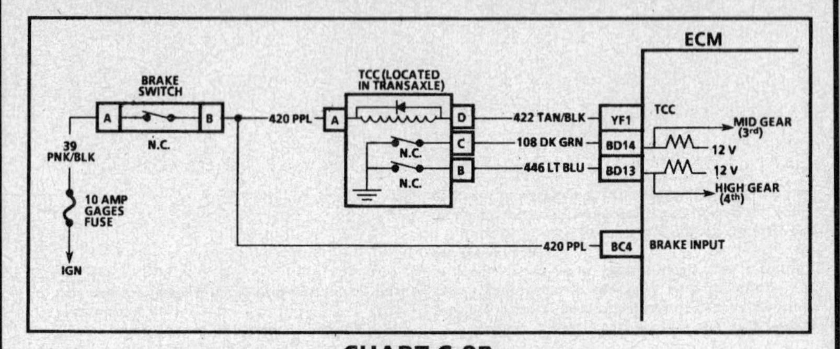

CHART C-8B

(Page 2 of 2)
TORQUE CONVERTER CLUTCH (TCC)
(ELECTRICAL DIAGNOSIS)
3300 (VIN N) "A" CARLINE (MFI)
4T60 TRANSAXLE

Circuit Description:

Each gear switch opens when the appropriate clutch is applied. All gear switches are open in fourth gear.

Test Description: Number(s) below refer to circled number(s) on the diagnostic chart.

1. Some scan tools display the state of these switches in different ways. Be familiar with the type of tool being used. All switches should be in the closed state during this test, and the tool should read the same for 3rd and 4th gear switches.
2. Determines whether the switch or signal circuit is open. The circuit can be checked for an open by measuring the voltage (with a voltmeter) at the Torque Converter Clutch (TCC) connector. Should be about 12 volts.
3. Because the switch(s) should be grounded in this step, disconnecting the TCC connector should cause the displayed switch states to change.
4. The switch state should change when the vehicle shifts into 3rd gear.

Diagnostic Aids:

If vehicle is road tested because of a TCC related problem, be sure the switch states do not change while in 4th gear because the TCC will disengage. If switches change state, carefully check wire routing and connections.

If the ECM, TCC solenoid, and control circuitry are functioning correctly and the TCC still does not apply properly, the problem may be hydraulic or mechanical.

3.3L (VIN N) ENGINE —COMPONENT DIAGNOSTIC CHART —CENTURY, CUTLASS CIERA/CRUISER, ACHIEVA, GRAND AM AND SKYLARK

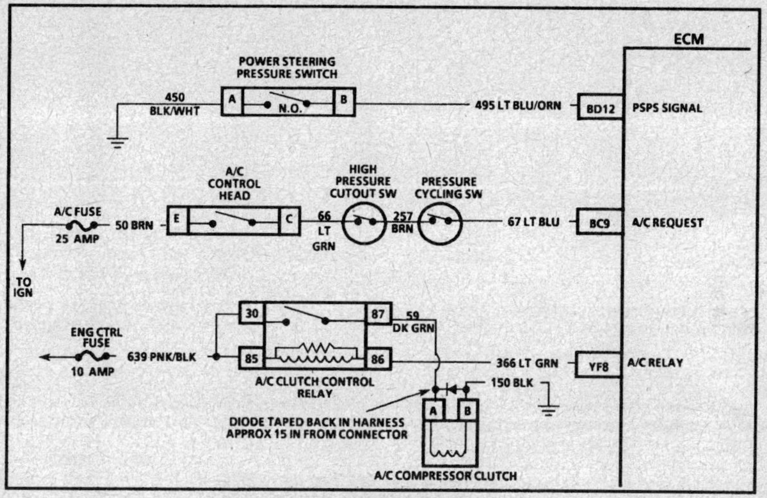

CHART C-10
(Page 1 of 2)
A/C CLUTCH CIRCUIT DIAGNOSIS
3300 (VIN N) "A" CARLINE (MFI)

Circuit Description:

The A/C relay is ECM controlled to delay A/C clutch engagement .3 second after A/C is turned "ON." This allows the IAC to adjust engine RPM before the A/C clutch engages. The ECM also causes the relay to disengage the A/C clutch during WOT operation. The A/C relay is energized when the ECM provides a ground path for CKT 366.

Test Description: Number(s) refer to circled number(s) on the diagnostic chart.
1. Checks A/C low pressure cutout switch.
2. Verifies that the A/C request signal is present at the ECM.

3.3L (VIN N) ENGINE —COMPONENT DIAGNOSTIC CHART —CENTURY, CUTLASS CIERA/CRUISER, ACHIEVA, GRAND AM AND SKYLARK

CHART C-8B
(Page 2 of 2)
TORQUE CONVERTER CLUTCH (TCC)
(ELECTRICAL DIAGNOSIS)
3300 (VIN N) "A" CARLINE (MFI)
4T60 TRANSAXLE

CHECKS MADE ON THIS PAGE WILL NOT PREVENT THE TCC FROM WORKING, BUT WILL AFFECT ENGAGEMENT OR DISENGAGEMENT POINTS.

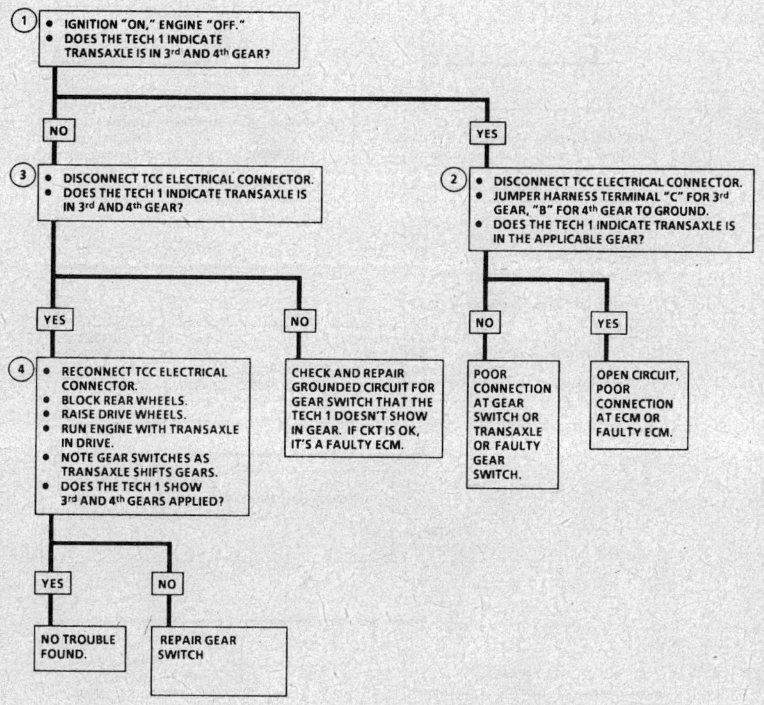

(1) IGNITION "ON," ENGINE "OFF."
- DOES THE TECH 1 INDICATE TRANSAXLE IS IN 3rd AND 4th GEAR?

NO →

(3) DISCONNECT TCC ELECTRICAL CONNECTOR.
- DOES THE TECH 1 INDICATE TRANSAXLE IS IN 3rd AND 4th GEAR?

YES →

(2) DISCONNECT TCC ELECTRICAL CONNECTOR. JUMPER HARNESS TERMINAL "C" FOR 3rd GEAR, "B" FOR 4th GEAR TO GROUND.
- DOES THE TECH 1 INDICATE TRANSAXLE IS IN THE APPLICABLE GEAR?

YES → OPEN CIRCUIT, POOR CONNECTION AT ECM OR FAULTY ECM.

NO → POOR CONNECTION AT GEAR SWITCH OR FAULTY TRANSAXLE GEAR SWITCH.

NO →

CHECK AND REPAIR GROUNDED CIRCUIT FOR GEAR SWITCH THAT THE TECH 1 DOESN'T SHOW IN GEAR. IF CKT IS OK, IT'S A FAULTY ECM.

YES →

(4) RECONNECT TCC ELECTRICAL CONNECTOR.
- BLOCK REAR WHEELS.
- RAISE DRIVE WHEELS.
- RUN ENGINE WITH TRANSAXLE IN DRIVE.
- NOTE GEAR SWITCHES AS TRANSAXLE SHIFTS GEARS.
- DOES THE TECH 1 SHOW 3rd AND 4th GEARS APPLIED?

NO → REPAIR GEAR SWITCH

YES → NO TROUBLE FOUND.

"AFTER REPAIRS," CONFIRM "CLOSED LOOP" OPERATION AND NO MIL (SERVICE ENGINE SOON).

3.3L (VIN N) ENGINE — COMPONENT DIAGNOSTIC CHART — CENTURY, CUTLASS CIERA/CRUISER, ACHIEVA, GRAND AM AND SKYLARK

CHART C-10

(Page 1 of 2)
A/C CLUTCH CIRCUIT DIAGNOSIS
3300 (VIN N) "A" CARLINE (MFI)

- INSTALL TECH 1.
- KEY "ON."
- A/C "ON."
- "SCAN" FOR "A/C REQUEST."
- "A/C REQUEST" SHOULD DISPLAY "YES."
DOES IT?

NO

(1)
- ENGINE IDLING.
- A/C "ON."
- DISCONNECT PRESSURE CYCLING SWITCH.
- JUMPER BETWEEN HARNESS CONNECTOR TERMINALS.
DOES A/C CLUTCH ENGAGE?

YES → SEE PAGE 2 OF THIS CHART.

NO
- PROBE CKT 257 AT PRESSURE CYCLING SWITCH CONNECTOR WITH A TEST LIGHT CONNECTED TO GROUND.
IS THE LIGHT "ON"?

YES
BASIC A/C SEALED SYSTEM PROBLEM
OR
POOR CONNECTION AT PRESSURE CYCLING SWITCH
OR
FAULTY PRESSURE CYCLING SWITCH.

YES

(2)
- RECONNECT PRESSURE CYCLING SWITCH.
- IGNITION "OFF."
- DISCONNECT BLACK ECM C-D CONNECTOR.
- IGNITION "ON."
- PROBE HARNESS TERMINAL "BC9" WITH A TEST LIGHT CONNECTED TO GROUND.
IS THE LIGHT "ON"?

YES
POOR CONNECTION AT ECM OR FAULTY ECM.

NO
OPEN CKT 67.

NO
- RECONNECT PRESSURE CYCLING SWITCH.
- DISCONNECT HIGH PRESSURE CUTOUT SWITCH.
- PROBE CKT 66 AT SWITCH CONNECTOR WITH A TEST LIGHT CONNECTED TO GROUND.
IS THE LIGHT "ON"?

YES
OPEN CKT 257,
OR
POOR CONNECTION AT HIGH PRESSURE CUTOUT SWITCH
OR
FAULTY HIGH PRESSURE CUTOUT SWITCH.

NO
OPEN CKT 66 OR 50
OR
POOR CONNECTION AT A/C CONTROL HEAD
OR
FAULTY CONTROL HEAD.

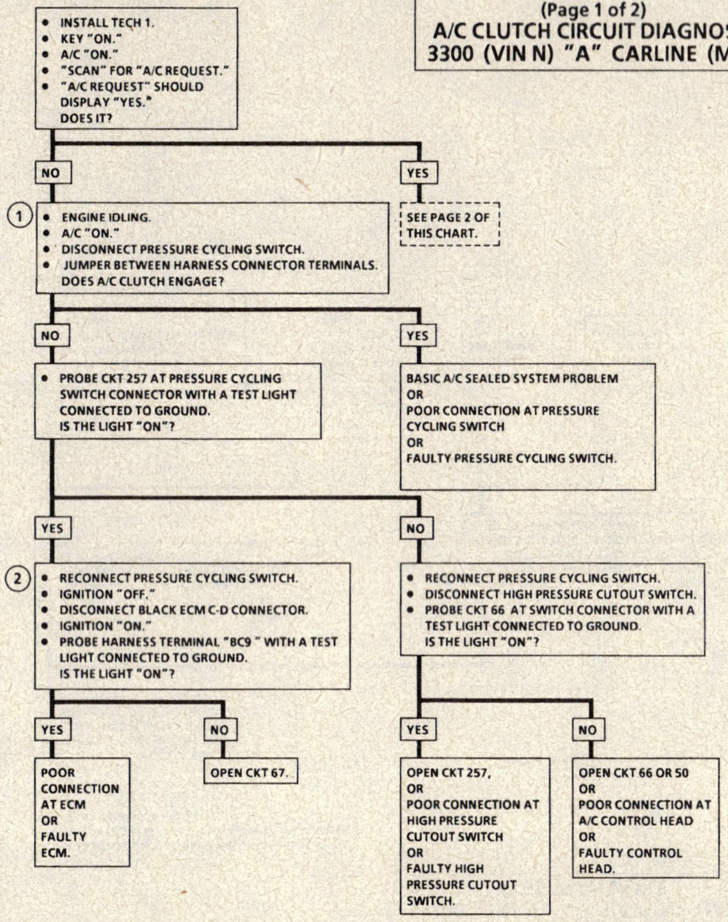

"AFTER REPAIRS," CONFIRM "CLOSED LOOP" OPERATION AND NO MIL (SERVICE ENGINE SOON).

3.3L (VIN N) ENGINE — COMPONENT DIAGNOSTIC CHART — CENTURY, CUTLASS CIERA/CRUISER, ACHIEVA, GRAND AM AND SKYLARK

ECM
POWER STEERING PRESSURE SWITCH
450 BLK/WHT
A N.O. B
495 LT BLU/ORN
BD12 PSPS SIGNAL
A/C FUSE 25 AMP
50 BRN
A/C CONTROL HEAD
E C
66 LT GRN
HIGH PRESSURE CUTOUT SW
257 BRN
PRESSURE CYCLING SW
67 LT BLU
BC9 A/C REQUEST
TO IGN
ENG CTRL FUSE 10 AMP
639 PNK/BLK
30 87 59 DK GRN
85 86
A/C CLUTCH CONTROL RELAY
366 LT GRN
150 BLK
YF8 A/C RELAY
DIODE TAPED BACK IN HARNESS APPROX 15 IN FROM CONNECTOR
A B
A/C COMPRESSOR CLUTCH

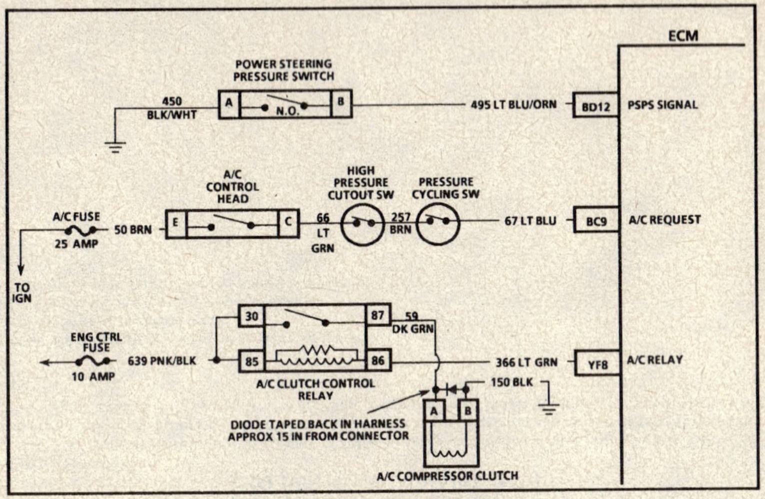

CHART C-10

(Page 2 of 2)
A/C CLUTCH CIRCUIT DIAGNOSIS
3300 (VIN N) "A" CARLINE (MFI)

Circuit Description:

The A/C relay is ECM controlled to delay A/C clutch engagement .3 second after A/C is turned "ON." This allows the IAC to adjust engine RPM before the A/C clutch engages. The ECM also causes the relay to disengage the A/C clutch during WOT operation. The A/C relay is energized when the ECM provides a ground path for CKT 366.

Test Description: Number(s) refer to circled number(s) on the diagnostic chart.

3. Test light "ON" verifies the integrity of CKT 366, 639, and QDM "B" in the ECM.

4. Test light "ON" verifies the integrity of CKT 639 to terminal "30" of the A/C relay.

5. Bypasses the relay to determine whether the relay or the compressor clutch or clutch wiring is faulty.

3.3L (VIN N) ENGINE — COMPONENT DIAGNOSTIC CHART — CENTURY, CUTLASS CIERA/CRUISER, ACHIEVA, GRAND AM AND SKYLARK

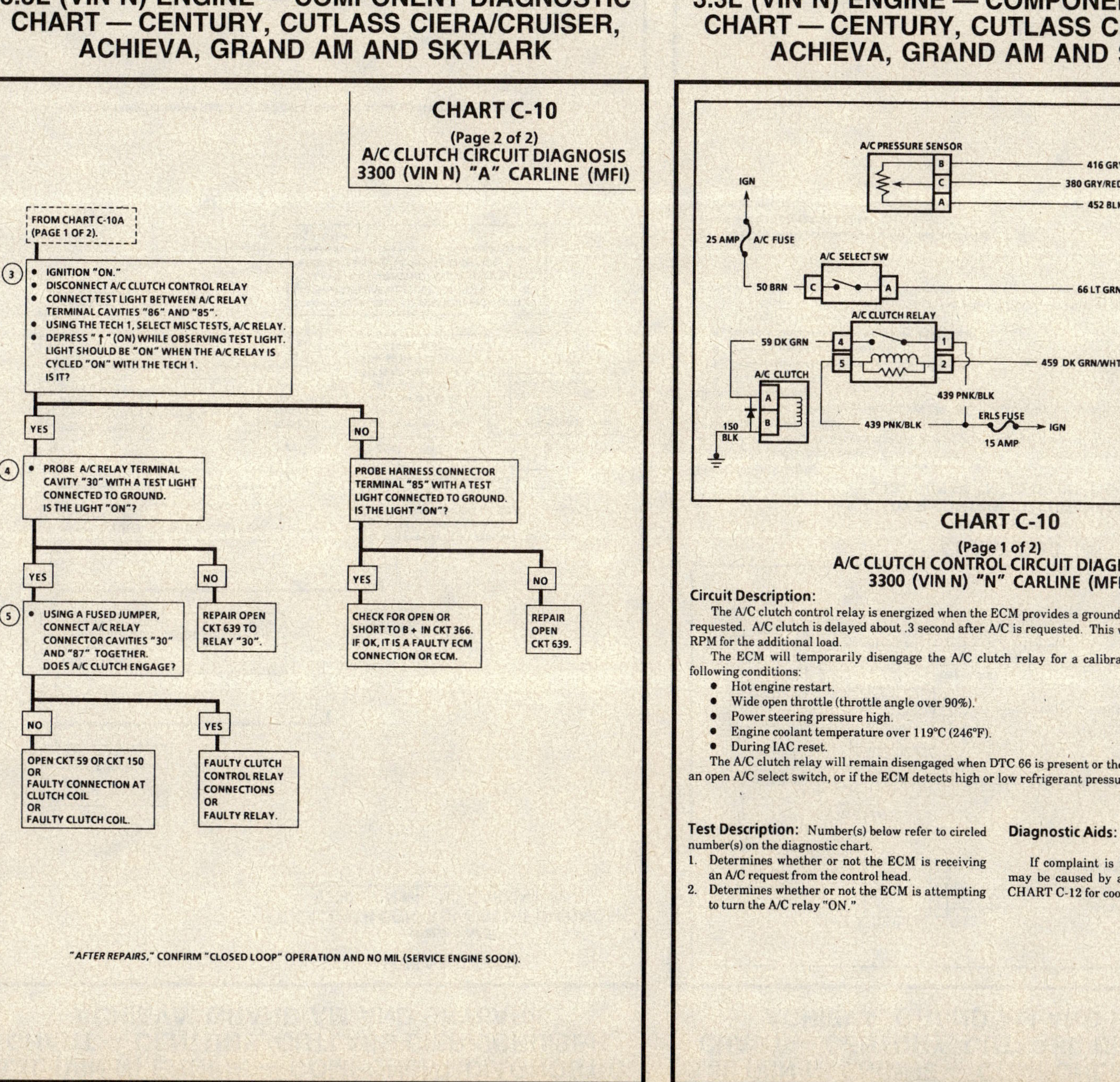

3.3L (VIN N) ENGINE — COMPONENT DIAGNOSTIC CHART — CENTURY, CUTLASS CIERA/CRUISER, ACHIEVA, GRAND AM AND SKYLARK

Circuit Description:

The A/C clutch control relay is energized when the ECM provides a ground path through CKT 459 and A/C is requested. A/C clutch is delayed about .3 second after A/C is requested. This will allow the IAC to adjust engine RPM for the additional load.

The ECM will temporarily disengage the A/C clutch relay for a calibrated time for one or more of the following conditions:
• Hot engine restart.
• Wide open throttle (throttle angle over 90%).
• Power steering pressure high.
• Engine coolant temperature over 119°C (246°F).
• During IAC reset.

The A/C clutch relay will remain disengaged when DTC 66 is present or there is no A/C request signal due to an open A/C select switch, or if the ECM detects high or low refrigerant pressure.

Test Description: Number(s) below refer to circled number(s) on the diagnostic chart.
1. Determines whether or not the ECM is receiving an A/C request from the control head.
2. Determines whether or not the ECM is attempting to turn the A/C relay "ON."

Diagnostic Aids:

If complaint is insufficient cooling, the problem may be caused by an inoperative cooling fan. See CHART C-12 for cooling fan diagnosis.

3.3L (VIN N) ENGINE — COMPONENT DIAGNOSTIC CHART — CENTURY, CUTLASS CIERA/CRUISER, ACHIEVA, GRAND AM AND SKYLARK

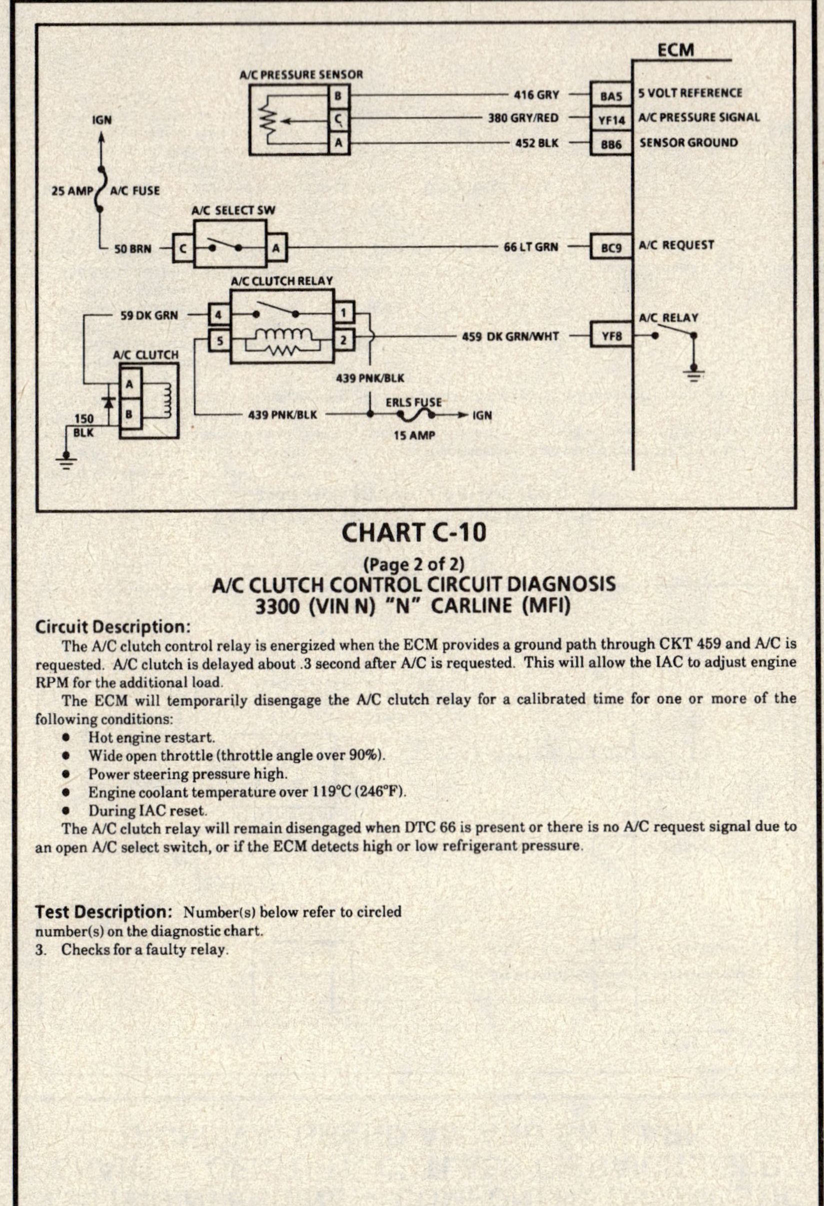

CHART C-10
(Page 2 of 2)
A/C CLUTCH CONTROL CIRCUIT DIAGNOSIS
3300 (VIN N) "N" CARLINE (MFI)

Circuit Description:

The A/C clutch control relay is energized when the ECM provides a ground path through CKT 459 and A/C is requested. A/C clutch is delayed about .3 second after A/C is requested. This will allow the IAC to adjust engine RPM for the additional load.

The ECM will temporarily disengage the A/C clutch relay for a calibrated time for one or more of the following conditions:
- Hot engine restart.
- Wide open throttle (throttle angle over 90%).
- Power steering pressure high.
- Engine coolant temperature over 119°C (246°F).
- During IAC reset.

The A/C clutch relay will remain disengaged when DTC 66 is present or there is no A/C request signal due to an open A/C select switch, or if the ECM detects high or low refrigerant pressure.

Test Description: Number(s) below refer to circled number(s) on the diagnostic chart.

3. Checks for a faulty relay.

3.3L (VIN N) ENGINE — COMPONENT DIAGNOSTIC CHART — CENTURY, CUTLASS CIERA/CRUISER, ACHIEVA, GRAND AM AND SKYLARK

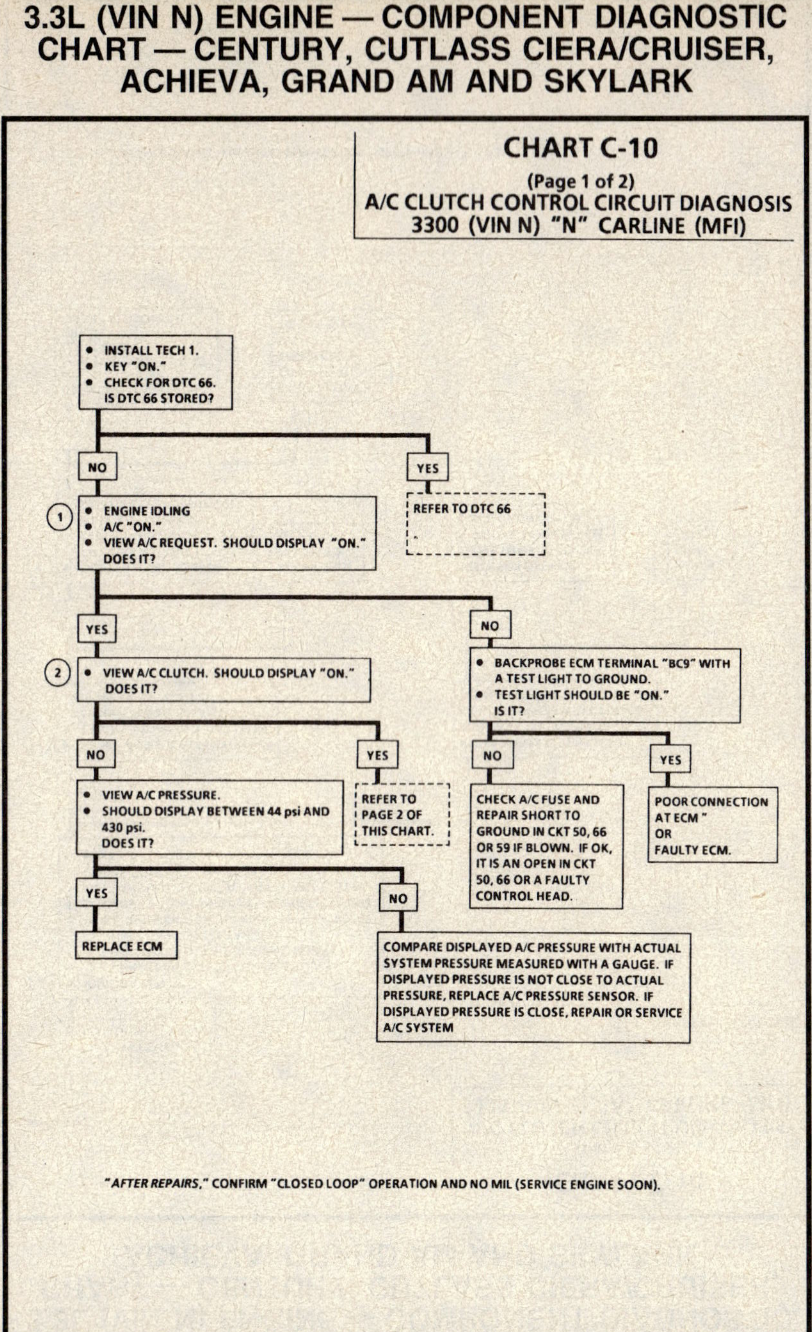

CHART C-10
(Page 1 of 2)
A/C CLUTCH CONTROL CIRCUIT DIAGNOSIS
3300 (VIN N) "N" CARLINE (MFI)

3.3L (VIN N) ENGINE — COMPONENT DIAGNOSTIC CHART — CENTURY, CUTLASS CIERA/CRUISER, ACHIEVA, GRAND AM AND SKYLARK

CHART C-10
(Page 2 of 2)
A/C CLUTCH CONTROL CIRCUIT DIAGNOSIS
3300 (VIN N) "N" CARLINE (MFI)

FROM CHART C-10, PAGE 1 OF 2.

(3)
- DISCONNECT A/C RELAY.
- INSTALL A FUSED JUMPER BETWEEN RELAY HARNESS TERMINALS "1" AND "4".
- A/C CLUTCH SHOULD ENGAGE. DOES IT?

NO:
- PROBE RELAY HARNESS TERMINAL "1" WITH A TEST LIGHT CONNECTED TO GROUND.
- TEST LIGHT SHOULD BE "ON," IS IT?

  YES:
  - DISCONNECT A/C CLUTCH ELECTRICAL CONNECTOR.
  - PROBE CONNECTOR TERMINAL "A" (CKT 59) WITH A TEST LIGHT CONNECTED TO GROUND.
  - TEST LIGHT SHOULD BE "ON," IS IT?

    YES:
    - CONNECT A TEST LIGHT BETWEEN A/C CLUTCH CONNECTOR TERMINALS "A" AND "B".
    - TEST LIGHT SHOULD BE "ON," IS IT?

      YES: FAULTY A/C CLUTCH

      NO: OPEN CKT 150 TO A/C CLUTCH

    NO: OPEN CKT 59

  NO: OPEN CKT 439 TO A/C RELAY TERMINAL "1".

YES:
- CONNECT A TEST LIGHT BETWEEN RELAY HARNESS TERMINALS "2" AND "5".
- TEST LIGHT SHOULD BE "ON," IS IT?

  YES: FAULTY A/C RELAY CONNECTION OR FAULTY A/C RELAY.

  NO: OPEN CKT 439 TO A/C RELAY TERMINAL "5" OR OPEN CKT 459 OR FAULTY ECM.

"AFTER REPAIRS," CONFIRM "CLOSED LOOP" OPERATION AND NO MIL (SERVICE ENGINE SOON).

3.3L (VIN N) ENGINE — COMPONENT DIAGNOSTIC CHART — CENTURY, CUTLASS CIERA/CRUISER, ACHIEVA, GRAND AM AND SKYLARK

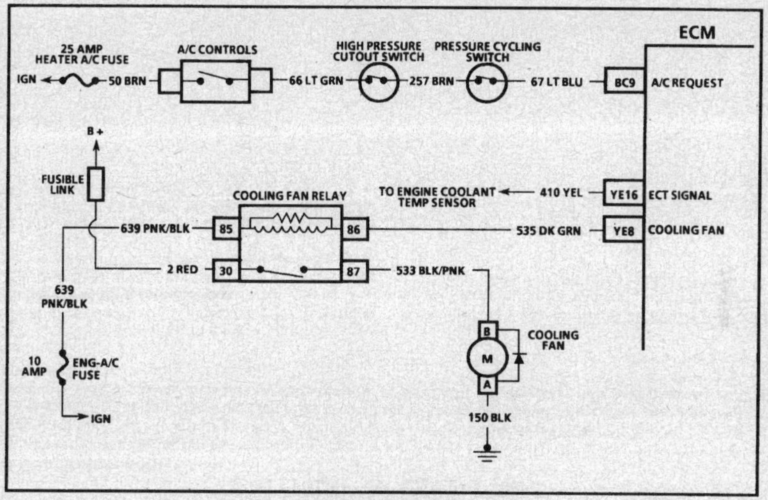

CHART C-12A

COOLING FAN FUNCTIONAL CHECK
3300 (VIN N) "A" CARLINE (MFI)

Circuit Description:

Power for the cooling fan motor comes through the fusible link to terminal "30" on the relay. The relay is energized when current flows to ground through the quad-driver module inside the ECM.

Cooling Fan Relay - The cooling fan relay is energized by the ECM. The ECM energizes the relay through terminal "YE8" when engine coolant temperature reaches 100°C (212°F). The fan relay is also energized when A/C is requested.

Test Description: Number(s) refer to circled number(s) on the diagnostic chart.

1. The Tech 1 output tests allows manual control of the cooling fan.
2. Fan should be "ON" when A/C is requested at idle.
3. Ensures that the ECM is receiving an A/C request.

Diagnostic Aids:

If fan operates normally but an overheat condition exists, refer to "Engine Coolant Temperature Sensor Temperature vs. Resistance Values" table on the DTC 15 chart, "Engine Components/Wiring Diagrams/Diagnostic Charts"
Replace sensor if mis-scaled. If not, refer to engine cooling and radiator diagnosis.

3.3L (VIN N) ENGINE — COMPONENT DIAGNOSTIC CHART — CENTURY, CUTLASS CIERA/CRUISER, ACHIEVA, GRAND AM AND SKYLARK

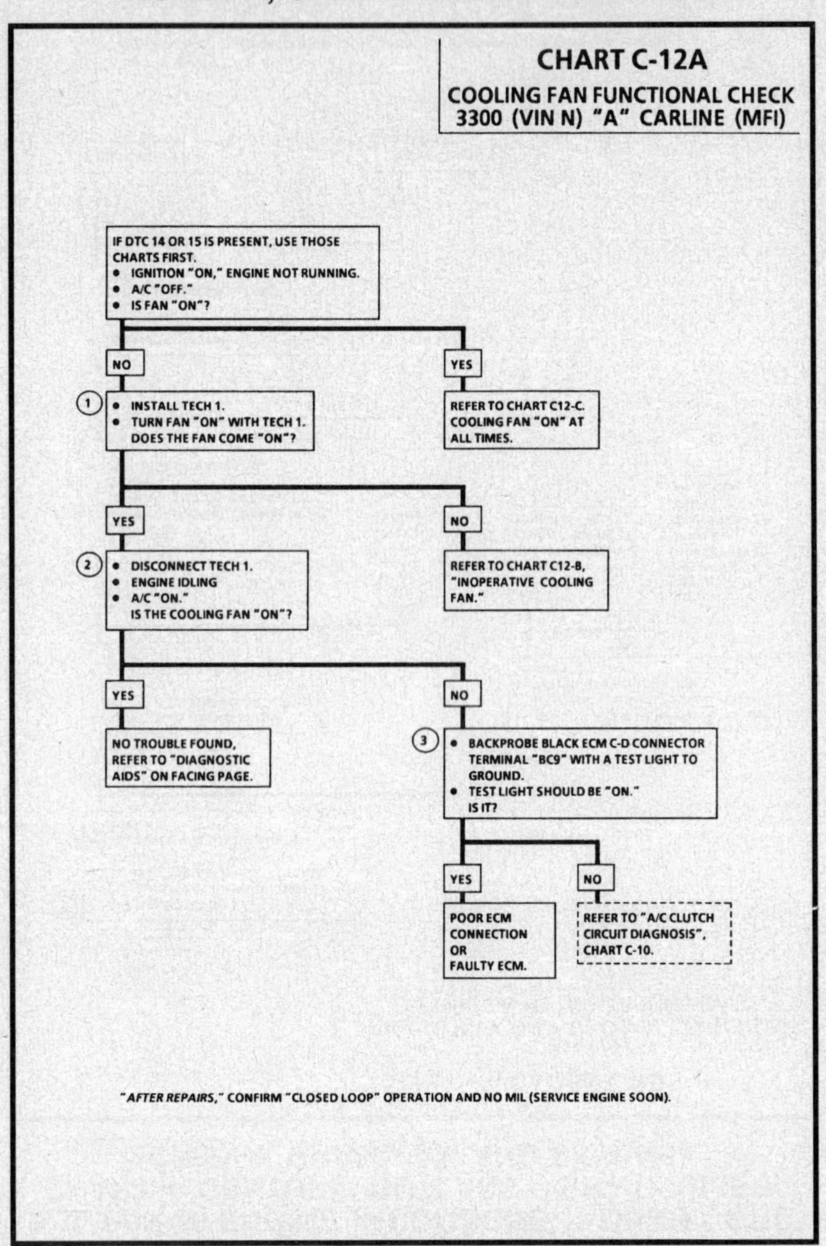

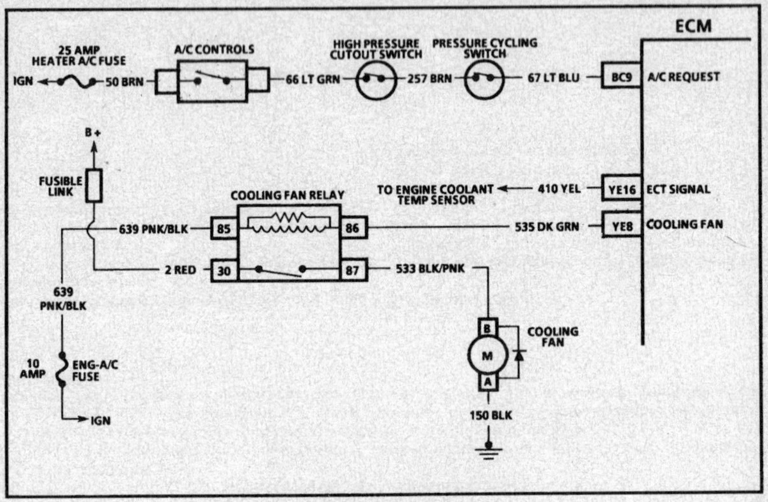

CHART C-12B
COOLING FAN INOPERATIVE
3300 (VIN N) "A" CARLINE (MFI)

Circuit Description:

Power for the cooling fan motor comes through the fusible link to terminal "30" on the relay. The relay is energized when current flows to ground through the quad-driver module inside the ECM.

Cooling Fan Relay - The cooling fan relay is energized by the ECM. The ECM energizes the relay through terminal "YE8" if engine coolant temperature reaches 100°C (212°F). The fan relay is also energized when A/C is requested.

Test Description: Number(s) refer to circled number(s) on the diagnostic chart.

1. Test light should be "ON" because the relay connector "85" terminal is fed directly from ignition.
2. The Tech 1 Misc. Test function allows manual control of the cooling fan relay. This test checks CKT 535 and the ECM.
3. Jumpering relay connector terminals "30" and "87" feeds the fan motor directly and it should run.
4. Checks for battery feed at the cooling fan relay.

3.3L (VIN N) ENGINE — COMPONENT DIAGNOSTIC CHART — CENTURY, CUTLASS CIERA/CRUISER, ACHIEVA, GRAND AM AND SKYLARK

3.3L (VIN N) ENGINE — COMPONENT DIAGNOSTIC CHART — CENTURY, CUTLASS CIERA/CRUISER, ACHIEVA, GRAND AM AND SKYLARK

CHART C-12B
COOLING FAN INOPERATIVE
3300 (VIN N) "A" CARLINE (MFI)

(1)
- PERFORM FUNCTIONAL TEST (CHART C-12A) FIRST.
- REMOVE COOLING FAN RELAY AND TURN IGNITION "ON."
- CONNECT A TEST LIGHT BETWEEN HARNESS TERMINAL "85 " (CKT 639) AND GROUND.
  TEST LIGHT SHOULD BE "ON."
  IS IT?

YES / NO

NO → CHECK 10 AMP ENG-A/C FUSE AND REPAIR SHORT TO GROUND IN CKT 639 IF BLOWN. IF OK, REPAIR OPEN CKT 639 TO RELAY.

(2)
- CONNECT TEST LIGHT BETWEEN HARNESS TERMINALS "85" AND "86".
- USING THE TECH 1, TURN THE FAN "ON."
  TEST LIGHT SHOULD BE "ON."
  IS IT?

YES / NO

NO → CHECK FOR OPEN CKT 535. IF CKT 535 IS OK, IT'S A POOR CONNECTION AT THE ECM OR FAULTY ECM.

(3)
- INSTALL A JUMPER BETWEEN HARNESS TERMINALS "30" AND "87".
  FAN SHOULD NOW BE "ON."
  IS IT?

NO / YES

YES → FAULTY FAN RELAY.

(4)
- CONNECT A TEST LIGHT BETWEEN HARNESS TERMINAL "30" AND GROUND.
  TEST LIGHT SHOULD BE "ON."
  IS IT?

YES / NO

YES → OPEN CKT 533 OR CKT 150, OR FAULTY FAN MOTOR.

NO → OPEN CKT 2 OR FUSIBLE LINK.

"AFTER REPAIRS," CONFIRM "CLOSED LOOP" OPERATION AND NO MIL (SERVICE ENGINE SOON).

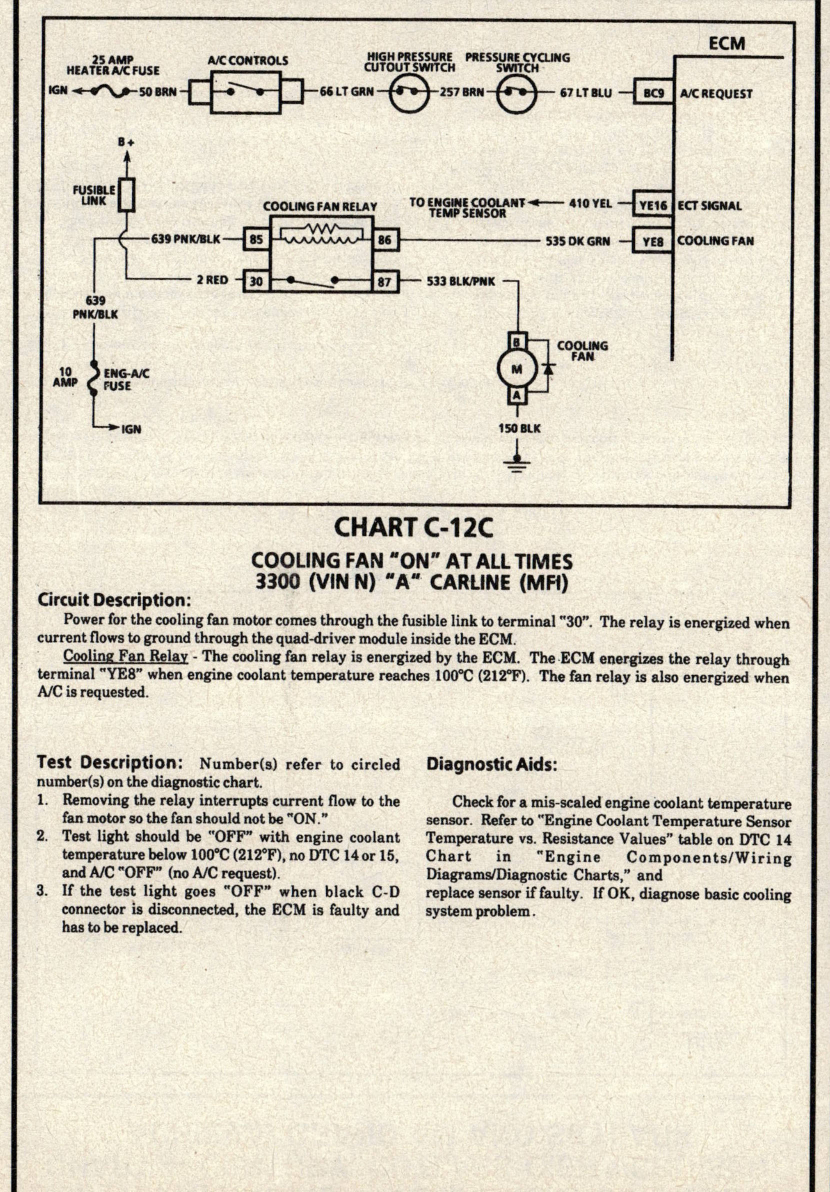

CHART C-12C
COOLING FAN "ON" AT ALL TIMES
3300 (VIN N) "A" CARLINE (MFI)

Circuit Description:
Power for the cooling fan motor comes through the fusible link to terminal "30". The relay is energized when current flows to ground through the quad-driver module inside the ECM.

Cooling Fan Relay - The cooling fan relay is energized by the ECM. The ECM energizes the relay through terminal "YE8" when engine coolant temperature reaches 100°C (212°F). The fan relay is also energized when A/C is requested.

Test Description: Number(s) refer to circled number(s) on the diagnostic chart.
1. Removing the relay interrupts current flow to the fan motor so the fan should not be "ON."
2. Test light should be "OFF" with engine coolant temperature below 100°C (212°F), no DTC 14 or 15, and A/C "OFF" (no A/C request).
3. If the test light goes "OFF" when black C-D connector is disconnected, the ECM is faulty and has to be replaced.

Diagnostic Aids:
Check for a mis-scaled engine coolant temperature sensor. Refer to "Engine Coolant Temperature Sensor Temperature vs. Resistance Values" table on DTC 14 Chart in "Engine Components/Wiring Diagrams/Diagnostic Charts," and replace sensor if faulty. If OK, diagnose basic cooling system problem.

3.3L (VIN N) ENGINE — COMPONENT DIAGNOSTIC CHART — CENTURY, CUTLASS CIERA/CRUISER, ACHIEVA, GRAND AM AND SKYLARK

3.3L (VIN N) ENGINE — COMPONENT DIAGNOSTIC CHART — CENTURY, CUTLASS CIERA/CRUISER, ACHIEVA, GRAND AM AND SKYLARK

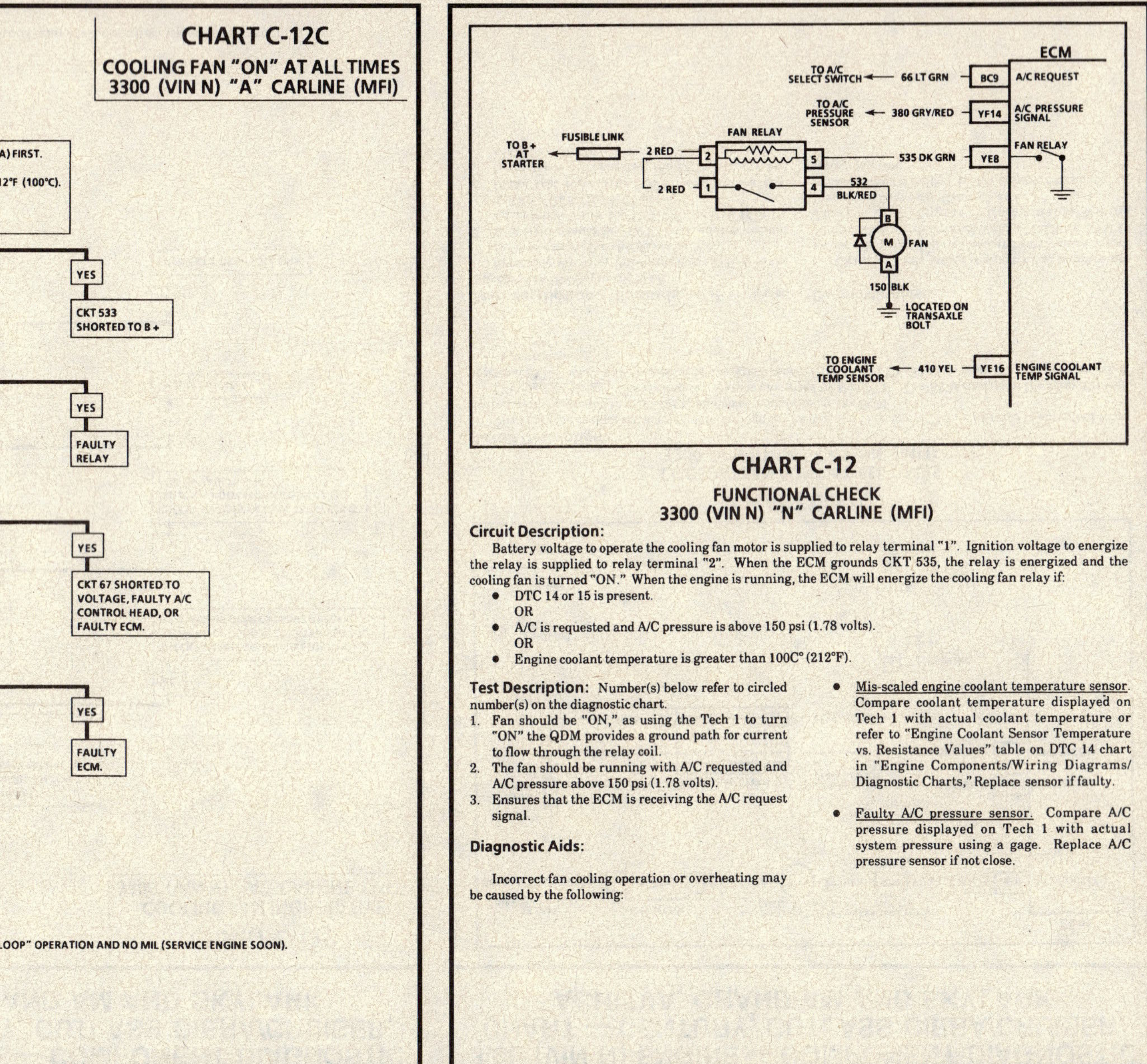

Circuit Description:

Battery voltage to operate the cooling fan motor is supplied to relay terminal "1". Ignition voltage to energize the relay is supplied to relay terminal "2". When the ECM grounds CKT 535, the relay is energized and the cooling fan is turned "ON." When the engine is running, the ECM will energize the cooling fan relay if:

- DTC 14 or 15 is present.
  OR
- A/C is requested and A/C pressure is above 150 psi (1.78 volts).
  OR
- Engine coolant temperature is greater than 100C° (212°F).

Test Description: Number(s) below refer to circled number(s) on the diagnostic chart.

1. Fan should be "ON," as using the Tech 1 to turn "ON" the QDM provides a ground path for current to flow through the relay coil.
2. The fan should be running with A/C requested and A/C pressure above 150 psi (1.78 volts).
3. Ensures that the ECM is receiving the A/C request signal.

Diagnostic Aids:

Incorrect fan cooling operation or overheating may be caused by the following:

- <u>Mis-scaled engine coolant temperature sensor</u>. Compare coolant temperature displayed on Tech 1 with actual coolant temperature or refer to "Engine Coolant Sensor Temperature vs. Resistance Values" table on DTC 14 chart in "Engine Components/Wiring Diagrams/ Diagnostic Charts," Replace sensor if faulty.

- <u>Faulty A/C pressure sensor</u>. Compare A/C pressure displayed on Tech 1 with actual system pressure using a gage. Replace A/C pressure sensor if not close.

3.3L (VIN N) ENGINE — COMPONENT DIAGNOSTIC CHART — CENTURY, CUTLASS CIERA/CRUISER, ACHIEVA, GRAND AM AND SKYLARK

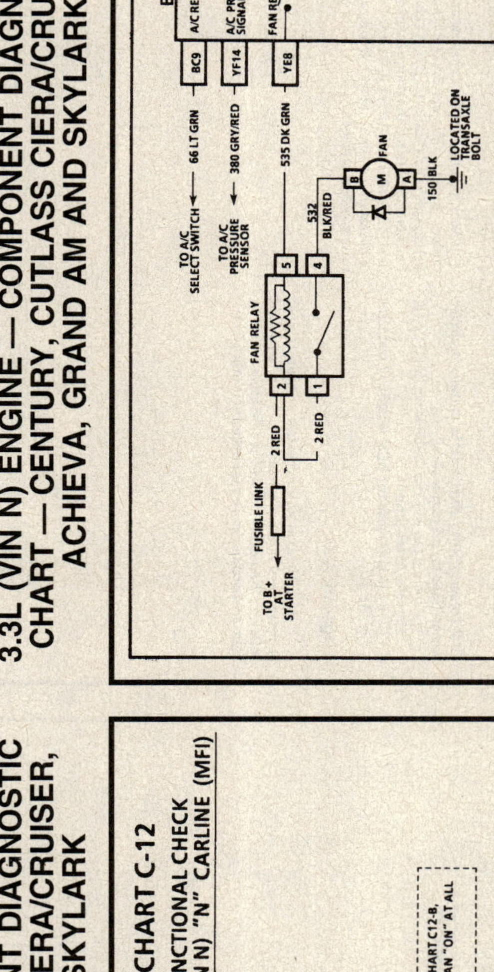

CHART C-12A
NO COOLING FAN
3300 (VIN N) "N" CARLINE (MFI)

Circuit Description:

Battery voltage to operate the cooling fan motor is supplied to relay terminal "1". Battery voltage to energize the relay is supplied to relay terminal "2". When the ECM grounds CKT 535, the relay is energized and the cooling fan is turned "ON." When the engine is running, the ECM will energize the cooling fan relay if an engine coolant temperature sensor DTC 14 or 15 has been set, or under the following conditions:

- A/C is requested and A/C pressure is above 150 psi (1.78 volts)
- Engine coolant temperature is greater than 100°C (212°F).

Test Description: Number(s) below refer to circled number(s) on the diagnostic chart.

1. Terminals "1" and "2" are connected directly to B+.
2. Test light should be "ON" as using the Tech 1 to turn the QDM "ON" allows current to flow through the fan relay coil.

Diagnostic Aids

If the vehicle has an overheating problem, it must be determined if the complaint was due to an actual boilover, the coolant temperature "temp" warning light, or the temperature gage indicated overheating.

If the gage or light indicates overheating but no boilover is detected, the gage circuit should be checked. The gage accuracy can be checked by comparing the engine coolant sensor reading using a scan tool with the gage reading.

If the engine is actually overheating and the gage indicates overheating, but scan values indicate a lower temperature and the cooling fan is not turning "ON," the engine coolant sensor has probably shifted out of calibration and should be replaced.

3.3L (VIN N) ENGINE — COMPONENT DIAGNOSTIC CHART — CENTURY, CUTLASS CIERA/CRUISER, ACHIEVA, GRAND AM AND SKYLARK

CHART C-12
FUNCTIONAL CHECK
3300 (VIN N) "N" CARLINE (MFI)

3.3L (VIN N) ENGINE — COMPONENT DIAGNOSTIC CHART — CENTURY, CUTLASS CIERA/CRUISER, ACHIEVA, GRAND AM AND SKYLARK

CHART C-12A

NO COOLING FAN
3300 (VIN N) "N" CARLINE (MFI)

1.
- PERFORM FUNCTIONAL TEST (CHART C-12A) FIRST.
- DISCONNECT FAN RELAY.
- USING A TEST LIGHT TO GROUND, CHECK HARNESS TERMINALS "1" AND "2".
- TEST LIGHT SHOULD BE "ON" FOR BOTH.
  IS IT?

YES → 2.
- IGNITION "ON."
- CONNECT TEST LIGHT BETWEEN TERMINALS "2" AND "5".
- INSTALL TECH 1.
- TURN FAN "ON" WITH TECH 1.
- TEST LIGHT SHOULD BE "ON."
  IS IT?

NO →
NO LIGHT ON HARNESS TERMINAL "2" - OPEN OR GROUNDED FUSE LINK OR CKT 2 TO RELAY. NO LIGHT ON "1" - OPEN OR GROUNDED CKT 2 BETWEEN HARNESS TERMINALS "2" AND "1".

YES →
- TURN FAN OFF WITH TECH 1.
- JUMPER HARNESS TERMINALS "1" AND "4".
- FAN SHOULD BE RUNNING.
  IS IT?

NO →
- CHECK FOR OPEN CKT 535. IF CIRCUIT IS OK, IT IS POOR CONNECTION AT ECM OR FAULTY ECM.

NO →
OPEN CKT 532
OR
OPEN CKT 150 TO FAN
OR
POOR CONNECTIONS AT FAN MOTOR
OR
FAULTY FAN MOTOR.

YES →
FAULTY FAN RELAY.

"AFTER REPAIRS," CONFIRM "CLOSED LOOP" OPERATION AND NO MIL (SERVICE ENGINE SOON).

3.3L (VIN N) ENGINE — COMPONENT DIAGNOSTIC CHART — CENTURY, CUTLASS CIERA/CRUISER, ACHIEVA, GRAND AM AND SKYLARK

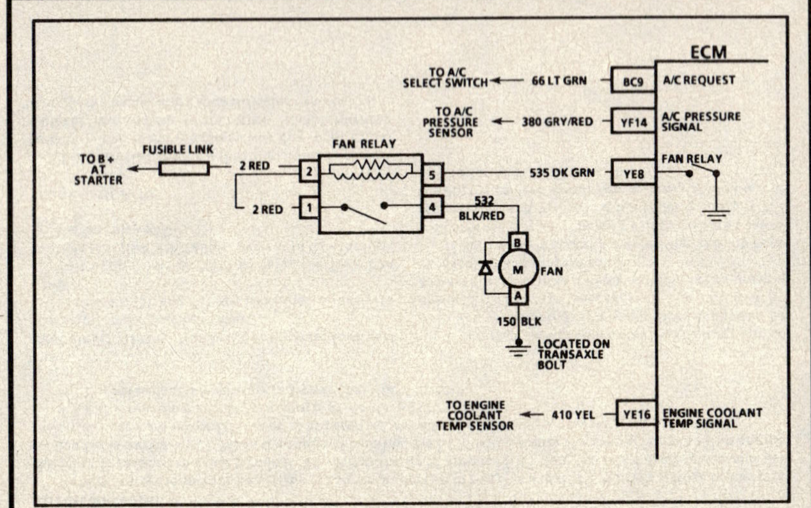

CHART C-12B

FAN "ON" AT ALL TIMES
3300 (VIN N) "N" CARLINE (MFI)

Circuit Description:

Battery voltage to operate the cooling fan motor is supplied to relay terminal "1". Battery voltage to energize the relay is supplied to relay terminal "2". When the ECM grounds CKT 535, the relay is energized and the cooling fan is turned "ON." When the engine is running, the ECM will energize the cooling fan relay if an engine coolant temperature sensor DTC 14 or 15 has been set, or under the following conditions:
- A/C is requested and A/C pressure is over 150 psi (1.78 volts).
- Engine coolant temperature is greater than 100°C (212°F).

Test Description: Number(s) below refer to circled number(s) on the diagnostic chart.
1. The cooling fan will run continuously if DTCs 14, 15 and/or 66 are present.
2. CKT 535 is grounded or ECM is faulty if test light is "ON."

Check for a mis-scaled engine coolant temperature sensor. Refer to "Engine Coolant Temperature vs. Resistance Values" table on DTC 14 chart in "Engine Components/Wiring Diagrams/Diagnostic Charts," Section and replace engine coolant temperature sensor if faulty. If OK, diagnose basic cooling system problem

Diagnostic Aids:

If the vehicle has an overheating problem, it must be determined if the complaint was due to an actual boilover, the engine coolant temperature warning light, or the temperature gage indicated overheating.

3.3L (VIN N) ENGINE — COMPONENT DIAGNOSTIC CHART — CENTURY, CUTLASS CIERA/CRUISER, ACHIEVA, GRAND AM AND SKYLARK

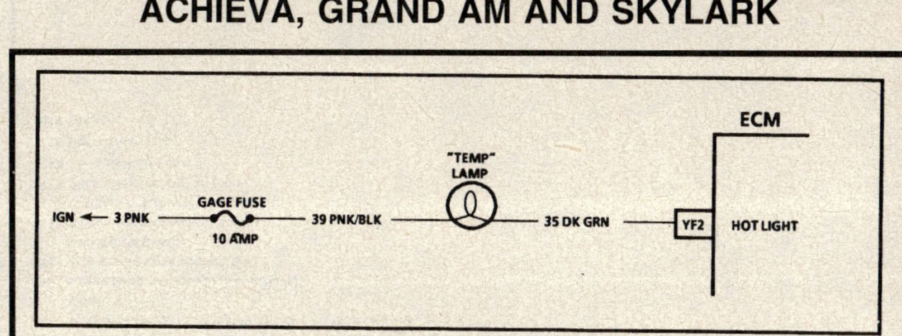

3.3L (VIN N) ENGINE — COMPONENT DIAGNOSTIC CHART — CENTURY, CUTLASS CIERA/CRUISER, ACHIEVA, GRAND AM AND SKYLARK

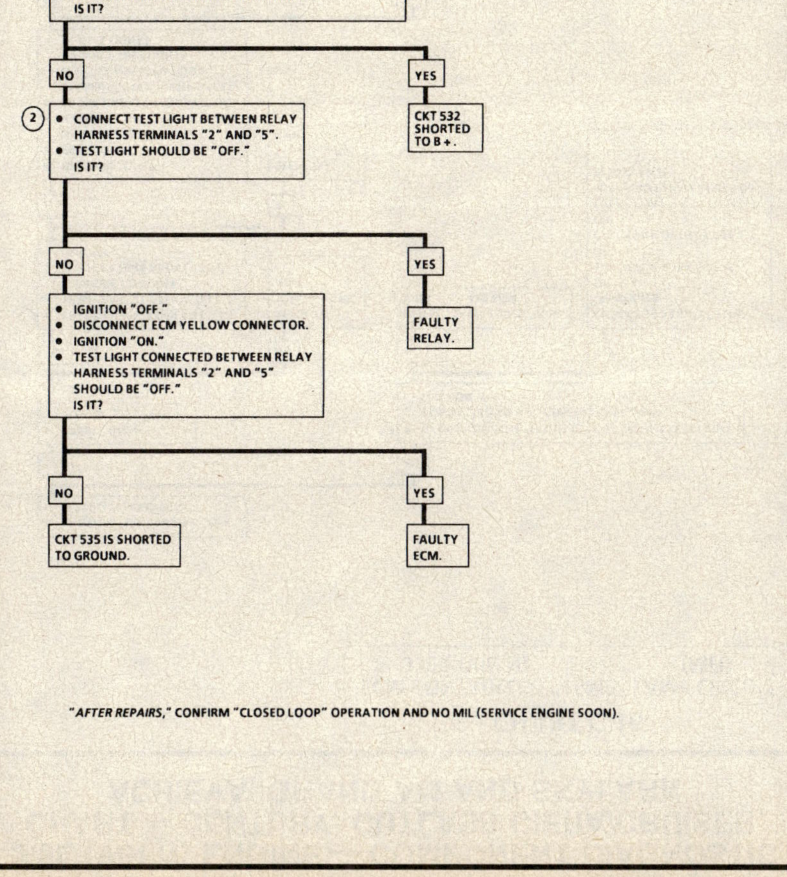

CHART C-16
ECM CONTROLLED "TEMP" LAMP CIRCUIT
3300 (VIN N)   (MFI)

Circuit Description:

The "Temp" lamp is powered through the gages fuse by ignition voltage. The ECM energizes the bulb by supplying a path to ground through quad-driver module 1 + 2 (A).

Test Description: Number(s) below refer to circled number(s) on the diagnostic chart.

1. With the ignition "ON" and engine "OFF," the ECM should be grounding CKT 35.
2. The "Temp" lamp should only be "ON" while the engine is running and the engine coolant temperature is 120°C (248°F) or higher.

Diagnostic Aids:

The engine coolant temperature sensor, in rare cases, may fail to indicate the correct engine coolant temperature without setting a diagnostic trouble code. This could result in turning "ON" the "Temp" lamp without having an overheating condition. It could also result in overheating without the "Temp" lamp being turned "ON." See "Engine Coolant Temperature Sensor Temperature vs. Resistance Values" diagnostic aids table on DTC 15 chart or compare coolant temperature reading on the Tech 1 with actual engine coolant temperature. Replace engine coolant temperature sensor if mis-scaled.

3.3L (VIN N) ENGINE — COMPONENT DIAGNOSTIC CHART — CENTURY, CUTLASS CIERA/CRUISER, ACHIEVA, GRAND AM AND SKYLARK

5.0L (VIN F) AND 5.7L (VIN 8) ENGINES — ENGINE COMPONENT LOCATION CHART — 1992 CAMARO

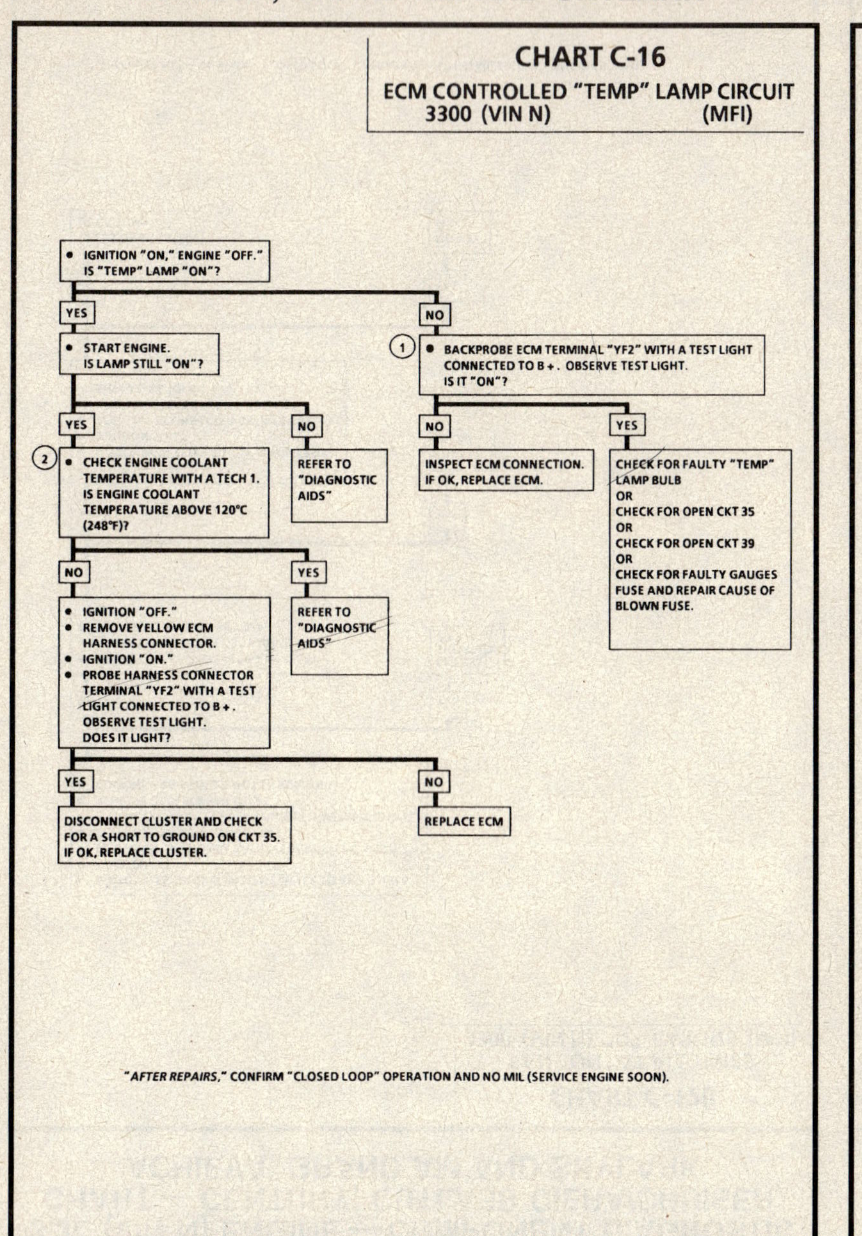

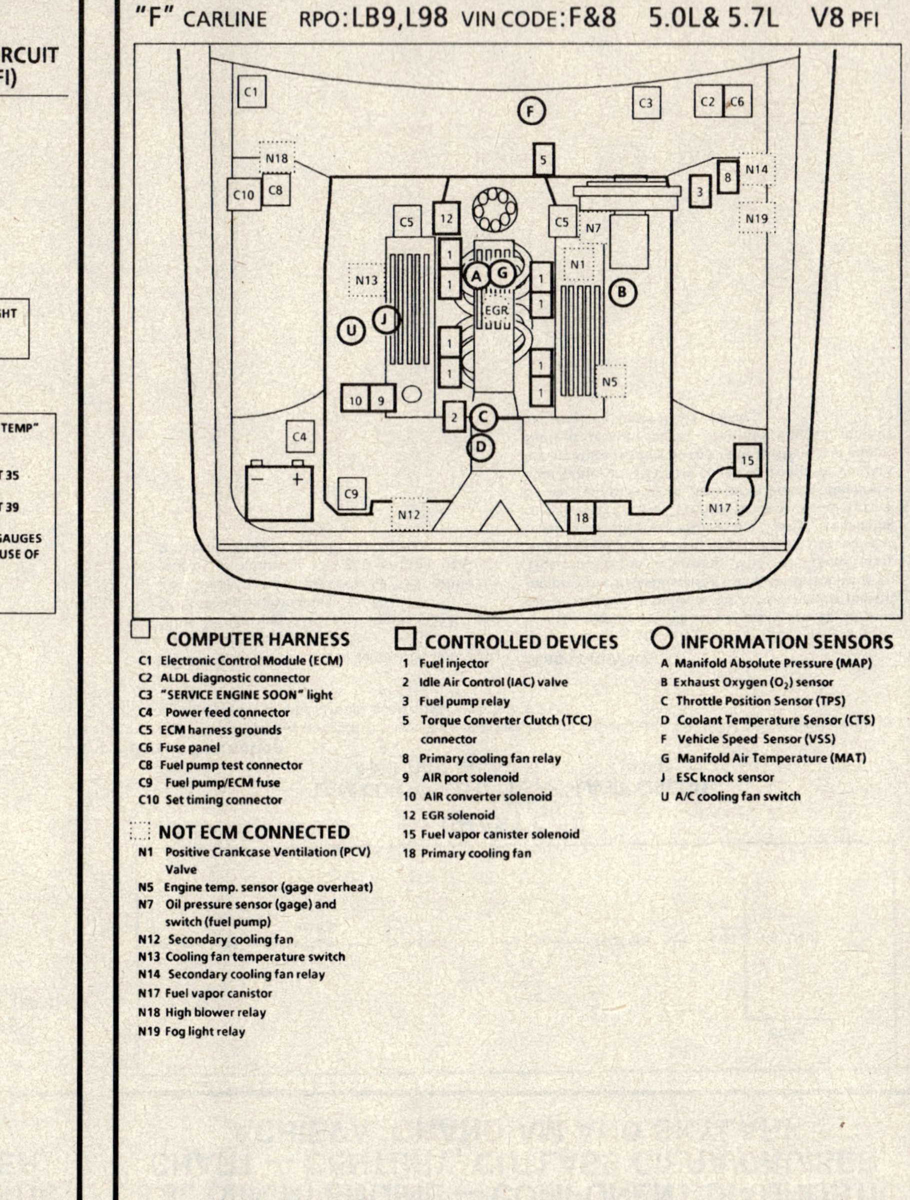

5.0L (VIN F) AND 5.7L (VIN 8) ENGINES — ENGINE COMPONENT LOCATION CHART — 1992 FIREBIRD

"F" CARLINE RPO: LB9, L98 VIN CODE: F&8 5.0L & 5.7L V8 PFI

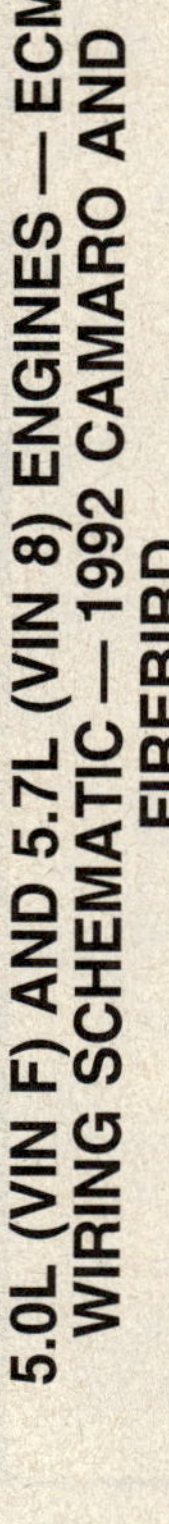

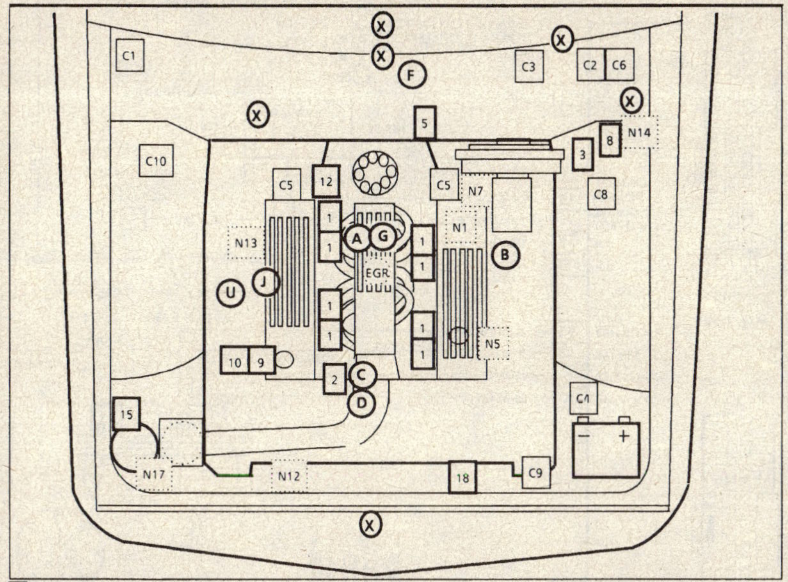

COMPUTER HARNESS

- C1 Electronic Control Module (ECM)
- C2 ALDL diagnostic connector
- C3 "SERVICE ENGINE SOON" light
- C4 Power feed connector
- C5 ECM harness grounds
- C6 Fuse panel
- C8 Fuel pump test connector
- C9 Fuel pump/ECM fuse
- C10 Set timing connector

NOT ECM CONNECTED

- N1 Positive Crankcase Ventilation (PCV) Valve
- N5 Engine temp. sensor (gage overheat)
- N7 Oil pressure sensor (gage) and switch (fuel pump)
- N12 Secondary cooling fan
- N13 Cooling fan temperature switch
- N14 Secondary cooling fan relay
- N17 Fuel vapor canister

CONTROLLED DEVICES

- 1 Fuel injector
- 2 Idle Air Control (IAC) valve
- 3 Primary cooling fan relay
- 5 Torque Converter Clutch (TCC) connector
- 8 Fuel pump relay
- 9 AIR port solenoid
- 10 AIR converter solenoid
- 12 EGR solenoid
- 15 Fuel vapor canister solenoid
- 18 Primary cooling fan

INFORMATION SENSORS

- A Manifold Absolute Pressure (MAP)
- B Exhaust Oxygen (O_2) sensor
- C Throttle Position Sensor (TPS)
- D Coolant Temperature Sensor (CTS)
- F Vehicle Speed Sensor (VSS)
- G Intake Air Temperature (IAT)
- J ESC knock sensor
- U A/C cooling fan switch

(X) SIR System Components. Refer to the Service Manual, for "Cautions" and information on SIR System Components.

5.0L (VIN F) AND 5.7L (VIN 8) ENGINES — ECM WIRING SCHEMATIC — 1992 CAMARO AND FIREBIRD

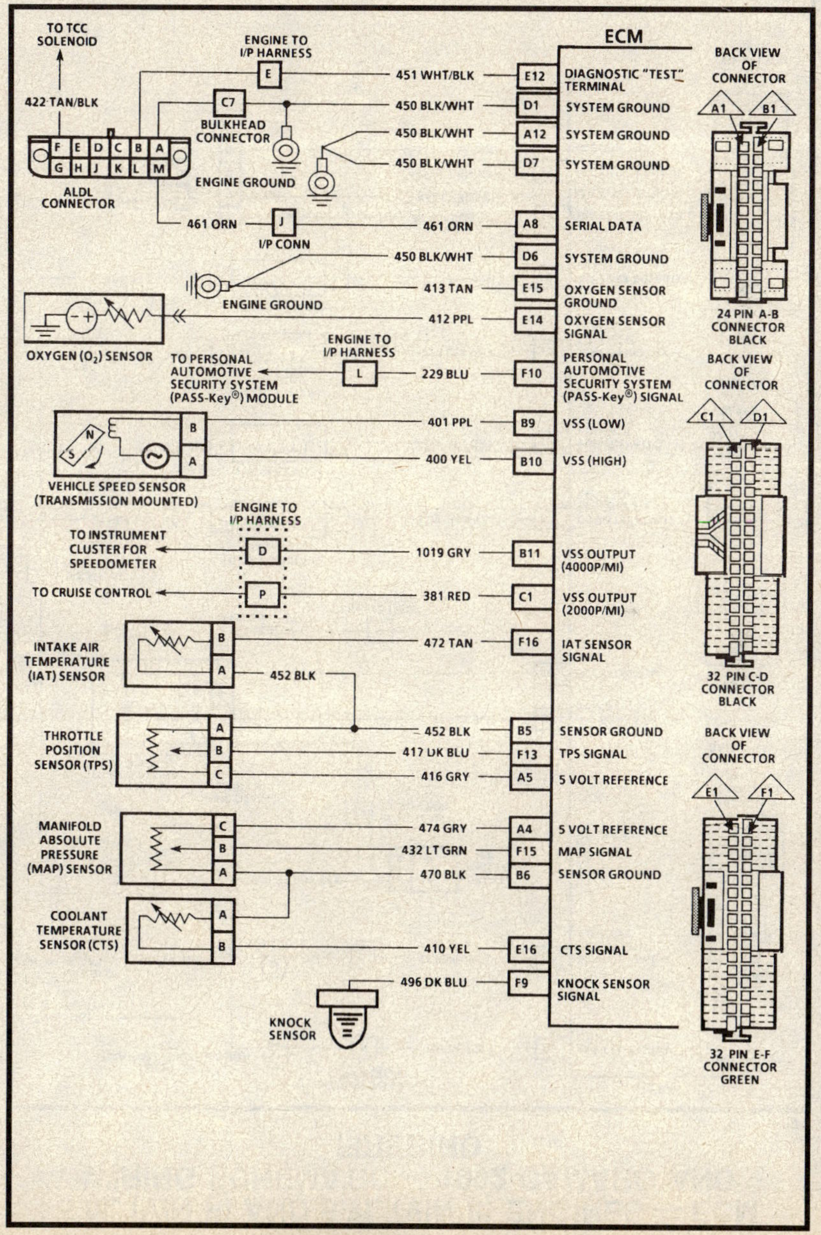

5.0L (VIN F) AND 5.7L (VIN 8) ENGINES — ECM WIRING SCHEMATIC — 1992 CAMARO AND FIREBIRD

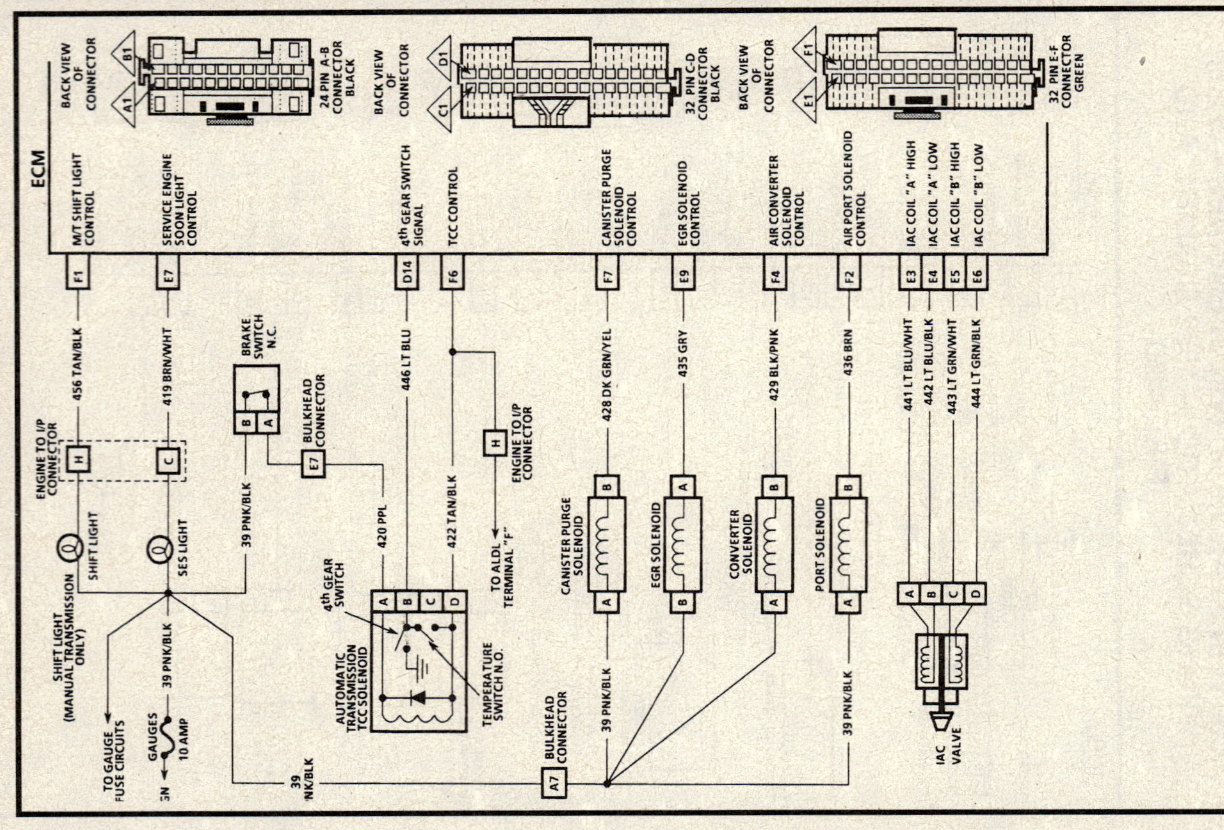

5.0L (VIN F) AND 5.7L (VIN 8) ENGINES — ECM WIRING SCHEMATIC — 1992 CAMARO AND FIREBIRD

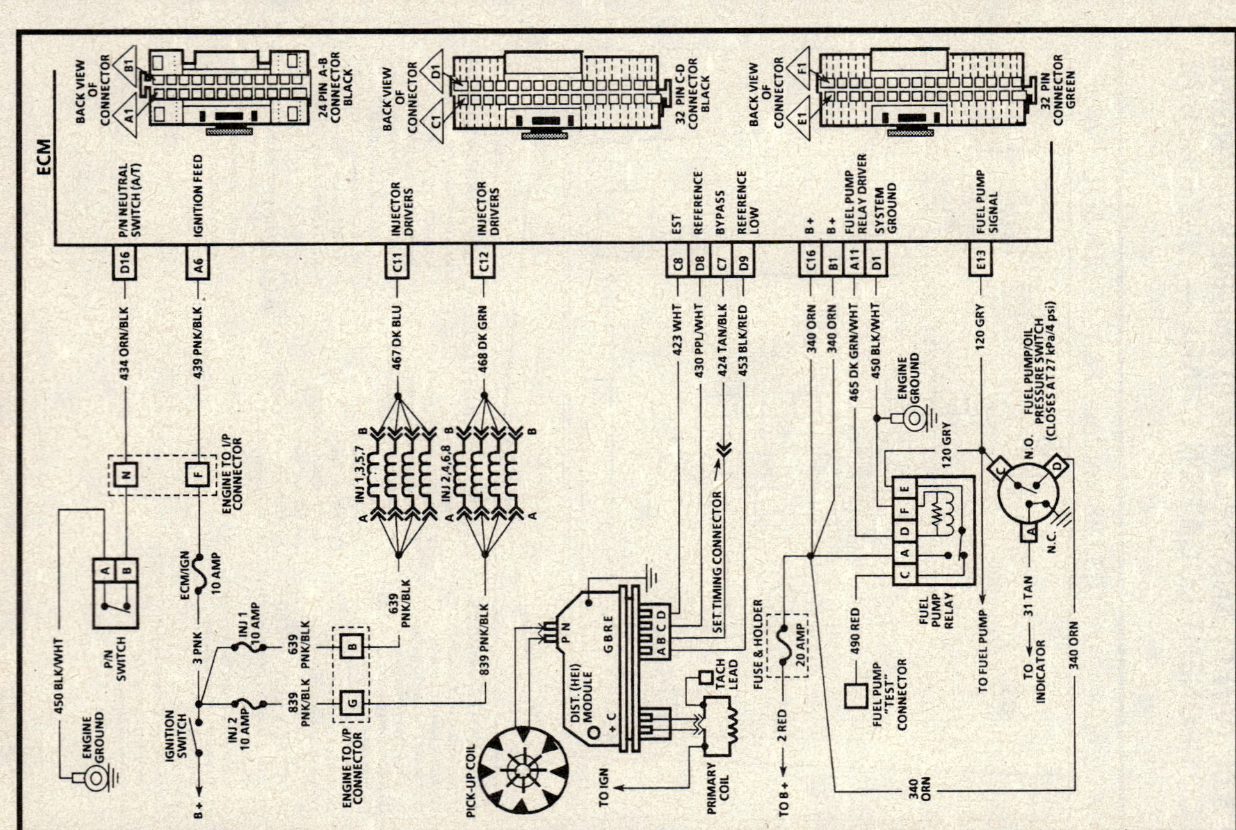

5.0L (VIN F) AND 5.7L (VIN 8) ENGINES — ECM WIRING SCHEMATIC — 1992 CAMARO AND FIREBIRD

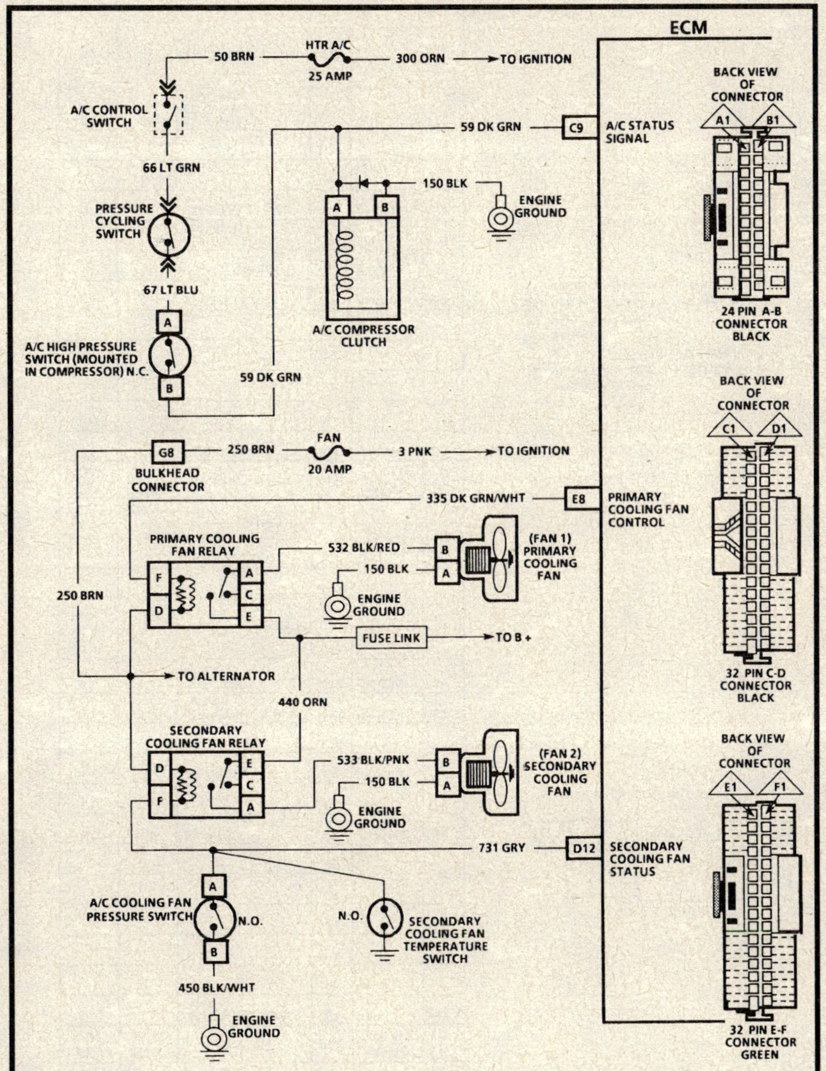

5.0L (VIN F) AND 5.7L (VIN 8) ENGINES — ECM CONNECTOR END VIEW — 1992 CAMARO AND FIREBIRD

PORT FUEL INJECTION ECM CONNECTOR IDENTIFICATION

This ECM voltage chart is for use with a digital voltmeter to further aid in diagnosis. The voltages you get may vary due to low battery charge or other reasons, but they should be very close.

THE FOLLOWING CONDITIONS MUST BE MET BEFORE TESTING:

● Engine at operating temperature ● Engine idling in closed loop (For "Engine Run" column) in park or neutral ● Diagnostic "Test" terminal not grounded ● ALDL "Scan" tool not installed

VOLTAGE

KEY "ON"	ENG. RUN	CIRCUIT	PIN	WIRE COLOR	CKT #
			A1		
			A2		
			A3		
5	5	+5V REFERENCE (MAP)	A4	GRY	474
5	5	+5V REFERENCE (TPS)	A5	GRY	416
B+	B+	IGNITION FEED	A6	PNK/BLK	439
			A7		
4.8 [②]	4.8	SERIAL DATA	A8	ORN	461
			A9		
			A10		
0* [④]	B+	FUEL PUMP RELAY DRIVE	A11	DK GRN/WHT	465
0*	0*	ECM GROUND	A12	BLK/WHT	450

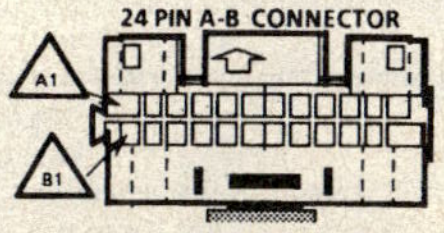

VOLTAGE

KEY "ON"	ENG. RUN	CIRCUIT	PIN	WIRE COLOR	CKT #
B+	B+	BATTERY FEED	B1	ORN	340
			B2		
			B3		
			B4		
0*	0*	TPS & IAT SENSOR GROUND	B5	BLK	452
0*	0*	CTS & MAP SENSOR GROUND	B6	BLK	470
			B7		
			B8		
0* [①]	0*	VSS (LOW)	B9	PPL	401
0* [①]	0*	VSS (HIGH)	B10	YEL	400
B+	B+	VSS TO I/P 4000 P/MI	B11	GRY	1019
			B12		

▽ Less than 1 volt.
* Less than .5 volt.
① Varies from .60 volt to Battery Voltage depending on position of drivewheels.
② Varies
③ Varies with temperature.
④ Battery Voltage for first two seconds with ignition "ON."

5.0L (VIN F) AND 5.7L (VIN 8) ENGINES — ECM CONNECTOR END VIEW — 1992 CAMARO AND FIREBIRD

VOLTAGE

Note	KEY "ON"	ENG. RUN	CIRCUIT	PIN	WIRE COLOR	CKT #
①			VSS OUTPUT 2000 P/MI	C1	RED	381
				C2		
				C3		
				C4		
				C5		
				C6		
	0*	4.6	BYPASS	C7	TAN/BLK	424
②	0*	1.3	EST	C8	WHT	423
"ON" / "OFF"	B+ / 0*	B+ / 0*	A/C REQUEST	C9	DK GRN	59
				C10		
	B+	B+	INJECTOR 1, 3, 5, 7 DRIVER	C11	DK BLU	467
	B+	B+	INJECTOR 2, 4, 6, 8 DRIVER	C12	DK GRN	468
				C13		
				C14		
				C15		
	B+	B+	BATTERY FEED	C16	ORN	340

VOLTAGE

Note	KEY "ON"	ENG. RUN	CIRCUIT	PIN	WIRE COLOR	CKT #
	0*	0*	ECM GROUND	D1	BLK/WHT	450
				D2		
				D3		
				D4		
				D5		
	0*	0*	ENGINE GROUND	D6	BLK/WHT	450
	0*	0*	ENGINE GROUND	D7	BLK/WHT	450
	0*	1.5	DISTRIBUTOR REFERENCE	D8	PPL/WHT	430
	0*	0*	REFERENCE LOW	D9	BLK/RED	453
				D10		
				D11		
"ON" / "OFF"	0* / B+	0* / B+	SECONDARY COOLING FAN CONTROL	D12	GRY	731
				D13		
	0*	0*	4TH GEAR SWITCH	D14	LT BLU	446
				D15		
⑤	0*	0*	PARK/NEUTRAL SWITCH SIGNAL	D16	ORN/BLK	434

▽ Less than 1 volt.
* Less than .5 volt.
① Varies from .60 volt to Battery Voltage depending on position of drivewheels.
② Varies
③ Varies with temperature.
④ Battery Voltage for first two seconds.
⑤ Reads battery voltage when in gear.

5.0L (VIN F) AND 5.7L (VIN 8) ENGINES — ECM CONNECTOR END VIEW — 1992 CAMARO AND FIREBIRD

VOLTAGE

Note	KEY "ON"	ENG. RUN	CIRCUIT	PIN	WIRE COLOR	CKT #
				E1		
				E2		
②			IAC "A" HIGH	E3	LT BLU/WHT	441
②			IAC "A" LOW	E4	LT BLU/BLK	442
②			IAC "B" HIGH	E5	LT GRN/WHT	443
②			IAC "B" LOW	E6	LT GRN/BLK	444
	0*	B+	"SERVICE ENGINE SOON LIGHT"	E7	BRN/WHT	419
"ON" / "OFF"	0* / B+	0* / B+	PRIMARY COOLING FAN CONTROL	E8	DK GRN/WHT	335
	B+	▽	EGR SOLENOID CONTROL	E9	GRY	435
				E10		
				E11		
	5	5	DIAGNOSTIC "TEST" TERMINAL	E12	WHT/BLK	451
④	0*	B+	FUEL PUMP SIGNAL	E13	GRY	120
②	.35 - .55	.1 - .9	OXYGEN (O$_2$) SENSOR SIGNAL	E14	PPL	412
	0*	0*	OXYGEN (O$_2$) SENSOR GROUND	E15	TAN	413
③	1.3	1.4	CTS SIGNAL	E16	YEL	410

VOLTAGE

Note	KEY "ON"	ENG. RUN	CIRCUIT	PIN	WIRE COLOR	CKT #
	0* / B+	0* / B+	M/T SHIFT LIGHT CONTROL	F1	TAN/BLK	456
	B+	B+	PORT (SWITCH) SOLENOID	F2	BRN	436
				F3		
	B+	▽	CONVERTER (DIVERT) SOLENOID	F4	BLK/PNK	429
				F5		
	0* / B+	0* / B+	TCC CONTROL A/T	F6	TAN/BLK	422
	B+	B+	CANISTER PURGE SOL CONTROL	F7	DK GRN/YEL	428
				F8		
	2.5	2.5	ESC KNOCK SENSOR SIGNAL	F9	DK BLU	496
	2.5	2.5	PASS-Key® SIGNAL	F10	DK BLU	229
				F11		
				F12		
	.65	.65	TPS SIGNAL	F13	DK BLU	417
				F14		
⑤	4.80	1.37	MAP SIGNAL	F15	LT GRN	432
③	3.9	2.6	IAT SIGNAL	F16	TAN	472

▽ Less than 1 volt.
* Less than .5 volt.
① Varies from .60 volt to Battery Voltage
② Varies
③ Varies with temperature.
④ Battery Voltage for first two seconds with ignition "ON."
⑤ Varies with altitude.

5.0L (VIN F) AND 5.7L (VIN 8) ENGINES — ON-BOARD DIAGNOSTIC SYSTEM CHART — 1992 CAMARO AND FIREBIRD

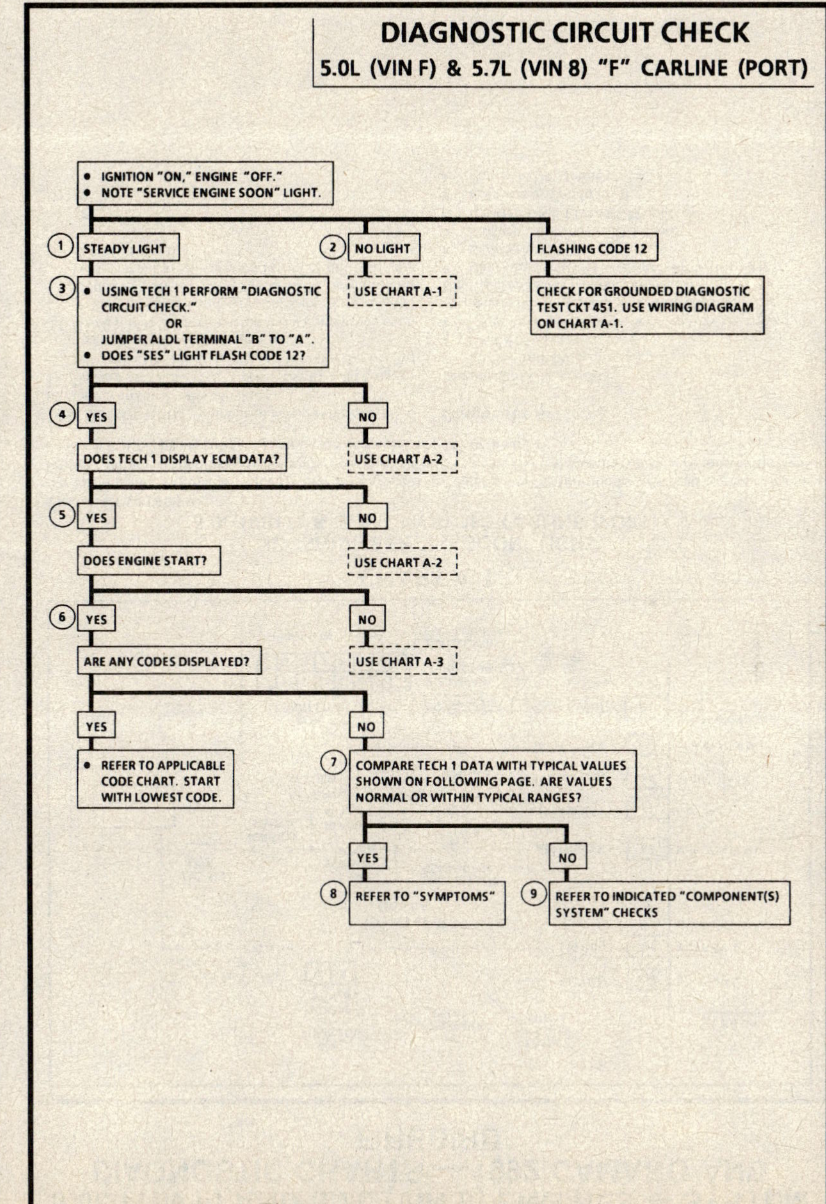

DIAGNOSTIC CIRCUIT CHECK
5.0L (VIN F) & 5.7L (VIN 8) "F" CARLINE (PORT)

Circuit Description:

The diagnostic circuit check is an organized approach to identifying a problem created by an Electronic Engine Control System malfunction. It must be the starting point for any driveability complaint diagnosis, because it directs the service technician to the next logical step in diagnosing the complaint. Understanding the chart and using it correctly will reduce diagnostic time and prevent the unnecessary replacement of good parts.

Test Description: Number(s) refer to circled number(s) on the diagnostic chart.

1. This step is a check for the proper operation of the "Service Engine Soon" light. The "SES" light should be "ON" steady.
2. No "SES" light at this point indicates that there is a problem with the "SES" light circuit or the ECM control of that circuit.
3. This test checks the ability of the ECM to control the "SES" light. With the diagnostic terminal grounded, the "SES" light should flash a Code 12 three times, followed by any trouble code stored in memory.
4. Most of the procedures use a Tech 1 to aid diagnosis, therefore, serial data must be available. If a MEM-CAL error is present, the ECM may have been able to flash Code 12/51, but not enable serial data.
5. Although the ECM is powered up, a "Cranks But Will Not Run" symptom could exist because of an ECM or system problem.
6. This step will isolate if the customer complaint is a "SES" light or a driveability problem with no "SES" light. An invalid code may be the result of a faulty "Scan" tool, MEM-CAL or ECM.
7. Comparison of actual control system data with the typical valves is a quick check to determine if any parameter is not within limits. Keep in mind that a base engine problem (i.e. advanced cam timing) may substantially alter sensor values.
8. Installation of a "Scan" tool will provide a good ground path for the ECM and may hide a driveability complaint due to poor ECM grounds.
9. If the actual data is not within the typical values established, the charts in "Components Systems," Section _____ will provide a functional check of the suspect component or system.

5.0L (VIN F) AND 5.7L (VIN 8) ENGINES — ON-BOARD DIAGNOSTIC SYSTEM CHART — 1992 CAMARO AND FIREBIRD

DIAGNOSTIC CIRCUIT CHECK
5.0L (VIN F) & 5.7L (VIN 8) "F" CARLINE (PORT)

5.0L (VIN F) AND 5.7L (VIN 8) ENGINES — ON-BOARD DIAGNOSTIC SYSTEM CHART — 1992 CAMARO AND FIREBIRD

If after completing the diagnostic circuit check and finding the Tech 1 diagnostics functioning properly and no trouble codes displayed, the "Typical Scan Values" may be used for comparison with values obtained on the vehicle being diagnosed. The "Typical Scan Values," are an average of display values recorded from normally operating vehicles and are intended to represent what a normally functioning system would display.

A "SCAN" TOOL THAT DISPLAYS FAULTY DATA SHOULD NOT BE USED, AND THE PROBLEM SHOULD BE REPORTED TO THE MANUFACTURER. THE USE OF A FAULTY "SCAN" TOOL CAN RESULT IN MISDIAGNOSIS AND UNNECESSARY PARTS REPLACEMENT.

Only the parameters listed are used in this manual for diagnosis. If a "Scan" tool reads the other parameters, the values are not recommended by General Motors for use in diagnosis. For more description on the values and use of the Tech 1 to diagnose ECM inputs, refer to the applicable diagnosis section in "Component Systems," If all values are within the range illustrated, refer to "Symptoms "

TYPICAL SCAN TOOL DATA VALUES

Idle / Upper Radiator Hose Hot / Closed Throttle / Park or Neutral / Closed Loop / Acc. off

"SCAN" Position	Units Displayed	Typical Data Values
Engine Speed	RPM	± 100 RPM from desired RPM in park (± 50 in drive)
Desired Idle	RPM	ECM idle command (700 rpm ± 100 rpm in Park)
Coolant Temp	C°/F°	85°C - 105°C (185°F - 221°F)
MAT/INT Air Temp	C°/F°	10°C - 90°C (50°F-194°F) (Depends on underhood temp.)
MAP	KPa/Volts	20 - 48/1 - 2 Volts (varies with vacuum and altitude)
Throt Position	Volts	.36 - 62
Throttle Angle	Percentage	0% at Idle
Oxygen Sensor	Millivolts	1 - 1000 and varying
Open/Close Loop	Open/Closed	Closed Loop (may go open with extended idle)
Block Learn Cell	Cell Number	4 (depends on Air Flow & RPM)
BLK Learn Enable	No/Yes	Yes (when BLM is enabled)
Fuel Integrator	Counts	Varies
Block Learn	Counts	118 - 138
Spark Advance	# of Degrees	Varies
INJ. Pulse Width	M/Sec	1 - 4 and varying
Knock Retard	Degrees of Retard	0
Knock Signal	Yes/No	No (Yes, when knock is detected)
Exhaust Recirc.	0 - 100%	0% at Idle
Purge Duty Cycle	0 - 100%	0% at Idle
Idle Air Control	Counts (Steps)	5 - 50
Learn IAC	Counts (Steps)	5 - 50
MPH/KPH	MPH/KPH	0 at Idle
Torque Conv CL	OFF/ON	"OFF" ("ON" with TCC commanded)
Battery Voltage	Volts	13.5 - 14.5
Fuel Pump Volts	Volts	13.5 - 14.5
Air Control Sol	Normal/Divert	Normal
Air Switch Sol	Port/Converter	Converter
A/C Request	Yes/No	No (Yes, with A/C Request)
Pass-Key®	Not OK/OK	OK
Fan 2 Request	Yes/No	No (Yes, with A/C High Pressure)
Fan 1	ON/OFF	"OFF" (below 226°F)
4th Gear (5.0L)	Yes/No	No (Yes, when in 4th gear)
4th Gear (5.7L)	No/Yes	No (Yes, when in 4th gear)
Park/Neutral	P-N--/R-DL	Park/Neutral
Shift Light	OFF/ON	OFF
Prom ID	0 - 9999	Prom ID Numbers (Varies)
Time From Start	Hrs/Min/sec	Varies (Time from start. Measures how long the engine has been running (if the engine stops, time from start will reset to 0:00:00)

5.0L (VIN F) AND 5.7L (VIN 8) ENGINES — SYSTEM DIAGNOSTIC CHARTS — 1992 CAMARO AND FIREBIRD

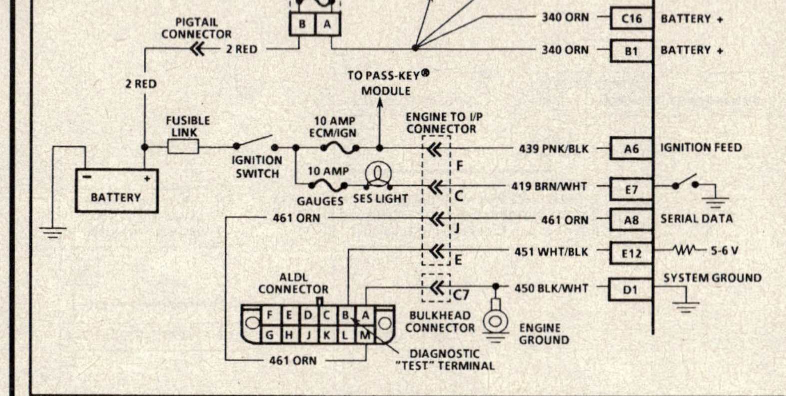

CHART A-1

NO "SERVICE ENGINE SOON" LIGHT
5.0L (VIN F) & 5.7L (VIN 8) "F" CARLINE (PORT)

Circuit Description:

There should always be a steady "Service Engine Soon" light when the ignition is "ON" and engine stopped. Ignition voltage is supplied directly to the light bulb. The Electronic Control Module (ECM) will control the light and turn it "ON" by providing a ground path through CKT 419 to the ECM.

Test Description: Number(s) below refer to circled number(s) on the diagnostic chart.

1. Battery feed CKT 340 is protected by a 20 amp in-line fuse. If this fuse was open, refer to wiring diagram on the facing page of Code 54.
2. Using a test light connected to 12 volts probe each of the system ground circuits to be sure a good ground is present. Refer to the ECM terminal end view in front of this section for ECM pin locations of ground circuits.

Diagnostic Aids:

Engine runs OK, check:
- Faulty light bulb.
- CKT 419 open.
- Gage fuse open, this will result in no oil or generator lights, seat belt reminder, etc.

Engine cranks but will not run, check:
- Continuous battery - fuse or fusible link open.
- ECM/Ignition fuse open.
- Battery CKT 340 to ECM open.
- Ignition CKT 439 to ECM open.
- Poor connection to ECM.
- Faulty ECM ground circuit(s).

5.0L (VIN F) AND 5.7L (VIN 8) ENGINES — SYSTEM DIAGNOSTIC CHARTS — 1992 CAMARO AND FIREBIRD

CHART A-1

NO "SERVICE ENGINE SOON" LIGHT
5.0L (VIN F) & 5.7L (VIN 8) "F" CARLINE (PORT)

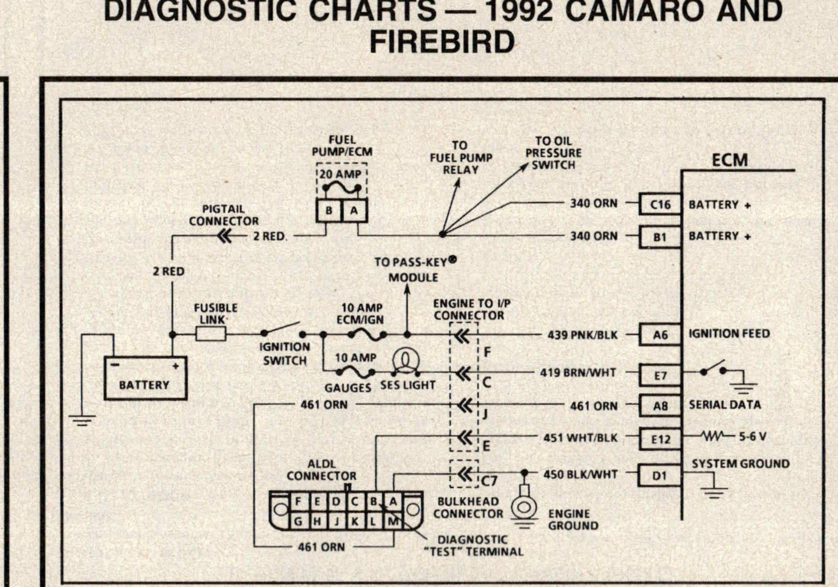

"AFTER REPAIRS," CONFIRM "CLOSED LOOP" OPERATION AND NO "SERVICE ENGINE SOON" LIGHT.

5.0L (VIN F) AND 5.7L (VIN 8) ENGINES — SYSTEM DIAGNOSTIC CHARTS — 1992 CAMARO AND FIREBIRD

CHART A-2

**NO ALDL DATA OR WILL NOT FLASH CODE 12
"SERVICE ENGINE SOON" LIGHT "ON" STEADY**
5.0L (VIN F) & 5.7L (VIN 8) "F" CARLINE (PORT)

Circuit Description:

There should always be a steady "Service Engine Soon" light when the ignition is "ON" and engine stopped. The Electronic Control Module (ECM) will turn the light "ON" by grounding CKT 419 in the ECM.

With the diagnostic "test" terminal grounded the light should flash a Code 12, followed by any trouble code(s) stored in memory.

A steady light suggests a short to ground in the light control CKT 419, or an open in diagnostic CKT 451.

Test Description: Number(s) below refer to circled number(s) on the diagnostic chart.

1. If there is a problem with the ECM that causes a "Scan" tool to not read Serial data, the ECM should not flash a Code 12. If Code 12 is flashing check for CKT 451 for being short to ground. If Code 12 does flash, make sure that the "Scan" tool is working properly on another vehicle. If the "Scan" tool is functioning properly and CKT 461 is OK, the MEM-CAL or ECM may be at fault.

2. If the light goes "OFF" when the ECM connector is disconnected, CKT 419 is not shorted to ground.

3. This step will check for an open diagnostic CKT 451.

4. At this point the "Service Engine Soon" light wiring is OK. The problem is a faulty ECM or MEM-CAL. If Code 12 does not flash, the ECM should be replaced using the original MEM-CAL. Replace the MEM-CAL only after trying an ECM, as a defective MEM-CAL is an unlikely cause of the problem.

5.0L (VIN F) AND 5.7L (VIN 8) ENGINES — SYSTEM DIAGNOSTIC CHARTS — 1992 CAMARO AND FIREBIRD

CHART A-2

NO ALDL DATA OR WILL NOT FLASH CODE 12
"SERVICE ENGINE SOON" LIGHT "ON" STEADY
5.0L (VIN F) & 5.7L (VIN 8) "F" CARLINE (PORT)

- IGNITION "ON," ENGINE "OFF." IS THE "SES" LIGHT "ON"?
  - **YES** → GROUND DIAGNOSTIC TERMINAL. DOES LIGHT FLASH CODE 12?
    - **NO** → (2) IGNITION "OFF". DISCONNECT ECM CONNECTORS. IGNITION "ON" AND NOTE "SERVICE ENGINE SOON" LIGHT.
      - (3) LIGHT "OFF" → IGNITION "OFF." RECONNECT ECM. IGNITION "ON," ENGINE "OFF," DIAGNOSTIC TERMINAL NOT GROUNDED. BACK PROBE ECM CKT 451 WITH TEST LIGHT TO GROUND. LEAVE CONNECTED AND WATCH "SES" LIGHT.
        - **NO CODE 12** → (4) CHECK MEM-CAL FOR PROPER INSTALLATION. IF OK, REPLACE ECM USING ORIGINAL MEM-CAL. RECHECK FOR CODE 12
          - **NO CODE 12** → REPLACE MEM-CAL
          - **CODE 12** → SYSTEM OK
        - **CODE 12** → CHECK FOR OPEN CKT 451 TO ECM. IF OK, CHECK FOR OPEN CIRCUIT BETWEEN ALDL TERMINAL "A" AND ECM.
      - LIGHT "ON" → REPAIR SHORT TO GROUND IN CKT 419.
    - **YES** → (1) IF PROBLEM WAS NO ALDL DATA: CHECK SERIAL DATA CKT 461 FOR OPEN OR SHORT TO GROUND BETWEEN ECM AND ALDL CONNECTOR. IF OK, IT IS A FAULTY ECM OR MEM-CAL.
  - **NO** → USE CHART A-1

"AFTER REPAIRS," CONFIRM "CLOSED LOOP" OPERATION AND NO "SERVICE ENGINE SOON" LIGHT.

5.0L (VIN F) AND 5.7L (VIN 8) ENGINES — SYSTEM DIAGNOSTIC CHARTS — 1992 CAMARO AND FIREBIRD

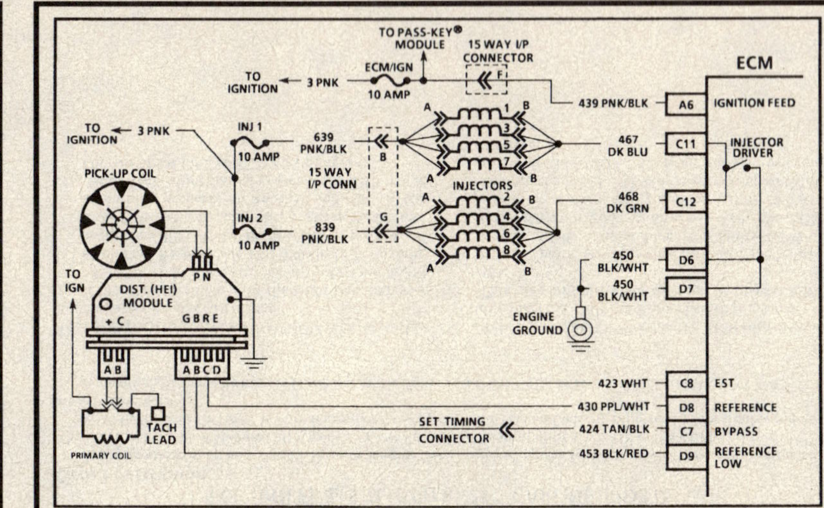

CHART A-3

(Page 1 of 2)
ENGINE CRANKS BUT WILL NOT RUN
5.0L (VIN F) & 5.7L (VIN 8) "F" CARLINE (PORT)

Circuit Description:
This chart assumes that battery condition and engine cranking speed are OK, and there is adequate fuel in the tank.

Test Description: Number(s) below refer to circled number(s) on the diagnostic chart.

1. A "Service Engine Soon" light "ON" is a basic test to determine if there is a 12 volt supply and ignition 12 volts to ECM. No ALDL may be due to an ECM problem and CHART A-2 will diagnose the ECM. If TPS is over 2.5 volts the engine may be in the clear flood mode which will cause starting problems. The engine will not start without reference pulses and therefore the "Scan" tool should read engine rpm (reference) during crank.
2. No spark may be caused by one of several components related to the ignition/EST system. CHART C-4 will address all problems related to the causes of a no spark condition.
3. The test light should blink, indicating the ECM is controlling the injectors OK. Use injector test light J 34730-3 or equivalent.
4. All injectors should be within 1.0 ohm of each other and should not be less than 10 ohms at 21°C (70°F). If an injector is suspected for a no start condition, unhook the suspected injector and try to start the engine.
5. Use fuel pressure gage J 34730-1 or equivalent. Wrap a shop towel around the fuel pressure tap to absorb any small amount of fuel leakage that may occur when installing the gage.

Diagnostic Aids:

- An EGR valve sticking open can cause a low air/fuel ratio during cranking.
- Unless engine enters "Clear Flood" at the first indication of a flooding condition, it can result in a no start.
- Check for fouled plugs.
- Water or foreign material in fuel line can cause a no start in freezing weather.
- Check for blinking light at injector harness, on both banks of the engine. If not OK, check injector fuses.

If above are all OK, refer to "Hard Start" in "Symptoms" Section

5.0L (VIN F) AND 5.7L (VIN 8) ENGINES — SYSTEM DIAGNOSTIC CHARTS — 1992 CAMARO AND FIREBIRD

CHART A-3

(Page 1 of 2)

ENGINE CRANKS BUT WILL NOT RUN

5.0L (VIN F) & 5.7L (VIN 8) "F" CARLINE (PORT)

① • PERFORM DIAGNOSTIC CIRCUIT CHECK.
CHECK THE FOLLOWING:
  • IF CODE 54 IS SET, REFER TO CODE 54 CHART.
  • CHECK PASS-KEY® STATUS. IF NOT OK, REFER TO CODE 46 CHART.
  • ACTUAL ENGINE TEMPERATURE AND CTS TEMPERATURE ON THE TECH 1 SHOULD BE CLOSE TO THE SAME, IF NOT REFER TO CODE 15.
  • TPS - IF OVER 2.5 VOLTS AT CLOSED THROTTLE, USE THE CODE 21 CHART.
  • IF TPS AND MAP VOLTAGES ARE OUT OF RANGE AND NO CODE IS SET REFER TO TPS AND MAP CODE CHARTS.
  IS RPM INDICATED DURING CRANKING?

YES
• USING A ST-125 (SPARK CHECKER) J 26792 OR EQUIVALENT, CHECK FOR SPARK WHILE CRANKING (CHECK TWO WIRES). IS SPARK PRESENT?

NO
• USING A ST-125 (SPARK CHECKER) J 26792 OR EQUIVALENT, CHECK FOR SPARK WHILE CRANKING (CHECK TWO WIRES). IS SPARK PRESENT?

YES →
③ • DISCONNECT ALL INJECTORS.
• CONNECT TEST LIGHT J 34730-2 OR EQUIVALENT TO INJECTOR HARNESS CONNECTOR. (TEST ONE INJECTOR HARNESS ON EACH SIDE OF THE ENGINE.)
• CHECK FOR BLINKING LIGHT WHILE CRANKING.

NO →
② BASIC HEI PROBLEM. REFER TO CHART C-4.

YES →
• IGNITION "OFF."
• DISCONNECT DISTRIBUTOR 4-WAY CONNECTOR.
• IGNITION "ON."
• MOMENTARILY TOUCH HARNESS CONNECTOR TERMINAL (CKT 430) WITH A TEST LIGHT TO 12 VOLTS.
• TECH 1 SHOULD INDICATE RPM WHEN TEST IS PERFORMED. DOES IT?

NO →
② CHECK FOR BATTERY VOLTAGE TO IGNITION SYSTEM. IF OK, THERE IS A BASIC HEI PROBLEM. REFER TO CHART C-4.

BLINKING LIGHT
④ • CHECK RESISTANCE ACROSS EACH INJECTOR AND COMPARE VALUES.

NO BLINKING LIGHT
USE CHART A-3 (2 OF 2)

YES
• FAULTY CONNECTION OR IGNITION MODULE.

NO
CKT 430 OPEN, SHORTED TO GROUND, OR FAULTY ECM.

OK
⑤ • IGNITION "OFF."
• INSTALL FUEL PRESSURE GAGE AND NOTE PRESSURE AFTER IGNITION "ON." SHOULD BE (283-325 kPa) 41-47 psi

NOT OK
REPLACE INJECTOR(S) THAT IS OUT OF RANGE.

OK
• REVIEW THE "DIAGNOSTIC AIDS" ON FACING PAGE FOR ADDITIONAL ITEMS TO CHECK. IF ALL ARE OK, EFI SYSTEM IS OK. REFER TO "HARD START "IN SECTION "B."

NOT OK
USE CHART A-7

"AFTER REPAIRS," CONFIRM "CLOSED LOOP" OPERATION AND NO "SERVICE ENGINE SOON" LIGHT.

5.0L (VIN F) AND 5.7L (VIN 8) ENGINES — SYSTEM DIAGNOSTIC CHARTS — 1992 CAMARO AND FIREBIRD

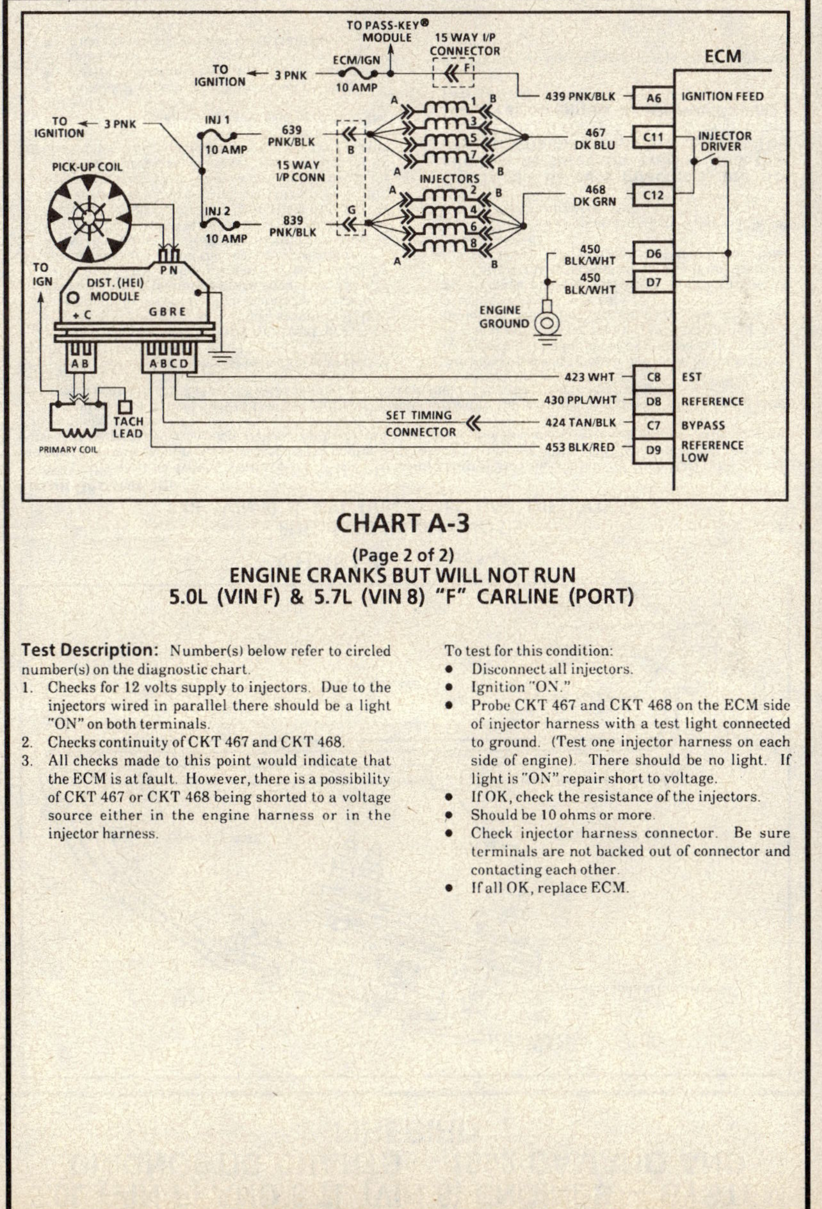

CHART A-3

(Page 2 of 2)

ENGINE CRANKS BUT WILL NOT RUN

5.0L (VIN F) & 5.7L (VIN 8) "F" CARLINE (PORT)

Test Description: Number(s) below refer to circled number(s) on the diagnostic chart.

1. Checks for 12 volts supply to injectors. Due to the injectors wired in parallel there should be a light "ON" on both terminals.
2. Checks continuity of CKT 467 and CKT 468.
3. All checks made to this point would indicate that the ECM is at fault. However, there is a possibility of CKT 467 or CKT 468 being shorted to a voltage source either in the engine harness or in the injector harness.

To test for this condition:
• Disconnect all injectors.
• Ignition "ON."
• Probe CKT 467 and CKT 468 on the ECM side of injector harness with a test light connected to ground. (Test one injector harness on each side of engine). There should be no light. If light is "ON" repair short to voltage.
• If OK, check the resistance of the injectors.
• Should be 10 ohms or more.
• Check injector harness connector. Be sure terminals are not backed out of connector and contacting each other.
• If all OK, replace ECM.

5.0L (VIN F) AND 5.7L (VIN 8) ENGINES — SYSTEM DIAGNOSTIC CHARTS — 1992 CAMARO AND FIREBIRD

CHART A-3

(Page 2 of 2)
ENGINE CRANKS BUT WILL NOT RUN
5.0L (VIN F) & 5.7L (VIN 8) "F" CARLINE (PORT)

FROM A-3 (1 OF 2)

NO BLINKING LIGHT AT INJECTOR

NO LIGHT

STEADY LIGHT

①
- IGNITION "ON."
- PROBE INJECTOR HARNESS TERMINALS WITH A TEST LIGHT TO GROUND.
- LIGHT SHOULD BE "ON" AT BOTH TERMINALS.

- CHECK INJECTOR DRIVER CIRCUIT WITH STEADY LIGHT FOR SHORT TO GROUND.
- IF CIRCUIT IS NOT SHORTED, CHECK RESISTANCE ACROSS EACH INJECTOR IN THE CIRCUIT.
- RESISTANCE SHOULD BE GREATER THAN 10 OHMS.

OK

NOT OK

FAULTY ECM

REPAIR SHORT TO GROUND OR REPLACE ANY INJECTOR THAT MEASURES UNDER 10 OHMS.

LIGHT "ON" BOTH

LIGHT "ON" ONE

LIGHT "OFF" BOTH

②
- RECONNECT INJECTOR(S)
- IGNITION "OFF."
- DISCONNECT ECM
- IGNITION "ON."
- PROBE TERMINALS "C11" AND "C12" WITH A TEST LIGHT TO GROUND.

DUE TO INJECTORS WIRED IN PARALLEL, THERE SHOULD BE A LIGHT ON BOTH TERMINALS.

IF NOT, THE PROBLEM IS AN OPEN IN THE HARNESS TO THE TESTED INJECTOR.

REPAIR OPEN IN INJECTOR FEED CIRCUIT.

LIGHT "ON"

LIGHT "OFF"

③ REFER TO FACING PAGE.

OPEN CKT 467 OR 468

"AFTER REPAIRS," CONFIRM "CLOSED LOOP" OPERATION AND NO "SERVICE ENGINE SOON" LIGHT.

5.0L (VIN F) AND 5.7L (VIN 8) ENGINES — SYSTEM DIAGNOSTIC CHARTS — 1992 CAMARO AND FIREBIRD

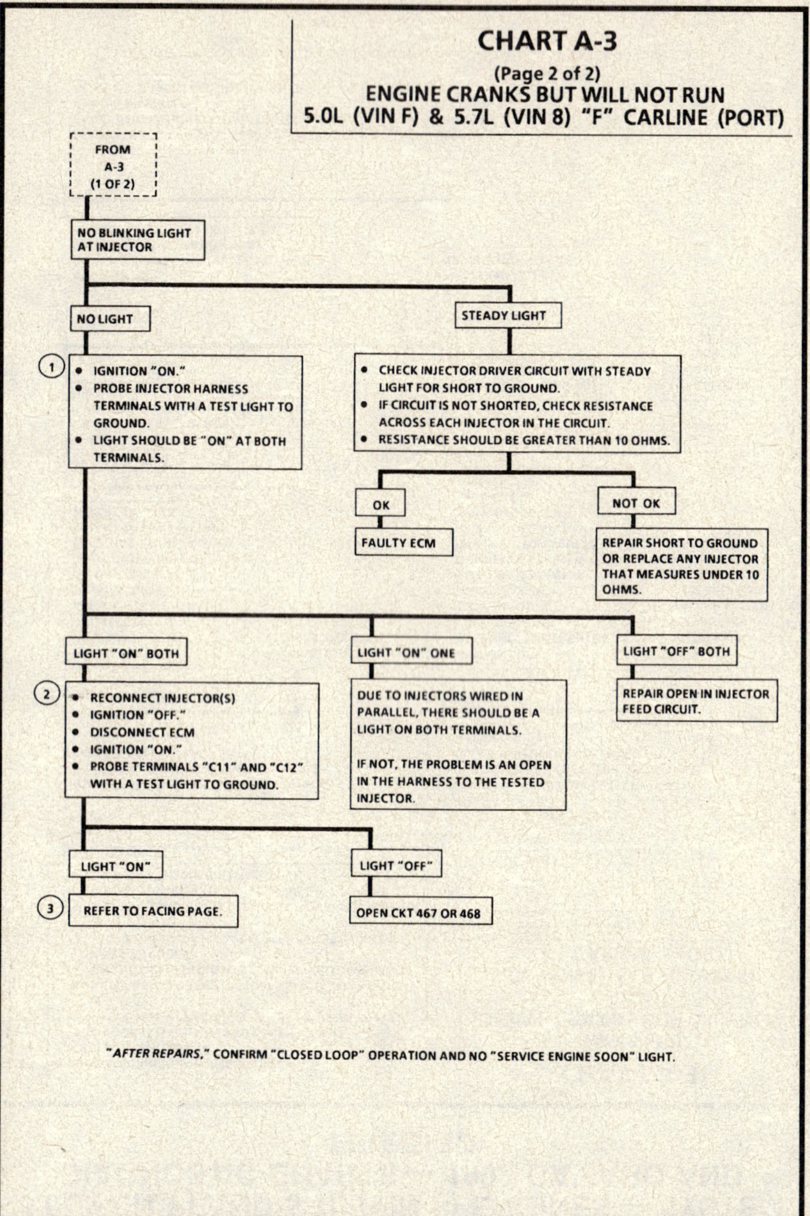

CHART A-7 (Page 1 of 2)

FUEL SYSTEM DIAGNOSIS
5.0L (VIN F) & 5.7L (VIN 8) "F" CARLINE (PORT)

Circuit Description:

When the ignition switch is turned "ON," the Electronic Control Module (ECM) will turn "ON" the in-tank fuel pump. It will remain "ON" as long as the engine is cranking or running, and the ECM is receiving reference pulses. If there are no reference pulses, the ECM will shut "OFF" the fuel pump within 2 seconds after ignition "ON" or engine stops.

The pump will deliver fuel to the fuel rail and injectors, then to the pressure regulator, where the system pressure is controlled at 234 to 325 kPa (34 to 47 psi). Excess fuel is then returned to the fuel tank.

Test Description: Number(s) below refer to circled number(s) on the diagnostic chart.

1. Wrap a shop towel around the fuel pressure connector to absorb any small amount of fuel leakage that may occur when installing the gage. Ignition "ON," pump pressure should be 280-325 kPa (41-47 psi). This pressure is controlled by spring pressure within the regulator assembly.
2. When the engine is idling, the manifold pressure is low (high vacuum) and is applied to the fuel regulator diaphragm. This will offset the spring and result in a lower fuel pressure. This idle pressure will vary somewhat depending on barometric pressure; however, the pressure idling should be less, indicating pressure regulator control.
3. Pressure that leaks down is caused by one of the following:
   - In-tank fuel pump check valve not holding.
   - Partially disconnected fuel pulse dampener (pulsator).
   - Fuel pressure regulator valve leaking.
   - Injector(s) sticking open.
4. An injector sticking open can best be determined by checking for a fouled or saturated spark plug(s). If a leaking injector cannot be determined by a fouled or saturated spark plug, the following procedure should be used:
   - Remove plenum, and remove fuel rail bolts. Follow the procedures in the Fuel Control Section of this manual, but leave fuel lines connected.
   - Lift fuel rail out just enough to observe injector nozzles in the ports.

CAUTION: Be sure injector(s) are not allowed to spray on engine and that injector retaining clips are intact. This should be carefully followed to prevent fuel spray on engine which would cause a fire hazard.

   - Pressurize the fuel system and observe injector nozzles.

5.0L (VIN F) AND 5.7L (VIN 8) ENGINES — SYSTEM DIAGNOSTIC CHARTS — 1992 CAMARO AND FIREBIRD

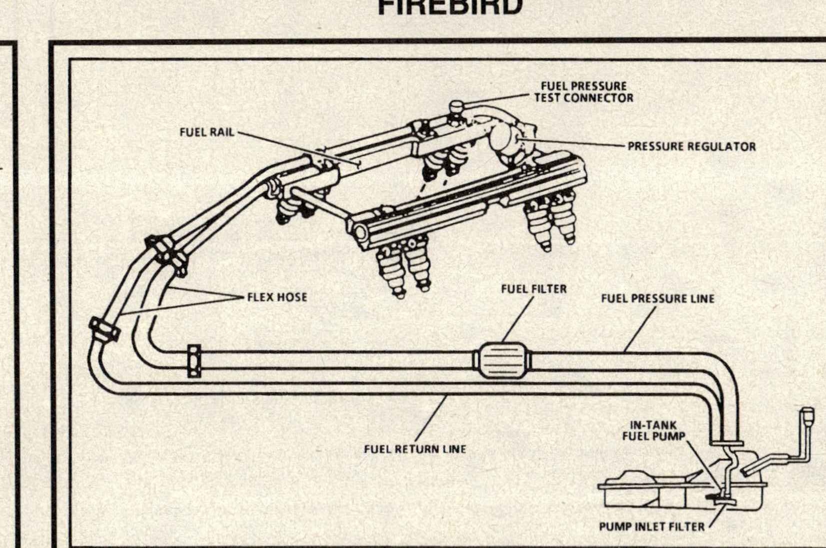

CHART A-7
(Page 2 of 2)
FUEL SYSTEM DIAGNOSIS
5.0L (VIN F) & 5.7L (VIN 8) "F" CARLINE (PORT)

Test Description: Number(s) below refer to circled number(s) on the diagnostic chart.

1. Fuel pressure less than 280 kPa (41 psi) falls into two areas:
   - Regulated pressure less than 280 kPa (41 psi). System will be lean and may set Code 44. Also, hard starting cold and overall poor performance.
   - Restricted flow causing pressure drop. Normally, a vehicle with a fuel pressure of less than 165 kPa (24 psi) at idle will not be driveable. However, if the pressure drop occurs only while driving, the engine will normally surge then stop running as pressure begins to drop rapidly. This is most likely caused by a restricted fuel line or plugged in-line filter.

2. Restricting the fuel return line allows the fuel pressure to build above regulated pressure. With battery voltage applied to the fuel pump test terminal, pressure should rise to 414 kPa (60 psi) as the fuel return hose is gradually pinched.

   NOTICE: Do not allow fuel pressure to exceed 414 kPa (60 psi), damage to the pressure regulator may result.

3. This test determines if the high fuel pressure is due to a restricted fuel return line or a pressure regulator problem.

CHART A-7
(Page 1 of 2)
FUEL SYSTEM DIAGNOSIS
5.0L (VIN F) & 5.7L (VIN 8) "F" CARLINE (PORT)

THIS CHART ASSUMES THERE IS NO CODE 54

FROM CHART A-3 PAGE 1

NOTE: THE IGNITION MAY HAVE TO BE CYCLED "ON" MORE THAN ONCE TO OBTAIN MAXIMUM PRESSURE. ALSO, IT IS NORMAL FOR THE PRESSURE TO DROP SLIGHTLY WHEN THE PUMP STOPS.

① INSTALL FUEL PRESSURE GAGE, J 34730-1 OR EQUIVALENT.
- IGNITION "OFF" FOR 10 SECONDS. A/C "OFF."
- IGNITION "ON." FUEL PUMP WILL RUN FOR ABOUT 2 SECONDS.
- NOTE FUEL PRESSURE, WITH PUMP RUNNING SHOULD BE 280-325 kPa (41-47 psi) AND HOLD STEADY WHEN PUMP STOPS.

② START AND IDLE ENGINE AT NORMAL OPERATING TEMPERATURE.
- PRESSURE SHOULD BE LOWER BY 21-69 kPa (3-10 psi).

USING AN EXTERNAL VACUUM SOURCE, APPLY 10 INCHES OF VACUUM TO FUEL PRESSURE REGULATOR.
- FUEL PRESSURE SHOULD DROP 21-69 kPa (3-10 psi).

NO TROUBLE FOUND. REVIEW "SYMPTOMS," SECTION "B."

REPAIR VACUUM SOURCE TO REGULATOR

REPLACE REGULATOR ASSEMBLY.

③ PRESSURE BUT NOT HOLDING — PRESSURE BELOW 280 kPa (41 psi) — PRESSURE ABOVE 325 kPa (47 psi) — NO PRESSURE

IGNITION "OFF" FOR 10 SECONDS. IGNITION "ON." BLOCK FUEL PRESSURE LINE BY PINCHING FLEX HOSE. PRESSURE SHOULD HOLD.

SEE CHART A-7 (2 of 2)

IGNITION "OFF" FOR 10 SECONDS. IGNITION "ON." APPLY 12 VOLTS TO FUEL PUMP TEST TERMINAL. LISTEN FOR FUEL PUMP RUNNING.

IGNITION "OFF" FOR 10 SECONDS. IGNITION "ON." BLOCK FUEL RETURN LINE BY PINCHING HOSE. RECHECK PRESSURE.

CHECK:
- LEAKING PUMP COUPLING HOSE OR PULSATOR.
- FAULTY IN-TANK PUMP.

④ LOCATE AND CORRECT LEAKING INJECTOR(S).

FAULTY FUEL PRESSURE REGULATOR

PUMP RUNS — PUMP NOT RUNNING

CHECK FOR:
- PLUGGED IN-LINE FILTER.
- PLUGGED PUMP INLET FILTER.
- RESTRICTED FUEL LINE.
- DISCONNECTED COUPLING HOSE OR PULSATOR.

CHECK FOR:
- OPEN WIRE IN CKT 120.
- OPEN PUMP. GROUND CKT 150.

REPLACE IN-TANK FUEL PUMP.

"AFTER REPAIRS," CONFIRM "CLOSED LOOP" OPERATION AND NO "SERVICE ENGINE SOON" LIGHT.

5.0L (VIN F) AND 5.7L (VIN 8) ENGINES — SYSTEM DIAGNOSTIC CHARTS — 1992 CAMARO AND FIREBIRD

5.0L (VIN F) AND 5.7L (VIN 8) ENGINES — DIAGNOSTIC TROUBLE CODE CHART — 1992 CAMARO AND FIREBIRD

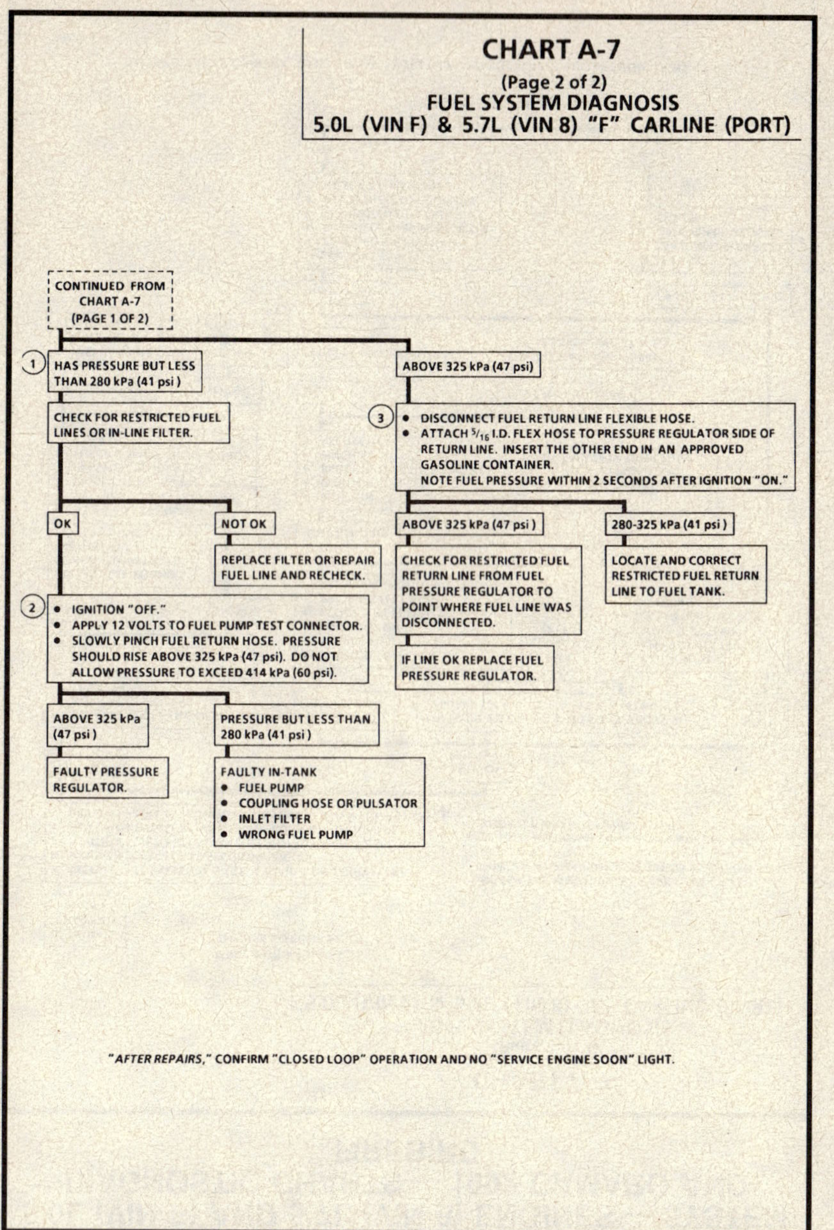

CODE 13
OXYGEN SENSOR CIRCUIT
(OPEN CIRCUIT)
5.0L (VIN F) & 5.7L (VIN 8) "F" CARLINE (PORT)

Circuit Description:

The ECM supplies a voltage of about 450 mV between terminals "E14" and "E15". (If measured with a 10 megohm digital voltmeter, this may read as low as 320 mV.) The O_2 sensor varies the voltage within a range of about 1000 mV if the exhaust is rich, down through about 10 mV if exhaust is lean.

The sensor is like an open circuit and produces no voltage when it is below 315°C (600°F). An open sensor circuit or cold sensor causes "Open Loop" operation.

Test Description: Number(s) below refer to circled number(s) on the diagnostic chart.

1. Code 13 will set:
   - Coolant temperature greater than 70°C (158°F).
   - At least 2 minutes engine time after start.
   - O_2 signal voltage steady between 350 mV and 550 mV.
   - Throttle position sensor signal above 5%. (about .3 volt above closed throttle voltage).
   - All conditions must be met for about 60 seconds.
   If the conditions for a Code 13 exist, the system will not go into "Closed Loop."
2. This will determine if the oxygen sensor is at fault or the wiring or ECM is the cause of the Code 13.
3. For this test use only a high impedance digital volt ohmmeter. This test checks the continuity of CKT 412 and CKT 413. If CKT 413 is open, the ECM voltage on CKT 412 will be over .6 volt (600 mV).

Diagnostic Aids:

Normal "Scan" tool voltage readings varies between 10 mV to 1000 mV (.1 and 1.0 volt) while in "Closed Loop." Code 13 sets in one minute if voltage remains between .35 and .55 volt, the system will go "Open Loop" in about 15 seconds.

Refer to "Intermittents" in "Symptoms," Section

5.0L (VIN F) AND 5.7L (VIN 8) ENGINES — DIAGNOSTIC TROUBLE CODE CHART — 1992 CAMARO AND FIREBIRD

CODE 13

OXYGEN SENSOR CIRCUIT
(OPEN CIRCUIT)
5.0L (VIN F) & 5.7L (VIN 8) "F" CARLINE (PORT)

"AFTER REPAIRS," REFER TO CODE CRITERIA ON FACING PAGE AND CONFIRM CODE DOES NOT RESET.

5.0L (VIN F) AND 5.7L (VIN 8) ENGINES — DIAGNOSTIC TROUBLE CODE CHART — 1992 CAMARO AND FIREBIRD

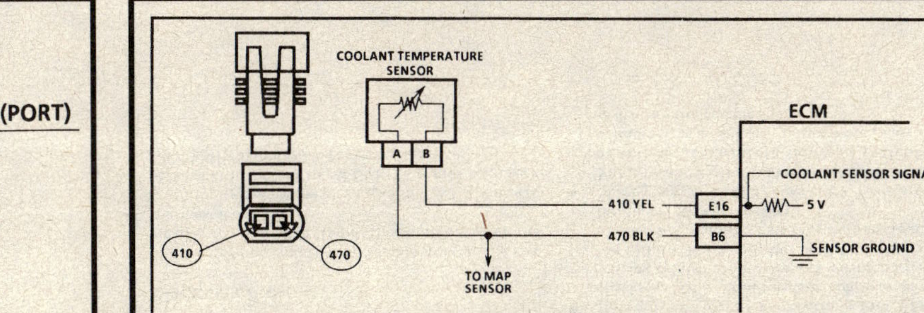

CODE 14

COOLANT TEMPERATURE SENSOR CIRCUIT
(HIGH TEMPERATURE INDICATED)
5.0L (VIN F) & 5.7L (VIN 8) "F" CARLINE (PORT)

Circuit Description:

The Coolant Temperature Sensor (CTS) uses a thermistor to control the signal voltage to the ECM. The ECM applies a voltage on CKT 410 to the sensor. When the engine coolant is cold, the sensor (thermistor) resistance is high, therefore, the ECM will sense high signal voltage.

As the engine coolant warms, the sensor resistance becomes less, and the voltage drops. At normal engine operating temperature (85°C - 95°C or 185°F - 203°F), the voltage will measure about 1.5 to 2.0 volts.

Test Description: Number(s) below refer to circled number(s) on the diagnostic chart.

1. Code 14 will set if:
   - Signal voltage indicates a coolant temperature above 130°C (266°F).
2. This test will determine if CKT 410 is shorted to ground which will cause the conditions for Code 14.

Diagnostic Aids:

Check harness routing for a potential short to ground in CKT 410.

"Scan" tool displays engine temperature in degrees celsius and fahrenheit. After engine is started, the temperature should rise steadily to about 90°C (194°F), then stabilize when thermostat opens.

Refer to "Intermittents" in "Symptoms " Section

Check for:

- An intermittent may be caused by a poor connection, rubbed through wire insulation or a wire broken inside the insulation.
- Poor Connection or Damaged Harness. Inspect ECM harness connectors for backed out terminals, improper mating, broken locks, improperly formed or damaged terminals, poor terminal to wire connection and damaged harness.
- Intermittent Test. If connections and harness checks out OK, "Scan" coolant temperature while moving related connectors and wiring harness. If the failure is induced, the "coolant temperature" display will change. This may help to isolate the location of the malfunction.
- Shifted Sensor. The "Temperature To Resistance Value" scale may be used to test the coolant sensor at various temperature levels to evaluate the possibility of a "shifted" sensor which may result in driveability complaints.

5.0L (VIN F) AND 5.7L (VIN 8) ENGINES — DIAGNOSTIC TROUBLE CODE CHART — 1992 CAMARO AND FIREBIRD

CODE 14
COOLANT TEMPERATURE SENSOR CIRCUIT
(HIGH TEMPERATURE INDICATED)
5.0L (VIN F) & 5.7L (VIN 8) "F" CARLINE (PORT)

① DOES TECH 1 DISPLAY COOLANT TEMPERATURE OF 130°C (266°F) OR HIGHER?

- YES
- NO

② • DISCONNECT COOLANT TEMPERATURE SENSOR. TECH 1 SHOULD DISPLAY COOLANT TEMPERATURE BELOW -30°C (-22°F). DOES IT?

CODE 14 IS INTERMITTENT. IF NO ADDITIONAL CODES WERE STORED, REFER TO "DIAGNOSTIC AIDS" ON FACING PAGE.

- YES
- NO

REPLACE COOLANT TEMPERATURE SENSOR.

CKT 410 SHORTED TO GROUND
OR
CKT 410 SHORTED TO SENSOR GROUND CIRCUIT
OR
FAULTY ECM.

DIAGNOSTIC AID

COOLANT SENSOR		
TEMPERATURE VS. RESISTANCE VALUES		
(APPROXIMATE)		
°C	°F	OHMS
100	212	177
90	194	241
80	176	332
70	158	467
60	140	667
50	122	973
45	113	1188
40	104	1459
35	95	1802
30	86	2238
25	77	2796
20	68	3520
15	59	4450
10	50	5670
5	41	7280
0	32	9420
-5	23	12300
-10	14	16180
-15	5	21450
-20	-4	28680
-30	-22	52700
-40	-40	100700

"AFTER REPAIRS," REFER TO CODE CRITERIA ON FACING PAGE AND CONFIRM CODE DOES NOT RESET.

5.0L (VIN F) AND 5.7L (VIN 8) ENGINES — DIAGNOSTIC TROUBLE CODE CHART — 1992 CAMARO AND FIREBIRD

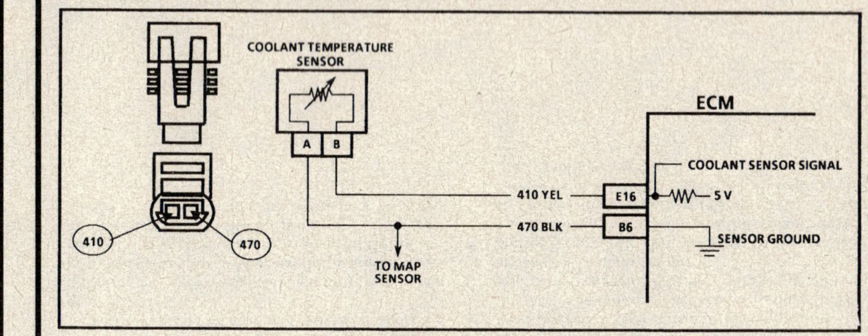

CODE 15
COOLANT TEMPERATURE SENSOR CIRCUIT
(LOW TEMPERATURE INDICATED)
5.0L (VIN F) & 5.7L (VIN 8) "F" CARLINE (PORT)

Circuit Description:
The Coolant Temperature Sensor (CTS) uses a thermistor to control the signal voltage to the ECM. The ECM applies a voltage on CKT 410 to the sensor. When the engine coolant is cold, the sensor (thermistor) resistance is high, therefore, the ECM will sense high signal voltage.

As the engine coolant warms, the sensor resistance becomes less, and the voltage drops. At normal engine operating temperature (85°C - 95°C or 185°F - 203°F), the voltage will measure about 1.5 to 2.0 volts.

Test Description: Number(s) below refer to circled number(s) on the diagnostic chart.
1. Code 15 will set if:
   - Signal voltage indicates an engine coolant temperature less than -49°C (-56°F).
2. This test simulates a Code 14. If the ECM recognizes the low signal voltage, (high temperature) and the "Scan" tool displays 130°C (266°F) or above, the ECM and wiring are OK.
3. This test will determine if CKT 410 is open. There should be 5 volts present at sensor connector if measured with a DVM.

Diagnostic Aids:

After engine is started the temperature should rise steadily to about 90°C (194°F) then stabilize when the thermostat opens.

A faulty connection or an open in CKT 410 or 470 will result in a Code 15. If Code 33 is also set, check CKT 470 for faulty wiring or connections.

Check terminals at sensor for good contact Refer to "Intermittents" in "Symptoms,"

Check for:
- An intermittent may be caused by a poor connection, rubbed through wire insulation or a wire broken inside the insulation.
- Poor Connection or Damaged Harness. Inspect ECM harness connectors for backed out terminals, improper mating, broken locks, improperly formed or damaged terminals, poor terminal to wire connection and damaged harness.
- Intermittent Test. If connections and harness checks out OK, "Scan" coolant temperature while moving related connectors and wiring harness. If the failure is induced, the "coolant temperature" display will change. This may help to isolate the location of the malfunction.
- Shifted Sensor. The "Temperature To Resistance Value" scale may be used to test the coolant sensor at various temperature levels to evaluate the possibility of a "shifted" sensor which may result in driveability complaints.

5.0L (VIN F) AND 5.7L (VIN 8) ENGINES — DIAGNOSTIC TROUBLE CODE CHART — 1992 CAMARO AND FIREBIRD

CODE 15
COOLANT TEMPERATURE SENSOR CIRCUIT
(LOW TEMPERATURE INDICATED)
5.0L (VIN F) & 5.7L (VIN 8) "F" CARLINE (PORT)

1. • DOES TECH 1 DISPLAY COOLANT TEMPERATURE OF -30°C (-22°F) OR LESS?

YES →

2. • DISCONNECT COOLANT TEMPERATURE SENSOR.
   • JUMPER HARNESS TERMINALS TOGETHER.
   • TECH 1 SHOULD DISPLAY 130°C (266°F) OR MORE. DOES IT?

NO →

CODE 15 IS INTERMITTENT. IF NO ADDITIONAL CODES WERE STORED, REFER TO "DIAGNOSTIC AIDS" ON FACING PAGE.

3. • JUMPER CKT 410 TO GROUND.
   • TECH 1 SHOULD DISPLAY OVER 130°C (266 °F). DOES IT?

YES →

FAULTY CONNECTION OR COOLANT TEMPERATURE SENSOR.

YES →

OPEN COOLANT TEMPERATURE SENSOR GROUND CIRCUIT, FAULTY CONNECTION OR FAULTY ECM.

NO →

OPEN CKT 410, FAULTY CONNECTION AT ECM, OR FAULTY ECM.

DIAGNOSTIC AID

COOLANT SENSOR TEMPERATURE VS. RESISTANCE VALUES (APPROXIMATE)		
°C	°F	OHMS
100	212	177
90	194	241
80	176	332
70	158	467
60	140	667
50	122	973
45	113	1188
40	104	1459
35	95	1802
30	86	2238
25	77	2796
20	68	3520
15	59	4450
10	50	5670
5	41	7280
0	32	9420
-5	23	12300
-10	14	16180
-15	5	21450
-20	-4	28680
-30	-22	52700
-40	-40	100700

"AFTER REPAIRS," REFER TO CODE CRITERIA ON FACING PAGE AND CONFIRM CODE DOES NOT RESET.

5.0L (VIN F) AND 5.7L (VIN 8) ENGINES — DIAGNOSTIC TROUBLE CODE CHART — 1992 CAMARO AND FIREBIRD

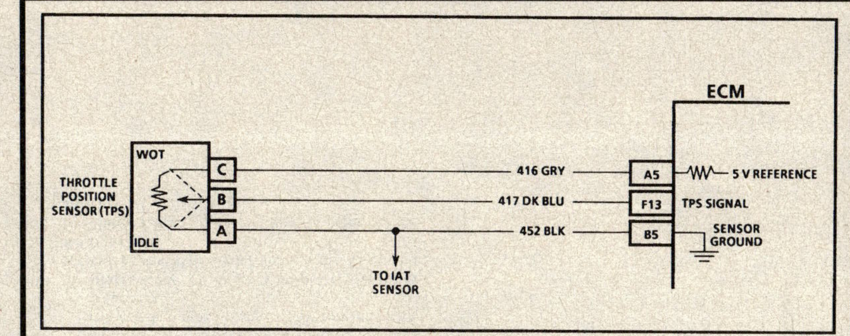

CODE 21
THROTTLE POSITION SENSOR (TPS) CIRCUIT
(SIGNAL VOLTAGE HIGH)
5.0L (VIN F) & 5.7L (VIN 8) "F" CARLINE (PORT)

Circuit Description:

The Throttle Position Sensor (TPS) provides a voltage signal that changes relative to the throttle blade position. Signal voltage will vary from about .5 at idle to about 5 volts at wide open throttle.

The TPS signal is one of the most important inputs used by the ECM for fuel control and for most of the ECM controlled outputs.

Test Description: Number(s) below refer to circled number(s) on the diagnostic chart.

1. Code 21 will set if:
   - TPS signal voltage is greater than 2.5 volts.
   - Engine is running.
   - Air flow is less than 15 GM/sec.
   - All conditions met for 3 seconds.
   OR
   - TPS signal voltage over about 4.8 volts with ignition "ON."

   With throttle closed, the TPS should read less than .96 volt. If it doesn't replace TPS sensor.

2. With the TPS sensor disconnected, the TPS voltage should go low if the ECM and wiring are OK.

3. Probing circuit with a test light to 12 volts checks the sensor ground CKT 452. A faulty sensor ground will cause a Code 21.

Diagnostic Aids:

A "Scan" tool displays throttle position in volts. With ignition "ON" or at idle, TPS signal voltage should display between .36 volt and .96 volt with the throttle closed and increase at a steady rate as throttle is moved toward WOT.

The Tech 1 will display throttle angle. Closed throttle equals 0% and wide open throttle equals 100%.

An open in CKT 452 will result in a Code 21.
Refer to "Intermittents" in "Symptoms," Section

5.0L (VIN F) AND 5.7L (VIN 8) ENGINES — DIAGNOSTIC TROUBLE CODE CHART — 1992 CAMARO AND FIREBIRD

CODE 21
THROTTLE POSITION SENSOR (TPS) CIRCUIT
(SIGNAL VOLTAGE HIGH)
5.0L (VIN F) & 5.7L (VIN 8) "F" CARLINE (PORT)

1. • THROTTLE CLOSED.
   DOES TECH 1 DISPLAY THROTTLE POSITION OVER 2.5 VOLTS?

   - YES
   2. • DISCONNECT THROTTLE POSITION SENSOR. TECH 1 SHOULD DISPLAY THROTTLE POSITION BELOW .2 VOLT (200mV). DOES IT?

      - YES
      3. • PROBE SENSOR GROUND CIRCUIT WITH A TEST LIGHT CONNECTED TO BATTERY VOLTAGE.

         - LIGHT "ON"
           FAULTY CONNECTION OR THROTTLE POSITION SENSOR.

         - LIGHT "OFF"
           OPEN SENSOR GROUND CIRCUIT OR FAULTY ECM.

      - NO
        TPS SIGNAL CIRCUIT SHORTED TO VOLTAGE OR FAULTY ECM.

   - NO
     CODE 21 IS INTERMITTENT. IF NO ADDITIONAL CODES WERE STORED, REFER TO "DIAGNOSTIC AIDS" ON FACING PAGE.

"AFTER REPAIRS," REFER TO CODE CRITERIA ON FACING PAGE AND CONFIRM CODE DOES NOT RESET.

5.0L (VIN F) AND 5.7L (VIN 8) ENGINES — DIAGNOSTIC TROUBLE CODE CHART — 1992 CAMARO AND FIREBIRD

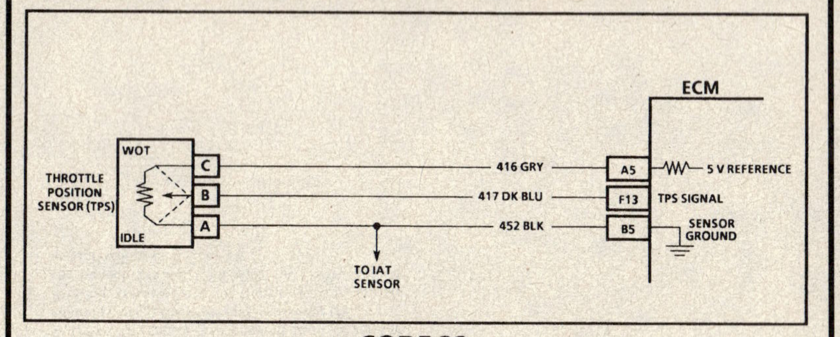

CODE 22
THROTTLE POSITION SENSOR (TPS) CIRCUIT
(SIGNAL VOLTAGE LOW)
5.0L (VIN F) & 5.7L (VIN 8) "F" CARLINE (PORT)

Circuit Description:

The Throttle Position Sensor (TPS) provides a voltage signal that changes relative to the throttle blade position. Signal voltage will vary from about .54 at idle to about 5 volts at Wide Open Throttle (WOT).

The TPS signal is one of the most important inputs used by the ECM for fuel control and for most of the ECM controlled outputs.

Test Description: Number(s) below refer to circled number(s) on the diagnostic chart.

1. Code 22 will set if:
   - Engine running.
   - TPS signal voltage is less than about .2 volt for 3 seconds.
   - The TPS has an auto zeroing feature. If the voltage reading is within the range of about .3 to 1.0 volts, the ECM will use that value as closed throttle. If the voltage reading is out of the auto zero range at closed throttle (about 1.5 volts or more), check for a binding throttle cable or damaged linkage, if OK, continue with diagnosis.
2. Simulates Code 21: (high voltage) If the ECM recognizes the high signal voltage the ECM and wiring are OK.
3. Simulates a high signal voltage to check for an open in CKT 417.

Diagnostic Aids:

A "Scan" tool displays throttle position in volts. With ignition "ON" or at idle, TPS signal voltage should display between .36 volt and .96 volt with the throttle closed and increase at a steady rate as throttle is moved toward WOT.

An open or short to ground in CKT 416 or CKT 417 will result in a Code 22.

Refer to "Intermittents" in "Symptoms," Section

5.0L (VIN F) AND 5.7L (VIN 8) ENGINES — DIAGNOSTIC TROUBLE CODE CHART — 1992 CAMARO AND FIREBIRD

CODE 22
THROTTLE POSITION SENSOR (TPS) CIRCUIT
(SIGNAL VOLTAGE LOW)
5.0L (VIN F) & 5.7L (VIN 8) "F" CARLINE (PORT)

1. • THROTTLE CLOSED.
   DOES "TECH 1" DISPLAY TPS .2 VOLT (200 mV) OR BELOW?

   YES →
   2. • DISCONNECT TPS SENSOR.
      • JUMPER CKTS 416 & 417 TOGETHER.
      "TECH 1" SHOULD DISPLAY TPS OVER 4.0 VOLTS (4000 mV).
      DOES IT?

   NO →
   • CODE 22 IS INTERMITTENT.
   IF NO ADDITIONAL CODES WERE STORED, REFER TO "DIAGNOSTIC AIDS" ON FACING PAGE.

   NO →
   3. • PROBE CKT 417 WITH A TEST LIGHT CONNECTED TO BATTERY VOLTAGE.
      "TECH 1" SHOULD DISPLAY TPS OVER 4.0 VOLTS (4000 mV).
      DOES IT?

   YES →
   FAULTY SENSOR CONNECTION
   OR
   FAULTY SENSOR.

   YES →
   CKT 416 OPEN OR SHORTED TO GROUND
   OR
   FAULTY CONNECTION
   OR
   FAULTY ECM.

   NO →
   CKT 417 OPEN OR SHORTED TO GROUND, OR SHORTED TO SENSOR GROUND CIRCUIT
   OR
   FAULTY ECM CONNECTION
   OR
   FAULTY ECM.

"AFTER REPAIRS," REFER TO CODE CRITERIA ON FACING PAGE AND CONFIRM CODE DOES NOT RESET.

5.0L (VIN F) AND 5.7L (VIN 8) ENGINES — DIAGNOSTIC TROUBLE CODE CHART — 1992 CAMARO AND FIREBIRD

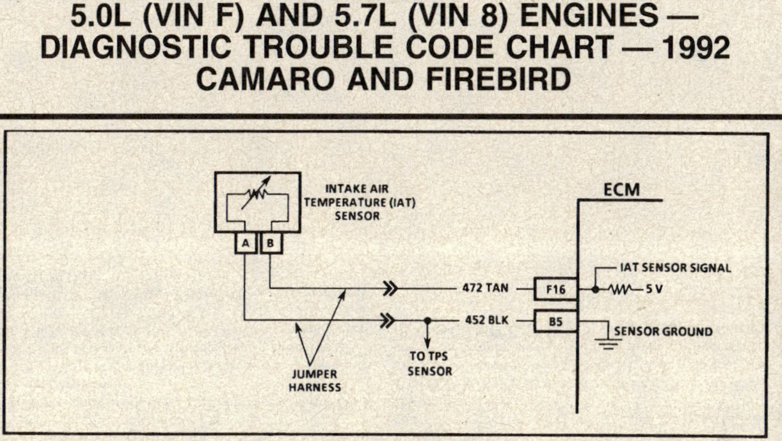

CODE 23
INTAKE AIR TEMPERATURE (IAT) SENSOR CIRCUIT
(LOW TEMPERATURE INDICATED)
5.0L (VIN F) & 5.7L (VIN 8) "F" CARLINE (PORT)

Circuit Description:

The Intake Air Temperature (IAT) sensor uses a thermistor to control the signal voltage to the ECM. The ECM applies a voltage (about 5 volts) on CKT 472 to the sensor. When the intake air is cold, the sensor (thermistor) resistance is high, therefore, the ECM will sense a high signal voltage. If the intake air is warm, the sensor (thermistor) resistance is low, therefore, the ECM will sense a low voltage.

Test Description: Number(s) below refer to circled number(s) on the diagnostic chart.
1. Code 23 will set if:
   • A signal voltage indicates an intake air temperature below -35°C (-31°F) for 12 seconds.
   • Time since engine start is 4 minutes or longer.
   • No VSS (vehicle not moving).
2. A Code 23 will set, due to an open sensor, wire or connection. This test will determine if the wiring and ECM are OK. The IAT sensor is difficult to reach and this test can be performed by disconnecting the IAT jumper harness connector. If the "Scan" indicates a temperature of over 130°C (266°F) the jumper harness to the sensor should be checked before replacing the sensor.
3. This will determine if the IAT sensor signal (CKT 472) or the IAT sensor ground (CKT 452) is open.

Diagnostic Aids:

A "Scan" tool displays temperature of the air entering the engine and should be close to ambient air temperature when engine is cold, and rises as underhood temperature increases. If the engine has been allowed to sit overnight, the intake air temperature and coolant temperature values should be within a few degrees of each other.

Carefully check harness and connections for possible open CKT 472 or CKT 452.

Refer to "Intermittents" in "Symptoms" Section

5.0L (VIN F) AND 5.7L (VIN 8) ENGINES — DIAGNOSTIC TROUBLE CODE CHART — 1992 CAMARO AND FIREBIRD

CODE 23
INTAKE AIR TEMPERATURE (IAT) SENSOR CIRCUIT
(LOW TEMPERATURE INDICATED)
5.0L (VIN F) & 5.7L (VIN 8) "F" CARLINE (PORT)

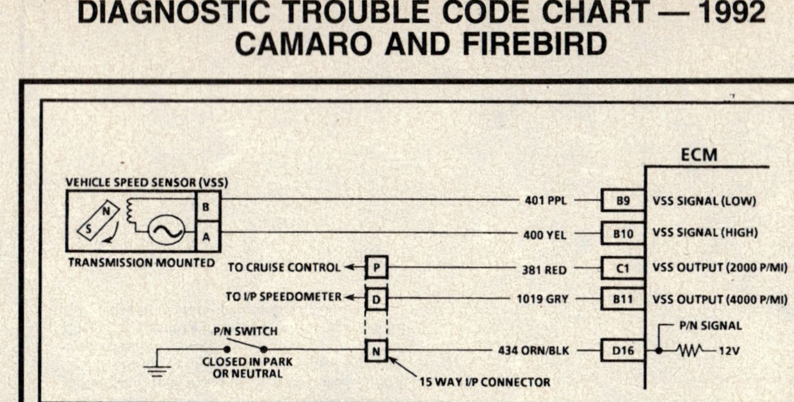

DIAGNOSTIC AID

IAT SENSOR		
TEMPERATURE VS. RESISTANCE VALUES (APPROXIMATE)		
°F	°C	OHMS
210	100	185
160	70	450
100	38	1,800
70	20	3,400
40	4	7,500
20	-7	13,500
0	-18	25,000
-40	-40	100,700

"AFTER REPAIRS," REFER TO CODE CRITERIA ON FACING PAGE AND CONFIRM CODE DOES NOT RESET.

5.0L (VIN F) AND 5.7L (VIN 8) ENGINES — DIAGNOSTIC TROUBLE CODE CHART — 1992 CAMARO AND FIREBIRD

CODE 24
VEHICLE SPEED SENSOR (VSS) CIRCUIT
5.0L (VIN F) & 5.7L (VIN 8) "F" CARLINE (PORT)

Circuit Description:

Vehicle speed information is provided to the ECM by the Vehicle Speed Sensor (VSS), which is a Permanent Magnet (PM) generator and it is mounted in the transmission. The PM generator produces a pulsing voltage whenever vehicle speed is over about 3 mph. The AC voltage level and the number of pulses increases with vehicle speed. The ECM then converts the pulsing voltage to mph which is used for calculations, and the mph can be displayed with a "Scan" tool.

The function of VSS buffer used in past model years has been incorporated into the ECM. The ECM then supplies the necessary signal for the instrument panel (4000 pulses per mile) for operating the speedometer and the odometer. If the vehicle is equipped with cruise control, the ECM also provides a signal (2000 pulses per mile) to the cruise control module.

Test Description: Number(s) below refer to circled number(s) on the diagnostic chart.

1. Code 24 will set if vehicle speed is less than 5 km/h (3 mph) when:
   - Coolant temperature is greater than 85°C (185°F).
   - Engine speed is between 900 and 3000 rpm.
   - TPS is less than 2%.
   - Low load condition (low air flow).
   - All conditions met for 2 seconds.

   These conditions are met during a road load deceleration.
   - The PM generator only produces a signal if drive wheels are turning greater than 3 mph.
2. CKTs 400, 401 and 1019 are OK if the speedometer works properly. Code 24 is being caused by a faulty ECM, faulty MEM-CAL or an incorrect MEM-CAL.

Diagnostic Aids:

The Tech 1 should indicate a vehicle speed whenever the drive wheels are turning greater than 5 km/h (3 mph).

A problem in CKT 1019 or CKT 381 will not affect the VSS input or the readings on a "Scan."

Check CKTs 400 and 401 for proper connections to be sure they're clean and tight and the harness is routed correctly. Refer to "Intermittents" in "Symptoms," Section

(A/T) A faulty or misadjusted Park/Neutral (P/N) switch can result in a false Code 24. Use a "Scan" and check for proper signal while in drive or overdrive. Refer to CHART C-1A for P/N switch diagnosis check.

5.0L (VIN F) AND 5.7L (VIN 8) ENGINES — DIAGNOSTIC TROUBLE CODE CHART — 1992 CAMARO AND FIREBIRD

CODE 24
VEHICLE SPEED SENSOR (VSS) CIRCUIT
5.0L (VIN F) & 5.7L (VIN 8) "F" CARLINE (PORT)

(1)
- RAISE DRIVE WHEELS
- WITH ENGINE IDLING IN GEAR, "SCAN" TECH 1 SHOULD DISPLAY VEHICLE SPEED ABOVE 0.
  DOES IT?

NO → DOES SPEEDOMETER WORK?

YES → CODE 24 IS INTERMITTENT. IF NO ADDITIONAL CODES WERE STORED, REFER TO "DIAGNOSTIC AIDS" ON FACING PAGE.

NO →
- IGNITION "OFF."
- DISCONNECT VSS HARNESS CONNECTOR AT TRANSMISSION.
- CONNECT SIGNAL GENERATOR TESTER J 33431-B OR EQUIVALENT TO VSS HARNESS CONNECTOR.
- IGNITION "ON," TOOL "ON" AND SET TO GENERATE A VSS SIGNAL.
- "SCAN" TOOL SHOULD DISPLAY VEHICLE SPEED ABOVE 0.
  DOES IT?

YES → (2) CHECK MEM-CAL FOR CORRECT APPLICATION. IF OK, REPLACE ECM.

NO → CKT 400 OR 401 OPEN, SHORTED TO GROUND, SHORTED TOGETHER, FAULTY CONNECTIONS, OR FAULTY ECM.

YES → REPLACE VEHICLE SPEED SENSOR.

"AFTER REPAIRS," CONFIRM "CLOSED LOOP" OPERATION AND NO "SERVICE ENGINE SOON" LIGHT.

5.0L (VIN F) AND 5.7L (VIN 8) ENGINES — DIAGNOSTIC TROUBLE CODE CHART — 1992 CAMARO AND FIREBIRD

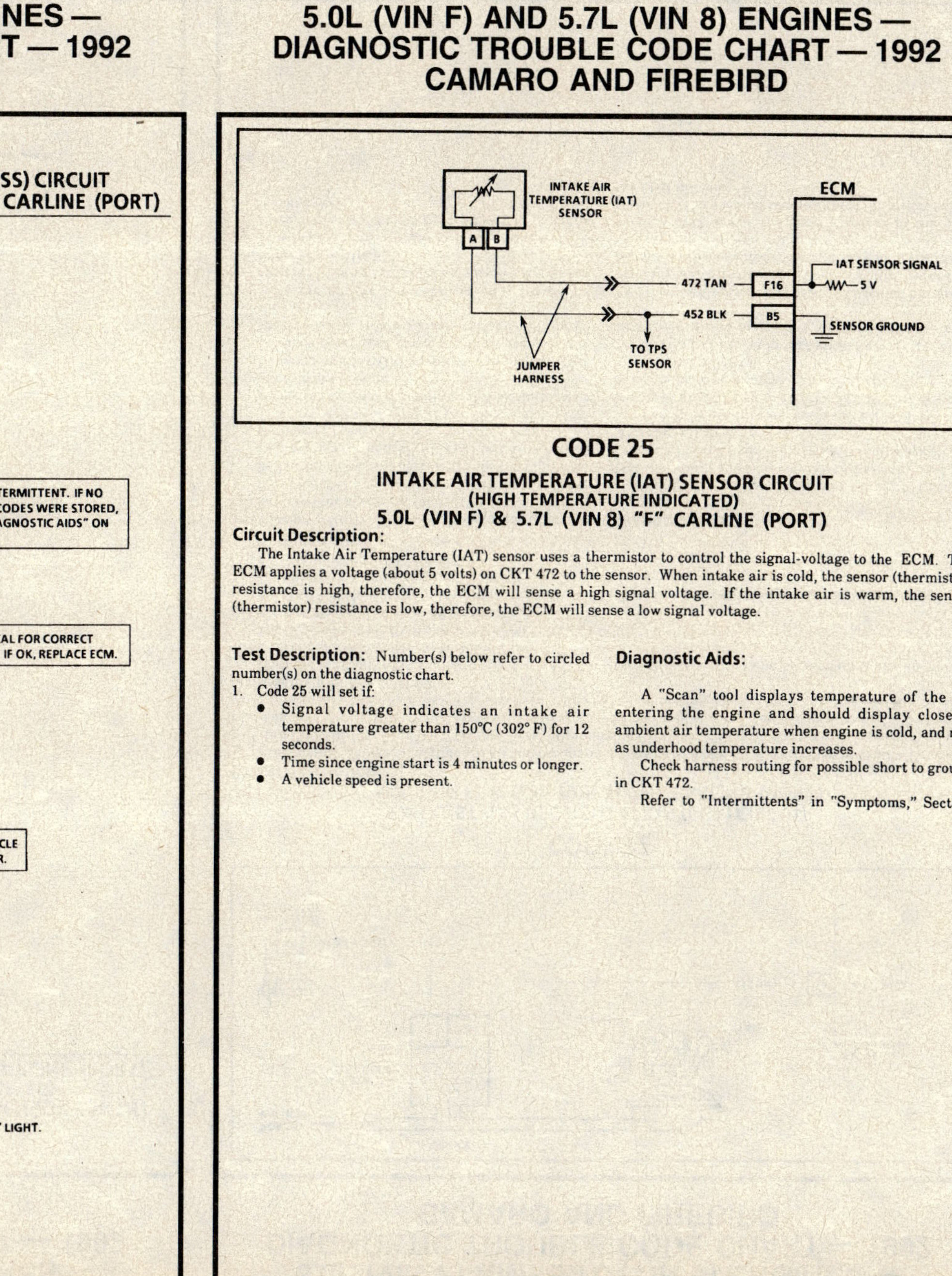

CODE 25
INTAKE AIR TEMPERATURE (IAT) SENSOR CIRCUIT
(HIGH TEMPERATURE INDICATED)
5.0L (VIN F) & 5.7L (VIN 8) "F" CARLINE (PORT)

Circuit Description:

The Intake Air Temperature (IAT) sensor uses a thermistor to control the signal-voltage to the ECM. The ECM applies a voltage (about 5 volts) on CKT 472 to the sensor. When intake air is cold, the sensor (thermistor) resistance is high, therefore, the ECM will sense a high signal voltage. If the intake air is warm, the sensor (thermistor) resistance is low, therefore, the ECM will sense a low signal voltage.

Test Description: Number(s) below refer to circled number(s) on the diagnostic chart.
1. Code 25 will set if:
- Signal voltage indicates an intake air temperature greater than 150°C (302° F) for 12 seconds.
- Time since engine start is 4 minutes or longer.
- A vehicle speed is present.

Diagnostic Aids:

A "Scan" tool displays temperature of the air entering the engine and should display close to ambient air temperature when engine is cold, and rise as underhood temperature increases.

Check harness routing for possible short to ground in CKT 472.

Refer to "Intermittents" in "Symptoms," Section

5.0L (VIN F) AND 5.7L (VIN 8) ENGINES — DIAGNOSTIC TROUBLE CODE CHART — 1992 CAMARO AND FIREBIRD

CODE 25

INTAKE AIR TEMPERATURE (IAT) SENSOR CIRCUIT
(HIGH TEMPERATURE INDICATED)
5.0L (VIN F) & 5.7L (VIN 8) "F" CARLINE (PORT)

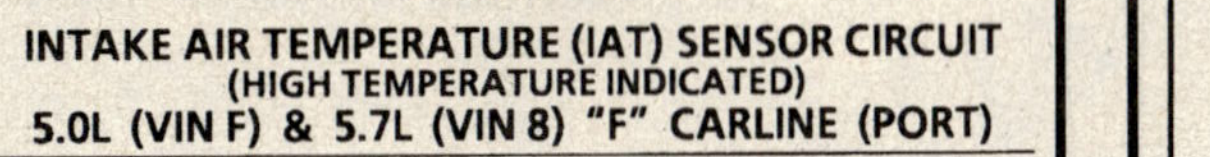

DIAGNOSTIC AID		
IAT SENSOR		
TEMPERATURE VS. RESISTANCE VALUES (APPROXIMATE)		
°F	°C	OHMS
210	100	185
160	70	450
100	38	1,800
70	20	3,400
40	4	7,500
20	-7	13,500
0	-18	25,000
-40	-40	100,700

"AFTER REPAIRS," REFER TO CODE CRITERIA ON FACING PAGE AND CONFIRM CODE DOES NOT RESET.

5.0L (VIN F) AND 5.7L (VIN 8) ENGINES — DIAGNOSTIC TROUBLE CODE CHART — 1992 CAMARO AND FIREBIRD

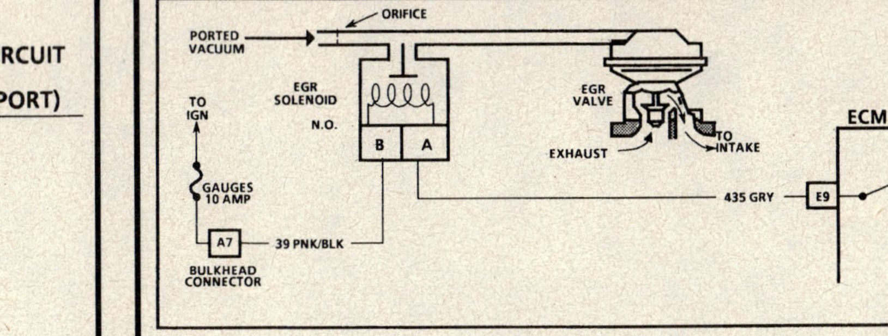

CODE 32

EXHAUST GAS RECIRCULATION (EGR) CIRCUIT
5.0L (VIN F) & 5.7L (VIN 8) "F" CARLINE (PORT)

Circuit Description:

The Exhaust Gas Recirculation (EGR) valve vacuum is controlled by an ECM controlled solenoid. The ECM will turn the EGR "ON" and "OFF" (Duty Cycle) by grounding CKT 435. The duty cycle is calculated by the ECM based on information from the coolant temperature, air flow and engine rpm. There should be no EGR when in park or neutral, TPS input below a specified value, TPS indicating Wide Open Throttle (WOT), or IAT temperature less than 15°C (59°F).

With the ignition "ON" engine stopped, the EGR solenoid is de-energized and by grounding the diagnostic "test" terminal, the solenoid is energized.

The ECM will check EGR operation when:
- Vehicle speed is above 48 km/h (30 mph).
- Manifold absolute pressure is between about: 30 - 85 at low altitudes (100 kPa BARO) or through 0 - 55 kPa at high altitudes (70 kPa BARO).
- TPS is between 5% and 30%.
- No more than .4% change in TPS.

Test Description: Number(s) below refer to circled number(s) on the diagnostic chart.

1. By grounding the diagnostic "test" terminal, the EGR solenoid should close and hold vacuum. The vacuum may bleed off slowly, but this is not considered a fault.
2. This test will determine if the electrical control part of the system is at fault, or if the connector or solenoid are at fault.
3. By plugging the EGR valve side and ungrounding the diagnostic "test" terminal, the solenoid valve should open and allow vacuum to bleed off through the vent.
4. With the engine not running and vacuum is applied to the valve, the valve should move to the fully open position.
5. This engine uses a negative backpressure EGR valve, the valve should close when the engine is cranked over.

Diagnostic Aids:

The ECM monitors integrator during a Code 32 test and must detect a change to pass the test. If integrator is at a fixed value (lean or rich) when a Code 32 is ran, a Code 32 may set.

The EGR circuit can be inoperative if the P/N switch is misadjusted or faulty. The EGR is disabled when in park or neutral. To check the P/N switch, refer to CHART C-1A.

Suction from shop exhaust hoses can alter backpressure and may affect the function of the EGR valve.

Thoroughly check that the EGR vacuum harness is not plugged or restricted. If the vacuum harness is pushed on the solenoid too far, it may plug the end of the vacuum line and cause a Code 32. Make sure vacuum lines are at the right locations of the EGR solenoid. The vacuum source goes to the orifice side of the EGR solenoid.

5.0L (VIN F) AND 5.7L (VIN 8) ENGINES — DIAGNOSTIC TROUBLE CODE CHART — 1992 CAMARO AND FIREBIRD

CODE 32

EXHAUST GAS RECIRCULATION (EGR) CIRCUIT
5.0L (VIN F) & 5.7L (VIN 8)
"F" CARLINE (PORT)

① • BEFORE USING THIS CHART, CHECK VACUUM HOSES FOR LEAKS, RESTRICTIONS, AND CHECK PORTED VACUUM SOURCE TO EGR SOLENOID, SHOULD HAVE AT LEAST 24 kPa (7" HG) AT 2000 RPM.
• IGNITION "ON," ENGINE STOPPED.
• GROUND DIAGNOSTIC TERMINAL.
• DISCONNECT EGR SOLENOID VACUUM HARNESS.
• APPLY 10" VACUUM TO MANIFOLD SIDE OF SOLENOID.
• SHOULD BE ABLE TO HOLD VACUUM.

UNABLE TO GET 33 kPa (10" HG) VACUUM

② • DISCONNECT EGR SOLENOID.
• CONNECT TEST LIGHT BETWEEN HARNESS CONNECTOR TERMS.

LIGHT "OFF"

PROBE EACH HARNESS CONNECTOR TERMINAL WITH A TEST LIGHT CONNECTED TO GROUND.

LIGHT "ON" ONE

OPEN CKT 435, FAULTY CONNECTION, OR FAULTY ECM.

LIGHT "OFF"

REPAIR OPEN IN IGNITION FEED CKT TO SOLENOID

LIGHT "ON" BOTH

REPAIR SHORT TO VOLTAGE IN CKT 435.

LIGHT "ON"

FAULTY SOLENOID CONNECTION OR SOLENOID.

ABLE TO GET 33 kPa (10" HG) VACUUM

③ • PLUG EGR SIDE OF SOLENOID.
• UNGROUND DIAGNOSTIC TERMINAL.
• NOTE VACUUM (SHOULD BLEED OFF). DOES IT?

YES

④ • IGNITION "OFF."
• CONNECT A VACUUM PUMP TO EGR VALVE.
• USING A MIRROR, OBSERVE EGR DIAPHRAGM WHILE APPLYING VACUUM.
• DIAPHRAGM SHOULD MOVE FREELY AND HOLD VACUUM FOR AT LEAST 20 SECONDS. DOES IT?

YES

⑤ • APPLY VACUUM TO EGR VALVE.
• START ENGINE AND IMMEDIATELY OBSERVE VACUUM GAGE ON VACUUM PUMP.
• VALVE IS GOOD IF DIAPHRAGM HAS MOVED TO SEATED POSITION (VALVE CLOSED) AND VACUUM DROPPED WHILE STARTING ENGINE.

VACUUM DROPPED

BE SURE VACUUM HOSE BETWEEN SOLENOID AND EGR VALVE IS OK. (NO LEAKS OR RESTRICTIONS) IF NO PROBLEM IS FOUND, THE EGR CIRCUIT IS OK.

NO VACUUM DROP

• REMOVE EGR VALVE.
• CHECK PASSAGES FOR BEING PLUGGED. IF NOT PLUGGED, REPLACE VALVE.

NO

• DISCONNECT SOLENOID ELECTRICAL CONNECTOR.
• NOTE VACUUM.

NO DROP

REPLACE SOLENOID

DROPS

REPAIR SHORT TO GROUND IN CKT 435. IF NOT SHORTED, IT IS A FAULTY ECM.

NO

REPLACE EGR VALVE.

"AFTER REPAIRS," CONFIRM "CLOSED LOOP" OPERATION AND NO "SERVICE ENGINE SOON" LIGHT.

5.0L (VIN F) AND 5.7L (VIN 8) ENGINES — DIAGNOSTIC TROUBLE CODE CHART — 1992 CAMARO AND FIREBIRD

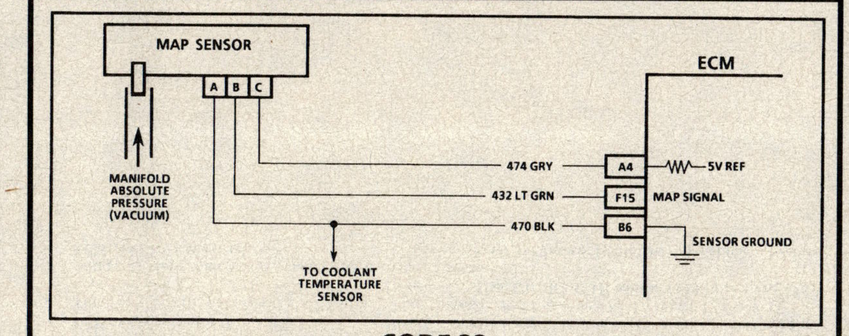

CODE 33

MANIFOLD ABSOLUTE PRESSURE (MAP) SENSOR CIRCUIT
(SIGNAL VOLTAGE HIGH - LOW VACUUM)
5.0L (VIN F) & 5.7L (VIN 8) "F" CARLINE (PORT)

Circuit Description:

The Manifold Absolute Pressure (MAP) sensor responds to changes in manifold pressure (vacuum). The ECM receives this information as a signal voltage that will vary from about 1-1.5 volts at idle to 4.0-4.8 volts at Wide Open Throttle (WOT).

The Tech 1 displays manifold pressure in kPa of pressure and voltage. Low pressure (high vacuum) reads a low voltage while a high pressure (low vacuum) reads a high voltage.

If the MAP sensor fails, the ECM will substitute a fixed MAP value and use the Throttle Position Sensor (TPS) to control fuel delivery.

Test Description: Number(s) below refer to circled number(s) on the diagnostic chart.

1. Code 33 will set when:
   • Signal is 85% of BARO for 1 second.
   • TPS less than 2%.
   Engine misfire or a low unstable idle may set Code 33. Disconnect MAP sensor and system will go into backup mode. If the misfire or idle condition remains, see "Symptoms," Section
2. If the ECM recognizes the low MAP signal, the ECM and wiring are OK.

Diagnostic Aids:

If the idle is rough or unstable, refer to "Symptoms," Section   for items which can cause an unstable idle.

An open in CKT 470 will result in a Code 33.

With the ignition "ON," and the engine "OFF," the manifold pressure is equal at atmospheric pressure and the signal voltage will be high. This information is used by the ECM as an indication of vehicle altitude and is referred to as BARO. Comparison of this BARO reading with a known good vehicle with the same sensor is a good way to check accuracy of a "suspect" sensor. Reading should be the same, ± .4 volt.

Also CHART C-1D can be used to test the MAP sensor.

Refer to "Intermittents" in "Symptoms," Section

5.0L (VIN F) AND 5.7L (VIN 8) ENGINES — DIAGNOSTIC TROUBLE CODE CHART — 1992 CAMARO AND FIREBIRD

CODE 33

MANIFOLD ABSOLUTE PRESSURE (MAP) SENSOR CIRCUIT
(SIGNAL VOLTAGE HIGH - LOW VACUUM)
5.0L (VIN F) & 5.7L (VIN 8) "F" CARLINE (PORT)

①
- IF ENGINE IDLE IS ROUGH, UNSTABLE, OR INCORRECT, CORRECT CONDITION BEFORE USING CHART. SEE "SYMPTOMS" IN SECTION "B".
- ENGINE IDLING.
- DOES TECH 1 DISPLAY A MAP VOLTAGE OF 4.0 VOLTS OR OVER?

YES → ②
- IGNITION "OFF."
- DISCONNECT MAP SENSOR ELECTRICAL CONNECTOR.
- IGNITION "ON."
- TECH 1 SHOULD READ A VOLTAGE OF 1 VOLT OR LESS. DOES IT?

NO → CODE 33 IS INTERMITTENT. IF NO ADDITIONAL CODES WERE STORED, REFER TO "DIAGNOSTIC AIDS"

YES →
- PROBE SENSOR GROUND CIRCUIT WITH A TEST LIGHT TO BATTERY VOLTAGE.
- TEST LIGHT SHOULD LIGHT. DOES IT?

NO → MAP SIGNAL CIRCUIT SHORTED TO VOLTAGE, SHORTED TO 5 VOLT REFERENCE CIRCUIT OR FAULTY ECM.

YES → PLUGGED OR LEAKING SENSOR VACUUM HOSE OR FAULTY MAP SENSOR.

NO → OPEN SENSOR GROUND CIRCUIT.

"AFTER REPAIRS," REFER TO CODE CRITERIA ON FACING PAGE AND CONFIRM CODE DOES NOT RESET.

5.0L (VIN F) AND 5.7L (VIN 8) ENGINES — DIAGNOSTIC TROUBLE CODE CHART — 1992 CAMARO AND FIREBIRD

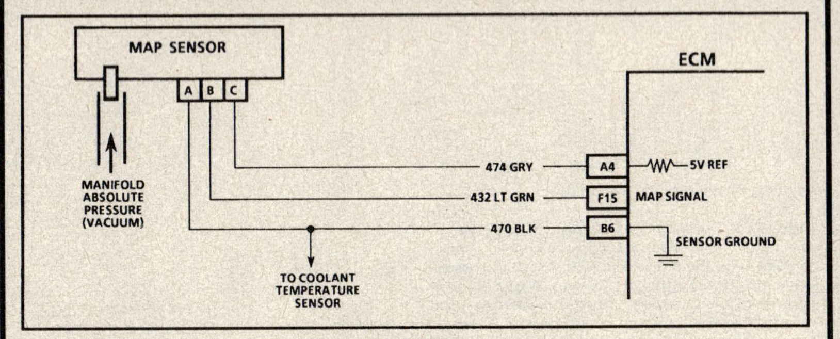

CODE 34

MANIFOLD ABSOLUTE PRESSURE (MAP) SENSOR CIRCUIT
(SIGNAL VOLTAGE LOW - HIGH VACUUM)
5.0L (VIN F) & 5.7L (VIN 8) "F" CARLINE (PORT)

Circuit Description:

The Manifold Absolute Pressure (MAP) sensor responds to changes in manifold pressure (vacuum). The ECM receives this information as a signal voltage that will vary from about 1-1.5 volts at idle to 4.0-4.8 volts at Wide Open Throttle (WOT).

The Tech 1 displays manifold pressure in kPa of pressure and voltage. Low pressure (high vacuum) reads a low voltage while a high pressure (low vacuum) reads a high voltage.

If the MAP sensor fails, the ECM will substitute a fixed MAP value and use the Throttle Position Sensor (TPS) to control fuel delivery.

Test Description: Number(s) below refer to circled number(s) on the diagnostic chart.

1. Code 34 will set when:
   - Engine running less than 1200 rpm.
   - Signal is 15% of BARO for 1 second.
     OR
   - Engine running greater than 1200 rpm.
   - Throttle position greater than 10%.
   - Signal is 15% of BARO for 1 second.
2. If the ECM recognizes the high MAP signal, the ECM and wiring are OK.
3. The Tech 1 will not display 12 volts. The important thing is that the ECM recognizes the voltage as more than 4 volts, indicating that the ECM and CKT 432 are OK.

Diagnostic Aids:

An intermittent open CKT 432 or CKT 416 will result in a Code 34.

With the ignition "ON" and engine "OFF," the manifold pressure is equal at atmospheric pressure and the signal voltage will be high. This information is used by the ECM as an indication of vehicle altitude and is referred to as BARO. Comparison of the BARO reading with a known good vehicle with the same sensor is a good way to check accuracy of a "suspect" sensor. Reading should be the same, ± .4 volt.

Also, CHART C-1D can be used to test the MAP sensor.

Refer to "Intermittents" in "Symptoms," Section

5.0L (VIN F) AND 5.7L (VIN 8) ENGINES — DIAGNOSTIC TROUBLE CODE CHART — 1992 CAMARO AND FIREBIRD

CODE 34
MANIFOLD ABSOLUTE PRESSURE (MAP) SENSOR CIRCUIT
(SIGNAL VOLTAGE LOW - HIGH VACUUM)
5.0L (VIN F) & 5.7L (VIN 8) "F" CARLINE (PORT)

(1)
- ENGINE IDLING.
- DOES TECH 1 DISPLAY MAP VOLTAGE BELOW .25 VOLT?

YES →

(2)
- IGNITION "OFF."
- DISCONNECT SENSOR ELECTRICAL CONNECTOR.
- JUMPER HARNESS TERMINALS "B" TO "C".
- IGNITION "ON."
- MAP VOLTAGE SHOULD READ OVER 4.7 VOLTS. DOES IT?

NO → CODE 34 IS INTERMITTENT. IF NO ADDITIONAL CODES WERE STORED, REFER TO "DIAGNOSTIC AIDS" ON FACING PAGE.

(3)
- IGNITION "OFF."
- REMOVE JUMPER WIRE.
- PROBE TERMINAL "B" (CKT 432) WITH A TEST LIGHT TO BATTERY VOLTAGE.
- IGNITION "ON."
- TECH 1 SHOULD READ OVER 4 VOLTS. DOES IT?

YES → FAULTY CONNECTION OR SENSOR.

NO →

5 VOLT REFERENCE CIRCUIT OPEN
OR
SHORTED TO GROUND
OR
FAULTY ECM.

CKT 432 OPEN
OR
CKT 432 SHORTED TO GROUND
OR
CKT 432 SHORTED TO SENSOR GROUND
OR
FAULTY ECM.

"AFTER REPAIRS," REFER TO CODE CRITERIA ON FACING PAGE AND CONFIRM CODE DOES NOT RESET.

5.0L (VIN F) AND 5.7L (VIN 8) ENGINES — DIAGNOSTIC TROUBLE CODE CHART — 1992 CAMARO AND FIREBIRD

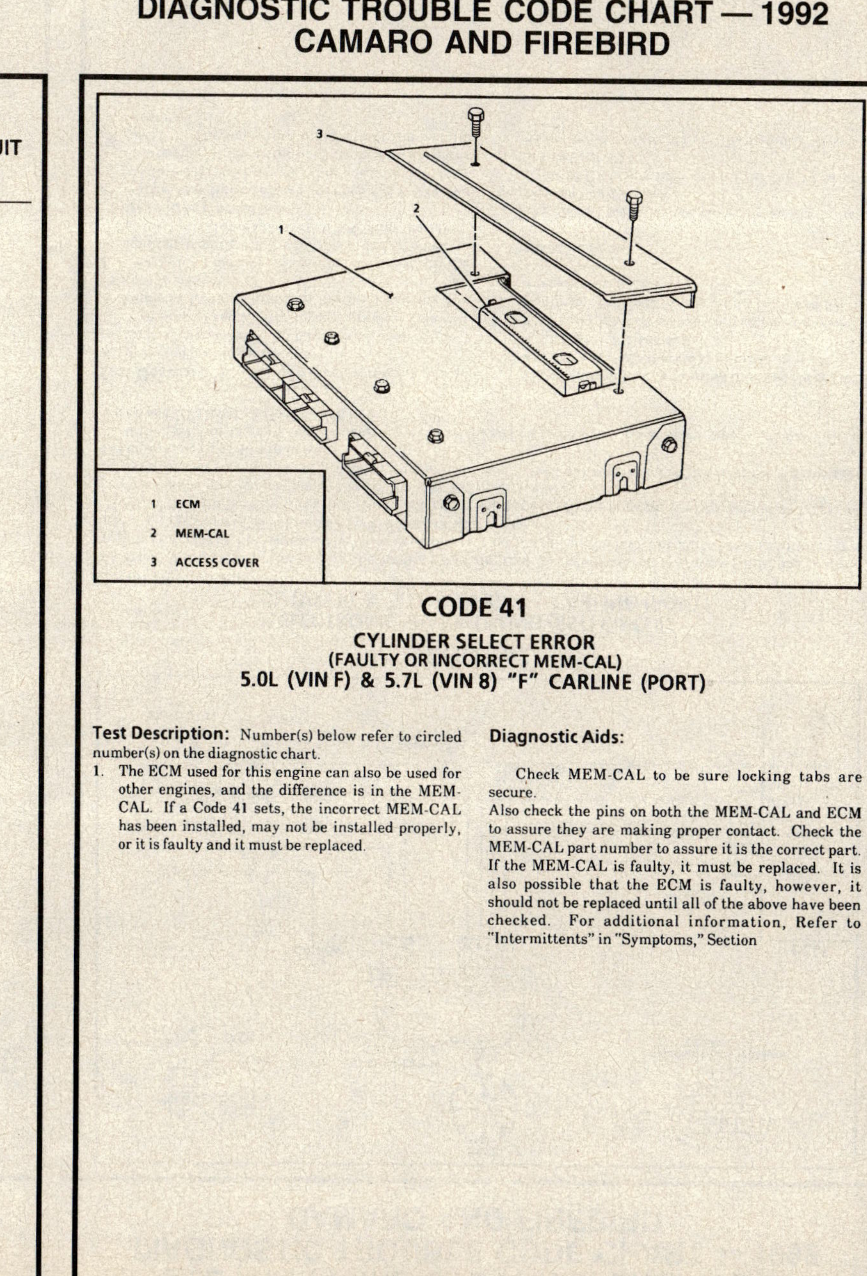

CODE 41
CYLINDER SELECT ERROR
(FAULTY OR INCORRECT MEM-CAL)
5.0L (VIN F) & 5.7L (VIN 8) "F" CARLINE (PORT)

Test Description: Number(s) below refer to circled number(s) on the diagnostic chart.

1. The ECM used for this engine can also be used for other engines, and the difference is in the MEM-CAL. If a Code 41 sets, the incorrect MEM-CAL has been installed, may not be installed properly, or it is faulty and it must be replaced.

Diagnostic Aids:

Check MEM-CAL to be sure locking tabs are secure.

Also check the pins on both the MEM-CAL and ECM to assure they are making proper contact. Check the MEM-CAL part number to assure it is the correct part. If the MEM-CAL is faulty, it must be replaced. It is also possible that the ECM is faulty, however, it should not be replaced until all of the above have been checked. For additional information, Refer to "Intermittents" in "Symptoms," Section

5.0L (VIN F) AND 5.7L (VIN 8) ENGINES — DIAGNOSTIC TROUBLE CODE CHART — 1992 CAMARO AND FIREBIRD

CODE 41

CYLINDER SELECT ERROR
(FAULTY OR INCORRECT MEM-CAL)
5.0L (VIN F) & 5.7L (VIN 8) "F" CARLINE (PORT)

① • IGNITION "OFF," CLEAR CODES.
 • START ENGINE AND RUN FOR 1 MINUTE OR UNTIL CODE 41 SETS. DOES CODE 41 SET?

YES → FAULTY CONNECTIONS DUE TO MEM-CAL NOT LOCKED IN PLACE, OR INCORRECT MEM-CAL INSTALLED.

NO → CODE 41 IS INTERMITTENT. REVIEW "DIAGNOSTIC AIDS" ON FACING PAGE.

"AFTER REPAIRS," REFER TO CODE CRITERIA ON FACING PAGE AND CONFIRM CODE DOES NOT RESET.

5.0L (VIN F) AND 5.7L (VIN 8) ENGINES — DIAGNOSTIC TROUBLE CODE CHART — 1992 CAMARO AND FIREBIRD

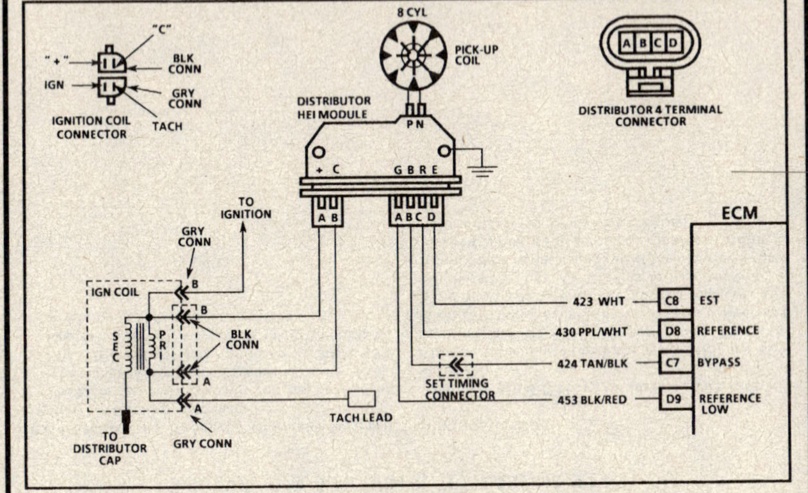

CODE 42

ELECTRONIC SPARK TIMING (EST) CIRCUIT
5.0L (VIN F) & 5.7L (VIN 8) "F" CARLINE (PORT)

Circuit Description:

When the system is running on the ignition module, that is, no voltage on the bypass line, the ignition module grounds the EST signal. The ECM expects to see no voltage on the EST line during this condition. If it sees a voltage, it sets Code 42 and will not go into the EST mode.

When the rpm for EST is reached (about 400 rpm), and bypass voltage applied, the EST should no longer be grounded in the ignition module, so the EST voltage should be varying.

If the bypass line is open or grounded, the ignition module will not switch to EST mode so the EST voltage will be low and Code 42 will be set.

If the EST line is grounded, the ignition module will switch to EST but, because the line is grounded there will be no EST signal. A Code 42 will be set.

Test Description: Number(s) below refer to circled number(s) on the diagnostic chart.

1. Code 42 means the ECM has seen an open or short to ground in the EST or bypass circuits. This test confirms Code 42 and that the fault causing the code is present.
2. Checks for a normal EST ground path through the ignition module. An EST CKT 423 shorted to ground will also read less than 500 ohms; however, this will be checked later.
3. As the test light voltage touches CKT 424 the module should switch. The ohmmeter may "overrange" if the meter is in the 1000-2000 ohms position. The important thing is that the module "switched."

4. The module did not switch and this step checks for:
 • EST CKT 423 shorted to ground.
 • Bypass CKT 424 open.
 • Faulty ignition module connection or module.
5. Confirms that Code 42 is a faulty ECM and not an intermittent in CKT 423 or CKT 424.

Diagnostic Aids:

The "Scan" tool does not have any ability to help diagnose a Code 42 problem.

A MEM-CAL not fully seated in the ECM can result in a Code 42.

Refer to "Intermittents" in "Symptoms," Section

5.0L (VIN F) AND 5.7L (VIN 8) ENGINES — DIAGNOSTIC TROUBLE CODE CHART — 1992 CAMARO AND FIREBIRD

CODE 42
ELECTRONIC SPARK TIMING (EST) CIRCUIT
5.0L (VIN F) & 5.7L (VIN 8)
"F" CARLINE (PORT)

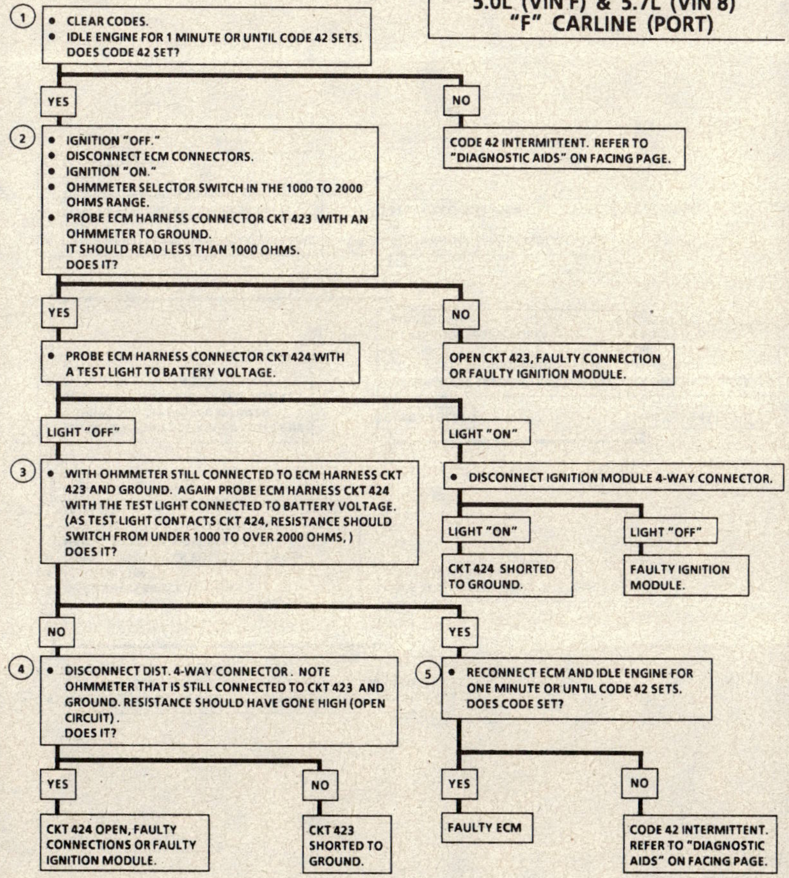

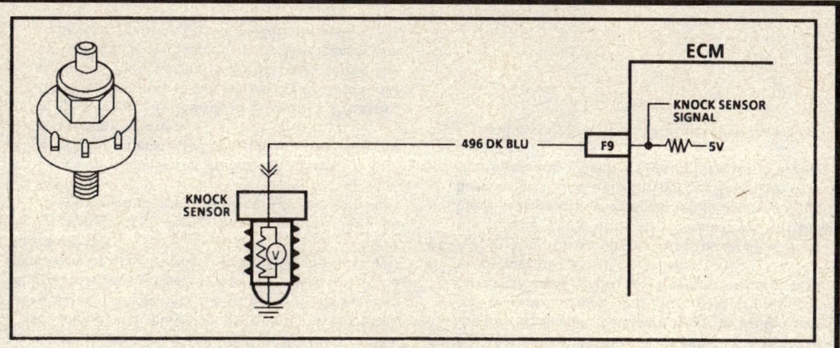

CLEAR CODES AND CONFIRM "CLOSED LOOP" OPERATION AND NO "SERVICE ENGINE SOON" LIGHT.

5.0L (VIN F) AND 5.7L (VIN 8) ENGINES — DIAGNOSTIC TROUBLE CODE CHART — 1992 CAMARO AND FIREBIRD

CODE 43
ELECTRONIC SPARK CONTROL (ESC) CIRCUIT
5.0L (VIN F) & 5.7L (VIN 8) "F" CARLINE (PORT)

Circuit Description:

The knock sensor is used to detect engine detonation and signal the ECM. The ECM will retard the electronic spark timing based on the signal being received. The circuitry within the knock sensor causes the ECM 5 volts to be pulled down so that under a no knock condition, CKT 496 would measure about 2.5 volts. The knock sensor produces an AC voltage. The AC voltage is then sent to the ECM which reduces the ECM signal voltage. The reduction of the ECM signal voltage triggers the ECM to retard timing. The amount of AC signal is determined by the amount of knock.

If CKT 496 becomes open or shorted to ground, the voltage will either go above 3.5 volts or below 1.5 volts. If either of these conditions are met for about 1/2 second, a Code 43 will be stored.

Test Description: Number(s) below refer to circled number(s) on the diagnostic chart.

1. This step determines if conditions for Code 43 still exist (voltage on CKT 496 above 3.5 volts or below 1.5 volts). The system is designed to retard the spark 15° if either condition exists.
2. The ECM has a 5 volt pullup resistor, which applies 5 volts to CKT 496. The 5 volt signal should be present at the knock sensor terminal during these test conditions.
3. This step determines if the knock sensor resistance is 3300 to 4500 ohms the sensor is OK.
4. Before replacing knock sensor, check threads in block for being clean. Do not apply thread sealer to sensor threads. Sensor is coated at factory and applying additional sealant will affect the sensors ability to detect detonation.

Diagnostic Aids:

If CKT 496 is not open or shorted to ground and the voltage reading is below 4 volts, the most likely cause is an open circuit in the ECM. It is possible that a faulty MEM-CAL could be drawing the 5 volt signal down, and it should be replaced, if a replacement ECM did not correct the problem.

Refer to "Intermittents" in "Symptoms," Section

5.0L (VIN F) AND 5.7L (VIN 8) ENGINES — DIAGNOSTIC TROUBLE CODE CHART — 1992 CAMARO AND FIREBIRD

5.0L (VIN F) AND 5.7L (VIN 8) ENGINES — DIAGNOSTIC TROUBLE CODE CHART — 1992 CAMARO AND FIREBIRD

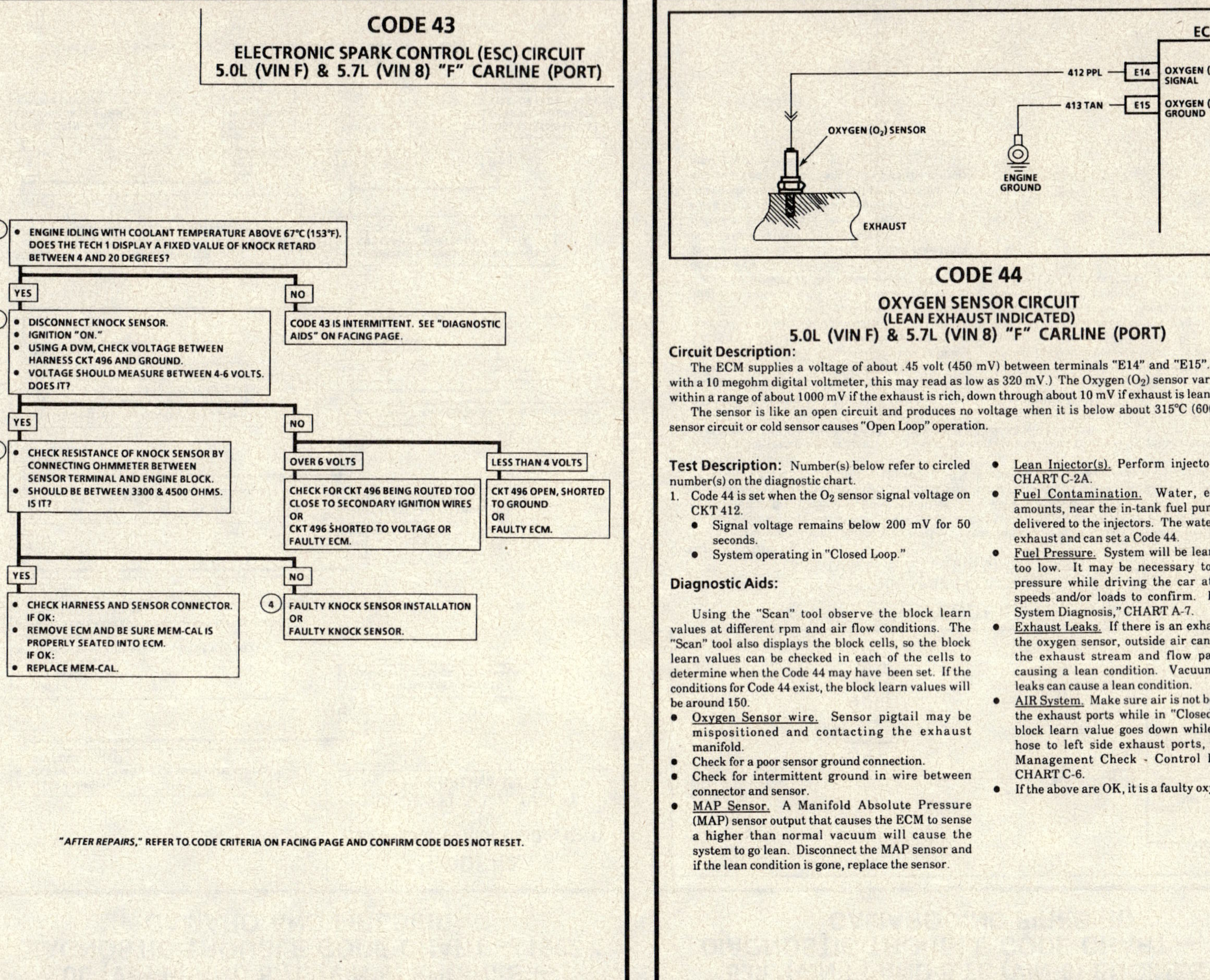

Circuit Description:

The ECM supplies a voltage of about .45 volt (450 mV) between terminals "E14" and "E15". (If measured with a 10 megohm digital voltmeter, this may read as low as 320 mV.) The Oxygen (O₂) sensor varies the voltage within a range of about 1000 mV if the exhaust is rich, down through about 10 mV if exhaust is lean.

The sensor is like an open circuit and produces no voltage when it is below about 315°C (600°F). An open sensor circuit or cold sensor causes "Open Loop" operation.

Test Description: Number(s) below refer to circled number(s) on the diagnostic chart.

1. Code 44 is set when the O₂ sensor signal voltage on CKT 412.
   - Signal voltage remains below 200 mV for 50 seconds.
   - System operating in "Closed Loop."

Diagnostic Aids:

Using the "Scan" tool observe the block learn values at different rpm and air flow conditions. The "Scan" tool also displays the block cells, so the block learn values can be checked in each of the cells to determine when the Code 44 may have been set. If the conditions for Code 44 exist, the block learn values will be around 150.

- <u>Oxygen Sensor wire.</u> Sensor pigtail may be mispositioned and contacting the exhaust manifold.
- Check for a poor sensor ground connection.
- Check for intermittent ground in wire between connector and sensor.
- <u>MAP Sensor.</u> A Manifold Absolute Pressure (MAP) sensor output that causes the ECM to sense a higher than normal vacuum will cause the system to go lean. Disconnect the MAP sensor and if the lean condition is gone, replace the sensor.

- <u>Lean Injector(s).</u> Perform injector balance test CHART C-2A.
- <u>Fuel Contamination.</u> Water, even in small amounts, near the in-tank fuel pump inlet can be delivered to the injectors. The water causes a lean exhaust and can set a Code 44.
- <u>Fuel Pressure.</u> System will be lean if pressure is too low. It may be necessary to monitor fuel pressure while driving the car at various road speeds and/or loads to confirm. Refer to "Fuel System Diagnosis," CHART A-7.
- <u>Exhaust Leaks.</u> If there is an exhaust leak above the oxygen sensor, outside air can be pulled into the exhaust stream and flow past the sensor causing a lean condition. Vacuum or crankcase leaks can cause a lean condition.
- <u>AIR System.</u> Make sure air is not being directed to the exhaust ports while in "Closed Loop." If the block learn value goes down while squeezing air hose to left side exhaust ports, Refer to "Air Management Check - Control Pedes Valve," CHART C-6.
- If the above are OK, it is a faulty oxygen sensor.

5.0L (VIN F) AND 5.7L (VIN 8) ENGINES — DIAGNOSTIC TROUBLE CODE CHART — 1992 CAMARO AND FIREBIRD

CODE 44

OXYGEN SENSOR CIRCUIT
(LEAN EXHAUST INDICATED)
5.0L (VIN F) & 5.7L (VIN 8) "F" CARLINE (PORT)

① • RUN WARM ENGINE (75°C/167°F TO 95°C/203°F) AT 1200 RPM.
• DOES TECH 1 INDICATE O₂ SENSOR VOLTAGE FIXED BELOW .35 VOLT (350 mV)?

YES
• DISCONNECT O₂ SENSOR.
• WITH ENGINE IDLING, TECH 1 SHOULD DISPLAY O₂ SENSOR VOLTAGE BETWEEN .35 VOLT AND .55 VOLT (350 mV AND 550 mV). DOES IT?

NO
CODE 44 IS INTERMITTENT. IF NO ADDITIONAL CODES WERE STORED, REFER TO "DIAGNOSTIC AIDS" ON FACING PAGE.

YES
REFER TO "DIAGNOSTIC AIDS" ON FACING PAGE.

NO
CKT 412 SHORTED TO GROUND OR FAULTY ECM.

"AFTER REPAIRS," REFER TO CODE CRITERIA ON FACING PAGE AND CONFIRM CODE DOES NOT RESET.

5.0L (VIN F) AND 5.7L (VIN 8) ENGINES — DIAGNOSTIC TROUBLE CODE CHART — 1992 CAMARO AND FIREBIRD

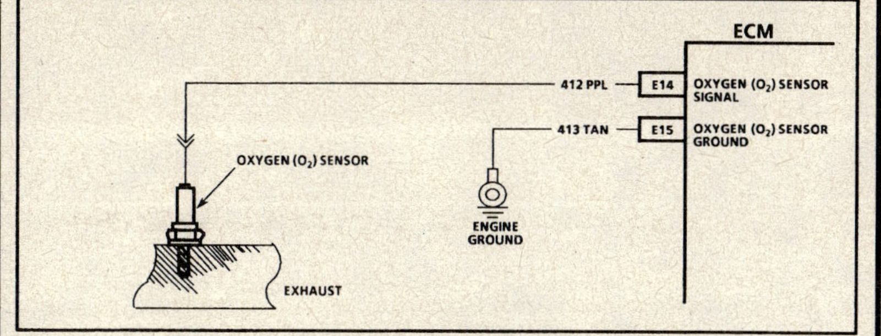

CODE 45

OXYGEN SENSOR CIRCUIT
(RICH EXHAUST INDICATED)
5.0L (VIN F) & 5.7L (VIN 8) "F" CARLINE (PORT)

Circuit Description:

The ECM supplies a voltage of about 450 mV between terminals "E14" and "E15". (If measured with a 10 megohm digital voltmeter, this may read as low as 320 mV). The O₂ sensor varies the voltage within a range of about 1000 mV if the exhaust is rich, down through about 10 mV if exhaust is lean.

The sensor is like an open circuit and produces no voltage when it is below about 315°C (600°F). An open sensor circuit or cold sensor causes "Open Loop" operation.

Test Description: Number(s) below refer to circled number(s) on the diagnostic chart.

1. Code 45 is set when the O₂ sensor signal voltage on CKT 412.
   • Signal voltage remains above 700 mV for 50 seconds.
   • Engine time after start is 1 minute or more.
   • Throttle angle greater than 2% (about .2 volt above idle voltage).
   • System operating in "Closed Loop."

Diagnostic Aids:

Using the "Scan" tool observe the block learn values at different rpm and air flow conditions. The "Scan" tool also displays the block cells, so the block learn values can be checked in each of the cells to determine when the Code 45 may have been set. If the conditions for Code 45 exists, the block learn values will be around 115.

• <u>Fuel Pressure.</u> System will go rich if pressure is too high. The ECM can compensate for some increase. However, if it gets too high, a Code 45 may be set.
Refer to "Fuel System Diagnosis," CHART A-7.
• <u>Rich injector.</u> Perform injector balance test CHART C-2A.
• <u>Leaking injector.</u> Refer to "Fuel System Diagnosis," CHART A-7.

• Check for fuel contaminated oil.
• <u>HEI Shielding.</u> An open ground CKT 453 (distributor ground, reference low) may result in Electromagnetic Interference (EMI), or induced electrical "noise" and may cause a rich condition.
• <u>Canister purge.</u> Check for fuel saturation. If full of fuel, check canister control and hoses. Refer to "Evaporative Emission Control System,"
• <u>MAP sensor.</u> An output that causes the ECM to sense a lower than normal vacuum can cause the system to go rich. Disconnecting the MAP sensor will allow the ECM to set a fixed value for the sensor. Substitute a different MAP sensor if the rich condition is gone while the sensor is disconnected.
• Check for leaking fuel pressure regulator diaphragm by checking vacuum line to regulator for fuel.
• <u>TPS.</u> An intermittent TPS output will cause the system to go rich, due to a false indication of the engine accelerating.
• <u>EGR.</u> An EGR staying open (especially at idle) will cause the O₂ sensor to indicate a rich exhaust, and this could result in a Code 45.

5.0L (VIN F) AND 5.7L (VIN 8) ENGS — DIAGNOSTIC TROUBLE CODE CHART — 1992 CAMARO AND FIREBIRD

CODE 45
OXYGEN SENSOR CIRCUIT
(RICH EXHAUST INDICATED)
5.0L (VIN F) & 5.7L (VIN 8) "F" CARLINE (PORT)

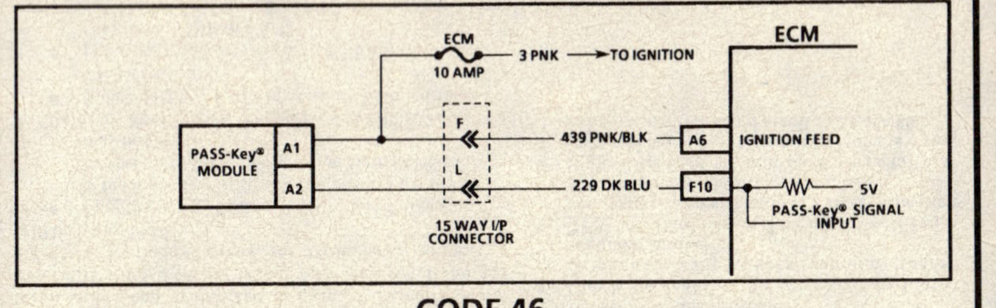

"AFTER REPAIRS," REFER TO CODE CRITERIA ON FACING PAGE AND CONFIRM CODE DOES NOT RESET.

5.0L (VIN F) AND 5.7L (VIN 8) ENGINES — DIAGNOSTIC TROUBLE CODE CHART — 1992 CAMARO AND FIREBIRD

CODE 46
PERSONAL AUTOMOTIVE SECURITY SYSTEM (PASS-Key®) CIRCUIT
5.0L (VIN F) & 5.7L (VIN 8) "F" CARLINE (PORT)

Circuit Description:

The PASS-Key® system is designed to disable vehicle operation if the incorrect key or starting procedure is used. The PASS-Key® decoder module sends a signal to the ECM if the correct key is being used. If the proper signal does not reach the ECM on CKT 229, the ECM will not pulse the injectors "ON" and thus not allow the vehicle to be started.

Code 46 will set, if the proper signal is not being received on CKT 229 by the ECM when the ignition is turned "ON." A PASS-Key® fault will store a Code 46 in the ECM memory.

Test Description: Number(s) below refer to circled number(s) on the diagnostic chart.

1. If the engine cranks, but doesn't start. It indicates that the portion of the module which generates the signal to the ECM is not operating or CKT 229 is open or shorted to ground.

2. If Code 46 is stored, and the engine will not crank, it indicates that there is a PASS-Key® problem or an incorrect key or starting procedure is being used.

5.0L (VIN F) AND 5.7L (VIN 8) ENGINES — DIAGNOSTIC TROUBLE CODE CHART — 1992 CAMARO AND FIREBIRD

CODE 46
**PERSONAL AUTOMOTIVE SECURITY SYSTEM (PASS-Key®) CIRCUIT
5.0L (VIN F) & 5.7L (VIN 8) "F" CARLINE (PORT)**

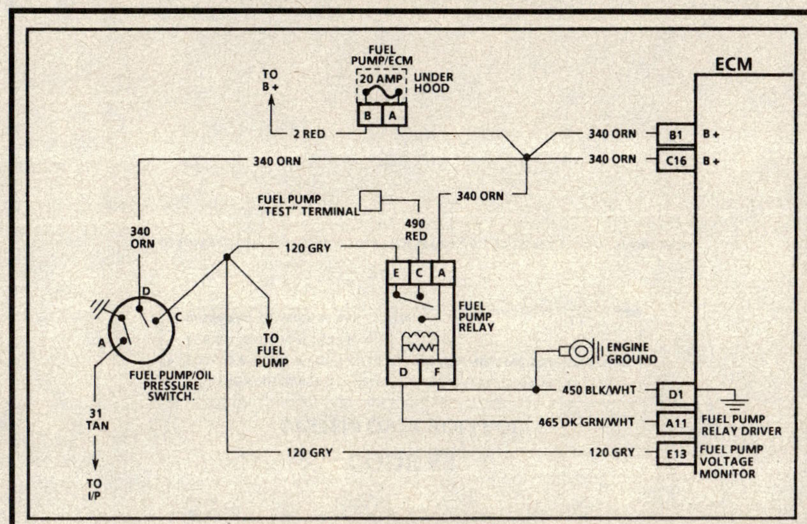

"AFTER REPAIRS," REFER TO CODE CRITERIA ON FACING PAGE AND CONFIRM CODE DOES NOT RESET.

5.0L (VIN F) AND 5.7L (VIN 8) ENGINES — DIAGNOSTIC TROUBLE CODE CHART — 1992 CAMARO AND FIREBIRD

CODE 54
**FUEL PUMP CIRCUIT
(LOW VOLTAGE)
5.0L (VIN F) & 5.7L (VIN 8) "F" CARLINE (PORT)**

Circuit Description:

The status of the fuel pump signal (CKT 120) is monitored for voltage by the ECM, and is used to compensate fuel delivery based on system voltage. This fuel pump signal is also used to store a trouble code if the fuel pump relay is defective or fuel pump voltage is lost while the engine is running. There should be about 12 volts on CKT 120 for 2 seconds after the ignition is turned "ON," or any time reference pulses are being received by the ECM.

Code 54 will set, if the voltage on CKT 120 is less than 2 volts for 1.5 seconds since the last reference pulse was received. This code is designed to detect a faulty relay, causing extended crank time, and the code will help the diagnosis of an engine that "CRANKS BUT WILL NOT RUN."

If a fault is detected during start-up, the "Service Engine Soon" light will stay "ON" until the ignition is cycled "OFF." However, if the voltage is detected below 2 volts, with the engine running, the light will only remain "ON" while the condition exists. The fuel pump test connector is located under the hood. Refer to "Component Location," Figure A-1.

5.0L (VIN F) AND 5.7L (VIN 8) ENGINES — DIAGNOSTIC TROUBLE CODE CHART — 1992 CAMARO AND FIREBIRD

CODE 54

FUEL PUMP CIRCUIT
(LOW VOLTAGE)
5.0L (VIN F) & 5.7L (VIN 8)
"F" CARLINE (PORT)

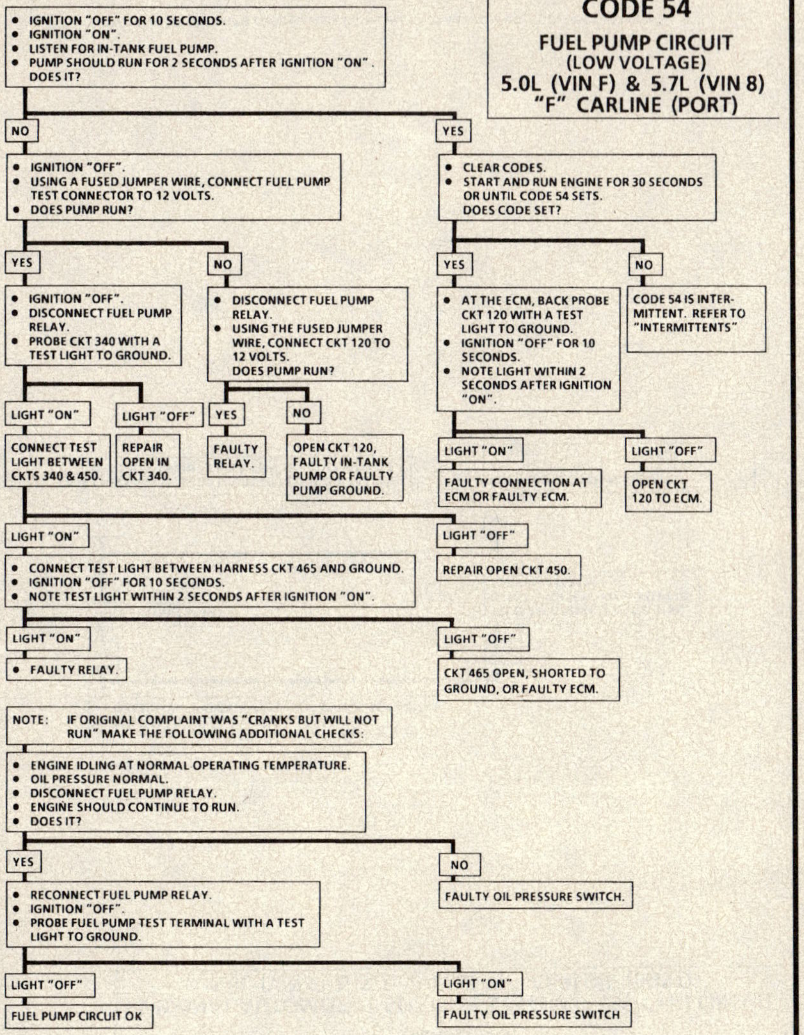

- IGNITION "OFF" FOR 10 SECONDS.
- IGNITION "ON".
- LISTEN FOR IN-TANK FUEL PUMP.
- PUMP SHOULD RUN FOR 2 SECONDS AFTER IGNITION "ON". DOES IT?

NO
- IGNITION "OFF".
- USING A FUSED JUMPER WIRE, CONNECT FUEL PUMP TEST CONNECTOR TO 12 VOLTS.
- DOES PUMP RUN?

YES
- IGNITION "OFF".
- DISCONNECT FUEL PUMP RELAY.
- PROBE CKT 340 WITH A TEST LIGHT TO GROUND.

NO
- DISCONNECT FUEL PUMP RELAY.
- USING THE FUSED JUMPER WIRE, CONNECT CKT 120 TO 12 VOLTS. DOES PUMP RUN?

LIGHT "ON" — CONNECT TEST LIGHT BETWEEN CKTS 340 & 450.

LIGHT "OFF" — REPAIR OPEN IN CKT 340.

YES — FAULTY RELAY.

NO — OPEN CKT 120, FAULTY IN-TANK PUMP OR FAULTY PUMP GROUND.

YES
- CLEAR CODES.
- START AND RUN ENGINE FOR 30 SECONDS OR UNTIL CODE 54 SETS. DOES CODE SET?

YES
- AT THE ECM, BACK PROBE CKT 120 WITH A TEST LIGHT TO GROUND.
- IGNITION "OFF" FOR 10 SECONDS.
- NOTE LIGHT WITHIN 2 SECONDS AFTER IGNITION "ON".

NO — CODE 54 IS INTER-MITTENT. REFER TO "INTERMITTENTS"

LIGHT "ON" — FAULTY CONNECTION AT ECM OR FAULTY ECM.

LIGHT "OFF" — OPEN CKT 120 TO ECM.

LIGHT "ON"
- CONNECT TEST LIGHT BETWEEN HARNESS CKT 465 AND GROUND.
- IGNITION "OFF" FOR 10 SECONDS.
- NOTE TEST LIGHT WITHIN 2 SECONDS AFTER IGNITION "ON".

LIGHT "OFF" — REPAIR OPEN CKT 450.

LIGHT "ON" — FAULTY RELAY.

LIGHT "OFF" — CKT 465 OPEN, SHORTED TO GROUND, OR FAULTY ECM.

NOTE: IF ORIGINAL COMPLAINT WAS "CRANKS BUT WILL NOT RUN" MAKE THE FOLLOWING ADDITIONAL CHECKS:

- ENGINE IDLING AT NORMAL OPERATING TEMPERATURE.
- OIL PRESSURE NORMAL.
- DISCONNECT FUEL PUMP RELAY.
- ENGINE SHOULD CONTINUE TO RUN.
- DOES IT?

YES
- RECONNECT FUEL PUMP RELAY.
- IGNITION "OFF".
- PROBE FUEL PUMP TEST TERMINAL WITH A TEST LIGHT TO GROUND.

NO — FAULTY OIL PRESSURE SWITCH.

LIGHT "OFF" — FUEL PUMP CIRCUIT OK

LIGHT "ON" — FAULTY OIL PRESSURE SWITCH

5.0L (VIN F) AND 5.7L (VIN 8) ENGINES — DIAGNOSTIC TROUBLE CODE CHART — 1992 CAMARO AND FIREBIRD

CODE 51
CODE 53

5.0L (VIN F) & 5.7L (VIN 8) "F" CARLINE (PORT)

CODE 51

MEM-CAL ERROR
(FAULTY OR INCORRECT MEM-CAL)

CHECK THAT ALL PINS ARE FULLY INSERTED IN THE SOCKET AND THAT MEM-CAL IS PROPERLY LATCHED. IF OK, REPLACE MEM/CAL, CLEAR MEMORY, AND RECHECK. IF CODE 51 REAPPEARS, REPLACE ECM.

"AFTER REPAIRS," CONFIRM "CLOSED LOOP" OPERATION AND NO "SERVICE ENGINE SOON" LIGHT.

CODE 53

SYSTEM OVER VOLTAGE

- THIS CODE INDICATES THERE IS A BASIC GENERATOR PROBLEM.
- CODE 53 WILL SET, IF VOLTAGE AT ECM IGNITION INPUT PIN IS GREATER THAN 16.9 VOLTS FOR 10 SECONDS.
- CHECK AND REPAIR CHARGING SYSTEM.

"AFTER REPAIRS," CONFIRM "CLOSED LOOP" OPERATION AND NO "SERVICE ENGINE SOON" LIGHT.

5.0L (VIN F) AND 5.7L (VIN 8) ENGINES — ECM SYMPTOM CHART "A" — 1992 CAMARO AND FIREBIRD

ECM PIN CONNECTOR "A" SYMPTOM CHART
BLACK 24 PIN CONNECTOR

PIN FUNCTION	PIN	WIRE COLOR	COMPONENT CONNECTOR CAVITY	CODES AFFECT	POSSIBLE SYMPTOMS FROM FAULTY CIRCUIT
A1					
A2					
A3					
A4 MAP SENSOR 5 VOLT REFERENCE	474	GRY	MAP SENSOR "C"	34 (3)	LACK OF POWER, ROUGH IDLE, SURGE, HESITATION, POOR PERFORMANCE.
A5 TPS 5 VOLT REFERENCE	416	GRY	TPS "C"	22 (3)	HIGH IDLE, SURGE, LACK OF PERFORMANCE.
A6 IGNITION FEED	439	PNK/BLK	ECM FUSE AND PASS-Key® MODULE "A1"		ENGINE CRANKS BUT WILL NOT START, NO SES LIGHT.
A7					
A8 SERIAL DATA	461	ORN	ALDL "M"		NO SERIAL DATA, "SCAN" TOOL INOPERATIVE.
A9					
A10					
A11 FUEL PUMP RELAY DRIVE	465	DK GRN/WHT	FUEL PUMP RELAY "D"		LONG CRANKING TIME, HARD TO START.
A12 ECM GROUND	450	BLK/WHT	ENGINE BLOCK		CRANKS, WILL NOT START. IF BOTH SYSTEM GROUNDS OPEN, CHECK GROUNDS AT ENGINE BLOCK.

(1) Open circuit.
(2) Grounded circuit.
(3) Open/grounded circuit.

5.0L (VIN F) AND 5.7L (VIN 8) ENGINES — ECM SYMPTOM CHART "B" — 1992 CAMARO AND FIREBIRD

ECM PIN CONNECTOR "B" SYMPTOM CHART
BLACK 24 PIN CONNECTOR

PIN FUNCTION	CKT #	WIRE COLOR	COMPONENT CONNECTOR CAVITY	CODES AFFECT	POSSIBLE SYMPTOMS FROM FAULTY CIRCUIT
B1 BATTERY FEED	340	ORN	FUEL PUMP RELAY "A" OPS "D	54	NO SES LIGHT, ENGINE CRANKS BUT WILL NOT START
B2					
B3					
B4					
B5 TPS AND IAT SENSOR GROUND	452	BLK	IAT "A" TPS "A"	21 (1) 23 (1)	HIGH IDLE
B6 CTS AND MAP SENSOR GROUND	470	BLK	CTS "A" MAP SENSOR "A"	15 (1) 33 (1)	ROUGH IDLE, LACK OF PERFORMANCE, EXHAUST ODOR
B7					
B8					
B9 VEHICLE SPEED SENSOR (VSS) SIGNAL LOW	401	PPL	VEHICLE SPEED SENSOR (VSS) "B"	24	NO VSS SIGNAL, INOPERATIVE SPEEDOMETER, INOPERATIVE CRUISE CONTROL
B10 VEHICLE SPEED SENSOR (VSS) SIGNAL HIGH	400	YEL	VEHICLE SPEED SENSOR (VSS) "A"	24	NO VSS SIGNAL, INOPERATIVE SPEEDOMETER, INOPERATIVE CRUISE CONTROL
B11 VSS OUTPUT 4000 P/MI	1019	GRY	I/P CLUSTER "C8"		INOPERATIVE SPEEDOMETER
B12					

(1) Open circuit
(2) Grounded circuit
(3) Open/grounded circuit

5.0L (VIN F) AND 5.7L (VIN 8) ENGINES — ECM SYMPTOM CHART "C" — 1992 CAMARO AND FIREBIRD

ECM PIN CONNECTOR "C" SYMPTOM CHART
BLACK 32 PIN CONNECTOR

PIN FUNCTION	CKT #	WIRE COLOR	COMPONENT CONNECTOR CAVITY	CODES AFFECT	POSSIBLE SYMPTOMS FROM FAULTY CIRCUIT
C1 2000 PULSE PER MILE SIGNAL	381	RED	I/P CONNECTOR "P"		
C2					
C3					
C4					
C5					
C6					
C7 BYPASS	424	TAN/BLK	DISTRIBUTOR 4 PIN CONNECTOR "B"	42 (3)	LACK OF POWER (SET TIMING CONNECTOR)
C8 ELECTRONIC SPARK TIMING (EST)	423	WHT	DISTRIBUTOR 4 PIN CONNECTOR "D"	42 (3)	LACK OF POWER, DETONATION, HARD START
C9 A/C REQUEST SIGNAL	59	DK GRN	A/C COMPRESSOR CLUTCH "A"		INOPERATIVE A/C, INCORRECT IDLE
C10					
C11 INJECTOR DRIVER 1, 3, 5, 7	467	DK BLU	INJECTOR CONNECTOR "B"		ROUGH IDLE, ENGINE CRANKS BUT WILL NOT START, LACK OF PERFORMANCE
C12 INJECTOR DRIVER 2, 4, 6, 8	468	DK GRN	INJECTOR CONNECTOR "B"		ROUGH IDLE, ENGINE CRANKS BUT WILL NOT START, LACK OF PERFORMANCE
C13					
C14					
C15					
C16 BATTERY FEED	340	ORN	FUEL PUMP RELAY "A" OPS "D"	54 (3)	NO SES LIGHT, ENGINE CRANKS BUT WILL NOT START

(1) Open circuit
(2) Grounded circuit
(3) Open/grounded circuit

5.0L (VIN F) AND 5.7L (VIN 8) ENGINES — ECM SYMPTOM CHART "D" — 1992 CAMARO AND FIREBIRD

ECM PIN CONNECTOR "D" SYMPTOM CHART
BLACK 32 PIN CONNECTOR

PIN FUNCTION	CKT #	WIRE COLOR	COMPONENT CONNECTOR CAVITY	CODES AFFECT	POSSIBLE SYMPTOMS FROM FAULTY CIRCUIT
D1 ECM GROUND	450	BLK/WHT	ENGINE BLOCK		ENGINE CRANKS BUT WILL NOT START, NO SES LIGHT
D2					
D3					
D4					
D5					
D6 INJECTOR GROUND	450	BLK/WHT	ENGINE BLOCK		ENGINE CRANKS BUT WILL NOT START
D7 INJECTOR GROUND	450	BLK/WHT	ENGINE BLOCK		ENGINE CRANKS BUT WILL NOT START
D8 REFERENCE	430	PPL/WHT	DISTRIBUTOR 4 PIN TERMINAL "C"		ENGINE CRANKS BUT WILL NOT START
D9 DISTRIBUTOR REFERENCE LOW	453	BLK/RED	DISTRIBUTOR 4 PIN TERMINAL "A"		LACK OF PERFORMANCE
D10					
D11					
D12 SECONDARY COOLING FAN STATUS	731	GRY	SECONDARY FAN (FAN 2) RELAY "F"		SECONDARY COOLING FAN INOPERATIVE, SECONDARY COOLING FAN "ON" AT ALL TIMES
D13					
D14 4th GEAR SWITCH SIGNAL	446	LT BLU	TCC CONNECTOR "B"		POOR FUEL ECONOMY
D15					
D16 PARK/NEUTRAL (P/N) SWITCH SIGNAL	434	ORN/BLK	PARK/NEUTRAL (P/N) SWITCH "B"	24	POOR FUEL ECONOMY, STALL, INCORRECT IDLE

(1) Open circuit
(2) Grounded circuit
(3) Open/grounded circuit

5.0L (VIN F) AND 5.7L (VIN 8) ENGINES — ECM SYMPTOM CHART "E" — 1992 CAMARO AND FIREBIRD

ECM PIN CONNECTOR "E" SYMPTOM CHART
GREEN 32 PIN CONNECTOR

	PIN FUNCTION	CKT #	WIRE COLOR	COMPONENT CONNECTOR CAVITY	CODES AFFECT	POSSIBLE SYMPTOMS FROM FAULTY CIRCUIT
E1						
E2						
E3	IAC COIL "A" HIGH	441	LT BLU/WHT	IAC VALVE "A"		INCORRECT IDLE, SURGE
E4	IAC COIL "A" LOW	442	LT BLU/BLK	IAC VALVE "B"		INCORRECT IDLE, SURGE
E5	IAC COIL "B" HIGH	443	LT GRN/WHT	IAC VALVE "C"		INCORRECT IDLE, SURGE
E6	IAC COIL "B" LOW	444	LT GRN/BLK	IAC VALVE "D"		INCORRECT IDLE, SURGE
E7	"SERVICE ENGINE SOON" LIGHT CONTROL	419	BRN/WHT	I/P PRINTED CIRCUIT CONNECTOR "C2"		NO SES LIGHT. SES LIGHT "ON" AT ALL TIMES.
E8	PRIMARY COOLING FAN CONTROL	335	DK GRN/WHT	PRIMARY FAN (FAN 1) RELAY "F"		INOPERATIVE FAN 1, FAN 1 RUNS ALL THE TIME
E9	EGR SOLENOID CONTROL	435	GRY	EGR SOLENOID "A"	32 (3)	DETONATION, STALL, ROUGH IDLE, POOR PERFORMANCE
E10						
E11						
E12	DIAGNOSTIC "TEST" TERMINAL	451	WHT/BLK	ALDL CONNECTOR "B"		NO "SCAN" TOOL DATA, WILL NOT FLASH CODE 12
E13	FUEL PUMP SIGNAL	120	GRY	FUEL PUMP RELAY "E" OPS "C"	54 (3)	ENGINE CRANKS BUT WILL NOT START
E14	OXYGEN (O₂) SENSOR SIGNAL	412	PPL	OXYGEN (O₂) SENSOR	13 (1) 44 (2)	EXHAUST ODOR, POOR PERFORMANCE
E15	OXYGEN (O₂) SENSOR GROUND	413	TAN	ENGINE GROUND	13 (1)	ROUGH IDLE, INCORRECT IDLE, LEAN EXHAUST, POOR PERFORMANCE
E16	CTS SIGNAL	410	YEL	CTS "B"	14 (2) 15 (1)	EXHAUST ODOR, ROUGH IDLE, LACK OF PERFORMANCE

(1) Open circuit
(2) Grounded circuit
(3) Open/grounded circuit

5.0L (VIN F) AND 5.7L (VIN 8) ENGINES — ECM SYMPTOM CHART "F" — 1992 CAMARO AND FIREBIRD

ECM PIN CONNECTOR "F" SYMPTOM CHART
GREEN 32 PIN CONNECTOR

	PIN FUNCTION	CKT #	WIRE COLOR	COMPONENT CONNECTOR CAVITY	CODES AFFECT	POSSIBLE SYMPTOMS FROM FAULTY CIRCUIT
F1	M/T SHIFT LIGHT CONTROL	456	TAN/BLK	I/P PRINTED CIRCUIT CONNECTOR "C10"		INOPERATIVE SHIFT LIGHT
F2	A.I.R. PORT (SWITCH) SOLENOID CONTROL	436	BRN	A.I.R. E.A.S. SOLENOID "B"		INCORRECT IDLE, ROUGH IDLE, FAIL EMISSIONS TEST
F3						
F4	A.I.R. CONVERTER (DIVERT) SOLENOID CONTROL	429	BLK/PNK	A.I.R. E.A.C. SOLENOID "B"		INCORRECT IDLE, ROUGH IDLE, BACKFIRE UPON DECELERATION
F5						
F6	A/T TCC CONTROL	422	TAN/BLK	TCC CONNECTOR "D" ALDL "F"		POOR FUEL ECONOMY, CHUGGLE
F7	CANISTER PURGE SOLENOID CONTROL	428	DK GRN/YEL	CANISTER PURGE SOLENOID "B"	45	FUEL ODOR, ROUGH IDLE
F8						
F9	ESC KNOCK SENSOR SIGNAL	496	DK BLU	KNOCK SENSOR	43 (3)	LACK OF PERFORMANCE
F10	PASS-Key® SIGNAL	229	DK BLU	PASS-Key® MODULE "A2"	46 (2)	ENGINE CRANKS BUT WILL NOT START
F11						
F12						
F13	TPS SIGNAL	417	DK BLU	TPS "B"	22 (3)	LACK OF PERFORMANCE, ROUGH IDLE
F14						
F15	MAP SIGNAL	432	LT GRN	MAP SENSOR "B"	34 (3)	LACK OF PERFORMANCE, ROUGH IDLE, SURGE
F16	IAT SIGNAL	472	TAN	IAT SENSOR "B"	23 (1) 25 (2)	POOR PERFORMANCE

(1) Open circuit
(2) Grounded circuit
(3) Open/grounded circuit

5.0L (VIN F) AND 5.7L (VIN 8) ENGINES — — RESTRICTED EXHAUST SYSTEM DIAGNOSTIC CHART — 1992 CAMARO AND FIREBIRD

CHART B-1

RESTRICTED EXHAUST SYSTEM CHECK
ALL ENGINES

Proper diagnosis for a restricted exhaust system is essential before any components are replaced. Either of the following procedures may be used for diagnosis, depending upon engine or tool used:

CHECK AT A. I. R. PIPE: OR **CHECK AT O_2 SENSOR:**

CHECK AT A. I. R. PIPE:
1. Remove the rubber hose at the exhaust manifold A.I.R. pipe check valve. Remove check valve.
2. Connect a fuel pump pressure gauge to a hose and nipple from a Propane Enrichment Device (J 26911) (see illustration).
3. Insert the nipple into the exhaust manifold A.I.R. pipe.

CHECK AT O_2 SENSOR:
1. Carefully remove O_2 sensor.
2. Install Borroughs exhaust backpressure tester (BT 8515 or BT 8603) or equivalent in place of O_2 sensor (see illustration).
3. After completing test described below, be sure to coat threads of O_2 sensor with anti-seize compound P/N 5613695 or equivalent prior to re-installation.

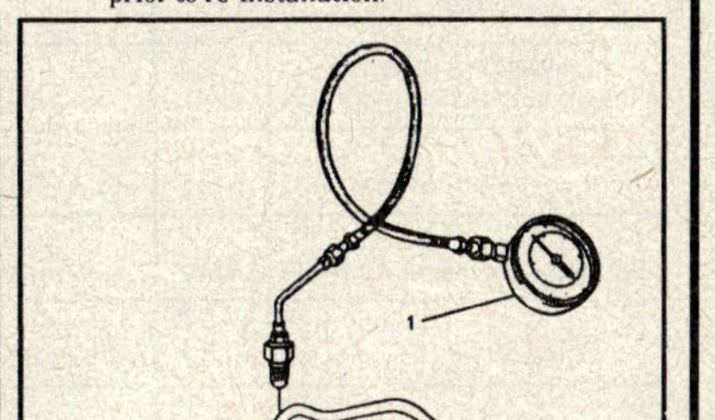

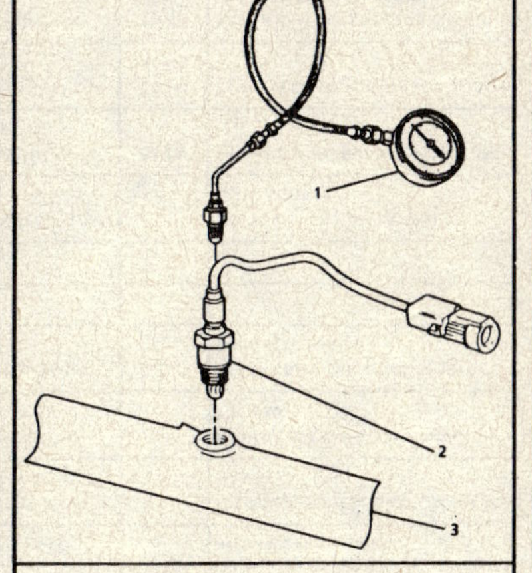

DIAGNOSIS:
1. With the engine idling at normal operating temperature, observe the exhaust system backpressure reading on the gage. Reading should not exceed 8.6 kPa (1.25 psi).
2. Increase engine speed to 2000 rpm and observe gage. Reading should not exceed 20.7 kPa (3 psi).
3. If the backpressure at either speed exceeds specification, a restricted exhaust system is indicated.
4. Inspect the entire exhaust system for a collapsed pipe, heat distress, or possible internal muffler failure.
5. If there are no obvious reasons for the excessive backpressure, the catalytic converter is suspected to be restricted and should be replaced using current recommended procedures.

5.0L (VIN F) AND 5.7L (VIN 8) ENGINES — COMPONENT DIAGNOSTIC CHART — 1992 CAMARO AND FIREBIRD

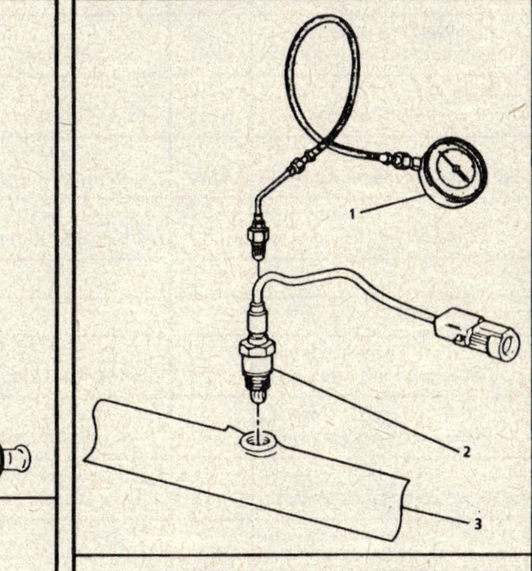

CHART C-1

PARK/NEUTRAL (P/N) SWITCH DIAGNOSIS
(AUTO TRANSMISSION ONLY)
5.0L (VIN F) & 5.7L (VIN 8) "F" CARLINE (PORT)

Circuit Description:
The Park/Neutral (P/N) switch contacts are a part of the neutral start switch, and are closed to ground in park or neutral and open in drive ranges.

The ECM supplies ignition voltage, through a current limiting resistor, to CKT 434 and senses a closed switch, when the voltage on CKT 434 drops to less than one volt.

The ECM uses the P/N signal as one of the inputs to control:
- Idle Air Control (IAC).
- Vehicle Speed Sensor (VSS) diagnostics.
- Exhaust Gas Recirculation (EGR).

If CKT 434 indicates P/N (grounded), while in drive range, the EGR would be inoperative, resulting in possible detonation.

If CKT 434 always indicates drive (open), a drop in the idle speed may exist when the gear selector is moved into drive range.

Test Description: Number(s) below refer to circled number(s) on the diagnostic chart.
1. Checks for a closed switch to ground in park position.
2. Checks for an open switch in drive range.
3. Make sure Tech 1 indicates drive, even while wiggling shifter to test for an intermittent or misadjusted switch in drive range.

5.0L (VIN F) AND 5.7L (VIN 8) ENGINES — COMPONENT DIAGNOSTIC CHART — 1992 CAMARO AND FIREBIRD

CHART C-1
PARK/NEUTRAL (P/N) SWITCH DIAGNOSIS
(AUTO TRANSMISSION ONLY)
5.0L (VIN F) & 5.7L (VIN 8) "F" CARLINE (PORT)

① • WITH TRANSMISSION IN PARK, TECH 1 SHOULD INDICATE PARK OR NEUTRAL. DOES IT?

YES → ③ • SHIFT TRANSMISSION INTO DRIVE. • TECH 1 SHOULD DISPLAY A CHANGE TO INDICATE DRIVE. DOES IT?

NO → ② • DISCONNECT PARK/NEUTRAL SWITCH CONNECTOR. • JUMPER HARNESS CONNECTOR TERMINALS "A" AND "B". • TECH 1 SHOULD INDICATE PARK OR NEUTRAL. DOES IT?

From ③:
NO → • DISCONNECT P/N SWITCH. • THIS SHOULD CAUSE TECH 1 TO DISPLAY DRIVE RANGE. DOES IT?

YES → NO TROUBLE FOUND. REFER TO "INTERMITTENTS" IN "SYMPTOM," SECTION "B".

From ②:
NO → • JUMPER HARNESS CONNECTOR (CKT 434) TO ENGINE GROUND. • TECH 1 SHOULD INDICATE PARK OR NEUTRAL. DOES IT?

YES → FAULTY P/N SWITCH CONNECTION OR P/N SWITCH MISADJUSTED OR FAULTY P/N SWITCH.

YES → FAULTY P/N SWITCH CONNECTION OR P/N SWITCH MISADJUSTED OR FAULTY P/N SWITCH.

NO → CKT 434 SHORTED TO GROUND OR FAULTY ECM.

YES → OPEN GROUND CIRCUIT.

NO → CKT 434 OPEN OR FAULTY ECM CONNECTION OR ECM.

"AFTER REPAIRS," CONFIRM "CLOSED LOOP" OPERATION AND NO "SERVICE ENGINE SOON" LIGHT.

5.0L (VIN F) AND 5.7L (VIN 8) ENGINES — COMPONENT DIAGNOSTIC CHART — 1992 CAMARO AND FIREBIRD

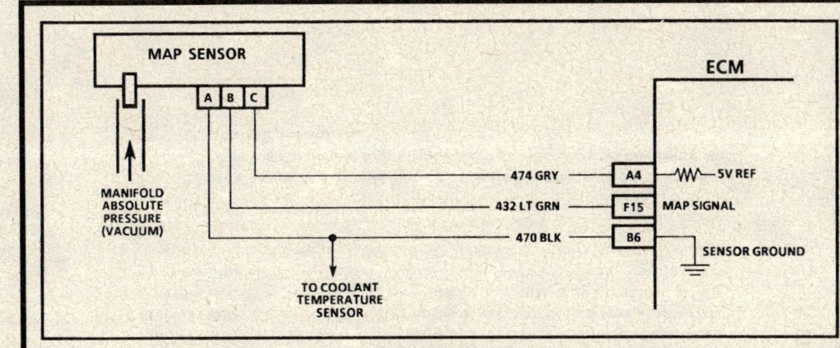

CHART C-1D
MANIFOLD ABSOLUTE PRESSURE (MAP) OUTPUT CHECK
5.0L (VIN F) & 5.7L (VIN 8) "F" CARLINE (PORT)

Circuit Description:

The Manifold Absolute Pressure (MAP) sensor measures the changes in the intake manifold pressure which result from engine load (intake manifold vacuum) and rpm changes: and converts theses into a voltage output. The ECM sends a 5 volt reference voltage to the MAP sensor. As the manifold pressure changed, the output voltage of the sensor also changes. By monitoring the sensor output voltage, the ECM knows the manifold pressure. A lower pressure (low voltage will be about 1 - 2 volts at idle. While higher pressure (high voltage) output voltage will be about 4 - 4.8 at Wide Open Throttle (WOT). The MAP sensor is also used, under certain conditions, to measure barometric pressure, allowing the ECM to make adjustments for different altitudes. The ECM used the MAP sensor to control fuel delivery and ignition timing.

Test Description: Number(s) below refer to circled number(s) on the diagnostic chart.

Important
• Be sure to use the same Diagnostic Test Equipment for all measurements.

1. When comparing Tech 1 reading to a known good vehicle, it is important to compare vehicles that use a sensor having the same color insert or having the same "Hot Stamped" number. See figures on facing page.
2. Applying 34 kPa (10" Hg) vacuum to the MAP sensor should cause the voltage to be 1.5 volts. Upon applying vacuum to the sensor, the change in voltage should be instantaneous. A slow voltage change indicates a faulty sensor.

3. Check vacuum hose to sensor for leaking or restriction. Be sure that no other vacuum devices are connected to the MAP hose.

Important
• Make sure electrical connector remains securely fastened.

4. Disconnect sensor from bracket and twist sensor by hand (only) to check for intermittent connection. Output changes greater than .10 volt indicates a bad sensor. If OK, replace sensor.

5.0L (VIN F) AND 5.7L (VIN 8) ENGINES — COMPONENT DIAGNOSTIC CHART — 1992 CAMARO AND FIREBIRD

CHART C-1D
MANIFOLD ABSOLUTE PRESSURE (MAP) OUTPUT CHECK
5.0L (VIN F) & 5.7L (VIN 8) "F" CARLINE (PORT)

NOTE: THIS CHART ONLY APPLIES TO MAP SENSORS HAVING GREEN OR BLACK COLOR KEY INSERT (SEE BELOW).

1.
   - IGNITION "ON," ENGINE "OFF."
   - TECH 1 SHOULD INDICATE A MAP SENSOR VOLTAGE.
   - COMPARE THIS READING WITH THE READING OF A KNOWN GOOD VEHICLE. SEE FACING PAGE TEST DESCRIPTION, STEP 1.
   - VOLTAGE READING SHOULD BE WITHIN, ± .4 VOLT.
   - IS IT?

   YES →

2.
   - DISCONNECT AND PLUG VACUUM SOURCE TO MAP SENSOR.
   - CONNECT A HAND VACUUM PUMP TO MAP SENSOR.
   - START ENGINE.
   - NOTE MAP SENSOR VOLTAGE.
   - APPLY 34 kPa (10" Hg) OF VACUUM AND NOTE VOLTAGE CHANGE. SUBTRACT SECOND READING FROM THE FIRST. VOLTAGE VALUE SHOULD BE GREATER THAN 1.5 VOLTS.
   - IS IT?

   YES →

3. NO TROUBLE FOUND. CHECK SENSOR VACUUM SOURCE FOR LEAKAGE OR RESTRICTION. BE SURE THIS SOURCE SUPPLIES VACUUM TO MAP SENSOR ONLY.

   NO (from 1) → REPLACE SENSOR.

4. CHECK SENSOR CONNECTION. IF OK, REPLACE SENSOR.

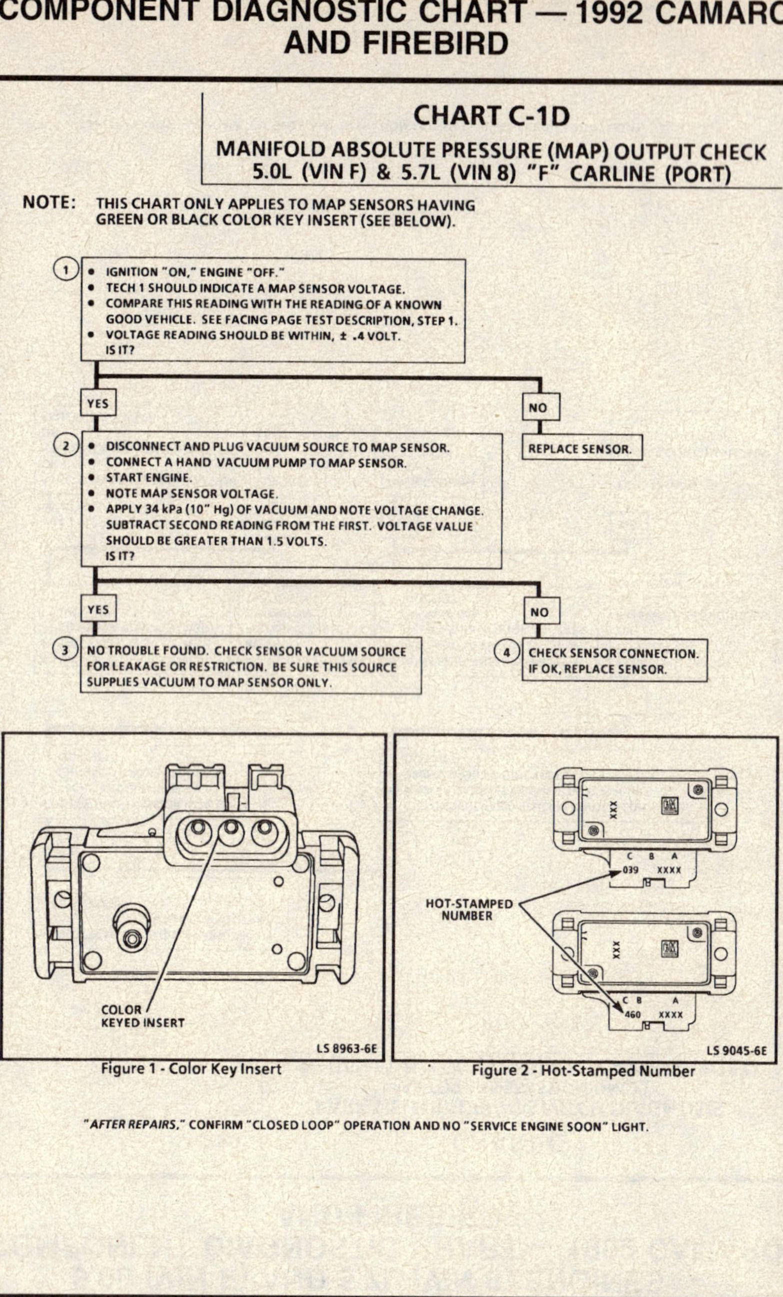

Figure 1 - Color Key Insert

Figure 2 - Hot-Stamped Number

"AFTER REPAIRS," CONFIRM "CLOSED LOOP" OPERATION AND NO "SERVICE ENGINE SOON" LIGHT.

5.0L (VIN F) AND 5.7L (VIN 8) ENGINES — COMPONENT DIAGNOSTIC CHART — 1992 CAMARO AND FIREBIRD

CHART C-2A
INJECTOR BALANCE TEST

The injector balance tester is a tool used to turn the injector on for a precise amount of time, thus spraying a measured amount of fuel into the manifold. This causes a drop in fuel rail pressure that we can record and compare between each injector. All injectors should have the same amount of pressure drop (± 10 kpa). Any injector with a pressure drop that is 10 kpa (or more) greater or less than the average drop of the other injectors should be considered faulty and replaced.

STEP 1

Engine "cool down" period (10 minutes) is necessary to avoid irregular readings due to "Hot Soak" fuel boiling. With ignition "OFF" connect fuel gauge J 347301 or equivalent to fuel pressure tap. Wrap a shop towel around fitting while connecting gage to avoid fuel spillage.

Disconnect harness connectors at all injectors, and connect injector tester J 34730-3, or equivalent, to one injector. Follow manufacturers instructions for use of adaptor harness. Ignition must be "OFF" at least 10 seconds to complete ECM shutdown cycle. Fuel pump should run about 2 seconds after ignition is turned "ON". At this point, insert clear tubing attached to vent valve into a suitable container and bleed air from gauge and hose to insure accurate gauge operation. Repeat this step until all air is bled from gauge.

STEP 2

Turn ignition "OFF" for 10 seconds and then "ON" again to get fuel pressure to its maximum. Record this initial pressure reading. Energize tester one time and note pressure drop at its lowest point (Disregard any slight pressure increase after drop hits low point). By subtracting this second pressure reading from the initial pressure, we have the actual amount of injector pressure drop.

STEP 3

Repeat Step 2 on each injector and compare the amount of drop. Usually, good injectors will have virtually the same drop. Retest any injector that has a pressure difference of 10 kPa, either more or less than the average of the other injectors on the engine. Replace any injector that also fails the retest. If the pressure drop of all injectors is within 10 kPa of this average, the injectors appear to be flowing properly. Reconnect them and review "Symptoms", Section

NOTE: *The entire test should __not__ be repeated more than once without running the engine to prevent flooding. (This includes any retest on faulty injectors).*

5.0L (VIN F) AND 5.7L (VIN 8) ENGINES — COMPONENT DIAGNOSTIC CHART — 1992 CAMARO AND FIREBIRD

NOTICE: The entire test should <u>NOT</u> be repeated more than once without running the engine to prevent flooding. (This includes any retest on faulty injectors.)

The fuel pressure test in Section "A" Chart A-7, should be completed prior to this test.

CHART C-2A

INJECTOR BALANCE TEST
5.0L (VIN F) & 5.7L (VIN 8)
"F" CARLINE (PORT)

Step 1. If engine is at operating temperature, allow a 10 minute "cool down" period then connect fuel pressure gauge and injector tester.
1. Ignition "OFF."
2. Connect fuel pressure gauge and injector tester.
3. Ignition "ON."
4. Bleed off air in gauge. Repeat until all air is bled from gauge.

Step 2. Run test:
1. Ignition "OFF" for 10 seconds.
2. Ignition "ON". Record gauge pressure. (Pressure must hold steady, if not see the Fuel System diagnosis, Chart A-7, in Section "A").
3. Turn injector on, by depressing button on injector tester, and note pressure at the instant the gauge needle stops.

Step 3.
1. Repeat step 2 on all injectors and record pressure drop on each. Retest injectors that appear faulty (Any injectors that have a 10 kPa (1.5 psi) difference, either more or less, in pressure from the average). If no problem is found, review "Symptoms" Section

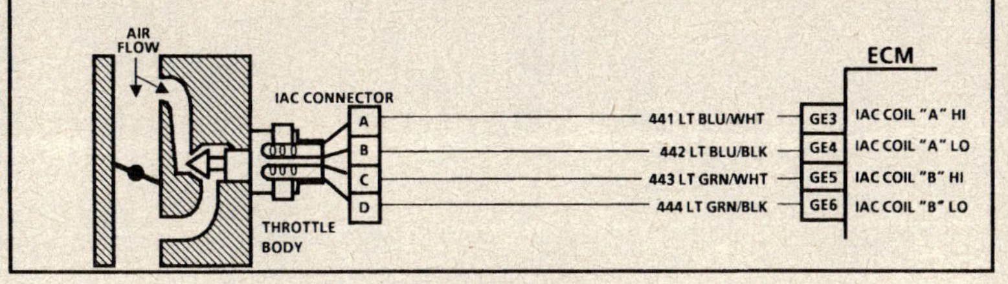

EXAMPLE

CYLINDER	1	2	3	4	5	6	7	8
1ST READING	293 kPa (43 psi)	293 kPa (43 psi)	293 kPa (43 psi)	293 kPa (43 psi)	293 kPa (43 psi)	293 kPa (43 psi)	293 kPa (43 psi)	293 kPa (43 psi)
2ND READING	131 kPa (19 psi)	131 kPa (19 psi)	131 kPa (19 psi)	115 kPa (17 psi)	131 kPa (19 psi)	145 kPa (21 psi)	131 kPa (19 psi)	131 kPa (19 psi)
AMOUNT OF DROP	162 kPa (24 psi)	162 kPa (24 psi)	162 kPa (24 psi)	178 kPa (26 psi)	162 kPa (24 psi)	148 kPa (21 psi)	162 kPa (24 psi)	162 kPa (24 psi)
	OK	OK	OK	FAULTY, RICH (TOO MUCH FUEL DROP)	OK	FAULTY, LEAN (TOO LITTLE FUEL DROP)	OK	OK

5.0L (VIN F) AND 5.7L (VIN 8) ENGINES — COMPONENT DIAGNOSTIC CHART — 1992 CAMARO AND FIREBIRD

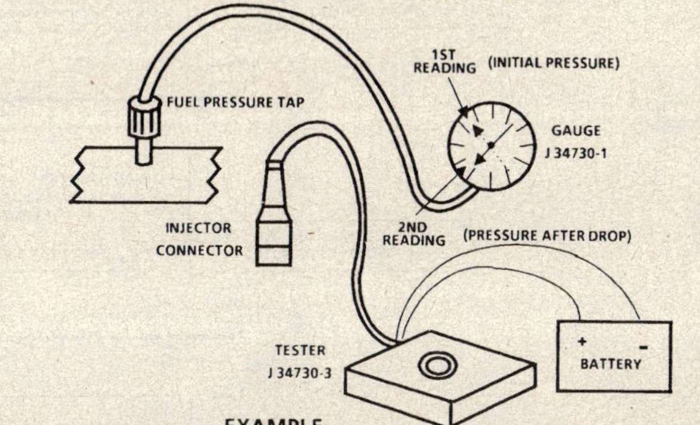

CHART C-2C

IDLE AIR CONTROL (IAC) SYSTEM CHECK
5.0L (VIN F) & 5.7L (VIN 8) "F" CARLINE (PORT)

Circuit Description:
The ECM controls idle rpm with the IAC valve. To increase idle rpm, the ECM moves the IAC valve out, allowing more air to bypass the throttle plate. To decrease rpm, it moves the IAC valve in, reducing air flow by-passing the throttle plate. A "Scan" tool will read the ECM commands to the IAC valve in counts. The higher the counts, the more air allowed (higher idle). The lower the counts, the less air allowed (lower idle).

Test Description: Number(s) below refer to circled number(s) on the diagnostic chart.
1. The IAC tester is used to extend and retract the IAC valve. Valve movement is verified by an engine speed change. If no change in engine speed occurs, the valve can be retested when removed from the throttle body.
2. This step checks the quality of the IAC movement in Step 1. Between 700 rpm and about 1500 rpm, the engine speed should change smoothly with each flash of the tester light in both extend and retract. If the IAC valve is retracted beyond the control range (about 1500 rpm), it may take many flashes in the extend position before engine speed will begin to drop. This is normal on certain engines, fully extending IAC may cause engine stall. This may be normal.
3. Steps 1 and 2 verified proper IAC valve operation while this step checks IAC circuits. Each lamp on the node light should flash red and green while the IAC valve is cycled. While the sequence of color is not important if either light is "OFF" or does not flash red and green, check the circuits for faults, beginning with poor terminal contacts.

Diagnostic Aids:

A slow, unstable, or fast idle may be caused by a non-IAC system problem that cannot be overcome by the IAC valve. Out of control range IAC "Scan" tool counts will be above 60 if idle is too low, and zero counts if idle is too high. The following checks should be made to repair a non-IAC system problem.

- <u>Vacuum Leak (High Idle)</u>
  If idle is too high, stop the engine. Fully extend (low) IAC with tester. Start engine. If idle speed is above 800 rpm, locate and correct vacuum leak including PCV system. Also check for binding of throttle blade or linkage.
- <u>System too lean (High Air/Fuel Ratio)</u>
  Idle speed may be too high or too low. Engine speed may vary up and down and disconnecting IAC does not help. Code 44 may be set. "Scan" O$_2$ voltage will be less than 300 mV (.3 volt). Check for low regulated fuel pressure, water in the fuel or a restricted injector.
- <u>System too rich (Low Air/Fuel Ratio)</u>
  The idle speed will be too low. "Scan" tool IAC counts will usually be above 80. System is obviously rich and may exhibit black smoke in exhaust.
  O$_2$ voltage will be fixed above 800 mV (.8 volt). Check for high fuel pressure, leaking or sticking injector. Silicone contaminated O$_2$ sensor will "Scan" an O$_2$ voltage slow to respond.
- <u>Throttle Body</u>
  Remove IAC and inspect bore for foreign material.
- <u>PCV Valve</u>
  An incorrect or faulty PCV valve may result in an incorrect idle speed.
- Refer to "Rough, Unstable, Incorrect Idle or Stalling" in "Symptoms," Section
- If intermittent poor driveability or idle symptoms are resolved by disconnecting the IAC, carefully recheck connections, valve terminal resistance, or replace IAC.

5.0L (VIN F) AND 5.7L (VIN 8) ENGINES — COMPONENT DIAGNOSTIC CHART — 1992 CAMARO AND FIREBIRD

CHART C-2C
IDLE AIR CONTROL (IAC) SYSTEM CHECK
5.0L (VIN F) & 5.7L (VIN 8) "F" CARLINE (PORT)

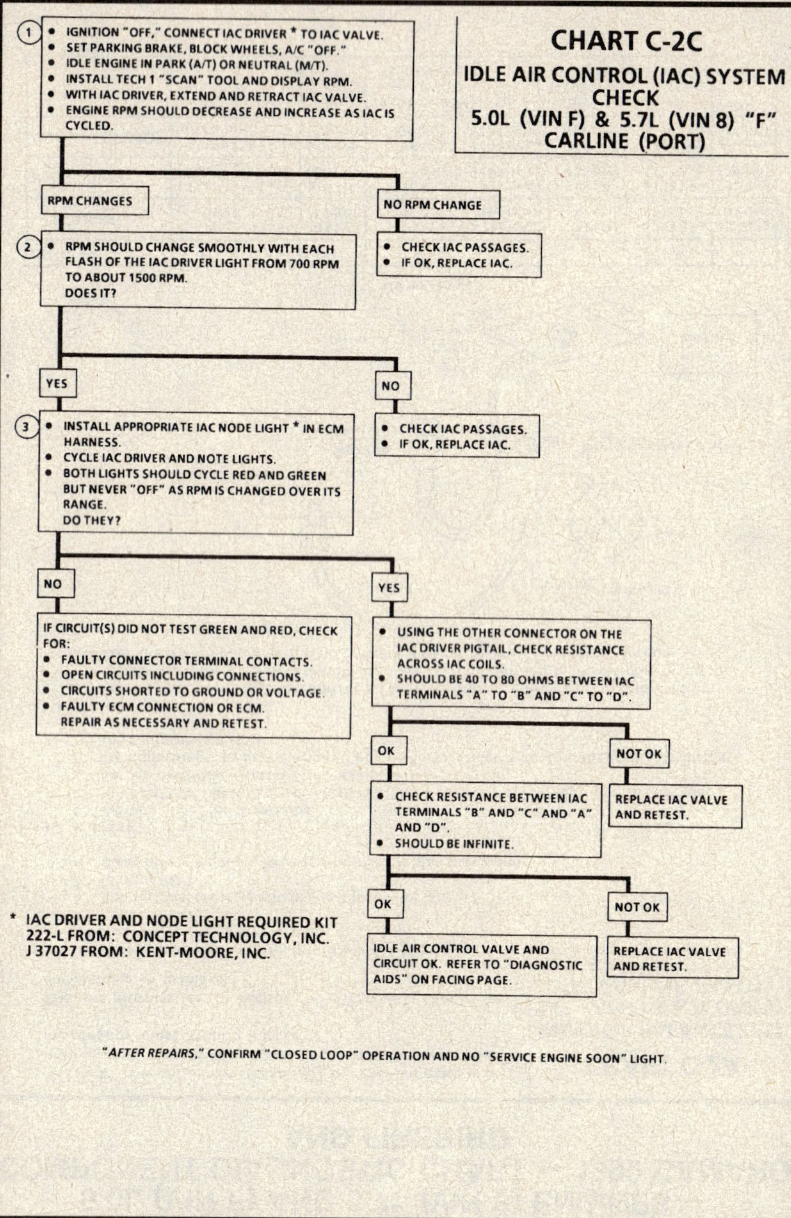

5.0L (VIN F) AND 5.7L (VIN 8) ENGINES — COMPONENT DIAGNOSTIC CHART — 1992 CAMARO AND FIREBIRD

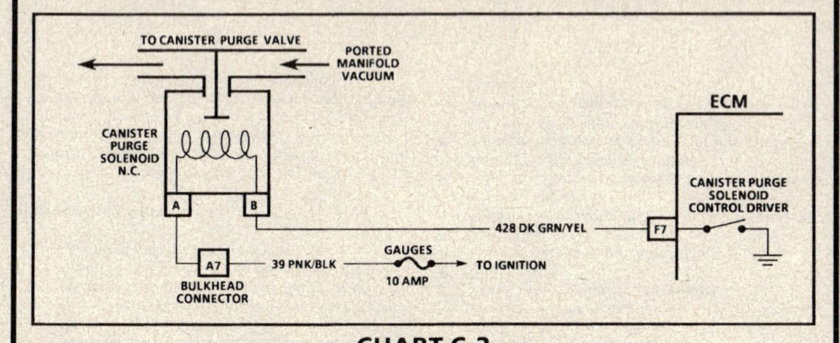

CHART C-3
CANISTER PURGE SOLENOID CHECK
5.0L (VIN F) & 5.7L (VIN 8) "F" CARLINE (PORT)

Circuit Description:

Canister purge is controlled by a solenoid that allows ported manifold vacuum to purge the fuel vapor canister when energized. The ECM supplies a ground to energize the solenoid (purge "ON").

If the diagnostic "test" terminal is grounded, with the engine stopped, or the following conditions are met with the engine running, the canister purge solenoid will be energized (purge "ON").

• Engine run time after start more than 1 minute.
• Coolant temperature above 75°C (167°F).
• Vehicle Speed Sensor (VSS) above 15 mph.
• Throttle Position Sensor (TPS) is above idle.

Test Description: Number(s) below refer to circled number(s) on the diagnostic chart.

1. The external vacuum source must be applied to the canister purge solenoid at the canister.

2. Grounding the diagnostic "test" terminal will energize the canister purge solenoid and allow vacuum to pass.

3. Some canister purge solenoids may have a large enough bleed built into them to appear to be operating incorrectly.

5.0L (VIN F) AND 5.7L (VIN 8) ENGINES — COMPONENT DIAGNOSTIC CHART — 1992 CAMARO AND FIREBIRD

CHART C-3
CANISTER PURGE SOLENOID CHECK
5.0L (VIN F) & 5.7L (VIN 8) "F" CARLINE (PORT)

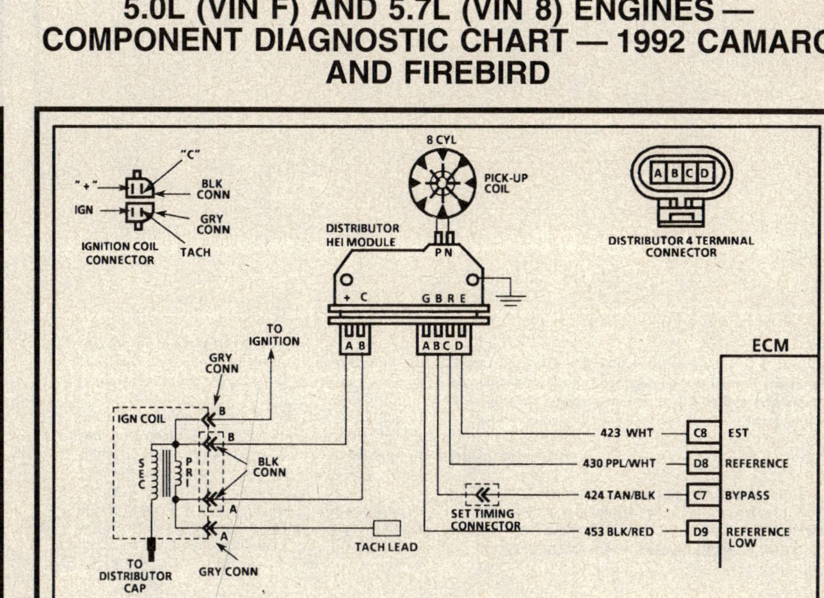

"AFTER REPAIRS," CONFIRM "CLOSED LOOP" OPERATION AND NO "SERVICE ENGINE SOON" LIGHT.

5.0L (VIN F) AND 5.7L (VIN 8) ENGINES — COMPONENT DIAGNOSTIC CHART — 1992 CAMARO AND FIREBIRD

CHART C-4
IGNITION SYSTEM CHECK
(REMOTE COIL/SEALED MODULE CONNECTOR DISTRIBUTOR)
5.0L (VIN F) & 5.7L (VIN 8) "F" CARLINE (PORT)

Test Description: Number(s) below refer to circled number(s) on the diagnostic chart.

1. A Tach filter or Tach circuit that is shorted to ground will result in a no spark condition. Two wires are checked, to ensure that an open is not present in a spark plug wire.

1A. If spark occurs with EST connector disconnected, pick-up coil output is too low for EST operation.

2. A spark indicates the problem must be the distributor cap, rotor, or coil output wire.

3. Normally, there should be battery voltage at the "C" and "+" terminals. Low voltage would indicate an open or a high resistance circuit from the distributor to the coil or ignition switch. If "C" terminal voltage was low, but "+" terminal voltage is 10 volts or more, circuit from "C" terminal to ignition coil or ignition coil primary winding is open.

4. Checks for a shorted module or grounded circuit from the ignition coil to the module. The distributor module should be turned "OFF," so normal voltage should be about 12 volts. If the module is turned "ON," the voltage would be low, but above 1 volt.

This could cause the ignition coil to fail from excessive heat. With an open ignition coil primary winding, a small amount of voltage will leak through the module from the "Batt" to the tach terminal.

5. Applying a voltage (1.35 to 1.5 volts) to module terminal "P" should turn the module "ON" and the tach terminal voltage should drop to about 7-9 volts. This test will determine whether the module or coil is faulty or if the pick-up coil is not generating the proper signal to turn the module "ON." This test can be performed by using a DC test battery with a rating of 1.35 to 1.5 volts. (Such as AA, C, or D cell). The test battery must be a known good battery with a voltage greater than 1.35 volts.

6. This should turn "OFF" the module and cause a spark. If no spark occurs, the fault is most likely in the ignition coil, because most module problems would have been found before this point in the procedure. A module tester (J 24642) could determine which is at fault.

5.0L (VIN F) AND 5.7L (VIN 8) ENGINES — COMPONENT DIAGNOSTIC CHART — 1992 CAMARO AND FIREBIRD

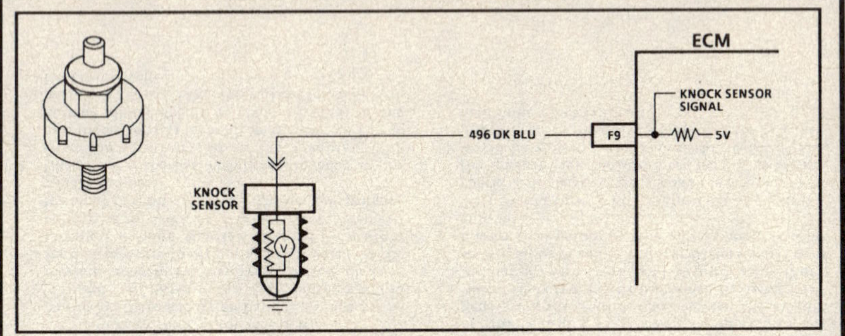

5.0L (VIN F) AND 5.7L (VIN 8) ENGINES — COMPONENT DIAGNOSTIC CHART — 1992 CAMARO AND FIREBIRD

CHART C-5
ELECTRONIC SPARK CONTROL (ESC)
5.0L (VIN F) & 5.7L (VIN 8) "F" CARLINE (PORT)

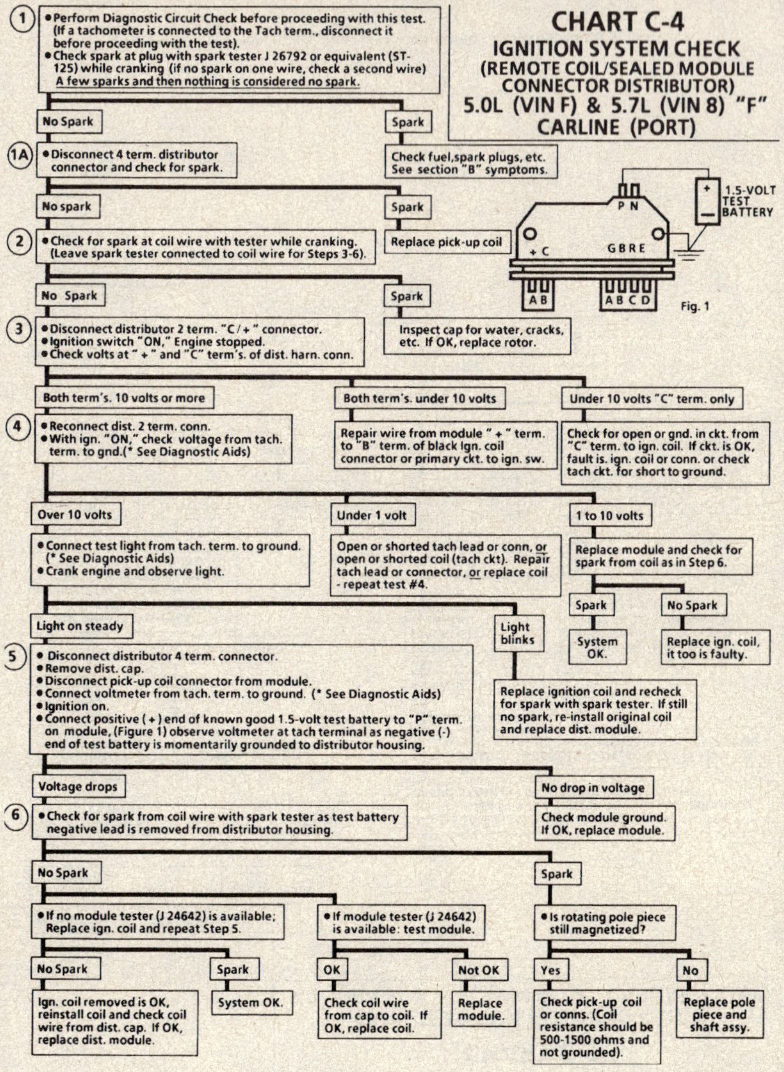

Circuit Description:

The knock sensor is used to detect engine detonation and the ECM will retard the electronic spark timing based on the signal being received. The circuitry within the knock sensor causes the ECMs 5 volts to be pulled down so that under a no knock condition, CKT 496 would measure about 2.5 volts. The knock sensor produces an AC voltage. The AC voltage is then sent to the ECM which reduces the ECM signal voltage. The reduction of the ECM signal voltage triggers the ECM to retard timing. The amount of AC signal is determined by the amount of knock.

The MEM-CAL used with this engine, contains the functions which were part of remotely mounted ESC modules used on other GM vehicles. The ESC portion of the MEM-CAL then sends a signal to other parts of the ECM which adjusts the spark timing to retard the spark and reduce the detonation.

Test Description: Number(s) below refer to circled number(s) on the diagnostic chart.

1. With engine idling, there should not be a knock signal present at the ECM because detonation is not likely under a no load condition.
2. Tapping on the engine should simulate a knock signal to determine if the sensor is capable of detecting detonation. If no knock is detected, try tapping on engine block closer to sensor before replacing sensor.
3. If the engine has an internal problem which is creating a knock, the knock sensor may be responding to the internal failure.
4. This test determines if the knock sensor is faulty or if the ESC portion of the MEM-CAL is faulty. If it is determined that the MEM-CAL is faulty, be sure that it is properly installed and latched into place. If not properly installed, repair and retest.

Diagnostic Aids:

While observing knock signal on the Tech 1 there should be an indication that knock is present when detonation can be heard. Detonation is most likely to occur under high engine load conditions.

5.0L (VIN F) AND 5.7L (VIN 8) ENGINES — COMPONENT DIAGNOSTIC CHART — 1992 CAMARO AND FIREBIRD

CHART C-5
ELECTRONIC SPARK CONTROL (ESC)
5.0L (VIN F) & 5.7L (VIN 8) "F" CARLINE (PORT)

1.
- IF CODE 43 IS SET, USE THE CODE CHART.
- ENGINE MUST BE IDLING AT NORMAL OPERATING TEMPERATURE.
- USE "SCAN" TOOL TO OBSERVE KNOCK SIGNAL. IS KNOCK INDICATED?

NO →

2.
- TAP ON RIGHT EXHAUST MANIFOLD WHILE OBSERVING KNOCK SIGNAL.
- "SCAN" TOOL SHOULD INDICATE KNOCK WHILE TAPPING ON MANIFOLD. DOES IT?

YES →

3.
- IF AN ENGINE KNOCK CAN BE HEARD, REPAIR THE BASIC ENGINE PROBLEM. IF NO AUDIBLE KNOCK IS HEARD, FOLLOW THE STEPS:
- DISCONNECT KNOCK SENSOR.
- CONNECT VOLTMETER TO KNOCK SENSOR AND ENGINE GROUND.
- SET VOLTMETER ON 2 VOLT AC SCALE. IS A SIGNAL INDICATED ON VOLTMETER?

NO →

4.
- DISCONNECT KNOCK SENSOR.
- CONNECT VOLTMETER TO KNOCK SENSOR AND ENGINE GROUND.
- SET VOLTMETER ON 2 VOLT AC SCALE.
- TAP ON ENGINE BLOCK NEAR SENSOR. IS A SIGNAL INDICATED ON VOLTMETER WHILE TAPPING ON ENGINE BLOCK?

NO → SYSTEM IS OPERATING PROPERLY. REFER TO "DIAGNOSTIC AIDS" ON FACING PAGE.

YES → REPLACE KNOCK SENSOR.

NO → CHECK CKT 496 FOR BEING NEAR A SPARK PLUG WIRE OR A FAULTY ECM CONNECTION OR FAULTY ECM OR MEM-CAL.

YES → REPLACE KNOCK SENSOR.

YES → REPLACE MEM-CAL OR ECM.

"AFTER REPAIRS," CONFIRM "CLOSED LOOP" OPERATION AND NO "SERVICE ENGINE SOON" LIGHT.

5.0L (VIN F) AND 5.7L (VIN 8) ENGINES — COMPONENT DIAGNOSTIC CHART — 1992 CAMARO AND FIREBIRD

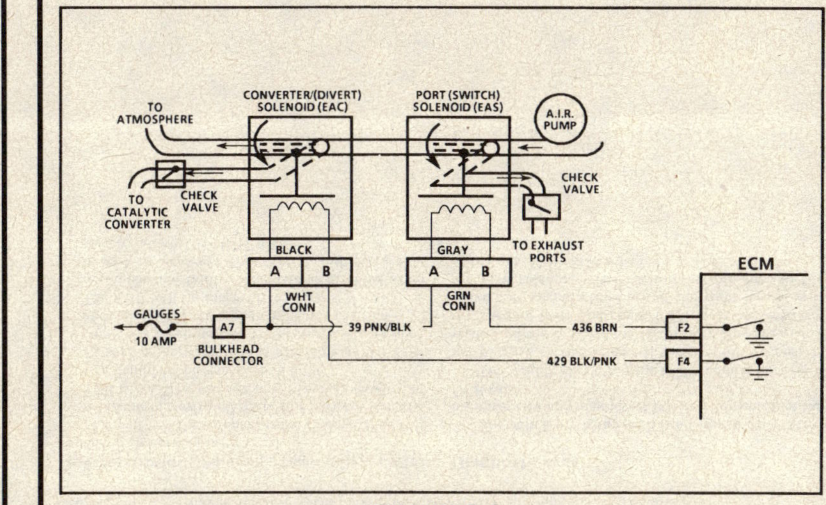

CHART C-6
AIR MANAGEMENT CHECK - CONTROL (PEDES) VALVE
(PRESSURE OPERATED ELECTRIC DIVERT/ELECTRIC SWITCHING)
5.0L (VIN F) & 5.7L (VIN 8) "F" CARLINE (PORT)

Circuit Description:

Air management is controlled by a pressure operated port valve and a converter valve, each with an ECM controlled solenoid. When the solenoid is grounded by the ECM, A.I.R. pump pressure will activate the valve and allow pump air to be directed as follows:

- Cold Mode - The port (switch) solenoid is energized which in turn opens the port valve and allows flow to the exhaust ports.
- Warm Mode - The port (switch) solenoid is de-energized and the converter (divert) solenoid energized which closes the port valve and keeps the converter valve seated, thus forcing flow past the converter valve and to the catalytic converter.
- Divert Mode - Both solenoids are de-energized, which opens the converter valve, allowing air to take the path of least resistance, i.e., out the divert/relief tube to atmosphere

Test Description: Number(s) below refer to circled number(s) on the diagnostic chart.

1. This is a system functional check. Air is directed to ports during "Open Loop" and all engines start in "Open Loop" even on a warm engine. Since the air to the ports may be very short, prepare to observe port air prior to engine start up. This can be done by squeezing a hose.
2. This should normally set a Code 22. When any code is set, the ECM opens the ground to the converter solenoid and allows air to divert. This checks for ECM response to a fault. A ground in the control valve circuit to the ECM would prevent divert action.
3. This checks for a grounded circuit to the ECM. Test light "OFF" is normal and would indicate the circuit is not grounded.
4. Checks for an open in the solenoid control circuits. Grounding the diagnostic "test" terminal should ground both solenoid circuits. Normally, the test light should be "ON," which indicates the problem is not in the ECM or wiring but at the solenoid connections or the control valve itself.
5. Checks for a grounded port (switch) solenoid circuit. Test light "OFF" would indicate the circuit is normal and fault is in the control valve.

5.0L (VIN F) AND 5.7L (VIN 8) ENGINES — COMPONENT DIAGNOSTIC CHART — 1992 CAMARO AND FIREBIRD

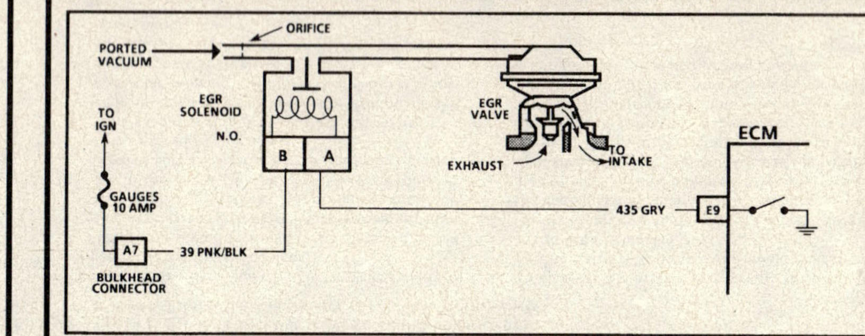

CHART C-6

AIR MANAGEMENT CHECK-CONTROL (PEDES) VALVE
(PRESSURE OPERATED ELECTRIC DIVERT/ ELECTRIC SWITCHING)
5.0L (VIN F) & 5.7L (VIN 8) "F" CARLINE (PORT)

1. • START ENGINE, RUNNING AT PART THROTTLE, BELOW 2000 RPM, AIR SHOULD BE FELT AT OUTLET TO EXHAUST PORTS DURING "OPEN LOOP" OPERATION (FROM 6 SECONDS TO 3 MINUTES ON WARM ENGINE DEPENDING ON APPLICATION) AND SWITCH TO CONVERTER WHEN SYSTEM GOES "CLOSED LOOP."

|OK|PORT, THEN DIVERT ONLY|CONSTANT PORT OR CONVERTER AIR|

2. • DISCONNECT TPS CONNECTOR AT IDLE AND AIR SHOULD DIVERT TO ATMOSPHERE AFTER A FEW SECONDS.

4. • IGNITION "ON," ENGINE STOPPED.
• INSTALL TECH 1.
• ENABLE AIR CONTROL SOLENOID.
• DISCONNECT CONVERTER VALVE SOLENOID.
• CONNECT TEST LIGHT BETWEEN HARNESS CONNECTOR TERMINALS AND NOTE LIGHT.

5. • IGNITION "ON," ENGINE STOPPED.
• DISCONNECT PORT SOLENOID ELECTRICAL CONNECTOR.
• CONNECT TEST LIGHT BETWEEN HARNESS CONNECTOR TERMINALS AND NOTE LIGHT.

|NO DIVERT|OK|

3. • IGNITION "ON," ENGINE STOPPED.
• DISCONNECT CONVERTER SOLENOID ELECTRICAL CONNECTOR.
• CONNECT TEST LIGHT BETWEEN HARNESS CONNECTOR TERMINALS.
• NOTE LIGHT.

NO TROUBLE FOUND. CLEAR MEMORY AND REVIEW SYMPTOMS SECTION "B".

LIGHT "OFF" → • CONNECT TEST LIGHT BETWEEN CKT 39 AND GROUND.

LIGHT "ON" → FAULT IS CONVERTER VALVE SOLENOID, CONNECTION, OR VALVE.

LIGHT "OFF" → • INSTALL TECH 1.
• ENABLE AIR SWITCH SOLENOID.
• NOTE TEST LIGHT.

LIGHT "ON" → GROUNDED CKT 436 OR FAULTY ECM.

LIGHT "OFF" → OPEN CKT 39

LIGHT "ON" → OPEN IN CKT 429. FAULTY ECM CONNECTION OR ECM.

LIGHT "OFF" → • CONNECT TEST LIGHT BETWEEN CKT 39 AND GROUND.

LIGHT "ON" → FAULTY CONNECTION AT PORT SOLENOID OR FAULTY PEDES VALVE.

LIGHT "ON" → CHECK CKT 429 FOR SHORT TO GROUND. IF NOT GROUNDED, IT IS A FAULTY ECM.

LIGHT "OFF" → REPLACE PEDES VALVE.

LIGHT "OFF" → REPAIR OPEN CKT 39

LIGHT "ON" → OPEN CKT 436. FAULTY CONNECTION OR ECM.

"AFTER REPAIRS," CONFIRM "CLOSED LOOP" OPERATION AND NO "SERVICE ENGINE SOON" LIGHT.

5.0L (VIN F) AND 5.7L (VIN 8) ENGINES — COMPONENT DIAGNOSTIC CHART — 1992 CAMARO AND FIREBIRD

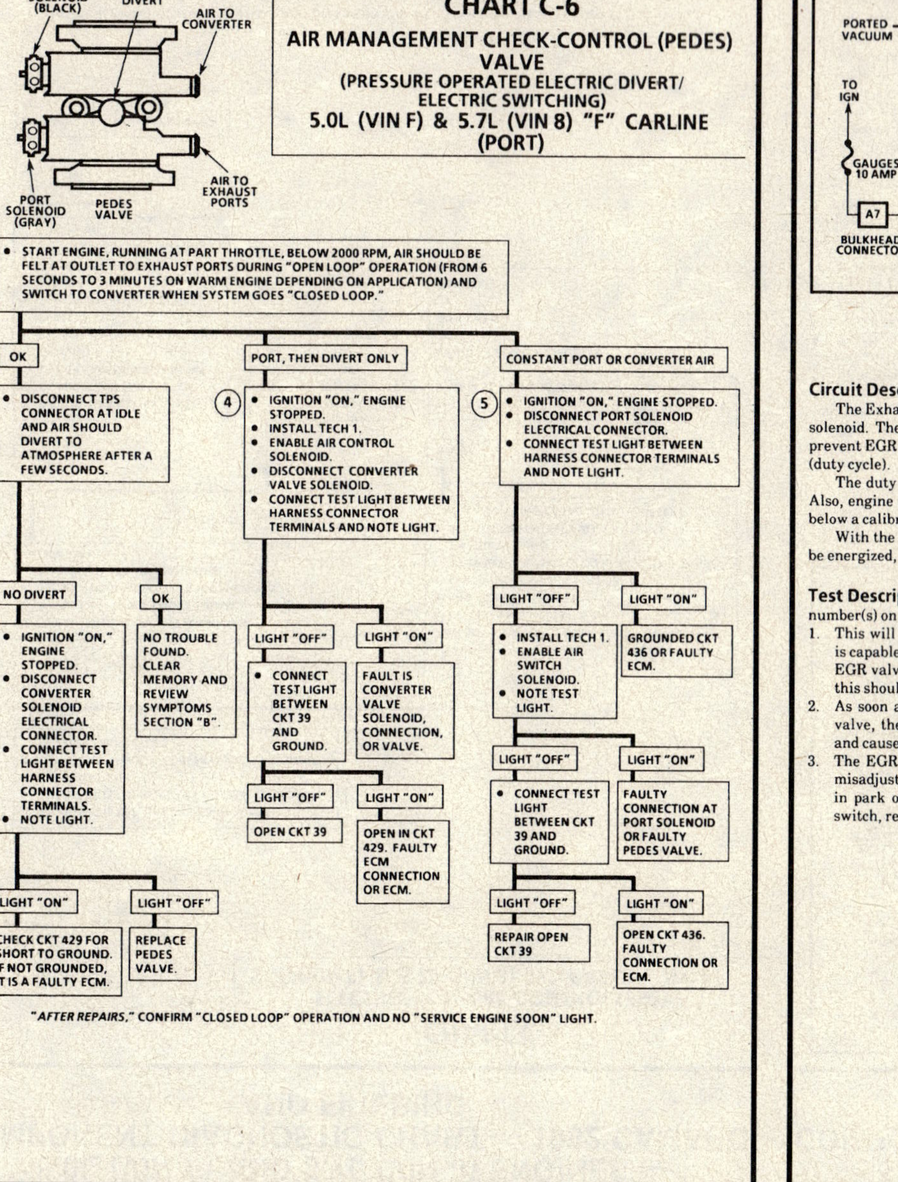

CHART C-7

EXHAUST GAS RECIRCULATION (EGR) CHECK
5.0L (VIN F) & 5.7L (VIN 8) "F" CARLINE (PORT)

Circuit Description:

The Exhaust Gas Recirculation (EGR) valve is controlled by a normally open Pulse Width Modulation (PWM) solenoid. The ECM turns the solenoid "OFF," to allow vacuum to pass to the EGR, and turns the solenoid "ON" to prevent EGR operation. When EGR is commanded, the solenoid is turned "ON" and "OFF" many times a second (duty cycle).

The duty cycle is calculation by the ECM based on information from the CTS, IAT, TPS, and MAP sensors. Also, engine rpm and the P/N switch input affect EGR. There should be no EGR when in park or neutral, TPS below a calibrated value, or TPS indicating WOT.

With the ignition "ON" and engine "OFF," the EGR solenoid is de-energized. The solenoid, however, should be energized, if the diagnostic "test" terminal is grounded with the ignition "ON" and engine "OFF."

Test Description: Number(s) below refer to circled number(s) on the diagnostic chart.

1. This will test the solenoid valve to determine if it is capable of closing off the ported vacuum from the EGR valve. The vacuum may bleed off slowly, but this should not be considered a fault.
2. As soon as backpressure is available at the EGR valve, the bleed portion in the valve should open and cause the valve to go to its seated position.
3. The EGR will be inoperative if the P/N switch is misadjusted or faulty. The EGR is disabled when in park or neutral. Use Tech 1 and check P/N switch, refer to CHART C-1A.

Diagnostic Aids:

Suction from shop exhaust hoses can alter backpressure and may affect the function check of the EGR valve.

Thoroughly check that the EGR vacuum harness if not plugged or restricted. If the vacuum harness is pushed on the solenoid too far, it may plug the end of the vacuum lines and cause EGR to be inoperative. Make sure vacuum lines are at the right locations of the EGR solenoid. The vacuum source goes to the orifice side of the EGR solenoid.

5.0L (VIN F) AND 5.7L (VIN 8) ENGINES — COMPONENT DIAGNOSTIC CHART — 1992 CAMARO AND FIREBIRD

CHART C-7
EXHAUST GAS RECIRCULATION (EGR) CHECK
5.0L (VIN F) & 5.7L (VIN 8) "F" CARLINE (PORT)

BEFORE USING THIS CHART, CHECK FOR PORTED VACUUM TO EGR SOLENOID, ALSO CHECK HOSES FOR LEAKS OR RESTRICTIONS. SHOULD BE AT LEAST 7" HG VACUUM AT 2000 RPM. THIS CHART ASSUMES THERE IS NO CODE 32.

Flow chart boxes:

(1)
- IGNITION "ON" ENGINE STOPPED.
- GROUND DIAGNOSTIC TERMINAL.
- DISCONNECT EGR SOLENOID VACUUM HARNESS.
- APPLY 10" VACUUM TO MANIFOLD SIDE OF SOLENOID.
- SHOULD BE ABLE TO HOLD VACUUM.

OK →
- UNGROUND DIAGNOSTIC TERMINAL.
- VACUUM SHOULD DROP.

DROPS →
- IGNITION "OFF."
- CONNECT A VACUUM PUMP TO EGR VALVE.
- USING A MIRROR, OBSERVE EGR DIAPHRAGM WHILE APPLYING VACUUM.
- DIAPHRAGM SHOULD MOVE FREELY AND HOLD VACUUM FOR AT LEAST 20 SECONDS.

OK →

(2)
- APPLY VACUUM TO EGR VALVE.
- START ENGINE AND IMMEDIATELY OBSERVE VACUUM.
- VALVE IS GOOD IF VALVE MOVES TO SEATED POSITION (VALVE CLOSED) AND VACUUM DROPPED WHILE STARTING ENGINE.

OK →

(3) IF AUTOMATIC TRANSMISSION, THEN CHECK P/N SWITCH.

IF OK, THERE IS NO TROUBLE FOUND.

NOT OK →
- REMOVE EGR VALVE. CHECK PASSAGES FOR BEING PLUGGED.
- IF NOT PLUGGED, REPLACE EGR VALVE.

NO DROP →
- DISCONNECT SOLENOID ELECTRICAL CONNECTOR
- NOTE VACUUM

NO DROP → REPLACE SOLENOID

DROPS → CKT 435 SHORTED TO GROUND OR FAULTY ECM

NOT OK → REPLACE EGR VALVE.

NOT OK →
- DISCONNECT EGR SOLENOID ELECTRICAL CONNECTOR.
- CONNECT A TEST LIGHT BETWEEN HARNESS TERMINALS.

LIGHT "ON" → REPLACE SOLENOID.

LIGHT "OFF" → PROBE EACH HARNESS CONNECTOR TERMINAL WITH A TEST LIGHT CONNECTED TO GROUND.

NO LIGHT → REPAIR OPEN CKT 39.

LIGHT "ON" ONE (CKT 39) → OPEN CKT 435. CKT 435 SHORTED TO VOLTAGE OR FAULTY ECM.

LIGHT "ON" BOTH → REPAIR SHORT TO VOLTAGE IN CKT 435.

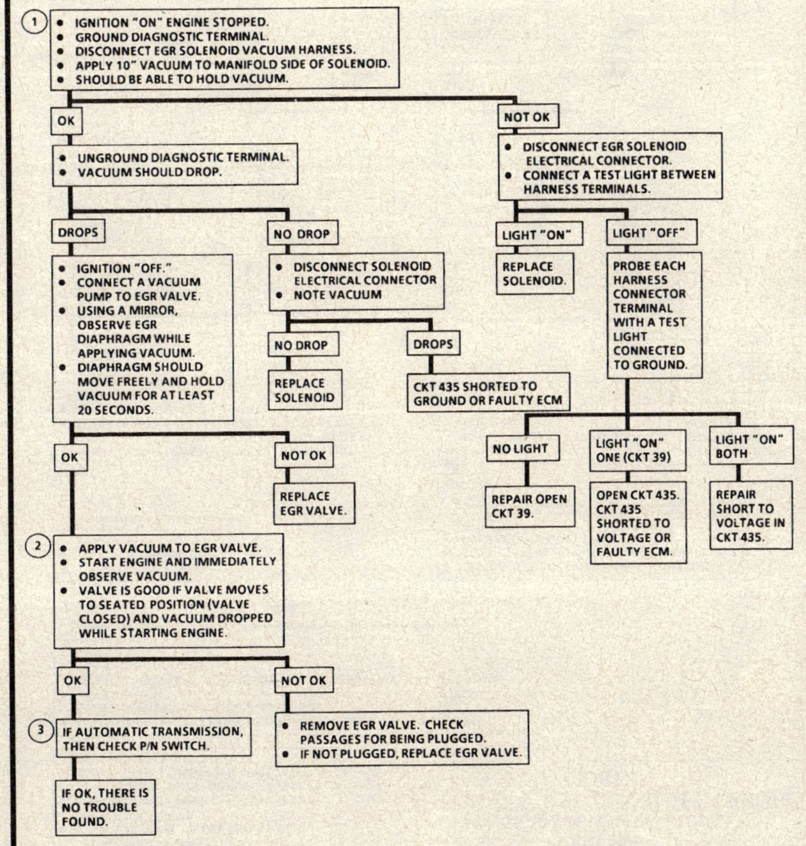

"AFTER REPAIRS," CONFIRM "CLOSED LOOP" OPERATION AND NO "SERVICE ENGINE SOON" LIGHT.

5.0L (VIN F) AND 5.7L (VIN 8) ENGINES — COMPONENT DIAGNOSTIC CHART — 1992 CAMARO AND FIREBIRD

Diagram labels (CHART C-8):

TO IGNITION
GAUGES 10 AMP
39 PNK/BLK
BRAKE SWITCH N.C.
A7 BULKHEAD CONNECTOR
420 PPL
4TH GEAR SWITCH (5.0L VIN F N.O.) (5.7L VIN 8 N.C.)
TRANSMISSION
TEMPERATURE SWITCH N.O.
TCC APPLY SOLENOID
A B C D
H 15 WAY I/P CONNECTOR
F E D C B A / G H J K L M
ALDL CONNECTOR
ECM
446 LT BLU D14 12 V
4TH GEAR SWITCH SIGNAL
422 TAN/BLK F6 TCC CONTROL

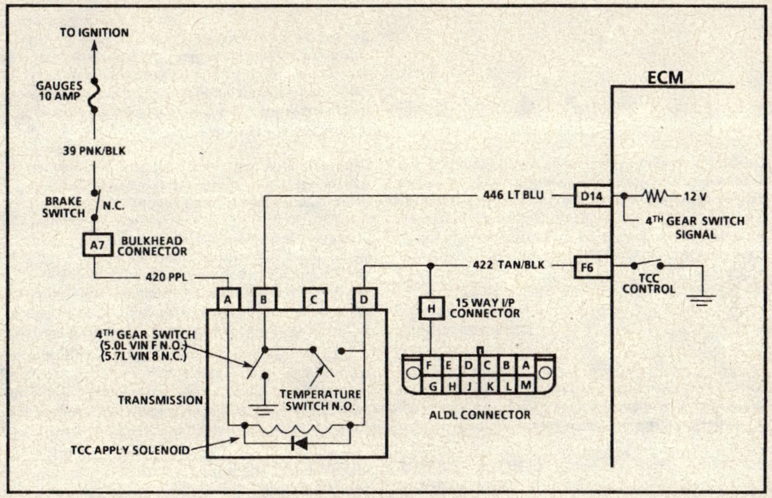

CHART C-8
(Page 1 of 3)
TORQUE CONVERTER CLUTCH (TCC) SYSTEM
(4th GEAR SWITCH DIAGNOSIS)
5.0L (VIN F) & 5.7L (VIN 8) "F" CARLINE (PORT)

Circuit Description:

The purpose of the automatic transmission Torque Converter Clutch (TCC) feature is to eliminate the power loss of the torque converter stage when the vehicle is in a cruise condition. This allows the convenience of the automatic transmission and the fuel economy of a manual transmission. The heart of the system is a solenoid, located inside the automatic transmission, which is controlled by the ECM.

When the solenoid coil is activated ("ON"), the Torque Converter Clutch (TCC) is applied which results in 100% mechanical coupling from the engine to transmission. When the transmission solenoid is deactivated, the Torque Converter Clutch (TCC) is released, which allows the torque converter to operate in the conventional manner (fluid coupling between engine and transmission).

The ECM turns "ON" the TCC when coolant temperature is above 50°C (122°F), TPS not changing, and vehicle speed above a specified value.

Test Description: Number(s) below refer to circled number(s) on the diagnostic chart.
1. When a test light is connected from ALDL terminal "F" to ground, a test light "ON" indicates battery voltage is OK and the TCC solenoid is disengaged.
2. When the diagnostic terminal is grounded, the ECM should energize the TCC solenoid and the test light should go out.
3. TCC solenoid coil resistance must measure more than 20 ohms. Less resistance will cause early failure of the ECM "Driver." Using an ohmmeter, check the solenoid coil resistance of all ECM controlled solenoids and relays before installing a replacement ECM. Replace any solenoid or relay that measures less than 20 ohms.

Diagnostic Aids:

A "Scan" tool only indicates when the ECM has turned "ON" the TCC driver (grounded CKT 422) but this does not confirm that the TCC has engaged. To determine if TCC is functioning properly, engine rpm should decrease when the "Scan" indicates the TCC driver has turned "ON." To determine if the 4th gear switch is functioning properly, perform the checks in CHART C-8 (Page 2 of 3) for 5.0L (VIN F) or (Page 3 of 3) for 5.7L (VIN 8). The switches will not prevent TCC from functioning but will affect TCC lock and unlock points. If the 4th gear switch circuit is always open/close depending on which TCC system is used, may engage as soon as sufficient oil pressure is reached.

5.0L (VIN F) AND 5.7L (VIN 8) ENGINES — COMPONENT DIAGNOSTIC CHART — 1992 CAMARO AND FIREBIRD

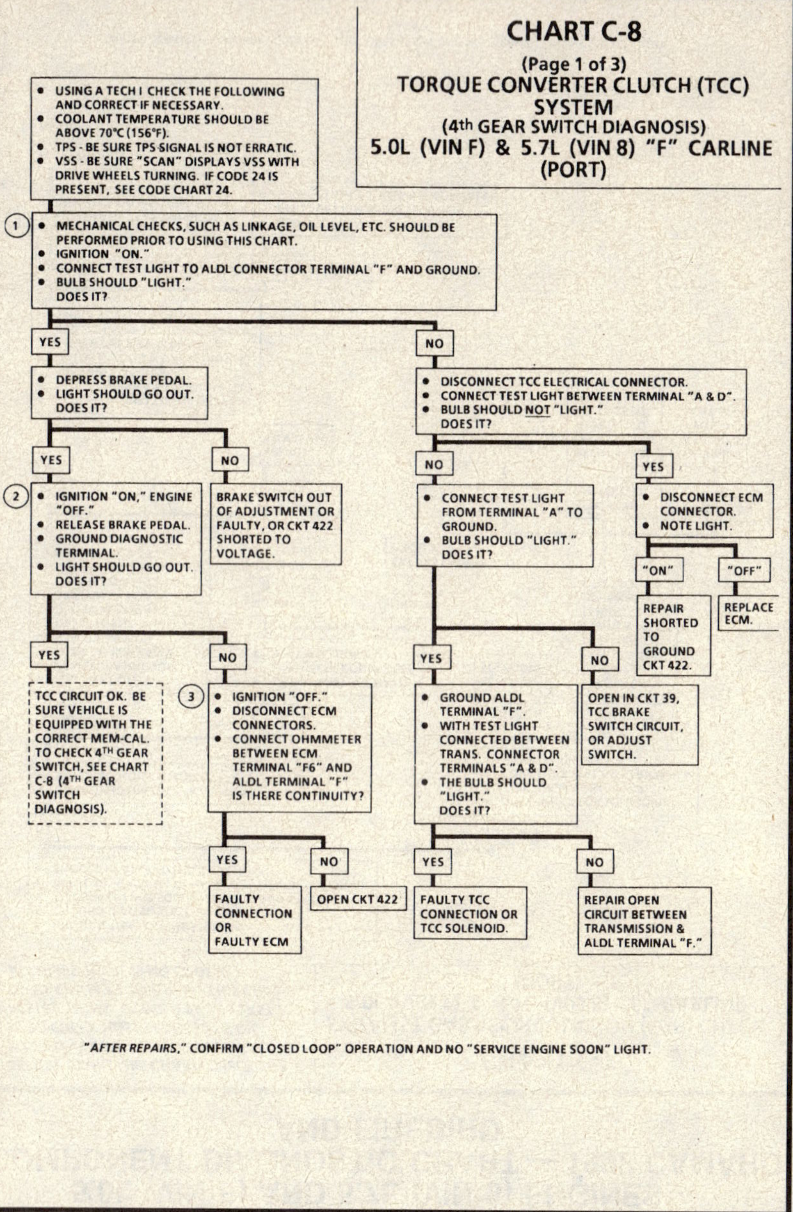

5.0L (VIN F) AND 5.7L (VIN 8) ENGINES — COMPONENT DIAGNOSTIC CHART — 1992 CAMARO AND FIREBIRD

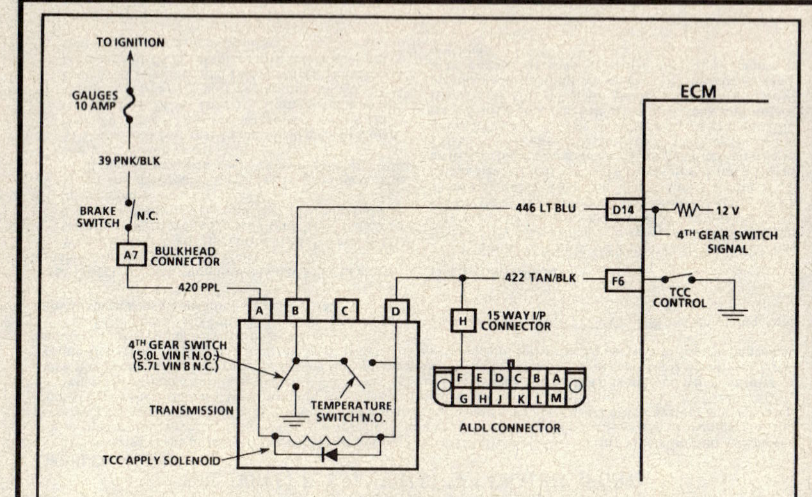

CHART C-8
(Page 2 of 3)
TORQUE CONVERTER CLUTCH (TCC) SYSTEM
(4th GEAR SWITCH DIAGNOSIS)
5.0L (VIN F) "F" CARLINE (PORT)

Circuit Description:

A 4th gear switch (mounted in the transmission) is normally open and closes when the transmission shifts into 4th gear, and this switch is used by the ECM to modify TCC lock and unlock points, when in a 4-3 downshift maneuver. Temperature switch is normally open. Transmission must be in 4th gear, for the temperature switch to apply TCC to cool transmission fluid. Temperature switch closes between 132.8°C and 140°C (271°F - 284°F) and opens between 121°C and 131°C (250°F - 267°F).

Test Description: Number(s) below refer to circled number(s) on the diagnostic chart.

1. Unless the switch is stuck closed or CKT 446 is shorted to ground. The "Scan" should display "NO," indicating the transmission is not in 4th gear. The 4th gear switch should only be closed while in 4th gear.
2. Checks the operation of the 4th gear switch. When the transmission shifts into 4th gear the switch should close and the "Scan" tool should display "YES."
3. Disconnecting the TCC connector simulates an open switch to determine if CKT 446 is shorted to ground or the problem is in the transmission.
4. This step determines if the ECM and wiring are OK. Grounding CKT 446 should cause the "Scan" tool to display "YES" indicating the transmission is in 4th gear.

Diagnostic Aids:

A road test may be necessary to verify the customer complaint. If the "Scan" tool indicates TCC is turning "ON" and "OFF" erratically, check the state of the 4th gear switch to be sure it is not changing states under a steady throttle position. If the switch is changing states, check connections and wire routing carefully. Also, if the 4th gear switch is always open the TCC may engage as soon as sufficient oil pressure is reached.

5.0L (VIN F) AND 5.7L (VIN 8) ENGINES — COMPONENT DIAGNOSTIC CHART — 1992 CAMARO AND FIREBIRD

CHART C-8

(Page 2 of 3)
TORQUE CONVERTER CLUTCH (TCC) SYSTEM
(4th GEAR SWITCH DIAGNOSIS)
5.0L (VIN F) "F" CARLINE (PORT)

CHECKS MADE ON THIS PAGE WILL NOT PREVENT THE TCC FROM WORKING, BUT WILL AFFECT ENGAGEMENT OR DISENGAGEMENT POINTS.

1.
- IGNITION "ON," ENGINE "OFF."
- DOES "SCAN" INDICATE TRANS. IS IN 4th GEAR?

NO →

2.
- RAISE DRIVE WHEELS.
- START ENGINE.
- SHIFT VEHICLE INTO OVERDRIVE.
- INCREASE SPEED SLOWLY UNTIL TRANS. SHIFTS INTO 4th GEAR.
- DOES "SCAN" INDICATE TRANS. IS IN 4th GEAR?

YES →
4th GEAR SWITCH OK. REFER TO "DIAGNOSTIC AIDS" ON FACING PAGE.

NO →

4.
- DISCONNECT TCC ELECTRICAL CONNECTOR.
- JUMPER HARNESS TERMINAL "B" (CKT 446) TO GROUND.
- DOES "SCAN" INDICATE TRANS. IS IN 4th GEAR?

YES →
FAULTY CONNECTION OR FAULTY 4th GEAR SWITCH IN TRANSMISSION.

NO →
OPEN CKT 446, FAULTY CONNECTION OR FAULTY ECM.

YES →

3.
- IGNITION "ON," ENGINE "OFF."
- DISCONNECT TRANS. ELECTRICAL CONNECTOR.
- DOES "SCAN" INDICATE TRANS. IS IN 4th GEAR?

YES →
CKT 446 SHORTED TO GROUND OR FAULTY ECM.

NO →
FAULTY 4th GEAR SWITCH OR CKT 446 GROUNDED INTERNALLY IN TRANSMISSION.

"AFTER REPAIRS," CONFIRM "CLOSED LOOP" OPERATION AND NO "SERVICE ENGINE SOON" LIGHT.

5.0L (VIN F) AND 5.7L (VIN 8) ENGINES — COMPONENT DIAGNOSTIC CHART — 1992 CAMARO AND FIREBIRD

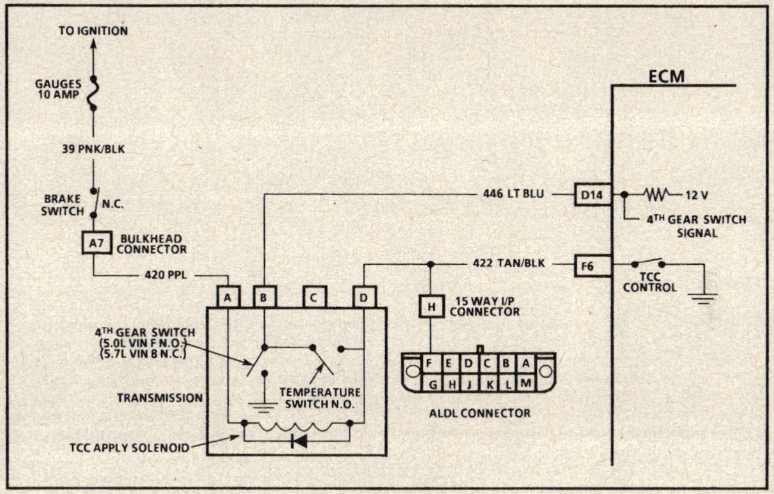

CHART C-8

(Page 3 of 3)
TORQUE CONVERTER CLUTCH (TCC) SYSTEM
(4th GEAR SWITCH DIAGNOSIS)
5.7L (VIN 8) "F" CARLINE (PORT)

Circuit Description:

A 4th gear switch (mounted in the transmission) is normally closed and opens when the transmission shifts into 4th gear, and this switch is used by the ECM to modify TCC lock and unlock points, when in a 4-3 downshift maneuver. Temperature switch is normally open. Transmission must be in 4th gear, for the temperature switch to apply TCC to cool transmission fluid. Temperature switch closes between 132.8°C and 140°C (271°F - 284°F) and opens between 121°C and 131°C (250°F - 267°F).

Test Description: Number(s) below refer to circled number(s) on the diagnostic chart.

1. Unless the switch is stuck open or CKT 446 is open, the "Scan" tool should display "NO," indicating the transmission is not in 4th gear. The 4th gear switch should only be open while in 4th gear.
2. This step determines if the ECM and wiring are OK. Grounding CKT 446 should cause the "Scan" tool to display "NO" indicating the transmission is not in 4th gear.
3. Checks the operation of the 4th gear switch. When the transmission shifts into 4th gear the switch should close and the "Scan" tool should display "YES."
4. Disconnecting the TCC connector simulates an open switch to determine if CKT 446 is shorted to ground or the problem is in the transmission.

Diagnostic Aids:

A road test may be necessary to verify the customer complaint. If the "Scan" tool indicates TCC is turning "ON" and "OFF" erratically, check the state of the 4th gear switch to be sure it is not changing states under a steady throttle position. If the switch is changing states, check connections and wire routing carefully. Also, if the 4th gear switch is always open the TCC may engage as soon as sufficient oil pressure is reached.

5.0L (VIN F) AND 5.7L (VIN 8) ENGINES — COMPONENT DIAGNOSTIC CHART — 1992 CAMARO AND FIREBIRD

CHART C-8
(Page 3 of 3)
TORQUE CONVERTER CLUTCH (TCC) SYSTEM
(4th GEAR SWITCH DIAGNOSIS)
5.7L (VIN 8) "F" CARLINE (PORT)

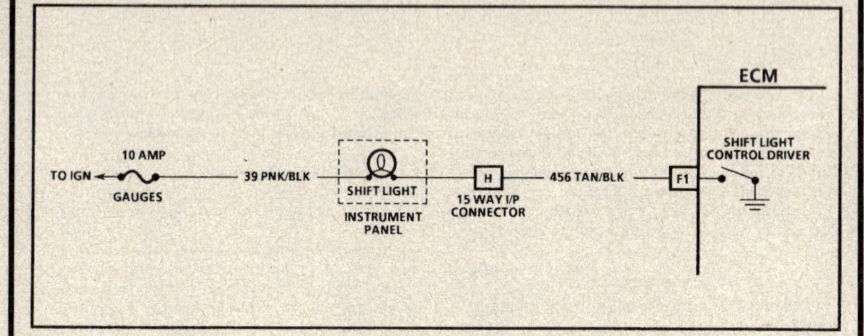

5.0L (VIN F) AND 5.7L (VIN 8) ENGINES — COMPONENT DIAGNOSTIC CHART — 1992 CAMARO AND FIREBIRD

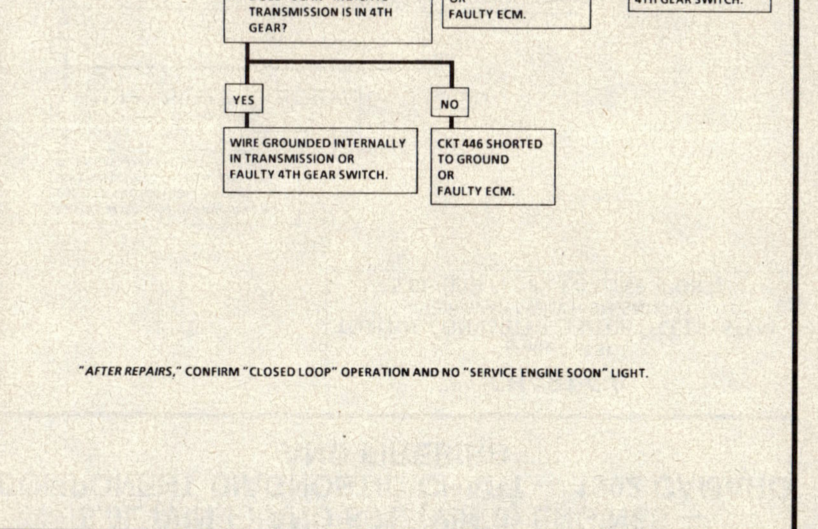

CHART C-8C
MANUAL TRANSMISSION SHIFT LIGHT DIAGNOSIS
5.0L (VIN F) "F" CARLINE (PORT)

Circuit Description:

The purpose of the shift light is to provide a display, to the driver, which indicates the best transmission shift point for maximum fuel economy based on engine speed and load. The light is controlled by the ECM and is turned "ON" by grounding CKT 456.

The ECM uses information from the following inputs to control the shift light:
- Coolant Temperature Sensor (CTS).
- Throttle Position Sensor (TPS).
- Vehicle Speed Sensor (VSS).
- Rpm.

The ECM uses the measured rpm and the vehicle speed to calculate what gear the vehicle is in. It's this calculation that determines when the shift light should be turned "ON."

Test Description: Number(s) below refer to circled number(s) on the diagnostic chart.
1. This should not turn "ON" the shift light. If the light is "ON," there is a short to ground in CKT 456 wiring or a fault in the ECM.
2. When the diagnostic terminal is grounded, the ECM should ground CKT 456 and the shift light should come "ON."
3. This checks the shift light circuit up to the ECM connector. If the shift light illuminates, then the ECM connector is faulty or the ECM does not have the ability to ground the circuit.

5.0L (VIN F) AND 5.7L (VIN 8) ENGINES — COMPONENT DIAGNOSTIC CHART — 1992 CAMARO AND FIREBIRD

CHART C-8C
MANUAL TRANSMISSION SHIFT LIGHT DIAGNOSIS
5.0L (VIN F) "F" CARLINE (PORT)

1. • IGNITION "ON," ENGINE "OFF."
   • OBSERVE SHIFT LIGHT.

LIGHT "OFF"

2. GROUND ALDL DIAGNOSTIC TERMINAL AND OBSERVE SHIFT LIGHT.

LIGHT "ON"

• IGNITION "OFF."
• DISCONNECT ECM CONNECTORS.
• TURN IGNITION "ON" AND OBSERVE SHIFT LIGHT.

LIGHT "ON"

REPAIR SHORT TO GROUND IN CKT 456.

LIGHT "OFF"

REPLACE ECM.

LIGHT "OFF"

3. • IGNITION "OFF."
   • DISCONNECT ECM CONNECTORS.
   • IGNITION "ON."
   • JUMPER CKT 456 TO GROUND AND OBSERVE SHIFT LIGHT.

LIGHT "ON"

CHECK FOR:
- CODE 24, (NO VSS).
- THERMOSTAT FAULTY OR HAS INCORRECT HEAT RANGE.
  IF OK, REVIEW SYMPTOMS IN SECTION "B".

LIGHT "OFF"

OPEN IGNITION CKT 39, OPEN CKT 456, OR FAULTY BULB.

LIGHT "ON"

POOR CONNECTION AT ECM OR FAULTY ECM.

"AFTER REPAIRS," CONFIRM "CLOSED LOOP" OPERATION AND NO "SERVICE ENGINE SOON" LIGHT.

5.0L (VIN F) AND 5.7L (VIN 8) ENGINES — COMPONENT DIAGNOSTIC CHART — 1992 CAMARO AND FIREBIRD

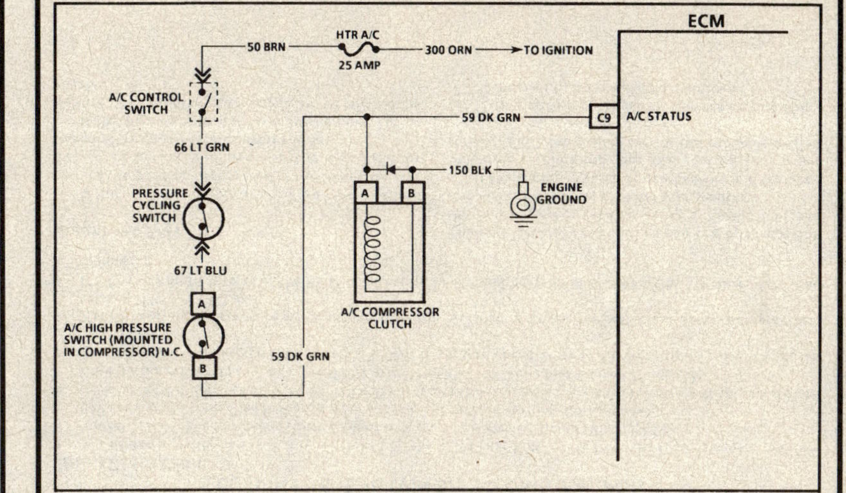

CHART C-10
A/C CLUTCH CIRCUIT DIAGNOSIS
5.0L (VIN F) & 5.7L (VIN 8) "F" CARLINE (PORT)

Circuit Description:
When the A/C selector control is switched to the "ON" position, ignition voltage is supplied to the pressure cycling switch through the control head. If there is sufficient A/C charge, the pressure cycling switch will be closed to complete the circuit to the ECM and to the A/C high pressure switch. If A/C head pressure is not too high, the circuit would be completed to the A/C clutch coil.

Test Description: Number(s) below refer to circled number(s) on the diagnostic chart.
1. Check for a short to ground in CKTs 59, 66, or 67 if fuse was open. Check for shorted compressor clutch coil.

2. Check to determine if the ECM is capable of detecting A/C status.

3. If A/C clutch engaged, check for 12 volt supply at ECM terminal "C9". If voltage is present then there is a faulty connection at the ECM or faulty ECM.

5.0L (VIN F) AND 5.7L (VIN 8) ENGINES — COMPONENT DIAGNOSTIC CHART — 1992 CAMARO AND FIREBIRD

CHART C-10

A/C CLUTCH CIRCUIT DIAGNOSIS
5.0L (VIN F) & 5.7L (VIN 8)
"F" CARLINE (PORT)

- CHECK HTR / A/C FUSE

OK →
- WITH ENGINE IDLING AT NORMAL OPERATING TEMPERATURE.
- A/C "ON."
IS A/C CLUTCH ENGAGED?

NOT OK →
(1) REFER TO ELECTRICAL DIAGNOSIS

NO →
- DISCONNECT CONNECTOR AT A/C COMPRESSOR.
- PROBE CKT 59 WITH A TEST LIGHT TO GROUND, SHOULD LIGHT, DOES IT?

YES →
(2) DOES SCAN TOOL INDICATE A/C REQUEST AS "YES"?

NO →
(3) OPEN CKT 59 FROM SPLICE TO ECM
OR
FAULTY ECM CONNECTION
OR
FAULTY ECM.

YES →
NO TROUBLE FOUND

NO →
- DISCONNECT HIGH PRESSURE SWITCH.
- PROBE CKT 67 WITH A TEST LIGHT TO GROUND. SHOULD LIGHT, DOES IT?

YES →
CONNECT TEST LIGHT BETWEEN A/C COMPRESSOR HARNESS CONNECTOR TERMINALS "A" AND "B."

LIGHT "ON" →
FAULTY CONNECTION AT CLUTCH COIL
OR
FAULTY A/C COMPRESSOR CLUTCH COIL.

LIGHT "OFF" →
OPEN CKT 150
OR
FAULTY GROUND CONNECTION

NO →
- RECONNECT HIGH PRESSURE SWITCH.
- DISCONNECT PRESSURE CYCLING SWITCH.
- PROBE CKT 66 WITH A TEST LIGHT TO GROUND. SHOULD LIGHT, DOES IT?

YES →
JUMPER CKTS 59 AND 67 TOGETHER AT HIGH PRESSURE SWITCH HARNESS CONNECTOR. DOES A/C CLUTCH ENGAGE?

YES →
FAULTY SWITCH CONNECTION
OR
FAULTY HIGH PRESSURE SWITCH
OR
HIGH A/C PRESSURE

NO →
OPEN CKT 59 FROM HIGH PRESSURE SWITCH TO A/C COMPRESSOR.

YES →
- JUMPER CKTS 66 AND 67 TOGETHER AT PRESSURE CYCLING SWITCH HARNESS. DOES A/C CLUTCH ENGAGE?

NO →
OPEN CKT 66
OR
FAULTY CONTROL HEAD CONNECTION
OR
FAULTY CONTROL HEAD. REFER TO MANUALLY CONTROLLED AIR CONDITIONING

YES →
FAULTY SWITCH CONNECTION
OR
FAULTY PRESSURE CYCLING SWITCH
OR
A/C SYSTEM LOW ON CHARGE

NO →
OPEN CKT 67

"AFTER REPAIRS," CONFIRM "CLOSED LOOP" OPERATION AND NO "SERVICE ENGINE SOON" LIGHT.

5.0L (VIN F) AND 5.7L (VIN 8) ENGINES — COMPONENT DIAGNOSTIC CHART — 1992 CAMARO AND FIREBIRD

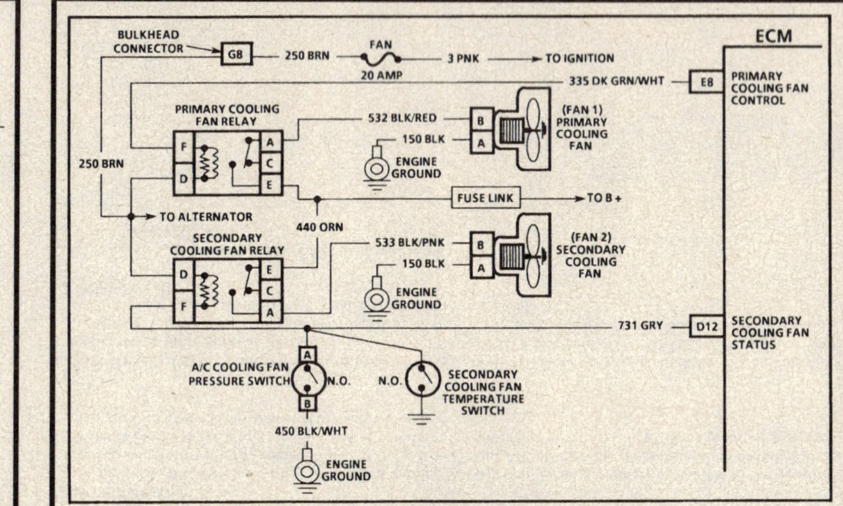

CHART C-12

(Page 1 of 2)
ELECTRIC COOLING FAN
5.0L (VIN F) & 5.7L (VIN 8) "F" CARLINE (PORT)

Circuit Description:

- The primary cooling fan is totally controlled by the ECM based on inputs from the coolant sensor and fan control switch. The fan should run, if coolant temperature is greater than 106°C (223°F).
- Battery voltage is supplied to the fan relay on terminal "E" and ignition voltage to terminal "D".
- Grounding CKT 335 (relay terminal "F") will energize the relay and supply battery voltage to the fan motor. Once the fan relay is energized by the ECM, it will remain "ON" for a minimum of 15 seconds.
- The ECM will remove the ground to CKT 335, if vehicle speed is over 40 mph, unless the engine is overheating.
- The A/C cooling fan pressure switch, mounted in the A/C high pressure line, will close when head pressure exceeds 1606 kPa (233 psi).
- If a Code 14 or 15 sets, or if the ECM is operating in the fuel back-up mode, the ECM will turn "ON" the cooling fan.

Diagnostic Aids:

If the owner complained of an overheating problem, it must be determined if the complaint was due to an actual boil over or the hot light or temperature gage indicated overheating.

If the gage or light indicates overheating, but no boilover is detected, the gage circuit should be checked.

The gage accuracy can also be checked by comparing the coolant sensor reading using a "Scan" tool and comparing its reading with the gage reading.

If the engine is actually overheating and the gage indicates overheating, but the cooling fan is not coming "ON", the coolant sensor has probably shifted out of calibration and should be replaced.

If the engine is overheating, and the cooling fan is "ON," the cooling system should be checked.

5.0L (VIN F) AND 5.7L (VIN 8) ENGINES — COMPONENT DIAGNOSTIC CHART — 1992 CAMARO AND FIREBIRD

CHART C-12
(Page 1 of 2)
ELECTRIC COOLING FAN
5.0L (VIN F) & 5.7L (VIN 8) "F" CARLINE (PORT)

- IGNITION "ON", ENGINE "OFF", A/C "OFF".
- ENGINE COOLANT TEMPERATURE BELOW 100°C (212°F).
  ARE BOTH COOLING FANS "OFF"?

YES →
- GROUND DIAGNOSTIC "TEST" TERMINAL.
- PRIMARY FAN SHOULD TURN "ON", DOES IT?

NO → (see PRIMARY / SECONDARY branch below)

PRIMARY:
- DISCONNECT PRIMARY FAN RELAY. FAN SHOULD STOP, DOES IT?
  - YES → PROBE CKT 335 WITH A TEST LIGHT TO B+
    - LIGHT "ON" → CKT 335 SHORTED TO GROUND OR FAULTY ECM
    - LIGHT "OFF" → FAULTY PRIMARY COOLING FAN RELAY
  - NO → CKT 532 SHORTED TO VOLTAGE

SECONDARY:
- DISCONNECT SECONDARY FAN RELAY. FAN SHOULD STOP, DOES IT?
  - YES → PROBE CKT 731 AT ECM WITH A DVOM TO GROUND. SHOULD BE BATTERY VOLTAGE, IS IT?
    - YES → FAULTY SECONDARY COOLING FAN RELAY
    - NO → CKT 731 SHORTED TO GROUND OR FAULTY COOLING FAN TEMPERATURE SWITCH OR FAULTY A/C COOLING FAN PRESSURE SWITCH
  - NO → CKT 533 SHORTED TO VOLTAGE

YES (from ground diagnostic) →
- UNGROUND DIAGNOSTIC "TEST" TERMINAL.
- PROBE CKT 731 AT THE ECM WITH A TEST LIGHT CONNECTED TO GROUND. SECONDARY COOLING FAN SHOULD TURN "ON", DOES IT?
  - NO → REFER TO PAGE (2 OF 2)

YES →
- START ENGINE, A/C "ON."
- ENGINE COOLANT TEMPERATURE BELOW 100°C (212°F).
- IF A/C INOP., SEE SECTION C-10.
- SECONDARY COOLING FAN SHOULD TURN "ON" WHEN A/C HEAD PRESSURE EXCEEDS ABOUT 233 psi (1606 kPa), DOES IT?
  - NO → REFER TO PAGE (2 OF 2)

YES →
- DISCONNECT PRIMARY COOLING FAN RELAY.
- ENGINE IDLING, A/C "OFF."
- WHEN COOLANT TEMPERATURE REACHES ABOUT 112°C (234°F) TO 114°C (238°F), SECONDARY COOLING FAN SHOULD TURN "ON", DOES IT?
  - YES → NO TROUBLE FOUND, REFER TO DIAGNOSTIC AIDS OF FACING PAGE
  - NO →
    - DISCONNECT SECONDARY COOLING FAN TEMPERATURE SWITCH.
    - WITH A FUSED JUMPER, GROUND CKT 731 AT THE SECONDARY COOLING FAN TEMPERATURE SWITCH HARNESS. SECONDARY COOLING FAN SHOULD TURN "ON", DOES IT?
      - YES → FAULTY SECONDARY COOLING FAN TEMPERATURE SWITCH
      - NO → OPEN CKT 731 FROM SECONDARY COOLING FAN TEMPERATURE SWITCH TO SECONDARY COOLING FAN RELAY

NO →
- DISCONNECT A/C COOLING FAN PRESSURE SWITCH.
- JUMPER TERMINALS "A" AND "B" TOGETHER OF A/C COOLING FAN PRESSURE SWITCH HARNESS. FAN SHOULD TURN "ON", DOES IT?
  - NO →
    - PROBE CKT 731 AT A/C PRESSURE SWITCH HARNESS WITH A DVOM TO GROUND.
    - VOLT METER SHOULD DISPLAY BATTERY VOLTAGE, DOES IT?
      - NO → CKT 731 OPEN FROM A/C COOLING FAN PRESSURE SWITCH TO SECONDARY COOLING FAN RELAY
      - YES → FAULTY CONNECTION AT A/C COOLING FAN PRESSURE SWITCH OR FAULTY SWITCH
  - YES → FAULTY CONNECTION AT A/C COOLING FAN PRESSURE SWITCH OR FAULTY SWITCH
    - YES → CKT 450 OPEN

"AFTER REPAIRS," CONFIRM "CLOSED LOOP" OPERATION AND NO "SERVICE ENGINE SOON" LIGHT.

5.0L (VIN F) AND 5.7L (VIN 8) ENGINES — COMPONENT DIAGNOSTIC CHART — 1992 CAMARO AND FIREBIRD

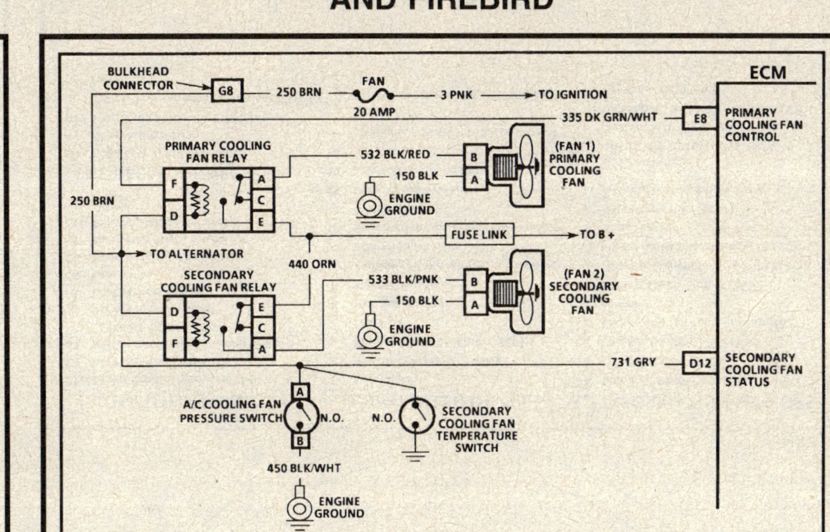

CHART C-12
(Page 2 of 2)
ELECTRIC COOLING FAN
5.0L (VIN F) & 5.7L (VIN 8) "F" CARLINE (PORT)

Circuit Description:

- The primary cooling fan is totally controlled by the ECM based on inputs from the coolant sensor and fan control switch. The fan should run, if coolant temperature is greater than 106°C (223°F).
- Battery voltage is supplied to the fan relay on terminal "E" and ignition voltage to terminal "D".
- Grounding CKT 335 (relay terminal "F") will energize the relay and supply battery voltage to the fan motor. Once the fan relay is energized by the ECM, it will remain "ON" for a minimum of 15 seconds.
- The ECM will remove the ground to CKT 335, if vehicle speed is over 40 mph, unless the engine is overheating.
- The A/C cooling fan switch, mounted in the A/C high pressure line, will close when head pressure exceeds 1606 kPa (233 psi).
- If a Code 14 or 15 sets, or if the ECM is operating in the fuel back-up mode, the ECM will turn "ON" the cooling fan.

Diagnostic Aids:

If the owner complained of an overheating problem, it must be determined if the complaint was due to an actual boil over or the hot light or temperature gage indicated overheating.

If the gage or light indicates overheating, but no boilover is detected, the gage circuit should be checked.

The gage accuracy can also be checked by comparing the coolant sensor reading using a "Scan" tool and comparing its reading with the gage reading.

If the engine is actually overheating and the gage indicates overheating, but the cooling fan is not coming "ON," the coolant sensor has probably shifted out of calibration and should be replaced.

If the engine is overheating, and the cooling fan is "ON," the cooling system should be checked.

5.0L (VIN F) AND 5.7L (VIN 8) ENGINES — COMPONENT DIAGNOSTIC CHART — 1992 CAMARO AND FIREBIRD

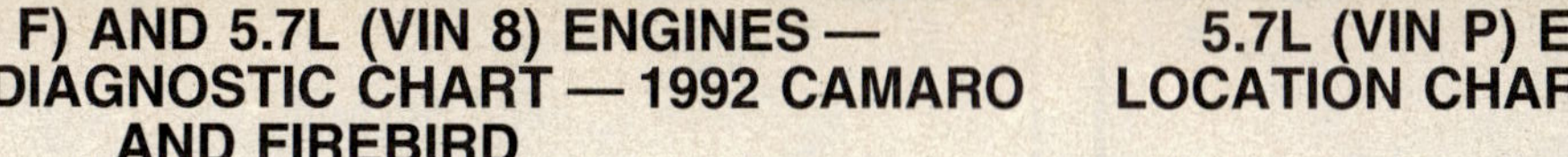

5.7L (VIN P) ENGINE — ENGINE COMPONENT LOCATION CHART — 1993 CAMARO AND FIREBIRD

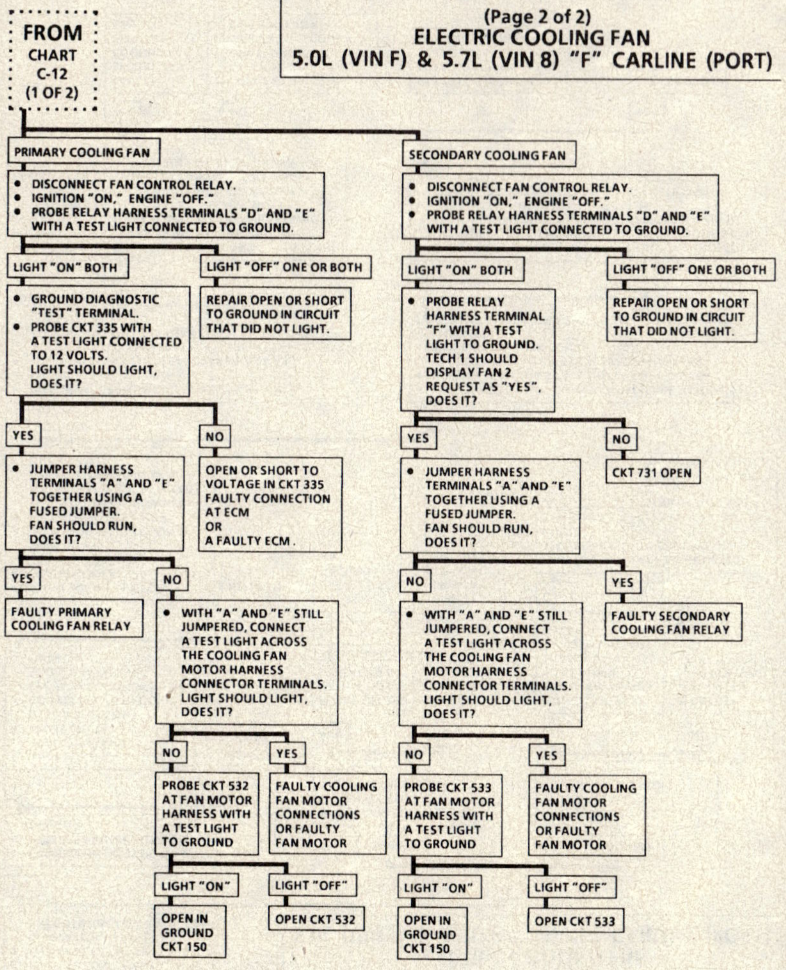

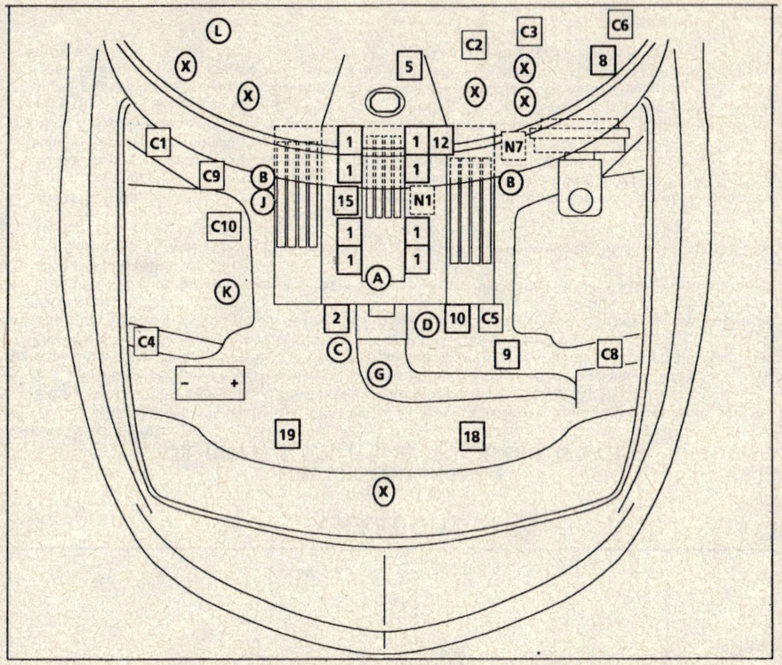

COMPUTER HARNESS

C1 Engine Control Module (ECM)
C2 Data Link Connector (DLC)
C3 Malfunction Indicator Lamp (MIL) "SERVICE ENGINE SOON"
C4 Battery junction block
C5 ECM harness grounds
C6 Fuse panel
C8 Underhood electrical center
C9 Fuel pump "test" connector
C10 A/C compressor low pressure cut-off switch

NOT ECM CONNECTED

N1 Crankcase ventilation valve
N7 Oil pressure sensor
N17 Evaporative emission canister (not shown), located in left rear, near filler tube

CONTROLLED DEVICES

1 Fuel injector
2 Idle Air Control (IAC) valve
5 Torque Converter Clutch (TCC) connector
8 Fuel pump relay
9 Electric air pump
10 Ignition coil and ignition coil module
12 EGR solenoid valve
15 Evaporative emission canister purge solenoid valve
18 Primary cooling fan
19 Secondary cooling fan
20 Reverse lockout solenoid (M/T only) (Transmission mounted, not shown)

Exhaust Gas Recirculation (EGR) valve

INFORMATION SENSORS

A Manifold Absolute Pressure (MAP)
B Oxygen sensor (O2S)
C Throttle Position (TP) sensor
D Engine Coolant Temperature (ECT) sensor
F Vehicle Speed Sensor (VSS) (Transmission mounted, not shown)
G Intake Air Temperature (IAT)
J Knock sensor
K A/C refrigerant pressure
L A/C evaporator temperature

X SIR system components. Refer to Section 9J of the service manual, for "Cautions" and information on SIR system components.

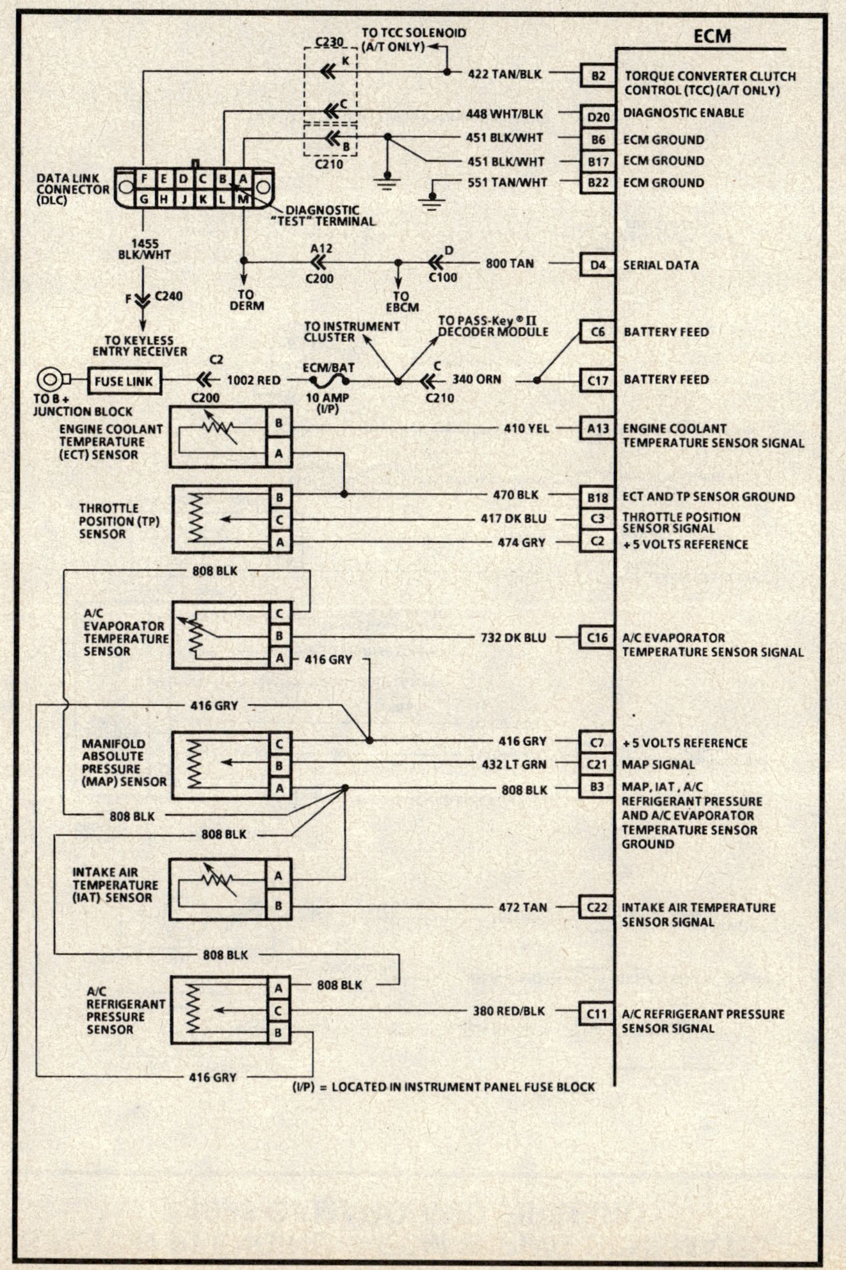

5.7L (VIN P) ENGINE — ECM WIRING SCHEMATIC — 1993 CAMARO AND FIREBIRD
ECM
C230
TO TCC SOLENOID (A/T ONLY)
422 TAN/BLK
B2
TORQUE CONVERTER CLUTCH CONTROL (TCC) (A/T ONLY)
448 WHT/BLK
D20
DIAGNOSTIC ENABLE
451 BLK/WHT
B6
ECM GROUND
451 BLK/WHT
B17
ECM GROUND
551 TAN/WHT
B22
ECM GROUND
DATA LINK CONNECTOR (DLC)
F E D C B A
G H J K L M
DIAGNOSTIC "TEST" TERMINAL
1455 BLK/WHT
A12
C200
800 TAN
D4
SERIAL DATA
C100
C240
F
TO DERM
TO EBCM
TO INSTRUMENT CLUSTER
TO PASS-Key®II DECODER MODULE
TO KEYLESS ENTRY RECEIVER
C6
BATTERY FEED
FUSE LINK
1002 RED
ECM/BAT 10 AMP (I/P)
C200
340 ORN
C210
C17
BATTERY FEED
TO B+ JUNCTION BLOCK
ENGINE COOLANT TEMPERATURE (ECT) SENSOR
410 YEL
A13
ENGINE COOLANT TEMPERATURE SENSOR SIGNAL
808 BLK
THROTTLE POSITION (TP) SENSOR
470 BLK
B18
ECT AND TP SENSOR GROUND
417 DK BLU
C3
THROTTLE POSITION SENSOR SIGNAL
474 GRY
C2
+5 VOLTS REFERENCE
808 BLK
A/C EVAPORATOR TEMPERATURE SENSOR
732 DK BLU
C16
A/C EVAPORATOR TEMPERATURE SENSOR SIGNAL
416 GRY
416 GRY
MANIFOLD ABSOLUTE PRESSURE (MAP) SENSOR
416 GRY
C7
+5 VOLTS REFERENCE
432 LT GRN
C21
MAP SIGNAL
808 BLK
B3
MAP, IAT, A/C REFRIGERANT PRESSURE AND A/C EVAPORATOR TEMPERATURE SENSOR GROUND
808 BLK
808 BLK
INTAKE AIR TEMPERATURE (IAT) SENSOR
472 TAN
C22
INTAKE AIR TEMPERATURE SENSOR SIGNAL
808 BLK
808 BLK
A/C REFRIGERANT PRESSURE SENSOR
380 RED/BLK
C11
A/C REFRIGERANT PRESSURE SENSOR SIGNAL
416 GRY
(I/P) = LOCATED IN INSTRUMENT PANEL FUSE BLOCK

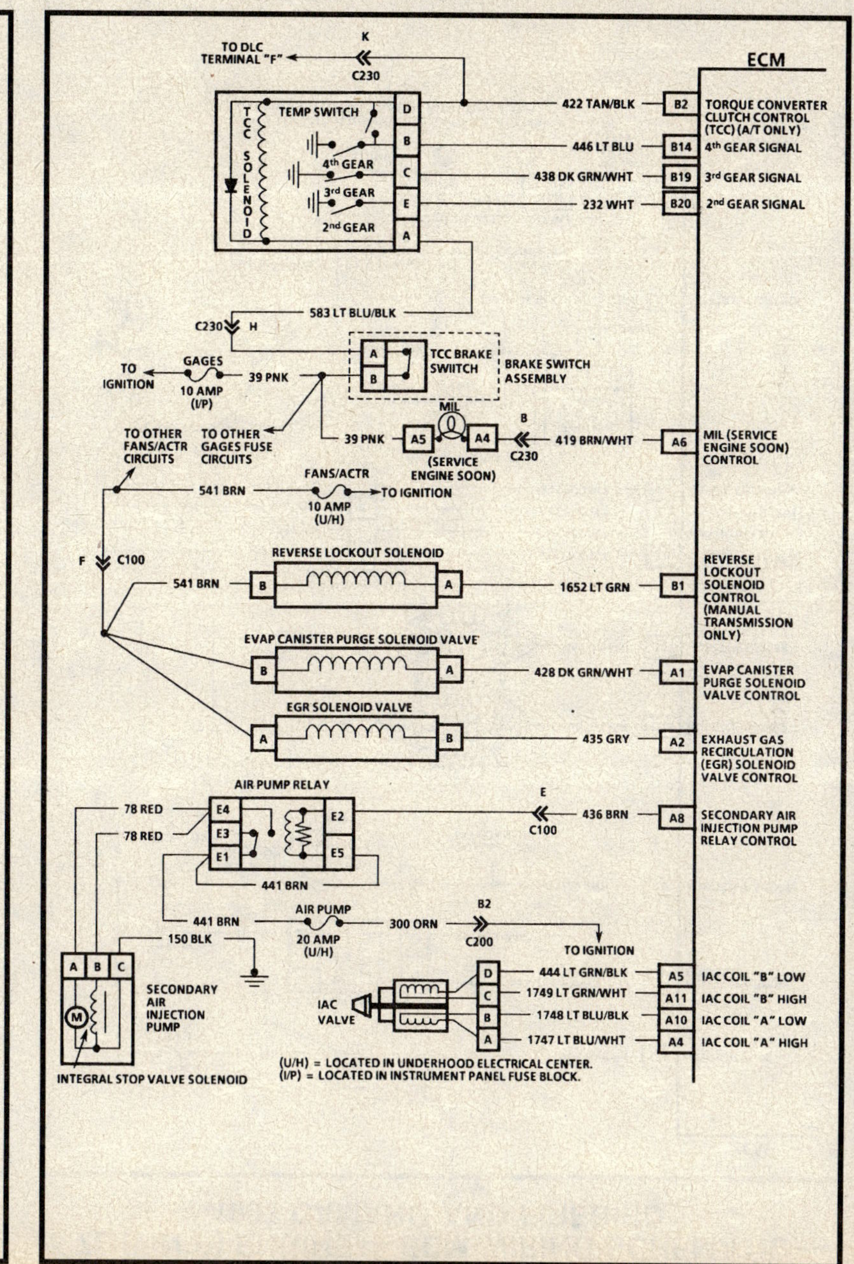

5.7L (VIN P) ENGINE — ECM WIRING SCHEMATIC — 1993 CAMARO AND FIREBIRD
ECM
TO DLC TERMINAL "F"
K
C230
TEMP SWITCH
TCC SOLENOID
4th GEAR
3rd GEAR
2nd GEAR
D
B
C
E
A
422 TAN/BLK
B2
TORQUE CONVERTER CLUTCH CONTROL (TCC) (A/T ONLY)
446 LT BLU
B14
4th GEAR SIGNAL
438 DK GRN/WHT
B19
3rd GEAR SIGNAL
232 WHT
B20
2nd GEAR SIGNAL
C230
H
583 LT BLU/BLK
TO IGNITION
GAGES 10 AMP (I/P)
39 PNK
A
B
TCC BRAKE SWIITCH
BRAKE SWITCH ASSEMBLY
TO OTHER FANS/ACTR CIRCUITS
TO OTHER GAGES FUSE CIRCUITS
39 PNK
A5
A4
MIL (SERVICE ENGINE SOON)
419 BRN/WHT
B
C230
A6
MIL (SERVICE ENGINE SOON) CONTROL
541 BRN
FANS/ACTR 10 AMP (U/H)
TO IGNITION
F
C100
REVERSE LOCKOUT SOLENOID
541 BRN
B
A
1652 LT GRN
B1
REVERSE LOCKOUT SOLENOID CONTROL (MANUAL TRANSMISSION ONLY)
EVAP CANISTER PURGE SOLENOID VALVE
B
A
428 DK GRN/WHT
A1
EVAP CANISTER PURGE SOLENOID VALVE CONTROL
EGR SOLENOID VALVE
A
B
435 GRY
A2
EXHAUST GAS RECIRCULATION (EGR) SOLENOID VALVE CONTROL
AIR PUMP RELAY
E4
E3
E1
E2
E5
436 BRN
E
C100
A8
SECONDARY AIR INJECTION PUMP RELAY CONTROL
78 RED
78 RED
441 BRN
441 BRN
150 BLK
AIR PUMP 20 AMP (U/H)
300 ORN
B2
C200
TO IGNITION
A
B
C
SECONDARY AIR INJECTION PUMP
INTEGRAL STOP VALVE SOLENOID
IAC VALVE
D
C
B
A
444 LT GRN/BLK
A5
IAC COIL "B" LOW
1749 LT GRN/WHT
A11
IAC COIL "B" HIGH
1748 LT BLU/BLK
A10
IAC COIL "A" LOW
1747 LT BLU/WHT
A4
IAC COIL "A" HIGH
(U/H) = LOCATED IN UNDERHOOD ELECTRICAL CENTER.
(I/P) = LOCATED IN INSTRUMENT PANEL FUSE BLOCK.

MULTIPORT FUEL INJECTION (MFI) SYSTEMS
EXCEPT LIGHT TRUCKS, VANS, GEO AND SATURN

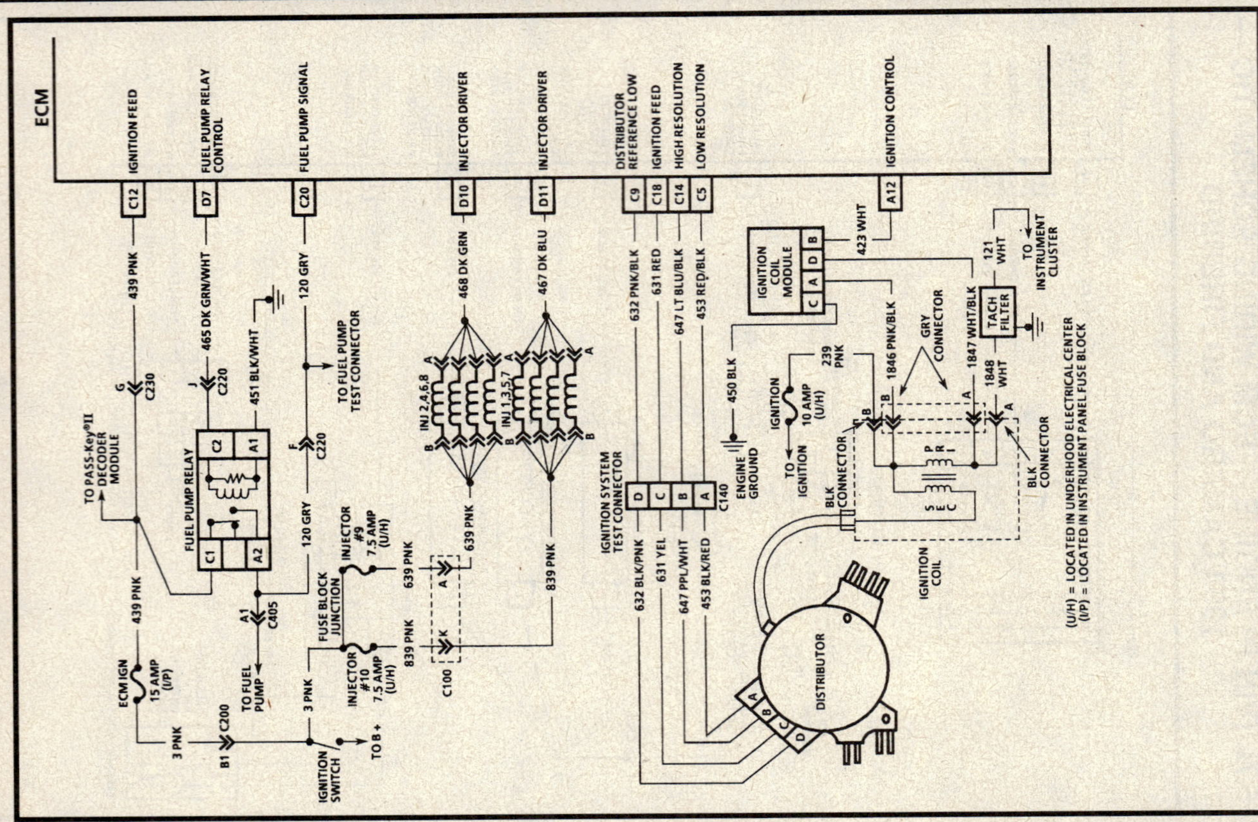

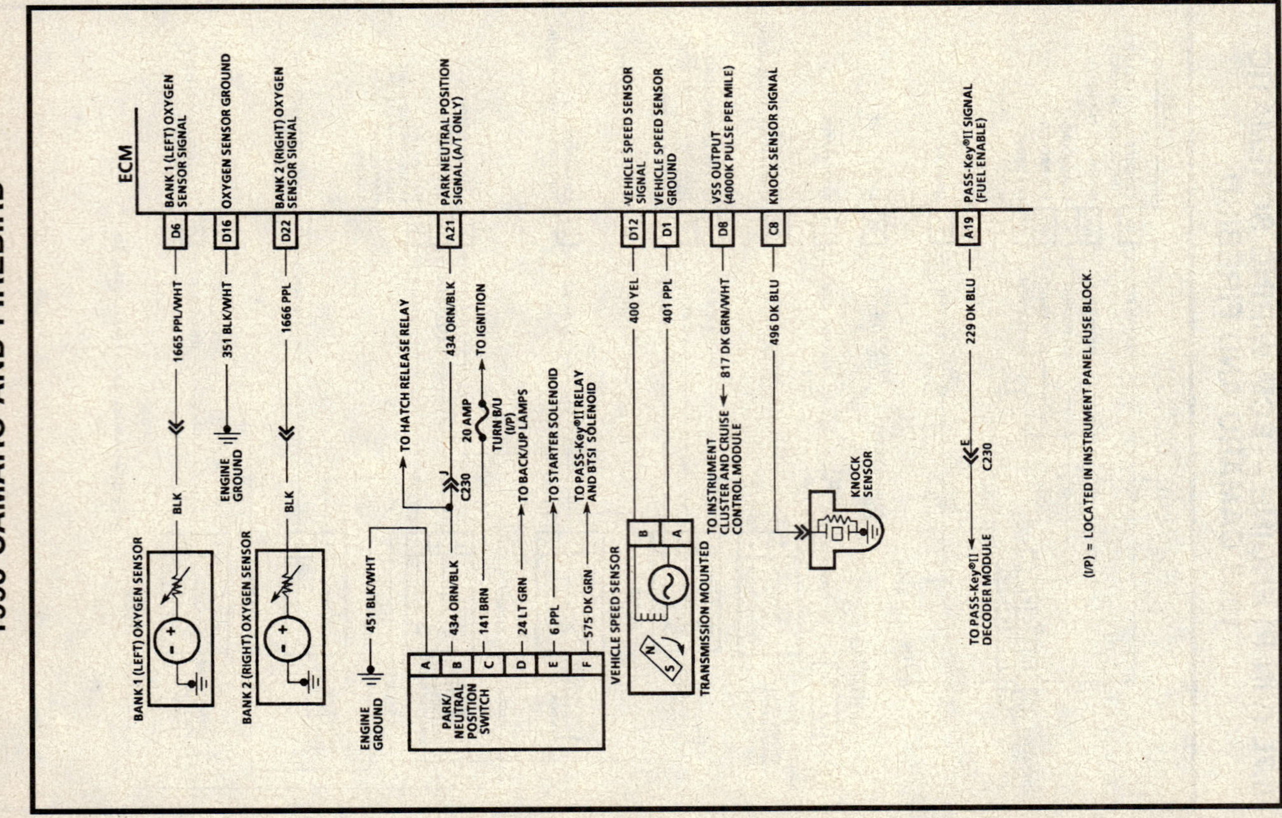

5.7L (VIN P) ENGINE — ECM WIRING SCHEMATIC — 1993 CAMARO AND FIREBIRD

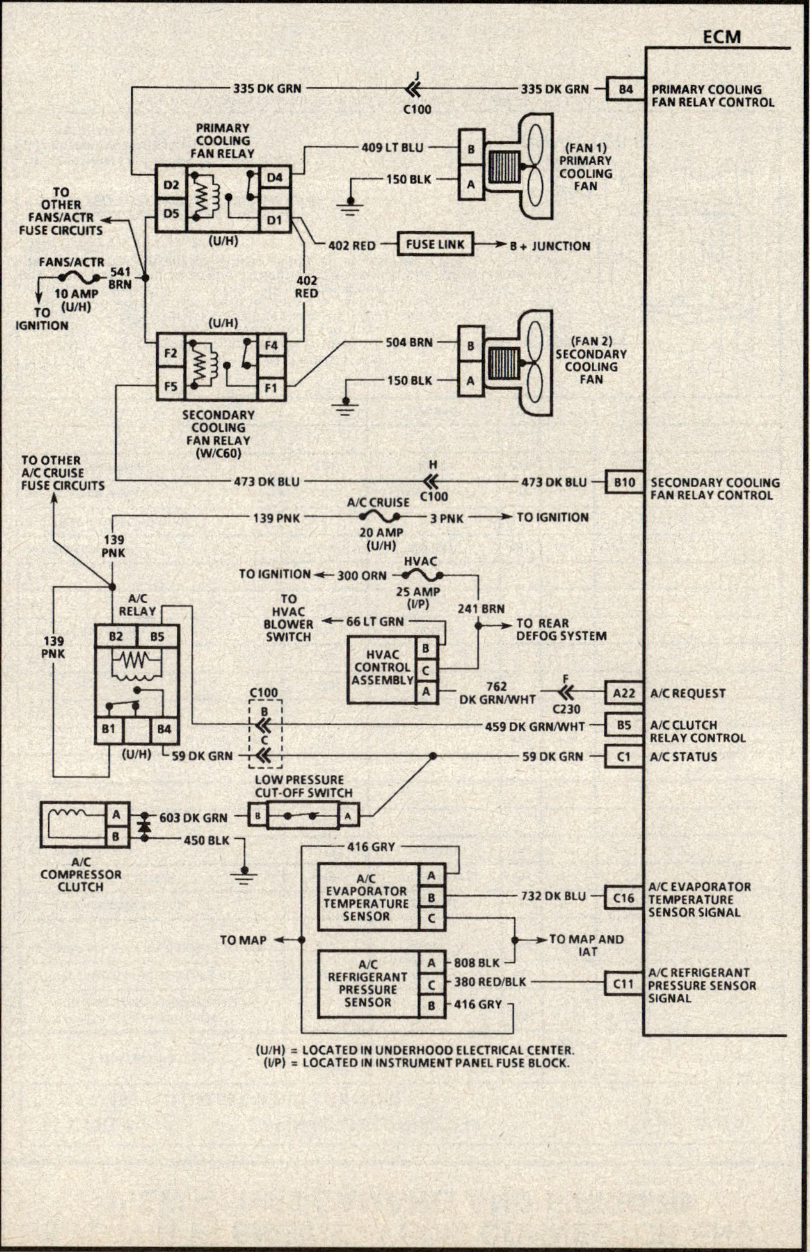

5.7L (VIN P) ENGINE — ECM CONNECTOR END VIEW — 1993 CAMARO AND FIREBIRD

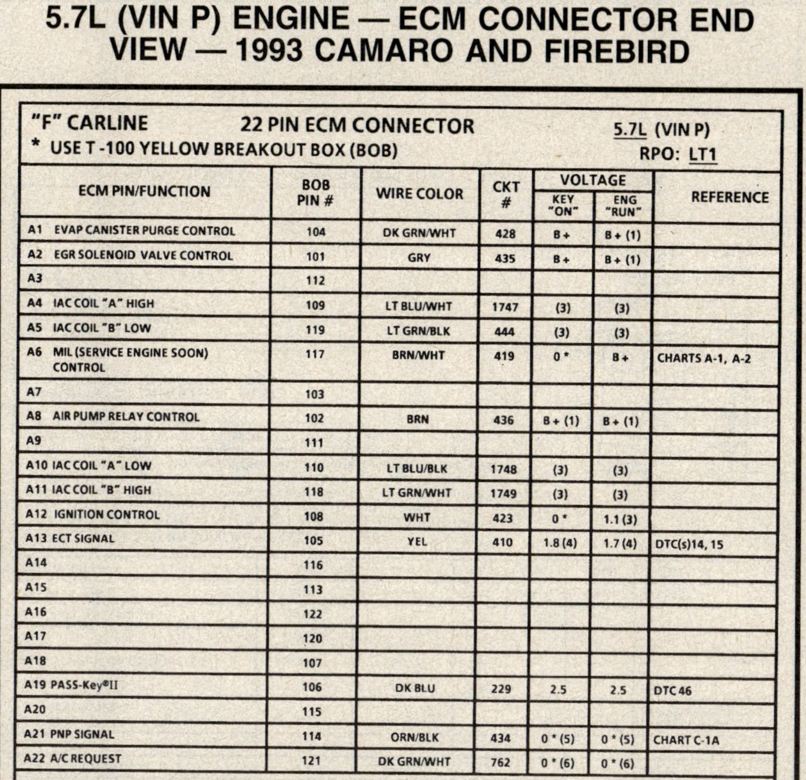

"F" CARLINE   22 PIN ECM CONNECTOR   5.7L (VIN P)
*** USE T-100 YELLOW BREAKOUT BOX (BOB)   RPO: LT1**

ECM PIN/FUNCTION	BOB PIN #	WIRE COLOR	CKT #	VOLTAGE KEY "ON"	VOLTAGE ENG "RUN"	REFERENCE
A1 EVAP CANISTER PURGE CONTROL	104	DK GRN/WHT	428	B+	B+ (1)	
A2 EGR SOLENOID VALVE CONTROL	101	GRY	435	B+	B+ (1)	
A3	112					
A4 IAC COIL "A" HIGH	109	LT BLU/WHT	1747	(3)	(3)	
A5 IAC COIL "B" LOW	119	LT GRN/BLK	444	(3)	(3)	
A6 MIL (SERVICE ENGINE SOON) CONTROL	117	BRN/WHT	419	0 *	B+	CHARTS A-1, A-2
A7	103					
A8 AIR PUMP RELAY CONTROL	102	BRN	436	B+ (1)	B+ (1)	
A9	111					
A10 IAC COIL "A" LOW	110	LT BLU/BLK	1748	(3)	(3)	
A11 IAC COIL "B" HIGH	118	LT GRN/WHT	1749	(3)	(3)	
A12 IGNITION CONTROL	108	WHT	423	0 *	1.1 (3)	
A13 ECT SIGNAL	105	YEL	410	1.8 (4)	1.7 (4)	DTC(s)14, 15
A14	116					
A15	113					
A16	122					
A17	120					
A18	107					
A19 PASS-Key®II	106	DK BLU	229	2.5	2.5	DTC 46
A20	115					
A21 PNP SIGNAL	114	ORN/BLK	434	0 * (5)	0 * (5)	CHART C-1A
A22 A/C REQUEST	121	DK GRN/WHT	762	0 * (6)	0 * (6)	

NOTICE: DO NOT BACKPROBE ECM CONNECTORS!
This Chart may be used in conjunction with the T-100 Yellow Breakout Box (48921) to obtain voltage present for each circuit listed. Install the BOB between the ECM harness connectors and the ECM, then probe the pin listed under "BOB PIN#". Voltage may vary due to low battery charge or other reasons, but should be very close. All voltages shown in the ENG "RUN" column are typical with engine at idle, closed throttle, normal operating temperature, park or neutral, system in "Closed Loop," all accessories "OFF," and scan tool not installed.

DVM NEGATIVE (BLACK) LEAD MUST BE CONNECTED TO A KNOWN GOOD GROUND.

(1) LESS THAN .5 VOLT WHEN SYSTEM ENABLED.
(2) 12 VOLTS FOR FIRST TWO SECONDS WITH IGNITION "ON."
(3) VARIES.
(4) VARIES WITH TEMPERATURE.
(5) BATTERY VOLTAGE WHEN IN GEAR.
(6) BATTERY VOLTAGE WHEN SYSTEM IS ENABLED.
*   LESS THAN .5 VOLT.

GRAY ECM CONNECTOR A

Figure A-7 - ECM Connector Terminal End View 5.7L (VIN P) "F" Carline (1 of 4)

5.7L (VIN P) ENGINE — ECM CONNECTOR END VIEW — 1993 CAMARO AND FIREBIRD

"F" CARLINE 22 PIN ECM CONNECTOR 5.7L (VIN P) RPO: LT1
* USE T-100 YELLOW BREAKOUT BOX (BOB)

ECM PIN/FUNCTION	BOB PIN #	WIRE COLOR	CKT #	VOLTAGE KEY "ON"	VOLTAGE ENG "RUN"	REFERENCE
B1 REVERSE LOCKOUT SOLENOID CONTROL (MANUAL TRANSMISSION)	204	LT GRN	1652	0 * (2)	0 * (2)	
B2 TCC (AUTOMATIC TRANSMISSION)	201	TAN/BLK	422	B+	B+ (1)	
B3 MAP, IAT, & A/C REFRIGERANT SENSOR GND	212	BLK	808	0 *	0 *	DTC(s) AFFECTED
B4 PRIMARY COOLING FAN	209	DK GRN	335	B+	B+ (1)	
B5 A/C CLUTCH CONTROL	219	DK GRN/WHT	459	B+	B+ (1)	
B6 ECM GROUND	217	BLK/WHT	451	0 *	0 *	
B7	203					
B8	202					
B9	211					
B10 SECONDARY COOLING FAN (WITH C60)	210	DK BLU	473	B+	B+ (1)	
B11	218					
B12	208					
B13	205					
B14 4th GEAR SIGNAL	216	LT BLU	446	B+	B+ (1)	
B15	213					
B16	222					
B17 ECM GROUND	220	BLK/WHT	451	0 *	0 *	
B18 TP SENSOR, ECT GROUND	207	BLK	470	0 *	0 *	DTC(s) AFFECTED
B19 3rd GEAR SIGNAL	206	DK GRN/WHT	438	0 *	0 *	
B20 2nd GEAR SIGNAL	215	WHT	232	0 *	0 *	
B21	214					
B22 ECM GROUND	221	TAN/WHT	551	0 *	0 *	

NOTICE: DO NOT BACKPROBE ECM CONNECTORS!
This Chart may be used in conjunction with the T-100 Yellow Breakout Box (48921) to obtain voltage present for each circuit listed. Install the BOB between the ECM harness connectors and the ECM, then probe the pin listed under "BOB PIN#". Voltage may vary due to low battery charge or other reasons, but should be very close. All voltages shown in the ENG "RUN" column are typical with engine at idle, closed throttle, normal operating temperature, park or neutral, system in "Closed Loop," all accessories "OFF," and scan tool not installed.

DVM NEGATIVE (BLACK) LEAD MUST BE CONNECTED TO A KNOWN GOOD GROUND.

(1) LESS THAN .5 VOLT WHEN SYSTEM IS ENABLED.
(2) BATTERY VOLTAGE ABOVE 5 MPH.
* LESS THAN .5 VOLT.

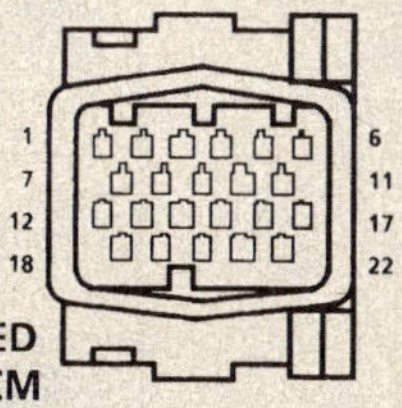

RED ECM CONNECTOR B

Figure A-8 - ECM Connector Terminal End View 5.7L (VIN P) "F" Carline (2 of 4)

5.7L (VIN P) ENGINE — ECM CONNECTOR END VIEW — 1993 CAMARO AND FIREBIRD

"F" CARLINE 22 PIN ECM CONNECTOR 5.7L (VIN P) RPO: LT1
* USE T-100 YELLOW BREAKOUT BOX (BOB)

ECM PIN/FUNCTION	BOB PIN #	WIRE COLOR	CKT #	VOLTAGE KEY "ON"	VOLTAGE ENG "RUN"	6E3 REFERENCE
C1 A/C STATUS	304	DK GRN	59	0	0 (3)	
C2 +5 VOLTS REFERENCE	301	GRY	474	5	5	DTC(s) AFFECTED
C3 TP SENSOR SIGNAL	312	DK BLU	417	.62	.62	DTC(s) 21, 22
C4	309					
C5 LOW RESOLUTION	319	RED/BLK	453	1.0 OR 5.0	1.0 (4)	DTC 16
C6 BATTERY FEED	317	ORN	340	B+	B+	CHART A-1
C7 +5 VOLTS REFERENCE	303	GRY	416	5	5	DTC(s) AFFECTED
C8 KNOCK SENSOR SIGNAL	302	DK BLU	496	2.5	2.5	DTC 43, SECTION C5
C9 REFERENCE LOW	311	PNK/BLK	632	0	0	
C10	310					
C11 A/C REFRIGERANT PRESSURE SENSOR SIGNAL	318	RED/BLK	380	.78	.6-1.0 (5)	DTC(s) 66, 67
C12 IGNITION FEED	308	PNK	439	B+	B+	CHART A-1
C13	305					
C14 HIGH RESOLUTION	316	LT BLU/BLK	647	0 *	2.5 (4)	DTC 36
C15	313					
C16 A/C EVAPORATOR TEMPERATURE SENSOR SIGNAL	322	DK BLU	732	1.4 (2)	1.4 (2)	CHART C-10 DTC 71
C17 BATTERY FEED	320	ORN	340	B+	B+	CHART A-1
C18 DISTRIBUTOR IGNITION FEED	307	RED	631	B+	B+	
C19	306					
C20 FUEL PUMP SIGNAL	315	GRY	120	0 * (6)	B+	
C21 MAP SIGNAL	314	LT GRN	432	4.8 (1)	1.2 (1)	DTC(s) 33, 34
C22 IAT SIGNAL	321	TAN	472	2.0 (2)	3.0 (2)	DTC(s) 23, 25

NOTICE: DO NOT BACKPROBE ECM CONNECTORS!
This Chart may be used in conjunction with the T-100 Yellow Breakout Box (48921) to obtain voltage present for each circuit listed. Install the BOB between the ECM harness connectors and the ECM, then probe the pin listed under "BOB PIN#". Voltage may vary due to low battery charge or other reasons, but should be very close. All voltages shown in the ENG "RUN" column are typical with engine at idle, closed throttle, normal operating temperature, park or neutral, system in "Closed Loop," all accessories "OFF," and scan tool not installed.

DVM NEGATIVE (BLACK) LEAD MUST BE CONNECTED TO A KNOWN GOOD GROUND.

(1) VARIES WITH ALTITUDE.
(2) VARIES WITH TEMPERATURE.
(3) B+ WHEN SYSTEM IS ENABLED.
(4) VARIES.
(5) WITH A/C "OFF"
(6) B+ WHEN FUEL PUMP IS ENABLED
* LESS THAN .5 VOLT.

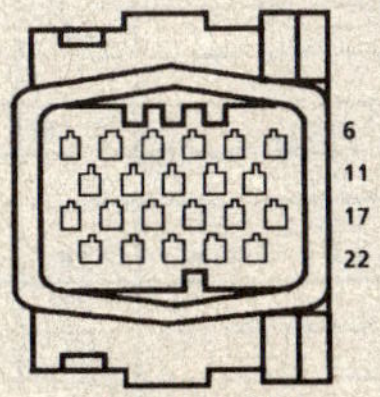

GREEN ECM CONNECTOR C

Figure A-9 - ECM Connector Terminal End View 5.7L (VIN P) "F" Carline (3 of 4)

5.7L (VIN P) ENGINE — ECM CONNECTOR END VIEW — 1993 CAMARO AND FIREBIRD

"F" CARLINE 22 PIN ECM CONNECTOR						5.7L (VIN P)
* USE T-100 YELLOW BREAKOUT BOX (BOB)						RPO: LT1

ECM PIN/FUNCTION	BOB PIN #	WIRE COLOR	CKT #	KEY "ON"	ENG "RUN"	6E3 REFERENCE
D1 VSS GROUND	404	PPL	401	O*	O*	DTC 24
D2	401					
D3	412					
D4 SERIAL DATA	409	TAN	800	(1)	(1)	CHART A-2
D5	419					
D6 BANK 1 (LEFT) O2S SIGNAL	417	PPL/WHT	1665	.38 (1)	.1-.9 (1)	DTC(s) 13, 44
D7 FUEL PUMP RELAY DRIVER	403	DK GRN/WHT	465	0 (2)	B+	CHART A-5
D8 VSS OUTPUT (4KPPM)	402	DK GRN/WHT	817	(3)	(3)	DTC 24
D9	411					
D10 INJECTOR DRIVER	410	DK GRN	468	B+	B+	CHART A-3
D11 INJECTOR DRIVER	418	DK BLU	467	B+	B+	CHART A-3
D12 VSS SIGNAL	408	YEL	400	(3)	(3)	DTC 24
D13	405					
D14	416					
D15	413					
D16 O2S GROUND	422	BLK/WHT	351	O*	O*	DTC(s) 13, 63
D17	420					
D18	407					
D19	406					
D20 DIAGNOSTIC ENABLE	415	WHT/BLK	448	5	5	CHART A-2
D21	414					
D22 BANK 2 (RIGHT) O2S SIGNAL	421	PPL	1666	.36	.1-.9 (1)	DTC(s) 63, 64

NOTICE: DO NOT BACKPROBE ECM CONNECTORS!
This Chart may be used in conjunction with the T-100 Yellow Breakout Box (48921) to obtain voltage present for each circuit listed. Install the BOB between the ECM harness connectors and the ECM, then probe the pin listed under "BOB PIN#". Voltage may vary due to low battery charge or other reasons, but should be very close. All voltages shown in the ENG "RUN" column are typical with engine at idle, closed throttle, normal operating temperature, park or neutral, system in "Closed Loop," all accessories "OFF," and scan tool not installed.

DVM NEGATIVE (BLACK) LEAD MUST BE CONNECTED TO A KNOWN GOOD GROUND.

**BROWN
ECM
CONNECTOR D**

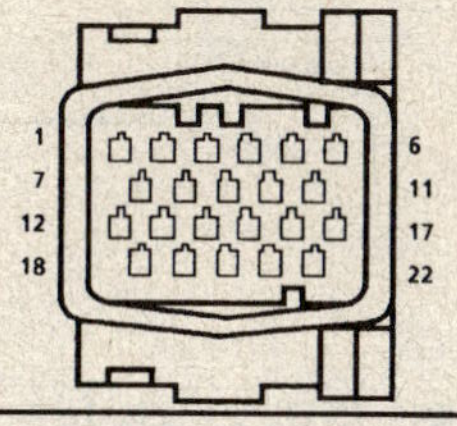

(1) VARIES.
(2) B+ FOR TWO SECONDS WITH IGNITION "ON."
(3) VARIES DEPENDING ON POSITION OF DRIVE WHEELS.
* LESS THAN .5 VOLT.

5.7L (VIN P) ENGINE — ON-BOARD DIAGNOSTIC SYSTEM CHART — 1993 CAMARO AND FIREBIRD

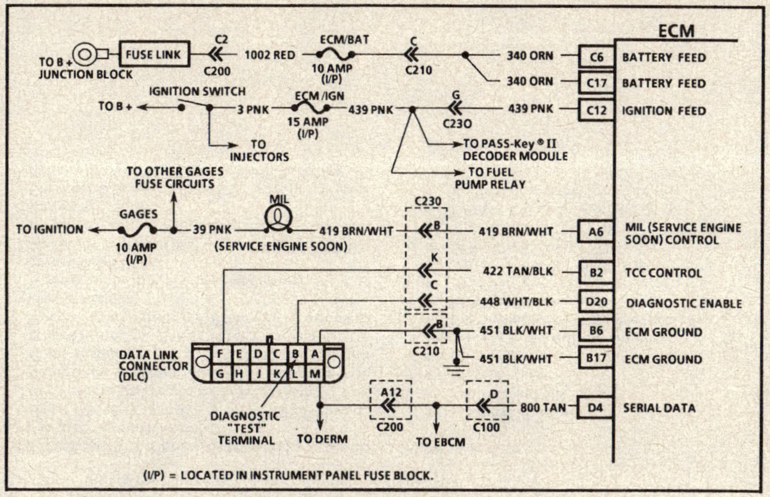

ON-BOARD DIAGNOSTIC (OBD) SYSTEM CHECK
5.7L (VIN P) "F" CARLINE (MFI)

Circuit Description:
The On-Board Diagnostic (OBD) system check is an organized approach to identifying a problem created by an engine control system malfunction. It must be the starting point for any driveability complaint diagnosis, because it directs the service technician to the next logical step in diagnosing the complaint. Understanding the chart and using it correctly will reduce diagnostic time and prevent the unnecessary replacement of good parts.

Test Description: Number(s) below refer to circled number(s) on the diagnostic chart.

1. This step is a check for the proper operation of the Malfunction Indicator Lamp (MIL) "Service Engine Soon." The MIL should be "ON" steady.
2. No MIL at this point indicates that there is a problem with the MIL circuit or the ECM control of that circuit.
3. This test checks the ability of the ECM to control the MIL. With the diagnostic terminal grounded, the MIL should flash a DTC 12 three times, followed by any Diagnostic Trouble Code (DTC) stored in memory.
4. Most of the   procedures use a Tech 1 to aid diagnosis, therefore, serial data must be available. If a PROM error is present, the ECM may have been able to flash DTC 12/51, but not enable serial data.
5. Although the ECM is powered up, a "Cranks But Will Not Run" symptom could exist because of an ECM or system problem.
6. This step will isolate if the customer complaint is a MIL or a driveability problem with no MIL. An invalid DTC may be the result of a faulty scan tool, PROM or ECM.
7. Comparison of actual control system data with the typical values is a quick check to determine if any parameter is not within limits. Keep in mind that a base engine problem (i.e. advanced cam timing) may substantially alter sensor values.
8. Installation of a scan tool will provide a good ground path for the ECM and may hide a driveability complaint due to poor ECM grounds.
9. If the actual data is not within the typical values established, the charts in "Components Systems," Section   will provide a functional check of the suspect component or system.

5.7L (VIN P) ENGINE — ON-BOARD DIAGNOSTIC SYSTEM CHART — 1993 CAMARO AND FIREBIRD

ON-BOARD DIAGNOSTIC (OBD) SYSTEM CHECK

5.7L (VIN P) "F" CARLINE (MFI)

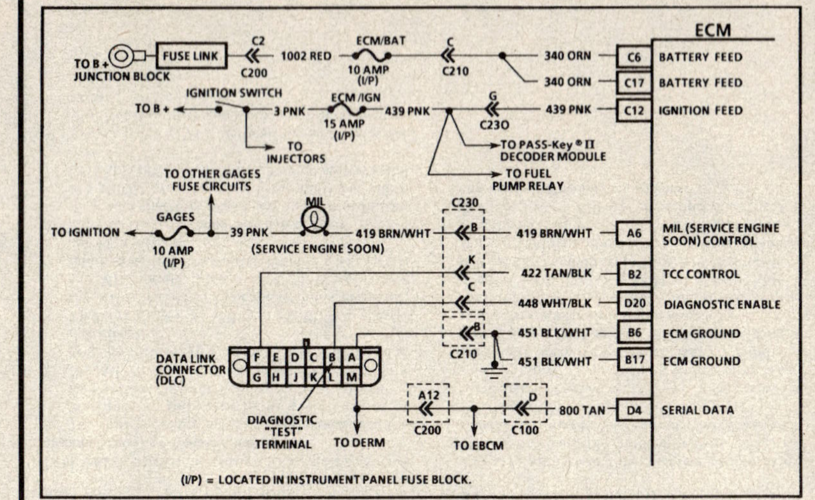

5.7L (VIN P) ENGINE — SYSTEM DIAGNOSTIC CHARTS — 1993 CAMARO AND FIREBIRD

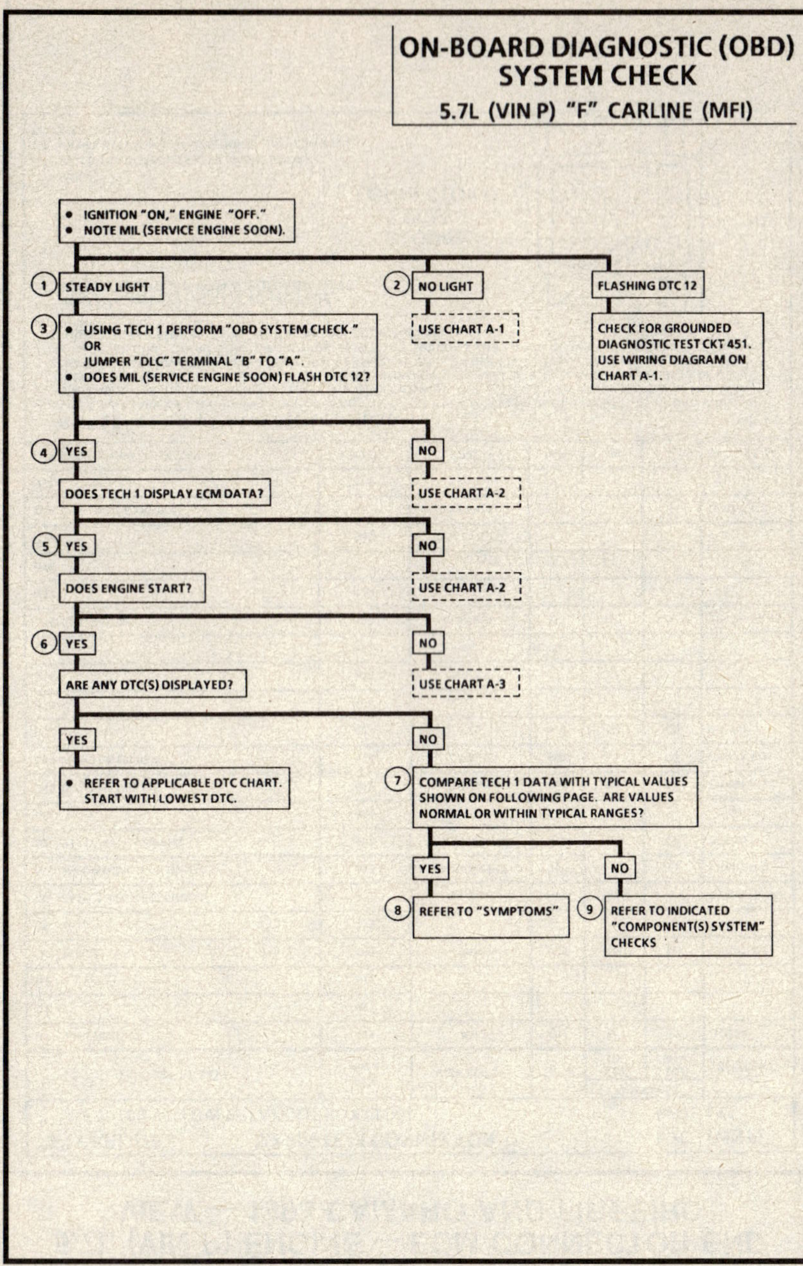

CHART A-1

NO MALFUNCTION INDICATOR LAMP (MIL)
5.7L (VIN P) "F" CARLINE (MFI)

Circuit Description:

There should always be a steady MIL (Service Engine Soon) when the ignition is "ON" and engine stopped. Ignition voltage is supplied directly to the light bulb. The Engine Control Module (ECM) will control the light and turn it "ON" by providing a ground path through CKT 419 to the ECM.

Test Description: Number(s) below refer to circled number(s) on the diagnostic chart.

1. Use J 35616-A tool kit to probe ECM harness terminals, so that you do not damage them.
2. Using a test light connected to 12 volts, probe each of the system ground circuits to be sure a good ground is present. See ECM terminal end view in front of this section for ECM pin locations of ground circuits.
3. If the fusible link is open, determine short to ground in CKT 1002.
4. Make sure to check CKT 439 if the ECM IGN fuse is open and both 340 circuits if the ECM/Batt fuse was open.
   Check all 439 circuits if the ECM IGN fuse is open. Refer to ECM wiring diagrams for information.

Diagnostic Aids:

Engine runs OK, check:
- Faulty light bulb.
- CKT 419 open.
- Open gages fuse.

Engine cranks but will not run, check:
- Continuous battery - fusible link open.
- ECM Batt fuse open.
- ECM IGN fuse open.
- Battery CKTs 340 to ECM open.
- Ignition CKT 439 to ECM open.
- Poor connection to ECM.
- Faulty ECM ground circuit(s).
- PASS-Key® II circuit problem.

5.7L (VIN P) ENGINE — SYSTEM DIAGNOSTIC CHARTS — 1993 CAMARO AND FIREBIRD

CHART A-1
NO MALFUNCTION INDICATOR LAMP (MIL)
5.7L (VIN P) "F" CARLINE (MFI)

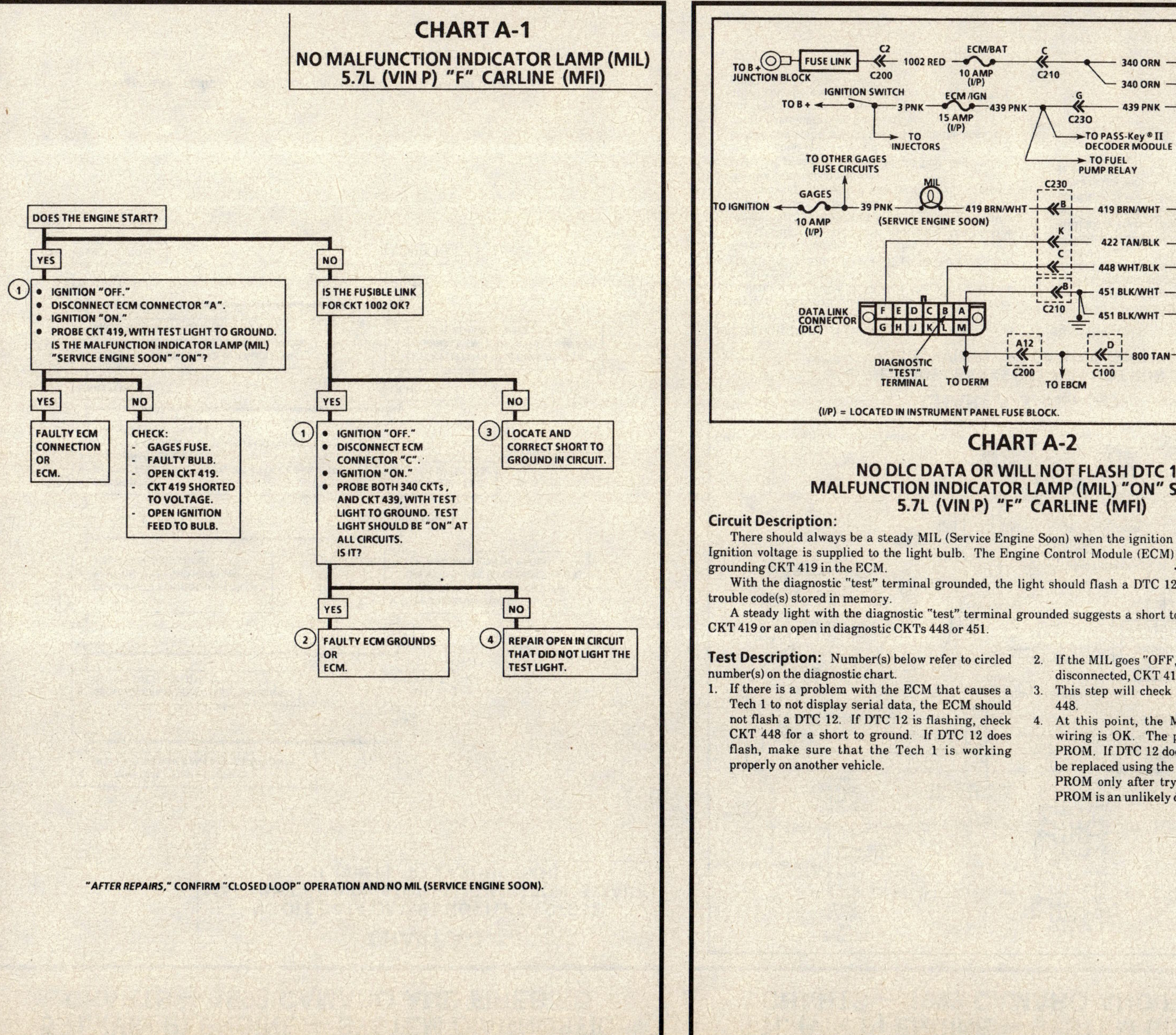

"AFTER REPAIRS," CONFIRM "CLOSED LOOP" OPERATION AND NO MIL (SERVICE ENGINE SOON).

5.7L (VIN P) ENGINE — SYSTEM DIAGNOSTIC CHARTS — 1993 CAMARO AND FIREBIRD

CHART A-2
NO DLC DATA OR WILL NOT FLASH DTC 12
MALFUNCTION INDICATOR LAMP (MIL) "ON" STEADY
5.7L (VIN P) "F" CARLINE (MFI)

Circuit Description:

There should always be a steady MIL (Service Engine Soon) when the ignition is "ON" and engine stopped. Ignition voltage is supplied to the light bulb. The Engine Control Module (ECM) will turn the light "ON" by grounding CKT 419 in the ECM.

With the diagnostic "test" terminal grounded, the light should flash a DTC 12, followed by any diagnostic trouble code(s) stored in memory.

A steady light with the diagnostic "test" terminal grounded suggests a short to ground in the light control CKT 419 or an open in diagnostic CKTs 448 or 451.

Test Description: Number(s) below refer to circled number(s) on the diagnostic chart.

1. If there is a problem with the ECM that causes a Tech 1 to not display serial data, the ECM should not flash a DTC 12. If DTC 12 is flashing, check CKT 448 for a short to ground. If DTC 12 does flash, make sure that the Tech 1 is working properly on another vehicle.

2. If the MIL goes "OFF," when the ECM connector is disconnected, CKT 419 is not shorted to ground.

3. This step will check for an open diagnostic CKT 448.

4. At this point, the MIL (Service Engine Soon) wiring is OK. The problem is a faulty ECM or PROM. If DTC 12 does not flash, the ECM should be replaced using the original PROM. Replace the PROM only after trying an ECM, as a defective PROM is an unlikely cause of the problem.

5.7L (VIN P) ENGINE — SYSTEM DIAGNOSTIC CHARTS — 1993 CAMARO AND FIREBIRD

CHART A-2

NO DLC DATA OR WILL NOT FLASH DTC 12
MALFUNCTION INDICATOR LAMP (MIL) "ON" STEADY
5.7L (VIN P) "F" CARLINE (MFI)

- IGNITION "ON," ENGINE "OFF."
- IS THE MIL (SERVICE ENGINE SOON) "ON"?

YES

- GROUND DIAGNOSTIC "TEST" TERMINAL. DOES LIGHT FLASH DTC 12?

NO → USE CHART A-1

(2)
- IGNITION "OFF."
- DISCONNECT ECM CONNECTOR "A".
- IGNITION "ON" AND NOTE MIL (SERVICE ENGINE SOON).

LIGHT "OFF"

(3)
- IGNITION "OFF."
- JUMPER TERMINALS "A" TO "B" AT DLC CONNECTOR
- DISCONNECT ECM CONNECTOR "D".
- CONNECT TEST LIGHT BETWEEN ECM CONNECTOR TERMINAL "D20" AND B +.

LIGHT "ON"

(4)
- CHECK PROM FOR PROPER INSTALLATION.
- IF OK, REPLACE ECM USING ORIGINAL PROM.
- RECHECK FOR DTC 12.

NO DTC 12 → REPLACE PROM.

DTC 12 → SYSTEM OK.

(1)
- IF PROBLEM WAS NO DLC DATA:
- CHECK SERIAL DATA 800 CKT FOR OPEN OR SHORT TO GROUND. IF OK, IT IS A FAULTY ECM OR PROM.

LIGHT "ON" → REPAIR SHORT TO GROUND IN CKT 419.

LIGHT "OFF" → CHECK FOR OPEN IN DLC DIAGNOSTIC TERMINALS "A" AND "B" (CKT 448 AND CKT 451), REPAIR AS NECESSARY.

"AFTER REPAIRS," CONFIRM "CLOSED LOOP" OPERATION AND NO MIL (SERVICE ENGINE SOON).

5.7L (VIN P) ENGINE — SYSTEM DIAGNOSTIC CHARTS — 1993 CAMARO AND FIREBIRD

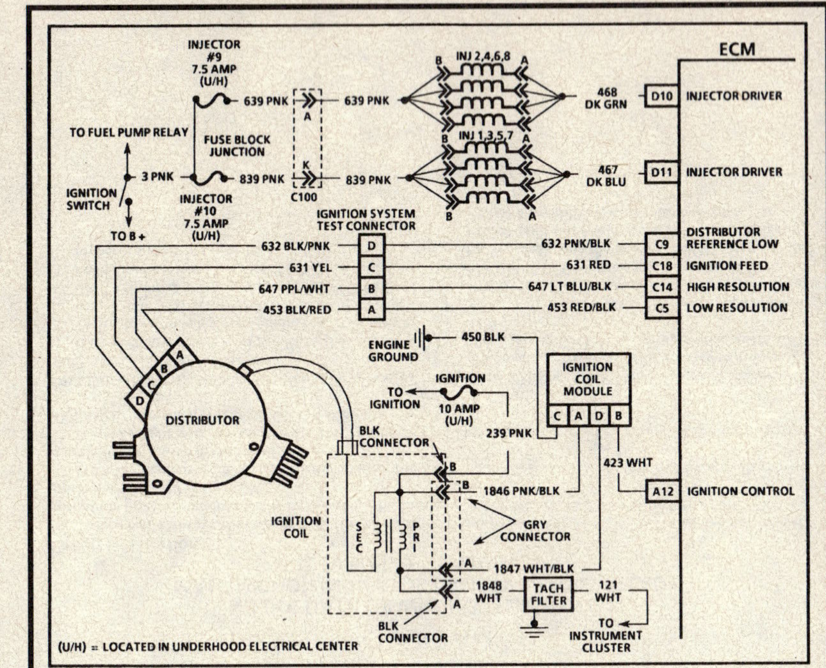

CHART A-3 (Page 1 of 2)

ENGINE CRANKS BUT WILL NOT RUN
5.7L (VIN P) "F" CARLINE (MFI)

Test Description: Number(s) below refer to circled number(s) on the diagnostic chart.

1. Perform On-Board Diagnostic (OBD) system check to determine if the ECM and the MIL (Service Engine Soon) are functioning.
2. CHART C-4 will address all problems related to the causes of a no spark condition.
3. The test light should blink, indicating the ECM is controlling the injectors OK. All lights should blink at the same brightness. If any injector light blinks dimly an injector may be shorted. Compare injector resistance values.
4. Use fuel pressure gage J 34730-1 or equivalent. Wrap a shop towel around the fuel pressure tap to absorb any small amount of fuel leakage that may occur when installing the gage.
5. All injectors should be within 1.0 ohm of each other and should not be less than 10 ohms at 21°C (70°F).

Diagnostic Aids:

- An EGR valve sticking open can cause a low air/fuel ratio during cranking.
- Unless engine enters "Clear Flood" at the first indication of a flooding condition, it can result in a no start.
- A defective MAP sensor may cause a no start or a stall after start. Disconnect the sensor. The ECM will use a default value for the sensor. If the condition is corrected and the connections are OK, replace the sensor.

If above are all OK, refer to "Symptoms," Section "Hard Start."

5.7L (VIN P) ENGINE — SYSTEM DIAGNOSTIC CHARTS — 1993 CAMARO AND FIREBIRD

CHART A-3
(Page 1 of 2)
ENGINE CRANKS BUT WILL NOT RUN
5.7L (VIN P) "F" CARLINE (MFI)

(1)
- PERFORM ON-BOARD DIAGNOSTIC SYSTEM CHECK. CHECK THE FOLLOWING:
  - IF ECM DTC(s) 16, 41, OR 42 ARE STORED, USE THOSE CHARTS FIRST.
  - CHECK PASS Key® STATUS. IF NOT OK, REFER TO ECM DTC 46 CHART FIRST.
  - ACTUAL ENGINE TEMPERATURE AND ECT TEMPERATURE ON THE TECH 1 SHOULD BE CLOSE TO THE SAME, IF NOT, REFER TO DTC 15.
  - TP SENSOR - IF OVER 2.5 VOLTS AT CLOSED THROTTLE, USE THE DTC 21 CHART.

  IS RPM INDICATED ON TECH 1 DURING CRANKING?

YES →
- USING A ST-125 (SPARK CHECKER), J 26792, CHECK FOR SPARK WHILE CRANKING (CHECK TWO WIRES). IS SPARK PRESENT?

NO →
- IGNITION "OFF."
- DISCONNECT DISTRIBUTOR ELECTRICAL CONNECTOR.
- IGNITION "ON."
- PROBE TERMINAL "C" WITH A TEST LIGHT TO GROUND.
- LIGHT SHOULD BE "ON."

IS IT?

YES →
- CHECK FOR CONTINUITY BETWEEN DISTRIBUTOR CONNECTOR TERMINAL "D" AND GROUND.
- CIRCUIT SHOULD HAVE CONTINUITY. DOES IT?

NO → FAULTY ECM CONNECTION OR OPEN OR GROUNDED IGNITION FEED CIRCUIT FROM DISTRIBUTOR TO ECM OR FAULTY ECM.

(From "IS SPARK PRESENT?") YES →
(3)
- DISCONNECT EACH INJECTOR ONE AT A TIME.
- CONNECT TEST LIGHT J 34730-2 TO EACH INJECTOR HARNESS CONNECTOR.
- CHECK FOR BLINKING LIGHT WHILE CRANKING.

(From "IS SPARK PRESENT?") NO →
(2) BASIC IGNITION SYSTEM PROBLEM. REFER TO CHART C-4.

(Continuity) YES → FAULTY DISTRIBUTOR CONNECTION OR FAULTY DISTRIBUTOR.

(Continuity) NO → FAULTY ECM CONNECTION OR OPEN GROUND CIRCUIT FROM DISTRIBUTOR TO ECM OR FAULTY ECM.

ALL LIGHTS BLINKING BRIGHTLY
- USING TECH 1, ENABLE FUEL PUMP. DOES FUEL PUMP OPERATE?

NO BLINKING LIGHT(S) → USE CHART A-3 (2 OF 2).

ANY LIGHTS BLINKING DIMLY
(5) SHORTED INJECTOR, COMPARE INJECTOR RESISTANCE VALUES.

YES →
(4)
- IGNITION "OFF."
- INSTALL FUEL PRESSURE GAGE AND NOTE PRESSURE AFTER IGNITION "ON." SHOULD BE (283-325 kPa) 41-47 psi.

NO → USE CHART A-5.

OK →
- REVIEW THE "DIAGNOSTIC AIDS" ON FACING PAGE FOR ADDITIONAL ITEMS TO CHECK. IF ALL ARE OK, MFI SYSTEM IS OK. REFER TO "HARD START" IN "SYMPTOMS," SECTION

NOT OK → USE CHART A-7.

"AFTER REPAIRS," CONFIRM "CLOSED LOOP" OPERATION AND NO MIL (SERVICE ENGINE SOON).

5.7L (VIN P) ENGINE — SYSTEM DIAGNOSTIC CHARTS — 1993 CAMARO AND FIREBIRD

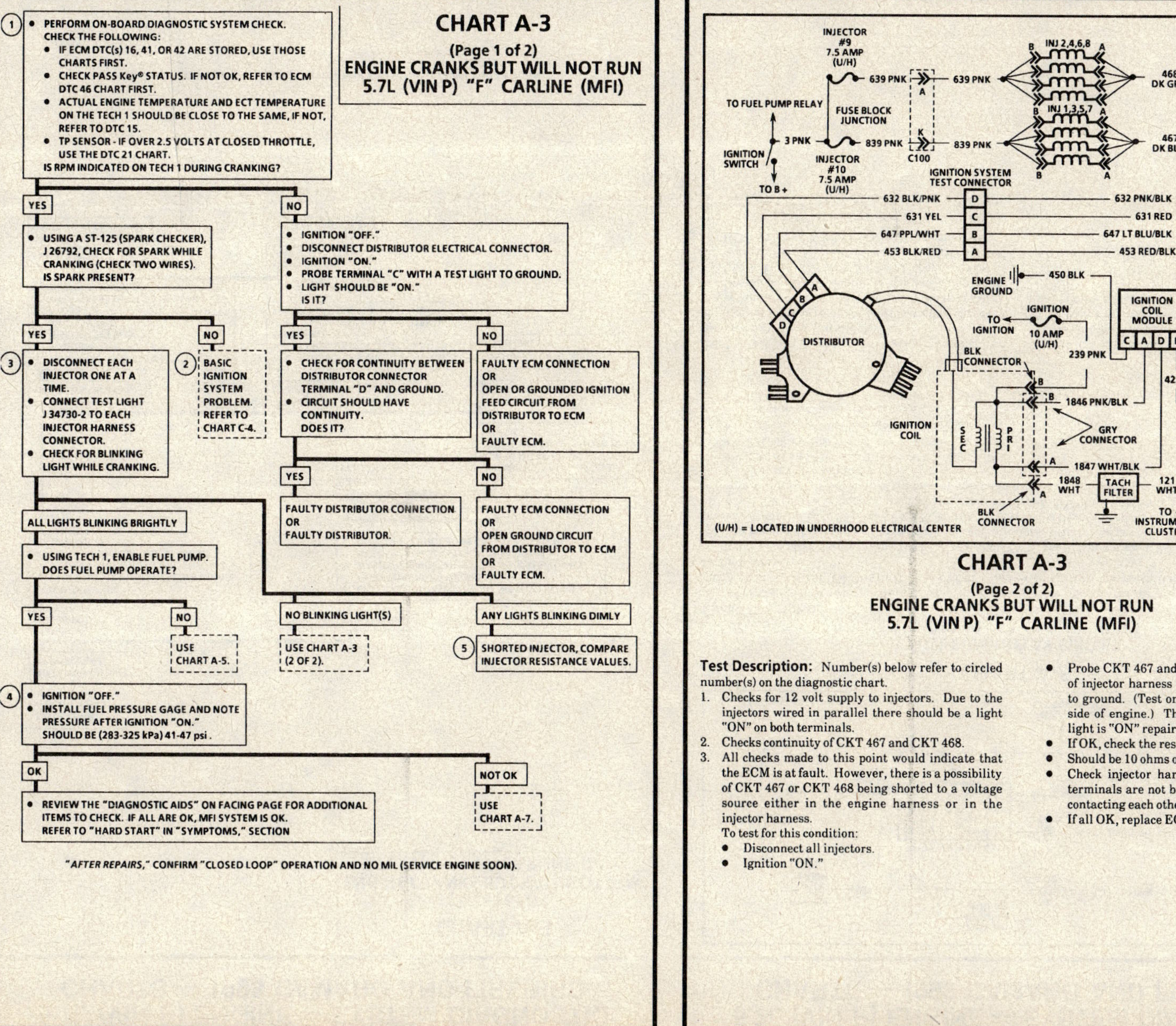

CHART A-3
(Page 2 of 2)
ENGINE CRANKS BUT WILL NOT RUN
5.7L (VIN P) "F" CARLINE (MFI)

Test Description: Number(s) below refer to circled number(s) on the diagnostic chart.

1. Checks for 12 volt supply to injectors. Due to the injectors wired in parallel there should be a light "ON" on both terminals.
2. Checks continuity of CKT 467 and CKT 468.
3. All checks made to this point would indicate that the ECM is at fault. However, there is a possibility of CKT 467 or CKT 468 being shorted to a voltage source either in the engine harness or in the injector harness.

To test for this condition:
- Disconnect all injectors.
- Ignition "ON."

- Probe CKT 467 and CKT 468 on the ECM side of injector harness with a test light connected to ground. (Test one injector harness on each side of engine.) There should be no light. If light is "ON" repair short to voltage.
- If OK, check the resistance of the injectors.
- Should be 10 ohms or more.
- Check injector harness connector. Be sure terminals are not backed out of connector and contacting each other.
- If all OK, replace ECM.

5.7L (VIN P) ENGINE — SYSTEM DIAGNOSTIC CHARTS — 1993 CAMARO AND FIREBIRD

CHART A-3
(Page 2 of 2)
ENGINE CRANKS BUT WILL NOT RUN
5.7L (VIN P) "F" CARLINE (MFI)

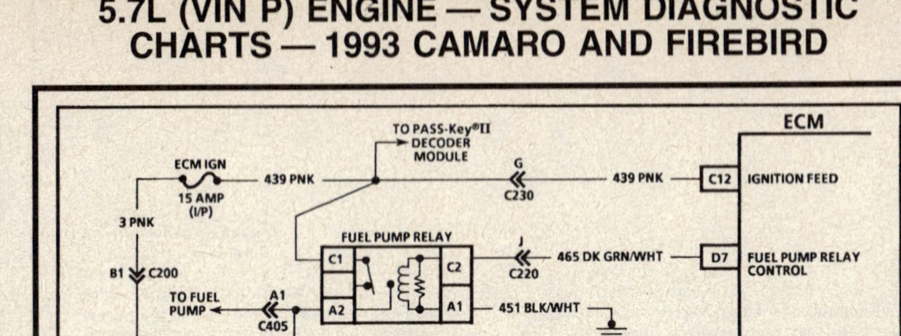

5.7L (VIN P) ENGINE — SYSTEM DIAGNOSTIC CHARTS — 1993 CAMARO AND FIREBIRD

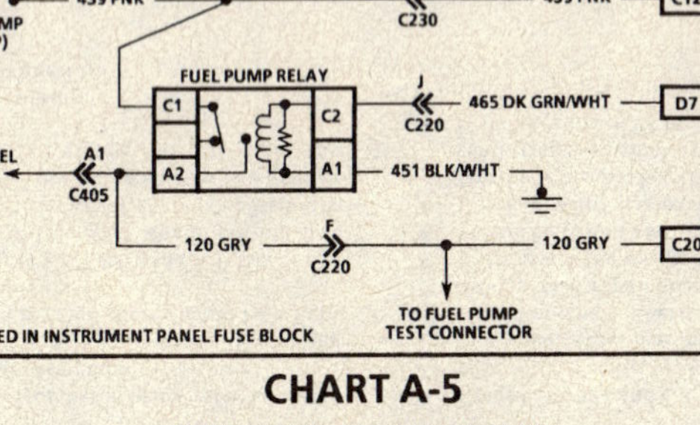

CHART A-5
FUEL PUMP RELAY CIRCUIT
5.7L (VIN P) "F" CARLINE (MFI)

Circuit Description:

When the ignition switch is turned "ON," the Engine Control Module (ECM) will turn "ON" the in-tank fuel pump. It will remain "ON" as long as the ECM is receiving ignition low resolution reference pulses from the ignition system.

If there are no reference pulses, the ECM will shut "OFF" the fuel pump about 2-3 seconds after key "ON."

The pump will deliver fuel to the fuel rail and injectors, then to the pressure regulator, where the system pressure is controlled to 284-325 kPa (41 - 47 psi). Excess fuel is then returned to the fuel tank.

When the engine is stopped, the fuel pump can be turned "ON" by using the Tech 1. Improper fuel system pressure will result in one or all of the following symptoms:

- Cranks but won't run.
- DTC 44 and 64.
- DTC 45 and 65.
- Cuts out, may feel like ignition problem.
- Poor fuel economy, loss of power.
- Hesitation.

Test Description: Number(s) below refer to circled number(s) on the diagnostic chart.

1. The fuel pump relay is located under the dead pedal, (foot rest) on the driver's side of the vehicle. Checks for 12 volt supply to the fuel pump relay. Check ECM IGN fuse for being open. If fuse is open, check CKT 439 for being grounded. If ECM IGN fuse was OK, check CKT 439 for being open.

2. The test light should illuminate for 2 seconds when the ignition was turned "ON."

   A Tech 1 can also be used to enable the fuel pump relay control circuit.

5.7L (VIN P) ENGINE—SYSTEM DIAGNOSTIC CHARTS—1993 CAMARO AND FIREBIRD

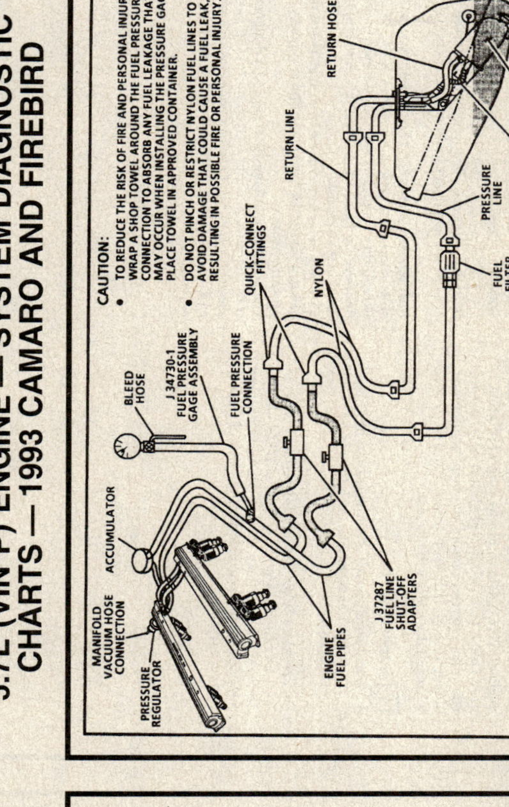

CHART A-7 (Page 1 of 3)
FUEL SYSTEM DIAGNOSIS
5.7L (VIN P) "F" CARLINE (MFI)

Circuit Description:

When the ignition switch is turned "ON," the Engine Control Module (ECM) will turn "ON" the in-tank fuel pump. It will remain "ON" as long as the engine is cranking or running, and the ECM is receiving reference pulses. If there are no reference pulses, the ECM will shut "OFF" the fuel pump within 2 seconds after ignition "ON" or engine stops.

Inside the fuel tank an electric fuel pump (within an integral reservoir), supplies fuel through an in-line filter to the fuel rail assembly. The pump is designed to provide fuel at a pressure above the regulated pressure needed by the injectors. A pressure regulator, attached to the fuel rail, keeps fuel available to the injectors at a regulated pressure. Unused fuel is returned to the fuel tank by a separate line.

Test Description: Number(s) below refer to circled number(s) on the diagnostic chart.

1. Connect fuel pressure gage as shown in illustration. Wrap a stop towel around the fuel pressure connection to absorb any small amount of fuel leakage that may occur when installing the gage. With ignition "ON" and fuel pump running, pressure should be 284-325 kPa (41-47 psi). This pressure is controlled by spring pressure within the regulator assembly.

2. When the engine is idling, manifold pressure is low (high vacuum) and is applied to the pressure regulator diaphragm. Vacuum will offset spring pressure and result in lower fuel pressure. Fuel pressure at idle will vary somewhat depending on barometric pressure but, should be less than pressure noted in Step (1).

3. A system that does not hold pressure is caused by one of the following:
   - Leaking fuel pump check ball.
   - Leaking fuel pump feed hose.
   - Leaking valve/seat within pressure regulator.
   - Leaking injector(s).

4. A leaking injector can best be determined by checking for a fouled or saturated spark plug(s). If a leaking injector can not be determined by a fouled or saturated spark plug, the following procedure should be used.
   - Remove fuel rail but leave fuel lines connected.
   - Lift fuel rail out just enough to leave injector nozzles in the ports.

CAUTION: To reduce the risk of fire and personal injury that may result from fuel spray on the engine, make sure fuel rail is positioned over injector ports and injector retaining clips are intact.

   - Pressurize the fuel system and observe injector nozzles.

5.7L (VIN P) ENGINE—SYSTEM DIAGNOSTIC CHARTS—1993 CAMARO AND FIREBIRD

CHART A-5
FUEL PUMP RELAY CIRCUIT
5.7L (VIN P) "F" CARLINE (MFI)

FROM CHART A-3 (1 OF 2).

①
- DISCONNECT FUEL PUMP RELAY.
- PROBE FUEL PUMP RELAY HARNESS TERMINAL "C1" WITH A TEST LIGHT TO GROUND.
- DOES TEST LIGHT ILLUMINATE?

YES
- USING A TEST LIGHT, PROBE FROM RELAY HARNESS TERMINAL "C1" TO "A1".
- DOES TEST LIGHT ILLUMINATE?

NO → FUEL PUMP RELAY IGNITION FEED CIRCUIT OPEN.

②
- IGNITION "OFF."
- PROBE FUEL PUMP RELAY HARNESS TERMINAL "C2" WITH A TEST LIGHT TO B+.
- OBSERVE TEST LIGHT AS IGNITION IS BEING TURNED "ON."
- DOES TEST LIGHT ILLUMINATE FOR 2 SECONDS WHEN IGNITION WAS TURNED "ON"?

NO → FUEL PUMP RELAY GROUND CIRCUIT OPEN.

YES
- USING A 15 AMP FUSED JUMPER WIRE, JUMP TERMINALS "C1" AND "A2" OF RELAY HARNESS. FUEL PUMP SHOULD RUN.
- DOES IT?

NO → CKT 465 OPEN OR FAULTY ECM CONNECTION OR FAULTY ECM.

- WITH "C1" AND "A2" STILL JUMPERED RAISE VEHICLE.
- DISCONNECT FUEL PUMP HARNESS CONNECTOR.
- PROBE CKT 120 OF FUEL PUMP HARNESS (RELAY SIDE) WITH A TEST LIGHT CONNECTED TO GROUND.
- DOES TEST LIGHT ILLUMINATE?

YES → FAULTY RELAY CONNECTION OR FAULTY RELAY.

- WITH "C1" AND "A2" STILL JUMPERED
- USING A TEST LIGHT PROBE CKT 120 TO FUEL PUMP GROUND CIRCUIT.
- DOES TEST LIGHT ILLUMINATE?

NO → CKT 120 OPEN FROM FUEL PUMP TO FUEL PUMP RELAY.

NO → OPEN FUEL PUMP GROUND CIRCUIT.

YES → OPEN FUEL PUMP HARNESS OR FAULTY INTANK FUEL PUMP.

5.7L (VIN P) ENGINE — SYSTEM DIAGNOSTIC CHARTS — 1993 CAMARO AND FIREBIRD

CHART A-7

(Page 1 of 3)
FUEL SYSTEM DIAGNOSIS
5.7L (VIN P) "F" CARLINE (MFI)

1. • INSTALL FUEL PRESSURE GAGE AS SHOWN ON FACING PAGE.
   • IGNITION "OFF" FOR 10 SECONDS. A/C "OFF."
   • IGNITION "ON." FUEL PUMP WILL RUN FOR ABOUT 2 SECONDS. IT MAY BE NECESSARY TO CYCLE THE IGNITION "ON" MORE THAN ONCE TO OBTAIN MAXIMUM PRESSURE.
   • NOTE FUEL PRESSURE WITH PUMP RUNNING, PRESSURE SHOULD BE 284-325 kPa (41-47 psi). WHEN PUMP STOPS, PRESSURE MAY VARY SLIGHTLY THEN SHOULD HOLD STEADY. IS PRESSURE CORRECT AND DOES IT HOLD?

YES → IF FUEL PRESSURE IS WITHIN NORMAL RANGE BUT IS SUSPECTED OF DROPPING OFF DURING ACCELERATION, CRUISE OR HARD CORNERING, SEE CHART A-7 (2 OF 3).

2. • START ENGINE, ALLOW IT TO IDLE AT NORMAL OPERATING TEMPERATURE.
   • FUEL PRESSURE NOTED IN STEP (1) SHOULD DROP APPROXIMATELY 21-69 kPa (3-10 psi). DOES IT?

YES → NO TROUBLE FOUND, REVIEW "SYMPTOMS."

NO → • DISCONNECT VACUUM HOSE FROM PRESSURE REGULATOR ASSEMBLY.
   • WITH ENGINE IDLING, APPLY 12-14 INCHES OF VACUUM TO PRESSURE REGULATOR. FUEL PRESSURE NOTED IN STEP (1) SHOULD DROP APPROXIMATELY 21-69 kPa (3-10 psi). DOES IT?

YES → LOCATE AND REPAIR LOSS OF VACUUM TO PRESSURE REGULATOR.

NO → PRESSURE REGULATOR IS FAULTY.

NO → FROM CHART A-3

3. FUEL PRESSURE WITHIN SPEC., BUT DOES NOT HOLD.

FUEL PRESSURE OUT OF SPEC. → SEE CHART A-7 (2 OF 3)

NO FUEL PRESSURE → USE CHART A-5 TO DIAGNOSE FUEL PUMP ELECTRICAL CIRCUIT.

• INSTALL J 37287 FUEL LINE SHUT-OFF ADAPTORS, REFER TO PAGES 3 OF 3 AND FACING PAGE ILLUSTRATION.
• MAKE SURE VALVES ARE OPEN.
• IGNITION "OFF."
• USING A 10 AMP FUSED JUMPER WIRE, CONNECT FUEL PUMP "TEST" CONNECTOR TO B+ AND WAIT FOR PRESSURE TO BUILD.
• DISCONNECT JUMPER AND CLOSE VALVE IN FUEL PRESSURE LINE. PRESSURE SHOULD HOLD. DOES IT?

IF OK → CHECK FOR:
• PLUGGED IN-LINE FILTER.
• RESTRICTED FUEL PRESSURE LINE.
• PLUGGED FUEL PUMP STRAINER.
• LEAKING FUEL PUMP FEED HOSE.

IF OK → FUEL PUMP IS FAULTY.

NO → • OPEN VALVE IN FUEL PRESSURE LINE.
• RECONNECT PUMP "TEST" JUMPER AND WAIT FOR PRESSURE TO BUILD.
• DISCONNECT JUMPER AND CLOSE VALVE IN FUEL RETURN LINE. PRESSURE SHOULD HOLD. DOES IT?

YES → CHECK FOR:
• LEAKING FUEL PUMP FEED HOSE.

IF OK → FUEL PUMP IS FAULTY. (LEAKING CHECK BALL INSIDE PUMP.)

NO → 4. LOCATE AND CORRECT LEAKING INJECTOR(S).

YES → PRESSURE REGULATOR IS FAULTY.

"AFTER REPAIRS," CONFIRM "CLOSED LOOP" OPERATION AND NO MIL (SERVICE ENGINE SOON).

5.7L (VIN P) ENGINE — SYSTEM DIAGNOSTIC CHARTS — 1993 CAMARO AND FIREBIRD

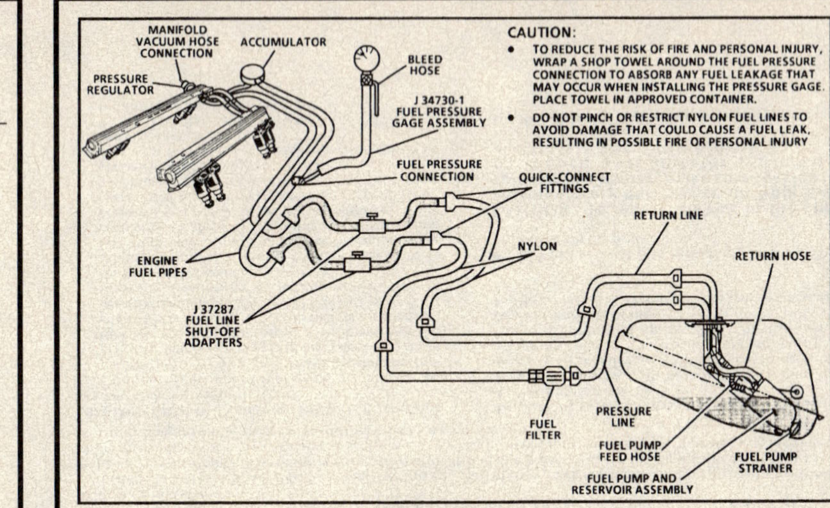

CHART A-7

(Page 2 of 3)
FUEL SYSTEM DIAGNOSIS
5.7L (VIN P) "F" CARLINE (MFI)

Test Description: Number(s) below refer to circled number(s) on the diagnostic chart.

5. Fuel pressure that drops off during acceleration, cruise or hard cornering may cause a lean condition and result in a loss of power, surging or misfire. This condition can be diagnosed using a Tech 1 scan tool. If the fuel system is very lean, one or both oxygen sensors will stop toggling and output voltage will drop below 300 mV. Also, injector pulse width will increase.

Important
• Make sure system is not operating at "Fuel-Cut-Off" which may cause false readings on the scan tool.

6. Fuel pressure below 284 kPa (41 psi) may cause a lean condition and may set a DTC 44/64. Driveability conditions can include hard starting cold, hesitation, poor driveability, lack of power, surging or misfire.

7. Restricting the fuel return line causes fuel pressure to build above regulated pressure. With battery voltage applied to the pump "test" connector, pressure should rise above 325 kPa (47 psi) as the valve in the return line is partially closed.

NOTICE: Do not allow pressure to exceed 414 kPa (60 psi) as damage to the regulator may result.

8. Fuel pressure above 325 kPa (47 psi) may cause a rich condition and may set a DTC 45/65. Driveability conditions can include hard starting (followed by black smoke) and a strong sulphur smell in the exhaust.

9. This test determines if the high fuel pressure is due to a restricted fuel return line or a faulty fuel pressure regulator.

10. The pressure regulator filter screen is designed to trap any contaminants introduced during engine assembly. If dirty, it can be removed with a small pick and discarded without potential harm to the regulator.

5.7L (VIN P) ENGINE — SYSTEM DIAGNOSTIC CHARTS — 1993 CAMARO AND FIREBIRD

CHART A-7
(Page 2 of 3)
FUEL SYSTEM DIAGNOSIS
5.7L (VIN P) "F" CARLINE (MFI)

FROM CHART A-7 (1 OF 3)

5. FUEL PRESSURE DROPS OFF DURING ACCELERATION, CRUISE OR HARD CORNERING.

CHECK FOR RESTRICTED IN-LINE FUEL FILTER OR FUEL PRESSURE LINE. IS THERE A RESTRICTION?

- NO
- YES → SERVICE AS REQUIRED AND RECHECK.

CHECK FOR:
- RESTRICTED FUEL PUMP STRAINER.
- LEAKING FUEL PUMP FEED HOSE.
- WRONG FUEL PUMP.

IF OK → FUEL PUMP IS FAULTY.

6. FUEL PRESSURE LESS THAN 284 kPa (41 psi).

7.
- INSTALL J 37287-2 FUEL RETURN LINE SHUT-OFF ADAPTER, REFER TO PAGES 3 OF 3 AND FACING PAGE ILLUSTRATION.
- MAKE SURE VALVE IS OPEN.
- IGNITION "OFF."
- USING A 10 AMP FUSED JUMPER WIRE, CONNECT FUEL PUMP "TEST" CONNECTOR TO B+.
- SLOWLY CLOSE SHUT-OFF VALVE. PRESSURE SHOULD RISE ABOVE 325 kPa (47 psi). DO NOT EXCEED 414 kPa (60 psi). DOES PRESSURE RISE ABOVE 325 kPa (47 psi)?

- NO → CHECK FOR: (see list above)
- YES → PRESSURE REGULATOR IS FAULTY.

8. FUEL PRESSURE ABOVE 325 kPa (47 psi).

9.
- DISCONNECT QUICK-CONNECT FITTING AT ENGINE FUEL RETURN PIPE, REFER TO PAGES 3 OF 3 AND FACING PAGE ILLUSTRATION.
- ATTACH A LENGTH OF FLEXIBLE FUEL HOSE TO ENGINE FUEL RETURN PIPE. PLACE OPEN END OF HOSE INTO AN APPROVED GASOLINE CONTAINER.
- IGNITION "OFF" FOR 10 SECONDS.
- IGNITION "ON." NOTE FUEL PRESSURE WITH PUMP RUNNING.
- PRESSURE SHOULD BE 284-325 kPa (41-47 psi). IS IT?

- NO → CHECK FOR RESTRICTED ENGINE FUEL RETURN PIPE. IS THERE A RESTRICTION?
  - NO → 10. REMOVE PRESSURE REGULATOR AND CHECK FOR RESTRICTED FILTER SCREEN (IF EQUIPPED).
    - IF OK → PRESSURE REGULATOR IS FAULTY.
  - YES → SERVICE AS REQUIRED AND RECHECK.
- YES → LOCATE AND CORRECT RESTRICTION IN FUEL RETURN LINE TO FUEL TANK.

"AFTER REPAIRS," CONFIRM "CLOSED LOOP" OPERATION AND NO MIL (SERVICE ENGINE SOON).

5.7L (VIN P) ENGINE — SYSTEM DIAGNOSTIC CHARTS — 1993 CAMARO AND FIREBIRD

CHART A-7
(Page 3 of 3)
FUEL SYSTEM PRESSURE TEST
5.7L (VIN P) "F" CARLINE (MFI)

FUEL SYSTEM PRESSURE RELIEF PROCEDURE

Engines With Fuel Pressure Connection

(Must Be Performed Before Disconnecting Fuel Line Fittings)

CAUTION:
- To reduce the risk of fire and personal injury, it is necessary to relieve fuel system pressure before disconnecting fuel line fittings.
- After relieving system pressure, a small amount of fuel may be released when disconnecting fuel line fittings. In order to reduce the chance of personal injury, cover fuel line fittings with a shop towel before disconnecting, to catch any fuel that may leak out. Place the towel in an approved container when disconnect is completed.

Tool Required: J 34730-1 Fuel Pressure Gage

1. Ignition "OFF."
2. Disconnect negative battery cable to avoid possible fuel discharge if an accidental attempt is made to start the engine.
3. Loosen fuel filler cap to relieve tank vapor pressure.
4. Connect gage J 34730-1 to fuel pressure connection. Wrap a shop towel around fitting while connecting gage to avoid spillage.
5. Install bleed hose into an approved container and open valve to bleed system pressure. Fuel line fittings are now safe for servicing.
6. Drain any fuel remaining in gage into an approved gasoline container.
7. Perform service required.
8. Tighten fuel filler cap.
9. Ignition "OFF."
10. Connect negative battery cable.
11. Cycle ignition "ON" and "OFF" twice, waiting ten seconds between cycles, then check for fuel leaks.

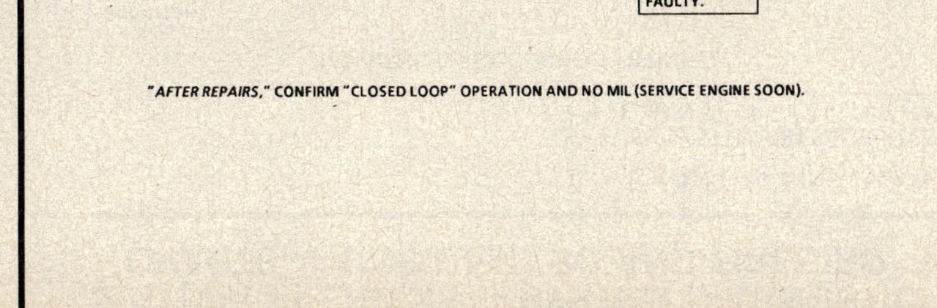

5.7L (VIN P) ENGINE — SYSTEM DIAGNOSTIC CHARTS — 1993 CAMARO AND FIREBIRD

CHART A-7 (Page 3 of 3)
FUEL SYSTEM PRESSURE TEST
5.7L (VIN P) "F" CARLINE (MFI)

SERVICING QUICK-CONNECT FITTINGS

Important
- In order to install fuel system diagnostic equipment on vehicles equipped with plastic quick-connect fittings, fuel line separator tools must be used to disconnect the fittings. Using the separator tools to release the fittings will cause the plastic retainer to remain inside the female connector allowing diagnostic equipment to be connected.

Tools required:
  J 37088-A tool set, fuel line quick-connect separator;
  J 39504 tool set, fuel line quick-connect separator (restricted access).

Remove or Disconnect
1. Relieve fuel system pressure (see "Fuel System Pressure Relief").
2. If equipped, slide dust cover back to access quick-connect fitting.
3. Grasp both sides of fitting. Twist female connector 1/4 turn in each direction to loosen any dirt within fitting.

CAUTION: Safety glasses must be worn when using compressed air, as flying dirt particles may cause eye injury.

4. Using compressed air, blow dirt out of fitting.
5. Choose correct tool from J 37088-A or J 39504 tool set for size of fitting. Insert tool into female connector, then push/pull inward to release locking tabs.
6. Pull connection apart.

Clean and Inspect

NOTICE: If it is necessary to remove rust or burrs from fuel pipe, use emery cloth in a radial motion with the pipe end to prevent damage to O-ring sealing surface.

- Using a clean shop towel, wipe off male pipe end.
- Inspect both ends of fitting for dirt and burrs. Clean or replace components/assemblies as required.

Install or Connect

CAUTION: To Reduce the Risk of Fire and Personal Injury:
- Before connecting fitting, always apply a few drops of clean engine oil to the male pipe end of engine fuel pipe, pressure gage adapter or fuel line shut-off adapter. This will ensure proper reconnection and prevent a possible fuel leak. (During normal operation, the O-rings located in the female connector will swell and may prevent proper reconnection if not lubricated.)

1. Apply a few drops of clean engine oil to the male pipe end of engine fuel pipe, pressure gage adapter or fuel line shut-off adapter.
2. Push both sides of fitting together to cause the retaining tabs/fingers to snap into place.
3. Once installed, pull on both sides of fitting to make sure connection is secure.
4. If equipped, reposition dust cover over quick-connect fitting.

5.7L (VIN P) ENGINE — DIAGNOSTIC TROUBLE CODE CHART — 1993 CAMARO AND FIREBIRD

DTC 13
BANK 1 (LEFT) OXYGEN SENSOR (O2S) CIRCUIT
(OPEN CIRCUIT)
5.7L (VIN P) "F" CARLINE (MFI)

Circuit Description:

The ECM supplies a voltage of about 450 mV between terminals "D6" and "D16". (If measured with a 10 megohm digital voltmeter, this may read as low as 320 mV). The oxygen sensor varies the voltage within a range of about 1000 mV if a rich O2S signal is indicated, down through about 10 mV if a lean O2S signal is indicated.

The sensor is like an open circuit and produces no voltage when it is below 315°C (600°F). An open sensor circuit or cold sensor causes "Open Loop" operation.

Test Description: Number(s) below refer to circled number(s) on the diagnostic chart.
1. DTC 13 will set:
   - Engine at normal operating temperature (above 70°C/158°F).
   - At least 2 minutes engine run time after start.
   - O2S signal voltage steady between 350 mV to 550 mV.
   - Throttle position sensor signal above 5% (about .3 volt above closed throttle voltage).
   - All conditions must be met for about 60 seconds.
   If the conditions for a DTC 13 exist, the system will not go into "Closed Loop."
2. This will determine if the oxygen sensor is at fault, or the wiring or ECM is the cause of the DTC 13.

3. For this test use only a high impedance digital volt ohmmeter (J 39200). This test checks the continuity of CKT 1665 and CKT 351. If CKT 351 is open, the ECM voltage on CKT 1665 will be over .6 volt (600 mV).

Diagnostic Aids:

Normal Tech 1 voltage readings vary between 10 mV to 1000 mV (.01 and 1.0 volt), while in "Closed Loop." The system will go into "Open Loop" operation when DTC 13 sets.

Refer to "Intermittents" in "Symptoms," Section

5.7L (VIN P) ENGINE — DIAGNOSTIC TROUBLE CODE CHART — 1993 CAMARO AND FIREBIRD

DTC 13
BANK 1 (LEFT) OXYGEN SENSOR (O2S) CIRCUIT
(OPEN CIRCUIT)
5.7L (VIN P) "F" CARLINE (MFI)

1. ENGINE AT NORMAL OPERATING TEMPERATURE (ABOVE 80°C/176°F).
   - RUN ENGINE ABOVE 1200 RPM FOR TWO MINUTES. DOES TECH 1 TOOL INDICATE "CLOSED LOOP"?

NO

2.
   - DISCONNECT BANK 1 O2S.
   - JUMPER HARNESS CKT 1665 (ECM SIDE) TO GROUND.
   - TECH 1 SHOULD DISPLAY O2S VOLTAGE BELOW .2 VOLT (200 mV) WITH ENGINE RUNNING. DOES IT?

YES

DTC 13 IS INTERMITTENT. IF NO ADDITIONAL DTC(s) WERE STORED, REFER TO "DIAGNOSTIC AIDS" ON FACING PAGE.

NO

3.
   - REMOVE JUMPER.
   - IGNITION "ON," ENGINE "OFF."
   - CHECK VOLTAGE OF CKT 1665 (ECM SIDE) AT O2S HARNESS CONNECTOR USING A DVM.

YES

FAULTY O2S CONNECTION OR O2S.

.3 - .6 VOLT (300 - 600 mV) → FAULTY ECM.

OVER .6 VOLT (600 mV) → OPEN CKT 351 OR FAULTY CONNECTION OR FAULTY ECM.

LESS THAN .3 VOLT (300 mV) → OPEN CKT 1665 OR FAULTY ECM CONNECTION OR FAULTY ECM.

"AFTER REPAIRS," REFER TO DTC CRITERIA ON FACING PAGE AND CONFIRM DTC DOES NOT RESET.

5.7L (VIN P) ENGINE — DIAGNOSTIC TROUBLE CODE CHART — 1993 CAMARO AND FIREBIRD

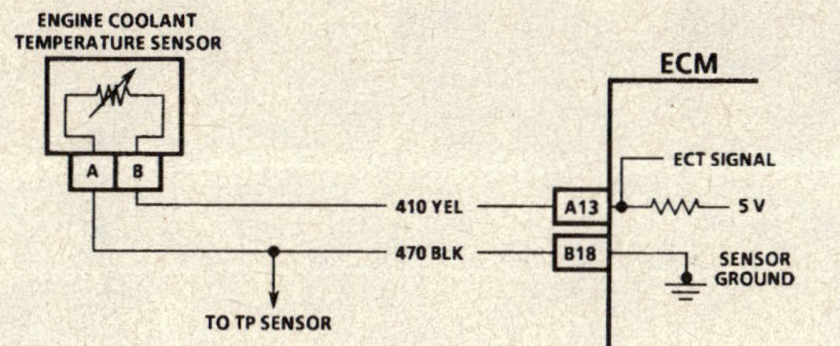

DTC 14
ENGINE COOLANT TEMPERATURE (ECT) SENSOR CIRCUIT
(HIGH TEMPERATURE INDICATED)
5.7L (VIN P) "F" CARLINE (MFI)

Circuit Description:

The Engine Coolant Temperature (ECT) sensor uses a thermistor to control a signal voltage to the ECM. The ECM applies a voltage on CKT 410 to the sensor. When the engine coolant is cold, the sensor (thermistor) resistance is high, therefore, the ECM will sense a high signal voltage. As the engine coolant warms, the sensor resistance becomes less, and the voltage drops. At normal engine operating temperature (85°C – 95°C or 185°F – 203°F) the voltage will measure about 1.5 to 2.0 volts.

Test Description: Number(s) below refer to circled number(s) on the diagnostic chart.

1. DTC 14 will set if:
   - Signal voltage indicates an engine coolant temperature above 130°C (266°F).
2. This test will determine if CKT 410 is shorted to ground, which will cause the condition for a DTC 14.

Diagnostic Aids:

Check harness routing for a potential short to ground in CKT 410.

Tech 1 displays engine coolant temperature in degrees celsius and fahrenheit. After engine is started, the temperature should rise steadily, reach normal operating temperature, and then stabilize when thermostat opens.

Refer to "Intermittents" in "Symptoms," Section

5.7L (VIN P) ENGINE — DIAGNOSTIC TROUBLE CODE CHART — 1993 CAMARO AND FIREBIRD

DTC 14

ENGINE COOLANT TEMPERATURE (ECT) SENSOR CIRCUIT
(HIGH TEMPERATURE INDICATED)
5.7L (VIN P) "F" CARLINE (MFI)

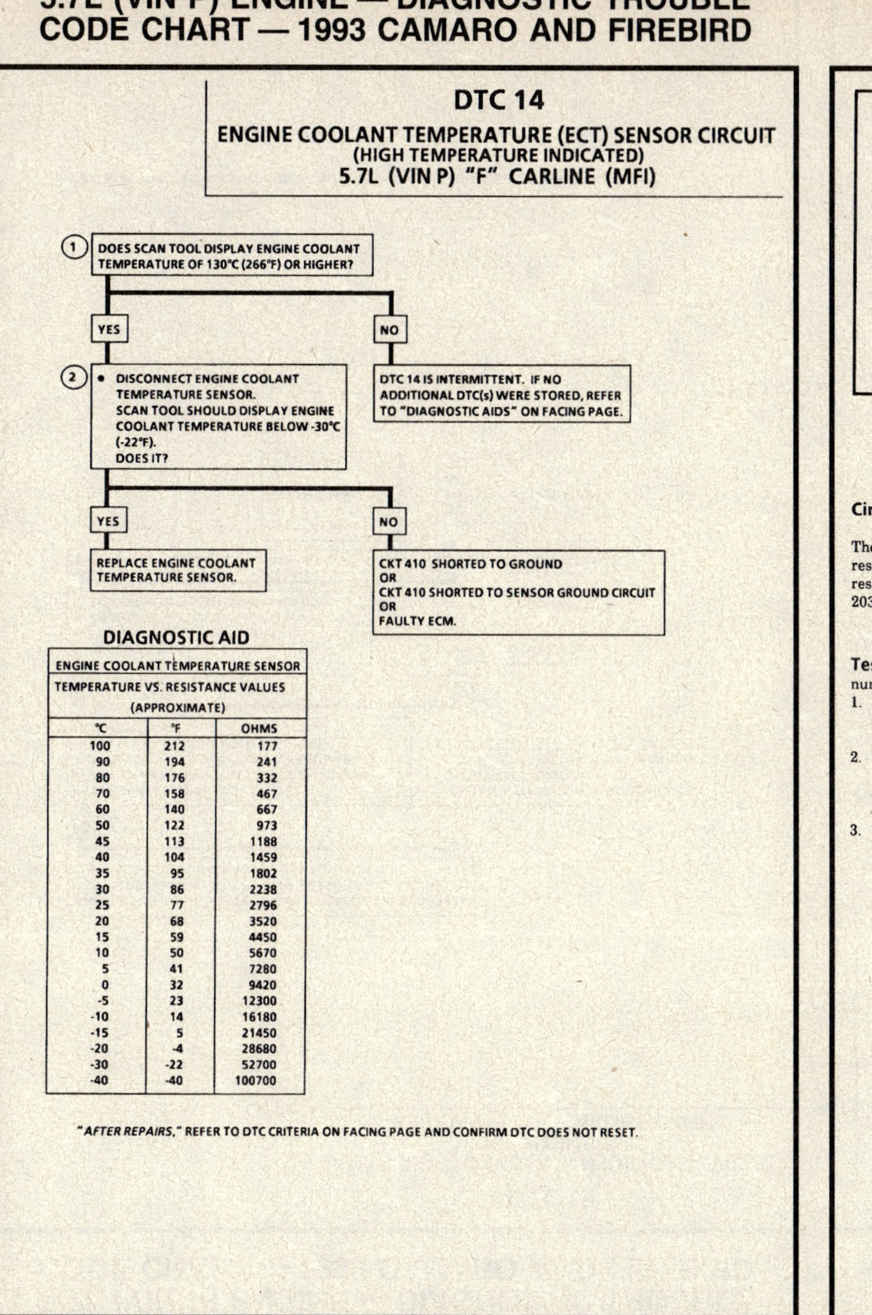

DIAGNOSTIC AID

ENGINE COOLANT TEMPERATURE SENSOR
TEMPERATURE VS. RESISTANCE VALUES
(APPROXIMATE)

°C	°F	OHMS
100	212	177
90	194	241
80	176	332
70	158	467
60	140	667
50	122	973
45	113	1188
40	104	1459
35	95	1802
30	86	2238
25	77	2796
20	68	3520
15	59	4450
10	50	5670
5	41	7280
0	32	9420
-5	23	12300
-10	14	16180
-15	5	21450
-20	-4	28680
-30	-22	52700
-40	-40	100700

"AFTER REPAIRS," REFER TO DTC CRITERIA ON FACING PAGE AND CONFIRM DTC DOES NOT RESET.

5.7L (VIN P) ENGINE — DIAGNOSTIC TROUBLE CODE CHART — 1993 CAMARO AND FIREBIRD

DTC 15

ENGINE COOLANT TEMPERATURE (ECT) SENSOR CIRCUIT
(LOW TEMPERATURE INDICATED)
5.7L (VIN P) "F" CARLINE (MFI)

Circuit Description:

The Engine Coolant Temperature (ECT) sensor uses a thermistor to control the signal voltage to the ECM. The ECM applies a voltage on CKT 410 to the sensor. When the engine coolant is cold, the sensor (thermistor) resistance is high, therefore the ECM will sense a high signal voltage. As the engine coolant warms, the sensor resistance becomes less, and the voltage drops. At normal engine operating temperature (85°C – 95°C or 185°F – 203°F), the voltage will measure about 1.5 to 2.0 volts at the ECM.

Test Description: Number(s) below refer to circled number(s) on the diagnostic chart.
1. DTC 15 will set if:
   - ECT signal voltage indicates an engine coolant temperature less than -49°C (-56.2°F).
2. This test simulates a DTC 14. If the ECM recognizes the low signal voltage, (high temperature) and the Tech 1 displays 130°C (266°F) or above, the ECM and wiring are OK.
3. This test will determine if CKT 410 is open. There should be 5 volts present at sensor connector, if measured with a DVM.

Diagnostic Aids:

A Tech 1 displays engine coolant temperature in degrees celsius and fahrenheit. After engine is started, the temperature should rise steadily, reach normal operating temperature, and then stabilize when the thermostat opens.

A faulty connection, or an open in CKT 410 or 470, will result in a DTC 15.

If DTC 21 is also set, check CKT 470 for faulty wiring or connections. Check terminals at sensor for good contact.

Refer to "Intermittents" in "Symptoms," Section

5.7L (VIN P) ENGINE — DIAGNOSTIC TROUBLE CODE CHART — 1993 CAMARO AND FIREBIRD

DTC 15

ENGINE COOLANT TEMPERATURE (ECT) SENSOR CIRCUIT
(LOW TEMPERATURE INDICATED)
5.7L (VIN P) "F" CARLINE (MFI)

1. • DOES TECH 1 SCAN TOOL DISPLAY ENGINE COOLANT TEMPERATURE OF -30°C (-22°F) OR LESS?

YES

NO

2. • DISCONNECT ENGINE COOLANT TEMPERATURE SENSOR.
   • JUMPER HARNESS TERMINALS TOGETHER.
   • TECH 1 SCAN TOOL SHOULD DISPLAY 130°C (266°F) OR MORE. DOES IT?

DTC 15 IS INTERMITTENT. IF NO ADDITIONAL DTC(S) WERE STORED, REFER TO "DIAGNOSTIC AIDS" ON FACING PAGE.

NO

YES

3. • JUMPER CKT 410 TO GROUND.
   • TECH 1 SCAN TOOL SHOULD DISPLAY OVER 130°C (266°F). DOES IT?

FAULTY CONNECTION OR ENGINE COOLANT TEMPERATURE SENSOR.

YES

NO

OPEN ENGINE COOLANT TEMPERATURE SENSOR GROUND CIRCUIT, FAULTY CONNECTION OR FAULTY ECM.

OPEN CKT 410, FAULTY CONNECTION AT ECM, OR FAULTY ECM.

DIAGNOSTIC AID

ENGINE COOLANT TEMPERATURE SENSOR		
TEMPERATURE VS. RESISTANCE VALUES (APPROXIMATE)		
°C	°F	OHMS
100	212	177
90	194	241
80	176	332
70	158	467
60	140	667
50	122	973
45	113	1188
40	104	1459
35	95	1802
30	86	2238
25	77	2796
20	68	3520
15	59	4450
10	50	5670
5	41	7280
0	32	9420
-5	23	12300
-10	14	16180
-15	5	21450
-20	-4	28680
-30	-22	52700
-40	-40	100700

"AFTER REPAIRS," REFER TO DTC CRITERIA ON FACING PAGE AND CONFIRM DTC DOES NOT RESET.

5.7L (VIN P) ENGINE — DIAGNOSTIC TROUBLE CODE CHART — 1993 CAMARO AND FIREBIRD

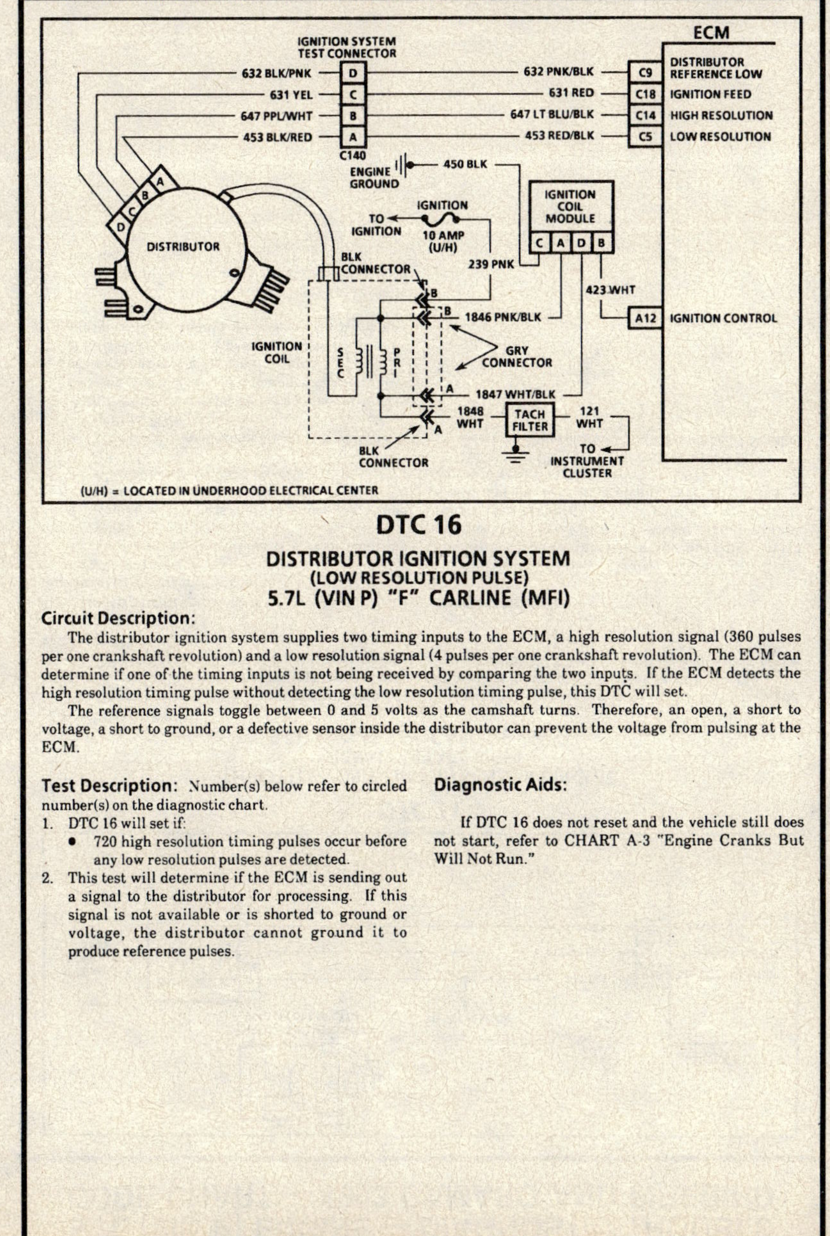

DTC 16

DISTRIBUTOR IGNITION SYSTEM
(LOW RESOLUTION PULSE)
5.7L (VIN P) "F" CARLINE (MFI)

Circuit Description:

The distributor ignition system supplies two timing inputs to the ECM, a high resolution signal (360 pulses per one crankshaft revolution) and a low resolution signal (4 pulses per one crankshaft revolution). The ECM can determine if one of the timing inputs is not being received by comparing the two inputs. If the ECM detects the high resolution timing pulse without detecting the low resolution timing pulse, this DTC will set.

The reference signals toggle between 0 and 5 volts as the camshaft turns. Therefore, an open, a short to voltage, a short to ground, or a defective sensor inside the distributor can prevent the voltage from pulsing at the ECM.

Test Description: Number(s) below refer to circled number(s) on the diagnostic chart.

1. DTC 16 will set if:
   • 720 high resolution timing pulses occur before any low resolution pulses are detected.
2. This test will determine if the ECM is sending out a signal to the distributor for processing. If this signal is not available or is shorted to ground or voltage, the distributor cannot ground it to produce reference pulses.

Diagnostic Aids:

If DTC 16 does not reset and the vehicle still does not start, refer to CHART A-3 "Engine Cranks But Will Not Run."

5.7L (VIN P) ENGINE — DIAGNOSTIC TROUBLE CODE CHART — 1993 CAMARO AND FIREBIRD

DTC 16
DISTRIBUTOR IGNITION SYSTEM
(LOW RESOLUTION PULSE)
5.7L (VIN P) "F" CARLINE (MFI)

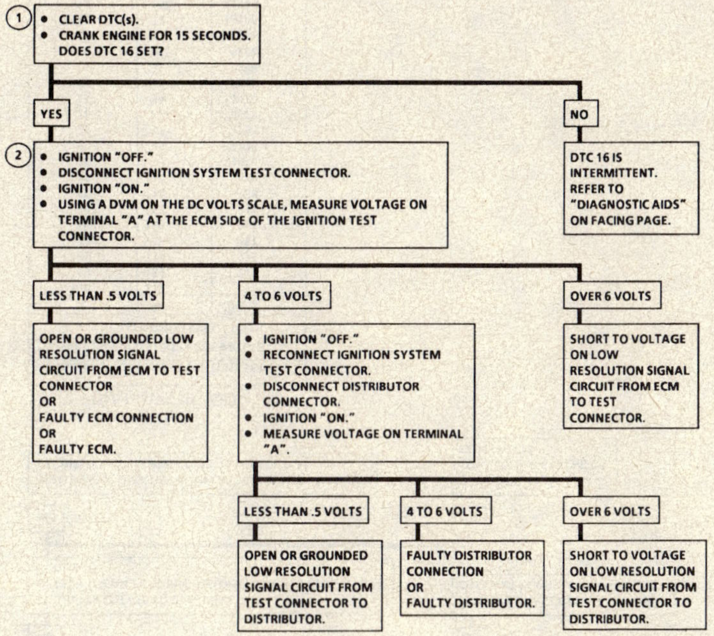

1.
- CLEAR DTC(s).
- CRANK ENGINE FOR 15 SECONDS. DOES DTC 16 SET?

YES

2.
- IGNITION "OFF."
- DISCONNECT IGNITION SYSTEM TEST CONNECTOR.
- IGNITION "ON."
- USING A DVM ON THE DC VOLTS SCALE, MEASURE VOLTAGE ON TERMINAL "A" AT THE ECM SIDE OF THE IGNITION TEST CONNECTOR.

NO

DTC 16 IS INTERMITTENT. REFER TO "DIAGNOSTIC AIDS" ON FACING PAGE.

LESS THAN .5 VOLTS

OPEN OR GROUNDED LOW RESOLUTION SIGNAL CIRCUIT FROM ECM TO TEST CONNECTOR OR FAULTY ECM CONNECTION OR FAULTY ECM.

4 TO 6 VOLTS

- IGNITION "OFF."
- RECONNECT IGNITION SYSTEM TEST CONNECTOR.
- DISCONNECT DISTRIBUTOR CONNECTOR.
- IGNITION "ON."
- MEASURE VOLTAGE ON TERMINAL "A."

OVER 6 VOLTS

SHORT TO VOLTAGE ON LOW RESOLUTION SIGNAL CIRCUIT FROM ECM TO TEST CONNECTOR.

LESS THAN .5 VOLTS

OPEN OR GROUNDED LOW RESOLUTION SIGNAL CIRCUIT FROM TEST CONNECTOR TO DISTRIBUTOR.

4 TO 6 VOLTS

FAULTY DISTRIBUTOR CONNECTION OR FAULTY DISTRIBUTOR.

OVER 6 VOLTS

SHORT TO VOLTAGE ON LOW RESOLUTION SIGNAL CIRCUIT FROM TEST CONNECTOR TO DISTRIBUTOR.

"AFTER REPAIRS," REFER TO DTC CRITERIA ON FACING PAGE AND CONFIRM DTC DOES NOT RESET.

5.7L (VIN P) ENGINE — DIAGNOSTIC TROUBLE CODE CHART — 1993 CAMARO AND FIREBIRD

THROTTLE POSITION (TP) SENSOR CONNECTOR

THROTTLE POSITION (TP) SENSOR

WOT

IDLE

TO ECT SENSOR

ECM

470 BLK — B18 — SENSOR GROUND
417 DK BLU — C3 — TP SENSOR SIGNAL
474 GRY — C2 — 5 V REFERENCE

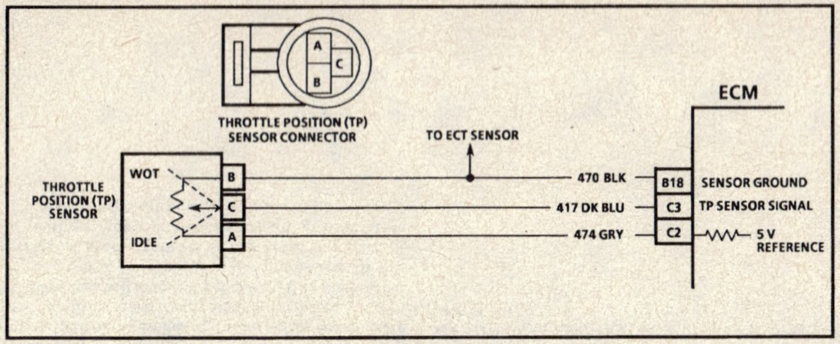

DTC 21
THROTTLE POSITION (TP) SENSOR CIRCUIT
(SIGNAL VOLTAGE HIGH)
5.7L (VIN P) "F" CARLINE (MFI)

Circuit Description:

The Throttle Position (TP) sensor provides a voltage signal that changes, relative to the throttle blade position. Signal voltage should vary from about .6 volt at idle to about 5 volts at Wide Open Throttle (WOT).

The TP sensor signal is one of the most important inputs used by the ECM for fuel control and for most of the ECM controlled outputs.

Test Description: Number(s) below refer to circled number(s) on the diagnostic chart.

1. DTC 21 will set if:
- TP sensor signal voltage is greater than 2.5 volts.
- Engine at idle (less than 12 grams per second calculated air flow).
- All conditions met for 3 seconds.
  OR
- TP sensor signal voltage over 4.8 volts with ignition "ON."

With throttle closed the TP sensor should display between .3 and .9 volt. If it does not, check throttle cable adjustment or for bent linkage.

2. With the TP sensor disconnected, the TP sensor voltage should go low, if the ECM and wiring are OK.
3. Probing CKT 470 with a test light to 12 volts checks the sensor ground. A faulty sensor ground will cause a DTC 21.

Diagnostic Aids:

The Tech 1 displays throttle position in voltage and percentage of throttle blade opening. With ignition "ON" or at idle, TP sensor signal voltage should read between .3 and .9 volt with the throttle closed and increase at a steady rate as throttle is moved toward Wide Open Throttle (WOT).

An open in CKT 470 will result in a DTC 21.

Refer to "Intermittents" in "Symptoms," Section

5.7L (VIN P) ENGINE — DIAGNOSTIC TROUBLE CODE CHART — 1993 CAMARO AND FIREBIRD

DTC 21
**THROTTLE POSITION (TP) SENSOR CIRCUIT
(SIGNAL VOLTAGE HIGH)
5.7L (VIN P) "F" CARLINE (MFI)**

1. • THROTTLE CLOSED.
   DOES TECH 1 SCAN TOOL DISPLAY
   THROTTLE POSITION OVER 2.5 VOLTS?

   YES → / NO →

   NO: DTC 21 IS INTERMITTENT. IF NO ADDITIONAL DTC(s) WERE STORED, REFER TO "DIAGNOSTIC AIDS" ON FACING PAGE.

2. • DISCONNECT THROTTLE POSITION SENSOR.
   TECH 1 SCAN TOOL SHOULD DISPLAY
   THROTTLE POSITION BELOW .2 VOLT (200 mV).
   DOES IT?

   YES → / NO →

   NO: TP SENSOR SIGNAL CIRCUIT SHORTED TO VOLTAGE
   OR
   FAULTY ECM.

3. • PROBE SENSOR GROUND CIRCUIT
   WITH A TEST LIGHT CONNECTED TO
   BATTERY VOLTAGE.

   LIGHT "ON" / LIGHT "OFF"

   LIGHT "ON":
   FAULTY CONNECTION
   OR
   THROTTLE POSITION SENSOR.

   LIGHT "OFF":
   OPEN SENSOR GROUND CIRCUIT
   OR
   FAULTY ECM.

"AFTER REPAIRS," REFER TO DTC CRITERIA ON FACING PAGE AND CONFIRM DTC DOES NOT RESET.

5.7L (VIN P) ENGINE — DIAGNOSTIC TROUBLE CODE CHART — 1993 CAMARO AND FIREBIRD

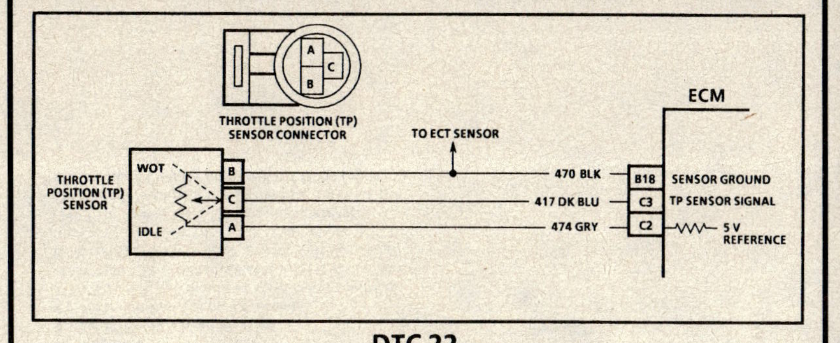

DTC 22
**THROTTLE POSITION (TP) SENSOR CIRCUIT
(SIGNAL VOLTAGE LOW)
5.7L (VIN P) "F" CARLINE (MFI)**

Circuit Description:
The Throttle Position (TP) sensor provides a voltage signal that changes, relative to the throttle blade position. Signal voltage should vary from about .6 volt at idle to about 5 volts at Wide Open Throttle (WOT).

The TP sensor signal is one of the most important inputs used by the ECM for fuel control and for most of the ECM controlled outputs.

Test Description: Number(s) below refer to circled number(s) on the diagnostic chart.
1. DTC 22, will set if:
   • TP sensor signal voltage is less than about .23 volt.
   The TP sensor has an auto zeroing feature. If the voltage reading is within the range of about .3 to .9 volt, the ECM will use that value as closed throttle. If the voltage reading is out of the auto zero range at closed throttle, check for a binding throttle cable or damaged linkage, if OK, continue with diagnosis.
2. Simulates DTC 21: (high voltage) If the ECM recognizes the high signal voltage, the ECM and wiring are OK.
3. This simulates a high signal voltage to check for an open in CKT 417.

Diagnostic Aids:
The Tech 1 displays throttle position in voltage and percentage of throttle blade opening. With ignition "ON" or at idle, TP sensor signal voltage should display between .3 and .9 volt with the throttle closed and increase at a steady rate as throttle is moved toward Wide Open-Throttle (WOT).

An open or short to ground in CKT 474 or CKT 417 will result in a DTC 22.

Refer to "Intermittents" in "Symptoms," Section

5.7L (VIN P) ENGINE — DIAGNOSTIC TROUBLE CODE CHART — 1993 CAMARO AND FIREBIRD

DTC 22
THROTTLE POSITION (TP) SENSOR CIRCUIT
(SIGNAL VOLTAGE LOW)
5.7L (VIN P) "F" CARLINE (MFI)

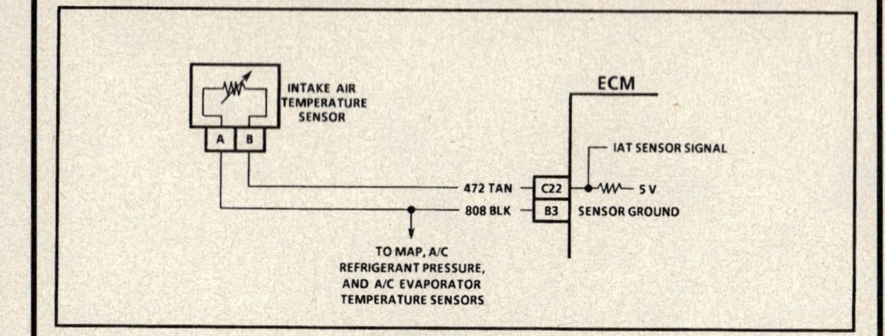

"AFTER REPAIRS," REFER TO DTC CRITERIA ON FACING PAGE AND CONFIRM DTC DOES NOT RESET.

5.7L (VIN P) ENGINE — DIAGNOSTIC TROUBLE CODE CHART — 1993 CAMARO AND FIREBIRD

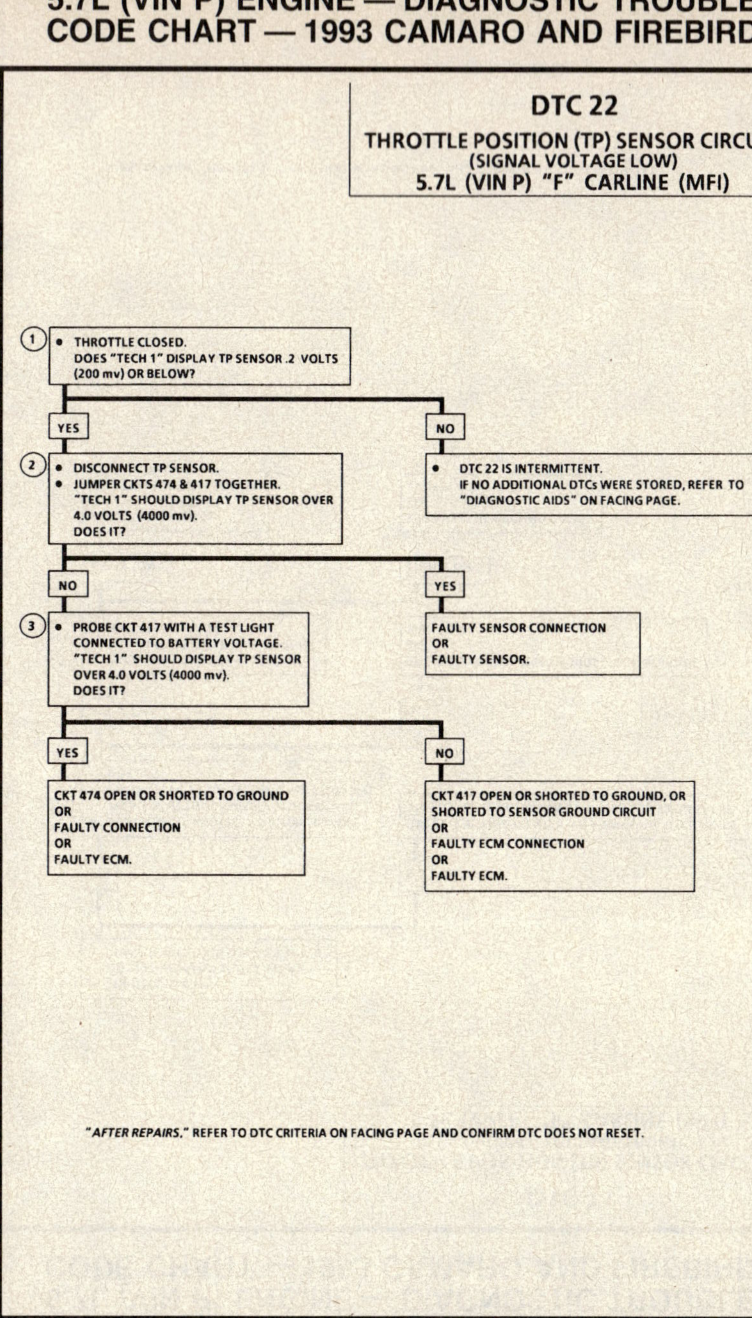

DTC 23
INTAKE AIR TEMPERATURE (IAT) SENSOR CIRCUIT
(LOW TEMPERATURE INDICATED)
5.7L (VIN P) "F" CARLINE (MFI)

Circuit Description:

The Intake Air Temperature (IAT) sensor uses a thermistor to control the signal voltage to the ECM. The ECM applies a voltage (about 5 volts) on CKT 472 to the sensor. When the intake air is cold, the sensor (thermistor) resistance is high, therefore, the ECM will sense a high signal voltage. If the intake air is warm, the sensor (thermistor) resistance is low, therefore, the ECM will sense a low voltage.

Test Description: Number(s) below refer to circled number(s) on the diagnostic chart.

1. DTC 23 will set if:
   - A signal voltage indicates an intake air temperature below -38°C (-36°F).
   - Engine coolant temperature above 29°C (85°F).
   - Engine idling (0% throttle).
   - No VSS (vehicle not moving).
   - All conditions met for 60 seconds,
   OR
   - A signal voltage indicates an intake air temperature below -38°C (-36°F).
   - DTC 14 or 15 present.
   - Engine run time greater than 60 seconds.
   - Engine idling (0% throttle).
   - No VSS (vehicle not moving)
   - All conditions met for 12 seconds.
2. A DTC 23 will set, due to an open sensor, wire, or connection. This test simulates a DTC 25. If the ECM recognizes the high signal voltage, (high temperature) and the Tech 1 displays 130°C (266°F) or above, the ECM and wiring are OK.
3. This will determine if the IAT sensor signal (CKT 472) or the IAT sensor ground (CKT 808) is open.

Diagnostic Aids:

The Tech 1 displays temperature of the air entering the engine and should be close to ambient air temperature when engine is cold, and rises as underhood temperature increases.

Carefully check harness and connections for possible open.

Refer to "Intermittents" in "Symptoms," Section

If the engine has been allowed to sit overnight, the intake air temperature, and engine coolant temperature values should be within a few degrees of each other.

5.7L (VIN P) ENGINE — DIAGNOSTIC TROUBLE CODE CHART — 1993 CAMARO AND FIREBIRD

DTC 23

INTAKE AIR TEMPERATURE (IAT) SENSOR CIRCUIT
(LOW TEMPERATURE INDICATED)
5.7L (VIN P) "F" CARLINE (MFI)

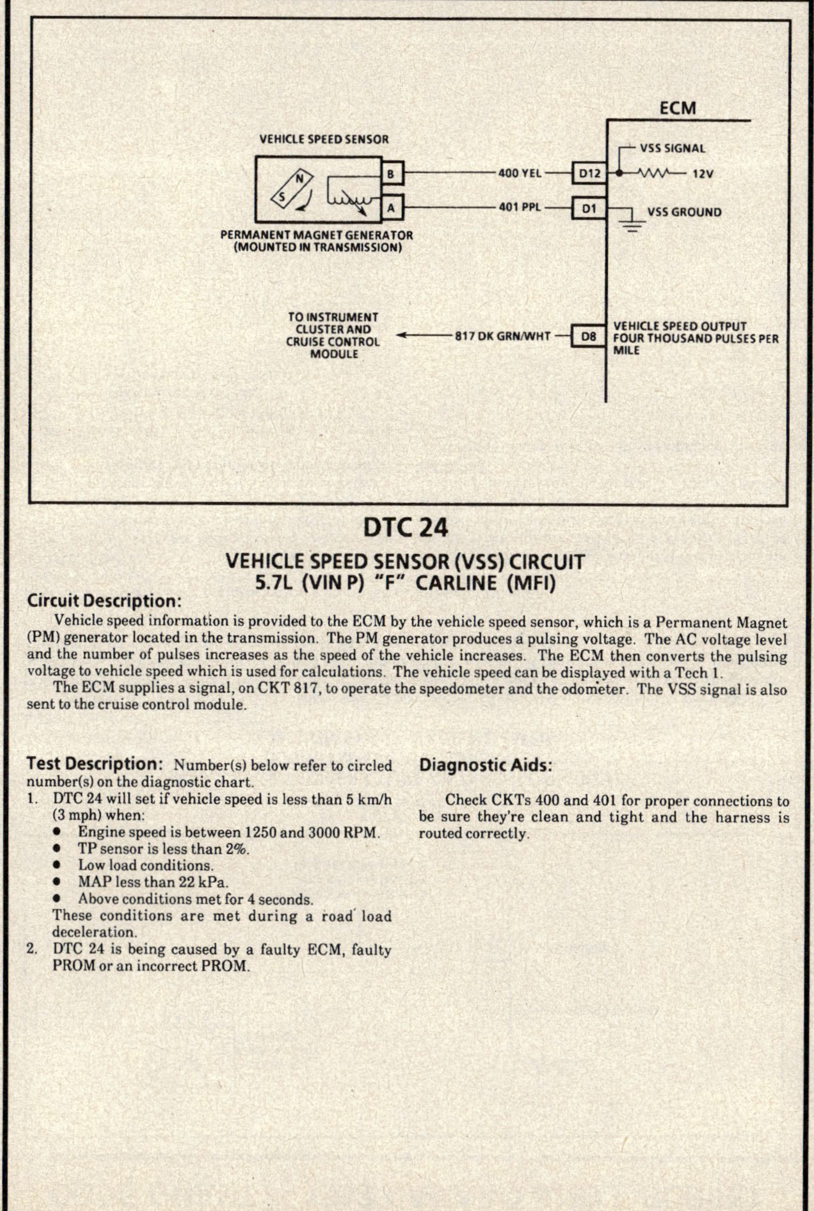

DIAGNOSTIC AID

INTAKE AIR TEMPERATURE SENSOR		
TEMPERATURE VS. RESISTANCE VALUES (APPROXIMATE)		
°C	°F	OHMS
100	212	177
90	194	241
80	176	332
70	158	467
60	140	667
50	122	973
45	113	1188
40	104	1459
35	95	1802
30	86	2238
25	77	2796
20	68	3520
15	59	4450
10	50	5670
5	41	7280
0	32	9420
-5	23	12300
-10	14	16180
-15	5	21450
-20	-4	28680
-30	-22	52700
-40	-40	100700

"AFTER REPAIRS," REFER TO DTC CRITERIA ON FACING PAGE AND CONFIRM DTC DOES NOT RESET.

5.7L (VIN P) ENGINE — DIAGNOSTIC TROUBLE CODE CHART — 1993 CAMARO AND FIREBIRD

DTC 24

VEHICLE SPEED SENSOR (VSS) CIRCUIT
5.7L (VIN P) "F" CARLINE (MFI)

Circuit Description:

Vehicle speed information is provided to the ECM by the vehicle speed sensor, which is a Permanent Magnet (PM) generator located in the transmission. The PM generator produces a pulsing voltage. The AC voltage level and the number of pulses increases as the speed of the vehicle increases. The ECM then converts the pulsing voltage to vehicle speed which is used for calculations. The vehicle speed can be displayed with a Tech 1.

The ECM supplies a signal, on CKT 817, to operate the speedometer and the odometer. The VSS signal is also sent to the cruise control module.

Test Description: Number(s) below refer to circled number(s) on the diagnostic chart.
1. DTC 24 will set if vehicle speed is less than 5 km/h (3 mph) when:
   - Engine speed is between 1250 and 3000 RPM.
   - TP sensor is less than 2%.
   - Low load conditions.
   - MAP less than 22 kPa.
   - Above conditions met for 4 seconds.
   These conditions are met during a road load deceleration.
2. DTC 24 is being caused by a faulty ECM, faulty PROM or an incorrect PROM.

Diagnostic Aids:

Check CKTs 400 and 401 for proper connections to be sure they're clean and tight and the harness is routed correctly.

5.7L (VIN P) ENGINE — DIAGNOSTIC TROUBLE CODE CHART — 1993 CAMARO AND FIREBIRD

DTC 24

VEHICLE SPEED SENSOR (VSS) CIRCUIT
5.7L (VIN P) "F" CARLINE (MFI)

① • IGNITION "OFF."
• RAISE DRIVE WHEELS.
• ENGINE IDLING.
• WITH ENGINE IDLING IN GEAR, TECH 1 SHOULD DISPLAY VEHICLE SPEED ABOVE 0.
DOES IT?

NO → DOES SPEEDOMETER WORK?

YES → DTC 24 IS INTERMITTENT. IF NO ADDITIONAL DTC(s) WERE STORED, REFER TO "DIAGNOSTIC AIDS" ON FACING PAGE.

NO →
• IGNITION "OFF."
• DISCONNECT VSS HARNESS CONNECTOR AT TRANSMISSION.
• CONNECT SIGNAL GENERATOR TESTER J 33431-B TO VSS HARNESS CONNECTOR.
• IGNITION "ON," TESTER "ON" AND SET TO GENERATE A VSS SIGNAL.
• THE TECH 1 SHOULD DISPLAY VEHICLE SPEED ABOVE 0.
DOES IT?

② CHECK PROM FOR CORRECT APPLICATION. IF OK, REPLACE ECM.

NO → CKT 400 OR 401 OPEN, SHORTED TO GROUND, SHORTED TOGETHER, FAULTY CONNECTIONS OR FAULTY ECM.

YES → REPLACE VEHICLE SPEED SENSOR.

"AFTER REPAIRS," CONFIRM "CLOSED LOOP" OPERATION AND NO MIL (SERVICE ENGINE SOON).

5.7L (VIN P) ENGINE — DIAGNOSTIC TROUBLE CODE CHART — 1993 CAMARO AND FIREBIRD

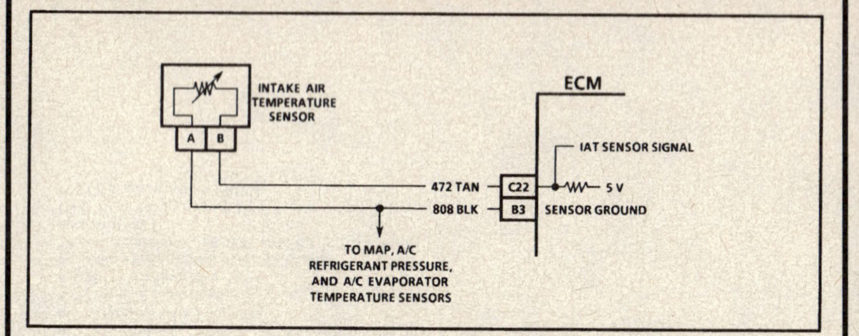

DTC 25

INTAKE AIR TEMPERATURE (IAT) SENSOR CIRCUIT
(HIGH TEMPERATURE INDICATED)
5.7L (VIN P) "F" CARLINE (MFI)

Circuit Description:

The Intake Air Temperature (IAT) sensor uses a thermistor to control the signal voltage to the ECM. The ECM applies a voltage (about 5 volts) on CKT 472 to the sensor. When intake air is cold, the sensor (thermistor) resistance is high, therefore, the ECM will sense a high signal voltage. If the intake air is warm, the sensor (thermistor) resistance is low, therefore, the ECM will sense a low signal voltage.

Test Description: Number(s) below refer to circled number(s) on the diagnostic chart.

1. DTC 25 will set if:
   • Signal voltage indicates an intake air temperature greater than 150°C (302°F) for 12 seconds.
   • Time since engine start is 4 minutes or longer.
   • Vehicle speed is greater than 8 km/h (5 mph).
   OR
   • Signal voltage indicates an intake air temperature greater than 150°C (302°F).
   • No VSS (vehicle not moving).
   • All conditions met for 60 seconds.

Diagnostic Aids:

The Tech 1 displays temperature of the air entering the engine and should display close to ambient air temperature, when engine is cold, and rise as underhood temperature increases.

Check harness routing for possible short to ground in CKT 472.

Refer to "Intermittents" in "Symptoms," Section

5.7L (VIN P) ENGINE — DIAGNOSTIC TROUBLE CODE CHART — 1993 CAMARO AND FIREBIRD

DTC 25
INTAKE AIR TEMPERATURE (IAT) SENSOR CIRCUIT
(HIGH TEMPERATURE INDICATED)
5.7L (VIN P) "F" CARLINE (MFI)

1. DOES TECH 1 SCAN TOOL DISPLAY IAT OF 145°C (293°F) OR HOTTER?

YES
- DISCONNECT SENSOR. TECH 1 SCAN TOOL SHOULD DISPLAY TEMPERATURE BELOW -30°C (-22°F). DOES IT?

YES
REPLACE SENSOR.

NO
DTC 25 IS INTERMITTENT. IF NO ADDITIONAL DTC(S) WERE STORED, REFER TO "DIAGNOSTIC AIDS" ON FACING PAGE.

NO
CKT 472 SHORTED TO GROUND
OR
TO SENSOR GROUND
OR
ECM IS FAULTY.

DIAGNOSTIC AID

INTAKE AIR TEMPERATURE SENSOR

TEMPERATURE VS. RESISTANCE VALUES (APPROXIMATE)

°C	°F	OHMS
100	212	177
90	194	241
80	176	332
70	158	467
60	140	667
50	122	973
45	113	1188
40	104	1459
35	95	1802
30	86	2238
25	77	2796
20	68	3520
15	59	4450
10	50	5670
5	41	7280
0	32	9420
-5	23	12300
-10	14	16180
-15	5	21450
-20	-4	28680
-30	-22	52700
-40	-40	100700

"AFTER REPAIRS," REFER TO DTC CRITERIA ON FACING PAGE AND CONFIRM DTC DOES NOT RESET.

5.7L (VIN P) ENGINE — DIAGNOSTIC TROUBLE CODE CHART — 1993 CAMARO AND FIREBIRD

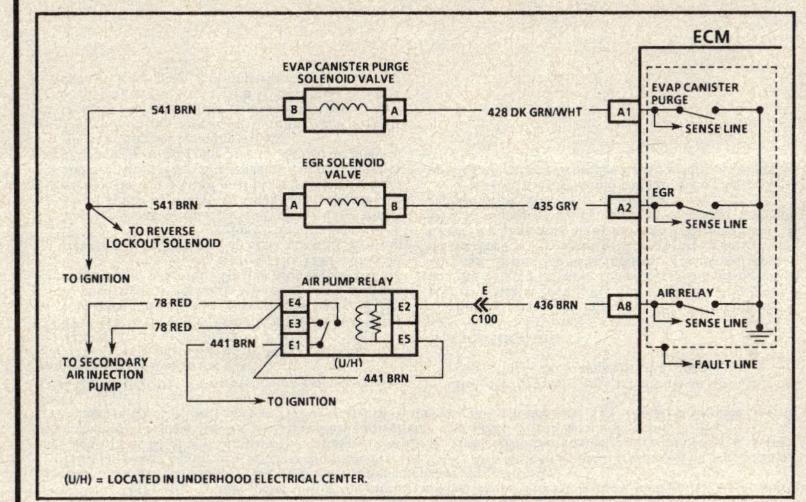

DTC 26
QUAD-DRIVER MODULE (QDM) #1 CIRCUIT
5.7L (VIN P) "F" CARLINE (MFI)

Circuit Description:

The ECM is used to control several components such as those illustrated above. The ECM controls these devices through the use of a Quad-Driver Module (QDM). When the ECM is commanding a component "ON," the voltage of the output circuit will be "low" (near 0 volts). When the ECM is commanding the output circuit to a component "OFF," the voltage potential of the circuit will be "high." (Near battery voltage.) The primary function of the QDM is to supply the ground for the component being controlled.

Each QDM has a fault line which is monitored by the ECM. The fault line signal status can be displayed on the Tech 1. The ECM will compare the voltage at the QDM. If the QDM fault detection circuit senses a voltage other than what is expected, the fault line status will change and a DTC 26 will set.

Test Description: Number(s) below refer to circled number(s) on the diagnostic chart.

1. DTC 26 will set if: the ECM detects the wrong voltage potential for 20 seconds.

   This test will begin to determine if the QDM connected AIR pump can be controlled by the ECM. If the relay appears to operate but the AIR pump does not turn "ON," refer to "Secondary Air Injection (AIR) System," for further AIR diagnosis.

2. This check can detect a partially shorted coil which would cause excessive current flow. Excessive current flow to a QDM will be detected as a fault and set this DTC. If excessive current flow is detected, a circuit check will be performed to isolate the device from the wiring.

3. The remaining checks will identify a circuit problem that has caused an excessive current flow or inoperative relay. If a QDM circuit check is done on a relay, it is important to identify and test the relay coil terminals of the harness connector to avoid improper diagnosis.

Diagnostic Aids:

Engine should be idling while monitoring QDM status. Using a Tech 1, monitor the QDM status while moving related harness connectors, including ECM harness. If the failure is induced, a fault will appear on the Tech 1. This can help locate the intermittent. Check for bent pins at ECM and ECM connector terminals. If DTC reoccurs with no apparent connector problem, replace ECM.

5.7L (VIN P) ENGINE — DIAGNOSTIC TROUBLE CODE CHART — 1993 CAMARO AND FIREBIRD

DTC 26
QUAD-DRIVER MODULE (QDM) #1 CIRCUIT
5.7L (VIN P) "F" CARLINE (MFI)

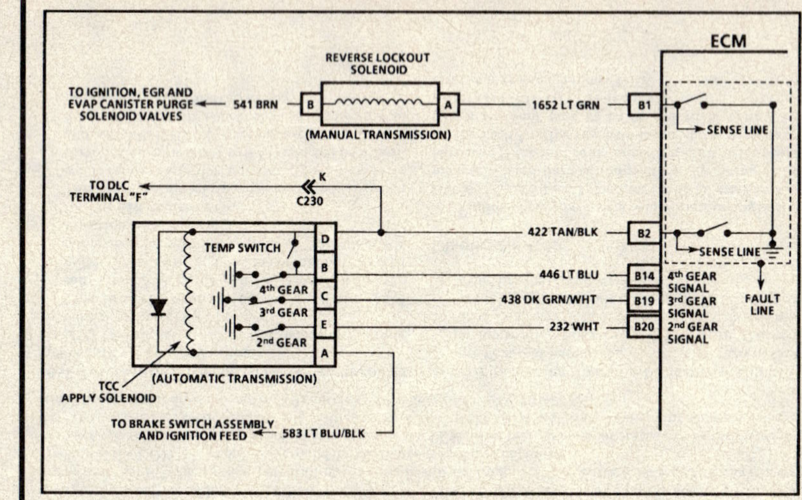

"AFTER REPAIRS," REFER TO DTC CRITERIA ON FACING PAGE AND CONFIRM DTC DOES NOT RESET.

5.7L (VIN P) ENGINE — DIAGNOSTIC TROUBLE CODE CHART — 1993 CAMARO AND FIREBIRD

DTC 27
QUAD-DRIVER MODULE (QDM) #2 CIRCUIT
5.7L (VIN P) "F" CARLINE (MFI)

Circuit Description:

The ECM is used to control several components such as those illustrated above. The ECM controls these devices through the use of a Quad-Driver Module (QDM). When the ECM is commanding a component "ON," the voltage of the output circuit will be "low" (near 0 volts). When the ECM is commanding the output circuit to a component "OFF," the voltage potential of the circuit will be "high." (Near battery voltage.) The primary function of the QDM is to supply the ground for the component being controlled.

Each QDM has a fault line which is monitored by the ECM. The fault line signal status can be displayed on the Tech 1. The ECM will compare the voltage at the QDM. If the QDM fault detection circuit senses a voltage other than what is expected, the fault line status will change and a DTC 27 will set if applicable.

A fault in QDM #2 will store a DTC 27 in the ECM memory but will not turn "ON" the MIL (Service Engine Soon).

Test Description: Number(s) below refer to circled number(s) on the diagnostic chart.

1. DTC 27 will set if the ECM detects an improper voltage potential for 26 seconds.
   This test will begin to determine if the QDM connected devices can be controlled by the ECM. If the reverse lockout solenoid or TCC are inoperative, refer to "Torque Converter Clutch (TCC) System (A/T), Automatic Transmission Gear Switch System, Reverse Lockout System (M/T)," for further diagnosis.
2. This check can detect a partially shorted coil which would cause excessive current flow. Excessive current flow to a QDM will be detected as a fault and set this DTC. If excessive current flow is detected, a circuit check will be performed to isolate the device from the wiring.
3. The remaining checks will identify a circuit problem that has caused an excessive current flow or inoperative solenoid. If a QDM circuit check is done on a solenoid, it is important to identify and test the solenoid coil terminals of the harness connector to avoid improper diagnosis.

Diagnostic Aids:

Engine should be idling while monitoring QDM status. Using a Tech 1, monitor the QDM status while moving related harness connectors, including ECM harness. If the failure is induced, a fault will appear on the Tech 1. This can help locate the intermittent. Check for bent pins at ECM and ECM terminals. If DTC 27 reoccurs with no apparent connector problem, replace ECM.

The QDM status should change when the brake is applied on vehicles equipped with a torque converter clutch.

5.7L (VIN P) ENGINE — DIAGNOSTIC TROUBLE CODE CHART — 1993 CAMARO AND FIREBIRD

DTC 27
QUAD-DRIVER MODULE (QDM) #2 CIRCUIT
5.7L (VIN P) "F" CARLINE (MFI)

(1)
- IF OTHER DIAGNOSTIC TROUBLE CODES (DTCs) ARE STORED USE THOSE CHARTS FIRST.
- IGNITION "ON," ENGINE "OFF."
- USING TECH 1 "MISC TEST," ACTIVATE REVERSE LOCKOUT SOLENOID (M/T) OR TCC SOLENOID (A/T).
  DOES SOLENOID OPERATE?

YES / NO

(2)
- KEY "OFF."
- DISCONNECT ECM CONNECTOR "B".
- KEY "ON."
- USING DVM ON AMP SCALE, MEASURE CURRENT FROM HARNESS TERMINAL "B2" OR "B1" TO GROUND.
- ECM TERMINAL "B1" (MANUAL TRANSMISSION) SHOULD MEASURE LESS THAN 1.5 AMPS (BUT NOT "0").
- ECM TERMINAL "B2" (AUTOMATIC TRANSMISSION) SHOULD MEASURE LESS THAN .75 AMPS (BUT NOT "0").
  DOES IT?

YES / NO

YES: DTC 27 IS INTERMITTENT. REFER TO "DIAGNOSTIC AIDS" ON FACING PAGE.

NO (MANUAL TRANSMISSION):

(3)
- IGNITION "OFF."
- RAISE VEHICLE.
- DISCONNECT REVERSE LOCKOUT SOLENOID.
- CONNECT TEST LIGHT BETWEEN SOLENOID HARNESS CONNECTOR TERMINALS.
- IGNITION "ON."
- LIGHT SHOULD BE "ON."
  IS IT?

NO / YES

NO:
- CONNECT TEST LIGHT FROM IGNITION FEED CIRCUIT OF SOLENOID HARNESS TO GROUND.
- LIGHT SHOULD BE "ON."
  IS IT?

YES: FAULTY ECM CONNECTION OR OPEN SOLENOID DRIVER CIRCUIT OR FAULTY ECM.

NO: OPEN SOLENOID IGNITION FEED CIRCUIT.

YES:
- GROUND DIAGNOSTIC "TEST" TERMINAL.
- LIGHT SHOULD BE "OFF."
  IS IT?

NO: DISCONNECT ECM CONNECTOR "B". IS LIGHT "ON"?

  NO: FAULTY ECM.

  YES: GROUNDED SOLENOID DRIVER CIRCUIT.

YES: FAULTY SOLENOID CONNECTION OR FAULTY SOLENOID.

NO (AUTOMATIC TRANSMISSION):

(3)
- IGNITION "OFF."
- RAISE VEHICLE.
- DISCONNECT TCC SOLENOID CONNECTOR.
- CONNECT TEST LIGHT BETWEEN SOLENOID HARNESS CONNECTOR TERMINALS.
- IGNITION "ON."
- GROUND DIAGNOSTIC TEST TERMINAL.
- LIGHT SHOULD BE "ON."
  IS IT?

NO / YES

NO:
- CONNECT TEST LIGHT FROM IGNITION FEED CIRCUIT OF SOLENOID TO GROUND.
- LIGHT SHOULD BE "ON."
  IS IT?

YES: FAULTY ECM CONNECTION OR OPEN SOLENOID DRIVER CIRCUIT OR FAULTY ECM.

NO: OPEN SOLENOID IGNITION FEED CIRCUIT.

YES:
- UNGROUND DIAGNOSTIC TEST TERMINAL.
- LIGHT SHOULD BE "OFF."
  IS IT?

NO: DISCONNECT ECM CONNECTOR "B". IS LIGHT "ON"?

  NO: FAULTY ECM.

  YES: GROUNDED SOLENOID DRIVER CIRCUIT.

YES: FAULTY SOLENOID CONNECTION OR FAULTY SOLENOID.

"AFTER REPAIRS," REFER TO DTC CRITERIA ON FACING PAGE AND CONFIRM DTC DOES NOT RESET.

5.7L (VIN P) ENGINE — DIAGNOSTIC TROUBLE CODE CHART — 1993 CAMARO AND FIREBIRD

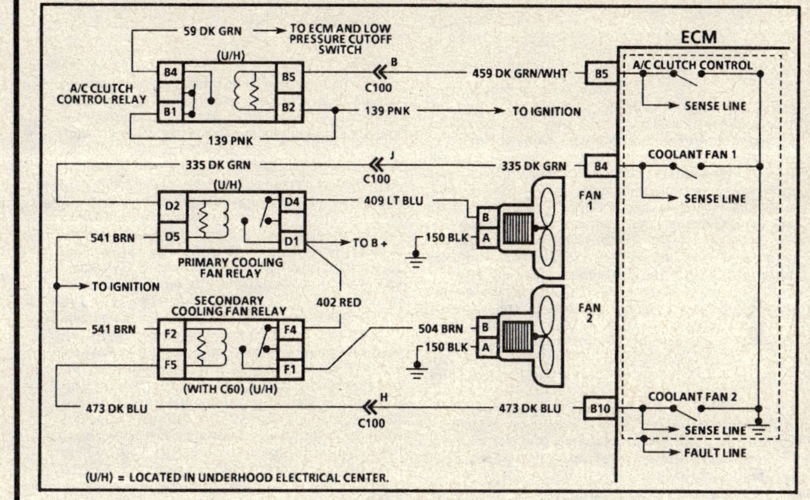

DTC 28
QUAD-DRIVER MODULE (QDM) #3 CIRCUIT
5.7L (VIN P) "F" CARLINE (MFI)

Circuit Description:

The ECM is used to control several components such as those illustrated above. The ECM controls these devices through the use of a Quad-Driver Module (QDM). When the ECM is commanding a component "ON," the voltage of the output circuit will be "low" (near 0 volts). When the ECM is commanding the output circuit to a component "OFF," the voltage potential of the circuit will be "high." (Near battery voltage.) The primary function of the QDM is to supply the ground for the component being controlled.

Each QDM has a fault line signal status which is monitored by the ECM. The fault line signal status can be displayed on the Tech 1. The ECM will compare the voltage at the QDM only when an A/C request signal is detected at the ECM. If the QDM fault detection circuit senses a voltage other than what is expected, the fault line status will change and a DTC 28 will set if applicable.

A QDM #3 fault will store a DTC 28 in the ECM memory but will not turn "ON" the MIL (Service Engine Soon).

Test Description: Number(s) below refer to circled number(s) on the diagnostic chart.

1. DTC 28 will set if the ECM detects an improper voltage potential for 26 seconds.
   These tests will begin to determine if the QDM connected devices can be controlled by the ECM. If a relay appears to operate but the device does not activate, refer to "Electric Cooling Fan," Section or "A/C Clutch Circuit Diagnosis," for further diagnosis.

2. This check can detect a partially shorted coil which would cause excessive current flow. Excessive current flow to a QDM will be detected as a fault and set this DTC. If excessive current flow is detected, a circuit check will be performed to isolate the device from the wiring

3. The remaining checks will identify a circuit problem that has caused an excessive current flow or inoperative relay. If a QDM circuit check is done on a relay, it is important to identify and test the relay coil terminals of the harness connector to avoid improper diagnosis.

Diagnostic Aids:

Engine should be idling while monitoring QDM status. Using a Tech 1, monitor the QDM status while moving related harness connectors, including ECM harness. If the failure is induced, a fault will appear on the Tech 1. This can help locate the intermittent. Check for bent pins at ECM and ECM connector terminals. If DTC reoccurs with no apparent connector problem, replace ECM.

5.7L (VIN P) ENGINE — DIAGNOSTIC TROUBLE CODE CHART — 1993 CAMARO AND FIREBIRD

DTC 28
QUAD-DRIVER MODULE (QDM) #3 CIRCUIT
5.7L (VIN P) "F" CARLINE (MFI)

①
- IF OTHER DTCs ARE STORED USE THOSE CHARTS FIRST.
- IGNITION "ON," ENGINE "OFF."
- USING TECH 1 "MISC TEST," ACTIVATE FAN 1 RELAY. DOES FAN 1 RELAY OPERATE?

YES / NO

- ACTIVATE FAN 2 RELAY. DOES FAN 2 RELAY OPERATE?

YES / NO

- ACTIVATE A/C RELAY. DOES A/C RELAY OPERATE?

YES / NO

②
- KEY "OFF."
- DISCONNECT ECM CONNECTOR "B".
- KEY "ON."
- USING DVM ON 2 AMP SCALE, MEASURE CURRENT FROM HARNESS TERMINALS "B4", "B5", "B10" TO GROUND.
- EACH TERMINAL SHOULD MEASURE LESS THAN .75 AMPS (BUT NOT "0"). DO THEY?

YES / NO

DTC 28 IS INTERMITTENT. REFER TO "DIAGNOSTIC AIDS" ON FACING PAGE.

③
- IGNITION "OFF."
- DISCONNECT AFFECTED RELAY CONNECTOR.
- CONNECT TEST LIGHT BETWEEN RELAY COIL CONNECTOR TERMINALS.
- IGNITION "ON."
- GROUND DIAGNOSTIC TEST TERMINAL.
- LIGHT SHOULD BE "ON." IS IT?

NO / YES

NO side:
- CONNECT TEST LIGHT FROM IGNITION FEED CIRCUIT OF RELAY/SOLENOID VALVE TO GROUND.
- LIGHT SHOULD BE "ON." IS IT?

YES → FAULTY ECM CONNECTION OR OPEN RELAY/SOLENOID VALVE DRIVER CIRCUIT OR FAULTY ECM.

NO → OPEN RELAY/SOLENOID VALVE IGNITION FEED CIRCUIT.

YES side:
- UNGROUND DIAGNOSTIC TEST TERMINAL.
- LIGHT SHOULD BE "OFF." IS IT?

NO → DISCONNECT ECM CONNECTOR "B". IS LIGHT "ON"?
- NO → FAULTY ECM.
- YES → GROUNDED RELAY/SOLENOID VALVE DRIVER CIRCUIT.

YES → FAULTY RELAY/SOLENOID VALVE CONNECTION OR FAULTY RELAY/SOLENOID VALVE.

"AFTER REPAIRS," REFER TO DTC CRITERIA ON FACING PAGE AND CONFIRM DTC DOES NOT RESET.

5.7L (VIN P) ENGINE — DIAGNOSTIC TROUBLE CODE CHART — 1993 CAMARO AND FIREBIRD

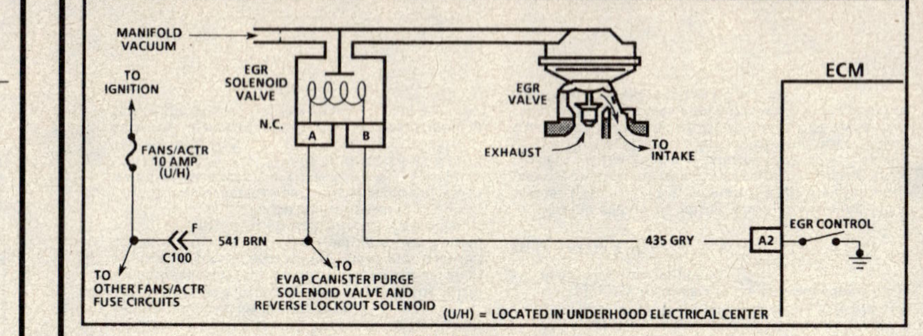

DTC 32 (Page 1 of 2)
EXHAUST GAS RECIRCULATION (EGR) CIRCUIT
5.7L (VIN P) "F" CARLINE (MFI)

Circuit Description:

The Exhaust Gas Recirculation (EGR) valve vacuum is controlled by an ECM controlled solenoid valve. The ECM will turn the EGR "ON" and "OFF" (Duty Cycle) by grounding CKT 435. The duty cycle is calculated by the ECM based on information from the engine coolant temperature, air flow and engine RPM. There should be no EGR when in park or neutral, TP sensor input below a specified value, or TP sensor indicating Wide Open Throttle (WOT).

With the ignition "ON" engine stopped, the EGR solenoid valve is de-energized and by grounding the diagnostic "test" terminal, the solenoid valve is energized. The ECM will check EGR operation when:

- Vehicle speed is above 50 mph.
- MAP is as listed in table (depends on altitude).
- Throttle angle is between 8% and 30%.
- No more than .4% change in throttle position.

	MAP WITH KEY "ON" ENGINE "OFF" (BARO)			
	100	90	80	70
MAP RANGE FOR TEST	30	20	10	0
	85	75	65	55

Test Description: Number(s) below refer to circled number(s) on the diagnostic chart.

1. **Intake Passage:** Shut "OFF" engine and remove the EGR valve from the manifold. Plug the exhaust side hole with a shop rag or suitable stopper. Leaving the intake side hole open, attempt to start the engine. If the engine runs at a very high idle (up to 3000 RPM is possible) or starts and stalls, the EGR passages are not restricted. If the engine starts and idles normally, the EGR intake side passage in the intake manifold is restricted.
Exhaust Passage: With EGR valve still removed, plug the intake side hole with a suitable stopper. With the exhaust side hole open, check for the presence of exhaust gas. If no exhaust gas is present, the EGR exhaust side passage in the intake manifold is restricted.
By grounding the diagnostic "test" terminal, the EGR solenoid valve should be energized and allow vacuum to be applied to the gage. The vacuum at the gage may or may not slowly bleed off. It is important that the gage is able to read the amount of vacuum being applied.

3. When the diagnostic "test" terminal is grounded, the vacuum gage should bleed off through a vent in the solenoid valve. The pump gage may or may not bleed off but this does not indicate a problem.

4. This test will determine if the electrical control part of the system is at fault or if the connector or solenoid valve is at fault.

Diagnostic Aids:

The EGR circuit can be inoperative if the PNP switch is misadjusted or faulty. The EGR is disabled when in park or neutral. To check the PNP switch, refer to CHART C-1A.

Suction from shop exhaust hoses can alter backpressure and may affect the operation of the EGR valve during install testing.

The ECM also monitors left side short term fuel trim during a DTC 32 test and must detect a change to pass the test. If short term fuel trim is at a fixed value (lean or rich) when a DTC 32 test is run, a DTC 32 may set.

Vacuum lines should be thoroughly checked for proper routing. Vacuum source goes to orifice side of the EGR solenoid valve. Refer to "Vehicle Emissions Control Information" label.

5.7L (VIN P) ENGINE — DIAGNOSTIC TROUBLE CODE CHART — 1993 CAMARO AND FIREBIRD

DTC 32

(Page 1 of 2)
EXHAUST GAS RECIRCULATION (EGR) CIRCUIT
5.7L (VIN P) "F" CARLINE (MFI)

BEFORE USING THIS CHART, CHECK FOR MANIFOLD VACUUM TO EGR SOLENOID VALVE, ALSO CHECK HOSES FOR LEAKS OR RESTRICTIONS. SHOULD BE AT LEAST 7" Hg VACUUM AT 2000 RPM.

- WITH ENGINE AT IDLE AND AT NORMAL OPERATING TEMPERATURE, MANUALLY LIFT EGR VALVE DIAPHRAGM.
- RPM SHOULD DECREASE OR ENGINE SHOULD STALL. DOES IT?

YES ⟶
- DISCONNECT EGR VACUUM HARNESS AT SOLENOID VALVE. INSTALL A VACUUM GAGE TO MANIFOLD SIDE OF HARNESS AND CHECK FOR VACUUM SUPPLY TO SOLENOID VALVE.
- VACUUM SHOULD BE AT LEAST 25 kPa (7" Hg) OF VACUUM AT 2000 RPM. IS IT?

NO ⟶ (1) CHECK FOR PLUGGED EGR PASSAGES.

(2) **YES** ⟶
- ROTATE VACUUM HARNESS AND REINSTALL ONLY THE EGR VALVE SIDE, TO THE SOLENOID VALVE.
- INSTALL A VACUUM GAGE IN PLACE OF EGR VALVE.
- INSTALL A HAND HELD VACUUM PUMP TO MANIFOLD SIDE OF EGR SOLENOID VALVE.
- IGNITION "ON," ENGINE STOPPED.
- GROUND DIAGNOSTIC "TEST" TERMINAL.
- APPLY 34 kPa (10" Hg) VACUUM AND OBSERVE GAGE.
- GAGE SHOULD READ VACUUM APPLIED BY PUMP. DOES IT?

NO ⟶ PLUGGED VACUUM PORT AT INTAKE MANIFOLD. LEAKING OR RESTRICTED VACUUM SUPPLY LINE TO EGR SOLENOID VALVE.

(3) **YES** ⟶
- APPLY 34 kPa (10" Hg) VACUUM.
- UNGROUND DIAGNOSTIC "TEST" TERMINAL.
- VACUUM SHOULD BLEED OFF COMPLETELY AT GAGE. DOES IT?

NO ⟶
- CONNECT VACUUM PUMP TO EGR VALVE SIDE OF HARNESS.
- APPLY VACUUM AND OBSERVE GAGE.
- GAGE SHOULD READ VACUUM APPLIED BY PUMP. DOES IT?

YES ⟶ REFER TO EGR CHART (2 OF 2).

NO ⟶
- DISCONNECT SOLENOID VALVE ELECTRICAL CONNECTOR. DOES VACUUM BLEED OFF RAPIDLY AT GAGE?
  - **YES** ⟶ ECM DRIVER CIRCUIT OPEN OR FAULTY ECM.
  - **NO** ⟶ REPLACE SOLENOID VALVE.

YES ⟶
- DISCONNECT EGR ELECTRICAL CONNECTOR.
- CONNECT TEST LIGHT FROM TERMINAL "B" OF EGR HARNESS CONNECTOR AND B +. IS LIGHT "ON"?

NO ⟶ (4)
- DISCONNECT EGR ELECTRICAL CONNECTOR.
- CONNECT TEST LIGHT BETWEEN HARNESS CONNECTOR TERMINALS.
- IGNITION "ON," ENGINE "OFF."
- DIAGNOSTIC "TEST" TERMINAL GROUNDED.
- TEST LIGHT SHOULD LIGHT. DOES IT?

  - **YES** ⟶ FAULTY SOLENOID VALVE CONNECTION OR FAULTY SOLENOID VALVE.
  - **NO** ⟶ CONNECT TEST LIGHT BETWEEN IGNITION FEED CIRCUIT AND GROUND.
    - **NO LIGHT** ⟶ REPAIR OPEN IGNITION FEED CIRCUIT.
    - **LIGHT** ⟶ OPEN ECM DRIVER CIRCUIT OR FAULTY ECM.

NO ⟶ FAULTY VACUUM SUPPLY LINE TO EGR VALVE.

YES ⟶ GROUNDED ECM DRIVER CIRCUIT.

"AFTER REPAIRS," CONFIRM "CLOSED LOOP" OPERATION AND NO MIL (SERVICE ENGINE SOON).

5.7L (VIN P) ENGINE — DIAGNOSTIC TROUBLE CODE CHART — 1993 CAMARO AND FIREBIRD

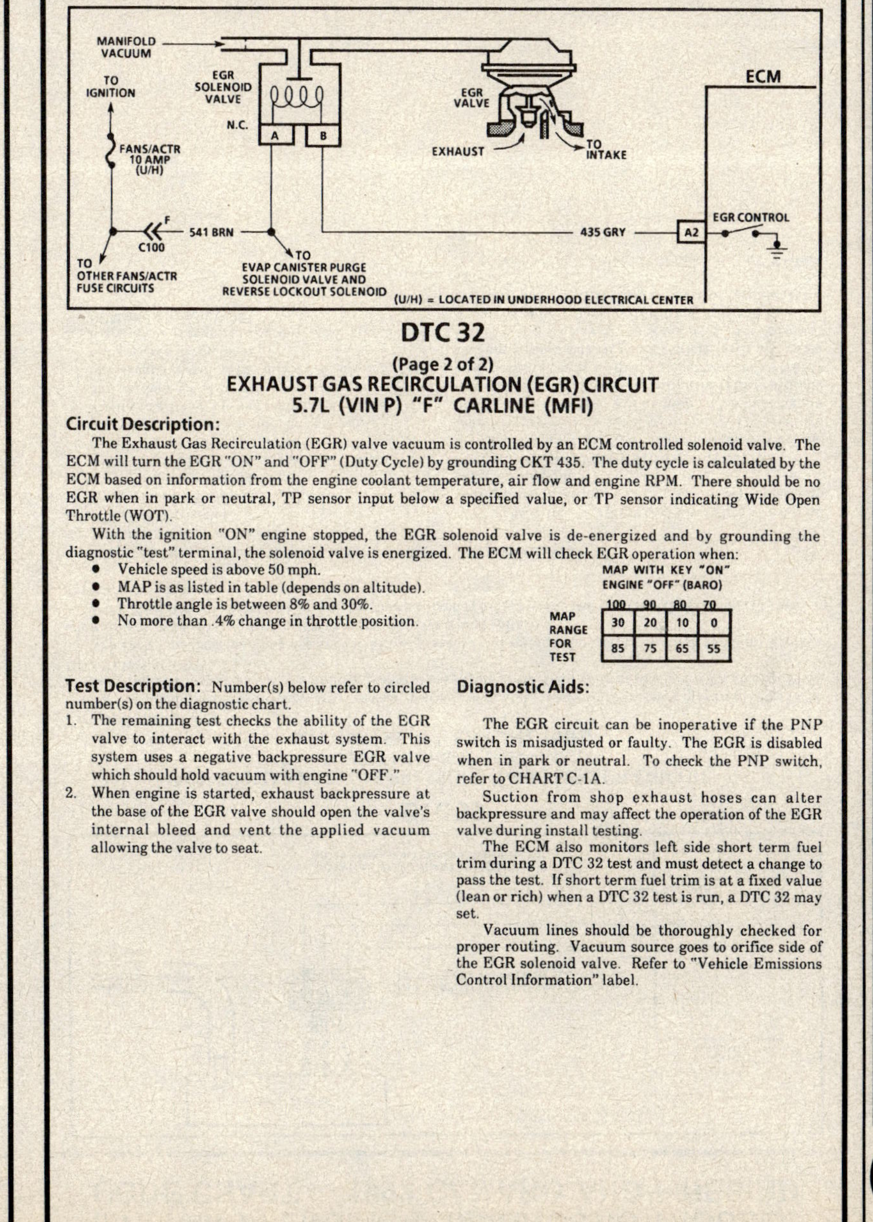

DTC 32

(Page 2 of 2)
EXHAUST GAS RECIRCULATION (EGR) CIRCUIT
5.7L (VIN P) "F" CARLINE (MFI)

Circuit Description:

The Exhaust Gas Recirculation (EGR) valve vacuum is controlled by an ECM controlled solenoid valve. The ECM will turn the EGR "ON" and "OFF" (Duty Cycle) by grounding CKT 435. The duty cycle is calculated by the ECM based on information from the engine coolant temperature, air flow and engine RPM. There should be no EGR when in park or neutral, TP sensor input below a specified value, or TP sensor indicating Wide Open Throttle (WOT).

With the ignition "ON" engine stopped, the EGR solenoid valve is de-energized and by grounding the diagnostic "test" terminal, the solenoid valve is energized. The ECM will check EGR operation when:

- Vehicle speed is above 50 mph.
- MAP is as listed in table (depends on altitude).
- Throttle angle is between 8% and 30%.
- No more than .4% change in throttle position.

MAP WITH KEY "ON" ENGINE "OFF" (BARO)

	100	90	80	70
MAP RANGE FOR TEST	30	20	10	0
	85	75	65	55

Test Description: Number(s) below refer to circled number(s) on the diagnostic chart.

1. The remaining test checks the ability of the EGR valve to interact with the exhaust system. This system uses a negative backpressure EGR valve which should hold vacuum with engine "OFF."
2. When engine is started, exhaust backpressure at the base of the EGR valve should open the valve's internal bleed and vent the applied vacuum allowing the valve to seat.

Diagnostic Aids:

The EGR circuit can be inoperative if the PNP switch is misadjusted or faulty. The EGR is disabled when in park or neutral. To check the PNP switch, refer to CHART C-1A.

Suction from shop exhaust hoses can alter backpressure and may affect the operation of the EGR valve during install testing.

The ECM also monitors left side short term fuel trim during a DTC 32 test and must detect a change to pass the test. If short term fuel trim is at a fixed value (lean or rich) when a DTC 32 test is run, a DTC 32 may set.

Vacuum lines should be thoroughly checked for proper routing. Vacuum source goes to orifice side of the EGR solenoid valve. Refer to "Vehicle Emissions Control Information" label.

5.7L (VIN P) ENGINE — DIAGNOSTIC TROUBLE CODE CHART — 1993 CAMARO AND FIREBIRD

DTC 32

(Page 2 of 2)

EXHAUST GAS RECIRCULATION (EGR) CIRCUIT
5.7L (VIN P) "F" CARLINE (MFI)

CONTINUED FROM EGR CHART (1 OF 2).

(1)
- IGNITION "OFF."
- CONNECT A VACUUM PUMP TO EGR VALVE.
- OBSERVE EGR DIAPHRAGM WHILE APPLYING VACUUM.
- DIAPHRAGM SHOULD MOVE FREELY AND HOLD VACUUM FOR AT LEAST 20 SECONDS.
DOES IT?

YES

(2)
- APPLY 34 kPa (10" Hg) VACUUM TO EGR VALVE.
- START ENGINE AND IMMEDIATELY OBSERVE GAGE ON VACUUM PUMP.
- EGR VALVE DIAPHRAGM SHOULD MOVE TO SEATED POSITION AND VACUUM SHOULD DROP FROM PUMP GAGE WHILE STARTING ENGINE.
DOES IT?

NO → REPLACE EGR VALVE.

NO
- REMOVE EGR VALVE.
- CHECK FOR PLUGGED OR RESTRICTED EXHAUST PASSAGES.

YES
EGR CIRCUIT IS OPERATING PROPERLY. REFER TO SYMPTOMS

PASSAGES OK → REPLACE EGR VALVE.

PASSAGES NOT OK
- CLEAN PASSAGES.
- RE-CHECK EGR VALVE.

"AFTER REPAIRS," CONFIRM "CLOSED LOOP" OPERATION AND NO MIL (SERVICE ENGINE SOON).

5.7L (VIN P) ENGINE — DIAGNOSTIC TROUBLE CODE CHART — 1993 CAMARO AND FIREBIRD

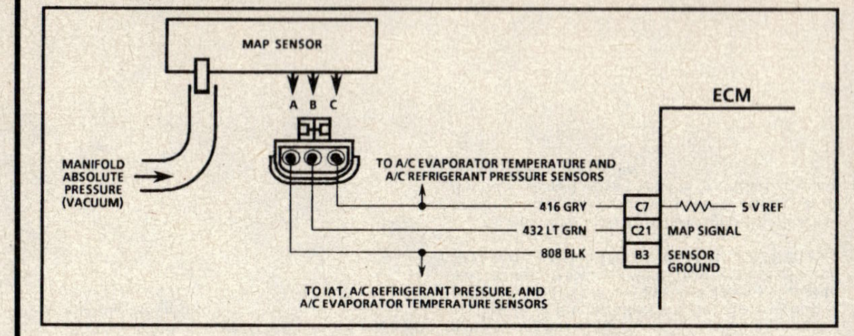

DTC 33

MANIFOLD ABSOLUTE PRESSURE (MAP) SENSOR CIRCUIT
(SIGNAL VOLTAGE HIGH - LOW VACUUM)
5.7L (VIN P) "F" CARLINE (MFI)

Circuit Description:

The Manifold Absolute Pressure (MAP) sensor responds to changes in manifold pressure (vacuum). The ECM receives this information as a signal voltage that will vary from about 1-1.5 volts at idle to 4.0-4.8 volts at Wide Open Throttle (WOT).

The Tech 1 displays manifold pressure in kPa of pressure and voltage. Low pressure (high vacuum) displays a low voltage, while high pressure (low vacuum) displays a high voltage.

If the MAP sensor fails, the ECM will substitute a fixed MAP value and use the Throttle Position (TP) sensor to control fuel delivery.

Test Description: Number(s) below refer to circled number(s) on the diagnostic chart.

1. DTC 33 will set when.
   - Less than 3% throttle.
   - MAP greater than 80 kPa.
   - MAP greater than 90 kPa with A/C "ON."
   - RPM greater than 500 RPM.
   - All above conditions met for 1 second.
   Engine misfire or a low unstable idle may set DTC 33. Disconnect MAP sensor and system will go into backup mode. If the misfire or idle condition remains, see "Symptoms,"
2. If the ECM recognizes the low MAP signal, the ECM and wiring are OK.

Diagnostic Aids:

If the idle is rough or unstable, refer to "Symptoms," Section   for items which can cause an unstable idle.

An open in CKT 808 will result in a DTC 33.

With the ignition "ON," and the engine "OFF," the manifold pressure is equal to atmospheric pressure and the signal voltage will be high. This information is used by the ECM as an indication of vehicle altitude and is referred to as BARO. Comparison of this BARO reading with a known good vehicle with the same sensor is a good way to check accuracy of a "suspect" sensor. Reading should be the same, ± .4 volt.

Also CHART C-1D can be used to test the MAP sensor.

Refer to "Intermittents" in "Symptoms," Section

5.7L (VIN P) ENGINE — DIAGNOSTIC TROUBLE CODE CHART — 1993 CAMARO AND FIREBIRD

DTC 33
MANIFOLD ABSOLUTE PRESSURE (MAP) SENSOR CIRCUIT
(SIGNAL VOLTAGE HIGH - LOW VACUUM)
5.7L (VIN P) "F" CARLINE (MFI)

1. • IF ENGINE IDLE IS ROUGH, UNSTABLE, OR INCORRECT, CORRECT CONDITION BEFORE USING CHART. SEE "SYMPTOMS"
   • ENGINE IDLING.
   • DOES SCAN TOOL DISPLAY A MAP VOLTAGE OF 4.0 VOLTS OR OVER?

YES → 2.
NO → DTC 33 IS INTERMITTENT. IF NO ADDITIONAL DTC(s) WERE STORED, REFER TO "DIAGNOSTIC AIDS" ON FACING PAGE.

2. • IGNITION "OFF."
   • DISCONNECT MAP SENSOR ELECTRICAL CONNECTOR.
   • IGNITION "ON."
   • SCAN TOOL SHOULD READ A VOLTAGE OF 1 VOLT OR LESS.
   DOES IT?

YES → • PROBE SENSOR GROUND CIRCUIT WITH A TEST LIGHT TO BATTERY VOLTAGE.
   • TEST LIGHT SHOULD LIGHT.
   DOES IT?

NO → MAP SIGNAL CIRCUIT SHORTED TO VOLTAGE, SHORTED TO 5 VOLT REFERENCE CIRCUIT OR FAULTY ECM.

YES → VACUUM LEAK AT MAP SENSOR SEAL OR FAULTY MAP SENSOR.

NO → OPEN SENSOR GROUND CIRCUIT.

"AFTER REPAIRS," REFER TO DTC CRITERIA ON FACING PAGE AND CONFIRM DTC DOES NOT RESET.

5.7L (VIN P) ENGINE — DIAGNOSTIC TROUBLE CODE CHART — 1993 CAMARO AND FIREBIRD

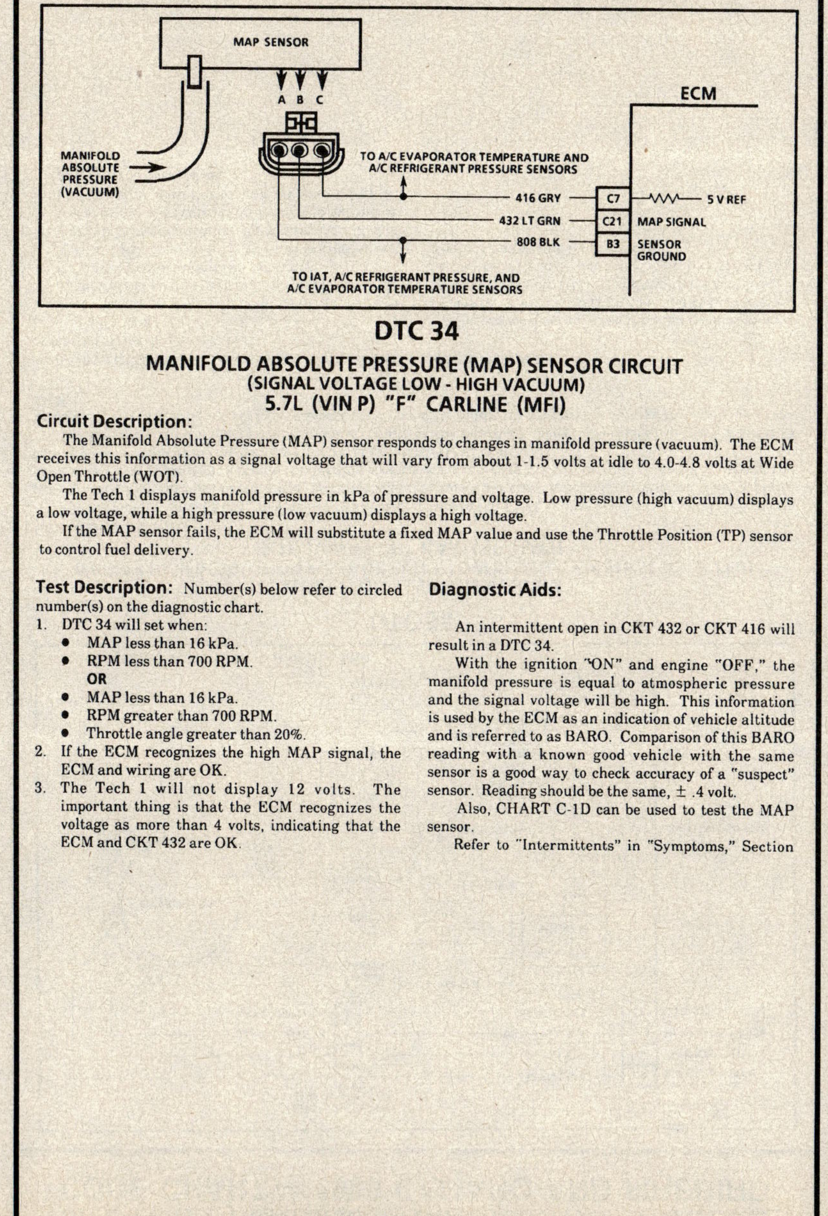

DTC 34
MANIFOLD ABSOLUTE PRESSURE (MAP) SENSOR CIRCUIT
(SIGNAL VOLTAGE LOW - HIGH VACUUM)
5.7L (VIN P) "F" CARLINE (MFI)

Circuit Description:

The Manifold Absolute Pressure (MAP) sensor responds to changes in manifold pressure (vacuum). The ECM receives this information as a signal voltage that will vary from about 1-1.5 volts at idle to 4.0-4.8 volts at Wide Open Throttle (WOT).

The Tech 1 displays manifold pressure in kPa of pressure and voltage. Low pressure (high vacuum) displays a low voltage, while a high pressure (low vacuum) displays a high voltage.

If the MAP sensor fails, the ECM will substitute a fixed MAP value and use the Throttle Position (TP) sensor to control fuel delivery.

Test Description: Number(s) below refer to circled number(s) on the diagnostic chart.

1. DTC 34 will set when:
   • MAP less than 16 kPa.
   • RPM less than 700 RPM.
   OR
   • MAP less than 16 kPa.
   • RPM greater than 700 RPM.
   • Throttle angle greater than 20%.
2. If the ECM recognizes the high MAP signal, the ECM and wiring are OK.
3. The Tech 1 will not display 12 volts. The important thing is that the ECM recognizes the voltage as more than 4 volts, indicating that the ECM and CKT 432 are OK.

Diagnostic Aids:

An intermittent open in CKT 432 or CKT 416 will result in a DTC 34.

With the ignition "ON" and engine "OFF," the manifold pressure is equal to atmospheric pressure and the signal voltage will be high. This information is used by the ECM as an indication of vehicle altitude and is referred to as BARO. Comparison of this BARO reading with a known good vehicle with the same sensor is a good way to check accuracy of a "suspect" sensor. Reading should be the same, ± .4 volt.

Also, CHART C-1D can be used to test the MAP sensor.

Refer to "Intermittents" in "Symptoms," Section

5.7L (VIN P) ENGINE — DIAGNOSTIC TROUBLE CODE CHART — 1993 CAMARO AND FIREBIRD

DTC 34

MANIFOLD ABSOLUTE PRESSURE (MAP) SENSOR CIRCUIT
(SIGNAL VOLTAGE LOW - HIGH VACUUM)
5.7L (VIN P) "F" CARLINE (MFI)

① • ENGINE IDLING.
• DOES SCAN TOOL DISPLAY MAP VOLTAGE BELOW .25 VOLT?

YES → ② • IGNITION "OFF."
• DISCONNECT SENSOR ELECTRICAL CONNECTOR.
• JUMPER HARNESS TERMINALS "B" TO "C".
• IGNITION "ON."
• MAP VOLTAGE SHOULD READ OVER 4.7 VOLTS.
DOES IT?

NO → DTC 34 IS INTERMITTENT. IF NO ADDITIONAL DTC(s) WERE STORED, REFER TO "DIAGNOSTIC AIDS" ON FACING PAGE.

② **NO** → ③ • IGNITION "OFF."
• REMOVE JUMPER WIRE.
• PROBE TERMINAL "B" (CKT 432) WITH A TEST LIGHT TO BATTERY VOLTAGE.
• IGNITION "ON."
• SCAN TOOL SHOULD READ OVER 4 VOLTS.
DOES IT?

② **YES** → FAULTY CONNECTION OR SENSOR.

③ **YES** → 5 VOLT REFERENCE CIRCUIT OPEN
OR
SHORTED TO GROUND
OR
FAULTY ECM.

③ **NO** → CKT 432 OPEN
OR
CKT 432 SHORTED TO GROUND
OR
CKT 432 SHORTED TO SENSOR GROUND
OR
FAULTY ECM.

"AFTER REPAIRS," REFER TO DTC CRITERIA ON FACING PAGE AND CONFIRM DTC DOES NOT RESET.

5.7L (VIN P) ENGINE — DIAGNOSTIC TROUBLE CODE CHART — 1993 CAMARO AND FIREBIRD

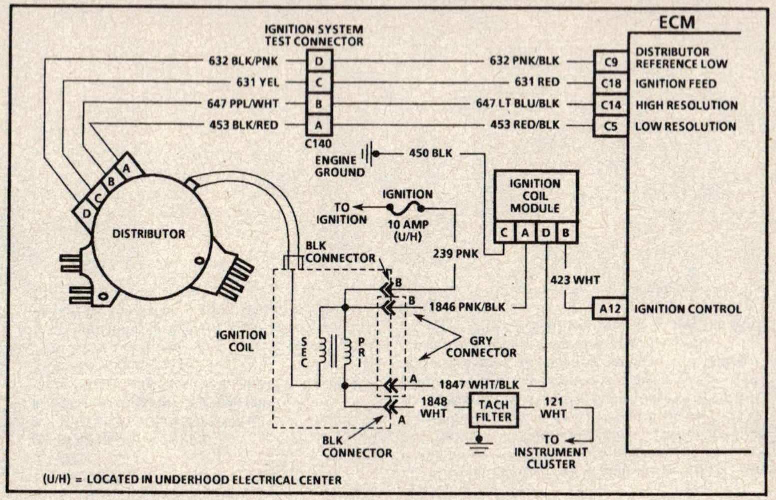

DTC 36

DISTRIBUTOR IGNITION SYSTEM
(FAULTY HIGH RESOLUTION PULSE OR EXTRA LOW RESOLUTION PULSE DETECTED)
5.7L (VIN P) "F" CARLINE (MFI)

Circuit Description:

The distributor ignition system supplies two timing inputs to the ECM, a high resolution signal (360 pulses per one crankshaft revolution) and a low resolution signal (4 pulses per one crankshaft revolution). The ECM can determine if one of the timing inputs is not being received by comparing the two inputs. If the ECM detects one timing pulse without detecting the other timing pulse, this DTC will set.

The reference signals toggle between 0 and 5 volts as the camshaft turns. Therefore, an open, a short to voltage, a short to ground, or a defective sensor inside the distributor can prevent the voltage from pulsing at the ECM.

Test Description: Number(s) below refer to circled number(s) on the diagnostic chart.
1. DTC 36 will set if:
   • If less than 60 high resolution pulses are detected between each low resolution pulse.
   • Failure occurs 5 consecutive times.
2. This test will determine if the ECM is sending out a signal to the distributor for processing. If this signal is not available, or is shorted to ground or voltage, the distributor cannot ground it to produce reference pulses.

Diagnostic Aids:

The engine does not need the high resolution pulse to operate. If a DTC 36 is present and the vehicle will not start, check for DTC 16 and use that chart first. If the engine will still not run, refer to CHART A-3 "Engine Cranks But Will Not Run."

5.7L (VIN P) ENGINE — DIAGNOSTIC TROUBLE CODE CHART — 1993 CAMARO AND FIREBIRD

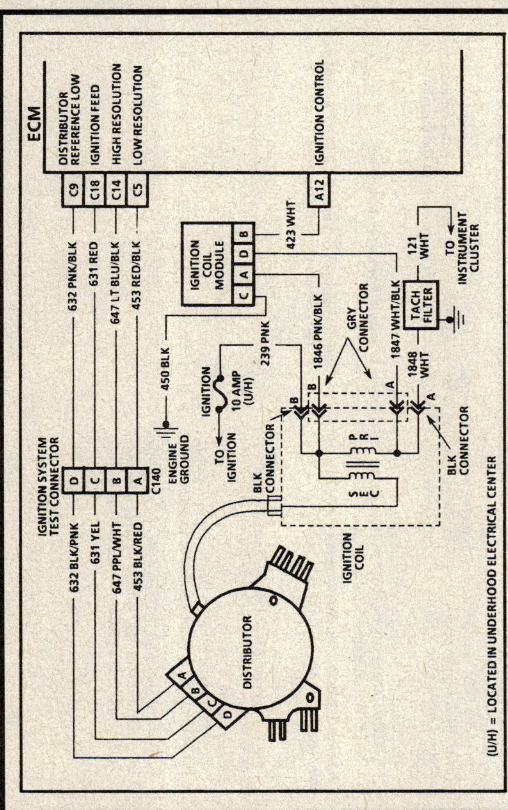

DTC 41

IGNITION CONTROL (IC) CIRCUIT
(OPEN OR SHORTED CIRCUIT)
5.7L (VIN P) "F" CARLINE (MFI)

Circuit Description:

The ignition system provides two timing inputs to the ECM, high resolution (360 pulses per crankshaft revolution) and low resolution (4 pulses per crankshaft revolution). The ECM uses these two reference pulses to determine individual ignition spark timing for each cylinder.

Once the ECM calculates ignition timing, the timing signal will be sent to the ignition coil module on the IC circuit. Each timing pulse received by the ignition coil module, on the IC circuit, will trigger the ignition coil module to operate the ignition coil. Secondary ignition voltage is induced and is then sent to the distributor for distribution to each spark plug.

The IC signal voltage ranges from about .5 volt to 4.5 volts.

If a DTC 41 is detected, the ECM will disable the fuel injectors to prevent flooding of the engine.

DTC 41 will set if the IC circuit problem occurs during cranking.

Test Description: Number(s) below refer to circled number(s) on the diagnostic chart.

1. DTC 41 will set if:
   - Voltage on CKT 423 exceeds 4.6 volts.
   - Engine is less than 1500 RPM.
2. Without the engine running or cranking there would be no voltage present on the IC line.
3. This check determines if the IC signal from the ECM is available at the ignition coil module. The IC circuit voltage should be between about .5 and 4.5 volts.
4. The remaining tests begin to check that the ignition coil module circuitry is OK.
5. If all wiring and connections are OK, check the ignition coil or the ignition coil voltage supply.

Diagnostic Aids:

Since the ignition coil module gets its power from the coil, check the ignition feed circuit to the ignition coil for opens.

5.7L (VIN P) ENGINE — DIAGNOSTIC TROUBLE CODE CHART — 1993 CAMARO AND FIREBIRD

DTC 36

DISTRIBUTOR IGNITION SYSTEM
(FAULTY HIGH RESOLUTION PULSE OR EXTRA/LOW RESOLUTION PULSE DETECTED)
5.7L (VIN P) "F" CARLINE (MFI)

(1)
- CLEAR DTC(s).
- RUN ENGINE FOR 30 SECONDS. DOES DTC 36 SET?

NO → DTC 36 IS INTERMITTENT. REFER TO "DIAGNOSTIC AIDS" ON FACING PAGE.

YES →

(2)
- IGNITION "OFF."
- DISCONNECT IGNITION "TEST" CONNECTOR.
- IGNITION "ON."
- USING A DVM (J 39200) ON THE DC VOLTS SCALE, MEASURE VOLTAGE ON TERMINAL "B" AT THE ECM SIDE OF THE IGNITION TEST CONNECTOR.

LESS THAN .5 VOLTS → OPEN OR GROUNDED HIGH RESOLUTION SIGNAL CIRCUIT FROM ECM TO TEST CONNECTOR OR FAULTY ECM CONNECTION OR FAULTY ECM.

4 TO 6 VOLTS →
- IGNITION "OFF."
- RECONNECT IGNITION SYSTEM TEST CONNECTOR.
- DISCONNECT DISTRIBUTOR CONNECTOR.
- IGNITION "ON."
- MEASURE VOLTAGE ON TERMINAL "A".

OVER 6 VOLTS → SHORT TO VOLTAGE ON HIGH RESOLUTION SIGNAL CIRCUIT FROM ECM TO TEST CONNECTOR.

From "MEASURE VOLTAGE ON TERMINAL A":

LESS THAN .5 VOLTS → OPEN OR GROUNDED HIGH RESOLUTION SIGNAL CIRCUIT FROM TEST CONNECTOR TO DISTRIBUTOR.

4 TO 6 VOLTS → FAULTY DISTRIBUTOR CONNECTION OR FAULTY DISTRIBUTOR.

OVER 6 VOLTS → SHORT TO VOLTAGE ON HIGH RESOLUTION SIGNAL CIRCUIT FROM TEST CONNECTOR TO DISTRIBUTOR.

"AFTER REPAIRS," REFER TO DTC CRITERIA ON FACING PAGE AND CONFIRM DTC DOES NOT RESET.

5.7L (VIN P) ENGINE — DIAGNOSTIC TROUBLE CODE CHART — 1993 CAMARO AND FIREBIRD

DTC 41
IGNITION CONTROL (IC) CIRCUIT
(OPEN OR SHORTED CIRCUIT)
5.7L (VIN P) "F" CARLINE (MFI)

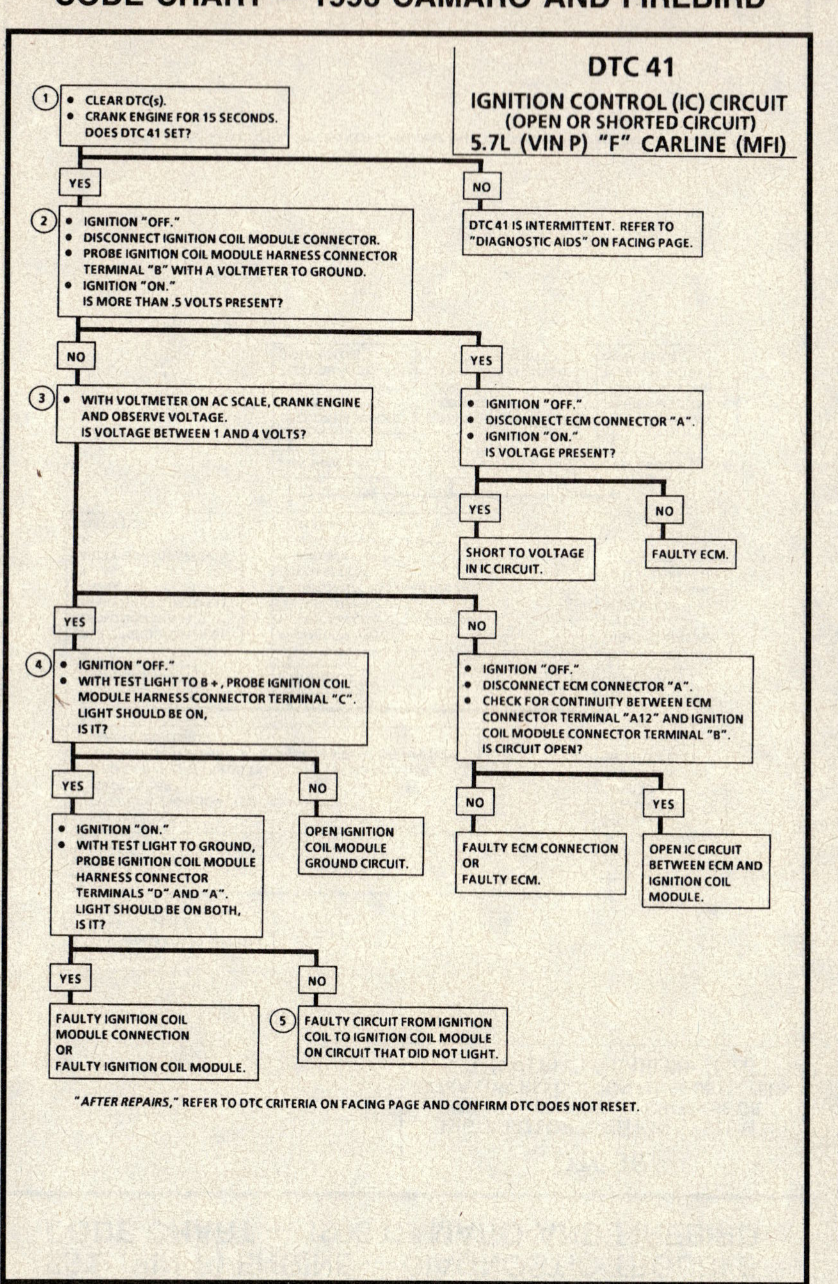

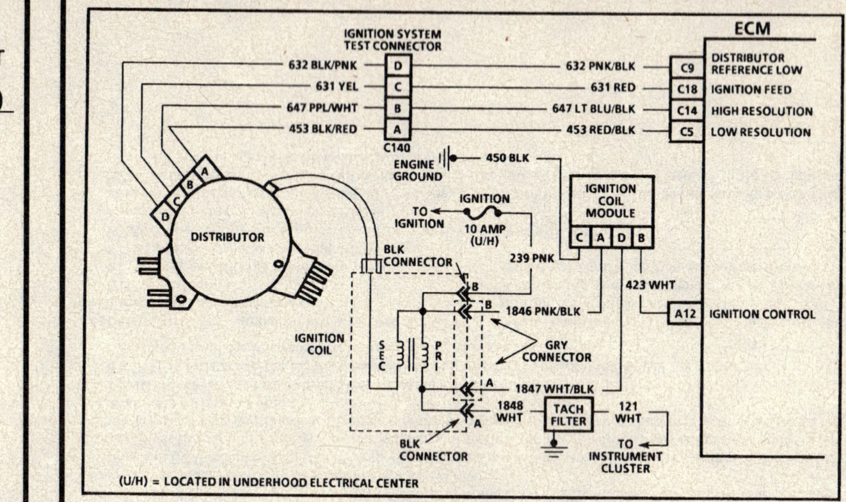

DTC 42
IGNITION CONTROL (IC) CIRCUIT
(GROUNDED CIRCUIT)
5.7L (VIN P) "F" CARLINE (MFI)

Circuit Description:

The ignition system provides two timing inputs to the ECM, high resolution (360 pulses per crankshaft revolution) and low resolution (4 pulses per crankshaft revolution). The ECM uses these two reference pulses to determine individual ignition spark timing for each cylinder.

Once the ECM calculates ignition timing, the timing signal will be sent to the ignition coil module on the IC circuit. Each timing pulse received by the ignition coil module on the IC circuit will trigger the ignition coil module to operate the ignition coil. Secondary ignition voltage is induced and is then sent to the distributor for distribution to each spark plug.

The IC signal voltage ranges from about .5 volt to 4.5 volts.

If DTC 42 is detected, the ECM will disable the fuel injectors to prevent flooding of the engine.

DTC 42 will set if the IC circuit problem occurs during cranking.

Test Description: Number(s) below refer to circled number(s) on the diagnostic chart.

1. DTC 42 will set if:
   - Engine speed below 3000 RPM.
   - Voltage on CKT 423 is below about .5 volt.

   If the engine starts at this point, DTC 42 is intermittent.

2. This check determines if the IC signal from the ECM is available at the ignition coil module.

3. The remaining tests begin to check that the ignition coil module circuitry is OK. If the ignition coil module loses its voltage source, secondary voltage will not be produced and a DTC 42 will set.

4. Since the ignition coil module gets its power from the coil, check the ignition feed circuit to the ignition coil for opens.

5. A DTC 42 will set if the high and low resolution CKTs are shorted to each other. Before replacing any components, make sure these circuits are not shorted together.

5.7L (VIN P) ENGINE — DIAGNOSTIC TROUBLE CODE CHART — 1993 CAMARO AND FIREBIRD

DTC 42

IGNITION CONTROL (IC) CIRCUIT
(GROUNDED CIRCUIT)
5.7L (VIN P) "F" CARLINE (MFI)

1. • CLEAR DTC(s).
   • CRANK ENGINE FOR 15 SECONDS. DOES DTC 42 SET?

YES →
2. • IGNITION "OFF."
   • DISCONNECT IGNITION COIL MODULE CONNECTOR.
   • WITH A VOLTMETER ON AC SCALE, PROBE IGNITION COIL MODULE CONNECTOR TERMINAL "B" TO GROUND.
   • CRANK ENGINE AND OBSERVE VOLTAGE. IS VOLTAGE BETWEEN 1 AND 4 VOLTS?

NO → DTC 42 IS INTERMITTENT. REFER TO "DIAGNOSTIC AIDS" ON FACING PAGE.

YES →
3. • IGNITION "OFF."
   • WITH TEST LIGHT TO B +, PROBE IGNITION COIL MODULE HARNESS CONNECTOR TERMINAL "C".
   • LIGHT SHOULD BE "ON." IS IT?

NO →
• IGNITION "OFF."
• DISCONNECT ECM CONNECTOR "A".
• WITH TEST LIGHT TO B +, PROBE ECM HARNESS TERMINAL "A12". IS TEST LIGHT ON?

YES →
• IGNITION "ON."
• WITH TEST LIGHT TO GROUND, PROBE IGNITION COIL MODULE HARNESS CONNECTOR TERMINALS "D" AND "A".
• LIGHT SHOULD BE "ON" BOTH. IS IT?

NO → OPEN IGNITION COIL MODULE GROUND CIRCUIT.

NO → FAULTY ECM CONNECTION OR FAULTY ECM.

YES → GROUNDED IC CIRCUIT BETWEEN ECM AND IGNITION COIL MODULE.

YES →
5. FAULTY IGNITION COIL MODULE CONNECTION OR FAULTY IGNITION COIL MODULE.

NO →
4. FAULTY CIRCUIT FROM COIL TO IGNITION COIL MODULE ON CIRCUIT THAT DID NOT LIGHT.

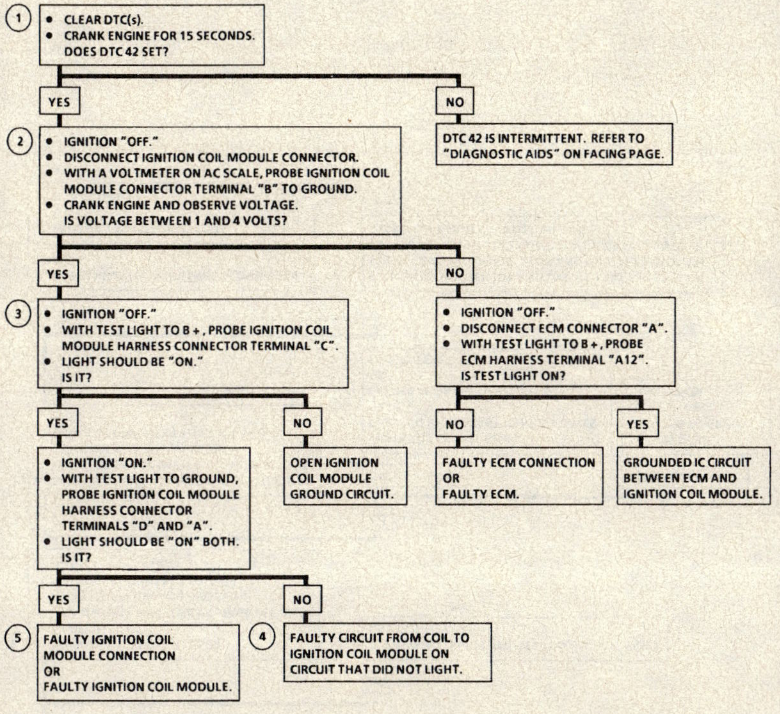

"AFTER REPAIRS," REFER TO DTC CRITERIA ON FACING PAGE AND CONFIRM DTC DOES NOT RESET.

5.7L (VIN P) ENGINE — DIAGNOSTIC TROUBLE CODE CHART — 1993 CAMARO AND FIREBIRD

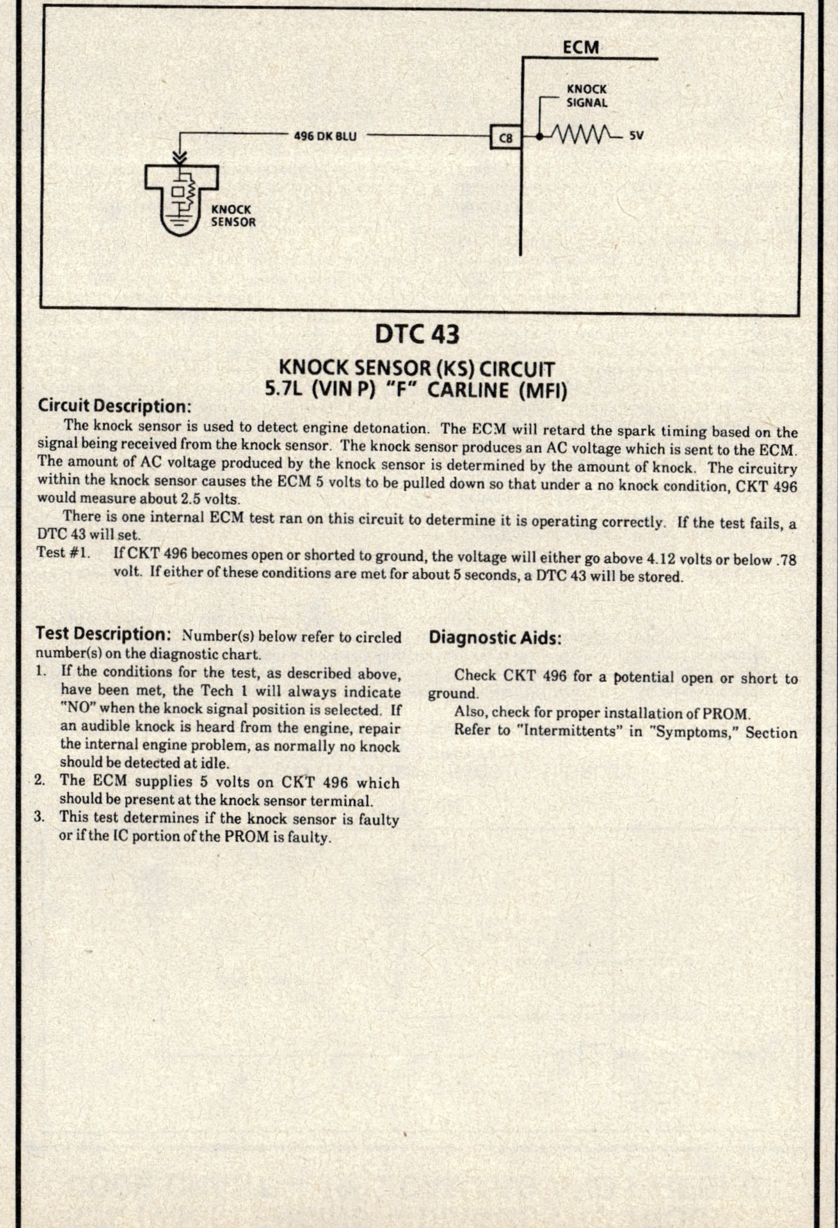

DTC 43

KNOCK SENSOR (KS) CIRCUIT
5.7L (VIN P) "F" CARLINE (MFI)

Circuit Description:

The knock sensor is used to detect engine detonation. The ECM will retard the spark timing based on the signal being received from the knock sensor. The knock sensor produces an AC voltage which is sent to the ECM. The amount of AC voltage produced by the knock sensor is determined by the amount of knock. The circuitry within the knock sensor causes the ECM 5 volts to be pulled down so that under a no knock condition, CKT 496 would measure about 2.5 volts.

There is one internal ECM test ran on this circuit to determine it is operating correctly. If the test fails, a DTC 43 will set.

Test #1.   If CKT 496 becomes open or shorted to ground, the voltage will either go above 4.12 volts or below .78 volt. If either of these conditions are met for about 5 seconds, a DTC 43 will be stored.

Test Description: Number(s) below refer to circled number(s) on the diagnostic chart.

1. If the conditions for the test, as described above, have been met, the Tech 1 will always indicate "NO" when the knock signal position is selected. If an audible knock is heard from the engine, repair the internal engine problem, as normally no knock should be detected at idle.
2. The ECM supplies 5 volts on CKT 496 which should be present at the knock sensor terminal.
3. This test determines if the knock sensor is faulty or if the IC portion of the PROM is faulty.

Diagnostic Aids:

Check CKT 496 for a potential open or short to ground.

Also, check for proper installation of PROM.

Refer to "Intermittents" in "Symptoms," Section

5.7L (VIN P) ENGINE — DIAGNOSTIC TROUBLE CODE CHART — 1993 CAMARO AND FIREBIRD

DTC 43

KNOCK SENSOR (KS) CIRCUIT
5.7L (VIN P) "F" CARLINE (MFI)

(1)
- INSTALL TECH 1.
- ENGINE IDLING, ENGINE COOLANT TEMP ABOVE 67°C. DOES TECH 1 INDICATE A FIXED VALUE OF KNOCK RETARD BETWEEN 4 AND 20 DEGREES?

YES → (2)

NO → DTC 43 IS INTERMITTENT. SEE "DIAGNOSTIC AIDS" ON FACING PAGE.

(2)
- DISCONNECT KNOCK SENSOR.
- IGNITION "ON."
- USING A DVM, MEASURE VOLTAGE BETWEEN HARNESS CKT 496 AND GROUND. VOLTAGE SHOULD BE 4-6 VOLTS. IS IT?

YES → (3)

NO →

OVER 6 VOLTS → CHECK FOR CKT 496 BEING ROUTED TOO CLOSE TO SECONDARY IGNITION WIRES OR CKT 496 SHORTED TO VOLTAGE OR FAULTY ECM.

LESS THAN 4 VOLTS → CKT 496 OPEN, SHORTED TO GROUND OR FAULTY ECM.

(3)
- MEASURE RESISTANCE OF KNOCK SENSOR BY CONNECTING OHMMETER BETWEEN SENSOR TERMINAL AND ENGINE BLOCK. SHOULD BE BETWEEN 3.3KΩ & 4.5KΩ. IS IT?

YES →
- CHECK HARNESS AND SENSOR CONNECTOR. IF OK:
- REMOVE ECM AND BE SURE PROM IS PROPERLY SEATED INTO ECM. IF OK:
- REPLACE PROM.

NO → REMOVE KNOCK SENSOR AND ENSURE THAT THREADS ON SENSOR AND BLOCK ARE CLEAN AND FREE OF CORROSION, TEFLON TAPE, ANTI-SEIZE COMPOUNDS, OR SEALANTS. IF OK, REPLACE FAULTY KNOCK SENSOR.

"AFTER REPAIRS," REFER TO DTC CRITERIA ON FACING PAGE AND CONFIRM DTC DOES NOT RESET.

5.7L (VIN P) ENGINE — DIAGNOSTIC TROUBLE CODE CHART — 1993 CAMARO AND FIREBIRD

DTC 44

BANK 1 (LEFT) OXYGEN SENSOR (O2S) CIRCUIT
(LEAN EXHAUST INDICATED)
5.7L (VIN P) "F" CARLINE (MFI)

Circuit Description:

The ECM supplies a voltage of about 450 mV between terminals "D6" and "D16". (If measured with a 10 megohm digital voltmeter, this may read as low as 320 mV.) The oxygen sensor varies the voltage within a range of about 1000 mV if a rich O2S signal is indicated, down through about 10 mV if a lean O2S signal is indicated.

The sensor is like an open circuit and produces no voltage when it is below 315°C (600°F). An open sensor circuit or cold sensor causes "Open Loop" operation.

Test Description: Number(s) below refer to circled number(s) on the diagnostic chart.

1. DTC 44 will set if:
- Signal voltage remains below 200 mV for 50 seconds.
- System is operating in "Closed Loop."

Diagnostic Aids:

Using the Tech 1, observe long term fuel trim values at different RPM and air flow conditions. The Tech 1 also displays the fuel trim cells, so the long term fuel trim values can be checked in each of the cells to determine when the DTC 44 may have been set. If the conditions for DTC 44 exist, the long term fuel trim values will be near 160.

- **Oxygen Sensor Wire.** Sensor pigtail may be mispositioned and contacting the exhaust manifold.
- Check for intermittent ground in wire between connector and sensor.
- **MAP Sensor.** A Manifold Absolute Pressure (MAP) sensor output that causes the ECM to sense a higher than normal vacuum, will cause the system to go lean. Disconnect the MAP sensor and if the lean condition is gone, replace the sensor.

- **Lean Injector(s).** Perform injector balance test CHART C-2A.
- **Fuel Contamination.** Water, even in small amounts, near the in-tank fuel pump inlet can be delivered to the injectors. The water causes a lean exhaust and can set a DTC 44.
- **Fuel Pressure.** System will be lean if pressure is too low. It may be necessary to monitor fuel pressure while driving the vehicle at various road speeds and/or loads to confirm. Refer to "Fuel System Diagnosis," CHART A-7.
- **Exhaust Leaks.** If there is an exhaust leak above the oxygen sensor, the engine can cause outside air to be pulled into the exhaust stream and flow past the sensor causing a lean condition. Vacuum or crankcase leaks can cause a lean condition.
- **AIR System.** Make sure air is not being directed to the exhaust ports while in "Closed Loop." If the long term fuel trim value goes down while squeezing air hose to left side exhaust ports, refer to CHART C-6.
- If the above are OK, the oxygen sensor may be at fault.

5.7L (VIN P) ENGINE — DIAGNOSTIC TROUBLE CODE CHART — 1993 CAMARO AND FIREBIRD

DTC 44
BANK 1 (LEFT) OXYGEN SENSOR (O2S) CIRCUIT
(LEAN EXHAUST INDICATED)
5.7L (VIN P) "F" CARLINE (MFI)

① • RUN WARM ENGINE (75°C/167°F TO 95°C/203°F) AT 1200 RPM. DOES TECH 1 INDICATE BANK1 O2S VOLTAGE FIXED BELOW .35 VOLT (350 mV)?

YES →
• DISCONNECT BANK1 O2S.
• WITH ENGINE IDLING, TECH 1 SHOULD DISPLAY BANK1 O2S VOLTAGE BETWEEN .35 VOLT AND .55 VOLT (350 mV AND 550 mV). DOES IT?

NO →
DTC 44 IS INTERMITTENT. IF NO ADDITIONAL DTC(s) WERE STORED, REFER TO "DIAGNOSTIC AIDS" ON FACING PAGE.

YES →
REFER TO "DIAGNOSTIC AIDS" ON FACING PAGE.

NO →
CKT 1665 SHORTED TO GROUND OR FAULTY ECM.

"AFTER REPAIRS," REFER TO DTC CRITERIA ON FACING PAGE AND CONFIRM DTC DOES NOT RESET.

5.7L (VIN P) ENGINE — DIAGNOSTIC TROUBLE CODE CHART — 1993 CAMARO AND FIREBIRD

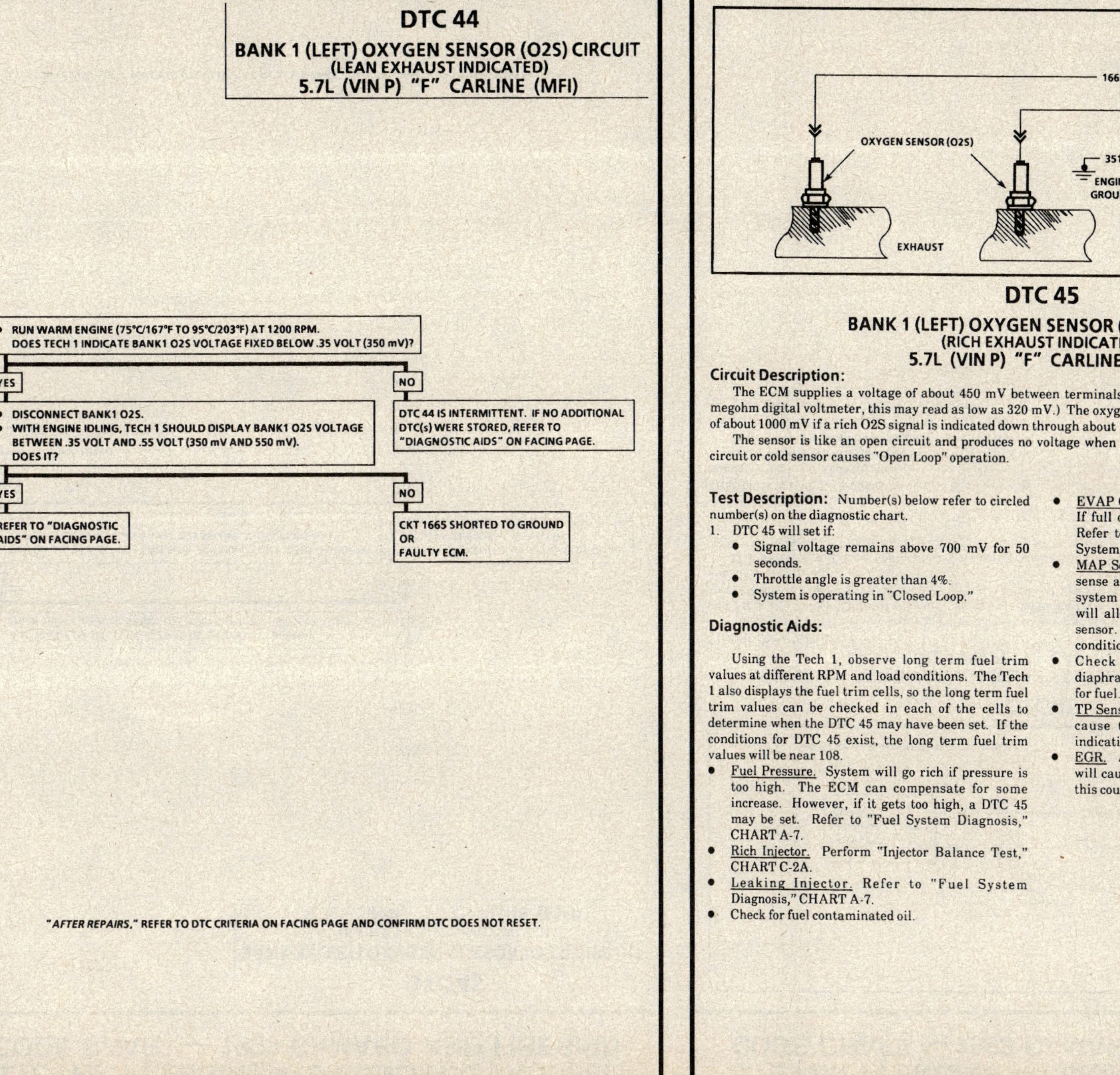

DTC 45
BANK 1 (LEFT) OXYGEN SENSOR (O2S) CIRCUIT
(RICH EXHAUST INDICATED)
5.7L (VIN P) "F" CARLINE (MFI)

Circuit Description:

The ECM supplies a voltage of about 450 mV between terminals "D6" and "D16". (If measured with a 10 megohm digital voltmeter, this may read as low as 320 mV.) The oxygen sensor varies the voltage within a range of about 1000 mV if a rich O2S signal is indicated down through about 10 mV if a lean O2S signal is indicated.

The sensor is like an open circuit and produces no voltage when it is below 315°C (600°F). An open sensor circuit or cold sensor causes "Open Loop" operation.

Test Description: Number(s) below refer to circled number(s) on the diagnostic chart.
1. DTC 45 will set if:
   • Signal voltage remains above 700 mV for 50 seconds.
   • Throttle angle is greater than 4%.
   • System is operating in "Closed Loop."

Diagnostic Aids:

Using the Tech 1, observe long term fuel trim values at different RPM and load conditions. The Tech 1 also displays the fuel trim cells, so the long term fuel trim values can be checked in each of the cells to determine when the DTC 45 may have been set. If the conditions for DTC 45 exist, the long term fuel trim values will be near 108.

• Fuel Pressure. System will go rich if pressure is too high. The ECM can compensate for some increase. However, if it gets too high, a DTC 45 may be set. Refer to "Fuel System Diagnosis," CHART A-7.
• Rich Injector. Perform "Injector Balance Test," CHART C-2A.
• Leaking Injector. Refer to "Fuel System Diagnosis," CHART A-7.
• Check for fuel contaminated oil.

• EVAP Canister Purge. Check for fuel saturation. If full of fuel, check canister control and hoses. Refer to "Evaporative Emission (EVAP) Control System,"
• MAP Sensor. An output that causes the ECM to sense a lower than normal vacuum can cause the system to go rich. Disconnecting the MAP sensor will allow the ECM to set a fixed value for the sensor. Substitute different MAP sensor if the rich condition is gone while the sensor is disconnected.
• Check for leaking fuel pressure regulator diaphragm by checking vacuum line to regulator for fuel.
• TP Sensor. An intermittent TP sensor output will cause the system to go rich, due to a false indication of the engine accelerating.
• EGR. An EGR staying open (especially at idle) will cause the O2S to indicate a rich exhaust, and this could result in a DTC 45.

5.7L (VIN P) ENGINE — DIAGNOSTIC TROUBLE CODE CHART — 1993 CAMARO AND FIREBIRD

DTC 45
BANK 1 (LEFT) OXYGEN SENSOR (O2S) CIRCUIT
(RICH EXHAUST INDICATED)
5.7L (VIN P) "F" CARLINE (MFI)

① • RUN WARM ENGINE (75°C/167°F TO 95°C/203°F) AT 1200 RPM.
  DOES TECH 1 DISPLAY BANK1 O2S VOLTAGE FIXED ABOVE .75 VOLT (750 mV)?

YES
• DISCONNECT BANK1 O2S AND JUMPER HARNESS CKT 1665 TO GROUND.
• TECH 1 SHOULD DISPLAY BANK1 O2S BELOW .35 VOLT (350 mV).
  DOES IT?

NO
DTC 45 IS INTERMITTENT. IF NO ADDITIONAL DTC(s) WERE STORED, REFER TO "DIAGNOSTIC AIDS" ON FACING PAGE.

YES
REFER TO "DIAGNOSTIC AIDS" ON FACING PAGE.

NO
REPLACE ECM.

"AFTER REPAIRS," REFER TO DTC CRITERIA ON FACING PAGE AND CONFIRM DTC DOES NOT RESET.

5.7L (VIN P) ENGINE — DIAGNOSTIC TROUBLE CODE CHART — 1993 CAMARO AND FIREBIRD

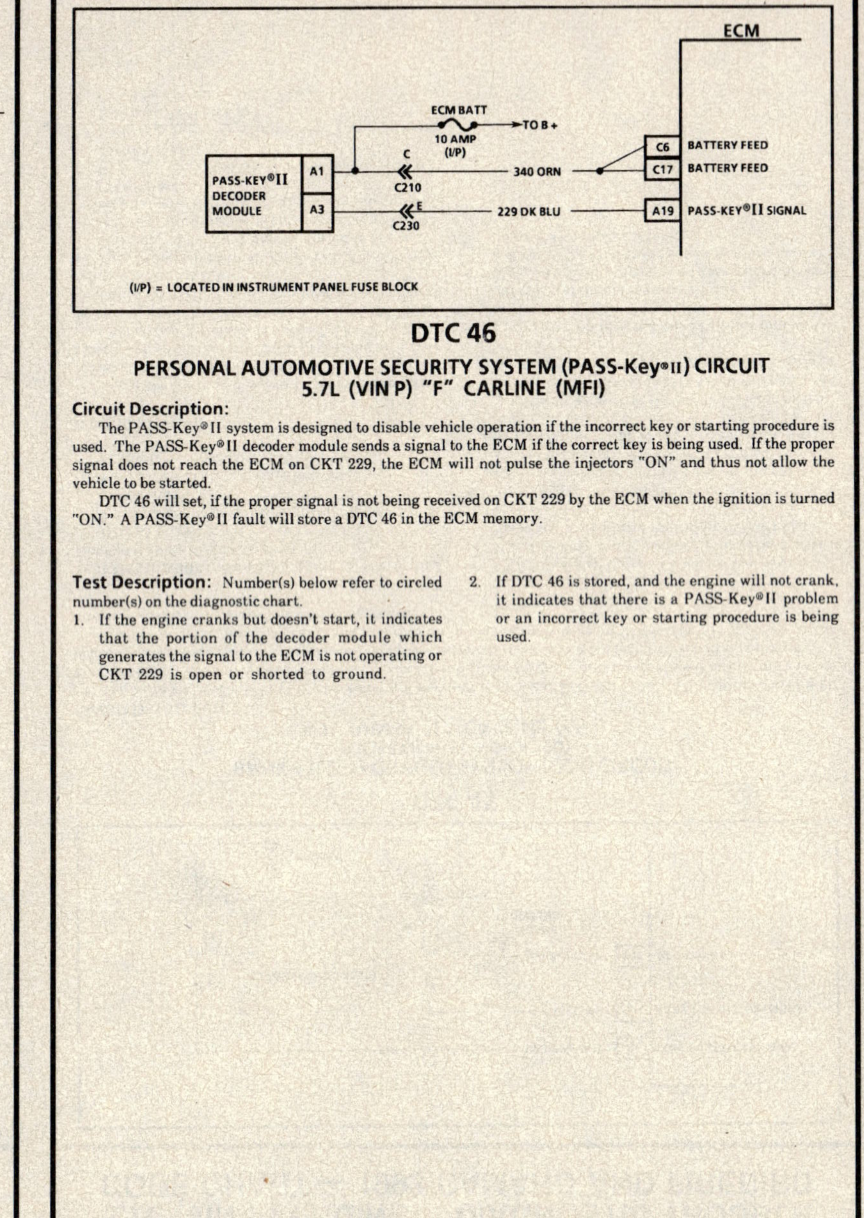

DTC 46
PERSONAL AUTOMOTIVE SECURITY SYSTEM (PASS-Key®II) CIRCUIT
5.7L (VIN P) "F" CARLINE (MFI)

Circuit Description:
The PASS-Key®II system is designed to disable vehicle operation if the incorrect key or starting procedure is used. The PASS-Key®II decoder module sends a signal to the ECM if the correct key is being used. If the proper signal does not reach the ECM on CKT 229, the ECM will not pulse the injectors "ON" and thus not allow the vehicle to be started.

DTC 46 will set, if the proper signal is not being received on CKT 229 by the ECM when the ignition is turned "ON." A PASS-Key®II fault will store a DTC 46 in the ECM memory.

Test Description: Number(s) below refer to circled number(s) on the diagnostic chart.

1. If the engine cranks but doesn't start, it indicates that the portion of the decoder module which generates the signal to the ECM is not operating or CKT 229 is open or shorted to ground.

2. If DTC 46 is stored, and the engine will not crank, it indicates that there is a PASS-Key®II problem or an incorrect key or starting procedure is being used.

5.7L (VIN P) ENGINE — DIAGNOSTIC TROUBLE CODE CHART — 1993 CAMARO AND FIREBIRD

DTC 46

PERSONAL AUTOMOTIVE SECURITY SYSTEM (PASS-Key®II) CIRCUIT
5.7L (VIN P) "F" CARLINE (MFI)

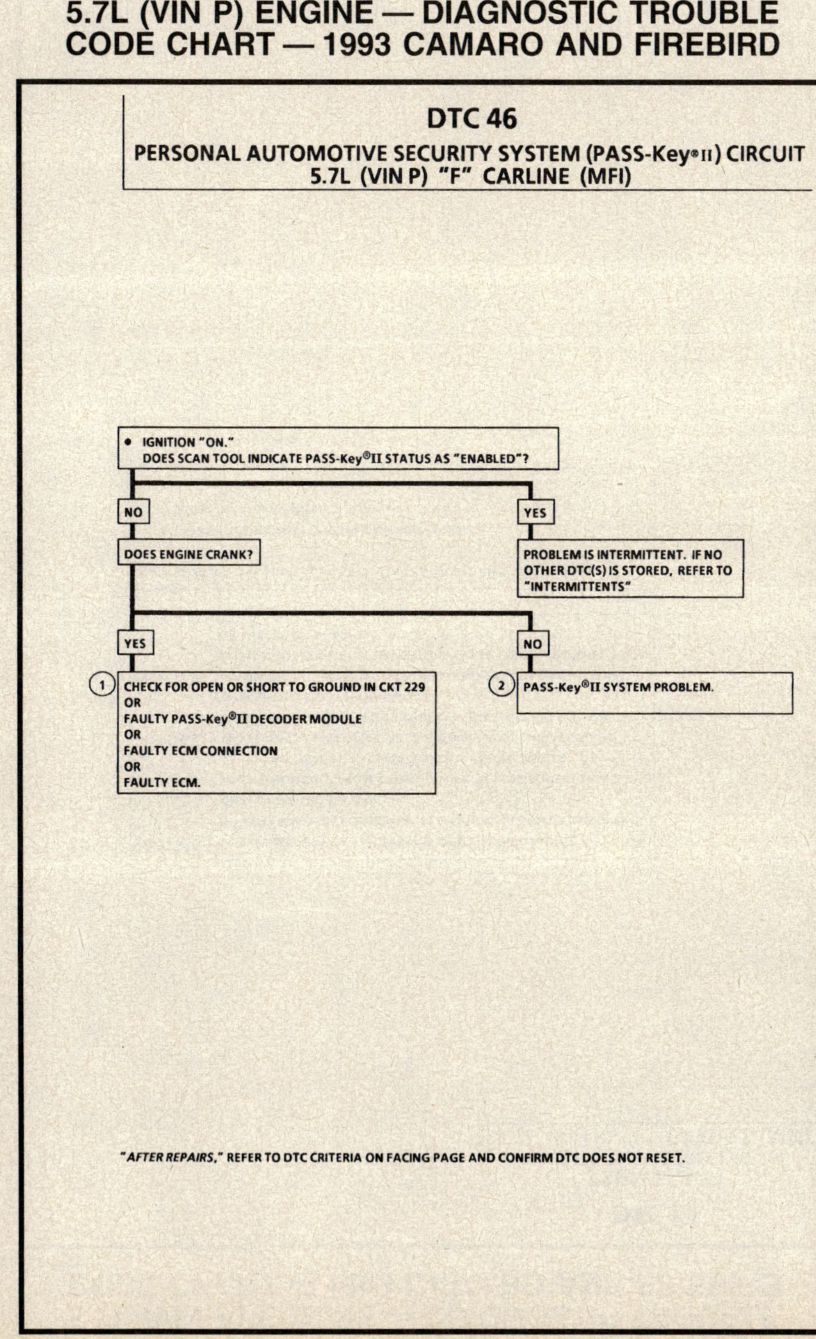

"AFTER REPAIRS," REFER TO DTC CRITERIA ON FACING PAGE AND CONFIRM DTC DOES NOT RESET.

5.7L (VIN P) ENGINE — DIAGNOSTIC TROUBLE CODE CHART — 1993 CAMARO AND FIREBIRD

DTC 51

PROM ERROR
(FAULTY OR INCORRECT PROM)
5.7L (VIN P) "F" CARLINE (MFI)

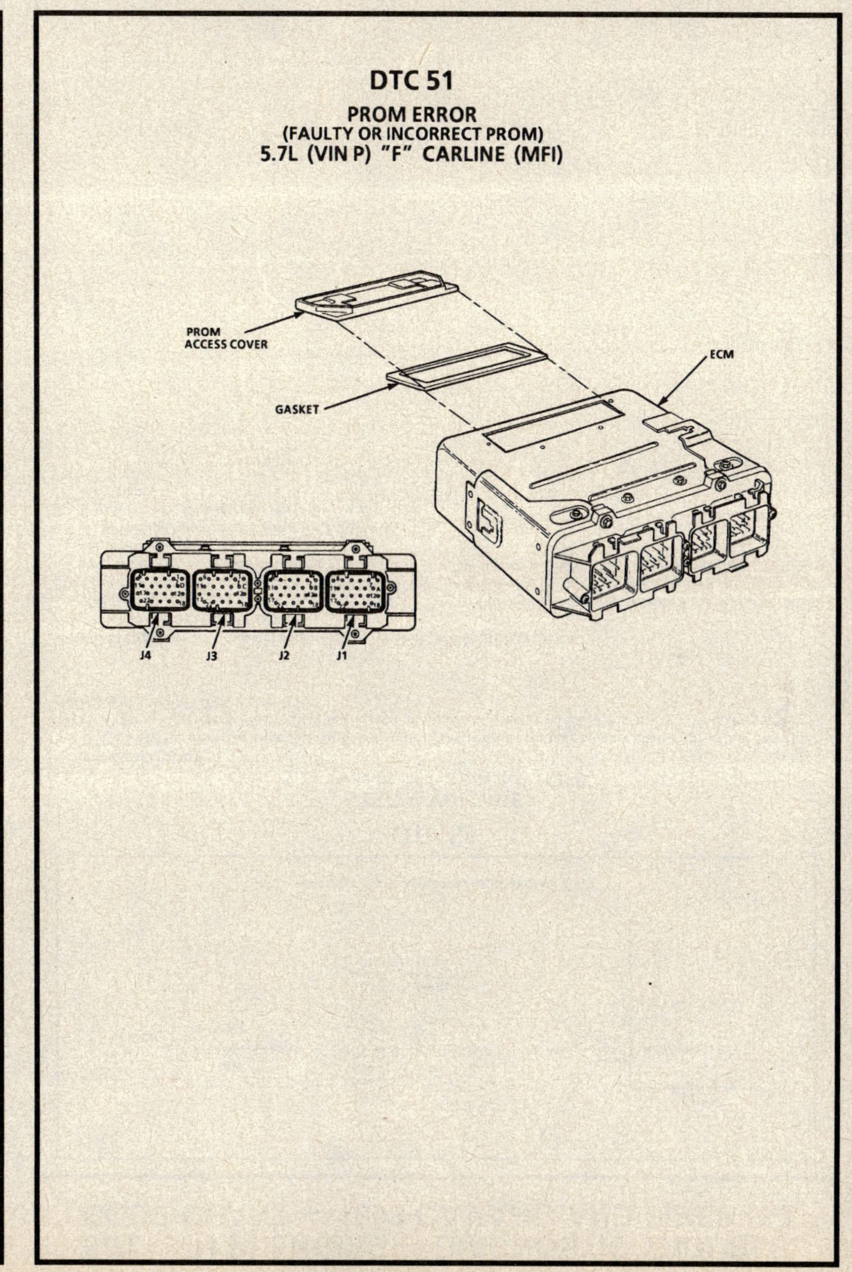

5.7L (VIN P) ENGINE — DIAGNOSTIC TROUBLE CODE CHART — 1993 CAMARO AND FIREBIRD

DTC 51

PROM ERROR
(FAULTY OR INCORRECT PROM)
5.7L (VIN P) "F" CARLINE (MFI)

NOTICE: TO PREVENT POSSIBLE ELECTROSTATIC DISCHARGE DAMAGE:
- DO NOT TOUCH THE ECM CONNECTOR PINS OR SOLDERED COMPONENTS ON THE ECM CIRCUIT BOARD.
- WHEN HANDLING A PROM, DO NOT TOUCH THE COMPONENT LEADS, AND DO NOT REMOVE INTEGRATED CIRCUIT FROM CARRIER.

NOTICE: TO PREVENT POSSIBLE ELECTROSTATIC DISCHARGE DAMAGE TO THE PROM, DO NOT TOUCH THE COMPONENT LEADS, AND DO NOT REMOVE INTEGRATED CIRCUIT FROM CARRIER.

NOTICE: TO PREVENT POSSIBLE ELECTROSTATIC DISCHARGE DAMAGE TO THE ECM, DO NOT TOUCH THE CONNECTOR PINS OR SOLDERED COMPONENTS ON THE CIRCUIT BOARD.

CHECK THAT ALL PINS ARE FULLY INSERTED IN THE SOCKET AND THAT PROM IS PROPERLY LATCHED.
IF OK, REPLACE PROM, CLEAR MEMORY, AND RECHECK.
IF DTC 51 REAPPEARS, REPLACE ECM.

5.7L (VIN P) ENGINE — DIAGNOSTIC TROUBLE CODE CHART — 1993 CAMARO AND FIREBIRD

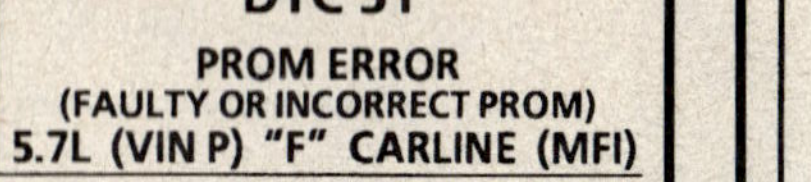

DTC 53

SYSTEM VOLTAGE
5.7L (VIN P) "F" CARLINE (MFI)

Circuit Description:
DTC 53 will set when the engine is running and the ECM senses a voltage of more than 17.1 volts or less than 8.0 volts on ECM terminal "C12". During the time the fault is present, all ECM outputs will be disengaged and setting of additional DTCs may result.

Test Description: Number(s) below refer to circled number(s) on the diagnostic chart.
1. Normal battery output is between 9.6 - 16.0 volts.
2. Checks to see if the high voltage reading is due to the generator or ECM. With engine running, check voltage at the battery.
   If the voltage is above 12.1 volts, the ECM is OK.
3. Checks to see if generator is faulty under load condition.

Diagnostic Aids:

During the time the failure is present, all ECM outputs will be disengaged to protect the hardware. The setting of additional DTCs may result.

5.7L (VIN P) ENGINE — DIAGNOSTIC TROUBLE CODE CHART — 1993 CAMARO AND FIREBIRD

DTC 53
SYSTEM VOLTAGE
5.7L (VIN P) "F" CARLINE (MFI)

- ① • ENGINE RUNNING ABOVE 800 RPM.
  • NOTE BATTERY VOLTAGE ON TECH 1.

- ABOVE 17.1 VOLTS
- ② • CHECK BATTERY VOLTAGE AT BATTERY.

  - ABOVE 17.1 VOLTS
    - • REMOVE GENERATOR FOR REPAIR.
  - BELOW 17.1 VOLTS
    - REPLACE ECM

- ABOVE 8.0 VOLTS AND BELOW 17.1 VOLTS
- ③ • RAISE ENGINE RPM TO 2000.
  • LOAD ELECTRICAL SYSTEM WITH HEADLAMPS AND HIGH BLOWER "ON."
  • NOTE BATTERY VOLTAGE VALUE.

  - ABOVE 8.0 VOLTS
    - • FAULT IS NOT PRESENT
    - • SEE "INTERMITTENTS"
  - BELOW 8.0 VOLTS
    - • REMOVE GENERATOR FOR REPAIR.

"AFTER REPAIRS," REFER TO DTC CRITERIA ON FACING PAGE AND CONFIRM DTC DOES NOT RESET.

5.7L (VIN P) ENGINE — DIAGNOSTIC TROUBLE CODE CHART — 1993 CAMARO AND FIREBIRD

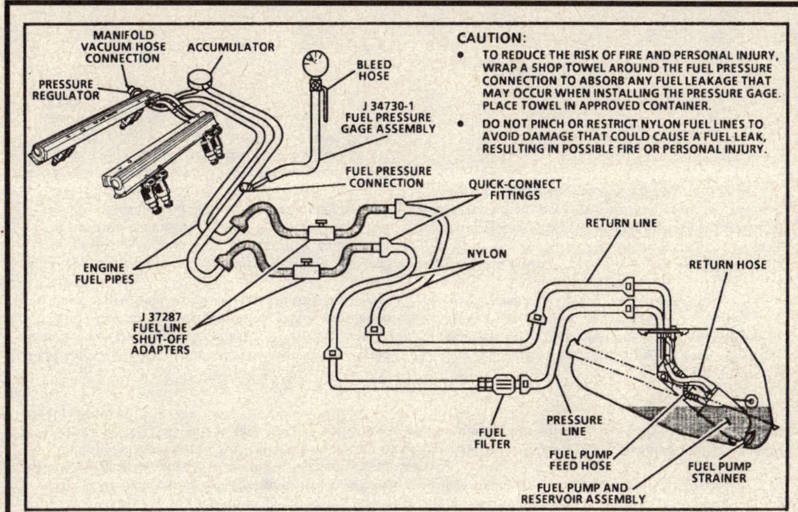

DTC 55
FUEL LEAN MONITOR
5.7L (VIN P) "F" CARLINE (MFI)

Circuit Description:

The ECMs internal circuitry can identify if the vehicle's fuel system is capable of supplying adequate amounts of fuel during heavy acceleration (power enrichment). When a power enrichment mode of operation is requested during "Closed Loop" operation (by heavy acceleration), the ECM will provide more fuel to the engine. Under these conditions, the ECM should detect a "rich" condition. If this "rich" exhaust is not detected at this time, a DTC 55 will set. A plugged fuel filter or restricted fuel line can prevent adequate amounts of fuel from being supplied during power enrichment mode.

Test Description: Number(s) below refer to circled number(s) on the diagnostic chart.
1. DTC 55 will set if:
   The ECM detects lean oxygen sensor voltage for 8 seconds during power enrichment modes of operation.
   Wrap a shop towel around the fuel pressure connector to absorb any small amount of fuel leakage that may occur when installing the gage. Ignition "ON," pump pressure should be 284-325 kPa (41-47 psi). This pressure is controlled by spring pressure within the regular assembly. Pressure below 284 kPa (41 psi) may cause a lean condition and may set a DTC 44 or 64. It could also cause hard starting cold and poor driveability. Low enough pressure will cause the engine not to run at all.

Restricted flow may allow the engine to run at idle, or low speeds, but may cause a surge and stall when more fuel is required, as when accelerating or driving at high speeds. Low fuel pressure under heavy acceleration conditions may set a DTC 55.
2. Restricting the fuel return line allows the fuel pressure to build above regulated pressure. With a Tech 1 enable the fuel pump, pressure should rise above 325 kPa (47 psi) as the valve in the return line is partially closed.

Diagnostic Aids:

A restricted fuel filter can supply adequate amounts of fuel at idle but may not be able to supply enough fuel during heavy acceleration. A vapor lock condition can cause a DTC 55.

5.7L (VIN P) ENGINE — DIAGNOSTIC TROUBLE CODE CHART — 1993 CAMARO AND FIREBIRD

DTC 55

FUEL LEAN MONITOR
5.7L (VIN P) "F" CARLINE (MFI)

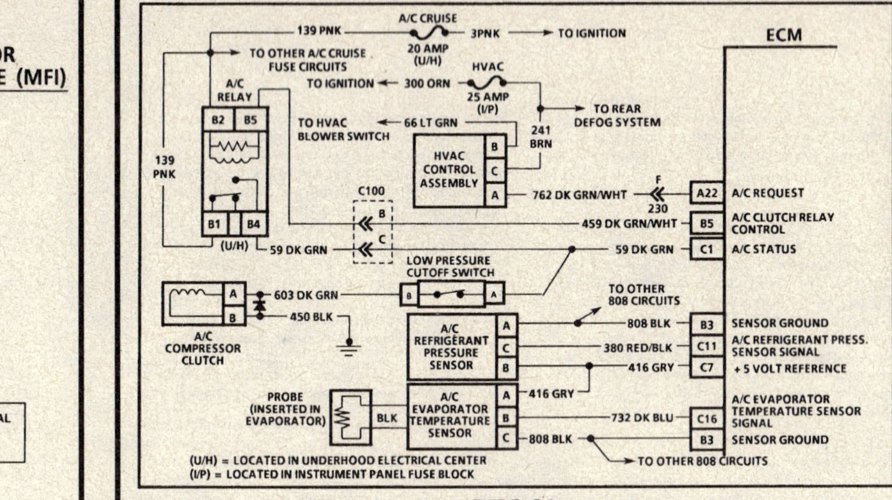

1. • IF DTC(s) 44 OR 64 ARE SET, REFER TO THOSE DTCs BEFORE PROCEEDING WITH THIS CHART.
   • INSTALL FUEL PRESSURE GAGE, J 34730-1 .
   • START AND IDLE ENGINE AT NORMAL OPERATING TEMPERATURE.
   • DISCONNECT VACUUM LINE GOING TO FUEL PRESSURE REGULATOR.
   • NOTE FUEL PRESSURE WITH ENGINE RUNNING SHOULD BE 284-325 kPa (41-47 psi).

NOT OK

CHECK FOR RESTRICTED FUEL LINES OR IN-LINE FILTER

OK

NO TROUBLE FOUND. IF NO ADDITIONAL DTCs WERE STORED, REVIEW CIRCUIT DESCRIPTION ON FACING PAGE.

OK

2. • IGNITION "OFF."
   • INSTALL J 37287-2 FUEL LINE ADAPTOR IN FUEL RETURN LINE. REFER TO FUEL SYSTEM PRESSURE TEST CHART A-7, PAGE 3 OF 3 FOR FUEL PRESSURE RELIEF PROCEDURE AND FOR SERVICING QUICK-CONNECT FITTINGS.
   • USING A TECH 1, ENABLE THE FUEL PUMP."
   • SLOWLY CLOSE VALVE IN RETURN LINE. PRESSURE SHOULD RISE ABOVE 325 kPa (47 psi). DO NOT EXCEED 414 kPa (60 psi).

NOT OK

REPLACE FILTER OR REPAIR FUEL LINE AND RECHECK.

ABOVE 325 kPa (47 psi)

DTC 55 IS INTERMITTENT.

PRESSURE BUT LESS THAN 284 kPa (41 psi)

CHECK FOR:
• RESTRICTED FUEL PUMP STRAINER.
• LEAKING FUEL PUMP FEED HOSE.
• FAULTY FUEL PUMP.
• INCORRECT FUEL PUMP.

"AFTER REPAIRS," REFER TO DTC CRITERIA ON FACING PAGE AND CONFIRM DTC DOES NOT RESET.

5.7L (VIN P) ENGINE — DIAGNOSTIC TROUBLE CODE CHART — 1993 CAMARO AND FIREBIRD

DTC 61

A/C SYSTEM PERFORMANCE
5.7L (VIN P) "F" CARLINE (MFI)

Circuit Description:

The ECM monitors A/C refrigerant pressure and A/C evaporator temperature to determine if the A/C system has adequate refrigerant (R134a) charge. The ECM will disable the A/C clutch whenever A/C refrigerant charge drops below a predetermined value, to reduce the chance of compressor damage.

The ECM determines A/C charge based on A/C refrigerant pressure and A/C evaporator temperature. The ECM uses these two inputs to calculate A/C refrigerant charge.

If the ECM determines A/C refrigerant charge to be low, the A/C clutch will be disabled and Diagnostic Trouble Code (DTC) 61 will be set. DTC 61 will store in ECM memory, but will not turn "ON" the MIL (Service Engine Soon).

Test Description: Number(s) below refer to circled number(s) on the diagnostic chart.

1. DTC 61 could be caused by other DTCs, so refer to those charts before proceeding with this diagnostic chart.
2. If the A/C refrigerant pressure sensor is "slewed" or "shifted" in value, the ECM could be detecting a lower than normal pressure. Therefore, the ECM would be sensing a low refrigerant change and DTC 61 could result.
3. The conditions to set a DTC 67 could cause DTC 61 to set. DTC 67 will set if the ECM does not detect an A/C refrigerant pressure change of more than 4 psi when the A/C clutch is cycled from "OFF" to "ON."
4. The A/C performance test <u>must</u> be performed, because A/C system hardware (A/C compressor, thermal expansion valve etc.) could cause a DTC 61.

Diagnostic Aids:

An A/C evaporator temperature sensor and/or an A/C refrigerant pressure sensor that is "slewed" (shifted in value) could cause a DTC 61.

5.7L (VIN P) ENGINE — DIAGNOSTIC TROUBLE CODE CHART — 1993 CAMARO AND FIREBIRD

5.7L (VIN P) ENGINE — DIAGNOSTIC TROUBLE CODE CHART — 1993 CAMARO AND FIREBIRD

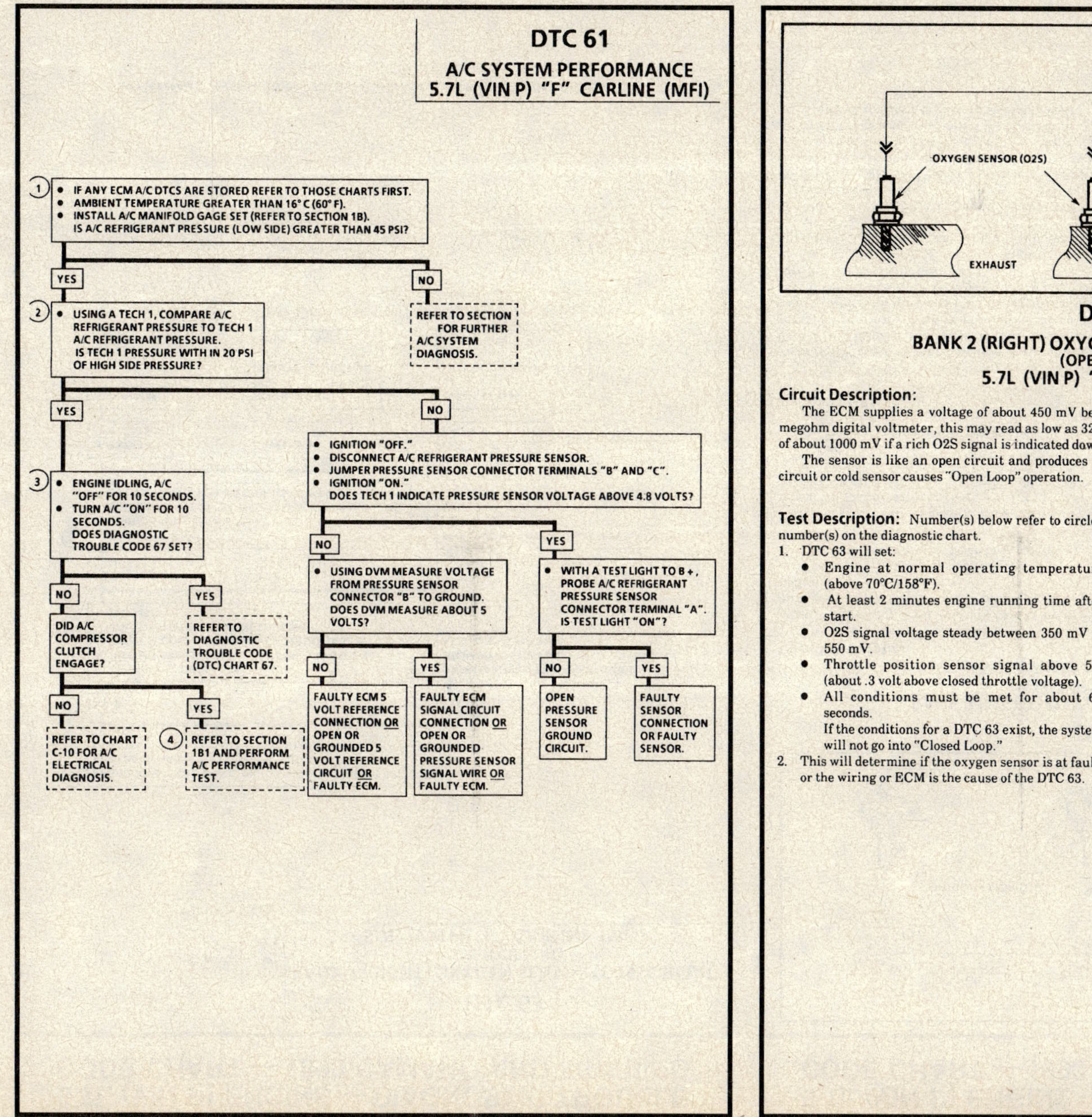

Circuit Description:

The ECM supplies a voltage of about 450 mV between terminals "D16" and "D22". (If measured with a 10 megohm digital voltmeter, this may read as low as 320 mV.) The oxygen sensor varies the voltage within a range of about 1000 mV if a rich O2S signal is indicated down through about 10 mV if a lean O2S signal is indicated.

The sensor is like an open circuit and produces no voltage when it is below 315°C (600°F). An open sensor circuit or cold sensor causes "Open Loop" operation.

Test Description: Number(s) below refer to circled number(s) on the diagnostic chart.

1. DTC 63 will set:
   • Engine at normal operating temperature (above 70°C/158°F).
   • At least 2 minutes engine running time after start.
   • O2S signal voltage steady between 350 mV to 550 mV.
   • Throttle position sensor signal above 5% (about .3 volt above closed throttle voltage).
   • All conditions must be met for about 60 seconds.
   If the conditions for a DTC 63 exist, the system will not go into "Closed Loop."

2. This will determine if the oxygen sensor is at fault, or the wiring or ECM is the cause of the DTC 63.

3. For this test use only a high impedance digital volt ohmmeter (J 39200). This test checks the continuity of CKT 1666 and CKT 351. If CKT 351 is open, the ECM voltage on CKT 1666 will be over .6 volt (600 mV).

Diagnostic Aids:

Normal Tech 1 voltage readings vary between 10 mV to 1000 mV (.01 and 1.0 volt), while in "Closed Loop." The system will go into "Open Loop" operation when DTC 63 sets.

Refer to "Intermittents" in "Symptoms," Section

5.7L (VIN P) ENGINE — DIAGNOSTIC TROUBLE CODE CHART — 1993 CAMARO AND FIREBIRD

DTC 63

BANK 2 (RIGHT) OXYGEN SENSOR (O2S) CIRCUIT
(OPEN CIRCUIT)
5.7L (VIN P) "F" CARLINE (MFI)

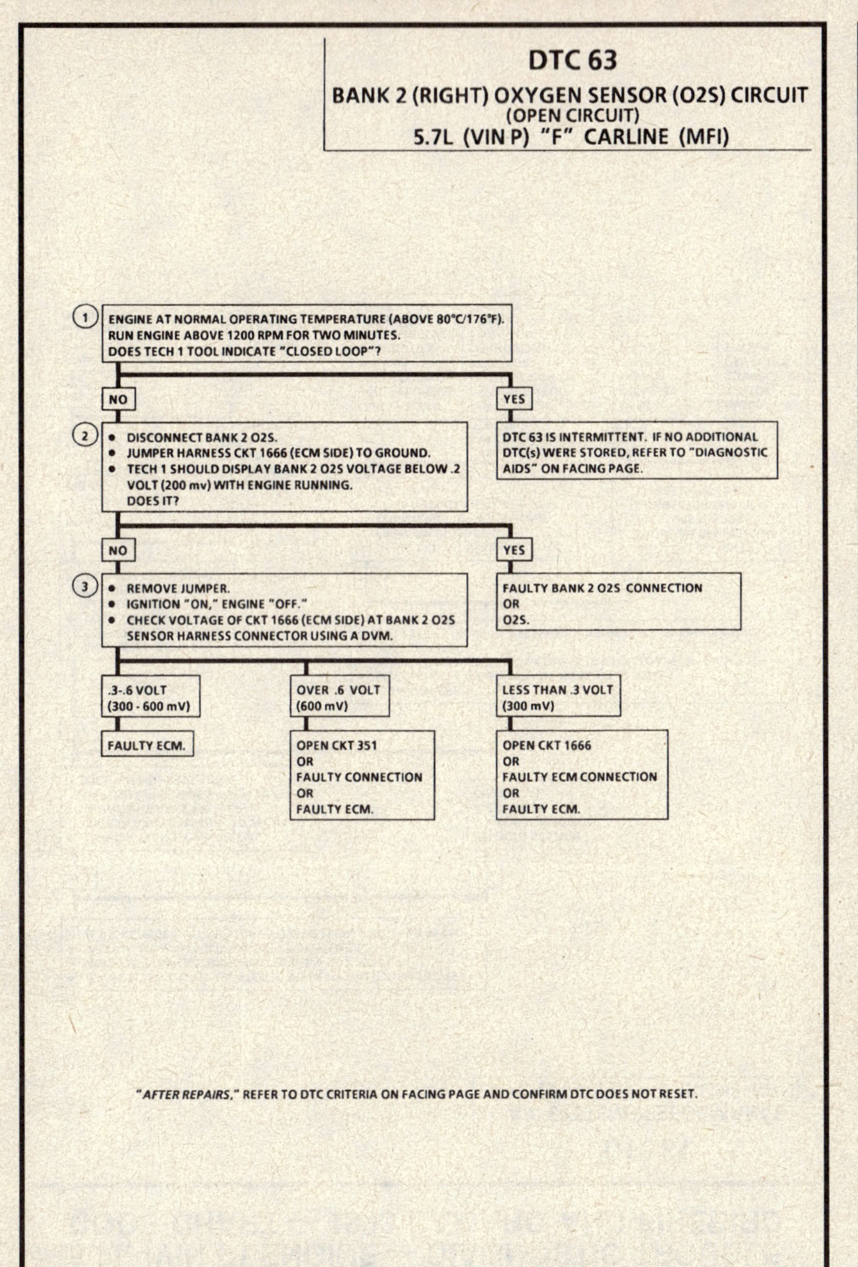

"AFTER REPAIRS," REFER TO DTC CRITERIA ON FACING PAGE AND CONFIRM DTC DOES NOT RESET.

DTC 64

BANK 2 (RIGHT) OXYGEN SENSOR (O2S) CIRCUIT
(LEAN EXHAUST INDICATED)
5.7L (VIN P) "F" CARLINE (MFI)

Circuit Description:

The ECM supplies a voltage of about 450 mV between terminals "D16" and "D22". (If measured with a 10 megohm digital voltmeter, this may read as low as 320 mV.) The Oxygen Sensor (O2S) varies the voltage within a range of about 1000 mV if a rich O2S signal is indicated down through about 10 mV if a lean O2S signal is indicated.

The sensor is like an open circuit and produces no voltage when it is below 315°C (600°F). An open sensor circuit or cold sensor causes "Open Loop" operation.

Test Description: Number(s) below refer to circled number(s) on the diagnostic chart.

1. DTC 64 will set if:
   - Signal voltage remains below 200 mV for 50 seconds.
   - System is operating in "Closed Loop."

Diagnostic Aids:

Using the Tech 1, observe long term fuel trim values at different RPM and air flow conditions. The Tech 1 also displays the fuel trim cells, so the long term fuel trim values can be checked in each of the cells to determine when the DTC 64 may have been set. If the conditions for DTC 64 exist, the long term fuel trim values will be near 160.

- **Oxygen Sensor Wire.** Sensor pigtail may be mispositioned and contacting the exhaust manifold.
- Check for intermittent ground in wire between connector and sensor.
- **MAP Sensor.** A Manifold Absolute Pressure (MAP) sensor output that causes the ECM to sense a higher than normal vacuum, will cause the system to go lean. Disconnect the MAP sensor and if the lean condition is gone, replace the sensor.
- **Lean Injector(s).** Perform injector balance test CHART C-2A.
- **Fuel Contamination.** Water, even in small amounts, near the in-tank fuel pump inlet can be delivered to the injectors. The water causes a lean exhaust and can set a DTC 64.
- **Fuel Pressure.** System will be lean if pressure is too low. It may be necessary to monitor fuel pressure while driving the vehicle at various road speeds and/or loads to confirm. Refer to "Fuel System Diagnosis," CHART A-7.
- **Exhaust Leaks.** If there is an exhaust leak above the oxygen sensor, the engine can cause outside air to be pulled into the exhaust stream and flow past the sensor causing a lean condition. Vacuum or crankcase leaks can cause a lean condition.
- **AIR System.** Make sure air is not being directed to the exhaust ports while in "Closed Loop." If the long term fuel trim value goes down while squeezing air hose to right side exhaust ports, refer to CHART C-6.
- If the above are OK, the oxygen sensor may be at fault.

5.7L (VIN P) ENGINE — DIAGNOSTIC TROUBLE CODE CHART — 1993 CAMARO AND FIREBIRD

DTC 64
BANK 2 (RIGHT) OXYGEN SENSOR (O2S) CIRCUIT
(LEAN EXHAUST INDICATED)
5.7L (VIN P) "F" CARLINE (MFI)

(1) • RUN WARM ENGINE (75°C/167°F TO 95°C/203°F) AT 1200 RPM. DOES TECH 1 INDICATE BANK 2 O2S VOLTAGE FIXED BELOW .35 VOLT (350 mV)?

YES
• DISCONNECT BANK 2 O2S.
• WITH ENGINE IDLING, TECH 1 SHOULD DISPLAY BANK 2 O2S VOLTAGE BETWEEN .35 VOLT AND .55 VOLT (350 mV AND 550 mV). DOES IT?

NO
DTC 64 IS INTERMITTENT. IF NO ADDITIONAL DTC(s) WERE STORED, REFER TO "DIAGNOSTIC AIDS" ON FACING PAGE.

YES
REFER TO "DIAGNOSTIC AIDS" ON FACING PAGE.

NO
CKT 1666 SHORTED TO GROUND OR FAULTY ECM.

"AFTER REPAIRS," REFER TO DTC CRITERIA ON FACING PAGE AND CONFIRM DTC DOES NOT RESET.

5.7L (VIN P) ENGINE — DIAGNOSTIC TROUBLE CODE CHART — 1993 CAMARO AND FIREBIRD

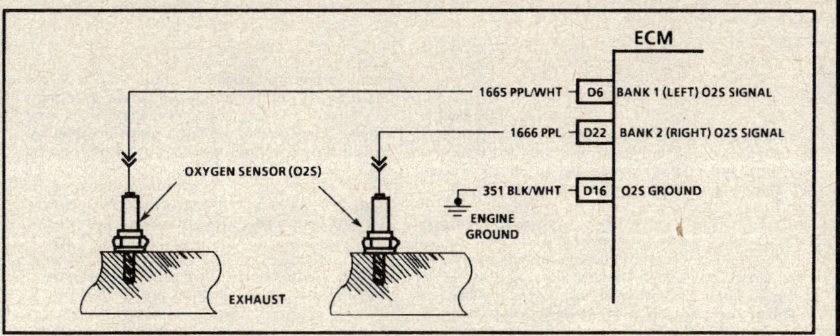

DTC 65
BANK 2 (RIGHT) OXYGEN SENSOR (O2S) CIRCUIT
(RICH EXHAUST INDICATED)
5.7L (VIN P) "F" CARLINE (MFI)

Circuit Description:

The ECM supplies a voltage of about 450 mV between terminals "D16" and "D22". (If measured with a 10 megohm digital voltmeter, this may read as low as 320 mV.) The oxygen sensor varies the voltage within a range of about 1000 mV if a rich O2S signal is indicated, down through about 10 mV if a lean O2S signal is indicated.

The sensor is like an open circuit and produces no voltage when it is below 315°C (600°F). An open sensor circuit or cold sensor causes "Open Loop" operation.

Test Description: Number(s) below refer to circled number(s) on the diagnostic chart.
1. DTC 65 will set if:
   • Signal voltage remains above 700 mV for 50 seconds.
   • Throttle angle is greater than 4%.
   • System is operating in "Closed Loop."

Diagnostic Aids:

Using the Tech 1, observe long term fuel trim values at different RPM and load conditions. The Tech 1 also displays the fuel trim cells, so the long term fuel trim values can be checked in each of the cells to determine when the DTC 65 may have been set. If the conditions for DTC 65 exist, the long term fuel trim values will be near 108.
• <u>Fuel Pressure.</u> System will go rich if pressure is too high. The ECM can compensate for some increase. However, if it gets too high, a DTC 65 may be set. Refer to "Fuel System Diagnosis," CHART A-7.
• <u>Rich Injector.</u> Perform "Injector Balance Test," CHART C-2A.
• <u>Leaking Injector.</u> See "Fuel System Diagnosis," CHART A-7.

• Check for fuel contaminated oil.
• <u>Canister Purge.</u> Check for fuel saturation. If full of fuel, check canister control and hoses. Refer to "Evaporative Emission (EVAP) Control System,"

• <u>MAP Sensor.</u> An output that causes the ECM to sense a lower than normal vacuum can cause the system to go rich. Disconnecting the MAP sensor will allow the ECM to set a fixed value for the sensor. Substitute different MAP sensor if the rich condition is gone while the sensor is disconnected.
• Check for leaking fuel pressure regulator diaphragm by checking vacuum line to regulator for fuel.
• <u>TP Sensor.</u> An intermittent TP sensor output will cause the system to go rich, due to a false indication of engine accelerating.
• <u>EGR.</u> An EGR staying open (especially at idle) will cause the O2S sensor to indicate a rich exhaust, and this could result in a DTC 65.

5.7L (VIN P) ENGINE — DIAGNOSTIC TROUBLE CODE CHART — 1993 CAMARO AND FIREBIRD

DTC 65
BANK 2 (RIGHT) OXYGEN SENSOR (O2S) CIRCUIT
(RICH EXHAUST INDICATED)
5.7L (VIN P) "F" CARLINE (MFI)

①
- RUN WARM ENGINE (75°C/167°F TO 95°C/203°F) AT 1200 RPM. DOES TECH 1 DISPLAY BANK 2 O2S VOLTAGE FIXED ABOVE .75 VOLT (750 mV)?

YES / **NO**

- DISCONNECT BANK 2 O2S AND JUMPER HARNESS CKT 1666 TO GROUND.
- TECH 1 SHOULD DISPLAY BANK 2 O2S BELOW .35 VOLT (350 mV). DOES IT?

DTC 65 IS INTERMITTENT. IF NO ADDITIONAL DTC(s) WERE STORED, REFER TO "DIAGNOSTIC AIDS" ON FACING PAGE.

YES / **NO**

REFER TO "DIAGNOSTIC AIDS" ON FACING PAGE.

REPLACE ECM.

"AFTER REPAIRS," REFER TO DTC CRITERIA ON FACING PAGE AND CONFIRM DTC DOES NOT RESET.

5.7L (VIN P) ENGINE — DIAGNOSTIC TROUBLE CODE CHART — 1993 CAMARO AND FIREBIRD

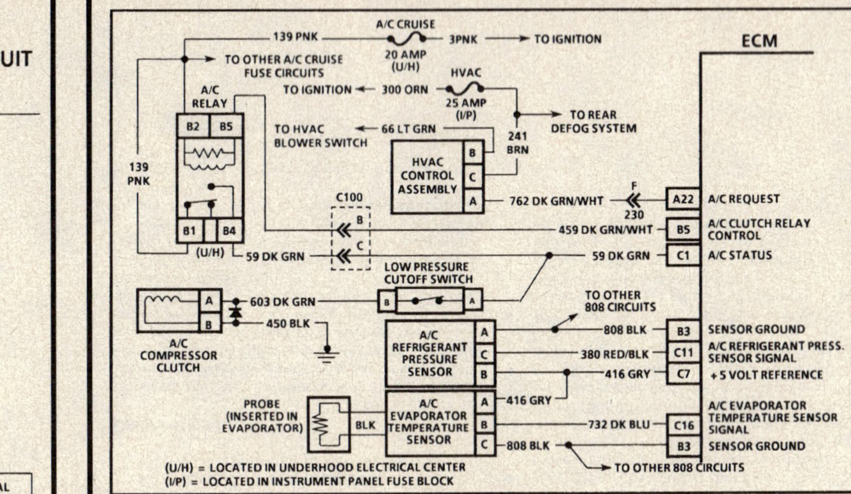

DTC 66
A/C REFRIGERANT PRESSURE SENSOR CIRCUIT
(OPEN OR SHORTED)
5.7L (VIN P) "F" CARLINE (MFI)

Circuit Description:

The A/C system uses an A/C refrigerant pressure sensor mounted in the high side of the A/C system to monitor A/C refrigerant pressure.

The ECM uses this information to turn "ON" the engine cooling fans when A/C pressure is high. The ECM can prevent compressor damage by disabling the compressor clutch if head pressure gets too high.

The A/C refrigerant pressure sensor operates much like other 3-wire sensors. A 5 volt reference is supplied to the sensor and returned to the ECM on the signal line. As the A/C refrigerant pressure increases or decreases, the resistance of the sensor changes and varies the amount of signal returning to the ECM. The amount of signal returned to the ECM is then converted to a pressure value which can be displayed on the Tech 1. A DTC 66 fault will store a DTC in the ECM memory but will not turn "ON" the MIL (Service Engine Soon).

Test Description: Number(s) below refer to circled number(s) on the diagnostic chart.
1. DTC 66 will set if:
   - A/C refrigerant pressure sensor indicates A/C refrigerant pressure is below -8 psi (.12 volt) for 10 seconds.
   OR
   - A/C refrigerant pressure sensor indicates A/C refrigerant pressure is above 448 psi (4.86 volts) for 10 seconds.
2. Checks to see if the high voltage signal is from a shorted sensor or a short to voltage in the circuit. Normally, disconnecting the sensor would make a normal circuit go to near zero volt.
3. Checks to see if low voltage signal is from the sensor or the circuit. Jumpering the sensor signal CKT 380 to 5 volts, checks the circuit, connections and ECM.
4. This step checks to see if the low voltage signal was due to an open in the sensor circuit or the 5 volts reference circuit since the prior step eliminated the pressure sensor.

Diagnostic Aids:

DTC 66 sets when signal voltage falls outside the normal range of the sensor and is not due to a refrigerant system problem. If problem is intermittent, check for opens or shorts in the harness, or poor connections.

5.7L (VIN P) ENGINE — DIAGNOSTIC TROUBLE CODE CHART — 1993 CAMARO AND FIREBIRD

DTC 66

A/C REFRIGERANT PRESSURE SENSOR CIRCUIT
(OPEN OR SHORTED)
5.7L (VIN P) "F" CARLINE (MFI)

5.7L (VIN P) ENGINE — DIAGNOSTIC TROUBLE CODE CHART — 1993 CAMARO AND FIREBIRD

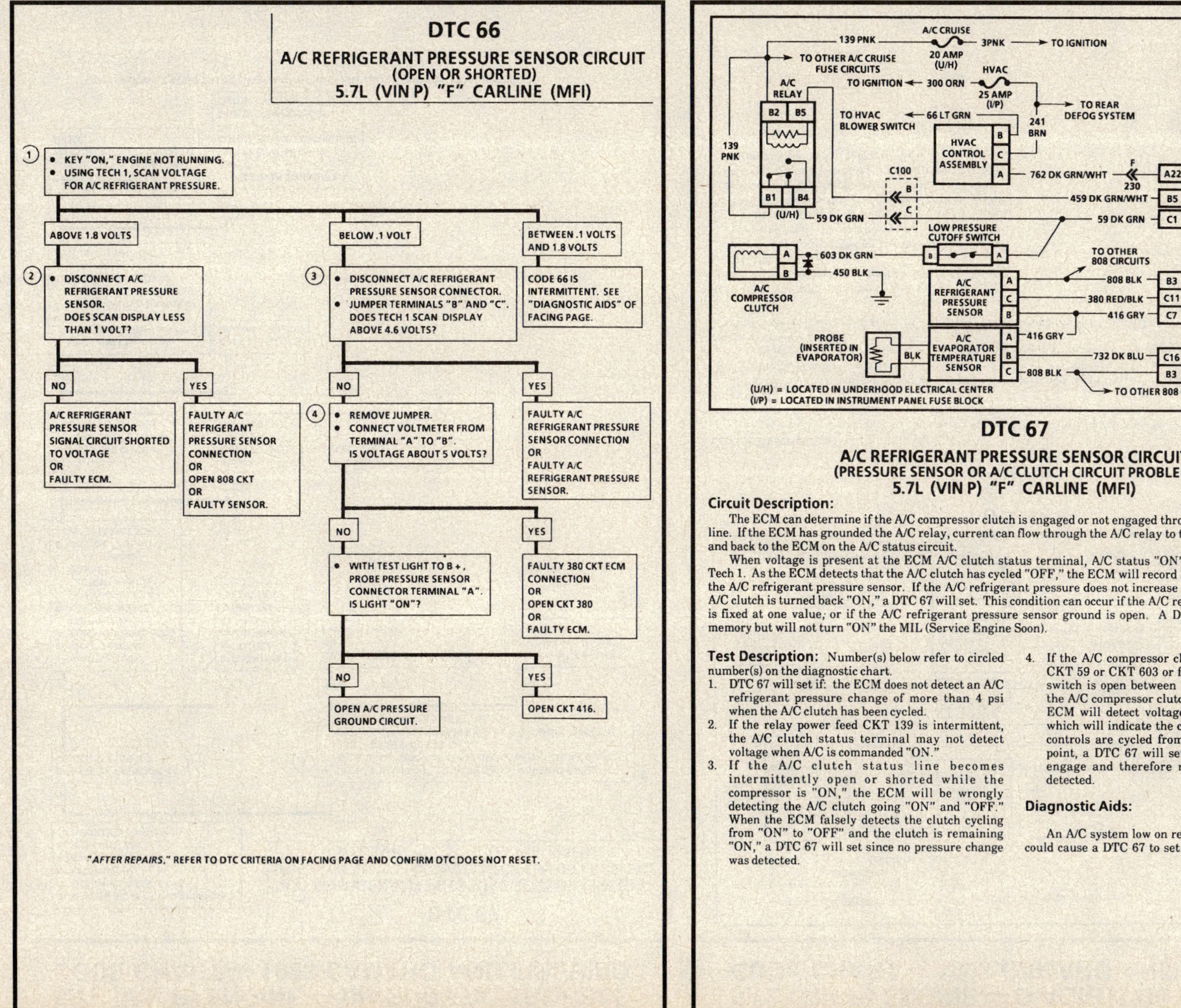

DTC 67

A/C REFRIGERANT PRESSURE SENSOR CIRCUIT
(PRESSURE SENSOR OR A/C CLUTCH CIRCUIT PROBLEM)
5.7L (VIN P) "F" CARLINE (MFI)

Circuit Description:

The ECM can determine if the A/C compressor clutch is engaged or not engaged through the A/C clutch status line. If the ECM has grounded the A/C relay, current can flow through the A/C relay to the A/C compressor clutch and back to the ECM on the A/C status circuit.

When voltage is present at the ECM A/C clutch status terminal, A/C status "ON" will be displayed on the Tech 1. As the ECM detects that the A/C clutch has cycled "OFF," the ECM will record A/C system pressure from the A/C refrigerant pressure sensor. If the A/C refrigerant pressure does not increase more than 4 psi when the A/C clutch is turned back "ON," a DTC 67 will set. This condition can occur if the A/C refrigerant pressure sensor is fixed at one value; or if the A/C refrigerant pressure sensor ground is open. A DTC 67 will store in ECM memory but will not turn "ON" the MIL (Service Engine Soon).

Test Description: Number(s) below refer to circled number(s) on the diagnostic chart.

1. DTC 67 will set if: the ECM does not detect an A/C refrigerant pressure change of more than 4 psi when the A/C clutch has been cycled.
2. If the relay power feed CKT 139 is intermittent, the A/C clutch status terminal may not detect voltage when A/C is commanded "ON."
3. If the A/C clutch status line becomes intermittently open or shorted while the compressor is "ON," the ECM will be wrongly detecting the A/C clutch going "ON" and "OFF." When the ECM falsely detects the clutch cycling from "ON" to "OFF" and the clutch is remaining "ON," a DTC 67 will set since no pressure change was detected.

4. If the A/C compressor clutch is disconnected or if CKT 59 or CKT 603 or faulty low pressure cut-off switch is open between the splice and the clutch, the A/C compressor clutch will not engage, but the ECM will detect voltage on the A/C status line which will indicate the clutch is "ON." If the A/C controls are cycled from "OFF" to "ON" at this point, a DTC 67 will set since the clutch did not engage and therefore no pressure change was detected.

Diagnostic Aids:

An A/C system low on refrigerant (R134a) charge, could cause a DTC 67 to set.

5.7L (VIN P) ENGINE — DIAGNOSTIC TROUBLE CODE CHART — 1993 CAMARO AND FIREBIRD

DTC 67

A/C REFRIGERANT PRESSURE SENSOR CIRCUIT
(PRESSURE SENSOR OR A/C CLUTCH CIRCUIT PROBLEM)
5.7L (VIN P) "F" CARLINE (MFI)

- IF DTC(s) 66, 69 OR 71 ARE PRESENT USE THOSE CHARTS FIRST.
- CLEAR DTC(s).
- START ENGINE.
- TURN A/C "ON."

DID CLUTCH ENGAGE?

YES → (1)
- USING A SCAN TOOL, OBSERVE A/C REFRIGERANT PRESSURE. A/C CLUTCH "OFF" FOR 10 SECONDS, THEN CYCLE A/C CLUTCH "ON" FOR 10 SECONDS.

DID A/C REFRIGERANT PRESSURE INCREASE MORE THAN 4 psi WHEN A/C CONTROLS WERE TURNED "ON"?

- **NO** → FAULTY SENSOR CONNECTION OR FAULTY SENSOR.
- **YES** → (3) DTC 67 IS INTERMITTENT. SEE "DIAGNOSTIC AIDS" ON FACING PAGE.

NO → DOES TECH 1 INDICATE A/C CLUTCH STATUS "ON"?

- **NO** → (2) INTERMITTENT A/C CLUTCH CIRCUIT FROM A/C RELAY TO SPLICE. SEE FACING PAGE.
- **YES** → (4)
  - IGNITION "OFF."
  - DISCONNECT A/C COMPRESSOR CLUTCH CONNECTOR.
  - CONNECT A TEST LIGHT BETWEEN A/C CLUTCH HARNESS CONNECTOR.
  - START ENGINE.
  - A/C "ON."

  IS TEST LIGHT "ON"?

  - **YES** → FAULTY A/C CLUTCH CONNECTION OR OPEN A/C CLUTCH COIL.
  - **NO** →
    - PROBE A/C CLUTCH TERMINAL "A" WITH TEST LIGHT TO GROUND.

    IS TEST LIGHT "ON"?

    - **YES** → OPEN A/C CLUTCH GROUND CIRCUIT.
    - **NO** →
      - DISCONNECT A/C LOW PRESSURE CUTOFF SWITCH.
      - PROBE A/C LOW PRESSURE CUTOFF SWITCH HARNESS TERMINAL "A" WITH A TEST LIGHT CONNECTED TO GROUND.

      IS TEST LIGHT "ON"?

      - **NO** → OPEN OR SHORTED A/C CLUTCH CIRCUIT FROM LOW PRESSURE CUTOFF SWITCH TO SPLICE.
      - **YES** → CKT 603 OPEN OR SHORTED OR FAULTY LOW PRESSURE CUTOFF SWITCH CONNECTION OR FAULTY LOW PRESSURE CUTOFF SWITCH.

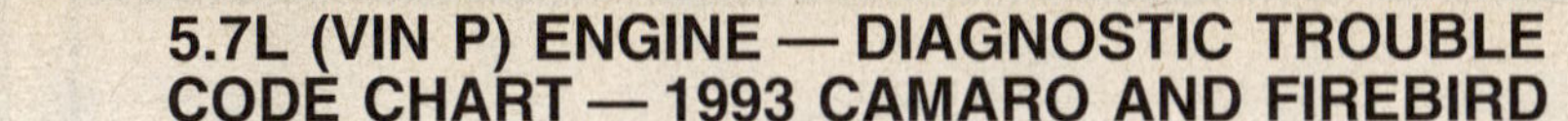

"AFTER REPAIRS," REFER TO DTC CRITERIA ON FACING PAGE AND CONFIRM DTC DOES NOT RESET.

5.7L (VIN P) ENGINE — DIAGNOSTIC TROUBLE CODE CHART — 1993 CAMARO AND FIREBIRD

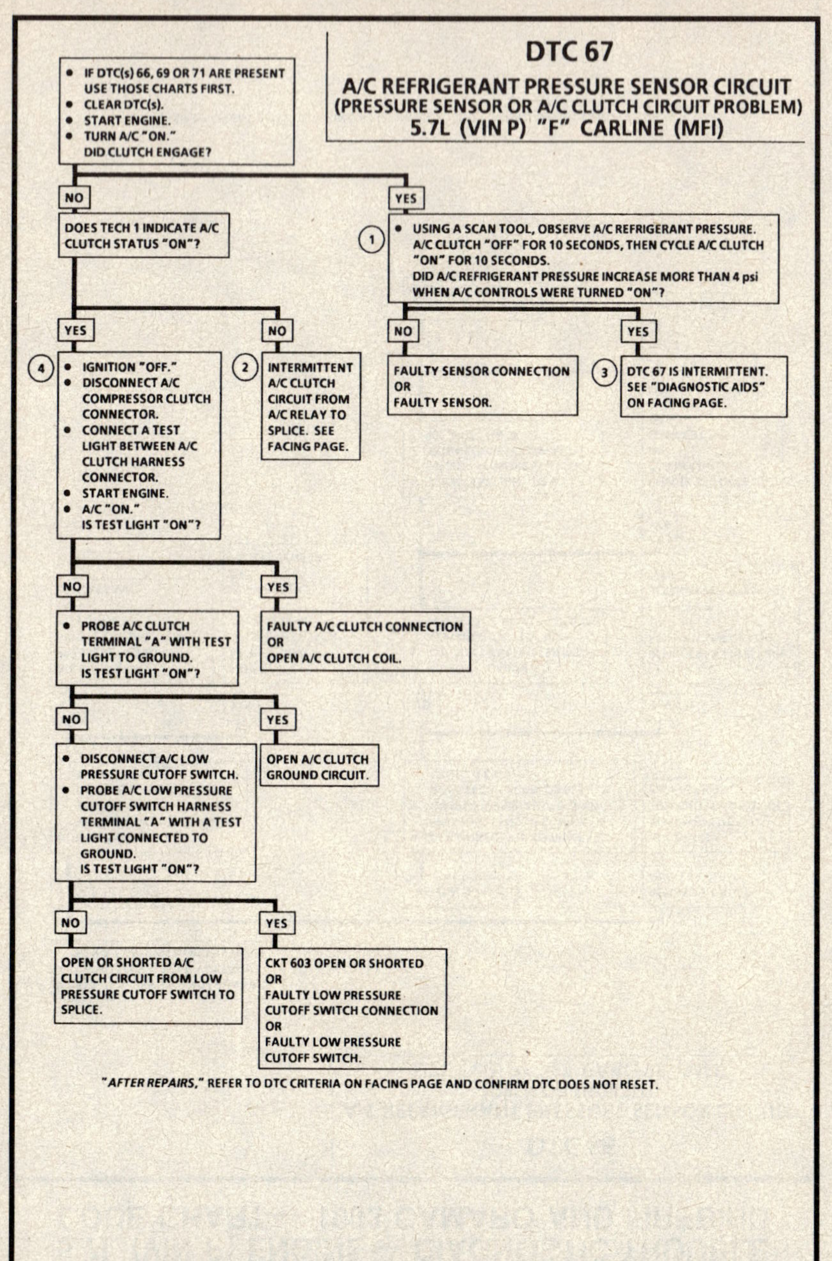

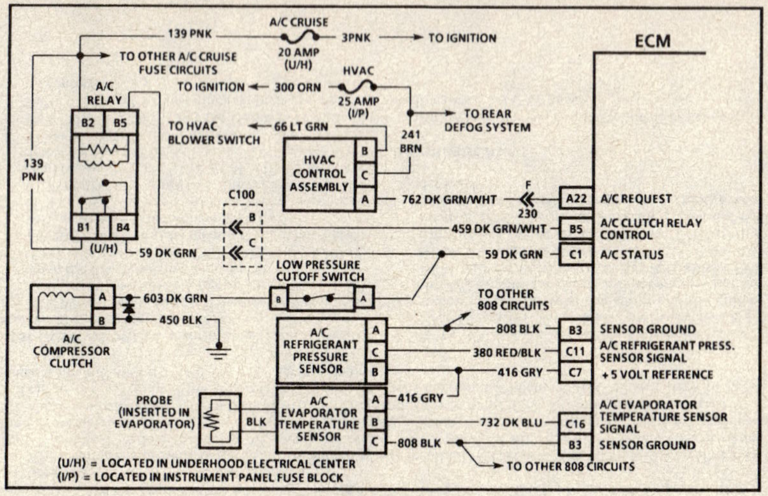

DTC 68

A/C RELAY CIRCUIT
(SHORTED CIRCUIT)
5.7L (VIN P) "F" CARLINE (MFI)

Circuit Description:

When the ECM detects that A/C has been requested, the ECM will activate the A/C clutch relay. When the relay is activated, voltage should be present at both the A/C compressor clutch and the A/C clutch status terminal at the ECM.

If the ECM detects voltage on the A/C clutch status terminal when the A/C relay has not been commanded "ON" by the ECM, a DTC 68 will set.

A short to voltage at any point in CKT 59 or CKT 603, a grounded relay control circuit or stuck A/C relay contacts can set a DTC 68.

A DTC 68 fault will be stored in ECM memory but will not turn "ON" the MIL (Service Engine Soon).

Test Description: Number(s) below refer to circled number(s) on the diagnostic chart.

1. DTC 68 will set if:
   - Voltage is detected on the A/C status line for more than 20 seconds after the ECM has disengaged the A/C relay.
2. This will isolate the ignition feed portion of the circuit from the clutch circuit.
3. If DTC 68 reappears at this step the ECM is internally shorted to voltage.

5.7L (VIN P) ENGINE — DIAGNOSTIC TROUBLE CODE CHART — 1993 CAMARO AND FIREBIRD

DTC 68
A/C RELAY CIRCUIT
(SHORTED CIRCUIT)
5.7L (VIN P) "F" CARLINE (MFI)

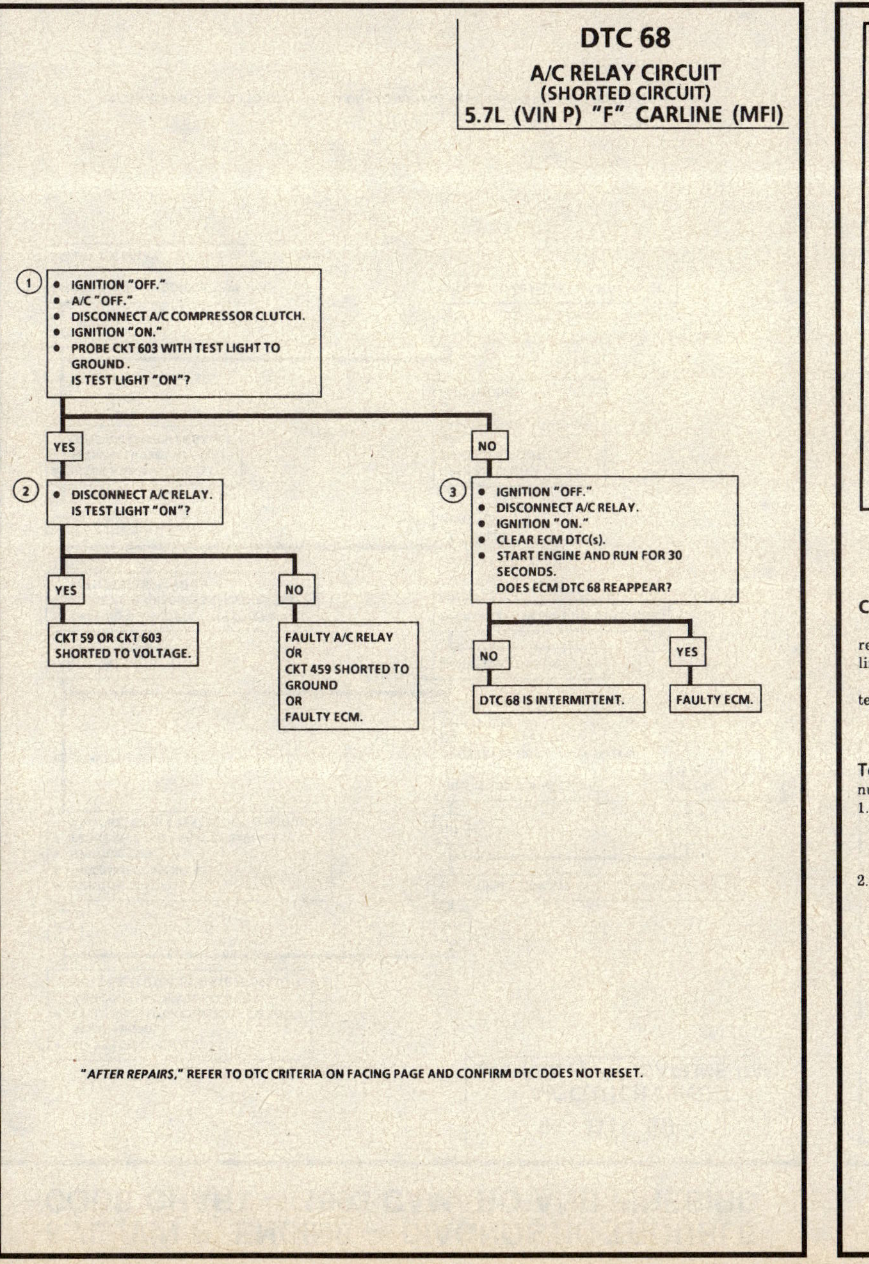

DTC 69
A/C CLUTCH CIRCUIT
5.7L (VIN P) "F" CARLINE (MFI)

Circuit Description:
When the ECM detects that A/C has been requested, the ECM will activate the A/C clutch relay. When the relay has been activated, voltage should be present at both the A/C compressor clutch and the A/C clutch status line terminal of the ECM.

If the ECM activates the A/C clutch relay but does not detect voltage present at the A/C clutch status terminal for more than 20 seconds, a DTC 69 will set.

An open or short to ground at any point in CKT 59 or CKT 603 will cause DTC 69 to set.

A DTC 69 fault will be stored in the ECM memory but will not turn "ON" the MIL (Service Engine Soon).

Test Description: Number(s) below refer to circled number(s) on the diagnostic chart.
1. DTC 69 will set if:
   - ECM has commanded A/C "ON."
   - No voltage detected on A/C status line for more than 20 seconds.
2. If the A/C clutch operates properly and the Tech 1 shows A/C clutch status "OFF," then an open exists between the ECM and the splice to the compressor.

"AFTER REPAIRS," REFER TO DTC CRITERIA ON FACING PAGE AND CONFIRM DTC DOES NOT RESET.

5.7L (VIN P) ENGINE — DIAGNOSTIC TROUBLE CODE CHART — 1993 CAMARO AND FIREBIRD

DTC 69
A/C CLUTCH CIRCUIT
5.7L (VIN P) "F" CARLINE (MFI)

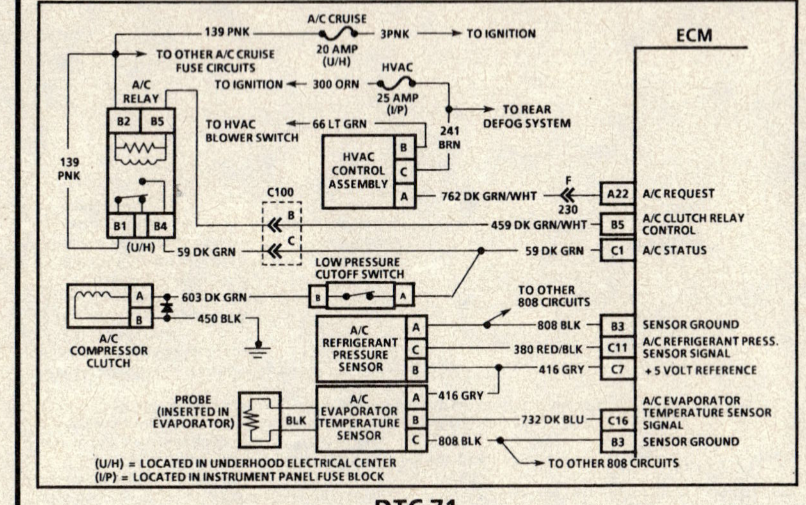

"AFTER REPAIRS," REFER TO DTC CRITERIA ON FACING PAGE AND CONFIRM DTC DOES NOT RESET.

5.7L (VIN P) ENGINE — DIAGNOSTIC TROUBLE CODE CHART — 1993 CAMARO AND FIREBIRD

DTC 71
A/C EVAPORATOR TEMPERATURE SENSOR CIRCUIT
(OPEN OR SHORTED)
5.7L (VIN P) "F" CARLINE (MFI)

Circuit Description:

The A/C system uses an A/C evaporator temperature sensor mounted in the A/C evaporator core to monitor A/C temperature for use by the ECM. The ECM uses this information to cycle the A/C compressor clutch "ON" and "OFF," so optimum cooling is attained. The A/C evaporator temperature sensor will also prevent A/C evaporator core freeze up. The ECM will disable the A/C compressor clutch whenever A/C evaporator temperature is below 2°C (36°F).

The A/C evaporator temperature sensor contains a two wire probe (thermistor) that is positioned in the center of the evaporator core. The two wire probe is part of the A/C evaporator temperature sensor interface module. The interface module works in conjunction with the ECM to determine A/C evaporator temperature. A DTC 71 fault will be stored in the ECM memory but will not turn on the MIL (Service Engine Soon).

Test Description: Number(s) below refer to circled number(s) on the diagnostic chart.

1. DTC 71 will set if:
   - A/C requested.
   - The ECM detects an open, short to ground or a short to voltage in the A/C evaporator temperature circuits.
   - All conditions met for 3 seconds.
2. With the A/C evaporator temperature sensor harness disconnected, normal voltage on CKT 732 is near 5 volts. If a short to voltage is suspected, reconnect sensor harness and back probe terminal "B" of sensor harness using a DVM. If 5 volts or more is indicated, then CKT 732 is shorted to voltage.

Diagnostic Aids:

The A/C evaporator temperature sensor and harness is located under the passenger side instrument panel.

5.7L (VIN P) ENGINE — DIAGNOSTIC TROUBLE CODE CHART — 1993 CAMARO AND FIREBIRD

DTC 71

A/C EVAPORATOR TEMPERATURE SENSOR CIRCUIT
(OPEN OR SHORTED)
5.7L (VIN P) "F" CARLINE (MFI)

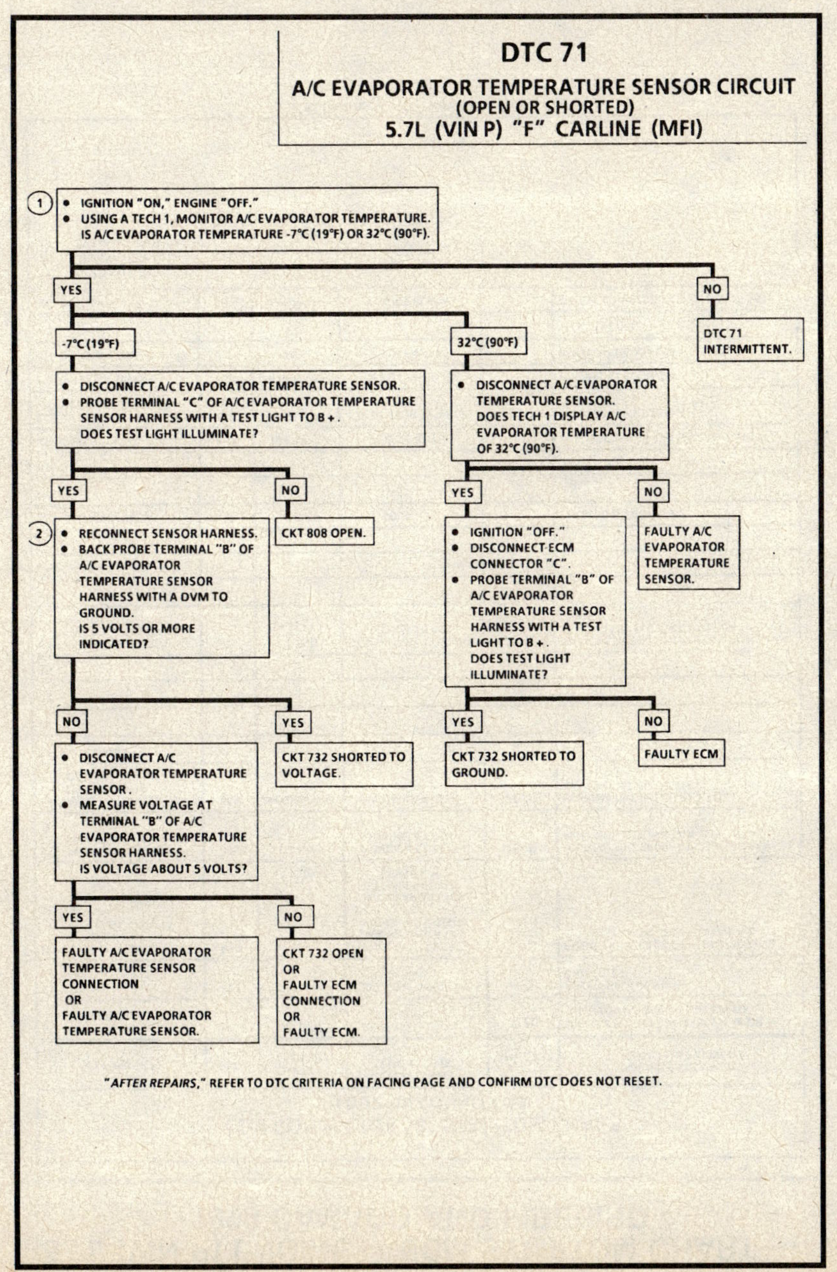

"AFTER REPAIRS," REFER TO DTC CRITERIA ON FACING PAGE AND CONFIRM DTC DOES NOT RESET.

5.7L (VIN P) ENGINE — ECM SYMPTOM CHART — 1993 CAMARO AND FIREBIRD

ECM PIN CONNECTOR "A" SYMPTOM CHART
GRAY 22 PIN CONNECTOR

PIN FUNCTION	CKT #	WIRE COLOR	COMPONENT CONNECTOR CAVITY	DTC(s) AFFECTED	POSSIBLE SYMPTOMS FROM FAULTY CIRCUIT
A1 EVAP CANISTER PURGE CONTROL	428	DK GRN/WHT	EVAP CANISTER SOLENOID VALVE "A"	26 (3) 45 (2)	FUEL ODOR, ROUGH IDLE.
A2 EGR SOLENOID VALVE CONTROL	435	GRY	EGR SOLENOID VALVE "B"	26 (3) 32 (3)	DETONATION, STALL, ROUGH IDLE, POOR PERFORMANCE.
A3					
A4 IAC COIL "A" HIGH	1747	LT BLU/WHT	IAC VALVE "A"	(3)	INCORRECT IDLE, SURGE, STALLING.
A5 IAC COIL "B" LOW	444	LT GRN/BLK	IAC VALVE "D"	(3)	INCORRECT IDLE, SURGE, STALLING.
A6 MIL (SERVICE ENGINE SOON) CONTROL	419	BRN/WHT			NO MIL MIL "ON" ALL THE TIME.
A7					
A8 AIR PUMP RELAY CONTROL	436	BRN	AIR RELAY "E2"	26 (3)	INOPERATIVE AIR PUMP MOTOR, AIR PUMP ON AT ALL TIMES.
A9					
A10 IAC COIL "A" LOW	1748	LT BLU/BLK	IAC VALVE "B"	(3)	INCORRECT IDLE, SURGE, STALLING.
A11 IAC COIL "B" HIGH	1749	LT GRN/WHT	IAC VALVE "C"	(3)	INCORRECT IDLE, SURGE, STALLING.
A12 IGNITION CONTROL	423	WHT	IGNITION COIL MODULE "B"	41, 42 (3)	ENGINE CRANKS, BUT WILL NOT START.
A13 ECT SIGNAL	410	YEL	ECT "B"	14 (2) 15 (1)	HARD STARTING EXHAUST ODOR, ROUGH IDLE, POOR FUEL ECONOMY, LACK OF PERFORMANCE.
A14					
A15					
A16					
A17					
A18					
A19 PASS-KEY®II	229	DK BLU	PASS-KEY®II MODULE "A3"		ENGINE CRANKS BUT WILL NOT START.
A20					
A21 PNP SIGNAL	434	ORN/BLK	PNP SWITCH "B"		POOR FUEL ECONOMY, STALL INCORRECT IDLE.
A22 A/C REQUEST	762	DK GRN/WHT	HVAC CONTROL ASSEMBLY "A"	(3)	A/C INOPERATIVE

(1) OPEN CIRCUIT.
(2) GROUNDED CIRCUIT.
(3) OPEN/GROUNDED CIRCUIT.

ECM Connector "A" Symptom Chart

5.7L (VIN P) ENGINE — ECM SYMPTOM CHART — 1993 CAMARO AND FIREBIRD

ECM CONNECTOR "B" SYMPTOM CHART
RED 22 PIN CONNECTOR

PIN FUNCTION	CKT #	WIRE COLOR	COMPONENT CONNECTOR CAVITY	DTC(s) AFFECTED	POSSIBLE SYMPTOMS FROM FAULTY CIRCUIT
B1 REVERSE LOCKOUT SOLENOID	1652	LT GRN	REVERSE LOCKOUT SOLENOID "A"	27 (3)	UNABLE TO SHIFT INTO REVERSE OR WILL GO INTO REVERSE AT ANY SPEED.
B2 TCC	422	TAN/BLK	TCC SOLENOID "D"	27 (3)	POOR FUEL ECONOMY. CHUGGLE.
B3 MAP, IAT, & A/C REFRIGERANT PRESSURE SENSOR GROUND	808	BLK	MAP "A", IAT "A", A/C REFRIGERANT PRESSURE SENSOR "A" A/C EVAPORATOR TEMPERATURE SENSOR "C"	23 (1) 33 (1) 67 (1) 71	ROUGH IDLE, LACK OF PERFORMANCE. EXHAUST ODOR, INOPERATIVE A/C.
B4 PRIMARY COOLING FAN CONTROL	335	DK GRN	PRIMARY COOLING FAN RELAY "D2"	28 (3)	INOPERATIVE PRIMARY COOLING FAN. PRIMARY COOLING FAN "ON" AT ALL TIMES.
B5 A/C CLUTCH RELAY CONTROL (C60 OPTION)	459	DK GRN/WHT	A/C RELAY "B5"	28 (3)	INOPERATIVE A/C CLUTCH (1). A/C "ON" AT ALL TIMES (2).
B6 ECM GROUND	451	BLK/WHT	ENGINE BLOCK		ENGINE CRANKS BUT WILL NOT START.
B7					
B8					
B9					
B10 SECONDARY COOLING FAN CONTROL (C60 OPTION)	473	DK BLU	SECONDARY COOLING FAN RELAY "F5"	28 (3)	INOPERATIVE SECONDARY COOLING FAN. SECONDARY COOLING FAN "ON" AT ALL TIMES.
B11					
B12					
B13					
B14 4th GEAR SIGNAL	446	LT BLU	TCC SOLENOID "B"		POOR FUEL ECONOMY.
B15					
B16					
B17 ECM GROUND	451	BLK/WHT	ENGINE BLOCK		ENGINE CRANKS BUT WILL NOT START.
B18 TP SENSOR, ECT GROUND	470	BLK	TP SENSOR "B", ECT "A"	15 (1) 21 (1)	LACK OF PERFORMANCE. ROUGH IDLE.
B19 3rd GEAR SIGNAL	438	DK GRN/WHT	TCC SOLENOID "C"		POOR FUEL ECONOMY.
B20 2nd GEAR SIGNAL	232	WHT	TCC SOLENOID "E"		POOR FUEL ECONOMY.
B21					
B22 ECM GROUND	551	TAN/WHT	ENGINE BLOCK		ENGINE CRANKS BUT WILL NOT START.

(1) OPEN CIRCUIT.
(2) GROUND CIRCUIT.
(3) OPEN/GROUND CIRCUIT.

ECM Connector "B" Symptom Chart

5.7L (VIN P) ENGINE — ECM SYMPTOM CHART — 1993 CAMARO AND FIREBIRD

ECM CONNECTOR "C" SYMPTOM CHART
GREEN 22 PIN CONNECTOR

PIN FUNCTION	CKT #	WIRE COLOR	COMPONENT CONNECTOR CAVITY	DTC(s) AFFECTED	POSSIBLE SYMPTOMS FROM FAULTY CIRCUIT
C1 A/C STATUS	59	DK GRN	A/C RELAY "B4" LOW PRESSURE SWITCH	69 (1)	A/C INOPERATIVE.
C2 +5 VOLTS REFERENCE	474	GRY	TP SENSOR "A"	22 (3)	HIGH IDLE, LACK OF PERFORMANCE.
C3 TP SENSOR SIGNAL	417	DK BLU	TP SENSOR "C"	22 (3)	HIGH IDLE, LACK OF PERFORMANCE.
C4					
C5 LOW RESOLUTION	453	RED/BLK	DISTRIBUTOR "A"	16 (3)	ENGINE CRANKS BUT WILL NOT START.
C6 BATTERY FEED	340	ORN			ENGINE CRANKS BUT WILL NOT START (1).
C7 +5 VOLTS REFERENCE	416	GRY	MAP "C", A/C REFRIG PRESSURE SENSOR "B", A/C EVAP TEMP SENSOR "A"	34 (3) 66 (3)	ROUGH IDLE, INOPERATIVE A/C.
C8 KNOCK SENSOR SIGNAL	496	DK BLU	KNOCK SENSOR	43 (3)	LACK OF PERFORMANCE.
C9 DISTRIBUTOR REFERENCE LOW	632	PNK/BLK	DISTRIBUTOR "D"		ENGINE CRANKS BUT WILL NOT START.
C10					
C11 A/C REFRIGERANT PRESSURE SIGNAL	380	RED/BLK	A/C REFRIGERANT PRESSURE SENSOR "C"	66 (3)	INOPERATIVE A/C.
C12 IGNITION FEED	439	PNK			ENGINE CRANKS BUT WILL NOT START.
C13					
C14 HIGH RESOLUTION	647	LT BLU/BLK	DISTRIBUTOR "B"	36 (3)	ENGINE SURGE, LACK OF PERFORMANCE.
C15					
C16 A/C EVAPORATOR TEMPERATURE SENSOR SIGNAL	732	DK BLU	A/C EVAPORATOR TEMPERATURE SENSOR "B"	71 (3)	INOPERATIVE A/C.
C17 BATTERY FEED	340	ORN			ENGINE CRANKS BUT WILL NOT START.
C18 DISTRIBUTOR IGNITION FEED	631	RED	DISTRIBUTOR "C"		ENGINE CRANKS BUT WILL NOT START.
C19					
C20 FUEL PUMP SIGNAL	120	GRY	FUEL PUMP RELAY "A2"		
C21 MAP SIGNAL	432	LT GRN	MAP "B"	34 (3)	ENGINE RUNS ROUGH, LACK OF POWER, POOR PERFORMANCE.
C22 IAT SIGNAL	472	TAN	IAT "B"	23 (1) 25 (2)	POOR FUEL ECONOMY.

(1) OPEN CIRCUIT.
(2) GROUNDED CIRCUIT.
(3) OPEN/GROUNDED CIRCUIT.

ECM Connector "C" Symptom Chart

5.7L (VIN P) ENGINE — ECM SYMPTOM CHART — 1993 CAMARO AND FIREBIRD

ECM PIN CONNECTOR "D" SYMPTOM CHART
BROWN 22 PIN CONNECTOR

PIN FUNCTION	CKT #	WIRE COLOR	COMPONENT CONNECTOR CAVITY	DTC(s) AFFECTED	POSSIBLE SYMPTOMS FROM FAULTY CIRCUIT
D1 VSS GROUND	401	~ PPL	VSS "A"	24 (1)	INOPERATIVE SPEEDOMETER. INOPERATIVE CRUISE CONTROL.
D2					
D3					
D4 SERIAL DATA	800	TAN			NO TECH 1 DATA.
D5					
D6 BANK 1 (LEFT) O2S SIGNAL	1665	PPL/WHT	O2S	13 (1) 44 (2)	EXHAUST ODOR, POOR PERFORMANCE. POOR FUEL ECONOMY, "OPEN LOOP."
D7 FUEL PUMP RELAY DRIVER	465	DK GRN/WHT	FUEL PUMP RELAY "C2"		LONG CRANK TIME BEFORE ENGINE STARTS.
D8 VSS OUTPUT (4KPPM)	817	DK GRN/WHT	I/P AND CRUISE MODULE		INOPERATIVE CRUISE CONTROL AND SPEEDOMETER.
D9					
D10 INJECTOR DRIVER	468	DK GRN	INJECTORS 2, 4, 6, 8 "A"	44 (1) 45 (2)	POSSIBLE ENGINE CRANKS BUT WILL NOT START, ROUGH IDLE, LACK OF PERFORMANCE.
D11 INJECTOR DRIVER	467	DK BLU	INJECTORS 1, 3, 5, 7 "A"	44 (1) 45 (2)	POSSIBLE ENGINE CRANKS BUT WILL NOT START, ROUGH IDLE, LACK OF PERFORMANCE.
D12 VSS SIGNAL	400	YEL	VSS "B"	24 (3)	INOPERATIVE SPEEDOMETER. INOPERATIVE CRUISE CONTROL.
D13					
D14					
D15					
D16 O2S GROUND	351	BLK/WHT	O2S	13 (1) 63 (1)	ROUGH IDLE, LEAN EXHAUST, POOR PERFORMANCE, POOR FUEL ECONOMY.
D17					
D18					
D19					
D20 DIAGNOSTIC ENABLE	448	WHT/BLK	DLC "B"		NO TECH 1 DATA. WILL NOT FLASH DTC 12.
D21					
D22 BANK 2 (RIGHT) O2S SIGNAL	1666	PPL	O2S	63 (1) 64 (2)	EXHAUST ODOR, POOR PERFORMANCE. POOR FUEL ECONOMY, "OPEN LOOP."

(1) OPEN CIRCUIT.
(2) GROUNDED CIRCUIT.
(3) OPEN/GROUNDED CIRCUIT.

ECM Connector "D" Symptom Chart

5.7L (VIN P) ENGINE — COMPONENT DIAGNOSTIC CHART — 1993 CAMARO AND FIREBIRD

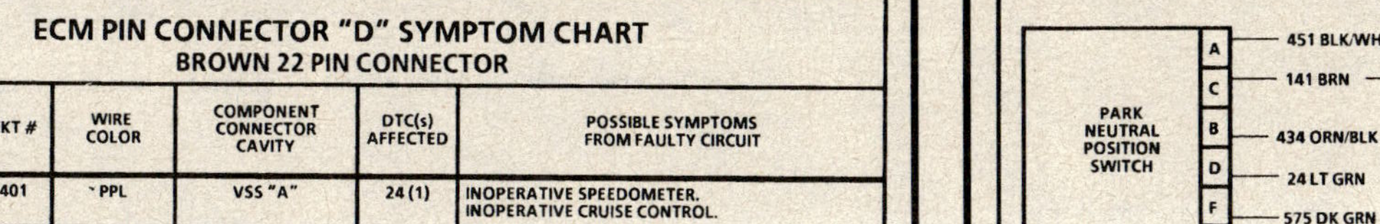

CHART C-1A

PARK/NEUTRAL POSITION (PNP) SWITCH DIAGNOSIS
(AUTOMATIC TRANSMISSION ONLY)
5.7L (VIN P) "F" CARLINE (MFI)

Circuit Description:

The Park/Neutral Position (PNP) switch contacts are a part of the neutral start switch, and are closed to ground in park or neutral and open in drive ranges.

The ECM supplies ignition voltage through a current limiting resistor to CKT 434 and senses a closed switch when the voltage on CKT 434 drops to less than one volt.

The ECM uses the PNP signal as one of the inputs to control:
- Idle Air Control (IAC).
- Vehicle Speed Sensor (VSS) diagnostics.
- Exhaust Gas Recirculation (EGR).

If CKT 434 indicates PNP (grounded), while in drive range, the EGR would be inoperative, resulting in possible detonation.

If CKT 434 always indicates drive (open), a drop in the idle speed may exist when the gear selector is moved into drive range.

Test Description: Number(s) below refer to circled number(s) on the diagnostic chart.
1. Checks for a closed switch to ground in park position.
2. Checks for an open switch in drive range.
3. Make sure the Tech 1 indicates drive, even while wiggling shifter to test for an intermittent or misadjusted switch in drive range.

5.7L (VIN P) ENGINE — COMPONENT DIAGNOSTIC CHART — 1993 CAMARO AND FIREBIRD

CHART C-1A

PARK/NEUTRAL POSITION (PNP) SWITCH DIAGNOSIS
(AUTOMATIC TRANSMISSION ONLY)
5.7L (VIN P) "F" CARLINE (MFI)

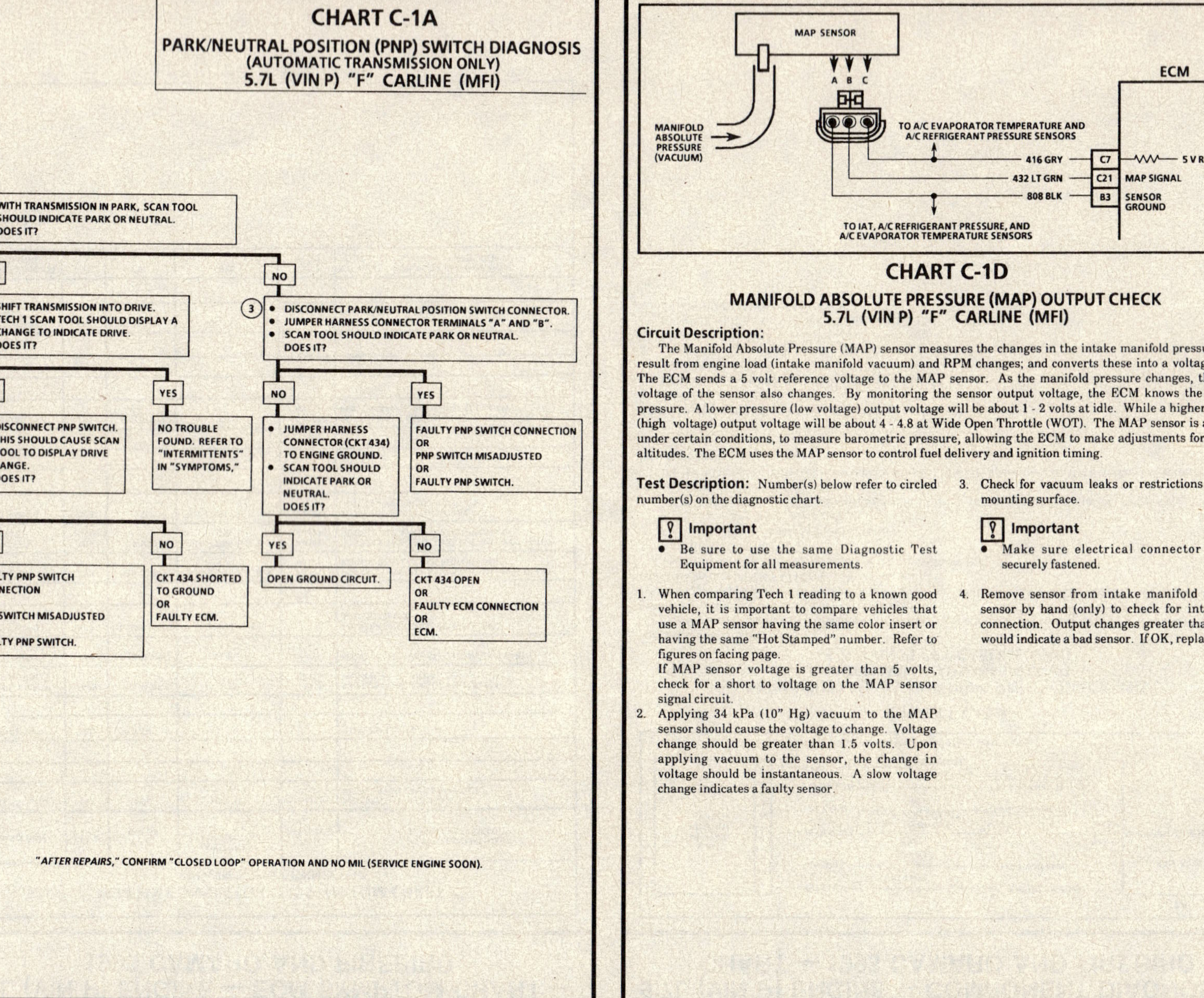

"AFTER REPAIRS," CONFIRM "CLOSED LOOP" OPERATION AND NO MIL (SERVICE ENGINE SOON).

5.7L (VIN P) ENGINE — COMPONENT DIAGNOSTIC CHART — 1993 CAMARO AND FIREBIRD

CHART C-1D

MANIFOLD ABSOLUTE PRESSURE (MAP) OUTPUT CHECK
5.7L (VIN P) "F" CARLINE (MFI)

Circuit Description:

The Manifold Absolute Pressure (MAP) sensor measures the changes in the intake manifold pressure which result from engine load (intake manifold vacuum) and RPM changes; and converts these into a voltage output. The ECM sends a 5 volt reference voltage to the MAP sensor. As the manifold pressure changes, the output voltage of the sensor also changes. By monitoring the sensor output voltage, the ECM knows the manifold pressure. A lower pressure (low voltage) output voltage will be about 1 - 2 volts at idle. While a higher pressure (high voltage) output voltage will be about 4 - 4.8 at Wide Open Throttle (WOT). The MAP sensor is also used, under certain conditions, to measure barometric pressure, allowing the ECM to make adjustments for different altitudes. The ECM uses the MAP sensor to control fuel delivery and ignition timing.

Test Description: Number(s) below refer to circled number(s) on the diagnostic chart.

> **Important**
> • Be sure to use the same Diagnostic Test Equipment for all measurements.

1. When comparing Tech 1 reading to a known good vehicle, it is important to compare vehicles that use a MAP sensor having the same color insert or having the same "Hot Stamped" number. Refer to figures on facing page.
   If MAP sensor voltage is greater than 5 volts, check for a short to voltage on the MAP sensor signal circuit.
2. Applying 34 kPa (10" Hg) vacuum to the MAP sensor should cause the voltage to change. Voltage change should be greater than 1.5 volts. Upon applying vacuum to the sensor, the change in voltage should be instantaneous. A slow voltage change indicates a faulty sensor.

3. Check for vacuum leaks or restrictions at intake mounting surface.

> **Important**
> • Make sure electrical connector remains securely fastened.

4. Remove sensor from intake manifold and twist sensor by hand (only) to check for intermittent connection. Output changes greater than .10 volt would indicate a bad sensor. If OK, replace sensor.

5.7L (VIN P) ENGINE — COMPONENT DIAGNOSTIC CHART — 1993 CAMARO AND FIREBIRD

CHART C-1D
MANIFOLD ABSOLUTE PRESSURE (MAP) OUTPUT CHECK
5.7L (VIN P) "F" CARLINE (MFI)

NOTICE: THIS CHART ONLY APPLIES TO MAP SENSORS HAVING GREEN OR BLACK COLOR KEY INSERT (SEE BELOW).

1.
- IF DTC 33 OR 34 IS SET, USE THOSE CHARTS FIRST.
- IGNITION "ON," ENGINE "OFF."
- SCAN TOOL SHOULD INDICATE A MAP SENSOR VOLTAGE.
- COMPARE THIS READING WITH THE READING OF A KNOWN GOOD VEHICLE. SEE FACING PAGE TEST DESCRIPTION, STEP 1. VOLTAGE READING SHOULD BE WITHIN ± .4 VOLT.
IS IT?

YES

NO — REPLACE MAP SENSOR.

2.
- DISCONNECT AND PLUG VACUUM SOURCE TO MAP SENSOR.
- CONNECT A HAND VACUUM PUMP TO MAP SENSOR.
- START ENGINE.
- NOTE MAP SENSOR VOLTAGE.
- APPLY 34 kPa (10" Hg) OF VACUUM AND NOTE VOLTAGE CHANGE. SUBTRACT SECOND READING FROM THE FIRST. VOLTAGE VALUE SHOULD BE GREATER THAN 1.5 VOLTS.
IS IT?

YES

3. NO TROUBLE FOUND. CHECK MAP SENSOR VACUUM SOURCE FOR LEAKAGE OR RESTRICTION. BE SURE THIS SOURCE SUPPLIES VACUUM TO MAP SENSOR ONLY.

NO

4. CHECK MAP SENSOR CONNECTION. IF OK, REPLACE MAP SENSOR.

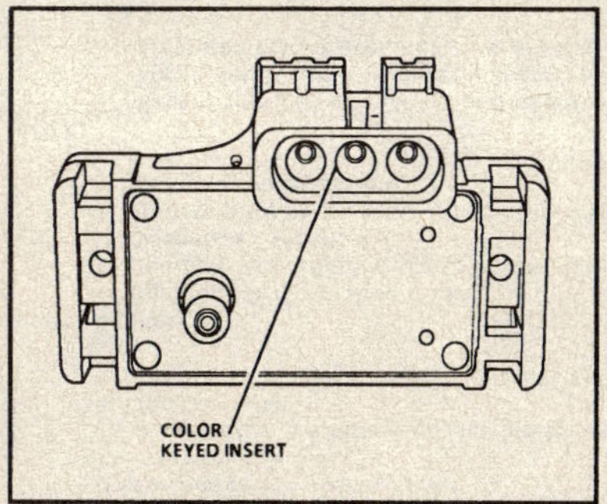

Figure 1 - Color Key Insert

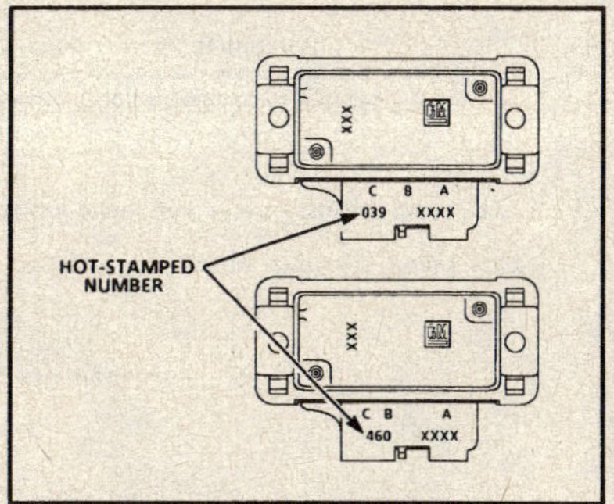

Figure 2 - Hot Stamped Number

"AFTER REPAIRS," CONFIRM "CLOSED LOOP" OPERATION AND NO MIL (SERVICE ENGINE SOON).

5.7L (VIN P) ENGINE — COMPONENT DIAGNOSTIC CHART — 1993 CAMARO AND FIREBIRD

CHART C-2A
INJECTOR BALANCE TEST

The injector balance tester is a tool used to turn the injector "ON" for a precise amount of time, thus spraying a measured amount of fuel into the manifold. This causes a drop in fuel rail pressure that we can record and compare between each injector. All injectors should have the same amount of pressure drop (± 10 kPa). Any injector with a pressure drop that is 10 kPa (or more) greater or less than the average drop of the other injectors should be considered faulty and replaced.

STEP 1

Engine "cool down" period (10 minutes) is necessary to avoid irregular readings due to "Hot Soak" fuel boiling. With ignition "OFF" connect fuel gauge J 3470-1 to fuel pressure tap. Wrap a shop towel around fitting while connecting gage to avoid fuel spillage.

Disconnect harness connectors at all injectors, and connect injector tester J 34730-3A to one injector. Follow manufacturers instructions for use of adaptor harness. Ignition must be "OFF" at least 10 seconds to complete ECM shutdown cycle. Fuel pump should run about 2 seconds after ignition is turned "ON." At this point, insert clear tubing attached to vent valve into a suitable container and bleed air from gauge and hose to insure accurate gauge operation. Repeat this step until all air is bled from gauge.

STEP 2

Turn ignition "OFF" for 10 seconds and then "ON" again to get fuel pressure to its maximum. Record this initial pressure reading. Energize tester one time and note pressure drop at its lowest point (disregard any slight pressure increase after drop hits low point). By subtracting this second pressure reading from the initial pressure, we have the actual amount of injector pressure drop.

STEP 3

Repeat Step 2 on each injector and compare the amount of drop. Usually, good injectors will have virtually the same drop. Retest any injector that has a pressure difference of 10 kPa, either more or less than the average of the other injectors on the engine. Replace any injector that also fails the retest. If the pressure drop of all injectors is within 10 kPa of this average, the injectors appear to be flowing properly. Reconnect them and review "Symptoms," Section "6E3-B".

NOTICE: *The entire test should not be repeated more than once without running the engine to prevent flooding. (This includes any retest on faulty injectors.)*

5.7L (VIN P) ENGINE — COMPONENT DIAGNOSTIC CHART — 1993 CAMARO AND FIREBIRD

NOTICE: The entire test should <u>NOT</u> be repeated more than once without running the engine to prevent flooding. (This includes any retest on faulty injectors.)

The fuel pressure test in Section "A" Chart A-7, should be completed prior to this test.

CHART C-2A
INJECTOR BALANCE TEST
5.7L (VIN P) "F" CARLINE (MFI)

Step 1. If engine is at operating temperature, allow a 10 minute "cool down" period then connect fuel pressure gage and injector tester.
1. Ignition "OFF."
2. Connect fuel pressure gage and injector tester.
3. Ignition "ON."
4. Bleed off air in gage. Repeat until all air is bled from gauge.

Step 2. Run test:
1. Ignition "OFF" for 10 seconds.
2. Ignition "ON." Record gage pressure. (Pressure must hold steady, if not, see "Fuel System Diagnosis," CHART A-7.
3. Turn injector on, by depressing button on injector tester, and note pressure at the instant the gauge needle stops.

Step 3.
1. Repeat Step 2 on all injectors and record pressure drop on each. Retest injectors that appear faulty (Any injectors that have a 10 kPa (1.5 psi) difference, either more or less, in pressure from the average). If no problem is found, review "Symptoms"

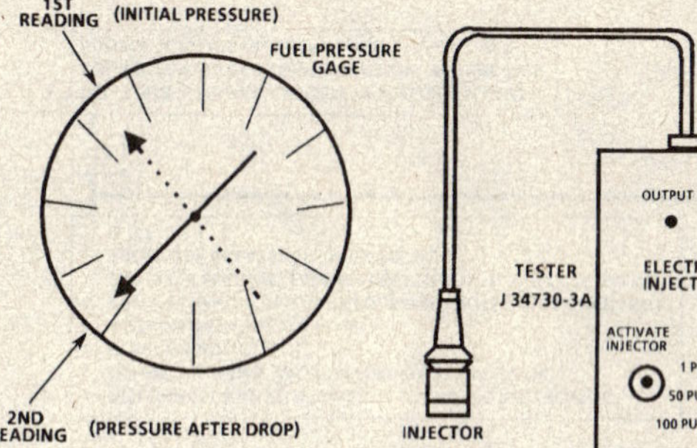

EXAMPLE

CYLINDER	1	2	3
1ST READING	293 kPa (43 psi)	293 kPa (43 psi)	293 kPa (43 psi)
2ND READING	131 kPa (19 psi)	115 kPa (17 psi)	145 kPa (21 psi)
AMOUNT OF DROP	162 kPa (24 psi)	178 kPa (26 psi)	148 kPa (21 psi)
	OK	FAULTY, RICH (TOO MUCH FUEL DROP)	FAULTY, LEAN (TOO LITTLE FUEL DROP)

5.7L (VIN P) ENGINE — COMPONENT DIAGNOSTIC CHART — 1993 CAMARO AND FIREBIRD

CHART C-2C
IDLE AIR CONTROL (IAC) VALVE CHECK
5.7L (VIN P) "F" CARLINE (MFI)

Circuit Description:

The ECM controls engine idle speed with the IAC valve. To increase idle speed, the ECM retracts the IAC valve pintle away from its seat, allowing more air to bypass the throttle bore. To decrease idle speed, it extends the IAC valve pintle towards its seat, reducing bypass air flow. The scan tool will display the ECM commands to the IAC valve in counts. Higher the counts indicate more air bypass (higher idle). The lower the counts indicate less air is allowed to bypass (lower idle).

Test Description: Number(s) below refer to circled number(s) on the diagnostic chart.
1. The IAC tester is used to extend and retract the IAC valve. Valve movement is verified by an engine speed change. If no change in engine speed occurs, the valve can be retested when removed from the throttle body.
2. This step checks the quality of the IAC movement in Step 1. Between 700 RPM and about 1500 RPM, the engine speed should change smoothly with each flash of the tester light in both extend and retract. If the IAC valve is retracted beyond the control range (about 1500 RPM), it may take many flashes in the extend position before engine speed will begin to drop. This is normal on certain engines, fully extending IAC may cause engine to stall. This may be normal.
3. Steps 1 and 2 verified proper IAC valve operation while this step checks the IAC circuits. Each lamp on the node light should flash red and green while the IAC valve is cycled. While the sequence of color is not important, if either light is "OFF" or does not flash red and green, check the circuits for faults, beginning with poor terminal contacts.

Diagnostic Aids:

A slow, unstable, or fast idle may be caused by a non-IAC system problem that cannot be overcome by the IAC valve. Out of control range IAC Tech 1 counts will be above 60 if idle is too low, and zero counts if idle is too high. The following checks should be made to repair a non-IAC system problem:
- Vacuum Leak (High Idle) - If idle is too high, stop the engine. Fully extend (low) IAC with tester. Start engine. If idle speed is above 800 RPM, locate and correct vacuum leaks including crankcase ventilation system. Also, check for binding of throttle blade or linkage.
- System too lean (High Air/Fuel Ratio) - The idle speed may be too high or too low. Engine speed may vary up and down and disconnecting the IAC valve does not help. DTC 44 or 64 may be set.
Tech 1 O2S voltage will be less than 300 mV (.3 volt). Check for low regulated fuel pressure, water in the fuel or a restricted injector(s).
- System too rich (Low Air/Fuel Ratio) - The idle speed will be too low. Tech 1 IAC counts will usually be above 80. System is obviously rich and may exhibit black smoke in exhaust.
Tech 1 O2S voltage will be fixed above 800 mV (.8 volt). DTC 45 or 65 may set.
Check for high fuel pressure, leaking or sticking injector(s). Silicone contaminated oxygen sensors will be slow to respond.
- Throttle Body - Remove IAC valve and inspect bore for foreign material.
- IAC Valve Electrical Connections - IAC valve connections should be carefully checked for proper contact.
- Crankcase Ventilation Valve - An incorrect or faulty crankcase ventilation valve may result in an incorrect idle speed.
- Refer to "Rough, Unstable, Incorrect Idle or Stalling" in "Symptoms,"
- If intermittent poor driveability or idle symptoms are resolved by disconnecting the IAC, carefully recheck connections, valve terminal resistance, or replace IAC.

5.7L (VIN P) ENGINE — COMPONENT DIAGNOSTIC CHART — 1993 CAMARO AND FIREBIRD

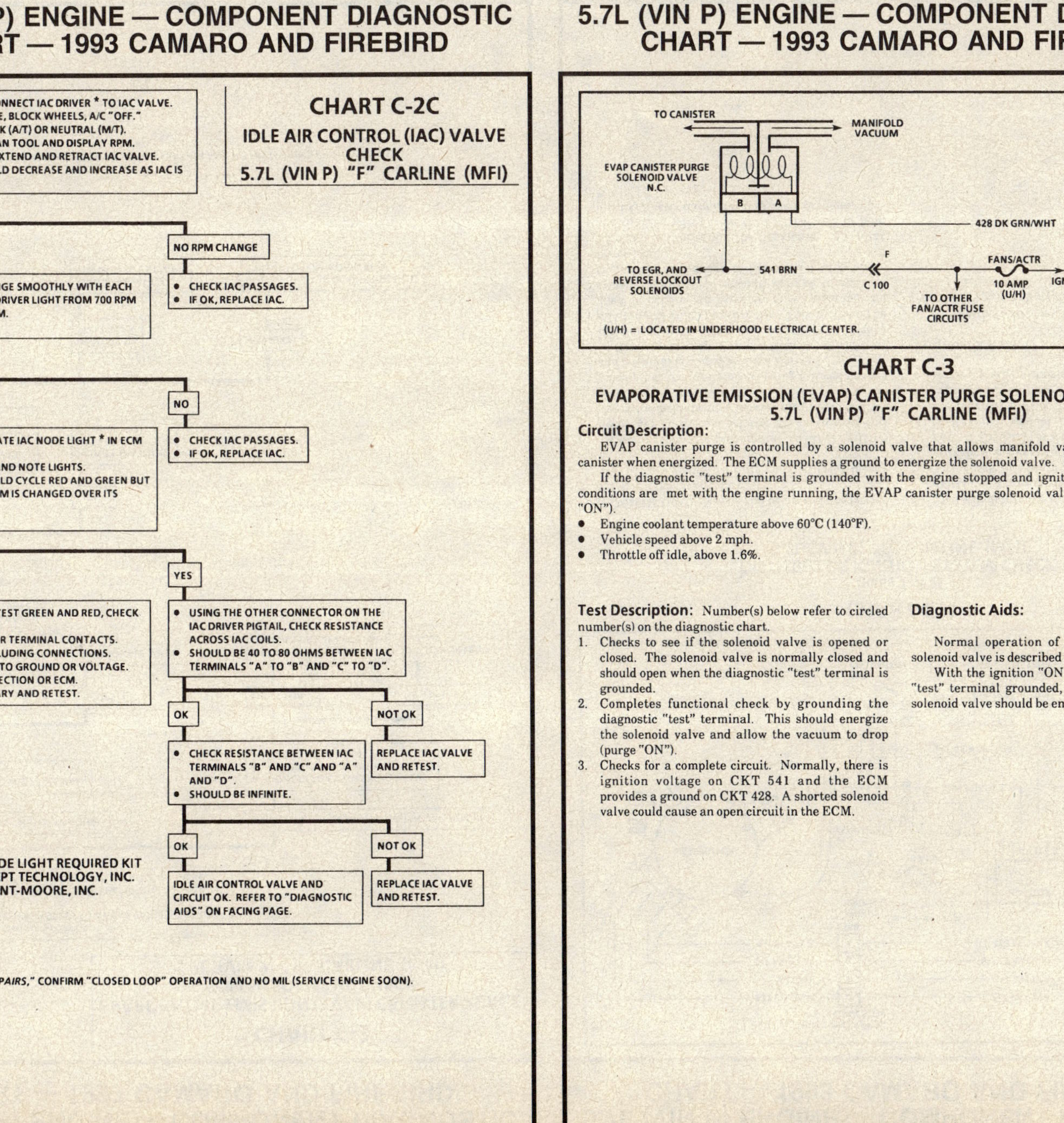

5.7L (VIN P) ENGINE — COMPONENT DIAGNOSTIC CHART — 1993 CAMARO AND FIREBIRD

CHART C-3

EVAPORATIVE EMISSION (EVAP) CANISTER PURGE SOLENOID VALVE CHECK
5.7L (VIN P) "F" CARLINE (MFI)

Circuit Description:

EVAP canister purge is controlled by a solenoid valve that allows manifold vacuum to purge the EVAP canister when energized. The ECM supplies a ground to energize the solenoid valve.

If the diagnostic "test" terminal is grounded with the engine stopped and ignition "ON," or the following conditions are met with the engine running, the EVAP canister purge solenoid valve will be energized (purge "ON").
* Engine coolant temperature above 60°C (140°F).
* Vehicle speed above 2 mph.
* Throttle off idle, above 1.6%.

Test Description: Number(s) below refer to circled number(s) on the diagnostic chart.
1. Checks to see if the solenoid valve is opened or closed. The solenoid valve is normally closed and should open when the diagnostic "test" terminal is grounded.
2. Completes functional check by grounding the diagnostic "test" terminal. This should energize the solenoid valve and allow the vacuum to drop (purge "ON").
3. Checks for a complete circuit. Normally, there is ignition voltage on CKT 541 and the ECM provides a ground on CKT 428. A shorted solenoid valve could cause an open circuit in the ECM.

Diagnostic Aids:

Normal operation of the EVAP canister purge solenoid valve is described as follows:

With the ignition "ON," engine "OFF," diagnostic "test" terminal grounded, the EVAP canister purge solenoid valve should be energized.

5.7L (VIN P) ENGINE — COMPONENT DIAGNOSTIC CHART — 1993 CAMARO AND FIREBIRD

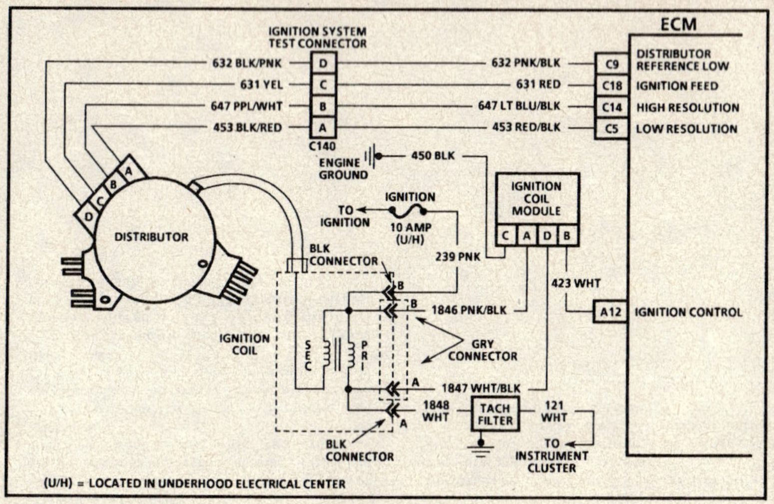

CHART C-4
(Page 1 of 2)
DISTRIBUTOR IGNITION SYSTEM CHECK
5.7L (VIN P) "F" CARLINE (MFI)

Test Description: Number(s) below refer to circled number(s) on the diagnostic chart.

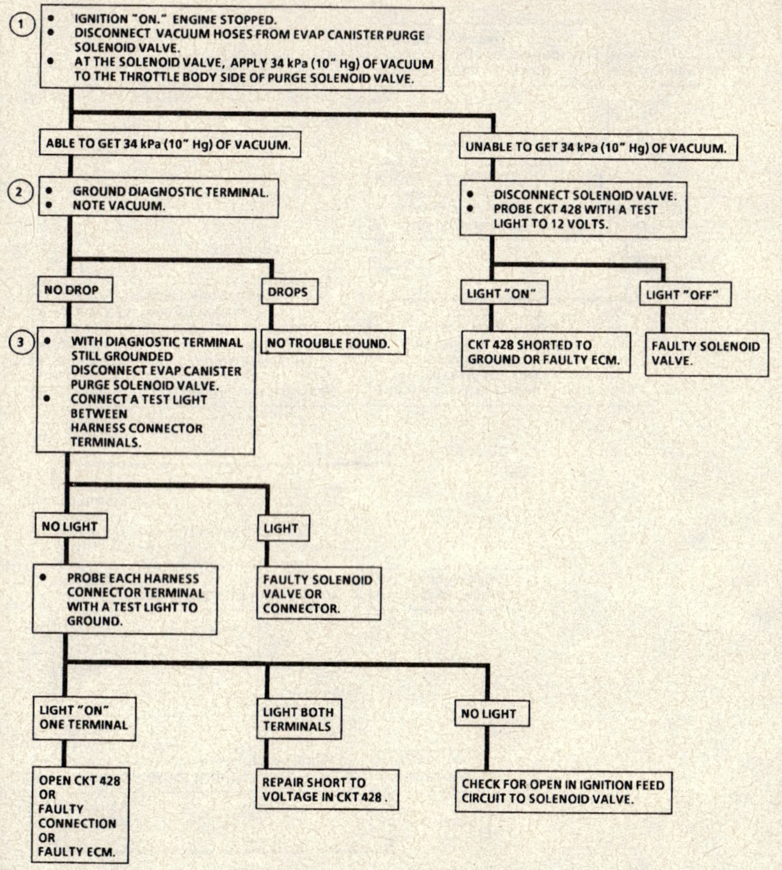

The battery should be fully charged prior to any tests.

1. To disconnect the tachometer lead, it will be necessary to disconnect the gray ignition coil connector from the ignition coil and jumper the "B" terminal of the ignition coil and jumper the "B" terminal of the harness connector to the "B" terminal of the ignition coil using a fused jumper and terminal test adapter kit J 35616-A.

   Checks for proper output from the distributor ignition system. The spark tester requires a minimum of 25,000 volts to fire. This check can be used in case of an ignition miss, because the system may provide enough voltage to run the engine but not enough to fire a spark plug under heavy load.

2. This test will separate the distributor cap, rotor and ignition wires from the ignition coil to help identify a secondary ignition system problem.

3. This will determine if the proper available voltage exists in the primary ignition circuit.

4. This check will begin to determine if the ECM is providing a signal to the ignition coil module or not. If the ECM is not providing a signal to the ignition coil module, the problem exists between the distributor and ECM.

5. A grounded coil circuit will cause the ignition fuse to open, and a no-start condition will occur.

Diagnostic Aids:

An ignition coil wire that is open or shorted to ground can cause a no start condition.

5.7L (VIN P) ENGINE — COMPONENT DIAGNOSTIC CHART — 1993 CAMARO AND FIREBIRD

CHART C-3
EVAPORATIVE EMISSION (EVAP) CANISTER PURGE SOLENOID VALVE CHECK
5.7L (VIN P) "F" CARLINE (MFI)

5.7L (VIN P) ENGINE — COMPONENT DIAGNOSTIC CHART — 1993 CAMARO AND FIREBIRD

CHART C-4
(Page 1 of 2)
DISTRIBUTOR IGNITION SYSTEM CHECK
5.7L (VIN P) "F" CARLINE (MFI)

1.
- PERFORM ON-BOARD DIAGNOSTIC (OBD) SYSTEM CHECK BEFORE PROCEEDING WITH THIS TEST. (DISCONNECT TACHOMETER BEFORE PROCEEDING WITH THE TEST.)
- CHECK SPARK AT PLUG WITH SPARK TESTER J 26792 (ST-125) WHILE CRANKING (IF NO SPARK ON ONE WIRE, CHECK A SECOND WIRE) A FEW SPARKS AND THEN NOTHING IS CONSIDERED NO SPARK.

NO SPARK

SPARK

2.
- CHECK FOR SPARK AT IGNITION COIL WIRE WITH TESTER WHILE CRANKING. (LEAVE SPARK TESTER CONNECTED TO IGNITION COIL WIRE FOR STEPS 3-6.)

CHECK FUEL, SPARK PLUGS, ETC. "SYMPTOMS."

NO SPARK

SPARK

3.
- DISCONNECT IGNITION COIL MODULE CONNECTOR.
- IGNITION "ON."
- WITH DVM, CHECK VOLTAGE FROM HARNESS TERMINALS "A" AND "D" TO GROUND.

INSPECT CAP FOR CRACKS, SECONDARY WIRES FOR OPENS; SHORTS; OR POOR CONNECTIONS. IF OK, REPLACE DISTRIBUTOR.

BOTH TERMINALS 10 VOLTS OR MORE

BOTH TERMINALS UNDER 10 VOLTS

EITHER TERMINAL UNDER 10 VOLTS

4.
- WITH DVM ON AC SCALE, CONNECTED FROM IGNITION COIL MODULE CONNECTOR TERMINAL "B" TO GROUND, CRANK ENGINE AND OBSERVE VOLTAGE. IS VOLTAGE BETWEEN 1 AND 4 VOLTS?

5.
FAULTY IGNITION FEED CIRCUIT TO IGNITION COIL
OR
GROUNDED EXTERNAL COIL CIRCUIT
OR
FAULTY IGNITION COIL.

FAULTY CIRCUIT FROM TERMINAL TO COIL
OR
FAULTY COIL CONNECTION
OR
FAULTY COIL.

YES

NO

PROBE IGNITION COIL MODULE CONNECTOR TERMINAL "C" WITH TEST LIGHT TO B +. IS LIGHT ON?

TO CHART C-4 2 OF 2.

NO

YES

OPEN IGNITION COIL MODULE GROUND CIRCUIT.

FAULTY IGNITION COIL MODULE CONNECTION
OR
FAULTY IGNITION COIL MODULE.

"AFTER REPAIRS," CONFIRM "CLOSED LOOP" OPERATION AND NO MIL (SERVICE ENGINE SOON).

5.7L (VIN P) ENGINE — COMPONENT DIAGNOSTIC CHART — 1993 CAMARO AND FIREBIRD

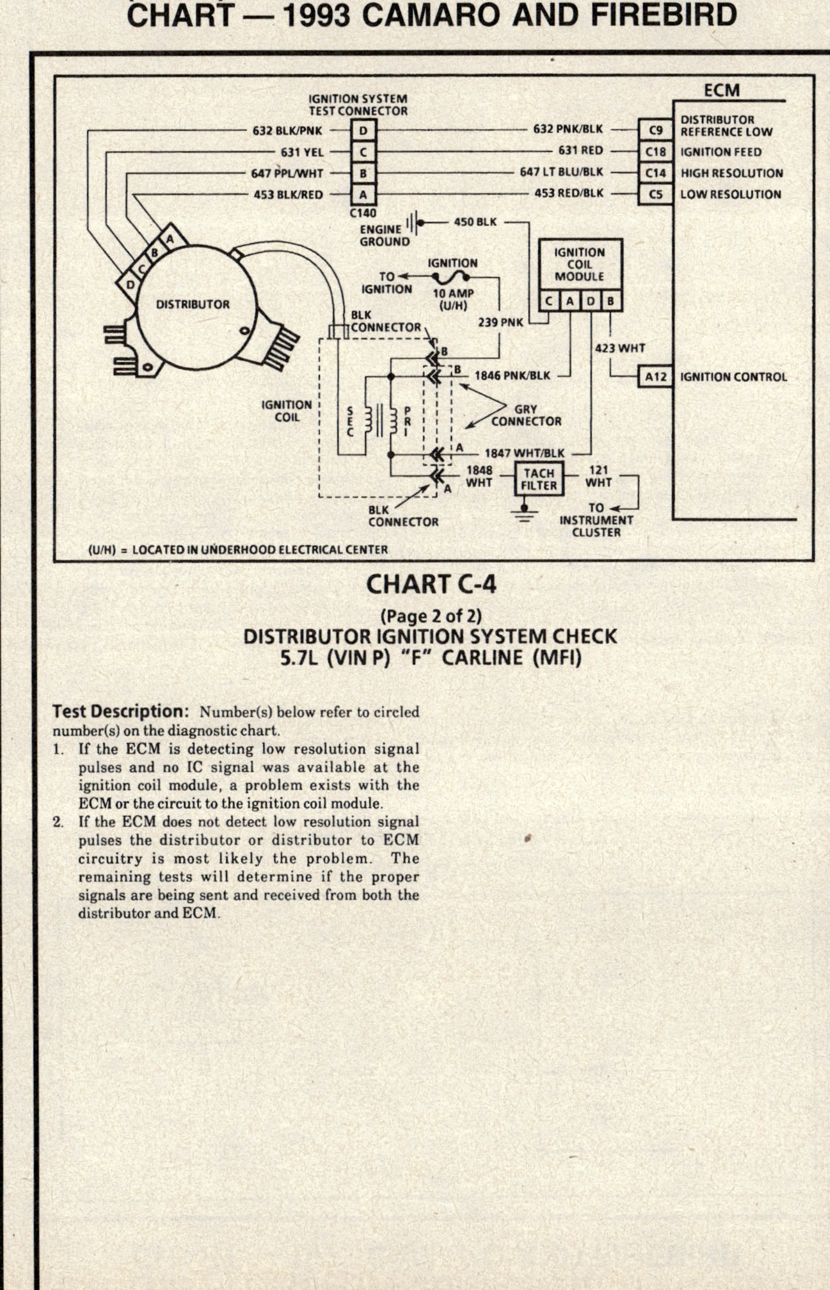

CHART C-4
(Page 2 of 2)
DISTRIBUTOR IGNITION SYSTEM CHECK
5.7L (VIN P) "F" CARLINE (MFI)

Test Description: Number(s) below refer to circled number(s) on the diagnostic chart.
1. If the ECM is detecting low resolution signal pulses and no IC signal was available at the ignition coil module, a problem exists with the ECM or the circuit to the ignition coil module.
2. If the ECM does not detect low resolution signal pulses the distributor or distributor to ECM circuitry is most likely the problem. The remaining tests will determine if the proper signals are being sent and received from both the distributor and ECM.

5.7L (VIN P) ENGINE — COMPONENT DIAGNOSTIC CHART — 1993 CAMARO AND FIREBIRD

CHART C-4
(Page 2 of 2)
DISTRIBUTOR IGNITION SYSTEM CHECK
5.7L (VIN P) "F" CARLINE (MFI)

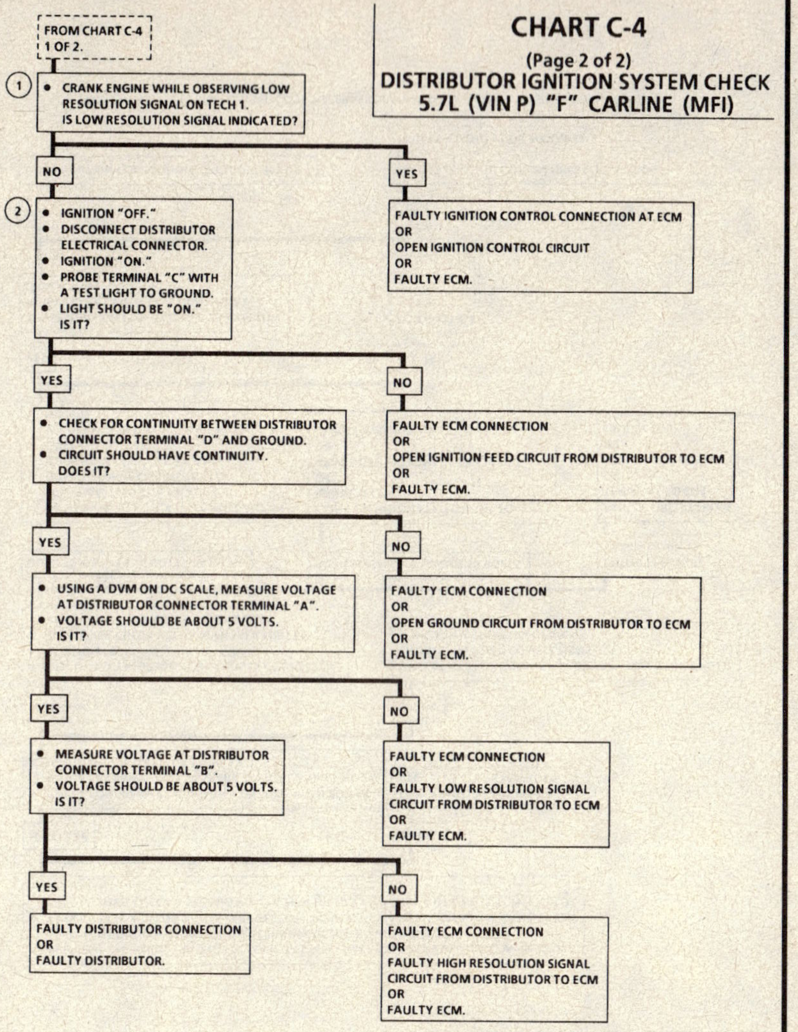

"AFTER REPAIRS," CONFIRM "CLOSED LOOP" OPERATION AND NO MIL (SERVICE ENGINE SOON).

5.7L (VIN P) ENGINE — COMPONENT DIAGNOSTIC CHART — 1993 CAMARO AND FIREBIRD

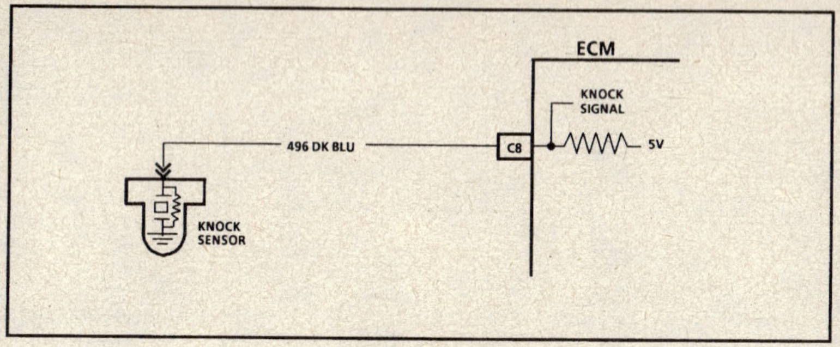

CHART C-5
KNOCK SENSOR (KS) SYSTEM CHECK
5.7L (VIN P) "F" CARLINE (MFI)

Circuit Description:

The knock sensor system is used to detect engine detonation. The ECM will retard the spark timing based on a signal being received from the knock sensor. The knock sensor produces an AC voltage which is sent to the ECM. The amount of AC voltage produced by the sensor is determined by the amount of engine knock. The circuitry within the knock sensor causes the ECMs 5 volts to be pulled down so that under a no engine knock condition, CKT 496 would measure about 2.5 volts.

Test Description: Number(s) below refer to circled number(s) on the diagnostic chart.

1. With engine idling, there should not be a knock signal present at the ECM, because detonation is not likely under a no load condition.

2. Tapping on the engines right exhaust manifold should simulate a knock signal to determine if the sensor is capable of detecting detonation. If no knock is detected, try tapping on engine block closer to sensor before replacing a sensor.

3. If the engine has an internal problem which is creating a knock, the knock sensor may be responding to the internal failure.

4. This test determines if the knock sensor is faulty or if the KS portion of the PROM is faulty. If it is determined that the PROM is faulty, be sure that it is properly installed and latched into place. If not properly installed, repair and retest.

Diagnostic Aids:

While observing knock signal on the Tech 1, there should be an indication that knock is present when detonation can be heard. Detonation is most likely to occur under high engine load conditions.

5.7L (VIN P) ENGINE — COMPONENT DIAGNOSTIC CHART — 1993 CAMARO AND FIREBIRD

CHART C-5
KNOCK SENSOR (KS) SYSTEM CHECK
5.7L (VIN P) "F" CARLINE (MFI)

1. • IF DTC 43 IS SET, USE THE DTC CHART.
   • ENGINE MUST BE IDLING AT NORMAL OPERATING TEMPERATURE.
   • USE SCAN TOOL TO OBSERVE KNOCK SIGNAL.
   IS KNOCK INDICATED?

NO →

2. • TAP ON RIGHT EXHAUST MANIFOLD WHILE OBSERVING KNOCK SIGNAL.
   • SCAN TOOL SHOULD INDICATE KNOCK WHILE TAPPING ON MANIFOLD.
   DOES IT?

YES →

3. IF AN ENGINE KNOCK CAN BE HEARD, REPAIR THE BASIC ENGINE PROBLEM. IF NO AUDIBLE KNOCK IS HEARD, FOLLOW THE STEPS:
   • DISCONNECT KNOCK SENSOR.
   • CONNECT VOLTMETER TO KNOCK SENSOR AND ENGINE GROUND.
   • VOLTMETER ON AC SCALE.
   IS A SIGNAL INDICATED ON VOLTMETER?

NO →

4. • DISCONNECT KNOCK SENSOR.
   • CONNECT VOLTMETER TO KNOCK SENSOR AND ENGINE GROUND.
   • VOLTMETER ON AC SCALE.
   • TAP ON ENGINE BLOCK NEAR SENSOR.
   IS A SIGNAL INDICATED ON VOLTMETER WHILE TAPPING ON ENGINE BLOCK?

| (2 NO) REPLACE PROM OR ECM. | (2 YES) REPLACE KNOCK SENSOR. |

(2 YES) SYSTEM IS OPERATING PROPERLY. REFER TO "DIAGNOSTIC AIDS" ON FACING PAGE.

(3 NO) CHECK CKT 496 FOR BEING NEAR A SPARK PLUG WIRE OR A FAULTY ECM CONNECTION OR FAULTY ECM OR PROM.

(3 YES) REPLACE KNOCK SENSOR.

"*AFTER REPAIRS," CONFIRM "CLOSED LOOP" OPERATION AND NO MIL (SERVICE ENGINE SOON).

5.7L (VIN P) ENGINE — COMPONENT DIAGNOSTIC CHART — 1993 CAMARO AND FIREBIRD

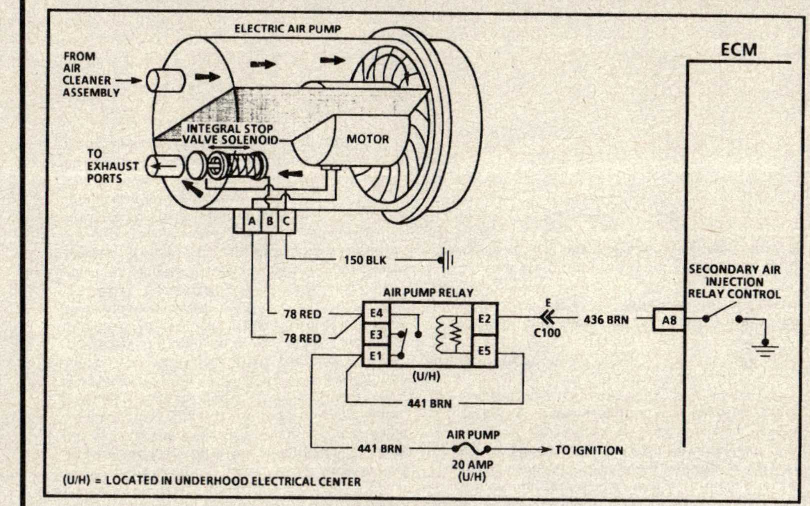

CHART C-6
SECONDARY AIR INJECTION (AIR) SYSTEM CHECK
5.7L (VIN P) "F" CARLINE (MFI)

Circuit Description:
    During cold starts, above 15℃ (59°F), the ECM completes the ground circuit to the air pump relay which enables the air pump and integral stop valve. Air is directed to the exhaust ports whenever the engine is started. Whenever the fuel system goes to "Closed Loop," or the air pump has been "ON" for greater than 240 seconds, the ECM opens the ground circuit to the air pump relay, and the air pump is de-energized and the integral stop valve closes.

Test Description: Number(s) below refer to circled number(s) on the diagnostic chart.
1. The ECM should energize the air pump when engine coolant temperature is greater than 15℃ (59°F) and fuel system not in "Closed Loop." If any Diagnostic Trouble Codes (DTCs) are present the ECM will not energize the AIR pump.
2. The ECM should de-energize the air pump relay when engine speed exceeds 2825 RPM.
3. Air pump relay is located in the underhood electrical center.

4. Selecting "Miscellaneous Test" then outputs and then AIR system on the Tech 1 with the ignition "ON" and the engine not running should enable the AIR pump and integral stop valve.
    There is a possibility that the AIR pump motor will be operating, but no air is being distributed to the AIR system. The integral stop valve may not be opening or a restriction in the AIR hoses or pipes may be the cause.

5.7L (VIN P) ENGINE — COMPONENT DIAGNOSTIC CHART — 1993 CAMARO AND FIREBIRD

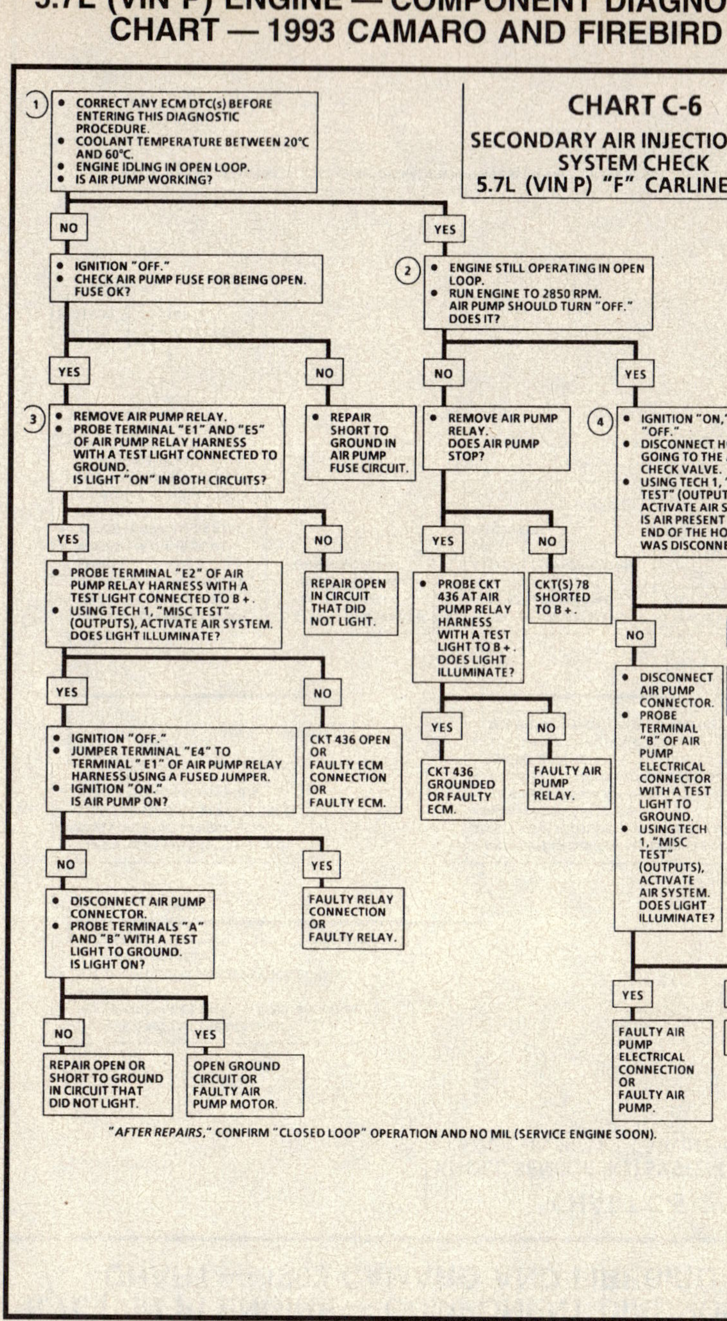

5.7L (VIN P) ENGINE — COMPONENT DIAGNOSTIC CHART — 1993 CAMARO AND FIREBIRD

CHART C-7

(Page 1 of 2)
EXHAUST GAS RECIRCULATION (EGR) SYSTEM CHECK
5.7L (VIN P) "F" CARLINE (MFI)

Circuit Description:

The Exhaust Gas Recirculation (EGR) valve is controlled by a normally closed solenoid valve. The ECM turns the solenoid valve "ON" to allow vacuum to pass to the EGR, and turns the solenoid valve "OFF" to prevent EGR operation.

The duty cycle is calculated by the ECM based on information from the ECT, IAT, TP sensor, and MAP sensors. Also, engine RPM and the PNP switch input affect EGR. There should be no EGR when in park or neutral, TP sensor below a calibrated value, or TP sensor indicating WOT.

With the ignition "ON" and engine "OFF," the EGR solenoid valve is de-energized. The solenoid valve, however, should be energized, if the diagnostic "test" terminal is grounded with the ignition "ON" and engine "OFF."

Test Description: Number(s) below refer to circled number(s) on the diagnostic chart.

1. **Intake Passage:** Shut "OFF" engine and remove the EGR valve from the manifold. Plug the exhaust side hole with a shop rag or suitable stopper. Leaving the intake side hole open, attempt to start the engine. If the engine runs at a very high idle (up to 3000 RPM is possible) or starts and stalls, the EGR passages are not restricted. If the engine starts and idles normally, the EGR intake side passage in the intake manifold is restricted.
Exhaust Passage: With EGR valve still removed, plug the intake side hole with a suitable stopper. With the exhaust side hole open, check for the presence of exhaust gas. If no exhaust gas is present, the EGR exhaust side passage in the intake manifold is restricted.

2. The vacuum at the gage may or may not <u>slowly</u> bleed off. It is important that the gage is able to read the amount of vacuum being applied.

3. When the diagnostic "test" terminal is grounded, the vacuum gage should bleed off through a vent in the solenoid valve. The pump gage may or may not bleed off but this does not indicate a problem.

4. This test will determine if the electrical control part of the system is at fault or if the connector or solenoid valve is at fault.

Diagnostic Aids:

Vacuum lines should be thoroughly checked for proper routing. Refer to "Vehicle Emission Control Information" label.

Suction from shop exhaust hoses can alter backpressure and may affect the functional check of the EGR valve.

5.7L (VIN P) ENGINE — COMPONENT DIAGNOSTIC CHART — 1993 CAMARO AND FIREBIRD

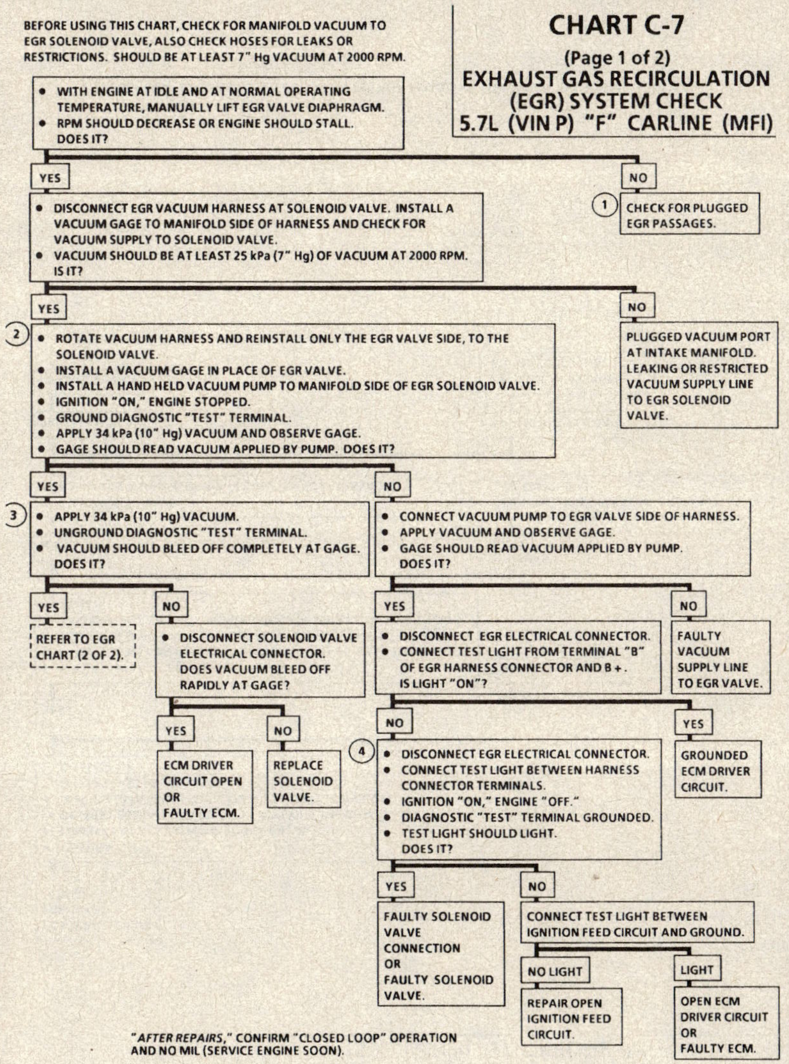
CHART C-7
(Page 1 of 2)
EXHAUST GAS RECIRCULATION (EGR) SYSTEM CHECK
5.7L (VIN P) "F" CARLINE (MFI)
BEFORE USING THIS CHART, CHECK FOR MANIFOLD VACUUM TO EGR SOLENOID VALVE, ALSO CHECK HOSES FOR LEAKS OR RESTRICTIONS. SHOULD BE AT LEAST 7" Hg VACUUM AT 2000 RPM.
• WITH ENGINE AT IDLE AND AT NORMAL OPERATING TEMPERATURE, MANUALLY LIFT EGR VALVE DIAPHRAGM.
• RPM SHOULD DECREASE OR ENGINE SHOULD STALL. DOES IT?
YES
NO
1
CHECK FOR PLUGGED EGR PASSAGES.
• DISCONNECT EGR VACUUM HARNESS AT SOLENOID VALVE. INSTALL A VACUUM GAGE TO MANIFOLD SIDE OF HARNESS AND CHECK FOR VACUUM SUPPLY TO SOLENOID VALVE.
• VACUUM SHOULD BE AT LEAST 25 kPa (7" Hg) OF VACUUM AT 2000 RPM. IS IT?
YES
NO
PLUGGED VACUUM PORT AT INTAKE MANIFOLD. LEAKING OR RESTRICTED VACUUM SUPPLY LINE TO EGR SOLENOID VALVE.
2
• ROTATE VACUUM HARNESS AND REINSTALL ONLY THE EGR VALVE SIDE, TO THE SOLENOID VALVE.
• INSTALL A VACUUM GAGE IN PLACE OF EGR VALVE.
• INSTALL A HAND HELD VACUUM PUMP TO MANIFOLD SIDE OF EGR SOLENOID VALVE.
• IGNITION "ON," ENGINE STOPPED.
• GROUND DIAGNOSTIC "TEST" TERMINAL.
• APPLY 34 kPa (10" Hg) VACUUM AND OBSERVE GAGE.
• GAGE SHOULD READ VACUUM APPLIED BY PUMP. DOES IT?
YES
NO
3
• APPLY 34 kPa (10" Hg) VACUUM.
• UNGROUND DIAGNOSTIC "TEST" TERMINAL.
• VACUUM SHOULD BLEED OFF COMPLETELY AT GAGE. DOES IT?
• CONNECT VACUUM PUMP TO EGR VALVE SIDE OF HARNESS.
• APPLY VACUUM AND OBSERVE GAGE.
• GAGE SHOULD READ VACUUM APPLIED BY PUMP. DOES IT?
YES
NO
REFER TO EGR CHART (2 OF 2).
• DISCONNECT SOLENOID VALVE ELECTRICAL CONNECTOR. DOES VACUUM BLEED OFF RAPIDLY AT GAGE?
YES
NO
ECM DRIVER CIRCUIT OPEN OR FAULTY ECM.
REPLACE SOLENOID VALVE.
YES
NO
• DISCONNECT EGR ELECTRICAL CONNECTOR.
• CONNECT TEST LIGHT FROM TERMINAL "B" OF EGR HARNESS CONNECTOR AND B+. IS LIGHT "ON"?
FAULTY VACUUM SUPPLY LINE TO EGR VALVE.
NO
YES
4
• DISCONNECT EGR ELECTRICAL CONNECTOR.
• CONNECT TEST LIGHT BETWEEN HARNESS CONNECTOR TERMINALS.
• IGNITION "ON," ENGINE "OFF."
• DIAGNOSTIC "TEST" TERMINAL GROUNDED.
• TEST LIGHT SHOULD LIGHT. DOES IT?
GROUNDED ECM DRIVER CIRCUIT.
YES
NO
FAULTY SOLENOID VALVE CONNECTION OR FAULTY SOLENOID VALVE.
CONNECT TEST LIGHT BETWEEN IGNITION FEED CIRCUIT AND GROUND.
NO LIGHT
LIGHT
REPAIR OPEN IGNITION FEED CIRCUIT.
OPEN ECM DRIVER CIRCUIT OR FAULTY ECM.

"AFTER REPAIRS," CONFIRM "CLOSED LOOP" OPERATION AND NO MIL (SERVICE ENGINE SOON).

5.7L (VIN P) ENGINE — COMPONENT DIAGNOSTIC CHART — 1993 CAMARO AND FIREBIRD

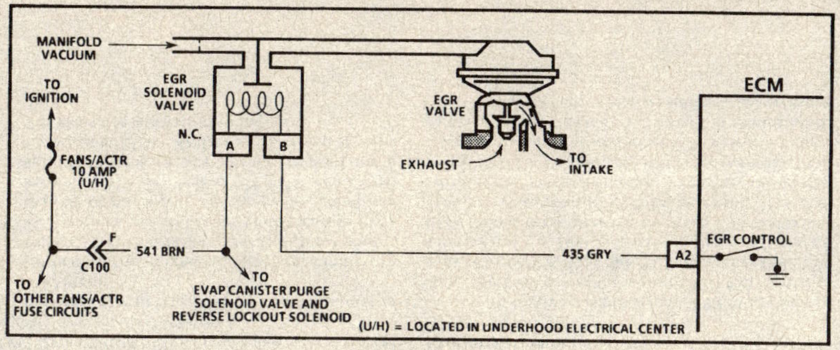
MANIFOLD VACUUM
TO IGNITION
FANS/ACTR 10 AMP (U/H)
EGR SOLENOID VALVE
N.C.
A
B
EGR VALVE
EXHAUST
TO INTAKE
ECM
F
C100
541 BRN
TO OTHER FANS/ACTR FUSE CIRCUITS
TO EVAP CANISTER PURGE SOLENOID VALVE AND REVERSE LOCKOUT SOLENOID
435 GRY
A2
EGR CONTROL
(U/H) = LOCATED IN UNDERHOOD ELECTRICAL CENTER

CHART C-7
(Page 2 of 2)

EXHAUST GAS RECIRCULATION (EGR) SYSTEM CHECK
5.7L (VIN P) "F" CARLINE (MFI)

Circuit Description:

The Exhaust Gas Recirculation (EGR) valve is controlled by a normally closed solenoid valve. The ECM turns the solenoid valve "ON" to allow vacuum to pass to the EGR, and turns the solenoid valve "OFF" to prevent EGR operation.

The duty cycle is calculated by the ECM based on information from the TP sensor, ECT, IAT, and MAP sensors. Also, engine RPM and the PNP switch input affect EGR. There should be no EGR when in park or neutral, TP sensor below a calibrated value, or TP sensor indicating WOT.

With the ignition "ON" and engine "OFF," the EGR solenoid valve is de-energized. The solenoid valve, however, should be energized, if the diagnostic "test" terminal is grounded with the ignition "ON" and engine "OFF."

Test Description: Number(s) below refer to circled number(s) on the diagnostic chart.

1. The remaining test checks the ability of the EGR valve to interact with the exhaust system. This system uses a negative backpressure EGR valve which should hold vacuum with engine "OFF."
2. When engine is started, exhaust backpressure at the base of the EGR valve should open the valve's internal bleed and vent the applied vacuum allowing the valve to seat.

Diagnostic Aids:

Suction from shop exhaust hoses can alter backpressure and may affect the functional check of the EGR valve.

During normal EGR valve operation, the movement of the EGR pintle is small. It is important to determine whether the valve pintle moves and not how much it moves.

5.7L (VIN P) ENGINE — COMPONENT DIAGNOSTIC CHART — 1993 CAMARO AND FIREBIRD

CHART C-7
(Page 2 of 2)
EXHAUST GAS RECIRCULATION (EGR) SYSTEM CHECK
5.7L (VIN P) "F" CARLINE (MFI)

CONTINUED FROM EGR CHART (1 OF 2).

1.
- IGNITION "OFF."
- CONNECT A VACUUM PUMP TO EGR VALVE.
- OBSERVE EGR DIAPHRAGM WHILE APPLYING VACUUM.
- DIAPHRAGM SHOULD MOVE FREELY AND HOLD VACUUM FOR AT LEAST 20 SECONDS.
DOES IT?

YES →

2.
- APPLY 34 kPa (10" Hg) VACUUM TO EGR VALVE.
- START ENGINE AND IMMEDIATELY OBSERVE GAGE ON VACUUM PUMP.
- EGR VALVE DIAPHRAGM SHOULD MOVE TO SEATED POSITION AND VACUUM SHOULD DROP FROM PUMP GAGE WHILE STARTING ENGINE.
DOES IT?

NO → REPLACE EGR VALVE.

NO →
- REMOVE EGR VALVE.
- CHECK FOR PLUGGED OR RESTRICTED EXHAUST PASSAGES.

PASSAGES OK → REPLACE EGR VALVE.

PASSAGES NOT OK →
- CLEAN PASSAGES.
- RE-CHECK EGR VALVE.

YES → EGR CIRCUIT IS OPERATING PROPERLY. REFER TO SYMPTOMS

"AFTER REPAIRS," CONFIRM "CLOSED LOOP" OPERATION AND NO MIL (SERVICE ENGINE SOON).

5.7L (VIN P) ENGINE — COMPONENT DIAGNOSTIC CHART — 1993 CAMARO AND FIREBIRD

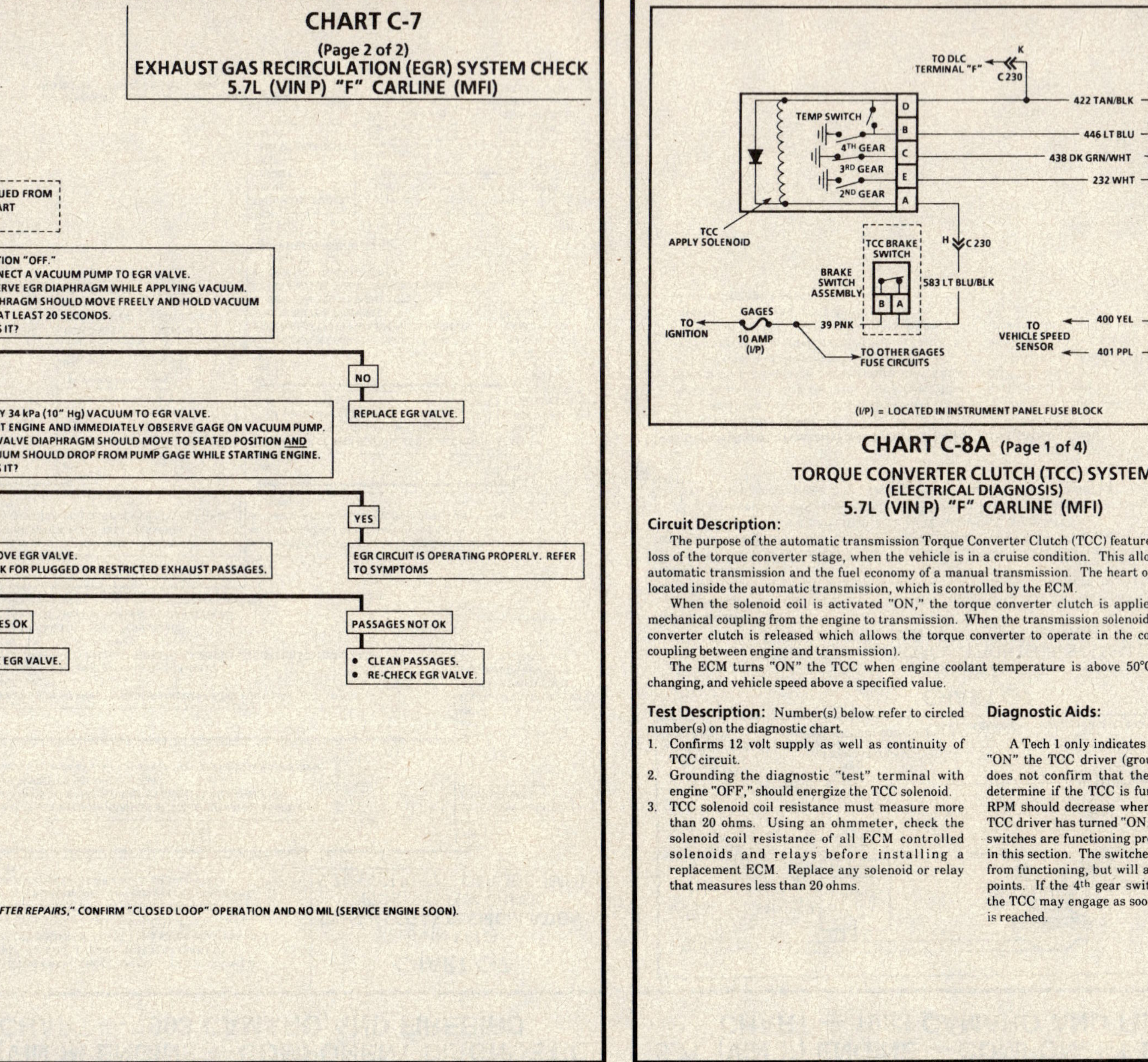

CHART C-8A (Page 1 of 4)
TORQUE CONVERTER CLUTCH (TCC) SYSTEM
(ELECTRICAL DIAGNOSIS)
5.7L (VIN P) "F" CARLINE (MFI)

Circuit Description:

The purpose of the automatic transmission Torque Converter Clutch (TCC) feature is to eliminate the power loss of the torque converter stage, when the vehicle is in a cruise condition. This allows the convenience of the automatic transmission and the fuel economy of a manual transmission. The heart of the system is a solenoid, located inside the automatic transmission, which is controlled by the ECM.

When the solenoid coil is activated "ON," the torque converter clutch is applied, which results in 100% mechanical coupling from the engine to transmission. When the transmission solenoid is deactivated, the torque converter clutch is released which allows the torque converter to operate in the conventional manner (fluid coupling between engine and transmission).

The ECM turns "ON" the TCC when engine coolant temperature is above 50°C (122°F), TP sensor not changing, and vehicle speed above a specified value.

Test Description: Number(s) below refer to circled number(s) on the diagnostic chart.

1. Confirms 12 volt supply as well as continuity of TCC circuit.
2. Grounding the diagnostic "test" terminal with engine "OFF," should energize the TCC solenoid.
3. TCC solenoid coil resistance must measure more than 20 ohms. Using an ohmmeter, check the solenoid coil resistance of all ECM controlled solenoids and relays before installing a replacement ECM. Replace any solenoid or relay that measures less than 20 ohms.

Diagnostic Aids:

A Tech 1 only indicates when the ECM has turned "ON" the TCC driver (grounded CKT 422) but this does not confirm that the TCC has engaged. To determine if the TCC is functioning properly, engine RPM should decrease when the Tech 1 indicates the TCC driver has turned "ON." To determine if the gear switches are functioning properly, perform the checks in this section. The switches will not prevent the TCC from functioning, but will affect TCC lock and unlock points. If the 4th gear switch circuit is always open, the TCC may engage as soon as sufficient oil pressure is reached.

5.7L (VIN P) ENGINE — COMPONENT DIAGNOSTIC CHART — 1993 CAMARO AND FIREBIRD

5.7L (VIN P) ENGINE — COMPONENT DIAGNOSTIC CHART — 1993 CAMARO AND FIREBIRD

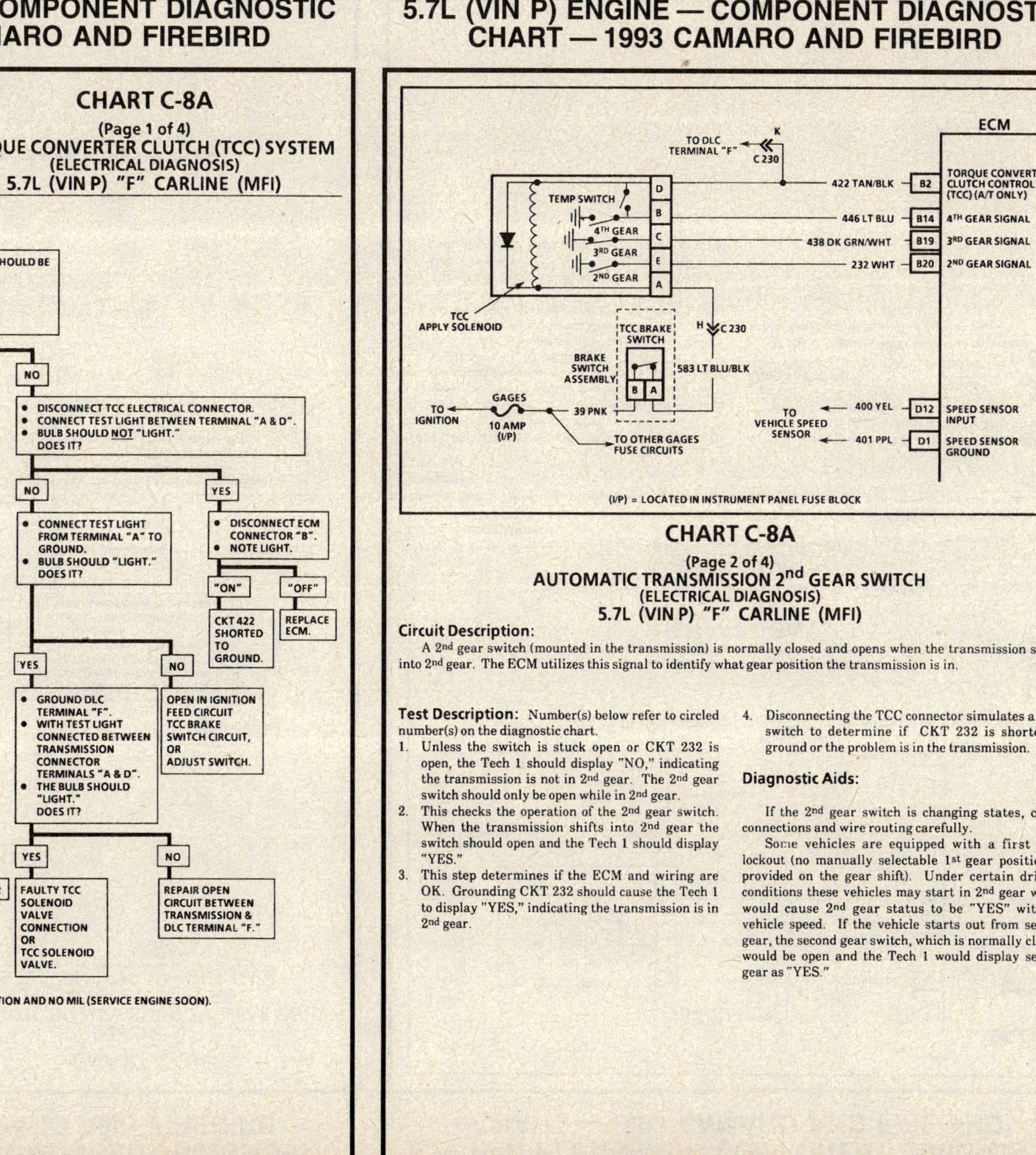

CHART C-8A

(Page 2 of 4)
AUTOMATIC TRANSMISSION 2nd GEAR SWITCH
(ELECTRICAL DIAGNOSIS)
5.7L (VIN P) "F" CARLINE (MFI)

Circuit Description:

A 2nd gear switch (mounted in the transmission) is normally closed and opens when the transmission shifts into 2nd gear. The ECM utilizes this signal to identify what gear position the transmission is in.

Test Description: Number(s) below refer to circled number(s) on the diagnostic chart.

1. Unless the switch is stuck open or CKT 232 is open, the Tech 1 should display "NO," indicating the transmission is not in 2nd gear. The 2nd gear switch should only be open while in 2nd gear.
2. This checks the operation of the 2nd gear switch. When the transmission shifts into 2nd gear the switch should open and the Tech 1 should display "YES."
3. This step determines if the ECM and wiring are OK. Grounding CKT 232 should cause the Tech 1 to display "YES," indicating the transmission is in 2nd gear.
4. Disconnecting the TCC connector simulates a open switch to determine if CKT 232 is shorted to ground or the problem is in the transmission.

Diagnostic Aids:

If the 2nd gear switch is changing states, check connections and wire routing carefully.

Some vehicles are equipped with a first gear lockout (no manually selectable 1st gear position is provided on the gear shift). Under certain driving conditions these vehicles may start in 2nd gear which would cause 2nd gear status to be "YES" with no vehicle speed. If the vehicle starts out from second gear, the second gear switch, which is normally closed, would be open and the Tech 1 would display second gear as "YES."

5.7L (VIN P) ENGINE — COMPONENT DIAGNOSTIC CHART — 1993 CAMARO AND FIREBIRD

CHART C-8A
(Page 2 of 4)
AUTOMATIC TRANSMISSION 2nd GEAR SWITCH
(ELECTRICAL DIAGNOSIS)
5.7L (VIN P) "F" CARLINE (MFI)

CHECKS MADE ON THIS PAGE WILL NOT PREVENT THE TCC FROM WORKING, BUT WILL AFFECT ENGAGEMENT OR DISENGAGEMENT POINTS.

(1)
- IGNITION "ON," ENGINE "OFF."
  DOES "SCAN" INDICATE TRANSMISSION IS IN 2nd GEAR?

NO →

YES →

(2)
- RAISE DRIVE WHEELS.
- START ENGINE.
- SHIFT VEHICLE INTO OVERDRIVE.
- INCREASE SPEED SLOWLY UNTIL TRANSMISSION SHIFTS INTO 2nd GEAR.
  DOES "SCAN" INDICATE TRANSMISSION IS IN 2nd GEAR?

(3)
- IGNITION "ON," ENGINE "OFF."
- DISCONNECT TRANSMISSION ELECTRICAL CONNECTOR.
- JUMPER HARNESS TERMINAL "E" TO GROUND.
  DOES "SCAN" INDICATE TRANSMISSION IS IN 2nd GEAR?

YES → 2nd GEAR SWITCH OK. REFER TO "DIAGNOSTIC AIDS" ON FACING PAGE.

NO →
(4)
- DISCONNECT TCC ELECTRICAL CONNECTOR.
  DOES "SCAN" INDICATE TRANSMISSION IS IN 2nd GEAR?

YES → FAULTY ECM CONNECTION OR CKT 232 OPEN OR FAULTY ECM.

NO → FAULTY 2nd GEAR SWITCH OR CKT 232 OPEN INTERNALLY IN TRANSMISSION.

YES → GROUNDED INTERNALLY IN TRANSMISSION OR FAULTY 2nd GEAR SWITCH IN TRANSMISSION.

NO → GROUNDED CKT 232 OR FAULTY ECM.

"AFTER REPAIRS," CONFIRM "CLOSED LOOP" OPERATION AND NO MIL (SERVICE ENGINE SOON).

5.7L (VIN P) ENGINE — COMPONENT DIAGNOSTIC CHART — 1993 CAMARO AND FIREBIRD

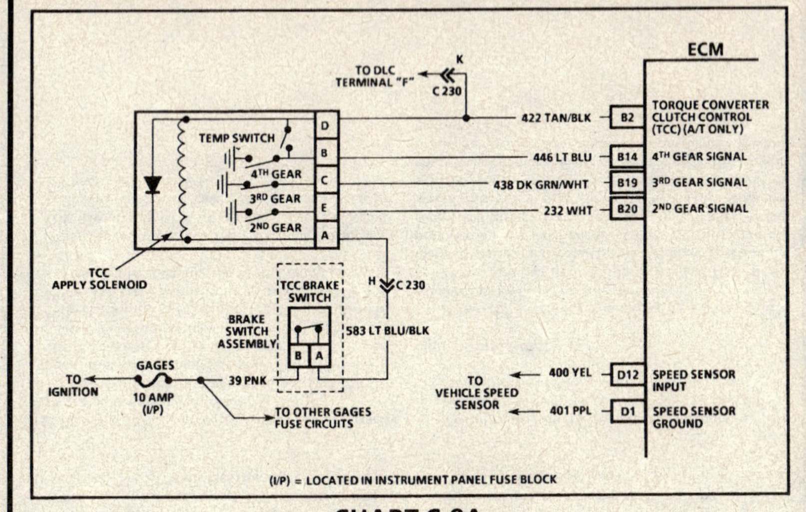

CHART C-8A
(Page 3 of 4)
AUTOMATIC TRANSMISSION 3rd GEAR SWITCH
(ELECTRICAL DIAGNOSIS)
5.7L (VIN P) "F" CARLINE (MFI)

Circuit Description:

A 3rd gear switch (mounted in the transmission) is normally closed and opens when the transmission shifts into 3rd gear. The ECM utilizes the 3rd gear switch signal for scheduling of the torque converter clutch engagement.

Test Description: Number(s) below refer to circled number(s) on the diagnostic chart.

1. Unless the switch is stuck open or CKT 438 is open, the Tech 1 should display "NO," indicating the transmission is not in 3rd gear. The 3rd gear switch should only be open while in 3rd gear.
2. This checks the operation of the 3rd gear switch. When the transmission shifts into 3rd gear the switch should open and the Tech 1 should display "YES."
3. This step determines if the ECM and wiring are OK. Grounding CKT 438 should cause the Tech 1 to display "NO," indicating the transmission is not in 3rd gear.
4. Disconnecting the TCC connector simulates an open switch to determine if CKT 438 is shorted to ground or the problem is in the transmission.

Diagnostic Aids:

If the 3rd gear switch is changing states, check connections and wire routing carefully.

5.7L (VIN P) ENGINE — COMPONENT DIAGNOSTIC CHART — 1993 CAMARO AND FIREBIRD

CHART C-8A

(Page 3 of 4)
AUTOMATIC TRANSMISSION 3rd GEAR SWITCH
(ELECTRICAL DIAGNOSIS)
5.7L (VIN P) "F" CARLINE (MFI)

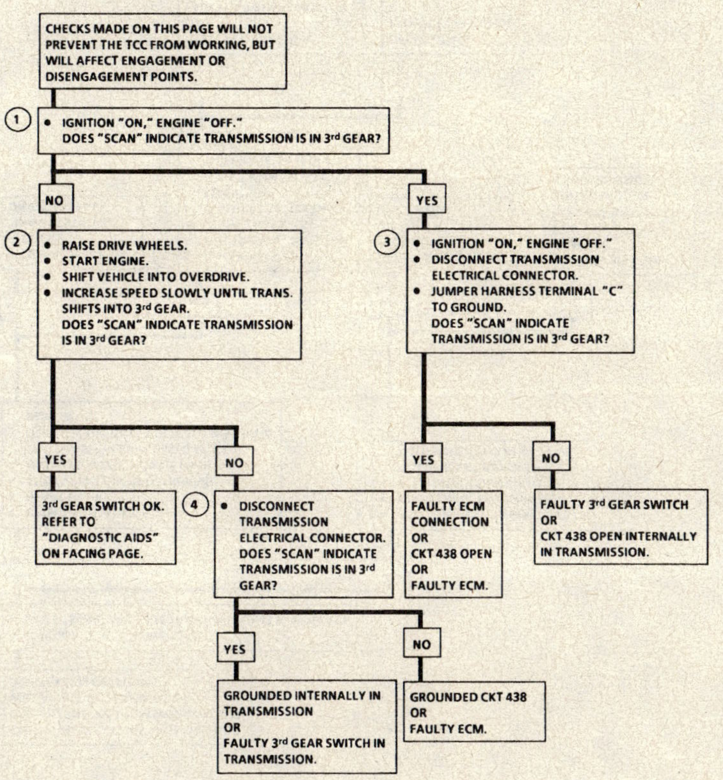

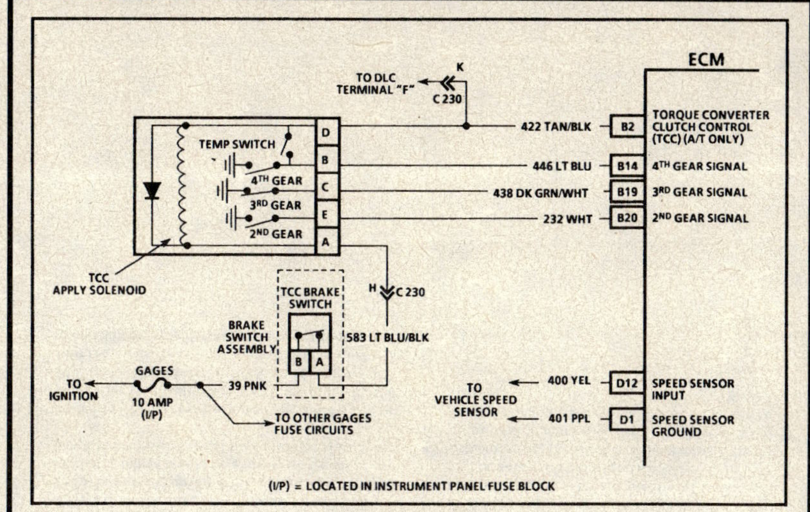

"AFTER REPAIRS," CONFIRM "CLOSED LOOP" OPERATION AND NO MIL (SERVICE ENGINE SOON).

5.7L (VIN P) ENGINE — COMPONENT DIAGNOSTIC CHART — 1993 CAMARO AND FIREBIRD

CHART C-8A

(Page 4 of 4)
AUTOMATIC TRANSMISSION 4th GEAR SWITCH
(ELECTRICAL DIAGNOSIS)
5.7L (VIN P) "F" CARLINE (MFI)

Circuit Description:

A 4th gear switch (mounted in the transmission) is normally open and closes when the transmission shifts into 4th gear. This switch is used by the ECM to modify TCC lock and unlock points, when in a 4-3 downshift maneuver. The temperature switch is normally open. Transmission must be in 4th gear for the temperature switch to apply TCC to cool transmission fluid. Temperature switch locks at about 155°C (279°F) and unlocks about 143°C (259°F).

Test Description: Number(s) below refer to circled number(s) on the diagnostic chart.

1. Unless the switch is stuck closed or CKT 446 is shorted to ground, the Tech 1 should display "NO," indicating the transmission is not in 4th gear. The 4th gear switch should only be closed while in 4th gear.

2. This checks the operation of the 4th gear switch. When the transmission shifts into 4th gear the switch should close and the Tech 1 should display "YES."

3. Disconnecting the TCC connector simulates an open switch to determine if CKT 446 is shorted to ground or the problem is in the transmission.

4. This step determines if the ECM and wiring are OK. Grounding CKT 446 should cause the Tech 1 to display "YES," indicating the transmission is in 4th gear.

Diagnostic Aids:

A road test may be necessary to verify the customer complaint. If the Tech 1 indicates TCC is turning "ON" and "OFF" erratically, check the state of the 4th gear switch to be sure they're not changing states under a steady throttle position. If the switch is changing states, check connections and wire routing carefully. Also, if the 4th gear switch is always closed, the TCC may engage as soon as sufficient oil pressure is reached.

5.7L (VIN P) ENGINE — COMPONENT DIAGNOSTIC CHART — 1993 CAMARO AND FIREBIRD

CHART C-8A

(Page 4 of 4)
AUTOMATIC TRANSMISSION 4th GEAR SWITCH
(ELECTRICAL DIAGNOSIS)
5.7L (VIN P) "F" CARLINE (MFI)

CHECKS MADE ON THIS PAGE WILL NOT PREVENT THE TCC FROM WORKING, BUT WILL AFFECT ENGAGEMENT OR DISENGAGEMENT POINTS.

1. • IGNITION "ON," ENGINE "OFF."
   DOES "SCAN" INDICATE TRANSMISSION IS IN 4th GEAR?

NO →

2. • RAISE DRIVE WHEELS.
   • START ENGINE.
   • SHIFT VEHICLE INTO OVERDRIVE.
   • INCREASE SPEED SLOWLY UNTIL TRANSMISSION SHIFTS INTO 4th GEAR.
   DOES "SCAN" INDICATE TRANSMISSION IS IN 4th GEAR?

YES →
4th GEAR SWITCH OK. REFER TO "DIAGNOSTIC AIDS" ON FACING PAGE.

NO →
4. • DISCONNECT TCC ELECTRICAL CONNECTOR.
   • JUMPER HARNESS TERMINAL "B" (CKT 446) TO GROUND.
   DOES "SCAN" INDICATE TRANSMISSION IS IN 4th GEAR?

YES →
FAULTY CONNECTION OR FAULTY 4th GEAR SWITCH IN TRANSMISSION.

NO →
OPEN CKT 446, FAULTY CONNECTION OR FAULTY ECM.

YES →

3. • IGNITION "ON," ENGINE "OFF."
   • DISCONNECT TRANSMISSION ELECTRICAL CONNECTOR
   DOES "SCAN" INDICATE TRANSMISSION IS IN 4th GEAR?

YES →
CKT 446 SHORTED TO GROUND OR FAULTY ECM.

NO →
FAULTY 4th GEAR SWITCH OR CKT 446 GROUNDED INTERNALLY IN TRANSMISSION.

"AFTER REPAIRS," CONFIRM "CLOSED LOOP" OPERATION AND NO MIL (SERVICE ENGINE SOON).

5.7L (VIN P) ENGINE — COMPONENT DIAGNOSTIC CHART — 1993 CAMARO AND FIREBIRD

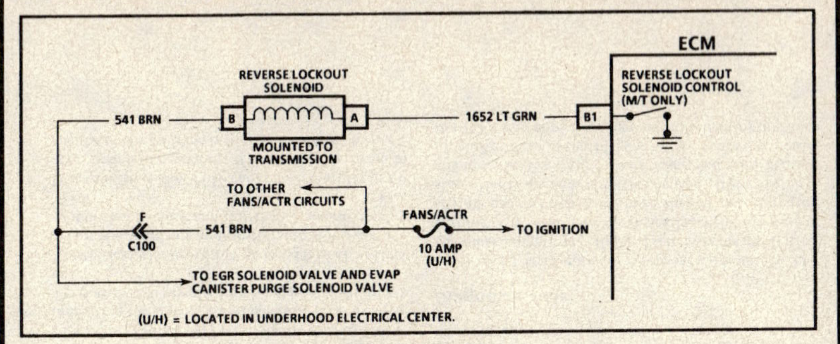

CHART C-8B

REVERSE LOCKOUT SYSTEM
(ELECTRICAL DIAGNOSIS)
5.7L (VIN P) "F" CARLINE (MFI)

Circuit Description:

The reverse lockout solenoid will be energized when the ECM grounds CKT 1652, which will allow the transmission to be shifted into reverse. The ECM will enable the reverse lockout solenoid whenever vehicle speed is below 4 mph. When vehicle speed is above 5 mph, the ECM will de-energize the reverse lockout solenoid, which will prohibit the transmission from being shifted into reverse.

Test Description: Number(s) below refer to circled number(s) on the diagnostic chart.

1. The reverse lockout solenoid will be enabled whenever vehicle speed is below 5 mph. If there is a fault in the vehicle speed sensor circuit and the ECM is not receiving VSS signal, the reverse lockout solenoid would be enabled at speeds above 5 mph (8 km/h) and the shifter could be placed into reverse.

2. Selecting "Field Service Mode" on the Tech 1 or grounding the diagnostic "test" terminal at the DLC with the ignition "ON" and engine not running, will de-energize the reverse lockout solenoid indicating normal system operation.

NOTICE: Do Not keep the diagnostic "test" terminal grounded for more than 1 minute, as damage could result to ECM driven solenoids.

3. Raise the vehicle enough to gain access to the reverse lockout solenoid electrical harness and still be able to turn the ignition "ON."

4. Disconnecting the ECM will determine if the ECM is internally shorted or CKT 1652 is shorted to ground causing the reverse lockout solenoid to be enabled.

5.7L (VIN P) ENGINE — COMPONENT DIAGNOSTIC CHART — 1993 CAMARO AND FIREBIRD

CHART C-8B

REVERSE LOCKOUT SYSTEM
(ELECTRICAL DIAGNOSIS)
5.7L (VIN P) "F" CARLINE (MFI)

① • IF DTC 24 IS PRESENT, USE THAT DTC CHART FIRST.
• IGNITION "OFF."
• FULLY DEPRESS CLUTCH PEDAL.
  CAN TRANSMISSION BE SHIFTED INTO REVERSE?

NO →
• IGNITION "ON," ENGINE "OFF."
• FULLY DEPRESS CLUTCH PEDAL.
  CAN TRANSMISSION BE SHIFTED INTO REVERSE?

YES →
FAULTY REVERSE LOCKOUT SOLENOID
OR
INTERNAL TRANSMISSION PROBLEM.

② • CLUTCH PEDAL DEPRESSED.
• SELECT "FIELD SERVICE" MODE ON TECH 1.
  CAN TRANSMISSION BE SHIFTED INTO REVERSE?

③ • IGNITION "OFF."
• RAISE VEHICLE.
• DISCONNECT ELECTRICAL HARNESS FROM REVERSE LOCKOUT SOLENOID.
• PROBE TERMINAL "B" OF REVERSE LOCKOUT SOLENOID ELECTRICAL HARNESS WITH A TEST LIGHT TO GROUND.
• IGNITION "ON."
  DOES TEST LIGHT ILLUMINATE?

③ • IGNITION "OFF."
• RAISE VEHICLE.
• DISCONNECT ELECTRICAL HARNESS FROM REVERSE LOCKOUT SOLENOID.
• PROBE TERMINAL "A" OF REVERSE LOCKOUT SOLENOID ELECTRICAL HARNESS WITH A TEST LIGHT TO B+.
• IGNITION "ON."
• SELECT "FIELD SERVICE" MODE ON TECH 1.
  DOES TEST LIGHT ILLUMINATE?

NORMAL SYSTEM OPERATION. EXIT "FIELD SERVICE" MODE ON TECH 1.

• PROBE TERMINAL "A" OF REVERSE LOCKOUT SOLENOID ELECTRICAL HARNESS WITH A TEST LIGHT CONNECTED TO B+.
  DOES TEST LIGHT ILLUMINATE?

NO → CKT 541 OPEN.

④ • IGNITION "OFF."
• DISCONNECT ECM CONNECTOR "B".
• IGNITION "ON."
• PROBE TERMINAL "A" OF REVERSE LOCKOUT SOLENOID WITH A TEST LIGHT CONNECTED TO B+.
  IS TEST LIGHT "ON"?

FAULTY REVERSE LOCKOUT SOLENOID OR INTERNAL TRANSMISSION PROBLEM.

FAULTY CONNECTION AT REVERSE LOCKOUT SOLENOID
OR
FAULTY REVERSE LOCKOUT SOLENOID.

CKT 1652 OPEN OR FAULTY ECM CONNECTION OR FAULTY ECM.

YES → CKT 1652 SHORTED TO GROUND.

NO → FAULTY ECM.

"AFTER REPAIRS," CONFIRM "CLOSED LOOP" OPERATION AND NO MIL (SERVICE ENGINE SOON).

5.7L (VIN P) ENGINE — COMPONENT DIAGNOSTIC CHART — 1993 CAMARO AND FIREBIRD

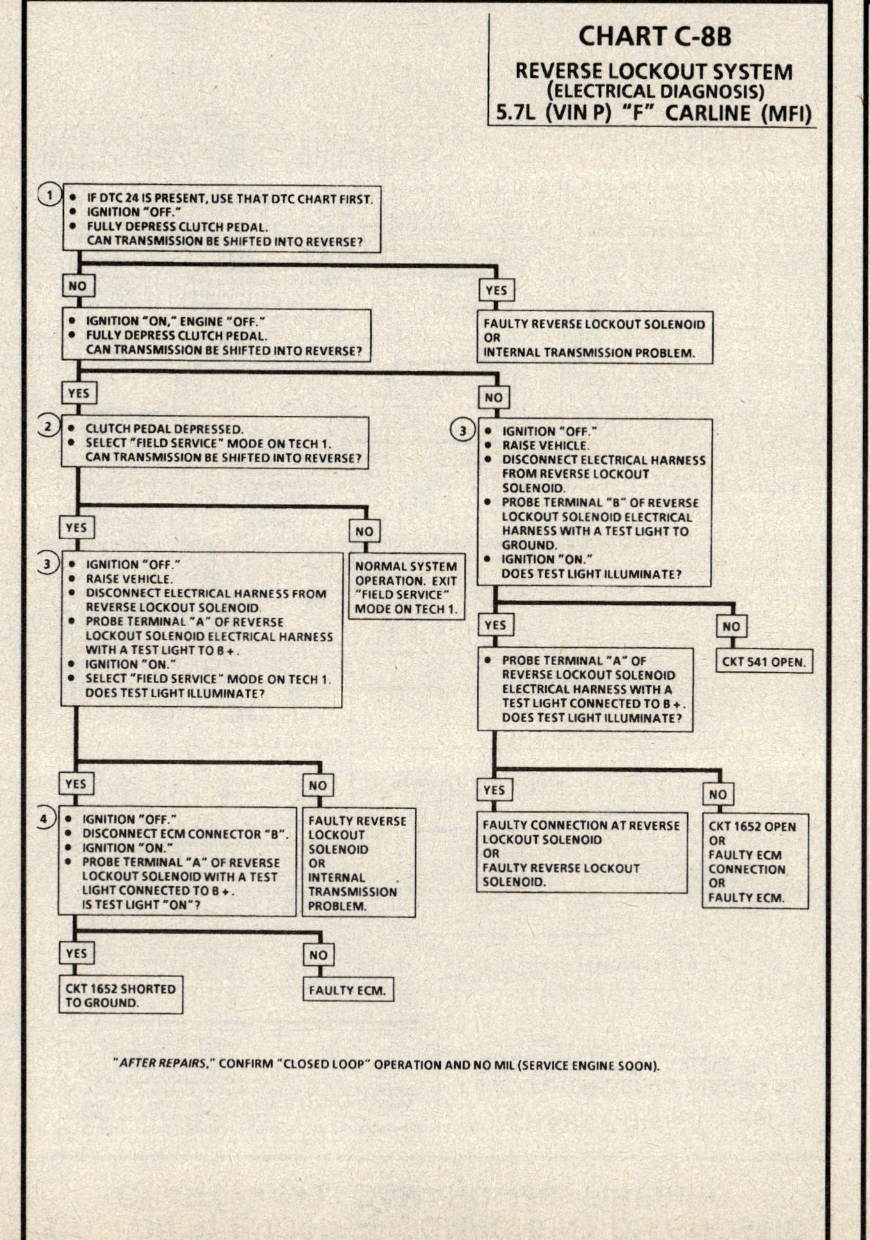

CHART C-10 (Page 1 of 3)

A/C CLUTCH CIRCUIT DIAGNOSIS
5.7L (VIN P) "F" CARLINE (MFI)

Circuit Description:

The A/C clutch control relay is ECM controlled to delay A/C clutch engagement after the A/C is turned "ON." This allows the ECM to adjust engine RPM before the A/C clutch engages.

The ECM will engage the A/C clutch any time A/C has been requested unless any of the following conditions exist:

- High coolant temperature.
- Low evaporator temperature.
- Low battery voltage.
- High engine RPM.
- High A/C system pressure.
- Low A/C system pressure.

When the HVAC control assembly is placed in the A/C mode, A 12 volt signal is sent to the ECM. When the ECM receives this signal the ECM will ground CKT 459 to energize the A/C relay. This will be displayed on the Tech 1 as A/C request "YES." A 3-wire A/C refrigerant pressure sensor is used to monitor A/C system pressure. If the A/C refrigerant pressure is greater than 414 psi or lower than 38 psi, the A/C clutch will not engage.

The A/C refrigerant pressure sensor uses ECM DTCs 66 and 67. If the A/C refrigerant pressure sensor signal wire becomes open, shorted to ground or shorted to voltage, DTC 66 will set. If the A/C clutch engages and no pressure change is detected, a DTC 67 will set.

The ECM also monitors A/C evaporator temperature. The ECM utilizes this information to cycle the A/C clutch. If A/C evaporator temperature is out of range (high or low), the ECM will disable the A/C clutch relay.

The A/C evaporator temperature sensor uses ECM DTC 71. If the A/C evaporator temperature sensor circuits becomes open, shorted to ground or shorted to voltage, DTC 71 will set.

When a request for A/C has been detected by the ECM, the ECM will ground the A/C clutch control relay driver circuit, the relay contacts will close, and current will flow through the relay to the A/C compressor clutch.

When A/C request has been detected by the ECM, the cooling fan(s) will be turned "ON" unless vehicle speed is too high.

Test Description: Number(s) below refer to circled number(s) on the diagnostic chart.
1. Checks the ECMs ability to control the A/C clutch control relay.
2. Checks for grounded CKT 459 to ECM.

Diagnostic Aids:

Before using CHART C-10, be sure no ECM DTC(s) are stored. The ECM will not activate the A/C clutch with a stored DTC.

5.7L (VIN P) ENGINE — COMPONENT DIAGNOSTIC CHART — 1993 CAMARO AND FIREBIRD

5.7L (VIN P) ENGINE — COMPONENT DIAGNOSTIC CHART — 1993 CAMARO AND FIREBIRD

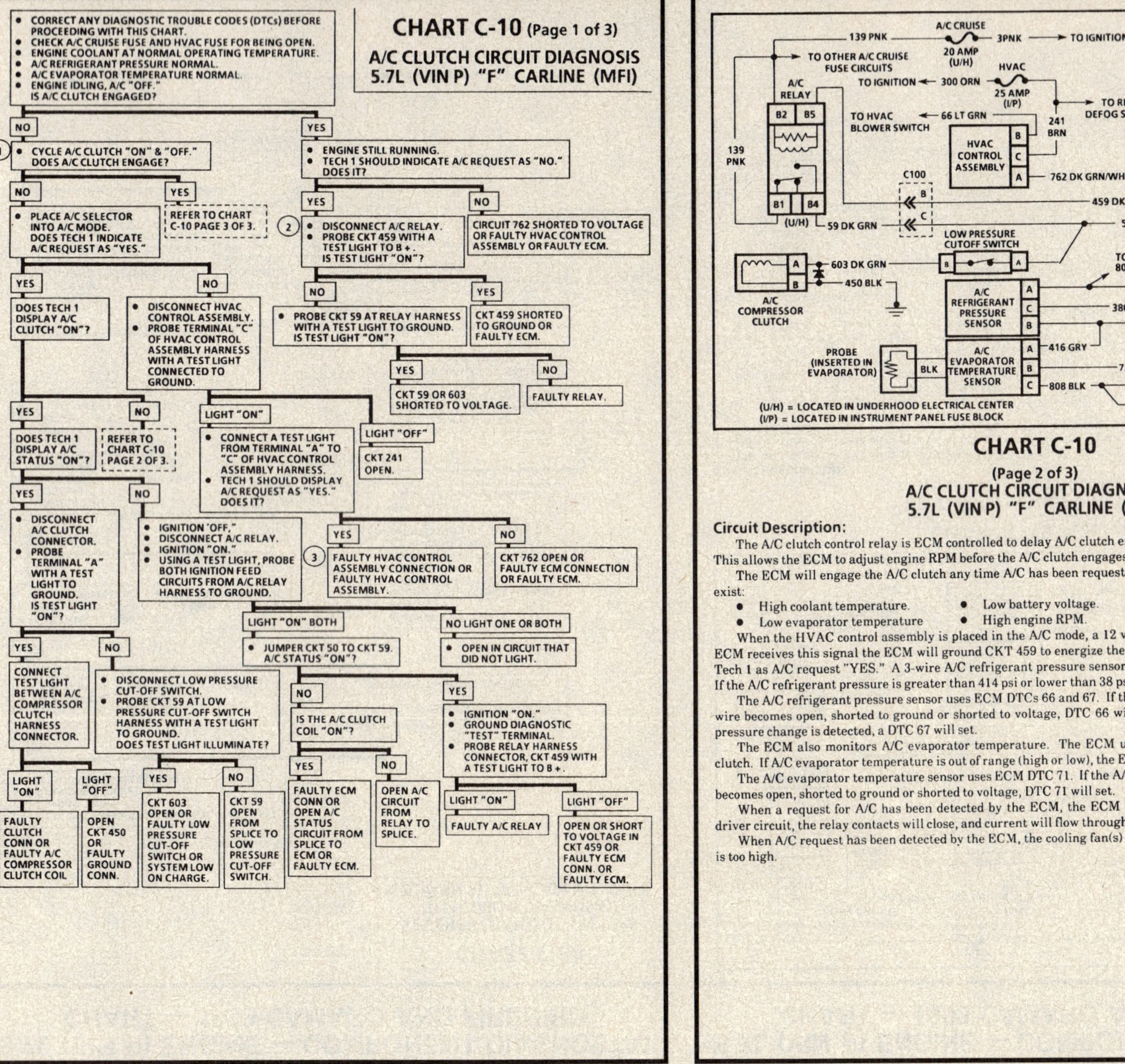

CHART C-10
(Page 2 of 3)
A/C CLUTCH CIRCUIT DIAGNOSIS
5.7L (VIN P) "F" CARLINE (MFI)

Circuit Description:

The A/C clutch control relay is ECM controlled to delay A/C clutch engagement after the A/C is turned "ON." This allows the ECM to adjust engine RPM before the A/C clutch engages.

The ECM will engage the A/C clutch any time A/C has been requested unless any of the following conditions exist:

- High coolant temperature.
- Low evaporator temperature
- Low battery voltage.
- High engine RPM.
- High A/C system pressure.
- Low A/C system pressure.

When the HVAC control assembly is placed in the A/C mode, a 12 volt signal is sent to the ECM. When the ECM receives this signal the ECM will ground CKT 459 to energize the A/C relay. This will be displayed on the Tech 1 as A/C request "YES." A 3-wire A/C refrigerant pressure sensor is used to monitor A/C system pressure. If the A/C refrigerant pressure is greater than 414 psi or lower than 38 psi, the A/C clutch will not engage.

The A/C refrigerant pressure sensor uses ECM DTCs 66 and 67. If the A/C refrigerant pressure sensor signal wire becomes open, shorted to ground or shorted to voltage, DTC 66 will set. If the A/C clutch engages and no pressure change is detected, a DTC 67 will set.

The ECM also monitors A/C evaporator temperature. The ECM utilizes this information to cycle the A/C clutch. If A/C evaporator temperature is out of range (high or low), the ECM will disable the A/C clutch relay.

The A/C evaporator temperature sensor uses ECM DTC 71. If the A/C evaporator temperature sensor circuits becomes open, shorted to ground or shorted to voltage, DTC 71 will set.

When a request for A/C has been detected by the ECM, the ECM will ground the A/C clutch control relay driver circuit, the relay contacts will close, and current will flow through the relay to the A/C compressor clutch.

When A/C request has been detected by the ECM, the cooling fan(s) will be turned "ON" unless vehicle speed is too high.

5.7L (VIN P) ENGINE — COMPONENT DIAGNOSTIC CHART — 1993 CAMARO AND FIREBIRD

CHART C-10
(Page 2 of 3)
A/C CLUTCH CIRCUIT DIAGNOSIS
5.7L (VIN P) "F" CARLINE (MFI)

5.7L (VIN P) ENGINE — COMPONENT DIAGNOSTIC CHART — 1993 CAMARO AND FIREBIRD

CHART C-10
(Page 3 of 3)
A/C CLUTCH CIRCUIT DIAGNOSIS
5.7L (VIN P) "F" CARLINE (MFI)

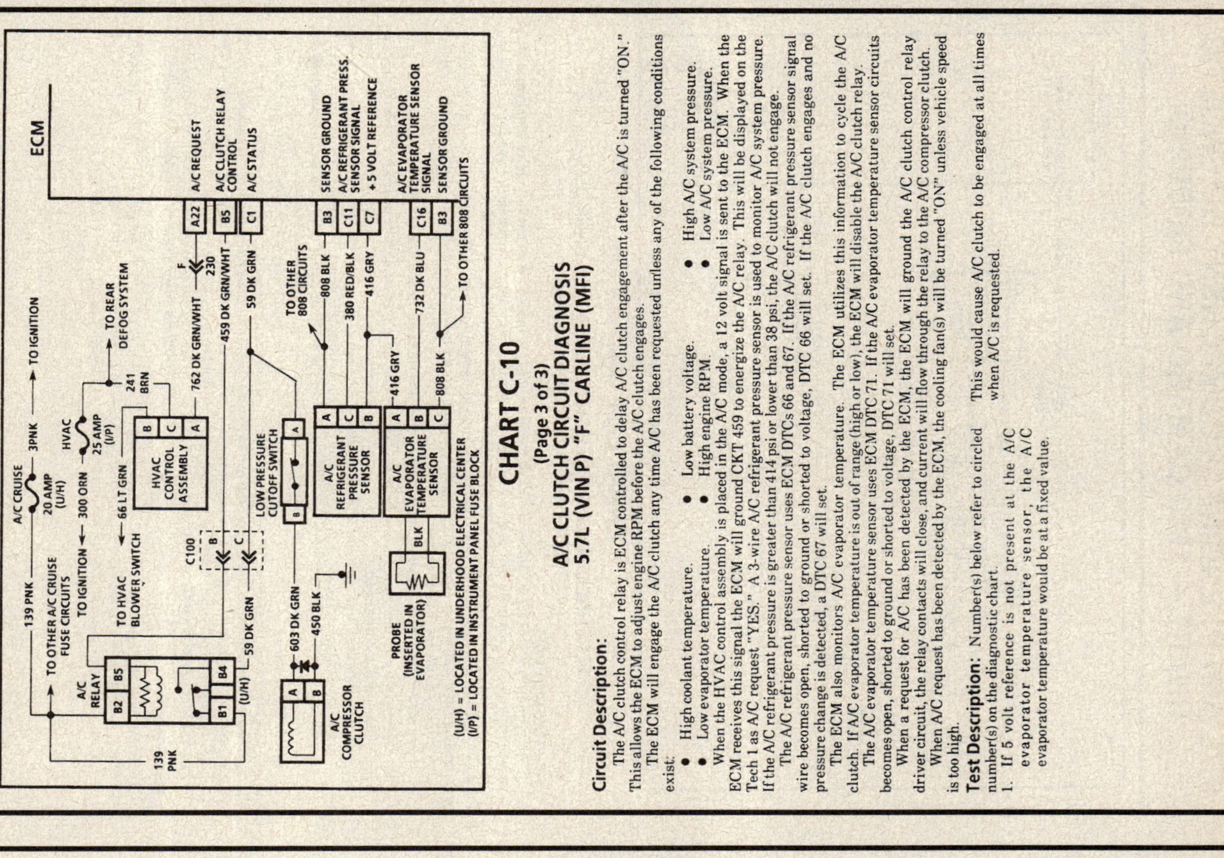

Circuit Description:

The A/C clutch control relay is ECM controlled to delay A/C clutch engagement after the A/C is turned "ON." This allows the ECM to adjust engine RPM before the A/C clutch engages.

The ECM will engage the A/C clutch any time A/C has been requested unless any of the following conditions exist:

- High coolant temperature.
- Low evaporator temperature.
- Low battery voltage.
- High engine RPM.
- High A/C system pressure.
- Low A/C system pressure.

When the HVAC control assembly is placed in the A/C mode, a 12 volt signal is sent to the ECM. When the ECM receives this signal the ECM will ground CKT 459 to energize the A/C relay. This will be displayed on the Tech 1 as A/C request "YES." A 3-wire A/C refrigerant pressure sensor is used to monitor A/C system pressure. If the A/C refrigerant pressure is greater than 414 psi or lower than 38 psi, the A/C clutch will not engage.

The A/C refrigerant pressure sensor uses ECM DTCs 66 and 67. If the A/C refrigerant pressure sensor signal wire becomes open, shorted to ground or shorted to voltage, DTC 66 will set. If the A/C clutch engages and no pressure change is detected, a DTC 67 will set.

The ECM also monitors A/C evaporator temperature. The ECM utilizes this information to cycle the A/C clutch. If A/C evaporator temperature is out of range (high or low), the ECM will disable the A/C clutch relay.

The A/C evaporator temperature sensor uses ECM DTC 71. If the A/C evaporator temperature sensor circuits becomes open, shorted to ground or shorted to voltage, DTC 71 will set.

When a request for A/C has been detected by the ECM, the ECM will ground the A/C clutch control relay driver circuit, the relay contacts will close, and current will flow through the relay to the A/C compressor clutch.

When A/C request has been detected by the ECM, the cooling fan(s) will be turned "ON" unless vehicle speed is too high.

Test Description: Number(s) below refer to circled number(s) on the diagnostic chart.

1. If 5 volt reference is not present at the A/C evaporator temperature sensor, the A/C evaporator temperature would be at a fixed value. This would cause A/C clutch to be engaged at all times when A/C is requested.

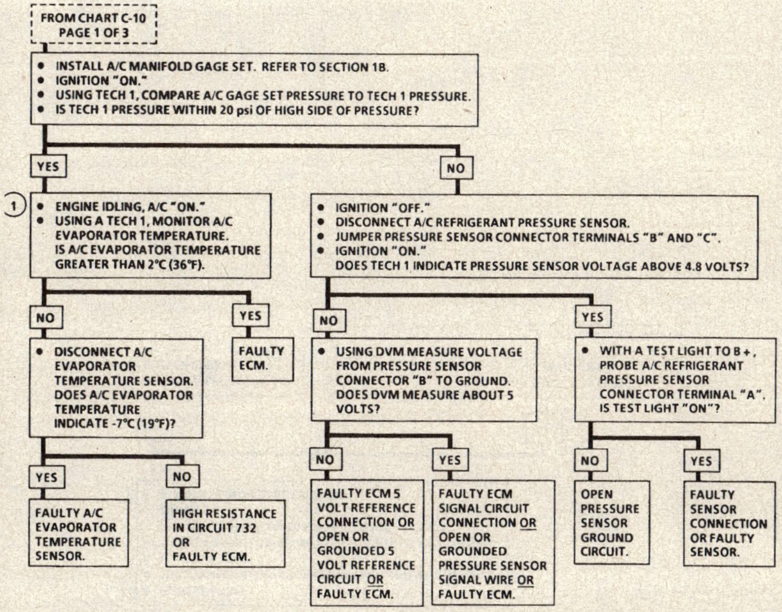

5.7L (VIN P) ENGINE — COMPONENT DIAGNOSTIC CHART — 1993 CAMARO AND FIREBIRD

CHART C-10

(Page 3 of 3)
A/C CLUTCH CIRCUIT DIAGNOSIS
5.7L (VIN P) "F" CARLINE (MFI)

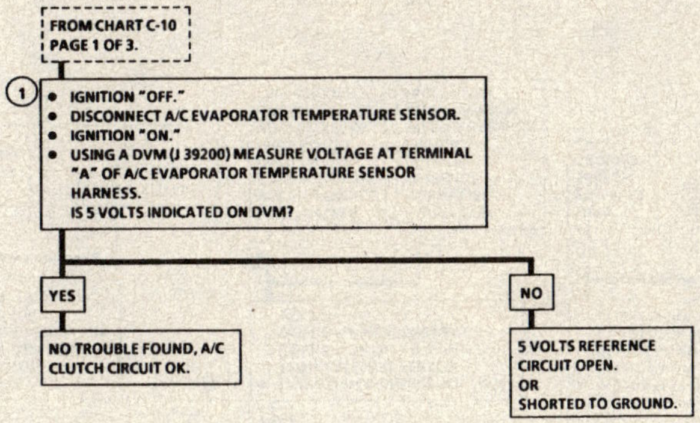

FROM CHART C-10 PAGE 1 OF 3.

1.
- IGNITION "OFF."
- DISCONNECT A/C EVAPORATOR TEMPERATURE SENSOR.
- IGNITION "ON."
- USING A DVM (J 39200) MEASURE VOLTAGE AT TERMINAL "A" OF A/C EVAPORATOR TEMPERATURE SENSOR HARNESS.
  IS 5 VOLTS INDICATED ON DVM?

YES — NO TROUBLE FOUND, A/C CLUTCH CIRCUIT OK.

NO — 5 VOLTS REFERENCE CIRCUIT OPEN. OR SHORTED TO GROUND.

5.7L (VIN P) ENGINE — COMPONENT DIAGNOSTIC CHART — 1993 CAMARO AND FIREBIRD

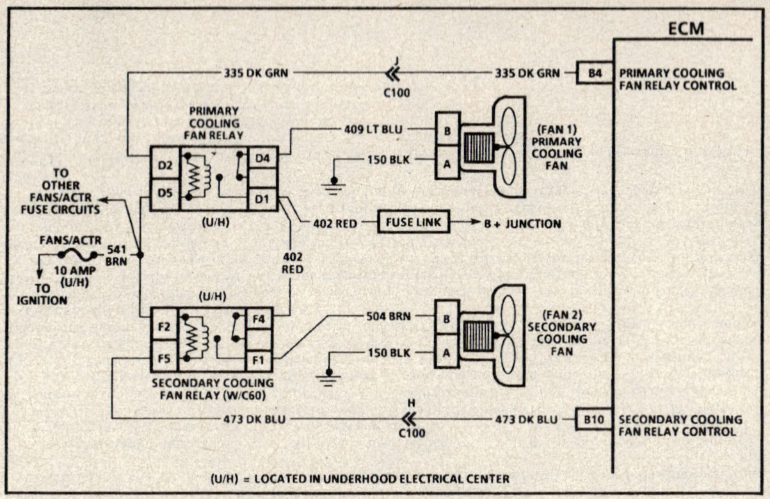

CHART C-12 (Page 1 of 2)

ELECTRIC COOLING FAN CONTROL CIRCUIT DIAGNOSIS
5.7L (VIN P) "F" CARLINE (MFI)

Circuit Description:

The cooling fans are controlled by the ECM based on various inputs. Battery voltage is supplied to the primary fan relay on terminals "D1" and "F4" of the secondary fan relay. Ignition voltage is supplied to terminal "D5" of the primary fan relay and "F2" of the secondary fan relay. Grounding CKT 335 (relay terminal "D2") will energize the primary cooling fan relay (Fan 1) and supply battery voltage to the primary cooling fan motor. Grounding CKT 473 (relay terminal "F5") will energize the secondary cooling fan relay (Fan 2) and supply battery voltage to the secondary fan motor.

When certain Diagnostic Trouble Codes (DTCs) are set, the ECM will enable the cooling fans.

Test Description: Number(s) below refer to circled number(s) on the diagnostic chart.

1. With the diagnostic "test" terminal grounded, the cooling fan control driver(s) will close, which should energize the fan control relay(s).
2. The cooling fans should come "ON" anytime A/C system is operating.
3. Comparing Tech 1 pressure and manifold gage set pressure will determine if the A/C refrigerant pressure sensor is out of range. An out of range A/C refrigerant pressure sensor can cause the cooling fans to operate at the wrong times.

light, or engine coolant temperature gage indicated overheating.

The gage accuracy can also be checked by comparing the Engine Coolant Temperature (ECT) sensor reading using a Tech 1 and comparing its reading with the gage reading.

If the engine is actually overheating and the gage indicated overheating, but the cooling fan is not coming "ON," the Engine Coolant Temperature (ECT) sensor has probably shifted out of calibration and should be replaced.

If the engine is overheating and the cooling fans are "ON," the cooling system should be checked

Diagnostic Aids:

If the owner complained of an overheating problem, it must be determined if the complaint was due to an actual boil over, or the warning indicator

The ECM will command fan 1 "ON" at 108°C (226°F) and "OFF" at 105°C (221°F) and, fan 2 "ON" at 113°C (235°F) and "OFF" at 110°C (230°F).

5.7L (VIN P) ENGINE — COMPONENT DIAGNOSTIC CHART — 1993 CAMARO AND FIREBIRD

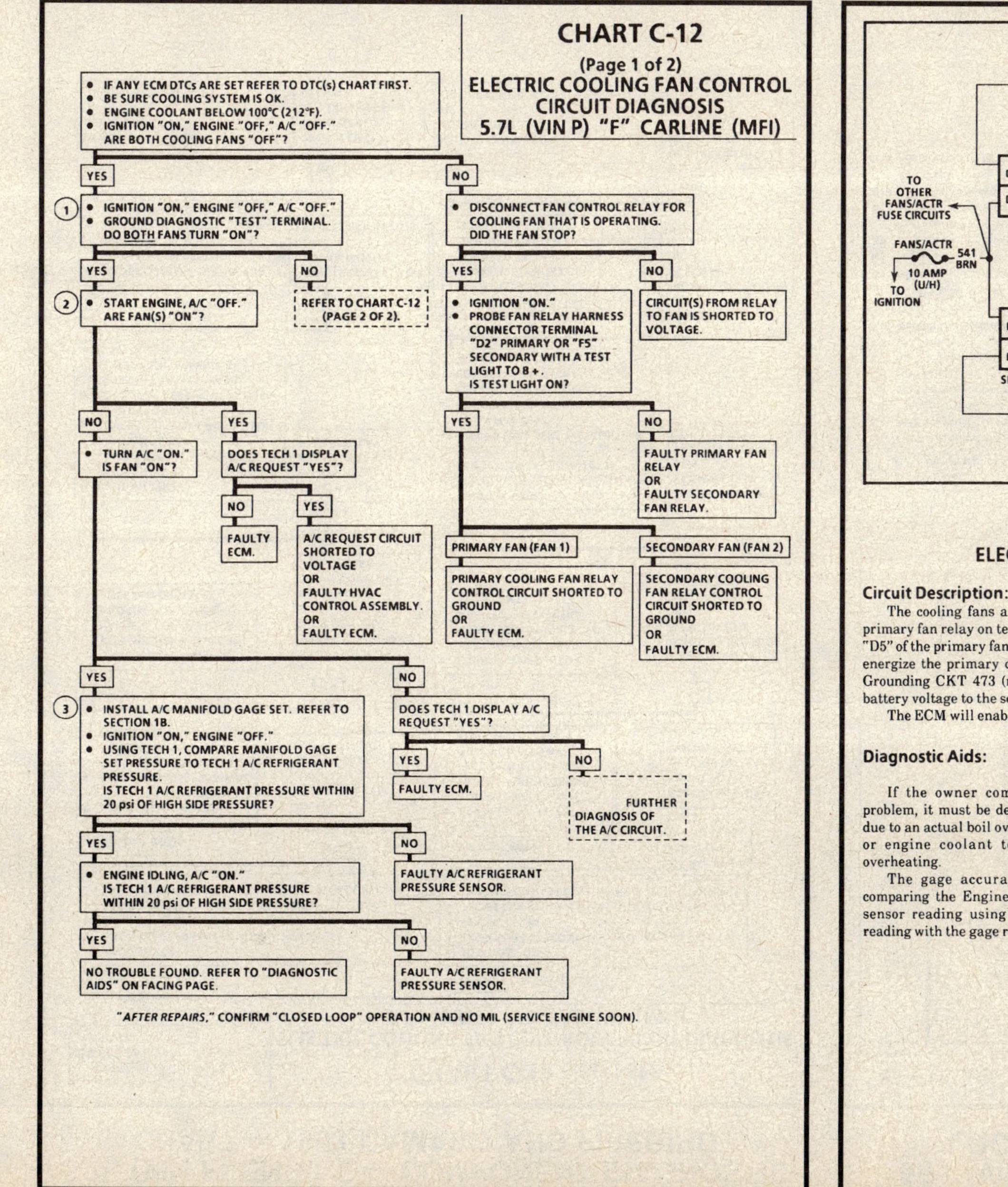

5.7L (VIN P) ENGINE — COMPONENT DIAGNOSTIC CHART — 1993 CAMARO AND FIREBIRD

CHART C-12
(Page 2 of 2)
ELECTRIC COOLING FAN CONTROL CIRCUIT DIAGNOSIS
5.7L (VIN P) "F" CARLINE (MFI)

Circuit Description:

The cooling fans are controlled by the ECM based on various inputs. Battery voltage is supplied to the primary fan relay on terminals "D1" and "F4" of the secondary fan relay. Ignition voltage is supplied to terminal "D5" of the primary fan relay and "F2" of the secondary fan relay. Grounding CKT 335 (relay terminal "D2") will energize the primary cooling fan relay (Fan 1) and supply battery voltage to the primary cooling fan motor. Grounding CKT 473 (relay terminal "F5") will energize the secondary cooling fan relay (Fan 2) and supply battery voltage to the secondary fan motor.

The ECM will enable the cooling fans, when certain Diagnostic Trouble Codes (DTCs) are set.

Diagnostic Aids:

If the owner complained of an overheating problem, it must be determined if the complaint was due to an actual boil over, the warning indicator light, or engine coolant temperature gage indicated overheating.

The gage accuracy can also be checked by comparing the Engine Coolant Temperature (ECT) sensor reading using a Tech 1 and comparing its reading with the gage reading.

If the engine is actually overheating and the gage indicated overheating, but the cooling fan is not coming "ON," the Engine Coolant Temperature (ECT) sensor has probably shifted out of calibration and should be replaced.

If the engine is overheating and the cooling fans are "ON," the cooling system should be checked.

The ECM will command fan 1 "ON" at 108°C (226°F) and "OFF" at 105°C (221°F) and an 2 "ON" at 113°C (235°F) and "OFF" at 110°C (230°F).

5.7L (VIN P) ENGINE — COMPONENT DIAGNOSTIC CHART — 1993 CAMARO AND FIREBIRD

CHART C-12 (Page 2 of 2)
ELECTRIC COOLING FAN CONTROL CIRCUIT DIAGNOSIS
5.7L (VIN P) "F" CARLINE (MFI)

FROM CHART C-12 (1 OF 2)

PRIMARY
- IGNITION "OFF."
- DISCONNECT PRIMARY COOLING FAN RELAY
- IGNITION "ON."
- PROBE FAN RELAY HARNESS TERMINALS "D1" AND "D5" WITH A TEST LIGHT CONNECTED TO GROUND.

LIGHT "ON" BOTH
- USING TECH 1 SELECT "FIELD SERVICE MODE."
- PROBE HARNESS TERMINAL "D2" WITH A TEST LIGHT CONNECTED TO B +. IS TEST LIGHT "ON"?

LIGHT "OFF" ONE OR BOTH
REPAIR OPEN OR SHORT TO GROUND IN CIRCUIT THAT DID NOT LIGHT.

YES
- USING A FUSED JUMPER WIRE, JUMPER FAN RELAY HARNESS TERMINALS "D1" AND "D4" TOGETHER. DOES FAN OPERATE?

NO
REPAIR OPEN OR SHORT TO VOLTAGE IN CIRCUIT 335. OR FAULTY ECM CONNECTION OR FAULTY ECM.

NO
- IGNITION "OFF."
- DISCONNECT PRIMARY COOLING FAN ELECTRICAL CONNECTOR
- IGNITION "ON."
- JUMPER WIRE STILL INSTALLED AT RELAY HARNESS.
- CONNECT A TEST LIGHT ACROSS COOLING FAN ELECTRICAL CONNECTOR TERMINALS. IS TEST LIGHT "ON".

YES
FAULTY RELAY CONNECTION OR FAULTY RELAY.

NO
- PROBE COOLING FAN MOTOR HARNESS CONNECTOR TERMINAL "B" WITH A TEST LIGHT CONNECTED TO GROUND. IS TEST LIGHT "ON"?

YES
FAULTY CONNECTION AT FAN MOTOR OR FAULTY FAN MOTOR.

YES
FAULTY FAN MOTOR CONNECTION OR CIRCUIT 150 OPEN.

NO
OPEN IN CIRCUIT 409 BETWEEN RELAY AND COOLING FAN MOTOR.

SECONDARY W/C60
- IGNITION "OFF."
- DISCONNECT SECONDARY COOLING FAN RELAY
- IGNITION "ON."
- PROBE FAN RELAY HARNESS TERMINALS "F2" AND "F4" WITH A TEST LIGHT CONNECTED TO GROUND.

LIGHT "ON" BOTH
- USING TECH 1 SELECT "FIELD SERVICE MODE."
- PROBE HARNESS TERMINAL "F5" WITH A TEST LIGHT CONNECTED TO B +. IS TEST LIGHT "ON"?

LIGHT "OFF" ONE OR BOTH
REPAIR OPEN OR SHORT TO GROUND IN CIRCUIT THAT DID NOT LIGHT.

YES
- USING A FUSED JUMPER WIRE, JUMPER FAN RELAY HARNESS TERMINALS "F4" AND "F1" TOGETHER. DOES FAN OPERATE?

NO
REPAIR OPEN OR SHORT TO VOLTAGE IN CIRCUIT 473. OR FAULTY ECM CONNECTION OR FAULTY ECM.

NO
- IGNITION "OFF."
- DISCONNECT SECONDARY COOLING FAN ELECTRICAL CONNECTOR
- IGNITION "ON."
- JUMPER WIRE STILL INSTALLED AT RELAY HARNESS.
- CONNECT A TEST LIGHT ACROSS COOLING FAN ELECTRICAL CONNECTOR TERMINALS. IS TEST LIGHT "ON".

YES
FAULTY RELAY CONNECTION OR FAULTY RELAY.

NO
- PROBE COOLING FAN MOTOR HARNESS CONNECTOR TERMINAL "B" WITH A TEST LIGHT CONNECTED TO GROUND. IS TEST LIGHT "ON"?

YES
FAULTY CONNECTION AT FAN MOTOR OR FAULTY FAN MOTOR.

YES
FAULTY FAN MOTOR CONNECTION OR CIRCUIT 150 OPEN.

NO
OPEN IN CIRCUIT 504 BETWEEN RELAY AND COOLING FAN MOTOR.

5.7L (VIN 8) ENGINE — ENGINE COMPONENT LOCATION CHART — 1992 CORVETTE

"Y" CARLINE   RPO: LT1   VIN CODE: P   5.7L V8

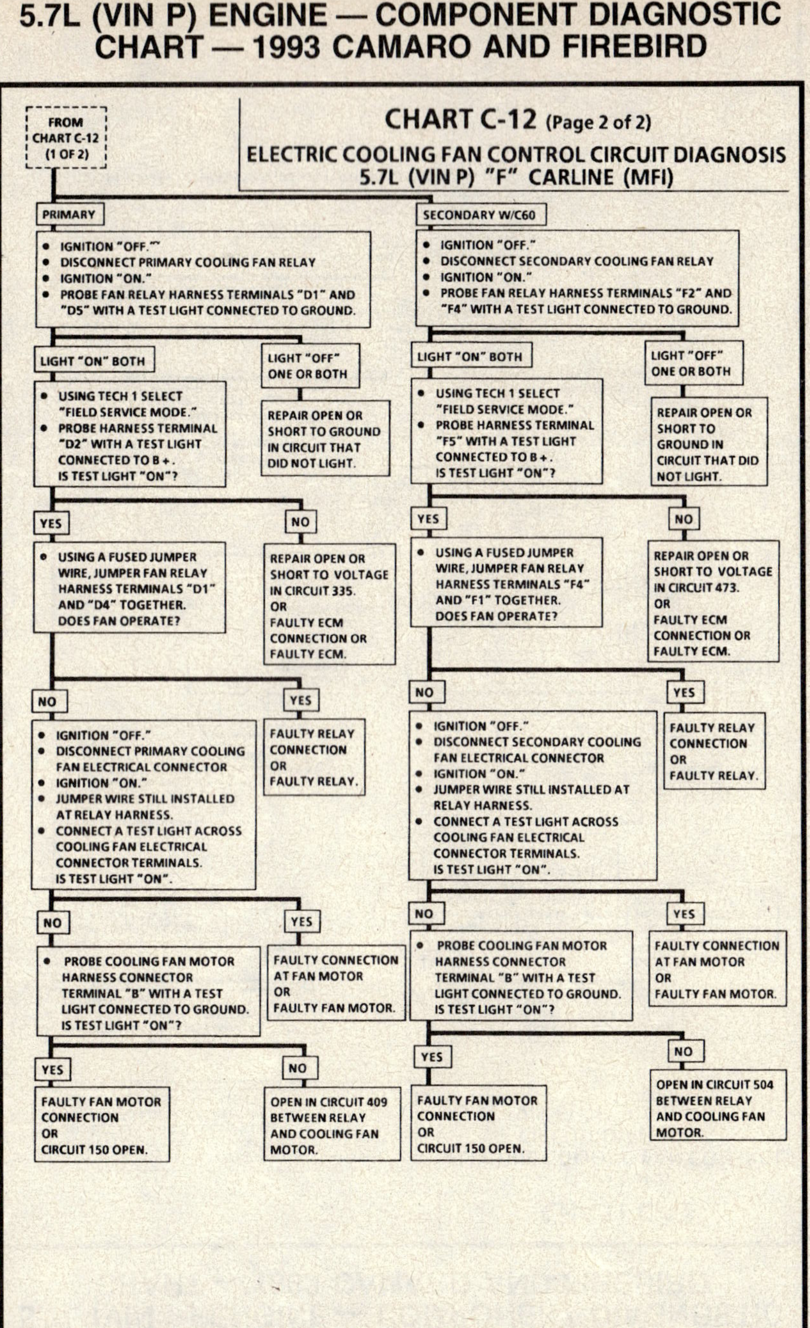

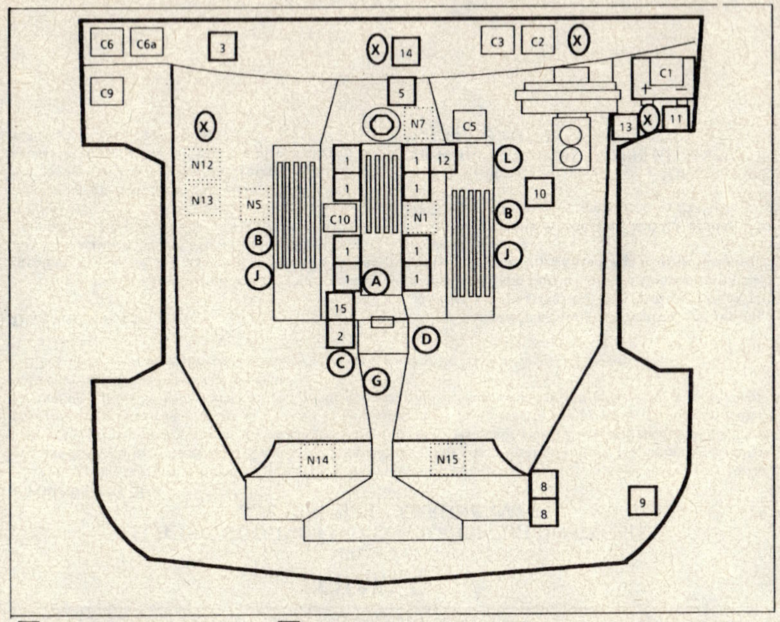

COMPUTER HARNESS
- C1 Electronic Control Module (ECM)
- C2 ALDL diagnostic connector
- C3 "Service Engine Soon" light
- C5 ECM harness grounds
- C6 Fuse panel
- C6a Auxiliary fuse panel
- C9 TPS Module
- C10 Ignition Test Connector

NOT ECM CONNECTED
- N1 Positive Crankcase Vent (PCV) valve
- N5 Engine temp. sensor (gage overheat)
- N7 Oil pressure sensor (gage) and switch (fuel pump)
- N12 A/C pressure cycling switch
- N13 A/C Pressure Sensor
- N14 Secondary cooling fan (FAN 2)
- N15 Primary cooling fan (FAN 1)

CONTROLLED DEVICES
- 1 Fuel injector
- 2 Idle Air Control (IAC) valve
- 3 Fuel pump relay
- 5 Torque Converter Clutch (TCC) connector
- 8 Cooling fan relay
- 9 Electric Air Pump
- 10 Electric switching valve solenoid
- 11 1-4 Upshift relay (M/T)
- 12 EGR solenoid
- 13 A/C clutch control relay
- 14 1-4 Upshift solenoid (M/T)
- 15 Fuel vapor canister solenoid

INFORMATION SENSORS
- A Manifold Absolute Pressure (MAP) sensor
- B Oxygen (O_2) sensor(s)
- C Throttle Position sensor (TPS)
- D Coolant Temperature Sensor (CTS)
- G Intake Air Temperature (IAT)
- J ESC knock sensor(s)
- L Engine Oil Temperature (EOT) sensor
- M Vehicle Speed Sensor (Mounted on transmission, not shown)

X

Exhaust Gas Recirculation (EGR) Valve

5.7L (VIN 8) ENGINE — ECM WIRING SCHEMATIC — 1992 CORVETTE

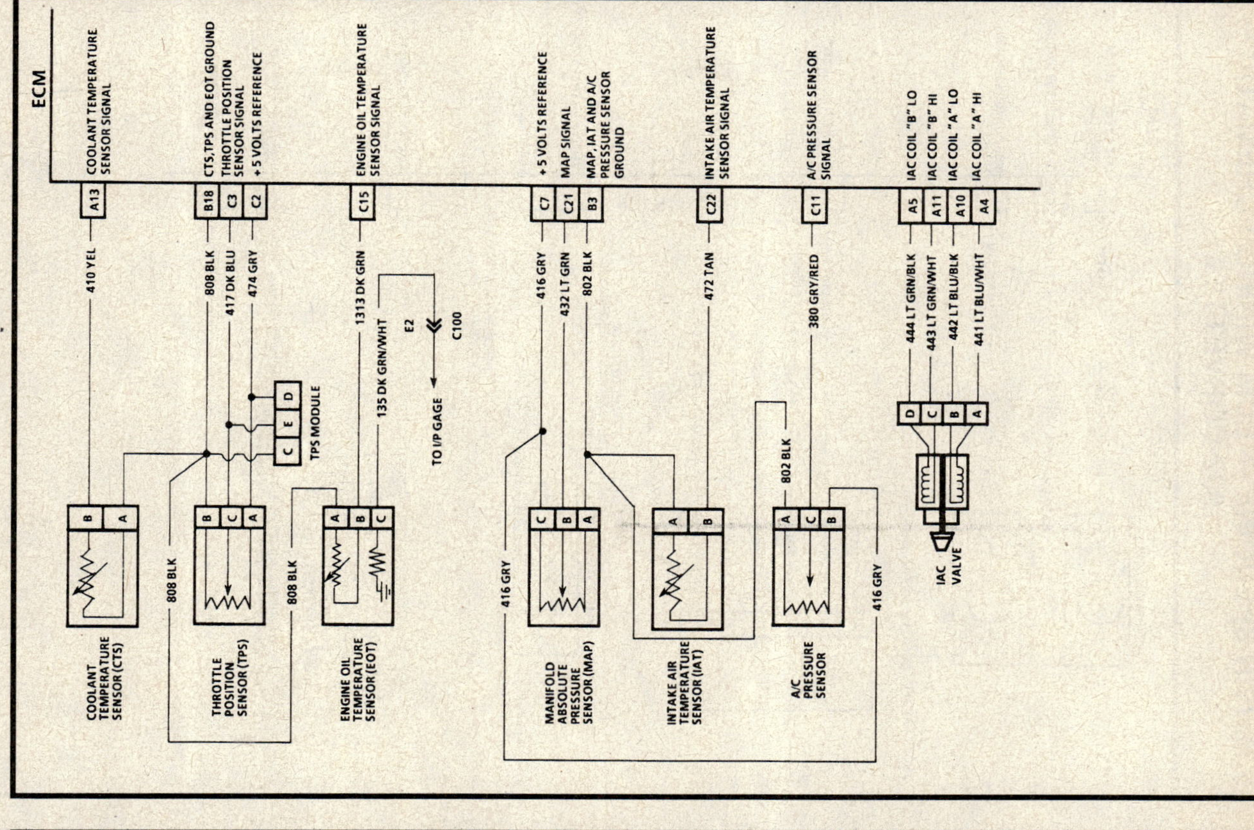

5.7L (VIN P) ENGINE — ENGINE COMPONENT LOCATION CHART — 1993 CORVETTE

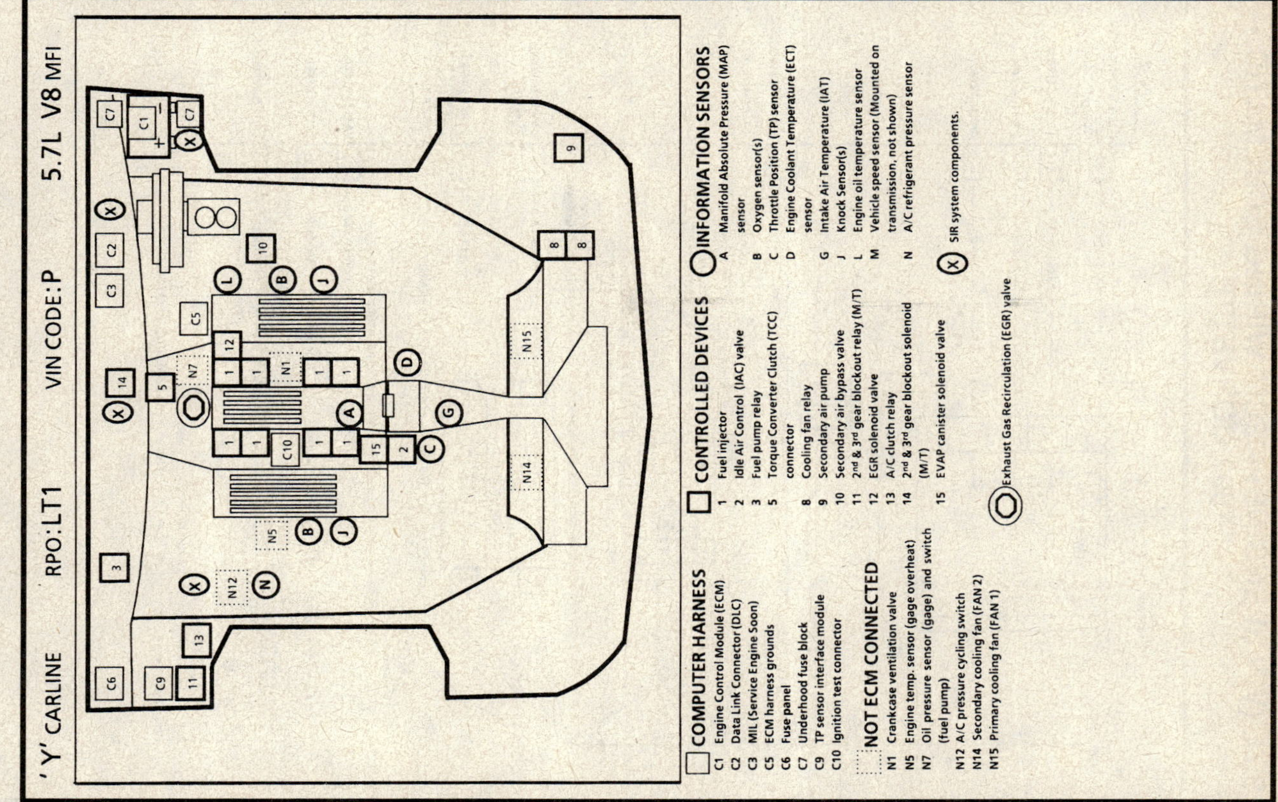

5.7L (VIN 8) ENGINE — ECM WIRING SCHEMATIC — 1992 CORVETTE

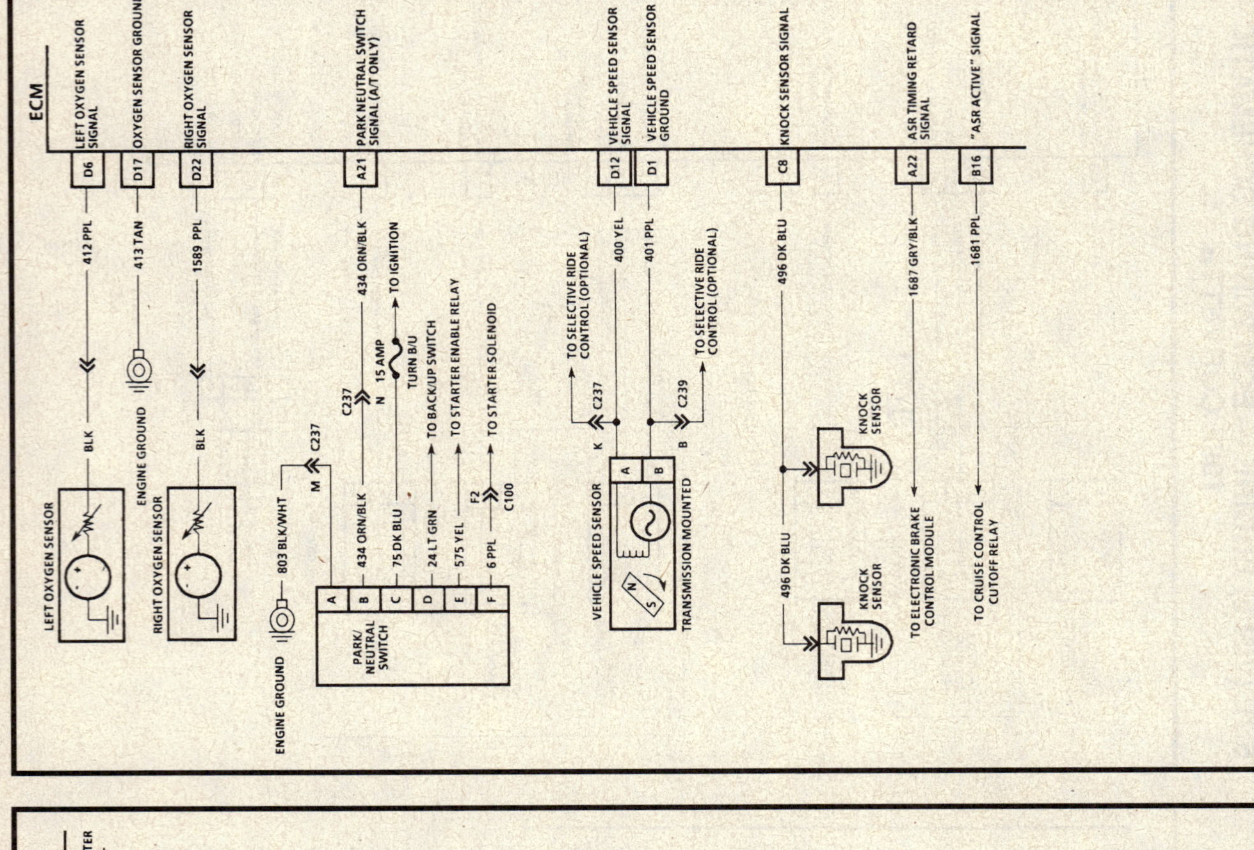

5.7L (VIN 8) ENGINE — ECM WIRING SCHEMATIC — 1992 CORVETTE

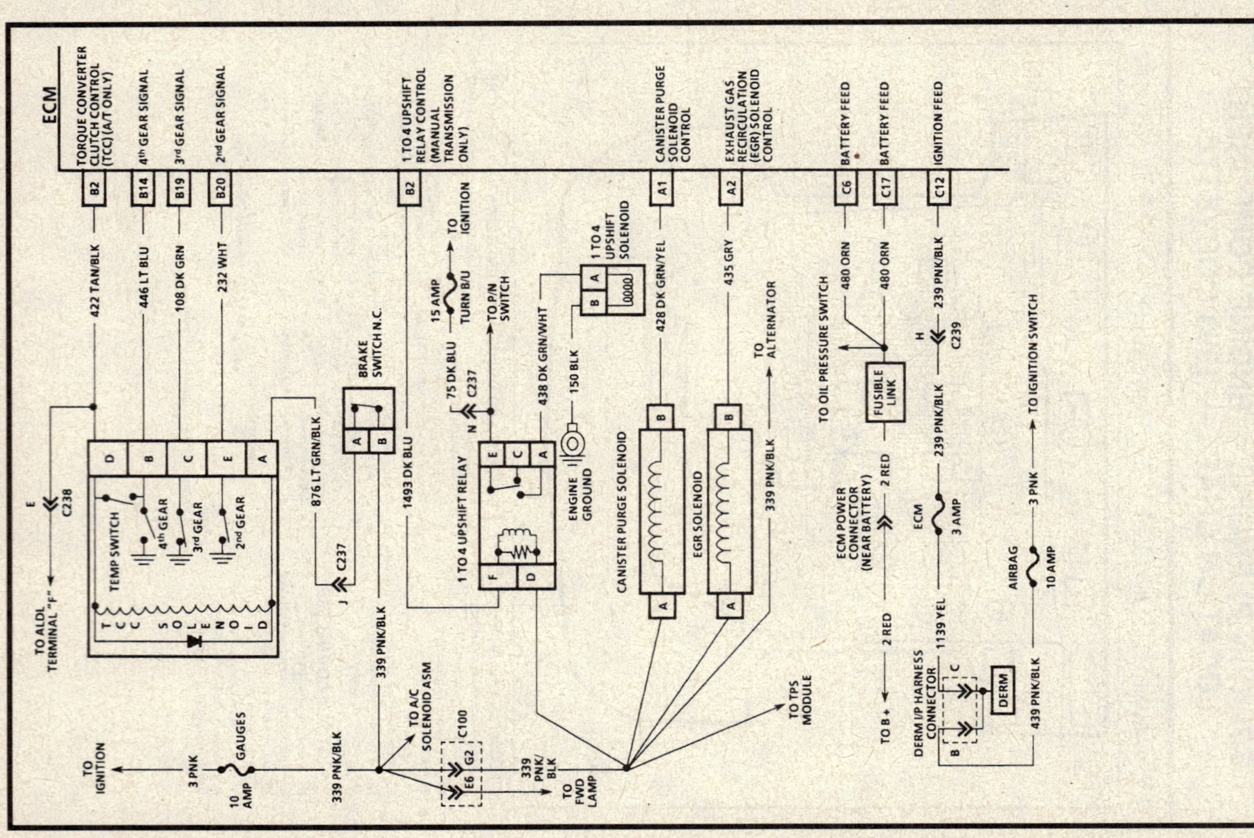

5.7L (VIN 8) ENGINE—ECM WIRING SCHEMATIC—1992 CORVETTE

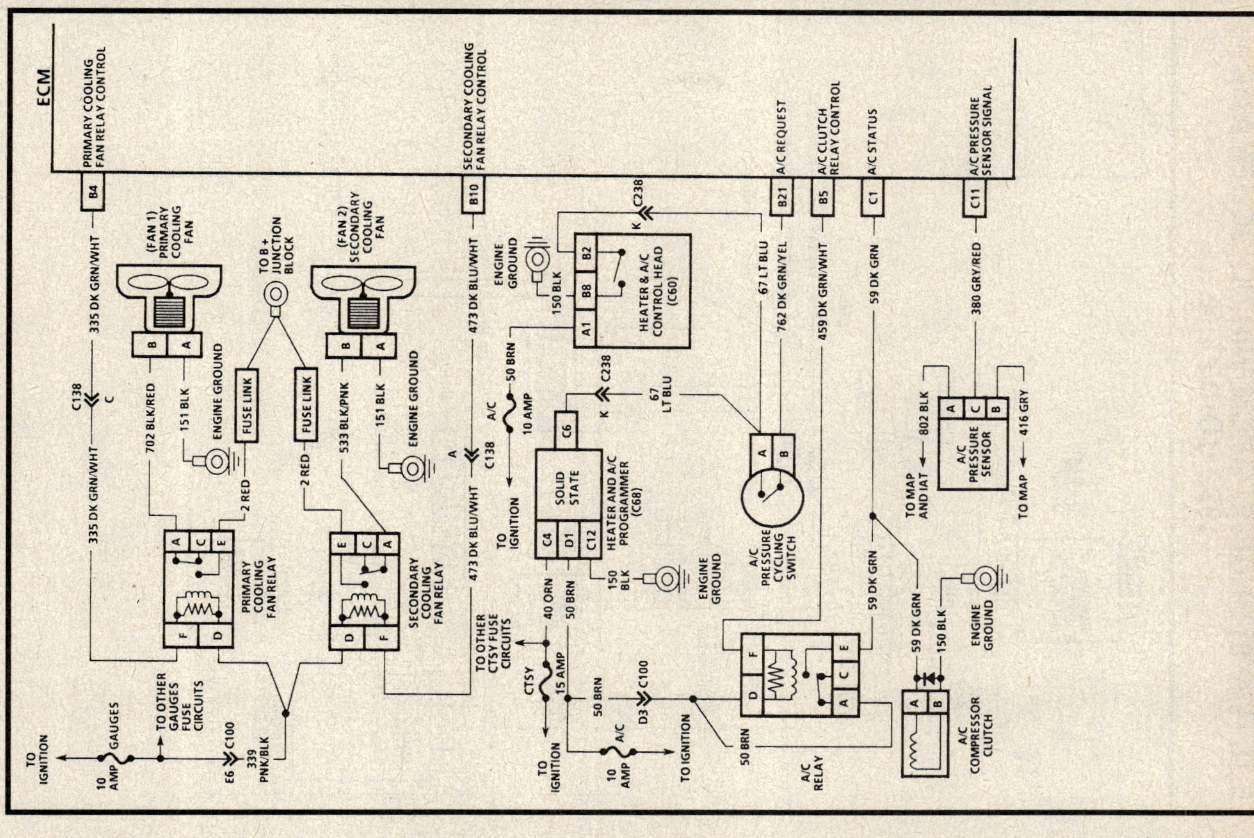

5.7L (VIN 8) ENGINE—ECM WIRING SCHEMATIC—1992 CORVETTE

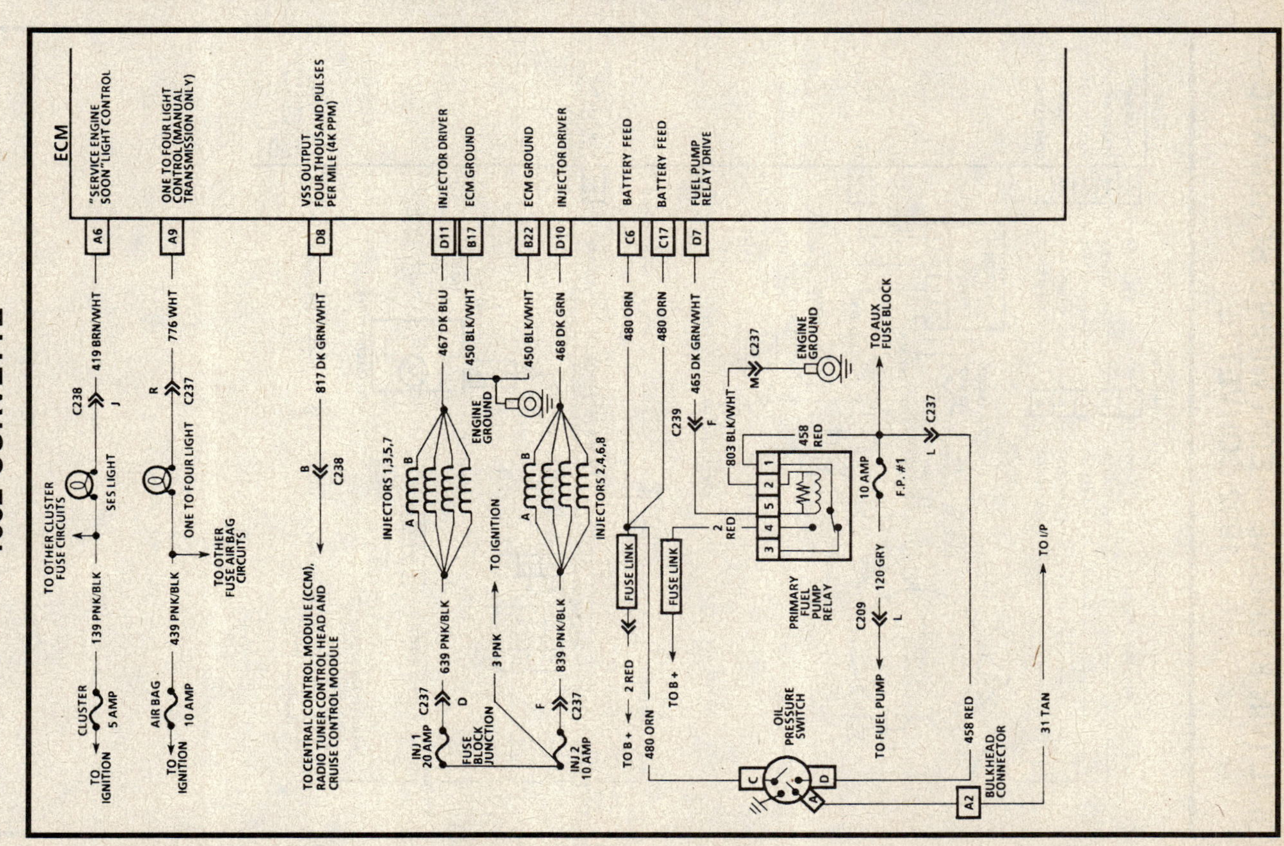
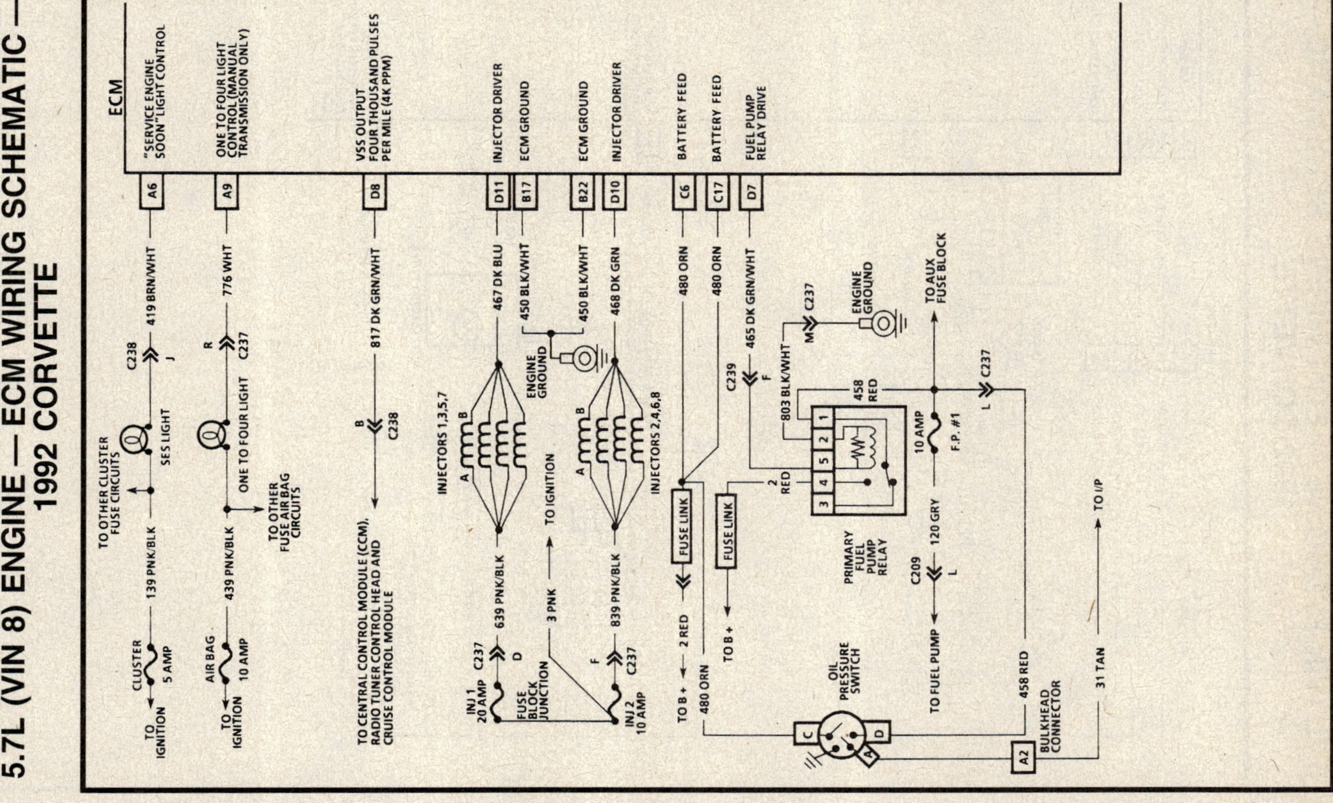

5.7L (VIN 8) ENGINE — ECM WIRING SCHEMATIC — 1992 CORVETTE

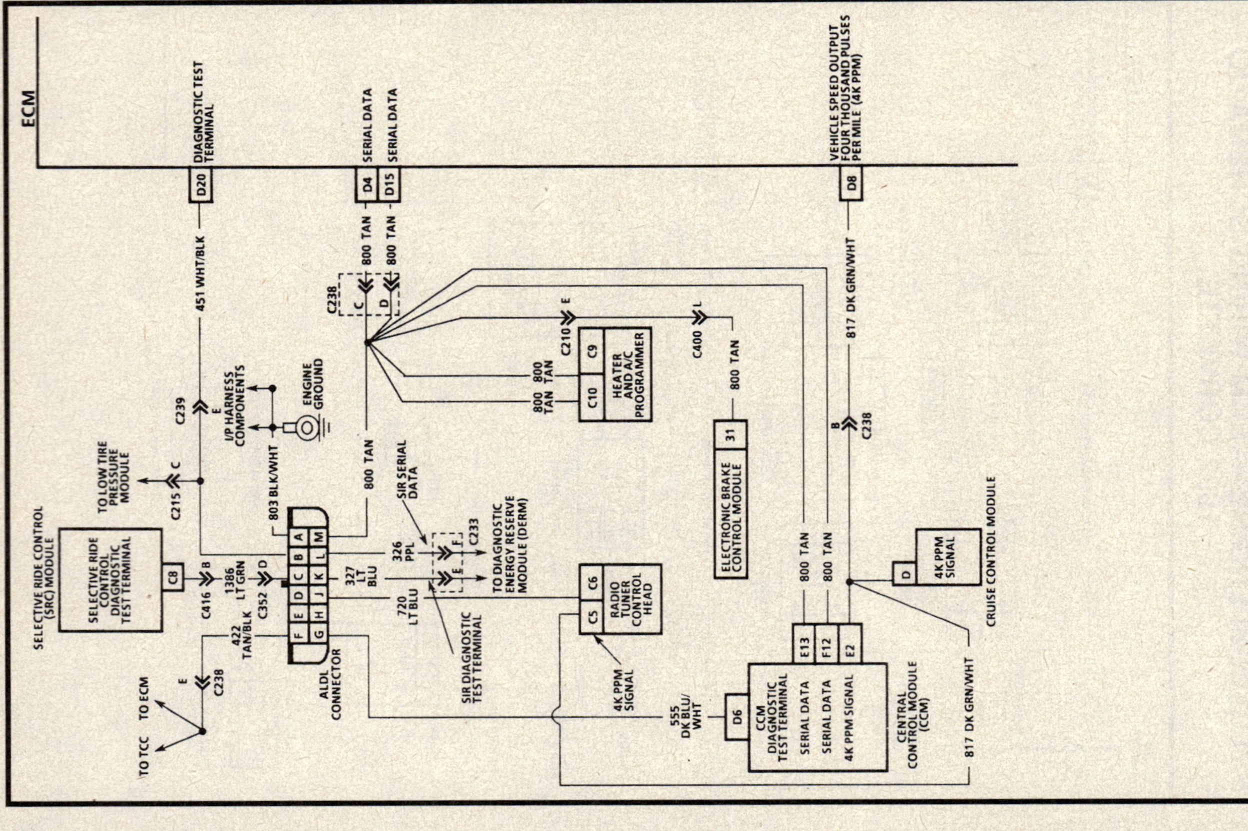

5.7L (VIN 8) ENGINE — ECM WIRING SCHEMATIC — 1992 CORVETTE

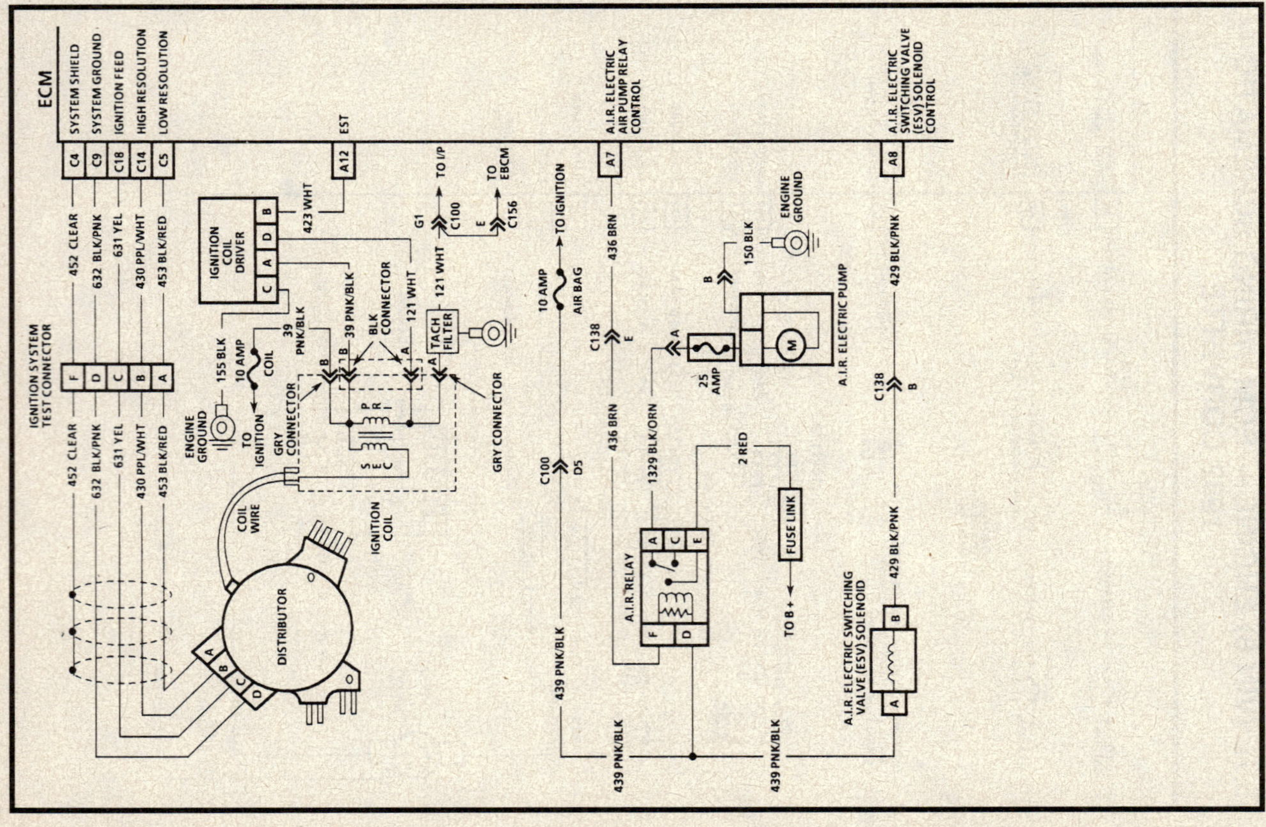

5.7L (VIN P) ENGINE — ECM WIRING SCHEMATIC — 1993 CORVETTE

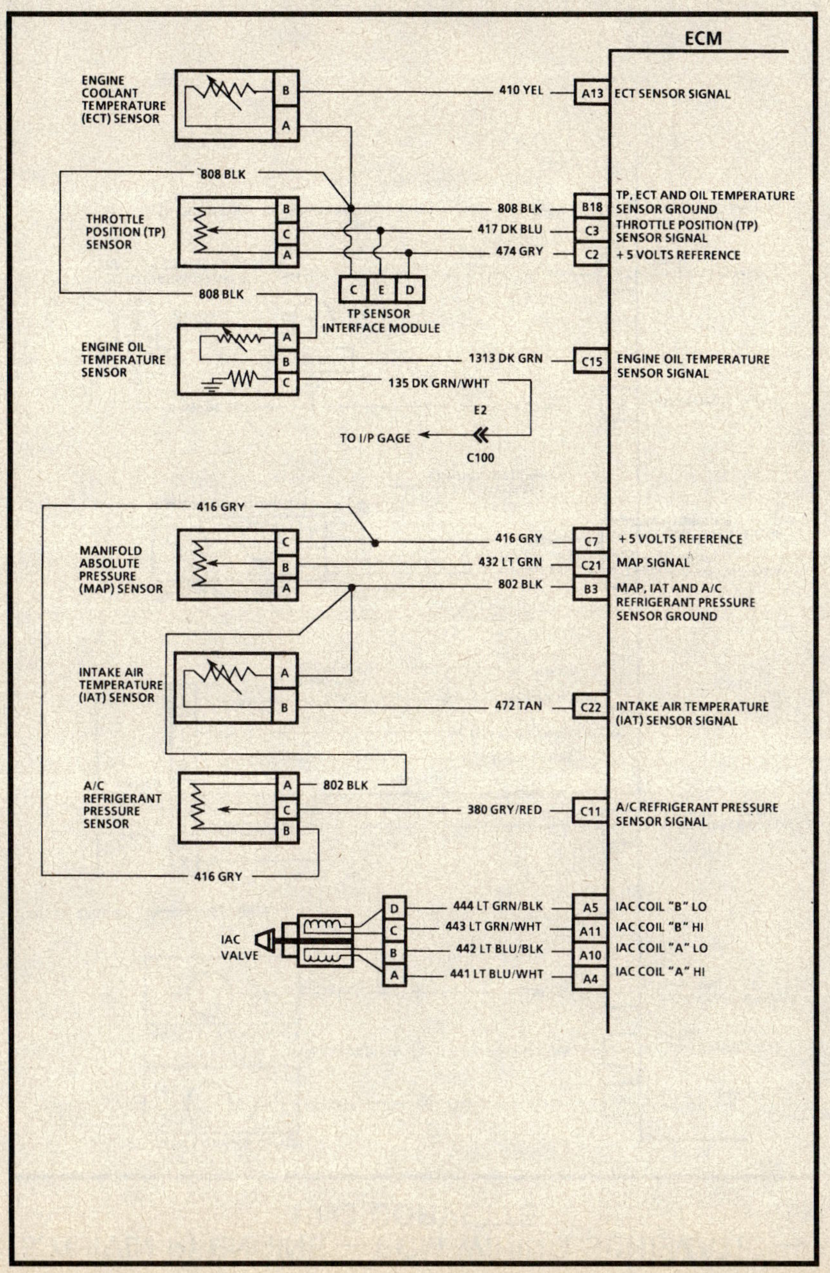

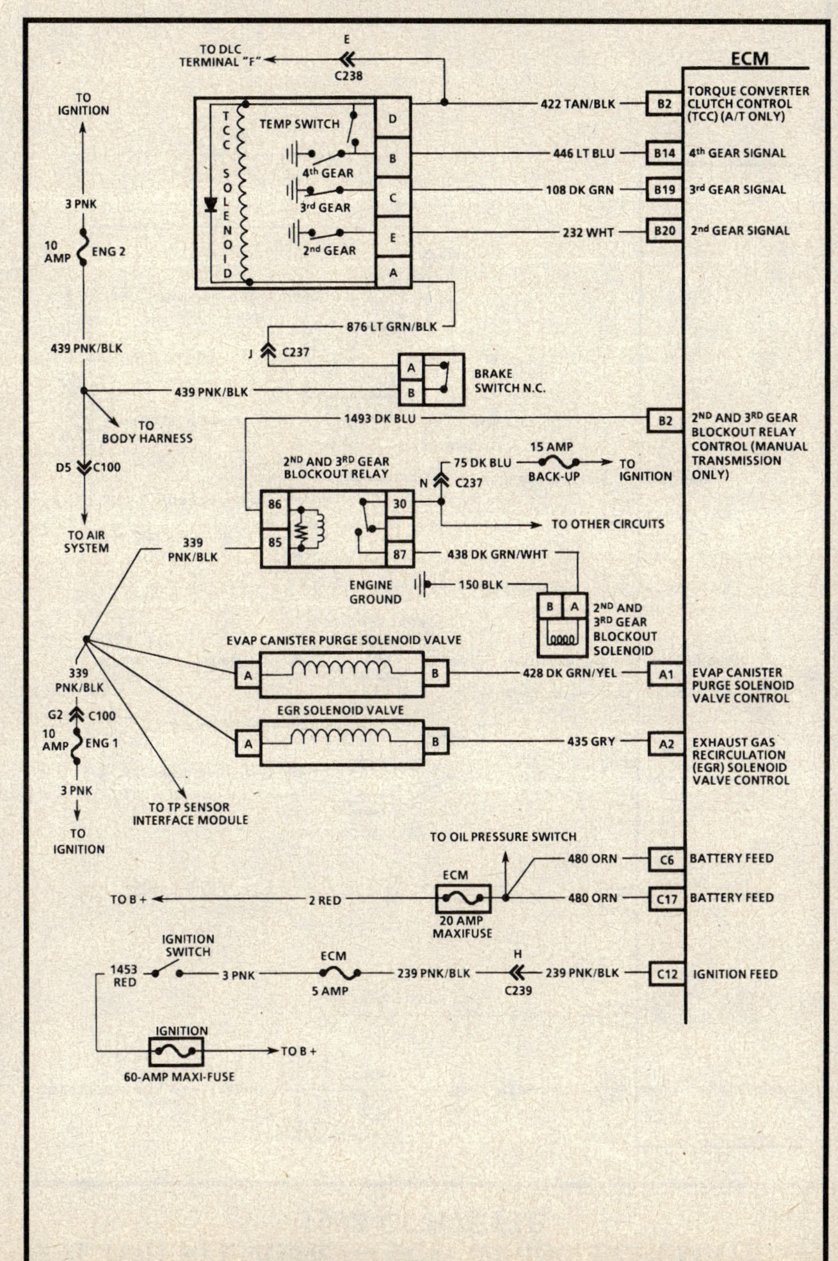

5.7L (VIN P) ENGINE — ECM WIRING SCHEMATIC — 1993 CORVETTE

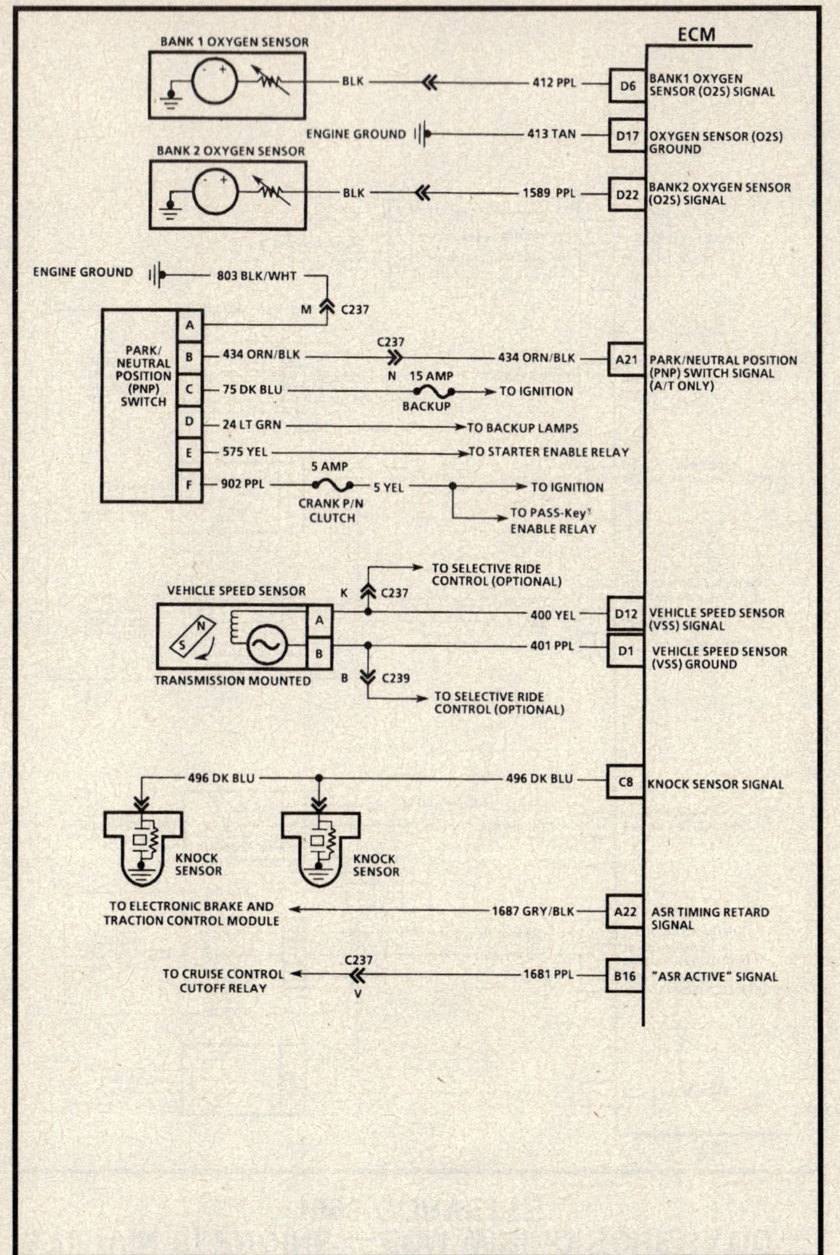

5.7L (VIN P) ENGINE — ECM WIRING SCHEMATIC — 1993 CORVETTE

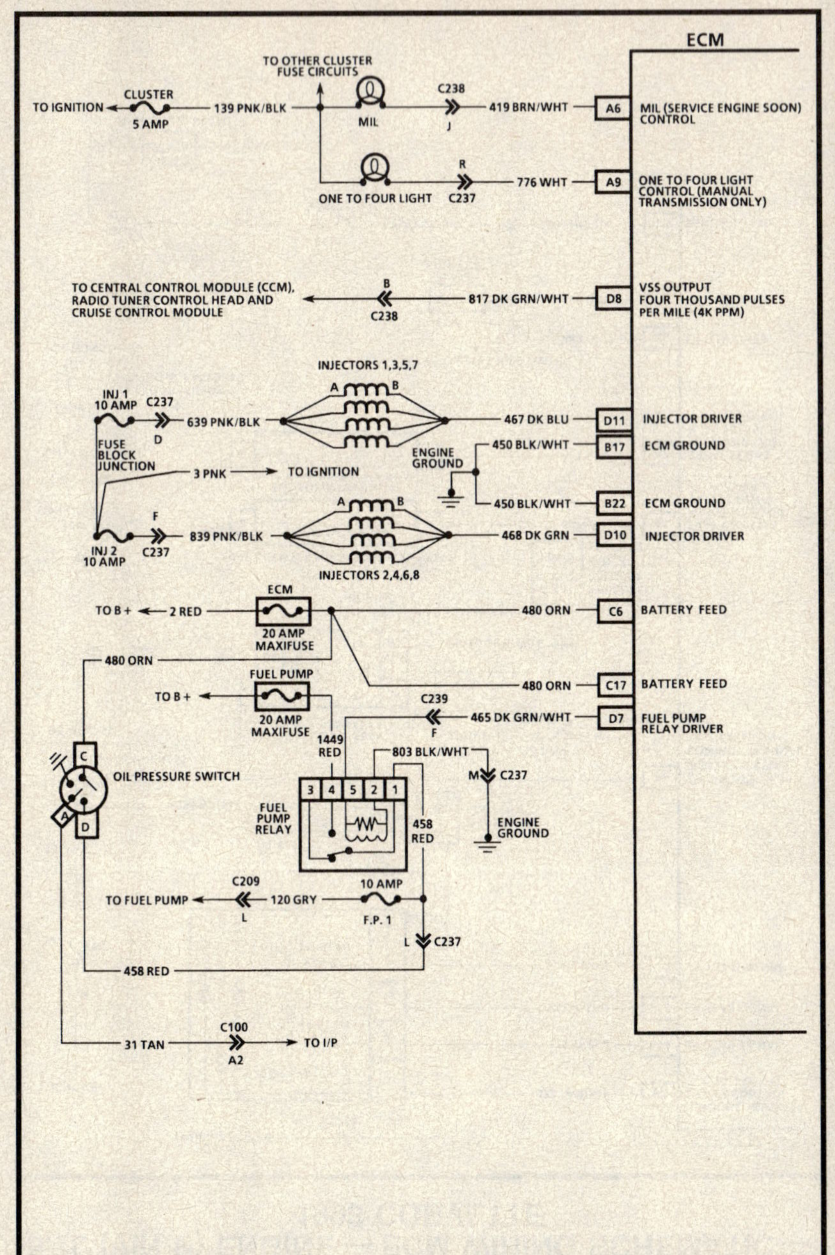

5.7L (VIN P) ENGINE — ECM WIRING SCHEMATIC — 1993 CORVETTE

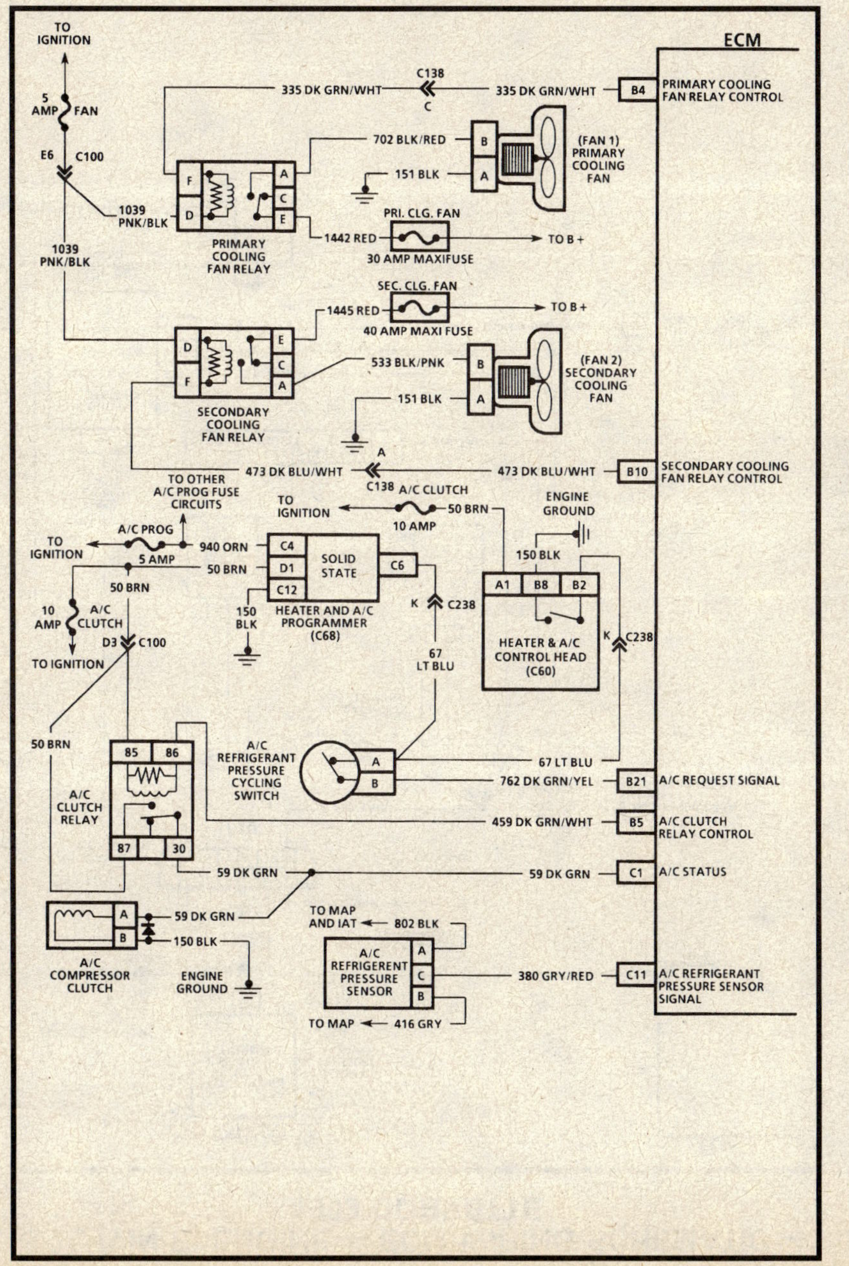

5.7L (VIN P) ENGINE — ECM WIRING SCHEMATIC — 1993 CORVETTE

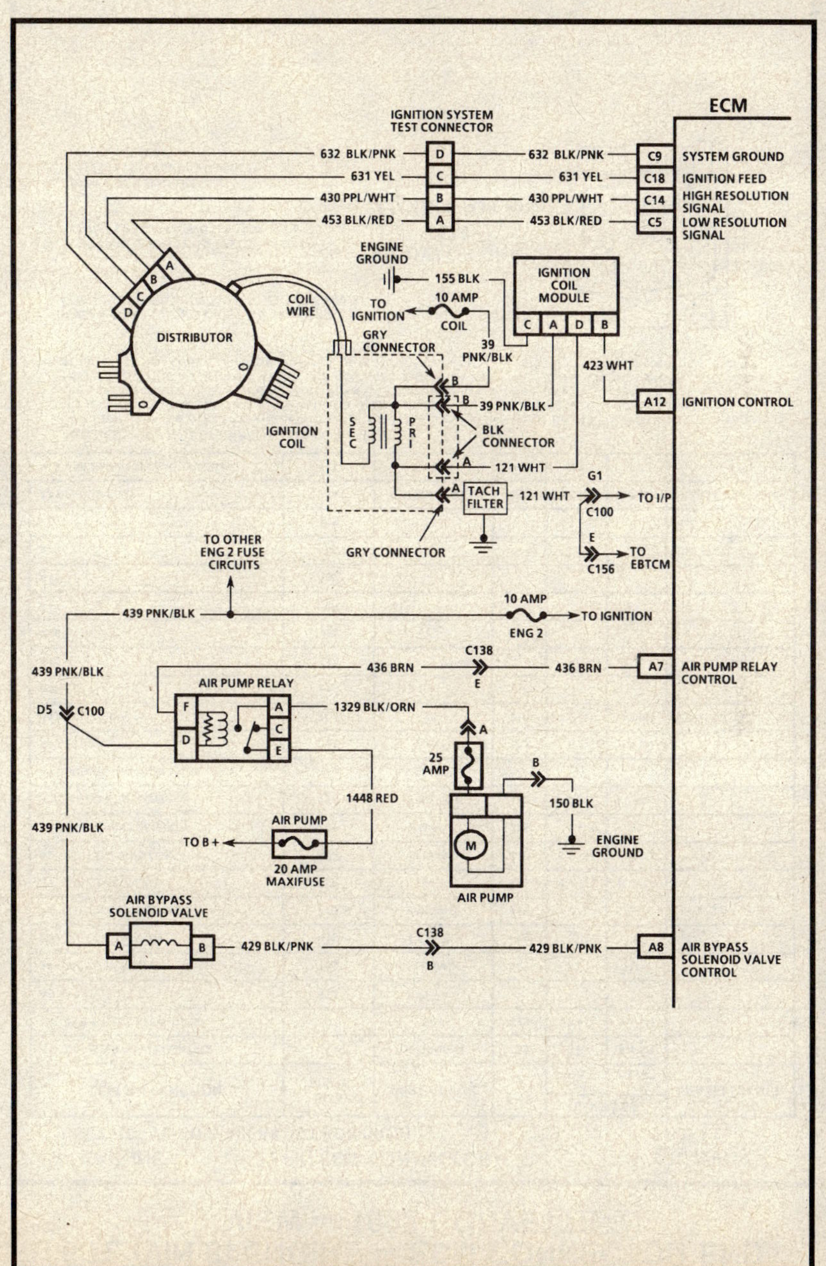

5.7L (VIN P) ENGINE — ECM WIRING SCHEMATIC — 1993 CORVETTE

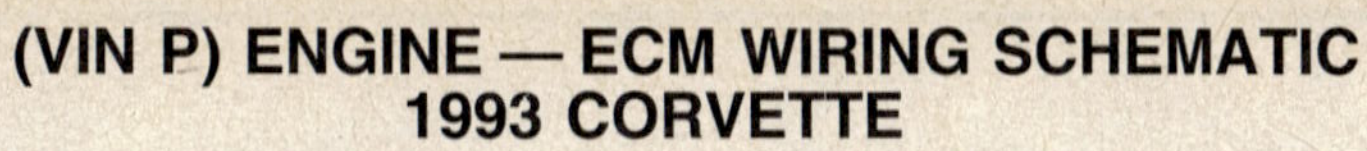

5.7L (VIN 8) ENGINE — ECM CONNECTOR END VIEW — 1992 CORVETTE

"Y" CARLINE — 22 PIN ECM CONNECTOR — 5.7L (VIN P)
*** USE T-100 YELLOW BREAKOUT BOX (BOB) — RPO: LT1**

ECM PIN/FUNCTION	BOB PIN #	WIRE COLOR	CKT #	VOLTAGE KEY "ON"	VOLTAGE ENG "RUN"	6E3 REFERENCE
A1 CANISTER PURGE CONTROL	104	DK GRN/YEL	428	B+	B+ (1)	
A2 EGR SOLENOID CONTROL	101	GRY	435	B+	B+ (1)	
A3	112					
A4 IAC COIL "A" HIGH	109	LT BLU/WHT	441	(3)	(3)	
A5 IAC COIL "B" LOW	119	LT GRN/BLK	444	(3)	(3)	
A6 "SES" LIGHT CONTROL	117	BRN/WHT	419	0 *	B+	CHARTS A-1, A-2
A7 AIR PUMP RELAY	103	BRN	436	B+ (1)	B+ (1)	
A8 ESV SOLENOID CONTROL	102	BLK/PNK	429	B+ (1)	B+ (1)	
A9 1 TO 4 LIGHT CONTROL	111	WHT	776	B+	B+	
A10 IAC COIL "A" LOW	110	LT BLU/BLK	442	(3)	(3)	
A11 IAC COIL "B" HIGH	118	LT GRN/WHT	443	(3)	(3)	
A12 EST	108	WHT	423	0 *	1.1 (3)	
A13 CTS SIGNAL	105	YEL	410	1.8 (4)	1.7 (4)	CODES 14, 15
A14	116					
A15	113					
A16	122					
A17	120					
A18	107					
A19	106					
A20	115					
A21 P/N SWITCH	114	ORN/BLK	434	0 * (5)	0 * (5)	CHART C-1A
A22 ASR TIMING RETARD SIGNAL	121	GRY/BLK	1687	0 *	0 *	

***NOTICE:** DO NOT BACKPROBE ECM CONNECTORS!
This Chart may be used in conjunction with the T-100 Yellow Breakout Box (48921) to obtain voltage present for each circuit listed. Install the BOB between the ECM harness connectors and the ECM, then probe the pin listed under "BOB PIN#". Voltage may vary due to low battery charge or other reasons, but should be very close. All voltages shown in the ENG "RUN" column are typical with engine at idle, closed throttle, normal operating temperature, park or neutral, system in "Closed Loop," all accessories "OFF," and scan tool not installed.

DVM NEGATIVE (BLACK) LEAD MUST BE CONNECTED TO A KNOWN GOOD GROUND.

(1) LESS THAN .5 VOLT WHEN SYSTEM ENABLED.
(2) 12 VOLTS FOR FIRST TWO SECONDS WITH IGNITION "ON."
(3) VARIES.
(4) VARIES WITH TEMPERATURE.
(5) BATTERY VOLTAGE WHEN IN GEAR.
* LESS THAN .5 VOLT.

5.7L (VIN 8) ENGINE — ECM CONNECTOR END VIEW — 1992 CORVETTE

"Y" CARLINE   **22 PIN ECM CONNECTOR**   **5.7L (VIN P) RPO: LT1**
* USE T-100 YELLOW BREAKOUT BOX (BOB)

ECM PIN/FUNCTION	BOB PIN #	WIRE COLOR	CKT #	VOLTAGE KEY "ON"	VOLTAGE ENG "RUN"	6E3 REFERENCE
B1	204					
B2 TCC OR 1 TO 4 UPSHIFT CONTROL	201	TAN/BLK OR DK BLU	422 1493	B+	B+ (1)	
B3 MAP, IAT, & A/C SENSOR GND	212	BLK	802	0*	0*	CODES AFFECTED
B4 PRIMARY COOLING FAN	209	DK GRN/WHT	335	B+	B+ (1)	
B5 A/C CLUTCH CONTROL	219	DK GRN/WHT	459	B+	B+ (1)	
B6 ECM GROUND	217	TAN/WHT	551	0*	0*	
B7	203					
B8	202					
B9	211					
B10 SECONDARY COOLING FAN	210	DK BLU/WHT	473	B+	B+ (1)	
B11	218					
B12	208					
B13	205					
B14 4th GEAR SIGNAL	216	LT BLU	446	B+	B+ (1)	
B15	213					
B16 "ASR ACTIVE" SIGNAL	222	PPL	1681	B+	B+	
B17 ECM GROUND	220	BLK/WHT	450	0*	0*	
B18 TPS, CTS, & EOT GROUND	207	BLK	808	0*	0*	CODES AFFECTED
B19 3rd GEAR SIGNAL	206	DK GRN	108	0*	0*	CODE 72
B20 2nd GEAR SIGNAL	215	WHT	232	0*	0*	CODE 72
B21 A/C REQUEST SIGNAL	214	DK GRN/YEL	762	B+	B+ (1)	
B22 ECM GROUND	221	BLK/WHT	450	0*	0*	

***NOTICE:** DO NOT BACKPROBE ECM CONNECTORS!
This Chart may be used in conjunction with the T-100 Yellow Breakout Box (48921) to obtain voltage present for each circuit listed. Install the BOB between the ECM harness connectors and the ECM, then probe the pin listed under "BOB PIN#." Voltage may vary due to low battery charge or other reasons, but should be very close. All voltages shown in the ENG "RUN" column are typical with engine at idle, closed throttle, normal operating temperature, park or neutral, system in "Closed Loop," all accessories "OFF," and scan tool not installed.

DVM NEGATIVE (BLACK) LEAD MUST BE CONNECTED TO A KNOWN GOOD GROUND.

(1) LESS THAN .5 VOLT WHEN SYSTEM IS ENABLED.
* LESS THAN .5 VOLT.

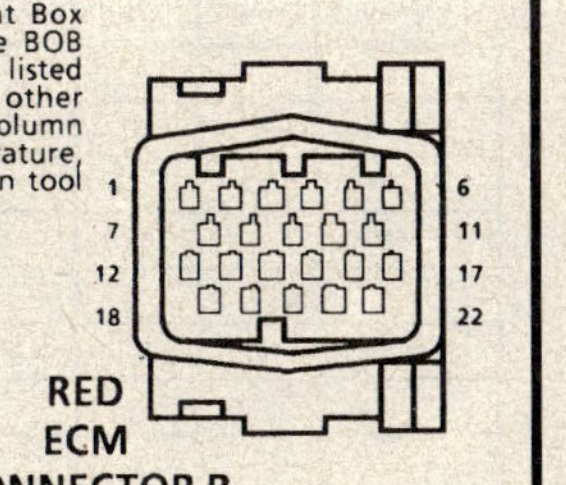

5.7L (VIN 8) ENGINE — ECM CONNECTOR END VIEW — 1992 CORVETTE

"Y" CARLINE   **22 PIN ECM CONNECTOR**   **5.7L (VIN P) RPO: LT1**
* USE T-100 YELLOW BREAKOUT BOX (BOB)

ECM PIN/FUNCTION	BOB PIN #	WIRE COLOR	CKT #	VOLTAGE KEY "ON"	VOLTAGE ENG "RUN"	6E3 REFERENCE
C1 A/C STATUS	304	DK GRN	59	0	0(3)	
C2 +5 VOLTS REFERENCE	301	GRY	474	5	5	CODES AFFECTED
C3 TPS SIGNAL	312	DK BLU	417	.62	.62	CODES 21, 22
C4 IGNITION SHIELD	309	CLEAR	452	0	0	
C5 LOW RESOLUTION	319	BLK/RED	453	1.0 OR 5.0	1.0(4)	CODE 16
C6 BATTERY FEED	317	ORN	480	B+	B+	CHART A-1
C7 +5 VOLTS REFERENCE	303	GRY	416	5	5	CODES AFFECTED
C8 KNOCK SENSOR SIGNAL	302	DK BLU	496	1.5	1.5	CODE 43
C9 REFERENCE LOW	311	BLK/PNK	632	0	0	
C10	310					
C11 A/C PRESSURE SIGNAL	318	GRY/RED	380	1.0	.6-1.0 (5)	CODES 66, 67
C12 IGNITION FEED	308	PNK/BLK	239	B+	B+	CHART A-1
C13	305					
C14 HIGH RESOLUTION	316	PPL/WHT	430	0*	2.5 (4)	CODE 36
C15 EOT SIGNAL	313	DK GRN	1313	1.8 (2)	1.6 (2)	CODES 52, 62
C16	322					
C17 BATTERY FEED	320	ORN	480	B+	B+	CHART A-1
C18 SYSTEM IGNITION FEED	307	YEL	631	B+	B+	
C19	306					
C20						
C21 MAP SIGNAL	314	LT GRN	432	4.8(1)	1.2(1)	CODES 33, 34
C22 IAT SIGNAL	321	TAN	472	2.0(2)	3.0(2)	CODES 23, 25

***NOTICE:** DO NOT BACKPROBE ECM CONNECTORS!
This Chart may be used in conjunction with the T-100 Yellow Breakout Box (48921) to obtain voltage present for each circuit listed. Install the BOB between the ECM harness connectors and the ECM, then probe the pin listed under "BOB PIN#." Voltage may vary due to low battery charge or other reasons, but should be very close. All voltages shown in the ENG "RUN" column are typical with engine at idle, closed throttle, normal operating temperature, park or neutral, system in "Closed Loop," all accessories "OFF," and scan tool not installed.

DVM NEGATIVE (BLACK) LEAD MUST BE CONNECTED TO A KNOWN GOOD GROUND.

(1) VARIES WITH ALTITUDE.
(2) VARIES WITH TEMPERATURE.
(3) B+ WHEN SYSTEM IS ENABLED.
(4) VARIES.
(5) WITH A/C "OFF"
* LESS THAN .5 VOLT.

5.7L (VIN P) ENGINE — ECM CONNECTOR END VIEW — 1993 CORVETTE

"Y" CARLINE   **22 PIN ECM CONNECTOR**   **5.7L (VIN P) RPO: LT1**
* USE T-100 YELLOW BREAKOUT BOX (BOB)

ECM PIN/FUNCTION	BOB PIN #	WIRE COLOR	CKT #	VOLTAGE KEY "ON"	VOLTAGE ENG "RUN"	REFERENCE
A1 EVAP CANISTER PURGE CONTROL	104	DK GRN/YEL	428	B+	B+ (1)	
A2 EGR SOLENOID VALVE CONTROL	101	GRY	435	B+	B+ (1)	
A3	112					
A4 IAC COIL "A" HIGH	109	LT BLU/WHT	441	(3)	(3)	
A5 IAC COIL "B" LOW	119	LT GRN/BLK	444	(3)	(3)	
A6 MIL (SERVICE ENGINE SOON) LIGHT CONTROL	117	BRN/WHT	419	0*	B+	CHARTS A-1, A-2
A7 AIR PUMP RELAY CONTROL	103	BRN	436	B+ (1)	B+ (1)	
A8 BYPASS SOLENOID VALVE CONTROL	102	BLK/PNK	429	B+ (1)	B+ (1)	
A9 1 TO 4 LIGHT CONTROL	111	WHT	776	B+	B+	
A10 IAC COIL "A" LOW	110	LT BLU/BLK	442	(3)	(3)	
A11 IAC COIL "B" HIGH	118	LT GRN/WHT	443	(3)	(3)	
A12 IGNITION CONTROL	108	WHT	423	0*	1.1 (3)	
A13 ECT SIGNAL	105	YEL	410	1.8(4)	1.7(4)	DTC(s)14, 15
A14	116					
A15	113					
A16	122					
A17	120					
A18	107					
A19	106					
A20	115					
A21 PNP SIGNAL	114	ORN/BLK	434	0* (5)	0* (5)	CHART C-1A
A22 ASR TIMING RETARD SIGNAL	121	GRY/BLK	1687	0*	0*	

*NOTICE: DO NOT BACKPROBE ECM CONNECTORS!
This Chart may be used in conjunction with the T-100 Yellow Breakout Box (48921) to obtain voltage present for each circuit listed. Install the BOB between the ECM harness connectors and the ECM, then probe the pin listed under "BOB PIN#". Voltage may vary due to low battery charge or other reasons, but should be very close. All voltages shown in the ENG "RUN" column are typical with engine at idle, closed throttle, normal operating temperature, park or neutral, system in "Closed Loop", all accessories "OFF", and scan tool not installed.

DVM NEGATIVE (BLACK) LEAD MUST BE CONNECTED TO A KNOWN GOOD GROUND.

(1) LESS THAN .5 VOLT WHEN SYSTEM ENABLED.
(2) 12 VOLTS FOR FIRST TWO SECONDS WITH IGNITION "ON".
(3) VARIES WITH TEMPERATURE.
(4) BATTERY VOLTAGE WHEN IN GEAR.
(5) LESS THAN .5 VOLT.

GRAY ECM CONNECTOR A

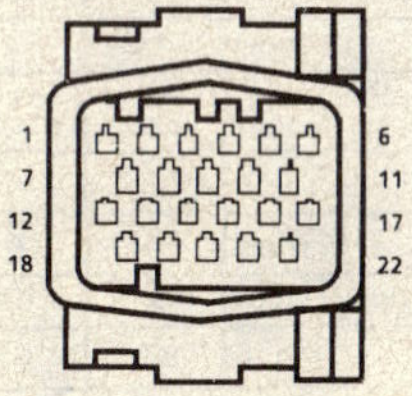

5.7L (VIN 8) ENGINE — ECM CONNECTOR END VIEW — 1992 CORVETTE

"Y" CARLINE   **22 PIN ECM CONNECTOR**   **5.7L (VIN P) RPO: LT1**
* USE T-100 YELLOW BREAKOUT BOX (BOB)

ECM PIN/FUNCTION	BOB PIN #	WIRE COLOR	CKT #	VOLTAGE KEY "ON"	VOLTAGE ENG "RUN"	REFERENCE
D1 VSS GROUND	404	PPL	401			CODE 24
D2	401					
D3	412					
D4 SERIAL DATA	409	TAN	800	(1)	(1)	CHART A-2
D5	419					
D6 LEFT O2 SENSOR SIGNAL	417	PPL	412	.38	.1-.9 (1)	CODES 13, 44
D7 FUEL PUMP RELAY DRIVER	403	DK GRN/WHT	465	0 (2)	B+	CHART A-5
D8 VSS OUTPUT (4KPPM)	402	DK GRN/WHT	817	(3)	(3)	CODE 24
D9	411					
D10 INJECTOR DRIVER	410	DK GRN	468	B+	B+	CHART A-3
D11 INJECTOR DRIVER	418	DK BLU	467	B+	B+	CHART A-3
D12 VSS SIGNAL	408	YEL	400	(3)	(3)	CODE 24
D13	405					
D14	416					
D15 SERIAL DATA	413	TAN	800	(1)	(1)	CHART A-2
D16	422					
D17 O2 GROUND	420	TAN	413	0*	0*	CODES 13, 63
D18	407					
D19	406					
D20 DIAGNOSTIC "TEST" TERM.	415	WHT/BLK	451	5	5	CHART A-2
D21	414					
D22 RIGHT O2 SENSOR SIGNAL	421	PPL	1589	.36	.1-.9 (1)	CODES 63, 64

*NOTICE: DO NOT BACKPROBE ECM CONNECTORS!
This Chart may be used in conjunction with the T-100 Yellow Breakout Box (48921) to obtain voltage present for each circuit listed. Install the BOB between the ECM harness connectors and the ECM, then probe the pin listed under "BOB PIN#". Voltage may vary due to low battery charge or other reasons, but should be very close. All voltages shown in the ENG "RUN" column are typical with engine at idle, closed throttle, normal operating temperature, park or neutral, system in "Closed Loop", all accessories "OFF", and scan tool not installed.

DVM NEGATIVE (BLACK) LEAD MUST BE CONNECTED TO A KNOWN GOOD GROUND.

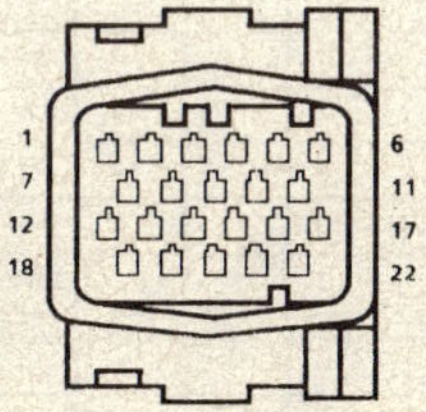

(1) VARIES
(2) B + FOR TWO SECONDS WITH IGNITION "ON"
(3) VARIES DEPENDING ON POSITION OF DRIVE WHEELS
* LESS THAN .5 VOLT.

BROWN ECM CONNECTOR D

5.7L (VIN P) ENGINE — ECM CONNECTOR END VIEW — 1993 CORVETTE

"Y" CARLINE — **22 PIN ECM CONNECTOR** — 5.7L (VIN P) RPO: LT1
* USE T-100 YELLOW BREAKOUT BOX (BOB)

ECM PIN/FUNCTION	BOB PIN #	WIRE COLOR	CKT #	VOLTAGE KEY "ON"	VOLTAGE ENG "RUN"	REFERENCE
B1	204					
B2 TCC OR 2nd & 3rd GEAR BLOCKOUT RELAY CONTROL	201	TAN/BLK OR DK BLU	422 1493	B+	B+ (1)	
B3 MAP, IAT, & A/C REFRIGERANT SENSOR GND	212	BLK	802	0 *	0 *	DTC(s) AFFECTED
B4 PRIMARY COOLING FAN	209	DK GRN/WHT	335	B+	B+ (1)	
B5 A/C CLUTCH CONTROL	219	DK GRN/WHT	459	B+	B+ (1)	
B6 ECM GROUND	217	TAN/WHT	551	0 *	0 *	
B7	203					
B8	202					
B9	211					
B10 SECONDARY COOLING FAN	210	DK BLU/WHT	473	B+	B+ (1)	
B11	218					
B12	208					
B13	205					
B14 4th GEAR SIGNAL	216	LT BLU	446	B+	B+ (1)	
B15	213					
B16 "ASR ACTIVE" SIGNAL	222	PPL	1681	B+	B+	
B17 ECM GROUND	220	BLK/WHT	450	0 *	0 *	
B18 TP SENSOR, ECT, & ENGINE OIL TEMPERATURE GROUND	207	BLK	808	0 *	0 *	DTC(s) AFFECTED
B19 3rd GEAR SIGNAL	206	DK GRN	108	0 *	0 *	DTC 72
B20 2nd GEAR SIGNAL	215	WHT	232	0 *	0 *	DTC 72
B21 A/C REQUEST SIGNAL	214	DK GRN/YEL	762	B+	B+ (1)	
B22 ECM GROUND	221	BLK/WHT	450	0 *	0 *	

***NOTICE:** DO NOT BACKPROBE ECM CONNECTORS!
This Chart may be used in conjunction with the T-100 Yellow Breakout Box (48921) to obtain voltage present for each circuit listed. Install the BOB between the ECM harness connectors and the ECM, then probe the pin listed under "BOB PIN#". Voltage may vary due to low battery charge or other reasons, but should be very close. All voltages shown in the ENG "RUN" column are typical with engine at idle, closed throttle, normal operating temperature, park or neutral, system in "Closed Loop," all accessories "OFF," and scan tool not installed.

DVM NEGATIVE (BLACK) LEAD MUST BE CONNECTED TO A KNOWN GOOD GROUND.

RED ECM CONNECTOR B

(1) LESS THAN .5 VOLT WHEN SYSTEM IS ENABLED.
* LESS THAN .5 VOLT.

5.7L (VIN P) ENGINE — ECM CONNECTOR END VIEW — 1993 CORVETTE

"Y" CARLINE — **22 PIN ECM CONNECTOR** — 5.7L (VIN P) RPO: LT1
* USE T-100 YELLOW BREAKOUT BOX (BOB)

ECM PIN/FUNCTION	BOB PIN #	WIRE COLOR	CKT #	VOLTAGE KEY "ON"	VOLTAGE ENG "RUN"	REFERENCE
C1 A/C STATUS	304	DK GRN	59	0	0 (3)	
C2 + 5 VOLTS REFERENCE	301	GRY	474	5	5	DTC(s) AFFECTED
C3 TP SENSOR SIGNAL	312	DK BLU	417	.62	.62	DTC(s) 21, 22
C4	309					
C5 LOW RESOLUTION SIGNAL	319	BLK/RED	453	1.0 OR 5.0	1.0(4)	DTC 16
C6 BATTERY FEED	317	ORN	480	B+	B+	CHART A-1
C7 + 5 VOLTS REFERENCE	303	GRY	416	5	5	DTC(s) AFFECTED
C8 KNOCK SENSOR SIGNAL	302	DK BLU	496	1.5	1.5	DTC 43
C9 REFERENCE LOW	311	BLK/PNK	632	0	0	
C10	310					
C11 A/C REFRIGERANT PRESSURE SIGNAL	318	GRY/RED	380	1.0	.6-1.0 (5)	DTC(s) 66, 67
C12 IGNITION FEED	308	PNK/BLK	239	B+	B+	CHART A-1
C13	305					
C14 HIGH RESOLUTION	316	PPL/WHT	430	0 *	2.5(4)	DTC 36
C15 ENGINE OIL TEMPERATURE SIGNAL	313	DK GRN	1313	1.8 (2)	1.6 (2)	DTC(s) 52, 62
C16	322					
C17 BATTERY FEED	320	ORN	480	B+	B+	CHART A-1
C18 DISTRIBUTOR IGNITION FEED	307	YEL	631	B+	B+	
C19	306					
C20	315					
C21 MAP SIGNAL	314	LT GRN	432	4.8 (1)	1.2 (1)	DTC(s) 33, 34
C22 IAT SIGNAL	321	TAN	472	2.0 (2)	3.0 (2)	DTC(s) 23, 25

***NOTICE:** DO NOT BACKPROBE ECM CONNECTORS!
This Chart may be used in conjunction with the T-100 Yellow Breakout Box (48921) to obtain voltage present for each circuit listed. Install the BOB between the ECM harness connectors and the ECM, then probe the pin listed under "BOB PIN#". Voltage may vary due to low battery charge or other reasons, but should be very close. All voltages shown in the ENG "RUN" column are typical with engine at idle, closed throttle, normal operating temperature, park or neutral, system in "Closed Loop," all accessories "OFF," and scan tool not installed.

DVM NEGATIVE (BLACK) LEAD MUST BE CONNECTED TO A KNOWN GOOD GROUND.

GREEN ECM CONNECTOR C

(1) VARIES WITH ALTITUDE.
(2) VARIES WITH TEMPERATURE.
(3) B + WHEN SYSTEM IS ENABLED.
(4) VARIES.
(5) WITH A/C "OFF"
* LESS THAN .5 VOLT.

5.7L (VIN P) ENGINE — ECM CONNECTOR END VIEW — 1993 CORVETTE

"Y" CARLINE   22 PIN ECM CONNECTOR   5.7L (VIN P)
* USE T-100 YELLOW BREAKOUT BOX (BOB)   RPO: LT1

ECM PIN/FUNCTION	BOB PIN #	WIRE COLOR	CKT #	VOLTAGE KEY "ON"	VOLTAGE ENG "RUN"	REFERENCE
D1 VSS GROUND	404	PPL	401			DTC 24
D2	401					
D3	412					
D4 SERIAL DATA	409	TAN	800	(1)	(1)	CHART A-2
D5	419					
D6 BANK1 (LEFT) OXYGEN SENSOR SIGNAL	417	PPL	412	.38	.1-.9 (1)	DTC(s) 13, 44
D7 FUEL PUMP RELAY DRIVER	403	DK GRN/WHT	465	0 (2)	B+	CHART A-5
D8 VSS OUTPUT (4K PPM)	402	DK GRN/WHT	817	(3)	(3)	DTC 24
D9	411					
D10 INJECTOR DRIVER	410	DK GRN	468	B+	B+	CHART A-3
D11 INJECTOR DRIVER	418	DK BLU	467	B+	B+	CHART A-3
D12 VSS SIGNAL	408	YEL	400	(3)	(3)	DTC 24
D13	405					
D14	416					
D15 SERIAL DATA	413	TAN	800	(1)	(1)	CHART A-2
D16	422					
D17 O2S GROUND	420	TAN	413	0 *	0 *	DTC(s) 13, 63
D18	407					
D19	406					
D20 DIAGNOSTIC "TEST" TERM.	415	WHT/BLK	451	5	5	CHART A-2
D21	414					
D22 BANK2 (RIGHT) OXYGEN SENSOR SIGNAL	421	PPL	1589	.36	.1-.9 (1)	DTC(s) 63, 64

***NOTICE: DO NOT BACKPROBE ECM CONNECTORS!**
This Chart may be used in conjunction with the T-100 Yellow Breakout Box (48921) to obtain voltage present for each circuit listed. Install the BOB between the ECM harness connectors and the ECM, then probe the pin listed under "BOB PIN#". Voltage may vary due to low battery charge or other reasons, but should be very close. All voltages shown in the ENG "RUN" column are typical with engine at idle, closed throttle, normal operating temperature, park or neutral, system in "Closed Loop," all accessories "OFF," and scan tool not installed.

DVM NEGATIVE (BLACK) LEAD MUST BE CONNECTED TO A KNOWN GOOD GROUND.

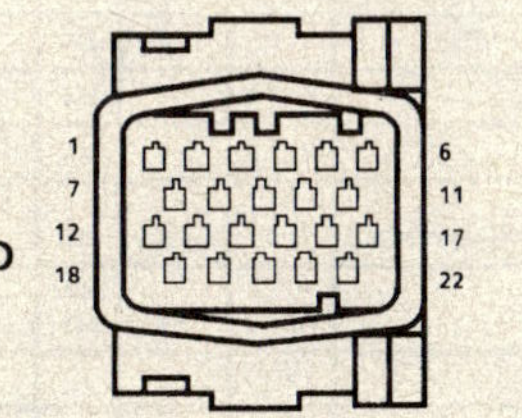

(1) VARIES.
(2) B + FOR TWO SECONDS WITH IGNITION "ON."
(3) VARIES DEPENDING ON POSITION OF DRIVE WHEELS.
* LESS THAN .5 VOLT.

5.7L (VIN 8) ENGINE — ON-BOARD DIAGNOSTIC SYSTEM CHART — 1992 CORVETTE

DIAGNOSTIC CIRCUIT CHECK
5.7L (VIN P) "Y" CARLINE

Circuit Description:

The diagnostic circuit check is an organized approach to identifying a problem created by an electronic engine control system malfunction. It must be the starting point for any driveability complaint diagnosis, because it directs the service technician to the next logical step in diagnosing the complaint. Understanding the chart and using it correctly will reduce diagnostic time and prevent the unnecessary replacement of good parts.

Test Description: Number(s) below refer to circled number(s) on the diagnostic chart.

1. This step is a check for the proper operation of the "Service Engine Soon" light. The "SES" light should be "ON" steady.
2. No "SES" light at this point indicates that there is a problem with the "SES" light circuit or the ECM control of that circuit.
3. This test checks the ability of the ECM to control the "SES" light. With the diagnostic terminal grounded, the "SES" light should flash a Code 12 three times, followed by any trouble code stored in memory.
4. Most of the   procedures use a Tech 1 to aid diagnosis, therefore, serial data must be available. If a PROM error is present, the ECM may have been able to flash Code 12/51, but not enable serial data.
5. Although the ECM is powered up, a "Cranks But Will Not Run" symptom could exist because of an ECM or system problem.
6. This step will isolate if the customer complaint is an "SES" light or a driveability problem with no "SES" light. An invalid code may be the result of a faulty "Scan" tool, PROM or ECM.
7. Comparison of actual control system data with the typical values is a quick check to determine if any parameter is not within limits. Keep in mind that a base engine problem (i.e. advanced cam timing) may substantially alter sensor values.
8. Installation of a "Scan" tool will provide a good ground path for the ECM and may hide a driveability complaint due to poor ECM grounds.
9. If the actual data is not within the typical values established, the charts in "Component Systems," Section   will provide a functional check of the suspect component or system.

5.7L (VIN 8) ENGINE — ON-BOARD DIAGNOSTIC SYSTEM CHART — 1992 CORVETTE

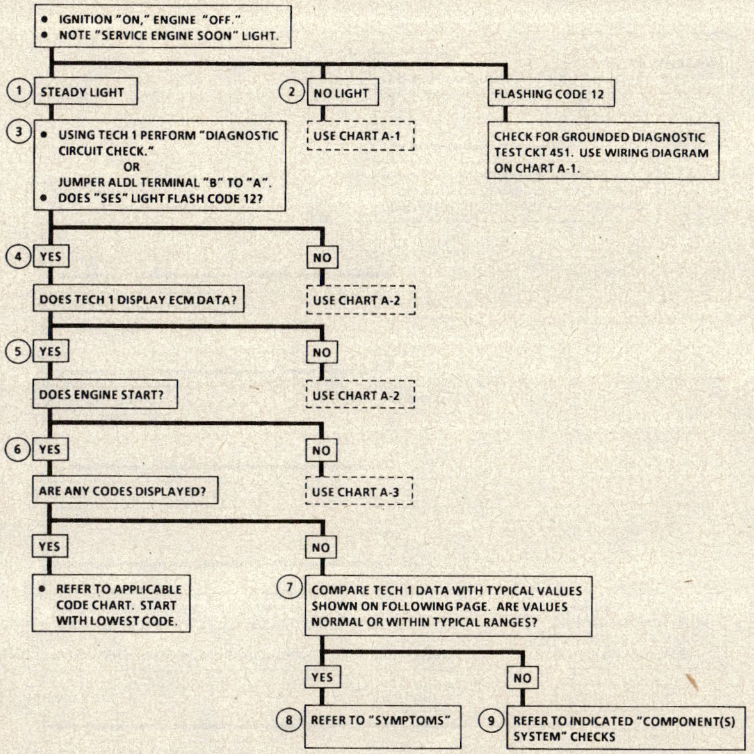

5.7L (VIN P) ENGINE — ON-BOARD DIAGNOSTIC SYSTEM CHART — 1993 CORVETTE

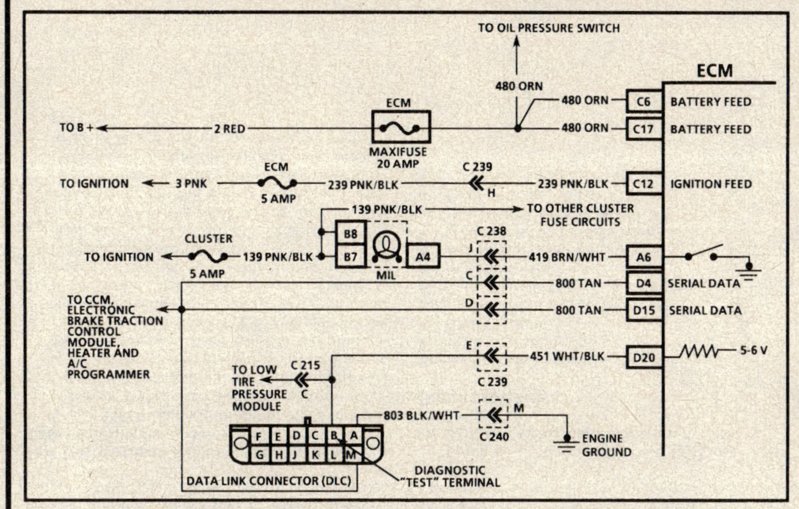

ON-BOARD DIAGNOSTIC (OBD) SYSTEM CHECK
5.7L (VIN P) "Y" CARLINE

Circuit Description:

The On-Board Diagnostic (OBD) system check is an organized approach to identifying a problem created by an electronic engine control system malfunction. It must be the starting point for any driveability complaint diagnosis, because it directs the service technician to the next logical step in diagnosing the complaint. Understanding the chart and using it correctly will reduce diagnostic time and prevent the unnecessary replacement of good parts.

Test Description: Number(s) below refer to circled number(s) on the diagnostic chart.

1. This step is a check for the proper operation of the MIL (Service Engine Soon). The MIL should be "ON" steady.
2. No MIL at this point indicates that there is a problem with the MIL circuit or the ECM control of that circuit.
3. This test checks the ability of the ECM to control the MIL. With the diagnostic terminal grounded, the MIL should flash a DTC 12 three times, followed by any Diagnostic Trouble Code (DTC) stored in memory.
4. Most of the 6E procedures use a Tech 1 to aid diagnosis, therefore, serial data must be available. If a PROM error is present, the ECM may have been able to flash DTCs 12/51, but not enable serial data.
5. Although the ECM is powered up, a "Cranks But Will Not Run" symptom could exist because of an ECM or system problem.
6. This step will isolate if the customer complaint is a MIL or a driveability problem with no MIL. An invalid DTC may be the result of a faulty scan tool, PROM or ECM.
7. Comparison of actual control system data with the typical values is a quick check to determine if any parameter is not within limits. Keep in mind that a base engine problem (i.e. advanced cam timing) may substantially alter sensor values.
8. Installation of a scan tool will provide a good ground path for the ECM and may hide a driveability complaint due to poor ECM grounds.
9. If the actual data is not within the typical values established, the charts in "Component Systems," will provide a functional check of the suspect component or system.

5.7L (VIN 8) ENGINE — SYSTEM DIAGNOSTIC CHARTS — 1992 CORVETTE

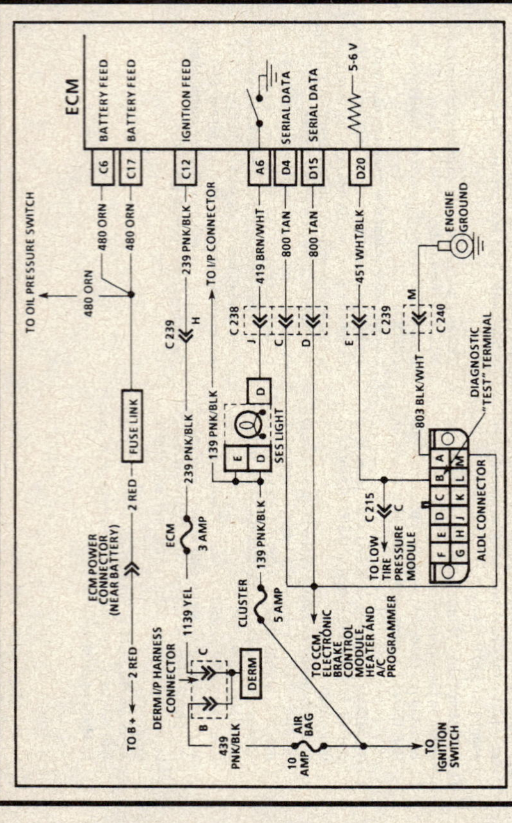

CHART A-1

NO "SERVICE ENGINE SOON" LIGHT
5.7L (VIN P) "Y" CARLINE

Circuit Description:

There should always be a steady "Service Engine Soon" light when the ignition is "ON" and engine stopped. Ignition voltage is supplied directly to the light bulb. The Electronic Control Module (ECM) will control the light and turn it "ON" by providing a ground path through CKT 419 to the ECM.

Test Description: Number(s) below refer to circled number(s) on the diagnostic chart.

1. Use J 35616 tool kit to probe ECM harness terminals, so that you do not damage them.
2. Using a test light connected to 12 volts, probe each of the system ground circuits to be sure a good ground is present. Refer to ECM terminal end view in front of this section for ECM pin locations of ground circuits.
3. If the fusible link is open, determine short to ground in CKT 480. CKT 480 also consists of the fuel pump oil pressure switch.
4. Make sure to check CKT 239 if the ECM fuse is open. Refer to the ECM wiring diagrams for information. The SIR system Diagnostic Energy Reserve Module (DERM) must be connected to the I/P harness or ignition voltage will not be supplied to the ECM and this will not allow the engine to start

Check all 439 circuits if the AIR BAG fuse is open. Refer to ECM wiring diagnosis for information.

Diagnostic Aids:

- Engine runs OK, check:
  - Faulty light bulb.
  - CKT 419 open.
- Engine cranks but will not run, check:
  - Continuous battery - fusible link open.
  - ECM fuse open.
  - Battery CKTs 480 to ECM open.
  - Ignition CKT 239 to ECM open.
  - Poor connection to ECM.
  - Faulty ECM ground circuit(s).
  - PASS-Key® circuit problem.

5.7L (VIN P) ENGINE — ON-BOARD DIAGNOSTIC SYSTEM CHART — 1993 CORVETTE

ON-BOARD DIAGNOSTIC (OBD) SYSTEM CHECK
5.7L (VIN P) "Y" CARLINE

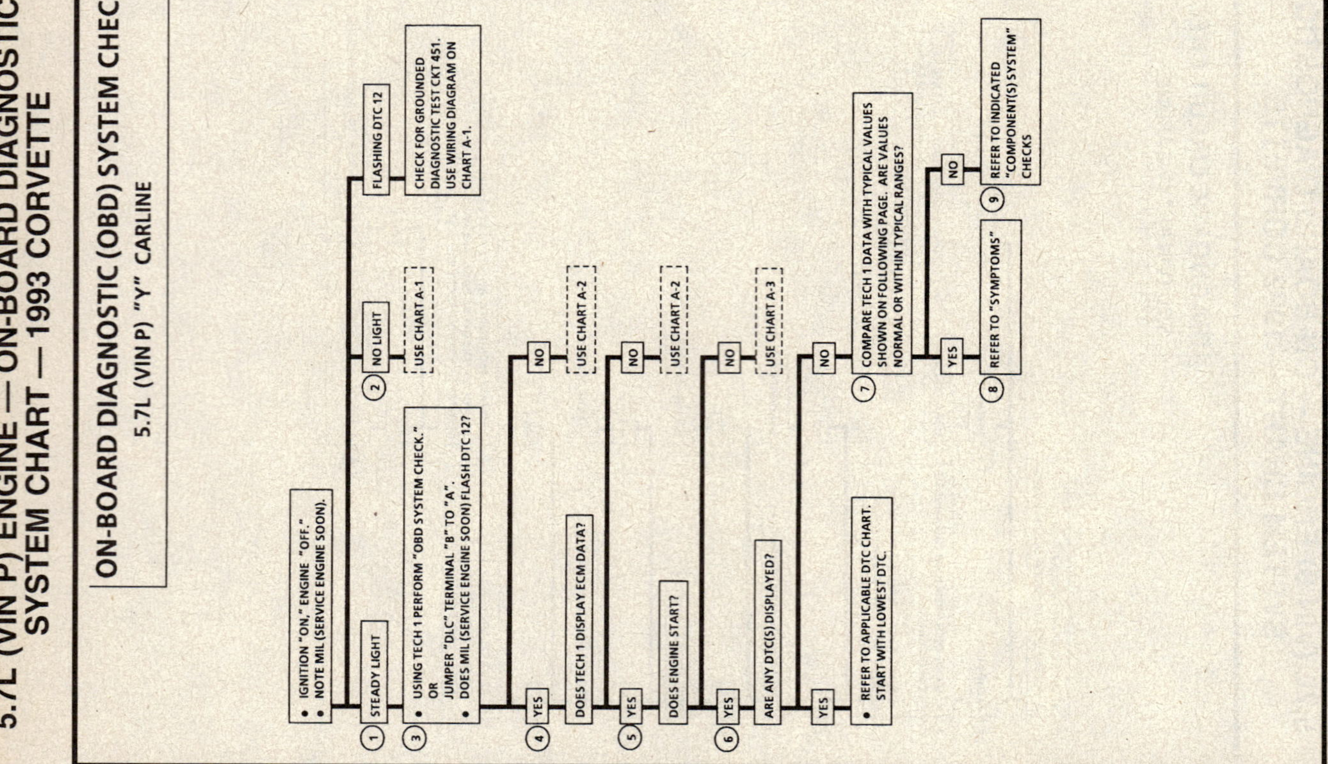

5.7L (VIN 8) ENGINE — SYSTEM DIAGNOSTIC CHARTS — 1992 CORVETTE

CHART A-1

NO "SERVICE ENGINE SOON" LIGHT
5.7L (VIN P) "Y" CARLINE

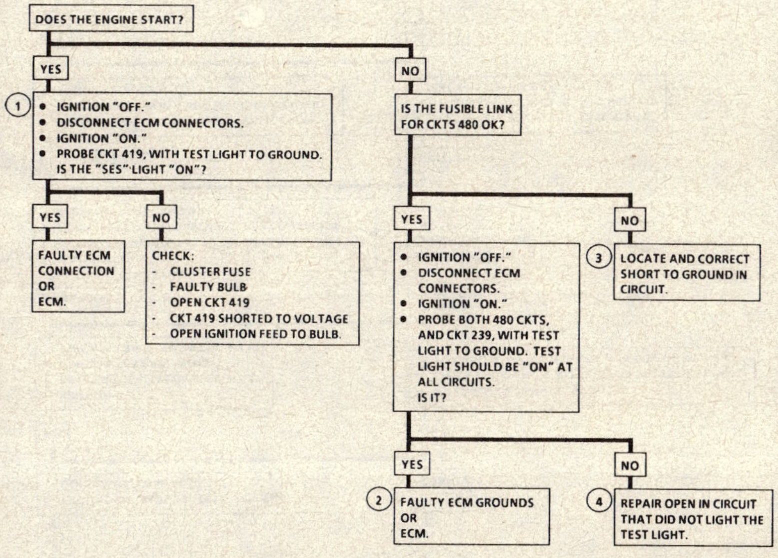

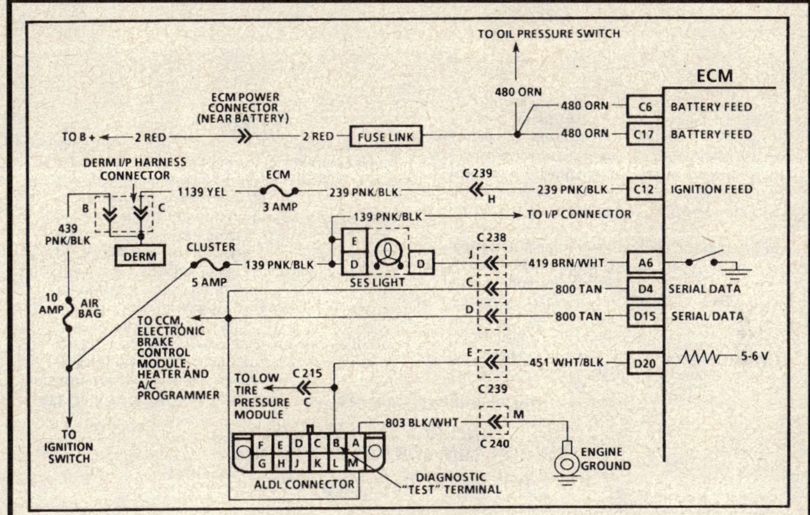

"AFTER REPAIRS," CONFIRM "CLOSED LOOP" OPERATION AND NO "SERVICE ENGINE SOON" LIGHT.

5.7L (VIN 8) ENGINE — SYSTEM DIAGNOSTIC CHARTS — 1992 CORVETTE

CHART A-2

NO ALDL DATA OR WILL NOT FLASH CODE 12
"SERVICE ENGINE SOON" LIGHT "ON" STEADY
5.7L (VIN P) "Y" CARLINE

Circuit Description:

There should always be a steady "Service Engine Soon" light when the ignition is "ON" and engine stopped. Ignition voltage is supplied to the light bulb. The Electronic Control Module (ECM) will turn the light "ON" by grounding CKT 419 in the ECM

With the diagnostic "test" terminal grounded, the light should flash a Code 12, followed by any trouble code(s) stored in memory.

A steady light with the diagnostic "test" terminal grounded suggests a short to ground in the light control CKT 419 or an open in diagnostic CKTs 451 or 803.

Test Description: Number(s) below refer to circled number(s) on the diagnostic chart.

1. If there is a problem with the ECM that causes a Tech 1 not to display serial data, the ECM should not flash a Code 12. If Code 12 is flashing, check CKT 451 for short to ground. If Code 12 does flash, make sure that the Tech 1 is working properly on another vehicle. If the Tech 1 is functioning properly and CKT 800 is OK, the MEM-CAL or ECM may be at fault for the no ALDL symptom. If the 800 circuit is short to ground or both 800 circuits are open, the Central Control Module (CCM) will set Code 41 (ECM serial data) and Code 54 (fuel enable failure) because the CCM did not receive ECM serial data

2. If the light goes "OFF," when the ECM connector is disconnected, CKT 419 is not shorted to ground.

3. This step will check for an open diagnostic CKT 451.

4. At this point, the "Service Engine Soon" light wiring is OK. The problem is a faulty ECM or MEM-CAL. If Code 12 does not flash, the ECM should be replaced using the original MEM-CAL. Replace the MEM-CAL only after trying an ECM, as a defective MEM-CAL is an unlikely cause of the problem.

5.7L (VIN 8) ENGINE — SYSTEM DIAGNOSTIC CHARTS — 1992 CORVETTE

CHART A-2
NO ALDL DATA OR WILL NOT FLASH CODE 12
"SERVICE ENGINE SOON" LIGHT "ON" STEADY
5.7L (VIN P) "Y" CARLINE

- IGNITION "ON," ENGINE "OFF."
  IS THE "SERVICE ENGINE SOON" LIGHT "ON"?

YES →
- GROUND DIAGNOSTIC "TEST" TERMINAL.
  DOES LIGHT FLASH CODE 12?

NO →
SEE CHART A-1

(2)
- IGNITION "OFF."
- DISCONNECT ECM CONNECTOR "A."
- IGNITION "ON" AND NOTE "SERVICE ENGINE SOON" LIGHT.

(1)
- IF PROBLEM WAS NO ALDL DATA:
- CHECK SERIAL DATA 800 CKTS FOR OPENS OR SHORTS TO GROUND. IF OK, IT IS A FAULTY ECM OR MEM-CAL.

LIGHT "OFF"

LIGHT "ON"

(3)
- IGNITION "OFF."
- JUMPER TERMINALS "A" TO "B" AT ALDL CONNECTOR
- CONNECT TEST LIGHT BETWEEN ECM CONNECTOR TERMINAL "D20" AND B +.

REPAIR SHORT TO GROUND IN CKT 419.

LIGHT "ON"

LIGHT "OFF"

(4)
- CHECK MEM-CAL FOR PROPER INSTALLATION.
- IF OK, REPLACE ECM USING ORIGINAL MEM-CAL.
- RECHECK FOR CODE 12.

- CHECK FOR OPEN IN ALDL DIAGNOSTIC TERMINALS "A" AND "B" (CKT 803 AND CKT 451), REPAIR AS NECESSARY.

NO CODE 12

CODE 12

REPLACE MEM-CAL.

SYSTEM OK.

"AFTER REPAIRS," CONFIRM "CLOSED LOOP" OPERATION AND NO "SERVICE ENGINE SOON" LIGHT.

5.7L (VIN 8) ENGINE — SYSTEM DIAGNOSTIC CHARTS — 1992 CORVETTE

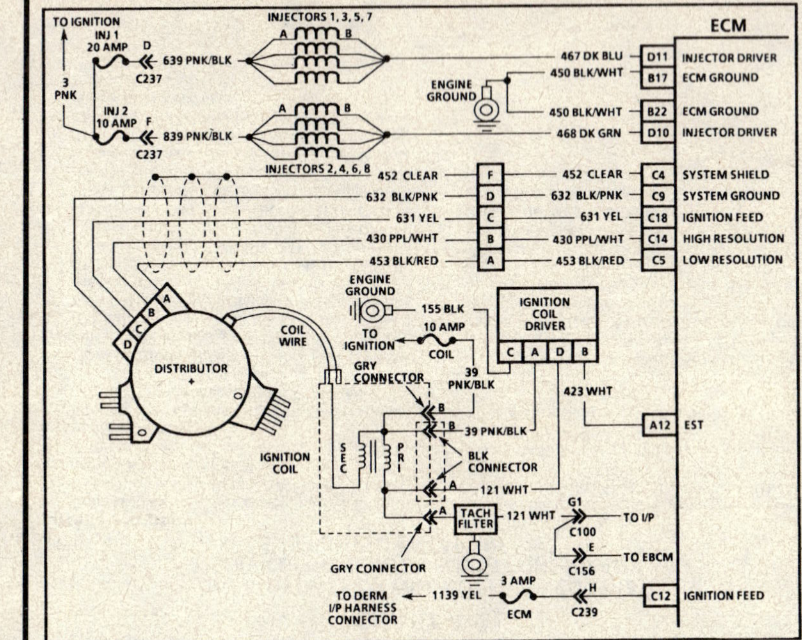

CHART A-3 (Page 1 of 2)
ENGINE CRANKS BUT WILL NOT RUN
5.7L (VIN P) "Y" CARLINE

Test Description: Number(s) below refer to circled number(s) on the diagnostic chart.

1. Perform diagnostic circuit check to determine if the ECM and the "Service Engine Soon" lights are functioning.
2. CHART C-4 will address all problems related to the causes of a no spark condition.
3. The test light should blink, indicating the ECM is controlling the injectors OK. All lights should blink at the same brightness. If any injector light blinks dimmly an injector may be shorted. Compare injector resistance values
4. Use fuel pressure gage J 34730-1 or equivalent. Wrap a shop towel around the fuel pressure tap to absorb any small amount of fuel leakage that may occur when installing the gage.

Diagnostic Aids:

- The SIR system will not allow the engine to start if the Diagnostic Energy Reserve Module (DERM) to I/P connector is disconnected

- An EGR valve sticking open can cause a low air/fuel ratio during cranking.
- Unless engine enters "Clear Flood" at the first indication of a flooding condition, it can result in a no start.
- A defective MAP sensor may cause a no start or a stall after start. Disconnect the sensor. The ECM will use a default value for the sensor. If the condition is corrected and the connections are OK, replace the sensor.
  If above are all OK, refer to "Symptoms," Section "Hard Start."

5.7L (VIN 8) ENGINE — SYSTEM DIAGNOSTIC CHARTS — 1992 CORVETTE

5.7L (VIN 8) ENGINE — SYSTEM DIAGNOSTIC CHARTS — 1992 CORVETTE

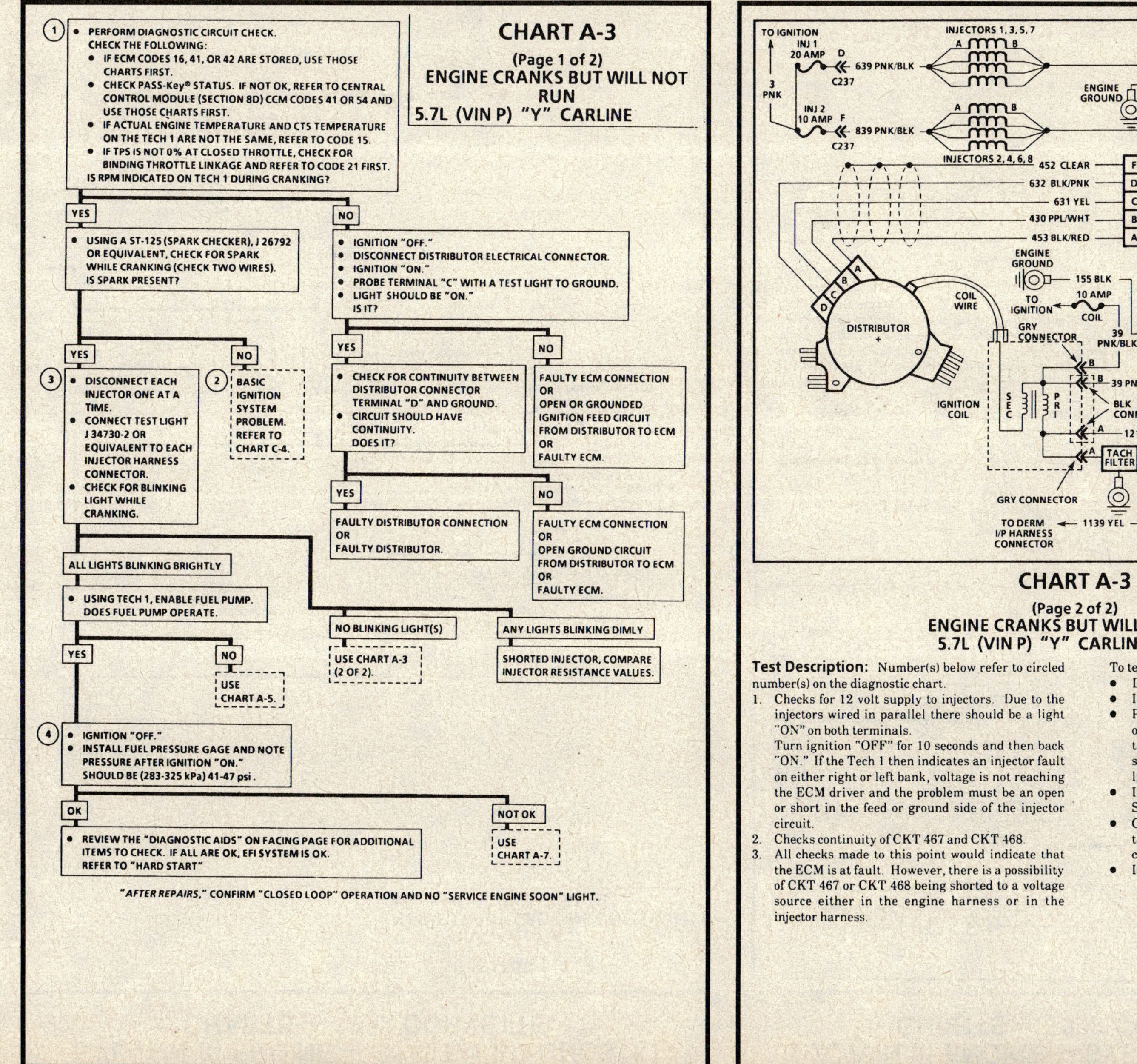

CHART A-3
(Page 2 of 2)
ENGINE CRANKS BUT WILL NOT RUN
5.7L (VIN P) "Y" CARLINE

Test Description: Number(s) below refer to circled number(s) on the diagnostic chart.

1. Checks for 12 volt supply to injectors. Due to the injectors wired in parallel there should be a light "ON" on both terminals.

   Turn ignition "OFF" for 10 seconds and then back "ON." If the Tech 1 then indicates an injector fault on either right or left bank, voltage is not reaching the ECM driver and the problem must be an open or short in the feed or ground side of the injector circuit.

2. Checks continuity of CKT 467 and CKT 468.

3. All checks made to this point would indicate that the ECM is at fault. However, there is a possibility of CKT 467 or CKT 468 being shorted to a voltage source either in the engine harness or in the injector harness.

To test for this condition:
• Disconnect all injectors.
• Ignition "ON."
• Probe CKT 467 and CKT 468 on the ECM side of injector harness with a test light connected to ground. (Test one injector harness on each side of engine.) There should be no light. If light is "ON" repair short to voltage.
• If OK, check the resistance of the injectors. Should be 10 ohms or more.
• Check injector harness connector. Be sure terminals are not backed out of connector and contacting each other.
• If all OK, replace ECM.

5.7L (VIN 8) ENGINE — SYSTEM DIAGNOSTIC CHARTS — 1992 CORVETTE

CHART A-3

(Page 2 of 2)
ENGINE CRANKS BUT WILL NOT RUN
5.7L (VIN P) "Y" CARLINE

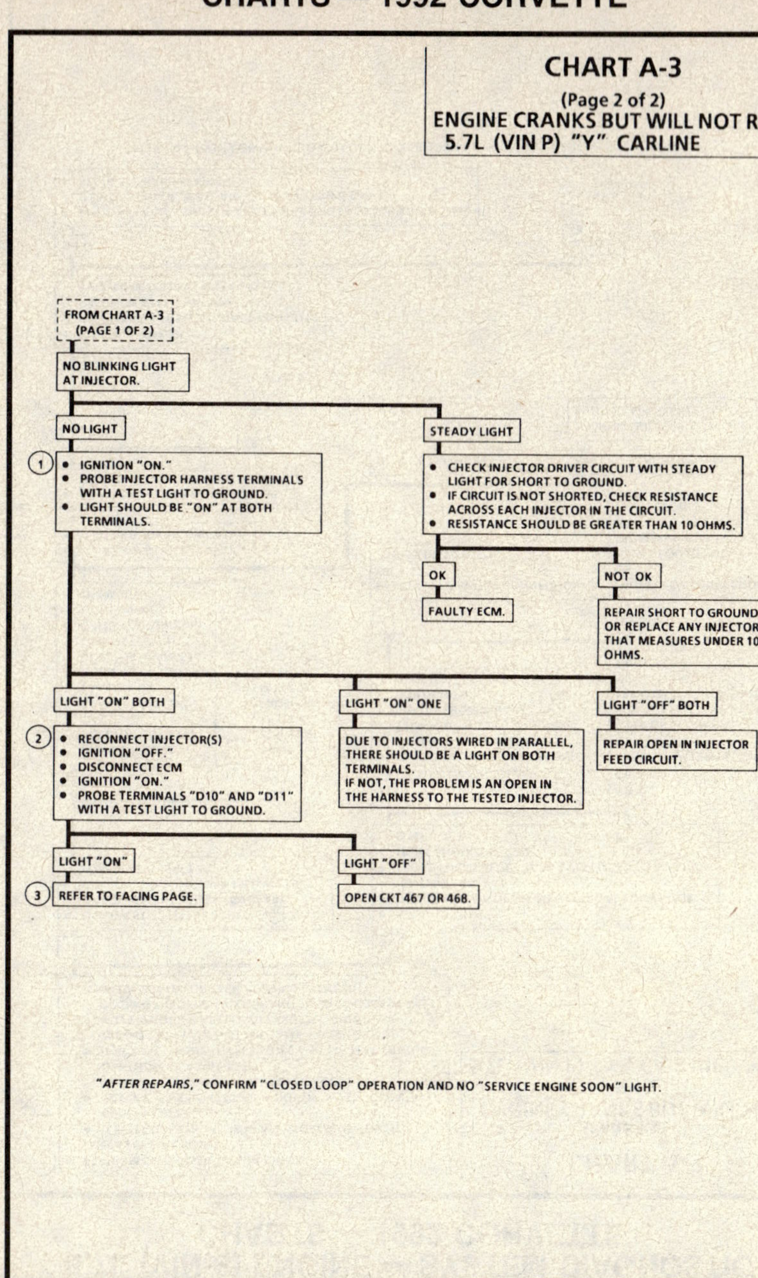

5.7L (VIN 8) ENGINE — SYSTEM DIAGNOSTIC CHARTS — 1992 CORVETTE

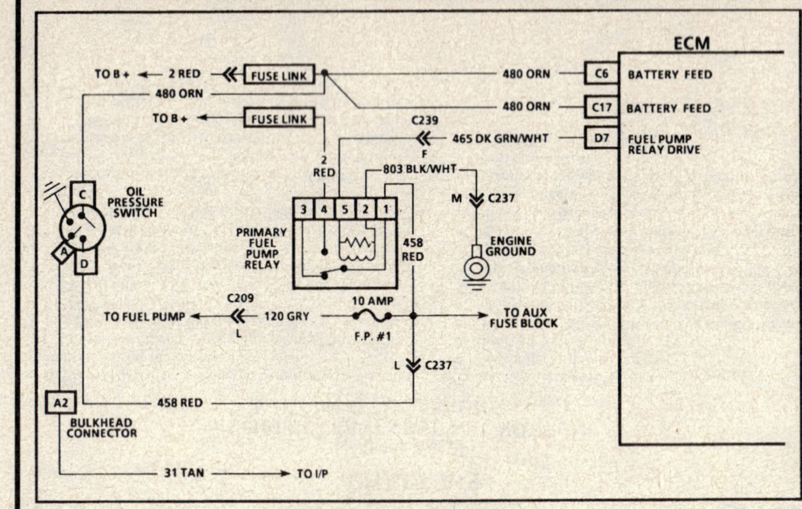

CHART A-5

FUEL PUMP RELAY CIRCUIT
5.7L (VIN P) "Y" CARLINE

Circuit Description:

When the ignition switch is turned "ON," the Electronic Control Module (ECM) will turn "ON" the in-tank fuel pump. It will remain "ON" as long as the ECM is receiving ignition low resolution reference pulses from the ignition system.

If there are no reference pulses, the ECM will shut "OFF" the fuel pump about 2-3 seconds after key "ON," or when engine stops. When sufficient oil pressure is present to close the oil pressure switch, the fuel pump will remain "ON."

The pump will deliver fuel to the fuel rail and injectors, then to the pressure regulator, where the system pressure is controlled to 284-325 kPa (41 - 47 psi). Excess fuel is then returned to the fuel tank.

When the engine is stopped, the fuel pump can be turned "ON" by using the Tech 1. Improper fuel system pressure will result in one or all of the following symptoms:

• Cranks but won't run.
• Code 44 and 64.
• Code 45 and 65.
• Cuts out, may feel like ignition problem.
• Poor fuel economy, loss of power.
• Hesitation.
• Code 55.

5.7L (VIN 8) ENGINE — SYSTEM DIAGNOSTIC CHARTS — 1992 CORVETTE

CHART A-5
FUEL PUMP RELAY CIRCUIT
5.7L (VIN P) "Y" CARLINE

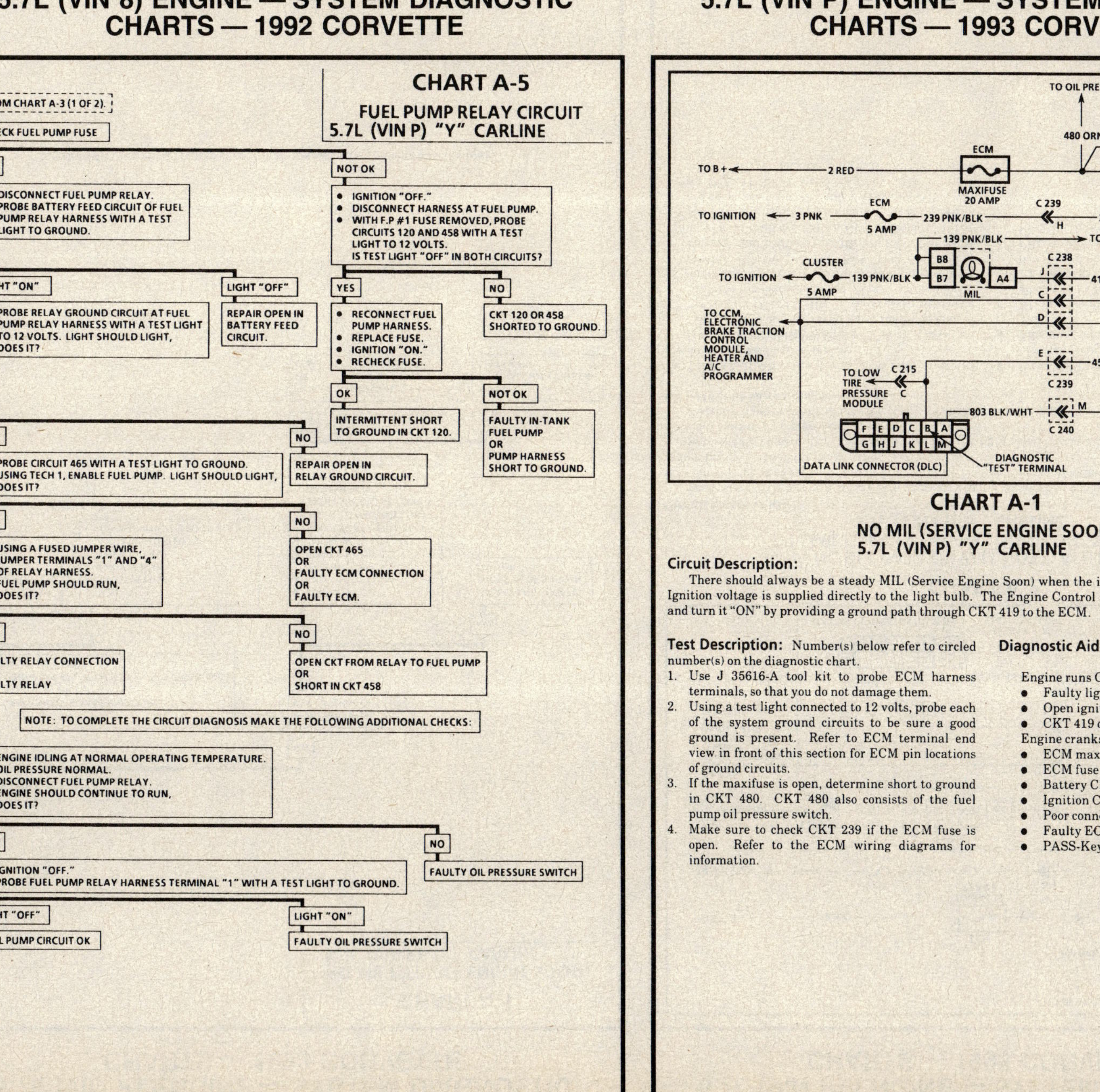

5.7L (VIN P) ENGINE — SYSTEM DIAGNOSTIC CHARTS — 1993 CORVETTE

CHART A-1
NO MIL (SERVICE ENGINE SOON)
5.7L (VIN P) "Y" CARLINE

Circuit Description:

There should always be a steady MIL (Service Engine Soon) when the ignition is "ON" and engine stopped. Ignition voltage is supplied directly to the light bulb. The Engine Control Module (ECM) will control the light and turn it "ON" by providing a ground path through CKT 419 to the ECM.

Test Description: Number(s) below refer to circled number(s) on the diagnostic chart.

1. Use J 35616-A tool kit to probe ECM harness terminals, so that you do not damage them.
2. Using a test light connected to 12 volts, probe each of the system ground circuits to be sure a good ground is present. Refer to ECM terminal end view in front of this section for ECM pin locations of ground circuits.
3. If the maxifuse is open, determine short to ground in CKT 480. CKT 480 also consists of the fuel pump oil pressure switch.
4. Make sure to check CKT 239 if the ECM fuse is open. Refer to the ECM wiring diagrams for information.

Diagnostic Aids:

Engine runs OK, check:
- Faulty light bulb.
- Open ignition feed to bulb.
- CKT 419 open.

Engine cranks but will not run, check:
- ECM maxifuse open.
- ECM fuse open.
- Battery CKTs 480 to ECM open.
- Ignition CKT 239 to ECM open.
- Poor connection to ECM.
- Faulty ECM ground circuit(s).
- PASS-Key® circuit problem.

5.7L (VIN P) ENGINE — SYSTEM DIAGNOSTIC CHARTS — 1993 CORVETTE

CHART A-1

NO MIL (SERVICE ENGINE SOON)
5.7L (VIN P) "Y" CARLINE

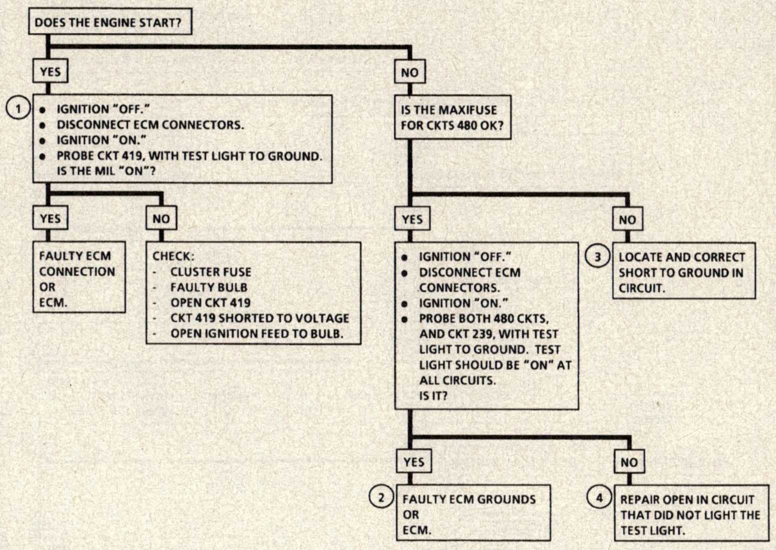

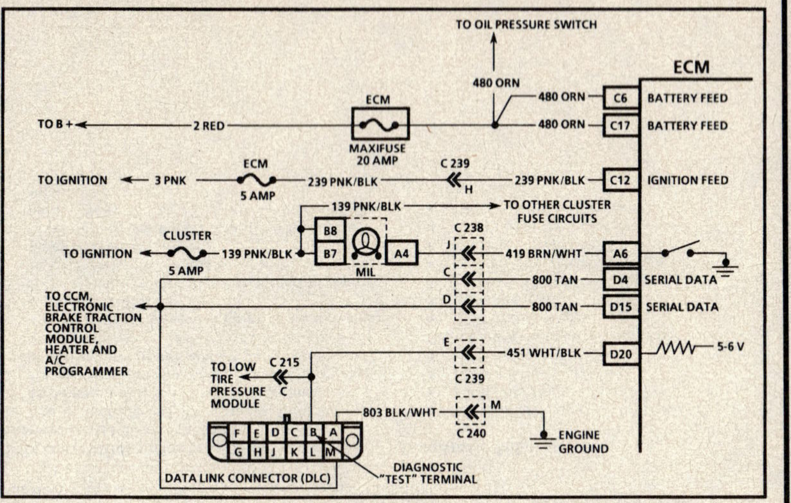

"AFTER REPAIRS," CONFIRM "CLOSED LOOP" OPERATION AND NO MIL (SERVICE ENGINE SOON).

5.7L (VIN P) ENGINE — SYSTEM DIAGNOSTIC CHARTS — 1993 CORVETTE

CHART A-2

NO DLC DATA OR WILL NOT FLASH DTC 12
MIL (SERVICE ENGINE SOON) "ON" STEADY
5.7L (VIN P) "Y" CARLINE

Circuit Description:

There should always be a steady MIL (Service Engine Soon) when the ignition is "ON" and engine stopped. Ignition voltage is supplied to the light bulb. The Engine Control Module (ECM) will turn the light "ON" by grounding CKT 419 in the ECM.

With the diagnostic "test" terminal grounded, the light should flash a DTC 12, followed by any Diagnostic Trouble Codes (DTCs) stored in memory.

A steady light with the diagnostic "test" terminal grounded suggests a short to ground in the light control CKT 419 or an open in diagnostic CKTs 451 or 803.

Test Description: Number(s) below refer to circled number(s) on the diagnostic chart.

1. If there is a problem with the ECM that causes a Tech 1 not to display serial data, the ECM should not flash a DTC 12. If DTC 12 is flashing, check CKT 451 for short to ground. If DTC 12 does flash, make sure that the Tech 1 is working properly on another vehicle. If the Tech 1 is functioning properly and CKT 800 is OK, the PROM or ECM may be at fault for the no DLC symptom. If the 800 circuit is short to ground or both 800 circuits are open, the Central Control Module (CCM) will set DTC 41 (ECM serial data) and DTC 54 (fuel enable failure) because the CCM did not receive ECM serial data.

2. If the light goes "OFF," when the ECM connector is disconnected, CKT 419 is not shorted to ground.

3. This step will check for an open diagnostic CKT 451.

4. At this point, the MIL (Service Engine Soon) wiring is OK. The problem is a faulty ECM or PROM. If DTC 12 does not flash, the ECM should be replaced using the original PROM. Replace the PROM only after trying an ECM, as a defective PROM is an unlikely cause of the problem.

5.7L (VIN P) ENGINE — SYSTEM DIAGNOSTIC CHARTS — 1993 CORVETTE

CHART A-2
NO DLC DATA OR WILL NOT FLASH DTC 12 MIL (SERVICE ENGINE SOON) "ON" STEADY 5.7L (VIN P) "Y" CARLINE

- IGNITION "ON," ENGINE "OFF." IS THE MIL "ON"?

YES →
- GROUND DIAGNOSTIC "TEST" TERMINAL. DOES LIGHT FLASH DTC 12?

NO →
(2)
- IGNITION "OFF."
- DISCONNECT ECM CONNECTOR "A".
- IGNITION "ON" AND NOTE MIL.

LIGHT "OFF" →
(3)
- IGNITION "OFF."
- JUMPER TERMINALS "A" TO "B" AT DATA LINK CONNECTOR
- CONNECT TEST LIGHT BETWEEN ECM CONNECTOR TERMINAL "D20" AND B +.

LIGHT "ON" →
(4)
- CHECK PROM FOR PROPER INSTALLATION.
- IF OK, REPLACE ECM USING ORIGINAL PROM.
- RECHECK FOR DTC 12.

NO DTC 12 → REPLACE PROM.

LIGHT "ON" → REPAIR SHORT TO GROUND IN CKT 419.

LIGHT "OFF" →
- CHECK FOR OPEN IN DLC DIAGNOSTIC TERMINALS "A" AND "B" (CKT 803 AND CKT 451), REPAIR AS NECESSARY.

DTC 12 → SYSTEM OK.

NO →
SEE CHART A-1

YES →
(1)
- IF PROBLEM WAS NO DLC DATA:
- CHECK SERIAL DATA 800 CKTS FOR OPENS OR SHORTS TO GROUND. IF OK, IT IS A FAULTY ECM OR PROM.

"AFTER REPAIRS," CONFIRM "CLOSED LOOP" OPERATION AND NO MIL (SERVICE ENGINE SOON).

5.7L (VIN P) ENGINE — SYSTEM DIAGNOSTIC CHARTS — 1993 CORVETTE

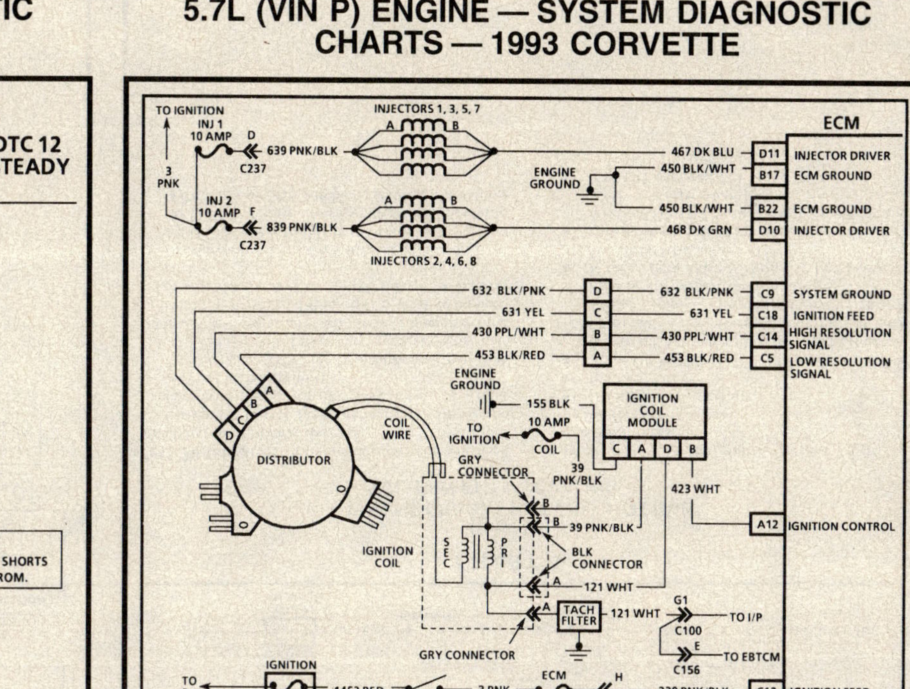

CHART A-3
(Page 1 of 2)
ENGINE CRANKS BUT WILL NOT RUN 5.7L (VIN P) "Y" CARLINE

Test Description: Number(s) below refer to circled number(s) on the diagnostic chart.

1. The OBD system check must be performed first, before proceeding with this diagnostic chart. The OBD system check will determine if the ECM and MIL are functioning.
2. CHART C-4 will address all problems related to the causes of a no spark condition.
3. The test light should blink, indicating the ECM is controlling the injectors OK. All lights should blink at the same brightness. If any injector light blinks dimly an injector may be shorted. Compare injector resistance values.
4. Use fuel pressure gage J 34730-1. Wrap a shop towel around the fuel pressure tap to absorb any small amount of fuel leakage that may occur when installing the gage.

Diagnostic Aids:

- An EGR valve sticking open can cause a low air/fuel ratio during cranking.
- Unless engine enters "Clear Flood" at the first indication of a flooding condition, it can result in a no start.
- A defective MAP sensor may cause a no start or a stall after start. Disconnect the sensor. The ECM will use a default value for the sensor. If the condition is corrected and the connections are OK, replace the sensor.

If above are all OK, refer to "Symptoms," Section "Hard Start."

5.7L (VIN P) ENGINE — SYSTEM DIAGNOSTIC CHARTS — 1993 CORVETTE

CHART A-3
(Page 1 of 2)
ENGINE CRANKS BUT WILL NOT RUN
5.7L (VIN P) "Y" CARLINE

1. PERFORM ON-BOARD DIAGNOSTIC (OBD) SYSTEM CHECK. CHECK THE FOLLOWING:
   - IF ECM DTC(s) 16, 41, OR 42 ARE STORED, USE THOSE CHARTS FIRST.
   - CHECK PASS-Key® STATUS. IF NOT OK, REFER TO CENTRAL CONTROL MODULE (SECTION 8D) CCM DTC(s) 41 OR 54 AND USE THOSE CHARTS FIRST.
   - IF ACTUAL ENGINE TEMPERATURE AND ECT SENSOR TEMPERATURE ON THE TECH 1 ARE NOT THE SAME, REFER TO DTC 15.
   - IF TP SENSOR IS NOT 0% AT CLOSED THROTTLE, CHECK FOR BINDING THROTTLE LINKAGE AND REFER TO DTC 21 FIRST.
   IS RPM INDICATED ON TECH 1 DURING CRANKING?

YES →
- USING A ST-125 (SPARK CHECKER), J 26792, CHECK FOR SPARK WHILE CRANKING (CHECK TWO WIRES). IS SPARK PRESENT?

NO →
- IGNITION "OFF."
- DISCONNECT DISTRIBUTOR ELECTRICAL CONNECTOR.
- IGNITION "ON."
- PROBE TERMINAL "C" WITH A TEST LIGHT TO GROUND.
- LIGHT SHOULD BE "ON." IS IT?

3.
   - DISCONNECT EACH INJECTOR ONE AT A TIME.
   - CONNECT TEST LIGHT J 34730-2C TO EACH INJECTOR HARNESS CONNECTOR.
   - CHECK FOR BLINKING LIGHT WHILE CRANKING.

NO → (2) BASIC IGNITION SYSTEM PROBLEM. REFER TO CHART C-4.

YES/NO → CHECK FOR CONTINUITY BETWEEN DISTRIBUTOR CONNECTOR TERMINAL "D" AND GROUND.
- CIRCUIT SHOULD HAVE CONTINUITY. DOES IT?

YES → FAULTY DISTRIBUTOR CONNECTION OR FAULTY DISTRIBUTOR.

NO → FAULTY ECM CONNECTION OR OPEN OR GROUNDED IGNITION FEED CIRCUIT FROM DISTRIBUTOR TO ECM OR FAULTY ECM.

FAULTY ECM CONNECTION OR OPEN GROUND CIRCUIT FROM DISTRIBUTOR TO ECM OR FAULTY ECM.

ALL LIGHTS BLINKING BRIGHTLY →
- USING TECH 1, ENABLE FUEL PUMP. DOES FUEL PUMP OPERATE.

NO BLINKING LIGHT(S) → USE CHART A-3 (2 OF 2).

ANY LIGHTS BLINKING DIMLY → SHORTED INJECTOR, COMPARE INJECTOR RESISTANCE VALUES.

YES → (to step 4)

NO → USE CHART A-5.

4.
   - IGNITION "OFF."
   - INSTALL FUEL PRESSURE GAGE AND NOTE PRESSURE AFTER IGNITION "ON." SHOULD BE (283-325 kPa) 41-47 psi.

OK →
- REVIEW THE "DIAGNOSTIC AIDS" ON FACING PAGE FOR ADDITIONAL ITEMS TO CHECK. IF ALL ARE OK, EFI SYSTEM IS OK. REFER TO "HARD START"

NOT OK → USE CHART A-7.

"AFTER REPAIRS," CONFIRM "CLOSED LOOP" OPERATION AND NO MIL (SERVICE ENGINE SOON).

5.7L (VIN P) ENGINE — SYSTEM DIAGNOSTIC CHARTS — 1993 CORVETTE

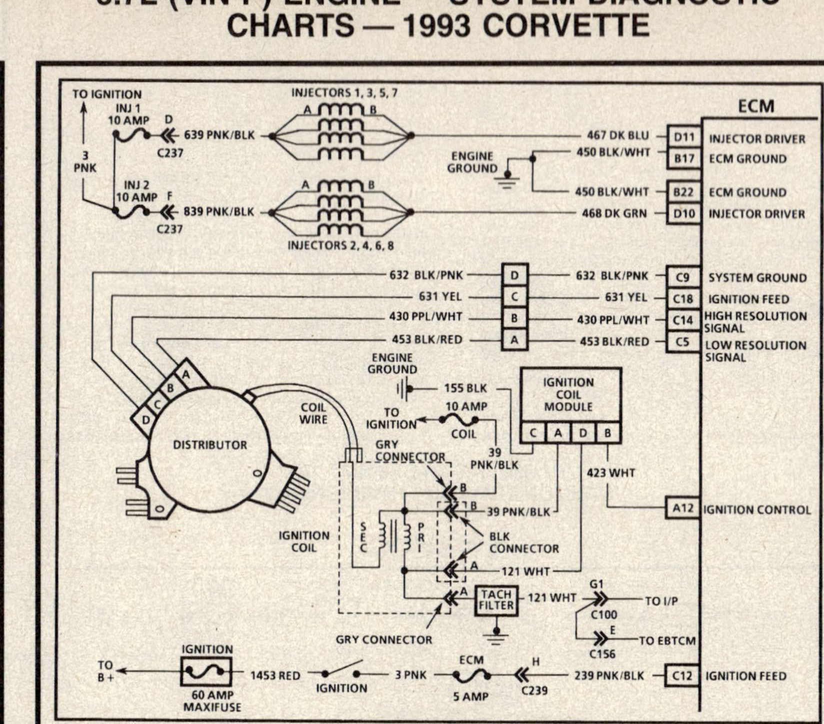

CHART A-3
(Page 2 of 2)
ENGINE CRANKS BUT WILL NOT RUN
5.7L (VIN P) "Y" CARLINE

Test Description: Number(s) below refer to circled number(s) on the diagnostic chart.

1. Checks for 12 volt supply to injectors. Due to the injectors wired in parallel there should be a light "ON" on both terminals.
   Turn ignition "OFF" for 10 seconds and then back "ON." If the Tech 1 then indicates an injector fault on either right or left bank, voltage is not reaching the ECM driver and the problem must be an open or short in the feed or ground side of the injector circuit.
2. Checks continuity of CKT 467 and CKT 468.
3. All checks made to this point would indicate that the ECM is at fault. However, there is a possibility of CKT 467 or CKT 468 being shorted to a voltage source either in the engine harness or in the injector harness.
   To test for this condition:
   - Disconnect all injectors.
   - Ignition "ON."
   - Probe CKT 467 and CKT 468 on the ECM side of injector harness with a test light connected to ground. (Test one injector harness on each side of engine.) There should be no light. If light is "ON" repair short to voltage.
   - If OK, check the resistance of the injectors. Should be 10 ohms or more.
   - Check injector harness connector. Be sure terminals are not backed out of connector and contacting each other.
   - If all OK, replace ECM.

5.7L (VIN P) ENGINE — SYSTEM DIAGNOSTIC CHARTS — 1993 CORVETTE

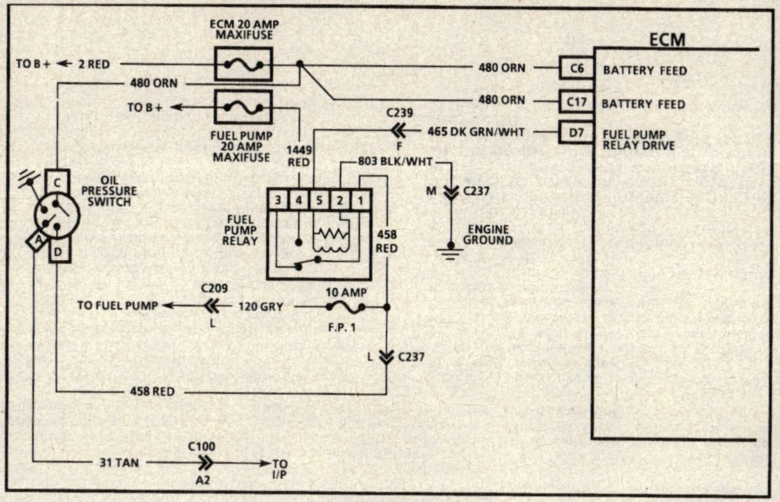

CHART A-5
FUEL PUMP RELAY CIRCUIT
5.7L (VIN P) "Y" CARLINE

Circuit Description:

When the ignition switch is turned "ON," the Engine Control Module (ECM) will turn "ON" the in-tank fuel pump. It will remain "ON" as long as the ECM is receiving ignition low resolution reference pulses from the distributor ignition system.

If there are no reference pulses, the ECM will shut "OFF" the fuel pump about 2-3 seconds after key "ON," or when engine stops. When sufficient oil pressure is present to close the oil pressure switch, the fuel pump will remain "ON."

The pump will deliver fuel to the fuel rail and injectors, then to the pressure regulator, where the system pressure is controlled to 284-325 kPa (41-47 psi). Excess fuel is then returned to the fuel tank.

When the engine is stopped, the fuel pump can be turned "ON" by using the Tech 1. Improper fuel system pressure will result in one or all of the following symptoms:

- Cranks but won't run.
- DTC 44 and 64.
- DTC 45 and 65.
- Cuts out, may feel like ignition problem.
- Poor fuel economy, loss of power.
- Hesitation.
- DTC 55.

5.7L (VIN P) ENGINE — SYSTEM DIAGNOSTIC CHARTS — 1993 CORVETTE

CHART A-3
(Page 2 of 2)
ENGINE CRANKS BUT WILL NOT RUN
5.7L (VIN P) "Y" CARLINE

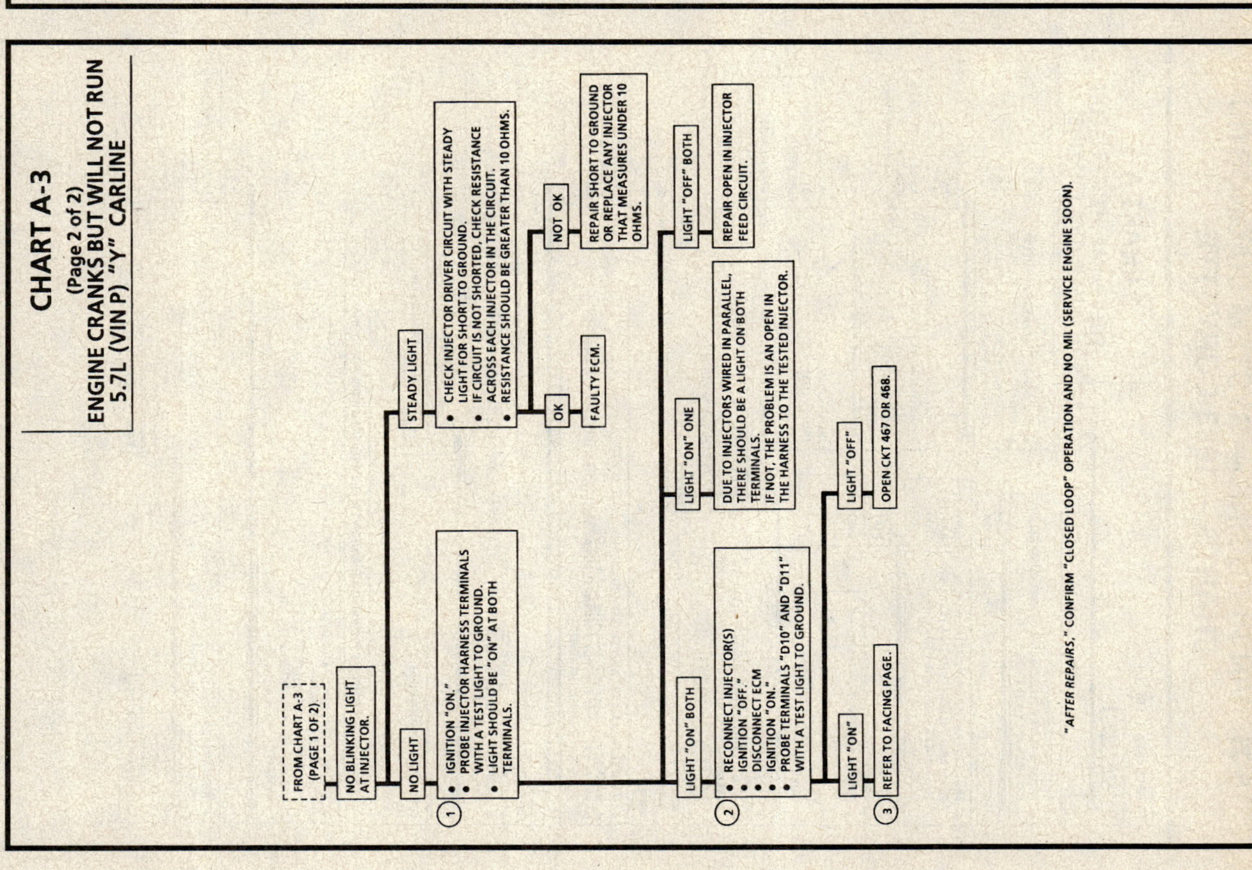

5.7L (VIN P) ENGINE — SYSTEM DIAGNOSTIC CHARTS — 1993 CORVETTE

CHART A-5
FUEL PUMP RELAY CIRCUIT
5.7L (VIN P) "Y" CARLINE

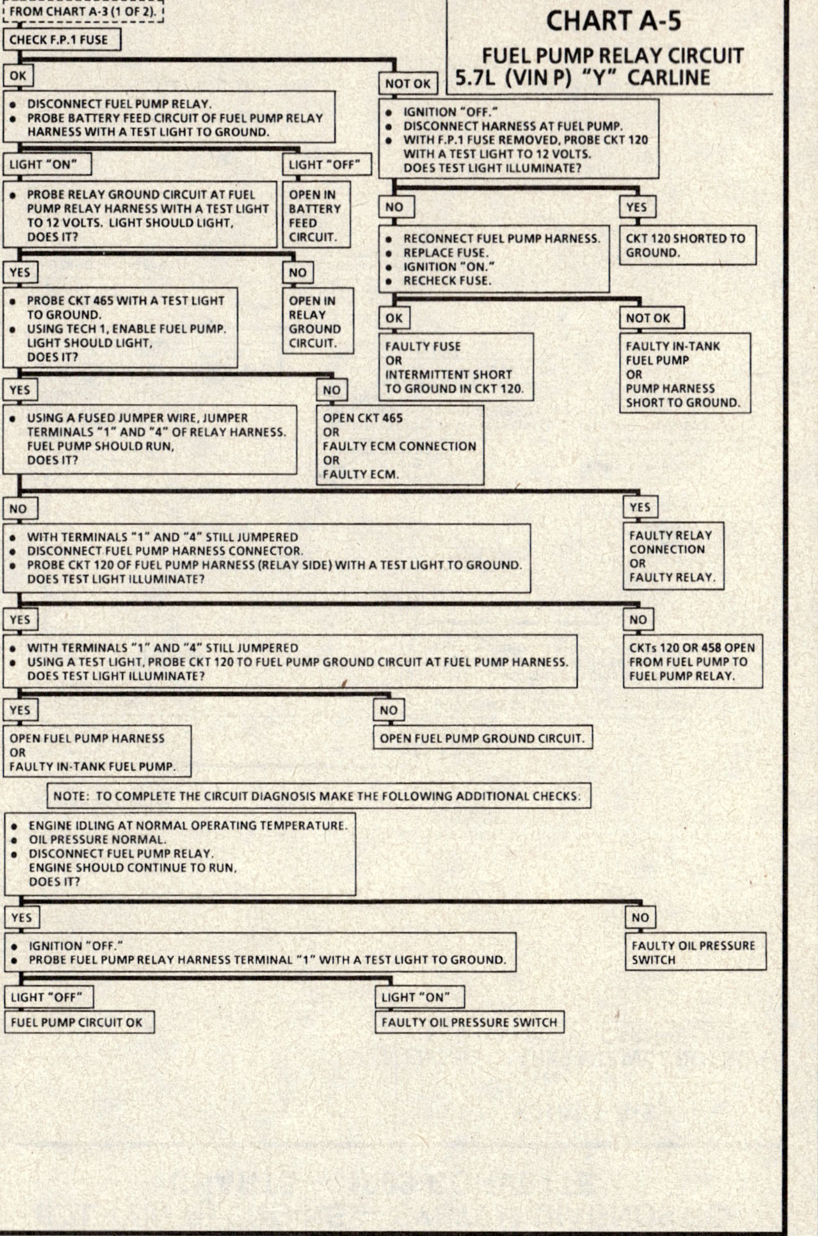

5.7L (VIN 8 & P) ENGINE — SYSTEM DIAGNOSTIC CHARTS — 1992–93 CORVETTE

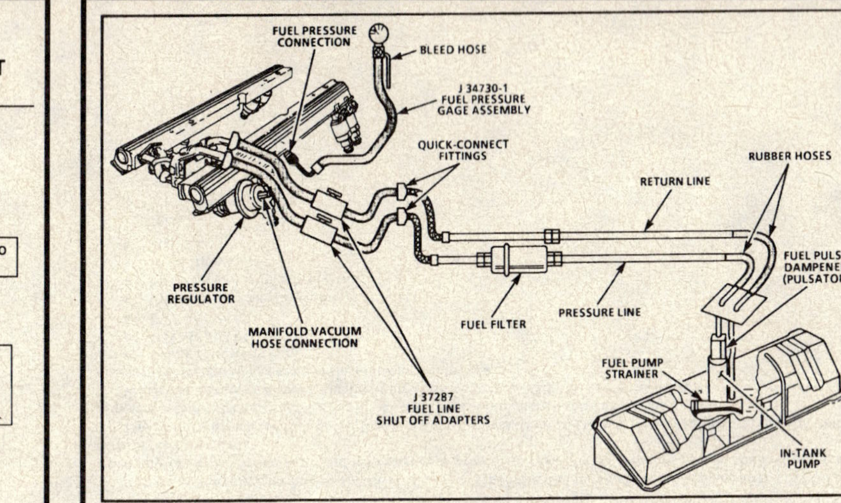

CHART A-7
(Page 1 of 3)
FUEL SYSTEM DIAGNOSIS
5.7L (VIN P) "Y" CARLINE (MFI)

Circuit Description:

When the ignition switch is turned "ON," the Electronic Control Module (ECM) will turn "ON" the in-tank fuel pump. It will remain "ON" as long as the engine is cranking or running, and the ECM is receiving reference pulses. If there are no reference pulses, the ECM will shut "OFF" the fuel pump within 2 seconds after ignition "ON" or engine stops.

An electric fuel pump, attached to the fuel sender assembly (inside the fuel tank) pumps fuel through an in-line filter to the fuel rail assembly. The pump is designed to provide fuel at a pressure above the regulated pressure needed by the injectors. A pressure regulator, attached to the fuel rail, keeps fuel available to the injectors at a regulated pressure. Unused fuel is returned to the fuel tank by a separate line.

Test Description: Number(s) below refer to circled number(s) on the diagnostic chart.

1. Wrap a shop towel around the fuel pressure connector to absorb any small amount of fuel leakage that may occur when installing the gage. Ignition "ON," pump pressure should be 284-325 kPa (41-47 psi). This pressure is controlled by spring pressure within the regulator assembly.
2. When the engine is idling, high vacuum is applied to the fuel regulator diaphragm. This will offset the spring and result in a lower fuel pressure. This idle pressure will vary somewhat depending on barometric pressure, however, the pressure idling should be less indicating pressure regulator control.
3. Pressure that leaks down is caused by one of the following:
   • In-tank fuel pump check valve not holding.
   • Partially disconnected fuel pulse damper (pulsator).
   • Fuel pressure regulator valve leaking.
   • Injector(s) sticking open.
4. An injector sticking open can best be determined by checking for a fouled or saturated spark plug(s). If a leaking injector can not be determined by a fouled or saturated spark plug, the following procedure should be used.
   • Remove fuel rail bolts. Follow the procedures in "Fuel Metering System," of this manual, but leave fuel lines connected.
   • Lift fuel rail out just enough to leave injector nozzles in the ports.

CAUTION: Be sure injector(s) are not allowed to spray on engine and that injector retaining clips are intact. This should be carefully followed to prevent fuel spray on engine which would cause a fire hazard.

   • Pressurize the fuel system and observe injector nozzles.

5.7L (VIN 8 & P) ENGINE — SYSTEM DIAGNOSTIC CHARTS — 1992–93 CORVETTE

CHART A-7
(Page 1 of 3)
FUEL SYSTEM DIAGNOSIS
5.7L (VIN P) "Y" CARLINE (MFI)

(1)
- INSTALL FUEL PRESSURE GAGE AS SHOWN ON FACING PAGE.
- IGNITION "OFF" FOR 10 SECONDS. A/C "OFF."
- IGNITION "ON." FUEL PUMP WILL RUN FOR ABOUT 2 SECONDS. IT MAY BE NECESSARY TO CYCLE THE IGNITION "ON" MORE THAN ONCE TO OBTAIN MAXIMUM PRESSURE.
- NOTE FUEL PRESSURE WITH PUMP RUNNING, PRESSURE SHOULD BE 284-325 kPa (41-47 psi). WHEN PUMP STOPS, PRESSURE MAY VARY SLIGHTLY THEN SHOULD HOLD STEADY. IS PRESSURE CORRECT AND DOES IT HOLD?

YES → IF FUEL PRESSURE IS WITHIN NORMAL RANGE BUT IS SUSPECTED OF DROPPING OFF DURING ACCELERATION, CRUISE OR HARD CORNERING, SEE CHART A-7 (2 OF 3).

(2)
- START ENGINE, ALLOW IT TO IDLE AT NORMAL OPERATING TEMPERATURE.
- FUEL PRESSURE NOTED IN STEP (1) SHOULD DROP APPROXIMATELY 21-69 kPa (3-10 psi). DOES IT?

YES → NO TROUBLE FOUND, REVIEW "SYMPTOMS,"

NO →
- DISCONNECT VACUUM HOSE FROM PRESSURE REGULATOR ASSEMBLY.
- WITH ENGINE IDLING, APPLY 12-14 INCHES OF VACUUM TO PRESSURE REGULATOR. FUEL PRESSURE NOTED IN STEP (1) SHOULD DROP APPROXIMATELY 21-69 kPa (3-10 psi). DOES IT?

  YES → LOCATE AND REPAIR LOSS OF VACUUM TO PRESSURE REGULATOR.

  NO → PRESSURE REGULATOR IS FAULTY.

NO → (from step 1) / FROM CHART A-3

(3) FUEL PRESSURE WITHIN SPEC., BUT DOES NOT HOLD.

FUEL PRESSURE OUT OF SPEC. — SEE CHART A-7 (2 OF 3)

NO FUEL PRESSURE. — USE CHART A-5 TO DIAGNOSE FUEL PUMP ELECTRICAL CIRCUIT.

(From "FUEL PRESSURE WITHIN SPEC., BUT DOES NOT HOLD":)
- INSTALL J 37287 FUEL LINE SHUT-OFF ADAPTORS, REFER TO PAGES 3 OF 3 AND FACING PAGE ILLUSTRATION.
- MAKE SURE VALVES ARE OPEN.
- USING TECH 1, ENABLE FUEL PUMP AND WAIT FOR PRESSURE TO BUILD.
- TURN PUMP "OFF."
- CLOSE VALVE IN FUEL PRESSURE LINE. PRESSURE SHOULD HOLD. DOES IT?

IF OK → CHECK FOR:
- PLUGGED IN-LINE FILTER.
- RESTRICTED FUEL PRESSURE LINE.
- PLUGGED FUEL PUMP STRAINER.
- LEAKING FUEL PULSE DAMPENER.

  IF OK → FUEL PUMP IS FAULTY.

NO →
- OPEN VALVE IN FUEL PRESSURE LINE.
- RE-ENABLE FUEL PUMP AND WAIT FOR PRESSURE TO BUILD.
- TURN PUMP "OFF."
- CLOSE VALVE IN FUEL RETURN LINE. PRESSURE SHOULD HOLD. DOES IT?

YES → CHECK FOR:
- LEAKING FUEL PULSE DAMPENER.

  IF OK → FUEL PUMP IS FAULTY. (LEAKING CHECK BALL INSIDE PUMP.)

NO → **(4)** LOCATE AND CORRECT LEAKING INJECTOR(S).

YES → PRESSURE REGULATOR IS FAULTY.

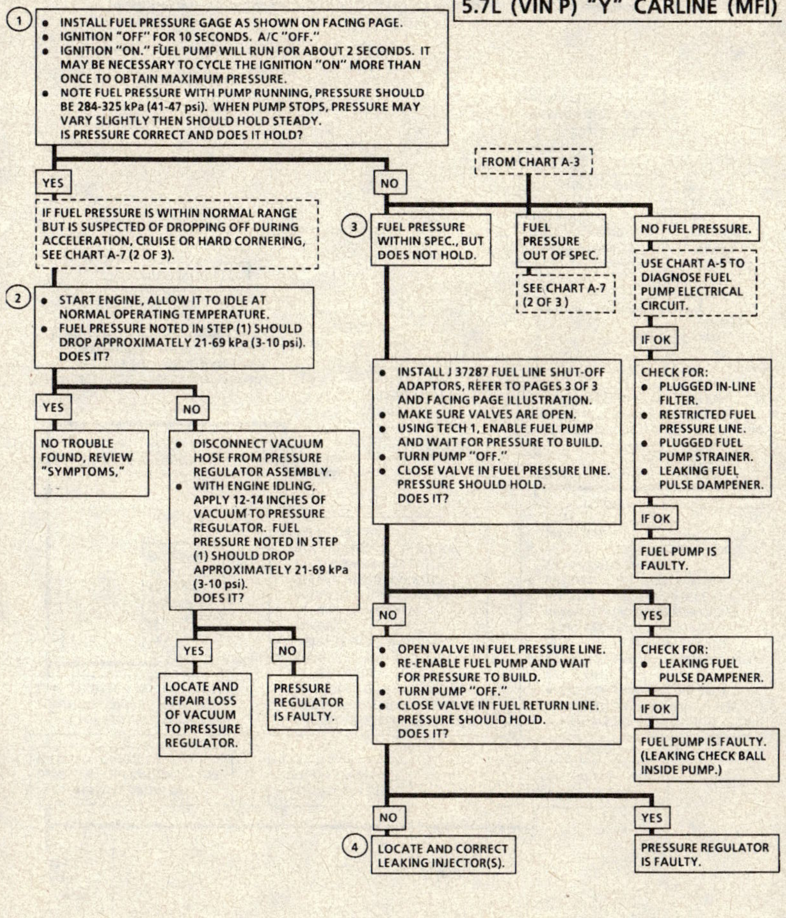

"AFTER REPAIRS," CONFIRM "CLOSED LOOP" OPERATION AND NO MIL (SERVICE ENGINE SOON).

5.7L (VIN 8 & P) ENGINE — SYSTEM DIAGNOSTIC CHARTS — 1992–93 CORVETTE

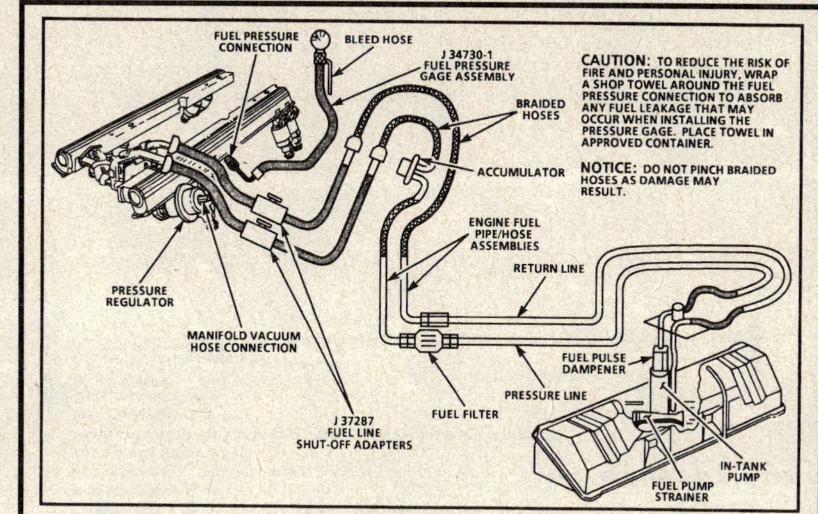

CHART A-7
(Page 2 of 3)
FUEL SYSTEM DIAGNOSIS
5.7L (VIN P) "Y" CARLINE (MFI)

Test Description: Number(s) below refer to circled number(s) on the diagnostic chart.

5. Fuel pressure that drops off during acceleration, cruise or hard cornering may cause a lean condition and result in a loss of power, surging or misfire. This condition can be diagnosed using a Tech 1 scan tool. If the fuel system is very lean, one or both oxygen sensors will stop toggling and output voltage will drop below 300 mV. Also, injector pulse width will increase.

⚠ Important
- Make sure system is not operating at "Fuel-Cut-Off" which may cause false readings on the scan tool.

6. Fuel pressure below 284 kPa (41 psi) may cause a lean condition and may set a DTC 44/64. Driveability conditions can include hard starting cold, hesitation, poor driveability, lack of power, surging or misfire.

7. Restricting the fuel return line causes fuel pressure to build above regulated pressure. Enable fuel pump using a Tech 1, pressure should rise above 324 kPa (47 psi) as the valve in the return line is partially closed.

NOTICE: Do not allow pressure to exceed 414 kPa (60 psi) as damage to the regulator may result.

8. Fuel pressure above 325 kPa (47 psi) may cause a rich condition and may set a DTC 45/65. Driveability conditions can include hard starting (followed by black smoke) and a strong sulphur smell in the exhaust.

9. This test determines if the high fuel pressure is due to a restricted fuel return line or a faulty fuel pressure regulator.

10. The pressure regulator filter screen is designed to trap any contaminants introduced during engine assembly. If dirty, it can be removed with a small pick and discarded without potential harm to the regulator.

5.7L (VIN 8 & P) ENGINE — SYSTEM DIAGNOSTIC CHARTS — 1992–93 CORVETTE

CHART A-7
(Page 2 of 3)
FUEL SYSTEM DIAGNOSIS
5.7L (VIN P) "Y" CARLINE (MFI)

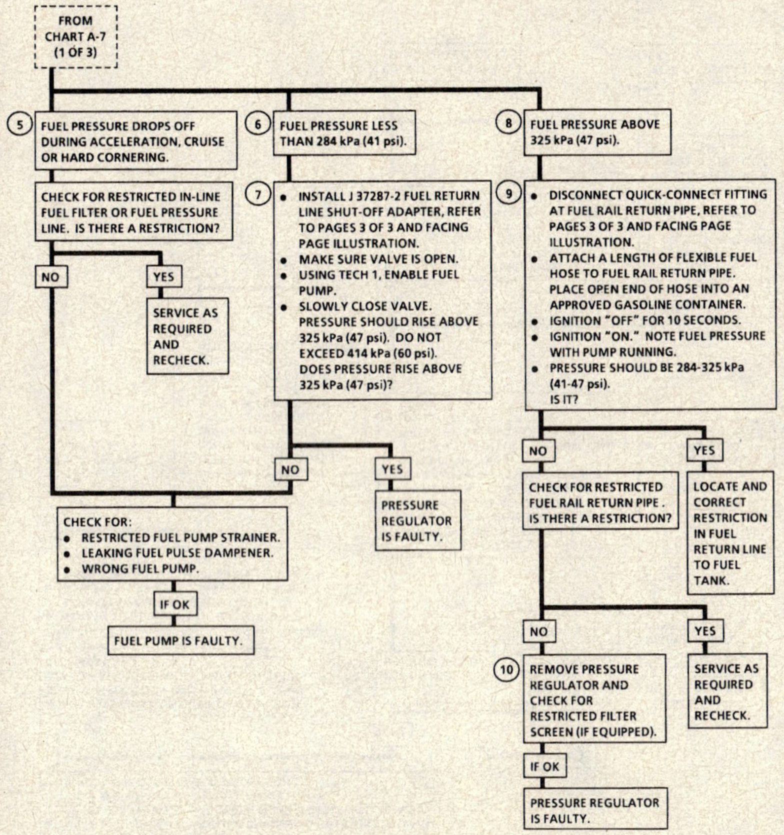

"AFTER REPAIRS," CONFIRM "CLOSED LOOP" OPERATION AND NO MIL (SERVICE ENGINE SOON).

5.7L (VIN 8 & P) ENGINE — SYSTEM DIAGNOSTIC CHARTS — 1992–93 CORVETTE

CHART A-7
(Page 3 of 3)
FUEL SYSTEM DIAGNOSIS
5.7L (VIN P) "Y" CARLINE (MFI)

FUEL SYSTEM PRESSURE RELIEF PROCEDURE

Engines With Fuel Pressure Connection

(Must Be Performed Before Disconnecting Fuel Line Fittings)

CAUTION:
- To reduce the risk of fire and personal injury, it is necessary to relieve fuel system pressure before disconnecting fuel line fittings.
- After relieving system pressure, a small amount of fuel may be released when disconnecting fuel line fittings. In order to reduce the chance of personal injury, cover fuel line fittings with a shop towel before disconnecting, to catch any fuel that may leak out. Place the towel in an approved container when disconnect is completed.

Tool Required: J 34730-1 Fuel Pressure Gage

1. Ignition "OFF."
2. Disconnect negative battery cable to avoid possible fuel discharge if an accidental attempt is made to start the engine.
3. Loosen fuel filler cap to relieve tank vapor pressure.
4. Connect gage J 34730-1 to fuel pressure connection. Wrap a shop towel around fitting while connecting gage to avoid spillage.
5. Install bleed hose into an approved container and open valve to bleed system pressure. Fuel line fittings are now safe for servicing.
6. Drain any fuel remaining in gage into an approved gasoline container.
7. Perform service required.
8. Tighten fuel filler cap.
9. Ignition "OFF."
10. Connect negative battery cable.
11. Cycle ignition "ON" and "OFF" twice, waiting ten seconds between cycles, then check for fuel leaks.

"AFTER REPAIRS," CONFIRM "CLOSED LOOP" OPERATION AND NO MIL (SERVICE ENGINE SOON).

5.7L (VIN 8 & P) ENGINE — SYSTEM DIAGNOSTIC CHARTS — 1992–93 CORVETTE

CHART A-7
(Page 3 of 3)
FUEL SYSTEM DIAGNOSIS
5.7L (VIN P) "Y" CARLINE (MFI)

SERVICING QUICK-CONNECT FITTINGS

Important

- In order to install fuel system diagnostic equipment on vehicles equipped with plastic quick-connect fittings, fuel line separator tools must be used to disconnect the fittings. Using the separator tools to release the fittings will cause the plastic retainer to remain inside the female connector allowing diagnostic equipment to be connected.

  Tools required:
  - J 37088-A tool set, fuel line quick-connect separator;
  - J 39504 tool set, fuel line quick-connect separator (restricted access).

Remove or Disconnect

1. Relieve fuel system pressure (see "Fuel System Pressure Relief").
2. If equipped, slide dust cover back to access quick-connect fitting.
3. Grasp both sides of fitting. Twist female connector 1/4 turn in each direction to loosen any dirt within fitting.

CAUTION: Safety glasses must be worn when using compressed air, as flying dirt particles may cause eye injury.

4. Using compressed air, blow dirt out of fitting.
5. Choose correct tool from J 37088-A or J 39504 tool set for size of fitting. Insert tool into female connector, then push/pull inward to release locking tabs.
6. Pull connection apart.

Clean and Inspect

NOTICE: If it is necessary to remove rust or burrs from fuel pipe, use emery cloth in a radial motion with the pipe end to prevent damage to O-ring sealing surface.

- Using a clean shop towel, wipe off male pipe end.
- Inspect both ends of fitting for dirt and burrs. Clean or replace components/assemblies as required.

Install or Connect

CAUTION: To Reduce the Risk of Fire and Personal Injury:
- Before connecting fitting, always apply a few drops of clean engine oil to the male pipe end of engine fuel pipe, pressure gage adapter or fuel line shut-off adapter. This will ensure proper reconnection and prevent a possible fuel leak. (During normal operation, the O-rings located in the female connector will swell and may prevent proper reconnection if not lubricated.)

1. Apply a few drops of clean engine oil to the male pipe end of engine fuel pipe, pressure gage adapter or fuel line shut-off adapter.
2. Push both sides of fitting together to cause the retaining tabs/fingers to snap into place.
3. Once installed, pull on both sides of fitting to make sure connection is secure.
4. If equipped, reposition dust cover over quick-connect fitting.

"AFTER REPAIRS," CONFIRM "CLOSED LOOP" OPERATION AND NO MIL (SERVICE ENGINE SOON).

5.7L (VIN 8 & P) ENGINE — DIAGNOSTIC TROUBLE CODE CHART — 1992–93 CORVETTE

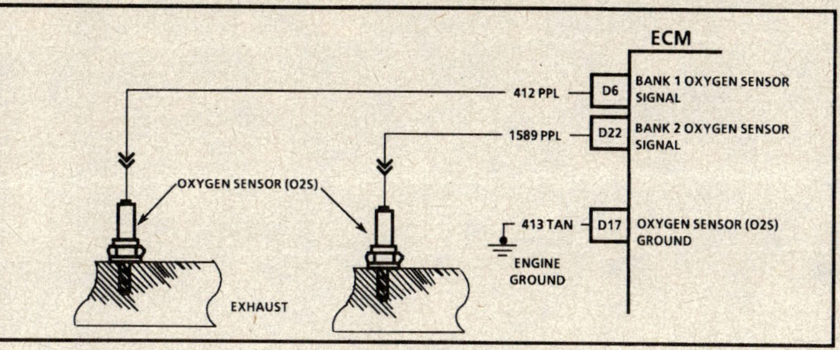

DTC 13
BANK 1 (LEFT) OXYGEN SENSOR (O2S) CIRCUIT
(OPEN CIRCUIT)
5.7L (VIN P) "Y" CARLINE (MFI)

Circuit Description:

The ECM supplies a voltage of about 450 mV between terminals "D6" and "D17". (If measured with a 10 megohm digital voltmeter, this may display as low as 320 mV.) The oxygen sensor varies the voltage within a range of about 1000 mV if the exhaust is rich, down through about 10 mV if exhaust is lean.

The sensor is like an open circuit and produces no voltage when it is below 315°C (600°F). An open sensor circuit or cold sensor causes "Open Loop" operation.

Test Description: Number(s) below refer to circled number(s) on the diagnostic chart.

1. DTC 13 will set if:
   - Engine at normal operating temperature (above 70°C/158°F).
   - At least 2 minutes engine run time after start.
   - Oxygen signal voltage steady between 350 mV to 550 mV.
   - Throttle position sensor signal above 5% (about .3 volt above closed throttle voltage).
   - All conditions must be met for about 60 seconds.

   If the conditions for a DTC 13 exist, the system will not go in "Closed Loop."
2. This will determine if the oxygen sensor is at fault, or the wiring or ECM is the cause of the DTC 13.
3. For this test use only a high impedance digital volt ohmmeter (J 39200). This test checks the continuity of CKT 412 and CKT 413. If CKT 413 is open, the ECM voltage on CKT 412 will be over .6 volt (600 mV).

Diagnostic Aids:

Normal Tech 1 voltage readings vary between 10 mV to 1000 mV (.01 and 1.0 volt), while in "Closed Loop." The system will go into "Open Loop" operation when DTC 13 sets.

Refer to "Intermittents" in "Symptoms,"

5.7L (VIN 8 & P) ENGINE — DIAGNOSTIC TROUBLE CODE CHART — 1992–93 CORVETTE

DTC 13

BANK 1 (LEFT) OXYGEN SENSOR (O2S) CIRCUIT
(OPEN CIRCUIT)
5.7L (VIN P) "Y" CARLINE (MFI)

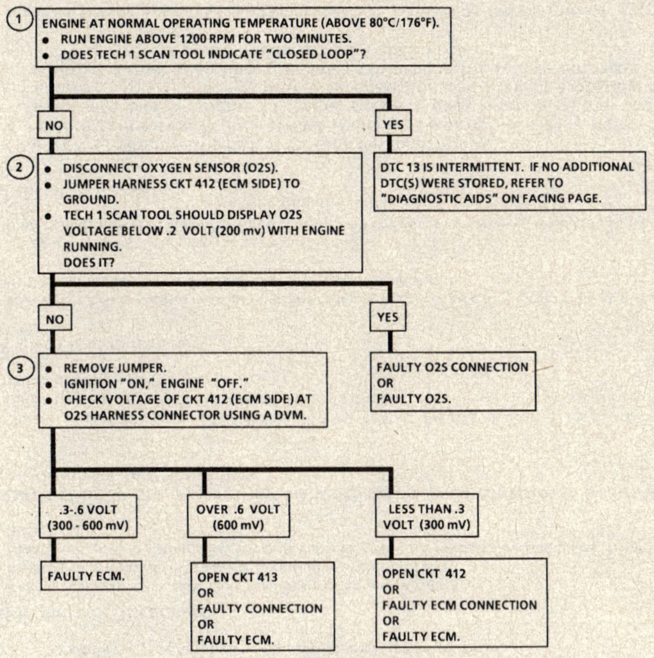
1
ENGINE AT NORMAL OPERATING TEMPERATURE (ABOVE 80°C/176°F).
• RUN ENGINE ABOVE 1200 RPM FOR TWO MINUTES.
• DOES TECH 1 SCAN TOOL INDICATE "CLOSED LOOP"?
NO
YES
DTC 13 IS INTERMITTENT. IF NO ADDITIONAL DTC(S) WERE STORED, REFER TO "DIAGNOSTIC AIDS" ON FACING PAGE.
2
• DISCONNECT OXYGEN SENSOR (O2S).
• JUMPER HARNESS CKT 412 (ECM SIDE) TO GROUND.
• TECH 1 SCAN TOOL SHOULD DISPLAY O2S VOLTAGE BELOW .2 VOLT (200 mv) WITH ENGINE RUNNING. DOES IT?
NO
YES
FAULTY O2S CONNECTION OR FAULTY O2S.
3
• REMOVE JUMPER.
• IGNITION "ON," ENGINE "OFF."
• CHECK VOLTAGE OF CKT 412 (ECM SIDE) AT O2S HARNESS CONNECTOR USING A DVM.
.3-.6 VOLT (300 - 600 mV)
OVER .6 VOLT (600 mV)
LESS THAN .3 VOLT (300 mV)
FAULTY ECM.
OPEN CKT 413 OR FAULTY CONNECTION OR FAULTY ECM.
OPEN CKT 412 OR FAULTY ECM CONNECTION OR FAULTY ECM.

"AFTER REPAIRS," REFER TO DTC CRITERIA ON FACING PAGE AND CONFIRM DTC DOES NOT RESET.

5.7L (VIN 8 & P) ENGINE — DIAGNOSTIC TROUBLE CODE CHART — 1992–93 CORVETTE

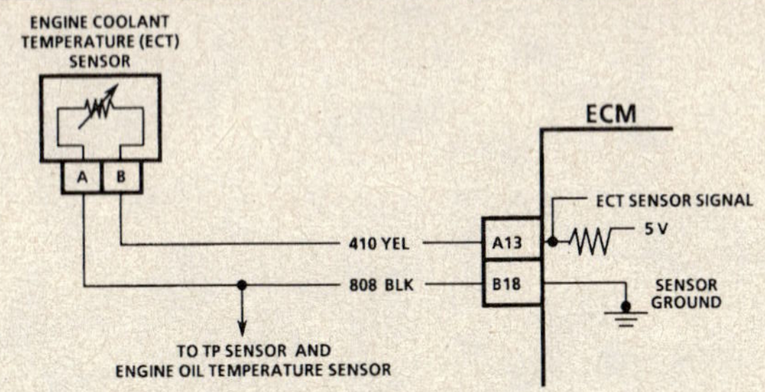
ENGINE COOLANT TEMPERATURE (ECT) SENSOR
A
B
ECM
ECT SENSOR SIGNAL
5 V
410 YEL
808 BLK
A13
B18
SENSOR GROUND
TO TP SENSOR AND ENGINE OIL TEMPERATURE SENSOR

DTC 14

ENGINE COOLANT TEMPERATURE (ECT) SENSOR CIRCUIT
(HIGH TEMPERATURE INDICATED)
5.7L (VIN P) "Y" CARLINE (MFI)

Circuit Description:

The Engine Coolant Temperature (ECT) sensor uses a thermistor to control a signal voltage to the ECM. The ECM applies a voltage on CKT 410 to the sensor. When the engine coolant is cold, the sensor (thermistor) resistance is high, therefore, the ECM signal voltage will be high. As the engine coolant warms, the sensor resistance becomes less, and the ECM voltage drops. At normal engine operating temperature (85°C - 95°C or 185°F - 203°F) the voltage should measure about 1.5 to 2.0 volts.

Test Description: Number(s) below refer to circled number(s) on the diagnostic chart.

1. DTC 14 will set if:
   - Signal voltage indicates a coolant temperature above 130°C (266°F).
2. This test will determine if CKT 410 is shorted to ground, which will cause the condition for DTC 14.

Diagnostic Aids:

Check harness routing for a potential short to ground in CKT 410.

Tech 1 displays engine temperature in degrees celsius and fahrenheit. After engine is started, the temperature should rise steadily, reach normal operating temperature, and then stabilize when thermostat opens.

Refer to "Intermittents" in "Symptoms,"

5.7L (VIN 8 & P) ENGINE — DIAGNOSTIC TROUBLE CODE CHART — 1992–93 CORVETTE

DTC 14
ENGINE COOLANT TEMPERATURE (ECT) SENSOR CIRCUIT
(HIGH TEMPERATURE INDICATED)
5.7L (VIN P) "Y" CARLINE (MFI)

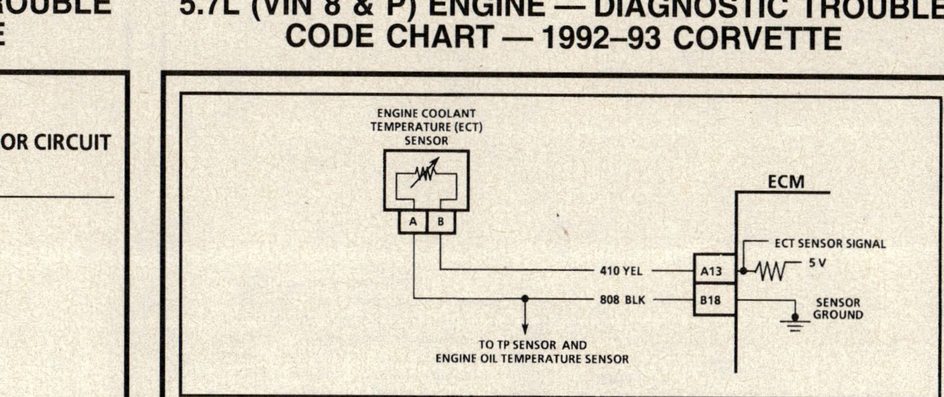

DIAGNOSTIC AID

ENGINE COOLANT TEMPERATURE SENSOR TEMPERATURE VS. RESISTANCE VALUES (APPROXIMATE)		
°C	°F	OHMS
100	212	177
90	194	241
80	176	332
70	158	467
60	140	667
50	122	973
45	113	1188
40	104	1459
35	95	1802
30	86	2238
25	77	2796
20	68	3520
15	59	4450
10	50	5670
5	41	7280
0	32	9420
-5	23	12300
-10	14	16180
-15	5	21450
-20	-4	28680
-30	-22	52700
-40	-40	100700

"AFTER REPAIRS," REFER TO DTC CRITERIA ON FACING PAGE AND CONFIRM DTC DOES NOT RESET.

5.7L (VIN 8 & P) ENGINE — DIAGNOSTIC TROUBLE CODE CHART — 1992–93 CORVETTE

DTC 15
ENGINE COOLANT TEMPERATURE (ECT) SENSOR CIRCUIT
(LOW TEMPERATURE INDICATED)
5.7L (VIN P) "Y" CARLINE (MFI)

Circuit Description:

The Engine Coolant Temperature (ECT) sensor uses a thermistor to control the signal voltage to the ECM. The ECM applies a voltage on CKT 410 to the sensor. When the engine coolant is cold, the sensor (thermistor) resistance is high, therefore the ECM signal voltage will be high. As the engine coolant warms, the sensor resistance becomes less, and the ECM voltage drops. At normal engine operating temperature (85°C – 95°C or 185°F – 203°F), the voltage should measure about 1.5 to 2.0 volts at the ECM.

Test Description: Number(s) below refer to circled number(s) on the diagnostic chart.
1. DTC 15 will set if:
   - ECT signal voltage indicates an engine coolant temperature less than -49°C (-56.2°F).
2. This test simulates a DTC 14. If the ECM recognizes the low signal voltage, (high temperature) and the Tech 1 displays 130°C (266°F) or above, the ECM and wiring are OK.
3. This test will determine if CKT 410 is open. There should be 5 volts present at sensor connector, if measured with a DVM.

Diagnostic Aids:

A Tech 1 displays engine coolant temperature in degrees celsius and fahrenheit. After engine is started, the temperature should rise steadily, reach normal operating temperature, and then stabilize when the thermostat opens.

A faulty connection, or an open in CKT 410 or 808, will result in a DTC 15.

If DTC 21 or 52 is also set, check CKT 808 for faulty wiring or connections. Check terminals at sensor for good contact.

Refer to "Intermittents" in "Symptoms,'

5.7L (VIN 8 & P) ENGINE — DIAGNOSTIC TROUBLE CODE CHART — 1992–93 CORVETTE

DTC 15
ENGINE COOLANT TEMPERATURE (ECT) SENSOR CIRCUIT
(LOW TEMPERATURE INDICATED)
5.7L (VIN P) "Y" CARLINE (MFI)

1. DOES TECH 1 SCAN TOOL DISPLAY ENGINE COOLANT TEMPERATURE OF -30°C (-22°F) OR LESS?

YES →

2.
- DISCONNECT ENGINE COOLANT TEMPERATURE SENSOR.
- JUMPER HARNESS TERMINALS TOGETHER.
- TECH 1 SCAN TOOL SHOULD DISPLAY 130°C (266°F) OR MORE. DOES IT?

NO → DTC 15 IS INTERMITTENT. IF NO ADDITIONAL DTC(S) WERE STORED, REFER TO "DIAGNOSTIC AIDS" ON FACING PAGE.

3.
- JUMPER CKT 410 TO GROUND.
- TECH 1 SCAN TOOL SHOULD DISPLAY OVER 130°C (266°F). DOES IT?

YES → FAULTY CONNECTION OR ENGINE COOLANT TEMPERATURE SENSOR.

YES → OPEN ENGINE COOLANT TEMPERATURE SENSOR GROUND CIRCUIT, FAULTY CONNECTION OR FAULTY ECM.

NO → OPEN CKT 410, FAULTY CONNECTION AT ECM, OR FAULTY ECM.

DIAGNOSTIC AID

ENGINE COOLANT TEMPERATURE SENSOR
TEMPERATURE VS. RESISTANCE VALUES (APPROXIMATE)

°C	°F	OHMS
100	212	177
90	194	241
80	176	332
70	158	467
60	140	667
50	122	973
45	113	1188
40	104	1459
35	95	1802
30	86	2238
25	77	2796
20	68	3520
15	59	4450
10	50	5670
5	41	7280
0	32	9420
-5	23	12300
-10	14	16180
-15	5	21450
-20	-4	28680
-30	-22	52700
-40	-40	100700

"AFTER REPAIRS," REFER TO DTC CRITERIA ON FACING PAGE AND CONFIRM DTC DOES NOT RESET.

5.7L (VIN 8 & P) ENGINE — DIAGNOSTIC TROUBLE CODE CHART — 1992–93 CORVETTE

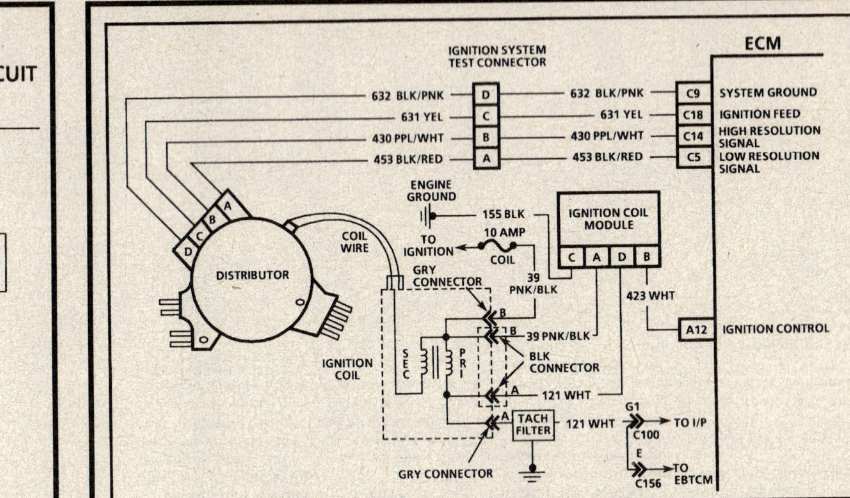

DTC 16
DISTRIBUTOR IGNITION SYSTEM
(LOW RESOLUTION PULSE)
5.7L (VIN P) "Y" CARLINE (MFI)

Circuit Description:

The distributor ignition system supplies two timing inputs to the ECM, a high resolution signal (360 pulses per one crankshaft revolution) and a low resolution signal (4 pulses per one crankshaft revolution). The ECM can determine if one of the timing inputs is not being received by comparing the two inputs. If the ECM detects the high resolution timing pulse without detecting the low resolution timing pulse, this DTC will set.

The reference signals toggle between 0 and 5 volts as the camshaft turns. Therefore, an open, a short to voltage, a short to ground, or a defective sensor inside the distributor can prevent the voltage from pulsing at the ECM.

Test Description: Number(s) below refer to circled number(s) on the diagnostic chart.

1. DTC 16 will set if:
- 720 high resolution timing pulses occur before any low resolution pulses are detected.
2. This test will determine if the ECM is sending out a signal to the distributor for processing. If this signal is not available or is shorted to ground or voltage, the distributor cannot ground it to produce reference pulses.

Diagnostic Aids:

If DTC 16 does not reset and the vehicle still does not start, refer to CHART A-3 "Engine Cranks But Will Not Run."

5.7L (VIN 8 & P) ENGINE — DIAGNOSTIC TROUBLE CODE CHART — 1992–93 CORVETTE

DTC 16

DISTRIBUTOR IGNITION SYSTEM
(LOW RESOLUTION PULSE)
5.7L (VIN P) "Y" CARLINE (MFI)

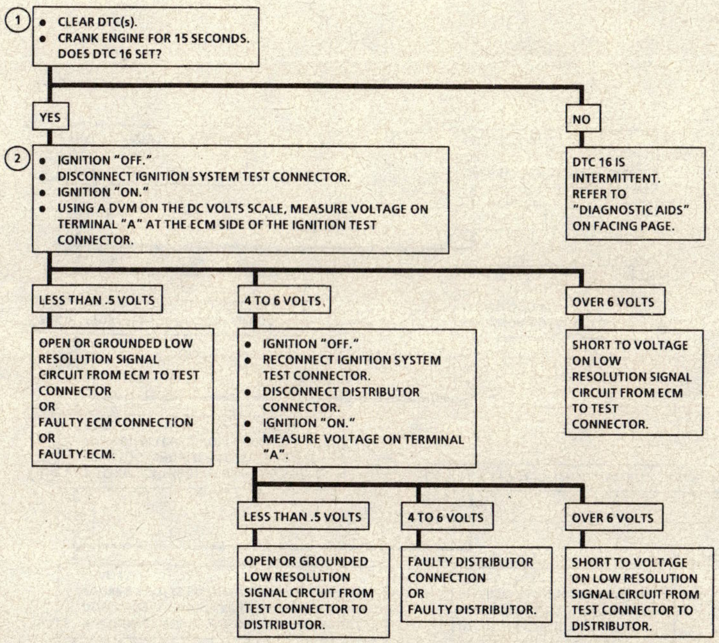

(1)
- CLEAR DTC(s).
- CRANK ENGINE FOR 15 SECONDS. DOES DTC 16 SET?

YES

(2)
- IGNITION "OFF."
- DISCONNECT IGNITION SYSTEM TEST CONNECTOR.
- IGNITION "ON."
- USING A DVM ON THE DC VOLTS SCALE, MEASURE VOLTAGE ON TERMINAL "A" AT THE ECM SIDE OF THE IGNITION TEST CONNECTOR.

NO

DTC 16 IS INTERMITTENT. REFER TO "DIAGNOSTIC AIDS" ON FACING PAGE.

LESS THAN .5 VOLTS

OPEN OR GROUNDED LOW RESOLUTION SIGNAL CIRCUIT FROM ECM TO TEST CONNECTOR OR FAULTY ECM CONNECTION OR FAULTY ECM.

4 TO 6 VOLTS

- IGNITION "OFF."
- RECONNECT IGNITION SYSTEM TEST CONNECTOR.
- DISCONNECT DISTRIBUTOR CONNECTOR.
- IGNITION "ON."
- MEASURE VOLTAGE ON TERMINAL "A".

OVER 6 VOLTS

SHORT TO VOLTAGE ON LOW RESOLUTION SIGNAL CIRCUIT FROM ECM TO TEST CONNECTOR.

LESS THAN .5 VOLTS

OPEN OR GROUNDED LOW RESOLUTION SIGNAL CIRCUIT FROM TEST CONNECTOR TO DISTRIBUTOR.

4 TO 6 VOLTS

FAULTY DISTRIBUTOR CONNECTION OR FAULTY DISTRIBUTOR.

OVER 6 VOLTS

SHORT TO VOLTAGE ON LOW RESOLUTION SIGNAL CIRCUIT FROM TEST CONNECTOR TO DISTRIBUTOR.

"AFTER REPAIRS," REFER TO DTC CRITERIA ON FACING PAGE AND CONFIRM DTC DOES NOT RESET.

5.7L (VIN 8 & P) ENGINE — DIAGNOSTIC TROUBLE CODE CHART — 1992–93 CORVETTE

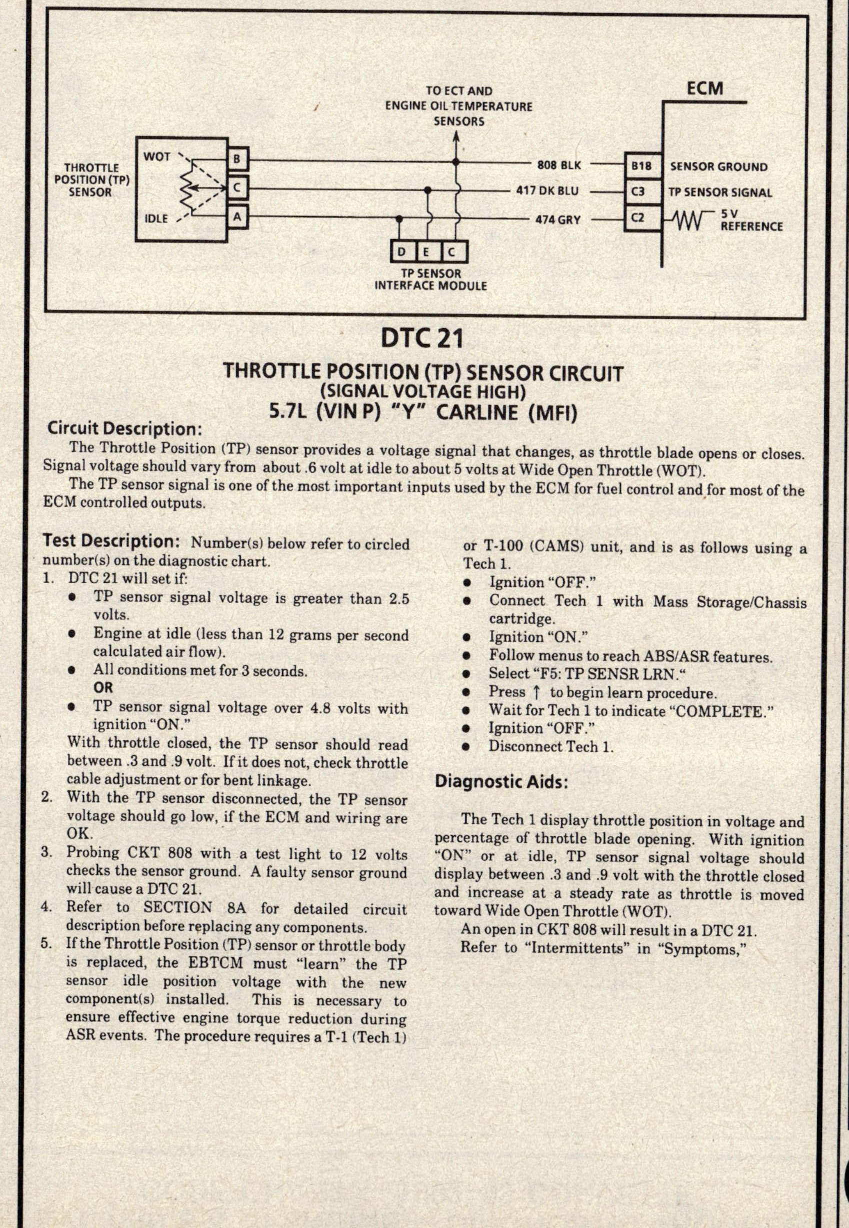

DTC 21

THROTTLE POSITION (TP) SENSOR CIRCUIT
(SIGNAL VOLTAGE HIGH)
5.7L (VIN P) "Y" CARLINE (MFI)

Circuit Description:

The Throttle Position (TP) sensor provides a voltage signal that changes, as throttle blade opens or closes. Signal voltage should vary from about .6 volt at idle to about 5 volts at Wide Open Throttle (WOT).

The TP sensor signal is one of the most important inputs used by the ECM for fuel control and for most of the ECM controlled outputs.

Test Description: Number(s) below refer to circled number(s) on the diagnostic chart.

1. DTC 21 will set if:
   - TP sensor signal voltage is greater than 2.5 volts.
   - Engine at idle (less than 12 grams per second calculated air flow).
   - All conditions met for 3 seconds.
   OR
   - TP sensor signal voltage over 4.8 volts with ignition "ON."
   With throttle closed, the TP sensor should read between .3 and .9 volt. If it does not, check throttle cable adjustment or for bent linkage.
2. With the TP sensor disconnected, the TP sensor voltage should go low, if the ECM and wiring are OK.
3. Probing CKT 808 with a test light to 12 volts checks the sensor ground. A faulty sensor ground will cause a DTC 21.
4. Refer to SECTION 8A for detailed circuit description before replacing any components.
5. If the Throttle Position (TP) sensor or throttle body is replaced, the EBTCM must "learn" the TP sensor idle position voltage with the new component(s) installed. This is necessary to ensure effective engine torque reduction during ASR events. The procedure requires a T-1 (Tech 1)

or T-100 (CAMS) unit, and is as follows using a Tech 1.
- Ignition "OFF."
- Connect Tech 1 with Mass Storage/Chassis cartridge.
- Ignition "ON."
- Follow menus to reach ABS/ASR features.
- Select "F5: TP SENSR LRN."
- Press ↑ to begin learn procedure.
- Wait for Tech 1 to indicate "COMPLETE."
- Ignition "OFF."
- Disconnect Tech 1.

Diagnostic Aids:

The Tech 1 display throttle position in voltage and percentage of throttle blade opening. With ignition "ON" or at idle, TP sensor signal voltage should display between .3 and .9 volt with the throttle closed and increase at a steady rate as throttle is moved toward Wide Open Throttle (WOT).

An open in CKT 808 will result in a DTC 21. Refer to "Intermittents" in "Symptoms,"

5.7L (VIN 8 & P) ENGINE — DIAGNOSTIC TROUBLE CODE CHART — 1992–93 CORVETTE

DTC 21
THROTTLE POSITION (TP) SENSOR CIRCUIT
(SIGNAL VOLTAGE HIGH)
5.7L (VIN P) "Y" CARLINE (MFI)

1. THROTTLE CLOSED. DOES SCAN TOOL DISPLAY THROTTLE POSITION (TP) SENSOR OVER 2.5 VOLTS OR THROTTLE ANGLE GREATER THAN 0%?

YES → DISCONNECT ASR/TP SENSOR INTERFACE MODULE. DOES SCAN TOOL DISPLAY THROTTLE VOLTAGE ABOVE 2.5 VOLTS OR THROTTLE ANGLE GREATER THAN 0%.

NO → DTC 21 IS INTERMITTENT. IF NO ADDITIONAL DTC(s) WERE STORED, REFER TO "DIAGNOSTIC AIDS" ON FACING PAGE.

YES →
2. DISCONNECT THROTTLE POSITION SENSOR. SCAN TOOL SHOULD DISPLAY THROTTLE POSITION BELOW .2 VOLT (200 mV). DOES IT?

NO →
4. FAULTY ASR/TP SENSOR INTERFACE MODULE.

YES →
3. PROBE SENSOR GROUND CIRCUIT WITH A TEST LIGHT CONNECTED TO BATTERY VOLTAGE.

NO → TP SENSOR SIGNAL CIRCUIT SHORTED TO VOLTAGE OR FAULTY ECM.

LIGHT "ON" →
5. FAULTY CONNECTION OR THROTTLE POSITION SENSOR.

LIGHT "OFF" → OPEN SENSOR GROUND CIRCUIT OR FAULTY ECM.

"AFTER REPAIRS," REFER TO DTC CRITERIA ON FACING PAGE AND CONFIRM DTC DOES NOT RESET.

5.7L (VIN 8 & P) ENGINE — DIAGNOSTIC TROUBLE CODE CHART — 1992–93 CORVETTE

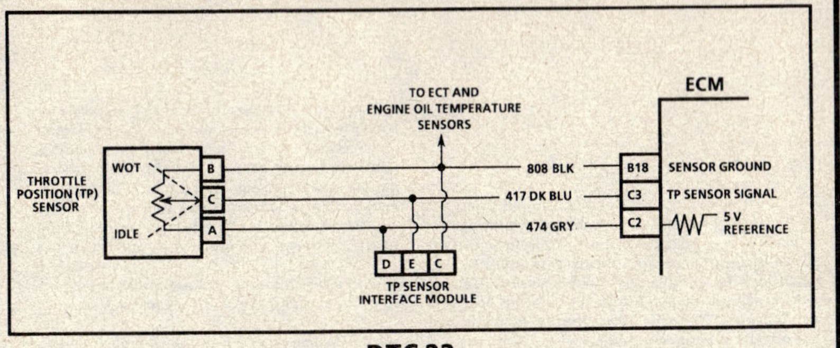

DTC 22
THROTTLE POSITION (TP) SENSOR CIRCUIT
(SIGNAL VOLTAGE LOW)
5.7L (VIN P) "Y" CARLINE (MFI)

Circuit Description:

The Throttle Position (TP) sensor provides a voltage signal that changes, relative to the throttle blade position. Signal voltage should vary from about .6 volt at idle to about 5 volts at Wide Open Throttle (WOT).

The TP sensor signal is one of the most important inputs used by the ECM for fuel control and for most of the ECM controlled outputs.

Test Description: Number(s) below refer to circled number(s) on the diagnostic chart.

1. DTC 22, will set if:
   - TP sensor signal voltage is less than about .23 volt.

   The TP sensor has an auto zeroing feature. If the voltage reading is within the range of about .3 to .9 volt, the ECM will use that value as closed throttle. If the voltage reading is out of the auto zero range at closed throttle, check for a binding throttle cable or damaged linkage, if OK, continue with diagnosis.

2. Simulates DTC 21: (high voltage) If the ECM recognizes the high signal voltage, the ECM and wiring are OK.

3. This simulates a high signal voltage to check for an open in CKT 417.

4. Refer to _______ detailed circuit description before replacing any components.

5. If the Throttle Position (TP) sensor or throttle body is replaced, the EBTCM must "learn" the TP sensor idle position voltage with the new component(s) installed. This is necessary to ensure effective engine torque reduction during ASR events. The procedure requires a T-1 (Tech 1) or T-100 (CAMS) unit, and is as follows using a Tech 1.
   - Ignition "OFF."
   - Connect Tech 1 with Mass Storage/Chassis cartridge.
   - Ignition "ON."
   - Follow menus to reach ABS/ASR features.
   - Select "F5: TP SENSR LRN."
   - Press ↑ to begin learn procedure.
   - Wait for Tech 1 to indicate "COMPLETE."
   - Ignition "OFF."
   - Disconnect Tech 1.

Diagnostic Aids:

The Tech 1 displays throttle position in voltage and percentage of throttle blade opening. With ignition "ON" or at idle, TP sensor signal voltage should display between .3 and .9 volt with the throttle closed and increase at a steady rate as throttle is moved toward Wide Open Throttle (WOT).

An open or short to ground in CKT 474 or CKT 417 will result in a DTC 22.

Refer to "Intermittents" in "Symptoms,"

5.7L (VIN 8 & P) ENGINE — DIAGNOSTIC TROUBLE CODE CHART — 1992–93 CORVETTE

DTC 22
THROTTLE POSITION (TP) SENSOR CIRCUIT
(SIGNAL VOLTAGE LOW)
5.7L (VIN P) "Y" CARLINE (MFI)

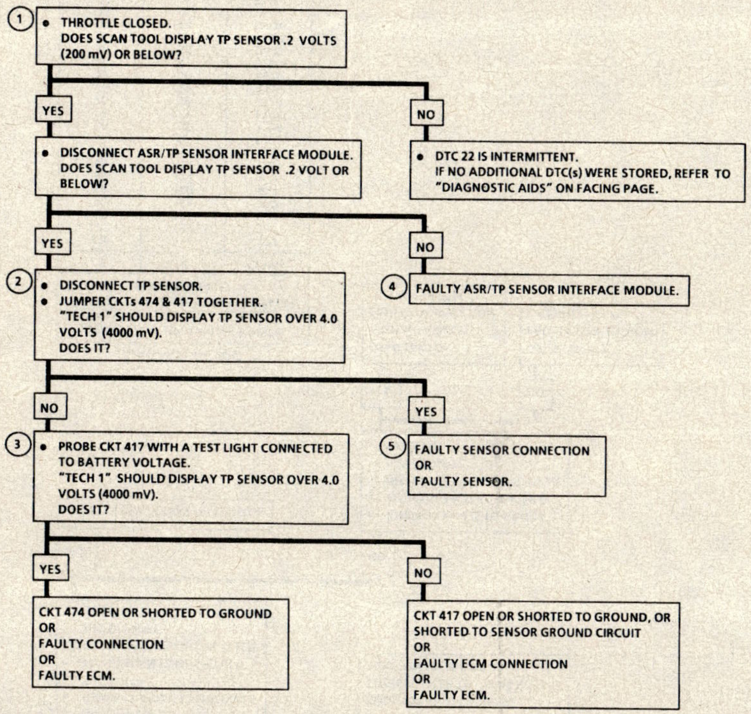

1
THROTTLE CLOSED.
DOES SCAN TOOL DISPLAY TP SENSOR .2 VOLTS (200 mV) OR BELOW?
YES
NO
DISCONNECT ASR/TP SENSOR INTERFACE MODULE.
DOES SCAN TOOL DISPLAY TP SENSOR .2 VOLT OR BELOW?
DTC 22 IS INTERMITTENT.
IF NO ADDITIONAL DTC(s) WERE STORED, REFER TO "DIAGNOSTIC AIDS" ON FACING PAGE.
YES
NO
2
DISCONNECT TP SENSOR.
JUMPER CKTs 474 & 417 TOGETHER.
"TECH 1" SHOULD DISPLAY TP SENSOR OVER 4.0 VOLTS (4000 mV).
DOES IT?
4
FAULTY ASR/TP SENSOR INTERFACE MODULE.
NO
YES
3
PROBE CKT 417 WITH A TEST LIGHT CONNECTED TO BATTERY VOLTAGE.
"TECH 1" SHOULD DISPLAY TP SENSOR OVER 4.0 VOLTS (4000 mV).
DOES IT?
5
FAULTY SENSOR CONNECTION
OR
FAULTY SENSOR.
YES
NO
CKT 474 OPEN OR SHORTED TO GROUND
OR
FAULTY CONNECTION
OR
FAULTY ECM.
CKT 417 OPEN OR SHORTED TO GROUND, OR SHORTED TO SENSOR GROUND CIRCUIT
OR
FAULTY ECM CONNECTION
OR
FAULTY ECM.

"AFTER REPAIRS," REFER TO DTC CRITERIA ON FACING PAGE AND CONFIRM DTC DOES NOT RESET.

5.7L (VIN 8 & P) ENGINE — DIAGNOSTIC TROUBLE CODE CHART — 1992–93 CORVETTE

INTAKE AIR TEMPERATURE SENSOR
A
B
ECM
IAT SENSOR SIGNAL
472 TAN
C22
5 V
802 BLK
B3
SENSOR GROUND
TO MAP AND A/C REFRIGERANT PRESSURE SENSORS

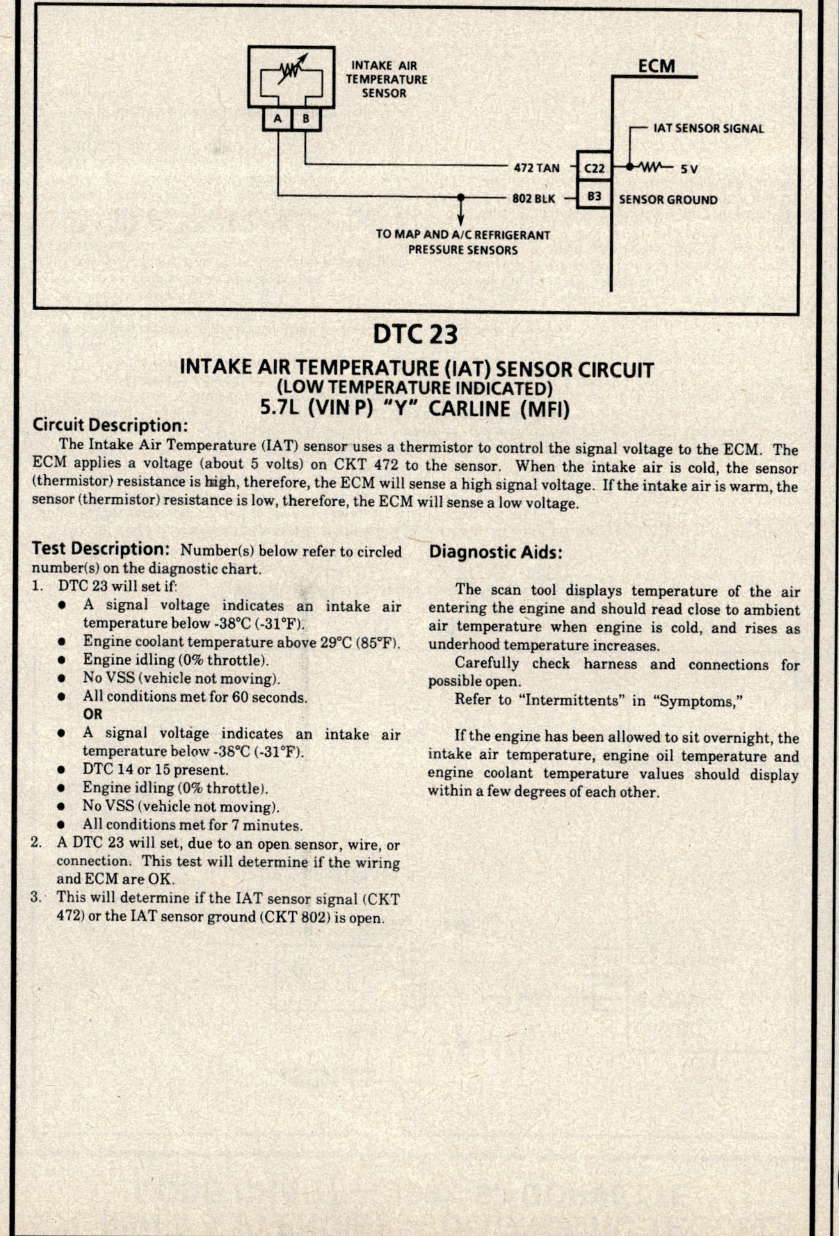

DTC 23
INTAKE AIR TEMPERATURE (IAT) SENSOR CIRCUIT
(LOW TEMPERATURE INDICATED)
5.7L (VIN P) "Y" CARLINE (MFI)

Circuit Description:

The Intake Air Temperature (IAT) sensor uses a thermistor to control the signal voltage to the ECM. The ECM applies a voltage (about 5 volts) on CKT 472 to the sensor. When the intake air is cold, the sensor (thermistor) resistance is high, therefore, the ECM will sense a high signal voltage. If the intake air is warm, the sensor (thermistor) resistance is low, therefore, the ECM will sense a low voltage.

Test Description: Number(s) below refer to circled number(s) on the diagnostic chart.

1. DTC 23 will set if:
   - A signal voltage indicates an intake air temperature below -38°C (-31°F).
   - Engine coolant temperature above 29°C (85°F).
   - Engine idling (0% throttle).
   - No VSS (vehicle not moving).
   - All conditions met for 60 seconds.

   OR
   - A signal voltage indicates an intake air temperature below -38°C (-31°F).
   - DTC 14 or 15 present.
   - Engine idling (0% throttle).
   - No VSS (vehicle not moving).
   - All conditions met for 7 minutes.
2. A DTC 23 will set, due to an open sensor, wire, or connection. This test will determine if the wiring and ECM are OK.
3. This will determine if the IAT sensor signal (CKT 472) or the IAT sensor ground (CKT 802) is open.

Diagnostic Aids:

The scan tool displays temperature of the air entering the engine and should read close to ambient air temperature when engine is cold, and rises as underhood temperature increases.

Carefully check harness and connections for possible open.

Refer to "Intermittents" in "Symptoms,"

If the engine has been allowed to sit overnight, the intake air temperature, engine oil temperature and engine coolant temperature values should display within a few degrees of each other.

5.7L (VIN 8 & P) ENGINE — DIAGNOSTIC TROUBLE CODE CHART — 1992–93 CORVETTE

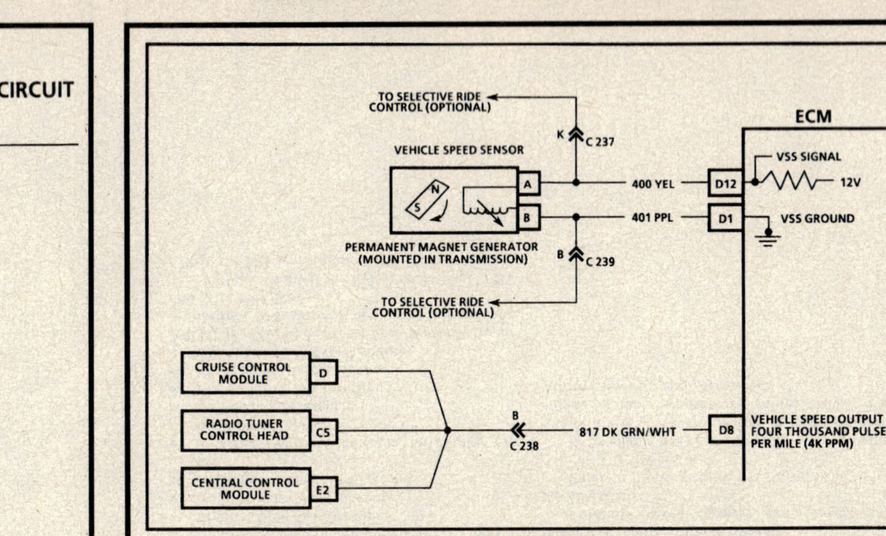

DTC 23
INTAKE AIR TEMPERATURE (IAT) SENSOR CIRCUIT
(LOW TEMPERATURE INDICATED)
5.7L (VIN P) "Y" CARLINE (MFI)

DIAGNOSTIC AID		
INTAKE AIR TEMPERATURE SENSOR		
TEMPERATURE VS. RESISTANCE VALUES (APPROXIMATE)		
°C	°F	OHMS
100	212	177
90	194	241
80	176	332
70	158	467
60	140	667
50	122	973
45	113	1188
40	104	1459
35	95	1802
30	86	2238
25	77	2796
20	68	3520
15	59	4450
10	50	5670
5	41	7280
0	32	9420
-5	23	12300
-10	14	16180
-15	5	21450
-20	-4	28680
-30	-22	52700
-40	-40	100700

"AFTER REPAIRS," REFER TO DTC CRITERIA ON FACING PAGE AND CONFIRM DTC DOES NOT RESET.

5.7L (VIN 8 & P) ENGINE — DIAGNOSTIC TROUBLE CODE CHART — 1992–93 CORVETTE

DTC 24
VEHICLE SPEED SENSOR (VSS) CIRCUIT
5.7L (VIN P) "Y" CARLINE (MFI)

Circuit Description:

Vehicle speed information is provided to the ECM by the Vehicle Speed Sensor (VSS), which is a Permanent Magnet (PM) generator located in the transmission. The PM generator produces a pulsing AC voltage. The AC voltage level and the number of pulses increases as the speed of the vehicle increases. The ECM then converts the pulsing voltage to vehicle speed which is used for calculations. The vehicle speed can be displayed with a Tech 1.

The ECM supplies a signal, on CKT 817, to the Central Control Module (CCM) for operating the speedometer and the odometer. The VSS signal is also sent to the cruise control module and may be sent to the radio control head, so that the volume of the radio will increase as the vehicle speed increases.

Test Description: Number(s) below refer to circled number(s) on the diagnostic chart.

1. DTC 24 will set if vehicle speed is less than 5 km/h (3 mph) when:
   - Engine speed is between 1250 and 3000 RPM.
   - TP sensor is less than 2%.
   - MAP less than 21 kPa.
   - Above conditions met for 4 seconds.

   These conditions can be met during a road load deceleration.

   The "ASR" system must be disabled when performing this step. Whenever the ignition key is cycled to the "OFF" position and then cycled back "ON," the ASR system will default "ON."
2. DTC 24 is being caused by a faulty ECM, faulty PROM or an incorrect PROM.
3. If the vehicle is equipped with selective ride, remove the selective ride control module connector and retest.

If the Tech 1 does not display vehicle speed, the fault is in the wiring or ECM.

Diagnostic Aids:

The Tech 1 should indicate a vehicle speed whenever the drive wheels are turning.

Check CKTs 400 and 401 for proper connections to be sure they're clean and tight and the harness is routed correctly.

A fault in CKT 400 or CKT 401 may cause a DTC 23 in the Selective Ride Control (SRC) module and the "selective ride control" light in the Driver Information Center (DIC) may be "ON."

5.7L (VIN 8 & P) ENGINE — DIAGNOSTIC TROUBLE CODE CHART — 1992–93 CORVETTE

DTC 24
VEHICLE SPEED SENSOR (VSS) CIRCUIT
5.7L (VIN P) "Y" CARLINE (MFI)

(1)
- IGNITION "OFF."
- RAISE DRIVE WHEELS.
- START ENGINE.
- DISABLE "ASR" SYSTEM.

NOTICE: DO NOT PERFORM THIS TEST WITHOUT SUPPORTING THE LOWER CONTROL ARMS SO THAT THE DRIVE AXLES ARE IN A NORMAL HORIZONTAL POSITION. RUNNING THE VEHICLE IN GEAR WITH THE WHEELS HANGING DOWN AT FULL TRAVEL MAY DAMAGE THE DRIVE AXLES.

- WITH ENGINE IDLING IN GEAR, TECH 1 SHOULD DISPLAY VEHICLE SPEED ABOVE 0.
  DOES IT?

NO → DOES SPEEDOMETER WORK?

YES →
- DTC 24 IS INTERMITTENT. IF NO ADDITIONAL DTC(S) WERE STORED, REFER TO "DIAGNOSTIC AIDS" ON FACING PAGE.

NO →
- IGNITION "OFF."
- DISCONNECT VSS HARNESS CONNECTOR AT TRANSMISSION.
- CONNECT SIGNAL GENERATOR TESTER J 33431-B TO VSS HARNESS CONNECTOR.
- IGNITION "ON," TESTER "ON" AND SET TO GENERATE A VSS SIGNAL.
- THE TECH 1 SHOULD DISPLAY VEHICLE SPEED ABOVE 0.
  DOES IT?

YES → (2)
- CHECK PROM FOR CORRECT APPLICATION. IF OK, REPLACE ECM.

NO → (3) CKT 400 OR 401 OPEN, SHORTED TO GROUND, SHORTED TOGETHER, FAULTY CONNECTIONS OR FAULTY ECM.

YES → REPLACE VEHICLE SPEED SENSOR.

"AFTER REPAIRS," REFER TO DTC CRITERIA ON FACING PAGE AND CONFIRM DTC DOES NOT RESET.

5.7L (VIN 8 & P) ENGINE — DIAGNOSTIC TROUBLE CODE CHART — 1992–93 CORVETTE

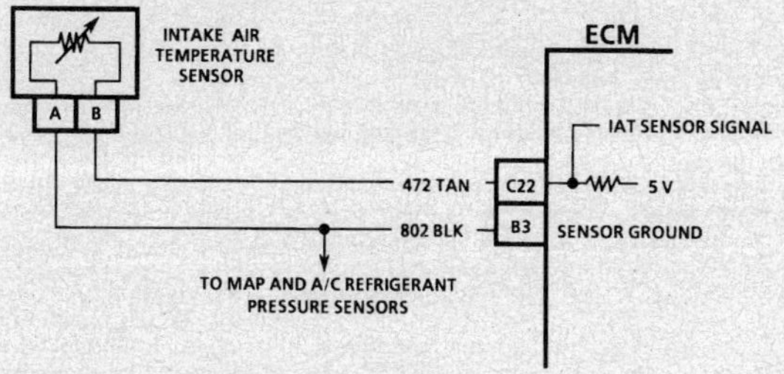

DTC 25
INTAKE AIR TEMPERATURE (IAT) SENSOR CIRCUIT
(HIGH TEMPERATURE INDICATED)
5.7L (VIN P) "Y" CARLINE (MFI)

Circuit Description:

The Intake Air Temperature (IAT) sensor uses a thermistor to control the signal voltage to the ECM. The ECM applies a voltage (about 5 volts) on CKT 472 to the sensor. When intake air is cold, the sensor (thermistor) resistance is high, therefore, the ECM will sense a high signal voltage. If the intake air is warm, the sensor (thermistor) resistance is low, therefore, the ECM will sense a low signal voltage.

Test Description: Number(s) below refer to circled number(s) on the diagnostic chart.

1. DTC 25 will set if:
   - Signal voltage indicates an intake air temperature greater than 150°C (302°F) for 12 seconds.
   - Time since engine start is 4 minutes or longer.
   - Vehicle speed is greater than 8 km/h (5 mph).
   OR
   - Signal voltage indicates a intake air temperature greater than 150°C (302°F).
   - No VSS (vehicle not moving).
   - All conditions met for 60 seconds.

Diagnostic Aids:

The Tech 1 displays temperature of the air entering the engine and should display close to ambient air temperature, when engine is cold, and rise as underhood temperature increases.

Check harness routing for possible short to ground in CKT 472.

Refer to "Intermittents" in "Symptoms,"

5.7L (VIN 8 & P) ENGINE — DIAGNOSTIC TROUBLE CODE CHART — 1992–93 CORVETTE

DTC 25
INTAKE AIR TEMPERATURE (IAT) SENSOR CIRCUIT
(HIGH TEMPERATURE INDICATED)
5.7L (VIN P) "Y" CARLINE (MFI)

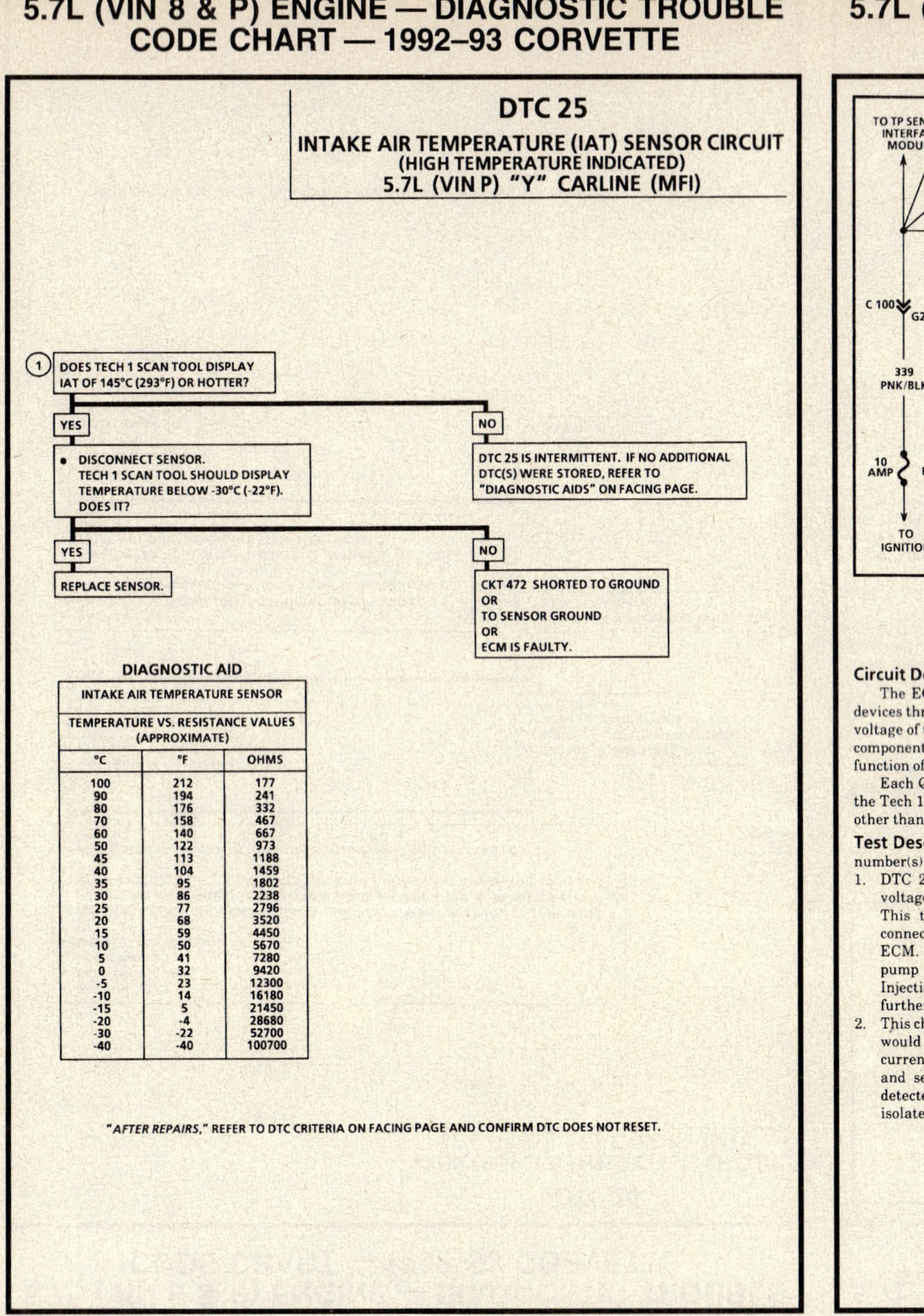

DIAGNOSTIC AID

INTAKE AIR TEMPERATURE SENSOR		
TEMPERATURE VS. RESISTANCE VALUES (APPROXIMATE)		
°C	°F	OHMS
100	212	177
90	194	241
80	176	332
70	158	467
60	140	667
50	122	973
45	113	1188
40	104	1459
35	95	1802
30	86	2238
25	77	2796
20	68	3520
15	59	4450
10	50	5670
5	41	7280
0	32	9420
-5	23	12300
-10	14	16180
-15	5	21450
-20	-4	28680
-30	-22	52700
-40	-40	100700

"AFTER REPAIRS," REFER TO DTC CRITERIA ON FACING PAGE AND CONFIRM DTC DOES NOT RESET.

5.7L (VIN 8 & P) ENGINE — DIAGNOSTIC TROUBLE CODE CHART — 1992–93 CORVETTE

DTC 26
QUAD-DRIVER MODULE (QDM) #1 CIRCUIT
5.7L (VIN P) "Y" CARLINE (MFI)

Circuit Description:

The ECM is used to control several components such as those illustrated above. The ECM controls these devices through the use of a Quad-Driver Module (QDM). When the ECM is commanding a component "ON," the voltage of the output circuit should be "low" (near 0 volts). When the ECM is commanding the output circuit to a component "OFF," the voltage potential of the circuit should be "high" (near battery voltage). The primary function of the QDM is to supply the ground for the component being controlled.

Each QDM has a fault line which is monitored by the ECM. The fault line signal status can be displayed on the Tech 1. The ECM will compare the voltage at the QDM. If the QDM fault detection circuit senses a voltage other than what is expected, the fault line status will change and a DTC 26 will set.

Test Description: Number(s) below refer to circled number(s) on the diagnostic chart.

1. DTC 26 will set if: the ECM detects the wrong voltage potential for 20 seconds.
   This test will begin to determine if the QDM connected AIR pump can be controlled by the ECM. If the relay appears to operate but the AIR pump does not turn on, refer to "Secondary Air Injection (AIR) System," for further AIR diagnosis.

2. This check can detect a partially shorted coil which would cause excessive current flow. Excessive current flow to a QDM will be detected as a fault and set this DTC. If excessive current flow is detected, a circuit check will be performed to isolate the device from the wiring.

3. The remaining checks will identify a circuit problem that has caused an excessive current flow or inoperative relay. If a QDM circuit check is done on a relay, it is important to identify and test the relay coil terminals of the harness connector to avoid improper diagnosis.

Diagnostic Aids:

Engine should be idling while monitoring QDM status. Using a Tech 1, monitor the QDM status while moving related harness connectors, including ECM harness. If the failure is induced, a fault will appear on the Tech 1. This can help locate the intermittent. Check for bent pins at ECM and ECM connector terminals. If DTC reoccurs with no apparent connector problem, replace ECM.

5.7L (VIN 8 & P) ENGINE — DIAGNOSTIC TROUBLE CODE CHART — 1992–93 CORVETTE

DTC 26
QUAD-DRIVER MODULE (QDM) #1 CIRCUIT
5.7L (VIN P) "Y" CARLINE (MFI)

(1)
- IF OTHER DTCs ARE STORED USE THOSE CHARTS FIRST.
- IGNITION "ON," ENGINE "OFF."
- USING TECH 1 "MISC TEST," ACTIVATE AIR PUMP RELAY.
- DOES AIR PUMP RELAY OPERATE?

YES → ACTIVATE AIR SOLENOID VALVE. DOES AIR SOLENOID VALVE OPERATE?
NO →

(3)
- IGNITION "OFF."
- DISCONNECT AFFECTED RELAY/SOLENOID VALVE CONNECTOR.
- CONNECT TEST LIGHT BETWEEN RELAY/SOLENOID COIL CONNECTOR TERMINALS.
- IGNITION "ON."
- GROUND DIAGNOSTIC TEST TERMINAL.
- LIGHT SHOULD BE "ON." IS IT?

YES → ACTIVATE EGR SOLENOID VALVE. DOES EGR SOLENOID VALVE OPERATE?
NO →

YES → ACTIVATE CANISTER PURGE SOLENOID VALVE. DOES CANISTER PURGE SOLENOID VALVE OPERATE?
NO →

NO:
- CONNECT TEST LIGHT FROM IGNITION FEED CIRCUIT OF RELAY/SOLENOID VALVE TO GROUND.
- LIGHT SHOULD BE "ON." IS IT?

YES:
- UNGROUND DIAGNOSTIC TEST TERMINAL.
- LIGHT SHOULD BE "OFF." IS IT?

YES →

(2)
- KEY "OFF."
- DISCONNECT ECM CONNECTOR "A."
- KEY "ON."
- USING DVM ON 2 AMP SCALE, MEASURE CURRENT FROM HARNESS TERMINALS "A7", "A8", "A2", "A1" TO GROUND.
- EACH TERMINAL SHOULD MEASURE LESS THAN .75 AMPS (BUT NOT "0").
- DO THEY?

YES → DTC 26 IS INTERMITTENT. REFER TO "DIAGNOSTIC AIDS" ON FACING PAGE.
NO →

YES → FAULTY ECM CONNECTION OR OPEN RELAY/SOLENOID VALVE DRIVER CIRCUIT OR FAULTY ECM.
NO → OPEN RELAY/SOLENOID VALVE IGNITION FEED CIRCUIT.

NO → DISCONNECT ECM CONNECTOR "A". IS LIGHT "ON"?
YES → FAULTY RELAY/SOLENOID VALVE CONNECTION OR FAULTY RELAY/SOLENOID VALVE.

NO → FAULTY ECM.
YES → GROUNDED RELAY/SOLENOID DRIVER CIRCUIT.

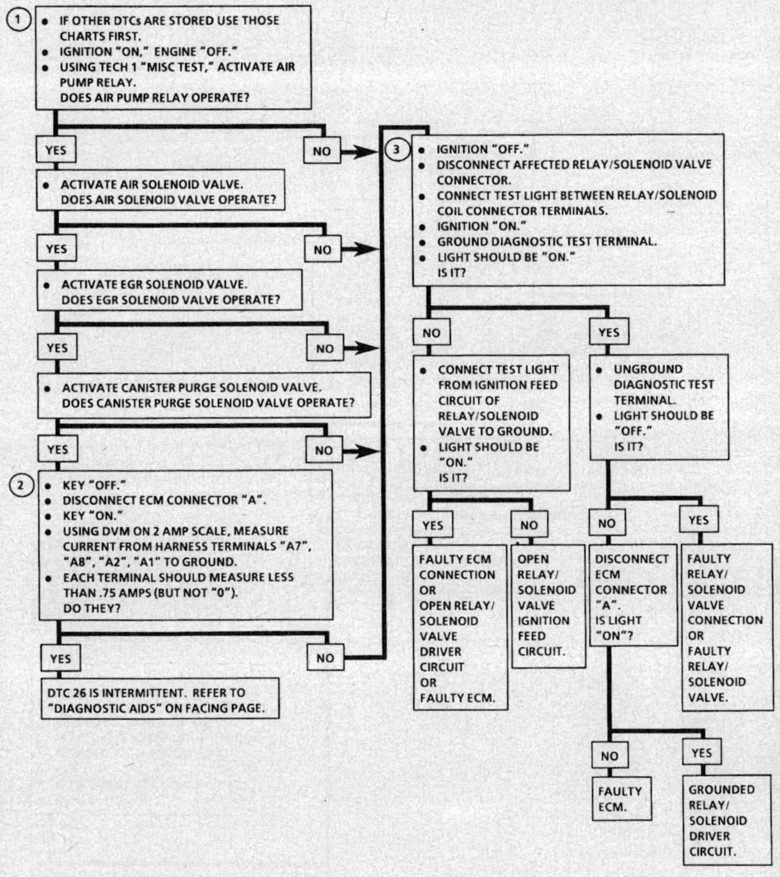

"AFTER REPAIRS," REFER TO DTC CRITERIA ON FACING PAGE AND CONFIRM DTC DOES NOT RESET.

5.7L (VIN 8 & P) ENGINE — DIAGNOSTIC TROUBLE CODE CHART — 1992–93 CORVETTE

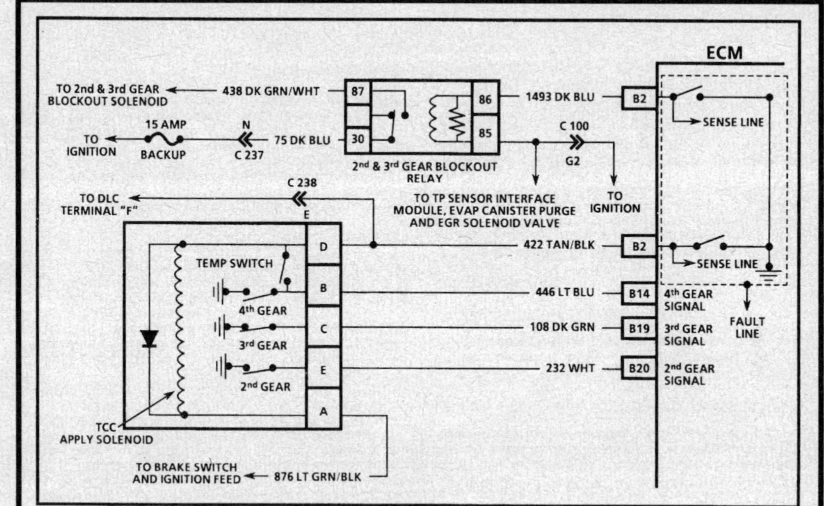

DTC 27
QUAD-DRIVER MODULE (QDM) #2 CIRCUIT
5.7L (VIN P) "Y" CARLINE (MFI)

Circuit Description:

The ECM is used to control several components such as those illustrated above. The ECM controls these devices through the use of a Quad-Driver Module (QDM). When the ECM is commanding a component "ON," the voltage of the output circuit will be "low" (near 0 volts). When the ECM is commanding the output circuit to a component "OFF," the voltage potential of the circuit will be "high" (near battery voltage). The primary function of the QDM is to supply the ground for the component being controlled.

Each QDM has a fault line which is monitored by the ECM. The fault line signal status can be displayed on the Tech 1. The ECM will compare the voltage at the QDM. If the QDM fault detection circuit senses a voltage other than what is expected, the fault line status will change and a DTC 27 will set if applicable.

A fault in QDM #2 will store a DTC 27 in the ECM memory but will not turn "ON" the MIL (Service Engine Soon).

Test Description: Number(s) below refer to circled number(s) on the diagnostic chart.

1. DTC 27 will set if the ECM detects an improper voltage potential for 25 seconds.
   This test will begin to determine if the QDM connected devices can be controlled by the ECM. If the 2nd & 3rd gear blockout relay (M/T) appears to operate but the 2nd & 3rd gear blockout solenoid does not, refer to "Torque Converter Clutch (TCC) System (A/T), Automatic Transmission Gear Switch System, 1-4 Upshift System (M/T),"

2. This check can detect a partially shorted coil which would cause excessive current flow. Excessive current flow to a QDM will be detected as a fault and set this DTC. If excessive current flow is detected, a circuit check will be performed to isolate the device from the wiring.

3. The remaining checks will identify a circuit problem that has caused an excessive current flow or inoperative relay. If a QDM circuit check is done on a relay, it is important to identify and test the relay coil terminals of the harness connector to avoid improper diagnosis.

Diagnostic Aids:

Engine should be idling while monitoring QDM status. Using a Tech 1, monitor the QDM status while moving related harness connectors, including ECM harness. If the failure is induced, a fault will appear on the Tech 1. This can help locate the intermittent. Check for bent pins at ECM and ECM terminals. If DTC reoccurs with no apparent connector problem, replace ECM.

The QDM status should change when the brake is applied on vehicles equipped with a torque converter clutch.

DTC 27

QUAD-DRIVER MODULE (QDM) #2 CIRCUIT
5.7L (VIN P) "Y" CARLINE (MFI)

(1)
- IF OTHER DTC(s) ARE STORED USE THOSE CHARTS FIRST.
- IGNITION "ON," ENGINE "OFF."
- USING TECH 1 "MISC TEST," ACTIVATE 2ND AND 3RD GEAR BLOCKOUT RELAY (M/T) OR TCC SOLENOID (A/T). DOES RELAY/SOLENOID OPERATE?

YES

(2)
- KEY "OFF."
- DISCONNECT ECM CONNECTOR "B".
- KEY "ON."
- USING DVM ON 2 AMP SCALE, MEASURE CURRENT FROM HARNESS TERMINAL "B2" TO GROUND.
- TERMINAL SHOULD MEASURE LESS THAN .75 AMPS (BUT NOT "0"). DOES IT?

YES

DTC 27 IS INTERMITTENT. REFER TO "DIAGNOSTIC AIDS" ON FACING PAGE.

NO

(3)
- IGNITION "OFF."
- DISCONNECT AFFECTED RELAY/SOLENOID CONNECTOR.
- CONNECT TEST LIGHT BETWEEN RELAY/SOLENOID COIL CONNECTOR TERMINALS.
- IGNITION "ON."
- GROUND DIAGNOSTIC TEST TERMINAL.
- LIGHT SHOULD BE "ON." IS IT?

NO
- CONNECT TEST LIGHT FROM IGNITION FEED CIRCUIT OF RELAY/SOLENOID TO GROUND.
- LIGHT SHOULD BE "ON." IS IT?

YES | NO

YES: FAULTY ECM CONNECTION OR OPEN RELAY/SOLENOID DRIVER CIRCUIT OR FAULTY ECM.

NO: OPEN RELAY/SOLENOID IGNITION FEED CIRCUIT.

YES
- UNGROUND DIAGNOSTIC TEST TERMINAL.
- LIGHT SHOULD BE "OFF." IS IT?

NO: DISCONNECT ECM CONNECTOR "B". IS LIGHT "ON"?

YES: FAULTY RELAY/SOLENOID CONNECTION OR FAULTY RELAY/SOLENOID.

NO: FAULTY ECM.

YES: GROUNDED RELAY/SOLENOID DRIVER CIRCUIT.

"AFTER REPAIRS," REFER TO DTC CRITERIA ON FACING PAGE AND CONFIRM DTC DOES NOT RESET.

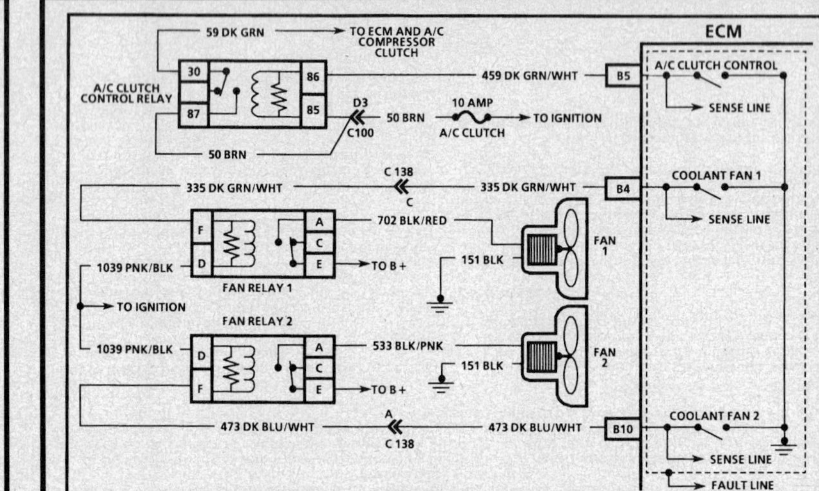

DTC 28

QUAD-DRIVER MODULE (QDM) #3 CIRCUIT
5.7L (VIN P) "Y" CARLINE (MFI)

Circuit Description:

The ECM is used to control several components such as those illustrated above. The ECM controls these devices through the use of a Quad-Driver Module (QDM). When the ECM is commanding a component "ON," the voltage of the output circuit will be "low" (near 0 volts). When the ECM is commanding the output circuit to a component "OFF," the voltage potential of the circuit will be "high" (near battery voltage). The primary function of the QDM is to supply the ground for the component being controlled.

Each QDM has a fault line which is monitored by the ECM. The fault line signal status can be displayed on the Tech 1. The ECM will compare the voltage at the QDM only when an A/C request signal is detected at the ECM. If the QDM fault detection circuit senses a voltage other than what is expected, the fault line status will change and a DTC 28 will set if applicable.

A QDM #3 fault will store a DTC 28 in the ECM memory but will not turn "ON" the MIL (Service Engine Soon).

Test Description: Number(s) below refer to circled number(s) on the diagnostic chart.

1. DTC 28 will set if the ECM detects an improper voltage potential for 25 seconds.
   These tests will begin to determine if the QDM connected devices can be controlled be the ECM. If a relay appears to operate but the device does not activate, refer to "Electric Cooling Fan, or "A/C Clutch Circuit Diagnosis," for further diagnosis.

2. This check can detect a partially shorted coil which would cause excessive current flow. Excessive current flow to a QDM will be detected as a fault and set this DTC. If excessive current flow is detected, a circuit check will be performed to isolate the device from the wiring.

3. The remaining checks will identify a circuit problem that has caused an excessive current flow or inoperative relay. If a QDM circuit check is done on a relay, it is important to identify and test the relay coil terminals of the harness connector to avoid improper diagnosis.

Diagnostic Aids:

Engine should be idling while monitoring QDM status. Using a Tech 1, monitor the QDM status while moving related harness connectors, including ECM harness. If the failure is induced, a fault will appear on the Tech 1. This can help locate the intermittent. Check for bent pins at ECM and ECM connector terminals. If DTC reoccurs with no apparent connector problem, replace ECM.

5.7L (VIN 8 & P) ENGINE — DIAGNOSTIC TROUBLE CODE CHART — 1992–93 CORVETTE

DTC 28
QUAD-DRIVER MODULE (QDM) #3 CIRCUIT
5.7L (VIN P) "Y" CARLINE (MFI)

① • IF OTHER DTCs ARE STORED USE THOSE CHARTS FIRST.
• IGNITION "ON," ENGINE "OFF."
• USING TECH 1 "MISC TEST," ACTIVATE FAN 1 RELAY.
DOES FAN 1 RELAY OPERATE?

YES → ACTIVATE FAN 2 RELAY.
DOES FAN 2 RELAY OPERATE?

YES → ACTIVATE A/C RELAY.
DOES A/C RELAY OPERATE?

② **YES** → • KEY "OFF."
• DISCONNECT ECM CONNECTOR "B".
• KEY "ON."
• USING DVM ON 2 AMP SCALE, MEASURE CURRENT FROM HARNESS TERMINALS "B4", "B5", "B10" TO GROUND.
• EACH TERMINAL SHOULD MEASURE LESS THAN .75 AMPS (BUT NOT "0").
DO THEY?

YES → DTC 28 IS INTERMITTENT. REFER TO "DIAGNOSTIC AIDS" ON FACING PAGE.

NO (from ①, ②, and A/C relay) →

③ • IGNITION "OFF."
• DISCONNECT AFFECTED RELAY CONNECTOR.
• CONNECT TEST LIGHT BETWEEN RELAY COIL CONNECTOR TERMINALS.
• IGNITION "ON."
• GROUND DIAGNOSTIC TEST TERMINAL.
• LIGHT SHOULD BE "ON."
IS IT?

NO → • CONNECT TEST LIGHT FROM IGNITION FEED CIRCUIT OF RELAY/SOLENOID VALVE TO GROUND.
• LIGHT SHOULD BE "ON."
IS IT?

  YES → OPEN RELAY/SOLENOID VALVE IGNITION FEED CIRCUIT.

  NO → FAULTY ECM CONNECTION OR OPEN RELAY/SOLENOID VALVE DRIVER CIRCUIT OR FAULTY ECM.

YES → • UNGROUND DIAGNOSTIC TEST TERMINAL.
• LIGHT SHOULD BE "OFF."
IS IT?

  YES → DISCONNECT ECM CONNECTOR "B". IS LIGHT "ON"?

    NO → FAULTY ECM.

    YES → GROUNDED RELAY/SOLENOID VALVE DRIVER CIRCUIT.

  NO → FAULTY RELAY/SOLENOID VALVE CONNECTION OR FAULTY RELAY/SOLENOID VALVE.

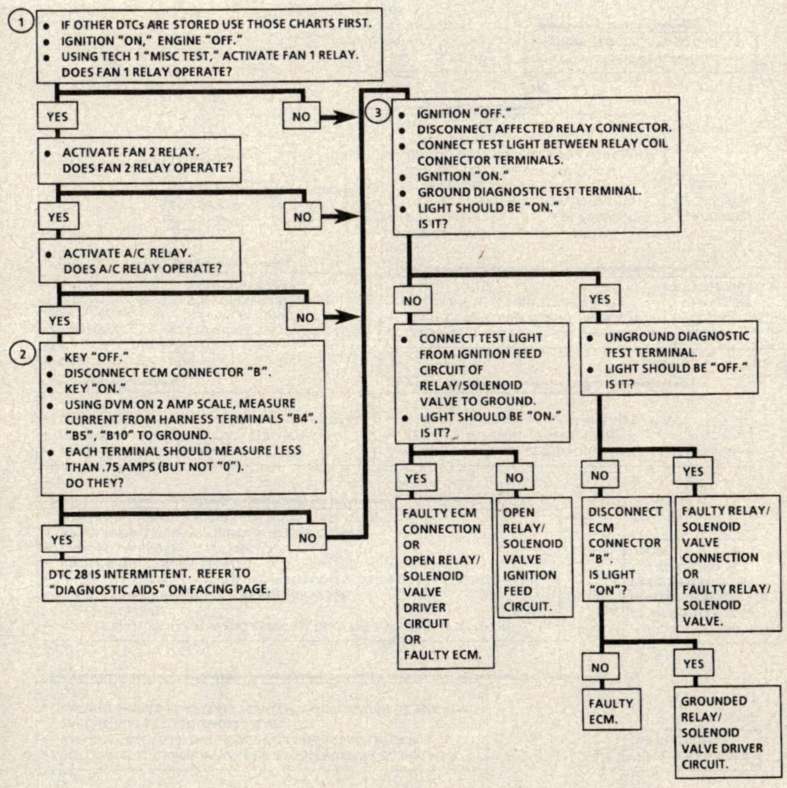

"AFTER REPAIRS," REFER TO DTC CRITERIA ON FACING PAGE AND CONFIRM DTC DOES NOT RESET.

5.7L (VIN 8 & P) ENGINE — DIAGNOSTIC TROUBLE CODE CHART — 1992–93 CORVETTE

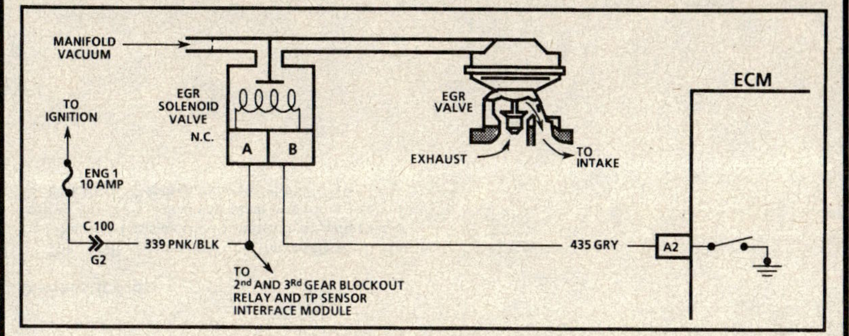

DTC 32
(Page 1 of 2)
EXHAUST GAS RECIRCULATION (EGR) CIRCUIT
5.7L (VIN P) "Y" CARLINE (MFI)

Circuit Description:

The Exhaust Gas Recirculation (EGR) valve vacuum is controlled by an ECM controlled solenoid valve. The ECM will turn the EGR "ON" and "OFF" (Duty Cycle) by grounding CKT 435. The duty cycle is calculated by the ECM based on information from the engine coolant temperature, air flow and engine RPM. There should be no EGR when in park or neutral, TP sensor input below a specified value, or TP sensor indicating Wide Open Throttle (WOT).

With the ignition "ON" engine stopped, the EGR solenoid valve is de-energized and by grounding the diagnostic "test" terminal, the solenoid valve is energized. The ECM will check EGR operation when:
• Vehicle speed is above 50 mph.
• MAP is as listed in table (depends on altitude).
• TP sensor is between 8% and 30%.
• No more than .4% change in TP sensor.

MAP WITH KEY "ON" ENGINE "OFF" (BARO)			
100	90	80	70
30	20	10	0

MAP RANGE FOR TEST			
85	75	65	55

Test Description: Number(s) below refer to circled number(s) on the diagnostic chart.

1. **Intake Passage:** Shut "OFF" engine and remove the EGR valve from the manifold. Plug the exhaust side hole with a suitable stopper. With the intake side hole open, attempt to start the engine. If the engine runs at a very high idle (up to 3000 RPM is possible) or starts and stalls, the EGR passages are not restricted. If the engine starts and idles normally, the EGR intake side passage in the intake manifold is restricted.
Exhaust Passage: With EGR valve still removed, plug the intake side hole with a suitable stopper. With the exhaust side hole open, check for the presence of exhaust gas. If no exhaust gas is present, the EGR exhaust side passage in the intake manifold is restricted.

2. By grounding the diagnostic "test" terminal, the EGR solenoid valve should be energized and allow vacuum to be applied to the gage. The vacuum at the gage may or may not <u>slowly</u> bleed off. It is important that the gage is able to read the amount of vacuum being applied.

3. When the diagnostic "test" terminal is grounded, the vacuum gage should bleed off through a vent in the solenoid valve. The pump gage may or may not bleed off but this does not indicate a problem.

4. This test will determine if the electrical control part of the system is at fault or if the connector or solenoid valve is at fault.

Diagnostic Aids:

Suction from shop exhaust hoses can alter backpressure and may affect the operation of the EGR valve during install testing.

The ECM also monitors left side short term fuel trim during a DTC 32 test and must detect a change to pass the test. If short term fuel trim is at a fixed value (lean or rich) when a DTC 32 test is ran, a DTC 32 may set.

Vacuum lines should be thoroughly checked for proper routing. Vacuum source goes to orifice side of the EGR solenoid valve. Refer to "Vehicle Emissions Control Information" Label.

5.7L (VIN 8 & P) ENGINE — DIAGNOSTIC TROUBLE CODE CHART — 1992–93 CORVETTE

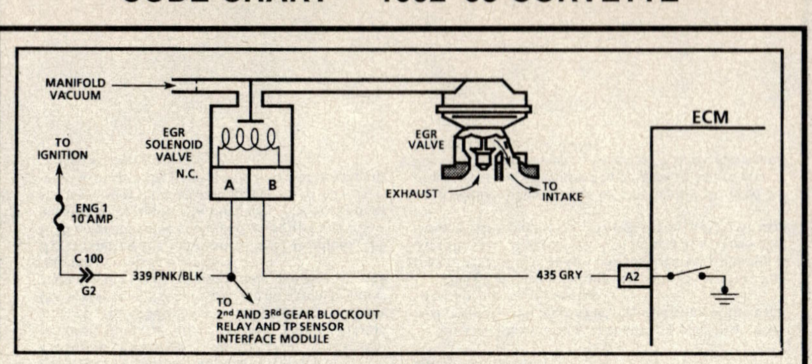

5.7L (VIN 8 & P) ENGINE — DIAGNOSTIC TROUBLE CODE CHART — 1992–93 CORVETTE

DTC 32
(Page 2 of 2)
EXHAUST GAS RECIRCULATION (EGR) CIRCUIT
5.7L (VIN P) "Y" CARLINE (MFI)

Circuit Description:

The Exhaust Gas Recirculation (EGR) valve vacuum is controlled by an ECM controlled solenoid valve. The ECM will turn the EGR "ON" and "OFF" (Duty Cycle) by grounding CKT 435. The duty cycle is calculated by the ECM based on information from the engine coolant temperature, air flow and engine RPM. There should be no EGR when in park or neutral, TP sensor input below a specified value, or TP sensor indicating Wide Open Throttle (WOT).

With the ignition "ON" engine stopped, the EGR solenoid valve is de-energized and by grounding the diagnostic "test" terminal, the solenoid valve is energized. The ECM will check EGR operation when:

- Vehicle speed is above 50 mph.
- MAP is as listed in table (depends on altitude).
- TP sensor is between 8% and 30%.
- No more than .4% change in TP sensor.

MAP RANGE FOR TEST	MAP WITH KEY "ON" ENGINE "OFF" (BARO)			
	100	90	80	70
	30	20	10	0
	85	75	65	55

Test Description: Number(s) below refer to circled number(s) on the diagnostic chart.

1. The remaining test checks the ability of the EGR valve to interact with the exhaust system. This system uses a negative backpressure EGR valve which should hold vacuum with engine "OFF."
2. When engine is started, exhaust backpressure at the base of the EGR valve should open the valve's internal bleed and vent the applied vacuum allowing the valve to seat.

During normal EGR valve operation, the movement of the EGR valve pintle is small. It is important to determine whether or not the valve pintle moves and not how much it moves.

Suction from shop exhaust hoses can alter backpressure and may affect the functional check of the EGR valve.

Diagnostic Aids:

The EGR circuit can be inoperative if the Park/Neutral Position (PNP) switch is misadjusted or faulty. The EGR is disabled when in park or neutral. To check the PNP switch, refer to CHART C-1A.

5.7L (VIN 8 & P) ENGINE — DIAGNOSTIC TROUBLE CODE CHART — 1992–93 CORVETTE

DTC 32
(Page 2 of 2)
EXHAUST GAS RECIRCULATION (EGR) CIRCUIT
5.7L (VIN P) "Y" CARLINE (MFI)

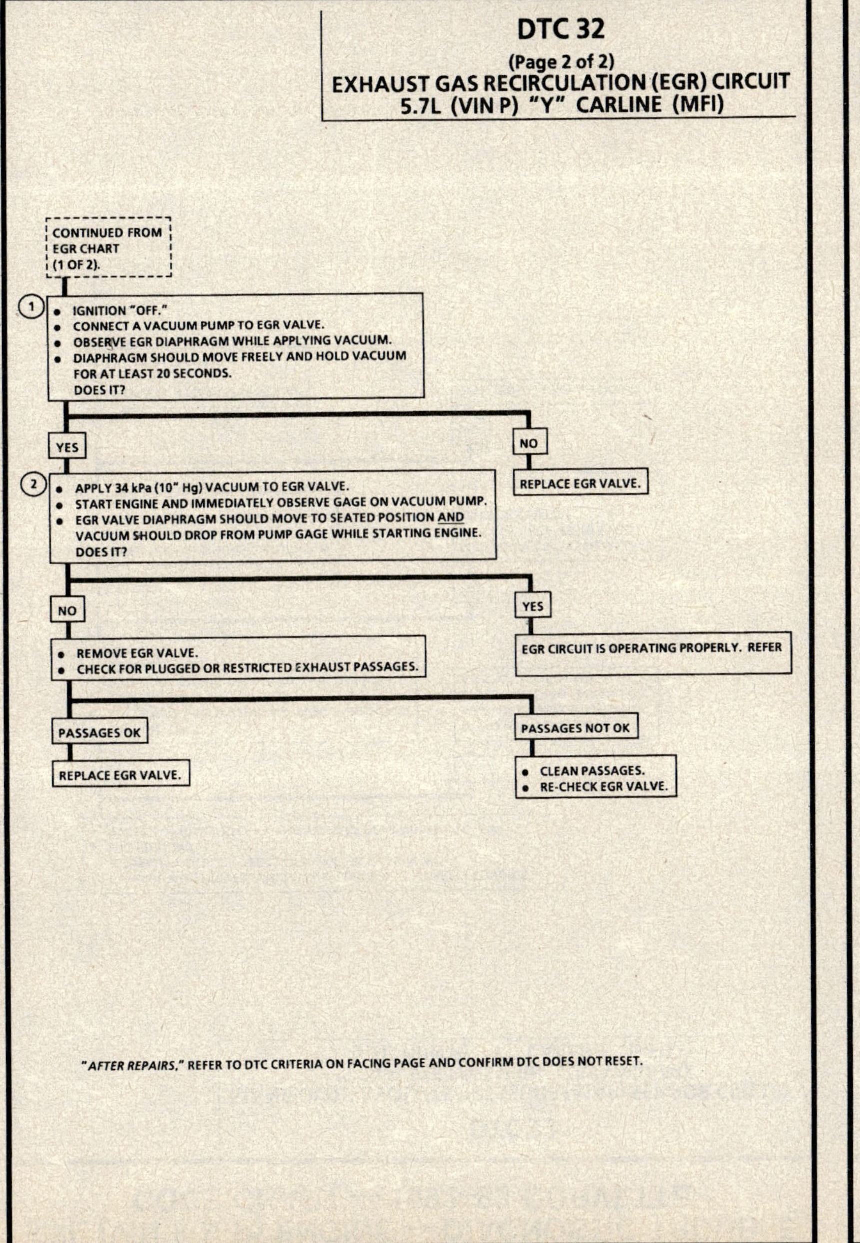

5.7L (VIN 8 & P) ENGINE — DIAGNOSTIC TROUBLE CODE CHART — 1992–93 CORVETTE

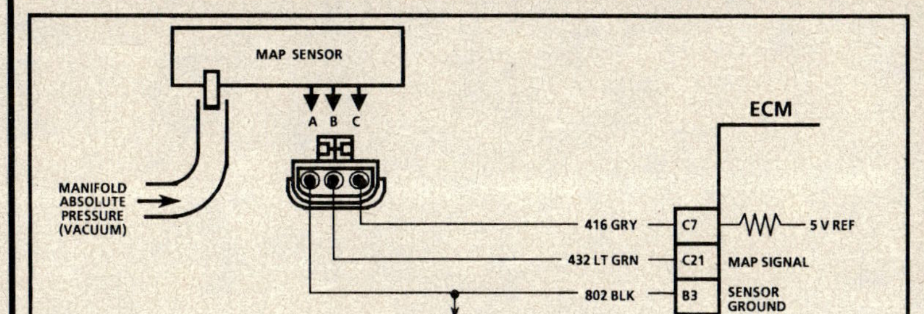

DTC 33
MANIFOLD ABSOLUTE PRESSURE (MAP) SENSOR CIRCUIT
(SIGNAL VOLTAGE HIGH - LOW VACUUM)
5.7L (VIN P) "Y" Carline (MFI)

Circuit Description:

The Manifold Absolute Pressure (MAP) sensor responds to changes in manifold pressure (vacuum). The ECM receives this information as a signal voltage that will vary from about 1 - 1.5 volts at idle to 4.0 - 4.8 volts at Wide Open Throttle (WOT).

The Tech 1 displays manifold pressure in kPa of pressure and voltage. Low pressure (high vacuum) indicates a low voltage while a high pressure (low vacuum) indicates a high voltage.

If the MAP sensor fails, the ECM will substitute a fixed MAP value and use the Throttle Position (TP) sensor to control fuel delivery.

Test Description: Number(s) below refer to circled number(s) on the diagnostic chart.

1. DTC 33 will set when:
   - Less than 3% throttle.
   - MAP greater than 80 kPa.
   - MAP greater than 90 kPa with A/C "ON."
   - RPM greater than 500 RPM.
   - All above conditions met for 1 second.

   Engine misfire or a low unstable idle may set DTC 33. Disconnect MAP sensor and system will go into backup mode. If the misfire or idle condition remains, refer to "Symptoms,"

2. If the ECM recognizes the low MAP signal, the ECM and wiring are OK.

Diagnostic Aids:

If the idle is rough or unstable, refer to "Symptoms," Section   for items which can cause an unstable idle. ·

An open in CKT 802 will result in a DTC 33.

With the ignition "ON," and the engine "OFF," the manifold pressure is equal to atmospheric pressure and the signal voltage will be high. This information is used by the ECM as an indication of vehicle altitude and is referred to as BARO. Comparison of this BARO reading with a known good vehicle with the same sensor is a good way to check accuracy of a "suspect" sensor. Reading should be the same, ± .4 volt.

Also, CHART C-1D can be used to test the MAP sensor.

Refer to "Intermittents" in "Symptoms,"

"AFTER REPAIRS," REFER TO DTC CRITERIA ON FACING PAGE AND CONFIRM DTC DOES NOT RESET.

5.7L (VIN 8 & P) ENGINE — DIAGNOSTIC TROUBLE CODE CHART — 1992–93 CORVETTE

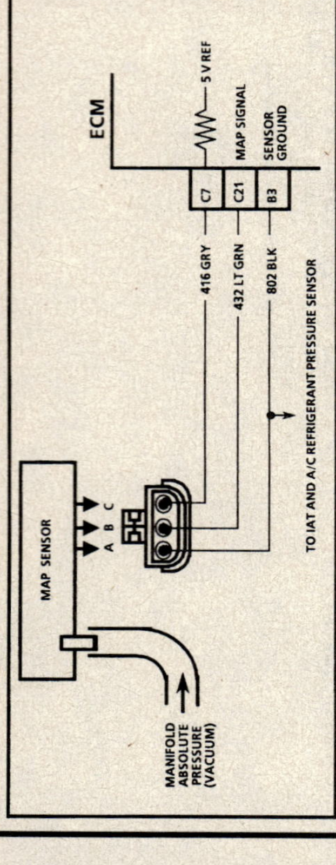

DTC 34

MANIFOLD ABSOLUTE PRESSURE (MAP) SENSOR CIRCUIT
(SIGNAL VOLTAGE LOW - HIGH VACUUM)
5.7L (VIN P) "Y" CARLINE (MFI)

Circuit Description:

The Manifold Absolute Pressure (MAP) sensor responds to changes in manifold pressure (vacuum). The ECM receives this information as a signal voltage that will vary from about 1 - 1.5 volts at idle to 4.0 - 4.8 volts at Wide Open Throttle (WOT).

The scan tool displays manifold pressure in kPa of pressure and voltage. Low pressure (high vacuum) indicates a low voltage while a high pressure (low vacuum) indicates a high voltage.

If the MAP sensor fails, the ECM will substitute a fixed MAP value and use the Throttle Position (TP) sensor to control fuel delivery.

Test Description: Number(s) below refer to circled number(s) on the diagnostic chart.

1. DTC 34 will set when:
   - MAP less than 20 kPa.
   - RPM less than 700 RPM.

   OR
   - MAP less than 20 kPa.
   - RPM greater than 700 RPM.
   - TP sensor greater than 20%.
2. If the ECM recognizes the high MAP signal, the ECM and wiring are OK.
3. The Tech 1 will not display 12 volts. The important thing is that the ECM recognizes the voltage as more than 4 volts, indicating that the ECM and CKT 416 are OK.

Diagnostic Aids:

An intermittent open in CKT 432 or CKT 416 will result in a DTC 34.

With the ignition "ON" and engine "OFF," the manifold pressure is equal to atmospheric pressure and the signal voltage will be high. This information is used by the ECM as an indication of vehicle altitude and is referred to as BARO. Comparison of this BARO reading with a known good vehicle with the same sensor is a good way to check accuracy of a "suspect" sensor. Reading should be the same, ± .4 volt.

Also, CHART C-1D can be used to test the MAP sensor.

Refer to "Intermittents" in "Symptoms."

5.7L (VIN 8 & P) ENGINE — DIAGNOSTIC TROUBLE CODE CHART — 1992–93 CORVETTE

DTC 33

MANIFOLD ABSOLUTE PRESSURE (MAP) SENSOR CIRCUIT
(SIGNAL VOLTAGE HIGH - LOW VACUUM)
5.7L (VIN P) "Y" CARLINE (MFI)

(1)
- IF ENGINE IDLE IS ROUGH, UNSTABLE, OR INCORRECT, CORRECT CONDITION BEFORE USING CHART. SEE "SYMPTOMS" IN SECTION "B".
- ENGINE IDLING.
- DOES SCAN TOOL DISPLAY A MAP VOLTAGE OF 4.0 VOLTS OR OVER?

NO → DTC 33 IS INTERMITTENT. IF NO ADDITIONAL DTC(s) WERE STORED, REFER TO "DIAGNOSTIC AIDS" ON FACING PAGE.

YES →

(2)
- IGNITION "OFF."
- DISCONNECT MAP SENSOR ELECTRICAL CONNECTOR.
- IGNITION "ON."
- SCAN TOOL SHOULD READ A VOLTAGE OF 1 VOLT OR LESS.
- DOES IT?

NO → MAP SIGNAL CIRCUIT SHORTED TO VOLTAGE, SHORTED TO 5 VOLT REFERENCE CIRCUIT OR FAULTY ECM.

YES →
- PROBE SENSOR GROUND CIRCUIT WITH A TEST LIGHT TO BATTERY VOLTAGE.
- TEST LIGHT SHOULD LIGHT.
- DOES IT?

NO → OPEN SENSOR GROUND CIRCUIT.

YES → VACUUM LEAK AT MAP SENSOR SEAL OR FAULTY MAP SENSOR.

"AFTER REPAIRS," REFER TO DTC CRITERIA ON FACING PAGE AND CONFIRM DTC DOES NOT RESET.

5.7L (VIN 8 & P) ENGINE — DIAGNOSTIC TROUBLE CODE CHART — 1992–93 CORVETTE

DTC 34
MANIFOLD ABSOLUTE PRESSURE (MAP) SENSOR CIRCUIT
(SIGNAL VOLTAGE LOW - HIGH VACUUM)
5.7L (VIN P) "Y" CARLINE (MFI)

① • ENGINE IDLING.
• DOES SCAN TOOL DISPLAY MAP VOLTAGE BELOW .25 VOLT?

YES

② • IGNITION "OFF."
• DISCONNECT SENSOR ELECTRICAL CONNECTOR.
• JUMPER HARNESS TERMINALS "B" TO "C".
• IGNITION "ON."
• MAP VOLTAGE SHOULD READ OVER 4.7 VOLTS. DOES IT?

NO

③ • IGNITION "OFF."
• REMOVE JUMPER WIRE.
• PROBE TERMINAL "B" (CKT 432) WITH A TEST LIGHT TO BATTERY VOLTAGE.
• IGNITION "ON."
• SCAN TOOL SHOULD READ OVER 4 VOLTS. DOES IT?

YES

5 VOLT REFERENCE CIRCUIT OPEN
OR
SHORTED TO GROUND
OR
FAULTY ECM.

NO

DTC 34 IS INTERMITTENT. IF NO ADDITIONAL DTC(s) WERE STORED, REFER TO "DIAGNOSTIC AIDS" ON FACING PAGE.

YES

FAULTY CONNECTION
OR
SENSOR.

NO

CKT 432 OPEN
OR
CKT 432 SHORTED TO GROUND
OR
CKT 432 SHORTED TO SENSOR GROUND
OR
FAULTY ECM.

"AFTER REPAIRS," REFER TO DTC CRITERIA ON FACING PAGE AND CONFIRM DTC DOES NOT RESET.

5.7L (VIN 8 & P) ENGINE — DIAGNOSTIC TROUBLE CODE CHART — 1992–93 CORVETTE

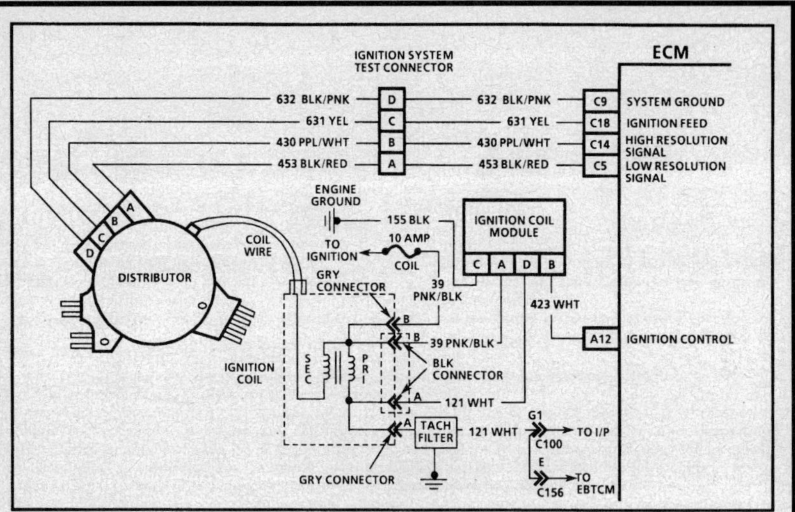

DTC 36
DISTRIBUTOR IGNITION SYSTEM
(FAULTY HIGH RESOLUTION PULSE OR EXTRA LOW RESOLUTION PULSE DETECTED)
5.7L (VIN P) "Y" CARLINE (MFI)

Circuit Description:

The distributor ignition system supplies two timing inputs to the ECM, a high resolution signal (360 pulses per one crankshaft revolution) and a low resolution signal (4 pulses per one crankshaft revolution). The ECM can determine if one of the timing inputs is not being received by comparing the two inputs. If the ECM detects the low resolution timing pulse without detecting the high resolution timing pulse, this DTC will set.

The reference signals toggle between 0 and 5 volts as the camshaft turns. Therefore, an open, a short to voltage, a short to ground, or a defective sensor inside the distributor can prevent the voltage from pulsing at the ECM.

Test Description: Number(s) below refer to circled number(s) on the diagnostic chart.
1. DTC 36 will set if:
   • If less than 60 high resolution pulses are detected between each low resolution pulse.
   • Failure occurs 5 times.
2. This test will determine if the ECM is sending out a signal to the distributor for processing. If this signal is not available, or is shorted to ground or voltage, the distributor cannot ground it to produce reference pulses.

Diagnostic Aids:

The engine does not need the high resolution pulse to operate. If a DTC 36 is present and the vehicle will not start, check for DTC 16 and use that chart first. If the engine will still not run, refer to CHART A-3 "Engine Cranks But Will Not Run."

5.7L (VIN 8 & P) ENGINE — DIAGNOSTIC TROUBLE CODE CHART — 1992–93 CORVETTE

DTC 36

DISTRIBUTOR IGNITION SYSTEM
(FAULTY HIGH RESOLUTION PULSE OR EXTRA LOW RESOLUTION PULSE DETECTED)
5.7L (VIN P) "Y" CARLINE (MFI)

(1)
- CLEAR DTC(s).
- RUN ENGINE FOR 30 SECONDS. DOES DTC 36 SET?

YES

(2)
- IGNITION "OFF."
- DISCONNECT IGNITION "TEST" CONNECTOR.
- IGNITION "ON."
- USING A DVM ON THE DC VOLTS SCALE, MEASURE VOLTAGE ON TERMINAL "B" AT THE ECM SIDE OF THE IGNITION TEST CONNECTOR.

NO

DTC 36 IS INTERMITTENT. REFER TO "DIAGNOSTIC AIDS" ON FACING PAGE.

LESS THAN .5 VOLTS

OPEN OR GROUNDED HIGH RESOLUTION SIGNAL CIRCUIT FROM ECM TO TEST CONNECTOR OR FAULTY ECM CONNECTION OR FAULTY ECM.

4 TO 6 VOLTS

- IGNITION "OFF."
- RECONNECT IGNITION SYSTEM TEST CONNECTOR.
- DISCONNECT DISTRIBUTOR CONNECTOR.
- IGNITION "ON."
- MEASURE VOLTAGE ON TERMINAL "A".

OVER 6 VOLTS

SHORT TO VOLTAGE ON HIGH RESOLUTION SIGNAL CIRCUIT FROM ECM TO TEST CONNECTOR.

LESS THAN .5 VOLTS

OPEN OR GROUNDED HIGH RESOLUTION SIGNAL CIRCUIT FROM TEST CONNECTOR TO DISTRIBUTOR.

4 TO 6 VOLTS

FAULTY DISTRIBUTOR CONNECTION OR FAULTY DISTRIBUTOR.

OVER 6 VOLTS

SHORT TO VOLTAGE ON HIGH RESOLUTION SIGNAL CIRCUIT FROM TEST CONNECTOR TO DISTRIBUTOR.

"AFTER REPAIRS," REFER TO DTC CRITERIA ON FACING PAGE AND CONFIRM DTC DOES NOT RESET.

5.7L (VIN 8 & P) ENGINE — DIAGNOSTIC TROUBLE CODE CHART — 1992–93 CORVETTE

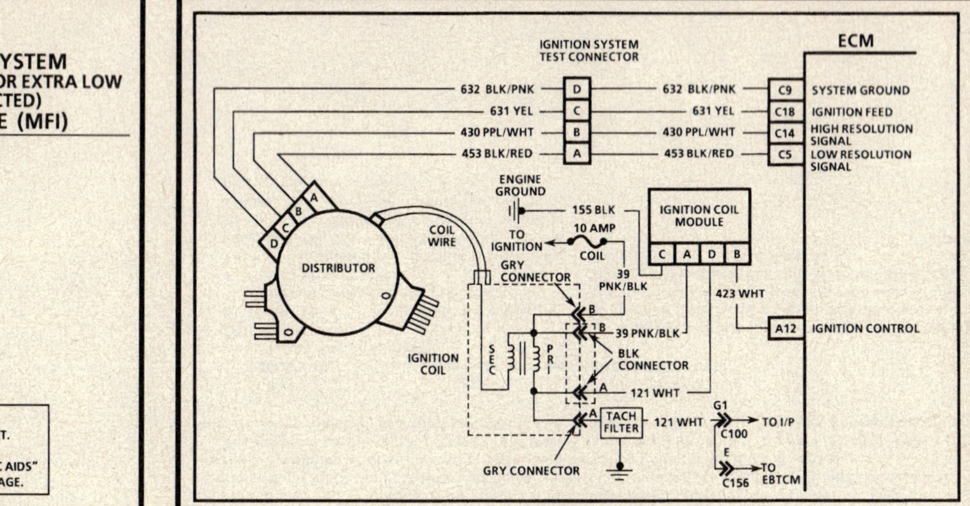

DTC 41

IGNITION CONTROL (IC) CIRCUIT
(OPEN OR SHORTED CIRCUIT)
5.7L (VIN P) "Y" CARLINE (MFI)

Circuit Description:

The distributor ignition system provides two timing inputs to the ECM, high resolution signal (360 pulses per crankshaft revolution) and low resolution signal (4 pulses per crankshaft revolution). The ECM uses these two reference pulses to determine individual ignition spark timing for each cylinder.

Once the ECM calculates ignition timing, the timing signal will be sent to the ignition coil module on the IC circuit. Each timing pulse received by the ignition coil module, on the IC circuit, will trigger the coil module to operate the ignition coil. Secondary ignition voltage is induced and is then sent to the distributor for distribution to each spark plug.

The IC signal voltage ranges from about .5 volt to 4.5 volts.

If a DTC 41 is detected, the ECM will disable the fuel injectors to prevent flooding of the engine.

DTC 41 will set if the IC circuit problem occurs during cranking.

Test Description: Number(s) below refer to circled number(s) on the diagnostic chart.

1. DTC 41 will set if:
   - Voltage on CKT 423 exceeds 4.6 volts.
   - Engine speed is less than 1500 RPM.
2. Without the engine running or cranking there would be no voltage present on the IC line.
3. This check determines if the IC signal from the ECM is available at the ignition coil module. The IC circuit voltage should be between about .5 and 4.5 volts.
4. The remaining tests begin to check that the coil driver circuitry is OK.
5. If all wiring and connections are OK, check the ignition coil or the ignition coil voltage supply. Check coil fuse for being open.

Diagnostic Aids:

Since the ignition coil module gets its power from the coil, check the ignition feed circuit to the ignition coil for opens.

5.7L (VIN 8 & P) ENGINE — DIAGNOSTIC TROUBLE CODE CHART — 1992–93 CORVETTE

DTC 41

IGNITION CONTROL (IC) CIRCUIT
(OPEN OR SHORTED CIRCUIT)
5.7L (VIN P) "Y" CARLINE (MFI)

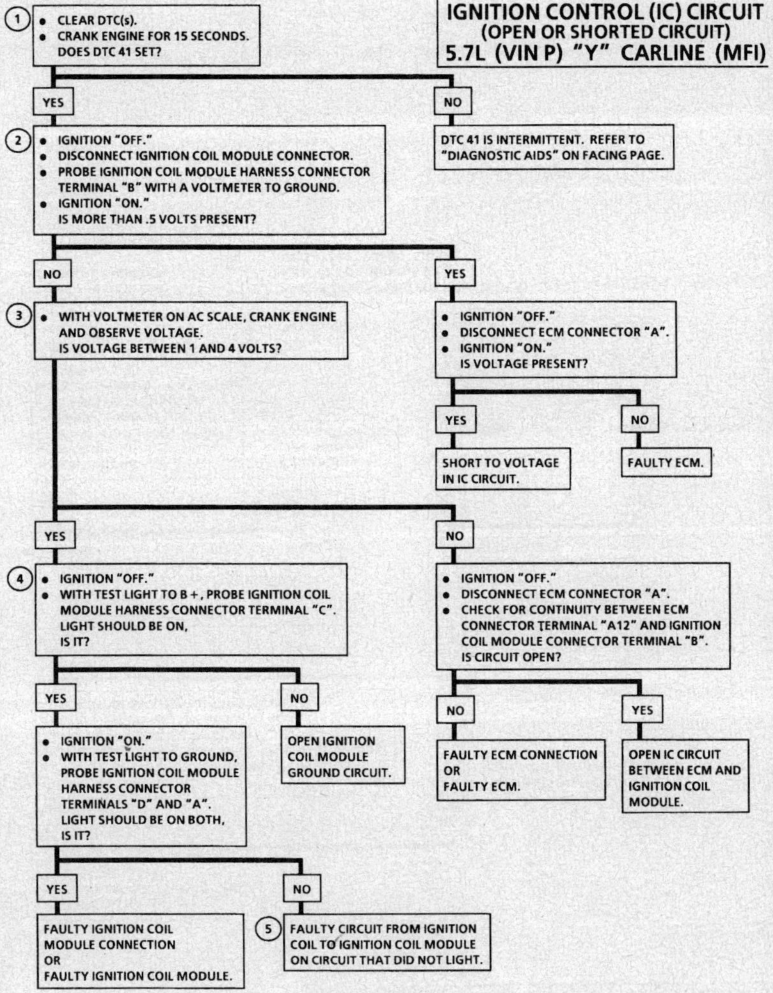

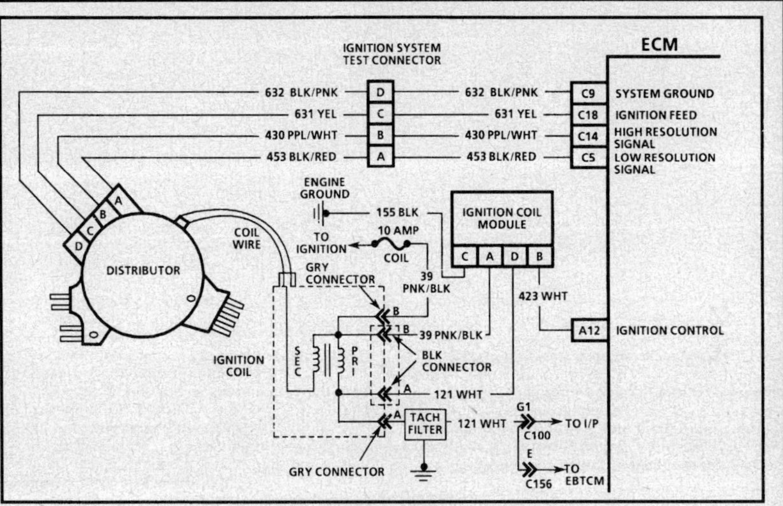

5.7L (VIN 8 & P) ENGINE — DIAGNOSTIC TROUBLE CODE CHART — 1992–93 CORVETTE

DTC 42

IGNITION CONTROL (IC) CIRCUIT
(GROUNDED CIRCUIT)
5.7L (VIN P) "Y" CARLINE (MFI)

Circuit Description:

The distributor ignition system provides two timing inputs to the ECM, high resolution signal (360 pulses per crankshaft revolution) and low resolution signal (4 pulses per crankshaft revolution). The ECM uses these two reference pulses to determine individual ignition spark timing for each cylinder.

Once the ECM calculates ignition timing, the timing signal will be sent to the ignition coil module on the IC circuit. Each timing pulse received by the ignition coil module on the IC circuit will trigger the ignition coil module to operate the ignition coil. Secondary ignition voltage is induced and is then sent to the distributor for distribution to each spark plug.

The IC signal voltage ranges from about .5 volt to 4.5 volts.

If DTC 42 is detected, the ECM will disable the fuel injectors to prevent flooding of the engine.

DTC 42 will set if the IC circuit problem occurs during cranking.

Test Description: Number(s) below refer to circled number(s) on the diagnostic chart.

1. DTC 42 will set if:
   - Engine speed below 3000 RPM.
   - Voltage on CKT 423 is below about .5 volt.
   If the engine starts at this point, DTC 42 is intermittent.
2. This check determines if the IC signal from the ECM is available at the ignition coil module.
3. The remaining tests begin to check that the ignition coil module circuitry is OK. If the ignition coil module loses its voltage source, secondary voltage will not be produced and a DTC 42 will set.

4. Since the ignition coil module gets its power from the coil, check the ignition feed circuit to the ignition coil for opens.
5. A DTC 42 will set if the high and low resolution signal CKTs are shorted to each other. Before replacing any components, make sure these circuits are not shorted together.

Diagnostic Aids:

Refer to "Intermittents" in "Symptoms," Section

5.7L (VIN 8 & P) ENGINE — DIAGNOSTIC TROUBLE CODE CHART — 1992–93 CORVETTE

DTC 42
IGNITION CONTROL (IC) CIRCUIT
(GROUNDED CIRCUIT)
5.7L (VIN P) "Y" CARLINE (MFI)

1. • CLEAR DTC(s).
   • CRANK ENGINE FOR 15 SECONDS. DOES DTC 42 SET?

YES →

2. • IGNITION "OFF."
   • DISCONNECT IGNITION COIL MODULE CONNECTOR.
   • WITH A VOLTMETER ON AC SCALE, PROBE IGNITION COIL MODULE CONNECTOR TERMINAL "B" TO GROUND.
   • CRANK ENGINE AND OBSERVE VOLTAGE. IS VOLTAGE BETWEEN 1 AND 4 VOLTS?

NO → DTC 42 IS INTERMITTENT. REFER TO "DIAGNOSTIC AIDS" ON FACING PAGE.

YES →

3. • IGNITION "OFF."
   • WITH TEST LIGHT TO B+, PROBE IGNITION COIL MODULE HARNESS CONNECTOR TERMINAL "C". LIGHT SHOULD BE "ON." IS IT?

NO →
   • IGNITION "OFF."
   • DISCONNECT ECM CONNECTOR "A".
   • WITH TEST LIGHT TO B+, PROBE ECM HARNESS TERMINAL "A12". IS TEST LIGHT ON?

YES →
   • IGNITION "ON."
   • WITH TEST LIGHT TO GROUND, PROBE IGNITION COIL MODULE HARNESS CONNECTOR TERMINALS "D" AND "A".
   • LIGHT SHOULD BE "ON" BOTH. IS IT?

NO → OPEN IGNITION COIL MODULE GROUND CIRCUIT.

NO → FAULTY ECM CONNECTION OR FAULTY ECM.

YES → GROUNDED IC CIRCUIT BETWEEN ECM AND IGNITION COIL MODULE.

YES →
5. FAULTY IGNITION COIL MODULE CONNECTION OR FAULTY IGNITION COIL MODULE.

NO →
4. FAULTY CIRCUIT FROM COIL TO IGNITION COIL MODULE ON CIRCUIT THAT DID NOT LIGHT.

"AFTER REPAIRS," REFER TO DTC CRITERIA ON FACING PAGE AND CONFIRM DTC DOES NOT RESET.

5.7L (VIN 8 & P) ENGINE — DIAGNOSTIC TROUBLE CODE CHART — 1992–93 CORVETTE

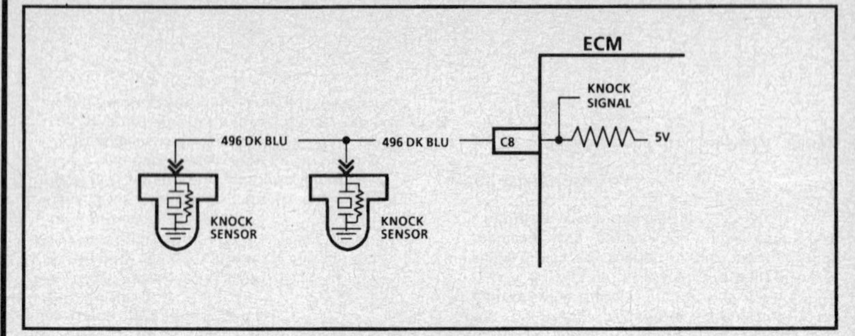

DTC 43
KNOCK SENSOR (KS) CIRCUIT
5.7L (VIN P) "Y" CARLINE (MFI)

Circuit Description:

The knock sensor system is used to detect engine detonation. The ECM will retard the spark timing based on a signal being received from the knock sensors. The knock sensor produces an AC voltage which is sent to the ECM. The amount of AC voltage produced by the sensors is determined by the amount of knock. The circuitry within each knock sensor causes the ECM's 5 volts to be pulled down so that under a no knock condition, CKT 496 would measure about 1.5 volts.

The internal ECM tests which are ran on this Knock Sensor (KS) circuit will determine if it is operating correctly. If the ECM detects that any of the following voltage conditions have been met for 5 seconds, DTC 43 will be set:
- One sensor being open - voltage between about 2.2 volts and 4.1 volts.
- Both knock sensors being open - voltage greater than 4.1 volts.
- Both knock sensors being grounded - voltage less than .78 volt.

Test Description: Number(s) below refer to circled number(s) on the diagnostic chart.

1. If an audible knock is heard from the engine, repair the internal engine problem, as normally no knock should be detected at idle.
   The ECM supplies 5 volts on CKT 496 which should be present at the knock sensor terminals when the sensors are disconnected.
2. This test determines if the knock sensor is faulty or if the KS portion of the PROM is faulty.
3. An improperly installed knock sensor can prevent the knock sensor from grounding to the block.

Diagnostic Aids:

The ECM has the ability to diagnose opens and shorts in the KS circuit. The Tech 1 will display if one or both sensor are open or if both are grounded when the ECM detects these conditions.

Check CKT 496 for a potential open or short to ground.

Also, check for proper installation of PROM.

Refer to "Intermittents" in "Symptoms," Section

5.7L (VIN 8 & P) ENGINE — DIAGNOSTIC TROUBLE CODE CHART — 1992–93 CORVETTE

DTC 43
KNOCK SENSOR (KS) CIRCUIT
5.7L (VIN P) "Y" CARLINE (MFI)

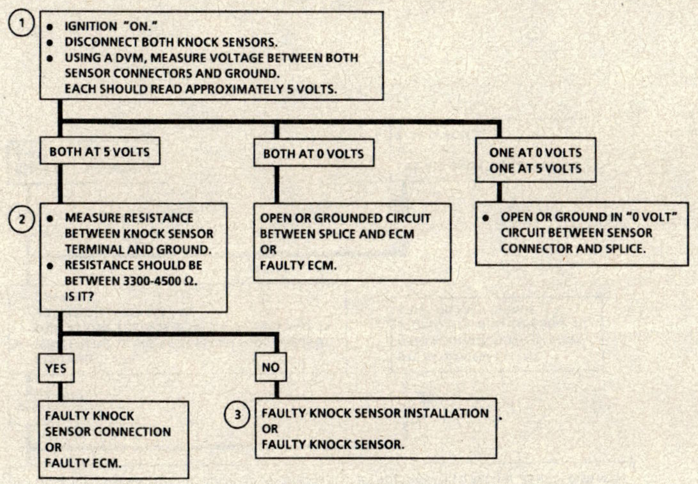

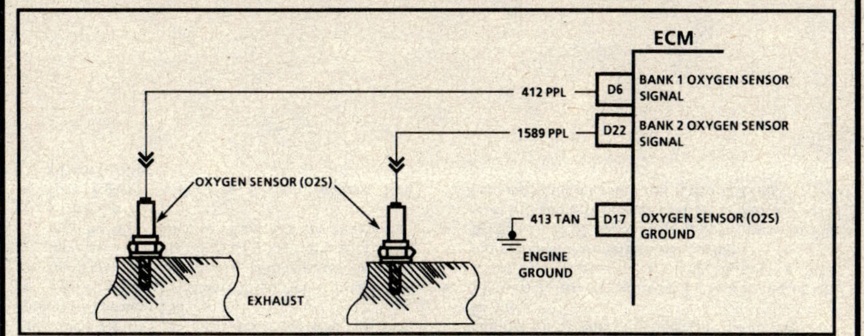

"AFTER REPAIRS," REFER TO DTC CRITERIA ON FACING PAGE AND CONFIRM DTC DOES NOT RESET.

5.7L (VIN 8 & P) ENGINE — DIAGNOSTIC TROUBLE CODE CHART — 1992–93 CORVETTE

DTC 44
BANK 1 (LEFT) OXYGEN SENSOR (O2S) CIRCUIT
(LEAN EXHAUST INDICATED)
5.7L (VIN P) "Y" CARLINE (MFI)

Circuit Description:

The ECM supplies a voltage of about 450 mV between terminals "D6" and "D17". (If measured with a 10 megohm digital voltmeter, this may display as low as 320 mV.) The Oxygen Sensor (O2S) varies the voltage within a range of about 1000 mV if the exhaust is rich, down through about 10 mV if exhaust is lean.

The sensor is like an open circuit and produces no voltage when it is below 315°C (600°F). An open sensor circuit or cold sensor causes "Open Loop" operation.

Test Description: Number(s) below refer to circled number(s) on the diagnostic chart.
1. DTC 44 will set if:
   • Signal voltage remains below 200 mV for 50 seconds.
   • System is operating in "Closed Loop."

Diagnostic Aids:

Using the Tech 1, observe the long term fuel trim values at different RPM and air flow conditions. The Tech 1 also displays the fuel trim cells so the long term fuel trim values can be checked in each of the cells to determine when the DTC 44 may have been set. If the conditions for DTC 44 exist, the long term fuel trim values will be near 160.

• <u>Oxygen Sensor (O2S) Wire.</u> Sensor pigtail may be mispositioned and contacting the exhaust manifold.
• Check for intermittent ground in wire between connector and sensor.
• <u>MAP Sensor.</u> A Manifold Absolute Pressure (MAP) sensor output that causes the ECM to sense a higher than normal vacuum, will cause the system to go lean. Disconnect the MAP sensor and if the lean condition is gone, replace the sensor.

• <u>Lean Injector(s).</u> Perform injector balance test CHART C-2A.
• <u>Fuel Contamination.</u> Water, even in small amounts, near the in-tank fuel pump inlet can be delivered to the injectors. The water causes a lean exhaust and can set a DTC 44.
• <u>Fuel Pressure.</u> System will be lean if pressure is too low. It may be necessary to monitor fuel pressure while driving the vehicle at various road speeds and/or loads to confirm. Refer to "Fuel System Diagnosis," CHART A-7.
• <u>Exhaust Leaks.</u> If there is an exhaust leak, above the oxygen sensor, the engine can cause outside air to be pulled into the exhaust stream and flow past the sensor causing lean condition. Vacuum or crankcase leaks can cause a lean condition.
• <u>AIR System.</u> Make sure the secondary air pump is not operating in "Closed Loop." While monitoring long term fuel trim, squeeze the air hose going to the exhaust ports. If long term fuel trim values go down, refer to CHART C-6.
• If the above are OK, the oxygen sensor may be at fault.

5.7L (VIN 8 & P) ENGINE — DIAGNOSTIC TROUBLE CODE CHART — 1992–93 CORVETTE

DTC 44

BANK 1 (LEFT) OXYGEN SENSOR (O2S) CIRCUIT
(LEAN EXHAUST INDICATED)
5.7L (VIN P) "Y" CARLINE (MFI)

① • RUN WARM ENGINE (75°C/167°F TO 95°C/203°F) AT 1200 RPM.
• DOES TECH 1 SCAN TOOL INDICATE OXYGEN SENSOR (O2S) VOLTAGE FIXED BELOW .35 VOLT (350 mV)?

YES

• DISCONNECT O2S.
• WITH ENGINE IDLING, TECH 1 SCAN TOOL SHOULD DISPLAY O2S VOLTAGE BETWEEN .35 VOLT AND .55 VOLT (350 mV AND 550 mV). DOES IT?

YES

REFER TO "DIAGNOSTIC AIDS" ON FACING PAGE.

NO

DTC 44 IS INTERMITTENT. IF NO ADDITIONAL DTC(S) WERE STORED, REFER TO "DIAGNOSTIC AIDS" ON FACING PAGE.

NO

CKT 412 SHORTED TO GROUND OR FAULTY ECM.

"AFTER REPAIRS," REFER TO DTC CRITERIA ON FACING PAGE AND CONFIRM DTC DOES NOT RESET.

5.7L (VIN 8 & P) ENGINE — DIAGNOSTIC TROUBLE CODE CHART — 1992–93 CORVETTE

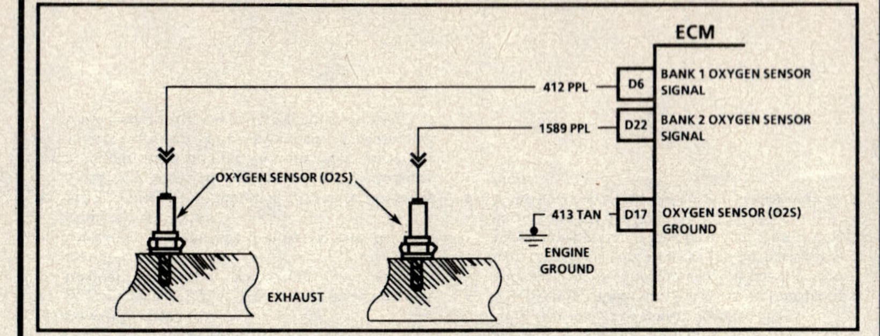

DTC 45

BANK 1 (LEFT) OXYGEN SENSOR (O2S) CIRCUIT
(RICH EXHAUST INDICATED)
5.7L (VIN P) "Y" CARLINE (MFI)

Circuit Description:

The ECM supplies a voltage of about 450 mV between terminals "D6" and "D17". (If measured with a 10 megohm digital voltmeter, this may read as low as 320 mV.) The Oxygen Sensor (O2S) varies the voltage within a range of about 1000 mV if the exhaust is rich down through about 10 mV if exhaust is lean.

The sensor is like an open circuit and produces no voltage when it is below 315°C (600°F). An open sensor circuit or cold sensor causes "Open Loop" operation.

Test Description: Number(s) below refer to circled number(s) on the diagnostic chart.
1. DTC 45 will set if:
   • Signal voltage remains above 700 mV for 50 seconds.
   • Throttle angle is greater than 4%.
   • System is operating in "Closed Loop."

Diagnostic Aids:

Using the Tech 1, observe the long term fuel trim values at different RPM and load conditions. The Tech 1 also displays the fuel trim cells, so the long term fuel trim values can be checked in each of the cells to determine when the DTC 45 may have been set. If the conditions for DTC 45 exist, the long term fuel trim values will be near 108.
• **Fuel Pressure.** System will go rich if pressure is too high. The ECM can compensate for some increase. However, if it gets too high, a DTC 45 may be set. Refer to "Fuel System Diagnosis," CHART A-7.
• **Rich Injector.** Perform injector balance test, CHART C-2A.
• **Leaking Injector.** Refer to "Fuel System Diagnosis," CHART A-7.
• Check for fuel contaminated oil.
• **Evaporative Emission (EVAP) Canister Purge.** Check for fuel saturation. If full of fuel, check canister control and hoses. Refer to "Evaporative Emission (EVAP) Control System," Section "6E3-C3".
• **MAP Sensor.** An output that causes the ECM to sense a lower than normal vacuum can cause the system to go rich. Disconnecting the MAP sensor will allow the ECM to set a fixed value for the sensor. Substitute different MAP sensor if the rich condition is gone while the sensor is disconnected. Check for leaking fuel pressure regulator diaphragm by checking vacuum line to regulator for fuel.
• **TP Sensor.** An intermittent TP sensor output will cause the system to go rich, due to a false indication of the engine accelerating.
• **EGR.** An EGR valve staying open (especially at idle) will cause the oxygen sensor to indicate a rich exhaust, and this could result in a DTC 45.

5.7L (VIN 8 & P) ENGINE — DIAGNOSTIC TROUBLE CODE CHART — 1992–93 CORVETTE

DTC 45

BANK 1 (LEFT) OXYGEN SENSOR (O2S) CIRCUIT
(RICH EXHAUST INDICATED)
5.7L (VIN P) "Y" CARLINE (MFI)

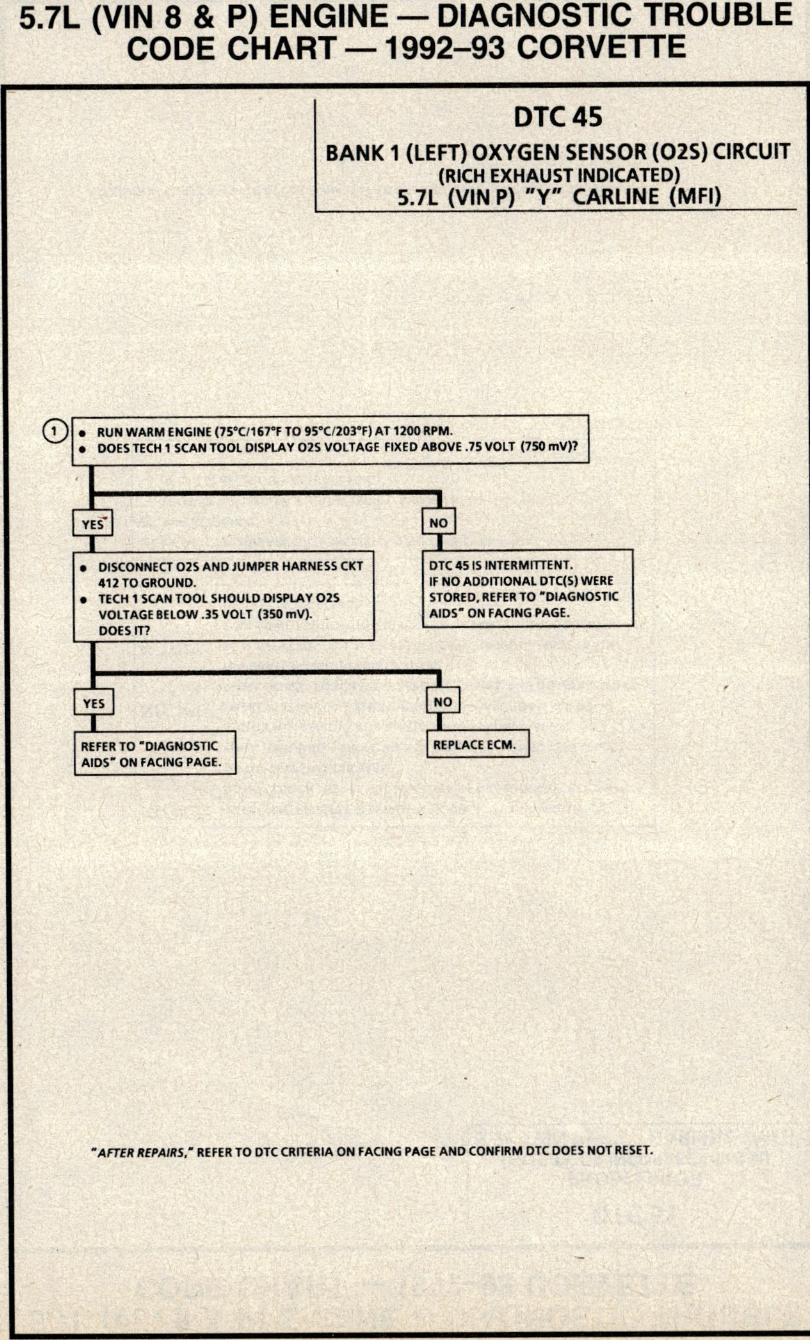

"AFTER REPAIRS," REFER TO DTC CRITERIA ON FACING PAGE AND CONFIRM DTC DOES NOT RESET.

5.7L (VIN 8 & P) ENGINE — DIAGNOSTIC TROUBLE CODE CHART — 1992–93 CORVETTE

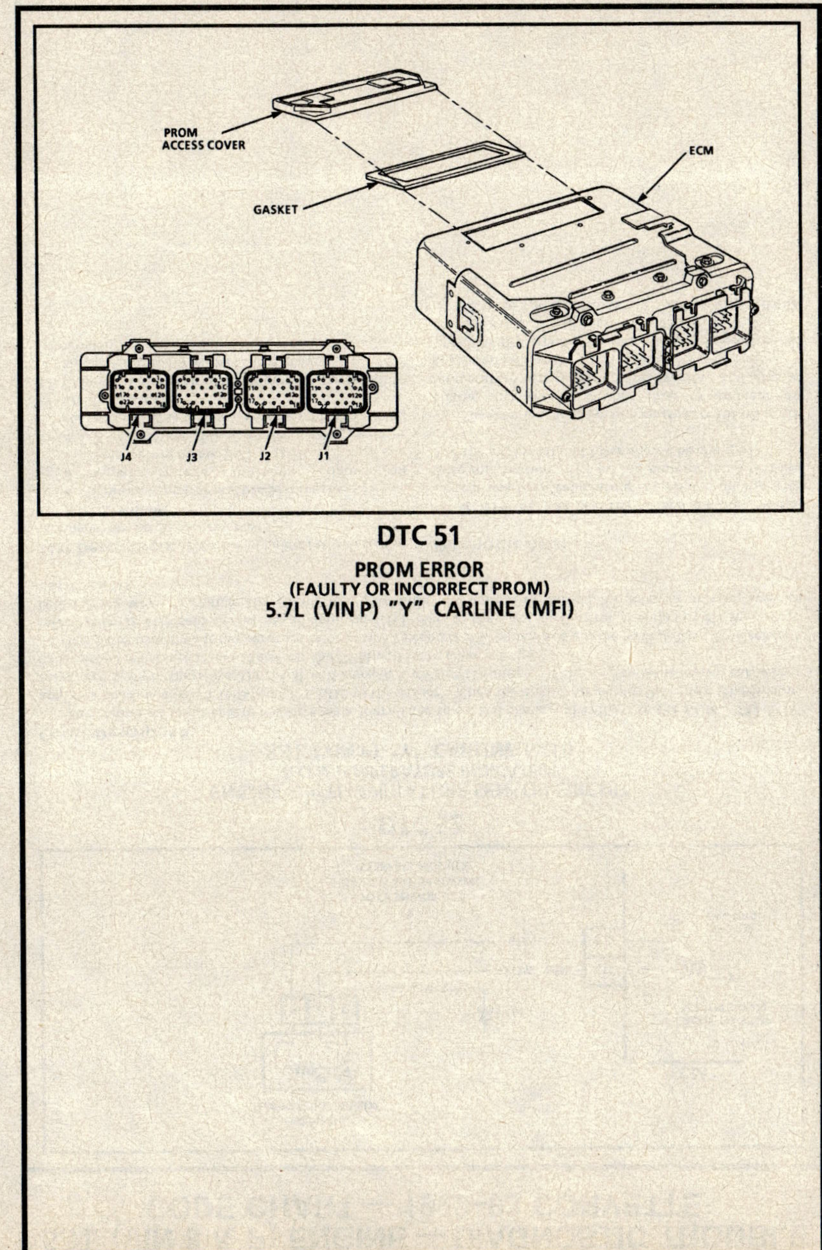

DTC 51

PROM ERROR
(FAULTY OR INCORRECT PROM)
5.7L (VIN P) "Y" CARLINE (MFI)

5.7L (VIN 8 & P) ENGINE — DIAGNOSTIC TROUBLE CODE CHART — 1992–93 CORVETTE

DTC 51

PROM ERROR
(FAULTY OR INCORRECT PROM)
5.7L (VIN P) "Y" CARLINE (MFI)

NOTICE: TO PREVENT POSSIBLE ELECTROSTATIC DISCHARGE DAMAGE:
- DO NOT TOUCH THE ECM CONNECTOR PINS OR SOLDERED COMPONENTS ON THE ECM CIRCUIT BOARD.
- WHEN HANDLING A PROM, DO NOT TOUCH THE COMPONENT LEADS, AND DO NOT REMOVE INTEGRATED CIRCUIT FROM CARRIER.

NOTICE: TO PREVENT POSSIBLE ELECTROSTATIC DISCHARGE DAMAGE TO THE PROM, DO NOT TOUCH THE COMPONENT LEADS, AND DO NOT REMOVE INTEGRATED CIRCUIT FROM CARRIER.

NOTICE: TO PREVENT POSSIBLE ELECTROSTATIC DISCHARGE DAMAGE TO THE ECM, DO NOT TOUCH THE CONNECTOR PINS OR SOLDERED COMPONENTS ON THE CIRCUIT BOARD.

CHECK THAT ALL PINS ARE FULLY INSERTED IN THE SOCKET AND THAT PROM IS PROPERLY LATCHED.

IF OK, REPLACE PROM, CLEAR MEMORY, AND RECHECK.

IF DTC 51 REAPPEARS, REPLACE ECM.

"AFTER REPAIRS," REFER TO DTC CRITERIA ON FACING PAGE AND CONFIRM DTC DOES NOT RESET.

5.7L (VIN 8 & P) ENGINE — DIAGNOSTIC TROUBLE CODE CHART — 1992–93 CORVETTE

DTC 52

ENGINE OIL TEMPERATURE SENSOR CIRCUIT
(LOW TEMPERATURE INDICATED)
5.7L (VIN P) "Y" CARLINE (MFI)

Circuit Description:

The engine oil temperature sensor uses a thermistor to control the signal voltage to the ECM. The ECM applies a voltage (about 5 volts) on CKT 1313 to the sensor. When the engine oil is cold, the sensor (thermistor) resistance is high, therefore, the ECM will sense a high signal voltage. If the engine oil is warm, the sensor (thermistor) resistance is low, therefore, the ECM will sense a low voltage.

The ECM uses the oil temperature information to control A/C clutch and engine cooling fan. This sensor's information is also sent across serial data, on CKT 800, to the Central Control Module (CCM) for use in determining when to change the engine oil. For more information on the oil life monitor system, refer to SECTION 8D.

Test Description: Number(s) below refer to circled number(s) on the diagnostic chart.
1. DTC 52 will set:
   - Time since engine start is 30 minutes.
   - Signal voltage indicates engine oil temperature below -35°C (-31°F).
2. A DTC 52 will set, due to an open sensor, wire or connection. This test will determine if the wiring and ECM are OK.
3. This will determine if the engine oil temperature signal (CKT 1313) or the sensor ground (CKT 808) is open.

Diagnostic Aids:

If DTC 52 was set; the CCM's oil life monitor system has been calculating the engine oil life with false information. The oil life monitor must be reset and the engine oil and filter should be changed.

The scan tool displays temperature of the oil in the engine and should display close to ambient air temperature when the engine is cold, and rise as engine oil temperature increases.

Carefully check harness and connections for possible open CKT 1313 or CKT 808.

Refer to "Intermittents" in "Symptoms," Section

5.7L (VIN 8 & P) ENGINE — DIAGNOSTIC TROUBLE CODE CHART — 1992–93 CORVETTE

DTC 52

ENGINE OIL TEMPERATURE SENSOR CIRCUIT
(LOW TEMPERATURE INDICATED)
5.7L (VIN P) "Y" CARLINE (MFI)

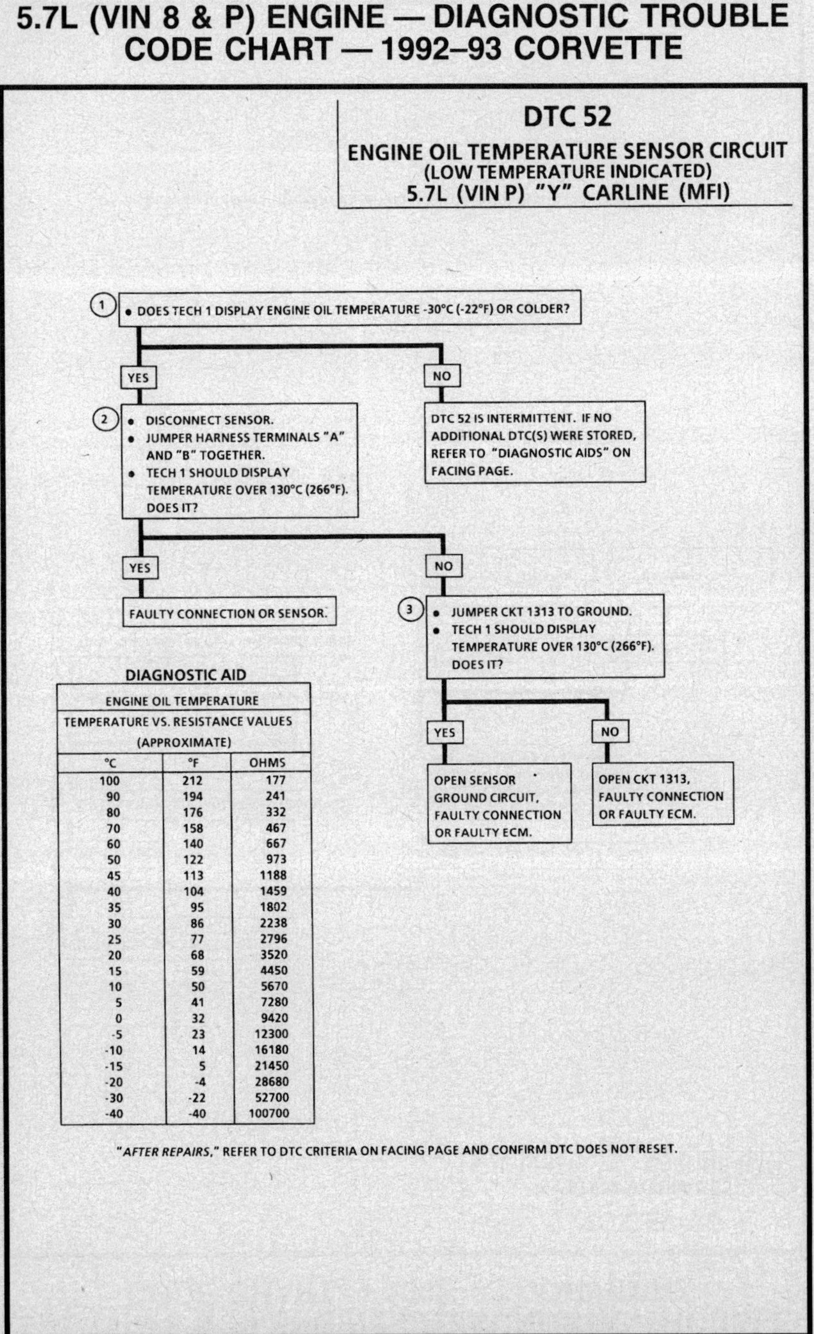

DIAGNOSTIC AID

ENGINE OIL TEMPERATURE		
TEMPERATURE VS. RESISTANCE VALUES		
(APPROXIMATE)		
°C	°F	OHMS
100	212	177
90	194	241
80	176	332
70	158	467
60	140	667
50	122	973
45	113	1188
40	104	1459
35	95	1802
30	86	2238
25	77	2796
20	68	3520
15	59	4450
10	50	5670
5	41	7280
0	32	9420
-5	23	12300
-10	14	16180
-15	5	21450
-20	-4	28680
-30	-22	52700
-40	-40	100700

"AFTER REPAIRS," REFER TO DTC CRITERIA ON FACING PAGE AND CONFIRM DTC DOES NOT RESET.

5.7L (VIN 8 & P) ENGINE — DIAGNOSTIC TROUBLE CODE CHART — 1992–93 CORVETTE

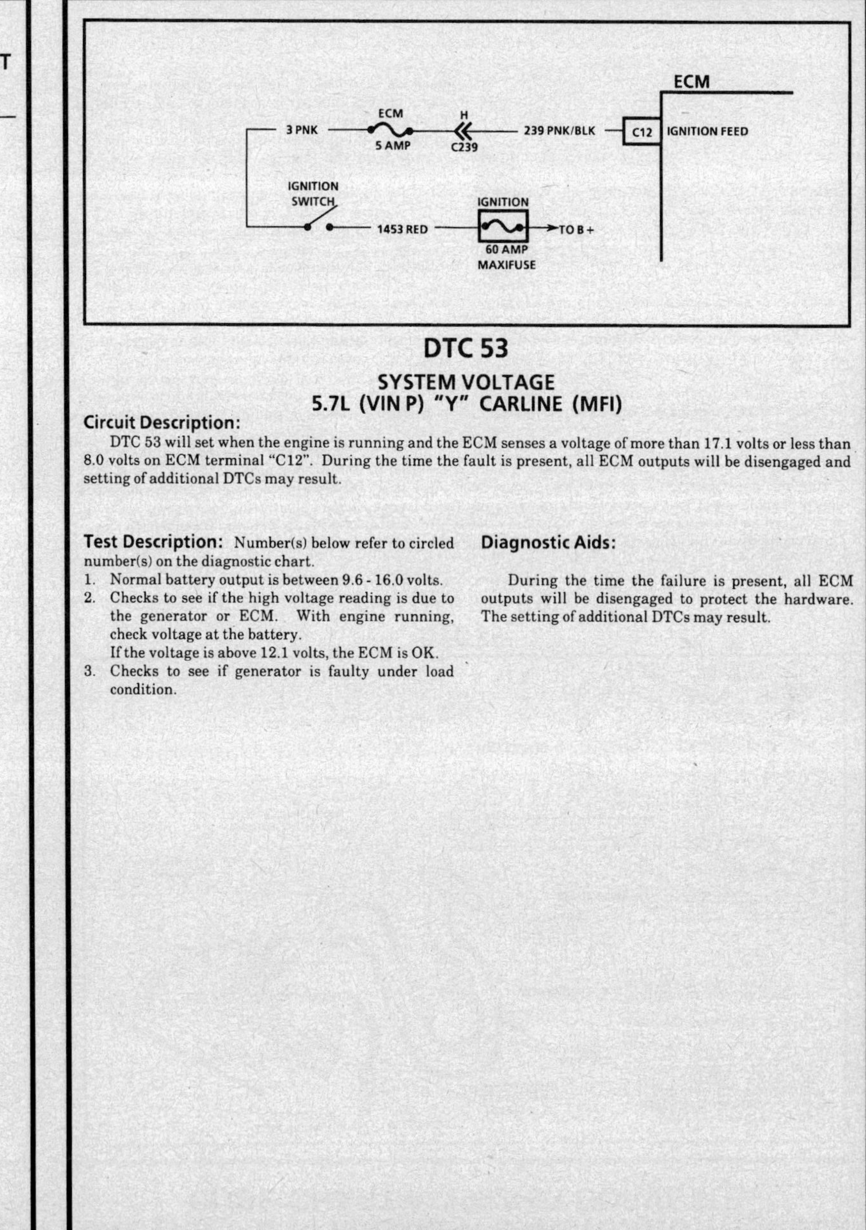

DTC 53

SYSTEM VOLTAGE
5.7L (VIN P) "Y" CARLINE (MFI)

Circuit Description:

DTC 53 will set when the engine is running and the ECM senses a voltage of more than 17.1 volts or less than 8.0 volts on ECM terminal "C12". During the time the fault is present, all ECM outputs will be disengaged and setting of additional DTCs may result.

Test Description: Number(s) below refer to circled number(s) on the diagnostic chart.

1. Normal battery output is between 9.6 - 16.0 volts.
2. Checks to see if the high voltage reading is due to the generator or ECM. With engine running, check voltage at the battery.
   If the voltage is above 12.1 volts, the ECM is OK.
3. Checks to see if generator is faulty under load condition.

Diagnostic Aids:

During the time the failure is present, all ECM outputs will be disengaged to protect the hardware. The setting of additional DTCs may result.

5.7L (VIN 8 & P) ENGINE — DIAGNOSTIC TROUBLE CODE CHART — 1992–93 CORVETTE

DTC 53
SYSTEM VOLTAGE
5.7L (VIN P) "Y" CARLINE (MFI)

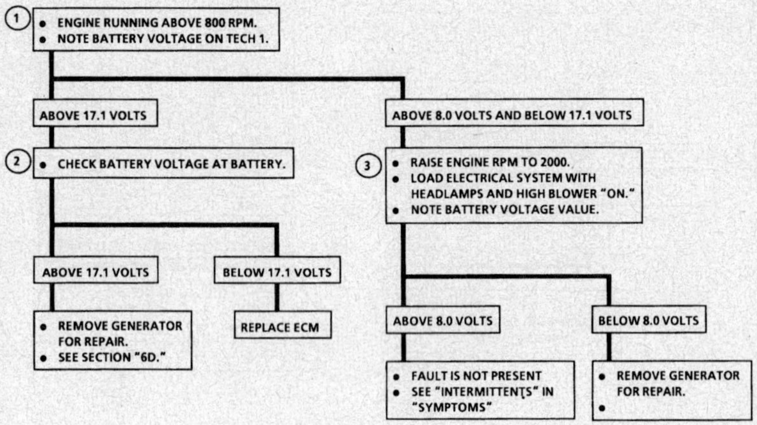

1. • ENGINE RUNNING ABOVE 800 RPM.
   • NOTE BATTERY VOLTAGE ON TECH 1.

ABOVE 17.1 VOLTS

ABOVE 8.0 VOLTS AND BELOW 17.1 VOLTS

2. • CHECK BATTERY VOLTAGE AT BATTERY.

3. • RAISE ENGINE RPM TO 2000.
   • LOAD ELECTRICAL SYSTEM WITH HEADLAMPS AND HIGH BLOWER "ON."
   • NOTE BATTERY VOLTAGE VALUE.

ABOVE 17.1 VOLTS

BELOW 17.1 VOLTS

ABOVE 8.0 VOLTS

BELOW 8.0 VOLTS

• REMOVE GENERATOR FOR REPAIR.
• SEE SECTION "6D."

REPLACE ECM

• FAULT IS NOT PRESENT
• SEE "INTERMITTENTS" IN "SYMPTOMS"
•

• REMOVE GENERATOR FOR REPAIR.
•

"AFTER REPAIRS," REFER TO DTC CRITERIA ON FACING PAGE AND CONFIRM DTC DOES NOT RESET.

5.7L (VIN 8 & P) ENGINE — DIAGNOSTIC TROUBLE CODE CHART — 1992–93 CORVETTE

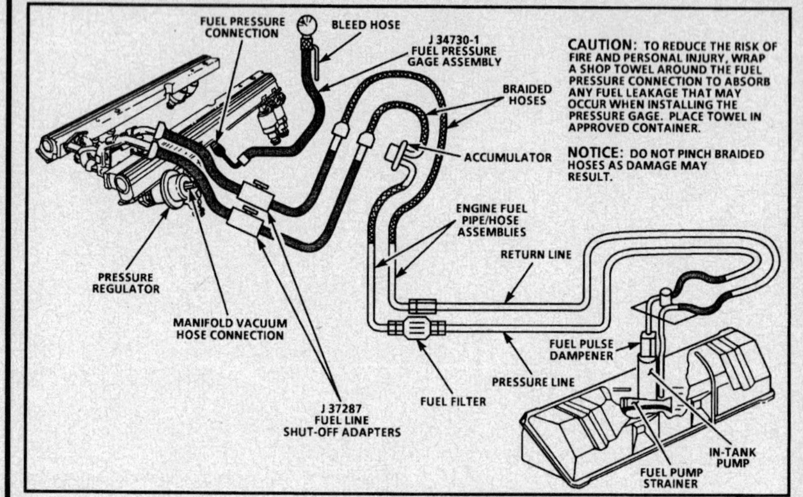

DTC 55
FUEL LEAN MONITOR
5.7L (VIN P) "Y" CARLINE (MFI)

Circuit Description:

The ECMs internal circuitry can identify if the vehicle fuel system is capable of supplying adequate amounts of fuel during heavy acceleration (power enrichment). When a power enrichment mode of operation is requested during "Closed Loop" operation (by heavy acceleration), the ECM will provide more fuel to the engine. Under these conditions, the ECM should detect a "rich" condition. If this "rich" exhaust is not detected at this time, a DTC 55 will set. A plugged fuel filter or restricted fuel line can prevent adequate amounts of fuel from being supplied during power enrichment mode.

Test Description: Number(s) below refer to circled number(s) on the diagnostic chart.

1. DTC 55 will set if:

   The ECM detects lean oxygen sensor voltage for 8 seconds during power enrichment modes of operation.

   Wrap a shop towel around the fuel pressure connector to absorb any small amount of fuel leakage that may occur when installing the gage. Ignition "ON," pump pressure should be 284-325 kPa (41-47 psi). This pressure is controlled by spring pressure within the regular assembly. Pressure below 284 kPa (41 psi) may cause a lean condition and may set a DTC 44 or 64. It could also cause hard starting cold and poor driveability. Low enough pressure will cause the engine not to run at all. Restricted flow may allow the engine to run at idle, or low speeds, but may cause a surge and stall when more fuel is required, as when

   accelerating or driving at high speeds. Low fuel pressure under heavy acceleration conditions may set a DTC 55.

2. Restricting the fuel return line allows the fuel pressure to build above regulated pressure. With a Tech 1 enable the fuel pump, pressure should rise above 325 kPa (47 psi) as the valve in the return line is partially closed.

Diagnostic Aids:

A restricted fuel filter can supply adequate amounts of fuel at idle but may not be able to supply enough fuel during heavy acceleration. A vapor lock condition can cause a DTC 55.

5.7L (VIN 8 & P) ENGINE — DIAGNOSTIC TROUBLE CODE CHART — 1992–93 CORVETTE

DTC 55
FUEL LEAN MONITOR
5.7L (VIN P) "Y" CARLINE (MFI)

① • IF DTC(s) 44 OR 64 ARE SET, REFER TO THOSE DTCs BEFORE PROCEEDING WITH THIS CHART.
• INSTALL FUEL PRESSURE GAGE, J 34730-1.
• START AND IDLE ENGINE AT NORMAL OPERATING TEMPERATURE.
• DISCONNECT VACUUM LINE GOING TO FUEL PRESSURE REGULATOR.
• NOTE FUEL PRESSURE WITH ENGINE RUNNING SHOULD BE 284-325 kPa (41-47 psi).

NOT OK →
CHECK FOR RESTRICTED FUEL LINES OR IN-LINE FILTER

OK →
NO TROUBLE FOUND. IF NO ADDITIONAL DTCs WERE STORED, REVIEW CIRCUIT DESCRIPTION ON FACING PAGE.

OK →
② • IGNITION "OFF."
• INSTALL J 37287-2 FUEL LINE ADAPTOR IN FUEL RETURN LINE. REFER TO FUEL SYSTEM PRESSURE TEST CHART A-7, PAGE 3 OF 3 FOR FUEL PRESSURE RELIEF PROCEDURE AND FOR SERVICING QUICK-CONNECT FITTINGS.
• USING A TECH 1, ENABLE THE FUEL PUMP."
• SLOWLY CLOSE VALVE IN RETURN LINE. PRESSURE SHOULD RISE ABOVE 325 kPa (47 psi). DO NOT EXCEED 414 kPa (60 psi).

NOT OK →
REPLACE FILTER OR REPAIR FUEL LINE AND RECHECK.

ABOVE 325 kPa (47 psi)
DTC 55 IS INTERMITTENT.

PRESSURE BUT LESS THAN 284 kPa (41 psi)
CHECK FOR:
• RESTRICTED FUEL PUMP STRAINER.
• LEAKING FUEL PUMP FEED HOSE.
• FAULTY FUEL PUMP.
• INCORRECT FUEL PUMP.

"AFTER REPAIRS," REFER TO DTC CRITERIA ON FACING PAGE AND CONFIRM DTC DOES NOT RESET.

5.7L (VIN 8 & P) ENGINE — DIAGNOSTIC TROUBLE CODE CHART — 1992–93 CORVETTE

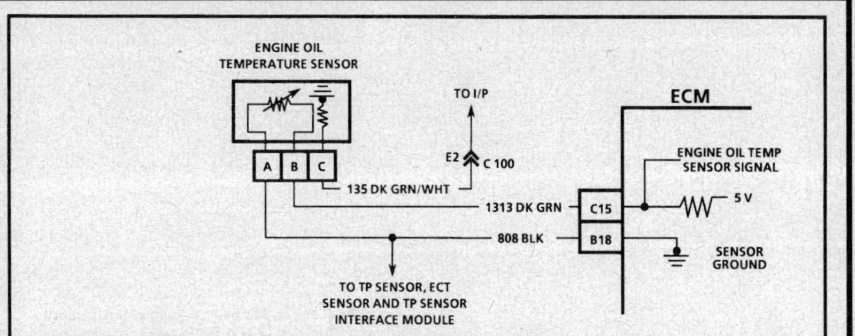

DTC 62
ENGINE OIL TEMPERATURE SENSOR CIRCUIT
(HIGH TEMPERATURE INDICATED)
5.7L (VIN P) "Y" CARLINE (MFI)

Circuit Description:

The engine oil temperature sensor uses a thermistor to control the signal voltage to the ECM. The ECM applies a voltage (about 5 volts) on CKT 1313 to the sensor. When the engine oil is cold, the sensor (thermistor) resistance is high, therefore, the ECM will sense a high signal voltage. If the engine oil is warm, the sensor (thermistor) resistance is low, therefore, the ECM will sense a low voltage.

The ECM uses the oil temperature information to control A/C clutch and engine cooling fan. This sensor's information is also sent across serial data, on CKT 800, to the Central Control Module (CCM) for use in determining when to change the engine oil. For more information on the oil life monitor system, refer to SECTION 8D.

Test Description: Number(s) below refer to circled number(s) on the diagnostic chart.

1. DTC 62 will set:
• Time since engine start is 30 minutes.
• Signal voltage indicates engine oil temperature greater than 140°C (284°F).

Diagnostic Aids:

If DTC 62 was set; the CCM's oil life monitor system has been calculating the engine oil life with false information. The oil life monitor must be reset and the engine oil and filter should be changed.

The scan tool displays temperature of the oil in the engine and should be close to ambient air temperature when the engine is cold, and rise as engine oil temperature increases.

Check harness routing for possible short to ground in CKT 1313.

Refer to "Intermittents" in "Symptoms," Section

5.7L (VIN 8 & P) ENGINE — DIAGNOSTIC TROUBLE CODE CHART — 1992–93 CORVETTE

DTC 62
ENGINE OIL TEMPERATURE SENSOR CIRCUIT
(HIGH TEMPERATURE INDICATED)
5.7L (VIN P) "Y" CARLINE (MFI)

(1) • DOES TECH 1 DISPLAY ENGINE OIL TEMPERATURE OF 140°C (284°F) OR HOTTER?

YES
- DISCONNECT ENGINE OIL TEMPERATURE SENSOR.
  TECH 1 SHOULD DISPLAY TEMPERATURE BELOW -30°C (-22°F).
  DOES IT?

YES
- REPLACE SENSOR.

NO
DTC 62 IS INTERMITTENT.
IF NO ADDITIONAL DTC(S) WERE STORED, REFER TO "DIAGNOSTIC AIDS" ON FACING PAGE.

NO
CKT 1313 SHORTED TO GROUND
OR
TO SENSOR GROUND
OR
ECM IS FAULTY.

DIAGNOSTIC AID

ENGINE OIL TEMPERATURE		
TEMPERATURE VS. RESISTANCE VALUES (APPROXIMATE)		
°C	°F	OHMS
100	212	177
90	194	241
80	176	332
70	158	467
60	140	667
50	122	973
45	113	1188
40	104	1459
35	95	1802
30	86	2238
25	77	2796
20	68	3520
15	59	4450
10	50	5670
5	41	7280
0	32	9420
-5	23	12300
-10	14	16180
-15	5	21450
-20	-4	28680
-30	-22	52700
-40	-40	100700

"AFTER REPAIRS," REFER TO DTC CRITERIA ON FACING PAGE AND CONFIRM DTC DOES NOT RESET.

5.7L (VIN 8 & P) ENGINE — DIAGNOSTIC TROUBLE CODE CHART — 1992–93 CORVETTE

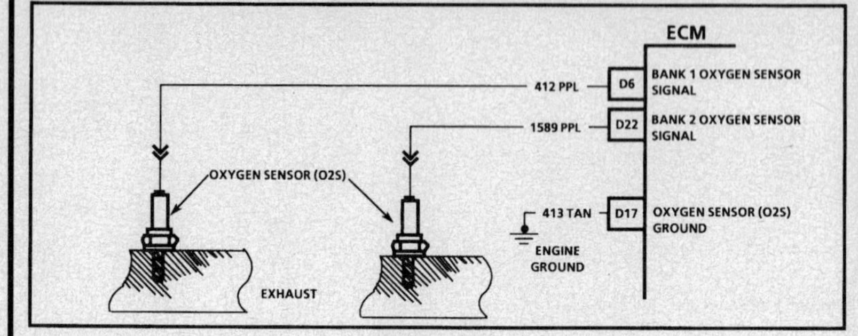

DTC 63
BANK 2 (RIGHT) OXYGEN SENSOR (O2S) CIRCUIT
(OPEN CIRCUIT)
5.7L (VIN P) "Y" CARLINE (MFI)

Circuit Description:

The ECM supplies a voltage of about 450 mV between terminals "D17" and "D22". (If measured with a 10 megohm digital voltmeter, this may read as low as 320 mV.) The Oxygen Sensor (O2S) varies the voltage within a range of about 1000 mV if the exhaust is rich, down through about 10 mV if exhaust is lean.

The sensor is like an open circuit and produces no voltage when it is below 315°C (600°F). An open sensor circuit or cold sensor causes "Open Loop" operation.

Test Description: Number(s) below refer to circled number(s) on the diagnostic chart.

1. DTC 63 will set:
   - Engine at normal operating temperature (above 70°C/158°F).
   - At least 2 minutes engine time after start.
   - Oxygen signal voltage steady between 350 mV to 550 mV.
   - Throttle position sensor signal above 5% (about .3 volt above closed throttle voltage).
   - All conditions must be met for about 60 seconds.
   If the conditions for a DTC 63 exist, the system will not go in "Closed Loop."
2. This will determine if the oxygen sensor is at fault, or the wiring or ECM is the cause of the DTC 63.

3. For this test use only a high impedance digital volt ohmmeter (J 39200). This test checks the continuity of CKT 1589 and CKT 413. If CKT 413 is open, the ECM voltage on CKT 1589 will be over .6 volt (600 mV).

Diagnostic Aids:

Normal Tech 1 voltage readings vary between 10 mV to 1000 mV (.01 and 1.0 volt), while in "Closed Loop." The system will go into "Open Loop" operation when DTC 63 sets.

Refer to "Intermittents" in "Symptoms," Section

5.7L (VIN 8 & P) ENGINE — DIAGNOSTIC TROUBLE CODE CHART — 1992–93 CORVETTE

DTC 63
BANK 2 (RIGHT) OXYGEN SENSOR (O2S) CIRCUIT
(OPEN CIRCUIT)
5.7L (VIN P) "Y" CARLINE (MFI)

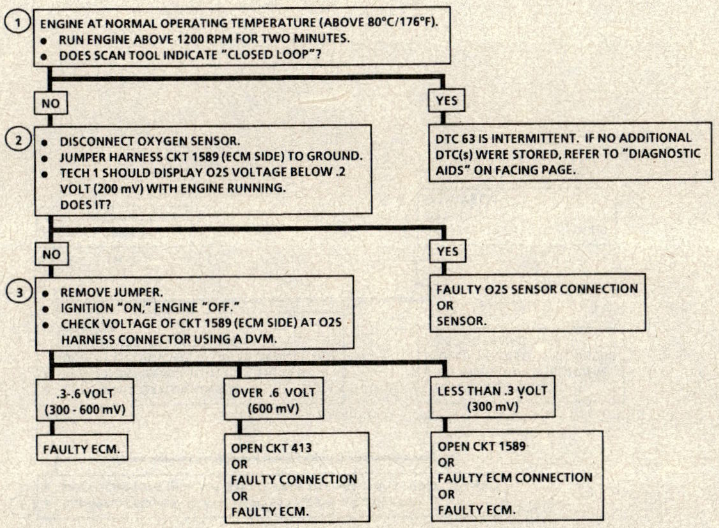

1. • ENGINE AT NORMAL OPERATING TEMPERATURE (ABOVE 80°C/176°F).
   • RUN ENGINE ABOVE 1200 RPM FOR TWO MINUTES.
   • DOES SCAN TOOL INDICATE "CLOSED LOOP"?

NO → 2. • DISCONNECT OXYGEN SENSOR.
   • JUMPER HARNESS CKT 1589 (ECM SIDE) TO GROUND.
   • TECH 1 SHOULD DISPLAY O2S VOLTAGE BELOW .2 VOLT (200 mV) WITH ENGINE RUNNING. DOES IT?

YES → DTC 63 IS INTERMITTENT. IF NO ADDITIONAL DTC(s) WERE STORED, REFER TO "DIAGNOSTIC AIDS" ON FACING PAGE.

NO → 3. • REMOVE JUMPER.
   • IGNITION "ON," ENGINE "OFF."
   • CHECK VOLTAGE OF CKT 1589 (ECM SIDE) AT O2S HARNESS CONNECTOR USING A DVM.

YES → FAULTY O2S SENSOR CONNECTION OR SENSOR.

.3 - .6 VOLT (300 - 600 mV) → FAULTY ECM.

OVER .6 VOLT (600 mV) → OPEN CKT 413 OR FAULTY CONNECTION OR FAULTY ECM.

LESS THAN .3 VOLT (300 mV) → OPEN CKT 1589 OR FAULTY ECM CONNECTION OR FAULTY ECM.

"AFTER REPAIRS," REFER TO DTC CRITERIA ON FACING PAGE AND CONFIRM DTC DOES NOT RESET.

5.7L (VIN 8 & P) ENGINE — DIAGNOSTIC TROUBLE CODE CHART — 1992–93 CORVETTE

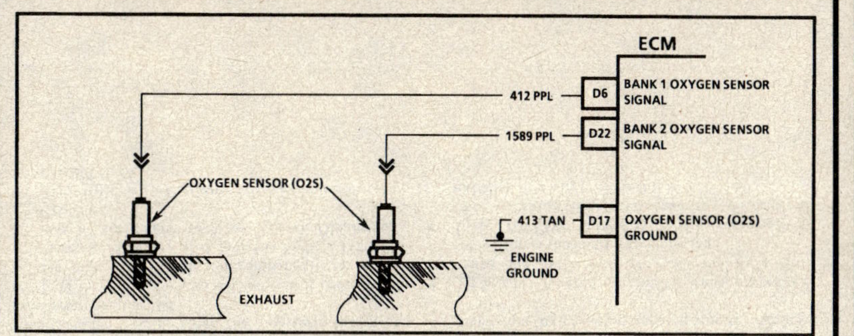

DTC 64
BANK 2 (RIGHT) OXYGEN SENSOR (O2S) CIRCUIT
(LEAN EXHAUST INDICATED)
5.7L (VIN P) "Y" CARLINE (MFI)

Circuit Description:

The ECM supplies a voltage of about 450 mV between terminals "D17" and "D22". (If measured with a 10 megohm digital voltmeter, this may read as low as 320 mV.) The Oxygen Sensor (O2S) varies the voltage within a range of about 1000 mV if the exhaust is rich, down through about 10 mV if exhaust is lean.

The sensor is like an open circuit and produces no voltage when it is below 315°C (600°F). An open sensor circuit or cold sensor causes "Open Loop" operation.

Test Description: Number(s) below refer to circled number(s) on the diagnostic chart.
1. DTC 64 will set if:
   • Signal voltage remains below 200 mV for 50 seconds.
   • System is operating in "Closed Loop."

Diagnostic Aids:

Using the Tech 1, observe the long term fuel trim values at different RPM and air flow conditions. The Tech 1 also, displays the fuel trim cells so the long term fuel trim values can be checked in each of the cells to determine when DTC 64 may have been set. If the conditions for DTC 64 exist, the long term fuel trim values will be near 160.

• Oxygen Sensor (O2S) Wire. Sensor pigtail may be mispositioned and contacting the exhaust manifold.
• Check for intermittent ground in wire between connector and sensor.
• MAP Sensor. A Manifold Absolute Pressure (MAP) sensor output that causes the ECM to sense a higher than normal vacuum, will cause the system to go lean. Disconnect the MAP sensor and if the lean condition is gone, replace the sensor.

• Lean Injector(s). Perform injector balance test CHART C-2A.
• Fuel Contamination. Water, even in small amounts, near the in-tank fuel pump inlet can be delivered to the injectors. The water causes a lean exhaust and can set a DTC 64.
• Fuel Pressure. System will be lean if pressure is too low. It may be necessary to monitor fuel pressure while driving the vehicle at various road speeds and/or loads to confirm. Refer to "Fuel System Diagnosis," CHART A-7.
• Exhaust Leaks. If there is an exhaust leak, above the oxygen sensor, the engine can cause outside air to be pulled into the exhaust stream and flow past the sensor causing lean condition. Vacuum or crankcase leaks can cause a lean condition.
• Air System. Make sure the secondary air pump is not operating in "Closed Loop." While monitoring long term fuel trim, squeeze the air hose going to the exhaust ports. If long term fuel trim values go down, refer to CHART C-6.
• If the above are OK, the oxygen sensor may be at fault.

5.7L (VIN 8 & P) ENGINE — DIAGNOSTIC TROUBLE CODE CHART — 1992–93 CORVETTE

DTC 64
BANK 2 (RIGHT) OXYGEN SENSOR (O2S) CIRCUIT
(LEAN EXHAUST INDICATED)
5.7L (VIN P) "Y" CARLINE (MFI)

① • RUN WARM ENGINE (75°C/167°F TO 95°C/203°F) AT 1200 RPM.
• DOES SCAN TOOL INDICATE O2S VOLTAGE FIXED BELOW .35 VOLT (350 mV)?

YES
• DISCONNECT OXYGEN SENSOR.
• WITH ENGINE IDLING, SCAN TOOL SHOULD DISPLAY O2S VOLTAGE BETWEEN .35 VOLT AND .55 VOLT (350 mV AND 550 mV). DOES IT?

NO
DTC 64 IS INTERMITTENT. IF NO ADDITIONAL DTC(s) WERE STORED, REFER TO "DIAGNOSTIC AIDS" ON FACING PAGE.

YES
REFER TO "DIAGNOSTIC AIDS" ON FACING PAGE.

NO
CKT 1589 SHORTED TO GROUND OR FAULTY ECM.

"AFTER REPAIRS," REFER TO DTC CRITERIA ON FACING PAGE AND CONFIRM DTC DOES NOT RESET.

5.7L (VIN 8 & P) ENGINE — DIAGNOSTIC TROUBLE CODE CHART — 1992–93 CORVETTE

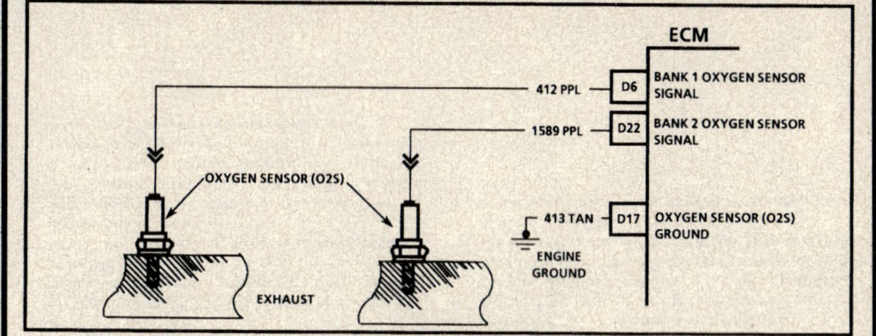

DTC 65
BANK 2 (RIGHT) OXYGEN SENSOR (O2S) CIRCUIT
(RICH EXHAUST INDICATED)
5.7L (VIN P) "Y" CARLINE (MFI)

Circuit Description:

The ECM supplies a voltage of about 450 mV between terminals "D22" and "D17". (If measured with a 10 megohm digital voltmeter, this may read as low as 320 mV.) The Oxygen Sensor (O2S) varies the voltage within a range of about 1000 mV if the exhaust is rich down through about 10 mV if exhaust is lean.

The sensor is like an open circuit and produces no voltage when it is below 315°C (600°F). An open sensor circuit or cold sensor causes "Open Loop" operation.

Test Description: Number(s) below refer to circled number(s) on the diagnostic chart.
1. DTC 65 will set if:
   • Signal voltage remains above 700 mV for 50 seconds.
   • Throttle angle is greater than 4%.
   • System is operating in "Closed Loop."

Diagnostic Aids:

Using the Tech 1, observe the long term fuel trim values at different RPM and load conditions. The Tech 1 also displays the fuel trim cells, so the long term fuel trim values can be checked in each of the cells to determine when the DTC 65 may have been set. If the conditions for DTC 65 exist, the long term fuel trim values will be near 108.
• <u>Fuel Pressure.</u> System will go rich if pressure is too high. The ECM can compensate for some increase. However, if it gets too high, a DTC 65 may be set. Refer to "Fuel System Diagnosis," CHART A-7.
• <u>Rich Injector.</u> Perform injector balance test, CHART C-2A.

• <u>Leaking Injector.</u> Refer to "Fuel System Diagnosis," CHART A-7.
• Check for fuel contaminated oil.
• <u>Evaporative Emission (EVAP) Canister Purge.</u> Check for fuel saturation. If full of fuel, check canister control and hoses. Refer to "Evaporative Emission (EVAP) Control System," Section "6E3-C3".
• <u>MAP Sensor.</u> An output that causes the ECM to sense a lower than normal vacuum can cause the system to go rich. Disconnecting the MAP sensor will allow the ECM to set a fixed value for the sensor. Substitute different MAP sensor if the rich condition is gone while the sensor is disconnected.
• Check for leaking fuel pressure regulator diaphragm by checking vacuum line to regulator for fuel.
• <u>TP Sensor.</u> An intermittent TP sensor output will cause the system to go rich, due to a false indication of the engine accelerating.
• <u>EGR.</u> An EGR valve staying open (especially at idle) will cause the oxygen sensor to indicate a rich exhaust, and this could result in a DTC 65.

5.7L (VIN 8 & P) ENGINE — DIAGNOSTIC TROUBLE CODE CHART — 1992–93 CORVETTE

DTC 65

BANK 2 (RIGHT) OXYGEN SENSOR (O2S) CIRCUIT
(RICH EXHAUST INDICATED)
5.7L (VIN P) "Y" CARLINE (MFI)

① • RUN WARM ENGINE (75°C/167°F TO 95°C/203°F) AT 1200 RPM.
• DOES TECH 1 DISPLAY O2S VOLTAGE FIXED ABOVE .75 VOLT (750 mV)?

YES → • DISCONNECT O2S SENSOR AND JUMPER HARNESS CKT 1589 TO GROUND.
• TECH 1 SHOULD DISPLAY O2S BELOW .35 VOLT (350 mV). DOES IT?

NO → DTC 65 IS INTERMITTENT. IF NO ADDITIONAL DTC(s) WERE STORED, REFER TO "DIAGNOSTIC AIDS" ON FACING PAGE.

YES → REFER TO "DIAGNOSTIC AIDS" ON FACING PAGE.

NO → REPLACE ECM.

"AFTER REPAIRS," REFER TO DTC CRITERIA ON FACING PAGE AND CONFIRM DTC DOES NOT RESET.

5.7L (VIN 8 & P) ENGINE — DIAGNOSTIC TROUBLE CODE CHART — 1992–93 CORVETTE

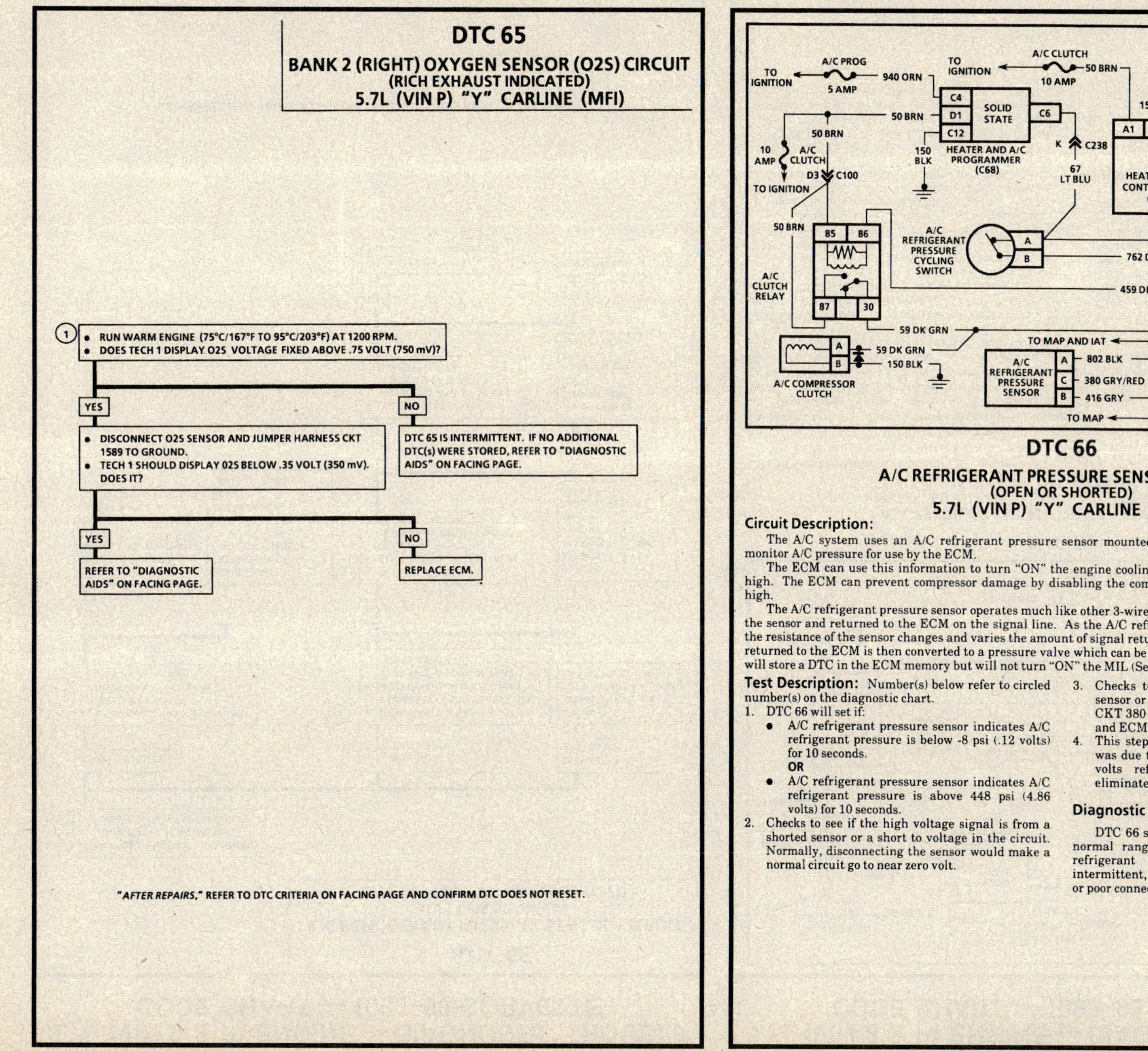

DTC 66

A/C REFRIGERANT PRESSURE SENSOR CIRCUIT
(OPEN OR SHORTED)
5.7L (VIN P) "Y" CARLINE (MFI)

Circuit Description:

The A/C system uses an A/C refrigerant pressure sensor mounted in the high side of the A/C system to monitor A/C pressure for use by the ECM.

The ECM can use this information to turn "ON" the engine cooling fans when A/C refrigerant pressure is high. The ECM can prevent compressor damage by disabling the compressor clutch if head pressure gets too high.

The A/C refrigerant pressure sensor operates much like other 3-wire sensors. A 5 volt reference is supplied to the sensor and returned to the ECM on the signal line. As the A/C refrigerant pressure increases or decreases, the resistance of the sensor changes and varies the amount of signal returning to the ECM. The amount of signal returned to the ECM is then converted to a pressure valve which can be displayed on the Tech 1. A DTC 66 fault will store a DTC in the ECM memory but will not turn "ON" the MIL (Service Engine Soon).

Test Description: Number(s) below refer to circled number(s) on the diagnostic chart.

1. DTC 66 will set if:
   • A/C refrigerant pressure sensor indicates A/C refrigerant pressure is below -8 psi (.12 volts) for 10 seconds.
   OR
   • A/C refrigerant pressure sensor indicates A/C refrigerant pressure is above 448 psi (4.86 volts) for 10 seconds.
2. Checks to see if the high voltage signal is from a shorted sensor or a short to voltage in the circuit. Normally, disconnecting the sensor would make a normal circuit go to near zero volt.
3. Checks to see if low voltage signal is from the sensor or the circuit. Jumpering the sensor signal CKT 380 to 5 volts, checks the circuit, connections and ECM.
4. This step checks to see if the low voltage signal was due to an open in the sensor circuit or the 5 volts reference circuit since the prior step eliminated the pressure sensor.

Diagnostic Aids:

DTC 66 sets when signal voltage falls outside the normal range of the sensor and is not due to a refrigerant system problem. If problem is intermittent, check for opens or shorts in the harness, or poor connections.

5.7L (VIN 8 & P) ENGINE — DIAGNOSTIC TROUBLE CODE CHART — 1992–93 CORVETTE

DTC 66
A/C REFRIGERANT PRESSURE SENSOR CIRCUIT
(OPEN OR SHORTED)
5.7L (VIN P) "Y" CARLINE (MFI)

5.7L (VIN 8 & P) ENGINE — DIAGNOSTIC TROUBLE CODE CHART — 1992–93 CORVETTE

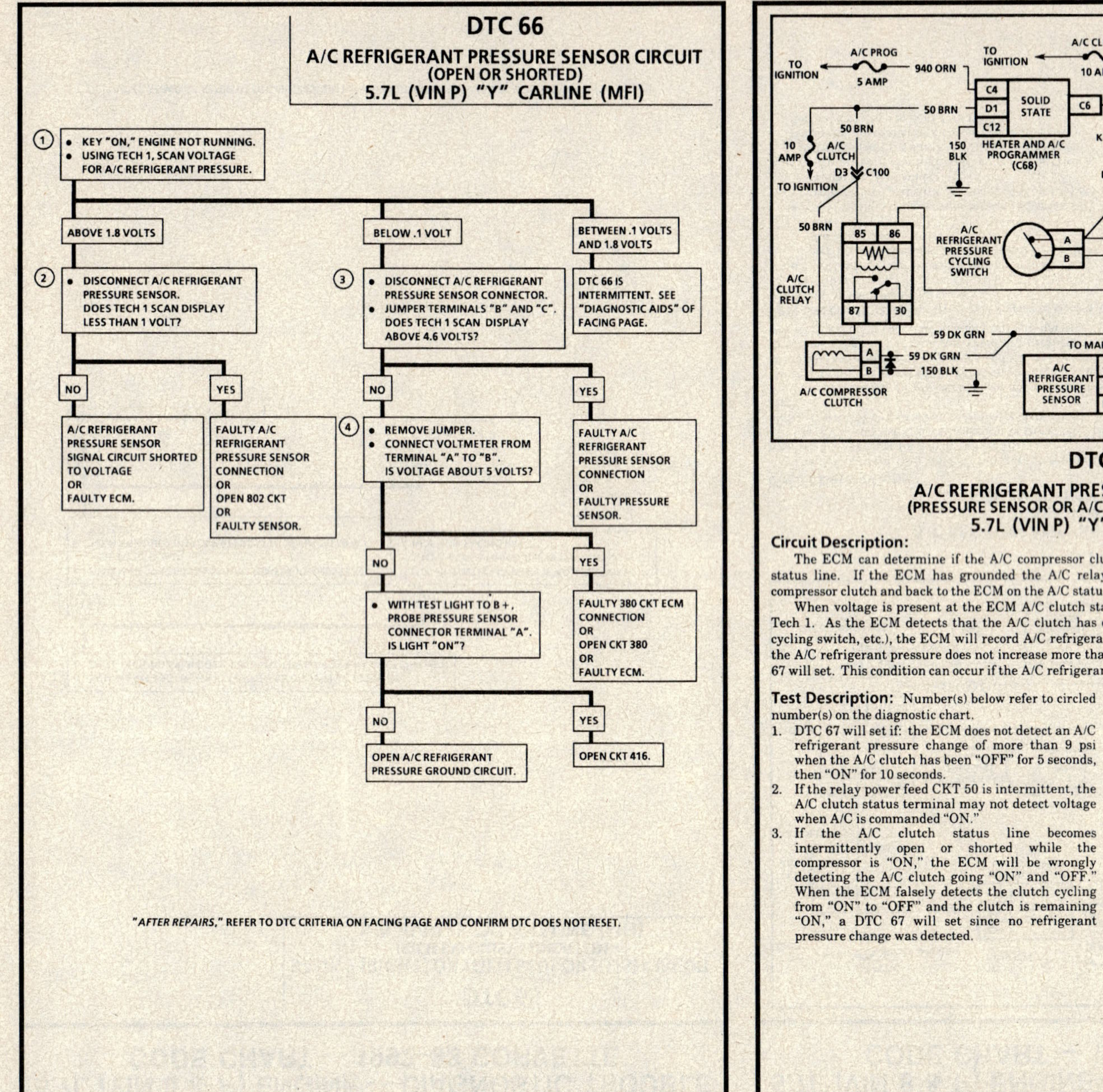

DTC 67
A/C REFRIGERANT PRESSURE SENSOR CIRCUIT
(PRESSURE SENSOR OR A/C CLUTCH CIRCUIT PROBLEM)
5.7L (VIN P) "Y" CARLINE (MFI)

Circuit Description:

The ECM can determine if the A/C compressor clutch is engaged or not engaged through the A/C clutch status line. If the ECM has grounded the A/C relay, current can flow through the A/C relay to the A/C compressor clutch and back to the ECM on the A/C status circuit.

When voltage is present at the ECM A/C clutch status terminal, A/C status "ON" will be displayed on the Tech 1. As the ECM detects that the A/C clutch has cycled "OFF" (through the A/C control switch, pressure cycling switch, etc.), the ECM will record A/C refrigerant pressure from the A/C refrigerant pressure sensor. If the A/C refrigerant pressure does not increase more than 9 psi when the A/C clutch is turned back "ON," a DTC 67 will set. This condition can occur if the A/C refrigerant pressure sensor is fixed at one value.

Test Description: Number(s) below refer to circled number(s) on the diagnostic chart.

1. DTC 67 will set if: the ECM does not detect an A/C refrigerant pressure change of more than 9 psi when the A/C clutch has been "OFF" for 5 seconds, then "ON" for 10 seconds.
2. If the relay power feed CKT 50 is intermittent, the A/C clutch status terminal may not detect voltage when A/C is commanded "ON."
3. If the A/C clutch status line becomes intermittently open or shorted while the compressor is "ON," the ECM will be wrongly detecting the A/C clutch going "ON" and "OFF." When the ECM falsely detects the clutch cycling from "ON" to "OFF" and the clutch is remaining "ON," a DTC 67 will set since no refrigerant pressure change was detected.
4. If the A/C compressor clutch is disconnected or if CKT 59 is open between the splice and the clutch, the A/C compressor clutch will not engage, but the ECM will detect voltage on the A/C status line which will indicate the clutch is "ON." If the A/C controls are cycled from "OFF" to "ON" at this point, a DTC 67 will set since the clutch did not engage and therefore no pressure change was detected.

Diagnostic Aids:

A DTC 67 will store in ECM memory but will not turn "ON" the MIL (Service Engine Soon). An A/C system low on charge could cause a DTC 67 to set.

"AFTER REPAIRS," REFER TO DTC CRITERIA ON FACING PAGE AND CONFIRM DTC DOES NOT RESET.

5.7L (VIN 8 & P) ENGINE — DIAGNOSTIC TROUBLE CODE CHART — 1992–93 CORVETTE

DTC 67

A/C REFRIGERANT PRESSURE SENSOR CIRCUIT
(PRESSURE SENSOR OR A/C CLUTCH CIRCUIT PROBLEM)
5.7L (VIN P) "Y" CARLINE (MFI)

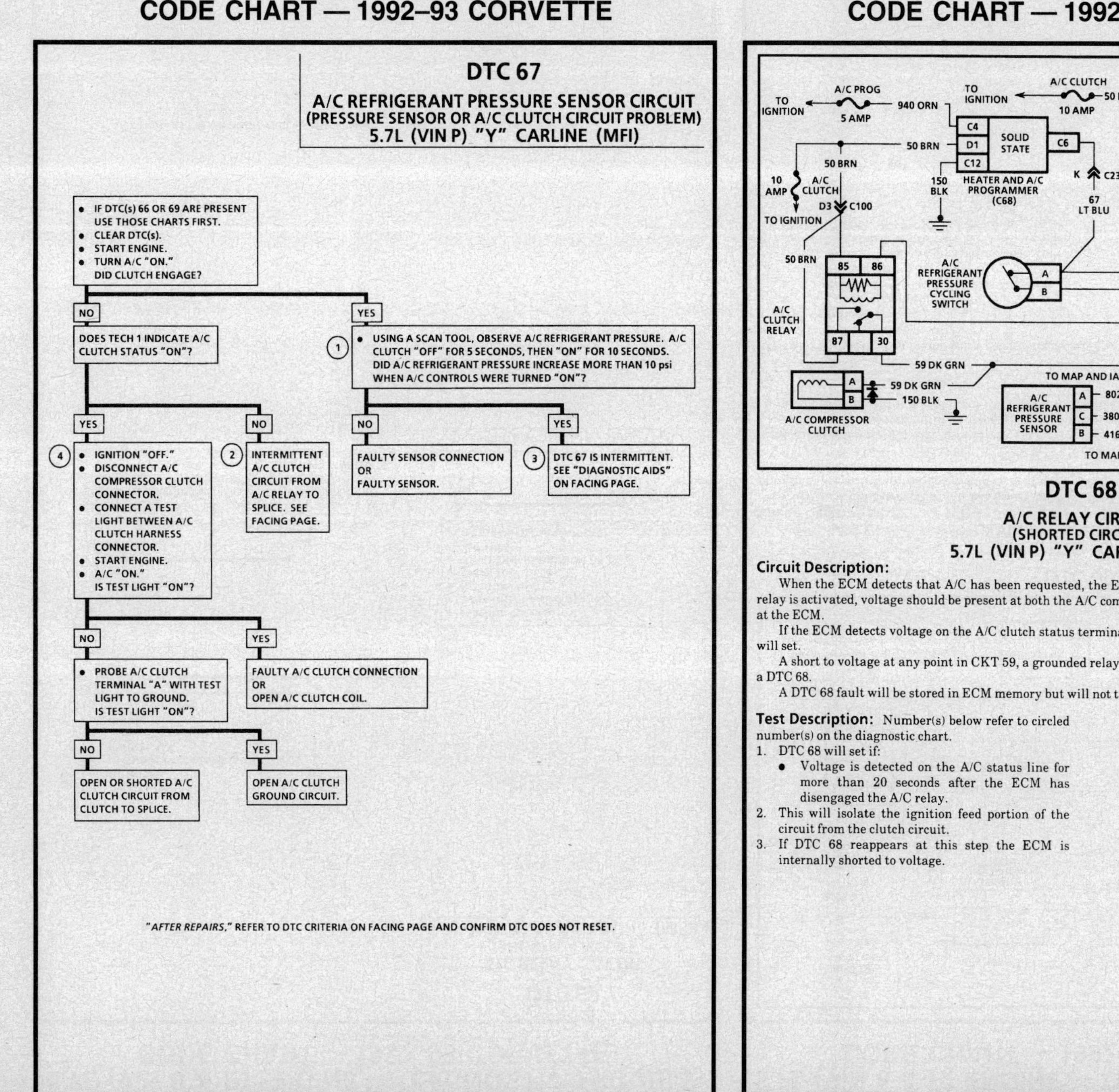

5.7L (VIN 8 & P) ENGINE — DIAGNOSTIC TROUBLE CODE CHART — 1992–93 CORVETTE

DTC 68

A/C RELAY CIRCUIT
(SHORTED CIRCUIT)
5.7L (VIN P) "Y" CARLINE (MFI)

Circuit Description:

When the ECM detects that A/C has been requested, the ECM will activate the A/C clutch relay. When the relay is activated, voltage should be present at both the A/C compressor clutch and the A/C clutch status terminal at the ECM.

If the ECM detects voltage on the A/C clutch status terminal when the A/C has not been requested a DTC 68 will set.

A short to voltage at any point in CKT 59, a grounded relay control circuit or stuck A/C relay contacts can set a DTC 68.

A DTC 68 fault will be stored in ECM memory but will not turn "ON" the MIL (Service Engine Soon).

Test Description: Number(s) below refer to circled number(s) on the diagnostic chart.

1. DTC 68 will set if:
   - Voltage is detected on the A/C status line for more than 20 seconds after the ECM has disengaged the A/C relay.
2. This will isolate the ignition feed portion of the circuit from the clutch circuit.
3. If DTC 68 reappears at this step the ECM is internally shorted to voltage.

5.7L (VIN 8 & P) ENGINE — DIAGNOSTIC TROUBLE CODE CHART — 1992–93 CORVETTE

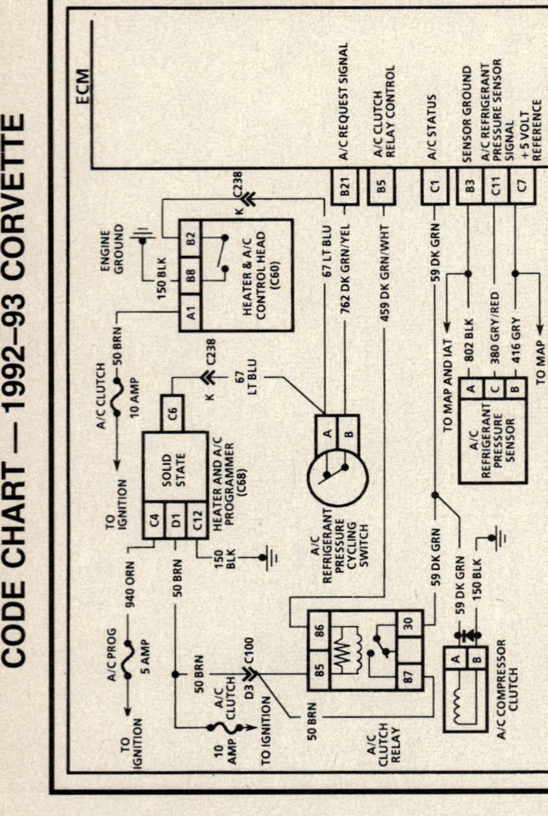

Circuit Description:

When the ECM detects that A/C has been requested, the ECM will activate the A/C clutch relay. When the relay has been activated, voltage should be present at both the A/C compressor clutch and the A/C clutch status line terminal of the ECM.

If the A/C has been requested but the ECM does not detect voltage present at the A/C clutch status terminal for more than 20 seconds, a DTC 69 will set.

An open or short to ground at any point in CKT 59 will cause DTC 69 to set.

A DTC 69 fault will be stored in the ECM memory but will not turn "ON" the MIL (Service Engine Soon).

Test Description: Number(s) below refer to circled number(s) on the diagnostic chart.
1. DTC 69 will set if:
   - ECM has commanded A/C "ON."
   - No voltage detected on A/C status line for more than 20 seconds.
2. If the A/C clutch operates properly and the Tech 1 shows A/C clutch status "OFF," then an open exists between the ECM and the splice to the compressor.

5.7L (VIN 8 & P) ENGINE — DIAGNOSTIC TROUBLE CODE CHART — 1992–93 CORVETTE

DTC 68
A/C RELAY CIRCUIT (SHORTED CIRCUIT)
5.7L (VIN P) "Y" CARLINE (MFI)

"AFTER REPAIRS," REFER TO DTC CRITERIA ON FACING PAGE AND CONFIRM DTC DOES NOT RESET.

5.7L (VIN 8 & P) ENGINE — DIAGNOSTIC TROUBLE CODE CHART — 1992–93 CORVETTE

DTC 69
A/C CLUTCH CIRCUIT
5.7L (VIN P) "Y" CARLINE (MFI)

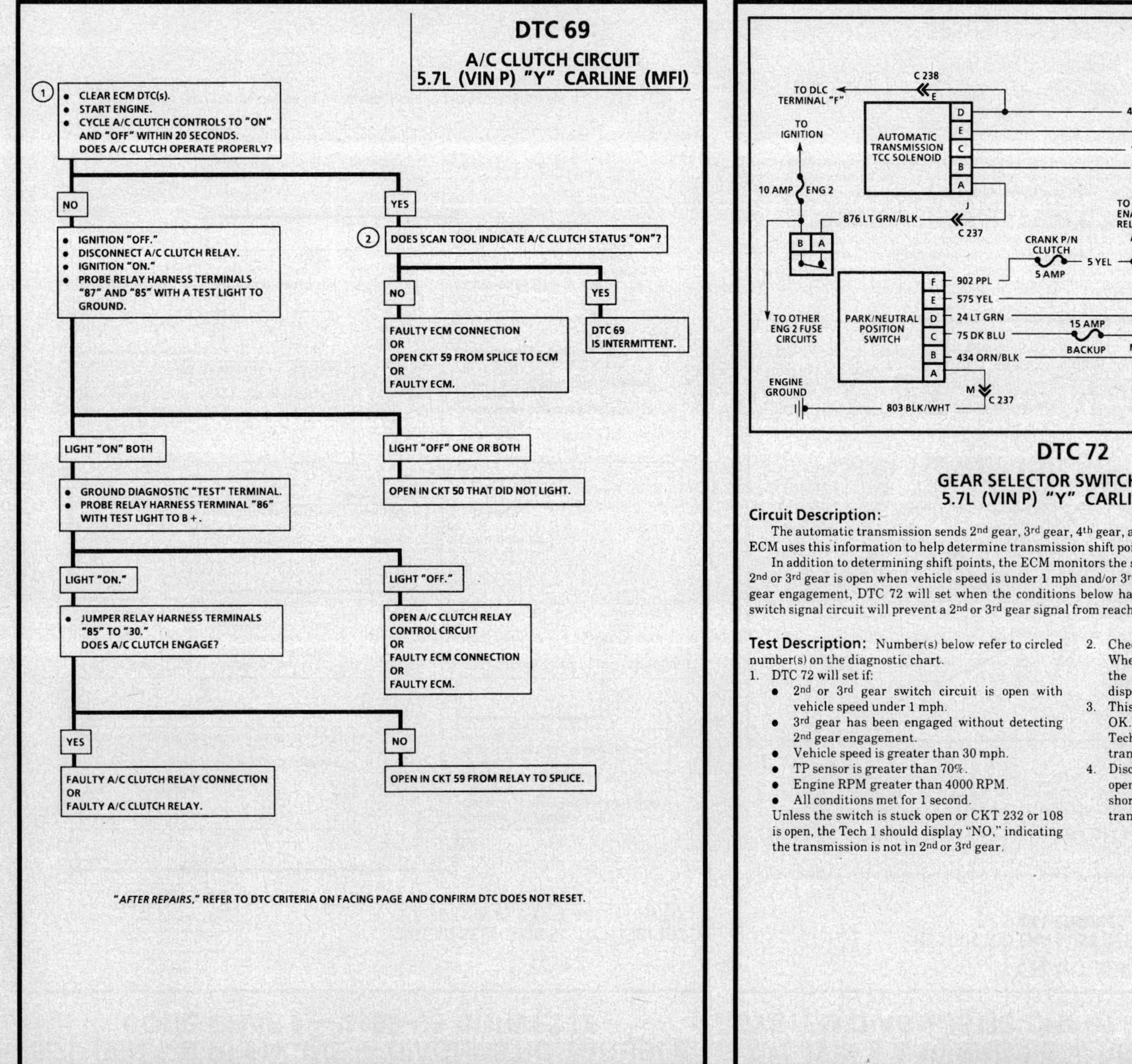

DTC 72
GEAR SELECTOR SWITCH CIRCUIT
5.7L (VIN P) "Y" CARLINE (MFI)

Circuit Description:
The automatic transmission sends 2nd gear, 3rd gear, 4th gear, and TCC information directly to the ECM. The ECM uses this information to help determine transmission shift points, as well as determining TCC apply times.

In addition to determining shift points, the ECM monitors the status of these inputs. If the ECM detects that 2nd or 3rd gear is open when vehicle speed is under 1 mph and/or 3rd gear has been engaged without detecting 2nd gear engagement, DTC 72 will set when the conditions below have been met. An open in the 2nd or 3rd gear switch signal circuit will prevent a 2nd or 3rd gear signal from reaching the ECM.

Test Description: Number(s) below refer to circled number(s) on the diagnostic chart.
1. DTC 72 will set if:
   - 2nd or 3rd gear switch circuit is open with vehicle speed under 1 mph.
   - 3rd gear has been engaged without detecting 2nd gear engagement.
   - Vehicle speed is greater than 30 mph.
   - TP sensor is greater than 70%.
   - Engine RPM greater than 4000 RPM.
   - All conditions met for 1 second.
   Unless the switch is stuck open or CKT 232 or 108 is open, the Tech 1 should display "NO," indicating the transmission is not in 2nd or 3rd gear.

2. Checks the operation of the 2nd or 3rd gear switch. When the transmission shifts into 2nd or 3rd gear, the switch should open and the Tech 1 should display "YES."
3. This step determines if the ECM and wiring are OK. Grounding CKT 232 or 108 should cause the Tech 1 to display "YES," indicating the transmission is in 2nd or 3rd gear.
4. Disconnecting the TCC connector simulates an open switch to determine if CKT 232 or 108 is shorted to ground or the problem is in the transmission.

"AFTER REPAIRS," REFER TO DTC CRITERIA ON FACING PAGE AND CONFIRM DTC DOES NOT RESET.

5.7L (VIN 8 & P) ENGINE — DIAGNOSTIC TROUBLE CODE CHART — 1992–93 CORVETTE

DTC 72

GEAR SELECTOR SWITCH CIRCUIT
5.7L (VIN P) "Y" CARLINE (MFI)

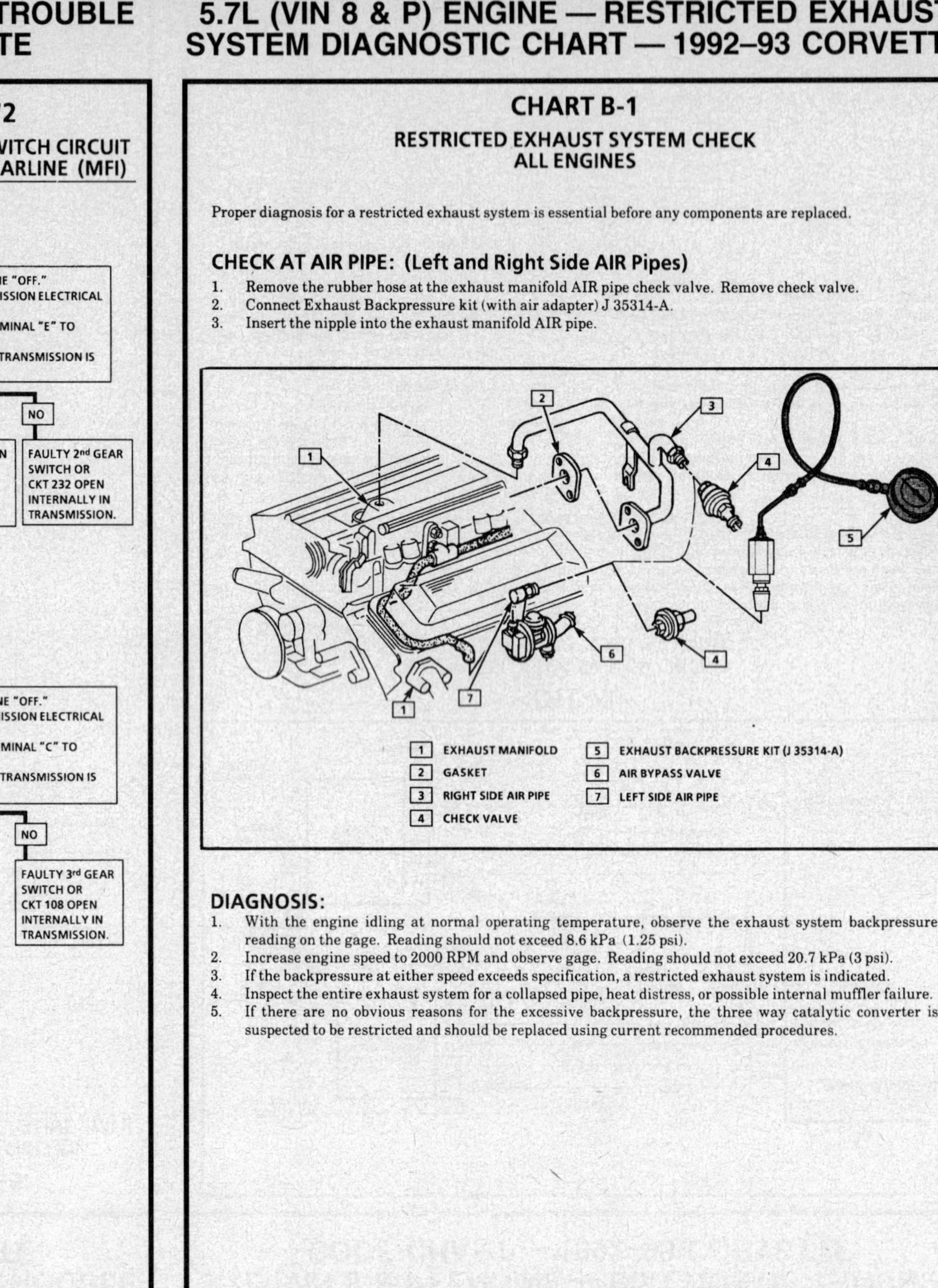

"AFTER REPAIRS," REFER TO DTC CRITERIA ON FACING PAGE AND CONFIRM DTC DOES NOT RESET.

5.7L (VIN 8 & P) ENGINE — RESTRICTED EXHAUST SYSTEM DIAGNOSTIC CHART — 1992–93 CORVETTE

CHART B-1

RESTRICTED EXHAUST SYSTEM CHECK
ALL ENGINES

Proper diagnosis for a restricted exhaust system is essential before any components are replaced.

CHECK AT AIR PIPE: (Left and Right Side AIR Pipes)

1. Remove the rubber hose at the exhaust manifold AIR pipe check valve. Remove check valve.
2. Connect Exhaust Backpressure kit (with air adapter) J 35314-A.
3. Insert the nipple into the exhaust manifold AIR pipe.

DIAGNOSIS:

1. With the engine idling at normal operating temperature, observe the exhaust system backpressure reading on the gage. Reading should not exceed 8.6 kPa (1.25 psi).
2. Increase engine speed to 2000 RPM and observe gage. Reading should not exceed 20.7 kPa (3 psi).
3. If the backpressure at either speed exceeds specification, a restricted exhaust system is indicated.
4. Inspect the entire exhaust system for a collapsed pipe, heat distress, or possible internal muffler failure.
5. If there are no obvious reasons for the excessive backpressure, the three way catalytic converter is suspected to be restricted and should be replaced using current recommended procedures.

5.7L (VIN 8 & P) ENGINE — COMPONENT DIAGNOSTIC CHART — 1992–93 CORVETTE

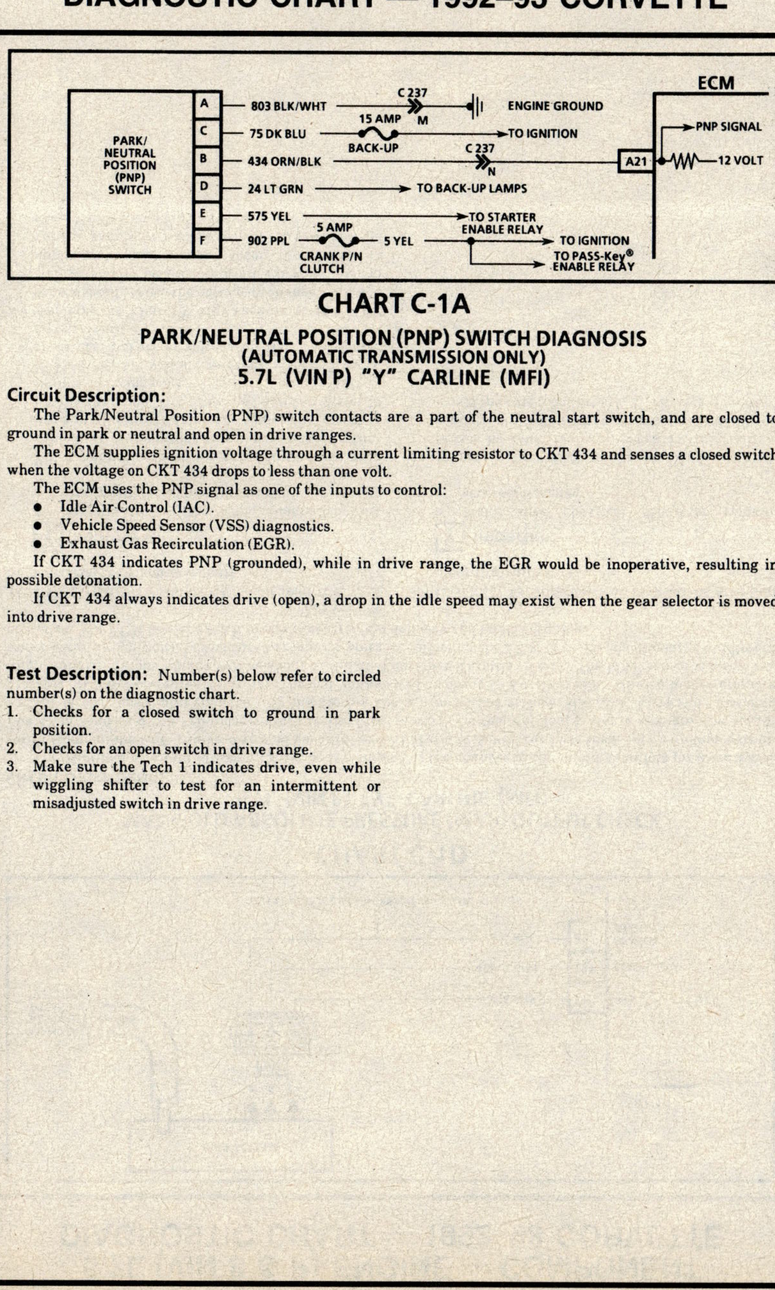

CHART C-1A

PARK/NEUTRAL POSITION (PNP) SWITCH DIAGNOSIS
(AUTOMATIC TRANSMISSION ONLY)
5.7L (VIN P) "Y" CARLINE (MFI)

Circuit Description:

The Park/Neutral Position (PNP) switch contacts are a part of the neutral start switch, and are closed to ground in park or neutral and open in drive ranges.

The ECM supplies ignition voltage through a current limiting resistor to CKT 434 and senses a closed switch when the voltage on CKT 434 drops to less than one volt.

The ECM uses the PNP signal as one of the inputs to control:

- Idle Air Control (IAC).
- Vehicle Speed Sensor (VSS) diagnostics.
- Exhaust Gas Recirculation (EGR).

If CKT 434 indicates PNP (grounded), while in drive range, the EGR would be inoperative, resulting in possible detonation.

If CKT 434 always indicates drive (open), a drop in the idle speed may exist when the gear selector is moved into drive range.

Test Description: Number(s) below refer to circled number(s) on the diagnostic chart.

1. Checks for a closed switch to ground in park position.
2. Checks for an open switch in drive range.
3. Make sure the Tech 1 indicates drive, even while wiggling shifter to test for an intermittent or misadjusted switch in drive range.

5.7L (VIN 8 & P) ENGINE — COMPONENT DIAGNOSTIC CHART — 1992–93 CORVETTE

CHART C-1A

PARK/NEUTRAL POSITION (PNP) SWITCH DIAGNOSIS
(AUTOMATIC TRANSMISSION ONLY)
5.7L (VIN P) "Y" CARLINE (MFI)

"AFTER REPAIRS," CONFIRM "CLOSED LOOP" OPERATION AND NO MIL (SERVICE ENGINE SOON).

5.7L (VIN 8 & P) ENGINE — COMPONENT DIAGNOSTIC CHART — 1992–93 CORVETTE

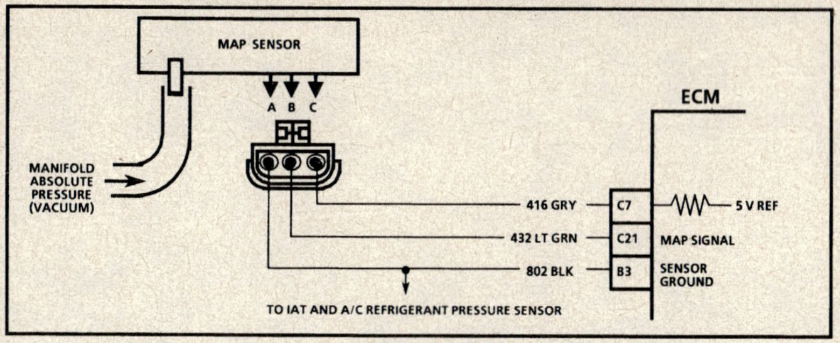

CHART C-1D
MANIFOLD ABSOLUTE PRESSURE (MAP) OUTPUT CHECK
5.7L (VIN P) "Y" CARLINE (MFI)

Circuit Description:

The Manifold Absolute Pressure (MAP) sensor measures the changes in the intake manifold pressure which result from engine load (intake manifold vacuum) and RPM changes; and converts these into a voltage output. The ECM sends a 5 volt reference voltage to the MAP sensor. As the manifold pressure changes, the output voltage of the sensor also changes. By monitoring the sensor output voltage, the ECM knows the manifold pressure. A lower pressure (low voltage) output voltage will be about 1 - 2 volts at idle. While a higher pressure (high voltage) output voltage will be about 4 - 4.8 at Wide Open Throttle (WOT). The MAP sensor is also used, under certain conditions, to measure barometric pressure, allowing the ECM to make adjustments for different altitudes. The ECM uses the MAP sensor to control fuel delivery and ignition timing.

Test Description: Number(s) below refer to circled number(s) on the diagnostic chart.

  Important
- Be sure to use the same Diagnostic Test Equipment for all measurements.

1. When comparing Tech 1 reading to a known good vehicle, it is important to compare vehicles that use a MAP sensor having the same color insert or having the same "Hot Stamped" number. Refer to figures on facing page.
   If MAP sensor voltage is greater than 5 volts, check for a short to voltage on the MAP sensor signal circuit.
2. Applying 34 kPa (10" Hg) vacuum to the MAP sensor should cause the voltage to change. Voltage change should be greater than 1.5 volts. Upon applying vacuum to the sensor, the change in voltage should be instantaneous. A slow voltage change indicates a faulty sensor.

3. Check for vacuum leaks or restrictions at intake mounting surface.

  Important
- Make sure electrical connector remains securely fastened.

4. Remove sensor from intake manifold and twist sensor by hand (only) to check for intermittent connection. Output changes greater than .10 volt would indicate a bad sensor. If OK, replace sensor.

5.7L (VIN 8 & P) ENGINE — COMPONENT DIAGNOSTIC CHART — 1992–93 CORVETTE

CHART C-1D
MANIFOLD ABSOLUTE PRESSURE (MAP) OUTPUT CHECK
5.7L (VIN P) "Y" CARLINE (MFI)

NOTICE: THIS CHART ONLY APPLIES TO MAP SENSORS HAVING GREEN OR BLACK COLOR KEY INSERT (SEE BELOW).

1.
   - IF DTC 33 OR 34 IS SET, USE THOSE CHARTS FIRST.
   - IGNITION "ON," ENGINE "OFF."
   - SCAN TOOL SHOULD INDICATE A MAP SENSOR VOLTAGE.
   - COMPARE THIS READING WITH THE READING OF A KNOWN GOOD VEHICLE. SEE FACING PAGE TEST DESCRIPTION, STEP 1. VOLTAGE READING SHOULD BE WITHIN ± .4 VOLT.
   IS IT?

  YES →   NO → REPLACE MAP SENSOR.

2.
   - DISCONNECT AND PLUG VACUUM SOURCE TO MAP SENSOR.
   - CONNECT A HAND VACUUM PUMP TO MAP SENSOR.
   - START ENGINE.
   - NOTE MAP SENSOR VOLTAGE.
   - APPLY 34 kPa (10" Hg) OF VACUUM AND NOTE VOLTAGE CHANGE. SUBTRACT SECOND READING FROM THE FIRST. VOLTAGE VALUE SHOULD BE GREATER THAN 1.5 VOLTS.
   IS IT?

  YES →   NO →

3. NO TROUBLE FOUND. CHECK MAP SENSOR VACUUM SOURCE FOR LEAKAGE OR RESTRICTION. BE SURE THIS SOURCE SUPPLIES VACUUM TO MAP SENSOR ONLY.

4. CHECK MAP SENSOR CONNECTION. IF OK, REPLACE MAP SENSOR.

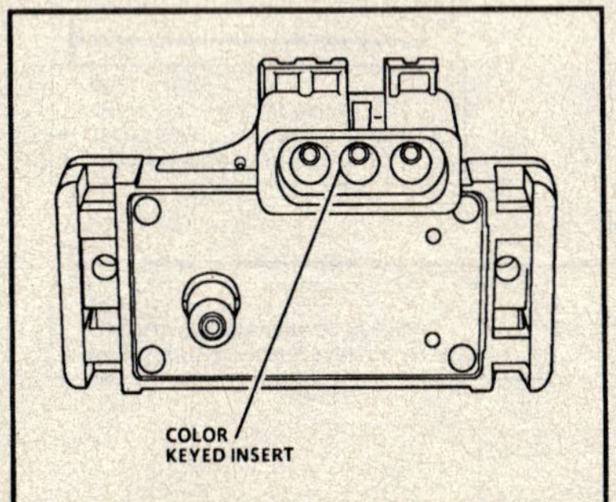

Figure 1 - Color Key Insert

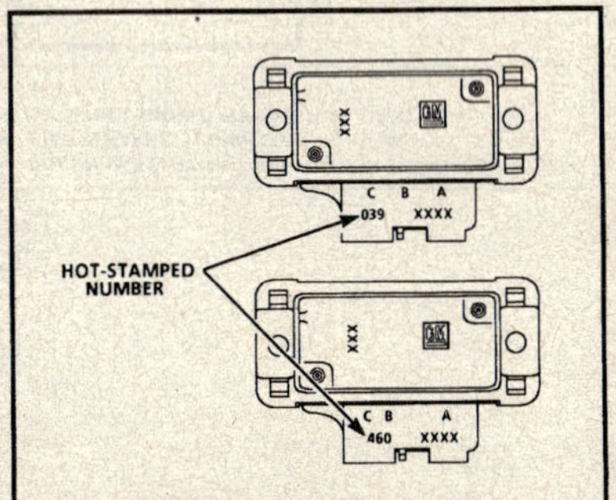

Figure 2 - Hot Stamped Number

"AFTER REPAIRS," CONFIRM "CLOSED LOOP" OPERATION AND NO MIL (SERVICE ENGINE SOON).

5.7L (VIN 8 & P) ENGINE — COMPONENT DIAGNOSTIC CHART — 1992–93 CORVETTE

CHART C-2A

INJECTOR BALANCE TEST
5.7L (VIN P) "Y" CARLINE (MFI)

The injector balance tester is a tool used to turn the injector "ON" for a precise amount of time, thus spraying a measured amount of fuel into the manifold. This causes a drop in fuel rail pressure that we can record and compare between each injector. All injectors should have the same amount of pressure drop ($\pm$ 10 kPa). Any injector with a pressure drop that is 10 kPa (or more) greater or less than the average drop of the other injectors should be considered faulty and replaced.

STEP 1

Engine "cool down" period (10 minutes) is necessary to avoid irregular readings due to "Hot Soak" fuel boiling. With ignition "OFF" connect fuel gauge J 3470-1 to fuel pressure tap. Wrap a shop towel around fitting while connecting gage to avoid fuel spillage.

Disconnect harness connectors at all injectors, and connect injector tester J 34730-3A to one injector. Follow manufacturers instructions for use of adaptor harness. Ignition must be "OFF" at least 10 seconds to complete ECM shutdown cycle. Fuel pump should run about 2 seconds after ignition is turned "ON." At this point, insert clear tubing attached to vent valve into a suitable container and bleed air from gauge and hose to insure accurate gauge operation. Repeat this step until all air is bled from gauge.

STEP 2

Turn ignition "OFF" for 10 seconds and then "ON" again to get fuel pressure to its maximum. Record this initial pressure reading. Energize tester one time and note pressure drop at its lowest point (disregard any slight pressure increase after drop hits low point). By subtracting this second pressure reading from the initial pressure, we have the actual amount of injector pressure drop.

STEP 3

Repeat Step 2 on each injector and compare the amount of drop. Usually, good injectors will have virtually the same drop. Retest any injector that has a pressure difference of 10 kPa, either more or less than the average of the other injectors on the engine. Replace any injector that also fails the retest. If the pressure drop of all injectors is within 10 kPa of this average, the injectors appear to be flowing properly. Reconnect them and review "Symptoms," Section

NOTICE: *The entire test should not be repeated more than once without running the engine to prevent flooding. (This includes any retest on faulty injectors.)*

5.7L (VIN 8 & P) ENGINE — COMPONENT DIAGNOSTIC CHART — 1992–93 CORVETTE

NOTICE: The entire test should NOT be repeated more than once without running the engine to prevent flooding. (This includes any retest on faulty injectors.)

CHART C-2A

INJECTOR BALANCE TEST
5.7L (VIN P) "Y" CARLINE (MFI)

The fuel pressure test in Section "A" Chart A-7, should be completed prior to this test.

Step 1. If engine is at operating temperature, allow a 10 minute "cool down" period then connect fuel pressure gage and injector tester.
1. Ignition "OFF."
2. Connect fuel pressure gage and injector tester.
3. Ignition "ON."
4. Bleed off air in gage. Repeat until all air is bled from gauge.

Step 2. Run test:
1. Ignition "OFF" for 10 seconds.
2. Ignition "ON." Record gage pressure. (Pressure must hold steady, if not, see "Fuel System Diagnosis," CHART A-7, in Section "A").
3. Turn injector on, by depressing button on injector tester, and note pressure at the instant the gauge needle stops.

Step 3.
1. Repeat Step 2 on all injectors and record pressure drop on each. Retest injectors that appear faulty (Any injectors that have a 10 kPa (1.5 psi) difference, either more or less, in pressure from the average). If no problem is found, review "Symptoms" Section

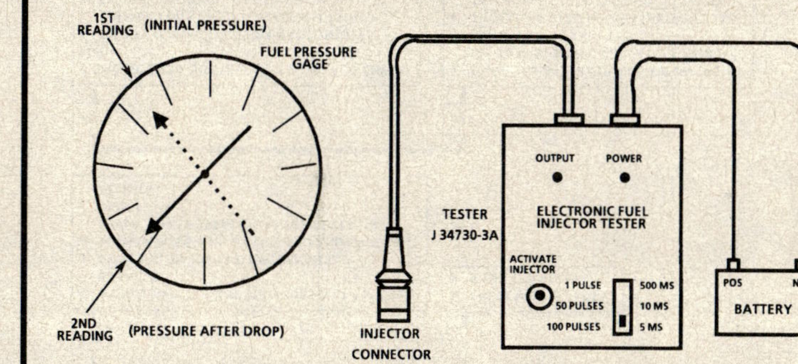

EXAMPLE

CYLINDER	1	2	3
1ST READING	293 kPa (43 psi)	293 kPa (43 psi)	293 kPa (43 psi)
2ND READING	131 kPa (19 psi)	115 kPa (17 psi)	145 kPa (21 psi)
AMOUNT OF DROP	162 kPa (24 psi)	178 kPa (26 psi)	148 kPa (21 psi)
	OK	FAULTY, RICH (TOO MUCH FUEL DROP)	FAULTY, LEAN (TOO LITTLE FUEL DROP)

5.7L (VIN 8 & P) ENGINE — COMPONENT DIAGNOSTIC CHART — 1992–93 CORVETTE

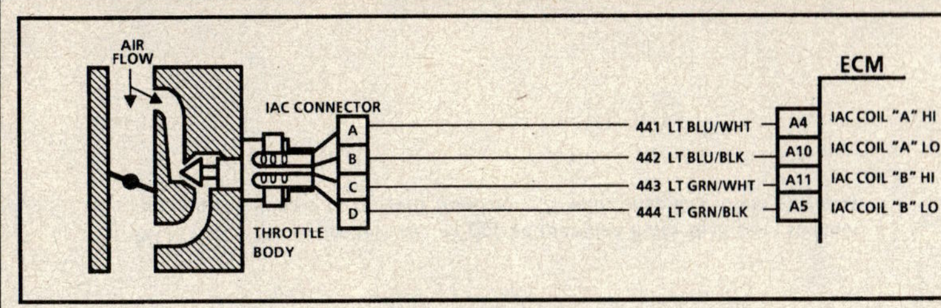

CHART C-2C
IDLE AIR CONTROL (IAC) VALVE CHECK
5.7L (VIN P) "Y" CARLINE (MFI)

Circuit Description:

The ECM controls engine idle speed with the IAC valve. To increase idle speed, the ECM retracts the IAC valve pintle away from its seat, allowing more air to bypass the throttle bore. To decrease idle speed, it extends the IAC valve pintle towards its seat, reducing bypass air flow. The scan tool will read the ECM commands to the IAC valve in counts. Higher the counts indicate more air bypass (higher idle). The lower the counts indicate less air is allowed to bypass (lower idle).

Test Description: Number(s) below refer to circled number(s) on the diagnostic chart.

1. The IAC tester is used to extend and retract the IAC valve. Valve movement is verified by an engine speed change. If no change in engine speed occurs, the valve can be retested when removed from the throttle body.
2. This step checks the quality of the IAC movement in Step 1. Between 700 RPM and about 1500 RPM, the engine speed should change smoothly with each flash of the tester light in both extend and retract. If the IAC valve is retracted beyond the control range (about 1500 RPM), it may take many flashes in the extend position before engine speed will begin to drop. This is normal on certain engines, fully extending IAC may cause engine stall. This may be normal.
3. Steps 1 and 2 verified proper IAC valve operation while this step checks the IAC circuits. Each lamp on the node light should flash red and green while the IAC valve is cycled. While the sequence of color is not important, if either light is "OFF" or does not flash red and green, check the circuits for faults, beginning with poor terminal contacts.

Diagnostic Aids:

A slow, unstable, or fast idle may be caused by a non-IAC system problem that cannot be overcome by the IAC valve. Out of control range IAC Tech 1 counts will be above 60 if idle is too low, and zero counts if idle is too high. The following checks should be made to repair a non-IAC system problem:

- <u>Vacuum Leak (High Idle)</u> - If idle is too high, stop the engine. Fully extend (low) IAC with tester. Start engine. If idle speed is above 800 RPM, locate and correct vacuum leak including crankcase ventilation system. Also, check for binding of throttle blade or linkage.
- <u>System too lean (High Air/Fuel Ratio)</u> - The idle speed may be too high or too low. Engine speed may vary up and down and disconnecting the IAC valve does not help. DTC 44 or 64 may be set. Tech 1 O2S voltage will be less than 300 mV (.3 volt). Check for low regulated fuel pressure, water in the fuel or a restricted injector.
- <u>System too rich (Low Air/Fuel Ratio)</u> - The idle speed will be too low. Tech 1 IAC counts will usually be above 80. System is obviously rich and may exhibit black smoke in exhaust. Tech 1 O2S voltage will be fixed above 800 mV (.8 volt). DTC 45 or 65 may set. Check for high fuel pressure, leaking or sticking injector. Silicone contaminated oxygen sensors will be slow to respond.
- <u>Throttle Body</u> - Remove IAC valve and inspect bore for foreign material.
- <u>IAC Valve Electrical Connections</u> - IAC valve connections should be carefully checked for proper contact.
- <u>Crankcase Ventilation Valve</u> - An incorrect or faulty crankcase ventilation valve may result in an incorrect idle speed.
- Refer to "Rough, Unstable, Incorrect Idle or Stalling" in "Symptoms," Section "6E3-B".
- If intermittent poor driveability or idle symptoms are resolved by disconnecting the IAC, carefully recheck connections, valve terminal resistance, or replace IAC.

5.7L (VIN 8 & P) ENGINE — COMPONENT DIAGNOSTIC CHART — 1992–93 CORVETTE

CHART C-2C
IDLE AIR CONTROL (IAC) VALVE CHECK
5.7L (VIN P) "Y" CARLINE (MFI)

①
- IGNITION "OFF," CONNECT IAC DRIVER * TO IAC VALVE.
- SET PARKING BRAKE, BLOCK WHEELS, A/C "OFF."
- IDLE ENGINE IN PARK (A/T) OR NEUTRAL (M/T).
- INSTALL TECH 1 SCAN TOOL AND DISPLAY RPM.
- WITH IAC DRIVER, EXTEND AND RETRACT IAC VALVE.
- ENGINE RPM SHOULD DECREASE AND INCREASE AS IAC IS CYCLED.

RPM CHANGES → **②**
- RPM SHOULD CHANGE SMOOTHLY WITH EACH FLASH OF THE IAC DRIVER LIGHT FROM 700 RPM TO ABOUT 1500 RPM. DOES IT?

NO RPM CHANGE
- CHECK IAC PASSAGES.
- IF OK, REPLACE IAC.

YES → **③**
- INSTALL APPROPRIATE IAC NODE LIGHT * IN ECM HARNESS.
- CYCLE IAC DRIVER AND NOTE LIGHTS.
- BOTH LIGHTS SHOULD CYCLE RED AND GREEN BUT NEVER "OFF" AS RPM IS CHANGED OVER ITS RANGE. DO THEY?

NO
- CHECK IAC PASSAGES.
- IF OK, REPLACE IAC.

NO
- IF CIRCUIT(S) DID NOT TEST GREEN AND RED, CHECK FOR:
- FAULTY CONNECTOR TERMINAL CONTACTS.
- OPEN CIRCUITS INCLUDING CONNECTIONS.
- CIRCUITS SHORTED TO GROUND OR VOLTAGE.
- FAULTY ECM CONNECTION OR ECM. REPAIR AS NECESSARY AND RETEST.

YES
- USING THE OTHER CONNECTOR ON THE IAC DRIVER PIGTAIL, CHECK RESISTANCE ACROSS IAC COILS.
- SHOULD BE 40 TO 80 OHMS BETWEEN IAC TERMINALS "A" TO "B" AND "C" TO "D".

OK
- CHECK RESISTANCE BETWEEN IAC TERMINALS "B" AND "C" AND "A" AND "D".
- SHOULD BE INFINITE.

NOT OK
- REPLACE IAC VALVE AND RETEST.

OK
- IDLE AIR CONTROL VALVE AND CIRCUIT OK. REFER TO "DIAGNOSTIC AIDS" ON FACING PAGE.

NOT OK
- REPLACE IAC VALVE AND RETEST.

* IAC DRIVER AND NODE LIGHT REQUIRED KIT 222-L FROM: CONCEPT TECHNOLOGY, INC. J 37027-A FROM: KENT-MOORE, INC.

"AFTER REPAIRS," CONFIRM "CLOSED LOOP" OPERATION AND NO MIL (SERVICE ENGINE SOON).

5.7L (VIN 8 & P) ENGINE — COMPONENT DIAGNOSTIC CHART — 1992–93 CORVETTE

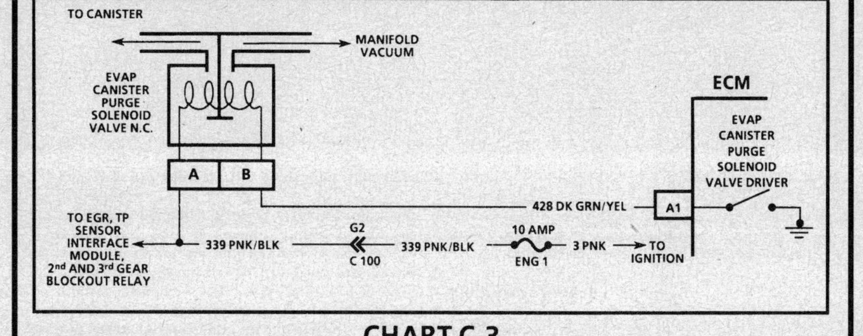

CHART C-3
EVAPORATIVE EMISSION (EVAP) CANISTER PURGE SOLENOID VALVE CHECK
5.7L (VIN P) "Y" CARLINE (MFI)

Circuit Description:

EVAP canister purge is controlled by a solenoid valve that allows manifold vacuum to purge the EVAP canister when energized. The ECM supplies a ground to energize the solenoid valve.

If the diagnostic "test" terminal is grounded with the engine stopped and ignition "ON," or the following conditions are met with the engine running, the EVAP canister purge solenoid valve will be energized (purge "ON").

- Engine run time after start more than 1 minute.
- Coolant temperature above 60°C (140°F).
- Vehicle speed above 2 mph.
- Throttle off idle, above 3%.
- IAT is less then 105°C (221°F).

Test Description: Number(s) below refer to circled number(s) on the diagnostic chart.

1. Checks to see if the solenoid valve is opened or closed. The solenoid valve is normally closed and should open when the diagnostic "test" terminal is grounded.
2. Completes functional check by grounding the diagnostic "test" terminal. This should energize the solenoid valve and allow the vacuum to drop (purge "ON").
3. Checks for a complete circuit. Normally, there is ignition voltage on CKT 339 and the ECM provides a ground on CKT 428. A shorted solenoid valve could cause an open circuit in the ECM.

Diagnostic Aids:

Normal operation of the EVAP canister purge solenoid valve is described as follows:

With the ignition "ON," engine "OFF," diagnostic "test" terminal grounded, the EVAP canister purge solenoid valve should be energized.

5.7L (VIN 8 & P) ENGINE — COMPONENT DIAGNOSTIC CHART — 1992–93 CORVETTE

CHART C-3
EVAPORATIVE EMISSION (EVAP) CANISTER PURGE SOLENOID VALVE CHECK
5.7L (VIN P) "Y" CARLINE (MFI)

1.
- IGNITION "ON." ENGINE STOPPED.
- DISCONNECT VACUUM HOSES FROM EVAP CANISTER PURGE SOLENOID VALVE.
- AT THE SOLENOID VALVE, APPLY 34 kPa (10" Hg) OF VACUUM TO THE THROTTLE BODY SIDE OF PURGE SOLENOID VALVE.

ABLE TO GET 34 kPa (10" Hg) OF VACUUM.

UNABLE TO GET 34 kPa (10" Hg) OF VACUUM.

2.
- GROUND DIAGNOSTIC TERMINAL.
- NOTE VACUUM.

- DISCONNECT SOLENOID VALVE.
- PROBE CKT 428 WITH A TEST LIGHT TO 12 VOLTS.

NO DROP

DROPS

LIGHT "ON"

LIGHT "OFF"

3.
- WITH DIAGNOSTIC TERMINAL STILL GROUNDED DISCONNECT EVAP CANISTER PURGE SOLENOID VALVE.
- CONNECT A TEST LIGHT BETWEEN HARNESS CONNECTOR TERMINALS.

NO TROUBLE FOUND.

CKT 428 SHORTED TO GROUND OR FAULTY ECM.

FAULTY SOLENOID VALVE.

NO LIGHT

LIGHT

- PROBE EACH HARNESS CONNECTOR TERMINAL WITH A TEST LIGHT TO GROUND.

FAULTY SOLENOID VALVE OR CONNECTOR.

LIGHT "ON" ONE TERMINAL

LIGHT BOTH TERMINALS

NO LIGHT

OPEN CKT 428 OR FAULTY CONNECTION OR FAULTY ECM.

REPAIR SHORT TO VOLTAGE IN CKT 428.

CHECK FOR OPEN IN IGNITION FEED CIRCUIT TO SOLENOID VALVE.

"AFTER REPAIRS," CONFIRM "CLOSED LOOP" OPERATION AND NO MIL (SERVICE ENGINE SOON).

5.7L (VIN 8 & P) ENGINE — COMPONENT DIAGNOSTIC CHART — 1992–93 CORVETTE

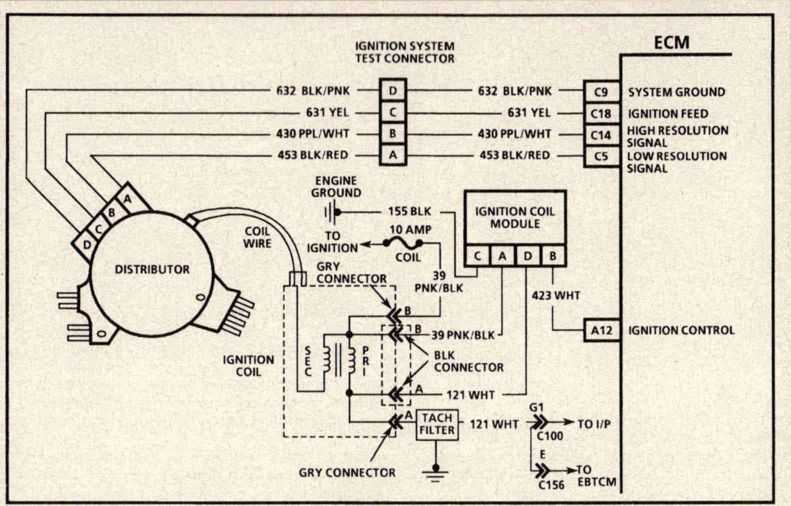

CHART C-4
(Page 1 of 2)
DISTRIBUTOR IGNITION SYSTEM CHECK
5.7L (VIN P) "Y" CARLINE (MFI)

Test Description: Number(s) below refer to circled number(s) on the diagnostic chart.

The battery should be fully charged prior to any tests.

1. To disconnect the tachometer lead, it will be necessary to disconnect the gray ignition coil connector from the ignition coil and jumper the "B" terminal of the harness connector to the "B" terminal of the ignition coil using a fused jumper and terminal test adapter kit J 35616-A.
Checks for proper output from the distributor ignition system. The spark tester requires a minimum of 25,000 volts to fire. This check can be used in case of an ignition miss, because the system may provide enough voltage to run the engine but not enough to fire a spark plug under heavy load.
2. This test will separate the distributor cap, rotor and ignition wires from the ignition coil to help identify a secondary ignition system problem.

3. This will determine if the proper available voltage exists in the primary ignition circuit.
4. This check will begin to determine if the ECM is providing a signal to the ignition coil module or not. If the ECM is not providing a signal to the ignition coil module, the problem exists between the distributor and ECM.
5. A grounded ignition coil circuit will cause the ignition coil fuse to open, and a no-start condition will occur.

Diagnostic Aids:

An ignition coil wire that is open or shorted to ground can cause a no start condition.

5.7L (VIN 8 & P) ENGINE — COMPONENT DIAGNOSTIC CHART — 1992–93 CORVETTE

CHART C-4
(Page 1 of 2)
DISTRIBUTOR IGNITION SYSTEM CHECK
5.7L (VIN P) "Y" CARLINE (MFI)

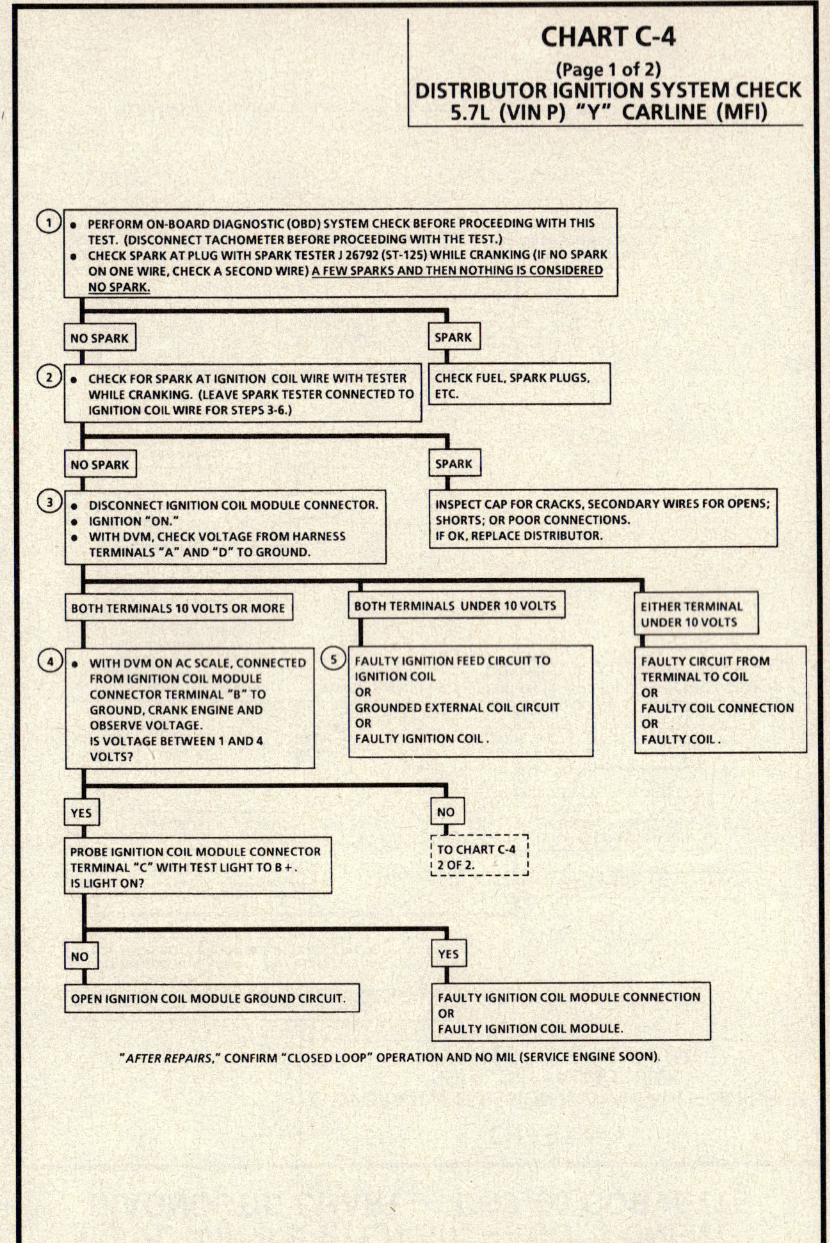

5.7L (VIN 8 & P) ENGINE — COMPONENT DIAGNOSTIC CHART — 1992–93 CORVETTE

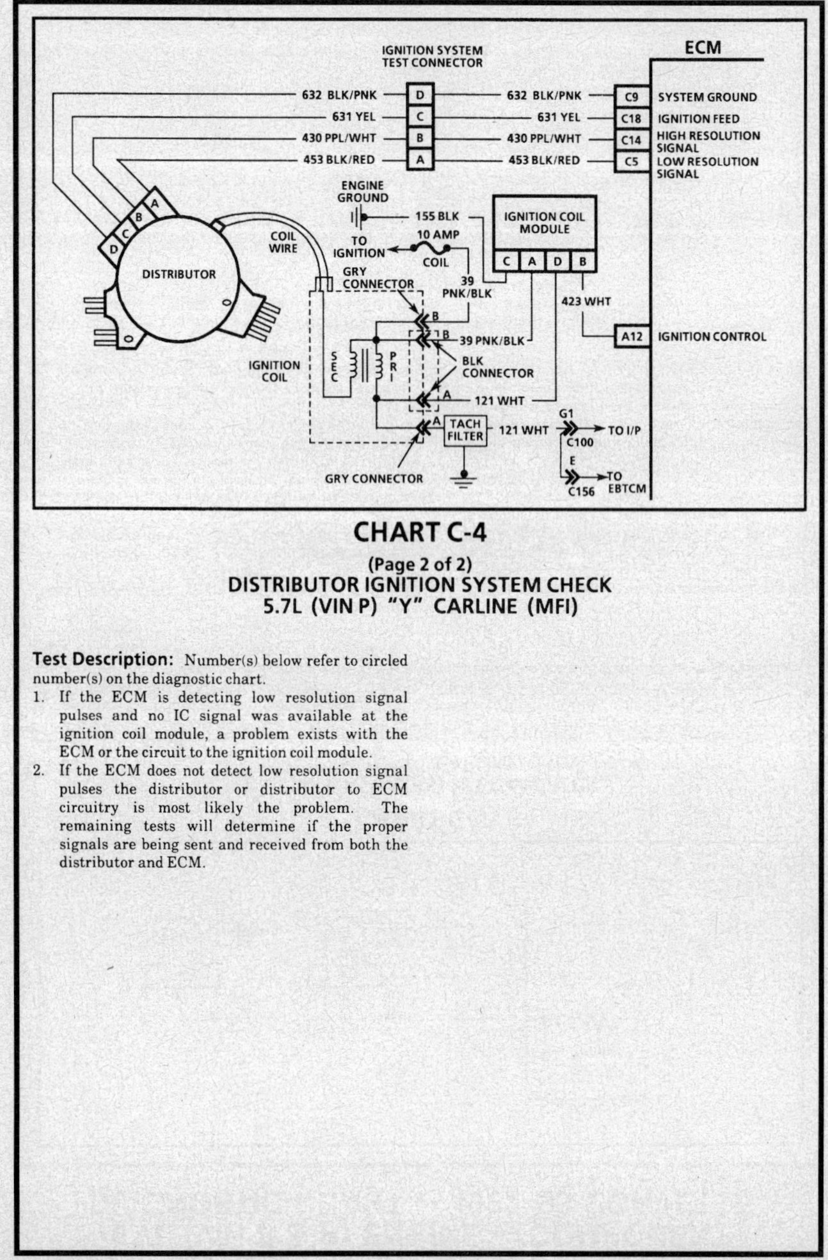

Test Description: Number(s) below refer to circled number(s) on the diagnostic chart.

1. If the ECM is detecting low resolution signal pulses and no IC signal was available at the ignition coil module, a problem exists with the ECM or the circuit to the ignition coil module.

2. If the ECM does not detect low resolution signal pulses the distributor or distributor to ECM circuitry is most likely the problem. The remaining tests will determine if the proper signals are being sent and received from both the distributor and ECM.

5.7L (VIN 8 & P) ENGINE — COMPONENT DIAGNOSTIC CHART — 1992–93 CORVETTE

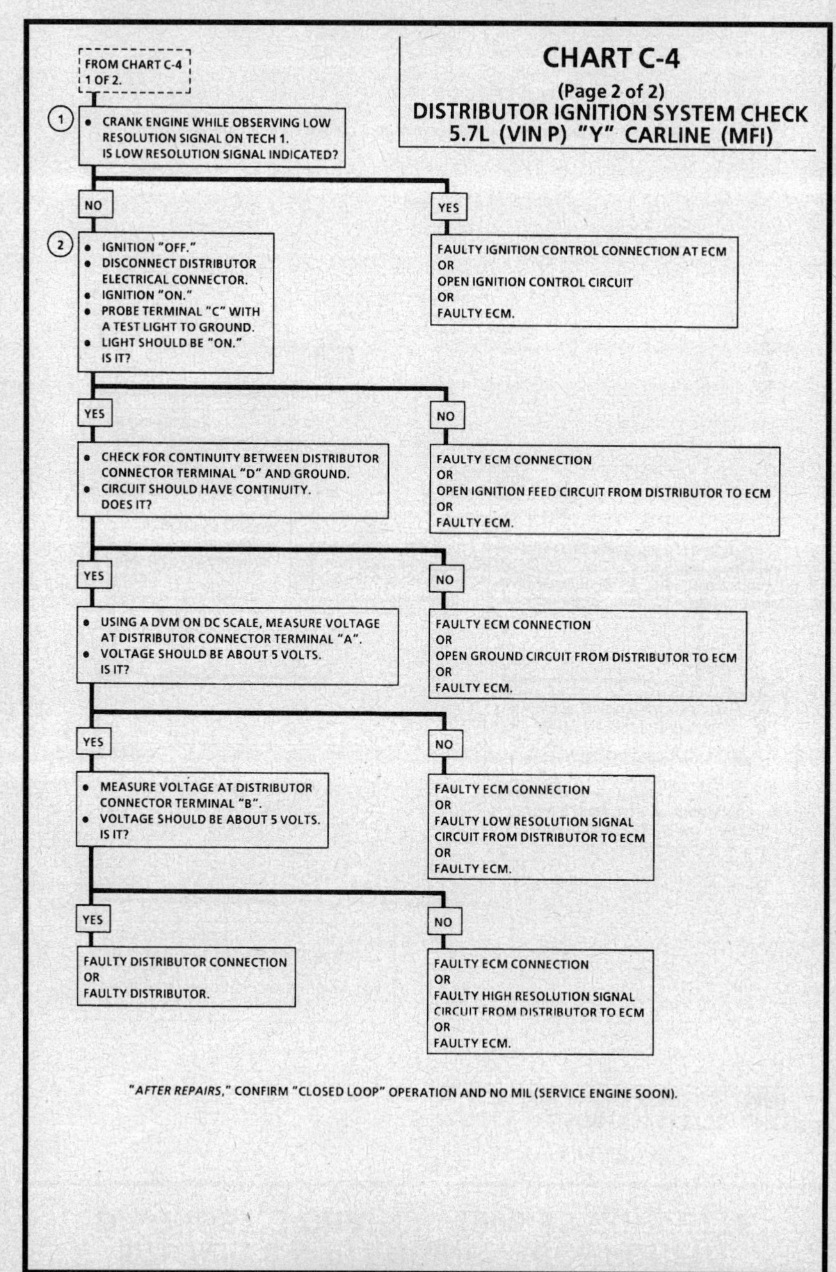

5.7L (VIN 8 & P) ENGINE — COMPONENT DIAGNOSTIC CHART — 1992–93 CORVETTE

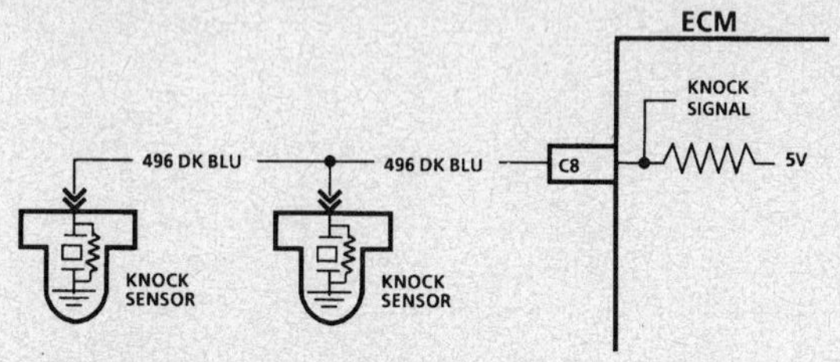

CHART C-5

KNOCK SENSOR (KS) SYSTEM CHECK
5.7L (VIN P) "Y" CARLINE (MFI)

Circuit Description:

The knock sensor system is used to detect engine detonation. The ECM will retard the spark timing based on a signal being received from the knock sensor. The knock sensors produce an AC voltage which is sent to the ECM. The amount of AC voltage produced by the sensors is determined by the amount of engine knock. The circuitry within each knock sensor causes the ECMs 5 volts to be pulled down so that under a no engine knock condition, CKT 496 would measure about 1.5 volts.

Test Description: Number(s) below refer to circled number(s) on the diagnostic chart.

1. With engine idling, there should not be a knock signal present at the ECM, because detonation is not likely under a no load condition.
2. Tapping on the engines right or left exhaust manifold should simulate a knock signal to determine if the sensors are capable of detecting detonation. If no knock is detected, try tapping on engine block closer to sensors before replacing a sensor.
3. If the engine has an internal problem which is creating a knock, the knock sensor may be responding to the internal failure.
4. This test determines if the knock sensor is faulty or if the KS portion of the PROM is faulty. If it is determined that the PROM is faulty, be sure that it is properly installed and latched into place. If not properly installed, repair and retest.

Diagnostic Aids:

While observing knock signal on the Tech 1, there should be an indication that knock is present when detonation can be heard. Detonation is most likely to occur under high engine load conditions.

5.7L (VIN 8 & P) ENGINE — COMPONENT DIAGNOSTIC CHART — 1992–93 CORVETTE

CHART C-5

KNOCK SENSOR (KS) SYSTEM CHECK
5.7L (VIN P) "Y" CARLINE (MFI)

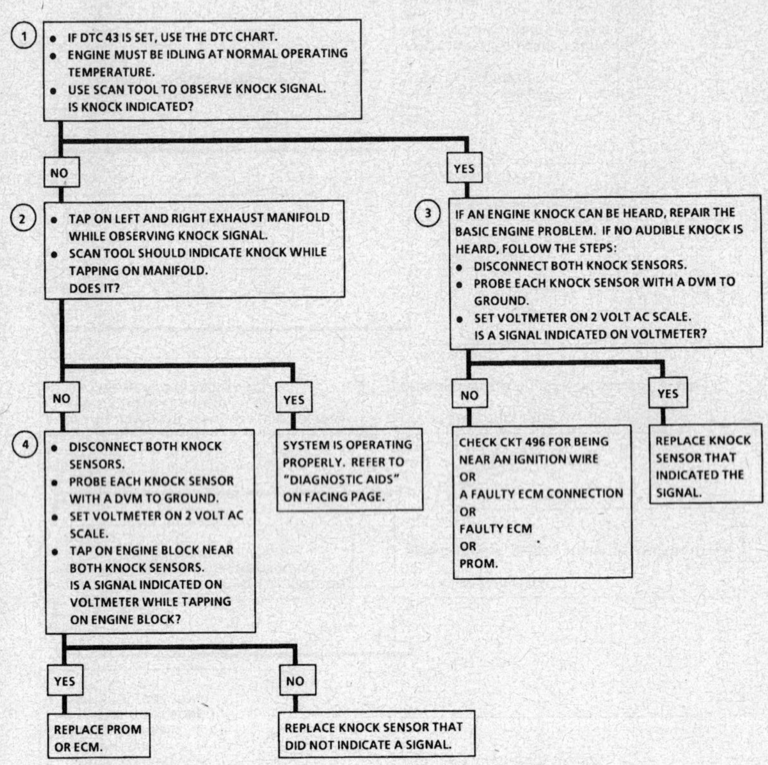

""*AFTER REPAIRS*," CONFIRM "CLOSED LOOP" OPERATION AND NO MIL (SERVICE ENGINE SOON).

5.7L (VIN 8 & P) ENGINE — COMPONENT DIAGNOSTIC CHART — 1992–93 CORVETTE

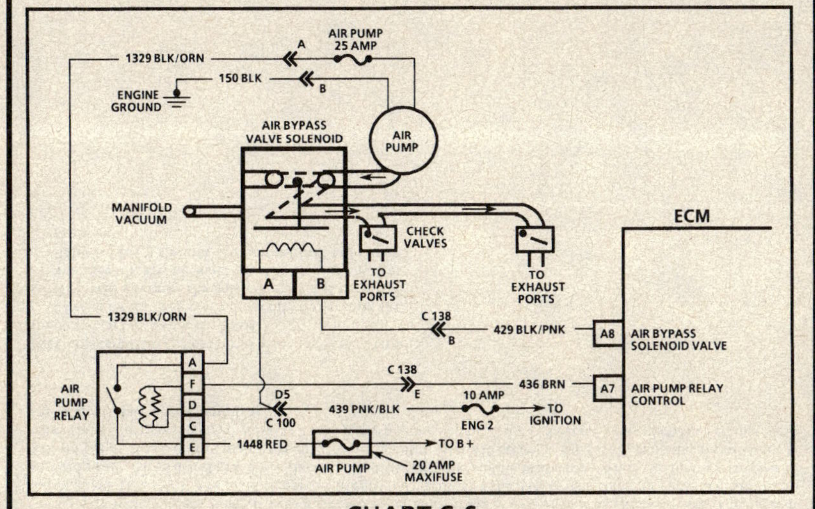

CHART C-6
(Page 1 of 2)
SECONDARY AIR INJECTION (AIR) SYSTEM CHECK
5.7L (VIN P) "Y" CARLINE (MFI)

Circuit Description:

An AIR bypass solenoid valve directs air into the exhaust ports or away from the exhaust ports. During cold starts, above 15°C (59°F), the ECM completes the ground circuit and the AIR bypass solenoid valve and AIR pump are energized. Air is directed to the exhaust ports whenever the engine is started. When the fuel system goes to "Closed Loop," the ECM opens the ground circuit. When the AIR bypass solenoid valve and AIR pump are de-energized, air is directed away from the exhaust ports until the AIR pump stops spinning.

Test Description: Number(s) below refer to circled number(s) on the diagnostic chart.

1. This is a system performance test. If both the AIR pump and AIR bypass valve are functioning properly, the amount of fresh air flowing passed the oxygen sensor will cause a false lean condition (low oxygen voltage). When the output test is enabled, it will only be "ON" for 5 seconds.
2. Test for a grounded AIR bypass valve circuit. The AIR bypass solenoid valve is normally closed and should hold vacuum.
3. Checks for voltage from battery through a fuse to the solenoid valve.
4. Checks for an open control circuit. Using the Tech 1 will energize the AIR bypass solenoid valve if the ECM and circuits are normal. In this step, if the test light is "ON," circuits are normal and the valve is at fault.

5.7L (VIN 8 & P) ENGINE — COMPONENT DIAGNOSTIC CHART — 1992–93 CORVETTE

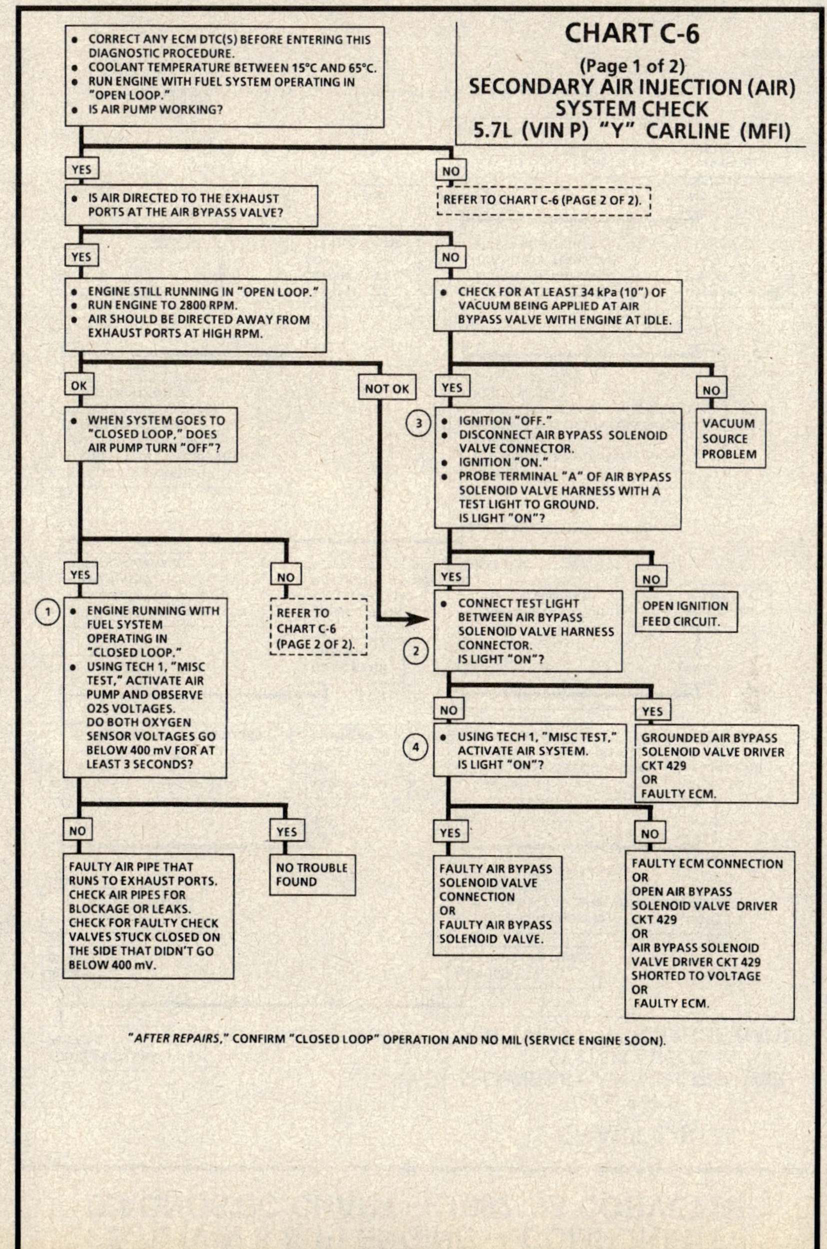

5.7L (VIN 8 & P) ENGINE — COMPONENT DIAGNOSTIC CHART — 1992–93 CORVETTE

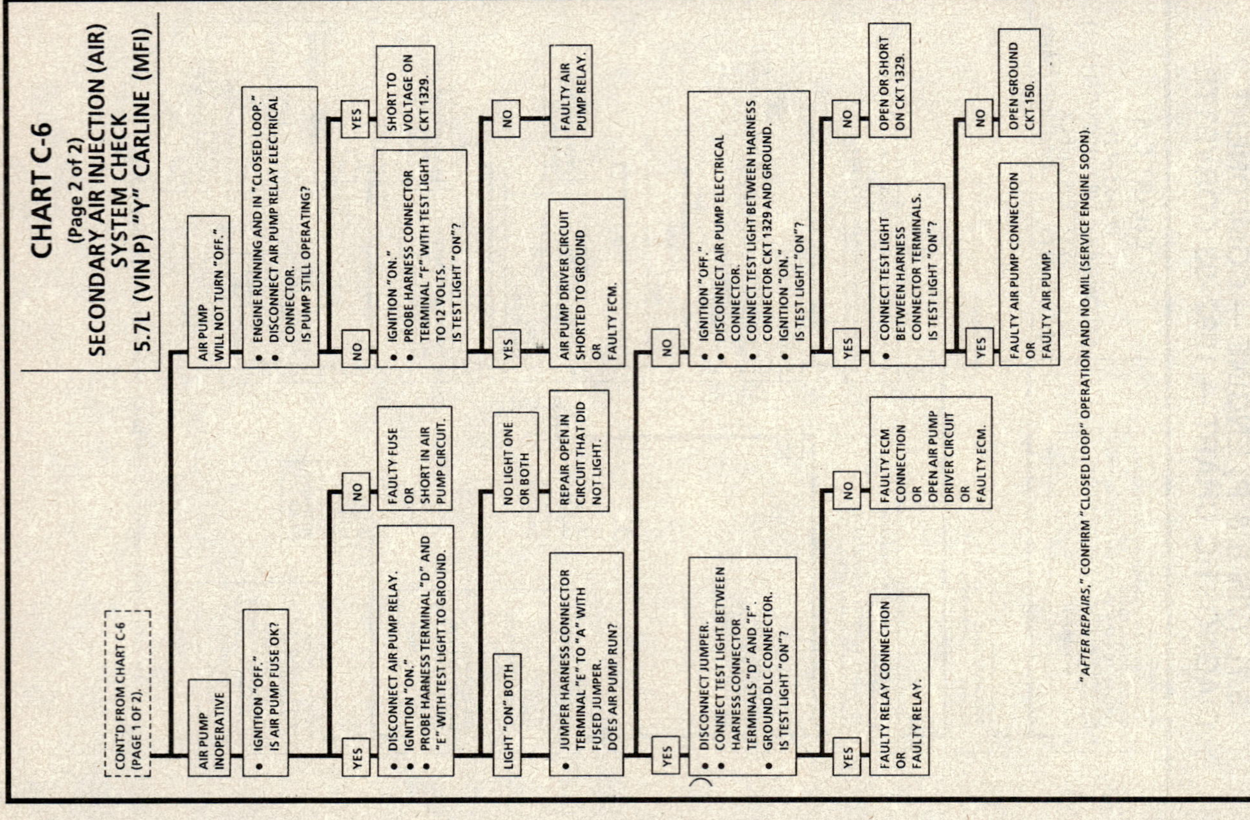

5.7L (VIN 8 & P) ENGINE — COMPONENT DIAGNOSTIC CHART — 1992–93 CORVETTE

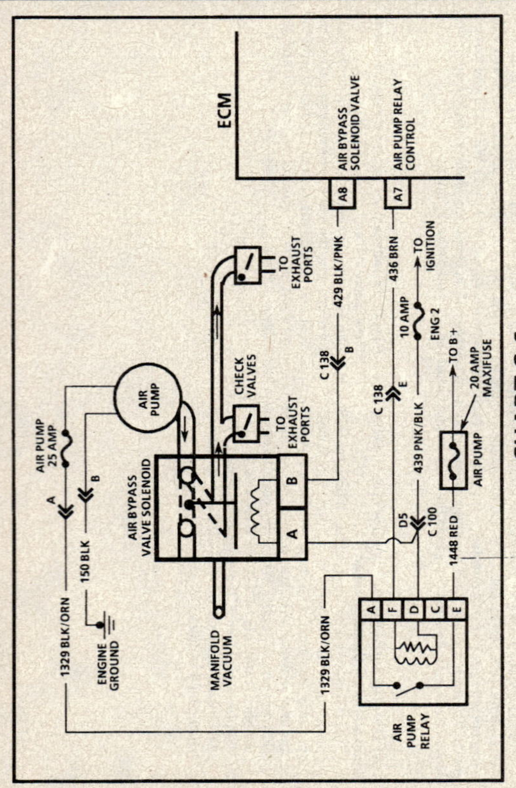

Circuit Description:

An AIR bypass solenoid valve directs air into the exhaust ports or away from the exhaust ports. During cold starts, above 15°C (59°F) the ECM completes the ground circuit and the AIR bypass solenoid valve and AIR pump are energized. Air is directed to the exhaust ports whenever the engine is started. When the fuel system goes to "Closed Loop," or the AIR pump has been "ON" for greater than 240 seconds, the ECM opens the ground circuit. When the AIR bypass solenoid valve and AIR pump are de-energized, air is directed away from the exhaust ports until the AIR pump stops spinning.

Test Description: Number(s) below refer to circled number(s) on the diagnostic chart.

1. Checks for an open control circuit. Grounding the DLC will energize the solenoid valve, if the ECM and circuits are normal. In this step, if the test light is "ON," circuits are normal and fault is in the relay.

5.7L (VIN 8 & P) ENGINE — COMPONENT DIAGNOSTIC CHART — 1992–93 CORVETTE

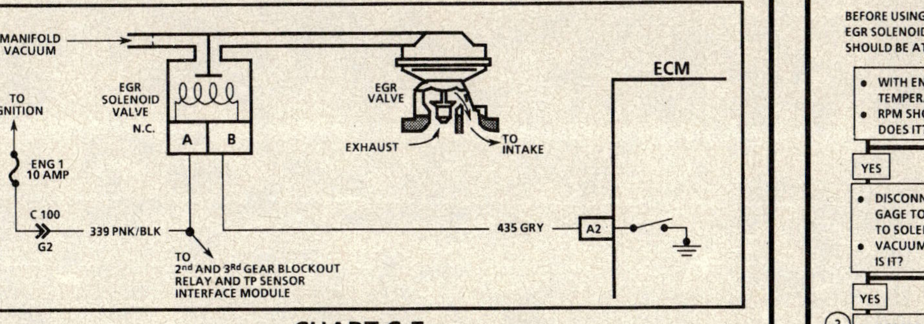

CHART C-7
(Page 1 of 2)
**EXHAUST GAS RECIRCULATION (EGR) SYSTEM CHECK
5.7L (VIN P) "Y" CARLINE (MFI)**

Circuit Description:

The Exhaust Gas Recirculation (EGR) valve is controlled by a normally closed solenoid valve. The ECM turns the solenoid "ON" to allow vacuum to pass to the EGR, and turns the solenoid "OFF" to prevent EGR operation.

The duty cycle is calculated by the ECM based on information from the ECT, IAT, TP sensor, and MAP sensors. Also, engine RPM and the PNP switch input affect EGR. There should be no EGR when in park or neutral, TP sensor below a calibrated value, or TP sensor indicating WOT.

With the ignition "ON" and engine "OFF," the EGR solenoid valve is de-energized. The solenoid valve, however, should be energized, if the diagnostic "test" terminal is grounded with the ignition "ON" and engine "OFF."

Test Description: Number(s) below refer to circled number(s) on the diagnostic chart.
1. **Intake Passage:** Shut "OFF" engine and remove the EGR valve from the manifold. Plug the exhaust side hole with a shop rag or suitable stopper. Leaving the intake side hole open, attempt to start the engine. If the engine runs at a very high idle (up to 3000 RPM is possible) or starts and stalls, the EGR passages are not restricted. If the engine starts and idles normally, the EGR intake side passage in the intake manifold is restricted.
  Exhaust Passage: With EGR valve still removed, plug the intake side hole with a suitable stopper. With the exhaust side hole open, check for the presence of exhaust gas. If no exhaust gas is present, the EGR exhaust side passage in the intake manifold is restricted.
2. The vacuum at the gage may or may not <u>slowly</u> bleed off. It is important that the gage is able to read the amount of vacuum being applied.
3. When the diagnostic "test" terminal is grounded, the vacuum gage should bleed off through a vent in the solenoid valve. The pump gage may or may not bleed off but this does not indicate a problem.
4. This test will determine if the electrical control part of the system is at fault or if the connector or solenoid valve is at fault.

Diagnostic Aids:

Vacuum lines should be thoroughly checked for proper routing. Refer to "Vehicle Emission Control Information" label.

Suction from shop exhaust hoses can alter backpressure and may affect the functional check of the EGR valve.

5.7L (VIN 8 & P) ENGINE — COMPONENT DIAGNOSTIC CHART — 1992–93 CORVETTE

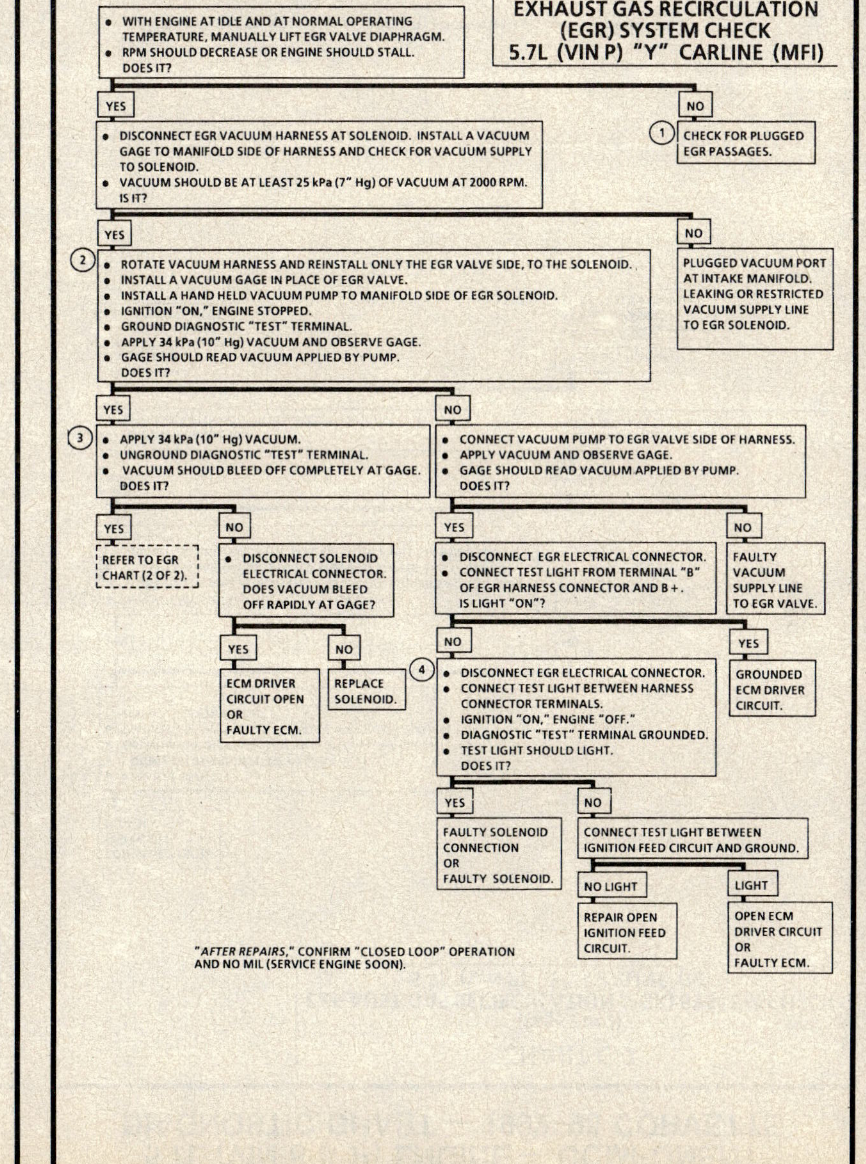

5.7L (VIN 8 & P) ENGINE — COMPONENT DIAGNOSTIC CHART — 1992–93 CORVETTE

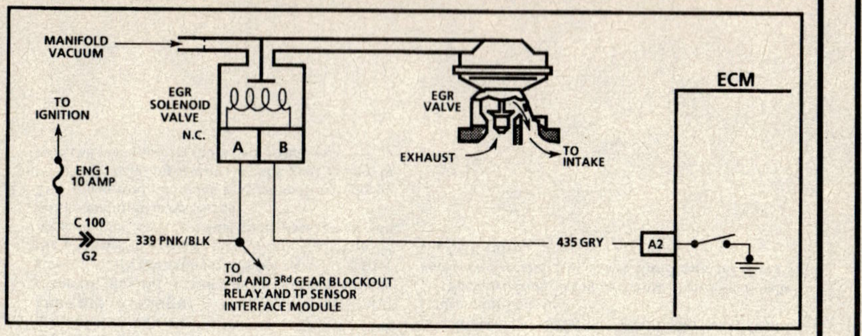

CHART C-7

(Page 2 of 2)
EXHAUST GAS RECIRCULATION (EGR) SYSTEM CHECK
5.7L (VIN P) "Y" CARLINE (MFI)

Circuit Description:

The Exhaust Gas Recirculation (EGR) valve is controlled by a normally closed solenoid valve. The ECM turns the solenoid "ON" to allow vacuum to pass to the EGR, and turns the solenoid "OFF" to prevent EGR operation.

The duty cycle is calculated by the ECM based on information from the TP sensor, ECT, IAT, and MAP sensors. Also, engine RPM and the PNP switch input affect EGR. There should be no EGR when in park or neutral, TP sensor below a calibrated value, or TP sensor indicating WOT.

With the ignition "ON" and engine "OFF," the EGR solenoid valve is de-energized. The solenoid valve, however, should be energized, if the diagnostic "test" terminal is grounded with the ignition "ON" and engine "OFF."

Test Description: Number(s) below refer to circled number(s) on the diagnostic chart.

1. The remaining test checks the ability of the EGR valve to interact with the exhaust system. This system uses a negative backpressure EGR valve which should hold vacuum with engine "OFF."
2. When engine is started, exhaust backpressure at the base of the EGR valve should open the valve's internal bleed and vent the applied vacuum allowing the valve to seat.

Diagnostic Aids:

Suction from shop exhaust hoses can alter backpressure and may affect the functional check of the EGR valve.

During normal EGR valve operation, the movement of the EGR pintle is small. It is important to determine whether the valve pintle moves and not how much it moves.

5.7L (VIN 8 & P) ENGINE — COMPONENT DIAGNOSTIC CHART — 1992–93 CORVETTE

CHART C-7

(Page 2 of 2)
EXHAUST GAS RECIRCULATION (EGR) SYSTEM CHECK
5.7L (VIN P) "Y" CARLINE (MFI)

CONTINUED FROM EGR CHART (1 OF 2).

1. • IGNITION "OFF."
   • CONNECT A VACUUM PUMP TO EGR VALVE.
   • OBSERVE EGR DIAPHRAGM WHILE APPLYING VACUUM.
   • DIAPHRAGM SHOULD MOVE FREELY AND HOLD VACUUM FOR AT LEAST 20 SECONDS.
   DOES IT?

→ YES

→ NO → REPLACE EGR VALVE.

2. • APPLY 34 kPa (10" Hg) VACUUM TO EGR VALVE.
   • START ENGINE AND IMMEDIATELY OBSERVE GAGE ON VACUUM PUMP.
   • EGR VALVE DIAPHRAGM SHOULD MOVE TO SEATED POSITION AND VACUUM SHOULD DROP FROM PUMP GAGE WHILE STARTING ENGINE.
   DOES IT?

→ NO
   • REMOVE EGR VALVE.
   • CHECK FOR PLUGGED OR RESTRICTED EXHAUST PASSAGES.

PASSAGES OK → REPLACE EGR VALVE.

→ YES
   EGR CIRCUIT IS OPERATING PROPERLY. REFER TO SYMPTOMS

PASSAGES NOT OK → • CLEAN PASSAGES.
   • RE-CHECK EGR VALVE.

"AFTER REPAIRS," CONFIRM "CLOSED LOOP" OPERATION AND NO MIL (SERVICE ENGINE SOON).

5.7L (VIN 8 & P) ENGINE — COMPONENT DIAGNOSTIC CHART — 1992–93 CORVETTE

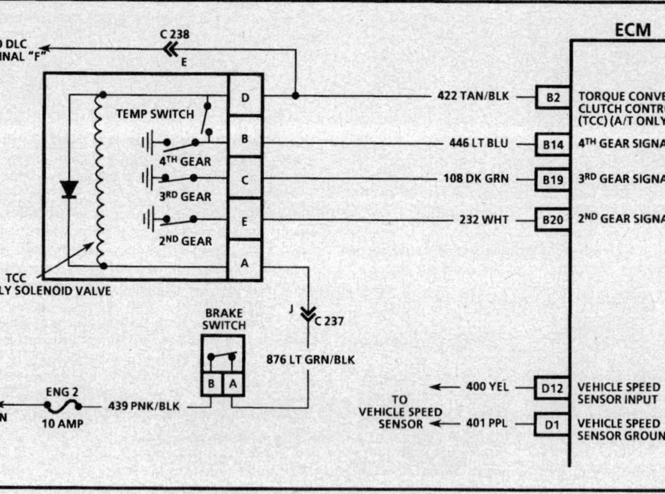

CHART C-8A (Page 1 of 4)
TORQUE CONVERTER CLUTCH (TCC) SYSTEM
(ELECTRICAL DIAGNOSIS)
5.7L (VIN P) "Y" CARLINE (MFI)

Circuit Description:

The purpose of the automatic transmission Torque Converter Clutch (TCC) feature is to eliminate the power loss of the torque converter stage, when the vehicle is in a cruise condition. This allows the convenience of the automatic transmission and the fuel economy of a manual transmission. The heart of the system is a solenoid, located inside the automatic transmission, which is controlled by the ECM.

When the solenoid coil is activated "ON," the torque converter clutch is applied, which results in 100% mechanical coupling from the engine to transmission. When the transmission solenoid is deactivated, the torque converter clutch is released which allows the torque converter to operate in the conventional manner (fluid coupling between engine and transmission).

The ECM turns "ON" the TCC when coolant temperature is above 65°C (149°F), TP sensor not changing, and vehicle speed above a specified value.

Test Description: Number(s) below refer to circled number(s) on the diagnostic chart.

1. Confirms 12 volt supply as well as continuity of TCC circuit.
2. Grounding the diagnostic "test" terminal with engine "OFF," should energize the TCC solenoid.
3. TCC solenoid coil resistance must measure more than 20 ohms. Using an ohmmeter, check the solenoid coil resistance of all ECM controlled solenoids and relays before installing a replacement ECM. Replace any solenoid or relay that measures less than 20 ohms.

Diagnostic Aids:

A Tech 1 only indicates when the ECM has turned "ON" the TCC driver (grounded CKT 422) but this does not confirm that the TCC has engaged. To determine if the TCC is functioning properly, engine RPM should decrease when the Tech 1 indicates the TCC driver has turned "ON." To determine if the gear switches are functioning properly, perform the checks in CHART C-8A. The switches will not prevent the TCC from functioning, but will affect TCC lock and unlock points. If the 4th gear switch circuit is always open, the TCC may engage as soon as sufficient oil pressure is reached.

5.7L (VIN 8 & P) ENGINE — COMPONENT DIAGNOSTIC CHART — 1992–93 CORVETTE

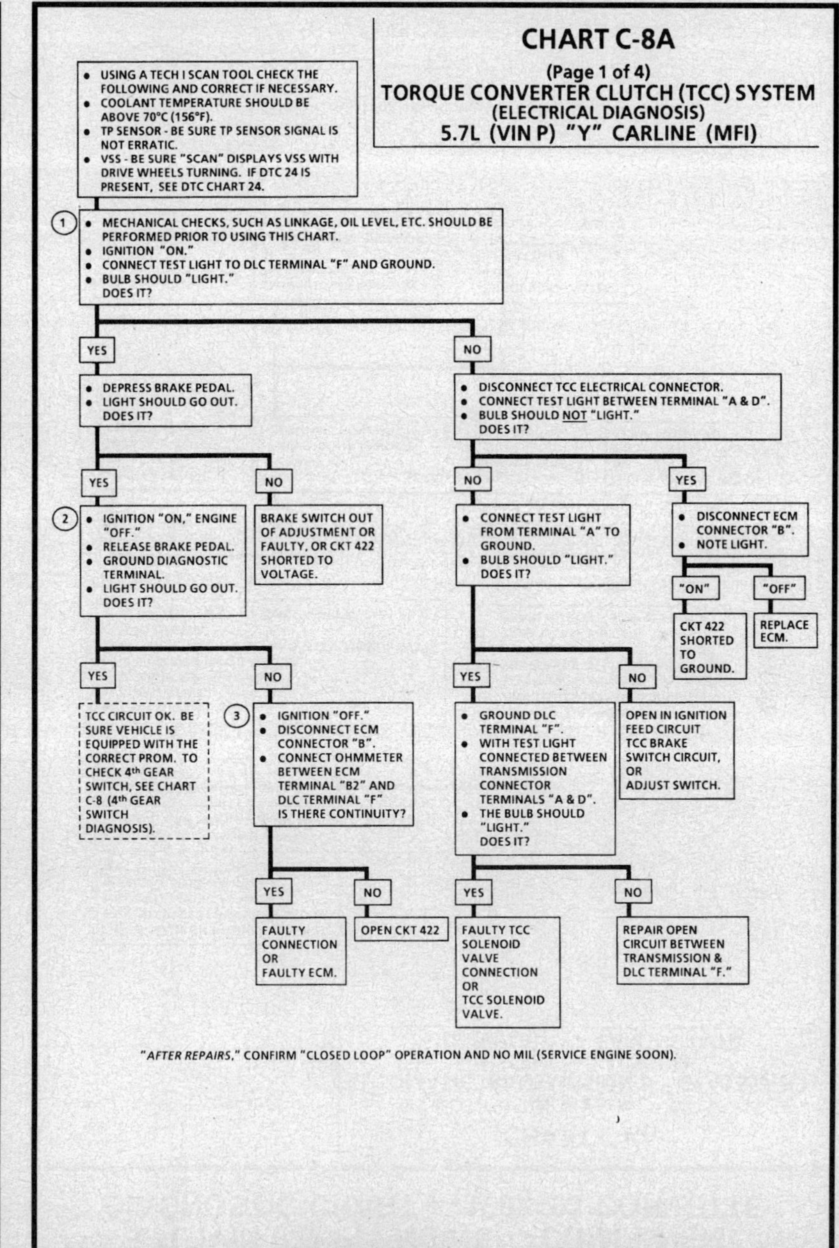

5.7L (VIN 8 & P) ENGINE — COMPONENT DIAGNOSTIC CHART — 1992–93 CORVETTE

CHART C-8A
(Page 2 of 4)
AUTOMATIC TRANSMISSION 2nd GEAR SWITCH
(ELECTRICAL DIAGNOSIS)
5.7L (VIN P) "Y" CARLINE (MFI)

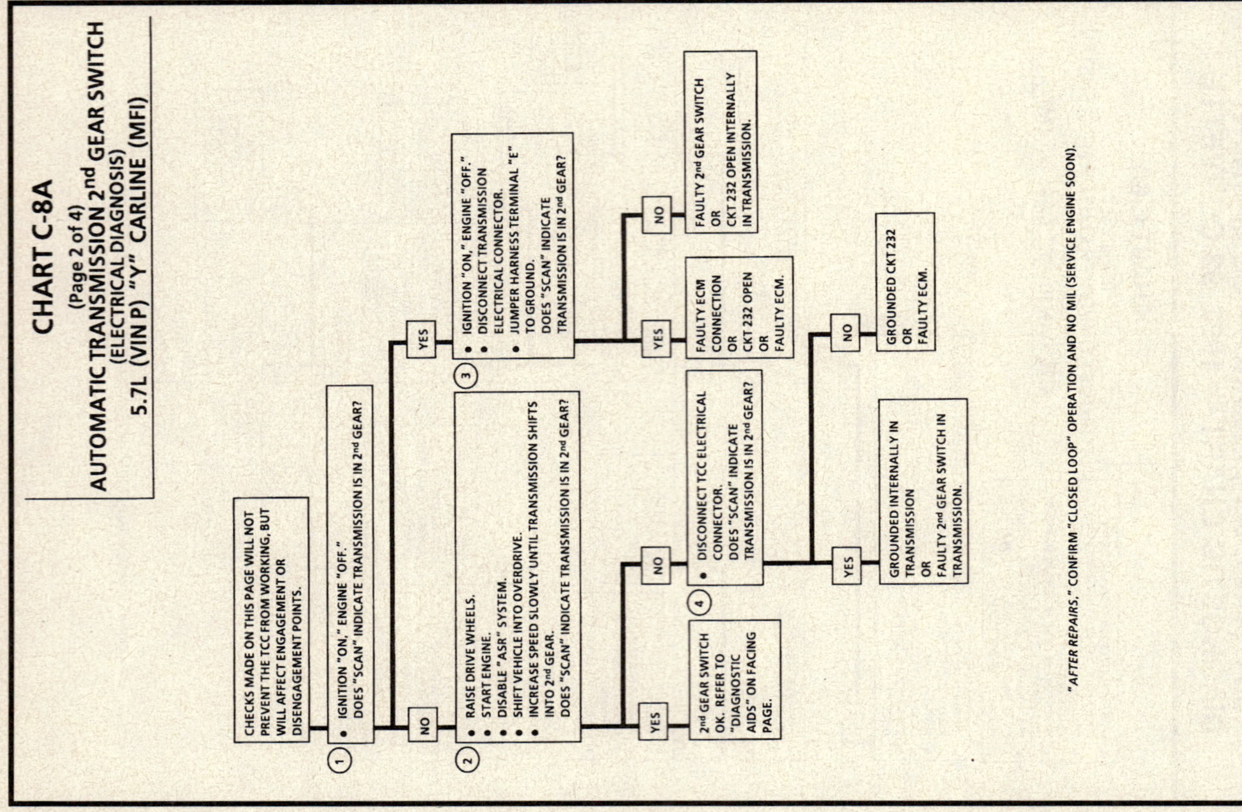

5.7L (VIN 8 & P) ENGINE — COMPONENT DIAGNOSTIC CHART — 1992–93 CORVETTE

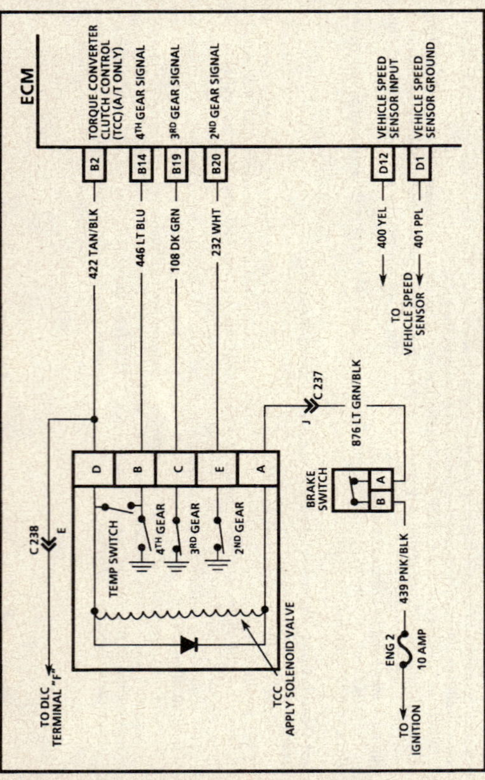

CHART C-8A
(Page 2 of 4)
AUTOMATIC TRANSMISSION 2nd GEAR SWITCH
(ELECTRICAL DIAGNOSIS)
5.7L (VIN P) "Y" CARLINE (MFI)

Circuit Description:

A 2nd gear switch (mounted in the transmission) is normally closed and opens when the transmission shifts into 2nd gear. The ECM uses gear switch signals to adjust ignition timing based on transmission shift points. Adjusting ignition timing after a transmission shift will allow for higher shift speeds.

Test Description: Number(s) below refer to circled number(s) on the diagnostic chart.

1. Unless the switch is stuck open or CKT 232 is open, the Tech 1 should display "NO," indicating the transmission is not in 2nd gear. The 2nd gear switch should only be open while in 2nd gear.

2. The "ASR" system must be disabled when performing this step. Whenever the ignition key is cycled to the "OFF" position and then cycled back "ON," the "ASR" system will default "ON." This checks the operation of the 2nd gear switch. When the transmission shifts into 2nd gear the switch should open and the Tech 1 should display "YES."

3. This step determines if the ECM and wiring are OK. Grounding CKT 232 should cause the Tech 1 to display "YES," indicating the transmission is in 2nd gear.

4. Disconnecting the TCC connector simulates a open switch to determine if CKT 232 is shorted to ground or the problem is in the transmission.

Diagnostic Aids:

A faulty 2nd or 3rd gear switch circuit can cause a DTC 72.

If the 2nd gear switch is changing states, check connections and wire routing carefully.

5.7L (VIN 8 & P) ENGINE — COMPONENT DIAGNOSTIC CHART — 1992–93 CORVETTE

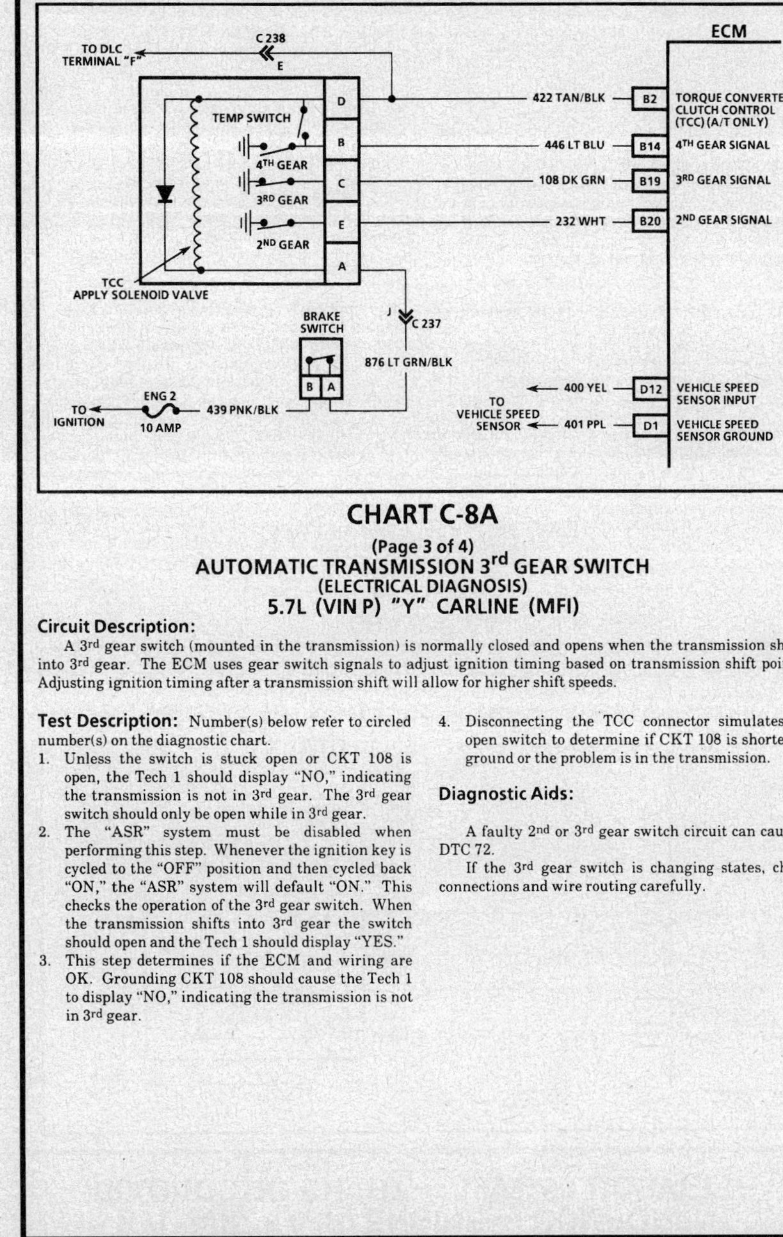

CHART C-8A

(Page 3 of 4)
AUTOMATIC TRANSMISSION 3rd GEAR SWITCH
(ELECTRICAL DIAGNOSIS)
5.7L (VIN P) "Y" CARLINE (MFI)

Circuit Description:

A 3rd gear switch (mounted in the transmission) is normally closed and opens when the transmission shifts into 3rd gear. The ECM uses gear switch signals to adjust ignition timing based on transmission shift points. Adjusting ignition timing after a transmission shift will allow for higher shift speeds.

Test Description: Number(s) below refer to circled number(s) on the diagnostic chart.

1. Unless the switch is stuck open or CKT 108 is open, the Tech 1 should display "NO," indicating the transmission is not in 3rd gear. The 3rd gear switch should only be open while in 3rd gear.
2. The "ASR" system must be disabled when performing this step. Whenever the ignition key is cycled to the "OFF" position and then cycled back "ON," the "ASR" system will default "ON." This checks the operation of the 3rd gear switch. When the transmission shifts into 3rd gear the switch should open and the Tech 1 should display "YES."
3. This step determines if the ECM and wiring are OK. Grounding CKT 108 should cause the Tech 1 to display "NO," indicating the transmission is not in 3rd gear.
4. Disconnecting the TCC connector simulates an open switch to determine if CKT 108 is shorted to ground or the problem is in the transmission.

Diagnostic Aids:

A faulty 2nd or 3rd gear switch circuit can cause a DTC 72.

If the 3rd gear switch is changing states, check connections and wire routing carefully.

5.7L (VIN 8 & P) ENGINE — COMPONENT DIAGNOSTIC CHART — 1992–93 CORVETTE

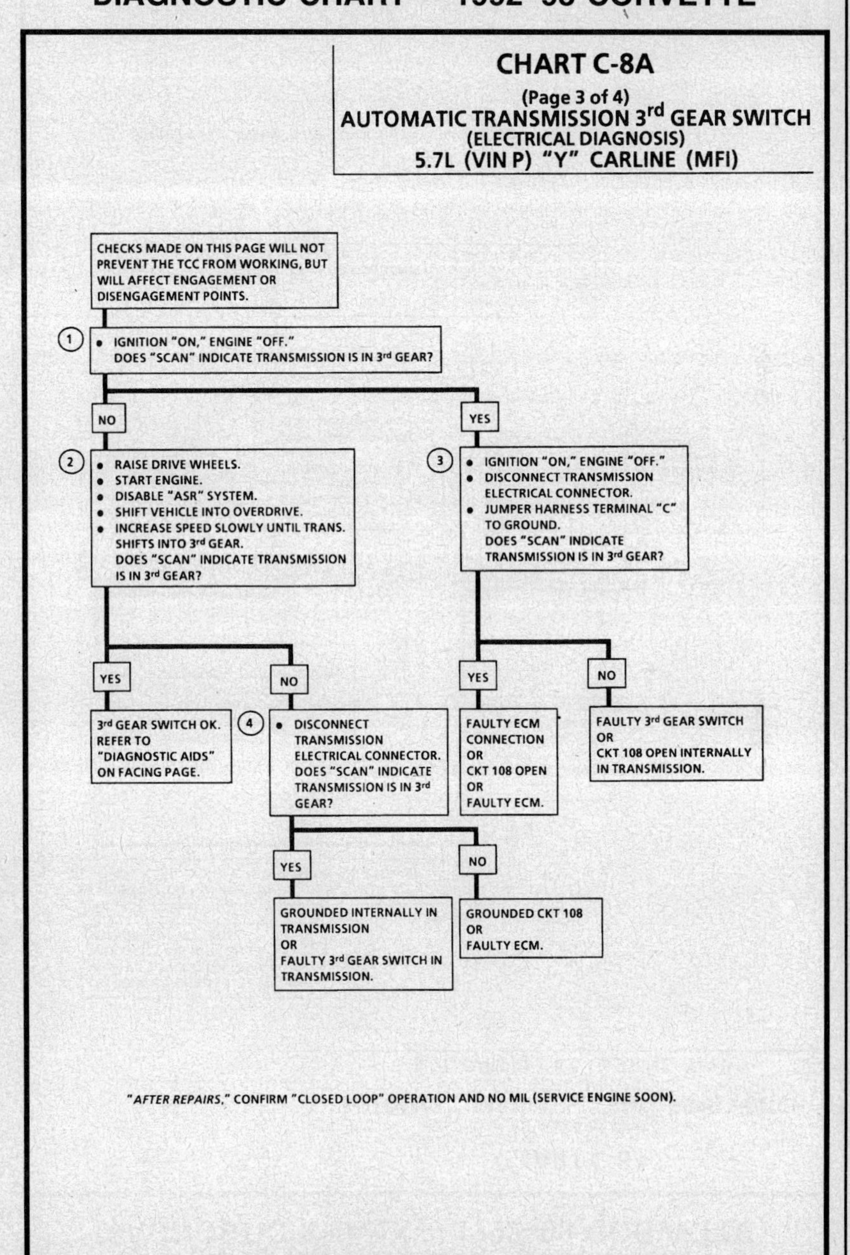

5.7L (VIN 8 & P) ENGINE — COMPONENT DIAGNOSTIC CHART — 1992-93 CORVETTE

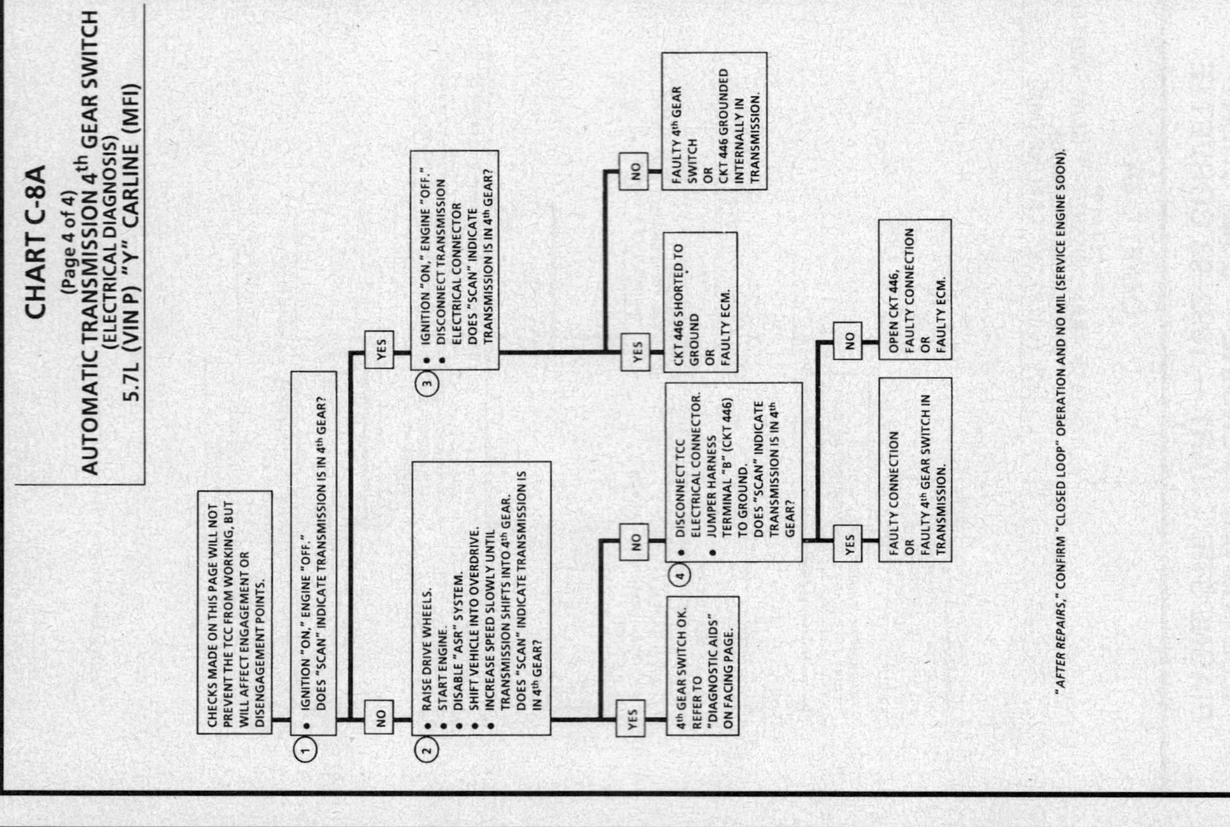

5.7L (VIN 8 & P) ENGINE — COMPONENT DIAGNOSTIC CHART — 1992-93 CORVETTE

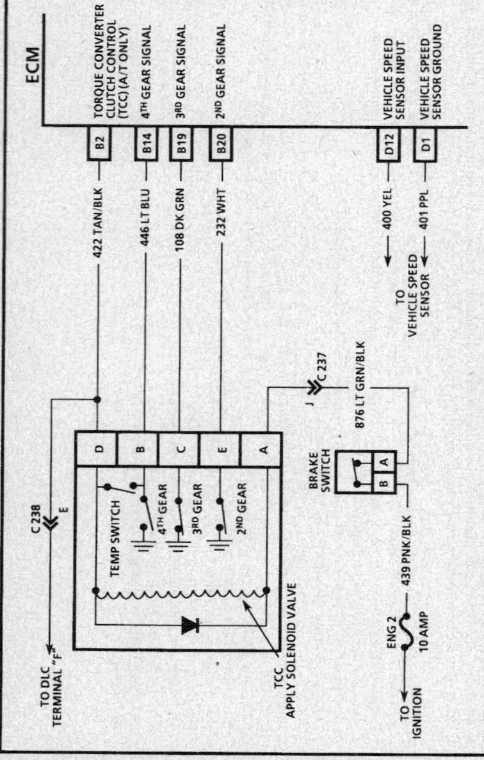

Circuit Description:

A 4th gear switch (mounted in the transmission) is normally open and closes when the transmission shifts into 4th gear. This switch is used by the ECM to modify TCC lock and unlock points, when in a 4-3 downshift maneuver. The temperature switch is normally open. Transmission must be in 4th gear for the temperature switch to apply TCC to cool transmission fluid. Temperature switch locks at about 155°C (279°F) and unlocks about 143°C (259°F).

Test Description: Number(s) below refer to circled number(s) on the diagnostic chart.

1. Unless the switch is stuck closed or CKT 446 is shorted to ground, the Tech 1 should display "NO," indicating the transmission is not in 4th gear. The 4th gear switch should only be closed while in 4th gear.

2. The "ASR" system must be disabled when performing this step. Whenever the ignition key is cycled to the "OFF" position and then cycled back "ON," the "ASR" system will default "ON." This checks the operation of the 4th gear switch. When the transmission shifts into 4th gear the switch should close and the Tech 1 should display "YES."

3. Disconnecting the TCC connector simulates an open switch to determine if CKT 446 is shorted to ground or the problem is in the transmission.

4. This step determines if the ECM and wiring are OK. Grounding CKT 446 should cause the Tech 1 to display "YES," indicating the transmission is in 4th gear.

Diagnostic Aids:

A road test may be necessary to verify the customer complaint. If the Tech 1 indicates TCC is turning "ON" and "OFF" erratically, check the state of the 4th gear switch to be sure they're not changing states under a steady throttle position. If the switch is changing states, check connections and wire routing carefully. Also, if the 4th gear switch is always closed, the TCC may engage as soon as sufficient oil pressure is reached.

5.7L (VIN 8 & P) ENGINE — COMPONENT DIAGNOSTIC CHART — 1992–93 CORVETTE

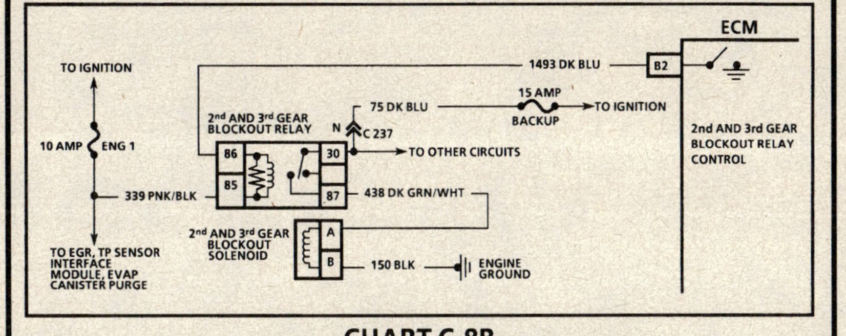

CHART C-8B
MANUAL TRANSMISSION 1-4 UPSHIFT SYSTEM
(ELECTRICAL DIAGNOSIS)
5.7L (VIN P) "Y" CARLINE (MFI)

Circuit Description:

The 2nd & 3rd gear blockout relay is energized when the ECM grounds CKT 1493, which provides battery voltage to the 2nd & 3rd gear blockout solenoid and activates it. When the 2nd & 3rd gear blockout solenoid (located in the transmission) is activated, it mechanically locks out the transmission linkage from being shifted from 1st gear to any gear other than 4th gear. Under certain low load conditions the vehicle can only be shifted from 1st to 4th gear to improve fuel economy.

The 2nd & 3rd gear blockout solenoid will be disabled when BARO pressure drops below 76 kPa and will not be re-enabled until BARO pressure is above 78 kPa.

Test Description: Number(s) below refer to circled number(s) on the diagnostic chart.
1. A mechanical transmission problem is indicated if the transmission can only be put in 1st or 4th gear in this step.
2. Selecting "Field Service Mode" on the Tech 1 or grounding the diagnostic "test" terminal with the ignition "ON" and engine not running, will energize the 2nd & 3rd gear blockout relay and solenoid indicating normal system operation.

NOTICE: Do Not keep the diagnostic "test" terminal grounded for more than 1 minute, as damage to the 2nd & 3rd gear blockout solenoid may occur.
3. If the fuse for CKT 75 is open, check CKT 438 for a short to ground.
4. Raise the vehicle enough to gain access to the 2nd & 3rd gear blockout solenoid harness connector and still be able to turn the ignition "ON."

5.7L (VIN 8 & P) ENGINE — COMPONENT DIAGNOSTIC CHART — 1992–93 CORVETTE

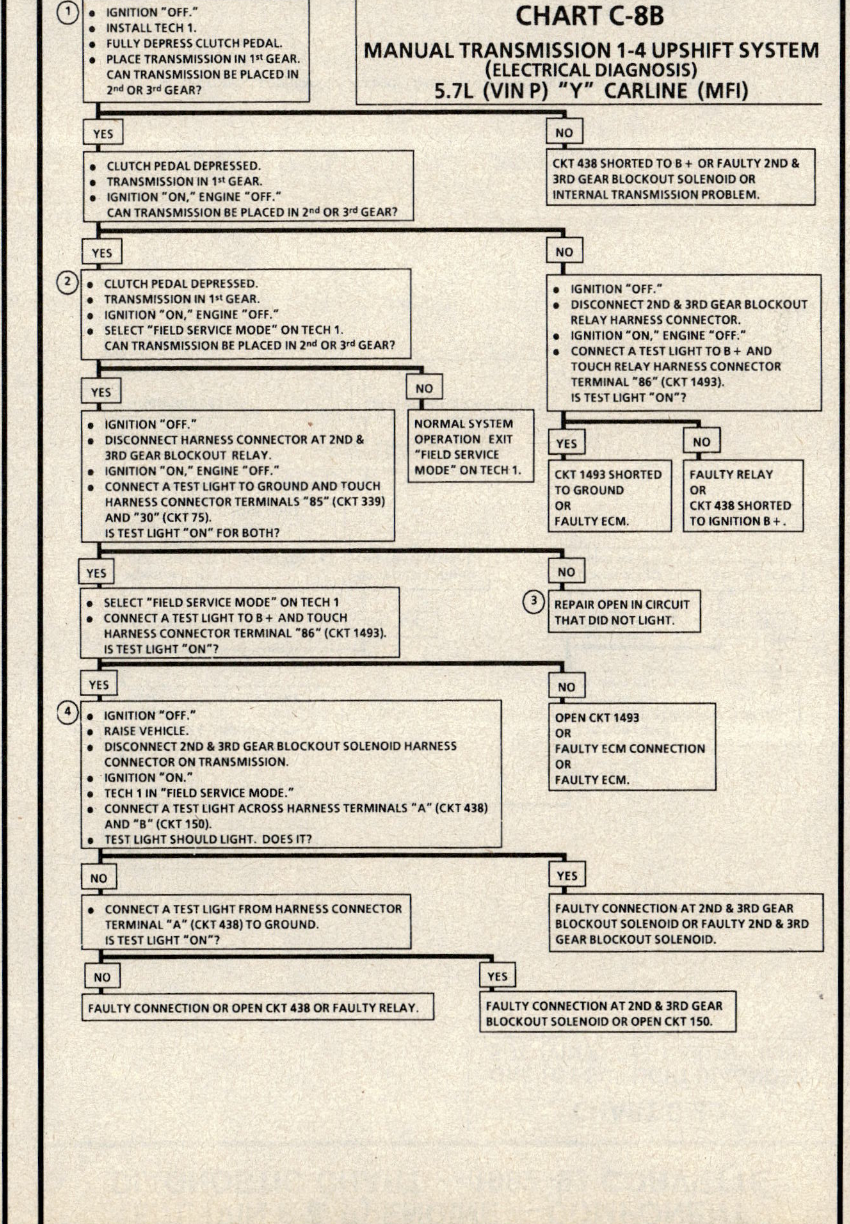

5.7L (VIN 8 & P) ENGINE — COMPONENT DIAGNOSTIC CHART — 1992–93 CORVETTE

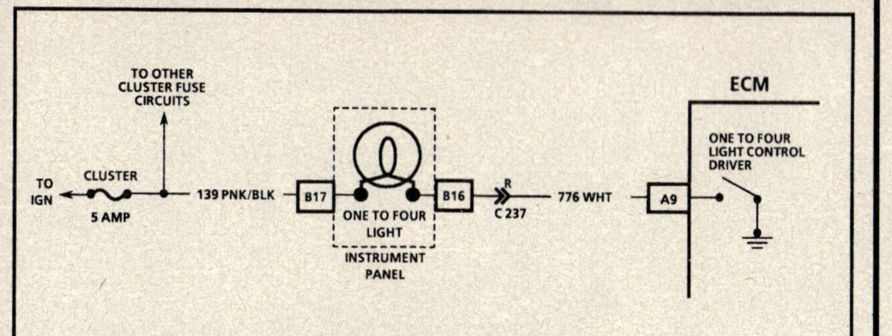

CHART C-8D

**ONE TO FOUR LIGHT DIAGNOSIS
5.7L (VIN P) "Y" CARLINE (MFI)**

Circuit Description:

The one to four light display is in the instrument panel. The purpose of the one to four light is to provide a display to the driver, which indicates the transmission should be shifted from one to four for maximum fuel economy. The light is controlled by the ECM and is turned "ON" by grounding CKT 776.

The one to four light will be enabled if all of the following conditions are met:

- Engine running.
- Vehicle speed greater than 12 mph but less than 19 mph.
- Engine speed less than 1200 RPM.
- Throttle position less than 35%.
- Engine coolant temperature greater than 50°C (122°F).

The ECM uses the measured RPM and the vehicle speed to calculate what gear the vehicle is in. It is this calculation that determines when the one to four light should be turned "ON."

Test Description: Number(s) below refer to circled number(s) on the diagnostic chart.

1. This should not turn "ON" the one to four light. If the light is "ON," there is a short to ground in CKT 776 wiring or a fault in the ECM.
2. When the diagnostic "test" terminal is grounded, the ECM should ground CKT 776 and the one to four light should come "ON."
3. This checks the one to four light circuit up to the ECM connector. If the one to four light illuminates, then the ECM connector is faulty or the ECM does not have the ability to ground the circuit.

5.7L (VIN 8 & P) ENGINE — COMPONENT DIAGNOSTIC CHART — 1992–93 CORVETTE

CHART C-8D

**ONE TO FOUR LIGHT DIAGNOSIS
5.7L (VIN P) "Y" CARLINE (MFI)**

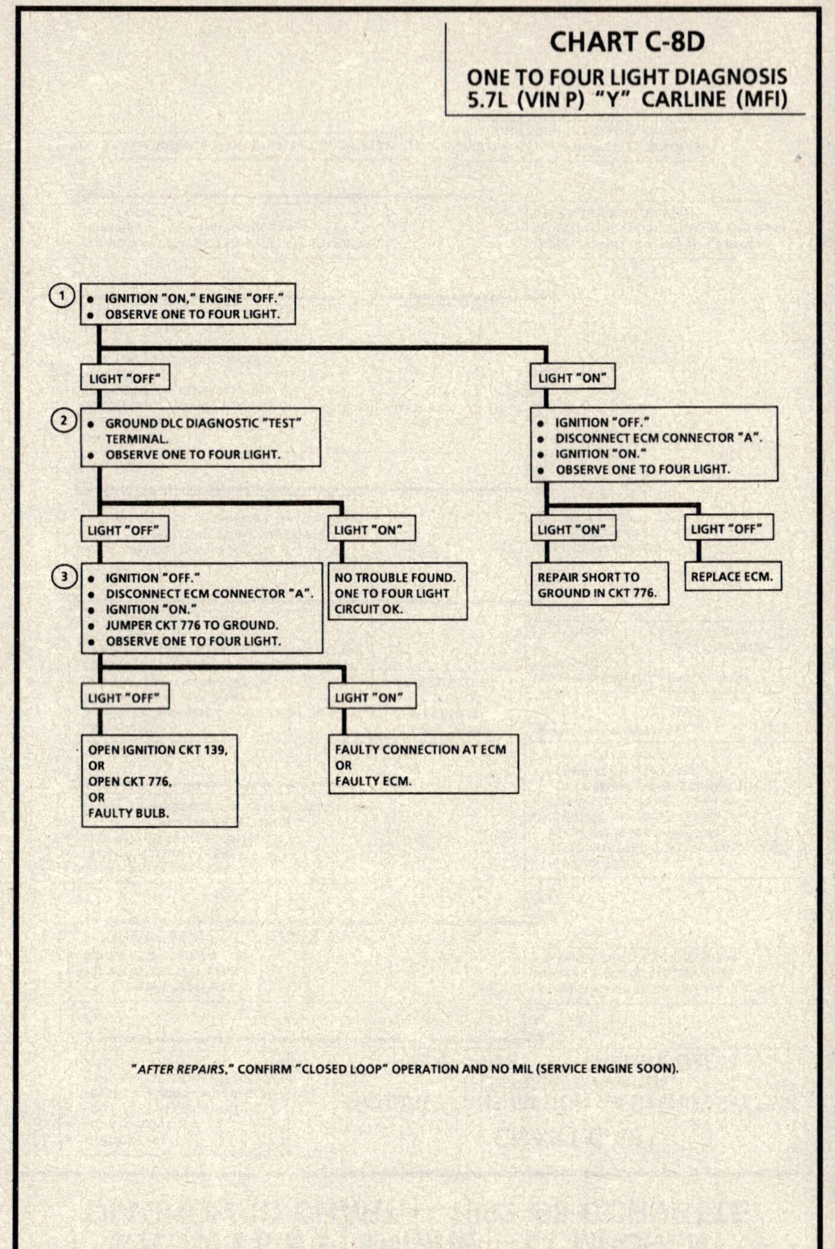

"AFTER REPAIRS," CONFIRM "CLOSED LOOP" OPERATION AND NO MIL (SERVICE ENGINE SOON).

5.7L (VIN 8 & P) ENGINE — COMPONENT DIAGNOSTIC CHART — 1993 CORVETTE

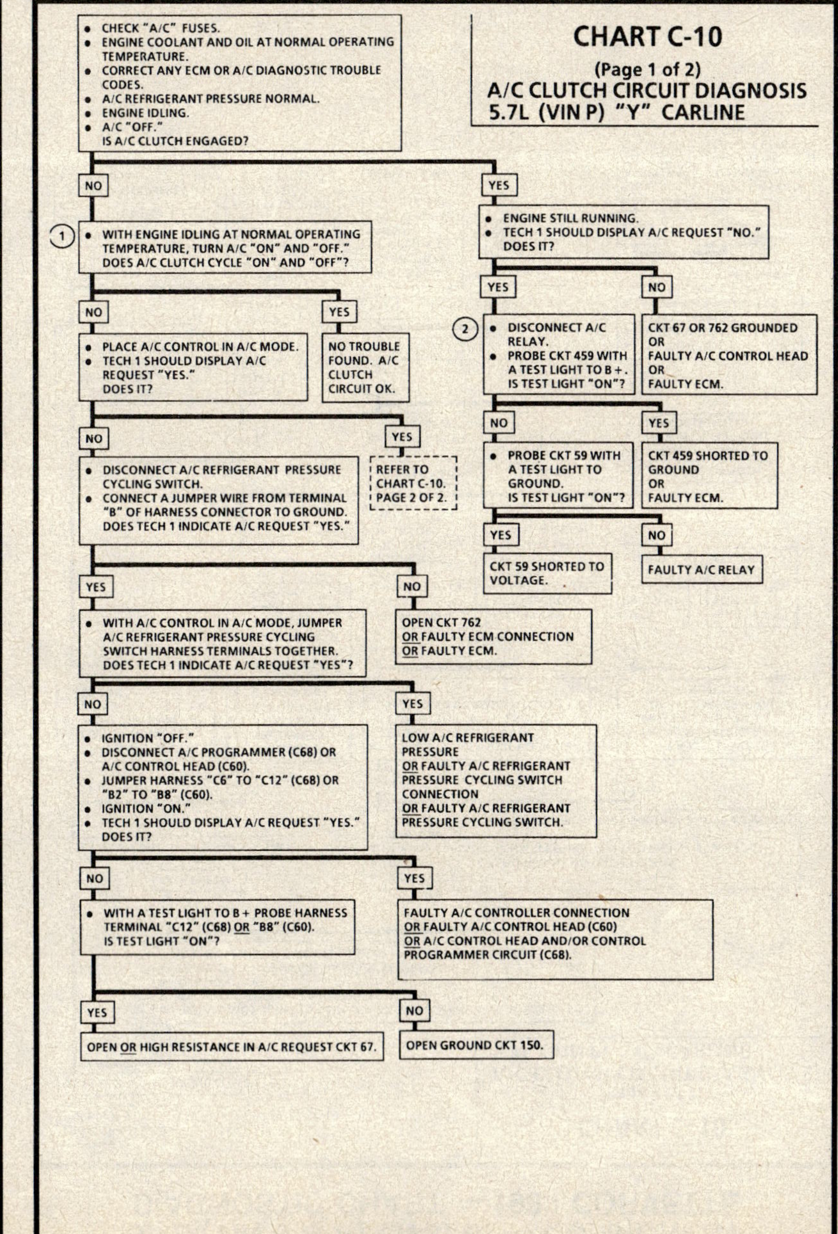

CHART C-10

(Page 1 of 2)
A/C CLUTCH CIRCUIT DIAGNOSIS
5.7L (VIN P) "Y" CARLINE

Circuit Description:

The A/C clutch relay is ECM controlled to delay A/C clutch engagement after the A/C is turned "ON." This allows the ECM to adjust engine RPM before the A/C clutch engages.

The ECM will engage the A/C clutch any time A/C has been requested unless any of the following conditions exist:

- High coolant temperature.
- Wide open throttle.
- High oil temperature.
- High engine RPM.
- High A/C system pressure.

The ECM can determine A/C has been requested by sending a voltage signal to the A/C control head. When the A/C control switch closes, the A/C request voltage signal is grounded. This is shown on the Tech 1 as A/C request "YES." A 3-wire A/C refrigerant pressure sensor is used to monitor A/C system pressure. If the A/C refrigerant pressure is greater than 429 psi or lower than 0 psi, the A/C clutch will not engage.

The A/C refrigerant pressure sensor uses ECM DTCs 66 and 67. If the A/C refrigerant pressure sensor signal wire becomes open, shorted to ground or shorted to voltage, DTC 66 will set. If the A/C clutch engages and no pressure change is detected, a DTC 67 will set.

When a request for A/C has been detected by the ECM, the ECM will ground the A/C clutch relay drive circuit, the relay contacts will close, and current will flow through the relay to the A/C compressor clutch.

When A/C request has been detected by the ECM, the cooling fan(s) will be turned "ON" unless vehicle speed is too high.

Test Description: Number(s) below refer to circled number(s) on the diagnostic chart.

1. Checks the ECMs ability to control the A/C clutch relay.
2. Checks for grounded CKT 459 to ECM.

Diagnostic Aids:

Before using CHART C-10, be sure no ECM or CCM DTC(s) are stored. The ECM will not activate the A/C clutch with a stored DTC.

5.7L (VIN 8 & P) ENGINE — COMPONENT DIAGNOSTIC CHART — 1993 CORVETTE

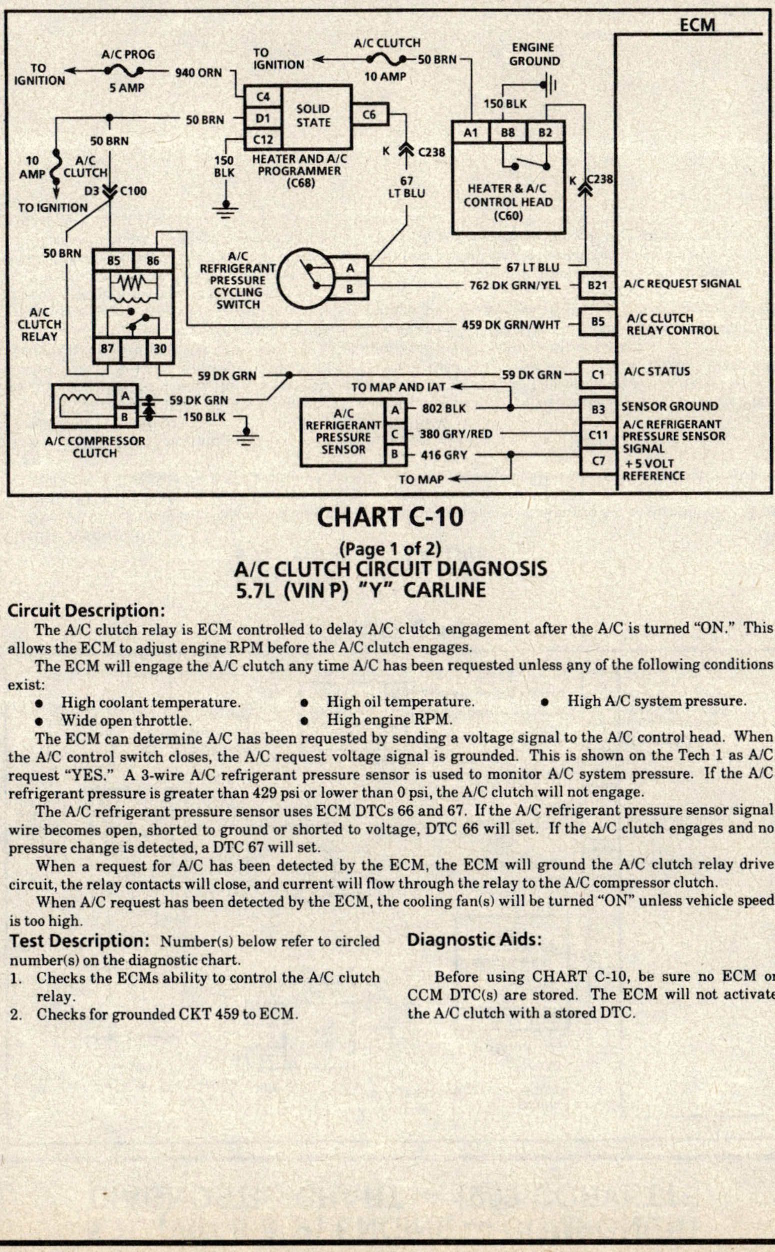

5.7L (VIN 8 & P) ENGINE — COMPONENT DIAGNOSTIC CHART — 1993 CORVETTE

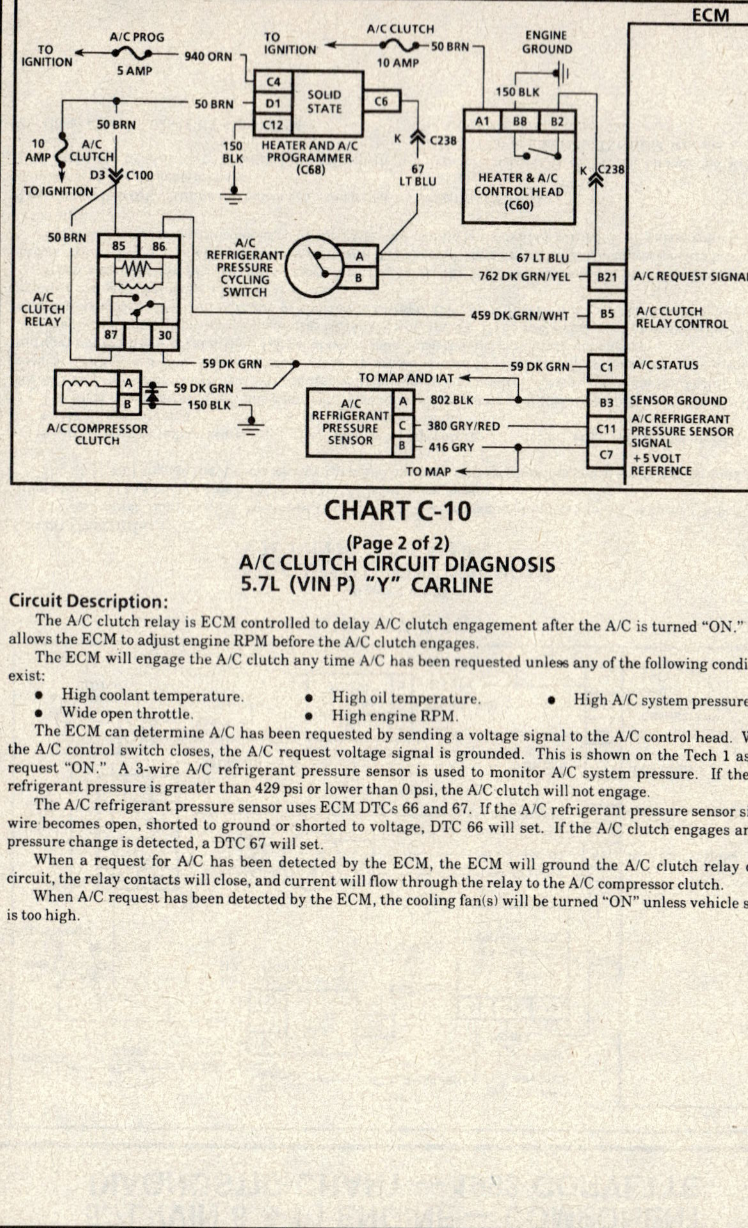

Circuit Description:

The A/C clutch relay is ECM controlled to delay A/C clutch engagement after the A/C is turned "ON." This allows the ECM to adjust engine RPM before the A/C clutch engages.

The ECM will engage the A/C clutch any time A/C has been requested unless any of the following conditions exist:

- High coolant temperature.
- Wide open throttle.
- High oil temperature.
- High engine RPM.
- High A/C system pressure.

The ECM can determine A/C has been requested by sending a voltage signal to the A/C control head. When the A/C control switch closes, the A/C request voltage signal is grounded. This is shown on the Tech 1 as A/C request "ON." A 3-wire A/C refrigerant pressure sensor is used to monitor A/C system pressure. If the A/C refrigerant pressure is greater than 429 psi or lower than 0 psi, the A/C clutch will not engage.

The A/C refrigerant pressure sensor uses ECM DTCs 66 and 67. If the A/C refrigerant pressure sensor signal wire becomes open, shorted to ground or shorted to voltage, DTC 66 will set. If the A/C clutch engages and no pressure change is detected, a DTC 67 will set.

When a request for A/C has been detected by the ECM, the ECM will ground the A/C clutch relay drive circuit, the relay contacts will close, and current will flow through the relay to the A/C compressor clutch.

When A/C request has been detected by the ECM, the cooling fan(s) will be turned "ON" unless vehicle speed is too high.

5.7L (VIN 8 & P) ENGINE — COMPONENT DIAGNOSTIC CHART — 1993 CORVETTE

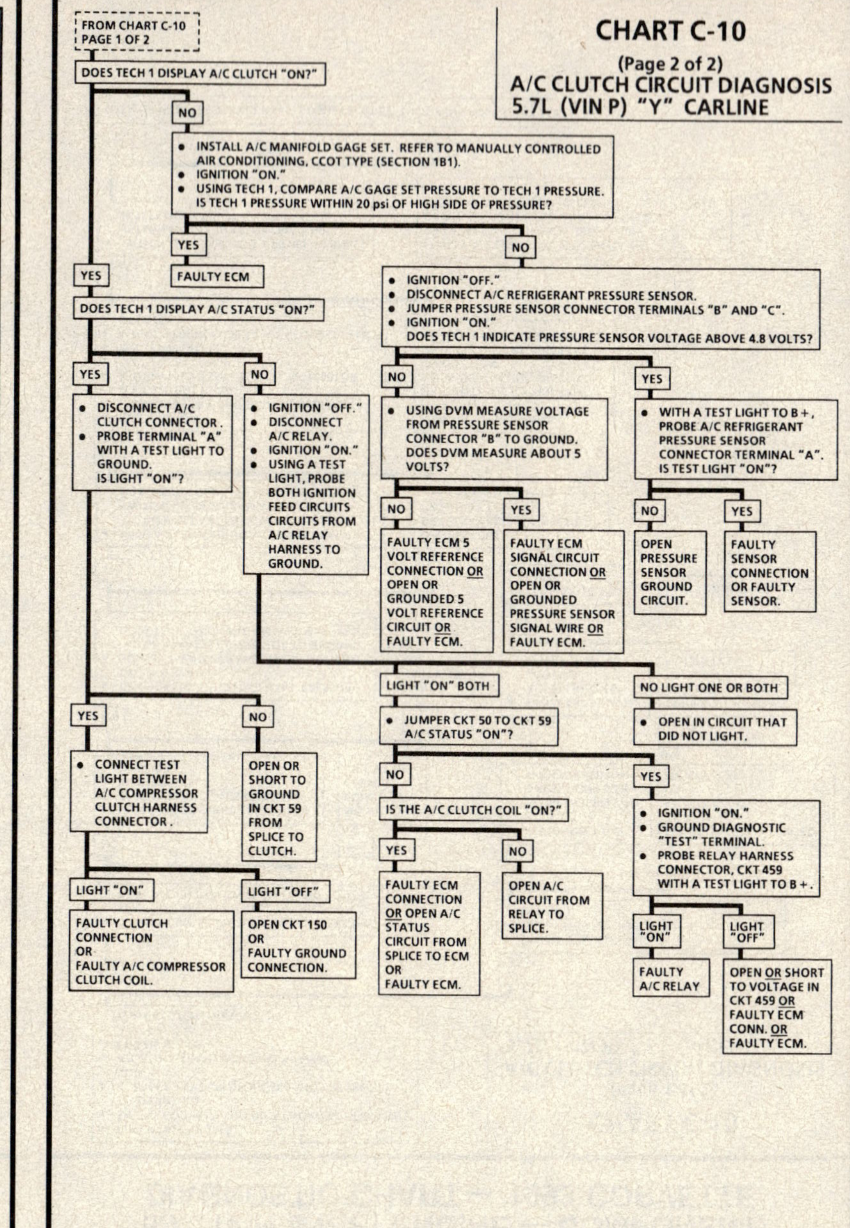

5.7L (VIN 8 & P) ENGINE — COMPONENT DIAGNOSTIC CHART — 1992 CORVETTE

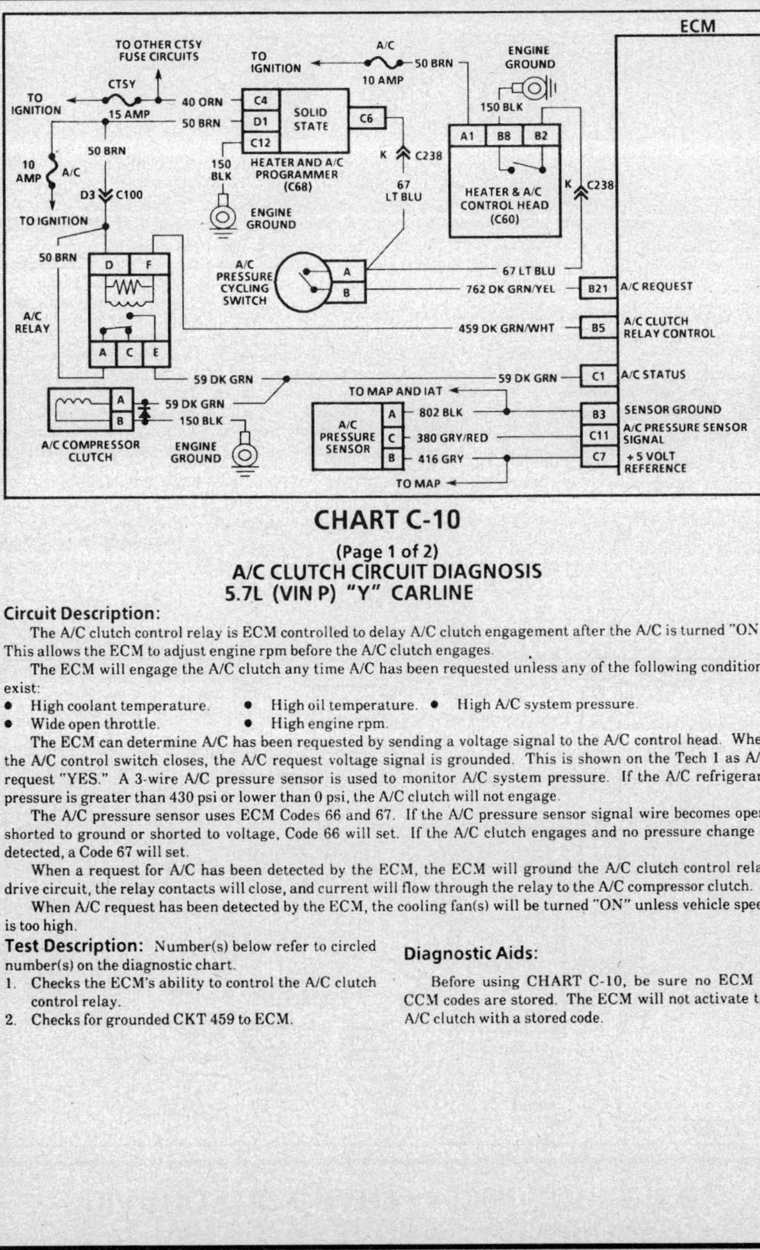

CHART C-10
(Page 1 of 2)
A/C CLUTCH CIRCUIT DIAGNOSIS
5.7L (VIN P) "Y" CARLINE

Circuit Description:

The A/C clutch control relay is ECM controlled to delay A/C clutch engagement after the A/C is turned "ON." This allows the ECM to adjust engine rpm before the A/C clutch engages.

The ECM will engage the A/C clutch any time A/C has been requested unless any of the following conditions exist:

- High coolant temperature.
- Wide open throttle.
- High oil temperature.
- High engine rpm.
- High A/C system pressure.

The ECM can determine A/C has been requested by sending a voltage signal to the A/C control head. When the A/C control switch closes, the A/C request voltage signal is grounded. This is shown on the Tech 1 as A/C request "YES." A 3-wire A/C pressure sensor is used to monitor A/C system pressure. If the A/C refrigerant pressure is greater than 430 psi or lower than 0 psi, the A/C clutch will not engage.

The A/C pressure sensor uses ECM Codes 66 and 67. If the A/C pressure sensor signal wire becomes open, shorted to ground or shorted to voltage, Code 66 will set. If the A/C clutch engages and no pressure change is detected, a Code 67 will set.

When a request for A/C has been detected by the ECM, the ECM will ground the A/C clutch control relay drive circuit, the relay contacts will close, and current will flow through the relay to the A/C compressor clutch.

When A/C request has been detected by the ECM, the cooling fan(s) will be turned "ON" unless vehicle speed is too high.

Test Description: Number(s) below refer to circled number(s) on the diagnostic chart.

1. Checks the ECM's ability to control the A/C clutch control relay.
2. Checks for grounded CKT 459 to ECM.

Diagnostic Aids:

Before using CHART C-10, be sure no ECM or CCM codes are stored. The ECM will not activate the A/C clutch with a stored code.

5.7L (VIN 8 & P) ENGINE — COMPONENT DIAGNOSTIC CHART — 1992 CORVETTE

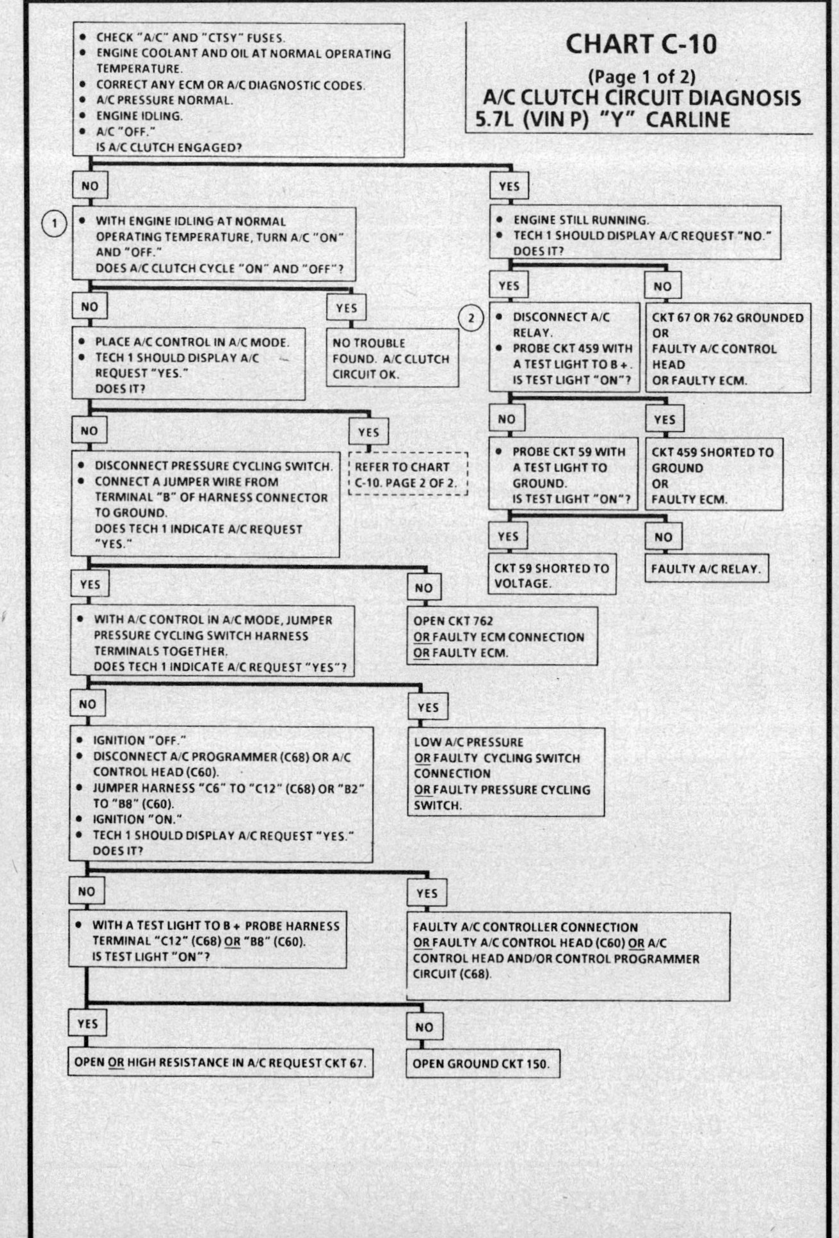

5.7L (VIN 8 & P) ENGINE — COMPONENT DIAGNOSTIC CHART — 1992 CORVETTE

5.7L (VIN 8 & P) ENGINE — COMPONENT DIAGNOSTIC CHART — 1992 CORVETTE

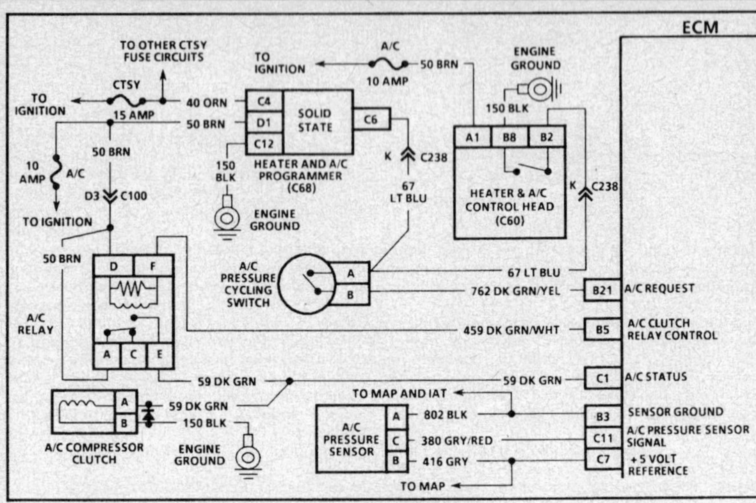

CHART C-10
(Page 2 of 2)
A/C CLUTCH CIRCUIT DIAGNOSIS
5.7L (VIN P) "Y" CARLINE

Circuit Description:

The A/C clutch control relay is ECM controlled to delay A/C clutch engagement after the A/C is turned "ON." This allows the ECM to adjust engine rpm before the A/C clutch engages.

The ECM will engage the A/C clutch any time A/C has been requested unless any of the following conditions exist.

- High coolant temperature.
- Wide open throttle.
- High oil temperature.
- High engine rpm.
- High A/C system pressure.

The ECM can determine A/C has been requested by sending a voltage signal to the A/C control head. When the A/C control switch closes, the A/C request voltage signal is grounded. This is shown on the Tech 1 as A/C request "ON." A 3-wire A/C pressure sensor is used to monitor A/C system pressure. If the A/C refrigerant pressure is greater than 430 psi or lower than 0 psi, the A/C clutch will not engage.

The A/C pressure sensor uses ECM Codes 66 and 67. If the A/C pressure sensor signal wire becomes open, shorted to ground or shorted to voltage, Code 66 will set. If the A/C clutch engages and no pressure change is detected, a Code 67 will set.

When a request for A/C has been detected by the ECM, the ECM will ground the A/C clutch control relay drive circuit, the relay contacts will close, and current will flow through the relay to the A/C compressor clutch.

When A/C request has been detected by the ECM, the cooling fan(s) will be turned "ON" unless vehicle speed is too high.

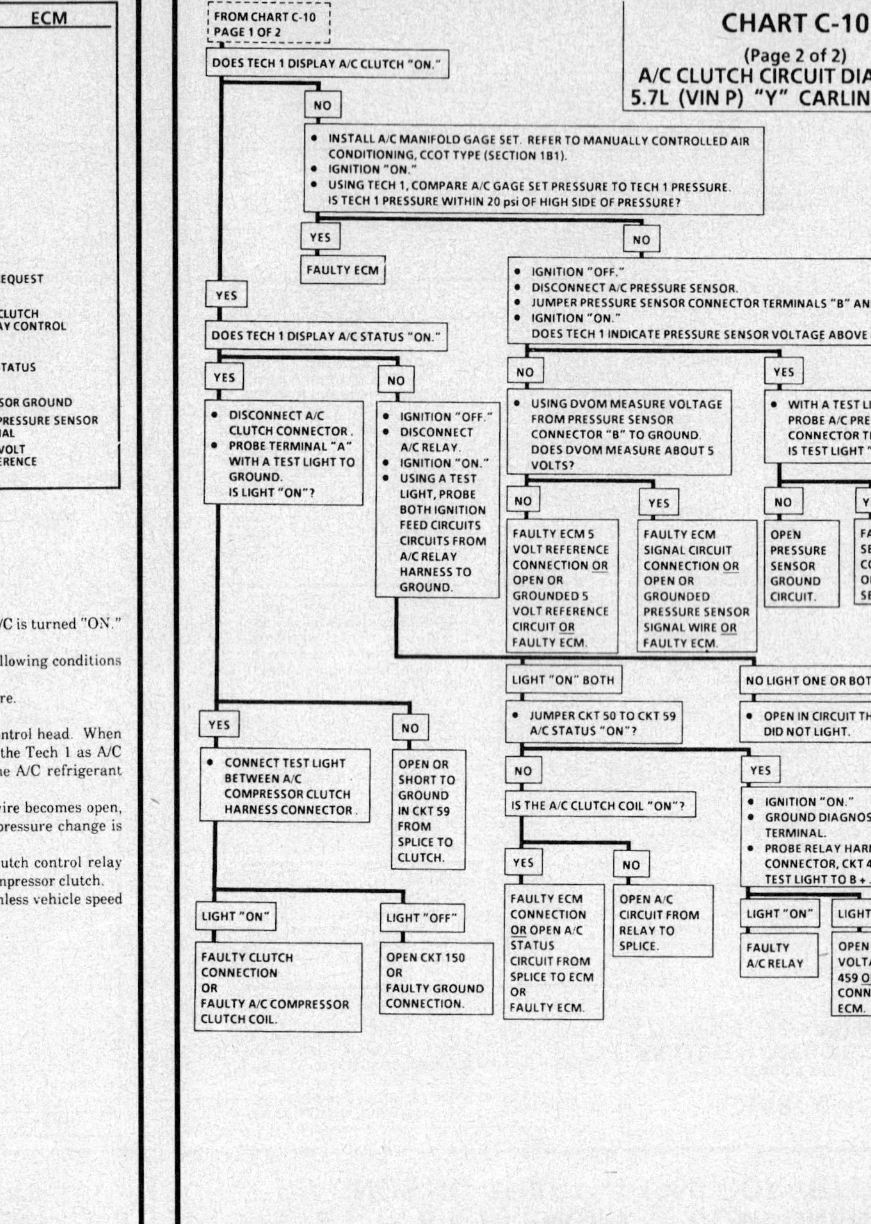

5.7L (VIN 8 & P) ENGINE — COMPONENT DIAGNOSTIC CHART — 1992–93 CORVETTE

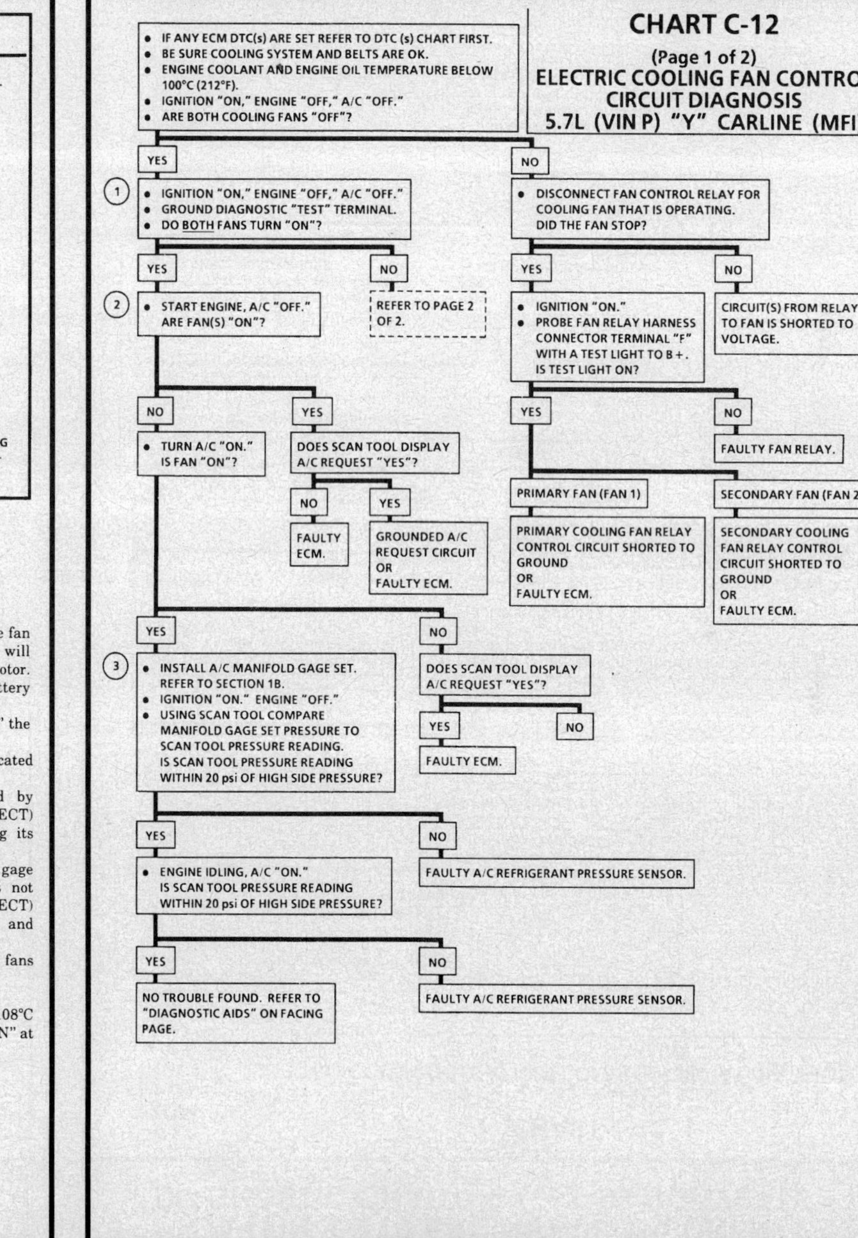

CHART C-12

(Page 1 of 2)

ELECTRIC COOLING FAN CONTROL CIRCUIT DIAGNOSIS
5.7L (VIN P) "Y" CARLINE (MFI)

Circuit Description:

The cooling fans are controlled by the ECM based on various inputs. Battery voltage is supplied to the fan relays on terminal "E" and ignition voltage to terminal "D". Grounding CKT 335 (relay terminal "F") will energize the primary cooling fan relay (Fan 1) and supply battery voltage to the primary cooling fan motor. Grounding CKT 473 (relay terminal "F") will energize the secondary cooling fan relay (Fan 2) and supply battery voltage to the secondary fan motor.

If any ECM DTC(s) are set or the ECM is operating in the fuel back-up mode, the ECM will turn "ON" the cooling fans.

Test Description: Number(s) below refer to circled number(s) on the diagnostic chart.

1. With the diagnostic "test" terminal grounded, the cooling fan control driver(s) will close, which should energize the fan control relay(s).
2. The cooling fans should come "ON" anytime A/C system is operating.
3. Comparing Tech 1 pressure and manifold gage set pressure will determine if the A/C refrigerant pressure sensor is out of range. An out of range A/C refrigerant pressure sensor can cause the cooling fans to operate at the wrong times.

Diagnostic Aids:

If the owner complained of an overheating problem, it must be determined if the complaint was due to an actual boil over, or the warning indicator light, or engine coolant temperature gage indicated overheating.

The gage accuracy can also be checked by comparing the Engine Coolant Temperature (ECT) sensor reading using a Tech 1 and comparing its reading with the gage reading.

If the engine is actually overheating and the gage indicated overheating, but the cooling fan is not coming "ON," the Engine Coolant Temperature (ECT) sensor has probably shifted out of calibration and should be replaced.

If the engine is overheating and the cooling fans are "ON," the cooling system should be checked

The ECM will command fan 1 "ON" at 108°C (226°F) and "OFF" at 105°C (221°F) and, fan 2 "ON" at 113°C (235°F) and "OFF" at 108°C (226°F).

5.7L (VIN 8 & P) ENGINE — COMPONENT DIAGNOSTIC CHART — 1992–93 CORVETTE

CHART C-12

(Page 1 of 2)

ELECTRIC COOLING FAN CONTROL CIRCUIT DIAGNOSIS
5.7L (VIN P) "Y" CARLINE (MFI)

- IF ANY ECM DTC(s) ARE SET REFER TO DTC (s) CHART FIRST.
- BE SURE COOLING SYSTEM AND BELTS ARE OK.
- ENGINE COOLANT AND ENGINE OIL TEMPERATURE BELOW 100°C (212°F).
- IGNITION "ON," ENGINE "OFF," A/C "OFF."
- ARE BOTH COOLING FANS "OFF"?

YES (1)
- IGNITION "ON," ENGINE "OFF," A/C "OFF."
- GROUND DIAGNOSTIC "TEST" TERMINAL.
- DO BOTH FANS TURN "ON"?

NO
- DISCONNECT FAN CONTROL RELAY FOR COOLING FAN THAT IS OPERATING. DID THE FAN STOP?

YES (2)
- START ENGINE, A/C "OFF." ARE FAN(S) "ON"?

NO
REFER TO PAGE 2 OF 2.

YES
- IGNITION "ON."
- PROBE FAN RELAY HARNESS CONNECTOR TERMINAL "F" WITH A TEST LIGHT TO B+. IS TEST LIGHT ON?

NO
CIRCUIT(S) FROM RELAY TO FAN IS SHORTED TO VOLTAGE.

NO
- TURN A/C "ON." IS FAN "ON"?

YES
DOES SCAN TOOL DISPLAY A/C REQUEST "YES"?

NO
FAULTY ECM.

YES
GROUNDED A/C REQUEST CIRCUIT OR FAULTY ECM.

YES
PRIMARY FAN (FAN 1)

PRIMARY COOLING FAN RELAY CONTROL CIRCUIT SHORTED TO GROUND OR FAULTY ECM.

NO
FAULTY FAN RELAY.

NO
SECONDARY FAN (FAN 2)

SECONDARY COOLING FAN RELAY CONTROL CIRCUIT SHORTED TO GROUND OR FAULTY ECM.

YES (3)
- INSTALL A/C MANIFOLD GAGE SET. REFER TO SECTION 1B.
- IGNITION "ON." ENGINE "OFF."
- USING SCAN TOOL COMPARE MANIFOLD GAGE SET PRESSURE TO SCAN TOOL PRESSURE READING. IS SCAN TOOL PRESSURE READING WITHIN 20 psi OF HIGH SIDE PRESSURE?

NO
DOES SCAN TOOL DISPLAY A/C REQUEST "YES"?

YES / **NO**

FAULTY ECM.

YES
- ENGINE IDLING, A/C "ON." IS SCAN TOOL PRESSURE READING WITHIN 20 psi OF HIGH SIDE PRESSURE?

NO
FAULTY A/C REFRIGERANT PRESSURE SENSOR.

YES
NO TROUBLE FOUND. REFER TO "DIAGNOSTIC AIDS" ON FACING PAGE.

NO
FAULTY A/C REFRIGERANT PRESSURE SENSOR.

5.7L (VIN 8 & P) ENGINE — COMPONENT DIAGNOSTIC CHART — 1992–93 CORVETTE

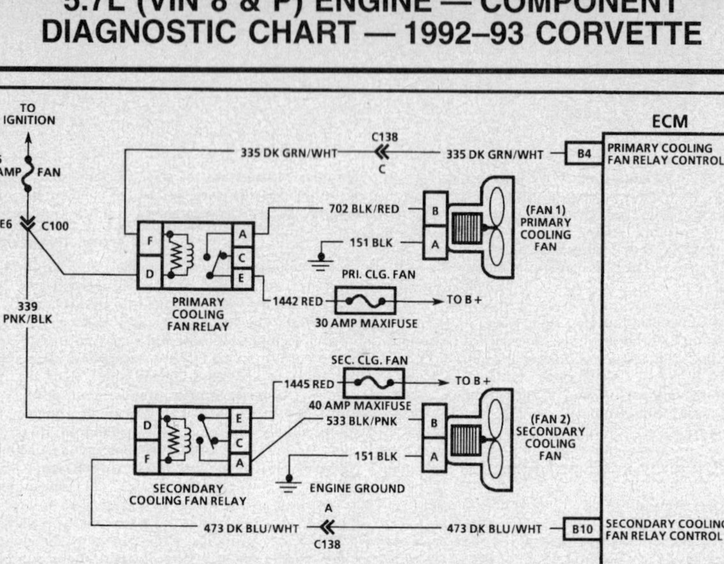

CHART C-12
(Page 2 of 2)
ELECTRIC COOLING FAN CONTROL CIRCUIT DIAGNOSIS
5.7L (VIN P) "Y" CARLINE (MFI)

Circuit Description:

The cooling fans are controlled by the ECM based on various inputs. Battery voltage is supplied to the fan relays on terminal "E" and ignition voltage to terminal "D". Grounding CKT 335 (relay terminal "F") will energize the primary cooling fan relay (Fan 1) and supply battery voltage to the primary cooling fan motor. Grounding CKT 473 (relay terminal "F") will energize the secondary cooling fan relay (Fan 2) and supply battery voltage to the secondary fan motor.

If any ECM DTC(s) are set or the ECM is operating in the fuel back-up mode, the ECM will turn "ON" the cooling fans.

Diagnostic Aids:

If the owner complained of an overheating problem, it must be determined if the complaint was due to an actual boil over, the warning indicator light, or engine coolant temperature gage indicated overheating.

The gage accuracy can also be checked by comparing the Engine Coolant Temperature (ECT) sensor reading using a Tech 1 and comparing its reading with the gage reading.

If the engine is actually overheating and the gage indicated overheating, but the cooling fan is not coming "ON," the Engine Coolant Temperature (ECT) sensor has probably shifted out of calibration and should be replaced.

If the engine is overheating and the cooling fans are "ON," the cooling system should be checked

The ECM will command fan 1 "ON" at 108°C (226°F) and "OFF" at 105°C (221°F) and, fan 2 "ON" at 113°C (235°F) and "OFF" at 108°C (226°F).

5.7L (VIN 8 & P) ENGINE — COMPONENT DIAGNOSTIC CHART — 1992–93 CORVETTE

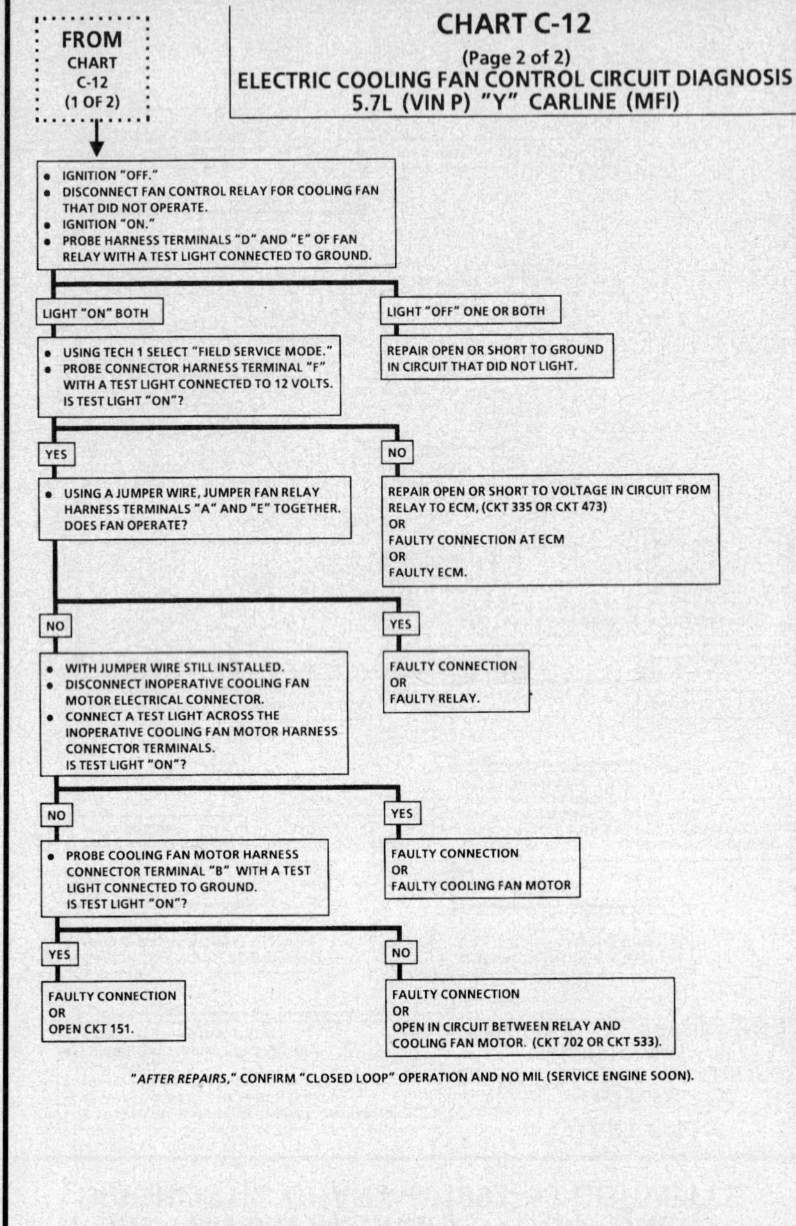

Component Replacement

NOTE: Many MFI repair procedures require the complete injection unit to removed from the engine. A part number is stamped on the throttle body. Refer to this number if servicing or part replacement is necessary.

SERVICE PRECAUTIONS

When working around any part of the fuel system, take the following precautionary steps to prevent fire and/or explosion:

- Disconnect negative terminal from battery (except when testing with battery voltage is required).
- Whenever possible, use a flashlight instead of a drop light.
- Keep all open flame and smoking material out of the area.
- Use a shop cloth or similar to catch fuel when opening a fuel system.
- Relieve fuel system pressure before servicing.
- Use eye protection.
- Always keep a dry chemical (class B) fire extinguisher near the area.

NOTE: Due to the amount of fuel pressure in the fuel lines, prior to performing any work to the fuel system, the fuel system should be de-pressurized.

CLEANING AND INSPECTION

All MPI component parts, with the exception of those noted below, should be cleaned in a cold immersion cleaner such as Carbon X (X–55) or equivalent. The throttle position sensor, idle air control valve, pressure regulator diaphragm assembly, fuel injectors or other components containing rubber should not be placed in a solvent or cleaner bath. A chemical reaction will cause these parts to swell, harden or distort. Do not soak the throttle body with the above parts attached. If the throttle body assembly requires cleaning, soak time in the cleaner should be kept to a minimum. Some vehicles have hidden throttle shaft dust seals that could lose their effectiveness by extended soaking.

1. Clean all parts thoroughly and blow dry with compressed air. Be sure that all fuel and air passages are free of dirt and burrs.
2. Inspect the gasket mating surfaces for damage that could affect proper gasket sealing.

THREAD LOCKING COMPOUND

Service repair kits are supplied with a small vial of thread locking compound with directions for use. If material is not available, use Loctite® 262 or GM part No. 10522624 or equivalent. In precoating the screws, do not use a higher strength locking compound than recommended, thus making removal of the screw extremely difficult, or result in damaging the screw head.

RELIEVING FUEL SYSTEM PRESSURE

Procedure

2.0L AND 2.2L ENGINES

1. Remove the fuel pump fuse from the fuse block.

2. Start the engine. It should run and then stall when the fuel in the lines is depleted. When the engine stops, crank the starter for about 3 seconds to make sure all pressure in the fuel lines is released.
3. Replace the fuel pump fuse.
4. On some vehicles a pressure relief valve is located on the fuel rail. To relive the fuel pressure using the relief valve use the following procedure.
5. Disconnect the negative battery cable to prevent accidental fuel system pressurization.

2.3L ENGINE

1. Loosen the fuel filler cap.
2. Raise and safely support the vehicle.
3. Disconnect the fuel pump wire connector at the tank.
4. Lower the vehicle and start the engine. It should run and then stall when the fuel in the lines consumed. When the engine stops, crank the starter for about 3 seconds to make sure all pressure in the fuel lines is released.
5. Raise and safely support the vehicle. Reconnect the fuel pump wire connector at the tank.
6. Disconnect the negative battery cable to prevent accidental fuel system pressurization.

3.1L, 3.3L, 3.4L, 5.0L AND 5.7L ENGINE

1. Disconnect the negative battery cable to avoid possible fuel discharge if an accidental attempt is made to start the engine.
2. Loosen the fuel filler cap to relieve fuel tank pressure.
3. Connect fuel gauge J–34730–1, or equivalent, to fuel pressure relief valve on the fuel rail. Wrap a rag around the pressure tap to absorb any leakage that may occur when installing the gauge.
4. Install bleed hose into a suitable container and open the valve to bleed off the fuel pressure in the fuel system.

ENGINE COOLANT TEMPERATURE (ECT) SENSOR

Removal and Installation

The coolant temperature sensor is usually located on the intake manifold water jacket or near (or in) the thermostat housing. It may be necessary to drain some of the coolant from the coolant system.

1. Disconnect the negative battery cable and disconnect the electrical connector at the sensor.
2. Remove the threaded temperature sensor from the engine.
3. Apply some pipe tape to the threaded sensor and install the senor. Torque the sensor to 22 ft. lbs. (30 Nm).
4. Reconnect the sensor and the negative battery cable.

CRANKSHAFT POSITION (CKP) SENSOR

Removal and Installation

2.0L ENGINE

1. Disconnect the negative battery cable.
2. Remove the serpentine belt from crankshaft pulley.
3. Remove the timing belt, alternator, power steering and bracket assembly.

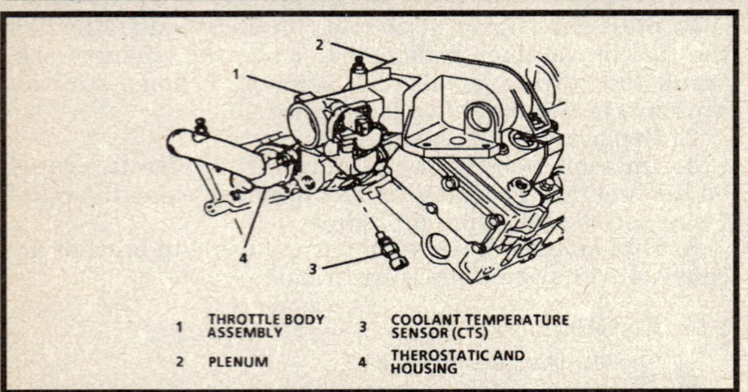

ECT sensor removal — 3.4L engine

4. Remove the sensor harness connector at the sensor. Remove the sensor to the block retaining bolt and remove the sensor from the block.

5. Installation is the reverse order of the removal procedure.

3.3L ENGINE

1. Disconnect the negative battery cable.
2. Remove the serpentine belt from crankshaft pulley.
3. Raise and support the vehicle safely.
4. Remove the right front tire and wheel assembly.
5. Remove the right inner fender access cover.
6. Using a 28mm socket and remove the crankshaft harmonic balancer retaining bolt.
7. Remove the harmonic balancer and disconnect the electrical connector. Remove the sensor and pedestal assembly from the mounting surface.
8. Remove the sensor from the pedestal.

To install:

9. Loosely install the crankshaft sensor on the pedestal. Position the sensor and pedestal assembly on special crankshaft sensor adjustment tool J–37089 or equivalent.
10. Position the special tool and pedestal assembly on the crankshaft. Install the retaining bolts to hold the pedestal to block. Torque the bolts to 14–28 ft. lbs. (20–40 Nm).
11. Torque the pedestal pinch bolt to 30–35 inch. lbs. (3–4 Nm). Remove special tool J–37089.
12. Place special tool J–37089 onto the harmonic balancer and turn it. If any vane of the harmonic balancer touches the tool, replace the balancer assembly.
13. Install the balancer onto the crankshaft and torque the bolt to 200–239 ft. lbs. (270–325 Nm).

14. Install the inner fender shield. Install the tire and wheel assembly. Lower the vehicle and install the serpentine belt and negative battery cable.

2.2L, 2.3L, 3.1L AND 3.4L ENGINES

1. Disconnect the negative battery cable.
2. Remove the sensor harness connector at the sensor.
3. Remove the sensor to the block retaining bolt and remove the sensor from the block.
4. Installation is the reverse order of the removal procedure and be sure to use a new O-ring on the sensor. Torque the sensor retaining bolt to 71–88 inch. lbs. (8–10 Nm).

ELECTRONIC CONTROL MODULE (ECM)

To prevent possible electrostatic discharge damage to the ECM, do not touch the connector pins or soldered components on the circuit board. Service of the ECM should normally consists of either replacement of the ECM or a PROM change.

If the diagnostic procedures call for the ECM to be replaced, the engine calibrator (Mem-Cal or PROM) and the ECM should be checked first to see if they are the correct parts. If they are, remove the Mem-Cal or PROM from the faulty ECM and install it in the new service ECM. The service ECM will not contain a new Mem-Cal or PROM.

The trouble Code 51 indicates that the Mem-Cal or PROM is installed improperly or has malfunctioned. When Code 51 is obtained, check the ECM installation for bent pins or pins not fully seated in the socket. If it is installed correctly and the Code 51 still appears, replace the Mem-Cal or PROM.

Removal and Installation

NOTE: When replacing the production ECM with a service ECM (controller), it is important to transfer the broadcast code and production ECM number to the service ECM label. Do not record the number on the ECM cover. This will allow positive identification of the ECM parts throughout the service life of the vehicle. To prevent internal ECM damage, the ignition must be OFF when disconnecting and reconnecting the power to the ECM (for example: the battery cable, ECM pigtail, ECM fuse, jumper cables, etc.).

1. Disconnect the negative battery cable.
2. Remove the glove box assembly and/or right hand side kick panel to gain access to the ECM, except on the

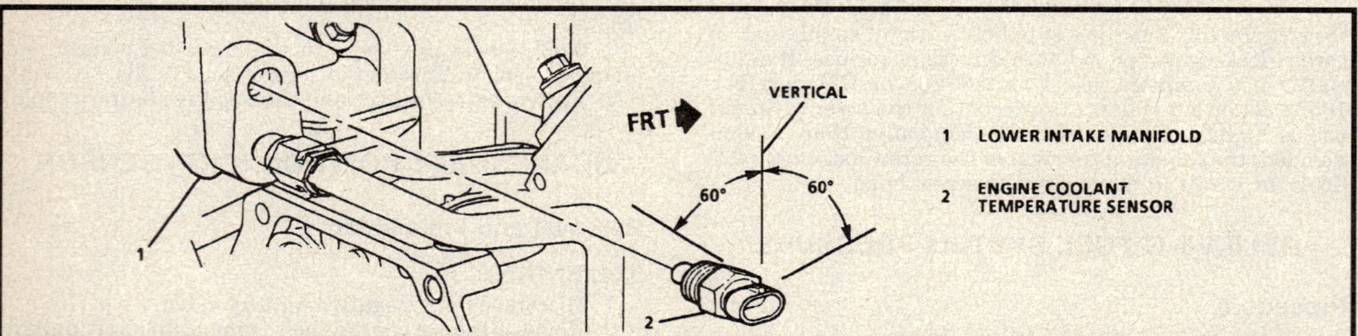

ECT sensor removal; Note the installed position of the sensor connector locking tab — 3.1L engine

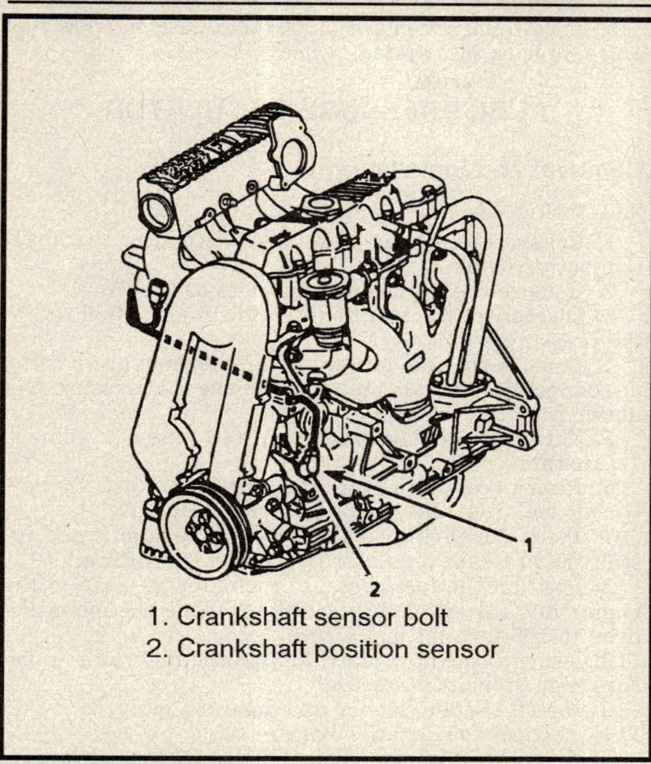

1. Crankshaft sensor bolt
2. Crankshaft position sensor

CKP sensor removal — 2.0L engine

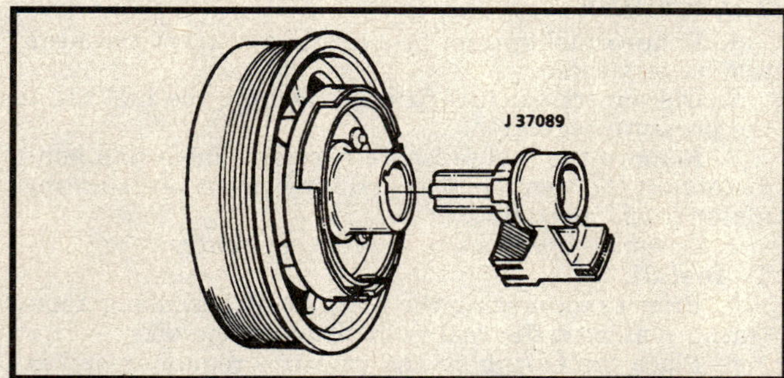

CKP sensor tool used to check for bent vanes on the harmonic balancer — 3.3L engine

Corvette and 1993–94 Camaro and Firebird. These vehicle have the ECM located in the engine compartment on either the left or right fenderwell.

3. Disconnect the connectors from the ECM.

4. Remove the ECM mounting hardware and remove the ECM from the passenger compartment.

5. Installation is the reverse order of the removal procedure. Be sure to remove the new ECM from its packaging and check the service number to make sure it is the same at the defective ECM.

PROM

Removal and Installation

1. Remove the ECM.

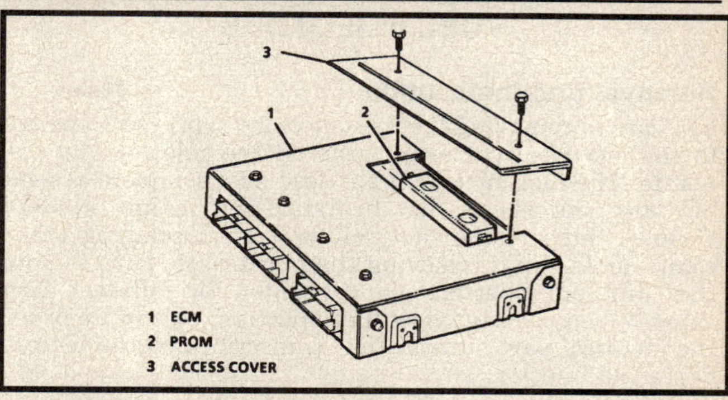

ECM and PROM assembly

2. Remove the PROM access panel.

3. Using the rocker type PROM removal tool, engage 1 end of the PROM carrier with the hook end of the tool. Press on the vertical bar end of the tool and rock the engaged end of the PROM carrier up as far as possible.

4. Engage the opposite end of the PROM carrier in the same manner and rock this end up as far as possible.

5. Repeat this process until the PROM carrier and PROM are free of the PROM socket. The PROM carrier with PROM in it should lift off of the PROM socket easily.

6. The PROM carrier should only be removed by using the special PROM removal tool. Other methods could cause damage to the PROM or PROM socket.

7. Before installing a new PROM, be sure the new PROM part number is the same as the old one or the as the updated number per a service bulletin.

8. Install the PROM with the small notch of the carrier aligned with the small notch in the socket. Press on the PROM carrier until the PROM is firmly seated, do not press on the PROM itself only on the PROM carrier.

9. Install the PROM access cover and reinstall the ECM.

Functional Check

1. Turn the ignition switch to the **ON** position.

2. Enter the diagnostic mode.

3. Allow a Code 12 to flash 4 times to verify that no other codes are present. This indicates that the PROM is installed properly.

4. If trouble Code 51 occurs or if the MIL in on constantly with no codes, the PROM is not fully seated, installed backwards, has bent pins or is defective.

5. If not fully seated press down firmly on the PROM carrier.

6. If installed backwards, replace the PROM.

NOTE: Any time the PROM is installed backwards and the ignition switch is turned ON, the PROM is destroyed.

7. If the pins are bent, remove the PROM straighten the pins and reinstall the PROM. If the bent pins break or crack during straightening, discard the PROM and replace with a new PROM.

NOTE: To prevent possible electrostatic discharge damage to the PROM or Cal-Pak, do not touch the component leads and do not remove the integrated circuit from the carrier.

FUEL INJECTORS

Removal and Installation

Use care in removing the fuel injectors to prevent damage to the electrical connector pins on the injector and the nozzle. The fuel injector is serviced as a complete assembly only and should not be immersed in any kind of cleaner. Support the fuel rail to avoid damaging other components while removing the injector. Be sure to note that different injectors are calibrated for different flow rates. When ordering new fuel injectors, be sure to order the identical part number that is inscribed on the bottom of the old injector.

EXCEPT 2.2L ENGINE

1. Relieve fuel system pressure. Disconnect the negative battery cable.
2. Remove the intake manifold plenum or throttle body assembly as required.
3. Remove the injector electrical connections. Remove the fuel rail assembly.
4. Remove the injector retaining clip, if equipped and remove the injector from the fuel rail.
5. Remove both injector seals from the injector and discard.

To install:

6. Prior to installing the injectors, coat the new injector O-ring seals with clean engine oil. Install the seals on the injector assembly.
7. Use new injector retainer clips on the injector assembly. Position the open end of the clip facing the injector electrical connector.
8. Install the injector into the fuel rail injector socket with the electrical connectors facing outward. Push the injector in firmly until it engages with the retainer clip locking it in place.
9. Install the fuel rail and injector assembly. Install the intake manifold, if equipped.
10. Connect the negative battery cable.
11. Turn the ignition switch **ON** and **OFF** to allow fuel pressure back into system. Check for leaks.

2.2L ENGINE

1. Relieve fuel system pressure. Disconnect the negative battery cable.
2. Remove the upper manifold plenum assembly.
3. Remove the fuel return line retaining bracket.
4. Remove the pressure regulator.
5. Remove the injector retaining bracket.
6. Remove the injector electrical connections.
7. Remove the injectors.
8. Remove both injector seals from the injector and discard. Ensure the lowers (smaller) O-ring does not remain in the lower manifold.

To install:

9. Prior to installing the injectors, coat the new injector O-ring seals with clean engine oil. Install the seals on the injector assembly.
10. Install the injectors into the lower manifold assembly. Position the injector with the electrical connector facing inward.
11. Install the injector retaining bracket. Connect the injector electrical connectors.
12. Install the pressure regulator assembly.
13. Install the fuel return line retaining bracket. Install the upper manifold assembly.
14. Connect the negative battery cable.

15. Turn the ignition switch **ON** and **OFF** to allow fuel pressure back into system. Check for leaks.

FUEL PRESSURE REGULATOR

Removal and Installation

2.0L ENGINE

1. Relieve fuel system pressure. Disconnect the negative battery cable.
2. Remove the air cleaner and duct assembly.
3. Disconnect the fuel fitting attaching the fuel rail to the pressure regulator.
4. Remove the fuel pressure regulator mounting bolts, disconnect the vacuum hose and remove the pressure regulator from the engine.
5. Cover all openings to prevent dirt entry.

To install:

6. Prior to connecting the fuel rail fitting, using a new O-ring seal, coat the seal with clean engine oil.
7. Place the O-ring on the pressure regulator and install the pressure regulator to the fuel rail fitting.
8. Position the fuel pressure regulator in place to the engine and install mounting bolts. Torque the mounting bolts to 105 inch lbs. (11.5 Nm).
9. Using a backup wrench tighten the fuel rail-to-pressure regulator fitting securely.
10. Install the air cleaner and duct assembly.
11. Connect the negative battery cable.
12. Turn the ignition switch **ON** and **OFF** to allow fuel pressure back into system. Check for leaks.

2.2L ENGINE

1. Relieve fuel system pressure. Disconnect the negative battery cable.
2. Disconnect the fuel fitting attaching the fuel rail to the pressure regulator.
3. Remove the fuel pressure regulator mounting bolts, disconnect the vacuum hose and remove the pressure regulator from the engine.
4. Cover all openings to prevent dirt entry.

To install:

5. Prior to connecting the fuel rail fitting, using a new O-ring seal, coat the seal with clean engine oil.
6. Place the O-ring on the pressure regulator and install the pressure regulator to the fuel rail fitting.
7. Position the fuel pressure regulator in place to the engine and install mounting bolts. Torque the mounting bolts to 31 inch lbs. (3.5 Nm).
8. Using a backup wrench tighten the fuel rail-to-pressure regulator fitting securely.
9. Connect the negative battery cable.
10. Turn the ignition switch **ON** and **OFF** to allow fuel pressure back into system. Check for leaks.

2.3L, 3.3L (VIN N) AND 3.4L ENGINES

1. Relieve fuel system pressure. Disconnect the negative battery cable.
2. On the 3.4L engine, remove the intake manifold.
3. Disconnect the fuel feed line and return line from the fuel rail assembly, be sure to use a backup wrench on the inlet fitting to prevent turning.
4. Remove the fuel rail assembly from the engine.
5. With the fuel rail assembly removed from the engine, remove the pressure regulator mounting screw.
6. Remove the pressure regulator from the rail assembly by twisting back and forth while pulling apart.

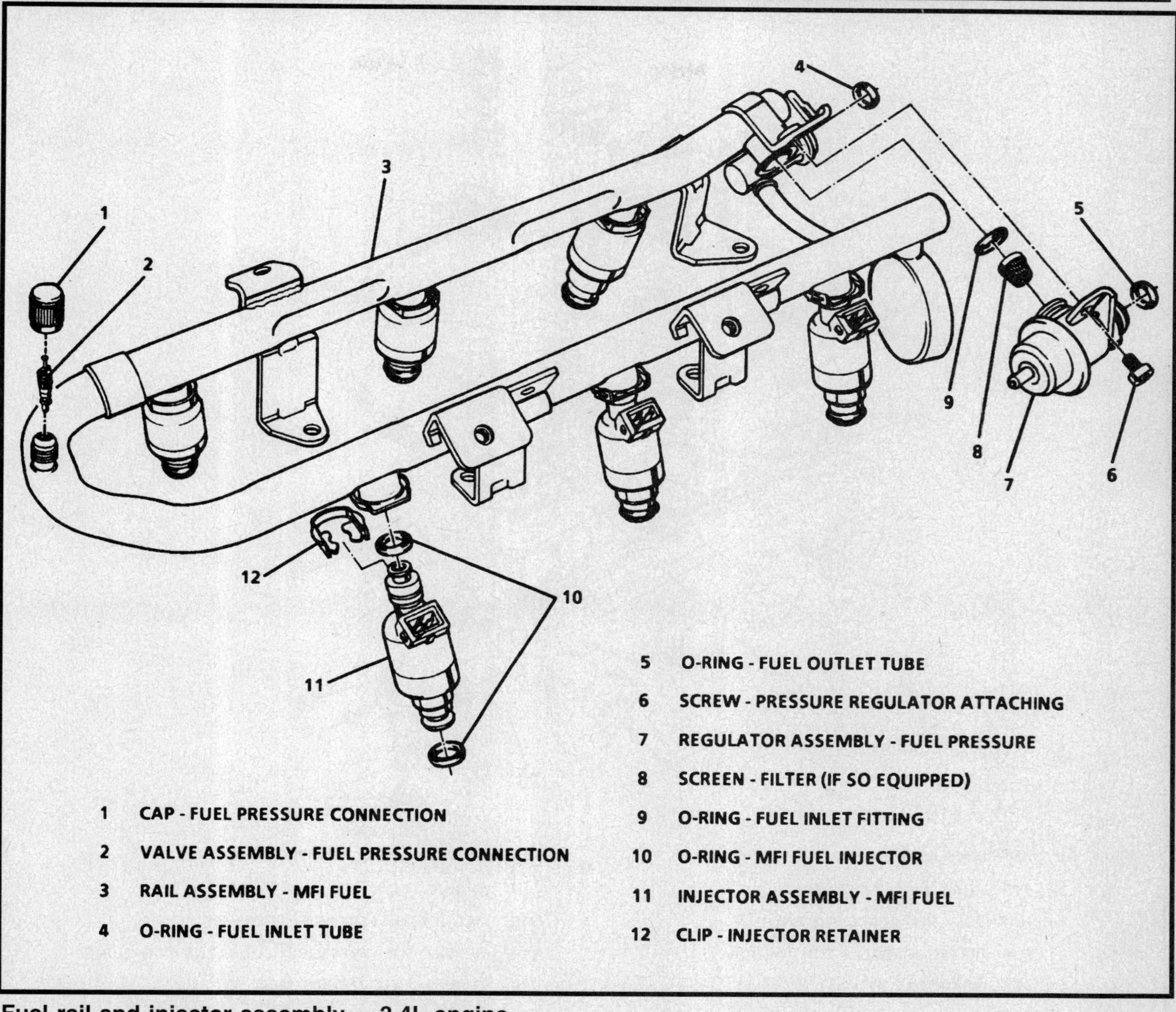

Fuel rail and injector assembly — 3.4L engine

To install:

7. Prior to assembling the pressure regulator to the fuel rail, lubricate the new rail-to-regulator O-ring seal with clean engine oil.

8. Place the O-ring on the pressure regulator and install the pressure regulator to the fuel rail.

9. Coat the regulator mounting screws with an approved tread locking compound and secure the pressure regulator in place. Torque the mounting screws to 102 inch lbs. (11.5 Nm).

10. Install the fuel rail assembly to the engine.

11. Connect the fuel feed line and return line to the fuel rail assembly, use a backup wrench on the inlet fitting to prevent turning.

12. Connect the negative battery cable.

13. Turn the ignition switch **ON** and **OFF** to allow fuel pressure back into system. Check for leaks.

3.1L ENGINE

NOTE: The fuel pressure regulator on the 1992 Firebird and Camaro is serviced only as an assembly with the fuel rail and therefore is not removable.

1. Relieve fuel system pressure. Disconnect the negative battery cable.

2. Remove the intake manifold plenum.

3. Disconnect the fuel feed line and return line from the fuel rail assembly, be sure to use a backup wrench on the inlet fitting to prevent turning.

4. Remove the fuel rail assembly from the engine.

5. With the fuel rail assembly removed from the engine, remove the pressure regulator mounting screw.

6. Disconnect the left and right hand fuel rails from the pressure regulator base.

MULTIPORT FUEL INJECTION (MFI) SYSTEMS
EXCEPT LIGHT TRUCKS, VANS, GEO AND SATURN

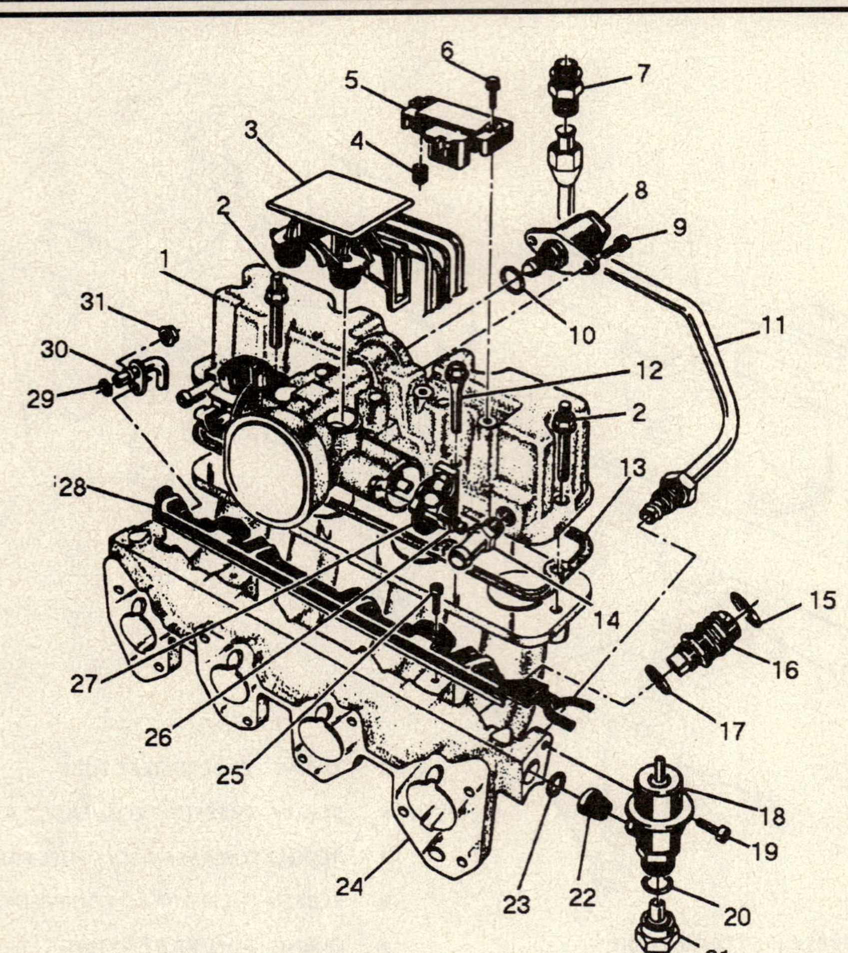

1	MANIFOLD ASM - UPPER	16	INJECTOR ASM - (BOTTOM FEED) MFI FUEL
2	STUD - UPPER INLET MANIFOLD	17	O-RING - LOWER
3	HARNESS ASM - EGR VALVE AND FUEL PRESSURE REGULATOR VACUUM	18	REGULATOR ASM - FUEL PRESSURE
4	SEAL - MAP SENSOR	19	SCREW - FUEL PRESSURE REGULATOR ATTACHING
5	SENSOR - MANIFOLD ABSOLUTE PRESSURE (MAP)	20	O-RING - FUEL RETURN LINE
6	BOLT - MAP SENSOR ATTACHING	21	PIPE ASM - FUEL INJECTOR FUEL RETURN
7	FITTING AND WASHER ASM	22	SCREEN - FILTER (IF SO EQUIPPED)
8	VALVE ASM - IDLE AIR CONTROL (IAC)	23	O-RING - FUEL INLET FITTING
9	SCREW - IAC VALVE ATTACHING	24	MANIFOLD ASM - LOWER
10	O-RING - IAC VALVE	25	SCREW - INJECTOR RETAINER ATTACHING
11	TUBE ASM - EGR TRANSPORT	26	SCREW - TP SENSOR ATTACHING
12	BOLT - UPPER INLET MANIFOLD	27	SENSOR - THROTTLE POSITION (TP)
13	GASKET - UPPER INLET MANIFOLD	28	RETAINER - INJECTOR
14	FITTING - POWER BRAKE	29	O-RING - FUEL FEED LINE
15	O-RING- FUEL INJECTOR	30	PIPE ASM - FUEL INJECTION FUEL FEED
		31	NUT - FUEL FEED

Fuel injection system components — 2.2L engine

7. Disconnect the regulator base to fuel rail connectors.

8. Remove the O-rings from the base and fuel rail connections.

9. Remove the pressure regulator from the assembly.

To install:

10. Prior to assembling the pressure regulator to the fuel rail, lubricate the new rail-to-regulator O-ring seal with clean engine oil.

11. Place the O-ring on the pressure regulator and install the pressure regulator to the fuel rail and base.

12. Coat the regulator mounting screws with an approved tread locking compound and secure the pressure regulator in place. Torque the mounting screws to 28 inch lbs. (3.2 Nm).

13. Install the fuel rail assembly to the engine.

14. Connect the fuel feed line and return line to the fuel rail assembly, use a backup wrench on the inlet fitting to prevent turning.

15. Install the intake manifold plenum.

16. Connect the negative battery cable.

17. Turn the ignition switch **ON** and **OFF** to allow fuel pressure back into system. Check for leaks.

5.0L AND 5.7L (VIN 8) ENGINES

NOTE: The fuel pressure regulator and base are serviced as a complete assembly. Do not remove the regulator cover or damage to the regulator may result.

1. Relieve fuel system pressure. Disconnect the negative battery cable.

2. Remove the intake manifold plenum and runners. Remove the fuel line bracket bolt.

3. Remove the fuel rail assembly.

4. Remove the rear crossover retainer-to-regulator base attaching screw and retainer.

5. Remove the crosssover tube and the O-ring from the pressure regulator base. Discard the O-ring.

6. Remove the pressure regulator bracket screws and bracket.

7. Remove the pressure regulator base-to-rail screw.

8. Remove the pressure regulator base assembly by separating the base from the fuel rail and disconnecting it from the fuel outlet tube.

9. Remove fuel outlet tube O-ring seal and discard it.

10. Remove the base-to-rail connector tube and remove the O-ring seals. Discard the seals.

To install:

11. Prior to assembly, lubricate the new base-to-rail connector tube O-rings and place them on the connector tube.

12. Position the base-to-rail connector tube into the regulator base.

13. Lubricate the new outlet tube O-ring and install it on the end of tube.

14. Connect the pressure regulator and base assembly between the fuel outlet tube and then to the fuel rail.

15. Install the base-to-rail retaining screw. Do not tighten the screw at this time.

16. Install the pressure regulator bracket with the retaining screws. Do not tighten at this time.

17. Lubricate the new rear crossover tube O-ring seal and install it on the end of the crossover tube.

18. Connect the crossover tube to the regulator base.

19. Install the crossover tube retainer with the retaining screw. Do not tighten at this time.

20. Torque all retaining bolts at this time to 44 inch lbs. (5.0 Nm).

21. Install the fuel rail assembly.

22. Temporarily connect the negative battery cable.

23. Turn the ignition to the **ON** position for 2 seconds, then turn it **OFF** for 10 seconds. Again turn the ignition to the **ON** position and check the system for fuel leaks.

24. Again, disconnect the negative battery cable.

25. Install the intake manifold plenum and runners.

26. Tighten the fuel tank cap. Connect the negative battery cable.

5.7L (VIN P) ENGINE

1. Relieve fuel system pressure.

2. Disconnect the negative battery cable.

3. Disconnect the fuel feed line and return line from the fuel rail assembly, be sure to use a backup wrench on the inlet fitting to prevent turning.

4. Remove the fuel rail assembly from the engine.

5. With the fuel rail assembly removed from the engine, remove the pressure regulator mounting screw.

6. Remove the pressure regulator from the rail assembly.

To install:

7. Prior to assembling the pressure regulator to the fuel rail, lubricate the new rail-to-regulator O-ring seal with clean engine oil.

8. Place the O-ring on the pressure regulator and install the pressure regulator to the fuel rail.

9. Coat the regulator mounting screws with an approved tread locking compound and secure the pressure regulator in place. Torque the mounting screws to 84 inch lbs. (9.5 Nm).

10. Install the fuel rail assembly to the engine.

11. Connect the fuel feed line and return line to the fuel rail assembly, use a backup wrench on the inlet fitting to prevent turning.

12. Connect the negative battery cable.

13. Turn the ignition switch **ON** and **OFF** to allow fuel pressure back into system. Check for leaks.

FUEL RAIL ASSEMBLY

When servicing the fuel rail assembly, be careful to prevent dirt and other contaminants from entering the fuel passages. Fitting should be capped and holes plugged during servicing. At any time the fuel system is opened for service, the O-ring seals and retainers used with related components should be replaced.

Before removing the fuel rail, the fuel rail assembly may be cleaned with a spray type cleaner, GM–30A or equivalent, following package instructions. Do not immerse fuel rails in liquid cleaning solvent.

There is an 8 digit number stamped on the under side of the fuel rail assembly on 4 cylinder engines and on the left hand fuel rail on dual rail assemblies (fueling even cylinders No. 2, 4, 6). Refer to this number if servicing or part replacement is required.

Removal and Installation

2.0L ENGINE

1. Relieve fuel system pressure. Disconnect the negative battery cable.

2. Remove the throttle body assembly.

3. Remove the crankcase ventilation oil/air separator.

4. Disconnect the electrical connector and vacuum lines from the canister purge solenoid, EGR control sole-

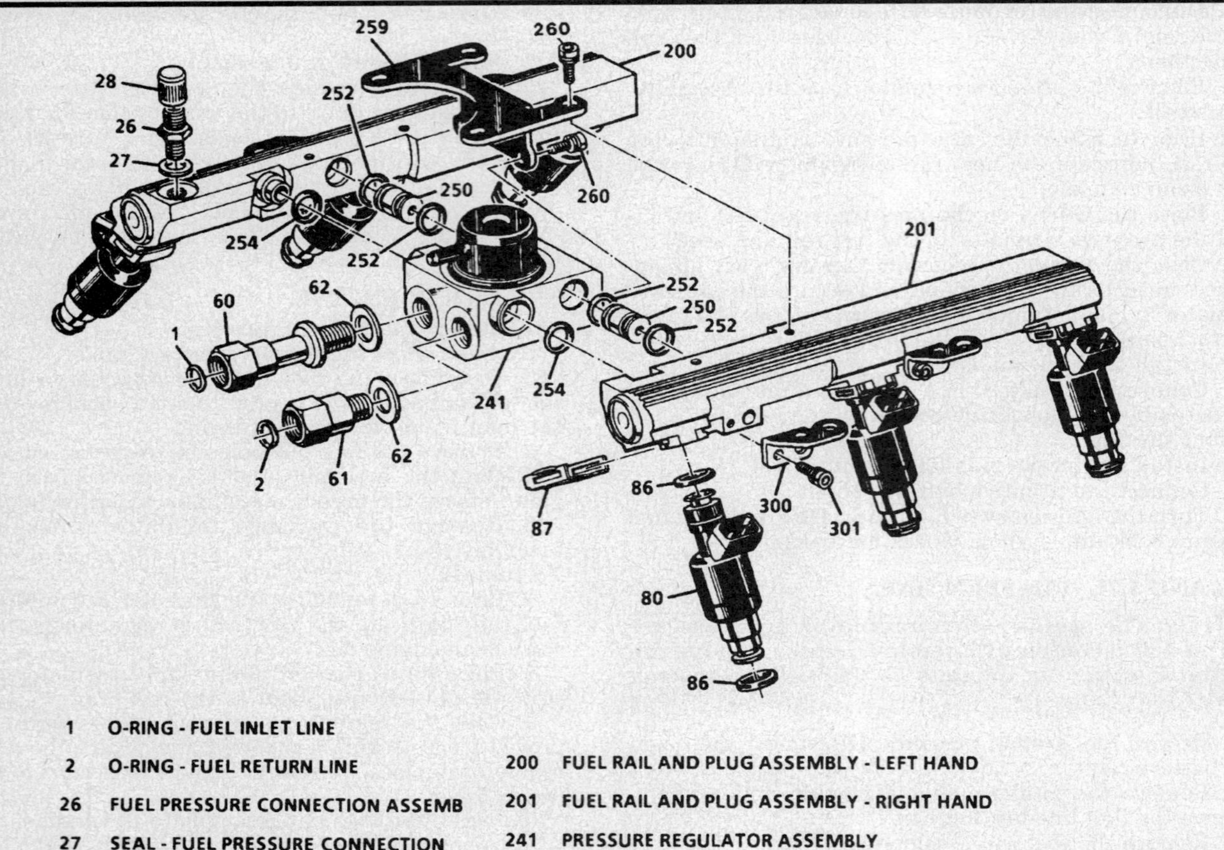

Fuel rail assembly with replaceable pressure regulator — 3.1L engine

noid and MAP sensor. Ensure lines are tagged for reinstallation.

5. Remove the canister purge solenoid, EGR solenoid, MAP sensor and brackets.

6. Push in on the fuel injector wire connector clip, while pulling the connector off.

7. Disconnect the fuel feed line and return line from the fuel rail assembly, be sure to use a backup wrench on the inlet fitting to prevent turning.

8. Remove the fuel rail assembly retaining bolts.

9. Remove the fuel rail assembly and cover all openings with masking tape to prevent dirt entry.

NOTE: If any injectors become separated from the fuel rail and remain in the intake manifold, both O-ring seals and injector retaining clip must be replaced. Use care in removing the fuel rail assembly, to prevent damage to the injector electrical connector terminals and the injector spray tips. When removed, support the fuel rail to avoid damaging its components. The fuel injector is serviced as a complete unit only. Since it is an electrical component, it should not be immersed in any type of cleaner.

To install:

10. Lubricate the new O-rings and seals with clean engine oil. Carefully push the injectors into the cylinder

head intake ports until the bolt holes on the fuel rail and manifold are aligned. Apply a coating of tread locking compound on the treads of the fittings. Torqe the fuel rail retaining bolts to 19 ft. lbs. (26 Nm).

11. Connect the fuel feed line and return line to the fuel rail assembly, be sure to use a backup wrench on the inlet fitting to prevent turning.

12. Reconnect the fuel injector wire connector.

13. Install the canister purge solenoid, EGR solenoid, MAP sensor and brackets.

14. Reconnect the electrical connector and vacuum lines to the canister purge solenoid, EGR control solenoid and MAP sensor.

15. Install the crankcase ventilation oil/air separator.

16. Install the thrrottle body assembly.

17. Connect the negative battery cable and pressurize the fuel system while checking for leaks.

2.2L ENGINE

The lower intake manifold replaces the fuel rail and is used to route the fuel and accommodate the injectors.

2.3L ENGINE

NOTE: Do not attempt to remove the fuel inlet fitting on the fuel rail. The fitting has been staked in place from the factory and any attempt made to remove it will result in damage to the fuel rail.

1. Relieve fuel system pressure.

2. Disconnect the negative battery cable.

3. Remove the crankcase ventilation oil/air separator and canister purge solenoid and position aside. Leave the hose connected to canister purge solenoid.

4. Disconnect the fuel pipe clamp bolt. Disconnect the vacuum hose at the pressure regulator.

5. Remove the fuel rail attaching bolts and remove the assembly from the cylinder head.

6. Push in on the fuel injector wire connector clip, while pulling the connector off.

7. Disconnect the fuel feed and return line being sure to use a backup wrench on the inlet fitting to prevent turning.

8. Remove the fuel rail assembly and cover all openings with masking tape to prevent dirt entry.

NOTE: If any injectors become separated from the fuel rail and remain in the intake manifold, both O-ring seals and injector retaining clip must be replaced. Use care in removing the fuel rail assembly, to prevent damage to the injector electrical connector terminals and the injector spray tips. When removed, support the fuel rail to avoid damaging its components. The fuel injector is serviced as a complete unit only. Since it is an electrical component, it should not be immersed in any type of cleaner.

To install:

9. Connect the fuel feed and return lines with new O-ring, being sure to use a backup wrench

10. Install the return pipe bracket retaining screw and torque to 53 inch lbs. (6 Nm).

11. Place the fuel rail over the cylinder head and connect the fuel injector wire connector.

12. Install the fuel rail attaching bolts and torque to 19 ft. lbs. (26 Nm).

13. Connect the vacuum hose at the pressure regulator.

14. Position the crankcase ventilation oil/air separator and canister purge solenoid in place.

15. Connect the negative battery cable and pressurize the fuel system while checking for leaks.

3.1L, 3.3L AND 3.4L ENGINES

1. Disconnect the negative battery cable. Relieve fuel system pressure.

2. Remove the intake manifold plenum. This is not necessary on 3.3L engine. Remove the fuel line bracket bolt.

3. Disconnect the fuel feed line and return line from the fuel rail assembly, be sure to use a backup wrench on the inlet fitting to prevent turning.

4. Remove the vacuum line at the pressure regulator. Remove the fuel rail assembly retaining bolts.

5. Push in the wire connector clip, while pulling the connector away from the injector.

6. Remove the fuel rail assembly and cover all openings with masking tape to prevent dirt entry.

NOTE: If any injectors become separated from the fuel rail and remain in the intake manifold, both O-ring seals and injector retaining clip must be replaced. Use care in removing the fuel rail assembly, to prevent damage to the injector electrical connector terminals and the injector spray tips. When removed, support the fuel rail to avoid damaging its components. The fuel injector is serviced as a complete unit only. Since it is an electrical component, it should not be immersed in any type of cleaner.

7. Installation is the reverse order of the removal procedure. Be sure to lubricate all the O-rings and seals with clean engine oil. Carefully tilt the fuel rail assembly and push the injectors into the cylinder head intake ports until the bolt holes on the fuel rail and manifold are aligned. Perform the following torque specifications:

    3.1L engine — Fuel rail retaining nuts to 88 inch. lbs. (10 Nm).

    3.1L engine — Fuel feed line nut to 17 ft. lbs. (23 Nm).

    3.3L engine — Fuel rail retaining nuts to 20 ft. lbs. (27 Nm).

    3.3L engine — Pressure regulator retaining screws–49 inch lbs. (5.5 Nm).

    3.3L engine — Fuel rail outlet fitting–20 ft. lbs. (27 Nm).

    3.4L engine — Fuel rail retaining nuts to 88 inch. lbs. (10 Nm).

    3.4L engine — Fuel feed line nut to 22 ft. lbs. (30 Nm).

8. Connect the negative battery cable. Energize the fuel pump and check for leaks.

5.0L (VIN F) AND 5.7L (VIN 8) ENGINES

1. Disconnect the negative battery cable. Relieve fuel system pressure.

2. Remove the intake manifold plenum and runners. Disconnect the fuel tube bracket retaining bolt.

3. Disconnect the fuel feed line and return line from the fuel rail assembly, be sure to use a backup wrench on the inlet fitting to prevent turning.

4. Disconnect the injector wire harness connectors.

5. Remove the vacuum line from the fuel pressure regulator. Remove the fuel rail assembly retaining bolts.

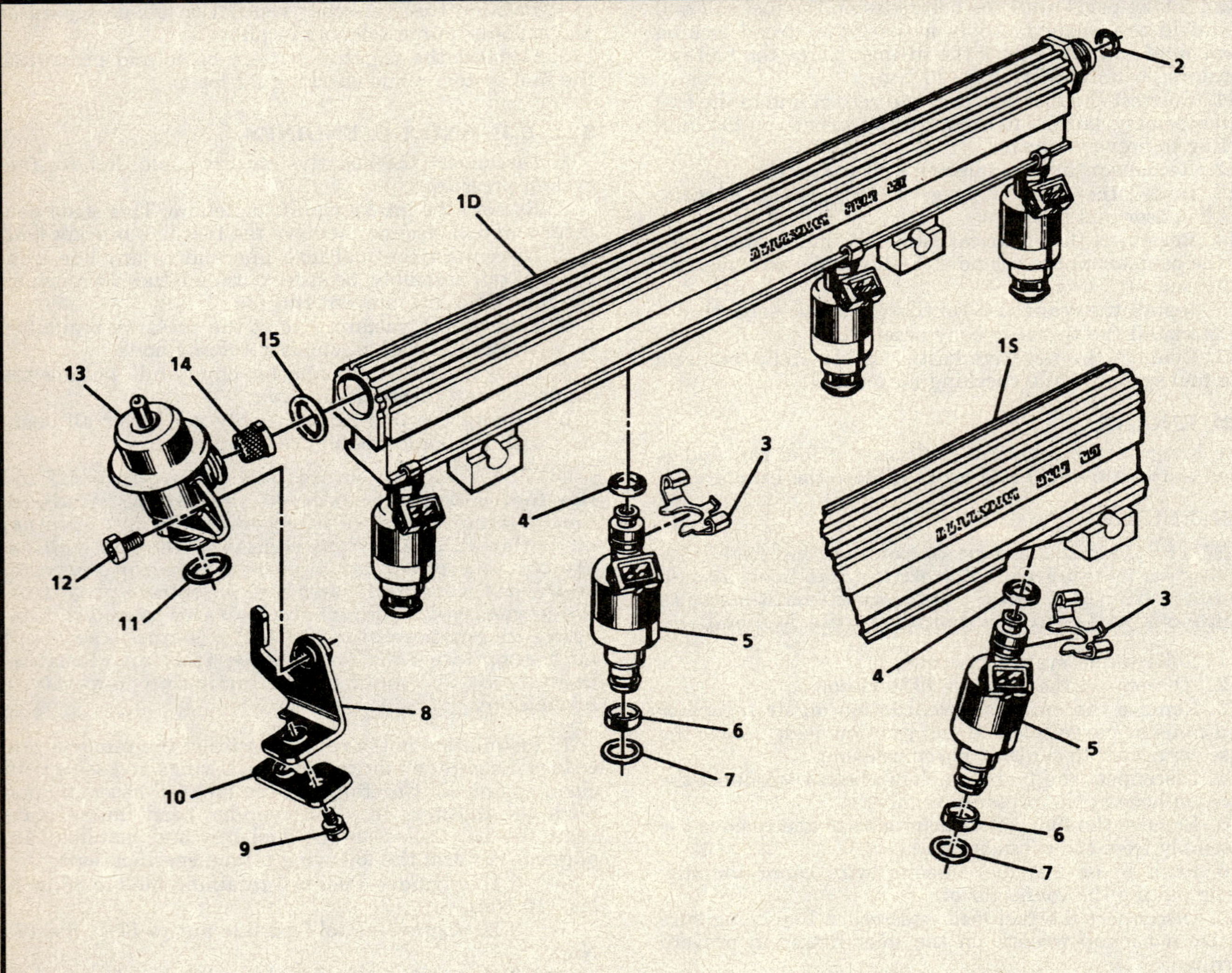

1	RAIL ASSEMBLY - MFI FUEL (S = SOHC; D = DOHC)	**9**	SCREW - BRACKET ATTACHING
2	O-RING - FUEL INLET LINE	**10**	BRACKET - RETURN LINE ATTACHING
3	CLIP - INJECTOR RETAINER	**11**	O-RING - FUEL RETURN LINE
4	O-RING - UPPER INJECTOR (BLACK)	**12**	SCREW - FUEL PRESSURE REGULATOR ATTACHING
5	INJECTOR ASSEMBLY - MFI FUEL	**13**	REGULATOR ASSEMBLY - FUEL PRESSURE
6	BACKUP - O-RING	**14**	SCREEN - FILTER (IF SO EQUIPPED)
7	O-RING - LOWER INJECTOR (BROWN)	**15**	O-RING - FUEL INLET FITTING
8	RETAINER AND SPACER ASSEMBLY		

Fuel rail assembly — 2.3L engine

6. Remove the fuel rail assembly and cover all openings with masking tape to prevent dirt entry.

NOTE: If any injectors become separated from the fuel rail and remain in the intake manifold, both O-ring seals and injector retaining clip must be replaced. Use care in removing the fuel rail assembly, to prevent damage to the injector electrical connector terminals and the injector spray tips. When removed, support the fuel rail to avoid damaging its components. The fuel injector is serviced as a complete unit only. Since it is an electrical component, it should not be immersed in any type of cleaner.

7. Installation is the reverse order of the removal procedure. Be sure to lubricate all the O-rings and seals with clean engine oil. Carefully tilt the fuel rail assembly and push the injectors into the cylinder head intake ports until the bolt holes on the fuel rail and manifold are aligned. Torque the fuel rail retaining bolts to 15 ft. lbs. (20 Nm), the fuel tube bracket bolt to 25 ft. lbs. (34 Nm), the fuel line fittings to 20 ft. lbs. (27 Nm).

8. Connect the negative battery cable. Energize the fuel pump and check for leaks.

5.7L ENGINE

1. Disconnect the negative battery cable. Relieve fuel system pressure.

2. Disconnect the inlet and return fuel lines using tool J–37088–A or equivalent.

3. Remove the vacuum line from the fuel pressure regulator and also the vacuum lines necessary tio gain access to the fuel rail. Remove the fuel rail assembly retaining bolts.

4. Disconnect the injector wire harness connectors.

5. Remove the fuel rail assembly and cover all openings with masking tape to prevent dirt entry.

NOTE: If any injectors become separated from the fuel rail and remain in the intake manifold, both O-ring seals and injector retaining clip must be replaced. Use care in removing the fuel rail assembly, to prevent damage to the injector electrical connector terminals and the injector spray tips. When removed, support the fuel rail to avoid damaging its components. The fuel injector is serviced as a complete unit only. Since it is an electrical component, it should not be immersed in any type of cleaner.

6. Install the fuel rail assembly into manifold. Be sure to lubricate all the O-rings and seals with clean engine oil. Carefully tilt the fuel rail assembly and push the injectors into the cylinder head intake ports until the bolt holes on the fuel rail and manifold are aligned. Torque the fuel rail retaining bolts to 15 ft. lbs. (20 Nm).

7. Connect the injector wire harness connectors and position the wire harness into the retaining clips on the mainfold.

8. Connect the vacuum lines to the fuel pressure regulator and to the intake manifold.

9. Connect the fuel feed and return lines. Apply a few drops of clean enngine oil to the ends of the fuel line; push the lines together until the retaining tabs on the quick connect fittings snap into place.

10. Connect the negative battery cable. Energize the fuel pump and check for leaks.

IDLE AIR CONTROL VALVE

Removal and Installation

EXCEPT 5.0L AND 5.7L ENGINES

1. Disconnect the negative battery cable.

2. Disconnect electrical connector from idle air control valve.

3. Remove the screws retaining the idle air control valve. Remove the idle air control valve from mounting position.

To install:

4. Prior to installing a new idle air control valve, measure the distance the valve plunger is extended. Measurement should be made from the edge of the valves mounting flange to the end of the cone. The distance should not exceed 1⅛ in. (28mm), or damage to the valve may occur when installed. If measuring distance is greater than specified above and the IAC valve is new, press on the valve firmly to retract it, using a slight side to side motion to help it retract easier. Do not attempt to push the pintle in on a used valve, otherwise the valve may become damaged.

5. Use a new gasket and install the idle air control valve in mounting position.

6. Install the retaining screws. Torque the retaining screws to 30 inch lbs. (3.4 Nm). Connect the electrical connector.

7. Reset the IAC valve position.

5.0L AND 5.7L ENGINES

1. Disconnect the negative battery cable. Remove the air inlet assembly, as required.

2. Removal electrical connector from idle air control valve.

3. Remove the idle air control valve using a 1¼ in. (32mm) wrench.

4. Installation is the reverse of removal. Before installing a new idle air control valve, measure the distance that the valve is extended. Measurement should be made from the motor housing to the end of the cone. The distance should not exceed 1⅛ in. (28mm), or damage to the valve may occur when installed. Use a new gasket and turn the ignition **ON** then **OFF** again to allow the ECM to reset the idle air control valve. Torque the IAC valve to 13 ft. lbs. (18 Nm). Reset the IAC valve position.

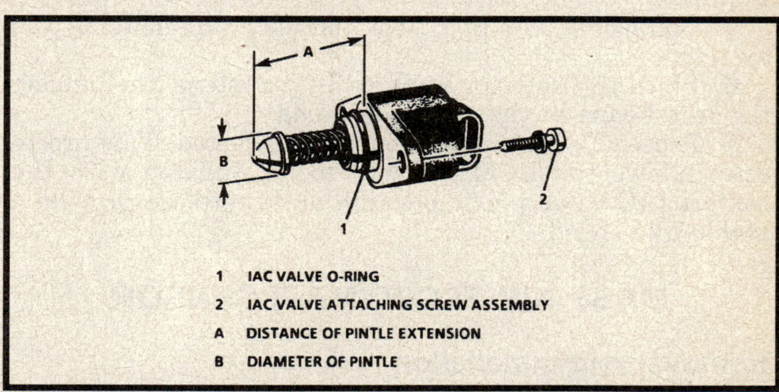

Measuring a new IAC valve pintle length — except 5.0L and 5.7L engines

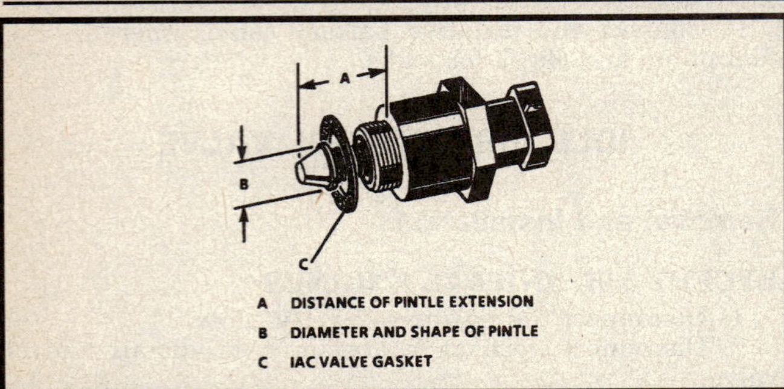

Measuring a new IAC valve pintle length — 5.0L and 5.7L engines

Resetting Procedure

2.0L ENGINE

Turn the ignition **ON** for 5 seconds, then **OFF** for 10 seconds. Start the engine. The ECM will then reset the IAC valve.

2.2L ENGINE

Using the appropriate scan tool and the "Idle Learn Procedure" as per instructions. If a scan tool is unavailable the ECM will reset the IAC automatically after at least 7 minutes of driving.

2.3L AND 3.3L ENGINES

1. Turn the ignition switch to the **ON** position, engine off.
2. Ground the diagnostic test terminal for 5 seconds, then remove the ground.
3. Turn the ignition **OFF** for 10 seconds.
4. Start the engine and check for proper idle operation.

3.1L, 3.4L, 5.0L AND 5.7L ENGINES

1. Turn the ignition switch to the **ON** position for 5 seconds, with the engine **OFF**.
2. Turn the ignition **OFF** for 5–10 seconds.
3. Start the engine and check for proper idle operation.

> NOTE: If the battery is disconnected for any reason, the programmed position of the IAC valve pintle is lost and replaced with a default value. This will cause the engine idle to be unstable for a period of approximately 7 minutes. To return the IAC valve pintle to the correct position, perform IAC resetting.

4. Re-establish a 12 volt battery supply source, as required.
5. Connect a TECH 1 scan tool or equivalent, to the system.
6. Place the selector knob to IAC System, then select the Idle Learn in the Misc test mode.
7. Proceed with the idle learn, as directed. This procedure allows the ECM memory to be updated with the correct IAC valve pintle position and therefore provide a stable idle speed.

MASS AIR FLOW (MAF) SENSOR

Removal and Installation

Only the 3.3L engine is equipped with a MAF sensor. The MAF is part of the throttle body assembly.
1. Disconnect the negative battery cable.

2. Disconnect the electrical connector to the MAF sensor.
3. Remove the air cleaner and duct assembly. Remove the sensor retaining bolts and remove the sensor.
4. Installation is the reverse of the removal procedure.

MANIFOLD ABSOLUTE PRESSURE (MAP) SENSOR

Removal and Installation

1. Disconnect the negative battery cable and remove the vacuum hose from the MAP sensor.
2. Disconnect the electrical connector to the MAP sensor.
3. Remove the MAP sensor retaining screws and remove the MAP sensor.
4. Installation is the reverse order of the removal procedure.

MANIFOLD/INTAKE AIR TEMPERATURE (MAT/IAT) SENSOR

Removal and Installation

1. Disconnect the negative battery cable.
2. Disconnect the electrical connector to the MAT/IAT sensor.
3. Remove the sensor retaining clip from the air cleaner or unscrew the MAT/IAT sensor from the intake manifold.
4. Installation is the reverse order of the removal procedure.

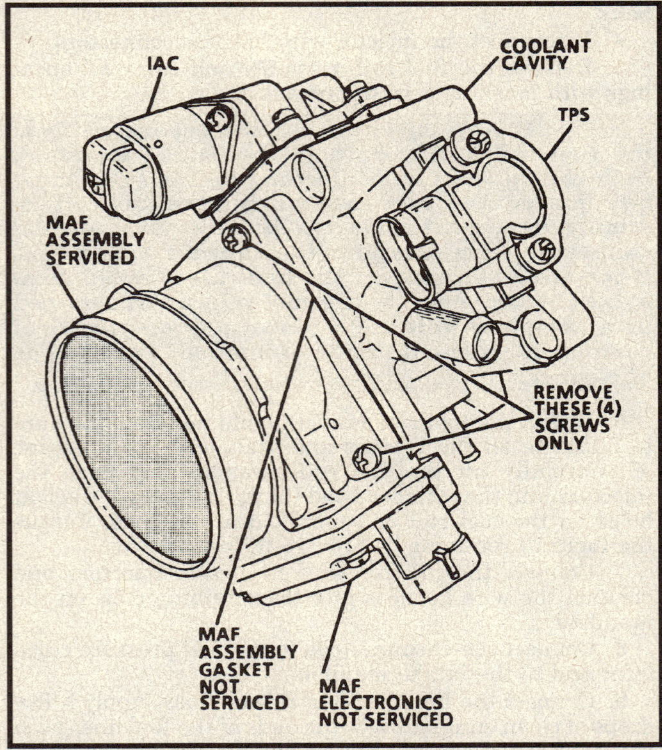

MAF and throttle body assembly — 3.3L engine

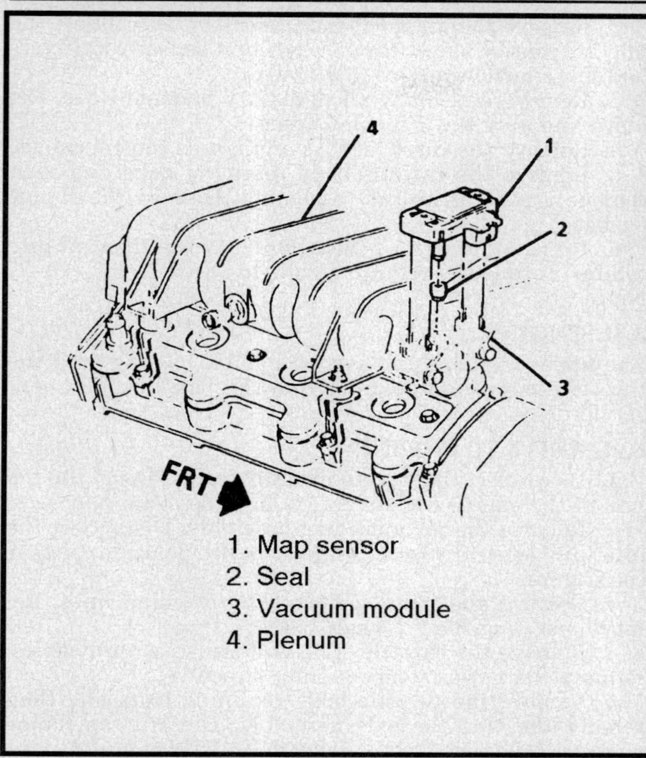

1. Map sensor
2. Seal
3. Vacuum module
4. Plenum

Removing MAP sensor — 3.4L engine

OXYGEN SENSOR

Removal and Installation

The oxygen sensor uses a permanently attached pigtail and connector. This pigtail should not be removed from the oxygen sensor. Damage or removal of the pigtail connector could effect the proper operation of the oxygen sensor. Use caution when handling the oxygen sensor. The in-line electrical connector and louvered end must be kept free of grease, dirt or other contaminants. Also avoid using cleaner solvent of any type. Do not drop or roughly handle the oxygen sensor. It is not recommended to clean or wire brush an oxygen sensor when in question, it is recommended that a new oxygen sensor be used.

NOTE: The oxygen sensor may be difficult to remove when the engine temperature is below 120°F (48°C). Excessive force may damage the threads in the exhaust manifold of exhaust pipe.

1. Start the engine and let it warm up to 120°F (48°C), stop the engine and disconnect the negative battery cable.
2. Disconnect the electrical connector from the oxygen sensor.

3. Using a special oxygen sensor socket, remove the sensor from the exhaust manifold or exhaust pipe.

NOTE: Anti-seize compound is used on the oxygen sensor threads. The compound consists of a liquid graphite and glass beads. The graphite will burn away, but the glass beads will remain, making the sensor easier to remove. New or service sensors will already have the compound applied to the threads. If a oxygen sensor is removed from the engine and if for any reason it is to be reinstalled, the threads must have anti-seize compound applied before installation.

To install:

4. Coat the threads of the oxygen sensor with anti-seize compound 5613695 or equivalent, if necessary.
5. Install the sensor and torque it to 30 ft. lbs (41 Nm).
6. Reconnect the electrical connector to the sensor and the negative battery cable.

POWER STEERING PRESSURE SWITCH

Removal and Installation

1. Disconnect the negative battery cable.
2. Disconnect the electrical connector to the switch and place a drip pan under the switch.
3. Unscrew switch and remove the switch from the power steering line.
4. Installation is the reverse order of the removal procedure. Be sure to replace the power steering fluid lost during removal of the switch.

VEHICLE SPEED SENSOR

Removal and Installation

1. Disconnect the negative battery cable. Remove the transmission crossmember on Corvette convertible.
2. Remove the vehicle speed sensor electrical lead from the transmission.
3. Remove the VSS sensor bolt and screw retainer.
4. Remove the vehicle speed sensor assembly and O-ring.
5. Installation is the reverse order of the removal procedure. Use a new O-ring and coat with transmission fluid.

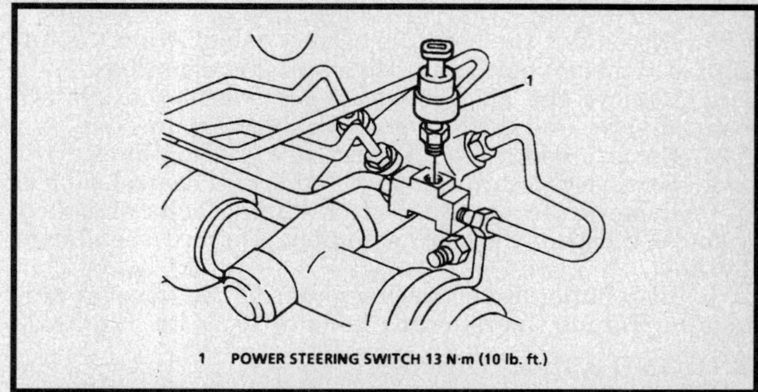

Power steering pressure switch — 3.1L engine

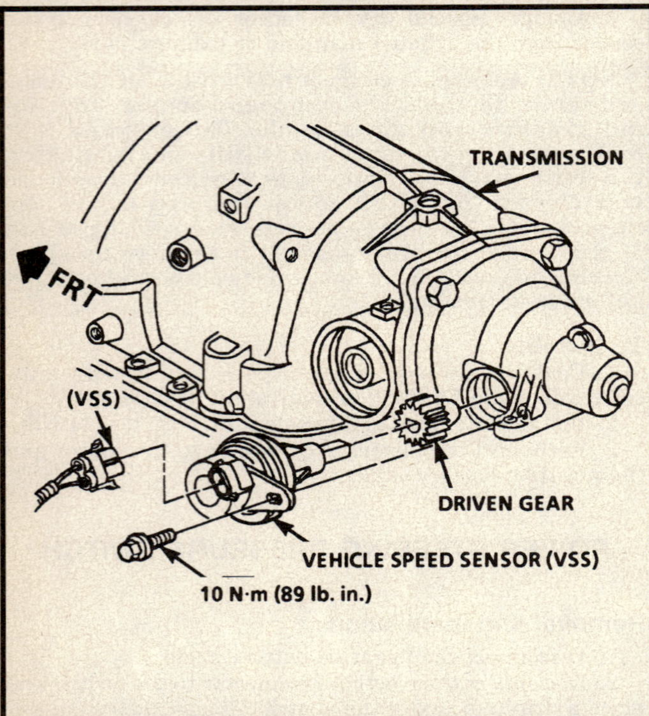

VSS with automatic transmission — Camaro, Firebird and Corvette

THROTTLE BODY

Removal and Installation

2.0L ENGINE

1. Disconnect the negative battery cable.
2. Remove the air intake duct.
3. Partially drain the antifreeze from the radiator and disconnect the coolant hoses from the throttle body assembly.
4. Disconnect the IAC and TPS electrical connectors.
5. Disconnect the vacuum lines, throttle valve and cruise control cables from the throttle body.
6. Remove the throttle body mounting bolts. Remove the throttle body and gasket.
7. Installation is the reverse order of the removal procedure. Use and new gasket. Torque the throttle body mounting bolts to 11 ft. lbs. (15 Nm).

2.3L ENGINE

1. Disconnect the negative battery cable. Drain the top half of the engine coolant into a suitable drain pan.
2. Remove the air inlet duct. Disconnect the idle air control valve and throttle position sensor connectors.
3. Remove and mark all necessary vacuum lines.
4. Remove the throttle, T.V. and cruise control cables.
5. Remove the throttle body retaining bolts and then remove the throttle body assembly. Discard the flange gasket.
6. Installation is the reverse order of the removal procedure. Torque the retaining bolts to 19 ft. lbs. (26 Nm).

3.1L AND 3.3L ENGINES

1. Disconnect the negative battery cable. Drain the top half of the engine coolant into a suitable drain pan.
2. Remove the air inlet duct assembly. Disconnect the idle air control valve, throttle position sensor connectors and mass airflow sensor (3.3L only).
3. Remove and mark all necessary vacuum lines. Remove and plug the 2 coolant hoses.
4. Remove the throttle, T.V. and cruise control cables.
5. Remove the throttle body retaining bolts and then remove the throttle body assembly. Discard the flange gasket.
6. Installation is the reverse order of the removal procedure. Torque the retaining bolts to 18–21 ft. lbs. (25–27 Nm).

3.4L ENGINE

The upper manifold consists of a single casting with the throttle body integrated; therefor the assembly is non-serviceable.

5.0L AND 5.7L ENGINES

1. Disconnect the negative battery cable. Drain the top half of the engine coolant into a suitable drain pan.
2. Remove the air inlet duct assembly. Disconnect the idle air control valve and throttle position sensor connectors.
3. Remove and mark all necessary vacuum lines. Remove and plug the 2 coolant hoses.
4. Remove the throttle, T.V. and cruise control cables. Remove the power steering pump brace.
5. Remove the throttle body retaining bolts and then remove the throttle body assembly. Discard the flange gasket.
6. Installation is the reverse order of the removal procedure. Torque the retaining bolts to 18 ft. lbs.

THROTTLE POSITION SENSOR

The throttle position sensor is non-adjustable. If the sensor is found to be out of specifications and the sensor is at fault, it must be replaced.

Removal and Installation

1. Disconnect the electrical connector from the sensor.
2. Remove the attaching screws, lock washers and retainers.
3. Remove the throttle position sensor. If necessary, remove the screw holding the actuator to the end of the throttle shaft.
4. With the throttle valve in the normal closed idle position, install the throttle position sensor on the throttle body assembly, making sure the sensor pickup lever is located above the tang on the throttle actuator lever.
5. Install the retainers, screws and lock washers using a thread locking compound. DO NOT tighten the screws until the throttle position switch is adjusted.

TORQUE CONVERTER CLUTCH (TCC) SOLENOID

Removal and Installation

1. Remove the negative battery cable. Raise and support the vehicle safely.
2. Drain the transmission fluid into a suitable drain pan. Remove the transmission pan.
3. Remove the secondary and third clutch switch as necessary.
4. Remove the TCC solenoid retaining screws and then remove the electrical connector, solenoid and check ball.

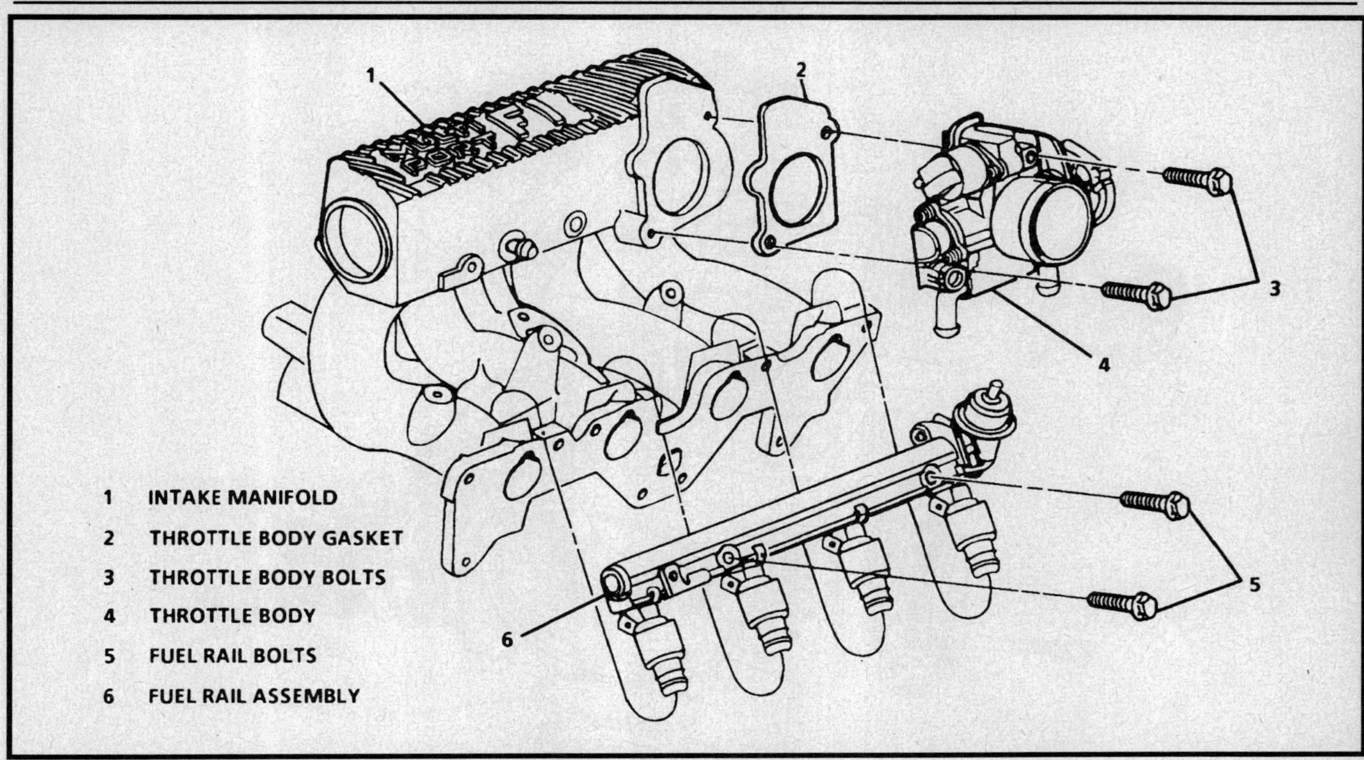

Fuel rail and throttle body assembly removal — 2.0L engine

5. Clean and inspect all parts. Replace defective parts as necessary.

To install:

6. Install the check ball, TCC solenoid and electrical connector. Install the solenoid retaining screws and torque them to 10 ft. lbs. (14 Nm).

7. Install the transmission pan with a new gasket and torque the pan retaining bolts to 10 ft. lbs. (14 Nm).

8. Lower the vehicle and refill the transmission with the proper amount of the recommended automatic transmission fluid.

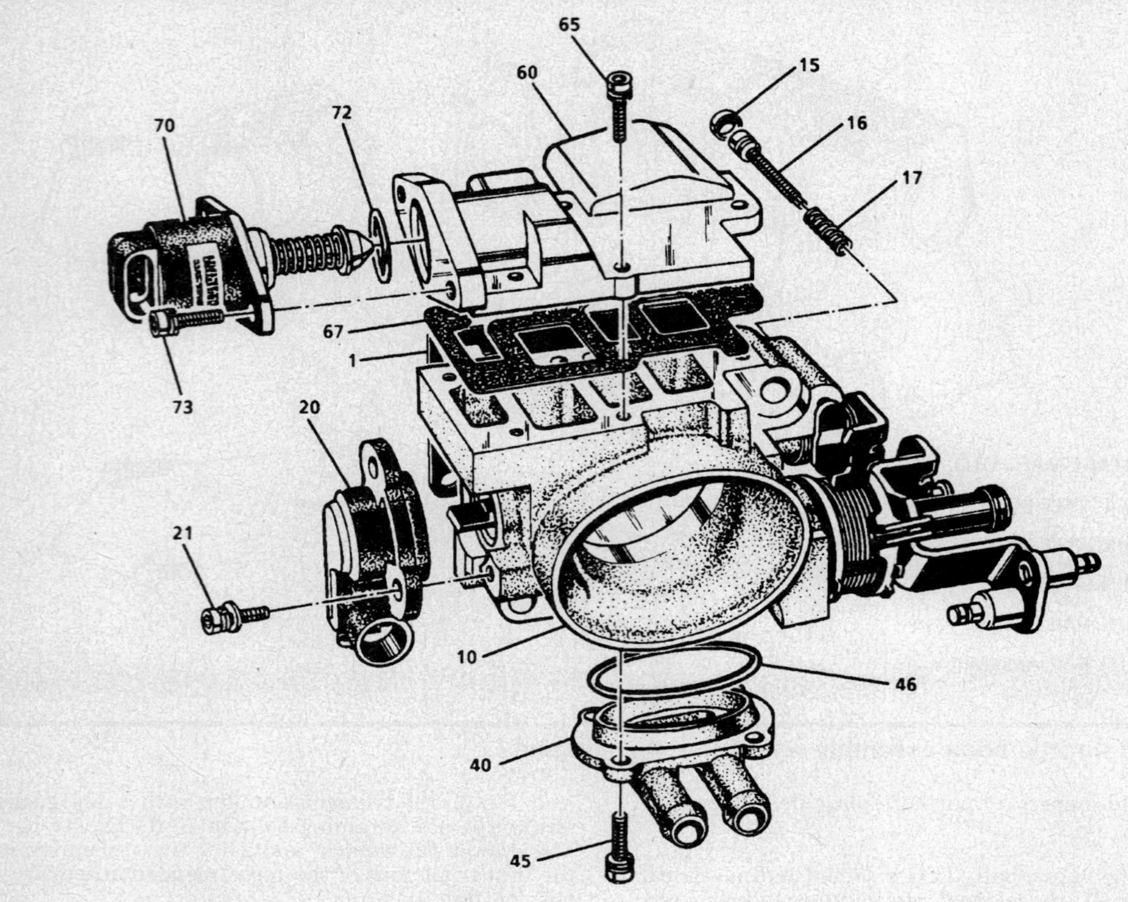

1	GASKET - FLANGE
10	THROTTLE BODY ASSEMBLY
15	PLUG - IDLE STOP SCREW
16	SCREW ASSEMBLY - IDLE STOP
17	SPRING - IDLE STOP SCREW ASSEMBLY
20	SENSOR - THROTTLE POSITION (TP)
21	SCREW ASSEMBLY - TP SENSOR ATTACHING
40	COVER - COOLANT CAVITY
45	SCREW ASSEMBLY - COOLANT COVER ATTACHING
46	O-RING - COOLANT COVER TO THROTTLE BODY
60	IDLE AIR/VACUUM SIGNAL HOUSING ASSEMBLY
65	SCREW ASSEMBLY - IDLE AIR/VACUUM SIGNAL ASSEMBLY
67	GASKET - IDLE AIR/VACUUM SIGNAL ASSEMBLY
70	VALVE ASSEMBLY - IDLE AIR CONTROL (IAC)
72	O-RING - IDLE AIR CONTROL VALVE
73	SCREW ASSEMBLY - IDLE AIR CONTROL VALVE ATTACHING

Throttle body assembly — 3.1L engine

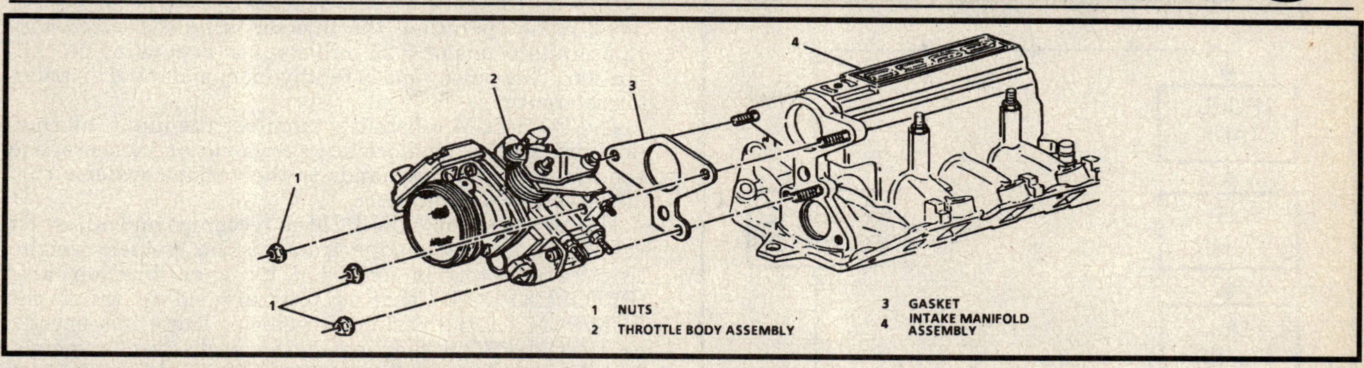

Throttle body assembly removal — 3.3L engine

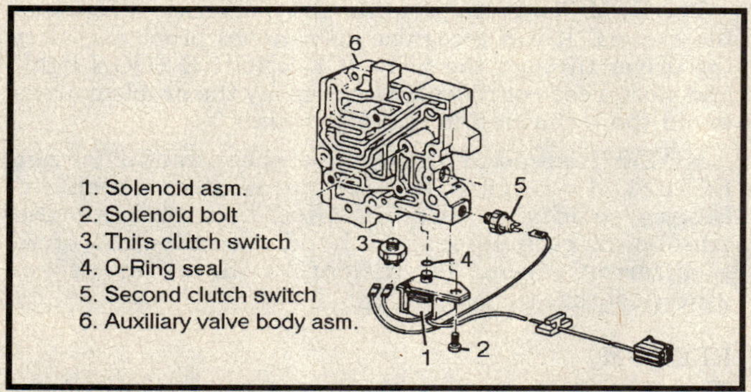

TCC solenoid assembly removal — Sunbird

LIGHT TRUCKS, VAN, GEO AND SATURN

General Description

There are several vehicle MFI systems covered in this section: the Chevrolet S/T series pick-ups with 2.2L (VIN 4), the GMC Syclone and Typhoon with the 4.3L (VIN Z) turbocharged engine, the Geo Prizm and Storm with either the 1.6L (VIN 5) engine which uses a mass air flow sensor or the 1.6L (VIN 6) and 1.8L (VIN 8) engines which use a manifold air pressure sensor and the Saturn equipped with 1.9L (VIN Z) engine.

The basic function of the fuel system is to control fuel delivery, by way of one individual fuel injectors per cylinder, to the engine in order to minimize exhaust emissions and maximize performance, while obtaining the 14.7:1 air/fuel ratio. The fuel system is constantly monitored by several sensors and in turn, adjusted by ECM controlled components, thus labeling it a "Closed Loop" system.

SYSTEM OPERATION

The Multiport Fuel Injection (MFI) system is controlled by an Electronic Control Module (ECM) which monitors engine operations and generates output signals to provide the correct air/fuel mixture, ignition timing and engine idle speed control. Input to the control unit is provided by an oxygen sensor, coolant temperature sensor, detonation sensor, mass air flow sensor (1.6L VIN 5), manifold absolute pressure sensor, manifold air temperature sensor and throttle position sensor. The ECM also receives information concerning engine rpm, road speed, transmission gear position, power steering and air conditioning.

The injectors are located, one at each intake port, rather than the single injector found on throttle body system. The injectors are mounted on a fuel rail and are activated by a signal from the electronic control module. The injector is a solenoid-operated valve which remains open depending on the width of the electronic pulses (length of the signal) from the ECM; the longer the open time, the more fuel is injected. In this manner, the air/fuel mixture can be precisely controlled for maximum performance with minimum emissions.

Fuel is pumped from the tank by a high pressure pump, located inside the fuel tank. It is a positive displacement roller vane pump. The impeller serves as a vapor separator and pre-charges the high pressure assembly. A pressure regulator maintains 28–36 psi (28–50 psi on turbocharged engines) in the fuel line to the injectors and the excess fuel is fed back to the tank. A fuel accumulator is used to dampen the hydraulic line hammer in the system created when all injectors open simultaneously.

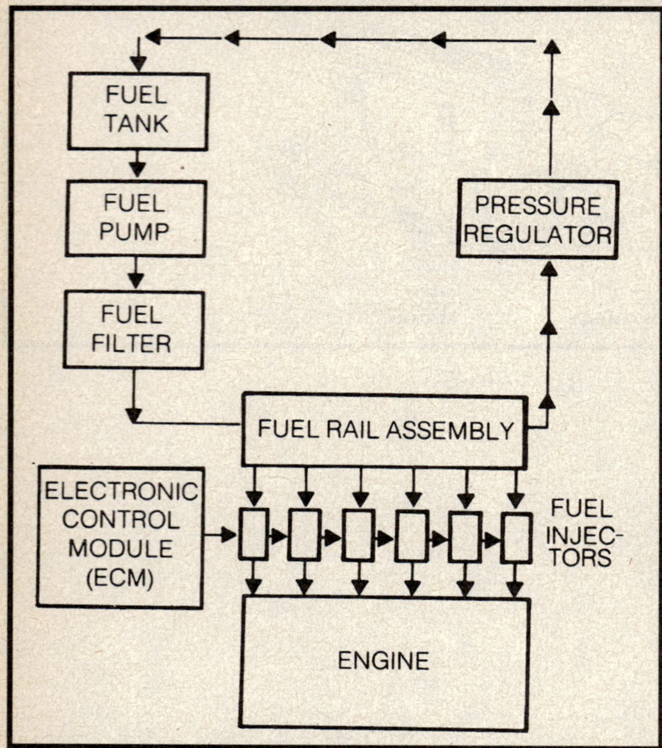

Typical fuel supply system

SYSTEM COMPONENTS

Electronic Control Module (ECM) or Powertrain Control Module

All GMC/Chevrolet truck use an ECM. All Saturn and Geo vehicles use an ECM if equipped with a manual transmission or a PCM if equipped with an automatic transaxle. The ECM/PCM consists of 2 parts; a Controller (the ECM without the PROM) and a Calibrator called a PROM (Programmable Read Only Memory). The Chevrolet 2.2L and Saturn 1.9L engine uses an ECM which is equipped with a Controller and an EEPROM (Electronically Eraseable Programmable Read Only Memory).

CONTROLLER

The fuel injection system is controlled by an on-board computer, the ECM/PCM or controller, usually located in the passenger compartment. The ECM/PCM monitors engine operations and environmental conditions (ambient temperature, barometric pressure, etc.) needed to calculate the fuel delivery time (pulse width/injector on-time) of the fuel injector. The fuel pulse may be modified by the ECM/PCM to account for special operating conditions, such as cranking, cold starting, altitude, acceleration and deceleration.

The control unit moderates the exhaust emissions by curving fuel delivery to achieve, as nearly as possible an air/fuel ratio of 14.7:1. The injector on-time is determined by the various sensor inputs to the ECM/PCM. By increasing the injector pulse, more fuel is delivered, enriching the air/fuel ratio. Pulses are sent to the injector in 2 different modes, synchronized and non-synchronized.

In synchronized mode operation, the injector is pulsed once for each distributor reference pulse. In non-synchro-

nized mode operation, the injector is pulsed once every 12.5 milliseconds or 6.25 milliseconds depending on calibration. This pulse time is totally independent of distributor reference.

The ECM/PCM constantly monitors the input information, processes this information from various sensors, and generates output commands to the various systems that affect vehicle performance.

The ability of the ECM/PCM to recognize and adjust for vehicle variations (engine transmission, vehicle weight, axle ratio, etc.) is provided by a calibration unit (PROM/EEPROM) that is programmed to tailor the ECM/PCM for the particular vehicle. There is a specific ECM/PCM PROM combination for each specific vehicle, and the combinations are not interchangeable with those of other vehicles.

The ECM/PCM also performs the diagnostic function of the system. It can recognize operational problems, alert the driver through the SERVICE ENGINE SOON light, and store a code or codes which identify the problem areas to aid the technician in making repairs.

NOTE: Instead of an edgeboard connector, newer ECM/PCM's have a header connector which attaches solidly to the ECM/PCM case. Like the edgeboard connectors, the header connectors have a different pinout identification for different engine designs.

EEPROM

THE EEPROM contains the preset calibrations of the vehicle according to the VIN. The EEPROM is non-repairable and can only be replaced with the controller assembly. The EEPROM will need to be reprogrammed using the manufacturers' equipment.

PROM

To allow one model of the ECM/PCM to be used for many different vehicles, a device called a Calibrator (or PROM) is used. The PROM is located inside the ECM/PCM and has information on the vehicle's weight, engine, transmission, axle ratio and other components.

While one ECM/PCM part number can be used by many different vehicles, a PROM is very specific and must be used for the right vehicle. For this reason, it is very important to check the latest parts book and or service bulletin information for the correct PROM part number when replacing the PROM.

An ECM/PCM used for service (called a controller) comes without a PROM. The PROM from the old ECM/PCM must be carefully removed and installed in the new ECM/PCM.

MEM-CAL

This assembly contains the functions of the PROM, Cal-Pak and the ESC module used on other GM applications. Like the PROM, it contains the calibrations needed for a specific vehicle as well as the back-up fuel control circuitry required if the rest of the ECM/PCM becomes damaged or faulty.

Starting Mode

When the engine is first turned **ON**, the ECM/PCM will turn on the fuel pump relay for 2 seconds and the fuel pump will build up pressure. The ECM/PCM then checks the coolant temperature sensor, throttle position sensor and crank sensor, then the ECM/PCM determines the proper air/fuel ratio for starting. This ranges from 1.5:1 at -33°F (-36°C) to 14.7:1 at 201°F (94°C).

The ECM/PCM controls the amount of fuel that is delivered in the Starting Mode by changing how long the injectors are turned on and off. This is done by pulsing the injectors for very short times.

Clear Flood Mode

If for some reason the engine should become flooded, provisions have been made to clear this condition. To clear the flood, the driver must depress the accelerator pedal to the wide-open throttle position. The ECM/PCM then completely turns off the fuel flow. The ECM/PCM holds this operational mode as long as the throttle stays in the wide open position and the engine rpm is below 600. If the throttle position becomes less than 62–70%, the ECM/PCM returns to the starting mode.

Run Mode

There are 2 different run modes. When the engine is first started and the rpm is above 400, the system goes into open loop operation. In open loop operation, the ECM/PCM will ignore the signal from the oxygen (O_2) sensor and calculate the injector on-time based upon inputs from the coolant sensor, MAF or MAP sensor and MAT sensors.

During open loop operation, the ECM/PCM analyzes the following items to determine when the system is ready to go to the closed loop mode.

1. The oxygen sensor varying voltage output. (This is dependent on temperature).

2. The coolant sensor must be above specified temperature.

3. A specific amount of time must elapse after starting the engine. These values are stored in the PROM.

When these conditions have been met, the system goes into closed loop operation. In closed loop operation, the ECM/PCM will calculate the air/fuel ratio (injector on time) based upon the signal from the oxygen sensor. The ECM/PCM will decrease the on-time if the air/fuel ratio is too rich, and will increase the on-time if the air/fuel ratio is too lean.

Acceleration Mode

When the accelerator pedal is depressed, the opening of the throttle valve(s) causes a rapid increase in airflow into the cylinders. The air flow is capable of moving much faster then the flow of fuel and therefore the engine may hesitate; to avoid this situation the ECM/PCM increases the fuel injector pulse width according to the TPS and MAF or MAP signals it receives.

Deceleration Mode

Upon deceleration, a leaner fuel mixture is required to reduce emission of hydrocarbons (HC) and carbon monoxide (CO). To adjust the injection on-time, the ECM/PCM uses the decrease in MAP and the decrease in throttle position to calculate a decrease in injector on time. To maintain an idle fuel ratio of 14.7:1, fuel output is momentarily reduced. If deceleration is very rapid or for long periods, the ECM/PCM can cut off the fuel completely to prevent damage to the catalytic converter.

Battery Voltage Correction Mode

The purpose of battery voltage correction is to compensate for variations in battery voltage to fuel pump and injector response. The ECM/PCM compensates by increasing the engine idle rpm.

Battery voltage correction takes place in all operating modes. When battery voltage is low, the spark delivered by the distributor may be low. To correct this low battery voltage problem, the ECM/PCM can do any or all of the following:

- Increase injector on time (increase fuel)
- Increase idle rpm
- Increase ignition dwell time

Fuel Cutoff Mode

Depending on the pre-programmed values, the ECM/PCM turns **OFF** the fuel flow to the injectors to prevent engine damage.

Converter Protection Mode

In this mode the ECM/PCM estimates the temperature of the catalytic converter and then modifies fuel delivery to protect the converter from high temperatures. When the ECM/PCM has determined that the converter may overheat, it will cause open loop operation and will enrichen the fuel delivery. A slightly richer mixture will then cause the converter temperature to be reduced.

Data Link Connector (DLC)

Formerly known as the Assembly Line Data Link (ALDL) or Assembly Line Communication Link (ALCL), the Data Link Connector (DLC) was renamed according to the new SAE J-1930 standards in order to universalize automotive terminology among manufacturers.

The diagnostic connector is located in the passenger compartment, usually under the instrument panel. The exception however, is Geo which has the DLC in the engine compartment.

The Diagnostic mode or the Field Service Mode may be entered on all except Geo by connecting or jumping terminal B, the diagnostic TEST terminal to terminal A, or ground circuit. On Geo Prizm and Storm, connect or jump terminal TE1, the diagnostic TEST terminal to terminal E1, or ground circuit. The DLC connector is a very useful tool in diagnosing performance and engine related ECM/PCM operating conditions. Information from the ECM/PCM is available at this terminal and can be read with one of the many popular scanner tools.

Fuel Control System

The fuel control system is made up of the following components:

1. Fuel supply system
2. Throttle body assembly
3. Fuel injectors
4. Fuel rail
5. Fuel pressure regulator
6. Idle Air Control (IAC)
7. Fuel pump
8. Fuel pump relay
9. In-line fuel filter

The fuel control system starts with the fuel in the fuel tank. An electric fuel pump, located in the fuel tank with the fuel gauge sending unit, pumps fuel to the fuel rail through an in-line fuel filter. The pump is designed to provide fuel at a pressure above the pressure needed by the injectors. A pressure regulator in the fuel rail keeps fuel available to the injectors at a constant pressure. Unused fuel is returned to the fuel tank by a separate line.

In order for the fuel injectors to supply a precise amount of fuel at the command of the ECM/PCM, the fuel supply system maintains a constant pressure (approximately 35 psi) drop across the injectors. As manifold vacuum changes, the fuel system pressure regulator controls the fuel supply pressure to compensate. The fuel pressure accumulator used on select models, isolates fuel line noise. The fuel rail is bolted rigidly to the engine and it provides the upper mount for the fuel injectors. It also contains a spring loaded pressure tap for testing the fuel system or relieving the fuel system pressure.

The injectors are controlled by the ECM/PCM. They deliver fuel in one of several modes as previously described. In order to properly control the fuel supply, the fuel pump is operated by the ECM/PCM through the fuel pump relay and oil pressure switch.

Throttle Body Unit

The throttle body unit has a throttle valve to control the amount of air delivered to the engine. The TPS and IAC valve are also mounted onto the throttle body. The throttle body contains vacuum ports located at, above or below the throttle valve. These vacuum ports generate the vacuum signals needed by various components.

On some vehicles, the engine coolant is directed through the coolant cavity at the bottom of the throttle body to warm the throttle valve and prevent icing.

Fuel Injector

A fuel injector is installed in the intake manifold at each cylinder. The fuel injector is a solenoid operated device controlled by the ECM. The ECM turns on the solenoid, which opens the valve to allow fuel delivery. The fuel, under pressure, is injected in a conical spray pattern at the opening of the intake valve. The regulated fuel, which is not used by the injectors is returned to the fuel tank.

An injector that is partly open, will cause loss of fuel pressure after the engine is shut down, so long crank time would be noticed on some engines. Also dieseling could occur because some fuel could be delivered after the ignition is turned **OFF**.

There are O-rings used to seal the injector at the intake manifold and fuel rail. The O-rings are lubricated and should be replaced whenever the injector is removed. Air leakage at the injector/intake area would create a lean cylinder and a possible driveability problem. The injectors are identified with an ID number cast on the injector near the top side.

Fuel Rail

The fuel rail is bolted rigidly to the engine and provides the upper mount for the fuel injectors. It distributes fuel to each injector. Fuel is delivered to the input end of the fuel rail by the fuel lines, goes through the rail, then to the fuel pressure regulator. The regulator keeps the fuel pressure to the injectors at a constant pressure. The remaining fuel is then returned to the fuel tank. The fuel rail also contains a spring loaded pressure tap for testing the fuel system or relieving the fuel system pressure.

Pressure Regulator

The fuel pressure regulator contains a pressure chamber separated by a diaphragm relief valve assembly with a calibrated spring in the vacuum chamber side. The fuel pressure is regulated when the pump pressure acting on the bottom spring of the diaphragm overcomes the force of the spring action on the top side.

The diaphragm relief valve moves, opening or closing an orifice in the fuel chamber to control the amount of fuel returned to the fuel tank. Vacuum acting on the top side of the diaphragm along with spring pressure controls the fuel pressure. A decrease in vacuum creates an increase in the fuel pressure. An increase in vacuum creates a decrease in fuel pressure.

An example of this is under heavy load conditions the engine requires more fuel flow. The vacuum decreases under a heavy load condition because of the throttle opening. A decrease in the vacuum allows more fuel pressure to the top side of the pressure relief valve, thus increasing the fuel pressure.

The pressure regulator is mounted on the fuel rail and serviced separately. If the pressure is too low, poor performance could result. If the pressure is too high, excessive odor and a DTC 45 on the 1.9L and 4.3L engines or a DTC 26 on the 1.6L and 1.8L engines may result.

Idle Air Control (IAC)

The purpose of the Idle Air Control (IAC) system is to control engine idle speeds while preventing stalls due to changes in engine load. The IAC assembly, mounted on the throttle body, controls bypass air around the throttle plate. By extending or retracting a conical valve, a controlled amount of air can move around the throttle plate. If rpm is too low, more air is diverted around the throttle plate to increase rpm.

During idle, the proper position of the IAC valve is calculated by the ECM/PCM based on battery voltage, coolant temperature, engine load, and engine rpm. If the rpm drops below a specified rate, the throttle plate is closed. The ECM/PCM will then calculate a new valve position.

Three different designs are used for the IAC conical valve. The first design used is single taper while the second design used is a dual taper. The third design is a blunt valve. Care should be taken to ensure use of the correct design when service replacement is required.

The IAC motor has 255 different positions or steps. The zero, or reference position, is the fully extended position at which the pintle is seated in the air bypass seat and no air is allowed to bypass the throttle plate. When the motor is fully retracted, maximum air is allowed to bypass the throttle plate. When the motor is fully retracted, maximum air is allowed to bypass the throttle plate.

The ECM/PCM always monitors how many steps it has extended or retracted the pintle from the zero or reference position; thus, it always calculates the exact position of the motor. Once the engine has started and the vehicle has reached approximately 40 mph, the ECM/PCM will extend the motor 255 steps from whatever position it is in. This will bottom out the pintle against the seat. The ECM/PCM will call this position **0** and thus keep its zero reference updated.

The IAC only affects the engine's idle characteristics. If it is stuck fully open, idle speed is too high (too much air enters the throttle bore) If it is stuck closed, idle speed is too low (not enough air entering). If it is stuck somewhere

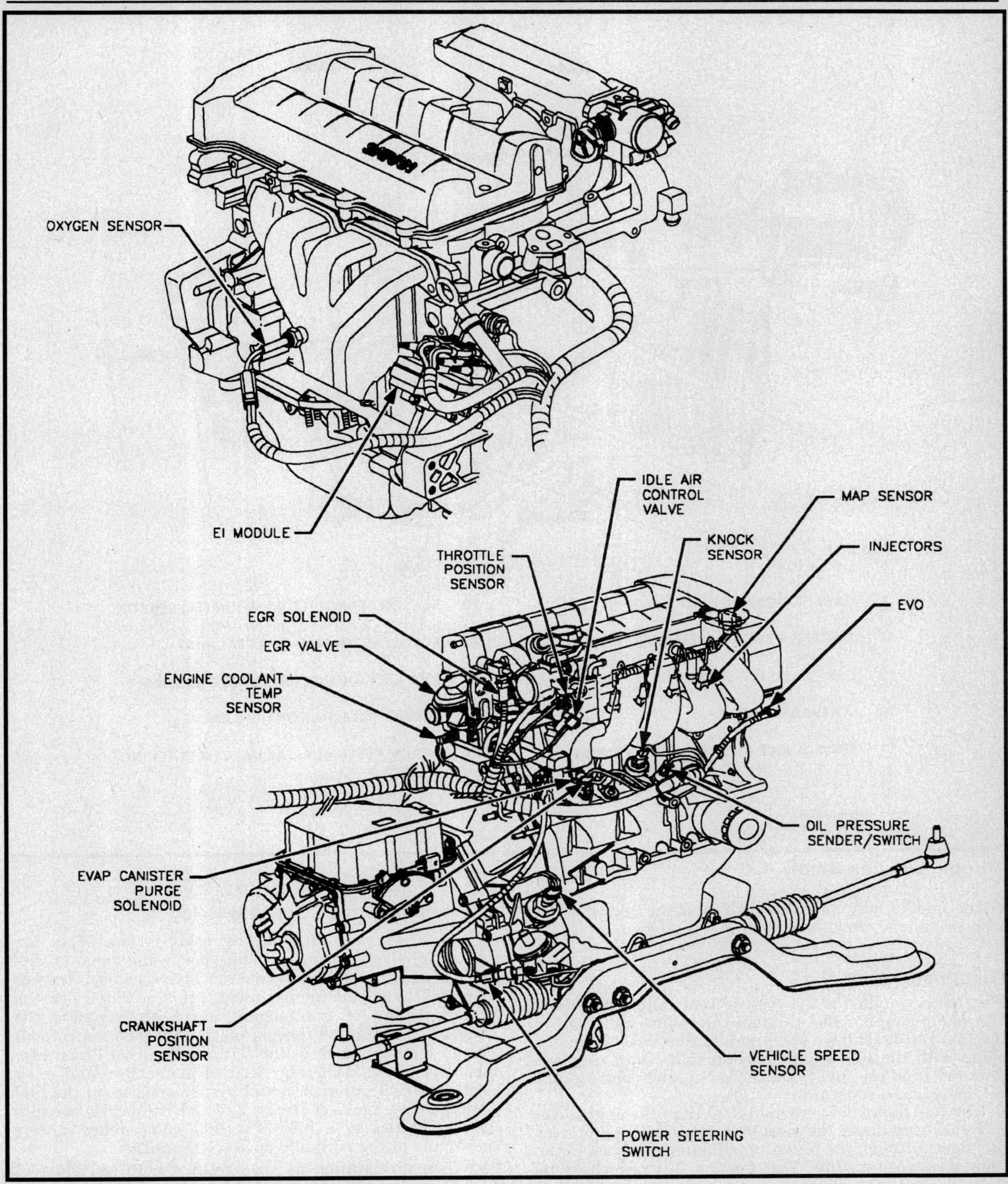

View of system components — Saturn

1	GASKET - FLANGE	20	SENSOR - THROTTLE POSITION (TP) SENSOR
10	THROTTLE BODY ASSEMBLY	21	SCREW ASSEMBLY - TP ATTACHING
13	CAP - IDLE STOP SCREW	70	IDLE AIR CONTROL (IAC) VALVE ASSEMBLY
16	SCREW ASSEMBLY - IDLE STOP	72	O-RING - IDLE AIR CONTROL VALVE
17	SPRING - IDLE STOP SCREW ASSEMBLY	73	SCREW ASSEMBLY - IAC VALVE ATTACHING

Throttle body assembly — Geo

in the middle, idle may be rough, and the engine won't respond to load changes.

Fuel Pump

The fuel is supplied to the system from an in-tank pump. The pump supplies fuel through the in-line fuel filter to the fuel rail assembly. The pump is removed for service along with the fuel gauge sending unit. Once they are removed from the fuel tank, the pump and sending unit can be serviced separately.

The fuel pump delivers more fuel than the engine can consume even under the most extreme conditions. Excess fuel flows through the pressure regulator and back to the tank via the return line. The constant flow of fuel means that the fuel system is always supplied with cool fuel, thereby preventing the formation of fuel-vapor bubbles (vapor lock).

Fuel Pump/Circuit Opening Relay

The fuel pump/circuit opening relay is located in the center of the instrument panel, below the radio on the Geo Prizm and in the fuse and relay box located at the left front of the engine compartment, near the battery on the Storm; on the left front engine side of the firewall on the GMC Syclone and Typhoon; on the left side instrument panel junction block in the Saturn. The Geo Prizm also uses an EFI/F-HTR relay located in the fuse and relay box. The fuel pump electrical system consists of the fuel pump relay, ignition circuit and the ECM/PCM circuits are protected by a fuse. The fuel pump relay contact switch is in the normally open (NO) position.

When the ignition is turned **ON** the ECM/PCM will supply voltage to the fuel pump relay coil, closing the open contact switch for 2 seconds. The ignition circuit fuse can now supply ignition voltage to the circuit which

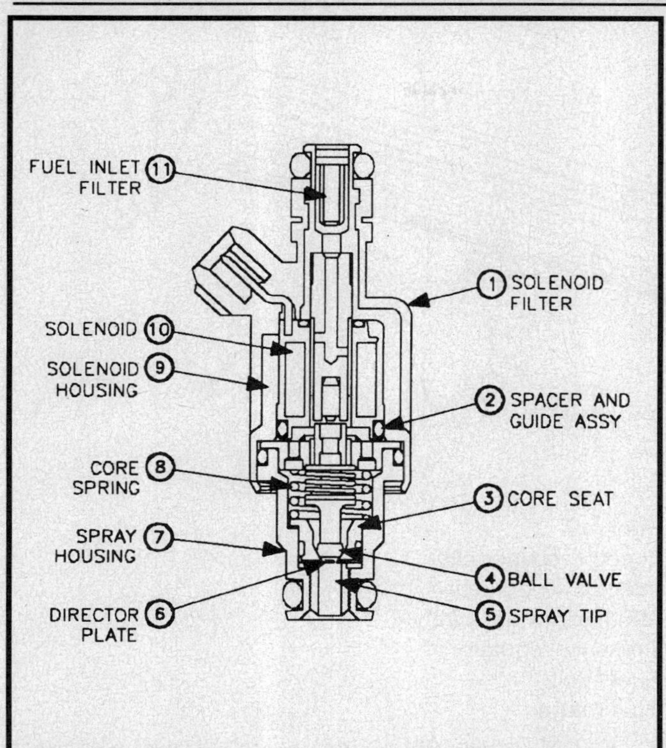

Exploded view of fuel injector

feeds the relay contact switch. With the relay contacts closed, ignition voltage is supplied to the fuel pump. The ECM/PCM will continue to supply voltage to the relay coil circuit as long as the ECM/PCM receives the rpm reference pulses from the ignition module.

In-Line Fuel Filter

The fuel filter is serviced only as a complete unit. O-rings are used at threaded connections to prevent fuel leakage. The threaded flex hoses connect the filter to the fuel tank feed line. The filter housing is constructed of steel with either threaded or quick-connect fittings. There is no service interval for the filter, only replace if restricted.

Cold Start Injector

The cold start injector is used to provide additional fuel during the crank mode to improve cold start-ups and is equipped on the 1.6L engine in 1992 only. This circuit is important when engine coolant temperature is low because the main injectors are not pulsed on long enough to provide the needed amount of fuel to start the engine.

During engine cranking, fuel is injected into the cylinder port through the individual fuel injectors and the required additional fuel is injected into a passage within the intake manifold by the cold start injector. The cold start injector is controlled by a thermal (Bosch) time switch and operates at engine temperatures below 95°F.

The circuit is activated only in the crank mode. The power is supplied directly from the starter solenoid and is protected by a fusible link. The system is control by a thermal time switch which provides a ground path for the cold start injector during cranking when the engine coolant is below 95°F.

The thermal switch is made of a bimetal material which opens at a specified coolant temperature. The bimetal is also heated by winding in the thermal switch which allows the injector to stay on for 8 seconds at 68°F coolant. The time the thermal switch will stay closed varies inversely with the coolant temperature. In other words, as the coolant temperature goes up, the cold start injector on-time goes down.

Data Sensors

A variety of sensors provide information to the ECM/PCM regarding engine operating characteristics. These sensors and their functions are described below. Be sure to take note that not every sensor described is used with every GM engine application.

Electronic Spark Timing (EST)

Electronic spark timing (EST) is used on the 4.3L engine with HEI distributor. The EST distributor contains no vacuum or centrifugal advance and uses a 7-terminal distributor module. It also has 4 wires going to a 4-terminal connector in addition to the connectors normally found on HEI distributors. A reference pulse, indicating both engine rpm and crankshaft position, is sent to the ECM/PCM. The ECM/PCM determines the proper spark advance for the engine operating conditions and sends an **EST** pulse to the distributor.

The EST system is designed to optimize spark timing for better control of exhaust emissions and for fuel economy improvements. The ECM/PCM monitors information from various engine sensors, computes the desired spark timing and changes the timing accordingly. A backup spark advance system is incorporated in the module in case of EST failure.

The basic function of the fuel control system is to control the fuel delivery to the engine. The fuel is delivered to the engine by individual fuel injectors mounted on the intake manifold near each cylinder.

The main control sensor is the oxygen sensor which is located in the exhaust manifold. The oxygen sensor tells the ECM/PCM how much oxygen is in the exhaust gas and the ECM/PCM changes the air/fuel ratio to the engine by controlling the fuel injectors. The best mixture (ratio) to minimize exhaust emissions is 14.7:1 which allows the catalytic converter to operate the most efficiently. Because of the constant measuring and adjusting of the air/fuel ratio, the fuel injection system is called a CLOSED LOOP system.

> **NOTE: When the term Electronic Control Module (ECM/PCM) is used in this manual it will refer to the engine control computer regardless that it may be a Powertrain Control Module (PCM) or Electronic Control Module (ECM/PCM).**

Electronic Spark Control (ESC)

ESC is used in conjunction with EST, ESC is used to reduce spark advance under conditions of detonation. A knock sensor signals a separate ESC controller to retard the timing when it senses knock. The ESC controller signals the ECM/PCM which reduces spark advance until no more signals are received from the knock sensor.

Engine Coolant Temperature

The coolant sensor is a thermistor (a resistor which changes value based on temperature) mounted on the en-

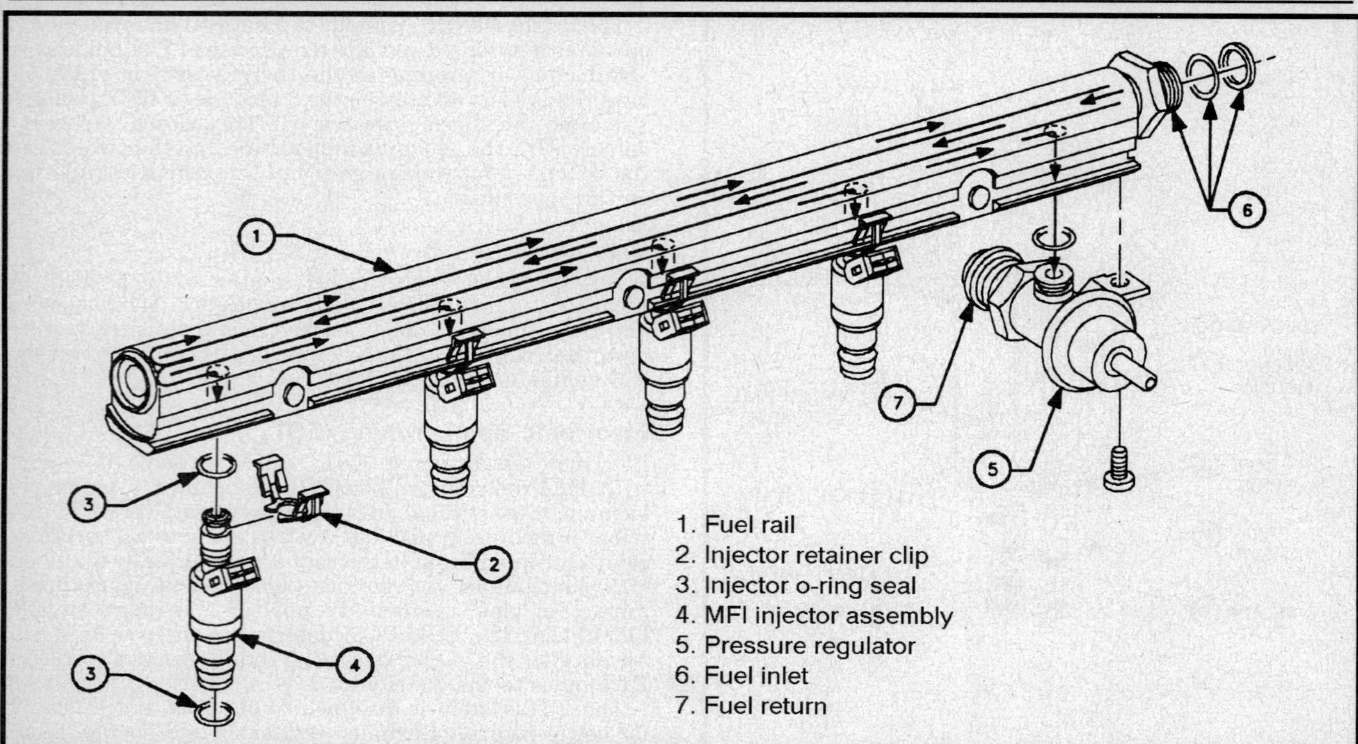

1. Fuel rail
2. Injector retainer clip
3. Injector o-ring seal
4. MFI injector assembly
5. Pressure regulator
6. Fuel inlet
7. Fuel return

Fuel rail and pressure regulator assembly — 1.9L engine

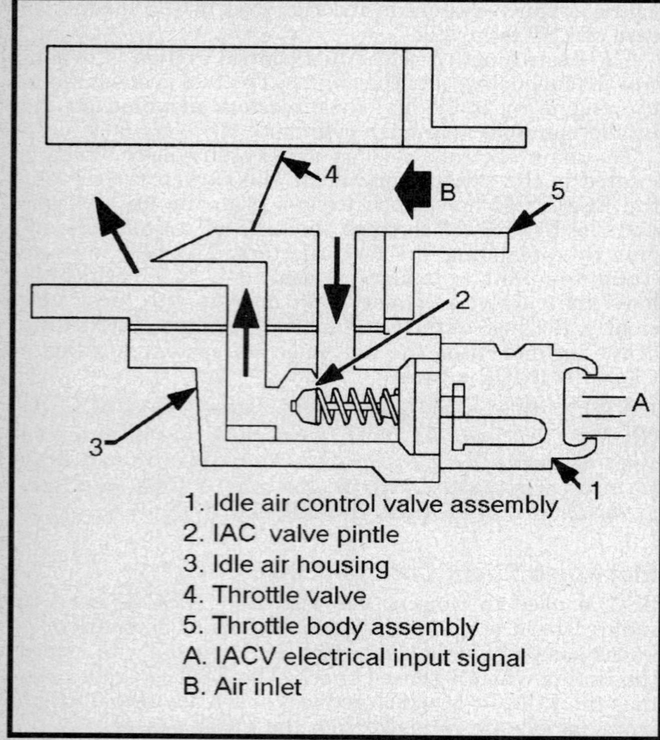

1. Idle air control valve assembly
2. IAC valve pintle
3. Idle air housing
4. Throttle valve
5. Throttle body assembly
A. IACV electrical input signal
B. Air inlet

IAC valve operating characteristics

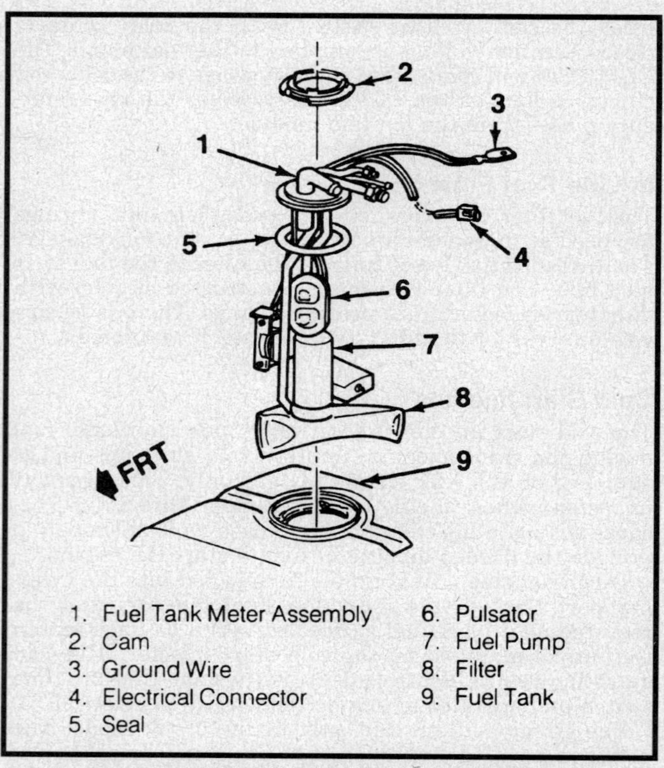

1. Fuel Tank Meter Assembly
2. Cam
3. Ground Wire
4. Electrical Connector
5. Seal
6. Pulsator
7. Fuel Pump
8. Filter
9. Fuel Tank

Fuel pump assembly — Syclone and Typhoon

gine coolant stream. As the temperature of the engine coolant changes, the resistance of the coolant sensor changes. Low coolant temperature produces a high resistance (approximately 93–100 kilo-ohms at -40°C/-40°F), while high temperature causes low resistance (approximately 65–70 ohms at 130°C/266°F).

The ECM/PCM supplies a 5 volt signal to the coolant sensor and measures the voltage that returns. By measuring the voltage change, the ECM/PCM determines the engine coolant temperature. The voltage will be high when the engine is cold and low when the engine is hot. This information is used to control fuel management, IAC, spark timing, EGR, canister purge and other engine operating conditions.

A failure in the coolant sensor circuit should either set a DTC 14 or 15 on GMC and Saturn vehicles; or a DTC 22 on Geo vehicles. These codes indicate a failure in the coolant temperature sensor circuit. Once the trouble code is set, the ECM/PCM will use a default value for engine coolant temperature.

Manifold/Intake Air Temperature (MAT/IAT) Sensor

The Manifold/Intake Air Temperature (MAT/IAT) sensor is a thermistor mounted in the intake manifold or air inlet duct. A thermistor is a resistor which changes resistance based on temperature. Low manifold air temperature produces a high resistance (approximately 93–100 kilo-ohms at -40°F/-40°C), while high temperature cause low resistance (65–70 ohms at 266°F/130°C).

The ECM supplies a 5 volt signal to the MAT/IAT sensor through a resistor in the ECM and monitors the voltage. The voltage will be high when the manifold air is cold and low when the air is hot. By monitoring the voltage, the ECM calculates the air temperature and uses this data to help determine the fuel delivery and spark advance. A failure in the MAT/IAT circuit should set either a DTC 23 or DTC 25 on GMC and Saturn vehicles or DTC 24 on Geo vehicles. Once the trouble code is set, the ECM will use an artificial default value for the MAT/IAT and some vehicle performance will return.

Oxygen Sensor

The exhaust oxygen sensor is mounted in the exhaust system where it can monitor the oxygen content of the exhaust gas stream. The oxygen content in the exhaust reacts with the oxygen sensor to produce a voltage output. This voltage ranges from approximately 100 millivolts (high oxygen — lean mixture) to 900 millivolts (low oxygen — rich mixture).

The sensor must reach a certain operating temperature prior to the system going into closed loop, therefor some vehicles are equipped with a heated sensor, thus reducing the amount of time needed by the sensor to reach operating temperature. The heated oxygen sensor may be recognized by 3 wires instead of a single wire on non-heated sensors.

By monitoring the voltage output of the oxygen sensor, the ECM/PCM will determine what fuel mixture command to give to the injector (lean mixture — low voltage — rich command, rich mixture — high voltage — lean command).

Remember that oxygen sensor indicates to the ECM/PCM the oxygen content in the exhaust. It does not cause things to happen. It is a gauge which indicates to the ECM/PCM: high oxygen content = lean mixture; low oxygen content = rich mixture. In turn the ECM/PCM adjust fuel injector pulse width to keep the system as close as possible to the stoichiometric ratio of 14.7:1.

The oxygen sensor, should set a DTC 13 or 21 if an open circuit is detected. A constant low voltage in the sensor circuit should set a DTC 44 or 25 while a constant high voltage in the circuit should set a DTC 45 or 26. DTCs 25, 26, 44 and 45 can also be set as a result of fuel system problems.

Manifold Absolute Pressure (MAP) Sensor

All engine except the 1.6L (VIN 5) engine use a MAP sensor. The Manifold Absolute Pressure (MAP) sensor measures the changes in the intake manifold pressure which result from engine load and speed changes. The pressure measured by the MAP sensor is the difference between barometric pressure (outside air) and manifold pressure (vacuum). A closed throttle engine coastdown would produce a relatively low MAP value (approximately 20–35 kPa), while wide-open throttle would produce a high value (100 kPa). This high value is produced when the pressure inside the manifold is the same as outside the manifold, and 100% of outside air (or 100 kPa) is being measured. This MAP output is the opposite of what you would measure on a vacuum gauge. The use of this sensor also allows the ECM/PCM to adjust automatically for different altitude.

The ECM/PCM sends a 5 volt reference signal to the MAP sensor. As the MAP changes, the electrical resistance of the sensor also changes. By monitoring the sensor output voltage the ECM/PCM can determine the manifold pressure. A higher pressure, lower vacuum (high voltage) requires more fuel, while a lower pressure, higher vacuum (low voltage) requires less fuel. The ECM/PCM uses the MAP sensor to control fuel delivery and ignition timing. A failure in the MAP sensor circuit should set a DTC 31, 33 or DTC 34.

Mass Air Flow (MAF) Sensor

The 1.6L (VIN 5) engine is the only engine which uses a MAF sensor. The Mass Air Flow (MAF) sensor measures the amount of air which passes through it. The ECM/PCM uses this information to determine the operating condition of the engine, to control fuel delivery. A large quantity of air indicates acceleration, while a small quantity indicates deceleration or idle.

The MAF sensor also contains a fuel pump control switch, which provides a ground to the circuit opening relay to run the fuel pump.

A scan tool will display air flow in terms of grams of air per second (gm/sec), with a range from 3gm/sec to 150 gm/sec. A DTC 25, 26 or 31 may be set if the sensor malfunctions.

Vehicle Speed Sensor (VSS)

NOTE: A vehicle equipped with a speed sensor, should not be driven with the speed sensor disconnected, as idle quality may be affected.

The Vehicle Speed Sensor (VSS) is mounted behind the speedometer in the instrument cluster on the Geo Storm or on the transmission/speedometer drive gear on all other vehicles. It provides electrical pulses to the ECM/PCM from the speedometer head. The pulses indicate the road speed. The ECM/PCM uses this information to operate the IAC, canister purge, and TCC. A failure in the VSS circuit should set a DTC 24 or 42.

Throttle Position Sensor (TPS)

The Throttle Position Sensor (TPS) is connected to the throttle shaft and is controlled by the throttle mechanism. A 5 volt reference signal is sent to the TPS from the ECM/PCM. As the throttle valve angle is changed (accelerator pedal moved), the resistance of the TPS also changes. At a closed throttle position, the resistance of the TPS is high, so the output voltage to the ECM/PCM will be low (approximately 0.5 volt). As the throttle plate opens, the resistance decreases so that, at wide open throttle, the output voltage should be approximately 5 volts. At closed throttle position, the voltage at the TPS should be less than 1.25 volts.

By monitoring the output voltage from the TPS, the ECM/PCM can determine fuel delivery based on throttle valve angle (driver demand). The TPS can either be misadjusted, shorted, open or loose. Misadjustment might result in poor idle or poor wide-open throttle performance. An open TPS signals the ECM/PCM that the throttle is always closed, resulting in poor performance. A shorted TPS gives the ECM/PCM a constant wide-open throttle signal. Any malfunction of the TP circuit should set a DTC 21, 22 or 41. A loose TPS indicates to the ECM/PCM that the throttle is moving. This causes intermittent bursts of fuel from the injector and an unstable idle. Once the trouble code is set, the ECM/PCM will use an artificial default value for the TBI and some vehicle performance will return.

Crankshaft Sensor

The Saturn 1.9L engine uses a magnetic crankshaft sensor, mounted remotely from the ignition module. The reluctor is a special wheel cast into the crankshaft with 7 slots machined into it, 6 of which are equally spaced (60 degrees apart). A seventh slot is spaced approximately 10 degrees from one of the other slots and serves to generate a Sync Pulse signal. As the reluctor rotates as part of the crankshaft, the slots change the magnetic field of the sensor, creating an induced voltage pulse.

Based on the crank sensor pulses, the ignition module sends reference signals to the ECM/PCM which are used to indicate crankshaft position and engine speed. The ignition module continues to send these reference pulses to the ECM/PCM at a rate of 1 per each 180 degrees of crankshaft rotation. This signal is called the 2X reference because it occurs 2 times per crankshaft revolution.

The ignition also sends a second, 1X reference signal to the ECM/PCM which occurs at the same time as the Sync Pulse from the crankshaft sensor. This signal is called the 1X reference because it occurs 1 time per crankshaft revolution. The 1X reference and the 2X reference signals are necessary for the ECM/PCM to determine when to activate the fuel injectors.

By comparing the time between pulses, the ignition module can recognize the pulse representing the seventh slot (sync pulse) which starts the calculation of the ignition coil sequencing.

Air Conditioning Request Signal

This signal indicates to the ECM/PCM that an air conditioning mode is selected at the switch and that the A/C low pressure switch is closed. When the A/C control switch is turned **ON** and the pressure sensor switch is closed, the A/C clutch relay sends a signal to the ECM/PCM. The ECM/PCM then supplies a ground to the A/C control relay, thus energizing the relay and sensing 12 volts to the compressor clutch.

The ECM/PCM also uses this input to calculate the A/C compressor load on the engine to help control the idle speed and turn the A/C **OFF** under certain engine operating conditions.

Detonation (Knock) Sensor

This sensor is a piezoelectric sensor located near the cylinder head (transmission end). It generates electrical impulses which are directly proportional to the frequency of the knock which is detected. A buffer then sorts these signals and eliminates all except for those frequency range of detonation. This information is passed to the module and then to the ECM/PCM, so that the ignition timing advance can be retarded until the detonation stops.

Park/Neutral Switch

NOTE: Vehicle should not be driven with the park/neutral switch disconnected as idle quality may be affected in park or neutral.

This switch indicates to the ECM/PCM when the transmission is in **P** or **N**. the information is used by the ECM/PCM for control on the torque converter clutch, EGR, and the idle air control valve operation.

Torque Converter Clutch Solenoid

The purpose of the torque converter clutch system is designed to eliminate power loss by the converter (slippage) to increase fuel economy. By locking the converter clutch, a more effective coupling to the flywheel is achieved. The converter clutch is operated by the ECM/PCM controlled torque converter clutch solenoid.

Power Steering Pressure (PSP) Switch

The power steering pressure switch is used so that the power steering oil pressure load will not effect the engine idle. Turning the steering wheel increases the power steering oil pressure and pump load on the engine, causing a possible stalling condition. The power steering pressure switch will close before the load can cause an idle problem. The ECM/PCM will turn the A/C clutch off when high power steering pressure is detected and spark timing is also fixed to prevent idle surging.

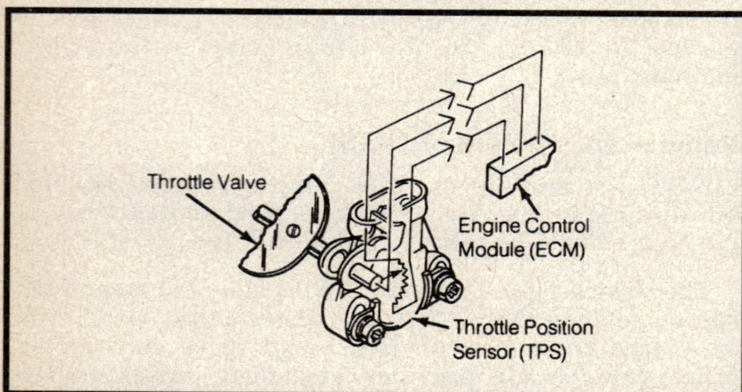

Throttle Position (TP) Sensor — 4.3L engine

Emission Control Systems

Various components are used to control exhaust emissions from a vehicle. These components are controlled by the ECM/PCM based on different engine operating conditions. These components are described in the following paragraphs. Not all components are used on all engines.

Exhaust Gas Recirculation (EGR) System

EGR is a oxides of nitrogen (NOx) control which recycles exhaust gases through the combustion cycle by admitting exhaust gases into the intake manifold. The amount of exhaust gas admitted is adjusted by a vacuum controlled valve in response to engine operating conditions. If the valve is open, the recirculated exhaust gas is released into the intake manifold to be drawn into the combustion chamber.

The integral exhaust pressure modulated EGR valve uses a transducer responsive to exhaust pressure to modulate the vacuum signal to the EGR valve. The vacuum signal is provided by an EGR vacuum port in the throttle body valve. Under conditions when exhaust pressure is lower than the control pressure, the EGR signal is reduced by an air bleed within the transducer. Under conditions when exhaust pressure is higher than the control pressure, the air bleed is closed and the EGR valve responds to an unmodified vacuum signal. Physical arrangement of the valve components will vary depending on whether the control pressure is positive or negative.

Positive Crankcase Ventilation (PCV) System

A Positive Crankcase Ventilation (PCV) system is used to provide more complete scavenging of crankcase vapors. Fresh air from the air cleaner is supplied to the crankcase, mixed with blow-by gases and then passed through a PCV valve into the induction system.

The primary mode of crankcase ventilation control is through the PCV valve which meters the mixture of fresh air and blow-by gases into the induction system at a rate dependent upon manifold vacuum.

To maintain the idle quality, the PCV valve restricts the ventilation system flow whenever intake manifold vacuum is designed to allow excessive amounts of blow-by gases to backflow through the breather assembly into the air cleaner and through the throttle body to be consumed by normal combustion.

Evaporative Emission Control System (EECS)

The basic evaporative emission control system used on all vehicles uses the carbon canister storage method. This method transfers fuel vapor to an activated carbon storage device for retention when the vehicle is not operating. A ported vacuum signal is used for purging vapors stored in the canister.

Saturn engine use the ECM/PCM to control a solenoid valve which regulates vacuum to the purge valve in the charcoal canister. In open loop, before a specified time has expired and below a specified rpm, the solenoid valve is energized and blocks vacuum to the purge valve. When the system is in closed loop, after a specified time and above a specified rpm, the solenoid valve is de-energized and vacuum can be applied to the purge valve. This releases the collected vapors into the intake manifold. On systems not using an ECM/PCM controlled solenoid, a Thermo Vacuum Valve (TVV) is used to control purge.

Catalytic Converter

Of all emission control devices available, the catalytic converter is the most effective in reducing tailpipe emissions. The major tailpipe pollutants are hydrocarbons (HC), carbon monoxide (CO), and oxides of nitrogen (NOx).

Diagnosis and Testing

SERVICE PRECAUTIONS

When working around any part of the fuel system, take precautionary steps to prevent fire and/or explosion:
- Disconnect negative terminal from battery (except when testing with battery voltage is required).
- Whenever possible, use a flashlight instead of a drop light.
- Keep all open flame and smoking material out of the area.
- Use a shop cloth or similar to catch fuel when opening a fuel system.
- Relieve fuel system pressure before servicing.
- Use eye protection.
- Always keep a dry chemical (class B) fire extinguisher near the area.

NOTE: Prior to performing any work on the fuel system, the fuel system must be de-pressurized.

Electrostatic Discharge Damage

Electronic components used in the control system are often design to carry very low voltage and are very susceptible to damage caused by electrostatic discharge. It is possible for less than 100 volts of static electricity to cause damage to some electronic components. By comparison it takes as much as 4000 volts for a person to even feel the zap of a static discharge.

There are several ways for a person to become statically charged. The most common methods of charging are by friction and induction. An example of charging by friction is a person sliding across a car seat, in which a charge as much as 25000 volts can build up. Charging by induction occurs when a person with well insulated shoes stands near a highly charged object and momentarily touches ground. Charges of the same polarity are drained off, leaving the person highly charged with the opposite polarity. Static charges of either type can cause damage, therefore, it is important to use care when handling and testing electronic components. Follow the guidelines listed below to avoid electrostatic discharge.
- To prevent possible electrostatic discharge damage to the ECM, do not touch the connector pins or soldered components on the circuit board.
- Always touch a known ground before handling any parts.
- When using a voltmeter, always connect the ground lead first.
- Do not remove any parts from there package until it is time to install them.
- Before removing the part from the package, touch the package to a good ground.

ECM/PCM LEARNING ABILITY

The ECM/PCM has a learning capability. If the battery is disconnected the learning process has to begin all over again. A change may be noted in the vehicle's performance. To teach the ECM/PCM, insure the vehicle is at operating temperature and drive at part throttle, with moderate acceleration and idle conditions, until performance returns.

DATA LINK CONNECTOR (DLC)

The DLC, previously known as the Assembly Line Diagnostic Link (ALDL), is a diagnostic connector located in the passenger compartment, usually under the instrument panel. The exception however, is Geo Prizm which incorporates the connector in the engine compartment. Geo Storm uses a white 3-pin connector which is located behind the passenger side kick panel.

The DLC may be used to enter the Diagnostic mode or the Field Service Mode by connecting or jumping terminal B (diagnostic TEST terminal) to terminal A (ground circuit) on the 1.9L and 4.3L engines. Diagnostic mode may be entered by connecting a jumper wire between terminals TE1 (diagnostic TEST terminal) to terminal E1 (ground circuit) on Geo Prizm. Diagnostic mode may be entered by jumping terminal 1 to terminal 3 on the Geo Storm. This connector is a very useful tool in diagnosing EFI engines. Important information from the ECM/PCM is available at this terminal and can be read with one of the many popular scan tools.

READING CODES

Diagnostic codes may be read by putting the ECM/PCM into the diagnostic or field service modes. When 0 resistance is between terminals A and B or 1 and 3 of the DLC connector, the diagnostic mode is entered. There are 2 positions to this mode. One with the engine **OFF**, but the ignition **ON**; the other is when the engine is running called Field Service Mode.

If the diagnostic mode is entered with the engine in the **OFF** position, trouble codes will flash and the idle air control motor will pulsate in and out. Also, the relays and solenoids are energized with the exception of the fuel pump and injector.

As a bulb and system check, the Malfunction Indicator Light (MIL) will come on with the ignition switch **ON** and the engine not running. When the engine is started, the MIL will turn off. If the MIL remains **ON**, the self-diagnostic system has detected a problem.

If the terminal B is then grounded with the ignition **ON**, engine not running, each trouble code will flash and repeat 3 times. If more than 1 problem has been detected, each trouble code will flash in numeric order (lowest num-

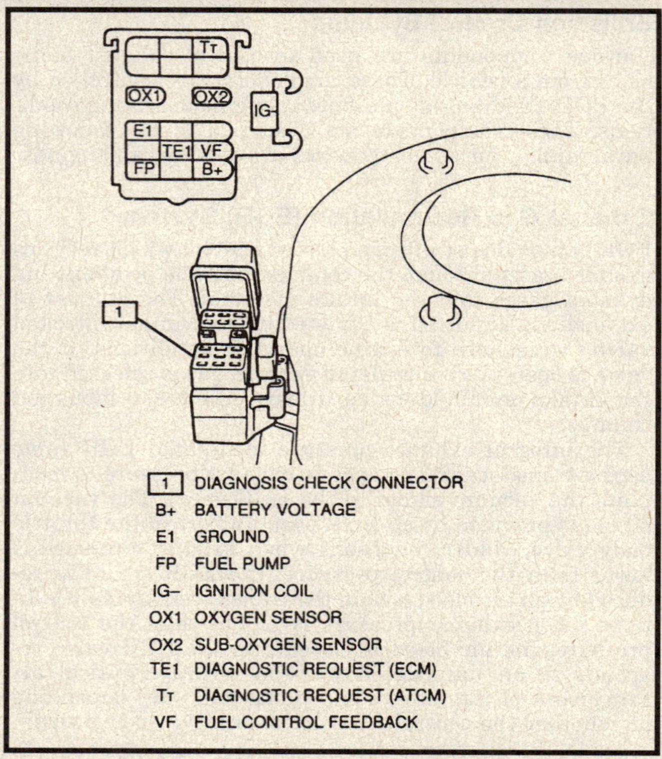

Data Link Connector (DLC) — 1992 Geo Prizm

ber first). The trouble code series will repeat as long as the B terminal is grounded.

A trouble code indicates a problem in a given circuit (DTC 14, for example, indicates a problem in the coolant sensor circuit or ignition circuit on Geo vehicles). The procedure for pinpointing the problem can be found in diagnosis. Similar charts are provided for each code.

This mode also allows checking the circuits which may be difficult to energize without driving the vehicle and being under particular operating conditions. The IAC valve will move to its fully extended position on most vehicles, block the idle air passage. This is useful in checking the minimum idle speed.

Scan Tester Information

A code retrieval unit (ALDL tester, scanner, monitor, etc), allows a technician to read the engine control system information from the DLC under the instrument panel or in the engine compartment. It can provide information faster than a digital voltmeter or ohmmeter can. The scan tool does not diagnose the exact location of the problem. The tool supplies information about the ECM/PCM, the information that it is receiving and the commands that it is sending plus special information such as integrator and block learn. To use a display tool you should understand thoroughly how an engine control system operates.

A scanner or monitor puts a fuel injection system into a special test mode. On vehicles with Electronic Spark Control (ESC), there will be a fixed spark, but it will be advanced. On vehicles with ESC, there might be a serious spark knock, this spark knock could be bad enough so as not being able to road test the vehicle in the test mode. Be sure to check the tool manufacturer for instructions on

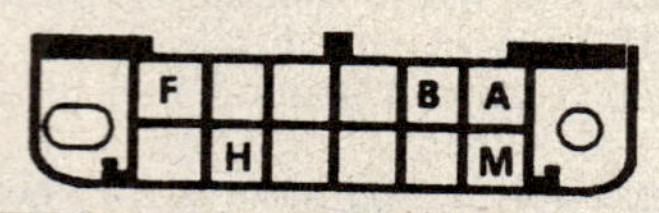

TERMINAL IDENTIFICATION

A. Ground
B. Diagnostic terminal
F. T.C.C. (if used)
H. Speed input from antilock module
M. Serial data

Data Link Connector (DLC) — GMC and Saturn

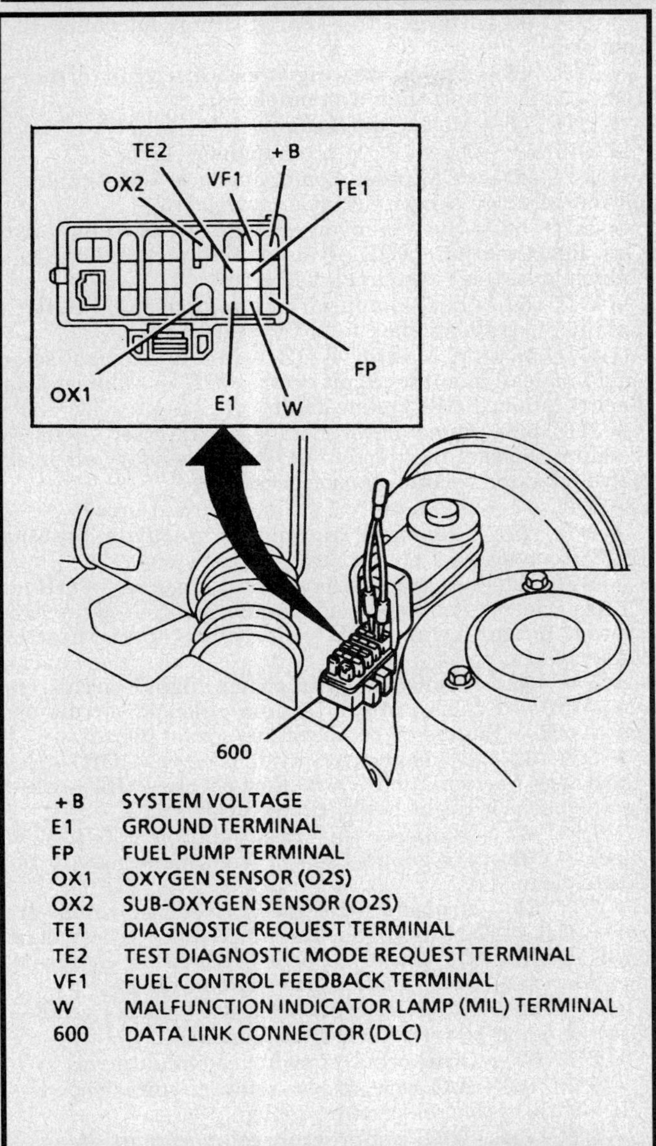

+B	SYSTEM VOLTAGE
E1	GROUND TERMINAL
FP	FUEL PUMP TERMINAL
OX1	OXYGEN SENSOR (O2S)
OX2	SUB-OXYGEN SENSOR (O2S)
TE1	DIAGNOSTIC REQUEST TERMINAL
TE2	TEST DIAGNOSTIC MODE REQUEST TERMINAL
VF1	FUEL CONTROL FEEDBACK TERMINAL
W	MALFUNCTION INDICATOR LAMP (MIL) TERMINAL
600	DATA LINK CONNECTOR (DLC)

Data Link Connector (DLC) — 1993 Geo Storm

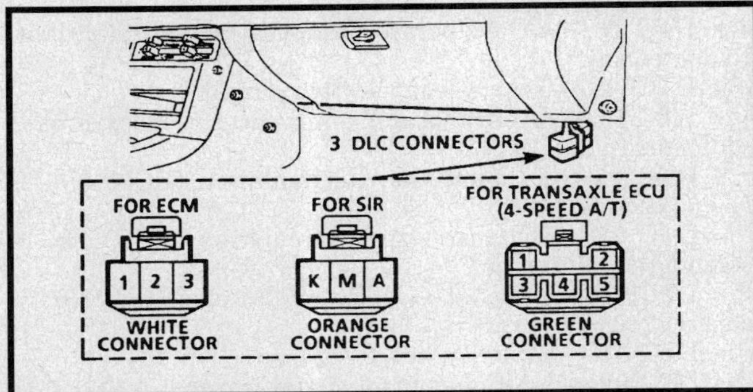

Data Link Connector (DLC) — Geo Storm

special test modes which should overcome these limitations.

When a tester is used with a fuel injected engine, it bypasses the timer that keeps the system in Open loop for a certain period of time. When all Closed loop conditions are met, the engine will go into Closed loop as soon as the vehicle is started.

These tools cannot diagnose everything. They do not tell the technician where a problem is located in a circuit. The diagnostic charts must still be used to pinpoint the problems.

Scan Tool for Intermittences

In some scan tool applications, the data update rate may make the tool less effective than a voltmeter, such as when trying to detect an intermittent problem which lasts for a very short time. Some scan tools have a snapshot function which stores several seconds or even minutes of operation to located an intermittent problem. Scan tools allow one to manipulate the wiring harness or components under the hood with the engine not running while observing the scan tool's readout.

The scan tool can be plugged in and observed while driving the vehicle under the condition when the MIL turns on momentarily or when the engine driveability is momentarily poor. If the problem seems to be related to certain parameters that can be checked on the scan tool, they should be checked while driving the vehicle. If there does not seem to be any correlation between the problem and any specific circuit, the scan tool can be checked on each position. Watching for a period of time to see if there is any change in the reading that indicates intermittent operation.

The scan tool is also an easy way to compare the operating parameters of a poorly operating engine with typical scan data for the vehicle being serviced or those of a known good engine. For example, a sensor may shift in value but not set a trouble code. Comparing the sensor's reading with those of a known good parameters may uncover the problem.

The scan tool has the ability to save time in diagnosis and prevent the replacement of good parts. The key to using the scan tool successfully for diagnosis lies in the technician's ability to understand the system he is trying to diagnose as well as an understanding of the scan tool's operation and limitations.

CLEARING TROUBLE CODES

Truck and Geo

When the ECM/PCM detects a problem with the system, the MIL will come on and a trouble code will be recorded in the ECM/PCM memory. If the problem is intermittent, the MIL will go out after 10 seconds, when the fault goes away. However the trouble code will stay in the ECM/PCM memory until the battery voltage to the ECM/PCM is removed. Removing the battery voltage for 30 seconds will clear all trouble codes. Do this by disconnecting the ECM/PCM harness from the positive battery terminal pigtail for 30 seconds with the key in the **OFF** position, or by removing the ECM/PCM fuse (EFI/F-HTR fuse on the Geo Storm) for 30 seconds with the key **OFF**.

NOTE: To prevent ECM/PCM damage, the key must be OFF when disconnecting and reconnecting ECM/PCM power.

Saturn

A diagnostic trouble code may be stored in either a General Information or Malfunction History table. Codes may be cleared from the **ON** and grounding DLC terminal A to B three times within 5 seconds. Malfunction History may only be cleared using a scan tool.

DIAGNOSTIC TROUBLE CODE

NOTE: The list of trouble codes below are a representation of all MFI vehicles, refer to the Troubleshooting Chart for the actual diagnostic codes per engine application.

Syclone, Typhoon, Saturn and Storm

- DTC **11** — Transmission diagnostic DTCs present
- DTC **12** — Diagnostic check only (flash code)
- DTC **13** — Oxygen (0_2) sensor circuit open
- DTC **14** — Engine Coolant Temperature (ECT) sensor circuit out of range (high/low)
- DTC **15** — Engine Coolant Temperature (ECT) sensor circuit out of range (low)
- DTC **16** — System voltage high or low — **OR** — Direct/electronic ignition system fault line circuit error
- DTC **17** — Camshaft position sensor/spark reference circuit error
- DTC **18** — Crankshaft/camshaft error
- DTC **19** — Ignition control module signal circuit error — **OR** — Crankshaft position sensor circuit
- DTC **21** — Throttle Position Sensor (TPS) circuit voltage out of range
- DTC **22** — Throttle Position Sensor (TPS) circuit voltage out of range
- DTC **23** — Intake/manifold air temperature sensor circuit voltage out of range
- DTC **24** — Vehicle speed sensor circuit signal error
- DTC **25** — Intake/manifold air temperature sensor circuit voltage out of range
- DTC **26** — Quad-driver module circuit error
- DTC **27** — Gear switch circuits error
- DTC **28** — Gear switch circuits error — **OR** — Quad-driver module control circuit
- DTC **29** — Gear switch circuits error
- DTC **31** — PRNDL/transmission neutral start/park-neutral position switch circuit error — **OR** — Turbo wastegate overboost — **OR** — Camshaft sensor circuit error
- DTC **32** — Exhaust Gas Recirculation (EGR) system fault
- DTC **33** — Mass air flow/manifold absolute pressure sensor circuit voltage out of range
- DTC **34** — Mass air flow/manifold absolute pressure sensor circuit voltage out of range
- DTC **35** — Idle speed/air/idle air control error
- DTC **36** — Transaxle shift problem — **OR** — Ignition control signal circuit error
- DTC **38** — Brake switch circuit error
- DTC **39** — Torque converter/transaxle clutch switch circuit error
- DTC **41** — Cam position sensor/cylinder select error
- DTC **42** — Electronic spark timing/ignition control circuit open or shorted
- DTC **43** — Knock sensor/electronic spark control circuit error
- DTC **44** — Oxygen (0_2) sensor circuit indicates system lean

- DTC **45** — Oxygen (0_2) sensor circuit indicates system rich
- DTC **46** — Power steering pressure circuit error — **OR** — Vehicle anti-theft system circuit
- DTC **48** — Misfire diagnosis
- DTC **51** — ECM, PCM prom memory error
- DTC **52** — PCM prom memory error — **OR** — Engine oil temperature sensor circuit indicates low
- DTC **53** — System over voltage — **OR** — Exhaust Gas Recirculation (EGR) system malfunction — **OR** — Vehicle anti-theft system circuit error
- DTC **54** — Fuel pump circuit voltage low — **OR** — Exhaust Gas Recirculation (EGR) system fault
- DTC **55** — PCM error — **OR** — Mixture control solenoid/fuel lean monitor circuit error — **OR** — Exhaust Gas Recirculation (EGR) system fault
- DTC **56** — Quad-driver module circuit error — **OR** — Vacuum sensor circuit error — **OR** — Secondary air inlet valve actuator vacuum sensor circuit error
- DTC **57** — Boost control solenoid circuit error
- DTC **58** — Personal automotive security system (PASS-key®II) fuel enable circuit error
- DTC **61** — Degraded oxygen (0_2) sensor — **OR** — Cruise control vent solenoid circuit error — **OR** — A/C system circuit error — **OR** — Secondary port throttle valve system error
- DTC **62** — Transaxle gear switch signal circuit error — **OR** — Cruise control vacuum solenoid circuit error — **OR** — Engine oil temperature sensor circuit
- DTC **63** — Cruise control system error — **OR** — Exhaust Gas Recirculation (EGR) flow check — **OR** — Oxygen (0_2) sensor (right bank) circuit open
- DTC **64** — Exhaust Gas Recirculation (EGR) flow check — **OR** — Oxygen (0_2) sensor (right bank) circuit indicates lean
- DTC **65** — Cruise control pressure sensor circuit error — **OR** — Exhaust Gas Recirculation (EGR) flow check — **OR** — Oxygen (0_2) sensor (right bank) circuit indicates rich — **OR** — Fuel injector circuit error
- DTC **66** — A/C refrigerant pressure sensor circuit — **OR** — Engine power switch circuit error
- DTC **67** — Cruise control switches circuit error
- DTC **68** — A/C compressor relay circuit shorted — **OR** — Cruise control system problem
- DTC **69** — A/C compressor relay circuit open — **OR** — A/C head pressure switch circuit error
- DTC **70** — A/C refrigerant pressure sensor circuit indicates high
- DTC **71** — A/C evaporator temperature sensor circuit indicates low
- DTC **72** — Gear selector switch circuit error
- DTC **73** — A/C evaporator temperature sensor circuit indicates high
- DTC **75** — Exhaust Gas Recirculation (EGR) No. 1 solenoid circuit error
- DTC **76** — Exhaust Gas Recirculation (EGR) No. 2 solenoid circuit error
- DTC **77** — Exhaust Gas Recirculation (EGR) No. 3 solenoid circuit error
- DTC **81** — Brake switch error
- DTC **82** — Ignition control signal error
- DTC **85** — PCM prom error
- DTC **86** — A/C multiplexer chip error
- DTC **87** — Electronically Erasable Programmable Read Only Memory (EEPROM) error

Except Geo Prizm

- DTC 12 — No RPM signal
- DTC 13 — No RPM signal
- DTC 14 — Ignition signal circuit (no signal)
- DTC 16 — PCM control signal (failure)
- DTC 21 — Oxygen (O_2) sensor circuit (open or shorted)
- DTC 22 — Engine Coolant Temperature (ECT) sensor circuit (open or shorted)
- DTC 24 — Intake Air Temperature (IAT) sensor circuit (open or shorted
- DTC 25 — Oxygen (O_2) sensor circuit indicates system lean
- DTC 26 — Oxygen (O_2) sensor circuit indicates system rich
- DTC 27 — Sub-Oxygen (O_2) sensor circuit (open or shorted) — California emissions
- DTC 31 — Mass Absolute Pressure (MAP) sensor circuit (open or shorted)
- DTC 41 — Throttle Position Sensor (TPS) circuit (open or shorted)
- DTC 42 — Vehicle speed sensor circuit (no input)
- DTC 43 — Crank signal (no signal)
- DTC 51 — Switch condition signal
- DTC 52 — Knock Sensor (KS) circuit (open or shorted)
- DTC 71 — Exhaust Gas Recirculation (EGR) system — California emissions

INTEGRATOR AND BLOCK LEARN

The integrator and block learn functions of the ECM/PCM are responsible for making minor adjustments to the air/fuel ratio on the fuel injected vehicles. These small adjustments are necessary to compensate for tiny air leaks and normal wear.

The integrator and block learn are 2 separate ECM/PCM memory functions which control fuel delivery. The integrator makes a temporary change and the block learn makes a more permanent change. Both of these functions apply only while the engine is in CLOSED LOOP. They represent the on-time of the injector. Also, integrator and block learn controls fuel delivery on the fuel injected engines as does the MC solenoid dwell on the CCC carbureted engines.

Integrator

Integrator is the term applied to a means of temporary change in fuel delivery. Integrator is displayed through the DLC and monitored with a scanner as a number between 0 and 255 with an average of 128. The integrator monitors the oxygen sensor output voltage and adds and subtracts fuel depending on the lean or rich condition of the oxygen sensor. When the integrator is displaying 128, it indicates a neutral condition. This means that the oxygen sensor is seeing results of the 14.7:1 air/fuel mixture burned in the cylinders.

> **NOTE: An air leak in the system (a lean condition) would cause the oxygen sensor voltage to decrease while the integrator would increase (add more fuel) to temporarily correct for the lean condition. If this happened the injector pulse width would increase.**

Block Learn

Although the integrator can correct fuel delivery over a wide range, it is only for a temporary correction. Therefore, another control called block learn was added. Although it cannot make as many corrections as the integrator, it does so for a longer period of time. It gets its name from the fact that the operating range of the engine for any given combinations of rpm and load is divided into 16 cell or blocks.

The computer has a given fuel delivery stored in each block. As the operating range gets into a given block the fuel delivery will be based on what value is stored in the memory in that block. Again, just like the integrator, the number represents the on-time of the injector. Also, just like the integrator, the number 128 represents no correction to the value that is stored in the cell or block. When the integrator increases or decreases, block learn which is also watching the integrator will make corrections in the same direction. As the block learn makes corrections, the integrator correction will be reduced until finally the integrator will return to 128 if the block learn has corrected the fuel delivery.

BLOCK LEARN MEMORY

Block learn operates on 1 of 2 types of memories depending on application, non-volatile and volatile. The non-volatile memories retain the value in the block learn cells even when the ignition switch is turned **OFF**. When the engine is restarted, the fuel delivery for a given block will be based on information stored in memory.

The volatile memories lose the numbers stored in the block learn cells when the ignition is turned to the **OFF** position. Upon restarting, the block learn starts at 128 in every block and corrects from that point as necessary.

INTEGRATOR/BLOCK LEARN LIMITS

Both the integrator and block learn have limits which will vary from engine to engine. If the mixture is off enough so that the block learn reaches the limit of its control and still cannot correct the condition, the integrator would also go to its limit of control in the same direction and the engine would then begin to run poorly. If the integrators and block learn are close to or at their limits of control, the engine hardware should be checked to determine the cause of the limits being reached, vacuum leaks, sticking injectors, etc.

If the integrator is lied to, for example, if the oxygen sensor lead was grounded (lean signal) the integrator and block learn would add fuel to the engine to cause it to run rich. However, with the oxygen sensor lead grounded, the ECM/PCM would continue seeing a lean condition eventually setting a DTC 44 and the fuel control system would change to open loop operations.

CLOSED LOOP FUEL CONTROL

The purpose of closed loop fuel control is to precisely maintain an air/fuel mixture 14.7:1. When the air/fuel mixture is maintained at 14.7:1, the catalytic converter is able to operate at maximum efficiency which results in lower emission levels.

Since the ECM/PCM controls the air/fuel mixture, it needs to check its output and correct the fuel mixture for deviations from the ideal ratio. The oxygen sensor feeds this output information back to the ECM/PCM.

ENGINE PERFORMANCE DIAGNOSIS

Engine performance diagnosis procedures are guides that will lead to the most probable causes of engine performance complaints. They consider the components of the fuel, ignition, and mechanical systems that could cause a particular complaint, and then outline repairs in a logical sequence.

It is important to determine if the MIL is **ON** or has illuminated for a short interval while driving. If the MIL has illuminated, the ECM/PCM should be checked for stored **TROUBLE CODES** which may indicate the cause for the performance complaint.

Basic Troubleshooting

Before suspecting the system or any of its components as faulty, check the ignition system including distributor, timing, spark plugs and wires. Check the engine compression, air cleaner, and emission control components not controlled by the ECM/PCM. Also check the intake manifold, vacuum hoses and hose connectors for leaks.

The following symptoms could indicate a possible problem with the system:
1. Detonation
2. Stalls or rough idle-cold
3. Stalls or rough idle-hot
4. Missing
5. Hesitation
6. Surges
7. Poor gasoline mileage
8. Sluggish or spongy performance
9. Hard starting-cold
10. Objectionable exhaust odors (that rotten egg smell)
11. Cuts out
12. Improper idle speed

INTERMITTENT MIL

An intermittent open in the ground circuit would cause loss of power through the ECM/PCM and intermittent MIL operation. When the ECM/PCM loses ground, distributor ignition is lost. An intermittent open in the ground circuit would be described as an engine miss.

Therefore, an intermittent MIL, no code stored and a driveability comment described as similar to a miss will require checking the grounding circuit and the ignition circuit as it originates at the ignition coil.

THROTTLE POSITION (TP) SENSOR PARAMETERS CHART

THROTTLE POSITION (TP SENSOR)

Engine	Volts
1.6L (VIN 5, 6)	0.30–1.00 V
1.8L (VIN 8)	0.30–1.00 V
1.9L (VIN 7)	0.40–0.55 V
4.3L (VIN 2)	0.40–1.25 V

V—Volts

ENGINE COOLANT TEMPERATURE (ECT) SENSOR PARAMETERS CHART

ENGINE COOLANT TEMPERATURE (ECT) SENSOR

Engine	Ohms	Degrees
1.6L (VIN 5, 6)	165–286	185–220°F (85–105°C)
1.8L (VIN 8)	165–286	185–220°F (85–105°C)
1.9L (VIN 7)	165–286	185–220°F (85–105°C)
4.3L (VIN 2)	165–286	185–220°F (85–105°C)

NOTE: Values are read via a scan tool.

IDLE AIR CONTROL (IAC) MOTOR PARAMETERS CHART

IDLE AIR CONTROL (IAC)

Engine	Key On, Engine Off	Key On, Engine Running
1.6L (VIN 5, 6)	above 50 counts	1–50 counts
1.8L (VIN 8)	above 50 counts	1–50 counts
1.9L (VIN 7)	N/A	N/A
4.3L (VIN 2)	above 50 counts	0 counts

NOTE: All specifications reflect those received via a scan tool.
N/A—Not available

INTAKE/MANIFOLD AIR TEMPERATURE (IAT/MAT) SENSOR PARAMETERS CHART

INTAKE/MANIFOLD AIR TEMPERATURE (IAT/MAT) SENSOR

Engine	Key On, Engine Off	Key On, Engine Running
1.6L (VIN 5, 6)	①	50–194°F (10–90°C)
1.8L (VIN 8)	①	50–194°F (10–90°C)
1.9L (VIN 7)	①	N/A
4.3L (VIN 2)	①	50–194°F (10–90°C)

NOTE: All specifications reflect those received via a scan tool.
N/A—Not available
① Either underhood ambient temperature or engine temperature depending where the sensor is located.

MANIFOLD AIR PRESSURE (MAP) SENSOR PARAMETERS CHART

MANIFOLD AIR PRESSURE (MAP) SENSOR

Engine	Key On, Engine Off	Key On, Engine Running
1.6L (VIN 5, 6)	above 4.0 V	1–2 V
1.8L (VIN 8)	above 4.0 V	1–2 V
1.9L (VIN 7)	above 4.0 V	0.80–1.5 V
4.3L (VIN 2)	above 4.0 V	1–2 V

NOTE: All specifications reflect those received via a scan tool.
V—Volts

OXYGEN (O₂S) SENSOR PARAMETERS CHART

OXYGEN (O₂S) SENSOR

Engine	Key On, Engine Off	Key On, Engine Running
1.6L (VIN 5, 6)	350–500 mv	1–1000 mv
1.8L (VIN 8)	350–500 mv	1–1000 mv
1.9L (VIN 7)	350–500 mv	1–1000 mv
4.3L (VIN 2)	350–500 mv	1–1000 mv

NOTE: All specifications reflect those received via a scan tool.
mv—milli-volts

FUEL SYSTEM

When the ignition switch is turned **ON**, the in-tank fuel pump is energized for as long as the engine is cranking or running and the control unit is receiving signals from the distributor or crank sensor. If there are no reference pulses, the control unit will shut off the fuel pump within 2 seconds. The pump will deliver fuel to the fuel rail and injectors; then the pressure regulator where the system pressure is maintained.

Fuel Pump

ELECTRICAL CIRCUIT TESTING

A quick test of the fuel pump electrical circuit may be accomplished on Saturn vehicles by interchanging one of the other relay located in the instrument panel junction block with the fuel pump relay.

The fuel pump electrical circuit may be tested on all Geo vehicles except the 1993 Storm by jumpering the FP and B+ terminals of the diagnostic connector (located in the engine compartment). The fuel pump electrical circuit may be tested on 1993 Geo Storm by connecting 12 volts through a 10 amp fused jumper wire to the pink/white lead of the ECM.

The GMC Syclone and Typhoon fuel pump may be tested by using a 10 amp fused jumper wire and applying 12 volts to the fuel pump test terminal (located on the center of the firewall), while listening for the pump to be running.

PRESSURE TESTING

1. Connect pressure gauge J–34730–1, or equivalent, to fuel pressure test point or the fuel rail. Wrap a rag around the pressure tap to absorb any leakage that may occur when installing the gauge.

2. Turn the ignition **ON** and listen for the pump to run for 2 seconds; then shut off. The pressure gauge should read within specification. The pressure is controlled by spring pressure within the regulator assembly.

3. Start the engine and allow it to idle. The fuel pressure should drop to slightly due to pressure regulator control.

NOTE: The idle pressure will vary somewhat depending on barometric pressure. Check for a drop in pressure indicating regulator control, rather than specific values.

4. If the fuel pressure drops, check the operation of the check valve, the pump coupling connection, fuel pressure regulator valve and the injectors. A restricted fuel line or filter may also cause a pressure drop. To check the fuel pump output, restrict the fuel return line and run 12 volts to the pump. The fuel pressure should rise approximately 20 psi with the return line restricted.

CAUTION
Before attempting to remove or service any fuel system component, it is necessary to relieve the fuel system pressure.

COLD START INJECTOR VALVE

Testing
PRIZM

1. Remove the screws holding the injector in the intake manifold. Do not disconnect the fuel lines or electrical connector.

2. Place the cold start injector in a container to catch fuel. Wrap a clean rag around the mouth of the container.

3. Operate the starter and note the injection time. injector should spray fuel for 1–8 seconds if the coolant temperature is lower than approximately 95°F (35°F). Above this temperature, no drip or spray should be noted.

4. If the cold start injector sprays continuously or drips, replace it.

5. If the cold start injector fails to function below 95°F (35°C), replace it.

NOTE: Perform this test as quickly as possible. Avoid energizing the injector for any length of time.

6. Disconnect the cold start injector and hook up a test light across its connector. Ground the ignition coil and run the starter. The light should glow for several seconds with the coolant temperature below 95°F (35°C) and then go out. If not, replace the thermo-time switch.

NOTE: No starts or poor cold starting can be caused by a malfunctioning cold start injector. Cranking a cold engine with the coil wire grounded should produce a cone-shaped spray from the cold start injector. Cranking a warm engine should produce no fuel; if the injector drips fuel, replace it.

SYMPTOMS — MULTI-PORT INJECTION SYSTEMS

IMPORTANT PRELIMINARY CHECKS

- Before using this section you should have performed the "Diagnostic Circuit Check."
- Verify the customer complaint, and locate the correct SYMPTOM below. Check the items indicated under that symptom.
- If the ENGINE CRANKS BUT WILL NOT RUN, use CHART A-3.
- Several of the following symptom procedures call for a careful visual/physical check.
  The importance of this step cannot be stressed too strongly - it can lead to correcting a problem without further checks and can save valuable time.

BEFORE STARTING

This check should include:
- Vacuum hoses for splits, kinks, and proper connections, as shown on Vehicle Emission Control Information label.
- Air leaks at throttle body mounting and intake manifold.
- Ignition wires for cracking, hardness, proper routing, and carbon tracking.
- Wiring for proper connections, pinches, and cuts.
- The following symptoms cover several engines.
To determine if a particular system or component is used, refer to the ECM wiring diagrams for application.

SYMPTOMS — MULTI-PORT INJECTION SYSTEMS

INTERMITTENTS
(Page 1 of 2)

Definition: Problem may or may not turn "ON" the "Service Engine Soon" light, or store a code.

PRELIMINARY CHECKS

- Perform the careful visual checks as described at start of "Symptoms,"

TROUBLE CODE CHARTS IN "ENGINE COMPONENTS/WIRING DIAGRAMS/DIAGNOSTIC CHARTS," SECTION "6E2-A"

- DO NOT use the Trouble Code Charts in for intermittent problems. The fault must be present to locate the problem. If a fault is intermittent, use of Trouble Code Charts may result in replacement of good parts.

FAULTY ELECTRICAL CONNECTIONS OR WIRING

- Most intermittent problems are caused by faulty electrical connections or wiring. Perform careful check of suspect circuits for:
  - Poor mating of the connector halves, or terminals, not fully seated in the connector body (backed out).
  - Improperly formed or damaged terminals. All connector terminals in problem circuit should be carefully reformed or replaced to insure proper contact tension.
  - Poor terminal to wire connection. This requires removing the terminal from the connector body to check.

ROAD TEST

- If a visual (physical) check does not find the cause of the problem, the vehicle can be driven with a voltmeter connected to a suspected circuit or a "Scan" tool may be used. An abnormal voltage or "Scan" reading, when the problem occurs, indicates the problem may be in that circuit. If the wiring and connectors check OK and a Trouble Code was stored for a circuit having a sensor, except for Codes 43, 44 and 45.

SYMPTOMS — MULTI-PORT INJECTION SYSTEMS

INTERMITTENTS
(Page 2 of 2)

Definition: Problem may or may not turn "ON" the "Service Engine Soon" light, or store a code.

PRELIMINARY CHECKS

- Perform the careful visual checks as described at start of "Symptoms."

INTERMITTENT "SERVICE ENGINE SOON LIGHT"

- An intermittent "Service Engine Soon" light, and No Trouble Codes, may be caused by:
  - Electrical system interference caused by a defective relay, ECM driven solenoid, or switch. They can cause a sharp electrical surge. Normally, the problem will occur when the faulty component is operated.
  - Improper installation of electrical options, such as lights, 2-way radios, cellular phones, etc.
  - EST wires should be routed away from spark plug wires, ignition system components, and generator. Wire for CKT 453 from ECM to ignition system should be a good ground.
  - Ignition secondary shorted to ground.
  - CKT 419 ("Service Engine Soon" light) or CKT 451 (Diagnostic "Test" Terminal) intermittently shorted to ground.
  - ECM power grounds.

LOSS OF TROUBLE CODE MEMORY

- To check, disconnect TPS and idle engine until "Service Engine Soon" light comes "ON." Code 22 should be stored, and kept in memory, when ignition is turned "OFF" for at least 10 seconds. If not, the ECM is faulty.

SYMPTOMS — MULTI-PORT INJECTION SYSTEMS

HARD START
(Page 1 of 2)

Definition: Engine cranks OK, but does not start for a long time. Does eventually run, or may start but immediately dies.

PRELIMINARY CHECKS

- Perform the careful visual checks as described at start of "Symptoms."
- Make sure the driver is using the correct starting procedure.

SENSORS

- **CHECK:** COOLANT TEMPERATURE SENSOR - Using a "Scan" tool compare coolant temperature with ambient temperature on cold engine.
  If coolant temperature reading is 5 degrees greater than or less than ambient air temperature, check for high resistance in coolant sensor circuit or sensor itself. Compare resistance value to Code 15 chart.

- **CHECK:** TPS - If a sticking throttle shaft or binding linkage causes a high TPS voltage (open throttle indication), the ECM will not control idle. Monitor TPS voltage. A "Scan" tool and/or voltmeter should display less than .95 volt with throttle closed.

FUEL SYSTEM

⚡ Important

- Fuel pump relay operation - pump should turn "ON" for 2 seconds when ignition is turned "ON." SEE CHART A-7 or Code 54.

- **CHECK:** Fuel pressure, CHART A-7.

- **CHECK:** Water contaminated fuel. For a faulty in-tank fuel pump check valve, which would allow the fuel in the lines to drain back to the tank after the engine is stopped. To check for this condition:
  1. Ignition "OFF."
  2. Disconnect fuel line at the filter.
  3. Remove the tank filler cap.
  4. Connect a radiator test pump to the fuel line and apply 103 kPa (15 psi) pressure. If the pressure will hold for 60 seconds, the check valve is OK.

SYMPTOMS — MULTI-PORT INJECTION SYSTEMS

HARD START
(Page 2 of 2)

Definition: Engine cranks OK, but does not start for a long time. Does eventually run, or may start but immediately dies.

PRELIMINARY CHECKS

- Perform the careful visual checks as described at start of "Symptoms."
- Make sure the driver is using the correct starting procedure.

IGNITION SYSTEM

- **CHECK:** Ignition system for:
  - Proper output with ST-125.
  - Worn distributor shaft.
  - Bare and shorted ignition wires.
  - Pickup coil resistance and connections.
  - Loose ignition coil connections.
  - Moisture in distributor cap.
  - Spark plugs, wet plugs, cracks, wear, improper gap, burned electrodes or heavy deposits.

Important
- If engine starts but then, immediately stalls, disconnect the set timing connector. If engine then starts, and runs OK, replace distributor pickup coil.

- **CHECK:** CKT 423 (EST) for short to ground.

ADDITIONAL CHECKS

- **CHECK:** IAC operation - see CHART C-2C.
- **CHECK:** No crank signal - see CHART C-1B.
- **CHECK:** EGR operation - see CHART C-7.

SYMPTOMS — MULTI-PORT INJECTION SYSTEMS

SURGES AND/OR CHUGGLES

Definition: Engine power variation, under steady throttle or cruise. Feels like the vehicle speeds up and slows down, with no change in the accelerator pedal.

PRELIMINARY CHECKS

- Perform the careful visual checks as described at start of "Symptoms."
- Be sure driver understands transmission/torque converter clutch and A/C compressor operation in owner's manual.
- Use a "Scan" tool to make sure reading of VSS matches vehicle speedometer.

SENSORS

- **CHECK:** Oxygen sensor for silicon contamination from fuel, or use of improper RTV sealant. The sensor may have a white, powdery coating and result in a high but false signal voltage (rich exhaust indication). The ECM will then reduce the amount of fuel delivered to the engine, causing a severe driveability problem.

FUEL SYSTEM

Important
- To determine if the condition is caused by a rich or lean system, the vehicle should be driven at the speed of the complaint. Monitoring block learn will help identify problem.
  Lean - Block learn near 150. Refer to "Diagnostic Aids" on facing page of Code 44.
  Rich - Block learn near 108. Refer to "Diagnostic Aids" on facing page of Code 45.

- **CHECK:** Fuel pressure while condition exists. See CHART A-7.
- **CHECK:** In-line fuel filter. Replace if dirty or plugged.

IGNITION SYSTEM

- **CHECK:** For proper output voltage using spark tester (ST-125) J 26792 or equivalent.
- **CHECK:** Spark plugs. After removing spark plugs, check for wet plugs, cracks, wear, improper gap, burned electrodes, or heavy deposits. Repair or replace as necessary.

ADDITIONAL CHECKS

- **CHECK:** Generator output voltage. Repair if less than 9 or more than 16 volts.
- **CHECK:** Vacuum lines for kinks or leaks.
- **CHECK:** For intermittent EGR at idle, use CHART C-7.
- **CHECK:** TCC operation. Use CHART C-8A.

SYMPTOMS — MULTI-PORT INJECTION SYSTEMS

LACK OF POWER, SLUGGISH, OR SPONGY
(Page 1 of 2)

Definition: Engine delivers less than expected power. Little or no increase in speed, when accelerator pedal is pushed down part way.

PRELIMINARY CHECKS

- Perform the careful visual checks as described at start of "Symptoms."

- Compare customer's vehicle to similar unit. Make sure the customer has an actual problem.
- Remove air filter and check air filter for dirt, or for being plugged, replace as necessary.
- If there is spray from only one injector, then there is a malfunction in the injector assembly, or in the signal to the injector assembly. The malfunction can be isolated, by switching the injector connectors. If the problem remains with the original injector, after switching the connector, the injector is defective. Replace the injector. If the problem moves with the injector connector, the problem is an improper signal in the injector circuits, use CHART A-3.

FUEL SYSTEM

CHECK: For restricted fuel filter, contaminated fuel or improper fuel pressure, use CHART A-7.

IGNITION SYSTEM

- **CHECK:** Proper output voltage with spark tester J 26792 or equivalent (ST-125).
- **CHECK:** Ignition timing. See "Vehicle Emission Control Information" label.
- **CHECK:** Proper operation of EST.

LACK OF POWER, SLUGGISH, OR SPONGY
(Page 2 of 2)

Definition: Engine delivers less than expected power. Little or no increase in speed, when accelerator pedal is pushed down part way.

EXHAUST SYSTEM

- **CHECK:** Exhaust system for possible restriction: See CHART B-1.
  Inspect exhaust system for damaged or collapsed pipes. Inspect muffler for heat distress or possible internal failure.
  1. With engine at normal operating temperature, connect a vacuum gage to any convenient vacuum port on intake manifold.
  2. Run engine at 1000 rpm and record vacuum reading.
  3. Increase rpm slowly to 2500 rpm. Note vacuum reading at steady 2500 rpm.
  4. If vacuum at 2500 rpm decreases more than 3", from reading at 1000 rpm, the exhaust system should be inspected for restrictions.
  5. Disconnect exhaust pipe from engine and repeat Steps 3 & 4. If vacuum still drops more than 3", with exhaust disconnected, check valve timing.

ADDITIONAL CHECKS

- **CHECK:** ECM power grounds, see wiring diagrams.
- **CHECK:** Engine valve timing and compression.
- **CHECK:** EGR operation for being open or partly open, all the time. Use CHART C-7.
- **CHECK:** Engine, for proper or worn camshaft.
- **CHECK:** Transmission torque converter operation.
- **CHECK:** Generator output voltage. Repair if less than 9 or more than 16 volts.

SYMPTOMS — MULTI-PORT INJECTION SYSTEMS

DETONATION/SPARK KNOCK
(Page 1 of 2)

Definition: A mild to severe ping, usually worse under acceleration. The engine makes sharp metallic knocks that change with throttle opening.

PRELIMINARY CHECKS

- Perform the careful visual checks as described at start of "Symptoms,"

- Make sure the customer has an actual problem.
- Remove air filter and check air filter for dirt, or for being plugged, replace as necessary.
- If there is spray from only one injector, then there is a malfunction in the injector assembly, or in the signal to the injector assembly. The malfunction can be isolated, by switching the injector connectors. If the problem remains with the original injector, after switching the connector, the injector is defective. Replace the injector. If the problem moves with the injector connector, the problem is an improper signal in the injector circuits, use CHART A-3.

COOLING SYSTEM

- **CHECK:** For obvious over heating problems.
- **CHECK:** Low coolant.
- **CHECK:** Loose water pump belt.
- **CHECK:** Restricted air flow to radiator, or restricted water flow thru radiator.
- **CHECK:** Faulty or incorrect thermostat.
- **CHECK:** Correct coolant solution - should be a 50/50 mix of GM #1052753 antifreeze coolant (or equivalent) and water.

SENSOR

- **CHECK:** Coolant Temperature Sensor (CTS), which has shifted in value.

SYMPTOMS — MULTI-PORT INJECTION SYSTEMS

DETONATION/SPARK KNOCK
(Page 2 of 2)

Definition: A mild to severe ping, usually worse under acceleration. The engine makes sharp metallic knocks that change with throttle opening.

FUEL SYSTEM

⚠ Important

- To determine if the condition is caused by a rich or lean system, the vehicle should be driven at the speed of the complaint. Monitoring block learn will help identify problem.
  Lean - Block learn near 150. Refer to "Diagnostic Aids" on facing page of Code 44.
  Rich - Block learn near 108. Refer to "Diagnostic Aids" on facing page of Code 45.

- **CHECK:** Fuel pressure, CHART A-7.
- **CHECK:** For poor fuel quality, proper octane rating.
- **CHECK:** If "Scan" tool readings are normal (see facing page of "Diagnostic Circuit Check") and there are no engine mechanical faults, fill fuel tank with a premium gasoline that has a minimum octane rating of 92 and re-evaluate vehicle performance.

IGNITION SYSTEM

- **CHECK:** ESC system operation see CHART C-5.
- **CHECK:** Ignition timing. See "Vehicle Emission "Control Information" label.

ENGINE MECHANICAL

- **CHECK:** For incorrect basic engine parts such as cam, heads, pistons, etc.
- **CHECK:** Excessive oil entering combustion chamber.

ADDITIONAL CHECKS

- **CHECK:** For carbon buildup. Remove carbon with top engine cleaner, follow instructions on can.
- **CHECK:** For proper transmission shift points.
- **CHECK:** TCC operation. Use CHART C-8.
- **CHECK:** For incorrect basic engine parts such as cam, heads, pistons, etc.
- **CHECK:** Excessive oil entering combustion chamber.
- **CHECK:** For correct PROM.
- **CHECK:** Spark plugs for correct heat range.
- **CHECK:** Proper operation of THERMAC.

HESITATION, SAG, STUMBLE

Definition: Momentary lack of response as the accelerator is pushed down. Can occur at all vehicle speeds. Usually more severe when first trying to make the vehicle move, as from a stop sign. May cause the engine to stall if severe enough.

PRELIMINARY CHECKS

- Perform the careful visual checks as described at start of "Symptoms."

FUEL SYSTEM

- **CHECK:** TPS - Check TPS for binding or sticking. Voltage should increase at a steady rate as throttle is moved toward Wide Open Throttle (WOT). TPS voltage should be less than .95 volt at idle.
- **CHECK:** MAP SENSOR - See CHART C-1D.

IGNITION SYSTEM

- **CHECK:** Spark plugs for being fouled, worn or cracked.
- **CHECK:** For open ignition system ground, CKT 453.
- **CHECK:** Ignition timing. See "Vehicle Emission Control Information" label.

ADDITIONAL CHECKS

- **CHECK:** Generator output voltage. Repair, if less than 9 or more than 16 volts.
- **CHECK:** EGR valve operation. Use CHART C-7

SYMPTOMS — MULTI-PORT INJECTION SYSTEMS

CUTS OUT, MISSES
(Page 1 of 2)

Definition: Steady pulsation or jerking that follows engine speed, usually more pronounced as engine load increases. The exhaust has a steady spitting sound at idle or low speed.

PRELIMINARY CHECKS

- Perform the careful visual checks as described at start of "Symptoms."
- If there is spray from only one injector, then there is a malfunction in the injector assembly, or in the signal to the injector assembly. The malfunction can be isolated, by switching the injector connectors. If the problem remains with the original injector, after switching the connector, the injector is defective. Replace the injector. If the problem moves with the injector connector, the problem is an improper signal in the injector circuits, use CHART A-3.

FUEL SYSTEM

- **CHECK:** Fuel pressure, CHART A-7.
- **CHECK:** Water contaminated fuel or restricted fuel filter.

IGNITION SYSTEM

- **CHECK:** For missing cylinder by:
1. Start engine, allow engine to stabilize then disconnect IAC motor. Remove one spark plug wire at a time, using insulated pliers.

   CAUTION: Do not perform this test for more than 2 minutes, as this may cause damage to the catalytic converter.

2. If there is an rpm drop, on all cylinders, (equal to within 50 rpm), go to ROUGH, UNSTABLE OR INCORRECT IDLE, STALLING symptom. Reconnect IAC motor.
3. If there is no rpm drop on one or more cylinders, or excessive variation in drop, check for spark, on the suspected cylinder(s) with J 26792 (ST-125) Spark Tester or equivalent. If no spark, use CHART C-4. If there is spark, remove spark plug(s) in these cylinders and check for:
   - Cracks.
   - Wear.
   - Improper Gap.
   - Burned Electrodes.
   - Heavy Deposits.
- **CHECK:** Ignition wire resistance (should not exceed 30,000 ohms), also, check rotor and distributor cap.

⚠ Important
- If the previous checks did not find the problem:
   - Visually inspect ignition system for moisture, dust, cracks, burns, etc. Spray plug wires with fine water mist to check for shorts.

SYMPTOMS — MULTI-PORT INJECTION SYSTEMS

CUTS OUT, MISSES
(Page 2 of 2)

Definition: Steady pulsation or jerking that follows engine speed, usually more pronounced as engine load increases. The exhaust has a steady spitting sound at idle or low speed.

ENGINE MECHANICAL

- **CHECK:** For proper valve timing. Remove rocker covers. Check for bent pushrods, worn rocker arms, broken or weak valve springs, worn camshaft lobes. Repair as necessary.
- **CHECK:** For casting flash in, intake and exhaust manifolds.
- **CHECK:** Low compression. Perform compression check.

SYMPTOMS — MULTI-PORT INJECTION SYSTEMS

POOR FUEL ECONOMY

Definition: Fuel economy, as measured by an actual road test, is noticeably lower than expected. Also, economy is noticeably lower than it was on this vehicle at one time, as previously shown by an actual road test.

PRELIMINARY CHECKS

- Perform the careful visual checks as described at start of "Symptoms."

- Perform "Diagnostic Circuit Check."
- Check owner's driving habits.
  - Is A/C "ON" full time (Defroster mode "ON")?
  - Are tires at correct pressure?
  - Are excessively heavy loads being carried?
  - Is acceleration too much, too often?
- Check air cleaner element (filter) for dirt or being plugged.
- Visually (physically) check: Vacuum hoses for splits, kinks, and proper connections as shown on "Vehicle Emission Control Information" label.

IGNITION SYSTEM

- **CHECK:** ESC operation.

- **CHECK:** Spark plugs. After removing spark plugs, check for wet plugs, cracks, wear, improper gap, burned electrodes, or heavy deposits. Repair or replace as necessary.
- **CHECK:** Ignition wires for cracking, hardness, and proper connections.
- **CHECK:** Ignition timing. See "Vehicle Emission Control Information" label.

COOLING SYSTEM

- **CHECK:** Engine thermostat for faulty part or for wrong heat range.

ADDITIONAL CHECKS

- **CHECK:** TCC operation - Check for proper operation. See CHART C-8. A "Scan" should indicate an rpm drop, when the TCC is commanded "ON."
- **CHECK:** For exhaust system restriction. See CHART B-1.
- **CHECK:** Fuel pressure. See CHART A-7
- **CHECK:** Compression.
- **CHECK:** For proper calibration of speedometer.
- **CHECK:** For dragging brakes.
- Suggest owner fill fuel tank and recheck fuel economy.
- Suggest driver read "Important Facts on Fuel Economy" in Owner's Manual.

SYMPTOMS — MULTI-PORT INJECTION SYSTEMS

STALLING, ROUGH, UNSTABLE OR INCORRECT IDLE
(Page 1 of 2)

Definition: The engine runs unevenly at idle. If bad enough, the vehicle may shake. Also, the idle may vary in rpm (called "hunting"). Either condition may be severe enough to cause stalling. Engine idles at incorrect speed.

PRELIMINARY CHECKS

- Perform the careful visual checks as described at start of "Symptoms,"

SENSORS

- **CHECK:** OXYGEN SENSOR - Inspect sensor for silicon contamination from fuel, or use of improper RTV sealant. The sensor will have a white, powdery coating, and will result in a high but false signal voltage (rich exhaust indication). The ECM will then reduce the amount of fuel delivered to the engine, causing a severe driveability problem.
- **CHECK:** TPS - If a sticking throttle shaft or binding linkage causes a high TPS voltage (open throttle indication), the ECM will not control idle. Monitor TPS voltage. A "Scan" tool and/or voltmeter should display less than .95 volt with throttle closed.
- **CHECK:** COOLANT TEMPERATURE SENSOR (CTS) - Using a "Scan" tool compare coolant temperature with ambient temperature on cold engine.
  - If coolant temperature reading is 5 degrees greater than or less than ambient air temperature. Check for high resistance in coolant sensor circuit or sensor itself. Compare resistance value to Code 15 chart.
- **CHECK:** MAP SENSOR - Refer to CHART C-1D MAP voltage output check.

FUEL SYSTEM

Important
- To determine if the condition is caused by a rich or lean system, the vehicle should be driven at the speed of the complaint. Monitoring block learn will help identify problem.
  Lean - Block learn near 150. Refer to "Diagnostic Aids" on facing page of Code 44.
  Rich - Block learn near 108. Refer to "Diagnostic Aids" on facing page of Code 45.

- **CHECK:** For fuel in pressure regulator hose. If present, replace regulator assembly.
- **CHECK:** Evaporative emission control system. CHART C-3.
- **CHECK:** Run a cylinder compression check.
- **CHECK:** For injector(s) leaking. Check fuel pressure, see CHART A-7.

SYMPTOMS — MULTI-PORT INJECTION SYSTEMS

STALLING, ROUGH, UNSTABLE OR INCORRECT IDLE
(Page 2 of 2)

Definition: The engine runs unevenly at idle. If bad enough, the vehicle may shake. Also, the idle may vary in rpm (called "hunting"). Either condition may be severe enough to cause stalling. Engine idles at incorrect speed.

IGNITION SYSTEM

- **CHECK:** Ignition system. Refer to "Ignition System (EST)
- **CHECK:** Ignition timing. See "Vehicle Emission Control Information" label.

ADDITIONAL CHECKS

- **CHECK:** IAC valve will not move, if system voltage is below 9 or greater than 16 volts.
- **CHECK:** IAC operation - See CHART C-2C.
- **CHECK:** ECM ground circuits.
- **CHECK:** P/N switch circuit. See CHART C-1A, or use "Scan" tool, and be sure tool indicates vehicle is in drive with gear selector is in drive.

Important
- Use "Scan" tool to determine if ECM is receiving A/C status signal.
  If problem exists with A/C "ON," check A/C system operation CHART C-10.

- **CHECK:** EGR "ON," while idling, will cause roughness, stalling, and hard starting. Use CHART C-7.
- **CHECK:** Battery cables and ground straps should be clean and secure. Erratic voltage will cause IAC to change its position, resulting in poor idle quality.
- **CHECK:** A/C refrigerant pressure too high.
- **CHECK:** For overcharge or faulty pressure switch.
- **CHECK:** PCV valve for proper operation by placing finger over inlet hole in valve end several times. Valve should snap back. If not, replace valve.

ENGINE MECHANICAL

- **CHECK:** Vacuum leaks can cause higher than normal idle.
- **CHECK:** For broken motor mounts.
- **CHECK:** For low compression.

SYMPTOMS — MULTI-PORT INJECTION SYSTEMS

EXCESSIVE EXHAUST EMISSIONS OR ODORS
(Page 1 of 2)

Definition: Vehicle fails an emission test. Vehicle has excessive "rotten egg" smell. Excessive odors do not necessarily indicate excessive emissions.

PRELIMINARY CHECKS

- Perform "Diagnostic Circuit Check."
- If Emission Test shows excessive CO and HC check items which cause vehicle to run RICH. Make sure engine is at normal operating temperature.
- If Emission Test shows excessive NO_x check items which can cause the vehicle to run LEAN or too hot.

SENSORS

⚠ **Important**
- If the "Scan" tool indicates a very high coolant temperature and the system is running LEAN:
  Check the coolant system and coolant fan for proper operation. Also check for vacuum leaks.

FUEL SYSTEM

⚠ **Important**
- If the system is running rich, (block learn near 108), refer to "Diagnostic Aids" on facing page of Code 45. If the system is running lean, (block learn near 150), refer to "Diagnostic Aids" on facing page of Code 44.

- **CHECK:** Fuel pressure. CHART A-7.

⚠ **Important**
- If test shows excessive NO_x, check items which cause vehicle to run LEAN or too hot.

- **CHECK:** Canister for fuel loading. See CHART C-3.

SYMPTOMS — MULTI-PORT INJECTION SYSTEMS

EXCESSIVE EXHAUST EMISSIONS OR ODORS
(Page 2 of 2)

Definition: Vehicle fails an emission test. Vehicle has excessive "rotten egg" smell. Excessive odors do not necessarily indicate excessive emissions.

COOLING SYSTEM

- **CHECK:** Coolant system and coolant fan for proper operation.

IGNITION SYSTEM

- **CHECK:** Ignition system.
- **CHECK:** For incorrect timing or excessive ignition timing advance. See "Vehicle Emission Control Information" label.
- **CHECK:** Spark plugs, plug wires, and ignition components.

ADDITIONAL CHECKS

- **CHECK:** For lead contamination of catalytic converter (look for the removal of fuel filler neck restrictor).
- **CHECK:** Remove carbon with top engine cleaner. Follow instructions on can.
- **CHECK:** EGR valve for not opening. See CHART C-7. Vacuum leaks.
- **CHECK:** PCV valve for being plugged, stuck, or blocked PCV hose, or fuel in the crankcase.

ENGINE MECHANICAL

- **CHECK:** EGR valve for not opening. See CHART C-7. Vacuum leaks.
- **CHECK:** PCV valve for being plugged, stuck, or blocked PCV hose, or fuel in the crankcase.

DIESELING, RUN-ON

Definition: Engine continues to run after key is turned "OFF," but runs very roughly. If engine runs smoothly, check ignition switch and adjustment.

PRELIMINARY CHECKS

- Perform the careful visual checks as described at start of "Symptoms." B"

FUEL SYSTEM

- **CHECK:** Injector(s) for leaking. Apply 12 volts to fuel pump "test" terminal (located underhood) to turn "ON" fuel pump and pressurize fuel system. Visually check injector(s) and TBI assembly for fuel leakage. Refer to CHART A-7, "Fuel System Diagnosis."

BACKFIRE

Definition: Fuel ignites in intake manifold, or in exhaust system, making a loud popping noise.

PRELIMINARY CHECKS

- Perform the careful visual checks as described at start of "Symptoms."

FUEL SYSTEM

- **CHECK:** Perform fuel system diagnosis check. Use CHART A-7.
- **CHECK:** For plugged fuel filter.

IGNITION SYSTEM

- **CHECK:** Proper output with spark tester J 26792 or equivalent (ST-125).
- **CHECK:** Spark plugs. After removing spark plugs, check for wet plugs, cracks, wear, improper gap, burned electrodes, or heavy deposits. Repair or replace as necessary.
- **CHECK:** Ignition system.
- **CHECK:** For crossfire between spark plugs (distributor cap, spark plug wires, and proper routing of plug wires.)
- **CHECK:** Ignition timing. See "Vehicle Emission Control Information" label.

ENGINE MECHANICAL

- **CHECK:** Compression. Perform a compression check - look for sticking or leaking valves.
    - For proper valve timing.
    - Broken or worn valve train parts.
- **CHECK:** Valve timing.
- **CHECK:** Intake manifold gasket for vacuum leaks.
- **CHECK:** Intake and exhaust manifolds for casting flash.
- **CHECK:** Faulty A.I.R. check valve. Use CHART C-6.
- **CHECK:** EGR operation for being open all the time. Use CHART C-7.

1.6L (VIN 5) ENGINE — ECM WIRING DIAGRAMS — 1992 PRIZM

1.6L (VIN 5) ENGINE — COMPONENT LOCATIONS — 1992 PRIZM

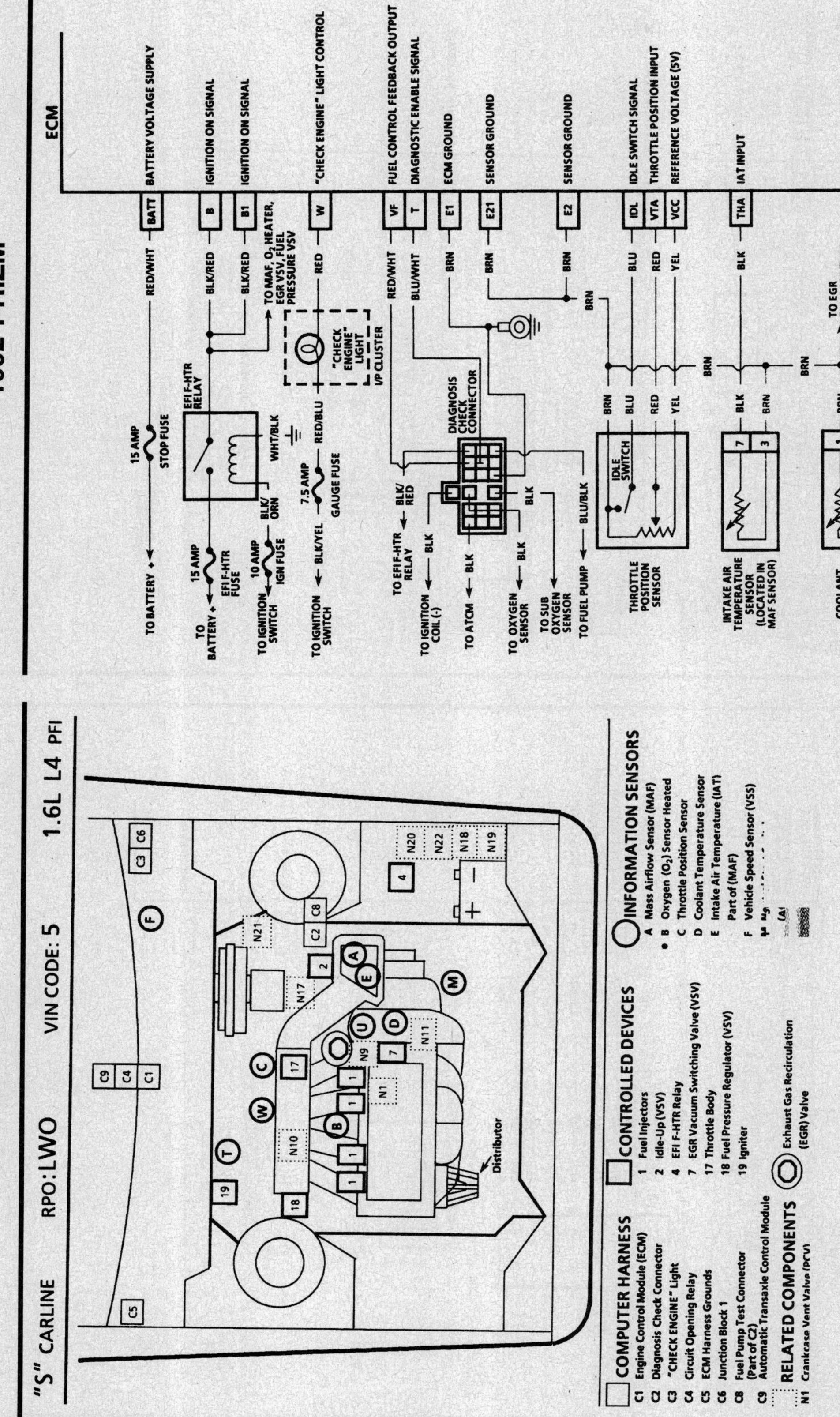

MULTIPORT FUEL INJECTION (MFI) SYSTEMS
LIGHT TRUCKS, VAN, GEO AND SATURN

1.6L (VIN 5) ENGINE —ECM WIRING DIAGRAMS — 1992 PRIZM

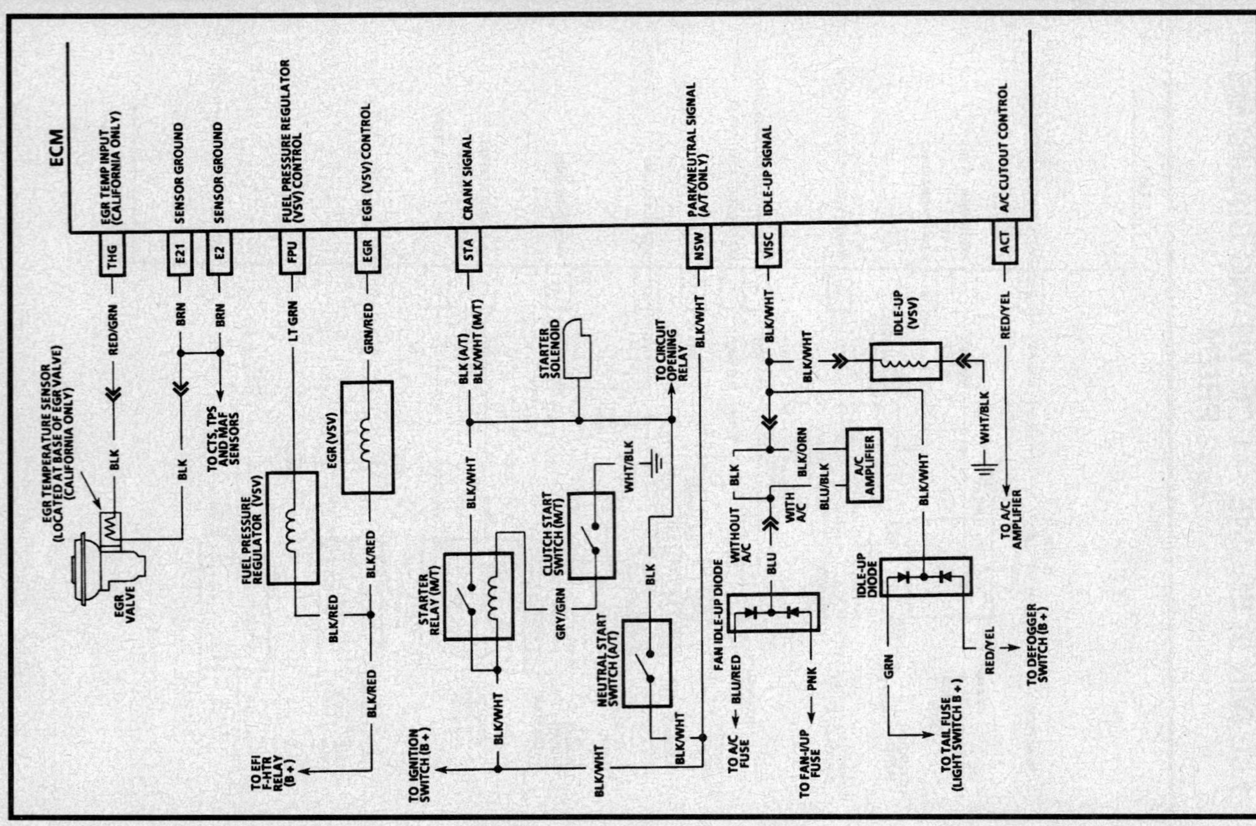

1.6L (VIN 5) ENGINE —ECM WIRING DIAGRAMS — 1992 PRIZM

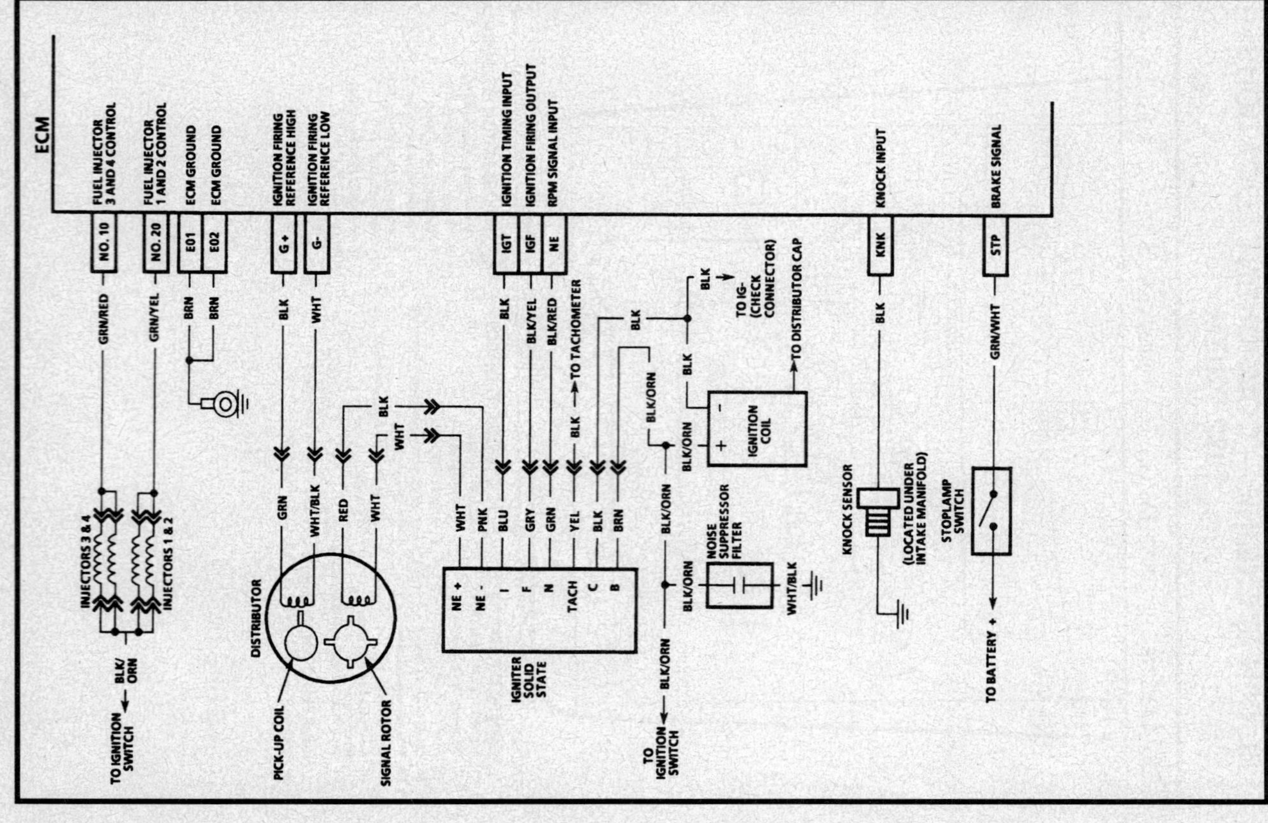

1.6L (VIN 5) ENGINE — CALIFORNIA ECM WIRING DIAGRAMS — 1992 PRIZM

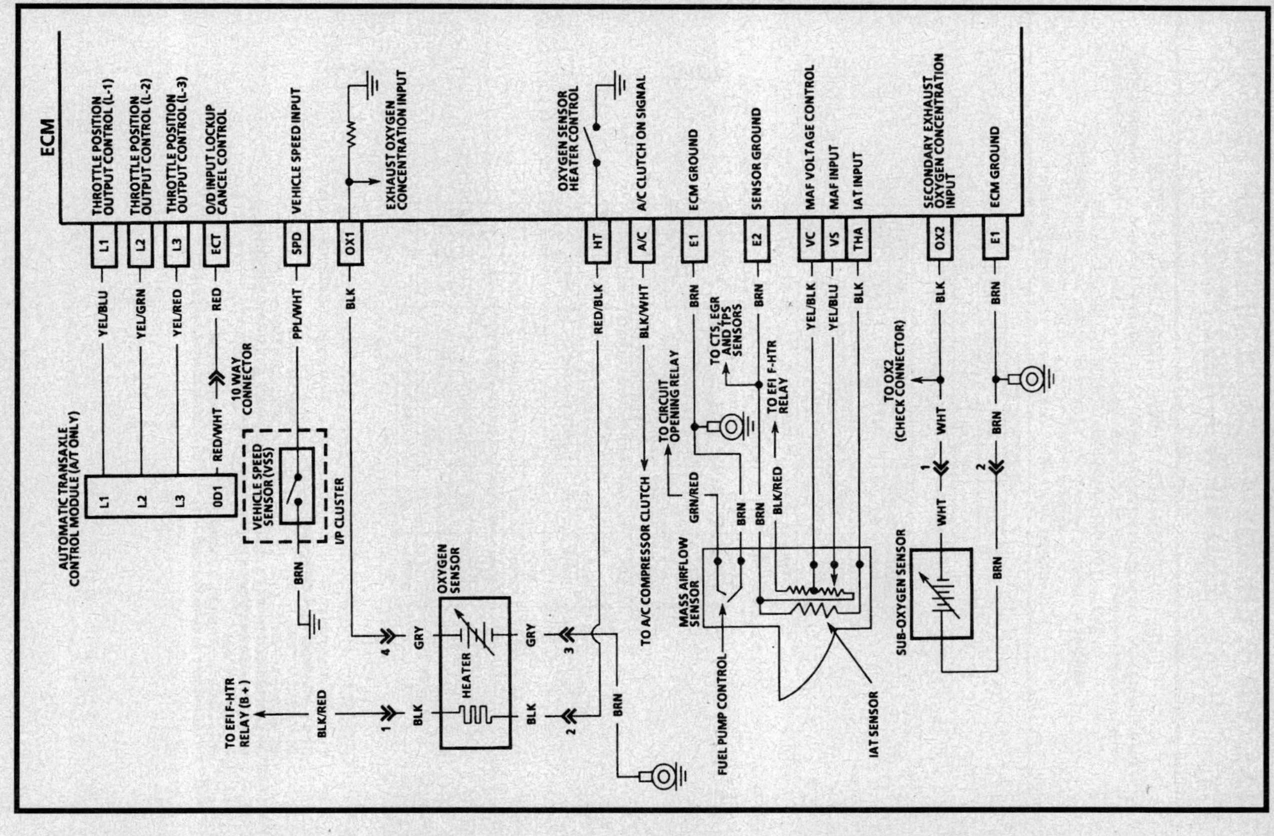

1.6L (VIN 5) ENGINE — FEDERAL ECM WIRING DIAGRAMS — 1992 PRIZM

1.6L (VIN 5) ENGINE — ECM CONNECTOR TERMINAL END VIEW — 1992 PRIZM

FUEL INJECTION ECM CONNECTOR IDENTIFICATION

The following ECM voltage charts are for use with a digital-voltmeter to further aid in diagnosis. The voltages you get may vary due to low battery charge or other reasons, but they should be very close.

THE FOLLOWING CONDITIONS MUST BE MET BEFORE TESTING:
- Engine at operating temperature
- Engine idling (for "Engine Run" column)
- Diagnostic Enable Terminal not grounded
- All voltages shown "B+" indicate battery or charging voltage

ECM Connector C3 — pin end view:

VF	OX+	OX1	OX2	THW	THA	VS	VC
E21	T	KNK	THG	IDL	VCC	VTA	E2

BACKVIEW OF ECM CONNECTOR C3

CAVITY/PIN	WIRE COLOR	CIRCUIT	VOLTAGE KEY "ON"	VOLTAGE ENG. RUN
1/VC	YEL/BLK	MAF VOLTAGE CONTROL	8V	8-9V
2/VS	YEL/BLU	MAF INPUT	2V	1-5V (10)
3/THA	BLK	IAT INPUT	2V (11)	2V (11)
4/THW	WHT	COOLANT TEMPERATURE INPUT	5V (11)	1-4V (11)
5/OX2 (★)	BLK	SECONDARY EXHAUST OXYGEN CONCENTRATION INPUT	0*	.1-.9V VARIES
6/OX1	BLK	EXHAUST OXYGEN CONCENTRATION INPUT	.1-.9V	.1-.9V VARIES
7/OX+ (●)	RED/GRN	EXHAUST OXYGEN CONCENTRATION INPUT	1V	1V
8/VF	RED/WHT	FUEL CONTROL FEEDBACK OUTPUT	0*	0-2.5V
9/E2	BRN	SENSOR GROUND	0*	0*
10/VTA	RED	THROTTLE POSITION INPUT	.60V	.6-4V (12)
11/VCC	YEL	REFERENCE VOLTAGE (5V)	5V	5V
12/IDL	BLU	IDLE SWITCH SIGNAL	0* (13)	0* (13)
13/THG (●)	RED/GRN	EGR TEMPERATURE INPUT	4.5V	.69V (11)
14/KNK	BLK	KNOCK INPUT	0*	0*
15/T	BLU/WHT	DIAGNOSTIC ENABLE SIGNAL	B+	B+
16/E21	BRN	SENSOR GROUND	0*	0*

0* LESS THAN 0.50 VOLTS
(1) B+ WITH BRAKE PEDAL DEPRESSED
(2) 0V WITH A/C "ON"
(3) B+ WITH A/C "ON"
(4) VARIES AS FRONT WHEELS ROTATE
(5) B+ WITH IGNITION SWITCH IN "START"
(6) 0V IN OVERDRIVE
(7) B+ WITH RADIATOR FAN, BLOWER MOTOR, HEADLIGHTS OR DEFOGGER "ON"
(8) 0V BELOW 4000 RPM, B+ ABOVE 4000 RPM
(9) B+ WITH MANUAL SELECTOR IN "R," "D," "2" OR "L"
(10) INCREASES WITH RPM
(11) VARIES WITH TEMPERATURE
(12) VARIES WITH THROTTLE POSITION
(13) B+ WHEN "OFF" IDLE
(★) CALIFORNIA ONLY
(●) FEDERAL AND CANADA ONLY
(X) AUTOMATIC TRANSAXLE ONLY

1.6L (VIN 5) ENGINE — ECM CONNECTOR TERMINAL END VIEW — 1992 PRIZM

FUEL INJECTION ECM CONNECTOR IDENTIFICATION

The following ECM voltage charts are for use with a digital-voltmeter to further aid in diagnosis. The voltages you get may vary due to low battery charge or other reasons, but they should be very close.

THE FOLLOWING CONDITIONS MUST BE MET BEFORE TESTING:
- Engine at operating temperature
- Engine idling (for "Engine Run" column)
- Diagnostic Enable Terminal not grounded
- All voltages shown "B+" indicate battery or charging voltage

ECM Connector C1 — pin end view:

ACT	STP	⊠	BATT	B1
SPD	A/C	⊠	W	B

ECM Connector C2 — pin end view (pins E01/E02, NO.10/NO.20, (★)HT/(●)HT, E1, VISC, IGT, L1, L2, L3, G-, NE, G+, EGR, NSW):

E01	NO.10	(★)HT	E1	VISC	IGT	L1	L2	L3
E02	NO.20	(●)HT	G-	NE	G+	EGR	NSW	

BACKVIEW OF ECM CONNECTOR C1

CAVITY/PIN	WIRE COLOR	CIRCUIT	VOLTAGE KEY "ON"	VOLTAGE ENG. RUN
1/B1	BLK/RED	IGNITION ON SIGNAL	B+	B+
2/BATT	RED/WHT	BATTERY VOLTAGE SUPPLY	B+	B+
3/-	-	NOT USED	-	-
4/-	-	NOT USED	-	-
5/STP	GRN/WHT	BRAKE SIGNAL	0* (1)	0* (1)
6/ACT	RED/YEL	A/C CUTOUT CONTROL	0*	B+ (2)
7/B	BLK/RED	IGNITION ON SIGNAL	B+	B+
8/W	RED	"CHECK ENGINE" LIGHT CONTROL	1-2V	B+
9/-	-	NOT USED	-	-
10/A/C	BLK/WHT	A/C CLUTCH ON SIGNAL	0* (3)	0* (3)
11/SPD	PPL/WHT	VEHICLE SPEED INPUT	0-B+ (4)	0-B+ (4)
12/-	-	NOT USED	-	-

BACKVIEW OF ECM CONNECTOR C2

CAVITY/PIN	WIRE COLOR	CIRCUIT	VOLTAGE KEY "ON"	VOLTAGE ENG. RUN
1/FPU	LT GRN	FUEL PRESSURE REGULATOR VSV CONTROL	B+	B+
2/STA	BLK (X)/BLK/WHT	CRANK SIGNAL	0* (5)	0* (5)
3/IGF	BLK/YEL	IGNITION FIRING OUTPUT	.25V	.6-.8V
4/NE	BLK/RED	RPM SIGNAL INPUT	B+	10V
5/G+	BLK	IGNITION FIRING REFERENCE HIGH	.60V	.60V
6/-	-	NOT USED	-	-
7/-	-	NOT USED	-	-
8/ECT (X)	RED	OVERDRIVE INPUT LOCKUP CANCEL CONTROL	0*	B+ (6)
9/VISC	BLK/WHT	IDLE-UP SIGNAL	0* (7)	0* (7)
10/-	-	NOT USED	-	-
11/HT (★)	RED/BLK	OXYGEN SENSOR HEATER CONTROL	0*	0* (8)
12/NO.10	GRN/RED	FUEL INJECTOR 3 & 4 CONTROL	B+	B+
13/E01	BRN	ECM GROUND	0*	0*
14/HT (●)	RED/BLK	OXYGEN SENSOR HEATER CONTROL	B+	.1-3V
15/NSW (X)	BLK/WHT	PARK/NEUTRAL SIGNAL	0* (9)	0* (9)
16/EGR	GRN/RED	EGR VSV CONTROL	B+	B+
17/G-	WHT	IGNITION FIRING REFERENCE LOW	.60V	.60V
18/-	-	NOT USED	-	-
19/L3	YEL RED	L3 THROTTLE POSITION OUTPUT CONTROL	9V	10V
20/L2 (X)	YEL/GRN	L2 THROTTLE POSITION OUTPUT CONTROL	9V	10V
21/L1 (X)	YEL/BLU	L1 THROTTLE POSITION OUTPUT CONTROL	9V	10V
22/IGT	BLK	IGNITION TIMING INPUT	0*	.76V
23/-	-	NOT USED	-	-
24/E1	BRN	ECM GROUND	0*	0*
25/NO.20	GRN/YEL	FUEL INJECTOR 1 & 2 CONTROL	B+	B+
26/E02	BRN	ECM GROUND	0*	0*

0* LESS THAN 0.50 VOLTS
(1) B+ WITH BRAKE PEDAL DEPRESSED
(2) 0V WITH A/C "ON"
(3) B+ WITH A/C "ON"
(4) VARIES AS FRONT WHEELS ROTATE
(5) B+ WITH IGNITION SWITCH IN "START"
(6) 0V IN OVERDRIVE
(7) B+ WITH RADIATOR FAN, BLOWER MOTOR, HEADLIGHTS OR DEFOGGER "ON"
(8) 0V BELOW 4000 RPM, B+ ABOVE 4000 RPM
(9) B+ WITH MANUAL SELECTOR IN "R," "D," "2" OR "L"
(10) INCREASES WITH RPM
(11) VARIES WITH TEMPERATURE
(12) VARIES WITH THROTTLE POSITION
(13) B+ WHEN "OFF" IDLE
(★) CALIFORNIA ONLY
(●) FEDERAL AND CANADA ONLY
(X) AUTOMATIC TRANSAXLE ONLY

1.6L (VIN 5) ENGINE — DIAGNOSTIC CIRCUIT CHECK — 1992 PRIZM

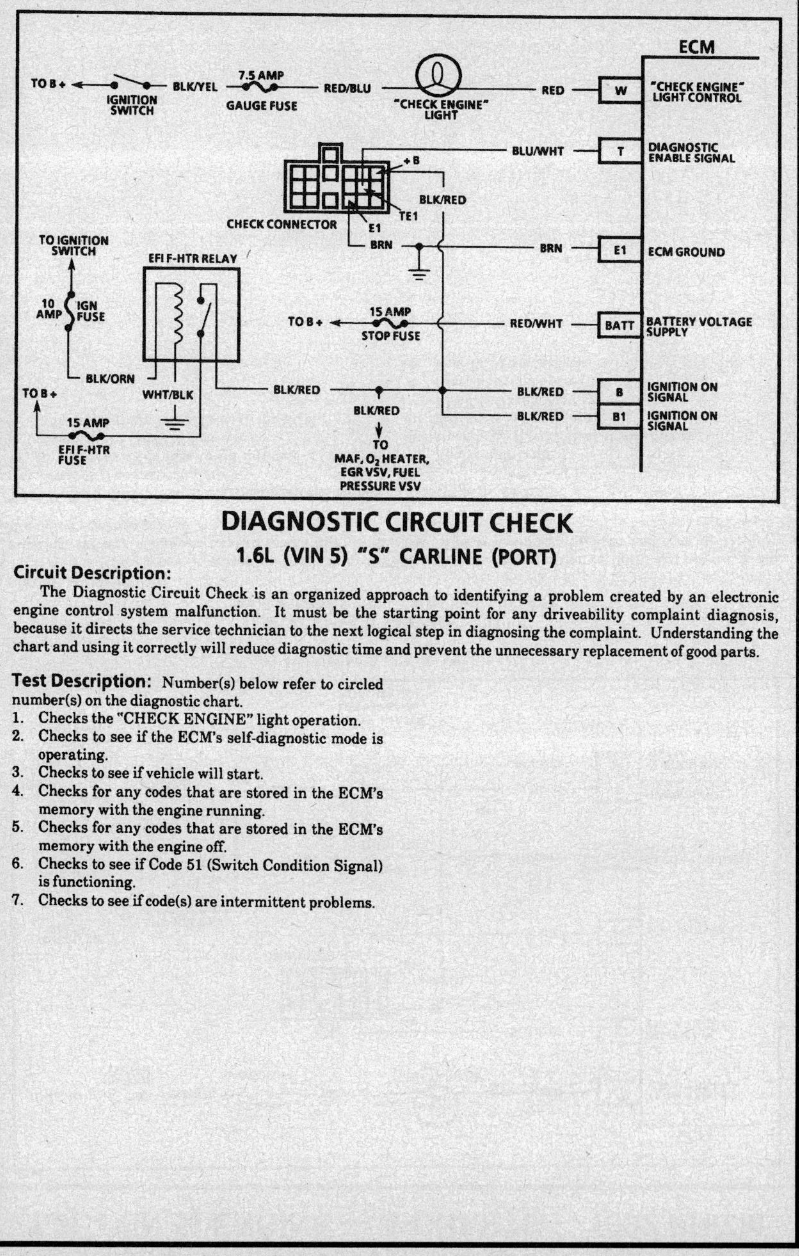

DIAGNOSTIC CIRCUIT CHECK
1.6L (VIN 5) "S" CARLINE (PORT)

Circuit Description:

The Diagnostic Circuit Check is an organized approach to identifying a problem created by an electronic engine control system malfunction. It must be the starting point for any driveability complaint diagnosis, because it directs the service technician to the next logical step in diagnosing the complaint. Understanding the chart and using it correctly will reduce diagnostic time and prevent the unnecessary replacement of good parts.

Test Description: Number(s) below refer to circled number(s) on the diagnostic chart.

1. Checks the "CHECK ENGINE" light operation.
2. Checks to see if the ECM's self-diagnostic mode is operating.
3. Checks to see if vehicle will start.
4. Checks for any codes that are stored in the ECM's memory with the engine running.
5. Checks for any codes that are stored in the ECM's memory with the engine off.
6. Checks to see if Code 51 (Switch Condition Signal) is functioning.
7. Checks to see if code(s) are intermittent problems.

1.6L (VIN 5) ENGINE — DIAGNOSTIC CIRCUIT CHECK — 1992 PRIZM

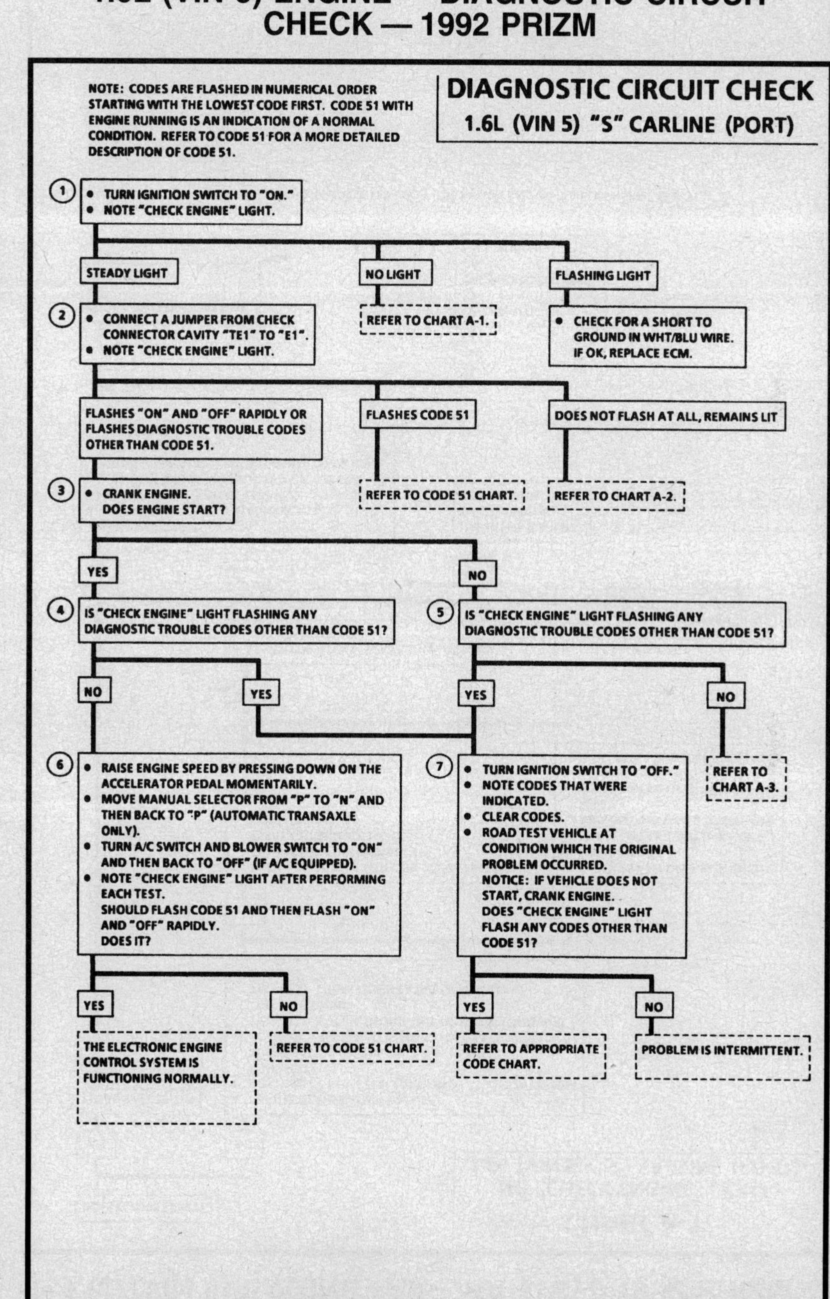

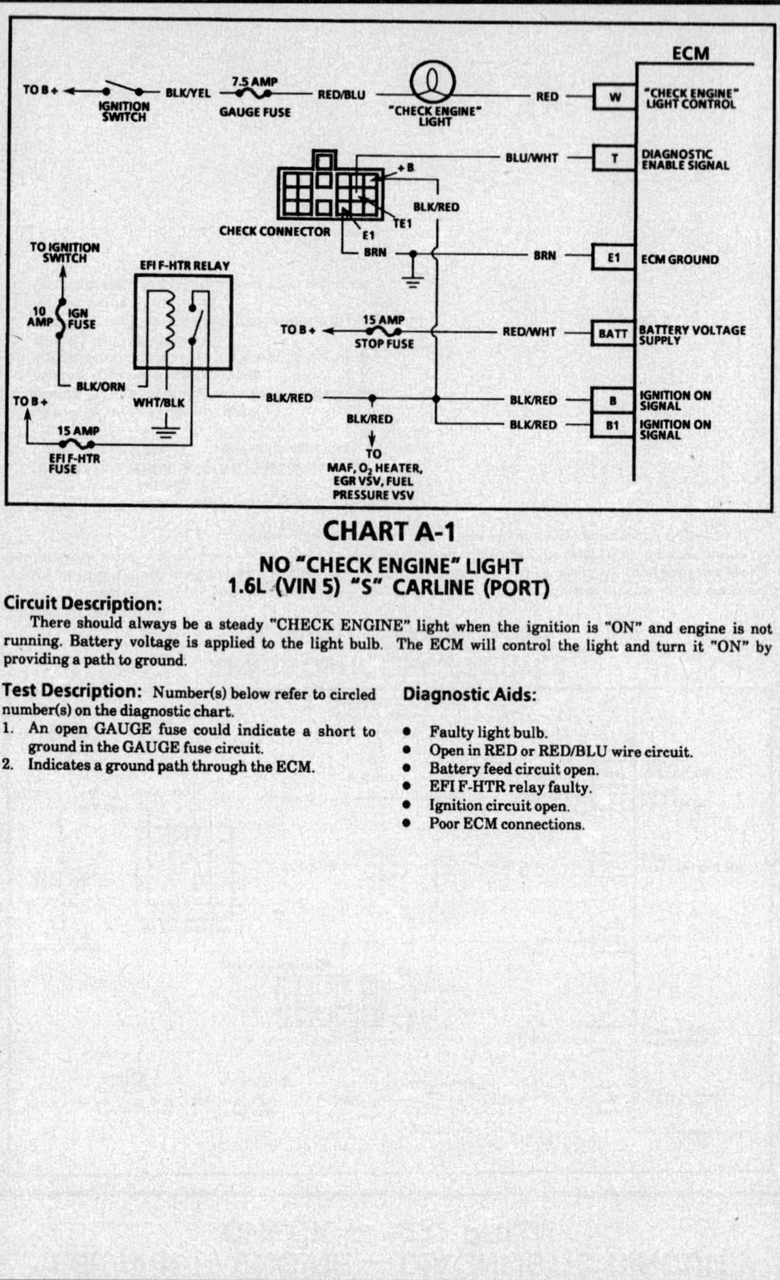

Circuit Description:
There should always be a steady "CHECK ENGINE" light when the ignition is "ON" and engine is not running. Battery voltage is applied to the light bulb. The ECM will control the light and turn it "ON" by providing a path to ground.

Test Description: Number(s) below refer to circled number(s) on the diagnostic chart.
1. An open GAUGE fuse could indicate a short to ground in the GAUGE fuse circuit.
2. Indicates a ground path through the ECM.

Diagnostic Aids:
• Faulty light bulb.
• Open in RED or RED/BLU wire circuit.
• Battery feed circuit open.
• EFI F-HTR relay faulty.
• Ignition circuit open.
• Poor ECM connections.

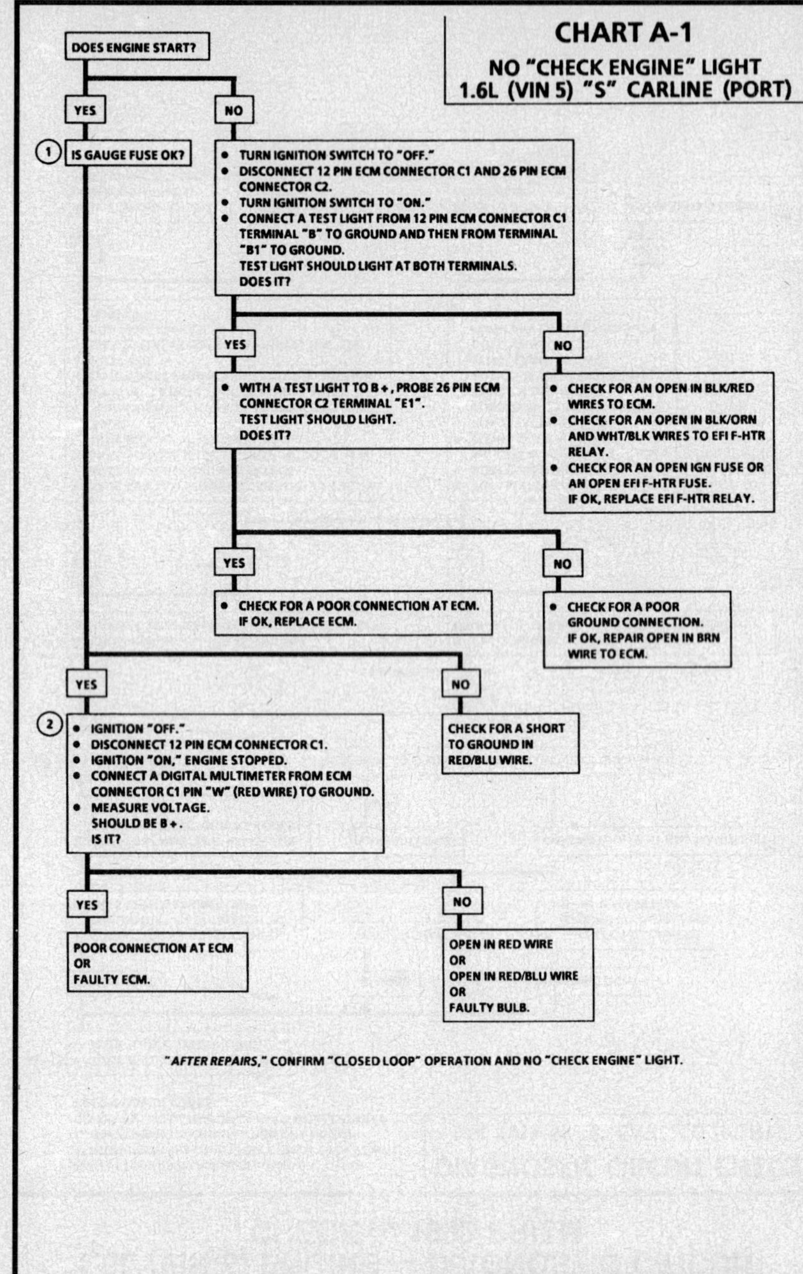

1.6L (VIN 5) ENGINE — A-CHARTS — 1992 PRIZM

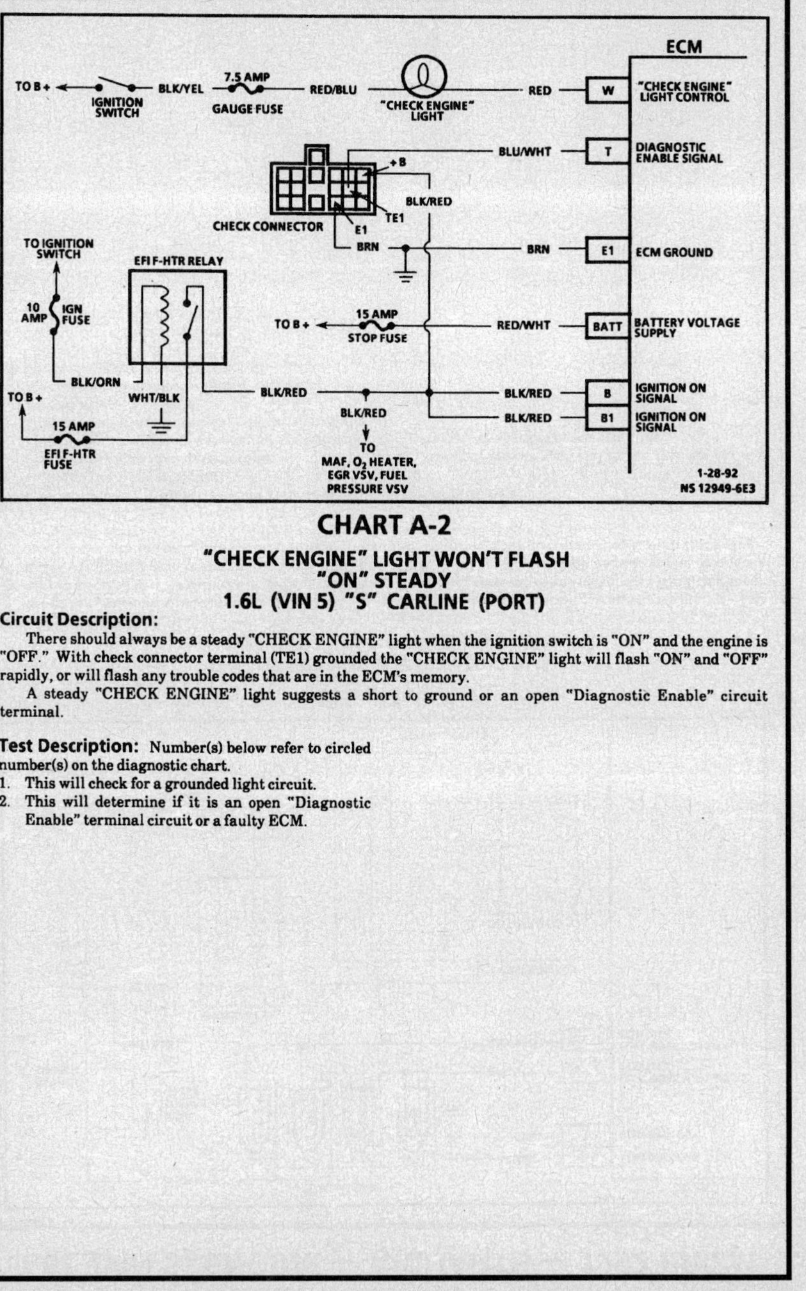

CHART A-2
"CHECK ENGINE" LIGHT WON'T FLASH "ON" STEADY
1.6L (VIN 5) "S" CARLINE (PORT)

Circuit Description:

There should always be a steady "CHECK ENGINE" light when the ignition switch is "ON" and the engine is "OFF." With check connector terminal (TE1) grounded the "CHECK ENGINE" light will flash "ON" and "OFF" rapidly, or will flash any trouble codes that are in the ECM's memory.

A steady "CHECK ENGINE" light suggests a short to ground or an open "Diagnostic Enable" circuit terminal.

Test Description: Number(s) below refer to circled number(s) on the diagnostic chart.

1. This will check for a grounded light circuit.
2. This will determine if it is an open "Diagnostic Enable" terminal circuit or a faulty ECM.

1.6L (VIN 5) ENGINE — A-CHARTS — 1992 PRIZM

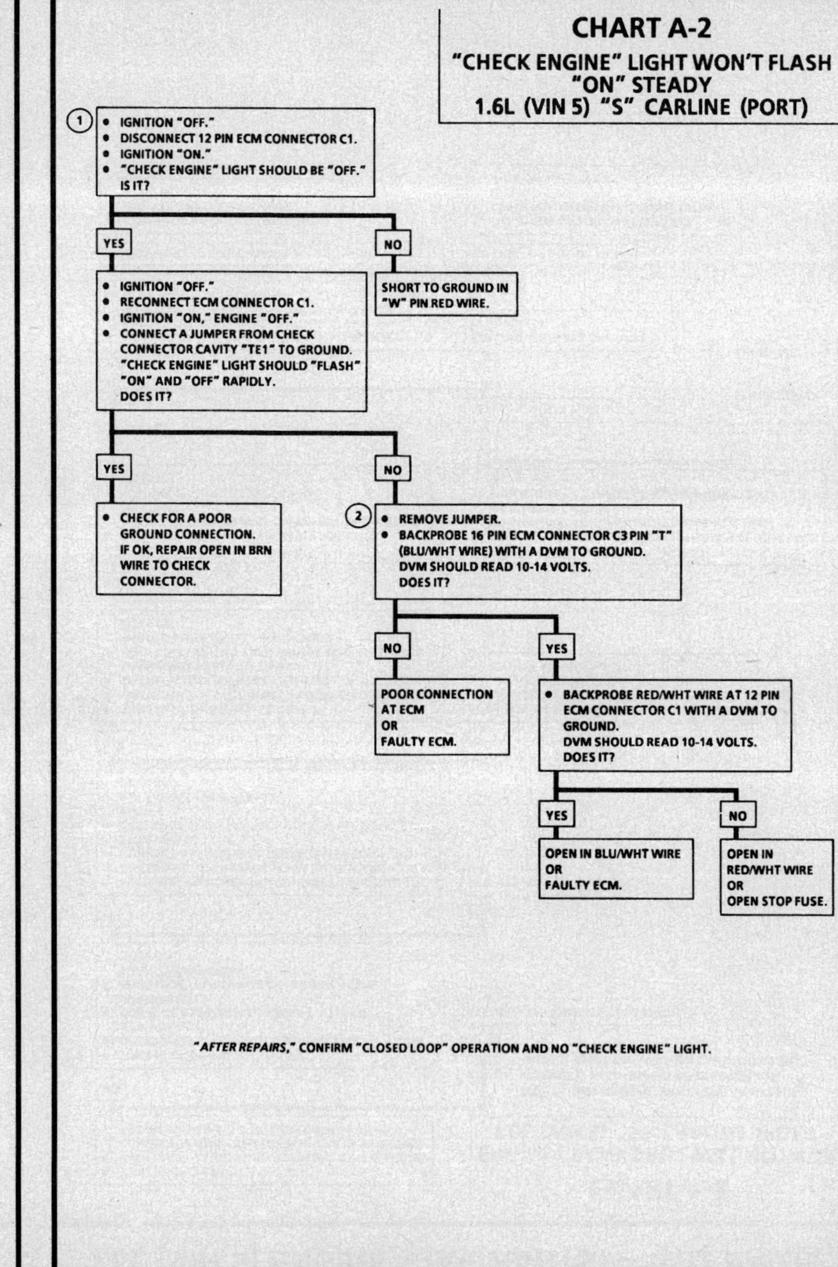

MULTIPORT FUEL INJECTION (MFI) SYSTEMS
LIGHT TRUCKS, VAN, GEO AND SATURN

1.6L (VIN 5) ENGINE — A-CHARTS — 1992 PRIZM

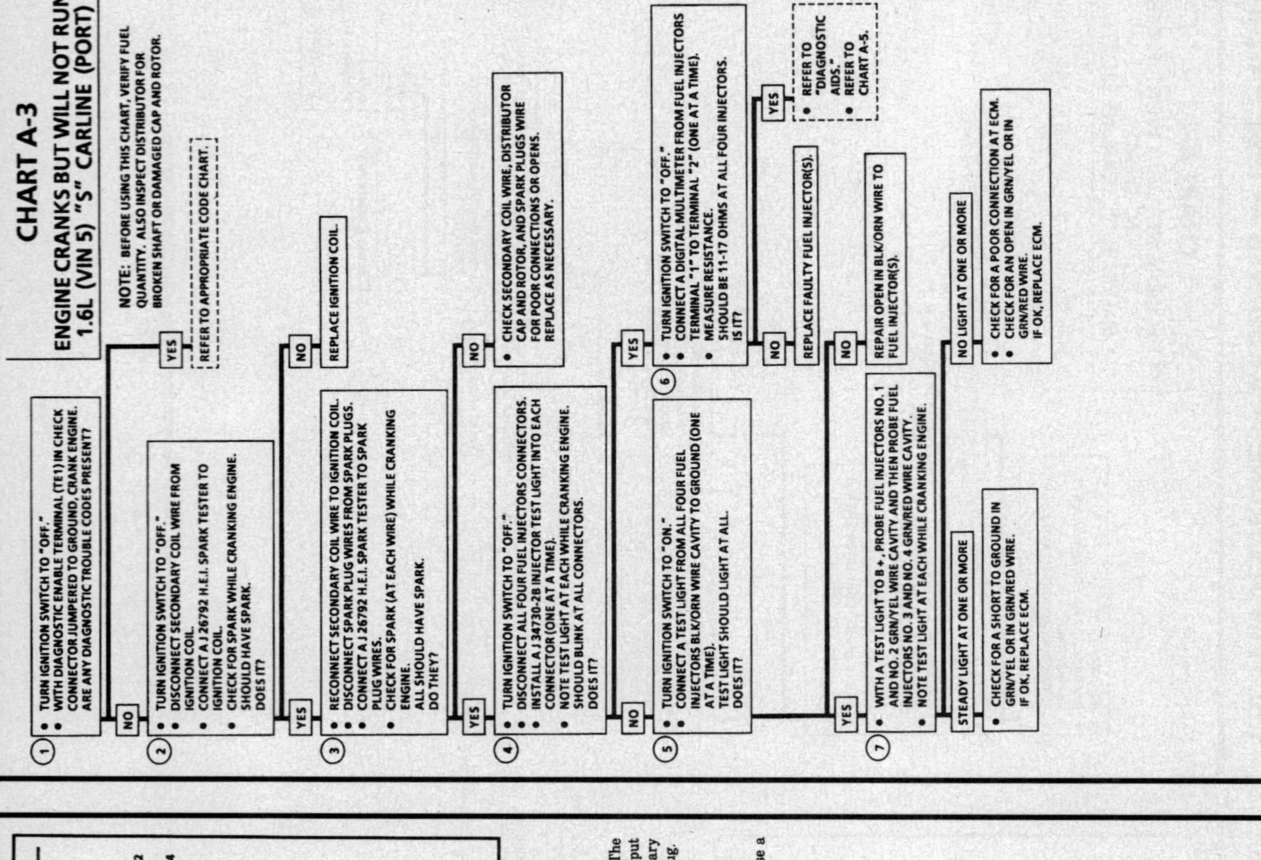

1.6L (VIN 5) ENGINE — A-CHARTS — 1992 PRIZM

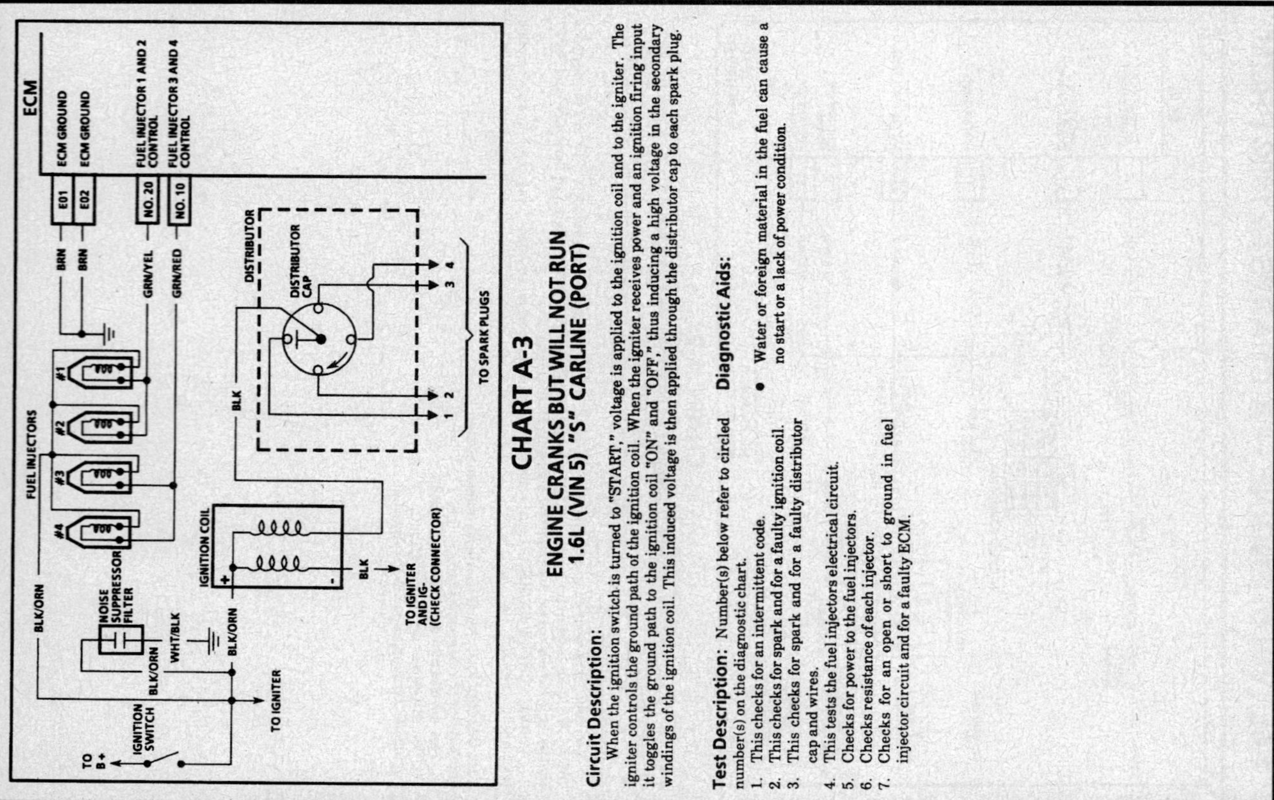

Circuit Description:

When the ignition switch is turned to "START," voltage is applied to the ignition coil and to the igniter. The igniter controls the ground path of the ignition coil. When the igniter receives power and an ignition firing input it toggles the ground path to the ignition coil "ON" and "OFF," thus inducing a high voltage in the secondary windings of the ignition coil. This induced voltage is then applied through the distributor cap to each spark plug.

Test Description: Number(s) below refer to circled number(s) on the diagnostic chart.

1. This checks for an intermittent code.
2. This checks for spark and for a faulty ignition coil.
3. This checks for spark and for a faulty distributor cap and wires.
4. This tests the fuel injectors electrical circuit.
5. Checks for power to the fuel injectors.
6. Checks resistance of each injector.
7. Checks for an open or short to ground in fuel injector circuit and for a faulty ECM.

Diagnostic Aids:

- Water or foreign material in the fuel can cause a no start or a lack of power condition.

1.6L (VIN 5) ENGINE — A-CHARTS — 1992 PRIZM

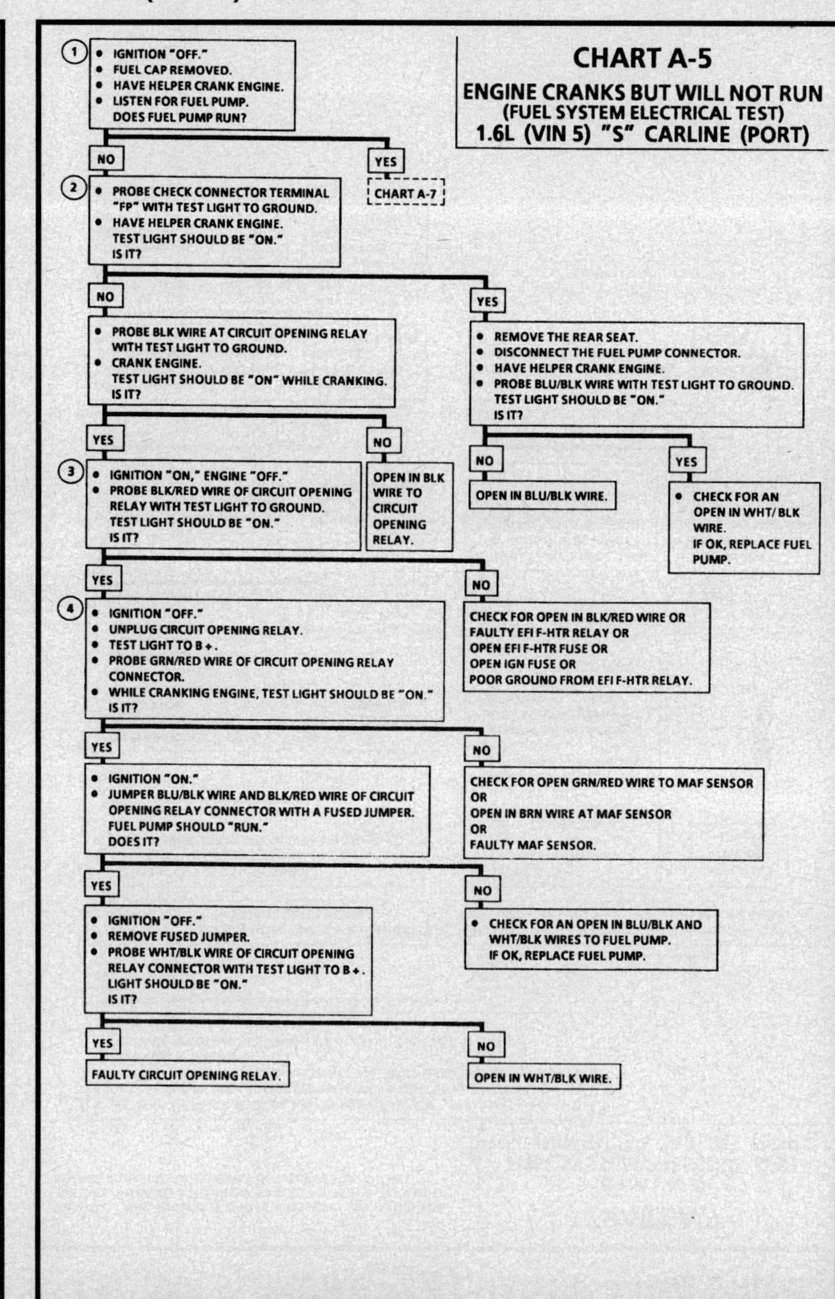

CHART A-5

ENGINE CRANKS BUT WILL NOT RUN
(FUEL SYSTEM ELECTRICAL TEST)
1.6L (VIN 5) "S" CARLINE (PORT)

Circuit Description:

When the engine is cranked the signal from starter relay (M/T) or neutral start switch (A/T) will turn "ON" the circuit opening relay to operate the fuel pump. Once the engine has started, the starter signal is removed from circuit opening relay and the fuel control switch in MAF sensor continues to run fuel pump. If the fuel control switch in the MAF sensor is nonfunctional, the engine will start but will not stay running.

Test Description: Number(s) below refer to circled number(s) on the diagnostic chart.
1. Fuel pump should run when ignition is cranked.
2. Test for B+ from circuit opening relay to fuel pump.
3. Checks for B+ signal to circuit opening relay.
4. Checks for proper MAF fuel pump switch operation.

Diagnostic Aids:

Visual inspection of wiring and connectors should be made if an intermittent problem exists.

1.6L (VIN 5) ENGINE — A-CHARTS — 1992 PRIZM

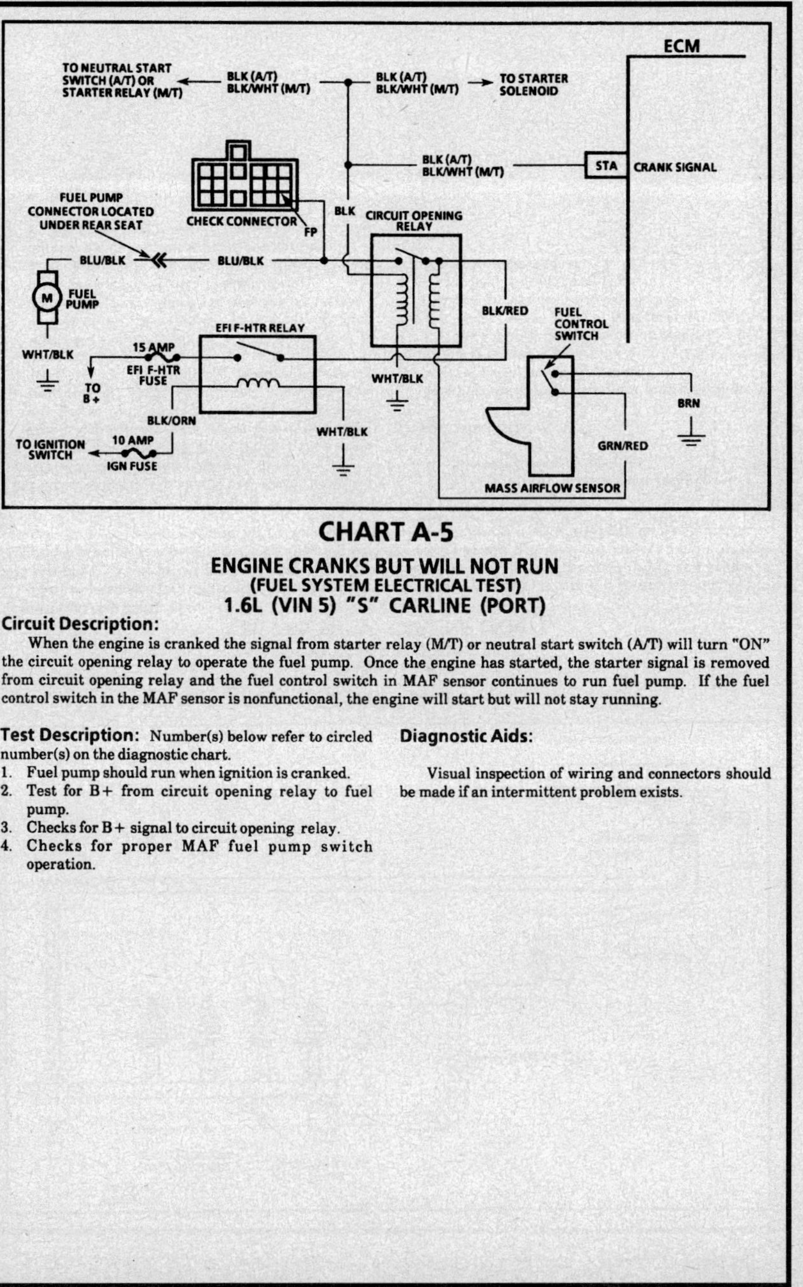

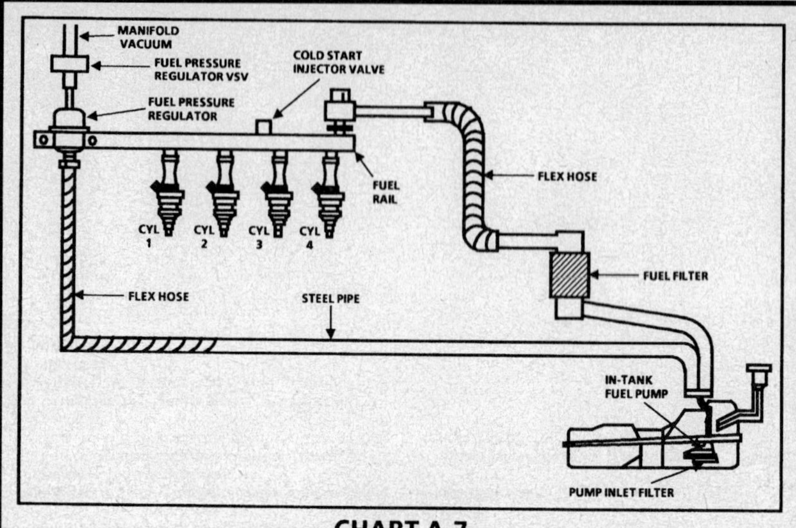

CHART A-7
(Page 1 of 3)
FUEL SYSTEM PRESSURE TEST
1.6L (VIN 5) "S" CARLINE (PORT)

Circuit Description:

The fuel pump delivers fuel to the fuel rail and injectors, where the system pressure is controlled from 265 to 304 kPa (38 to 44 psi) by the pressure regulator. Excess fuel is returned to the fuel tank. When the engine is stopped, the pump can be energized by applying battery voltage to the test terminal located in the engine compartment. With engine turned "OFF," fuel pressure should remain above 147 kPa (21 psi) for five minutes.

Test Description: Number(s) below refer to circled number(s) on the diagnostic chart.

1. Use pressure gage J 34730-1. Wrap a shop towel around the fuel pressure tap to absorb any small amount of fuel leakage that may occur when installing the gage. (The pressure will not leak down after the fuel pump is stopped on a correctly functioning system.)
2. While the engine is idling, manifold pressure is low (vacuum). When applied to the fuel regulator diaphragm, the pressure will result in a lower fuel pressure of about 207-241 kPa (30-35 psi).
3. The application of vacuum to the pressure regulator should result in a fuel pressure drop.
4. Pressure leak-down may be caused by one of the following:
   - In-tank fuel pump check valve not holding.
   - Pump coupling hose leaking.

- Fuel pressure regulator valve leaking.
- Injector sticking open.

Diagnostic Aids:

Improper fuel system pressure may contribute to one or all of the following symptoms:
- Cranks but won't run.
- Cutting out (may feel like ignition problem).
- Hesitation, loss of power or poor fuel economy. Refer to "Symptoms," Section "6E3-B".
- Check for basic engine problem.
- Check for water in fuel.

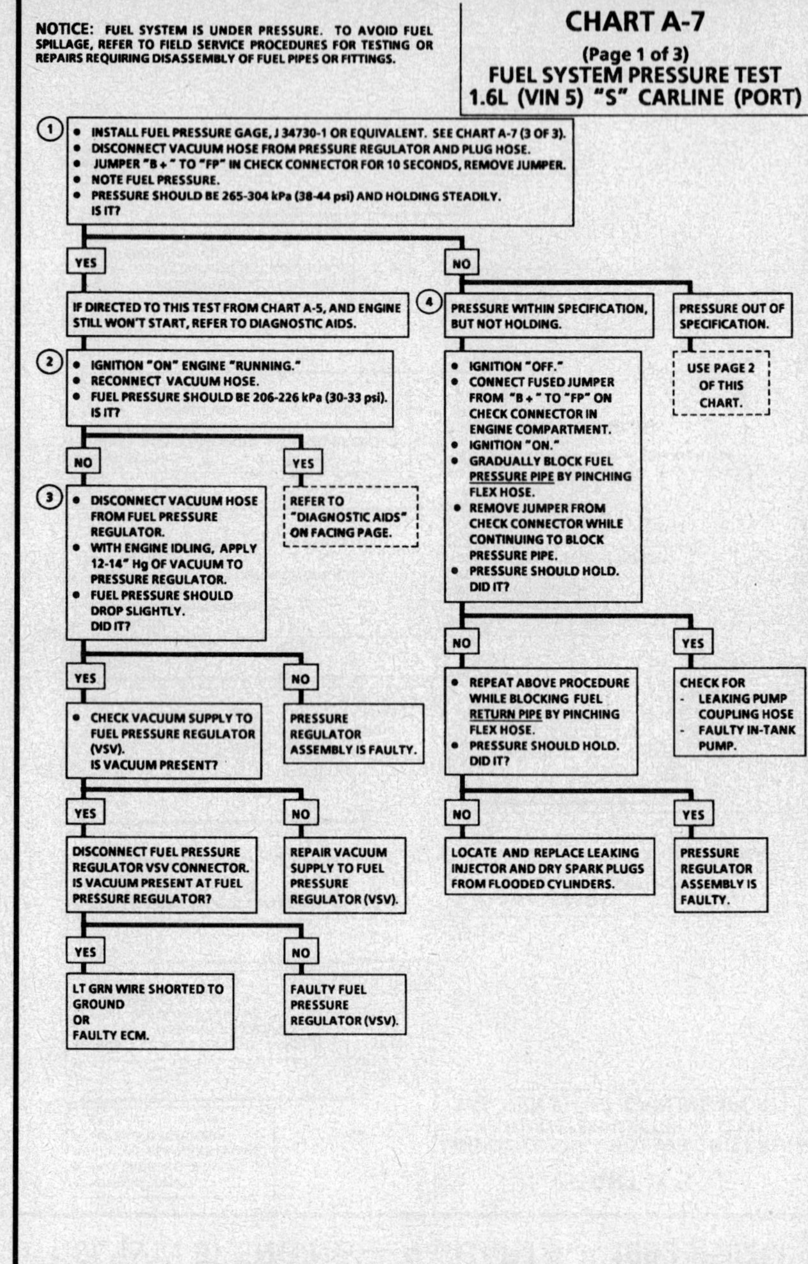

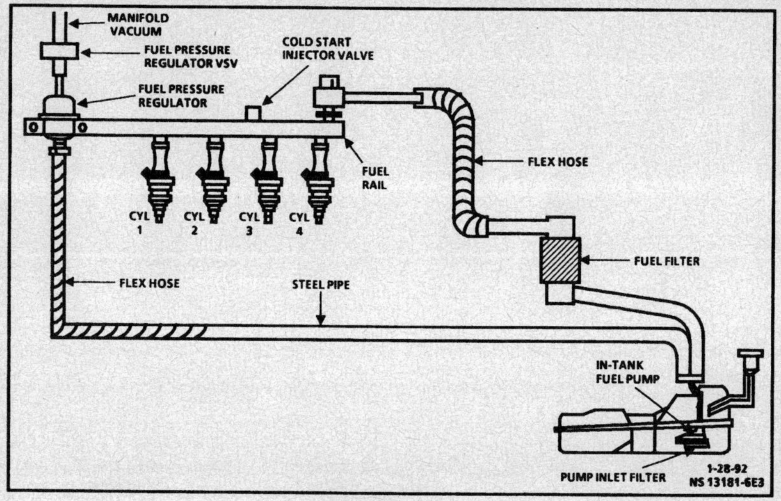

1.6L (VIN 5) ENGINE — A-CHARTS — 1992 PRIZM

CHART A-7
(Page 2 of 3)
FUEL SYSTEM PRESSURE TEST
1.6L (VIN 5) "S" CARLINE (PORT)

Circuit Description:

The fuel pump delivers fuel to the fuel rail and injectors, where the system pressure is controlled from 265 to 304 kPa (38 to 44 psi) by the pressure regulator. Excess fuel is returned to the fuel tank. When the engine is stopped, the pump can be energized by applying battery voltage to the test terminal located in the engine compartment. With engine turned "OFF," fuel pressure should remain above 21 psi (147 kPa) for five minutes.

Test Description: Number(s) below refer to circled number(s) on the diagnostic chart.

5. Pressure less than 265 kPa (38 psi) may be caused by one of two problems:
   - The regulated fuel pressure is too low. The system will be running lean and may set Code 25. Also, hard cold starting and overall poor performance is possible.
   - Restricted flow is causing a pressure drop. Normally, a vehicle with a fuel pressure loss at idle will not be drivable. However, if the pressure drop occurs only while driving, the engine will surge and then stop as pressure begins to drop rapidly.

6. Restricting the fuel return line allows the fuel pump to build above regulated pressure. When battery voltage is applied to "FP" on the check connector, pressure should be above 450 kPa (65 psi).

7. This test determines if the high fuel pressure is due to a restricted fuel return pipe or a pressure regulator problem.

1.6L (VIN 5) ENGINE — A-CHARTS — 1992 PRIZM

NOTICE: FUEL SYSTEM IS UNDER PRESSURE. TO AVOID FUEL SPILLAGE, REFER TO FIELD SERVICE PROCEDURES FOR TESTING OR REPAIRS REQUIRING DISASSEMBLY OF FUEL PIPES OR FITTINGS.

CHART A-7
(Page 2 of 3)
FUEL SYSTEM PRESSURE TEST
1.6L (VIN 5) "S" CARLINE (PORT)

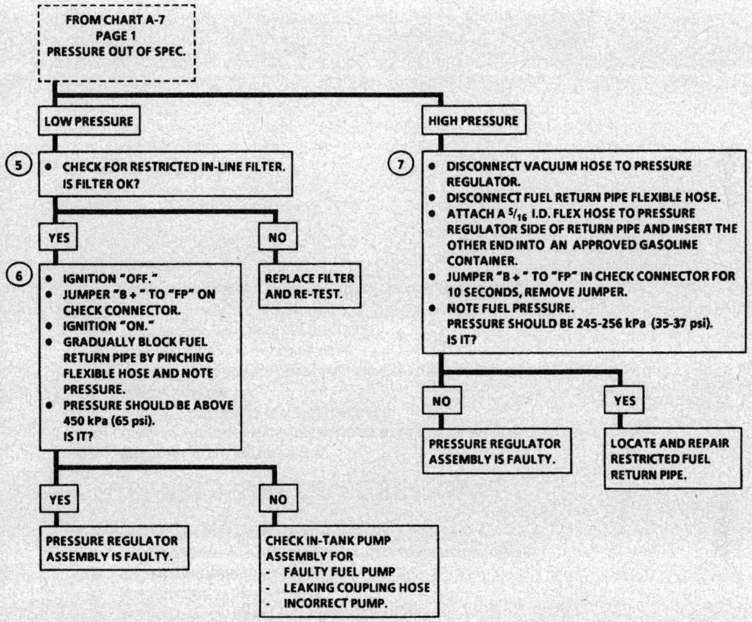

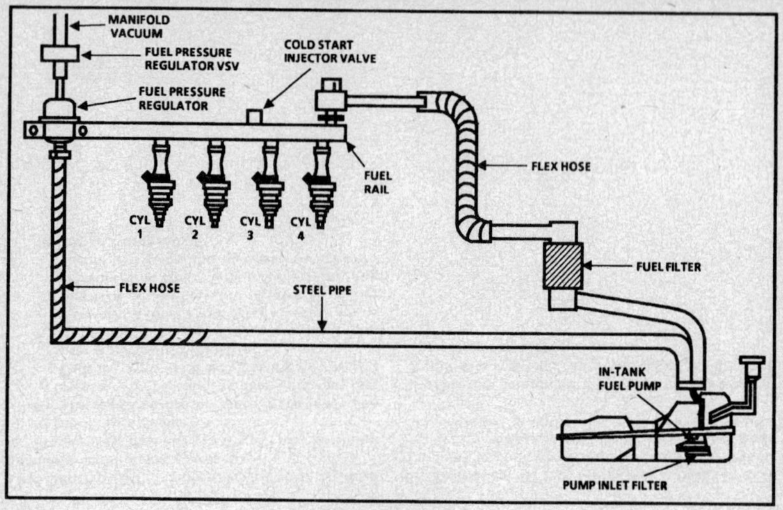

CHART A-7
(Page 3 of 3)
FUEL SYSTEM PRESSURE TEST
1.6L (VIN 5) "S" CARLINE (PORT)

CHART A-7
(Page 3 of 3)
FUEL SYSTEM PRESSURE TEST
1.6L (VIN 5) "S" CARLINE (PORT)

The following procedure outlines the installation and removal of the fuel pressure gage assembly. Make certain to observe all "Cautions" while performing this procedure.

Required Tools:

Fuel Pressure Gage	J 34730-1
Fuel Pressure Gage Adapter	J 38347

CAUTION: TO REDUCE THE RISK OF FIRE AND PERSONAL INJURY, RELIEVE FUEL SYSTEM PRESSURE BEFORE DISCONNECTING FUEL PIPES.

- LOOSEN FUEL FILLER CAP TO RELIEVE TANK PRESSURE.
- REMOVE CIRCUIT OPENING RELAY, LOCATED UNDER CONSOLE BELOW RADIO.
- CRANK ENGINE AND ALLOW TO RUN UNTIL IT STALLS.
- CRANK ENGINE FOR AN ADDITIONAL 3 SECONDS.
- USE SHOP TOWEL TO COLLECT FUEL DURING FUEL PIPE DISCONNECT, PLACE TOWEL IN APPROVED CONTAINER.

FUEL PRESSURE GAGE INSTALLATION

1. IGNITION "OFF." DISCONNECT NEGATIVE BATTERY CABLE.
2. CAREFULLY, REMOVE COLD START INJECTOR VALVE FUEL PIPE HOLD DOWN BOLT AT THE FUEL RAIL.
3. INSTALL ADAPTER J 38347 AND TIGHTEN SECURELY.
4. INSTALL FUEL PRESSURE GAGE J 34730-1 AND TIGHTEN SECURELY.
5. RECONNECT NEGATIVE BATTERY CABLE.
6. REINSTALL CIRCUIT OPENING RELAY.
7. START ENGINE *AND CHECK FOR LEAKS.
8. PURGE AIR FROM FUEL PRESSURE GAGE INTO AN APPROVED CONTAINER.
9. PROCEED TO FUEL SYSTEM PRESSURE TEST CHART A-7 (1 OF 3).

FUEL PRESSURE GAGE REMOVAL

1. REMOVE NEGATIVE BATTERY CABLE.
2. PURGE FUEL PRESSURE FROM FUEL PRESSURE GAGE INTO AN APPROVED CONTAINER.
3. CAREFULLY DISCONNECT FUEL PRESSURE GAGE.
4. REMOVE ADAPTER.
5. REINSTALL FUEL PIPE HOLD DOWN BOLT. TIGHTEN TO 15 N·m (11 LB. FT.).
6. RECONNECT NEGATIVE BATTERY CABLE.
7. START ENGINE *AND CHECK FOR FUEL LEAKS.

* IF ENGINE WILL NOT START, JUMPER B + TO "FP" ON CHECK CONNECTOR.

1.6L (VIN 5) ENGINE — A-CHARTS — 1992 PRIZM

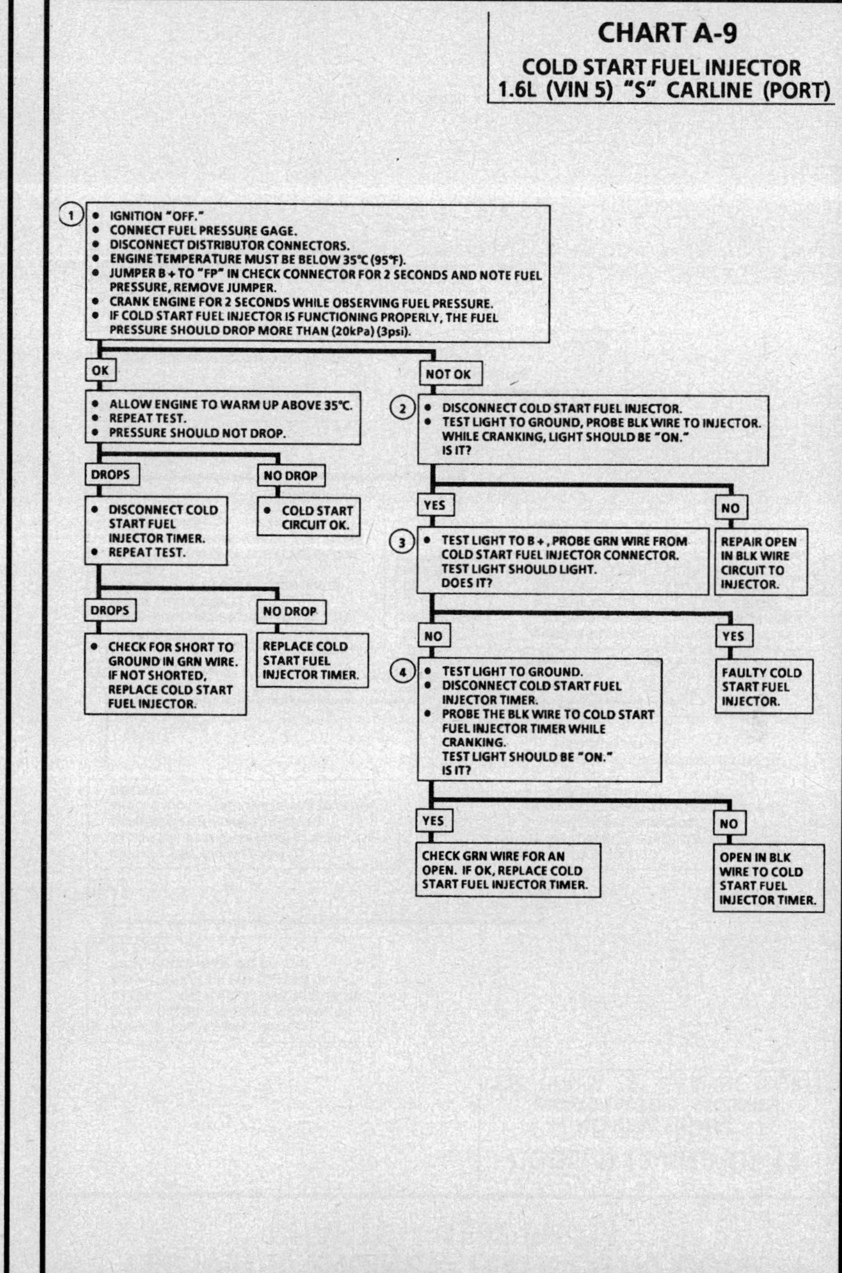

CHART A-9
COLD START FUEL INJECTOR
1.6L (VIN 5) "S" CARLINE (PORT)

Circuit Description:

The cold start fuel injector is used to provide additional fuel during the crank mode to improve cold start-ups. This circuit is important when engine coolant temperature is low because the other injectors are not pulsed "ON" long enough to provide the needed amount of fuel to start.

The circuit is activated only in the crank mode. The power is supplied directly from the starter solenoid circuit and is protected by a fuse. The system is controlled by a cold start fuel injector timer which provides a ground path for the injector during cranking when engine coolant is below 35°C (95°F).

The timer contains a bimetal material which opens at a specified coolant temperature. This bimetal is also heated by the winding in the thermal switch which allows the valve to stay "ON" for 8 seconds at -20°C (-5°F) coolant. The time the switch will stay closed varies inversely with coolant temperature. In other words, as the coolant temperature goes up, the cold start fuel injector "ON" time goes down.

Test Description: Number(s) below refer to circled number(s) on the diagnostic chart.

1. Disconnecting the two distributor connectors will disable the other injectors. The amount of pressure drop depends on the temperature of the engine.
2. This checks for power at the cold start fuel injector.
3. This checks for ground source at the cold start fuel injector.
4. This checks for power at the cold start fuel injector timer.

1.6L (VIN 5) ENGINE — A-CHARTS — 1992 PRIZM

CHART A-9
COLD START FUEL INJECTOR
1.6L (VIN 5) "S" CARLINE (PORT)

(1)
- IGNITION "OFF."
- CONNECT FUEL PRESSURE GAGE.
- DISCONNECT DISTRIBUTOR CONNECTORS.
- ENGINE TEMPERATURE MUST BE BELOW 35°C (95°F).
- JUMPER B+ TO "FP" IN CHECK CONNECTOR FOR 2 SECONDS AND NOTE FUEL PRESSURE, REMOVE JUMPER.
- CRANK ENGINE FOR 2 SECONDS WHILE OBSERVING FUEL PRESSURE.
- IF COLD START FUEL INJECTOR IS FUNCTIONING PROPERLY, THE FUEL PRESSURE SHOULD DROP MORE THAN (20kPa) (3psi).

OK
- ALLOW ENGINE TO WARM UP ABOVE 35°C.
- REPEAT TEST.
- PRESSURE SHOULD NOT DROP.

DROPS
- DISCONNECT COLD START FUEL INJECTOR TIMER.
- REPEAT TEST.

NO DROP
- COLD START CIRCUIT OK.

DROPS
- CHECK FOR SHORT TO GROUND IN GRN WIRE. IF NOT SHORTED, REPLACE COLD START FUEL INJECTOR.

NO DROP
- REPLACE COLD START FUEL INJECTOR TIMER.

NOT OK

(2)
- DISCONNECT COLD START FUEL INJECTOR.
- TEST LIGHT TO GROUND, PROBE BLK WIRE TO INJECTOR. WHILE CRANKING, LIGHT SHOULD BE "ON." IS IT?

YES

(3)
- TEST LIGHT TO B+, PROBE GRN WIRE FROM COLD START FUEL INJECTOR CONNECTOR. TEST LIGHT SHOULD LIGHT. DOES IT?

NO
- REPAIR OPEN IN BLK WIRE CIRCUIT TO INJECTOR.

NO

(4)
- TEST LIGHT TO GROUND.
- DISCONNECT COLD START FUEL INJECTOR TIMER.
- PROBE THE BLK WIRE TO COLD START FUEL INJECTOR TIMER WHILE CRANKING. TEST LIGHT SHOULD BE "ON." IS IT?

YES
- FAULTY COLD START FUEL INJECTOR.

YES
CHECK GRN WIRE FOR AN OPEN. IF OK, REPLACE COLD START FUEL INJECTOR TIMER.

NO
OPEN IN BLK WIRE TO COLD START FUEL INJECTOR TIMER.

1.6L (VIN 5) ENGINE — DIAGNOSTIC CODE CHARTS — 1992 PRIZM

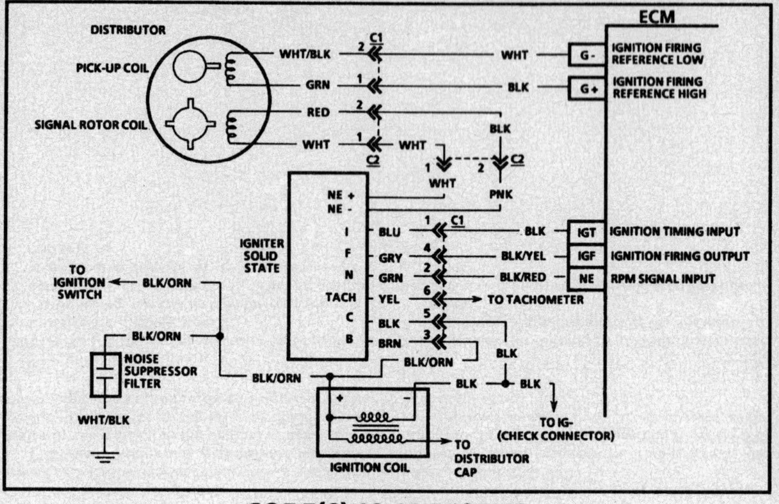

CODE(S) 12 AND/OR 13
NO RPM SIGNAL
(NO SIGNAL FOR 2 SECONDS)
1.6L (VIN 5) "S" CARLINE (PORT)

Circuit Description:

The ECM supplies a voltage signal to the pick-up coil. As the stator inside the distributor passes the pick-up coil the voltage signal is pulled low. This low voltage (signal) is then used to sequence the fuel injectors. If this signal is not present for 2 seconds, Code(s) 12 and/or 13 will set.

Test Description: Number(s) below refer to circled number(s) on the diagnostic chart.
1. Voltage is supplied by the ECM to the pick-up coil.
2. Determines if rpm signal is present at the ECM.
3. Checks for proper resistance of the pick-up coil.
4. Checks for proper resistance of the signal rotor coil.

Diagnostic Aids:

An intermittent may be caused by a poor connection, rubbed through wire insulation or a wire broken inside the insulation. Inspect ECM harness connectors for backed out terminals, "NE," "G+," "G–," "IGT," "IGF," improper mating, broken locks, improperly formed or damaged terminals, poor terminal-to-wire connection and damaged harness.

1.6L (VIN 5) ENGINE — DIAGNOSTIC CODE CHARTS — 1992 PRIZM

CODE(S) 12 AND/OR 13
NO RPM SIGNAL
(NO SIGNAL FOR 2 SECONDS)
1.6L (VIN 5) "S" CARLINE (PORT)

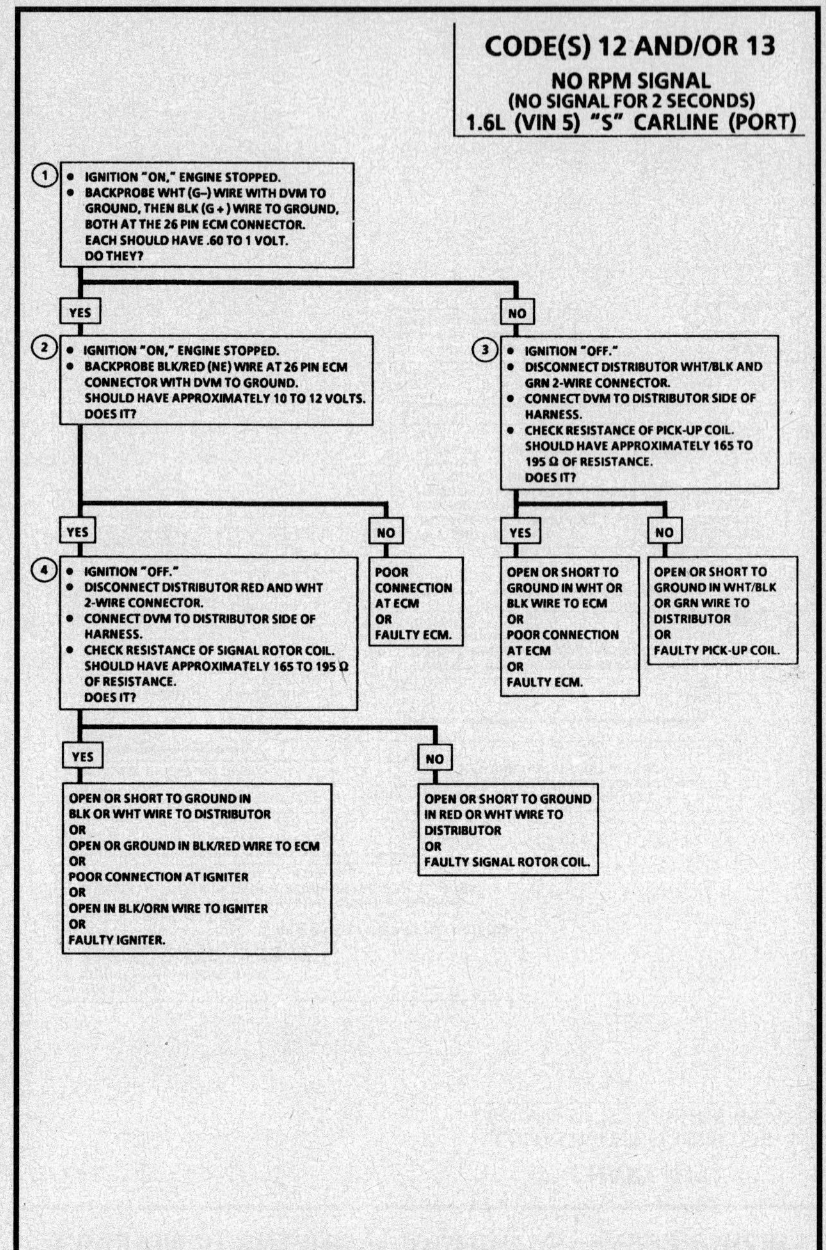

1.6L (VIN 5) ENGINE — DIAGNOSTIC CODE CHARTS — 1992 PRIZM

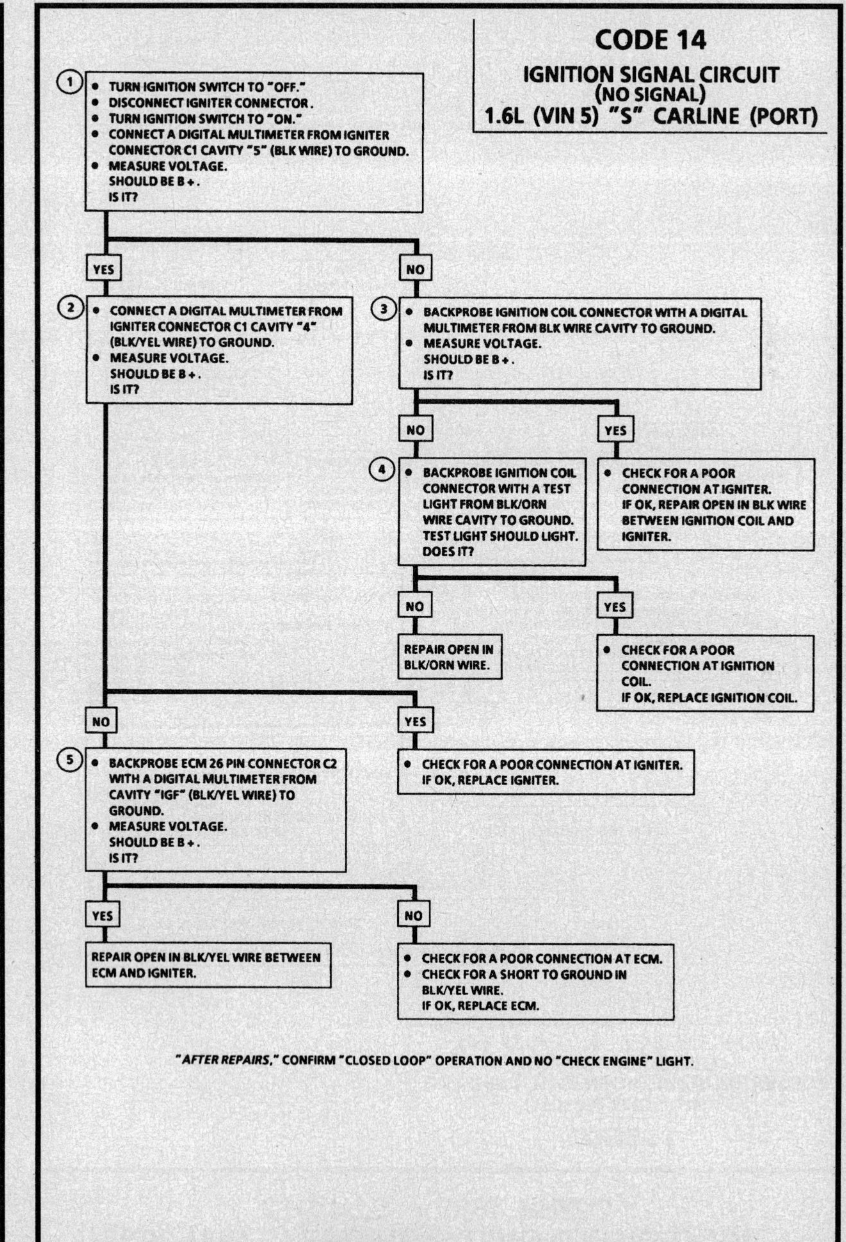

CODE 14
IGNITION SIGNAL CIRCUIT
(NO SIGNAL)
1.6L (VIN 5) "S" CARLINE (PORT)

Circuit Description:

The ECM sends a voltage signal to the igniter on terminal "IGF" of ECM. If this signal is not present four consecutive times during cranking, the Code 14 will set. The igniter uses this signal to pulse the injector.

Test Description: Number(s) below refer to circled number(s) on the diagnostic chart.
1. Checks for voltage from ignition coil.
2. Checks for voltage from ECM.
3. Checks for an open BLK wire between ignition coil and igniter.
4. Checks for power to ignition coil or for a faulty ignition coil.
5. Checks for an open or short to ground in BLK/YEL wire or for a faulty ECM.

Diagnostic Aids:

An intermittent may be caused by a poor connection, rubbed through wire insulation or a wire broken inside the insulation. Inspect ECM and igniter harness connectors for backed out terminals, improper mating, broken locks, improperly formed or damaged terminals, poor terminal-to-wire connection and damaged harness. Inspect tachometer in I/P for proper operation. If inoperative, check for poor connection, or open or ground in BLK (IGT) wire.

1.6L (VIN 5) ENGINE — DIAGNOSTIC CODE CHARTS — 1992 PRIZM

CODE 14
IGNITION SIGNAL CIRCUIT
(NO SIGNAL)
1.6L (VIN 5) "S" CARLINE (PORT)

①
- TURN IGNITION SWITCH TO "OFF."
- DISCONNECT IGNITER CONNECTOR.
- TURN IGNITION SWITCH TO "ON."
- CONNECT A DIGITAL MULTIMETER FROM IGNITER CONNECTOR C1 CAVITY "5" (BLK WIRE) TO GROUND.
- MEASURE VOLTAGE.
  SHOULD BE B+.
  IS IT?

YES →

②
- CONNECT A DIGITAL MULTIMETER FROM IGNITER CONNECTOR C1 CAVITY "4" (BLK/YEL WIRE) TO GROUND.
- MEASURE VOLTAGE.
  SHOULD BE B+.
  IS IT?

NO →

③
- BACKPROBE IGNITION COIL CONNECTOR WITH A DIGITAL MULTIMETER FROM BLK WIRE CAVITY TO GROUND.
- MEASURE VOLTAGE.
  SHOULD BE B+.
  IS IT?

NO →

④
- BACKPROBE IGNITION COIL CONNECTOR WITH A TEST LIGHT FROM BLK/ORN WIRE CAVITY TO GROUND. TEST LIGHT SHOULD LIGHT. DOES IT?

YES (from ③) →
- CHECK FOR A POOR CONNECTION AT IGNITER. IF OK, REPAIR OPEN IN BLK WIRE BETWEEN IGNITION COIL AND IGNITER.

NO (from ④) →
REPAIR OPEN IN BLK/ORN WIRE.

YES (from ④) →
- CHECK FOR A POOR CONNECTION AT IGNITION COIL. IF OK, REPLACE IGNITION COIL.

⑤
- BACKPROBE ECM 26 PIN CONNECTOR C2 WITH A DIGITAL MULTIMETER FROM CAVITY "IGF" (BLK/YEL WIRE) TO GROUND.
- MEASURE VOLTAGE.
  SHOULD BE B+.
  IS IT?

NO (from ②) →
- CHECK FOR A POOR CONNECTION AT IGNITER. IF OK, REPLACE IGNITER.

YES (from ⑤) →
REPAIR OPEN IN BLK/YEL WIRE BETWEEN ECM AND IGNITER.

NO (from ⑤) →
- CHECK FOR A POOR CONNECTION AT ECM.
- CHECK FOR A SHORT TO GROUND IN BLK/YEL WIRE. IF OK, REPLACE ECM.

"AFTER REPAIRS," CONFIRM "CLOSED LOOP" OPERATION AND NO "CHECK ENGINE" LIGHT.

1.6L (VIN 5) ENGINE — DIAGNOSTIC CODE CHARTS — 1992 PRIZM

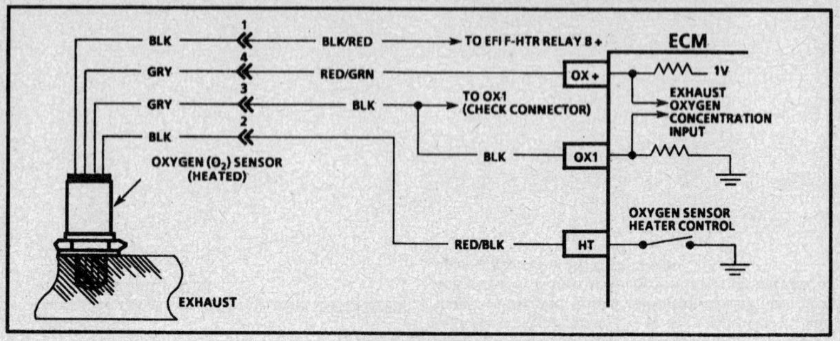

CODE 21
(FEDERAL EMISSIONS)
OXYGEN (O₂) SENSOR CIRCUIT (HEATED)
(OPEN/SHORT CIRCUIT)
1.6L (VIN 5) "S" CARLINE (PORT)

Circuit Description:

The oxygen sensor is located in the exhaust system. It indirectly determines whether the fuel mixture is rich or lean by detecting the concentration of oxygen present in the exhaust gas. The oxygen sensor is provided with a heater. The heater is controlled by the ECM. When the intake air volume is low (the temperature of the exhaust gas is low) current flows to the heater to heat the sensor for accurate detection. The oxygen sensor heater used on a Federal emissions vehicle is a pulse width modulated type. The ECM provides a modulated ground path to the heater in relation to oxygen sensor temperature. When the sensor heater is cold, the energization time of the heater is long and the oxygen sensor warms up. When the sensor heater becomes hot, the energization time is shortened and the oxygen sensor temperature is reduced.

The Federal emission vehicle oxygen sensor is provided with a 1 volt signal on the "OX+" wire from the ECM. The oxygen sensor varies this 1 volt signal and sends this varying voltage back to the ECM on the "OX1" wire.

Test Description: Number(s) below refer to circled number(s) on the diagnostic chart.

1. This checks for an intermittent Code 21. Code 21 will set under the following conditions:
   - Vehicle must be driven above 40 mph.
   - Turn ignition "OFF" then "ON" and run engine above 1900 rpm for 1 minute. If fault is present, Code 21 will be logged.
2. This test determines if the O₂ sensor is varying voltage from .1 to 1.0 volt.
3. This test determines if the heater circuit of the O₂ sensor is working.
4. This test checks for a 1 volt reference voltage from the ECM.

Diagnostic Aids:

Normal O₂ sensor voltage varies between 0 mV and 999 mV (.1 and 1.0 volt) while at normal operating temperature. Code 21 sets in one minute on a restart if the fault is still present.

Verify a clean, tight ground connection at engine block for all ECM ground wires. Open or shorted RED/BLK or RED/GRN wire will result in a Code 21. If Code 21 is intermittent, refer to "Symptoms."

1.6L (VIN 5) ENGINE — DIAGNOSTIC CODE CHARTS — 1992 PRIZM

CODE 21
(FEDERAL EMISSIONS)
OXYGEN (O₂) SENSOR CIRCUIT (HEATED)
(OPEN/SHORT CIRCUIT)
1.6L (VIN 5) "S" CARLINE (PORT)

(1)
- CLEAR CODE AND ROAD TEST VEHICLE ABOVE 40 MPH FOR TWO MINUTES.
- TURN IGNITION "OFF" THEN "ON."
- RUN ENGINE ABOVE 1900 RPM FOR 1 MINUTE. IS CODE 21 PRESENT?

YES →

(2)
- IGNITION "ON," ENGINE RUNNING.
- USING A DVM, PROBE THE OX1 TERMINAL OF THE CHECK CONNECTOR.
- RUN ENGINE RPM TO 1900. DVM SHOULD INDICATE VARYING VOLTAGE FROM .1 V TO 1.0 V. DOES IT?

NO → CODE 21 IS INTERMITTENT. REFER TO "DIAGNOSTIC AIDS".

From (2): **NO** →

(4)
- IGNITION "OFF."
- UNPLUG OXYGEN SENSOR.
- IGNITION "ON," ENGINE "OFF."
- USING DVM, PROBE OX + RED/GRN WIRE HARNESS SIDE OF CONNECTOR. DVM SHOULD INDICATE APPROXIMATELY 1 VOLT. DOES IT?

From (2): **YES** →

(3)
- IGNITION "OFF."
- UNPLUG OXYGEN SENSOR.
- INSTALL DVM BETWEEN THE RED/BLK HT WIRE AND THE BLK/RED WIRE HARNESS SIDE OF CONNECTOR.
- IGNITION "ON," ENGINE RUNNING. DVM SHOULD INDICATE A VARYING VOLTAGE. DOES IT?

From (4): **YES** →
- IGNITION "ON," ENGINE "OFF."
- JUMPER OX + RED/GRN WIRE TO OX1 BLK WIRE AT HARNESS SIDE OF O₂ SENSOR CONNECTOR.
- USING DVM, PROBE THE OX1 BLK WIRE OF THE 16 PIN ECM CONNECTOR. DVM SHOULD INDICATE APPROXIMATELY 1 VOLT. DOES IT?

From (4): **NO** → OPEN/SHORT IN RED/GRN WIRE OR FAULTY ECM.

From (3): **YES** → POOR CONNECTION AT OXYGEN SENSOR OR FAULTY OXYGEN SENSOR.

From (3): **NO** → OPEN IN BLK/RED WIRE OR OPEN/SHORT IN RED/BLK WIRE OR FAULTY ECM.

YES → POOR CONNECTION AT OXYGEN SENSOR OR FAULTY OXYGEN SENSOR.

NO → OPEN/SHORT IN OX1 BLK WIRE.

"AFTER REPAIRS," CONFIRM "CLOSED LOOP" OPERATION AND NO "CHECK ENGINE" LIGHT.

1.6L (VIN 5) ENGINE — DIAGNOSTIC CODE CHARTS — 1992 PRIZM

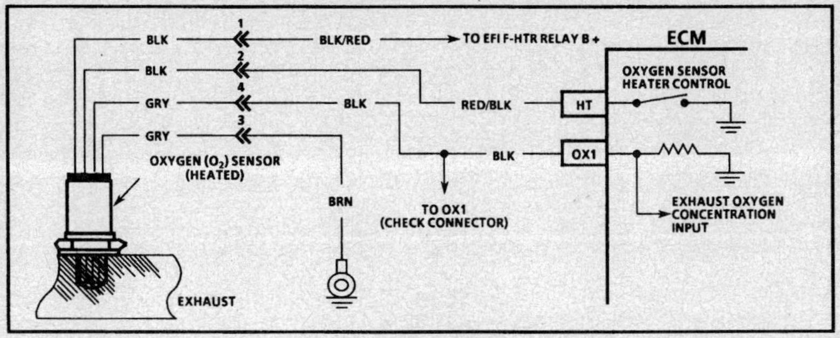

CODE 21

(CALIFORNIA EMISSIONS)
OXYGEN (O₂) SENSOR CIRCUIT (HEATED)
(OPEN/SHORT CIRCUIT)
1.6L (VIN 5) "S" CARLINE (PORT)

Circuit Description:

The oxygen sensor is located in the exhaust system. It indirectly determines whether the fuel mixture is rich or lean by detecting the concentration of oxygen present in the exhaust gases. The main oxygen sensor is provided with a heater. The heater is controlled by the ECM. When the intake air volume is low (the temperature of the exhaust gas is low) current flows to the heater to heat the sensor for accurate detection. The oxygen sensor heater on the California emissions vehicle is an "ON" or "OFF" signal. When the engine rpm is below 4000, the ECM turns "ON" a ground path and turns "ON" the sensor heater. When the rpm is above 4000, the ECM turns "OFF" the ground path to the heater.

On the California emissions vehicles, the OX+ line is not used to this oxygen sensor. A ground wire is used in place of the OX+ 1 volt line. This oxygen sensor produces its own varying voltage from .1 to 1.0 volt when heated and with oxygen concentration in exhaust, and sends this varying voltage back to the ECM on the OX1 BLK wire.

Test Description: Number(s) below refer to circled number(s) on the diagnostic chart.

1. This checks for an intermittent Code 21. Code 21 will set under the following conditions:
   - Vehicle must be driven above 40 mph.
   - Turn ignition "OFF" then "ON" and run engine above 1900 rpm for 1 minute. If fault is present, Code 21 will be logged.
2. This test determines if the oxygen sensor is varying voltage from .1 to 1.0 volt.
3. This test determines if the heater circuit wiring is working properly.

Diagnostic Aids:

Normal oxygen sensor voltage varies between 0 mV and 999 mV (.1 and 1.0 volt) while at normal operating temperature. Code 21 sets in one minute on a restart if a fault is present in either the oxygen signal voltage line or in the heater HT line.

Verify a clean, tight ECM ground connection. Open or shorted RED/BLK or BLK wire will result in a Code 21. If Code 21 is intermittent, refer to "Symptoms."

1.6L (VIN 5) ENGINE — DIAGNOSTIC CODE CHARTS — 1992 PRIZM

CODE 21

(CALIFORNIA EMISSIONS)
OXYGEN (O₂) SENSOR CIRCUIT (HEATED)
(OPEN/SHORT CIRCUIT)
1.6L (VIN 5) "S" CARLINE (PORT)

1.6L (VIN 5) ENGINE — DIAGNOSTIC CODE CHARTS — 1992 PRIZM

CODE 22
COOLANT TEMPERATURE SENSOR (CTS) CIRCUIT
(OPEN OR SHORTED CIRCUIT)
1.6L (VIN 5) "S" CARLINE (PORT)

1
- IGNITION "OFF," DISCONNECT COOLANT TEMPERATURE SENSOR.
- IGNITION "ON," ENGINE STOPPED.
- USING DVM, PROBE COOLANT TEMPERATURE SENSOR CONNECTOR WHT WIRE. SHOULD HAVE 4-5 VOLTS.
DOES IT?

YES →

2
- CONNECT A TEST LIGHT TO B+.
- PROBE COOLANT TEMPERATURE SENSOR CONNECTOR BRN WIRE. TEST LIGHT SHOULD BE "ON."
IS IT?

YES → POOR CONNECTION AT COOLANT TEMPERATURE SENSOR OR FAULTY COOLANT TEMPERATURE SENSOR

NO →

3
- CONNECT A TEST LIGHT TO B+.
- BACKPROBE THE ECM 16 PIN CONNECTOR BRN WIRE. LIGHT SHOULD BE "ON."
IS IT?

YES → OPEN IN BRN WIRE

NO → FAULTY ECM

NO (from 1) →

4
- USING DVM, BACKPROBE THE ECM 16 PIN CONNECTOR AT WHT WIRE. SHOULD HAVE 4-5 VOLTS. DOES IT?

YES → OPEN IN WHT WIRE

NO → SHORT TO GROUND IN WHT WIRE OR POOR CONNECTION AT ECM OR FAULTY ECM.

"AFTER REPAIRS," CONFIRM "CLOSED LOOP" OPERATION AND NO "CHECK ENGINE" LIGHT.

1.6L (VIN 5) ENGINE — DIAGNOSTIC CODE CHARTS — 1992 PRIZM

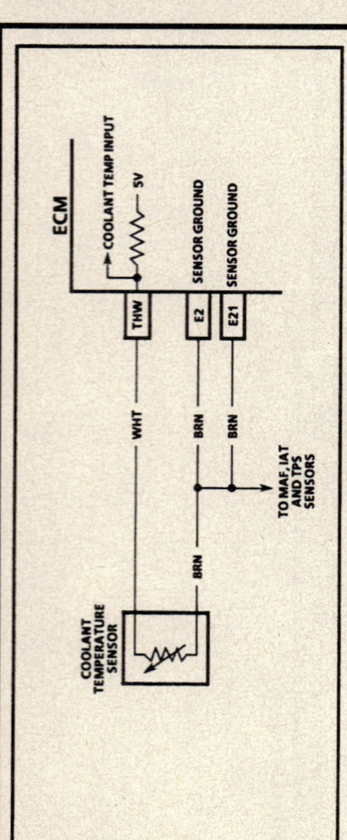

CODE 22
COOLANT TEMPERATURE SENSOR (CTS) CIRCUIT
(OPEN OR SHORTED CIRCUIT)
1.6L (VIN 5) "S" CARLINE (PORT)

Circuit Description:

The coolant temperature sensor (CTS) uses a thermistor to control the signal voltage to the ECM. The ECM applies a voltage on wire to the sensor. When the engine is cold, the sensor (thermistor) resistance is high, therefore, ECM will see high signal voltage.

As the engine warms, the sensor resistance becomes less, and the signal voltage drops. At normal engine operating temperature 80°C to 95°C (176° to 203°F), the signal voltage will measure about 1.5 to 2.0 volts at the ECM.

Test Description: Number(s) below refer to circled number(s) on the diagnostic chart.

1. Verifies that 4-5 volts is present at the CTS connector.
2. This test checks the ground side of the CTS circuit.
3. Determines if the problem is caused by an open BRN wire or the ECM. Be sure to check the ECM connections for proper fit.
4. Determines if the circuit or ECM is the problem.

Diagnostic Aids:

After engine is started the temperature should rise steadily to about 95°C (203°F) then stabilize when thermostat opens.

A faulty connection, or an open in the WHT wire or BRN wire can result in a Code 22.

Refer to "Intermittents" in "Symptoms."

1.6L (VIN 5) ENGINE — DIAGNOSTIC CODE CHARTS — 1992 PRIZM

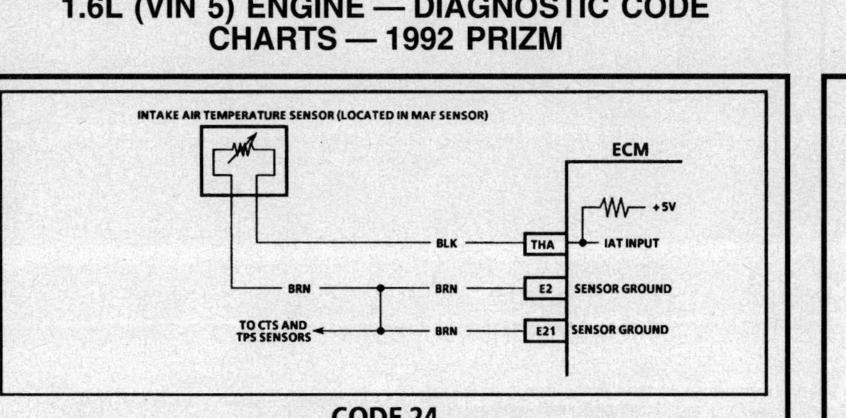

CODE 24
INTAKE AIR TEMPERATURE (IAT) SENSOR CIRCUIT
(OPEN OR SHORTED CIRCUIT)
1.6L (VIN 5) "S" CARLINE (PORT)

Circuit Description:

The intake air temperature (IAT) sensor uses a thermistor to control the signal voltage at the ECM. The ECM applies a voltage (4-6 volts) to the sensor. When the air is cold, the sensor (thermistor) resistance is high, therefore, the ECM will see a high signal voltage. If the air is warm, the sensor resistance is low, therefore, the ECM will see a low signal voltage.

Test Description: Number(s) below refer to circled number(s) on the diagnostic chart.

1. Verifies that 4-5 volts is present at the IAT sensor connector.
2. This test checks the ground side of the IAT sensor circuit.
3. This test determines if the problem is caused by an open circuit or the ECM.
4. This test determines if the BLK wire is open or shorted or if the ECM is defective.

Diagnostic Aids:

An intermittent may be caused by a poor connection, rubbed through wire insulation or a wire broken inside the insulation. Inspect ECM harness connectors for backed out terminals "THA," "E2" or "E21," improper mating, broken locks, improperly formed or damaged terminals, poor terminal-to-wire connection and damaged harness. Also check for damaged MAF housing.

1.6L (VIN 5) ENGINE — DIAGNOSTIC CODE CHARTS — 1992 PRIZM

CODE 24
INTAKE AIR TEMPERATURE (IAT) SENSOR CIRCUIT
(OPEN OR SHORTED CIRCUIT)
1.6L (VIN 5) "S" CARLINE (PORT)

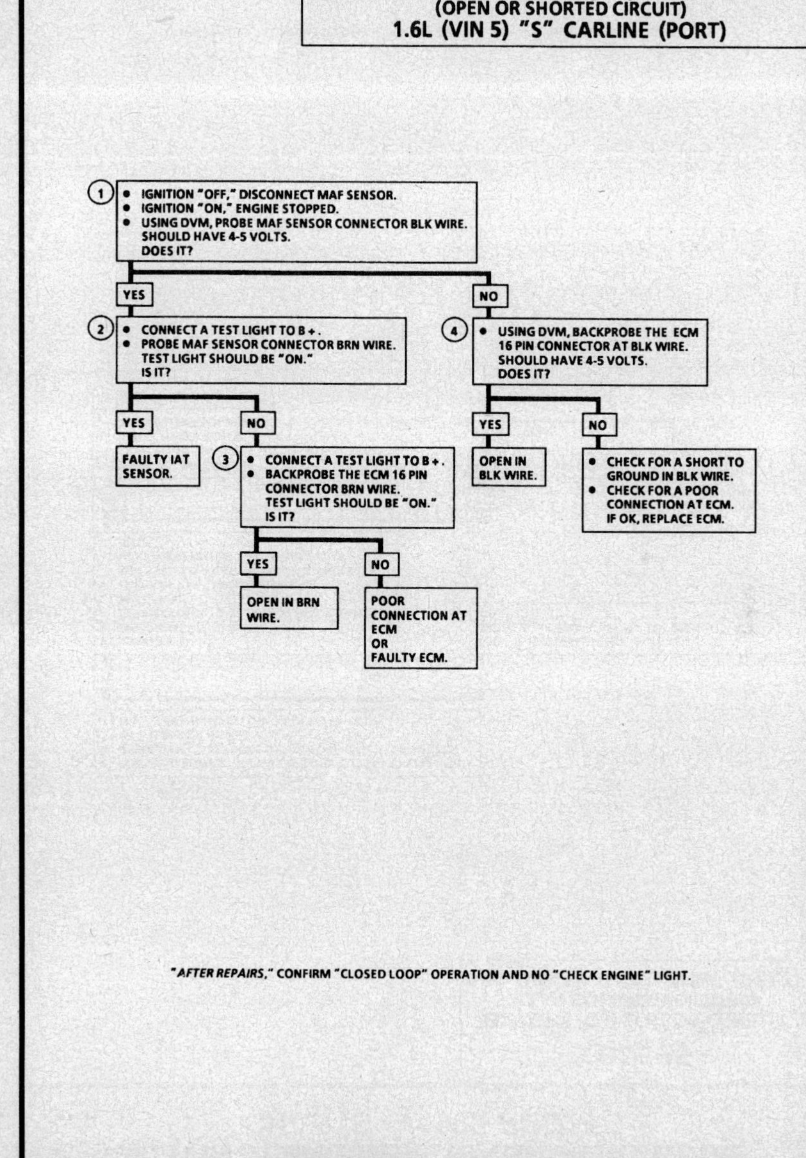

"AFTER REPAIRS," CONFIRM "CLOSED LOOP" OPERATION AND NO "CHECK ENGINE" LIGHT.

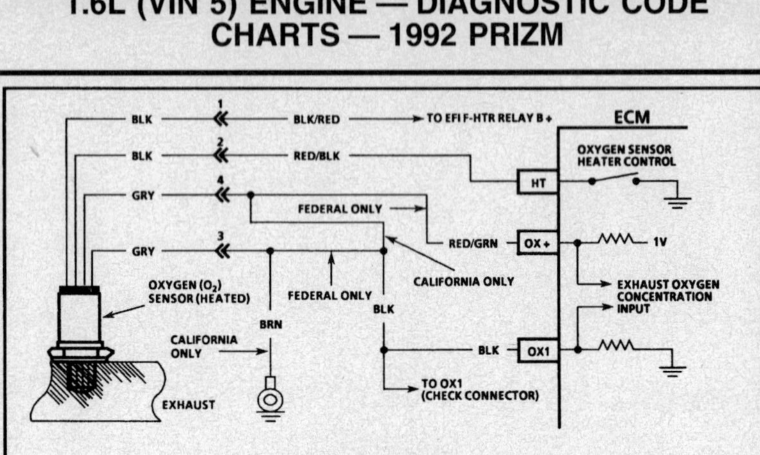

CODE 25

OXYGEN (O_2) SENSOR CIRCUIT
(LEAN EXHAUST INDICATED)
1.6L (VIN 5) "S" CARLINE (PORT)

Circuit Description:

When the O_2 sensor reaches operating temperature, it varies a voltage from about .1 volt (exhaust is lean) to about 1.0 volt (exhaust is rich).

The sensor is like an open circuit and produces no voltage when it is below 360°C (600°F). An open sensor circuit, or cold sensor, causes "Open Loop" operation.

Test Description: Number(s) below refer to circled number(s) on the diagnostic chart.

1. This checks for an intermittent Code 25. Code 25 is set when the O_2 sensor signal voltage on the OX1 wire remains below .2 volt for 60 seconds or more and the system is operating in "Closed Loop."

Diagnostic Aids:

- O_2 Sensor Wire. Sensor pigtail may be mispositioned and contacting the exhaust manifold.
  Check for ground in wires between connector and sensor.

- Fuel Contamination. Water, even in small amounts, near the in-tank fuel pump inlet can be delivered to the injector. The water causes a lean exhaust and can set a Code 25.
- Fuel Pressure. System will be lean if fuel pressure is too low. It may be necessary to monitor fuel pressure while driving the car at various road speeds and/or loads to confirm. See "Fuel System Diagnosis," CHART A-7.
- Exhaust Leaks. If there is an exhaust leak, the engine can cause outside air to be pulled into the exhaust and past the sensor. Vacuum or crankcase leaks can cause a lean condition.
- If Code 25 is intermittent, refer to "Symptoms."

CODE 25

OXYGEN (O_2) SENSOR CIRCUIT
(LEAN EXHAUST INDICATED)
1.6L (VIN 5) "S" CARLINE (PORT)

(1)
- RUN WARM ENGINE (75°C/167°F TO 95°C/203°F) AT 1200 RPM.
- USING DVM, BACKPROBE UNDERHOOD CHECK CONNECTOR OX1 TERMINAL WIRE.
  IS VOLTAGE FIXED BELOW .2 VOLT (200 mV)?

YES
- IGNITION "OFF."
- DISCONNECT O_2 SENSOR.
- CONNECT DVM TO O_2 SENSOR SIGNAL WIRE AT THE SENSOR.
- IGNITION "ON," ENGINE RUNNING.
  NOTE SENSOR VOLTAGE ON DVM, SHOULD BE VARYING BETWEEN 0 VOLT AND 1 VOLT.
  IS IT?

NO
CODE 25 IS INTERMITTENT. IF NO ADDITIONAL CODES WERE STORED, REFER TO "DIAGNOSTIC AIDS".

NO
- REFER TO "DIAGNOSTIC AIDS" ON FACING PAGE.
  IF ALL CHECKS ARE OK, REPLACE OXYGEN SENSOR.

YES
OX1 WIRE SHORTED TO GROUND
OR
FAULTY ECM.

"AFTER REPAIRS," CONFIRM "CLOSED LOOP" OPERATION AND NO "CHECK ENGINE" LIGHT.

1.6L (VIN 5) ENGINE — DIAGNOSTIC CODE CHARTS — 1992 PRIZM

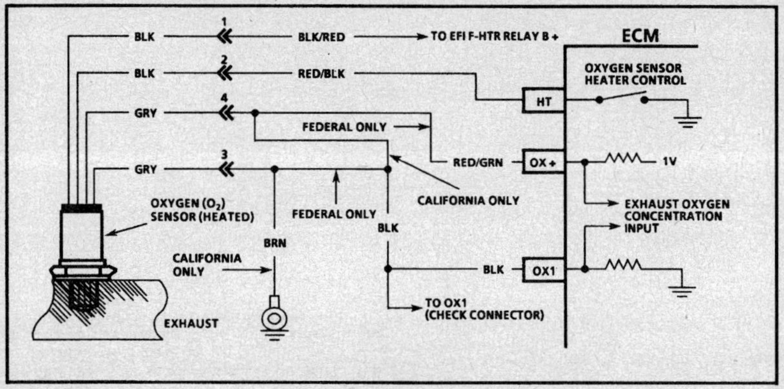

CODE 26
OXYGEN (O₂) SENSOR CIRCUIT
(RICH EXHAUST INDICATED)
1.6L (VIN 5) "S" CARLINE (PORT)

Circuit Description:

When the O_2 sensor reaches operating temperature, it varies a voltage from about .1 volt (exhaust is lean) to about 1.0 volt (exhaust is rich).

The sensor is like an open circuit and produces no voltage when it is below 360°C (600°F).

Test Description: Number(s) below refer to circled number(s) on the diagnostic chart.

1. This checks for an intermittent Code 26. Code 26 is set when the O_2 sensor signal voltage on the OX1 wire remains above .7 volt for 60 seconds or more and the system is operating in "Closed Loop."

Diagnostic Aids:

Code 26, or rich exhaust, is most likely caused by one of the following:

- Fuel Pressure. System will go rich, if pressure is too high. The ECM can compensate for some increase. However, if it gets too high, a Code 26 will be set. See "Fuel System Diagnosis," CHART A-7
- Leaking Injector. See CHART A-7.
- Canister Purge. Check for fuel saturation. If full of fuel, check canister control and hoses.

- MAF Sensor. An output that causes the ECM to sense a higher than normal air flow can cause the system to go rich. Disconnecting the MAF sensor will allow the ECM to set a fixed value for the sensor. Substitute a different MAF sensor if the rich condition is gone while the sensor is disconnected.
- TPS. An intermittent throttle position sensor output will cause the system to operate richly due to a false indication of the engine accelerating.
- O₂ Sensor Contamination. Inspect oxygen sensor for silicone contamination from fuel, or use of improper RTV sealant. The sensor may have a white, powdery coating and result in a high but false signal voltage (rich exhaust indication). The ECM will then reduce the amount of fuel delivered to the engine causing a severe surge driveability problem.
- EGR Valve. EGR sticking open at idle is usually accompanied by a rough idle and/or stall condition. If Code 26 is intermittent, refer to "Symptoms."

1.6L (VIN 5) ENGINE — DIAGNOSTIC CODE CHARTS — 1992 PRIZM

CODE 26
OXYGEN (O₂) SENSOR CIRCUIT
(RICH EXHAUST INDICATED)
1.6L (VIN 5) "S" CARLINE (PORT)

"AFTER REPAIRS," CONFIRM "CLOSED LOOP" OPERATION AND NO "CHECK ENGINE" LIGHT.

1.6L (VIN 5) ENGINE — DIAGNOSTIC CODE CHARTS — 1992 PRIZM

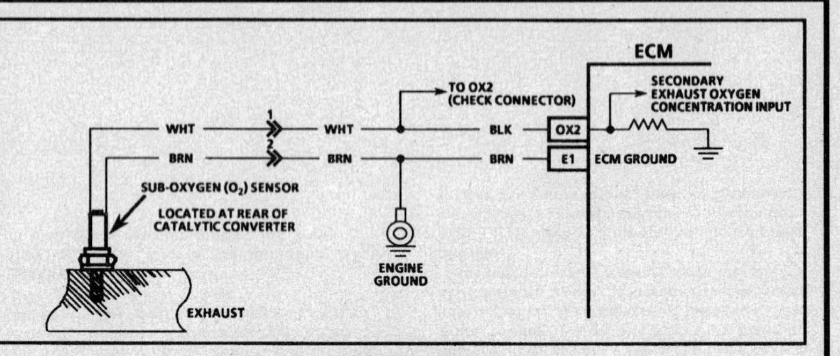

CODE 27
SUB OXYGEN (O$_2$) SENSOR CIRCUIT
(OPEN/SHORTED CIRCUIT)
(CALIFORNIA ONLY)
1.6L (VIN 5) "S" CARLINE (PORT)

Circuit Description:

When the sub O$_2$ sensor reaches operating temperature, it varies a voltage from about .1 volt (exhaust is lean) to about 1.0 volt (exhaust is rich).

The sensor is like an open circuit and produces no voltage when it is below 360°C (600°F). An open sensor circuit, or cold sensor, causes "Open Loop" operation.

Test Description: Number(s) below refer to circled number(s) on the diagnostic chart.

1. This checks for a poor connection or for a faulty ECM. Code 27 will set under the following conditions:
   - Engine at normal operating temperature.
   - At least 2 minutes have elapsed since engine start-up.
   - O$_2$ signal voltage is steady between 1 volt or 0 volt.
   - All above conditions are met for about 60 seconds.
2. This checks for an open circuit in the WHT, BLK or BRN wire to the ECM.
3. This test determines if the sub O$_2$ sensor is the problem or if the ECM and wiring are at fault.

Diagnostic Aids:

Normal sub O$_2$ sensor voltage varies between 0 mV and 999 mV (.1 and 1.0 volt) while at normal operating temperature. Code 27 sets in one minute if sensor signal voltage remains between 0 volt or 1 volt.

Verify a clean, tight ground connection for BRN wire. Open BRN wire, WHT wire or BLK wire will result in a Code 27. If Code 27 is intermittent, refer to "Symptoms."

1.6L (VIN 5) ENGINE — DIAGNOSTIC CODE CHARTS — 1992 PRIZM

CODE 27
SUB OXYGEN (O$_2$) SENSOR CIRCUIT
(OPEN/SHORTED CIRCUIT)
(CALIFORNIA ONLY)
1.6L (VIN 5) "S" CARLINE (PORT)

(1)
- ENGINE AT NORMAL OPERATING TEMPERATURE (ABOVE 80°C/176°F).
- RUN ENGINE ABOVE 1200 RPM FOR TWO MINUTES.
- USING A DVM TO GROUND, BACKPROBE ECM CONNECTOR C3 TERMINAL "OX2" BLK WIRE.
- MEASURE VOLTAGE.
  VOLTAGE SHOULD VARY BETWEEN 0 VOLT AND 1 VOLT.
  DOES IT?

NO → **(2)**

YES → CHECK FOR POOR CONNECTION AT ECM. IF OK, REPLACE ECM.

(2)
- BACKPROBE SUB O$_2$ SENSOR CONNECTOR CAVITY "1" (WHT WIRE) WITH A DVM TO GROUND.
- MEASURE VOLTAGE.
  VOLTAGE SHOULD VARY BETWEEN 0 VOLT AND 1 VOLT.
  DOES IT?

NO → **(3)**

YES → REPAIR OPEN IN WHT, BLK OR BRN WIRE.

(3)
- IGNITION "OFF."
- DISCONNECT SUB O$_2$ SENSOR CONNECTOR.
- IGNITION "ON," ENGINE RUNNING.
- CHECK VOLTAGE OF WHT WIRE (SENSOR SIDE) AT SUB O$_2$ SENSOR CONNECTOR USING A DVM. VOLTAGE SHOULD VARY BETWEEN 0 VOLT AND 1 VOLT.
  DOES IT?

NO → REPLACE SUB OXYGEN SENSOR.

YES → CHECK FOR A SHORT TO GROUND IN WHT OR BLK WIRE. IF OK, REPLACE ECM.

"*AFTER REPAIRS*," CONFIRM "CLOSED LOOP" OPERATION AND NO "CHECK ENGINE" LIGHT.

1.6L (VIN 5) ENGINE — DIAGNOSTIC CODE CHARTS — 1992 PRIZM

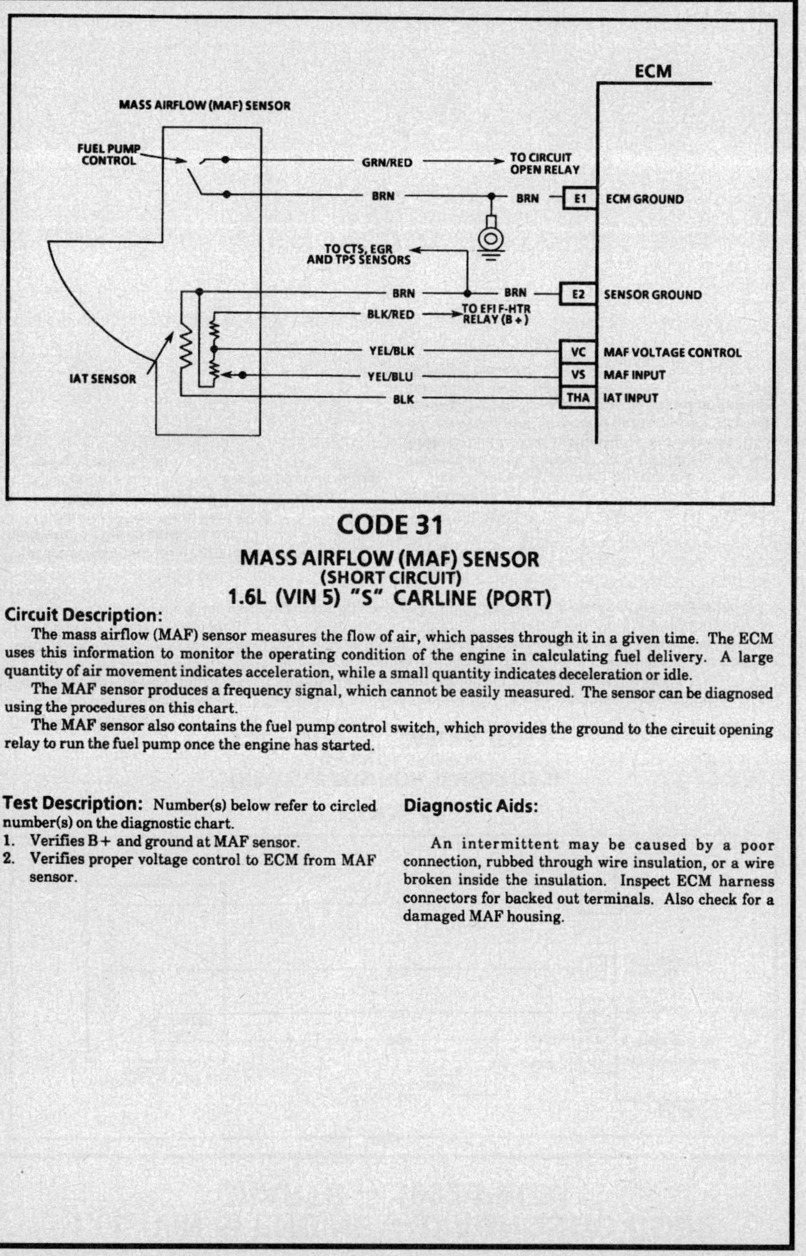

CODE 31
MASS AIRFLOW (MAF) SENSOR
(SHORT CIRCUIT)
1.6L (VIN 5) "S" CARLINE (PORT)

Circuit Description:

The mass airflow (MAF) sensor measures the flow of air, which passes through it in a given time. The ECM uses this information to monitor the operating condition of the engine in calculating fuel delivery. A large quantity of air movement indicates acceleration, while a small quantity indicates deceleration or idle.

The MAF sensor produces a frequency signal, which cannot be easily measured. The sensor can be diagnosed using the procedures on this chart.

The MAF sensor also contains the fuel pump control switch, which provides the ground to the circuit opening relay to run the fuel pump once the engine has started.

Test Description: Number(s) below refer to circled number(s) on the diagnostic chart.
1. Verifies B+ and ground at MAF sensor.
2. Verifies proper voltage control to ECM from MAF sensor.

Diagnostic Aids:

An intermittent may be caused by a poor connection, rubbed through wire insulation, or a wire broken inside the insulation. Inspect ECM harness connectors for backed out terminals. Also check for a damaged MAF housing.

1.6L (VIN 5) ENGINE — DIAGNOSTIC CODE CHARTS — 1992 PRIZM

CODE 31
MASS AIRFLOW (MAF) SENSOR
(SHORT CIRCUIT)
1.6L (VIN 5) "S" CARLINE (PORT)

(1)
- IGNITION "ON," ENGINE "OFF."
- PEEL BACK RUBBER BOOT FROM BACK OF MAF CONNECTOR. MAF SENSOR CONNECTED.
- CONNECT DVM BETWEEN MAF CONNECTOR "E2" BRN WIRE AND "VB" BLK/RED WIRE. DVM SHOULD HAVE 10 VOLTS OR MORE. DOES IT?

YES → (2)
- CONNECT DVM BETWEEN "VC" YEL/BLK WIRE AND "E2" BRN WIRE AT MAF CONNECTOR. DVM SHOULD READ A STEADY VALUE BETWEEN 6-10 VOLTS. DOES IT?

YES
- TURN IGNITION SWITCH TO "OFF."
- DISCONNECT 16 PIN ECM CONNECTOR C3.
- CONNECT A DVM FROM ECM CONNECTOR C3 "VC" YEL/BLK WIRE CAVITY TO GROUND.
- MEASURE VOLTAGE. SHOULD BE B+. IS IT?

NO
- ABOVE 10 VOLTS: YEL/BLK WIRE SHORTED TO VOLTAGE OR FAULTY MAF SENSOR.
- BELOW 6 VOLTS: YEL/BLK WIRE SHORTED TO GROUND OR FAULTY MAF.

NO
- CONNECT TEST LIGHT TO B+.
- BACKPROBE "E2" BRN WIRE AT MAF CONNECTOR. TEST LIGHT SHOULD BE "ON." IS IT?

YES
- OPEN IN BLK/RED WIRE FROM EFI F-HTR RELAY TO MAF SENSOR OR FAULTY EFI F-HTR RELAY.

NO
- OPEN IN "E2" BRN WIRE SENSOR GROUND OR FAULTY ECM.

YES
- CONNECT A DVM FROM ECM CONNECTOR C3 "VS" YEL/BLU WIRE CAVITY TO GROUND.
- MEASURE VOLTAGE. SHOULD BE B+. IS IT?

NO
- REPAIR OPEN IN YEL/BLK WIRE.

YES
- CHECK FOR A SHORT TO VOLTAGE IN YEL/BLU WIRE.
- CHECK FOR A POOR CONNECTION AT ECM. IF OK, REPLACE ECM.

NO
- CHECK FOR AN OPEN OR SHORT TO GROUND IN YEL/BLU WIRE.
- CHECK FOR A POOR CONNECTION AT MAF SENSOR. IF OK, REPLACE MAF SENSOR.

"AFTER REPAIRS," CONFIRM "CLOSED LOOP" OPERATION AND NO "CHECK ENGINE" LIGHT.

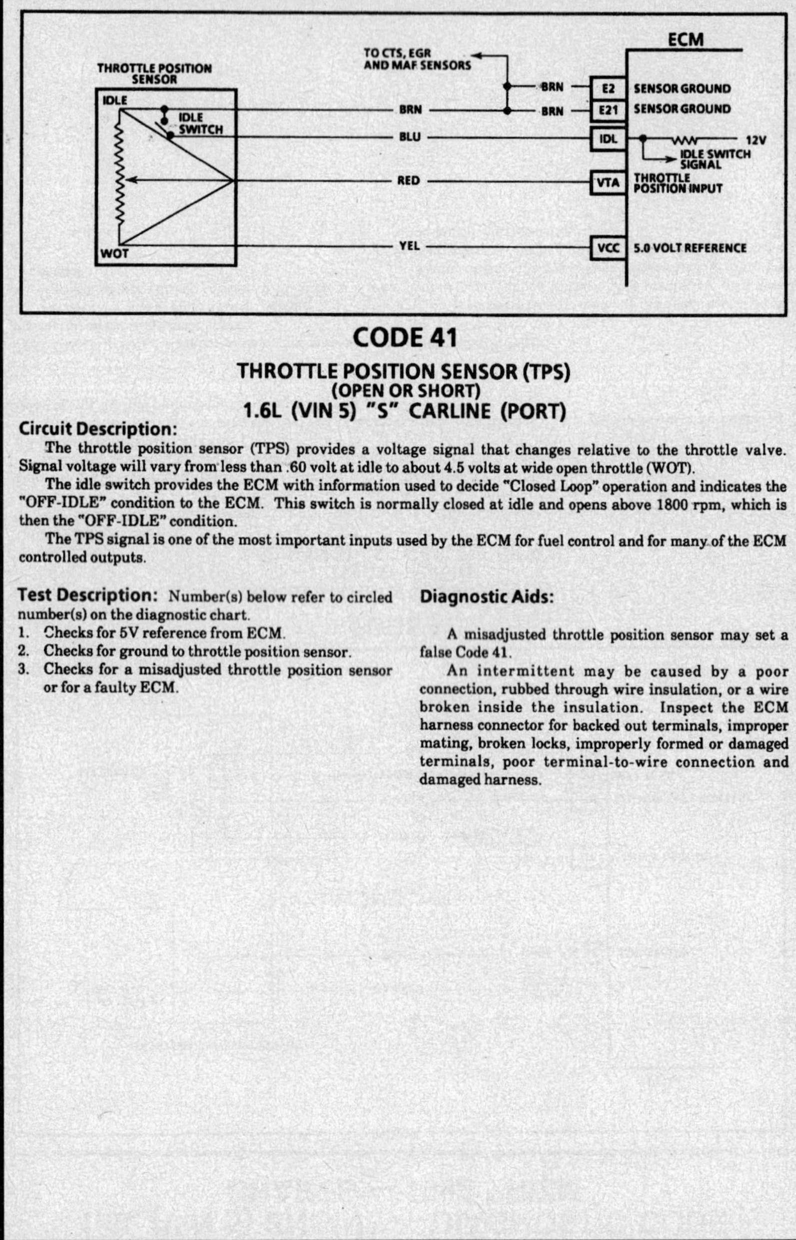

CODE 41
THROTTLE POSITION SENSOR (TPS)
(OPEN OR SHORT)
1.6L (VIN 5) "S" CARLINE (PORT)

Circuit Description:

The throttle position sensor (TPS) provides a voltage signal that changes relative to the throttle valve. Signal voltage will vary from less than .60 volt at idle to about 4.5 volts at wide open throttle (WOT).

The idle switch provides the ECM with information used to decide "Closed Loop" operation and indicates the "OFF-IDLE" condition to the ECM. This switch is normally closed at idle and opens above 1800 rpm, which is then the "OFF-IDLE" condition.

The TPS signal is one of the most important inputs used by the ECM for fuel control and for many of the ECM controlled outputs.

Test Description: Number(s) below refer to circled number(s) on the diagnostic chart.
1. Checks for 5V reference from ECM.
2. Checks for ground to throttle position sensor.
3. Checks for a misadjusted throttle position sensor or for a faulty ECM.

Diagnostic Aids:

A misadjusted throttle position sensor may set a false Code 41.

An intermittent may be caused by a poor connection, rubbed through wire insulation, or a wire broken inside the insulation. Inspect the ECM harness connector for backed out terminals, improper mating, broken locks, improperly formed or damaged terminals, poor terminal-to-wire connection and damaged harness.

CODE 41
THROTTLE POSITION SENSOR (TPS)
(OPEN OR SHORT)
1.6L (VIN 5) "S" CARLINE (PORT)

①
- TURN IGNITION SWITCH TO "OFF."
- DISCONNECT THROTTLE POSITION SENSOR CONNECTOR.
- TURN IGNITION SWITCH TO "ON."
- CONNECT A DIGITAL MULTIMETER FROM THROTTLE POSITION SENSOR CONNECTOR YEL WIRE CAVITY TO GROUND.
- MEASURE VOLTAGE.
  SHOULD BE 4-6 VOLTS.
  IS IT?

YES → **②**
- CONNECT A DIGITAL MULTIMETER FROM THROTTLE POSITION SENSOR CONNECTOR YEL WIRE CAVITY TO BRN WIRE CAVITY.
- MEASURE VOLTAGE.
  SHOULD BE 4-6 VOLTS.
  IS IT?

NO
- CHECK FOR AN OPEN OR SHORT TO GROUND IN YEL WIRE.
- CHECK FOR A POOR CONNECTION AT ECM.
  IF OK, REPLACE ECM.

YES → **③**
- RECONNECT THROTTLE POSITION SENSOR CONNECTOR.
- BACKPROBE ECM 16 PIN CONNECTOR WITH A DIGITAL MULTIMETER FROM "VTA" RED WIRE CAVITY TO GROUND.
- MEASURE VOLTAGE WITH THROTTLE LEVER AT "WOT" POSITION.
  SHOULD BE 3.5-4.5 VOLTS.
  IS IT?

NO
- CHECK FOR AN OPEN IN BRN WIRE.
- CHECK FOR A POOR CONNECTION AT ECM.
  IF OK, REPLACE ECM.

NO
- CHECK FOR AN OPEN OR SHORT TO GROUND IN RED WIRE.
  IF OK, ADJUST/REPLACE THROTTLE POSITION SENSOR.

YES
- CHECK FOR A POOR CONNECTION AT ECM.
  IF OK, REPLACE ECM.

"AFTER REPAIRS," CONFIRM "CLOSED LOOP" OPERATION AND NO "CHECK ENGINE" LIGHT.

1.6L (VIN 5) ENGINE — DIAGNOSTIC CODE CHARTS — 1992 PRIZM

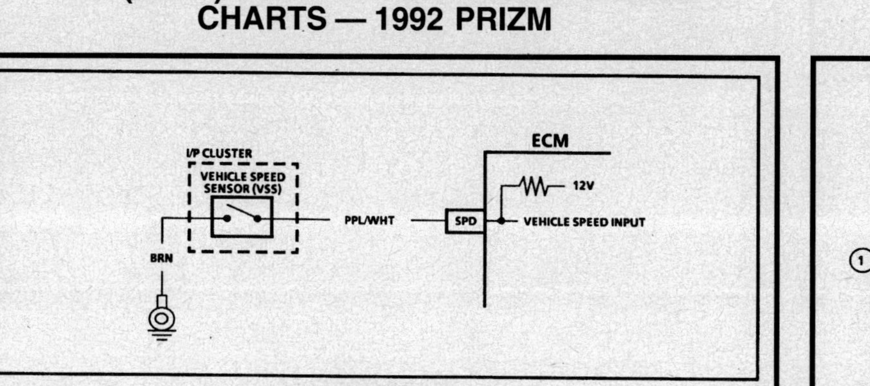

CODE 42
VEHICLE SPEED SENSOR (VSS) CIRCUIT
1.6L (VIN 5) "S" CARLINE (PORT)

Circuit Description:

The ECM provides a 12 volt signal to the vehicle speed sensor (VSS). The VSS will toggle the signal low indicating the vehicle is moving. Code 42 will set if the signal line stays "High" or "Low" for 4 seconds or more.

Test Description: Number(s) below refer to circled number(s) on the diagnostic chart.

1. If speedometer does not work, it could be the cause of Code 42.
2. Checks to see if the vehicle speed sensor on the back of the instrument panel will toggle the B+ supplied by the ECM.

Diagnostic Aids:

An intermittent may be caused by a poor connection, rubbed through wire insulation or a wire broken inside the insulation. Inspect ECM harness connectors for backed out terminals, improper mating, broken locks, improperly formed or damaged terminals, poor terminal-to-wire connection and damaged harness. If no trouble found and Code 42 resets, replace ECM.

1.6L (VIN 5) ENGINE — DIAGNOSTIC CODE CHARTS — 1992 PRIZM

CODE 42
VEHICLE SPEED SENSOR (VSS) CIRCUIT
1.6L (VIN 5) "S" CARLINE (PORT)

(1)
- RAISE FRONT OF VEHICLE AND SUPPORT DRIVE AXLES.
- IGNITION "ON," ENGINE RUNNING.
- VEHICLE IN DRIVE OR GEAR.
- NOTE SPEEDOMETER.
- DOES IT INDICATE 5 MPH OR MORE?

NO → CHECK FOR MECHANICAL PROBLEM.

YES → (2)
- IGNITION "ON," ENGINE STOPPED.
- BACKPROBE PPL/WHT WIRE ECM TERMINAL "SPD" WITH A DIGITAL MULTIMETER.
- MEASURE VOLTAGE
- NOTE: TURN DRIVE WHEELS BY HAND WHILE CHECKING.

0 V → DISCONNECT 12 WAY ECM CONNECTOR. WITH A TEST LIGHT TO B+, PROBE PPL/WHT WIRE TO ECM "SPD" TERMINAL. NOTE: TURN DRIVE WHEELS BY HAND WHILE CHECKING. TEST LIGHT SHOULD BLINK. DOES IT?

- **NO** → CHECK FOR A SHORT TO GROUND IN PPL/WHT WIRE. IF OK, REPLACE VSS.
- **YES** → CHECK FOR POOR CONNECTION AT ECM. IF OK, REPLACE ECM.

12 V → CHECK FOR AN OPEN IN BRN OR PPL/WHT WIRES. IF OK, REPLACE VSS.

VARIES BETWEEN 0 AND 12 V → PROBLEM IS INTERMITTENT. REFER TO "DIAGNOSTIC AIDS."

"AFTER REPAIRS," CONFIRM "CLOSED LOOP" OPERATION AND NO "CHECK ENGINE" LIGHT.

1.6L (VIN 5) ENGINE — DIAGNOSTIC CODE CHARTS — 1992 PRIZM

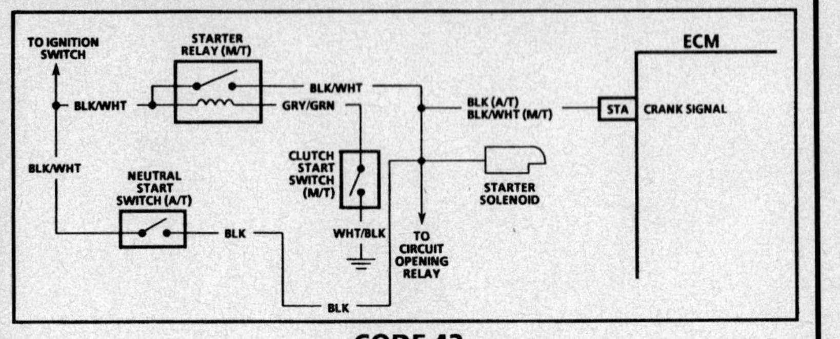

CODE 43
**NO CRANK SIGNAL
(NO SIGNAL FROM STARTER)
1.6L (VIN 5) "S" CARLINE (PORT)**

Circuit Description:

The ECM receives a B+ signal when the starter is in the cranking position. If the signal is not present, and engine speed is higher than 800 rpm, Code 43 will then set.

Test Description: Number(s) below refer to circled number(s) on the diagnostic chart.

1. Battery voltage should be present on the BLK/WHT wire while the engine is cranking.

Diagnostic Aids:

An intermittent may be caused by a poor connection, rubbed through wire insulation or a wire broken inside the insulation. Inspect the ECM harness connectors for backed out terminals, improper mating, broken locks, improperly formed or damaged terminals, poor terminal-to-wire connection and damaged harness.

1.6L (VIN 5) ENGINE — DIAGNOSTIC CODE CHARTS — 1992 PRIZM

CODE 43
**NO CRANK SIGNAL
(NO SIGNAL FROM STARTER)
1.6L (VIN 5) "S" CARLINE (PORT)**

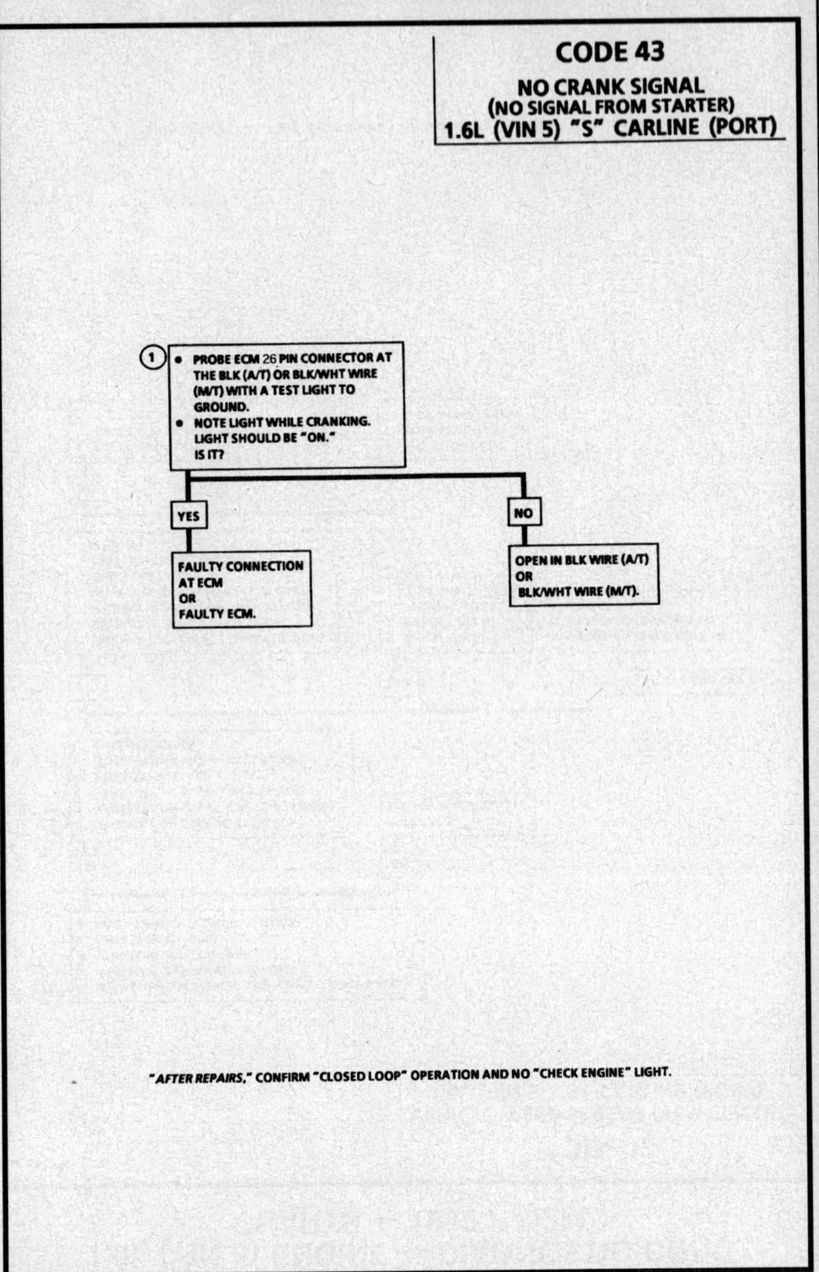

"AFTER REPAIRS," CONFIRM "CLOSED LOOP" OPERATION AND NO "CHECK ENGINE" LIGHT.

1.6L (VIN 5) ENGINE — DIAGNOSTIC CODE CHARTS — 1992 PRIZM

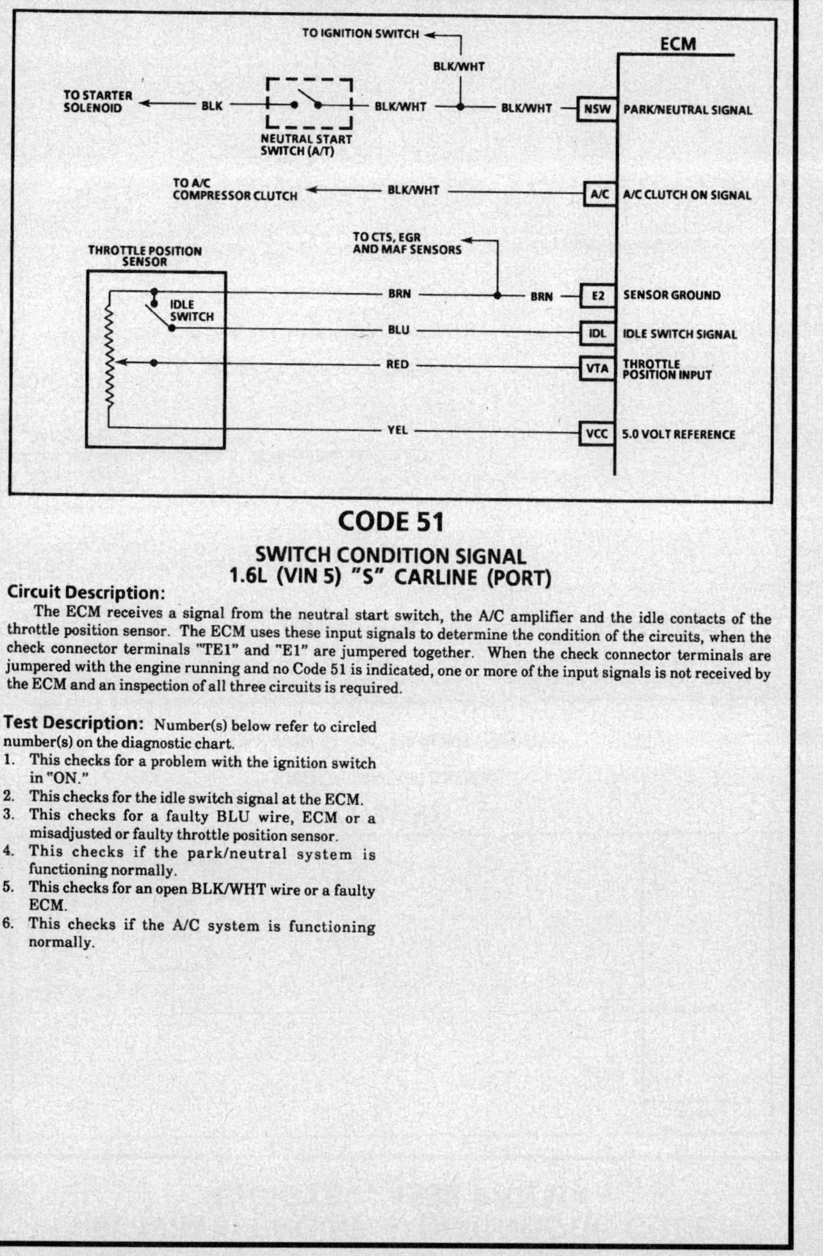

CODE 51
SWITCH CONDITION SIGNAL
1.6L (VIN 5) "S" CARLINE (PORT)

Circuit Description:

The ECM receives a signal from the neutral start switch, the A/C amplifier and the idle contacts of the throttle position sensor. The ECM uses these input signals to determine the condition of the circuits, when the check connector terminals "TE1" and "E1" are jumpered together. When the check connector terminals are jumpered with the engine running and no Code 51 is indicated, one or more of the input signals is not received by the ECM and an inspection of all three circuits is required.

Test Description: Number(s) below refer to circled number(s) on the diagnostic chart.

1. This checks for a problem with the ignition switch in "ON."
2. This checks for the idle switch signal at the ECM.
3. This checks for a faulty BLU wire, ECM or a misadjusted or faulty throttle position sensor.
4. This checks if the park/neutral system is functioning normally.
5. This checks for an open BLK/WHT wire or a faulty ECM.
6. This checks if the A/C system is functioning normally.

1.6L (VIN 5) ENGINE — DIAGNOSTIC CODE CHARTS — 1992 PRIZM

CODE 51
SWITCH CONDITION SIGNAL
1.6L (VIN 5) "S" CARLINE (PORT)

(1)
- A. CLEAR ANY CODES FROM ECM.
- B. CONNECT A JUMPER BETWEEN CHECK CONNECTOR TERMINALS TE1 AND E1"
- C. TURN IGNITION SWITCH TO "ON."
- D. NOTE "CHECK ENGINE" LIGHT. SHOULD NOT FLASH CODE 51. DOES IT?

NO → (2)
- START ENGINE AND RUN AT IDLE.
- MOMENTARILY DEPRESS ACCELERATOR PEDAL.
- NOTE "CHECK ENGINE" LIGHT. SHOULD FLASH CODE 51 ONCE AND THEN FLASH "ON" AND "OFF" RAPIDLY. DOES IT?

YES →
- CHECK FOR AN OPEN IN BLU WIRE FROM THROTTLE POSITION SENSOR.
- CHECK FOR A SHORT TO VOLTAGE IN BLK/WHT WIRE FROM A/C AMPLIFIER (IF EQUIPPED).
- CHECK FOR AN OPEN IN BLK/WHT WIRE FROM NEUTRAL START SWITCH (AUTOMATIC TRANSAXLE ONLY).
- CHECK FOR A POOR CONNECTION AT ECM. IF OK, REPLACE ECM.

(2) **YES →**
- IF VEHICLE IS EQUIPPED WITH AN AUTOMATIC TRANSAXLE PROCEED TO AUTOMATIC TRANSAXLE TEST.
- IF VEHICLE IS NOT EQUIPPED WITH AN AUTOMATIC TRANSAXLE, BUT IS EQUIPPED WITH AIR CONDITIONING PROCEED TO A/C TEST.
- IF VEHICLE IS NOT EQUIPPED WITH EITHER AN AUTOMATIC TRANSAXLE OR AIR CONDITIONING, SWITCH CONDITION SIGNAL SYSTEM IS FUNCTIONING NORMALLY.

NO → (3)
- TURN IGNITION SWITCH TO "OFF."
- DISCONNECT THROTTLE POSITION SENSOR CONNECTOR.
- CONNECT A DIGITAL MULTIMETER FROM THROTTLE POSITION SENSOR CONNECTOR BLU WIRE CAVITY TO GROUND.
- TURN IGNITION SWITCH TO "ON."
- MEASURE VOLTAGE. SHOULD BE B+. IS IT?

AUTOMATIC TRANSAXLE TEST

(4)
- TURN IGNITION SWITCH TO "OFF."
- ENGAGE PARKING BRAKE.
- TURN IGNITION SWITCH TO "ON."
- DEPRESS BRAKE PEDAL AND MOVE MANUAL SELECTOR FROM "P" TO "N" AND THEN BACK TO "P".
- NOTE "CHECK ENGINE" LIGHT. SHOULD FLASH CODE 51 ONCE AND THEN FLASH "ON" AND "OFF" RAPIDLY. DOES IT?

(3) **NO →**
- CHECK FOR AN OPEN OR SHORT TO GROUND IN BLU WIRE TO ECM.
- CHECK FOR A POOR CONNECTION AT ECM. IF OK, REPLACE ECM.

(3) **YES →** ADJUST/REPLACE THROTTLE POSITION SENSOR.

(4) **YES →**
- IF VEHICLE IS EQUIPPED WITH AIR CONDITIONING, START AND RUN ENGINE AT IDLE AND PROCEED TO A/C TEST.
- IF VEHICLE IS NOT EQUIPPED WITH AIR CONDITIONING, SWITCH CONDITION SIGNAL SYSTEM IS FUNCTIONING NORMALLY.

(4) **NO →** (5)
- TURN IGNITION SWITCH TO "OFF."
- BACKPROBE 26 PIN ECM CONNECTOR C2 WITH A DIGITAL MULTIMETER FROM NSW BLK/WHT WIRE CAVITY TO GROUND.
- TURN IGNITION SWITCH TO "ON."
- MEASURE VOLTAGE WITH MANUAL SELECTOR IN "R". SHOULD BE B+. IS IT?

A/C TEST

(6)
- TURN A/C SWITCH AND BLOWER SWITCH TO "ON."
- NOTE "CHECK ENGINE" LIGHT. SHOULD FLASH CODE 51 WHILE A/C COMPRESSOR CLUTCH IS ENGAGED. DOES IT?

(5) **YES →** REPAIR OPEN IN BLK/WHT WIRE TO NEUTRAL START SWITCH.

(5) **NO →** REPLACE ECM.

(6) **NO →** DID A/C COMPRESSOR CLUTCH ENGAGE?

(6) **YES →** SWITCH CONDITION SIGNAL SYSTEM IS FUNCTIONING NORMALLY.

DID A/C COMPRESSOR CLUTCH ENGAGE? **YES →**
- CHECK FOR AN OPEN IN BLK/WHT WIRE TO A/C AMPLIFIER.
- CHECK FOR A POOR CONNECTION AT ECM. IF OK, REPLACE ECM.

NO → REFER TO A/C COMPRESSOR CONTROLS ELECTRICAL DIAGNOSIS.

"AFTER REPAIRS" CONFIRM "CLOSED LOOP" OPERATION AND NO "CHECK ENGINE" LIGHT.

1.6L (VIN 5) ENGINE — DIAGNOSTIC CODE CHARTS — 1992 PRIZM

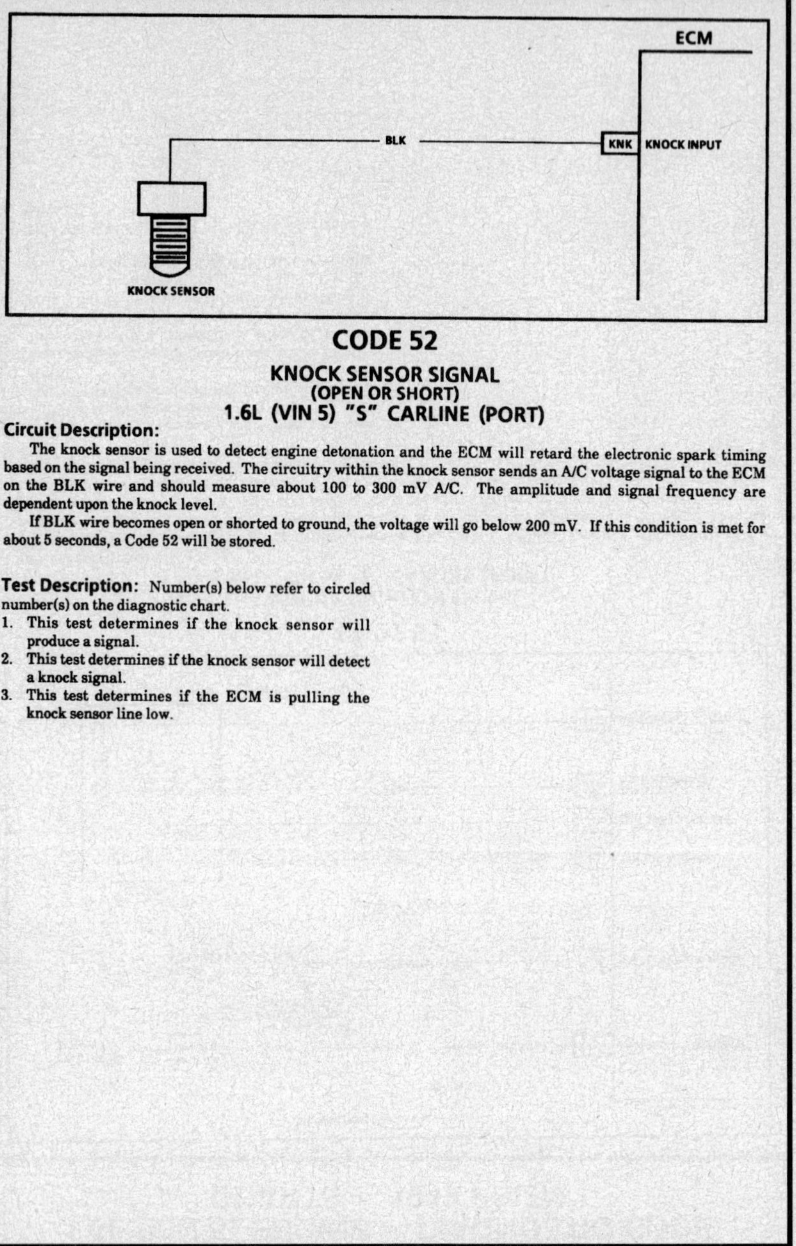

CODE 52
KNOCK SENSOR SIGNAL
(OPEN OR SHORT)
1.6L (VIN 5) "S" CARLINE (PORT)

Circuit Description:

The knock sensor is used to detect engine detonation and the ECM will retard the electronic spark timing based on the signal being received. The circuitry within the knock sensor sends an A/C voltage signal to the ECM on the BLK wire and should measure about 100 to 300 mV A/C. The amplitude and signal frequency are dependent upon the knock level.

If BLK wire becomes open or shorted to ground, the voltage will go below 200 mV. If this condition is met for about 5 seconds, a Code 52 will be stored.

Test Description: Number(s) below refer to circled number(s) on the diagnostic chart.

1. This test determines if the knock sensor will produce a signal.
2. This test determines if the knock sensor will detect a knock signal.
3. This test determines if the ECM is pulling the knock sensor line low.

1.6L (VIN 5) ENGINE — DIAGNOSTIC CODE CHARTS — 1992 PRIZM

CODE 52
KNOCK SENSOR SIGNAL
(OPEN OR SHORT)
1.6L (VIN 5) "S" CARLINE (PORT)

(1)
- IGNITION "OFF."
- USING DVM SET TO 2 VOLT A/C SCALE, BACKPROBE THE KNK BLK WIRE AT THE 16 PIN ECM CONNECTOR TO GROUND.
- TAP ON ENGINE, LIFT BRACKET.
- DVM SHOULD INDICATE 100 mV TO 300 mV ON THE A/C SCALE. DOES IT?

YES

(2)
- IGNITION "ON," ENGINE RUNNING.
- WITH DVM STILL CONNECTED TO KNK BLK WIRE, IS KNOCK SIGNAL INDICATED BETWEEN 100 mV AND 300 mV ON THE A/C SCALE?

YES — POSSIBLE CAUSES OF ENGINE KNOCK. MUST BE CORRECTED: (INTERNAL ENGINE PROBLEMS), BELT DRIVEN COMPONENTS. ETC.

NO — FAULTY KNOCK SENSOR.

NO

(3)
- DISCONNECT ECM 16 PIN CONNECTOR AND REPEAT PREVIOUS TEST. IS SIGNAL INDICATED ON DVM?

YES — FAULTY ECM.

NO — OPEN OR GROUND IN BLK WIRE TO SENSOR OR FAULTY KNOCK SENSOR.

"AFTER REPAIRS," CONFIRM "CLOSED LOOP" OPERATION AND NO "CHECK ENGINE" LIGHT.

1.6L (VIN 5) ENGINE — DIAGNOSTIC CODE CHARTS — 1992 PRIZM

CODE 53

KNOCK CONTROL SIGNAL
(FAULTY ECM)
1.6L (VIN 5) "S" CARLINE (PORT)

CHECK THAT ALL ECM CONNECTORS ARE FULLY SEATED INTO THE ECM.
IF OK, REPLACE ECM. RUN ENGINE AND RECHECK FOR CODE.

"AFTER REPAIRS," CONFIRM "CLOSED LOOP" OPERATION AND NO "CHECK ENGINE" LIGHT.

1.6L (VIN 5) ENGINE — DIAGNOSTIC CODE CHARTS — 1992 PRIZM

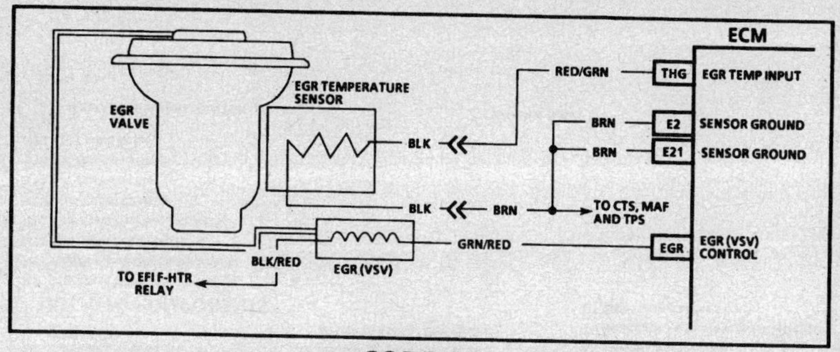

CODE 71

EGR TEMPERATURE SENSOR CIRCUIT
(CALIFORNIA ONLY)
1.6L (VIN 5) "S" CARLINE (PORT)

Circuit Description:

Code 71 indicates that the EGR temperature sensor has signaled the ECM of a temperature above or below the calibrated range. The EGR temperature sensor is used by the ECM to detect one of the following:

- EGR valve opened when not commanded by the ECM.
- EGR valve did not open when commanded by the ECM.
- EGR valve stuck in the opened position.

The EGR temperature sensor is a thermistor whose resistance changes relative to temperature change in the EGR valve exhaust gas passage of the intake manifold.

Test Description: Number(s) below refer to circled number(s) on the diagnostic chart.

1. The ECM should provide 4-5 volts to the EGR temperature sensor.
2. The ECM provides ground for the EGR temperature sensor.
3. Checks EGR temperature sensor resistance.

1.6L (VIN 5) ENGINE — DIAGNOSTIC CODE CHARTS — 1992 PRIZM

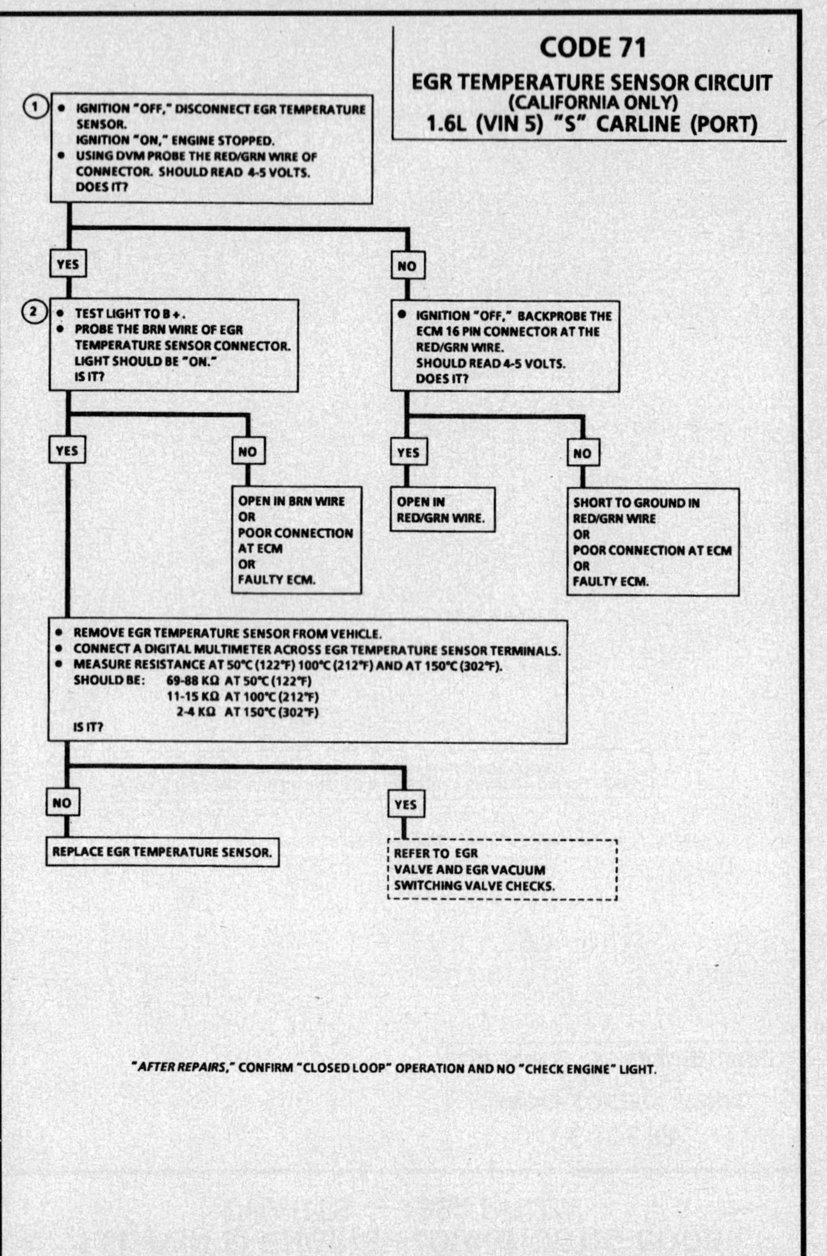

1.6L (VIN 6) ENGINE — COMPONENT LOCATIONS — 1992 PRIZM

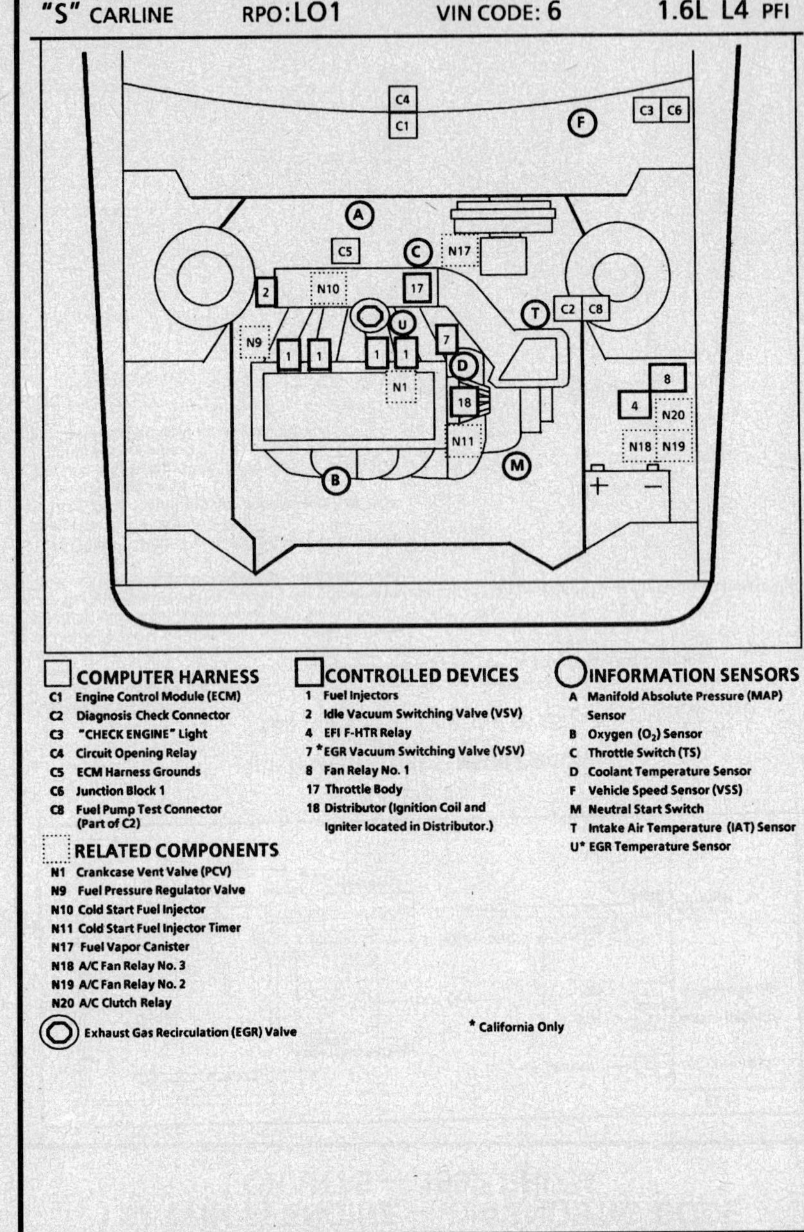

1.6L (VIN 6) ENGINE — ECM WIRING DIAGRAMS — 1992 PRIZM

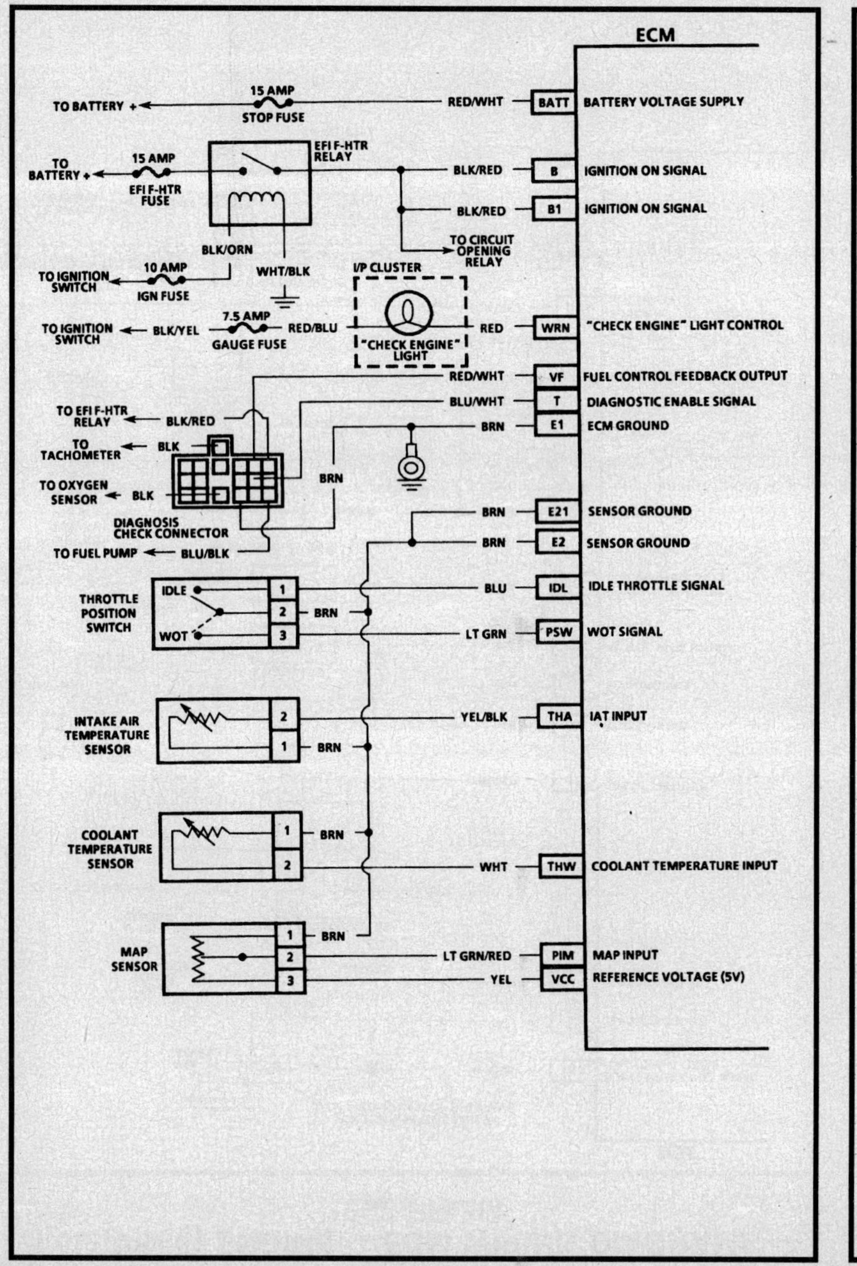

1.6L (VIN 6) ENGINE — ECM WIRING DIAGRAMS — 1992 PRIZM

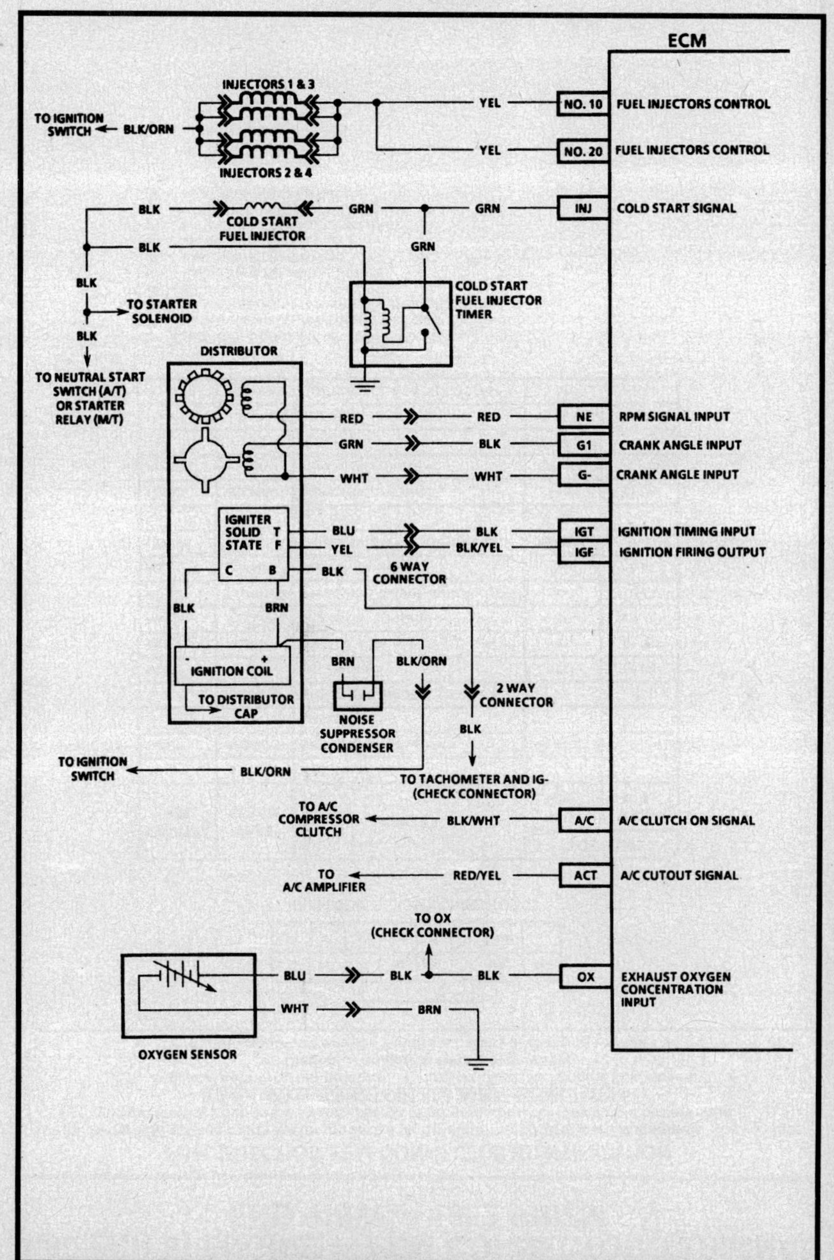

1.6L (VIN 6) ENGINE — ECM WIRING DIAGRAMS — 1992 PRIZM

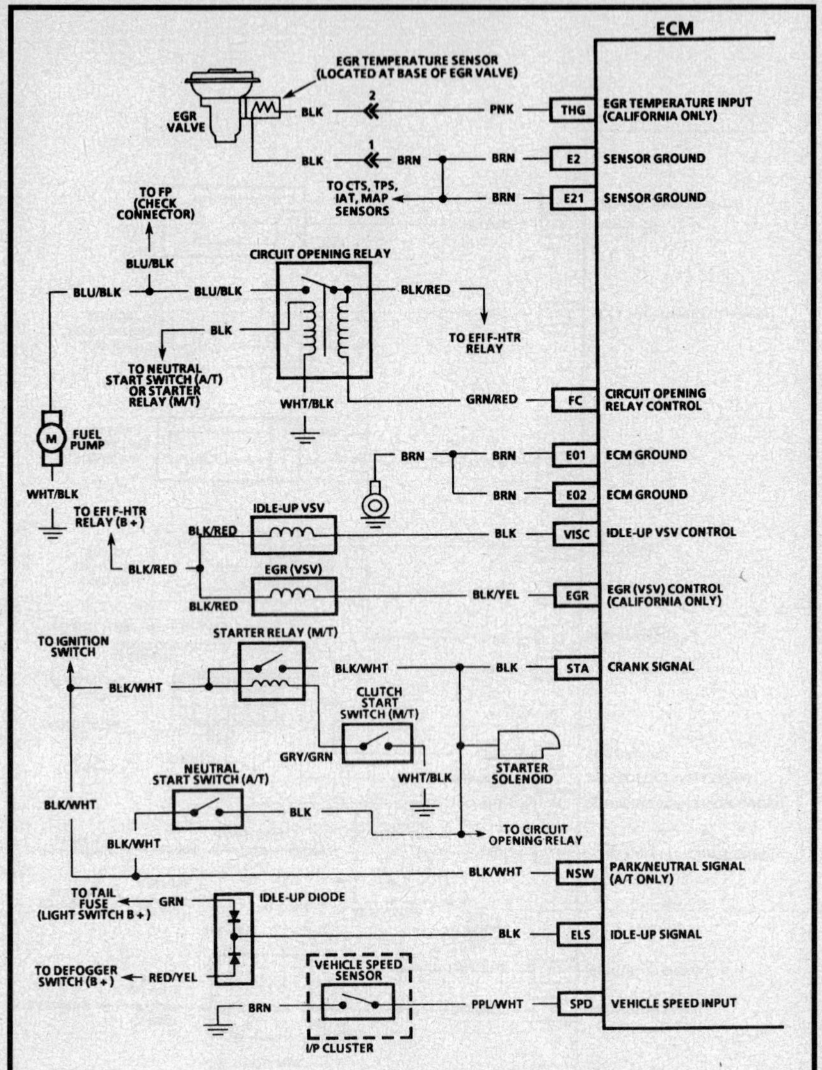

1.6L (VIN 6) ENGINE — ECM CONNECTOR TERMINAL END VIEW — 1992 PRIZM

FUEL INJECTION ECM CONNECTOR IDENTIFICATION

The following ECM voltage charts are for use with a digital voltmeter to further aid in diagnosis. The voltages you get may vary due to low battery charge or other reasons, but they should be very close.
THE FOLLOWING CONDITIONS JUST BE MET BEFORE TESTING:
- Engine at operating temperature   • Engine idling (for "Engine Run" column)
- Diagnostic enable terminal not grounded
- All voltages shown "B +" indicate battery or charging voltage

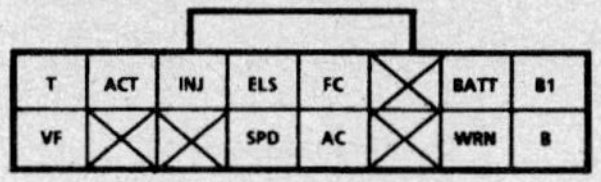

CAVITY/ PIN	WIRE COLOR	CIRCUIT	VOLTAGE	
			KEY "ON"	ENG. RUN
1/B1	BLK/RED	IGNITION ON SIGNAL	B+	B+
2/BATT	RED/WHT	BATTERY VOLTAGE SUPPLY	B+	B+
3/–	–	NOT USED	–	–
4/FC	GRN/RED	CIRCUIT OPENING RELAY CONTROL	B+	0*
5/ELS	BLK	IDLE-UP SIGNAL	0* (1)	0* (1)
6/INJ	GRN	COLD START SIGNAL	0*	0* (2)
7/ACT	RED/YEL	A/C CUTOUT CONTROL	0*	B + (3)
8/T	BLU/WHT	DIAGNOSTIC ENABLE SIGNAL	B+	B+
9/B	BLK/RED	IGNITION ON SIGNAL	B+	B+
10/WRN	RED	"CHECK ENGINE" LIGHT CONTROL	.8-1.5V	B+
11/–	–	NOT USED	–	–
12/A/C	BLK/WHT	A/C CLUTCH ON SIGNAL	0* (3)	0* (3)
13/SPD	PPL/WHT	VEHICLE SPEED INPUT	0-B + (4)	0-B + (4)
14/–	–	NOT USED	–	–
15/–	–	NOT USED	–	–
16/VF	RED/WHT	FUEL CONTROL FEEDBACK OUTPUT	0*	0-2.5V

0*	LESS THAN 0.50 VOLTS
(1)	B + WITH HEADLIGHTS OR DEFOGGER "ON"
(2)	B + WITH COOLANT TEMPERATURE BELOW 35°C (95°F)
(3)	B + WITH A/C "ON," 0V WITH A/C "OFF"
(4)	VARIES AS FRONT WHEELS ROTATE
(5)	VARIES WITH TEMPERATURE
(6)	VARIES WITH ENGINE LOAD
(7)	B + WITH IGNITION SWITCH IN "START"
(8)	VARIES WITH THROTTLE POSITION
(9)	0V AT WIDE OPEN THROTTLE
(10)	B + OFF IDLE
(★)	CALIFORNIA ONLY
(X)	AUTOMATIC TRANSAXLE ONLY

1.6L (VIN 6) ENGINE — ECM CONNECTOR TERMINAL END VIEW — 1992 PRIZM

FUEL INJECTION ECM CONNECTOR IDENTIFICATION

The following ECM voltage charts are for use with a digital voltmeter to further aid in diagnosis. The voltages you get may vary due to low battery charge or other reasons, but they should be very close.

THE FOLLOWING CONDITIONS JUST BE MET BEFORE TESTING:
- Engine at operating temperature • Engine idling (for "Engine Run" column)
- Diagnostic enable terminal not grounded
- All voltages shown "B +" indicate battery or charging voltage

E01	NO. 10	STA	OX	G-	G1	IGF	IGT	THA	PIM	THW	NSW	EGR
E02	NO. 20	E1	[X]	E21	NE	THG	IDL	VCC	PSW	E2	O/D	VISC

BACKVIEW OF ECM CONNECTOR C2

CAVITY/ PIN	WIRE COLOR	CIRCUIT	VOLTAGE KEY "ON"	VOLTAGE ENG. RUN
1/EGR (★)	BLK/YEL	EGR VSV CONTROL	B+	B+
2/NSW (X)	BLK/WHT	PARK/NEUTRAL SIGNAL	B+	B+
3/THW	WHT	COOLANT TEMPERATURE INPUT	5V	1-4V (5)
4/PIM	LT GRN/ RED	MAP INPUT	3.5V	1-3V (6)
5/THA	YEL/BLK	IAT INPUT	1-3V (5)	1-3V (5)
6/IGT	BLK	IGNITION TIMING INPUT	0*	.69V
7/IGF	BLK/YEL	IGNITION FIRING OUTPUT	.5-1V	1V
8/G1	BLK	CRANK ANGLE INPUT	.69V	.67V
9/G-	WHT	CRANK ANGLE INPUT	.69V	.67V
10/0X	BLK	EXHAUST OXYGEN CONCENTRATION INPUT	0*	.1-.9V VARIES
11/STA	BLK	CRANK SIGNAL	0*	0* (7)
12/NO. 10	YEL	FUEL INJECTORS CONTROL	B+	B+
13E01	BRN	ECM GROUND	0*	0*

CAVITY/ PIN	WIRE COLOR	CIRCUIT	VOLTAGE KEY "ON"	VOLTAGE ENG. RUN
14/VISC	BLK	IDLE-UP VSV CONTROL	1.4V	B + (8)
15/O/D	RED/WHT	NOT USED	-	-
16/E2	BRN	SENSOR GROUND	0*	0*
17/PSW	LT GRN	WIDE OPEN THROTTLE SIGNAL	B + (9)	B + (9)
18/VCC	YEL	REFERENCE VOLTAGE (5V)	5V	5V
19/IDL	BLU	IDLE THROTTLE SIGNAL	0* (10)	0* (10)
20/THG (★)	PNK	EGR TEMPERATURE INPUT	5V (5)	.69V (5)
21/NE	RED	RPM INPUT	.69V	.67V
22/E21	BRN	SENSOR GROUND	0*	0*
23/-	-	NOT USED	-	-
24/E1	BRN	ECM GROUND	0*	0*
25/NO. 20	YEL	FUEL INJECTORS CONTROL	B +	B +
26/E02	BRN	ECM GROUND	0*	0*

0* LESS THAN 0.50 VOLTS
(1) B+ WITH HEADLIGHTS OR DEFOGGER "ON"
(2) B+ WITH COOLANT TEMPERATURE BELOW 35°C (95°F)
(3) B+ WITH A/C "ON," 0V WITH A/C "OFF"
(4) VARIES AS FRONT WHEELS ROTATE
(5) VARIES WITH TEMPERATURE
(6) VARIES WITH ENGINE LOAD
(7) B + WITH IGNITION SWITCH IN "START"
(8) VARIES WITH THROTTLE POSITION
(9) 0V AT WIDE OPEN THROTTLE
(10) B + OFF IDLE
(★) CALIFORNIA ONLY
(X) AUTOMATIC TRANSAXLE ONLY

1.6L (VIN 6) ENGINE — DIAGNOSTIC CIRCUIT CHECK — 1992 PRIZM

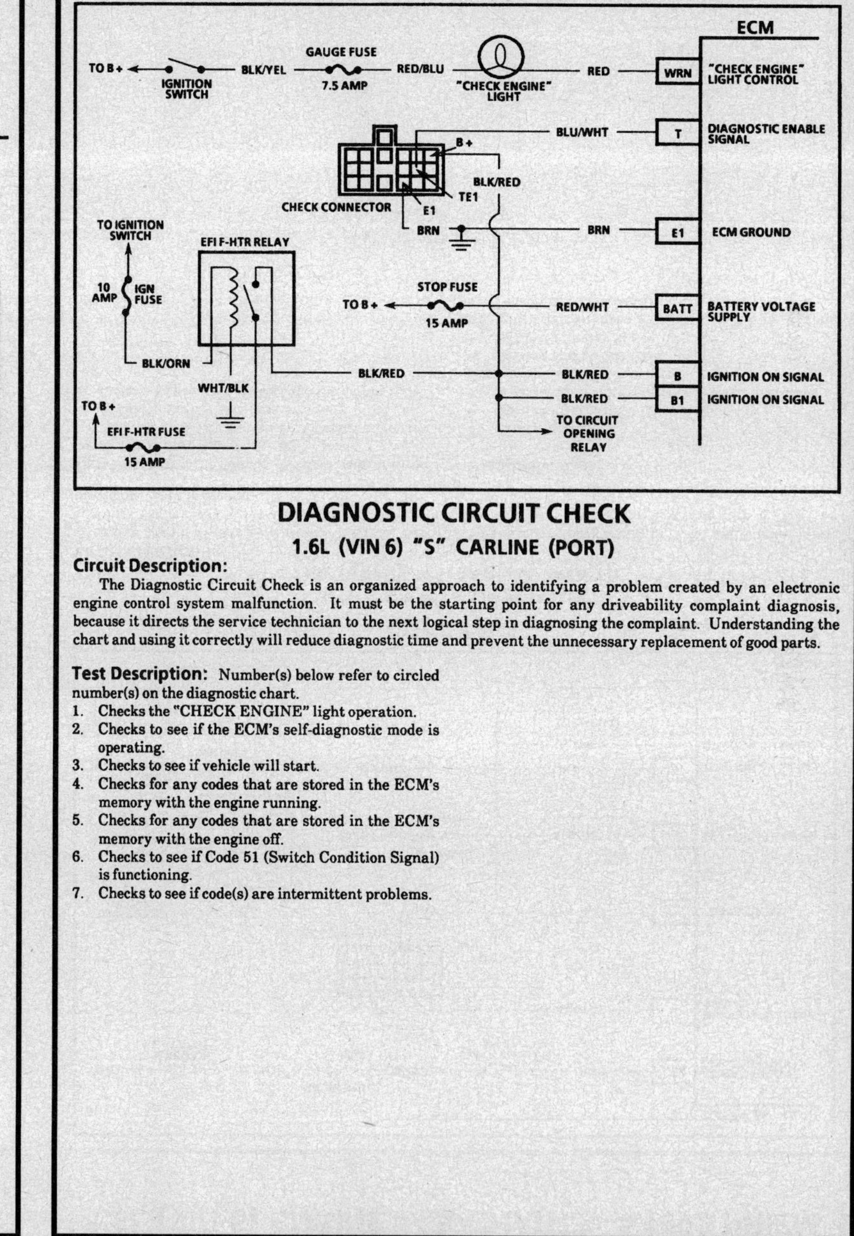

DIAGNOSTIC CIRCUIT CHECK
1.6L (VIN 6) "S" CARLINE (PORT)

Circuit Description:

The Diagnostic Circuit Check is an organized approach to identifying a problem created by an electronic engine control system malfunction. It must be the starting point for any driveability complaint diagnosis, because it directs the service technician to the next logical step in diagnosing the complaint. Understanding the chart and using it correctly will reduce diagnostic time and prevent the unnecessary replacement of good parts.

Test Description: Number(s) below refer to circled number(s) on the diagnostic chart.
1. Checks the "CHECK ENGINE" light operation.
2. Checks to see if the ECM's self-diagnostic mode is operating.
3. Checks to see if vehicle will start.
4. Checks for any codes that are stored in the ECM's memory with the engine running.
5. Checks for any codes that are stored in the ECM's memory with the engine off.
6. Checks to see if Code 51 (Switch Condition Signal) is functioning.
7. Checks to see if code(s) are intermittent problems.

1.6L (VIN 6) ENGINE — DIAGNOSTIC CIRCUIT CHECK — 1992 PRIZM

DIAGNOSTIC CIRCUIT CHECK
1.6L (VIN 6) "S" CARLINE (PORT)

NOTE: CODES ARE FLASHED IN NUMERICAL ORDER STARTING WITH THE LOWEST CODE FIRST. CODE 51 WITH ENGINE RUNNING IS AN INDICATION OF A NORMAL CONDITION. REFER TO CODE 51 FOR A MORE DETAILED DESCRIPTION OF CODE 51.

1.
- TURN IGNITION SWITCH TO "ON."
- NOTE "CHECK ENGINE" LIGHT.

STEADY LIGHT | **NO LIGHT** | **FLASHING LIGHT**

2.
- CONNECT A JUMPER FROM CHECK CONNECTOR CAVITY "TE1" TO "E1".
- NOTE "CHECK ENGINE" LIGHT.

NO LIGHT → REFER TO CHART A-1

FLASHING LIGHT →
- CHECK FOR A SHORT TO GROUND IN WHT/BLU WIRE. IF OK, REPLACE ECM.

FLASHES "ON" AND "OFF" RAPIDLY OR FLASHES DIAGNOSTIC TROUBLE CODES OTHER THAN CODE 51. | **FLASHES CODE 51** | **DOES NOT FLASH AT ALL, REMAINS LIT**

FLASHES CODE 51 → REFER TO CODE 51 CHART.

DOES NOT FLASH AT ALL, REMAINS LIT → REFER TO CHART A-2.

3.
- CRANK ENGINE.
- DOES ENGINE START?

YES / NO

4.
- IS "CHECK ENGINE" LIGHT FLASHING ANY DIAGNOSTIC TROUBLE CODES OTHER THAN CODE 51?

NO / YES

5.
- IS "CHECK ENGINE" LIGHT FLASHING ANY DIAGNOSTIC TROUBLE CODES OTHER THAN CODE 51?

YES / NO

NO → REFER TO CHART A-3.

6.
- RAISE ENGINE SPEED BY PRESSING DOWN ON THE ACCELERATOR PEDAL MOMENTARILY.
- MOVE MANUAL SELECTOR FROM "P" TO "N" AND THEN BACK TO "P" (AUTOMATIC TRANSAXLE ONLY).
- TURN A/C SWITCH AND BLOWER SWITCH TO "ON" AND THEN BACK TO "OFF" (IF A/C EQUIPPED).
- NOTE "CHECK ENGINE" LIGHT AFTER PERFORMING EACH TEST. SHOULD FLASH CODE 51 AND THEN FLASH "ON" AND "OFF" RAPIDLY. DOES IT?

7.
- TURN IGNITION SWITCH TO "OFF."
- NOTE CODES THAT WERE INDICATED.
- CLEAR CODES.
- ROAD TEST VEHICLE AT CONDITION WHICH THE ORIGINAL PROBLEM OCCURRED. NOTICE: IF VEHICLE DOES NOT START, CRANK ENGINE. DOES "CHECK ENGINE" LIGHT FLASH ANY CODES OTHER THAN CODE 51?

YES → THE ELECTRONIC ENGINE CONTROL SYSTEM IS FUNCTIONING NORMALLY.

NO → REFER TO CODE 51 CHART.

YES → REFER TO APPROPRIATE CODE CHART.

NO → PROBLEM IS INTERMITTENT.

1.6L (VIN 6) ENGINE — A-CHARTS — 1992 PRIZM

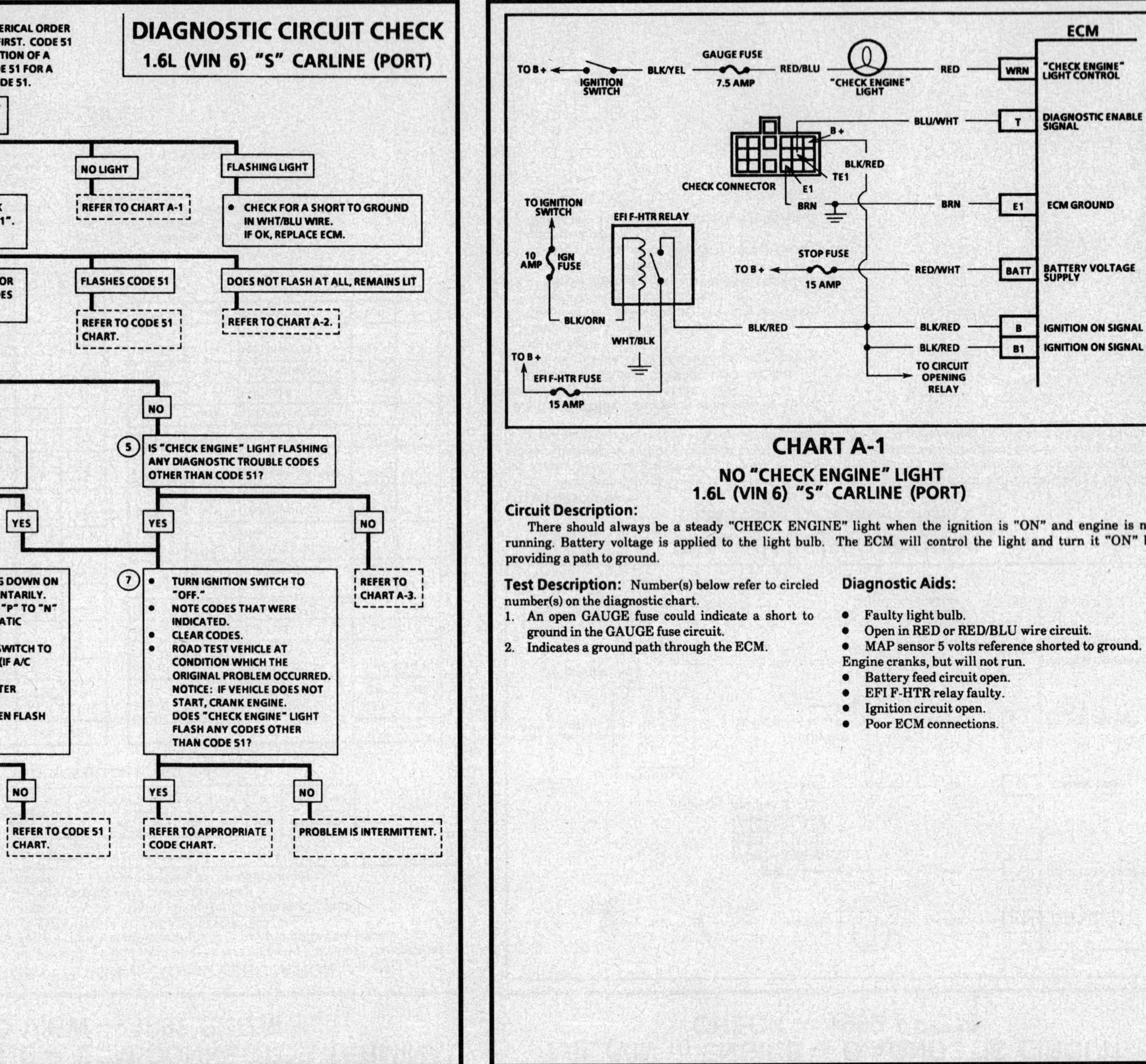

CHART A-1
NO "CHECK ENGINE" LIGHT
1.6L (VIN 6) "S" CARLINE (PORT)

Circuit Description:

There should always be a steady "CHECK ENGINE" light when the ignition is "ON" and engine is not running. Battery voltage is applied to the light bulb. The ECM will control the light and turn it "ON" by providing a path to ground.

Test Description: Number(s) below refer to circled number(s) on the diagnostic chart.

1. An open GAUGE fuse could indicate a short to ground in the GAUGE fuse circuit.
2. Indicates a ground path through the ECM.

Diagnostic Aids:

- Faulty light bulb.
- Open in RED or RED/BLU wire circuit.
- MAP sensor 5 volts reference shorted to ground. Engine cranks, but will not run.
- Battery feed circuit open.
- EFI F-HTR relay faulty.
- Ignition circuit open.
- Poor ECM connections.

1.6L (VIN 6) ENGINE — A-CHARTS — 1992 PRIZM

CHART A-1

**NO "CHECK ENGINE" LIGHT
1.6L (VIN 6) "S" CARLINE (PORT)**

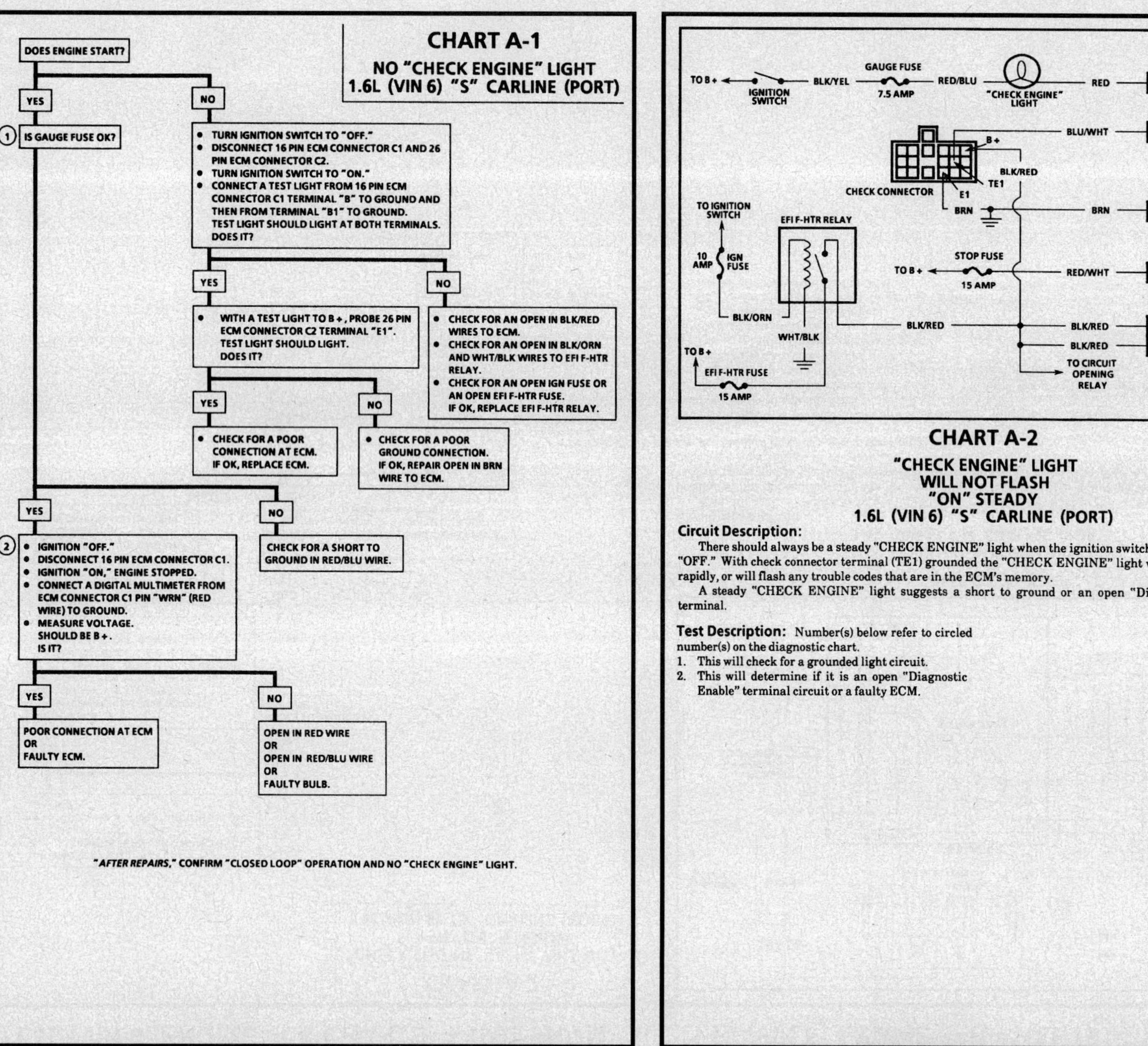

"AFTER REPAIRS," CONFIRM "CLOSED LOOP" OPERATION AND NO "CHECK ENGINE" LIGHT.

1.6L (VIN 6) ENGINE — A-CHARTS — 1992 PRIZM

CHART A-2

**"CHECK ENGINE" LIGHT
WILL NOT FLASH
"ON" STEADY
1.6L (VIN 6) "S" CARLINE (PORT)**

Circuit Description:

There should always be a steady "CHECK ENGINE" light when the ignition switch is "ON" and the engine is "OFF." With check connector terminal (TE1) grounded the "CHECK ENGINE" light will flash "ON" and "OFF" rapidly, or will flash any trouble codes that are in the ECM's memory.

A steady "CHECK ENGINE" light suggests a short to ground or an open "Diagnostic Enable" circuit terminal.

Test Description: Number(s) below refer to circled number(s) on the diagnostic chart.
1. This will check for a grounded light circuit.
2. This will determine if it is an open "Diagnostic Enable" terminal circuit or a faulty ECM.

CHART A-2

"CHECK ENGINE" LIGHT WILL NOT FLASH "ON" STEADY
1.6L (VIN 6) "S" CARLINE (PORT)

①
- IGNITION "OFF."
- DISCONNECT 16 PIN ECM CONNECTOR C1.
- IGNITION "ON."
- "CHECK ENGINE" LIGHT SHOULD BE "OFF." IS IT?

YES →
- IGNITION "OFF."
- RECONNECT ECM CONNECTOR C1.
- IGNITION "ON," ENGINE "OFF."
- CONNECT A JUMPER FROM CHECK CONNECTOR CAVITY "TE1" TO GROUND.
- "CHECK ENGINE" LIGHT SHOULD "FLASH" "ON" AND "OFF" RAPIDLY. DOES IT?

NO → SHORT TO GROUND IN RED WIRE.

YES →
- CHECK FOR POOR GROUND CONNECTION. IF OK, REPAIR OPEN IN BRN WIRE TO CHECK CONNECTOR.

NO →
②
- REMOVE JUMPER.
- BACKPROBE 16 PIN ECM CONNECTOR C1 PIN "T" (BLU/WHT WIRE) WITH A DVM TO GROUND. DVM SHOULD READ 10-14 VOLTS. DOES IT?

NO → POOR CONNECTION AT ECM OR FAULTY ECM.

YES →
- BACKPROBE RED/WHT WIRE AT 16 PIN ECM CONNECTOR C1 WITH A DVM TO GROUND. DVM SHOULD READ 10-14 VOLTS DOES IT?

NO → OPEN IN RED/WHT WIRE OR OPEN IN STOP FUSE.

YES → OPEN IN BLU/WHT WIRE OR FAULTY ECM.

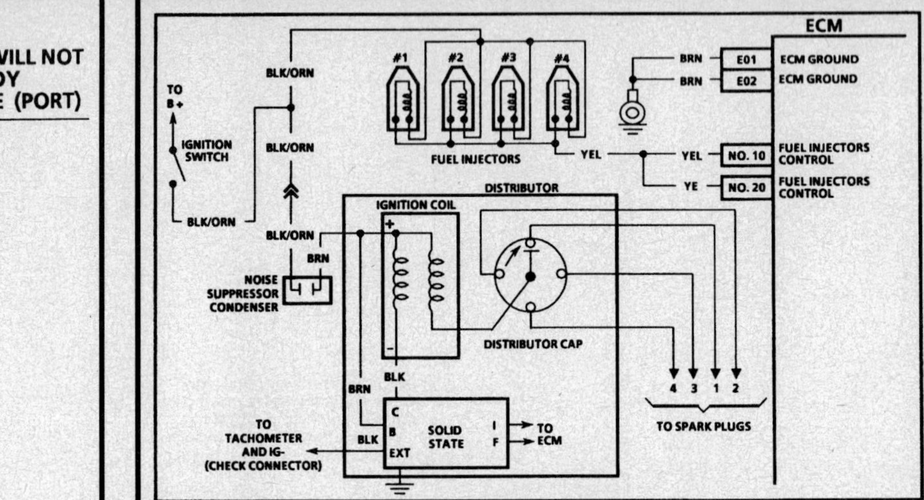

CHART A-3

ENGINE CRANKS BUT WILL NOT RUN
1.6L (VIN 6) "S" CARLINE (PORT)

Circuit Description:

When the ignition switch is turned to "START," voltage is applied to the ignition coil and to the igniter (located within the distributor). The igniter controls the ground path of the ignition coil. When the igniter receives power and an ignition firing input it toggles the ground path to the ignition coil "ON" and "OFF," thus inducing a high voltage in the secondary windings of the ignition coil. This induced voltage is then applied through the distributor cap to each spark plug.

Test Description: Number(s) below refer to circled number(s) on the diagnostic chart.

1. This checks for an intermittent code.
2. This checks for spark and for a faulty ignition coil and a faulty distributor cap and wires.
3. This tests the fuel injectors electrical circuit.
4. Checks for power to the fuel injectors.
5. Checks resistance of each injector.
6. Checks for an open or short to ground in fuel injector circuit and for a faulty ECM.

Diagnostic Aids:

- Water or foreign material in the fuel can cause a no start or a lack of power condition.

1.6L (VIN 6) ENGINE — A-CHARTS — 1992 PRIZM

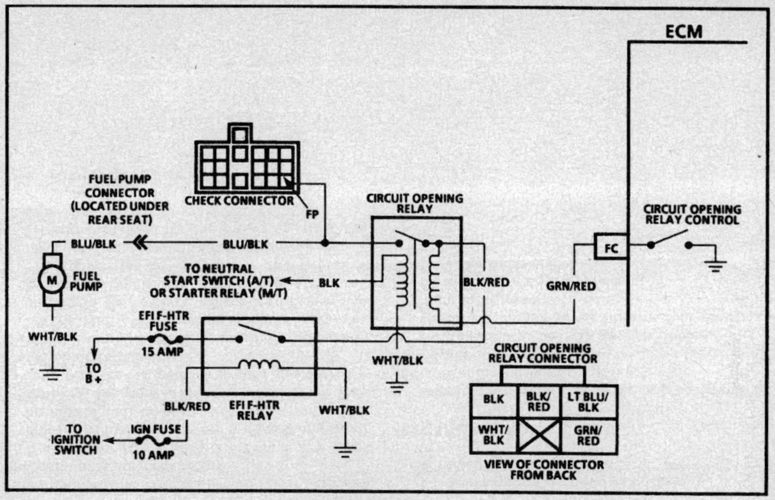

CHART A-5

ENGINE CRANKS BUT WILL NOT RUN
(FUEL SYSTEM ELECTRICAL TEST)
1.6L (VIN 6) "S" CARLINE (PORT)

Circuit Description:

When the engine is cranked the ECM will turn "ON" the circuit opening relay to operate the fuel pump. If the ECM receives ignition reference pulses the fuel pump will continue to run. If the ECM receives no reference pulses from the ignition circuit it will shut "OFF" the fuel pump.

Test Description: Number(s) below refer to circled number(s) on the diagnostic chart.

1. Fuel pump should run when ignition is cranked.
2. Test for B+ from circuit opening relay to fuel pump.
3. Checks for crank signal to circuit opening relay.
4. This will determine if ECM fuel pump driver circuit inside ECM is faulty.

Diagnostic Aids:

Visual inspection of wiring and connectors should be made if an intermittent problem exists.

1.6L (VIN 6) ENGINE — A-CHARTS — 1992 PRIZM

CHART A-3

ENGINE CRANKS BUT WILL NOT RUN
1.6L (VIN 6) "S" CARLINE (PORT)

NOTE: BEFORE USING THIS CHART, VERIFY FUEL QUANTITY. ALSO INSPECT DISTRIBUTOR FOR BROKEN SHAFT OR DAMAGED CAP AND ROTOR.

1.
- TURN IGNITION SWITCH TO "OFF."
- WITH DIAGNOSTIC ENABLE TERMINAL (TE1) IN CHECK CONNECTOR JUMPERED TO GROUND, CRANK ENGINE.
- ARE ANY DIAGNOSTIC TROUBLE CODES PRESENT?

YES → REFER TO APPROPRIATE CODE CHART.

NO ↓

2.
- TURN IGNITION SWITCH TO "OFF."
- DISCONNECT SPARK PLUG WIRES FROM SPARK PLUGS.
- CONNECT A J 26792 H.E.I. SPARK TESTER TO SPARK PLUG WIRES.
- CHECK FOR SPARK (AT EACH WIRE) WHILE CRANKING ENGINE.
- ALL SHOULD HAVE SPARK.
- DO THEY?

NO **(5)** → CHECK DISTRIBUTOR CAP AND ROTOR AND SPARK PLUG WIRES FOR POOR CONNECTIONS OR OPENS. CHECK FOR A POOR CONNECTION AT IGNITION COIL (INSIDE DISTRIBUTOR). IF OK, REPLACE IGNITION COIL.

YES ↓

3.
- TURN IGNITION SWITCH TO "OFF."
- DISCONNECT ALL FOUR FUEL INJECTORS CONNECTORS.
- INSTALL A J34730-2B INJECTOR TEST LIGHT INTO EACH CONNECTOR (ONE AT A TIME).
- NOTE TEST LIGHT AT EACH WHILE CRANKING ENGINE.
- SHOULD BLINK AT ALL CONNECTORS.
- DOES IT?

NO **(5)** → TURN IGNITION SWITCH TO "OFF." CONNECT A DIGITAL MULTIMETER FROM FUEL INJECTORS TERMINAL "1" TO TERMINAL "2" (ONE AT A TIME). MEASURE RESISTANCE. SHOULD BE 11-17 OHMS AT ALL FOUR INJECTORS. IS IT?
  - NO → REPLACE FAULTY FUEL INJECTOR(S).
  - YES → NO → REFER TO "DIAGNOSTIC AIDS." REFER TO CHART A-5.

YES ↓

4.
- TURN IGNITION SWITCH TO "ON."
- CONNECT A TEST LIGHT FROM ALL FOUR FUEL INJECTORS BLK/ORN WIRE CAVITY TO GROUND (ONE AT A TIME).
- TEST LIGHT SHOULD LIGHT AT ALL.
- DOES IT?

NO → REPAIR OPEN IN BLK/ORN WIRE TO FUEL INJECTOR(S).

YES ↓

6.
- WITH A TEST LIGHT TO B + PROBE ALL FOUR FUEL INJECTORS YEL WIRE CAVITY.
- NOTE TEST LIGHT AT EACH WHILE CRANKING ENGINE.

STEADY LIGHT AT ONE OR MORE → CHECK FOR A SHORT TO GROUND IN YEL WIRE. IF OK, REPLACE ECM.

NO LIGHT AT ONE OR MORE →
- CHECK FOR A POOR CONNECTION AT ECM.
- CHECK FOR AN OPEN IN YEL WIRE.
- IF OK, REPLACE ECM.

CHART A-5

ENGINE CRANKS BUT WILL NOT RUN
(FUEL SYSTEM ELECTRICAL TEST)
1.6L (VIN 6) "S" CARLINE (PORT)

(1)
- IGNITION "OFF."
- FUEL CAP REMOVED.
- HAVE HELPER CRANK ENGINE.
- LISTEN FOR FUEL PUMP.
- DOES FUEL PUMP RUN?

NO → YES → CHART A-7

(2)
- PROBE CHECK CONNECTOR TERMINAL "FP" WITH TEST LIGHT TO GROUND.
- HAVE HELPER CRANK ENGINE.
- TEST LIGHT SHOULD BE "ON." IS IT?

NO → YES

- PROBE BLK WIRE AT CIRCUIT OPENING RELAY WITH TEST LIGHT TO GROUND.
- CRANK ENGINE.
- TEST LIGHT SHOULD BE "ON" WHILE CRANKING. IS IT?

- REMOVE THE REAR SEAT.
- DISCONNECT THE FUEL PUMP CONNECTOR.
- HAVE HELPER CRANK ENGINE.
- PROBE BLU/BLK WIRE WITH TEST LIGHT TO GROUND. TEST LIGHT SHOULD BE "ON." IS IT?

YES → NO → OPEN IN BLK WIRE TO CIRCUIT OPENING RELAY.

NO → OPEN IN BLU/BLK WIRE.

YES
- CHECK FOR AN OPEN IN WHT/BLK WIRE. IF OK, REPLACE FUEL PUMP.

(3)
- IGNITION "ON," ENGINE "OFF."
- PROBE BLK/RED WIRE OF CIRCUIT OPENING RELAY WITH TEST LIGHT TO GROUND. TEST LIGHT SHOULD BE "ON." IS IT?

YES → NO

CHECK FOR OPEN IN BLK/RED WIRE OR
FAULTY EFI F-HTR RELAY OR
OPEN EFI F-HTR FUSE OR
OPEN IGN FUSE OR
POOR GROUND FROM EFI F-HTR RELAY.

(4)
- IGNITION "OFF."
- UNPLUG CIRCUIT OPENING RELAY.
- TEST LIGHT TO B +.
- PROBE GRN/RED WIRE OF CIRCUIT OPENING RELAY CONNECTOR.
- WHILE CRANKING ENGINE, TEST LIGHT SHOULD BE "ON." IS IT?

YES → NO

CHECK FOR OPEN GRN/RED WIRE TO ECM
OR
POOR CONNECTION AT ECM
OR
FAULTY ECM.

- IGNITION "ON."
- JUMPER BLU/BLK WIRE AND BLK/RED WIRE OF CIRCUIT OPENING RELAY CONNECTOR WITH A FUSED JUMPER.
- FUEL PUMP SHOULD "RUN." DOES IT?

YES → NO

- CHECK FOR AN OPEN IN BLU/BLK AND WHT/BLK WIRES TO FUEL PUMP. IF OK, REPLACE FUEL PUMP.

- IGNITION "OFF."
- REMOVE FUSED JUMPER.
- PROBE WHT/BLK WIRE OF CIRCUIT OPENING RELAY CONNECTOR WITH TEST LIGHT TO B +. LIGHT SHOULD BE "ON." IS IT?

YES → NO

FAULTY CIRCUIT OPENING RELAY.

OPEN IN WHT/BLK WIRE.

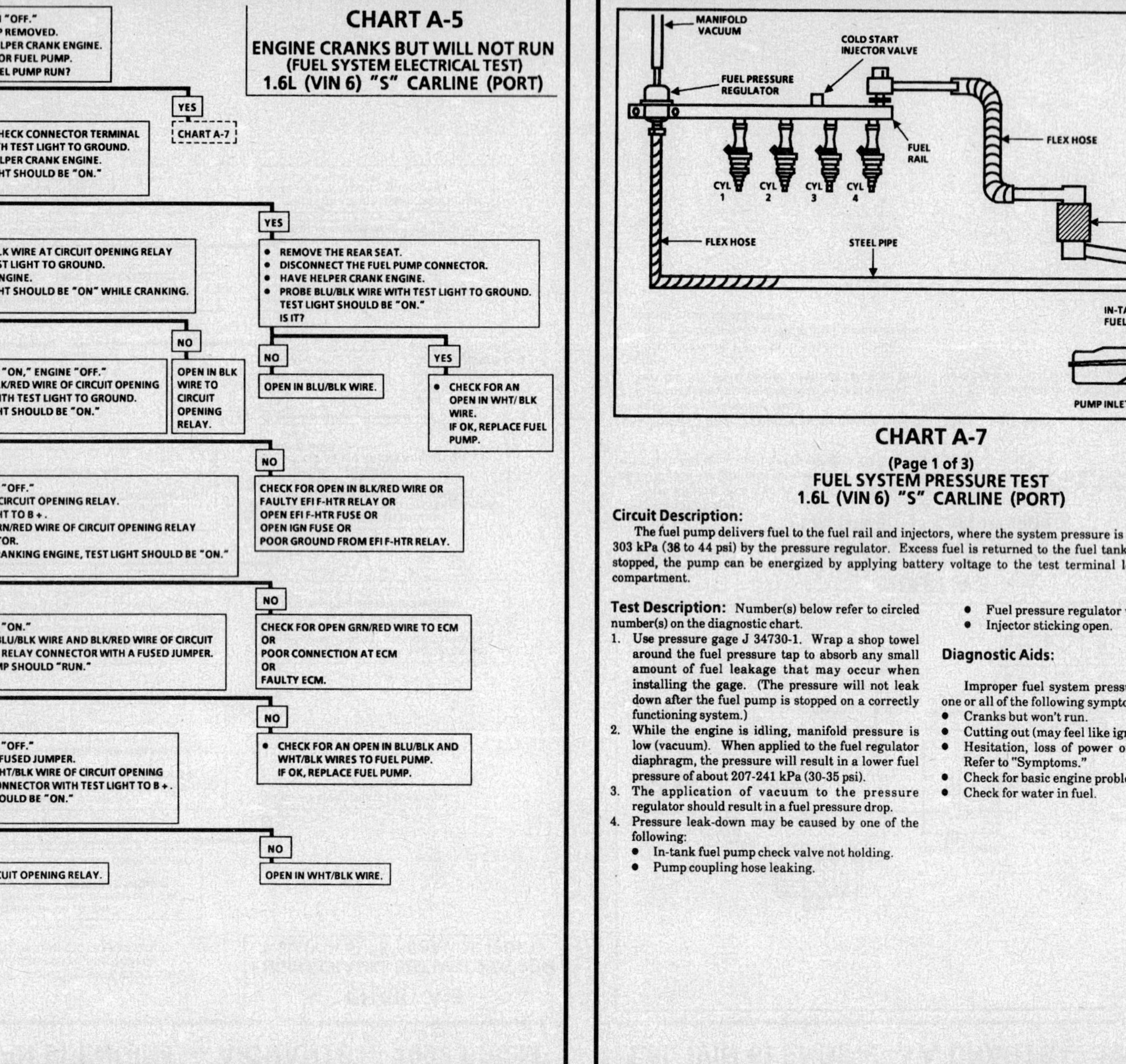

CHART A-7

(Page 1 of 3)
FUEL SYSTEM PRESSURE TEST
1.6L (VIN 6) "S" CARLINE (PORT)

Circuit Description:

The fuel pump delivers fuel to the fuel rail and injectors, where the system pressure is controlled from 262 to 303 kPa (38 to 44 psi) by the pressure regulator. Excess fuel is returned to the fuel tank. When the engine is stopped, the pump can be energized by applying battery voltage to the test terminal located in the engine compartment.

Test Description: Number(s) below refer to circled number(s) on the diagnostic chart.

1. Use pressure gage J 34730-1. Wrap a shop towel around the fuel pressure tap to absorb any small amount of fuel leakage that may occur when installing the gage. (The pressure will not leak down after the fuel pump is stopped on a correctly functioning system.)
2. While the engine is idling, manifold pressure is low (vacuum). When applied to the fuel regulator diaphragm, the pressure will result in a lower fuel pressure of about 207-241 kPa (30-35 psi).
3. The application of vacuum to the pressure regulator should result in a fuel pressure drop.
4. Pressure leak-down may be caused by one of the following:
   - In-tank fuel pump check valve not holding.
   - Pump coupling hose leaking.

- Fuel pressure regulator valve leaking.
- Injector sticking open.

Diagnostic Aids:

Improper fuel system pressure may contribute to one or all of the following symptoms:
- Cranks but won't run.
- Cutting out (may feel like ignition problem).
- Hesitation, loss of power or poor fuel economy. Refer to "Symptoms."
- Check for basic engine problem.
- Check for water in fuel.

1.6L (VIN 6) ENGINE — A-CHARTS — 1992 PRIZM

CHART A-7
(Page 1 of 3)
FUEL SYSTEM PRESSURE TEST
1.6L (VIN 6) "S" CARLINE (PORT)

NOTICE: FUEL SYSTEM IS UNDER PRESSURE. TO AVOID FUEL SPILLAGE, REFER TO FIELD SERVICE PROCEDURES FOR TESTING OR REPAIRS REQUIRING DISASSEMBLY OF FUEL PIPES OR FITTINGS.

(1)
- INSTALL FUEL PRESSURE GAGE, J 34730-1 OR EQUIVALENT. SEE CHART A-7 (3 OF 3).
- DISCONNECT VACUUM HOSE FROM PRESSURE REGULATOR AND PLUG HOSE.
- JUMPER "B +" TO "FP" IN CHECK CONNECTOR FOR 10 SECONDS, REMOVE JUMPER.
- NOTE FUEL PRESSURE.
- PRESSURE SHOULD BE 265-304 kPa (38-44 psi) AND HOLDING STEADILY.
  IS IT?

YES → IF DIRECTED TO THIS TEST FROM CHART A-5, AND ENGINE STILL WON'T START, REFER TO DIAGNOSTIC AIDS.

(2)
- IGNITION "ON" ENGINE "RUNNING."
- RECONNECT VACUUM HOSE.
- FUEL PRESSURE SHOULD BE 206-226 kPa (30-33 psi).
  IS IT?

NO →
(3)
- DISCONNECT VACUUM HOSE FROM FUEL PRESSURE REGULATOR.
- WITH ENGINE IDLING, APPLY 12-14" Hg OF VACUUM TO PRESSURE REGULATOR.
- FUEL PRESSURE SHOULD DROP SLIGHTLY.
  DID IT?

YES ← REFER TO "DIAGNOSTIC AIDS"

YES → REPAIR VACUUM SUPPLY TO FUEL PRESSURE REGULATOR.

NO → PRESSURE REGULATOR ASSEMBLY IS FAULTY.

NO →
(4) PRESSURE WITHIN SPECIFICATION, BUT NOT HOLDING.

PRESSURE OUT OF SPECIFICATION.

- IGNITION "OFF."
- CONNECT FUSED JUMPER FROM "B +" TO "FP" ON CHECK CONNECTOR IN ENGINE COMPARTMENT.
- IGNITION "ON."
- GRADUALLY BLOCK FUEL PRESSURE PIPE BY PINCHING FLEX HOSE.
- REMOVE JUMPER FROM CHECK CONNECTOR WHILE CONTINUING TO BLOCK PRESSURE PIPE.
- PRESSURE SHOULD HOLD.
  DID IT?

USE PAGE 2 OF THIS CHART.

NO →
- REPEAT ABOVE PROCEDURE WHILE BLOCKING FUEL RETURN PIPE BY PINCHING FLEX HOSE.
- PRESSURE SHOULD HOLD.
  DID IT?

YES → CHECK FOR
- LEAKING PUMP COUPLING HOSE.
- FAULTY IN-TANK PUMP.

NO → LOCATE AND REPLACE LEAKING INJECTOR AND DRY SPARK PLUGS FROM FLOODED CYLINDERS.

YES → PRESSURE REGULATOR ASSEMBLY IS FAULTY.

1.6L (VIN 6) ENGINE — A-CHARTS — 1992 PRIZM

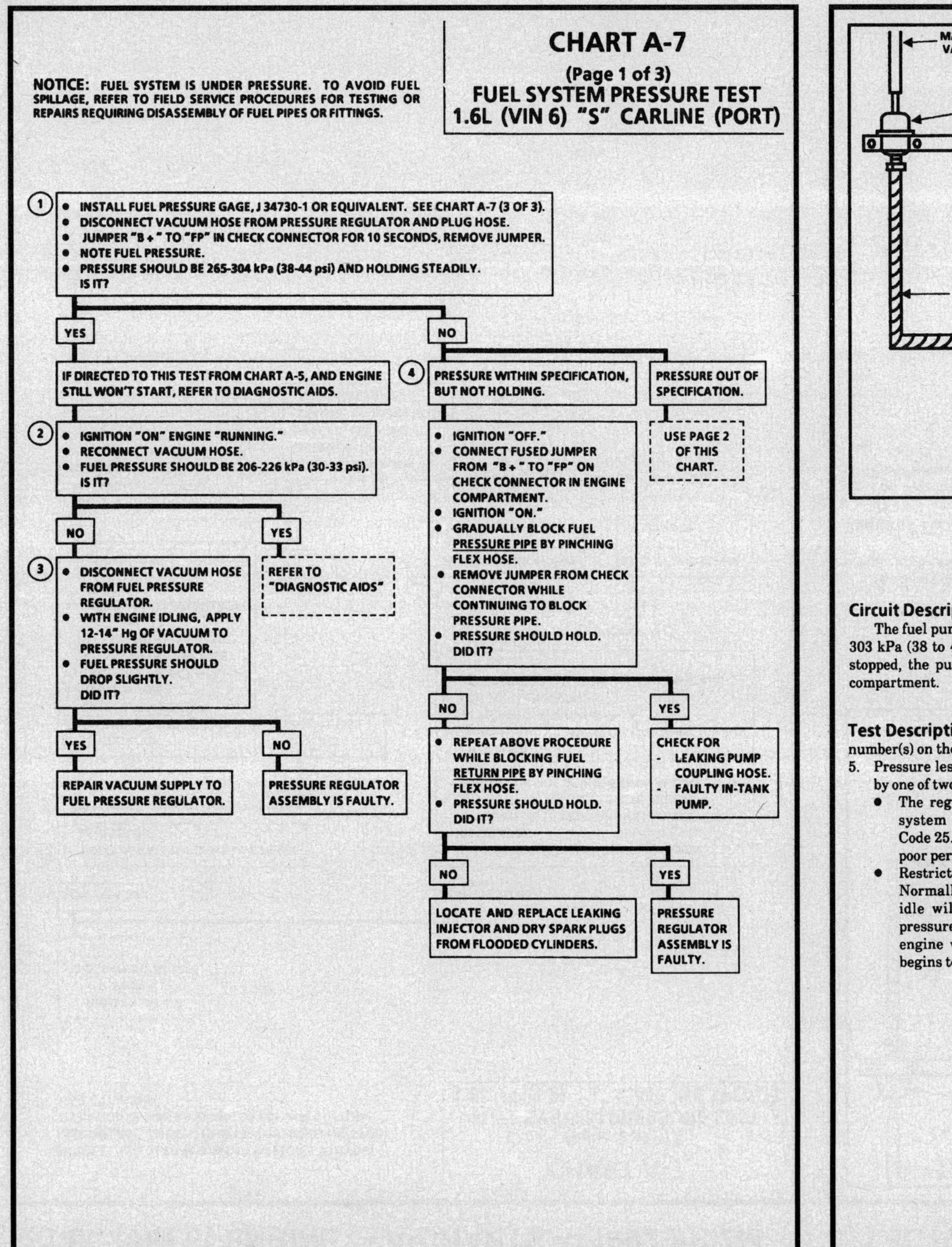

CHART A-7
(Page 2 of 3)
FUEL SYSTEM PRESSURE TEST
1.6L (VIN 6) "S" CARLINE (PORT)

Circuit Description:

The fuel pump delivers fuel to the fuel rail and injectors, where the system pressure is controlled from 262 to 303 kPa (38 to 44 psi) by the pressure regulator. Excess fuel is returned to the fuel tank. When the engine is stopped, the pump can be energized by applying battery voltage to the test terminal located in the engine compartment.

Test Description: Number(s) below refer to circled number(s) on the diagnostic chart.

5. Pressure less than 262 kPa (38 psi) may be caused by one of two problems:
   - The regulated fuel pressure is too low. The system will be running lean and may set Code 25. Also, hard cold starting and overall poor performance is possible.
   - Restricted flow is causing a pressure drop. Normally, a vehicle with a fuel pressure loss at idle will not be drivable. However, if the pressure drop occurs only while driving, the engine will surge and then stop as pressure begins to drop rapidly.

6. Restricting the fuel return line allows the fuel pump to build above regulated pressure. When battery voltage is applied to "FP" on the check connector, pressure should be above 450 kPa (65 psi).

7. This test determines if the high fuel pressure is due to a restricted fuel return pipe or a pressure regulator problem.

MULTIPORT FUEL INJECTION (MFI) SYSTEMS
LIGHT TRUCKS, VAN, GEO AND SATURN

1.6L (VIN 6) ENGINE — A-CHARTS — 1992 PRIZM

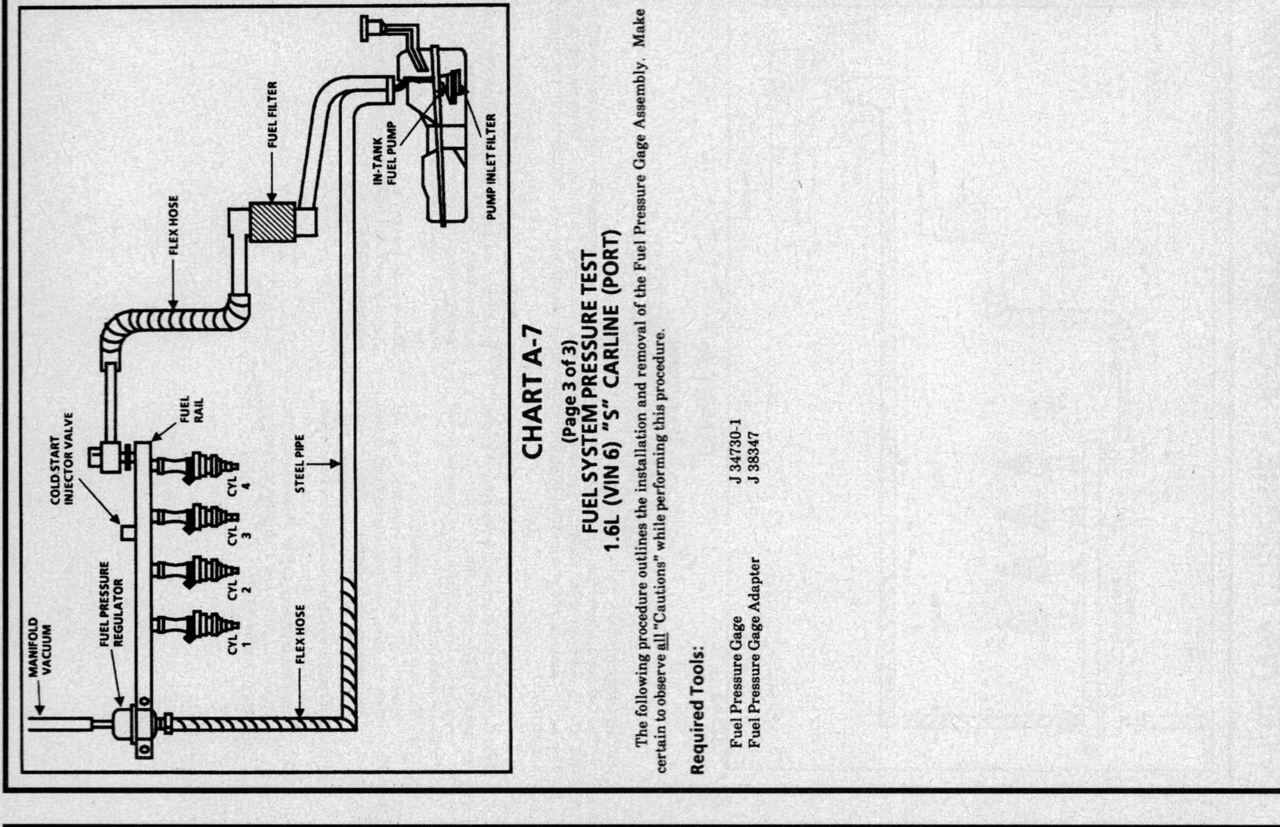

1.6L (VIN 6) ENGINE — A-CHARTS — 1992 PRIZM

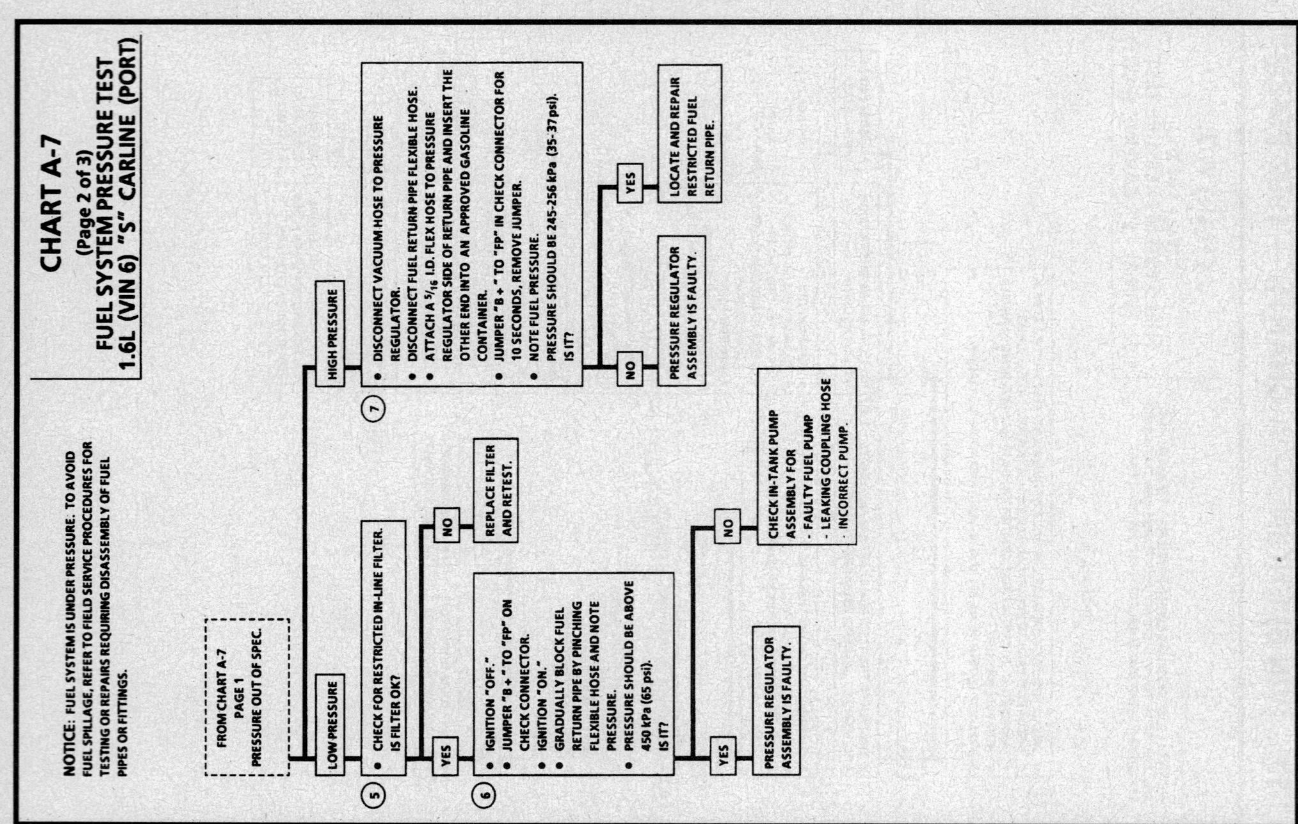

1.6L (VIN 6) ENGINE — A-CHARTS — 1992 PRIZM

CAUTION: TO REDUCE THE RISK OF FIRE AND PERSONAL INJURY, RELIEVE FUEL SYSTEM PRESSURE BEFORE DISCONNECTING FUEL PIPES.

- LOOSEN FUEL FILLER CAP TO RELIEVE TANK PRESSURE.
- REMOVE CIRCUIT OPENING RELAY, LOCATED UNDER CONSOLE BELOW RADIO.
- CRANK ENGINE AND ALLOW TO RUN UNTIL IT STALLS.
- CRANK ENGINE FOR AN ADDITIONAL 3 SECONDS.
- USE SHOP TOWEL TO COLLECT FUEL DURING FUEL PIPE DISCONNECT, PLACE TOWEL IN APPROVED CONTAINER.

CHART A-7
(Page 3 of 3)
FUEL SYSTEM PRESSURE TEST
1.6L (VIN 6) "S" CARLINE (PORT)

FUEL PRESSURE GAGE INSTALLATION

1. IGNITION "OFF." DISCONNECT NEGATIVE BATTERY CABLE.
2. CAREFULLY, REMOVE COLD START INJECTOR VALVE FUEL PIPE HOLD DOWN BOLT AT THE FUEL RAIL.
3. INSTALL ADAPTER J 38347 AND TIGHTEN SECURELY.
4. INSTALL FUEL PRESSURE GAGE J 34730-1 AND TIGHTEN SECURELY.
5. RECONNECT NEGATIVE BATTERY CABLE.
6. REINSTALL CIRCUIT OPENING RELAY.
7. START ENGINE *AND CHECK FOR LEAKS.
8. PURGE AIR FROM FUEL PRESSURE GAGE INTO AN APPROVED CONTAINER.
9. PROCEED TO FUEL SYSTEM PRESSURE TEST CHART A-7 (1 OF 3).

FUEL PRESSURE GAGE REMOVAL

1. REMOVE NEGATIVE BATTERY CABLE.
2. PURGE FUEL PRESSURE FROM FUEL PRESSURE GAGE INTO AN APPROVED CONTAINER.
3. CAREFULLY DISCONNECT FUEL PRESSURE GAGE.
4. REMOVE ADAPTER.
5. REINSTALL FUEL PIPE HOLD DOWN BOLT. TIGHTEN TO 15 N·m (11 LB. FT.).
6. RECONNECT NEGATIVE BATTERY CABLE.
7. START ENGINE *AND CHECK FOR FUEL LEAKS.

* IF ENGINE WILL NOT START, JUMPER B + TO "FP" ON CHECK CONNECTOR.

1.6L (VIN 6) ENGINE — A-CHARTS — 1992 PRIZM

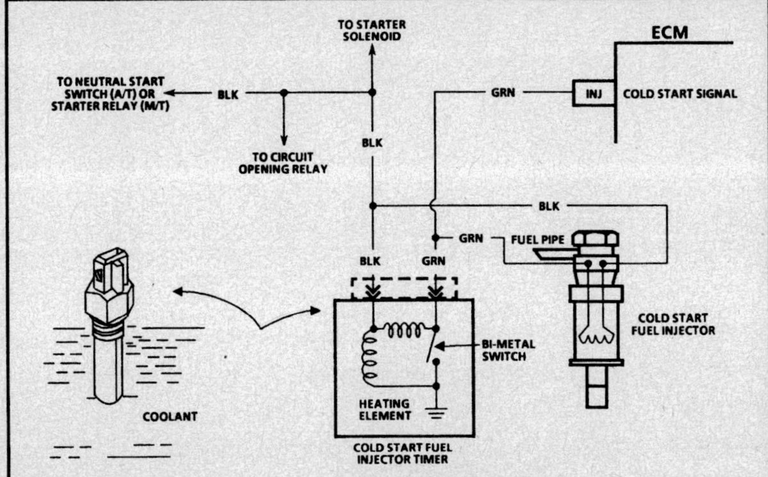

CHART A-9
COLD START FUEL INJECTOR
1.6L (VIN 6) "S" CARLINE (PORT)

Circuit Description:

The cold start fuel injector is used to provide additional fuel during the crank mode to improve cold start-ups. This circuit is important when engine coolant temperature is low because the other injectors are not pulsed "ON" long enough to provide the needed amount of fuel to start.

The circuit is activated only in the crank mode. The power is supplied directly from the starter solenoid circuit and is protected by a fuse. The system is controlled by a cold start fuel injector timer which provides a ground path for the injector during cranking when engine coolant is below 35°C (95°F).

The timer contains a bimetal material which opens at a specified coolant temperature. This bimetal is also heated by the winding in the thermal switch which allows the valve to stay "ON" for 8 seconds at -20°C (-5°F) coolant. The time the switch will stay closed varies inversely with coolant temperature. In other words, as the coolant temperature goes up, the cold start fuel injector "ON" time goes down. The ECM receives a signal when the cold start fuel injector timer switch is closed. The ECM uses this information to control the "ON" time of the other fuel injectors.

Test Description: Number(s) below refer to circled number(s) on the diagnostic chart.

1. Disconnecting the distributor connectors will disable the other injectors. The amount of pressure drop depends on the temperature of the engine.
2. This checks for power at the cold start fuel injector.
3. This checks for ground source at the cold start fuel injector.
4. This checks for power at the cold start fuel injector timer.

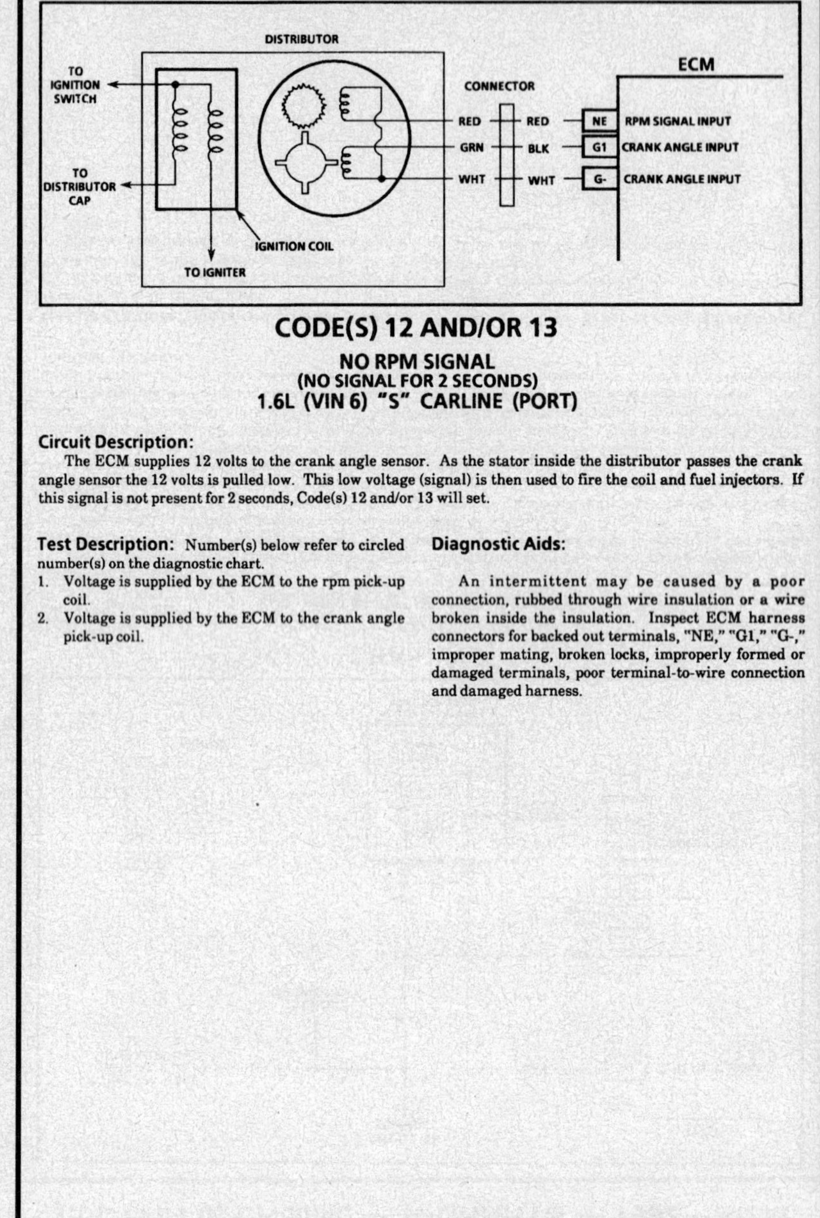

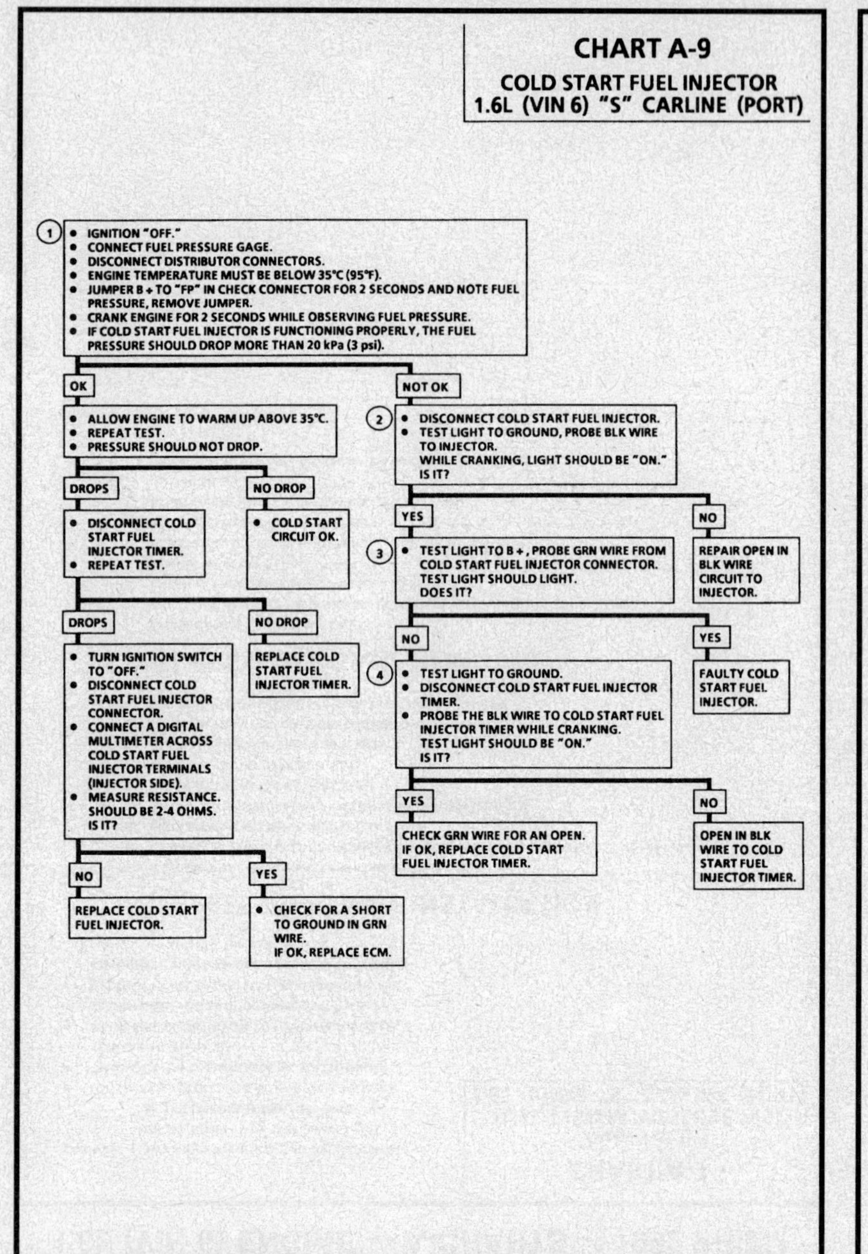

CODE(S) 12 AND/OR 13

NO RPM SIGNAL
(NO SIGNAL FOR 2 SECONDS)
1.6L (VIN 6) "S" CARLINE (PORT)

Circuit Description:
The ECM supplies 12 volts to the crank angle sensor. As the stator inside the distributor passes the crank angle sensor the 12 volts is pulled low. This low voltage (signal) is then used to fire the coil and fuel injectors. If this signal is not present for 2 seconds, Code(s) 12 and/or 13 will set.

Test Description: Number(s) below refer to circled number(s) on the diagnostic chart.
1. Voltage is supplied by the ECM to the rpm pick-up coil.
2. Voltage is supplied by the ECM to the crank angle pick-up coil.

Diagnostic Aids:
An intermittent may be caused by a poor connection, rubbed through wire insulation or a wire broken inside the insulation. Inspect ECM harness connectors for backed out terminals, "NE," "G1," "G-," improper mating, broken locks, improperly formed or damaged terminals, poor terminal-to-wire connection and damaged harness.

1.6L (VIN 6) ENGINE — DIAGNOSTIC CODE CHARTS — 1992 PRIZM

CODE(S) 12 AND/OR 13

NO RPM SIGNAL
(NO SIGNAL FOR 2 SECONDS)
1.6L (VIN 6) "S" CARLINE (PORT)

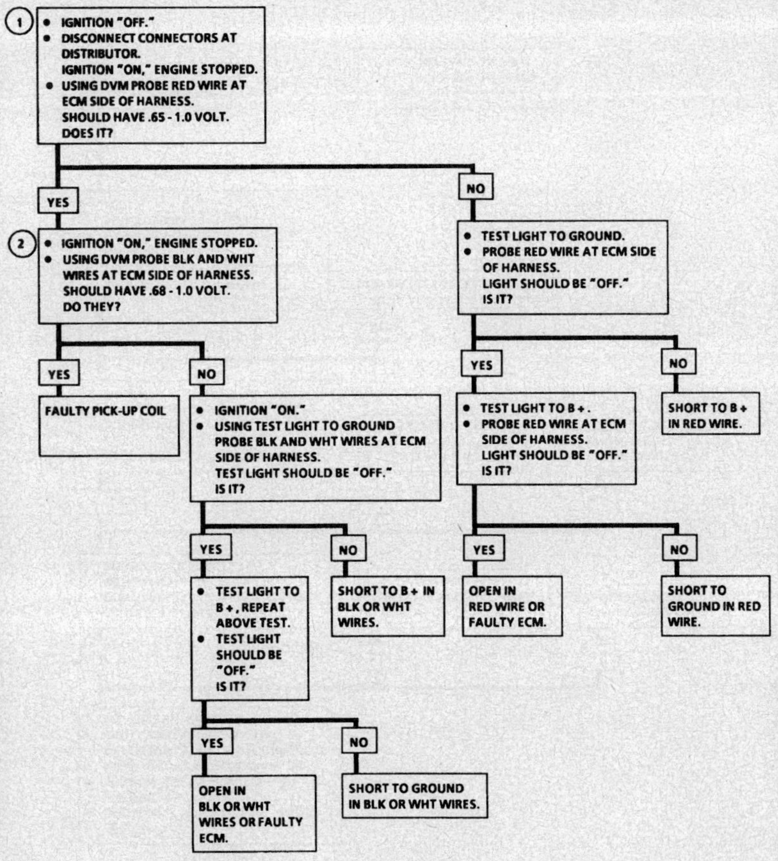

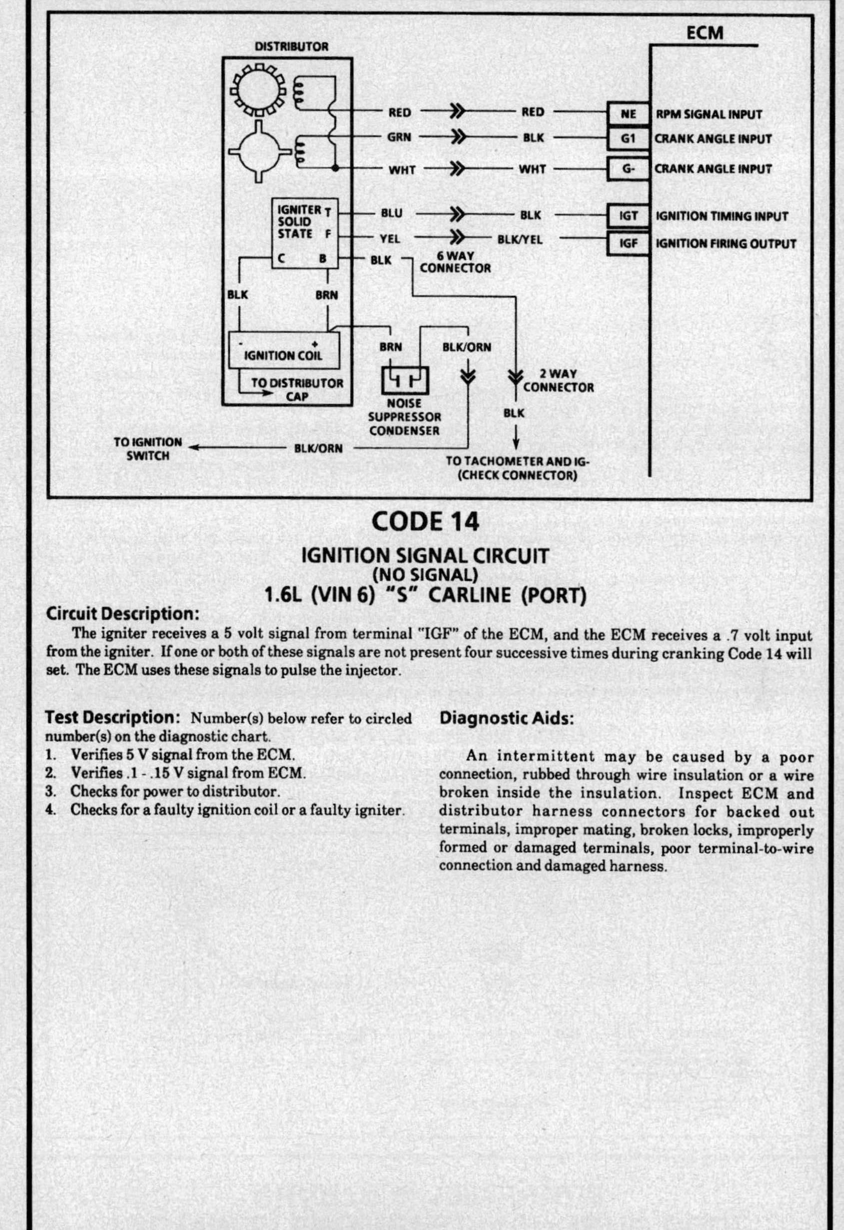

"AFTER REPAIRS," CONFIRM "CLOSED LOOP" OPERATION AND NO "CHECK ENGINE" LIGHT.

1.6L (VIN 6) ENGINE — DIAGNOSTIC CODE CHARTS — 1992 PRIZM

CODE 14

IGNITION SIGNAL CIRCUIT
(NO SIGNAL)
1.6L (VIN 6) "S" CARLINE (PORT)

Circuit Description:

The igniter receives a 5 volt signal from terminal "IGF" of the ECM, and the ECM receives a .7 volt input from the igniter. If one or both of these signals are not present four successive times during cranking Code 14 will set. The ECM uses these signals to pulse the injector.

Test Description: Number(s) below refer to circled number(s) on the diagnostic chart.

1. Verifies 5 V signal from the ECM.
2. Verifies .1 - .15 V signal from ECM.
3. Checks for power to distributor.
4. Checks for a faulty ignition coil or a faulty igniter.

Diagnostic Aids:

An intermittent may be caused by a poor connection, rubbed through wire insulation or a wire broken inside the insulation. Inspect ECM and distributor harness connectors for backed out terminals, improper mating, broken locks, improperly formed or damaged terminals, poor terminal-to-wire connection and damaged harness.

1.6L (VIN 6) ENGINE — DIAGNOSTIC CODE CHARTS — 1992 PRIZM

CODE 14

IGNITION SIGNAL CIRCUIT
(NO SIGNAL)
1.6L (VIN 6) "S" CARLINE (PORT)

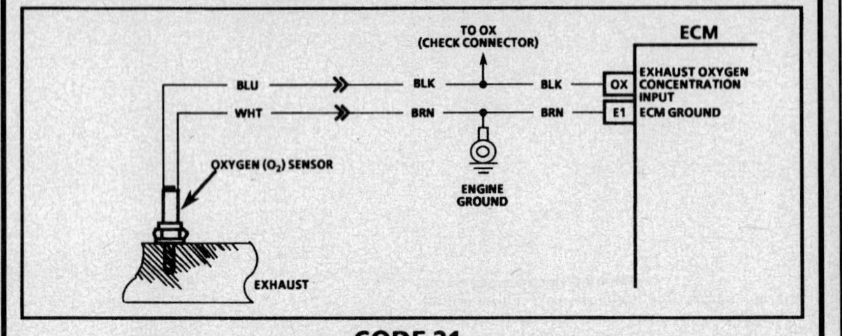

"AFTER REPAIRS," CONFIRM "CLOSED LOOP" OPERATION AND NO "CHECK ENGINE" LIGHT.

1.6L (VIN 6) ENGINE — DIAGNOSTIC CODE CHARTS — 1992 PRIZM

CODE 21

OXYGEN (O₂) SENSOR CIRCUIT
(OPEN/SHORTED CIRCUIT)
1.6L (VIN 6) "S" CARLINE (PORT)

Circuit Description:

When the O_2 sensor reaches operating temperature, it has a varying voltage from about .1 volt (exhaust is lean) to about 1.0 volt (exhaust is rich). The ECM uses this varying voltage signal to make fuel correction to maintain the optimum air fuel ratio.

The sensor is like an open circuit and produces no voltage when it is below 360°C (600°F). An open sensor circuit, or cold sensor, causes "Open Loop" operation.

Test Description: Number(s) below refer to circled number(s) on the diagnostic chart.

1. This checks for a poor connection or for a faulty ECM. Code 21 will set under the following conditions:
   - Engine at normal operating temperature.
   - At least 2 minutes have elapsed since engine start-up.
   - O_2 signal voltage is steady between 1 volt or 0 volt.
   - All above conditions are met for about 60 seconds.
2. This test determines if the O_2 sensor is the problem or if the ECM and wiring are at fault.

Diagnostic Aids:

Normal O_2 sensor voltage varies between 0 mV and 999 mV (.1 and 1.0 volt) while at normal operating temperature. Code 21 sets in one minute if sensor signal voltage remains constant between 0 volt and 1 volt.

Verify a clean, tight ground connection for BRN wire. Open BLK wire or BRN wire will result in a Code 21. If Code 21 is intermittent, refer to "Symptoms."

1.6L (VIN 6) ENGINE — DIAGNOSTIC CODE CHARTS — 1992 PRIZM

CODE 21
OXYGEN (O₂) SENSOR CIRCUIT
(OPEN/SHORTED CIRCUIT)
1.6L (VIN 6) "S" CARLINE (PORT)

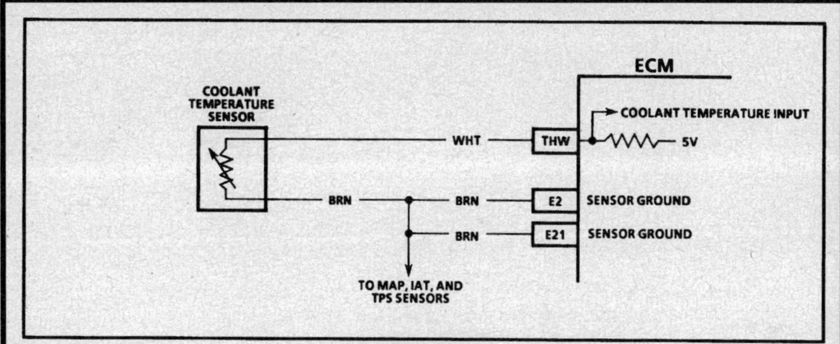

"AFTER REPAIRS," CONFIRM "CLOSED LOOP" OPERATION AND NO "CHECK ENGINE" LIGHT.

1.6L (VIN 6) ENGINE — DIAGNOSTIC CODE CHARTS — 1992 PRIZM

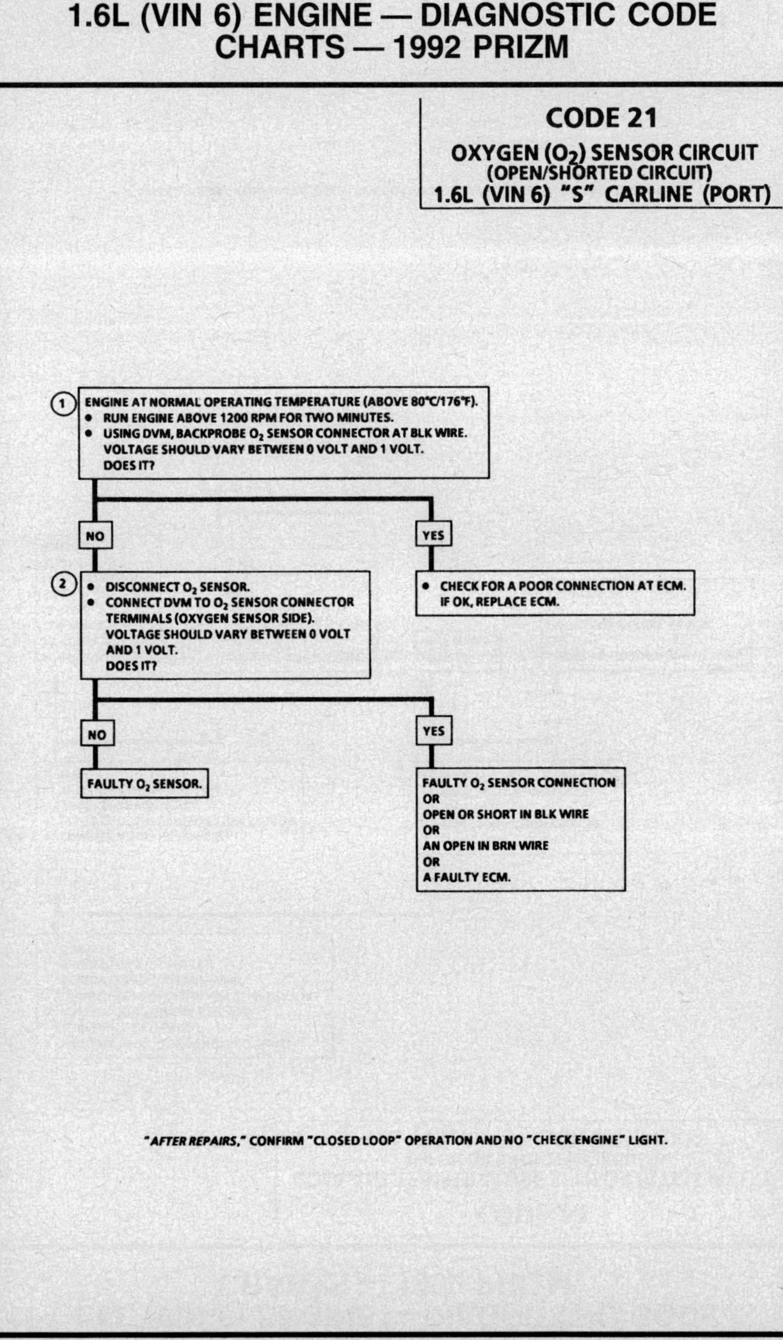

CODE 22
COOLANT TEMPERATURE SENSOR (CTS) CIRCUIT
(OPEN OR SHORTED CIRCUIT)
1.6L (VIN 6) "S" CARLINE (PORT)

Circuit Description:

The coolant temperature sensor (CTS) uses a thermistor to control the signal voltage to the ECM. The ECM applies a voltage on wire to the sensor. When the engine is cold, the sensor (thermistor) resistance is high, therefore, ECM will see high signal voltage.

As the engine warms, the sensor resistance becomes less, and the signal voltage drops. At normal engine operating temperature 80°C to 95°C (176°F to 203°F), the signal voltage will measure about 1.5 to 2.0 volts at the ECM.

Test Description: Number(s) below refer to circled number(s) on the diagnostic chart.
1. Verifies that 4-5 volts is present at the CTS connector.
2. This test checks the ground side of the CTS circuit.
3. Determines if the problem is caused by an open BRN wire or the ECM. Be sure to check the ECM connections for proper fit.
4. Determines if the circuit or ECM is the problem.

Diagnostic Aids:

After engine is started the temperature should rise steadily to about 95°C (203°F) then stabilize when thermostat opens.

A faulty connection, or an open in the WHT wire or BRN wire can result in a Code 22.

Refer to "Intermittents" in "Symptoms."

1.6L (VIN 6) ENGINE — DIAGNOSTIC CODE CHARTS — 1992 PRIZM

CODE 22
COOLANT TEMPERATURE SENSOR (CTS) CIRCUIT
(OPEN OR SHORTED CIRCUIT)
1.6L (VIN 6) "S" CARLINE (PORT)

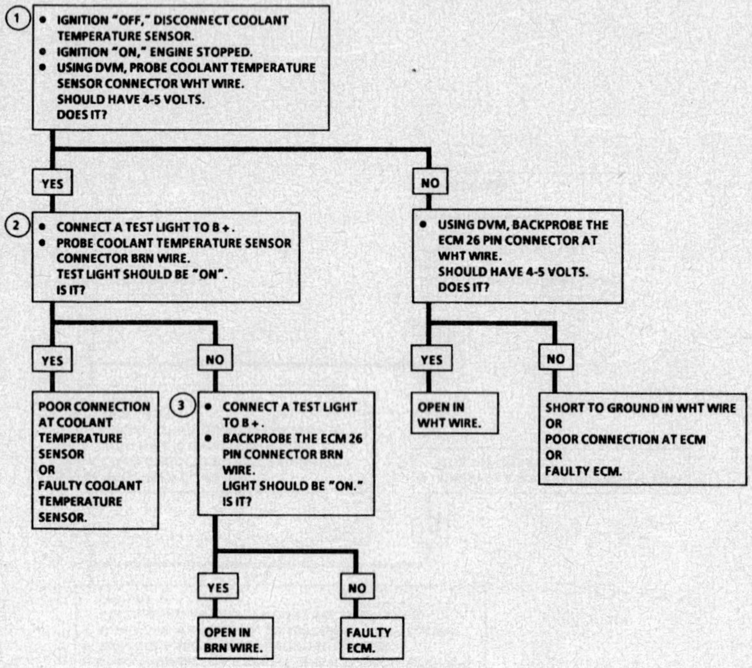

1.6L (VIN 6) ENGINE — DIAGNOSTIC CODE CHARTS — 1992 PRIZM

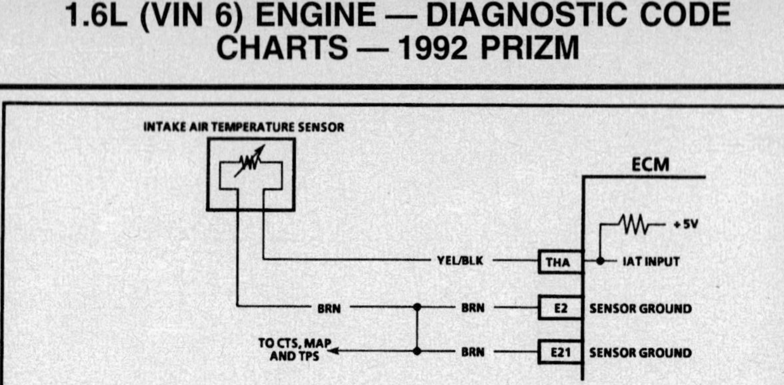

CODE 24
INTAKE AIR TEMPERATURE (IAT) SENSOR CIRCUIT
(OPEN OR SHORTED CIRCUIT)
1.6L (VIN 6) "S" CARLINE (PORT)

Circuit Description:

The intake air temperature (IAT) sensor uses a thermistor to control the signal voltage at the ECM. The ECM applies a voltage (4-6 volts) to the sensor. When the air is cold, the sensor (thermistor) resistance is high, therefore, the ECM will see a high signal voltage. If the air is warm, the sensor resistance is low, therefore, the ECM will see a low signal voltage.

Test Description: Number(s) below refer to circled number(s) on the diagnostic chart.
1. Verifies that 4-5 volts is present at the IAT sensor connector.
2. This test checks the ground side of the IAT sensor circuit.
3. This test determines if the problem is caused by an open circuit or the ECM.
4. This test determines if the YEL/BLK wire is open or shorted or if the ECM is defective.

Diagnostic Aids:

An intermittent may be caused by a poor connection, rubbed through wire insulation or a wire broken inside the insulation. Inspect ECM harness connectors for backed out terminals "THA," "E2" or "E21," improper mating, broken locks, improperly formed or damaged terminals, poor terminal-to-wire connection and damaged harness.

1.6L (VIN 6) ENGINE — DIAGNOSTIC CODE CHARTS — 1992 PRIZM

CODE 24
INTAKE AIR TEMPERATURE (IAT) SENSOR CIRCUIT
(OPEN OR SHORTED CIRCUIT)
1.6L (VIN 6) "S" CARLINE (PORT)

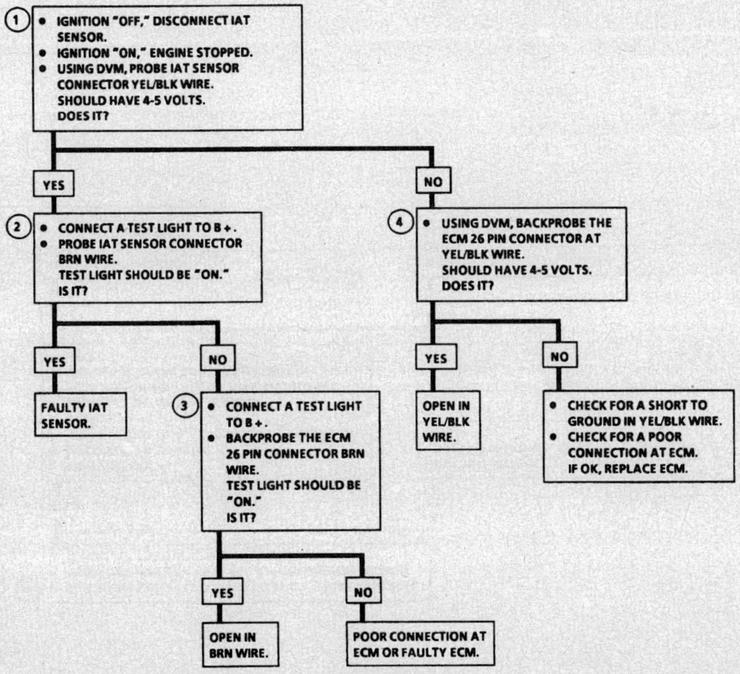

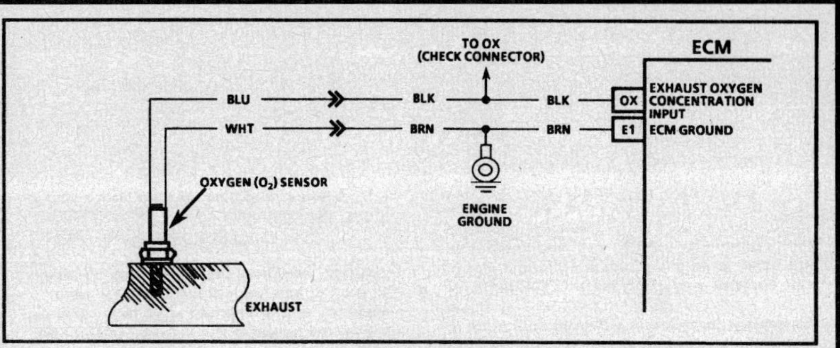

"AFTER REPAIRS," CONFIRM "CLOSED LOOP" OPERATION AND NO "CHECK ENGINE" LIGHT.

1.6L (VIN 6) ENGINE — DIAGNOSTIC CODE CHARTS — 1992 PRIZM

CODE 25
OXYGEN (O₂) SENSOR CIRCUIT
(LEAN EXHAUST INDICATED)
1.6L (VIN 6) "S" CARLINE (PORT)

Circuit Description:

When the O_2 sensor reaches operating temperature, it varies voltage from about .1 volt (exhaust is lean) to about .9 volt (exhaust is rich).

The sensor is like an open circuit and produces no voltage when it is below 360°C (600°F). An open sensor circuit, or cold sensor, causes "Open Loop" operation.

Test Description: Number(s) below refer to circled number(s) on the diagnostic chart.

1. This checks for an intermittent Code 25. Code 25 is set when the O_2 sensor signal voltage on BLK wire remains below .2 volt for 60 seconds or more and the system is operating in "Closed Loop."

Diagnostic Aids:

- O_2 Sensor Wire. Sensor pigtail may be mispositioned and contacting the exhaust manifold.
  Check for ground in wires between connector and sensor.

- Fuel Contamination. Water, even in small amounts, near the in-tank fuel pump inlet can be delivered to the injector. The water causes a lean exhaust and can set a Code 25.

- Fuel Pressure. System will be lean if fuel pressure is too low. It may be necessary to monitor fuel pressure while driving the car at various road speeds and/or loads to confirm. See "Fuel System Diagnosis," CHART A-7.

- Exhaust Leaks. If there is an exhaust leak, the engine can cause outside air to be pulled into the exhaust and past the sensor. Vacuum or crankcase leaks can cause a lean condition.

- If Code 25 is intermittent, refer to "Symptoms."

1.6L (VIN 6) ENGINE — DIAGNOSTIC CODE CHARTS — 1992 PRIZM

CODE 25

OXYGEN (O_2) SENSOR CIRCUIT
(LEAN EXHAUST INDICATED)
1.6L (VIN 6) "S" CARLINE (PORT)

(1)
- RUN WARM ENGINE (75°C/167°F TO 95°C/203°F) AT 1200 RPM.
- USING DVM, BACKPROBE O_2 SENSOR CONNECTOR BLK WIRE. IS VOLTAGE FIXED BELOW .2 VOLT (200 mV)?

YES
- IGNITION "OFF."
- DISCONNECT O_2 SENSOR.
- CONNECT DVM TO O_2 SENSOR SIGNAL WIRE AT THE SENSOR.
- IGNITION "ON," ENGINE RUNNING. NOTE SENSOR VOLTAGE ON DVM, SHOULD BE VARYING BETWEEN 0 VOLT AND 1 VOLT. IS IT?

NO
CODE 25 IS INTERMITTENT. IF NO ADDITIONAL CODES WERE STORED, REFER TO "DIAGNOSTIC AIDS"

NO
REFER TO "DIAGNOSTIC AIDS" IF ALL CHECKS ARE OK, REPLACE OXYGEN SENSOR.

YES
BLK WIRE SHORTED TO GROUND OR FAULTY ECM.

"AFTER REPAIRS," CONFIRM "CLOSED LOOP" OPERATION AND NO "CHECK ENGINE" LIGHT.

1.6L (VIN 6) ENGINE — DIAGNOSTIC CODE CHARTS — 1992 PRIZM

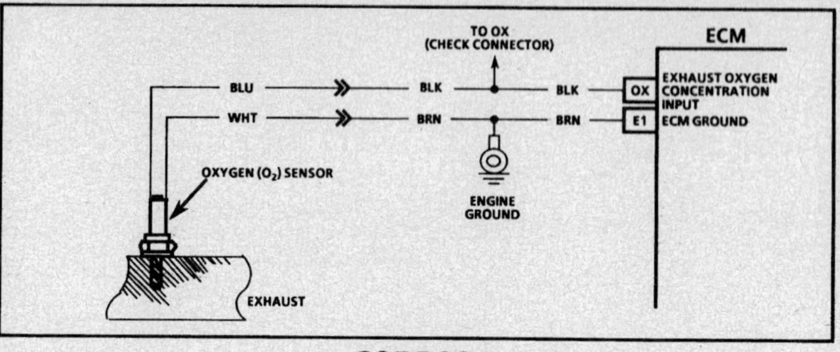

CODE 26

OXYGEN (O_2) SENSOR CIRCUIT
(RICH EXHAUST INDICATED)
1.6L (VIN 6) "S" CARLINE (PORT)

Circuit Description:

When the O_2 sensor reaches operating temperature, it varies this voltage from about .1 volt (exhaust is lean) to about 1.0 volt (exhaust is rich).

The sensor is like an open circuit and produces no voltage when it is below 360°C (600°F).

Test Description: Number(s) below refer to circled number(s) on the diagnostic chart.

1. This checks for an intermittent Code 26. Code 26 is set when the O_2 sensor signal voltage on BLK wire remains above .7 volt under the following conditions:
   - 30 seconds or more.
   - System is operating in normal operating temperature 60C°(176°F).
   - Engine run time after start is 1 minute or more.
   - Throttle angle is between 3% and 45%.

Diagnostic Aids:

Code 26, or rich exhaust, is most likely caused by one of the following:
- Fuel Pressure. System will go rich, if pressure is too high. The ECM can compensate for some increase. However, if it gets too high, a Code 26 will be set. See "Fuel System Diagnosis," CHART A-7
- Leaking Injector. See CHART A-7.
- Canister Purge. Check for fuel saturation. If full of fuel, check canister control and hoses.

- MAP Sensor. An output that causes the ECM to sense a higher than normal manifold pressure (low vacuum) can cause the system to go rich. Disconnecting the MAP sensor will allow the ECM to set a fixed value for the MAP sensor. Substitute a different MAP sensor if the rich condition is gone, while the sensor is disconnected.
- TS. An intermittent throttle switch output will cause the system to operate richly due to a false indication of the engine accelerating.
- O_2 Sensor Contamination. Inspect oxygen sensor for silicone contamination from fuel, or use of improper RTV sealant. The sensor may have a white, powdery coating and result in a high but false signal voltage (rich exhaust indication). The ECM will then reduce the amount of fuel delivered to the engine causing a severe surge driveability problem.
- EGR Valve (California Only). EGR sticking open at idle is usually accompanied by a rough idle and/or stall condition. If Code 26 is intermittent, refer to "Symptoms."

1.6L (VIN 6) ENGINE — DIAGNOSTIC CODE CHARTS — 1992 PRIZM

CODE 26

OXYGEN (O₂) SENSOR CIRCUIT
(RICH EXHAUST INDICATED)
1.6L (VIN 6) "S" CARLINE (PORT)

① • RUN WARM ENGINE (75°C/167°F TO 95°C/203°F) AT 1200 RPM.
• USING DVM, BACKPROBE O₂ SENSOR CONNECTOR BLK WIRE.
IS VOLTAGE FIXED ABOVE .75 VOLT (750mV)?

YES
• IGNITION "OFF."
• DISCONNECT O₂ SENSOR.
• CONNECT DVM TO O₂ SENSOR SIGNAL WIRE AT THE SENSOR.
• IGNITION "ON," ENGINE RUNNING.
NOTE SENSOR VOLTAGE ON DVM, SHOULD BE VARYING BETWEEN 0 VOLT AND 1 VOLT.
IS IT?

NO
CODE 26 IS INTERMITTENT. IF NO ADDITIONAL CODES WERE STORED, REFER TO "DIAGNOSTIC AIDS"

NO
REFER TO "DIAGNOSTIC AIDS."
IF ALL CHECKS ARE OK, REPLACE OXYGEN SENSOR.

YES
SHORT TO VOLTAGE IN BLK WIRE TO ECM PIN "OX"
OR
FAULTY ECM.

"AFTER REPAIRS," CONFIRM "CLOSED LOOP" OPERATION AND NO "CHECK ENGINE" LIGHT.

1.6L (VIN 6) ENGINE — DIAGNOSTIC CODE CHARTS — 1992 PRIZM

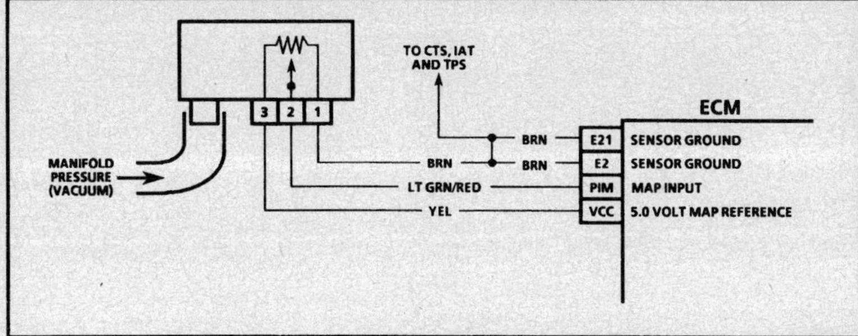

CODE 31

MANIFOLD ABSOLUTE PRESSURE (MAP) SENSOR
(OPEN OR SHORTED CIRCUIT)
1.6L (VIN 6) "S" CARLINE (PORT)

Circuit Description:
The manifold absolute pressure (MAP) sensor responds to changes in manifold pressure (vacuum). The ECM receives this information as a signal voltage that will vary from about 1 to 1.5 volts at closed throttle idle, to 4 - 4.5 volts at wide open throttle.

If the MAP sensor fails, the ECM will substitute a fixed MAP value and use the throttle position switch (TPS) to control fuel delivery.

Test Description: Number(s) below refer to circled number(s) on the diagnostic chart.
1. This step verifies that 5 volts reference is available from the ECM.
2. This step verifies the ECM is grounding the circuit.

Diagnostic Aids:

An intermittent may be caused by a poor connection, rubbed through wire insulation or a wire broken inside the insulation. Inspect ECM harness connectors for backed out terminals, improper mating, broken locks, improperly formed or damaged terminals, poor terminal-to-wire connection and damaged harness.

1.6L (VIN 6) ENGINE — DIAGNOSTIC CODE CHARTS — 1992 PRIZM

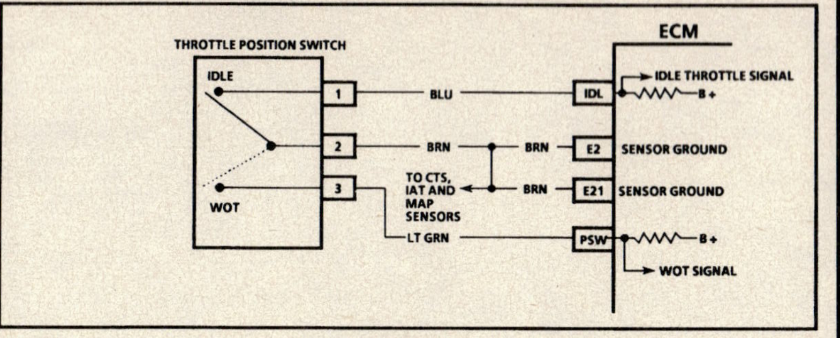

1.6L (VIN 6) ENGINE — DIAGNOSTIC CODE CHARTS — 1992 PRIZM

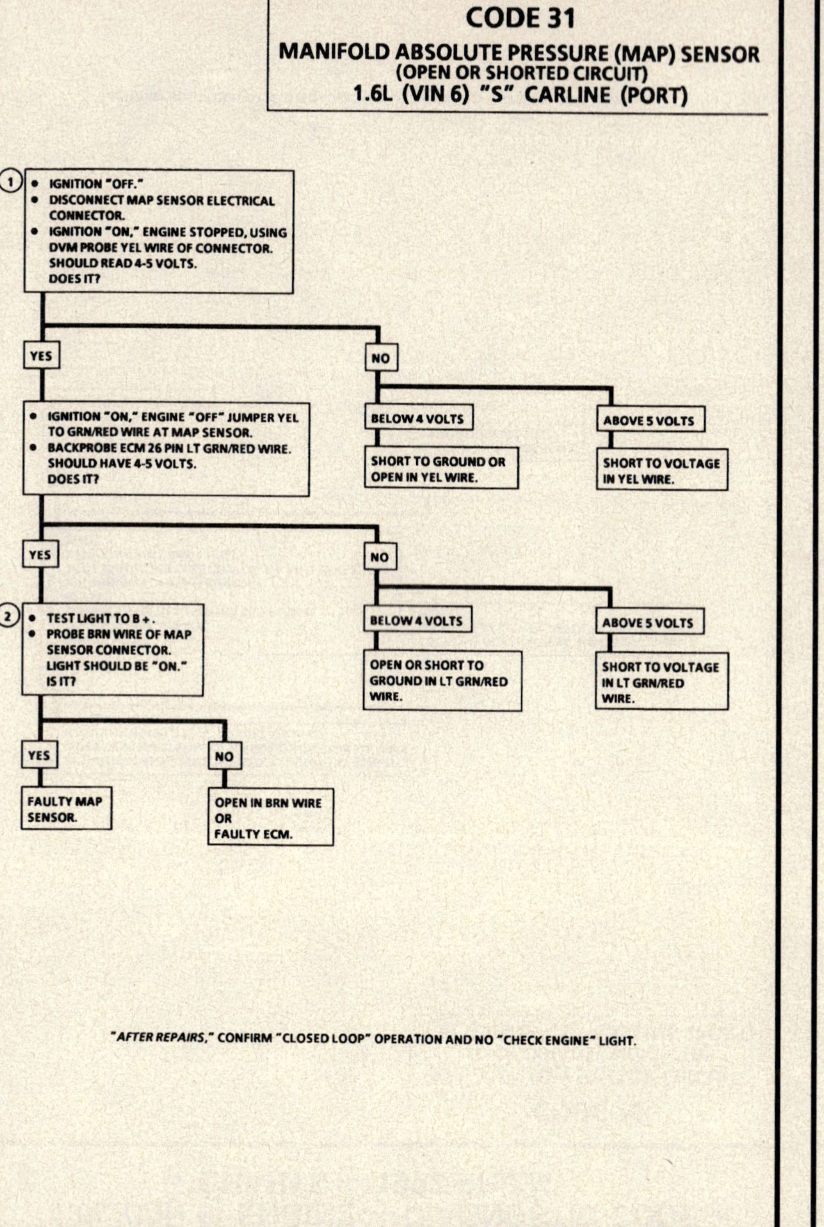

CODE 41
THROTTLE POSITION SWITCH CIRCUIT
1.6L (VIN 6) "S" CARLINE (PORT)

Circuit Description:

The ECM supplies B+ to a throttle switch on the throttle body. When engine is idling the ECM senses a low voltage signal at the idle signal input, slight throttle the ECM will sense a high voltage signal. At wide open throttle the signal voltage, at the WOT signal input is low. Code 41 will set if the ECM sees a low signal all the time or a high voltage signal.

Test Description: Number(s) below refer to circled number(s) on the diagnostic chart.

1. The ECM supplies battery voltage to the throttle switch. Poor connection or a grounded circuit will indicate a faulty signal.
2. This will check for a misadjusted throttle switch.

Diagnostic Aids:

An intermittent may be caused by a poor connection, rubbed through wire insulation or a wire broken inside the insulation. Inspect ECM harness connectors for backed out terminals, improper mating, broken locks, improperly formed or damaged terminals, poor terminal-to-wire connection and damaged harness. Refer to "Intermittents," in "Symptoms," for additional information.

1.6L (VIN 6) ENGINE — DIAGNOSTIC CODE CHARTS — 1992 PRIZM

CODE 41
THROTTLE POSITION SWITCH CIRCUIT
1.6L (VIN 6) "S" CARLINE (PORT)

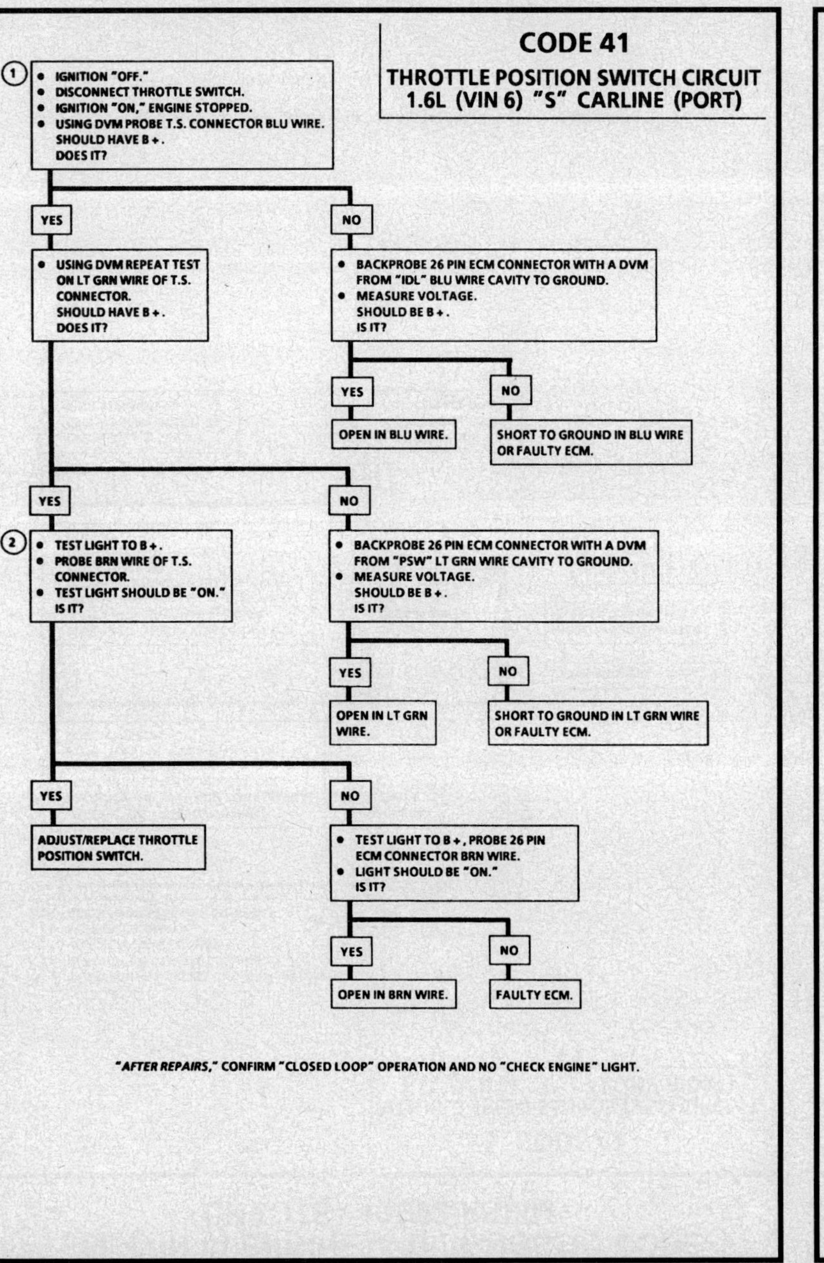

1.6L (VIN 6) ENGINE — DIAGNOSTIC CODE CHARTS — 1992 PRIZM

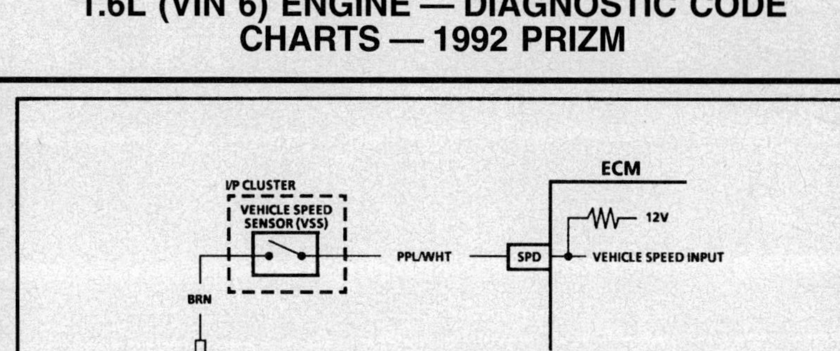

CODE 42
VEHICLE SPEED SENSOR (VSS) CIRCUIT
1.6L (VIN 6) "S" CARLINE (PORT)

Circuit Description:
The ECM provides a 12 volt signal to the vehicle speed sensor (VSS). The VSS will toggle the signal low indicating the vehicle is moving. Code 42 will set if the signal line stays "High" or "Low" for 4 seconds or more.

Test Description: Number(s) below refer to circled number(s) on the diagnostic chart.
1. If speedometer does not work, it could be the cause of Code 42.
2. Checks to see if the vehicle speed sensor on the back of the instrument panel will toggle the 12 volts supplied by the ECM.

Diagnostic Aids:
An intermittent may be caused by a poor connection, rubbed through wire insulation or a wire broken inside the insulation. Inspect ECM harness connectors for backed out terminals, improper mating, broken locks, improperly formed or damaged terminals, poor terminal-to-wire connection and damaged harness. If no trouble found and Code 42 resets, replace ECM.

1.6L (VIN 6) ENGINE — DIAGNOSTIC CODE CHARTS — 1992 PRIZM

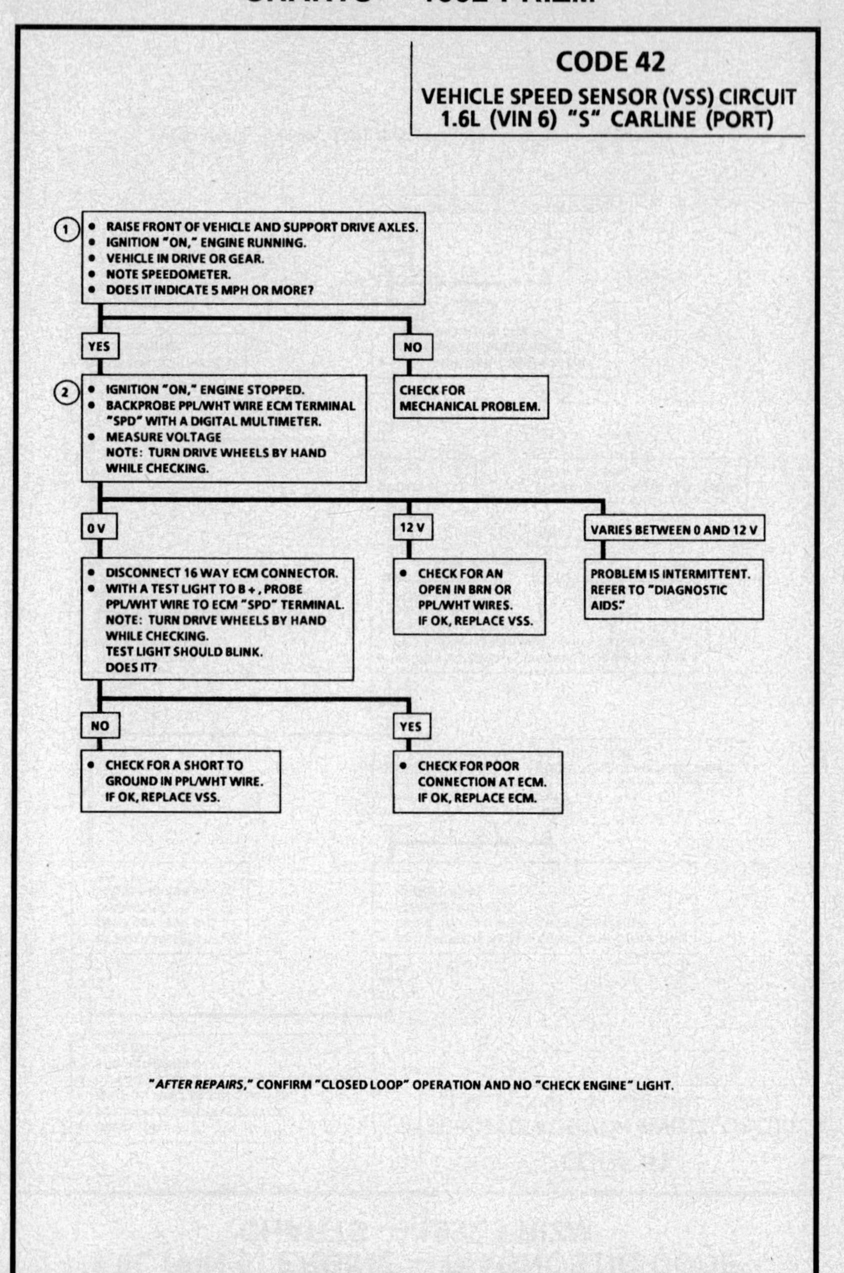

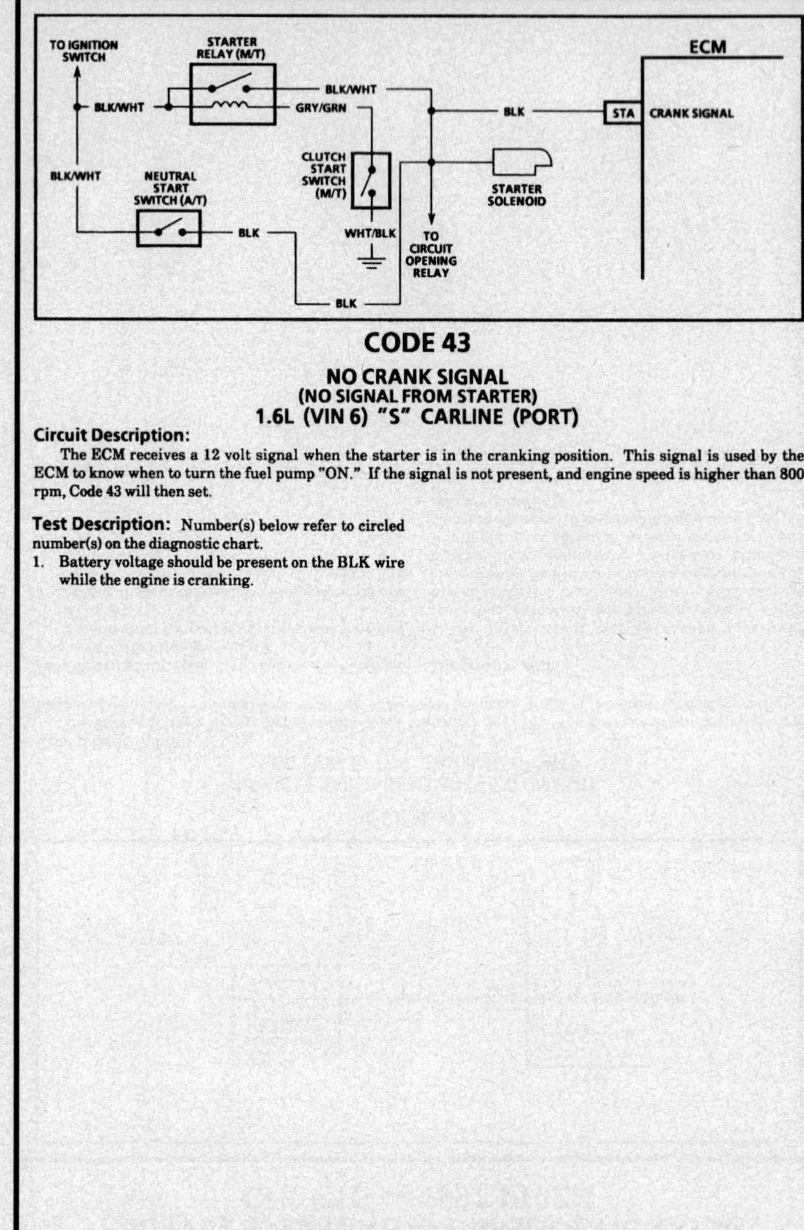

CODE 43

NO CRANK SIGNAL
(NO SIGNAL FROM STARTER)
1.6L (VIN 6) "S" CARLINE (PORT)

Circuit Description:
The ECM receives a 12 volt signal when the starter is in the cranking position. This signal is used by the ECM to know when to turn the fuel pump "ON." If the signal is not present, and engine speed is higher than 800 rpm, Code 43 will then set.

Test Description: Number(s) below refer to circled number(s) on the diagnostic chart.
1. Battery voltage should be present on the BLK wire while the engine is cranking.

1.6L (VIN 6) ENGINE — DIAGNOSTIC CODE CHARTS — 1992 PRIZM

CODE 43

NO CRANK SIGNAL
(NO SIGNAL FROM STARTER)
1.6L (VIN 6) "S" CARLINE (PORT)

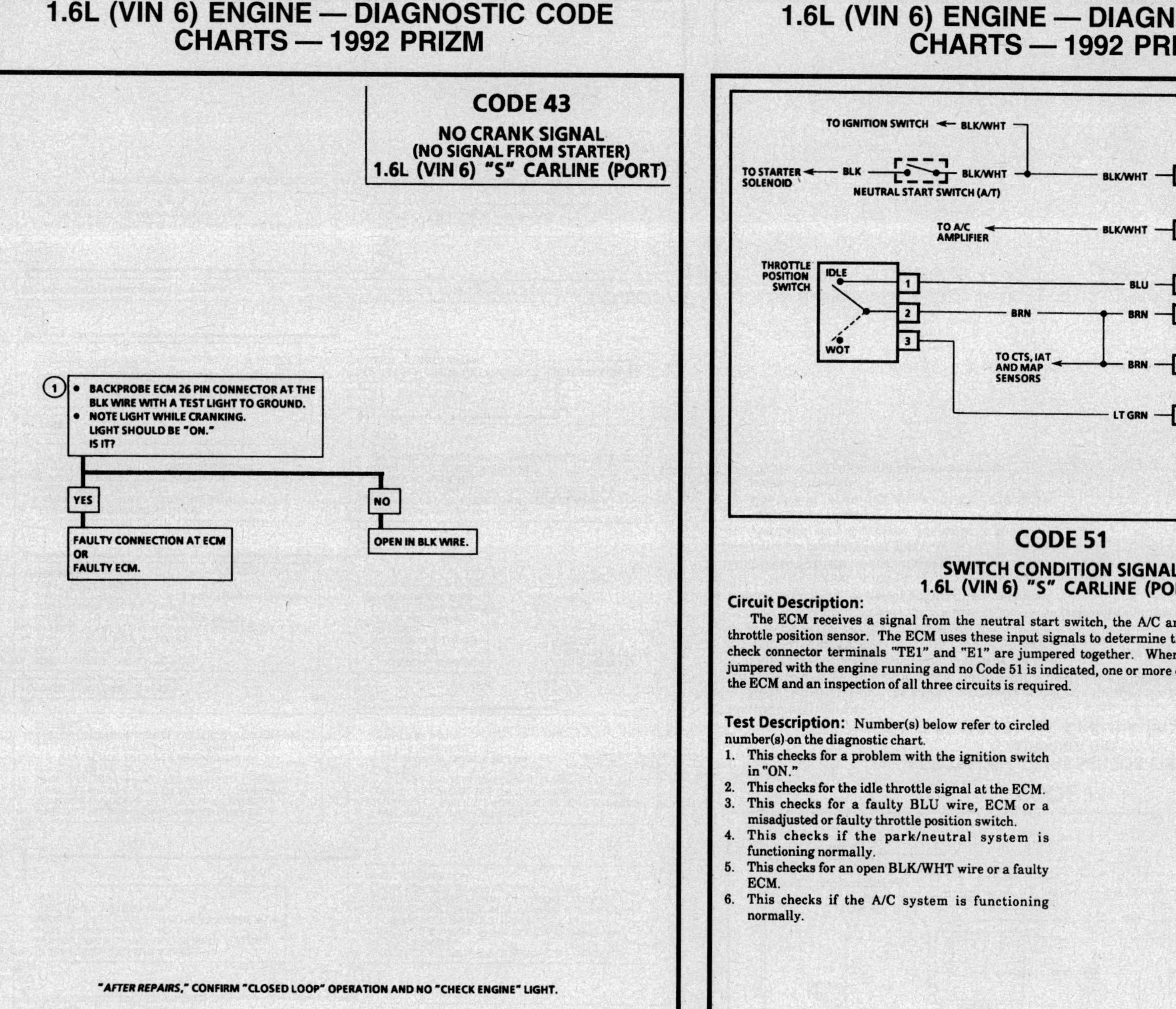

"AFTER REPAIRS," CONFIRM "CLOSED LOOP" OPERATION AND NO "CHECK ENGINE" LIGHT.

1.6L (VIN 6) ENGINE — DIAGNOSTIC CODE CHARTS — 1992 PRIZM

CODE 51

SWITCH CONDITION SIGNAL
1.6L (VIN 6) "S" CARLINE (PORT)

Circuit Description:
The ECM receives a signal from the neutral start switch, the A/C amplifier and the idle contacts of the throttle position sensor. The ECM uses these input signals to determine the condition of the circuits when the check connector terminals "TE1" and "E1" are jumpered together. When the check connecter terminals are jumpered with the engine running and no Code 51 is indicated, one or more of the input signals is not received by the ECM and an inspection of all three circuits is required.

Test Description: Number(s) below refer to circled number(s) on the diagnostic chart.
1. This checks for a problem with the ignition switch in "ON."
2. This checks for the idle throttle signal at the ECM.
3. This checks for a faulty BLU wire, ECM or a misadjusted or faulty throttle position switch.
4. This checks if the park/neutral system is functioning normally.
5. This checks for an open BLK/WHT wire or a faulty ECM.
6. This checks if the A/C system is functioning normally.

1.6L (VIN 6) ENGINE — DIAGNOSTIC CODE CHARTS — 1992 PRIZM

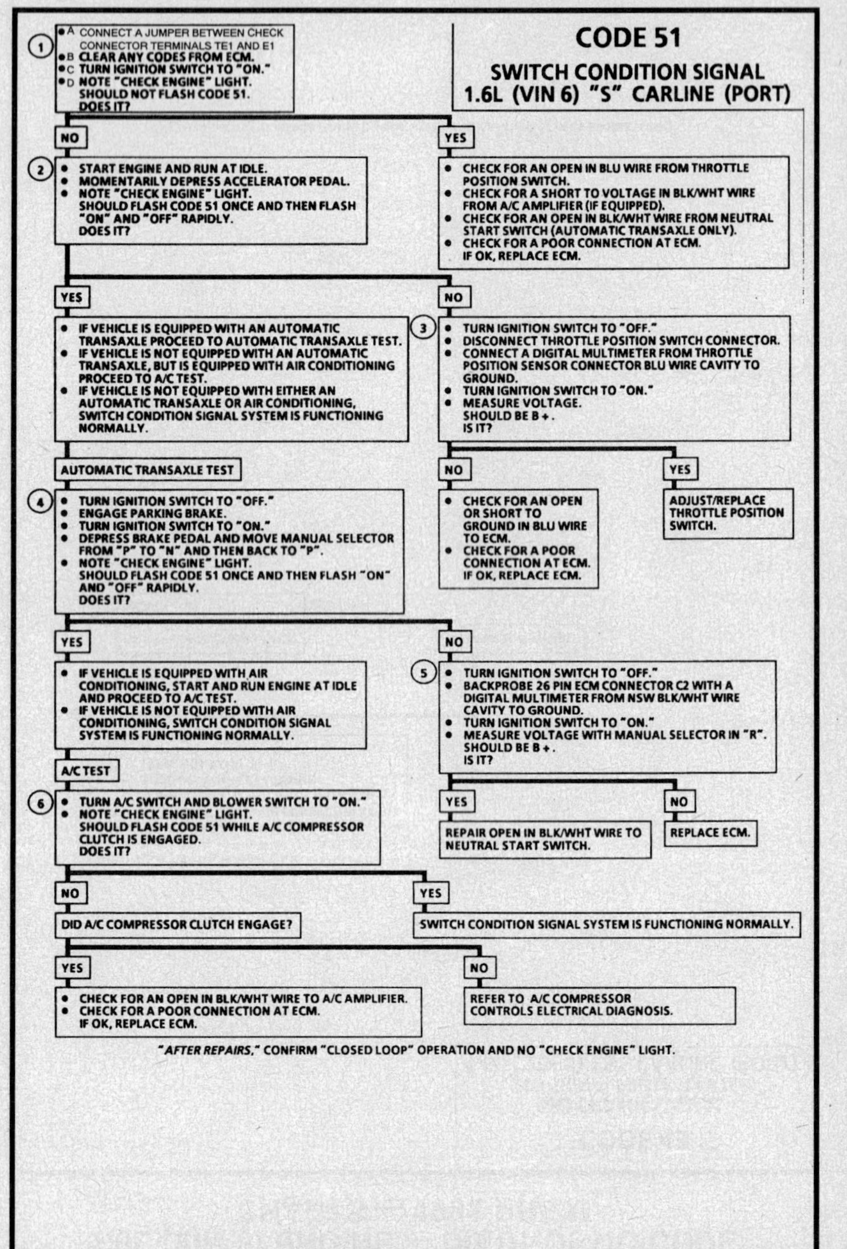

1.6L (VIN 6) ENGINE — DIAGNOSTIC CODE CHARTS — 1992 PRIZM

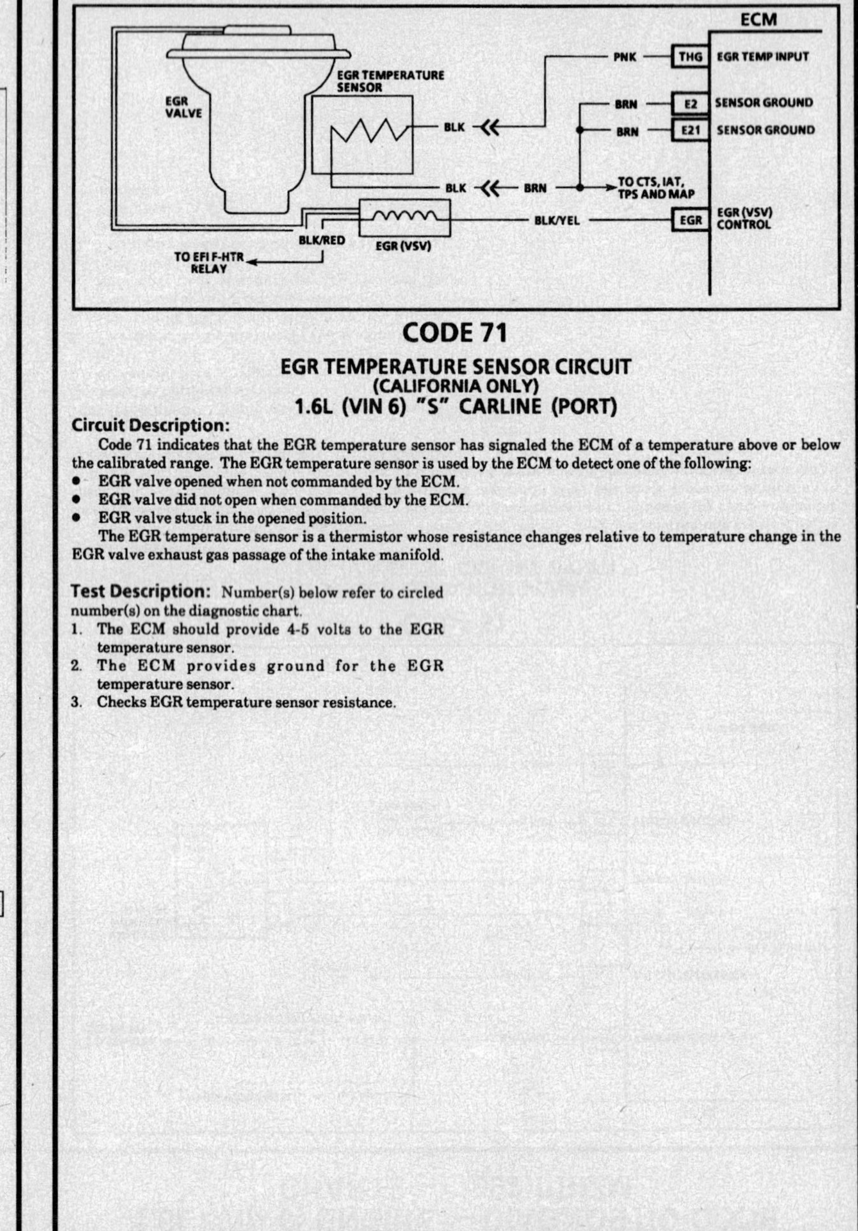

CODE 71
EGR TEMPERATURE SENSOR CIRCUIT
(CALIFORNIA ONLY)
1.6L (VIN 6) "S" CARLINE (PORT)

Circuit Description:

Code 71 indicates that the EGR temperature sensor has signaled the ECM of a temperature above or below the calibrated range. The EGR temperature sensor is used by the ECM to detect one of the following:

- EGR valve opened when not commanded by the ECM.
- EGR valve did not open when commanded by the ECM.
- EGR valve stuck in the opened position.

The EGR temperature sensor is a thermistor whose resistance changes relative to temperature change in the EGR valve exhaust gas passage of the intake manifold.

Test Description: Number(s) below refer to circled number(s) on the diagnostic chart.

1. The ECM should provide 4-5 volts to the EGR temperature sensor.
2. The ECM provides ground for the EGR temperature sensor.
3. Checks EGR temperature sensor resistance.

1.6L (VIN 6) ENGINE — DIAGNOSTIC CODE CHARTS — 1992 PRIZM

1.6L (VIN 6) ENGINE — C-CHARTS — 1992 PRIZM

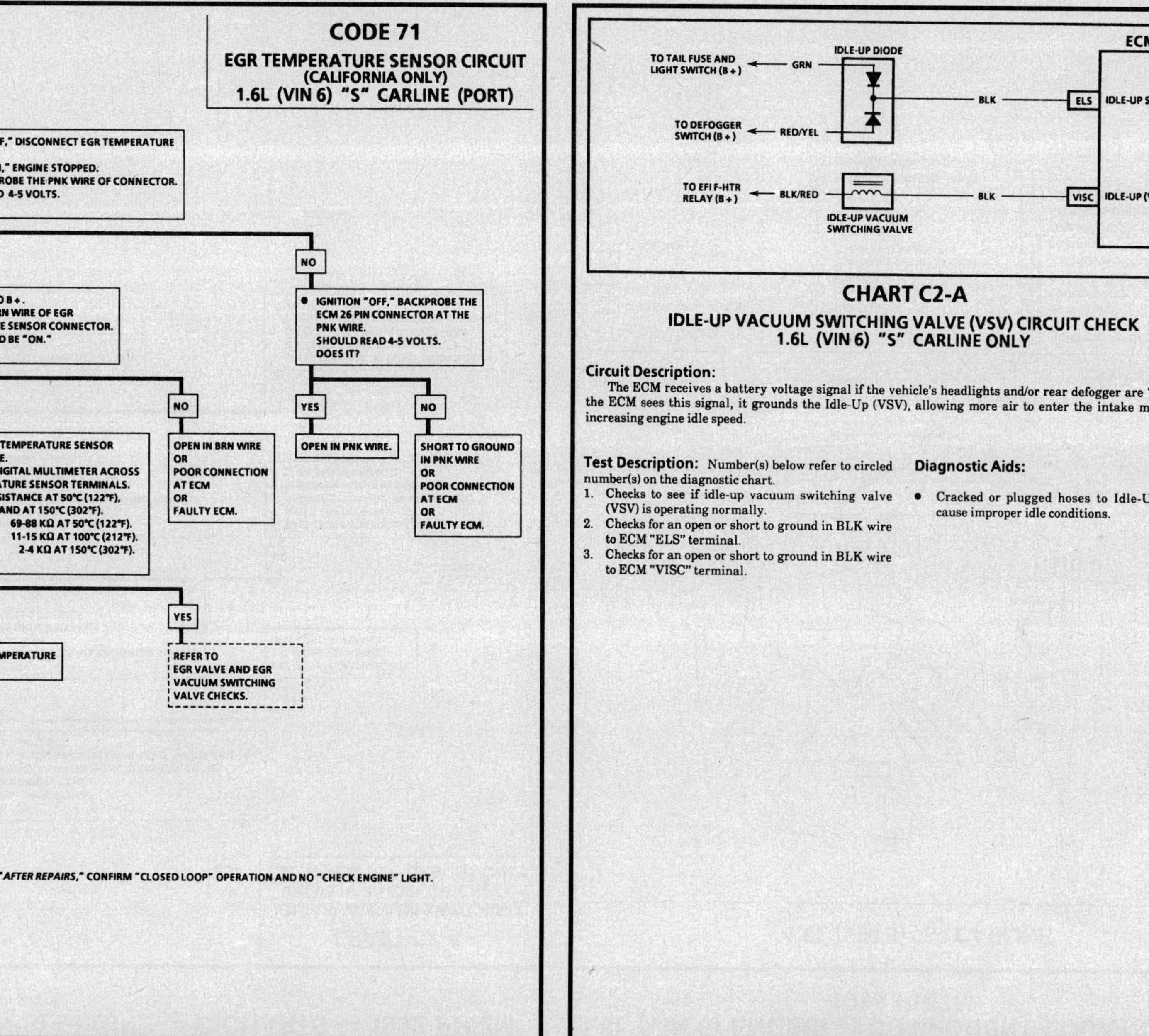

CHART C2-A

IDLE-UP VACUUM SWITCHING VALVE (VSV) CIRCUIT CHECK
1.6L (VIN 6) "S" CARLINE ONLY

Circuit Description:

The ECM receives a battery voltage signal if the vehicle's headlights and/or rear defogger are "ON." When the ECM sees this signal, it grounds the Idle-Up (VSV), allowing more air to enter the intake manifold, thus increasing engine idle speed.

Test Description: Number(s) below refer to circled number(s) on the diagnostic chart.

1. Checks to see if idle-up vacuum switching valve (VSV) is operating normally.
2. Checks for an open or short to ground in BLK wire to ECM "ELS" terminal.
3. Checks for an open or short to ground in BLK wire to ECM "VISC" terminal.

Diagnostic Aids:

• Cracked or plugged hoses to Idle-Up VSV may cause improper idle conditions.

CHART C2-A

IDLE-UP VACUUM SWITCHING VALVE (VSV) CIRCUIT CHECK 1.6L (VIN 6) "S" CARLINE ONLY

1
- LIGHT SWITCH "OFF."
- REAR DEFOGGER SWITCH "OFF."
- START AND RUN ENGINE AT IDLE.
- TURN LIGHT SWITCH AND REAR DEFOGGER SWITCH TO "ON."
  DOES ENGINE IDLE SPEED INCREASE WITH LIGHT SWITCH AND REAR DEFOGGER SWITCH "ON"?

NO →

2
- SHUT ENGINE "OFF."
- WITH A TEST LIGHT TO GROUND, BACKPROBE ECM BLK "ELS" TERMINAL WIRE.
- TURN IGNITION SWITCH TO "ON."
  TEST LIGHT SHOULD BE "ON."
  IS IT?

YES → IDLE-UP VACUUM SWITCHING VALVE (VSV) SYSTEM IS OPERATING NORMALLY.

YES →

3
- TURN IGNITION SWITCH TO "OFF."
- DISCONNECT 26 PIN ECM CONNECTOR.
- WITH A DVM TO GROUND, PROBE BLK "VISC" TERMINAL WIRE.
- TURN IGNITION SWITCH TO "ON."
- MEASURE VOLTAGE
  SHOULD BE B+.
  IS IT?

NO →
- CHECK FOR POOR CONNECTION AT ECM.
- CHECK FOR OPEN OR SHORT TO GROUND IN BLK "ELS" TERMINAL WIRE.
- CHECK FOR FAULTY IDLE-UP DIODE.
  IF OK, REPLACE ECM.

YES →
- CHECK FOR A POOR CONNECTION AT ECM. IF OK, REPLACE ECM.

NO →
- CHECK FOR OPEN OR SHORT TO GROUND IN BLK "VISC" TERMINAL WIRE.
- CHECK FOR AN OPEN IN BLK/RED WIRE. IF OK, REPLACE IDLE-UP VACUUM SWITCHING VALVE.

1.6L (VIN 6) "S" CARLINE

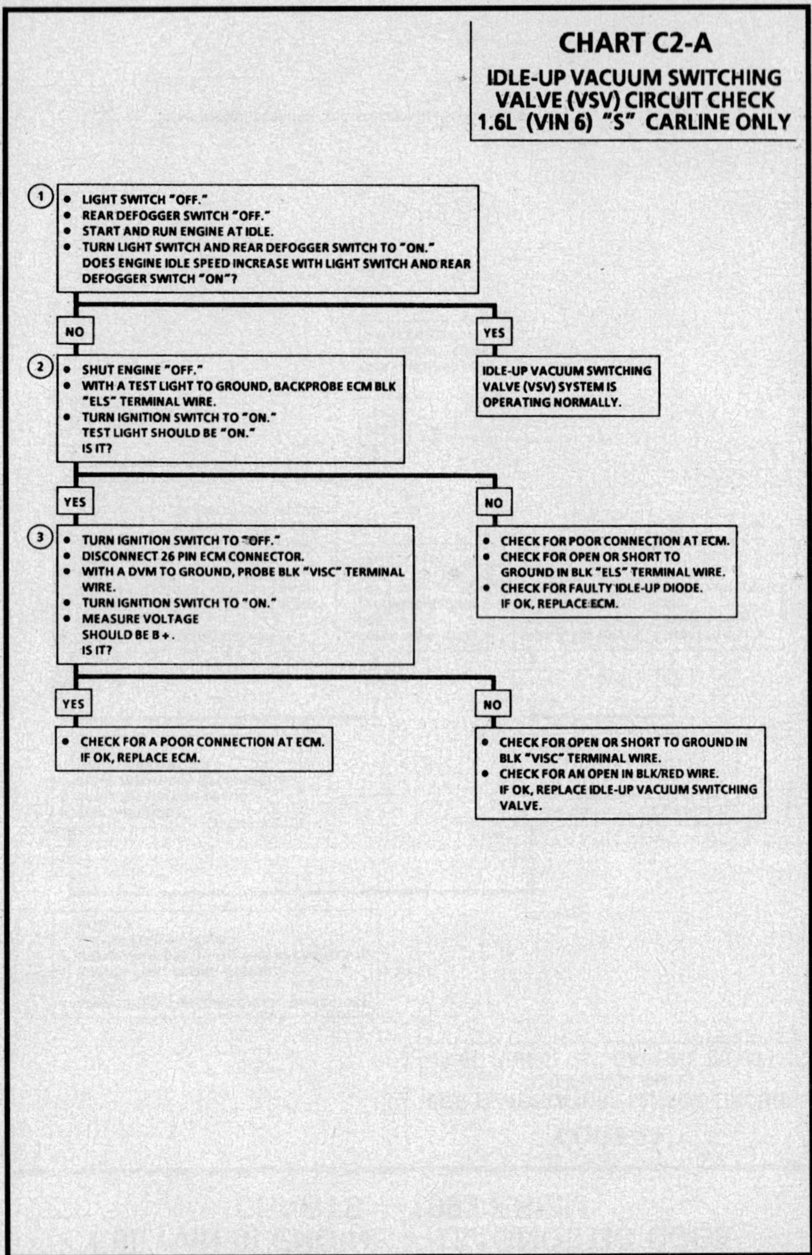

1.6L (VIN 6) ENGINE — ECM WIRING DIAGRAMS — 1993 PRIZM
ENGINE CONTROL MODULE (ECM)
ELECTRICAL LOAD IDLE-UP SIGNAL
A/C ON SIGNAL
A/C CUTOUT CONTROL
A12
A10
A6
AUDIO ALARM MODULE
IGNITION POWER
JUNCTION BLOCK 1
JUNCTION BLOCK 3
ELECTRICAL LOAD IDLE-UP DIODE
REAR DEFOGGER SWITCH
A/C AMPLIFIER
RED/BLU
BLK
YEL/BLU
RED/YEL
C207
JUNCTION CONNECTOR 1
G200
WHT/BLK
IGN
A/C ON OUTPUT CONTROL
A/C CUTOUT SIGNAL
RED/WHT
GAUGE FUSE 10 AMP
DEFOGGER RELAY
DEF. I/UP FUSE 7.5 AMP
MOMENTARY SWITCH
TIMER ENABLE/DISABLE SIGNAL
SOLID STATE
GROUND
WHT/BLK
JUNCTION BLOCK 3
JUNCTION CONNECTOR 3
TO IGNITION SWITCH (B+)
BLK/RED
TO DEFOGGER
TO DEFOGGER INDICATOR
BLK
RED/BLU
BLK/BLU
GRN
DEFOGGER RELAY CONTROL
IGNITION POWER
COMBINATION SWITCH
WHT/BLK
WHT — TO B+
FUSES
TAIL LAMP RELAY
DEF FUSE 30 AMP
FUSE (NOT USED)
TAIL FUSE 15 AMP
EXTERIOR LIGHTS
JUNCTION CONNECTOR 2
LT GRN
LIGHT/TURN SIGNAL SWITCH
OFF
ON
PARK
HEAD
WHT
WHT/BLK

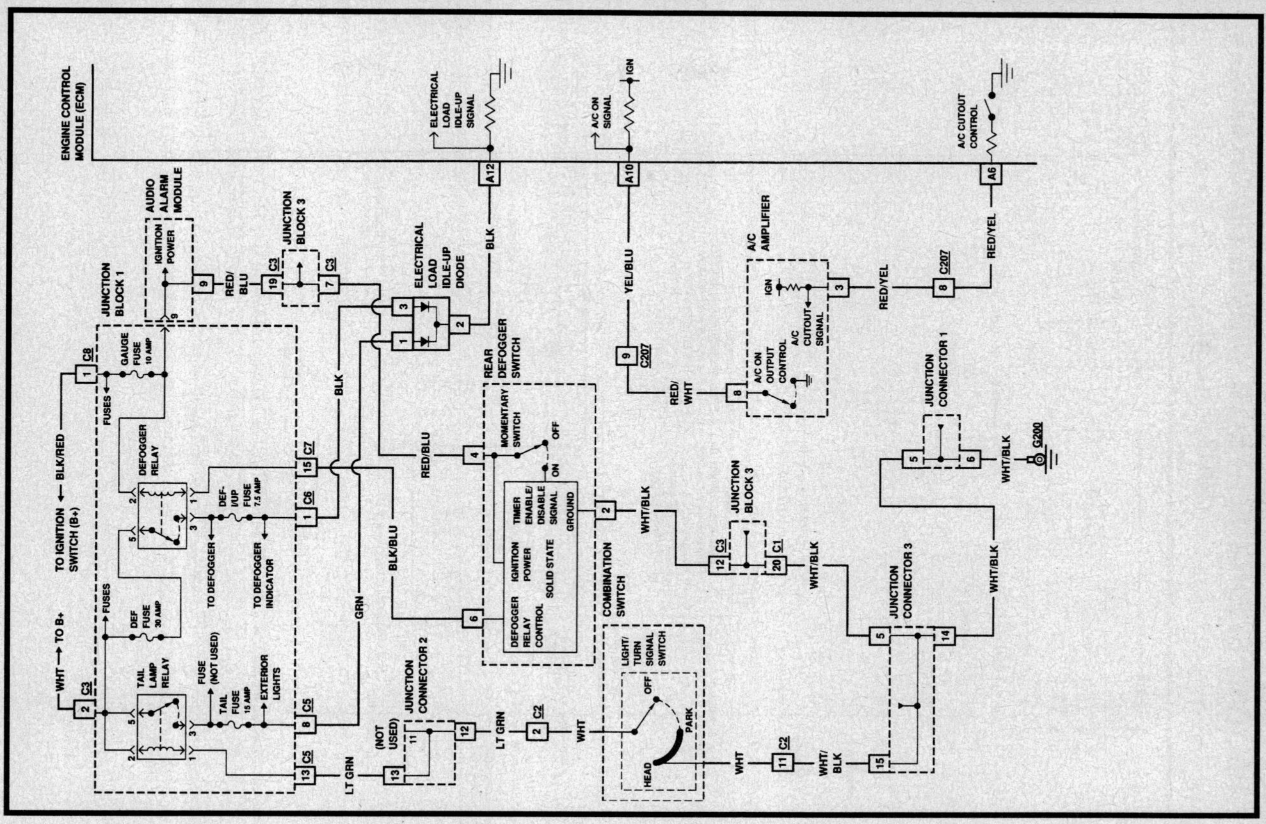

1.6L (VIN 6) ENGINE — ECM WIRING DIAGRAMS — 1993 PRIZM
ENGINE CONTROL MODULE (ECM)
MEMORY POWER
IGNITION POWER
IGNITION POWER
TEST DIAGNOSTIC MODE SIGNAL
DIAGNOSTIC REQUEST SIGNAL
5V
FUEL CONTROL FEEDBACK OUTPUT
MIL CONTROL
GROUND
GROUND
GROUND
A2
A1
A7
B7
B15
B8
A8
C15
C13
C26
BLK/ORN — TO IGNITION SWITCH (B+)
JUNCTION BLOCK 1
RED/WHT
BLK/RED
BLK/RED
GRN/WHT
BLU/WHT
RED/WHT
RED/YEL
BRN
BRN
IGN FUSE 10 AMP
DATA LINK CONNECTOR (DLC)
MANUAL TRANSAXLE ONLY
BLK — TO B+
FUSES
FUSE AND RELAY BOX
EFI F-HTR RELAY
RADIATOR FAN MOTOR AND ENGINE MAIN RELAY
EFI F-HTR FUSE 15 AMP
BLK/ORN
BLK/ORN
BLK/ORN
BLK/RED
TO IAC VALVE, A/C SV VALVE AND EGR SV VALVE
TO CIRCUIT OPENING RELAY
BLK/RED
RED/YEL
RED/YEL
RED/YEL
INSTRUMENT CLUSTER
MALFUNCTION INDICATOR LAMP (MIL)
BRN
BRN
G106
WHT/BLK
G102
TO IGNITION SWITCH (B+)
JUNCTION BLOCK 1
BLK/YEL
GAUGE FUSE 10 AMP
AUDIO ALARM MODULE
IGNITION POWER
RED/BLU
JUNCTION BLOCK 3
RED/YEL
RED/BLU
TO O2S, SUB O2S AND SHIELDS
TO O2S, SUB O2S AND SHIELDS AND ECM

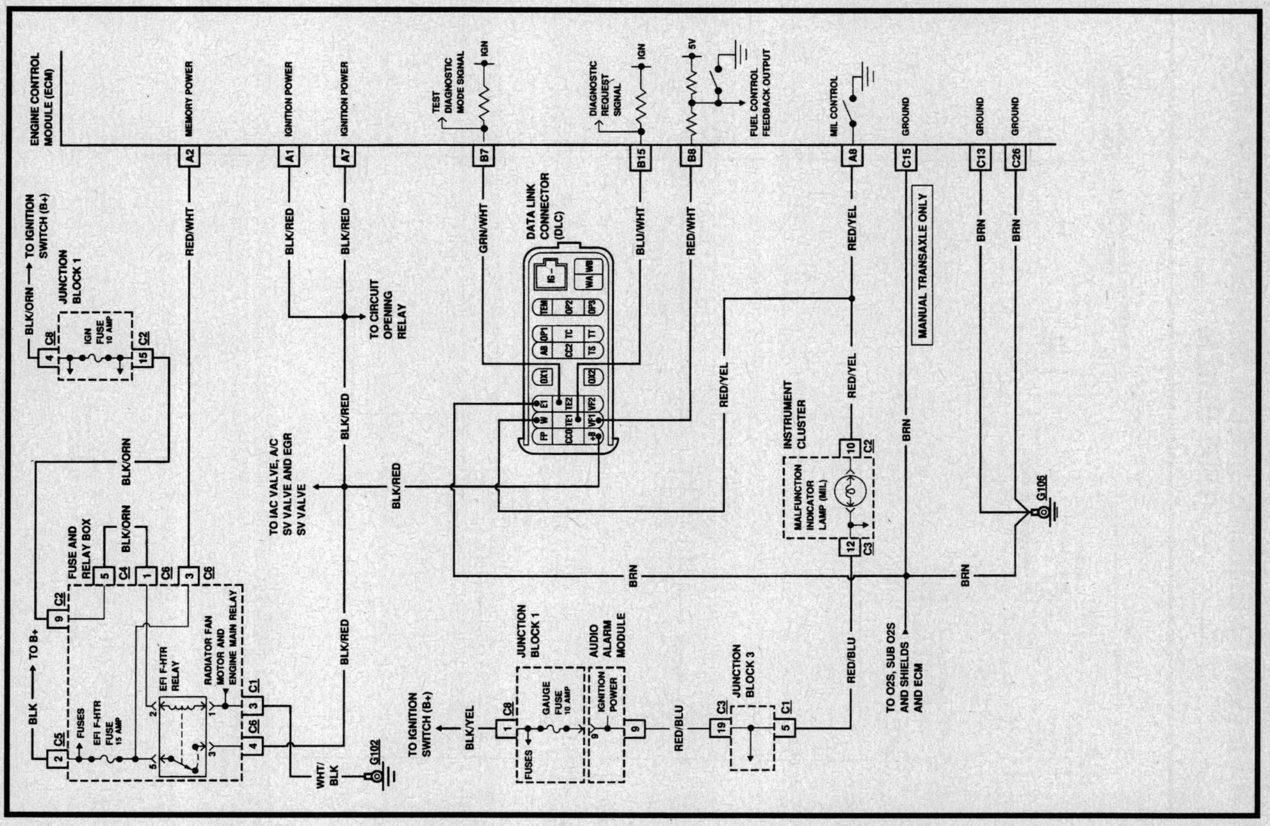

1.6L (VIN 6) ENGINE — ECM WIRING DIAGRAMS — 1993 PRIZM

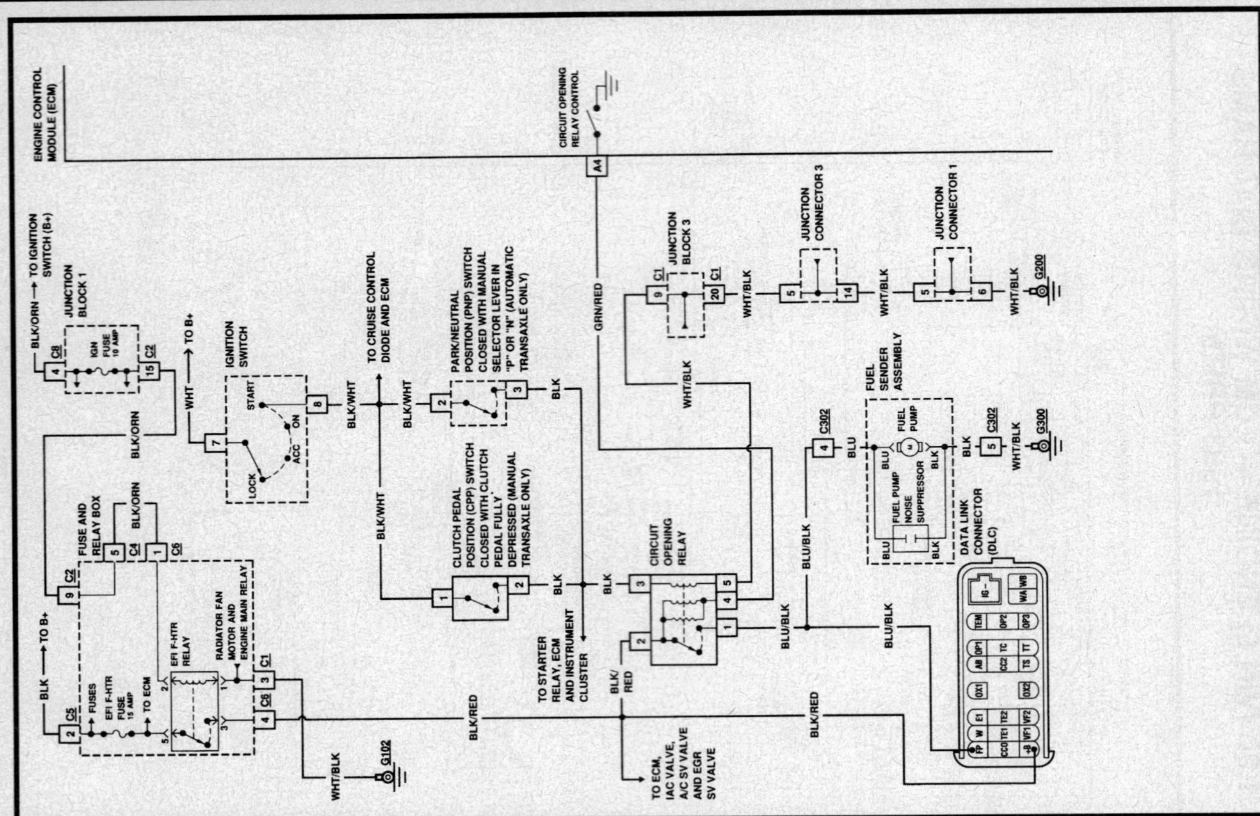

1.6L (VIN 6) ENGINE — ECM WIRING DIAGRAMS — 1993 PRIZM

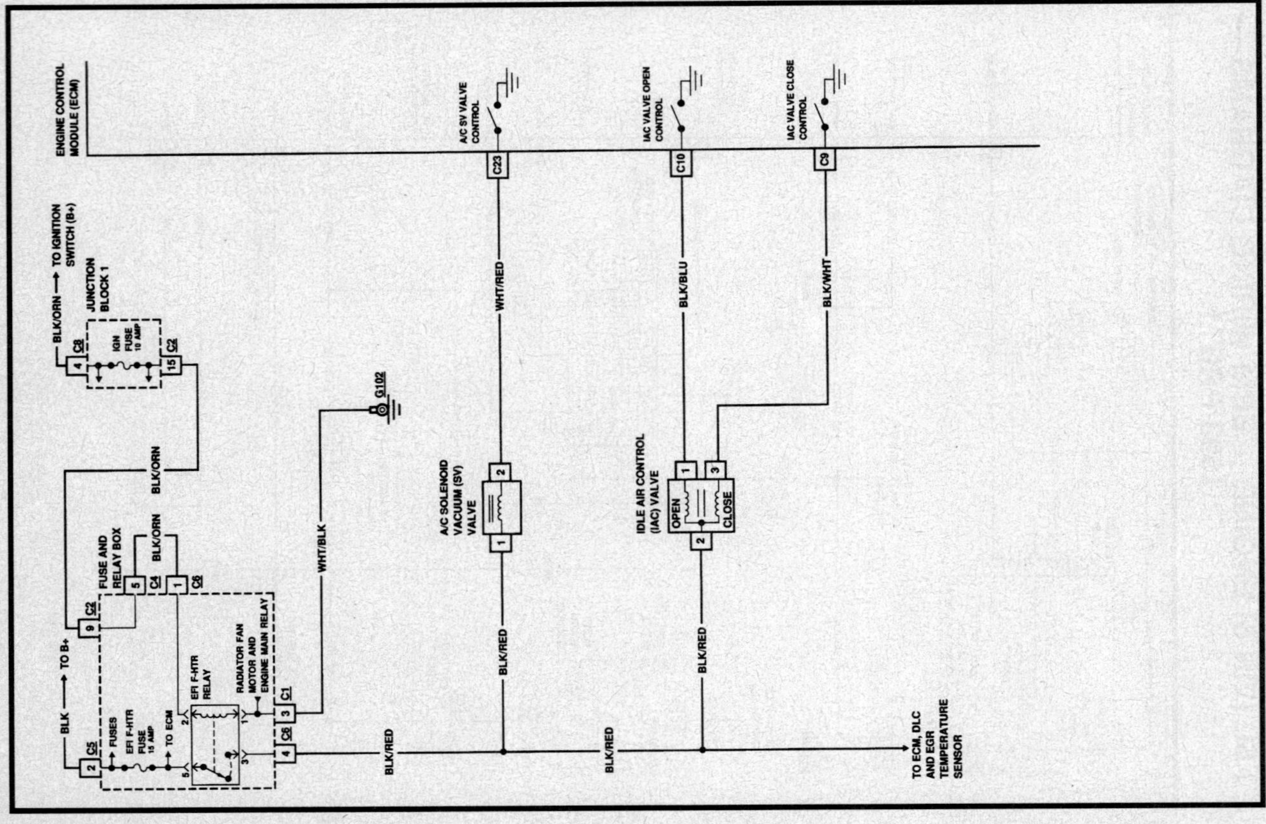

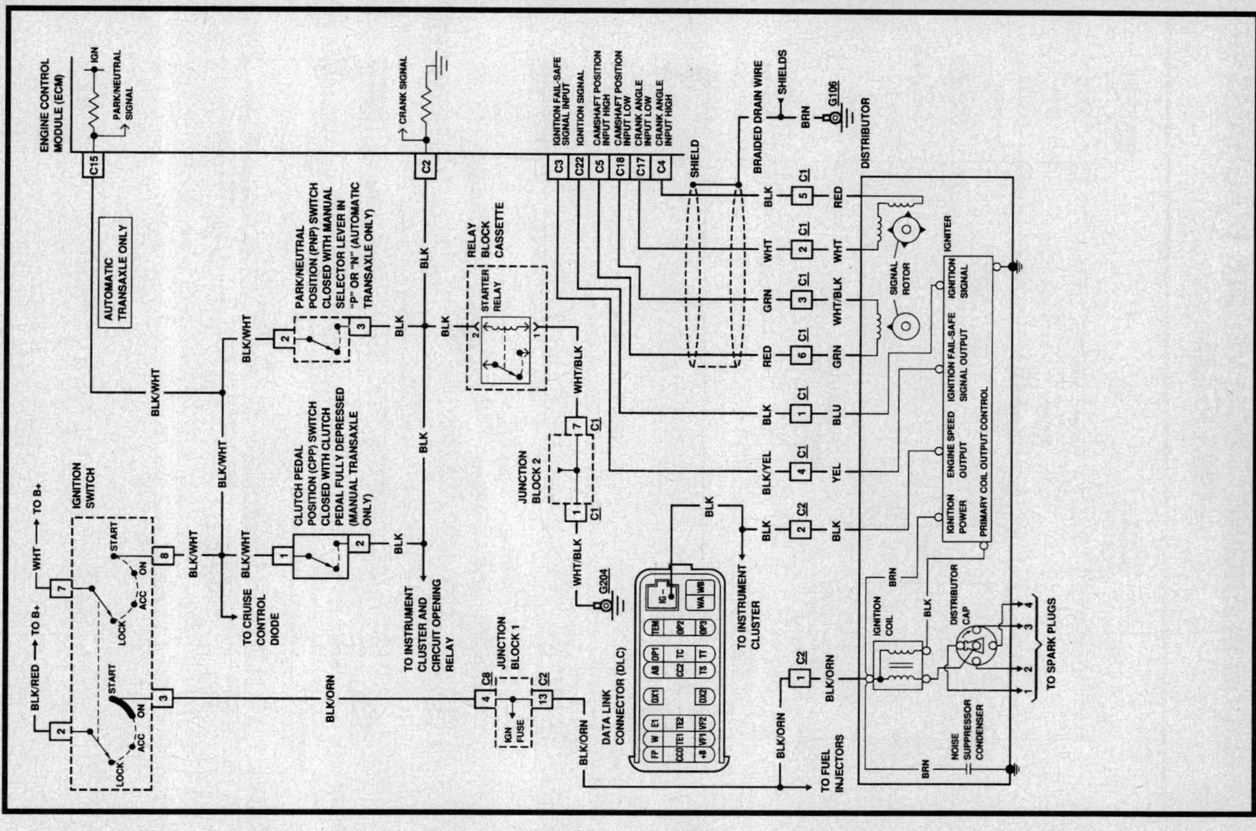
1.6L (VIN 6) ENGINE — ECM WIRING DIAGRAMS — 1993 PRIZM

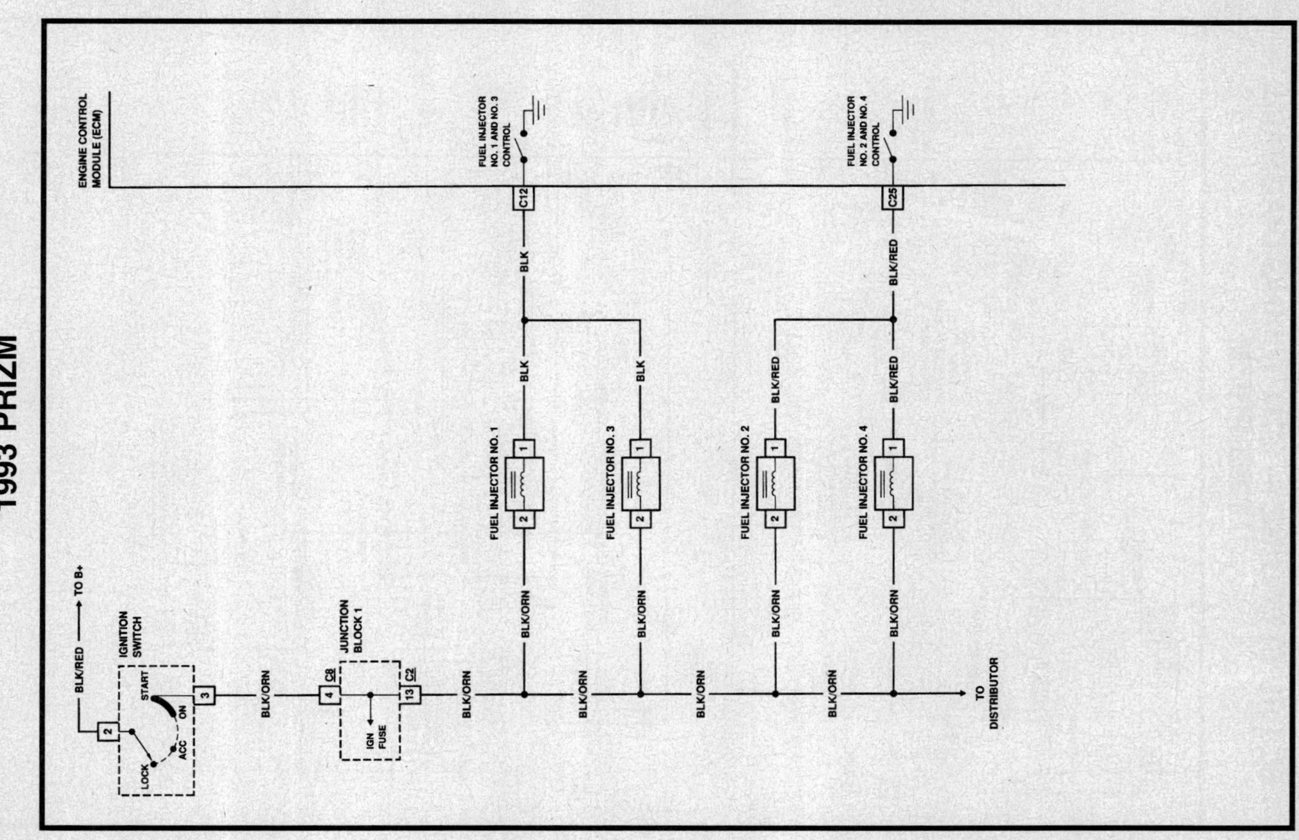
1.6L (VIN 6) ENGINE — ECM WIRING DIAGRAMS — 1993 PRIZM

1.6L (VIN 6) ENGINE — ECM WIRING DIAGRAMS — 1993 PRIZM

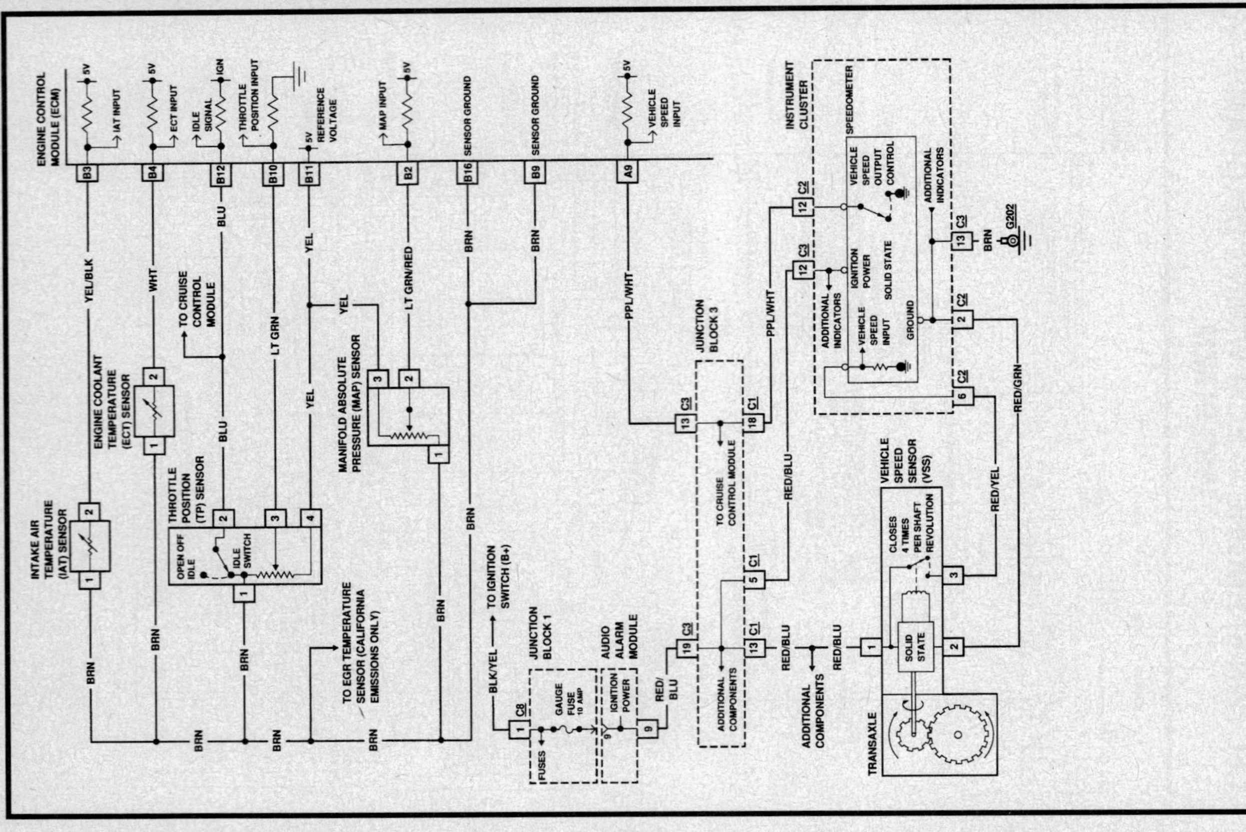

1.6L (VIN 6) ENGINE — ECM WIRING DIAGRAMS — 1993 PRIZM

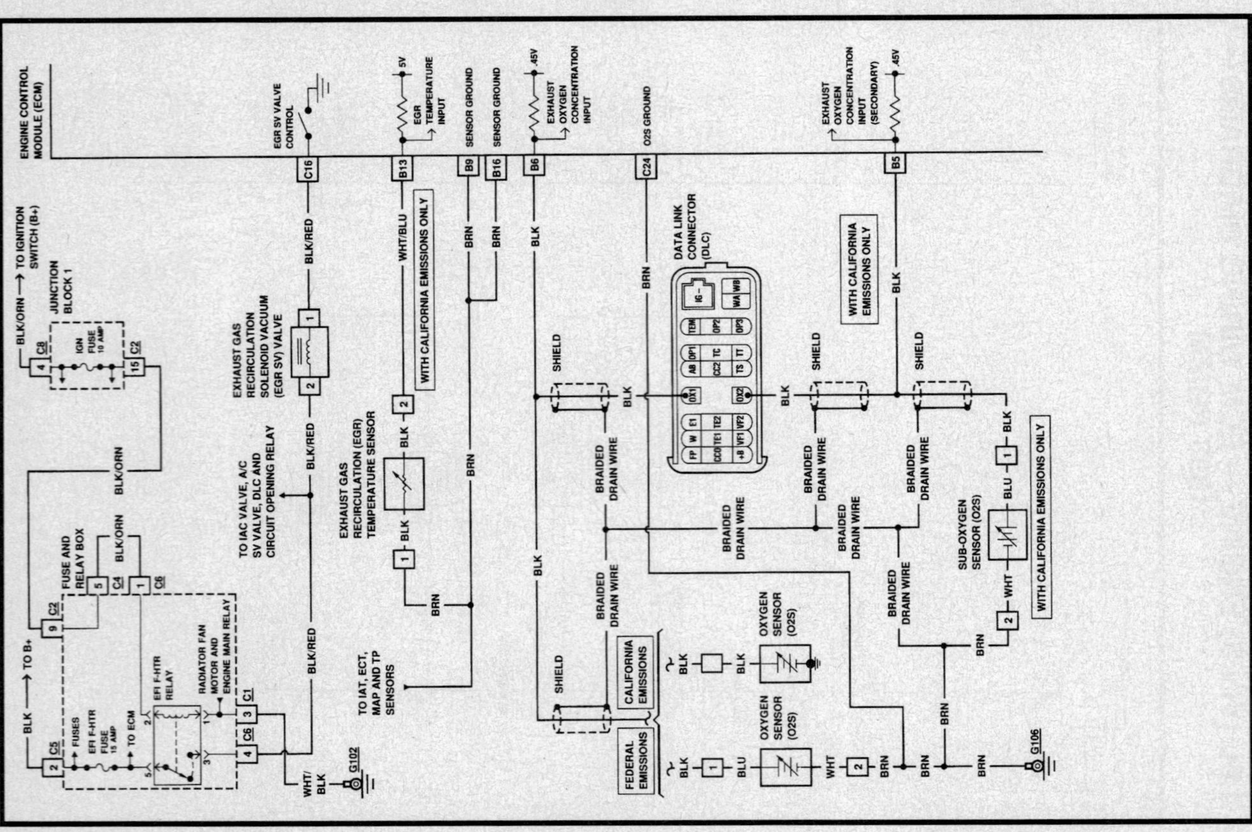

1.6L (VIN 6) ENGINE — ECM CONNECTOR TERMINAL END VIEW — 1993 PRIZM

ECM CONNECTOR IDENTIFICATION

The following ECM voltage charts are for use with a digital multimeter to further aid in diagnosis. The voltages you get may vary due to low battery charge or other reasons, but they should be very close.

THE FOLLOWING CONDITIONS MUST BE MET BEFORE TESTING:
- Engine at operating temperature   • Engine idling (for "ENG. RUN" column)
- Test terminal not grounded
- All voltages shown "B+" indicate system voltage

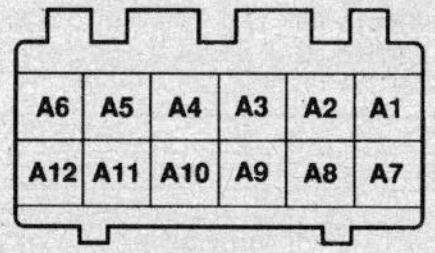

A6	A5	A4	A3	A2	A1
A12	A11	A10	A9	A8	A7

BACK VIEW OF ECM CONNECTOR C1 (GRAY)

CAVITY/PIN	WIRE COLOR	CIRCUIT	VOLTAGE KEY "ON"	VOLTAGE ENG. RUN
1/A1	BLK/RED	IGNITION POWER	B+	B+
2/A2	RED/WHT	MEMORY POWER	B+	B+
3/A3	—	NOT USED	—	—
4/A4	GRN/RED	CIRCUIT OPENING RELAY CONTROL	(1)	0*
5/A5	—	NOT USED	—	—
6/A6	RED/YEL	A/C CUTOUT CONTROL	0*	0* (2)
7/A7	BLK/RED	IGNITION POWER	B+	B+
8/A8	RED/YEL	MIL CONTROL	0-1V	B+
9/A9	—	NOT USED	—	—
10/A10	YEL/BLU	A/C ON SIGNAL	B+(3)	B+(3)
11/A11	PPL/WHT	VEHICLE SPEED INPUT	0* (4)	2-3V VARIES
12/A12	BLK	ELECTRICAL LOAD IDLE-UP SIGNAL	0* (5)	0* (5)

0* LESS THAN 0.50 VOLTS
(1) 0 VOLTS FOR 3 SECONDS, THEN B+
(2) 5 TO 6 VOLTS WITH A/C SWITCH DEPRESSED
(3) 0 VOLTS WITH A/C SWITCH DEPRESSED
(4) VARIES FROM 0 TO 5V WITH FRONT WHEELS ROTATING
(5) B+ WITH LIGHT/TURN SIGNAL SWITCH IN PARK OR HEAD OR WITH DEFOGGER ON
(6) VARIES WITH INTAKE AIR TEMPERATURE
(7) VARIES WITH ENGINE COOLANT TEMPERATURE
(8) VARIES WITH THROTTLE LEVER POSITION
(9) 0 VOLTS AT IDLE; B+ OFF IDLE
(10) VARIES WITH EXHAUST GAS TEMPERATURE
(11) B+ WITH IGNITION SWITCH IN "START"
(12) 0 VOLTS IN "P" OR "N"; B+ IN "R", "D", "2" OR "L"
(13) VARIES CONSTANTLY WITH SYSTEM IN CLOSED LOOP

1.6L (VIN 6) ENGINE — ECM CONNECTOR TERMINAL END VIEW — 1993 PRIZM

ECM CONNECTOR IDENTIFICATION

The following ECM voltage charts are for use with a digital multimeter to further aid in diagnosis. The voltages you get may vary due to low battery charge or other reasons, but they should be very close.

THE FOLLOWING CONDITIONS MUST BE MET BEFORE TESTING:
- Engine at operating temperature   • Engine idling (for "ENG. RUN" column)
- Test terminal not grounded
- All voltages shown "B+" indicate system voltage

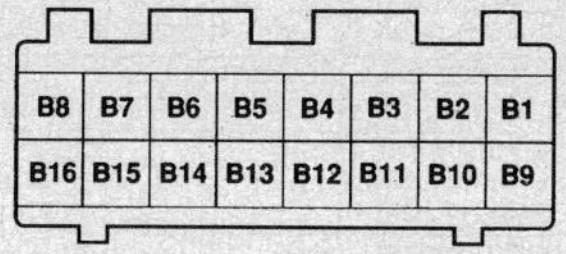

B8	B7	B6	B5	B4	B3	B2	B1
B16	B15	B14	B13	B12	B11	B10	B9

BACK VIEW OF ECM CONNECTOR C2 (GRAY)

CAVITY/PIN	WIRE COLOR	CIRCUIT	VOLTAGE KEY "ON"	VOLTAGE ENG. RUN
1/B1	—	NOT USED	—	—
2/B2	LT GRN/RED	MAP INPUT	3-4V	1.1-1.8V
3/B3	YEL/BLK	IAT INPUT	1-3.5V (6)	1.5-2.5V (6)
4/B4	WHT	ECT INPUT	.5-3.5V (7)	1-2V (7)
5/B5	BLK	SECONDARY) (CALIFORNIA ONLY) EXHAUST OXYGEN CONCENTRATION INPUT	0*	0-1V (13)
6/B6	BLK	TEST DIAGNOSTIC MODE SIGNAL	0*	0-1V (13)
7/B7	GRN/WHT	FUEL CONTROL FEEDBACK OUTPUT	B+	B+
8/B8	RED/WHT	SENSOR GROUND	0*	1.5-3.5V
9/B9	BRN	THROTTLE POSITION INPUT	0*	0*
10/B10	LT GRN	REFERENCE VOLTAGE (5V)	.5-4.5V (8)	.5-4.5V (8)
11/B11	YEL	IDLE SIGNAL	5V	5V
12/B12	BLU	EGR TEMPERATURE INPUT (CALIFORNIA ONLY)	(9)	(9)
13/B13	WHT/BLU	NOT USED	1-3.5V (10)	1.5-2.5V(10)
14/B14	—	DIAGNOSTIC REQUEST SIGNAL	—	—
15/B15	BLU/WHT	SENSOR GROUND	B+	B+
16/B16	BRN		0*	0*

0* LESS THAN 0.50 VOLTS
(1) 0 VOLTS FOR 3 SECONDS, THEN B+
(2) 5 TO 6 VOLTS WITH A/C SWITCH DEPRESSED
(3) 0 VOLTS WITH A/C SWITCH DEPRESSED
(4) VARIES FROM 0 TO 5V WITH FRONT WHEELS ROTATING
(5) B+ WITH LIGHT/TURN SIGNAL SWITCH IN PARK OR HEAD OR WITH DEFOGGER ON
(6) VARIES WITH INTAKE AIR TEMPERATURE
(7) VARIES WITH ENGINE COOLANT TEMPERATURE
(8) VARIES WITH THROTTLE LEVER POSITION
(9) 0 VOLTS AT IDLE; B+ OFF IDLE
(10) VARIES WITH EXHAUST GAS TEMPERATURE
(11) B+ WITH IGNITION SWITCH IN "START"
(12) 0 VOLTS IN "P" OR "N"; B+ IN "R", "D", "2" OR "L"
(13) VARIES CONSTANTLY WITH SYSTEM IN CLOSED LOOP

1.6L (VIN 6) ENGINE — ECM CONNECTOR TERMINAL END VIEW — 1993 PRIZM

1.6L (VIN 6) ENGINE — DIAGNOSTIC TROUBLE CODE (DTC) TABLE — 1993 PRIZM

ECM CONNECTOR IDENTIFICATION

The following ECM voltage charts are for use with a digital multimeter to further aid in diagnosis. The voltages you get may vary due to low battery charge or other reasons, but they should be very close.

__THE FOLLOWING CONDITIONS MUST BE MET BEFORE TESTING:__
- Engine at operating temperature   • Engine idling (for "ENG. RUN" column)
- Test terminal not grounded
- All voltages shown "B+" indicate system voltage

BACK VIEW OF ECM CONNECTOR C3 (GRAY)

CAVITY/ PIN	WIRE COLOR	CIRCUIT	VOLTAGE KEY "ON"	VOLTAGE ENG. RUN
1/C1	—	NOT USED	—	—
2/C2	BLK	CRANK SIGNAL	0* (11)	0* (11)
3/C3	BLK/YEL	IGNITION FAIL-SAFE SIGNAL INPUT	.56V	1.5V
4/C4	BLK	CRANK ANGLE INPUT HIGH	.69V	.69V
5/C5	RED	CAMSHAFT POSITION INPUT HIGH	.69V	.69V
6/C6	—	NOT USED	—	—
7/C7	—	NOT USED	—	—
8/C8	—	NOT USED	—	—
9/C9	BLK/WHT	IAC VALVE CLOSE CONTROL	.3-.7V	5-7V
10/C10	BLK/BLU	IAC VALVE OPEN CONTROL	B+	8-10V
11/C11	—	NOT USED	—	—
12/C12	BLK	FUEL INJECTOR NO. 1 AND NO. 3 CONTROL	B+	B+
13/C13	BRN	GROUND	0*	0*
14/C14	—	NOT USED	—	—
15/C15	BRN (M/T)	GROUND (M/T ONLY)	0*	0*
	BLK/WHT (A/T)	PARK/NEUTRAL POSITION SIGNAL (A/T ONLY)	(12)	(12)
16/C16	BLK/RED	EGR SV VALVE CONTROL (CALIFORNIA ONLY)	B+	B+
17/C17	WHT	CRANK ANGLE INPUT LOW	.69V	.69V
18/C18	GRN	CAMSHAFT POSITION INPUT LOW	.69V	.69V
19/C19	—	NOT USED	—	—
20/C20	—	NOT USED	—	—
21/C21	—	NOT USED	—	—
22/C22	BLK	IGNITION SIGNAL	0*	0-2V (8)
23/C23	WHT/RED	A/C SV VALVE CONTROL	B+	B+ (3)
24/C24	BRN	O2S GROUND	0*	0*
25/C25	BLK/RED	FUEL INJECTOR NO. 2 AND NO. 4 CONTROL	B+	B+
26/C26	BRN	GROUND	0*	0*

0* LESS THAN 0.50 VOLTS
(1) 0 VOLTS FOR 3 SECONDS, THEN B+
(2) 5 TO 6 VOLTS WITH A/C SWITCH DEPRESSED
(3) 0 VOLTS WITH A/C SWITCH DEPRESSED
(4) VARIES FROM 0 TO 5V WITH FRONT WHEELS ROTATING
(5) B+ WITH LIGHT/TURN SIGNAL SWITCH IN PARK
OR HEAD OR WITH DEFOGGER ON
(6) VARIES WITH INTAKE AIR TEMPERATURE
(7) VARIES WITH ENGINE COOLANT TEMPERATURE
(8) VARIES WITH THROTTLE LEVER POSITION
(9) 0 VOLTS AT IDLE; B+ OFF IDLE
(10) VARIES WITH EXHAUST GAS TEMPERATURE
(11) B+ WITH IGNITION SWITCH IN "START"
(12) 0 VOLTS IN "P" OR "N"; B+ IN "R", "D", "2" OR "L"
(13) VARIES CONSTANTLY WITH SYSTEM IN CLOSED LOOP

DIAGNOSTIC TROUBLE CODE (DTC) TABLE
1.6L (VIN 6) "S" CARLINE
(PAGE 1 OF 2)

EXAMPLE: RPM SIGNAL DTC 13 AND MANIFOLD ABSOLUTE PRESSURE (MAP) SENSOR INPUT DTC 31

DIAGNOSTIC TROUBLE CODE (DTC) NO.	MODE	SYSTEM	CONDITIONS THAT MUST BE MET BEFORE DTC WILL SET
—		NORMAL	• INDICATES THAT ELECTRONIC ENGINE CONTROL SYSTEM IS FUNCTIONING NORMALLY.
12		RPM SIGNAL	(1) NO CRANK ANGLE INPUT DETECTED AT ECM FOR 2 SECONDS OR MORE WHILE CRANKING ENGINE. (2) NO CAMSHAFT POSITION INPUT DETECTED AT ECM FOR 3 SECONDS OR MORE WITH ENGINE SPEED BETWEEN 600 AND 4,000 RPM. • THIS DTC CAN ONLY BE ACCESSED WHEN IN THE NORMAL DIAGNOSTIC MODE.
13		RPM SIGNAL	• NO CRANK ANGLE INPUT DETECTED AT ECM FOR 0.30 SECONDS OR MORE WITH ENGINE SPEED ABOVE 1,500 RPM.
14		IGNITION SIGNAL	• NO IGNITION FAIL-SAFE SIGNAL INPUT DETECTED AT ECM FOR 4 CONSECUTIVE IGNITION CYCLES. • THIS DTC CAN ONLY BE ACCESSED WHEN IN THE NORMAL DIAGNOSTIC MODE.
21		OXYGEN SENSOR (O2S) INPUT	• OXYGEN SENSOR VOLTAGE INPUT AT ECM CONSTANT BETWEEN 0.35 AND 0.70 VOLTS FOR 60 SECONDS OR MORE WITH ENGINE SPEED ABOVE 1,500 RPM.
22		ENGINE COOLANT TEMPERATURE (ECT) SENSOR	• ECT INPUT AT ECM OPEN OR SHORTED FOR 0.50 SECONDS OR MORE.
24		INTAKE AIR TEMPERATURE (IAT) SENSOR	• IAT INPUT AT ECM OPEN OR SHORTED FOR 0.50 SECONDS OR MORE.

1.6L (VIN 6) ENGINE — DIAGNOSTIC TROUBLE CODE (DTC) TABLE — 1993 PRIZM

DIAGNOSTIC TROUBLE CODE (DTC) TABLE
1.6L (VIN 6) "S" CARLINE
(PAGE 2 OF 2)

DIAGNOSTIC TROUBLE CODE (DTC) NO.	MODE	SYSTEM	CONDITIONS THAT MUST BE MET BEFORE DTC WILL SET
25		AIR/FUEL RATIO LEAN	(1) OXYGEN SENSOR INPUT AT ECM IS LESS THAN 0.45 VOLTS FOR 90 SECONDS OR MORE WITH OXYGEN SENSOR WARM AND ENGINE SPEED ABOVE 1,500 RPM. (2) WHEN ENGINE SPEED VARIES BY MORE THAN 50 RPM OVER THE PRECEDING CRANK ANGLE PERIOD DURING A PERIOD OF 50 SECONDS WHILE IDLING WITH THE ECT ABOVE 50°C (122°F). • EITHER CONDITION MUST BE DETECTED DURING 2 CONSECUTIVE IGNITION CYCLES FOR DTC TO BE STORED IN ECM MEMORY.
26		AIR/FUEL RATIO RICH	• WHEN ENGINE SPEED VARIES BY MORE THAN 50 RPM OVER THE PRECEDING CRANK ANGLE PERIOD DURING A PERIOD OF 50 SECONDS WHILE IDLING WITH THE ECT ABOVE 50°C (122°F). • CONDITION MUST BE DETECTED DURING 2 CONSECUTIVE IGNITION CYCLES FOR DTC TO BE STORED IN ECM MEMORY.
27		SUB-OXYGEN SENSOR (O2S) INPUT	• WHEN SUB-OXYGEN SENSOR IS WARMED UP AND ENGINE IS OFF IDLE FOR 2 SECONDS OR MORE AND OXYGEN SENSOR INPUT AT PCM IS 0.45 VOLTS OR MORE AND SUB-OXYGEN SENSOR INPUT AT ECM IS 0.45 VOLTS OR LESS. • CONDITION MUST BE DETECTED DURING 2 CONSECUTIVE IGNITION CYCLES FOR DTC TO BE STORED IN ECM MEMORY.
31		MANIFOLD ABSOLUTE PRESSURE (MAP) SENSOR INPUT	• MAP INPUT AT ECM OPEN OR SHORTED FOR 0.50 SECONDS OR MORE.
41		THROTTLE POSITION (TP) SENSOR INPUT	• THROTTLE POSITION INPUT AT ECM OPEN OR SHORTED FOR 0.50 SECONDS OR MORE.
42		VEHICLE SPEED INPUT	• VEHICLE SPEED INPUT NOT DETECTED AT ECM FOR 8 SECONDS OR MORE WITH ENGINE SPEED BETWEEN 3,000 AND 5,000 RPM AND TRANSAXLE IN GEAR.
43		CRANK SIGNAL	• CRANK SIGNAL NOT DETECTED AT ECM DURING IGNITION. • THIS DTC CAN ONLY BE DETECTED WHEN IN THE TEST DIAGNOSTIC MODE.
51		SWITCH CONDITION SIGNAL	• DISPLAYED WHEN A/C IS ON, ACCELERATOR PEDAL IS DEPRESSED OR WHEN MANUAL SELECTOR IS IN "R", "D", "2" OR "L" POSITION WITH DLC TERMINAL "TE1" AND "E1" CONNECTED. • THIS DTC CAN ONLY BE ACCESSED WHEN IN THE TEST DIAGNOSTIC MODE AND INDICATES THE ABOVE MENTIONED SYSTEMS ARE FUNCTIONING NORMALLY. IF DTC 51 IS NOT INDICATED WHEN THE CONDITIONS ARE MET, A FAULT IS PRESENT, AND DTC 51's DIAGNOSTIC CHART MUST BE USED.
71		EGR SYSTEM	• WITH ECT ABOVE 60°C (140°F) 50 SECONDS FROM START OF EGR OPERATION. THE EGR TEMPERATURE IS LESS THAN 80°C (176°F) AND THE EGR TEMPERATURE HAS RISEN LESS THAN 10°C (18°F) DURING THE 50 SECONDS. • CONDITION MUST BE DETECTED DURING 2 CONSECUTIVE IGNITION CYCLES FOR DTC TO BE STORED IN ECM MEMORY.

1.6L (VIN 6) ENGINE — ON-BOARD DIAGNOSTIC (OBD) SYSTEM CHECK — 1993 PRIZM

ON-BOARD DIAGNOSTIC (OBD) SYSTEM CHECK
(Page 1 of 2)
1.6L (VIN 6) "S" CARLINE

Circuit Description

The OBD System Check is an organized approach to identifying a malfunction created by the electronic engine control system. It must be the starting point for any driveability complaint diagnosis; because it directs the service technician to the next logical step in diagnosing the complaint. Understanding the chart and using it correctly will reduce diagnostic time and prevent the unnecessary replacement of good parts.

Test Description: Number(s) below refer to circled number(s) on the diagnostic chart.

1. Checks the MIL operation with the ignition switch in the "ON" position.
2. Checks to see if the ECM's normal diagnostic mode is functioning.
3. Checks to see if the ECM's test diagnostic mode is functioning.
4. Checks to see if the vehicle will start.
5. Checks for any DTCs that are stored in the ECM's memory with the engine running.
6. Checks for any DTCs that are stored in the ECM's memory with the engine off.

1.6L (VIN 6) ENGINE — ON-BOARD DIAGNOSTIC (OBD) SYSTEM CHECK — 1993 PRIZM

ON-BOARD DIAGNOSTIC (OBD) SYSTEM CHECK
(Page 1 of 2)
1.6L (VIN 6) "S" CARLINE

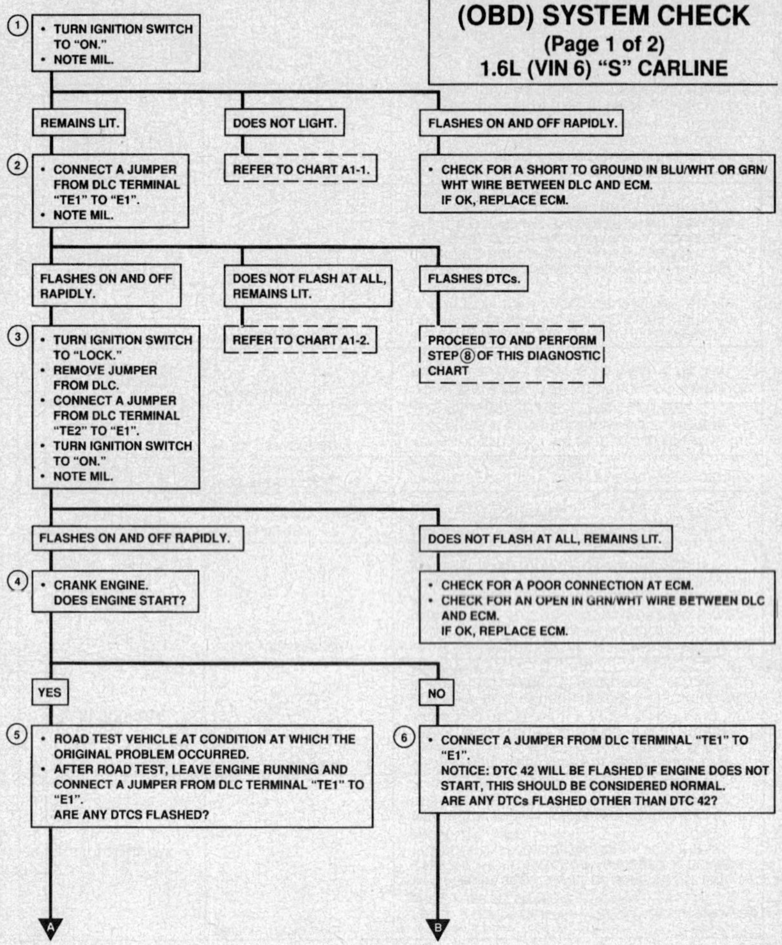
1 • TURN IGNITION SWITCH TO "ON." • NOTE MIL.
REMAINS LIT.
DOES NOT LIGHT.
FLASHES ON AND OFF RAPIDLY.
2 • CONNECT A JUMPER FROM DLC TERMINAL "TE1" TO "E1". • NOTE MIL.
REFER TO CHART A1-1.
• CHECK FOR A SHORT TO GROUND IN BLU/WHT OR GRN/WHT WIRE BETWEEN DLC AND ECM. IF OK, REPLACE ECM.
FLASHES ON AND OFF RAPIDLY.
DOES NOT FLASH AT ALL, REMAINS LIT.
FLASHES DTCs.
3 • TURN IGNITION SWITCH TO "LOCK." • REMOVE JUMPER FROM DLC. • CONNECT A JUMPER FROM DLC TERMINAL "TE2" TO "E1". • TURN IGNITION SWITCH TO "ON." • NOTE MIL.
REFER TO CHART A1-2.
PROCEED TO AND PERFORM STEP 8 OF THIS DIAGNOSTIC CHART
FLASHES ON AND OFF RAPIDLY.
DOES NOT FLASH AT ALL, REMAINS LIT.
4 • CRANK ENGINE. DOES ENGINE START?
• CHECK FOR A POOR CONNECTION AT ECM. • CHECK FOR AN OPEN IN GRN/WHT WIRE BETWEEN DLC AND ECM. IF OK, REPLACE ECM.
YES
NO
5 • ROAD TEST VEHICLE AT CONDITION AT WHICH THE ORIGINAL PROBLEM OCCURRED. • AFTER ROAD TEST, LEAVE ENGINE RUNNING AND CONNECT A JUMPER FROM DLC TERMINAL "TE1" TO "E1". ARE ANY DTCS FLASHED?
6 • CONNECT A JUMPER FROM DLC TERMINAL "TE1" TO "E1". NOTICE: DTC 42 WILL BE FLASHED IF ENGINE DOES NOT START, THIS SHOULD BE CONSIDERED NORMAL. ARE ANY DTCs FLASHED OTHER THAN DTC 42?
A
B

1.6L (VIN 6) ENGINE — ON-BOARD DIAGNOSTIC (OBD) SYSTEM CHECK — 1993 PRIZM

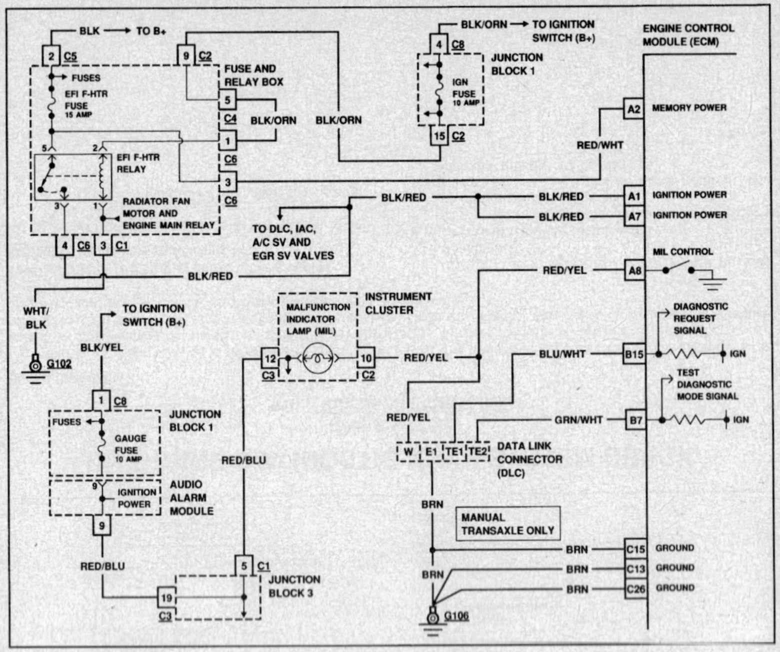
BLK → TO B+
C5
FUSES
EFI F-HTR FUSE 15 AMP
EFI F-HTR RELAY
RADIATOR FAN MOTOR AND ENGINE MAIN RELAY
FUSE AND RELAY BOX
BLK/ORN
BLK/ORN → TO IGNITION SWITCH (B+)
JUNCTION BLOCK 1
IGN FUSE 10 AMP
ENGINE CONTROL MODULE (ECM)
MEMORY POWER
RED/WHT
BLK/RED
IGNITION POWER
TO DLC, IAC, A/C SV AND EGR SV VALVES
MIL CONTROL
RED/YEL
WHT/BLK
G102
TO IGNITION SWITCH (B+)
BLK/YEL
MALFUNCTION INDICATOR LAMP (MIL)
INSTRUMENT CLUSTER
DIAGNOSTIC REQUEST SIGNAL
BLU/WHT
IGN
TEST DIAGNOSTIC MODE SIGNAL
GRN/WHT
GAUGE FUSE 10 AMP
JUNCTION BLOCK 1
RED/BLU
DATA LINK CONNECTOR (DLC)
IGNITION POWER
AUDIO ALARM MODULE
BRN
MANUAL TRANSAXLE ONLY
GROUND
JUNCTION BLOCK 3
G106

ON-BOARD DIAGNOSTIC (OBD) SYSTEM CHECK
(Page 2 of 2)
1.6L (VIN 6) "S" CARLINE

Circuit Description

The OBD System Check is an organized approach to identifying a malfunction created by the electronic engine control system. It must be the starting point for any driveability complaint diagnosis; because it directs the service technician to the next logical step in diagnosing the complaint. Understanding the chart and using it correctly will reduce diagnostic time and prevent the unnecessary replacement of good parts.

Test Description: Number(s) below refer to circled number(s) on the diagnostic chart.

7. Checks to see if DTC 51 is functioning normally.
8. Checks to see if DTCs that were stored are intermittent malfunctions.

1.6L (VIN 6) ENGINE — ON-BOARD DIAGNOSTIC (OBD) SYSTEM CHECK — 1993 PRIZM

1.6L (VIN 6) ENGINE — A-CHARTS — 1993 PRIZM

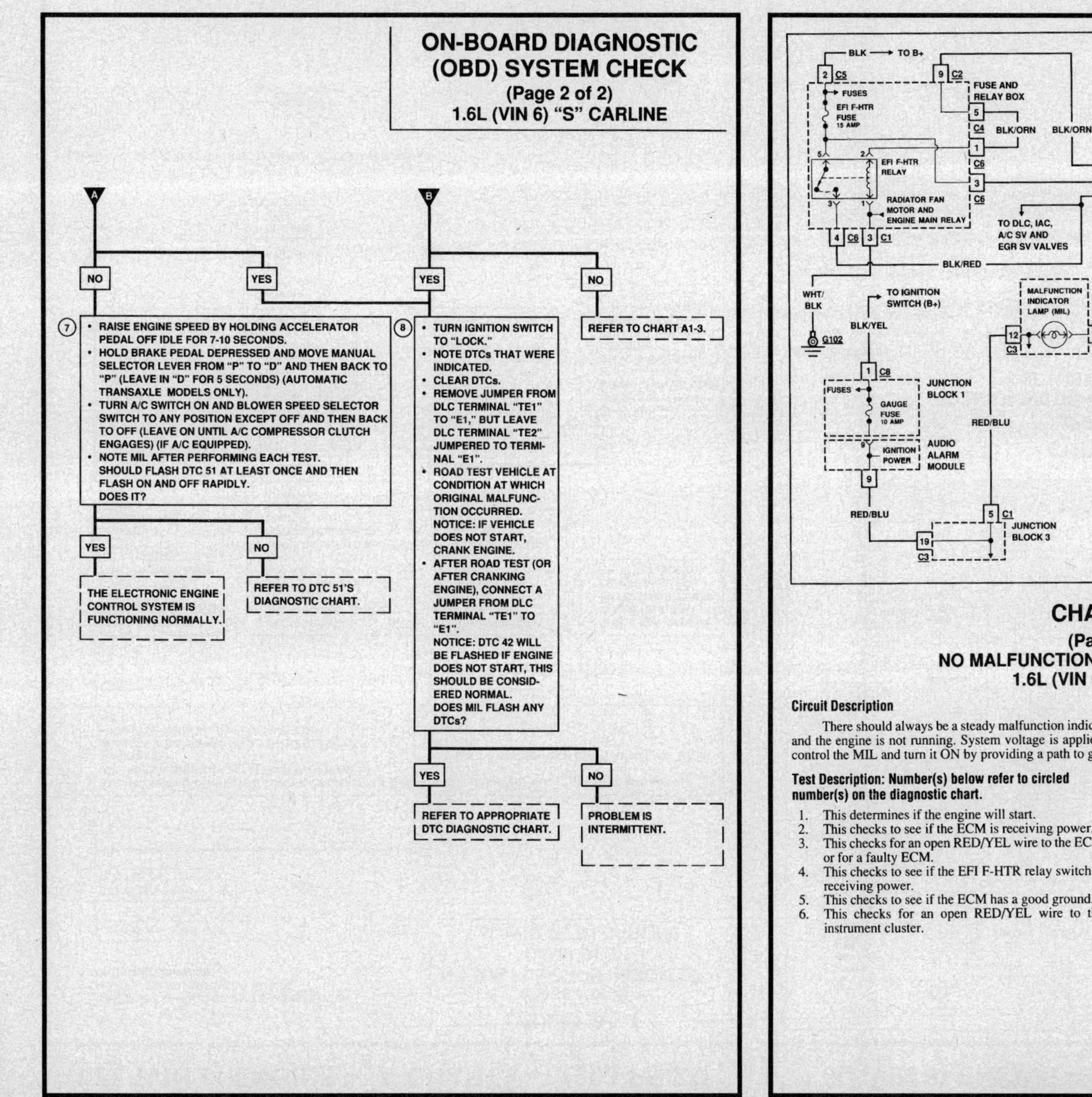

Circuit Description

There should always be a steady malfunction indicator lamp (MIL) when the ignition switch is in the "ON" position and the engine is not running. System voltage is applied to the indicator bulb. The engine control module (ECM) will control the MIL and turn it ON by providing a path to ground through the RED/YEL wire.

Test Description: Number(s) below refer to circled number(s) on the diagnostic chart.

1. This determines if the engine will start.
2. This checks to see if the ECM is receiving power.
3. This checks for an open RED/YEL wire to the ECM or for a faulty ECM.
4. This checks to see if the EFI F-HTR relay switch is receiving power.
5. This checks to see if the ECM has a good ground.
6. This checks for an open RED/YEL wire to the instrument cluster.
7. This checks to see if the MIL is receiving power.

Diagnostic Aids:

Check EFI F-HTR, IGN and GAUGE fuses for opens and make sure EFI F-HTR Relay is securely mounted into the fuse and relay box.

An intermittent may be caused by a poor connection, rubbed through wire insulation, or a wire broken inside the insulation. Inspect harness connectors for backed out terminals, improper mating, broken locks, improperly formed or damaged terminals and poor terminal-to-wire connections before component replacement.

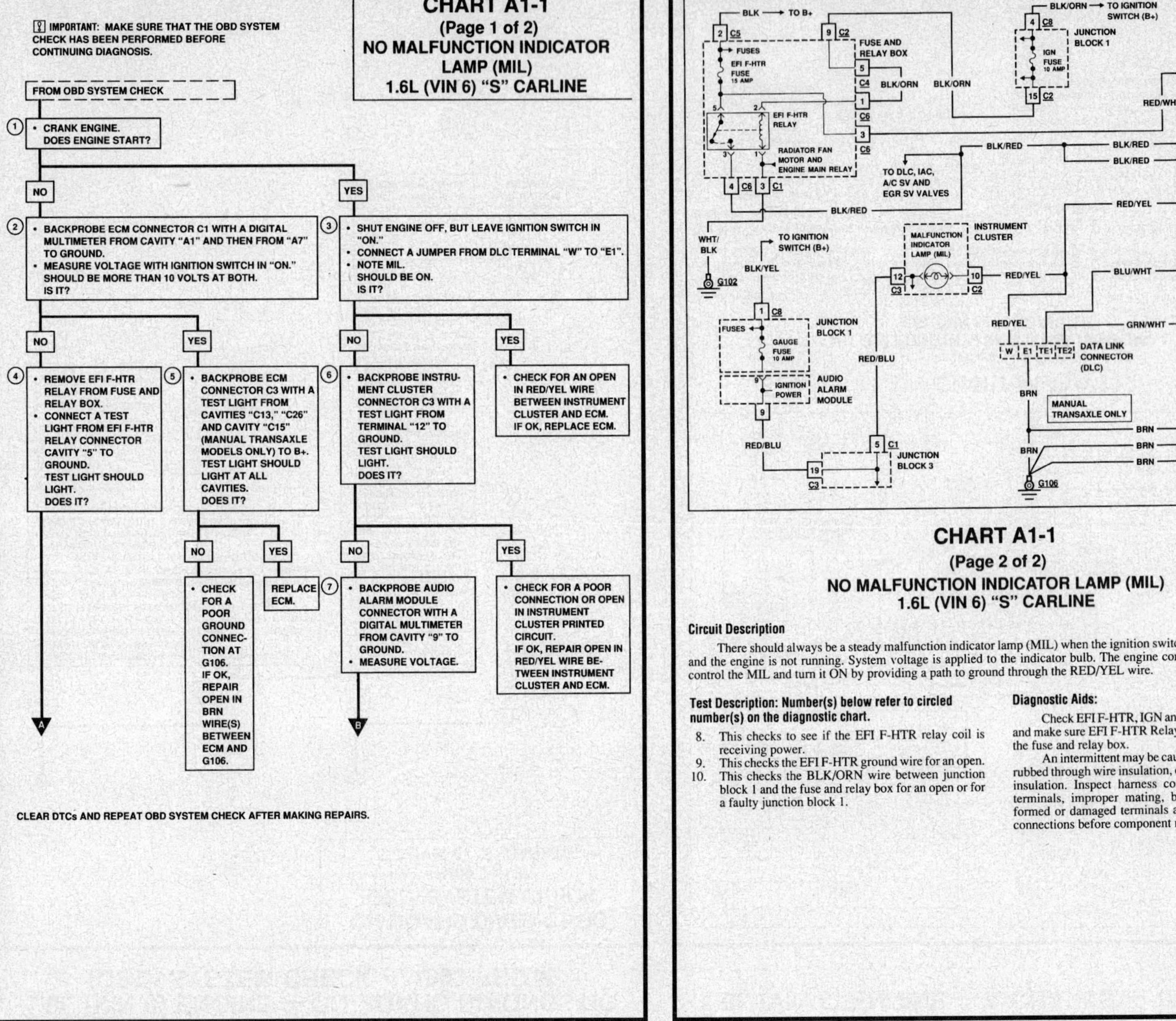

Circuit Description

There should always be a steady malfunction indicator lamp (MIL) when the ignition switch is in the "ON" position and the engine is not running. System voltage is applied to the indicator bulb. The engine control module (ECM) will control the MIL and turn it ON by providing a path to ground through the RED/YEL wire.

Test Description: Number(s) below refer to circled number(s) on the diagnostic chart.

8. This checks to see if the EFI F-HTR relay coil is receiving power.
9. This checks the EFI F-HTR ground wire for an open.
10. This checks the BLK/ORN wire between junction block 1 and the fuse and relay box for an open or for a faulty junction block 1.

Diagnostic Aids:

Check EFI F-HTR, IGN and GAUGE fuses for opens and make sure EFI F-HTR Relay is securely mounted into the fuse and relay box.

An intermittent may be caused by a poor connection, rubbed through wire insulation, or a wire broken inside the insulation. Inspect harness connectors for backed out terminals, improper mating, broken locks, improperly formed or damaged terminals and poor terminal-to-wire connections before component replacement.

1.6L (VIN 6) ENGINE — A-CHARTS — 1993 PRIZM

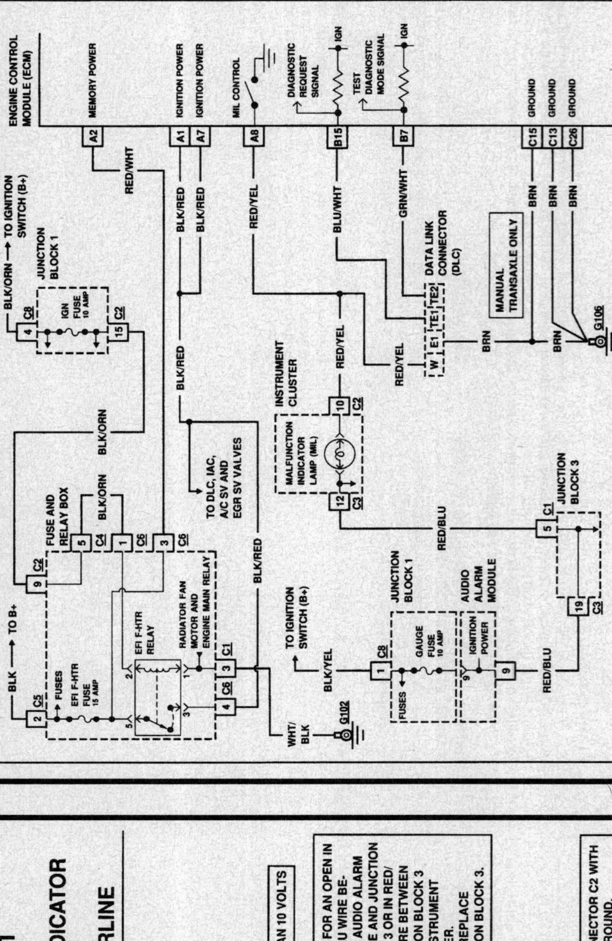

1.6L (VIN 6) ENGINE — A-CHARTS — 1993 PRIZM

CHART A1-2
MALFUNCTION INDICATOR LAMP (MIL) WILL NOT FLASH ANY DTCS, ON STEADY
1.6L (VIN 6) "S" CARLINE

Circuit Description:

The malfunction indicator lamp (MIL) will flash diagnostic trouble codes (DTCs) that are in the engine control modules (ECM's) memory when the data link connector (DLC) terminals "TE1" and "E1" are jumpered.

Test Description: Number(s) below refer to circled number(s) on the diagnostic chart.

1. This checks for a short to ground in RED/YEL wire.
2. This checks for an open BLU/WHT (diagnostic request signal) wire.
3. This checks for an open in the BRN ground wire to the DLC or for a faulty ECM.

Diagnostic Aids:

An intermittent may be caused by a poor connection, rubbed through wire insulation, or a wire broken inside the insulation. Inspect harness connectors for backed out terminals, improper mating, broken locks, improperly formed or damaged terminals and poor terminal-to-wire connections before component replacement.

CHART A1-1
(Page 2 of 2)
NO MALFUNCTION INDICATOR LAMP (MIL)
1.6L (VIN 6) "S" CARLINE

(8) CONNECT A TEST LIGHT FROM EFI F-HTR RELAY CONNECTOR CAVITY "2" TO GROUND. TEST LIGHT SHOULD LIGHT. DOES IT?

- **NO** → CHECK FOR AN OPEN IN BLK WIRE BETWEEN IGNITION SWITCH AND FUSE AND RELAY BOX. IF OK, REPLACE FUSE AND RELAY BOX.
- **YES** ↓

(9) CONNECT A TEST LIGHT FROM EFI F-HTR RELAY CONNECTOR CAVITY "1" TO B+. TEST LIGHT SHOULD LIGHT. DOES IT?

- **NO** → CHECK FOR A POOR CONNECTION AT G102. IF OK, REPAIR OPEN IN WHT/BLK WIRE BETWEEN FUSE AND RELAY BOX AND G102.
- **YES** → CHECK FOR AN OPEN IN BLK/RED WIRE BETWEEN FUSE AND ECM. IF OK, REPLACE EFI F-HTR RELAY.

B

- **LESS THAN 10 VOLTS** → CHECK FOR AN OPEN IN BLK/YEL WIRE BETWEEN IGNITION SWITCH AND JUNCTION BLOCK 1. IF OK, REPLACE AUDIO ALARM MODULE AND REPEAT OBD SYSTEM CHECK. IF MIL REMAINS INOPERATIVE, REPLACE JUNCTION BLOCK 1.
- **MORE THAN 10 VOLTS** → CHECK FOR AN OPEN IN RED/BLU WIRE BETWEEN AUDIO ALARM MODULE AND JUNCTION BLOCK 3 OR IN RED/BLU WIRE BETWEEN JUNCTION BLOCK 3 AND INSTRUMENT CLUSTER. IF OK, REPLACE JUNCTION BLOCK 3.

(10) BACKPROBE JUNCTION BLOCK 1 CONNECTOR C2 WITH A TEST LIGHT FROM CAVITY "15" TO GROUND. TEST LIGHT SHOULD LIGHT. DOES IT?

- **NO** → CHECK FOR AN OPEN IN BLK/ORN WIRE BETWEEN IGNITION SWITCH AND JUNCTION BLOCK 1. IF OK, REPLACE JUNCTION BLOCK 1.
- **YES** → CHECK FOR AN OPEN IN BLK/ORN WIRE BETWEEN JUNCTION BLOCK 1 AND FUSE AND RELAY BOX OR IN BLK/ORN WIRE BETWEEN FUSE AND RELAY BOX CONNECTORS C4 AND C6. IF OK, REPLACE JUNCTION BLOCK 1.

CLEAR DTCs AND REPEAT OBD SYSTEM CHECK AFTER MAKING REPAIRS.

CHART A1-2
MALFUNCTION INDICATOR LAMP (MIL) WILL NOT FLASH ANY DTCs, ON STEADY
1.6L (VIN 6) "S" CARLINE

IMPORTANT: MAKE SURE THAT THE OBD SYSTEM CHECK HAS BEEN PERFORMED BEFORE CONTINUING DIAGNOSIS.

FROM OBD SYSTEM CHECK

1.
- TURN IGNITION SWITCH TO "LOCK."
- REMOVE JUMPER FROM DLC.
- DISCONNECT ECM CONNECTOR C1.
- TURN IGNITION SWITCH TO "ON."
- NOTE MIL.
  SHOULD BE OFF.
  IS IT?

YES

NO → REPAIR SHORT TO GROUND IN RED/YEL WIRE BETWEEN INSTRUMENT CLUSTER, DLC AND ECM.

2.
- TURN IGNITION SWITCH TO "LOCK."
- RECONNECT ECM CONNECTOR C1.
- TURN IGNITION SWITCH TO "ON."
- CONNECT A DIGITAL MULTIMETER FROM DLC TERMINAL "TE1" TO GROUND.
- MEASURE VOLTAGE.

MORE THAN 10 VOLTS

LESS THAN 10 VOLTS → CHECK FOR AN OPEN IN BLU/WHT WIRE BETWEEN ECM AND DLC. IF OK, REPLACE ECM.

3.
- CONNECT A TEST LIGHT FROM DLC TERMINAL "E1" TO B+.
  TEST LIGHT SHOULD LIGHT.
  DOES IT?

NO → CHECK FOR A POOR GROUND CONNECTION AT G106. IF OK, REPAIR OPEN IN BRN WIRE BETWEEN DLC AND G106.

YES → REPLACE ECM.

CLEAR DTCs AND REPEAT OBD SYSTEM CHECK AFTER MAKING REPAIRS.

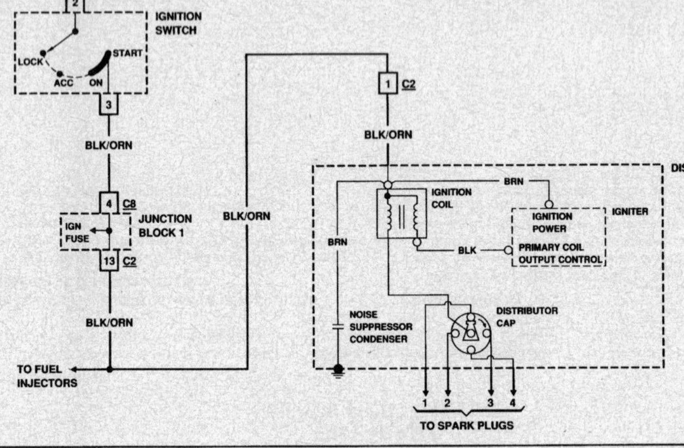

CHART A1-3
ENGINE CRANKS BUT WILL NOT RUN
1.6L (VIN 6) "S" CARLINE

Circuit Description:

When the ignition switch is turned to the "ON" or "START" positions, voltage is applied to the ignition coil and igniter. When the igniter receives an ignition signal from the engine control module (ECM), it toggles the primary coil windings of the ignition coil ON and OFF. By toggling these windings ON and OFF, a high voltage is produced in the secondary coil windings. The high voltage induced is forced to find a ground path through the distributor cap and each secondary spark plug wire to each spark plug.

Test Description: Number(s) below refer to circled number(s) on the diagnostic chart.

1. This checks for spark at the distributor cap.
2. This checks for faulty spark plug wires or for a mis-adjusted signal rotor air gap.

Diagnostic Aids:

Water or foreign material in the fuel can cause a no start or lack of power condition. Low fuel pressure, a restricted fuel filter or fuel injectors may also cause a no start condition. Refer to chart A1-5 for fuel injector circuit test.

An intermittent may be caused by a poor connection, rubbed through wire insulation, or a wire broken inside the insulation. Inspect harness connectors for backed out terminals, improper mating, broken locks, improperly formed or damaged terminals and poor terminal-to-wire connections before component replacement.

1.6L (VIN 6) ENGINE — A-CHARTS — 1993 PRIZM

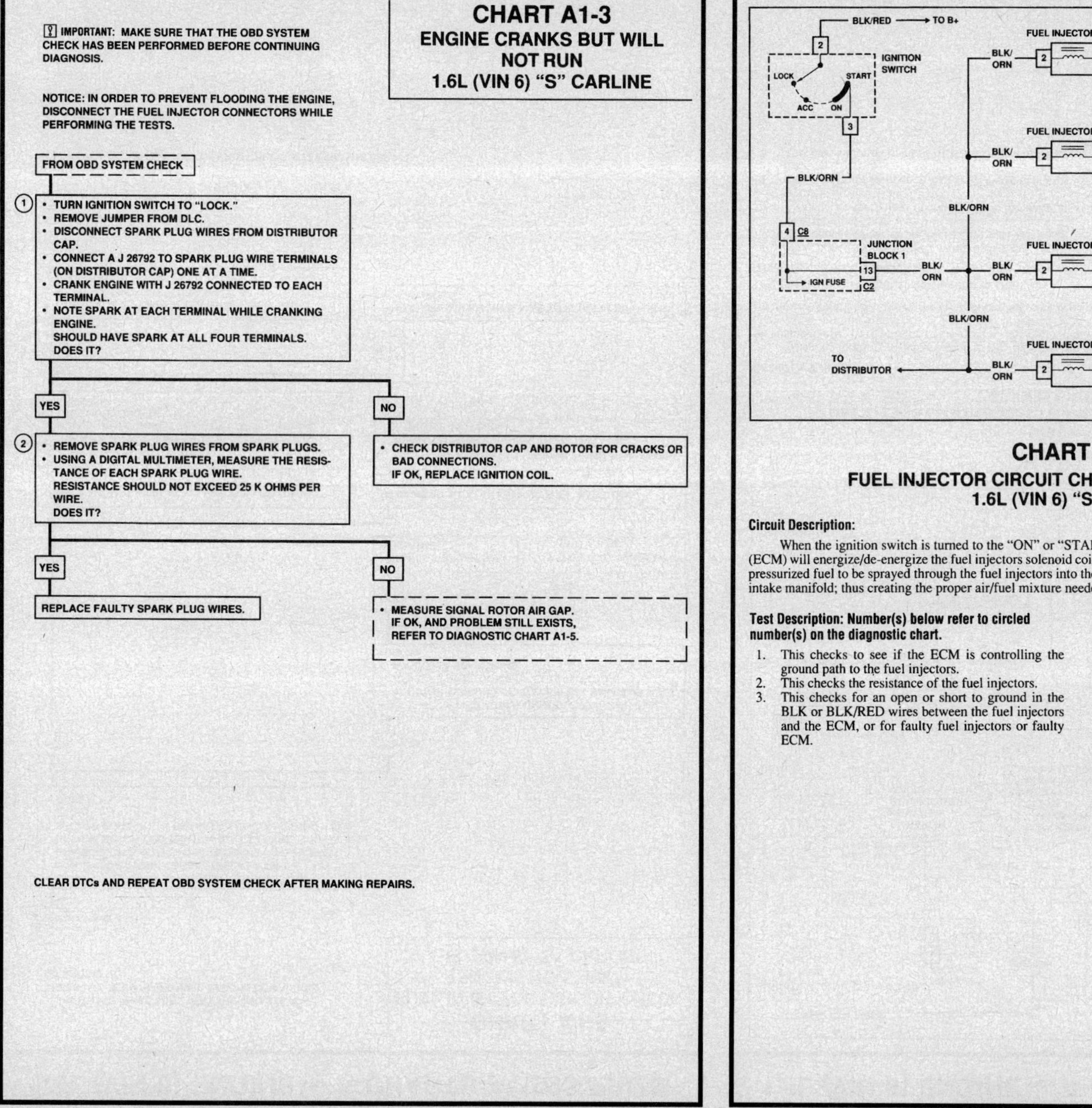

CHART A1-5
FUEL INJECTOR CIRCUIT CHECK (ENGINE NO-START)
1.6L (VIN 6) "S" CARLINE

Circuit Description:

When the ignition switch is turned to the "ON" or "START" position (engine running), the engine control module (ECM) will energize/de-energize the fuel injectors solenoid coil. With this coil energized, a plunger is activated, allowing pressurized fuel to be sprayed through the fuel injectors into the combustion chamber where it is mixed with air from the intake manifold; thus creating the proper air/fuel mixture needed for combustion.

Test Description: Number(s) below refer to circled number(s) on the diagnostic chart.

1. This checks to see if the ECM is controlling the ground path to the fuel injectors.
2. This checks the resistance of the fuel injectors.
3. This checks for an open or short to ground in the BLK or BLK/RED wires between the fuel injectors and the ECM, or for faulty fuel injectors or faulty ECM.

Diagnostic Aids:

There may be fuel spray at the fuel injectors, but it may not be enough to start the engine. If the fuel injectors and their circuits are OK, and fuel spray is detected, the fuel injector nozzle may be partly blocked or restricted, refer to Chart A1-7A for circuit opening relay circuit test.

An intermittent may be caused by a poor connection, rubbed through wire insulation, or a wire broken inside the insulation. Inspect harness connectors for backed out terminals, improper mating, broken locks, improperly formed or damaged terminals and poor terminal-to-wire connections before component replacement.

1.6L (VIN 6) ENGINE — A-CHARTS — 1993 PRIZM

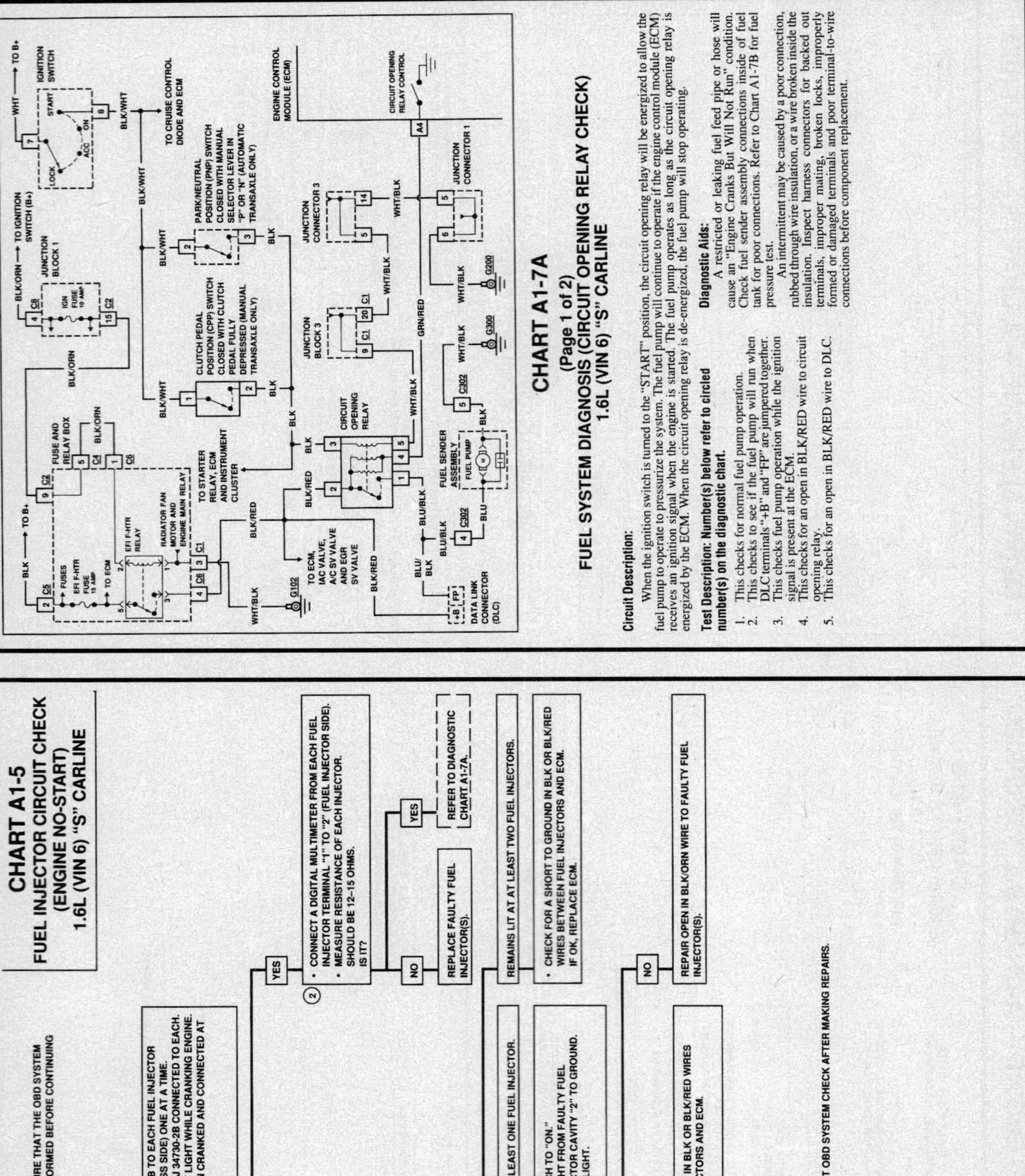

1.6L (VIN 6) ENGINE — A-CHARTS — 1993 PRIZM

CHART A1-7A
(Page 1 of 2)
FUEL SYSTEM DIAGNOSIS
(CIRCUIT OPENING RELAY CHECK)
1.6L (VIN 6) "S" CARLINE

IMPORTANT: MAKE SURE THAT THE OBD SYSTEM CHECK HAS BEEN PERFORMED BEFORE CONTINUING DIAGNOSIS.

FROM CHART A1-5

(1) • TURN IGNITION SWITCH TO "ON" AND MOMENTARILY BACKPROBE ECM CONNECTOR C1 WITH A FUSED JUMPER FROM CAVITY "A4" TO GROUND.
• LISTEN FOR FUEL PUMP TO RUN.
NOTICE: FUEL PUMP SHOULD RUN WHILE FUSED JUMPER IS CONNECTED.
DOES IT?

NO → (2) • MOMENTARILY CONNECT A JUMPER FROM DLC TERMINAL "+B" TO "FP".
• LISTEN FOR FUEL PUMP TO RUN.
DOES IT?

YES → (3) • CONNECT A DIGITAL MULTIMETER FROM DLC TERMINAL "FP" TO GROUND.
• MEASURE VOLTAGE WHILE CRANKING ENGINE.
SHOULD BE MORE THAN 10 VOLTS WHILE CRANKING.
IS IT?

NO → • CHECK FOR AN OPEN IN BLK WIRE BETWEEN CPP SWITCH/PNP SWITCH AND CIRCUIT OPENING RELAY OR FOR AN OPEN IN WHT/BLK WIRE BETWEEN CIRCUIT OPENING RELAY AND G200.
IF OK, REPLACE CIRCUIT OPENING RELAY AND RETEST SYSTEM. IF SYSTEM REMAINS INOPERATIVE, REPLACE ECM.

YES → REFER TO CHART A1-7B.

(4) • REMOVE JUMPER FROM DLC.
• BACKPROBE CIRCUIT OPENING RELAY CONNECTOR WITH A TEST LIGHT FROM CAVITY "2" TO GROUND.
TEST LIGHT SHOULD LIGHT.
DOES IT?

(5) • REMOVE JUMPER FROM DLC.
• CONNECT A TEST LIGHT FROM DLC TERMINAL "+B" TO GROUND.
TEST LIGHT SHOULD LIGHT.
DOES IT?

A

B

CLEAR DTCs AND REPEAT OBD SYSTEM CHECK AFTER MAKING REPAIRS.

1.6L (VIN 6) ENGINE — A-CHARTS — 1993 PRIZM

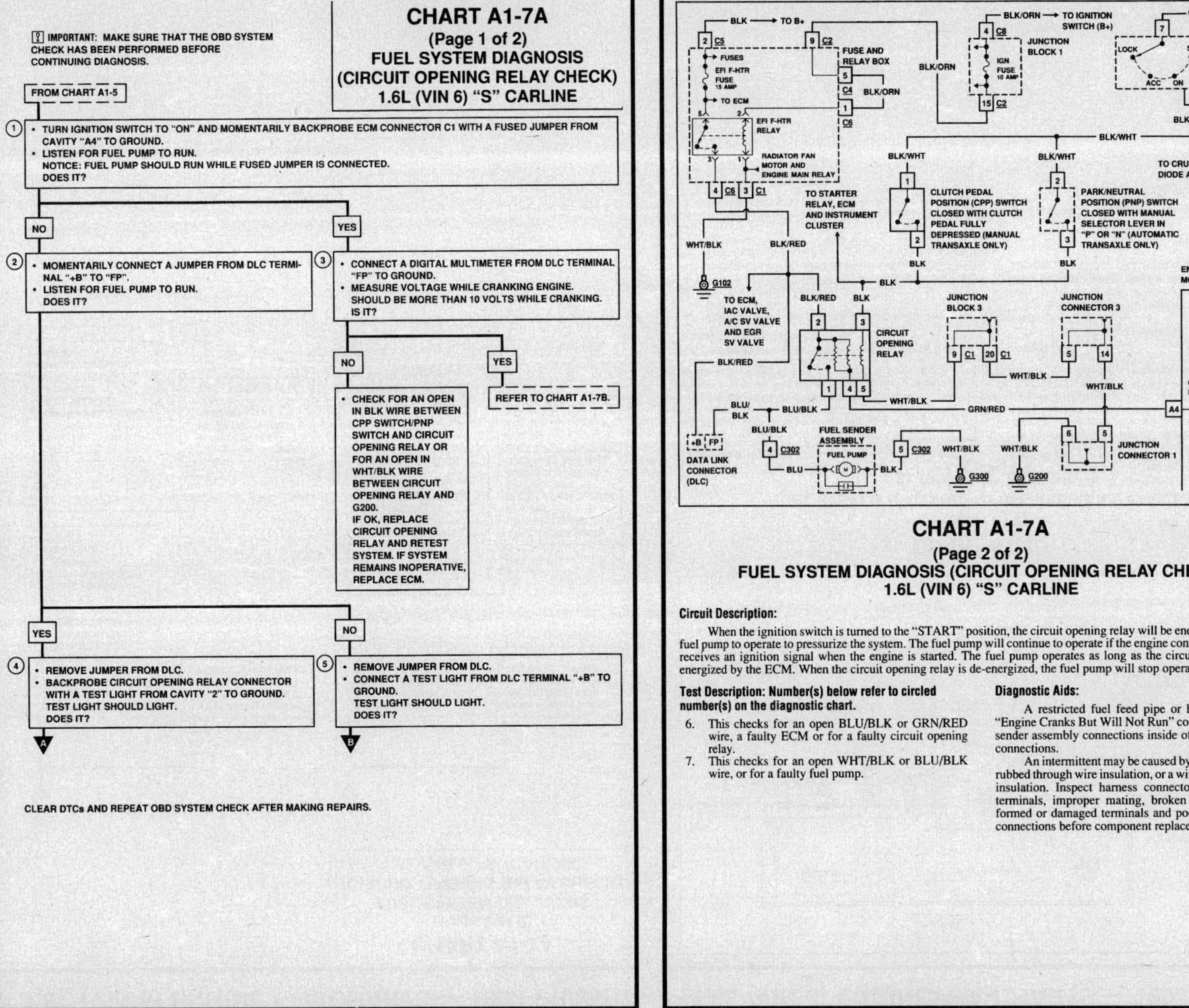

CHART A1-7A
(Page 2 of 2)
FUEL SYSTEM DIAGNOSIS (CIRCUIT OPENING RELAY CHECK)
1.6L (VIN 6) "S" CARLINE

Circuit Description:

When the ignition switch is turned to the "START" position, the circuit opening relay will be energized to allow the fuel pump to operate to pressurize the system. The fuel pump will continue to operate if the engine control module (ECM) receives an ignition signal when the engine is started. The fuel pump operates as long as the circuit opening relay is energized by the ECM. When the circuit opening relay is de-energized, the fuel pump will stop operating.

Test Description: Number(s) below refer to circled number(s) on the diagnostic chart.

6. This checks for an open BLU/BLK or GRN/RED wire, a faulty ECM or for a faulty circuit opening relay.
7. This checks for an open WHT/BLK or BLU/BLK wire, or for a faulty fuel pump.

Diagnostic Aids:

A restricted fuel feed pipe or hose will cause an "Engine Cranks But Will Not Run" condition. Check fuel sender assembly connections inside of fuel tank for poor connections.

An intermittent may be caused by a poor connection, rubbed through wire insulation, or a wire broken inside the insulation. Inspect harness connectors for backed out terminals, improper mating, broken locks, improperly formed or damaged terminals and poor terminal-to-wire connections before component replacement.

CHART A1-7A
(Page 2 of 2)
FUEL SYSTEM DIAGNOSIS
(CIRCUIT OPENING RELAY CHECK)
1.6L (VIN 6) "S" CARLINE

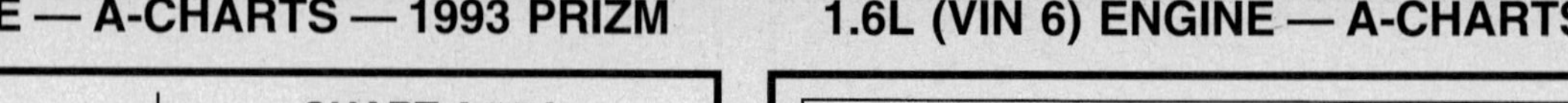

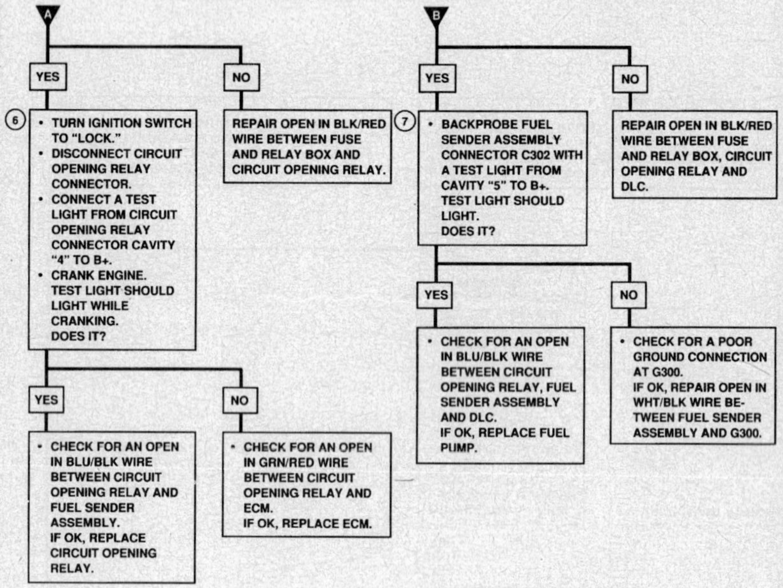

CLEAR DTCS AND REPEAT OBD SYSTEM CHECK AFTER MAKING REPAIRS.

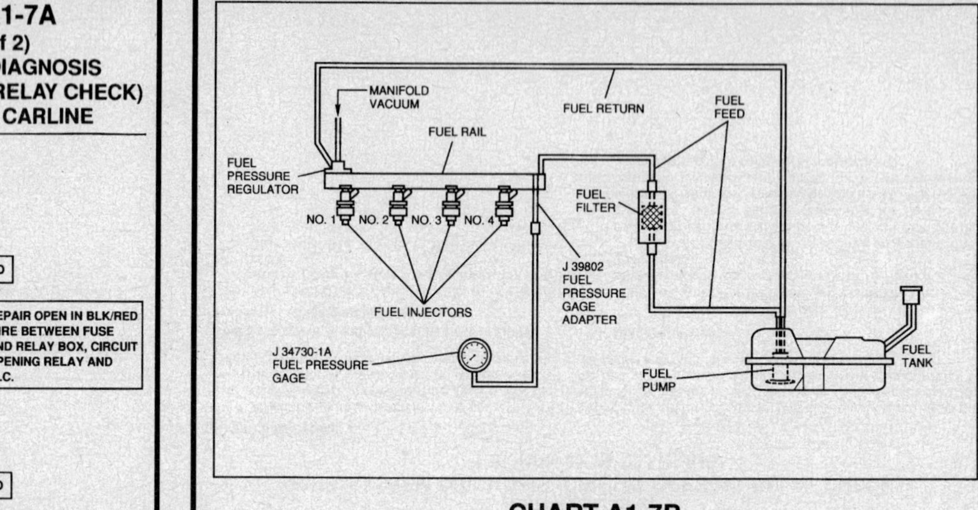

CHART A1-7B
(Page 1 of 2)
FUEL SYSTEM DIAGNOSIS (FUEL PRESSURE CHECK)
1.6L (VIN 6) "S" CARLINE

Circuit Description:

When the ignition switch is turned to the "START" position, the circuit opening relay will be energized to allow the fuel pump to operate to pressurize the system. The fuel pump will continue to operate if the engine control module (ECM) receives an ignition signal when the engine is started. The fuel pump operates as long as the circuit opening relay is energized by the ECM. When the circuit opening relay is de-energized, the fuel pump will stop operating. The fuel pump delivers fuel to the fuel rail where the pressure is maintained by the fuel pressure regulator at 284 kPa (41 psi) higher than the pressure inside the intake manifold. Excess fuel is returned to the fuel tank through a return pipe and hoses.

Test Description: Number(s) below refer to the circled number(s) on the diagnostic chart.

1. This measures the fuel pressure with the ignition switch in "ON."
2. This determines if the fuel pressure is normal.
3. This checks for a leaking fuel pressure regulator.
4. This checks for plugged fuel injectors.
5. This checks for leaking fuel injectors or a leaking fuel pump.

Diagnostic Aids:

Check for leakage around all fuel hose/pipe connections.

A plugged fuel injector could be the cause of a hard/no-start condition or loss of power/poor fuel economy. Refer to for mechanical checks and inspections if fuel spray and pressure is normal.

1.6L (VIN 6) ENGINE — A-CHARTS — 1993 PRIZM

1.6L (VIN 6) ENGINE — A-CHARTS — 1993 PRIZM

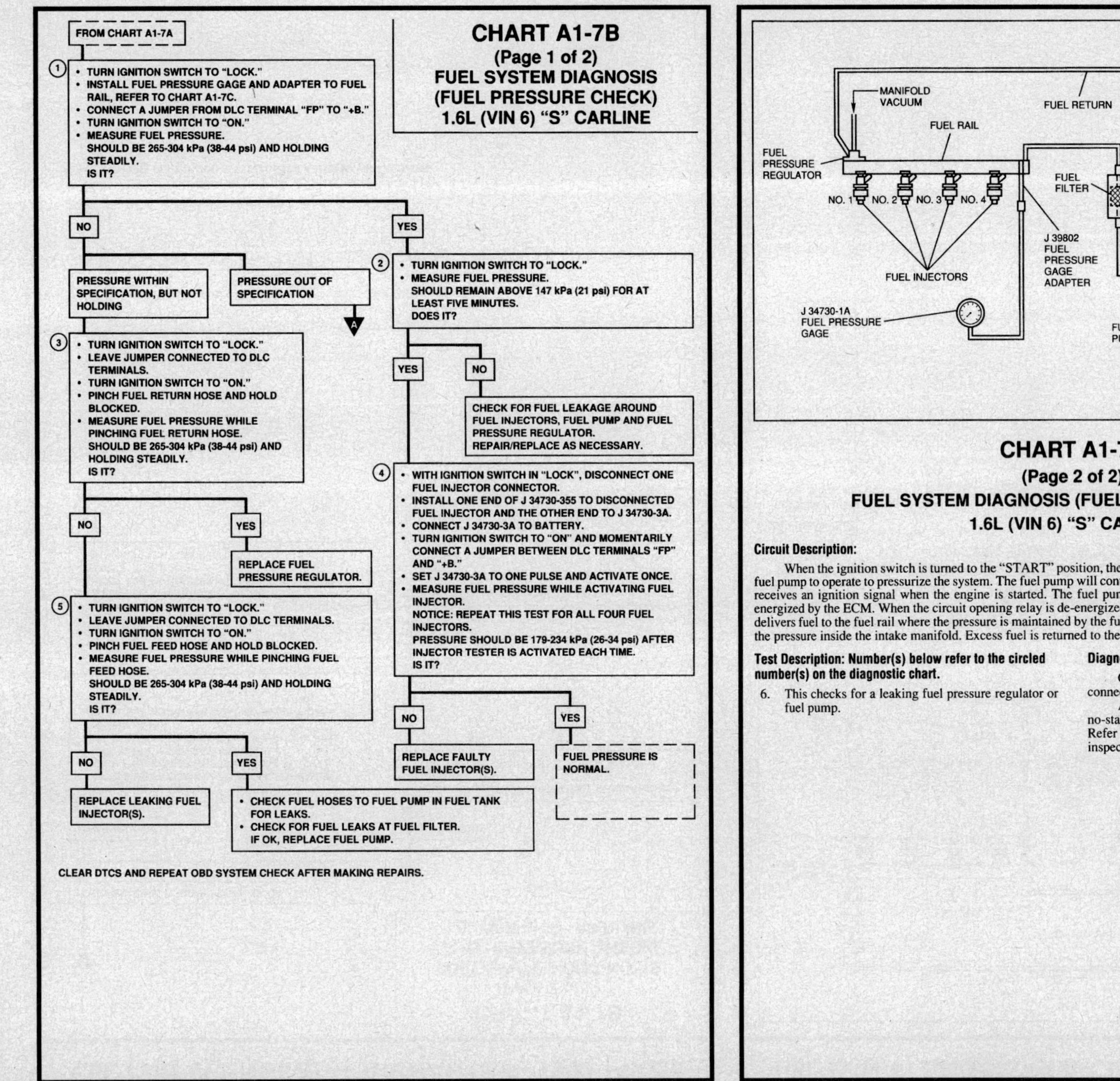

Circuit Description:

When the ignition switch is turned to the "START" position, the circuit opening relay will be energized to allow the fuel pump to operate to pressurize the system. The fuel pump will continue to operate if the engine control module (ECM) receives an ignition signal when the engine is started. The fuel pump operates as long as the circuit opening relay is energized by the ECM. When the circuit opening relay is de-energized, the fuel pump will stop operating. The fuel pump delivers fuel to the fuel rail where the pressure is maintained by the fuel pressure regulator at 284 kPa (41 psi) higher than the pressure inside the intake manifold. Excess fuel is returned to the fuel tank through a return pipe and hoses.

Test Description: Number(s) below refer to the circled number(s) on the diagnostic chart.

6. This checks for a leaking fuel pressure regulator or fuel pump.

Diagnostic Aids:

Check for leakage around all fuel hose/pipe connections.

A plugged fuel injector could be the cause of a hard/no-start condition or loss of power/poor fuel economy. Refer to for mechanical checks and inspections if fuel spray and pressure is normal.

CHART A1-7B
(Page 2 of 2)
FUEL SYSTEM DIAGNOSIS
(FUEL PRESSURE CHECK)
1.6L (VIN 6) "S" CARLINE

A

FUEL PRESSURE OUT OF SPECIFICATION

LOW PRESSURE

HIGH PRESSURE

6
- TURN IGNITION SWITCH TO "LOCK."
- LEAVE JUMPER CONNECTED TO DLC TERMINALS.
- PINCH FUEL RETURN HOSE AND HOLD BLOCKED.
- TURN IGNITION SWITCH TO "ON."
- MEASURE FUEL PRESSURE WHILE PINCHING FUEL RETURN HOSE.
  SHOULD BE 265-304 kPa (38-44 psi) AND HOLDING STEADILY.
  DOES IT?

- CHECK FOR RESTRICTIONS IN FUEL RETURN HOSES AND PIPES.
  IF OK, REPLACE FUEL PRESSURE REGULATOR.

NO

YES

- CHECK FOR A RESTRICTED FUEL FILTER.
- CHECK FOR RESTRICTED FUEL FEED HOSES AND PIPES.
  IF OK, REPLACE FUEL PUMP.

REPLACE FUEL PRESSURE REGULATOR.

CLEAR DTCs AND REPEAT OBD SYSTEM CHECK AFTER MAKING REPAIRS.

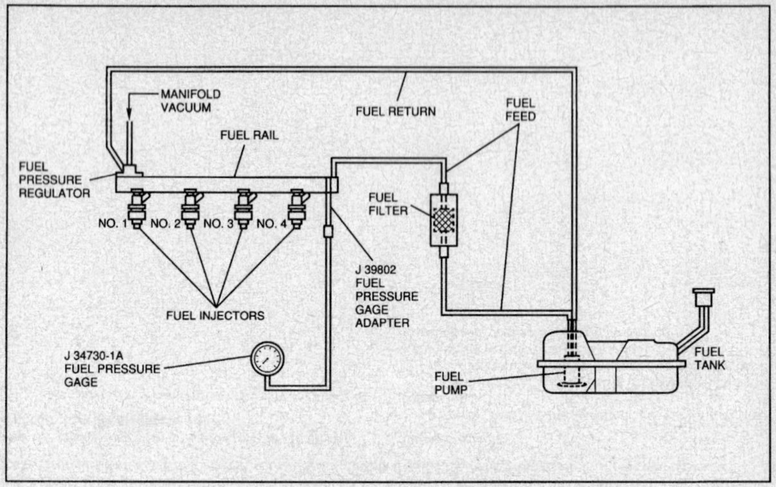

CHART A1-7C
FUEL SYSTEM DIAGNOSIS (FUEL PRESSURE GAGE)
1.6L (VIN 6) "S" CARLINE

Circuit Description:

The procedure on the right must be followed in order to install/remove the fuel pressure gage and adapter correctly. Make sure to observe all CAUTIONS while performing this procedure.

1.6L (VIN 6) ENGINE — A-CHARTS — 1993 PRIZM

1.6L (VIN 6) ENGINE — DIAGNOSTIC TROUBLE CODE CHARTS — 1993 PRIZM

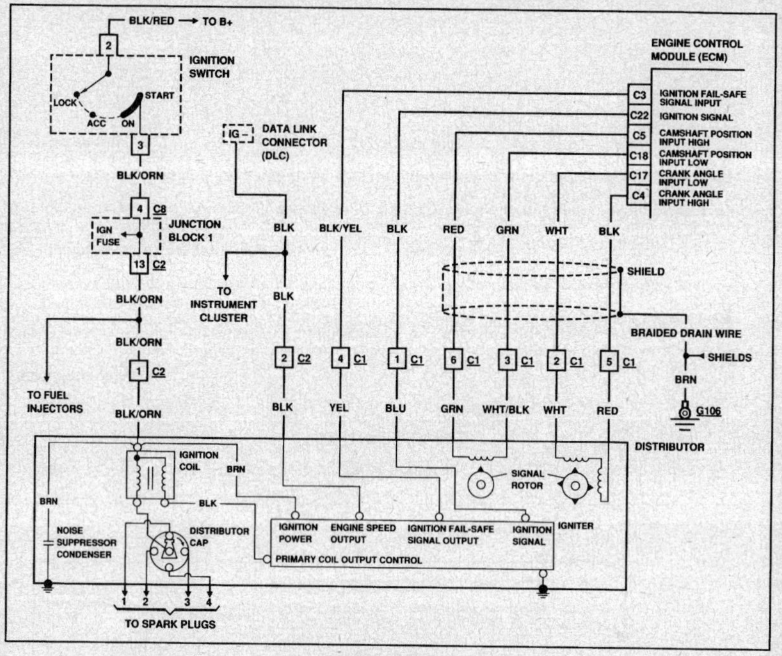

CHART A1-7C
FUEL SYSTEM DIAGNOSIS
(FUEL PRESSURE GAGE)
1.6L (VIN 6) "S" CARLINE

Tools Required:
J 34730-1A Fuel Pressure Gage
J 39802 Fuel Pressure Gage Adapter

CAUTION

- To reduce the risk of fire and personal injury, it is necessary to relieve the fuel system pressure before servicing the fuel system.
- After relieving fuel system pressure, a small amount of fuel may be released when servicing fuel pipe or hose connections. To reduce the chance of personal injury and to catch any fuel that may leak out, cover fuel pipe fittings with a shop towel before disconnecting them. Place the shop towel in an approved container when disconnect is completed.

FUEL PRESSURE GAGE INSTALLATION:

1. Relieve fuel system pressure.
   A. Loosen fuel filler cap to relieve tank pressure.
   B. Remove center trim bezel from center console by gently prying around edges to disconnect clips.
   C. Remove four screws and radio from center console (if equipped).
   D. Disconnect circuit opening relay electrical connector.
   E. Crank engine and allow to stall. Crank engine for an additional 3 seconds to assure relief of any remaining fuel pressure.
   F. Remove negative (-) battery cable.
   G. Reconnect circuit opening relay electrical connector.
   H. Tighten fuel filler cap.
2. Remove one bolt and fuel feed hose bracket from bottom on intake manifold.
3. Remove one bolt and fuel feed hose from fuel rail.
4. Install J 39802 with two gaskets and fuel feed hose to fuel rail.

   Tighten
   - J 39802 bolt to 29 N·m (22 lb.ft.).
5. Connect J 34730-1A to J 39802.
6. Reconnect negative (-) battery cable.

   Tighten
   - Negative (-) battery cable-to-negative (-) battery cable retainer to 15 N·m (11 lb.ft.).

FUEL PRESSURE GAGE REMOVAL

1. Relieve fuel system pressure.
   A. Loosen fuel filler cap to relieve tank pressure.
   B. Disconnect circuit opening relay electrical connector.
   C. Crank engine and allow to stall. Crank engine for an additional 3 seconds to assure relief of any remaining fuel pressure.
   D. Remove negative (-) battery cable.
   E. Reconnect circuit opening relay electrical connector.
   F. Install radio into center console (if equipped); secure with four screws.
   G. Install center trim bezel to center console, pushing in until clips engage.
   H. Tighten fuel filler cap.
2. Remove J 34730-1A from J 39802 and J 39802 from fuel rail.
3. Connect fuel feed hose with new gaskets to fuel rail.

   Tighten
   - Fuel feed hose bolt to 29 N·m (22 lb.ft.).
4. Connect fuel feed hose bracket to bottom of intake manifold, secure with one bolt.

   Tighten
   - Fuel feed hose bracket bolt to 10 N·m (89 lb. in.)
5. Reconnect negative (-) battery cable.

   Tighten
   - Negative (-) battery cable-to-negative (-) battery cable retainer to 15 N·m (11 lb.ft.).
6. Turn ignition switch to "ON" and momentarily connect a jumper between DLC terminals "+B" and "Fp" to pressurize the fuel system. Check for fuel leaks.

DTC 12
NO RPM SIGNAL
1.6L (VIN 6) "S" CARLINE

Circuit Description:

The engine control module (ECM) monitors the crank angle and the camshaft position. The ECM applies 0.70 volts to the crank angle and the camshaft position (CMP) sensors. This voltage is toggled high and low by the rotation of the signal rotor. It is by the toggled voltage input that the ECM can determine the crank angle and camshaft positions.
DTC 12 will set if either of the following conditions are met:
(1) No crank angle input detected at the ECM for 2 seconds or more while cranking engine.
(2) No camshaft position input detected at the ECM for 3 seconds or more with engine speed between 600 and 4,000 rpm.

Test Description: Number(s) below refer to circled number(s) on the diagnostic chart.
1. This checks the crank angle input low circuitry and ECM for faults.
2. This checks the crank angle input high circuitry and ECM for faults.
3. This checks the camshaft position input low circuitry and ECM for faults.
4. This checks the camshaft position input high circuitry and the ECM for faults and also for a misadjusted signal rotor air gap.

Diagnostic Aids:

Check distributor cap and rotor for cracks. Also check that signal rotor air gap is adjusted properly.
An intermittent may be caused by a poor connection, rubbed through wire insulation, or a wire broken inside the insulation. Inspect harness connectors for backed out terminals, improper mating, broken locks, improperly formed or damaged terminals and poor terminal-to-wire connections before component replacement.

DTC 12
NO RPM SIGNAL
1.6L (VIN 6) "S" CARLINE

IMPORTANT: MAKE SURE THAT THE OBD SYSTEM CHECK HAS BEEN PERFORMED BEFORE CONTINUING DIAGNOSIS.

FROM OBD SYSTEM CHECK

(1)
- TURN IGNITION SWITCH TO "LOCK."
- DISCONNECT DISTRIBUTOR CONNECTOR C1.
- CONNECT A DIGITAL MULTIMETER FROM DISTRIBUTOR CONNECTOR C1 CAVITY "2" TO GROUND.
- TURN IGNITION SWITCH TO "ON."
- MEASURE VOLTAGE. SHOULD BE 0.60–0.80 VOLTS. IS IT?

YES / NO

NO:
- CHECK FOR A POOR CONNECTION AT ECM.
- CHECK FOR AN OPEN, A SHORT TO GROUND OR A SHORT TO VOLTAGE IN WHT WIRE BETWEEN DISTRIBUTOR AND ECM. IF OK, REPLACE ECM.

(2)
- CONNECT A DIGITAL MULTIMETER FROM DISTRIBUTOR CONNECTOR C1 CAVITY "5" TO GROUND.
- MEASURE VOLTAGE. SHOULD BE 0.60–0.80 VOLTS. IS IT?

YES / NO

NO:
- CHECK FOR A POOR CONNECTION AT ECM.
- CHECK FOR AN OPEN, A SHORT TO GROUND OR A SHORT TO VOLTAGE IN BLK WIRE BETWEEN DISTRIBUTOR AND ECM. IF OK, REPLACE ECM.

(3)
- CONNECT A DIGITAL MULTIMETER FROM DISTRIBUTOR CONNECTOR C1 CAVITY "3" TO GROUND.
- MEASURE VOLTAGE. SHOULD BE 0.60–0.80 VOLTS. IS IT?

YES / NO

NO:
- CHECK FOR A POOR CONNECTION AT ECM.
- CHECK FOR AN OPEN, A SHORT TO GROUND OR A SHORT TO VOLTAGE IN GRN WIRE BETWEEN DISTRIBUTOR AND ECM. IF OK, REPLACE ECM.

(4)
- CONNECT A DIGITAL MULTIMETER FROM DISTRIBUTOR CONNECTOR C1 CAVITY "6" TO GROUND.
- MEASURE VOLTAGE. SHOULD BE 0.60–0.80 VOLTS. IS IT?

YES / NO

NO:
- CHECK FOR A POOR CONNECTION AT ECM.
- CHECK FOR AN OPEN, A SHORT TO GROUND OR A SHORT TO VOLTAGE IN RED WIRE BETWEEN DISTRIBUTOR AND ECM. IF OK, REPLACE ECM.

- CHECK SIGNAL ROTOR AIR GAP. IF OK, REPLACE DISTRIBUTOR HOUSING.

CLEAR DTCs AND REPEAT OBD SYSTEM CHECK AFTER MAKING REPAIRS.

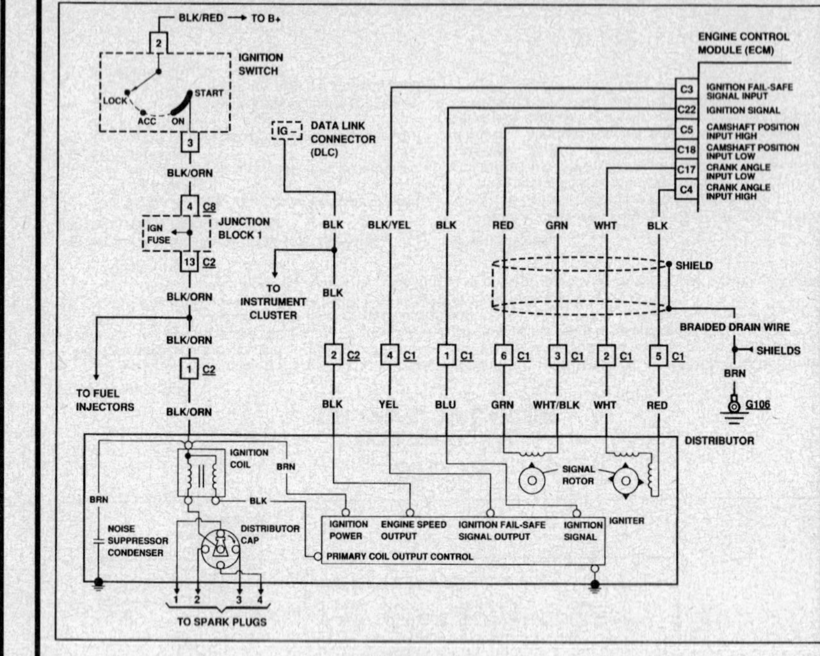

DTC 13
NO RPM SIGNAL
1.6L (VIN 6) "S" CARLINE

Circuit Description:

The engine control module (ECM) monitors the crank angle and the camshaft position. The ECM applies 0.70 volts to the crank angle and the camshaft position (CMP) sensors. This voltage is toggled high and low by the rotation of the signal rotor. It is by the toggled voltage input that the ECM can determine the crank angle and camshaft positions.

DTC 13 will set if the following condition is met:
- No crank angle input detected at the ECM for 0.30 seconds or more with engine speed above 1,500 rpm.

Test Description: Number(s) below refer to circled number(s) on the diagnostic chart.

1. This checks the crank angle input low circuitry and ECM for faults.
2. This checks the crank angle input high circuitry and ECM for faults and also for a misadjusted signal rotor air gap.

Diagnostic Aids:

Check distributor cap and rotor for cracks. Also check that signal rotor air gap is adjusted properly.

An intermittent may be caused by a poor connection, rubbed through wire insulation, or a wire broken inside the insulation. Inspect harness connectors for backed out terminals, improper mating, broken locks, improperly formed or damaged terminals and poor terminal-to-wire connections before component replacement.

1.6L (VIN 6) ENGINE — DIAGNOSTIC TROUBLE CODE CHARTS — 1993 PRIZM

1.6L (VIN 6) ENGINE — DIAGNOSTIC TROUBLE CODE CHARTS — 1993 PRIZM

DTC 13
NO RPM SIGNAL
1.6L (VIN 6) "S" CARLINE

[T] IMPORTANT: MAKE SURE THAT THE OBD SYSTEM CHECK HAS BEEN PERFORMED BEFORE CONTINUING DIAGNOSIS.

FROM OBD SYSTEM CHECK

(1)
- TURN IGNITION SWITCH TO "LOCK."
- DISCONNECT DISTRIBUTOR CONNECTOR C1.
- CONNECT A DIGITAL MULTIMETER FROM DISTRIBUTOR CONNECTOR C1 CAVITY "2" TO GROUND.
- TURN IGNITION SWITCH TO "ON."
- MEASURE VOLTAGE. SHOULD BE 0.60–0.80 VOLTS. IS IT?

YES

NO
- CHECK FOR A POOR CONNECTION AT ECM.
- CHECK FOR AN OPEN, A SHORT TO GROUND OR A SHORT TO VOLTAGE IN WHT WIRE BETWEEN DISTRIBUTOR AND ECM. IF OK, REPLACE ECM.

(2)
- CONNECT A DIGITAL MULTIMETER FROM DISTRIBUTOR CONNECTOR C1 CAVITY "5" TO GROUND.
- MEASURE VOLTAGE. SHOULD BE 0.60–0.80 VOLTS. IS IT?

YES

NO
- CHECK FOR A POOR CONNECTION AT ECM.
- CHECK FOR AN OPEN, A SHORT TO GROUND OR A SHORT TO VOLTAGE IN BLK WIRE BETWEEN DISTRIBUTOR AND ECM. IF OK, REPLACE ECM.

- CHECK SIGNAL ROTOR AIR GAP. REFER TO SECTION 6D4. IF OK, REPLACE DISTRIBUTOR HOUSING.

CLEAR DTCs AND REPEAT OBD SYSTEM CHECK AFTER MAKING REPAIRS.

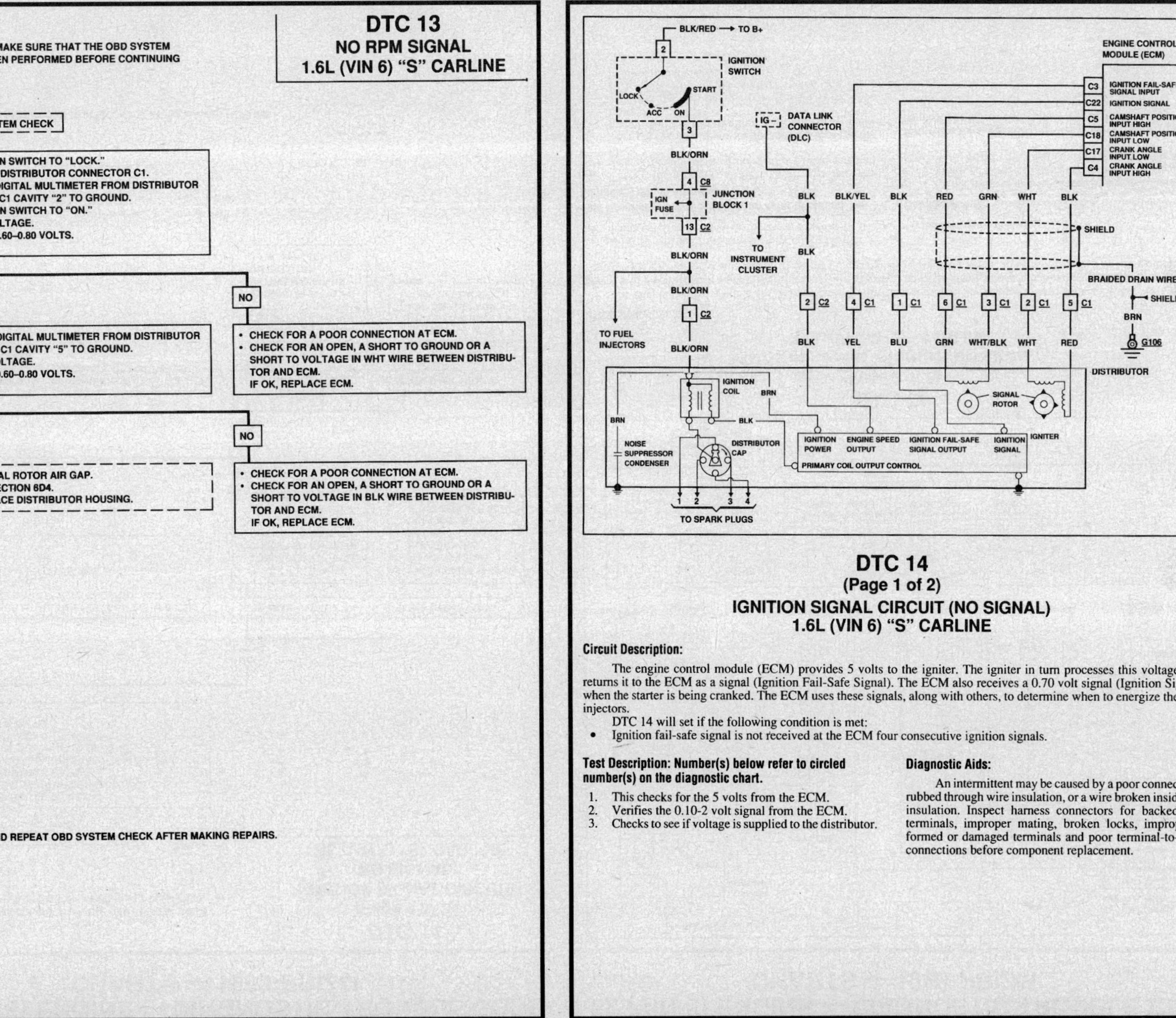

DTC 14
(Page 1 of 2)
IGNITION SIGNAL CIRCUIT (NO SIGNAL)
1.6L (VIN 6) "S" CARLINE

Circuit Description:

The engine control module (ECM) provides 5 volts to the igniter. The igniter in turn processes this voltage and returns it to the ECM as a signal (Ignition Fail-Safe Signal). The ECM also receives a 0.70 volt signal (Ignition Signal) when the starter is being cranked. The ECM uses these signals, along with others, to determine when to energize the fuel injectors.

DTC 14 will set if the following condition is met:
- Ignition fail-safe signal is not received at the ECM four consecutive ignition signals.

Test Description: Number(s) below refer to circled number(s) on the diagnostic chart.

1. This checks for the 5 volts from the ECM.
2. Verifies the 0.10-2 volt signal from the ECM.
3. Checks to see if voltage is supplied to the distributor.

Diagnostic Aids:

An intermittent may be caused by a poor connection, rubbed through wire insulation, or a wire broken inside the insulation. Inspect harness connectors for backed out terminals, improper mating, broken locks, improperly formed or damaged terminals and poor terminal-to-wire connections before component replacement.

DTC 14
(Page 1 of 2)
IGNITION SIGNAL CIRCUIT (NO SIGNAL)
1.6L (VIN 6) "S" CARLINE

IMPORTANT: MAKE SURE THAT THE OBD SYSTEM CHECK HAS BEEN PERFORMED BEFORE CONTINUING DIAGNOSIS.

FROM OBD SYSTEM CHECK

(1)
- TURN IGNITION SWITCH TO "LOCK."
- DISCONNECT DISTRIBUTOR CONNECTOR C1.
- TURN IGNITION SWITCH TO "ON."
- CONNECT A DIGITAL MULTIMETER FROM DISTRIBUTOR CONNECTOR C1 CAVITY "4" TO GROUND.
- MEASURE VOLTAGE. SHOULD BE 4–6 VOLTS. IS IT?

YES → (2)
NO → LESS THAN 4 VOLTS / MORE THAN 6 VOLTS

(2)
- CONNECT A DIGITAL MULTIMETER FROM DISTRIBUTOR CONNECTOR C1 CAVITY "1" TO GROUND.
- MEASURE VOLTAGE. SHOULD BE 0.10–2 VOLTS (VARIES WITH THROTTLE POSITION). IS IT?

LESS THAN 4 VOLTS
- CHECK FOR A POOR CONNECTION AT ECM.
- CHECK FOR AN OPEN OR FOR A SHORT TO GROUND IN BLK/YEL WIRE BETWEEN ECM AND DISTRIBUTOR. IF OK, REPLACE ECM.

MORE THAN 6 VOLTS
- CHECK FOR A SHORT TO VOLTAGE IN BLK/YEL WIRE BETWEEN ECM AND DISTRIBUTOR. IF OK, REPLACE ECM.

YES → (3)
NO → LESS THAN 0.10 VOLTS / MORE THAN 2 VOLTS

(3)
- BACKPROBE DISTRIBUTOR CONNECTOR C2 WITH A DIGITAL MULTIMETER FROM CAVITY "1" TO GROUND.
- MEASURE VOLTAGE. SHOULD BE MORE THAN 10 VOLTS. IS IT?

LESS THAN 0.10 VOLTS
- CHECK FOR A POOR CONNECTION AT ECM.
- CHECK FOR AN OPEN OR FOR A SHORT TO GROUND IN BLK WIRE BETWEEN ECM AND DISTRIBUTOR. IF OK, REPLACE ECM.

MORE THAN 2 VOLTS
- CHECK FOR A SHORT TO VOLTAGE IN BLK WIRE BETWEEN ECM AND DISTRIBUTOR. IF OK, REPLACE ECM.

A

CLEAR DTCs AND REPEAT OBD SYSTEM CHECK AFTER MAKING REPAIRS.

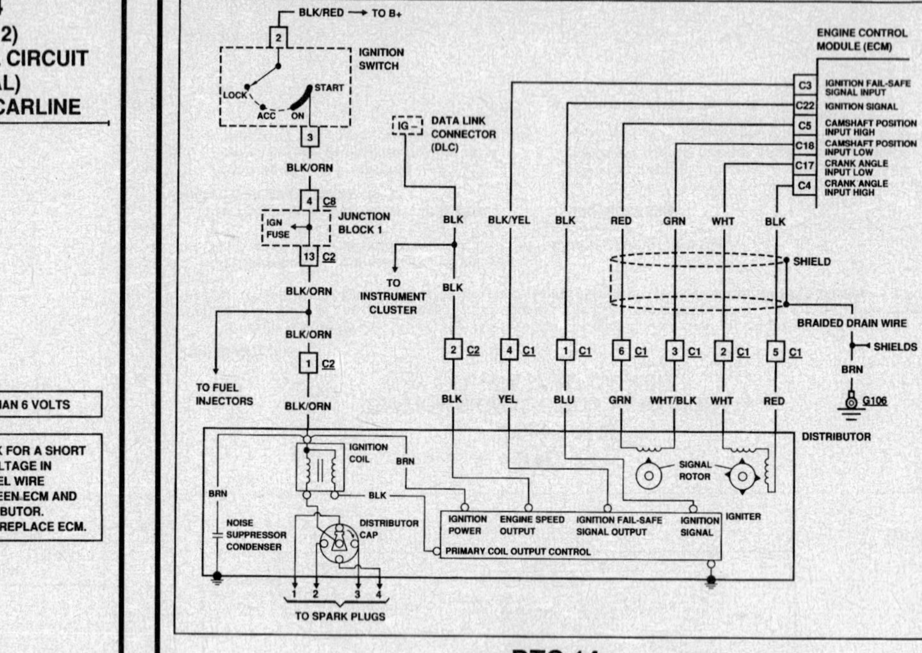

DTC 14
(Page 2 of 2)
IGNITION SIGNAL CIRCUIT (NO SIGNAL)
1.6L (VIN 6) "S" CARLINE

Circuit Description:

The engine control module (ECM) provides 5 volts to the igniter. The igniter in turn processes this voltage and returns it to the ECM as a signal (Ignition Fail-Safe Signal). The ECM also receives a 0.70 volt signal (Ignition Signal) when the starter is being cranked. The ECM uses these signals, along with others, to determine when to energize the fuel injectors.

DTC 14 will set if the following condition is met:
- Ignition fail-safe signal is not received at the ECM four consecutive ignition signals.

Test Description: Number(s) below refer to circled number(s) on the diagnostic chart.

4. Checks for an open in BLK/ORN wire between junction block 1 and distributor or for a faulty junction block 1.
5. This checks for opens in the power feed to the igniter, for a faulty ignition coil and for a faulty igniter.
6. This checks for an open in the RED/BLK or BLK/ORN wire to the ignition switch and for a faulty ignition switch.

Diagnostic Aids:

An intermittent may be caused by a poor connection, rubbed through wire insulation, or a wire broken inside the insulation. Inspect harness connectors for backed out terminals, improper mating, broken locks, improperly formed or damaged terminals and poor terminal-to-wire connections before component replacement.

1.6L (VIN 6) ENGINE — DIAGNOSTIC TROUBLE CODE CHARTS — 1993 PRIZM

DTC 14
(Page 2 of 2)
IGNITION SIGNAL CIRCUIT
(NO SIGNAL)
1.6L (VIN 6) "S" CARLINE

(4) • BACKPROBE JUNCTION BLOCK 1 CONNECTOR C8 WITH A TEST LIGHT FROM CAVITY "4" TO GROUND. TEST LIGHT SHOULD LIGHT. DOES IT?

(5) • REMOVE DISTRIBUTOR CAP FROM DISTRIBUTOR.
• CONNECT A DIGITAL MULTIMETER FROM IGNITION COIL BLK WIRE TERMINAL TO GROUND.
• MEASURE VOLTAGE. SHOULD BE MORE THEN 10 VOLTS. IS IT?

• CHECK BRN AND BLK WIRES BETWEEN IGNITION COIL AND IGNITER FOR OPENS. IF OK, REPLACE IGNITER.

REPLACE IGNITION COIL.

(6) • BACKPROBE IGNITION SWITCH CONNECTOR WITH A TEST LIGHT FROM CAVITY "2" TO GROUND. TEST LIGHT SHOULD LIGHT. DOES IT?

• CHECK FOR A POOR CONNECTION AT JUNCTION BLOCK 1.
• CHECK FOR AN OPEN IN BLK/ORN WIRE BETWEEN JUNCTION BLOCK 1 AND DISTRIBUTOR. IF OK, REPLACE JUNCTION BLOCK 1.

• CHECK FOR A POOR CONNECTION AT IGNITION SWITCH.
• CHECK FOR AN OPEN IN BLK/ORN WIRE BETWEEN IGNITION SWITCH AND JUNCTION BLOCK 1. IF OK, REPLACE IGNITION SWITCH.

REPAIR OPEN IN BLK/RED WIRE TO IGNITION SWITCH.

CLEAR DTCs AND REPEAT OBD SYSTEM CHECK AFTER MAKING REPAIRS.

1.6L (VIN 6) ENGINE — DIAGNOSTIC TROUBLE CODE CHARTS — 1993 PRIZM

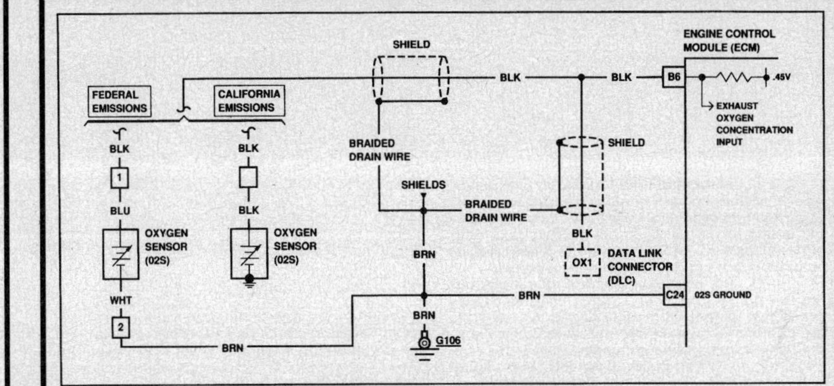

DTC 21
OXYGEN SENSOR (O2S) CIRCUIT
(OPEN/SHORTED CIRCUIT)
1.6L (VIN 6) "S" CARLINE

Circuit Description:

When the oxygen sensor (O2S) reaches operating temperature, it produces a varying voltage between 0.10 volts (exhaust is lean) and 1.0 volts (exhaust is rich). The engine control module (ECM) monitors this voltage and determines the concentration of gases in the exhaust. The ECM then uses this input to make fuel corrections to maintain the optimum air/fuel ratio.

DTC 21 will set if the following condition is met:
- O2S voltage input at the ECM is constant between 0.35 and 0.70 volts for 60 seconds or more with engine speed above 1,500 rpm and engine coolant temperature (ECT) above 80° C (176° F).

Test Description: Number(s) below refer to circled number(s) on the diagnostic chart.

1. This checks to see if the O2S is faulty.
2. This checks for an open in the BLK wire between the ECM and O2S.
3. This checks for an open in the BRN wire between the ECM and G106.
4. This checks for a short to ground in the BLK wire between the ECM and O2S, an open in the BRN wire to the O2S on federal emissions equipped vehicles and for a faulty ECM.

Diagnostic Aids:

Make sure that G106 is clean and tight.

An intermittent may be caused by a poor connection, rubbed through wire insulation, or a wire broken inside the insulation. Inspect harness connectors for backed out terminals, improper mating, broken locks, improperly formed or damaged terminals and poor terminal-to-wire connections before component replacement.

DTC 21
OXYGEN SENSOR (O2S) CIRCUIT
(OPEN/SHORTED CIRCUIT)
1.6L (VIN 6) "S" CARLINE

IMPORTANT: MAKE SURE THAT THE OBD SYSTEM CHECK HAS BEEN PERFORMED BEFORE CONTINUING DIAGNOSIS.

FROM OBD SYSTEM CHECK

(1)
- DISCONNECT O2S CONNECTOR.
- START AND RUN ENGINE AT 1,500 RPM FOR 1 MINUTE OR UNTIL NORMAL OPERATING TEMPERATURE IS ATTAINED.
- CONNECT A DIGITAL MULTIMETER FROM O2S CONNECTOR TERMINAL "1" (O2S SIDE) TO GROUND.
- MEASURE VOLTAGE.
SHOULD BE BETWEEN 0.80 AND 0.95 VOLTS.
IS IT?

YES → (2)
NO → REPLACE O2S.

(2)
- SHUT ENGINE OFF.
- DISCONNECT ECM CONNECTOR C2 AND C3.
- CONNECT A DIGITAL MULTIMETER FROM ECM CONNECTOR C2 CAVITY "B6" TO O2S CONNECTOR CAVITY "1" (HARNESS SIDE).
- MEASURE RESISTANCE.
SHOULD BE LESS THAN 0.50 OHMS.
IS IT?

YES → (3)
NO → REPAIR OPEN IN BLK WIRE BETWEEN ECM AND O2S.

(3)
- CONNECT A TEST LIGHT FROM ECM CONNECTOR C3 CAVITY "C24" TO B+.
TEST LIGHT SHOULD LIGHT.
DOES IT?

YES → (4)
NO → REPAIR OPEN IN BRN WIRE BETWEEN ECM AND G106.

(4)
- CONNECT A TEST LIGHT FROM ECM CONNECTOR C2 CAVITY "B6" TO B+.
TEST LIGHT SHOULD NOT LIGHT.
DOES IT?

NO → CHECK FOR AN OPEN IN BRN WIRE BETWEEN O2S AND G106 (FEDERAL EMISSIONS ONLY).
- CHECK FOR A POOR CONNECTION AT ECM.
IF OK, REPLACE ECM.

YES → REPAIR SHORT TO GROUND IN BLK WIRE BETWEEN ECM AND O2S.

CLEAR DTCs AND REPEAT OBD SYSTEM CHECK AFTER MAKING REPAIRS.

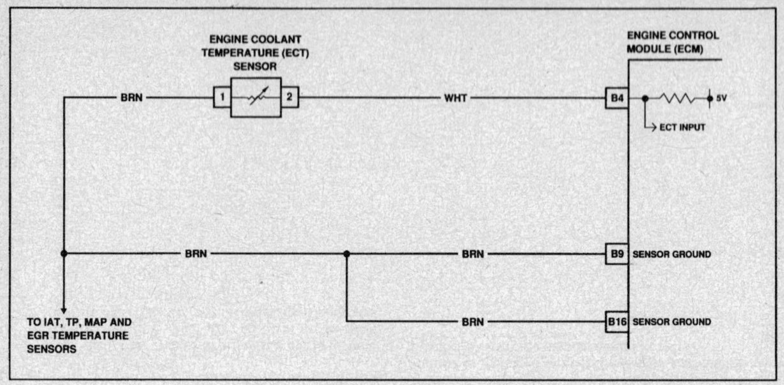

DTC 22
ENGINE COOLANT TEMPERATURE (ECT) SENSOR CIRCUIT
(OPEN/SHORTED CIRCUIT)
1.6L (VIN 6) "S" CARLINE

Circuit Description:

The engine coolant temperature (ECT) sensor is a thermistor (a variable resistor that changes along with ECT changes) in series with a fixed resistor within the engine control module (ECM). The ECM applies 5 volts to the sensor. The ECM monitors the voltage across the ECT sensor and converts it into a temperature reading. When the engine is cold the ECT sensor resistance is high, and when the engine is warm the ECT sensor resistance is low. Therefore, when the engine is cold the ECM will receive a high voltage input, and when the engine is warm the ECM will receive a low voltage input.

DTC 22 will set if either of the following conditions are met for at least 0.50 seconds:
(1) Voltage input at the ECM indicates a very low ECT.
(2) Voltage input at the ECM indicates a very high ECT.

Test Description: Number(s) below refer to circled number(s) on the diagnostic chart.

1. This checks for an open or short to ground in the WHT wire between the ECM and ECT sensor and for a faulty ECM.
2. This checks for an open in the BRN wire between the ECM and the ECT sensor and for a faulty ECM.
3. This determines if the problem is a faulty ECT sensor or ECM.

Diagnostic Aids:

Verify that the engine is not overheating and has not been subjected to conditions which could create an overheating condition (ie., overload, trailer towing, hilly terrain, heavy stop and go traffic, etc.).

A "shifted" (mis-scaled) sensor could result in poor driveability complaints. Measure the resistance of the ECT sensor according to the diagnostic aids chart on the diagnostic chart. If DTCs 24, 31 and 41 are also set, problem is open sensor ground circuit.

An intermittent may be caused by a poor connection, rubbed through wire insulation, or a wire broken inside the insulation. Inspect harness connectors for backed out terminals, improper mating, broken locks, improperly formed or damaged terminals and poor terminal-to-wire connections before component replacement.

1.6L (VIN 6) ENGINE — DIAGNOSTIC TROUBLE CODE CHARTS — 1993 PRIZM

[ⓘ] IMPORTANT: MAKE SURE THAT THE OBD SYSTEM CHECK HAS BEEN PERFORMED BEFORE CONTINUING DIAGNOSIS.

DTC 22
ENGINE COOLANT TEMPERATURE (ECT) SENSOR CIRCUIT
(OPEN/SHORTED CIRCUIT)
1.6L (VIN 6) "S" CARLINE

FROM OBD SYSTEM CHECK

① • TURN IGNITION SWITCH TO "LOCK."
• DISCONNECT ECT SENSOR CONNECTOR.
• TURN IGNITION SWITCH TO "ON."
• CONNECT A DIGITAL MULTIMETER FROM ECT SENSOR CONNECTOR CAVITY "2" TO GROUND.
• MEASURE VOLTAGE.
SHOULD BE 4–6 VOLTS.
IS IT?

YES / NO

NO → • CHECK FOR A POOR CONNECTION AT ECM AND AT ECT SENSOR.
• CHECK FOR AN OPEN OR SHORT TO GROUND IN WHT WIRE BETWEEN ECM AND ECT SENSOR.
IF OK, REPLACE ECM.

② • CONNECT A TEST LIGHT FROM ECT SENSOR CONNECTOR CAVITY "1" TO B+.
TEST LIGHT SHOULD LIGHT.
DOES IT?

YES / NO

NO → • CHECK FOR A POOR CONNECTION AT ECM AND AT ECT SENSOR.
• CHECK FOR AN OPEN IN BRN WIRE BETWEEN ECM AND ECT SENSOR.
IF OK, REPLACE ECM.

③ • TURN IGNITION SWITCH TO "LOCK."
• REMOVE ECT SENSOR.
• MEASURE RESISTANCE ACCORDING TO CHART BELOW. RESISTANCE SHOULD BE WITHIN SPECIFIED RANGES.
IS IT?

NO → REPLACE ECT SENSOR.

YES → REPLACE ECM.

DIAGNOSTIC AID		
ECT SENSOR		
TEMPERATURE TO RESISTANCE VALUES		
TEMPERATURE		RESISTANCE
-20°C	-4°F	10 – 20 K OHMS
0°C	32°F	4 – 7 K OHMS
20°C	68°F	2 – 3 K OHMS
40°C	104°F	0.9 – 1.3 K OHMS
60°C	140°F	0.4 – 0.7 K OHMS
80°C	176°F	0.2 – 0.4 K OHMS

CLEAR DTCs AND REPEAT OBD SYSTEM CHECK AFTER MAKING REPAIRS.

1.6L (VIN 6) ENGINE — DIAGNOSTIC TROUBLE CODE CHARTS — 1993 PRIZM

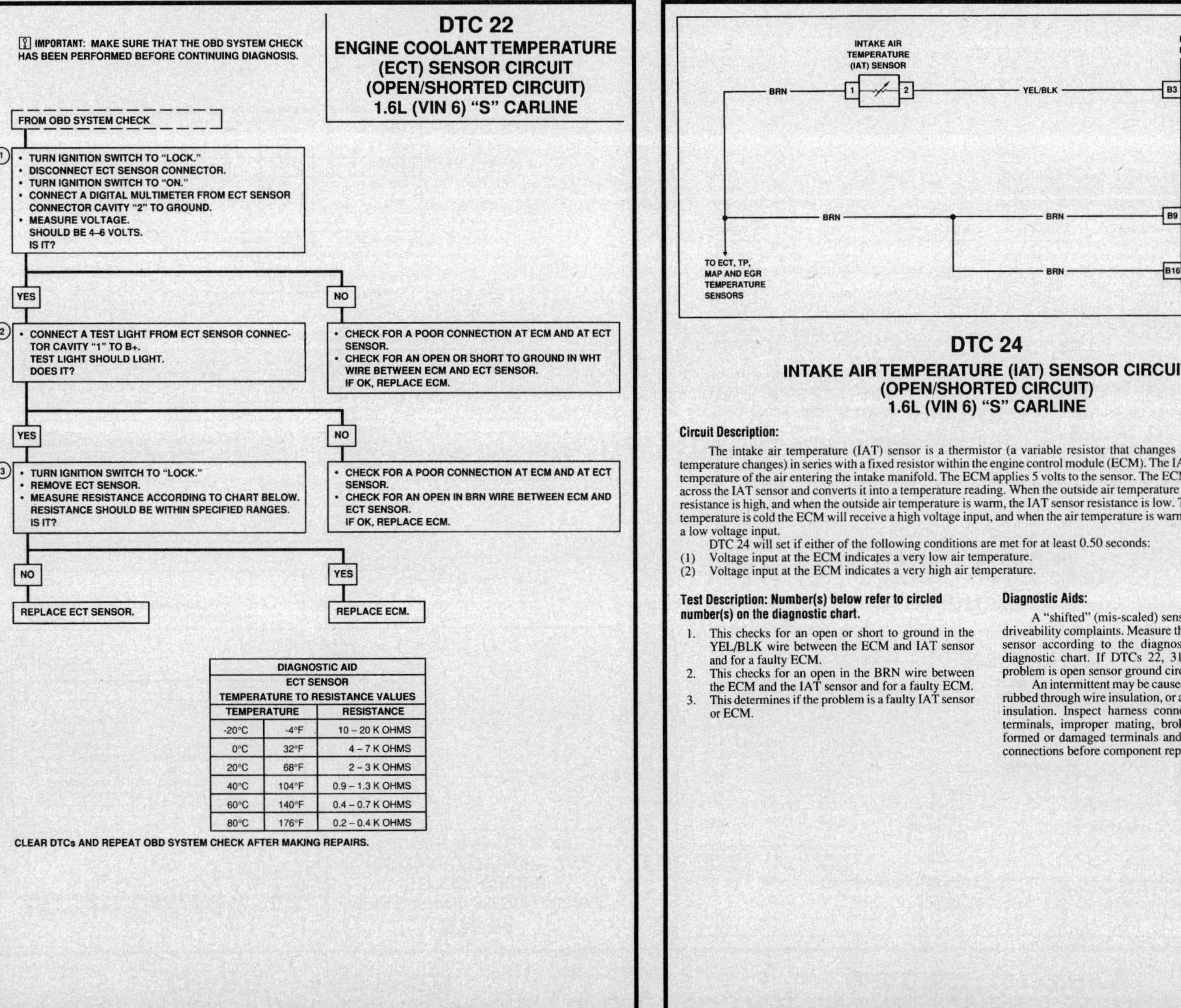

DTC 24
INTAKE AIR TEMPERATURE (IAT) SENSOR CIRCUIT
(OPEN/SHORTED CIRCUIT)
1.6L (VIN 6) "S" CARLINE

Circuit Description:

The intake air temperature (IAT) sensor is a thermistor (a variable resistor that changes along with outside air temperature changes) in series with a fixed resistor within the engine control module (ECM). The IAT sensor measures the temperature of the air entering the intake manifold. The ECM applies 5 volts to the sensor. The ECM monitors the voltage across the IAT sensor and converts it into a temperature reading. When the outside air temperature is cold, the IAT sensor resistance is high, and when the outside air temperature is warm, the IAT sensor resistance is low. Therefore, when the air temperature is cold the ECM will receive a high voltage input, and when the air temperature is warm the ECM will receive a low voltage input.

DTC 24 will set if either of the following conditions are met for at least 0.50 seconds:
(1) Voltage input at the ECM indicates a very low air temperature.
(2) Voltage input at the ECM indicates a very high air temperature.

Test Description: Number(s) below refer to circled number(s) on the diagnostic chart.

1. This checks for an open or short to ground in the YEL/BLK wire between the ECM and IAT sensor and for a faulty ECM.
2. This checks for an open in the BRN wire between the ECM and the IAT sensor and for a faulty ECM.
3. This determines if the problem is a faulty IAT sensor or ECM.

Diagnostic Aids:

A "shifted" (mis-scaled) sensor could result in poor driveability complaints. Measure the resistance of the IAT sensor according to the diagnostic aids chart on the diagnostic chart. If DTCs 22, 31 and 41 are also set, problem is open sensor ground circuit.

An intermittent may be caused by a poor connection, rubbed through wire insulation, or a wire broken inside the insulation. Inspect harness connectors for backed out terminals, improper mating, broken locks, improperly formed or damaged terminals and poor terminal-to-wire connections before component replacement.

1.6L (VIN 6) ENGINE — DIAGNOSTIC TROUBLE CODE CHARTS — 1993 PRIZM

DTC 24
INTAKE AIR TEMPERATURE (IAT) SENSOR CIRCUIT
(OPEN/SHORTED CIRCUIT)
1.6L (VIN 6) "S" CARLINE

? IMPORTANT: MAKE SURE THAT THE OBD SYSTEM CHECK HAS BEEN PERFORMED BEFORE CONTINUING DIAGNOSIS.

FROM OBD SYSTEM CHECK

(1)
- TURN IGNITION SWITCH TO "LOCK."
- DISCONNECT IAT SENSOR CONNECTOR.
- TURN IGNITION SWITCH TO "ON."
- CONNECT A DIGITAL MULTIMETER FROM IAT SENSOR CONNECTOR CAVITY "2" TO GROUND.
- MEASURE VOLTAGE. SHOULD BE 4–6 VOLTS. IS IT?

YES → (2)
- CONNECT A TEST LIGHT FROM IAT SENSOR CONNECTOR CAVITY "1" TO B+. TEST LIGHT SHOULD LIGHT. DOES IT?

NO →
- CHECK FOR A POOR CONNECTION AT ECM AND AT IAT SENSOR.
- CHECK FOR AN OPEN OR SHORT TO GROUND IN YEL/BLK WIRE BETWEEN ECM AND IAT SENSOR. IF OK, REPLACE ECM.

YES → (3)
- TURN IGNITION SWITCH TO "LOCK."
- REMOVE IAT SENSOR.
- MEASURE RESISTANCE ACCORDING TO CHART BELOW. RESISTANCE SHOULD BE WITHIN SPECIFIED RANGES. IS IT?

NO →
- CHECK FOR A POOR CONNECTION AT ECM AND AT IAT SENSOR.
- CHECK FOR AN OPEN IN BRN WIRE BETWEEN ECM AND IAT SENSOR. IF OK, REPLACE ECM.

NO → REPLACE IAT SENSOR.

YES → REPLACE ECM.

DIAGNOSTIC AID		
IAT SENSOR		
TEMPERATURE TO RESISTANCE VALUES		
TEMPERATURE		RESISTANCE
-20°C	-4°F	10 – 20 K OHMS
0°C	32°F	4 – 7 K OHMS
20°C	68°F	2 – 3 K OHMS
40°C	104°F	0.9 – 1.3 K OHMS
60°C	140°F	0.4 – 0.7 K OHMS
80°C	176°F	0.2 – 0.4 K OHMS

CLEAR DTCs AND REPEAT OBD SYSTEM CHECK AFTER MAKING REPAIRS.

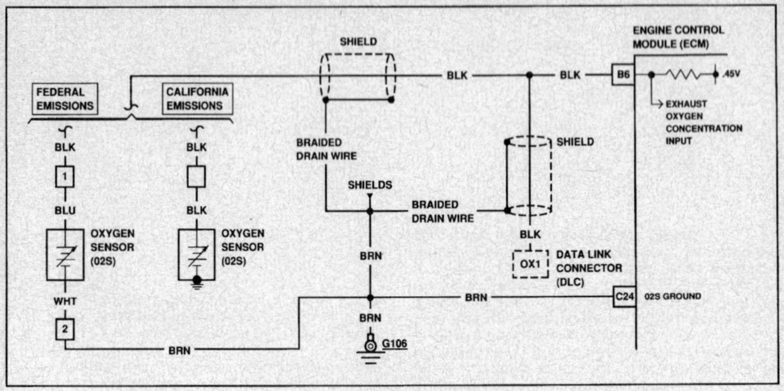

1.6L (VIN 6) ENGINE — DIAGNOSTIC TROUBLE CODE CHARTS — 1993 PRIZM

DTC 25
OXYGEN SENSOR (O2S) CIRCUIT
(AIR/FUEL RATIO LEAN)
1.6L (VIN 6) "S" CARLINE

Circuit Description:

When the oxygen sensor (O2S) reaches operating temperature, it produces a varying voltage between 0.10 volts (exhaust is lean) and 1.0 volts (exhaust is rich). The engine control module (ECM) monitors this voltage and determines the concentration of gases in the exhaust. The ECM then uses this input to make fuel corrections to maintain the optimum air/fuel ratio.

DTC 25 will set if either of the following condition are met:

(1) O2S voltage input at the ECM is less than 0.45 volts for 90 seconds or more with the O2S warm and engine speed above 1,500 rpm, vehicle speed below 100 km/h (60 mph), and with engine coolant temperature (ECT) above 50° C (122° F).

(2) When the engine speed varies by more than 15 rpm over the preceding crank angle period during a period of 30 seconds while idling with the ECT above 50°C (122°F).

Test Description: Number(s) below refer to circled number(s) on the diagnostic chart.

1. This checks to see if the O2S is faulty.
2. This checks for an open in the BLK wire between the ECM and O2S.
3. This checks for an open in the BRN wire between the ECM and G106.
4. This checks for a short to ground in the BLK wire between the ECM and O2S, an open in the BRN wire to the O2S on federal emissions equipped vehicles and for a faulty ECM.

Diagnostic Aids:

Make sure that G106 is clean and tight.

A lean air/fuel condition may also be caused by contaminated fuel (water in the fuel), low fuel pressure (leakage somewhere in the fuel system, restricted fuel filter or injectors or a faulty fuel pressure regulator) or even vacuum and exhaust leaks.

An intermittent may be caused by a poor connection, rubbed through wire insulation, or a wire broken inside the insulation. Inspect harness connectors for backed out terminals, improper mating, broken locks, improperly formed or damaged terminals and poor terminal-to-wire connections before component replacement.

1.6L (VIN 6) ENGINE — DIAGNOSTIC TROUBLE CODE CHARTS — 1993 PRIZM

IMPORTANT: MAKE SURE THAT THE OBD SYSTEM CHECK HAS BEEN PERFORMED BEFORE CONTINUING DIAGNOSIS.

DTC 25
OXYGEN SENSOR (O2S) CIRCUIT
(AIR/FUEL RATIO LEAN)
1.6L (VIN 6) "S" CARLINE

FROM OBD SYSTEM CHECK

1. • DISCONNECT O2S CONNECTOR.
   • START AND RUN ENGINE AT 1,500 RPM FOR 1 MINUTE OR UNTIL NORMAL OPERATING TEMPERATURE IS ATTAINED.
   • CONNECT A DIGITAL MULTIMETER FROM O2S CONNECTOR TERMINAL "1" (O2S SIDE) TO GROUND.
   • MEASURE VOLTAGE.
   SHOULD BE BETWEEN 0.80 AND 0.95 VOLTS.
   IS IT?

   YES → | NO → REPLACE O2S.

2. • SHUT ENGINE OFF.
   • DISCONNECT ECM CONNECTOR C2 AND C3.
   • CONNECT A DIGITAL MULTIMETER FROM ECM CONNECTOR C2 CAVITY "B6" TO O2S CONNECTOR CAVITY "1" (HARNESS SIDE).
   • MEASURE RESISTANCE.
   SHOULD BE LESS THAN 0.50 OHMS.
   IS IT?

   YES → | NO → REPAIR OPEN IN BLK WIRE BETWEEN ECM AND O2S.

3. • CONNECT A TEST LIGHT FROM ECM CONNECTOR C3 CAVITY "C24" TO B+.
   TEST LIGHT SHOULD LIGHT.
   DOES IT?

   YES → | NO → REPAIR OPEN IN BRN WIRE BETWEEN ECM AND G106.

4. • CONNECT A TEST LIGHT FROM ECM CONNECTOR C2 CAVITY "B6" TO B+.
   TEST LIGHT SHOULD NOT LIGHT.
   DOES IT?

   NO → CHECK FOR AN OPEN IN BRN WIRE BETWEEN O2S AND G106 (FEDERAL EMISSIONS ONLY).
   • CHECK FOR A POOR CONNECTION AT ECM. IF OK, REPLACE ECM.

   YES → REPAIR SHORT TO GROUND IN BLK WIRE BETWEEN ECM AND O2S.

CLEAR DTCs AND REPEAT OBD SYSTEM CHECK AFTER MAKING REPAIRS.

1.6L (VIN 6) ENGINE — DIAGNOSTIC TROUBLE CODE CHARTS — 1993 PRIZM

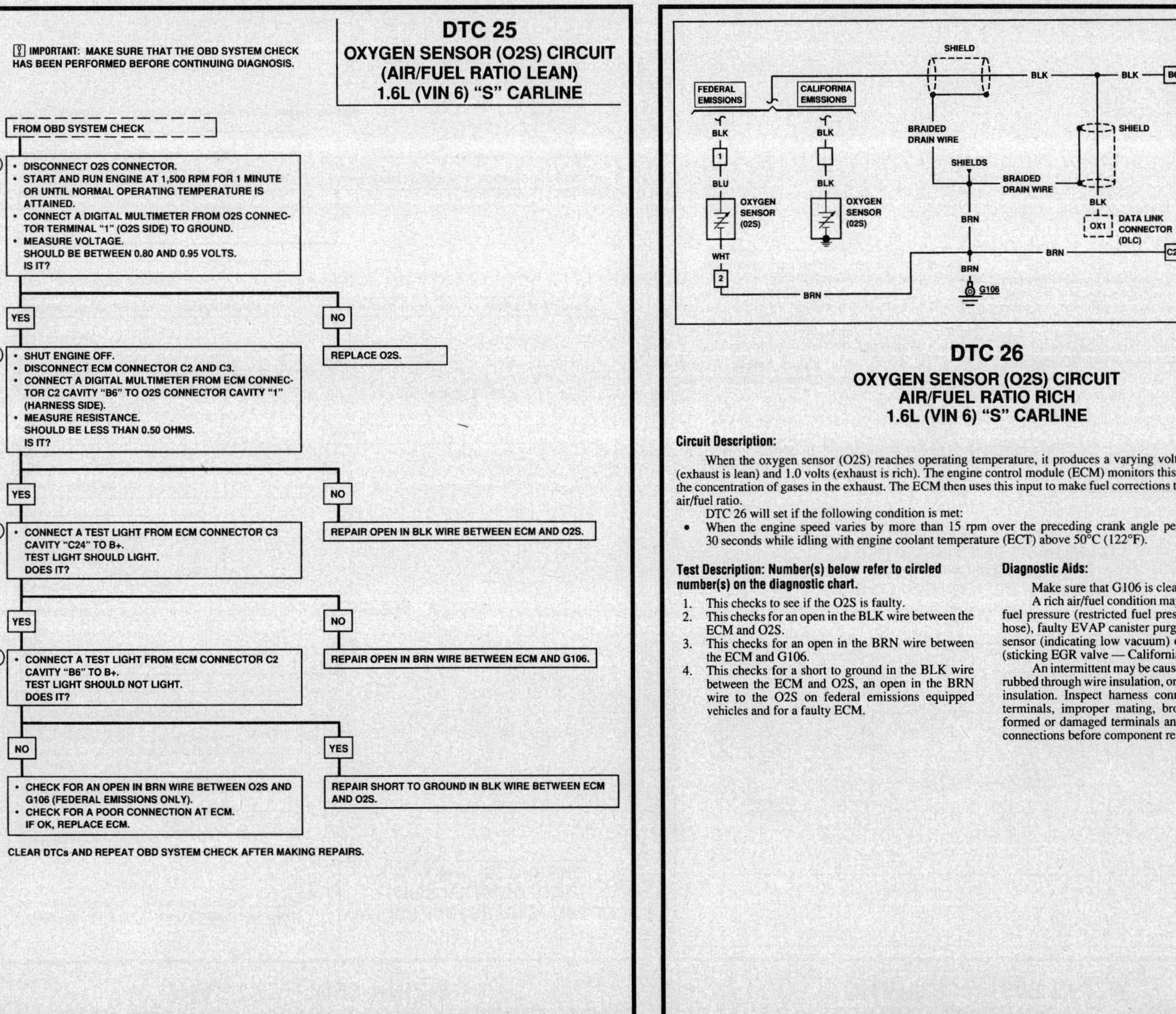

DTC 26
OXYGEN SENSOR (O2S) CIRCUIT
AIR/FUEL RATIO RICH
1.6L (VIN 6) "S" CARLINE

Circuit Description:

When the oxygen sensor (O2S) reaches operating temperature, it produces a varying voltage between 0.10 volts (exhaust is lean) and 1.0 volts (exhaust is rich). The engine control module (ECM) monitors this voltage and determines the concentration of gases in the exhaust. The ECM then uses this input to make fuel corrections to maintain the optimum air/fuel ratio.

DTC 26 will set if the following condition is met:
• When the engine speed varies by more than 15 rpm over the preceding crank angle period during a period of 30 seconds while idling with engine coolant temperature (ECT) above 50°C (122°F).

Test Description: Number(s) below refer to circled number(s) on the diagnostic chart.

1. This checks to see if the O2S is faulty.
2. This checks for an open in the BLK wire between the ECM and O2S.
3. This checks for an open in the BRN wire between the ECM and G106.
4. This checks for a short to ground in the BLK wire between the ECM and O2S, an open in the BRN wire to the O2S on federal emissions equipped vehicles and for a faulty ECM.

Diagnostic Aids:

Make sure that G106 is clean and tight.

A rich air/fuel condition may also be caused by high fuel pressure (restricted fuel pressure regulator or return hose), faulty EVAP canister purge system, a faulty MAP sensor (indicating low vacuum) or a faulty EGR system (sticking EGR valve — California emissions only).

An intermittent may be caused by a poor connection, rubbed through wire insulation, or a wire broken inside the insulation. Inspect harness connectors for backed out terminals, improper mating, broken locks, improperly formed or damaged terminals and poor terminal-to-wire connections before component replacement.

IMPORTANT: MAKE SURE THAT THE OBD SYSTEM CHECK HAS BEEN PERFORMED BEFORE CONTINUING DIAGNOSIS.

FROM OBD SYSTEM CHECK

① • DISCONNECT O2S CONNECTOR.
• START AND RUN ENGINE AT 1,500 RPM FOR 1 MINUTE OR UNTIL NORMAL OPERATING TEMPERATURE IS ATTAINED.
• CONNECT A DIGITAL MULTIMETER FROM O2S CONNECTOR TERMINAL "1" (O2S SIDE) TO GROUND.
• MEASURE VOLTAGE.
SHOULD BE BETWEEN 0.80 AND 0.95 VOLTS.
IS IT?

YES →

NO → REPLACE O2S.

② • SHUT ENGINE OFF.
• DISCONNECT ECM CONNECTOR C2 AND C3.
• CONNECT A DIGITAL MULTIMETER FROM ECM CONNECTOR C2 CAVITY "B6" TO O2S CONNECTOR CAVITY "1" (HARNESS SIDE).
• MEASURE RESISTANCE.
SHOULD BE LESS THAN 0.50 OHMS.
IS IT?

YES →

NO →

③ • CONNECT A TEST LIGHT FROM ECM CONNECTOR C3 CAVITY "C24" TO B+.
TEST LIGHT SHOULD LIGHT.
DOES IT?

REPAIR OPEN IN BLK WIRE BETWEEN ECM AND O2S.

YES →

NO →

④ • CONNECT A TEST LIGHT FROM ECM CONNECTOR C2 CAVITY "B6" TO B+.
TEST LIGHT SHOULD NOT LIGHT.
DOES IT?

REPAIR OPEN IN BRN WIRE BETWEEN ECM AND G106.

NO →

YES →

• CHECK FOR AN OPEN IN BRN WIRE BETWEEN O2S AND G106 (FEDERAL EMISSIONS ONLY).
• CHECK FOR A POOR CONNECTION AT ECM.
IF OK, REPLACE ECM.

REPAIR SHORT TO GROUND IN BLK WIRE BETWEEN ECM AND O2S.

CLEAR DTCs AND REPEAT OBD SYSTEM CHECK AFTER MAKING REPAIRS.

DTC 26
OXYGEN SENSOR (O2S) CIRCUIT
(AIR/FUEL RATIO RICH)
1.6L (VIN 6) "S" CARLINE

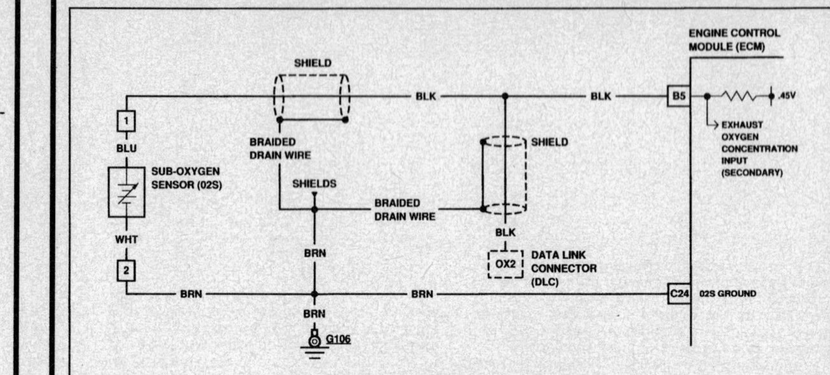

DTC 27
SUB-OXYGEN SENSOR (O2S) CIRCUIT
(OPEN/SHORTED CIRCUIT)
(CALIFORNIA EMISSIONS ONLY)
1.6L (VIN 6) "S" CARLINE

Circuit Description:

When the sub-oxygen sensor (O2S) reaches operating temperature, it produces a varying voltage between 0.10 volts (exhaust is lean) and 1.0 volts (exhaust is rich). The engine control module (ECM) monitors this voltage and determines the concentration of gases in the exhaust after the converter. The ECM then uses this input along with the oxygen sensor (O2S) input to make fuel corrections to maintain the optimum air/fuel ratio.

DTC 27 will set if the following condition is met:
• When the engine is warmed up and the engine is off idle for 2 seconds or more with the oxygen sensor input at the ECM 0.45 volts or more and the sub-oxygen sensor input at the ECM 0.45 volts or less.

Test Description: Number(s) below refer to circled number(s) on the diagnostic chart.

1. This checks to see if the sub-O2S is faulty.
2. This checks for an open in the BLK wire between the ECM and sub-O2S.
3. This checks for an open in the BRN wire between the ECM and G106.
4. This checks for a short to ground in the BLK wire between the ECM and sub-O2S, an open in the BRN wire to the sub-O2S and for a faulty ECM.

Diagnostic Aids:

Make sure that G106 is clean and tight.

An intermittent may be caused by a poor connection, rubbed through wire insulation, or a wire broken inside the insulation. Inspect harness connectors for backed out terminals, improper mating, broken locks, improperly formed or damaged terminals and poor terminal-to-wire connections before component replacement.

1.6L (VIN 6) ENGINE — DIAGNOSTIC TROUBLE CODE CHARTS — 1993 PRIZM

DTC 27
SUB-OXYGEN SENSOR (O2S) CIRCUIT
(OPEN/SHORTED CIRCUIT)
(CALIFORNIA EMISSIONS ONLY)
1.6L (VIN 6) "S" CARLINE

FROM OBD SYSTEM CHECK

⚠ IMPORTANT: MAKE SURE THAT THE OBD SYSTEM CHECK HAS BEEN PERFORMED BEFORE CONTINUING DIAGNOSIS.

(1)
- START AND RUN ENGINE AT 1,500 RPM FOR 1 MINUTE OR UNTIL NORMAL OPERATING TEMPERATURE IS ATTAINED.
- RAISE AND SUITABLY SUPPORT VEHICLE.
- DISCONNECT SUB-O2S CONNECTOR.
- CONNECT A DIGITAL MULTIMETER FROM SUB-O2S CONNECTOR TERMINAL "1" (SUB-O2S SIDE) TO GROUND.
- MEASURE VOLTAGE. SHOULD BE BETWEEN 0.80 AND 0.95 VOLTS. IS IT?

YES → (2)
NO → REPLACE SUB-O2S.

(2)
- LOWER VEHICLE.
- SHUT ENGINE OFF.
- DISCONNECT ECM CONNECTOR C2 AND C3.
- CONNECT A DIGITAL MULTIMETER FROM ECM CONNECTOR C2 CAVITY "B5" TO SUB-O2S CONNECTOR CAVITY "1" (HARNESS SIDE).
- MEASURE RESISTANCE. SHOULD BE LESS THAN 0.50 OHMS. IS IT?

YES → (3)
NO → REPAIR OPEN IN BLK WIRE BETWEEN ECM AND SUB-O2S.

(3)
- CONNECT A TEST LIGHT FROM ECM CONNECTOR C3 CAVITY "C24" TO B+. TEST LIGHT SHOULD LIGHT. DOES IT?

YES → (4)
NO → REPAIR OPEN IN BRN WIRE BETWEEN ECM AND G106.

(4)
- CONNECT A TEST LIGHT FROM ECM CONNECTOR C2 CAVITY "B5" TO B+. TEST LIGHT SHOULD NOT LIGHT. DOES IT?

NO →
- CHECK FOR AN OPEN IN BRN WIRE BETWEEN SUB-O2S AND G106.
- CHECK FOR A POOR CONNECTION AT ECM. IF OK, REPLACE ECM.

YES → REPAIR SHORT TO GROUND IN BLK WIRE BETWEEN ECM AND SUB-O2S.

CLEAR DTCs AND REPEAT OBD SYSTEM CHECK AFTER MAKING REPAIRS.

1.6L (VIN 6) ENGINE — DIAGNOSTIC TROUBLE CODE CHARTS — 1993 PRIZM

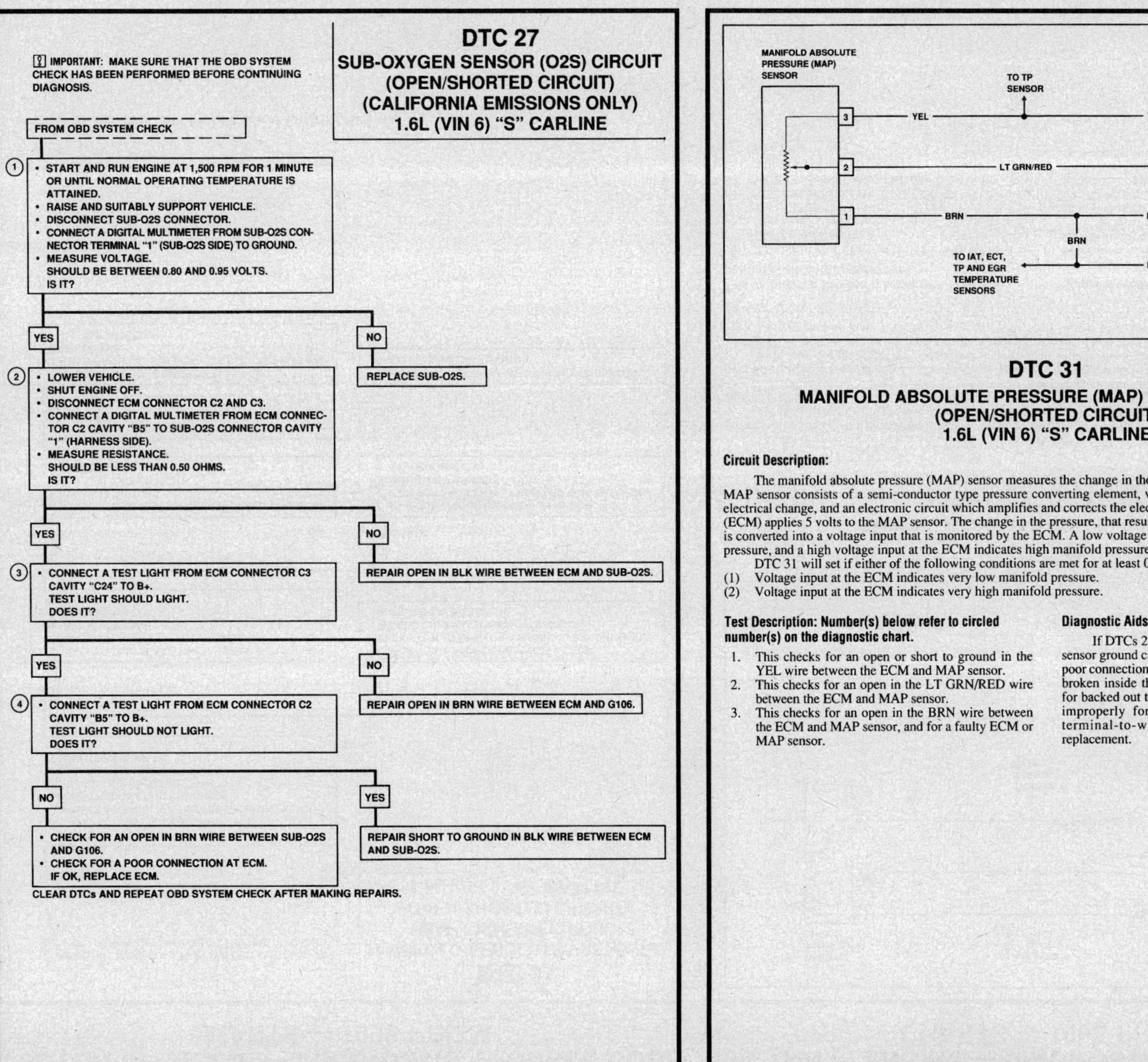

DTC 31
MANIFOLD ABSOLUTE PRESSURE (MAP) SENSOR CIRCUIT
(OPEN/SHORTED CIRCUIT)
1.6L (VIN 6) "S" CARLINE

Circuit Description:

The manifold absolute pressure (MAP) sensor measures the change in the intake manifold pressure (vacuum). The MAP sensor consists of a semi-conductor type pressure converting element, which converts a pressure change into an electrical change, and an electronic circuit which amplifies and corrects the electrical change. The engine control module (ECM) applies 5 volts to the MAP sensor. The change in the pressure, that results from the engine load and rpm changes, is converted into a voltage input that is monitored by the ECM. A low voltage input at the ECM indicates low manifold pressure, and a high voltage input at the ECM indicates high manifold pressure.

DTC 31 will set if either of the following conditions are met for at least 0.50 seconds:
(1) Voltage input at the ECM indicates very low manifold pressure.
(2) Voltage input at the ECM indicates very high manifold pressure.

Test Description: Number(s) below refer to circled number(s) on the diagnostic chart.

1. This checks for an open or short to ground in the YEL wire between the ECM and MAP sensor.
2. This checks for an open in the LT GRN/RED wire between the ECM and MAP sensor.
3. This checks for an open in the BRN wire between the ECM and MAP sensor, and for a faulty ECM or MAP sensor.

Diagnostic Aids:

If DTCs 22, 24 and 41 are also set, problem is open sensor ground circuit. An intermittent may be caused by a poor connection, rubbed through wire insulation, or a wire broken inside the insulation. Inspect harness connectors for backed out terminals, improper mating, broken locks, improperly formed or damaged terminals and poor terminal-to-wire connections before component replacement.

1.6L (VIN 6) ENGINE — DIAGNOSTIC TROUBLE CODE CHARTS — 1993 PRIZM

⚠ **IMPORTANT:** MAKE SURE THAT THE OBD SYSTEM CHECK HAS BEEN PERFORMED BEFORE CONTINUING DIAGNOSIS.

DTC 31
MANIFOLD ABSOLUTE PRESSURE (MAP) SENSOR CIRCUIT
(OPEN/SHORTED CIRCUIT)
1.6L (VIN 6) "S" CARLINE

FROM OBD SYSTEM CHECK

(1)
- TURN IGNITION SWITCH TO "LOCK."
- DISCONNECT MAP SENSOR CONNECTOR.
- CONNECT A DIGITAL MULTIMETER FROM MAP SENSOR CONNECTOR CAVITY "3" TO GROUND.
- TURN IGNITION SWITCH TO "ON."
- MEASURE VOLTAGE. SHOULD BE 4–6 VOLTS. IS IT?

YES → (2)
NO →
- CHECK FOR A POOR CONNECTION AT ECM.
- CHECK FOR AN OPEN OR A SHORT TO GROUND IN YEL WIRE BETWEEN ECM AND MAP SENSOR. IF OK, REPLACE ECM.

(2)
- CONNECT A DIGITAL MULTIMETER FROM MAP SENSOR CONNECTOR CAVITY "2" TO GROUND.
- MEASURE VOLTAGE. SHOULD BE 4–5 VOLTS. IS IT?

YES → (3)
NO →
- CHECK FOR A POOR CONNECTION AT ECM.
- CHECK FOR AN OPEN IN LT GRN/RED WIRE BETWEEN ECM AND MAP SENSOR. IF OK, REPLACE ECM.

(3)
- CONNECT A TEST LIGHT FROM MAP SENSOR CONNECTOR CAVITY "1" TO B+.
- TEST LIGHT SHOULD LIGHT. DOES IT?

NO →
- CHECK FOR A POOR CONNECTION AT ECM.
- CHECK FOR AN OPEN IN BRN WIRE BETWEEN ECM AND MAP SENSOR. IF OK, REPLACE ECM.

YES →
- CHECK FOR A POOR CONNECTION AT MAP SENSOR. IF OK, REPLACE MAP SENSOR.

CLEAR DTCs AND REPEAT OBD SYSTEM CHECK AFTER MAKING REPAIRS.

1.6L (VIN 6) ENGINE — DIAGNOSTIC TROUBLE CODE CHARTS — 1993 PRIZM

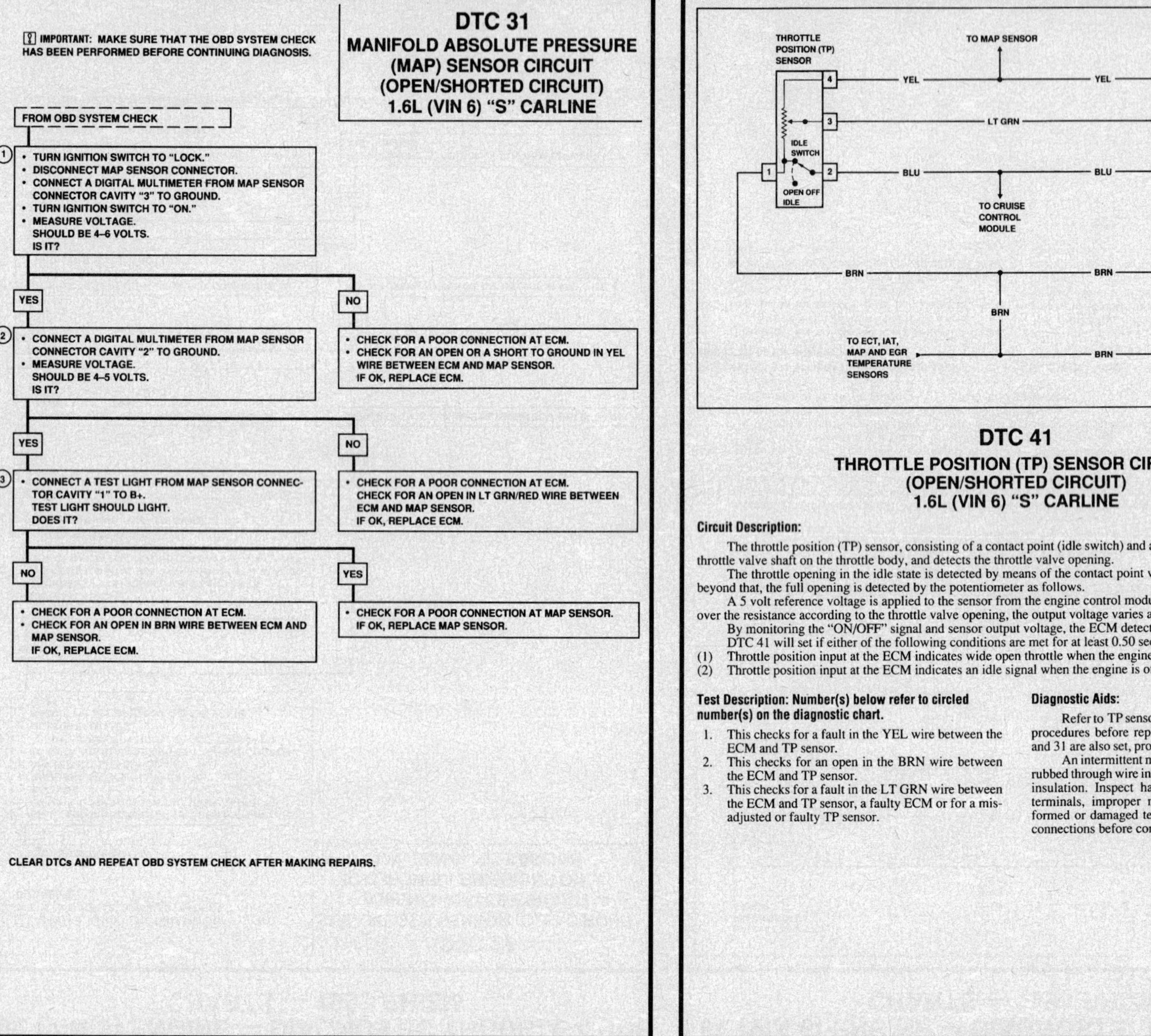

DTC 41
THROTTLE POSITION (TP) SENSOR CIRCUIT
(OPEN/SHORTED CIRCUIT)
1.6L (VIN 6) "S" CARLINE

Circuit Description:

The throttle position (TP) sensor, consisting of a contact point (idle switch) and a potentiometer, is connected to the throttle valve shaft on the throttle body, and detects the throttle valve opening.

The throttle opening in the idle state is detected by means of the contact point which turns "ON" in that state. But beyond that, the full opening is detected by the potentiometer as follows.

A 5 volt reference voltage is applied to the sensor from the engine control module (ECM), and, as its brush moves over the resistance according to the throttle valve opening, the output voltage varies accordingly.

By monitoring the "ON/OFF" signal and sensor output voltage, the ECM detects the throttle valve opening. DTC 41 will set if either of the following conditions are met for at least 0.50 seconds:

(1) Throttle position input at the ECM indicates wide open throttle when the engine is idling.
(2) Throttle position input at the ECM indicates an idle signal when the engine is off idle.

Test Description: Number(s) below refer to circled number(s) on the diagnostic chart.

1. This checks for a fault in the YEL wire between the ECM and TP sensor.
2. This checks for an open in the BRN wire between the ECM and TP sensor.
3. This checks for a fault in the LT GRN wire between the ECM and TP sensor, a faulty ECM or for a misadjusted or faulty TP sensor.

Diagnostic Aids:

Refer to TP sensor adjustment procedures before replacing TP sensor. If DTCs 22, 24 and 31 are also set, problem is open sensor ground circuit.

An intermittent may be caused by a poor connection, rubbed through wire insulation, or a wire broken inside the insulation. Inspect harness connectors for backed out terminals, improper mating, broken locks, improperly formed or damaged terminals and poor terminal-to-wire connections before component replacement.

1.6L (VIN 6) ENGINE — DIAGNOSTIC TROUBLE CODE CHARTS — 1993 PRIZM

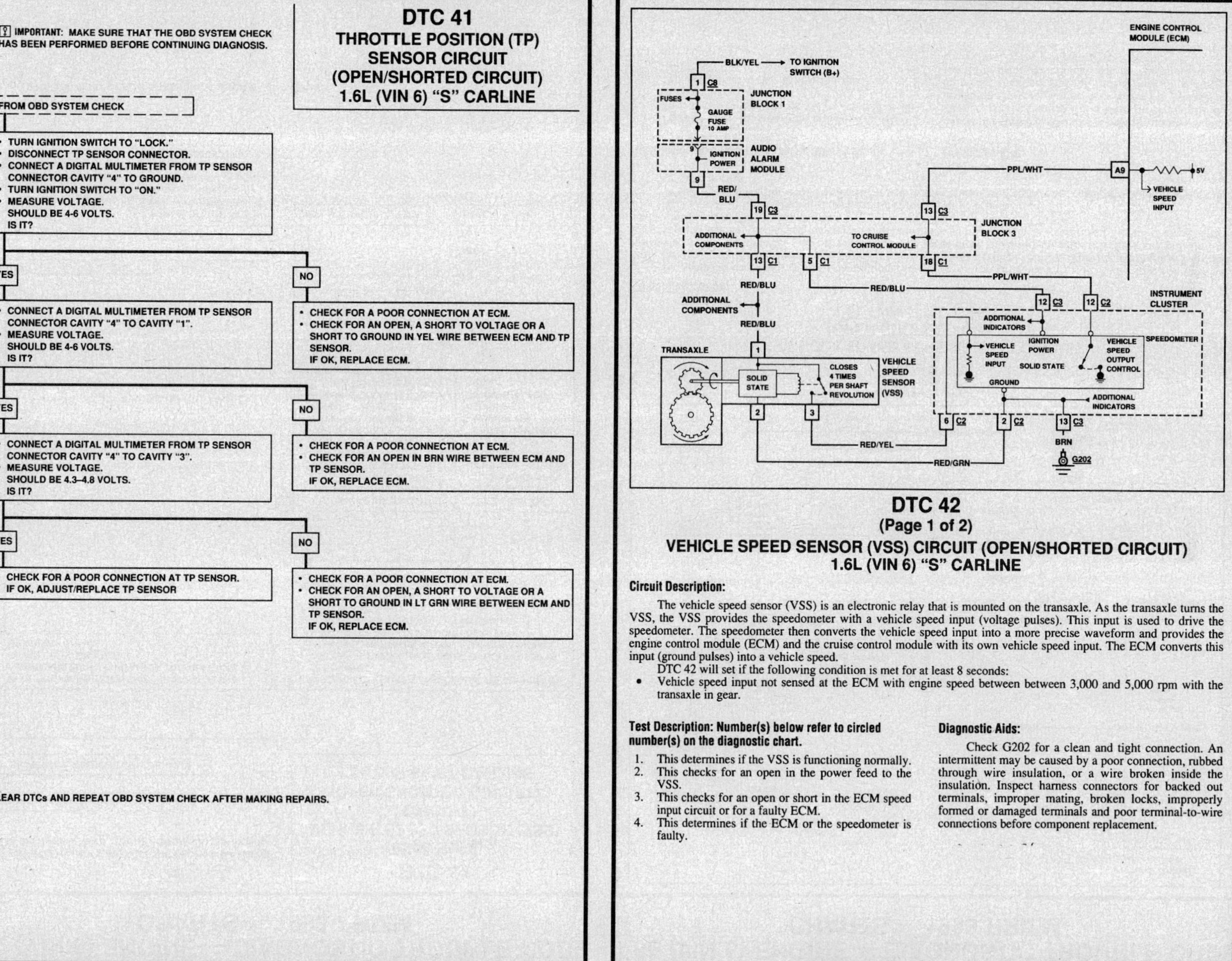

DTC 42
(Page 1 of 2)
VEHICLE SPEED SENSOR (VSS) CIRCUIT (OPEN/SHORTED CIRCUIT)
1.6L (VIN 6) "S" CARLINE

Circuit Description:

The vehicle speed sensor (VSS) is an electronic relay that is mounted on the transaxle. As the transaxle turns the VSS, the VSS provides the speedometer with a vehicle speed input (voltage pulses). This input is used to drive the speedometer. The speedometer then converts the vehicle speed input into a more precise waveform and provides the engine control module (ECM) and the cruise control module with its own vehicle speed input. The ECM converts this input (ground pulses) into a vehicle speed.

DTC 42 will set if the following condition is met for at least 8 seconds:

• Vehicle speed input not sensed at the ECM with engine speed between between 3,000 and 5,000 rpm with the transaxle in gear.

Test Description: Number(s) below refer to circled number(s) on the diagnostic chart.

1. This determines if the VSS is functioning normally.
2. This checks for an open in the power feed to the VSS.
3. This checks for an open or short in the ECM speed input circuit or for a faulty ECM.
4. This determines if the ECM or the speedometer is faulty.

Diagnostic Aids:

Check G202 for a clean and tight connection. An intermittent may be caused by a poor connection, rubbed through wire insulation, or a wire broken inside the insulation. Inspect harness connectors for backed out terminals, improper mating, broken locks, improperly formed or damaged terminals and poor terminal-to-wire connections before component replacement.

1.6L (VIN 6) ENGINE — DIAGNOSTIC TROUBLE CODE CHARTS — 1993 PRIZM

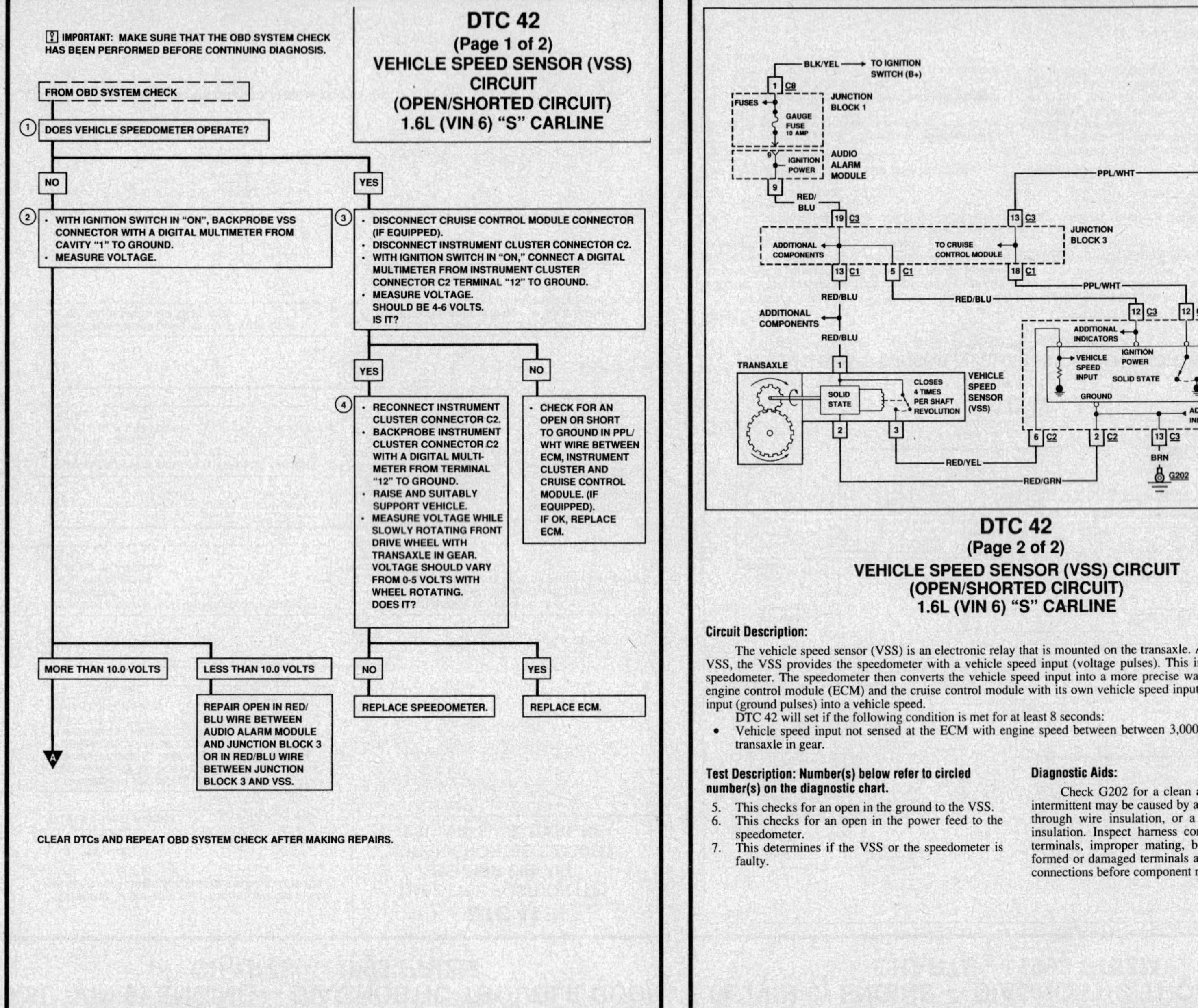

DTC 42 (Page 2 of 2)
VEHICLE SPEED SENSOR (VSS) CIRCUIT (OPEN/SHORTED CIRCUIT)
1.6L (VIN 6) "S" CARLINE

Circuit Description:

The vehicle speed sensor (VSS) is an electronic relay that is mounted on the transaxle. As the transaxle turns the VSS, the VSS provides the speedometer with a vehicle speed input (voltage pulses). This input is used to drive the speedometer. The speedometer then converts the vehicle speed input into a more precise waveform and provides the engine control module (ECM) and the cruise control module with its own vehicle speed input. The ECM converts this input (ground pulses) into a vehicle speed.

DTC 42 will set if the following condition is met for at least 8 seconds:
· Vehicle speed input not sensed at the ECM with engine speed between between 3,000 and 5,000 rpm with the transaxle in gear.

Test Description: Number(s) below refer to circled number(s) on the diagnostic chart.

5. This checks for an open in the ground to the VSS.
6. This checks for an open in the power feed to the speedometer.
7. This determines if the VSS or the speedometer is faulty.

Diagnostic Aids:

Check G202 for a clean and tight connection. An intermittent may be caused by a poor connection, rubbed through wire insulation, or a wire broken inside the insulation. Inspect harness connectors for backed out terminals, improper mating, broken locks, improperly formed or damaged terminals and poor terminal-to-wire connections before component replacement.

1.6L (VIN 6) ENGINE — DIAGNOSTIC TROUBLE CODE CHARTS — 1993 PRIZM

1.6L (VIN 6) ENGINE — DIAGNOSTIC TROUBLE CODE CHARTS — 1993 PRIZM

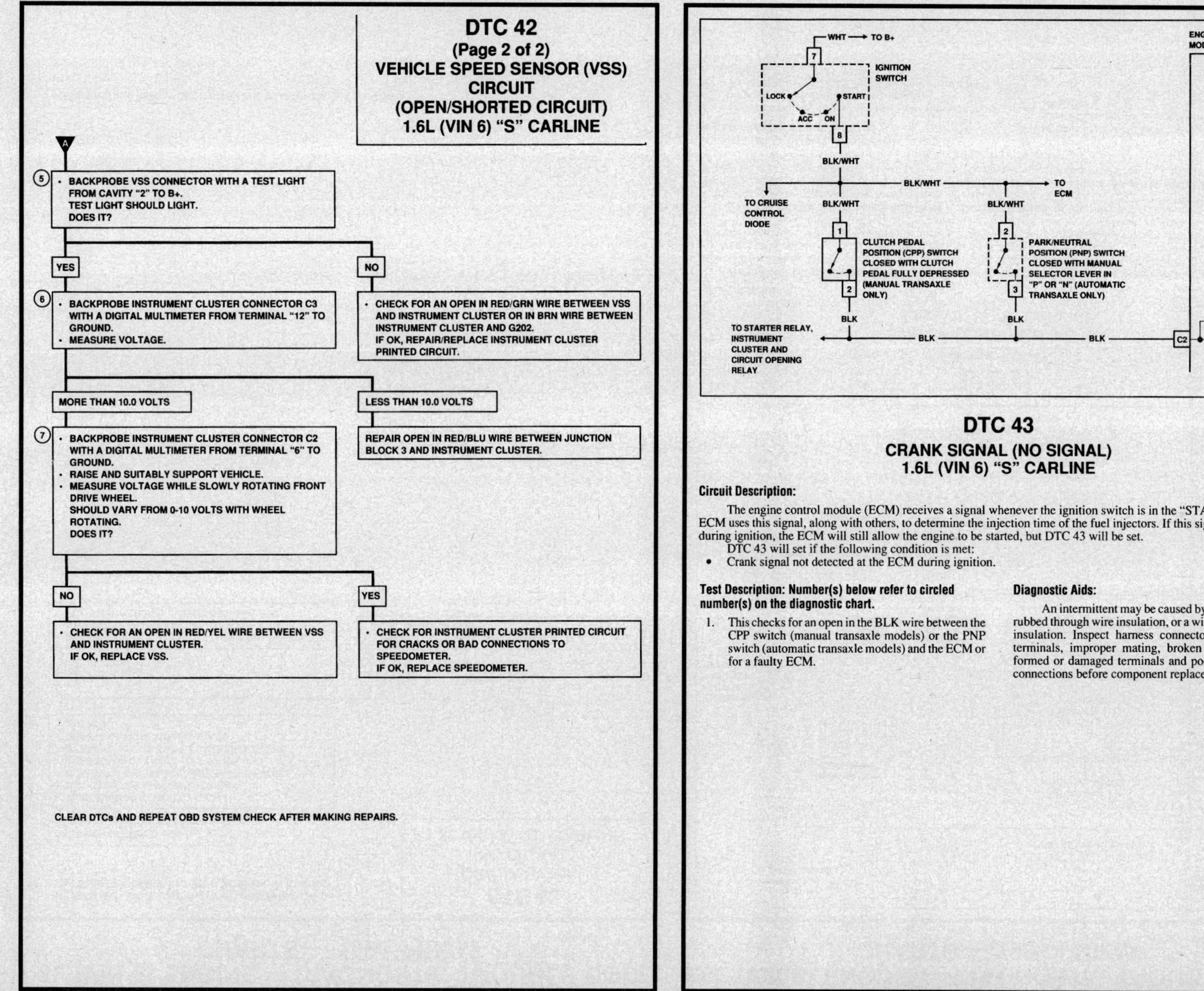

DTC 43
CRANK SIGNAL (NO SIGNAL)
1.6L (VIN 6) "S" CARLINE

Circuit Description:

The engine control module (ECM) receives a signal whenever the ignition switch is in the "START" position. The ECM uses this signal, along with others, to determine the injection time of the fuel injectors. If this signal is not detected during ignition, the ECM will still allow the engine to be started, but DTC 43 will be set.

DTC 43 will set if the following condition is met:
• Crank signal not detected at the ECM during ignition.

Test Description: Number(s) below refer to circled number(s) on the diagnostic chart.

1. This checks for an open in the BLK wire between the CPP switch (manual transaxle models) or the PNP switch (automatic transaxle models) and the ECM or for a faulty ECM.

Diagnostic Aids:

An intermittent may be caused by a poor connection, rubbed through wire insulation, or a wire broken inside the insulation. Inspect harness connectors for backed out terminals, improper mating, broken locks, improperly formed or damaged terminals and poor terminal-to-wire connections before component replacement.

1.6L (VIN 6) ENGINE — DIAGNOSTIC TROUBLE CODE CHARTS — 1993 PRIZM

☒ IMPORTANT: MAKE SURE THAT THE OBD SYSTEM CHECK HAS BEEN PERFORMED BEFORE CONTINUING DIAGNOSIS.

FROM OBD SYSTEM CHECK

① • TURN IGNITION SWITCH TO "LOCK."
• DISCONNECT ECM CONNECTOR C3.
• CONNECT A DIGITAL MULTIMETER FROM ECM CONNECTOR C3 CAVITY "C2" TO GROUND.
• MEASURE VOLTAGE WHILE CRANKING ENGINE. SHOULD BE MORE THAN 10 VOLTS. IS IT?

NO

• REPAIR OPEN IN BLK WIRE BETWEEN CPP SWITCH (MANUAL TRANSAXLE MODELS) OR PNP SWITCH (AUTOMATIC TRANSAXLE MODELS) AND ECM.

YES

• CHECK FOR A POOR CONNECTION AT ECM. IF OK, REPLACE ECM.

CLEAR DTCs AND REPEAT OBD SYSTEM CHECK AFTER MAKING REPAIRS.

DTC 43
CRANK SIGNAL
(NO SIGNAL)
1.6L (VIN 6) "S" CARLINE

1.6L (VIN 6) ENGINE — DIAGNOSTIC TROUBLE CODE CHARTS — 1993 PRIZM

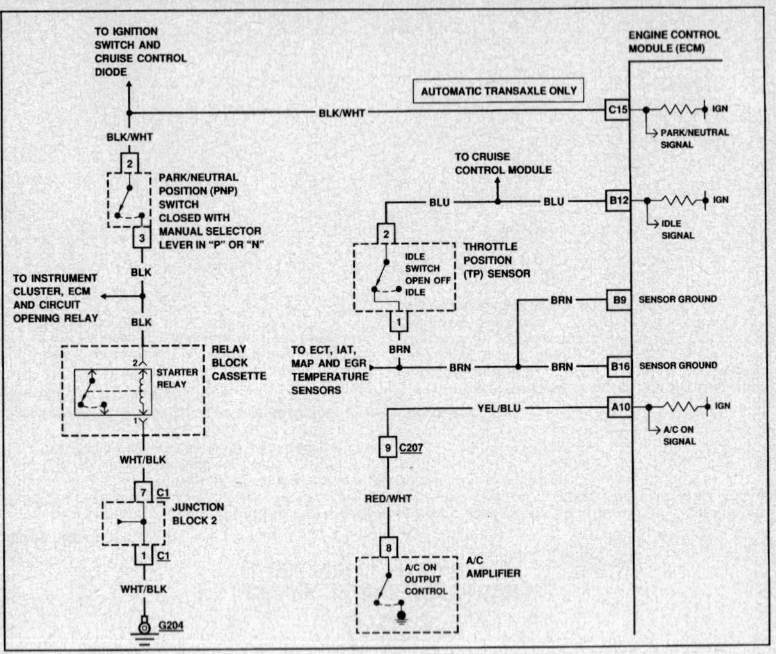

DTC 51
(Page 1 of 4)
SWITCH CONDITION SIGNAL
1.6L (VIN 6) "S" CARLINE

Circuit Description:

The engine control module (ECM) receives a signal whenever the following occurs: (1) the accelerator pedal is depressed (indicating that the throttle lever is in the off idle position (2) if the vehicle is equipped with air conditioning; the A/C switch is depressed with the blower speed selector switch in any position except "OFF" (indicating that A/C operation has been requested (3) or if the vehicle is equipped with an automatic transaxle; the manual selector lever is in the "R," "D," "2" or "L" positions (indicating that the transaxle is in another gear besides park or neutral).

DTC 51 indicates that the above mentioned systems are functioning normally. If DTC 51 is not indicated when the accelerator pedal is depressed, A/C is ON or with the manual selector lever out of "P" or "N," with DLC terminals "TE2" and "TE1" jumpered to ground a fault is present. If DTC 51 is indicated without the accelerator pedal being depressed, without the A/C switch or blower switch ON or with the manual selector lever in "P" or "N," a fault is also present.

Test Description: Number(s) below refer to circled number(s) on the diagnostic chart.

1. This determines if the systems are functioning normally in the test diagnostic mode.
2. This checks to see if the TP sensor circuit is operating normally.
3. This checks for a short to ground in the A/C circuit, an open in the PNP switch circuit, an open in the TP sensor circuit or for a faulty ECM or for a faulty A/C amplifier.

1.6L (VIN 6) ENGINE — DIAGNOSTIC TROUBLE CODE CHARTS — 1993 PRIZM

IMPORTANT: MAKE SURE THAT THE OBD SYSTEM CHECK HAS BEEN PERFORMED BEFORE CONTINUING DIAGNOSIS.

DTC 51
(Page 1 of 4)
SWITCH CONDITION SIGNAL
1.6L (VIN 6) "S" CARLINE

FROM OBD SYSTEM CHECK

(1)
- TURN IGNITION SWITCH TO "LOCK."
- CLEAR DTCs.
- CONNECT A JUMPER FROM DLC TERMINAL "TE2" TO TERMINAL "E1".
- TURN IGNITION SWITCH TO "ON." MIL SHOULD BE FLASHING ON AND OFF RAPIDLY.
- START AND ROAD TEST VEHICLE FOR 1 MINUTE ABOVE 16 KPH (10 MPH).
- LEAVE ENGINE RUNNING AT IDLE.
- CONNECT A JUMPER FROM DLC TERMINAL "TE1" TO TERMINAL "E1".
- NOTE MIL. SHOULD FLASH ON AND OFF RAPIDLY, NOT DTC 51. DOES IT?

YES

(2)
- HOLD ACCELERATOR PEDAL DEPRESSED OFF IDLE FOR 7-10 SECONDS.
- NOTE MIL. SHOULD FLASH DTC 51 AT LEAST ONCE AND THEN FLASH ON AND OFF RAPIDLY. DOES IT?

NO, FLASHES DTC 51

(3)
- SHUT ENGINE OFF, BUT LEAVE IGNITION SWITCH IN "ON."
- DISCONNECT TP SENSOR CONNECTOR.
- CONNECT A DIGITAL MULTIMETER FROM TP SENSOR CONNECTOR CAVITY "2" TO GROUND.
- MEASURE VOLTAGE.

MORE THAN 10 VOLTS

- CONNECT A DIGITAL MULTIMETER FROM TP SENSOR TERMINAL "1" TO TERMINAL "2" (TP SENSOR SIDE).
- MEASURE RESISTANCE.

LESS THAN 10 VOLTS

- CHECK FOR A POOR CONNECTION AT ECM.
- CHECK FOR AN OPEN IN BLU WIRE BETWEEN TP SENSOR AND ECM. IF OK, REPLACE ECM.

LESS THAN INFINITE

INFINITE

ADJUST/REPLACE TP SENSOR.

A
TO PAGE 6E3-A1-73

B
TO PAGE 6E3-A1-73

CLEAR DTCs AND REPEAT OBD SYSTEM CHECK AFTER MAKING REPAIRS.

1.6L (VIN 6) ENGINE — DIAGNOSTIC TROUBLE CODE CHARTS — 1993 PRIZM

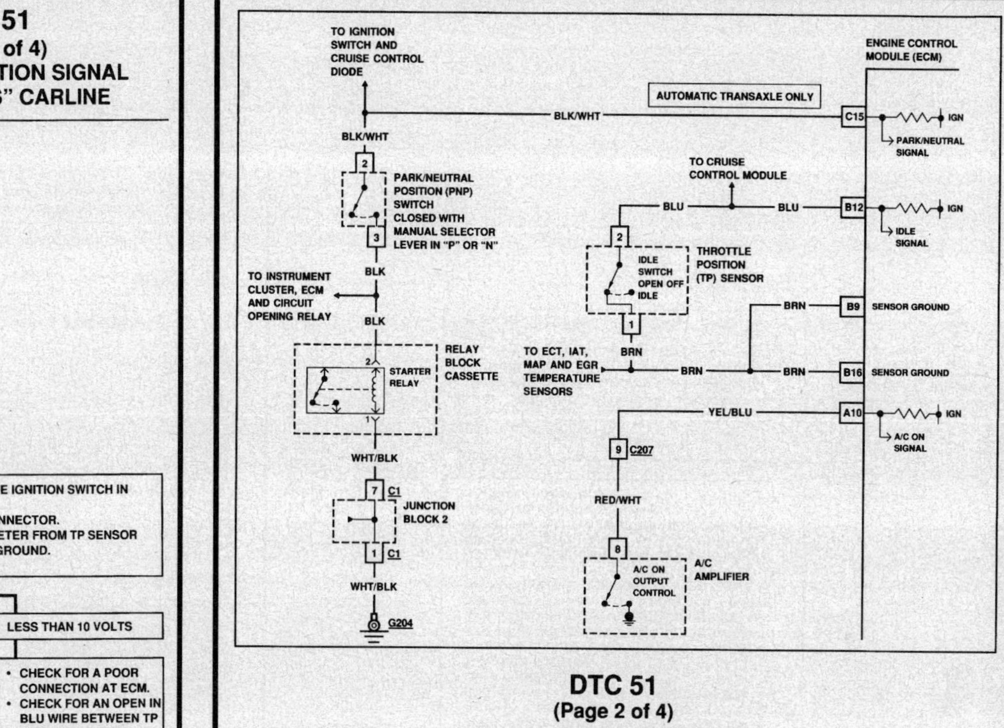

DTC 51
(Page 2 of 4)
SWITCH CONDITION SIGNAL
1.6L (VIN 6) "S" CARLINE

Circuit Description:

The engine control module (ECM) receives a signal whenever the following occurs: (1) the accelerator pedal is depressed (indicating that the throttle lever is in the off idle position (2) if the vehicle is equipped with air conditioning; the A/C switch is depressed with the blower speed selector switch in any position except "OFF" (indicating that A/C operation has been requested (3) or if the vehicle is equipped with an automatic transaxle; the manual selector lever is in the "R," "D," "2" or "L" positions (indicating that the transaxle is in another gear besides park or neutral).

DTC 51 indicates that the above mentioned systems are functioning normally. If DTC 51 is not indicated when the accelerator pedal is depressed, A/C is ON or with the manual selector lever out of "P" or "N," with DLC terminals "TE2" and "TE1" jumpered to ground a fault is present. If DTC 51 is indicated without the accelerator pedal being depressed, without the A/C switch or blower switch ON or with the manual selector lever in "P" or "N," a fault is also present.

1.6L (VIN 6) ENGINE — DIAGNOSTIC TROUBLE CODE CHARTS — 1993 PRIZM

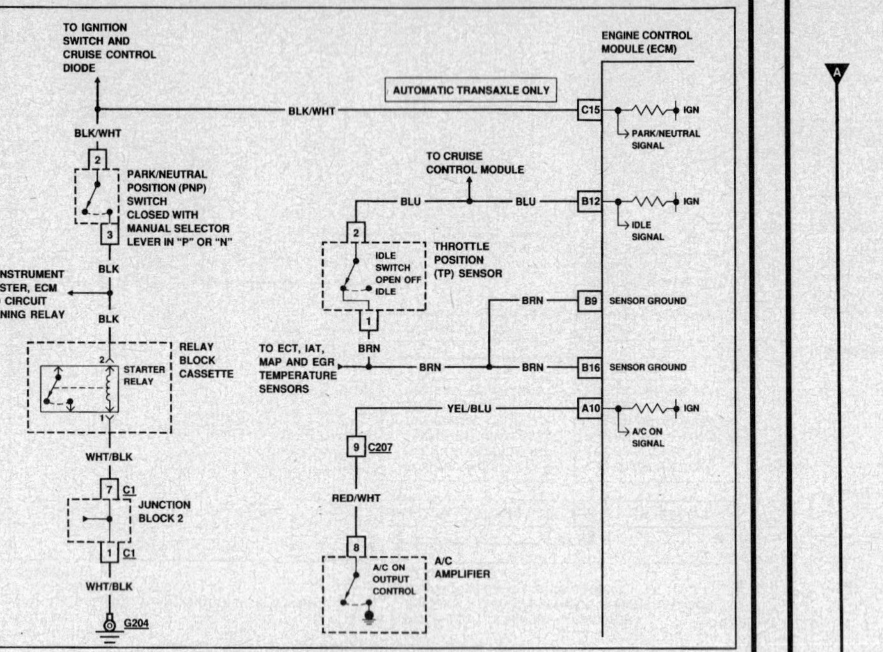

DTC 51
(Page 3 of 4)
SWITCH CONDITION SIGNAL
1.6L (VIN 6) "S" CARLINE

Circuit Description:

The engine control module (ECM) receives a signal whenever the following occurs: (1) the accelerator pedal is depressed (indicating that the throttle lever is in the off idle position (2) if the vehicle is equipped with air conditioning; the A/C switch is depressed with the blower speed selector switch in any position except "OFF" (indicating that A/C operation has been requested (3) or if the vehicle is equipped with an automatic transaxle; the manual selector lever is in the "R," "D," "2" or "L" positions (indicating that the transaxle is in another gear besides park or neutral).

DTC 51 indicates that the above mentioned systems are functioning normally. If DTC 51 is not indicated when the accelerator pedal is depressed, A/C is ON or with the manual selector lever out of "P" or "N," with DLC terminals "TE2" and "TE1" jumpered to ground a fault is present. If DTC 51 is indicated without the accelerator pedal being depressed, without the A/C switch or blower switch ON or with the manual selector lever in "P" or "N," a fault is also present.

Test Description: Number(s) below refer to circled number(s) on the diagnostic chart.

4. This checks for a short to ground in the TP sensor circuit or for a faulty TP sensor or ECM.

5. This checks for a short to ground in the PNP switch circuit.

6. This determines whether the system is functioning normally or if there is a problem in the A/C circuit.

1.6L (VIN 6) ENGINE — DIAGNOSTIC TROUBLE CODE CHARTS — 1993 PRIZM

DTC 51
(Page 2 of 4)
SWITCH CONDITION SIGNAL
1.6L (VIN 6) "S" CARLINE

A

B

IS VEHICLE EQUIPPED WITH A/C OR AUTOMATIC TRANSAXLE?

YES

NO → REPLACE ECM.

- IF VEHICLE IS EQUIPPED WITH AN AUTOMATIC TRANSAXLE ONLY, CHECK FOR A POOR CONNECTION AT ECM, AND FOR AN OPEN IN BLK/WHT WIRE BETWEEN ECM AND PNP SWITCH. IF OK, REPLACE ECM.
- IF VEHICLE IS EQUIPPED WITH A/C ONLY, DISCONNECT A/C AMPLIFIER CONNECTOR. CONNECT A DIGITAL MULTIMETER FROM A/C AMPLIFIER CONNECTOR CAVITY "8" TO GROUND. MEASURE VOLTAGE. IF VOLTAGE IS MORE THAN 10 VOLTS, REPLACE A/C AMPLIFIER. IF VOLTAGE IS LESS THAN 10 VOLTS, CHECK FOR A SHORT TO GROUND IN YEL/BLU OR RED/WHT BETWEEN ECM AND A/C AMPLIFIER. IF OK, REPLACE ECM.
- IF VEHICLE IS EQUIPPED WITH BOTH A/C AND AN AUTOMATIC TRANSAXLE, TURN IGNITION SWITCH TO "LOCK." RECONNECT TP SENSOR CONNECTOR AND DISCONNECT A/C AMPLIFIER CONNECTOR. REMOVE JUMPER FROM DLC TERMINAL "TE1" BUT LEAVE TERMINAL "TE2" JUMPERED. TURN IGNITION SWITCH TO "ON" AND THEN CONNECT A JUMPER FROM DLC TERMINAL "TE1" TO "E1." NOTICE: MIL WILL FLASH DTC 42 AND DTC 43, THIS SHOULD BE CONSIDERED NORMAL. DOES MIL FLASH DTC 51?

YES

NO → REPLACE A/C AMPLIFIER

- CHECK FOR AN OPEN IN BLK/WHT WIRE BETWEEN ECM AND PNP SWITCH OR FOR A SHORT TO GROUND IN YEL/BLU AND RED/WHT WIRE BETWEEN ECM AND A/C AMPLIFIER. IF OK, REPLACE ECM.

A

CLEAR DTCs AND REPEAT OBD SYSTEM CHECK AFTER MAKING REPAIRS.

1.6L (VIN 6) ENGINE — DIAGNOSTIC TROUBLE CODE CHARTS — 1993 PRIZM

DTC 51
(Page 3 of 4)
SWITCH CONDITION SIGNAL
1.6L (VIN 6) "S" CARLINE

A

YES

- IF THE VEHICLE IS EQUIPPED WITH AN AUTOMATIC TRANSAXLE, PROCEED TO THE AUTOMATIC TRANSAXLE TEST.
- IF THE VEHICLE IS NOT EQUIPPED WITH AN AUTOMATIC TRANSAXLE, BUT IS EQUIPPED WITH AIR CONDITIONING, PROCEED TO THE A/C TEST.
- IF THE VEHICLE IS NOT EQUIPPED WITH EITHER AN AUTOMATIC TRANSAXLE OR AIR CONDITIONING, THE SWITCH CONDITION SIGNAL SYSTEM IS FUNCTIONING NORMALLY.

AUTOMATIC TRANSAXLE TEST

(5)
- ENGAGE PARKING BRAKE.
- MOVE MANUAL SELECTOR LEVER FROM "P" TO "L" AND THEN BACK TO "P" (LEAVE IN "L" FOR 5 SECONDS).
- NOTE MIL.
  SHOULD FLASH DTC 51 AT LEAST ONCE AND THEN FLASH ON AND OFF RAPIDLY.
  DOES IT?

YES

- IF VEHICLE IS EQUIPPED WITH AIR CONDITIONING, PROCEED TO THE A/C TEST.
- IF VEHICLE IS NOT EQUIPPED WITH AIR CONDITIONING, THE SWITCH CONDITION SIGNAL SYSTEM IS FUNCTIONING NORMALLY.

A/C TEST

(6)
- DEPRESS A/C SWITCH AND TURN BLOWER SPEED SELECTOR SWITCH TO ANY POSITION EXCEPT "OFF."
- NOTE MIL.
  SHOULD FLASH DTC 51 AS LONG AS A/C SWITCH IS DEPRESSED WITH BLOWER SPEED SELECTOR SWITCH IN ANY POSITION EXCEPT "OFF."
  DOES IT?

C

NO

(4)
- TURN IGNITION SWITCH TO "LOCK."
- DISCONNECT TP SENSOR CONNECTOR.
- CONNECT A DIGITAL MULTIMETER FROM TP SENSOR CONNECTOR CAVITY "2" TO GROUND.
- TURN IGNITION SWITCH TO "ON."
- MEASURE VOLTAGE.

LESS THAN 10 VOLTS

- CHECK FOR A SHORT TO GROUND IN BLU WIRE BETWEEN ECM AND TP SENSOR. IF OK, REPLACE ECM.

MORE THAN 10 VOLTS

- ADJUST/REPLACE TP SENSOR.

NO

- CHECK FOR A SHORT TO GROUND IN BLK/WHT WIRE BETWEEN ECM AND PNP SWITCH. IF OK, REPLACE ECM.

CLEAR DTCs AND REPEAT OBD SYSTEM CHECK AFTER MAKING REPAIRS.

1.6L (VIN 6) ENGINE — DIAGNOSTIC TROUBLE CODE CHARTS — 1993 PRIZM

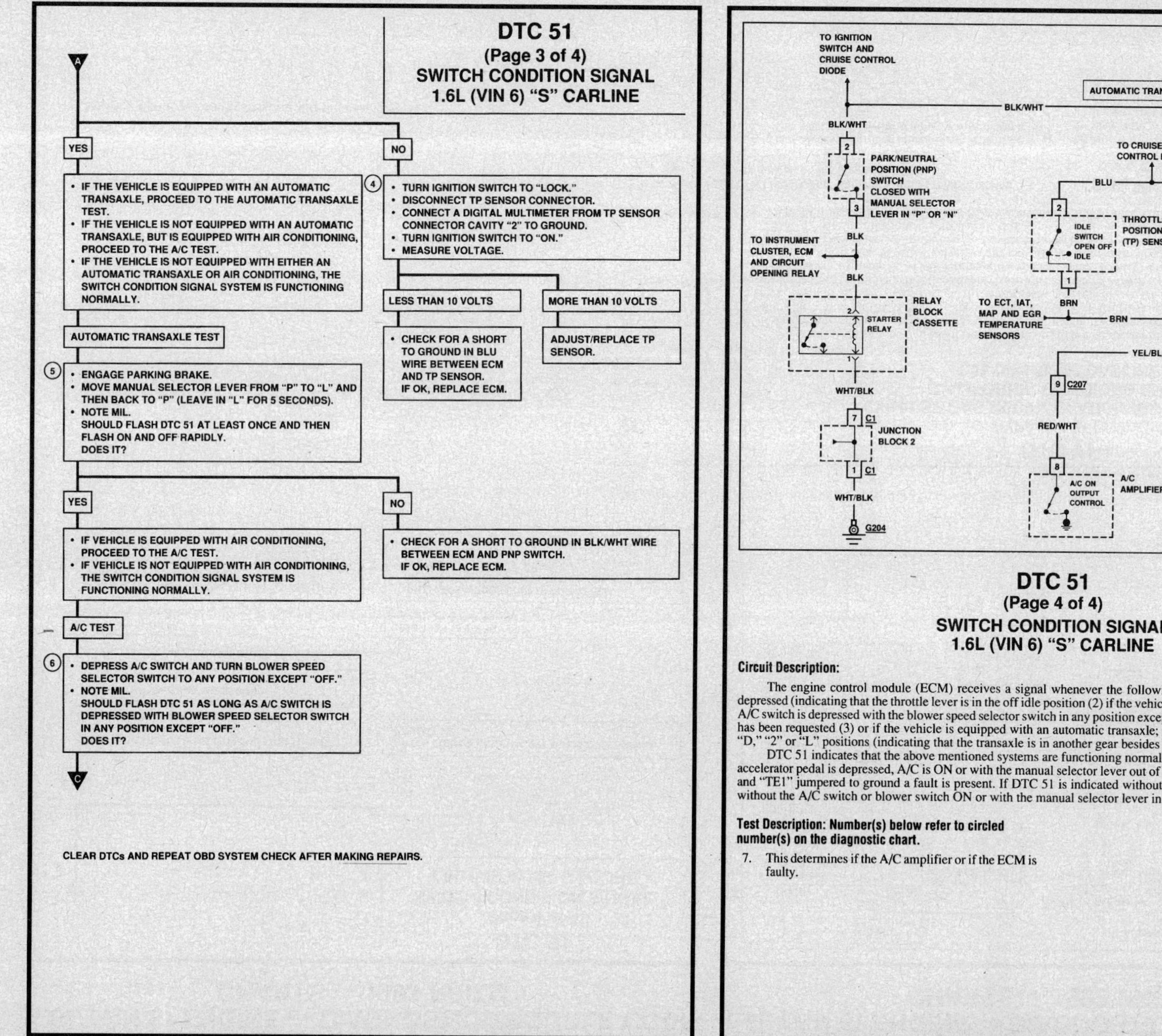

DTC 51
(Page 4 of 4)
SWITCH CONDITION SIGNAL
1.6L (VIN 6) "S" CARLINE

Circuit Description:

The engine control module (ECM) receives a signal whenever the following occurs: (1) the accelerator pedal is depressed (indicating that the throttle lever is in the off idle position (2) if the vehicle is equipped with air conditioning; the A/C switch is depressed with the blower speed selector switch in any position except "OFF" (indicating that A/C operation has been requested (3) or if the vehicle is equipped with an automatic transaxle; the manual selector lever is in the "R," "D," "2" or "L" positions (indicating that the transaxle is in another gear besides park or neutral).

DTC 51 indicates that the above mentioned systems are functioning normally. If DTC 51 is not indicated when the accelerator pedal is depressed, A/C is ON or with the manual selector lever out of "P" or "N," with DLC terminals "TE2" and "TE1" jumpered to ground a fault is present. If DTC 51 is indicated without the accelerator pedal being depressed, without the A/C switch or blower switch ON or with the manual selector lever in "P" or "N," a fault is also present.

Test Description: Number(s) below refer to circled number(s) on the diagnostic chart.

7. This determines if the A/C amplifier or if the ECM is faulty.

1.6L (VIN 6) ENGINE — DIAGNOSTIC TROUBLE CODE CHARTS — 1993 PRIZM

DTC 51
(Page 4 of 4)
SWITCH CONDITION SIGNAL
1.6L (VIN 6) "S" CARLINE

C

NO

(7) • TURN BLOWER SPEED SELECTOR SWITCH TO "OFF" AND A/C SWITCH TO OFF.
• DISCONNECT A/C AMPLIFIER CONNECTOR.
• CONNECT A DIGITAL MULTIMETER FROM A/C AMPLIFIER CONNECTOR CAVITY "8" TO GROUND.
• MEASURE VOLTAGE.

MORE THAN 10 VOLTS

• CHECK FOR A POOR CONNECTION AT A/C AMPLIFIER. IF OK, REPLACE A/C AMPLIFIER.

YES

SWITCH CONDITION SIGNAL SYSTEM IS FUNCTIONING NORMALLY.

LESS THAN 10 VOLTS

• CHECK FOR A POOR CONNECTION AT ECM.
• CHECK FOR AN OPEN IN YEL/BLU OR RED/WHT WIRE BETWEEN ECM AND A/C AMPLIFIER. IF OK, REPLACE ECM.

CLEAR DTCs AND REPEAT OBD SYSTEM CHECK AFTER MAKING REPAIRS.

1.6L (VIN 6) ENGINE — DIAGNOSTIC TROUBLE CODE CHARTS — 1993 PRIZM

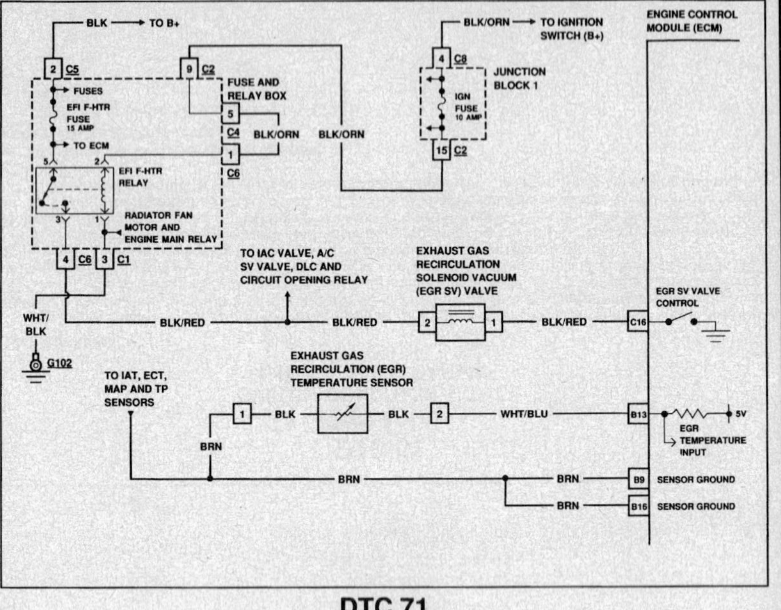

DTC 71
(Page 1 of 2)
EXHAUST GAS RECIRCULATION (EGR) SYSTEM
(CALIFORNIA EMISSIONS ONLY)
1.6L (VIN 6) "S" CARLINE

Circuit Description:

The exhaust gas recirculation (EGR) temperature sensor is a thermistor (a variable resistor that changes along with EGR temperature changes) in series with a fixed resistor within the engine control module (ECM). The ECM applies 5 volts to the sensor. The ECM monitors the voltage across the EGR temperature sensor and converts it into a temperature reading. When the exhaust gases are cold the sensor resistance is high, and when the exhaust gases are warm the sensor resistance is low. Therefore, when the exhaust gases are cold the ECM will receive a high voltage input, and when the exhaust gases are warm the ECM will receive a low voltage input. The ECM also maintains control over a EGR solenoid vacuum (SV) valve. The EGR SV valve permits exhaust gas to be recirculated and incorporated into the combustion process. The ECM will determine the best time to energize the EGR SV valve by monitoring the exhaust gas temperature.
DTC 71 will set if the following condition is met:
• With ECT above 60°C (140°F) with EGR temperature less than 70°C (158°F) for at least 50 seconds.

Test Description: Number(s) below refer to circled number(s) on the diagnostic chart.

1. This checks the power feed to the sensor.
2. This checks the ground circuit to the sensor.
3. This checks the EGR temperature resistance vs. temperature values.
4. This checks the power feed circuit to the EGR SV valve.

Diagnostic Aids:

Check EGR vacuum modulator and EGR valve for sticking open or closed. An intermittent may be caused by a poor connection, rubbed through wire insulation, or a wire broken inside the insulation. Inspect harness connectors for backed out terminals, improper mating, broken locks, improperly formed or damaged terminals and poor terminal-to-wire connections before component replacement.

1.6L (VIN 6) ENGINE — DIAGNOSTIC TROUBLE CODE CHARTS — 1993 PRIZM

⚠ IMPORTANT: MAKE SURE THAT THE OBD SYSTEM CHECK HAS BEEN PERFORMED BEFORE CONTINUING DIAGNOSIS.

FROM OBD SYSTEM CHECK

DTC 71
(Page 1of 2)
EXHAUST GAS RECIRCULATION (EGR) SYSTEM
(CALIFORNIA EMISSIONS ONLY)
1.6L (VIN 6) "S" CARLINE

1.
- TURN IGNITION SWITCH TO "LOCK."
- DISCONNECT EGR TEMPERATURE SENSOR CONNECTOR.
- CONNECT A DIGITAL MULTIMETER FROM EGR TEMPERATURE SENSOR CONNECTOR CAVITY "2" TO GROUND.
- TURN IGNITION SWITCH TO "ON."
- MEASURE VOLTAGE.
  SHOULD BE 4–6 VOLTS.
  IS IT?

YES → 2 / NO →

NO:
- CHECK FOR A POOR CONNECTION AT ECM.
- CHECK FOR AN OPEN, A SHORT TO VOLTAGE OR SHORT TO GROUND IN WHT/BLU WIRE BETWEEN ECM AND EGR TEMPERATURE SENSOR.
  IF OK, REPLACE ECM.

2.
- CONNECT A TEST LIGHT FROM EGR TEMPERATURE SENSOR CONNECTOR CAVITY "1" TO B+.
  TEST LIGHT SHOULD LIGHT.
  DOES IT?

YES → 3 / NO →

NO: REPAIR OPEN IN BRN WIRE BETWEEN EGR TEMPERATURE SENSOR AND ECM.

3.
- REMOVE EGR TEMPERATURE SENSOR FROM EGR VALVE.
- MEASURE RESISTANCE ACROSS SENSOR.
  RESISTANCE SHOULD BE:69–88 K OHMS AT 50°C (122°F)
  11–15 K OHMS AT 100°C (212°F)
  2–4 K OHMS AT 150°C (302°F)
  IS IT?

YES → 4 / NO →

NO: REPLACE EGR TEMPERATURE SENSOR.

4.
- BACKPROBE EGR SV VALVE CONNECTOR WITH A DIGITAL MULTIMETER FROM CAVITY "2" TO GROUND.
- MEASURE VOLTAGE.
  SHOULD BE MORE THAN 10 VOLTS.
  IS IT?

A

CLEAR DTCs AND REPEAT OBD SYSTEM CHECK AFTER MAKING REPAIRS.

1.6L (VIN 6) ENGINE — DIAGNOSTIC TROUBLE CODE CHARTS — 1993 PRIZM

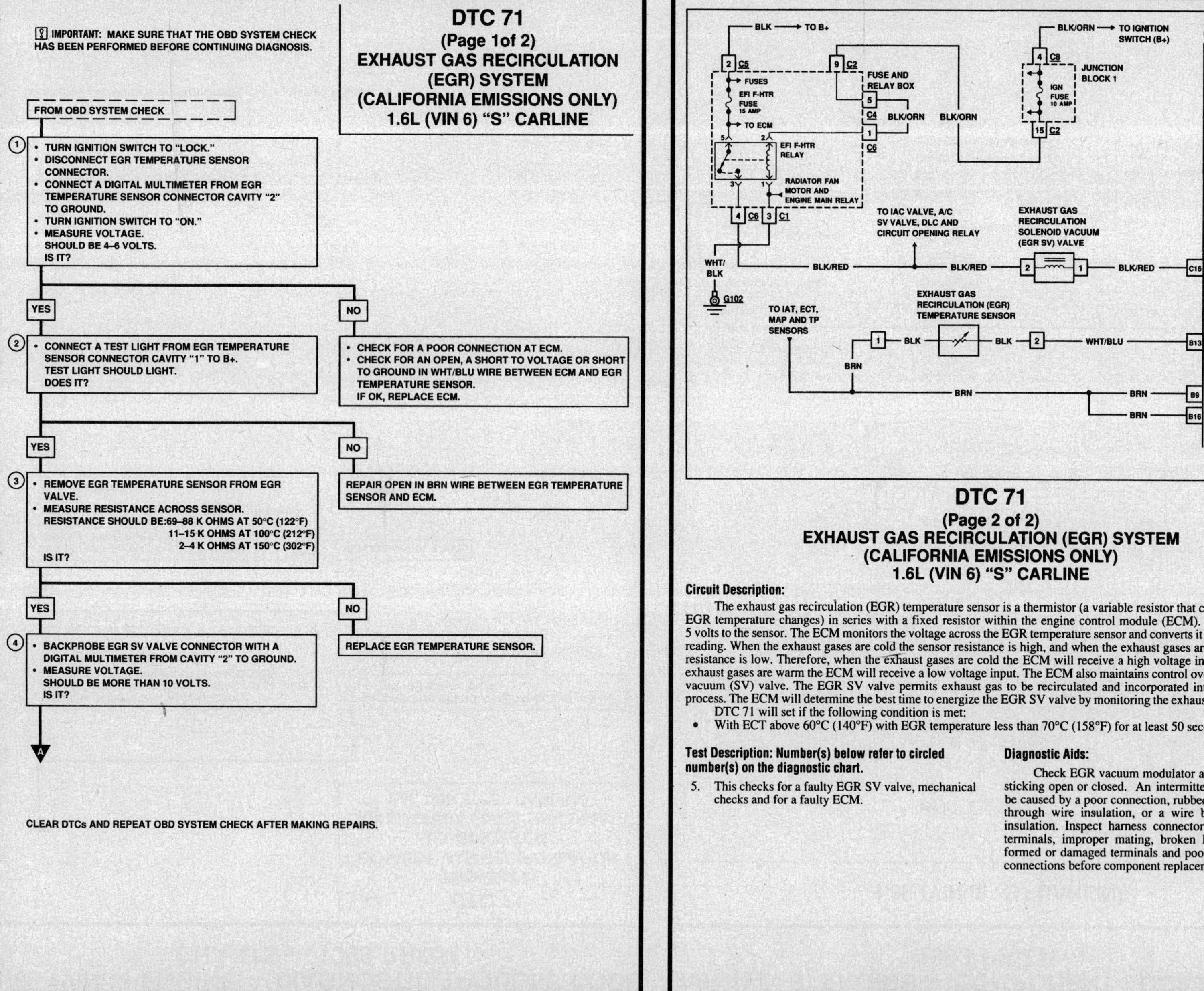

DTC 71
(Page 2 of 2)
EXHAUST GAS RECIRCULATION (EGR) SYSTEM
(CALIFORNIA EMISSIONS ONLY)
1.6L (VIN 6) "S" CARLINE

Circuit Description:

The exhaust gas recirculation (EGR) temperature sensor is a thermistor (a variable resistor that changes along with EGR temperature changes) in series with a fixed resistor within the engine control module (ECM). The ECM applies 5 volts to the sensor. The ECM monitors the voltage across the EGR temperature sensor and converts it into a temperature reading. When the exhaust gases are cold the sensor resistance is high, and when the exhaust gases are warm the sensor resistance is low. Therefore, when the exhaust gases are cold the ECM will receive a high voltage input, and when the exhaust gases are warm the ECM will receive a low voltage input. The ECM also maintains control over a EGR solenoid vacuum (SV) valve. The EGR SV valve permits exhaust gas to be recirculated and incorporated into the combustion process. The ECM will determine the best time to energize the EGR SV valve by monitoring the exhaust gas temperature.

DTC 71 will set if the following condition is met:
- With ECT above 60°C (140°F) with EGR temperature less than 70°C (158°F) for at least 50 seconds.

Test Description: Number(s) below refer to circled number(s) on the diagnostic chart.

5. This checks for a faulty EGR SV valve, mechanical checks and for a faulty ECM.

Diagnostic Aids:

Check EGR vacuum modulator and EGR valve for sticking open or closed. An intermittent may be caused by a poor connection, rubbed through wire insulation, or a wire broken inside the insulation. Inspect harness connectors for backed out terminals, improper mating, broken locks, improperly formed or damaged terminals and poor terminal-to-wire connections before component replacement.

1.6L (VIN 6) ENGINE — DIAGNOSTIC TROUBLE CODE CHARTS — 1993 PRIZM

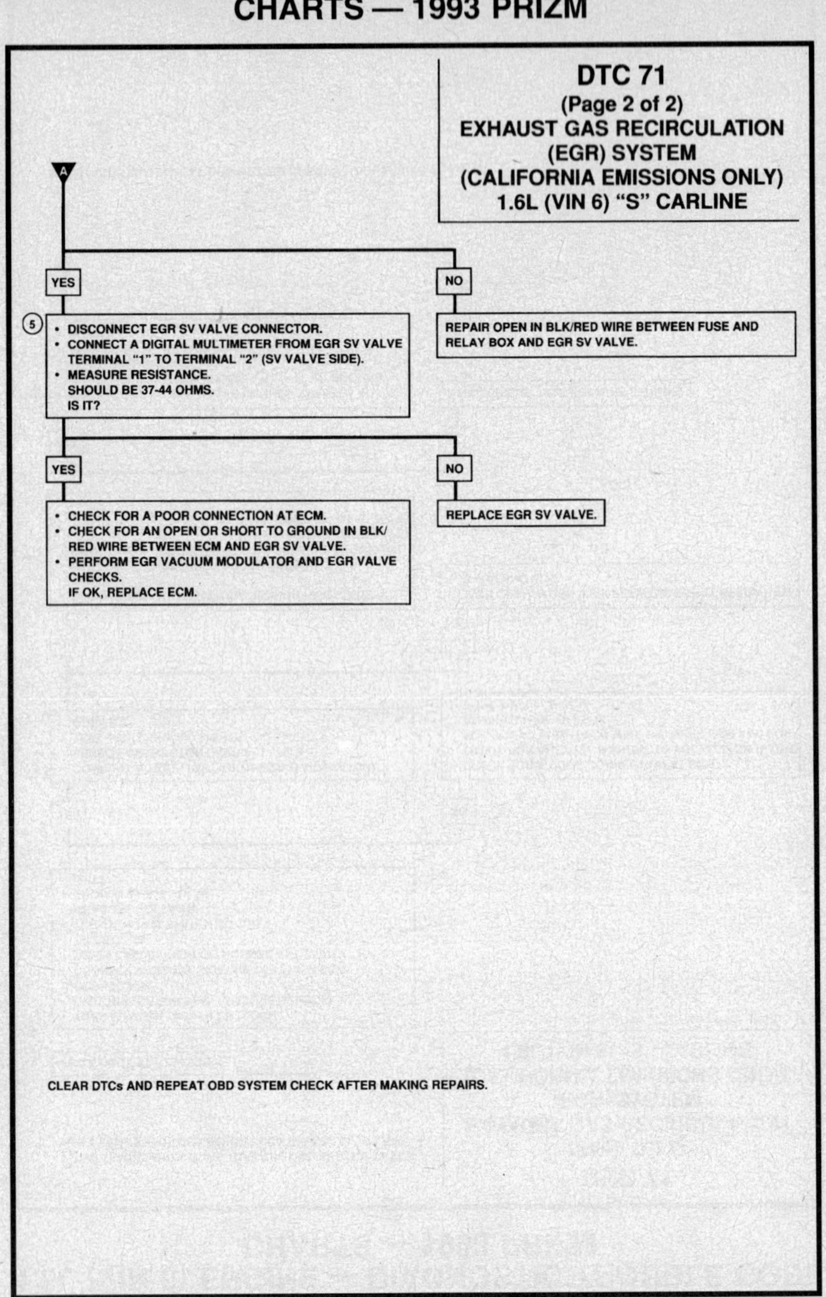

1.8L (VIN 8) ENGINE — COMPONENT LOCATIONS — 1993 PRIZM

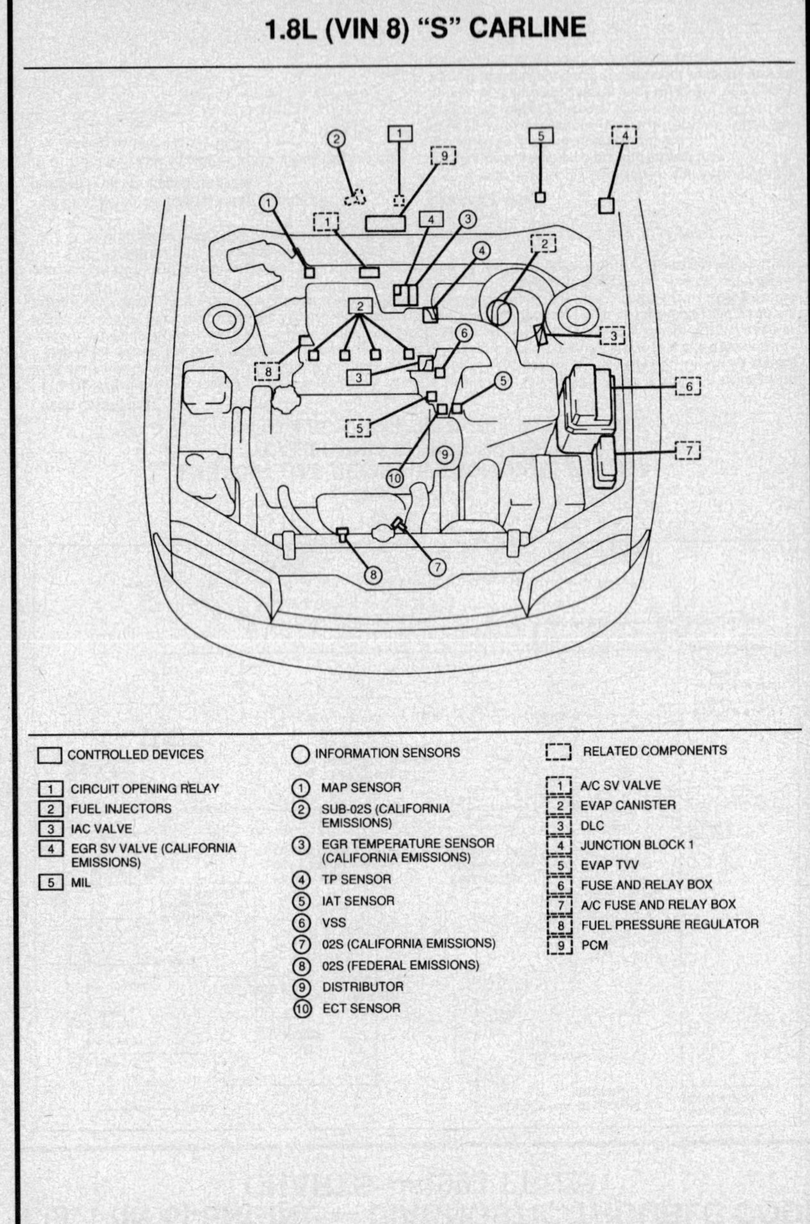

1.8L (VIN 8) ENGINE — PCM WIRING DIAGRAMS — 1993 PRIZM

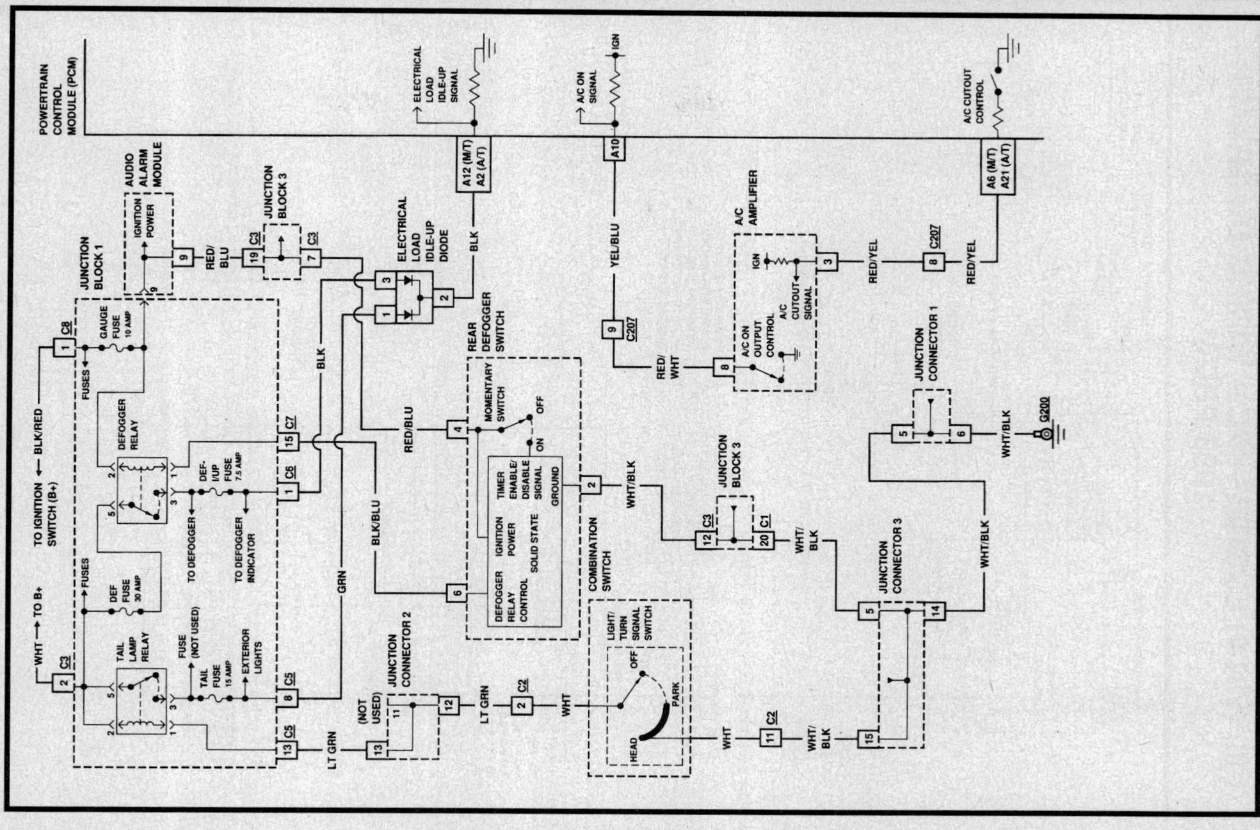

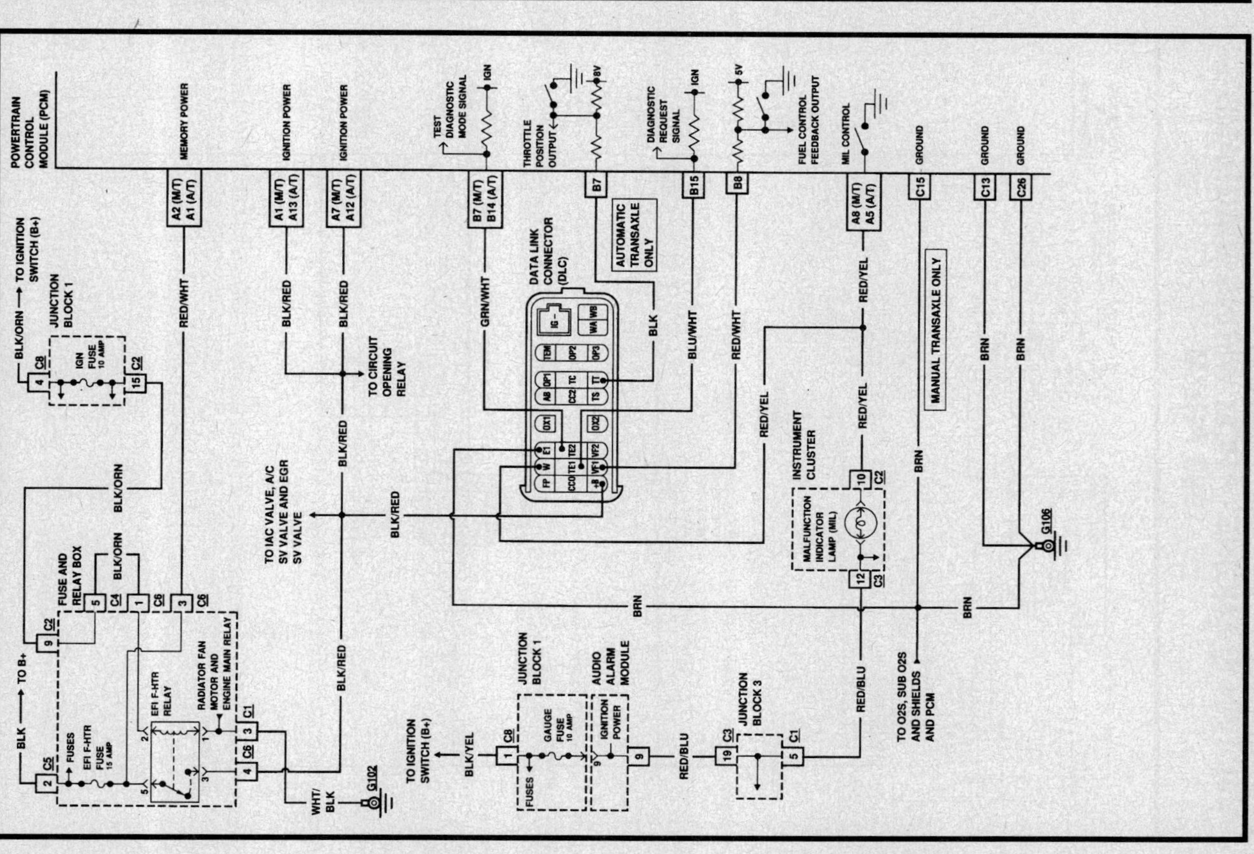

1.8L (VIN 8) ENGINE — PCM WIRING DIAGRAMS — 1993 PRIZM

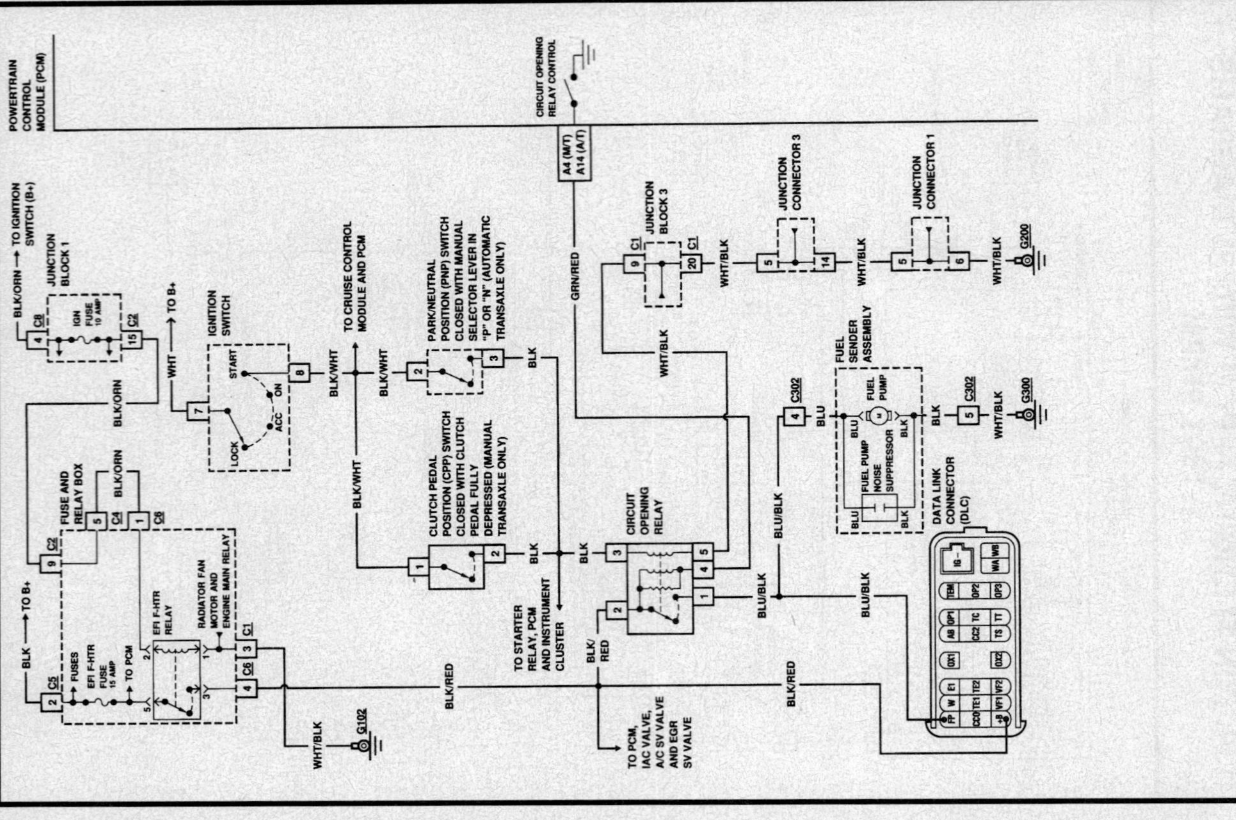

1.8L (VIN 8) ENGINE — PCM WIRING DIAGRAMS — 1993 PRIZM

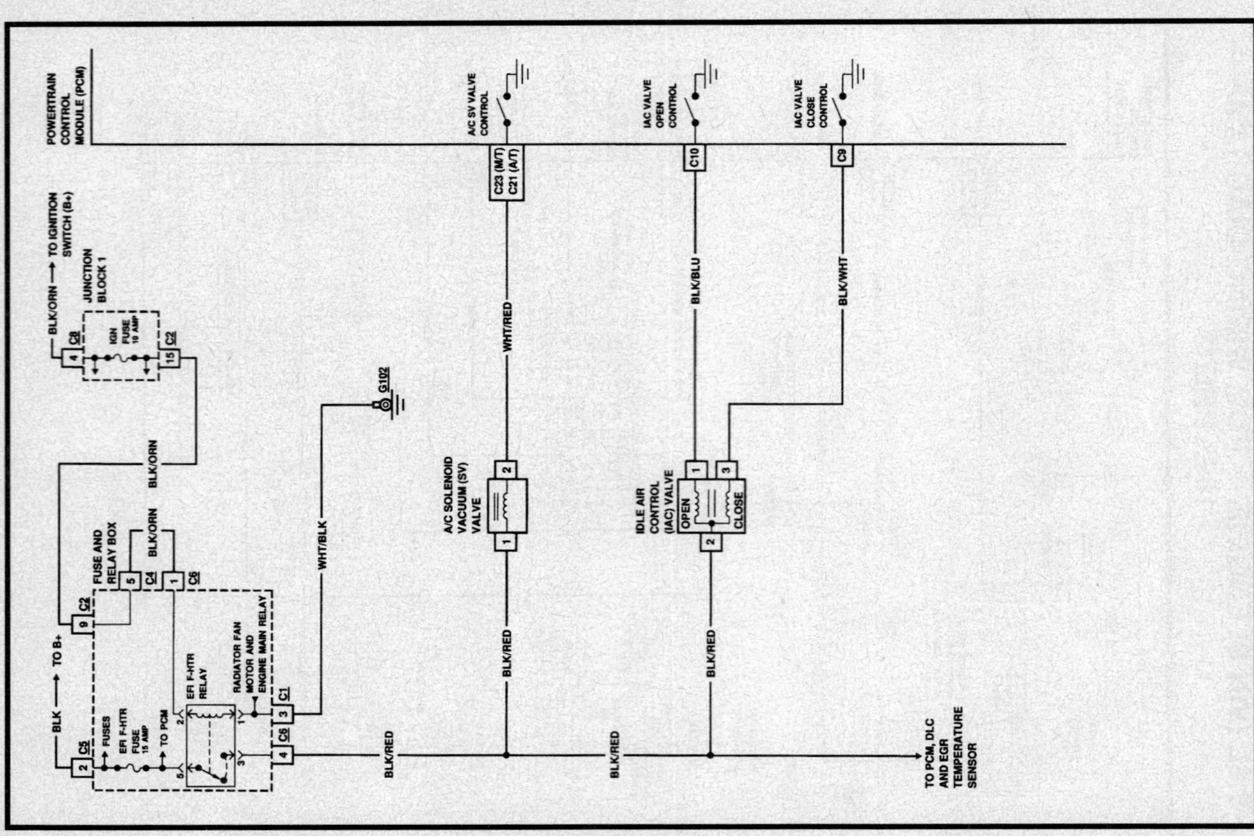

1.8L (VIN 8) ENGINE — PCM WIRING DIAGRAMS — 1993 PRIZM

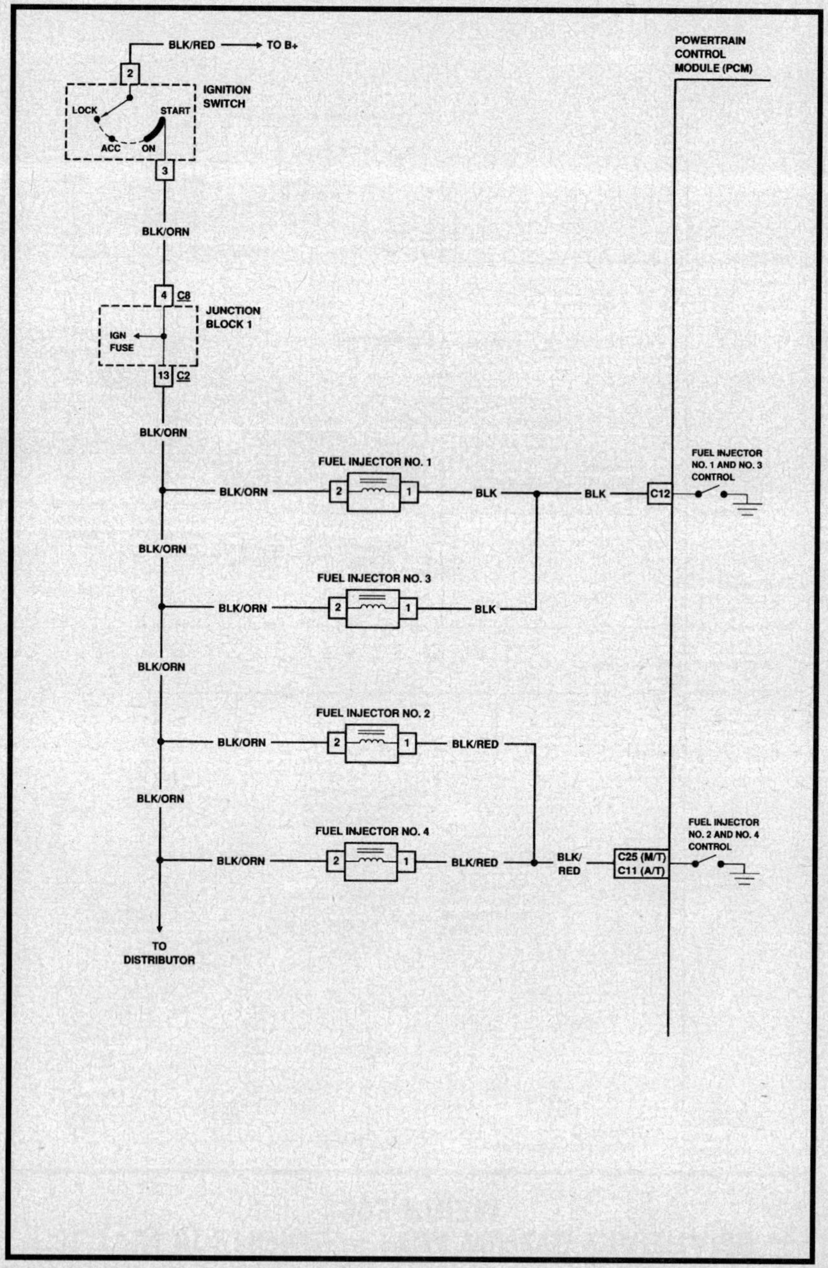

1.8L (VIN 8) ENGINE — PCM WIRING DIAGRAMS — 1993 PRIZM

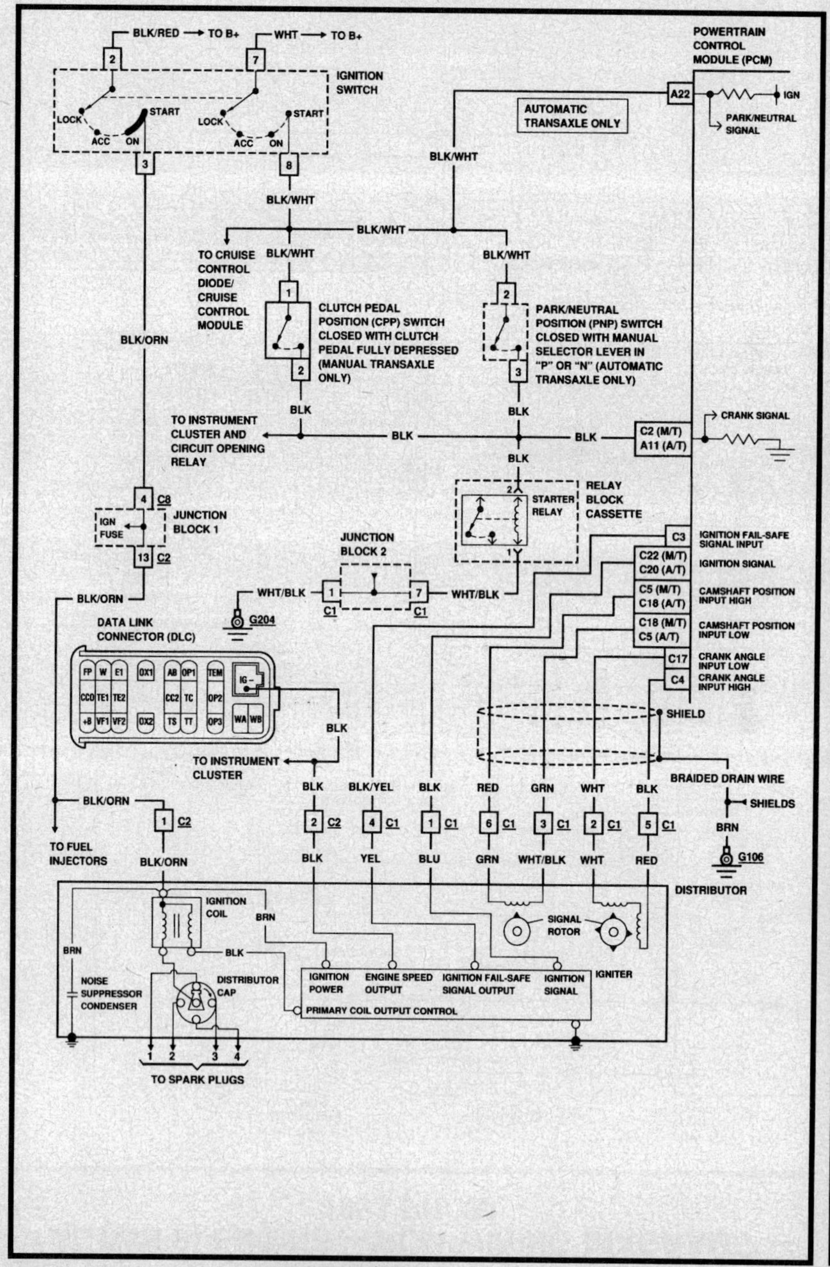

1.8L (VIN 8) ENGINE—PCM WIRING DIAGRAMS—1993 PRIZM

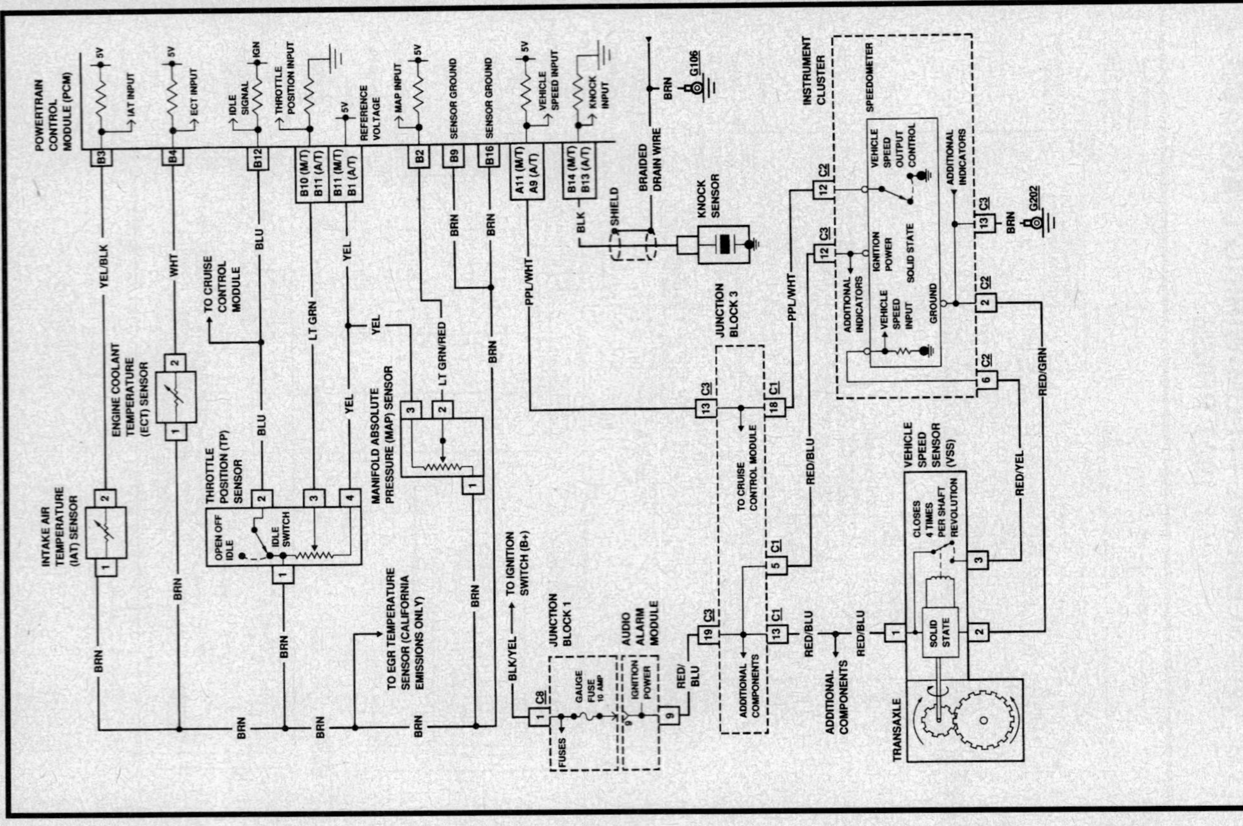

1.8L (VIN 8) ENGINE—PCM WIRING DIAGRAMS—1993 PRIZM

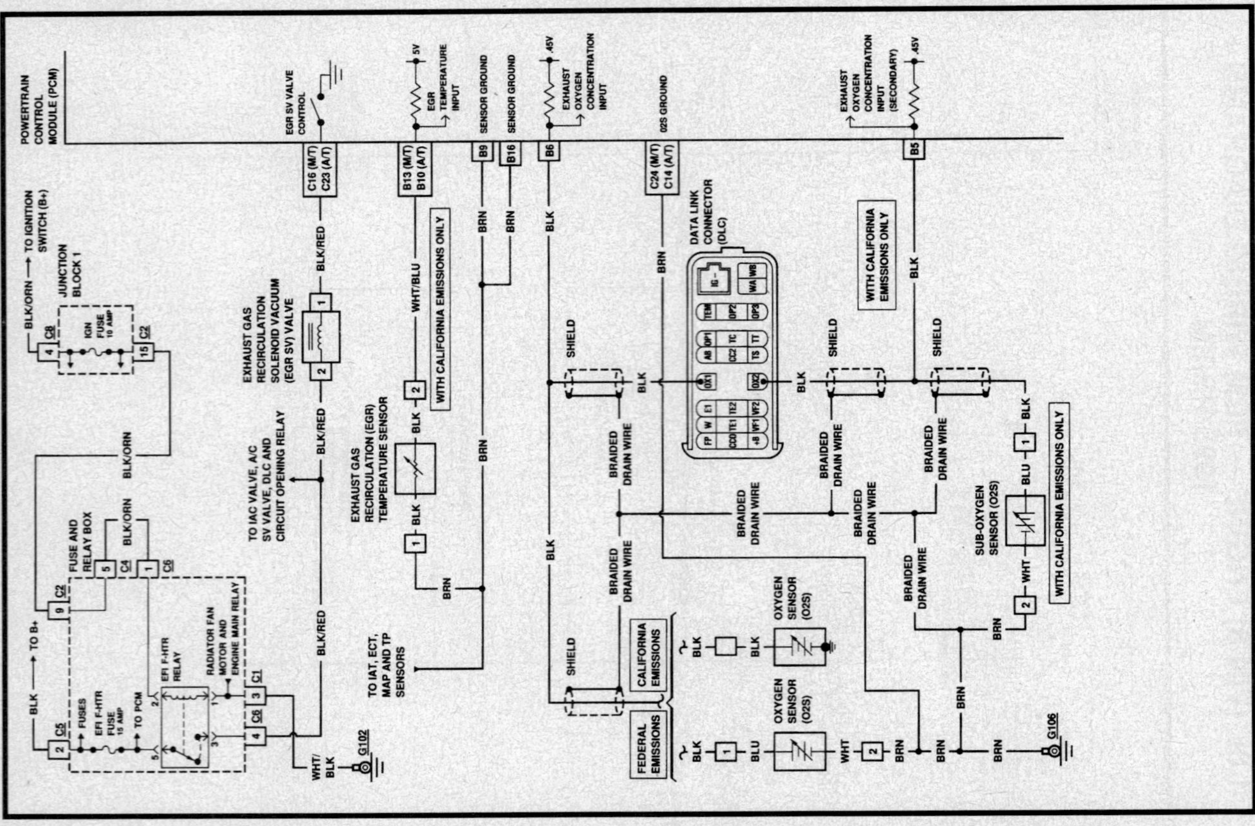

1.8L (VIN 8) ENGINE — PCM WIRING DIAGRAMS — 1993 PRIZM

1.8L (VIN 8) ENGINE — PCM CONNECTOR TERMINAL END VIEW — 1993 PRIZM

PCM CONNECTOR IDENTIFICATION

The following PCM voltage charts are for use with a digital multimeter to further aid in diagnosis. The voltages you get may vary due to low battery charge or other reasons, but they should be very close.

THE FOLLOWING CONDITIONS MUST BE MET BEFORE TESTING:
- Engine at operating temperature • Engine idling (for "ENG. RUN" column)
- Test terminal not grounded
- All voltages shown "B+" indicate system voltage

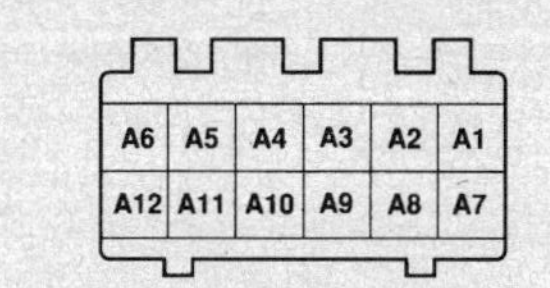

BACK VIEW OF PCM CONNECTOR C1 (MANUAL TRANSAXLE) (GRAY)

CAVITY/ PIN	WIRE COLOR	CIRCUIT	VOLTAGE KEY "ON"	VOLTAGE ENG. RUN
1/A1	BLK/RED	IGNITION POWER	B+	B+
2/A2	RED/WHT	MEMORY POWER	B+	B+
3/A3	—	NOT USED	—	—
4/A4	GRN/RED	CIRCUIT OPENING RELAY CONTROL	(7)	0*
5/A5	—	NOT USED	—	—
6/A6	RED/YEL	A/C CUTOUT CONTROL	0*	0* (8)
7/A7	BLK/RED	IGNITION POWER	B+	B+
8/A8	RED/YEL	MIL CONTROL	0-1V	B+
9/A9	—	NOT USED	—	—
10/A10	YEL/BLU	A/C ON SIGNAL	B+(5)	B+(5)
11/A11	PPL/WHT	VEHICLE SPEED INPUT	0* (4)	2-3V VARIES
12/A12	BLK	ELECTRICAL LOAD IDLE-UP SIGNAL	0* (1)	0* (1)

0* LESS THAN 0.50 VOLTS
(1) B+ WITH LIGHT/TURN SIGNAL SWITCH IN PARK OR HEAD OR WITH DEFOGGER ON
(2) B+ WITH BRAKE PEDAL DEPRESSED
(3) 0 VOLTS WITH OVERDRIVE SWITCH RELEASED; B+ WITH OVERDRIVE SWITCH DEPRESSED
(4) VARIES FROM 0 TO 5V WITH FRONT WHEELS ROTATING
(5) 0 VOLTS WITH A/C SWITCH DEPRESSED
(6) B+ WITH IGNITION SWITCH IN "START"
(7) 0 VOLTS FOR 3 SECONDS, THEN B+
(8) 5 TO 6 VOLTS WITH A/C SWITCH DEPRESSED
(9) 0 VOLTS IN "P" OR "N"; B+ IN "R", "D", "2" OR "L"
(10) VARIES WITH INTAKE AIR TEMPERATURE
(11) VARIES WITH ENGINE COOLANT TEMPERATURE
(12) VARIES WITH EXHAUST GAS TEMPERATURE
(13) VARIES WITH THROTTLE LEVER POSITION
(14) 0 VOLTS AT IDLE; B+ OFF IDLE
(15) B+ WITH MANUAL SELECTOR LEVER IN "2"
(16) B+ WITH MANUAL SELECTOR LEVER IN "L"
(17) B+ WITH MANUAL SELECTOR LEVER IN "R"
(18) VARIES CONSTANTLY WITH SYSTEM IN CLOSED LOOP
(19) B+ WITH VEHICLE IN TCC LOCKUP CONDITION
(20) B+ WITH TRANSAXLE IN 1st OR 2nd GEAR
(21) B+ WITH TRANSAXLE IN 2nd OR 3rd GEAR

1.8L (VIN 8) ENGINE — PCM CONNECTOR TERMINAL END VIEW — 1993 PRIZM

PCM CONNECTOR IDENTIFICATION

The following PCM voltage charts are for use with a digital multimeter to further aid in diagnosis. The voltages you get may vary due to low battery charge or other reasons, but they should be very close.

THE FOLLOWING CONDITIONS MUST BE MET BEFORE TESTING:
- Engine at operating temperature
- Engine idling (for "ENG. RUN" column)
- Test terminal not grounded
- All voltages shown "B+" indicate system voltage

A11	A10	A9	A8	A7	A6	A5	A4	A3	A2	A1
A22	A21	A20	A19	A18	A17	A16	A15	A14	A13	A12

BACK VIEW OF PCM CONNECTOR C1 (AUTOMATIC TRANSAXLE) (GRAY)

CAVITY/PIN	WIRE COLOR	CIRCUIT	VOLTAGE KEY "ON"	VOLTAGE ENG. RUN
1/A1	RED/WHT	MEMORY POWER	B+	B+
2/A2	BLK	ELECTRICAL LOAD IDLE-UP SIGNAL	0* (1)	0* (1)
3/A3	—	NOT USED		
4/A4	GRN/WHT	BRAKE SIGNAL	0* (2)	0* (2)
5/A5	RED/YEL	MIL CONTROL	0-1V	B+
6/A6	—	NOT USED		
7/A7	LT GRN	O/D ON/OFF SIGNAL/"O/D OFF" DIAGNOSTIC OUTPUT CONTROL	(3)	(3)
8/A8	—	NOT USED		
9/A9	PPL/WHT	VEHICLE SPEED INPUT	0* (4)	2-3V VARIES
10/A10	YEL/BLU	A/C ON SIGNAL	B+ (5)	B+ (5)
11/A11	BLK	CRANK SIGNAL	0* (6)	0* (6)
12/A12	BLK/RED	IGNITION POWER	B+	B+
13/A13	BLK/RED	IGNITION POWER	B+	B+
14/A14	GRN/RED	CIRCUIT OPENING RELAY CONTROL	(7)	0*
15/A15	—	NOT USED	—	—
16/A16	—	NOT USED	—	—
17/A17	—	NOT USED	—	—
18/A18	—	NOT USED	—	—
19/A19	—	NOT USED	—	—
20/A20	RED	O/D CUT SIGNAL	B+	B+
21/A21	RED/YEL	A/C CUTOUT CONTROL	0*	0* (8)
22/A22	BLK/YEL	PARK/NEUTRAL POSITION SIGNAL	(9)	(9)

0* LESS THAN 0.50 VOLTS
(1) B+ WITH LIGHT/TURN SIGNAL SWITCH IN PARK OR HEAD OR WITH DEFOGGER ON
(2) B+ WITH BRAKE PEDAL DEPRESSED
(3) 0 VOLTS WITH OVERDRIVE SWITCH RELEASED; B+ WITH OVERDRIVE SWITCH DEPRESSED
(4) VARIES FROM 0 TO 5V WITH FRONT WHEELS ROTATING
(5) 0 VOLTS WITH A/C SWITCH DEPRESSED
(6) B+ WITH IGNITION SWITCH IN "START"
(7) 0 VOLTS FOR 3 SECONDS, THEN B+
(8) 5 TO 6 VOLTS WITH A/C SWITCH DEPRESSED
(9) 0 VOLTS IN "P" OR "N"; B+ IN "R", "D", "2" OR "L"
(10) VARIES WITH INTAKE AIR TEMPERATURE
(11) VARIES WITH ENGINE COOLANT TEMPERATURE
(12) VARIES WITH EXHAUST GAS TEMPERATURE
(13) VARIES WITH THROTTLE LEVER POSITION
(14) 0 VOLTS AT IDLE; B+ OFF IDLE
(15) B+ WITH MANUAL SELECTOR LEVER IN "2"
(16) B+ WITH MANUAL SELECTOR LEVER IN "L"
(17) B+ WITH MANUAL SELECTOR LEVER IN "R"
(18) VARIES CONSTANTLY WITH SYSTEM IN CLOSED LOOP
(19) B+ WITH VEHICLE IN TCC LOCKUP CONDITION
(20) B+ WITH TRANSAXLE IN 1st OR 2nd GEAR
(21) B+ WITH TRANSAXLE IN 2nd OR 3rd GEAR

1.8L (VIN 8) ENGINE — PCM CONNECTOR TERMINAL END VIEW — 1993 PRIZM

PCM CONNECTOR IDENTIFICATION

The following PCM voltage charts are for use with a digital multimeter to further aid in diagnosis. The voltages you get may vary due to low battery charge or other reasons, but they should be very close.

THE FOLLOWING CONDITIONS MUST BE MET BEFORE TESTING:
- Engine at operating temperature
- Engine idling (for "ENG. RUN" column)
- Test terminal not grounded
- All voltages shown "B+" indicate system voltage

B8	B7	B6	B5	B4	B3	B2	B1
B16	B15	B14	B13	B12	B11	B10	B9

BACK VIEW OF PCM CONNECTOR C2 (GRAY)

CAVITY/PIN	WIRE COLOR	CIRCUIT	VOLTAGE KEY "ON"	VOLTAGE ENG. RUN
1/B1	YEL	REFERENCE VOLTAGE (5V) (A/T ONLY)	4-6V	4-6V
2/B2	LT GRN/RED	MAP INPUT	3-4V	1.1-1.8V
3/B3	YEL/BLK	IAT INPUT	1-3.5V (10)	1.5-2.5V (10)
4/B4	WHT	ECT INPUT	.5-3.5V (11)	1-2V (11)
5/B5	BLK	EXHAUST OXYGEN CONCENTRATION INPUT (SECONDARY) (CALIFORNIA ONLY)	0*	0-1V (18)
6/B6	BLK	EXHAUST OXYGEN CONCENTRATION INPUT	0*	0-1V (18)
7/B7	GRN/WHT (M/T)	TEST DIAGNOSTIC MODE SIGNAL (M/T ONLY)	B+	B+
	BLK (A/T)	THROTTLE POSITION OUTPUT (A/T ONLY)	0*	(13)
8/B8	RED/WHT	FUEL CONTROL FEEDBACK OUTPUT	0*	1.5-3.5V
9/B9	BRN	SENSOR GROUND	0*	0*
10/B10	LT GRN (M/T)	THROTTLE POSITION INPUT (M/T ONLY)	.5-4.5V (13)	.5-4.5V (13)
	WHT/BLU (A/T)	EGR TEMPERATURE INPUT (CALIFORNIA A/T ONLY)	1-3.5V (12)	1.5-2.5V (12)
11/B11	YEL (M/T)	REFERENCE VOLTAGE (5V) (M/T ONLY)	4-6V	4-6V
	LT GRN (A/T)	THROTTLE POSITION INPUT (A/T ONLY)	.5-4.5V (13)	.5-4.5V (13)
12/B12	BLU	IDLE SIGNAL	(14)	(14)
13/B13	WHT/BLU (M/T)	EGR TEMPERATURE INPUT (CALIFORNIA M/T ONLY)	1-3.5V (12)	1.5-2.5V (12)
	BLK (A/T)	KNOCK INPUT	0*	0-.2V
14/B14	BLK (M/T)	KNOCK INPUT	0*	0-.2V
	GRN/WHT (A/T)	TEST DIAGNOSTIC MODE SIGNAL (A/T ONLY)	B+	B+
15/B15	BLU/WHT	DIAGNOSTIC REQUEST SIGNAL	B+	B+
16/B16	BRN	SENSOR GROUND	0*	0*

0* LESS THAN 0.50 VOLTS
(1) B+ WITH LIGHT/TURN SIGNAL SWITCH IN PARK OR HEAD OR WITH DEFOGGER ON
(2) B+ WITH BRAKE PEDAL DEPRESSED
(3) 0 VOLTS WITH OVERDRIVE SWITCH RELEASED; B+ WITH OVERDRIVE SWITCH DEPRESSED
(4) VARIES FROM 0 TO 5V WITH FRONT WHEELS ROTATING
(5) 0 VOLTS WITH A/C SWITCH DEPRESSED
(6) B+ WITH IGNITION SWITCH IN "START"
(7) 0 VOLTS FOR 3 SECONDS, THEN B+
(8) 5 TO 6 VOLTS WITH A/C SWITCH DEPRESSED
(9) 0 VOLTS IN "P" OR "N"; B+ IN "R", "D", "2" OR "L"
(10) VARIES WITH INTAKE AIR TEMPERATURE
(11) VARIES WITH ENGINE COOLANT TEMPERATURE
(12) VARIES WITH EXHAUST GAS TEMPERATURE
(13) VARIES WITH THROTTLE LEVER POSITION
(14) 0 VOLTS AT IDLE; B+ OFF IDLE
(15) B+ WITH MANUAL SELECTOR LEVER IN "2"
(16) B+ WITH MANUAL SELECTOR LEVER IN "L"
(17) B+ WITH MANUAL SELECTOR LEVER IN "R"
(18) VARIES CONSTANTLY WITH SYSTEM IN CLOSED LOOP
(19) B+ WITH VEHICLE IN TCC LOCKUP CONDITION
(20) B+ WITH TRANSAXLE IN 1st OR 2nd GEAR
(21) B+ WITH TRANSAXLE IN 2nd OR 3rd GEAR

1.8L (VIN 8) ENGINE — PCM CONNECTOR TERMINAL END VIEW — 1993 PRIZM

1.8L (VIN 8) ENGINE — DIAGNOSTIC TROUBLE CODE (DTC) TABLE — 1993 PRIZM

PCM CONNECTOR IDENTIFICATION

The following PCM voltage charts are for use with a digital multimeter to further aid in diagnosis. The voltages you get may vary due to low battery charge or other reasons, but they should be very close.

THE FOLLOWING CONDITIONS MUST BE MET BEFORE TESTING:
- Engine at operating temperature • Engine idling (for "ENG. RUN" column)
- Test terminal not grounded
- All voltages shown "B+" indicate system voltage

BACK VIEW OF PCM CONNECTOR C3 (GRAY)

CAVITY/PIN	WIRE COLOR	CIRCUIT	VOLTAGE KEY "ON"	VOLTAGE ENG. RUN
1/C1	BLU/YEL	TCC SOLENOID CONTROL (A/T ONLY)	0*	0* (19)
2/C2	BLK (M/T)	CRANK SIGNAL (M/T ONLY)	0* (6)	0* (6)
	PPL (A/T)	SHIFT SOLENOID NO. 1 CONTROL (A/T ONLY)	0*	B+ (20)
3/C3	BLK/YEL	IGNITION FAIL-SAFE SIGNAL INPUT	.56V	1.5V
4/C4	BLK	CRANK ANGLE INPUT HIGH	.69V	.69V
5/C5	RED (M/T)	CAMSHAFT POSITION INPUT HIGH (M/T ONLY)	.69V	.69V
	GRN (A/T)	CAMSHAFT POSITION INPUT LOW (A/T ONLY)	.69V	.69V
6/C6	LT GRN/RED	2 POSITION SIGNAL (A/T ONLY)	0* (15)	0* (15)
7/C7	—	NOT USED	—	—
8/C8	—	NOT USED	—	—
9/C9	BLK/WHT	IAC VALVE CLOSE CONTROL	.3-.7V	5-7V
10/C10	BLK/BLU	IAC VALVE OPEN CONTROL	B+	8-10V
11/C11	BLK/RED	FUEL INJECTOR NO. 2 AND NO. 4 CONTROL (A/T ONLY)	B+	B+
12/C12	BLK	FUEL INJECTOR NO. 1 AND NO. 3 CONTROL	B+	B+
13/C13	BRN	GROUND	0*	0*
14/C14	BRN	O2S GROUND (A/T ONLY)	0*	0*
15/C15	BRN (M/T)	GROUND (M/T ONLY)	0*	0*
	BRN/YEL (A/T)	SHIFT SOLENOID NO. 2 CONTROL (A/T ONLY)	0*	0* (21)
16/C16	BLK/RED	EGR SV VALVE CONTROL (CALIFORNIA M/T ONLY)	B+	B+
17/C17	WHT	CRANK ANGLE INPUT LOW	.69V	.69V
18/C18	GRN (M/T)	CAMSHAFT POSITION INPUT LOW (M/T ONLY)	.69V	.69V
	RED (A/T)	CAMSHAFT POSITION INPUT HIGH (A/T ONLY)	.69V	.69V
19/C19	LT GRN	L POSITION SIGNAL (A/T ONLY)	0* (16)	0* (16)
20/C20	BLK	IGNITION SIGNAL (A/T ONLY)	0*	0-2V (13)
21/C21	WHT/RED	A/C SV VALVE CONTROL (A/T ONLY)	B+	B+ (5)
22/C22	BLK (M/T)	IGNITION SIGNAL (M/T ONLY)	0*	0-2V (13)
	RED/BLK (A/T)	R POSITION SIGNAL (A/T ONLY)	0* (17)	0* (17)
23/C23	WHT/RED (M/T)	A/C SV VALVE CONTROL (M/T ONLY)	B+	B+(5)
	BLK/RED (A/T)	EGR SV VALVE CONTROL (CALIFORNIA A/T ONLY)	B+	B+
24/C24	BRN	O2S GROUND (M/T ONLY)	0*	0*
25/C25	BLK/RED	FUEL INJECTOR NO. 2 AND NO. 4 CONTROL (M/T ONLY)	B+	B+
26/C26	BRN	GROUND	0*	0*

0* LESS THAN 0.50 VOLTS
(1) B+ WITH LIGHT/TURN SIGNAL SWITCH IN PARK OR HEAD OR WITH DEFOGGER ON
(2) B+ WITH BRAKE PEDAL DEPRESSED
(3) 0 VOLTS WITH OVERDRIVE SWITCH RELEASED; B+ WITH OVERDRIVE SWITCH DEPRESSED
(4) VARIES FROM 0 TO 5V WITH FRONT WHEELS ROTATING
(5) 0 VOLTS WITH A/C SWITCH DEPRESSED
(6) B+ WITH IGNITION SWITCH IN "START"
(7) 0 VOLTS FOR 3 SECONDS, THEN B+
(8) 5 TO 6 VOLTS WITH A/C SWITCH DEPRESSED
(9) 0 VOLTS IN "P" OR "N"; B+ IN "R", "D", "2" OR "L"
(10) VARIES WITH INTAKE AIR TEMPERATURE
(11) VARIES WITH ENGINE COOLANT TEMPERATURE
(12) VARIES WITH EXHAUST GAS TEMPERATURE
(13) VARIES WITH THROTTLE LEVER POSITION
(14) 0 VOLTS AT IDLE; B+ OFF IDLE
(15) B+ WITH MANUAL SELECTOR LEVER IN "2"
(16) B+ WITH MANUAL SELECTOR LEVER IN "L"
(17) B+ WITH MANUAL SELECTOR LEVER IN "R"
(18) VARIES CONSTANTLY WITH SYSTEM IN CLOSED LOOP
(19) B+ WITH VEHICLE IN TCC LOCKUP CONDITION
(20) B+ WITH TRANSAXLE IN 1st OR 2nd GEAR
(21) B+ WITH TRANSAXLE IN 2nd OR 3rd GEAR

DIAGNOSTIC TROUBLE CODE (DTC) TABLE
1.8L (VIN 8) "S" CARLINE
(PAGE 1 OF 2)

EXAMPLE: RPM SIGNAL DTC 13 AND MANIFOLD ABSOLUTE PRESSURE (MAP) SENSOR INPUT DTC 31

ON / OFF — DTC 13 — DTC 31 — 4.5 — 0.52 — 1.5 — 2.5 — 4.5 — (SECONDS)

DIAGNOSTIC TROUBLE CODE (DTC) NO.	MODE	SYSTEM	CONDITIONS THAT MUST BE MET BEFORE DTC WILL SET
—		NORMAL	• INDICATES THAT THE ELECTRONIC ENGINE CONTROL SYSTEM IS FUNCTIONING NORMALLY.
12		RPM SIGNAL	(1) NO CRANK ANGLE INPUT DETECTED AT PCM FOR 2 SECONDS OR MORE WHILE CRANKING ENGINE. (2) NO CAMSHAFT POSITION INPUT DETECTED AT PCM FOR 3 SECONDS OR MORE WITH ENGINE SPEED BETWEEN 600 AND 4,000 RPM. • THIS DTC CAN ONLY BE ACCESSED WHEN IN THE NORMAL DIAGNOSTIC MODE.
13		RPM SIGNAL	• NO CRANK ANGLE INPUT DETECTED AT PCM FOR 0.30 SECONDS OR MORE WITH ENGINE SPEED ABOVE 1,500 RPM.
14		IGNITION SIGNAL	• NO IGNITION FAIL-SAFE SIGNAL INPUT DETECTED AT PCM FOR 4 CONSECUTIVE IGNITION CYCLES. • THIS DTC CAN ONLY BE ACCESSED WHEN IN THE NORMAL DIAGNOSTIC MODE.
16		PCM CONTROL SIGNAL	• COMMUNICATION FAILURE BETWEEN PCM'S ELECTRONIC ENGINE AND ELECTRONIC TRANSAXLE CIRCUITRY. • THIS DTC CAN ONLY BE ACCESSED WHEN IN THE NORMAL DIAGNOSTIC MODE.
21		OXYGEN SENSOR (O2S) INPUT	• OXYGEN SENSOR VOLTAGE INPUT AT PCM CONSTANT BETWEEN 0.35 AND 0.70 VOLTS FOR 60 SECONDS OR MORE WITH ENGINE SPEED ABOVE 1,500 RPM.
22		ENGINE COOLANT TEMPERATURE (ECT) SENSOR	• ECT INPUT AT PCM OPEN OR SHORTED FOR 0.50 SECONDS OR MORE.
24		INTAKE AIR TEMPERATURE (IAT) SENSOR	• IAT INPUT AT PCM OPEN OR SHORTED FOR 0.50 SECONDS OR MORE.
25		AIR/FUEL RATIO LEAN	(1) OXYGEN SENSOR INPUT AT PCM IS LESS THAN 0.45 VOLTS FOR 90 SECONDS OR MORE WITH OXYGEN SENSOR WARM AND ENGINE SPEED ABOVE 1,500 RPM. (2) WHEN ENGINE SPEED VARIES BY MORE THAN 50 RPM OVER THE PRECEDING CRANK ANGLE PERIOD DURING A PERIOD OF 50 SECONDS WHILE IDLING WITH THE ECT ABOVE 50°C (122°F). • EITHER CONDITION MUST BE DETECTED DURING 2 CONSECUTIVE IGNITION CYCLES FOR DTC TO BE STORED IN PCM MEMORY.

1.8L (VIN 8) ENGINE — DIAGNOSTIC TROUBLE CODE (DTC) TABLE — 1993 PRIZM

DIAGNOSTIC TROUBLE CODE (DTC) TABLE
1.8L (VIN 8) "S" CARLINE
(PAGE 2 OF 2)

DIAGNOSTIC TROUBLE CODE (DTC) NO.	MODE	SYSTEM	CONDITIONS THAT MUST BE MET BEFORE DTC WILL SET
26		AIR/FUEL RATIO RICH	• WHEN ENGINE SPEED VARIES BY MORE THAN 50 RPM OVER THE PRECEDING CRANK ANGLE PERIOD DURING A PERIOD OF 50 SECONDS WHILE IDLING WITH THE ECT ABOVE 50°C (122°F). • CONDITION MUST BE DETECTED DURING 2 CONSECUTIVE IGNITION CYCLES FOR DTC TO BE STORED IN PCM MEMORY.
27		SUB-OXYGEN SENSOR (O2S) INPUT	• WHEN SUB-OXYGEN SENSOR IS WARMED UP AND ENGINE IS OFF IDLE FOR 2 SECONDS OR MORE AND OXYGEN SENSOR INPUT AT PCM IS 0.45 VOLTS OR MORE AND SUB-OXYGEN SENSOR INPUT AT PCM IS 0.45 VOLTS OR LESS. • CONDITION MUST BE DETECTED DURING 2 CONSECUTIVE IGNITION CYCLES FOR DTC TO BE STORED IN PCM MEMORY.
31		MANIFOLD ABSOLUTE PRESSURE (MAP) SENSOR INPUT	• MAP INPUT AT PCM OPEN OR SHORTED FOR 0.50 SECONDS OR MORE.
41		THROTTLE POSITION (TP) SENSOR INPUT	• THROTTLE POSITION INPUT AT PCM OPEN OR SHORTED FOR 0.50 SECONDS OR MORE.
42		VEHICLE SPEED INPUT	• VEHICLE SPEED INPUT NOT DETECTED AT PCM FOR 8 SECONDS OR MORE WITH ENGINE SPEED BETWEEN 3,000 AND 5,000 RPM AND TRANSAXLE IN GEAR.
43		CRANK SIGNAL	• CRANK SIGNAL NOT DETECTED AT PCM DURING IGNITION. • THIS DTC CAN ONLY BE DETECTED WHEN IN THE TEST DIAGNOSTIC MODE.
51		SWITCH CONDITION SIGNAL	• DISPLAYED WHEN A/C IS ON, ACCELERATOR PEDAL IS DEPRESSED OR WHEN MANUAL SELECTOR IS IN "R", "D", "2" OR "L" POSITION WITH DLC TERMINAL "TE1" AND "E1" CONNECTED. • THIS DTC CAN ONLY BE ACCESSED WHEN IN THE TEST DIAGNOSTIC MODE AND INDICATES THE ABOVE MENTIONED SYSTEMS ARE FUNCTIONING NORMALLY. IF DTC 51 IS NOT INDICATED WHEN THE CONDITIONS ARE MET, A FAULT IS PRESENT, AND DTC 51's DIAGNOSTIC CHART MUST BE USED.
52		KNOCK SENSOR (KS) INPUT	• KNOCK SENSOR INPUT AT PCM OPEN OR SHORTED WITH ENGINE SPEED BETWEEN 1,200 RPM AND 6,000 RPM. • THIS DTC CAN ONLY BE DETECTED WHEN IN THE NORMAL DIAGNOSTIC MODE.
71		EGR SYSTEM	• WITH ECT ABOVE 60°C (140°F) 50 SECONDS FROM START OF EGR OPERATION, THE EGR TEMPERATURE IS LESS THAN 80°C (176°F) AND THE EGR TEMPERATURE HAS RISEN LESS THAN 10°C (18°F) DURING THE 50 SECONDS. • CONDITION MUST BE DETECTED DURING 2 CONSECUTIVE IGNITION CYCLES FOR DTC TO BE STORED IN PCM MEMORY.

1.8L (VIN 8) ENGINE — ON-BOARD DIAGNOSTIC (OBD) SYSTEM CHECK — 1993 PRIZM

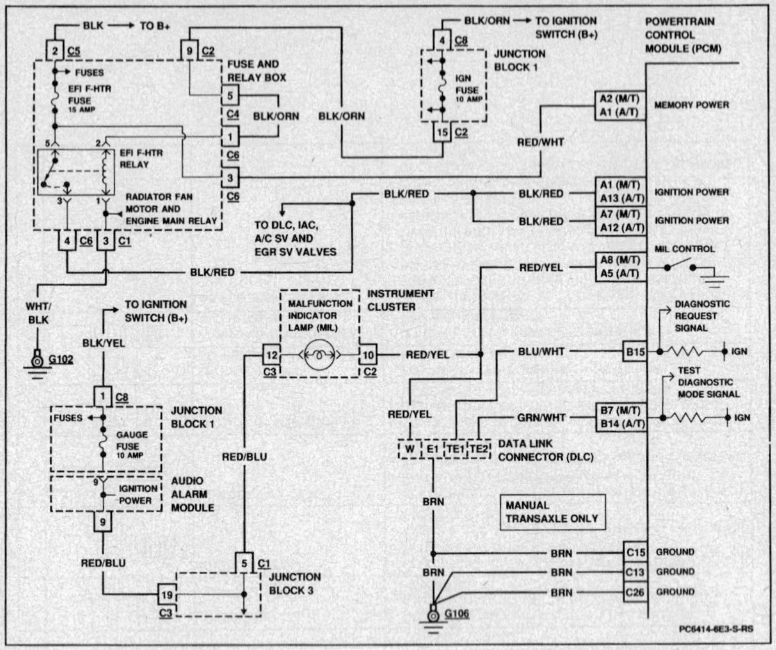

ON-BOARD DIAGNOSTIC (OBD) SYSTEM CHECK
(Page 1 of 2)
1.8L (VIN 8) "S" CARLINE

Circuit Description

The OBD System Check is an organized approach to identifying a malfunction created by the electronic engine control system. It must be the starting point for any driveability complaint diagnosis; because it directs the service technician to the next logical step in diagnosing the complaint. Understanding the chart and using it correctly will reduce diagnostic time and prevent the unnecessary replacement of good parts.

Test Description: Number(s) below refer to circled number(s) on the diagnostic chart.

1. Checks the MIL operation with the ignition switch in the "ON" position.
2. Checks to see if the PCM's normal diagnostic mode is functioning.
3. Checks to see if the PCM's test diagnostic mode is functioning.
4. Checks to see if the vehicle will start.
5. Checks for any DTCs that are stored in the PCM's memory with the engine running.
6. Checks for any DTCs that are stored in the PCM's memory with the engine off.

1.8L (VIN 8) ENGINE — ON-BOARD DIAGNOSTIC (OBD) SYSTEM CHECK — 1993 PRIZM

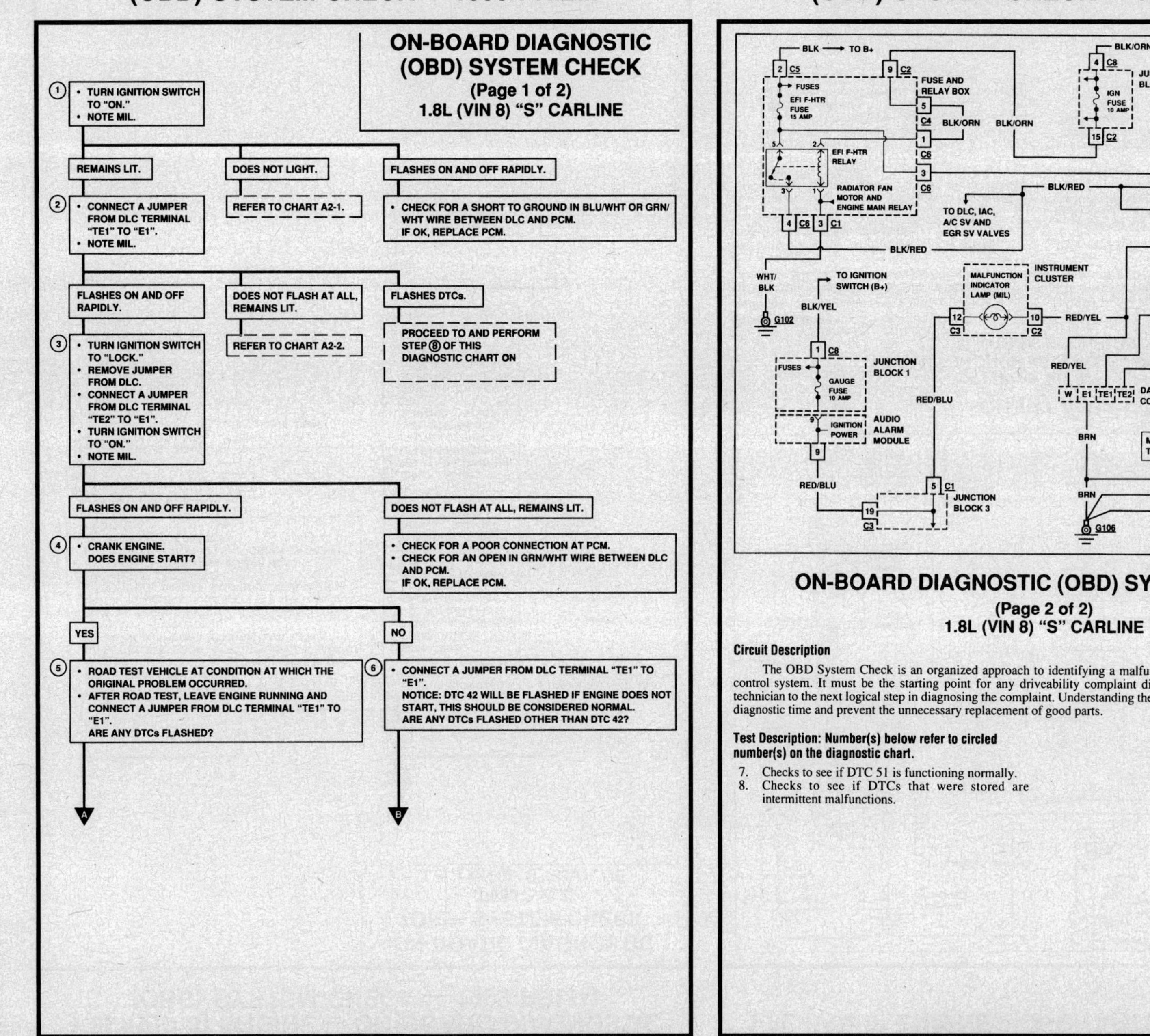

ON-BOARD DIAGNOSTIC (OBD) SYSTEM CHECK
(Page 2 of 2)
1.8L (VIN 8) "S" CARLINE

Circuit Description

The OBD System Check is an organized approach to identifying a malfunction created by the electronic engine control system. It must be the starting point for any driveability complaint diagnosis; because it directs the service technician to the next logical step in diagnosing the complaint. Understanding the chart and using it correctly will reduce diagnostic time and prevent the unnecessary replacement of good parts.

Test Description: Number(s) below refer to circled number(s) on the diagnostic chart.

7. Checks to see if DTC 51 is functioning normally.
8. Checks to see if DTCs that were stored are intermittent malfunctions.

ON-BOARD DIAGNOSTIC (OBD) SYSTEM CHECK
(Page 2 of 2)
1.8L (VIN 8) "S" CARLINE

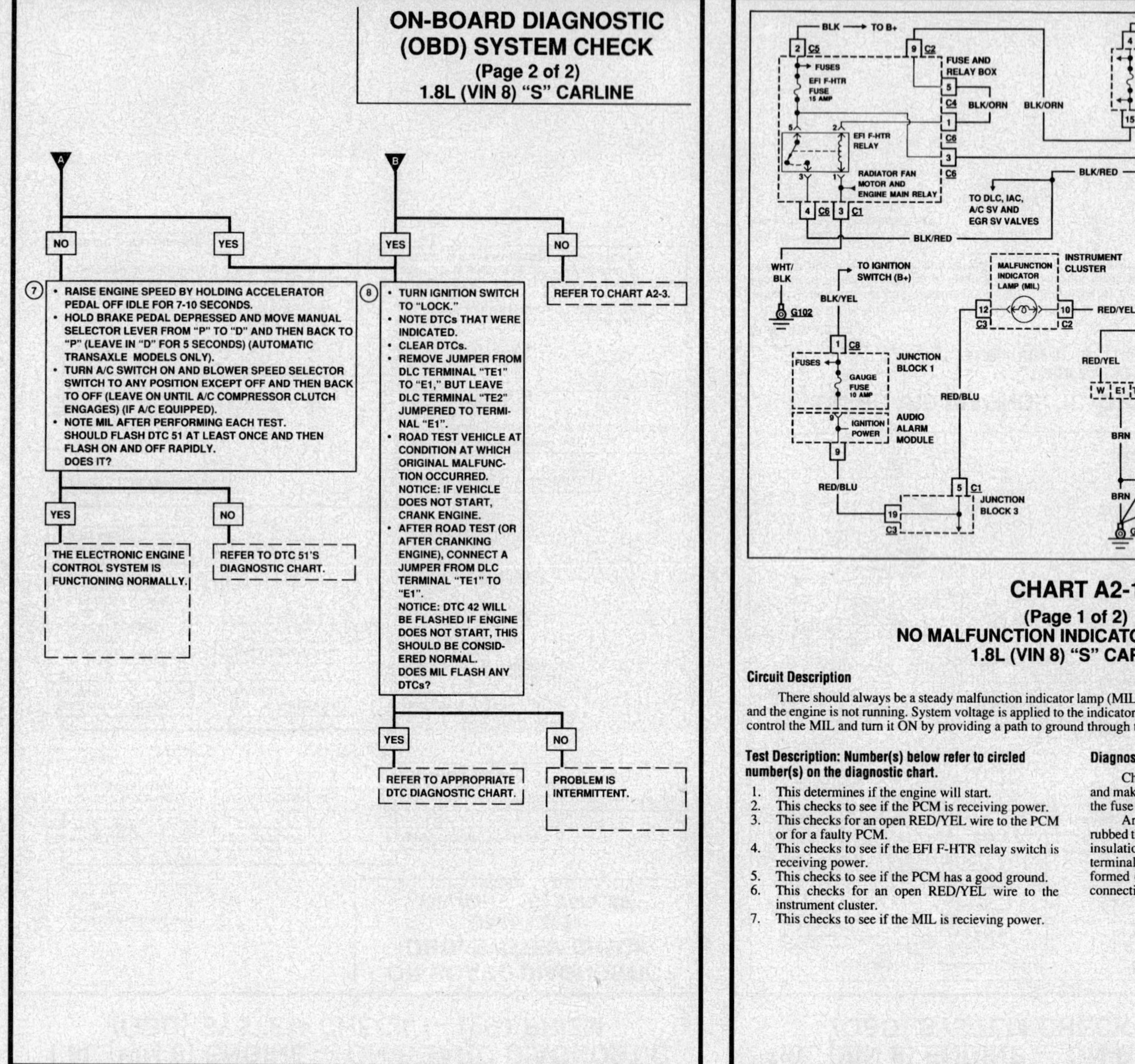

CHART A2-1
(Page 1 of 2)
NO MALFUNCTION INDICATOR LAMP (MIL)
1.8L (VIN 8) "S" CARLINE

Circuit Description

There should always be a steady malfunction indicator lamp (MIL) when the ignition switch is in the "ON" position and the engine is not running. System voltage is applied to the indicator bulb. The powertrain control module (PCM) will control the MIL and turn it ON by providing a path to ground through the RED/YEL wire.

Test Description: Number(s) below refer to circled number(s) on the diagnostic chart.

1. This determines if the engine will start.
2. This checks to see if the PCM is receiving power.
3. This checks for an open RED/YEL wire to the PCM or for a faulty PCM.
4. This checks to see if the EFI F-HTR relay switch is receiving power.
5. This checks to see if the PCM has a good ground.
6. This checks for an open RED/YEL wire to the instrument cluster.
7. This checks to see if the MIL is recieving power.

Diagnostic Aids:

Check EFI F-HTR, IGN and GAUGE fuses for opens and make sure EFI F-HTR Relay is securely mounted into the fuse and relay box.

An intermittent may be caused by a poor connection, rubbed through wire insulation, or a wire broken inside the insulation. Inspect harness connectors for backed out terminals, improper mating, broken locks, improperly formed or damaged terminals and poor terminal-to-wire connections before component replacement.

1.8L (VIN 8) ENGINE — A-CHARTS — 1993 PRIZM

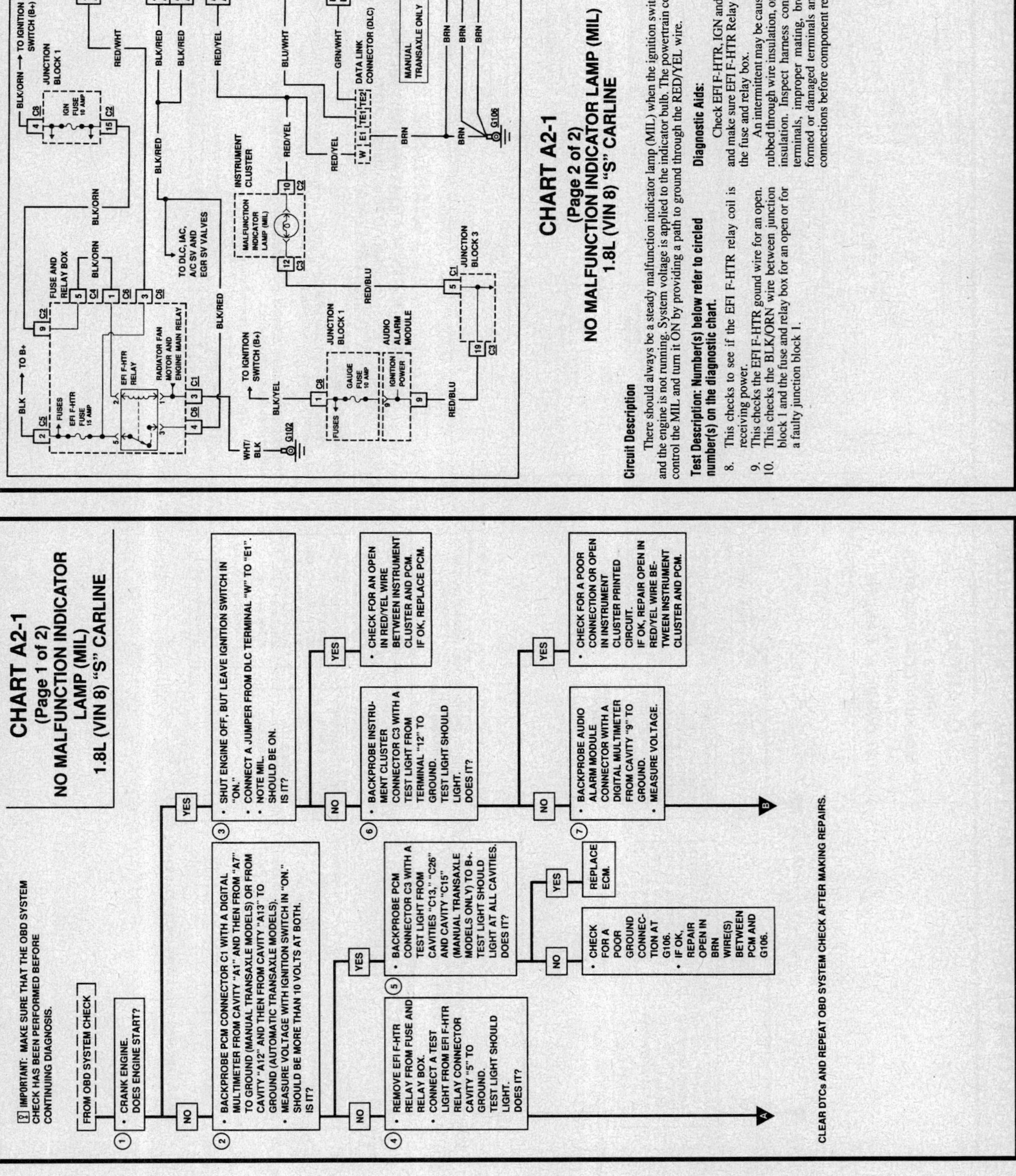

Circuit Description

There should always be a steady malfunction indicator lamp (MIL) when the ignition switch is in the "ON" position and the engine is not running. System voltage is applied to the indicator bulb. The powertrain control module (PCM) will control the MIL and turn it ON by providing a path to ground through the RED/YEL wire.

Test Description: Number(s) below refer to circled number(s) on the diagnostic chart.

8. This checks to see if the EFI F-HTR relay coil is receiving power.
9. This checks the EFI F-HTR gound wire for an open.
10. This checks the BLK/ORN wire between junction block 1 and the fuse and relay box for an open or for a faulty junction block 1.

Diagnostic Aids:

Check EFI F-HTR, IGN and GAUGE fuses for opens and make sure EFI F-HTR Relay is securely mounted into the fuse and relay box.

An intermittent may be caused by a poor connection, rubbed through wire insulation, or a wire broken inside the insulation. Inspect harness connectors for backed out terminals, improper mating, broken locks, improperly formed or damaged terminals and poor terminal-to-wire connections before component replacement.

1.8L (VIN 8) ENGINE — A-CHARTS — 1993 PRIZM

CHART A2-1
(Page 1 of 2)
NO MALFUNCTION INDICATOR
LAMP (MIL)
1.8L (VIN 8) "S" CARLINE

[i] IMPORTANT: MAKE SURE THAT THE OBD SYSTEM CHECK HAS BEEN PERFORMED BEFORE CONTINUING DIAGNOSIS.

FROM OBD SYSTEM CHECK

(1) • CRANK ENGINE.
• DOES ENGINE START?

(2) • BACKPROBE PCM CONNECTOR C1 WITH A DIGITAL MULTIMETER FROM CAVITY "A1" AND THEN FROM "A7" (MANUAL TRANSAXLE MODELS) OR FROM CAVITY "A12" AND THEN FROM CAVITY "A13" TO GROUND (AUTOMATIC TRANSAXLE MODELS).
• MEASURE VOLTAGE WITH IGNITION SWITCH IN "ON."
SHOULD BE MORE THAN 10 VOLTS AT BOTH.
IS IT?

(3) • SHUT ENGINE OFF, BUT LEAVE IGNITION SWITCH IN "ON."
• CONNECT A JUMPER FROM DLC TERMINAL "W" TO "E1".
• NOTE MIL.
SHOULD BE ON.
IS IT?

• CHECK FOR AN OPEN IN RED/YEL WIRE BETWEEN INSTRUMENT CLUSTER AND PCM. IF OK, REPLACE PCM.

(6) • BACKPROBE INSTRU-MENT CLUSTER CONNECTOR C3 WITH A TEST LIGHT FROM TERMINAL "12" TO GROUND.
TEST LIGHT SHOULD LIGHT.
DOES IT?

• CHECK FOR A POOR CONNECTION OR OPEN IN INSTRUMENT CLUSTER PRINTED CIRCUIT. IF OK, REPAIR OPEN IN RED/YEL WIRE BE-TWEEN INSTRUMENT CLUSTER AND PCM.

(7) • BACKPROBE AUDIO ALARM MODULE CONNECTOR WITH A DIGITAL MULTIMETER FROM CAVITY "9" TO GROUND.
• MEASURE VOLTAGE.

(4) • REMOVE EFI F-HTR RELAY FROM FUSE AND RELAY BOX.
• CONNECT A TEST LIGHT FROM EFI F-HTR RELAY CONNECTOR CAVITY "5" TO GROUND.
TEST LIGHT SHOULD LIGHT.
DOES IT?

(5) • BACKPROBE PCM CONNECTOR C3 WITH A TEST LIGHT FROM CAVITIES "C13," "C26" AND CAVITY "C15" (MANUAL TRANSAXLE MODELS ONLY) TO B+.
TEST LIGHT SHOULD LIGHT AT ALL CAVITIES.
DOES IT?

• CHECK FOR A POOR GROUND CONNEC-TION AT G106.
• IF OK, REPAIR OPEN IN BRN WIRE(S) BETWEEN PCM AND G106.

REPLACE ECM.

B →

A →

CLEAR DTCs AND REPEAT OBD SYSTEM CHECK AFTER MAKING REPAIRS.

CHART A2-1

(Page 2 of 2)
NO MALFUNCTION INDICATOR LAMP (MIL)
1.8L (VIN 8) "S" CARLINE

A → YES / NO

- **8** CONNECT A TEST LIGHT FROM EFI F-HTR RELAY CONNECTOR CAVITY "2" TO GROUND. TEST LIGHT SHOULD LIGHT. DOES IT?

  - (NO) CHECK FOR AN OPEN IN BLK WIRE BETWEEN IGNITION SWITCH AND FUSE AND RELAY BOX. IF OK, REPLACE FUSE AND RELAY BOX.

B → LESS THAN 10 VOLTS / MORE THAN 10 VOLTS

- (LESS THAN 10 VOLTS) CHECK FOR AN OPEN IN BLK/YEL WIRE BETWEEN IGNITION SWITCH AND JUNCTION BLOCK 1. IF OK, REPLACE AUDIO ALARM MODULE AND REPEAT OBD SYSTEM CHECK. IF MIL REMAINS INOPERATIVE, REPLACE JUNCTION BLOCK 1.

- (MORE THAN 10 VOLTS) CHECK FOR AN OPEN IN RED/BLU WIRE BETWEEN AUDIO ALARM MODULE AND JUNCTION BLOCK 3 OR IN RED/BLU WIRE BETWEEN JUNCTION BLOCK 3 AND INSTRUMENT CLUSTER. IF OK, REPLACE JUNCTION BLOCK 3.

(YES) →

- **9** CONNECT A TEST LIGHT FROM EFI F-HTR RELAY CONNECTOR CAVITY "1" TO B+. TEST LIGHT SHOULD LIGHT. DOES IT?

  - (YES) CHECK FOR AN OPEN IN BLK/RED WIRE BETWEEN FUSE AND RELAY BOX AND PCM. IF OK, REPLACE EFI F-HTR RELAY.
  - (NO) CHECK FOR A POOR GROUND CONNECTION AT G102. IF OK, REPAIR OPEN IN WHT/BLK WIRE BETWEEN FUSE AND RELAY BOX AND G102.

(NO) →

- **10** BACKPROBE JUNCTION BLOCK 1 CONNECTOR C2 WITH A TEST LIGHT FROM CAVITY "15" TO GROUND. TEST LIGHT SHOULD LIGHT. DOES IT?

  - (NO) CHECK FOR AN OPEN IN BLK/ORN WIRE BETWEEN IGNITION SWITCH AND JUNCTION BLOCK 1. IF OK, REPLACE JUNCTION BLOCK 1.
  - (YES) CHECK FOR AN OPEN IN BLK/ORN WIRE BETWEEN JUNCTION BLOCK 1 AND FUSE AND RELAY BOX OR IN BLK/ORN WIRE BETWEEN JUNCTION BLOCK 1 AND RELAY BOX CONNECTORS C4 AND C6. IF OK, REPLACE JUNCTION BLOCK 1.

CLEAR DTCs AND REPEAT OBD SYSTEM CHECK AFTER MAKING REPAIRS.

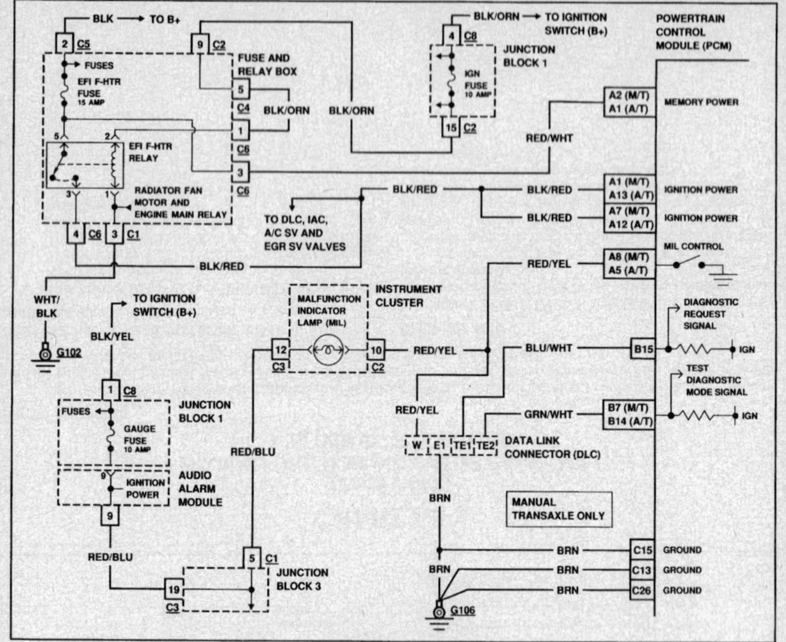

CHART A2-2

MALFUNCTION INDICATOR LAMP (MIL) WILL NOT FLASH ANY DTCs, ON STEADY
1.8L (VIN 8) "S" CARLINE

Circuit Description:

The malfunction indicator lamp (MIL) will flash diagnostic trouble codes (DTCs) that are in the powertrain control modules (PCM's) memory when the data link connector (DLC) terminals "TE1" and "E1" are jumpered.

Test Description: Number(s) below refer to circled number(s) on the diagnostic chart.

1. This checks for a short to ground in RED/YEL wire.
2. This checks for an open BLU/WHT (diagnostic request signal) wire.
3. This checks for an open in the BRN ground wire to the DLC or for a faulty PCM.

Diagnostic Aids:

An intermittent may be caused by a poor connection, rubbed through wire insulation, or a wire broken inside the insulation. Inspect harness connectors for backed out terminals, improper mating, broken locks, improperly formed or damaged terminals and poor terminal-to-wire connections before component replacement.

1.8L (VIN 8) ENGINE — A-CHARTS — 1993 PRIZM

CHART A2-2
MALFUNCTION INDICATOR LAMP (MIL) WILL NOT FLASH ANY DTCs, ON STEADY
1.8L (VIN 8) "S" CARLINE

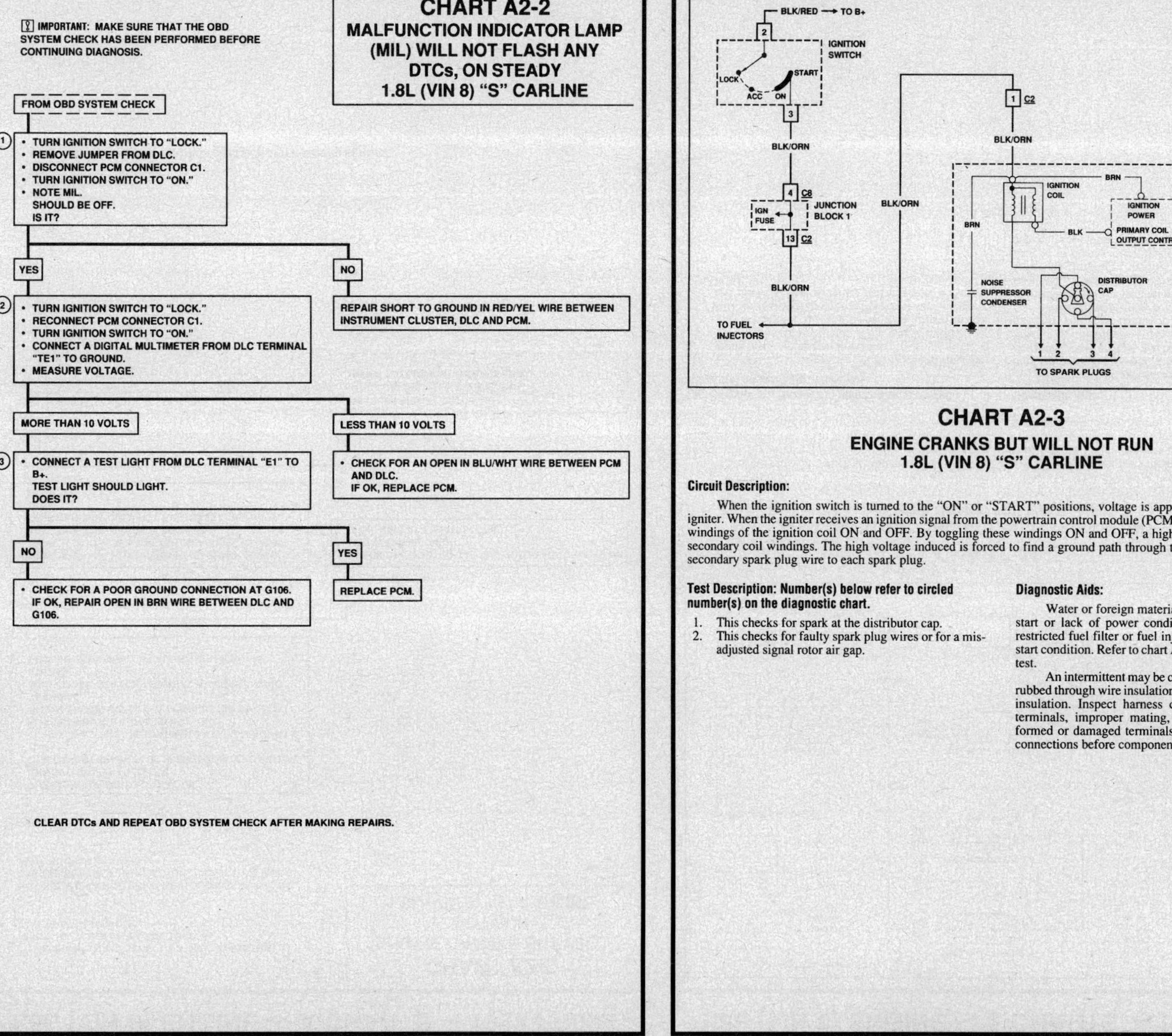

CHART A2-3
ENGINE CRANKS BUT WILL NOT RUN
1.8L (VIN 8) "S" CARLINE

Circuit Description:

When the ignition switch is turned to the "ON" or "START" positions, voltage is applied to the ignition coil and igniter. When the igniter receives an ignition signal from the powertrain control module (PCM), it toggles the primary coil windings of the ignition coil ON and OFF. By toggling these windings ON and OFF, a high voltage is produced in the secondary coil windings. The high voltage induced is forced to find a ground path through the distributor cap and each secondary spark plug wire to each spark plug.

Test Description: Number(s) below refer to circled number(s) on the diagnostic chart.

1. This checks for spark at the distributor cap.
2. This checks for faulty spark plug wires or for a mis-adjusted signal rotor air gap.

Diagnostic Aids:

Water or foreign material in the fuel can cause a no start or lack of power condition. Low fuel pressure, a restricted fuel filter or fuel injectors may also cause a no start condition. Refer to chart A2-5 for fuel injector circuit test.

An intermittent may be caused by a poor connection, rubbed through wire insulation, or a wire broken inside the insulation. Inspect harness connectors for backed out terminals, improper mating, broken locks, improperly formed or damaged terminals and poor terminal-to-wire connections before component replacement.

CHART A2-3
ENGINE CRANKS BUT WILL NOT RUN
1.8L (VIN 8) "S" CARLINE

[!] IMPORTANT: MAKE SURE THAT THE OBD SYSTEM CHECK HAS BEEN PERFORMED BEFORE CONTINUING DIAGNOSIS.

NOTICE: IN ORDER TO PREVENT FLOODING THE ENGINE, DISCONNECT THE FUEL INJECTOR CONNECTORS WHILE PERFORMING THE TESTS.

FROM OBD SYSTEM CHECK

1. • TURN IGNITION SWITCH TO "LOCK."
   • REMOVE JUMPER FROM DLC.
   • DISCONNECT SPARK PLUG WIRES FROM DISTRIBUTOR CAP.
   • CONNECT A J 26792 TO SPARK PLUG WIRE TERMINALS (ON DISTRIBUTOR CAP) ONE AT A TIME.
   • CRANK ENGINE WITH J 26792 CONNECTED TO EACH TERMINAL.
   • NOTE SPARK AT EACH TERMINAL WHILE CRANKING ENGINE.
   SHOULD HAVE SPARK AT ALL FOUR TERMINALS.
   DOES IT?

YES

NO → • CHECK DISTRIBUTOR CAP AND ROTOR FOR CRACKS OR BAD CONNECTIONS.
   IF OK, REPLACE IGNITION COIL.

2. • REMOVE SPARK PLUG WIRES FROM SPARK PLUGS.
   • USING A DIGITAL MULTIMETER, MEASURE THE RESISTANCE OF EACH SPARK PLUG WIRE.
   RESISTANCE SHOULD NOT EXCEED 25 K OHMS PER WIRE.
   DOES IT?

YES

NO → • MEASURE SIGNAL ROTOR AIR GAP.
   IF OK, AND PROBLEM STILL EXISTS, REFER TO DIAGNOSTIC CHART A2-5.

REPLACE FAULTY SPARK PLUG WIRES.

CLEAR DTCs AND REPEAT OBD SYSTEM CHECK AFTER MAKING REPAIRS.

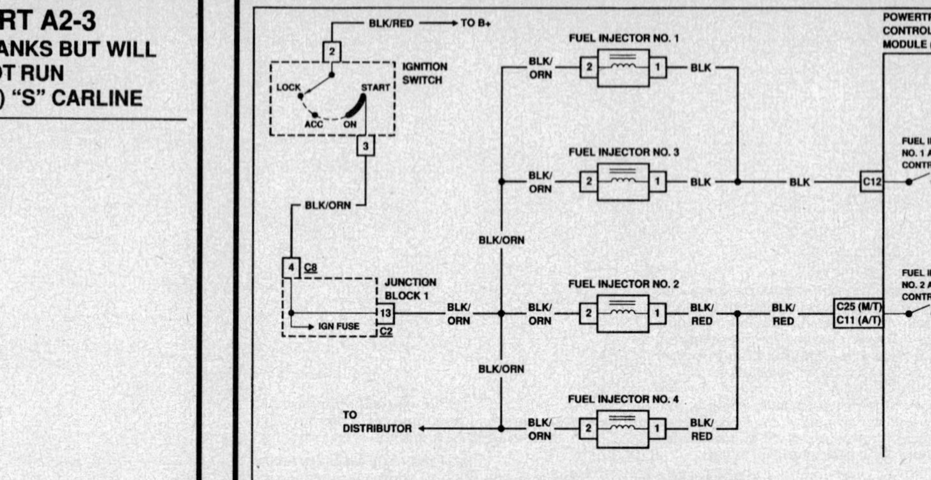

CHART A2-5
FUEL INJECTOR CIRCUIT CHECK (ENGINE NO-START)
1.8L (VIN 8) "S" CARLINE

Circuit Description:

When the ignition switch is turned to the "ON" or "START" position (engine running), the powertrain control module (PCM) will energize/de-energize the fuel injectors solenoid coil. With this coil energized, a plunger is activated, allowing pressurized fuel to be sprayed through the fuel injectors into the combustion chamber where it is mixed with air from the intake manifold; thus creating the proper air/fuel mixture needed for combustion.

Test Description: Number(s) below refer to circled number(s) on the diagnostic chart.

1. This checks to see if the PCM is controlling the ground path to the fuel injectors.
2. This checks the resistance of the fuel injectors.
3. This checks for an open or short to ground in the BLK or BLK/RED wires between the fuel injectors and the PCM, or for faulty fuel injectors or faulty PCM.

Diagnostic Aids:

There may be fuel spray at the fuel injectors, but it may not be enough to start the engine. If the fuel injectors and their circuit are OK, and fuel spray is detected, the fuel injector nozzle may be partly blocked or restricted. Refer to Chart A2-7A for circuit opening relay circuit test.

An intermittent may be caused by a poor connection, rubbed through wire insulation, or a wire broken inside the insulation. Inspect harness connectors for backed out terminals, improper mating, broken locks, improperly formed or damaged terminals and poor terminal-to-wire connections before component replacement.

1.8L (VIN 8) ENGINE — A-CHARTS — 1993 PRIZM

CHART A2-5
FUEL INJECTOR CIRCUIT CHECK
(ENGINE NO-START)
1.8L (VIN 8) "S" CARLINE

⚠ IMPORTANT: MAKE SURE THAT THE OBD SYSTEM CHECK HAS BEEN PERFORMED BEFORE CONTINUING DIAGNOSIS.

FROM CHART A2-3

① • CONNECT A J 34730-2B TO EACH FUEL INJECTOR CONNECTOR (HARNESS SIDE) ONE AT A TIME.
• CRANK ENGINE WITH J 34730-2B CONNECTED TO EACH.
• NOTE INJECTOR TEST LIGHT WHILE CRANKING ENGINE. SHOULD FLASH WHEN CRANKED AND CONNECTED AT EACH.
DOES IT?

NO

YES

② • CONNECT A DIGITAL MULTIMETER FROM EACH FUEL INJECTOR TERMINAL "1" TO "2" (FUEL INJECTOR SIDE).
• MEASURE RESISTANCE OF EACH INJECTOR. SHOULD BE 12–15 OHMS.
IS IT?

NO

YES

REPLACE FAULTY FUEL INJECTOR(S).

REFER TO DIAGNOSTIC CHART A2-7A.

DOES NOT LIGHT AT AT LEAST ONE FUEL INJECTOR.

REMAINS LIT AT AT LEAST TWO FUEL INJECTORS.

③ • TURN IGNITION SWITCH TO "ON."
• CONNECT A TEST LIGHT FROM FAULTY FUEL INJECTOR(S) CONNECTOR CAVITY "2" TO GROUND. TEST LIGHT SHOULD LIGHT.
DOES IT?

• CHECK FOR A SHORT TO GROUND IN BLK OR BLK/RED WIRES BETWEEN FUEL INJECTORS AND PCM. IF OK, REPLACE PCM.

YES

NO

• CHECK FOR AN OPEN IN BLK OR BLK/RED WIRES BETWEEN FUEL INJECTORS AND PCM. IF OK, REPLACE PCM.

REPAIR OPEN IN BLK/ORN WIRE TO FAULTY FUEL INJECTOR(S).

CLEAR DTCs AND REPEAT OBD SYSTEM CHECK AFTER MAKING REPAIRS.

1.8L (VIN 8) ENGINE — A-CHARTS — 1993 PRIZM

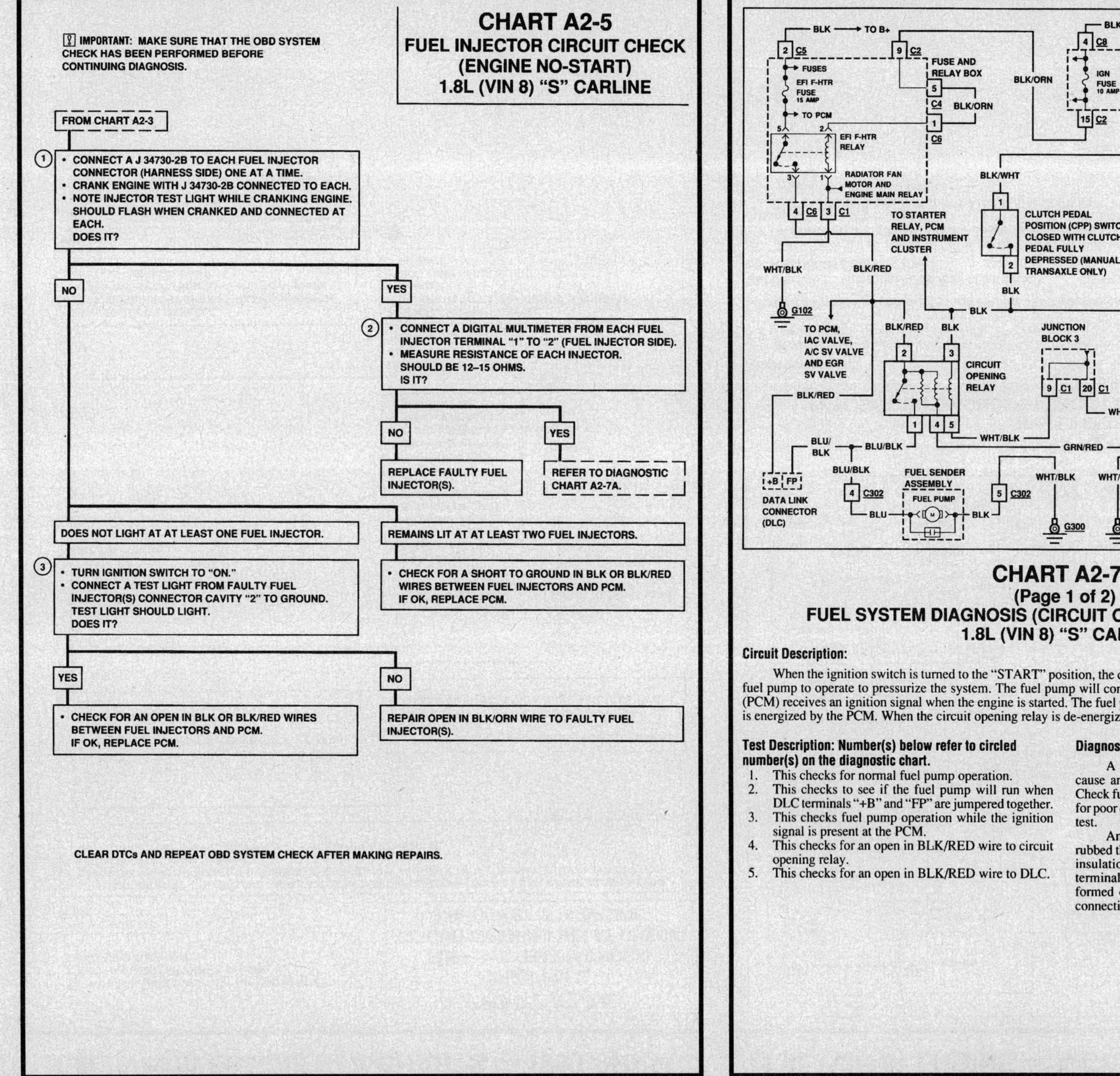

CHART A2-7A
(Page 1 of 2)
FUEL SYSTEM DIAGNOSIS (CIRCUIT OPENING RELAY CHECK)
1.8L (VIN 8) "S" CARLINE

Circuit Description:

When the ignition switch is turned to the "START" position, the circuit opening relay will be energized to allow the fuel pump to operate to pressurize the system. The fuel pump will continue to operate if the powertrain control module (PCM) receives an ignition signal when the engine is started. The fuel pump operates as long as the circuit opening relay is energized by the PCM. When the circuit opening relay is de-energized, the fuel pump will stop operating.

Test Description: Number(s) below refer to circled number(s) on the diagnostic chart.
1. This checks for normal fuel pump operation.
2. This checks to see if the fuel pump will run when DLC terminals "+B" and "FP" are jumpered together.
3. This checks fuel pump operation while the ignition signal is present at the PCM.
4. This checks for an open in BLK/RED wire to circuit opening relay.
5. This checks for an open in BLK/RED wire to DLC.

Diagnostic Aids:

A restricted or leaking fuel feed pipe or hose will cause an "Engine Cranks But Will Not Run" condition. Check fuel sender assembly connections inside of fuel tank for poor connections. Refer to Chart A2-7B for fuel pressure test.

An intermittent may be caused by a poor connection, rubbed through wire insulation, or a wire broken inside the insulation. Inspect harness connectors for backed out terminals, improper mating, broken locks, improperly formed or damaged terminals and poor terminal-to-wire connections before component replacement.

1.8L (VIN 8) ENGINE — A-CHARTS — 1993 PRIZM

CHART A2-7A (Page 1 of 2) — FUEL SYSTEM DIAGNOSIS (CIRCUIT OPENING RELAY CHECK) — 1.8L (VIN 8) "S" CARLINE

IMPORTANT: MAKE SURE THAT THE OBD SYSTEM CHECK HAS BEEN PERFORMED BEFORE CONTINUING DIAGNOSIS.

FROM CHART A2-5

1. • TURN IGNITION SWITCH TO "ON" AND MOMENTARILY BACKPROBE PCM CONNECTOR C1 WITH A FUSED JUMPER FROM CAVITY "A4" (MANUAL TRANSAXLE MODELS) OR FROM CAVITY "A14" (AUTOMATIC TRANSAXLE MODELS) TO GROUND.
   • LISTEN FOR FUEL PUMP TO RUN.
   NOTICE: FUEL PUMP SHOULD RUN WHILE FUSED JUMPER IS CONNECTED.
   DOES IT?

   NO →

2. • MOMENTARILY CONNECT A JUMPER FROM DLC TERMINAL "+B" TO "FP".
   • LISTEN FOR FUEL PUMP TO RUN.
   DOES IT?

   YES →

3. • CONNECT A DIGITAL MULTIMETER FROM DLC TERMINAL "FP" TO GROUND.
   • MEASURE VOLTAGE WHILE CRANKING ENGINE. SHOULD BE MORE THAN 10 VOLTS WHILE CRANKING.
   IS IT?

   NO → CHECK FOR AN OPEN IN BLK WIRE BETWEEN CPP SWITCH/PNP SWITCH AND CIRCUIT OPENING RELAY OR FOR AN OPEN IN WHT/BLK WIRE BETWEEN CIRCUIT OPENING RELAY AND G200. IF OK, REPLACE CIRCUIT OPENING RELAY AND RETEST SYSTEM. IF SYSTEM REMAINS INOPERATIVE, REPLACE PCM.

   YES → REFER TO CHART A2-7B.

4. (YES) • REMOVE JUMPER FROM DLC.
   • BACKPROBE CIRCUIT OPENING RELAY CONNECTOR WITH A TEST LIGHT FROM CAVITY "2" TO GROUND.
   TEST LIGHT SHOULD LIGHT.
   DOES IT?
   → A

5. (NO) • REMOVE JUMPER FROM DLC.
   • CONNECT A TEST LIGHT FROM DLC TERMINAL "+B" TO GROUND.
   TEST LIGHT SHOULD LIGHT.
   DOES IT?
   → B

CLEAR DTCs AND REPEAT OBD SYSTEM CHECK AFTER MAKING REPAIRS.

1.8L (VIN 8) ENGINE — A-CHARTS — 1993 PRIZM

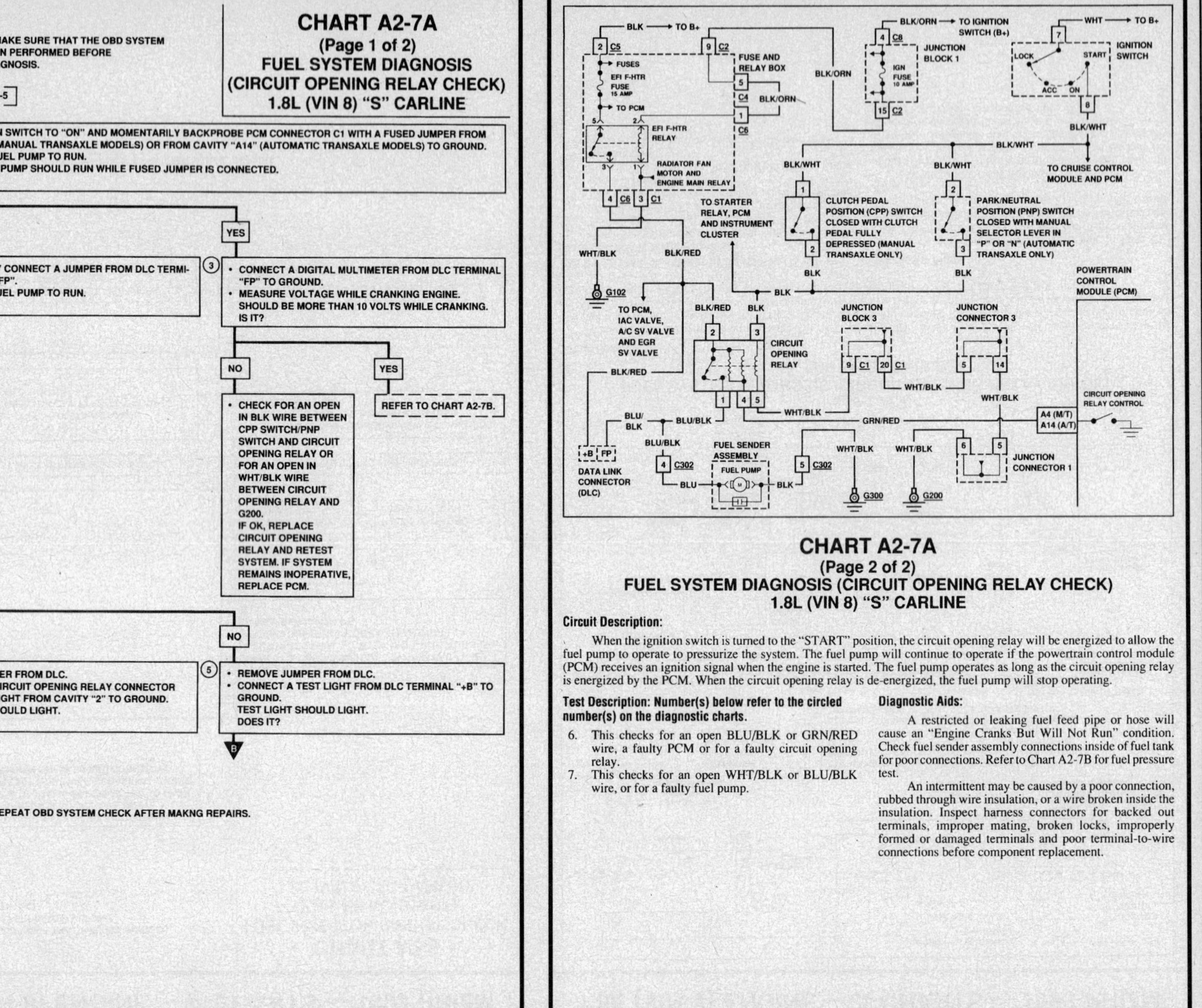

CHART A2-7A (Page 2 of 2) — FUEL SYSTEM DIAGNOSIS (CIRCUIT OPENING RELAY CHECK) — 1.8L (VIN 8) "S" CARLINE

Circuit Description:

When the ignition switch is turned to the "START" position, the circuit opening relay will be energized to allow the fuel pump to operate to pressurize the system. The fuel pump will continue to operate if the powertrain control module (PCM) receives an ignition signal when the engine is started. The fuel pump operates as long as the circuit opening relay is energized by the PCM. When the circuit opening relay is de-energized, the fuel pump will stop operating.

Test Description: Number(s) below refer to the circled number(s) on the diagnostic charts.

6. This checks for an open BLU/BLK or GRN/RED wire, a faulty PCM or for a faulty circuit opening relay.

7. This checks for an open WHT/BLK or BLU/BLK wire, or for a faulty fuel pump.

Diagnostic Aids:

A restricted or leaking fuel feed pipe or hose will cause an "Engine Cranks But Will Not Run" condition. Check fuel sender assembly connections inside of fuel tank for poor connections. Refer to Chart A2-7B for fuel pressure test.

An intermittent may be caused by a poor connection, rubbed through wire insulation, or a wire broken inside the insulation. Inspect harness connectors for backed out terminals, improper mating, broken locks, improperly formed or damaged terminals and poor terminal-to-wire connections before component replacement.

1.8L (VIN 8) ENGINE — A-CHARTS — 1993 PRIZM

CHART A2-7A
(Page 2 of 2)
FUEL SYSTEM DIAGNOSIS
(CIRCUIT OPENING RELAY CHECK)
1.8L (VIN 8) "S" CARLINE

A

YES / **NO**

6. • TURN IGNITION SWITCH TO "LOCK."
   • DISCONNECT CIRCUIT OPENING RELAY CONNECTOR.
   • CONNECT A TEST LIGHT FROM CIRCUIT OPENING RELAY CONNECTOR CAVITY "4" TO B+.
   • CRANK ENGINE. TEST LIGHT SHOULD LIGHT WHILE CRANKING. DOES IT?

NO: • REPAIR OPEN IN BLK/RED WIRE BETWEEN FUSE AND RELAY BOX AND CIRCUIT OPENING RELAY.

YES / **NO**

YES: • CHECK FOR AN OPEN IN BLU/BLK WIRE BETWEEN CIRCUIT OPENING RELAY AND FUEL SENDER ASSEMBLY. IF OK, REPLACE CIRCUIT OPENING RELAY.

NO: • CHECK FOR AN OPEN IN GRN/RED WIRE BETWEEN CIRCUIT OPENING RELAY AND PCM. IF OK, REPLACE PCM.

B

YES / **NO**

7. • BACKPROBE FUEL SENDER ASSEMBLY CONNECTOR C302 WITH A TEST LIGHT FROM CAVITY "5" TO B+. TEST LIGHT SHOULD LIGHT. DOES IT?

NO: • REPAIR OPEN IN BLK/RED WIRE BETWEEN FUSE AND RELAY BOX, CIRCUIT OPENING RELAY AND DLC.

YES / **NO**

YES: • CHECK FOR AN OPEN IN BLU/BLK WIRE BETWEEN CIRCUIT OPENING RELAY, FUEL SENDER ASSEMBLY AND DLC. IF OK, REPLACE FUEL PUMP.

NO: • CHECK FOR A POOR GROUND CONNECTION AT G300. IF OK, REPAIR OPEN IN WHT/BLK WIRE BETWEEN FUEL SENDER ASSEMBLY AND G300.

CLEAR DTCS AND REPEAT OBD SYSTEM CHECK AFTER MAKING REPAIRS.

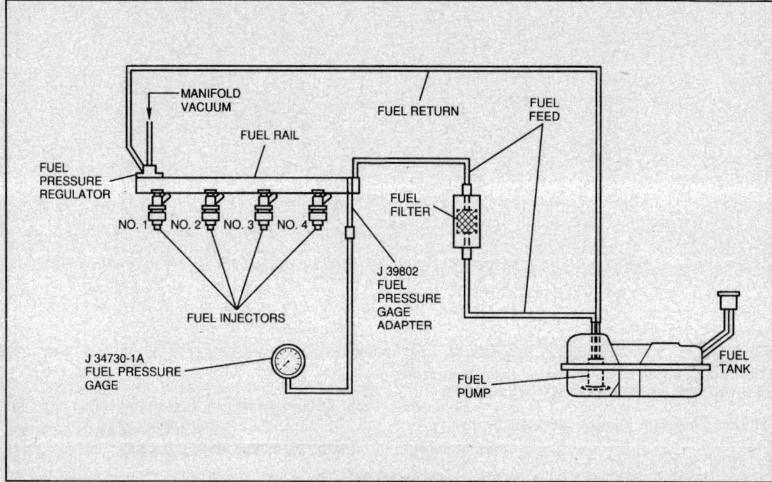

1.8L (VIN 8) ENGINE — A-CHARTS — 1993 PRIZM

CHART A2-7B
(Page 1 of 2)
FUEL SYSTEM DIAGNOSIS (FUEL PRESSURE CHECK)
1.8L (VIN 8) "S" CARLINE

Circuit Description:

When the ignition switch is turned to the "START" position, the circuit opening relay will be energized to allow the fuel pump to operate to pressurize the system. The fuel pump will continue to operate if the powertrain control module (PCM) receives an ignition signal when the engine is started. The fuel pump operates as long as the circuit opening relay is energized by the PCM. When the circuit opening relay is de-energized, the fuel pump will stop operating. The fuel pump delivers fuel to the fuel rail where the pressure is maintained by the fuel pressure regulator at 284 kPa (41 psi) higher than the pressure inside the intake manifold. Excess fuel is returned to the fuel tank through a return pipe and hoses.

Test Description: Number(s) below refer to the circled number(s) on the diagnostic chart.

1. This measures the fuel pressure with the ignition switch in "ON."
2. This determines if the fuel pressure is normal.
3. This checks for a leaking fuel pressure regulator.
4. This checks for plugged fuel injectors.
5. This checks for leaking fuel injectors or a leaking fuel pump.

Diagnostic Aids:

Check for leakage around all fuel hose/pipe connections.

A plugged fuel injector could be the cause of a hard/no-start condition or loss of power/poor fuel economy. Refer to mechanical checks and inspections if fuel spray and pressure is normal.

CHART A2-7B
(Page 1 of 2)
FUEL SYSTEM DIAGNOSIS
(FUEL PRESSURE CHECK)
1.8L (VIN 8) "S" CARLINE

FROM CHART A2-7A

① • TURN IGNITION SWITCH TO "LOCK."
• INSTALL FUEL PRESSURE GAGE AND ADAPTER TO FUEL RAIL, REFER TO CHART A1-7C.
• CONNECT A JUMPER FROM DLC TERMINAL "FP" TO "+B."
• TURN IGNITION SWITCH TO "ON."
• MEASURE FUEL PRESSURE. SHOULD BE 265-304 kPa (38-44 psi) AND HOLDING STEADILY. IS IT?

NO → PRESSURE WITHIN SPECIFICATION, BUT NOT HOLDING

YES → PRESSURE OUT OF SPECIFICATION ▲A

② • TURN IGNITION SWITCH TO "LOCK."
• MEASURE FUEL PRESSURE. SHOULD REMAIN ABOVE 147 kPa (21 psi) FOR AT LEAST FIVE MINUTES. DOES IT?

③ • TURN IGNITION SWITCH TO "LOCK."
• LEAVE JUMPER CONNECTED TO DLC TERMINALS.
• TURN IGNITION SWITCH TO "ON."
• PINCH FUEL RETURN HOSE AND HOLD BLOCKED.
• MEASURE FUEL PRESSURE WHILE PINCHING FUEL RETURN HOSE. SHOULD BE 265-304 kPa (38-44 psi) AND HOLDING STEADILY. IS IT?

YES | NO

NO → CHECK FOR FUEL LEAKAGE AROUND FUEL INJECTORS, FUEL PUMP AND FUEL PRESSURE REGULATOR. REPAIR/REPLACE AS NECESSARY.

YES → REPLACE FUEL PRESSURE REGULATOR.

④ • WITH IGNITION SWITCH IN "LOCK", DISCONNECT ONE FUEL INJECTOR CONNECTOR.
• INSTALL ONE END OF J 34730-355 TO DISCONNECTED FUEL INJECTOR AND THE OTHER END TO J 34730-3A.
• CONNECT J 34730-3A TO BATTERY.
• TURN IGNITION SWITCH TO "ON" AND MOMENTARILY CONNECT A JUMPER BETWEEN DLC TERMINALS "FP" AND "+B."
• SET J 34730-3A TO ONE PULSE AND ACTIVATE ONCE.
• MEASURE FUEL PRESSURE WHILE ACTIVATING FUEL INJECTOR. NOTICE: REPEAT THIS TEST FOR ALL FOUR FUEL INJECTORS. PRESSURE SHOULD BE 179-234 kPa (26-34 PSI) AFTER INJECTOR TESTER IS ACTIVATED EACH TIME. IS IT?

⑤ • TURN IGNITION SWITCH TO "LOCK."
• LEAVE JUMPER CONNECTED TO DLC TERMINALS.
• TURN IGNITION SWITCH TO "ON."
• PINCH FUEL FEED HOSE AND HOLD BLOCKED.
• MEASURE FUEL PRESSURE WHILE PINCHING FUEL FEED HOSE. SHOULD BE 265-304 kPa (38-44 psi) AND HOLDING STEADILY. IS IT?

NO → REPLACE LEAKING FUEL INJECTOR(S).

YES → • CHECK FUEL HOSES TO FUEL PUMP IN FUEL TANK FOR LEAKS.
• CHECK FOR FUEL LEAKS AT FUEL FILTER. IF OK, REPLACE FUEL PUMP.

NO → REPLACE FAULTY FUEL INJECTOR(S).

YES → FUEL PRESSURE IS NORMAL.

CLEAR DTCS AND REPEAT OBD SYSTEM CHECK AFTER MAKING REPAIRS.

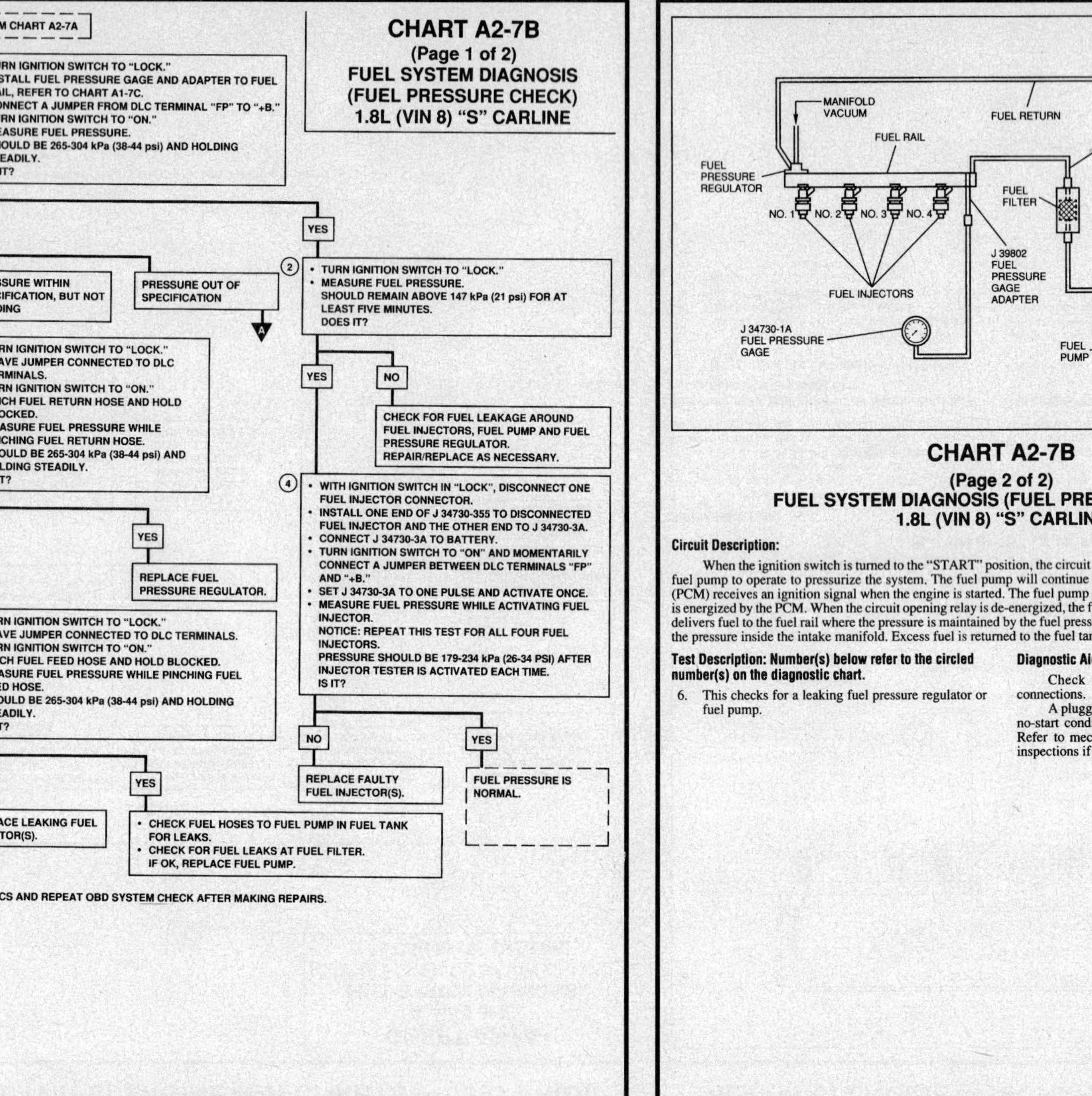

CHART A2-7B
(Page 2 of 2)
FUEL SYSTEM DIAGNOSIS (FUEL PRESSURE CHECK)
1.8L (VIN 8) "S" CARLINE

Circuit Description:

When the ignition switch is turned to the "START" position, the circuit opening relay will be energized to allow the fuel pump to operate to pressurize the system. The fuel pump will continue to operate if the powertrain control module (PCM) receives an ignition signal when the engine is started. The fuel pump operates as long as the circuit opening relay is energized by the PCM. When the circuit opening relay is de-energized, the fuel pump will stop operating. The fuel pump delivers fuel to the fuel rail where the pressure is maintained by the fuel pressure regulator at 284 kPa (41 psi) higher than the pressure inside the intake manifold. Excess fuel is returned to the fuel tank through a return pipe and hoses.

Test Description: Number(s) below refer to the circled number(s) on the diagnostic chart.

6. This checks for a leaking fuel pressure regulator or fuel pump.

Diagnostic Aids:

Check for leakage around all fuel hose/pipe connections.

A plugged fuel injector could be the cause of a hard/no-start condition or loss of power/poor fuel economy. Refer to mechanical checks and inspections if fuel spray and pressure is normal.

1.8L (VIN 8) ENGINE — A-CHARTS — 1993 PRIZM

CHART A2-7B
(Page 2 of 2)
FUEL SYSTEM DIAGNOSIS
(FUEL PRESSURE CHECK)
1.8L (VIN 8) "S" CARLINE

A

FUEL PRESSURE OUT OF SPECIFICATION

LOW PRESSURE

HIGH PRESSURE

- TURN IGNITION SWITCH TO "LOCK."
- LEAVE JUMPER CONNECTED TO DLC TERMINALS.
- PINCH FUEL RETURN HOSE AND HOLD BLOCKED.
- TURN IGNITION SWITCH TO "ON."
- MEASURE FUEL PRESSURE WHILE PINCHING FUEL RETURN HOSE.
  SHOULD BE 265-304 kPa (38-44 psi) AND HOLDING STEADILY.
  DOES IT?

- CHECK FOR RESTRICTIONS IN FUEL RETURN HOSES AND PIPES.
  IF OK, REPLACE FUEL PRESSURE REGULATOR.

NO

YES

- CHECK FOR A RESTRICTED FUEL FILTER.
- CHECK FOR RESTRICTED FUEL FEED HOSES AND PIPES.
  IF OK, REPLACE FUEL PUMP.

REPLACE FUEL PRESSURE REGULATOR.

CLEAR DTCs AND REPEAT OBD SYSTEM CHECK AFTER MAKING REPAIRS.

1.8L (VIN 8) ENGINE — A-CHARTS — 1993 PRIZM

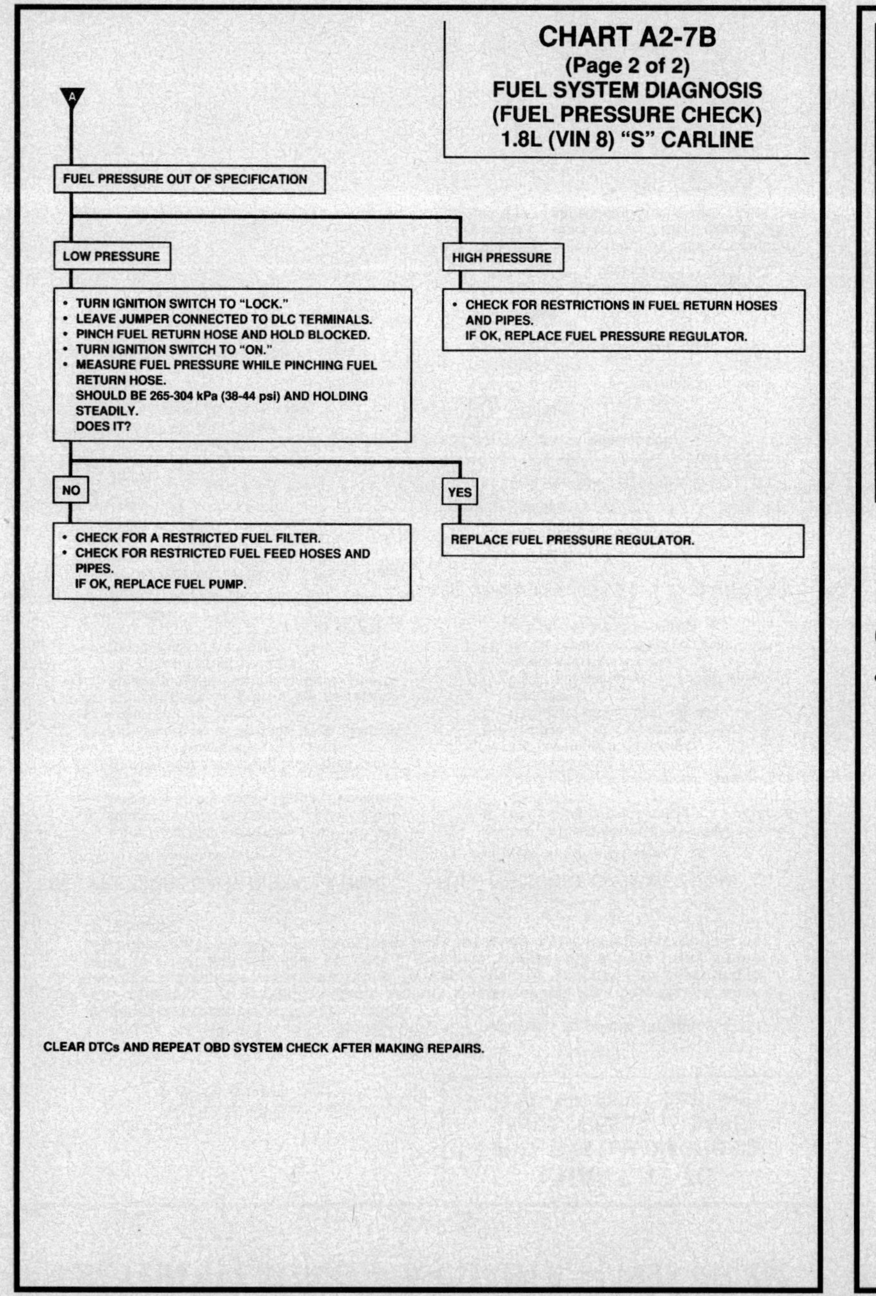

CHART A2-7C
FUEL SYSTEM DIAGNOSIS
(FUEL PRESSURE GAGE)
1.8L (VIN 8) "S" CARLINE

Circuit Description:

The procedure on the   must be followed in order to install/remove the fuel pressure gage and adapter correctly. Make sure to observe all CAUTIONS while performing this procedure.

CHART A2-7C
FUEL SYSTEM DIAGNOSIS
(FUEL PRESSURE GAGE)
1.8L (VIN 8) "S" CARLINE

Tools Required:
J 34730-1A Fuel Pressure Gage
J 39802 Fuel Pressure Gage Adapter

CAUTION

- To reduce the risk of fire and personal injury, it is necessary to relieve the fuel system pressure before servicing the fuel system.
- After relieving fuel system pressure, a small amount of fuel may be released when servicing fuel pipe or hose connections. To reduce the chance of personal injury and to catch any fuel that may leak out, cover fuel pipe fittings with a shop towel before disconnecting them. Place the shop towel in an approved container when disconnect is completed.

FUEL PRESSURE GAGE INSTALLATION:

1. Relieve fuel system pressure.
   A. Loosen fuel filler cap to relieve tank pressure.
   B. Remove center trim bezel from center console by gently prying around edges to disconnect clips.
   C. Remove four screws and radio from center console (if equipped).
   D. Disconnect circuit opening relay electrical connector.
   E. Crank engine and allow to stall. Crank engine for an additional 3 seconds to assure relief of any remaining fuel pressure.
   F. Remove negative (-) battery cable.
   G. Reconnect circuit opening relay electrical connector.
   H. Tighten fuel filler cap.
2. Remove one bolt and fuel feed hose bracket from bottom on intake manifold.
3. Remove one bolt and fuel feed hose from fuel rail.
4. Install J 39802 with two gaskets and fuel feed hose to fuel rail.

   ⊠ **Tighten**
   - J 39802 bolt to 29 N·m (22 lb.ft.).
5. Connect J 34730-1A to J 39802.
6. Reconnect negative (-) battery cable.

   ⊠ **Tighten**
   - Negative (-) battery cable-to-negative (-) battery cable retainer to 15 N·m (11 lb.ft.).

FUEL PRESSURE GAGE REMOVAL

1. Relieve fuel system pressure.
   A. Loosen fuel filler cap to relieve tank pressure.
   B. Disconnect circuit opening relay electrical connector.
   C. Crank engine and allow to stall. Crank engine for an additional 3 seconds to assure relief of any remaining fuel pressure.
   D. Remove negative (-) battery cable.
   E. Reconnect circuit opening relay electrical connector.
   F. Install radio into center console (if equipped); secure with four screws.
   G. Install center trim bezel to center console, pushing in until clips engage.
   H. Tighten fuel filler cap.
2. Remove J 34730-1A from J 39802 and J 39802 from fuel rail.
3. Connect fuel feed hose with new gaskets to fuel rail.

   ⊠ **Tighten**
   - Fuel feed hose bolt to 29 N·m (22 lb.ft.).
4. Connect fuel feed hose bracket to bottom of intake manifold, secure with one bolt.

   ⊠ **Tighten**
   - Fuel feed hose bracket bolt to 10 N·m (89 lb. in.)
5. Reconnect negative (-) battery cable.

   ⊠ **Tighten**
   - Negative (-) battery cable-to-negative (-) battery cable retainer to 15 N·m (11 lb.ft.).
6. Turn ignition switch to "ON" and momentarily connect a jumper between DLC terminals "+B" and "Fp" to pressurize the fuel system. Check for fuel leaks.

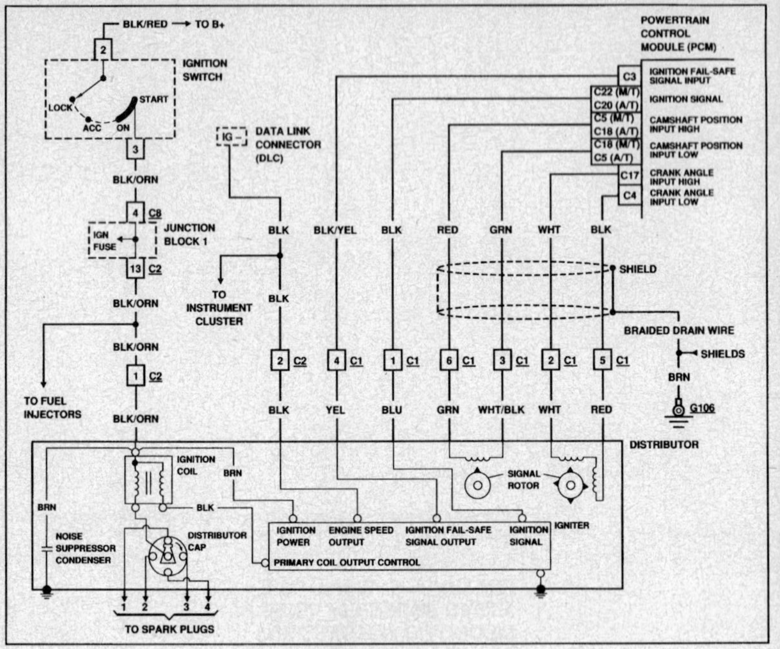

DTC 12
NO RPM SIGNAL
1.8L (VIN 8) "S" CARLINE

Circuit Description:

The powertrain control module (PCM) monitors the crank angle and the camshaft position. The PCM applies 0.70 volts to the crank angle and the camshaft position (CMP) sensors. This voltage is toggled high and low by the rotation of the signal rotor. It is by the toggled voltage input that the PCM can determine the crank angle and camshaft positions.
DTC 12 will set if either of the following conditions are met:
(1) No crank angle input detected at the PCM for 2 seconds or more while cranking engine.
(2) No camshaft position input detected at the PCM for 3 seconds or more with engine speed between 600 and 4,000 rpm.

Test Description: Number(s) below refer to circled number(s) on the diagnostic chart.

1. This checks the crank angle input low circuitry and PCM for faults.
2. This checks the crank angle input high circuitry and PCM for faults.
3. This checks the camshaft position input low circuitry and PCM for faults.
4. This checks the camshaft position input high circuitry and the PCM for faults and also for a misadjusted signal rotor air gap.

Diagnostic Aids:

Check distributor cap and rotor for cracks. Also check that signal rotor air gap is adjusted properly.

An intermittent may be caused by a poor connection, rubbed through wire insulation, or a wire broken inside the insulation. Inspect harness connectors for backed out terminals, improper mating, broken locks, improperly formed or damaged terminals and poor terminal-to-wire connections before component replacement.

1.8L (VIN 8) ENGINE — DIAGNOSTIC TROUBLE CODE CHARTS — 1993 PRIZM

DTC 12
NO RPM SIGNAL
1.8L (VIN 8) "S" CARLINE

[i] IMPORTANT: MAKE SURE THAT THE OBD SYSTEM CHECK HAS BEEN PERFORMED BEFORE CONTINUING DIAGNOSIS.

FROM OBD SYSTEM CHECK

1.
- TURN IGNITION SWITCH TO "LOCK."
- DISCONNECT DISTRIBUTOR CONNECTOR C1.
- CONNECT A DIGITAL MULTIMETER FROM DISTRIBUTOR CONNECTOR C1 CAVITY "2" TO GROUND.
- TURN IGNITION SWITCH TO "ON."
- MEASURE VOLTAGE. SHOULD BE 0.60–0.80 VOLTS. IS IT?

YES / NO

NO →
- CHECK FOR A POOR CONNECTION AT PCM.
- CHECK FOR AN OPEN, A SHORT TO GROUND OR A SHORT TO VOLTAGE IN WHT WIRE BETWEEN DISTRIBUTOR AND PCM. IF OK, REPLACE PCM.

2.
- CONNECT A DIGITAL MULTIMETER FROM DISTRIBUTOR CONNECTOR C1 CAVITY "5" TO GROUND.
- MEASURE VOLTAGE. SHOULD BE 0.60–0.80 VOLTS. IS IT?

YES / NO

NO →
- CHECK FOR A POOR CONNECTION AT PCM.
- CHECK FOR AN OPEN, A SHORT TO GROUND OR A SHORT TO VOLTAGE IN BLK WIRE BETWEEN DISTRIBUTOR AND PCM. IF OK, REPLACE PCM.

3.
- CONNECT A DIGITAL MULTIMETER FROM DISTRIBUTOR CONNECTOR C1 CAVITY "3" TO GROUND.
- MEASURE VOLTAGE. SHOULD BE 0.60–0.80 VOLTS. IS IT?

YES / NO

NO →
- CHECK FOR A POOR CONNECTION AT PCM.
- CHECK FOR AN OPEN, A SHORT TO GROUND OR A SHORT TO VOLTAGE IN GRN WIRE BETWEEN DISTRIBUTOR AND PCM. IF OK, REPLACE PCM.

4.
- CONNECT A DIGITAL MULTIMETER FROM DISTRIBUTOR CONNECTOR C1 CAVITY "6" TO GROUND.
- MEASURE VOLTAGE. SHOULD BE 0.60–0.80 VOLTS. IS IT?

YES / NO

NO →
- CHECK FOR A POOR CONNECTION AT PCM.
- CHECK FOR AN OPEN, A SHORT TO GROUND OR A SHORT TO VOLTAGE IN RED WIRE BETWEEN DISTRIBUTOR AND PCM. IF OK, REPLACE PCM.

- CHECK SIGNAL ROTOR AIR GAP. REFER TO SECTION 6D4. IF OK, REPLACE DISTRIBUTOR HOUSING.

CLEAR DTCs AND REPEAT OBD SYSTEM CHECK AFTER MAKING REPAIRS.

1.8L (VIN 8) ENGINE — DIAGNOSTIC TROUBLE CODE CHARTS — 1993 PRIZM

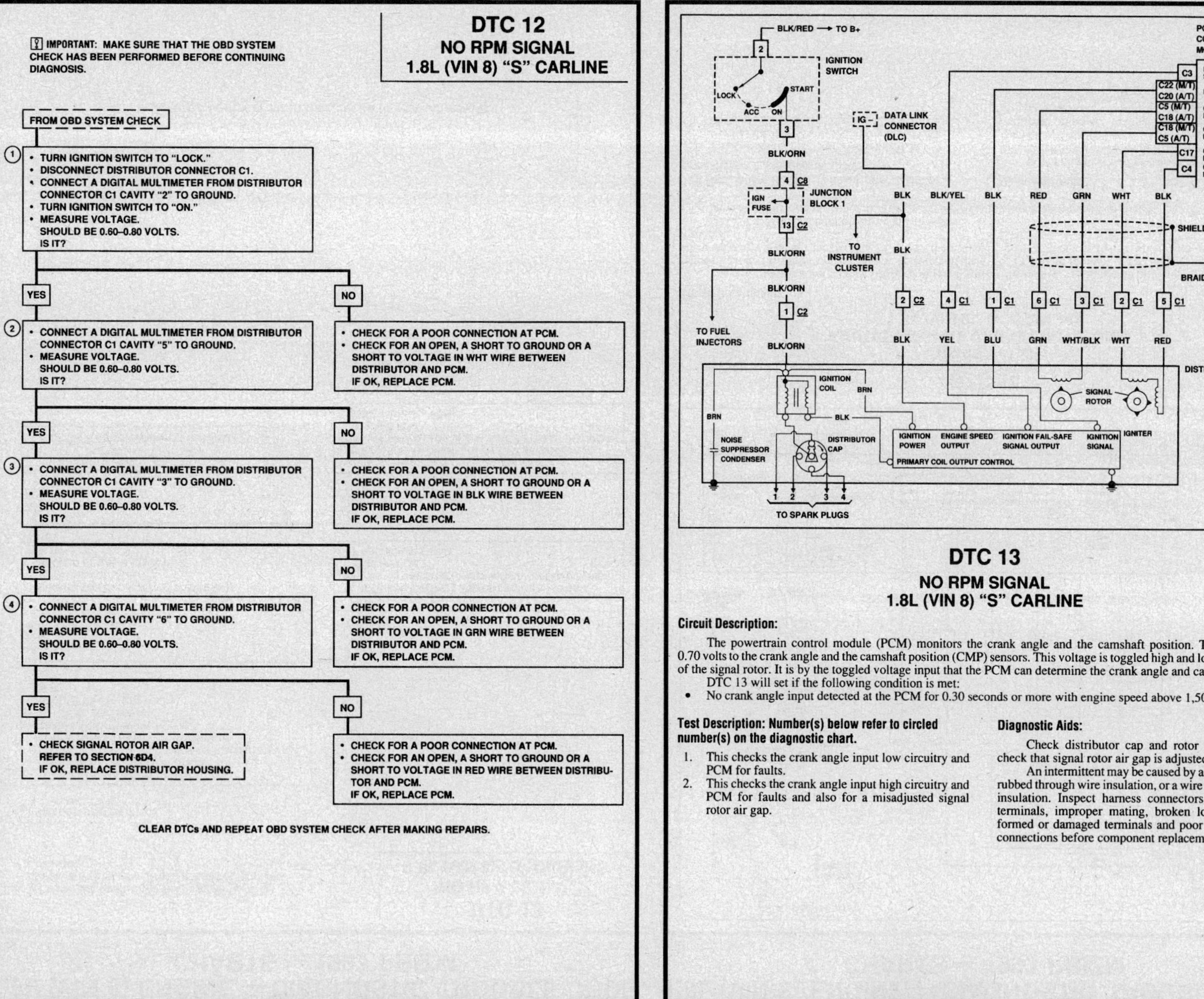

DTC 13
NO RPM SIGNAL
1.8L (VIN 8) "S" CARLINE

Circuit Description:

The powertrain control module (PCM) monitors the crank angle and the camshaft position. The PCM applies 0.70 volts to the crank angle and the camshaft position (CMP) sensors. This voltage is toggled high and low by the rotation of the signal rotor. It is by the toggled voltage input that the PCM can determine the crank angle and camshaft positions. DTC 13 will set if the following condition is met:
- No crank angle input detected at the PCM for 0.30 seconds or more with engine speed above 1,500 rpm.

Test Description: Number(s) below refer to circled number(s) on the diagnostic chart.

1. This checks the crank angle input low circuitry and PCM for faults.
2. This checks the crank angle input high circuitry and PCM for faults and also for a misadjusted signal rotor air gap.

Diagnostic Aids:

Check distributor cap and rotor for cracks. Also check that signal rotor air gap is adjusted properly.

An intermittent may be caused by a poor connection, rubbed through wire insulation, or a wire broken inside the insulation. Inspect harness connectors for backed out terminals, improper mating, broken locks, improperly formed or damaged terminals and poor terminal-to-wire connections before component replacement.

1.8L (VIN 8) ENGINE — DIAGNOSTIC TROUBLE CODE CHARTS — 1993 PRIZM

DTC 13
NO RPM SIGNAL
1.8L (VIN 8) "S" CARLINE

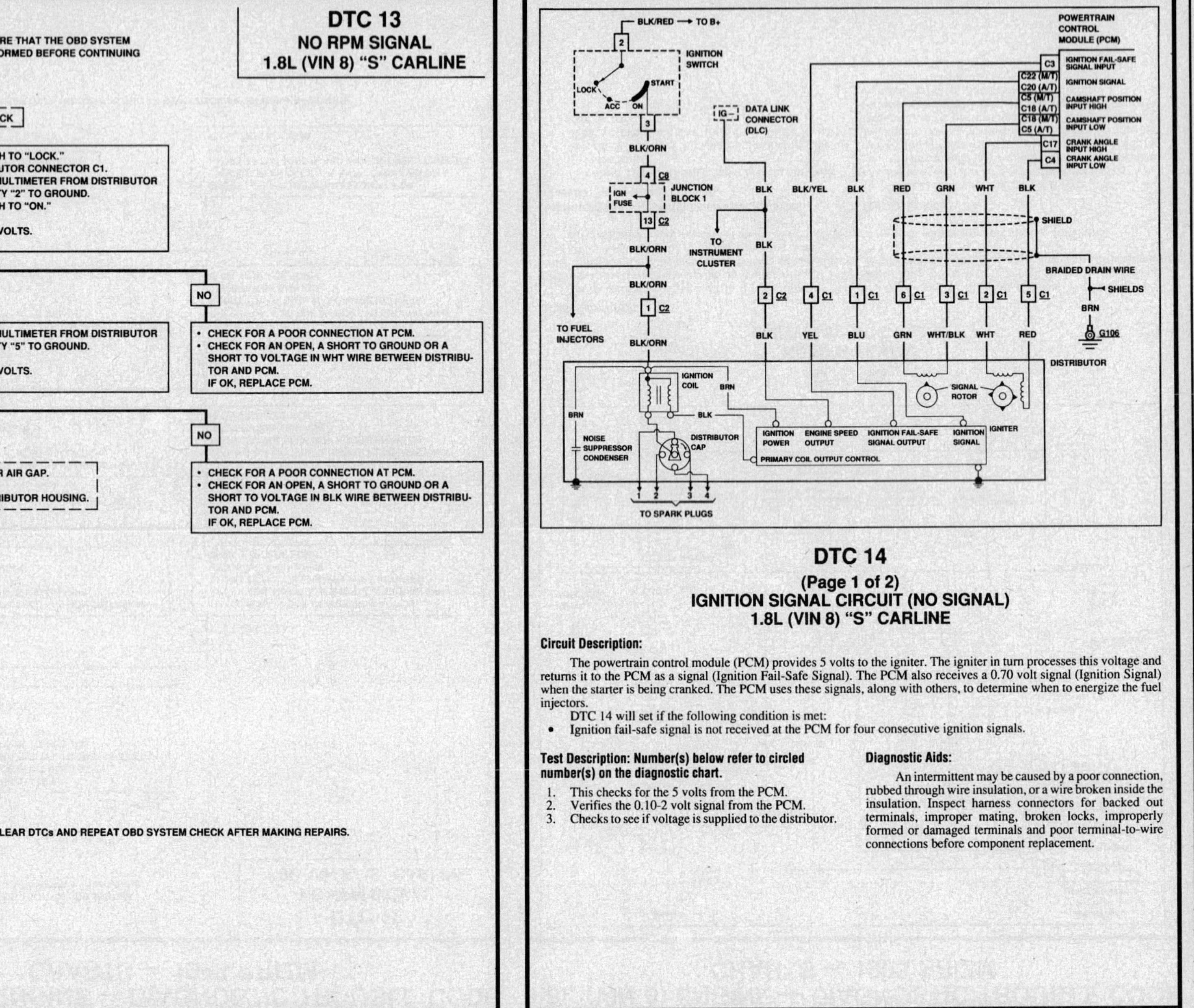

DTC 14
(Page 1 of 2)
IGNITION SIGNAL CIRCUIT (NO SIGNAL)
1.8L (VIN 8) "S" CARLINE

Circuit Description:

The powertrain control module (PCM) provides 5 volts to the igniter. The igniter in turn processes this voltage and returns it to the PCM as a signal (Ignition Fail-Safe Signal). The PCM also receives a 0.70 volt signal (Ignition Signal) when the starter is being cranked. The PCM uses these signals, along with others, to determine when to energize the fuel injectors.

DTC 14 will set if the following condition is met:
• Ignition fail-safe signal is not received at the PCM for four consecutive ignition signals.

Test Description: Number(s) below refer to circled number(s) on the diagnostic chart.

1. This checks for the 5 volts from the PCM.
2. Verifies the 0.10-2 volt signal from the PCM.
3. Checks to see if voltage is supplied to the distributor.

Diagnostic Aids:

An intermittent may be caused by a poor connection, rubbed through wire insulation, or a wire broken inside the insulation. Inspect harness connectors for backed out terminals, improper mating, broken locks, improperly formed or damaged terminals and poor terminal-to-wire connections before component replacement.

1.8L (VIN 8) ENGINE — DIAGNOSTIC TROUBLE CODE CHARTS — 1993 PRIZM

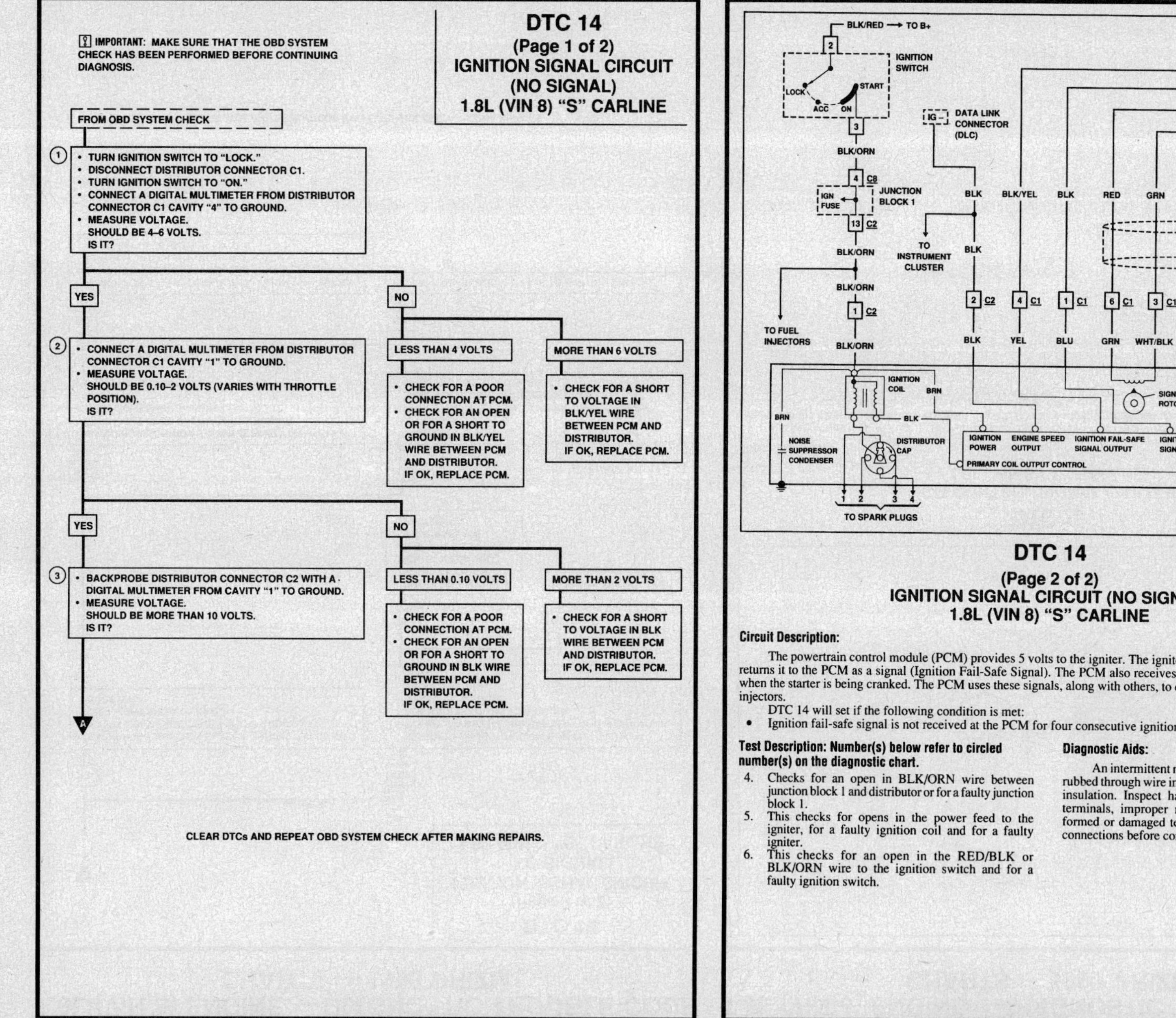

DTC 14
(Page 2 of 2)
IGNITION SIGNAL CIRCUIT (NO SIGNAL)
1.8L (VIN 8) "S" CARLINE

Circuit Description:

The powertrain control module (PCM) provides 5 volts to the igniter. The igniter in turn processes this voltage and returns it to the PCM as a signal (Ignition Fail-Safe Signal). The PCM also receives a 0.70 volt signal (Ignition Signal) when the starter is being cranked. The PCM uses these signals, along with others, to determine when to energize the fuel injectors.

DTC 14 will set if the following condition is met:
• Ignition fail-safe signal is not received at the PCM for four consecutive ignition signals.

Test Description: Number(s) below refer to circled number(s) on the diagnostic chart.

4. Checks for an open in BLK/ORN wire between junction block 1 and distributor or for a faulty junction block 1.
5. This checks for opens in the power feed to the igniter, for a faulty ignition coil and for a faulty igniter.
6. This checks for an open in the RED/BLK or BLK/ORN wire to the ignition switch and for a faulty ignition switch.

Diagnostic Aids:

An intermittent may be caused by a poor connection, rubbed through wire insulation, or a wire broken inside the insulation. Inspect harness connectors for backed out terminals, improper mating, broken locks, improperly formed or damaged terminals and poor terminal-to-wire connections before component replacements.

1.8L (VIN 8) ENGINE — DIAGNOSTIC TROUBLE CODE CHARTS — 1993 PRIZM

DTC 14
(Page 2 of 2)
IGNITION SIGNAL CIRCUIT
(NO SIGNAL)
1.8L (VIN 8) "S" CARLINE

A

NO

④ • BACKPROBE JUNCTION BLOCK 1 CONNECTOR C8 WITH A TEST LIGHT FROM CAVITY "4" TO GROUND.
TEST LIGHT SHOULD LIGHT.
DOES IT?

YES

⑤ • REMOVE DISTRIBUTOR CAP FROM DISTRIBUTOR.
• CONNECT A DIGITAL MULTIMETER FROM IGNITION COIL BLK WIRE TERMINAL TO GROUND.
• MEASURE VOLTAGE.
SHOULD BE B+.
IS IT?

YES

• CHECK BRN AND BLK WIRES BETWEEN IGNITION COIL AND IGNITER FOR OPENS.
IF OK, REPLACE IGNITER.

NO

REPLACE IGNITION COIL.

NO

⑥ • BACKPROBE IGNITION SWITCH CONNECTOR WITH A TEST LIGHT FROM CAVITY "2" TO GROUND.
TEST LIGHT SHOULD LIGHT.
DOES IT?

YES

• CHECK FOR A POOR CONNECTION AT JUNCTION BLOCK 1.
• CHECK FOR AN OPEN IN BLK/ORN WIRE BETWEEN JUNCTION BLOCK 1 AND DISTRIBUTOR.
IF OK, REPLACE JUNCTION BLOCK 1.

YES

• CHECK FOR A POOR CONNECTION AT IGNITION SWITCH.
• CHECK FOR AN OPEN IN BLK/ORN WIRE BETWEEN IGNITION SWITCH AND JUNCTION BLOCK 1.
IF OK, REPLACE IGNITION SWITCH.

NO

REPAIR OPEN IN BLK/RED WIRE TO IGNITION SWITCH.

CLEAR DTCs AND REPEAT OBD SYSTEM CHECK AFTER MAKING REPAIRS.

1.8L (VIN 8) ENGINE — DIAGNOSTIC TROUBLE CODE CHARTS — 1993 PRIZM

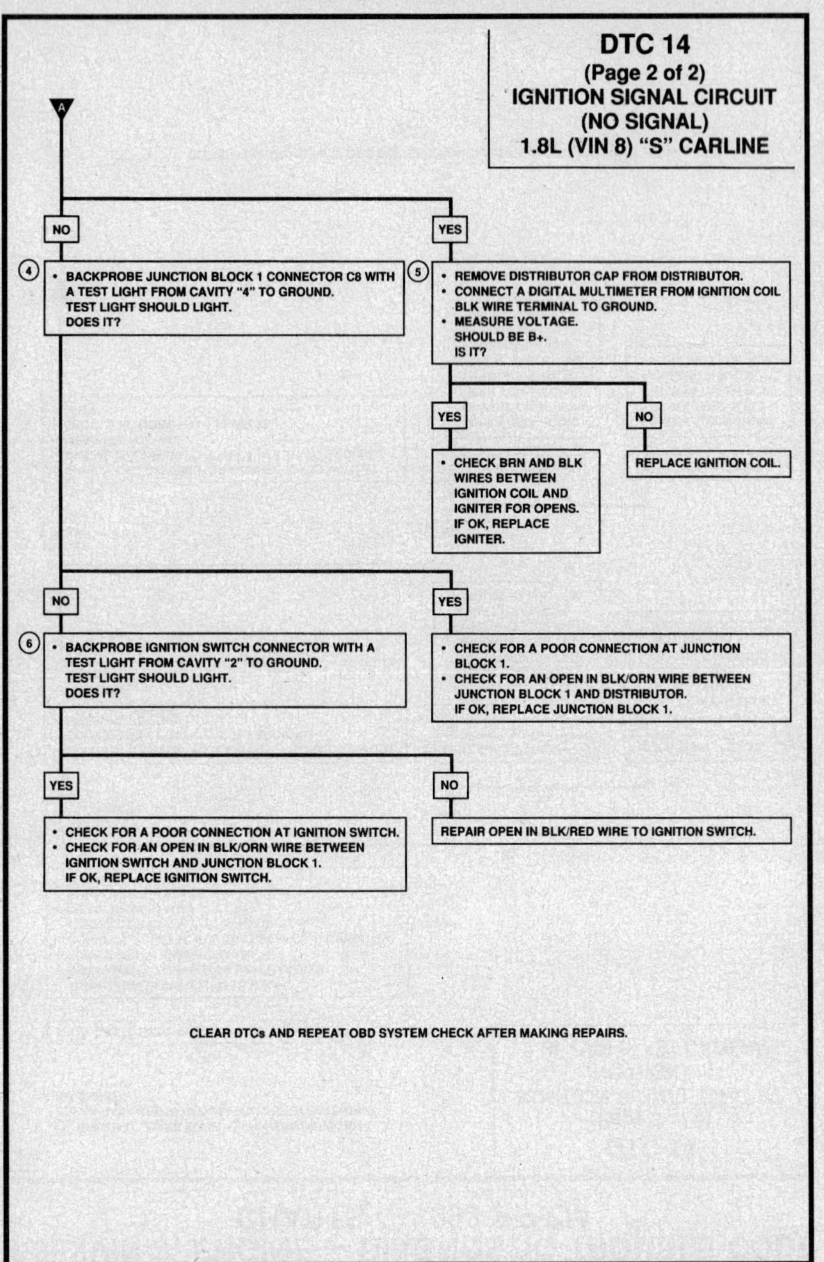

DTC 16
PCM CONTROL SIGNAL (FAILURE)
1.8L (VIN 8) "S" CARLINE

Circuit Description:

The powertrain control module (PCM) has detected a failure in the communication circuitry between the electronic engine and the electronic transaxle controls. The PCM will no longer allow the transaxle to go into a torque converter clutch (TCC) lockup condition and must be replaced if this DTC is set.

1.8L (VIN 8) ENGINE — DIAGNOSTIC TROUBLE CODE CHARTS — 1993 PRIZM

[!] IMPORTANT: MAKE SURE THAT THE OBD SYSTEM CHECK HAS BEEN PERFORMED BEFORE CONTINUING DIAGNOSIS.

FROM OBD SYSTEM CHECK

- PCM COMMUNICATION CIRCUITRY BETWEEN ELECTRONIC ENGINE AND ELECTRONIC TRANSAXLE CONTROLS HAS DETECTED A FAILURE.
- REPLACE PCM.

DTC 16
PCM CONTROL SIGNAL (FAILURE)
1.8L (VIN 8) "S" CARLINE

CLEAR DTCs AND REPEAT OBD SYSTEM CHECK AFTER MAKING REPAIRS.

1.8L (VIN 8) ENGINE — DIAGNOSTIC TROUBLE CODE CHARTS — 1993 PRIZM

DTC 21
OXYGEN SENSOR (O2S) CIRCUIT
(OPEN/SHORTED CIRCUIT)
1.8L (VIN 8) "S" CARLINE

Circuit Description:

When the oxygen sensor (O2S) reaches operating temperature, it produces a varying voltage between 0.10 volts (exhaust is lean) and 1.0 volts (exhaust is rich). The powertrain control module (PCM) monitors this voltage and determines the concentration of gases in the exhaust.

The PCM then uses this input to make fuel corrections to maintain the optimum air/fuel ratio.

DTC 21 will set if the following condition is met:

- O2S voltage input at the PCM is constant between 0.35 and 0.70 volts for 60 seconds or more with engine speed above 1,500 rpm and engine coolant temperature (ECT) above 80°C (176°F).

Test Description: Number(s) below refer to circled number(s) on the diagnostic chart.

1. This checks to see if the O2S is faulty.
2. This checks for an open in the BLK wire between the PCM and O2S.
3. This checks for an open in the BRN wire between the PCM and G106.
4. This checks for a short to ground in the BLK wire between the PCM and O2S, an open in the BRN wire to the O2S on federal emissions equipped vehicles and for a faulty PCM.

Diagnostic Aids:

Make sure that G106 is clean and tight.

An intermittent may be caused by a poor connection, rubbed through wire insulation, or a wire broken inside the insulation. Inspect harness connectors for backed out terminals, improper mating, broken locks, improperly formed or damaged terminals and poor terminal-to-wire connections before component replacement.

1.8L (VIN 8) ENGINE — DIAGNOSTIC TROUBLE CODE CHARTS — 1993 PRIZM

[i] IMPORTANT: MAKE SURE THAT THE OBD SYSTEM CHECK HAS BEEN PERFORMED BEFORE CONTINUING DIAGNOSIS.

DTC 21
OXYGEN SENSOR (O2S) CIRCUIT
(OPEN/SHORTED CIRCUIT)
1.8L (VIN 8) "S" CARLINE

FROM OBD SYSTEM CHECK

(1)
- DISCONNECT O2S CONNECTOR.
- START AND RUN ENGINE AT 1,500 RPM FOR 1 MINUTE OR UNTIL NORMAL OPERATING TEMPERATURE IS ATTAINED.
- CONNECT A DIGITAL MULTIMETER FROM O2S CONNECTOR TERMINAL "1" (O2S SIDE) TO GROUND.
- MEASURE VOLTAGE.
  SHOULD BE BETWEEN 0.80 AND 0.95 VOLTS.
  IS IT?

YES → (2) | NO → REPLACE O2S.

(2)
- SHUT ENGINE OFF.
- DISCONNECT PCM CONNECTOR C2 AND C3.
- CONNECT A DIGITAL MULTIMETER FROM PCM CONNECTOR C2 CAVITY "B6" TO O2S CONNECTOR CAVITY "1" (HARNESS SIDE).
- MEASURE RESISTANCE.
  SHOULD BE LESS THAN 0.50 OHMS.
  IS IT?

YES → (3) | NO → REPAIR OPEN IN BLK WIRE BETWEEN PCM AND O2S.

(3)
- CONNECT A TEST LIGHT FROM PCM CONNECTOR C3 CAVITY "C24" (MANUAL TRANSAXLE MODELS) OR FROM CAVITY "C14" (AUTOMATIC TRANSAXLE MODELS) TO B+.
  TEST LIGHT SHOULD LIGHT.
  DOES IT?

YES → (4) | NO → REPAIR OPEN IN BRN WIRE BETWEEN PCM AND G106.

(4)
- CONNECT A TEST LIGHT FROM PCM CONNECTOR C2 CAVITY "B6" TO B+.
  TEST LIGHT SHOULD NOT LIGHT.
  DOES IT?

NO | YES → REPAIR SHORT TO GROUND IN BLK WIRE BETWEEN PCM AND O2S.

- CHECK FOR AN OPEN IN BRN WIRE BETWEEN O2S AND G106 (FEDERAL EMISSIONS ONLY).
- CHECK FOR A POOR CONNECTION AT PCM.
  IF OK, REPLACE PCM.

CLEAR DTCs AND REPEAT OBD SYSTEM CHECK AFTER MAKING REPAIRS.

1.8L (VIN 8) ENGINE — DIAGNOSTIC TROUBLE CODE CHARTS — 1993 PRIZM

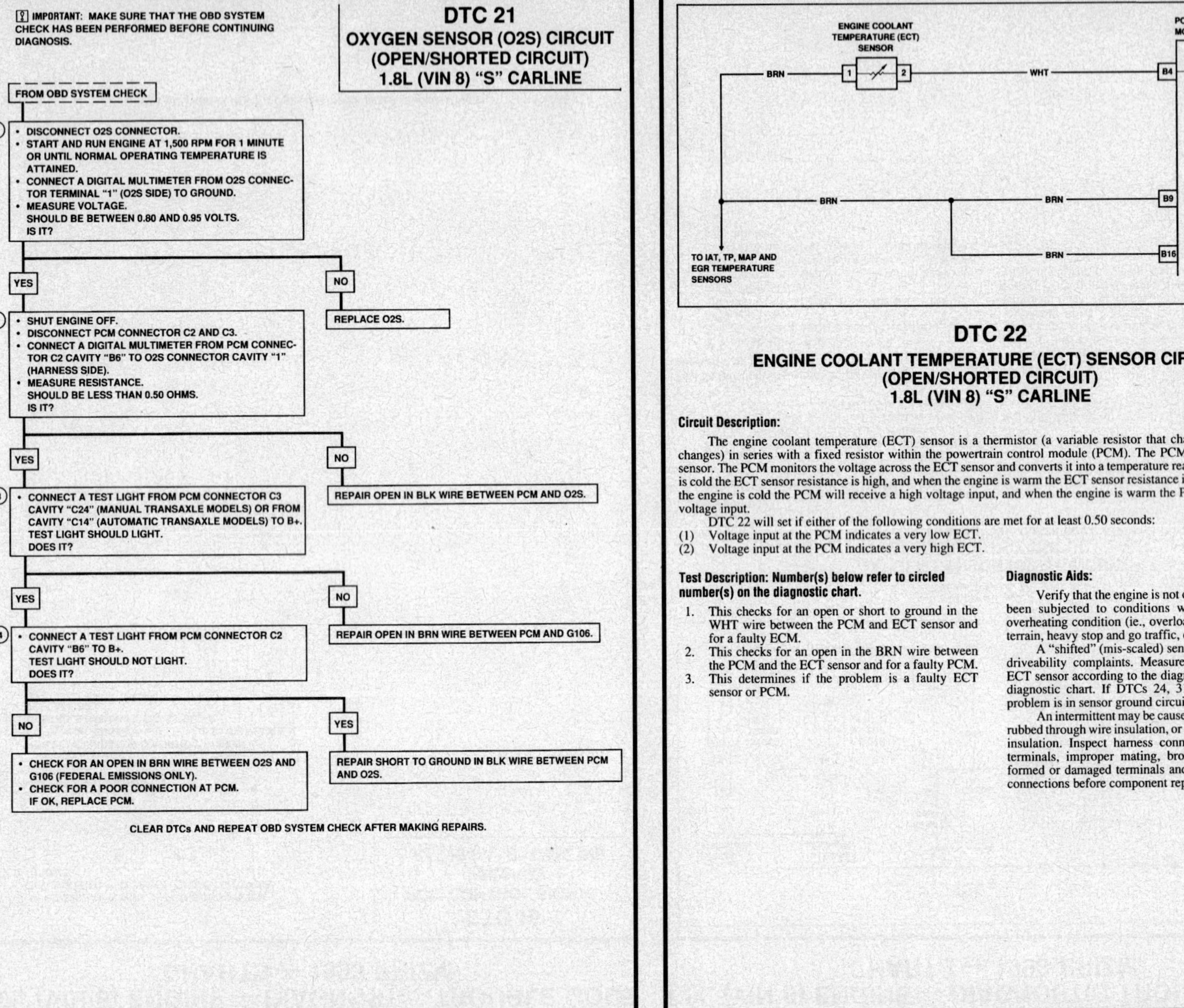

DTC 22
ENGINE COOLANT TEMPERATURE (ECT) SENSOR CIRCUIT
(OPEN/SHORTED CIRCUIT)
1.8L (VIN 8) "S" CARLINE

Circuit Description:

The engine coolant temperature (ECT) sensor is a thermistor (a variable resistor that changes along with ECT changes) in series with a fixed resistor within the powertrain control module (PCM). The PCM applies 5 volts to the sensor. The PCM monitors the voltage across the ECT sensor and converts it into a temperature reading. When the engine is cold the ECT sensor resistance is high, and when the engine is warm the ECT sensor resistance is low. Therefore, when the engine is cold the PCM will receive a high voltage input, and when the engine is warm the PCM will receive a low voltage input.

DTC 22 will set if either of the following conditions are met for at least 0.50 seconds:

(1) Voltage input at the PCM indicates a very low ECT.
(2) Voltage input at the PCM indicates a very high ECT.

Test Description: Number(s) below refer to circled number(s) on the diagnostic chart.

1. This checks for an open or short to ground in the WHT wire between the PCM and ECT sensor and for a faulty ECM.
2. This checks for an open in the BRN wire between the PCM and the ECT sensor and for a faulty PCM.
3. This determines if the problem is a faulty ECT sensor or PCM.

Diagnostic Aids:

Verify that the engine is not overheating and has not been subjected to conditions which could create an overheating condition (ie., overload, trailer towing, hilly terrain, heavy stop and go traffic, etc.).

A "shifted" (mis-scaled) sensor could result in poor driveability complaints. Measure the resistance of the ECT sensor according to the diagnostic aids chart on the diagnostic chart. If DTCs 24, 31 and 41 are also set, problem is in sensor ground circuit.

An intermittent may be caused by a poor connection, rubbed through wire insulation, or a wire broken inside the insulation. Inspect harness connectors for backed out terminals, improper mating, broken locks, improperly formed or damaged terminals and poor terminal-to-wire connections before component replacement.

1.8L (VIN 8) ENGINE — DIAGNOSTIC TROUBLE CODE CHARTS — 1993 PRIZM

DTC 22
ENGINE COOLANT TEMPERATURE (ECT) SENSOR CIRCUIT (OPEN/SHORTED CIRCUIT)
1.8L (VIN 8) "S" CARLINE

DIAGNOSTIC AID		
ECT SENSOR		
TEMPERATURE TO RESISTANCE VALUES		
TEMPERATURE		RESISTANCE
-20°C	-4°F	10 – 20 K OHMS
0°C	32°F	4 – 7 K OHMS
20°C	68°F	2 – 3 K OHMS
40°C	104°F	0.9 – 1.3 K OHMS
60°C	140°F	0.4 – 0.7 K OHMS
80°C	176°F	0.2 – 0.4 K OHMS

CLEAR DTCs AND REPEAT OBD SYSTEM CHECK AFTER MAKING REPAIRS.

1.8L (VIN 8) ENGINE — DIAGNOSTIC TROUBLE CODE CHARTS — 1993 PRIZM

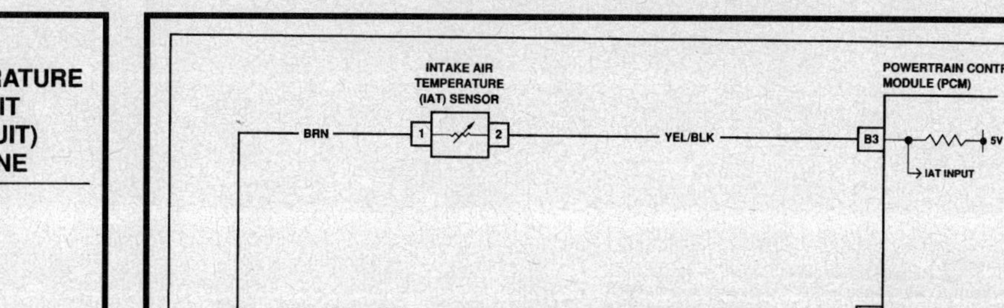

DTC 24
INTAKE AIR TEMPERATURE (IAT) SENSOR CIRCUIT (OPEN/SHORTED CIRCUIT)
1.8L (VIN 8) "S" CARLINE

Circuit Description:

The intake air temperature (IAT) sensor is a thermistor (a variable resistor that changes along with outside air temperature changes) in series with a fixed resistor within the powertrain control module (PCM). The IAT sensor measures the temperature of the air entering the intake manifold. The PCM applies 5 volts to the sensor. The PCM monitors the voltage across the IAT sensor and converts it into a temperature reading. When the outside air temperature is cold, the IAT sensor resistance is high, and when the outside air temperature is warm, the IAT sensor resistance is low. Therefore, when the air temperature is cold the PCM will receive a high voltage input, and when the air temperature is warm the PCM will receive a low voltage input.

DTC 24 will set if either of the following conditions are met for at least 0.50 seconds:

(1) Voltage input at the PCM indicates a very low air temperature.
(2) Voltage input at the PCM indicates a very high air temperature.

Test Description: Number(s) below refer to circled number(s) on the diagnostic chart.

1. This checks for an open or short to ground in the YEL/BLK wire between the PCM and IAT sensor and for a faulty PCM.
2. This checks for an open in the BRN wire between the PCM and the IAT sensor and for a faulty PCM.
3. This determines if the problem is a faulty IAT sensor or PCM.

Diagnostic Aids:

A "shifted" (mis-scaled) sensor could result in poor driveability complaints. Measure the resistance of the IAT sensor according to the diagnostic aids chart on the diagnostic chart. If DTCs 22, 31 and 41 are also set, problem is in sensor ground circuit.

An intermittent may be caused by a poor connection, rubbed through wire insulation, or a wire broken inside the insulation. Inspect harness connectors for backed out terminals, improper mating, broken locks, improperly formed or damaged terminals and poor terminal-to-wire connections before component replacement.

1.8L (VIN 8) ENGINE — DIAGNOSTIC TROUBLE CODE CHARTS — 1993 PRIZM

② IMPORTANT: MAKE SURE THAT THE OBD SYSTEM CHECK HAS BEEN PERFORMED BEFORE CONTINUING DIAGNOSIS.

FROM OBD SYSTEM CHECK

① • TURN IGNITION SWITCH TO "LOCK."
• DISCONNECT IAT SENSOR CONNECTOR.
• TURN IGNITION SWITCH TO "ON."
• CONNECT A DIGITAL MULTIMETER FROM IAT SENSOR CONNECTOR CAVITY "2" TO GROUND.
• MEASURE VOLTAGE.
SHOULD BE 4–6 VOLTS.
IS IT?

YES

② • CONNECT A TEST LIGHT FROM IAT SENSOR CONNECTOR CAVITY "1" TO B+.
TEST LIGHT SHOULD LIGHT.
DOES IT?

NO
• CHECK FOR A POOR CONNECTION AT PCM AND AT IAT SENSOR.
• CHECK FOR AN OPEN OR SHORT TO GROUND IN YEL/BLK WIRE BETWEEN PCM AND IAT SENSOR.
IF OK, REPLACE PCM.

YES

③ • TURN IGNITION SWITCH TO "LOCK."
• REMOVE IAT SENSOR.
• MEASURE RESISTANCE ACCORDING TO CHART BELOW.
RESISTANCE SHOULD BE WITHIN SPECIFIED RANGES.
IS IT?

NO
• CHECK FOR A POOR CONNECTION AT PCM AND AT IAT SENSOR.
• CHECK FOR AN OPEN IN BRN WIRE BETWEEN PCM AND IAT SENSOR.
IF OK, REPLACE PCM.

NO

REPLACE IAT SENSOR.

YES

REPLACE PCM.

DTC 24
INTAKE AIR TEMPERATURE (IAT) SENSOR CIRCUIT
(OPEN/SHORTED CIRCUIT)
1.8L (VIN 8) "S" CARLINE

DIAGNOSTIC AID		
IAT SENSOR		
TEMPERATURE TO RESISTANCE VALUES		
TEMPERATURE		RESISTANCE
-20°C	-4°F	10 – 20 K OHMS
0°C	32°F	4 – 7 K OHMS
20°C	68°F	2 – 3 K OHMS
40°C	104°F	0.9 – 1.3 K OHMS
60°C	140°F	0.4 – 0.7 K OHMS
80°C	176°F	0.2 – 0.4 K OHMS

CLEAR DTCs AND REPEAT OBD SYSTEM CHECK AFTER MAKING REPAIRS.

1.8L (VIN 8) ENGINE — DIAGNOSTIC TROUBLE CODE CHARTS — 1993 PRIZM

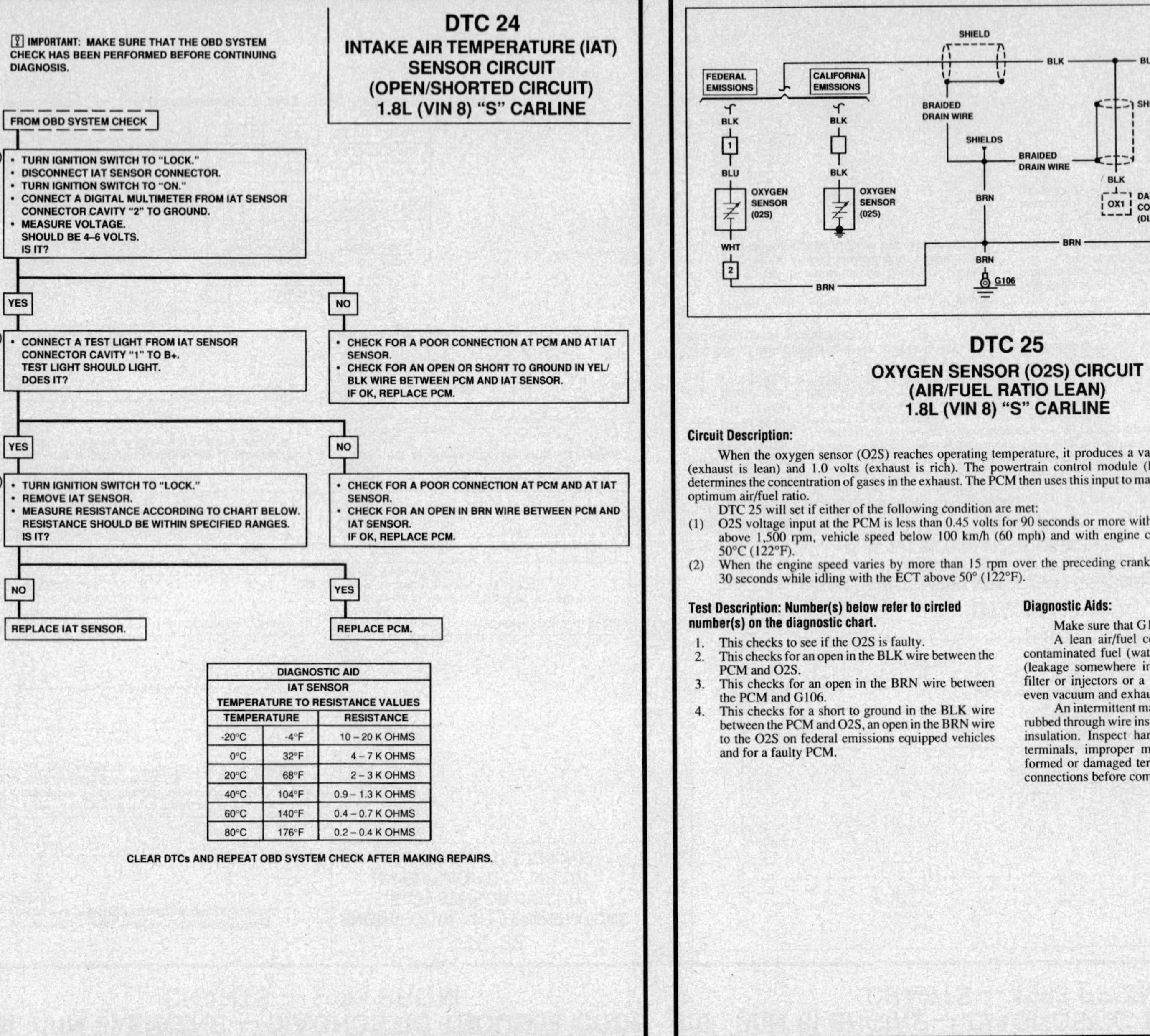

DTC 25
OXYGEN SENSOR (O2S) CIRCUIT
(AIR/FUEL RATIO LEAN)
1.8L (VIN 8) "S" CARLINE

Circuit Description:

When the oxygen sensor (O2S) reaches operating temperature, it produces a varying voltage between 0.10 volts (exhaust is lean) and 1.0 volts (exhaust is rich). The powertrain control module (PCM) monitors this voltage and determines the concentration of gases in the exhaust. The PCM then uses this input to make fuel corrections to maintain the optimum air/fuel ratio.

DTC 25 will set if either of the following condition are met:
(1) O2S voltage input at the PCM is less than 0.45 volts for 90 seconds or more with the O2S warm and engine speed above 1,500 rpm, vehicle speed below 100 km/h (60 mph) and with engine coolant temperature (ECT) above 50°C (122°F).
(2) When the engine speed varies by more than 15 rpm over the preceding crank angle period during a period of 30 seconds while idling with the ECT above 50° (122°F).

Test Description: Number(s) below refer to circled number(s) on the diagnostic chart.

1. This checks to see if the O2S is faulty.
2. This checks for an open in the BLK wire between the PCM and O2S.
3. This checks for an open in the BRN wire between the PCM and G106.
4. This checks for a short to ground in the BLK wire between the PCM and O2S, an open in the BRN wire to the O2S on federal emissions equipped vehicles and for a faulty PCM.

Diagnostic Aids:

Make sure that G106 is clean and tight.

A lean air/fuel condition may also be caused by contaminated fuel (water in the fuel), low fuel pressure (leakage somewhere in the fuel system, restricted fuel filter or injectors or a faulty fuel pressure regulator) or even vacuum and exhaust leaks.

An intermittent may be caused by a poor connection, rubbed through wire insulation, or a wire broken inside the insulation. Inspect harness connectors for backed out terminals, improper mating, broken locks, improperly formed or damaged terminals and poor terminal-to-wire connections before component replacement.

1.8L (VIN 8) ENGINE — DIAGNOSTIC TROUBLE CODE CHARTS — 1993 PRIZM

▣ IMPORTANT: MAKE SURE THAT THE OBD SYSTEM CHECK HAS BEEN PERFORMED BEFORE CONTINUING DIAGNOSIS.

FROM OBD SYSTEM CHECK

① • DISCONNECT O2S CONNECTOR.
• START AND RUN ENGINE AT 1,500 RPM FOR 1 MINUTE OR UNTIL NORMAL OPERATING TEMPERATURE IS ATTAINED.
• CONNECT A DIGITAL MULTIMETER FROM O2S CONNECTOR TERMINAL "1" (O2S SIDE) TO GROUND.
• MEASURE VOLTAGE.
SHOULD BE BETWEEN 0.80 AND 0.95 VOLTS.
IS IT?

YES — NO

NO → **REPLACE O2S.**

② • SHUT ENGINE OFF.
• DISCONNECT PCM CONNECTOR C2 AND C3.
• CONNECT A DIGITAL MULTIMETER FROM PCM CONNECTOR C2 CAVITY "B6" TO O2S CONNECTOR CAVITY "1" (HARNESS SIDE).
• MEASURE RESISTANCE.
SHOULD BE LESS THAN 0.50 OHMS.
IS IT?

YES — NO

NO → **REPAIR OPEN IN BLK WIRE BETWEEN PCM AND O2S.**

③ • CONNECT A TEST LIGHT FROM PCM CONNECTOR C3 CAVITY "C24" (MANUAL TRANSAXLE MODELS) OR FROM CAVITY "C14" (AUTOMATIC TRANSAXLE MODELS) TO B+.
TEST LIGHT SHOULD LIGHT.
DOES IT?

YES — NO

NO → **REPAIR OPEN IN BRN WIRE BETWEEN PCM AND G106.**

④ • CONNECT A TEST LIGHT FROM PCM CONNECTOR C2 CAVITY "B6" TO B+.
TEST LIGHT SHOULD NOT LIGHT.
DOES IT?

NO — YES

YES → **REPAIR SHORT TO GROUND IN BLK WIRE BETWEEN PCM AND O2S.**

NO → • CHECK FOR AN OPEN IN BRN WIRE BETWEEN O2S AND G106 (FEDERAL EMISSIONS ONLY).
• CHECK FOR A POOR CONNECTION AT PCM.
IF OK, REPLACE PCM.

CLEAR DTCs AND REPEAT OBD SYSTEM CHECK AFTER MAKING REPAIRS.

DTC 25
OXYGEN SENSOR (O2S) CIRCUIT (AIR/FUEL RATIO LEAN) 1.8L (VIN 8) "S" CARLINE

1.8L (VIN 8) ENGINE — DIAGNOSTIC TROUBLE CODE CHARTS — 1993 PRIZM

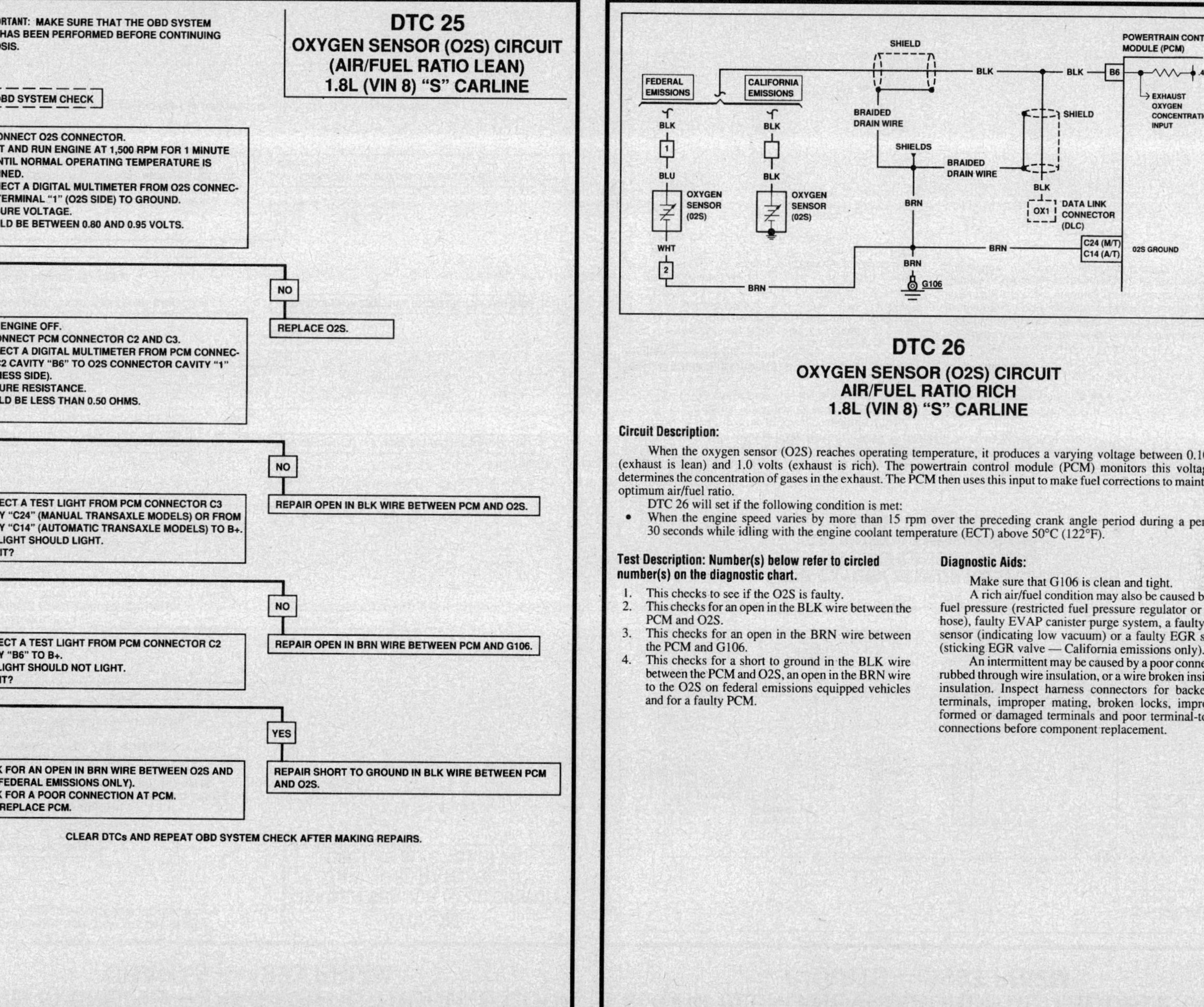

DTC 26
OXYGEN SENSOR (O2S) CIRCUIT AIR/FUEL RATIO RICH 1.8L (VIN 8) "S" CARLINE

Circuit Description:

When the oxygen sensor (O2S) reaches operating temperature, it produces a varying voltage between 0.10 volts (exhaust is lean) and 1.0 volts (exhaust is rich). The powertrain control module (PCM) monitors this voltage and determines the concentration of gases in the exhaust. The PCM then uses this input to make fuel corrections to maintain the optimum air/fuel ratio.

DTC 26 will set if the following condition is met:
• When the engine speed varies by more than 15 rpm over the preceding crank angle period during a period of 30 seconds while idling with the engine coolant temperature (ECT) above 50°C (122°F).

Test Description: Number(s) below refer to circled number(s) on the diagnostic chart.

1. This checks to see if the O2S is faulty.
2. This checks for an open in the BLK wire between the PCM and O2S.
3. This checks for an open in the BRN wire between the PCM and G106.
4. This checks for a short to ground in the BLK wire between the PCM and O2S, an open in the BRN wire to the O2S on federal emissions equipped vehicles and for a faulty PCM.

Diagnostic Aids:

Make sure that G106 is clean and tight.

A rich air/fuel condition may also be caused by high fuel pressure (restricted fuel pressure regulator or return hose), faulty EVAP canister purge system, a faulty MAP sensor (indicating low vacuum) or a faulty EGR system (sticking EGR valve — California emissions only).

An intermittent may be caused by a poor connection, rubbed through wire insulation, or a wire broken inside the insulation. Inspect harness connectors for backed out terminals, improper mating, broken locks, improperly formed or damaged terminals and poor terminal-to-wire connections before component replacement.

1.8L (VIN 8) ENGINE — DIAGNOSTIC TROUBLE CODE CHARTS — 1993 PRIZM

IMPORTANT: MAKE SURE THAT THE OBD SYSTEM CHECK HAS BEEN PERFORMED BEFORE CONTINUING DIAGNOSIS.

FROM OBD SYSTEM CHECK

1.
- DISCONNECT O2S CONNECTOR.
- START AND RUN ENGINE AT 1,500 RPM FOR 1 MINUTE OR UNTIL NORMAL OPERATING TEMPERATURE IS ATTAINED.
- CONNECT A DIGITAL MULTIMETER FROM O2S CONNECTOR TERMINAL "1" (O2S SIDE) TO GROUND.
- MEASURE VOLTAGE. SHOULD BE BETWEEN 0.80 AND 0.95 VOLTS. IS IT?

YES / NO → REPLACE O2S.

2.
- SHUT ENGINE OFF.
- DISCONNECT PCM CONNECTOR C2 AND C3.
- CONNECT A DIGITAL MULTIMETER FROM PCM CONNECTOR C2 CAVITY "B6" TO O2S CONNECTOR CAVITY "1" (HARNESS SIDE).
- MEASURE RESISTANCE. SHOULD BE LESS THAN 0.50 OHMS. IS IT?

YES / NO → REPAIR OPEN IN BLK WIRE BETWEEN PCM AND O2S.

3.
- CONNECT A TEST LIGHT FROM PCM CONNECTOR C3 CAVITY "C24" (MANUAL TRANSAXLE MODELS) OR FROM CAVITY "C14" (AUTOMATIC TRANSAXLE MODELS) TO B+. TEST LIGHT SHOULD LIGHT. DOES IT?

YES / NO → REPAIR OPEN IN BRN WIRE BETWEEN PCM AND G106.

4.
- CONNECT A TEST LIGHT FROM PCM CONNECTOR C2 CAVITY "B6" TO B+. TEST LIGHT SHOULD NOT LIGHT. DOES IT?

NO / YES → REPAIR SHORT TO GROUND IN BLK WIRE BETWEEN PCM AND O2S.

- CHECK FOR AN OPEN IN BRN WIRE BETWEEN O2S AND G106 (FEDERAL EMISSIONS ONLY).
- CHECK FOR A POOR CONNECTION AT PCM. IF OK, REPLACE PCM.

CLEAR DTCs AND REPEAT OBD SYSTEM CHECK AFTER MAKING REPAIRS.

DTC 26
OXYGEN SENSOR (O2S) CIRCUIT
(AIR/FUEL RATIO RICH)
1.8L (VIN 8) "S" CARLINE

1.8L (VIN 8) ENGINE — DIAGNOSTIC TROUBLE CODE CHARTS — 1993 PRIZM

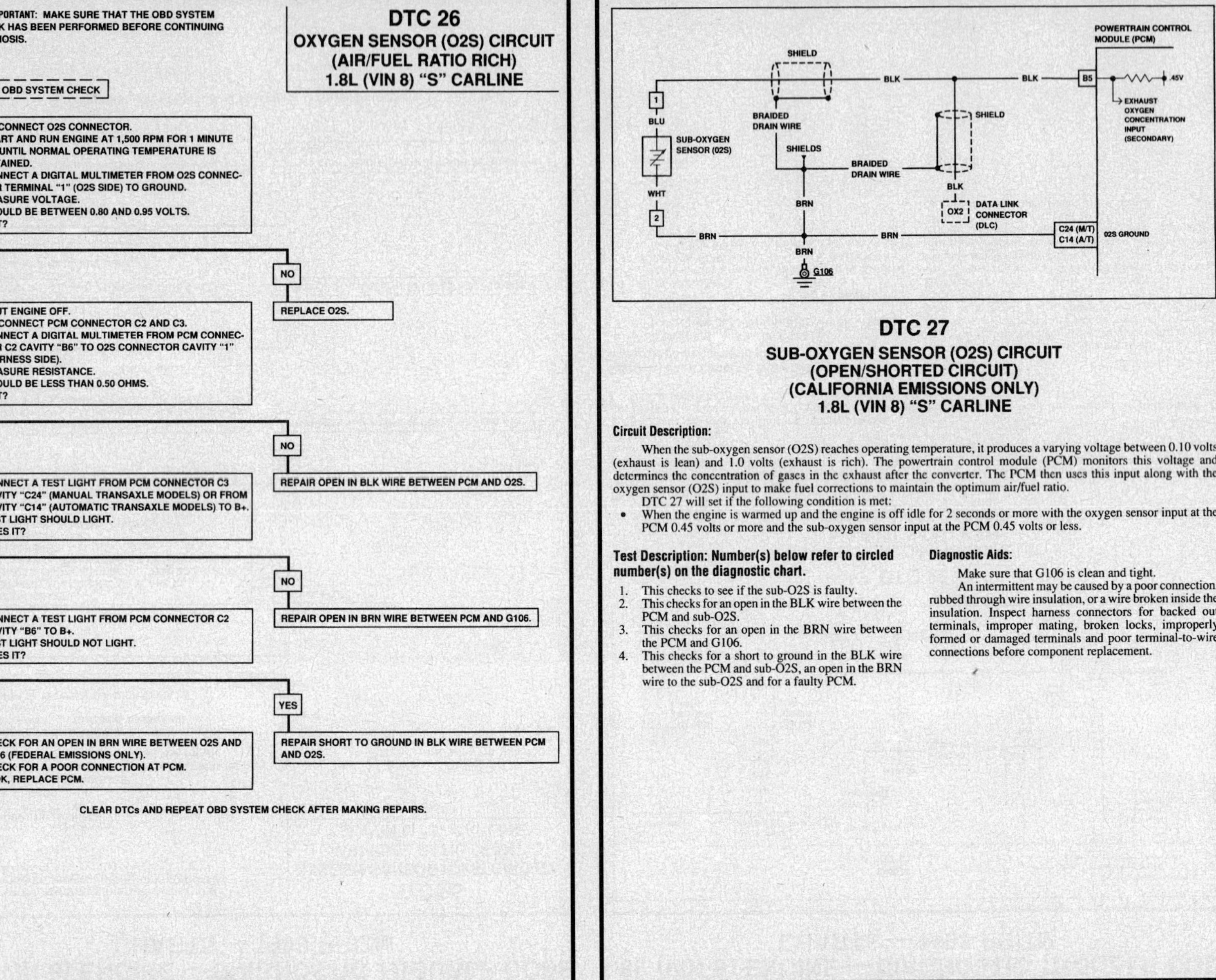

DTC 27
SUB-OXYGEN SENSOR (O2S) CIRCUIT
(OPEN/SHORTED CIRCUIT)
(CALIFORNIA EMISSIONS ONLY)
1.8L (VIN 8) "S" CARLINE

Circuit Description:

When the sub-oxygen sensor (O2S) reaches operating temperature, it produces a varying voltage between 0.10 volts (exhaust is lean) and 1.0 volts (exhaust is rich). The powertrain control module (PCM) monitors this voltage and determines the concentration of gases in the exhaust after the converter. The PCM then uses this input along with the oxygen sensor (O2S) input to make fuel corrections to maintain the optimum air/fuel ratio.

DTC 27 will set if the following condition is met:
- When the engine is warmed up and the engine is off idle for 2 seconds or more with the oxygen sensor input at the PCM 0.45 volts or more and the sub-oxygen sensor input at the PCM 0.45 volts or less.

Test Description: Number(s) below refer to circled number(s) on the diagnostic chart.

1. This checks to see if the sub-O2S is faulty.
2. This checks for an open in the BLK wire between the PCM and sub-O2S.
3. This checks for an open in the BRN wire between the PCM and G106.
4. This checks for a short to ground in the BLK wire between the PCM and sub-O2S, an open in the BRN wire to the sub-O2S and for a faulty PCM.

Diagnostic Aids:

Make sure that G106 is clean and tight.

An intermittent may be caused by a poor connection, rubbed through wire insulation, or a wire broken inside the insulation. Inspect harness connectors for backed out terminals, improper mating, broken locks, improperly formed or damaged terminals and poor terminal-to-wire connections before component replacement.

1.8L (VIN 8) ENGINE — DIAGNOSTIC TROUBLE CODE CHARTS — 1993 PRIZM

IMPORTANT: MAKE SURE THAT THE DIAGNOSTIC CIRCUIT CHECK HAS BEEN PERFORMED BEFORE CONTINUING DIAGNOSIS.

DTC 27
SUB-OXYGEN SENSOR (O2S) CIRCUIT
(OPEN/SHORTED CIRCUIT)
(CALIFORNIA EMISSIONS ONLY)
1.8L (VIN 8) "S" CARLINE

FROM DIAGNOSTIC CIRCUIT CHECK

(1) • START AND RUN ENGINE AT 1,500 RPM FOR 1 MINUTE OR UNTIL NORMAL OPERATING TEMPERATURE IS ATTAINED.
• RAISE AND SUITABLY SUPPORT VEHICLE.
• DISCONNECT SUB-O2S CONNECTOR.
• CONNECT A DIGITAL MULTIMETER FROM SUB-O2S CONNECTOR TERMINAL "1" (SUB-O2S SIDE) TO GROUND.
• MEASURE VOLTAGE.
SHOULD BE BETWEEN 0.80 AND 0.95 VOLTS.
IS IT?

YES → (2)
NO → REPLACE SUB-O2S.

(2) • LOWER VEHICLE.
• SHUT ENGINE OFF.
• DISCONNECT PCM CONNECTOR C2 AND C3.
• CONNECT A DIGITAL MULTIMETER FROM PCM CONNECTOR C2 CAVITY "B5" TO SUB-O2S CONNECTOR CAVITY "1" (HARNESS SIDE).
• MEASURE RESISTANCE.
SHOULD BE LESS THAN 0.50 OHMS.
IS IT?

YES → (3)
NO → REPAIR OPEN IN BLK WIRE BETWEEN PCM AND SUB-O2S.

(3) • CONNECT A TEST LIGHT FROM PCM CONNECTOR C3 CAVITY "C24" (MANUAL TRANSAXLE MODELS) OR FROM "C14" (AUTOMATIC TRANSXLE MODELS) TO B+.
TEST LIGHT SHOULD LIGHT.
DOES IT?

YES → (4)
NO → REPAIR OPEN IN BRN WIRE BETWEEN PCM AND G106.

(4) • CONNECT A TEST LIGHT FROM PCM CONNECTOR C2 CAVITY "B5" TO B+.
TEST LIGHT SHOULD NOT LIGHT.
DOES IT?

NO → CHECK FOR AN OPEN IN BRN WIRE BETWEEN SUB-O2S AND G106.
• CHECK FOR A POOR CONNECTION AT PCM.
IF OK, REPLACE PCM.

YES → REPAIR SHORT TO GROUND IN BLK WIRE BETWEEN PCM AND SUB-O2S.

CLEAR DTCs AND REPEAT OBD SYSTEM CHECK AFTER MAKING REPAIRS.

1.8L (VIN 8) ENGINE — DIAGNOSTIC TROUBLE CODE CHARTS — 1993 PRIZM

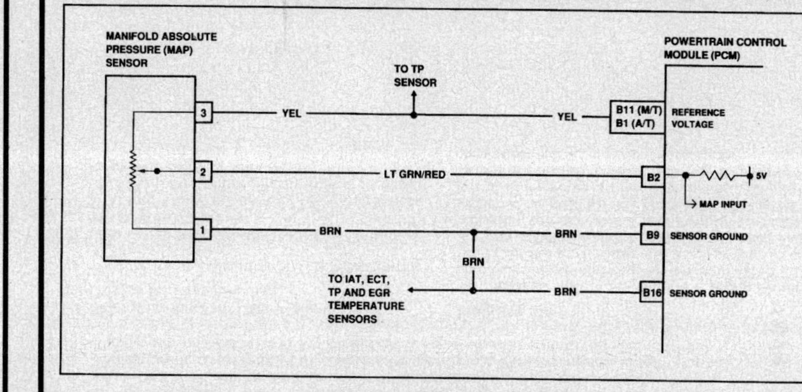

DTC 31
MANIFOLD ABSOLUTE PRESSURE (MAP) SENSOR CIRCUIT
(OPEN/SHORTED CIRCUIT)
1.8L (VIN 8) "S" CARLINE

Circuit Description:

The manifold absolute pressure (MAP) sensor measures the change in the intake manifold pressure (vacuum). The MAP sensor consists of a semi-conductor type pressure converting element, which converts a pressure change into an electrical change, and an electronic circuit which amplifies and corrects the electrical change. The powertrain control module (PCM) applies 5 volts to the MAP sensor. The change in the pressure, that results from the engine load and rpm changes, is converted into a voltage input that is monitored by the PCM. A low voltage input at the PCM indicates low manifold pressure, and a high voltage input at the PCM indicates high manifold pressure.

DTC 31 will set if either of the following conditions are met for at least 0.50 seconds:
(1) Voltage input at the PCM indicates very low manifold pressure.
(2) Voltage input at the PCM indicates very high manifold pressure.

Test Description: Number(s) below refer to circled number(s) on the diagnostic chart.

1. This checks for an open or short to ground in the YEL wire between the PCM and MAP sensor.
2. This checks for an open in the LT GRN/RED wire between the PCM and MAP sensor.
3. This checks for an open in the BRN wire between the PCM and MAP sensor, and for a faulty PCM or MAP sensor.

Diagnostic Aids:

If DTCs 22, 24 and 41 are also set, problem is in sensor ground circuit.

An intermittent may be caused by a poor connection, rubbed through wire insulation, or a wire broken inside the insulation. Inspect harness connectors for backed out terminals, improper mating, broken locks, improperly formed or damaged terminals and poor terminal-to-wire connections before component replacement.

1.8L (VIN 8) ENGINE — DIAGNOSTIC TROUBLE CODE CHARTS — 1993 PRIZM

[!] IMPORTANT: MAKE SURE THAT THE OBD SYSTEM CHECK HAS BEEN PERFORMED BEFORE CONTINUING DIAGNOSIS.

FROM OBD SYSTEM CHECK

(1)
- TURN IGNITION SWITCH TO "LOCK."
- DISCONNECT MAP SENSOR CONNECTOR.
- CONNECT A DIGITAL MULTIMETER FROM MAP SENSOR CONNECTOR CAVITY "3" TO GROUND.
- TURN IGNITION SWITCH TO "ON."
- MEASURE VOLTAGE. SHOULD BE 4–6 VOLTS. IS IT?

YES / NO

(2)
- CONNECT A DIGITAL MULTIMETER FROM MAP SENSOR CONNECTOR CAVITY "2" TO GROUND.
- MEASURE VOLTAGE. SHOULD BE 4–5 VOLTS. IS IT?

NO:
- CHECK FOR A POOR CONNECTION AT PCM.
- CHECK FOR AN OPEN OR A SHORT TO GROUND IN YEL WIRE BETWEEN PCM AND MAP SENSOR. IF OK, REPLACE PCM.

YES / NO

(3)
- CONNECT A TEST LIGHT FROM MAP SENSOR CONNECTOR CAVITY "1" TO B+.
- TEST LIGHT SHOULD LIGHT. DOES IT?

NO:
- CHECK FOR A POOR CONNECTION AT PCM. CHECK FOR AN OPEN IN LT GRN/RED WIRE BETWEEN PCM AND MAP SENSOR. IF OK, REPLACE PCM.

NO / YES

NO:
- CHECK FOR A POOR CONNECTION AT PCM.
- CHECK FOR AN OPEN IN BRN WIRE BETWEEN PCM AND MAP SENSOR. IF OK, REPLACE PCM.

YES:
- CHECK FOR A POOR CONNECTION AT MAP SENSOR. IF OK, REPLACE MAP SENSOR.

CLEAR DTCs AND REPEAT OBD SYSTEM CHECK AFTER MAKING REPAIRS.

DTC 31
MANIFOLD ABSOLUTE PRESSURE (MAP) SENSOR CIRCUIT (OPEN/SHORTED CIRCUIT) 1.8L (VIN 8) "S" CARLINE

1.8L (VIN 8) ENGINE — DIAGNOSTIC TROUBLE CODE CHARTS — 1993 PRIZM

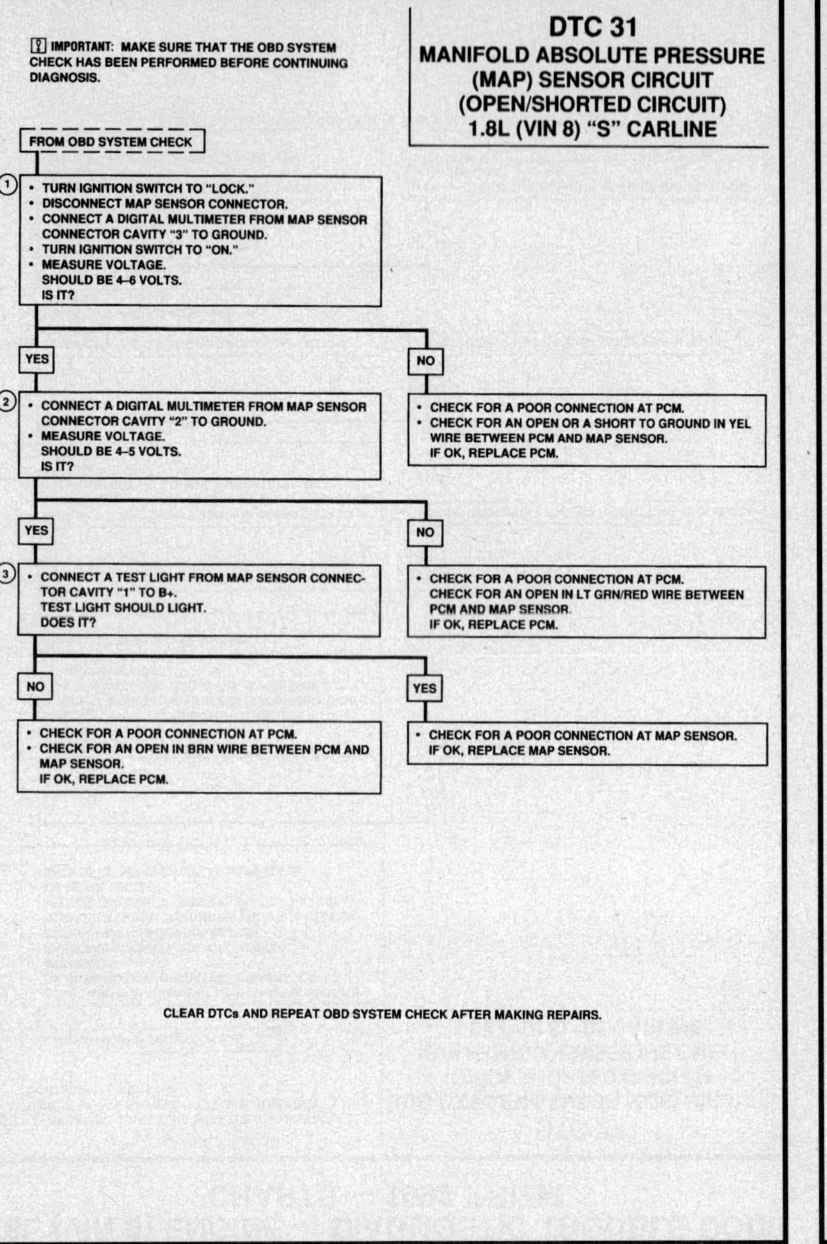

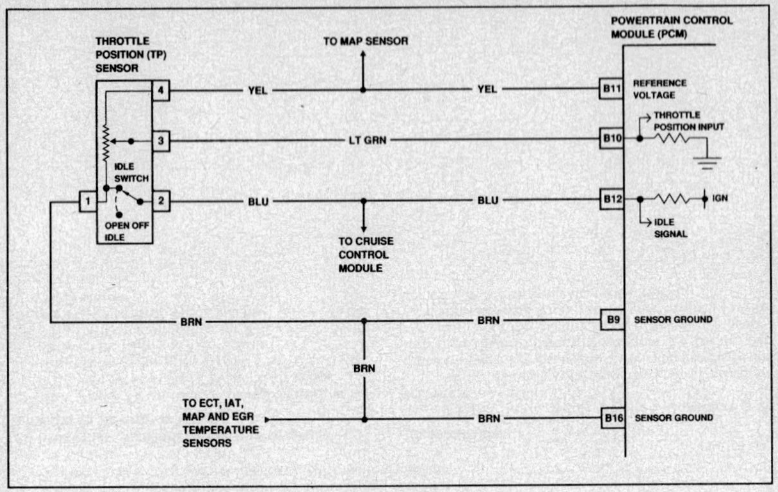

DTC 41
THROTTLE POSITION (TP) SENSOR CIRCUIT (OPEN/SHORTED CIRCUIT) 1.8L (VIN 8) "S" CARLINE

Circuit Description:

The throttle position (TP) sensor, consisting of a contact point (idle switch) and a potentiometer, is connected to the throttle valve shaft on the throttle body, and detects the throttle valve opening.

The throttle opening in the idle state is detected by means of the contact point which turns "ON" in that state. But beyond that, the full opening is detected by the potentiometer as follows.

A 5 volt reference voltage is applied to the sensor from the powertrain control module (PCM), and, as its brush moves over the resistance according to the throttle valve opening, the output voltage varies accordingly.

By monitoring the "ON/OFF" signal and sensor output voltage, the PCM detects the throttle valve opening. DTC 41 will set if either of the following conditions are met for at least 0.50 seconds:

(1) Throttle position input at the PCM indicates wide open throttle when the engine is idling.
(2) Throttle position input at the PCM indicates an idle signal when the engine is off idle.

Test Description: Number(s) below refer to circled number(s) on the diagnostic chart.

1. This checks for a fault in the YEL wire between the PCM and TP sensor.
2. This checks for an open in the BRN wire between the PCM and TP sensor.
3. This checks for a fault in the LT GRN wire between the PCM and TP sensor, a faulty PCM or for a misadjusted or faulty TP sensor.

Diagnostic Aids:

Refer to sensor adjustment procedures before replacing TP sensor. If DTCs 22, 24 and 31 are also set, problem is in sensor ground circuit.

An intermittent may be caused by a poor connection, rubbed through wire insulation, or a wire broken inside the insulation. Inspect harness connectors for backed out terminals, improper mating, broken locks, improperly formed or damaged terminals and poor terminal-to-wire connections before component replacement.

1.8L (VIN 8) ENGINE — DIAGNOSTIC TROUBLE CODE CHARTS — 1993 PRIZM

DTC 41
THROTTLE POSITION (TP) SENSOR CIRCUIT (OPEN/SHORTED CIRCUIT)
1.8L (VIN 8) "S" CARLINE

[!] **IMPORTANT:** MAKE SURE THAT THE OBD SYSTEM CHECK HAS BEEN PERFORMED BEFORE CONTINUING DIAGNOSIS.

FROM OBD SYSTEM CHECK

(1)
- TURN IGNITION SWITCH TO "LOCK."
- DISCONNECT TP SENSOR CONNECTOR.
- CONNECT A DIGITAL MULTIMETER FROM TP SENSOR CONNECTOR CAVITY "4" TO GROUND.
- TURN IGNITION SWITCH TO "ON."
- MEASURE VOLTAGE. SHOULD BE 4–6 VOLTS. IS IT?

YES ↓   **NO** →

NO:
- CHECK FOR A POOR CONNECTION AT PCM.
- CHECK FOR AN OPEN, A SHORT TO VOLTAGE OR A SHORT TO GROUND IN YEL WIRE BETWEEN PCM AND TP SENSOR. IF OK, REPLACE PCM.

(2)
- CONNECT A DIGITAL MULTIMETER FROM TP SENSOR CONNECTOR CAVITY "4" TO CAVITY "1".
- MEASURE VOLTAGE. SHOULD BE 4–6 VOLTS. IS IT?

YES ↓   **NO** →

NO:
- CHECK FOR A POOR CONNECTION AT PCM.
- CHECK FOR AN OPEN IN BRN WIRE BETWEEN PCM AND TP SENSOR. IF OK, REPLACE PCM.

(3)
- CONNECT A DIGITAL MULTIMETER FROM TP SENSOR CONNECTOR CAVITY "4" TO CAVITY "3".
- MEASURE VOLTAGE. SHOULD BE 4.3–4.8 VOLTS. IS IT?

YES ↓   **NO** →

YES:
- CHECK FOR A POOR CONNECTION AT TP SENSOR. IF OK, ADJUST/REPLACE TP SENSOR.

NO:
- CHECK FOR A POOR CONNECTION AT PCM.
- CHECK FOR AN OPEN, A SHORT TO VOLTAGE OR A SHORT TO GROUND IN LT GRN WIRE BETWEEN PCM AND TP SENSOR. IF OK, REPLACE PCM.

CLEAR DTCs AND REPEAT OBD SYSTEM CHECK AFTER MAKING REPAIRS.

1.8L (VIN 8) ENGINE — DIAGNOSTIC TROUBLE CODE CHARTS — 1993 PRIZM

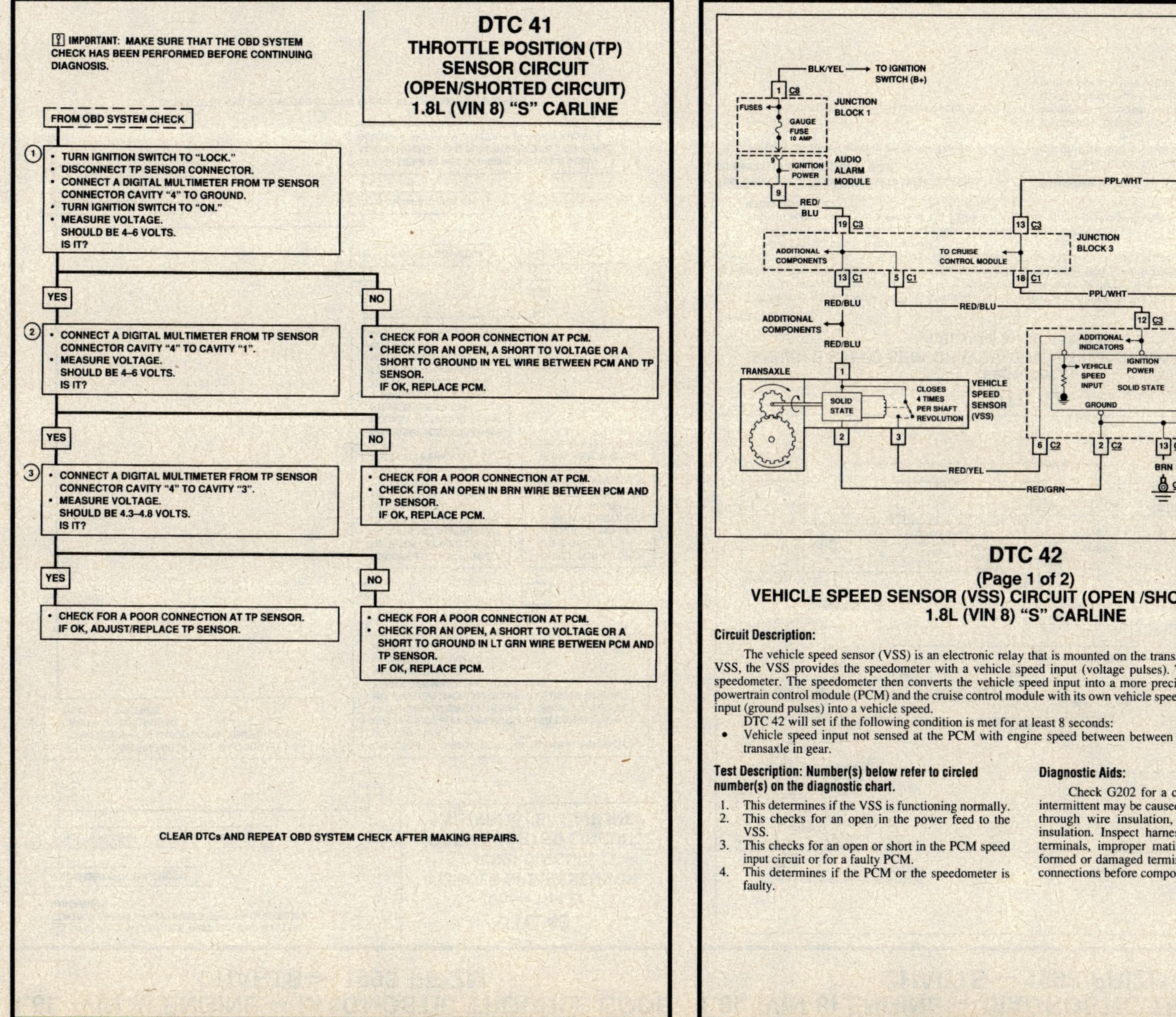

DTC 42
(Page 1 of 2)
VEHICLE SPEED SENSOR (VSS) CIRCUIT (OPEN /SHORTED CIRCUIT)
1.8L (VIN 8) "S" CARLINE

Circuit Description:

The vehicle speed sensor (VSS) is an electronic relay that is mounted on the transaxle. As the transaxle turns the VSS, the VSS provides the speedometer with a vehicle speed input (voltage pulses). This input is used to drive the speedometer. The speedometer then converts the vehicle speed input into a more precise waveform and provides the powertrain control module (PCM) and the cruise control module with its own vehicle speed input. The PCM converts this input (ground pulses) into a vehicle speed.

DTC 42 will set if the following condition is met for at least 8 seconds:
- Vehicle speed input not sensed at the PCM with engine speed between between 3,000 and 5,000 rpm with the transaxle in gear.

Test Description: Number(s) below refer to circled number(s) on the diagnostic chart.

1. This determines if the VSS is functioning normally.
2. This checks for an open in the power feed to the VSS.
3. This checks for an open or short in the PCM speed input circuit or for a faulty PCM.
4. This determines if the PCM or the speedometer is faulty.

Diagnostic Aids:

Check G202 for a clean and tight connection. An intermittent may be caused by a poor connection, rubbed through wire insulation, or a wire broken inside the insulation. Inspect harness connectors for backed out terminals, improper mating, broken locks, improperly formed or damaged terminals and poor terminal-to-wire connections before component replacement.

MULTIPORT FUEL INJECTION (MFI) SYSTEMS
LIGHT TRUCKS, VAN, GEO AND SATURN

1.8L (VIN 8) ENGINE—DIAGNOSTIC TROUBLE CODE CHARTS—1993 PRIZM

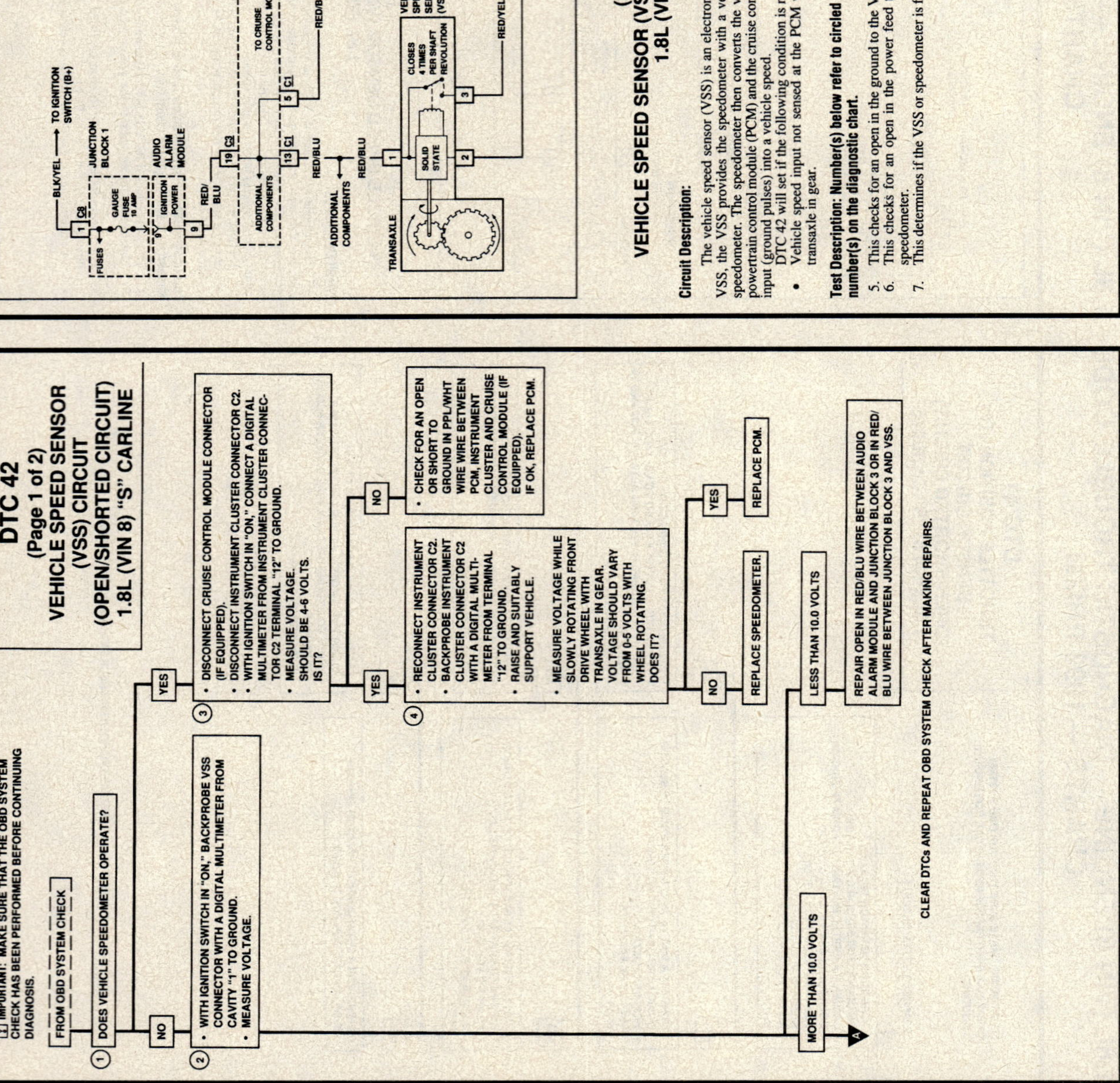

DTC 42
(Page 2 of 2)
VEHICLE SPEED SENSOR (VSS) CIRCUIT (OPEN/SHORTED CIRCUIT)
1.8L (VIN 8) "S" CARLINE

Circuit Description:

The vehicle speed sensor (VSS) is an electronic relay that is mounted on the transaxle. As the transaxle turns the VSS, the VSS provides the speedometer with a vehicle speed input (voltage pulses). This input is used to drive the speedometer. The speedometer then converts the vehicle speed input into a more precise waveform and provides the powertrain control module (PCM) and the cruise control module with its own vehicle speed input. The PCM converts this input (ground pulses) into a vehicle speed.

DTC 42 will set if the following condition is met for at least 8 seconds:
- Vehicle speed input not sensed at the PCM with engine speed between 3,000 and 5,000 rpm with the transaxle in gear.

Test Description: Number(s) below refer to circled number(s) on the diagnostic chart.

5. This checks for an open in the ground to the VSS.
6. This checks for an open in the power feed to the speedometer.
7. This determines if the VSS or speedometer is faulty.

Diagnostic Aids:

Check G202 for a clean and tight connection. An intermittent may be caused by a poor connection, rubbed through wire insulation, or a wire broken inside the insulation. Inspect harness connectors for backed out terminals, improper mating, broken locks, improperly formed or damaged terminals and poor terminal-to-wire connections before component replacement.

1.8L (VIN 8) ENGINE—DIAGNOSTIC TROUBLE CODE CHARTS—1993 PRIZM

DTC 42
(Page 1 of 2)
VEHICLE SPEED SENSOR (OPEN/SHORTED CIRCUIT) CIRCUIT
1.8L (VIN 8) "S" CARLINE

☒ IMPORTANT: MAKE SURE THAT THE OBD SYSTEM CHECK HAS BEEN PERFORMED BEFORE CONTINUING DIAGNOSIS.

FROM OBD SYSTEM CHECK

(1) DOES VEHICLE SPEEDOMETER OPERATE?
NO / YES

(2) WITH IGNITION SWITCH IN "ON," BACKPROBE VSS CONNECTOR WITH A DIGITAL MULTIMETER FROM CAVITY "1" TO GROUND. MEASURE VOLTAGE.

(3) • DISCONNECT CRUISE CONTROL MODULE CONNECTOR (IF EQUIPPED).
• DISCONNECT INSTRUMENT CLUSTER CONNECTOR C2.
• WITH IGNITION SWITCH IN "ON," CONNECT A DIGITAL MULTIMETER FROM INSTRUMENT CLUSTER CONNECTOR C2 TERMINAL "12" TO GROUND.
• MEASURE VOLTAGE. SHOULD BE 4-6 VOLTS. IS IT?
YES / NO

NO → CHECK FOR AN OPEN OR SHORT TO GROUND IN PPL/WHT WIRE BETWEEN PCM, INSTRUMENT CLUSTER AND CRUISE CONTROL MODULE (IF EQUIPPED). IF OK, REPLACE PCM.

(4) • RECONNECT INSTRUMENT CLUSTER CONNECTOR C2.
• BACKPROBE INSTRUMENT CLUSTER CONNECTOR C2 WITH A DIGITAL MULTIMETER FROM TERMINAL "12" TO GROUND.
• RAISE AND SUITABLY SUPPORT VEHICLE.
• MEASURE VOLTAGE WHILE SLOWLY ROTATING FRONT DRIVE WHEEL WITH TRANSAXLE IN GEAR. VOLTAGE SHOULD VARY FROM 0-5 VOLTS WITH WHEEL ROTATING. DOES IT?
YES / NO

YES → REPLACE PCM.
NO → REPLACE SPEEDOMETER.

MORE THAN 10.0 VOLTS ▷
LESS THAN 10.0 VOLTS → REPAIR OPEN IN RED/BLU WIRE BETWEEN AUDIO ALARM MODULE AND JUNCTION BLOCK 3 OR IN RED/BLU WIRE BETWEEN JUNCTION BLOCK 3 AND VSS.

CLEAR DTCs AND REPEAT OBD SYSTEM CHECK AFTER MAKING REPAIRS.

1.8L (VIN 8) ENGINE — DIAGNOSTIC TROUBLE CODE CHARTS — 1993 PRIZM

1.8L (VIN 8) ENGINE — DIAGNOSTIC TROUBLE CODE CHARTS — 1993 PRIZM

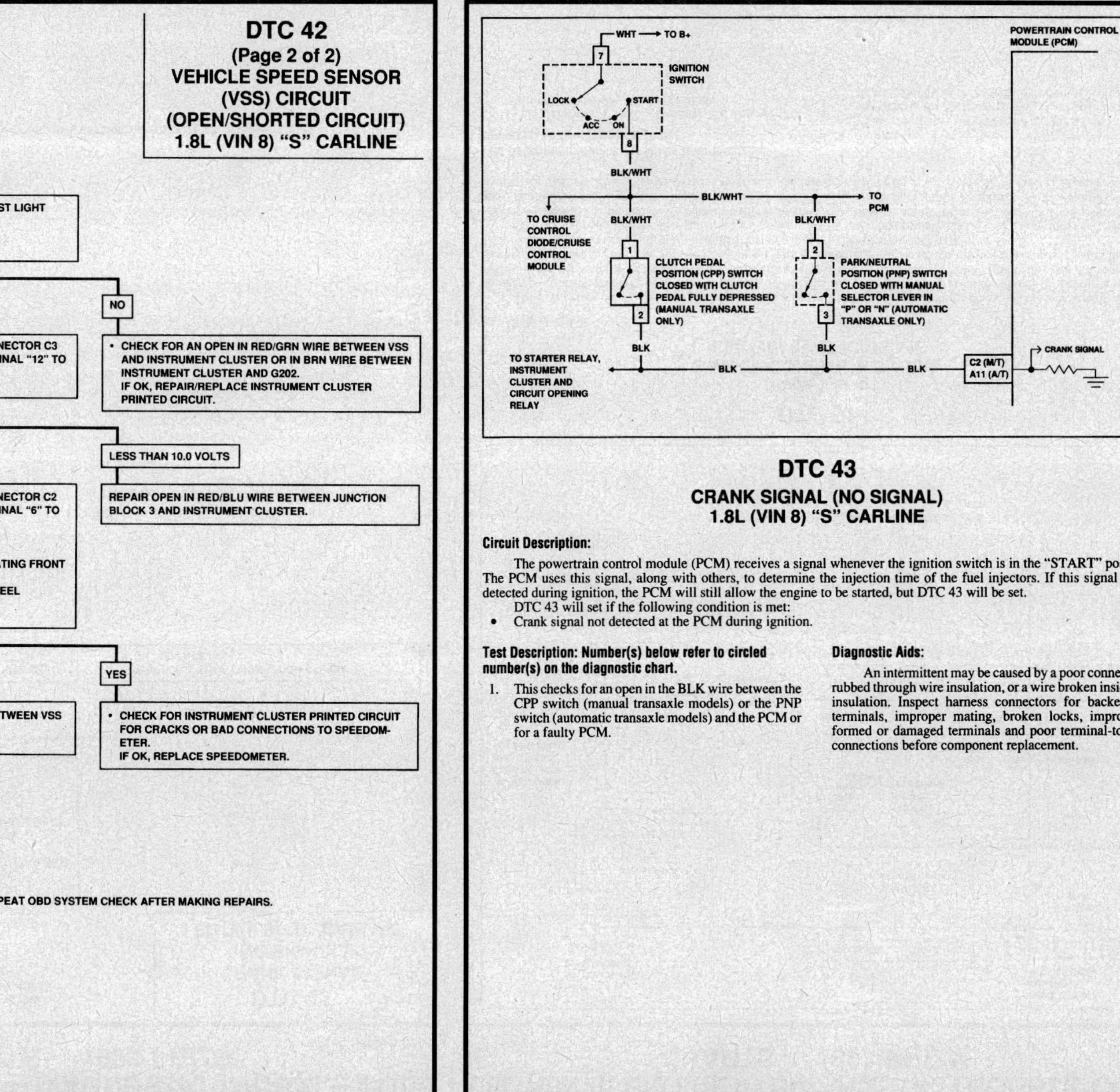

DTC 43
CRANK SIGNAL (NO SIGNAL)
1.8L (VIN 8) "S" CARLINE

Circuit Description:

The powertrain control module (PCM) receives a signal whenever the ignition switch is in the "START" position. The PCM uses this signal, along with others, to determine the injection time of the fuel injectors. If this signal is not detected during ignition, the PCM will still allow the engine to be started, but DTC 43 will be set.

DTC 43 will set if the following condition is met:
• Crank signal not detected at the PCM during ignition.

Test Description: Number(s) below refer to circled number(s) on the diagnostic chart.

1. This checks for an open in the BLK wire between the CPP switch (manual transaxle models) or the PNP switch (automatic transaxle models) and the PCM or for a faulty PCM.

Diagnostic Aids:

An intermittent may be caused by a poor connection, rubbed through wire insulation, or a wire broken inside the insulation. Inspect harness connectors for backed out terminals, improper mating, broken locks, improperly formed or damaged terminals and poor terminal-to-wire connections before component replacement.

1.8L (VIN 8) ENGINE — DIAGNOSTIC TROUBLE CODE CHARTS — 1993 PRIZM

1.8L (VIN 8) ENGINE — DIAGNOSTIC TROUBLE CODE CHARTS — 1993 PRIZM

Ⓘ IMPORTANT: MAKE SURE THAT THE OBD SYSTEM CHECK HAS BEEN PERFORMED BEFORE CONTINUING DIAGNOSIS.

DTC 43
CRANK SIGNAL
(NO SIGNAL)
1.8L (VIN 8) "S" CARLINE

FROM OBD SYSTEM CHECK

① • TURN IGNITION SWITCH TO "LOCK."
• DISCONNECT PCM CONNECTOR C3 (MANUAL TRANSAXLE MODELS) OR CONNECTOR C1 (AUTOMATIC TRANSAXLE MODELS).
• CONNECT A DIGITAL MULTIMETER FROM PCM CONNECTOR C3 CAVITY "C2" OR PCM CONNECTOR C1 CAVITY "A11" TO GROUND.
• MEASURE VOLTAGE WHILE CRANKING ENGINE. SHOULD BE MORE THAN 10 VOLTS. IS IT?

NO

YES

• REPAIR OPEN IN BLK WIRE BETWEEN CPP SWITCH (MANUAL TRANSAXLE MODELS) OR PNP SWITCH (AUTOMATIC TRANSAXLE MODELS) AND PCM.

• CHECK FOR A POOR CONNECTION AT PCM. IF OK, REPLACE PCM.

CLEAR DTCs AND REPEAT OBD SYSTEM CHECK AFTER MAKING REPAIRS.

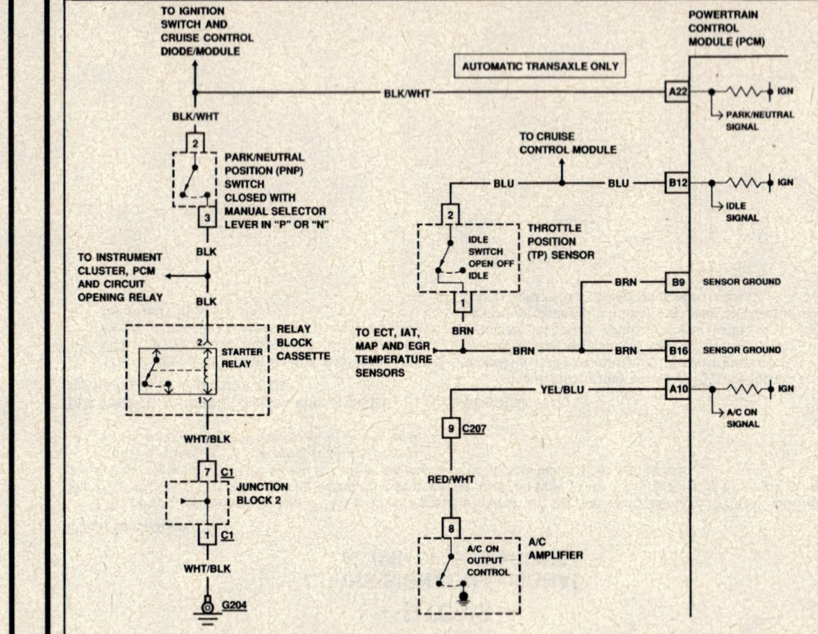

DTC 51
(Page 1 of 4)
SWITCH CONDITION SIGNAL
1.8L (VIN 8) "S" CARLINE

Circuit Description:

The powertrain control module (PCM) receives a signal whenever the following occurs: (1) the accelerator pedal is depressed (indicating that the throttle lever is in the off idle position (2) if the vehicle is equipped with air conditioning; the A/C switch is depressed with the blower speed selector switch in any position except "OFF" (indicating that A/C operation has been requested (3) or if the vehicle is equipped with an automatic transaxle; the manual selector lever is in the "R," "D," "2" or "L" positions (indicating that the transaxle is in another gear besides park or neutral).

DTC 51 indicates that the above mentioned systems are functioning normally. If DTC 51 is not indicated when the accelerator pedal is depressed, A/C is ON or with the manual selector lever out of "P" or "N," with DLC terminals "TE2" and "TE1" jumpered to ground a fault is present. If DTC 51 is indicated without the accelerator pedal being depressed, without the A/C switch or blower switch ON or with the manual selector lever in "P" or "N," a fault is also present.

Test Description: Number(s) below refer to circled number(s) on the diagnostic chart.

1. This determines if the systems are functioning normally in the test diagnostic mode.

2. This checks to see if the TP sensor circuit is operating normally.

3. This checks for a short to ground in the A/C circuit, an open in the PNP switch circuit, an open in the TP sensor circuit or for a faulty PCM or A/C amplifier.

1.8L (VIN 8) ENGINE — DIAGNOSTIC TROUBLE CODE CHARTS — 1993 PRIZM

DTC 51
(Page 1 of 4)
SWITCH CONDITION SIGNAL
1.8L (VIN 8) "S" CARLINE

[!] IMPORTANT: MAKE SURE THAT THE OBD SYSTEM CHECK HAS BEEN PERFORMED BEFORE CONTINUING DIAGNOSIS.

FROM OBD SYSTEM CHECK

(1)
- TURN IGNITION SWITCH TO "LOCK."
- CLEAR DTCs.
- CONNECT A JUMPER FROM DLC TERMINAL "TE2" TO TERMINAL "E1".
- TURN IGNITION SWITCH TO "ON."
  MIL SHOULD BE FLASHING ON AND OFF RAPIDLY.
- START AND ROAD TEST VEHICLE FOR 1 MINUTE ABOVE 16 KPH (10 MPH).
- LEAVE ENGINE RUNNING AT IDLE.
- CONNECT A JUMPER FROM DLC TERMINAL "TE1" TO TERMINAL "E1".
- NOTE MIL.
  SHOULD FLASH ON AND OFF RAPIDLY, NOT DTC 51.
  DOES IT?

YES

NO, FLASHES DTC 51

(2)
- HOLD ACCELERATOR PEDAL DEPRESSED OFF IDLE FOR 7–10 SECONDS.
- NOTE MIL.
  SHOULD FLASH DTC 51 AT LEAST ONCE AND THEN FLASH ON AND OFF RAPIDLY.
  DOES IT?

(3)
- SHUT ENGINE OFF, BUT LEAVE IGNITION SWITCH IN "ON."
- DISCONNECT TP SENSOR CONNECTOR.
- CONNECT A DIGITAL MULTIMETER FROM TP SENSOR CONNECTOR CAVITY "2" TO GROUND.
- MEASURE VOLTAGE.

MORE THAN 10 VOLTS

LESS THAN 10 VOLTS

- CONNECT A DIGITAL MULTIMETER FROM TP SENSOR TERMINAL "1" TO TERMINAL "2" (TP SENSOR SIDE).
- MEASURE RESISTANCE.

- CHECK FOR A POOR CONNECTION AT PCM.
- CHECK FOR AN OPEN IN BLU WIRE BETWEEN TP SENSOR AND PCM. IF OK, REPLACE PCM.

LESS THAN INFINITE

INFINITE

ADJUST/REPLACE TP SENSOR.

A

B

CLEAR DTCs AND REPEAT OBD SYSTEM CHECK AFTER MAKING REPAIRS.

1.8L (VIN 8) ENGINE — DIAGNOSTIC TROUBLE CODE CHARTS — 1993 PRIZM

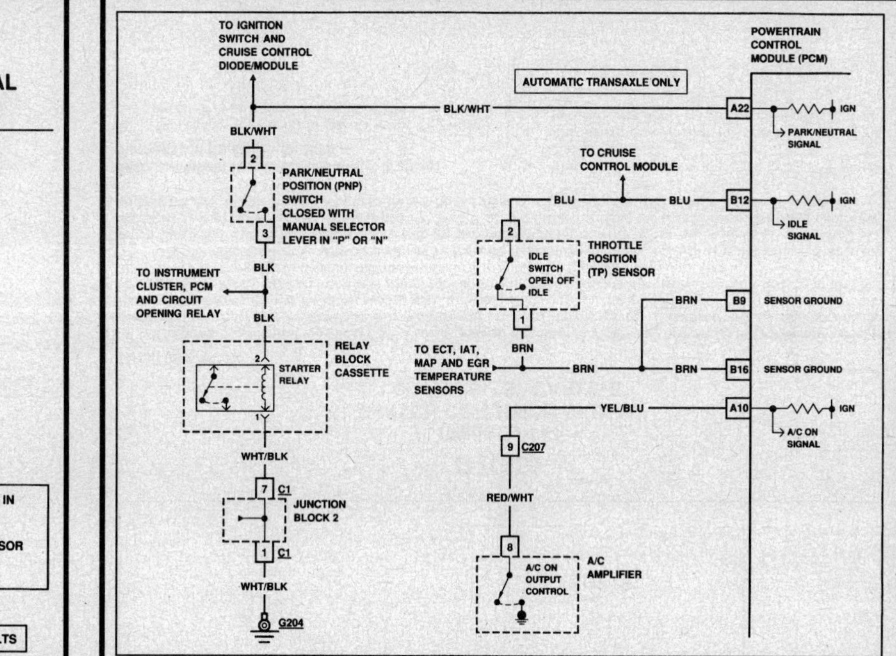

DTC 51
(Page 2 of 4)
SWITCH CONDITION SIGNAL
1.8L (VIN 8) "S" CARLINE

Circuit Description:

The powertrain control module (PCM) receives a signal whenever the following occurs: (1) the accelerator pedal is depressed (indicating that the throttle lever is in the off idle position (2) if the vehicle is equipped with air conditioning; the A/C switch is depressed with the blower speed selector switch in any position except "OFF" (indicating that A/C operation has been requested (3) or if the vehicle is equipped with an automatic transaxle; the manual selector lever is in the "R," "D," "2" or "L" positions (indicating that the transaxle is in another gear besides park or neutral).

DTC 51 indicates that the above mentioned systems are functioning normally. If DTC 51 is not indicated when the accelerator pedal is depressed, A/C is ON or with the manual selector lever out of "P" or "N," with DLC terminals "TE2" and "TE1" jumpered to ground a fault is present. If DTC 51 is indicated without the accelerator pedal being depressed, without the A/C switch or blower switch ON or with the manual selector lever in "P" or "N," a fault is also present.

DTC 51
(Page 2 of 4)
SWITCH CONDITION SIGNAL
1.8L (VIN 8) "S" CARLINE

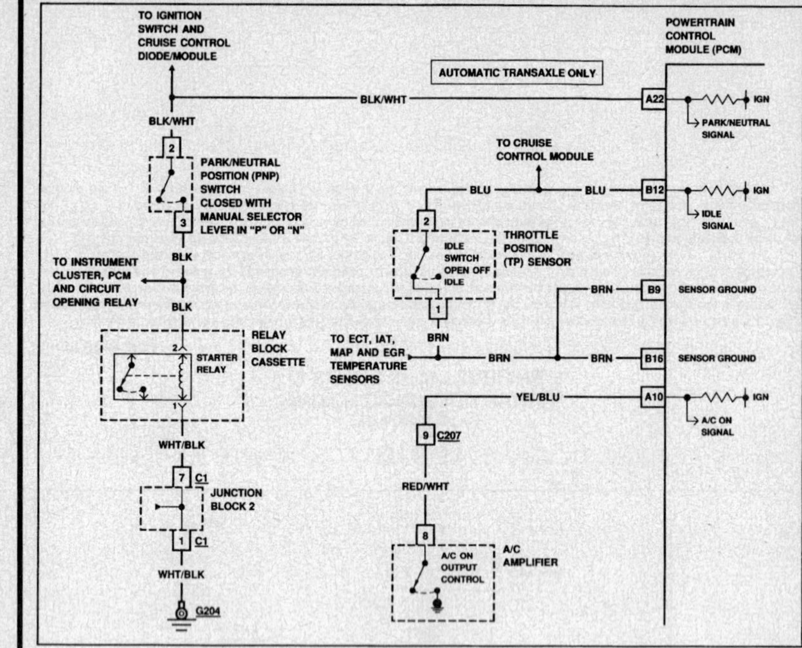

DTC 51
(Page 3 of 4)
SWITCH CONDITION SIGNAL
1.8L (VIN 8) "S" CARLINE

Circuit Description:

The powertrain control module (PCM) receives a signal whenever the following occurs: (1) the accelerator pedal is depressed (indicating that the throttle lever is in the off idle position (2) if the vehicle is equipped with air conditioning; the A/C switch is depressed with the blower speed selector switch in any position except "OFF" (indicating that A/C operation has been requested (3) or if the vehicle is equipped with an automatic transaxle; the manual selector lever is in the "R," "D," "2" or "L" positions (indicating that the transaxle is in another gear besides park or neutral).

DTC 51 indicates that the above mentioned systems are functioning normally. If DTC 51 is not indicated when the accelerator pedal is depressed, A/C is ON or with the manual selector lever out of "P" or "N," with DLC terminals "TE2" and "TE1" jumpered to ground a fault is present. If DTC 51 is indicated without the accelerator pedal being depressed, without the A/C switch or blower switch ON or with the manual selector lever in "P" or "N," a fault is also present.

Test Description: Number(s) below refer to circled number(s) on the diagnostic chart.

4. This checks for a short to ground in the TP sensor circuit or for a faulty TP sensor or PCM.

5. This checks for a short to ground in the PNP switch circuit.

6. This determines whether the system is functioning normally or if there is a problem in the A/C circuit.

CLEAR DTCs AND REPEAT OBD SYSTEM CHECK AFTER MAKING REPAIRS.

1.8L (VIN 8) ENGINE — DIAGNOSTIC TROUBLE CODE CHARTS — 1993 PRIZM

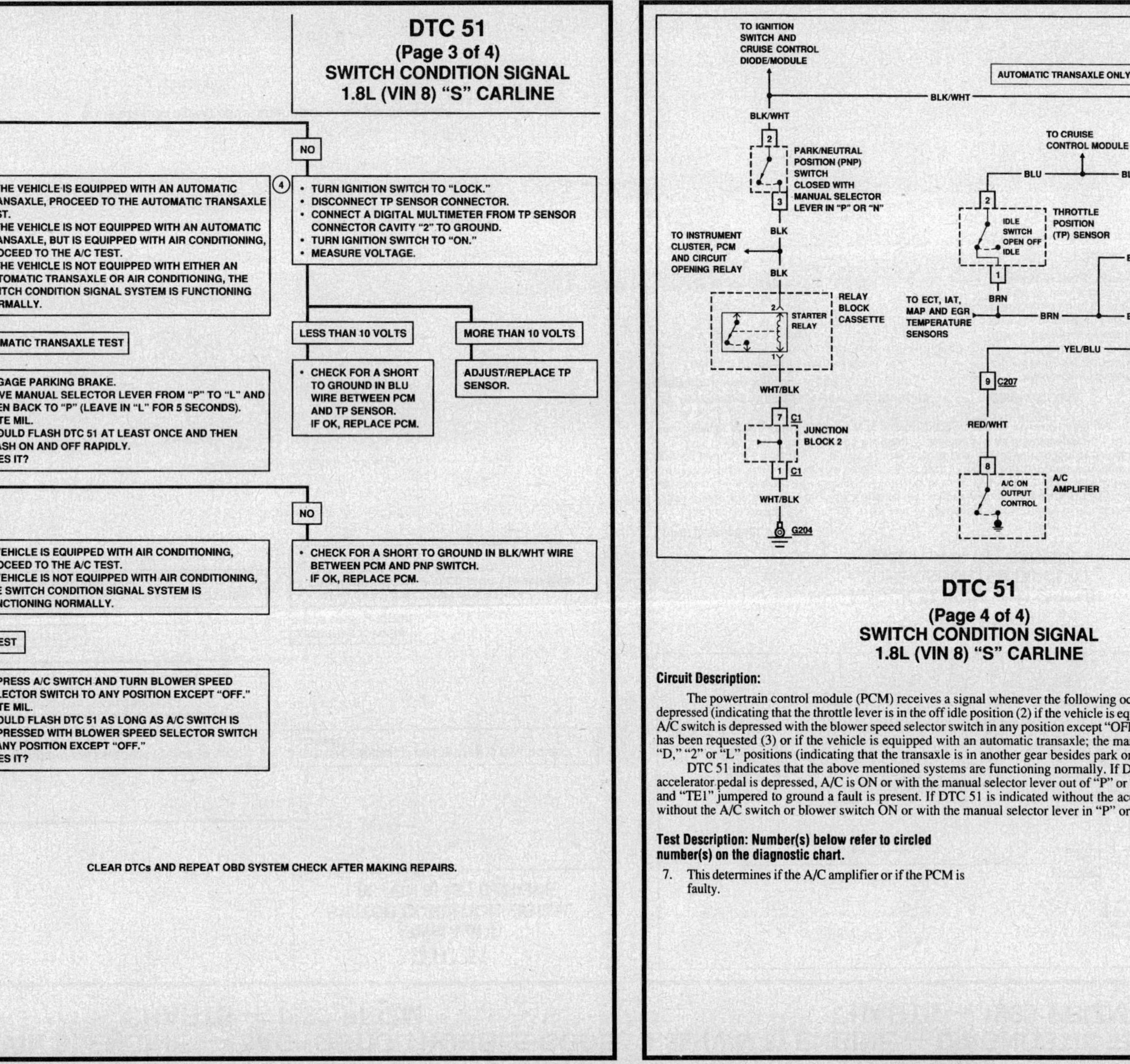

Circuit Description:

The powertrain control module (PCM) receives a signal whenever the following occurs: (1) the accelerator pedal is depressed (indicating that the throttle lever is in the off idle position (2) if the vehicle is equipped with air conditioning; the A/C switch is depressed with the blower speed selector switch in any position except "OFF" (indicating that A/C operation has been requested (3) or if the vehicle is equipped with an automatic transaxle; the manual selector lever is in the "R," "D," "2" or "L" positions (indicating that the transaxle is in another gear besides park or neutral).

DTC 51 indicates that the above mentioned systems are functioning normally. If DTC 51 is not indicated when the accelerator pedal is depressed, A/C is ON or with the manual selector lever out of "P" or "N," with DLC terminals "TE2" and "TE1" jumpered to ground a fault is present. If DTC 51 is indicated without the accelerator pedal being depressed, without the A/C switch or blower switch ON or with the manual selector lever in "P" or "N," a fault is also present.

Test Description: Number(s) below refer to circled number(s) on the diagnostic chart.

7. This determines if the A/C amplifier or if the PCM is faulty.

DTC 51
(Page 4 of 4)
SWITCH CONDITION SIGNAL
1.8L (VIN 8) "S" CARLINE

NO

(7) • TURN BLOWER SPEED SELECTOR SWITCH TO "OFF" AND A/C SWITCH TO OFF.
• DISCONNECT A/C AMPLIFIER CONNECTOR.
• CONNECT A DIGITAL MULTIMETER FROM A/C AMPLIFIER CONNECTOR CAVITY "8" TO GROUND.
• MEASURE VOLTAGE.

YES

SWITCH CONDITION SIGNAL SYSTEM IS FUNCTIONING NORMALLY.

MORE THAN 10 VOLTS

• CHECK FOR A POOR CONNECTION AT A/C AMPLIFIER. IF OK, REPLACE A/C AMPLIFIER.

LESS THAN 10 VOLTS

• CHECK FOR A POOR CONNECTION AT PCM.
• CHECK FOR AN OPEN IN YEL/BLU OR RED/WHT WIRE BETWEEN PCM AND A/C AMPLIFIER. IF OK, REPLACE PCM.

CLEAR DTCs AND REPEAT OBD SYSTEM CHECK AFTER MAKING REPAIRS.

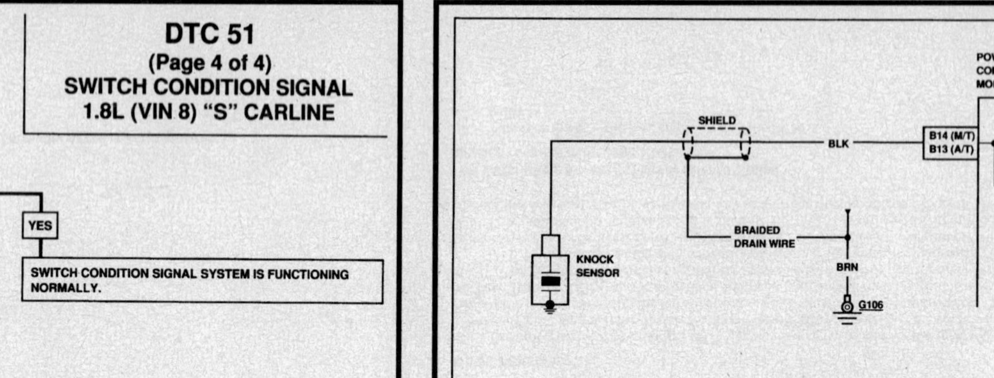

DTC 52
KNOCK SENSOR (KS) CIRCUIT
(OPEN/SHORTED CIRCUIT)
1.8L (VIN 8) "S" CARLINE

Circuit Description:

The knock sensor (KS), located in the rear of the engine block, is used to detect engine knock (abnormal vibration). When engine knock is present, the KS signals the powertrain control module (PCM) to retard ignition timing. The KS is constructed of a piezoelectric element which generates an AC voltage when it becomes deformed (vibration or engine knock detected). The KS sends this voltage signal back to the powertrain control module (PCM), which in turn retards the ignition timing. By reducing the ignition timing, the engine knock will no longer be detected.

DTC 52 will set if the following condition is met for at least 0.50 seconds:
• Open or short in the KS circuit with engine speed between 1,200 and 6,000 rpm.

Test Description: Number(s) below refer to circled number(s) on the diagnostic chart.

1. Checks KS for circuit for possible shorts.
2. Determines if KS is faulty, BLK wire is open/shorted or if the PCM is faulty.

Diagnostic Aids:

Make sure that there are no mechanical failures that are inducing engine knock.

An intermittent may be caused by a poor connection, rubbed through wire insulation, or a wire broken inside the insulation. Inspect harness connectors for backed out terminals, improper mating, broken locks, improperly formed or damaged terminals and poor terminal-to-wire connections before component replacement.

1.8L (VIN 8) ENGINE — DIAGNOSTIC TROUBLE CODE CHARTS — 1993 PRIZM

DTC 52
KNOCK SENSOR (KS) CIRCUIT (OPEN/SHORTED CIRCUIT)
1.8L (VIN 8) "S" CARLINE

[i] IMPORTANT: MAKE SURE THAT THE OBD SYSTEM CHECK HAS BEEN PERFORMED BEFORE CONTINUING DIAGNOSIS.

FROM OBD SYSTEM CHECK

① • TURN IGNITION SWITCH TO "LOCK" AND DISCONNECT PCM CONNECTOR C2.
• CONNECT A DIGITAL MULTIMETER FROM PCM CONNECTOR C2 CAVITY "B14" (MANUAL TRANSAXLE EQUIPPED VEHICLES) OR FROM CAVITY "B13" (AUTOMATIC TRANSAXLE EQUIPPED VEHICLES) TO GROUND.
• MEASURE RESISTANCE. SHOULD BE 1 M OHM OR HIGHER. IS IT?

NO — YES

② • DISCONNECT KS CONNECTOR.
• CONNECT A DIGITAL MULTIMETER FROM KS TERMINAL (KS SIDE) TO GROUND.
• MEASURE RESISTANCE. SHOULD BE 1 M OHM OR HIGHER. IS IT?

NO — YES

REPLACE KS.

• CHECK FOR AN OPEN, A SHORT TO GROUND OR A SHORT TO VOLTAGE IN BLK WIRE BETWEEN PCM AND KS.
IF OK, REPLACE KS AND REPEAT OBD SYSTEM CHECK. IF DTC 52 RESETS, REPLACE PCM.

CLEAR DTCs AND REPEAT OBD SYSTEM CHECK AFTER MAKING REPAIRS.

1.8L (VIN 8) ENGINE — DIAGNOSTIC TROUBLE CODE CHARTS — 1993 PRIZM

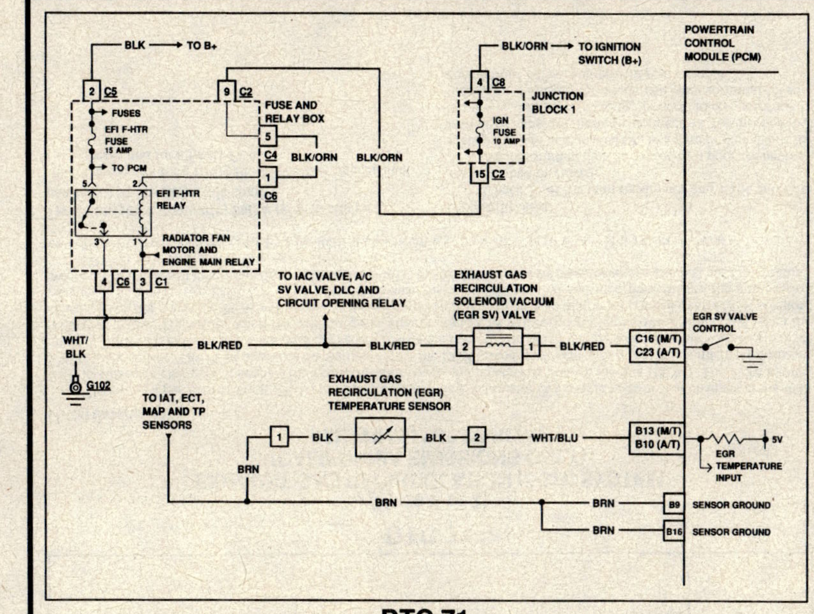

DTC 71
(Page 1 of 2)
EXHAUST GAS RECIRCULATION (EGR) SYSTEM (CALIFORNIA EMISSIONS ONLY)
1.8L (VIN 8) "S" CARLINE

Circuit Description:

The exhaust gas recirculation (EGR) temperature sensor is a thermistor (a variable resistor that changes along with EGR temperature changes) in series with a fixed resistor within the powertrain control module (PCM). The PCM applies 5 volts to the sensor. The PCM monitors the voltage across the EGR temperature sensor and converts it into a temperature reading. When the exhaust gases are cold the sensor resistance is high, and when the exhaust gases are warm the sensor resistance is low. Therefore, when the exhaust gases are cold the PCM will receive a high voltage input, and when the exhaust gases are warm the PCM will receive a low voltage input. The PCM also maintains control over a EGR solenoid vacuum (SV) valve. The EGR SV valve permits exhaust gas to be recirculated and incorporated into the combustion process. The PCM will determine the best time to energize the EGR SV valve by monitoring the exhaust gas temperature.
DTC 71 will set if the following condition is met:
• With ECT above 60°C (140°F) with EGR temperature less than 70°C (158°F) for at least 50 seconds.

Test Description: Number(s) below refer to circled number(s) on the diagnostic chart.
1. This checks the power feed to the sensor.
2. This checks the ground circuit to the sensor.
3. This checks the EGR temperature resistance vs. temperature values.
4. This checks the power feed circuit to the EGR SV valve.

Diagnostic Aids:

Check EGR vacuum modulator and EGR valve for sticking open or closed.
An intermittent may be caused by a poor connection, rubbed through wire insulation, or a wire broken inside the insulation. Inspect harness connectors for backed out terminals, improper mating, broken locks, improperly formed or damaged terminals and poor terminal-to-wire connections before component replacement.

1.8L (VIN 8) ENGINE — DIAGNOSTIC TROUBLE CODE CHARTS — 1993 PRIZM

1.8L (VIN 8) ENGINE — DIAGNOSTIC TROUBLE CODE CHARTS — 1993 PRIZM

DTC 71
(Page 1 of 2)
EXHAUST GAS RECIRCULATION (EGR) SYSTEM
(CALIFORNIA EMISSIONS ONLY)
1.8L (VIN 8) "S" CARLINE

⚠ IMPORTANT: MAKE SURE THAT THE OBD SYSTEM CHECK HAS BEEN PERFORMED BEFORE CONTINUING DIAGNOSIS.

FROM OBD SYSTEM CHECK

① • TURN IGNITION SWITCH TO "LOCK."
• DISCONNECT EGR TEMPERATURE SENSOR CONNECTOR.
• CONNECT A DIGITAL MULTIMETER FROM EGR TEMPERATURE SENSOR CONNECTOR CAVITY "2" TO GROUND.
• TURN IGNITION SWITCH TO "ON."
• MEASURE VOLTAGE.
SHOULD BE 4–6 VOLTS.
IS IT?

YES / NO

NO → • CHECK FOR A POOR CONNECTION AT PCM.
• CHECK FOR AN OPEN, A SHORT TO VOLTAGE OR SHORT TO GROUND IN WHT/BLU WIRE BETWEEN PCM AND EGR TEMPERATURE SENSOR.
IF OK, REPLACE PCM.

② • CONNECT A TEST LIGHT FROM EGR TEMPERATURE SENSOR CONNECTOR CAVITY "1" TO B+.
TEST LIGHT SHOULD LIGHT.
DOES IT?

YES / NO

NO → REPAIR OPEN IN BRN WIRE BETWEEN EGR TEMPERATURE SENSOR AND PCM.

③ • REMOVE EGR TEMPERATURE SENSOR FROM EGR VALVE.
• MEASURE RESISTANCE ACROSS SENSOR.
RESISTANCE SHOULD BE: 69–88 K OHMS AT 50°C (122°F)
11–15 K OHMS AT 100°C (212°F)
2–4 K OHMS AT 150°C (302°F)
IS IT?

YES / NO

NO → REPLACE EGR TEMPERATURE SENSOR.

④ • BACKPROBE EGR SV VALVE CONNECTOR WITH A DIGITAL MULTIMETER FROM CAVITY "2" TO GROUND.
• MEASURE VOLTAGE.
SHOULD BE MORE THAN 10 VOLTS.
IS IT?

A

CLEAR DTCs AND REPEAT OBD SYSTEM CHECK AFTER MAKING REPAIRS.

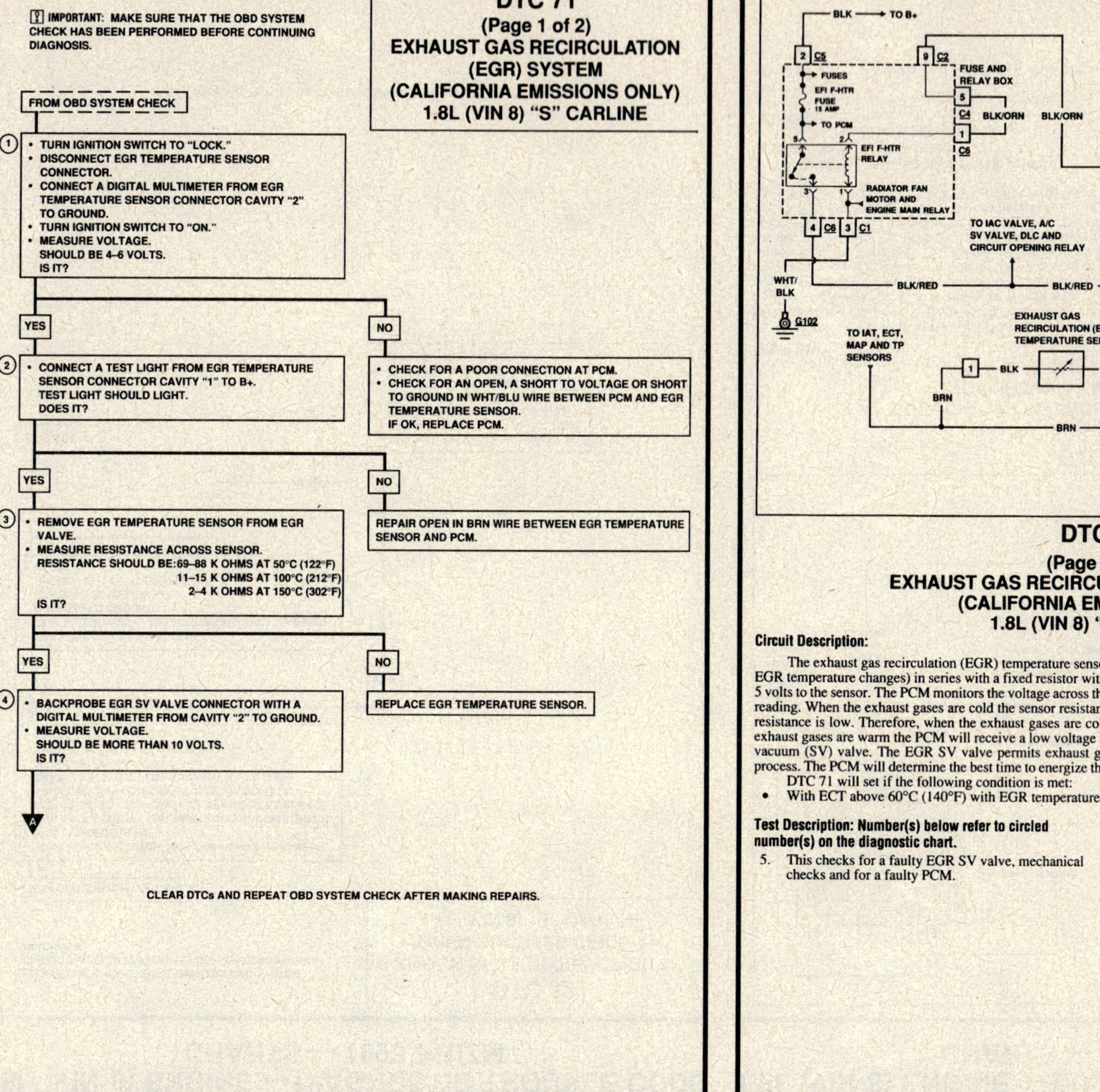

DTC 71
(Page 2 of 2)
EXHAUST GAS RECIRCULATION (EGR) SYSTEM
(CALIFORNIA EMISSIONS ONLY)
1.8L (VIN 8) "S" CARLINE

Circuit Description:

The exhaust gas recirculation (EGR) temperature sensor is a thermistor (a variable resistor that changes along with EGR temperature changes) in series with a fixed resistor within the powertrain control module (PCM). The PCM applies 5 volts to the sensor. The PCM monitors the voltage across the EGR temperature sensor and converts it into a temperature reading. When the exhaust gases are cold the sensor resistance is high, and when the exhaust gases are warm the sensor resistance is low. Therefore, when the exhaust gases are cold the PCM will receive a high voltage input, and when the exhaust gases are warm the PCM will receive a low voltage input. The PCM also maintains control over a EGR solenoid vacuum (SV) valve. The EGR SV valve permits exhaust gas to be recirculated and incorporated into the combustion process. The PCM will determine the best time to energize the EGR SV valve by monitoring the exhaust gas temperature.
DTC 71 will set if the following condition is met:
• With ECT above 60°C (140°F) with EGR temperature less than 70°C (158°F) for at least 50 seconds.

Test Description: Number(s) below refer to circled number(s) on the diagnostic chart.
5. This checks for a faulty EGR SV valve, mechanical checks and for a faulty PCM.

Diagnostic Aids:

Check EGR vacuum modulator and EGR valve for sticking open or closed.

An intermittent may be caused by a poor connection, rubbed through wire insulation, or a wire broken inside the insulation. Inspect harness connectors for backed out terminals, improper mating, broken locks, improperly formed or damaged terminals and poor terminal-to-wire connections before component replacement.

1.8L (VIN 8) ENGINE — DIAGNOSTIC TROUBLE CODE CHARTS — 1993 PRIZM

1.8L (VIN 8) ENGINE — C-CHARTS — 1993 PRIZM

DTC 71
(Page 2 of 2)
EXHAUST GAS RECIRCULATION
(EGR) SYSTEM
(CALIFORNIA EMISSION ONLY)
1.8L (VIN 8) "S" CARLINE

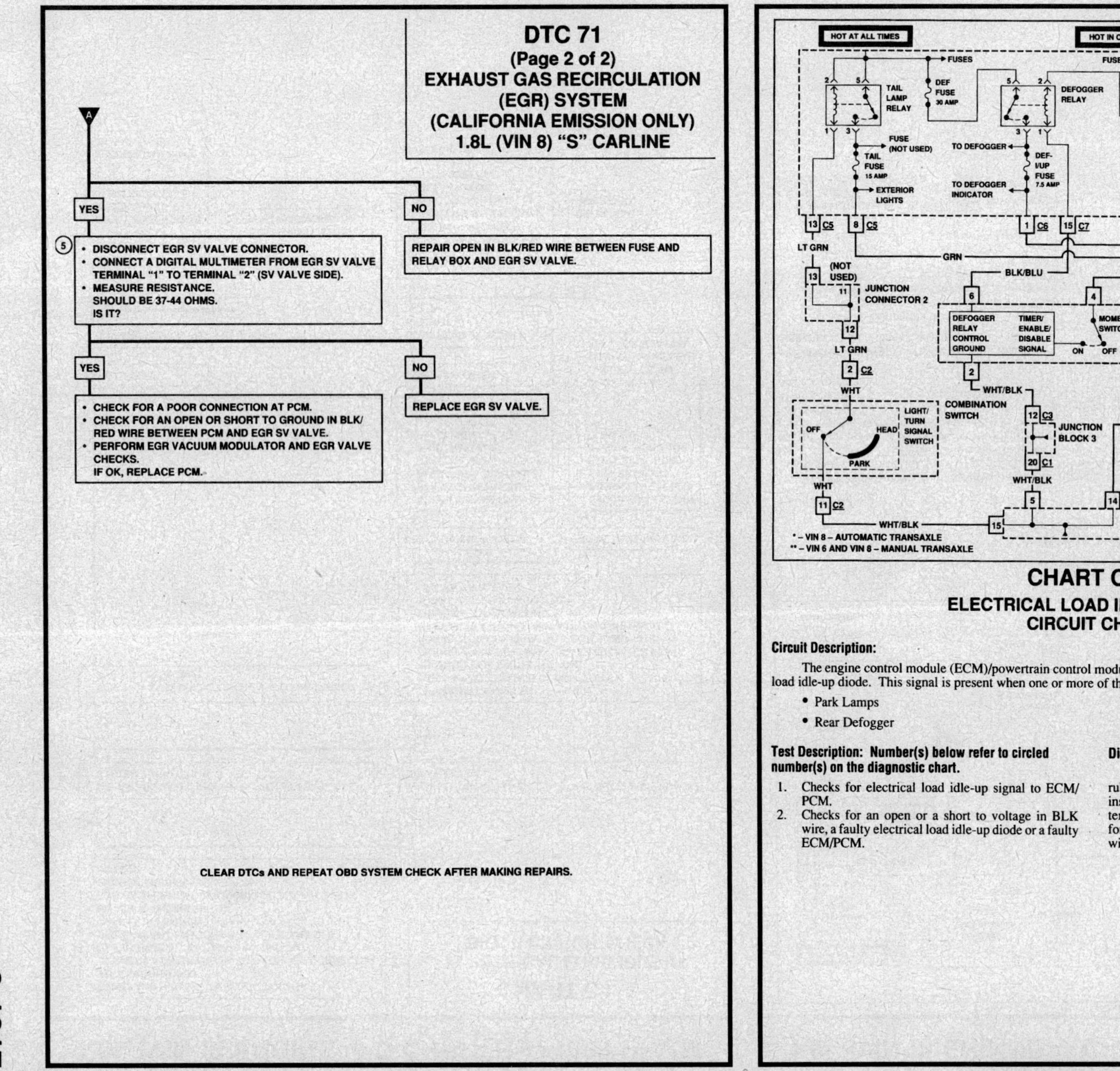

CHART C1-A
ELECTRICAL LOAD IDLE-UP DIODE
CIRCUIT CHECK

Circuit Description:

The engine control module (ECM)/powertrain control module (PCM) receives an idle-up signal from the electrical load idle-up diode. This signal is present when one or more of the following systems is turned on or operated.

• Park Lamps
• Rear Defogger

Test Description: Number(s) below refer to circled number(s) on the diagnostic chart.

1. Checks for electrical load idle-up signal to ECM/ PCM.
2. Checks for an open or a short to voltage in BLK wire, a faulty electrical load idle-up diode or a faulty ECM/PCM.

Diagnostic Aids:

An intermittent may be caused by a poor connection rubbed through wire insulation or a wire broken inside the insulation. Inspect the harness connector for backed out terminals, improper mating, broken lock, improperly formed or damaged terminals and for poor terminal-to-wire connection before component replacement.

1.8L (VIN 8) ENGINE — C-CHARTS — 1993 PRIZM

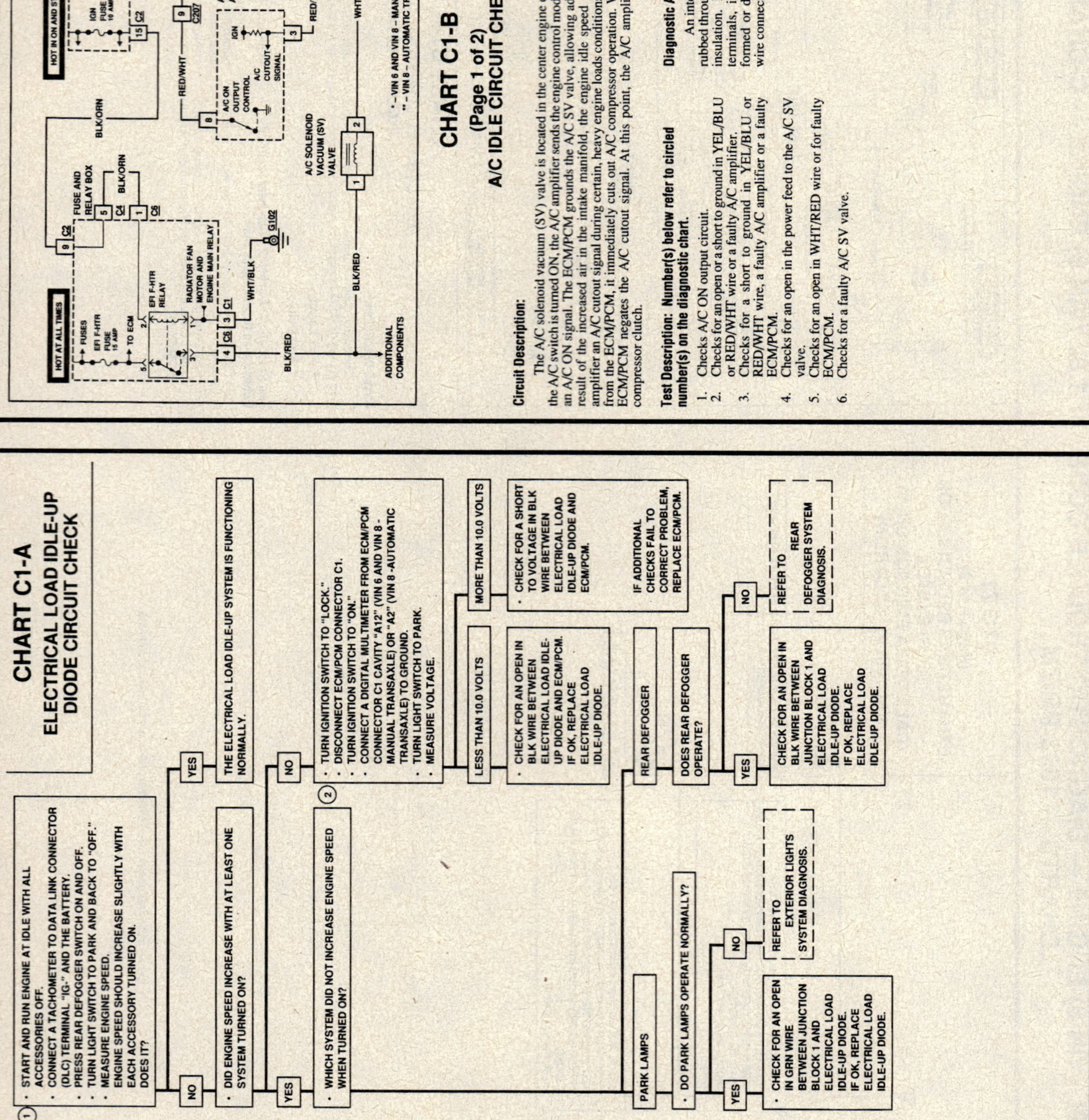

CHART C1-B
(Page 1 of 2)
A/C IDLE CIRCUIT CHECK

Circuit Description:

The A/C solenoid vacuum (SV) valve is located in the center engine compartment mounted to the bulkhead. When the A/C switch is turned ON, the A/C amplifier sends the engine control module (ECM)/powertrain control module (PCM) an A/C ON signal. The ECM/PCM grounds the A/C SV valve, allowing additional air to enter the intake manifold. As a result of the increased air in the intake manifold, the engine idle speed is increased. The ECM/PCM sends the A/C amplifier an A/C cutout signal during certain, heavy engine loads conditions. When the A/C amplifier receives this signal from the ECM/PCM, it immediately cuts out A/C compressor operation. When the engine returns to a lighter load, the ECM/PCM negates the A/C cutout signal. At this point, the A/C amplifier can resume normal cycling of the A/C compressor clutch.

Test Description: Number(s) below refer to circled number(s) on the diagnostic chart.

1. Checks A/C ON output circuit.
2. Checks for an open or a short to ground in YEL/BLU or RED/WHT wire or a faulty A/C amplifier.
3. Checks for a short to ground in YEL/BLU or RED/WHT wire, a faulty A/C amplifier or a faulty ECM/PCM.
4. Checks for an open in the power feed to the A/C SV valve.
5. Checks for an open in WHT/RED wire or for faulty ECM/PCM.
6. Checks for a faulty A/C SV valve.

Diagnostic Aids:

An intermittent may be caused by a poor connection rubbed through wire insulation or a wire broken inside the insulation. Inspect the harness connector for backed out terminals, improper mating, broken lock, improperly formed or damaged terminals and for poor terminal-to-wire connection before component replacement.

1.8L (VIN 8) ENGINE — C-CHARTS — 1993 PRIZM

CHART C1-A
ELECTRICAL LOAD IDLE-UP DIODE CIRCUIT CHECK

(1)
- START AND RUN ENGINE AT IDLE WITH ALL ACCESSORIES OFF.
- CONNECT A TACHOMETER TO DATA LINK CONNECTOR (DLC) TERMINAL "IG-" AND THE BATTERY.
- PRESS REAR DEFOGGER SWITCH ON AND OFF.
- TURN LIGHT SWITCH TO PARK AND BACK TO "OFF."
- MEASURE ENGINE SPEED.
- ENGINE SPEED SHOULD INCREASE SLIGHTLY WITH EACH ACCESSORY TURNED ON.

DOES IT?

YES → THE ELECTRICAL LOAD IDLE-UP SYSTEM IS FUNCTIONING NORMALLY.

NO → DID ENGINE SPEED INCREASE WITH AT LEAST ONE SYSTEM TURNED ON?

- **YES** → WHICH SYSTEM DID NOT INCREASE ENGINE SPEED WHEN TURNED ON?
- **NO** → (2)

(2)
- TURN IGNITION SWITCH TO "LOCK."
- DISCONNECT ECM/PCM CONNECTOR C1.
- TURN IGNITION SWITCH TO "ON."
- CONNECT A DIGITAL MULTIMETER FROM ECM/PCM CONNECTOR C1 CAVITY "A12" (VIN 6 AND VIN 8 - MANUAL TRANSAXLE) OR "A2" (VIN 8 -AUTOMATIC TRANSAXLE) TO GROUND.
- TURN LIGHT SWITCH TO PARK.
- MEASURE VOLTAGE.

LESS THAN 10.0 VOLTS →
- CHECK FOR AN OPEN IN BLK WIRE BETWEEN ELECTRICAL LOAD IDLE-UP DIODE AND ECM/PCM. IF OK, REPLACE ELECTRICAL LOAD IDLE-UP DIODE.

MORE THAN 10.0 VOLTS →
- CHECK FOR A SHORT TO VOLTAGE IN BLK WIRE BETWEEN ELECTRICAL LOAD IDLE-UP DIODE AND ECM/PCM.

IF ADDITIONAL CHECKS FAIL TO CORRECT PROBLEM, REPLACE ECM/PCM.

REAR DEFOGGER → DOES REAR DEFOGGER OPERATE?
- **YES** → CHECK FOR AN OPEN IN BLK WIRE BETWEEN JUNCTION BLOCK 1 AND ELECTRICAL LOAD IDLE-UP DIODE. IF OK, REPLACE ELECTRICAL LOAD IDLE-UP DIODE.
- **NO** → REFER TO REAR DEFOGGER SYSTEM DIAGNOSIS.

PARK LAMPS → DO PARK LAMPS OPERATE NORMALLY?
- **YES** → CHECK FOR AN OPEN IN GRN WIRE BETWEEN JUNCTION BLOCK 1 AND ELECTRICAL LOAD IDLE-UP DIODE. IF OK, REPLACE ELECTRICAL LOAD IDLE-UP DIODE.
- **NO** → REFER TO EXTERIOR LIGHTS SYSTEM DIAGNOSIS.

1.8L (VIN 8) ENGINE — C-CHARTS — 1993 PRIZM

CHART C1-B
(Page 1 of 2)
A/C IDLE CIRCUIT CHECK

①
- TURN IGNITION SWITCH TO "ON."
- BACKPROBE ECM/PCM CONNECTOR C1 WITH A DIGITAL MULTIMETER FROM CAVITY "A10" TO GROUND.
- MEASURE VOLTAGE.

MORE THAN 10.0 VOLTS

②
- START ENGINE.
- TURN BLOWER SPEED SELECTOR SWITCH TO "LOW."
- PRESS A/C SWITCH TO ON.
- BACKPROBE ECM/PCM CONNECTOR C1 WITH A DIGITAL MULTIMETER FROM CAVITY "A10" TO GROUND.
- MEASURE VOLTAGE.

LESS THAN 10.0 VOLTS

③
- TURN IGNITION SWITCH TO "LOCK."
- DISCONNECT ECM/PCM CONNECTOR C1.
- CONNECT A DIGITAL MULTIMETER FROM ECM/PCM CONNECTOR C1 CAVITY "A10" TO B+.
- MEASURE VOLTAGE.

MORE THAN 10.0 VOLTS
- CHECK FOR A SHORT TO GROUND IN YEL/BLU OR RED/WHT WIRE BETWEEN ECM/PCM AND A/C AMPLIFIER. IF OK, REPLACE A/C AMPLIFIER.

LESS THAN 10.0 VOLTS
- REPLACE ECM/PCM.

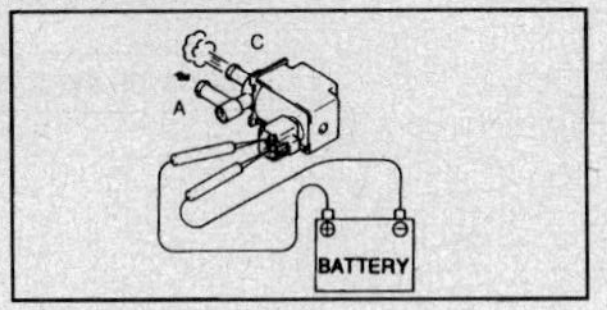

LESS THAN 10.0 VOLTS

④
- BACKPROBE A/C SV VALVE CONNECTOR WITH A DIGITAL MULTIMETER FROM CAVITY "1" TO GROUND.
- MEASURE VOLTAGE.

MORE THAN 10.0 VOLTS
- CHECK FOR AN OPEN IN YEL/BLU OR RED/WHT WIRE BETWEEN A/C AMPLIFIER AND ECM/PCM. IF OK, REPLACE A/C AMPLIFIER.

MORE THAN 10.0 VOLTS

⑤
- DISCONNECT A/C SV VALVE CONNECTOR.
- CONNECT A DIGITAL MULTIMETER FROM CONNECTOR CAVITY "2" TO B+.
- MEASURE VOLTAGE.

LESS THAN 10.0 VOLTS
- REPAIR OPEN IN BLK/RED WIRE BETWEEN FUSE AND RELAY BOX AND A/C SV VALVE.

LESS THAN 10.0 VOLTS

⑥
- TURN IGNITION SWITCH TO "LOCK."
- DISCONNECT A/C SV VALVE CONNECTOR.
- REMOVE A/C SV VALVE.
- CONNECT A FUSED JUMPER FROM A/C SV VALVE TERMINAL "1" TO B+.
- CONNECT A FUSED JUMPER FROM A/C SV VALVE TERMINAL "2" TO GROUND.
- BLOW INTO PIPE "A" AND MAKE SURE AIR COMES OUT OF PIPE "C" (REFER TO FIGURE A) DOES IT?

MORE THAN 10.0 VOLTS
- CHECK FOR AN OPEN IN WHT/RED WIRE BETWEEN A/C SV VALVE AND ECM/PCM. IF OK, REPLACE ECM/PCM.

▲ A

Figure A A/C SV Valve Check

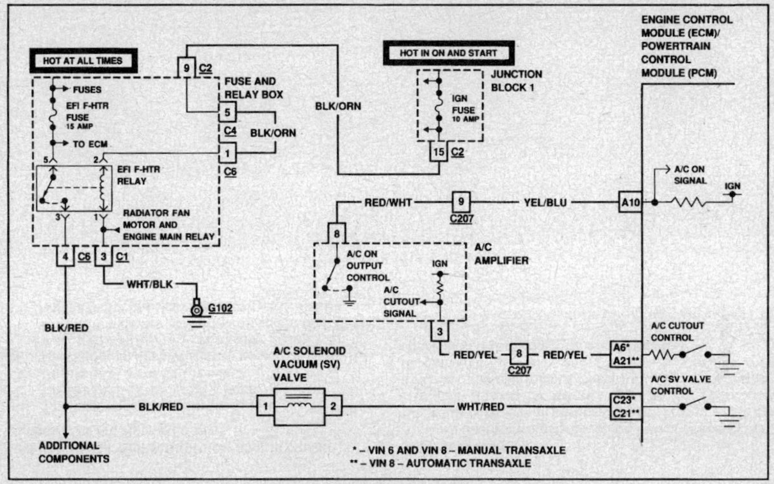

1.8L (VIN 8) ENGINE — C-CHARTS — 1993 PRIZM

CHART C1-B
(Page 2 of 2)
A/C IDLE CIRCUIT CHECK

Circuit Description:

The A/C solenoid vacuum (SV) valve is located in the center engine compartment mounted to the bulkhead. When the A/C switch is turned ON, the A/C amplifier sends the engine control module (ECM)/powertrain control module (PCM) an A/C ON signal. The ECM/PCM grounds the A/C SV valve, allowing additional air to enter the intake manifold. As a result of the increased air in the intake manifold, the engine idle speed is increased. The ECM/PCM sends the A/C amplifier an A/C cutout signal during certain, heavy engine loads conditions. When the A/C amplifier receives this signal from the ECM/PCM, it immediately cuts out A/C compressor operation. When the engine returns to a lighter load, the ECM/PCM negates the A/C cutout signal. At this point, the A/C amplifier can resume normal cycling of the A/C compressor clutch.

Test Description: Number(s) below refer to circled number(s) on the diagnostic chart.

7. Checks A/C cutout circuit.
8. Checks for an open in RED/YEL wire.
9. Checks for an open or a short to ground in RED/YEL wire, a faulty A/C amplifier or a faulty ECM/PCM.
10. Checks for a faulty ECM/PCM or a faulty A/C amplifier.

Diagnostic Aids:

An intermittent may be caused by a poor connection rubbed through wire insulation or a wire broken inside the insulation. Inspect the harness connector for backed out terminals, improper mating, broken lock, improperly formed or damaged terminals and for poor terminal-to-wire connection before component replacement.

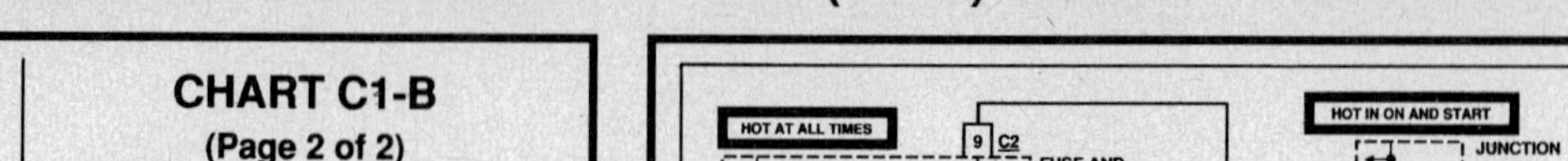

CHART C1-B
(Page 2 of 2)
A/C IDLE CIRCUIT CHECK

A

YES

NO

(7)
- RECONNECT A/C SV VALVE IN VEHICLE.
- TURN IGNITION SWITCH TO "ON."
- BACKPROBE A/C AMPLIFIER CONNECTOR WITH A DIGITAL MULTIMETER FROM CAVITY "3" TO GROUND.
- MEASURE VOLTAGE.

REPLACE A/C SV VALVE.

MORE THAN 10.0 VOLTS

LESS THAN 10.0 VOLTS

(8)
- BACKPROBE ECM/PCM CONNECTOR C1 WITH A DIGITAL MULTIMETER FROM CAVITY "A6" (VIN 6 AND VIN 8 – MANUAL TRANSAXLE) OR "A21" (VIN 8 – AUTOMATIC TRANSAXLE) TO GROUND.
- MEASURE VOLTAGE.

(9)
- TURN IGNITION SWITCH TO "LOCK."
- DISCONNECT ECM/PCM CONNECTOR C1.
- TURN IGNITION SWITCH TO "ON."
- CONNECT A DIGITAL MULTIMETER FROM ECM/PCM CONNECTOR C1 CAVITY "A6" (VIN 6 AND VIN 8 – MANUAL TRANSAXLE) OR "A26" (VIN 8 – AUTOMATIC TRANSAXLE) TO GROUND.
- MEASURE VOLTAGE.

MORE THAN 10.0 VOLTS

LESS THAN 10.0 VOLTS

REPLACE ECM/PCM.

- CHECK FOR AN OPEN OR A SHORT TO GROUND IN RED/YEL WIRE BETWEEN A/C AMPLIFIER AND ECM/PCM.
  IF OK, REPLACE A/C AMPLIFIER.

MORE THAN 10.0 VOLTS

LESS THAN 10.0 VOLTS

(10)
- LEAVE MULTIMETER CONNECTED.
- ROAD TEST VEHICLE.
- DEPRESS ACCELERATOR TO WIDE OPEN THROTTLE (WOT).
- NOTICE: ONLY PERFORM THIS TEST MOMENTARILY SO VEHICLE SPEED DOES NOT EXCEED SPEED LIMIT.
- MEASURE VOLTAGE.

REPAIR OPEN IN RED/YEL WIRE BETWEEN A/C AMPLIFIER AND ECM/PCM.

MORE THAN 10.0 VOLTS

LESS THAN 10.0 VOLTS

IF ADDITIONAL CHECKS FAIL TO CORRECT THE PROBLEM, REPLACE ECM/PCM.

A/C IDLE CIRCUIT IS FUNCTIONING NORMALLY.

IF ADDITIONAL CHECKS FAIL TO CORRECT THE PROBLEM, REPLACE A/C AMPLIFIER.

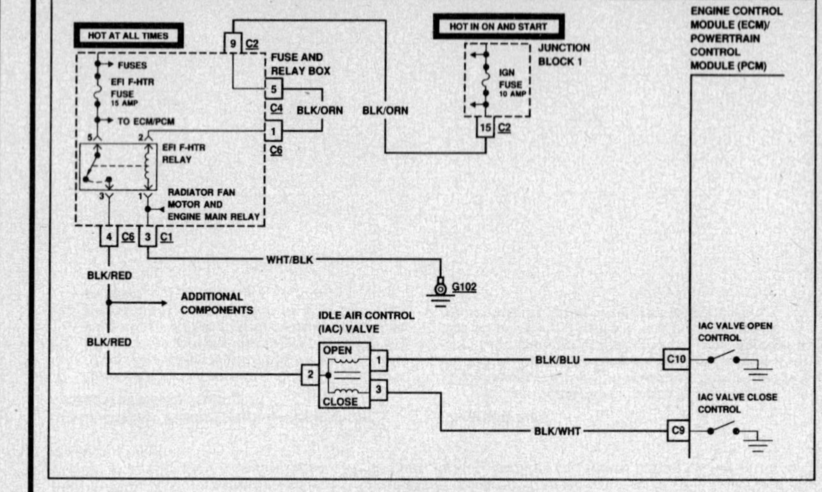

CHART C2-A
(Page 1 of 2)
IDLE AIR CONTROL (IAC) VALVE CHECK

Circuit Description:

The engine control module (ECM)/powertrain control module (PCM) controls engine idle speed with the idle air control (IAC) valve. To increase idle speed, the ECM/PCM grounds the OPEN solenoid within the IAC valve. This action opens a plate within the IAC valve, allowing air to bypass the throttle plate. The ECM/PCM regulates (fast) idle speed via the IAC valve by increasing the opening of the air bypass port. To decrease idle speed, the ECM/PCM grounds the CLOSE solenoid within the IAC valve, thus allowing less air to bypass the throttle plate. The ECM/PCM regulates (slow) idle speed via the IAC valve by decreasing the opening of the air bypass port.

Test Description: Number(s) below refer to circled number(s) on the diagnostic chart.

1. Checks for open in power feed to IAC valve.
2. Checks IAC valve CLOSE circuit.
3. Checks IAC valve OPEN circuit.
4. Checks for an open or start to ground in BLK/BLU wire, a faulty IAC valve or a faulty ECM/PCM.
5. Checks for an open or short to ground in BLK/WHT wire, a faulty IAC valve or a faulty ECM/PCM.

Diagnostic Aids:

An intermittent may be caused by a poor connection, rubbed through wire insulation or a wire broken inside the insulation. Inspect the harness connector for backed out terminals, improper mating, broken lock, improperly formed or damaged terminals and for poor terminal-to-wire connection before component replacement.

1.8L (VIN 8) ENGINE — C-CHARTS — 1993 PRIZM

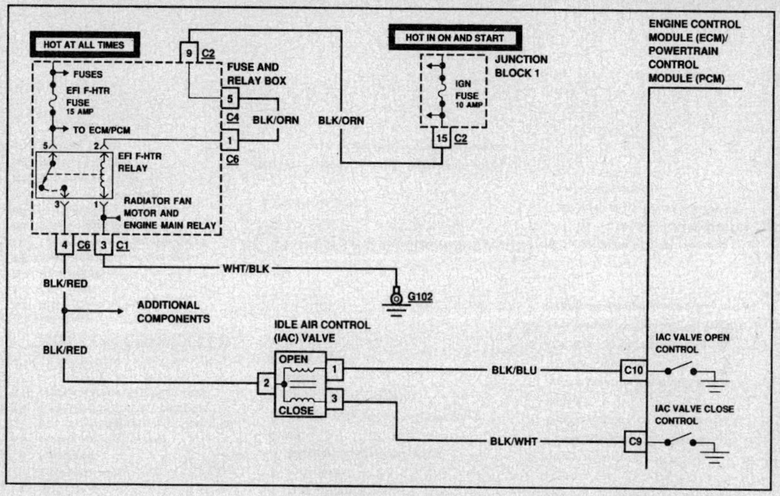

CHART C2-A
(Page 2 of 2)
IDLE AIR CONTROL (IAC) VALVE CHECK

Circuit Description:

The engine control module (ECM)/powertrain control module (PCM) controls engine idle speed with the idle air control (IAC) valve. To increase idle speed, the ECM/PCM grounds the OPEN solenoid within the IAC valve. This action opens a plate within the IAC valve, allowing air to bypass the throttle plate. The ECM/PCM regulates (fast) idle speed via the IAC valve by increasing the opening of the air bypass port. To decrease idle speed, the ECM/PCM grounds the CLOSE solenoid within the IAC valve, thus allowing less air to bypass the throttle plate. The ECM/PCM regulates (slow) idle speed via the IAC valve by decreasing the opening of the air bypass port.

Test Description: Number(s) below refer to circled number(s) on the diagnostic chart.

6. Checks that IAC valve closes.
7. Checks that IAC valve opens.

Diagnostic Aids:

An intermittent may be caused by a poor connection, rubbed through wire insulation or a wire broken inside the insulation. Inspect the harness connector for backed out terminals, improper mating, broken lock, improperly formed or damaged terminals and for poor terminal-to-wire connection before component replacement.

Diagram labels: ENGINE CONTROL MODULE (ECM)/POWERTRAIN CONTROL MODULE (PCM); IAC VALVE OPEN CONTROL; IAC VALVE CLOSE CONTROL; C10; C9; BLK/BLU; BLK/WHT; HOT IN ON AND START; JUNCTION BLOCK 1; IGN FUSE 10 AMP; 15 C2; BLK/ORN; HOT AT ALL TIMES; FUSE AND RELAY BOX; 9 C2; 5; C4; 1; C6; BLK/ORN; BLK/ORN; G102; FUSES; EFI F-HTR FUSE 15 AMP; TO ECM/PCM; EFI F-HTR RELAY; RADIATOR FAN MOTOR AND ENGINE MAIN RELAY; 4 C6; 3 C1; BLK/RED; BLK/RED; WHT/BLK; IDLE AIR CONTROL (IAC) VALVE; OPEN; CLOSE; 1; 2; 3; ADDITIONAL COMPONENTS.

1.8L (VIN 8) ENGINE — C-CHARTS — 1993 PRIZM

CHART C2-A
(Page 1 of 2)
IDLE AIR CONTROL (IAC) VALVE CHECK

(1)
- TURN IGNITION SWITCH TO "ON."
- BACKPROBE IDLE AIR CONTROL (IAC) VALVE WITH A TEST LIGHT FROM CAVITY "2" TO GROUND.

TEST LIGHT SHOULD LIGHT. DOES IT?

NO → REPAIR OPEN IN BLK/RED WIRE BETWEEN FUSE AND RELAY BOX AND IAC VALVE.

YES →

(2)
- BACKPROBE ECM/PCM CONNECTOR C3 WITH A DIGITAL MULTIMETER FROM CAVITY "C10" TO GROUND.
- MEASURE VOLTAGE.

MORE THAN 10.0 VOLTS →

(3)
- BACKPROBE ECM/PCM CONNECTOR C3 WITH A DIGITAL MULTIMETER FROM CAVITY "C9" TO GROUND.
- MEASURE VOLTAGE.

SHOULD BE APPROXIMATELY 0.5 VOLTS. IS IT?

(4)
- TURN IGNITION SWITCH TO "LOCK."
- DISCONNECT ECM/PCM CONNECTOR C3.
- TURN IGNITION SWITCH TO "ON."
- CONNECT A DIGITAL MULTIMETER FROM ECM/PCM CONNECTOR C3 CAVITY "C10" TO GROUND.
- MEASURE VOLTAGE.

LESS THAN 10.0 VOLTS →
- CHECK FOR AN OPEN OR A SHORT TO GROUND IN BLK/BLU WIRE BETWEEN ECM/PCM AND IAC VALVE. IF OK, REPLACE IAC VALVE.

MORE THAN 10.0 VOLTS → REPLACE ECM/PCM.

YES → (A)

NO →

(5)
- TURN IGNITION SWITCH TO "LOCK."
- DISCONNECT ECM/PCM CONNECTOR C3.
- TURN IGNITION SWITCH TO "ON."
- CONNECT A DIGITAL MULTIMETER FROM ECM/PCM CONNECTOR C3 CAVITY "C9" TO GROUND.
- MEASURE VOLTAGE.

LESS THAN 10.0 VOLTS →
- CHECK FOR AN OPEN OR A SHORT TO GROUND IN BLK/WHT WIRE BETWEEN ECM/PCM AND IAC VALVE. IF OK, REPLACE IAC VALVE.

MORE THAN 10.0 VOLTS → REPLACE ECM/PCM.

1.8L (VIN 8) ENGINE — C-CHARTS — 1993 PRIZM

1.6L (VIN 6) ENGINE — COMPONENT LOCATIONS — 1992–93 STORM

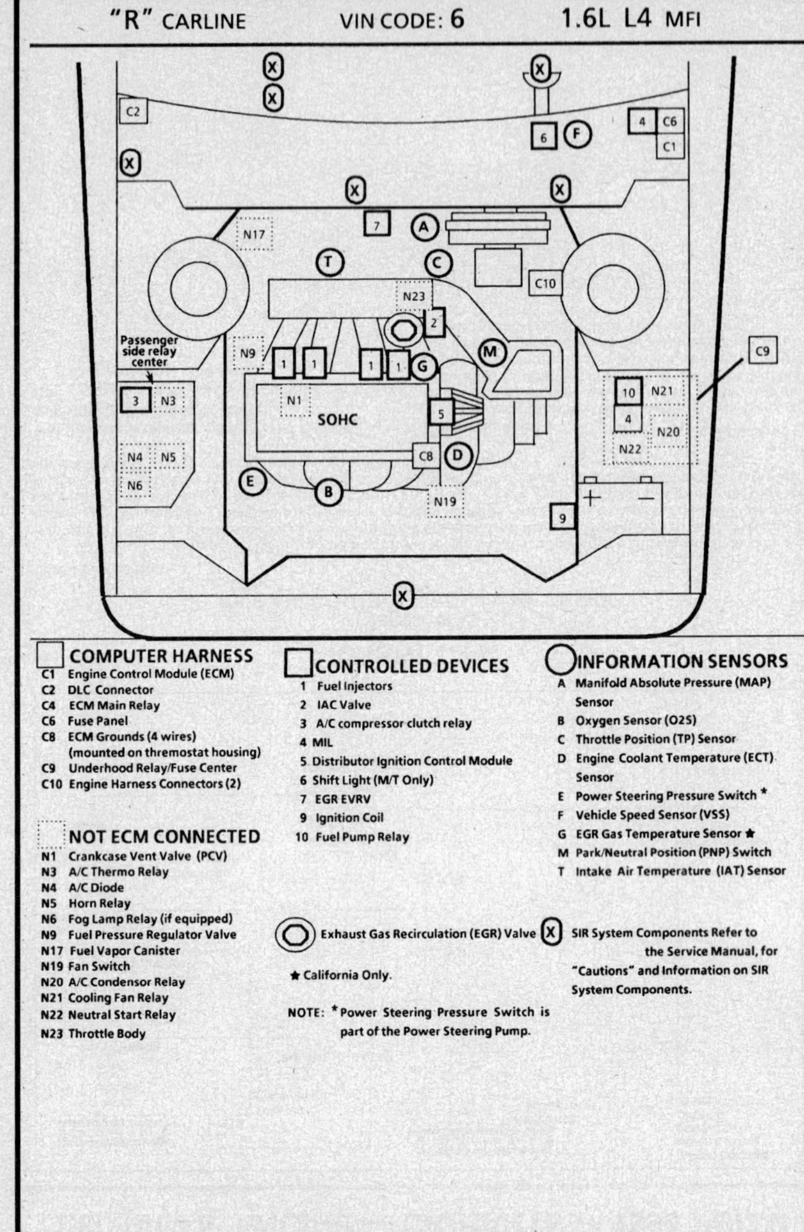

1.6L (VIN 6) ENGINE — ECM WIRING DIAGRAMS — 1992–93 STORM

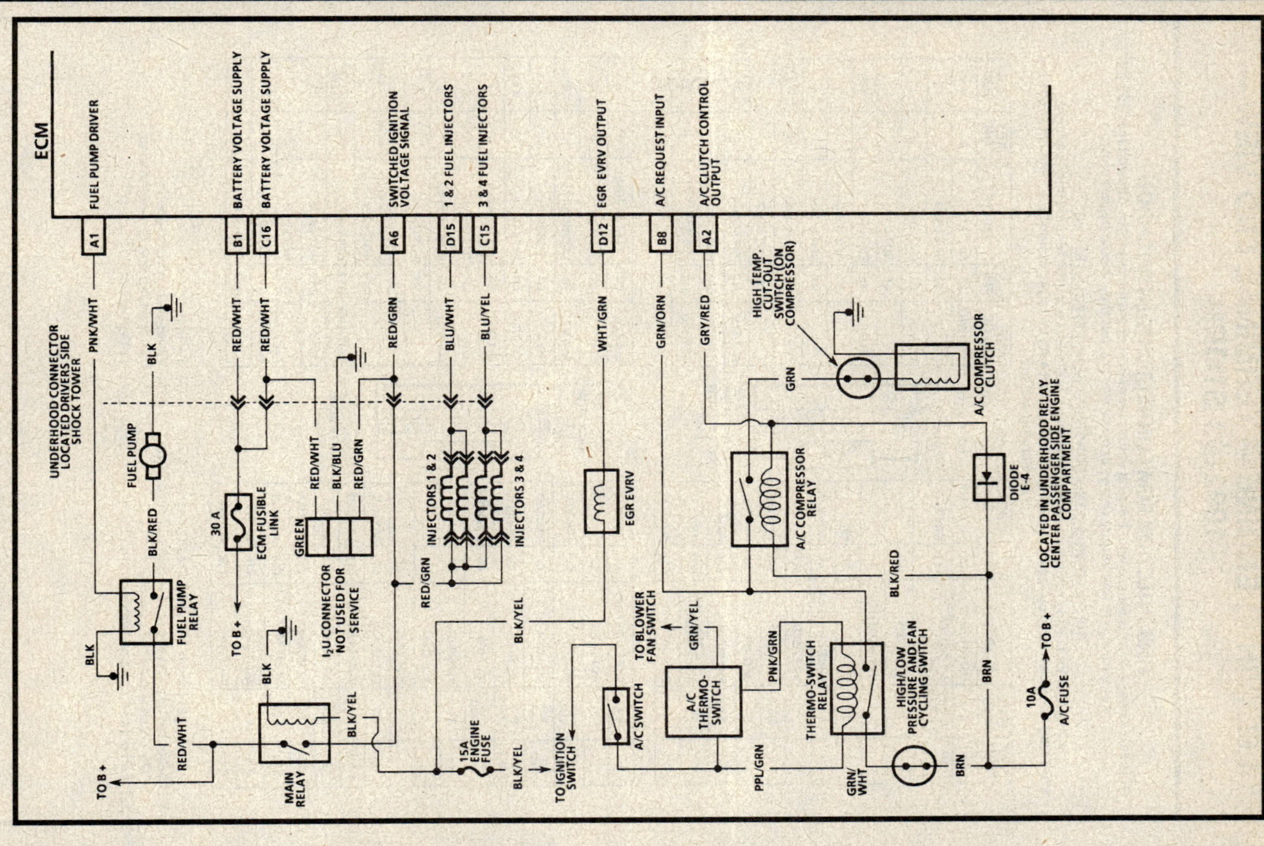

1.6L (VIN 6) ENGINE — ECM WIRING DIAGRAMS — 1992–93 STORM

1.6L (VIN 6) ENGINE — ECM WIRING DIAGRAMS — 1992–93 STORM

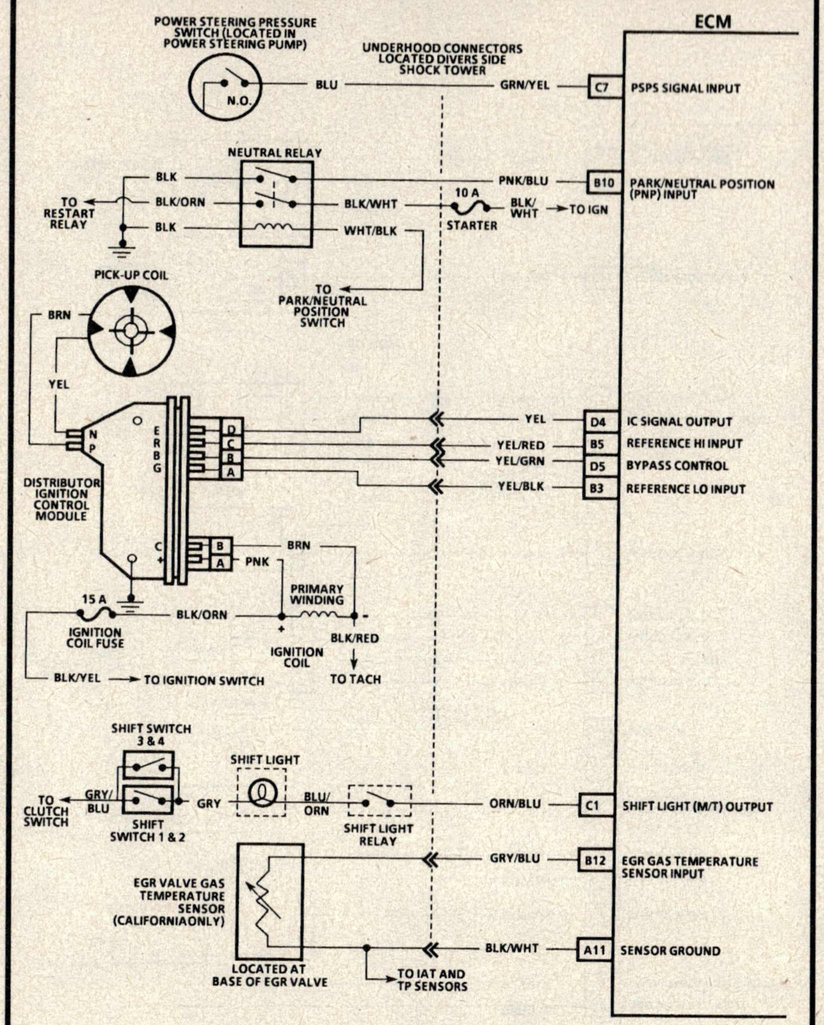

Wiring diagram labels:

POWER STEERING PRESSURE SWITCH (LOCATED IN POWER STEERING PUMP) — N.O. — BLU — GRN/YEL — C7 PSPS SIGNAL INPUT

UNDERHOOD CONNECTORS LOCATED DIVERS SIDE SHOCK TOWER — ECM

NEUTRAL RELAY — BLK — BLK/ORN — BLK — TO RESTART RELAY — BLK/WHT — WHT/BLK — PNK/BLU — 10 A — BLK/WHT TO IGN — STARTER — B10 PARK/NEUTRAL POSITION (PNP) INPUT — TO PARK/NEUTRAL POSITION SWITCH

PICK-UP COIL — BRN — YEL — DISTRIBUTOR IGNITION CONTROL MODULE — E R B G / D C B A

YEL — D4 IC SIGNAL OUTPUT
YEL/RED — B5 REFERENCE HI INPUT
YEL/GRN — D5 BYPASS CONTROL
YEL/BLK — B3 REFERENCE LO INPUT

PNK — BRN — PRIMARY WINDING — 15 A — BLK/ORN — IGNITION COIL FUSE — IGNITION COIL — BLK/RED TO TACH — BLK/YEL TO IGNITION SWITCH

SHIFT SWITCH 3 & 4 — SHIFT LIGHT — GRY/BLU — GRY — BLU/ORN — SHIFT SWITCH 1 & 2 — SHIFT LIGHT RELAY — TO CLUTCH SWITCH — ORN/BLU — C1 SHIFT LIGHT (M/T) OUTPUT

GRY/BLU — B12 EGR GAS TEMPERATURE SENSOR INPUT

EGR VALVE GAS TEMPERATURE SENSOR (CALIFORNIA ONLY) — LOCATED AT BASE OF EGR VALVE — BLK/WHT — A11 SENSOR GROUND — TO IAT AND TP SENSORS

1.6L (VIN 6) ENGINE — TERMINAL END VIEW — 1992–93 STORM

FUEL INJECTION ECM CONNECTOR IDENTIFICATION

This ECM voltage chart is for use with a digital voltmeter to further aid in diagnosis. The voltages you get may vary due to low battery charge or other reasons, but they should be very close.

__THE FOLLOWING CONDITIONS MUST BE MET BEFORE TESTING:__

● Engine at operating temperature   ● Engine idling in Closed Loop (for "Engine Run" column)   ● DLC OBD terminal not grounded ● "Scan" tool not installed ● All voltages shown "B +" indicates battery or charging voltage.

KEY "ON"	ENG. RUN	CIRCUIT	PIN	WIRE COLOR
B +	B +	FUEL PUMP	A1	PNK/WHT
0	0-ON B + -OFF	A/C CLUTCH CONTROL OUTPUT	A2	GRY/RED
		NOT USED	A3	
		NOT USED	A4	
0*	B +	MIL	A5	PPL
B +	B +	IGNITION SIGNAL	A6	RED/GRN
		NOT USED	A7	
5.0V	5.0V	SERIAL DATA	A8	ORN/BLK
B +	B +	OBD REQUEST	A9	ORN/YEL
0* - B +	B +	VSS INPUT	A10	WHT
0*	0*	ECT & MAP, EGR SENSOR GND	A11	GRY
0*	0*	ECM GROUND	A12	BLK/BLU

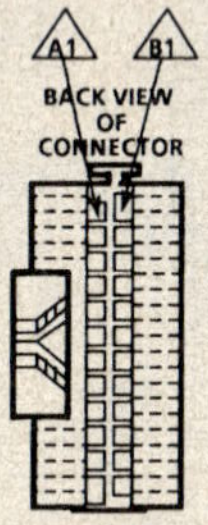

(1) marks the A1 row; (4) marks the A10 row.

KEY "ON"	ENG. RUN	CIRCUIT	PIN	WIRE COLOR
B +	B +	SHIFT LIGHT M/T	C1	ORN/BLU
		NOT USED	C2	
NOT USABLE	NOT USABLE	IAC LOW	C3	BRN/YEL
NOT USABLE	NOT USABLE	IAC HIGH	C4	BRN/RED
NOT USABLE	NOT USABLE	IAC HIGH	C5	BRN/BLK
NOT USABLE	NOT USABLE	IAC LOW	C6	BRN/WHT
B +	B +	PSPS SIGNAL	C7	GRN/YEL
		NOT USED	C8	
		NOT USED	C9	
(2)	(2)	ECT SIGNAL	C10	GRY/BLK
4.79V	(5)	MAP SIGNAL	C11	GRY/RED
(2)	(2)	IAT SIGNAL	C12	BLU/BLK
.50	.53	TP SIGNAL	C13	YEL/BLU
5.0V	5.0V	SV REF MAP & TP	C14	BLU/ORN
B +	B +	INJECTOR DRIVER 1&3	C15	BLU/YEL
B +	B +	BATTERY	C16	RED/WHT

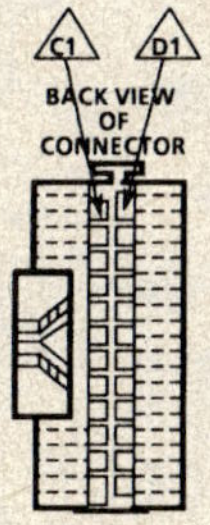

24 PIN BLACK CONNECTOR WITH YELLOW INSERT — BACK VIEW OF CONNECTOR (A1, B1)

WIRE COLOR	PIN	CIRCUIT	KEY "ON"	ENG. RUN
RED/WHT	B1	BATTERY	B +	B +
	B2	NOT USED		
YEL/BLK	B3	REF LOW	0*	0*
	B4	NOT USED		
YEL/RED	B5	REF HIGH	0*	1.0V
	B6	NOT USED		
	B7	NOT USED		
GRN/ORN	B8	A/C REQUEST	0*	8-10 ON B + OFF
	B9	NOT USED		
PNK/BLU	B10	PNP SIGNAL	0*	0* (3)
	B11	NOT USED		
GRY/BLU	B12	EGR GAS TEMP. SIGNAL	(2)	(2) (6)

24 PIN BLACK CONNECTOR WITH BLUE INSERT — BACK VIEW OF CONNECTOR (C1, D1)

WIRE COLOR	PIN	CIRCUIT	KEY "ON"	ENG. RUN
BLK/BLU	D1	ECM GROUND	0*	0*
BLK	D2	TP & IAT SENSOR GROUND	0*	0*
BLK/BLU	D3	ENGINE GROUND	0*	0*
YEL	D4	IC SIGNAL	0*	1.15
YEL/GRN	D5	BYPASS	0*	4.6V
BLK/BLU	D6	O2S GROUND	0*	0*
BLU	D7	O2S SIGNAL	.43V	.1-.9
BLK/BLU	D8	ECM GROUND	0*	0*
	D9	NOT USED		
	D10	NOT USED		
	D11	NOT USED		
WHT/GRN	D12	EGR EVRV	B +	B + -OFF 0-ON
	D13	NOT USED		
	D14	NOT USED		
BLU/WHT	D15	INJ DRIVER 2&4	B +	B +
	D16	NOT USED		

*   All voltages shown "0" should read less than .5 volt.
(1) Reads battery voltage for 2 seconds after ignition "ON" then should read 0 volts.
(2) Varies depending on temperature.
(3) B + in R, D or L.
(4) Varies with vehicle speed.
(5) Varies depending on engine load.
(6) California only.

1.8L (VIN 8) ENGINE — COMPONENT LOCATIONS —

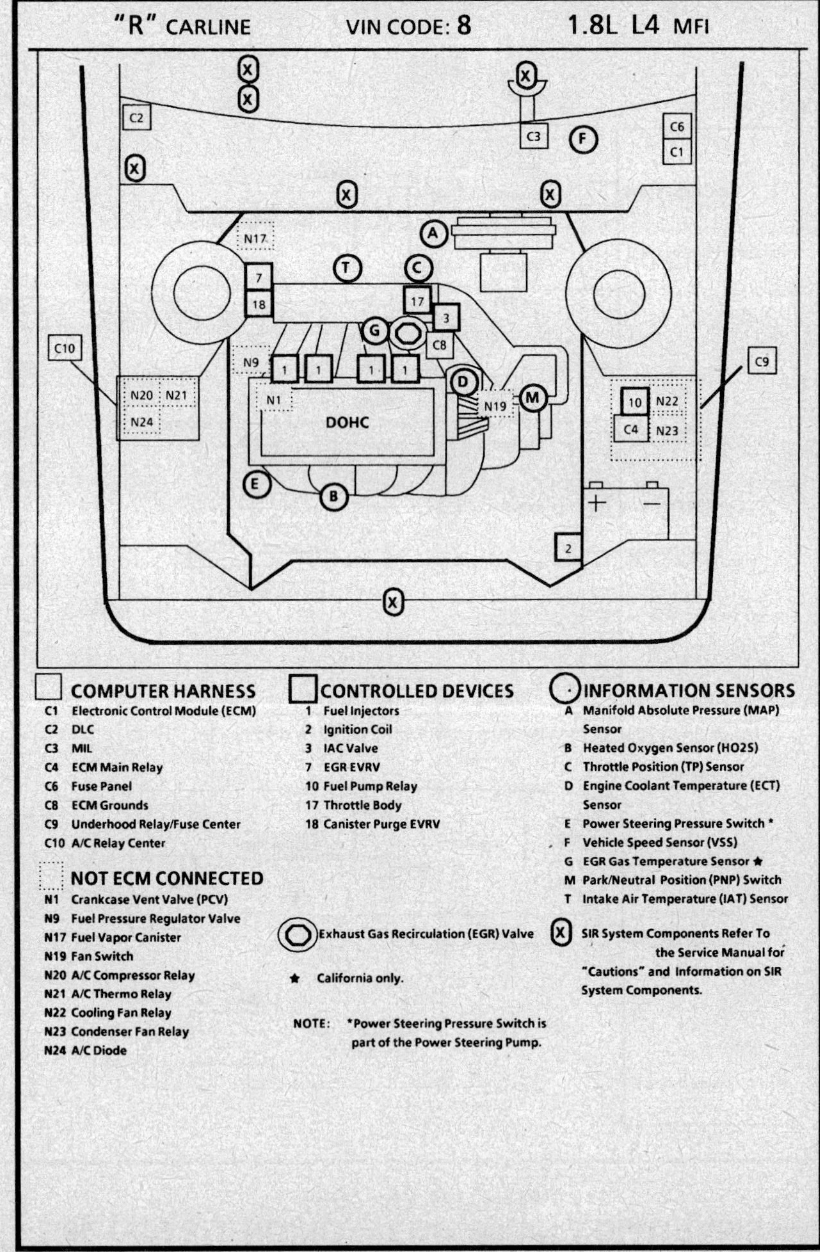

1.8L (VIN 8) ENGINE — ECM WIRING DIAGRAMS — 1992–93 STORM

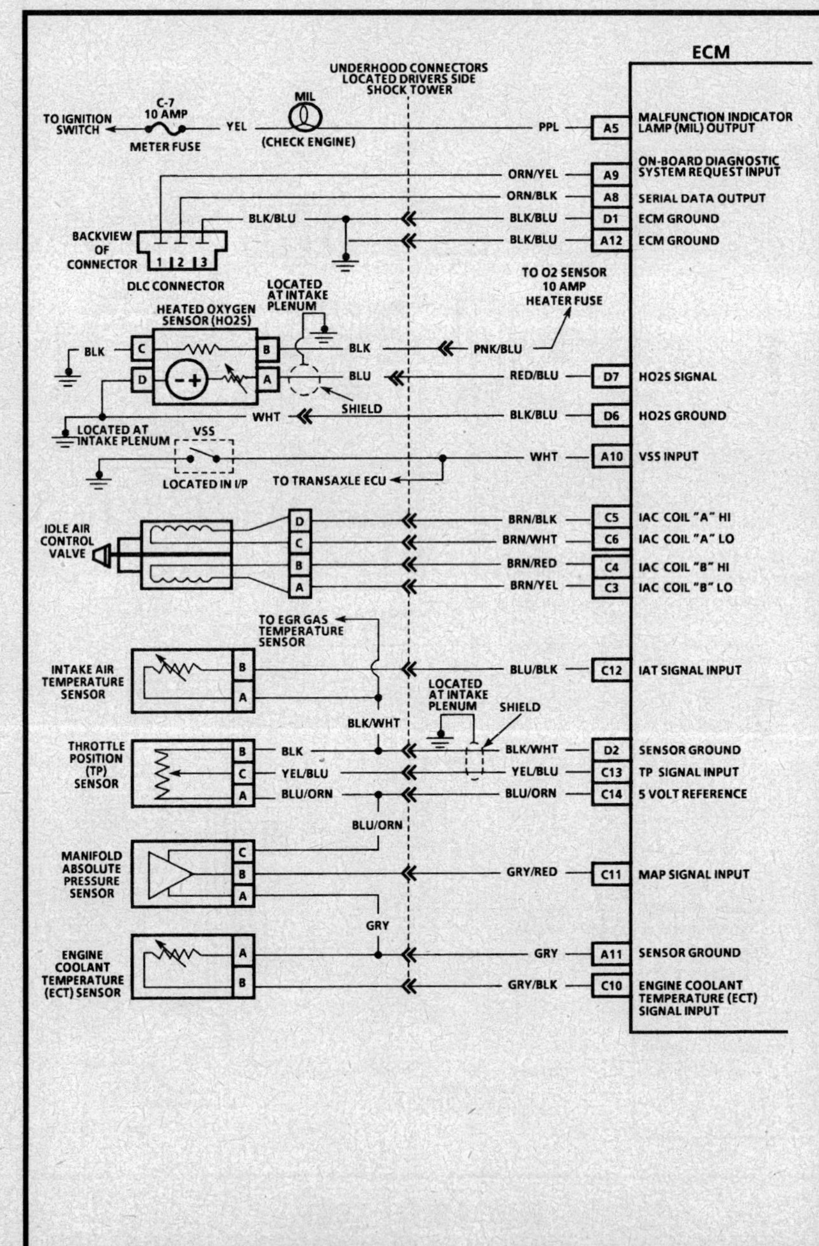

1.8L (VIN 8) ENGINE—ECM WIRING DIAGRAMS—1992–93 STORM

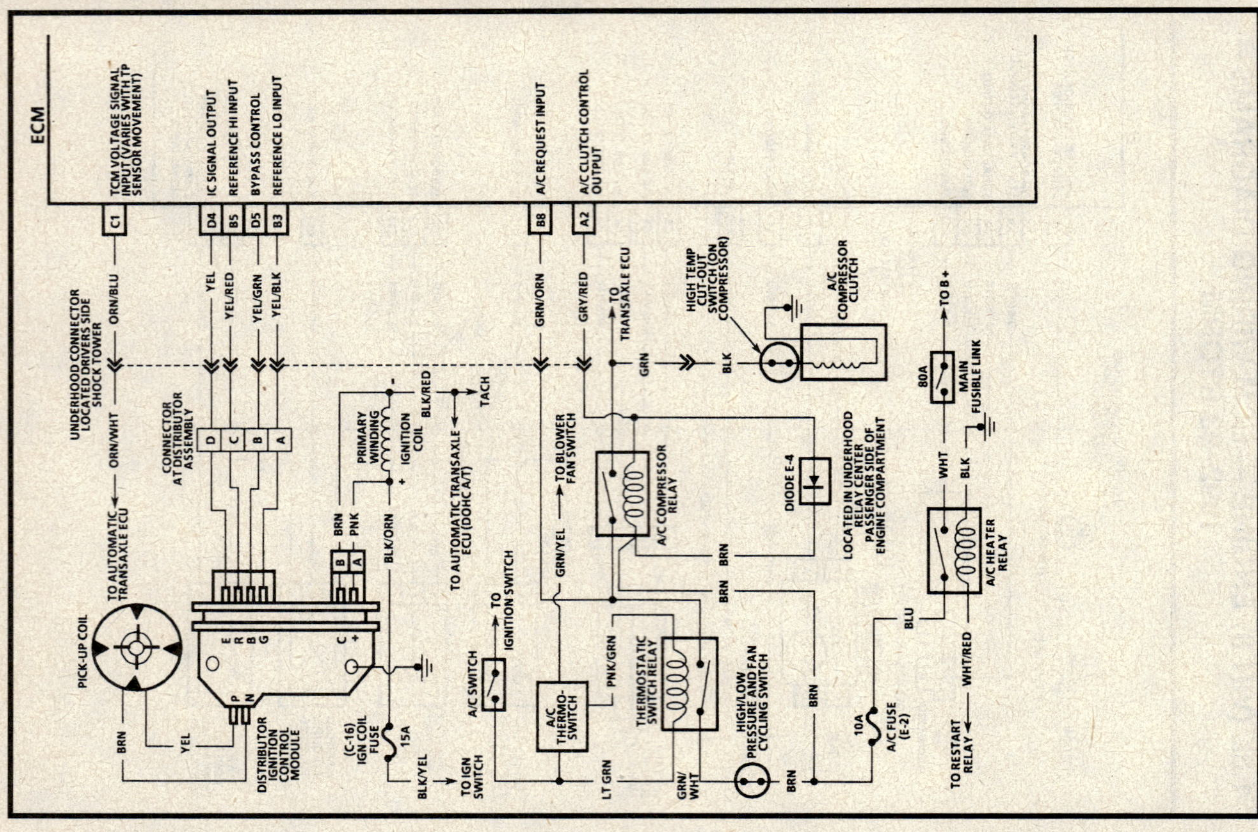

1.8L (VIN 8) ENGINE—ECM WIRING DIAGRAMS—1992–93 STORM

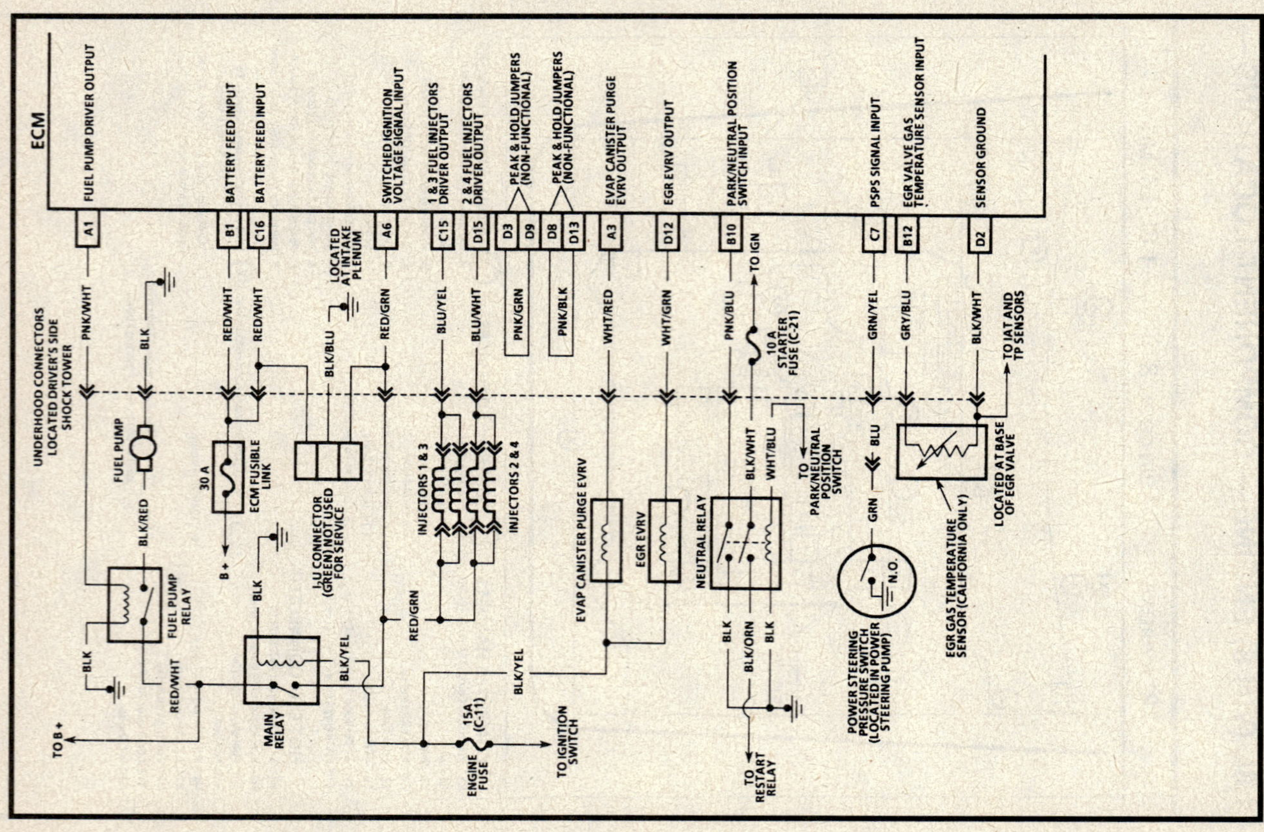

1.8L (VIN 8) ENGINE — ECM TERMINAL END VIEW —

FUEL INJECTION ECM CONNECTOR IDENTIFICATION

This ECM voltage chart is for use with a digital voltmeter to further aid in diagnosis. The voltages you get may vary due to low battery charge or other reasons, but they should be very close.

THE FOLLOWING CONDITIONS MUST BE MET BEFORE TESTING:

• Engine at operating temperature • Engine idling in "Closed Loop" (for "Engine Run" column)
• ALDL Diagnostic terminal not grounded • "Scan" tool not installed • All voltages shown "B + " indicates battery or charging voltage.

VOLTAGE

KEY "ON"	ENG. RUN	CIRCUIT	PIN	WIRE COLOR
B + (1)	B +	FUEL PUMP	A1	PNK/WHT
0*	B + OFF 0* ON	A/C CLUTCH CONTROL RELAY	A2	GRY/RED
B +	B +	CANISTER PURGE VSV	A3	WHT/RED
		NOT USED	A4	
0*	B +	CHECK ENGINE LIGHT	A5	PPL
B +	B +	IGNITION VOLTAGE SIGNAL	A6	RED/GRN
		NOT USED	A7	
(3) 4.7V	(3) 4.7V	SERIAL DATA	A8	ORN/BLK
B +	B +	DIAGNOSTIC ALDL REQUEST	A9	ORN/YEL
0* (5)	0*	VSS INPUT	A10	WHT
0*	0*	CTS & MAP, EGR SENSOR GROUND	A11	GRY
0*	0*	ECM GROUND	A12	BLK/BLU

VOLTAGE

WIRE COLOR	PIN	CIRCUIT	KEY "ON"	ENG. RUN
RED/WHT	B1	BATTERY	B +	B +
	B2	NOT USED		
YEL/BLK OR YEL/PPL	B3	REF LOW	0*	0*
	B4	NOT USED		
YEL/RED	B5	REF HIGH	0*	.9-1.1V
	B6	NOT USED		
	B7	NOT USED		
GRN/ORN	B8	A/C SIGNAL	0*	0* OFF B + ON
	B9	NOT USED		
PNK/BLU	B10	P/N SIGNAL (A/T)	0*	0* (4)
	B11	NOT USED		
GRY/BLU	B12	EGR GAS TEMP. SIGNAL	5.0V	(2) (7)

24 PIN A-B CONNECTOR

BACK VIEW OF CONNECTOR

6-8V (6)	6-8V (6)	SIG. (A/T ECU)	C1	ORN/BLU
		NOT USED	C2	
NOT USEABLE	NOT USEABLE	IAC LOW	C3	BRN/YEL
NOT USEABLE	NOT USEABLE	IAC HIGH	C4	BRN/RED
NOT USEABLE	NOT USEABLE	IAC HIGH	C5	BRN/BLK
NOT USEABLE	NOT USEABLE	IAC LOW	C6	BRN/WHT
B +	B +	PSPS SIGNAL	C7	GRN/YEL
		NOT USED	C8	
		NOT USED	C9	
(2)	(2)	CTS SIGNAL	C10	GRY/BLK
4.7V	1.9V	MAP SIGNAL	C11	GRY/RED
(2)	(2)	IAT SIGNAL	C12	BLU/BLK
.40-.56V	.40-.56V	TPS SIGNAL	C13	YEL/BLU
5.0V	5.0V	5V REF MAP & TPS	C14	BLU/ORN
B +	B +	INJECTOR DRIVER 1 & 3	C15	BLU/YEL
B +	B +	BATTERY	C16	RED/WHT

WIRE COLOR	PIN	CIRCUIT	KEY "ON"	ENG. RUN
BLK/BLU	D1	ECM GROUND	0*	0*
BLK/WHT	D2	TPS, IAT & EGR GAS TEMP GROUND	0*	0*
PNK/GRN	D3	PEAK & HOLD	**	**
YEL	D4.	EST SIGNAL	0*	1.1V
YEL/GRN	D5	BYPASS	0*	4.6V
BLK/BLU	B6	O$_2$ GROUND	0*	0*
RED/BLU	D7	O$_2$ SIGNAL	.380	VARIES
PNK/BLK	D8	PEAK & HOLD	**	**
PNK/GRN	D9	PEAK & HOLD	**	**
	D10	NOT USED		
	D11	NOT USED		
WHT/GRN	D12	EGR VSV	B +	B +
PNK/BLK	D13	PEAK & HOLD	**	**
	D14	NOT USED		
BLU/WHT	D15	INJ DRIVER 2 & 4	B +	B +
	D16	NOT USED		

32 PIN C-D CONNECTOR

BACK VIEW OF CONNECTOR

* All voltages shown "0" should read less than .5 volt.
** Non-functional.
(1) Reads battery voltage for 2 seconds after ignition "ON" then should read 0 volts.
(2) Varies depending on temperature.
(3) Scan Tool Disconnected.
(4) B + in R, D or L.
(5) Varies between 0-B + with vehicle speed.
(6) At idle & varies with TPS signal.
(7) California only.

1.6L (VIN 6) AND 1.8L (VIN 8) ENGINES — ON-BOARD DIAGNOSTIC (OBD) SYSTEM CHECK — 1992–93 STORM

ECM
I₄U CONNECTOR (GREEN) NOT FOR USE IN SERVICE
RED/WHT
BLK/BLU
RED/GRN
ECM 30A
FUSIBLE LINK
TO B +
RED/WHT   C16   BATTERY FEED INPUT
RED/WHT   B1   BATTERY FEED INPUT
TO FUEL INJECTORS
TO MAIN RELAY
RED/GRN   A6   SWITCHED IGNITION VOLTAGE SIGNAL INPUT
TO IGNITION SWITCH
10A
METER FUSE (C-7)
MALFUNCTION INDICATOR LAMP (MIL)
PPL   A5   MIL DRIVER OUTPUT
ORN/YEL   A9   DLC OBD SYSTEM CHECK REQUEST INPUT
ORN/BLK   A8   SERIAL DATA INPUT/OUTPUT
BLK/BLU   D1   ECM GROUND
BLK/BLU   A12   ECM GROUND
DLC CONNECTOR (WHITE) LOCATED BEHIND PASSENGER KICK PANEL
LOCATED AT INTAKE PLENUM
1 2 3

ON-BOARD DIAGNOSTIC (OBD) SYSTEM CHECK

1.6L SOHC & 1.8L DOHC (VIN 6 & 8) "R" CARLINE (MFI)

Circuit Description:

The OBD system check is an organized approach to identifying a problem created by a malfunction in the engine control system. It must be the starting point for any driveability complaint diagnosis, because it directs the service technician to the next logical step in diagnosing the complaint. Understanding the chart and using it correctly will reduce diagnostic time and prevent the unnecessary replacement of good parts.

Test Description: Number(s) below refer to circled number(s) on the diagnostic chart.

1. This step is a check for the proper operation of the malfunction indication lamp. The MIL should be "ON" steady.
2. No MIL at this point indicates that there is a problem with the MIL circuit or the ECM control of that circuit.
3. This test checks the ability of the ECM to control the MIL. With the DLC grounded, the MIL should flash a DTC 12 three times, followed by any DTC stored in memory. Depending upon the type of ECM, an ECM error may result in the inability to flash DTC 12.
4. Most of the   procedures use a scan tool to aid diagnosis, therefore, serial data must be available. If an ECM or EEPROM error is present, the ECM may have been able to flash DTC 12/51, but not enable serial data.
5. Although the ECM is powered up, a "Cranks But Will Not Run" symptom could exist because of an ECM or system problem.
6. This step will isolate if the customer complaint is a MIL or a driveability problem with no MIL. Refer to diagnostic trouble code   for valid DTC(s). An invalid DTC may be the result of a faulty scan tool or ECM.
7. Comparison of actual control system data with the typical values is a quick check to determine if any parameter is not within limits. Keep in mind that a base engine problem (i.e. advanced cam timing) may substantially alter sensor values.
8. Installation of a scan tool will provide a good ground path for the ECM and may hide a driveability complaint due to poor ECM grounds.
9. If the actual data is not within the typical values established, the charts in "Symptoms,"   will provide a functional check of the suspect component or system.

1.6L (VIN 6) AND 1.8L (VIN 8) ENGINES — ON-BOARD DIAGNOSTIC (OBD) SYSTEM CHECK — 1992–93 STORM

1.6L (VIN 6) AND 1.8L (VIN 8) ENGINES — SCAN TOOL DATA VALUES — 1992–93 STORM

ON-BOARD DIAGNOSTIC (OBD) SYSTEM CHECK
1.6L SOHC & 1.8L DOHC (VIN 6 & 8) "R" CARLINE (MFI)

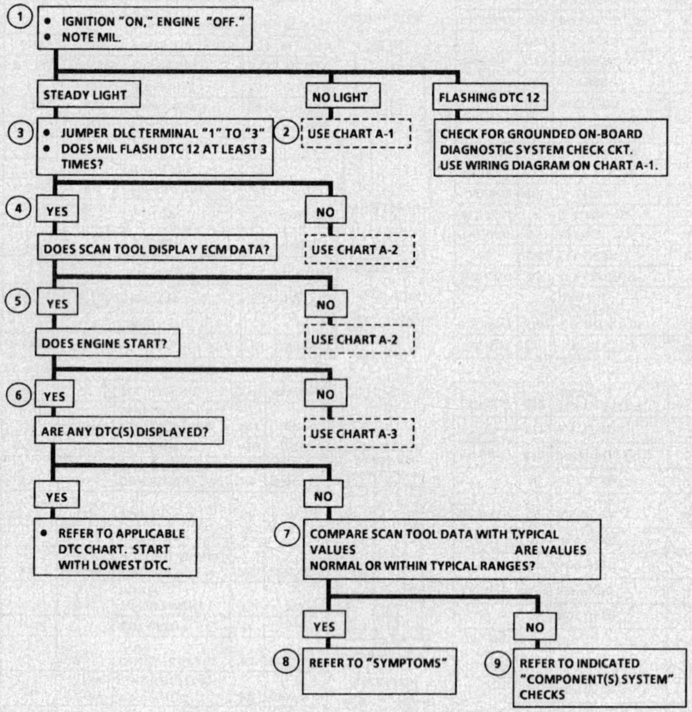

1. IGNITION "ON," ENGINE "OFF."
   - NOTE MIL.

STEADY LIGHT — NO LIGHT — FLASHING DTC 12

3. JUMPER DLC TERMINAL "1" TO "3"
   - DOES MIL FLASH DTC 12 AT LEAST 3 TIMES?

2. USE CHART A-1

CHECK FOR GROUNDED ON-BOARD DIAGNOSTIC SYSTEM CHECK CKT. USE WIRING DIAGRAM ON CHART A-1.

4. YES — NO

DOES SCAN TOOL DISPLAY ECM DATA? — USE CHART A-2

5. YES — NO

DOES ENGINE START? — USE CHART A-2

6. YES — NO

ARE ANY DTC(S) DISPLAYED? — USE CHART A-3

YES — NO

- REFER TO APPLICABLE DTC CHART. START WITH LOWEST DTC.

7. COMPARE SCAN TOOL DATA WITH TYPICAL VALUES ARE VALUES NORMAL OR WITHIN TYPICAL RANGES?

YES — NO

8. REFER TO "SYMPTOMS"

9. REFER TO INDICATED "COMPONENT(S) SYSTEM" CHECKS

If after completing the on-board diagnostic system check and finding scan tool diagnostics functioning properly and no diagnostic trouble codes displayed, the "Typical Scan Values" may be used for comparison with values obtained on the vehicle being diagnosed. The "Typical Scan Values," are an average of display values recorded from normally operating vehicles and are intended to represent what a normally functioning system would display.

A SCAN TOOL THAT DISPLAYS FAULTY DATA SHOULD NOT BE USED, AND THE PROBLEM SHOULD BE REPORTED TO THE MANUFACTURER. THE USE OF A FAULTY SCAN TOOL CAN RESULT IN MISDIAGNOSIS AND UNNECESSARY PARTS REPLACEMENT.

Only the parameters listed are used in this manual for diagnosis. If a scan tool reads other parameters, the values are not recommended for use in diagnosis. If all values are within the range illustrated, refer to "Symptom."

TYPICAL TECH 1 DATA VALUES
Idle / Upper Radiator Hose Hot / Closed Throttle / Park or Neutral / Closed Loop / Acc. off

SCAN Position	Units Displayed	Typical Data Value
Engine Speed	RPM	± 100 RPM from desired RPM (± 50 in drive)
Desired Idle	RPM	ECM idle command (varies with temperature)
Engine Coolant Temp	C°/F°	85°C - 105°C (185°F - 221°F)
Intake Air Temp	C°/F°	10°C - 90°C (50°F - 194°F) (Depends on underhood temp)
MAP	kPa/Volts	20-48 kPa/1.0-2.0 Volts (Varies with altitude)
BARO	kPa/Volts	Varies with altitude
Throttle Position	Volts	.3 - 1.0
Throttle Angle	Percentage	0%
Injector Pulse Width	Milliseconds	.8-1.2
Oxygen Sensor	Millivolts	1 - 1000 and varying
Closed Throttle	No/Yes	Yes
Fuel Trim Enable	No/Yes	Yes
Injector Pulse Width	Milliseconds	Varies
St Fuel Trim	Counts	Varies 110-145
Lt Fuel Trim	Counts	Varies 115-138
Open/Closed Loop	Open/Closed	Closed Loop (may enter Open Loop with extended idle SOHC ONLY)
Fuel Trim Cell	Cell Number	Cell number varies with engine load
VSS	MPH/KPH	0 varies with vehicle movement
Engine Speed	RPM	± 100 RPM from desired RPM (± 50 in drive)
Spark Advance	Degrees	Varies
Ignition Control (EST)	No/Yes	Yes
EGR EVRV	"ON/OFF"	"OFF"
Fuel EVAP Purge	"ON/OFF"	"OFF"
Air Fuel Ratio	Ratio	14.7:1
System Voltage	Volts	13-14.5 Volts
Idle Air Control	Steps	5-29
Park/Neutral Position	PRDL	PRDL
A/C Request	YES/NO	No & Yes when requested
A/C Clutch	"ON/OFF"	"OFF" & "ON" when requested
Power Steering	Normal/High pressure	Normal
Time from start	0:00	

1.6L (VIN 6) AND 1.8L (VIN 8) ENGINES — SCAN TOOL DATA DEFINITIONS — 1992–93 STORM

TYPICAL SCAN TOOL DATA DEFINITIONS

ECM DATA DESCRIPTION

A list of explanations for each data message displayed on the scan tool begins below.

This information will assist in tracking down emission or driveability problems, since the displays can be viewed while the vehicle is being driven. Refer to the "OBD System Check" for additional information.

ENGINE SPEED - Range 0-6375 RPM - Engine speed is computed by the ECM from the fuel control reference input. It should remain close to desired idle under various engine loads with engine idling.

DESIRED IDLE - Range 0-3187 RPM - The idle speed that is commanded by the ECM. The ECM will compensate for various engine loads to keep the engine at the desired idle speed.

ENGINE COOLANT TEMP - Range -40°C to 151°C, -40° to 304°F - The Engine Coolant Temperature (ECT) sensor is mounted in the coolant stream and sends engine temperature information to the ECM. The ECM supplies 5 volts to the engine coolant temperature sensor circuit. The sensor is a thermistor which changes internal resistance as temperature changes. When the sensor is cold (internal resistance high), the ECM monitors a high signal voltage which it interprets as a cold engine. As the sensor warms (internal resistance decreases), the voltage signal will decrease and the ECM will interpret the lower voltage as a warm engine.

INTAKE AIR TEMP - Range -38°C to 199°C, -36°F to 390°F - The ECM converts the resistance of the intake air temperature sensor to degrees. Intake Air Temperature (IAT) is used by the ECM to adjust fuel delivery and spark timing according to incoming air density.

MAP - Range 11-105 kPa/0-5.10 Volts - The Manifold Absolute Pressure (MAP) sensor measures the changes in the intake manifold pressure which result from engine load and speed changes, varying the signal voltage. The MAP sensor is also used to measure barometric pressure at start up and other certain conditions, which allows the ECM to automatically adjust for different altitudes.

BARO - Range 11-105 kPa/0.00-5.10 Volts - The BARO reading displayed is measured from the MAP sensor at key "ON," engine "OFF" and WOT conditions. The BARO reading displayed represents barometric pressure and is used to compensate for altitude.

THROT POSITION - Range 0-5.10 Volts - Used by the ECM to determine the amount of throttle demanded by the driver. Should read .33 - 1.33 volts at idle to above 4 volts at wide open throttle.

THROTTLE ANGLE - Range 0-100% - Computed by the ECM from TP sensor voltage (throt position) should read 0% at idle and 100% at wide open throttle.

OXYGEN SENSOR - Range 0-1132 mV - Represents the exhaust oxygen sensor output voltage. Should fluctuate constantly within a range between 100 mV (lean exhaust) and 1000 mV (rich exhaust) when operating in "Closed Loop."

INJECTOR PULSE WIDTH - Range 0-499 mS - Indicates the "ON" time of the fuel injector(s) in milliseconds. When engine load is increased, injector pulse width will increase.

SPARK ADVANCE - Range -90° to 90° - This is a display of the spark advance (IC) calculation which the ECM is programming into the ignition system. It computes the desired spark advance using data such as engine temperature, RPM, load, vehicle speed and operating mode.

SHORT TERM FUEL TRIM - Range 0-255 - ST fuel trim (formerly Fuel Integrator) represents a short-term correction to fuel delivery by the ECM in response to the amount of time the oxygen sensor voltage spends above or below the 450 mV threshold. If the oxygen sensor voltage has mainly been below 450 mV, indicating a lean air/fuel mixture, ST fuel trim will increase to tell the ECM to add fuel. If the oxygen sensor voltage stays mainly above the threshold, the ECM will reduce fuel delivery to compensate for the indicated rich condition. Under certain conditions such as extended idle and high ambient temperatures, canister purge may cause short term fuel trim to read less than 100 counts.

LONG TERM FUEL TRIM - Range 0-255 - LT Fuel Trim (formerly Block Learn) is derived from the ST fuel trim (fuel integrator) value and is used for long-term correction of fuel delivery. A value of 128 counts indicates that fuel delivery requires no compensation to maintain a 14.7:1 air/fuel ratio. A value below 128 counts means that the fuel system is too rich and fuel delivery is being reduced (decreased injector pulse width). A value above 128 counts indicates that a lean condition exists and the ECM is compensating by adding fuel (increased injector pulse width).

LT fuel trim tends to follow ST fuel trim; a value of less than 100 counts due to canister purge at idle should not be considered unusual.

OPEN/CLOSED LOOP - Scan Tool Displays "Open" LP or "Closed" LP - The ECM controls various functions of the engine in two different modes. When in "Open Loop" a preset of fixed calibration. When the oxygen sensor begins to toggle between 100 mV and 900 mV and engine temperature is above approximately 50°C the ECM switches to the "Closed Loop" mode and controls fuel delivery based on the oxygen sensor.

FUEL TRIM CELL (FTC) - Range 0-4 - Fuel trim cell (formerly Block Learn Cell) is selecting between one of four (FTC) cells for maintaining proper fuel control under any and all types of engine load conditions, is made on the basis of throttle position, vacuum load and engine speed. If the engine is at closed throttle and at idle, FTC cell 0 will be used. If the engine speed is not at idle with a closed throttle position, than the FTC cell 1 will be used. When the engine speed is "OFF" idle and the throttle is open (but not at WOT), FTC cell 2 is used. FTC cell 3 will be used when the engine speed is at Wide Open Throttle (WOT).

IDLE AIR CONTROL - Range 0-255 - Displays the commanded position of the idle air control pintle in counts. The higher the number of counts, the greater the commanded idle speed. Idle air control should respond fairly quickly to changes in engine load to maintain desired idle RPM.

PARK/NEUTRAL POSITION - Scan Tool Displays "P-N" or "R-D-L" - "P-N" displayed indicates that the gear select lever is in park or neutral.

MPH/km/h - Range 0-255 mph, 0-255 kph - The vehicle speed sensor signal is converted into km/h and mph for display.

TCC - Scan Tool Displays "OFF" or "ON" - The scan tool only indicates when the ECM has enabled the TCC driver, but this does not confirm that the TCC is engaged. To determine if TCC is operating properly, engine RPM should decrease when TCC is enabled.

SHIFT LIGHT - Scan Tool Displays "ON" or "OFF" - The shift light is used to indicate the best point for upshifting the manual transaxle. When the shift light (if equipped) is "ON," the scan tool should display "ON."

IDLE LEARN - Scan Tool Displays "Yes" or "No"

1.6L (VIN 6) AND 1.8L (VIN 8) ENGINES — SCAN TOOL DATA DEFINITIONS — 1992–93 STORM

A/C REQUEST - Scan Tool Displays "YES" or "NO" - When air conditioning is selected and the A/C low pressure switch is closed (sufficient refrigerant pressure) the scan tool will display "YES." Displays "NO" when A/C is turned "OFF" or other non-A/C functions are selected.

A/C RELAY - Scan Tool Displays "ON" or "OFF" - Represents the commanded state of the A/C clutch control relay. Clutch should be engaged when "ON" is displayed.

EGR EVRV - Scan Tool Displays "ON" OR "OFF" - The exhaust gas recirculation system is controlled by the ECM using a vacuum solenoid. When EGR is needed, the ECM grounds the solenoid thus turning "ON" a vacuum source to the EGR valve.

FUEL EVAP PURGE - This represents the canister purge EVRV. The ECM commands the canister purge "ON" under several conditions. The scan tool will display "ON"/"OFF." If the vehicle is at operation temperature the canister purge EVRV should be "OFF" if at idle.

NOTICE: When using a scan tool on this vehicle, the power connection for the scan tool should be done at the battery using a suitable adapter. The cigarette lighter(s) are connected to the ignition switch and have power to them when the key is "ON." Power is lost during cranking.

1.6L (VIN 6) AND 1.8L (VIN 8) ENGINES — A-CHARTS — 1992–93 STORM

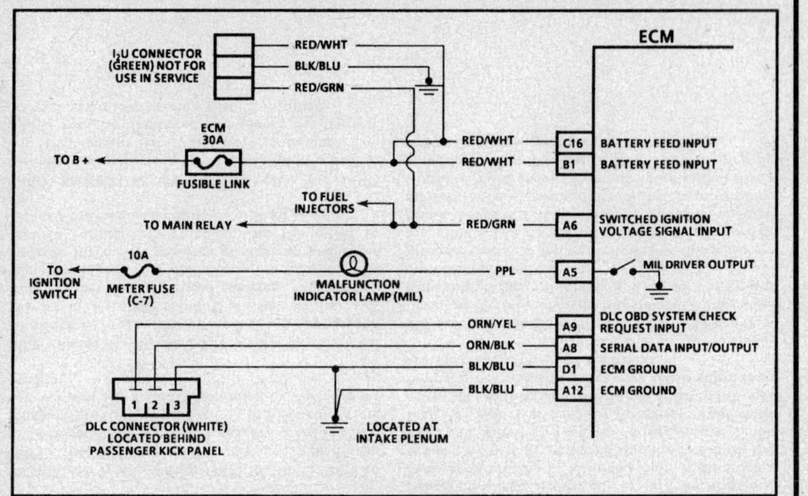

CHART A-1
WON'T FLASH DTC 12
NO MALFUNCTION INDICATOR LAMP (MIL)
1.6L SOHC & 1.8L DOHC (VIN 6 & 8) "R" CARLINE (MFI)

Circuit Description:

There should always be a steady Malfunction Indicator Lamp (MIL), when the ignition is "ON" and engine stopped. Battery voltage is supplied directly to the light bulb. The Engine Control Module (ECM) will control the MIL and turn it "ON" by providing a ground path through the MIL circuit from the ECM to the bulb.

Test Description: Number(s) below refer to circled number(s) on the diagnostic chart.

1. Battery feed terminals "C16" and "B1" are protected by a fusible link in relay/fuse center under the hood near the battery.
2. If terminals "C16" and "B1" have voltage, the ECM grounds or the ECM are faulty.
3. Using a test light connected to 12 volts, probe each of the system ground circuits to be sure a good ground is present. Refer to the ECM terminal end view in front of this section for ECM pin locations of ground circuits.

Diagnostic Aids:

Engine runs OK, check:
- Faulty light bulb.
- PPL wire open.
- Meter fuse blown. This will result in no oil or generator lights, seat belt reminder, etc.

Engine cranks, but will not run.
- Continuous battery - fuse or fusible link open.
- ECM ignition fuse open.
- Battery RED/WHT wire to ECM open.
- Ignition RED/GRN wire to ECM open.
- Poor connection to ECM.

1.6L (VIN 6) AND 1.8L (VIN 8) ENGINES — A-CHARTS — 1992–93 STORM

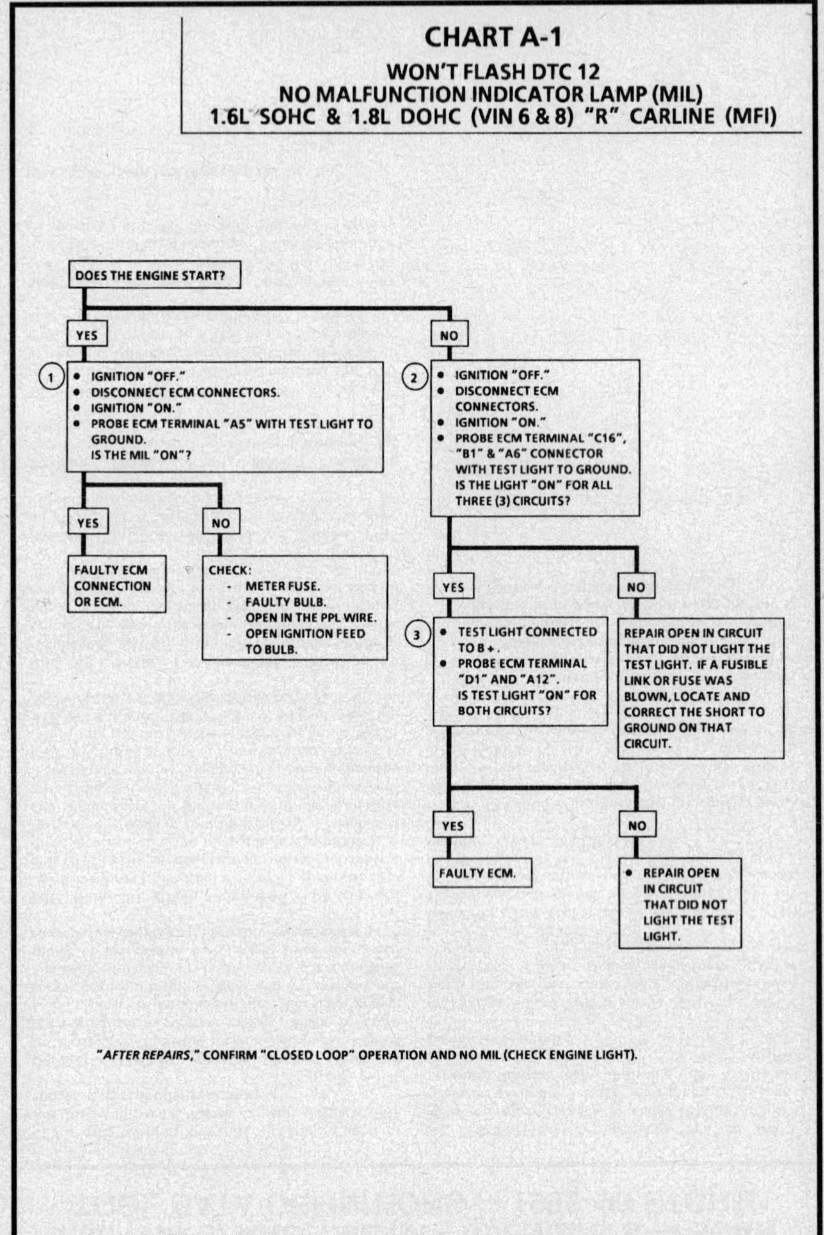

1.6L (VIN 6) AND 1.8L (VIN 8) ENGINES — A-CHARTS — 1992–93 STORM

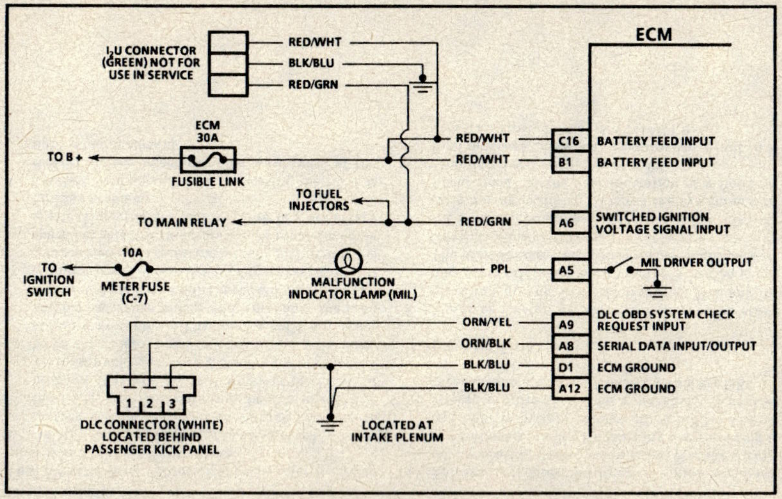

CHART A-2

**NO DLC DATA OR WON'T FLASH DTC 12
MALFUNCTION INDICATOR LAMP (MIL) "ON" STEADY
1.6L SOHC & 1.8L DOHC (VIN 6 & 8) "R" CARLINE (MFI)**

Circuit Description:

There should always be a steady Malfunction Indicator Lamp (MIL) when the ignition is "ON" and engine stopped. Battery voltage is supplied directly to the light bulb. The Engine Control Module (ECM) will turn the light "ON" by grounding MIL circuit at the ECM.

With the on-board diagnostic terminal grounded, the light should flash a DTC 12, followed by any trouble DTC(S) stored in memory.

A steady light suggests a short to ground in the MIL circuit, or an open in OBD system check request circuit.

Test Description: Number(s) below refer to circled number(s) on the diagnostic chart.

1. If there is a problem with the ECM that causes a scan tool to not read Serial data, then the ECM should not flash a DTC 12. If DTC 12 does flash, be sure that the scan tool is working properly on another vehicle. If the scan tool is functioning properly and OBD system request circuit is OK, the ECM may be at fault for the NO DLC symptom.
2. If the light goes "OFF" when the ECM connector is disconnected, then the MIL circuit is not shorted to ground.
3. This step will check for an open OBD system check request circuit.
4. At this point, the MIL wiring is OK. The problem is a faulty ECM. If DTC 12 does not flash, the ECM should be replaced.

Diagnostic Aid:

If the Malfunction Indicator Lamp (MIL) starts to flash DTC 12 but does not flash DTC at least 3 times, the cause of this condition could be a short in one of the QDM circuits. Check the following circuits for a possible short:

- A/C clutch circuit.
- Canister purge circuit.
- EGR valve EVRV circuit.
- Shift light circuit.

If no problem is found the ECM is the cause of the problem.

1.6L (VIN 6) AND 1.8L (VIN 8) ENGINES — A-CHARTS — 1992–93 STORM

CHART A-2

**NO DLC DATA OR WON'T FLASH DTC 12 MALFUNCTION INIDCATOR LAMP (MIL) "ON" STEADY
1.6L SOHC & 1.8L DOHC (VIN 6 & 8) "R" CARLINE (MFI)**

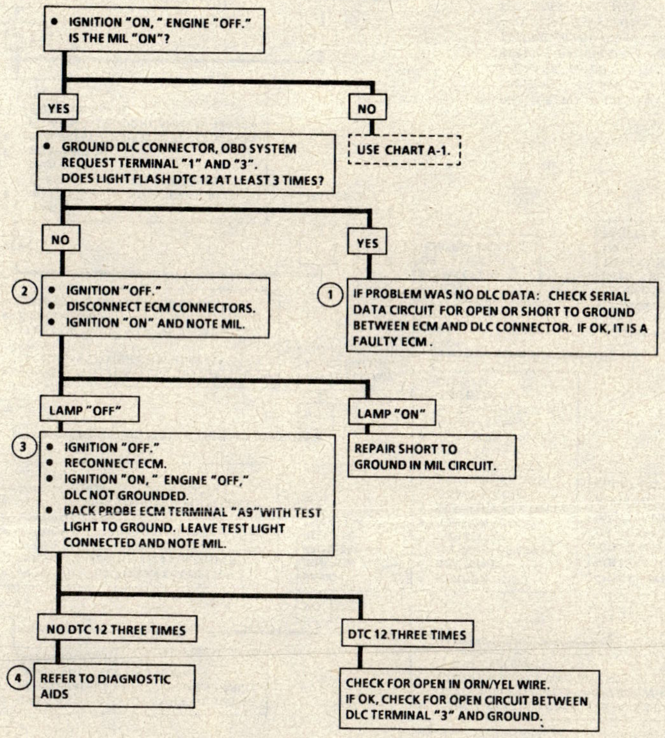

"AFTER REPAIRS," CONFIRM "CLOSED LOOP" OPERATION AND NO MIL (CHECK ENGINE LIGHT).

1.6L (VIN 6) AND 1.8L (VIN 8) ENGINES — A-CHARTS — 1992–93 STORM

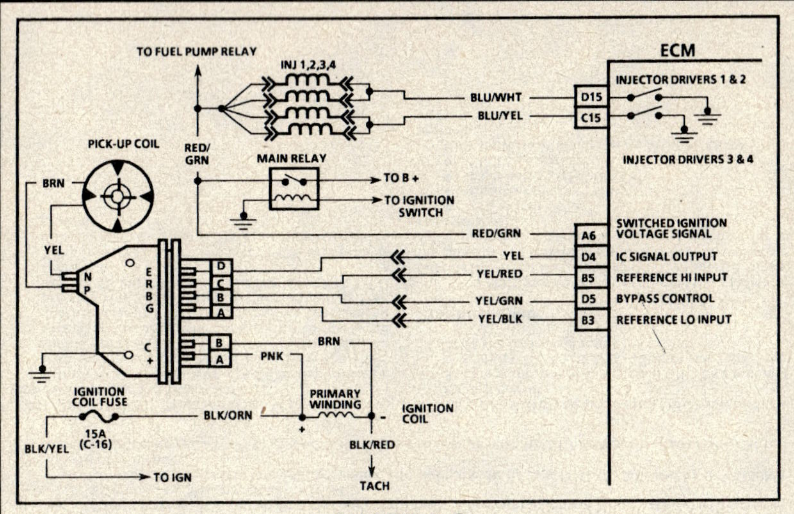

CHART A-3
(Page 1 of 2)
ENGINE CRANKS BUT WON'T RUN
1.6L SOHC (VIN 6) "R" CARLINE (MFI)

Circuit Description:
Before using this chart, battery condition, engine cranking speed, and fuel quantity should be checked and verified as being OK.

Test Description: Number(s) below refer to circled number(s) on the diagnostic chart.

1. An MIL "ON" is a basic test to determine if there is battery voltage and ignition voltage supplied to the ECM. No DLC data may be due to an ECM problem. CHART A-2 will diagnose the ECM. If TP sensor voltage is over 2.5 volts, the engine may be in the "clear flood" mode, which will cause starting problems. The engine will not start without reference pulses and, therefore, the scan tool should indicate engine speed during cranking.

2. If engine speed was indicated during crank, the DI control module is receiving a crank signal, but "no spark" at this test indicates the DI control module is not triggering the coil or there is a secondary ignition problem.

3. The test light should flash, indicating the ECM is controlling the injectors. The brightness of the light is not important.

4. This test will determine if the DI control module is not generating the reference pulse, or if the wiring or ECM are at fault. By touching and removing a test light to battery voltage on YEL/RED wire, a reference pulse should be generated. If engine speed is indicated, the ECM and wiring are OK.

Diagnostic Aids:

- Water or foreign material can cause a no start condition during freezing weather. The engine may start after 5 or 6 minutes in a heated shop. The problem may recur after an overnight park in freezing temperatures.
- An EGR sticking open can cause a low air/fuel ratio during cranking. Unless the system enters "clear flood" at the first indication of a flooding condition it can result in a no start.
- Fuel pressure: Low fuel pressure can result in a very lean air/fuel ratio. See CHART A-7.

1.6L (VIN 6) AND 1.8L (VIN 8) ENGINES — A-CHARTS — 1992–93 STORM

CHART A-3
(Page 1 of 2)
ENGINE CRANKS BUT WON'T RUN
1.6L SOHC (VIN 6)
"R" CARLINE (MFI)

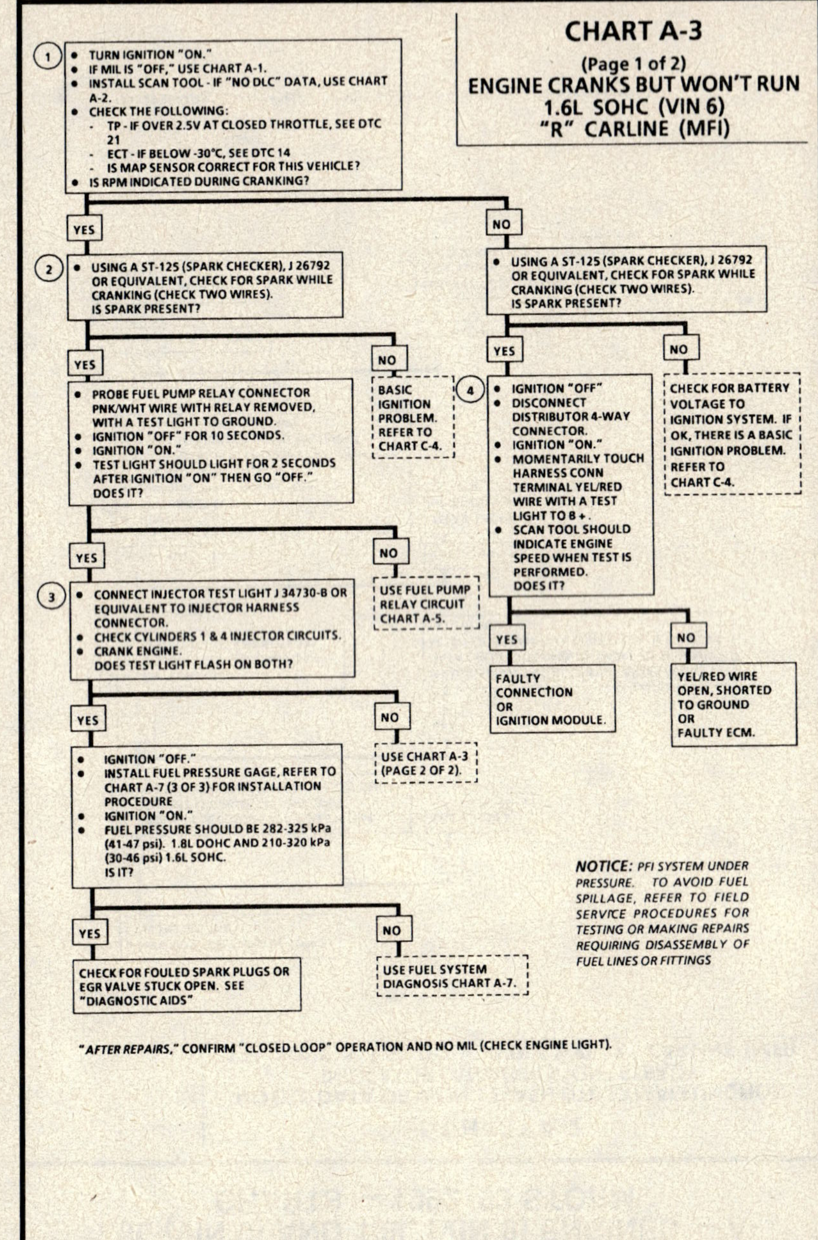

"*AFTER REPAIRS*," CONFIRM "CLOSED LOOP" OPERATION AND NO MIL (CHECK ENGINE LIGHT).

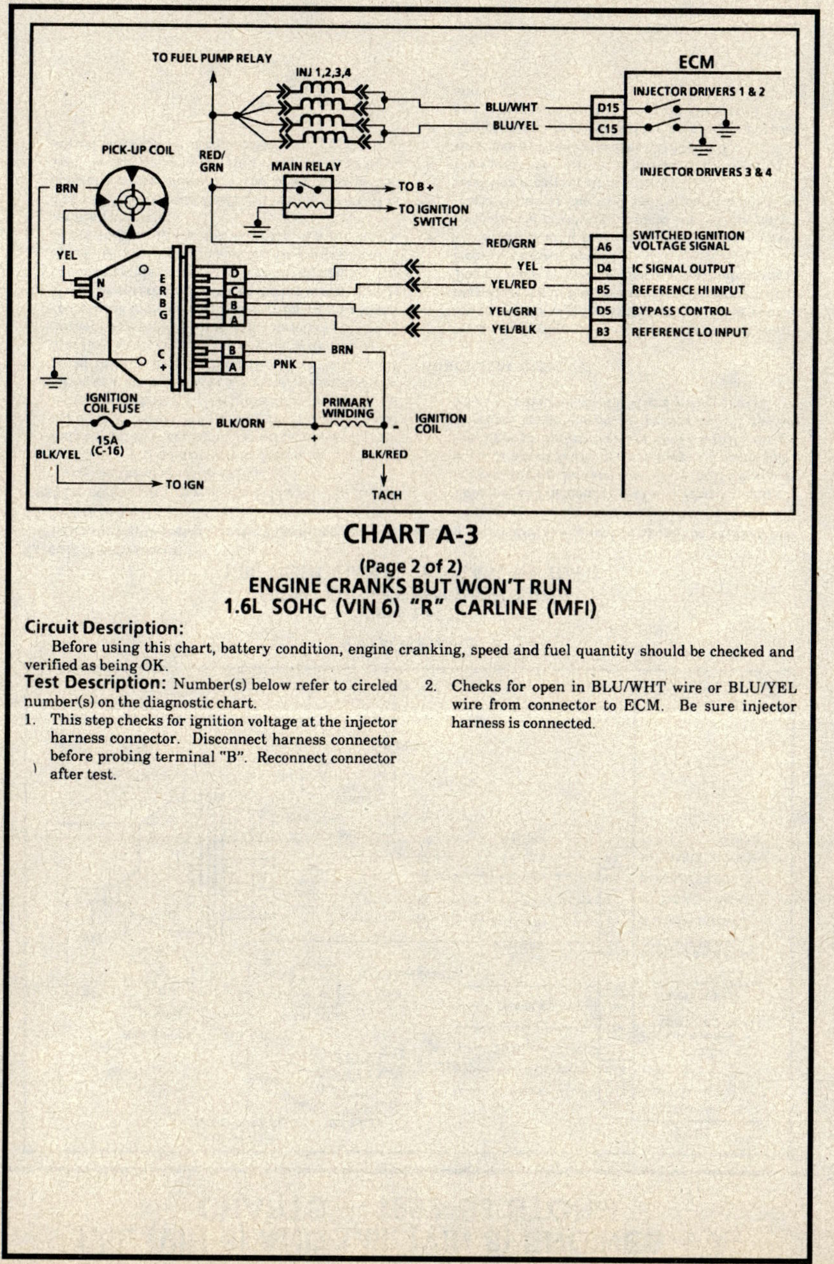

Circuit Description:
Before using this chart, battery condition, engine cranking, speed and fuel quantity should be checked and verified as being OK.

Test Description: Number(s) below refer to circled number(s) on the diagnostic chart.

1. This step checks for ignition voltage at the injector harness connector. Disconnect harness connector before probing terminal "B". Reconnect connector after test.

2. Checks for open in BLU/WHT wire or BLU/YEL wire from connector to ECM. Be sure injector harness is connected.

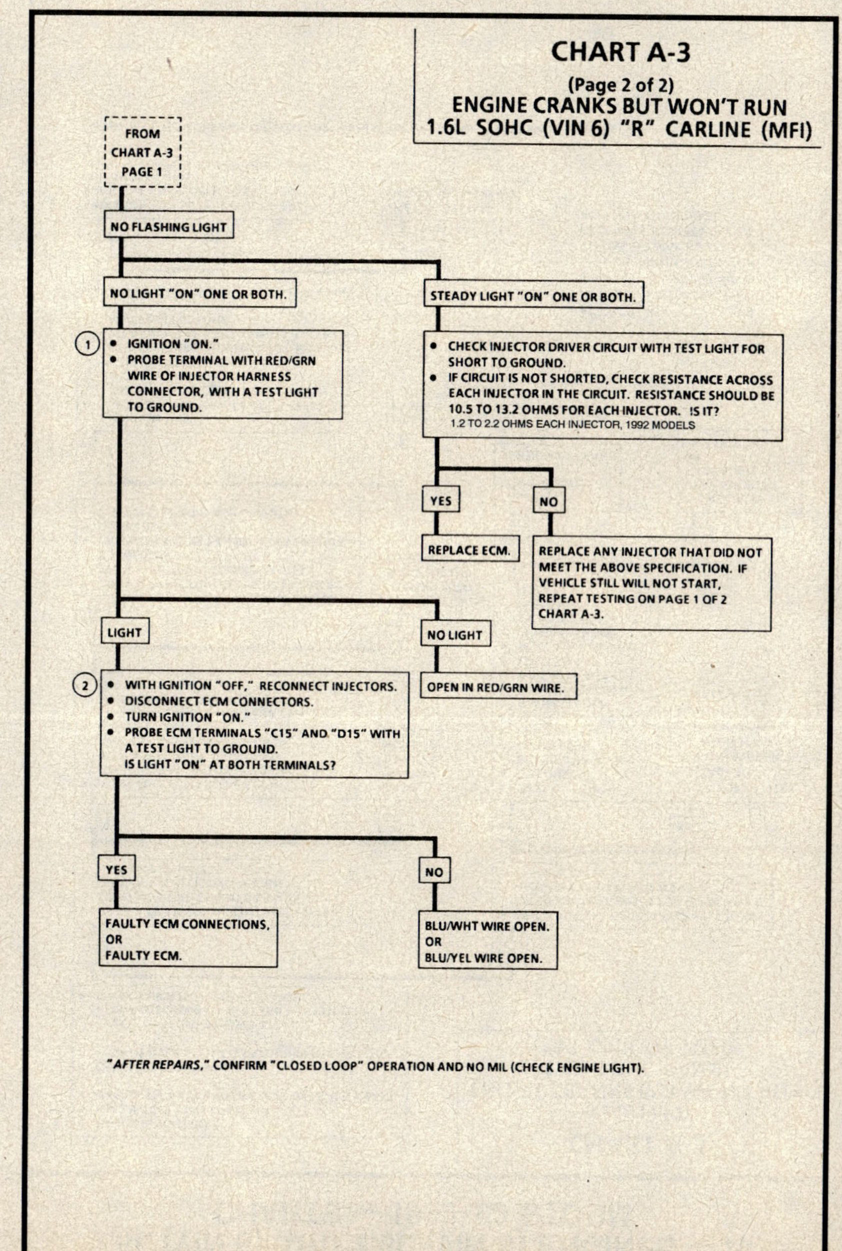

1.6L (VIN 6) AND 1.8L (VIN 8) ENGINES — A-CHARTS — 1992–93 STORM

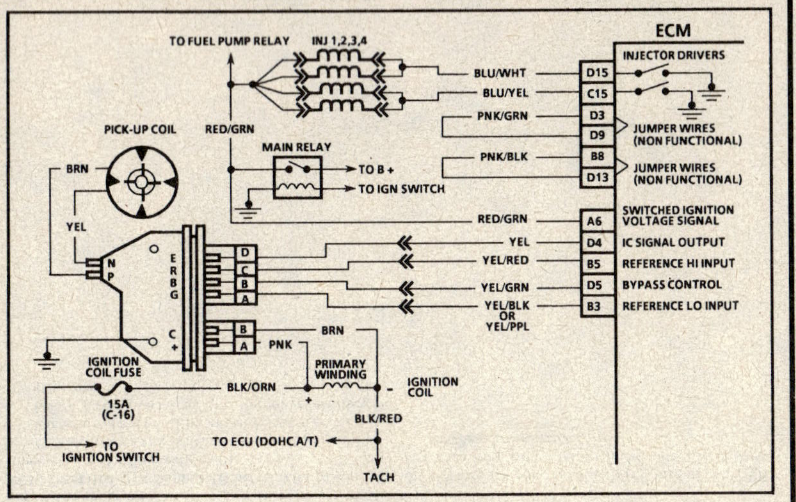

CHART A-3
(Page 1 of 2)
**ENGINE CRANKS BUT WON'T RUN
1.8L DOHC (VIN 8) "R" CARLINE (MFI)**

Circuit Description:
Before using this chart, battery condition, engine cranking speed, and fuel quantity should be checked and verified as being OK.

Test Description: Number(s) below refer to circled number(s) on the diagnostic chart.

1. A MIL "ON" is a basic test to determine if there is battery voltage and ignition voltage supplied to the ECM. No DLC data may be due to an ECM problem. CHART A-2 will diagnose the ECM. If TP sensor voltage is over 2.5 volts, the engine may be in the "clear flood" mode, which will cause starting problems. The engine will not start without reference pulses and, therefore, the scan tool should indicate engine speed during cranking.

2. If engine speed was indicated during crank, the DI control module is receiving a crank signal, but "no spark" at this test indicates the DI control module is not triggering the coil or there is a secondary ignition problem.

3. The test light should flash, indicating the ECM is controlling the injectors. The brightness of the light is not important. However, the test light should be a J 34730 or equivalent.

4. This test will determine if the DI control module is not generating the reference pulse, or if the wiring or ECM are at fault. By touching and removing a test light to battery voltage on YEL/RED wire, a reference pulse should be generated. If engine speed is indicated, the ECM and wiring are OK.

Diagnostic Aids:

- Water or foreign material can cause a no start condition during freezing weather. The engine may start after 5 or 6 minutes in a heated shop. The problem may recur after an overnight park in freezing temperatures.
- An EGR sticking open can cause a low air/fuel ratio during cranking. Unless the system enters "clear flood" at the first indication of a flooding condition it can result in a no start.
- Fuel pressure: Low fuel pressure can result in a very lean air/fuel ratio. See CHART A-7.
- The peak and hold wires have no useable purpose and will not affect the injector circuits if open and/or shorted.

1.6L (VIN 6) AND 1.8L (VIN 8) ENGINES — A-CHARTS — 1992–93 STORM

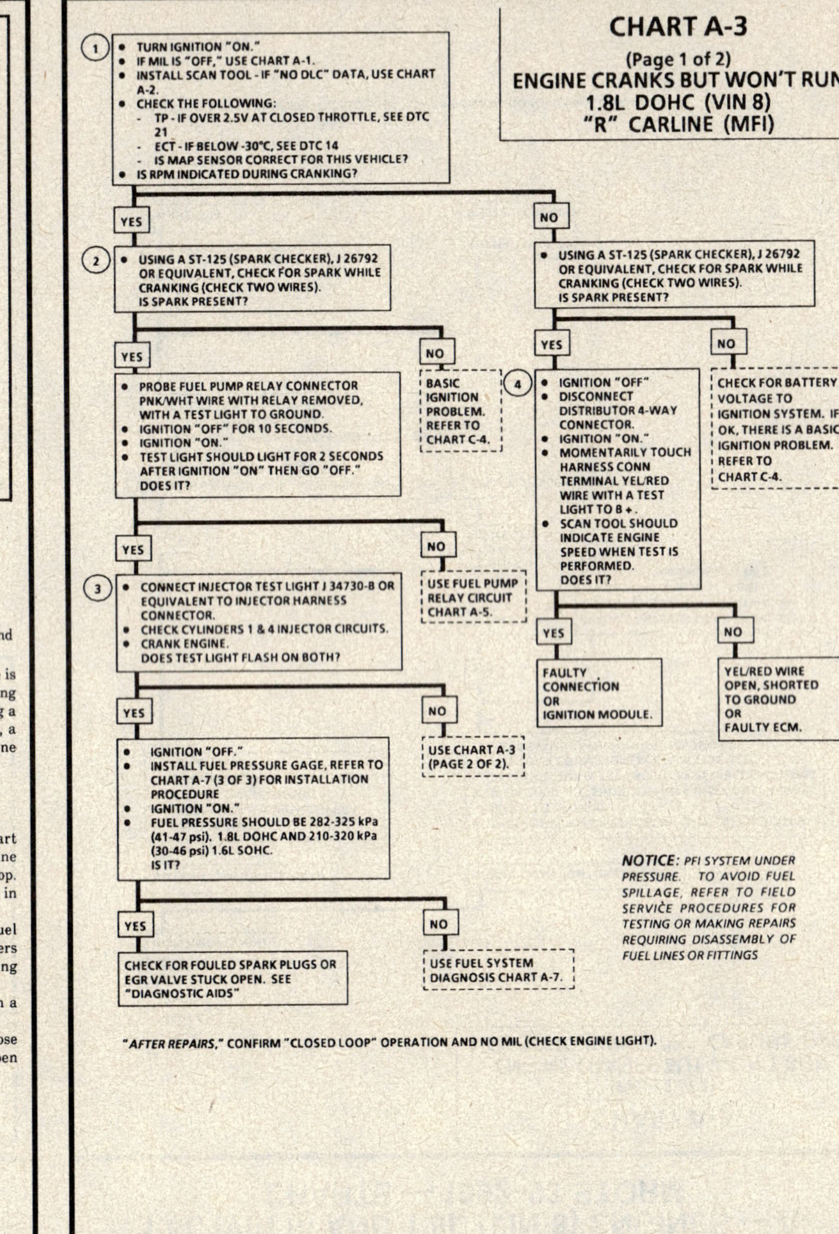

1.6L (VIN 6) AND 1.8L (VIN 8) ENGINES — A- CHARTS — 1992–93 STORM

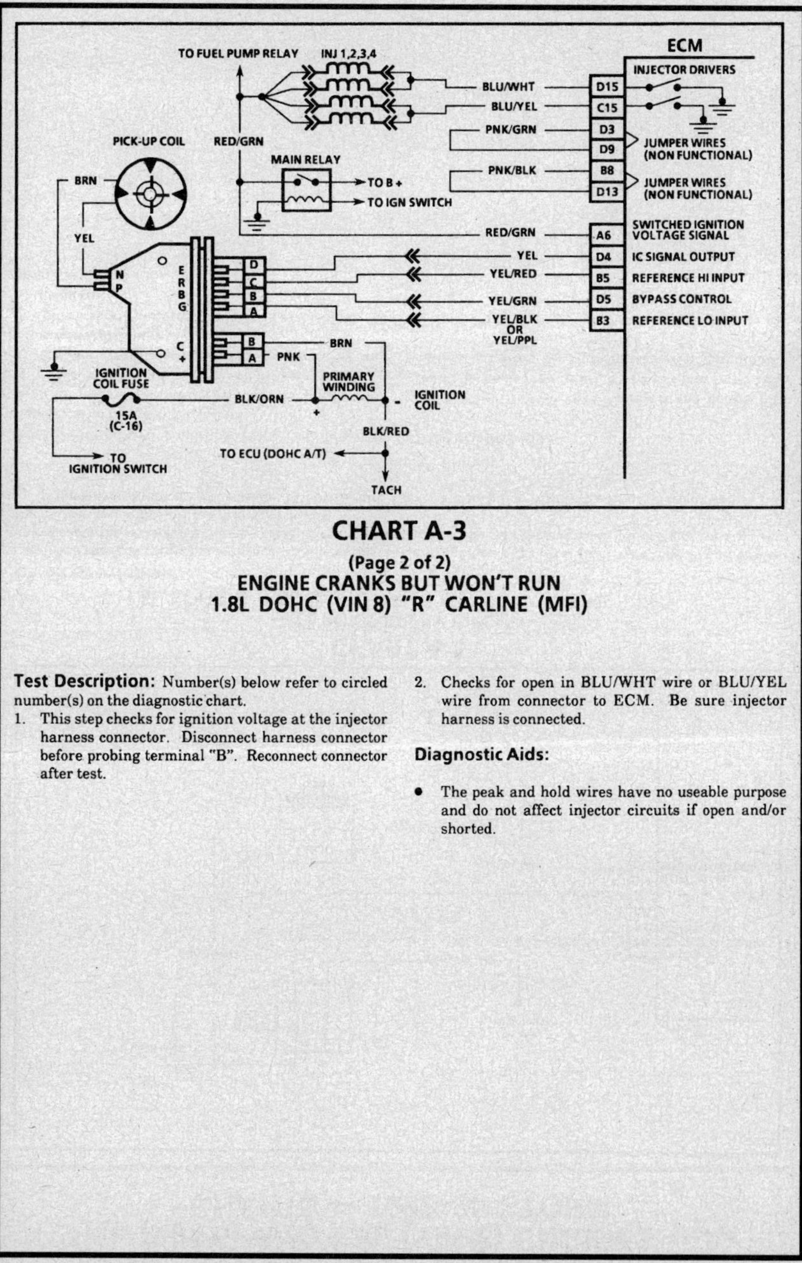

CHART A-3
(Page 2 of 2)
ENGINE CRANKS BUT WON'T RUN
1.8L DOHC (VIN 8) "R" CARLINE (MFI)

Test Description: Number(s) below refer to circled number(s) on the diagnostic chart.

1. This step checks for ignition voltage at the injector harness connector. Disconnect harness connector before probing terminal "B". Reconnect connector after test.

2. Checks for open in BLU/WHT wire or BLU/YEL wire from connector to ECM. Be sure injector harness is connected.

Diagnostic Aids:

- The peak and hold wires have no useable purpose and do not affect injector circuits if open and/or shorted.

1.6L (VIN 6) AND 1.8L (VIN 8) ENGINES — A- CHARTS — 1992–93 STORM

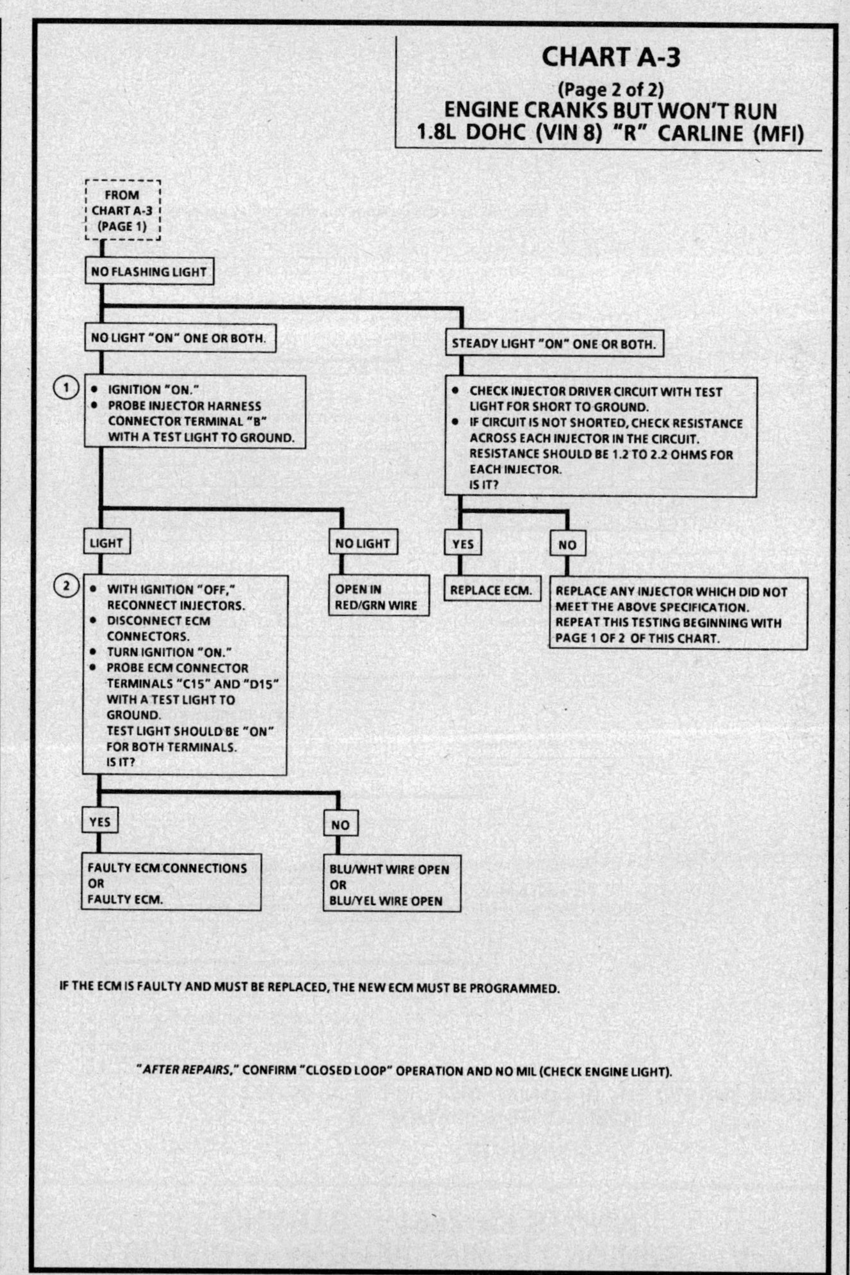

1.6L (VIN 6) AND 1.8L (VIN 8) ENGINES — A-CHARTS — 1992–93 STORM

1.6L (VIN 6) AND 1.8L (VIN 8) ENGINES — A-CHARTS — 1992–93 STORM

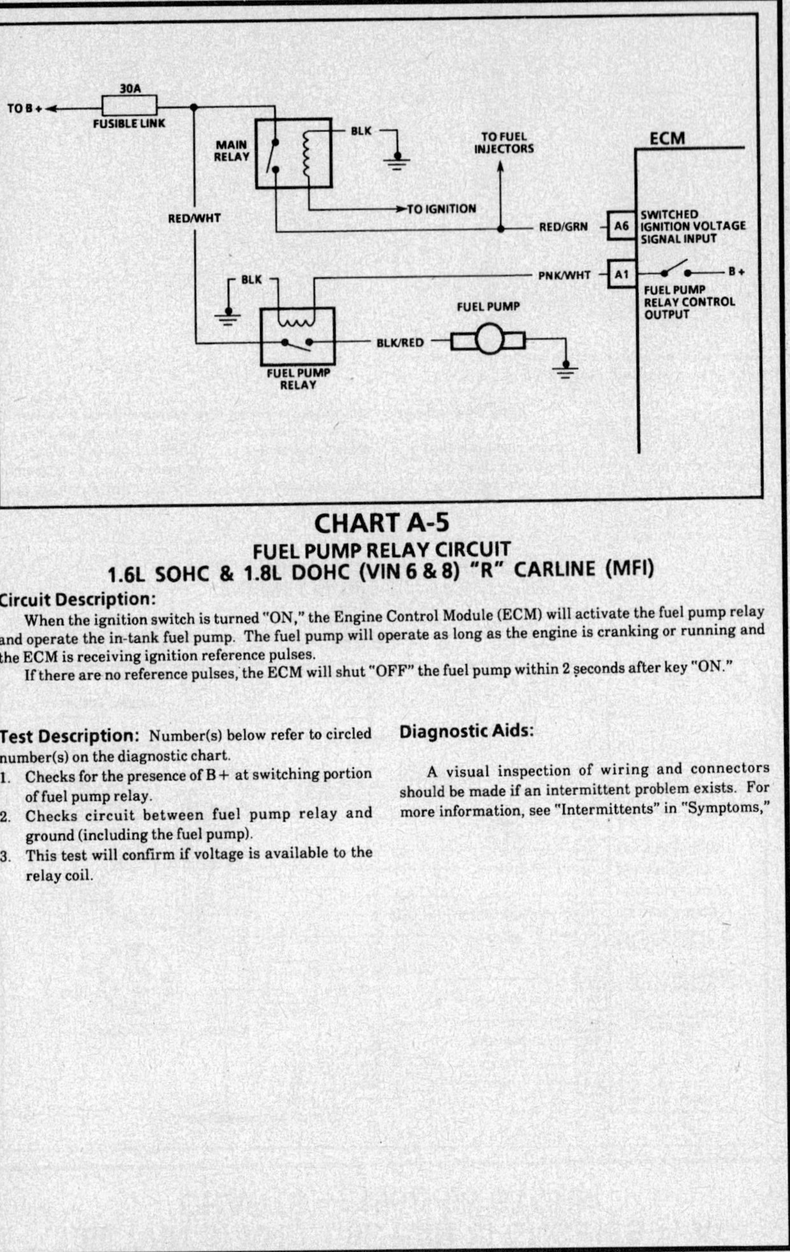

CHART A-5
FUEL PUMP RELAY CIRCUIT
1.6L SOHC & 1.8L DOHC (VIN 6 & 8) "R" CARLINE (MFI)

Circuit Description:

When the ignition switch is turned "ON," the Engine Control Module (ECM) will activate the fuel pump relay and operate the in-tank fuel pump. The fuel pump will operate as long as the engine is cranking or running and the ECM is receiving ignition reference pulses.

If there are no reference pulses, the ECM will shut "OFF" the fuel pump within 2 seconds after key "ON."

Test Description: Number(s) below refer to circled number(s) on the diagnostic chart.

1. Checks for the presence of B+ at switching portion of fuel pump relay.
2. Checks circuit between fuel pump relay and ground (including the fuel pump).
3. This test will confirm if voltage is available to the relay coil.

Diagnostic Aids:

A visual inspection of wiring and connectors should be made if an intermittent problem exists. For more information, see "Intermittents" in "Symptoms,"

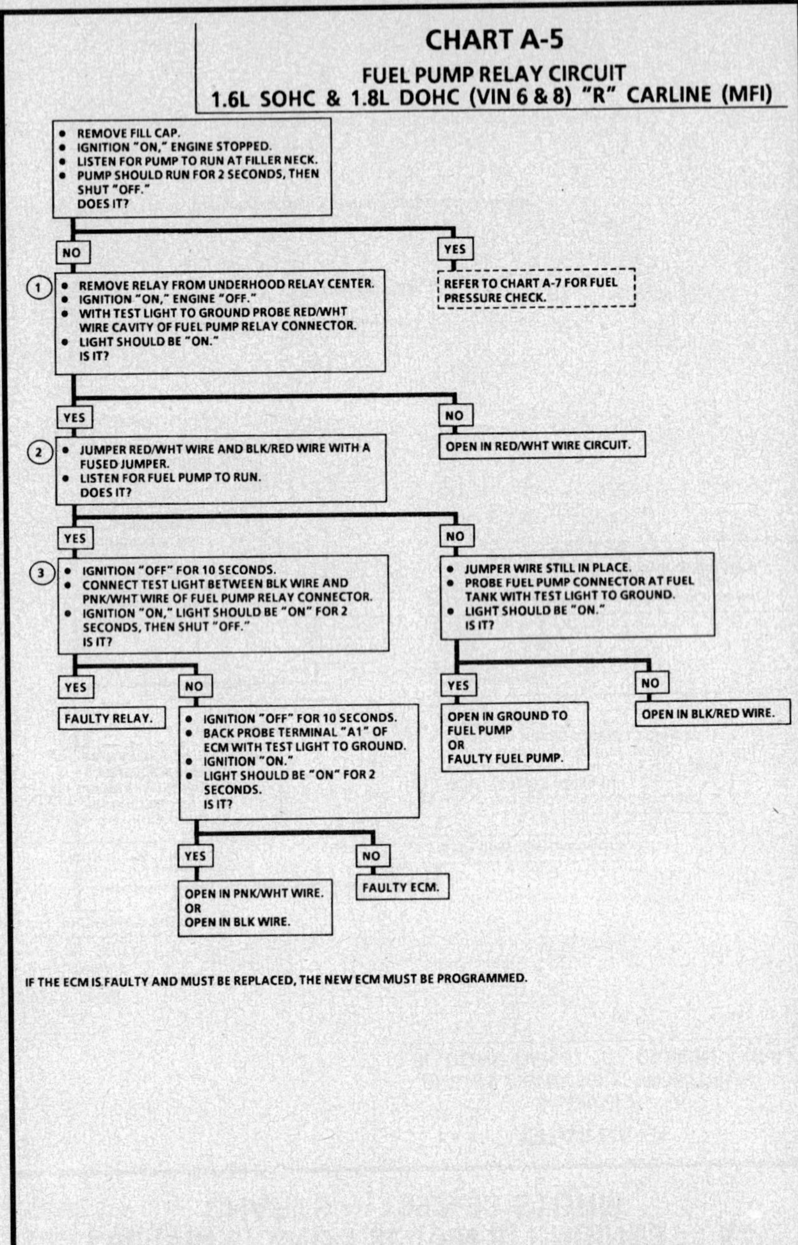

1.6L (VIN 6) AND 1.8L (VIN 8) ENGINES — A- CHARTS — 1992–93 STORM

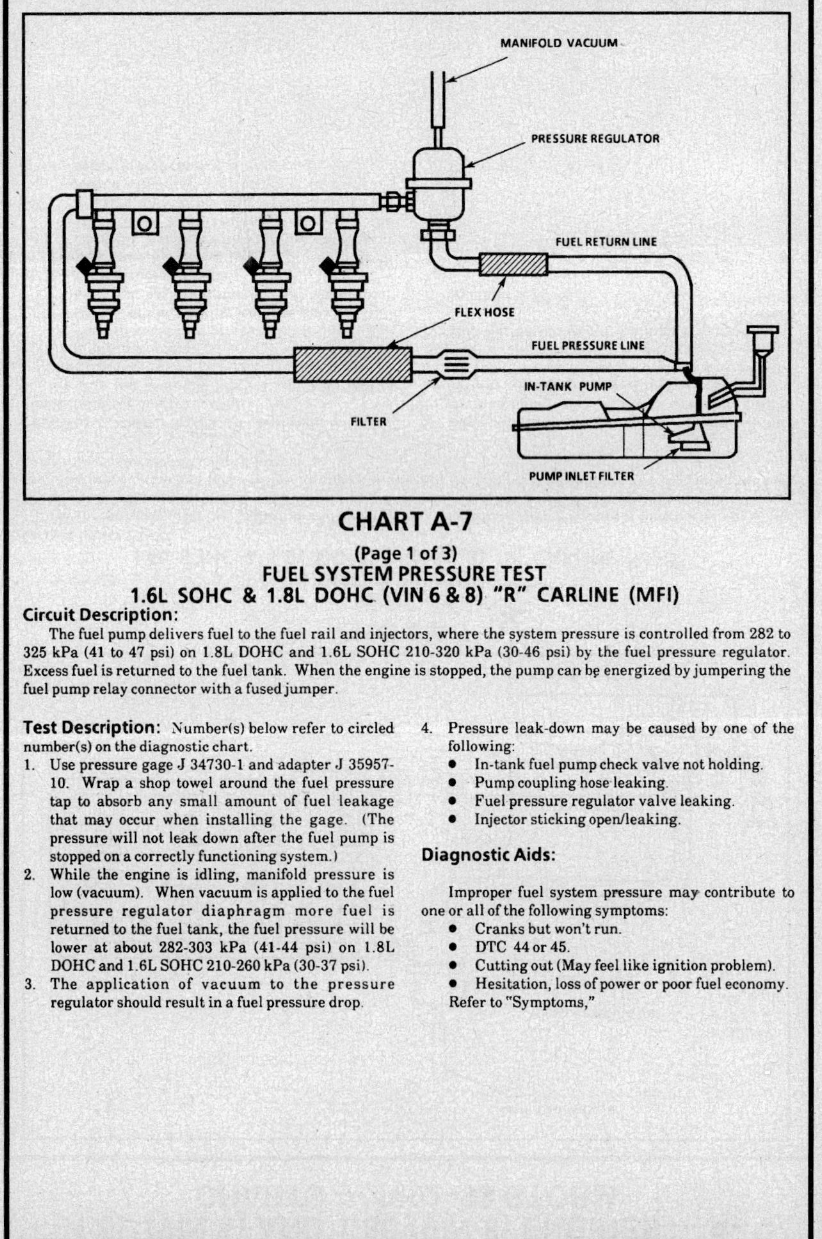

CHART A-7
(Page 1 of 3)
FUEL SYSTEM PRESSURE TEST
1.6L SOHC & 1.8L DOHC (VIN 6 & 8) "R" CARLINE (MFI)

Circuit Description:

The fuel pump delivers fuel to the fuel rail and injectors, where the system pressure is controlled from 282 to 325 kPa (41 to 47 psi) on 1.8L DOHC and 1.6L SOHC 210-320 kPa (30-46 psi) by the fuel pressure regulator. Excess fuel is returned to the fuel tank. When the engine is stopped, the pump can be energized by jumpering the fuel pump relay connector with a fused jumper.

Test Description: Number(s) below refer to circled number(s) on the diagnostic chart.

1. Use pressure gage J 34730-1 and adapter J 35957-10. Wrap a shop towel around the fuel pressure tap to absorb any small amount of fuel leakage that may occur when installing the gage. (The pressure will not leak down after the fuel pump is stopped on a correctly functioning system.)
2. While the engine is idling, manifold pressure is low (vacuum). When vacuum is applied to the fuel pressure regulator diaphragm more fuel is returned to the fuel tank, the fuel pressure will be lower at about 282-303 kPa (41-44 psi) on 1.8L DOHC and 1.6L SOHC 210-260 kPa (30-37 psi).
3. The application of vacuum to the pressure regulator should result in a fuel pressure drop.
4. Pressure leak-down may be caused by one of the following:
   - In-tank fuel pump check valve not holding.
   - Pump coupling hose leaking.
   - Fuel pressure regulator valve leaking.
   - Injector sticking open/leaking.

Diagnostic Aids:

Improper fuel system pressure may contribute to one or all of the following symptoms:
- Cranks but won't run.
- DTC 44 or 45.
- Cutting out (May feel like ignition problem).
- Hesitation, loss of power or poor fuel economy. Refer to "Symptoms,"

1.6L (VIN 6) AND 1.8L (VIN 8) ENGINES — A- CHARTS — 1992–93 STORM

NOTICE: FUEL SYSTEM IS UNDER PRESSURE. TO AVOID FUEL SPILLAGE, REFER TO FIELD SERVICE PROCEDURES FOR TESTING OR REPAIRS REQUIRING DISASSEMBLY OF FUEL LINES OR FITTINGS. REFER TO PAGE 3 OF 3 OF THIS CHART.

CHART A-7
(Page 1 of 3)
FUEL SYSTEM PRESSURE TEST
1.6L SOHC & 1.8L DOHC (VIN 6 & 8) "R" CARLINE (MFI)

(1)
- INSTALL FUEL PRESSURE GAGE, J 34730-1 OR EQUIVALENT, SEE CHART A-7 (PAGE 3 OF 3) FOR INSTALLATION PROCEDURE.
- DISCONNECT VACUUM HOSE FROM PRESSURE REGULATOR.
- IGNITION MUST BE "OFF" FOR 10 SECONDS.
- TURN IGNITION "ON." FUEL PUMP SHOULD RUN FOR ABOUT 2 SECONDS.
- NOTE FUEL PRESSURE AFTER PUMP STOPS.
- PRESSURE SHOULD BE 282-325 kPa (41-47 PSI) 1.8L DOHC AND 210-320 kPa (30-46 psi) ON 1.6L SOHC AND HOLDING STEADY. IS IT?

YES →

(2)
- START AND IDLE ENGINE AT NORMAL OPERATING TEMPERATURE.
- RECONNECT VACUUM LINE.
- FUEL PRESSURE SHOULD HAVE DROPPED SLIGHTLY. DID IT?

NO →

(3)
- DISCONNECT VACUUM HOSE FROM FUEL PRESSURE REGULATOR.
- WITH ENGINE IDLING, APPLY 12-14" Hg OF VACUUM TO PRESSURE REGULATOR.
- FUEL PRESSURE SHOULD DROP SLIGHTLY. DID IT?

YES →

NO TROUBLE FOUND. REFER TO "DIAGNOSTIC AIDS"

YES (from 3): LOCATE AND CORRECT CAUSE OF INTERRUPTED MANIFOLD PRESSURE SUPPLY TO REGULATOR.

NO (from 3): PRESSURE REGULATOR ASSEMBLY IS FAULTY.

NO (from 1) →

(4)
PRESSURE WITHIN SPECIFICATION, BUT NOT HOLDING.

PRESSURE OUT OF SPECIFICATION. → USE PAGE 2 OF THIS CHART.

NO PRESSURE → USE CHART A-5.

(from 4):
- IGNITION "OFF."
- CONNECT FUSED JUMPER TO FUEL PUMP RELAY CONNECTOR BETWEEN THE BLK/RED WIRE AND RED/WHT WIRE IN ENGINE COMPARTMENT.
- GRADUALLY BLOCK FUEL PRESSURE LINE BY PINCHING FLEX HOSE.
- REMOVE JUMPER FROM RELAY CONNECTOR WHILE CONTINUING TO BLOCK PRESSURE LINE.
- PRESSURE SHOULD HOLD. DID IT?

NO →
- REPEAT ABOVE PROCEDURE WHILE BLOCKING FUEL RETURN LINE INSTEAD OF PRESSURE LINE BY PINCHING FLEX HOSE.
- PRESSURE SHOULD HOLD. DID IT?

NO: LOCATE AND REPLACE LEAKING INJECTOR AND DRY SPARK PLUGS FROM FLOODED CYLINDERS.

YES: PRESSURE REGULATOR ASSEMBLY IS FAULTY.

YES (from first 4 block): CHECK FOR
- LEAKING PUMP COUPLING HOSE
- FAULTY IN-TANK PUMP.

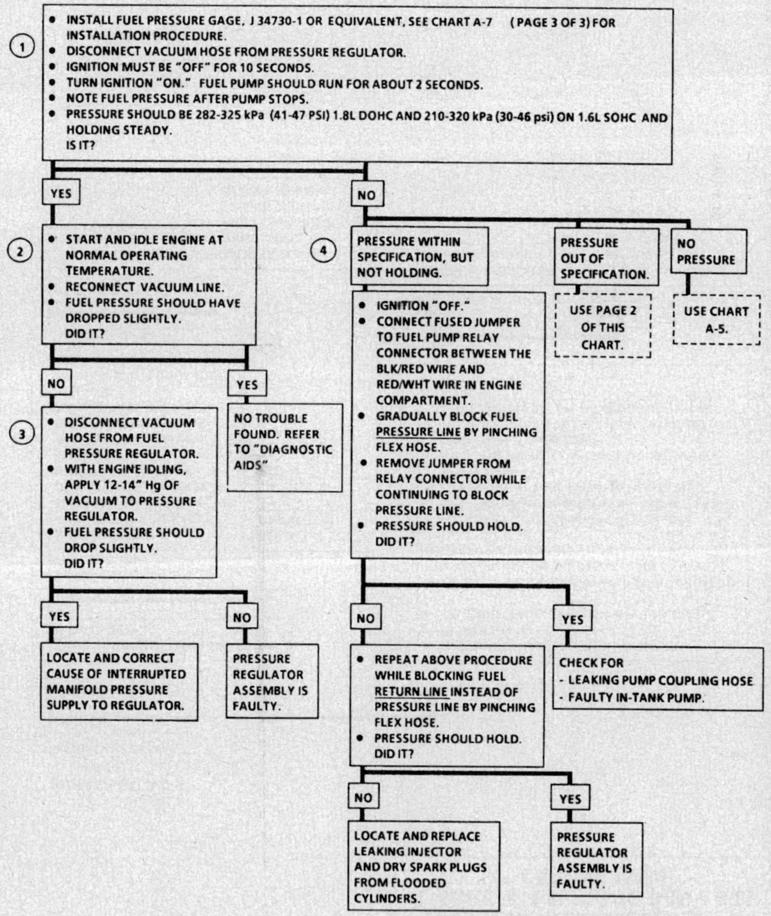

1.6L (VIN 6) AND 1.8L (VIN 8) ENGINES — A-CHARTS — 1992–93 STORM

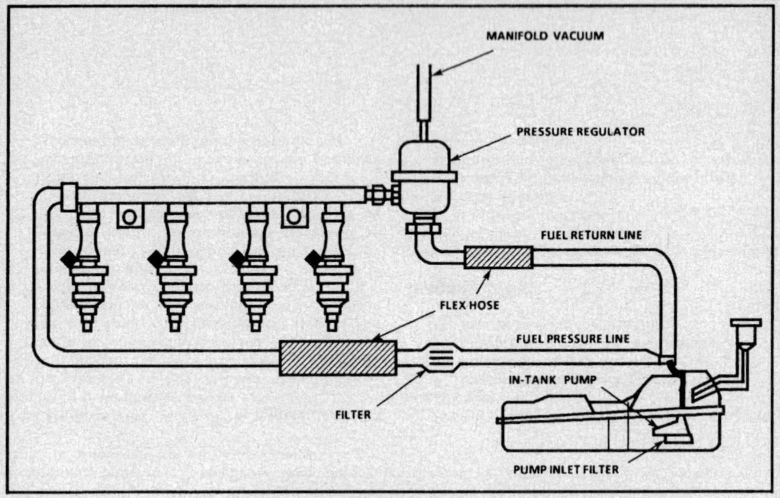

CHART A-7
(Page 2 of 3)
FUEL SYSTEM PRESSURE TEST
1.6L SOHC & 1.8L DOHC (VIN 6 & 8) "R" CARLINE (MFI)

Circuit Description:

The fuel pump delivers fuel to the fuel rail and injectors, where the system pressure is controlled from 282 to 325 kPa (41 to 47 psi) 1.8L DOHC and 210-320 kPa (30-46 psi) 1.6L SOHC by the fuel pressure regulator. Excess fuel is returned to the fuel tank. When the engine is stopped, the pump can be energized by jumpering the fuel pump relay connector with a fused jumper.

Test Description: Number(s) below refer to circled number(s) on the diagnostic chart.

5. Pressure less than 282 kPa (41 psi) 1.8L DOHC and 210 kPa (30 psi) 1.6L SOHC may be caused by one of two problems.
   - The regulated fuel pressure is too low. The system will be running lean and may set DTC 44. Also, hard cold starting and overall poor performance is possible.
   - Restricted flow is causing a pressure drop. Normally, a vehicle with a fuel pressure loss at idle will not be driveable. However, if the pressure drop occurs only while driving, the engine will surge and then stop as pressure begins to drop rapidly.

6. Restricting the fuel return line allows the fuel pump to build above regulated pressure. When battery voltage is applied to the pump test terminal, pressure should be above 413 kPa (60 psi).

7. This test determines if the high fuel pressure is due to a restricted fuel return line or a pressure regulator problem.

1.6L (VIN 6) AND 1.8L (VIN 8) ENGINES — A-CHARTS — 1992–93 STORM

NOTICE: FUEL SYSTEM IS UNDER PRESSURE. TO AVOID FUEL SPILLAGE, REFER TO FIELD SERVICE PROCEDURES FOR TESTING OR REPAIRS REQUIRING DISASSEMBLY OF FUEL LINES OR FITTING. REFER TO PAGE 3 OF 3 OF THIS CHART.

CHART A-7
(Page 2 of 3)
FUEL SYSTEM PRESSURE TEST
1.6L SOHC & 1.8L DOHC (VIN 6 & 8)
"R" CARLINE (MFI)

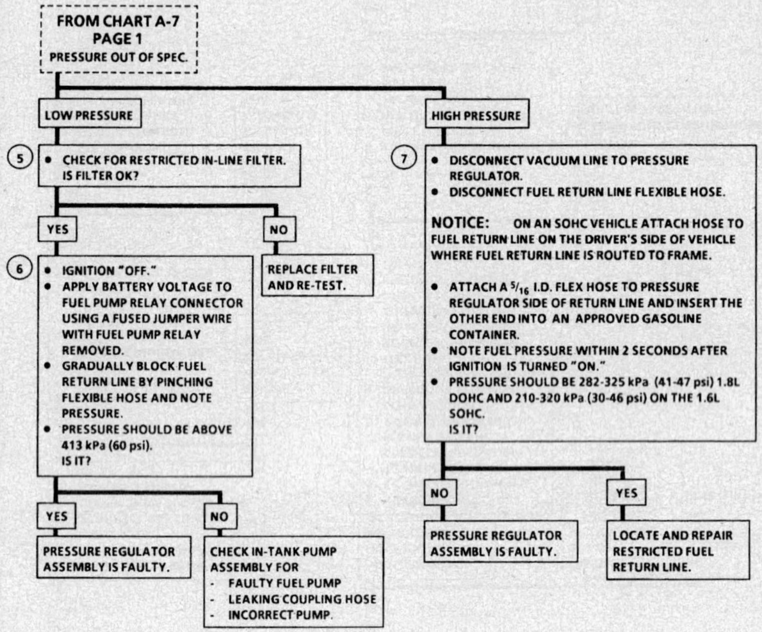

1.6L (VIN 6) AND 1.8L (VIN 8) ENGINES — A-CHARTS — 1992–93 STORM

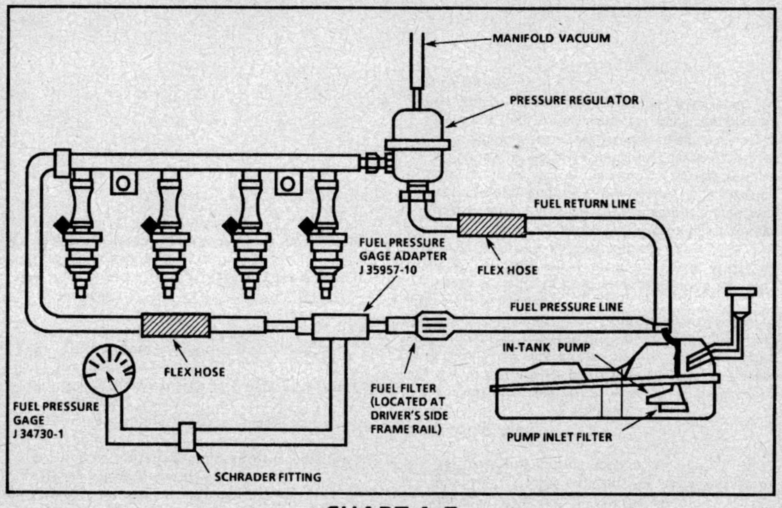

CHART A-7

(Page 3 of 3)
FUEL SYSTEM PRESSURE TEST
1.6L SOHC & 1.8L DOHC (VIN 6 & 8) "R" CARLINE (MFI)

Circuit Description:

The fuel rail is not equipped with a fuel system test fitting. In order to install fuel pressure gage J 34730-1, gage adapter J 23597-10 must be installed as shown.

1.6L (VIN 6) AND 1.8L (VIN 8) ENGINES — A-CHARTS — 1992–93 STORM

CHART A-7

(Page 3 of 3)
FUEL SYSTEM PRESSURE TEST
1.6L SOHC & 1.8L DOHC (VIN 6 & 8) "R" CARLINE (MFI)

Tools Required: J 34730 - 1 FUEL PRESSURE GAGE

J 35957 - 10 FUEL PRESSURE GAGE ADAPTER

CAUTION: TO REDUCE THE RISK OF FIRE AND PERSONAL INJURY:
- IT IS NECESSARY TO RELIEVE FUEL SYSTEM PRESSURE BEFORE CONNECTING A FUEL PRESSURE GAGE.
- A SMALL AMOUNT OF FUEL MAY BE RELEASED WHEN DISCONNECTING THE FUEL LINES. COVER FUEL LINE FITTINGS WITH A SHOP TOWEL BEFORE DISCONNECTING, TO CATCH ANY FUEL THAT MAY LEAK OUT. PLACE TOWEL IN APPROVED CONTAINER WHEN DISCONNECT IS COMPLETED.

FUEL PRESSURE RELIEF PROCEDURE
1. REMOVE FUEL CAP.
2. REMOVE FUEL PUMP RELAY FROM UNDERHOOD RELAY CENTER.
3. START ENGINE AND ALLOW TO STALL.
4. CRANK ENGINE FOR ADDITIONAL 30 SECONDS.
5. REMOVE NEGATIVE BATTERY CABLE.

FUEL GAGE ADAPTER INSTALLATION
1. DISCONNECT FUEL PRESSURE LINE AT THE FUEL FILTER. USE A SHOP TOWEL TO CATCH ANY ADDITIONAL FUEL THAT WAS LEFT IN THE LINE.
2. INSTALL FUEL GAGE ADAPTER J 35957 - 10 BETWEEN FUEL PRESSURE LINE AND FUEL FILTER AT FRAME RAIL.
3. INSTALL FUEL GAGE J 34730 - 1 TO ADAPTER.
4. REINSTALL FUEL PUMP RELAY.
5. CONNECT NEGATIVE BATTERY CABLE.
6. INSTALL FUEL CAP.
7. START ENGINE AND CHECK FOR LEAKS.

NOTE: REVERSE THE PROCEDURE ABOVE TO REMOVE FUEL GAGE ADAPTER.

1.6L (VIN 6) AND 1.8L (VIN 8) ENGINES — DIAGNOSTIC TROUBLE CODE CHARTS — 1992–93 STORM

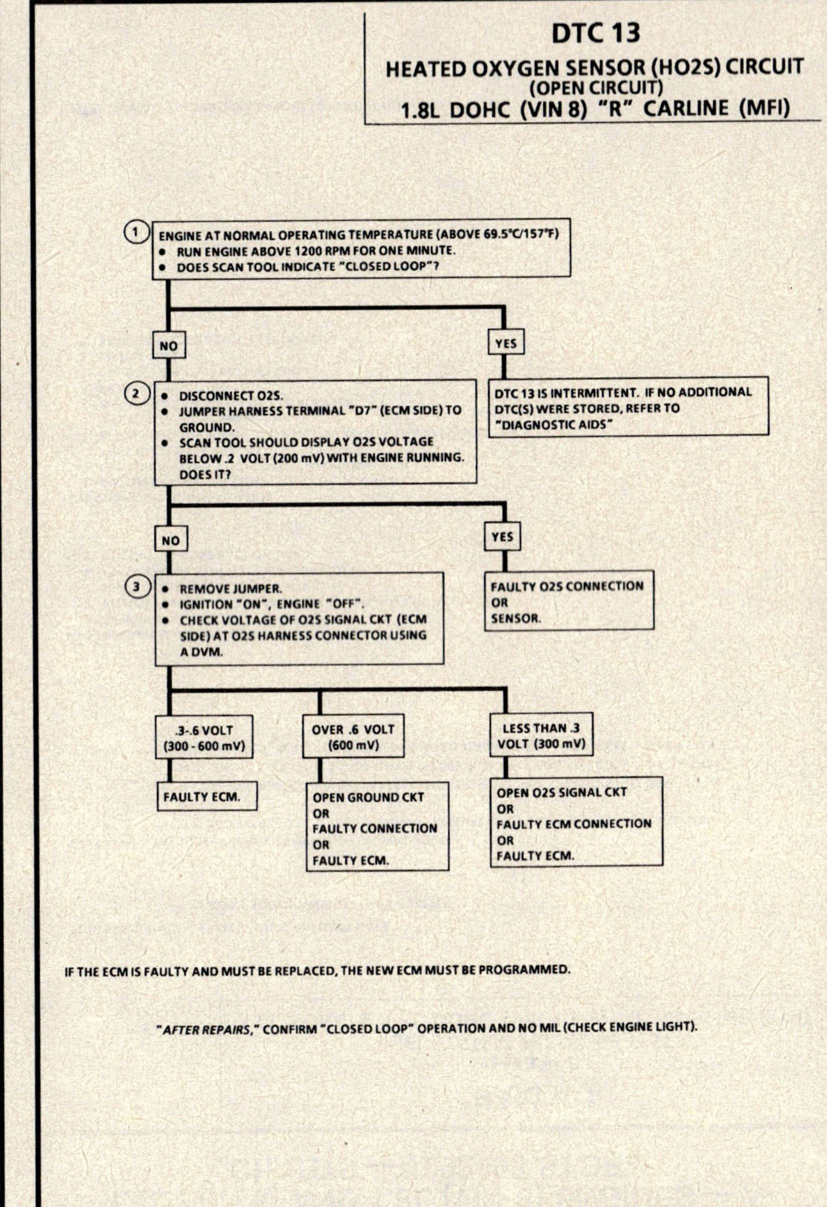

DTC 13
HEATED OXYGEN SENSOR (HO2S) CIRCUIT
(OPEN CIRCUIT)
1.8L DOHC (VIN 8) "R" CARLINE (MFI)

Circuit Description:

The ECM supplies a voltage of about .45 volt between terminals "D7" and "D6". (If measured with a 10 megohm digital voltmeter, this may read as low as .32 volt.)

The heater circuit of the HO2S is turned "ON" when the engine is running. The HO2S as it heats up will begin to send a varying voltage signal to the ECM. This voltage will vary from about .1 volt (lean exhaust) to about 1.0 volts (rich exhaust) depending on the condition of the exhaust gases in the exhaust manifold.

The sensor is like an open circuit and produces no voltage when it is below 315°C (600°F). An open sensor circuit causes "Open Loop" operation.

Test Description: Number(s) below refer to circled number(s) on the diagnostic chart.

1. DTC 13 will set under if:
   - Engine operating temperature above 69.5°C (183°F).
   - At least 2 minutes have elapsed since engine start-up.
   - HO2S signal voltage is steady between .347 volt and .547 volt.
   - Vehicle speed > 3 mph.
   - Throttle angle is above 5%.
   - All above conditions are met for about 40 seconds.

   If the conditions for a DTC 13 exist, the system will not go "Closed Loop."

2. This test determines if the HO2S is the problem or if the ECM and wiring are at fault.

3. Use only a 10 megohm digital when performing this test. This test checks the continuity of "D7" and "D6". If "D6" is open, the ECM voltage on "D7" will be over .6 volt (600 mV).

Diagnostic Aids:

Normal scan tool HO2S voltage varies between 100 mV to 999 mV (.1 and 1.0 volt) while in "Closed Loop." DTC 13 sets in one minute if sensor signal voltage remains between .347 volt and .547 volt, but the system will go to "Open Loop" in about 15 seconds after DTC 13 has set.

Verify a clean, tight ground connection for "D6". Open "D7" or "D6" will result in a DTC 13. If DTC 13 is intermittent, refer to "Symptoms,"

- Check the oxygen sensor wire for induced voltage (EMI). Disconnect the oxygen sensor signal wire at the sensor and the ECM. Using DVM measure the wire with the engine running for any voltage. If 300 mV or more are indicated, check the signal wire shield and/or reroute the signal wire away from any high voltage sources. Such source could be ignition wires, battery cables, Alternator "B" circuit, ect.

1.6L (VIN 6) AND 1.8L (VIN 8) ENGINES — DIAGNOSTIC TROUBLE CODE CHARTS — 1992–93 STORM

DTC 13
HEATED OXYGEN SENSOR (HO2S) CIRCUIT
(OPEN CIRCUIT)
1.8L DOHC (VIN 8) "R" CARLINE (MFI)

IF THE ECM IS FAULTY AND MUST BE REPLACED, THE NEW ECM MUST BE PROGRAMMED.

"AFTER REPAIRS," CONFIRM "CLOSED LOOP" OPERATION AND NO MIL (CHECK ENGINE LIGHT).

1.6L (VIN 6) AND 1.8L (VIN 8) ENGINES — DIAGNOSTIC TROUBLE CODE CHARTS — 1992–93 STORM

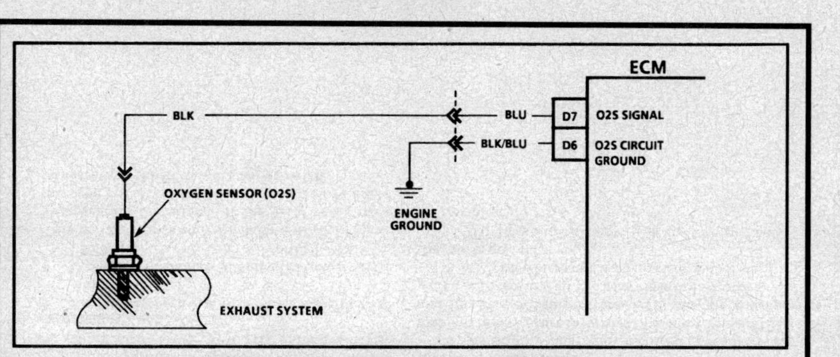

DTC 13
OXYGEN SENSOR (O2S) CIRCUIT
(OPEN CIRCUIT)
1.6L SOHC (VIN 6) "R" CARLINE (MFI)

Circuit Description:

The ECM supplies a voltage of about of .45 volt between terminals "D7" and "D6". (If measured with 10 megohm digital voltmeter, this may read as low as .32 volt).

The oxygen sensor varies the voltage within a range of about 1.0 volts, if the exhaust is rich, down to about .1 volt, if exhaust is lean.

The sensor is like an open circuit and produces no voltage, when it is below 315°C (600°F). An open sensor circuit causes "Open Loop" operation.

Test Description: Number(s) below refer to circled number(s) on the diagnostic chart.
1. DTC 13 will set if:
   - Engine operating temperature above 69.5°C (183°F).
   - At least 50 seconds engine run time after start.
   - O2S signal voltage steady between .347 volt and .547 volt.
   - Throttle angle above 5%.
   - Vehicle speed > 3 mph.
   - All conditions must be met for about 60 seconds.

   If the conditions for a DTC 13 exist, the system will not go "Closed Loop."
2. This test determines if the oxygen sensor is the problem, or if the ECM/wiring are at fault.
3. Use only a high impedance digital volt ohmmeter when preforming this test. This test checks the continuity of "D7" and "D6". If "D6" is open, the ECM voltage on "D7" will be over .6 volt (600 mV).

Diagnostic Aids:

Normal scan voltage varies between 100 mV to 999 mV (.1 and 1.0 volt), while in "Closed Loop." DTC 13 sets within one minute, if voltage remains between .347 volt and .547 volt, but the system will go "Open Loop" in about 15 seconds after DTC 13 has set.

Verify a clean, tight ground connection for "D6". Open "D7" or "D6" will result in a DTC 13.

If DTC 13 is intermittent, refer to "Intermittents," in "Symptoms,"
- Check the oxygen sensor wire for induced voltage (EMI). Disconnect the oxygen sensor signal wire at the sensor and the ECM. Using DVM measure the wire with the engine running for any voltage. If 300 mV or more are indicated, check the signal wire shield and/or reroute the signal wire away from any high voltage sources. Such source could be ignition wires, battery cables, Alternator "B" circuit, ect.

1.6L (VIN 6) AND 1.8L (VIN 8) ENGINES — DIAGNOSTIC TROUBLE CODE CHARTS — 1992–93 STORM

DTC 13
OXYGEN SENSOR (O2S) CIRCUIT
(OPEN CIRCUIT)
1.6L SOHC (VIN 6) "R" CARLINE (MFI)

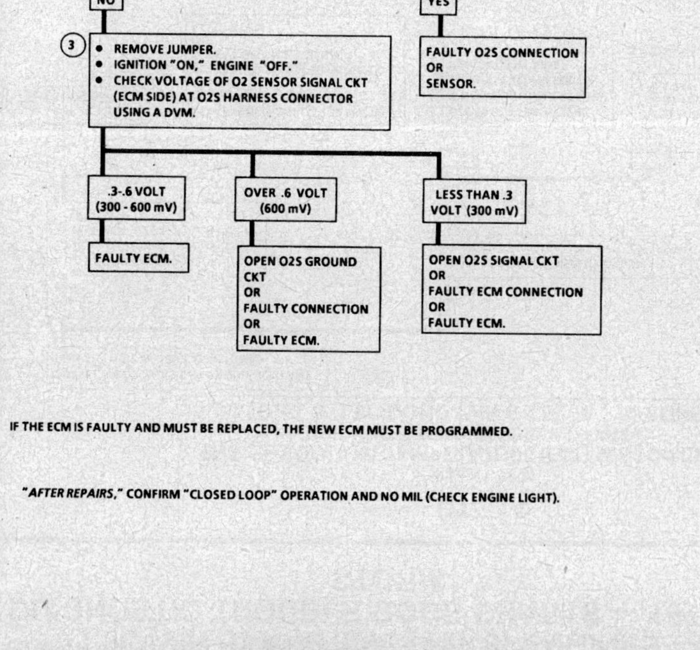

IF THE ECM IS FAULTY AND MUST BE REPLACED, THE NEW ECM MUST BE PROGRAMMED.

"AFTER REPAIRS," CONFIRM "CLOSED LOOP" OPERATION AND NO MIL (CHECK ENGINE LIGHT).

1.6L (VIN 6) AND 1.8L (VIN 8) ENGINES — DIAGNOSTIC TROUBLE CODE CHARTS — 1992–93 STORM

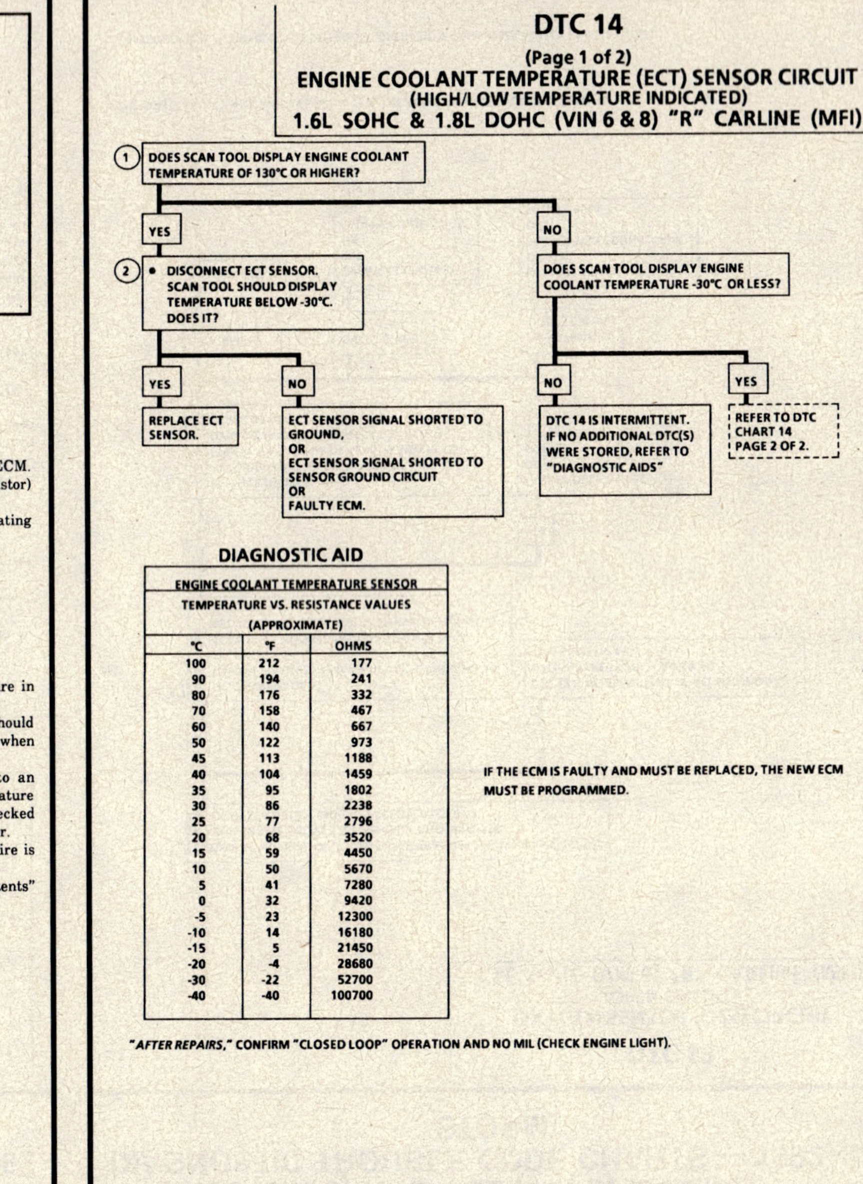

DTC 14
(Page 1 of 2)
ENGINE COOLANT TEMPERATURE (ECT) SENSOR CIRCUIT
(HIGH/LOW TEMPERATURE INDICATED)
1.6L SOHC & 1.8L DOHC (VIN 6 & 8) "R" CARLINE (MFI)

Circuit Description:

The Engine Coolant Temperature (ECT) sensor uses a thermistor to control the signal voltage from the ECM. The ECM applies a reference voltage on GRY wire to the sensor. When the engine is cold, the sensor (thermistor) resistance is high. The ECM will then sense a high signal voltage.

As the engine warms up, the sensor resistance decreases and the voltage drops. At normal engine operating temperature, the voltage will measure about 1.5 to 2.0 volts at ECM terminal.

Engine coolant temperature is one of the inputs used to control the following:

- Fuel Delivery.
- Ignition Control (IC).
- Idle Air Control (IAC).

Test Description: Number(s) below refer to circled number(s) on the diagnostic chart.

1. Checks to see if a DTC was set as result of hard failure or intermittent condition.
   DTC 14 will set if:
   <u>High Temperature Criteria:</u>
   - Signal voltage indicates an engine coolant temperature above 213°C (412°F) for 3 seconds.
   <u>Low Temperature Criteria:</u>
   - Engine has been running for at least two (2) minutes.
   - Coolant temperature is indicated below -37°C (-35°F).
2. This test simulates conditions for a DTC 14 (Low Temperature Indicated). If the ECM recognizes the open circuit (high voltage) and displays a low temperature, the ECM and wiring are OK.

Diagnostic Aids:

A scan tool reads engine coolant temperature in degrees celsius and degrees of fahrenheit.

After the engine is started, the temperature should rise steadily to about 90°C (220°F), then stabilize when the thermostat opens.

If the engine has been allowed to cool to an ambient temperature (overnight), coolant temperature and Intake Air Temperature (IAT) may be checked with a scan tool and should read close to each other.

A DTC 14 will result if the sensor signal wire is shorted to ground.

If DTC 14 is intermittent, refer to "Intermittents" in "Symptoms,"

1.6L (VIN 6) AND 1.8L (VIN 8) ENGINES — DIAGNOSTIC TROUBLE CODE CHARTS — 1992–93 STORM

DTC 14
(Page 1 of 2)
ENGINE COOLANT TEMPERATURE (ECT) SENSOR CIRCUIT
(HIGH/LOW TEMPERATURE INDICATED)
1.6L SOHC & 1.8L DOHC (VIN 6 & 8) "R" CARLINE (MFI)

DIAGNOSTIC AID

ENGINE COOLANT TEMPERATURE SENSOR		
TEMPERATURE VS. RESISTANCE VALUES (APPROXIMATE)		
°C	°F	OHMS
100	212	177
90	194	241
80	176	332
70	158	467
60	140	667
50	122	973
45	113	1188
40	104	1459
35	95	1802
30	86	2238
25	77	2796
20	68	3520
15	59	4450
10	50	5670
5	41	7280
0	32	9420
-5	23	12300
-10	14	16180
-15	5	21450
-20	-4	28680
-30	-22	52700
-40	-40	100700

IF THE ECM IS FAULTY AND MUST BE REPLACED, THE NEW ECM MUST BE PROGRAMMED.

"AFTER REPAIRS," CONFIRM "CLOSED LOOP" OPERATION AND NO MIL (CHECK ENGINE LIGHT).

1.6L (VIN 6) AND 1.8L (VIN 8) ENGINES — DIAGNOSTIC TROUBLE CODE CHARTS — 1992–93 STORM

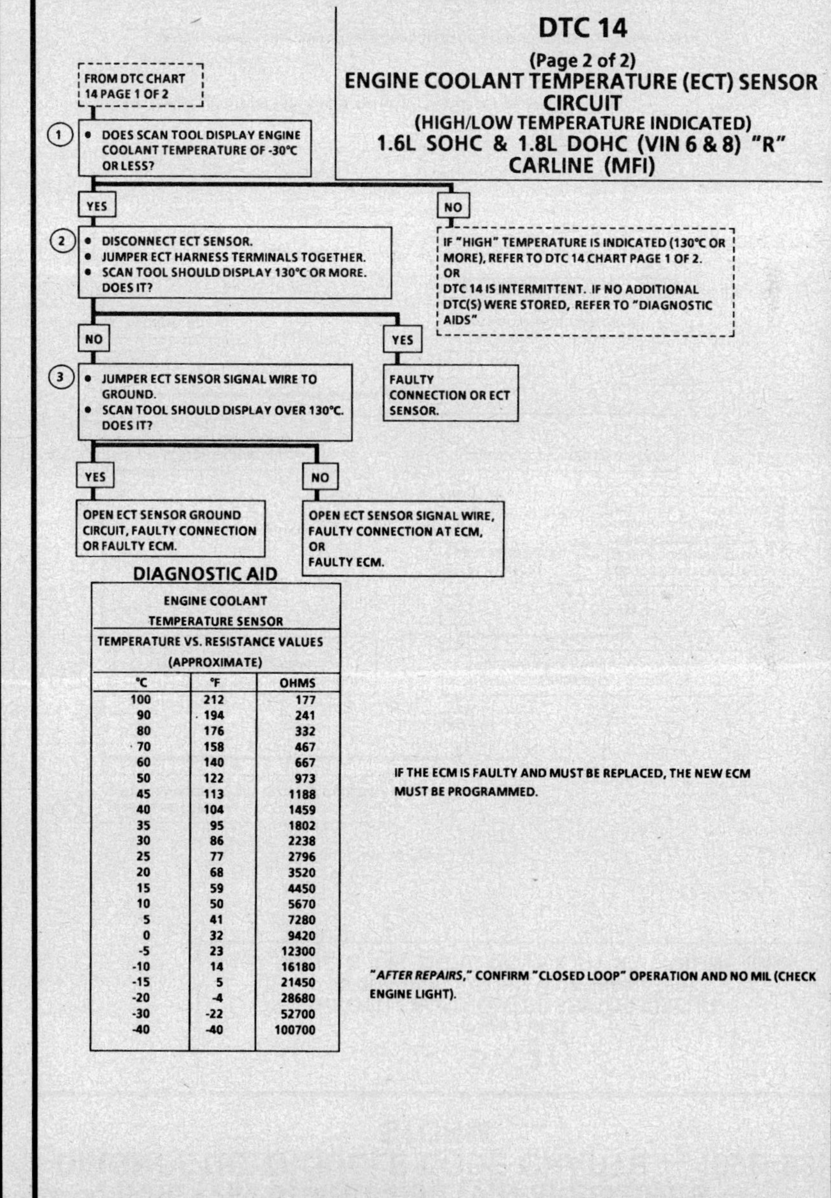

DTC 14
(Page 2 of 2)
ENGINE COOLANT TEMPERATURE (ECT) SENSOR CIRCUIT
(HIGH/LOW TEMPERATURE INDICATED)
1.6L SOHC & 1.8L DOHC (VIN 6 & 8) "R" CARLINE (MFI)

Circuit Description:

The Engine Coolant Temperature (ECT) sensor uses a thermistor to control the signal voltage to the ECM. The ECM applies a voltage on the sensor signal wire to the sensor. When the engine is cold, the sensor (thermistor) resistance is high, therefore, the ECM will see high signal voltage.

As the engine warms, the sensor resistance becomes less and the voltage drops. At normal engine operating temperature, the voltage will measure about 1.5 to 2.0 volts at the ECM terminal.

Engine coolant temperature is one of the inputs used to control:

- Fuel Delivery.
- Ignition Control (IC).
- Engine Idle Speed (IAC).

Test Description: Number(s) below refer to circled number(s) on the diagnostic chart.

1. Checks to see if code was set as a result of hard failure or intermittent condition.
2. This test simulates conditions for a DTC 14 (High Temperature Indicated). If the ECM recognizes the grounded circuit (low voltage), and displays a high temperature, the ECM and wiring are OK.
3. This test will determine if there is a wiring problem or a faulty ECM.

Diagnostic Aids:

A scan tool reads engine temperature in degrees centigrade and degrees in fahrenheit. After the engine is started, the temperature should rise steadily to about 90°C (220°F), then stabilize when the thermostat opens.

If the engine has been allowed to cool to an ambient temperature (overnight), coolant and IAT temperatures may be checked with a scan tool and should read close to each other.

A DTC 14 will result if the sensor ground wire or the sensor signal wire are open.

If DTC 14 is intermittent, refer to "Intermittents" in "Symptoms,"

DIAGNOSTIC AID

ENGINE COOLANT TEMPERATURE SENSOR TEMPERATURE VS. RESISTANCE VALUES (APPROXIMATE)		
°C	°F	OHMS
100	212	177
90	194	241
80	176	332
70	158	467
60	140	667
50	122	973
45	113	1188
40	104	1459
35	95	1802
30	86	2238
25	77	2796
20	68	3520
15	59	4450
10	50	5670
5	41	7280
0	32	9420
-5	23	12300
-10	14	16180
-15	5	21450
-20	-4	28680
-30	-22	52700
-40	-40	100700

1.6L (VIN 6) AND 1.8L (VIN 8) ENGINES — DIAGNOSTIC TROUBLE CODE CHARTS — 1992–93 STORM

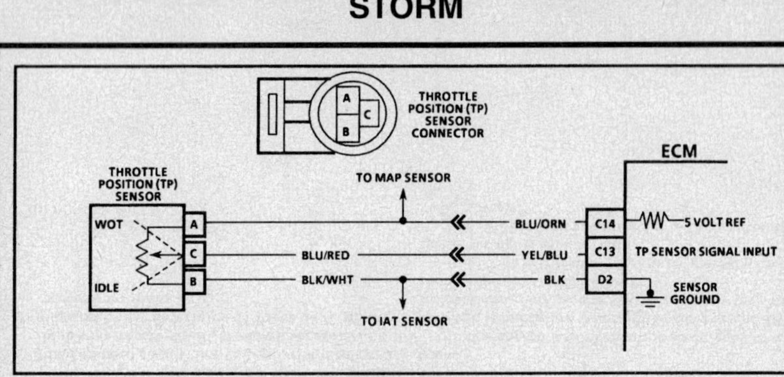

DTC 21

(Page 1 of 2)
THROTTLE POSITION (TP) SENSOR CIRCUIT
(SIGNAL VOLTAGE HIGH/LOW INDICATED)
1.6L SOHC & 1.8L DOHC (VIN 6 & 8) "R" CARLINE (MFI)

Circuit Description:

The Throttle Position (TP) sensor provides a voltage signal that changes relative to the throttle valve position. Signal voltage will vary from less than 1.25 volts at idle to about 5 volts at Wide Open Throttle (WOT).

The TP sensor signal is one of the most important inputs used by the ECM for fuel control and for many of the ECM controlled outputs.

Test Description: Number(s) below refer to circled number(s) on the diagnostic chart.

1. This step checks to see if DTC 21 is the result of a hard failure or an intermittent condition.
   A DTC 21 will set if:
   High TP Sensor Voltage Criteria:
   - TP sensor reading above 2.5 volts.
   - Engine speed less than 1800 RPM.
   - MAP reading below 60 kPa.
   - All of the above conditions present for 2 seconds.
   Low TP Sensor Voltage Criteria:
   - Engine running.
   - TP sensor voltage below .2 volt (200 mV).
2. This step simulates conditions for a DTC 21 (Signal Voltage Low). If the ECM recognizes the change, the ECM, and the signal circuit wire and the 5 volt reference wire are OK.

3. This step isolates a faulty sensor ECM or an open BLK/WHT wire. If the BLK/WHT wire is open, there may also be a DTC 23.

Diagnostic Aids:

A scan tool displays throttle position in volts. Closed throttle voltage should be less than 1.25 volts. TP sensor voltage should increase at a steady rate as throttle is moved to WOT.

A DTC 21 will result if the sensor ground wire is open or the signal circuit wire is shorted to voltage. If DTC 21 is intermittent, refer to "Intermittents" in "Symptoms,"

NOTICE: On the 1.8L DOHC vehicles with A/T transaxles, the "ECONO" light will also be flashing when DTC 21 is logged in the ECM's memory or the circuit fault is still present. An incorrectly programmed EEPROM may also cause the "ECONO" light to be flashing.

1.6L (VIN 6) AND 1.8L (VIN 8) ENGINES — DIAGNOSTIC TROUBLE CODE CHARTS — 1992–93 STORM

DTC 21

(Page 1 of 2)
THROTTLE POSITION (TP) SENSOR CIRCUIT
(SIGNAL VOLTAGE HIGH/LOW INDICATED)
1.6L SOHC & 1.8L DOHC (VIN 6 & 8) "R" CARLINE (MFI)

① • THROTTLE CLOSED.
DOES SCAN TOOL DISPLAY TP OVER 2.5 VOLTS?

YES → ② • DISCONNECT TP SENSOR. SCAN TOOL SHOULD DISPLAY TP BELOW .2 VOLT (200mV). DOES IT?

NO → • THROTTLE AT WIDE OPEN POSITION DOES SCAN TOOL DISPLAY TP AT 4.0 VOLTS OR MORE?

NO → REFER TO DTC 21 CHART PAGE 2 OF 2

YES → CODE 21 IS INTERMITTENT. IF NO ADDITIONAL DTC(S) WERE STORED, REFER TO "DIAGNOSTIC AIDS"

YES → ③ • PROBE TP SENSOR GROUND CIRCUIT WITH A TEST LIGHT CONNECTED TO BATTERY VOLTAGE.

NO → TP SIGNAL CKT SHORTED TO VOLTAGE OR FAULTY ECM.

LIGHT "ON" → FAULTY CONNECTION OR TP SENSOR.

LIGHT "OFF" → OPEN TP SENSOR GROUND CIRCUIT OR FAULTY ECM.

IF THE ECM IS FAULTY AND MUST BE REPLACED, THE NEW ECM MUST BE PROGRAMMED.

"AFTER REPAIRS," CONFIRM "CLOSED LOOP" OPERATION AND NO MIL (CHECK ENGINE LIGHT).

1.6L (VIN 6) AND 1.8L (VIN 8) ENGINES — DIAGNOSTIC TROUBLE CODE CHARTS — 1992–93 STORM

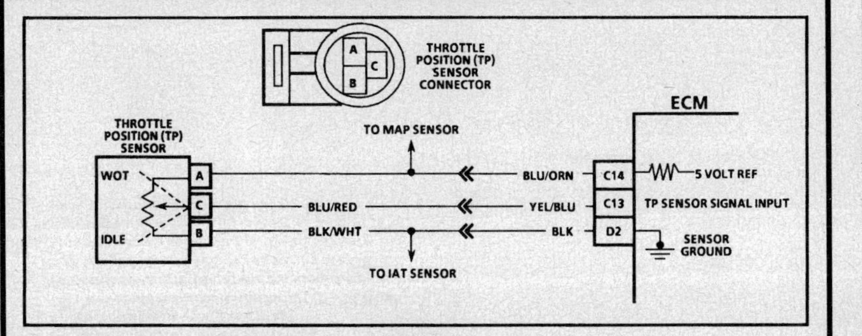

DTC 21
(Page 2 of 2)
THROTTLE POSITION (TP) SENSOR CIRCUIT
(SIGNAL VOLTAGE HIGH/LOW INDICATED)
1.6L SOHC & 1.8L DOHC (VIN 6 & 8) "R" CARLINE (MFI)

Circuit Description:

The Throttle Position (TP) sensor provides a voltage signal that changes, relative to the throttle valve position. Signal voltage will vary from less than 1.25 volts at idle to about 5 volts at Wide Open Throttle (WOT).

The TP sensor signal is one of the most important inputs used by the ECM for fuel control and for many of the ECM controlled outputs.

Test Description: Number(s) below refer to circled number(s) on the diagnostic chart.

1. This step checks to see if DTC 21 is the result of a hard failure or an intermittent condition.
2. This step simulates conditions for a DTC 21 (Signal Voltage High). If a DTC 21 (Signal Voltage High) is set, or the scan tool displays over 4 volts, the ECM and wiring are OK.
3. The scan tool may not display 12 volts. The important thing is that the ECM recognizes the voltage as over 4 volts, indicating that the signal circuit wire and the ECM are OK.
4. If the 5 volt reference wire is open or shorted to ground, there may also be a stored DTC 33.

Diagnostic Aids:

A scan tool displays throttle position in volts. Closed throttle voltage should be less than 1.25 volts. TP sensor voltage should increase at a steady rate as throttle is moved to WOT.

An open or grounded 5 volt reference wire or signal circuit wire will result in a DTC 21.

If DTC 21 is intermittent, refer to "Symptoms,"

NOTICE: On the 1.8L DOHC vehicles with A/T transaxles, the "ECONO" light will also be flashing when DTC 21 is logged in the ECM's memory or the circuit fault is still present. An incorrectly programmed EEPROM may also cause the "ECONO" light to be flashing.

1.6L (VIN 6) AND 1.8L (VIN 8) ENGINES — DIAGNOSTIC TROUBLE CODE CHARTS — 1992–93 STORM

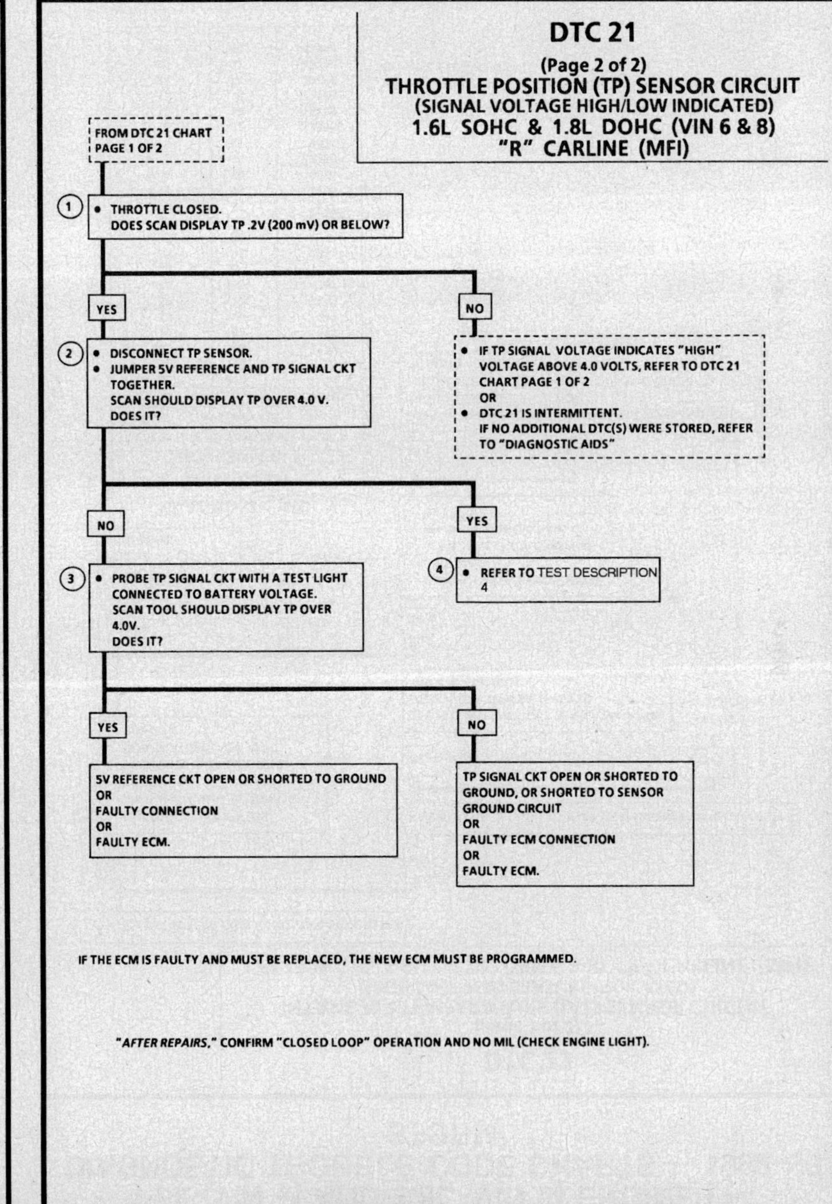

1.6L (VIN 6) AND 1.8L (VIN 8) ENGINES — DIAGNOSTIC TROUBLE CODE CHARTS — 1992–93 STORM

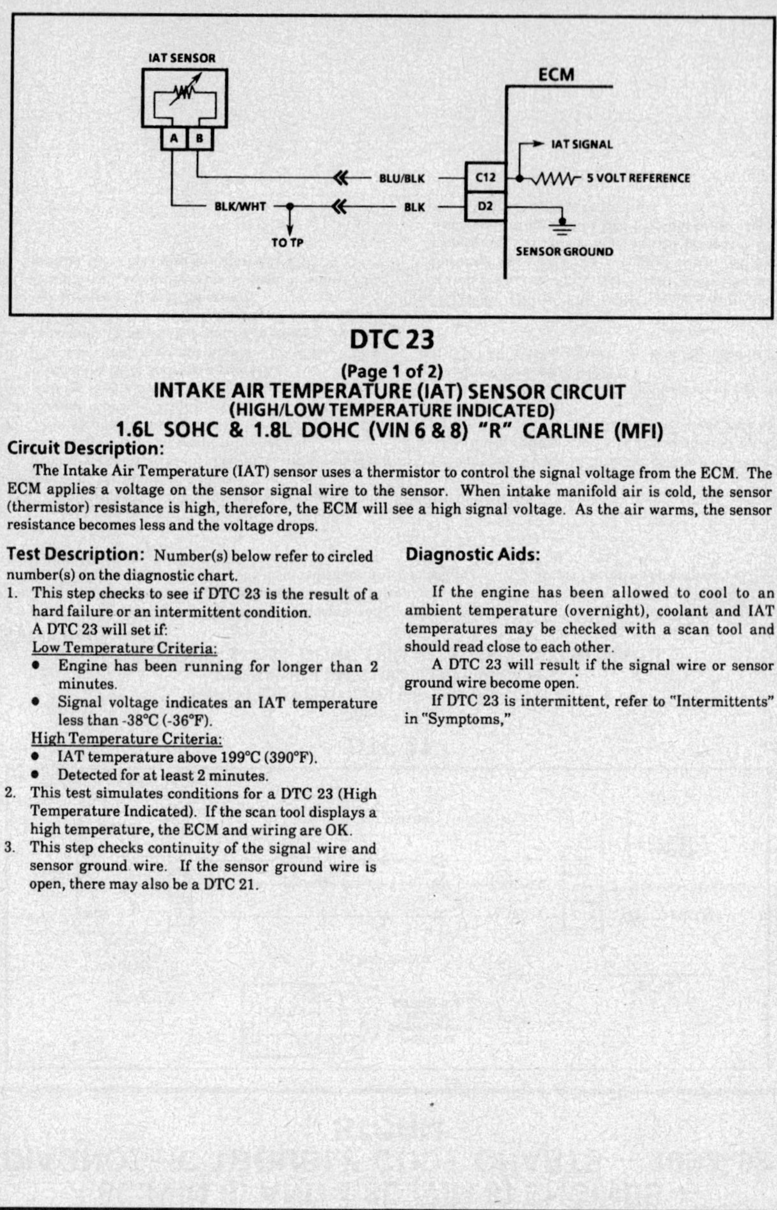

DTC 23
(Page 1 of 2)
INTAKE AIR TEMPERATURE (IAT) SENSOR CIRCUIT
(HIGH/LOW TEMPERATURE INDICATED)
1.6L SOHC & 1.8L DOHC (VIN 6 & 8) "R" CARLINE (MFI)

Circuit Description:

The Intake Air Temperature (IAT) sensor uses a thermistor to control the signal voltage from the ECM. The ECM applies a voltage on the sensor signal wire to the sensor. When intake manifold air is cold, the sensor (thermistor) resistance is high, therefore, the ECM will see a high signal voltage. As the air warms, the sensor resistance becomes less and the voltage drops.

Test Description: Number(s) below refer to circled number(s) on the diagnostic chart.

1. This step checks to see if DTC 23 is the result of a hard failure or an intermittent condition.
   A DTC 23 will set if:
   Low Temperature Criteria:
   - Engine has been running for longer than 2 minutes.
   - Signal voltage indicates an IAT temperature less than -38°C (-36°F).
   High Temperature Criteria:
   - IAT temperature above 199°C (390°F).
   - Detected for at least 2 minutes.
2. This test simulates conditions for a DTC 23 (High Temperature Indicated). If the scan tool displays a high temperature, the ECM and wiring are OK.
3. This step checks continuity of the signal wire and sensor ground wire. If the sensor ground wire is open, there may also be a DTC 21.

Diagnostic Aids:

If the engine has been allowed to cool to an ambient temperature (overnight), coolant and IAT temperatures may be checked with a scan tool and should read close to each other.

A DTC 23 will result if the signal wire or sensor ground wire become open.

If DTC 23 is intermittent, refer to "Intermittents" in "Symptoms,"

1.6L (VIN 6) AND 1.8L (VIN 8) ENGINES — DIAGNOSTIC TROUBLE CODE CHARTS — 1992–93 STORM

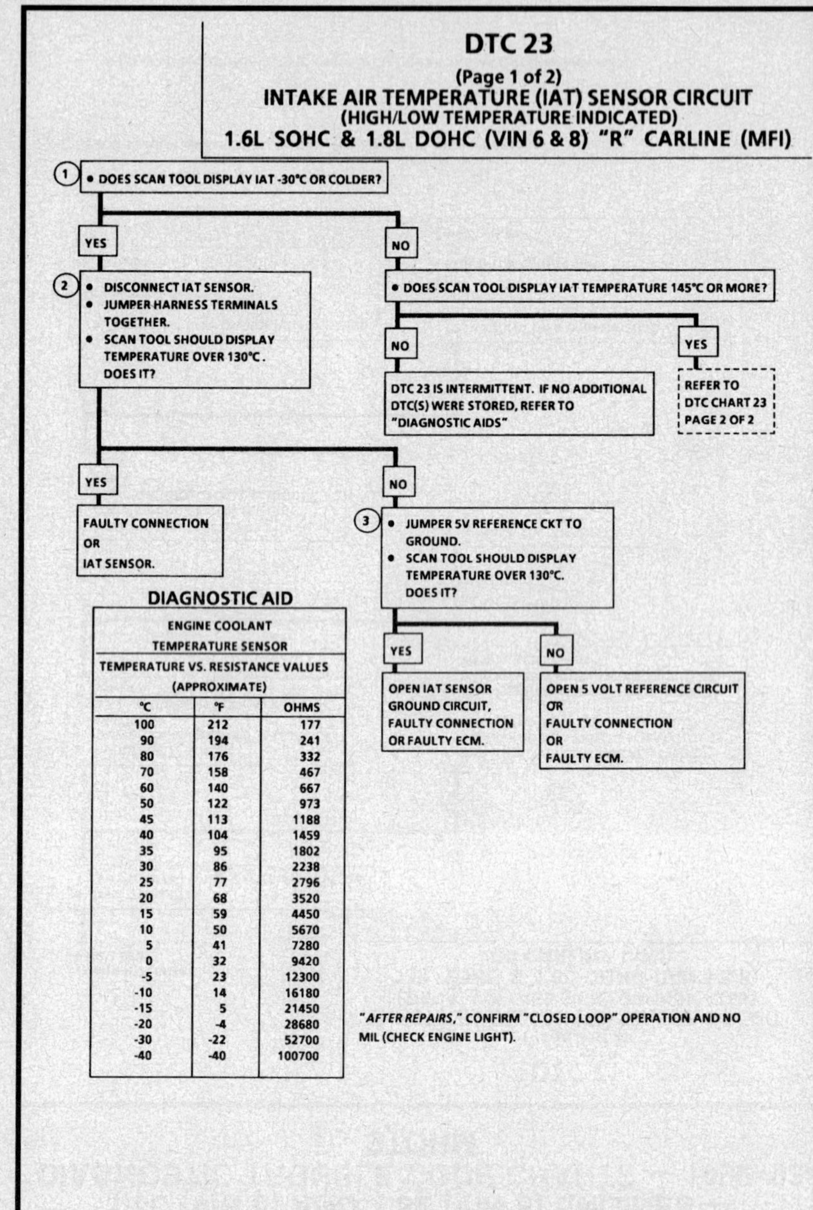

DIAGNOSTIC AID

ENGINE COOLANT TEMPERATURE SENSOR

TEMPERATURE VS. RESISTANCE VALUES (APPROXIMATE)

°C	°F	OHMS
100	212	177
90	194	241
80	176	332
70	158	467
60	140	667
50	122	973
45	113	1188
40	104	1459
35	95	1802
30	86	2238
25	77	2796
20	68	3520
15	59	4450
10	50	5670
5	41	7280
0	32	9420
-5	23	12300
-10	14	16180
-15	5	21450
-20	-4	28680
-30	-22	52700
-40	-40	100700

1.6L (VIN 6) AND 1.8L (VIN 8) ENGINES — DIAGNOSTIC TROUBLE CODE CHARTS — 1992–93 STORM

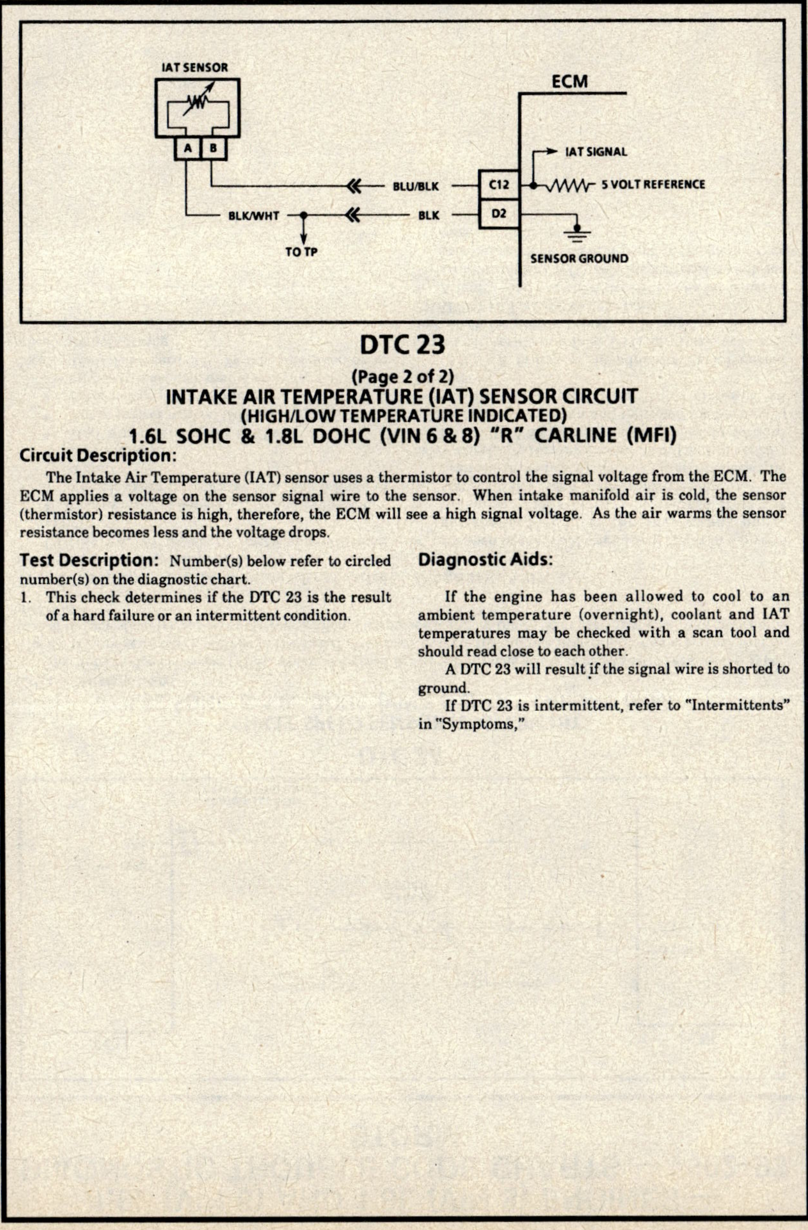

DTC 23
(Page 2 of 2)
INTAKE AIR TEMPERATURE (IAT) SENSOR CIRCUIT
(HIGH/LOW TEMPERATURE INDICATED)
1.6L SOHC & 1.8L DOHC (VIN 6 & 8) "R" CARLINE (MFI)

Circuit Description:

The Intake Air Temperature (IAT) sensor uses a thermistor to control the signal voltage from the ECM. The ECM applies a voltage on the sensor signal wire to the sensor. When intake manifold air is cold, the sensor (thermistor) resistance is high, therefore, the ECM will see a high signal voltage. As the air warms the sensor resistance becomes less and the voltage drops.

Test Description: Number(s) below refer to circled number(s) on the diagnostic chart.

1. This check determines if the DTC 23 is the result of a hard failure or an intermittent condition.

Diagnostic Aids:

If the engine has been allowed to cool to an ambient temperature (overnight), coolant and IAT temperatures may be checked with a scan tool and should read close to each other.

A DTC 23 will result if the signal wire is shorted to ground.

If DTC 23 is intermittent, refer to "Intermittents" in "Symptoms,"

1.6L (VIN 6) AND 1.8L (VIN 8) ENGINES — DIAGNOSTIC TROUBLE CODE CHARTS — 1992–93 STORM

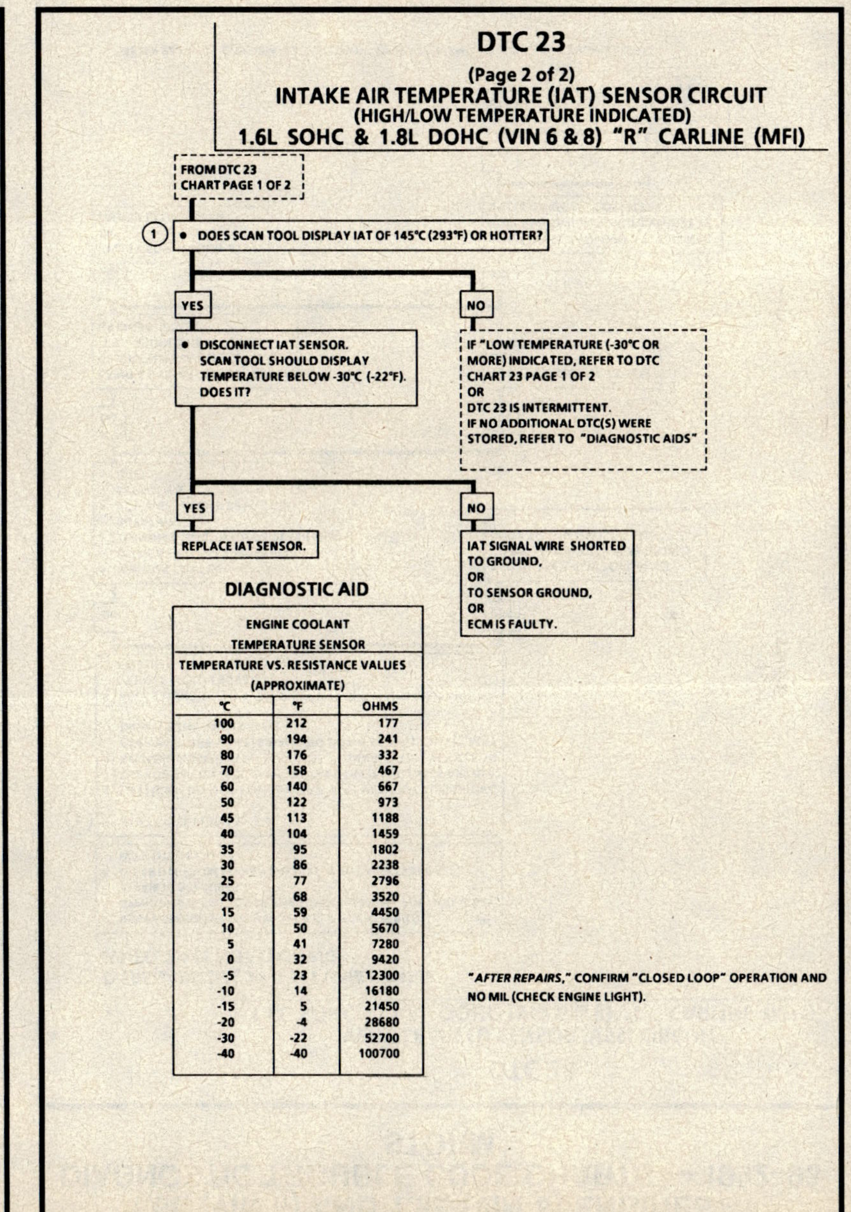

DIAGNOSTIC AID

ENGINE COOLANT TEMPERATURE SENSOR		
TEMPERATURE VS. RESISTANCE VALUES (APPROXIMATE)		
°C	°F	OHMS
100	212	177
90	194	241
80	176	332
70	158	467
60	140	667
50	122	973
45	113	1188
40	104	1459
35	95	1802
30	86	2238
25	77	2796
20	68	3520
15	59	4450
10	50	5670
5	41	7280
0	32	9420
-5	23	12300
-10	14	16180
-15	5	21450
-20	-4	28680
-30	-22	52700
-40	-40	100700

1.6L (VIN 6) AND 1.8L (VIN 8) ENGINES — DIAGNOSTIC TROUBLE CODE CHARTS — 1992–93 STORM

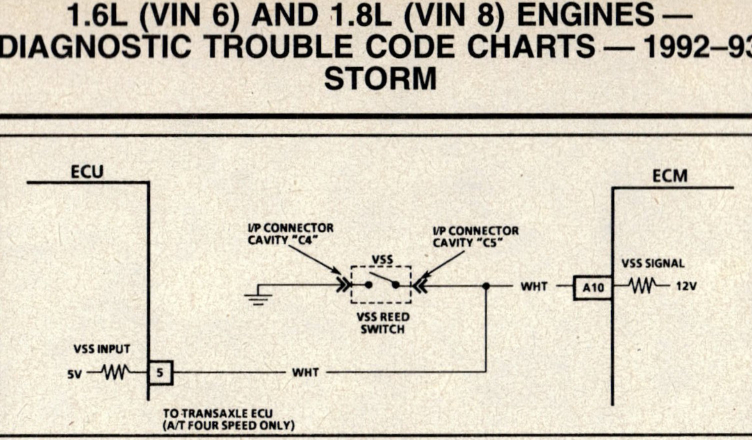

DTC 24
VEHICLE SPEED SENSOR (VSS) CIRCUIT
1.6L SOHC & 1.8L DOHC (VIN 6 & 8) "R" Carline (MFI)

Circuit Description:

The Vehicle Speed Sensor (VSS), which is located in the instrument panel is a magnetic reed switch type. The ECM provides battery voltage to the switch, which will toggle the voltage high and low as a magnet that is part of the speedometer, passed by the reed switch. This pulsed signal to the ECM is used by the ECM to determine mph/km/h.

Test Description: Number(s) below refer to circled number(s) on the diagnostic chart.

1. DTC 24 will set if vehicle speed equals 0 mph when:
   - Engine speed is between 1500 and 4400 RPM.
   - Vehicle speed is indicated less than 5 km/h (3 mph).
   - No DTC 33 set.
   - MAP sensor is less than 43 kPa.
   - Not in park or neutral (A/T).
   - All conditions met for 5 seconds.

   These conditions are met during a road load deceleration. Disregard DTC 24 that sets when drive wheels are not turning.

Diagnostic Aids:

- An open in the ECM circuit will result in 4-5 volts at terminal "C5" of the instrument panel connector. This voltage is sourced from the transaxle ECU for 4 speed transaxle only.

  Scan should indicate a vehicle speed whenever the drive wheels are turning greater than 3 mph (5 km/h).

  Check the VSS signal wire for proper connections to be sure they are clean and tight and the harness is routed correctly. Refer to "Intermittents" in "Symptoms,"

  (A/T) - A faulty or misadjusted Park/Neutral switch can result in a false DTC 24. Use a scan tool and check for proper signal while in drive. Refer to CHART C-1A for PNP switch check.

- A faulty reed switch in the I/P can be the cause of an intermittent DTC 24. Inspect reed switch for poor connections and/or a crack in the glass vapor tube.

1.6L (VIN 6) AND 1.8L (VIN 8) ENGINES — DIAGNOSTIC TROUBLE CODE CHARTS — 1992–93 STORM

DTC 24
VEHICLE SPEED SENSOR (VSS) CIRCUIT
1.6L SOHC & 1.8L DOHC (VIN 6 & 8) "R" CARLINE (MFI)

DISREGARD DTC 24 IF SET WHILE DRIVE WHEELS ARE NOT TURNING.

- VERIFY SPEEDOMETER IS WORKING, IF NOT REFER TO APPROPRIATE SECTION OF THE SERVICE MANUAL FOR NECESSARY REPAIR PROCEDURES.
- CLEAR DTC FROM ECM AND RETEST. IF DTC RESETS PROCEED WITH DTC 24 CHART.

(1)
- RAISE DRIVE WHEELS.
- **NOTICE:** DO NOT PERFORM THIS TEST WITHOUT SUPPORTING THE LOWER CONTROL ARMS SO THAT THE DRIVE AXLES ARE IN A NORMAL HORIZONTAL POSITION. RUNNING THE VEHICLE IN GEAR WITH THE WHEELS HANGING DOWN AT FULL TRAVEL MAY DAMAGE THE DRIVE AXLES.
- WITH ENGINE IDLING IN GEAR, SCAN TOOL SHOULD DISPLAY VEHICLE SPEED ABOVE 0. DOES IT?

NO →
- IGNITION "ON" ENGINE "OFF."
- USING A DVM BACKPROBE ECM TERMINAL "A10" WITH DRIVE WHEELS ROTATING.
- VOLTAGE SHOULD BE VARYING BETWEEN 0-12 VOLTS. IS IT?

YES → DTC 24 IS INTERMITTENT. IF NO ADDITIONAL DTC(S) WERE STORED, REFER TO "DIAGNOSTIC AIDS"

NO → USING A DVM BACKPROBE THE I/P CONNECTOR CAVITIES "C4" AND "C5". SHOULD HAVE 12 VOLTS OR MORE. DOES IT?

YES → POOR CONNECTION AT ECM OR FAULTY ECM.

YES → FAULTY CONNECTION AT I/P OR FAULTY REED SWITCH.

NO → REPEAT TEST WITH DVM AT TERMINAL "C5" AND GROUND. SHOULD HAVE 12 VOLTS OR MORE. DOES IT?

YES → FAULTY I/P GROUND.

NO → OPEN CIRCUIT OR FAULTY ECM.

"AFTER REPAIRS," CONFIRM "CLOSED LOOP" OPERATION AND NO MIL (CHECK ENGINE LIGHT).

1.6L (VIN 6) AND 1.8L (VIN 8) ENGINES — DIAGNOSTIC TROUBLE CODE CHARTS — 1992–93 STORM

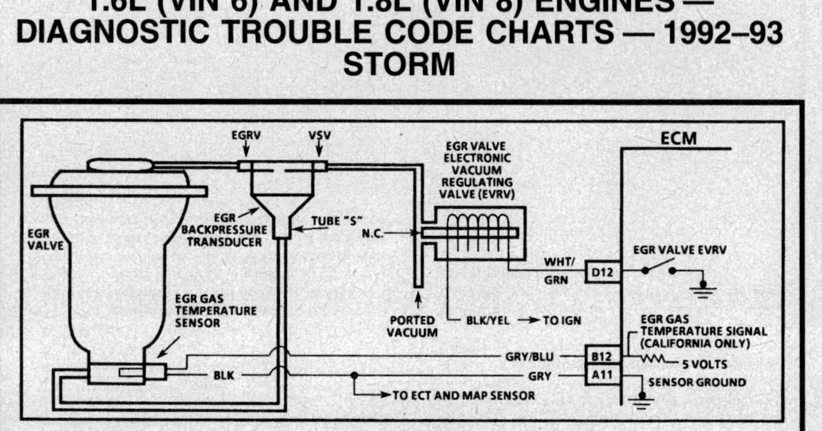

DTC 32
EXHAUST GAS RECIRCULATION (EGR) SYSTEM ERROR
(CALIFORNIA ONLY)
1.6L SOHC & 1.8L DOHC (VIN 6 & 8) "R" CARLINE (MFI)

Circuit Description:

EGR valve operation is ECM controlled. A driver inside the ECM provides a ground source to the EGR Electronic Vacuum Regulating Valve (EVRV). By turning the EVRV "ON" and "OFF," the ECM controls the vacuum source to the EGR back-pressure transducer. The EGR back-pressure transducer then regulates vacuum to the EGR valve using back pressure provided by the exhaust system.

The ECM uses an EGR gas temperature sensor mounted in the base of the EGR valve for diagnosing if the EGR system is functioning properly. This temperature sensor contains a thermistor. The ECM provides a 5 volt reference and a ground for the sensor. The ECM monitors a change in the exhaust gas temperature to determine if the EGR valve is working when the EGR is commanded "ON." If the EGR valve fails to open, the gas temperature sensor signal will indicate "Low" Temperature to the ECM. If the EGR valve is stuck open, the gas temperature sensor signal will indicate "High" Temperature to the ECM. As a result, DTC 32 will be logged in the ECM's memory, and the "Check Engine" light will come "ON." A failure in the EGR gas temperature circuit will also set a DTC 32.

DTC 32 will set if any of the following conditions exist:
- "High" Temperature indicated at idle.
- "Low" Temperature indicated when EGR is commanded "ON."
- Open or short in the EGR gas temperature sensor circuit.
- One of the above conditions exist for 120 seconds.

EGR is not enabled if coolant temperature is below 69°C (182°F).

Test Description: Number(s) below refer to circled number(s) on the diagnostic chart.
1. Checks for the gas temperature sensors 5 volts reference.
2. Looking for a poor or open ground wire.
3. Checks for short to voltage or ground in sensor 5 volts reference circuit.
4. Checking the EGR gas temperatures sensors resistance.
5. A non-functioning EGR valve can cause DTC 32 to set.

Diagnostic Aids:

A poor connection or "shifted" EGR gas temperature sensor could cause a DTC 32 to set. To check for a "shifted" sensor, monitor the sensor resistance at ambient temperature with a DVM connected to GRY/BLU wire and BLK wire. Start and run the engine noting the resistance change in the sensor as engine temperature increases. Refer to Temperature to Resistance chart on diagnostic chart. If the sensor fails the check, replace the sensor.

1.6L (VIN 6) AND 1.8L (VIN 8) ENGINES — DIAGNOSTIC TROUBLE CODE CHARTS — 1992–93 STORM

DTC 32
EXHAUST GAS RECIRCULATION (EGR) SYSTEM ERROR
(CALIFORNIA ONLY)
1.6L SOHC & 1.8L DOHC (VIN 6 & 8) "R" CARLINE (MFI)

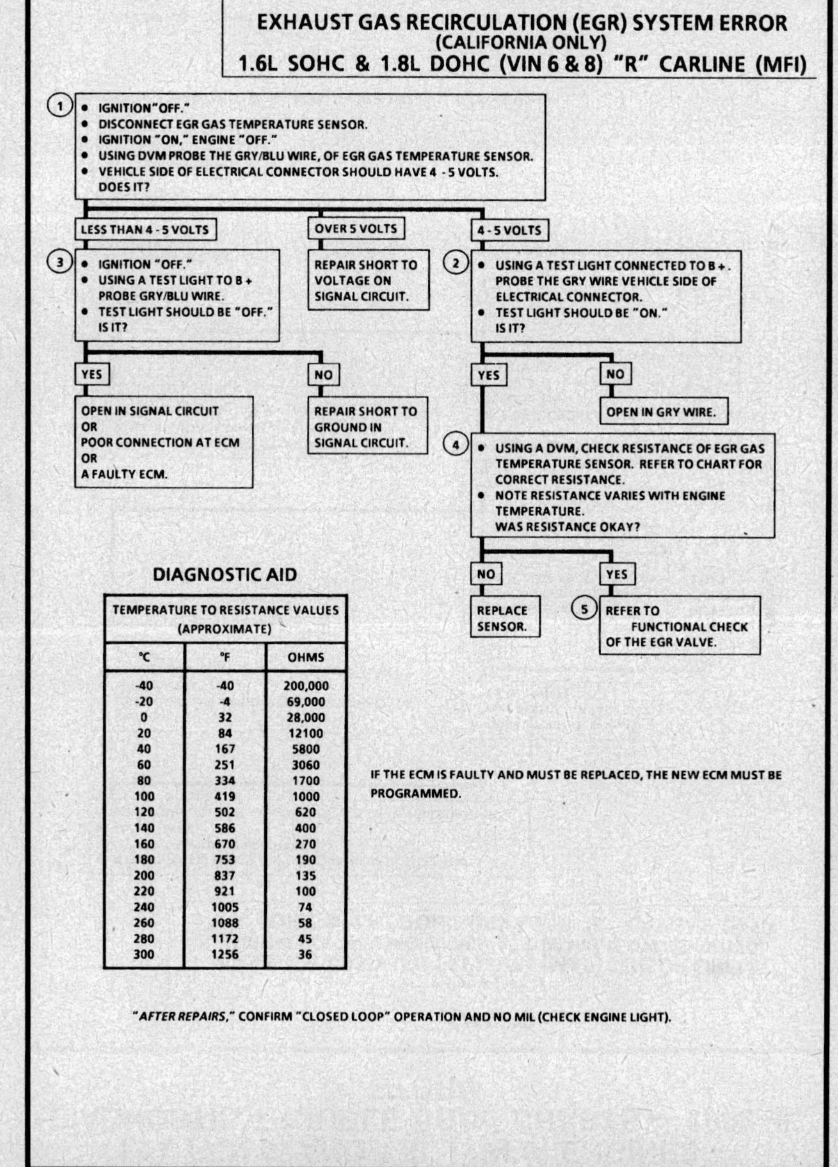

DIAGNOSTIC AID

TEMPERATURE TO RESISTANCE VALUES (APPROXIMATE)		
°C	°F	OHMS
-40	-40	200,000
-20	-4	69,000
0	32	28,000
20	84	12100
40	167	5800
60	251	3060
80	334	1700
100	419	1000
120	502	620
140	586	400
160	670	270
180	753	190
200	837	135
220	921	100
240	1005	74
260	1088	58
280	1172	45
300	1256	36

IF THE ECM IS FAULTY AND MUST BE REPLACED, THE NEW ECM MUST BE PROGRAMMED.

"AFTER REPAIRS," CONFIRM "CLOSED LOOP" OPERATION AND NO MIL (CHECK ENGINE LIGHT).

1.6L (VIN 6) AND 1.8L (VIN 8) ENGINES — DIAGNOSTIC TROUBLE CODE CHARTS — 1992–93 STORM

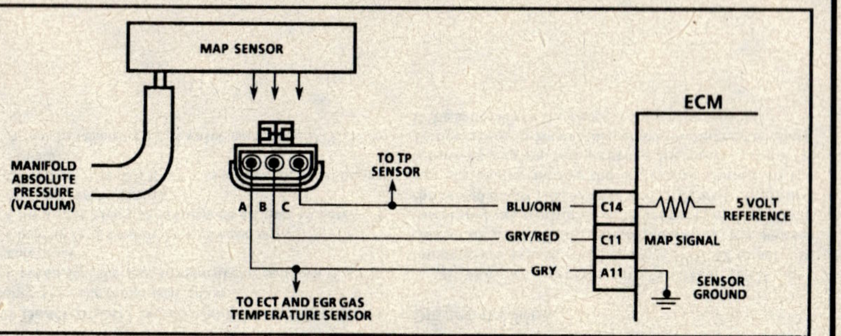

DTC 33
(Page 1 of 2)
MANIFOLD ABSOLUTE PRESSURE (MAP) SENSOR CIRCUIT
(SIGNAL VOLTAGE HIGH/LOW-VACUUM HIGH/LOW INDICATED)
1.6L SOHC & 1.8L DOHC (VIN 6 & 8) "R" CARLINE (MFI)

Circuit Description:

The Manifold Absolute Pressure (MAP) sensor responds to changes in manifold pressure (vacuum). The ECM receives this information as a signal voltage that will vary from about 1 to 1.5 volts, at closed throttle idle, to 4 - 4.5 volts at Wide Open Throttle (low vacuum).

If the MAP sensor fails, the ECM will substitute a fixed MAP value and use the Throttle Position (TP) sensor to control fuel delivery.

Test Description: Number(s) below refer to circled number(s) on the diagnostic chart.

1. This step will determine if DTC 33 is the result of a hard failure or an intermittent condition.
A DTC 33 will set if:
High Voltage-Low Vacuum Criteria:
- MAP signal indicates greater than 84 kPa (low vacuum).
- No DTC 21 is set.
- TP sensor less than 5%.
- These conditions are present longer than 5 seconds.
Low Voltage-High Vacuum Criteria:
- Engine RPM > 1200 RPM.
- No DTC 21 is set.
- TP sensor is > 15%.
- MAP voltage indicated too low.
2. This step simulates conditions for a DTC 33 (Signal Voltage Low-High Vacuum). If the ECM recognizes the change, the ECM and sensor ground wire and signal circuit wire are OK. If the 5 volt reference wire is open, there may also be a stored DTC 23.

Diagnostic Aids:

With the ignition "ON" and the engine stopped, the manifold pressure is equal to atmospheric pressure and the signal voltage will be high. This information is used by the ECM as an indication of vehicle altitude and is referred to as BARO. Comparison of this BARO reading with a known good vehicle with the same sensor is a good way to check accuracy of a "suspect" sensor. Readings should be the same ± .4 volt.

A DTC 33 will result if the ground wire is open, or if the signal wire is shorted to voltage or to the 5 volt reference wire.

If DTC 33 is intermittent, refer to "Intermittents" in "Symptoms,"

NOTICE: If MAP sensor is mounted below intake manifold, you may set a false DTC 33. Be sure sensor is mounted in the correct location.

1.6L (VIN 6) AND 1.8L (VIN 8) ENGINES — DIAGNOSTIC TROUBLE CODE CHARTS — 1992–93 STORM

DTC 33
(Page 1 of 2)
MANIFOLD ABSOLUTE PRESSURE (MAP) SENSOR CIRCUIT
(SIGNAL VOLTAGE HIGH/LOW-VACUUM HIGH/LOW INDICATED)
1.6L SOHC & 1.8L DOHC (VIN 6 & 8) "R" CARLINE (MFI)

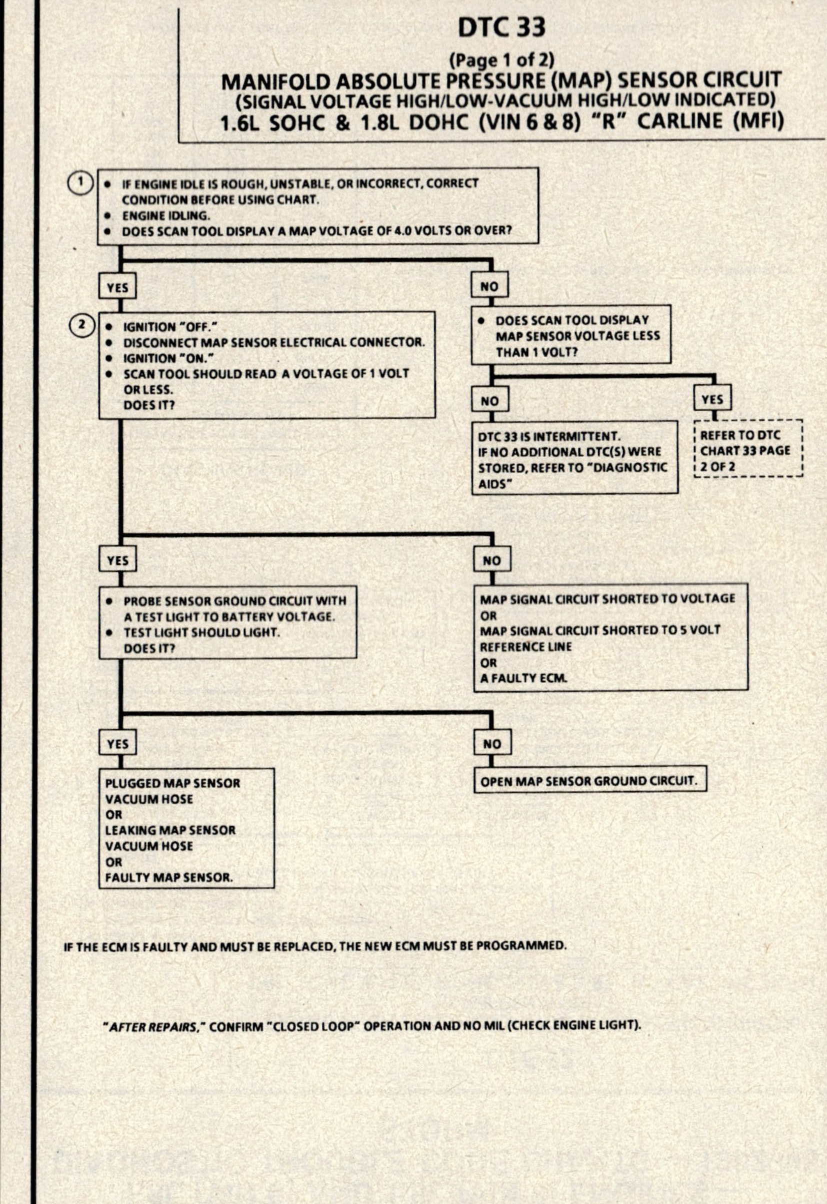

1.6L (VIN 6) AND 1.8L (VIN 8) ENGINES — DIAGNOSTIC TROUBLE CODE CHARTS — 1992–93 STORM

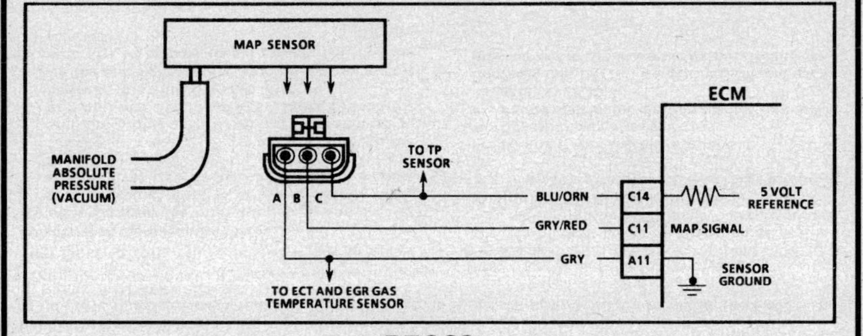

DTC 33
(Page 2 of 2)
MANIFOLD ABSOLUTE PRESSURE (MAP) SENSOR CIRCUIT
(SIGNAL VOLTAGE HIGH/LOW-VACUUM HIGH/LOW INDICATED)
1.6L SOHC & 1.8L DOHC (VIN 6 & 8) "R" CARLINE (MFI)

Circuit Description:

The Manifold Absolute Pressure (MAP) sensor responds to changes in manifold pressure (vacuum). The ECM receives this information as a signal voltage that will vary from about 1 to 1.5 volts at closed throttle idle, to 4 - 4.5 volts at Wide Open Throttle (WOT).

If the MAP sensor fails, the ECM will substitute a fixed MAP value and use the Throttle Position (TP) sensor to control fuel delivery.

Test Description: Number(s) below refer to circled number(s) on the diagnostic chart.
1. This step determines if DTC 33 is the result of a hard failure or an intermittent condition.
2. Jumpering harness terminals "B" to "C", 5 volts to signal will determine if the sensor is at fault or if there is a problem with the ECM or wiring.
3. The scan tool may not display 12 volts. The important thing is that the ECM recognizes the voltage as more than 4 volts, indicating that the ECM and the signal circuit wire are OK.

Diagnostic Aids:

With the ignition "ON" and the engine stopped, the manifold pressure is equal to atmospheric pressure and the signal voltage will be high. This information is used by the ECM as an indication of vehicle altitude and is referred to as BARO. Comparison of this BARO reading with a known good vehicle with the same sensor is a good way to check accuracy of a "suspect" sensor. Readings should be the same ± .4 volt.

A DTC 33 will result if the 5 volt reference wire or the signal wire are open or shorted to ground.

If DTC 33 is intermittent, refer to "Intermittents" in "Symptoms,"

NOTICE: If MAP sensor is mounted below intake manifold, you may set a false DTC 33. Be sure sensor is mounted in the correct location.

1.6L (VIN 6) AND 1.8L (VIN 8) ENGINES — DIAGNOSTIC TROUBLE CODE CHARTS — 1992–93 STORM

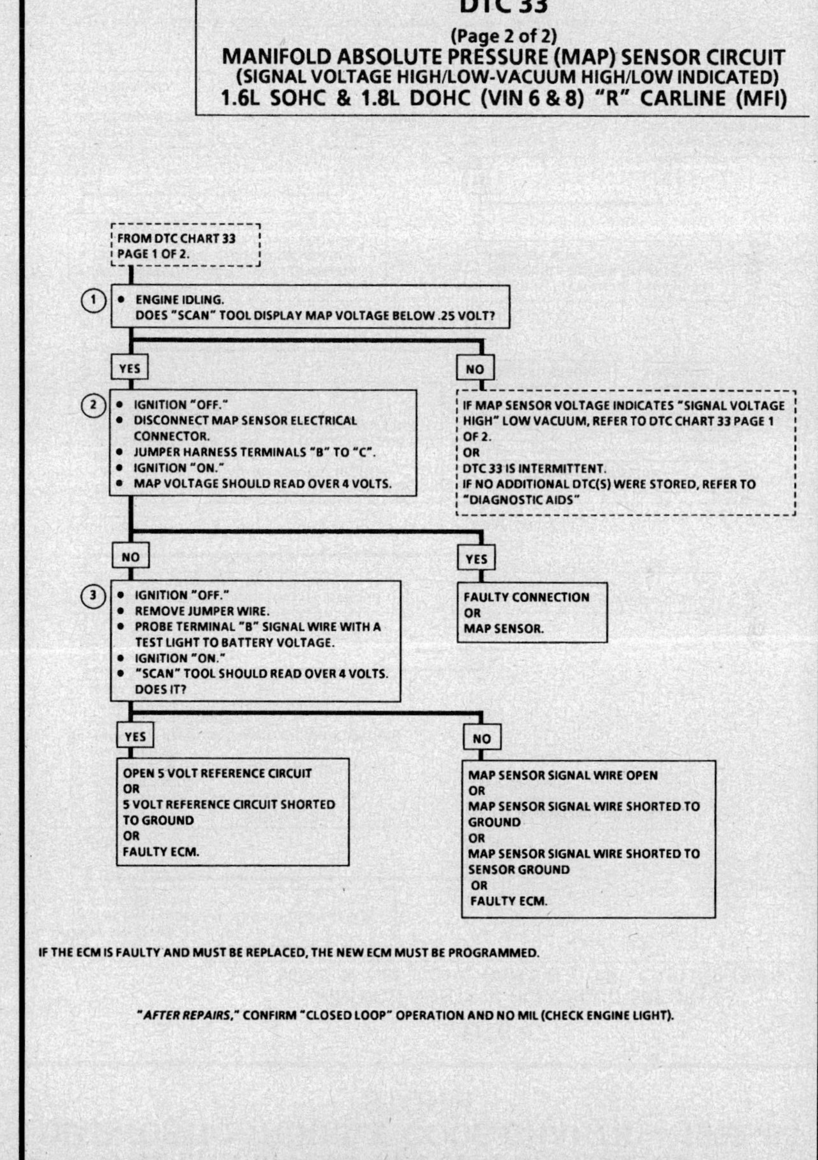

1.6L (VIN 6) AND 1.8L (VIN 8) ENGINES — DIAGNOSTIC TROUBLE CODE CHARTS — 1992–93 STORM

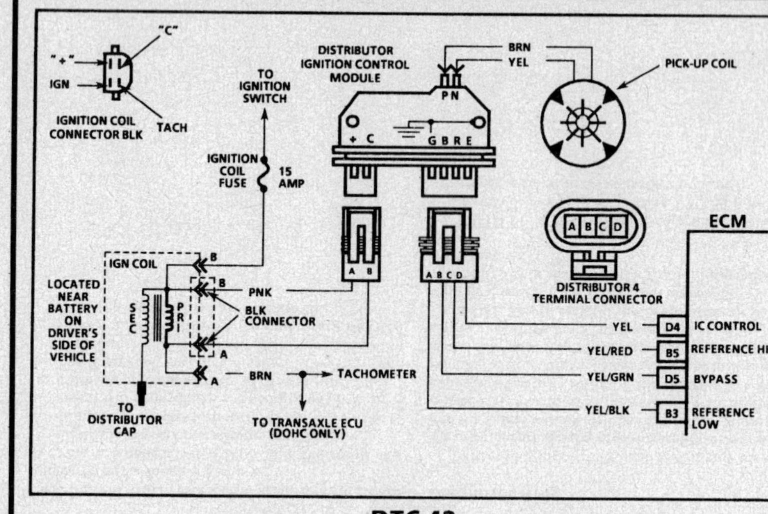

DTC 42

IGNITION CONTROL (IC) CIRCUIT ERROR
1.6L SOHC & 1.8L DOHC (VIN 6 & 8) "R" CARLINE (MFI)

Circuit Description:

The Distributor Ignition (DI) module sends a reference signal (YEL/RED wire) to the ECM, when the engine is cranking. While the engine speed is under 400 RPM, the DI module will control ignition timing. When the engine speed exceeds 400 RPM, the ECM applies 5 volts to the bypass line to switch the timing to ECM control.

When the system is running on the DI module, that is, no voltage on the bypass line, the DI module grounds the IC circuit. The ECM expects to sense or detect no voltage on the IC line during this condition. If it senses or detects a voltage, it sets DTC 42 and will not go into the IC mode.

When the RPM for IC is reached (about 400 RPM), voltage will be applied to the bypass line. The IC should no longer be grounded in the DI module, so the IC voltage should be varying.

If the bypass line is open or grounded, the DI module will not switch to IC mode, so the IC voltage will be low and DTC 42 will be set.

If the IC line is grounded, the DI module will switch to IC but, because the line is grounded, there will be no IC signal. A DTC 42 will be set.

Test Description: Number(s) below refer to circled number(s) on the diagnostic chart.

1. DTC 42 means the ECM has detected an open or short to ground in the IC or bypass circuits. This test confirms DTC 42 and that the fault causing the DTC is present.
2. Checks for a normal IC ground path through the ignition module. An IC circuit wire shorted to ground, will also read less than 500 ohms; however, this will be checked later.
3. As the test light voltage touches bypass circuit wire, the DI module should switch, causing the ohmmeter to

"overrange" if the meter is in the 1000-2000 ohms position. Selecting the 10-20,000 ohms position will indicate above 5000 ohms. The important thing is that the DI module "switched."
4. The DI module did not switch and this step checks for:
   - IC circuit wire shorted to ground.
   - Bypass circuit wire open.
   - Faulty distributor ignition module connection or faulty DI module.
5. Confirms that DTC 42 is a faulty ECM and not an intermittent in circuit wire or bypass circuit wire.

1.6L (VIN 6) AND 1.8L (VIN 8) ENGINES — DIAGNOSTIC TROUBLE CODE CHARTS — 1992–93 STORM

DTC 42

IGNITION CONTROL (IC) CIRCUIT ERROR
1.6L SOHC & 1.8L DOHC (VIN 6 & 8) "R" CARLINE (MFI)

(1)
- CLEAR DTC(S).
- IDLE ENGINE FOR 1 MINUTE OR UNTIL DTC 42 SETS.
  DOES DTC 42 SET?

YES →

(2)
- IGNITION "OFF."
- DISCONNECT ECM CONNECTORS.
- IGNITION "ON."
- OHMMETER SELECTOR SWITCH IN THE 1000 TO 2000 OHMS RANGE.
- PROBE ECM HARNESS CONNECTOR IC CIRCUIT WITH AN OHMMETER TO GROUND.
  IT SHOULD READ LESS THAN 1000 OHMS.
  DOES IT?

NO → DTC 42 INTERMITTENT. REFER TO "SYMPTOMS," "INTERMITTENTS."

YES →

- PROBE ECM HARNESS CONNECTOR BYPASS CIRCUIT WITH A TEST LIGHT TO BATTERY VOLTAGE.

NO → OPEN IC CIRCUIT, FAULTY CONNECTION OR FAULTY EI MODULE.

LIGHT "OFF"

(3)
- WITH OHMMETER STILL CONNECTED TO ECM HARNESS IC CIRCUIT AND GROUND, AGAIN PROBE ECM HARNESS BYPASS CIRCUIT WITH THE TEST LIGHT CONNECTED TO BATTERY VOLTAGE. (AS TEST LIGHT CONTACTS BYPASS CIRCUIT, RESISTANCE SHOULD SWITCH FROM UNDER 1000 TO OVER 2000 OHMS.)
  DOES IT?

LIGHT "ON" →
- DISCONNECT ELECTRONIC IGNITION SYSTEM MODULE 4-WAY CONNECTOR.

  LIGHT "ON" → BYPASS CIRCUIT SHORTED TO GROUND.

  LIGHT "OFF" → FAULTY EI MODULE.

NO →

(4)
- DISCONNECT 4-WAY CONNECTOR AT ELECTRONIC IGNITION SYSTEM MODULE. NOTE OHMMETER THAT IS STILL CONNECTED TO IC CIRCUIT AND GROUND. RESISTANCE SHOULD HAVE GONE HIGH (OPEN CIRCUIT).

YES → BYPASS CIRCUIT OPEN, FAULTY CONNECTIONS, OR FAULTY EI MODULE.

NO → IC CIRCUIT SHORTED TO GROUND.

YES →

(5)
- RECONNECT ECM AND IDLE ENGINE FOR ONE MINUTE OR UNTIL DTC 42 SETS.
  DOES DTC SET?

YES → FAULTY ECM

NO → DTC 42 INTERMITTENT. REFER TO "SYMPTOMS," "INTERMITTENTS."

"AFTER REPAIRS," CONFIRM "CLOSED LOOP" OPERATION AND NO MIL (CHECK ENGINE LIGHT).

1.6L (VIN 6) AND 1.8L (VIN 8) ENGINES — DIAGNOSTIC TROUBLE CODE CHARTS — 1992–93 STORM

DTC 44
HEATED OXYGEN SENSOR (HO2S) CIRCUIT
(LEAN EXHAUST INDICATED)
1.8L DOHC (VIN 8) "R" CARLINE (MFI)

Circuit Description:

The ECM supplies a voltage of about .45 volt between terminals "D7" and "D6". (If measured with a 10 megohm digital voltmeter, this may read as low as .32 volt.)

The heater circuit of the oxygen sensor is turned "ON" when the engine is running. The oxygen sensor as it heats up will begin to send a varying voltage signal to the ECM. This voltage will vary from about .1 volt (lean exhaust) to about 1.0 volts (rich exhaust) depending on the condition of the exhaust gases in the exhaust manifold.

The sensor is like an open circuit and produces no voltage when it is below 315°C (600°F). An open sensor circuit causes "Open Loop" operation.

Test Description: Number(s) below refer to circled number(s) on the diagnostic chart.
1. DTC 44 is set when the O2 sensor signal voltage is fixed above 280 mV) and the following:
   - TP sensor greater than 3%.
   - No DTC 21 or 33.
   - System is operating in "Closed Loop."
   - Engine Coolant Temperature (ECT) above 69.5°C (183°F).
   - Short term fuel trim not at 128.
   - VSS > 3 mph.

Diagnostic Aids:

Using the scan tool, observe the long term fuel trim value at different RPMs. The scan tool also displays the long term fuel trim cells, so the long term fuel trim values can be checked in each of the cells, to determine when the DTC 44 may have been set. If the conditions for DTC 44 exist, the long term fuel trim values will be around 150.
- HO2 Sensor Wire - Sensor pigtail may be mispositioned and contacting the exhaust manifold.
- Check for ground in wire between connector and sensor.

- Fuel Contamination - Water, even in small amounts, near the in-tank fuel pump inlet can be delivered to the injector. The water causes a lean exhaust and can set a DTC 44.
- Fuel Pressure - System will be lean if pressure is too low. It may be necessary to monitor fuel pressure, while driving the vehicle at various road speeds and/or loads to confirm. See "Fuel System Diagnosis," CHART A-7.
- Exhaust Leaks - If there is an exhaust leak, the engine can cause outside air to be pulled into the exhaust and past the sensor. Vacuum or crankcase leaks can cause a lean condition.
- If DTC 44 intermittent, refer to "Symptoms,"
- If the heater part of the O2 sensor fails, this may cause a DTC 44 or DTC 45 to set.

- Check the oxygen sensor wire for induced voltage (EMI). Disconnect the oxygen sensor signal wire at the sensor and the ECM. Using DVM measure the wire with the engine running for any voltage. If 300 mV or more are indicated, check the signal wire shield and/or reroute the signal wire away from any high voltage sources. Such source could be ignition wires, battery cables, Alternator "B" circuit, etc.

1.6L (VIN 6) AND 1.8L (VIN 8) ENGINES — DIAGNOSTIC TROUBLE CODE CHARTS — 1992–93 STORM

DTC 44
HEATED OXYGEN SENSOR (HO2S) CIRCUIT
(LEAN EXHAUST INDICATED)
1.8L DOHC (VIN 8) "R" CARLINE (MFI)

IF THE ECM IS FAULTY AND MUST BE REPLACED, THE NEW ECM MUST BE PROGRAMMED.

"AFTER REPAIRS," CONFIRM "CLOSED LOOP" OPERATION AND NO MIL (CHECK ENGINE LIGHT).

1.6L (VIN 6) AND 1.8L (VIN 8) ENGINES — DIAGNOSTIC TROUBLE CODE CHARTS — 1992–93 STORM

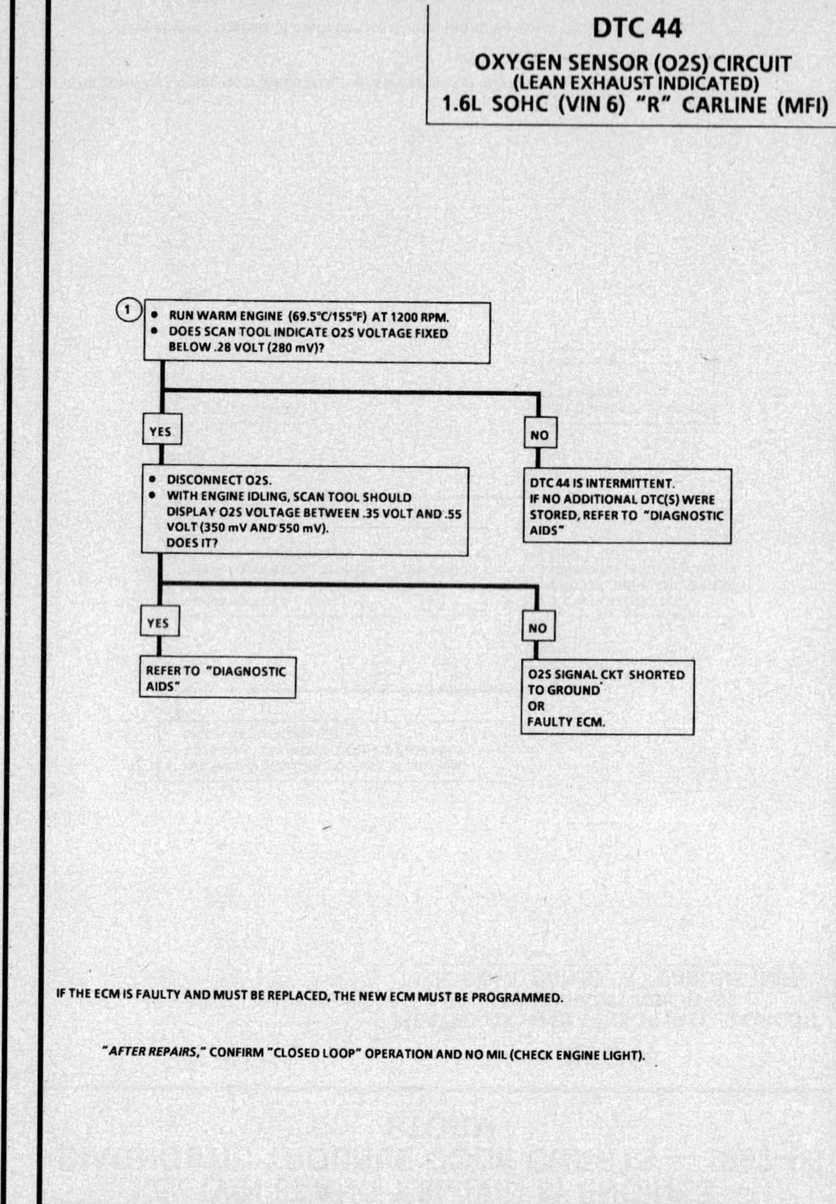

DTC 44
OXYGEN SENSOR (O2S) CIRCUIT
(LEAN EXHAUST INDICATED)
1.6L SOHC (VIN 6) "R" CARLINE (MFI)

Circuit Description:

The ECM supplies a voltage of about .43 volt between terminals "D7" and "D6". (If measured with a 10 megohm digital voltmeter, this may read as low as .32 volt.) The oxygen sensor varies the voltage within a range of about 1.0 volt, if the exhaust is rich, down to about .10 volt, if exhaust is lean.

The sensor is like an open circuit and produces no voltage, when it is below about 315°C (600°F). An open sensor circuit causes "Open Loop" operation.

Test Description: Number(s) below refer to circled number(s) on the diagnostic chart.

1. DTC 44 will set when the following occurs:
   - No DTC 21 or DTC 33.
   - Engine running for 50 seconds or more.
   - O2S voltage below 278 mV for 99 seconds.
   - System is in "Closed Loop."
   - Engine coolant temperature above 69.5°C (183°F).
   - TP sensor is above 3%.
   - Short term fuel trim not at 128.
   - VSS > 3 mph.

Diagnostic Aids:

Using the scan tool, observe the long term fuel trim value at different RPMs. The scan tool also displays the long term fuel trim cells, so the long term fuel trim values can be checked in each of the cells, to determine when the DTC 44 may have been set. If the conditions for DTC 44 exist, the long term fuel trim values will be around 150.

- <u>Oxygen Sensor Wire</u> - Sensor pigtail may be mispositioned and contacting the exhaust manifold.
- Check for ground in wire between connector and sensor.

- <u>Fuel Contamination</u> - Water, even in small amounts, near the in-tank fuel pump inlet can be delivered to the injector. The water causes a lean exhaust and can set a DTC 44.
- <u>Fuel Pressure</u> - System will be lean if pressure is too low. It may be necessary to monitor fuel pressure, while driving the vehicle at various road speeds and/or loads to confirm. See "Fuel System Diagnosis," CHART A-7.
- <u>Exhaust Leaks</u> - If there is an exhaust leak, the engine can cause outside air to be pulled into the exhaust and past the sensor. Vacuum or crankcase leaks can cause a lean condition.
- If DTC 44 intermittent, refer to "Symptoms,"
- <u>Fuel Injectors</u> - The wrong fuel injector(s) could be the cause of a lean exhaust condition. Verify that the correct fuel injector(s) are installed in the vehicle. Refer to the service parts manual for correct part and part number.
- Check the oxygen sensor wire for induced voltage (EMI). Disconnect the oxygen sensor signal wire at the sensor and the ECM. Using DVM, measure the wire with the engine running for any voltage. If 300 mV or more are indicated, check the signal wire shield and/or reroute the signal wire away from any high voltage sources. Such source could be ignition wires, battery cables, Alternator "B" circuit, etc.

1.6L (VIN 6) AND 1.8L (VIN 8) ENGINES — DIAGNOSTIC TROUBLE CODE CHARTS — 1992–93 STORM

DTC 44
OXYGEN SENSOR (O2S) CIRCUIT
(LEAN EXHAUST INDICATED)
1.6L SOHC (VIN 6) "R" CARLINE (MFI)

IF THE ECM IS FAULTY AND MUST BE REPLACED, THE NEW ECM MUST BE PROGRAMMED.

"AFTER REPAIRS," CONFIRM "CLOSED LOOP" OPERATION AND NO MIL (CHECK ENGINE LIGHT).

1.6L (VIN 6) AND 1.8L (VIN 8) ENGINES — DIAGNOSTIC TROUBLE CODE CHARTS — 1992–93 STORM

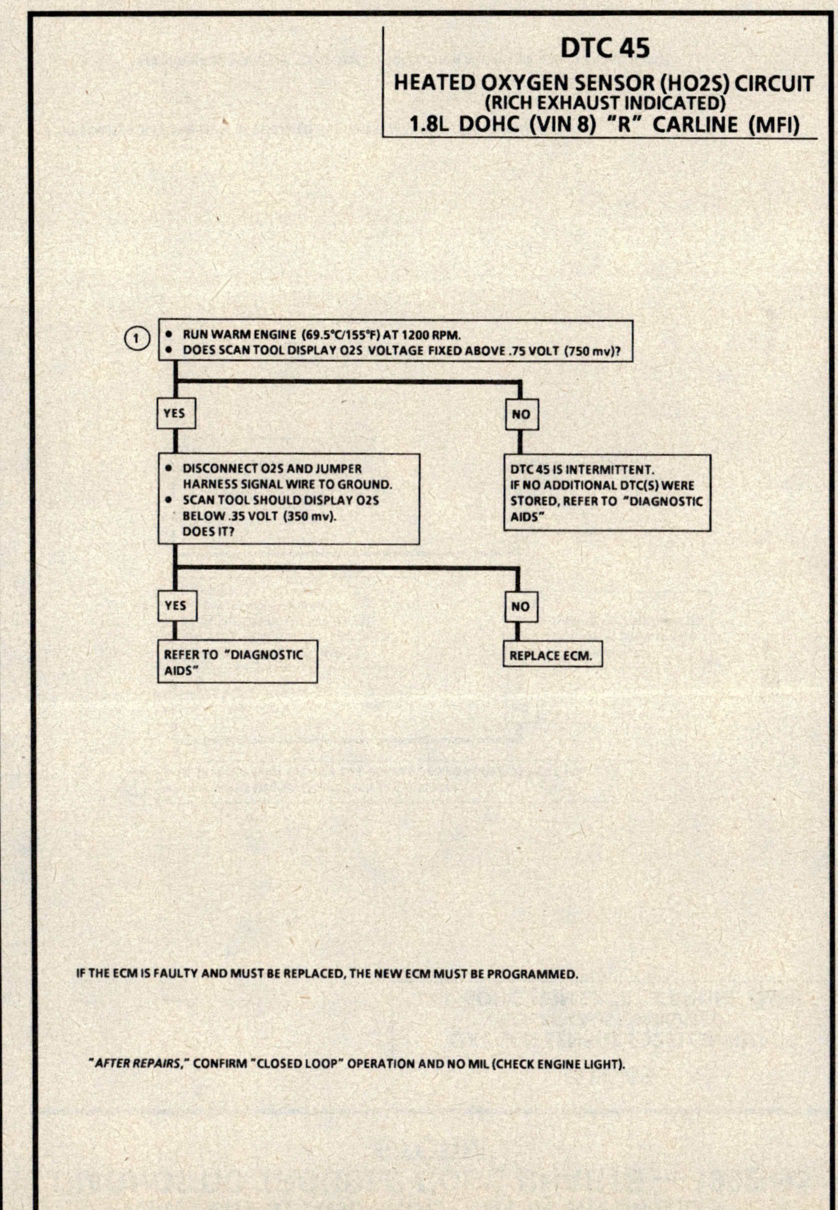

DTC 45
HEATED OXYGEN SENSOR (HO2S) CIRCUIT
(RICH EXHAUST INDICATED)
1.8L DOHC (VIN 8) "R" CARLINE (MFI)

Circuit Description:

The ECM supplies a voltage of about .45 volt between terminals "D7" and "D6" (If measured with a 10 megohm digital voltmeter, this may read as low as .32 volt.)

The heater circuit of the oxygen sensor is turned "ON" when the engine is running. The oxygen sensor as it heats up will begin to send a varying voltage signal to the ECM. This voltage will vary from about .1 volt (lean exhaust) to about 1.0 volts (rich exhaust) depending on the condition of the exhaust gases in the exhaust manifold.

The sensor is like an open circuit and produces no voltage when it is below 315°C (600°F). An open sensor circuit or cold sensor causes "Open Loop" operation.

Test Description: Number(s) below refer to circled number(s) on the diagnostic chart.

1. DTC 45 is set, when the O2 sensor signal voltage is fixed above .75 (750 mV) and the following:
   - TP sensor greater than 3%.
   - No DTC 21 or 33.
   - System is operating in "Closed Loop."
   - Engine Coolant Temperature (ECT) above 69.5°C (183°F).
   - Short term fuel trim not at 128.
   - VSS > 3 mph.

Diagnostic Aids:

The DTC 45, or rich exhaust, is most likely caused by one of the following:

- Fuel Pressure - System will go rich if pressure is too high. The ECM can compensate for some increase. However, if it gets too high, a DTC 45 will be set. See "Fuel System Diagnosis" CHART A-7.
- HEI Shielding - An open ignition ref. low YEL/BLK wire may result in EMI, or induced electrical "noise." The ECM looks at this "noise" as reference pulses. The additional pulses result in a higher than actual engine speed signal. The ECM then delivers too much fuel, causing system to go rich. Engine tachometer will also show higher than actual engine speed, which can help in diagnosing this problem. Refer to ECM wiring schematic for ignition system wiring.

- Canister Purge - Check for fuel saturation. If full of fuel, check canister control and hoses. See "Evaporative Emission (EVAP) Control System,"
- MAP Sensor - An output that causes the ECM to sense a higher than normal manifold pressure (low vacuum) can cause the system to go rich. Disconnecting the MAP sensor will allow the ECM to set a fixed value for the MAP sensor. Substitute a different MAP sensor if the rich condition is gone, while the sensor is disconnected.
- TP Sensor - An intermittent TP sensor output will cause the system to go rich, due to a false indication of the engine accelerating.
- Oxygen Sensor Contamination - Inspect oxygen sensor for silicone contamination from fuel, or use of improper RTV sealant. The sensor may have a white, powdery coating and result in a high, but false signal voltage (rich exhaust indication). The ECM will then reduce the amount of fuel delivered to the engine, causing a severe surge driveability problem.
- EGR - Valve sticking open at idle, usually accompanied by a rough idle, stall complaint. If DTC 45 is intermittent, refer to "Symptoms,"

1.6L (VIN 6) AND 1.8L (VIN 8) ENGINES — DIAGNOSTIC TROUBLE CODE CHARTS — 1992–93 STORM

DTC 45
HEATED OXYGEN SENSOR (HO2S) CIRCUIT
(RICH EXHAUST INDICATED)
1.8L DOHC (VIN 8) "R" CARLINE (MFI)

IF THE ECM IS FAULTY AND MUST BE REPLACED, THE NEW ECM MUST BE PROGRAMMED.

"AFTER REPAIRS," CONFIRM "CLOSED LOOP" OPERATION AND NO MIL (CHECK ENGINE LIGHT).

1.6L (VIN 6) AND 1.8L (VIN 8) ENGINES — DIAGNOSTIC TROUBLE CODE CHARTS — 1992–93 STORM

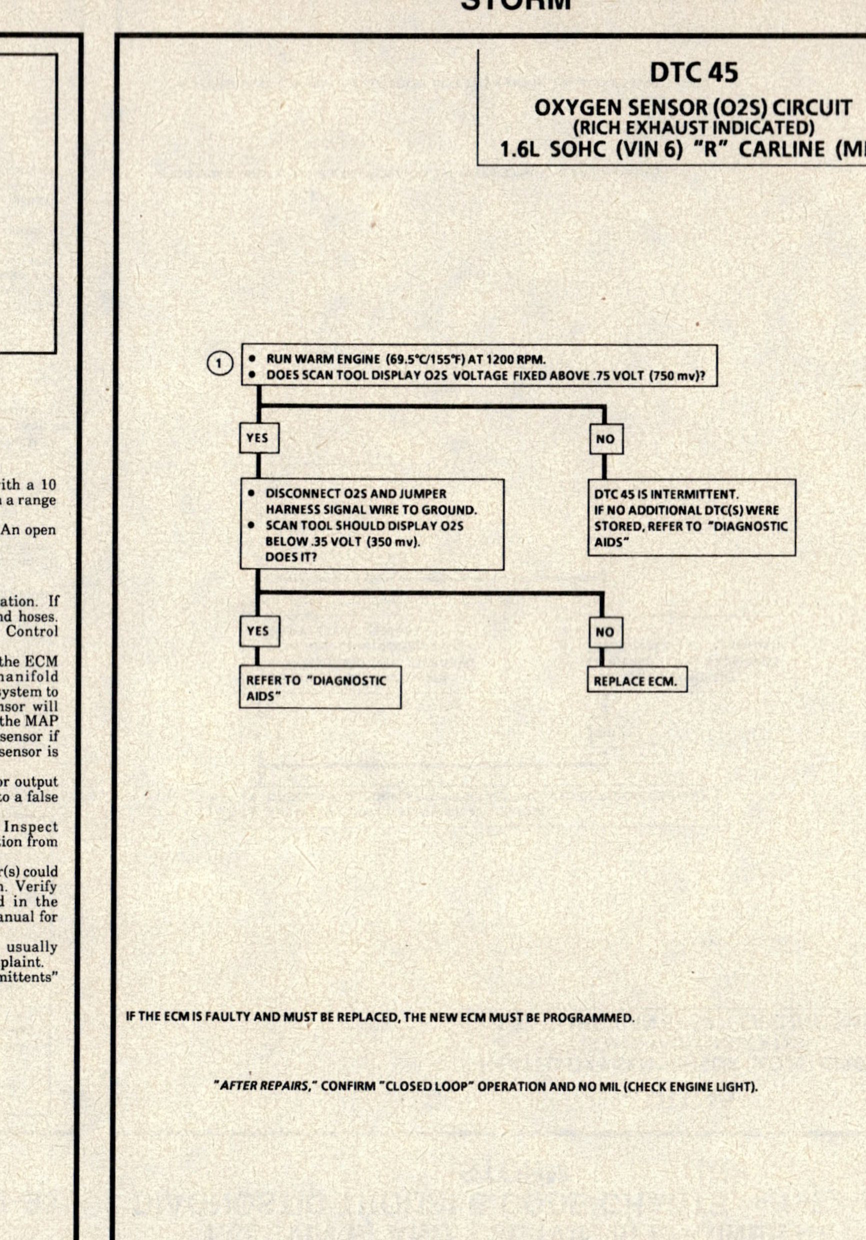

DTC 45

OXYGEN SENSOR (O2S) CIRCUIT
(RICH EXHAUST INDICATED)
1.6L SOHC (VIN 6) "R" CARLINE (MFI)

Circuit Description:

The ECM supplies a voltage of about .43 volt between terminals "D7" and "D6". (If measured with a 10 megohm digital voltmeter, this may read as low as .32 volt.) The oxygen sensor varies the voltage within a range of about 1.0 volt, if the exhaust is rich, down to about .10 volt, if exhaust is lean.

The sensor is like an open circuit and produces no voltage, when it is below about 315°C (600°F). An open sensor circuit causes "Open Loop" operation.

Test Description: Number(s) below refer to circled number(s) on the diagnostic chart.

1. DTC 45 is set, when the O2 sensor signal voltage is fixed above .75 (750 mV) and the following:
   - TP sensor greater than 3%.
   - No DTC 21 or 33.
   - System is operating in "Closed Loop."
   - Engine Coolant Temperature (ECT) above 69.5°C (183°F).
   - Short term fuel trim not at 128.
   - VSS > 3 mph.

Diagnostic Aids:

The DTC 45, or rich exhaust, is most likely caused by one of the following:

- Fuel Pressure - System will go rich if pressure is too high. The ECM can compensate for some increase. However, if it gets too high, a DTC 45 will be set. See "Fuel System Diagnosis" CHART A-7.
- HEI Shielding - An open ignition ref low wire may result in EMI, or induced electrical "noise." The ECM looks at this "noise" as reference pulses. The additional pulses result in a higher than actual engine speed signal. Refer to

ECM wiring schematics for ignition system wiring. The ECM then delivers too much fuel, causing system to go rich. Engine tachometer will also show higher than actual engine speed, which can help in diagnosing this problem.

- Canister Purge - Check for fuel saturation. If full of fuel, check canister control and hoses. See "Evaporative Emission (EVAP) Control System,"
- MAP Sensor - An output that causes the ECM to sense a higher than normal manifold pressure (low vacuum) can cause the system to go rich. Disconnecting the MAP sensor will allow the ECM to set a fixed value for the MAP sensor. Substitute a different MAP sensor if the rich condition is gone, while the sensor is disconnected.
- TP Sensor - An intermittent TP sensor output will cause the system to go rich, due to a false indication of the engine accelerating.
- Oxygen Sensor Contamination - Inspect oxygen sensor for silicone contamination from fuel, or use of improper RTV sealant.
- Fuel Injectors - The wrong fuel injector(s) could be the cause of a rich exhaust condition. Verify the correct injector(s) are installed in the vehicle. Refer to the service parts manual for the correct part and part number.
- EGR - Valve sticking open at idle, usually accompanied by a rough idle, stall complaint.

If DTC 45 is intermittent, refer to "Intermittents" in "Symptoms,"

1.6L (VIN 6) AND 1.8L (VIN 8) ENGINES — DIAGNOSTIC TROUBLE CODE CHARTS — 1992–93 STORM

DTC 45

OXYGEN SENSOR (O2S) CIRCUIT
(RICH EXHAUST INDICATED)
1.6L SOHC (VIN 6) "R" CARLINE (MFI)

IF THE ECM IS FAULTY AND MUST BE REPLACED, THE NEW ECM MUST BE PROGRAMMED.

"AFTER REPAIRS," CONFIRM "CLOSED LOOP" OPERATION AND NO MIL (CHECK ENGINE LIGHT).

1.6L (VIN 6) AND 1.8L (VIN 8) ENGINES — DIAGNOSTIC TROUBLE CODE CHARTS — 1992–93 STORM

DTC 51
1.6L SOHC & 1.8L DOHC (VIN 6 & 8) "R" CARLINE (MFI)

DTC 51

ECM FAILURE
(ECM FAILED OR EEPROM FAILURE)

CHECK THAT ALL ECM CONNECTIONS ARE GOOD.
IF OK, CLEAR MEMORY AND RECHECK ECM.
IF DTC 51 REAPPEARS, REPLACE ECM AND REPROGRAM.

IF THE ECM IS FAULTY AND MUST BE REPLACED, THE NEW ECM MUST BE PROGRAMMED.

"AFTER REPAIRS," CONFIRM "CLOSED LOOP" OPERATION AND NO MIL (CHECK ENGINE LIGHT).

1.6L (VIN 6) AND 1.8L (VIN 8) ENGINES — B-CHARTS — 1992–93 STORM

CHART B-1
RESTRICTED EXHAUST SYSTEM CHECK
1.6L SOHC & 1.8L DOHC (VIN 6 & 8)
"R" CARLINE (MFI)

Proper diagnosis for a restricted exhaust system is essential before any components are replaced.

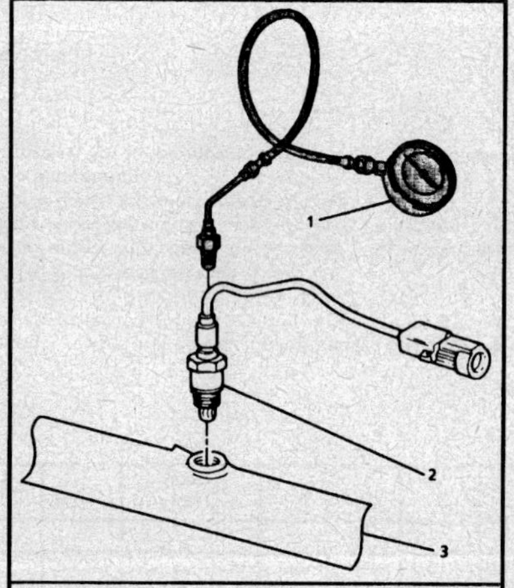

1   BACK PRESSURE GAGE
2   OXYGEN SENSOR (O2S)
3   EXHAUST MANIFOLD

DIAGNOSIS:

1. With the engine idling at normal operating temperature, observe the exhaust system backpressure reading on the gage. Reading should not exceed 7.8 kPa (1.13 psi) for Federal and 9.5 kPa (1.42 psi) for California.
2. Increase engine speed to 2000 RPM and observe gage. Reading should not exceed 14.6 kPa (2.11 psi) for Federal and 11.8 kPa (1.71 psi) for California.
3. If the backpressure at either speed exceeds specification, a restricted exhaust system is indicated.
4. Inspect the entire exhaust system for a collapsed pipe, heat distress, or possible internal muffler failure.
5. If there are no obvious reasons for the excessive backpressure, the catalytic converter is suspected to be restricted and should be replaced using current recommended procedures.

1.6L (VIN 6) AND 1.8L (VIN 8) ENGINES — ECM SYMPTOMS CHARTS — 1992–93 STORM

"A" 24 PIN ECM CONNECTOR

PIN FUNCTION	CKT #	WIRE COLOR	COMPONENT CONNECTOR CAVITY	NORMAL VOLTAGE KEY "ON"	NORMAL VOLTAGE ENG RUN**	DTC AFFECTED	POSSIBLE SYMPTOMS FROM FAULTY CIRCUIT
A1 FUEL PUMP		PNK/WHT	FUEL PUMP RELAY	(6)	B+		(5) NO START.
A2 A/C CLUTCH CONTROL OUTPUT		GRY/RED	A/C COMPRESSOR RELAY	*0	(7)		(5) INOPERATIVE A/C. A/C OPERATES AT ALL TIMES.
A3 EVAP CANISTER PURGE EVRV		WHT/RED	EVAP EVRV	B+	B+	45	POOR FUEL ECONOMY, ROUGH IDLE
A4 NOT USED							
A5 MALFUNCTION INDICATOR LAMP (MIL)		PPL	I/P C1-2	0*	B+		(4)
A6 IGNITION VOLTAGE SIGNAL		RED/GRN	MAIN RELAY	B+	B+		(3) NO START. (4) NO START. BLOWN ECM FUSE.
A7 NOT USED							
A8 SERIAL DATA		ORN/BLK	DLC "1"	4.7	4.7		(5) NO SCAN TOOL DATA.
A9 OBD SYSTEM REQUEST		ORN/YEL	DLC "2"	B+	B+		(4) FIELD SERVICE MODE ACTIVE, MIL FLASHES RICH/LEAN.
A10 VEHICLE SPEED SIGNAL (VSS)		WHT	VEHICLE SPEED SENSOR	(2)	(2)	24	REFER TO ELECTRICAL DIAGNOSIS
A11 ECT AND MAP, EGR GAS TEMP GROUND		GRY	MAP SENSOR "A" ECT "A" EGR GAS TEMP "A"	* *	* *	33 14 32	INCORRECT IDLE HESITATION. WITH "SCAN" TOOL, MAP READS 5V, ECT READS -39°C.
A12 ECM SYSTEM GROUND		BLK/BLU	INT. PLENUM (1.8L) THERMOSTAT HOUSING (1.6L)	*	*		NO EFFECT. IF "D2" IS ALSO OPEN, NO START.

NOTICE: The voltages you get may vary due to low battery charge or other reasons, but they should be very close.

* All voltages shown "0" should read less than .5 volt.
** All voltages shown are typical with engine at idle, closed throttle, normal operating engine temperature, park or neutral and "Closed Loop." All accessories "OFF."
(A) A/C select switch "OFF."
(B) Varies depending on temperature.
(1) Changes with IAC valve activity (when moving throttle slightly up and down).
(2) Varies.
(3) Open circuit.
(4) Grounded circuit.
(5) Open or grounded circuit.
(6) Reads B + for 2 seconds after ignition "ON," then should read 0 volts.
(7) A/C "OFF" B +, A/C "ON" 0 volt.
(8) B + above 5000 rpm.

ECM Symptoms Chart (1 of 4)

1.6L (VIN 6) AND 1.8L (VIN 8) ENGINES — ECM SYMPTOMS CHARTS — 1992–93 STORM

"B" 24 PIN ECM CONNECTOR

PIN FUNCTION	CKT #	WIRE COLOR	COMPONENT CONNECTOR CAVITY	NORMAL VOLTAGE KEY "ON"	NORMAL VOLTAGE ENG RUN**	CODES AFFECT	POSSIBLE SYMPTOMS FROM FAULTY CIRCUIT
B1 BATTERY FEED		RED/WHT	30 AMP MAIN FUSE	B+	B+		(3) NO EFFECT. IF C16 IS ALSO OPEN, NO START. (4) BLOWN FUSE LINK.
B2 NOT USED							
B3 REF LOW		YEL/BLK	IGN MODULE "A"	*	*		(5) NO EFFECT.
B4 NOT USED							
B5 REFERENCE "HI"		YEL/RED	IGNITION MODULE	0*	1.1V		(5) NO START.
B6 NOT USED							
B7 NOT USED							
B8 A/C REQUEST INPUT		GRN/ORN	THERMO SWITCH RELAY	0*	(A)		(5) LOW IDLE WITH A/C "ON." NO A/C.
B9 NOT USED							
B10 PNP SWITCH		PNK/PPL	PARK/NEUTRAL	(6)	(6)		(5) INCORRECT IDLE, STALLING.
B11 NOT USED							
B12 EGR GAS TEMP (CALIFORNIA ONLY)		GRY/BLU	EGR GAS TEMP SENSOR	(2)	(2)	32	NO EFFECT.

NOTICE: The voltages you get may vary due to low battery charge or other reasons, but they should be very close.

* All voltages shown "0" should read less than .5 volt.
** All voltages shown are typical with engine at idle, closed throttle, normal operating temperature, park or neutral and "Closed Loop." All accessories "OFF."
(A) A/C select switch "OFF" 0 volt, A/C select switch "ON" B +.
(B) Varies depending on temperature.
(1) Changes with IAC valve activity (when moving throttle slightly up and down).
(2) Varies.
(3) Open circuit.
(4) Grounded circuit.
(5) Open or grounded circuit.
(6) Reads B + in gear.

ECM Symptoms Chart (2 of 4)

1.6L (VIN 6) AND 1.8L (VIN 8) ENGINES — ECM SYMPTOMS CHARTS — 1992–93 STORM

"C" 32 PIN ECM CONNECTOR							
PIN FUNCTION	CKT #	WIRE COLOR	COMPONENT CONNECTOR CAVITY	NORMAL VOLTAGE KEY "ON"	NORMAL VOLTAGE ENG RUN**	DTC AFFECTED	POSSIBLE SYMPTOMS FROM FAULTY CIRCUIT
C1 SHIFT LIGHT M/T (SOHC) TCM SIGNAL A/T (DOHC)		ORN/BLU	SHIFT LIGHT A/T TCM	B+ 4.0	B+ 3.3		(3) INOPERATIVE SHIFT A/T SHIFT PROBLEM, ECONO LIGHT FLASHING.
C2 NOT USED							
C3 IAC COIL "B" LO		BRN/YEL	IAC VALVE "D"	NOT USEABLE	NOT USEABLE		(5) UNSTABLE, INCORRECT IDLE.
C4 IAC COIL "B" HI		BRN/RED	IAC VALVE "C"	NOT USEABLE	NOT USEABLE		(5) UNSTABLE, INCORRECT IDLE.
C5 IAC COIL "A" HI		BRN/BLK	IAC VALVE "A"	NOT USEABLE	NOT USEABLE		(5) UNSTABLE, INCORRECT IDLE.
C6 IAC COIL "A" LO		BRN/WHT	IAC VALVE "B"	NOT USEABLE	NOT USEABLE		(5) UNSTABLE, INCORRECT IDLE.
C7 PSPS		GRN/YEL	PSPS	B+	B+		(5) STALL, HIGH IDLE, LACK OF POWER.
C8 NOT USED							
C9 NOT USED							
C10 ECT SIGNAL		GRY/BLK	ECT "B"	1.9V (2)	1.9V (2)	14 (4) (3)	(3) INCORRECT IDLE.
C11 MAP SIGNAL		GRY/RED	MAP SENSOR "B"	4.8V	1.1V	33	(5) INCORRECT IDLE, CHUGGLE, POOR PERFORMANCE.
C12 IAT SIGNAL		BLU/BLK	IAT SENSOR "A"	(2)	(2)	23	(5) POOR PERFORMANCE.
C13 TP SENSOR SIGNAL		YEL/BLU	TP "B"	.6V	.6V IDLE	21	(5) POOR PERFORMANCE, BACKFIRE, HESITATION.
C14 5 VOLT REFERENCE		BLU/ORN	MAP "C" TP "C"	5.0V	5.0V	21 (5) 34 (5)	(5) STUMBLES, HESITATES, EXHAUST ODOR, FUEL INTEGRATOR REMOVING FUEL (LOW COUNTS).
C15 INJECTOR DRIVER		BLU/YEL	INJECTOR	B+	B+		(3) ROUGH IDLE, NO POWER.
C16 BATTERY FEED		RED/WHT	30 AMP MAIN FUSE	B+	B+		(3) NO EFFECT. IF "B1" IS ALSO OPEN, NO START.

NOTICE: The voltages you get may vary due to low battery charge or other reasons, but they should be very close.

* All voltages shown "0" should read less than .5 volt.
** All voltages shown are typical with engine at idle, closed throttle, normal operating temperature, park or neutral and "Closed Loop." All accessories "OFF."
(1) Changes with IAC valve activity (when moving throttle slightly up and down).
(2) Varies.
(3) Open circuit.
(4) Grounded circuit.
(5) Open or grounded circuit.
(6) Reads B + for 2 seconds after ignition "ON," then should read 0 volts.

ECM Symptoms Chart (3 of 4)

1.6L (VIN 6) AND 1.8L (VIN 8) ENGINES — ECM SYMPTOMS CHARTS — 1992–93 STORM

"D" 32 PIN ECM CONNECTOR							
PIN FUNCTION	CKT #	WIRE COLOR	COMPONENT CONNECTOR CAVITY	NORMAL VOLTAGE KEY "ON"	NORMAL VOLTAGE ENG RUN**	DTC AFFECTED	POSSIBLE SYMPTOMS FROM FAULTY CIRCUIT
D1 ECM SYSTEM GROUND		BLK/BLU	INTAKE PLENUM	0*	0*		(3) NO EFFECT.
D2 TP AND IAT GROUND		BLK	TP "A" IAT "B"	0*	0*	21	(3) HESITATION, STUMBLE, UNSTABLE IDLE.
D3 SOHC ECM SYSTEM GROUND		BLK/BLU	THERMOSTAT HOUSING	0*	0*		NO EFFECT.
D3 NOT USED (7)							
D4 IC CIRCUIT		YEL	ELECTRONIC IGNITION MODULE "D"	0*	1.3V	42	(5) STUMBLES, UNSTABLE IDLE.
D5 BYPASS		YEL/GRN	DISTRIBUTOR MODULE "B"	0*	4.6V	42	(5) POOR PERFORMANCE.
D6 OXYGEN SENSOR (O2S) GROUND		BLK/BLU	INTAKE PLENUM	0*	0*	13	(3) "OPEN LOOP" "SCAN" TOOL INDICATES O2S VOLTAGE AT 400-500 mV.
D7 OXYGEN SENSOR (O2S) SIGNAL		BLU	O2 SENSOR	.33 - .55V	.1 - .9V	13 (3) 44 (4)	(5) "OPEN LOOP."
D8 ECM GROUND SYSTEM SOHC		BLK/BLU	ENGINE BLOCK	0*	0*		NO EFFECT.
D8 NOT USED (7)							
D9 NOT USED (7)							
D10 NOT USED							
D11 NOT USED							
D12 EGR EVRV		WHT/GRN	EGR EVRV	0*	B+	32	(3) INOPERATIVE EGR, SPARK KNOCK.
D13 NOT USED (7)							
D14 NOT USED							
D15 INJECTOR DRIVER SOHC		BLU/WHT	INJECTOR	B+	B+		(3) ROUGH IDLE, NO POWER.
D15 INJECTOR DRIVER DOHC		BLU/WHT	INJECTOR	B+	B+		(3) ROUGH IDLE, NO POWER.
D16 NOT USED							

NOTICE: The voltages you get may vary due to low battery charge or other reasons, but they should be very close.

* All voltages shown "0" should read less than .5 volt.
** All voltages shown are typical with engine at idle, closed throttle, normal operating temperature, park or neutral and "Closed Loop." All accessories "OFF."
(1) Changes with IAC valve activity (when moving throttle slightly up and down).
(2) Varies.
(3) Open circuit.
(4) Grounded circuit.
(5) Open or grounded circuit.
(6) Reads B + for 2 seconds after ignition "ON," then should read 0 volts.
(7) On the DOHC a wire may be located in this cavity, it is non-functional.

ECM Symptoms Chart (4 of 4)

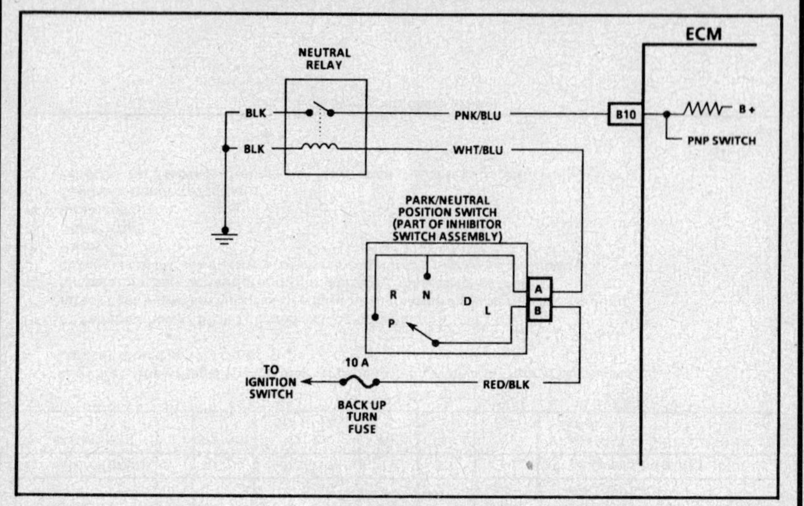

CHART C-1A

PARK/NEUTRAL POSITION (PNP) SWITCH DIAGNOSIS
(AUTO TRANSAXLE ONLY)
1.6L SOHC & 1.8L DOHC (VIN 6 & 8) "R" CARLINE (MFI)

Circuit Description:

The Park/Neutral Position (PNP) switch contacts are a part of the inhibitor switch assembly. When the PNP is in park or neutral the neutral relay is "Closed" indicating to the ECM the vehicle is in park/neutral (low voltage at ECM terminal "B10"). When the gear selector is moved to reverse or drive the neutral relay is "Open" (high voltage at ECM terminal "B10") the ECM will raise the engine RPM for the additional load the transaxle put on the engine.

The ECM uses the PNP signal as one of the inputs to control idle air, VSS diagnostics, and EGR.

If signal wire indicates PNP (grounded), while in drive range, the EGR would be inoperative, resulting in possible detonation.

If PNP signal wire indicates drive (closed), a dip in the idle may exist when the gear selector is moved into drive range and a harsh bump may be felt. If the ECM thinks the vehicle is in drive range all of the time, a "High" idle condition may exist.

Test Description: Number(s) below refer to circled number(s) on the diagnostic chart.

1. Checks for a closed switch to ground in park position. Different makes of scan tools will read PNP differently. Refer to tool operator's manual for type of display used for a specific tool.

2. Checks for an open switch in drive range.

3. Be sure scan indicates drive, even while wiggling shifter, to test for an intermittent or misadjusted switch in drive or overdrive range.

4. Checking for an open in the power feed circuit.

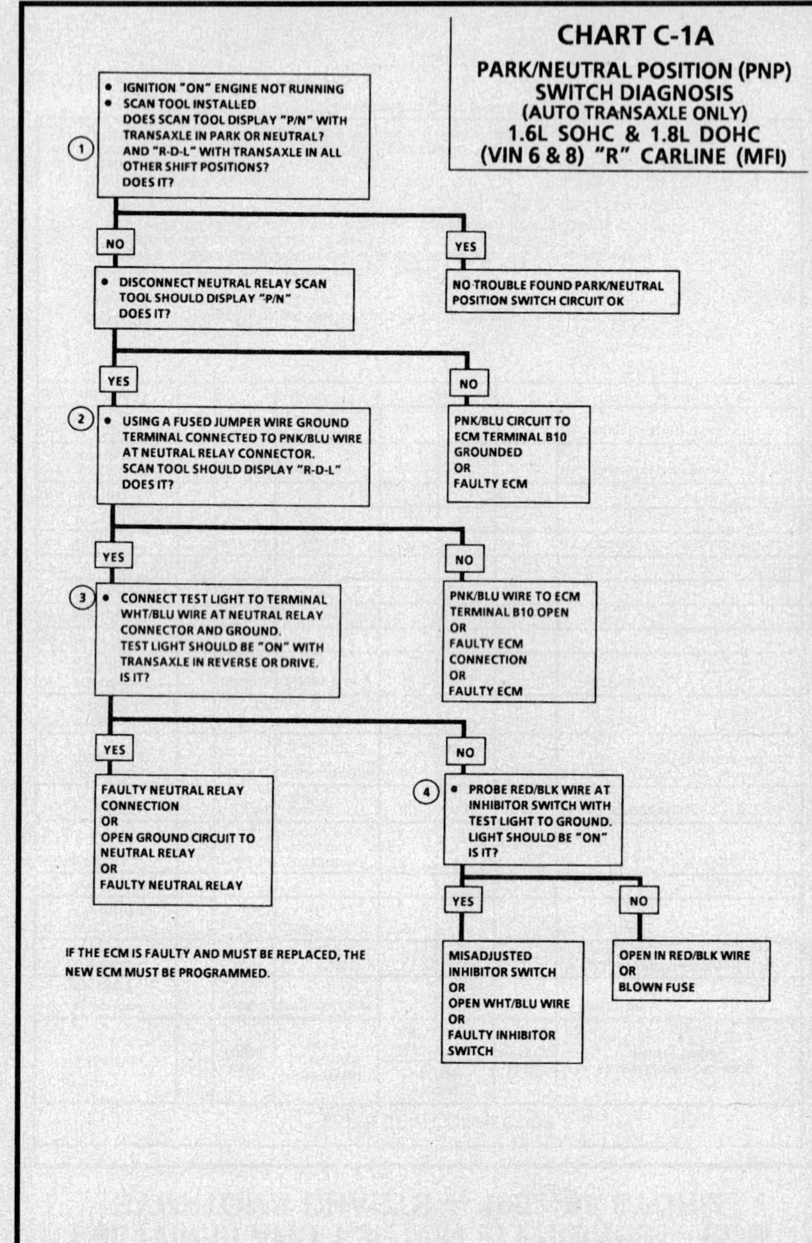

1.6L (VIN 6) AND 1.8L (VIN 8) ENGINES — C-CHARTS — 1992–93 STORM

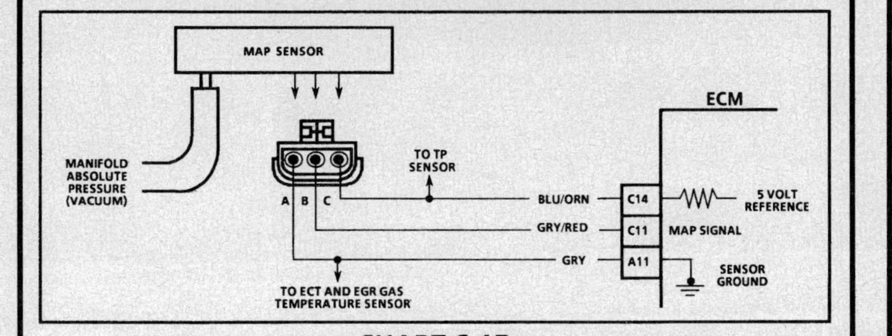

CHART C-1D
MANIFOLD ABSOLUTE PRESSURE (MAP) OUTPUT CHECK
1.6L SOHC & 1.8L DOHC (VIN 6 & 8) "R" CARLINE (MFI)

Circuit Description:

Manifold Absolute Pressure (MAP) sensor measures the changes in the intake manifold pressure which result from engine load (intake manifold vacuum) and rpm changes; and converts these into a voltage output. The ECM sends a 5-volt reference voltage to the MAP sensor. As the manifold pressure changes, the output voltage of the sensor also changes. By monitoring the sensor output voltage, the ECM knows the manifold pressure. A lower pressure (low voltage) output voltage will be about 1-2 volts at idle. While higher pressure (high voltage) output voltage will be about 4-4.8 at Wide Open Throttle (WOT). The MAP sensor is also used, under certain conditions, to measure barometric pressure, allowing the ECM to make adjustments for different altitudes. The ECM uses the MAP sensor to control fuel delivery and distributor ignition timing.

Test Description: Number(s) below refer to circled number(s) on the diagnostic chart.

Important
- Be sure to use the same Diagnostic Test Equipment for all measurements.

1. When comparing scan tool readings to a known good vehicle, it is important to compare vehicles that use a MAP sensor having the same color insert or having the same "Hot Stamped" number. See figures on facing page.
2. Applying 34 kPa (10" Hg) vacuum to the MAP sensor should cause the voltage to be 1.5 to 2.1 volts less than the voltage at Step 1. Upon applying vacuum to the sensor, the change in voltage should be instantaneous. A slow voltage change indicates a faulty sensor.

3. Check vacuum hose to sensor for leaking or restriction. Be sure than no other vacuum devices are connected to the MAP hose.

   NOTICE: Make sure electrical connector remains securely fastened.

4. Disconnect sensor from bracket and twist sensor by hand (only) to check for intermittent connection. Output changes greater than .10 volt indicate a bad sensor.

   NOTICE: Make sure the MAP sensor is not hanging below the intake manifold, a false engine load could be the result of an improperly located MAP sensor.

1.6L (VIN 6) AND 1.8L (VIN 8) ENGINES — C-CHARTS — 1992–93 STORM

CHART C-1D
MANIFOLD ABSOLUTE PRESSURE (MAP) OUTPUT CHECK
1.6L SOHC & 1.8L DOHC (VIN 6 & 8) "R" CARLINE (MFI)

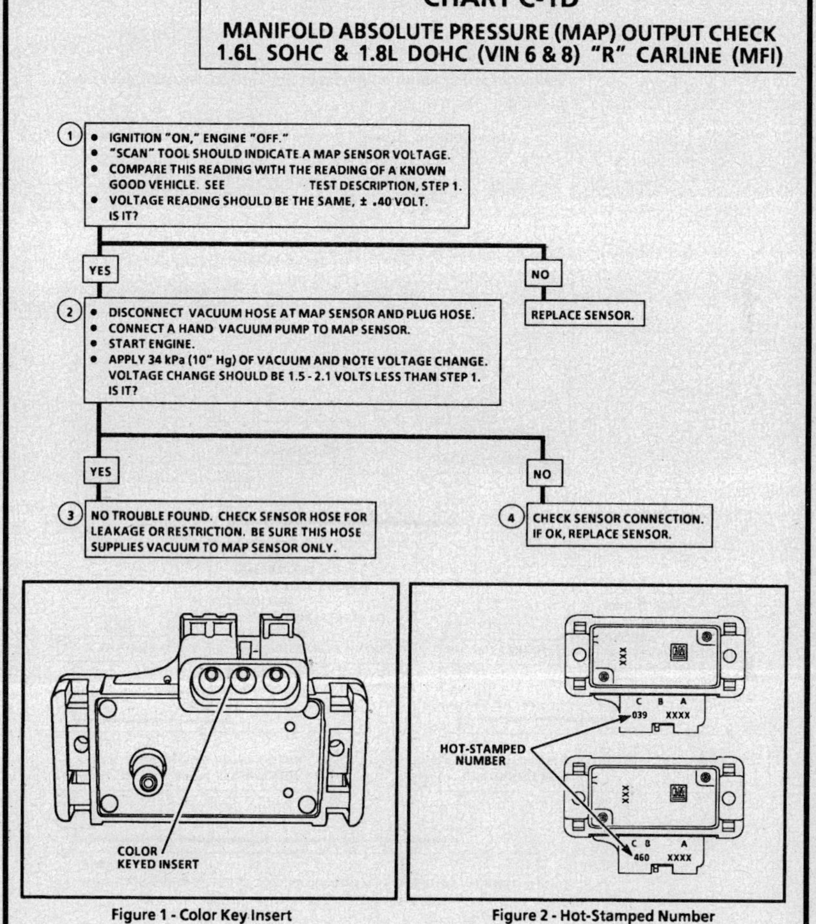

Figure 1 - Color Key Insert

Figure 2 - Hot-Stamped Number

"AFTER REPAIRS," CONFIRM "CLOSED LOOP" OPERATION AND NO MIL (CHECK ENGINE LIGHT).

1.6L (VIN 6) AND 1.8L (VIN 8) ENGINES — C-CHARTS — 1992–93 STORM

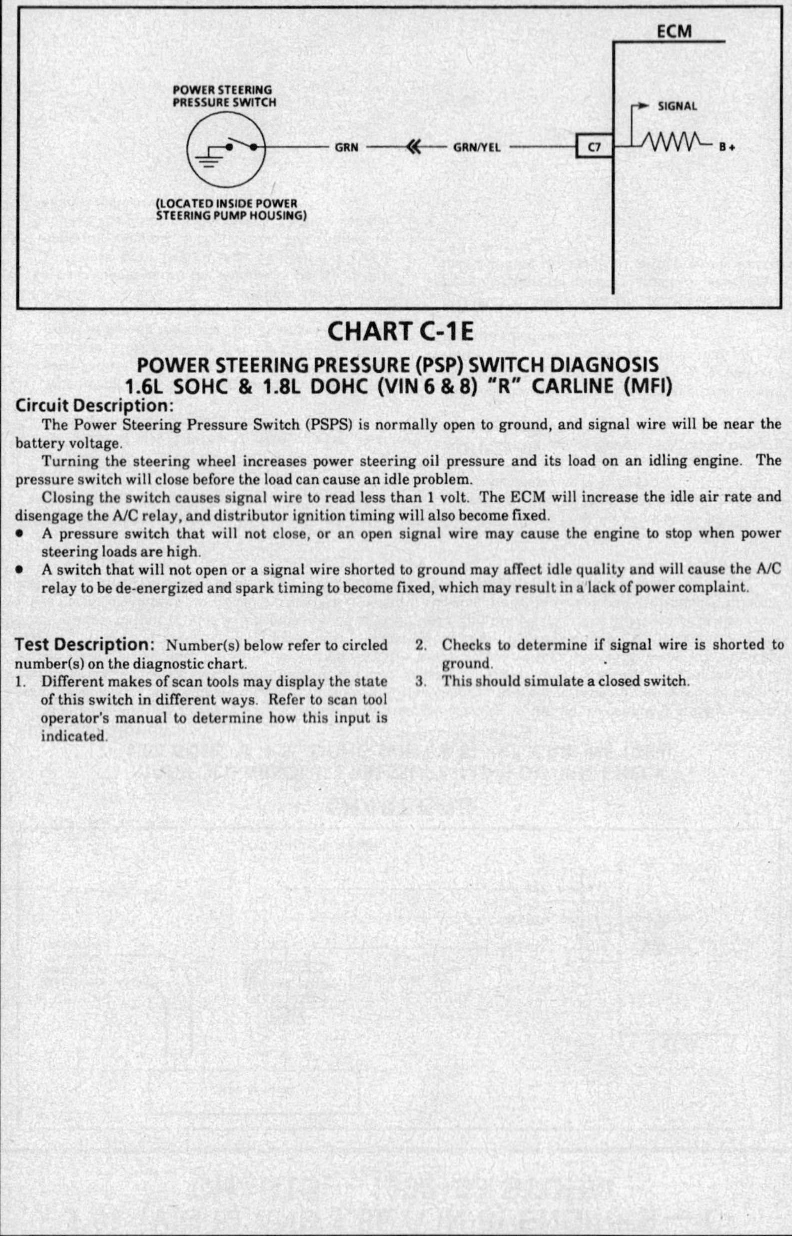

CHART C-1E
POWER STEERING PRESSURE (PSP) SWITCH DIAGNOSIS
1.6L SOHC & 1.8L DOHC (VIN 6 & 8) "R" CARLINE (MFI)

Circuit Description:

The Power Steering Pressure Switch (PSPS) is normally open to ground, and signal wire will be near the battery voltage.

Turning the steering wheel increases power steering oil pressure and its load on an idling engine. The pressure switch will close before the load can cause an idle problem.

Closing the switch causes signal wire to read less than 1 volt. The ECM will increase the idle air rate and disengage the A/C relay, and distributor ignition timing will also become fixed.

- A pressure switch that will not close, or an open signal wire may cause the engine to stop when power steering loads are high.
- A switch that will not open or a signal wire shorted to ground may affect idle quality and will cause the A/C relay to be de-energized and spark timing to become fixed, which may result in a lack of power complaint.

Test Description: Number(s) below refer to circled number(s) on the diagnostic chart.

1. Different makes of scan tools may display the state of this switch in different ways. Refer to scan tool operator's manual to determine how this input is indicated.

2. Checks to determine if signal wire is shorted to ground.

3. This should simulate a closed switch.

1.6L (VIN 6) AND 1.8L (VIN 8) ENGINES — C-CHARTS — 1992–93 STORM

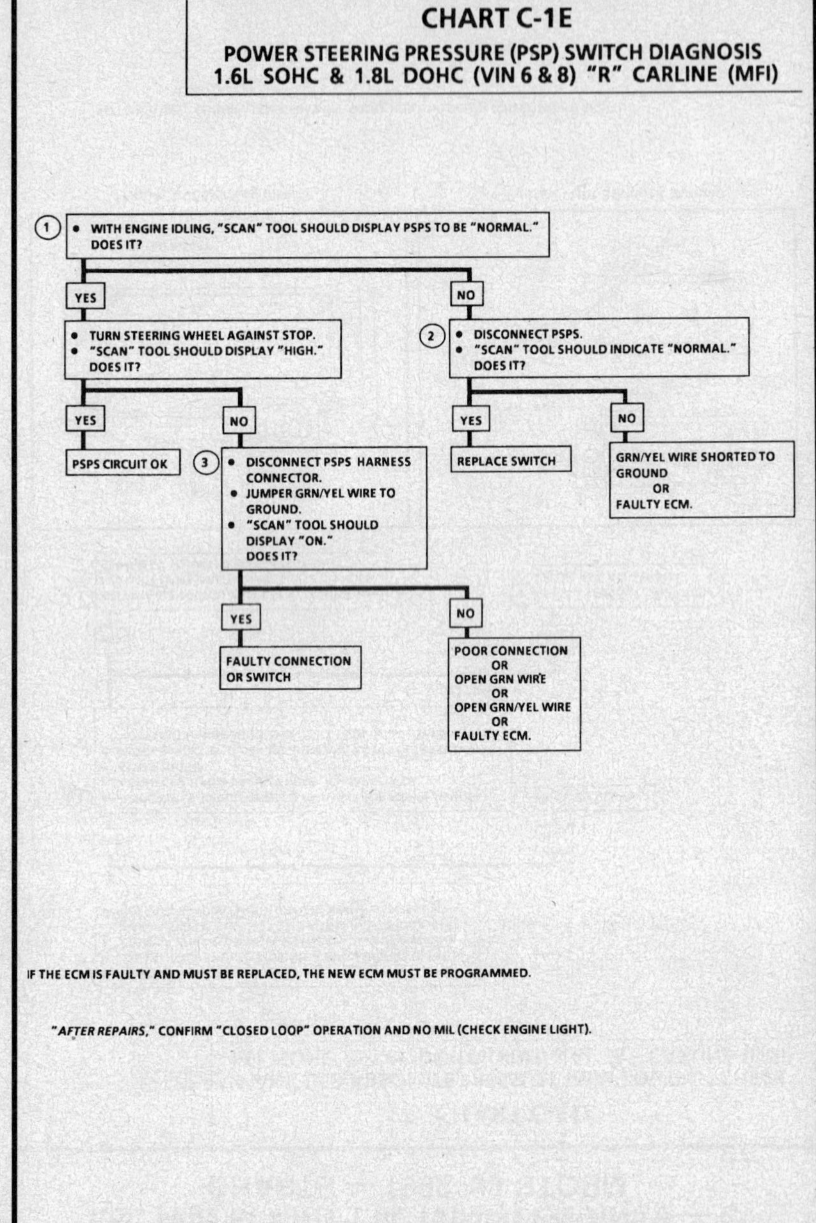

1.6L (VIN 6) AND 1.8L (VIN 8) ENGINES — C-CHARTS — 1992–93 STORM

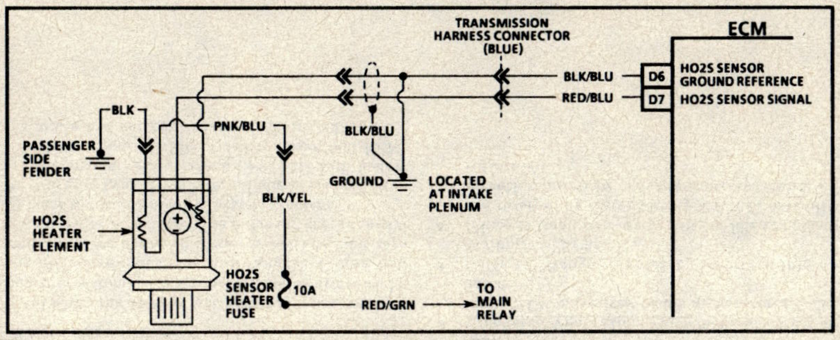

CHART C-1F
HEATED OXYGEN SENSOR (HO2S) CHECK
1.8L DOHC (VIN 8) "R" CARLINE (MFI)

Circuit Description:

The generator charging circuit supplies battery voltage to the heater part of the heated oxygen sensor. The heater warms the HO2S to allow a varying voltage to be produced by the HO2S more rapidly instead of waiting for the HO2S to heat-up on its own.

Test Description: Number(s) below refer to circled number(s) on the diagnostic chart.

The following test procedures are for the heater part of the heated oxygen sensor. No DTC(s) will be logged if the heater fails to function properly.

1. This will verify that battery voltage is present for the heater.
2. This check will test the ground circuit to the heater.
3. This tests the heater element resistance. Resistance will vary with the temperature of the sensor. Normal resistance is between 3.5 - 14.2 ohms.

Diagnostic Aids:

- A poor or loose connection at the heated oxygen sensor could result in a false DTC being set or the heated oxygen heater not functioning properly.
- A shorted HO2S heater could cause a DTC 13 to set and/or DTC 44 and DTC 45.

1.6L (VIN 6) AND 1.8L (VIN 8) ENGINES — C-CHARTS — 1992–93 STORM

CHART C-1F
HEATED OXYGEN SENSOR (HO2S) CHECK
1.8L DOHC (VIN 8) "R" CARLINE (MFI)

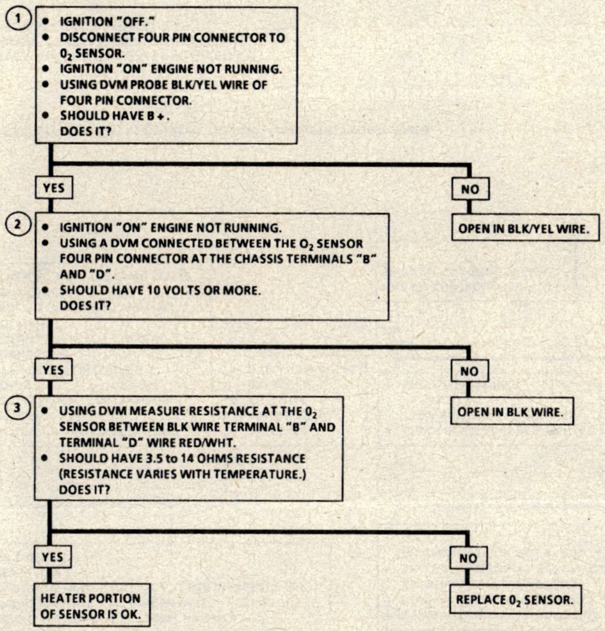

1.6L (VIN 6) AND 1.8L (VIN 8) ENGINES — C-CHARTS — 1992–93 STORM

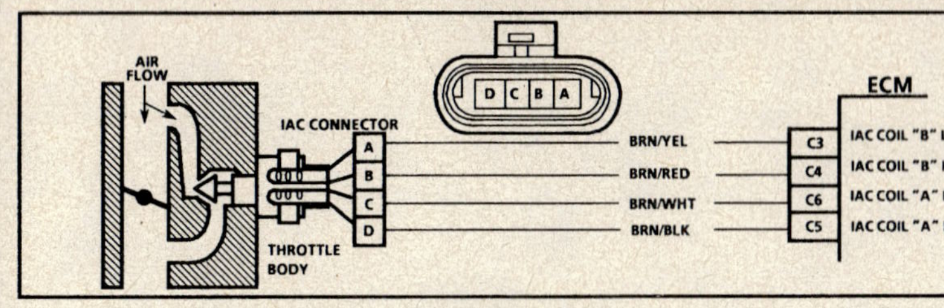

CHART C-2C
IDLE AIR CONTROL (IAC) SYSTEM CHECK
1.6L SOHC & 1.8L DOHC (VIN 6 & 8) "R" CARLINE (MFI)

Circuit Description:

The ECM controls engine idle speed with the IAC valve. To increase idle speed, the ECM retracts the IAC valve pintle away from its seat, allowing more air to bypass the throttle bore. To decrease idle speed, it extends the IAC valve pintle towards its seat, reducing bypass air flow. A scan tool will read the ECM commands to the IAC valve in counts. Higher the counts indicate more air bypass (higher idle). The lower the counts indicate less air allowed to bypass (lower idle).

Test Description: Number(s) below refer to circled number(s) on the diagnostic chart.

1. The Tech 1 RPM output control mode is used to extend and retract the IAC valve. The valve should move smoothly within the specified range. If the idle speed is commanded (IAC extended) too low (below 700 RPM), the engine may stall. This may be normal and would not indicate a problem. Retracting the IAC beyond its controlled range (above 1500 RPM) will cause a delay before the RPM's start dropping. This too is normal.
2. This test uses the Tech 1 to command the IAC controlled idle speed. The ECM issues commands to obtain commanded idle speed. The node lights each should flash red and green to indicate a good circuit as the ECM issues commands. While the sequence of color is not important if either light is "OFF" or does not flash red and green, check the circuits for faults, beginning with poor terminal contacts.

Diagnostic Aids:

A slow, unstable, or fast idle may be caused by a non-IAC system problem that cannot be overcome by the IAC valve. Out of control range IAC scan tool counts will be above 60 if idle is too low, and zero counts if idle is too high. The following checks should be made to repair a non-IAC system problem:

- **Vacuum Leak (High Idle)** - If idle is too high, stop the engine. Fully extend (low) IAC with tester. Start engine. If Idle speed is above 800 RPM, locate and correct vacuum leak including PCV system. Also check for binding of throttle blade or linkage.
- **System too lean (High Air/Fuel Ratio)** - Idle speed may be too high or too low. Engine speed may vary up and down and disconnecting IAC does not help. DTC 44 may be set. Scan O2S voltage will be less than 300 mV (.3 volt). Check for low regulated fuel pressure, water in the fuel or a restricted injector.
- **System too rich (Low Air/Fuel Ratio)** - The idle speed will be too low. Scan tool IAC counts will usually be above 80. System is obviously rich and may exhibit black smoke exhaust. Scan tool O2S voltage will be fixed above 800 mV (.8 volt). Check for high fuel pressure, leaking or sticking injector. Silicone contaminated O2 sensor will scan an O2S voltage slow to respond.
- **Throttle Body** - Remove IAC and inspect bore for foreign material.
- **IAC Valve Electrical Connections** - IAC valve connections should be carefully checked for proper contact.
- **Crankcase Ventilation Valve** - An incorrect or faulty PCV valve may result in an incorrect idle speed.
- Refer to "Rough, Unstable, Incorrect Idle or Stalling" in "Symptoms."
- If intermittent, poor driveability or idle symptoms are resolved by disconnecting the IAC, carefully recheck connections, valve terminal resistance, or replace IAC.

1.6L (VIN 6) AND 1.8L (VIN 8) ENGINES — C-CHARTS — 1992–93 STORM

CHART C-2C
IDLE AIR CONTROL (IAC) SYSTEM CHECK
1.6L SOHC & 1.8L DOHC (VIN 6 & 8) "R" CARLINE (MFI)

(1)
- INSTALL TECH 1 SCAN TOOL.
- ENGINE AT NORMAL OPERATING TEMPERATURE IN PARK/NEUTRAL WITH PARKING BRAKE SET.
- A/C "OFF."
- SELECT RPM CONTROL. (MISC. TESTS)
- CYCLE IAC THROUGH ITS RANGE FROM 700 RPM UP TO 1500 RPM.
- RPM SHOULD CHANGE SMOOTHLY.

DOES IT?

NO →

(2)
- INSTALL IAC NODE LIGHT * IN IAC HARNESS.
- ENGINE RUNNING. CYCLE IAC WITH TECH 1 SCAN TOOL.
- EACH NODE LIGHT SHOULD CYCLE RED AND GREEN BUT NEVER "OFF."

DO THEY?

YES →
- USING THE IAC DRIVER * OR OTHER CONVENIENT CONNECTOR, CHECK RESISTANCE ACROSS IAC COILS.
- SHOULD BE 40 TO 80 OHMS BETWEEN IAC TERMINALS "A" TO "B" AND "C" TO "D".

OK →
- CHECK RESISTANCE BETWEEN IAC TERMINALS "B" AND "C" AND "A" AND "D".
- SHOULD BE INFINITE.

NOT OK → REPLACE IAC VALVE AND RETEST.

OK → IDLE AIR CONTROL CIRCUIT OK. REFER TO "DIAGNOSTIC AIDS"

NOT OK → REPLACE IAC VALVE AND RETEST.

(Box 2) NO → IF CIRCUIT(S) DID NOT TEST RED AND GREEN, CHECK FOR:
- FAULTY CONNECTOR TERMINAL CONTACTS.
- OPEN CIRCUITS INCLUDING CONNECTORS.
- CIRCUITS SHORTED TO GROUND OR VOLTAGE.
- FAULTY ECM CONNECTIONS OR REPLACE ECM.
REPAIR AS NECESSARY AND RETEST.

(Box 2) YES →
- CHECK IAC CONNECTIONS.
- CHECK IAC PASSAGES.
- IF OK, REPLACE IAC.

IF THE ECM IS FAULTY AND MUST BE REPLACED, THE NEW ECM MUST BE PROGRAMMED.

* IAC DRIVER AND NODE LIGHT REQUIRED KIT
222-L FROM: CONCEPT TECHNOLOGY, INC.
J 37027 FROM: KENT-MOORE, INC.

1.6L (VIN 6) AND 1.8L (VIN 8) ENGINES — C-CHARTS — 1992–93 STORM

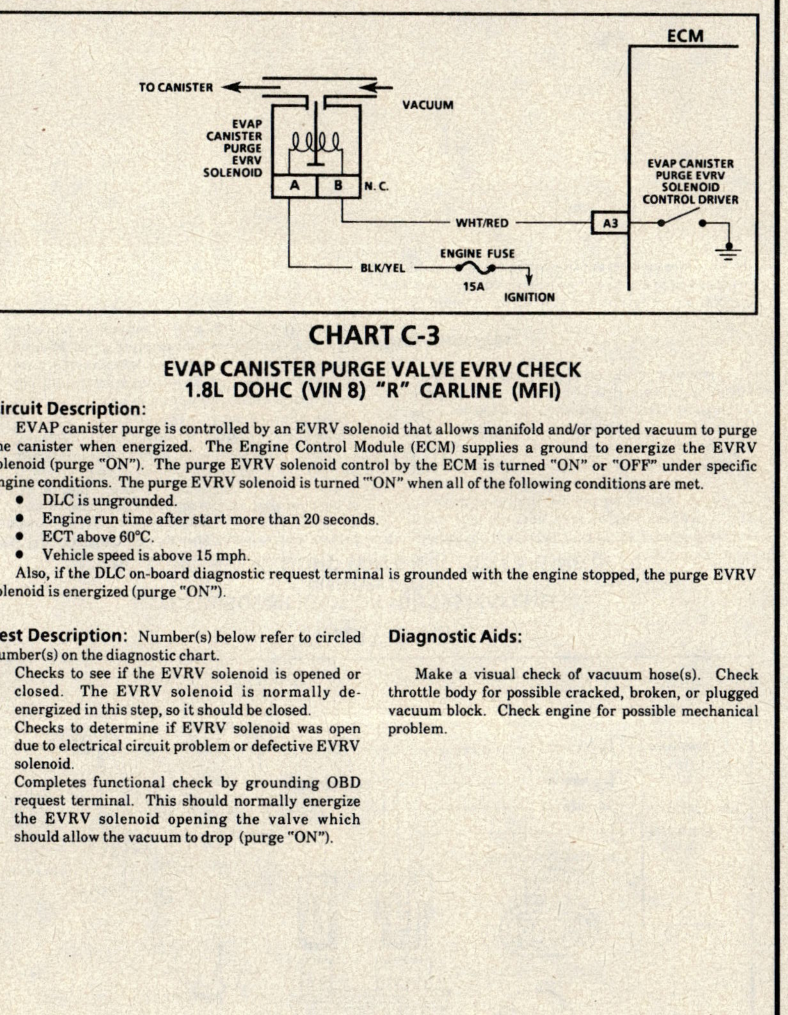

CHART C-3
EVAP CANISTER PURGE VALVE EVRV CHECK
1.8L DOHC (VIN 8) "R" CARLINE (MFI)

Circuit Description:

EVAP canister purge is controlled by an EVRV solenoid that allows manifold and/or ported vacuum to purge the canister when energized. The Engine Control Module (ECM) supplies a ground to energize the EVRV solenoid (purge "ON"). The purge EVRV solenoid control by the ECM is turned "ON" or "OFF" under specific engine conditions. The purge EVRV solenoid is turned '"ON" when all of the following conditions are met.

- DLC is ungrounded.
- Engine run time after start more than 20 seconds.
- ECT above 60°C.
- Vehicle speed is above 15 mph.

Also, if the DLC on-board diagnostic request terminal is grounded with the engine stopped, the purge EVRV solenoid is energized (purge "ON").

Test Description: Number(s) below refer to circled number(s) on the diagnostic chart.

1. Checks to see if the EVRV solenoid is opened or closed. The EVRV solenoid is normally de-energized in this step, so it should be closed.
2. Checks to determine if EVRV solenoid was open due to electrical circuit problem or defective EVRV solenoid.
3. Completes functional check by grounding OBD request terminal. This should normally energize the EVRV solenoid opening the valve which should allow the vacuum to drop (purge "ON").

Diagnostic Aids:

Make a visual check of vacuum hose(s). Check throttle body for possible cracked, broken, or plugged vacuum block. Check engine for possible mechanical problem.

1.6L (VIN 6) AND 1.8L (VIN 8) ENGINES — C-CHARTS — 1992–93 STORM

CHART C-3
CANISTER PURGE VALVE EVRV CHECK
1.8L DOHC (VIN 8) "R" CARLINE (MFI)

(1)
- IGNITION "OFF."
- IGNITION "ON" ENGINE STOPPED. (DO NOT CRANK ENGINE BEFORE FOLLOWING STEPS BELOW.)
- DISCONNECT THROTTLE BODY TO CANISTER PURGE EVRV SOLENOID VACUUM HOSE FROM EVRV SOLENOID.
- APPLY VACUUM (10" Hg OR 34 kPa) TO EVRV SOLENOID. EVRV SOLENOID SHOULD HOLD VACUUM. DOES IT?

YES →

*** (3)**
- IGNITION "ON" ENGINE RUNNING.
- CONNECT A TECH 1 SCAN TOOL TO THE DLC CONNECTOR.
- SELECT "MISC." TEST (F4), THEN SELECT "OUTPUT" TEST (FO).
- SELECT FUEL EVAP PURGE (F4).
- APPLY 10" OF VACUUM TO THE CANISTER PURGE EVRV PORTED VACUUM HOSE TO THE EVRV.
- SELECT THE "UP" ARROW ON THE TECH 1.
- VACUUM GAUGE SHOULD DROP TO ZERO "0". DOES IT?

YES →
- INSTALL VACUUM GAGE TO HOSE PREVIOUSLY DISCONNECTED FROM EVRV SOLENOID.
- START ENGINE.
- STABILIZE ENGINE RPM AT ABOUT 2500.
- MOMENTARILY SNAP THROTTLE OPEN AND LET RETURN TO IDLE. VERIFY THAT ABOUT 10" Hg (34 kPa) OF VACUUM IS AVAILABLE AT CANISTER PURGE EVRV SOLENOID, IS IT?

YES → FAULTY PURGE EVRV SOLENOID.

NO → SEE "DIAGNOSTIC AIDS"

NO →
- ENGINE STILL RUNNING.
- DISCONNECT EVRV SOLENOID ELECTRICAL CONNECTOR.
- CONNECT TEST LIGHT BETWEEN HARNESS TERMINALS.
- SELECT "UP" ARROW AGAIN, TEST LIGHT SHOULD LIGHT. DOES IT?

YES → FAULTY PURGE EVRV SOLENOID.

NO → PROBE EACH HARNESS TERMINAL WITH A TEST LIGHT TO GROUND.

LIGHT "ON" ONE → OPEN WHT/RED WIRE OR FAULTY ECM.

LIGHT "ON" BOTH → REPAIR SHORT TO VOLTAGE IN WHT/RED WIRE.

NO LIGHT → OPEN BLK/YEL.

NO →

(2)
- DISCONNECT EVRV SOLENOID ELECTRICAL CONNECTOR.
- APPLY VACUUM TO EVRV SOLENOID AS IN STEP 1. EVRV SOLENOID SHOULD HOLD VACUUM. DOES IT?

YES → CHECK FOR SHORT TO GROUND IN WHT/RED WIRE. IF CIRCUIT IS OK, ECM IS FAULTY.

NO → FAULTY EVRV SOLENOID.

(bottom box, YES/NO) NO PROBLEM FOUND.

* 92 CAR MAY/MAY NOT REQUIRE THIS STEP IF NOT, PROCEED TO NEXT STEP

1.6L (VIN 6) AND 1.8L (VIN 8) ENGINES — C-CHARTS — 1992–93 STORM

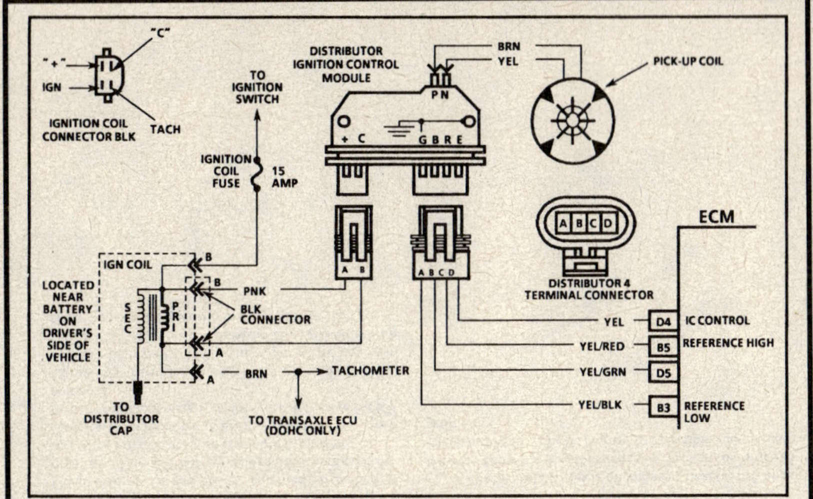

CHART C-4C

DISTRIBUTOR IGNITION (DI) SYSTEM CHECK
(REMOTE COIL)
1.6L SOHC & 1.8L DOHC (VIN 6 & 8) "R" CARLINE (MFI)

Test Description: Number(s) below refer to circled number(s) on the diagnostic chart.

1. Two wires are checked, to ensure that an open is not present in a spark plug wire.
2. A spark indicates the problem must be the distributor cap, rotor or coil output wire.
3. Normally, there should be battery voltage at the "C" and "+" terminals. Low voltage would indicate an open or a high resistance circuit from the distributor to the coil or ignition switch. If "C" terminal voltage was low, but "+" terminal voltage is 10 volts or more, circuit from "C" terminal to ignition coil or ignition coil primary winding is open.
4. Checks for a shorted module or grounded circuit from the ignition coil to the module. The distributor module should be turned "OFF," so normal voltage should be about 12 volts. If the module is turned "ON," the voltage would be low, but above 1 volt. This could cause the ignition coil to fail from excessive heat. With an open ignition coil primary winding, a small amount of voltage will leak through the module from the "batt" to the "tach" terminal.

Diagnostic Aids:

In order to do the test procedures at the ignition coil, use of the Kent-Moore J 35616-A will be needed to jumper the engine harness at the ignition coil.

1.6L (VIN 6) AND 1.8L (VIN 8) ENGINES — C-CHARTS — 1992–93 STORM

CHART C-4C
DISTRIBUTOR IGNITION (DI) SYSTEM CHECK (REMOTE COIL)
1.6L SOHC & 1.8L DOHC (VIN 6 & 8) "R" CARLINE (MFI)

(1)
- CHECK SPARK PLUG WIRES FOR OPEN CIRCUITS, CRACKS IN INSULATION OR IMPROPER SEATING OF TERMINALS AT SPARK PLUGS, DISTRIBUTOR, AND IGNITION COIL BEFORE PROCEEDING WITH THIS TEST.
- CHECK SPARK AT PLUG WITH SPARK TESTER J 26792 OR EQUIVALENT ST 125 WHILE CRANKING (IF NO SPARK ON ONE WIRE, CHECK A SECOND WIRE) A FEW SPARKS AND THEN NOTHING IS CONSIDERED NO SPARK.

NO SPARK →
SPARK → CHECK FUEL, SPARK PLUGS, ETC. REFER TO SYMPTOMS

(2)
- DISCONNECT 4 TERMINAL DISTRIBUTOR CONNECTOR. CHECK FOR SPARK AT IGNITION COIL TOWER WITH TESTER J 26792 OR EQUIVALENT ST 125 WHILE CRANKING. (LEAVE TESTER CONNECTED TO COIL TOWER FOR STEPS 3-6.)

NO SPARK →
SPARK → CHECK CAP, ROTOR AND COIL WIRES FOR OPENS, DAMAGE, OR WEAR. REPLACE AS NECESSARY.

(3)
- DISCONNECT DISTRIBUTOR 2 TERMINAL "C/+" CONNECTOR.
- IGNITION SWITCH "ON" ENGINE STOPPED.
- CHECK VOLTAGE AT "+" AND "C" TERMINALS OF DISTRIBUTOR HARNESS CONNECTOR.

BOTH TERMINALS 10 VOLTS OR MORE | BOTH TERMINALS UNDER 10 VOLTS. | UNDER 10 VOLTS "C" TERMINAL ONLY.

(4)
- RECONNECT DISTRIBUTOR 2 TERMINAL CONNECTOR.
- INSTALL JUMPERS AT IGNITION COIL USING J 35616-A OR EQUIVALENT.
- WITH IGNITION "ON," CHECK VOLTAGE FROM TACH TERMINAL TO GROUND. (TERMINAL MAY BE TAPED BACK IN HARNESS.)

REPAIR WIRE FROM IGNITION CONTROL MODULE "+" TERMINAL TO "B" TERMINAL OF BLACK IGNITION COIL CONNECTOR OR PRIMARY CIRCUIT TO IGNITION SWITCH.

CHECK FOR OPEN OR GROUND IN CIRCUIT FROM "C" TERMINAL TO IGNITION COIL. IF CIRCUIT IS OK, CHECK IGNITION COIL OR CONNECTION.

GREATER THAN 10 VOLTS | LESS THAN 1 VOLT | 1 TO 10 VOLTS

- CONNECT TEST LIGHT FROM TACHOMETER TERMINAL TO GROUND.
- CRANK ENGINE AND OBSERVE LIGHT.

REPAIR OPEN TACHOMETER LEAD OR CONNECTOR. REPEAT TEST #4.

REPLACE IGNITION CONTROL MODULE AND CHECK FOR SPARK FROM COIL AS IN STEP 6.

REFER TO IGNITION SYSTEM CHECK (2 OF 2).

SPARK → SYSTEM OK
NO SPARK → REPLACE IGNITION COIL.

1.6L (VIN 6) AND 1.8L (VIN 8) ENGINES — C-CHARTS — 1992–93 STORM

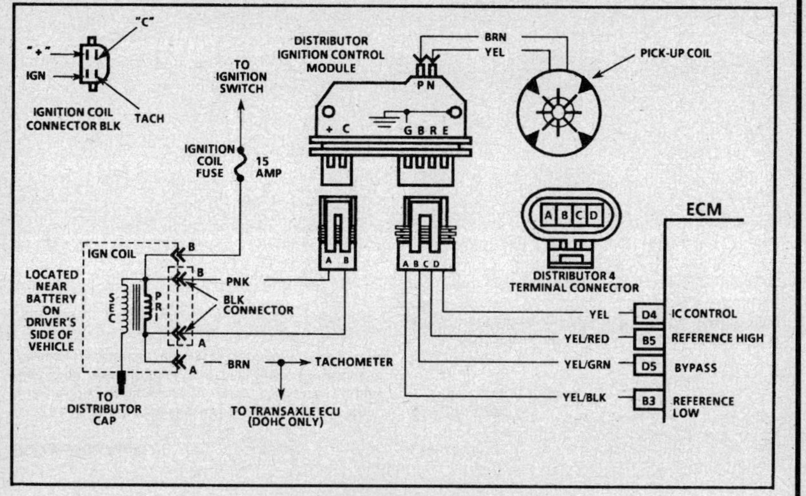

CHART C-4C
DISTRIBUTOR IGNITION (DI) SYSTEM CHECK
(REMOTE COIL)
1.6L SOHC & 1.8L DOHC (VIN 6 & 8) "R" CARLINE (MFI)

Test Description: Number(s) below refer to circled number(s) on the diagnostic chart.

5. Applying a voltage (1.35 to 1.50 volts) to module terminal "P" should turn the module "ON" and the "tach" terminal voltage should drop to about 7-9 volts. This test will determine whether the module or coil is faulty or if the pick-up coil is not

generating the proper signal to turn the module "ON." This test can be performed by using a DC test battery with a rating of 1.5 volts. (Such as AA, C, or D cell.) The battery must be a known good battery with a voltage of over 1.35 volts.

6. This should turn "OFF" the module and cause a spark. If no spark occurs, the fault is most likely in the ignition coil because most module problems would have been found before this point in the procedure. A distributor ignition control module tester (J 24642) could determine which is at fault.

Diagnostic Aids:

In order to do the test procedures at the ignition coil, use of the Kent-Moore J 35616-A will be needed to jumper the engine harness at the ignition coil.

1.6L (VIN 6) AND 1.8L (VIN 8) ENGINES — C-CHARTS — 1992–93 STORM

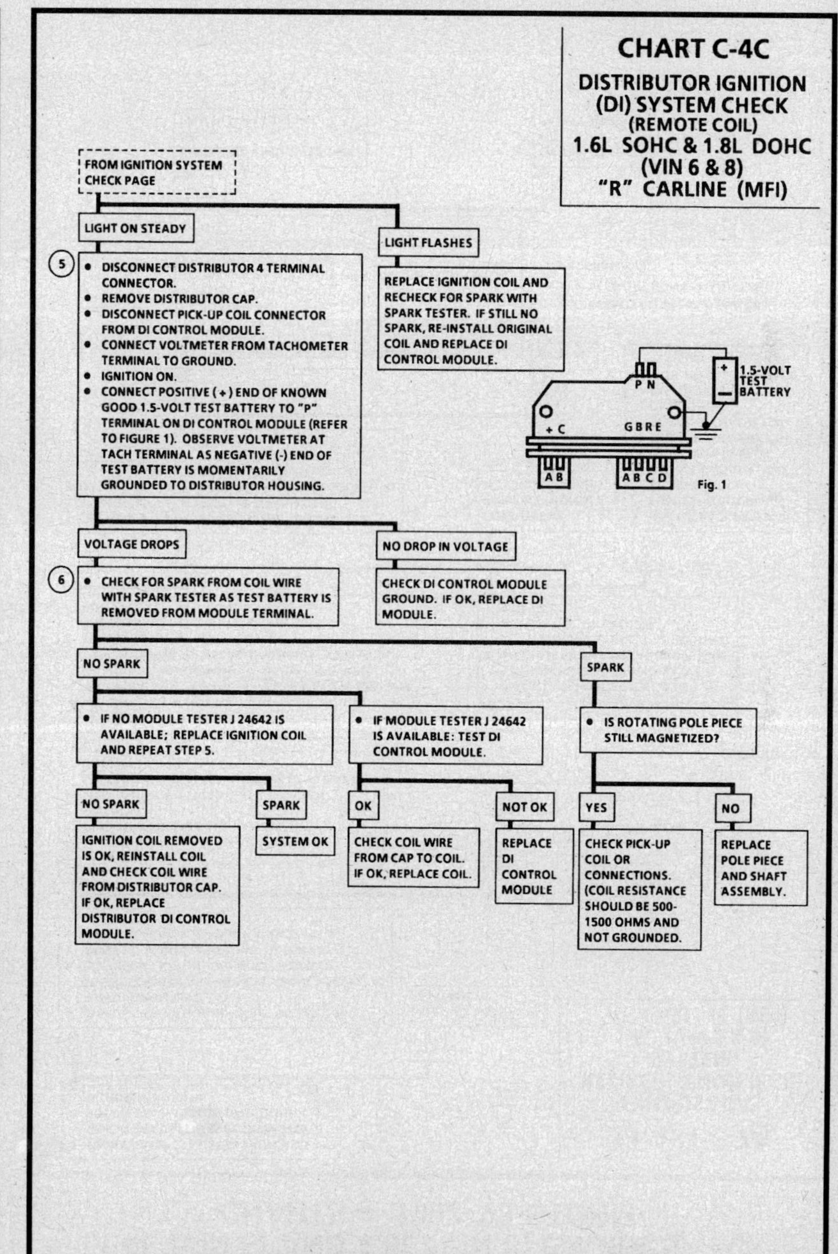

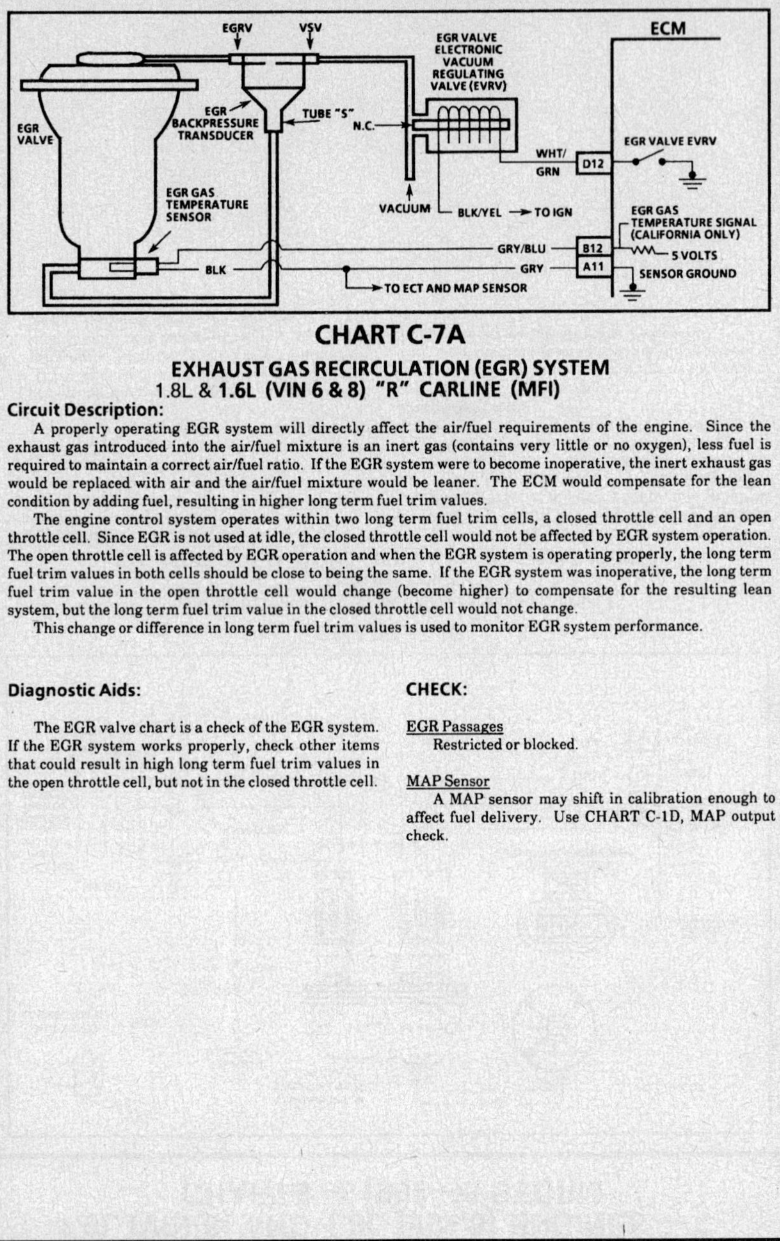

CHART C-7A

EXHAUST GAS RECIRCULATION (EGR) SYSTEM
1.8L & 1.6L (VIN 6 & 8) "R" CARLINE (MFI)

Circuit Description:

A properly operating EGR system will directly affect the air/fuel requirements of the engine. Since the exhaust gas introduced into the air/fuel mixture is an inert gas (contains very little or no oxygen), less fuel is required to maintain a correct air/fuel ratio. If the EGR system were to become inoperative, the inert exhaust gas would be replaced with air and the air/fuel mixture would be leaner. The ECM would compensate for the lean condition by adding fuel, resulting in higher long term fuel trim values.

The engine control system operates within two long term fuel trim cells, a closed throttle cell and an open throttle cell. Since EGR is not used at idle, the closed throttle cell would not be affected by EGR system operation. The open throttle cell is affected by EGR operation and when the EGR system is operating properly, the long term fuel trim values in both cells should be close to being the same. If the EGR system was inoperative, the long term fuel trim value in the open throttle cell would change (become higher) to compensate for the resulting lean system, but the long term fuel trim value in the closed throttle cell would not change.

This change or difference in long term fuel trim values is used to monitor EGR system performance.

Diagnostic Aids:

The EGR valve chart is a check of the EGR system. If the EGR system works properly, check other items that could result in high long term fuel trim values in the open throttle cell, but not in the closed throttle cell.

CHECK:

EGR Passages
Restricted or blocked.

MAP Sensor
A MAP sensor may shift in calibration enough to affect fuel delivery. Use CHART C-1D, MAP output check.

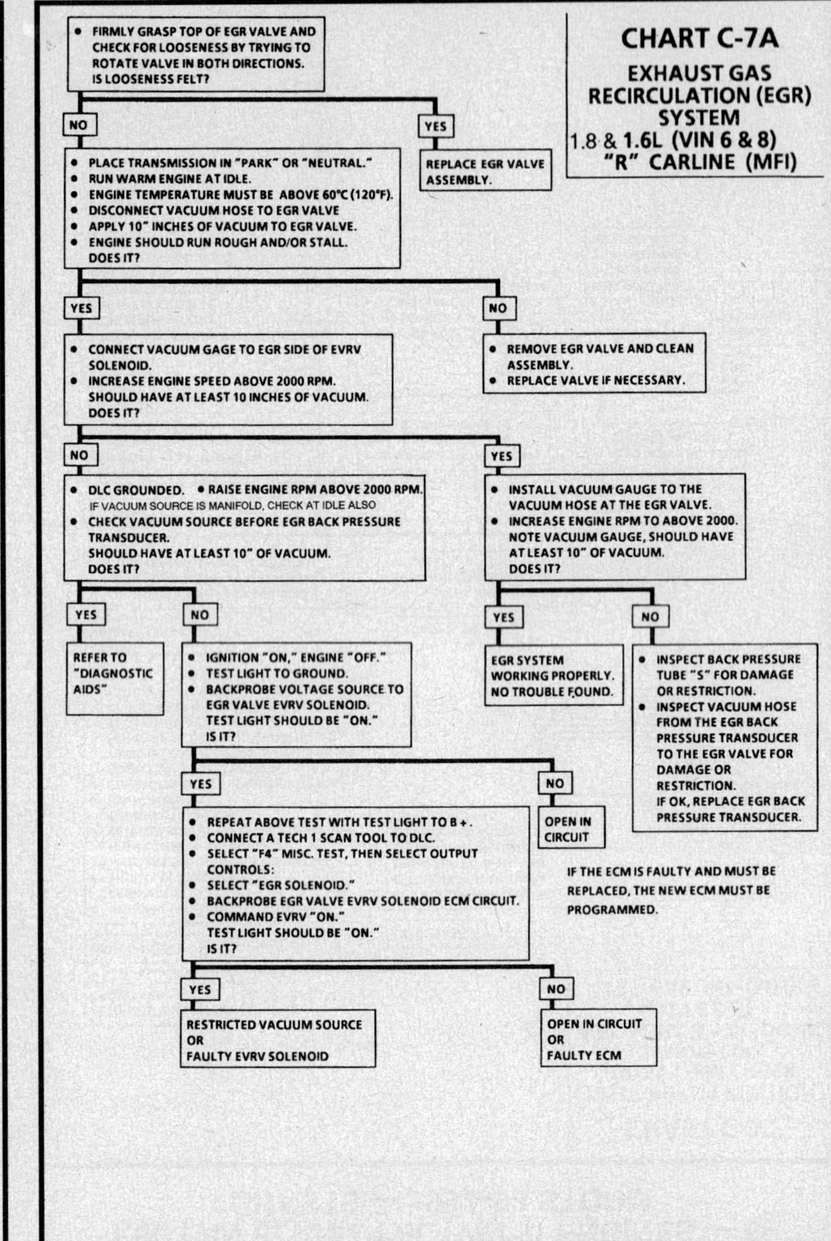

1.6L (VIN 6) AND 1.8L (VIN 8) ENGINES — C-CHARTS — 1992–93 STORM

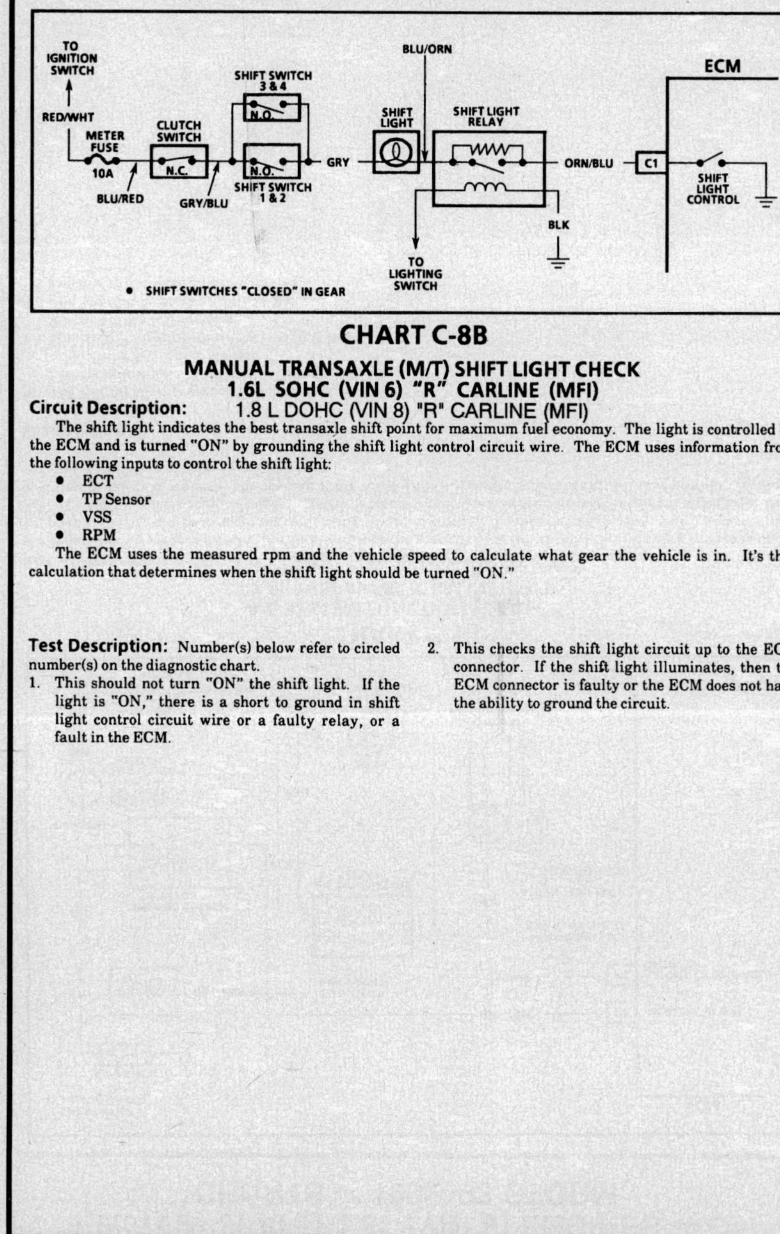

CHART C-8B

MANUAL TRANSAXLE (M/T) SHIFT LIGHT CHECK
1.6L SOHC (VIN 6) "R" CARLINE (MFI)
1.8 L DOHC (VIN 8) "R" CARLINE (MFI)

Circuit Description:
The shift light indicates the best transaxle shift point for maximum fuel economy. The light is controlled by the ECM and is turned "ON" by grounding the shift light control circuit wire. The ECM uses information from the following inputs to control the shift light:

- ECT
- TP Sensor
- VSS
- RPM

The ECM uses the measured rpm and the vehicle speed to calculate what gear the vehicle is in. It's this calculation that determines when the shift light should be turned "ON."

Test Description: Number(s) below refer to circled number(s) on the diagnostic chart.
1. This should not turn "ON" the shift light. If the light is "ON," there is a short to ground in shift light control circuit wire or a faulty relay, or a fault in the ECM.
2. This checks the shift light circuit up to the ECM connector. If the shift light illuminates, then the ECM connector is faulty or the ECM does not have the ability to ground the circuit.

1.6L (VIN 6) AND 1.8L (VIN 8) ENGINES — C-CHARTS — 1992–93 STORM

CHART C-8B

MANUAL TRANSAXLE (M/T) SHIFT LIGHT CHECK
1.6L SOHC (VIN 6) "R" CARLINE (MFI)
1.8 L DOHC (VIN 8) "R" CARLINE (MFI)

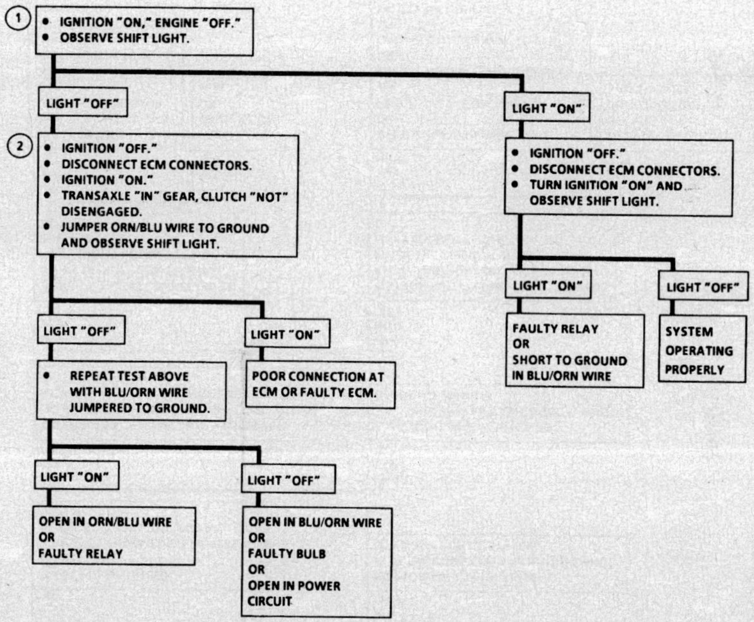

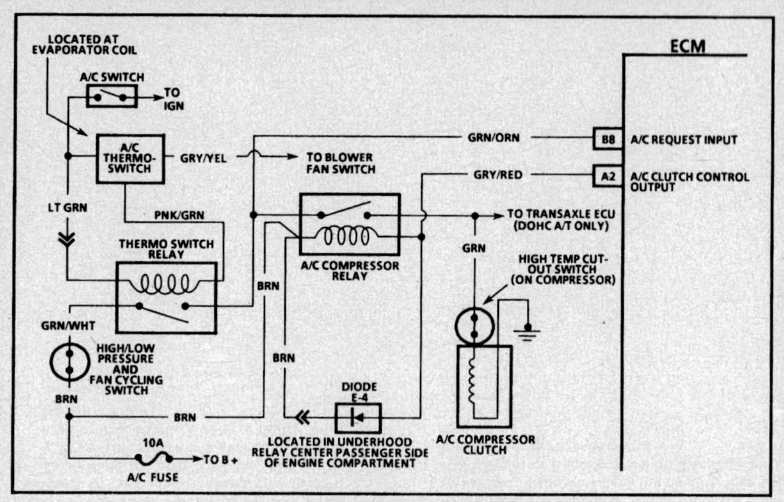

CHART C-10
A/C SYSTEM FUNCTIONAL CHECK
1.6L (VIN 6) "R" CARLINE (MFI)
1.8 L DOHC (VIN 8) "R" CARLINE (MFI)

Circuit Description:
When the A/C mode is selected the ECM will supply a ground to the A/C clutch relay. Battery voltage is then supplied to the A/C clutch through the completing of circuit through the thermo switch relay, which is part of the A/C "ON/OFF" switch circuit. If the A/C system exhibits high/low pressure or an overheating condition exist, the A/C system will be disabled by the opening of the A/C high/low pressure switch or the A/C thermo switch, or A/C compressor HIGH temperature cutout switch.

Diagnostic Aids:

- The A/C high/low pressure switch is located in the high pressure side A/C hose. This switch also controls the A/C condenser fan.
- The HIGH temperature cutout switch is located in the back of the A/C compressor.
- The thermo cut-out switch is part of the A/C evaporator mounted under the dash.

CHART C-10
A/C SYSTEM FUNCTIONAL CHECK
1.6L (VIN 6) "R" CARLINE (PORT)
1.8 L (VIN 8) "R" CARLINE (PORT)

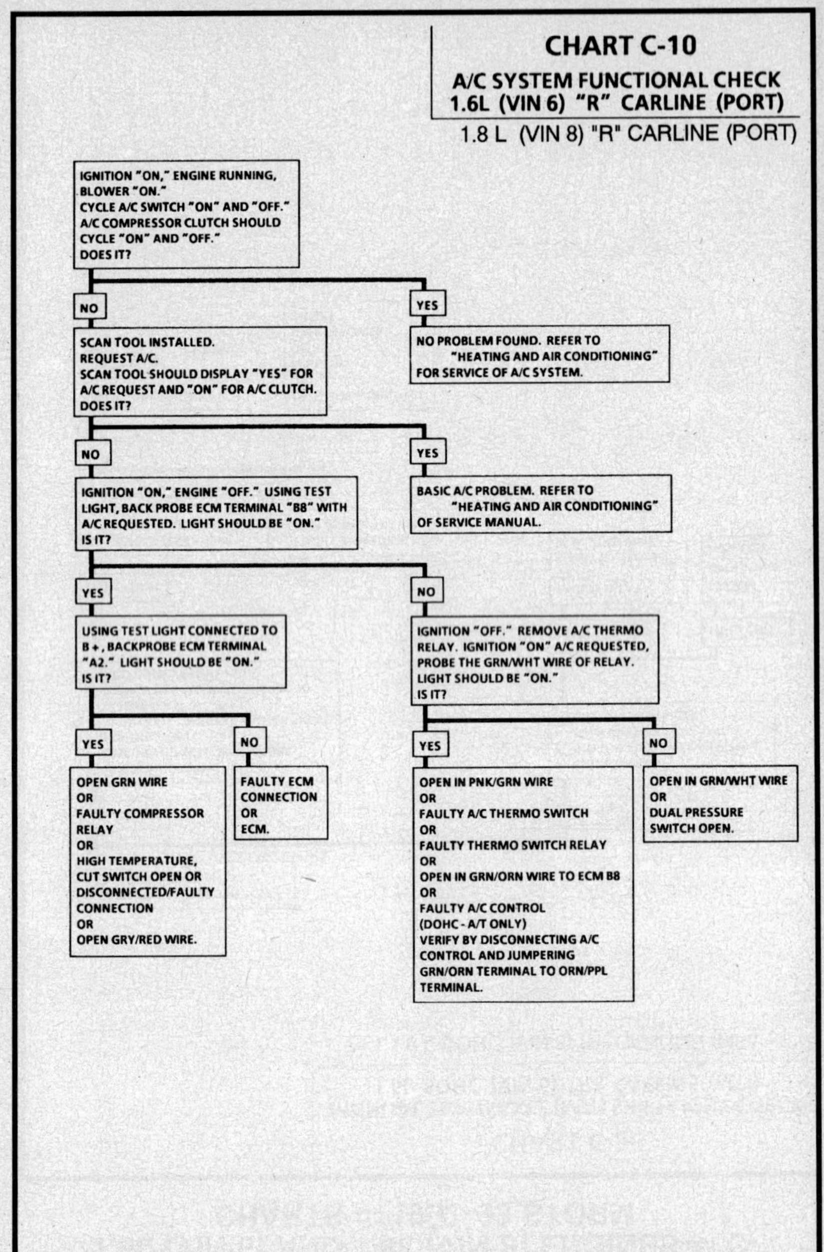

1.9L (VIN 7) ENGINE — COMPONENT LOCATIONS — SATURN

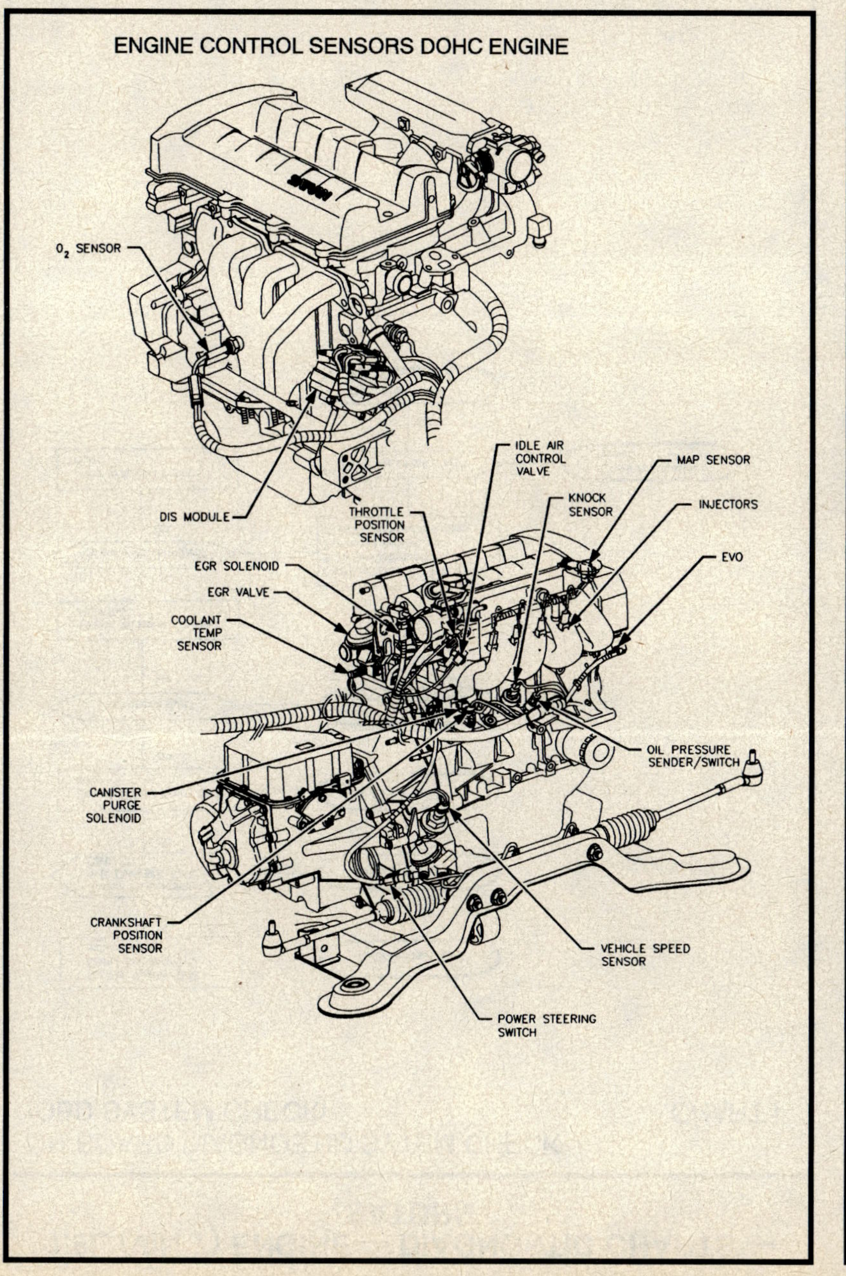

1.9L (VIN 7) ENGINE — DIAGNOSTIC CHARTS — SATURN

DIAGNOSTIC I

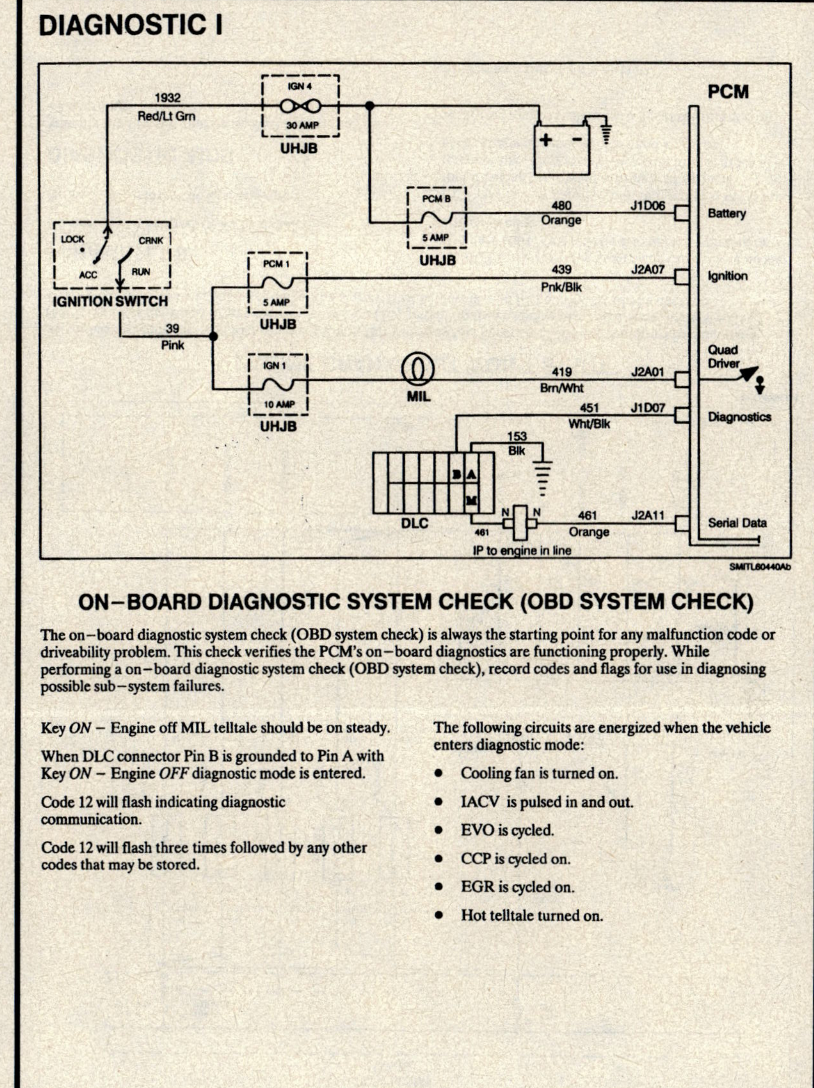

ON—BOARD DIAGNOSTIC SYSTEM CHECK (OBD SYSTEM CHECK)

The on—board diagnostic system check (OBD system check) is always the starting point for any malfunction code or driveability problem. This check verifies the PCM's on—board diagnostics are functioning properly. While performing a on—board diagnostic system check (OBD system check), record codes and flags for use in diagnosing possible sub—system failures.

Key *ON* — Engine off MIL telltale should be on steady.

When DLC connector Pin B is grounded to Pin A with Key *ON* — Engine *OFF* diagnostic mode is entered.

Code 12 will flash indicating diagnostic communication.

Code 12 will flash three times followed by any other codes that may be stored.

The following circuits are energized when the vehicle enters diagnostic mode:

- Cooling fan is turned on.
- IACV is pulsed in and out.
- EVO is cycled.
- CCP is cycled on.
- EGR is cycled on.
- Hot telltale turned on.

1.9L (VIN 7) ENGINE — DIAGNOSTIC CHARTS — SATURN

ON-BOARD DIAGNOSTIC SYTEM CHECK (OBD SYSTEM CHECK)

CHART I

Key *On*, Engine *off*. Is there a malfunction indicator lamp. — No → Refer to Chart III
↓ Yes
Jumper DLC terminal *A* to *B*. Does MIL flash Code 12? — No → Refer to Chart IV
↓ Yes
Does Scan Tool display PCM data? — No → Refer to Chart IV
↓ Yes
Does engine start? — No → Refer to Chart II or Diagnostic matrix charts.
↓ Yes
Engine *off*, key *on*. Connect Scan Tool.
↓
Are any codes displayed? — No → Compare Scan Tool data with Scan Tool Information Parameter, located after PCM/EC Information Flags.
↓ Yes
Refer to applicable code chart.

Compare Scan Tool data with Scan Tool Information Parameter, located after PCM/EC Information Flags. → Are values within range? — No → Check appropriate sensor.
↓ Yes
Diagnostic system functioning properly.

1.9L (VIN 7) ENGINE — DIAGNOSTIC CHARTS — SATURN

DIAGNOSTIC II

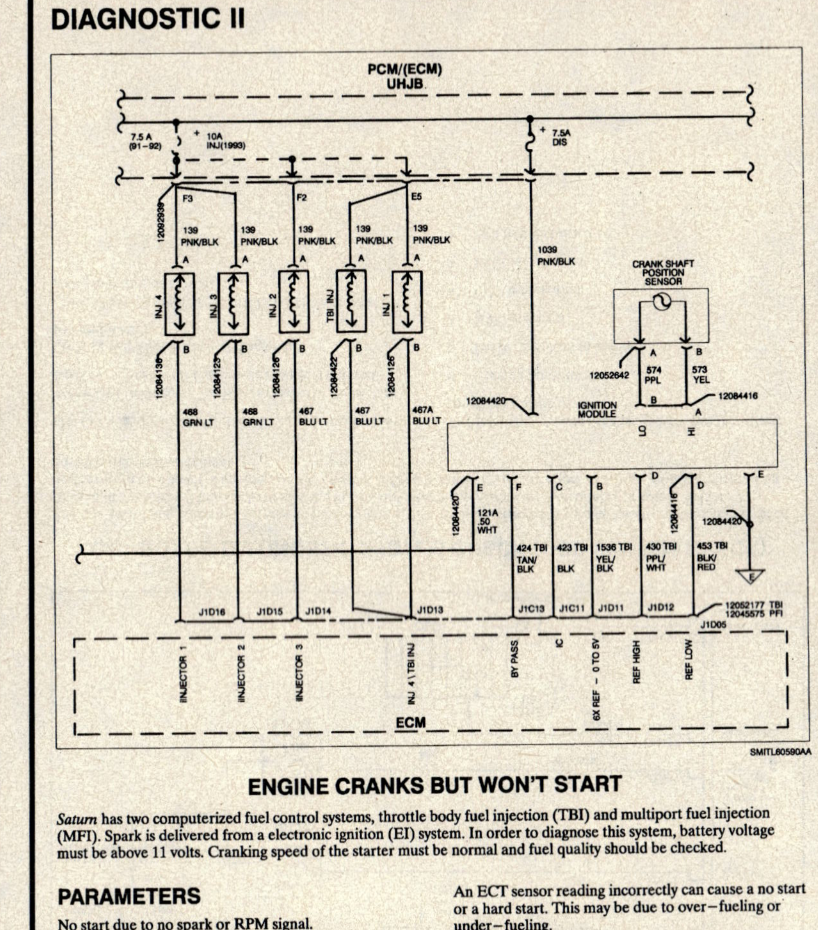

ENGINE CRANKS BUT WON'T START

Saturn has two computerized fuel control systems, throttle body fuel injection (TBI) and multiport fuel injection (MFI). Spark is delivered from a electronic ignition (EI) system. In order to diagnose this system, battery voltage must be above 11 volts. Cranking speed of the starter must be normal and fuel quality should be checked.

PARAMETERS

No start due to no spark or RPM signal.

No start due to low or no fuel pressure.

DIAGNOSTIC AIDS

Water in the fuel can cause a no start during cold weather starts.

An ECT sensor reading incorrectly can cause a no start or a hard start. This may be due to over−fueling or under−fueling.

ECT may be compared to IAT temperature with scan tool if engine has cooled overnight at ambient temperature. They should be within four degrees of each other with ignition "On" and key "Off".

A shorted injector will cause a no start condition on both TBI and MFI.

MAP sensor can cause a no start.

1.9L (VIN 7) ENGINE — DIAGNOSTIC CHARTS — SATURN

ENGINE CRANKS BUT WON'T START — CHART II

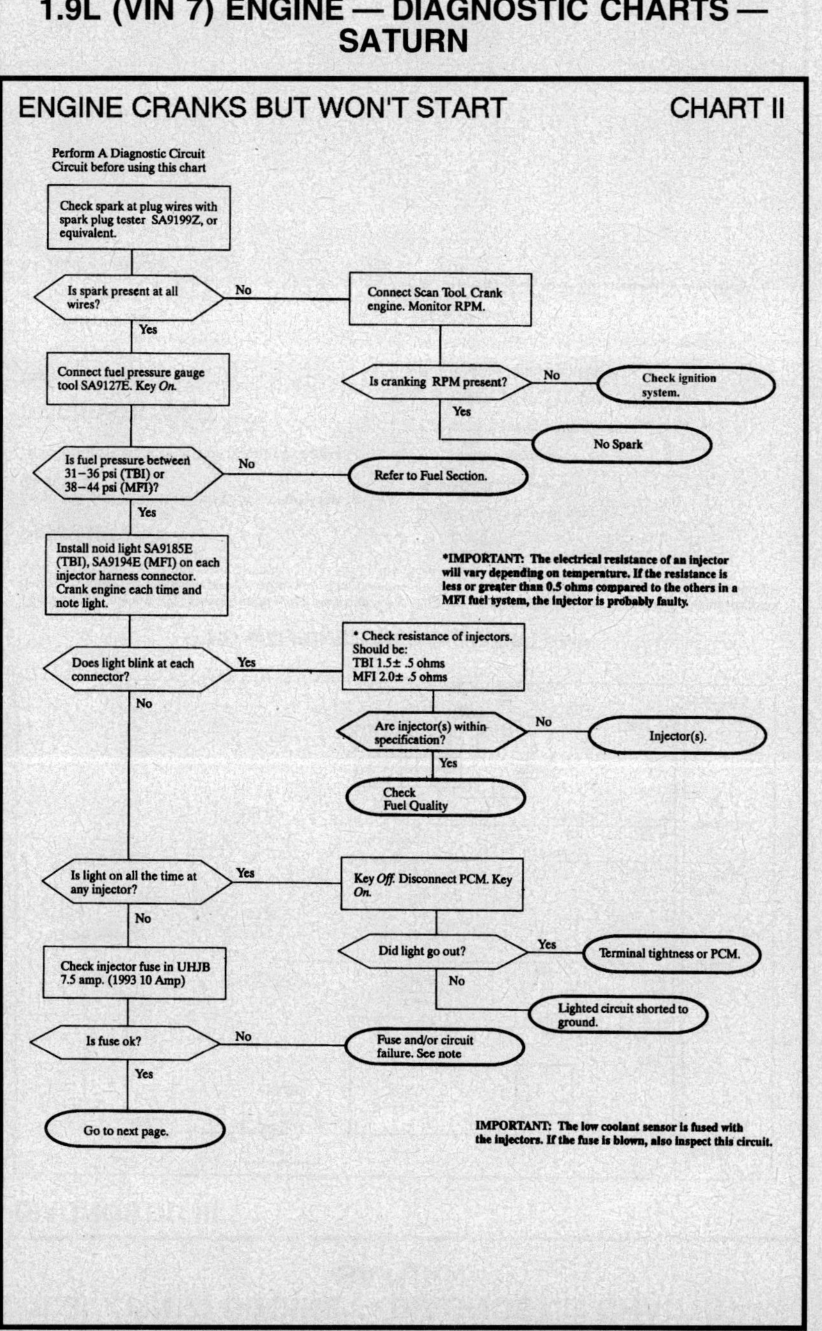

1.9L (VIN 7) ENGINE — DIAGNOSTIC CHARTS — SATURN

DIAGNOSTIC III

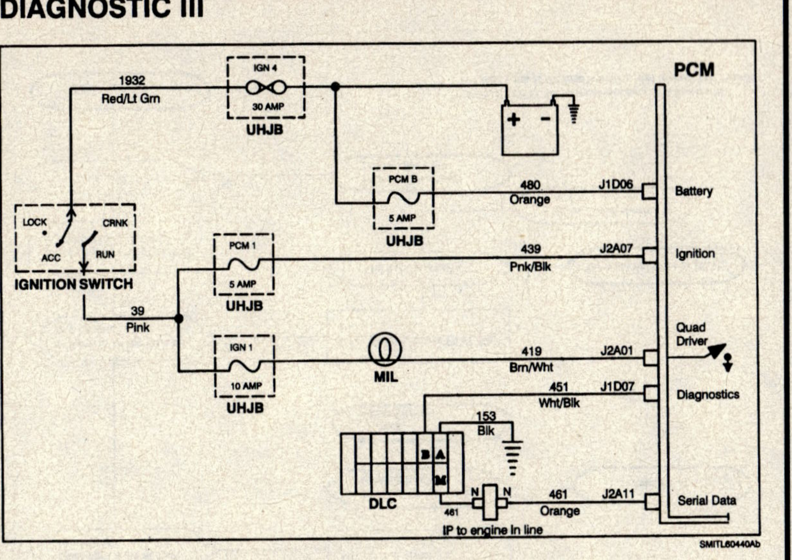

NO MALFUNCTION INDICATOR LAMP

There should always be a steady *malfunction indicator lamp* illuminated when the ignition is *On* and engine stopped. Battery voltage is supplied directly to the bulb. The powertrain control module (PCM) will control the lamp and turn it *On* by providing a ground path through Circuit 419 to the PCM.

PARAMETERS

Lamp will not illuminate if power or ground is lost to the PCM or bulb.

Lamp will not illuminate if bulb is shorted or open.

DIAGNOSTIC AIDS

Engine runs ok, check:

- Faulty bulb.
- Circuit 419 open.
- Ignition 1 fuse blown.
- Faulty Quad driver.

Engine cranks but will not run:

- PCM B fuse blown
- PCM 1 fuse blown
- Ignition 4 Maxi fuse blown.
- Circuits 480 or 439 open.
- Terminal tightness at PCM.

PCM Grounds

- Circuits 551A and 450E from J1C02 and J1C01.

1.9L (VIN 7) ENGINE — DIAGNOSTIC CHARTS — SATURN

NO MALFUNCTION INDICATOR LAMP   CHART III

1.9L (VIN 7) ENGINE — DIAGNOSTIC CHARTS — SATURN

DIAGNOSTIC IV

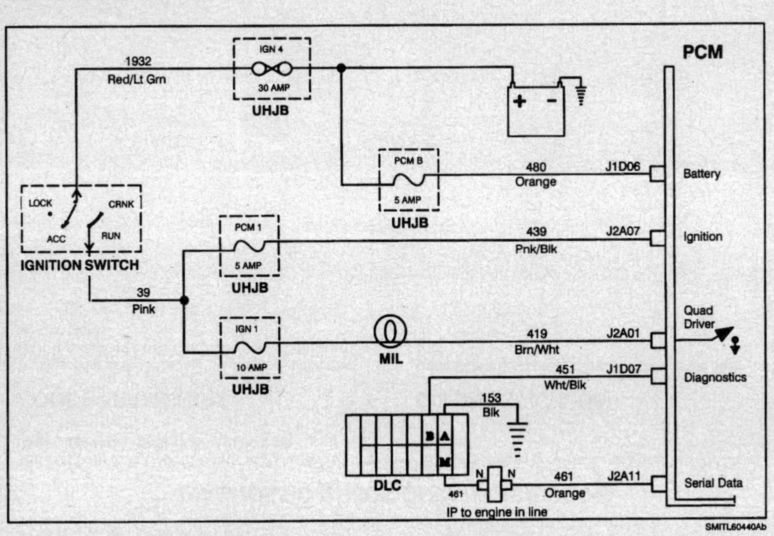

NO SCAN TOOL DATA OR WON'T FLASH CODE 12

There should always be a steady *malfunction indicator lamp* when the ignition is *On* and engine stopped. Battery voltage is supplied directly to the lamp bulb. The powertrain control module (PCM) will control the lamp and turn it *On* by providing a ground path through Circuit 419 to the PCM. With the diagnostic terminal grounded, the lamp should flash a Code 12, followed by any trouble code(s) stored in memory.

DIAGNOSTIC AIDS

- If there is a problem with the PCM that causes a scan tool not to read serial data, the PCM should not flash a Code 12.

- If Code 12 does flash, make sure Scan Tool will work properly on another vehicle. If Code 81 is flashed, refer to the code chart.

- A steady light suggests a short to ground in the lamp control Circuit 419, or an open in diagnostic Circuit 451.

- Check the tightness of the female terminals with a spare male terminal.

1.9L (VIN 7) ENGINE — DIAGNOSTIC CHARTS — SATURN

NO SCAN DATA/WON'T FLASH CODE 12

CHART IV

1.9L (VIN 7) ENGINE — PCM/EC DIAGNOSTIC CHARTS — SATURN

PCM/EC CODE 11

TRANSMISSION DIAGNOSTIC CODES PRESENT

The powertrain control module, used on vehicles equipped with automatic transaxles, contains an engine control module (EC) and a transaxle control module (TC). These control modules communicate internally over a serial peripheral interface (SPI) bus. When the transaxle control module (TC) detects a malfunction in the automatic transaxle, a code is stored in the transaxle control module's memory and an indication is sent to the engine control module to turn on the *malfunction indicator lamp*. If the DLC is grounded, the transaxle codes flash on the *SHIFT TO D2* telltale or in 1993 the HOT light telltale located in the instrument cluster.

CODE PARAMETERS

- Code 11 will set if the PCM/TC has set a transaxle code.

DIAGNOSTIC AIDS

- Search the scan tool for stored transaxle codes. *Select TCM trouble codes.

When the PCM/TC codes are cleared, Code 11 will also be cleared from general information but not from Malf history.

- Code 11 will not set if transaxle flags are present.

PCM/EC CODE 12

DIAGNOSTIC CHECK ONLY (FLASH CODE)

The PCM will flash Code 12 on the *malfunction indicator lamp* when it is able to perform diagnostics. The DLC must be grounded with the key *On* to flash codes.

CODE PARAMETERS

If diagnostics in the PCM are functioning, when the DLC is grounded Code 12 will flash on the *malfunction indicator lamp*.

DIAGNOSTIC AIDS

Code 12 is a flash code. It is not shown on the Scan Tool.

1.9L (VIN 7) ENGINE — PCM/EC DIAGNOSTIC CHARTS — SATURN

PCM/EC CODE 13

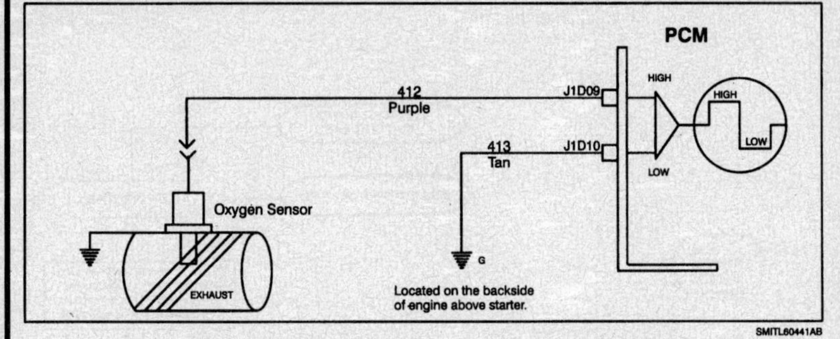

OXYGEN SENSOR CIRCUIT — OPEN/NOT READY

The oxygen sensor consists of a zirconia electrolyte between two platinum plates. When the sensor reaches approximately 318°C (600°F) it becomes an electrical source that responds to the oxygen content in the exhaust. The PCM produces a bias voltage of approximately 470 millivolts on the oxygen sensor circuit. When the sensor is cold its internal resistance is extremely high, therefore, the PCM recognizes the bias as an open circuit. As the sensor warms up, the internal resistance decreases. When the sensor reaches approximately 318°C (600°F), it starts producing a voltage based on the oxygen content in the exhaust stream. This voltage is used by the PCM to determine a rich or lean signal and adjust the fuel mixture accordingly.

CODE PARAMETERS

- The oxygen sensor voltage is inactive (not reading outside 350−550 mV) for 20 seconds.

When:

- The engine is at normal operating temperature (the ECT reads over 60°C [140°F]).

- The engine has been running at least one minute.

- The TP sensor reads over 6.5% (the engine is off idle).

- Code 21 and 22 are not present (TP sensor codes).

DIAGNOSTIC AIDS

The PCM will not go into closed loop if a Code 13 is set.

When attempting to diagnose an intermittent problem, use the Scan Tool* to review supplemental diagnostic information. The supplemental data can be used to duplicate a problem.

* Select MALF History from Scan Tool EC menu.

Intermittents or opens suspected to be at the connectors can be detected by using Diagnostic Service Probe. Voltage can be read on wires without disconnecting any connectors.

Check the tightness of the female terminal grip with a spare male terminal.

Normal Scan Tool readings in closed loop will show the oxygen sensor voltage varying between 10 mV and 999 mV.

A Code 13 can be set if the vehicle runs out of fuel or stalls while the vehicle is in motion.

Code 13 can set or cause false oxygen sensor readings if the oxygen sensor ground becomes loose. Refer to PCM and Engine Grounds in this manual.

1.9L (VIN 7) ENGINE — PCM/EC DIAGNOSTIC CHARTS — SATURN

OXYGEN SENSOR CIRCUIT - OPEN/NOT READY

PCM/EC CODE 13

Engine running time over five minutes. Connect Scan Tool. ECT above 60°C (140°F)?

Is oxygen sensor voltage varying over 550 mV to under 350 mV? — **Yes** → Problem Intermittent: see Diagnostic Aids.

No

Engine *Off*. Key *On*. Disconnect oxygen sensor at sensor connector. Jumper Ckt. 412 to ground at PCM harness end.

Does Scan Tool read less than 100 mV? — **Yes** → Terminal Tightness or oxygen sensor

No

Remove jumper. Key *Off*. Disconnect PCM. Check continuity of Circuit 412.

Is Circuit 412 ok? — **No** → Circuit 412 open.

Yes

Back probe Circuit 413 at PCM harness. Measure resistance to ground.

Is resistance below 22 ohms? — **Yes** → Terminal Tightness or PCM

No → Circuit 413 open.

1.9L (VIN 7) ENGINE — PCM/EC DIAGNOSTIC CHARTS — SATURN

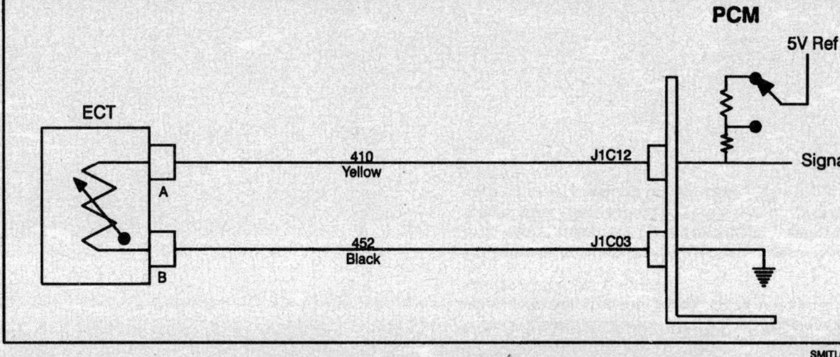

ENGINE COOLANT TEMPERATURE CIRCUIT — TEMPERATURE OUT OF RANGE HIGH

The engine coolant temperature sensor is a thermistor located in the lower coolant passage of the engine's cylinder head. When coolant temperature is cold the sensor has a high resistance, as temperature increases the resistance of the sensor decreases. The PCM provides a five volt signal to the coolant sensor, which is also connected to PCM ground. The PCM reads the voltage drop on the signal line to determine coolant temperature.

CODE PARAMETERS

Code 14 will set if:

- The ECT reads over 140°C (284°F).
- The engine has been running over 10 seconds.

DIAGNOSTIC AIDS

Refer to Description and Operation in this manual for Temperature vs. Resistance Chart.

When attempting to diagnose an intermittent problem, use the Scan Tool* to review supplemental diagnostic information. The supplemental data can be used to duplicate a problem.

* Select MALF History from Scan Tool EC menu.

Check the tightness of the female terminal grip with a spare male terminal.

Start engine and observe ECT reading on Scan Tool. Normal operation is for the ECT to rise smoothly to approximately 88°C (190°F) thermostat opens then stabilize.

The PCM will turn the coolant fan *On* if a Code 14 is set.

Code 14 can be set if there is a problem with the engine cooling system.

ECT may be compared to IAT temperature with Scan Tool if engine has cooled overnight at ambient temperature. They should be within four degrees of each other with ignition *On* and engine *Off*.

Engine cooling fan also comes on if ECT exceeds 106°C.

ENGINE COOLANT TEMPERATURE CIRCUIT (HIGH) — PCM/EC CODE 14

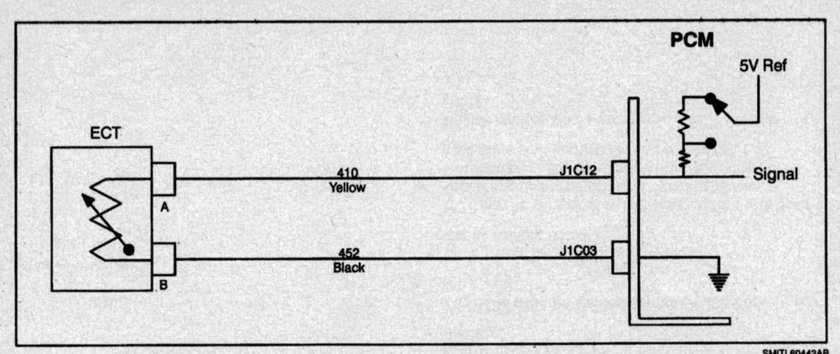

PCM/EC CODE 15

ECT CIRCUIT — TEMPERATURE OUT OF RANGE LOW

The engine coolant temperature sensor (ECT) is a thermistor located in a coolant passage. When coolant temperature is cold the sensor has a high resistance, as temperature increases the resistance of the sensor decreases. The PCM provides a five volt signal to the coolant sensor, which is also connected to PCM ground. The PCM reads the voltage drop on the signal line to determine coolant temperature.

CODE PARAMETERS

Code 15 will set if:

- The ECT reads below −35°C (−31°F).
- The engine has been running longer than five minutes with a temperature less than −35°C (−31°F).

DIAGNOSTIC AIDS

Refer to Description and Operation in this manual for Temperature vs. Resistance Chart.

When attempting to diagnose an intermittent problem, use the Scan Tool* to review supplemental diagnostic information. The supplemental data can be used to duplicate a problem.

* Select MALF History from Scan Tool EC menu.

Check the tightness of the female terminal grip with a spare male terminal.

Start engine and observe ECT reading on Scan Tool. Normal operation is for the ECT to rise smoothly to approximately 88°C (190°F) as thermostat opens then stabilize.

Visually inspect ECT wires for damage. Wiggle wires while observing scan tool ECT reading and watch for a sudden shift in temperature.

ECT may be compared to IAT temperature, with a Scan Tool if engine has cooled overnight at ambient temperature. They should be within four degrees of each other with ignition On − engine Off.

1.9L (VIN 7) ENGINE — PCM/EC DIAGNOSTIC CHARTS — SATURN

IECT CIRCUIT (LOW) PCM/EC CODE 15

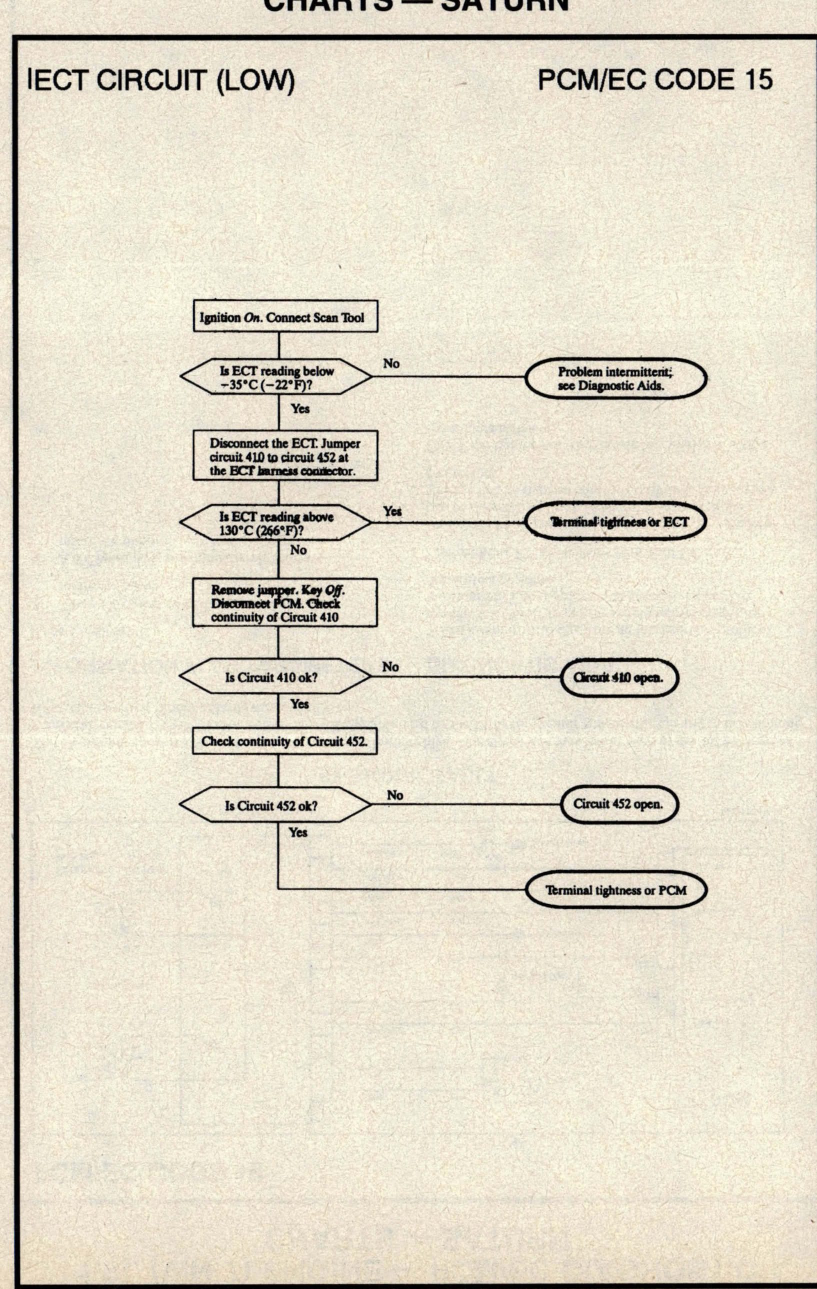

1.9L (VIN 7) ENGINE — PCM/EC DIAGNOSTIC CHARTS — SATURN

PCM/EC CODE 17

PCM FAULT — PULL-UP RESISTOR

The PCM uses a pull-up resistor to increase resolution throughout the entire range of engine operating temperatures. When the coolant temperature is less than 40°C (104°F) the 4K ohm resistor is used. When temperature is above 40°C (104°F) the PCM switches to the 348 ohm resistor. If the pull−up resistor does not switch, Code 17 will set.

CODE PARAMETERS

Code 17 will set if:

- The pull-up resistor inside the PCM switches and there is no change in the coolant temperatures signal.

DIAGNOSTIC AIDS

Code 17 is an internal fault within the PCM. The PCM must be replaced.

1.9L (VIN 7) ENGINE — PCM/EC DIAGNOSTIC CHARTS — SATURN

PCM/EC CODE 19

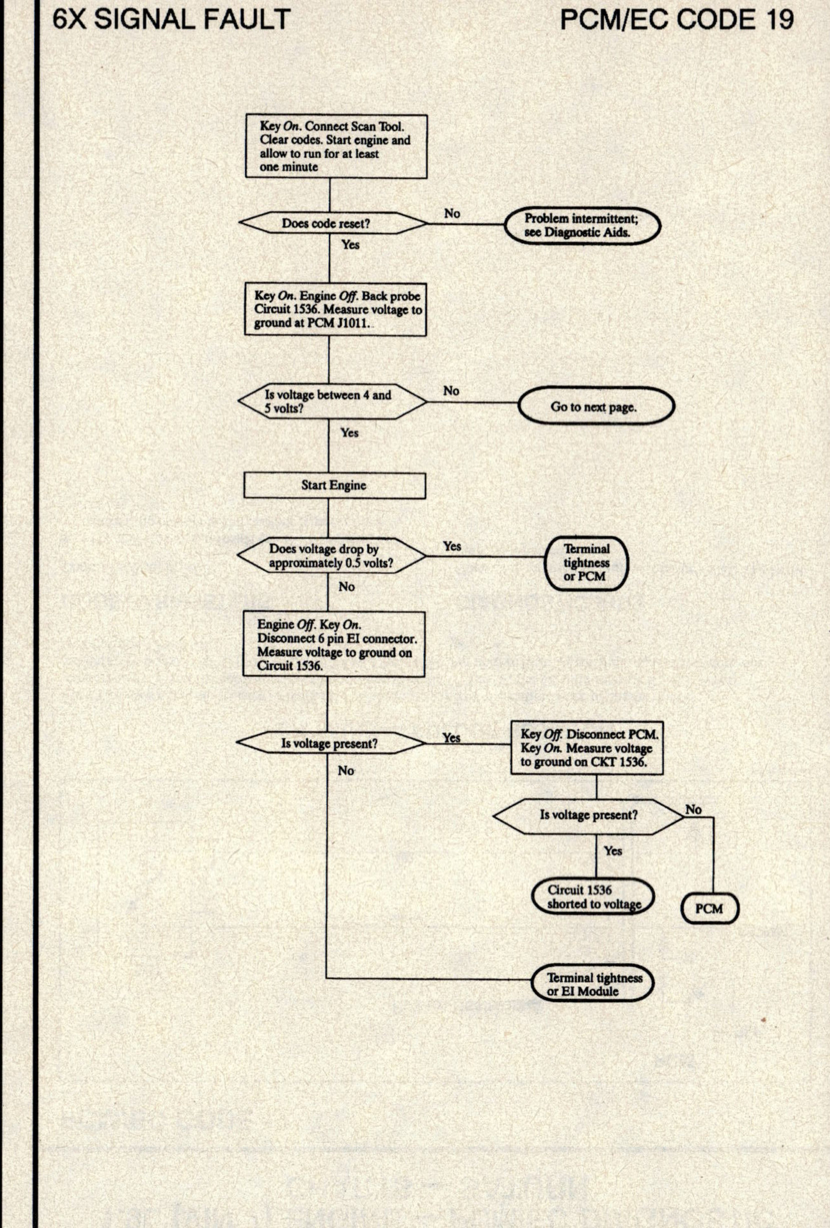

6X SIGNAL FAULT

The 6X output signal is a five volt square wave that is initially high. The signal switches low at each of the six, 60 degrees crankshaft position (CKP) sensor pulses. The 6X signal is used by the PCM for greater rpm resolution and in determining KS circuit (system) (knock retard) windowing.

INFORMATION FLAG PARAMETERS

Code 19 will set if:

- Three 6X pulses do not occur between each reference pulse.

- A 6X pulse does not immediately follow a reference pulse.

DIAGNOSTIC AIDS

When attempting to diagnose an intermittent problem, use the Scan Tool* to review supplemental diagnostic information. The supplemental data can be used to duplicate a problem.

* Select MALF History from Scan Tool EC menu.

Intermittents or opens suspected to be at the connector can be detected by using Diagnostic Service Probe. Voltage can be read on wires without disconnecting any connectors.

Check the tightness of the female terminal grip with a spare male terminal.

1.9L (VIN 7) ENGINE — PCM/EC DIAGNOSTIC CHARTS — SATURN

6X SIGNAL FAULT

PCM/EC CODE 19

1.9L (VIN 7) ENGINE — PCM/EC DIAGNOSTIC CHARTS — SATURN

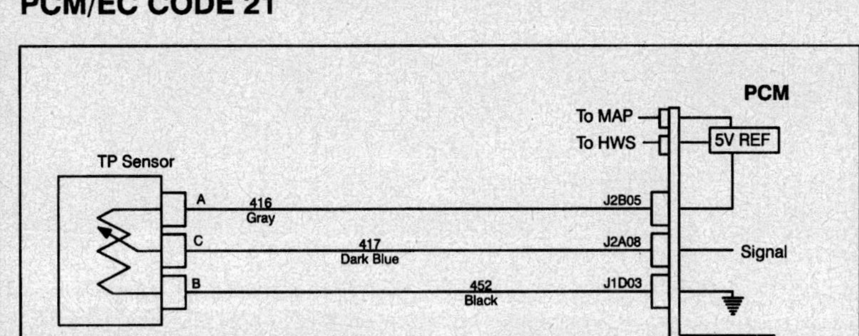

1.9L (VIN 7) ENGINE — PCM/EC DIAGNOSTIC CHARTS — SATURN

PCM/EC CODE 21

TP SENSOR CIRCUIT — VOLTAGE OUT OF RANGE HIGH

The throttle position (TP) sensor is a variable resistor that is connected to a five volt reference source, a ground, and an input signal at the PCM. When the throttle blade is closed, the sensor input voltage will be low and increase as the throttle is opened. The TP sensor is mounted on the throttle body and connected to the throttle shaft which is controlled by accelerator movement.

CODE PARAMETERS

Code 21 will set if:

- The PCM sees TP sensor voltage over 4.9 volts.

DIAGNOSTIC AIDS

When attempting to diagnose an intermittent problem, use the Scan Tool* to review supplemental diagnostic information. The supplemental data can be used to duplicate a problem.

* Select MALF History from Scan Tool EC menu.

Intermittents or opens suspected to be at the connector can be detected by using Diagnostic Service Probe. Voltage can be read on wires without disconnecting any connectors.

Check the tightness of the female terminal grip with a spare male terminal.

Normal voltage readings should vary smoothly from 0.4 volts to 4.7 volts ± 0.2 volts as the throttle is moved from closed to wide open position.

1.9L (VIN 7) ENGINE — PCM/EC DIAGNOSTIC CHARTS — SATURN

PCM/EC CODE 21

TP SENSOR CIRCUIT (HIGH)

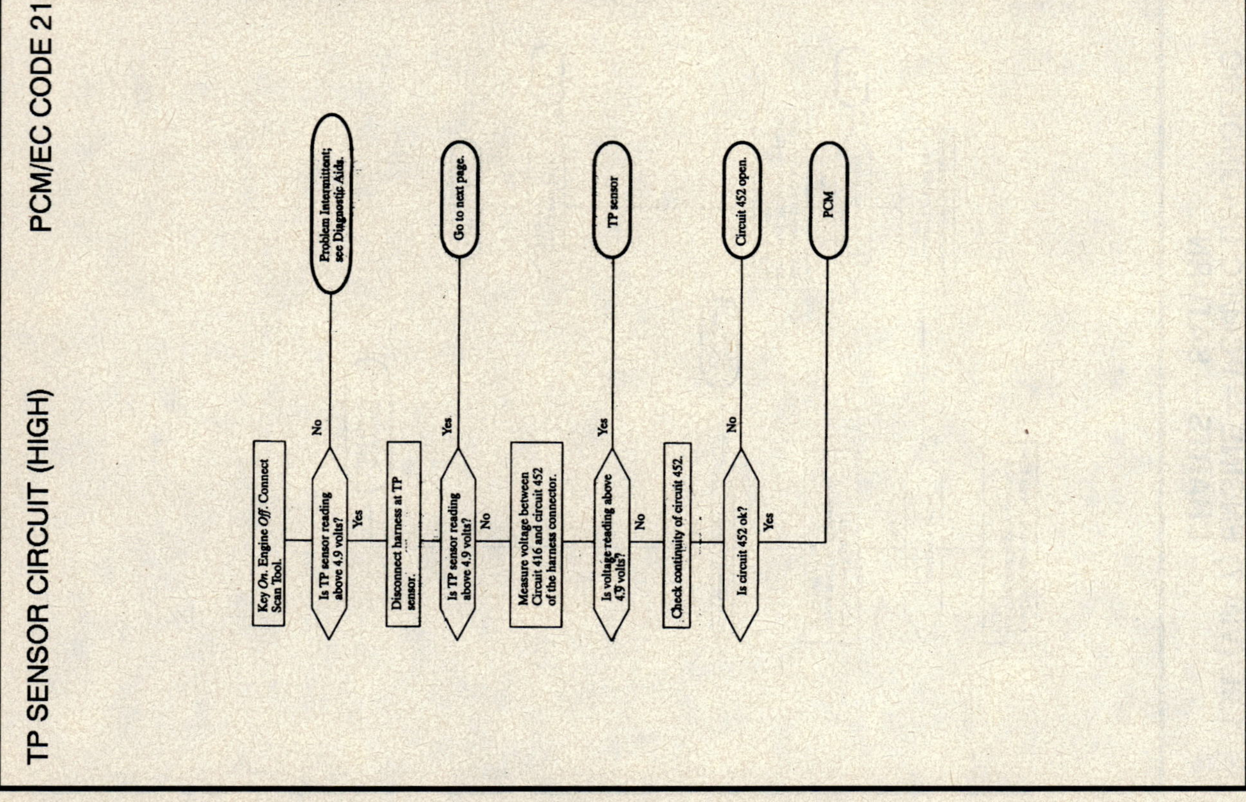

1.9L (VIN 7) ENGINE — PCM/EC DIAGNOSTIC CHARTS — SATURN

PCM/EC CODE 21

TP SENSOR CIRCUIT (HIGH)

1.9L (VIN 7) ENGINE — PCM/EC DIAGNOSTIC CHARTS — SATURN

PCM/EC CODE 22

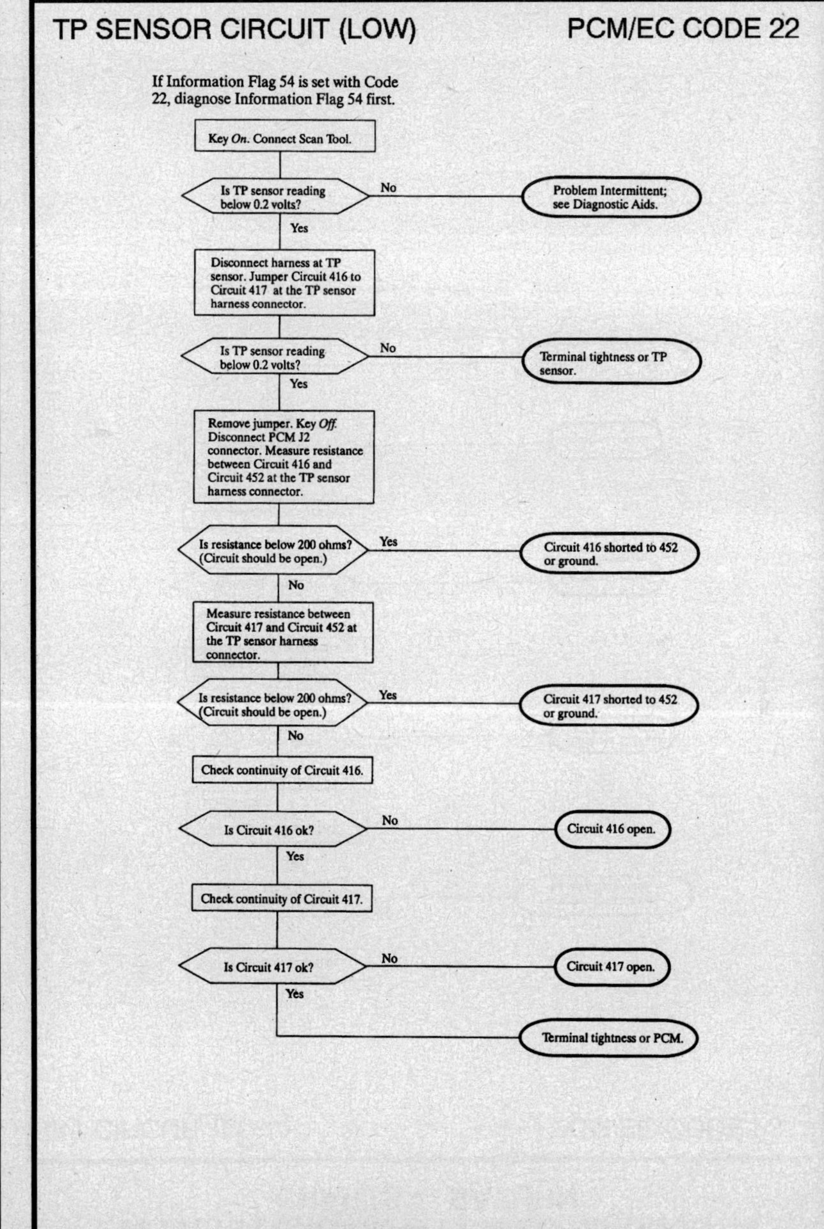

TP SENSOR CIRCUIT — VOLTAGE OUT OF RANGE LOW

The throttle position (TP) sensor is a variable resistor that is connected to a five volt reference source, a ground, and an input signal at the PCM. When the throttle blade is closed, the sensor input voltage will be low and increase as the throttle is opened. The TP sensor is mounted on the throttle body and connected to the throttle shaft which is controlled by accelerator movement.

CODE PARAMETERS

Code 22 will set if:

- TP sensor voltage less than 0.2 volts.

DIAGNOSTIC AIDS

When attempting to diagnose an intermittent problem, use the Scan Tool* to review supplemental diagnostic information. The supplemental data can be used to duplicate a problem.

* Select MALF History from Scan Tool EC menu.

Intermittents or opens suspected to be at the connector can be detected by using Diagnostic Service Probe. Voltage can be read on wires without disconnecting any connectors.

Check the tightness of the female terminal grip with a spare male terminal.

Normal voltage readings should vary smoothly from 0.4 volts to 4.7 volts ± 0.2 volts as the throttle is moved from closed to wide open position.

If the 5V Ref line to MAP or HWS is shorted or open, a Code 22 may set.

1.9L (VIN 7) ENGINE — PCM/EC DIAGNOSTIC CHARTS — SATURN

TP SENSOR CIRCUIT (LOW) — PCM/EC CODE 22

If Information Flag 54 is set with Code 22, diagnose Information Flag 54 first.

1.9L (VIN 7) ENGINE — PCM/EC DIAGNOSTIC CHARTS — SATURN

PCM/EC CODE 23

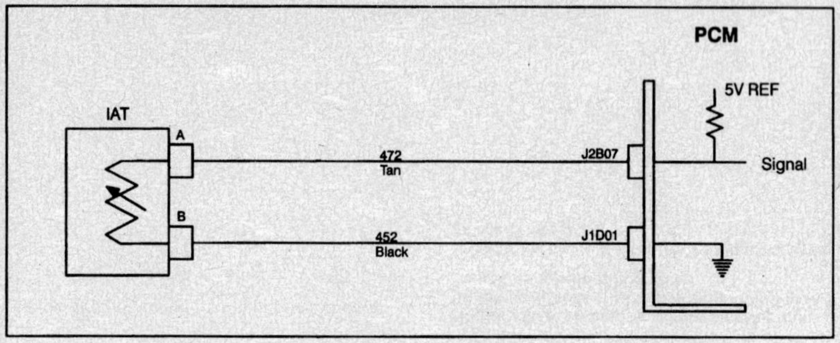

IAT CIRCUIT — TEMPERATURE OUT OF RANGE LOW

The inlet air temperature (IAT) sensor is a thermistor located in the air intake passage of the air induction system on the engine. When the sensor is cold it has a high resistance, the resistance decreases as the temperature of the sensor increases. The PCM supplies a five volt signal to the sensor which is also connected to ground. The PCM reads the voltage drop on the sensor signal line to determine air temperature.

CODE PARAMETERS

Code 23 will set if:

- The PCM sees IAT temperature less than −30°C (−22°F).

DIAGNOSTIC AIDS

Refer to Description and Operation in this manual for Temperature vs. Resistance chart.

When attempting to diagnose an intermittent problem, use the Scan Tool* to review supplemental diagnostic information. The supplemental data can be used to duplicate a problem.

* Select MALF History from Scan Tool EC menu.

Check the tightness of the female terminal grip with a spare male terminal.

- This code may be set if the vehicle has been in extremely cold ambients. Ask the customer if this has occurred.

- Code 23 will set if the IAT is disconnected and the ignition is cycled on.

- IAT may be compared to ECT temperature, with a Scan Tool, if engine has been cooled overnight at ambient temperature. They should be within four degrees of each other with ignition *On* and engine *Off*.

- The IAT and ECT are the same part and can be used interchangeably.

1.9L (VIN 7) ENGINE — PCM/EC DIAGNOSTIC CHARTS — SATURN

IAT CIRCUIT (LOW)   PCM/EC CODE 23

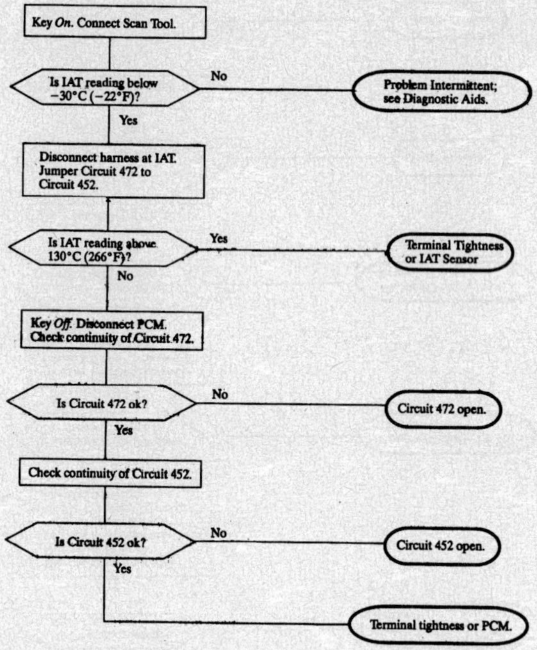

1.9L (VIN 7) ENGINE — PCM/EC DIAGNOSTIC CHARTS — SATURN

PCM/EC CODE 24

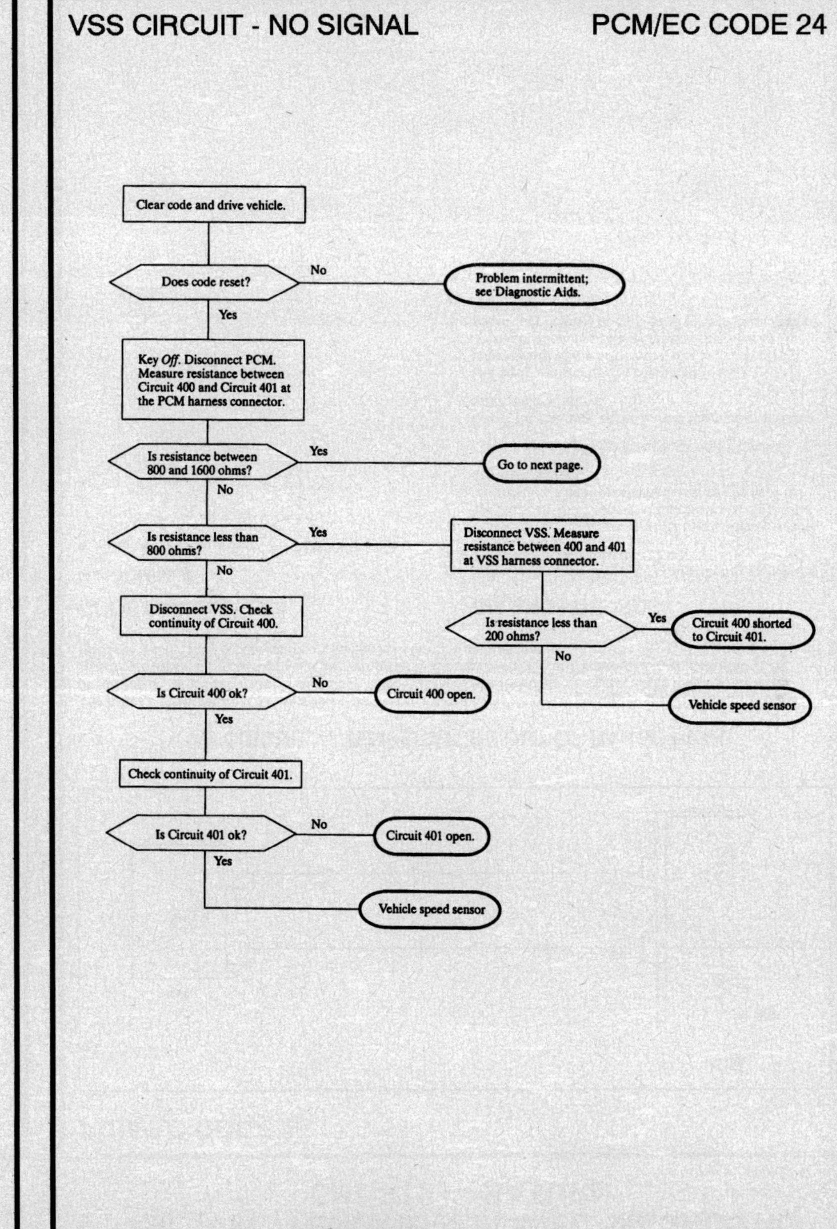

VSS CIRCUIT — NO SIGNAL

Vehicle speed values are provided by the vehicle speed sensor (VSS). The sensor is a permanent magnet variable reluctance sensor (VRS) mounted on the differential area of the transaxle converter housing. The PM generator produces a pulsing voltage whenever vehicle speed is above 5 KPH (3 MPH). These pulses occur 16 times per each revolution of the differential housing and are translated to provide vehicle speed in miles per hour.

CODE PARAMETERS

Code 24 will set if:

- The VSS indication is less than 1.6 KPH (1 MPH).
- The engine speed is above idle. The MAP indication is less than .5 V for greater than 5 sec.
- Vehicle not in park/neutral (Automatic Only)

DIAGNOSTIC AIDS

When attempting to diagnose an intermittent problem, use the Scan Tool* to review supplemental diagnostic information. The supplemental data can be used to duplicate a problem.

* Select MALF History from Scan Tool EC menu.

Intermittents or opens suspected to be at the connector can be detected by using Diagnostic Service Probe. Voltage can be read on wires without disconnecting any connectors.

Check the tightness of the female terminal grip with a spare male terminal.

- The speedometer will not work if the VSS subsystem has failed.
- If VSS reading is accurate and speedometer is incorrect, refer to Instrument Cluster Section.
- VSS resistance 700 – 900 ohms.
- If problem is intermittent wiggle wires while monitoring VSS signal on scan tool and watch for signal to jump or for loss of signal.
- The PCM will substitute VSS based on RPM if code 24 is set.

1.9L (VIN 7) ENGINE — PCM/EC DIAGNOSTIC CHARTS — SATURN

VSS CIRCUIT - NO SIGNAL — PCM/EC CODE 24

1.9L (VIN 7) ENGINE — PCM/EC DIAGNOSTIC CHARTS — SATURN

VSS CIRCUIT - NO SIGNAL — PCM/EC CODE 24

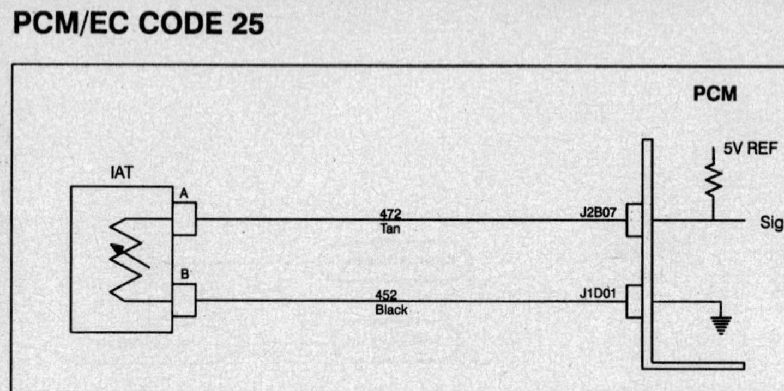

1.9L (VIN 7) ENGINE — PCM/EC DIAGNOSTIC CHARTS — SATURN

PCM/EC CODE 25

IAT CIRCUIT — TEMPERATURE OUT OF RANGE HIGH

The inlet air temperature (IAT) sensor is a thermistor located in the air intake passage of the air induction system on the engine. When the sensor is cold it has a high resistance, the resistance decreases as the temperature of the sensor increases. The PCM supplies a five volt signal to the sensor which is also connected to ground. The PCM reads the voltage drop on the sensor signal line to determine air temperature.

CODE PARAMETERS

Code 25 will set if:

- The IAT reading is over 125°C (257°F).

DIAGNOSTIC AIDS

Refer to Description and Operation in this manual for Temperature vs. Resistance chart.

When attempting to diagnose an intermittent problem, use the Scan Tool* to review supplemental diagnostic information. The supplemental data can be used to duplicate a problem.

* Select MALF History from Scan Tool EC menu.

Check the tightness of the female terminal grip with a spare male terminal.

IAT temperature may be compared to ECT temperature with a Scan tool if engine has cooled overnight at ambient temperature. They should be within four degrees of each other with ignition *On* and engine *Off*.

1.9L (VIN 7) ENGINE — PCM/EC DIAGNOSTIC CHARTS — SATURN

IAT CIRCUIT (HIGH) PCM/EC CODE 25

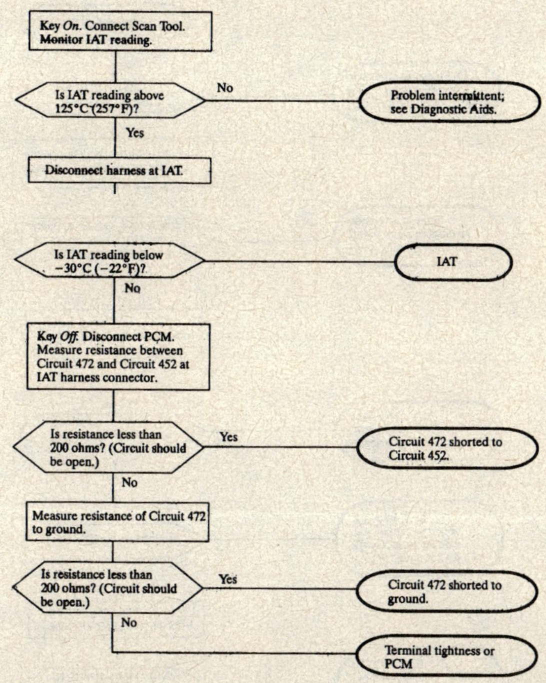

Key *On.* Connect Scan Tool. Monitor IAT reading.

Is IAT reading above 125°C (257°F)? — No → Problem intermittent; see Diagnostic Aids.

Yes

Disconnect harness at IAT.

Is IAT reading below −30°C (−22°F)? — No → IAT

Key *Off.* Disconnect PCM. Measure resistance between Circuit 472 and Circuit 452 at IAT harness connector.

Is resistance less than 200 ohms? (Circuit should be open.) — Yes → Circuit 472 shorted to Circuit 452.

No

Measure resistance of Circuit 472 to ground.

Is resistance less than 200 ohms? (Circuit should be open.) — Yes → Circuit 472 shorted to ground.

No

Terminal tightness or PCM

1.9L (VIN 7) ENGINE — PCM/EC DIAGNOSTIC CHARTS — SATURN

PCM/EC CODE 26

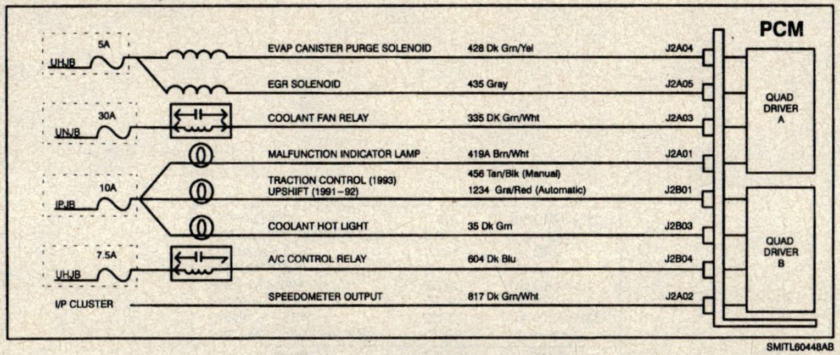

Fuse		Load	Circuit	Terminal
5A UH/JB		EVAP CANISTER PURGE SOLENOID	428 Dk Grn/Yel	J2A04
		EGR SOLENOID	435 Gray	J2A05
30A UH/JB		COOLANT FAN RELAY	335 DK Grn/Wht	J2A03
		MALFUNCTION INDICATOR LAMP	419A Brn/Wht	J2A01
10A IP/JB		TRACTION CONTROL (1993) UPSHIFT (1991–92)	456 Tan/Blk (Manual) 1234 Gra/Red (Automatic)	J2B01
		COOLANT HOT LIGHT	35 Dk Grn	J2B03
7.5A UH/JB		A/C CONTROL RELAY	604 Dk Blu	J2B04
I/P CLUSTER		SPEEDOMETER OUTPUT	817 Dk Grn/Wht	J2A02

PCM — QUAD DRIVER A — QUAD DRIVER B

SMITL60448AB

QUAD DRIVER OUTPUT FAULT

A quad drive module is a semiconductor device capable of controlling four separate outputs. Each output of the QDM is an open collector driver which when turned on pulls the output to ground. A load is connected between 12 volts and the driver. When the output is turn on, current flows from the battery, through the load, through the driver to ground. Each QDM has a fault line feedback. A comparison of the driver input and output states is performed. If the input and output are in the same state, a fault will be detected. Each QDM has only one fault line that will detect a fault on any of the four outputs.

CODE PARAMETERS

Code 26 will set if:

There is an open or short on any of the QDM output circuits.

Fault detected for 25 seconds.

DIAGNOSTIC AIDS

- EGR Solenoid . . . . 37–44 ohms
- EVAP Canister Purge Solenoid . . . . 32 ohms
- Coolant Fan Relay . . . . 80 ohms
- A/C Relay . . . . 80 ohms

  * all resistance are ± 10 ohms

- A shorted or open solenoid, relay or bulb can cause a Code 26.
- If one output is bad the PCM will shut down only that QDM output and not the entire quad driver unless the Quad driver reaches its thermal limit (short to voltage). At this time it will turn all four Quad driver outputs off. When the Quad driver cools down it will turn back on and this process will start over.

- If any corresponding codes have been set (example Code 32) diagnose that QDM circuit first.

Check the tightness of the female terminal grip with the spare male terminal.

Use the Saturn Service Stall (if available) to diagnose the Quad Driver circuits.

When viewing "QDM FAULT" (using a scan tool, Dynamic Display or MIL telltale), you may notice the fault appears to be intermittent. This may be due to it being an intermittent fault, but more likely is due to the way the fault detection logic works.

For example, if the EGR solenoid is off, the PCM expects to see 12 volts on the sense line. If the circuit is open, 0 volts will bee seen, and the fault will be displayed. However, if the EGR solenoid is turned on (for instance at cruising speeds) the PCM expects to see 0 volts on the sense line, and will not detect a fault. In this example, the display would show "QDM FAULT A" at idle, but if the RPM is raised, the display will show "QDM FAULT NONE". When returned to idle, the fault will again be displayed.

- Check circuit 817 to Passive Restraint Module for an open or short to ground.

1.9L (VIN 7) ENGINE — PCM/EC DIAGNOSTIC CHARTS — SATURN

QUAD DRIVER OUTPUT FAULT — PCM/EC CODE 26

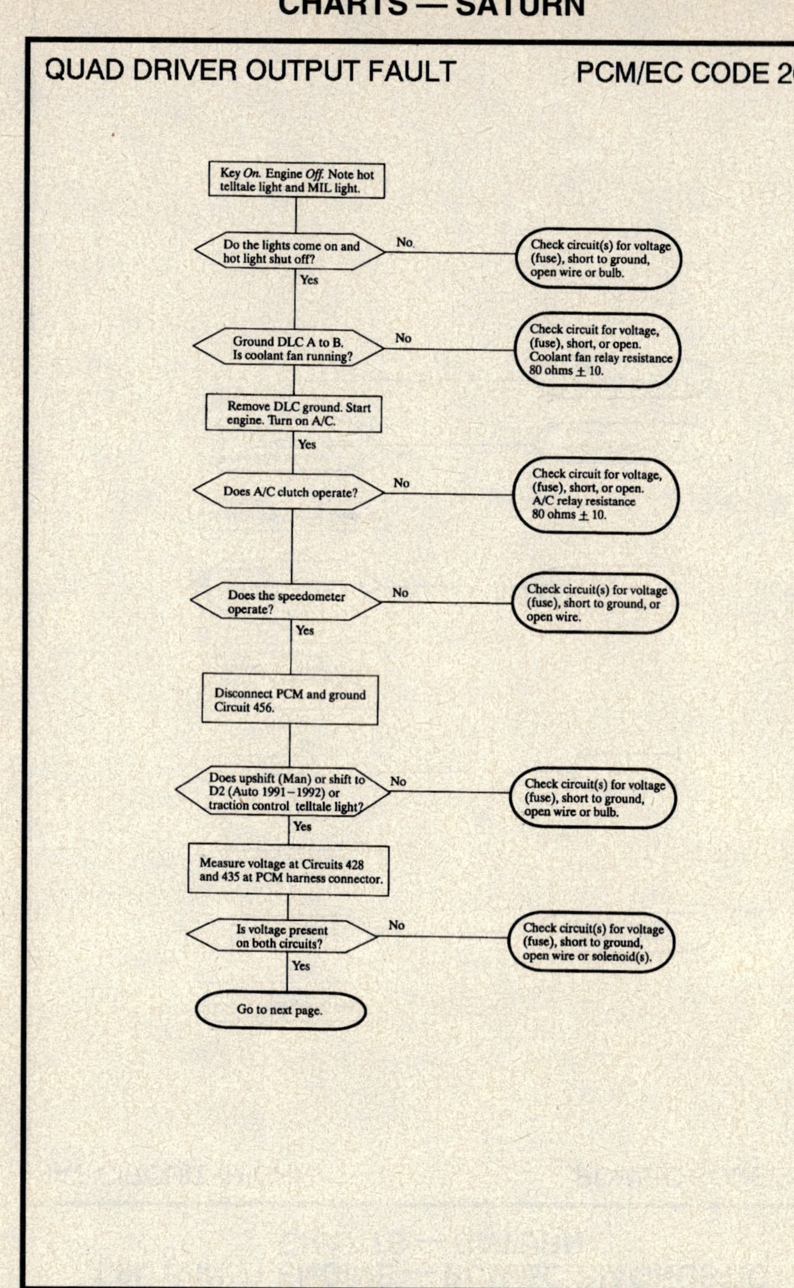

1.9L (VIN 7) ENGINE — PCM/EC DIAGNOSTIC CHARTS — SATURN

QUAD DRIVER OUTPUT FAULT — PCM/EC CODE 26

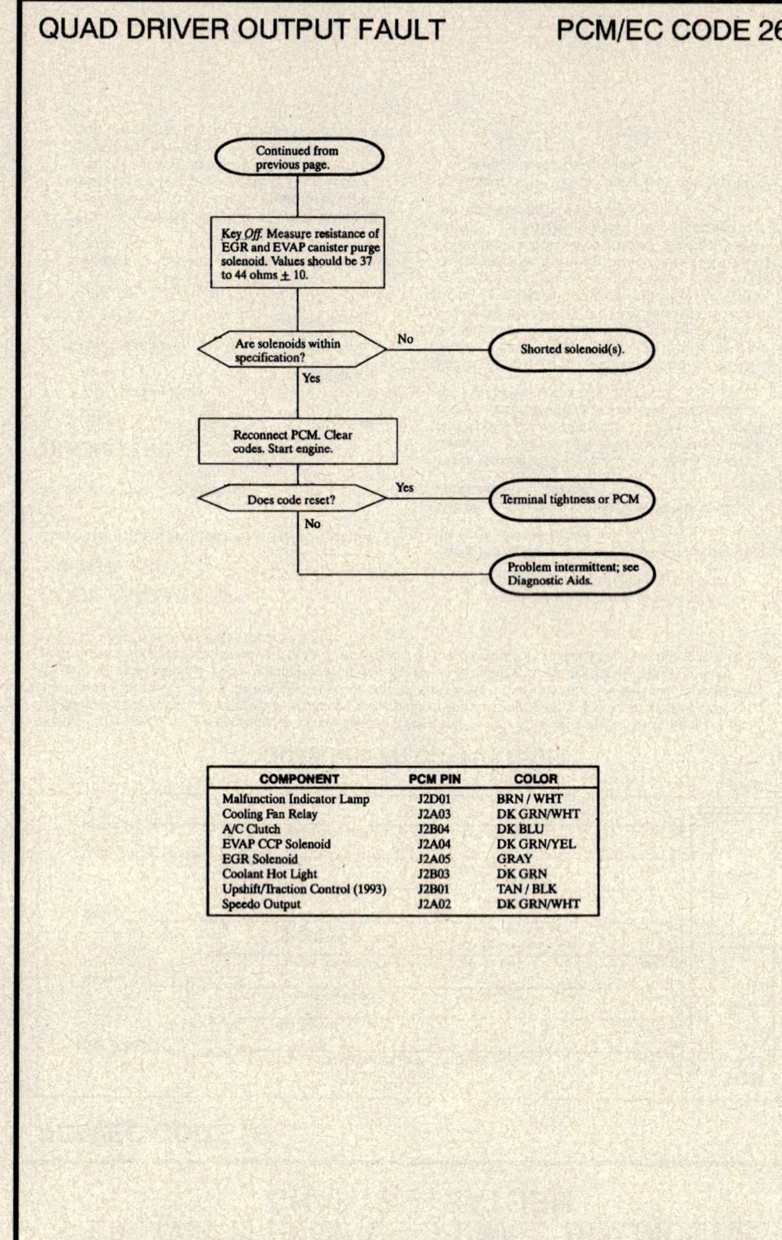

COMPONENT	PCM PIN	COLOR
Malfunction Indicator Lamp	J2D01	BRN / WHT
Cooling Fan Relay	J2A03	DK GRN/WHT
A/C Clutch	J2B04	DK BLU
EVAP CCP Solenoid	J2A04	DK GRN/YEL
EGR Solenoid	J2A05	GRAY
Coolant Hot Light	J2B03	DK GRN
Upshift/Traction Control (1993)	J2B01	TAN / BLK
Speedo Output	J2A02	DK GRN/WHT

1.9L (VIN 7) ENGINE — PCM/EC DIAGNOSTIC CHARTS — SATURN

PCM/EC CODE 32

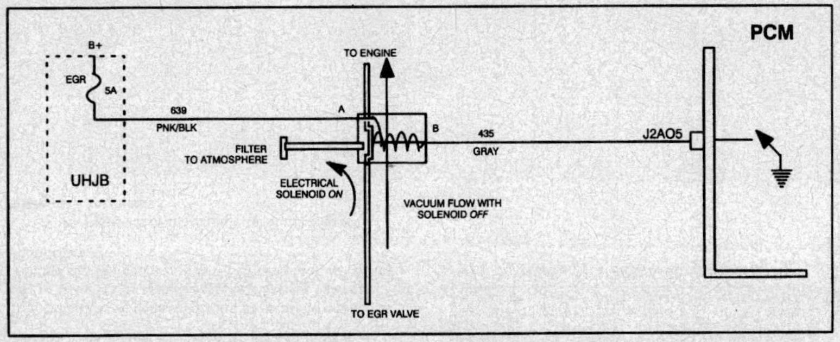

EGR SYSTEM FAULT

The exhaust gas recirculation (EGR) solenoid is connected to voltage through the ignition switch and to a ground through a QDM output. The QDM output is pulled low in order to *turn on* the solenoid and *turn off* the EGR valve. The EGR solenoid defaults to the open position to prevent the cylinder from becoming too hot in the event of an EGR system failure. The PCM checks the EGR system by grounding Ckt. 435 (no EGR) and monitoring the change in the short term fuel trim (integrator) or oxygen sensor.

CODE PARAMETERS

Code 32 will set if:

- Vacuum between .25v to 2.6v (60 and 25 kpa).
- TP sensor is between 5 and 70 percent.
- Throttle is constant.
- Vehicle speed is greater than 25 KPH (15 MPH).
- Coolant temperature above 70°C (158°F).
- Minimum engine run time of 15 sec.
- short term fuel trim (integrator) does not move at least 10 counts in 5 seconds.

DIAGNOSTIC AIDS

IMPORTANT: If Code 26 is set in combination with Code 32, diagnose Code 26 first. Then clear codes and test drive the vehicle to see if Code 32 reset. If it does reset, continue with the Code 32 trouble tree. If it does not reset, is can be assumed that Code 32 was set because of faulty Quad driver circuitry.

When attempting to diagnose an intermittent problem use the Scan Tool* to review supplemental diagnostic information. The supplemental data can be used to duplicate a problem.

* Select MALF History from Scan Tool EC menu.

Intermittents or opens suspected to be at the connector can be detected by using Diagnostic Service Probe. Voltage can be read on both wires without disconnecting any connectors.

Check the tightness of the female terminal grip with a spare male terminal.

EGR passages (cylinder head and intake manifold) restricted or blocked could also cause a Code 32.

EGR is ported vacuum. In the event the EGR solenoid fails you will still get EGR.

Check for vacuum present at EGR solenoid connection and insure hose is not plugged.

1.9L (VIN 7) ENGINE — PCM/EC DIAGNOSTIC CHARTS — SATURN

EGR SYSTEM FAULT   PCM/EC CODE 32

IMPORTANT: If Code 26 is set along with Code 32, diagnose Code 26 first.

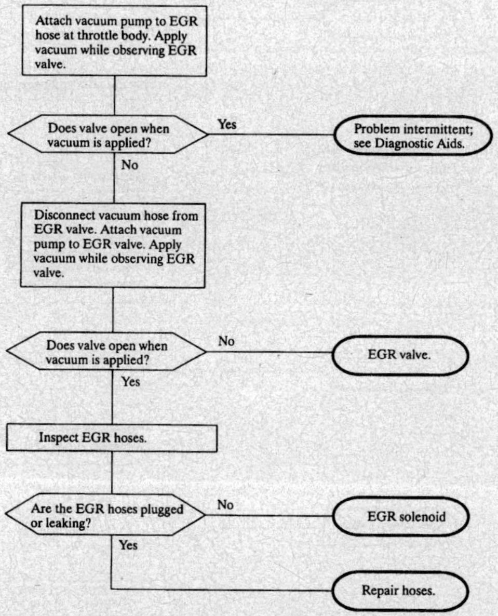

1.9L (VIN 7) ENGINE — PCM/EC DIAGNOSTIC CHARTS — SATURN

PCM/EC CODE 33

MAP CIRCUIT – VOLTAGE OUT OF RANGE HIGH

The manifold absolute pressure (MAP) sensor is a variable resistor that responds to changes in manifold pressure (vacuum). The PCM measures the voltage drop across the MAP sensor. The voltage ranges from 1–1.5 volts at idle (high vacuum) to 4–4.5 volts at wide open throttle (low vacuum).

CODE PARAMETERS

Code 33 will set if:

- MAP is greater than 4.2 volts or (90 kpa)
- TP sensor is less than 5 percent.
- Code 21 and 22 are not present.

DIAGNOSTIC AIDS

When attempting to diagnose an intermittent problem use the Scan Tool* to review supplemental diagnostic information. The supplemental data can be used to duplicate a problem.

* Select MALF History from Scan Tool EC menu.

Intermittents or opens suspected to be at the connector can be detected by using Diagnostic Service Probe. Voltage can be read on wires without disconnecting any connectors.

Check the tightness of the female terminal grip with a spare male terminal.

With engine *Off* and key *On* the map reading should be the same as barometric because manifold pressure is equal to atmospheric pressure (no vacuum — high voltage). Comparing this reading to a known good vehicle with the same sensor is a good way to verify accuracy of a *suspect* sensor. Readings should be within 0.4 volts.

ALTITUDE		VOLTAGE RANGE	kPa
Meters	Feet		
Below 305	Below 1,000	3.8 – 5.5V	98 – 100 kPa
305 – 610	1,000 – 2,000	3.6 – 5.3V	98 – 95 kPa
610 – 914	2,000 – 3,000	3.5 – 5.1V	95 – 92 kPa
914 – 1219	3,000 – 4,000	3.3 – 5.0V	92 – 89 kPa
1219 – 1524	4,000 – 5,000	3.2 – 4.8V	89 – 86 kPa
1524 – 1829	5,000 – 6,000	3.0 – 4.6V	86 – 83 kPa
1829 – 2133	6,000 – 7,000	2.9 – 4.5V	83 – 80 kPa
2133 – 2438	7,000 – 8,000	2.8 – 4.3V	80 – 77 kPa
2438 – 2743	8,000 – 9,000	2.6 – 4.2V	77 – 74 kPa
2743 – 3948	9,000 – 10,000	2.5 – 4.0V	74 – 71 kPa

IGNITION ON ENGINE OFF VOLTAGE

LOW ALTITUDE = HIGH PRESSURE = HIGH VOLTAGE

1.9L (VIN 7) ENGINE — PCM/EC DIAGNOSTIC CHARTS — SATURN

MAP CIRCUIT - RANGE HIGH   PCM/EC CODE 33

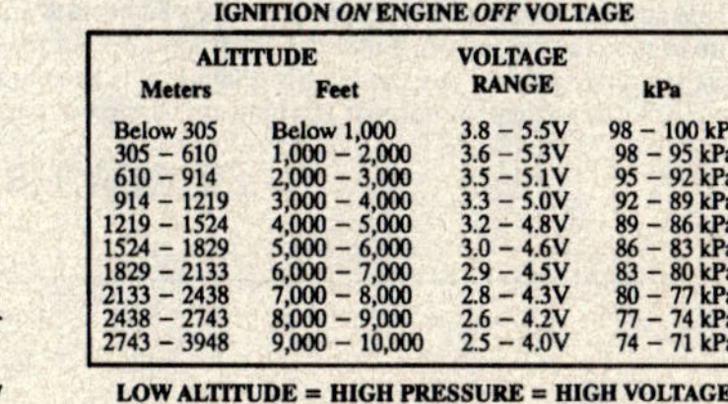

1.9L (VIN 7) ENGINE — PCM/EC DIAGNOSTIC CHARTS — SATURN

MAP CIRCUIT- RANGE HIGH — PCM/EC CODE 33

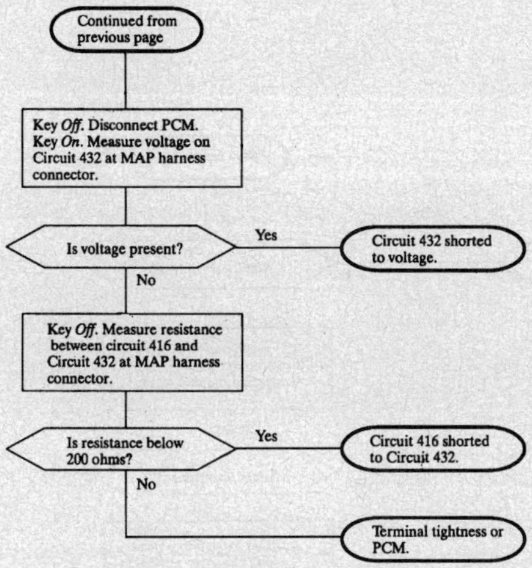

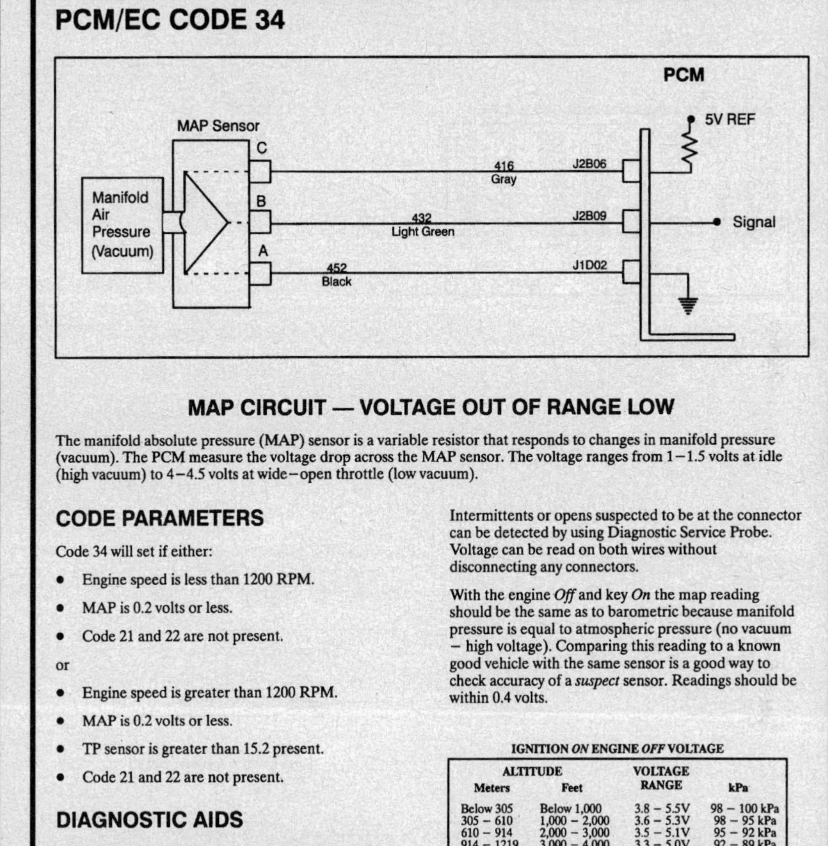

IGNITION *ON* ENGINE *OFF* VOLTAGE			
ALTITUDE		VOLTAGE	
Meters	Feet	RANGE	kPa
Below 305	Below 1,000	3.8 − 5.5V	98 − 100 kPa
305 − 610	1,000 − 2,000	3.6 − 5.3V	98 − 95 kPa
610 − 914	2,000 − 3,000	3.5 − 5.1V	95 − 92 kPa
914 − 1219	3,000 − 4,000	3.3 − 5.0V	92 − 89 kPa
1219 − 1524	4,000 − 5,000	3.2 − 4.8V	89 − 86 kPa
1524 − 1829	5,000 − 6,000	3.0 − 4.6V	86 − 83 kPa
1829 − 2133	6,000 − 7,000	2.9 − 4.5V	83 − 80 kPa
2133 − 2438	7,000 − 8,000	2.8 − 4.3V	80 − 77 kPa
2438 − 2743	8,000 − 9,000	2.6 − 4.2V	77 − 74 kPa
2743 − 3948	9,000 − 10,000	2.5 − 4.0V	74 − 71 kPa

LOW ALTITUDE = HIGH PRESSURE = HIGH VOLTAGE

1.9L (VIN 7) ENGINE — PCM/EC DIAGNOSTIC CHARTS — SATURN

PCM/EC CODE 34

MAP CIRCUIT — VOLTAGE OUT OF RANGE LOW

The manifold absolute pressure (MAP) sensor is a variable resistor that responds to changes in manifold pressure (vacuum). The PCM measure the voltage drop across the MAP sensor. The voltage ranges from 1−1.5 volts at idle (high vacuum) to 4−4.5 volts at wide−open throttle (low vacuum).

CODE PARAMETERS

Code 34 will set if either:

- Engine speed is less than 1200 RPM.
- MAP is 0.2 volts or less.
- Code 21 and 22 are not present.

or

- Engine speed is greater than 1200 RPM.
- MAP is 0.2 volts or less.
- TP sensor is greater than 15.2 present.
- Code 21 and 22 are not present.

DIAGNOSTIC AIDS

When attempting to diagnose an intermittent problem use the Scan Tool* to review supplemental diagnostic information. The supplemental data can be used to duplicate a problem.

* Select MALF History from Scan Tool EC menu.

Intermittents or opens suspected to be at the connector can be detected by using Diagnostic Service Probe. Voltage can be read on both wires without disconnecting any connectors.

With the engine *Off* and key *On* the map reading should be the same as to barometric because manifold pressure is equal to atmospheric pressure (no vacuum − high voltage). Comparing this reading to a known good vehicle with the same sensor is a good way to check accuracy of a *suspect* sensor. Readings should be within 0.4 volts.

IGNITION *ON* ENGINE *OFF* VOLTAGE			
ALTITUDE		VOLTAGE	
Meters	Feet	RANGE	kPa
Below 305	Below 1,000	3.8 − 5.5V	98 − 100 kPa
305 − 610	1,000 − 2,000	3.6 − 5.3V	98 − 95 kPa
610 − 914	2,000 − 3,000	3.5 − 5.1V	95 − 92 kPa
914 − 1219	3,000 − 4,000	3.3 − 5.0V	92 − 89 kPa
1219 − 1524	4,000 − 5,000	3.2 − 4.8V	89 − 86 kPa
1524 − 1829	5,000 − 6,000	3.0 − 4.6V	86 − 83 kPa
1829 − 2133	6,000 − 7,000	2.9 − 4.5V	83 − 80 kPa
2133 − 2438	7,000 − 8,000	2.8 − 4.3V	80 − 77 kPa
2438 − 2743	8,000 − 9,000	2.6 − 4.2V	77 − 74 kPa
2743 − 3948	9,000 − 10,000	2.5 − 4.0V	74 − 71 kPa

LOW ALTITUDE = HIGH PRESSURE = HIGH VOLTAGE

1.9L (VIN 7) ENGINE — PCM/EC DIAGNOSTIC CHARTS — SATURN

PCM/EC CODE 34

MAP CIRCUIT - RANGE LOW

IGNITION *ON* ENGINE *OFF* VOLTAGE

ALTITUDE		VOLTAGE RANGE	
Meters	Feet		kPa
Below 305	Below 1,000	3.8 – 5.5V	98 – 100 kPa
305 – 610	1,000 – 2,000	3.6 – 5.3V	98 – 95 kPa
610 – 914	2,000 – 3,000	3.5 – 5.1V	95 – 92 kPa
914 – 1219	3,000 – 4,000	3.3 – 5.0V	92 – 89 kPa
1219 – 1524	4,000 – 5,000	3.2 – 4.8V	89 – 86 kPa
1524 – 1829	5,000 – 6,000	3.0 – 4.6V	86 – 83 kPa
1829 – 2133	6,000 – 7,000	2.9 – 4.5V	83 – 80 kPa
2133 – 2438	7,000 – 8,000	2.8 – 4.3V	80 – 77 kPa
2438 – 2743	8,000 – 9,000	2.6 – 4.2V	77 – 74 kPa
2743 – 3948	9,000 – 10,000	2.5 – 4.0V	74 – 71 kPa

LOW ALTITUDE = HIGH PRESSURE = HIGH VOLTAGE

1.9L (VIN 7) ENGINE — PCM/EC DIAGNOSTIC CHARTS — SATURN

PCM/EC CODE 34

MAP CIRCUIT - RANGE LOW

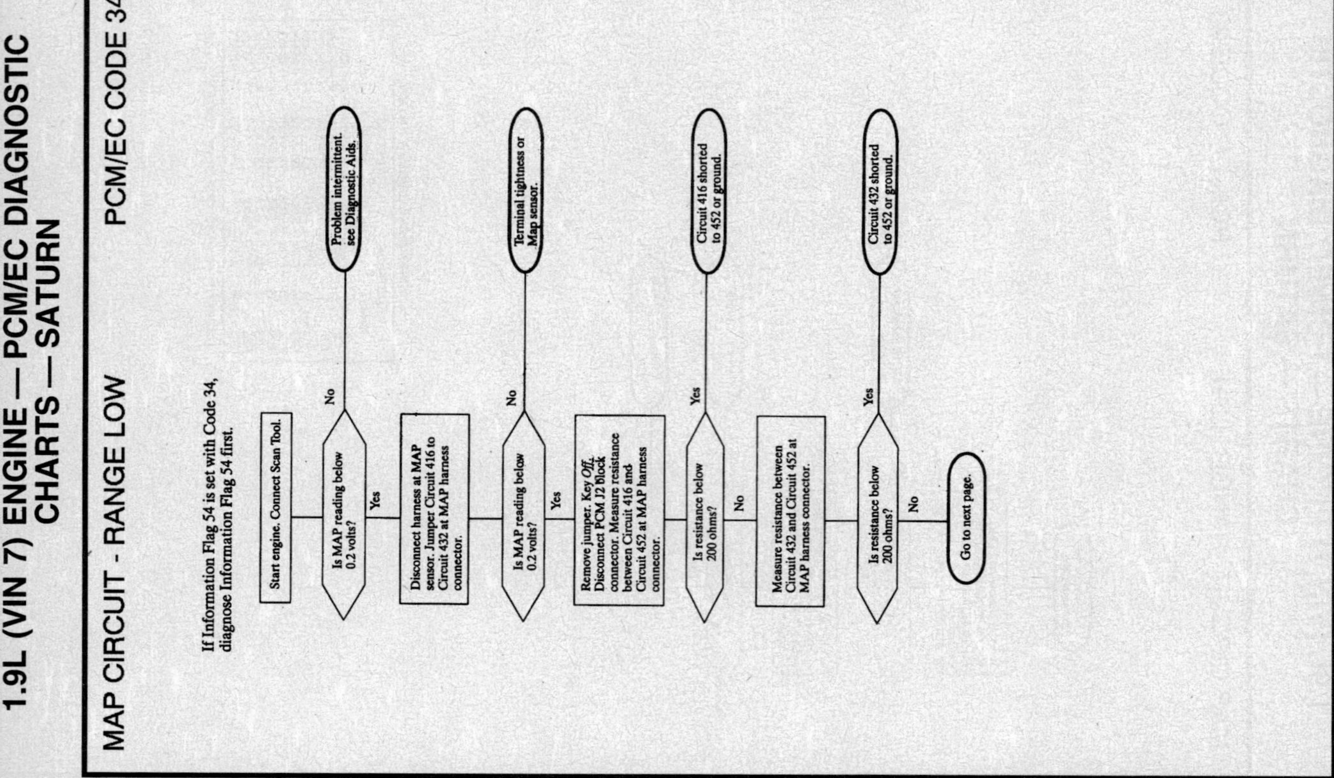

1.9L (VIN 7) ENGINE — PCM/EC DIAGNOSTIC CHARTS — SATURN

PCM/EC CODE 35

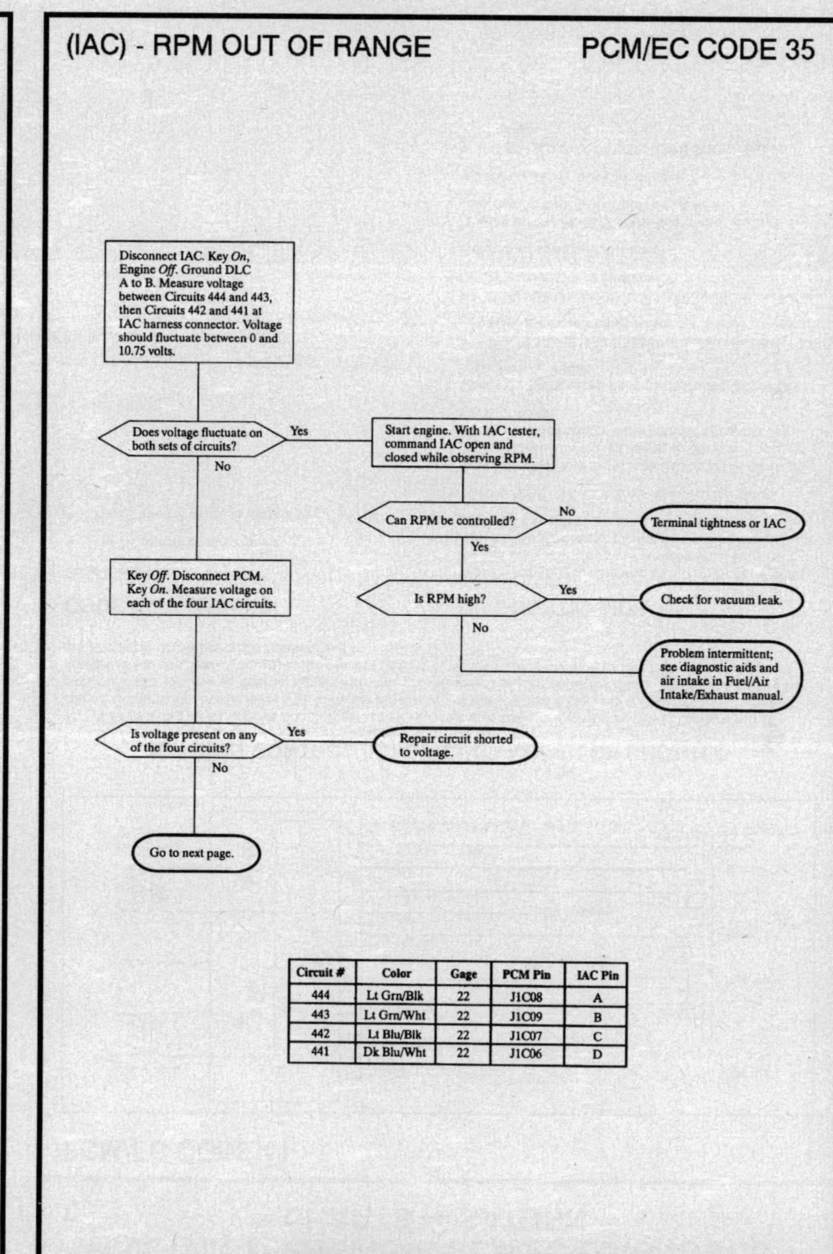

IDLE AIR CONTROL (IAC) — RPM OUT OF RANGE

The idle air control (IAC) consists of a two coil stepper motor controlling a valve. The motor is used to move the IAC valve pintle in and out of a seat to control the amount of air that enters the intake manifold at idle.

CODE PARAMETERS

Code 35 will set if:

- Idle speed is greater than desired after PCM commanding the IAC motor closed.
- Idle speed is less than desired after PCM commanding the IAC motor wide open.

DIAGNOSTIC AIDS

When attempting to diagnose an intermittent problem use the Scan Tool* to review supplemental diagnostic information. The supplemental data can be used to duplicate a problem.

* Select MALF History from Scan Tool EC menu.

Intermittents or opens suspected to be at the connector can be detected by using Diagnostic Service Probe. Voltage can be read on both wires without disconnecting any connectors.

Check the tightness of the female terminal grip with a spare male terminal.

- Desired idle is based on a look–up table but anything that requires an idle speed boost will affect desired idle.
- The desired idle speed can be found in either the Scan Tool or the *Saturn* Service Stall.
- IAC is controlled by the PCM with four pulse width modulated (PWM) outputs.
- Cruise cable misadjustment may cause a high idle due to throttle plates being held open.
- Inspect PCV valve and system.
- Incorrect minimum idle air will cause a Code 35.

1.9L (VIN 7) ENGINE — PCM/EC DIAGNOSTIC CHARTS — SATURN

(IAC) - RPM OUT OF RANGE PCM/EC CODE 35

Disconnect IAC. Key *On*, Engine *Off*. Ground DLC A to B. Measure voltage between Circuits 444 and 443, then Circuits 442 and 441 at IAC harness connector. Voltage should fluctuate between 0 and 10.75 volts.

- Does voltage fluctuate on both sets of circuits? — Yes → Start engine. With IAC tester, command IAC open and closed while observing RPM.
  - No
- Can RPM be controlled? — No → Terminal tightness or IAC
  - Yes
- Key *Off*. Disconnect PCM. Key *On*. Measure voltage on each of the four IAC circuits.
- Is RPM high? — Yes → Check for vacuum leak.
  - No → Problem intermittent; see diagnostic aids and air intake in Fuel/Air Intake/Exhaust manual.
- Is voltage present on any of the four circuits? — Yes → Repair circuit shorted to voltage.
  - No
- Go to next page.

Circuit #	Color	Gage	PCM Pin	IAC Pin
444	Lt Grn/Blk	22	J1C08	A
443	Lt Grn/Wht	22	J1C09	B
442	Lt Blu/Blk	22	J1C07	C
441	Dk Blu/Wht	22	J1C06	D

1.9L (VIN 7) ENGINE — PCM/EC DIAGNOSTIC CHARTS — SATURN

(IAC) - RPM OUT OF RANGE — PCM/EC CODE 35

Circuit #	Color	Gage	PCM Pin	IAC Pin
444	Lt Grn/Blk	22	J1C08	A
443	Lt Grn/Wht	22	J1C09	B
442	Lt Blu/Blk	22	J1C07	C
441	Dk Blu/Wht	22	J1C06	D

1.9L (VIN 7) ENGINE — PCM/EC DIAGNOSTIC CHARTS — SATURN

PCM/EC CODE 41

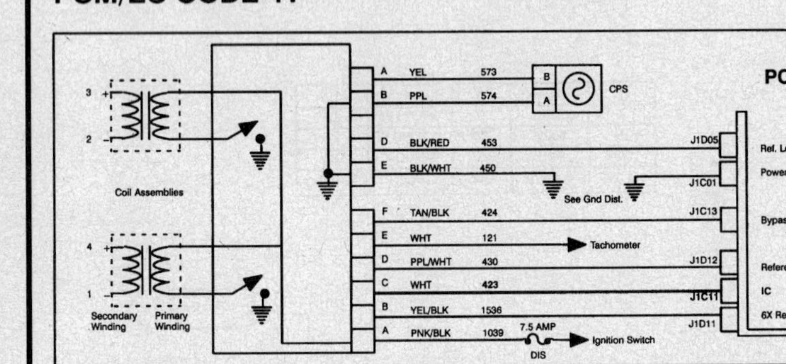

IGNITION CONTROL (IC) CIRCUIT — OPEN OR SHORTED

The PCM provides the IC control circuit signal. The EI module uses this signal to fire the ignition coils, this is IC control circuit mode (bypass line high). In bypass mode, (bypass line low) the EI module will ground the IC control circuit line and generate an internal signal to fire the ignition coils. The EI module has the ability to force the bypass line low (Back–up Spark Control) in the event of missing IC control circuit pulses. The PCM monitors the state of the IC control circuit line while cranking and running.

CODE PARAMETERS

Code 41 will set if:

- The IC control circuit is open.
- The IC control circuit is shorted to ground or voltage.

DIAGNOSTIC AIDS

When attempting to diagnose an intermittent problem use the Scan Tool* to review supplemental diagnostic information. The supplemental data can be used to duplicate a problem.

* Select MALF History from Scan Tool EC menu.

Intermittents or opens suspected to be at the connector can be detected by using Diagnostic Service Probe. Voltage can be read on wires without disconnecting any connectors.

Check the tightness of the female terminal grip with a spare male terminal.

- The PCM will hold IC control circuit low when there are no reference pulses.
- PCM resistance with no IC control circuit output is approximately 14 M ohms.

Refer to EI section of this manual.

If using the *Saturn* PDT at the end of the General Information table will display a 41A or 41B.

- 41A – IC control Circuit open
- 41B – IC control Circuit shorted to ground or voltage.

1.9L (VIN 7) ENGINE — PCM/EC DIAGNOSTIC CHARTS — SATURN

IC CONTROL CIRCUIT OPEN OR SHORTED

PCM/EC CODE 41

IMPORTANT: If Code 41 is set along with Code 42, refer to Chart Codes 41–42.

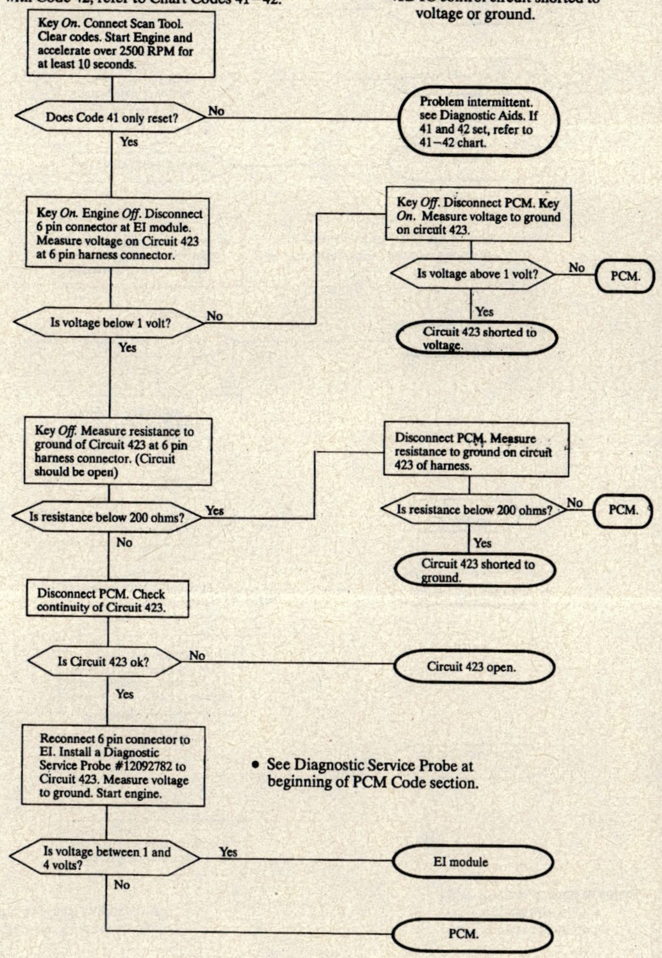

41A IC Control circuit open
41B IC control circuit shorted to voltage or ground.

Key On. Connect Scan Tool. Clear codes. Start Engine and accelerate over 2500 RPM for at least 10 seconds.

Does Code 41 only reset? — No → Problem intermittent. see Diagnostic Aids. If 41 and 42 set, refer to 41–42 chart.

Yes

Key On. Engine Off. Disconnect 6 pin connector at EI module. Measure voltage on Circuit 423 at 6 pin harness connector.

Key Off. Disconnect PCM. Key On. Measure voltage to ground on circuit 423.

Is voltage above 1 volt? — No → PCM.

Yes

Circuit 423 shorted to voltage.

Is voltage below 1 volt? — No

Yes

Key Off. Measure resistance to ground of Circuit 423 at 6 pin harness connector. (Circuit should be open.)

Disconnect PCM. Measure resistance to ground on circuit 423 of harness.

Is resistance below 200 ohms? — No → PCM.

Yes

Circuit 423 shorted to ground.

Is resistance below 200 ohms? — Yes

No

Disconnect PCM. Check continuity of Circuit 423.

Is Circuit 423 ok? — No → Circuit 423 open.

Yes

Reconnect 6 pin connector to EI. Install a Diagnostic Service Probe #12092782 to Circuit 423. Measure voltage to ground. Start engine.

• See Diagnostic Service Probe at beginning of PCM Code section.

Is voltage between 1 and 4 volts? — Yes → EI module

No

PCM.

1.9L (VIN 7) ENGINE — PCM/EC DIAGNOSTIC CHARTS — SATURN

PCM/EC CODE 42

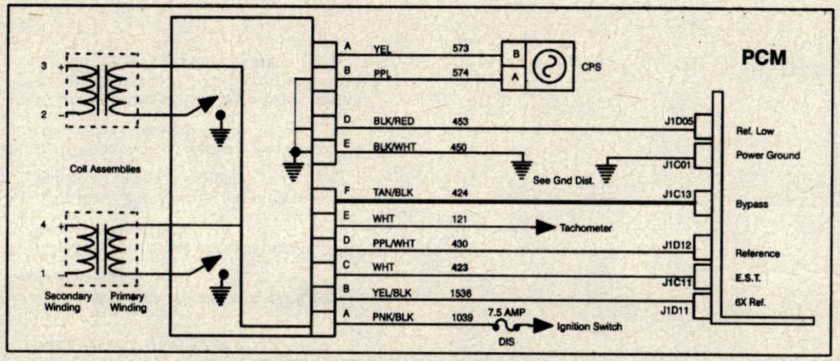

BYPASS CIRCUIT — OPEN OR SHORTED

The PCM/EC enables the IC control circuit mode by switching the bypass line high. When the bypass line is low the EI module controls secondary coil firing. To enable IC control circuit, the PCM/EC applies 5 volts on the bypass line (at approximately 400 RPM) indicating IC control circuit Mode to the EI module. In IC control circuit mode the EI module uses the IC control circuit signal from the PCM/EC to fire the secondary coils. The PCM/EC has internal feedback to monitor the state of the bypass line.

CODE PARAMETERS

Code 42 will set if:

- The bypass line is open.
- The bypass line is grounded.
- The bypass line is shorted to voltage.

DIAGNOSTIC AIDS

When attempting to diagnose an intermittent problem use the Scan Tool* to review supplemental diagnostic information. The supplemental data can be used to duplicate a problem.

* Select MALF History from Scan Tool EC menu.

Intermittents or opens suspected to be at the connector can be detected by using Diagnostic Service Probe. Voltage can be read on wires without disconnecting any connectors.

Check the tightness of the female terminal grip with a spare male terminal.

Refer to EI section of this manual.

If using the *Saturn* PDT, at the end of the General Information table it will display a 42A, 42B, or 42C.

- 42A – Bypass line grounded
- 42B – Bypass line open
- 42C – Bypass line shorted to voltage

1.9L (VIN 7) ENGINE — PCM/EC DIAGNOSTIC CHARTS — SATURN

BYPASS CIRCUIT - OPEN OR SHORTED

PCM/EC CODE 42

IMPORTANT: If Code 41 is set along with Code 42, refer to Chart Codes 41–42.

42A – Bypass Grounded
42B – Bypass Open
42C – Bypass shorted to voltage.

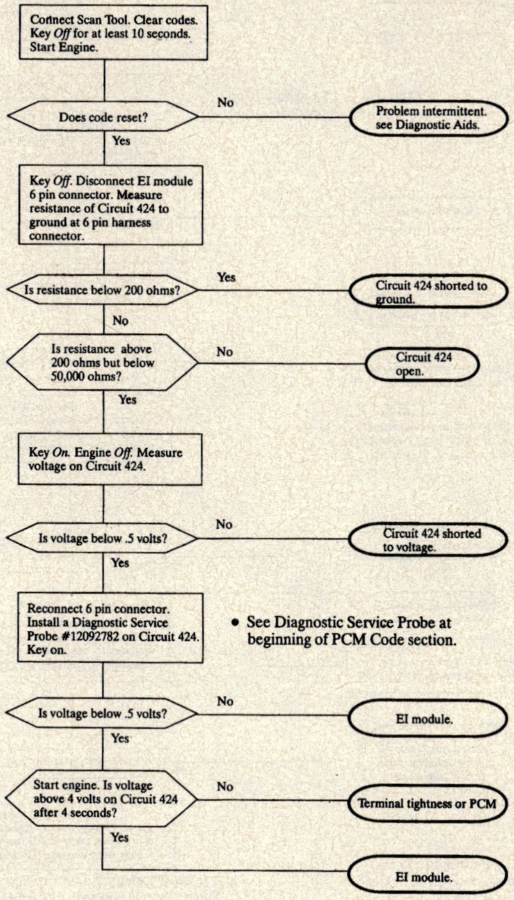

Connect Scan Tool. Clear codes. Key *Off* for at least 10 seconds. Start Engine.

Does code reset? — No → Problem intermittent. see Diagnostic Aids.
Yes ↓

Key *Off*. Disconnect EI module 6 pin connector. Measure resistance of Circuit 424 to ground at 6 pin harness connector.

Is resistance below 200 ohms? — Yes → Circuit 424 shorted to ground.
No ↓

Is resistance above 200 ohms but below 50,000 ohms? — No → Circuit 424 open.
Yes ↓

Key *On.* Engine *Off.* Measure voltage on Circuit 424.

Is voltage below .5 volts? — Yes → Circuit 424 shorted to voltage.
Yes ↓

Reconnect 6 pin connector. Install a Diagnostic Service Probe #12092782 on Circuit 424. Key on.

- See Diagnostic Service Probe at beginning of PCM Code section.

Is voltage below .5 volts? — No → EI module.
Yes ↓

Start engine. Is voltage above 4 volts on Circuit 424 after 4 seconds? — No → Terminal tightness or PCM
Yes ↓

EI module.

1.9L (VIN 7) ENGINE — PCM/EC DIAGNOSTIC CHARTS — SATURN

PCM/EC CODE 41 AND 42

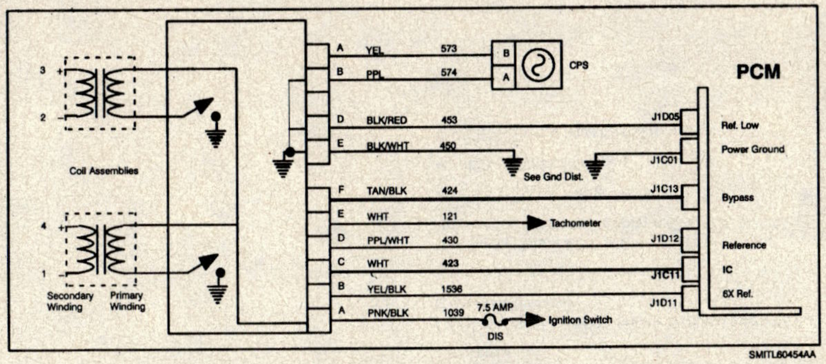

IC CONTROL CIRCUIT GROUNDED/BYPASS OPEN

The EI module looks for IC control circuit pulses from the PCM/EC. Until it recognizes one IC control circuit pulse it will not let the bypass line go high. When both codes are present (41 and 42) this indicates an open bypass or a grounded IC control circuit. If the bypass is open the PCM has received no IC control circuit feedback pulses this ignition cycle.

CODE PARAMETERS

Set 41 IC control circuit grounded with Code 42 Engine Running:

- IC control circuit signal sent from PCM/EC bypass line commanded high.

- EI does not see IC control circuit pulses from PCM/EC. Pulls bypass line low (module timing) Code 41 sets.

- IC control circuit enabled again and no feedback pulse seen by the PCM for 100 milliseconds and bypass monitor is high Code 42 sets.

Set 42 IC control circuit bypass open with Code 41 Engine Running:

- IC control circuit enabled (signal sent from PCM) bypass command high.

- More than eight reference pulses have occurred with no IC control circuit pulse.

- No IC control circuit feedback seen from PCM/EC this ignition cycle Code 42 sets.

- IC control circuit enabled again. No IC control circuit feedback to PCM/EC occurred in 100 milliseconds (bypass high) Code 41 sets.

DIAGNOSTIC AIDS

Intermittent:

- Visually inspect Circuit 423 for a chaffed wire shorting to voltage ground.

- Visually inspect Circuit 424 for a cut in wire or deformed wire insulation. Also check wire terminal connection at EI module and PCM for good connections.

- Wiggle wires while monitoring Scan tool to see if codes will reset.

1.9L (VIN 7) ENGINE — PCM/EC DIAGNOSTIC CHARTS — SATURN

IC CONTROL CIRCUIT/BYPASS GROUNDED OR OPEN — PCM/EC CODES 41-42

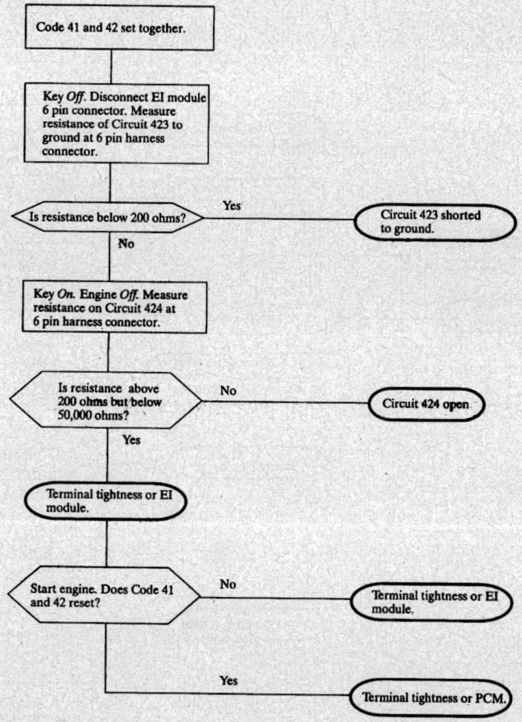

1.9L (VIN 7) ENGINE — PCM/EC DIAGNOSTIC CHARTS — SATURN

PCM/EC CODE 43

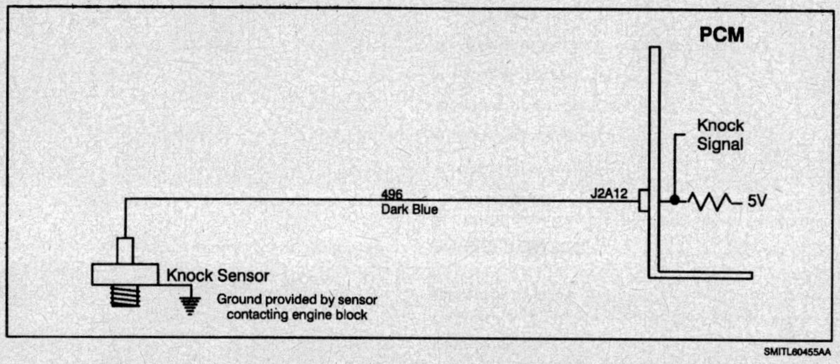

KNOCK SENSOR CIRCUIT (KS) — OPEN OR SHORTED

The knock sensor is a piezoelectric device that produces an output voltage based on mechanical engine vibration. The PCM will retard the electronic spark timing on the signal being received. The circuitry within the knock sensor causes the PCM/EC five volts reference to be pulled down so that under a no knock condition, Circuit 496 would measure 2.5 volts. The knock sensor produces an AC signal. The amplitude and signal frequency is dependant upon the knock level. If Circuit 496 becomes open or shorted to ground the voltage will either go above 3.5 volts or below 1.5 volts. If either of these conditions are met for about 10 seconds, a Code 43 will be set.

CODE PARAMETERS

- Code 43 will set if the KS circuit (system) voltage reading is out of range either high or low for at least 10 seconds.

DIAGNOSTIC AIDS

When attempting to diagnose an intermittent problem use the Scan Tool* to review supplemental diagnostic information. The supplemental data can be used to duplicate a problem.

* Select MALF History from Scan Tool EC menu.

Intermittents or opens suspected to be at the connector can be detected by using Diagnostic Service Probe. Voltage can be read on wire without disconnecting any connectors.

The resistance to ground through the knock sensor should be between 3300 and 4500 ohms.

Loose knock sensor could set PCM/EC Code 43.

Torque: 15 N•m (133 in−lbs)

1.9L (VIN 7) ENGINE — PCM/EC DIAGNOSTIC CHARTS — SATURN

KS CIRCUIT (SYSTEM) — OPEN OR SHORTED

PCM/EC CODE 43

Connect Scan tool, clear codes, start engine.

Does code reset? — No → Problem Intermittent See Diagnostic Aids

Yes

Key *On*. Back probe Circuit 496 of the PCM. Measure voltage.

Is voltage between 1.5 and 3.5 volts? — Yes → Terminal tightness or PCM

No

Key *Off*. Disconnect PCM. Back probe Circuit 496. Measure resistance to ground.

Is resistance between 3300 and 4500 ohms? — Yes → Terminal tightness or PCM.

No

Is resistance more than 4500 ohms? — Yes → Disconnect KS circuit (system) harness. Check continuity of Circuit 496 to ground.

No

Disconnect KS circuit (system) harness from knock sensor. Measure resistance of wire (Circuit 496) to ground.

Is Circuit 496 ok? — No → Circuit 496 open.

Yes → Terminal tightness or knock sensor.

Is resistance less than 200 ohms? (Circuit should be open.) — Yes → Circuit 496 shorted to ground.

No

Knock Sensor

1.9L (VIN 7) ENGINE — PCM/EC DIAGNOSTIC CHARTS — SATURN

PCM/EC CODE 44

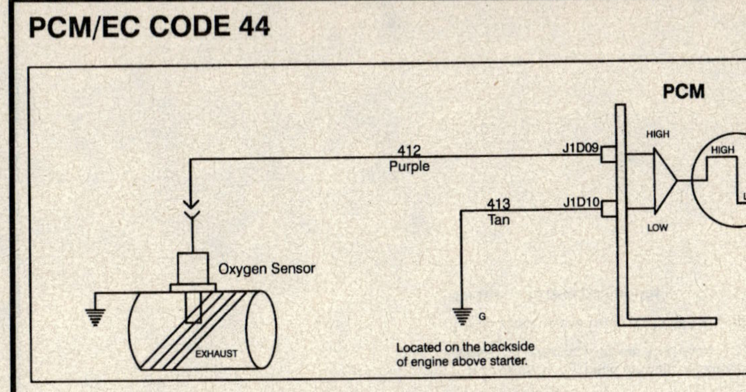

LEAN OXYGEN SENSOR SIGNAL

The oxygen sensor (O2S) consists of a zirconia electrolyte between two platinum plates. When the sensor reaches approximately 318°C (600°F) it becomes an electrical source that responds to the oxygen content in the exhaust. The PCM produces a bias voltage of approximately 450 millivolts on the oxygen sensor circuit. When the sensor is cold its internal resistance is extremely high, therefore the PCM recognizes the bias as an open circuit. As the sensor warms up, the internal resistance decreases. When the sensor reaches approximately 318°C (600°F), it starts producing a voltage based on the oxygen content in the exhaust stream. This voltage is used by the PCM to determine a rich or lean oxygen sensor signal and adjusts the fuel mixture accordingly.

CODE PARAMETERS

Code 44 will set if:

- The system is in *closed loop* mode.
- Oxygen sensor is less than 300 mV for more than 3 minutes.
- Codes 33 – 34 not present.

DIAGNOSTIC AIDS

When attempting to diagnose an intermittent problem use the Scan Tool* to review supplemental diagnostic information. The supplemental data can be used to duplicate a problem.

* Select MALF History from Scan Tool EC menu.

Intermittents or opens suspected to be at the connector can be detected by using Diagnostic Service Probe. Voltage can be read on wires without disconnecting any connectors.

Check the tightness of the female terminal grip with a spare male terminal.

POSSIBLE CAUSES:

- Misfiring cylinders (see Ignition System section of this manual).
- Running out of fuel.
- Low fuel system pressure.
- Water on sensor.
- Plugged injector.
- Engine stalling while vehicle is moving

1.9L (VIN 7) ENGINE — PCM/EC DIAGNOSTIC CHARTS — SATURN

OXYGEN SENSOR INDICATES SYSTEM LEAN

PCM/EC CODE 44

Engine at idle for over one minute. Connect Scan Tool ECT reading above 60°C (140°F).

Is oxygen sensor reading constantly below 350 mV? — No → Problem intermittent; see Diagnostic Aids.

Yes ↓

Disconnect oxygen sensor.

Is oxygen sensor reading between 350−550 mV? — Yes → System lean; see Diagnostic Aids;

No ↓

Ignition *Off*. Disconnect PCM. Measure resistance of Circuit 412 to ground.

Is resistance less than 200 ohms? — Yes → Circuit 412 shorted to 413 or ground.

No ↓

PCM

1.9L (VIN 7) ENGINE — PCM/EC DIAGNOSTIC CHARTS — SATURN

PCM/EC CODE 45

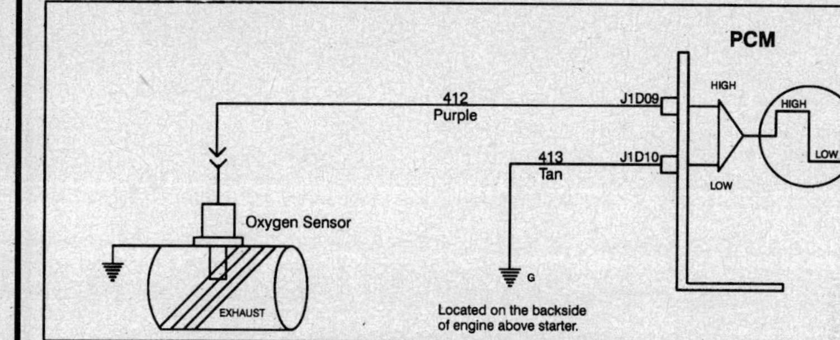

RICH OXYGEN SENSOR SIGNAL

The oxygen sensor (O2S) consists of a zirconia electrolyte between two platinum plates. When the sensor reaches approximately 318°C (600°F) it becomes an electrical source that responds to the oxygen content in the exhaust. The PCM produces a bias voltage of approximately 450 millivolts on the oxygen sensor circuit. When the sensor is cold its internal resistance is extremely high, therefore, the PCM recognizes the bias as an open circuit. As the sensor warms up, the internal resistance decreases. When the sensor reaches approximately 318°C (600°F), it starts producing a voltage based on the oxygen content in the exhaust stream. This voltage is used by the PCM to determine a rich or lean oxygen sensor signal and adjusts the fuel mixture accordingly.

CODE PARAMETERS

Code 46 will set if:

- The system is in *closed loop* mode.
- TP sensor in a 2−20% range.
- Oxygen sensor is greater than 750 mV for 30 seconds.
- Codes 33 − 34 are not present.

DIAGNOSTIC AIDS

When attempting to diagnose an intermittent problem use the Scan Tool* to review supplemental diagnostic information. The supplemental data can be used to duplicate a problem.

* Select MALF History from Scan Tool EC menu.

Intermittents or opens suspected to be at the connector can be detected by using Diagnostic Service Probe. Voltage can be read on both wires without disconnecting any connectors.

POSSIBLE CAUSES:

If system is commanding lean, the oxygen sensor signal is around 900 mV and short term fuel trim (integrator) is in the 60's, the system is nearing or at maximum correction. Check for fuel leaking into cylinders or manifold.

- Fuel pressure high
- Leaking injector
- EVAP canister purge
- Oxygen sensor contamination
- EGR valve
- Engine oil contamination
- Shorted injector
- Vacuum leak to map sensor

1.9L (VIN 7) ENGINE — PCM/EC DIAGNOSTIC CHARTS — SATURN

OXYGEN SENSOR INDICATES SYSTEM RICH

PCM/EC CODE 45

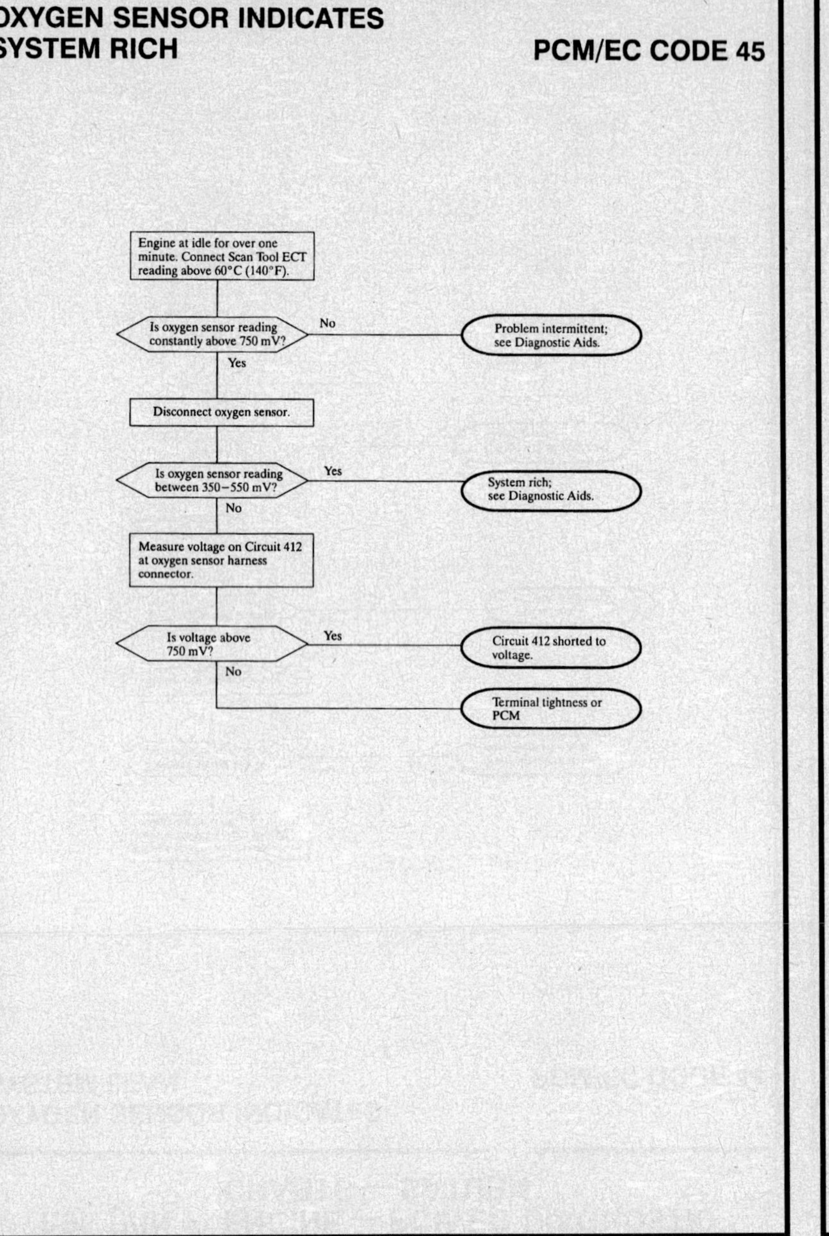

1.9L (VIN 7) ENGINE — PCM/EC DIAGNOSTIC CHARTS — SATURN

PCM/EC CODE 49

RPM — HIGH IDLE (VACUUM LEAK)

The vacuum lines include the MAP sensor junction, PCV hoses, PCV valve, brake booster hose, the fuel pressure regulator hose, throttle body, intake manifold gasket and throttle body minimum idle adjustment. Code 49 detects a leak in any of these lines which may cause the idle speed to rise above the control of the IAC system. Code 49 is detected when engine idle cannot be reduced by the IAC motor and the IAC is determined to be working.

CODE PARAMETERS

Code 49 will set if:

- The engine is idling higher than desired RPM.
- The IAC is working properly.

DIAGNOSTIC AIDS

- Inspect the hoses for cracks or splits.
- If all the hoses and connections are good, inspect the manifold gasket and MAP sensor seal for leaks.
- Check minimum idle speed.
- A short to ground in one of the IAC circuits may cause a Code 49 to set.
- Check PCV valve. Refer to Fuel and Emissions Section.

1.9L (VIN 7) ENGINE — PCM/EC DIAGNOSTIC CHARTS — SATURN

RPM — HIGH IDLE (VACUUM LEAK)

PCM/EC CODE 49

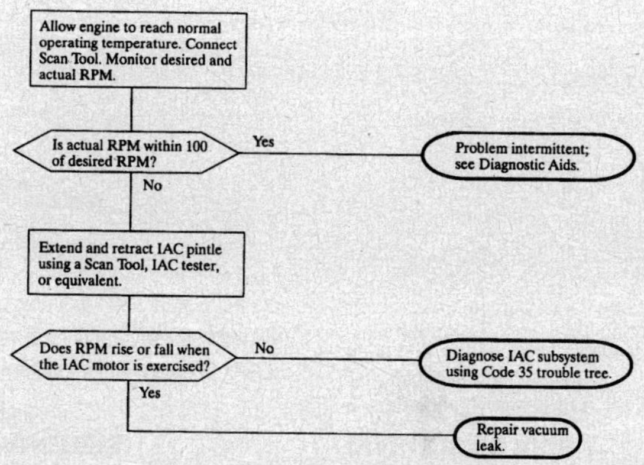

Allow engine to reach normal operating temperature. Connect Scan Tool. Monitor desired and actual RPM.

↓

Is actual RPM within 100 of desired RPM? — Yes → Problem intermittent; see Diagnostic Aids.

No ↓

Extend and retract IAC pintle using a Scan Tool, IAC tester, or equivalent.

↓

Does RPM rise or fall when the IAC motor is exercised? — No → Diagnose IAC subsystem using Code 35 trouble tree.

Yes ↓

Repair vacuum leak.

1.9L (VIN 7) ENGINE — PCM/EC DIAGNOSTIC CHARTS — SATURN

PCM/EC CODE 51

PCM MEMORY ERROR

Each time the PCM is powered up, several terminal tests are performed on the random access memory (RAM), electronically programmable read only memory (EPROM) and electronically erasable/programmable read−only memory (EEPROM). If there is a fault in one or more of these, Code 51 will set.

CODE PARAMETERS

Code 51 will set if:

The PCM memory is malfunctioning.

DIAGNOSTIC AIDS

Code 51 indicates an internal PCM/EC malfunction. The PCM should be replaced..

Tampering with the PCM/EC will cause a code 51.

PCM/EC CODE 55

A/D ERROR

The PCM receives analog voltage signals from various sensors and converts them to digital square waves. If there is a problem with the conversion process, Code 55 will set.

CODE PARAMETERS

Code 55 will set if:

- The analog to digital conversion takes too long.
- The low voltage of a square wave is greater than 0.1 volt.
- The high voltage of a square wave is less than 4.88 volts.

DIAGNOSTIC AIDS

Code 55 indicates an internal PCM malfunction. The PCM should be replaced.

1.9L (VIN 7) ENGINE — PCM/EC DIAGNOSTIC CHARTS — SATURN

PCM/EC CODE 81 – 1993 VEHICLES

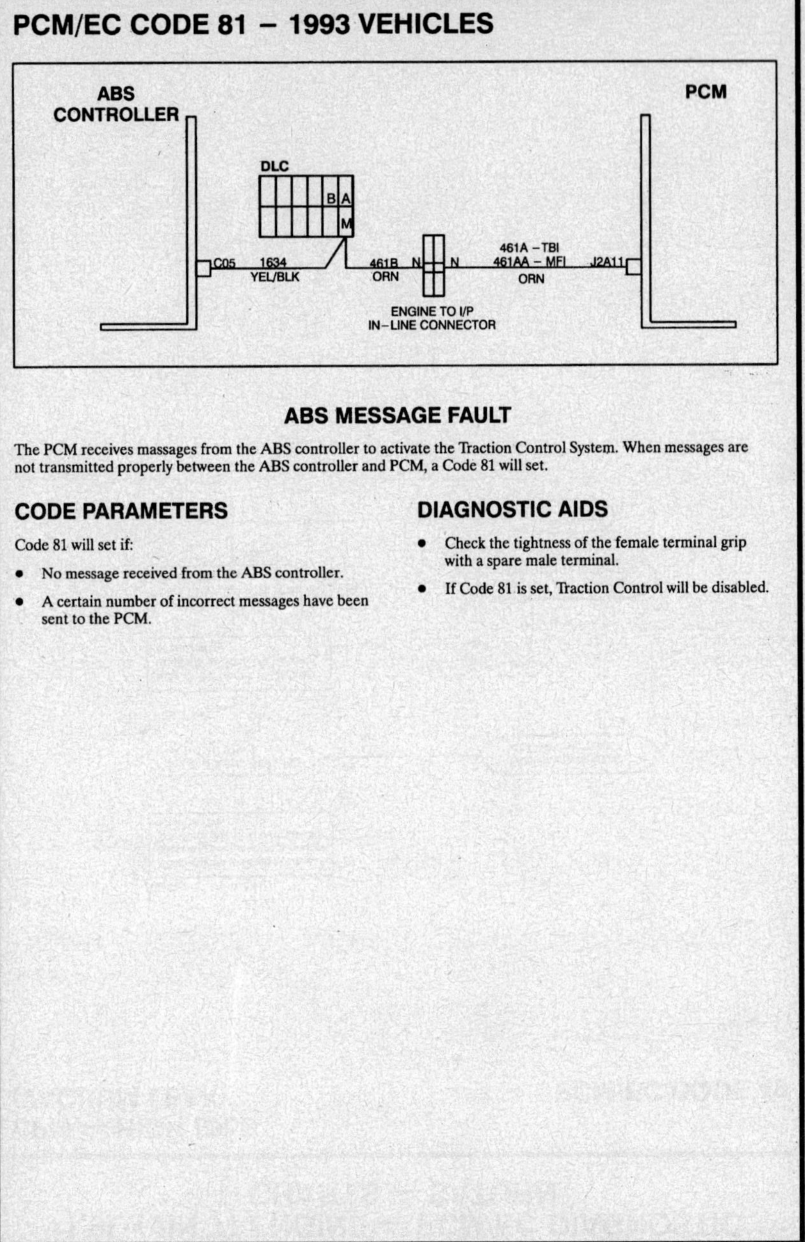

ABS MESSAGE FAULT

The PCM receives massages from the ABS controller to activate the Traction Control System. When messages are not transmitted properly between the ABS controller and PCM, a Code 81 will set.

CODE PARAMETERS

Code 81 will set if:
- No message received from the ABS controller.
- A certain number of incorrect messages have been sent to the PCM.

DIAGNOSTIC AIDS

- Check the tightness of the female terminal grip with a spare male terminal.
- If Code 81 is set, Traction Control will be disabled.

1.9L (VIN 7) ENGINE — PCM/EC DIAGNOSTIC CHARTS — SATURN

ABS MESSAGE FAULT — 1993 VEHICLES

PCM/EC CODE 81

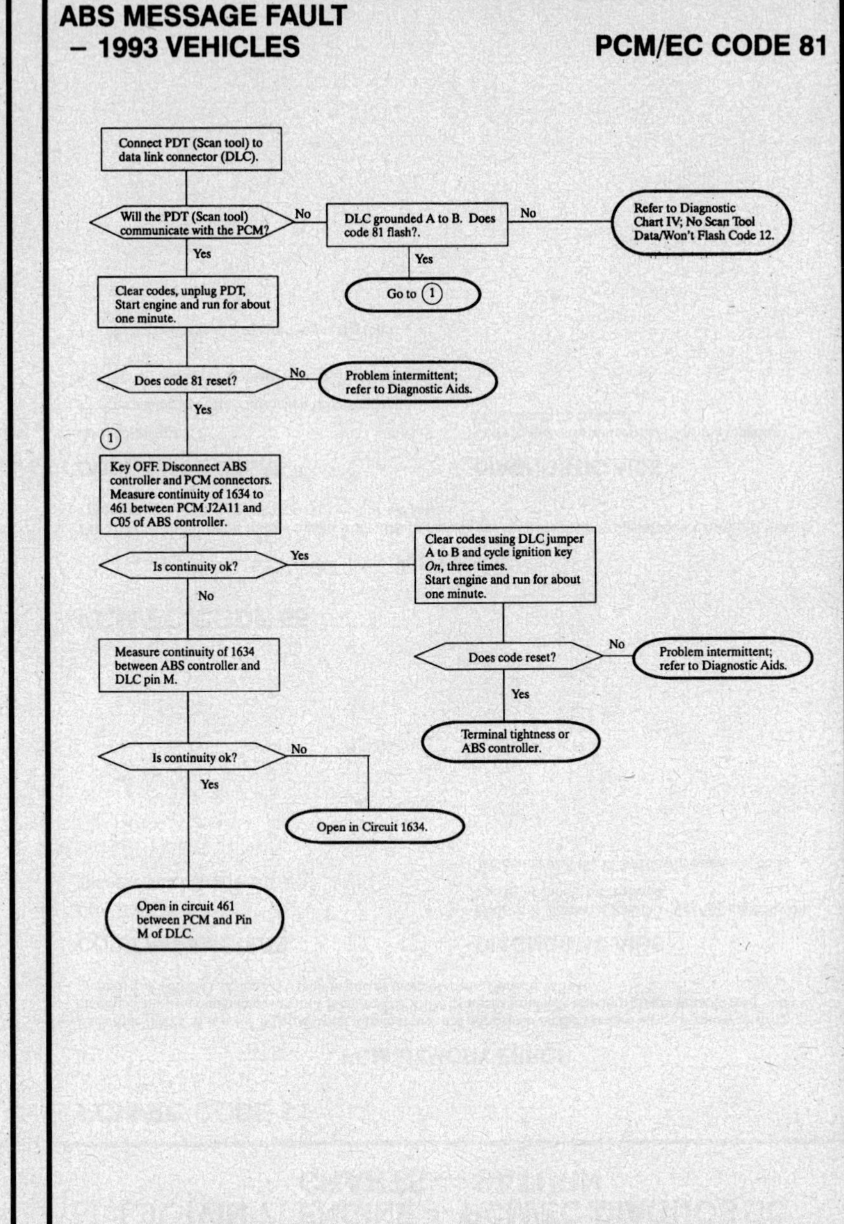

1.9L (VIN 7) ENGINE — FLAG CODE DIAGNOSTIC CHARTS — SATURN

PCM/EC CODE 82

PCM — INTERNAL COMMUNICATION FAULT

The PCM has diagnostic features that monitor communications between the engine control module and transaxle control module within the PCM. When messages are not properly transmitted between the two, Code 82 will be set.

CODE PARAMETERS

Code 82 will set if:

- A certain number of incorrect messages have been sent between the engine control module and transaxle control module within the PCM.

DIAGNOSTIC AIDS

Clear the code and recalibrate the PCM

Cycle the ignition. If Code 82 resets, replace the PCM.

1.9L (VIN 7) ENGINE — FLAG CODE DIAGNOSTIC CHARTS — SATURN

INFORMATION FLAG 16

ELECTRONIC VARIABLE ORIFICE (EVO) FAULT

The electronic variable orifice (EVO) signal is pulse width modulated to control the power steering pressure/flow from the power steering pump. The PCM controls the ground and source voltage. When the vehicle speed is zero, current flow is zero milliamps. When vehicle speed reaches 19 KPH (12 MPH), current starts to flow and steering effort begins to increase.

INFORMATION FLAG PARAMETERS

Information Flag 16 will set if:

- The PCM desired current does not match with feedback current.

- In 1992 the fault must be present for 2.5 seconds before flag is set.

DIAGNOSTIC AIDS

When the vehicle speed is zero MPH the current from the PCM will be zero milliamps (maximum power assist). Once the vehicle speed reaches 19 KPH (12 MPH) the current will gradually increase from 120 milliamps to 650 milliamps as the vehicle speed increases from 19 to 89 KPH (12 to 55 MPH). The increased current will provide decreasing amounts of power assist as speed increases.

NOTICE: Information Flags should be used for diagnostic purposes only. They do not necessarily indicate component malfunctions/failure.

- Flag 16 may be set by loss of ignition when the vehicle is still moving. This is a normal condition.

- The Scan Tool displays EVO output and feedback values to assist in EVO diagnosis. The two values have an inverse relation when output is 100% (at idle). Feedback should be at 0% and as vehicle speed increases above 19 KPH (12 MPH). Output should drop and feedback should rise indicating a good EVO subsystem.

IMPORTANT: The ignition must be cycled after a repair or Information Flag 16 will reset.

MULTIPORT FUEL INJECTION (MFI) SYSTEMS
LIGHT TRUCKS, VAN, GEO AND SATURN

1.9L (VIN 7) ENGINE — FLAG CODE DIAGNOSTIC CHARTS — SATURN
ELECTRONIC VARIABLE ORIFICE FAULT
INFORMATION FLAG 16

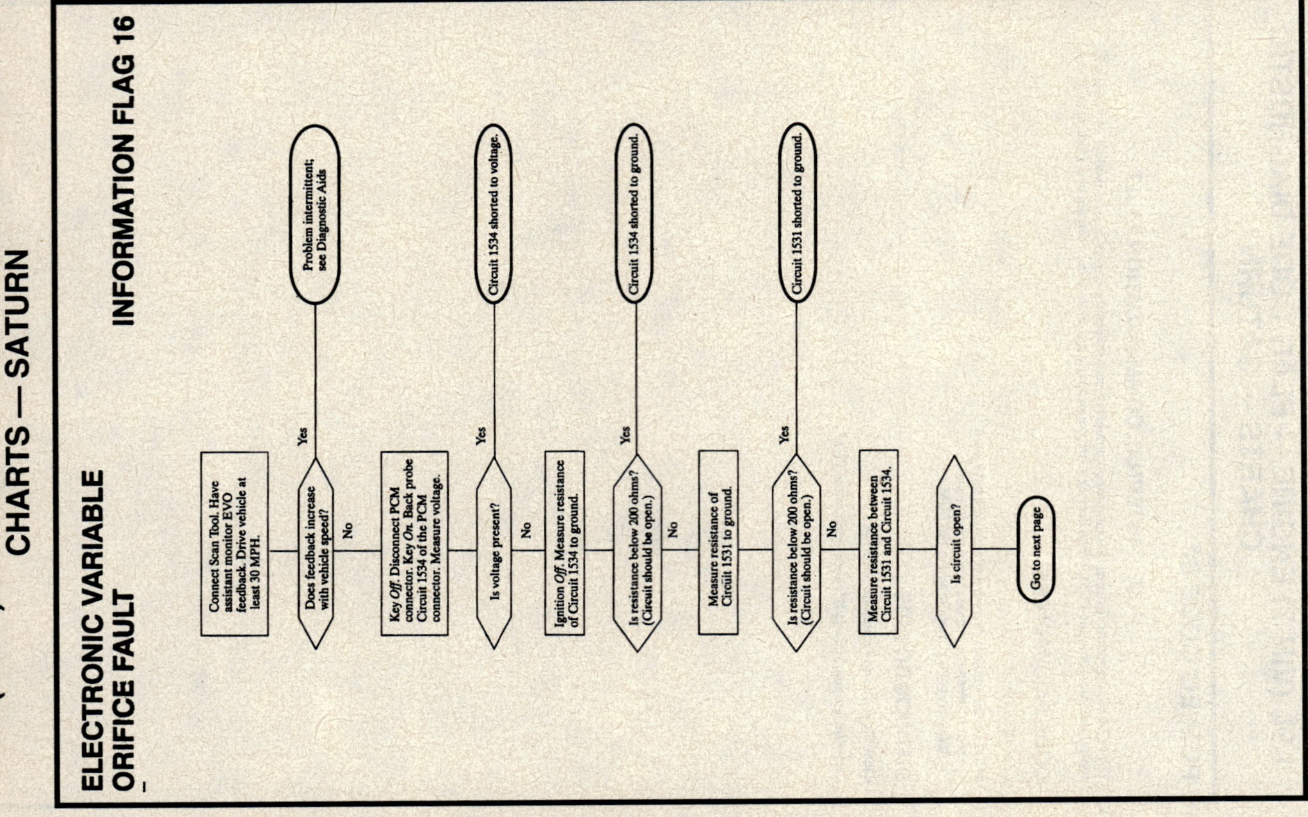

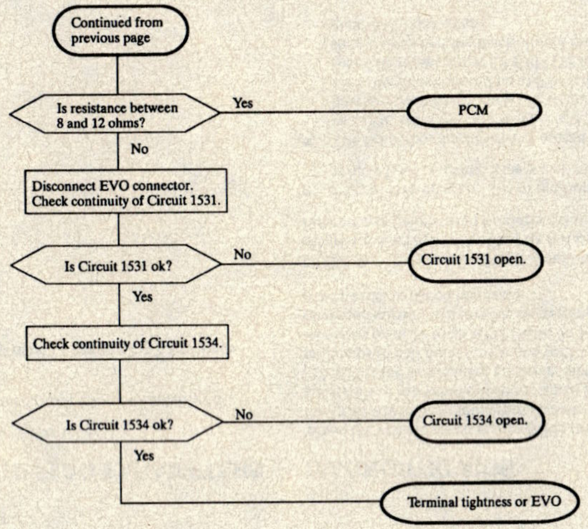

1.9L (VIN 7) ENGINE — FLAG CODE DIAGNOSTIC CHARTS — SATURN

INFORMATION FLAG 27 - EXCEPT 1992

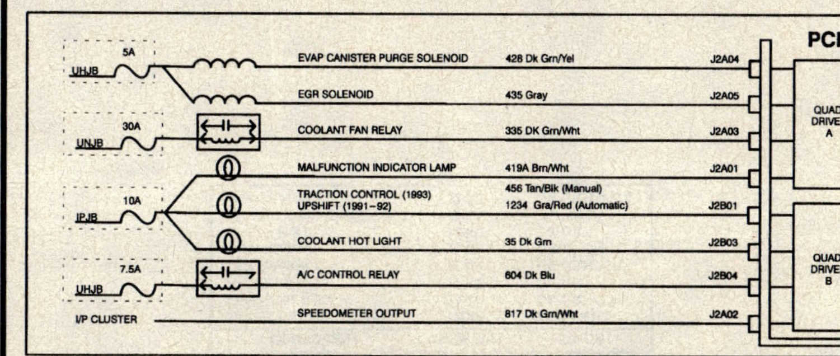

QUICK QUAD DRIVER OUTPUT FAULT

A quad driver module is a semiconductor device capable of controlling four separate outputs. Each output of the QDM is an open collector driver which when turned on pulls the output to ground. A load is connected between 12 volts and the driver. When the output is turned on, current flows from the battery, through the load, through the driver to ground. Each QDM has a fault line feedback. A comparison of the driver input and output states is performed. If the input and output are in the same state, a fault will be detected. Each QDM has only one fault line that will detect a fault on any of the four outputs.

INFORMATION FLAG PARAMETERS

Information Flag 27 will set if:

There is an open or short on any of the QDM output circuits. Fault is detected for 5 seconds.

DIAGNOSTIC AIDS

- Code 27 is used to aid in finding the problem area with an intermittent circuit.
- If any corresponding codes have been set (example Code 32) diagnose that QDM first.
- If only quick QDM 27 is set, this indicates an intermittent problem. Carefully check the tightness of the female terminal grips with the spare male terminal.

Components Resistances:

- EGR Solenoid . . . . 37–44 ohms
- Evaporative Emission Solenoid . . . . . 32 ohms
- Coolant Fan Relay . . . . 80 ohms
- A/C Relay . . . . 80 ohms

  * all resistance are ± 10 ohms

- A shorted or open solenoid, relay or bulb can cause a Code 27.
- If any corresponding codes have been set (example Code 32) diagnose that QDM circuit first.
- When viewing "QDM FAULT" (using a scan tool or Dynamic Display), you may notice the fault appears to be intermittent. This may be due to it being an intermittent fault, but more likely is due to the way the fault detection logic works. For example, if the EGR solenoid is off, the PCM expects to see 12 volts on the sense line. If the circuit is open, 0 volts will bee seen, and the fault will be displayed. However, if the EGR solenoid is turned on (for instance at cruising speeds) the PCM expects to see 0 volts on the sense line, and will not detect a fault. In this example, the display would show "QDM FAULT A" at idle, but if the RPM is raised, the display will show "QDM FAULT NONE". When returned to idle, the fault will again be displayed.
- Use Saturn Service Stall (if available) to diagnose the quad driver circuits.
- Check circuit 817 to Passive Restraint Module for an open or short to ground.

1.9L (VIN 7) ENGINE — FLAG CODE DIAGNOSTIC CHARTS — SATURN

QUICK QUAD DRIVER OUTPUT FAULT - EXCEPT 1992   INFORMATION FLAG 27

1.9L (VIN 7) ENGINE — FLAG CODE DIAGNOSTIC CHARTS — SATURN

QUICK QUAD DRIVER
OUTPUT FAULT - EXCEPT 1992   INFORMATION FLAG 27

COMPONENT	PCM PIN	COLOR
SERVICE ENGINE SOON	J2D01	BRN / WHT
Cooling Fan Relay	J2A03	DK GRN/WHT
A/C Clutch	J2B04	DK BLU
CCP Solenoid	J2A04	DK GRN/YEL
EGR Solenoid	J2A05	GRAY
Coolant Hot Light	J2B03	DK GRN
Upshift/Traction Control (1993)	J2B01	TAN / BLK
Speedo Output	J2A02	DK GRN/WHT

1.9L (VIN 7) ENGINE — FLAG CODE DIAGNOSTIC CHARTS — SATURN

INFORMATION FLAG 48

REFERENCE INPUT — INTERMITTENT OR NOISY

The crankshaft position sensor (CKP) defines engine position by generating an electrical pulse each time it passes by a notch of the crankshaft reluctor wheel. The reference pulse is a 5 volt square wave that switches low two times per crankshaft revolution. The PCM uses reference pulses for fuel delivery and spark timing.

INFORMATION FLAG PARAMETERS

Information Flag 48 will set if:

- Extra or missing reference pulses are detected.
- Time between reference pulses is too long or too short.

DIAGNOSTIC AIDS

When attempting to diagnose an intermittent problem, use the Scan Tool* to review supplemental diagnostic information. The supplemental data can be used to duplicate a problem.

* Select MALF History from Scan Tool EC menu.

Intermittents or opens suspected to be at the connector can be detected by using Diagnostic Service Probe. Voltage can be read on wires without disconnecting any connectors.

Check the tightness of the female terminal grip with a spare male terminal.

Verify that the secondary leads are not in close proximity to the signal wires (EST, REF, etc.).

NOTICE: Information flags should be used for diagnostic purposes only. They do not necessarily indicate component malfunction/failure.

1.9L (VIN 7) ENGINE — FLAG CODE DIAGNOSTIC CHARTS — SATURN

REFERENCE INPUT — INTERMITTENT OR NOISY

INFORMATION FLAG 48

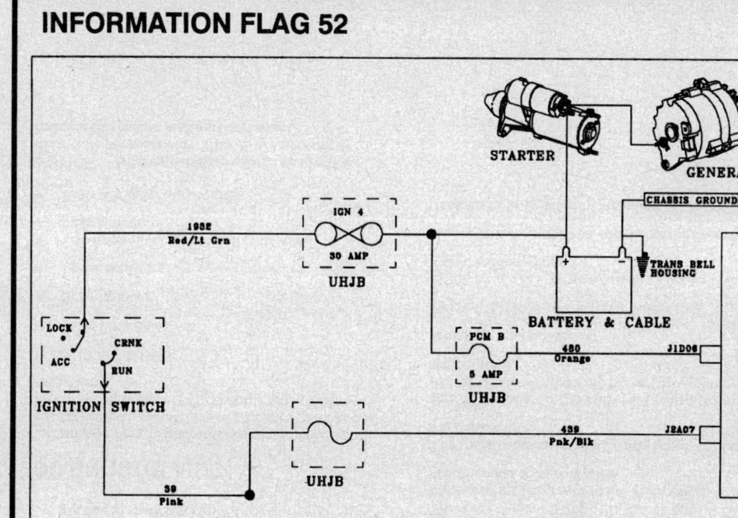

1.9L (VIN 7) ENGINE — FLAG CODE DIAGNOSTIC CHARTS — SATURN

INFORMATION FLAG 52

BATTERY VOLTAGE OUT OF RANGE

The PCM internally monitors ignition voltage on Circuit 439.

INFORMATION FLAG PARAMETERS

Information Flag 52 will set if:

- Battery voltage at the PCM is greater than 17 volts.
- Battery voltage at the PCM is less than 11 volts.

DIAGNOSTIC AIDS

When attempting to diagnose an intermittent problem, use the Scan Tool* to review supplemental diagnostic information. The supplemental data can be used to duplicate a problem.

* Select MALF History from Scan Tool EC menu.

Intermittents or opens suspected to be at the connector can be detected by using Diagnostic Service Probe. Voltage can be read on wires without disconnecting any connectors.

Check the tightness of the female terminal grip with a spare male terminal.

The engine must meet the flag setting conditions in order to diagnose the code.

NOTICE: Information flags should be used for diagnostic purposes only. They do not necessarily indicate component malfunction/failure.

Information Flag 52 can set if all accessories are on for extend idling periods. This does not indicate a problem.

1.9L (VIN 7) ENGINE — FLAG CODE DIAGNOSTIC CHARTS — SATURN

BATTERY VOLTAGE OUT OF RANGE

INFORMATION FLAG 52

- Connect Scan Tool. Key *On*. Read battery voltage.
- Is voltage between 11 and 17 volts? — Yes → Problem intermittent; see Diagnostic Aids.
- No
- Is voltage below 11 volts? — No → Check charging* system for over charge.
- Yes
- Check for:
  - Discharged battery
  - High resistance in Circuit 439.
  - Poor PCM grounds.
  - Charging system problems.

* Flag 52 could be set with battery charger.

1.9L (VIN 7) ENGINE — FLAG CODE DIAGNOSTIC CHARTS — SATURN

INFORMATION FLAG 53

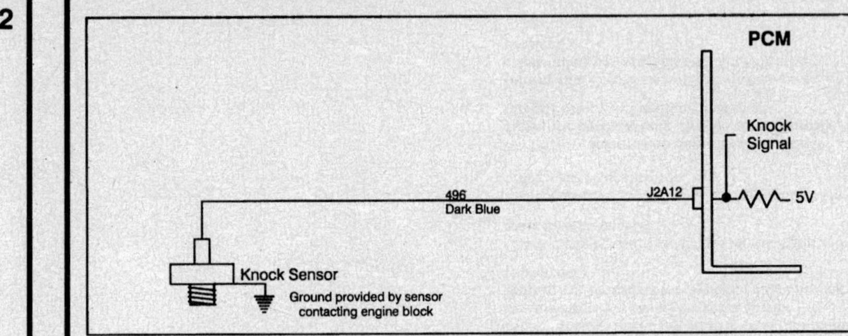

KNOCK SENSOR CIRCUIT — KNOCK PRESENT

The purpose of this flag is to indicate excessive engine noise. The knock sensor cannot differentiate between spark knock and other engine noise. When the knock sensor cannot eliminate noise by retarding timing, Flag 53 will be set.

INFORMATION FLAG PARAMETERS

Information Flag 53 will set if:

- If the PCM cannot reduce engine noise by retarding timing (i.e. the engine noise is most likely not due to spark knock).

DIAGNOSTIC AIDS

Information Flag 53 indicates excessive engine noise usually caused by something else besides spark knock. It does not indicate a faulty knock sensor circuit subsystem.

POSSIBLE CAUSES:

- Valve lifter noise
- Belt tensioner
- Loose bracket
- Low octane fuel in high ambients while pulling loads
- Piston/Cylinder bore scuffed

NOTICE: Information flags should be used for diagnostic purposes only. They do not necessarily indicate component malfunction/failure.

The PCM has the ability to learn spark retard. If knock is present long enough to fill the mid, hi and low spark compensating cells a low power condition may result due to maximum spark retard. The spark compensating cells will eventually learn back (0) when knock condition is corrected, however, this is a slow process. The spark compensating cells may be quickly cleared by disconnecting the power to the PCM when the knock condition has been corrected.

The spark compensating cells may be viewed on the PDT spark table.

In 1993 we adjust the stored spark retard (hi, mid, & low table values) according to start up coolant temperature.

If coolant temperature is:

Below 8°C (46°F) we set table values to 0.

Below 56°C (133°F) we set table values to 75% of stored earned values.

Above 104°C (219°F) we use the stored values.

Temperatures in between are adjusted on a percentage value.

1.9L (VIN 7) ENGINE — FLAG CODE DIAGNOSTIC CHARTS — SATURN

INFORMATION FLAG 54

FIVE VOLT REFERENCE GROUNDED

The five volt reference circuit provides a signal voltage to the handwheel sensor (91 & 92 only), MAP sensor, and TP sensor. The return line is grounded through the PCM. Flag 54 detects a short to ground in any of the three, five volt reference circuits or inside the PCM.

INFORMATION FLAG PARAMETERS

Information Flag 54 will set if:

- MAP sensor signal is zero.
- Handwheel sensor signal is zero (if equipped).
- TP sensor signal is zero.

DIAGNOSTIC AIDS

When attempting to diagnose an intermittent problem, use the Scan Tool* to review supplemental diagnostic information. The supplemental data can be used to duplicate a problem.

* Select MALF History from Scan Tool EC menu.

Intermittents or opens suspected to be at the connector can be detected by using Diagnostic Service Probe. Voltage can be read on both wires without disconnecting any connectors.

Check the tightness of the female terminal grip with a spare male terminal.

The engine must meet the flag setting conditions in order to diagnose the code.

NOTICE: Information flags should be used for diagnostic purposes only. They do not necessarily indicate component malfunction/failure.

IMPORTANT: 1992 Vehicles without handwheel sensors are still equipped with the wiring for the sensor. Except for the base model ("ZF" 4th and 5th digit of VIN).

1.9L (VIN 7) ENGINE — FLAG CODE DIAGNOSTIC CHARTS — SATURN

FIVE VOLT REFERENCE GROUND

INFORMATION FLAG 54

1.9L (VIN 7) ENGINE — FLAG CODE DIAGNOSTIC CHARTS — SATURN

INFORMATION FLAG 58

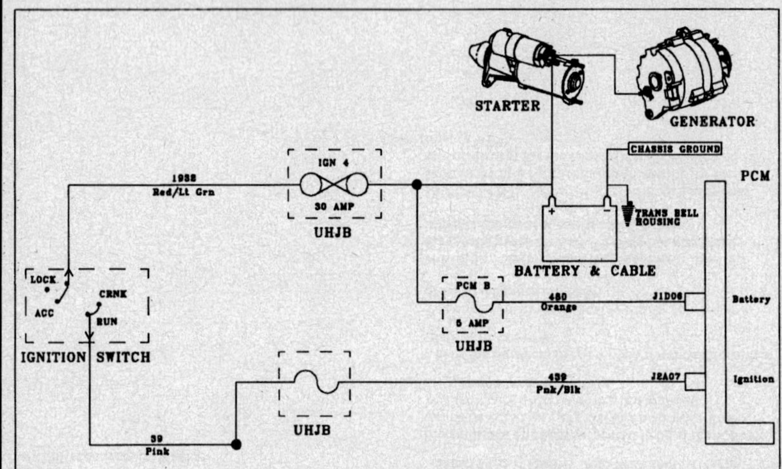

BATTERY VOLTAGE UNSTABLE

The PCM monitors ignition voltage on Ckt. 439 when powered up. The purpose of flag 58 is to detect an unstable ignition voltage to the PCM/EC.

INFORMATION FLAG PARAMETERS

Information Flag 54 will set if:

- Battery voltage changes more than three volts instantaneously.

DIAGNOSTIC AIDS

When attempting to diagnose an intermittent problem, use the Scan Tool* to review supplemental diagnostic information. The supplemental data can be used to duplicate a problem.

* Select MALF History from Scan Tool EC menu.

Wiggle the PCM connector. Watch Scan Tool to see if battery voltage reading changes sharply.

Check PCM connector. Check the tightness of the female terminal grip with a spare male terminal.

Wiggle and tug the harness at the ignition switch. Watch Scan Tool to see if battery voltage changes sharply. Lightly tap the PCM. Watch for changes in the ECT reading

NOTICE: Information flags should be used for diagnostic purposes only. They do not necessarily indicate component malfunction/failure.

1.9L (VIN 7) ENGINE — FLAG CODE DIAGNOSTIC CHARTS — SATURN

INFORMATION FLAG 63

OPTION CHECK SUM ERROR

Each time the PCM is powered up, a comparison of tire size and options are performed. Tire size and options must compare to valid combination of options stored in electronically erasable programmable read only memory (EEPROM).

CODE PARAMETERS

Information Flag 63 will set if:

- Invalid combinations of tire size and/or options.

DIAGNOSTIC AIDS

If Flag 63 is set, recalibrate PCM. If Flag 63 reoccurs, replace PCM.

1.9L (VIN 7) ENGINE — FLAG CODE DIAGNOSTIC CHARTS — SATURN

INFORMATION FLAG 67 – 1992

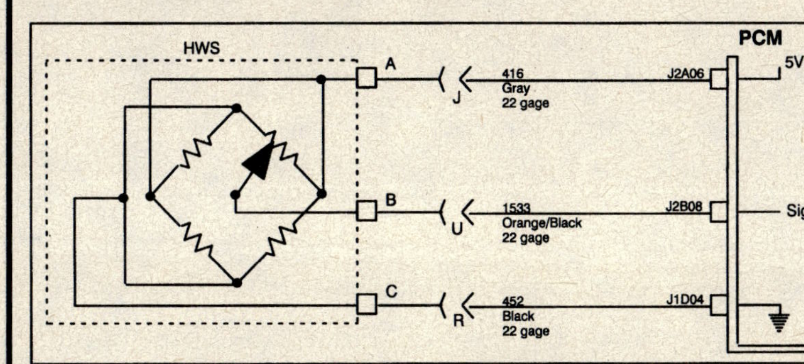

HANDWHEEL SENSOR CIRCUIT FAULT

The handwheel sensor (HWS) is used to determine the rate at which the steering wheel is being turned. The PCM calculates the rate of change in the sensor signal to determine the control cycle of the steering orifice solenoid. The duty cycle applied to the electronic variable orifice steering (EVO) solenoid controls the amount of power assist applied to the steering gear.

INFORMATION FLAG PARAMETERS

Information Flag 67 will set if:

- Vehicle speed is 8–48 KPH (5–30 MPH).
- Handwheel reads greater than 4.8 volts.
- Handwheel reads less than 0.2 volts.

DIAGNOSTIC AIDS

If Information Flag 54 is set with Flag 67, diagnose Flag 54 first.

When attempting to diagnose an intermittent problem, use the Scan Tool* to review supplemental diagnostic information. The supplemental data can be used to duplicate a problem.

* Select MALF History from Scan Tool EC menu.

Intermittents or opens suspected to be at the connector can be detected by using Diagnostic Service Probe.

Voltage can be read on both wires without disconnecting any connectors.

Check the tightness of the female terminal grip with a spare male terminal.

To disconnect HWS:

4. Locate HWS on steering column (under dash, approximately five inches from bulkhead).
5. Remove HWS from support bracket by pressing both locking tabs.
6. Rotate HWS 180 degrees so the connector is on bottom.
7. Push connector locking tab and disconnect harness.

NOTICE: Information flags should be used for diagnostic purposes only. They do not necessarily indicate component malfunction/failure.

- Flag 67 may set if the drive wheels are operated while vehicle is on a hoist.

1.9L (VIN 7) ENGINE — FLAG CODE DIAGNOSTIC CHARTS — SATURN

HANDWHEEL SENSOR CIRCUIT FAULT

INFORMATION FLAG 67

1.9L (VIN 7) ENGINE — FLAG CODE DIAGNOSTIC CHARTS — SATURN

HANDWHEEL SENSOR CIRCUIT FAULT

INFORMATION FLAG 67

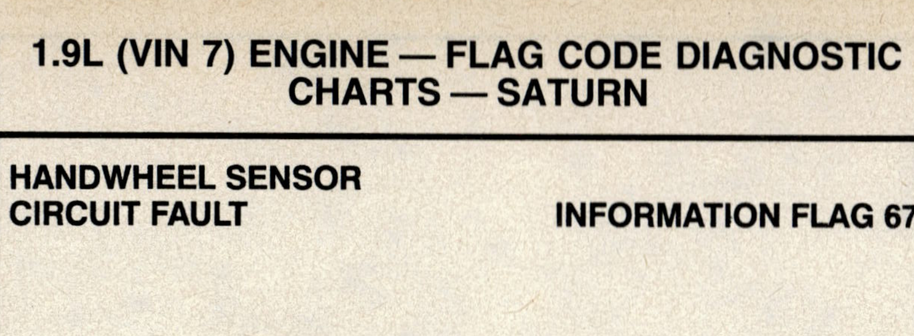

1.9L (VIN 7) ENGINE — FLAG CODE DIAGNOSTIC CHARTS — SATURN

HANDWHEEL SENSOR— LOW VOLTAGE

INFORMATION FLAG 67

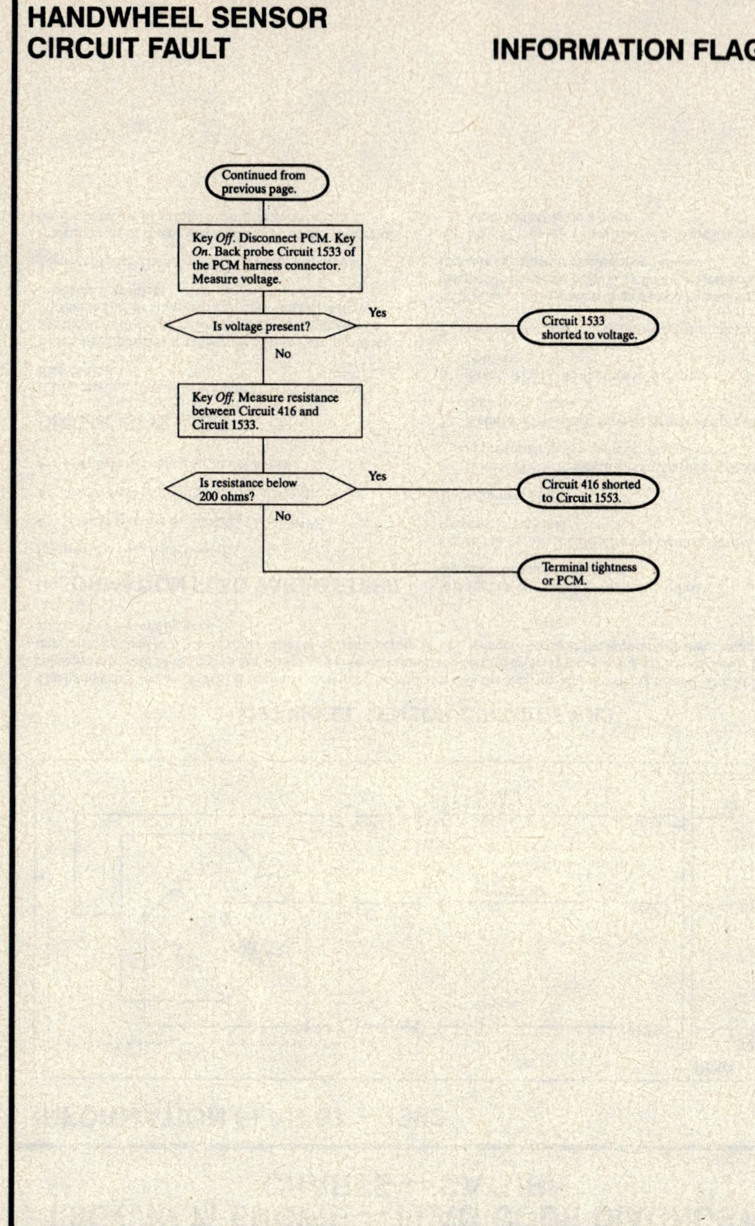

1.9L (VIN 7) ENGINE — FLAG CODE DIAGNOSTIC CHARTS — SATURN

INFORMATION FLAG 71

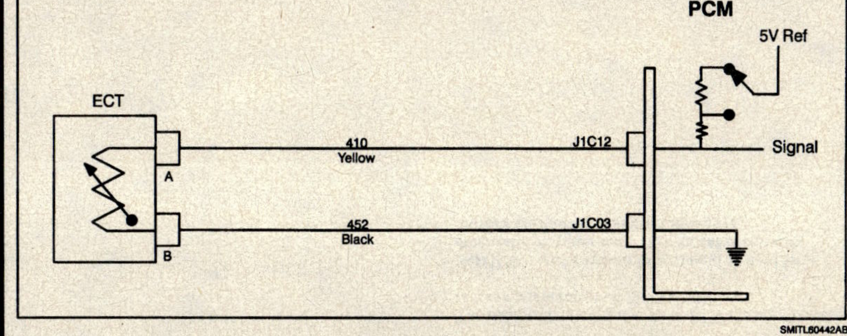

ENGINE COOLING SYSTEM — TEMPERATURE HIGH

The engine coolant temperature (ECT)sensor is a thermistor located in a coolant passage on the engine. When coolant temperature is cold the sensor has a high resistance, as temperature increases the resistance of the sensor decrease. The PCM provides a five volt signal to the coolant sensor, which is also connect to PCM ground. The PCM reads the voltage drop on the signal line to determine coolant temperature.

INFORMATION FLAG PARAMETERS

Information Flag 71 will set if:

- Engine coolant is above 118°C (244°F).

DIAGNOSTIC AIDS

Information Flag 71 indicates higher than normal coolant temperatures. It does not indicate a faulty ECT subsystem.

Check engine cooling system for proper operation. Check for correct operation of fan and coolant level.

Possible causes of high coolant temperature:

- Low coolant
- Thermostat stuck closed
- Reservoir cap damage
- Hoses collapsing
- Water pump damaged
- Plugged radiator

NOTICE: Information flags should be used for diagnostic purposes only. They do not necessarily indicate component malfunction/failure.

NOTICE: The A/C will shut off at 118°C (244°F) to protect the engine.

NOTICE: Trailer grade loads may cause this temperature without system problems.

NOTICE: Extended high ambient idle conditions may set Flag 71 without a system problem.

1.9L (VIN 7) ENGINE — FLAG CODE DIAGNOSTIC CHARTS — SATURN

INFORMATION FLAG 72

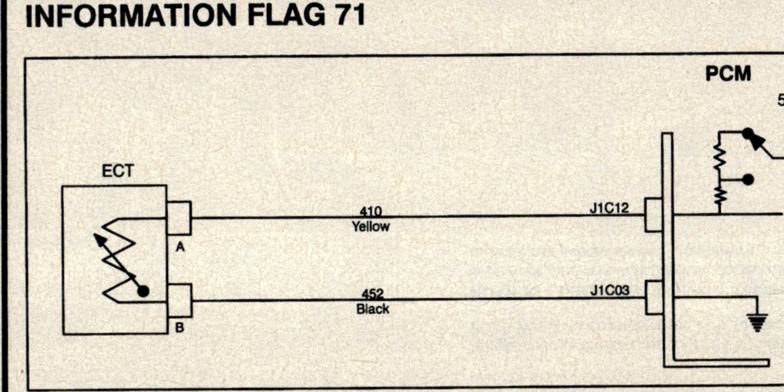

ENGINE COOLING SYSTEM — TEMPERATURE LOW

The engine coolant temperature (ECT)sensor is a thermistor located in a coolant passage on the engine. When coolant temperature is cold the sensor has a high resistance, as temperature increases the resistance of the sensor decrease. The PCM provides a five volt signal to the coolant sensor, which is also connect to PCM ground. The PCM reads the voltage drop on the signal line to determine coolant temperature.

INFORMATION FLAG PARAMETERS

Information Flag 72 will set if:

- The coolant temperature is below 0°C (32°F) after the engine has been running for more than five minutes.

DIAGNOSTIC AIDS

Information Flag 72 indicates lower than normal coolant temperatures and does not indicate a faulty ECT subsystem.

Possible causes of low coolant temperature

- Thermostat stuck open
- Cooling fan always on

NOTICE: Information flags should be used for diagnostic purposes only. They do not necessarily indicate component malfunction/failure.

1.9L (VIN 7) ENGINE — FLAG CODE DIAGNOSTIC CHARTS — SATURN

INFORMATION FLAG 73

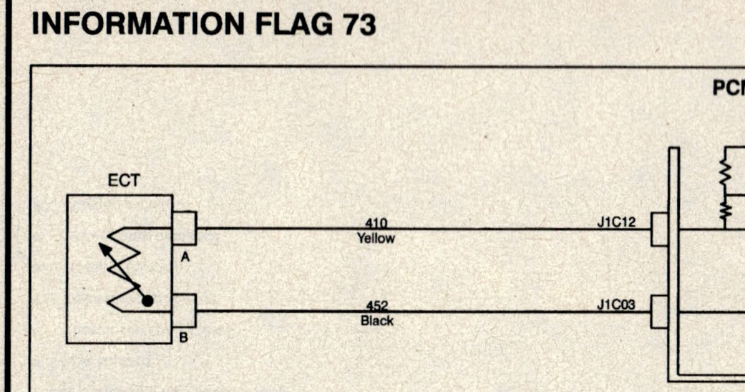

ECT UNSTABLE

The engine coolant temperature (ECT) sensor is a thermistor located in a coolant passage on the engine. When coolant temperature is cold the sensor has a high resistance, as temperature increase the resistance of the sensor decreases. The PCM provides a five volt signal to the coolant sensor, which is also connected to PCM ground. The PCM reads the voltage drop on the signal line to determine coolant temperature.

INFORMATION FLAG PARAMETERS

Information Flag 73 will set if:

- ECT reading changes more than 15°C (59°F) in 100 milli seconds.

DIAGNOSTIC AIDS

When attempting to diagnose an intermittent problem, use the Scan Tool* to review supplemental diagnostic information. The supplemental data can be used to duplicate a problem.

* Select MALF History from Scan Tool EC menu.

Wiggle the ECT connector. Watch Scan Tool to see if ECT reading changes falls sharply.

Check ECT connector. Check the tightness of the female terminal grip with a spare male terminal.

Wiggle and tug the harness. Watch Scan Tool to see if reading change sharply.

Wiggle and tug the harness at the PCM. Lightly tap the PCM. Watch for changes in the ECT reading.

NOTICE: Information flags should be used for diagnostic purposes only. They do not necessarily indicate component malfunction/failure.

1.9L (VIN 7) ENGINE — FLAG CODE DIAGNOSTIC CHARTS — SATURN

INFORMATION FLAG 74

ECT/TRANS TEMP SENSOR RATIO ERROR

The engine coolant temperature (ECT) sensor is a thermistor located in a coolant passage on the engine. When coolant temperature is cold the sensor has a high resistance, as temperature increase the resistance of the sensor decreases. The PCM provides a five volt signal to the coolant sensor, which is also connected to PCM ground. The PCM reads the voltage drop on the signal line to determine coolant temperature.

INFORMATION FLAG PARAMETERS

Information Flag 74 will set if:

- The transmission temperature sensor is functioning properly.
- The ECT reading is less than Transaxle Temperature Sensor (TTS) reading.

DIAGNOSTIC AIDS

Information Flag 74 is used to indicate a degrading ECT.

Remove the ECT and inspect for contamination.

Measure resistance through the ECT by attaching female terminals to the two ECT pins.

The resistance reading should be 2786 ohms ± 40 ohms at room temperature of 25°C (77°F).

NOTICE: Information flags should be used for diagnostic purposes only. They do not necessarily indicate component malfunction/failure.

1.9L (VIN 7) ENGINE — FLAG CODE DIAGNOSTIC CHARTS — SATURN

INFORMATION FLAG 75

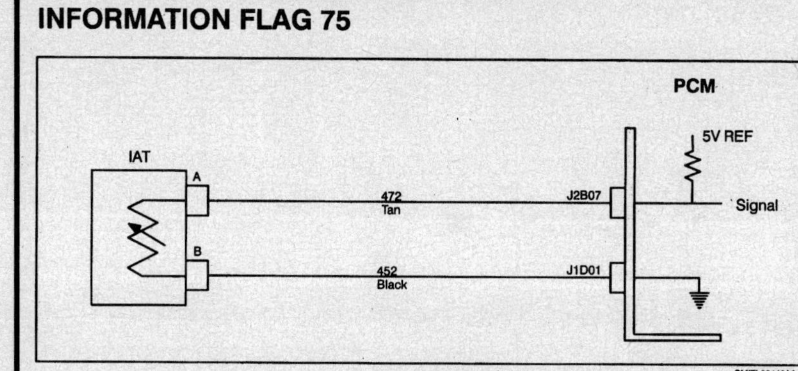

IAT UNSTABLE

The inlet air temperature (IAT) sensor is a thermistor located in the air intake passage of the air induction system on the engine. When the sensor is cold it has a high resistance, the resistance decreases as the temperature of the sensor increases. The PCM supplies a five volt signal to the sensor which is also connected to ground. The PCM reads the voltage drop on the sensor signal line to determine air temperature.

INFORMATION FLAG PARAMETERS

Information Flag 75 will set if:

- IAT reading changes more than 15°C (59°F) in 100 m seconds.

DIAGNOSTIC AIDS

When attempting to diagnose an intermittent problem, use the Scan Tool* to review supplemental diagnostic information. The supplemental data can be used to duplicate a problem.

* Select MALF History from Scan Tool EC menu.

Check the tightness of the female terminal grip with a spare male terminal.

Wiggle the IAT connector. Watch Scan Tool to see if IAT reading changes falls sharply.

Check IAT connector. Check for shorted terminals.

Wiggle and tug the harness. Watch Scan Tool to see if IAT reading rises or drops sharply.

Wiggle and tug the harness at the PCM. Lightly tap the PCM. Watch for changes in the IAT reading.

Check the PCM connector. Check for shorted terminals.

NOTICE: Information flags should be used for diagnostic purposes only. They do not necessarily indicate component malfunction/failure.

1.9L (VIN 7) ENGINE — FLAG CODE DIAGNOSTIC CHARTS — SATURN

INFORMATION FLAG 76

TP SENSOR VOLTAGE VS MAP VOLTAGE — OUT OF RANGE

The PCM has an internal table that looks at MAP sensor values with respect to TP sensor values. When their relationship to each other falls outside normal operating ranges (example: high vacuum during wide open throttle) Flag 76 will be set.

INFORMATION FLAG PARAMETERS

Information Flag 76 will set if:

- The TP sensor vs MAP sensor relationship is not consistent with normal engine operation.

DIAGNOSTIC AIDS

Information Flag 76 indicates that either the TP sensor or MAP sensor is skewed outside its normal operating range.

Check TP sensor and MAP sensor connectors for corrosion.

Check PCM connector for corrosion.

Move throttle and check TP sensor voltage readings on Scan Tool.

At a closed throttle position, the output of the TP sensor is low (approximately 0.4 volts). As the throttle plate opens, the output increases so that, at wide−open throttle, the output voltage should be approximately 4.9 volts.

When attempting to diagnose an intermittent problem, use the Scan Tool* to review supplemental diagnostic information. The supplemental data can be used to duplicate a problem.

* Select MALF History from Scan Tool EC menu.

Check the tightness of the female terminal grip with a spare male terminal.

NOTICE: Information flags should be used for diagnostic purposes only. They do not necessarily indicate component malfunction/failure.

1.9L (VIN 7) ENGINE — FLAG CODE DIAGNOSTIC CHARTS — SATURN

INFORMATION FLAG 83 – 1993 ONLY

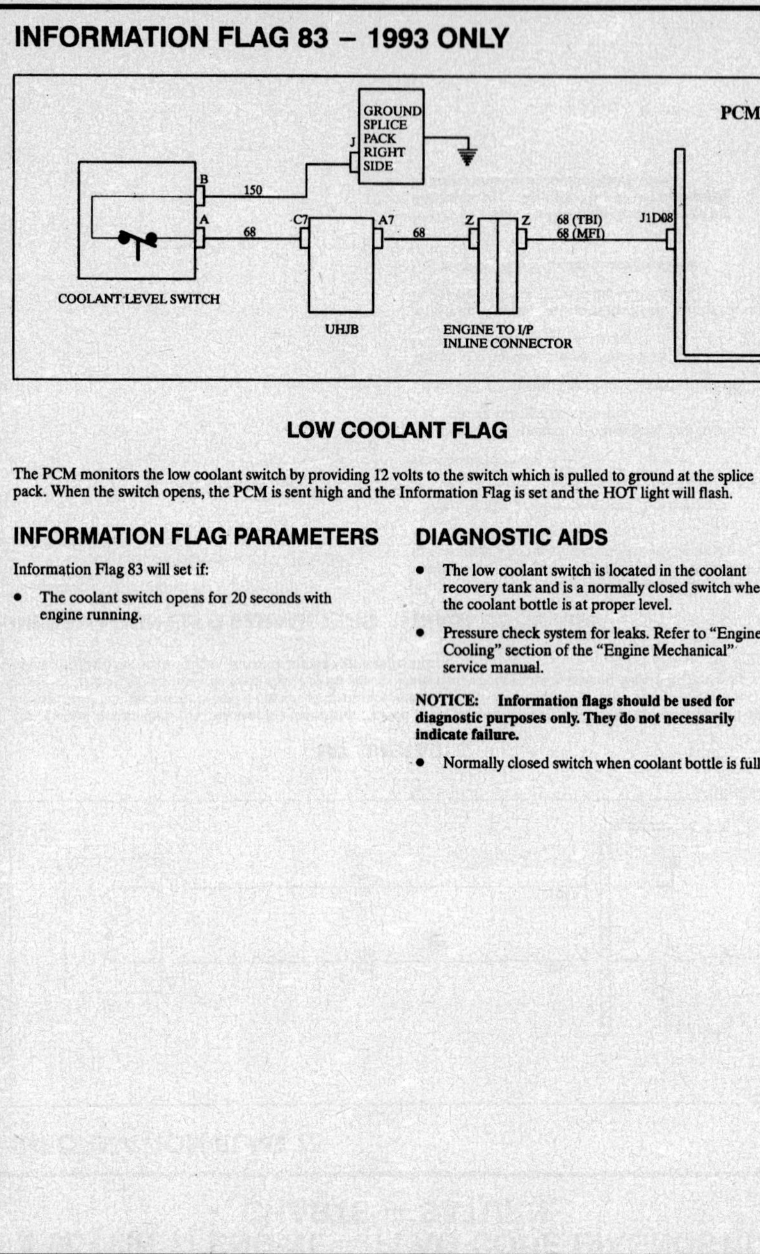

LOW COOLANT FLAG

The PCM monitors the low coolant switch by providing 12 volts to the switch which is pulled to ground at the splice pack. When the switch opens, the PCM is sent high and the Information Flag is set and the HOT light will flash.

INFORMATION FLAG PARAMETERS

Information Flag 83 will set if:

- The coolant switch opens for 20 seconds with engine running.

DIAGNOSTIC AIDS

- The low coolant switch is located in the coolant recovery tank and is a normally closed switch when the coolant bottle is at proper level.

- Pressure check system for leaks. Refer to "Engine Cooling" section of the "Engine Mechanical" service manual.

NOTICE: Information flags should be used for diagnostic purposes only. They do not necessarily indicate failure.

- Normally closed switch when coolant bottle is full.

1.9L (VIN 7) ENGINE — FLAG CODE DIAGNOSTIC CHARTS — SATURN

LOW COOLANT FLAG – 1993 ONLY

INFORMATION FLAG 83

1.9L (VIN 7) ENGINE — FLAG CODE DIAGNOSTIC CHARTS — SATURN

INFORMATION FLAG 84

PCM – INTERNAL COMMUNICATION FAILURE

The PCM/EC communicates to the PCM/TC by an internal serial peripheral interface (SPI) line. When SPI communications fail, the Information Flag 84 is set and the PCM/EC will continue communicating with the PCM/TC by way of the serial communication interface (SCI) line, with no performance impacts.

INFORMATION FLAG PARAMETERS

Information Flag 84 will set if:

- A certain number of incorrect or missing messages have been sent between the engine controller and the transaxle controller within the PCM.

DIAGNOSTIC AIDS

If Information Flag 84 is active and you plug in the PDT or Scan tool, a code 82 will set. This is due to the PDT or Scan tool also communicating over the SCI line which interrupts the PCM/EC to TC communication. This condition will cause the PCM/TC to go into back–up.

- Clear all codes and information flags.

- Test drive vehicle. If Information Flag 84 resets, replace PCM.

- If Information Flag 84 does not reset, do not replace the PCM.

When a code is active it means the conditions for the code are currently being met, and if it is cleared, it will reset immediately.

1.9L (VIN 7) ENGINE — SCAN TOOL INFORMATION — SATURN

SATURN SCAN TOOL PARAMETER INFORMATION

PARAMETER	DESCRIPTION
T–CODES/FLAG:	Displays a "YES" when codes or flags are present in the PCM.
ENGINE RPM:	The RPM signal is generated from the Crankshaft Position (CKP) sensor, and provided to the PCM by the EI module thru the reference signal circuit. RPM is active during crank but the PDT update rate must be set fast enough to pick it up.
DESIRED IDLE RPM:	The target RPM that the PCM is commanding the engine to. Desired idle RPM is based on coolant temperature, P/N, A/C– on/off, and low battery voltage inputs.
COOLANT TEMP.:	Displays coolant temperature in degrees, input received from the Engine Coolant Temperature (ECT) sensor. Normal operating temperature is 88 Degrees C to 106 Degrees C (176 Degrees to 212 Degrees F). Thermostat opens at 88 degrees C and the fan turns on at 106 degrees C.
START–UP COOLANT:	Displays engine coolant temperature as seen by the PCM when ignition was cycle on and stored until engine is turned off.
O$_2$ SENSOR: (OXYGEN SENSOR)	Displays oxygen sensor signal voltage in millivolts (mv), The PCM puts out a signal voltage of 450 mv (approx.), this bias voltage will be shown when sensor is cool. A lean exhaust will cause oxygen sensor to reduce signal voltage toward 050 mv, a rich exhaust will cause oxygen sensor to raise signal voltage toward 950 mv.
ATS:	Displays underhood temperature from the Inlet Air Temperature sensor (IAT) located in the engine air inlet duct. The PCM uses IAT temperature to taylor spark timing at idle for cold or hot weather, fuel calculation and EGR operation.
TPS VOLTS: (TP SENSOR VOLTS)	Displays throttle position as reported by Throttle Position (TP) Sensor in volts. With .40 to .55 volts being closed throttle and 4.80 to 4.90 being WOT. Throttle position is used for fuel control, and some specific idle to WOT spark controls.
TPS %: (TP Sensor %)	Displays of throttle position in percent, percent converted from TP sensor voltage by PCM (0% equals closed throttle and 100% equals wide open throttle).
TPS COUNTS: (TP SENSOR COUNTS)	At idle TP sensor counts should not vary more than 4 counts on DOHC and no more than 3 counts on SOHC.
MAP SENSOR:	Displays manifold absolute pressure which is used as a measure of engine load reported in volts. The readings will vary slightly due to altitude changes. At idle the readings should be between .8 to 1.5 volts and increase uniformly to 4.5 to 5.0 volts at wide open throttle. (WOT).
BARO PRESSURE:	Displays what BARO was at start–up and is used primarily for EGR– Fuel compensation corrections. BARO is updated after start–up based on RPM and TPS values close to WOT. At key ON, engine OFF, BARO and MAP should be the same value around 5.0 volts.
HWS:	Displays the Handwheel Sensor (HWS) reported voltage, will vary as steering wheel position is changed. The PCM uses this parameter to taylor EVO operation on HWS equipped vehicles (early 1992 coupes).
SPEED KPH (MPH):	Displays vehicle speed as calculated by the PCM. The Vehicle Speed Sensor (VSS) located in transaxle provides the raw speed signal to PCM.
INTEGRATOR: (SHORT TERM FUEL TRIM)	Displays in counts the temporary changes being made to the fuel delivery. Ideal is 128, with an operating range from typically 64 to 196. As short term fuel trim moves from 128 to 40 fuel is being subtracted (oxygen sensor indicating a rich mixture) as short term fuel trim moves from 128 to 240 fuel is being added (oxygen sensor indicated a lean mixture). The short term fuel trim value is used by long term fuel trim to keep fuel delivery within limits.

1.9L (VIN 7) ENGINE — SCAN TOOL INFORMATION — SATURN

Parameter	Description
BLOCK LEARN: (LONG TERM FUEL TRIM)	Displays in counts the average short term fuel trim learned variations to fuel delivery over a wide range of driving conditions. Changes to fuel control is represented by a value from 108 to 148, with ideal being 128. Long term fuel trim is used to maintain an ideal air/fuel mixture of 14.7 to 1. Without long term fuel trim, this mixture would be effected by build variations in the vehicle. High long term fuel trim counts indicate the PCM is adding fuel to compensate for a lean exhaust. Low long term fuel trim indicates the PCM is reducing fuel to compensate for a rich condition.
INJECTOR BPW:	Displays the PCM commanded on time of the injectors in milliseconds. The injector on time varies according to inputs from ECT, MAP, TP (TPS), long term fuel trim, and IAT.
ENGINE ON TIME:	Displays the actual time the engine has been running per ignition cycle, resets to 0 time when ignition switch is turned switch off.
BATTERY VOLTS:	Displays battery voltage at the PCM, provided thru the ignition switch.
HIGH BATTERY:	Displays "ON" when the battery voltage at the PCM exceeds 17 volts.
LOW BATTERY:	Displays "ON" if the PCM detects a voltage below 11 volts on the ignition circuit.
SPARK ADVANCE DEGREES:	Displays in degrees the commanded spark advance by the PCM. There are two modes of spark control, bypass mode where spark is controlled by the EI module (Back–up timing) and EST mode where the PCM controls spark timing. The vehicle is started in bypass mode.
KNOCK RETARD DEGREES:	Displays in degrees the PCM commanded spark retard from normal spark advance. The PCM retards timing as a result of knock detected from the knock sensor. There are two modes of spark retard, low RPM throttle tip– in, and steady state driving conditions. Knock windowing is used, which looks only at the knock signal during the firing event when actual spark knock occurs. This helps to discriminate knock from other engine noises.
EST ENABLE:	Displays "ON" when the PCM is controlling spark timing. To enter EST mode, PCM drives the bypass signal circuit high and controls the timing from the 2X signal generated by the ignition module and sent to the PCM.
LEARN CONTROL:	Displays "ON" when PCM allows block learn cells to be updated. Learn control is on after closed loop and engine temperature have been met.
A/C REQUEST:	Displays "YES" when the PCM has received input from the A/C control panel and IAT is above 10 degrees C and ECT is less than 118 degrees C. The request is only received when Fan switch, A/C switch, low pressure switch and high pressure switch all are closed.
A/C CLUTCH:	Displays "ON" when the PCM commands the A/C relay on, providing battery positive to A/C clutch. The A/C will be temporarily turned off during a tip–in launch, wide open throttle, some trans. shifts, and when power steering is cramped or RPM drops below 600.
A/C CTS DELAY:	Displays "YES" if A/C is no longer being keep off by the PCM due to coolant temperature timer having expired after engine startup.
CCP STATUS:	Displays "ON" when the PCM commands the EVAP canister purge solenoid off and purge is allowed. The decision to purge the vapors depends on ECT (CTS), VSS and TP (TPS). EVAP canister purge solenoid fails closed.
COOLANT FAN REQ:	Displays "ON" when the PCM commands the coolant fan relay on. The fan is turned on when engine cooling temperature reaches approximately 106 degrees C (215 Degrees F) or the A/C is on. Above 70 MPH the fan is turned off.
EGR SOLENOID:	Displays "ON" when the PCM commands the EGR solenoid off (vacuum passage open), allowing the EGR valve to open. EGR is enabled depending on MAP, TP (TPS), ECT (CTS), IAT, VSS and run time. EGR solenoid fails closed.
EVO STATUS:	Displays "ON" when the PCM is monitoring power steering effort during Electronic Variable Orifice (EVO) operation.
EVO OUTPUT DC :	Displays the duty cycle (100–0) of the EVO commanded by the PCM. At idle When full assist is needed 100% duty cycle is commanded; full power steering assist.
EVO FEEDBACK DC :	Displays the duty cycle reported back to the PCM. Current is used to monitor and control EVO. The PCM converts the current reading to a duty cycle (0 at idle).

1.9L (VIN 7) ENGINE — SCAN TOOL INFORMATION — SATURN

Parameter	Description
PRNDL SWITCH:	Displays "P–N" or "R–D–L" status as reported by the PCM/TC to the PCM/EC.
ESC ACTIVE:	Displays "YES" when the PCM is receiving a spark knock active signal from the knock sensor.
ESC SEVERE KNOCK:	Displays "ON" when the PCM receive a spark knock signal which indicates spark knock is over active, indicating a possible false knock.
ASYNC FLAG:	Displays "ON" when the PCM enters asynchronous fuel delivery mode. Asynchronous fuel delivery occurs when the BPW is too small to deliver synchronously (decel condition) or the BPW is not providing enough fuel (acceleration). In ASYNC mode fuel is delivered every 12.5 millisecond in 1992 in 1993 6.25 ms.
DECEL MODE:	Displays "ACTIVE" when the PCM is commanding only a small amount of fuel to be delivery during deceleration.
EXHAUST GASES:	Displays "RICH" or "LEAN" status of exhaust, as translated by the PCM. Exhaust status is received from oxygen sensor signal.
LOOP STATUS:	Displays "OPEN" or "CLOSED" loop status. Open loop indicates the system is running in a fixed fuel calibration. Closed loop indicates that the PCM is using the oxygen sensor signal to control fuel delivery. Closed loop will not be entered after start–up until TP Sensor is seen.
DECEL FUEL OFF:	Displays "YES" when PCM has turned off fuel during deceleration.
FUEL CUTOFF P–N:	Displays "ON" wher the PCM stops delivering fuel to prevent powertrain damage (approx 4000 RPM) at high engine RPM's in park or neutral.
FUEL CUTOFF DRIVE:	Displays "ON" when the PCM stops delivering fuel to prevent tire damage (All tires are rated for a specific MPH) at high vehicle speeds.
FUEL CUTOFF REVERSE:	Displays "ON" when the PCM stops delivering fuel to prevent transaxle damage at high vehicle speeds in reverse.
SES LIGHT:	Displays "ON" when the PCM has commanded the malfunction indicator lamp on, light should be illuminated.
HOT LIGHT:	Displays "ON" when the PCM has commanded the hot light on, light should be illuminated. In 1993 the hot light has 3 functions. It will come on solid when engine temperature is above 121°C (250°F) or when transaxle temperature is above 140°C (248°F) degrees C. The Hot light will also flash when coolant level is below normal.
SHIFT LIGHT:	Displays "ON" when the PCM has commanded the shift light on, light should be illuminated. The indicates that the transaxle should be shifted to the next highest gear to obtain maximum fuel economy.
QDM FAULT:	Displays status of Quad Driver Modules (QDM) faults. There are two QDM'S in the PCM which run 4 outputs each and are displayed as QDM A and QDM B. "A" indicates there is a QDM fault on one of A's outputs. "B" indicates there is a QDM fault on one of B's outputs. "BOTH" indicates a failure on both QDM's. "NONE" indicates there is presently no QDM failures.
LOW COOLANT:	Displays "YES" when the PCM has commanded the hot light on due to low coolant switch, light should be flashing (93 vehicles only).
TCM REQ HOT LIGHT:	Displays "YES" when the PCM has commanded the hot light on due to transaxle over temperature, light should be illuminated (except '92).
FAN RUN ON:	Displays "ON" when the PCM has commanded the coolant fan to continue running, coolant fan should be running.
TCC IS LOCKED:	Displays "YES" when the PCM/TC has commanded the TCC to be locked up.
STALL SAVER A/C:	DISPLAYS "ON" when the PCM has commanded the A/C clutch off to prevent engine from stalling.
FUEL PUMP:	Displays PCM's ability to turn the fuel pump on, not actually reporting that the fuel pump is running.

1.9L (VIN 7) ENGINE — SCAN TOOL INFORMATION — SATURN

O2 CROSSCOUNTS: (OXYGEN SENSOR CROSSCOUNTS)	Displays number of times oxygen sensor voltage has crosses a PCM voltage check point (approx. 440 mV).
O2 SENSOR READY: (OXYGEN SENSOR CROSSCOUNTS)	Displays "YES" when the PCM has seen oxygen sensor voltage fluctuating high and low.
BLM IDLE CEL 0: (LONG TERM FUEL TRIM ACCE: CEL 0))	Displays in counts idle fuel adjustments, learned changes to fuel control. Ideal is 128 with an operating range from 108 to 148. Long term fuel trim cells are used to compensate for variation due to changes in the engine, driving conditions. High block learn counts indicate the PCM is adding fuel to compensating for a lean exhaust. Low long term fuel trim indicates the PCM is reducing fuel to compensate for a rich condition.
BLM ACCEL CEL 1: (LONG TERM FUEL TRIM ACCE: CEL 1))	Displays in counts acceleration fuel adjustments, learned changes to fuel control. Ideal is 128 with an operating range from 108 to 148 (except '92).
BLM CRUISE CEL 1: (LONG TERM FUEL TRIM CRUISE CEL 1)	Displays in counts cruise fuel adjustments, learned changes to fuel control. Ideal is 128 with an operating range from 108 to 148.
BLM DECEL CEL 2: (LONG TERM FUEL TRIM DECEL CEL 2)	Displays in counts deceleration fuel adjustments, learned changes to fuel control. Ideal is 128 with an operating range from 108 to 148.
BLM ACCEL CEL 3: (LONG TERM FUEL TRIM ACCEL CEL 3)	Displays in counts acceleration fuel adjustments, learned changes to fuel control. Ideal is 128 with an operating range from 108 to 148.
AIR FUEL RATIO:	Displays the number of parts of air to 1 part of fuel as calculate by the PCM.
CRANK FUEL:	Displays "ON" when PCM is commanding fuel pump to run during engine cranking.
LOW SPARK COMP:	Low RPM adaptive spark retard value, used to control engine spark knock. In 1992 values are learned out during low RPM driving and no spark knock present. In 1993 depending on coolant temperature we will clear adaptives (cold) or use only a percentage of the current adaptives.
MID SPARK COMP:	Mid RPM adaptive spark retard value, used to control engine spark knock. In 1992 values are learned out during mid RPM driving and no spark knock present. In 1993 depending on coolant temperature we will clear adaptives (cold) or use only a percentage of the current adaptives.
HIGH SPARK COMP:	High RPM adaptive spark retard value, used to control engine spark knock. In 1992 values are learned out during high RPM driving and no spark knock present. In 1993 depending on coolant temperature we will clear adaptives (cold) or use only a percentage of the current adaptives.
IDLE SPARK CONTROL:	Displays commanded spark modification so engine RPM equals desired.
KNOCK SIGNAL:	Displays engine knock in counts as reported to the PCM by the knock sensor. The Knock signal is the amount of detonation (or noise) during a fixed time.
IAC POSITION:	Displays in counts IAC position as calculated by the PCM.
DES IAC POS:	Displays in counts the desired IAC position, commanded by the PCM.
ENGINE LOAD:	Displays in percent the PCM calculated engine load, determined by comparing MAP vs TP Sensor.
CURRENT GEAR:	Displays current gear being commanded by PCM/TC (except '92).
ABS R SENS FAULT:	Displays "NO" if the ABS control module hasn't reported rear wheel sensor faults to the PCM, used for traction control operation (except '92).
BRAKE SWITCH:	Displays "ON" when brake pedal is depressed. This input is read by the PCM/TC and sent to the PCM/EC.

4.3L (VIN Z) TURBOCHARGED ENGINE — COMPONENT LOCATIONS — SYCLONE AND TYPHOON

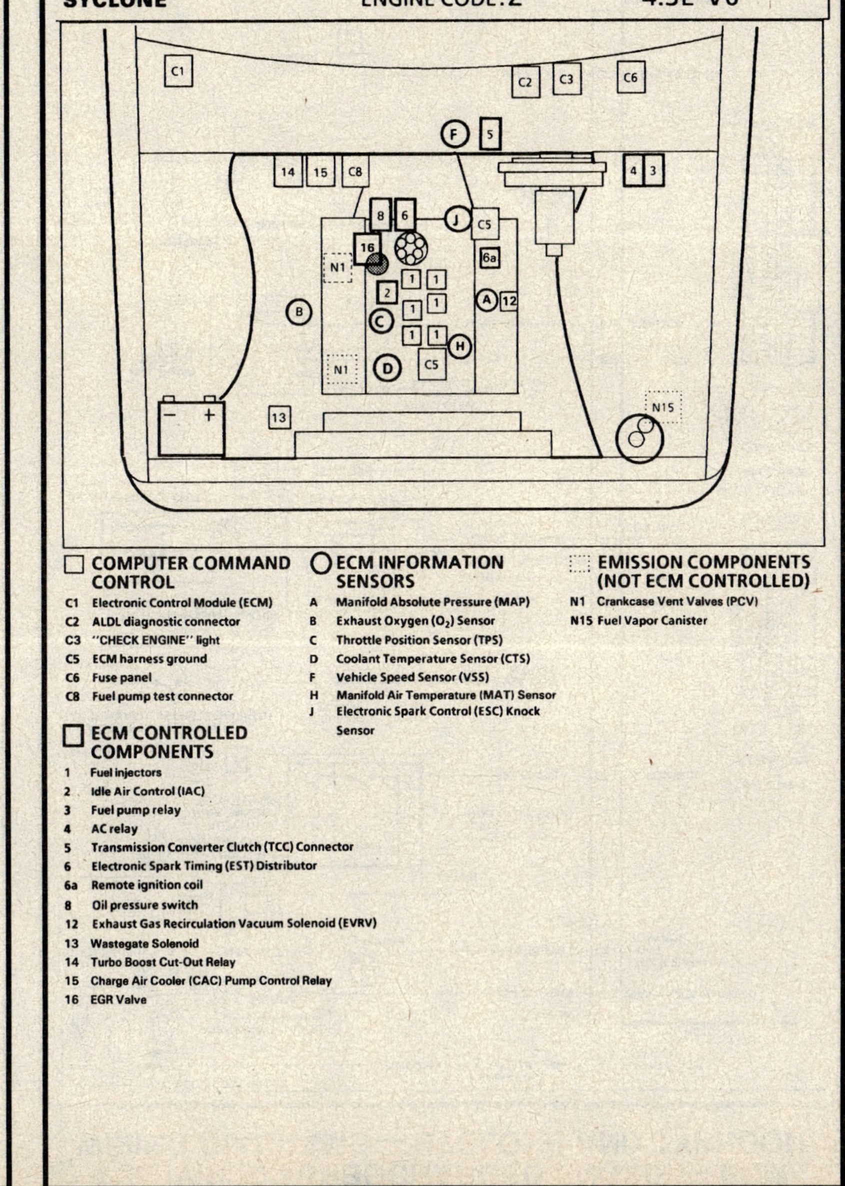

□ COMPUTER COMMAND CONTROL

C1 Electronic Control Module (ECM)
C2 ALDL diagnostic connector
C3 "CHECK ENGINE" light
C5 ECM harness ground
C6 Fuse panel
C8 Fuel pump test connector

□ ECM CONTROLLED COMPONENTS

1 Fuel injectors
2 Idle Air Control (IAC)
3 Fuel pump relay
4 AC relay
5 Transmission Converter Clutch (TCC) Connector
6 Electronic Spark Timing (EST) Distributor
6a Remote ignition coil
8 Oil pressure switch
12 Exhaust Gas Recirculation Vacuum Solenoid (EVRV)
13 Wastegate Solenoid
14 Turbo Boost Cut-Out Relay
15 Charge Air Cooler (CAC) Pump Control Relay
16 EGR Valve

○ ECM INFORMATION SENSORS

A Manifold Absolute Pressure (MAP)
B Exhaust Oxygen (O₂) Sensor
C Throttle Position Sensor (TPS)
D Coolant Temperature Sensor (CTS)
F Vehicle Speed Sensor (VSS)
H Manifold Air Temperature (MAT) Sensor
J Electronic Spark Control (ESC) Knock Sensor

▦ EMISSION COMPONENTS (NOT ECM CONTROLLED)

N1 Crankcase Vent Valves (PCV)
N15 Fuel Vapor Canister

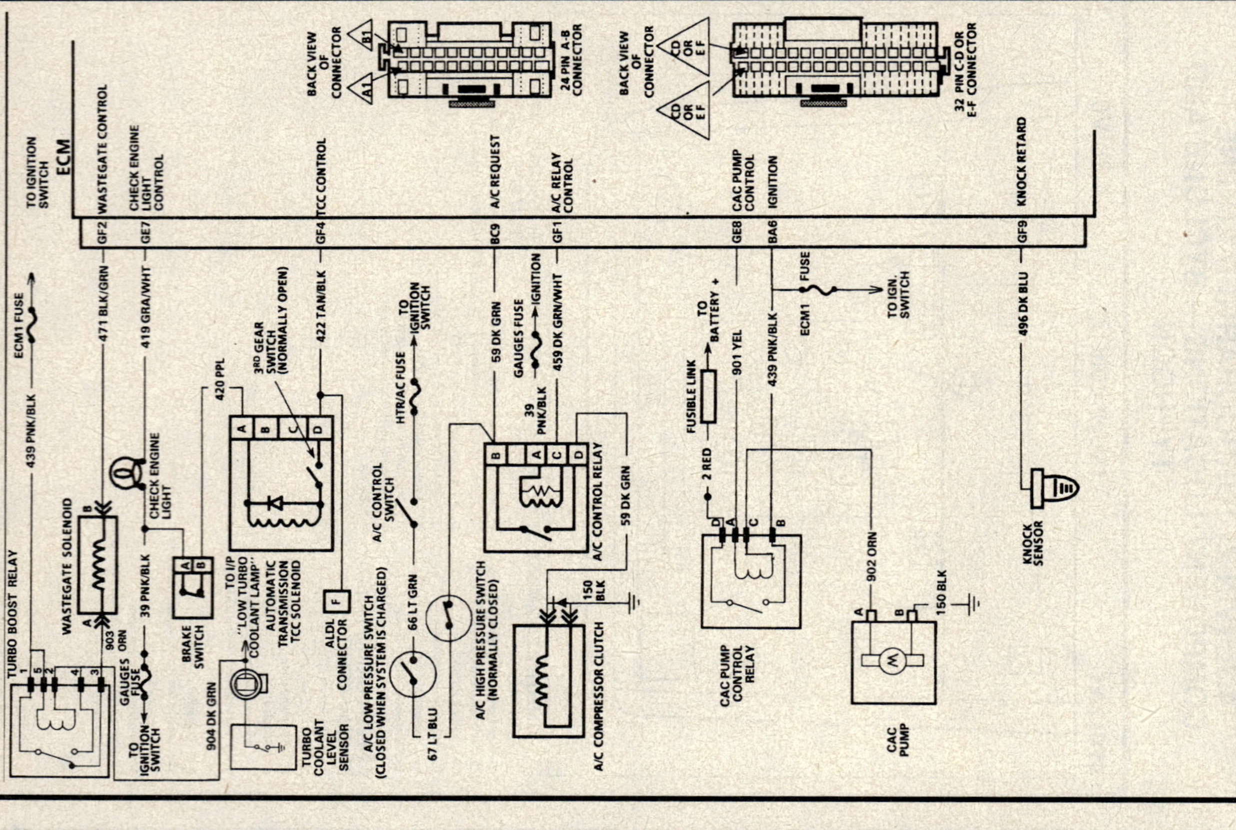
4.3L (VIN Z) TURBOCHARGED ENGINE — ECM
WIRING DIAGRAMS — SYCLONE AND TYPHOON
ECM
BACK VIEW OF CONNECTOR
24 PIN A-B CONNECTOR
BACK VIEW OF CONNECTOR
32 PIN C-D OR E-F CONNECTOR
TO IGNITION SWITCH
WASTEGATE CONTROL
CHECK ENGINE LIGHT CONTROL
A/C REQUEST
A/C RELAY CONTROL
CAC PUMP CONTROL
IGNITION
KNOCK RETARD
GF2
GE7
GF4 TCC CONTROL
BC9
GF1
GE8
BA6
GF9
ECM1 FUSE
439 PNK/BLK
471 BLK/GRN
419 GRA/WHT
420 PPL
422 TAN/BLK
59 DK GRN
459 DK GRN/WHT
ECM1 FUSE
439 PNK/BLK
496 DK BLU
GAUGES ORN
39 PNK/BLK
3RD GEAR SWITCH (NORMALLY OPEN)
A B C D
TO I/P "LOW TURBO COOLANT LAMP"
AUTOMATIC TRANSMISSION TCC SOLENOID
HTR/AC FUSE
TO IGNITION SWITCH
39 PNK/BLK
GAUGES FUSE
TO IGNITION SWITCH
FUSIBLE LINK
TO BATTERY +
2 RED
901 YEL
ECM1 FUSE
TO IGN. SWITCH
439 PNK/BLK
WASTEGATE SOLENOID
903 ORN
CHECK ENGINE LIGHT
BRAKE SWITCH
ALDL CONNECTOR
66 LT GRN
67 LT BLU
A/C CONTROL SWITCH
A/C LOW PRESSURE SWITCH (CLOSED WHEN SYSTEM IS CHARGED)
A/C HIGH PRESSURE SWITCH (NORMALLY CLOSED)
150 BLK
A/C COMPRESSOR CLUTCH
A/C CONTROL RELAY
59 DK GRN
CAC PUMP CONTROL RELAY
902 ORN
150 BLK
CAC PUMP
KNOCK SENSOR
TURBO BOOST RELAY
TO IGNITION SWITCH
904 DK GRN
TURBO COOLANT LEVEL SENSOR

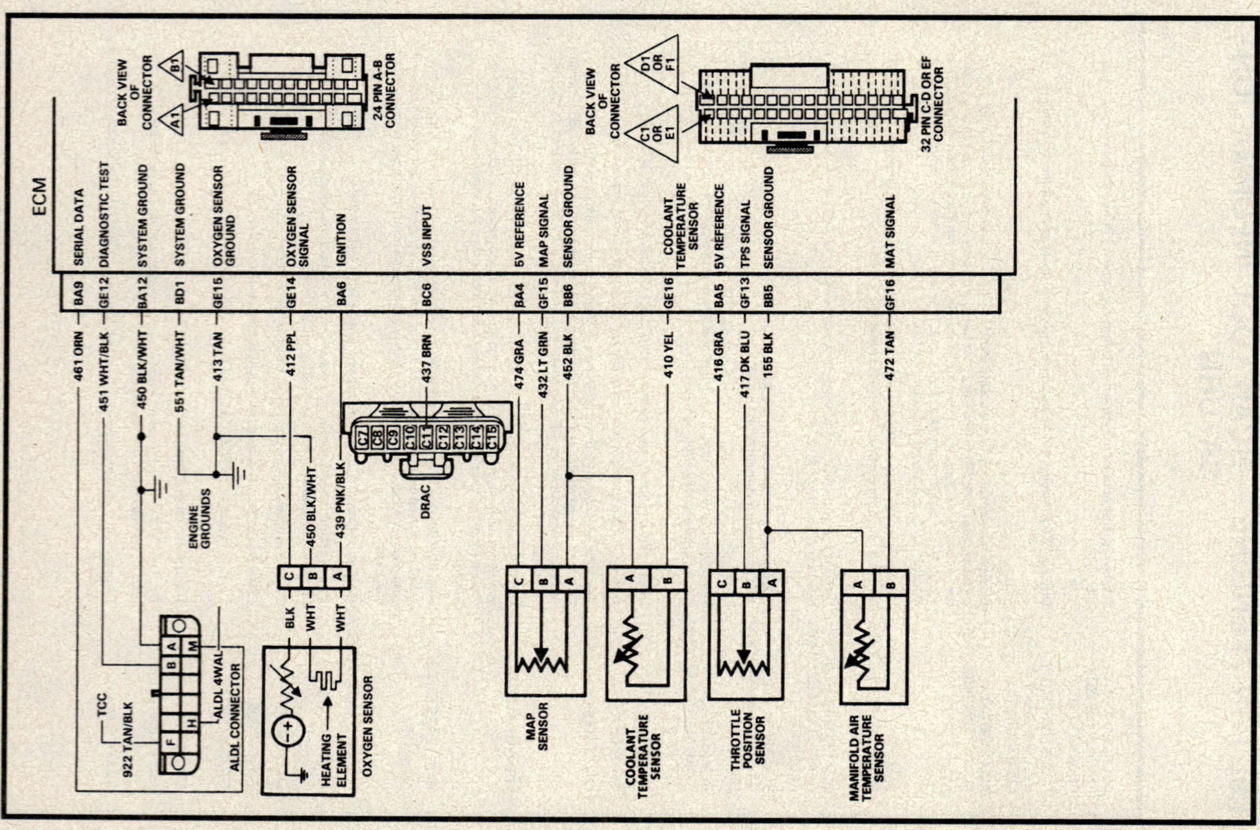
4.3L (VIN Z) TURBOCHARGED ENGINE — ECM
WIRING DIAGRAMS — SYCLONE AND TYPHOON
BACK VIEW OF CONNECTOR
24 PIN A-B CONNECTOR
BACK VIEW OF CONNECTOR
32 PIN C-D OR EF CONNECTOR
ECM
SERIAL DATA
DIAGNOSTIC TEST
SYSTEM GROUND
SYSTEM GROUND
OXYGEN SENSOR GROUND
OXYGEN SENSOR SIGNAL
IGNITION
VSS INPUT
5V REFERENCE
MAP SIGNAL
SENSOR GROUND
COOLANT TEMPERATURE SENSOR
5V REFERENCE
TPS SIGNAL
SENSOR GROUND
MAT SIGNAL
BA9
GE12
BA12
BD1
GE15
GE14
BA6
BC6
BA4
GF15
BB6
GE16
BA5
GF13
BB5
GF16
461 ORN
451 WHT/BLK
450 BLK/WHT
551 TAN/WHT
413 TAN
412 PPL
437 BRN
474 GRA
432 LT GRN
452 BLK
410 YEL
416 GRA
417 DK BLU
155 BLK
472 TAN
ENGINE GROUNDS
450 BLK/WHT
439 PNK/BLK
DRAC
922 TAN/BLK
TCC
ALDL 4WAL
ALDL CONNECTOR
BLK
WHT
WHT
OXYGEN SENSOR
HEATING ELEMENT
MAP SENSOR
COOLANT TEMPERATURE SENSOR
THROTTLE POSITION SENSOR
MANIFOLD AIR TEMPERATURE SENSOR

4.3L (VIN Z) TURBOCHARGED ENGINE — ECM WIRING DIAGRAMS — SYCLONE AND TYPHOON

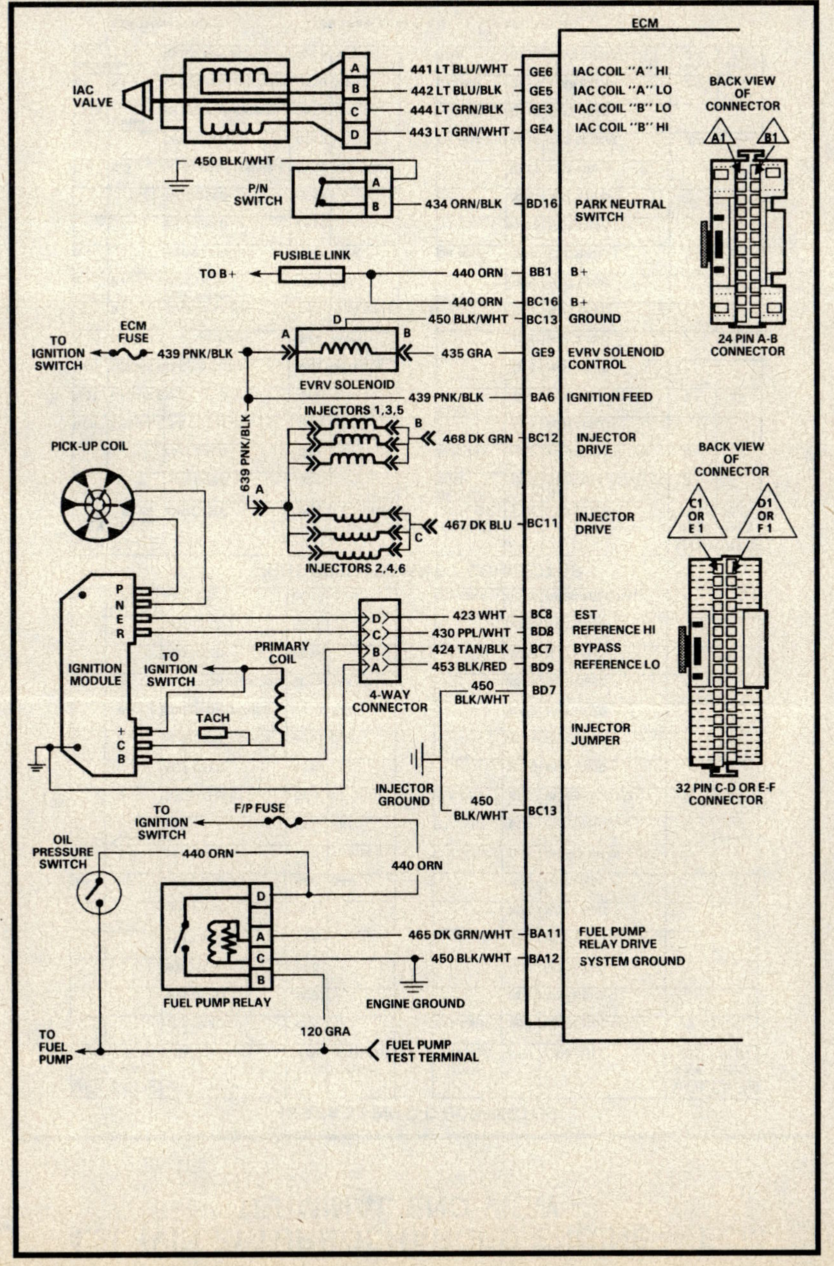

4.3L (VIN Z) TURBOCHARGED ENGINE — ECM TERMINAL END VIEW —

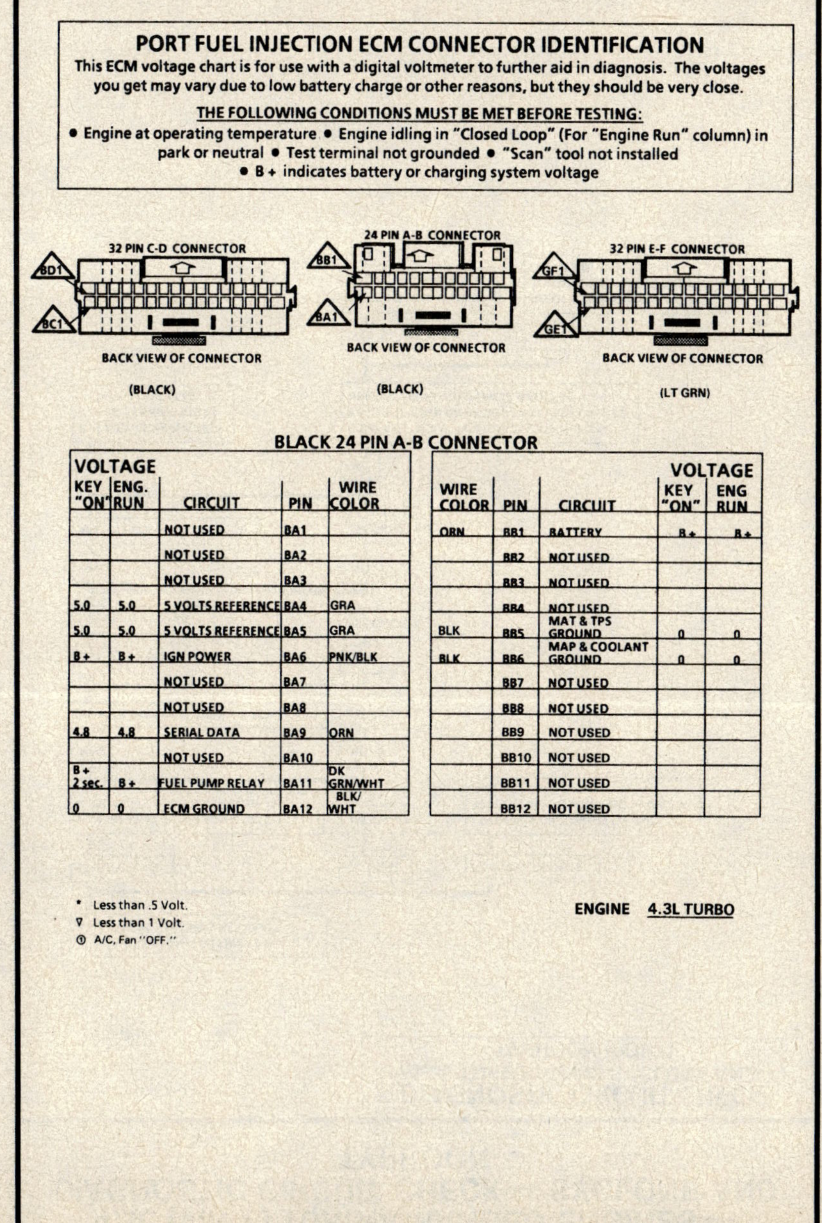

BLACK 24 PIN A-B CONNECTOR

VOLTAGE KEY "ON"	VOLTAGE ENG. RUN	CIRCUIT	PIN	WIRE COLOR
		NOT USED	BA1	
		NOT USED	BA2	
		NOT USED	BA3	
5.0	5.0	5 VOLTS REFERENCE	BA4	GRA
5.0	5.0	5 VOLTS REFERENCE	BA5	GRA
B+	B+	IGN POWER	BA6	PNK/BLK
		NOT USED	BA7	
		NOT USED	BA8	
4.8	4.8	SERIAL DATA	BA9	ORN
		NOT USED	BA10	
B+ 2 sec.	B+	FUEL PUMP RELAY	BA11	DK GRN/WHT
0	0	ECM GROUND	BA12	BLK/WHT

WIRE COLOR	PIN	CIRCUIT	VOLTAGE KEY "ON"	VOLTAGE ENG. RUN
ORN	BB1	BATTERY	B+	B+
	BB2	NOT USED		
	BB3	NOT USED		
	BB4	NOT USED		
BLK	BB5	MAT & TPS GROUND	0	0
BLK	BB6	MAP & COOLANT GROUND	0	0
	BB7	NOT USED		
	BB8	NOT USED		
	BB9	NOT USED		
	BB10	NOT USED		
	BB11	NOT USED		
	BB12	NOT USED		

4.3L (VIN Z) TURBOCHARGED ENGINE — ECM TERMINAL END VIEW —

BLACK 32 PIN C-D CONNECTOR

VOLTAGE KEY "ON"	ENG. RUN	CIRCUIT	PIN	WIRE COLOR
		NOT USED	BC1	
		NOT USED	BC2	
		NOT USED	BC3	
		NOT USED	BC4	
		NOT USED	BC5	
B+	B+	DRAC SIGNAL	BC6	BRN
0*	4.7	BYPASS	BC7	TAN/BLK
0*	varies 1.2	EST	BC8	WHT
(1) 0	0	A/C REQUEST	BC9	DK GRN
		NOT USED	BC10	
B+	B+	INJECTOR DRIVER	BC11	BLU
B+	B+	INJECTOR DRIVER	BC12	GRN
0*	0*	INJECTOR GROUND	BC13	BLK/WHT
		NOT USED	BC14	
		NOT USED	BC15	
B+	B+	BATTERY	BC16	ORN

WIRE COLOR	PIN	CIRCUIT	VOLTAGE KEY "ON"	ENG RUN
TAN/WHT	BD1	ECM GROUND	0	0
	BD2	NOT USED		
	BD3	NOT USED		
	BD4	NOT USED		
	BD5	NOT USED		
	BD6	NOT USED		
BLK/WHT	BD7	INJECTOR GROUND	0	0
PPL/WHT	BD8	DIST. REFERENCE HI	0*	varies 1.0
BLK/RED	BD9	DIST. REFERENCE LO	0*	0*
	BD10	NOT USED		
	BD11	NOT USED		
	BD12	NOT USED		
	BD13	NOT USED		
	BD14	NOT USED		
	BD15	NOT USED		
ORN/BLK	BD16	PARK/NEUTRAL	0*	0*

LIGHT GREEN 32 PIN E-F CONNECTOR

VOLTAGE KEY "ON"	ENG. RUN	CIRCUIT	PIN	WIRE COLOR
		NOT USED	GE1	
		NOT USED	GE2	
NOT	USEABLE	IAC "B" LOW	GE3	LT GRN/WHT
NOT	USEABLE	IAC "B" HIGH	GE4	LT GRN/BLK
NOT	USEABLE	IAC "A" LOW	GE5	BLU/BLK
NOT	USEABLE	IAC "A" HIGH	GE6	LT BLU/WHT
.1	B+	CHECK ENGINE LAMP	GE7	GRA/WHT
B+	B+	CAC PUMP CONTROL	GE8	YEL
B+	.1	EVRV CONTROL	GE9	GRA
		NOT USED	GE10	
		NOT USED	GE11	
5.0	5.0	ALDL DIAG. ENABLE	GE12	WHT/BLK
		NOT USED	GE13	
.1-.3	varies .1-.9	O² SENSOR SIGNAL	GE14	PPL
0*	0*	O² SENSOR GROUND	GE15	TAN
varies 1.5	varies 1.5	COOLANT SIGNAL	GE16	YEL

WIRE COLOR	PIN	CIRCUIT	VOLTAGE KEY "ON"	ENG RUN
DK GRN/WHT	GF1	A/C CONTROL RELAY	B+	B+ (1)
BLK/GRN	GF2	WASTEGATE SOL.	B+	.5
	GF3	NOT USED		
TAN/BLK	GF4	TCC SOLENOID	B+	B+
	GF5	NOT USED		
	GF6	NOT USED		
	GF7	NOT USED		
	GF8	NOT USED		
DK BLU	GF9	ESC SIGNAL	2.4	2.4
	GF10	NOT USED		
	GF11	NOT USED		
	GF12	NOT USED		
DK BLU	GF13	TPS SIGNAL	.84	.84
	GF14	NOT USED		
LT GRN	GF15	MAP SIGNAL	2.2	varies .8
TAN	GF16	MAT SIGNAL	varies 2.4	varies 2.2

* LESS THAN .5 VOLT   ▽ LESS THAN 1 VOLT   (1) A/C, FAN OFF

4.3L (VIN Z) TURBOCHARGED ENGINES — DIAGNOSTIC CIRCUIT CHECK — SYCLONE AND TYPHOON

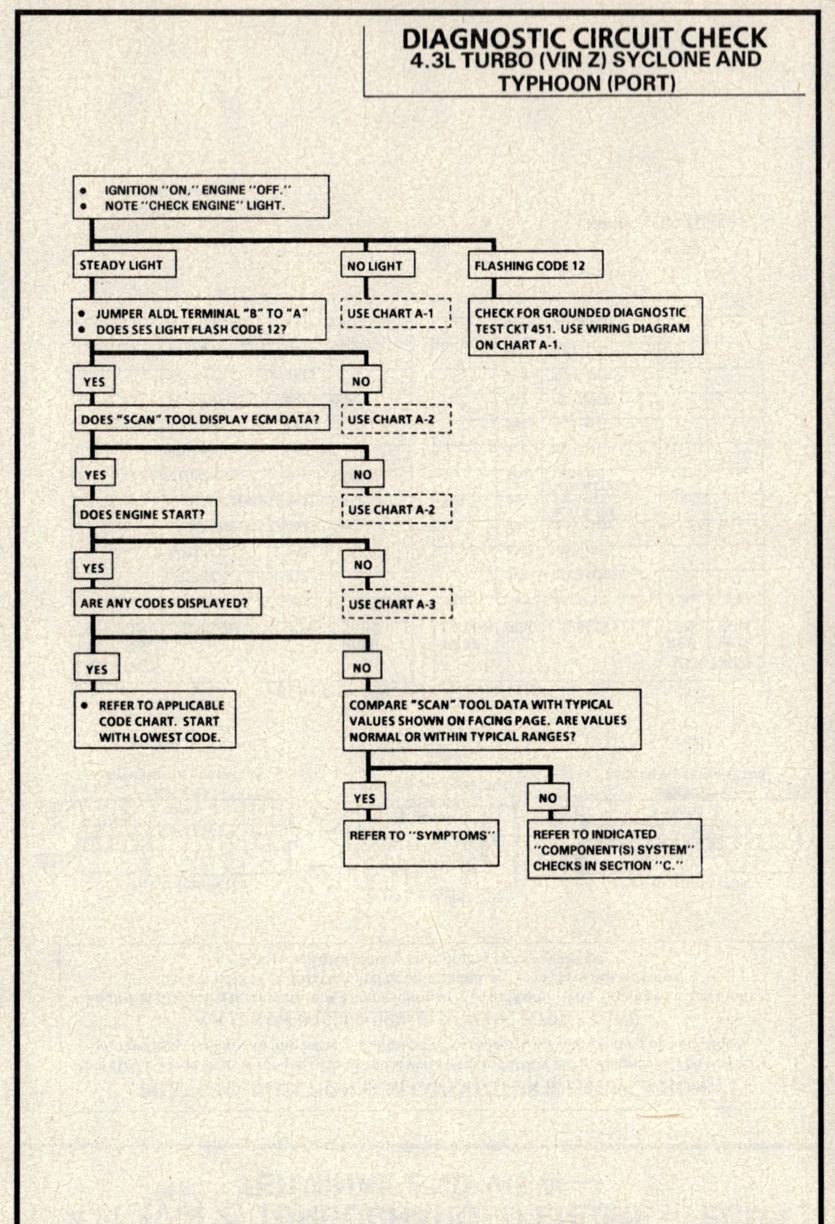

4.3L (VIN Z) TURBOCHARGED ENGINES — A-CHARTS — SYCLONE AND TYPHOON

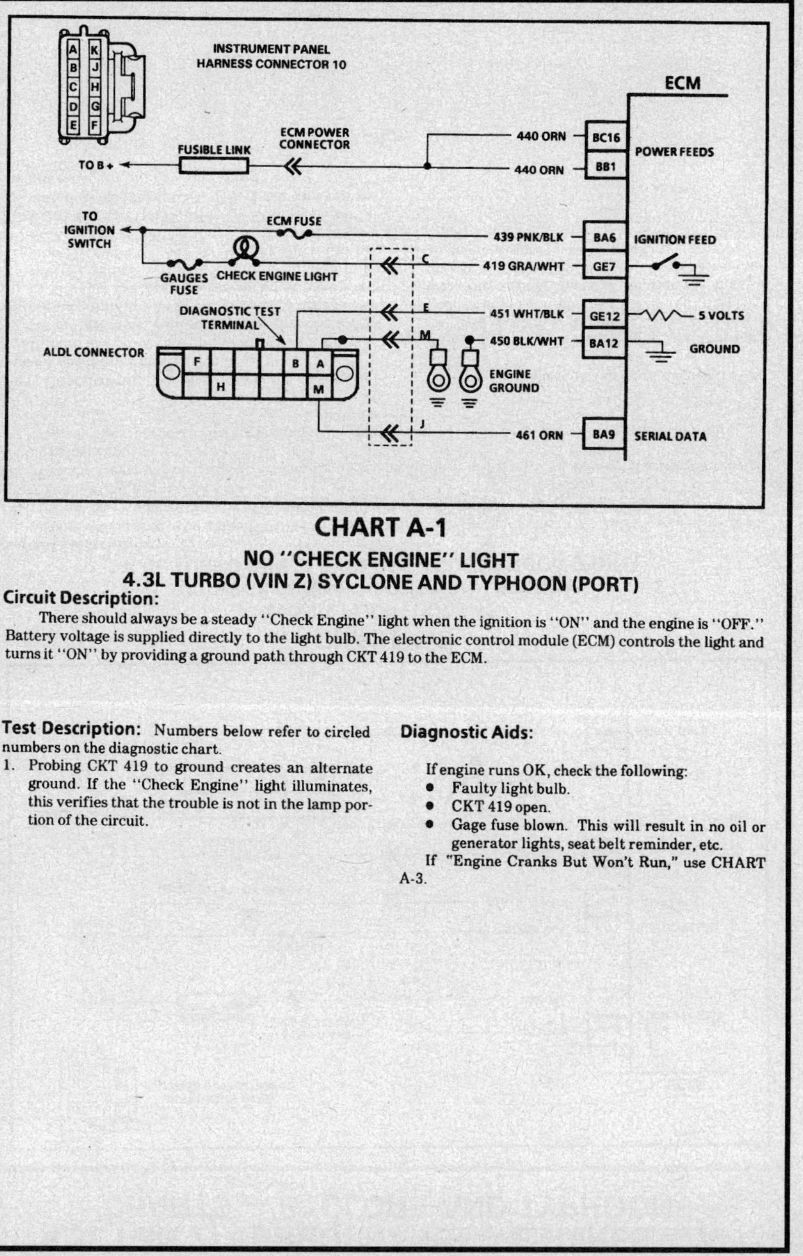

CHART A-1
NO "CHECK ENGINE" LIGHT
4.3L TURBO (VIN Z) SYCLONE AND TYPHOON (PORT)

Circuit Description:

There should always be a steady "Check Engine" light when the ignition is "ON" and the engine is "OFF." Battery voltage is supplied directly to the light bulb. The electronic control module (ECM) controls the light and turns it "ON" by providing a ground path through CKT 419 to the ECM.

Test Description: Numbers below refer to circled numbers on the diagnostic chart.

1. Probing CKT 419 to ground creates an alternate ground. If the "Check Engine" light illuminates, this verifies that the trouble is not in the lamp portion of the circuit.

Diagnostic Aids:

If engine runs OK, check the following:
- Faulty light bulb.
- CKT 419 open.
- Gage fuse blown. This will result in no oil or generator lights, seat belt reminder, etc.

If "Engine Cranks But Won't Run," use CHART A-3.

4.3L (VIN Z) TURBOCHARGED ENGINES — A-CHARTS — SYCLONE AND TYPHOON

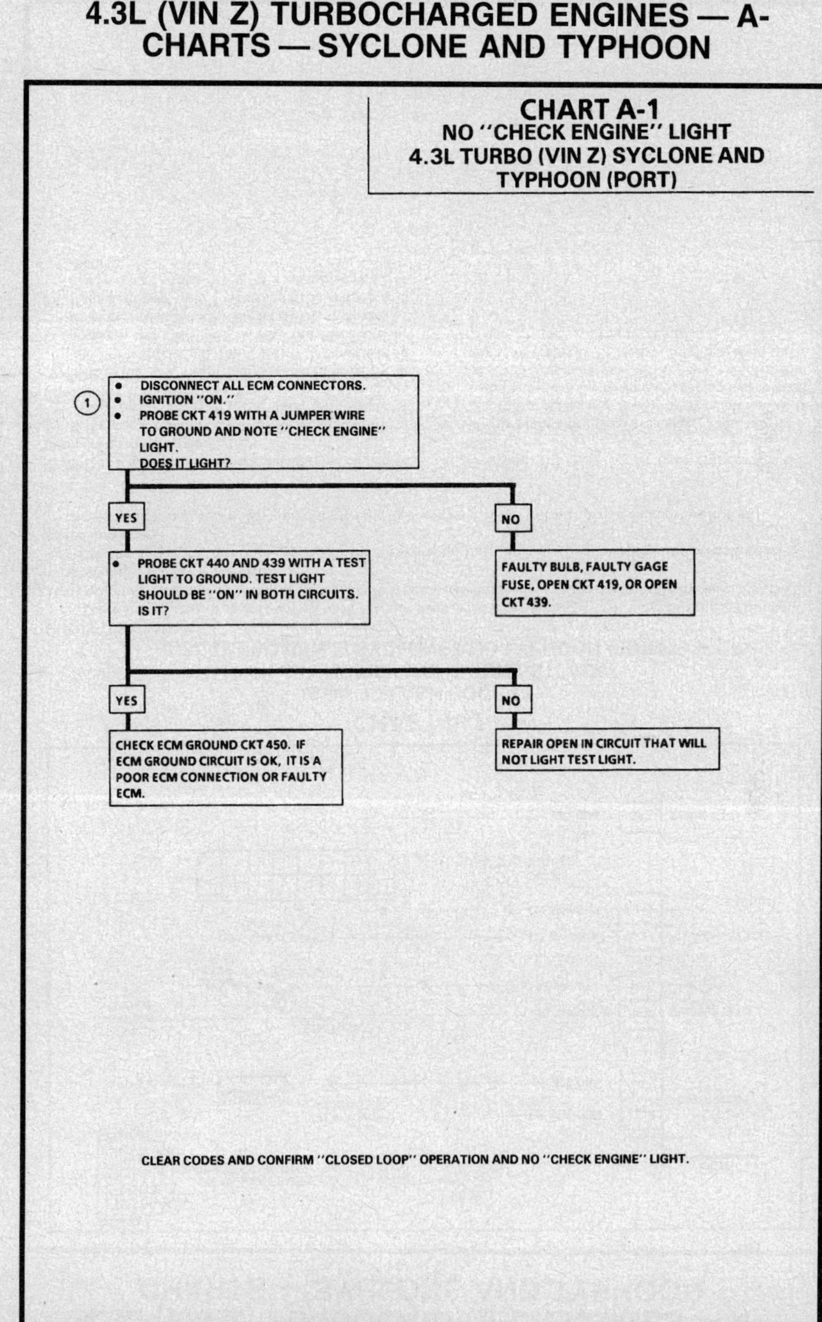

4.3L (VIN Z) TURBOCHARGED ENGINES — A-CHARTS — SYCLONE AND TYPHOON

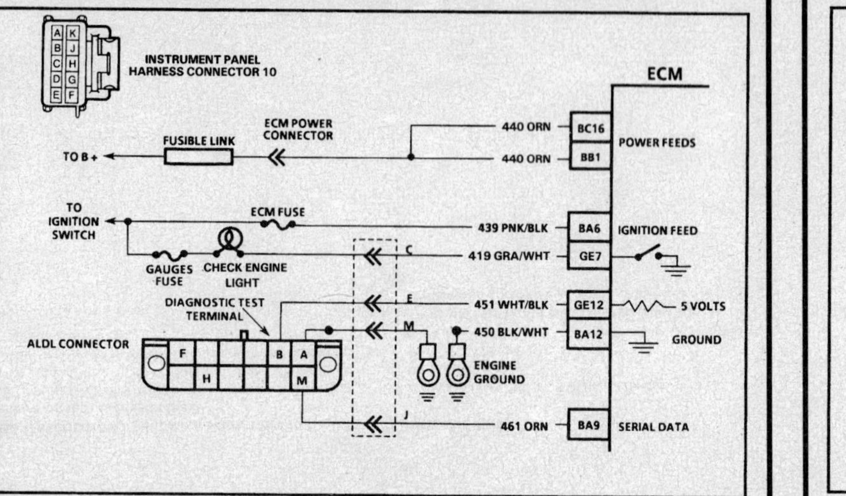

CHART A-2
WON'T FLASH CODE 12
"CHECK ENGINE" LIGHT "ON" STEADY
4.3L TURBO (VIN Z) SYCLONE AND TYPHOON (PORT)

Circuit Description:

There should always be a steady "Check Engine" light when the ignition is "ON" and the engine stopped. Battery voltage is supplied directly to the light bulb. The electronic control module (ECM) will turn the light "ON" by grounding CKT 419 at the ECM.

With the diagnostic terminal grounded, the light should flash a Code 12 followed by any trouble code(s) stored in memory.

A steady light suggests a short to ground in the light control CKT 419 or an open in diagnostic CKT 451.

Test Description: Numbers below refer to circled numbers on the diagnostic chart.

1. If there is a problem with the ECM that causes a "Scan" tool to not read serial data, then the ECM should not flash a Code 12. If Code 12 does flash, be sure that the "Scan" tool is working properly on another vehicle. If the "Scan" is functioning properly and CKT 461 is OK, the Mem-Cal or ECM may be at fault for the no ALDL symptom.

2. If the light goes "OFF" when the ECM connector is disconnected, then CKT 419 is not shorted to ground.

3. This step will check for an open diagnostic CKT 451.

4. At this point the "Check Engine" light wiring is OK. The problem is a faulty ECM or Mem-Cal. If Code 12 does not flash, the ECM should be replaced using the original Mem-Cal. Replace the Mem-Cal only after trying an ECM, as a defective Mem-Cal is an unlikely cause of the problem.

4.3L (VIN Z) TURBOCHARGED ENGINES — A-CHARTS — SYCLONE AND TYPHOON

CHART A-2
WON'T FLASH CODE 12
"CHECK ENGINE" LIGHT "ON" STEADY
4.3L TURBO (VIN Z) SYCLONE AND TYPHOON (PORT)

Circuit Description:

There should always be a steady "Check Engine" light when the ignition is "ON" and the engine stopped. Battery voltage is supplied directly to the light bulb. The electronic control module (ECM) will turn the light "ON" by grounding CKT 419 at the ECM.

With the diagnostic terminal grounded, the light should flash a Code 12 followed by any trouble code(s) stored in memory.

A steady light suggests a short to ground in the light control CKT 419 or an open in diagnostic CKT 451.

Test Description: Numbers below refer to circled numbers on the diagnostic chart.

1. If there is a problem with the ECM that causes a "Scan" tool to not read serial data, then the ECM should not flash a Code 12. If Code 12 does flash, be sure that the "Scan" tool is working properly on another vehicle. If the "Scan" is functioning properly and CKT 461 is OK, the Mem-Cal or ECM may be at fault for the no ALDL symptom.

2. If the light goes "OFF" when the ECM connector is disconnected, then CKT 419 is not shorted to ground.

3. This step will check for an open diagnostic CKT 451.

4. At this point the "Check Engine" light wiring is OK. The problem is a faulty ECM or Mem-Cal. If Code 12 does not flash, the ECM should be replaced using the original Mem-Cal. Replace the Mem-Cal only after trying an ECM, as a defective Mem-Cal is an unlikely cause of the problem.

4.3L (VIN Z) TURBOCHARGED ENGINES — A-CHARTS — SYCLONE AND TYPHOON

CHART A-2
WON'T FLASH CODE 12
"CHECK ENGINE" LIGHT "ON" STEADY
4.3L TURBO (VIN Z) SYCLONE AND TYPHOON (PORT)

4.3L (VIN Z) TURBOCHARGED ENGINES — A-CHARTS — SYCLONE AND TYPHOON

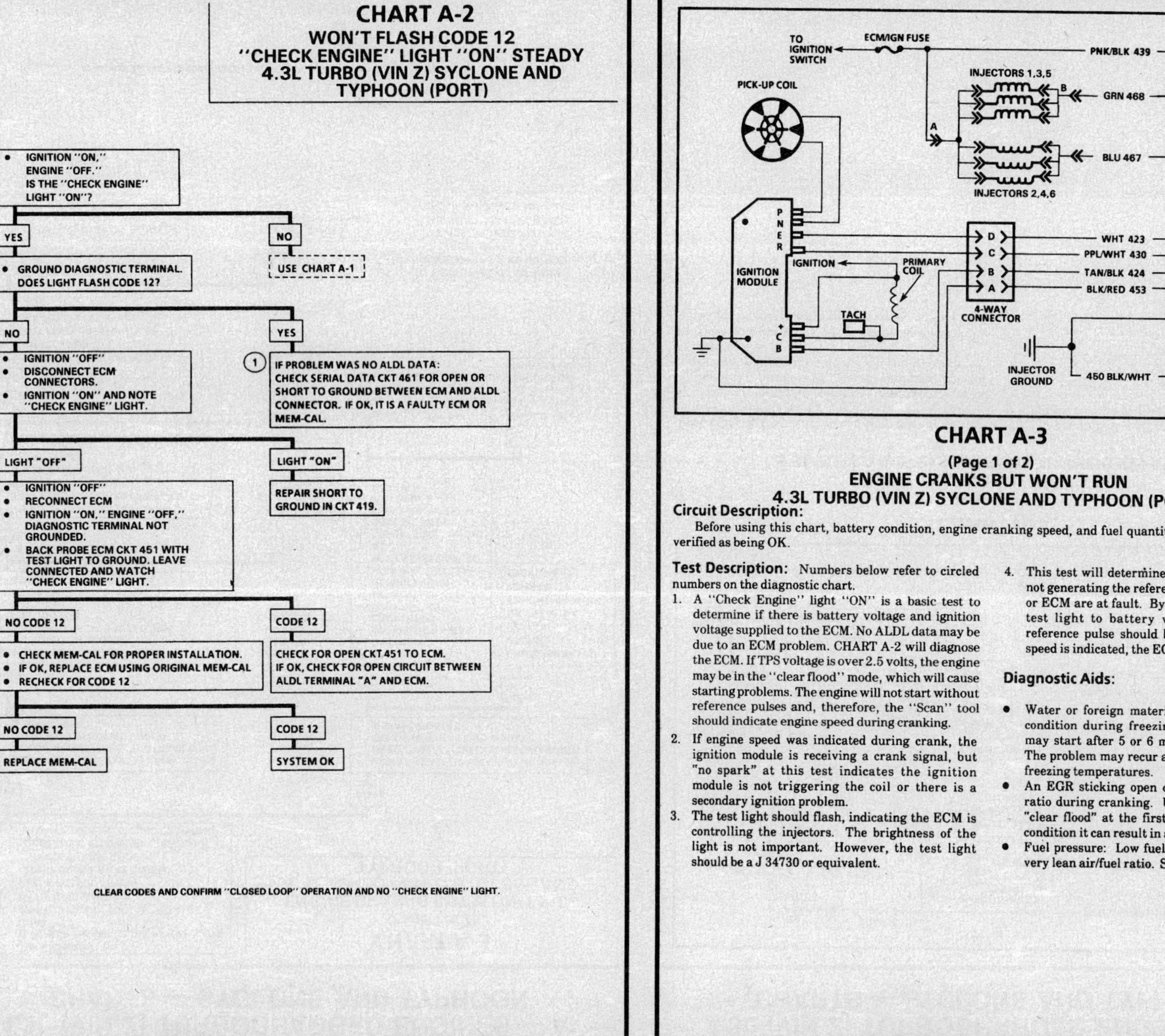

CHART A-3
(Page 1 of 2)
ENGINE CRANKS BUT WON'T RUN
4.3L TURBO (VIN Z) SYCLONE AND TYPHOON (PORT)

Circuit Description:

Before using this chart, battery condition, engine cranking speed, and fuel quantity should be checked and verified as being OK.

Test Description: Numbers below refer to circled numbers on the diagnostic chart.

1. A "Check Engine" light "ON" is a basic test to determine if there is battery voltage and ignition voltage supplied to the ECM. No ALDL data may be due to an ECM problem. CHART A-2 will diagnose the ECM. If TPS voltage is over 2.5 volts, the engine may be in the "clear flood" mode, which will cause starting problems. The engine will not start without reference pulses and, therefore, the "Scan" tool should indicate engine speed during cranking.

2. If engine speed was indicated during crank, the ignition module is receiving a crank signal, but "no spark" at this test indicates the ignition module is not triggering the coil or there is a secondary ignition problem.

3. The test light should flash, indicating the ECM is controlling the injectors. The brightness of the light is not important. However, the test light should be a J 34730 or equivalent.

4. This test will determine if the ignition module is not generating the reference pulse, or if the wiring or ECM are at fault. By touching and removing a test light to battery voltage on CKT 430, a reference pulse should be generated. If engine speed is indicated, the ECM and wiring are OK.

Diagnostic Aids:

- Water or foreign material can cause a no start condition during freezing weather. The engine may start after 5 or 6 minutes in a heated shop. The problem may recur after an overnight park in freezing temperatures.
- An EGR sticking open can cause a low air/fuel ratio during cranking. Unless the system enters "clear flood" at the first indication of a flooding condition it can result in a no start.
- Fuel pressure: Low fuel pressure can result in a very lean air/fuel ratio. See CHART A-7.

4.3L (VIN Z) TURBOCHARGED ENGINES — A-CHARTS — SYCLONE AND TYPHOON

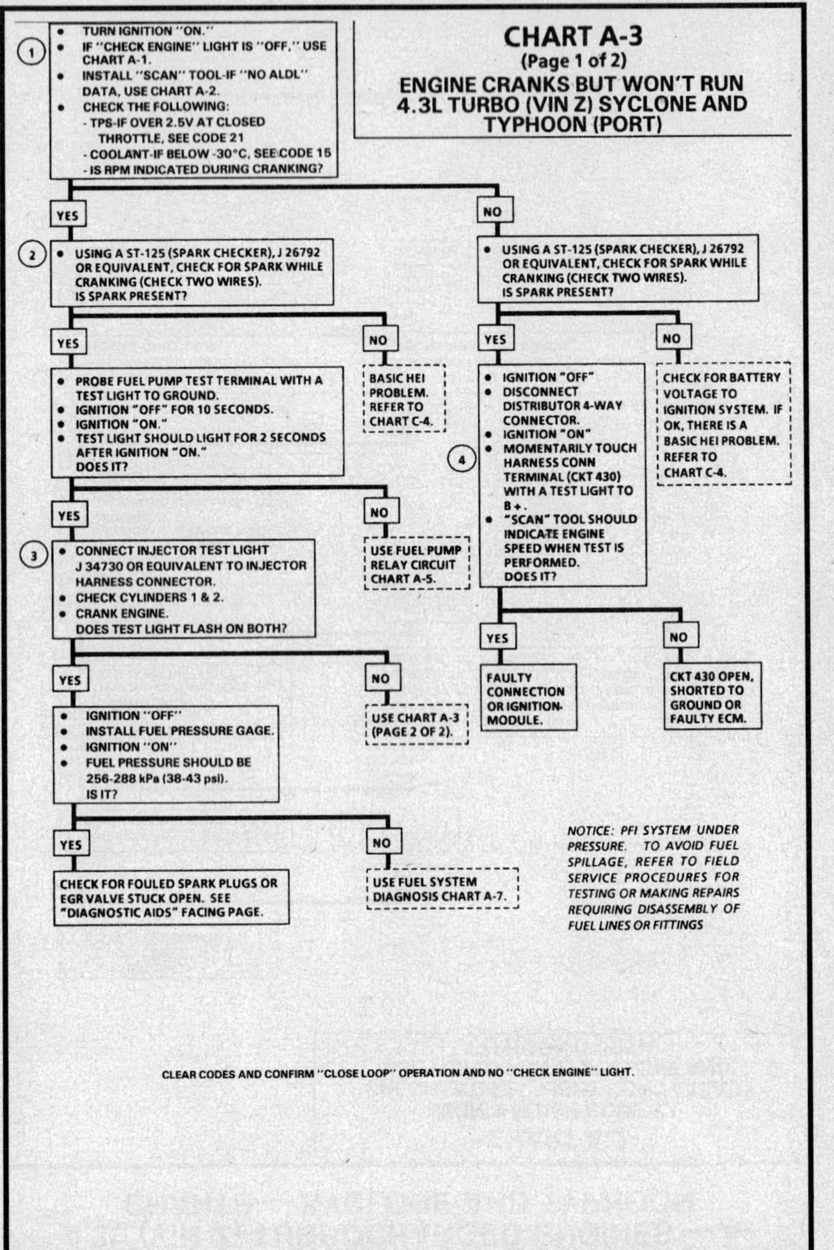

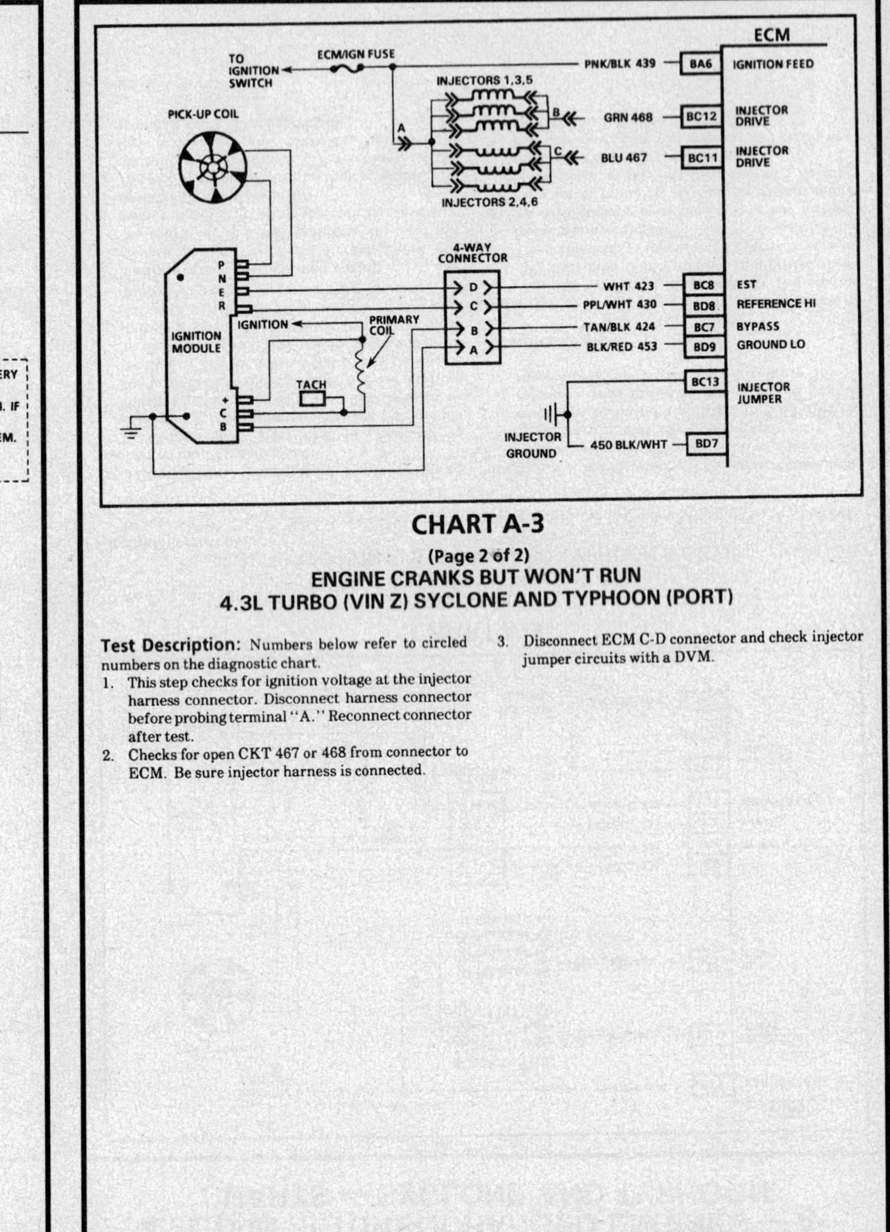

Test Description: Numbers below refer to circled numbers on the diagnostic chart.

1. This step checks for ignition voltage at the injector harness connector. Disconnect harness connector before probing terminal "A." Reconnect connector after test.

2. Checks for open CKT 467 or 468 from connector to ECM. Be sure injector harness is connected.

3. Disconnect ECM C-D connector and check injector jumper circuits with a DVM.

4.3L (VIN Z) TURBOCHARGED ENGINES — A-CHARTS — SYCLONE AND TYPHOON

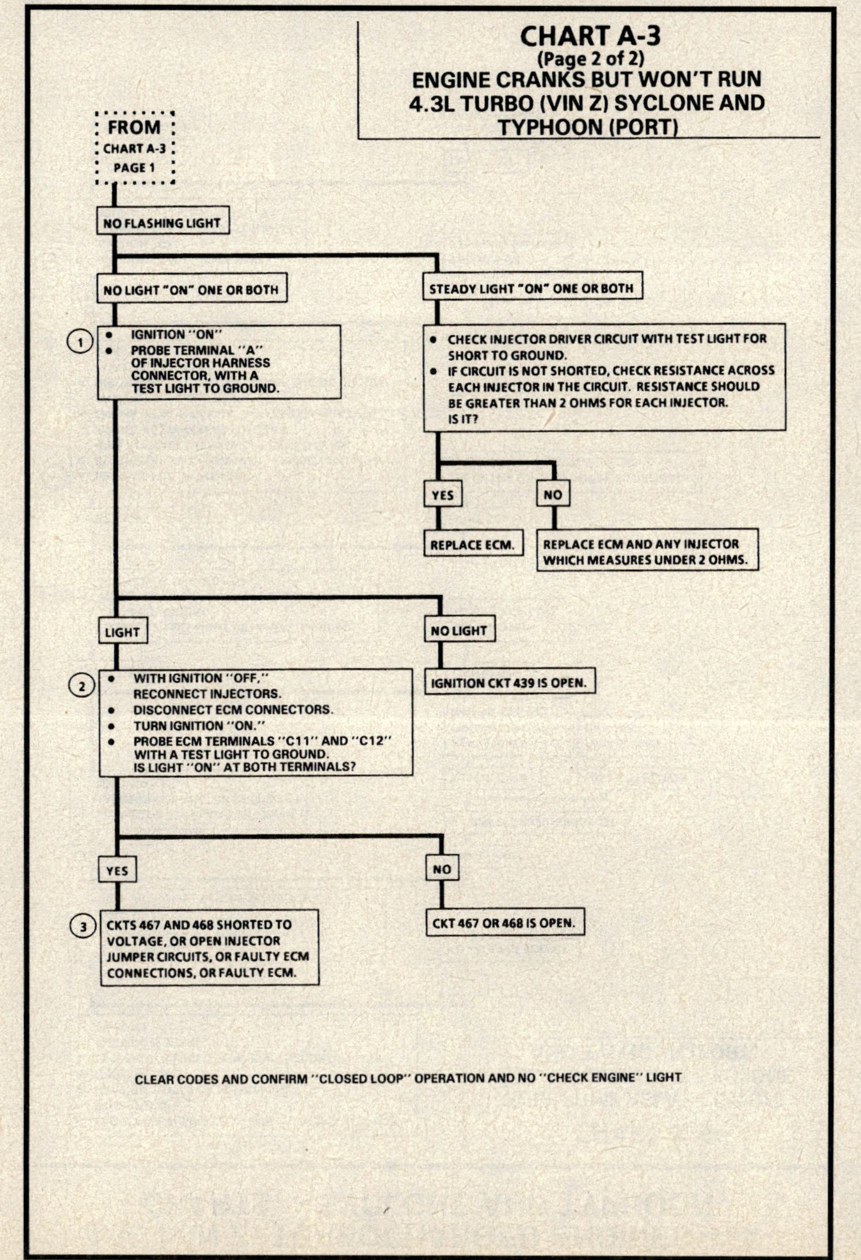

4.3L (VIN Z) TURBOCHARGED ENGINES — A-CHARTS — SYCLONE AND TYPHOON

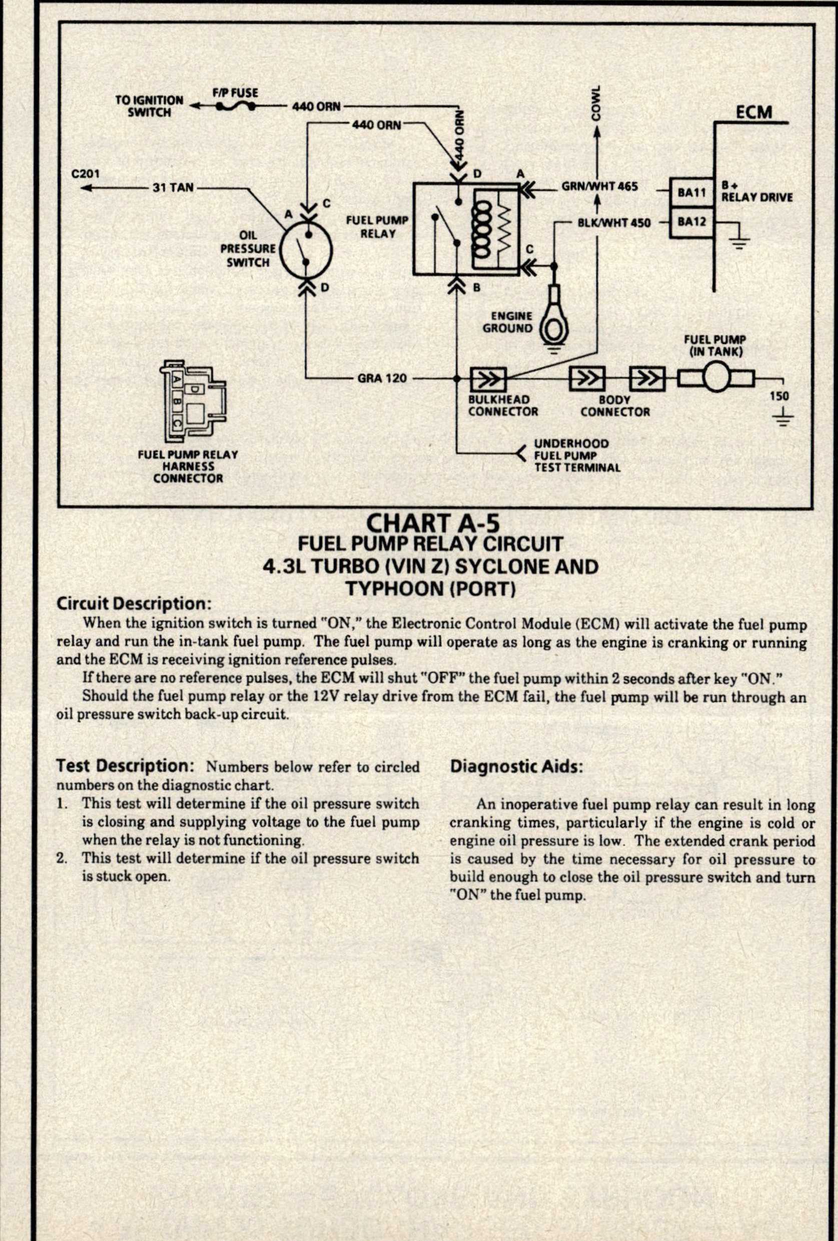

CHART A-5
FUEL PUMP RELAY CIRCUIT
4.3L TURBO (VIN Z) SYCLONE AND TYPHOON (PORT)

Circuit Description:
When the ignition switch is turned "ON," the Electronic Control Module (ECM) will activate the fuel pump relay and run the in-tank fuel pump. The fuel pump will operate as long as the engine is cranking or running and the ECM is receiving ignition reference pulses.

If there are no reference pulses, the ECM will shut "OFF" the fuel pump within 2 seconds after key "ON."

Should the fuel pump relay or the 12V relay drive from the ECM fail, the fuel pump will be run through an oil pressure switch back-up circuit.

Test Description: Numbers below refer to circled numbers on the diagnostic chart.
1. This test will determine if the oil pressure switch is closing and supplying voltage to the fuel pump when the relay is not functioning.
2. This test will determine if the oil pressure switch is stuck open.

Diagnostic Aids:

An inoperative fuel pump relay can result in long cranking times, particularly if the engine is cold or engine oil pressure is low. The extended crank period is caused by the time necessary for oil pressure to build enough to close the oil pressure switch and turn "ON" the fuel pump.

4.3L (VIN Z) TURBOCHARGED ENGINES — A-CHARTS — SYCLONE AND TYPHOON

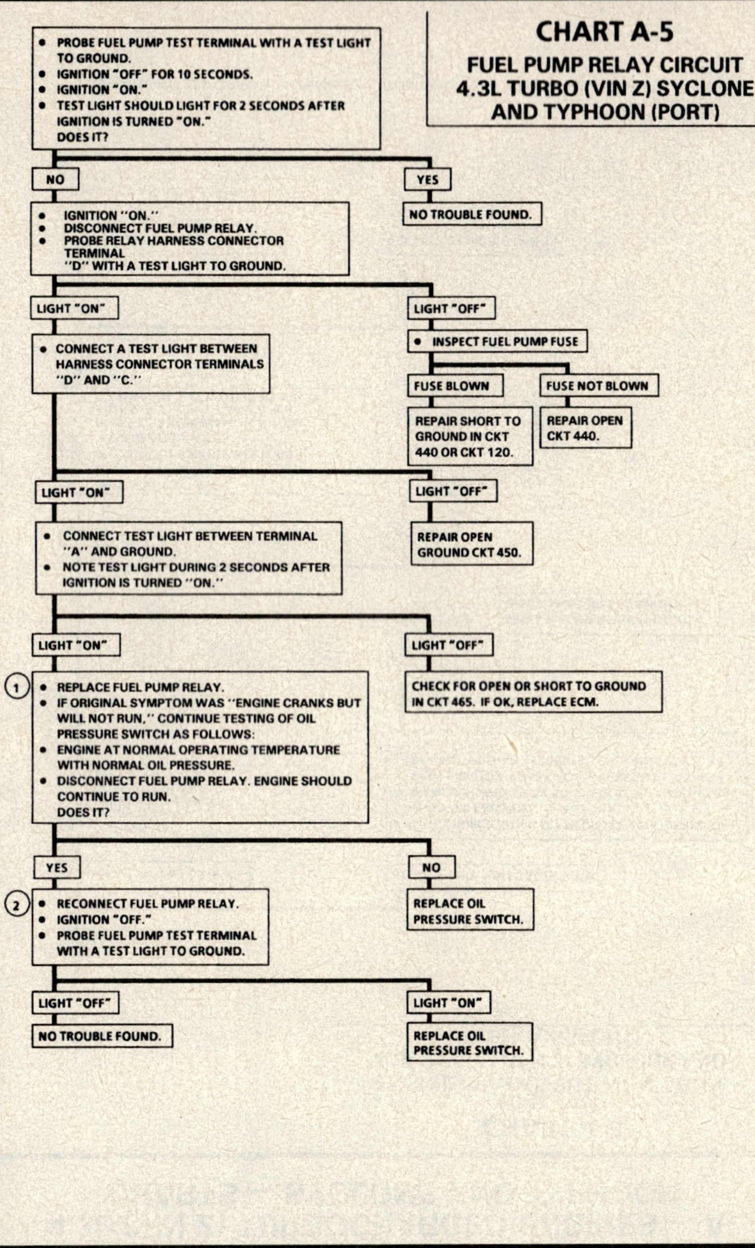

4.3L (VIN Z) TURBOCHARGED ENGINES — A-CHARTS — SYCLONE AND TYPHOON

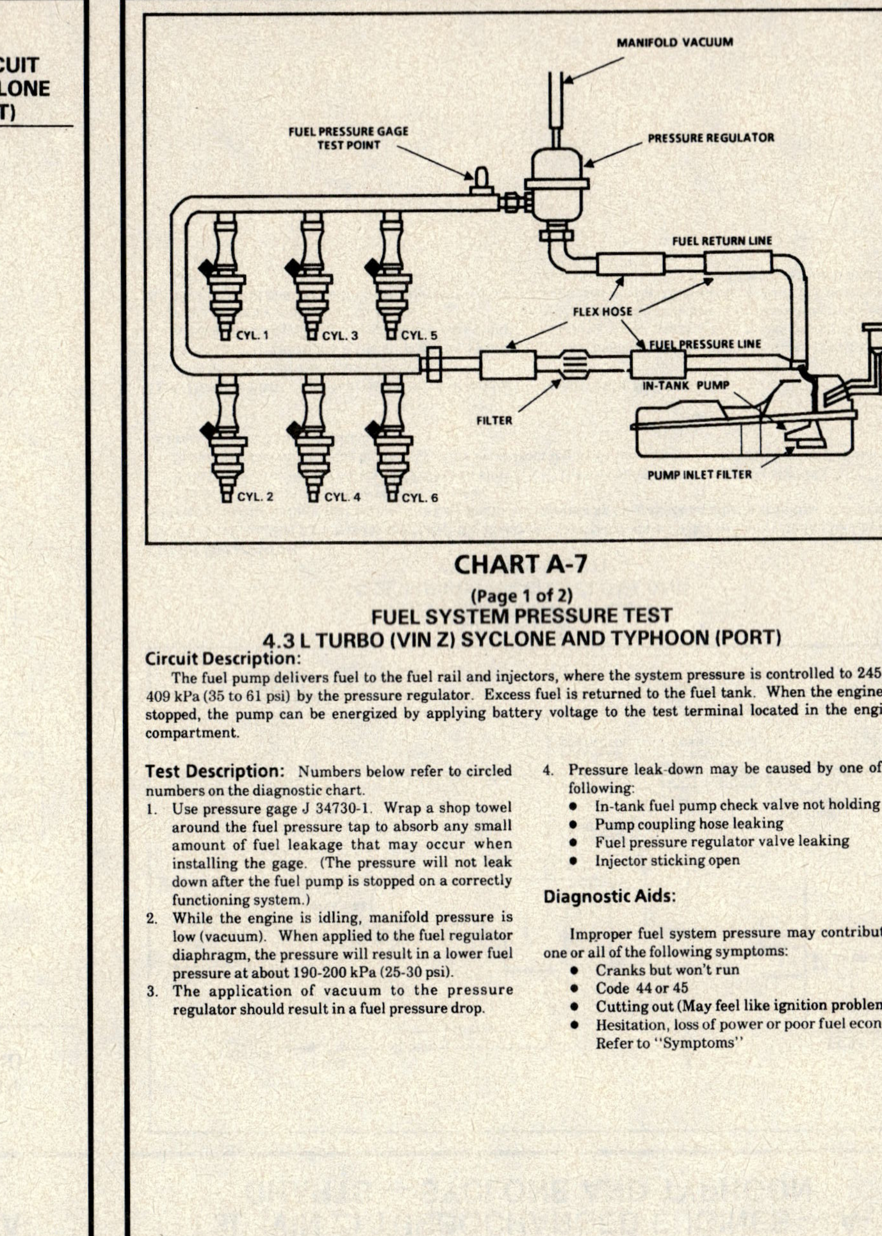

CHART A-7
(Page 1 of 2)
FUEL SYSTEM PRESSURE TEST
4.3 L TURBO (VIN Z) SYCLONE AND TYPHOON (PORT)

Circuit Description:

The fuel pump delivers fuel to the fuel rail and injectors, where the system pressure is controlled to 245 to 409 kPa (35 to 61 psi) by the pressure regulator. Excess fuel is returned to the fuel tank. When the engine is stopped, the pump can be energized by applying battery voltage to the test terminal located in the engine compartment.

Test Description: Numbers below refer to circled numbers on the diagnostic chart.

1. Use pressure gage J 34730-1. Wrap a shop towel around the fuel pressure tap to absorb any small amount of fuel leakage that may occur when installing the gage. (The pressure will not leak down after the fuel pump is stopped on a correctly functioning system.)
2. While the engine is idling, manifold pressure is low (vacuum). When applied to the fuel regulator diaphragm, the pressure will result in a lower fuel pressure at about 190-200 kPa (25-30 psi).
3. The application of vacuum to the pressure regulator should result in a fuel pressure drop.

4. Pressure leak-down may be caused by one of the following:
   • In-tank fuel pump check valve not holding
   • Pump coupling hose leaking
   • Fuel pressure regulator valve leaking
   • Injector sticking open

Diagnostic Aids:

Improper fuel system pressure may contribute to one or all of the following symptoms:
   • Cranks but won't run
   • Code 44 or 45
   • Cutting out (May feel like ignition problem)
   • Hesitation, loss of power or poor fuel economy Refer to "Symptoms"

4.3L (VIN Z) TURBOCHARGED ENGINES — A-CHARTS — SYCLONE AND TYPHOON

CHART A-7
(Page 1 of 2)
FUEL SYSTEM PRESSURE TEST
4.3L TURBO (VIN Z) SYCLONE AND TYPHOON (PORT)

NOTICE: FUEL SYSTEM IS UNDER PRESSURE. TO AVOID FUEL SPILLAGE, REFER TO FIELD SERVICE PROCEDURES FOR TESTING OR REPAIRS REQUIRING DISASSEMBLY OF FUEL LINES OR FITTINGS.

(1)
- INSTALL FUEL PRESSURE GAGE, J 34730-1 OR EQUIVALENT.
- DISCONNECT VACUUM HOSE FROM PRESSURE REGULATOR.
- IGNITION MUST BE "OFF" FOR 10 SECONDS AND ENSURE THAT A/C IS "OFF."
- TURN IGNITION "ON." FUEL PUMP SHOULD RUN FOR ABOUT 2 SECONDS.
- NOTE FUEL PRESSURE AFTER PUMP STOPS.
- PRESSURE SHOULD BE 256-288 kPA (38-43 PSI) AND HOLDING STEADY. IS IT?

YES →

(2)
- START AND IDLE ENGINE AT NORMAL OPERATING TEMPERATURE.
- RECONNECT VACUUM LINE.
- FUEL PRESSURE SHOULD HAVE DROPPED SLIGHTLY. DID IT?

NO →

(3)
- DISCONNECT VACUUM HOSE FROM FUEL PRESSURE REGULATOR.
- WITH ENGINE IDLING, APPLY 12-14" Hg OF VACUUM TO PRESSURE REGULATOR.
- FUEL PRESSURE SHOULD DROP SLIGHTLY. DID IT?

YES →
LOCATE AND CORRECT CAUSE OF INTERRUPTED MANIFOLD PRESSURE SUPPLY TO REGULATOR.

NO →
PRESSURE REGULATOR ASSEMBLY IS FAULTY.

YES (from 2) →
NO TROUBLE FOUND. REFER TO "DIAGNOSTIC AIDS" ON FACING PAGE.

NO (from 1) →

(4)
PRESSURE WITHIN SPECIFICATION, BUT NOT HOLDING.

- IGNITION "OFF."
- CONNECT FUSED JUMPER FROM BATTERY VOLTAGE TO FUEL PUMP TEST CONNECTOR IN ENGINE COMPARTMENT.
- GRADUALLY BLOCK FUEL PRESSURE LINE BY PINCHING FLEX HOSE.
- REMOVE JUMPER FROM TEST CONNECTOR WHILE CONTINUING TO BLOCK PRESSURE LINE.
- PRESSURE SHOULD HOLD. DID IT?

NO →
- REPEAT ABOVE PROCEDURE WHILE BLOCKING FUEL RETURN LINE INSTEAD OF PRESSURE LINE BY PINCHING FLEX HOSE.
- PRESSURE SHOULD HOLD. DID IT?

YES →
CHECK FOR
- LEAKING PUMP COUPLING HOSE
- FAULTY IN-TANK PUMP.

NO →
LOCATE AND REPLACE LEAKING INJECTOR AND DRY SPARK PLUGS FROM FLOODED CYLINDERS.

YES →
PRESSURE REGULATOR ASSEMBLY IS FAULTY.

PRESSURE OUT OF SPECIFICATION.
USE PAGE 2 OF THIS CHART.

NO PRESSURE
USE CHART A-5.

CLEAR CODES AND CONFIRM "CLOSED LOOP" OPERATION AND NO "CHECK ENGINE" LIGHT.

4.3L (VIN Z) TURBOCHARGED ENGINES — A-CHARTS — SYCLONE AND TYPHOON

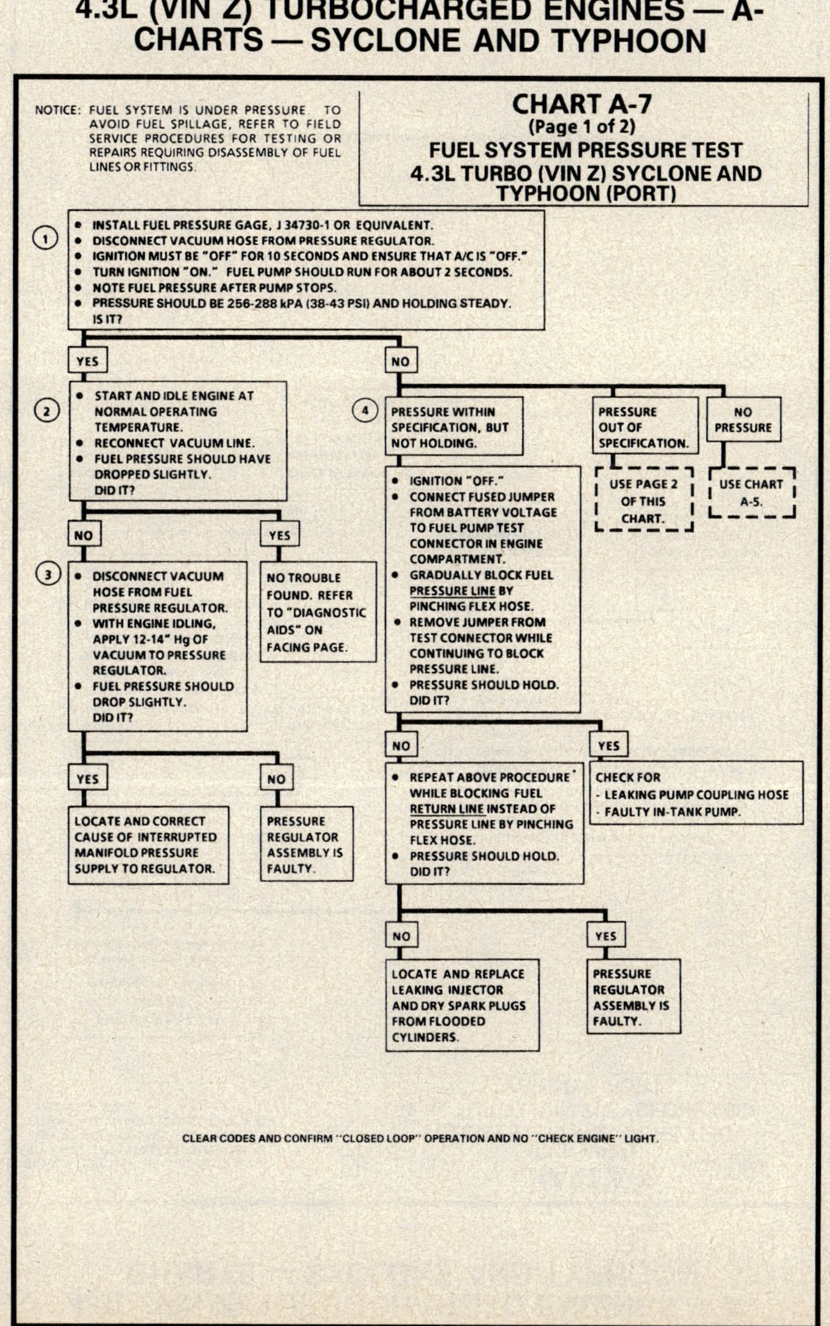

CHART A-7
(Page 2 of 2)
FUEL SYSTEM PRESSURE TEST
4.3 L TURBO (VIN Z) SYCLONE AND TYPHOON (PORT)

Circuit Description:

The fuel pump delivers fuel to the fuel rail and injectors, where the system pressure is controlled to 245 to 409 kPa (35 to 61 psi) by the pressure regulator. Excess fuel is returned to the fuel tank. When the engine is stopped, the pump can be energized by applying battery voltage to the test terminal located in the engine compartment.

Test Description: Numbers below refer to circled numbers on the diagnostic chart.

5. Pressure less than 245 kPa (35 psi) may be caused by one of two problems.
   - The regulated fuel pressure is too low. The system will be running lean and may set Code 44. Also, hard cold starting and overall poor performance is possible.
   - Restricted flow is causing a pressure drop. Normally, a vehicle with a fuel pressure loss at idle will not be driveable. However, if the pressure drop occurs only while driving, the engine will surge and then stop as pressure begins to drop rapidly.

6. Restricting the fuel return line allows the fuel pump to build above regulated pressure. When battery voltage is applied to the pump test terminal, pressure should be above 450 kPa (65 psi).

7. This test determines if the high fuel pressure is due to a restricted fuel return line or a pressure regulator problem.

4.3L (VIN Z) TURBOCHARGED ENGINES — A-CHARTS — SYCLONE AND TYPHOON

4.3L (VIN Z) TURBOCHARGED ENGINES — DIAGNOSTIC CODE CHARTS — SYCLONE AND TYPHOON

NOTICE; FUEL SYSTEM IS UNDER PRESSURE. TO AVOID FUEL SPILLAGE, REFER TO FIELD SERVICE PROCEDURES FOR TESTING OR REPAIRS REQUIRING DISASSEMBLY OF FUEL LINES OR FITTINGS.

CHART A-7
(Page 2 of 2)
FUEL SYSTEM PRESSURE TEST
4.3L TURBO (VIN Z) SYCLONE AND TYPHOON (PORT)

FROM CHART A-7
PAGE 1
PRESSURE OUT OF SPEC.
LESS THAN 256 kPa (38 PSI)
GREATER THAN 288 kPa (43 PSI)

LOW PRESSURE

(5) • CHECK FOR RESTRICTED IN-LINE FILTER. IS FILTER OK?

YES — NO

(6) • IGNITION "OFF."
• APPLY BATTERY VOLTAGE TO FUEL PUMP TEST CONNECTOR.
• GRADUALLY BLOCK FUEL RETURN LINE BY PINCHING FLEXIBLE HOSE AND NOTE PRESSURE.
• PRESSURE SHOULD BE ABOVE 450 kPa (65 psi). IS IT?

REPLACE FILTER AND RE-TEST.

YES — NO

PRESSURE REGULATOR ASSEMBLY IS FAULTY.

CHECK IN-TANK PUMP ASSEMBLY FOR
- FAULTY FUEL PUMP
- LEAKING COUPLING HOSE
- INCORRECT PUMP.

HIGH PRESSURE

(7) • DISCONNECT VACUUM LINE TO PRESSURE REGULATOR.
• DISCONNECT FUEL RETURN LINE FLEXIBLE HOSE.
• ATTACH A 5/16 I.D. FLEX HOSE TO PRESSURE REGULATOR SIDE OF RETURN LINE AND INSERT THE OTHER END INTO AN APPROVED GASOLINE CONTAINER.
• NOTE FUEL PRESSURE WITHIN 2 SECONDS AFTER IGNITION IS TURNED "ON".
• PRESSURE SHOULD BE 256-288 kPa (38-43 PSI). IS IT?

NO — YES

PRESSURE REGULATOR ASSEMBLY IS FAULTY.

LOCATE AND REPAIR RESTRICTED FUEL RETURN LINE.

CLEAR CODES AND CONFIRM "CLOSED LOOP" OPERATION AND NO "CHECK ENGINE" LIGHT.

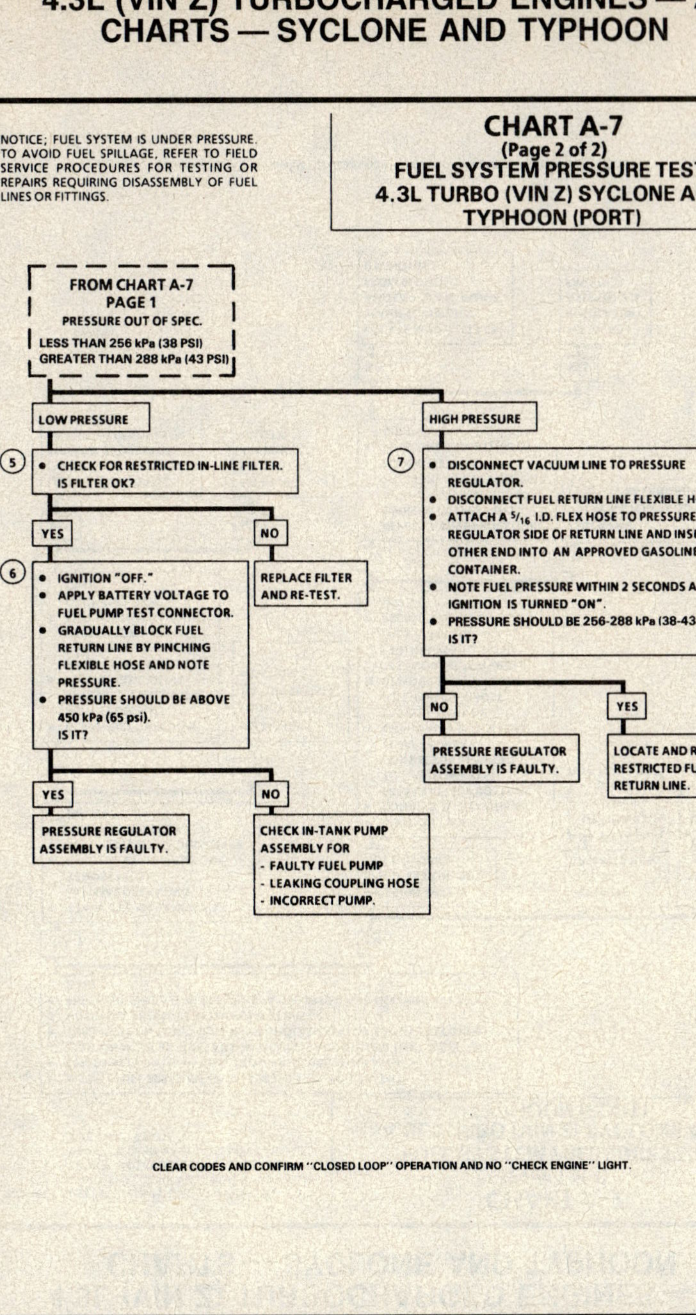

CODE 13
OXYGEN (O₂) SENSOR CIRCUIT
(OPEN CIRCUIT)
4.3L TURBO (VIN Z) SYCLONE AND TYPHOON (PORT)

Circuit Description:

The ECM applies a bias voltage of approximately 450 millivolts (350-550 mV is normal bias voltage) between terminals "GE14" and "GE15." The oxygen (O_2) sensor varies the voltage within a range of about 1 volt if the exhaust is rich, down through about .10 volt if exhaust is lean. Code 13 is set when the voltage does not vary on CKT 412 within a predetermined amount of time.

The sensor is like an open circuit and produces no voltage when it is below 360°C (680°F). An open sensor circuit causes "Open Loop" operation.

Test Description: Numbers below refer to circled numbers on the diagnostic chart.

1. Code 13 will set:
   • Engine coolant temperature greater than 70°C (168°F).
   • At least 2 minutes engine run time after start.
   • O_2 signal voltage steady between .35 and .55 volt.
   • Throttle position sensor signal above 5% (about .3 volt above closed throttle voltage).
   • All conditions must be met for about 60 seconds.
   If the conditions for a Code 13 exist, the system will not go "Closed Loop."
2. This test checks the oxygen sensor's heating element. The heating element resistance should be 3.5 ohms at 20°C (68°F) or 14 ohms at 350°C (562°F).
3. This will determine if the sensor is at fault or the wiring or ECM is the cause of the Code 13.

4. For this test use only a high impedance digital volt ohmmeter (DVM). This test checks the continuity of CKT 412 and CKT 413. If CKT 413 is open, the ECM voltage on CKT 412 will be over .6 volt (600 mV).
5. If the A/C fuse was blown check the A/C control circuit or generator for short circuits.

Diagnostic Aids:

Both the oxygen sensor and heater must be working properly to enable "Closed Loop" operation.

Normal "Scan" voltage varies between 100 mV to 999 mV (.1 and 1.0 volt), while in "Closed Loop." Code 13 sets in one minute, if voltage remains between .35 and .55 volt, but the system will go "Open Loop" in about 15 seconds.

Refer to "Intermittents"

4.3L (VIN Z) TURBOCHARGED ENGINES — DIAGNOSTIC CODE CHARTS — SYCLONE AND TYPHOON

CODE 13
OXYGEN (O_2) SENSOR CIRCUIT (OPEN CIRCUIT)
4.3L TURBO (VIN Z) SYCLONE AND TYPHOON (PORT)

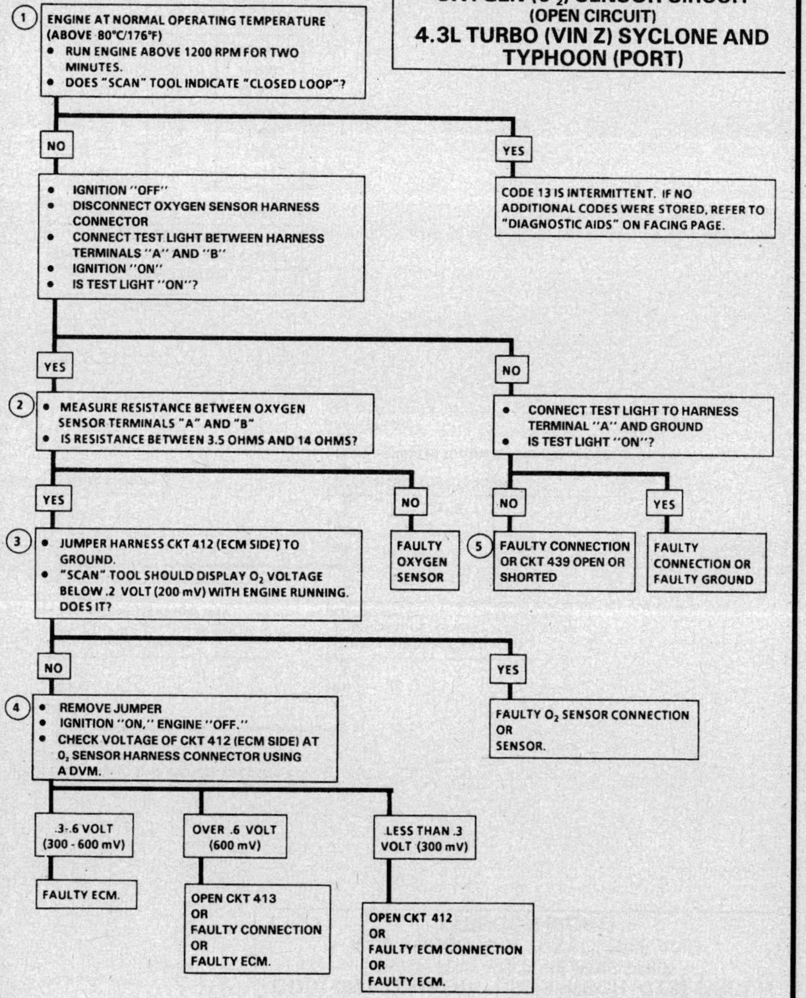

1 ENGINE AT NORMAL OPERATING TEMPERATURE (ABOVE 80°C/176°F)
• RUN ENGINE ABOVE 1200 RPM FOR TWO MINUTES.
• DOES "SCAN" TOOL INDICATE "CLOSED LOOP"?
NO
YES
CODE 13 IS INTERMITTENT. IF NO ADDITIONAL CODES WERE STORED, REFER TO "DIAGNOSTIC AIDS" ON FACING PAGE.
• IGNITION "OFF"
• DISCONNECT OXYGEN SENSOR HARNESS CONNECTOR
• CONNECT TEST LIGHT BETWEEN HARNESS TERMINALS "A" AND "B"
• IGNITION "ON"
• IS TEST LIGHT "ON"?
YES
NO
2 • MEASURE RESISTANCE BETWEEN OXYGEN SENSOR TERMINALS "A" AND "B"
• IS RESISTANCE BETWEEN 3.5 OHMS AND 14 OHMS?
• CONNECT TEST LIGHT TO HARNESS TERMINAL "A" AND GROUND
• IS TEST LIGHT "ON"?
YES
NO
FAULTY OXYGEN SENSOR
NO
YES
5 FAULTY CONNECTION OR CKT 439 OPEN OR SHORTED
FAULTY CONNECTION OR FAULTY GROUND
3 • JUMPER HARNESS CKT 412 (ECM SIDE) TO GROUND.
• "SCAN" TOOL SHOULD DISPLAY O2 VOLTAGE BELOW .2 VOLT (200 mV) WITH ENGINE RUNNING. DOES IT?
NO
YES
FAULTY O2 SENSOR CONNECTION OR SENSOR.
4 • REMOVE JUMPER
• IGNITION "ON," ENGINE "OFF."
• CHECK VOLTAGE OF CKT 412 (ECM SIDE) AT O2 SENSOR HARNESS CONNECTOR USING A DVM.
.3-.6 VOLT (300 - 600 mV)
OVER .6 VOLT (600 mV)
LESS THAN .3 VOLT (300 mV)
FAULTY ECM.
OPEN CKT 413 OR FAULTY CONNECTION OR FAULTY ECM.
OPEN CKT 412 OR FAULTY ECM CONNECTION OR FAULTY ECM.

4.3L (VIN Z) TURBOCHARGED ENGINES — DIAGNOSTIC CODE CHARTS — SYCLONE AND TYPHOON

COOLANT TEMPERATURE SENSOR
A B
452
410
410 YEL
452 BLK
GE16
BB6
ECM
COOLANT SENSOR SIGNAL
5 VOLTS REFERENCE
SENSOR GROUND
TO MAP SENSOR

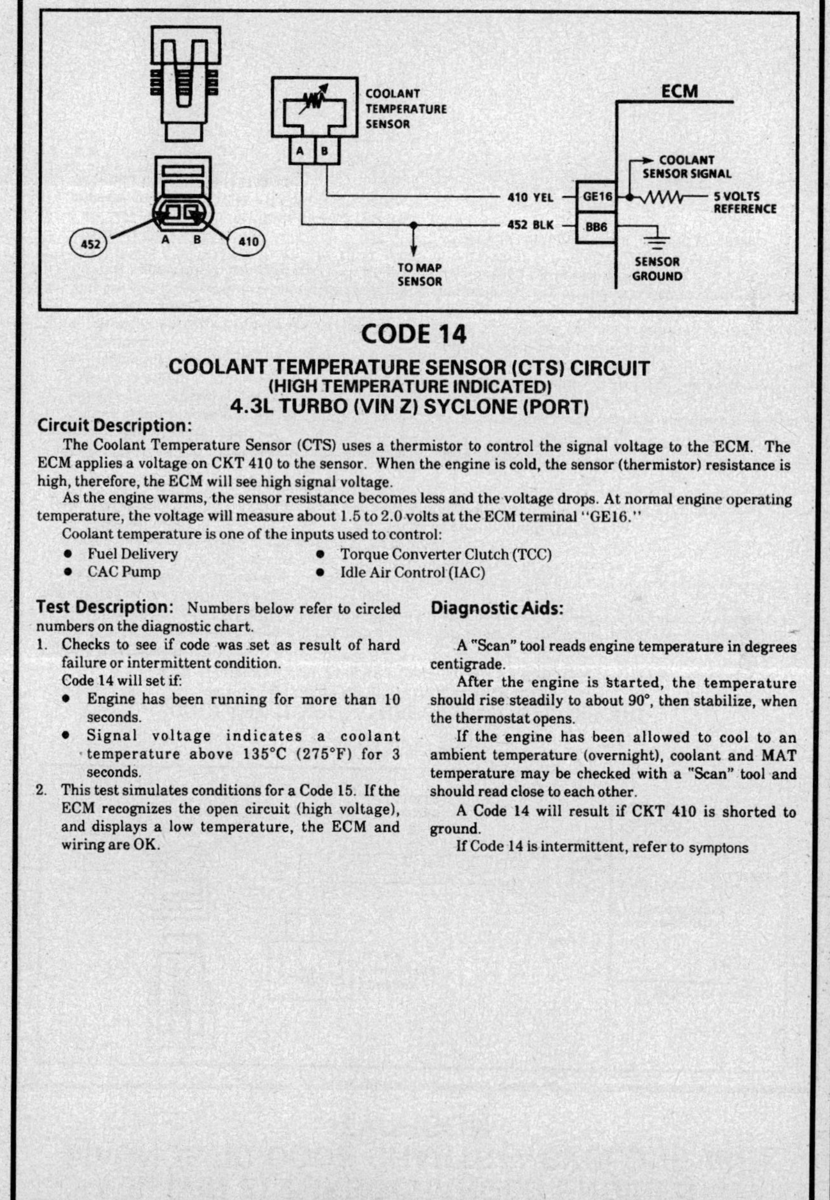

CODE 14
COOLANT TEMPERATURE SENSOR (CTS) CIRCUIT (HIGH TEMPERATURE INDICATED)
4.3L TURBO (VIN Z) SYCLONE (PORT)

Circuit Description:

The Coolant Temperature Sensor (CTS) uses a thermistor to control the signal voltage to the ECM. The ECM applies a voltage on CKT 410 to the sensor. When the engine is cold, the sensor (thermistor) resistance is high, therefore, the ECM will see high signal voltage.

As the engine warms, the sensor resistance becomes less and the voltage drops. At normal engine operating temperature, the voltage will measure about 1.5 to 2.0 volts at the ECM terminal "GE16."

Coolant temperature is one of the inputs used to control:

- Fuel Delivery
- CAC Pump
- Torque Converter Clutch (TCC)
- Idle Air Control (IAC)

Test Description: Numbers below refer to circled numbers on the diagnostic chart.

1. Checks to see if code was set as result of hard failure or intermittent condition.
   Code 14 will set if:
   - Engine has been running for more than 10 seconds.
   - Signal voltage indicates a coolant temperature above 135°C (275°F) for 3 seconds.
2. This test simulates conditions for a Code 15. If the ECM recognizes the open circuit (high voltage), and displays a low temperature, the ECM and wiring are OK.

Diagnostic Aids:

A "Scan" tool reads engine temperature in degrees centigrade.

After the engine is started, the temperature should rise steadily to about 90°, then stabilize, when the thermostat opens.

If the engine has been allowed to cool to an ambient temperature (overnight), coolant and MAT temperature may be checked with a "Scan" tool and should read close to each other.

A Code 14 will result if CKT 410 is shorted to ground.

If Code 14 is intermittent, refer to symptons

4.3L (VIN Z) TURBOCHARGED ENGINES — DIAGNOSTIC CODE CHARTS — SYCLONE AND TYPHOON

CODE 14
COOLANT TEMPERATURE SENSOR (CTS) CIRCUIT
(HIGH TEMPERATURE INDICATED)
4.3L TURBO (VIN Z) SYCLONE AND TYPHOON (PORT)

1. DOES "SCAN" TOOL DISPLAY COOLANT TEMPERATURE OF 130°C OR HIGHER?

— YES

2. • DISCONNECT SENSOR. "SCAN" TOOL SHOULD DISPLAY TEMPERATURE BELOW -30°C. DOES IT?

— NO: CODE 14 IS INTERMITTENT. IF NO ADDITIONAL CODES WERE STORED, REFER TO "DIAGNOSTIC AIDS" ON FACING PAGE.

— YES: REPLACE SENSOR.

— NO: CKT 410 SHORTED TO GROUND, OR CKT 410 SHORTED TO SENSOR GROUND CIRCUIT OR FAULTY ECM.

DIAGNOSTIC AID

COOLANT SENSOR		
TEMPERATURE VS. RESISTANCE VALUES (APPROXIMATE)		
°F	°C	OHMS
210	100	185
160	70	450
100	38	1,800
70	20	3,400
40	4	7,500
20	-7	13,500
0	-18	25,000
-40	-40	100,700

CLEAR CODES AND CONFIRM "CLOSED LOOP" OPERATION AND NO "CHECK ENGINE" LIGHT.

4.3L (VIN Z) TURBOCHARGED ENGINES — DIAGNOSTIC CODE CHARTS — SYCLONE AND TYPHOON

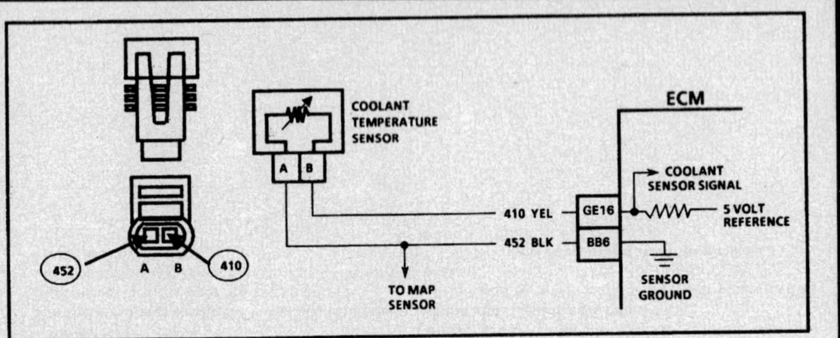

CODE 15
COOLANT TEMPERATURE SENSOR (CTS) CIRCUIT
(LOW TEMPERATURE INDICATED)
4.3L TURBO (VIN Z) SYCLONE AND TYPHOON (PORT)

Circuit Description:
The Coolant Temperature Sensor (CTS) uses a thermistor to control the signal voltage to the ECM. The ECM applies a voltage on CKT 410 to the sensor. When the engine is cold, the sensor (thermistor) resistance is high, therefore, the ECM will see high signal voltage.

As the engine warms, the sensor resistance becomes less and the voltage drops. At normal engine operating temperature, the voltage will measure about 1.5 volts to 2.0 volts at the ECM terminal "GE16."

Coolant temperature is one of the inputs used to control:
- Fuel Delivery
- CAC Pump
- Torque Converter Clutch (TCC)
- Idle Air Control (IAC)

Test Description: Numbers below refer to circled numbers on the diagnostic chart.
1. Checks to see if code was set as result of hard failure or intermittent condition.
   Code 15 will set if:
   - Engine has been running for more than 50 seconds.
   - Signal voltage indicates a coolant temperature below -30°C (-22°F).
2. This test simulates conditions for a Code 14. If the ECM recognizes the grounded circuit (low voltage) and displays a high temperature, the ECM and wiring are OK.
3. This test will determine if there is a wiring problem or a faulty ECM. If CKT 452 is open, there may also be a Code 21 stored.

Diagnostic Aids:

A "Scan" tool reads engine temperature in degrees centigrade.

After the engine is started, the temperature should rise steadily to about 90°C (194°F), then stabilize when the thermostat opens.

If the engine has been allowed to cool to an ambient temperature (overnight), coolant and MAT temperature may be checked with a "Scan" tool and should read close to each other.

A Code 15 will result if CKTs 410 or 452 are open. If Code 15 is intermittent, refer to symptons.

4.3L (VIN Z) TURBOCHARGED ENGINES — DIAGNOSTIC CODE CHARTS — SYCLONE AND TYPHOON

CODE 15
**COOLANT TEMPERATURE SENSOR (CTS) CIRCUIT
(LOW TEMPERATURE INDICATED)
4.3L TURBO (VIN Z) SYCLONE AND
TYPHOON (PORT)**

① • DOES "SCAN" TOOL DISPLAY COOLANT TEMPERATURE OF -30°C OR LESS?

YES

② • DISCONNECT SENSOR.
• JUMPER HARNESS TERMINALS TOGETHER.
• "SCAN" TOOL SHOULD DISPLAY 130°C OR MORE. DOES IT?

NO
CODE 15 IS INTERMITTENT. IF NO ADDITIONAL CODES WERE STORED, REFER TO "DIAGNOSTIC AIDS" ON FACING PAGE.

NO

③ • JUMPER CKT 410 TO GROUND.
• "SCAN" TOOL SHOULD DISPLAY OVER 130°C. DOES IT?

YES
FAULTY CONNECTION OR SENSOR.

YES
OPEN SENSOR GROUND CIRCUIT, FAULTY CONNECTION OR FAULTY ECM.

NO
OPEN CKT 410, FAULTY CONNECTION AT ECM, OR FAULTY ECM.

DIAGNOSTIC AID

COOLANT SENSOR		
TEMPERATURE TO RESISTANCE VALUES (APPROXIMATE)		
°F	°C	OHMS
210	100	185
160	70	450
100	38	1,800
70	20	3,400
40	4	7,500
20	-7	13,500
0	-18	25,000
-40	-40	100,700

CLEAR CODES AND CONFIRM "CLOSED LOOP" OPERATION AND NO "CHECK ENGINE" LIGHT.

4.3L (VIN Z) TURBOCHARGED ENGINES — DIAGNOSTIC CODE CHARTS — SYCLONE AND TYPHOON

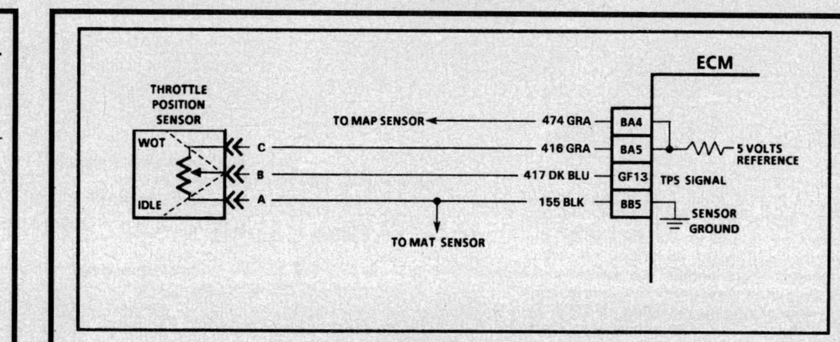

CODE 21

**THROTTLE POSITION SENSOR (TPS) CIRCUIT
(SIGNAL VOLTAGE HIGH)
4.3L TURBO (VIN Z) SYCLONE AND TYPHOON (PORT)**

Circuit Description:

The Throttle Position Sensor (TPS) provides a voltage signal that changes relative to the throttle valve. Signal voltage will vary from less than 1.0 volt at idle to about 4.6 volts at wide open throttle (WOT).

The TPS signal is one of the most important inputs used by the ECM for fuel control and for many of the ECM controlled outputs.

Test Description: Numbers below refer to circled numbers on the diagnostic chart.
1. This step checks to see if Code 21 is the result of a hard failure or an intermittent condition.
   A Code 21 will set if:
   • TPS reading above 2.5 volts
   • MAP reading below 70 kPa (M/T) or 81 kPa (A/T)
   • Engine speed less than 1300 rpm
   • All of the above conditions are present for 10 seconds
2. This step simulates conditions for a Code 22. If the ECM recognizes the change of state, the ECM and CKTs 416 and 417 are OK.
3. This step isolates a faulty sensor, ECM, or an open CKT 155.

Diagnostic Aids:

A "Scan" tool displays throttle position in volts. Closed throttle voltage should be less than 1.0 volt. TPS voltage should increase at a steady rate as throttle is moved to WOT.

A Code 21 will result if CKT 155 is open or CKT 417 is shorted to voltage. If Code 21 is intermittent, refer to symptons.

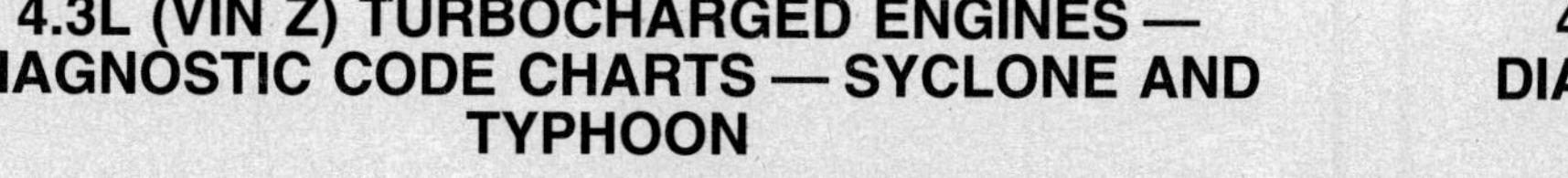

4.3L (VIN Z) TURBOCHARGED ENGINES — DIAGNOSTIC CODE CHARTS — SYCLONE AND TYPHOON

CODE 21
THROTTLE POSITION SENSOR (TPS) CIRCUIT
(SIGNAL VOLTAGE HIGH)
4.3L TURBO (VIN Z) SYCLONE AND TYPHOON (PORT)

① • THROTTLE CLOSED.
   DOES "SCAN" TOOL DISPLAY TPS OVER 2.5 VOLTS?

YES → ② • DISCONNECT SENSOR. "SCAN" TOOL SHOULD DISPLAY TPS BELOW .2 VOLT (200mV). DOES IT?

NO → CODE 21 IS INTERMITTENT. IF NO ADDITIONAL CODES WERE STORED, REFER TO "DIAGNOSTIC AIDS" ON FACING PAGE.

YES → ③ • PROBE SENSOR GROUND CIRCUIT WITH A TEST LIGHT CONNECTED TO BATTERY VOLTAGE.

NO → CKT 417 SHORTED TO VOLTAGE OR FAULTY ECM.

LIGHT "ON" → FAULTY CONNECTION OR SENSOR.

LIGHT "OFF" → OPEN SENSOR GROUND CIRCUIT OR FAULTY ECM.

CLEAR CODES AND CONFIRM "CLOSED LOOP" OPERATION AND NO "CHECK ENGINE" LIGHT.

4.3L (VIN Z) TURBOCHARGED ENGINES — DIAGNOSTIC CODE CHARTS — SYCLONE AND TYPHOON

CODE 22
THROTTLE POSITION SENSOR (TPS) CIRCUIT
(SIGNAL VOLTAGE LOW)
4.3L TURBO (VIN Z) SYCLONE AND TYPHOON (PORT)

Circuit Description:

The Throttle Position Sensor (TPS) provides a voltage signal that changes relative to the throttle valve. Signal voltage will vary from less than 1.0 volt at idle to about 4.5 volts at wide open throttle (WOT).

The TPS signal is one of the most important inputs used by the ECM for fuel control and for many of the ECM controlled outputs.

Test Description: Numbers below refer to circled numbers on the diagnostic chart.

1. Code 22 will set if:
   - Engine is running
   - TPS signal voltage is less than .2 volt for 4 seconds
2. Simulates Code 21: (high voltage). If ECM recognizes the high signal voltage the ECM and wiring are OK.
3. With closed throttle, ignition "ON" or at idle, voltage at "GF13" should be .36-.44 volt. If not, replace the TPS.

4. Simulates a high signal voltage. Checks CKT 417 for an open.

Diagnostic Aids:

A "Scan" tool reads throttle position in volts. Voltage should increase at a steady rate as throttle is moved toward WOT.

Also some "Scan" tools will read throttle angle 0% = closed throttle, 100% = WOT.

An open or short to ground in CKTs 416 or 417 will result in a Code 22.

- **Poor Connection or Damaged Harness.** Inspect ECM harness connectors for backed out terminal "GF13," improper mating, broken locks, improperly formed or damaged terminals, poor terminal to wire connection, and damaged harness.
- **Intermittent Test.** If connections and harness check OK, monitor TPS voltage display while moving related connectors and wiring harness. If the failure is induced, the display will change. This may help to isolate the location of the malfunction.
- **TPS Scaling.** Observe TPS voltage display while depressing accelerator pedal with engine stopped and ignition "ON." Display should vary from closed throttle TPS voltage when throttle was closed, to over 4.5 volts (4500 mV) when throttle is held at wide open throttle position.

4.3L (VIN Z) TURBOCHARGED ENGINES — DIAGNOSTIC CODE CHARTS — SYCLONE AND TYPHOON

CODE 22
THROTTLE POSITION SENSOR (TPS) CIRCUIT
(SIGNAL VOLTAGE LOW)
4.3L TURBO (VIN Z) SYCLONE AND TYPHOON (PORT)

1. • THROTTLE CLOSED.
   • DOES "SCAN" DISPLAY TPS .2V (200 mV) OR BELOW?

YES

2. • DISCONNECT TPS SENSOR.
   • JUMPER CKTS 416 & 417 TOGETHER.
   • "SCAN" SHOULD DISPLAY TPS OVER 4.0 V (4000 mV). DOES IT?

NO
• CODE 22 IS INTERMITTENT.
  IF NO ADDITIONAL CODES WERE STORED, REFER TO "DIAGNOSTIC AIDS" ON FACING PAGE.

NO

4. • PROBE CKT 417 WITH A TEST LIGHT CONNECTED TO BATTERY VOLTAGE.
   • "SCAN" TOOL SHOULD DISPLAY TPS OVER 4.0V (4000 mV). DOES IT?

YES
3. • REFER TO FACING PAGE FOR SPECIFIC INSTRUCTIONS.

YES
CKT 416 OPEN OR SHORTED TO GROUND
OR
FAULTY CONNECTION
OR
FAULTY ECM.

NO
CKT 417 OPEN OR SHORTED TO GROUND, OR SHORTED TO SENSOR GROUND CIRCUIT
OR
FAULTY ECM CONNECTION
OR
FAULTY ECM.

CLEAR CODES AND CONFIRM "CLOSED LOOP" OPERATION AND NO "CHECK ENGINE" LIGHT.

4.3L (VIN Z) TURBOCHARGED ENGINES — DIAGNOSTIC CODE CHARTS — SYCLONE AND TYPHOON

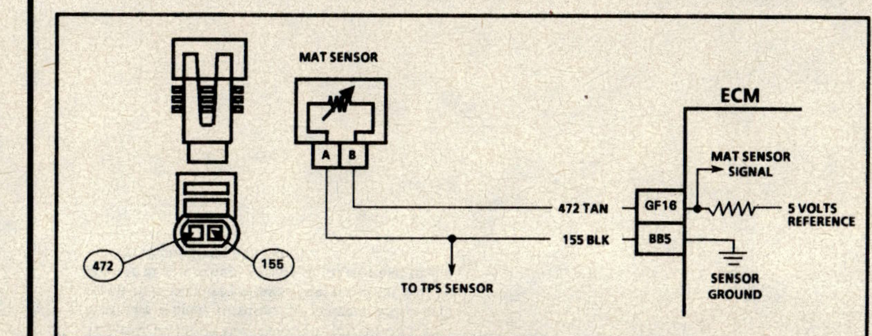

CODE 23
MANIFOLD AIR TEMPERATURE (MAT) SENSOR CIRCUIT
(LOW TEMPERATURE INDICATED)
4.3L TURBO (VIN Z) SYCLONE AND TYPHOON (PORT)

Circuit Description:
The Manifold Air Temperature (MAT) sensor uses a thermistor to control the signal voltage to the ECM. The ECM applies a voltage (about 5 volts) on CKT 472 to the sensor. When the air is cold the sensor (thermistor) resistance is high, therefore the ECM will see a high signal voltage. If the air is warm, the sensor resistance is low, therefore, the ECM will detect a low voltage.

Test Description: Numbers below refer to circled numbers on the diagnostic chart.

Code 23 will set if:
- A signal voltage indicates a manifold air temperature below −30°C (−22°F).
- Boost conditions are present.

Due to the conditions necessary to set a Code 23, the "Check Engine" light will only stay "ON" while the fault is present.

1. A "Scan" tool may not be used to diagnose this fault, due to the ECM transmitting "default" (substitute) values while the fault is present. A Code 23 will set due to an open sensor, wire, or connection. This test will determine if the wiring and ECM are OK.

2. If the resistance is greater than 25,000 ohms, replace the sensor.

4.3L (VIN Z) TURBOCHARGED ENGINES — DIAGNOSTIC CODE CHARTS — SYCLONE AND TYPHOON

CODE 23

MANIFOLD AIR TEMPERATURE (MAT) SENSOR CIRCUIT
(LOW TEMPERATURE INDICATED)
4.3L TURBO (VIN Z) SYCLONE AND TYPHOON (PORT)

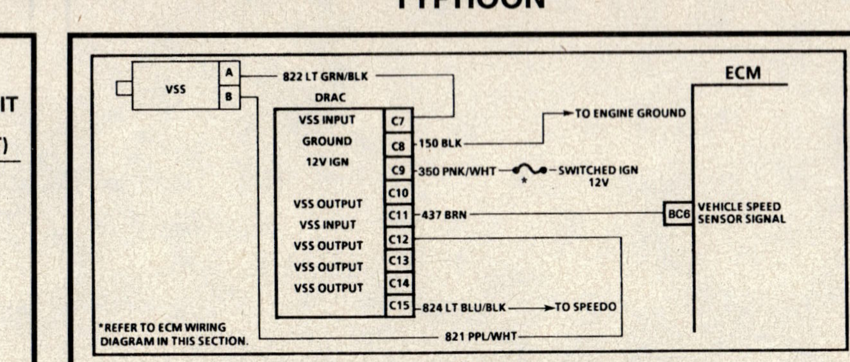

MAT SENSOR		
TEMPERATURE TO RESISTANCE VALUES (APPROXIMATE)		
°F	°C	OHMS
210	100	185
160	70	450
100	38	1,800
70	20	3,400
40	4	7,500
20	-7	13,500
0	-18	25,000
-40	-40	100,700

CLEAR CODES AND CONFIRM "CLOSED LOOP" OPERATION AND NO "CHECK ENGINE" LIGHT.

4.3L (VIN Z) TURBOCHARGED ENGINES — DIAGNOSTIC CODE CHARTS — SYCLONE AND TYPHOON

CODE 24

VEHICLE SPEED SENSOR (VSS) CIRCUIT
4.3L TURBO (VIN Z) SYCLONE AND TYPHOON (PORT)

Circuit Description:

The ECM applies and monitors 12 volts on CKT 437. CKT 437 connects to the DRAC, which alternately grounds CKT 437, when receiving voltage pulses from Vehicle Speed Sensor (VSS) when drive wheels are turning. This pulsing action takes place about 2000 times per mile and the ECM will calculate vehicle speed based on the time between "pulses."

A "Scan" tool reading should closely match the speedometer reading with drive wheels turning.

Test Description: Number(s) below refer to circled number(s) on the diagnostic chart.

1. Code 24 will set if:
   - CKT 437 voltage is constant.
   - Engine speed is more than 1200 rpm.
   - Vehicle speed signal indicates less than 2 mph (3 km/h) on Tech 1.
   - Not in park or neutral.
   - All conditions must be met for 5 seconds.
   These conditions are met during a road load deceleration.
2. This test determines if the DRAC is receiving the A/C signal from the VSS.
3. This test monitors the DRAC voltage on CKT 437. With the wheels turning, the pulsing action will result in a varying voltage. The variation will be greater at low wheel speeds to an average of 4-6 volts at about 20 mph (32 km/h).

Diagnostic Aids:

1. "Scan" reading should closely match speedometer reading, with drive wheels turning.
2. Check Park/Neutral (P/N) switch diagnosis chart.
3. If Park/Neutral (P/N) switch is OK, refer to "Intermittents" in "Symptoms,"

4.3L (VIN Z) TURBOCHARGED ENGINES — DIAGNOSTIC CODE CHARTS — SYCLONE AND TYPHOON

CODE 24
VEHICLE SPEED SENSOR (VSS) CIRCUIT
4.3L TURBO (VIN Z) SYCLONE AND TYPHOON (PORT)

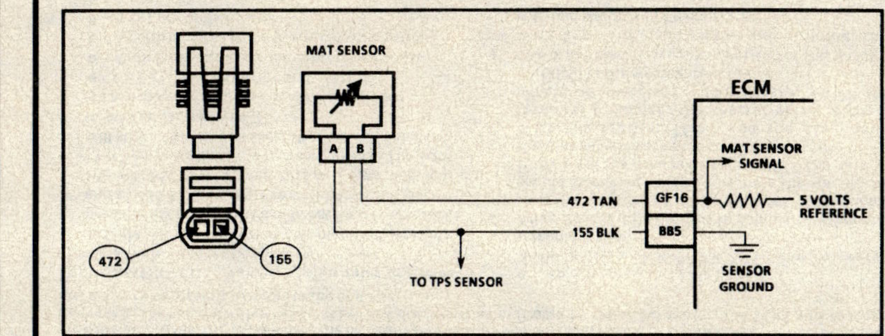

"AFTER REPAIRS," REFER TO CODE CRITERIA ON FACING PAGE AND CONFIRM CODE DOES NOT RESET.

4.3L (VIN Z) TURBOCHARGED ENGINES — DIAGNOSTIC CODE CHARTS — SYCLONE AND TYPHOON

CODE 25
MANIFOLD AIR TEMPERATURE (MAT) SENSOR CIRCUIT
(HIGH TEMPERATURE INDICATED)
4.3L TURBO (VIN Z) SYCLONE AND TYPHOON (PORT)

Circuit Description:

The Manifold Air Temperature (MAT) sensor uses a thermistor to control the signal voltage to the ECM. The ECM applies a voltage (4-6 volts) on CKT 472 to the sensor. When manifold air is cold, the sensor (thermistor) resistance is high, therefore the ECM will detect a high signal voltage. If the air warms, the sensor resistance becomes less, and the voltage drops.

Test Description: Numbers below refer to circled numbers on the diagnostic chart.

Code 25 will set if the engine is not experiencing turbocharger boost and the following conditions are met:

- Signal voltage indicates a manifold air temperature greater than 135°C (275°F).
- The above requirement is met for at least 30 seconds.

Due to the conditions necessary to set a Code 25, the "Check Engine" light will only stay "ON" while the fault is present.

1. A "Scan" tool may not be used to diagnose this fault due to the ECM transmitting "default" (substitute) values while the fault is present. If voltage is above 4 volts, the ECM and wiring are OK.
2. If the resistance is less than 100 ohms, replace the sensor.

4.3L (VIN Z) TURBOCHARGED ENGINES — DIAGNOSTIC CODE CHARTS — SYCLONE AND TYPHOON

CODE 25
MANIFOLD AIR TEMPERATURE (MAT) SENSOR CIRCUIT
(HIGH TEMPERATURE INDICATED)
4.3L TURBO (VIN Z) SYCLONE AND TYPHOON (PORT)

NOTE: A "SCAN" TOOL MAY NOT BE USED TO DIAGNOSE THIS FAULT, DUE TO THE ECM TRANSMITTING "DEFAULT" (SUBSTITUTE) VALUES WHEN THE FAULT IS PRESENT.

① • DISCONNECT MAT SENSOR.
• IGNITION "ON," ENGINE STOPPED.
• CHECK VOLTAGE BETWEEN HARNESS CONNECTOR TERMINALS.

4 VOLTS OR OVER

② CHECK RESISTANCE ACROSS MAT SENSOR TERMINALS. SHOULD BE MORE THAN 185 OHMS. SEE TABLE FOR APPROXIMATE TEMPERATURE TO RESISTANCE VALUES.

BELOW 4 VOLTS

CKT 472 SHORTED TO GROUND.
OR
CKT 472 SHORTED TO SENSOR GROUND CIRCUIT.
OR
FAULTY ECM.

OK

INTERMITTENT FAULT IN SENSOR CIRCUIT OR CONNECTOR. IF ADDITIONAL CODES WERE STORED, SEE APPLICABLE CHART. IF NO CODES, REFER TO "INTERMITTENTS."

NOT OK

REPLACE SENSOR

MAT SENSOR		
TEMPERATURE TO RESISTANCE VALUES (APPROXIMATE)		
°F	°C	OHMS
210	100	185
160	70	450
100	38	1,600
70	20	3,400
40	4	7,500
20	-7	13,500
0	-18	25,000
-40	-40	100,700

CLEAR CODES AND CONFIRM "CLOSED LOOP" OPERATION AND NO "CHECK ENGINE" LIGHT.

4.3L (VIN Z) TURBOCHARGED ENGINES — DIAGNOSTIC CODE CHARTS — SYCLONE AND TYPHOON

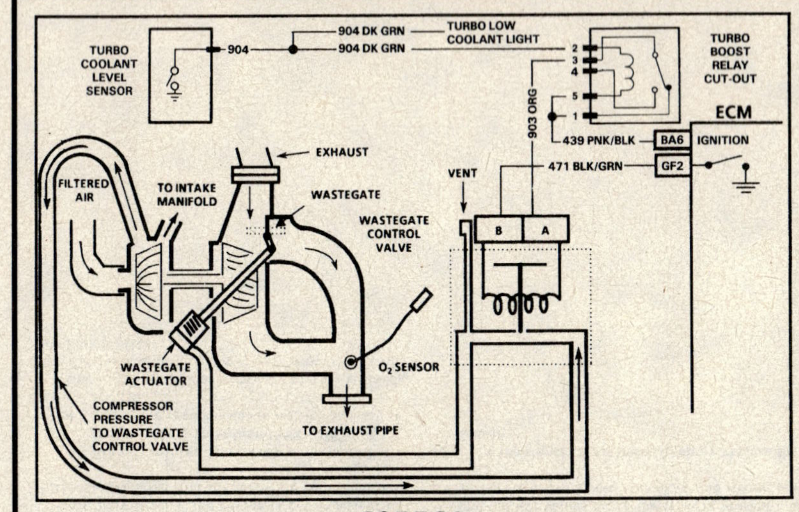

CODE 31
(Page 1 of 2)
TURBO WASTEGATE OVERBOOST
4.3L TURBO (VIN Z) SYCLONE AND TYPHOON (PORT)

Circuit Description:
On turbo charged engines, the exhaust gases pass from the exhaust manifold through the turbocharger, turning the turbine blades. The compressor side of the turbocharger also turns, pulling air through the air filter and pushing the air into the intake manifold, pressurizing the intake manifold.

The wastegate is normally closed, but opens to bypass exhaust gas to prevent an overboost condition. The wastegate will open when pressure is applied to the actuator, and is controlled by a wastegate control solenoid valve pulsed "ON" and "OFF" by the ECM. Under normal driving conditions, the control solenoid is energized all the time which closes "OFF" the manifold pressure to the wastegate actuator. This allows for a rapid increase in boost pressure. A boost increase will be detected by the MAP sensor, and the ECM will pulse the wastegate control valve. Manifold pressure will then be allowed to pass to the wastegate actuator, and the actuator will open the wastegate. This will prevent an overboost condition on heavy acceleration. As boost pressure decreases, the ECM closes the control valve and the wastegate actuator pressure bleeds "OFF" through the vent in the control valve. If an overboost does exist as indicated by the MAP sensor, the ECM will reduce fuel delivery to prevent damage to the engine.

Test Description: Numbers below refer to circled numbers on the diagnostic chart.

1. A Code 31 will set when the manifold absolute pressure exceeds 205 kPa of boost for two seconds, and a Code 33 has not previously been set. Code 31 will set, but the "Check Engine" light will stay "ON" only while the overboost exists. The light will stay "ON" for 10 seconds after the condition exists and then go "OUT."

An overboost condition could be caused by:
• CKT 471 shorted to ground
• A sticking wastegate actuator or wastegate
• A control valve stuck in the closed position
• A cut or pinched hose
• A faulty ECM

• An extremely dirty air filter
With ignition shut "OFF," the control valve solenoid is open.

2. After the 103 kPa (15 psi) is applied to valve and then the pressure source is removed, the actuator should slowly move back and close the wastegate. If the pressure does not bleed "OFF," the vent in the control valve solenoid could be plugged.

3. With the ignition "ON" and the diagnostic terminal grounded, the control valve solenoid should be energized. This closes "OFF" the manifold to the wastegate actuator.

4. Check the electrical control portion of the system. With key "ON" and engine not running, the solenoid should not be energized.

4.3L (VIN Z) TURBOCHARGED ENGINES — DIAGNOSTIC CODE CHARTS — SYCLONE AND TYPHOON

CODE 31
(Page 1 of 2)
TURBO WASTEGATE OVERBOOST
4.3L TURBO (VIN Z) SYCLONE AND TYPHOON (PORT)

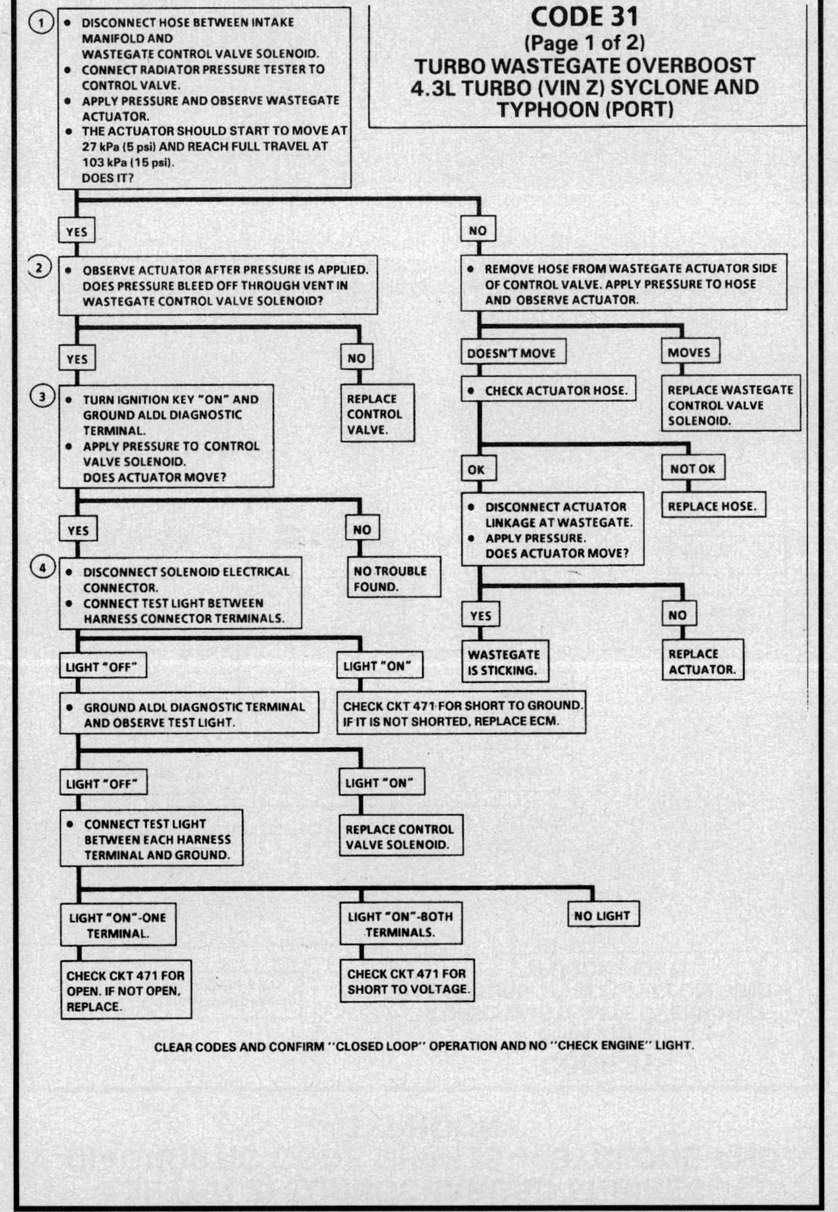

CLEAR CODES AND CONFIRM "CLOSED LOOP" OPERATION AND NO "CHECK ENGINE" LIGHT.

4.3L (VIN Z) TURBOCHARGED ENGINES — DIAGNOSTIC CODE CHARTS — SYCLONE AND TYPHOON

CODE 31
(Page 2 of 2)
TURBO WASTEGATE OVERBOOST
4.3L TURBO (VIN Z) SYCLONE AND TYPHOON (PORT)

Circuit Description:

If the level of coolant in the charge air cooler (CAC) is low, the turbo boost cut-out relay will shut off all power to the wastegate solenoid. This will open the wastegate to bypass exhaust gas and lower the boost. The "LOW TURBO COOLANT" lamp will remain "ON" until coolant is added.

4.3L (VIN Z) TURBOCHARGED ENGINES — DIAGNOSTIC CODE CHARTS — SYCLONE AND TYPHOON

CODE 31
(Page 2 of 2)
TURBO WASTEGATE OVERBOOST
4.3L TURBO (VIN Z) SYCLONE AND TYPHOON (PORT)

4.3L (VIN Z) TURBOCHARGED ENGINES — DIAGNOSTIC CODE CHARTS — SYCLONE AND TYPHOON

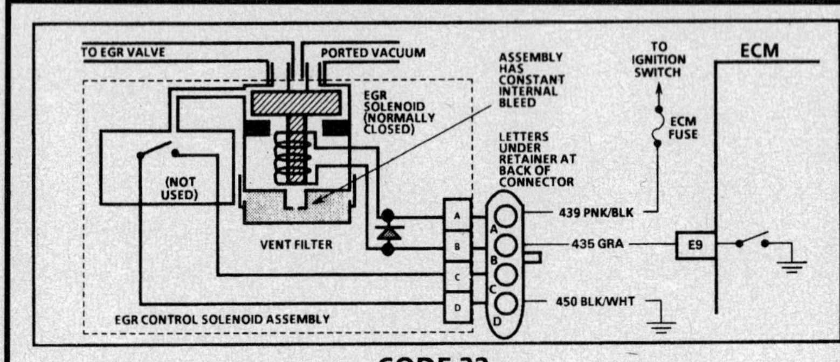

CODE 32

EXHAUST GAS RECIRCULATION (EGR) CIRCUIT
4.3L TURBO (VIN Z) SYCLONE AND TYPHOON (PORT)

Circuit Description:

The ECM operates a vacuum solenoid to control the Exhaust Gas Recirculation (EGR) valve. This solenoid is normally closed. By providing a ground path, the ECM energizes the solenoid, allowing vacuum to reach the EGR valve.

Under certain conditions, when the EGR valve is normally open, the ECM tests the EGR function by de-energizing the EGR control solenoid, blocking vacuum to the EGR valve diaphragm. Without EGR, the system will sense a lean condition and will increase the fuel integrator rate in response. The ECM monitors the amount of fuel delivery increase. If the increase is below a specified value, the ECM will interpret that the test was failed. The failure indicates that closing the EGR valve when it would normally be open does not make a significant change, indicating a problem in the EGR system.

Test Description: Numbers below refer to circled numbers on the diagnostic chart.

The diagnostic chart covers checks for the entire EGR system. If no trouble is found but Code 32 was set, an intermittent electrical condition or a sticky EGR valve is at fault.

Diagnostic Aids:

The vacuum switch in the EGR solenoid assembly is not used.

An EGR valve stuck open will cause a rough idle.

A plugged EGR solenoid vent filter could cause the EGR valve to remain open or to close slowly.

An inoperative check valve in the ported vacuum line will result in faulty EGR system operation.

4.3L (VIN Z) TURBOCHARGED ENGINES — DIAGNOSTIC CODE CHARTS — SYCLONE AND TYPHOON

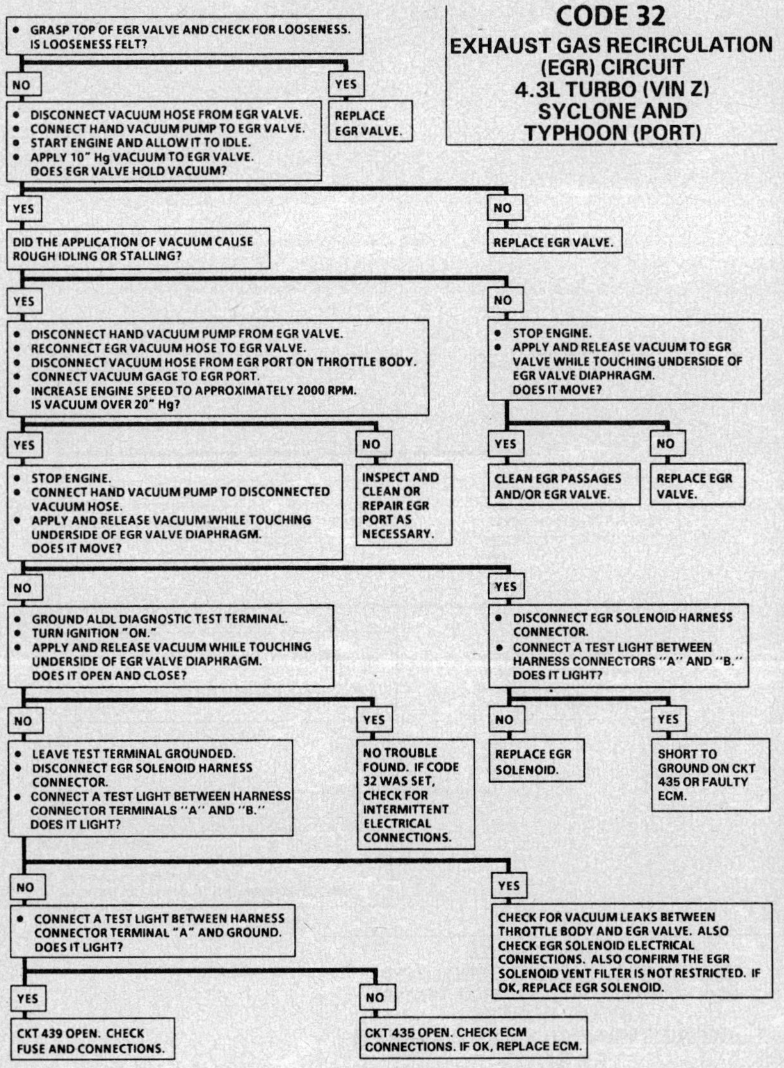

GRASP TOP OF EGR VALVE AND CHECK FOR LOOSENESS. IS LOOSENESS FELT?

- NO →
  - DISCONNECT VACUUM HOSE FROM EGR VALVE.
  - CONNECT HAND VACUUM PUMP TO EGR VALVE.
  - START ENGINE AND ALLOW IT TO IDLE.
  - APPLY 10" Hg VACUUM TO EGR VALVE. DOES EGR VALVE HOLD VACUUM?
- YES → REPLACE EGR VALVE.

DID THE APPLICATION OF VACUUM CAUSE ROUGH IDLING OR STALLING? (YES)

- NO → REPLACE EGR VALVE.

YES →
- DISCONNECT HAND VACUUM PUMP FROM EGR VALVE.
- RECONNECT EGR VACUUM HOSE TO EGR VALVE.
- DISCONNECT VACUUM HOSE FROM EGR PORT ON THROTTLE BODY.
- CONNECT VACUUM GAGE TO EGR PORT.
- INCREASE ENGINE SPEED TO APPROXIMATELY 2000 RPM. IS VACUUM OVER 20" Hg?

- YES →
  - STOP ENGINE.
  - CONNECT HAND VACUUM PUMP TO DISCONNECTED VACUUM HOSE.
  - APPLY AND RELEASE VACUUM WHILE TOUCHING UNDERSIDE OF EGR VALVE DIAPHRAGM. DOES IT MOVE?
- NO → INSPECT AND CLEAN OR REPAIR EGR PORT AS NECESSARY.

(from STOP ENGINE branch:)
- NO →
  - GROUND ALDL DIAGNOSTIC TEST TERMINAL.
  - TURN IGNITION "ON."
  - APPLY AND RELEASE VACUUM WHILE TOUCHING UNDERSIDE OF EGR VALVE DIAPHRAGM. DOES IT OPEN AND CLOSE?
- YES →
  - STOP ENGINE.
  - APPLY AND RELEASE VACUUM TO EGR VALVE WHILE TOUCHING UNDERSIDE OF EGR VALVE DIAPHRAGM. DOES IT MOVE?
    - YES → CLEAN EGR PASSAGES AND/OR EGR VALVE.
    - NO → REPLACE EGR VALVE.

- NO →
  - LEAVE TEST TERMINAL GROUNDED.
  - DISCONNECT EGR SOLENOID HARNESS CONNECTOR.
  - CONNECT A TEST LIGHT BETWEEN HARNESS CONNECTOR TERMINALS "A" AND "B." DOES IT LIGHT?
- YES → NO TROUBLE FOUND. IF CODE 32 WAS SET, CHECK FOR INTERMITTENT ELECTRICAL CONNECTIONS.

YES (from 2000 RPM branch) →
- DISCONNECT EGR SOLENOID HARNESS CONNECTOR.
- CONNECT A TEST LIGHT BETWEEN HARNESS CONNECTORS "A" AND "B." DOES IT LIGHT?
  - NO → REPLACE EGR SOLENOID.
  - YES → SHORT TO GROUND ON CKT 435 OR FAULTY ECM.

- NO →
  - CONNECT A TEST LIGHT BETWEEN HARNESS CONNECTOR TERMINAL "A" AND GROUND. DOES IT LIGHT?
- YES → CHECK FOR VACUUM LEAKS BETWEEN THROTTLE BODY AND EGR VALVE. ALSO CHECK EGR SOLENOID ELECTRICAL CONNECTIONS. ALSO CONFIRM THE EGR SOLENOID VENT FILTER IS NOT RESTRICTED. IF OK, REPLACE EGR SOLENOID.

- YES → CKT 439 OPEN. CHECK FUSE AND CONNECTIONS.
- NO → CKT 435 OPEN. CHECK ECM CONNECTIONS. IF OK, REPLACE ECM.

CODE 32
EXHAUST GAS RECIRCULATION (EGR) CIRCUIT
4.3L TURBO (VIN Z) SYCLONE AND TYPHOON (PORT)

4.3L (VIN Z) TURBOCHARGED ENGINES — DIAGNOSTIC CODE CHARTS — SYCLONE AND TYPHOON

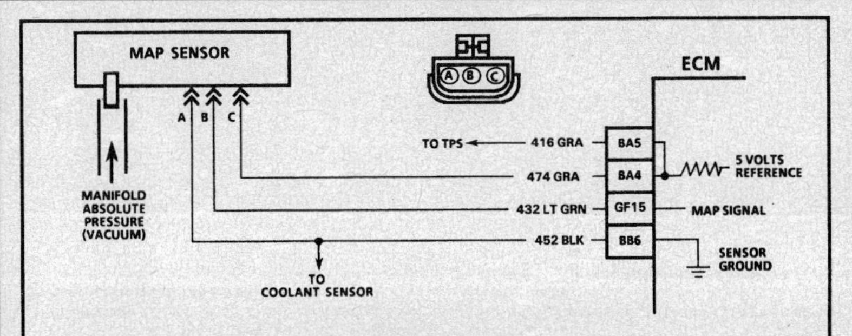

CODE 33
MANIFOLD ABSOLUTE PRESSURE (MAP) SENSOR CIRCUIT
(SIGNAL VOLTAGE HIGH - LOW VACUUM)
4.3L TURBO (VIN Z) SYCLONE AND TYPHOON (PORT)

Circuit Description:

The Manifold Absolute Pressure (MAP) sensor responds to changes in manifold pressure (vacuum). The ECM receives this information as a signal voltage that will vary from less than 1.0 volt at closed throttle idle, to 4-4.5 volts at wide open throttle (low vacuum or boost).

If the MAP sensor fails, the ECM will substitute a fixed MAP value and use the Throttle Position Sensor (TPS) to control fuel delivery.

Test Description: Numbers below refer to circled numbers on the diagnostic chart.

1. This step will determine if Code 33 is the result of a hard failure or an intermittent condition.
   A Code 33 will set if:
   - MAP signal voltage is too high (low vacuum)
   - TPS less than 2%
   - These conditions for a time longer than 5 seconds

2. This step simulates conditions for a Code 34. If the ECM recognizes the change, the ECM and CKTs 474 and 432 are OK. If CKT 452 is open, there may also be a stored Code 23.

Diagnostic Aids:

With the ignition "ON" and the engine stopped, the manifold pressure is equal to atmospheric pressure and the signal voltage will be approximately 2.5 volts. This information is used by the ECM as an indication of vehicle altitude and is referred to as BARO. Comparison of this BARO reading with a known good vehicle with the same sensor is a good way to check accuracy of a "suspect" sensor. Reading should be the same ± .4 volt.

A Code 33 will result if CKT 452 is open, or if CKT 432 is shorted to voltage or to CKT 474.

If Code 33 is intermittent, refer to symptons.

4.3L (VIN Z) TURBOCHARGED ENGINES — DIAGNOSTIC CODE CHARTS — SYCLONE AND TYPHOON

CODE 33
MANIFOLD ABSOLUTE PRESSURE (MAP) SENSOR CIRCUIT
(SIGNAL VOLTAGE HIGH - LOW VACUUM)
4.3L TURBO (VIN Z) SYCLONE AND TYPHOON (PORT)

①
- IF ENGINE IDLE IS ROUGH, UNSTABLE OR INCORRECT, CORRECT BEFORE USING CHART. SEE SYMPTOMS IN SECTION B.
- ENGINE IDLING
- DOES "SCAN" DISPLAY A MAP OF 2.0 VOLTS OR OVER?

YES → ②
- IGNITION "OFF"
- DISCONNECT MAP SENSOR ELECTRICAL CONNECTOR.
- IGNITION "ON"
- "SCAN" SHOULD READ A VOLTAGE OF 1 VOLT OR LESS. DOES IT?

NO → CODE 33 IS INTERMITTENT. IF NO ADDITIONAL CODES WERE STORED, REFER TO DIAGNOSTIC AIDS ON FACING PAGE.

YES →
- PROBE CKT 452 WITH A TEST LIGHT TO BATTERY VOLTAGE.
- TEST LIGHT SHOULD LIGHT. DOES IT?

NO → CKT 432 SHORTED TO VOLTAGE, SHORTED TO CKT 474, OR FAULTY ECM

YES → PLUGGED OR LEAKING SENSOR VACUUM HOSE OR FAULTY MAP SENSOR

NO → OPEN CIRCUIT 452

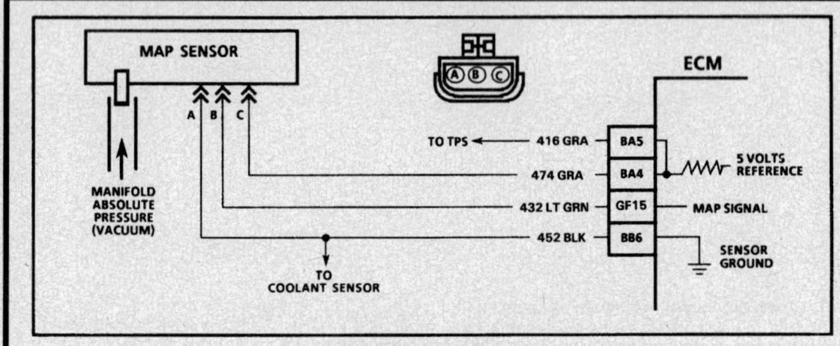

CODE 34
MANIFOLD ABSOLUTE PRESSURE (MAP) SENSOR CIRCUIT
(SIGNAL VOLTAGE LOW - HIGH VACUUM)
4.3L TURBO (VIN Z) SYCLONE AND TYPHOON (PORT)

Circuit Description:

The Manifold Absolute Pressure (MAP) sensor responds to changes in manifold pressure (vacuum). The ECM receives this information as a signal voltage that will vary from less than 1.0 volt at closed throttle idle, to 4-4.5 volts at wide open throttle.

If the MAP sensor fails, the ECM will substitute a fixed MAP value and use the Throttle Position Sensor (TPS) to control fuel delivery.

Test Description: Numbers below refer to circled numbers on the diagnostic chart.

1. This step determines if Code 34 is the result of a hard failure or an intermittent condition.
   A Code 34 will set when:
   - MAP signal voltage is too low
   - Engine speed below 1200 rpm and/or TPS greater than 20%
2. Jumpering harness terminals "B" to "C", 5 volts to signal, will determine if the sensor is at fault or if there is a problem with the ECM or wiring.
3. The "Scan" tool may not display battery voltage. The important thing is that the ECM recognizes the voltage as more than 4 volts, indicating that the ECM and CKT 432 are OK.

Diagnostic Aids:

With the ignition "ON" and the engine stopped, the manifold pressure is equal to atmospheric pressure and the signal voltage will be approximately 2.5 volts. This information is used by the ECM as an indication of vehicle altitude and is referred to as BARO. Comparison of this BARO reading with a known good vehicle with the same sensor is a good way to check accuracy of a "suspect" sensor. Reading should be the same ± .4 volt.

A Code 34 will result if CKTs 474 or 432 are open or shorted to ground.

If CKT 416 is shorted to ground, there may also be a stored Code 22.

If Code 34 is intermittent, refer to symptons.

4.3L (VIN Z) TURBOCHARGED ENGINES — DIAGNOSTIC CODE CHARTS — SYCLONE AND TYPHOON

CODE 34

MANIFOLD ABSOLUTE PRESSURE (MAP) SENSOR CIRCUIT
(SIGNAL VOLTAGE LOW - HIGH VACUUM)
4.3L TURBO (VIN Z) SYCLONE AND TYPHOON (PORT)

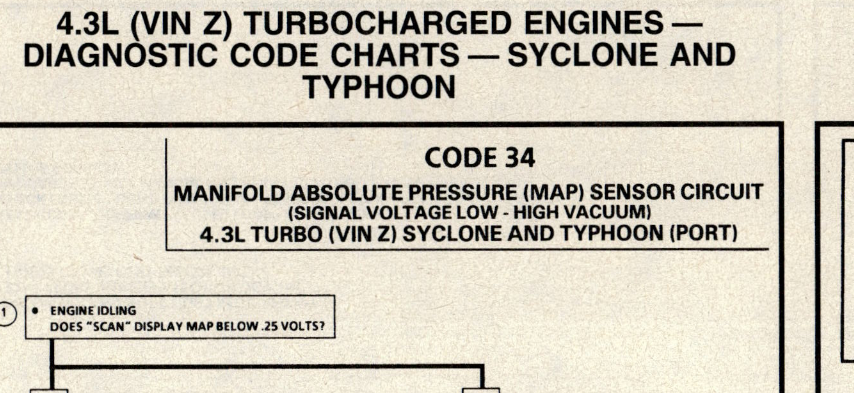

4.3L (VIN Z) TURBOCHARGED ENGINES — DIAGNOSTIC CODE CHARTS — SYCLONE AND TYPHOON

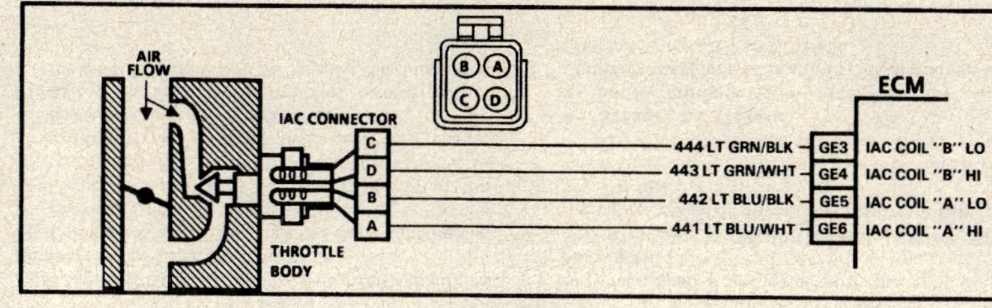

CODE 35

IDLE SPEED ERROR
4.3L TURBO (VIN Z) SYCLONE AND TYPHOON (PORT)

Circuit Description:

The ECM controls engine idle speed with the IAC valve. To increase idle speed, the ECM retracts the IAC valve pintle away from its seat, allowing more air to bypass the throttle bore. To decrease idle speed, it extends the IAC valve pintle towards its seat, reducing bypass air flow. A "Scan" tool will read the ECM commands to the IAC valve in counts. Higher counts indicate more air bypass (higher idle). The lower the counts indicate less air is allowed to bypass (lower idle). Code 35 will set when the closed throttle engine speed is 225 rpm above or below the desired (commanded) idle speed for 20 seconds. Review the general description of the IAC operation in section "C2."

Test Description: Numbers below refer to circled numbers on the diagnostic chart.

1. The Tech 1 rpm control mode is used to extend and retract the IAC valve. The valve should move smoothly within the specified range. If the idle speed is commanded (IAC extended) too low (below 700 rpm), the engine may stall. This may be normal and would not indicate a problem. Retracting the IAC beyond its controlled range (above 1500 rpm) will cause a delay before the rpm's start dropping. This too is normal.

2. This test uses the Tech 1 to command the IAC controlled idle speed. The ECM issues commands to obtain commended idle speed. The node lights each should flash red and green to indicate a good circuit as the ECM issues commands. While the sequence of color is not important if either light is "OFF" or does not flash red and green, check the circuits for faults, beginning with poor terminal contacts.

Diagnostic Aids:

A slow, unstable, or fast idle may be caused by a non--IAC system problem that cannot be overcome by the IAC valve. Out of control range IAC "Scan" tool counts will be above 60 if idle is too low, and zero counts if idle is too high. The following checks should be made to repair a non-IAC system problem:
- Vacuum Leak (High Idle)
  If idle is too high, stop the engine. Fully extend (low) IAC with tester. Start engine. If idle speed is above 800 rpm, locate and correct vacuum leak including PCV system. Also check for binding of throttle blade or linkage.

- System too lean (High Air/Fuel Ratio)
  The idle speed may be too high or too low. Engine speed may vary up and down and disconnecting the IAC valve does not help. Code 44 may be set. "Scan O_2 voltage will be less than 300 mV (.3 volts). Check for low regulated fuel pressure, water in the fuel or a restricted injector.
- System too rich (Low Air/Fuel Ratio)
  The idle speed will be too low. "Scan" tool IAC Counts will usually be above 80. System is obviously rich and may exhibit black smoke in exhaust.

"Scan" tool O_2 voltage will be fixed above 800 m V (.8 volt).

Check for high fuel pressure, leaking or sticking injector. Silicone contaminated O_2 "Scan" voltage will be slow to respond.
- Throttle Body
  Remove IAC valve and inspect bore for foreign material.
- IAC Valve Electrical Connections
  IAC valve connections should be carefully checked for proper contact.
- PCV Valve
  An incorrect or faulty PCV valve may result in an incorrect idle speed.
- Refer to "Rough, Unstable, Incorrect Idle or Stalling" in "Symptoms," in
- If intermittent poor driveability or idle symptoms are resolved by disconnecting the IAC, carefully recheck connections, valve terminal resistance, or replace IAC.

4.3L (VIN Z) TURBOCHARGED ENGINES — DIAGNOSTIC CODE CHARTS — SYCLONE AND TYPHOON

CODE 35
IDLE SPEED ERROR
4.3L TURBO (VIN Z) SYCLONE AND TYPHOON (PORT)

(1)
- ENGINE AT NORMAL OPERATING TEMPERATURE IN PARK/NEUTRAL WITH PARKING BRAKE SET.
- A/C "OFF."
- SELECT RPM CONTROL. (MISC. TESTS)
- CYCLE IAC THROUGH ITS RANGE FROM 700 RPM UP TO 1500 RPM USING SCAN TOOL.
- RPM SHOULD CHANGE SMOOTHLY. DOES IT?

NO → (2)
- INSTALL IAC NODE LIGHT * IN IAC HARNESS.
- ENGINE RUNNING. CYCLE IAC WITH TECH 1.
- EACH NODE LIGHT SHOULD CYCLE RED AND GREEN BUT NEVER "OFF." DO THEY?

NO →
IF CIRCUIT(S) DID NOT TEST RED AND GREEN, CHECK FOR:
- FAULTY CONNECTOR TERMINAL CONTACTS.
- OPEN CIRCUITS INCLUDING CONNECTORS.
- CIRCUITS SHORTED TO GROUND OR VOLTAGE.
- FAULTY ECM CONNECTIONS OR REPLACE ECM. REPAIR AS NECESSARY AND RETEST.

YES →
- CHECK IAC CONNECTIONS.
- CHECK IAC PASSAGES.
- IF OK, REPLACE IAC.

YES →
- USING THE IAC DRIVER * OR OTHER CONVENIENT CONNECTOR, CHECK RESISTANCE ACROSS IAC COILS.
- SHOULD BE 40 TO 80 OHMS BETWEEN IAC TERMINALS "A" TO "B" AND "C" TO "D."

OK →
- CHECK RESISTANCE BETWEEN IAC TERMINALS "B" AND "D" AND "A" AND "C."
- SHOULD BE INFINITE.

NOT OK → REPLACE IAC VALVE AND RETEST.

OK → IDLE AIR CONTROL CIRCUIT OK. REFER TO "DIAGNOSTIC AIDS" ON FACING PAGE.

NOT OK → REPLACE IAC VALVE AND RETEST.

* IAC DRIVER AND NODE LIGHT REQUIRED KIT
222-L FROM: CONCEPT TECHNOLOGY, INC.
J 37027 FROM: KENT-MOORE, INC.

CLEAR CODES, CONFIRM "CLOSED LOOP" OPERATION, NO "CHECK ENGINE" LIGHT, PERFORM IAC RESET PROCEDURE PER APPLICABLE SERVICE MANUAL AND VERIFY CONTROLLED IDLE SPEED IS CORRECT.

4.3L (VIN Z) TURBOCHARGED ENGINES — DIAGNOSTIC CODE CHARTS — SYCLONE AND TYPHOON

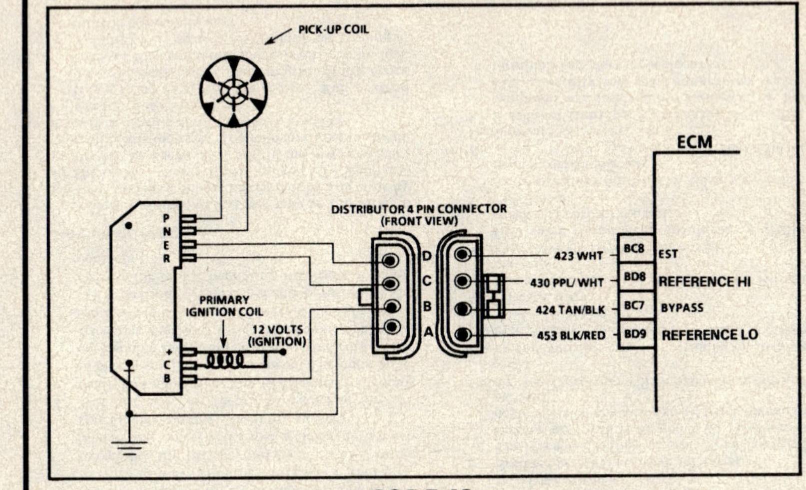

CODE 42
ELECTRONIC SPARK TIMING (EST) CIRCUIT
4.3L TURBO (VIN Z) SYCLONE AND TYPHOON (PORT)

Circuit Description:

The ignition module sends a reference signal (CKT 430) to the ECM when the engine is cranking. While the engine speed is under 400 rpm, the ignition module will control ignition timing. When the engine speed exceeds 400 rpm, the ECM applies 5 volts to the bypass line (CKT 424) to switch the timing to ECM control (EST CKT 423).

When the system is running "ON" the ignition module, that is, no voltage on the bypass line, the ignition module grounds the EST signal. The ECM expects to see no voltage on the EST line during this condition. If it sees a voltage, it sets Code 42 and will not go into the EST mode.

When the rpm for EST is reached (about 400 rpm), voltage will be applied to the bypass line, the EST should no longer be grounded in the ignition module, so the EST voltage should be varying.

If the bypass line is open or grounded, the ignition module will not switch to EST mode, so the EST voltage will be low and Code 42 will be set.

If the EST line is grounded, the ignition module will switch to EST but, because the line is grounded, there will be no EST signal. A Code 42 will be set.

Test Description: Numbers below refer to circled numbers on the diagnostic chart.

1. Confirms Code 42 and that the fault causing the code is present.
2. Checks for a normal EST ground path through the ignition module. An EST CKT 423 shorted to ground will also read less than 500 ohms, however, this will be checked later.
3. As the test light voltage touches terminal "C7," the module should switch, causing the ohmmeter to "overrange" if the meter is in the 1000-2000 ohms position. Selecting the 10-20,000 ohms position will indicate above 5000 ohms. The important thing is that the module "switched."
4. The module did not switch and this step checks for:
   - EST CKT 423 shorted to ground
   - Bypass CKT 424 open
   - Faulty ignition module connection or module
5. Confirms that Code 42 is a faulty ECM and not an intermittent in CKTs 423 or 424.

4.3L (VIN Z) TURBOCHARGED ENGINES — DIAGNOSTIC CODE CHARTS — SYCLONE AND TYPHOON

CODE 42
ELECTRONIC SPARK TIMING (EST) CIRCUIT
4.3L TURBO (VIN Z) SYCLONE AND TYPHOON (PORT)

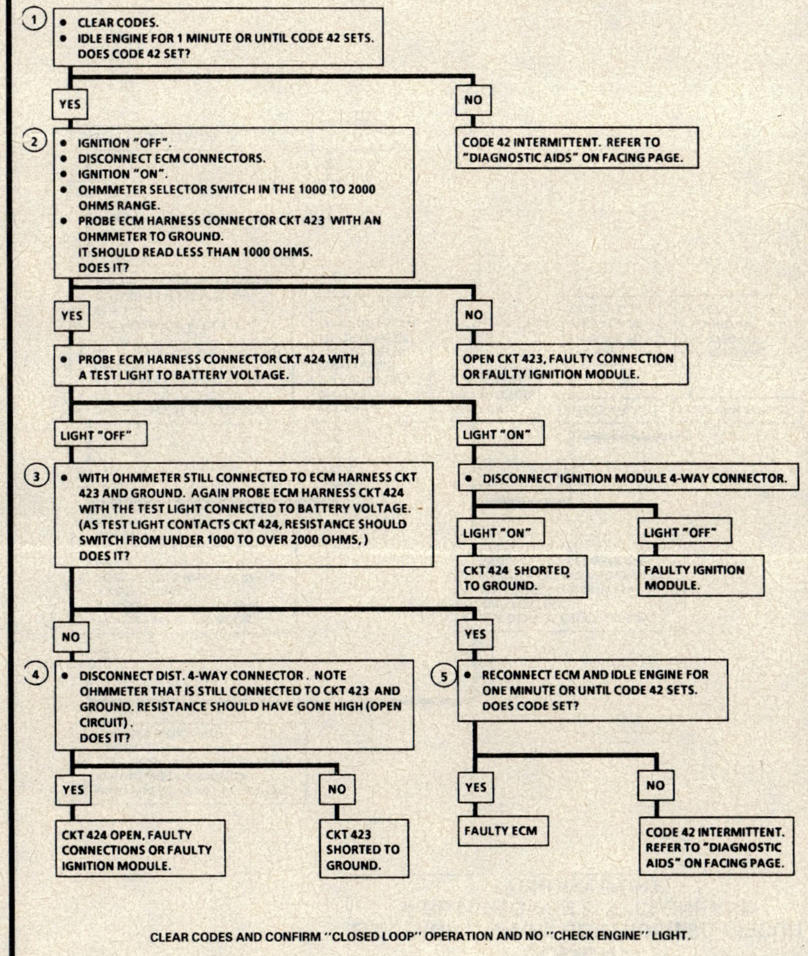

1. • CLEAR CODES.
   • IDLE ENGINE FOR 1 MINUTE OR UNTIL CODE 42 SETS.
   DOES CODE 42 SET?

   YES →

2. • IGNITION "OFF".
   • DISCONNECT ECM CONNECTORS.
   • IGNITION "ON".
   • OHMMETER SELECTOR SWITCH IN THE 1000 TO 2000 OHMS RANGE.
   • PROBE ECM HARNESS CONNECTOR CKT 423 WITH AN OHMMETER TO GROUND.
   IT SHOULD READ LESS THAN 1000 OHMS.
   DOES IT?

   YES →

   • PROBE ECM HARNESS CONNECTOR CKT 424 WITH A TEST LIGHT TO BATTERY VOLTAGE.

   LIGHT "OFF" →

3. • WITH OHMMETER STILL CONNECTED TO ECM HARNESS CKT 423 AND GROUND. AGAIN PROBE ECM HARNESS CKT 424 WITH THE TEST LIGHT CONNECTED TO BATTERY VOLTAGE. (AS TEST LIGHT CONTACTS CKT 424, RESISTANCE SHOULD SWITCH FROM UNDER 1000 TO OVER 2000 OHMS.)
   DOES IT?

   NO →

4. • DISCONNECT DIST. 4-WAY CONNECTOR. NOTE OHMMETER THAT IS STILL CONNECTED TO CKT 423 AND GROUND. RESISTANCE SHOULD HAVE GONE HIGH (OPEN CIRCUIT).
   DOES IT?

   YES → CKT 424 OPEN, FAULTY CONNECTIONS OR FAULTY IGNITION MODULE.

   NO → CKT 423 SHORTED TO GROUND.

NO branches:
- CODE 42 INTERMITTENT. REFER TO "DIAGNOSTIC AIDS" ON FACING PAGE.
- OPEN CKT 423, FAULTY CONNECTION OR FAULTY IGNITION MODULE.

LIGHT "ON" →
• DISCONNECT IGNITION MODULE 4-WAY CONNECTOR.

- **LIGHT "ON"** → CKT 424 SHORTED TO GROUND.
- **LIGHT "OFF"** → FAULTY IGNITION MODULE.

YES →

5. • RECONNECT ECM AND IDLE ENGINE FOR ONE MINUTE OR UNTIL CODE 42 SETS.
   DOES CODE SET?

   YES → FAULTY ECM

   NO → CODE 42 INTERMITTENT. REFER TO "DIAGNOSTIC AIDS" ON FACING PAGE.

CLEAR CODES AND CONFIRM "CLOSED LOOP" OPERATION AND NO "CHECK ENGINE" LIGHT.

4.3L (VIN Z) TURBOCHARGED ENGINES — DIAGNOSTIC CODE CHARTS — SYCLONE AND TYPHOON

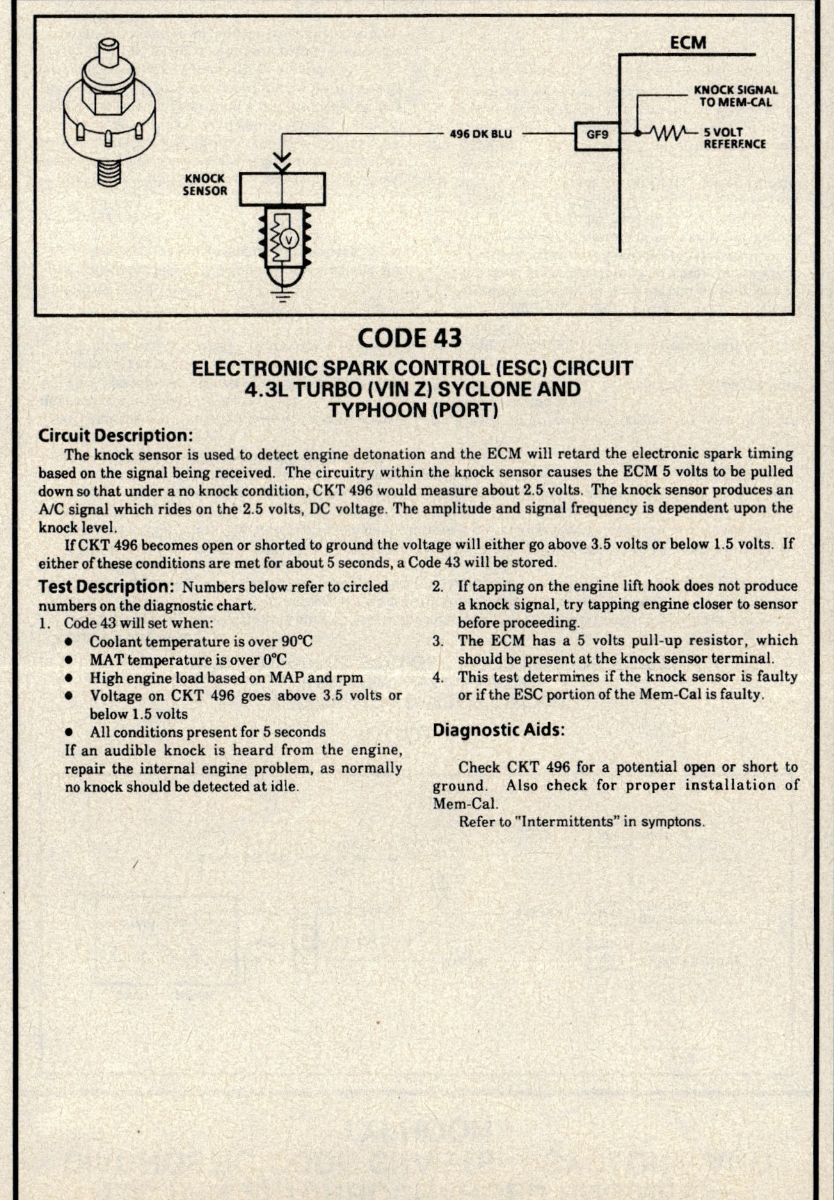

CODE 43
ELECTRONIC SPARK CONTROL (ESC) CIRCUIT
4.3L TURBO (VIN Z) SYCLONE AND TYPHOON (PORT)

Circuit Description:

The knock sensor is used to detect engine detonation and the ECM will retard the electronic spark timing based on the signal being received. The circuitry within the knock sensor causes the ECM 5 volts to be pulled down so that under a no knock condition, CKT 496 would measure about 2.5 volts. The knock sensor produces an A/C signal which rides on the 2.5 volts, DC voltage. The amplitude and signal frequency is dependent upon the knock level.

If CKT 496 becomes open or shorted to ground the voltage will either go above 3.5 volts or below 1.5 volts. If either of these conditions are met for about 5 seconds, a Code 43 will be stored.

Test Description: Numbers below refer to circled numbers on the diagnostic chart.

1. Code 43 will set when:
   • Coolant temperature is over 90°C
   • MAT temperature is over 0°C
   • High engine load based on MAP and rpm
   • Voltage on CKT 496 goes above 3.5 volts or below 1.5 volts
   • All conditions present for 5 seconds

   If an audible knock is heard from the engine, repair the internal engine problem, as normally no knock should be detected at idle.

2. If tapping on the engine lift hook does not produce a knock signal, try tapping engine closer to sensor before proceeding.
3. The ECM has a 5 volts pull-up resistor, which should be present at the knock sensor terminal.
4. This test determines if the knock sensor is faulty or if the ESC portion of the Mem-Cal is faulty.

Diagnostic Aids:

Check CKT 496 for a potential open or short to ground. Also check for proper installation of Mem-Cal.

Refer to "Intermittents" in symptons.

4.3L (VIN Z) TURBOCHARGED ENGINES — DIAGNOSTIC CODE CHARTS — SYCLONE AND TYPHOON

CODE 43
ELECTRONIC SPARK CONTROL (ESC) CIRCUIT
4.3L TURBO (VIN Z) SYCLONE AND TYPHOON (PORT)

1. • WITH ENGINE IDLING AT NORMAL OPERATING TEMPERATURE.
   • USE "SCAN" TOOL AND OBSERVE KNOCK SIGNAL.
   • IS KNOCK INDICATED?

2. • TAP ON ENGINE LIFT HOOK BRACKET WHILE OBSERVING KNOCK SIGNAL.
   • TOOL SHOULD INDICATE KNOCK WHILE TAPPING ON BRACKET. (SEE NOTE ON FACING PAGE). DOES IT?

3. • DISCONNECT KNOCK SENSOR.
   • IGNITION "ON".
   • CHECK VOLTAGE BETWEEN HARNESS CKT 496 AND GROUND. SHOULD READ BETWEEN 4-6 VOLTS. DOES IT?

4. • DISCONNECT KNOCK SENSOR.
   • CONNECT VOLTMETER TO KNOCK SENSOR AND ENGINE GROUND.
   • SET VOLTMETER ON 2 VOLT AC SCALE.
   • TAP ON ENGINE BLOCK NEAR SENSOR.
   • IS A SIGNAL INDICATED ON VOLTMETER WHILE TAPPING ON BLOCK?

- CODE 43 IS INTERMITTENT. IF NO ADDITIONAL CODES WERE STORED, REFER TO "DIAGNOSIC AIDS" ON FACING PAGE.
- FAULTY CONNECTOR OR SENSOR
- OVER 6 VOLTS → CKT 496 SHORTED TO VOLTAGE OR FAULTY ECM.
- LESS THAN 4 VOLTS → CKT 496 OPEN, SHORTED TO GROUND, OR FAULTY ECM.
- FAULTY KNOCK SENSOR.
- FAULTY MEM-CAL

CLEAR CODES AND CONFIRM "CLOSED LOOP" OPERATION AND NO "CHECK ENGINE" LIGHT.

4.3L (VIN Z) TURBOCHARGED ENGINES — DIAGNOSTIC CODE CHARTS — SYCLONE AND TYPHOON

CODE 44
OXYGEN (O₂) SENSOR CIRCUIT
(LEAN EXHAUST INDICATED)
4.3L TURBO (VIN Z) SYCLONE AND TYPHOON (PORT)

Circuit Description:

The ECM applies a bias voltage of approximately 450 millivolts (350-550 m V is normal bias voltage) between terminals "GE14" and "GE15" (if measured with a 10 megohm digital voltmeter, this may read as low as .32 volt). The oxygen (O_2) sensor varies the voltage within a range of about 1 volt, if the exhaust is rich, down through about .10 volt, if exhaust is lean.

A lean exhaust condition will cause the oxygen sensor to output a low voltage, which will pull the bias voltage from CKT 412 low. The ECM is programmed to interpret any voltage less than 500 m V as a "lean exhaust condition."

The sensor is like an open circuit and produces no voltage when it is below about 316°C (600°F). An open sensor circuit causes "Open Loop" operation. The heating element in the 0² sensor causes the sensor to heat up quickly, allowing for quicker closed loop operation.

Test Description: Numbers below refer to circled numbers on the diagnostic chart.

1. This test checks the oxygen sensor's heating element. The heating element resistance should be 3.5 ohms at 20°C (68°F) or 14 ohms at 350°C (562°F).
2. Code 44 is set when the oxygen (O_2) sensor signal voltage on CKT 412;
   - Remains below .2 volt for 2 minutes and the fuel system is operating in "Closed Loop."

Diagnostic Aids:

Using the Tech I diagnostic computer "Scan" tool, observe the block learn values at different rpm and air flow conditions. The "Scan" tool also displays the block cells, so the block learn values can be checked in each of the cells to determine when the Code 44 may have been set. If the conditions for Code 44 or Code 64 exist, the block learn values will be around 150.

- O_2 Sensor Wire: Sensor pigtail may be mispositioned and contacting the exhaust manifold.
- Check for intermittent ground in wire between connector and sensor.
- Poor connection at oxygen (O_2) sensor ground wire.
- Lean Injector(s): Perform injector balance test CHART C-2A.
- Fuel Contamination: Water, even in small amounts, near the in-tank fuel pump inlet can be delivered to the injectors. The water causes a lean exhaust and can set a Code 44 or Code 64.
- Fuel Pressure: System will be lean if pressure is too low. It may be necessary to monitor fuel pressure while driving the car at various road speeds and/or loads to confirm. Refer to Fuel System Diagnosis CHART A-7.
- Exhaust Leaks: If there is an exhaust leak, the engine can cause outside air to be pulled into the exhaust and past the sensor. Vacuum or crankcase leaks can cause a lean condition.
- If the above are OK, it is a faulty oxygen sensor.

4.3L (VIN Z) TURBOCHARGED ENGINES — DIAGNOSTIC CODE CHARTS — SYCLONE AND TYPHOON

CODE 44
OXYGEN (O$_2$) SENSOR CIRCUIT
(LEAN EXHAUST INDICATED)
4.3L TURBO (VIN Z) SYCLONE AND TYPHOON (PORT)

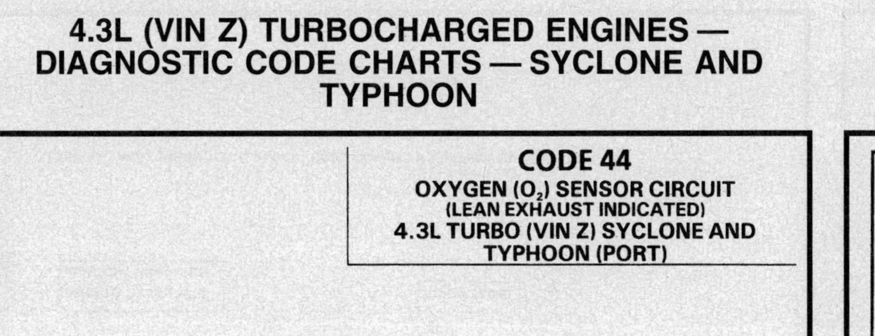

4.3L (VIN Z) TURBOCHARGED ENGINES — DIAGNOSTIC CODE CHARTS — SYCLONE AND TYPHOON

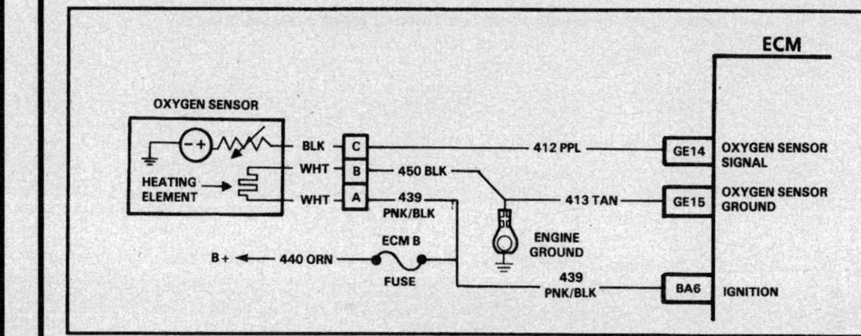

CODE 45
OXYGEN (O$_2$) SENSOR CIRCUIT
(RICH EXHAUST INDICATED)
4.3L TURBO (VIN Z) SYCLONE AND TYPHOON (PORT)

Circuit Description:

The ECM applies a bias voltage of approximately 450 millivolts (350-550 mV is normal bias voltage) between terminals "GE14" and "GE15" (if measured with a 10 megohm digital voltmeter, this may read as low as .32 volt). The O$_2$ sensor varies the voltage within a range of about 1 volt if the exhaust is rich, down through about .10 volt if exhaust is lean.

The sensor is like an open circuit and produces no voltage when it is below about 316°C (600°F). An open sensor circuit causes "Open Loop" operation. The heating element in the O$_2$ sensor causes the sensor to heat up quickly, allowing for quicker closed-loop operation.

Test Description: Numbers below refer to circled numbers on the diagnostic chart.
1. Code 45 is set when the O$_2$ sensor voltage:
   • Remains above .75 volt for 50 seconds; and the system is in "Closed Loop."

Diagnostic Aids:

Using the "Scan," observe the block learn values at different rpm and air flow conditions. The "Scan" also displays the block cells, so the block learn values can be checked in each of the cells to determine when the Code 45 may have been set. If the conditions for Code 45 exists, the block learn values will be around 115.

• <u>Fuel Pressure.</u> System will go rich if pressure is too high. The ECM can compensate for some increase. However, if it gets too high, a Code 45 may be set. See Fuel System diagnosis CHART A-7.
• <u>Rich Injector.</u> Perform injector balance test CHART C-2A.
• <u>Leaking Injector.</u> See CHART A-7.
• Check for fuel contaminated oil.
• <u>HEI Shielding.</u> An open ignition ground CKT 453 may result in EMI, or induced electrical "noise." The ECM looks at this "noise" as reference pulses.

The additional pulses result in a higher than actual engine speed signal. The ECM then delivers too much fuel, causing system to go rich. Engine tachometer will also show higher than actual engine speed, which can help in diagnosing this problem.

• <u>Canister purge.</u> Check canister for fuel saturation. If full of fuel, check canister control and hoses. See "Purge Valve Operation" Section "C3."
• <u>MAP Sensor.</u> An output that causes the ECM to sense a higher than normal manifold pressure can cause the system to go rich. Disconnecting the MAP sensor will allow the ECM to set a fixed value for the sensor. Substitute a different MAP sensor if the rich condition is gone while the sensor is disconnected.
• Check for leaking fuel pressure regulator diaphragm by checking vacuum line to regulator for fuel.
• <u>TPS.</u> An intermittent TPS output will cause the system to go rich, due to a false indication of the engine accelerating.
• <u>EGR.</u> An EGR staying open (especially at idle) will cause the O$_2$ sensor to indicate a rich exhaust, and this could result in a Code 45.

4.3L (VIN Z) TURBOCHARGED ENGINES — DIAGNOSTIC CODE CHARTS — SYCLONE AND TYPHOON

CODE 45
OXYGEN (O₂) SENSOR CIRCUIT
(RICH EXHAUST INDICATED)
4.3L TURBO (VIN Z) SYCLONE
AND TYPHOON (PORT)

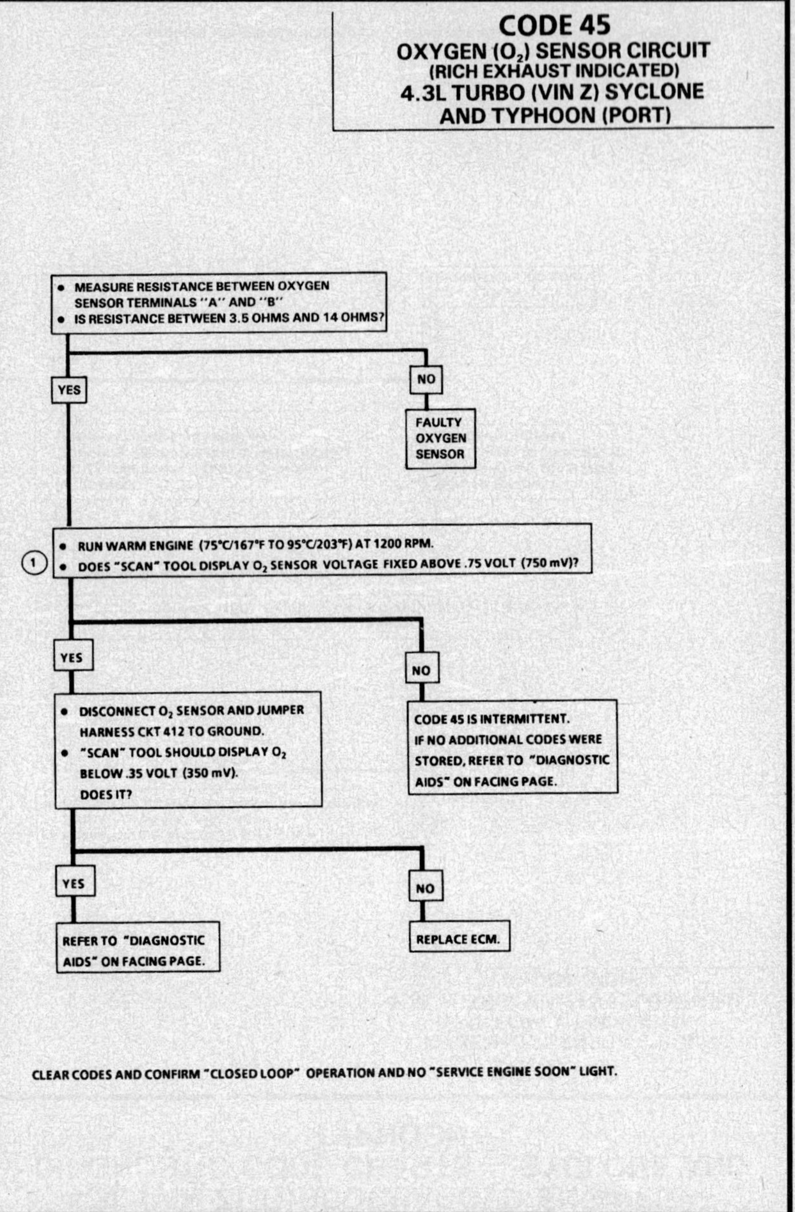

CLEAR CODES AND CONFIRM "CLOSED LOOP" OPERATION AND NO "SERVICE ENGINE SOON" LIGHT.

4.3L (VIN Z) TURBOCHARGED ENGINES — DIAGNOSTIC CODE CHARTS — SYCLONE AND TYPHOON

CODE 51
PROM ERROR
(FAULTY OR INCORRECT PROM)
4.3L TURBO (VIN Z) SYCLONE AND
TYPHOON (PORT)

CHECK THAT ALL PINS ARE FULLY INSERTED IN THE SOCKET AND THAT PROM IS PROPERLY SEATED. IF OK, REPLACE PROM, CLEAR MEMORY, AND RECHECK. IF CODE 51 REAPPEARS, REPLACE ECM.

CLEAR CODES AND CONFIRM "CLOSED LOOP" OPERATION AND NO "CHECK ENGINE" LIGHT.

4.3L (VIN Z) TURBOCHARGED ENGINES — C-CHARTS — SYCLONE AND TYPHOON

CHART C-1A

PARK/NEUTRAL SWITCH DIAGNOSIS
4.3L TURBO (VIN Z) SYCLONE AND TYPHOON (PORT)

Circuit Description:

The Park/Neutral switch contacts are a part of the neutral start switch. They are closed to ground in park or neutral and open in drive ranges.

The ECM supplies voltage through a current limiting resistor to CKT 434 and senses a closed switch when the voltage on CKT 434 drops to less than one volt.

The ECM uses the P/N signal as one of the inputs to control idle air and EGR.

If CKT 434 indicates P/N (grounded), while in drive range, the EGR would be inoperative, resulting in possible detonation.

If CKT 434 indicates drive (open), a dip in the idle may exist when the gear selector is moved into drive range.

Test Description: Numbers below refer to circled numbers on the diagnostic chart.

1. Checks for a closed switch to ground in park position. Different makes of "Scan" tools will read P/N differently. Refer to tool operator's manual for type of display used for a specific tool.
2. Checks for an open switch in drive range.
3. Be sure "Scan" indicates drive, even while wiggling shifter, to test for an intermittent or misadjusted switch in drive or overdrive range.

4.3L (VIN Z) TURBOCHARGED ENGINES — C-CHARTS — SYCLONE AND TYPHOON

CHART C-1A

PARK/NEUTRAL SWITCH DIAGNOSIS
4.3L TURBO (VIN Z) SYCLONE AND TYPHOON (PORT)

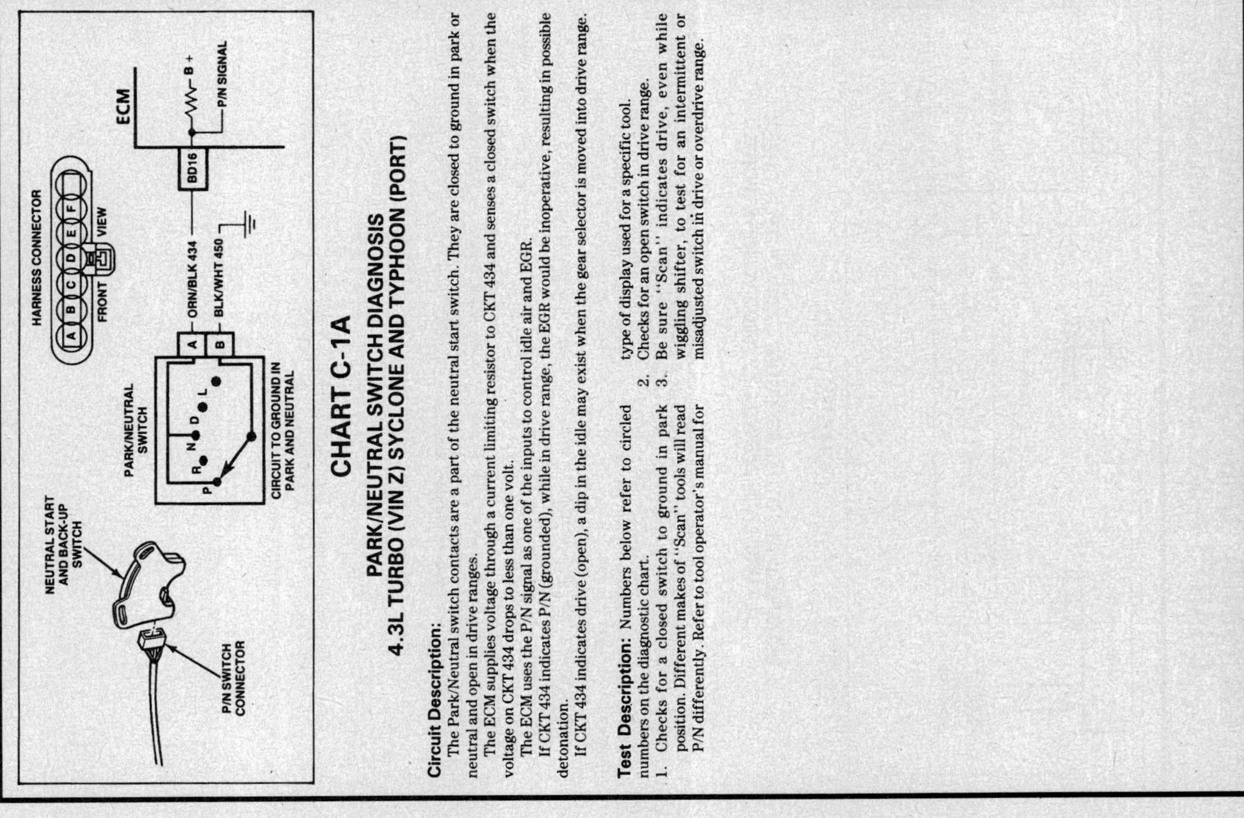

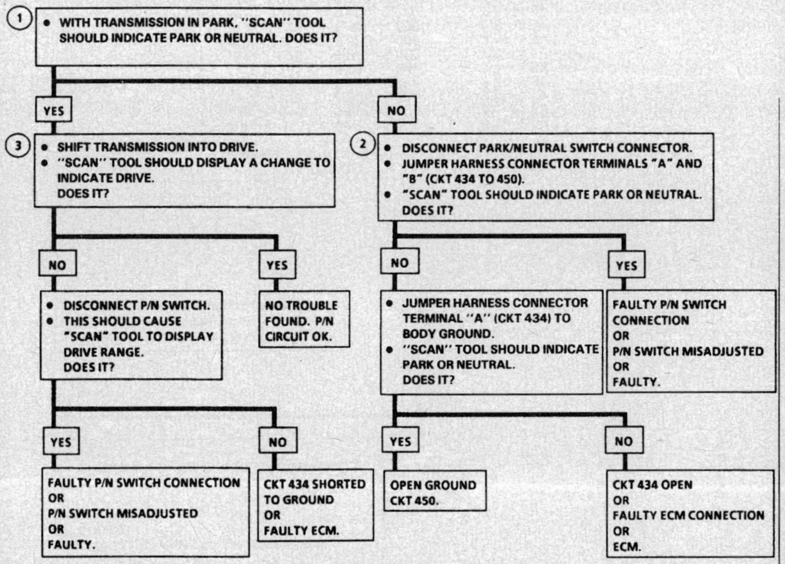

4.3L (VIN Z) TURBOCHARGED ENGINES — C-CHARTS — SYCLONE AND TYPHOON

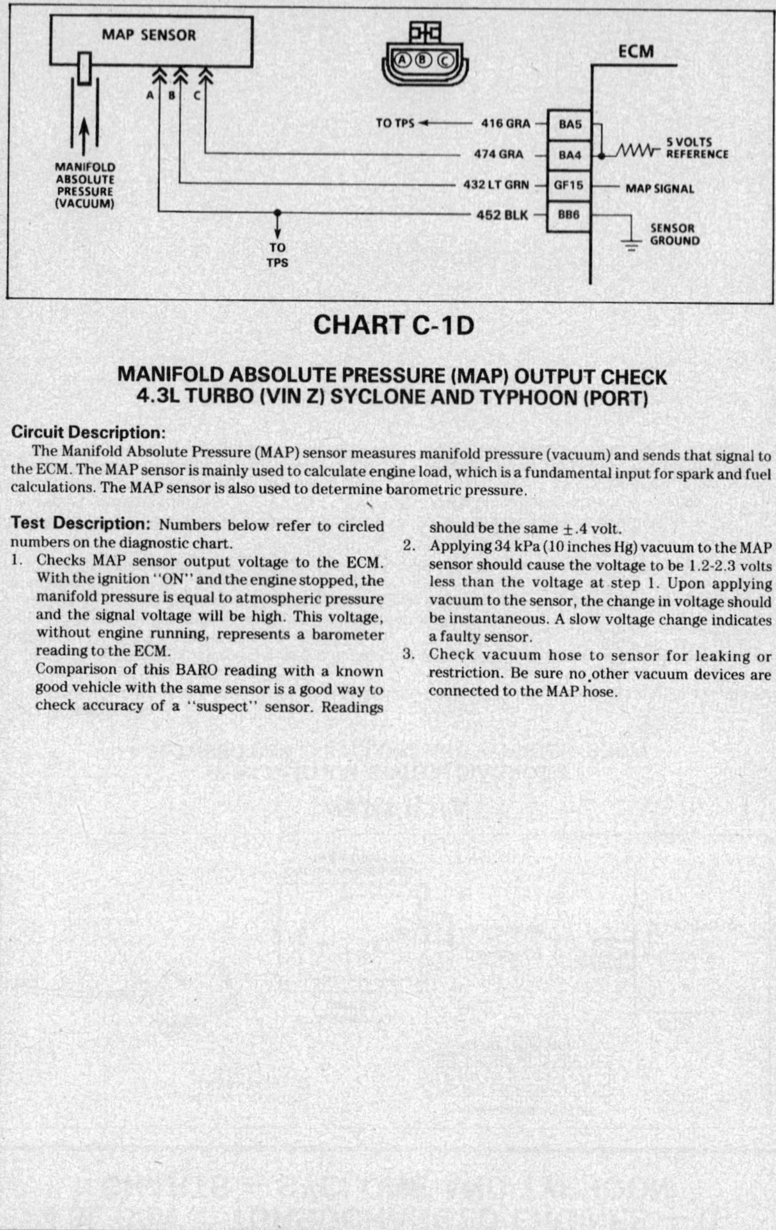

CHART C-1D

MANIFOLD ABSOLUTE PRESSURE (MAP) OUTPUT CHECK
4.3L TURBO (VIN Z) SYCLONE AND TYPHOON (PORT)

Circuit Description:

The Manifold Absolute Pressure (MAP) sensor measures manifold pressure (vacuum) and sends that signal to the ECM. The MAP sensor is mainly used to calculate engine load, which is a fundamental input for spark and fuel calculations. The MAP sensor is also used to determine barometric pressure.

Test Description: Numbers below refer to circled numbers on the diagnostic chart.

1. Checks MAP sensor output voltage to the ECM. With the ignition "ON" and the engine stopped, the manifold pressure is equal to atmospheric pressure and the signal voltage will be high. This voltage, without engine running, represents a barometer reading to the ECM.

   Comparison of this BARO reading with a known good vehicle with the same sensor is a good way to check accuracy of a "suspect" sensor. Readings should be the same ±.4 volt.

2. Applying 34 kPa (10 inches Hg) vacuum to the MAP sensor should cause the voltage to be 1.2-2.3 volts less than the voltage at step 1. Upon applying vacuum to the sensor, the change in voltage should be instantaneous. A slow voltage change indicates a faulty sensor.

3. Check vacuum hose to sensor for leaking or restriction. Be sure no other vacuum devices are connected to the MAP hose.

4.3L (VIN Z) TURBOCHARGED ENGINES — C-CHARTS — SYCLONE AND TYPHOON

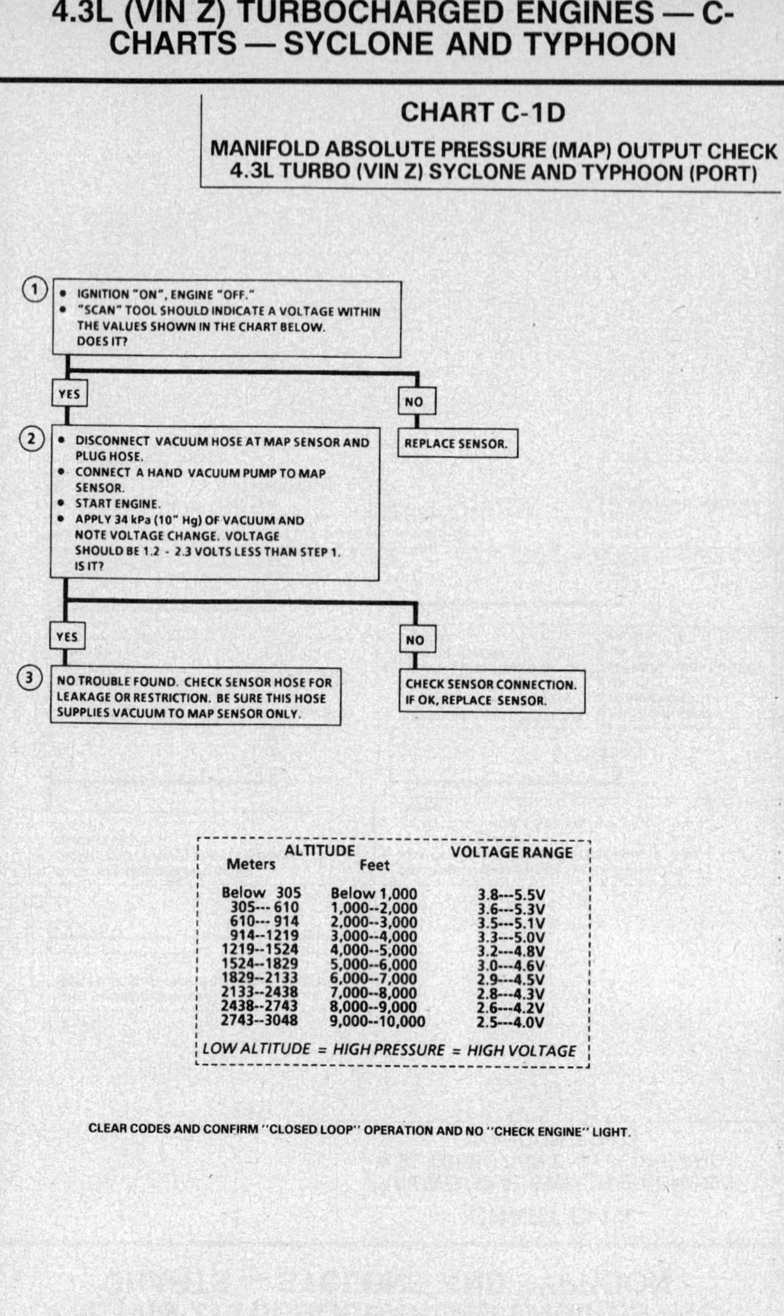

CHART C-1D

MANIFOLD ABSOLUTE PRESSURE (MAP) OUTPUT CHECK
4.3L TURBO (VIN Z) SYCLONE AND TYPHOON (PORT)

ALTITUDE		VOLTAGE RANGE
Meters	Feet	
Below 305	Below 1,000	3.8---5.5V
305--- 610	1,000--2,000	3.6---5.3V
610--- 914	2,000--3,000	3.5---5.1V
914--1219	3,000--4,000	3.3---5.0V
1219--1524	4,000--5,000	3.2---4.8V
1524--1829	5,000--6,000	3.0---4.6V
1829--2133	6,000--7,000	2.9---4.5V
2133--2438	7,000--8,000	2.8---4.3V
2438--2743	8,000--9,000	2.6---4.2V
2743--3048	9,000--10,000	2.5---4.0V

LOW ALTITUDE = HIGH PRESSURE = HIGH VOLTAGE

CLEAR CODES AND CONFIRM "CLOSED LOOP" OPERATION AND NO "CHECK ENGINE" LIGHT.

4.3L (VIN Z) TURBOCHARGED ENGINES — C-CHARTS — SYCLONE AND TYPHOON

CHART C-2A

INJECTOR BALANCE TEST

The injector balance tester is a tool used to turn the injector on for a precise amount of time, thus spraying a measured amount of fuel into the manifold. This causes a drop in fuel rail pressure that we can record and compare between each injector. All injectors should have the same amount of pressure drop ($\pm$ 10 kPa). Any injector with a pressure drop that is 10 kPa (or more) greater or less than the average drop of the other injectors should be considered faulty and replaced.

STEP 1

Engine "cool down" period (10 minutes) is necessary to avoid irregular readings due to "Hot Soak" fuel boiling. With ignition "OFF" connect fuel gage J 34730-1 or equivalent to fuel pressure tap. Wrap a shop towel around fitting while connecting gage to avoid fuel spillage.

Disconnect harness connectors at all injectors, and connect injector tester J 34730-3, or equivalent, to one injector. Ignition must be "OFF" at least 10 seconds to complete ECM shutdown cycle. Fuel pump should run about 2 seconds after ignition is turned "ON." At this point, insert clear tubing attached to vent valve into a suitable container and bleed air from gage and hose to insure accurate gage operation. Repeat this step until all air is bled from gage.

STEP 2

Turn ignition "OFF" for 10 seconds and then "ON" again to get fuel pressure to its maximum. Record this initial pressure reading. Energize tester one time and note pressure drop at its lowest point (Disregard any slight pressure increase after drop hits low point). By subtracting this second pressure reading from the initial pressure, we have the actual amount of injector pressure drop.

STEP 3

Repeat step 2 on each injector and compare the amount of drop. Usually, good injectors will have virtually the same drop. Retest any injector that has a pressure difference of 10 kPa, either more or less than the average of the other injectors on the engine. Replace any injector that also fails the retest. If the pressure drop of all injectors is within 10 kPa of this average, the injectors appear to be flowing properly. Reconnect them and review "Symptoms,"

NOTE: *The entire test should not be repeated more than once without running the engine to prevent flooding. (This includes any retest on faulty injectors).*

4.3L (VIN Z) TURBOCHARGED ENGINES — C-CHARTS — SYCLONE AND TYPHOON

CHART C-2A
INJECTOR BALANCE TEST
4.3L TURBO (VIN Z) SYCLONE AND TYPHOON (PORT)

NOTE: If injectors are suspected of being dirty, they should be cleaned using an approved tool and procedure prior to performing this test. The fuel pressure test in Section "A," Chart A-7, should be completed prior to this test.

Step 1. If engine is at operating temperature, allow a 10 minute "cool down" period then connect fuel pressure gage and injector tester.
1. Ignition "OFF."
2. Connect fuel pressure gage and injector tester.
3. Ignition "ON."
4. Bleed off air in gage. Repeat until all air is bled from gage.

Step 2. Run test:
1. Ignition "OFF" for 10 seconds.
2. Ignition "ON." Record gage pressure. (Pressure must hold steady, if not see the Fuel System diagnosis, Chart A-7, in Section "A".)
3. Turn injector on, by depressing button on injector tester, and note pressure at the instant the gage needle stops.

Step 3.
1. Repeat step 2 on all injectors and record pressure drop on each.
Retest injectors that appear faulty (any injectors that have a 10 kPa difference, either more or less, in pressure from the average). If no problem is found, review "Symptoms" Section

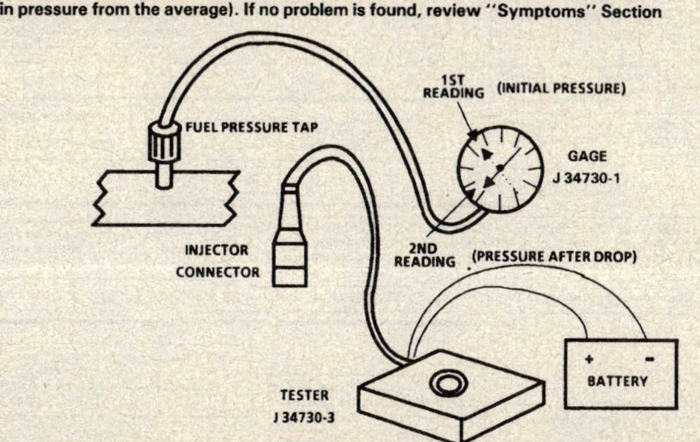

— EXAMPLE —

CYLINDER	1	2	3	4	5	6
1ST READING	225	225	225	225	225	225
2ND READING	100	100	100	90	100	115
AMOUNT OF DROP	125	125	125	135	125	110
	OK	OK	OK	FAULTY, RICH (TOO MUCH) (FUEL DROP)	OK	FAULTY, LEAN (TOO LITTLE) (FUEL DROP)

4.3L (VIN Z) TURBOCHARGED ENGINES — C-CHARTS — SYCLONE AND TYPHOON

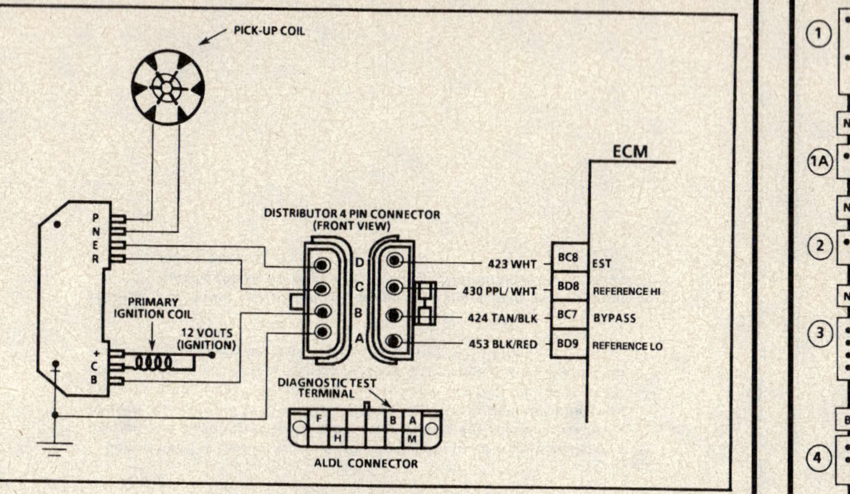

CHART C-4C
IGNITION SYSTEM CHECK
(REMOTE COIL)
4.3L TURBO (VIN Z) SYCLONE AND TYPHOON (PORT)

Test Description: Numbers below refer to circled numbers on the diagnostic chart.

1. Two wires are checked to ensure that an open is not present in a spark plug wire.

1A. If spark occurs with 4 terminal distributor connector disconnected, pick-up coil output is too low for EST operation.

2. A spark indicates the problem must be the distributor cap or rotor.

3. Normally, there should be battery voltage at the "C" and "+" terminals. Low voltage would indicate an open or a high resistance circuit from the distributor to the coil or ignition switch. If "C" terminal voltage was low, but "+" terminal voltage is 10 volts or more, circuit from "C" terminal to ignition coil or ignition coil primary winding is open.

4. Check for a shorted module or grounded circuit from the ignition coil to the module. The distributor module should be turned "OFF." Normal voltage should be about 12 volts.

   If the module is turned "ON," the voltage would be low, but above 1 volt. This could cause the ignition coil to fail from excessive heat.

   With an open ignition coil primary winding, a small amount of voltage will leak through the mod-ule from the "BATT" to the "TACH" terminal.

5. Applying voltage (1.5 to 8 volts) to module terminal "P" should turn the module "ON" and the tachometer terminal voltage should drop to about 7-9 volts. This test will determine whether the module or coil is faulty or if the pick-up coil is not generating the proper signal to turn the module "ON." This test can be performed by using a DC battery with a rating of 1.5 to 8 volts. The use of the test light is mainly to allow the "P" terminal to be probed more easily.

   Some digital multi-meters can also be used to trigger the module by selecting ohms, usually the diode position. In this position the meter may have a voltage across its terminals which can be used to trigger the module. The voltage in the ohms position can be checked by using a second meter or by checking the manufacturer's specification of the tool being used.

6. This should turn "OFF" the module and cause a spark. If no spark occurs, the fault is most likely in the ignition coil because most module problems would have been found before this point in the procedure. A module tester (J 24642-F) could determine which is at fault.

4.3L (VIN Z) TURBOCHARGED ENGINES — C-CHARTS — SYCLONE AND TYPHOON

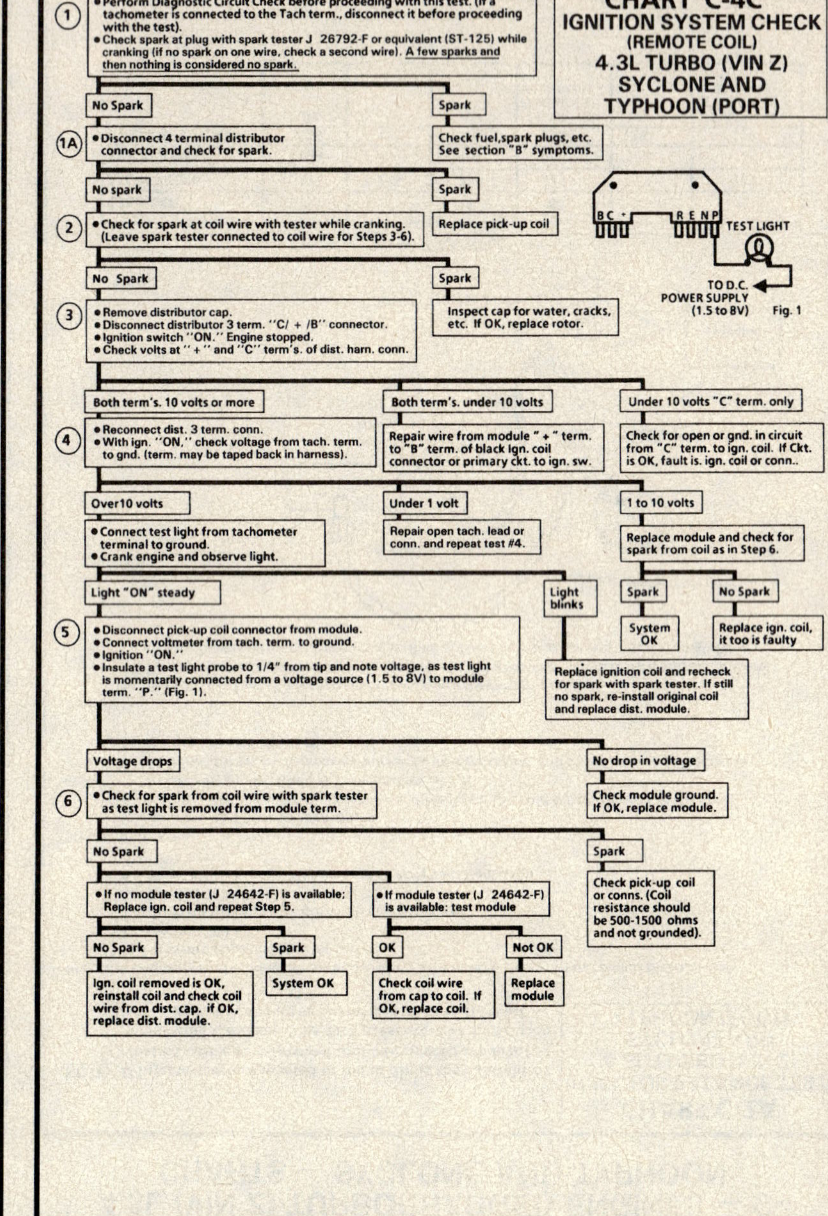

4.3L (VIN Z) TURBOCHARGED ENGINES — C-CHARTS — SYCLONE AND TYPHOON

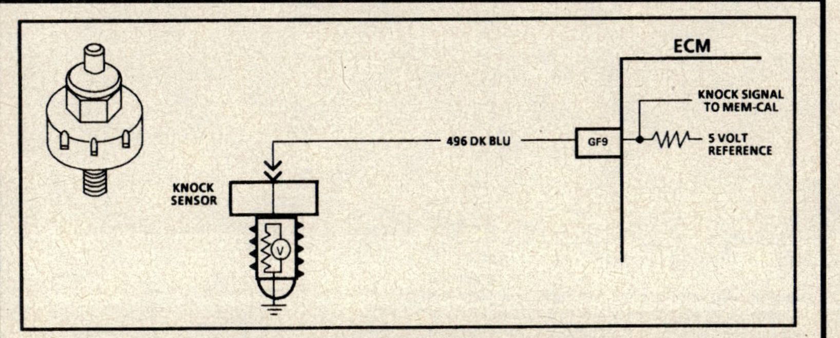

CHART C-5
ELECTRONIC SPARK CONTROL (ESC) SYSTEM CHECK
4.3L TURBO (VIN Z) SYCLONE AND TYPHOON (PORT)

Circuit Description:

The ESC knock sensor is used to detect engine detonation and the ECM will retard the electronic spark timing based on the signal being received. The circuitry, within the knock sensor, causes the ECM's 5 volts to be pulled down so that under a no knock condition, CKT 496 would measure about 2.5 volts. The knock sensor produces an A/C signal, which rides on the 2.5 volts DC voltage. The amplitude and frequency are dependent upon the knock level.

The Mem-Cal, used with this engine, contains the functions which were part of remotely mounted ESC modules used on other GM vehicles. The ESC portion of the Mem-Cal then sends a signal to other parts of the ECM which adjusts the spark timing to retard the spark and reduce the detonation.

Test Description: Numbers below refer to circled numbers on the diagnostic chart.

1. With engine idling, there should not be a knock signal present at the ECM, because detonation is not likely under a no load condition.
2. Tapping on the engine lift hood bracket should simulate a knock signal to determine if the sensor is capable of detecting detonation. If no knock is detected, try tapping on engine block closer to sensor before replacing sensor.
3. If the engine has an internal problem, which is creating a knock, the knock sensor may be responding to the internal failure.
4. This test determines if the knock sensor is faulty, or, if the ESC portion of the Mem-Cal is faulty. If it is determined that the Mem-Cal is faulty, be sure that it is properly installed and latched into place. If not properly installed, repair and retest.

Diagnostic Aids:

While observing knock signal on the "Scan," there should be an indication that knock is present, when detonation can be heard. Detonation is most likely to occur under high engine load conditions.

4.3L (VIN Z) TURBOCHARGED ENGINES — C-CHARTS — SYCLONE AND TYPHOON

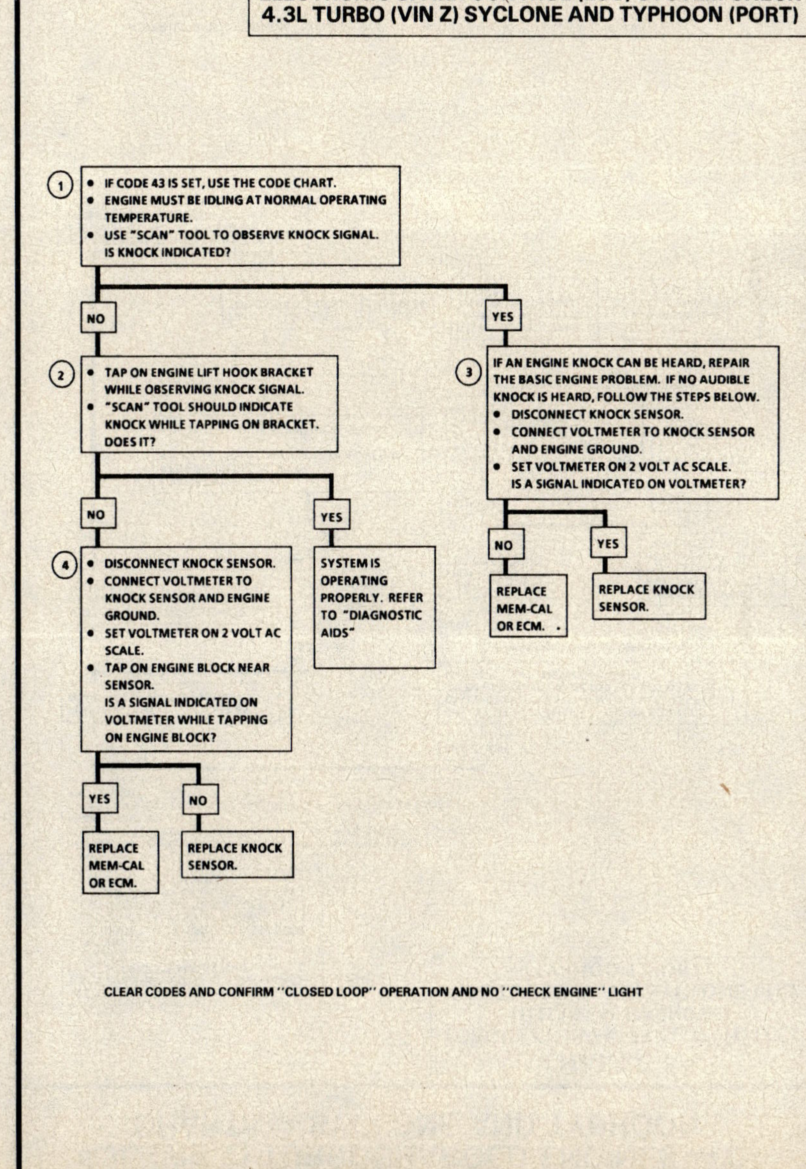

4.3L (VIN Z) TURBOCHARGED ENGINES — C-CHARTS — SYCLONE AND TYPHOON

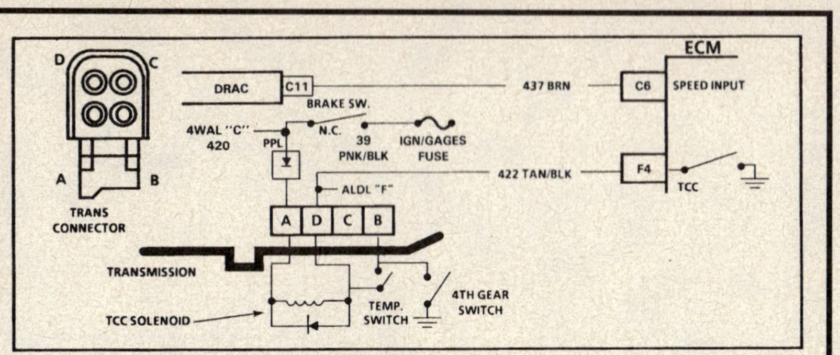

CHART C-8A
TORQUE CONVERTER CLUTCH (TCC)
(ELECTRICAL DIAGNOSIS)
4.3L TURBO (VIN Z) SYCLONE AND TYPHOON (PORT)

Circuit Description:

The purpose of the automatic transmission Torque Converter Clutch (TCC) feature is to eliminate the power loss of the torque converter stage when the vehicle is in a cruise condition. This allows the convenience of the automatic transmission and the fuel economy of a manual transmission.

Fused battery ignition is supplied to the TCC solenoid through the TCC brake switch.

The ECM will engage TCC by grounding CKT 422 to energize the solenoid.

TCC will engage when:

- Vehicle speed above 30 mph (48 km/h).
- Engine at normal operating temperature (above 65°C/149°F)
- Throttle Position Sensor (TPS) output not changing, indicating a steady road speed.
- Brake switch closed.
- 3rd or 4th gears.

Test Description: Number(s) below refer to circled number(s) on the diagnostic chart.

1. A test light "ON" indicates battery voltage and continuity through the TCC solenoid is OK.
2. Checks for vehicle speed sensor signal to ECM using a "Scan" tool.

Diagnostic Aids:

Solenoid coil resistance must measure more than 20 ohms. Less resistance will cause early failure of the ECM "Driver." Refer to "ECM QDR" check CHART C-1A in "Electronic Control Module (ECM) and Sensors," Section "6E3-C1". Using an ohmmeter, check the solenoid coil resistance of all ECM controlled solenoids and relays before installing a replacement ECM. Replace any solenoid or relay that measures less than 20 ohms resistance.

To prevent TCC over heat condition TCC, temperature switch closes at 279°F ± 7° and reopens at 259°F ± 9°.

4.3L (VIN Z) TURBOCHARGED ENGINES — C-CHARTS — SYCLONE AND TYPHOON

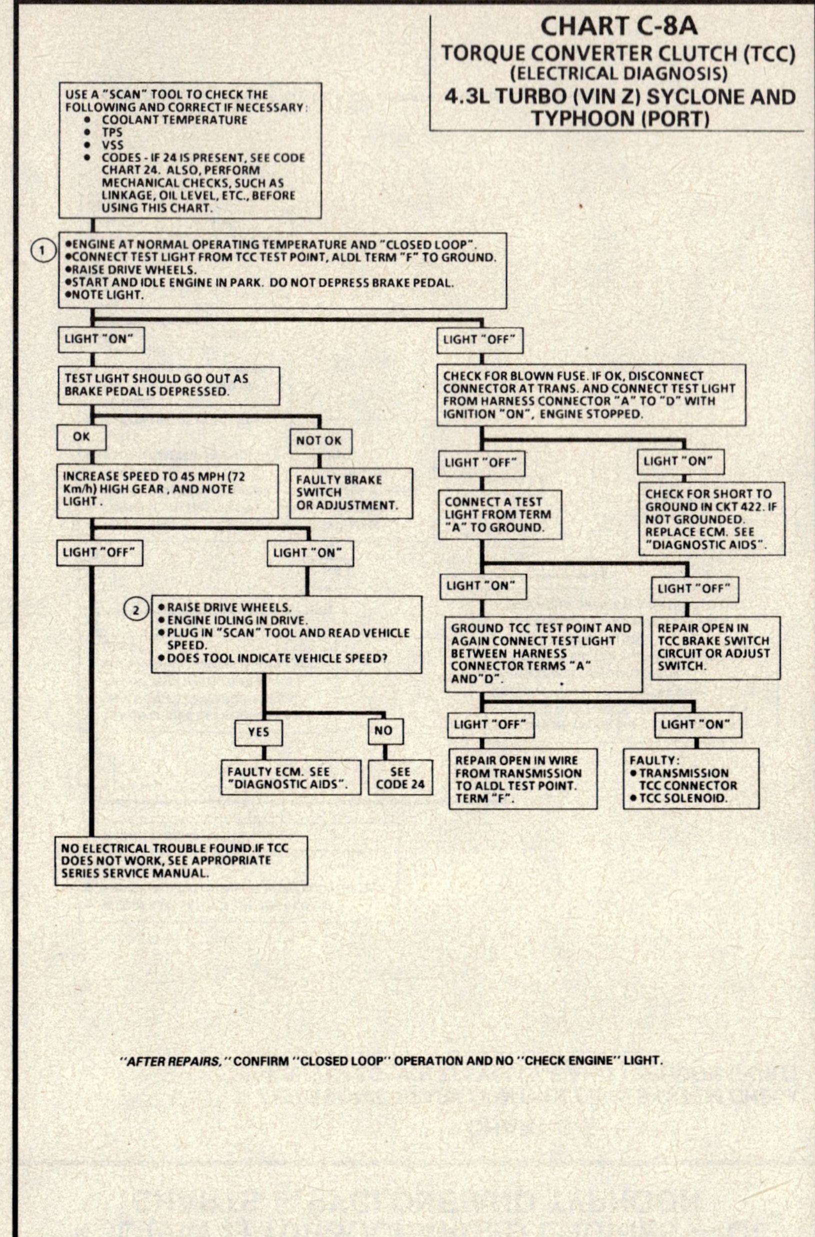

4.3L (VIN Z) TURBOCHARGED ENGINES — C-CHARTS — SYCLONE AND TYPHOON

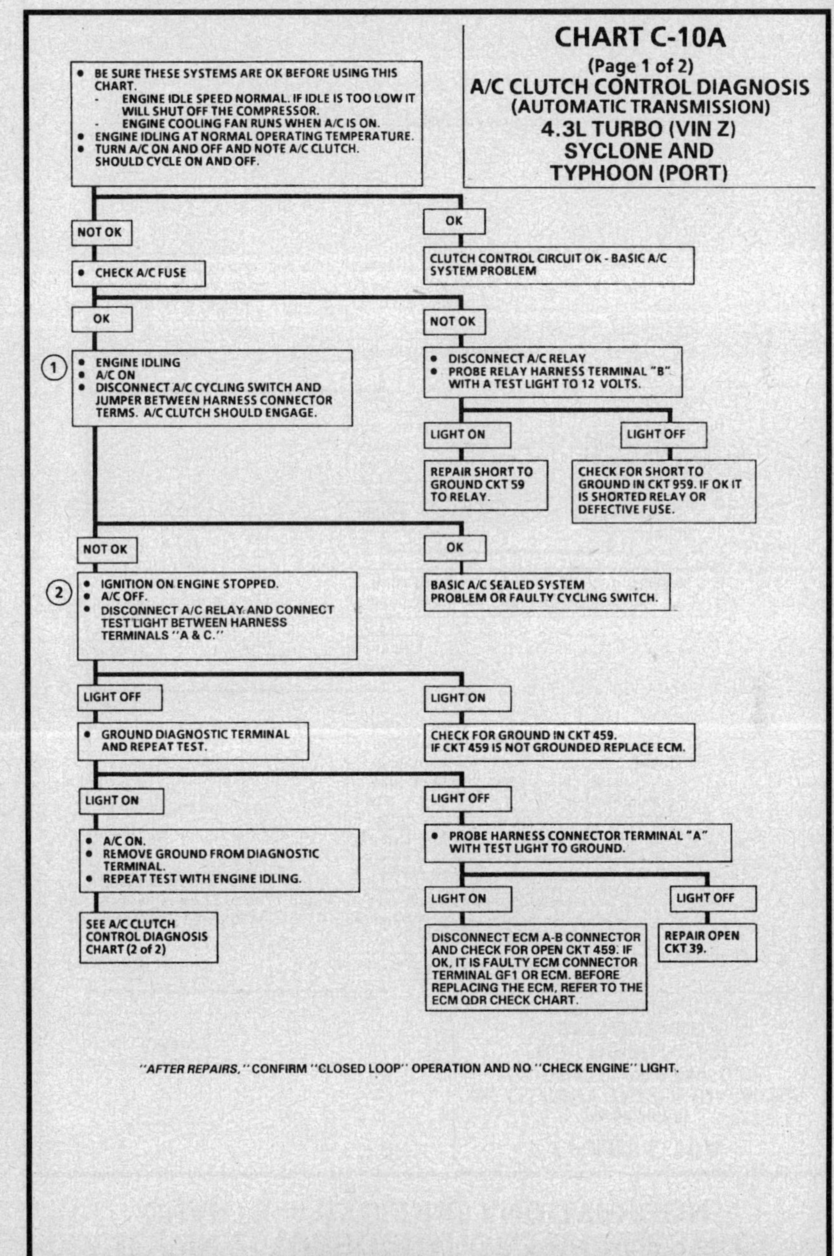

CHART C-10A
(Page 1 of 2)
A/C CLUTCH CONTROL DIAGNOSIS
(AUTOMATIC TRANSMISSION)
4.3L TURBO (VIN Z) SYCLONE AND TYPHOON (PORT)

Circuit Description:
ECM control of the A/C clutch improves idle quality and performance by:
- Delaying clutch apply until the idle air rate is increased.
- Releasing clutch when idle speed is too low.
- Releasing clutch at wide open throttle.
- Smooths cycling of the compressor by providing additional fuel at the instant clutch is applied.

Turning on air conditioning supplies CKT 59 battery voltage to the clutch control relay and terminal "BC9" of the ECM connector. After a time delay of about 1/2 second the ECM will ground terminal "GF1" of the ECM connector, CKT 459, and close the control relay. A/C compressor clutch will engage.

Test Description: Number(s) below refer to circled number(s) on the diagnostic chart.
1. Checks for low refrigerant as cause for no A/C.
2. This and following tests check for faulty A/C control relay.

4.3L (VIN Z) TURBOCHARGED ENGINES — C-CHARTS — SYCLONE AND TYPHOON

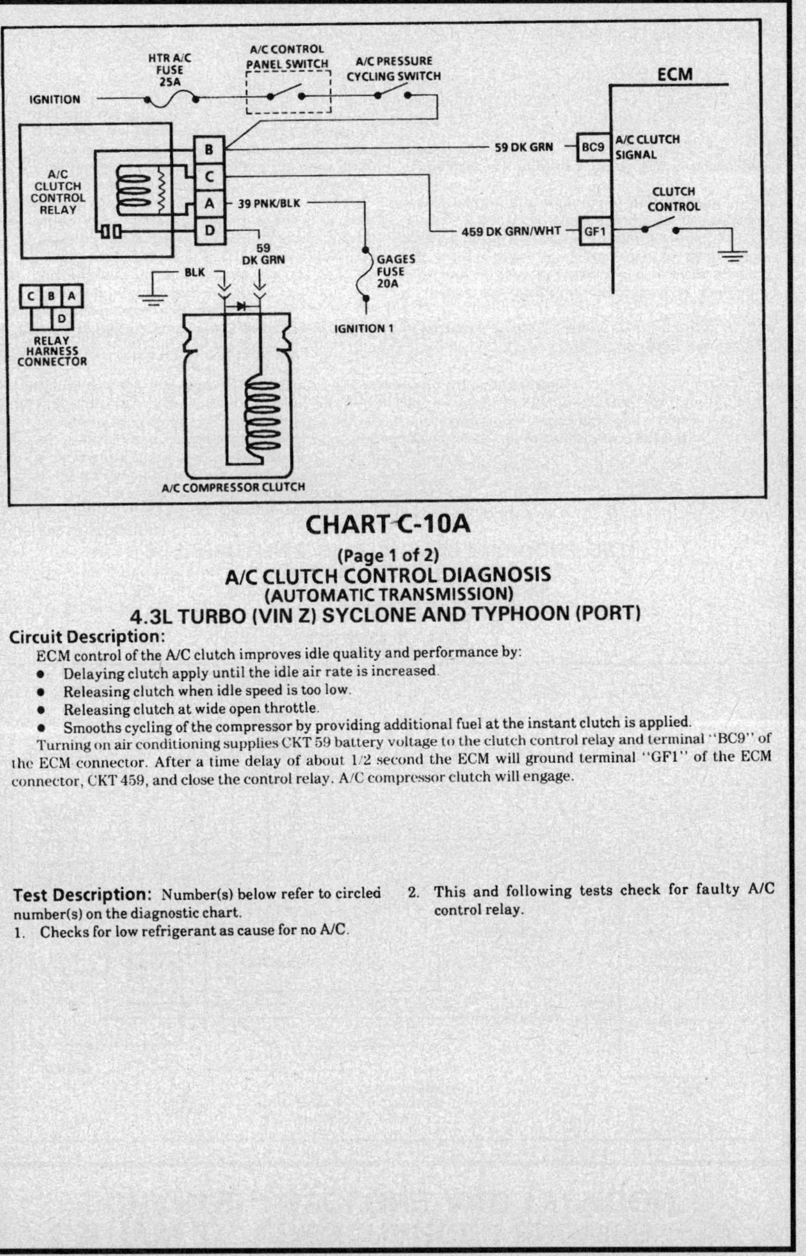

4.3L (VIN Z) TURBOCHARGED ENGINES — C-CHARTS — SYCLONE AND TYPHOON

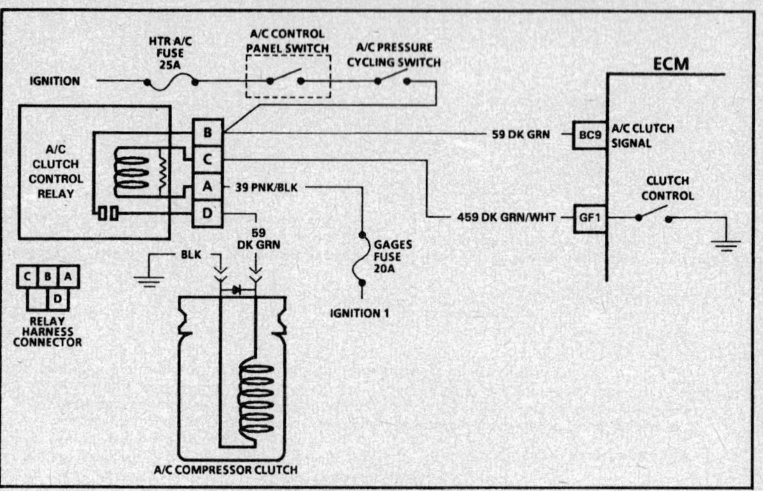

CHART C-10A
(Page 2 of 2)
A/C CLUTCH CONTROL DIAGNOSIS
(AUTOMATIC TRANSMISSION)
4.3L TURBO (VIN Z) SYCLONE AND TYPHOON (PORT)

Circuit Description:
ECM control of the A/C clutch improves idle quality and performance by:
- Delaying clutch apply until the idle air rate is increased.
- Releasing clutch when idle speed is too low.
- Releasing clutch at wide open throttle.
- Smooths cycling of the compressor by providing additional fuel at the instant clutch is applied.

Turning on air conditioning supplies CKT 59 battery voltage to the clutch control relay and terminal "BC9" of the ECM connector. After a time delay of about 1/2 second the ECM will ground terminal "GF1" of the ECM connector, CKT 459, and close the control relay. A/C compressor clutch will engage.

Test Description: Number(s) below refer to circled number(s) on the diagnostic chart.
3. Checks for faulty cycling switch.
- Solenoids and relays are turned "ON" or "OFF" by the ECM, using internal electronic switches called "drivers." Each driver is part of a group of four, called "Quad-Drivers." Failure of one driver can damage any other driver in the set.

Solenoid and relay coil resistance must measure more than 20 ohms. Less resistance will cause early failure of the ECM "driver." Using an ohmmeter, check the coil resistance of the A/C relay before replacing the ECM.

Diagnostic Aids:

Before replacing ECM, use ohmmeter and check resistance of each ECM controlled relay or solenoid coil. Refer to ECM QDR Check, CHART C-1A in "Electronic Control Module (ECM) and Sensors."

See ECM wiring diagram for coil terminal identification for solenoid(s) and relay(s) to be checked.

Replace any relay or solenoid that measures less than 20 ohms.

4.3L (VIN Z) TURBOCHARGED ENGINES — C-CHARTS — SYCLONE AND TYPHOON

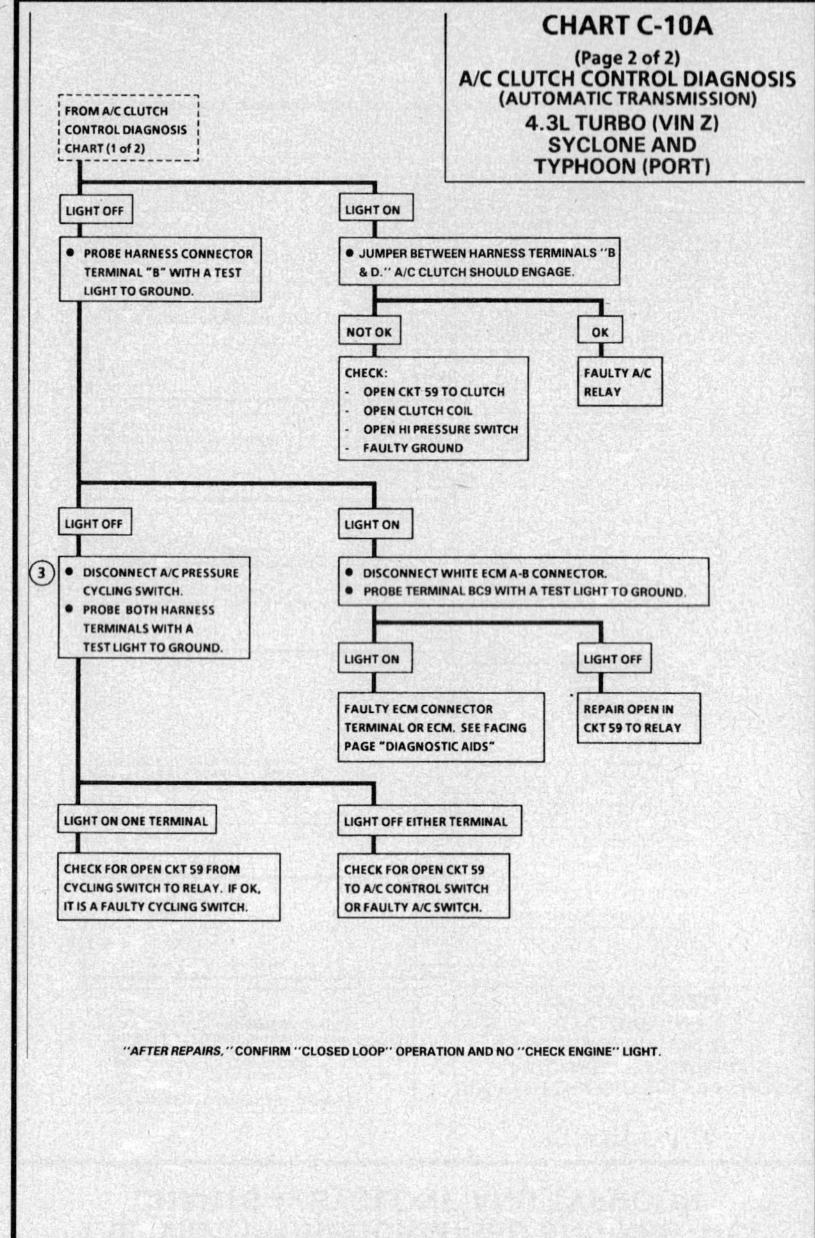

4.3L (VIN Z) TURBOCHARGED ENGINES — C-CHARTS — SYCLONE AND TYPHOON

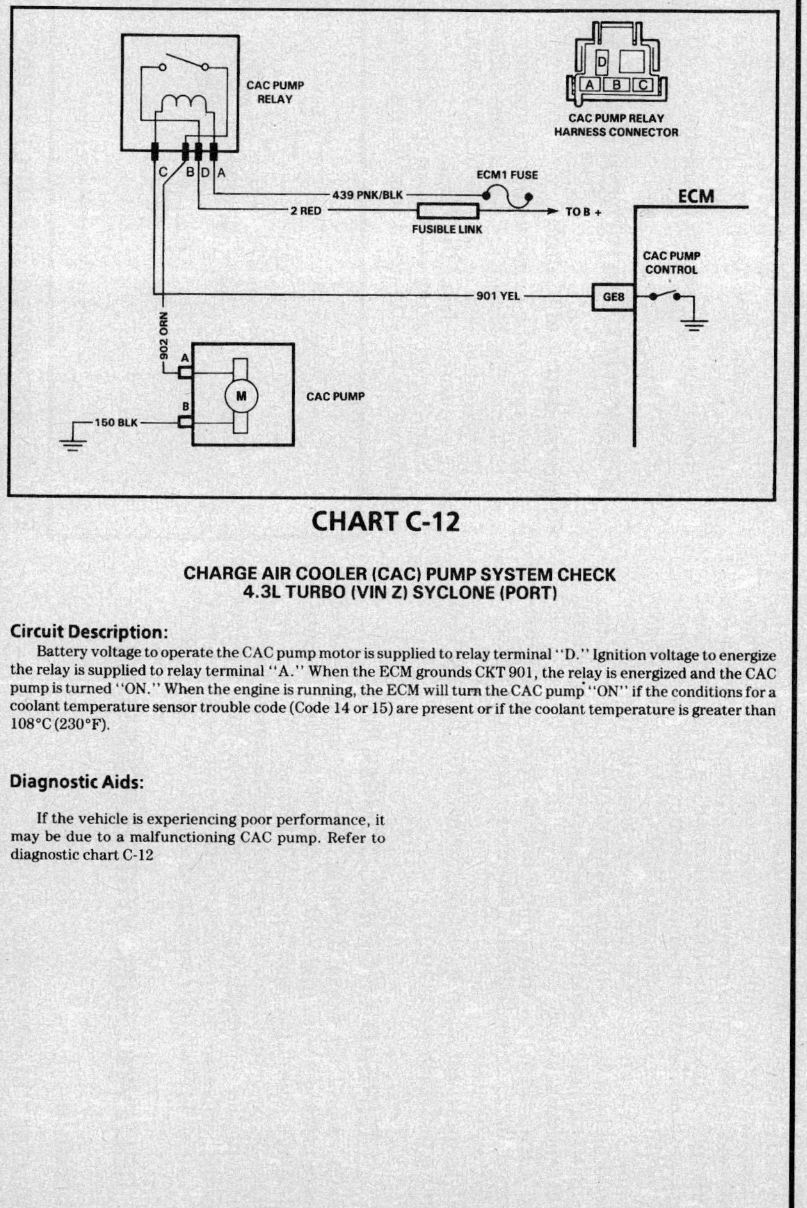

CHART C-12

CHARGE AIR COOLER (CAC) PUMP SYSTEM CHECK
4.3L TURBO (VIN Z) SYCLONE (PORT)

Circuit Description:

Battery voltage to operate the CAC pump motor is supplied to relay terminal "D." Ignition voltage to energize the relay is supplied to relay terminal "A." When the ECM grounds CKT 901, the relay is energized and the CAC pump is turned "ON." When the engine is running, the ECM will turn the CAC pump "ON" if the conditions for a coolant temperature sensor trouble code (Code 14 or 15) are present or if the coolant temperature is greater than 108°C (230°F).

Diagnostic Aids:

If the vehicle is experiencing poor performance, it may be due to a malfunctioning CAC pump. Refer to diagnostic chart C-12

4.3L (VIN Z) TURBOCHARGED ENGINES — C-CHARTS — SYCLONE AND TYPHOON

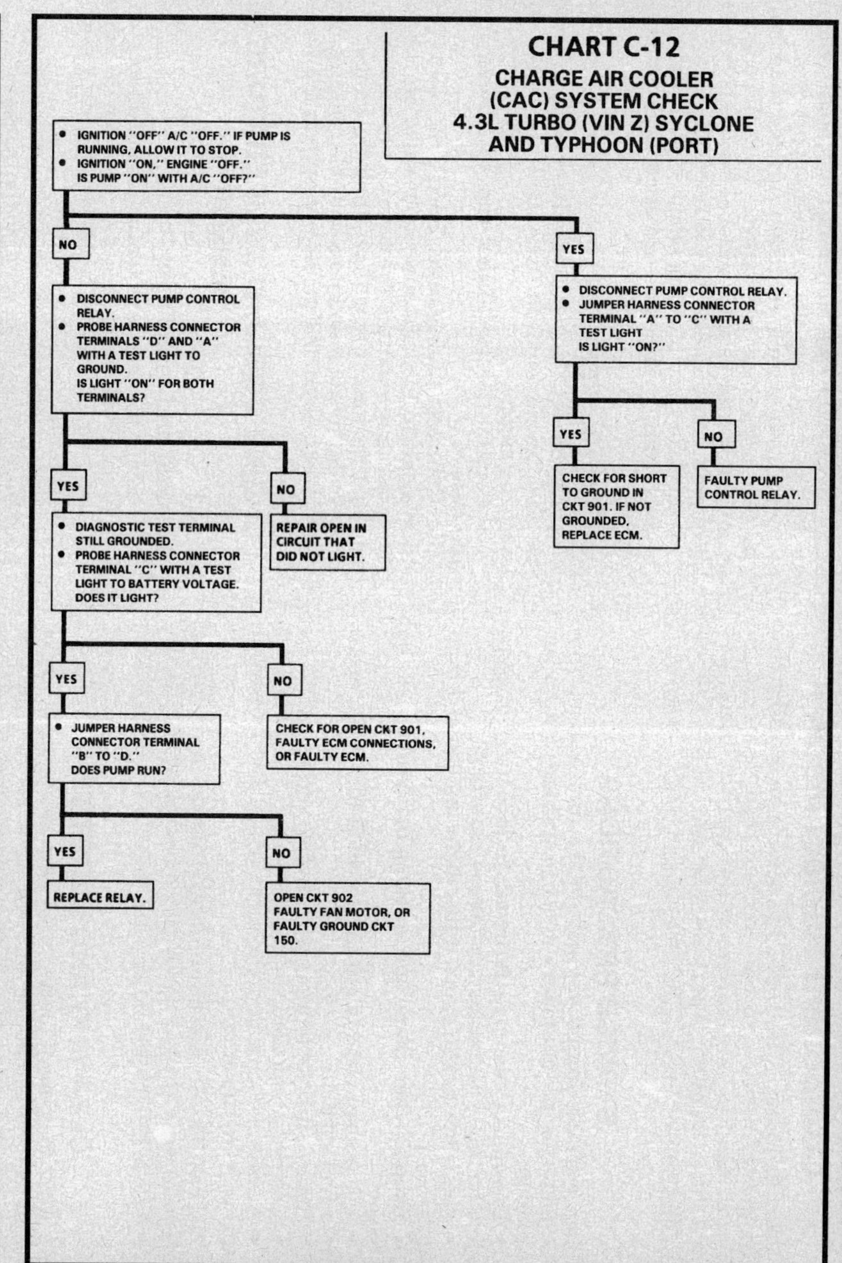

Component Replacement

NOTE: Many MPI repair procedures require the complete injection to be removed from the engine. A part number is stamped on the throttle body. Refer to this number if servicing or part replacement is necessary.

SERVICE PRECAUTIONS

When working around any part of the fuel system, take the following precautionary steps to prevent fire and/or explosion:

- Disconnect negative terminal from battery (except when testing with battery voltage is required).
- Whenever possible, use a flashlight instead of a drop light.
- Keep all open flame and smoking material out of the area.
- Use a shop cloth or similar to catch fuel when opening a fuel system.
- Relieve fuel system pressure before servicing.
- Use eye protection.
- Always keep a dry chemical (class B) fire extinguisher near the area.

NOTE: The fuel system should be de-pressurized prior to servicing, due to the amount of fuel pressure in the fuel lines.

CLEANING AND INSPECTION

All MFI component parts, with the exception of those noted below, should be cleaned in a cold immersion cleaner such as Carbon X (X–55) or equivalent. The throttle position sensor, idle air control valve, pressure regulator diaphragm assembly, fuel injectors or other components containing rubber should not be placed in a solvent or cleaner bath. A chemical reaction will cause these parts to swell, harden or distort. Do not soak the throttle body with the above parts attached. If the throttle body assembly requires cleaning, soak time in the cleaner should be kept to a minimum. Some vehicles have hidden throttle shaft dust seals that could lose their effectiveness by extended soaking.

1. Clean all parts thoroughly and blow dry with compressed air. Be sure that all fuel and air passages are free of dirt and burrs.

2. Inspect the mating casting surfaces for damage that could affect gasket sealing.

THREAD LOCKING COMPOUND

Service repair kits are supplied with a small vial of thread locking compound with directions for use. If material is not available, use Loctite® 262 or GM part No. 10522624 or equivalent. In precoating the screws, do not use a higher strength locking compound than recommended, since to do so could make removing the screw extremely difficult, or result in damaging the screw head.

RELIEVING FUEL SYSTEM PRESSURE

Procedure

1.6L AND 1.8L ENGINES
1. Loosen the fuel filler cap.

2. Remove the circuit opening (fuel pump) relay located under the dash or in the engine compartment.

3. Start the engine. It should run and then stall when the fuel in the lines is depleted. When the engine stops, crank the starter for an additional 3 seconds to make sure all pressure in the fuel lines is released.

4. Replace the circuit opening relay and disconnect the negative battery cable to avoid accidental cranking of the engine.

1.9L AND 4.3L ENGINE
1. Disconnect the negative battery cable to avoid possible fuel discharge if an accidental attempt is made to start the engine.

2. Loosen the fuel filler cap to relieve fuel tank pressure.

3. Connect fuel gauge SA9127E, J–34730–1 or equivalent, to fuel pressure relief valve on the fuel rail. Wrap a rag around the pressure tap to absorb any leakage that may occur when installing the gauge.

4. Install bleed hose into a suitable container and open the valve to bleed off the fuel pressure in the fuel system.

ENGINE COOLANT TEMPERATURE (ECT) SENSOR

Removal and Installation

The coolant temperature sensor is usually located on the intake manifold water jacket or near (or in) the thermostat housing. It may be necessary to drain some of the coolant from the cooling system.

1. Disconnect the negative battery cable and disconnect the electrical connector at the sensor.

2. Remove the threaded temperature sensor from the engine.

3. Apply some pipe tape to the threaded sensor and install the sensor.

4. Reconnect the sensor connector and the negative battery cable.

CRANKSHAFT POSITION (CKP) SENSOR

Removal and Installation

1. Disconnect the negative battery cable.
2. Remove the sensor harness connector at the sensor.
3. Remove the sensor to the block retaining bolt and remove the sensor from the block.
4. Installation is the reverse order of the removal procedure and be sure to use a new O-ring on the sensor.

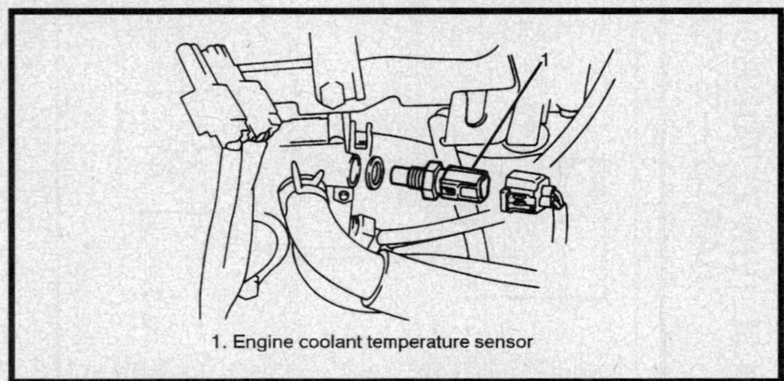

ECT sensor removal — Prizm 1.6L and 1.8L engine

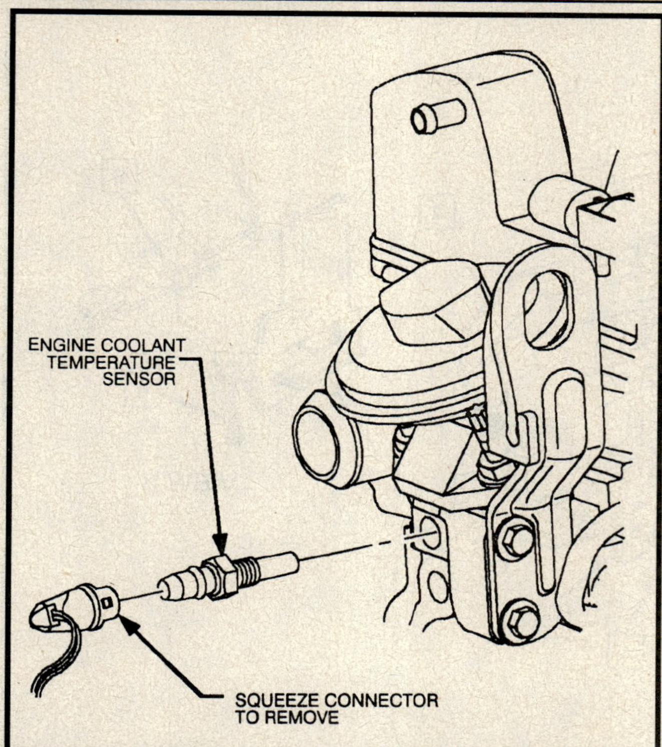

ECT sensor removal — 1.9L engine

ELECTRONIC/POWERTRAIN CONTROL MODULE (ECM/PCM)

To prevent possible electrostatic discharge damage to the ECM/PCM, do not touch the connector pins or soldered components on the circuit board. Service of the ECM/PCM should normally consists of either replacement of the ECM/PCM or a Mem-Cal (PROM) change.

If the diagnostic procedures call for the ECM/PCM to be replaced, the engine calibrator (Mem-Cal or PROM) and the ECM/PCM should be checked first to see if they are the correct parts. If they are, remove the Mem-Cal or PROM from the faulty ECM/PCM and install it in the new service ECM/PCM. The service ECM/PCM will not contain a new Mem-Cal or PROM.

The trouble DTC 51 indicates that the Mem-Cal or PROM is installed improperly or has malfunctioned. When DTC 16 on Prizm or DTC 51 on all other vehicles is obtained, check the ECM/PCM installation for bent pins or pins not fully seated in the socket. If it is installed correctly and the DTC 16 or 51 still appears, replace the Mem-Cal or PROM.

Some vehicles are equipped with an EEPROM. The EEPROM contains the preset calibrations of the vehicle according to the VIN. The EEPROM is non-repairable and can only be replaced with the controller assembly. The EEPROM will also to be reprogrammed using GM Techline or equivalent scan tool equipment.

Removal and Installation

> **NOTE: When replacing the production ECM/PCM with a service ECM/PCM (controller), it is important to transfer the broadcast code and production ECM/PCM number to the service ECM/PCM label. Do not record the number on the ECM/PCM cover. This will allow positive identification of the ECM/PCM parts throughout the service life of the vehicle. To prevent internal ECM/PCM damage, the ignition must be OFF when disconnecting and reconnecting the power the ECM/PCM (for example: the battery cable, ECM/PCM pigtail, ECM/PCM fuse, jumper cables, etc.).**

1992 PRIZM

1. Disconnect the negative battery cable.
2. Remove the center console assembly.
3. Disconnect the connectors from the ECM/PCM.
4. Remove the ECM/PCM mounting hardware and remove the ECM/PCM from the passenger compartment.
5. Installation is the reverse order of the removal procedure. Be sure to remove the new ECM/PCM from its packaging and check the service number to make sure it is the same at the defective ECM/PCM.

1993–94 PRIZM

1. Disconnect the negative battery cable.
2. Remove the carpet retainer from the carpet retaining bracket and pull the carpet back from the floor under the glove box.
3. Remove the ECM/PCM mounting hardware and remove the ECM/PCM from the passenger compartment.
4. Disconnect the connectors from the ECM/PCM.
5. Installation is the reverse order of the removal procedure. Be sure to remove the new ECM/PCM from its packaging and check the service number to make sure it is the same at the defective ECM/PCM.

STORM

1. Disconnect the negative battery cable.
2. Remove the driver's side sound insulator.
3. Disconnect the connectors from the ECM/PCM.
4. Remove the ECM/PCM mounting hardware and remove the ECM/PCM from the passenger compartment.
5. Installation is the reverse order of the removal procedure. Be sure to remove the new ECM/PCM from its packaging and check the service number to make sure it is the same at the defective ECM/PCM.

SYCLONE AND TYPHOON

1. Disconnect the negative battery cable.
2. Remove the glove right hand side kick panel to gain access to the ECM.
3. Disconnect the connectors from the ECM.
4. Remove the ECM mounting hardware and remove the ECM from the passenger compartment.
5. Installation is the reverse order of the removal procedure. Be sure to remove the new ECM from its packaging and check the service number to make sure it is the same at the defective ECM.

CAL-PAK

Removal and Installation

1. Remove the ECM/PCM.
2. Remove the Cal-Pak access panel.

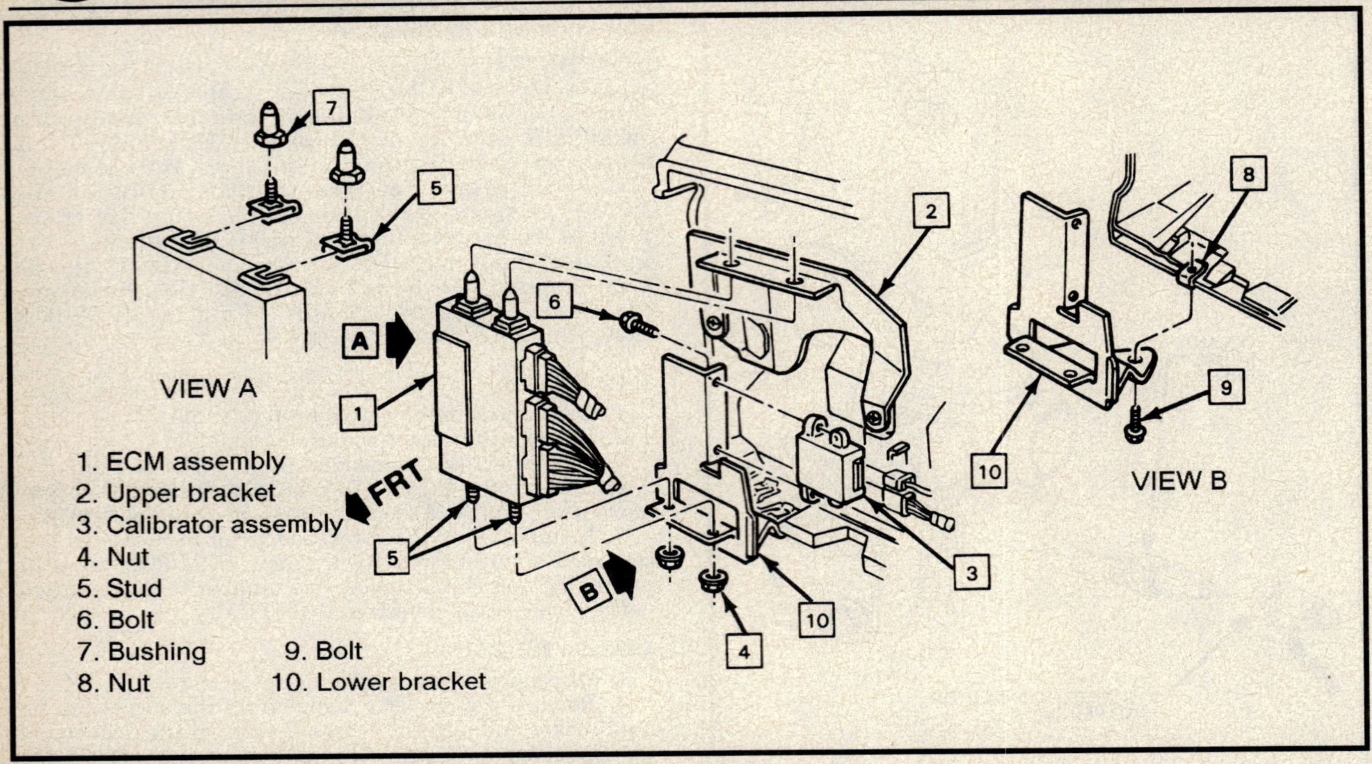

1. ECM assembly
2. Upper bracket
3. Calibrator assembly
4. Nut
5. Stud
6. Bolt
7. Bushing
8. Nut
9. Bolt
10. Lower bracket

ECM assembly removal — Syclone and Typhoon

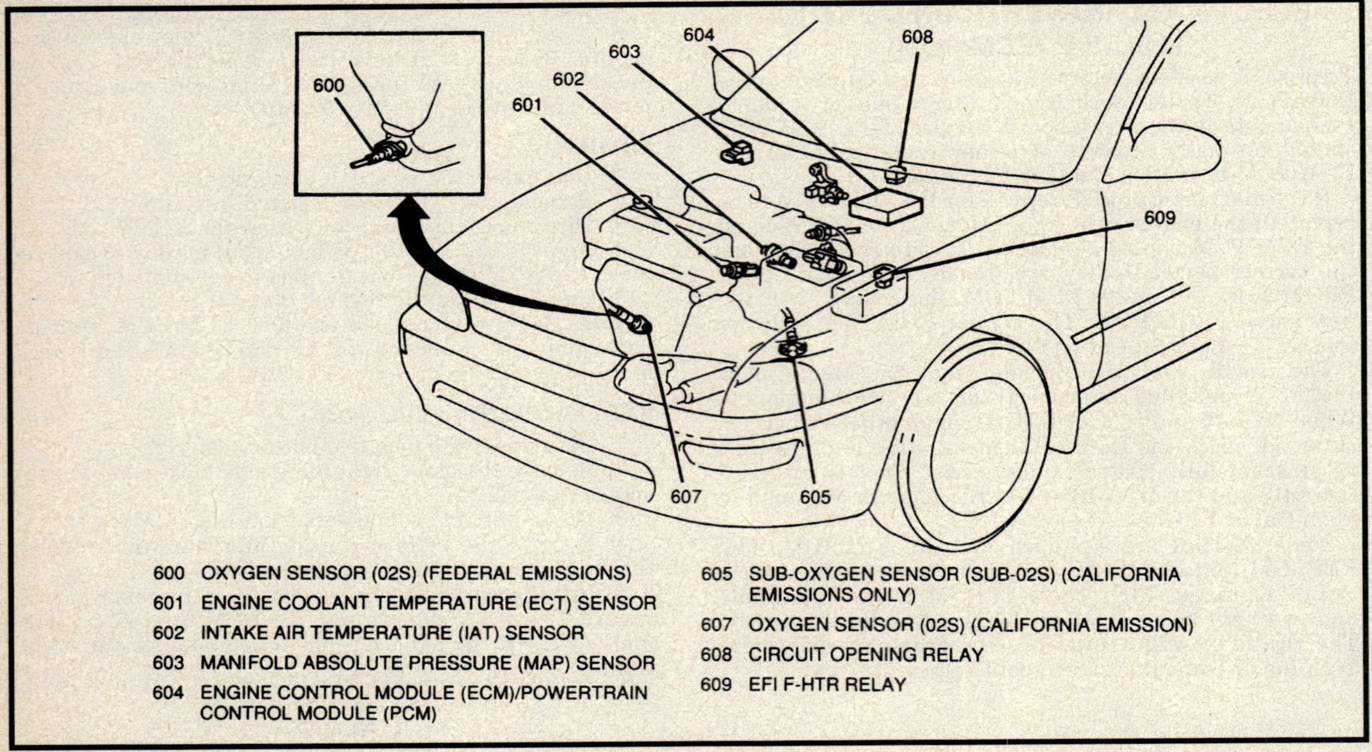

600 OXYGEN SENSOR (02S) (FEDERAL EMISSIONS)
601 ENGINE COOLANT TEMPERATURE (ECT) SENSOR
602 INTAKE AIR TEMPERATURE (IAT) SENSOR
603 MANIFOLD ABSOLUTE PRESSURE (MAP) SENSOR
604 ENGINE CONTROL MODULE (ECM)/POWERTRAIN CONTROL MODULE (PCM)
605 SUB-OXYGEN SENSOR (SUB-02S) (CALIFORNIA EMISSIONS ONLY)
607 OXYGEN SENSOR (02S) (CALIFORNIA EMISSION)
608 CIRCUIT OPENING RELAY
609 EFI F-HTR RELAY

Component locations — Prizm

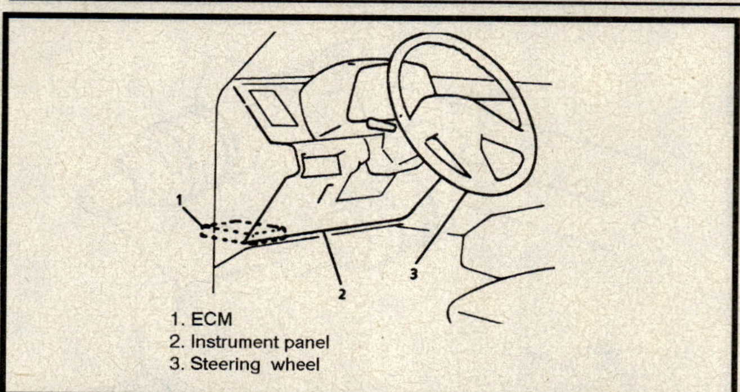

ECM location — Storm

3. Using the special Cal-Pak removal tool, grasp the Cal-Pak carrier at the narrow end and gently rock back and fourth while applying a firm upward force.

4. Inspect the reference end of the Cal-Pak carrier and set it aside. Do not remove the Cal-Pak from the carrier to confirm the Cal-Pak correctness.

5. Before installing a new Cal-Pak, be sure the new Cal-Pak part number is the same as the old one or the as the updated number per a service bulletin.

6. Install the Cal-Pak with the small notch of the carrier aligned with the small notch in the socket. Press on the Cal-Pak carrier until the Cal-Pak is firmly seated, do not press on the Cal-Pak itself only on the carrier.

7. Install the Cal-Pak access cover and reinstall the ECM/PCM.

MEM-CAL

Removal and Installation

1. Remove the ECM/PCM.
2. Remove the Mem-Cal access panel.
3. Using 2 fingers, push both retaining clips back away from the Mem-Cal. At the same time, grasp it at both ends and lift it out of the socket. Do not remove the cover of the Mem-Cal. Use of an unapproved Mem-Cal removal procedures may cause damage to the Mem-Cal or the socket.

4. Install the new Mem-Cal by pressing only the ends of the Mem-Cal. Small notches in the Mem-Cal must be aligned with the small notches in the Mem-Cal socket.

5. Press on the ends of the Mem-Cal until the retaining clips snap into the ends of the Mem-Cal. Do not press on the middle of the Mem-Cal, only on the ends.

6. Install the Mem-Cal access cover and reinstall the ECM/PCM.

Functional Check

1. Turn the ignition switch to the **ON** position.
2. Enter the diagnostic mode.
3. Allow a DTC 12 to flash 4 times to verify that no other codes are present. This indicates that the Mem-Cal is installed properly and the ECM/PCM is working correctly.
4. If trouble DTCs 41, 42, 43, 51 or 55 occurs or if the MIL in on constantly with no codes, the Mem-Cal is not fully seated or is defective.

5. If not fully seated press down firmly on the Mem-Cal carrier If found to be defective, replace it with a new one.

NOTE: To prevent possible electrostatic discharge damage to the Mem-Cal, do not touch the component leads and do not remove the integrated circuit from the carrier.

PROM

Removal and Installation

1. Remove the ECM/PCM.
2. Remove the PROM access panel.
3. Using the rocker type PROM removal tool, engage 1 end of the PROM carrier with the hook end of the tool. Press on the vertical bar end of the tool and rock the engaged end of the PROM carrier up as far as possible.
4. Engage the opposite end of the PROM carrier in the same manner and rock this end up as far as possible.
5. Repeat this process until the PROM carrier and PROM are free of the PROM socket. The PROM carrier with PROM in it should lift off of the PROM socket easily.
6. The PROM carrier should only be removed by using the special PROM removal tool. Other methods could cause damage to the PROM or PROM socket.
7. Before installing a new PROM, be sure the new PROM part number is the same as the old one or the as the updated number per a service bulletin.
8. Install the PROM with the small notch of the carrier aligned with the small notch in the socket. Press on the PROM carrier until the PROM is firmly seated, do not press on the PROM itself only on the PROM carrier.
9. Install the PROM access cover and reinstall the ECM/PCM.

Functional Check

1. Turn the ignition switch to the **ON** position.
2. Enter the diagnostic mode.
3. Allow a DTC 12 to flash 4 times to verify that no other codes are present. This indicates that the PROM is installed properly.
4. If trouble DTC 51 occurs or if the MIL in on constantly with no codes, the PROM is not fully seated, installed backwards, has bent pins or is defective.
5. If not fully seated press down firmly on the PROM carrier.
6. If installed backwards, replace the PROM.

NOTE: Any time the PROM is installed backwards and the ignition switch is turned ON, the PROM is destroyed.

7. If the pins are bent, remove the PROM straighten the pins and reinstall the PROM. If the bent pins break or crack during straightening, discard the PROM and replace with a new PROM.

NOTE: To prevent possible electrostatic discharge damage to the PROM or Cal-Pak, do not touch the component leads and do not remove the integrated circuit from the carrier.

FUEL INJECTORS

Removal and Installation

Use care in removing the fuel injectors to prevent damage to the electrical connector pins on the injector and the nozzle. The fuel injector is serviced as a complete assembly only and should not be immersed in any kind of cleaner. Support the fuel rail to avoid damaging other components while removing the injector. Be sure to note that different injectors are calibrated for different flow rates. When ordering new fuel injectors, be sure to order the identical part number that is inscribed on the bottom of the old injector.

1. Relieve fuel system pressure. Disconnect the negative battery cable.

2. Remove the intake manifold plenum or throttle body assembly as required.

3. Remove the injector electrical connections. Remove the fuel rail assembly.

4. Remove the injector retaining clip, if equipped and remove the injector from the fuel rail.

5. Remove both injector seals from the injector and discard.

To install:

6. Prior to installing the injectors, coat the new injector O-ring seals with clean engine oil. Install the seals on the injector assembly.

7. Use new injector retainer clips on the injector assembly, if equipped and position the open end of the clip facing the injector electrical connector.

8. Install the injector into the fuel rail injector socket with the electrical connectors facing outward. Push the injector in firmly until it engages with the retainer clip locking it in place.

9. Install the fuel rail and injector assembly. Install the intake manifold plenum, if equipped.

10. Connect the negative battery cable.

11. Turn the ignition switch **ON** and **OFF** to allow fuel pressure back into system. Check for leaks.

COLD START INJECTOR

Removal and Installation

1. Relieve fuel system pressure.

2. Disconnect the negative battery cable.

3. Disconnect the electrical connector from the cold start injector and clean off any dirt or grease from the injector.

4. Provide a clean container to catch any fuel or wrap a clean rag around the fuel connection.

5. Disconnect the fuel line from the cold start injector.

6. Remove injector retaining screws. Remove the cold start injector from the intake manifold.

7. Installation is the reverse order of the removal procedure. Replace the rubber sealing ring or gasket and hose clamp is necessary. Torque the retaining screws to 82 inch lbs. (9.3 Nm).

NOTE: The fuel injections systems are very susceptible to dirt in the system. Be sure that all components are clean and free from dirt and grease before reinstalling them.

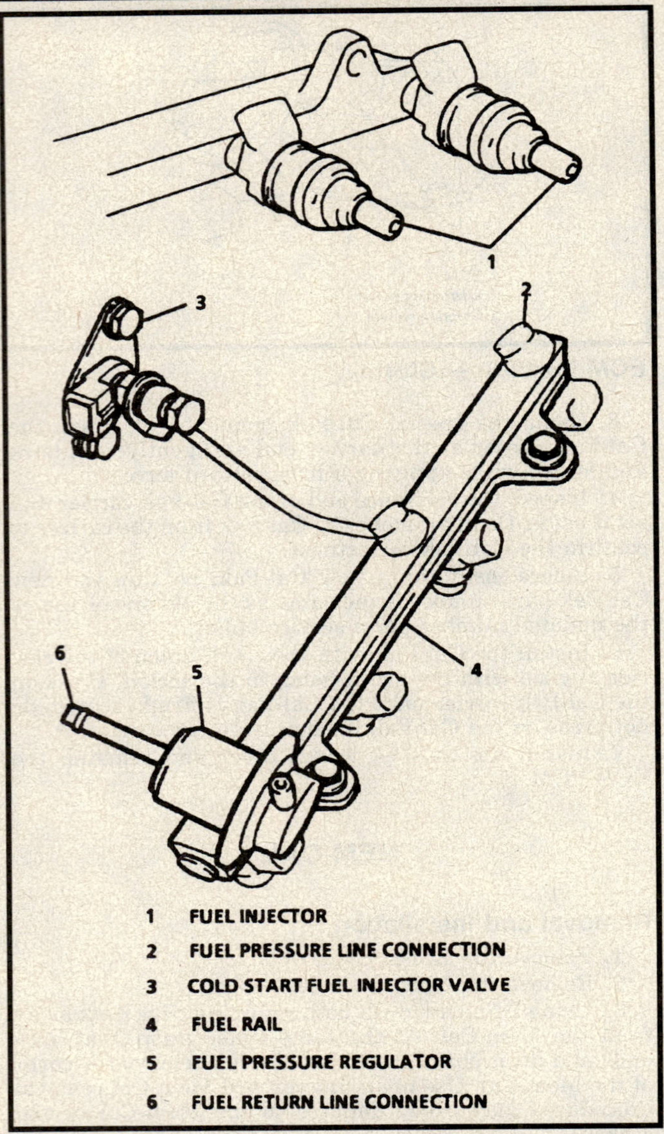

Fuel rail and cold start injector assembly — 1992 1.6L engine

FUEL PRESSURE REGULATOR

Removal and Installation

1.6L AND 1.8L ENGINES

1. Relieve fuel system pressure. Disconnect the negative battery cable.

2. Disconnect the fuel fitting attaching the fuel rail to the pressure regulator.

3. Remove the fuel pressure regulator mounting bolts, disconnect the vacuum hose and remove the pressure regulator from the engine.

4. Cover all openings with masking tape to prevent dirt entry.

To install:

5. Prior to connecting the fuel rail fitting, using a new O-ring seal, coat the seal with clean engine oil.

6. Place the O-ring on the pressure regulator and install the pressure regulator to the fuel rail fitting.

7. Position the fuel pressure regulator in place to the engine and install mounting bolts. Torque the mounting bolts to 14–19 ft. lbs. (20–27 Nm).

8. Using a backup wrench tighten the fuel rail-to-pressure regulator fitting securely.

9. Connect the negative battery cable.

10. Turn the ignition switch **ON** and **OFF** to allow fuel pressure back into system. Check for leaks.

1.9L ENGINE

1. Remove the air intake tube/resonator.

2. Relieve fuel system pressure. Disconnect the negative battery cable.

3. Remove the bolt from the fuel line clip.

4. Disconnect the fuel inlet and return lines from the pressure regulator.

5. Remove the fuel pressure regulator mounting bolts, disconnect the vacuum hose and remove the pressure regulator from the engine.

6. Cover all openings with masking tape to prevent dirt entry.

To install:

7. Prior to connecting the regulator, install a new fuel inlet O-ring seal, coat the seal with clean engine oil. Do not replace the fuel outlet O-ring unless damaged. The outlet O-ring must be replaced using a special brass seal removal and installation tool.

8. Place the O-ring on the pressure regulator and install the pressure regulator to the fuel rail fitting. Install the mounting bolts using Loctite® 242 or equivalent. Torque the mounting bolts to 106 inch lbs. (12 Nm).

9. Using a backup wrench tighten the fuel rail-to-pressure regulator fitting securely.

10. Install the air intake tube and resonator. Connect the negative battery cable.

11. Turn the ignition switch **ON** and **OFF** to allow fuel pressure back into system. Check for leaks.

4.3L ENGINE

1. Properly relieve the fuel system pressure and disconnect the negative battery cable.

2. Remove the regulator retaining bolts.

3. Disconnect the vacuum line from the regulator.

4. Remove the regulator from the vehicle.

5. Installation is the reverse of the removal procedure.

FUEL RAIL ASSEMBLY

When servicing the fuel rail assembly, be careful to prevent dirt and other contaminants from entering the fuel passages. Fitting should be capped and holes plugged during servicing. At any time the fuel system is opened for service, the O-ring seals and retainers used with related components should be replaced.

Before removing the fuel rail, the fuel rail assembly may be cleaned with a spray type cleaner, GM–30A or equivalent, following package instructions. Do not immerse fuel rails in liquid cleaning solvent. Be sure to always use new O-rings and seals when reinstalling the fuel rail assemblies.

There is an 8 digit number stamped on the under side of the fuel rail assembly on 4 cylinder engines and on the left hand fuel rail on dual rail assemblies (fueling even cylinders No. 2, 4, 6). Refer to this number if servicing or part replacement is required.

Removal and Installation

1992 PRIZM

1. Disconnect the negative battery cable. Relieve fuel system pressure.

2. Remove the PCV hose from the valve cover.

3. Disconnect the fuel feed line and return line from the fuel rail assembly, be sure to use a backup wrench on the inlet fitting to prevent turning.

4. Remove the vacuum line at the pressure regulator. Remove the return line from pressure regulator.

5. Disconnect the electrical connectors from each fuel injector.

6. Disconnect the electrical connector and fuel line at the cold start injector.

7. Remove the fuel rail assembly retaining bolts. Remove the fuel rail assembly and cover all openings with masking tape to prevent dirt entry.

NOTE: If any injectors become separated from the fuel rail and remain in the intake manifold, both O-ring seals and injector retaining clip must be replaced. Use care in removing the fuel rail assembly, to prevent damage to the injector electrical connector terminals and the injector spray tips. When removed, support the fuel rail to avoid damaging its components. The fuel injector is serviced as a complete unit only. Since it is an electrical component, it should not be immersed in any type of cleaner.

To install:

8. Be sure to lubricate all the O-rings and seals with clean engine oil. Carefully push the injectors into the cylinder head intake ports until the bolt holes on the fuel rail and manifold are aligned.

9. Apply a coating of tread locking compound on the treads of the fittings. Torque the fuel rail retaining bolts to 11 ft. lbs. (15 Nm).

10. Connect the cold start injector and fuel rail feed and return line.

11. Connect the fuel injector electrical connectors.

12. Connect the vacuum hose to the regulator.

13. Connect the PCV hose to the valve cover.

14. Connect the negative battery cable.

15. Energize the fuel pump and check for leaks.

1993–94 PRIZM

1. Disconnect the negative battery cable. Relieve fuel system pressure.

2. Drain the engine coolant into a suitable container.

3. Remove the accelerator and cruise control cable (if equipped) from the air cleaner clip.

4. Disconnect the intake air temperature sensor electrical connector.

5. Loosen the air cleaner hose clamp bolt on the throttle.

6. Remove the 4 cap clips from the air cleaner.

7. Remove the air cleaner hose and cap from the throttle body.

8. Disconnect the throttle valve cable at the throttle body and remove it from the bracket, if equipped with automatic transaxle.

9. Disconnect the 2 EGR and 1 evaporative emission vacuum hoses from the throttle body, if equipped with California emissions.

10. Disconnect the electrical connectors at the TP and IAC sensors.

11. Remove the throttle body retaining bolts and position the throttle body aside.

12. Remove the 2 bolts securing the fuel feed and return line bracket to the intake manifold.

13. Remove the EGR hoses from the EGR valve and pipe, if equipped with California emissions.

14. Remove the engine hanger and EGR vacuum modulator, if equipped with California emissions, from the intake manifold.

15. Remove the EGR temperature sensor electrical connector, if equipped with California emissions.

16. Remove the EGR valve and pipe, if equipped with California emissions.

17. Remove the vacuum hoses at the pressure regulator and the PCV hose from the intake plenum.

18. Remove the air intake chamber cover and gasket.

19. Disconnect the fuel injector electrical connectors.

20. Disconnect the fuel feed line at the fuel rail assembly and the return line, from the pressure regulator, be sure to use a backup wrench on the inlet fitting to prevent turning.

21. Remove the injectors and fuel rail assembly from the intake plenum.

NOTE: If any injectors become separated from the fuel rail and remain in the intake manifold, both O-ring seals and injector retaining clip must be replaced. Use care in removing the fuel rail assembly, to prevent damage to the injector electrical connector terminals and the injector spray tips. When removed, support the fuel rail to avoid damaging its components. The fuel injector is serviced as a complete unit only. Since it is an electrical component, it should not be immersed in any type of cleaner.

To install:

22. Be sure to lubricate all the O-rings and seals with clean engine oil. Carefully push the injectors into the fuel rail.

23. Install new insulators and 2 spacers in position on intake manifold.

24. Install the fuel rail, securing loosely with 2 bolts and ensure the injectors rotate smoothly. If they do not rotate smoothly, remove the injectors and inspect the O-rings. Torque the fuel rail retaining bolts to 11 ft. lbs. (15 Nm).

25. Connect the fuel feed line at the fuel rail assembly and the return line to the pressure regulator; be sure to use a backup wrench on the inlet fitting to prevent turning.

26. Reconnect the fuel injector electrical connectors.

27. Install the air intake chamber cover and gasket. Torque all bolts to 14 ft. lbs. (19 Nm).

28. Connect the vacuum hoses at the pressure regulator and the PCV hose onto the intake plenum.

29. Install the EGR valve and pipe, if equipped with California emissions. Torque the nuts to 115 inch lbs. (13 Nm).

30. Connect the EGR temperature sensor electrical connector, if equipped with California emissions.

31. Install the engine hanger and EGR vacuum modulator, if equipped with California emissions, from the intake manifold. Torque the engine hanger assembly nut and bolt to 21 ft. lbs. (28 Nm).

32. Connect the EGR hoses to the EGR valve and pipe, if equipped with California emissions.

33. Install the 2 bolts securing the fuel feed and return line bracket to the intake manifold. Torque the bolts to 89 inch lbs. (10 Nm).

34. Install the throttle body assembly with new gaskets and torque the retaining bolts 16 ft. lbs. (22 Nm).

35. Reconnect the electrical connectors at the TP and IAC sensors.

36. Connect the 2 EGR and 1 evaporative emission vacuum hoses to the throttle body, if equipped with California emissions.

37. Connect the throttle valve and accelerator cables at the throttle body, if equipped with automatic transaxle. Secure with nuts and adjust the throttle valve cable.

38. Install the air cleaner hose and cap to the throttle body.

39. Install the 4 cap clips onto the air cleaner.

40. Tighten the air cleaner hose clamp bolt at the throttle body to 11 ft. lbs. (15 Nm).

41. Connect the intake air temperature sensor electrical connector.

42. Connect the accelerator and cruise control cable (if equipped) to the air cleaner clip.

43. Fill the cooling system with antifreeze.

44. Connect the negative battery cable.

45. Turn the ignition switch **ON** and momentarily connect a jumper wire between the DLC terminals +B and FP to pressurize the fuel system. Inspect for fuel leaks.

STORM 1.6L ENGINE

1. Disconnect the negative battery cable. Relieve fuel system pressure.

2. Remove the air cleaner duct assembly.

3. Remove the vacuum hoses and the PCV hose from the intake plenum.

4. Disconnect the fuel feed line at the retainer bracket and the return line at the pressure regulator, be sure to use a backup wrench on the inlet fitting to prevent turning.

5. Disconnect the fuel pressure regulator vacuum hose.

6. Disconnect the fuel injector electrical connectors.

7. Disconnect the electrical connectors at the IAT sensor and position harness away from the fuel rail.

8. Remove the plenum support bracket and fasteners at the passenger side of the plenum.

9. Remove the two 12mm bolts with spacers securing the fuel rail to the plenum.

10. Remove the injectors from the manifold and fuel rail. Discard the fuel injector seals.

11. Remove the fuel rail assembly.

To install:

12. Be sure to lubricate all the O-rings and seals with clean engine oil. Carefully push the injectors into the cylinder head intake ports until the bolt holes on the fuel rail and manifold are aligned. Apply a coating of tread locking compound on the treads of the fittings. Torque the fuel rail retaining bolts to 28 ft. lbs. (32 Nm).

13. Connect the fuel line at the pressure regulator. Be sure to use a backup wrench on the inlet fitting to prevent turning.

14. Install the fuel pressure regulator vacuum hose.

15. Connect the vacuum hoses and the PCV hose to the intake plenum.

16. Connect the fuel feed line, be sure to use a backup wrench on the inlet fitting to prevent turning.

17. Connect the electrical connectors at the IAT sensor.

18. Install the plenum support bracket and fasteners at the passenger side of the plenum.

19. Install the air cleaner duct assembly.
20. Connect the negative battery cable. Energize the fuel pump and check for leaks.

STORM 1.8L ENGINE

1. Disconnect the negative battery cable. Relieve fuel system pressure.
2. Remove the air cleaner duct assembly.
3. Disconnect the throttle cable at the throttle body and remove it from the bracket.
4. Disconnect the electrical connectors at the TPS, IAC and IAT sensors.
5. Label and disconnect all vacuum and coolant hoses from the throttle body.
6. Remove the throttle body retaining bolts and remove the throttle body.
7. Remove the vacuum hoses and the PCV hose from the intake plenum.
8. Disconnect the ECM/PCM ground connectors from the intake plenum.
9. Remove the engine hook bracket and the plenum support brackets.
10. Remove plenum retaining bolts and remove the plenum assembly.
11. Disconnect the fuel injector electrical connectors.
12. Disconnect the fuel feed line at the fuel rail assembly and the return line, from the pressure regulator, be sure to use a backup wrench on the inlet fitting to prevent turning.
13. Remove the fuel rail assembly retaining bolts. Remove the fuel rail assembly and cover all openings with masking tape to prevent dirt entry.

> **NOTE: If any injectors become separated from the fuel rail and remain in the intake manifold, both O-ring seals and injector retaining clip must be replaced. Use care in removing the fuel rail assembly, to prevent damage to the injector electrical connector terminals and the injector tips. When removed, support the fuel rail to avoid damaging its components. The fuel injector is serviced as a complete unit only. Since it is an electrical component, it should not be immersed in any type of cleaner.**

To install:

14. Be sure to lubricate all the O-rings and seals with clean engine oil. Carefully push the injectors into the cylinder head intake ports until the bolt holes on the fuel rail and manifold are aligned. Apply a coating of tread locking compound on the treads of the fittings. Torque the fuel rail retaining bolts to 28 ft. lbs. (32 Nm).
15. Connect the fuel feed line at the fuel rail assembly and the return line to the pressure regulator. Be sure to use a backup wrench on the inlet fitting to prevent turning.
16. Install plenum assembly and tighten the retaining bolts.
17. Connect the fuel injector electrical connectors.
18. Install the engine hook bracket and the plenum support brackets.
19. Connect the ECM/PCM ground connectors to the intake plenum.
20. Install the vacuum hoses and the PCV hose at the intake plenum.
21. Install the throttle body using a new gasket. Torque the throttle body retaining bolts to 16 ft. lbs. (22 Nm).
22. Connect all vacuum and coolant hoses to the throttle body.

23. Connect the electrical connectors at the TPS, IAC and IAT sensors.
24. Connect the throttle cable at the throttle body and to the bracket.
25. Install the air cleaner duct assembly.
26. Connect the negative battery cable. Energize the fuel pump and check for leaks.

1.9L ENGINE

1. Disconnect the negative battery cable. Relieve fuel system pressure.
2. Remove the air intake duct, resonator and fresh air tube assembly.
3. Remove the fuel line bracket bolt. Remove the fuel supply and return lines using a backup wrench to prevent damage to the fuel lines.

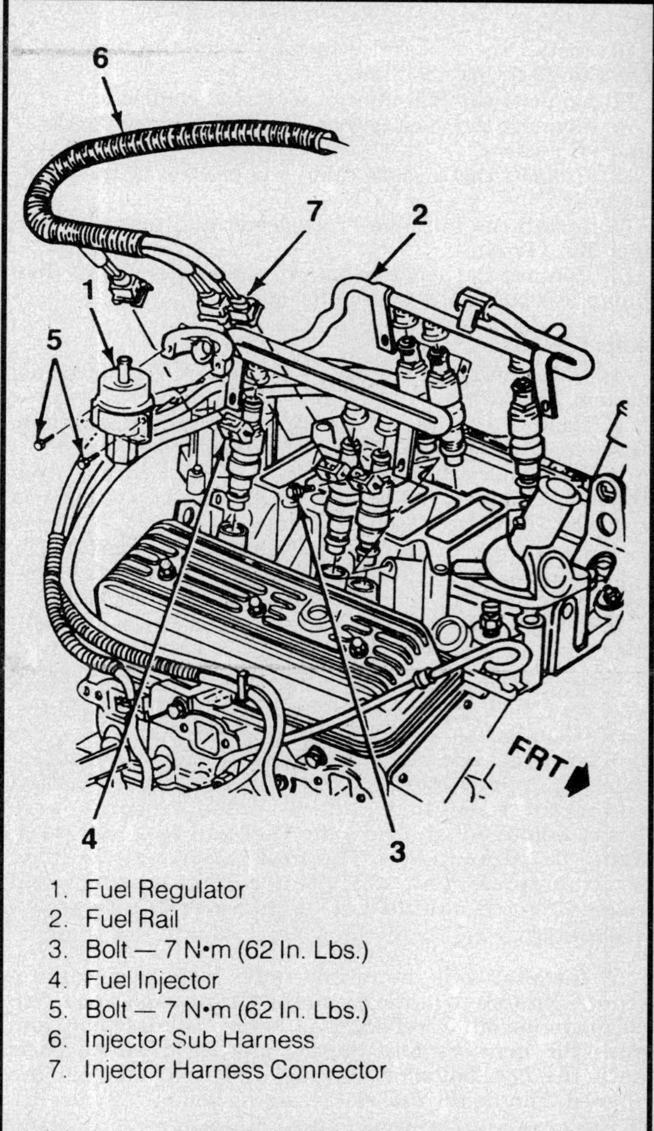

1. Fuel Regulator
2. Fuel Rail
3. Bolt — 7 N•m (62 In. Lbs.)
4. Fuel Injector
5. Bolt — 7 N•m (62 In. Lbs.)
6. Injector Sub Harness
7. Injector Harness Connector

Fuel rail and injector assembly — 4.3L engine

4. Disconnect the throttle cable at the throttle body and remove it from the bracket.

NOTE: It may be necessary to remove the fuel rail attaching bolts and rotate the fuel rail in order to gain access for the wrench.

5. Disconnect the vacuum hose at the pressure regulator and PCV valve.

6. Remove the throttle cable bracket bolts, disconnect the throttle cable from the throttle lever and lay over the intake manifold.

7. Disconnect the fuel injector electrical connectors.

8. Remove the fuel rail attaching bolts and remove the assembly by rotating the throttle lever.

To install:

9. Be sure to lubricate all the O-rings and seals with clean engine oil. Carefully push the injectors into the cylinder head intake ports until the bolt holes on the fuel rail and manifold are aligned. Connect the fuel feed and return lines.

10. Install the fuel rail attaching bolts and torque the bolts to 22 ft. lbs. (30 Nm).

11. Connect the fuel injector electrical connectors.

12. Connect the vacuum hose at the pressure regulator and PCV valve.

13. Connect the throttle cable and bracket at the throttle body.

14. Install the fuel line bracket bolt and torque to 106 inch lbs. (12 Nm).

15. Connect the negative battery cable, energize the fuel pump and check the system for leaks.

4.3L ENGINE

1. Disconnect the negative battery cable. Relieve fuel system pressure.

2. Remove the intake manifold plenum and runners. Disconnect the fuel tube bracket retaining bolt.

3. Disconnect the fuel feed line and return line from the fuel rail assembly, be sure to use a backup wrench on the inlet fitting to prevent turning.

4. Disconnect the injector wire harness connectors.

5. Remove the vacuum line from the fuel pressure regulator. Remove the fuel rail assembly retaining bolts.

6. Remove the fuel rail assembly and cover all openings with masking tape to prevent dirt entry.

NOTE: If any injectors become separated from the fuel rail and remain in the intake manifold, both O-ring seals and injector retaining clip must be replaced. Use care in removing the fuel rail assembly, to prevent damage to the injector electrical connector terminals and the injector spray tips. When removed, support the fuel rail to avoid damaging its components. The fuel injector is serviced as a complete unit only. Since it is an electrical component, it should not be immersed in any type of cleaner.

7. Installation is the reverse order of the removal procedure. Be sure to lubricate all the O-rings and seals with clean engine oil. Carefully tilt the fuel rail assembly and push the injectors into the cylinder head intake ports until the bolt holes on the fuel rail and manifold are aligned. Torque the fuel rail retaining bolts to 62 inch lbs. (7 Nm).

8. Connect the negative battery cable. Energize the fuel pump and check for leaks.

IDLE AIR CONTROL VALVE

Removal and Installation

PRIZM

1. Disconnect the negative battery cable.

2. Remove the throttle body from the vehicle.

3. Remove the 4 screws retaining the idle air control valve. Remove the idle air control valve from mounting position.

To install:

4. Use a new gasket and install the idle air control valve in mounting position.

5. Install the retaining screws.

SATURN AND STORM

1. Disconnect the negative battery cable.

2. Disconnect electrical connector from idle air control valve.

3. Remove the screws retaining the idle air control valve. Remove the idle air control valve from mounting position.

To install:

4. Prior to installing the idle air control valve, measure the distance the valve plunger is extended. Measurement should be made from the edge of the valves mounting flange to the end of the cone. The distance should not exceed 1⅛ in. (28mm). If a new valve is being installed and the measured distance is greater than specified above, press on the valve firmly to retract it.

5. Use a new gasket and install the idle air control valve in mounting position.

6. Install the retaining screws. Torque the retaining screws to 27 inch lbs. (3 Nm). Connect the electrical connector.

7. Reset the IAC valve position.

SYCLONE AND TYPHOON

1. Disconnect the negative battery cable. Removal electrical connector from idle air control valve.

2. Remove the idle air control valve using a 1¼ in. (32mm) wrench.

3. Installation is the reverse of removal. Before installing a new idle air control valve, measure the distance that the valve is extended. Measurement should be made from the motor housing to the end of the cone. The distance should not exceed 1⅛ in. (28mm), or damage to the valve may occur when installed. If the length of the valve is too long, press on valve firmly to retract it, using a slight side to side motion to help it retract easier.

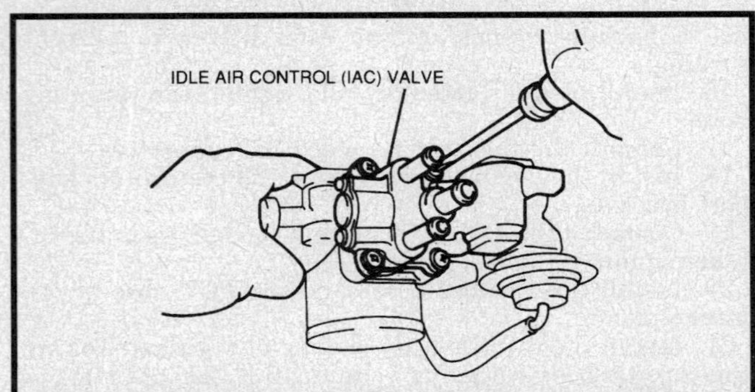

Removing IAC valve — Prizm

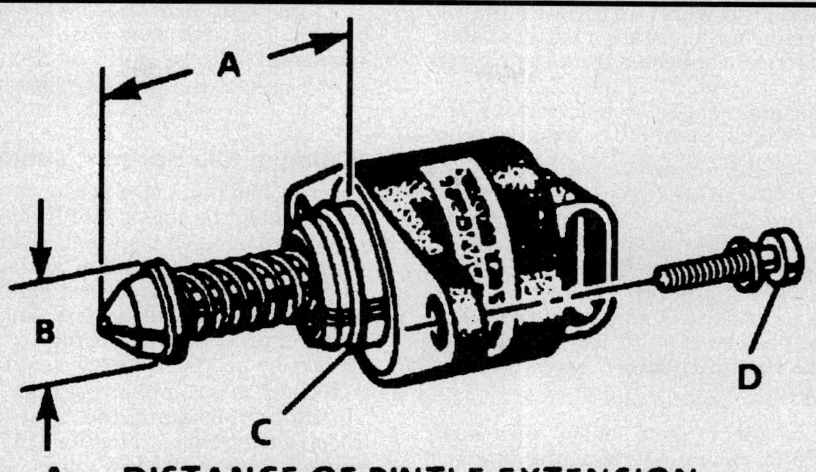

Measuring new IAC valve length — Storm and Saturn

4. Use a new gasket and install the valve. Torque the IAC valve to 13 ft. lbs. (18 Nm).

5. Start the vehicle and allow the engine to reach normal operating temperature, then drive the vehicle above 35 mph and the ECM will automatically reset the the IAC position.

MASS AIR FLOW (MAF) SENSOR

Removal and Installation

1. Disconnect the negative battery cable.

2. Disconnect the electrical connector to the MAF sensor.

3. Remove the air cleaner cover. Remove the sensor retaining bolts and remove the sensor.

4. Installation is the reverse order of the removal procedure.

MANIFOLD ABSOLUTE PRESSURE(MAP) SENSOR

Removal and Installation

1. Disconnect the negative battery cable and remove the vacuum hose from the MAP sensor, except Saturn.

2. Disconnect the electrical connector to the MAP sensor.

3. Remove the MAP sensor retaining screws and remove the MAP sensor.

4. Installation is the reverse order of the removal procedure.

INTAKE/MANIFOLD AIR TEMPERATURE (IAT/MAT) SENSOR

Removal and Installation

1. Disconnect the negative battery cable.

2. Disconnect the electrical connector to the MAT/IAT sensor.

3. Remove the sensor retaining clip from the air cleaner duct or unscrew the MAT/IAT sensor from the intake manifold.

4. Installation is the reverse order of the removal procedure.

OXYGEN SENSOR

Removal and Installation

The oxygen sensor uses a permanently attached pigtail and connector. This pigtail should not be removed from the oxygen sensor. Damage or removal of the pigtail connector could effect the proper operation of the oxygen sensor. Use caution when handling the oxygen sensor. The in-line electrical connector and louvered end must be kept free of grease, dirt or other contaminants. Also avoid using cleaner solvent of any type. Do not drop or roughly handle the oxygen sensor. It is not recommended to clean or wire brush an oxygen sensor when in question, it is recommended that a new oxygen sensor be used.

NOTE: The oxygen sensor may be difficult to remove when the engine temperature is below 120°F (48°C). Excessive force may damage the threads in the exhaust manifold of exhaust pipe.

1. Start the engine and let it warm up to 120°F (48°C), stop the engine and disconnect the negative battery cable.

2. Disconnect the electrical connector from the oxygen sensor.

3. Using a special oxygen sensor socket, remove the oxygen sensor from the exhaust manifold or exhaust pipe.

NOTE: A special anti-seize compound is used on the oxygen sensor threads. The compound consists of a liquid graphite and glass beads. The graphite will burn away, but the glass beads will remain, making the sensor easier to remove. New or service sensors will already have the compound applied to the threads. If a oxygen sensor is removed from the engine and if for any reason it is to be reinstalled, the threads must have this anti-seize compound applied before installation.

4. Coat the threads of the oxygen sensor with anti-seize compound 5613695 or equivalent, if necessary.

5. Install the sensor and torque it to 18 ft. lbs. (25 Nm) on Saturn or 30 ft. lbs (41 Nm) on all other vehicles.

6. Reconnect the electrical connector to the sensor and the negative battery cable.

POWER STEERING PRESSURE SWITCH

Removal and Installation

1. Disconnect the negative battery cable.

2. Remove the belt and remove the pump leaving the pressure and return lines attached.

3. Disconnect the electrical connector to the switch and place a drip pan under the switch.

4. Unscrew switch and remove the switch from the power steering line.

5. Installation is the reverse order of the removal procedure. Be sure to replace the power steering fluid lost during removal of the switch.

VEHICLE SPEED SENSOR

Removal and Installation

1. Disconnect the negative battery cable.

2. Remove the vehicle speed sensor electrical lead from the transmission.

3. Remove the VSS sensor bolt and screw retainer or unscrew the sensor from the transmission/transaxle.

4. Remove the vehicle speed sensor assembly and O-ring.

5. Installation is the reverse order of the removal procedure. Use a new O-ring and coat with transmission fluid.

THROTTLE BODY

Removal and Installation

1. Disconnect the negative battery cable.

2. Remove the air cleaner duct assembly.

3. Disconnect the IAC and TPS electrical connectors.

4. Disconnect the vacuum lines, throttle valve and cruise control cables from the throttle body.

5. Partially drain engine coolant to allow coolant hoses at throttle body to be removed, Geo Storm only.

6. Remove the throttle body mounting bolts. Remove the throttle body and gasket.

7. Installation is the reverse order of the removal procedure. Use and new gasket. Torque the throttle body mounting bolts to 10–15 ft. lbs. (14–20 Nm).

8. Connect the negative battery cable.

Minimum Idle Speed Adjustment

The throttle plate stop screw or minimum air rate adjustment should be considered the minimum idle speed as on other fuel injected engine. Low internal friction and provision for slight production variations and various operation altitudes resulted in a calibrated minimum air rate which is too low allow most engines to idle. The adjustment is preset at the factory and no further adjustment should be necessary, except on the 1992 Prizm.

If there is a complaint of high idle speed, vacuum leaks should be considered the most likely cause. Because the Electronic Control Module (ECM/PCM) learns idle air control, it is even less likely that a stalling complaint would be due to incorrect minimum air rate.

If it is determined that the minimum air rate is suspect, be sure the IAC valve is not lost (not actually at the location indicated by current IAC counts). The IAC valve could be lost if the ECM/PCM power has been interrupted with the ignition in the **ON** position or the IAC valve has been disconnected with the engine running since the last reset. The minimum air rate may be checked using the following procedures:

EXCEPT PRIZM

1. Block the drive wheels and apply the parking brake.

2. Connect a suitable scan tool the ALDL.

3. Start the engine and allow it to reach normal operating temperature and go into the **CLOSED LOOP** mode.

4. On vehicles equipped with automatic transaxles, shift to **D** and then back to the **N** position.

5. With the A/C and all electrical accessories **OFF** and the transaxle in the **N** position, scan the power steering pressure input. It should be **OFF** or normal. If it is not repair the fault in the power steering pump circuit and allow the engine idle to stabilize.

6. Scan the IAC counts, if the counts are between 1–50 counts, the throttle plate stop screw adjustment is acceptable. It is important to allow the engine idle speed to stabilize to assure correct counts are determined.

7. If the counts are below specifications, check the intake manifold for vacuum leaks at the hoses, throttle body and intake manifold or damaged throttle lever and correct as necessary.

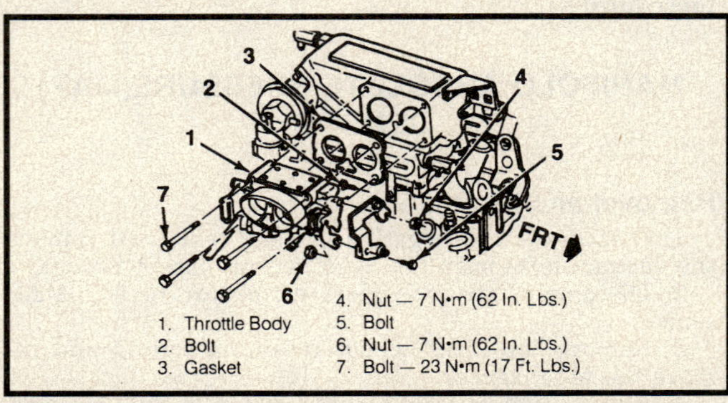

Throttle body assembly removal — 4.3L engine

8. If no vacuum leaks or other causes of excessive air into the intake are found, the throttle body will have to be replaced.

If the counts are too high, check for damaged throttle lever or airflow restriction by the throttle plate. If there is no problem evident, remove the air inlet air duct at the throttle body and clean the residue from the inside of the throttle bore and from the edges of the throttle plate. Use a suitable solvent, but do not use a solvent containing methyl ethyl ketone. Recheck the IAC counts, if the counts are still too high, the throttle body will have to be replaced.

PRISM

This procedure is capable of being used on 1992 Geo Prizm only. There is no procedure available for 1993–94 vehicles. Before adjusting the idle speed, ensure that the air cleaner is installed, all air intake hoses and electrical connectors are connected.

1. Start the engine and allow it to reach normal operating temperature.
2. Turn all accessories and air conditioning controls **OFF**.
3. Connect a tachometer to the IG terminal of the check connector.
4. Connect a jumper between terminals TE1 to E1 of the check connector on 1.6L (VIN 6) engine only.
5. Set the idle speed by adjusting the idle speed screw.
   VIN 5 with manual transaxle — 800 rpm
   VIN 5 with automatic transaxle — 800 rpm
   VIN 6 with manual transaxle — 700 rpm
   VIN 6 with manual transaxle — 700 rpm
6. Turn the air conditioning switch ON, the engine idle should be 900–1000 rpm.
7. If the air conditioning idle is incorrect, check the air conditioning Vacuum Switching Valve (VSV).
8. Remove the tachometer and jumper from the check connector.

THROTTLE POSITION SENSOR

Removal and Installation

1. Disconnect the electrical connector from the sensor.
2. Remove the attaching screws, lock washers and retainers.
3. Remove the throttle position sensor. If necessary, remove the screw holding the actuator to the end of the throttle shaft.
4. With the throttle valve in the normal closed idle position, install the throttle position sensor on the throt-

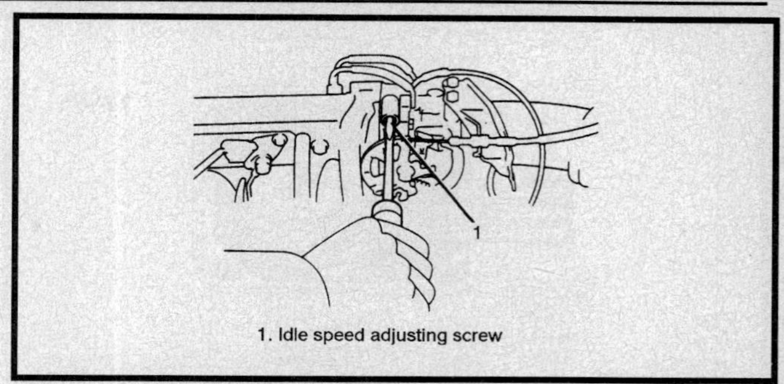

Minimum idle speed adjusting screw — 1.6L (VIN 6) engine

tle body assembly, making sure the sensor pickup lever is located above the tang on the throttle actuator lever.
5. Install the retainers, screws and lock washers using a thread locking compound.

Adjustment

The throttle position sensor is only adjustable on the Geo Prizm. If the sensor is found to be at fault and cannot be adjusted, it should be replaced.

1992 PRISM 1.6L (VIN 5) ENGINE

1. Remove the electrical connector from the throttle position sensor.
2. Loosen the throttle position sensor mounting bolts.
3. Insert a 0. 014 in. (0.35mm) feeler gauge tool between the throttle stop screw and the throttle lever.
4. Using DVM check for continuity between terminals E2 and IDL of the throttle switch. Continuity should be present.
5. Insert a 0.023 in (0.59mm) feeler gauge tool. Continuity should not be present.
6. Rotate the throttle position sensor to obtain the desired adjustment. Tighten the retaining screws.
7. Reconnect the electrical connector to the sensor.

1992 PRISM 1.6L (VIN 6) ENGINE

1. Remove the electrical connector from the throttle position sensor.
2. Loosen the throttle position sensor mounting bolts.
3. Insert a 0.028 in. (0.70mm) feeler gauge tool between the throttle stop screw and the throttle lever.
4. Using DVM check for continuity between terminals E2 and IDL of the throttle switch. Continuity should be present.

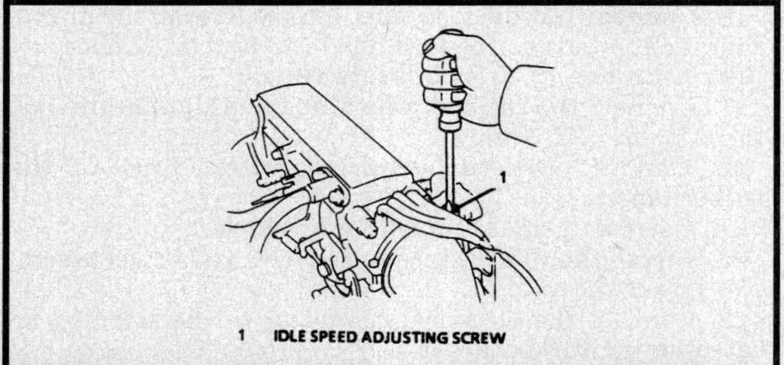

Minimum idle speed adjusting screw — 1.6L (VIN 5) engine

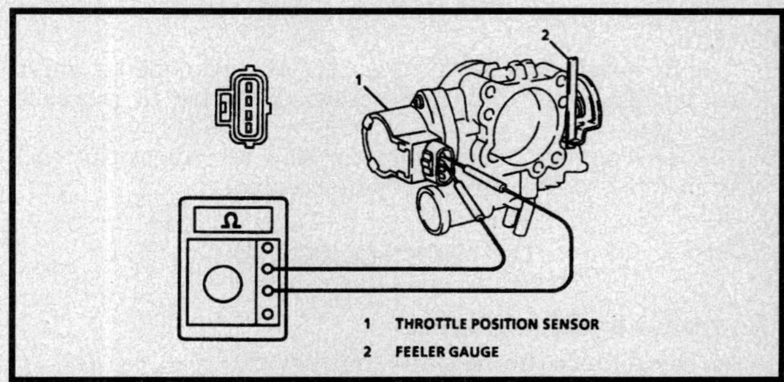

Throttle Position Sensor — 1992 Prizm 1.6L (VIN 5) engine

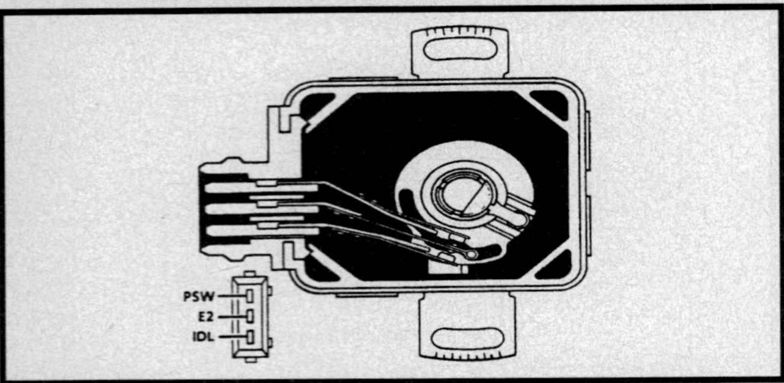

Throttle Position Sensor — 1992 Prizm 1.6L (VIN 6) engine

5. Insert a 0.032 in (0.80mm) feeler gauge tool. Continuity should not be present.

6. Rotate the throttle position sensor to obtain the desired adjustment. Tighten the retaining screws.

7. Reconnect the electrical connector to the sensor.

1993–94 PRISM

1. Remove the electrical connector from the throttle position sensor.

2. Loosen the throttle position sensor mounting bolts.

3. If equipped with California emissions, disconnect and apply vacuum to the throttle opener valve at the throttle body.

4. Insert a 0.016 in. (0.40) feeler gauge tool between the throttle stop screw and the throttle lever.

5. Using DVM check for continuity between terminals E2 and IDL of the throttle switch. Continuity should be present.

6. Insert a 0.035 in (0.90mm) feeler gauge tool. Continuity should not be present.

7. Rotate the throttle position sensor to obtain the desired adjustment. Tighten the retaining screws.

8. Reconnect the electrical connector to the sensor and disconnect the vacuum supply at the throttle opener and reconnect the vacuum line to the valve.

9. If equipped California emissions, adjust the throttle opener valve as follows:

   a. Start the engine and run until normal operating temperature is achieved.

   b. Connect a tachometer to the battery and IG terminal of the DLC connector.

   c. Disconnect and plug the vacuum hose at the throttle opener valve.

   d. Maintain engine speed at 2500 rpm for approximately 15 seconds. Release the throttle and verify the rpm is between 1300–1500 rpm with the radiator fan **OFF**.

   e. If not as specified, adjust the throttle opener valve by turning the adjusting screw clockwise to increase the rpm.

   f. Disconnect the tachometer and reconnect the vacuum hose to the throttle opener valve.

TURBOCHARGER

Removal and Installation

1. Disconnect the negative battery cable.
2. Drain the coolant from the radiator.
3. Remove the air intake duct.
4. Remove the turbocharger air intake duct.

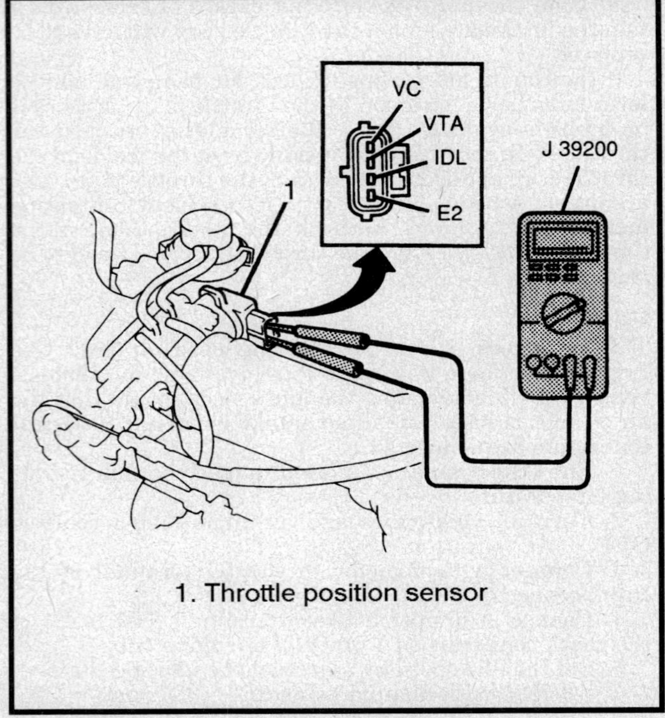

1. Throttle position sensor

Throttle Position Sensor terminal identification — 1993–94 Prizm

5. Remove the battery, battery tray and vacuum tank.
6. Disconnect the turbocharger oil feed line.
7. Disconnect the electrical connector from the solenoid located on the turbocharger.
8. Disconnect the turbocharger coolant return line.
9. Raise and safely support the vehicle.
10. Remove the right tire and wheelhouse panel.
11. Remove the turbocharger outlet pipe.
12. Remove the turbocharger oil return pipe.
13. Remove the turbocharger mounting nuts.
14. Remove the turbocharger coolant feed lines.
15. Remove the turbocharger assembly from the engine.

To install:

16. Install the turbocharger to the engine and tighten the turbocharger to exhaust manifold nuts to 33 ft. lbs. (45 Nm).

17. Connect the turbocharger coolant feed line and torque to 26 ft. lbs. (35 Nm).

18. Connect the oil feed and return line at the turbocharger and tighten the feed line to 13 ft. lbs. (17 Nm) and the return line to 35 inch lbs. (4 Nm).

19. Connect the coolant return line and tighten the line to 16 ft. lbs. (22 Nm).

20. Connect the turbocharger outlet pipe to the turbocharger.

21. Raise and safely support the vehicle.

22. Install the right wheelhouse panel and wheel assembly. Lower the vehicle.

23. Connect the electrical connector to the solenoid located on the turbocharger.

24. Install the vacuum tank, battery tray and battery.

25. Install the air intake duct and turbocharger air intake duct.

26. Fill the cooling system with antifreeze and check system for leaks.

TORQUE CONVERTER CLUTCH (TCC) SOLENOID

Removal and Installation

1. Remove the negative battery cable. Raise and support the vehicle safely.

2. Drain the transmission fluid into a suitable drain pan. Remove the transmission pan.

3. Remove the TCC solenoid retaining screws and then remove the electrical connector, solenoid and check ball.

4. Clean and inspect all parts. Replace defective parts as necessary.

5. Install the check ball, TCC solenoid and electrical connector. Install the solenoid retaining screws and torque them to 10 ft. lbs. (14 Nm).

6. Install the transmission pan with a new gasket and torque the pan retaining bolts to 10 ft. lbs. (14 Nm).

7. Lower the vehicle and refill the transmission with the proper amount of the recommended automatic transmission fluid.

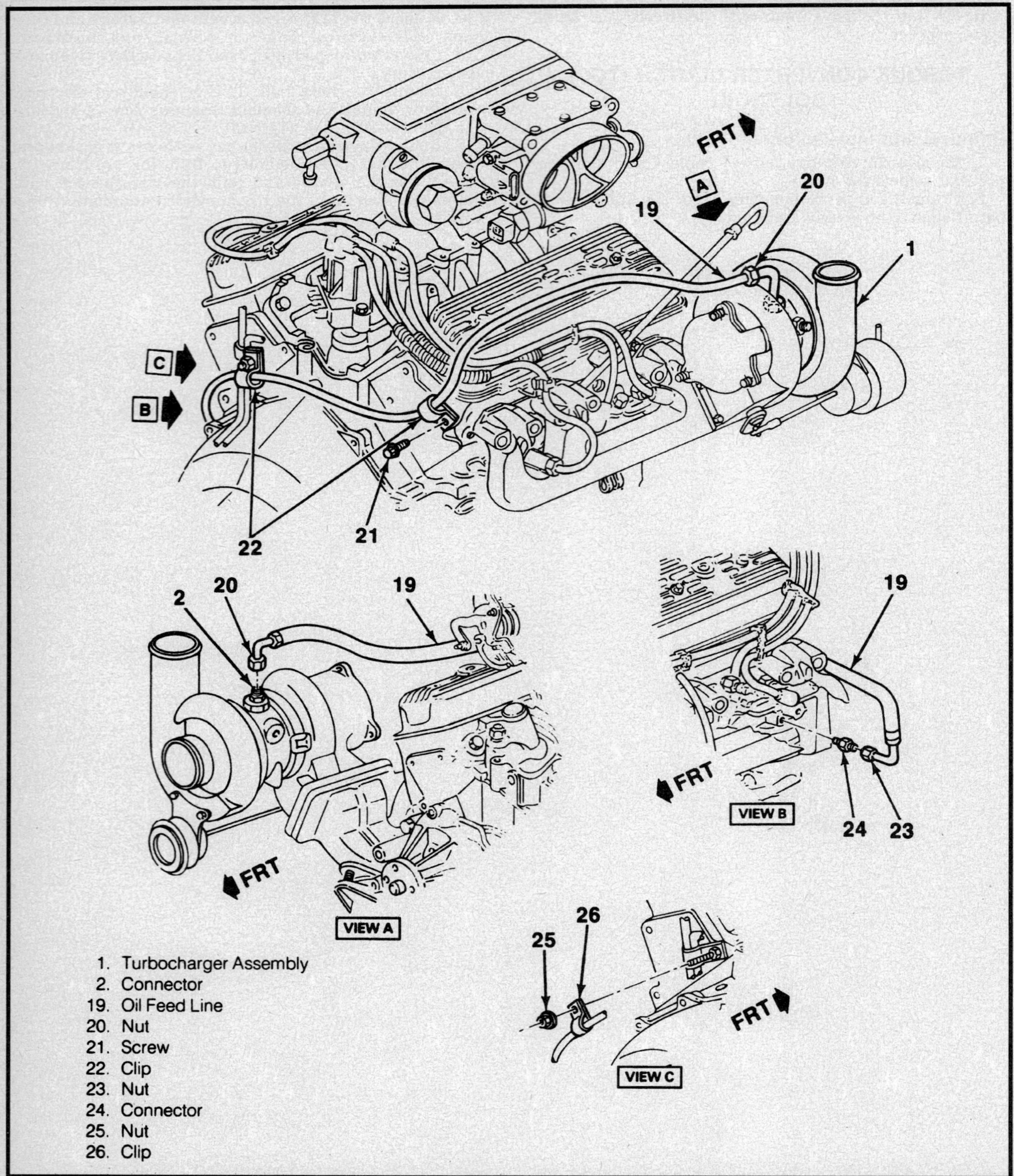

1. Turbocharger Assembly
2. Connector
19. Oil Feed Line
20. Nut
21. Screw
22. Clip
23. Nut
24. Connector
25. Nut
26. Clip

Turbocharger oil feed line assembly — 4.3L engine

CENTRAL PORT FUEL INJECTION (CPI) SYSTEM

LIGHT TRUCK AND VAN

APPLICATIONS CHART — LIGHT TRUCKS AND VANS

GM LIGHT TRUCKS, VANS AND APVS					
Model	**Body VIN**	**Engine Liter**	**Engine VIN**	**Ignition Type ①**	**Fuel System ②**
Astro/Safari	L,M	4.3	W	C³I	CPI
Light Truck	S,T	4.3	W	C³I	CPI

C³I—Computer Control Command Ignition
SFI—Sequential Port Fuel Injection

General Description

The 4.3L VIN W engine is used in the compact truck (S/T) and mid size van (M/L). The Central Port Fuel Injection (CPI) fuel metering system is designed to improve engine torque at both the low and high ends of the power band. The system consists of a CPI assembly (contained within the intake manifold), fuel pump, fuel regulator, Engine Control Module (ECM) and connecting wires and lines.

Pressurized fuel is delivered to the CPI assembly which distributes the fuel though nylon hoses (located under the intake manifold) to six poppet nozzles (located at each cylinder).

SYSTEM OPERATION

The fuel system has an electric fuel pump located in the fuel tank on the gauge sending unit. It pumps fuel to the CPI unit through an in-line fuel filter and fuel supply line. The pump provides fuel at a pressure above the regulated pressure needed by CPI injector.

A pressure regulator in the CPI unit keeps fuel available to the injector at a constant pressure. Fuel in excess of injector needs is returned to the fuel tank via a separate line.

The ECM controls the injector that is located in the Central Port Injector (CPI) assembly. The CPI deliveries fuel in response to the pulse width mandated signal from the ECM in one of several modes.

In order to properly control the fuel supply, the fuel pump is operated by the ECM through the fuel pump relay and oil pressure switch.

SYSTEM COMPONENTS

Electronic Control Module (ECM)

All vehicles use an ECM which consists of 2 parts; a controller (the ECM without the PROM) and a calibrator called a PROM (Programmable Read Only Memory) or MEMCAL (Memory Calibration).

The Powertrain Control Module (PCM) is used when vehicles are equipped with electronically controlled transmissions. Its function is identical to that of the ECM with the added features of controlling shift points in the transmission, as well as cruise control operation and diagnostics.

NOTE: When the term Electronic Control Module (ECM) is used, it refers to the engine control computer regardless of whether it is a Powertrain Control Module (PCM) or not.

CONTROLLER

The fuel injection system is controlled by an on-board computer, the Electronic Control Module (ECM) or controller, located in the passenger compartment. The ECM monitors engine operations and environmental conditions (ambient temperature, barometric pressure, etc.) needed to calculate the fuel delivery time (pulse width/injector on-time) of the fuel injector. The fuel pulse may be modified by the ECM to account for special operating conditions, such as cranking, cold starting, altitude, acceleration and deceleration.

The control unit moderates the exhaust emissions by curving fuel delivery to achieve, as nearly as possible an air/fuel ratio of 14.7:1. The injector on-time is determined by the various sensor inputs to the ECM. By increasing the injector pulse, more fuel is delivered, enriching the air/fuel ratio.

PROM

The ability of the ECM to recognize and adjust for vehicle variations (engine transmission, vehicle weight, axle ratio, etc.) is provided by a calibration unit (PROM or MEMCAL) that is programmed to tailor the ECM for the particular vehicle. There is a specific ECM/PROM combination for each specific vehicle, and the combinations are not interchangeable with those of other vehicles.

The PROM is located inside the ECM and has information on the vehicle's weight, engine, transmission, axle ratio and other components.

An ECM used for service (called a controller) comes without a PROM. The PROM from the old ECM must be carefully removed and installed in the new ECM.

ECM MODES OF OPERATION

The ECM operates in several modes:

Starting Mode

When the engine is first turned **ON**, the ECM will turn ON the fuel pump relay for 2 seconds and the fuel pump will pressurize the fuel system. The ECM then checks the engine coolant temperature sensor, throttle position sensor and crank sensor, then the ECM determines the proper air/fuel ratio for starting. This ranges from 1.5:1 at -33°F (-36°C) to 14.7:1 at 201°F (94°C).

The ECM controls the amount of fuel that is delivered in the Starting Mode by changing how long the injectors are turned ON and OFF. This is done by pulsing the injectors for very short times.

Clear Flood Mode

If for some reason the engine should become flooded, provisions have been made to clear this condition. To clear the flood, the driver must depress the accelerator pedal to the wide-open throttle position. The ECM then completely turns off the fuel flow. The ECM holds this operational mode as long as the throttle stays in the wide open position and the engine rpm is below 600. If the throttle position becomes less than 62–70%, the ECM returns to the starting mode.

Run Mode

There are 2 different run modes. When the engine is first started and the rpm is above 400, the system goes into open loop operation. In open loop operation, the ECM will ignore the signal from the oxygen (O_2) sensor and calculate the injector on-time based upon inputs from the ECT, MAP and IAT sensors.

During open loop operation, the ECM analyzes the following items to determine when the system is ready to go to the closed loop mode.

1. The oxygen sensor varying voltage output. (This is dependent on temperature).

2. The engine coolant temperature sensor must be above specified temperature.

3. A specific amount of time must elapse after starting the engine. These values are stored in the PROM.

When these conditions have been met, the system goes into closed loop operation. In closed loop operation, the ECM will calculate the air/fuel ratio (injector on-time) based upon the signal from the oxygen sensor. The ECM will decrease the on-time if the air/fuel ratio is too rich, and will increase the on-time if the air/fuel ratio is too lean.

Acceleration Mode

When the accelerator pedal is depressed, the opening of the throttle valve(s) causes a rapid increase in airflow into the cylinders. The air flow is capable of moving much faster then the flow of fuel and therefore the engine may hesitate; to avoid this situation the ECM increases the fuel injector pulse width according to the TPS and MAP signals it receives.

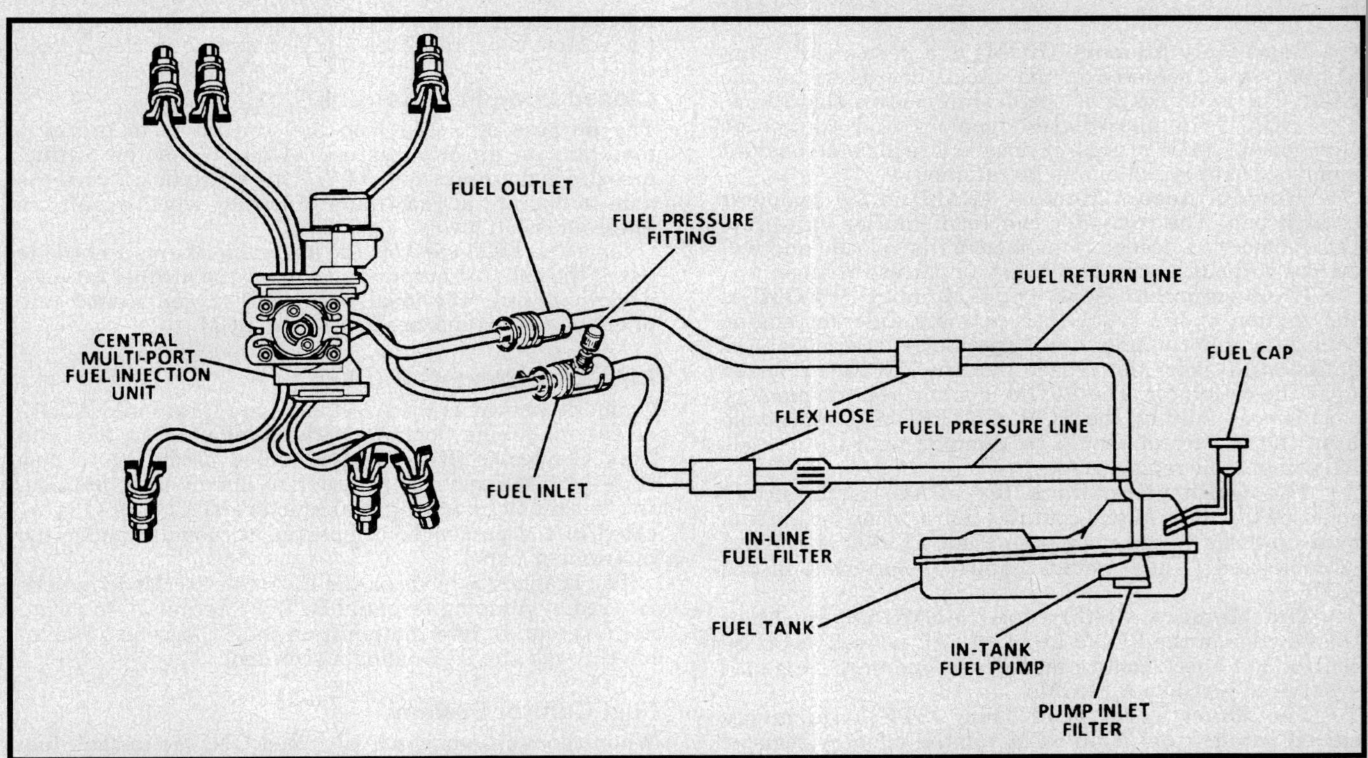

CPI fuel supply system

Deceleration Mode

Upon deceleration, a leaner fuel mixture is required to reduce emission of hydrocarbons (HC) and carbon monoxide (CO). To adjust the injection on-time, the ECM uses the decrease in MAP and the decrease in throttle position to calculate a decrease in injector on-time. To maintain an idle fuel ratio of 14.7:1, fuel output is momentarily reduced. If deceleration is very rapid or for long periods, the ECM can cut off the fuel completely to prevent damage to the catalytic converter.

Battery Voltage Correction Mode

The purpose of battery voltage correction is to compensate for variations in battery voltage to fuel pump and injector response. The ECM compensates by increasing the engine idle rpm.

Battery voltage correction takes place in all operating modes. When battery voltage is low, the spark delivered by the distributor may be low. To correct this low battery voltage problem, the ECM can do any or all of the following:
- Increase injector on-time (increase fuel)
- Increase idle rpm
- Increase ignition dwell time

Fuel Cutoff Mode
If the sensors exceed a pre-programmed value, the ECM turns OFF the fuel flow to the injectors to prevent engine damage.

Converter Protection Mode
In this mode, the ECM estimates the temperature of the catalytic converter and then modifies fuel delivery to protect the converter from high temperatures. When the ECM has determined that the converter may overheat, it will cause open loop operation and will enrichen the fuel delivery. A slightly richer mixture will then cause the converter temperature to be reduced.

Memory

NOTE: Not every type of memory control is used in each vehicle application.

- **Read Only Memory (ROM)** is a permanent memory physically soldered to the circuit boards within the ECM. The ROM contains the overall control algorithms. The ROM is a non-volatile memory and cannot be changed. Because of this, it does not require a constant supply of battery voltage to be retained.
- **Random Access Memory (RAM)** is the computer scratch pad. The computer can read from or write into this memory as needed. This memory is volatile and will be lost without a constant supply of battery voltage.
- **Programmable Read Only Memory (PROM)** is the portion of the ECM that contains different engine calibration information that is specific to year, model and emissions. It is for this reason that the PROM is referred to as the calibrator. The PROM is a non-volatile memory that is read only by the ECM. The PROM is removable from the ECM and should be retained with the vehicle following ECM replacement.
- **The Calibration Pack (CALPAC)** is physically soldered to the ECM and contains information for the fuel back-up mode of the computer. The CALPAC is not serviceable and failure would require replacement of the ECM.
- **The Memory Calibration (MEMCAL)** combines the function of the PROM and CALPAC as well as cruise control and knock control modules in one unit. This unit is serviced just like a PROM.
- **The Short Term Fuel Trim (STFT) (formerly called integrator)** is an ECM volatile memory register that will contain a number between 0 and 255. The neutral value for STFT is 128; any deviation from this value indicates a change in the injector pulse width as commanded from STFT. This function is only active in the closed loop mode of operation. As the ECM monitors the oxygen sensor voltage input, it is constantly varying the STFT value. For example, if a vacuum leak were to occur causing a lean condition (O_2 voltage low), the STFT would respond by increasing injector pulse width. This increase would be seen on a scan tool as a number greater than 128. Conversely, if the air filter became plugged causing a rich condition (O_2 voltage high), the STFT would respond with a decrease in injector pulse width; a number less than 128. Because this value is updated very quickly, the STFT only corrects for short term mixture trends.
- **The Long Term Fuel Trim (LTFT) (formerly called block learn)** is a matrix of cells arranged by rpm and load. As engine operating conditions change, the ECM will switch from cell to cell to determine which LTFT factor is appropriate. While in any given block, the ECM also monitors the STFT. If the STFT is far enough from 128 in either direction, the ECM will change the LTFT value for that given cell. Once the LTFT value is changed the STFT value should revert back to 128, this represents a neutral condition. If the mixture is still not correct (as judged by the O_2 sensor), the process of short and long term fuel correction will continue until they reach the limits of their control. These limits are programmed into the PROM and would be different for each vehicle. At this point of maximum correction a trouble code 44 or 45 would be registered in the computer's RAM. As with STFT the long term fuel trim is only active in closed loop operation. The LTFT represents the computer's learning capability as these values can be tailored to the specific driving habits of the operator. Once battery power is lost however, the LTFT values will revert back to default settings, and a loss of performance may be noticed. Original performance should return after the vehicle is operated for a period of time in closed loop.

Closed Loop Fuel Control
The purpose of closed loop fuel control is to precisely maintain an air/fuel mixture 14.7:1. When the air/fuel mixture is maintained at 14.7:1, the catalytic converter is able to operate at maximum efficiency which results in lower emission levels.

Since the ECM controls the air/fuel mixture, it needs to check the exhaust output and correct the air/fuel ratio for deviations from the ideal ratio. The oxygen sensor supplies the output information to the ECM.

Data Link Connector (DLC)
Formerly known as the Assembly Line Data Link (ALDL) or Assembly Line Communication Link (ALCL), the Data Link Connector (DLC) was renamed according to new SAE J-2008 standards in order to universalize automotive terminology among manufacturers. The DLC is located in the passenger compartment, usually under the instrument panel.

The Diagnostic mode or the Field Service Mode may be entered by jumping terminal B (TEST terminal) to terminal A (ground). Information from the ECM is also available through the DLC using a scan tool.

Fuel Control System
When the ignition switch is turned **ON**, the in-tank fuel pump is energized for as long as the engine is cranking or running and the control unit is receiving signals from the HEI distributor. If there are no reference pulses, the control unit will shut off the fuel pump within 2 seconds. The pump will deliver fuel to the CPI assembly where the system pressure range is 55–61 psi (380–420 kPa).

The fuel control system is made up of the following components:
1. Fuel supply system (tank, fuel lines and filter)
2. CPI assembly
3. Intake Manifold Tuning Valve (IMTV)

4. Accelerator control components (IAC system)
5. Fuel pump
6. Fuel pump relay circuit
7. Emission control system

CPI Assembly

The CPI assembly is a sealed unit consisting of a fuel meter body, gasket seal, fuel pressure regulator, fuel injector and 6 poppet nozzles with fuel tubes. It is located in the middle of the lower intake manifold.

The fuel meter body portion of the CPI assembly is connected by nylon tubes to 6 poppet valves that feed fuel into the runners just before the individual cylinders. The ECM controls the flow of fuel by modulating the pulse-width via the injector solenoid. The poppet valves allow fuel from the injector solenoid to flow into the runners when pressure exceeds 37–43 psi (254–296 kPa).

Intake Manifold Tuning Valve (IMTV)

The Intake Manifold Tuning Valve (IMTV) is a rotary solenoid that streamlines air into the plenum during mid to high range rpm, heavy throttle applications. It is located in the center of the upper intake manifold.

During normal vehicle acceleration and at cruising speeds (low torque conditions), the IMTV is closed and the intake manifold supplies a more conservative air flow pattern to the cylinders. When the torque requirements change (high torque conditions), the ECM opens the IMTV and redirects air flows within the manifold to utilize all the manifold tubes and increase high rpm torque.

Idle Air Control (IAC)

The purpose of the Idle Air Control (IAC) system is to control engine idle speeds while preventing stalls due to changes in engine load. The IAC assembly, mounted on the throttle body, controls bypass air around the throttle plate. The ECM send a signal to the IAC motor to move the shaft and pintle IN or OUT, allowing a controlled amount of air to move around the throttle plate. If rpm is too low, more air is diverted around the throttle plate to increase rpm.

During idle, the proper position of the IAC valve is calculated by the ECM based on battery voltage, coolant temperature, engine load and engine rpm. If the rpm drops below a specified rate and the throttle plate is closed, the ECM will calculate a new valve position.

The IAC motor has 255 different positions, which are counted in steps. The zero, or reference position, is the fully extended position at which the pintle is seated in the air bypass seat and no air is allowed to bypass the throttle plate. When the motor is fully retracted, maximum air is allowed to bypass the throttle plate. When the motor is fully retracted, maximum air is allowed to bypass the throttle plate. The ECM always monitors how many steps it has extended or retracted the pintle from the zero or reference position; thus, it always calculates the exact position of the motor.

If the IAC valve is replaced or disconnected or battery voltage to the ECM is lost for any reason, the valve will need to be reset by following the Idle Learn Procedure.

The IAC system uses the information from the following sensors to determine idle speed:
1. Park/Neutral Switch
2. Power Steering Pressure Switch (PPS)
3. Air Conditioning (A/C) Pressure Sensor
4. Intake Air Temperature (IAT) Sensor
5. Throttle Position Sensor (TPS)

Park/Neutral Switch

NOTE: Vehicle should not be driven with the park/neutral switch disconnected as idle quality may be affected in park or neutral and a Code 24 (VSS) may be set.

This switch indicates to the ECM when the transmission is in **P** or **N**. The information is used by the ECM for control of the torque converter clutch, EGR and the idle air control valve operation.

Power Steering Pressure Switch (PPS)

The power steering pressure switch is used so that the power steering oil pressure pump load will not effect the engine idle. Turning the steering wheel increase the power steering oil pressure and pump load on the engine. The power steering pressure switch will close before the load can cause an idle problem. The ECM will also turn the A/C clutch off when high power steering pressure is detected.

Air Conditioning (A/C) Pressure Sensor

The air conditioning (A/C) pressure sensor provides a signal to the ECM which indicates varying high side refrigerant pressure between approximately 0–450 psi. The ECM uses the A/C compressor load on the engine to help control the idle speed with the IAC valve.

Intake Air Temperature (IAT) Sensor

The Intake Air Temperature (IAT) sensor is a thermistor (a resistor which changes resistance based on temperature) mounted in the intake manifold.

The ECM supplies a 5 volt signal to the IAT sensor through a resistor in the ECM and monitors the voltage. The voltage will be high when the manifold air is cold and low when the air is hot. By monitoring the voltage, the ECM calculates the air temperature and uses this data to help determine fuel delivery and spark advance.

Throttle Position Sensor (TPS)

The Throttle Position Sensor (TPS) is connected to the throttle shaft and is controlled by the throttle mechanism. A 5 volt reference signal is sent to the TPS from the ECM. As the throttle valve angle is changed (accelerator pedal moved), the resistance of the TPS also changes. At a closed throttle position, the resistance of the TPS is high, so the output voltage to the ECM will be low (approximately 0.5 volt). As the throttle plate opens, the resistance decreases so that, at wide open throttle, the output voltage should be approximately 5 volts. At closed throttle position, the voltage at the TPS should be less than 1.25 volts. By monitoring the output voltage from the TPS, the ECM can determine fuel delivery based on throttle valve angle (driver demand).

Fuel Pump

The fuel is supplied to the system from an in-tank pump. The pump supplies fuel through the in-line fuel filter to the fuel rail assembly. The pump is removed for service along with the fuel gauge sending unit. Once they are removed from the fuel tank, the pump and sending unit can be serviced separately.

The fuel pump delivers more fuel than the engine can consume even under the most extreme conditions. Excess fuel flows through the pressure regulator and back to the tank via the return line. The constant flow of fuel means that the fuel system is always supplied with cool fuel, thereby preventing the formation of fuel-vapor bubbles (vapor lock).

Fuel Pump Relay Circuit

The fuel pump relay is located on the engine side of the firewall (center cowl). The fuel pump electrical system consists of the fuel pump relay, ignition circuit and the ECM circuits. The fuel pump relay contact switch is in the normally open position, with no voltage flowing to the pump.

When the ignition is turned **ON** the ECM will for 2 seconds, supply voltage to the fuel pump relay coil, closing the contact switch. The ignition circuit fuse can now supply ignition voltage to the circuit which feeds the relay contact switch. With the relay contacts closed, ignition voltage is supplied to the fuel pump. The ECM will continue to supply voltage to the relay coil circuit as long as the ECM receives the rpm reference pulses from the ignition module.

The fuel pump control circuit also includes an engine oil pressure switch with a set of normally open contacts. The switch closes at approximately 4–6.5 psi and provides a secondary battery feed path to the fuel pump. If the relay fails, the pump will continue to run using the battery feed supplied by the closed oil pressure switch. A failed fuel pump relay will result in extended engine crank times in order to build up enough oil pressure to close the switch and turn on the fuel pump.

Emission Control Systems

Various components are used to control the exhaust emissions from a vehicle. These components are controlled by the ECM based on different engine operating conditions.
1. Exhaust Gas Recirculation (EGR) System.
2. Positive Crankcase Ventilation (PCV) System.
3. Evaporative Emission Control (EEC) Systems.
4. Catalytic convertor.

Exhaust Gas Recirculation (EGR) System

The EGR is used to lower oxides of nitrogen (NOx) emission levels caused by high combustion temperatures. The EGR meters the flow of exhaust gases into the intake manifold. The amount of exhaust gas admitted is adjusted by the ECM in response to information from the RPM, TPS, ECT, VSS, MAP and PPS sensors. When the valve is opened, the metered amount of exhaust gas is released into the intake manifold and then drawn into the combustion chamber to be burned again, thus lowering the combustion temperature.

Positive Crankcase Ventilation (PCV) System

A closed Positive Crankcase Ventilation (PCV) system is used to recapture crankcase vapors. Fresh air from the air cleaner is supplied to the crankcase, mixed with blow-by gases and then passed through a PCV valve into the induction system.

The primary mode of crankcase ventilation control is through the PCV valve which meters the mixture of fresh air and blow-by gases into the induction system at a rate dependent upon manifold vacuum.

To maintain the idle quality, the PCV valve restricts the ventilation system flow and redirects excessive blow-by gases through the breather assembly into the air cleaner and through the throttle body to be consumed by normal combustion.

Evaporative Emission Control (EEC) System

The evaporative emission control system uses a carbon canister storage method. This method transfers fuel vapor to an activated carbon storage device for retention when the vehicle is not operating. When the engine is started, a ported vacuum signal is used to purge the vapors from the canister into the intake manifold.

The ECM controls a solenoid valve which controls vacuum to the purge valve in the charcoal canister. In open loop, before a specified time has expired and below a specified rpm, the solenoid valve is energized and blocks vacuum to the purge valve. When the system is in closed loop, after a specified time and above a specified rpm, the solenoid valve is de-energized and vacuum can be applied to the purge valve. This releases the collected vapors into the intake manifold.

Catalytic Converter

Of all emission control devices available, the catalytic converter is the most effective in reducing tail pipe emissions. The convertor uses heat and oxygen to change the chemical makeup of the exhaust pollutants and reduce the output of hydrocabons (HC), carbon monoxide (CO), and oxides of nitrogen (NOx).

Data Sensors

A variety of sensors provide information to the ECM regarding engine operating characteristics. These sensors and their functions are described below.
1. Engine Coolant Temperature (ECT) Sensor
2. Oxygen (O_2) Sensor
3. Manifold Absolute Pressure (MAP) Sensor
4. Vehicle Speed Sensor (VSS)
5. Knock (Detonation) Sensor (KS)
6. Oil Pressure Switch

Engine Coolant Temperature (ECT) Sensor

The coolant sensor is a thermistor (a resistor which changes value based on temperature) mounted in the front of the intake manifold in the engine coolant stream. As the temperature of the engine coolant changes, the resistance of the coolant sensor changes. Low coolant temperature produces a high resistance, while high temperature causes low resistance.

The ECM supplies a 5 volt signal to the coolant sensor and measures the voltage that returns. By measuring the voltage change, the ECM determines the engine coolant temperature. The voltage will be high when the engine is cold and low when the engine is hot. This information is used to control fuel management, IAC, spark timing, EGR, canister purge and other engine operating conditions.

Oxygen (O_2) Sensor

The oxygen sensor is mounted in the exhaust crossover, before the catalytic convertor, where it can monitor the oxygen content of the exhaust gas stream. The oxygen content in the exhaust reacts with the oxygen sensor to produce a voltage output. By monitoring the voltage output of the oxygen sensor, the ECM will determine what fuel mixture command to give to the injector (lean mixture — low voltage — rich command, rich mixture — high voltage — lean command).

The oxygen sensor only monitors the exhaust. It does not make any changes to the system.

Manifold Absolute Pressure (MAP) Sensor

The Manifold Absolute Pressure (MAP) sensor measures the changes in the intake manifold pressure which result from engine load and speed changes. The pressure measured by the MAP sensor is the difference between barometric pressure (outside air) and manifold pressure (vacuum). A closed throttle engine coast-down would produce a relatively low MAP value (approximately 20–35 kPa), while wide-open throttle would produce a high value (100 kPa). This high value is produced when the pressure inside the manifold is the same as outside the manifold, and 100% of outside air (or 100 kPa) is being measured. This MAP output is the opposite of what would be measured on a vacuum gauge. The use of this sensor also allows the ECM to adjust for changing altitudes.

The ECM sends a 5 volt reference signal to the MAP sensor. As the MAP changes, the electrical resistance of the sensor also changes. By monitoring the sensor output voltage the ECM can determine the manifold pressure. A higher pressure, lower vacuum requires more fuel, while a lower pressure, higher vacuum requires less fuel. The ECM uses the MAP sensor to control fuel delivery and ignition timing.

Vehicle Speed Sensor (VSS)

NOTE: A vehicle equipped with a speed sensor, should not be driven without the speed sensor connected, as idle quality may be affected.

The vehicle speed sensor (VSS) is mounted behind the speedometer in the instrument cluster or on the transmission/speedometer drive gear. The sensor is a Permanent Magnet (PM), which generates a small AC voltage and transmits the signal to the ECM. The pulses indicate the road speed. The ECM uses this information to operate the IAC, canister purge and TCC.

Knock (Detonation) Sensors

The CPI system uses 2 knock sensors. On S/T models, the sensors are located in the side of the left head and the rear of the right head. On L/M models, the sensors are located in the rear of the left head and the side of the right head. They generate an AC signal which varies with the severity of the knock. This information is passed to the ESC module and then to the ECM, so ignition timing advance can be retarded until the detonation stops.

Oil Pressure Switch

The oil pressure switch is usually mounted on the back of the engine, just below the intake manifold. The CPI system uses the oil pressure switch as a parallel power supply with the fuel pump relay and will provide voltage to the fuel pump after approximately 4 psi (28 kPa) of oil pressure is reached. This switch will also help prevent engine seizure by shutting off the power to the fuel pump and causing the engine to stop when the oil pressure is lower than 4 psi.

Diagnosis and Testing

SERVICE PRECAUTIONS

When working around any part of the fuel system, take precautionary steps to prevent fire and/or explosion:

- Disconnect negative terminal from battery (except when testing where battery voltage is required).
- Whenever possible, use a flashlight instead of a drop light.
- Keep all open flame and smoking material out of the area.
- Use a shop cloth or similar to absorb fuel when opening a fuel system.
- Properly relieve fuel system pressure before servicing.
- Always wear suitable eye protection.
- Always keep a dry chemical (class B) fire extinguisher near the area.

Handling Methanol (M85) or Ethanol (E85) Fuels

- Both methanol and ethanol, like all fuels, are poisonous and should be handled with great care. These fuels can be absorbed into your body through the skin much faster than gasoline and can cause headaches, blindness, loss of memory and even death.
- Avoid spills. If either fuel comes in contact with your skin, wash the exposed area immediately with ample amounts of soap and water. If any fuel contacts your clothing, change clothes right away and wash your skin immediately. Allow contaminated clothes to air dry first, then machine wash.
- Always wear methanol/ethanol resistant gloves, eye or face protection and NEVER try to siphon these fuels by mouth. Drinking a mouthful of these fuels can cause death. Additionally, work in a well ventilated area, as these fuels give off harmful vapors which can cause serious bodily harm if ingested over a long period of time. If fumes become prominent, move to an area of fresh air immediately.
- Partial combustion of methanol yields formaldehyde which is suspected to be a cancer causing agent. A byproduct of ethanol fuel combustion is acetaldehyde, which causes burning of the eyes, nose and throat. When working in an enclosed space with the engine running, it is absolutely essential to use exhaust hoses.

CAUTION

Before attempting to remove or service any fuel system component, it is necessary to properly relieve the fuel system pressure.

Electrostatic Discharge Damage

Electronic components used in the control system are often designed to carry very low voltage and are very susceptible to damage caused by electrostatic discharge (ESD). It is possible for less than 100 volts of static electricity to cause damage to some electronic components. By comparison it takes as much as 4,000 volts for a person to feel the zap of a static discharge.

There are several ways for a person to become statically charged. The most common methods of charging are by friction and induction. An example of charging by friction is a person sliding across a car seat, in which a charge as much as 25,000 volts can build up. Charging by induction occurs when a person with well insulated shoes stands near a highly charged object and momentarily touches ground. Charges of the same polarity are drained off, leaving the person highly charged with the opposite polarity. Static charges of either type can cause damage, therefore, it is important to use great care when handling and testing electronic components. Follow the guidelines listed below to avoid electrostatic discharge.

- To prevent possible electrostatic discharge damage to the ECM, do not touch the connector pins or soldered components on the circuit board.
- Always touch a known ground before handling any parts.
- When using a voltmeter, always connect the ground lead first.
- Do not remove any parts from there package until it is time to install them.
- Before removing the part from the package, touch the package to a good ground.

ECM Learning Ability

The ECM has a learning capability. If the battery is disconnected the learning process has to begin all over again. A change may be noted in the vehicle's performance. To teach the ECM, insure the vehicle is at operating temperature and drive at part throttle, with moderate acceleration and idle conditions, until performance returns.

READING CODES

The Data Link Connector (DLC), previously known as the ALDL connector, is used for communicating with the ECM. It is usually located under the instrument panel and is sometimes covered by a plastic cover labeled DIAGNOSTIC CONNECTOR. Codes stored in the ECM's memory can be read through a handheld diagnostic scanner plugged into the DLC connector. Codes can also be read by connecting terminals A and B of the DLC connector, turning the ignition switch **ON**, and counting the number of flashes of the SERVICE ENGINE SOON light.

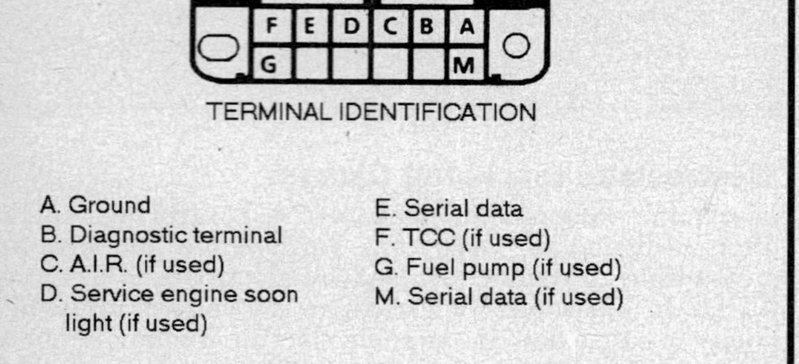

Data link connector (DLC)

CLEARING CODES

To clear codes from the ECM memory, the ECM power feed must be disconnected for at least 30 seconds. Depending on the vehicle, the ECM power feed can be disconnected at the positive battery terminal pigtail, the in-line fuseholder that originates at the positive connection at the battery or the ECM fuse in the fuse block. The negative battery cable may also be disconnected, however, other on-board memory data such as preset radio tuning will also be lost.

DIAGNOSTIC TROUBLE CODES (DTC)

- DTC **12** — Diagnostic check only (flash code)
- DTC **13** — Oxygen (O_2) sensor circuit open
- DTC **14** — Engine Coolant Temperature (ECT) sensor circuit out of range (high)
- DTC **15** — Engine Coolant Temperature (ECT) sensor circuit out of range (low)
- DTC **16** — Transmission output speed low
- DTC **21** — Throttle Position Sensor (TPS) circuit voltage out of range
- DTC **22** — Throttle Position Sensor (TPS) circuit voltage out of range
- DTC **24** — Vehicle speed sensor circuit signal error
- DTC **25** — Intake air temperature high indicated
- DTC **28** — Fluid pressure switch assembly
- DTC **32** — Exhaust Gas Recirculation (EGR) system fault
- DTC **33** — Manifold absolute pressure sensor circuit voltage out of range
- DTC **34** — Manifold absolute pressure sensor circuit voltage out of range
- DTC **35** — Idle air control error
- DTC **37** — Brake switch circuit error, stuck ON
- DTC **38** — Brake switch circuit error, stuck OFF
- DTC **41** — Fuel moding error
- DTC **42** — Ignition control circuit open or shorted
- DTC **43** — Knock sensor/electronic spark control circuit error
- DTC **44** — Oxygen (O_2) sensor circuit indicates system lean
- DTC **45** — Oxygen (O_2) sensor circuit indicates system rich
- DTC **51** — PROM memory error
- DTC **52** — CALPAC missing
- DTC **53** — System over voltage
- DTC **54** — Fuel pump circuit voltage low
- DTC **55** — PCM error
- DTC **58** — Transmission fluid temperature high
- DTC **59** — Transmission fluid temperature low
- DTC **66** — Control solenoid circuit error
- DTC **67** — Torque converter clutch circuit error
- DTC **69** — Torque converter clutch stuck ON
- DTC **72** — Vehicle speed sensor loss
- DTC **73** — Pressure control solenoid
- DTC **79** — Transmission fluid over temperature
- DTC **81** — Shift solenoid 2–3 circuit fault.
- DTC **82** — Shift solenoid 1–2 circuit fault.

COOLANT, AIR AND OIL TEMPERATURE SENSOR COMPONENT PARAMETERS

COOLANT, AIR AND OIL TEMPERATURE SENSOR RESISTANCE

Engine	Degrees F	Degrees C	Ohms
All	212	100	177
	194	90	241
	176	80	332
	158	70	467
	140	60	667
	122	50	973
	113	45	1188
	104	40	1459
	95	35	1802
	86	30	2238
	77	25	2796
	68	20	3520
	59	15	4450
	50	10	5670
	41	5	7280
	32	0	9420
	23	−5	12300
	14	−10	16180
	5	−15	21450
	−4	−20	28680
	−22	−30	52700
	−40	−40	100700

MAP SENSOR COMPONENT PARAMETERS

MAP SENSOR

Engine	Signal Voltage	
	Idle	WOT
3.8L (VIN L)	1–1.5	4–4.6
4.3L (VIN W)	1–1.5	4–4.6

OXYGEN SENSOR COMPONENT PARAMETERS

OXYGEN SENSOR

Engine	Signal Voltage	
	Lean	Rich
3.8L (VIN L)	0.1	1.0
4.3L (VIN W)	0.1	1.0

THROTTLE POSITION SENSOR COMPONENT PARAMETERS

THROTTLE POSITION SENSOR		
	Signal Voltage	
Engine	Idle	WOT
3.8L (VIN L)	0.5	4.0
4.3L (VIN W)	0.5	4.0

ENGINE PERFORMANCE DIAGNOSIS

Engine performance diagnosis procedures are guides that will lead to the most probable causes of engine performance complaints. They consider the components of the fuel, ignition and mechanical systems that could cause a particular complaint, then outline repairs in a logical sequence.

It is important to determine if the SERVICE ENGINE SOON light is ON or has come ON for a short interval while driving. If the SERVICE ENGINE SOON light has come ON, the computer should be checked for stored TROUBLE CODES which may indicate the cause for the performance complaint.

All of the symptoms can be caused by worn out or defective parts such as spark plugs, ignition wiring, etc. If time and/or mileage indicate that parts should be replaced, it is recommended that it be done.

NOTE: Before checking any system controlled by the computer, the Diagnostic Circuit Check must be performed or misdiagnosis may occur. If the complaint involves the SERVICE ENGINE SOON light, go directly to the Diagnostic Circuit Check.

Basic Troubleshooting

NOTE: The following explains how to activate the trouble code signal light in the instrument cluster. This is not a full system troubleshooting and isolation procedure.

Before suspecting the system or any of its components as faulty, check the ignition system including distributor, timing, spark plugs and wires. Check the engine compression, air cleaner, and emission control components not controlled by the ECM. Also check the intake manifold, vacuum hoses and hose connectors for leaks.

The following symptoms could indicate a possible problem with the system:
- Detonation
- Stalls or rough idle-cold
- Stalls or rough idle-hot
- Missing
- Hesitation
- Surges
- Poor gasoline mileage
- Sluggish or spongy performance
- Hard starting-cold
- Objectionable exhaust odors (rotten egg smell) or black smoke
- Cuts out
- Improper idle speed

As a bulb and system check, the SERVICE ENGINE SOON light will come ON when the ignition switch is turned to the **ON** position but the engine is not started. The SERVICE ENGINE SOON light will also produce the trouble code or codes by a series of flashes. When the diagnostic test terminal under the dash is grounded, with the ignition in the **ON** position and the engine not running, the SERVICE ENGINE SOON light will flash once, pause, then flash twice in rapid succession. This is a Code 12, which indicates that the diagnostic system is working. After a long pause, the Code 12 will repeat itself 2 more times. The cycle will then repeat itself until the engine is started or the ignition is turned **OFF**.

When the engine is started, the SERVICE ENGINE SOON light will remain ON for a few seconds, then turn OFF. If the SERVICE ENGINE SOON light remains ON, the self-diagnostic system has detected a problem. If the test terminal is then grounded, the trouble code will flash 3 times. If more than 1 problem is found, each trouble code will flash 3 times. Trouble codes will flash in numerical order (lowest code number to highest). The trouble codes series will repeat as long as the test terminal is grounded.

A trouble code indicates a problem with a given circuit. For example, Code 14 indicates a problem in the cooling sensor circuit. This includes the coolant sensor, its electrical harness and the ECM. Since the self-diagnostic system cannot diagnose every possible fault in the system, the absence of a trouble code does not mean the system is trouble-free. To determine problems within the system which do not activate a trouble code, a system performance check must be made.

In the case of an intermittent fault in the system, the SERVICE ENGINE SOON light will go out when the fault goes away, but the trouble code will remain in the memory of the ECM. Therefore, if a trouble code can be obtained even though the SERVICE ENGINE SOON light is not ON, the trouble code must be evaluated. It must be determined if the fault is intermittent or if the engine must be at certain operating conditions (under load, etc.) before the SERVICE ENGINE SOON light will come ON. Some trouble codes will not be recorded in the ECM until the engine has been operated at part throttle for about 5–18 minutes.

INTERMITTENT SERVICE ENGINE SOON LIGHT

An intermittent open in the ground circuit would cause loss of power through the ECM and intermittent SERVICE ENGINE SOON light operation. When the ECM loses ground, distributor ignition is lost. An intermittent open in the ground circuit would be described as an engine miss.

Therefore, an intermittent SERVICE ENGINE SOON light, no code stored and a driveability comment described as similar to a miss will require checking the grounding circuit and the Code 12 circuit as it originates at the ignition coil.

UNDERVOLTAGE TO THE ECM

A system voltage below 9 volts will cause the SERVICE ENGINE SOON light to come ON as long as the condition exist. Therefore, an intermittent SERVICE ENGINE SOON light, no code stored and a driveability comment described as similar to a miss will require checking the grounding circuit, Code 12 circuit and the ignition feed circuit to terminal C of the ECM. This does not eliminate the necessity of checking the normal vehicle electrical system for possible cause such as a loose battery cable.

OVERVOLTAGE TO THE ECM

The ECM will also shut off when the power supply rises above 16 volts. The overvoltage condition will also cause the SERVICE ENGINE SOON light to come ON and stay ON as long as this condition exist.

A momentary voltage surge in a vehicle's electrical system is a common occurrence. These voltage surges do not present a problem because the entire electrical system acts as a shock absorber until the surge is dissipated. Voltage surges or spikes in the vehicle's electrical system have been known, on occasion, to exceed 100 volts.

The system is a low voltage (between 9 and 16 volts) system and will not tolerate these surges. The ECM will be shut OFF by any surge in excess of 16 volts and will come back ON only after the surge has dissipated sufficiently to bring the voltage under 16 volts.

A surge will usually occur when an accessory requiring a high voltage supply is turned OFF or DOWN. The voltage regulator in the vehicle's charging system cannot react to the changes in the voltage demands quickly enough and a surge occurs. The driver should be questioned to determine which accessory circuit was turned OFF when the SERVICE ENGINE SOON light came ON.

Therefore, intermittent SERVICE ENGINE SOON light operation, with no trouble code stored, may require installation of a diode in the appropriate accessory circuit.

THROTTLE POSITION SENSOR (TPS)

Testing

WITH SCAN TOOL

1. Connect a suitable scan tool to read the TPS voltage.
2. With the ignition switch **ON** and the engine not running, the TPS voltage should be less than 1.25 volts or throttle angle 0%.
3. If the voltage or throttle angle reading is higher than specified, replace the throttle position sensor.

4. Depress the accelerator pedal. The throttle position sensor reading should increase steadily to over 4.0 volts or 100% throttle angle at Wide Open Throttle (WOT).

WITHOUT SCAN TOOL

1. Remove air cleaner. Disconnect the TPS harness from the TPS.
2. Using suitable jumper wires, connect a digital voltmeter J–29125–A or equivalent to the correct TPS terminals A and B.
3. With the ignition **ON** and the engine running, The TPS voltage should be 0.3–1.0 volts at base idle to approximately 4.5 volts at Wide Open Throttle (WOT).
4. If the reading on the TPS is out of specification, check the minimum idle speed before replacing the TPS.
5. If the voltage reading is correct, turn the engine off, remove the voltmeter and jumper wires and reconnect the TPS connector to the sensor.
6. Reinstall the air cleaner.

FUEL PUMP RELAY

Testing

The fuel pump relay is located on the right or left side of the engine compartment, on the firewall. The fuel pump test terminal is located near the relay or along the PCM harness. The fuel pump may be tested by using a 10 amp fused jumper wire and applying 12 volts to the fuel pump test terminal while listening for the pump to operate.

1. Disconnect the negative battery cable.
2. Disconnect the fuel pump (5 terminal) relay.
3. Connect an ohmmeter between terminals E and B of the relay. Verify there is continuity. Check the resistance between terminals E and B of the relay. There should be at least 20 ω resistance.
4. Next, connect a 12 volt source to terminal A and ground terminal C.
5. Verify the resistance is greater than 10 kilo-ohms between terminals D and E.
6. Connect an ohmmeter between terminals D and B of the relay with the 12 volt source still connected.
7. Verify there is continuity.
8. If not as specified, replace the relay.

FUEL PRESSURE

Testing

CAUTION
Before attempting to remove or service any fuel system component, it is necessary to properly relieve the fuel system pressure.

---

1. Connect pressure gauge J–34730–1A, or equivalent, to fuel pressure test point on the intake manifold fuel inlet tube. Wrap a rag around the pressure tap to absorb any leakage that may occur when installing the gauge.
2. Turn the ignition **ON** and check that pump pressure is 54–64 psi (370–440 kPa) after 2 seconds. The pressure should not bleed down even after the pump stops operating.
3. If pressure if OK:
   a. Start the engine and allow it to idle. The fuel pressure should drop 5–10 psi. Open the throttle rapidly and observe the pressure reading. The pressure should increase as manifold pressure increases. If not,

the pressure regulator may not be functioning properly. Replace the CPI unit.

4. If pressure drops:

a. Turn the ignition **OFF** for 10 seconds. Raise and safely support the vehicle.

b. Turn the ignition **ON** and then pinch the fuel pressure line. If pressure does not bleed down, then there may be a poor connection at the pulse dampener coupling or a faulty fuel pump check valve. If pressure continues to bleed down, remove the upper intake manifold and inspect for fuel leaks. If no leaks can be found, replace the CPI unit.

4.3L (VIN W) ENGINE — COMPONENT LOCATIONS — MID SIZE VAN (M/L)

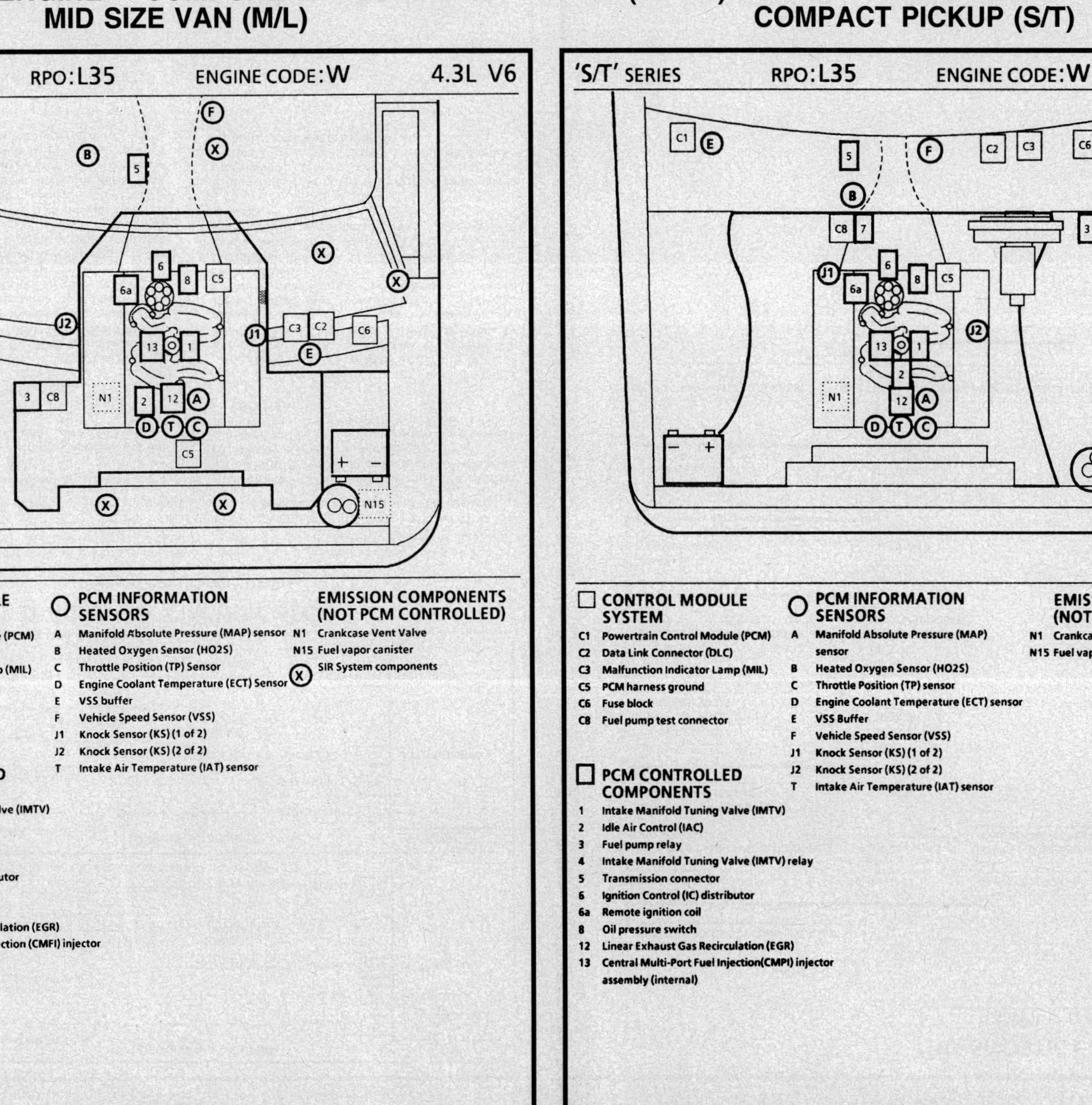

CONTROL MODULE SYSTEM

C1 Powertrain Control Module (PCM)
C2 Data Link Connector (DLC)
C3 Malfunction Indicator Lamp (MIL)
C5 PCM harness ground
C6 Fuse block
C8 Fuel pump test connector

PCM CONTROLLED COMPONENTS

1 Intake Manifold Tuning Valve (IMTV)
2 Idle Air Control (IAC)
3 Fuel pump relay
5 Transmission connector
6 Ignition Control (IC) distributor
6a Remote ignition coil
8 Oil pressure switch
12 Linear Exhaust Gas Recirculation (EGR)
13 Central Multi-Port Fuel Injection (CMFI) injector assembly (internal)

PCM INFORMATION SENSORS

A Manifold Absolute Pressure (MAP) sensor
B Heated Oxygen Sensor (HO2S)
C Throttle Position (TP) Sensor
D Engine Coolant Temperature (ECT) Sensor
E VSS buffer
F Vehicle Speed Sensor (VSS)
J1 Knock Sensor (KS) (1 of 2)
J2 Knock Sensor (KS) (2 of 2)
T Intake Air Temperature (IAT) sensor

EMISSION COMPONENTS (NOT PCM CONTROLLED)

N1 Crankcase Vent Valve
N15 Fuel vapor canister
X SIR System components

4.3L (VIN W) ENGINE — COMPONENT LOCATIONS — COMPACT PICKUP (S/T)

CONTROL MODULE SYSTEM

C1 Powertrain Control Module (PCM)
C2 Data Link Connector (DLC)
C3 Malfunction Indicator Lamp (MIL)
C5 PCM harness ground
C6 Fuse block
C8 Fuel pump test connector

PCM CONTROLLED COMPONENTS

1 Intake Manifold Tuning Valve (IMTV)
2 Idle Air Control (IAC)
3 Fuel pump relay
4 Intake Manifold Tuning Valve (IMTV) relay
5 Transmission connector
6 Ignition Control (IC) distributor
6a Remote ignition coil
8 Oil pressure switch
12 Linear Exhaust Gas Recirculation (EGR)
13 Central Multi-Port Fuel Injection(CMPI) injector assembly (internal)

PCM INFORMATION SENSORS

A Manifold Absolute Pressure (MAP) sensor
B Heated Oxygen Sensor (HO2S)
C Throttle Position (TP) sensor
D Engine Coolant Temperature (ECT) sensor
F Vehicle Speed Sensor (VSS)
J1 Knock Sensor (KS) (1 of 2)
J2 Knock Sensor (KS) (2 of 2)
T Intake Air Temperature (IAT) sensor

EMISSION COMPONENTS (NOT PCM CONTROLLED)

N1 Crankcase vent valve
N15 Fuel vapor canister

4.3L (VIN W) ENGINE — DIAGNOSTIC CIRCUIT CHECK

4.3L (VIN W) ENGINE — DIAGNOSTIC CIRCUIT CHECK

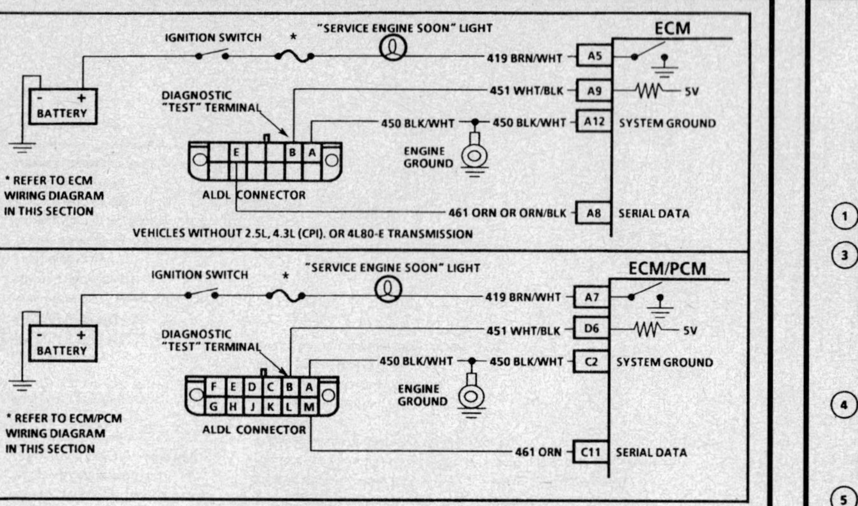

DIAGNOSTIC CIRCUIT CHECK
(Using A Tech 1 "Scan" Tool)

The diagnostic circuit check is an organized approach to identifying a problem created by an Computer Command Control System malfunction. It must be the starting point for any driveability complaint diagnosis, because it directs the service technician to the next logical step in diagnosing the complaint. Understanding the chart and using it correctly will reduce diagnostic time and prevent the unnecessary replacement of good parts.

Test Description: Number(s) below refer to circled number(s) on the diagnostic chart.

1. This step is a check for the proper operation of the "Service engine Soon" light. The "SES" light should be "ON" steady.
2. No "SES" light at this point indicates that there is a problem with the "SES" light circuit or the ECM/PCM control of that circuit.
3. This test checks the ability of the ECM/PCM to control the "SES" light. With the diagnostic terminal grounded, the "SES" light should flash a Code 12 three times, followed by any trouble code stored in memory. Depending upon the type of ECM/PCM, PROM or MEM-CAL error may result in the inability to flash Code 12.
4. This test checks the "Quad-Driver" circuit. If the "SES" light flashes less than three times, indicates that there is a problem with the "SES" light circuit or the ECM/PCM control of that circuit. Refer to "ECM/PCM QDR Check Procedure."
5. Use Tech 1 to aid diagnosis, therefore, serial data must b available. If a PROM or MEM-CAL error is present, the ECM/PCM may have been able to flash Code 12/51, but not enable serial data.

6. Although the ECM is powered up, a "Cranks But Will Not Run" symptom could exist because of an ECM/PCM or system problem.
7. This step will isolate if the customer complaint is a "SES" light or a driveability problem with no "SES" light.

An invalid code may be the result of a faulty "Scan" tool, MEM-CAL/PROM or ECM/PCM.
8. Comparison of actual control system data with the Typical Tech 1 Data Values is a quick check to determine if any parameter is not within limits. Keep in mind that a base engine problem (i.e. advanced cam timing) may substantially alter sensor values.
9. Installation of a Tech 1 "Scan" tool will provide a good ground path for the ECM/PCM and may hide a driveability complaint due to poor ECM/PCM grounds.
10. If the actual data is not within the typical values established, refer to the applicable diagnosis section to provide a functional check for the suspect component or system.

DIAGNOSTIC CIRCUIT CHECK
(Using A Tech 1 "Scan" Tool)

- IGNITION "ON," ENGINE "OFF."
- NOTE "SERVICE ENGINE SOON" (SES) LIGHT.

(1) STEADY LIGHT

(2) NO LIGHT → FLASHING CODE 12 → CHECK FOR GROUNDED DIAGNOSTIC TEST CKT 451.

(3)
- USING TECH 1 PERFORM "DIAGNOSTIC CIRCUIT CHECK." OR
- JUMPER ALDL TERMINAL "B" TO "A".
- DOES SES LIGHT FLASH CODE 12?

YES / NO → USE CHART A-1

(4) DOES SES LIGHT FLASH CODE 12 AT LEAST THREE TIMES?
YES / NO → USE CHART A-2

(5) DOES TECH 1 DISPLAY ENGINE DATA?
YES / NO → USE ECM/PCM QDR CHECK PROCEDURE.

(6) DOES ENGINE START?
YES / NO → USE CHART A-2.

(7) ARE ANY CODES DISPLAYED?
YES / NO → USE CHART A-3.

YES → REFER TO APPLICABLE CODE CHART. START WITH LOWEST CODE.

NO →
(8) COMPARE TECH 1 DATA WITH TYPICAL TECH 1 DATA VALUES. ARE VALUES NORMAL OR WITHIN TYPICAL RANGES?
YES / NO →
(10) REFER TO THE APPLICABLE SYSTEM SECTION.

4.3L (VIN W) ENGINE — SYMPTOMS — MID SIZE VAN (M/L) AND COMPACT PICKUP (S/T)

IMPORTANT PRELIMINARY CHECKS

BEFORE USING THIS SECTION

Before using this section you should have performed the "On-Board Diagnostic System Check" and determined that:
1. The control module and MIL (Service Engine Soon) are operating correctly.
2. There are no diagnostic trouble codes stored, or there is a diagnostic code but no MIL (Service Engine Soon).
- Several of the following symptom procedures call for a careful visual/physical check. The importance of visual/physical checks cannot be over stressed, because they can lead to correcting a problem without further checks and can save valuable time.

BEFORE STARTING

This check should include:

- Control module grounds for being clean, tight, and in their proper location.
- Vacuum hoses for splits, kinks and proper connections, as shown on "Vehicle Emission Control Information" label. Check thoroughly for any type of leak or restriction.
- Air leaks at all mounting areas of the intake manifold sealing surfaces.
- Ignition wires for cracking, hardness, proper routing, or carbon tracking.
- Wiring for proper connections, pinches, and cuts.
- The following symptom charts contain groups of possible causes for each symptom and cover several engines. **These procedures are not necessarily meant to be done in consecutive order.** If Tech 1 scan tool readings do not indicate the problems, then proceed in a logical order, easiest to check or most likely cause first.

SYMPTOM

Verify the customer complaint, and locate the correct symptom in the table of contents. Check the items indicated under that symptom.

4.3L (VIN W) ENGINE — SYMPTOMS — MID SIZE VAN (M/L) AND COMPACT PICKUP (S/T)

INTERMITTENTS

(Page 1 of 2)

Definition: Problem may or may not turn "ON" the Malfunction Indicator Lamp (MIL) or store a Diagnostic Trouble Code (DTC).

PRELIMINARY CHECKS

- Refer to "Important Preliminary Checks."
- DO NOT use the Diagnostic Trouble Code (DTC) charts for intermittent problems. The fault must be present to locate the problem. If a fault is intermittent, use of diagnostic trouble code charts may result in replacement of good parts.

FAULTY ELECTRICAL CONNECTIONS OR WIRING

- Most intermittent problems are caused by faulty electrical connections or wiring. Perform careful check of suspect circuits for:
  - Poor mating of the connector halves, or terminals, not fully seated in the connector body (backed out).
  - Improperly formed or damaged terminals. All connector terminals in problem circuit should be carefully reformed or replaced to insure proper contact tension.
  - Poor terminal to wire connection. This requires removing the terminal from the connector body to check. Refer to "On-Vehicle Service".

ROAD TEST

- If a visual/physical check does not locate the cause of the problem, the vehicle can be driven with a voltmeter connected to a suspected circuit or a Tech 1 scan tool may be used. An abnormal voltage or scan reading, when the problem occurs, indicates the problem may be in that circuit. If the wiring and connectors check OK, and a diagnostic trouble code was stored for a circuit having a sensor, except for DTCs 44 and 45, replace the sensor.

4.3L (VIN W) ENGINE — SYMPTOMS — MID SIZE VAN (M/L) AND COMPACT PICKUP (S/T)

INTERMITTENTS

(Page 2 of 2)

Definition: Problem may or may not turn "ON" the Malfunction Indicator Lamp (MIL), or store a Diagnostic Trouble Code (DTC).

INTERMITTENT "MALFUNCTION INDICATOR LAMP (MIL)"

- An intermittent MIL and No Diagnostic Trouble Codes (DTCs), may be caused by:
  - Electrical system interference caused by a defective relay, control module driven solenoid, or switch. They can cause a sharp electrical surge. Normally, the problem will occur when the faulty component is operated.
  - Improper installation of electrical devices, such as lights, 2-way radios, electric motors, etc.
  - Ignition Control (IC) wires should be routed away from spark plug wires, ignition system components and generator. Wire for CKT 453 from control module to ignition system should be a good ground.
  - Ignition secondary shorted to ground.
  - CKT 419 Malfunction Indicator Lamp (MIL) or CKT 451 (diagnostic "test" terminal) intermittently shorted to ground.
  - Control module grounds.

LOSS OF DIAGNOSTIC TROUBLE CODE MEMORY

- To check, disconnect Throttle Position (TP) sensor and idle engine until the "Malfunction Indicator Lamp" comes "ON." DTC 22 should be stored, and kept in memory when the ignition is turned "OFF" for at least 10 seconds. If not, the control module is faulty.

4.3L (VIN W) ENGINE — SYMPTOMS — MID SIZE VAN (M/L) AND COMPACT PICKUP (S/T)

HARD START

(Page 1 of 2)

Definition: Engine cranks OK, but does not start for a long time. Does eventually run, or may start but immediately dies.

PRELIMINARY CHECKS

- Refer to "Important Preliminary Checks."
- Make sure the driver is using the correct starting procedure.

SENSORS

- **CHECK:** Engine Coolant Temperature (ECT) sensor - Using a scan tool, compare engine coolant temperature with ambient temperature on a cold engine.
  If coolant temperature reading is 5 degrees greater than or less than ambient air temperature on a cold engine, check for high resistance in coolant sensor circuit or sensor itself. Refer to "DTC 15" and compare resistance values.
- **CHECK:** Throttle Position (TP) sensor - If a sticking throttle shaft or binding linkage causes a high TP sensor voltage (open throttle indication), the control module will not control idle. Monitoring TP sensor voltage. Scan tool and/or voltmeter should display less than 1.25 volts with throttle closed.

FUEL SYSTEM

- **CHECK:** Fuel pump relay operation. Fuel pump should operate for 20 seconds on vehicles with a fuel module and for 2 seconds on all other vehicles when ignition is turned "ON." Refer to "CHART A-5".
- **CHECK:** Fuel Pressure. Refer to "CHART A-6".
- **CHECK:** For water contaminated fuel.
- **CHECK:** For a faulty in-tank fuel pump check valve, which would allow the fuel in the lines to drain back to the tank after the engine is stopped. To check for this condition:
  1. Ignition "OFF."
  2. Disconnect fuel line at the filter.
  3. Remove the tank filler cap.
  4. Connect a radiator test pump to the fuel line and apply 103 kPa (15 psi) pressure. If the pressure will hold for 60 seconds, the check valve is OK.
- **CHECK:** Fuel pump relay - Connect test light between fuel pump "test" terminal and ground. Test light should be "ON" for 20 seconds on vehicles with a fuel module and for 2 seconds on all other vehicles following ignition "ON." If not, refer to "CHART A-5"

4.3L (VIN W) ENGINE — SYMPTOMS — MID SIZE VAN (M/L) AND COMPACT PICKUP (S/T)

HARD START

(Page 2 of 2)

Definition: Engine cranks OK, but does not start for a long time. Does eventually operate, or may start but immediately stalls.

IGNITION SYSTEM

- **CHECK:** Ignition system for:
  - Proper ignition voltage output with spark tester J 26792 or equivalent (ST-125).
  - Spark plugs; wet plugs, cracks, wear, improper gap, burned electrodes or heavy deposits.
  - Bare and shorted wires.
  - Moisture in distributor cap.
  - Worn distributor shaft.
  - Pickup coil resistance and connections.
  - Loose ignition coil connections.
- **CHECK:** If engine starts, but then immediately stalls, disconnect the set timing connector. If engine then starts, and operates OK, replace distributor pickup coil.
- **CHECK:** CKT 423 (Ignition Control) for short to ground.

ADDITIONAL CHECKS

- **CHECK:** IAC operation - refer to "DTC 35", and "Diagnosis"
- **CHECK:** No crank signal .
- **CHECK:** EGR operation .
- **CHECK:** Service Bulletins for PROM updates.

4.3L (VIN W) ENGINE — SYMPTOMS — MID SIZE VAN (M/L) AND COMPACT PICKUP (S/T)

SURGES AND/OR CHUGGLES

Definition: Engine power variation, under steady throttle or cruise. Feels like the vehicle speeds up and slows down, with no change in the accelerator pedal.

PRELIMINARY CHECKS

- Refer to "Important Preliminary Checks."
- Be sure driver understands Torque Converter Clutch (TCC) and A/C compressor operation in owner's manual.
- Use a scan tool to make sure reading of VSS matches vehicle speedometer except vehicles with electronic transmissions where some variation between VSS and speedometer is normal. Refer to "DTC 24 Diagnostic Aids".

SENSORS

- **CHECK:** Oxygen Sensor (O2S) for silicon contamination from fuel, or use of improper sealant (never use any sealant on waterproof connectors). The sensor may have a white, powdery coating and result in a high but false signal voltage (rich exhaust indication). The control module will then reduce the amount of fuel delivered to the engine, causing a severe driveability problem.

FUEL SYSTEM

- **CHECK:** To determine if the condition is caused by a rich or lean system, the vehicle should be driven at the speed of the complaint. Monitoring fuel trim will help identify a problem.
  Lean - Long term fuel trim greater than 150. Refer to "DTC 44 Diagnostic Aids"
  Rich - Long term fuel trim less than 115. Refer to "DTC 45 Diagnostic Aids"
- **CHECK:** Fuel pressure while condition exists, refer to "CHART A-6".

IGNITION SYSTEM

- **CHECK:** For proper ignition voltage output voltage using spark tester J 26792 or equivalent (ST-125).
- **CHECK:** Spark plugs. Remove spark plugs, check for wet plugs, cracks, wear, improper gap, burned electrodes, or heavy deposits. Repair or replace as necessary.
- **CHECK:** Ignition timing.

ADDITIONAL CHECKS

- **CHECK:** Control module grounds for being clean, tight, and in their proper locations.
- **CHECK:** Generator output voltage. Repair if less than 9 or more than 16 volts.
- **CHECK:** Vacuum lines for kinks or leaks.
- **CHECK:** For intermittent EGR.
- **CHECK:** TCC operation.

4.3L (VIN W) ENGINE — SYMPTOMS — MID SIZE VAN (M/L) AND COMPACT PICKUP (S/T)

LACK OF POWER, SLUGGISH, OR SPONGY

(Page 1 of 2)

Definition: Engine delivers less than expected power. Little or no increase in speed, when accelerator pedal is partially applied.

PRELIMINARY CHECKS

- Refer to "Important Preliminary Checks."
- Compare customer's vehicle with a similar unit. Make sure the customer has an actual problem.
- Remove air filter and check for dirt, or for being plugged, replace as necessary.
- Transmission shift pattern and down shift operation.
- If there is spray from only one injector, then, there may be a malfunction in the injector assembly, or in the signal to the injector assembly. The malfunction can be isolated, by switching the injector connectors. If the problem remains with the original injector, after switching the connector, the injector is defective. Replace the injector. If the problem moves with the injector connector, the problem is an improper signal in the injector circuits. Refer to "CHART A-3"

FUEL SYSTEM

- **CHECK:** For contaminated fuel.
- **CHECK:** For restricted fuel filter, contaminated fuel or improper fuel pressure. Refer to "CHART A-6"

IGNITION SYSTEM

- **CHECK:** Proper ignition voltage output with spark tester J 26792 or equivalent (ST-125).
- **CHECK:** Ignition timing and proper operation of ignition control.

EXHAUST SYSTEM

- **CHECK:** Exhaust system for possible restriction:

  Inspect exhaust system for damaged or collapsed pipes. Inspect muffler for heat distress or possible internal failure.
  1. With engine at normal operating temperature, connect a vacuum gage to any convenient vacuum port on intake manifold.
  2. Operate engine at 1000 RPM and record vacuum reading.
  3. Increase RPM slowly to 2500 RPM. Note vacuum reading at a steady 2500 RPM.
  4. If vacuum at 2500 RPM decreases more than 10 kPa (3" Hg) from reading at 1000 RPM, the exhaust system should be inspected for restrictions.
  5. Disconnect exhaust pipe from engine and repeat Steps 3 & 4. If vacuum still drops more than 10 kPa (3" Hg) with exhaust disconnected, check for exhaust manifold restriction and valve timing.

4.3L (VIN W) ENGINE — SYMPTOMS — MID SIZE VAN (M/L) AND COMPACT PICKUP (S/T)

LACK OF POWER, SLUGGISH, OR SPONGY

(Page 2 of 2)

Definition: Engine delivers less than expected power. Little or no increase in speed, when accelerator pedal is pushed down part way.

ADDITIONAL CHECKS

- **CHECK:** Control module grounds for being clean, tight, and in their proper location.
- **CHECK:** EGR operation for being open or partly open all the time.
- **CHECK:** Torque Converter Clutch (TCC) operation.
- **CHECK:** A/C operation. Refer to "A/C Chart"
- **CHECK:** Generator output voltage. Repair if less than 9 or more than 16 volts.

ENGINE MECHANICAL

- **CHECK:** Engine compression, valve timing, and for proper or worn camshaft.

4.3L (VIN W) ENGINE — SYMPTOMS — MID SIZE VAN (M/L) AND COMPACT PICKUP (S/T)

DETONATION/SPARK KNOCK

(Page 1 of 2)

Definition: A mild to severe ping, usually worse under acceleration. The engine makes sharp metallic knocks that change with throttle opening.

PRELIMINARY CHECKS

- Refer to "Important Preliminary Checks."
- Make sure the customer has an actual problem.
- If there is spray from only one injector, there man be a malfunction in the injector assembly, or in the signal to the injector assembly. The malfunction can be isolated, by switching the injector connectors. If the problem remains with the original injector, after switching the connector, the injector is defective. Replace the injector. If the problem moves with the injector connector, the problem is an improper signal in the injector circuits. Refer to "CHART A-3"
- Park/Neutral Position (PNP) switch or manual valve position pressure switch operation.

COOLING SYSTEM

- **CHECK:** For obvious over heating problems.
- **CHECK:** Low engine coolant.
- **CHECK:** Loose water pump belt
- **CHECK:** Restricted air flow to radiator, or restricted coolant flow.
- **CHECK:** Faulty or incorrect thermostat.
- **CHECK:** Correct coolant solution.

SENSOR

- **CHECK:** Engine Coolant Temperature (ECT) sensor, which has shifted in value. Refer to "DTC 15 Diagnostic Aids"

FUEL SYSTEM

- **CHECK:** To determine if the condition is caused by a lean system, the vehicle should be driven at the speed of the complaint. Monitoring fuel trim will help identify the problem. Lean - Long term fuel trim greater than 150. Refer to "DTC 44 Diagnostic Aids"
- **CHECK:** Fuel pressure. Refer to "CHART A-6"
- **CHECK:** For poor fuel quality and proper octane rating.
- If scan tool readings are normal and there are no engine mechanical faults, fill fuel tank with a premium gasoline that has a minimum octane rating of 92 and re-evaluate vehicle performance.

4.3L (VIN W) ENGINE — SYMPTOMS — MID SIZE VAN (M/L) AND COMPACT PICKUP (S/T)

DETONATION/SPARK KNOCK

(Page 2 of 2)

Definition: A mild to severe ping, usually worse under acceleration. The engine makes sharp metallic knocks that change with throttle opening.

IGNITION SYSTEM

- **CHECK:** Spark plugs for proper heat range and gap.
- **CHECK:** Knock Sensor (KS) system operation.
- **CHECK:** Ignition timing.

ENGINE MECHANICAL

- **CHECK:** For carbon buildup. Remove carbon with top engine cleaner. Follow instructions on can.
- **CHECK:** For incorrect basic engine parts such as cam, heads, pistons, etc.
- **CHECK:** For excessive oil entering combustion chamber.

ADDITIONAL CHECKS

- **CHECK:** Proper operation of EGR valve.
- **CHECK:** Proper operation of Thermac.
- **CHECK:** For proper transmission shift points. manual.
- **CHECK:** Torque Converter Clutch (TCC) operation.
- **CHECK:** Service Bulletins for PROM updates.

4.3L (VIN W) ENGINE — SYMPTOMS — MID SIZE VAN (M/L) AND COMPACT PICKUP (S/T)

HESITATION, SAG, STUMBLE

Definition: Momentary lack of response as the accelerator is pushed down. Can occur at any vehicle speed. Usually most severe when first trying to make the vehicle move, as from a stop sign. May cause the engine to stall if severe enough.

PRELIMINARY CHECKS

- Refer to "Important Preliminary Checks."

FUEL SYSTEM

- **CHECK:** Fuel pressure.
- **CHECK:** Throttle Position (TP) sensor - Check TP sensor for binding or sticking. Voltage should increase at a steady rate as throttle is moved toward Wide Open Throttle (WOT).
- **CHECK:** MAP sensor response and accuracy. Refer to "Manifold Absolute Pressure (MAP) Output Check"
- **CHECK:** Water/Contaminated fuel.
- **CHECK:** Canister purge system for proper operation.

IGNITION SYSTEM

- **CHECK:** Spark plug wires for being faulty.
- **CHECK:** Spark plugs for being fouled.
- **CHECK:** Open ignition system ground CKT 453 or ignition timing.

ADDITIONAL CHECKS

- **CHECK:** Service Bulletins for PROM updates.
- **CHECK:** Generator output voltage. Repair if less than 9 or more than 16 volts.
- **CHECK:** EGR valve operation.

4.3L (VIN W) ENGINE — SYMPTOMS — MID SIZE VAN (M/L) AND COMPACT PICKUP (S/T)

CUTS OUT, MISSES

Definition: Steady pulsation or jerking that follows engine speed, usually more pronounced as engine load increases. The exhaust has a steady spitting sound at idle or low speed.

PRELIMINARY CHECKS

- Refer to "Important Preliminary Checks."
- If there is spray from only one injector, then, there may be a malfunction in the injector assembly, or in the signal to the injector assembly. The malfunction can be isolated, by switching the injector connectors. If the problem remains with the original injector, after switching the connector, the injector is defective. Replace the injector. If the problem moves with the injector connector, the problem is an improper signal in the injector circuits. Refer to "CHART A-3"

IGNITION SYSTEM

- **CHECK:** For cylinder miss by:
  1. Start engine, allow engine to stabilize then disconnect IAC motor. Remove one spark plug wire at a time, using insulated pliers.

  NOTICE: Do not perform this test for more than 2 minutes, as this may cause damage to the catalytic converter.

  2. If there is an RPM drop, on all cylinders, (equal to within 50 RPM), go to "Rough, Unstable Or Incorrect Idle, Stalling" symptom. Reconnect IAC motor with ignition "OFF."
  3. If there is no RPM drop on one or more cylinders, or excessive variation in drop, check for spark, on the suspected cylinder(s) with J 26792 (ST-125) Spark Tester or equivalent.
  4. If no spark, refer to "CHART A-3"
  5. If there is spark, remove spark plug(s) in these cylinders and check for:
     - Insulation Cracks.
     - Wear.
     - Improper Gap.
     - Burned Electrodes.
     - Heavy Deposits.
- **CHECK:** Spark plug wire resistance (should not exceed 30,000 ohms), also, check rotor and distributor cap.
- **CHECK:** If the previous checks did not find the problem:
  - Visually inspect ignition system for moisture, dust, cracks, burns, etc. With engine running, spray plug wires with fine water mist to check for shorts.

FUEL SYSTEM

- **CHECK:** Fuel pressure. Refer to "CHART A-6"
- **CHECK:** For contaminated fuel or restricted fuel filter.

ENGINE MECHANICAL

- **CHECK:** For proper valve timing, low compression, bent pushrods, worn rocker arms, broken or weak valve springs, worn camshaft lobes.
- **CHECK:** Intake and exhaust manifold passages for casting flash.

4.3L (VIN W) ENGINE — SYMPTOMS — MID SIZE VAN (M/L) AND COMPACT PICKUP (S/T)

POOR FUEL ECONOMY

Definition: Fuel economy, as measured by an actual road test, is noticeably lower than expected. Also, economy is noticeably lower than it was on this vehicle at one time, as previously shown by an actual road test.

PRELIMINARY CHECKS

- Refer to "Important Preliminary Checks."
- Check air cleaner element (filter) for dirt or being plugged and operation of thermostatic air cleaner.
- Visually (physically) check: Vacuum hoses for splits, kinks, and proper connections as shown on "Vehicle Emission Control Information" label.
- Perform "On-Board Diagnostic System Check."
- Check owner's driving habits.
  - Is A/C "ON" full time (Defroster mode "ON")?
  - Are tires at correct pressure?
  - Are excessively heavy loads being carried?
  - Is acceleration too much, too often?
- Suggest owner fill fuel tank and recheck fuel economy.
- Suggest driver read "Important Facts on Fuel Economy" in Owner's Manual.

FUEL SYSTEM

- **CHECK:** Fuel type, quality and alcohol content.
- **CHECK:** Fuel pressure. Refer to "CHART A-6"

IGNITION SYSTEM

- **CHECK:** Spark plugs. Remove spark plugs, check for wet plugs, cracks, wear, improper gap, burned electrodes, or heavy deposits. Repair or replace as necessary.
- **CHECK:** Ignition wires for cracking, hardness, and proper connections.
- **CHECK:** Operation.
- **CHECK:** Ignition timing.

COOLING SYSTEM

- **CHECK:** Engine coolant level.
- **CHECK:** Engine thermostat for faulty part (always open) or for wrong heat range.

ADDITIONAL CHECKS

- **CHECK:** Transmission shift pattern.
- **CHECK:** TCC operation -
  A scan tool should indicate an rpm drop when the TCC is commanded "ON."
- **CHECK:** For proper calibration of speedometer.
- **CHECK:** For dragging brakes.

4.3L (VIN W) ENGINE — SYMPTOMS — MID SIZE VAN (M/L) AND COMPACT PICKUP (S/T)

ROUGH, UNSTABLE OR INCORRECT IDLE, STALLING

(Page 1 of 2)

Definition: The engine runs unevenly at idle. If bad enough, the vehicle may shake. Also, the idle may vary in RPM (called "hunting"). Either condition may be severe enough to cause stalling. Engine idles at incorrect speed.

PRELIMINARY CHECKS

- Refer to "Important Preliminary Checks."

SENSORS

- **CHECK:** Oxygen Sensor (O2S) - Inspect sensor for silicon contamination from fuel, or use of improper sealant (never use sealant on any waterproof connectors). The sensor will have a white, powdery coating, and will result in a high but false signal voltage (rich exhaust indication). The control module will then reduce the amount of fuel delivered to the engine, causing a severe driveability problem.
- **CHECK:** Throttle Position (TP) sensor - If a sticking throttle shaft or binding linkage causes a high TP sensor voltage open throttle indication, the control module will not control idle. Monitor TP sensor voltage. A scan tool and/or voltmeter should read less than 1.25 volts with throttle closed.
- **CHECK:** Engine Coolant Temperature (ECT) sensor - Using a scan tool, compare engine coolant temperature with ambient temperature on a cold engine.
  - If engine coolant temperature reads 5 degrees greater than or less than ambient air temperature. Check for high resistance in coolant sensor circuit or sensor itself. Refer to "DTC 15 Diagnostic Aids"
- **CHECK:** MAP sensor response and accuracy - Refer to "Manifold Absolute Pressure (MAP) Output Check"

FUEL SYSTEM

- **CHECK:** To determine if the condition is caused by a rich or lean system, the vehicle should be driven at the speed of the complaint. Monitoring fuel trim will help identify problem.
  Lean - Long term fuel trim greater than 150. Refer to "DTC 44 Diagnostic Aids"
  Rich - Long term fuel trim less than 115. Refer to "DTC 45 Diagnostic Aids"
- **CHECK:** Evaporative emission control system.
- **CHECK:** Perform a cylinder compression check.
- **CHECK:** For injector(s) leaking. Check fuel pressure. Refer to "CHART A-6"

IGNITION SYSTEM

- **CHECK:** Ignition system and ignition timing.

ROUGH, UNSTABLE OR INCORRECT IDLE, STALLING

(Page 2 of 2)

Definition: The engine runs unevenly at idle. If bad enough, the vehicle may shake. Also, the idle may vary in RPM (called "hunting"). Either condition may be severe enough to cause stalling. Engine idles at incorrect speed.

ADDITIONAL CHECKS

- **CHECK:** Vacuum leaks can cause higher than normal idle and low IAC counts.
- **CHECK:** IAC operation - refer to "DTC 35" and "Diagnosis"
- **CHECK:** Control module grounds for clean, tight, and proper routing.
- **CHECK:** Park/Neutral Position (PNP) switch or Pressure Switch Manifold (PSM) operation.
- **CHECK:** Use scan tool to determine if control module is receiving A/C request signal. Whenever A/C is selected If problem exists with A/C "ON," check A/C system operation.
- **CHECK:** EGR "ON," while idling, will cause roughness, stalling and hard starting.
- **CHECK:** Battery cables and ground straps should be clean and secure. Erratic voltage will cause IAC to change its position, resulting in poor idle quality.
- **CHECK:** IAC valve will not move if system voltage is below 9 or greater than 16 volts.
- **CHECK:** A/C refrigerant pressure too high or faulty high pressure switch.
- **CHECK:** Crankcase ventilation valve for proper operation by placing finger over inlet hole in valve end several times. Valve should snap back. If not, replace valve.
- **CHECK:** Air system.

ENGINE MECHANICAL

- **CHECK:** For broken motor mounts or low compression.

EXCESSIVE EXHAUST EMISSIONS OR ODORS

Definition: Vehicle fails an emission test. Vehicle has excessive "rotten egg" smell. Excessive odors do not necessarily indicate excessive emissions.

PRELIMINARY CHECKS

- Refer to "Important Preliminary Checks."
- Perform "Diagnostic Circuit Check."
- Make sure engine is at normal operating temperature.
- If EMISSION TEST shows excessive CO and HC check items which cause vehicle to run RICH. Long Term Fuel Trim Memory less than 115 refer to "DTC 45 Diagnostic Aids"
- If EMISSION TEST shows excessive NOx check items which cause engine to operate LEAN or too hot.

SENSORS

- **CHECK:** If the scan tool indicates a very high coolant temperature and the system is running LEAN: check the cooling system and cooling fan for proper operation.

FUEL SYSTEM

- **CHECK:** If the system is running rich, (long term fuel trim less than 115), refer to "DTC 45 Diagnostic Aids" or if the system is running lean, (long term fuel trim greater than 150), refer to "DTC 44 Diagnostic Aids"
- **CHECK:** For properly installed fuel cap.
- **CHECK:** Fuel pressure. Refer to "CHART A-6"
- **CHECK:** If test shows excessive NOx, check items which cause engine to operate LEAN or too hot.
- **CHECK:** Canister for fuel loading.

IGNITION SYSTEM

- **CHECK:** Ignition system, incorrect timing, or excessive advance.
- **CHECK:** Spark plugs, plug wires, and ignition components.

ADDITIONAL CHECKS

- **CHECK:** For vacuum leaks.
- **CHECK:** For lead contamination of catalytic converter (look for the removal of fuel filler neck restrictor).
- **CHECK:** Carbon build-up. Remove carbon with top engine cleaner. Follow instructions on can.
- **CHECK:** EGR valve for not opening.
- **CHECK:** Crankcase ventilation valve for being plugged, stuck, or blocked PCV hose.

4.3L (VIN W) ENGINE — SYMPTOMS — MID SIZE VAN (M/L) AND COMPACT PICKUP (S/T)

DIESELING, RUN-ON

Definition: Engine continues to run after key is turned "OFF," but runs very roughly. If engine runs smoothly, check ignition switch and adjustment.

PRELIMINARY CHECKS

- Refer to "Important Preliminary Checks."

FUEL SYSTEM

- **CHECK:** Evaporative system and fuel tank venting.
- **CHECK:** Injector(s) on TBI unit for leakage. Apply 12 volts to fuel pump "test" terminal to turn "ON" fuel pump and pressurize fuel system. Visually check injector(s) for fuel leakage.

4.3L (VIN W) ENGINE — SYMPTOMS — MID SIZE VAN (M/L) AND COMPACT PICKUP (S/T)

BACKFIRE

Definition: Fuel ignites in intake manifold, or in exhaust system, making a loud popping noise.

PRELIMINARY CHECKS

- Refer to "Important Preliminary Checks."

IGNITION SYSTEM

- **CHECK:** Proper ignition coil output voltage with spark tester J 26792 or equivalent (ST-125).
- **CHECK:** Spark plugs. Remove spark plugs, check for wet plugs, cracks, wear, improper gap, burned electrodes, or heavy deposits. Repair or replace as necessary.
- **CHECK:** Ignition system and ignition timing.
- **CHECK:** For crossfire between spark plugs (distributor cap, spark plug wires, and proper routing of plug wires).

ENGINE MECHANICAL

- **CHECK:** Compression, valve timing, manifold gaskets, sticking or leaking valves.
- **CHECK:** Faulty A.I.R. check valve.
- **CHECK:** EGR operation for being open all the time.
- **CHECK:** Intake and exhaust system for casting flash or other restrictions.

4.3L (VIN W) ENGINE — A CHARTS — MID SIZE VAN (M/L) AND COMPACT PICKUP (S/T)

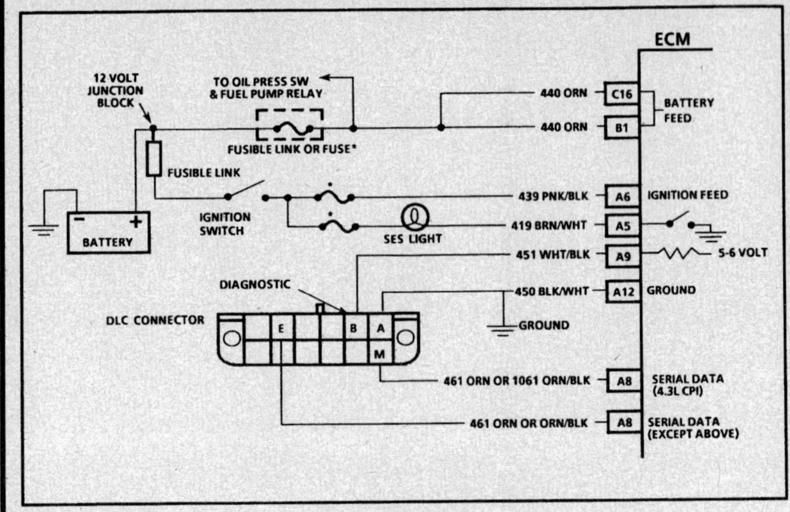

CHART A-1

NO "SERVICE ENGINE SOON" LIGHT
ALL VEHICLES WITHOUT 4L80-E TRANSMISSION

Circuit Description:

There should always be a steady "Service Engine Soon" light when the ignition is "ON" and engine stopped. Battery voltage is supplied to the light bulb. The ECM will control the light and turn it "ON" by providing a ground path through CKT 419 to the ECM.

Test Description: Number(s) below refer to circled number(s) on the diagnostic chart.
1. This step will isolate the problem in CKT 419.
2. If CKT's 440 and 439 have voltage, the ECM connections, grounds, or ECM is faulty.
3. Using a test light connected to 12 volts, probe each of the system ground circuits to be sure a good ground is present. Refer to ECM terminal end view in this section for ECM pin locations of ground circuits.
4. If the fuse/fusible link is blown, refer to ECM wiring diagram for complete circuit.

Diagnostic Aids:

If the engine runs OK, check:
- Faulty light bulb.
- CKT 419 open, shorted to ground or blown fuse.

If the engine cranks but will not run, check:
- Continuous battery-fuse or fusible link open.
- ECM ignition fuse open.
- Battery CKT 440 to ECM open.
- Ignition CKT 439 to ECM open.
- Poor connection to ECM.

4.3L (VIN W) ENGINE — A CHARTS — MID SIZE VAN (M/L) AND COMPACT PICKUP (S/T)

CHART A-1

NO "SERVICE ENGINE SOON" LIGHT
ALL VEHICLES WITHOUT 4L80-E TRANSMISSION

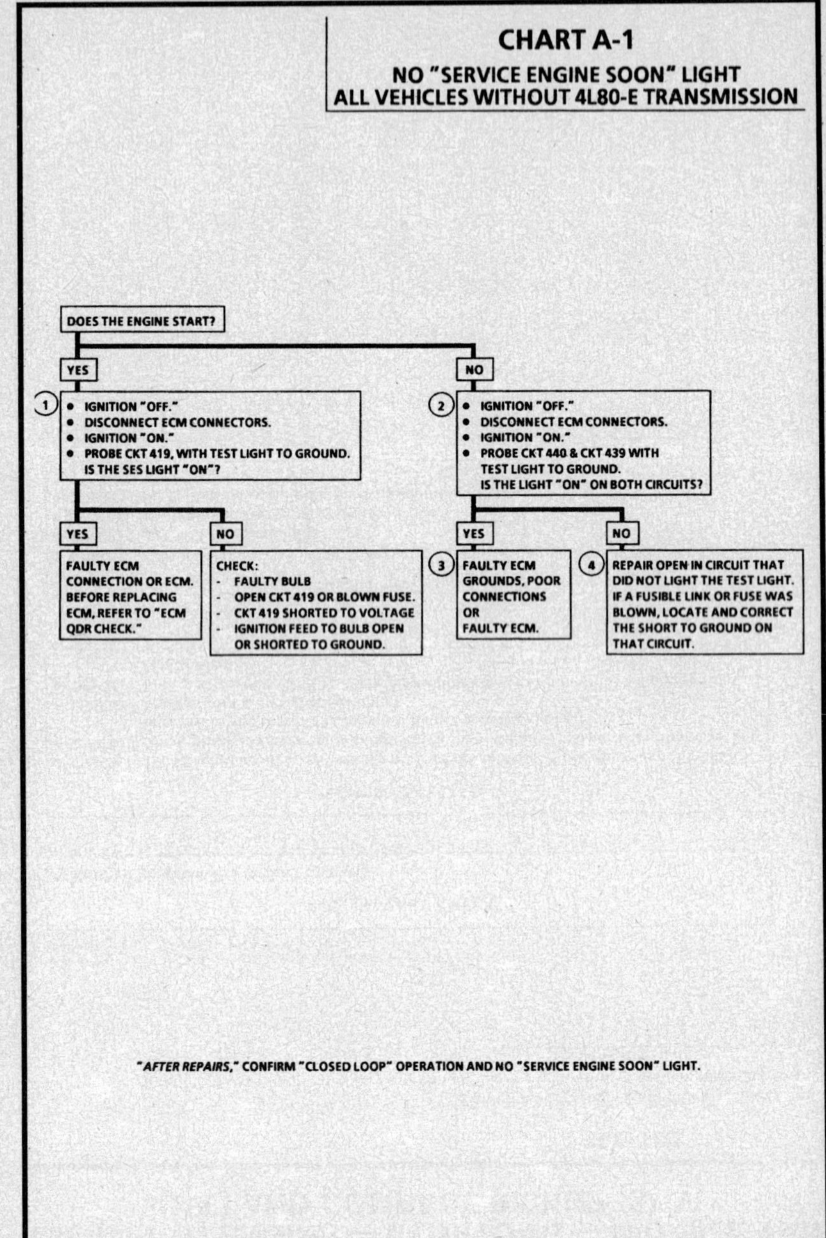

"AFTER REPAIRS," CONFIRM "CLOSED LOOP" OPERATION AND NO "SERVICE ENGINE SOON" LIGHT.

4.3L (VIN W) ENGINE — A CHARTS — MID SIZE VAN (M/L) AND COMPACT PICKUP (S/T)

4.3L (VIN W) ENGINE — A CHARTS — MID SIZE VAN (M/L) AND COMPACT PICKUP (S/T)

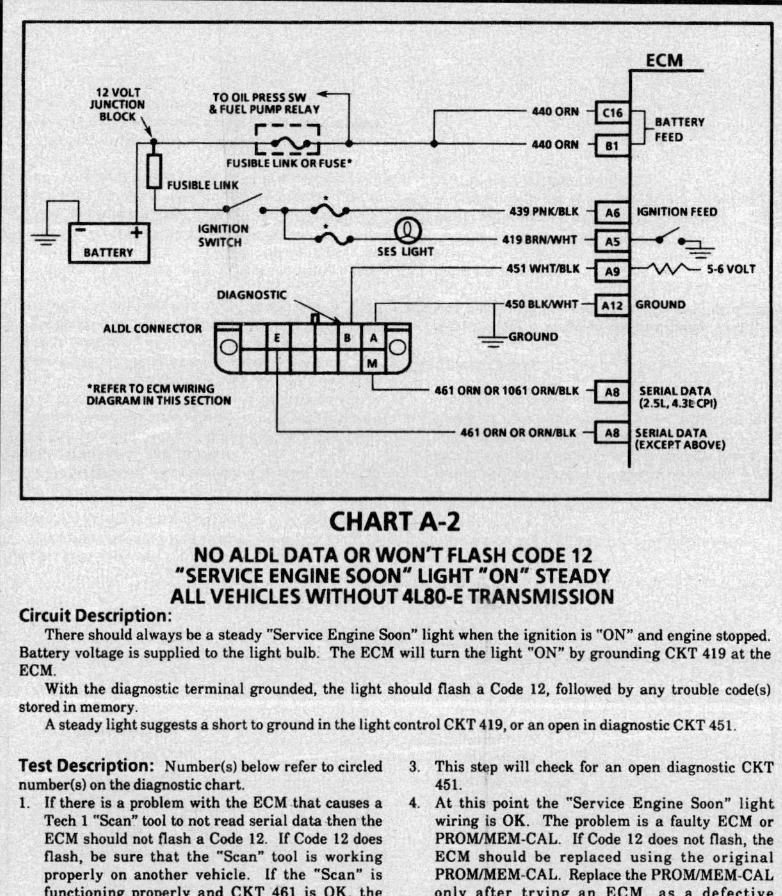

CHART A-2
NO ALDL DATA OR WON'T FLASH CODE 12
"SERVICE ENGINE SOON" LIGHT "ON" STEADY
ALL VEHICLES WITHOUT 4L80-E TRANSMISSION

Circuit Description:

There should always be a steady "Service Engine Soon" light when the ignition is "ON" and engine stopped. Battery voltage is supplied to the light bulb. The ECM will turn the light "ON" by grounding CKT 419 at the ECM.

With the diagnostic terminal grounded, the light should flash a Code 12, followed by any trouble code(s) stored in memory.

A steady light suggests a short to ground in the light control CKT 419, or an open in diagnostic CKT 451.

Test Description: Number(s) below refer to circled number(s) on the diagnostic chart.

1. If there is a problem with the ECM that causes a Tech 1 "Scan" tool to not read serial data then the ECM should not flash a Code 12. If Code 12 does flash, be sure that the "Scan" tool is working properly on another vehicle. If the "Scan" is functioning properly and CKT 461 is OK, the PROM/MEM-CAL or ECM may be at fault for the No ALDL symptom.

2. If the light goes "OFF" when the ECM connector is disconnected, then CKT 419 is not shorted to ground.

3. This step will check for an open diagnostic CKT 451.

4. At this point the "Service Engine Soon" light wiring is OK. The problem is a faulty ECM or PROM/MEM-CAL. If Code 12 does not flash, the ECM should be replaced using the original PROM/MEM-CAL. Replace the PROM/MEM-CAL only after trying an ECM, as a defective PROM/MEM-CAL is an unlikely cause of the problem.

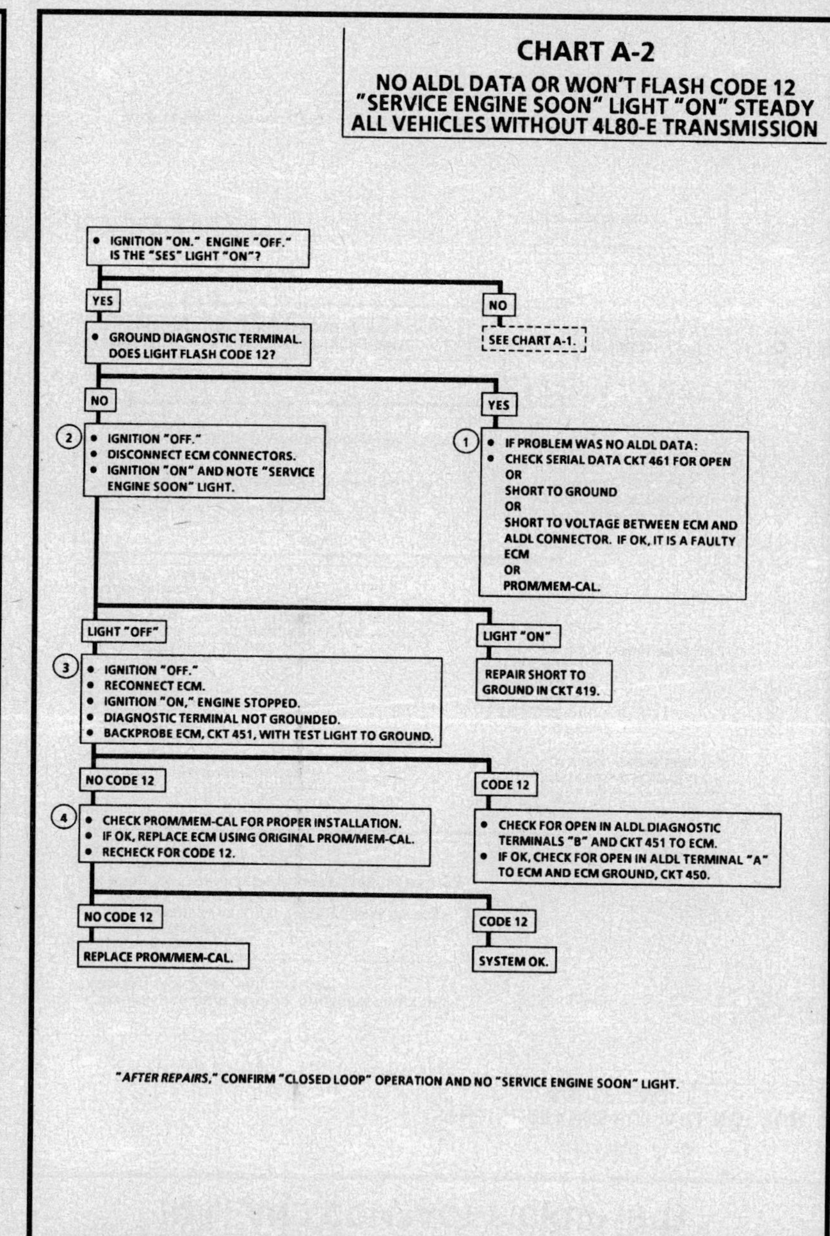

4.3L (VIN W) ENGINE — A CHARTS — MID SIZE VAN (M/L) AND COMPACT PICKUP (S/T)

4.3L (VIN W) ENGINE — A CHARTS — MID SIZE VAN (M/L) AND COMPACT PICKUP (S/T)

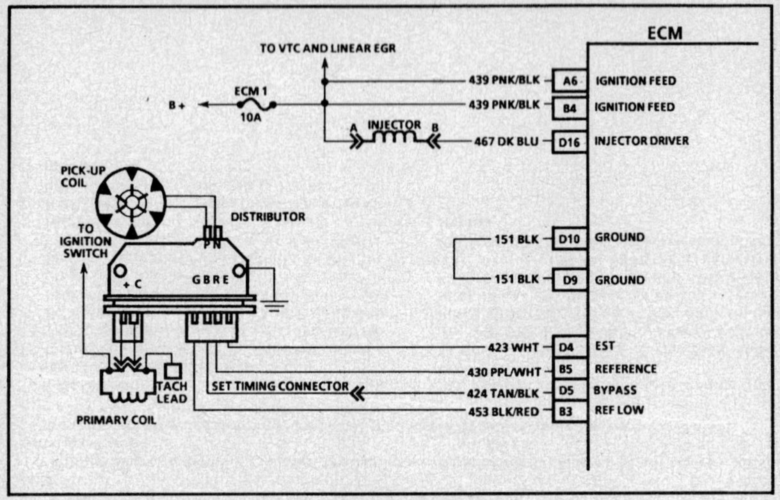

CHART A-3
ENGINE CRANKS BUT WILL NOT RUN
4.3L CPI ENGINE

Circuit Description:

This chart assumes that battery condition and engine cranking speed are OK, and there is adequate fuel in the tank. This chart should be used on 4.3L engines with CPI.

Test Description: Number(s) below refer to circled number(s) on the diagnostic chart.

1. A "Service Engine Soon" light "ON" is a basic test to determine if there is a 12 volt supply and ignition 12 volts to ECM. No ALDL may be due to an ECM problem and CHART A-2 will diagnose the ECM. If TPS is over 2.5 volts, the engine may be in the clear flood mode which will cause starting problems. If coolant sensor is below -30°C, the ECM will provide fuel for this extremely cold temperature which will severely flood the engine.
2. Voltage at the spark plug is checked using spark tester tool ST-125 (J 26792) or equivalent. No spark indicates a basic ignition problem.
3. Use an injector test light like J 34730-2B, BT-8329A or equivalent, to test injector circuit. A blinking light indicates the ECM is controlling the injector.
4. This test will determine if there is fuel pressure at the CPI unit and holding steady. Use fuel pressure gage J 34730-1 or equivalent.

Wrap a shop towel around the fuel pressure tap to absorb any small amount of fuel leakage that may occur when installing the gage.
5. Checking resistance of the injector will determine if the injector is at fault.

Diagnostic Aids:

If no trouble is found in the fuel pump circuit or ignition system and the cause of a "Engine Cranks But Will Not Run" has not been found, check for:
- Fouled spark plugs.
- EGR valve stuck open.
- Low fuel pressure. Refer to CHART A-6.
- Water or foreign material in the fuel system.
- A ground CKT 423 (EST) may cause a "No Start" or a "Start then Stall" condition.
- Basic engine problem.

CHART A-3
ENGINE CRANKS BUT WILL NOT RUN
4.3L CPI ENGINE

BEFORE USING THIS CHART, REFER TO "DIAGNOSTIC CIRCUIT CHECK." IF CODE 54 IS STORED, USE THAT CHART FIRST.

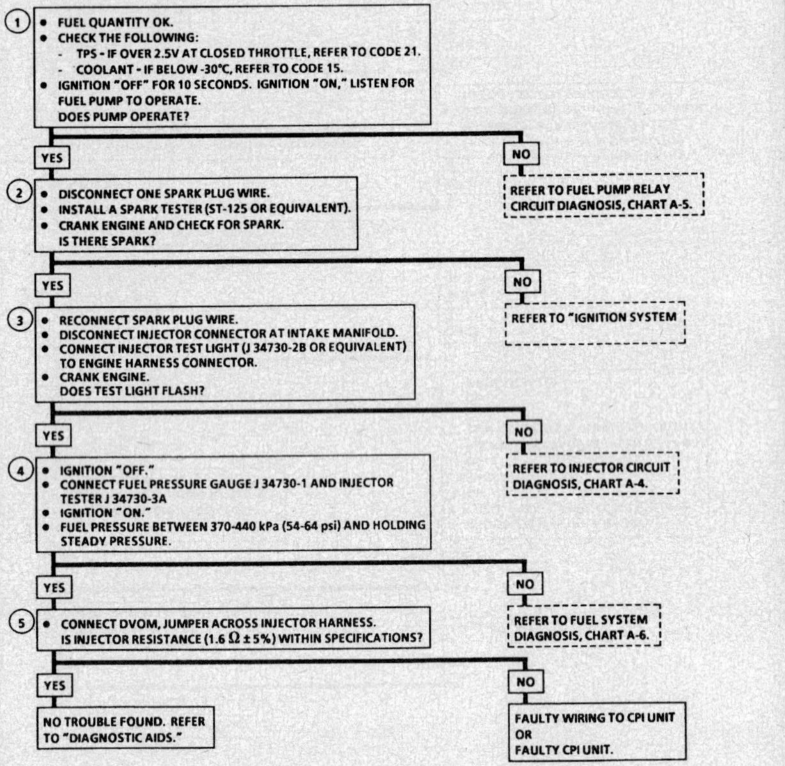

①
- FUEL QUANTITY OK.
- CHECK THE FOLLOWING:
  - TPS - IF OVER 2.5V AT CLOSED THROTTLE, REFER TO CODE 21.
  - COOLANT - IF BELOW -30°C, REFER TO CODE 15.
- IGNITION "OFF" FOR 10 SECONDS. IGNITION "ON," LISTEN FOR FUEL PUMP TO OPERATE.
- DOES PUMP OPERATE?

YES → ② · NO → REFER TO FUEL PUMP RELAY CIRCUIT DIAGNOSIS, CHART A-5.

②
- DISCONNECT ONE SPARK PLUG WIRE.
- INSTALL A SPARK TESTER (ST-125 OR EQUIVALENT).
- CRANK ENGINE AND CHECK FOR SPARK.
- IS THERE SPARK?

YES → ③ · NO → REFER TO "IGNITION SYSTEM"

③
- RECONNECT SPARK PLUG WIRE.
- DISCONNECT INJECTOR CONNECTOR AT INTAKE MANIFOLD.
- CONNECT INJECTOR TEST LIGHT (J 34730-2B OR EQUIVALENT) TO ENGINE HARNESS CONNECTOR.
- CRANK ENGINE.
- DOES TEST LIGHT FLASH?

YES → ④ · NO → REFER TO INJECTOR CIRCUIT DIAGNOSIS, CHART A-4.

④
- IGNITION "OFF."
- CONNECT FUEL PRESSURE GAUGE J 34730-1 AND INJECTOR TESTER J 34730-3A
- IGNITION "ON."
- FUEL PRESSURE BETWEEN 370-440 kPa (54-64 psi) AND HOLDING STEADY PRESSURE.

YES → ⑤ · NO → REFER TO FUEL SYSTEM DIAGNOSIS, CHART A-6.

⑤
- CONNECT DVOM, JUMPER ACROSS INJECTOR HARNESS. IS INJECTOR RESISTANCE (1.6 Ω ± 5%) WITHIN SPECIFICATIONS?

YES → NO TROUBLE FOUND. REFER TO "DIAGNOSTIC AIDS." · NO → FAULTY WIRING TO CPI UNIT OR FAULTY CPI UNIT.

"AFTER REPAIRS," CONFIRM "CLOSED LOOP" OPERATION AND NO "SERVICE ENGINE SOON" LIGHT.

4.3L (VIN W) ENGINE — A CHARTS — MID SIZE VAN (M/L) AND COMPACT PICKUP (S/T)

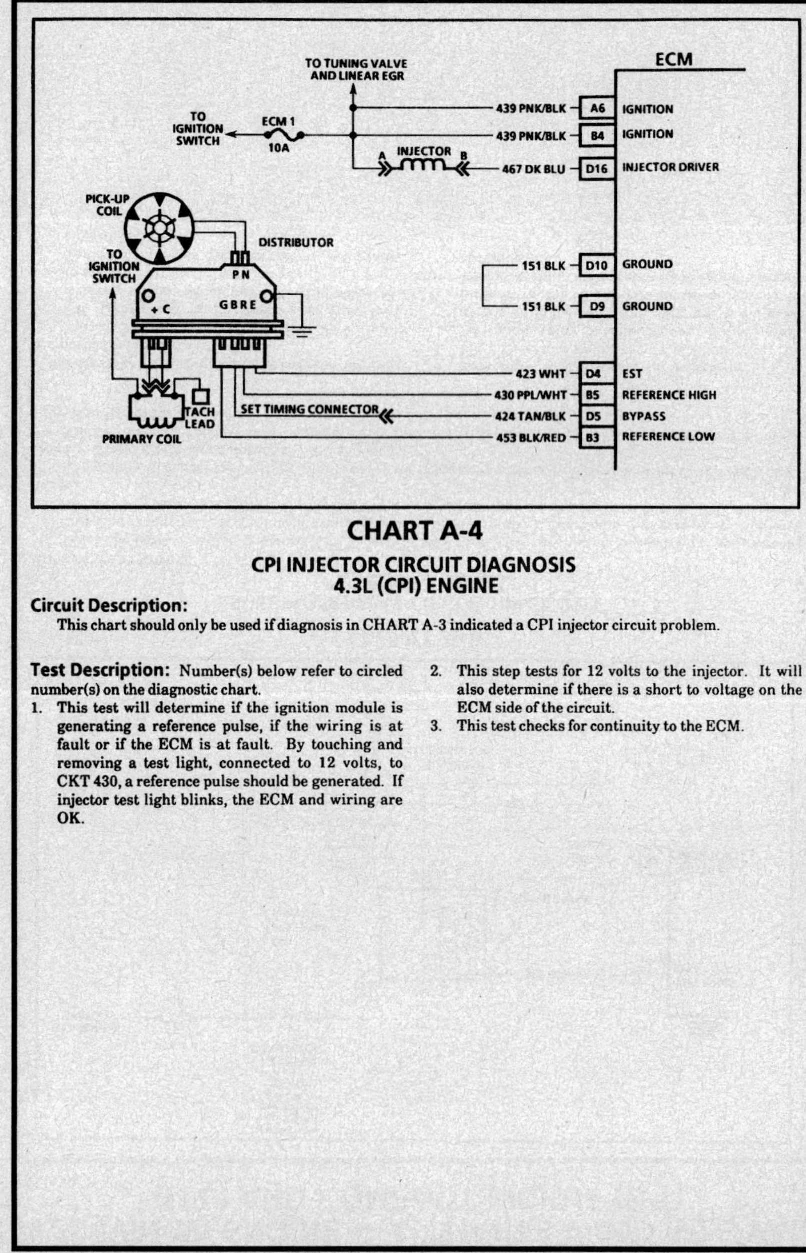

CHART A-4
CPI INJECTOR CIRCUIT DIAGNOSIS
4.3L (CPI) ENGINE

Circuit Description:

This chart should only be used if diagnosis in CHART A-3 indicated a CPI injector circuit problem.

Test Description: Number(s) below refer to circled number(s) on the diagnostic chart.

1. This test will determine if the ignition module is generating a reference pulse, if the wiring is at fault or if the ECM is at fault. By touching and removing a test light, connected to 12 volts, to CKT 430, a reference pulse should be generated. If injector test light blinks, the ECM and wiring are OK.

2. This step tests for 12 volts to the injector. It will also determine if there is a short to voltage on the ECM side of the circuit.

3. This test checks for continuity to the ECM.

4.3L (VIN W) ENGINE — A CHARTS — MID SIZE VAN (M/L) AND COMPACT PICKUP (S/T)

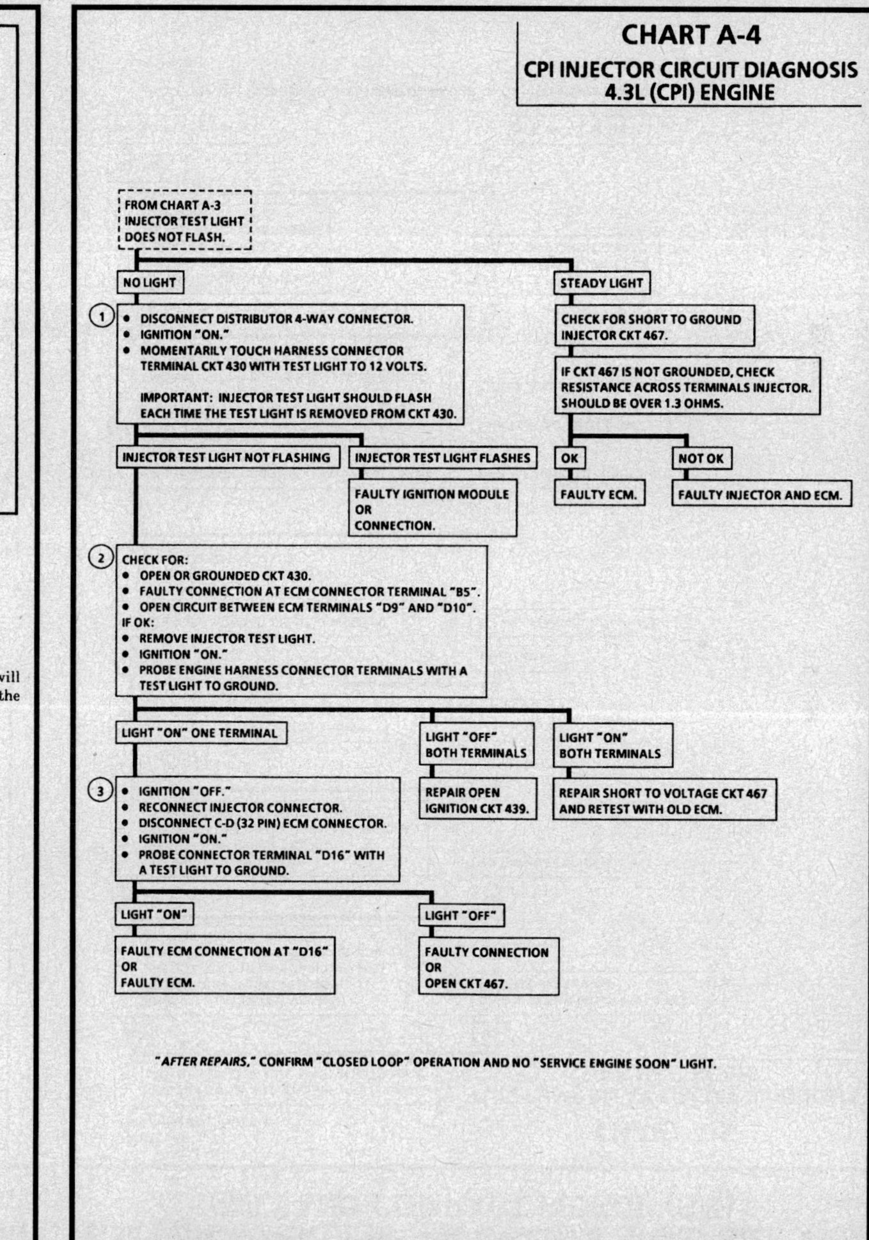

4.3L (VIN W) ENGINE — A CHARTS — MID SIZE VAN (M/L) AND COMPACT PICKUP (S/T)

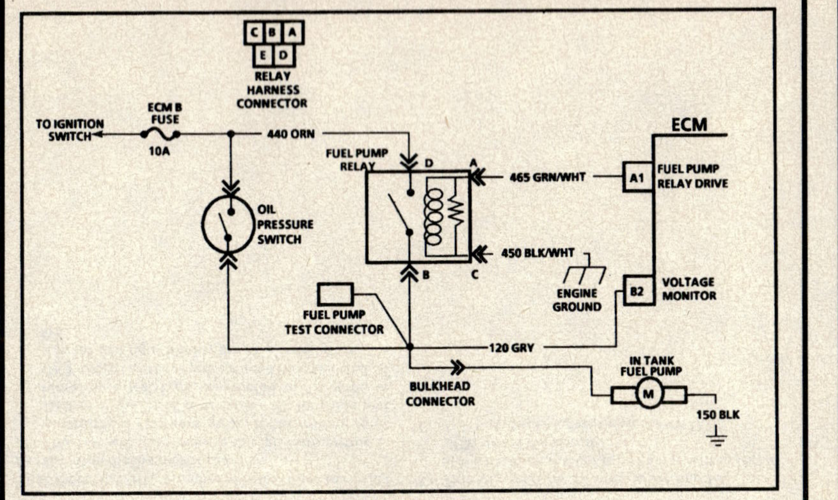

CHART A-5
FUEL PUMP RELAY CIRCUIT DIAGNOSIS
"S/T & M/L" SERIES

Circuit Description:

When the ignition switch is turned "ON," the ECM will turn "ON" the in-tank fuel pump. It will remain "ON" as long as the engine is cranking or running, and the ECM is receiving distributor reference pulses. If there are no reference pulses, the ECM will shut "OFF" the fuel pump within 2 seconds after ignition "ON" or engine stops.

The pump will deliver fuel to the TBI unit where the system pressure is controlled to about 62 to 90 kPa (9 to 13 psi). Excess fuel is then returned to the fuel tank.

The CPI pump will deliver system pressure that is controlled to about 370-440 kPa (54-64 psi). Excess fuel is then returned to the fuel tank.

Test Description: Number(s) below refer to circled number(s) on the diagnostic chart.

1. Turns "ON" the fuel pump if CKT 120 wiring is OK. If the pump runs, it maybe a fuel pump relay circuit problem, which the following steps will locate.
2. The next two steps check for power and ground circuits to the relay.
3. Determines if ECM can control the relay.
4. The oil pressure switch serves as a backup for the fuel pump relay to help prevent a "no start" situation. If the fuel pump relay was found to be inoperative, the oil pressure switch circuit should also be tested to determine why it did not operate the fuel pump.

4.3L (VIN W) ENGINE — A CHARTS — MID SIZE VAN (M/L) AND COMPACT PICKUP (S/T)

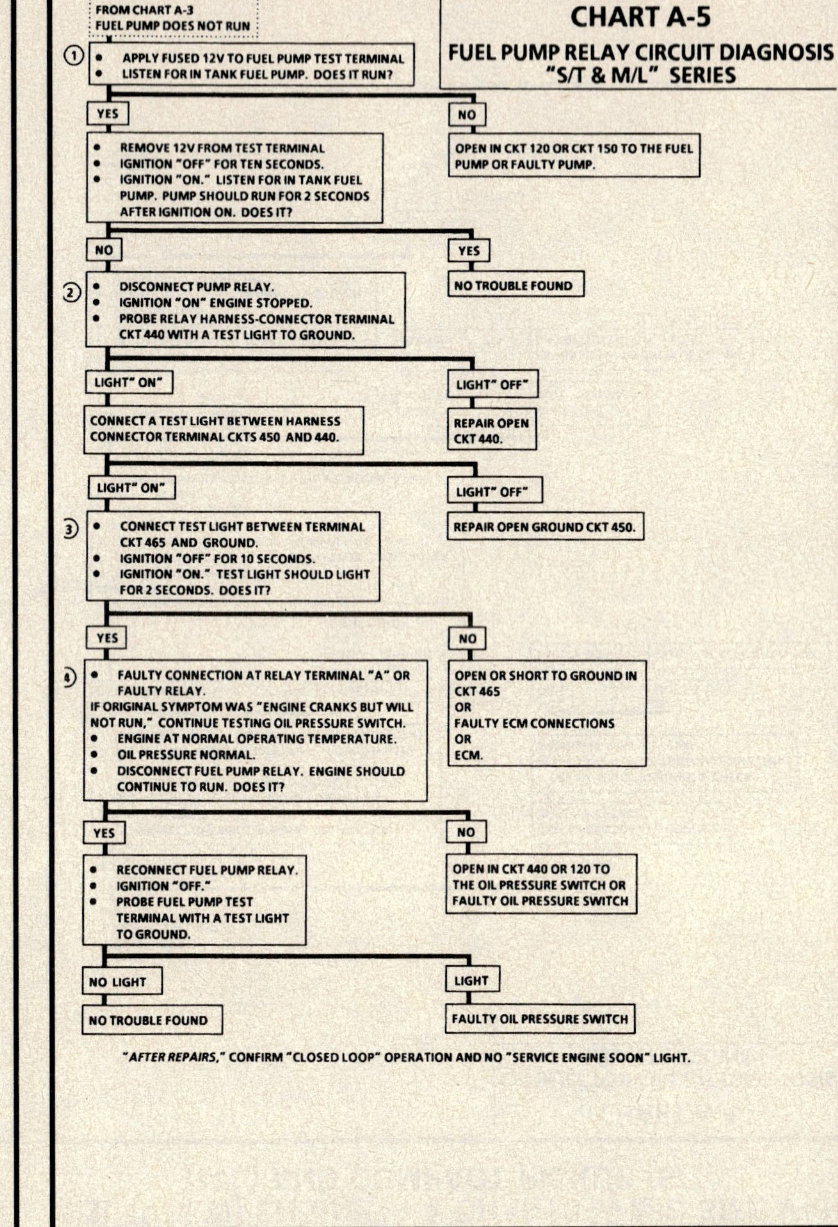

4.3L (VIN W) ENGINE — A CHARTS — MID SIZE VAN (M/L) AND COMPACT PICKUP (S/T)

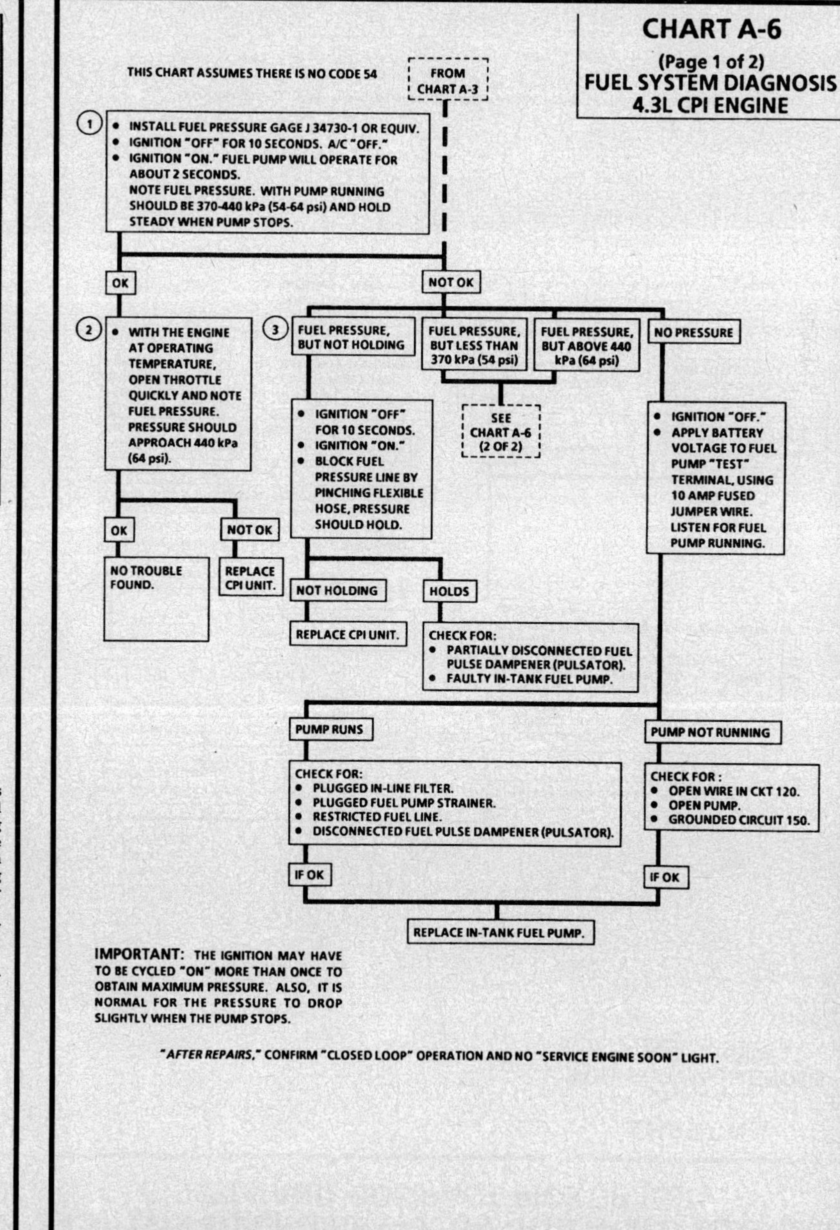

CHART A-6
(Page 1 of 2)
FUEL SYSTEM DIAGNOSIS
4.3L CPI ENGINE

Circuit Description:

When the ignition switch is turned "ON," the Electronic Control Module (ECM) will turn "ON" the in-tank fuel pump. It will remain "ON" as long as the engine is cranking or running, and the ECM is receiving reference pulses. If there are no reference pulses, the ECM will shut "OFF" the fuel pump within 2 seconds after ignition "ON" or engine stops.

An electric fuel pump, attached to the fuel level meter assembly (inside the fuel tank) pumps fuel through an in-line filter to the CPI unit. The pump is designed to provide fuel at a pressure above the regulated pressure needed by the injector. A pressure regulator, an integral part of the CPI unit, keeps fuel available to the injector at a regulated pressure. Unused fuel is returned to the fuel tank by a separate line.

Tools Required:

J 34730-1A Fuel Pressure Gage
or
J 34730-1 Fuel Pressure Gage with J 34730-250 Fuel Pressure Adapter Kit.

Test Description: Numbers below refer to circled numbers on the diagnostic chart.

1. Wrap a shop towel around the fuel pressure connection to absorb any small amount of fuel leakage that may occur when installing the gage. Ignition "ON" pump pressure should be 370-440 kPa (54-64 psi). This pressure is controlled by spring pressure within the regulator assembly.

2. When the engine is idling, the manifold pressure is low (high vacuum) and is applied to the fuel regulator diaphragm. This will offset the spring and result in a lower fuel pressure. This idle pressure will vary somewhat depending on barometric pressure, however, the pressure idling should be less, indicating pressure regulator control.

3. Fuel pressure that continues to fall is caused by one of the following:
   • In-tank fuel pump check valve not holding.
   • A partially disconnected fuel pulse dampener (pulsator).
   • Fuel pressure regulator valve leaking.
   • CPI injector and poppet valve(s) leaking.

4.3L (VIN W) ENGINE — A CHARTS — MID SIZE VAN (M/L) AND COMPACT PICKUP (S/T)

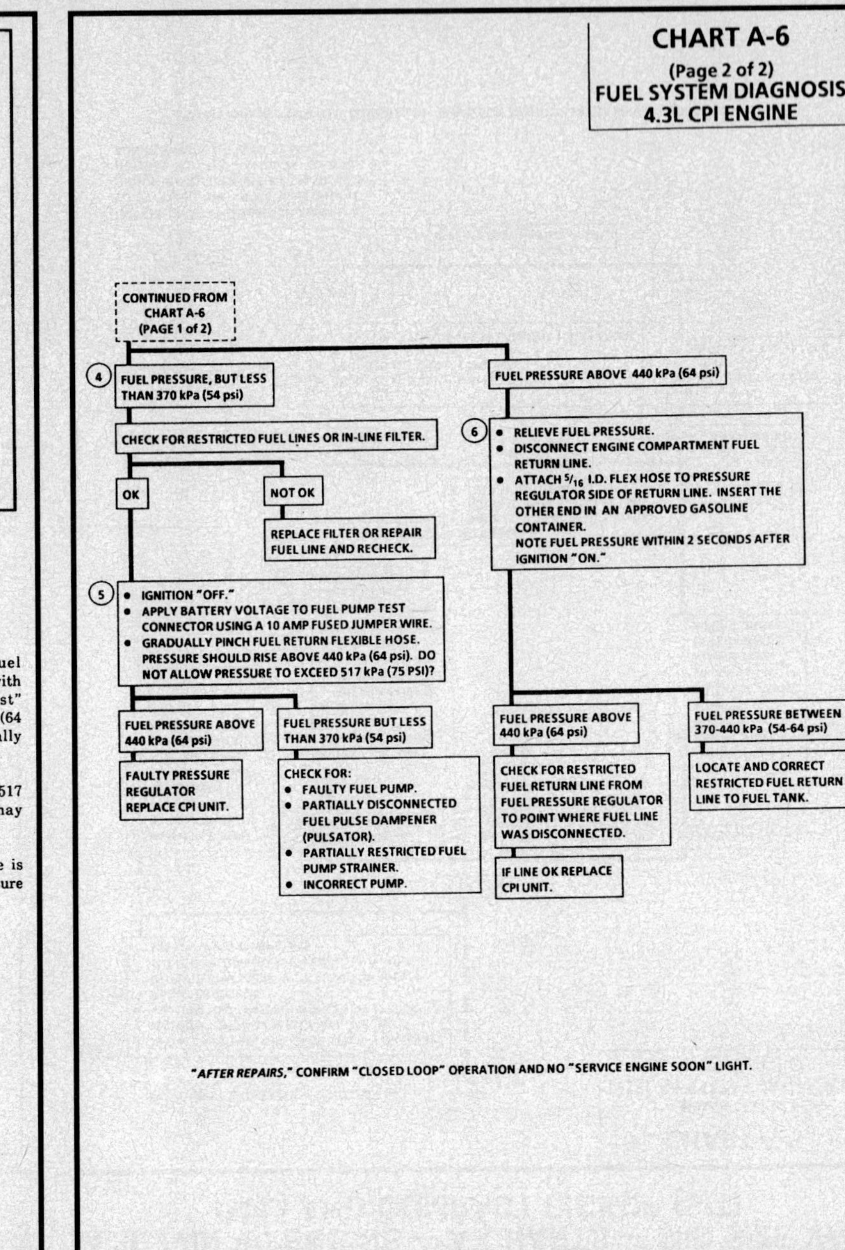

CHART A-6
(Page 2 of 2)
FUEL SYSTEM DIAGNOSIS
4.3L CPI ENGINE

Test Description: Number(s) below refer to circled number(s) on the diagnostic chart.

4. Fuel pressure but less than 370 kPa (54 psi) falls into three areas:
   - Regulated pressure but less than 370 kPa (54 psi). Amount of fuel to injector OK, but pressure is too low. System will be lean running and may set Code 44. Also, hard starting cold and overall poor performance or may not start at all (refer to CHART A3).
   - Restricted flow causing pressure drop - Normally, a vehicle with a fuel pressure of less than 300 kPa (44 psi) at idle will not be driveable. However, if the pressure drop occurs only while driving, the engine will normally surge then stop running as pressure begins to drop rapidly. This is most likely caused by a restricted fuel line or plugged filter.
   - Leaking or contaminated pressure regulator valve/seat interface may not allow regulated pressure to be achieved. Refer to Step 3.

5. Restricting the fuel return line allows fuel pressure to build above regulated pressure, with battery voltage applied to the pump "test" terminal, pressure should rise above 440 kPa (64 psi) as the valve in the return line is partially closed.

NOTICE: Do Not allow pressure to exceed 517 kPa (75 psi), as damage to the regulator may result.

6. This test determines if the high fuel pressure is due to a restricted fuel return line or a pressure regulator problem.

4.3L (VIN W) ENGINE — A CHARTS — MID SIZE VAN (M/L) AND COMPACT PICKUP (S/T)

CHART A-6
(Page 2 of 2)
FUEL SYSTEM DIAGNOSIS
4.3L CPI ENGINE

CONTINUED FROM
CHART A-6
(PAGE 1 of 2)

(4) FUEL PRESSURE, BUT LESS THAN 370 kPa (54 psi)

CHECK FOR RESTRICTED FUEL LINES OR IN-LINE FILTER.

OK

NOT OK

REPLACE FILTER OR REPAIR FUEL LINE AND RECHECK.

(5)
- IGNITION "OFF."
- APPLY BATTERY VOLTAGE TO FUEL PUMP TEST CONNECTOR USING A 10 AMP FUSED JUMPER WIRE.
- GRADUALLY PINCH FUEL RETURN FLEXIBLE HOSE. PRESSURE SHOULD RISE ABOVE 440 kPa (64 psi). DO NOT ALLOW PRESSURE TO EXCEED 517 kPa (75 PSI)?

FUEL PRESSURE ABOVE 440 kPa (64 psi)

FAULTY PRESSURE REGULATOR REPLACE CPI UNIT.

FUEL PRESSURE BUT LESS THAN 370 kPa (54 psi)

CHECK FOR:
- FAULTY FUEL PUMP.
- PARTIALLY DISCONNECTED FUEL PULSE DAMPENER (PULSATOR).
- PARTIALLY RESTRICTED FUEL PUMP STRAINER.
- INCORRECT PUMP.

FUEL PRESSURE ABOVE 440 kPa (64 psi)

(6)
- RELIEVE FUEL PRESSURE.
- DISCONNECT ENGINE COMPARTMENT FUEL RETURN LINE.
- ATTACH 5/16 I.D. FLEX HOSE TO PRESSURE REGULATOR SIDE OF RETURN LINE. INSERT THE OTHER END IN AN APPROVED GASOLINE CONTAINER. NOTE FUEL PRESSURE WITHIN 2 SECONDS AFTER IGNITION "ON."

FUEL PRESSURE ABOVE 440 kPa (64 psi)

CHECK FOR RESTRICTED FUEL RETURN LINE FROM FUEL PRESSURE REGULATOR TO POINT WHERE FUEL LINE WAS DISCONNECTED.

IF LINE OK REPLACE CPI UNIT.

FUEL PRESSURE BETWEEN 370-440 kPa (54-64 psi)

LOCATE AND CORRECT RESTRICTED FUEL RETURN LINE TO FUEL TANK.

"AFTER REPAIRS," CONFIRM "CLOSED LOOP" OPERATION AND NO "SERVICE ENGINE SOON" LIGHT.

4.3L (VIN W) ENGINE — A CHARTS — MID SIZE VAN (M/L) AND COMPACT PICKUP (S/T)

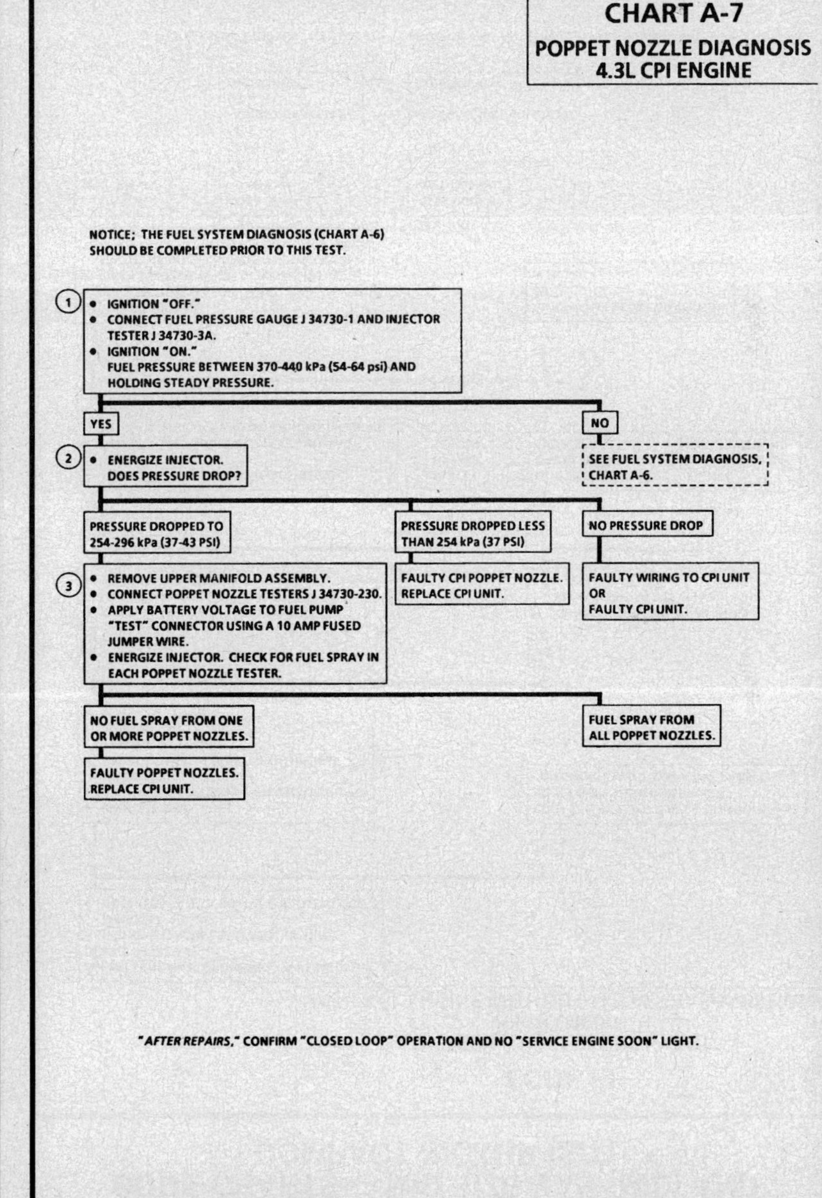

CHART A-7
POPPET NOZZLE DIAGNOSIS
4.3L CPI ENGINE

NOTICE: Poppet nozzles cannot be cleaned.

The injector tester is a tool used to turn "ON" the injector and open the poppet nozzles for a precise amount of time, thus spraying a measured amount of fuel into the manifold. This causes a drop in fuel pressure that we can record and compare to the pressure drop specification. A injector/poppet nozzle, that drops less than 275 kPa (39 psi), is considered faulty.

Test Description: Number(s) below refer to circled number(s) on the diagnostic chart.

1. Engine "cool down" period (10 minutes) is necessary to avoid irregular readings due to "Hot Soak" fuel boiling. With ignition "OFF," connect fuel gauge J 34730-1 or equivalent to fuel pressure tap. Wrap a shop towel around fitting while connecting gage to avoid fuel spillage.
Disconnect harness connector at intake plenum and connect injector tester J 34730-3, or equivalent to injector harness. Ignition must be "OFF" at least 10 seconds to complete ECM shutdown cycle. Fuel pump should run about 2 seconds after ignition is turned "ON." At this point, insert clear tubing attached to vent valve into a suitable container and bleed air from gauge and hose to insure accurate gage operation. Repeat this step until all air is bled from gauge.
2. Turn ignition "OFF" for 10 seconds and then "ON" again to get fuel pressure to its maximum. Record this initial pressure reading. Energize injector by depressing button on injector tester one time and note pressure drop at its lowest point (disregard any slight pressure increase after drop hits low point). This will determine the pressure drop, and check for a poppet nozzle stuck open.
3. Remove upper manifold assembly (with throttle body).
Remove the poppet nozzles and install poppet nozzle testers J 34730-230 to poppet nozzles. Connect nozzle testers to lower manifold assembly. Apply battery voltage to fuel pump "test" connector, using a 10 amp fused jumper wire to energize fuel pump. Pressure should be 370-440 kPa (54-64 psi).
Energize injector, this will verity if the poppet nozzle opens by checking for presence of fuel spray in each tester. This will determine if you have a faulty poppet nozzle or basic engine problem.

4.3L (VIN W) ENGINE — A CHARTS — MID SIZE VAN (M/L) AND COMPACT PICKUP (S/T)

CHART A-7
POPPET NOZZLE DIAGNOSIS
4.3L CPI ENGINE

NOTICE: THE FUEL SYSTEM DIAGNOSIS (CHART A-6) SHOULD BE COMPLETED PRIOR TO THIS TEST.

1.
- IGNITION "OFF."
- CONNECT FUEL PRESSURE GAUGE J 34730-1 AND INJECTOR TESTER J 34730-3A.
- IGNITION "ON."
- FUEL PRESSURE BETWEEN 370-440 kPa (54-64 psi) AND HOLDING STEADY PRESSURE.

YES → 2.
- ENERGIZE INJECTOR. DOES PRESSURE DROP?

NO → SEE FUEL SYSTEM DIAGNOSIS, CHART A-6.

- PRESSURE DROPPED TO 254-296 kPa (37-43 PSI)
- PRESSURE DROPPED LESS THAN 254 kPa (37 PSI)
- NO PRESSURE DROP

3.
- REMOVE UPPER MANIFOLD ASSEMBLY.
- CONNECT POPPET NOZZLE TESTERS J 34730-230.
- APPLY BATTERY VOLTAGE TO FUEL PUMP "TEST" CONNECTOR USING A 10 AMP FUSED JUMPER WIRE.
- ENERGIZE INJECTOR. CHECK FOR FUEL SPRAY IN EACH POPPET NOZZLE TESTER.

FAULTY CPI POPPET NOZZLE. REPLACE CPI UNIT.

FAULTY WIRING TO CPI UNIT OR FAULTY CPI UNIT.

- NO FUEL SPRAY FROM ONE OR MORE POPPET NOZZLES.
- FUEL SPRAY FROM ALL POPPET NOZZLES.

FAULTY POPPET NOZZLES. REPLACE CPI UNIT.

"AFTER REPAIRS," CONFIRM "CLOSED LOOP" OPERATION AND NO "SERVICE ENGINE SOON" LIGHT.

4.3L (VIN W) ENGINE — DIAGNOSTIC TROUBLE CODE CHARTS — MID SIZE VAN (M/L) AND COMPACT PICKUP (S/T)

4.3L (VIN W) ENGINE — DIAGNOSTIC TROUBLE CODE CHARTS — MID SIZE VAN (M/L) AND COMPACT PICKUP (S/T)

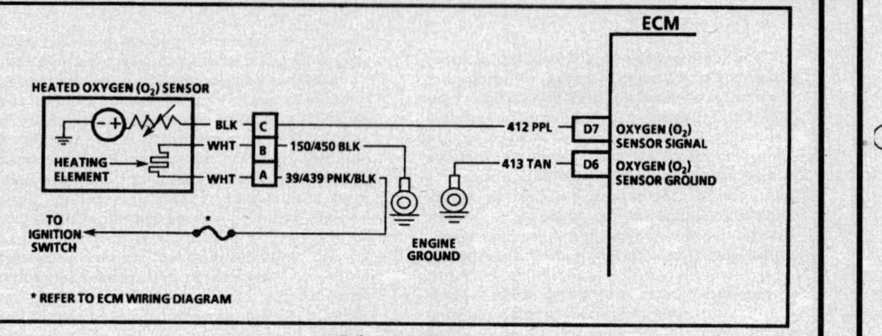

CODE 13
OXYGEN (O₂) SENSOR CIRCUIT
(OPEN CIRCUIT)
WITH 4.3L ENGINE WITHOUT 4L80-E TRANSMISSION

Circuit Description:

The ECM applies a bias voltage of approximately 450 millivolts (350-550 mV is normal bias voltage) between terminals "D7" and "D6". The Oxygen (O₂) sensor varies the voltage within a range of about 1 volt if the exhaust is rich, and down through about .010 volt if exhaust is lean. Code 13 is set when the voltage does not vary on CKT 412 within a predetermined amount of time.

The sensor is like an open circuit and produces no voltage when it is below 315°C (600°F). An open sensor circuit, or a cold sensor causes "Open Loop" operation.

Test Description:Test Description: Number(s) below refer to circled number(s) on the diagnostic chart.

1. Code 13 will set:
   - Engine has been running longer than 2 minutes.
   - Engine coolant temperature greater than 70°C (158°F).
   - Throttle position sensor signal has been above 5% for 60 seconds or longer.
   - O₂ signal voltage steady between .35 and .55 volt.
   If the conditions for a Code 13 or Code 63 exist, the system will not go into "Closed Loop."
2. This test checks the oxygen sensor's heating element. The heating element resistance should be 3.5 ohms at 20°C (68°F) or 14 ohms at 350°C (662°F).
3. This will determine if the sensor is at fault.
4. For this test use only a high impedance Digital Volt Ohmmeter (DVM). This test checks the continuity of CKT 412 and CKT 413. If CKT 413 is open, the ECM voltage on CKT 412 will be over .6 volt (600 mV).

5. If the fuse was open, check the ECM wiring diagram for complete wiring circuit.

Diagnostic Aids:

If the oxygen sensor heaters are not operating properly, system may go into "Open Loop" after extended idle.

Normal Tech 1 "Scan" tool voltage varies between 100 mV to 1000 mV (.1 and 1.0 volt), while in "Closed Loop." Code 13 sets in one minute if voltage remains between .35 and .55 volt, however the system will go "Open Loop" in about 15 seconds.

Refer to "Intermittents" in "Driveability Symptoms," Section "2".

An oxygen supply inside the O₂ sensor is necessary for proper O₂ sensor operation. This supply of oxygen is supplied through the O₂ sensor wires. All O₂ sensor wires and connections should be inspected for breaks or contamination that could prevent reference oxygen from reaching the O₂ sensor.

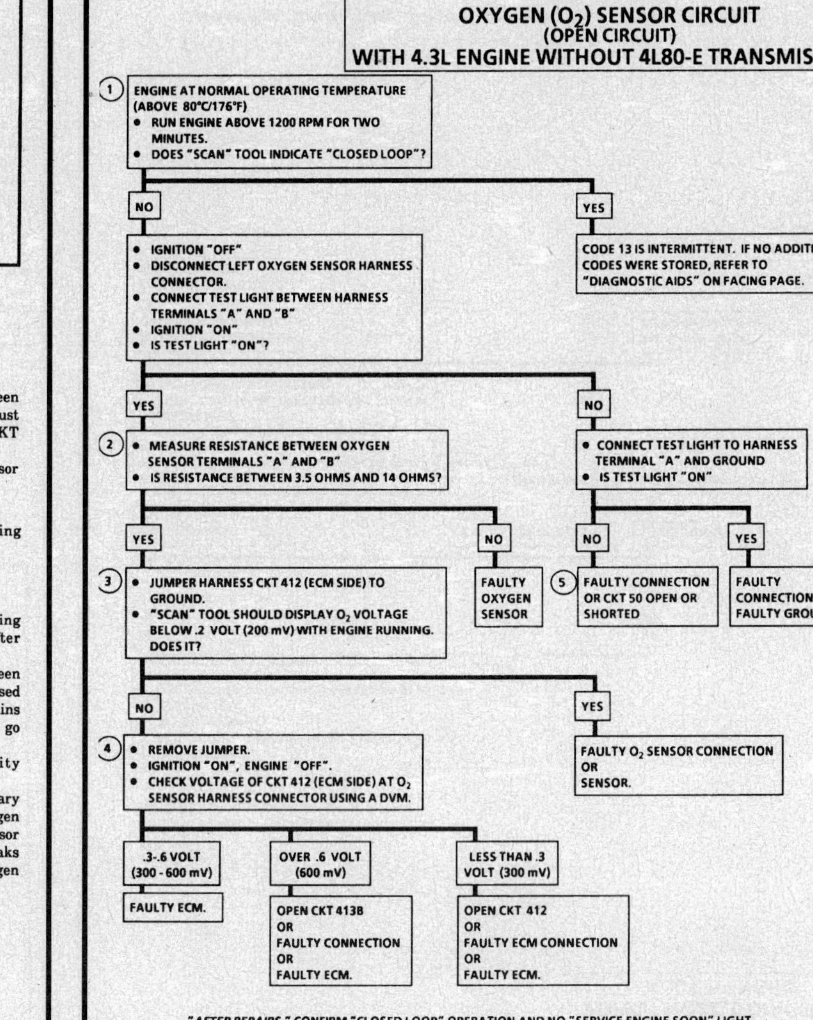

4.3L (VIN W) ENGINE — DIAGNOSTIC TROUBLE CODE CHARTS — MID SIZE VAN (M/L) AND COMPACT PICKUP (S/T)

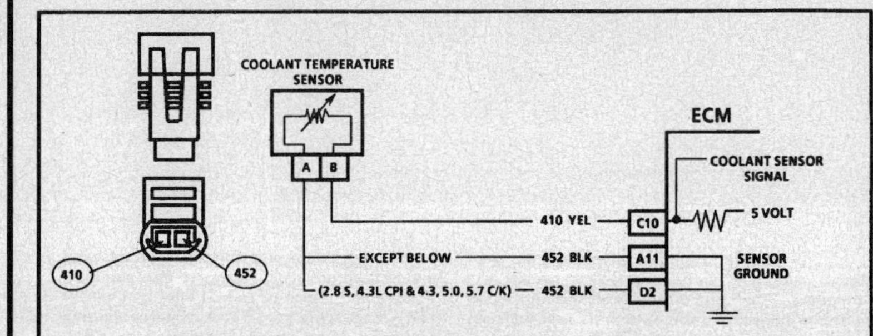

CODE 14
COOLANT TEMPERATURE SENSOR (CTS) CIRCUIT
(HIGH TEMPERATURE INDICATED)
WITHOUT 4L80-E TRANSMISSION

Circuit Description:

The Coolant Temperature Sensor (CTS) is a thermistor that controls the signal voltage to the ECM. The ECM applies a voltage on CKT 410 to the sensor. When the engine is cold, the sensor (thermistor) resistance is high, therefore the ECM will see high signal voltage.

As the engine warms, the sensor resistance becomes less and the voltage drops. At normal engine operating temperature (85°C to 95°C), the voltage will measure about 1.5 to 2.0 volts.

Test Description: Number(s) below refer to circled number(s) on the diagnostic chart.

1. Checks to see if code was set as result of a hard failure or intermittent condition.
   - Code 14 will set if:
     Signal voltage indicates a coolant temperature above 135°C (270°F) for 6 seconds.
2. This test simulates conditions for a Code 15. If the ECM recognizes the open circuit (high voltage) and displays a low temperature, the ECM and wiring are OK.

Diagnostic Aids:

Check harness routing for a potential short to ground in CKT 410.

Tech 1 "Scan" tool displays engine temperature in degrees centigrade. After engine is started, the temperature should rise steadily to about 90°C then stabilize when thermostat opens.

The "Temperature to Resistance Value" scale at the right may be used to test the coolant sensor at various temperature levels to evaluate the possibility of a "skewed" (mis-scaled) sensor. A "skewed" sensor could result in poor driveability complaints.

4.3L (VIN W) ENGINE — DIAGNOSTIC TROUBLE CODE CHARTS — MID SIZE VAN (M/L) AND COMPACT PICKUP (S/T)

CODE 14
COOLANT TEMPERATURE SENSOR (CTS) CIRCUIT
(HIGH TEMPERATURE INDICATED)
WITHOUT 4L80-E TRANSMISSION

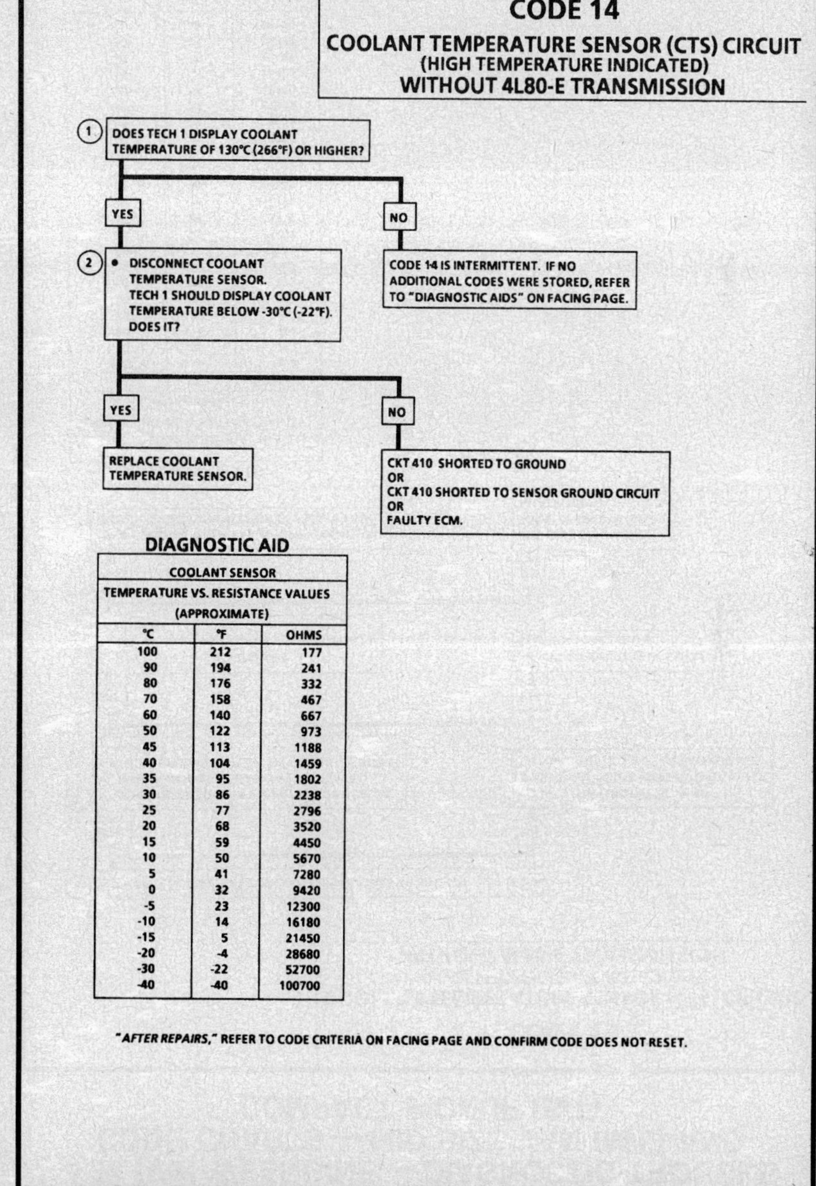

DIAGNOSTIC AID

COOLANT SENSOR		
TEMPERATURE VS. RESISTANCE VALUES (APPROXIMATE)		
°C	°F	OHMS
100	212	177
90	194	241
80	176	332
70	158	467
60	140	667
50	122	973
45	113	1188
40	104	1459
35	95	1802
30	86	2238
25	77	2796
20	68	3520
15	59	4450
10	50	5670
5	41	7280
0	32	9420
-5	23	12300
-10	14	16180
-15	5	21450
-20	-4	28680
-30	-22	52700
-40	-40	100700

"AFTER REPAIRS," REFER TO CODE CRITERIA ON FACING PAGE AND CONFIRM CODE DOES NOT RESET.

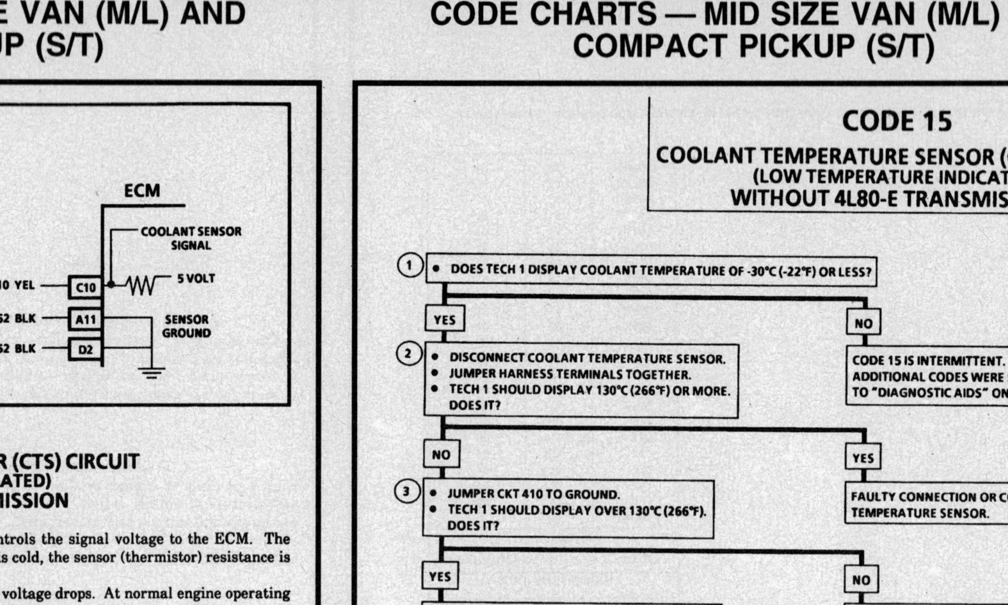

CODE 15
COOLANT TEMPERATURE SENSOR (CTS) CIRCUIT
(LOW TEMPERATURE INDICATED)
WITHOUT 4L80-E TRANSMISSION

Circuit Description:

The Coolant Temperature Sensor (CTS) is a thermistor that controls the signal voltage to the ECM. The ECM applies a voltage on CKT 410 to the sensor. When the engine is cold, the sensor (thermistor) resistance is high, therefore the ECM will see high signal voltage.

As the engine warms, the sensor resistance becomes less and the voltage drops. At normal engine operating temperature (85°C to 95°C), the voltage will measure about 1.5 to 2.0 volts.

Test Description: Number(s) below refer to circled number(s) on the diagnostic chart.

1. Checks to see if code was set as result of a hard failure or intermittent condition.
   - Code 15 will set if:
     Signal voltage indicates a coolant temperature less than -33°C (-27°F) for 30 seconds.
2. This test simulates a Code 14. If the ECM recognizes the low signal voltage (high temperature), and the Tech 1 "Scan" reads 130°C (266°F) or above, the ECM and wiring are OK.
3. This test will determine if CKT 410 is open. There should be 5 volts present at sensor connector if measured with a DVOM. This will determine if there is a wiring problem or a faulty ECM.

Diagnostic Aids:

A Tech 1 "Scan" tool reads engine temperature in degrees centigrade. After engine is started, the temperature should rise steadily to about 90°C then stabilize when thermostat opens.

A faulty connection, or an open in CKTs 410 or 452 will result in a Code 15.

The "Temperature To Resistance Value" scale at the right may be used to test the coolant sensor at various temperature levels to evaluate the possibility of a "skewed" (mis-scaled) sensor. A "skewed" sensor could result in poor driveability complaints.

CODE 15
COOLANT TEMPERATURE SENSOR (CTS) CIRCUIT
(LOW TEMPERATURE INDICATED)
WITHOUT 4L80-E TRANSMISSION

(1) • DOES TECH 1 DISPLAY COOLANT TEMPERATURE OF -30°C (-22°F) OR LESS?

YES →

(2) • DISCONNECT COOLANT TEMPERATURE SENSOR.
• JUMPER HARNESS TERMINALS TOGETHER.
• TECH 1 SHOULD DISPLAY 130°C (266°F) OR MORE. DOES IT?

NO →

(3) • JUMPER CKT 410 TO GROUND.
• TECH 1 SHOULD DISPLAY OVER 130°C (266°F). DOES IT?

YES →

OPEN COOLANT TEMPERATURE SENSOR GROUND CIRCUIT, FAULTY CONNECTION OR FAULTY ECM.

NO →

CODE 15 IS INTERMITTENT. IF NO ADDITIONAL CODES WERE STORED, REFER TO "DIAGNOSTIC AIDS" ON FACING PAGE.

YES →

FAULTY CONNECTION OR COOLANT TEMPERATURE SENSOR.

NO →

OPEN CKT 410, FAULTY CONNECTION AT ECM, OR FAULTY ECM.

DIAGNOSTIC AID

COOLANT SENSOR		
TEMPERATURE VS. RESISTANCE VALUES (APPROXIMATE)		
°C	°F	OHMS
100	212	177
90	194	241
80	176	332
70	158	467
60	140	667
50	122	973
45	113	1188
40	104	1459
35	95	1802
30	86	2238
25	77	2796
20	68	3520
15	59	4450
10	50	5670
5	41	7280
0	32	9420
-5	23	12300
-10	14	16180
-15	5	21450
-20	-4	28680
-30	-22	52700
-40	-40	100700

"AFTER REPAIRS," REFER TO CODE CRITERIA ON FACING PAGE AND CONFIRM CODE DOES NOT RESET.

4.3L (VIN W) ENGINE — DIAGNOSTIC TROUBLE CODE CHARTS — MID SIZE VAN (M/L) AND COMPACT PICKUP (S/T)

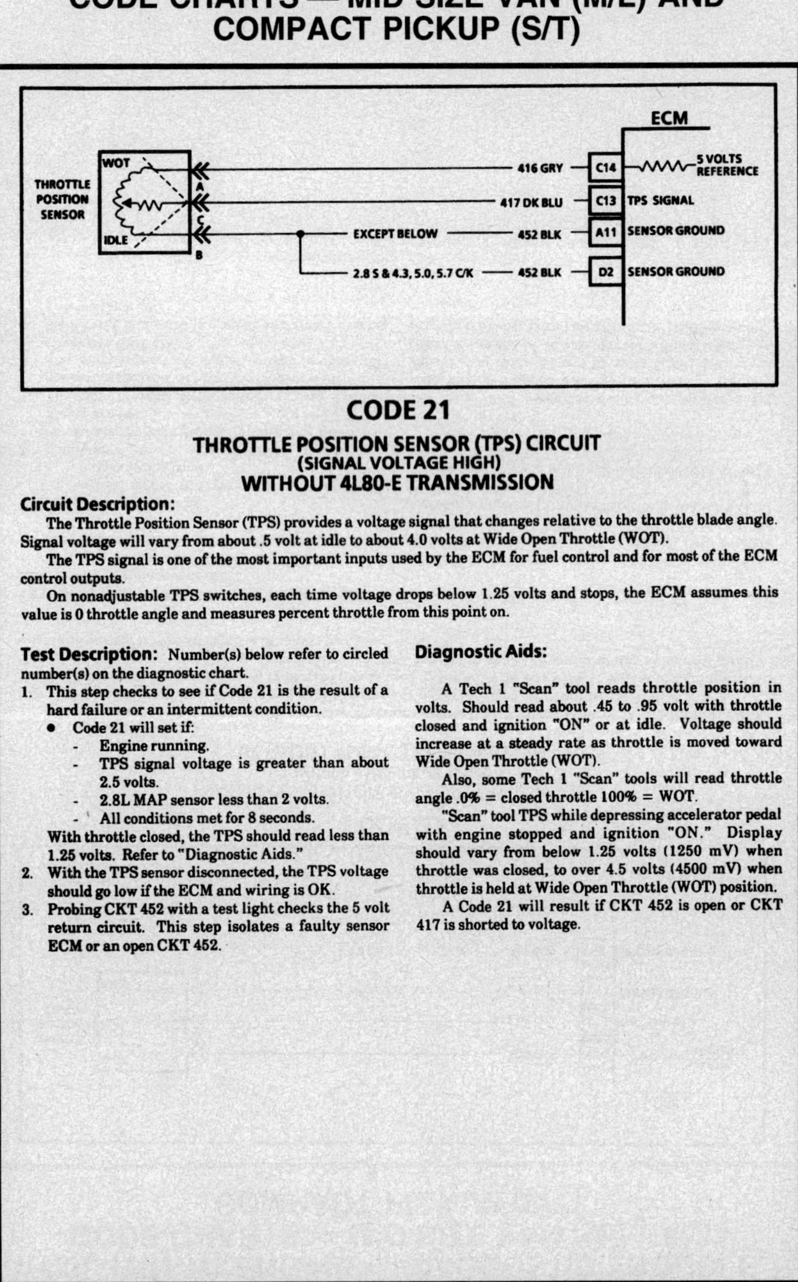

CODE 21
THROTTLE POSITION SENSOR (TPS) CIRCUIT
(SIGNAL VOLTAGE HIGH)
WITHOUT 4L80-E TRANSMISSION

Circuit Description:

The Throttle Position Sensor (TPS) provides a voltage signal that changes relative to the throttle blade angle. Signal voltage will vary from about .5 volt at idle to about 4.0 volts at Wide Open Throttle (WOT).

The TPS signal is one of the most important inputs used by the ECM for fuel control and for most of the ECM control outputs.

On nonadjustable TPS switches, each time voltage drops below 1.25 volts and stops, the ECM assumes this value is 0 throttle angle and measures percent throttle from this point on.

Test Description: Number(s) below refer to circled number(s) on the diagnostic chart.

1. This step checks to see if Code 21 is the result of a hard failure or an intermittent condition.
   - Code 21 will set if:
     - Engine running.
     - TPS signal voltage is greater than about 2.5 volts.
     - 2.8L MAP sensor less than 2 volts.
     - All conditions met for 8 seconds.

   With throttle closed, the TPS should read less than 1.25 volts. Refer to "Diagnostic Aids."
2. With the TPS sensor disconnected, the TPS voltage should go low if the ECM and wiring is OK.
3. Probing CKT 452 with a test light checks the 5 volt return circuit. This step isolates a faulty sensor ECM or an open CKT 452.

Diagnostic Aids:

A Tech 1 "Scan" tool reads throttle position in volts. Should read about .45 to .95 volt with throttle closed and ignition "ON" or at idle. Voltage should increase at a steady rate as throttle is moved toward Wide Open Throttle (WOT).

Also, some Tech 1 "Scan" tools will read throttle angle .0% = closed throttle 100% = WOT.

"Scan" tool TPS while depressing accelerator pedal with engine stopped and ignition "ON." Display should vary from below 1.25 volts (1250 mV) when throttle was closed, to over 4.5 volts (4500 mV) when throttle is held at Wide Open Throttle (WOT) position.

A Code 21 will result if CKT 452 is open or CKT 417 is shorted to voltage.

4.3L (VIN W) ENGINE — DIAGNOSTIC TROUBLE CODE CHARTS — MID SIZE VAN (M/L) AND COMPACT PICKUP (S/T)

CODE 21
THROTTLE POSITION SENSOR (TPS) CIRCUIT
(SIGNAL VOLTAGE HIGH)
WITHOUT 4L80-E TRANSMISSION

"AFTER REPAIRS," REFER TO CODE CRITERIA ON FACING PAGE AND CONFIRM CODE DOES NOT RESET.

4.3L (VIN W) ENGINE — DIAGNOSTIC TROUBLE CODE CHARTS — MID SIZE VAN (M/L) AND COMPACT PICKUP (S/T)

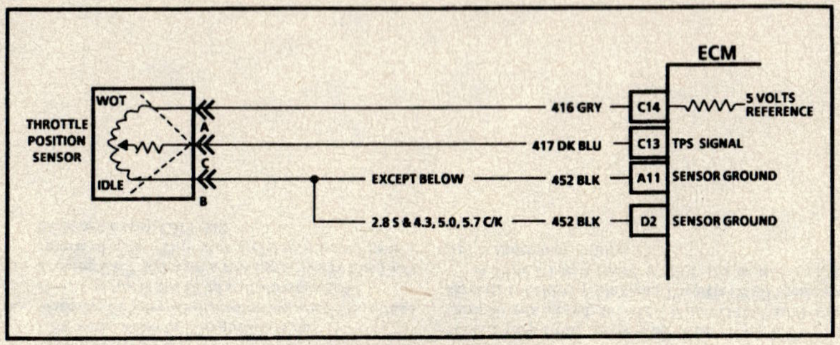

CODE 22

THROTTLE POSITION SENSOR (TPS) CIRCUIT
(SIGNAL VOLTAGE LOW)
WITHOUT 4L80-E TRANSMISSION

Circuit Description:

The Throttle Position sensor (TPS) provides a voltage signal that changes relative to the throttle blade. Signal voltage will vary from about .5 volt at idle to about 4.0 volts at Wide Open Throttle (WOT).

The TPS signal is one of the most important inputs used by the ECM for fuel control and for most of the ECM control outputs every time voltage drops below 1.25 volts and stops, the ECM assumes this valve is 0 throttle angle and measures percent throttle from this point on.

Test Description: Number(s) below refer to circled number(s) on the diagnostic chart.

1. This step checks to see if Code 22 is the result of a hard failure or an intermittent condition.
   - Code 22 will set if:
     - Engine running.
     - TPS signal voltage is less than about .2 volt for 2 seconds.
2. Simulates Code 21: (high voltage) - If the ECM recognizes the high signal voltage, the ECM and wiring are OK.
3. The ECM recognizes the voltage as over 4 volts, indicating the CKT 417 and the ECM are OK.
4. This simulates a high signal voltage to check for an open in CKT 417.
5. If CKT 416 is shorted to ground, there may also be a stored Code 34.

Diagnostic Aids:

A Tech 1 "Scan" tool reads throttle position in volts. Should read about .45 to .95 volt with throttle closed and ignition "ON" or at idle. Voltage should increase at a steady rate as throttle is moved toward Wide Open Throttle (WOT).

An open or short to ground in CKTs 416 or 417 will result in a Code 22.

"Scan" TPS while depressing accelerator pedal with engine stopped and ignition "ON". Display should vary from below 1.25 volts (1250 mV) when throttle was closed, to over 4.5 volts (4500 mV) when throttle is held at Wide Open Throttle (WOT) position.

4.3L (VIN W) ENGINE — DIAGNOSTIC TROUBLE CODE CHARTS — MID SIZE VAN (M/L) AND COMPACT PICKUP (S/T)

CODE 22

THROTTLE POSITION SENSOR (TPS) CIRCUIT
(SIGNAL VOLTAGE LOW)
WITHOUT 4L80-E TRANSMISSION

(1) • THROTTLE CLOSED.
DOES TECH 1 DISPLAY THROTTLE POSITION .2V (200 mV) OR BELOW?

- YES → (2)
- NO → • CODE 22 IS INTERMITTENT. IF NO ADDITIONAL CODES WERE STORED, REFER TO "DIAGNOSTIC AIDS" ON FACING PAGE.

(2) • DISCONNECT TPS SENSOR.
• JUMPER CKTs 416 & 417 TOGETHER.
TECH 1 SHOULD DISPLAY THROTTLE POSITION OVER 4.0V (4000 mV).
DOES IT?

- NO → (4)
- YES → (3) REPLACE TPS.

(4) • PROBE CKT 417 WITH A TEST LIGHT CONNECTED TO BATTERY VOLTAGE.
TECH 1 SHOULD DISPLAY THROTTLE POSITION OVER 4.0V (4000 mV).
DOES IT?

- YES → (5) CKT 416 OPEN OR SHORTED TO GROUND OR FAULTY CONNECTION OR FAULTY ECM.
- NO → CKT 417 OPEN OR SHORTED TO GROUND OR SHORTED TO THROTTLE POSITION SENSOR GROUND CIRCUIT OR FAULTY ECM CONNECTION OR FAULTY ECM.

"AFTER REPAIRS," REFER TO CODE CRITERIA ON FACING PAGE AND CONFIRM CODE DOES NOT RESET.

4.3L (VIN W) ENGINE — DIAGNOSTIC TROUBLE CODE CHARTS — MID SIZE VAN (M/L) AND COMPACT PICKUP (S/T)

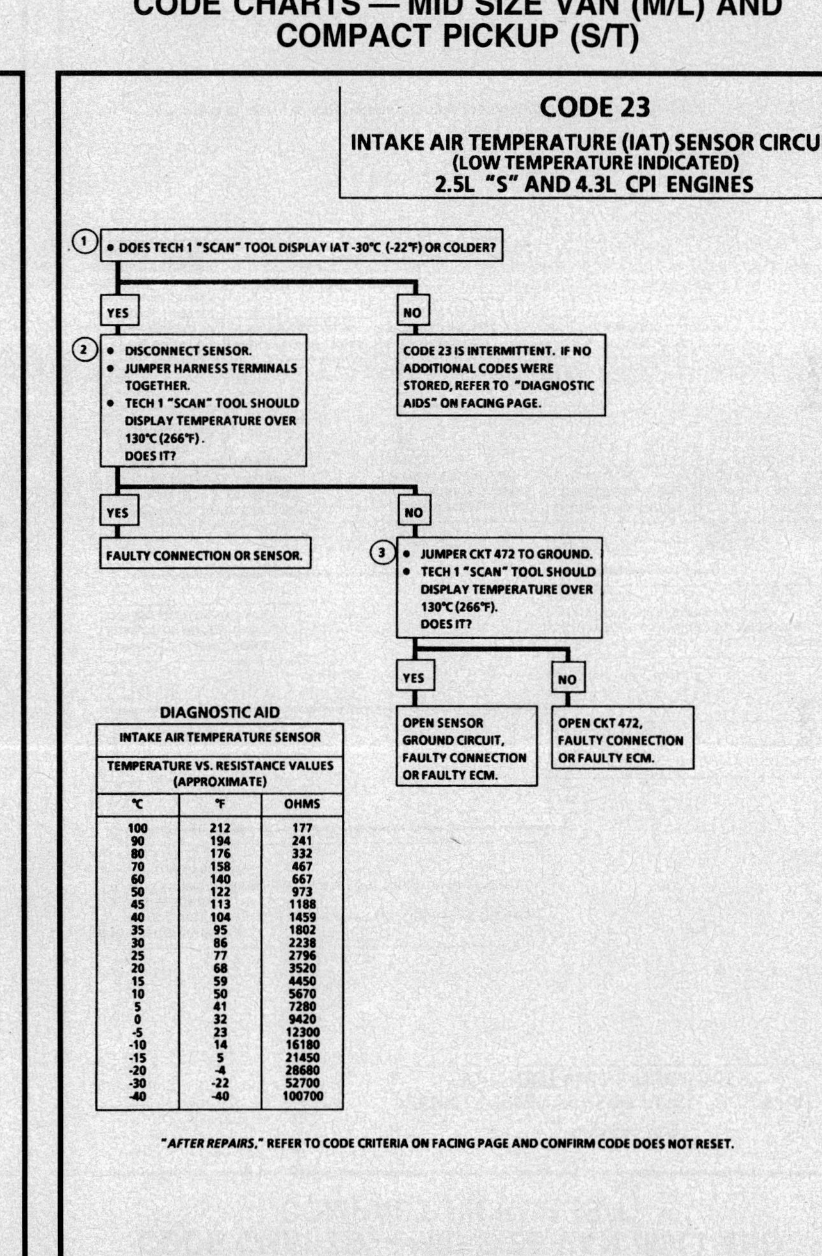

CODE 23
INTAKE AIR TEMPERATURE (IAT) SENSOR CIRCUIT
(LOW TEMPERATURE INDICATED)
2.5L "S" AND 4.3L CPI ENGINES

Circuit Description:

The Intake Air Temperature (IAT) sensor is a thermistor that controls the signal voltage to the ECM. The ECM applies a voltage (4-6 volts) on CKT 472 to the sensor. When the air is cold, the sensor (thermistor) resistance is high, therefore, the ECM will see a high signal voltage. If the air is warm, the sensor resistance is low, therefore, the ECM will see a low voltage.

Test Description: Number(s) below refer to circled number(s) on the diagnostic chart.
1. This step will determine if Code 33 is the result of a hard failure or an intermittent condition.
   - Code 23 will set if all conditions are met:
     - A signal voltage indicates a Intake Air Temperature (IAT) below -30°C (-22°F) for 12 seconds.
     - Time since engine start is 1 minute or longer.
     - No VSS (vehicle not moving)
2. A Code 23 will set, due to an open sensor, wire, or connection. This test will determine if the wiring and ECM are OK.
3. This will determine if the signal CKT 472 or the 5 volt return CKT 469/452 is open.

Diagnostic Aids:

A Tech 1 "Scan" tool reads temperature of the air entering the engine and should read close to ambient air temperature when engine is cold, and rises as underhood temperature increases.

Carefully check harness and connections for possible open CKT 472 or 469/452.

The "Temperature to Resistance Value" scale at the right may be used to test the IAT sensor at various temperature levels to evaluate the possibility of a "skewed" (mis-scaled) sensor. A "skewed" sensor could result in poor driveability complaints.

4.3L (VIN W) ENGINE — DIAGNOSTIC TROUBLE CODE CHARTS — MID SIZE VAN (M/L) AND COMPACT PICKUP (S/T)

CODE 23
INTAKE AIR TEMPERATURE (IAT) SENSOR CIRCUIT
(LOW TEMPERATURE INDICATED)
2.5L "S" AND 4.3L CPI ENGINES

DIAGNOSTIC AID

INTAKE AIR TEMPERATURE SENSOR		
TEMPERATURE VS. RESISTANCE VALUES (APPROXIMATE)		
°C	°F	OHMS
100	212	177
90	194	241
80	176	332
70	158	467
60	140	667
50	122	973
45	113	1188
40	104	1459
35	95	1802
30	86	2238
25	77	2796
20	68	3520
15	59	4450
10	50	5670
5	41	7280
0	32	9420
-5	23	12300
-10	14	16180
-15	5	21450
-20	-4	28680
-30	-22	52700
-40	-40	100700

"AFTER REPAIRS," REFER TO CODE CRITERIA ON FACING PAGE AND CONFIRM CODE DOES NOT RESET.

4.3L (VIN W) ENGINE — DIAGNOSTIC TROUBLE CODE CHARTS — MID SIZE VAN (M/L) AND COMPACT PICKUP (S/T)

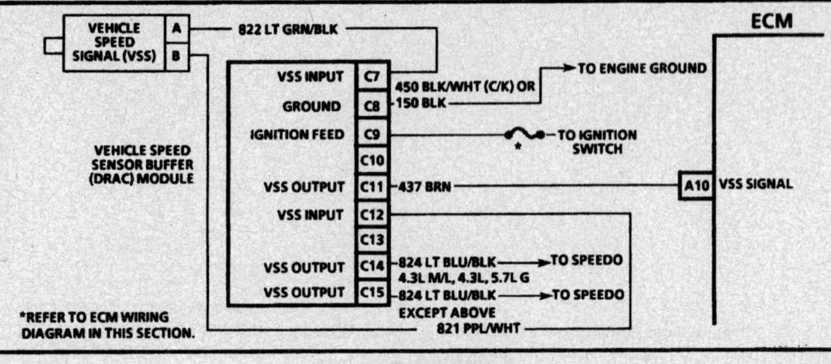

CODE 24

VEHICLE SPEED SENSOR (VSS) CIRCUIT FAULT
WITHOUT 4L80-E TRANSMISSION

Circuit Description:

The ECM applies and monitors 12 volts on CKT 437. CKT 437 connects to the Vehicle Speed Sensor (VSS) buffer (DRAC) module, which alternately grounds CKT 437, when receiving voltage pulses from Vehicle Speed Sensor (VSS) when drive wheels are turning. This pulsing action takes place about 2000 times per mile and the ECM will calculate vehicle speed based on the time between "pulses."

A Tech 1 "Scan" tool reading should closely match the speedometer reading with drive wheels turning.

Test Description: Number(s) below refer to circled number(s) on the diagnostic chart.

1. Code 24 will set if:
   - CKT 437 voltage is constant.
   - Engine speed is more than 1200 rpm.
   - Vehicle speed signal indicates less than 2 mph (3 km/h) on Tech 1.
   - Automatic transmission in drive.
   - All conditions must be met for 5 seconds.

   These conditions are met during a road load deceleration except 2.8L which sets on acceleration or at highway speed.

2. This test determines if the DRAC is receiving the A/C signal from the VSS.

3. This test monitors the DRAC voltage on CKT 437. With the wheels turning, the pulsing action will result in a varying voltage. The variation will be greater at low wheel speeds to an average of 4-6 volts at about 20 mph (32 km/h).

Diagnostic Aids:

1. Tech 1 "Scan" tool reading should closely match speedometer reading, with drive wheels turning.
2. Check Park/Neutral (P/N) switch diagnosis chart if vehicle equipped with automatic transmission.
3. If Park/Neutral (P/N) switch is OK, refer to "ECM Intermittent Codes or Performance"

4.3L (VIN W) ENGINE — DIAGNOSTIC TROUBLE CODE CHARTS — MID SIZE VAN (M/L) AND COMPACT PICKUP (S/T)

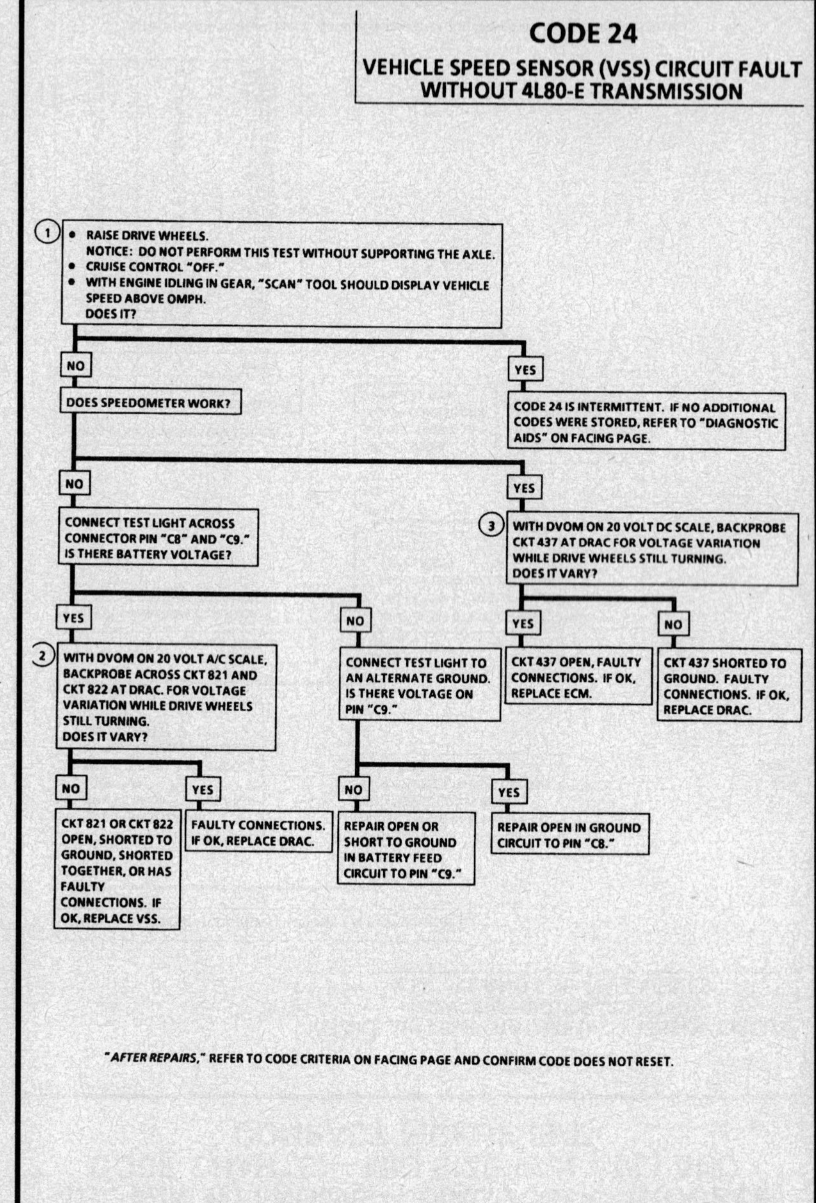

4.3L (VIN W) ENGINE — DIAGNOSTIC TROUBLE CODE CHARTS — MID SIZE VAN (M/L) AND COMPACT PICKUP (S/T)

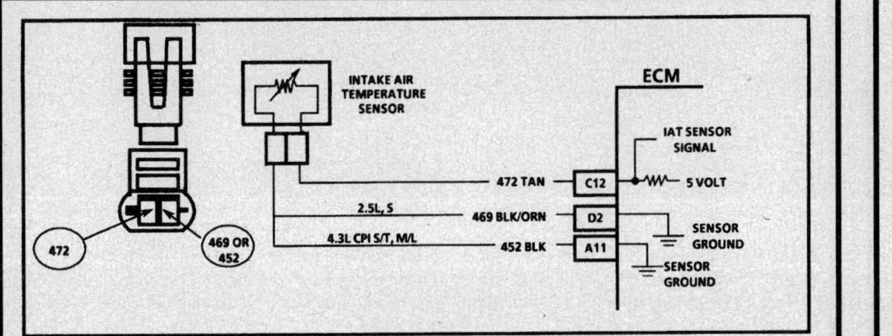

CODE 25

INTAKE AIR TEMPERATURE (IAT) SENSOR CIRCUIT
(HIGH TEMPERATURE INDICATED)
2.5L AND 4.3L CPI ENGINES

Circuit Description:

The Intake Air Temperature (IAT) sensor is a thermistor that controls the signal voltage to the ECM. The ECM applies a voltage (4-6 volts) on CKT 472 to the sensor. When the air is cold, the sensor (thermistor) resistance is high, therefore, the ECM will see a high signal voltage. As the air warms, the sensor resistance is low, therefore, the ECM will see a low voltage.

Test Description: Number(s) below refer to circled number(s) on the diagnostic chart.

1. Code 25 will set if:
   - Signal voltage indicates a Intake Air Temperature (IAT) above 150°C (302°F) for 2 seconds.
   - Time since engine start is 1 minute or longer.
   - A vehicle speed is present.

Diagnostic Aids:

A Tech 1 "Scan" tool reads temperature of the air entering the engine and should read close to ambient air temperature when engine is cold, and rises as underhood temperature increases.

Check harness routing for possible short to ground in CKT 472.

The "Temperature to Resistance Value" scale at the right may be used to test the IAT sensor at various temperature levels to evaluate the possibility of a "skewed" (mis-scaled) sensor. A "skewed" sensor could result in poor driveability complaints.

4.3L (VIN W) ENGINE — DIAGNOSTIC TROUBLE CODE CHARTS — MID SIZE VAN (M/L) AND COMPACT PICKUP (S/T)

CODE 25

INTAKE AIR TEMPERATURE (IAT) SENSOR CIRCUIT
(HIGH TEMPERATURE INDICATED)
2.5L AND 4.3L CPI ENGINES

1. DOES TECH 1 "SCAN" TOOL DISPLAY IAT OF 145°C (293°F) OR HOTTER?

YES
- DISCONNECT SENSOR. TECH 1 "SCAN" TOOL SHOULD DISPLAY TEMPERATURE BELOW -30°C (-22°F). DOES IT?

NO
CODE 25 IS INTERMITTENT. IF NO ADDITIONAL CODES WERE STORED, REFER TO "DIAGNOSTIC AIDS" ON FACING PAGE.

YES
REPLACE SENSOR.

NO
CKT 472 SHORTED TO GROUND
OR
TO SENSOR GROUND
OR
ECM IS FAULTY.

DIAGNOSTIC AID

INTAKE AIR TEMPERATURE SENSOR		
TEMPERATURE VS. RESISTANCE VALUES (APPROXIMATE)		
°C	°F	OHMS
100	212	177
90	194	241
80	176	332
70	158	467
60	140	667
50	122	973
45	113	1188
40	104	1459
35	95	1802
30	86	2238
25	77	2796
20	68	3520
15	59	4450
10	50	5670
5	41	7280
0	32	9420
-5	23	12300
-10	14	16180
-15	5	21450
-20	-4	28680
-30	-22	52700
-40	-40	100700

"AFTER REPAIRS," REFER TO CODE CRITERIA ON FACING PAGE AND CONFIRM CODE DOES NOT RESET.

4.3L (VIN W) ENGINE — DIAGNOSTIC TROUBLE CODE CHARTS — MID SIZE VAN (M/L) AND COMPACT PICKUP (S/T)

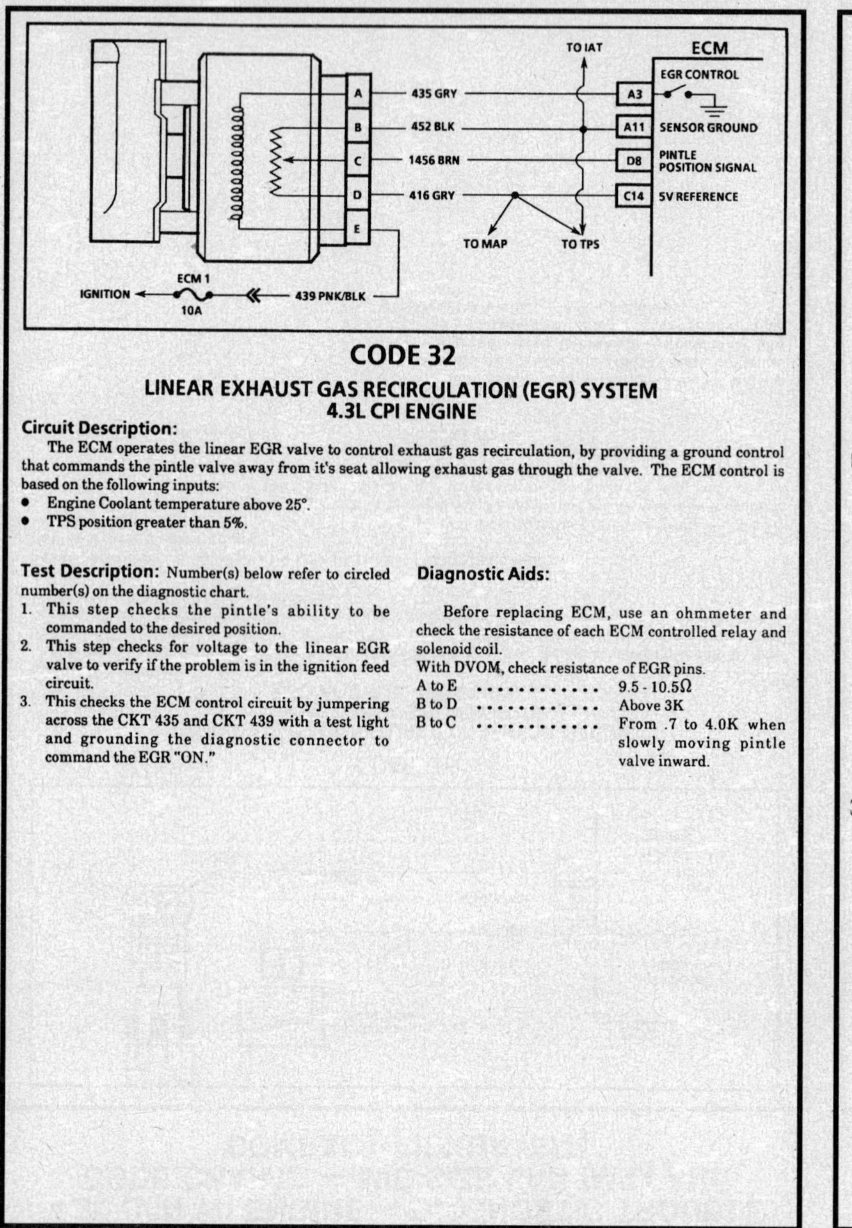

CODE 32
LINEAR EXHAUST GAS RECIRCULATION (EGR) SYSTEM
4.3L CPI ENGINE

Circuit Description:

The ECM operates the linear EGR valve to control exhaust gas recirculation, by providing a ground control that commands the pintle valve away from it's seat allowing exhaust gas through the valve. The ECM control is based on the following inputs:

- Engine Coolant temperature above 25°.
- TPS position greater than 5%.

Test Description: Number(s) below refer to circled number(s) on the diagnostic chart.

1. This step checks the pintle's ability to be commanded to the desired position.
2. This step checks for voltage to the linear EGR valve to verify if the problem is in the ignition feed circuit.
3. This checks the ECM control circuit by jumpering across the CKT 435 and CKT 439 with a test light and grounding the diagnostic connector to command the EGR "ON."

Diagnostic Aids:

Before replacing ECM, use an ohmmeter and check the resistance of each ECM controlled relay and solenoid coil.

With DVOM, check resistance of EGR pins.

- A to E .... 9.5 - 10.5Ω
- B to D .... Above 3K
- B to C .... From .7 to 4.0K when slowly moving pintle valve inward.

4.3L (VIN W) ENGINE — DIAGNOSTIC TROUBLE CODE CHARTS — MID SIZE VAN (M/L) AND COMPACT PICKUP (S/T)

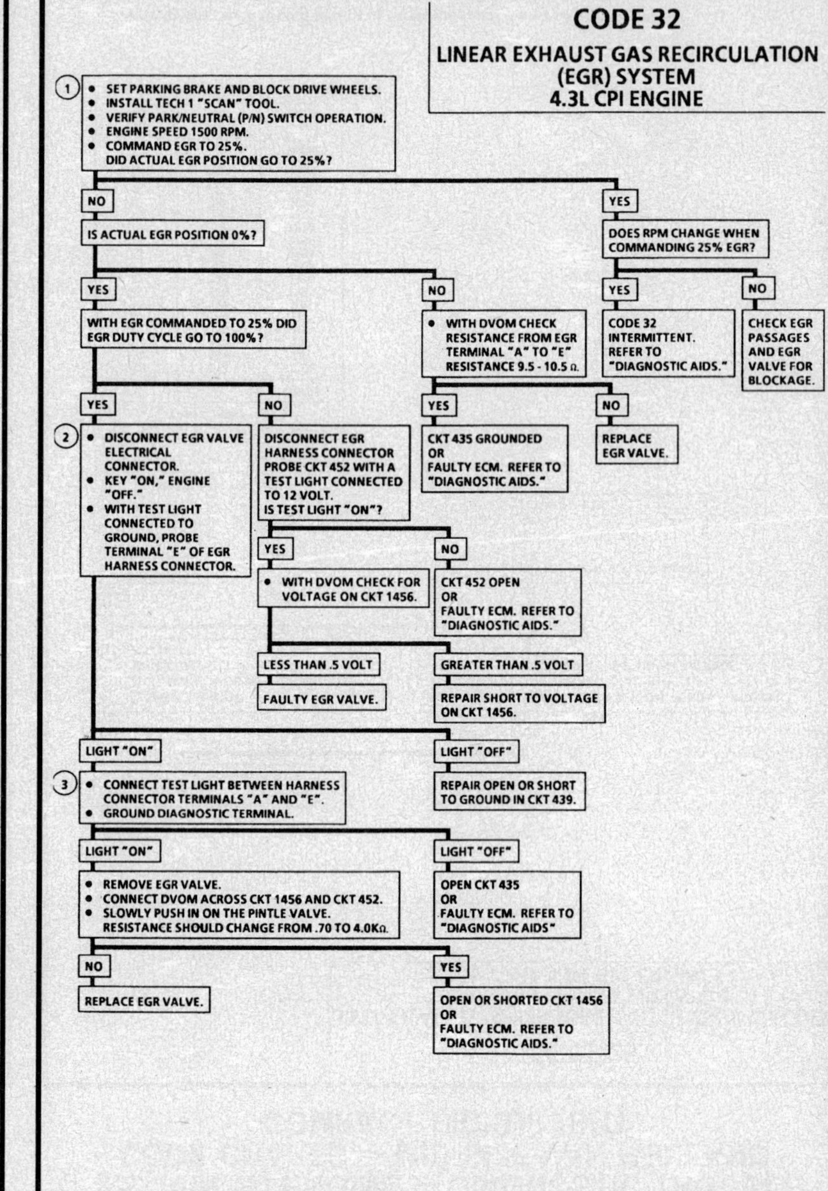

4.3L (VIN W) ENGINE — DIAGNOSTIC TROUBLE CODE CHARTS — MID SIZE VAN (M/L) AND COMPACT PICKUP (S/T)

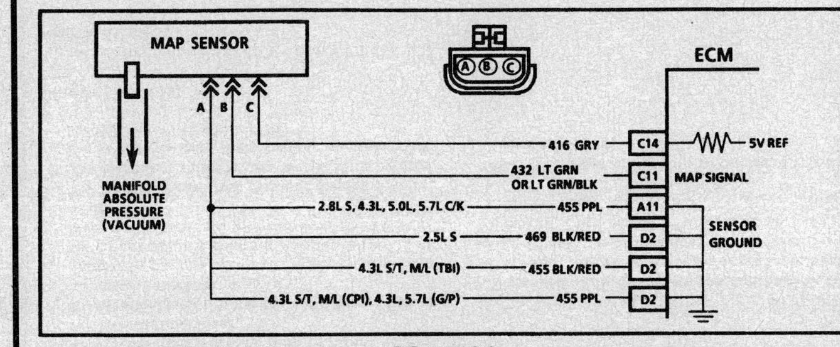

CODE 33
MANIFOLD ABSOLUTE PRESSURE (MAP) SENSOR CIRCUIT
(SIGNAL VOLTAGE HIGH - LOW VACUUM)
ALL VEHICLES WITHOUT 4L80-E TRANSMISSION

Circuit Description:

The Manifold Absolute Pressure (MAP) sensor responds to changes in manifold pressure (vacuum). The ECM receives this information as a signal voltage that will vary from about 1-1.5 volts at closed throttle (idle) to 4-4.6 volts at Wide Open Throttle (WOT) (low vacuum).

If the MAP sensor fails, the ECM will substitute a fixed MAP value and use the Throttle Position Sensor (TPS) to control fuel delivery.

Test Description: Number(s) below refer to circled number(s) on the diagnostic chart.

1. This step will determine if Code 33 is the result of a hard failure or an intermittent condition.
   A Code 33 will set under the following conditions:
   - MAP signal voltage is too high (low vacuum).
   - TPS less than 4%.
   - These conditions exist longer than 5 seconds.
   - Engine misfire or a low unstable idle may set Code 33.
2. This step simulates conditions for a Code 34. If the ECM recognizes the change, the ECM and CKT 416 and CKT 432 are OK. If CKT 455/469 is open, there may also be a stored Code 23.

A Code 33 will result if CKT 455/469 is open or if CKT 432 is shorted to voltage or to CKT 416.

- Check all connections.
- Disconnect sensor from bracket and twist sensor connections. Output changes greater than .1 volt indicates a bad connector or connection. If OK, replace sensor.

NOTICE: Make sure electrical connector remains securely fastened.

Diagnostic Aids:

With the ignition "ON" and the engine stopped, the manifold pressure is equal to atmospheric pressure and the signal voltage will be high. This information is used by the ECM as an indication of vehicle altitude. Comparison of this reading with a known good vehicle with the same sensor is a good way to check accuracy of a "suspect" sensor. Readings should be the same ± .4 volt.

4.3L (VIN W) ENGINE — DIAGNOSTIC TROUBLE CODE CHARTS — MID SIZE VAN (M/L) AND COMPACT PICKUP (S/T)

CODE 33
MANIFOLD ABSOLUTE PRESSURE (MAP) SENSOR CIRCUIT
(SIGNAL VOLTAGE HIGH - LOW VACUUM)
ALL VEHICLES WITHOUT 4L80-E TRANSMISSION

(1)
- IF ENGINE IDLE IS ROUGH, UNSTABLE OR INCORRECT OR IF MANIFOLD VACUUM, AT IDLE, IS BELOW 15", CORRECT BEFORE USING CHART. SEE "SYMPTOMS" IN SECTION "2."
- ENGINE IDLING.
- DOES "SCAN" DISPLAY A MAP OF 4.0 VOLTS OR OVER?

YES →

(2)
- IGNITION "OFF."
- DISCONNECT MAP SENSOR ELECTRICAL CONNECTOR.
- IGNITION "ON."
- "SCAN" SHOULD READ A VOLTAGE OF 1 VOLT OR LESS. DOES IT?

NO → CODE 33 IS INTERMITTENT. IF NO ADDITIONAL CODES WERE STORED, REFER TO "DIAGNOSTIC AIDS."

YES →
- PROBE CKT 455/469 WITH A TEST LIGHT TO 12 VOLT.
- TEST LIGHT SHOULD LIGHT. DOES IT?

NO → CKT 432 SHORTED TO VOLTAGE OR SHORTED TO CKT 416 OR FAULTY ECM.

YES → PLUGGED OR LEAKING SENSOR VACUUM HOSE OR FAULTY MAP SENSOR. REFER TO "DIAGNOSTIC AIDS."

NO → OPEN CIRCUIT 455/469.

"AFTER REPAIRS," REFER TO CODE CRITERIA ON FACING PAGE AND CONFIRM CODE DOES NOT RESET.

4.3L (VIN W) ENGINE — DIAGNOSTIC TROUBLE CODE CHARTS — MID SIZE VAN (M/L) AND COMPACT PICKUP (S/T)

CODE 34

MANIFOLD ABSOLUTE PRESSURE (MAP) SENSOR CIRCUIT
(SIGNAL VOLTAGE LOW - HIGH VACUUM)
ALL VEHICLES WITHOUT 4L80-E TRANSMISSION

Circuit Description:

The Manifold Absolute Pressure (MAP) sensor responds to changes in manifold pressure (vacuum). The ECM receives this information as a signal voltage that will vary from about 1-1.5 volts at idle to 4-4.6 volts at Wide Open Throttle (WOT).

A Tech 1 "Scan" tool displays manifold pressure in volts. Low pressure (high vacuum) reads a low voltage while a high pressure (low vacuum) reads a high voltage.

If the MAP sensor fails the ECM will substitute a fixed MAP value and use the Throttle Position Sensor (TPS) to control fuel delivery.

Test Description: Number(s) below refer to circled number(s) on the diagnostic chart.

1. This step determines if Code 34 is the result of a hard failure or an intermittent condition. A Code 34 will set when:
   - When engine is less than 1200 rpm.
   - Manifold pressure reading less than 14 kPa, conditions met for 1 second.
   OR
   - Engine speed is greater than 1200 rpm.
   - Throttle angle over 21%.
   - Manifold pressure less than 14 kPa, conditions met for 1 second.
2. Jumpering harness terminals "B" to "C" (5 volts to signal circuit) will determine if the sensor is at fault, or if there is a problem with the ECM or wiring.
3. The Tech 1 "Scan" tool may not display 5 volts. The important thing is that the ECM recognized the voltage as more than 4 volts, indicating that the ECM and CKT 432 are OK.

With the ignition "ON" and the engine "OFF," the manifold pressure is equal to atmospheric pressure and the signal voltage will be high. This information is used by the ECM as an indication of vehicle altitude.

Comparison of this reading with a known good vehicle with the same sensor is a good way to check accuracy of a "suspect" sensor. Reading should be the same ± .4 volt. Also, MAP output check in "Computer Command Control,"

- Disconnect sensor from bracket and twist sensor by hand (only) to check for intermittent connections. Output changes greater than .1 volt indicates a bad connector or connection. If OK, replace sensor.

NOTICE: Make sure electrical connector remains securely fastened.

Diagnostic Aids:

An intermittent open in CKT 432 or CKT 416 will result in a Code 34.

CODE 34

MANIFOLD ABSOLUTE PRESSURE (MAP) SENSOR CIRCUIT
(SIGNAL VOLTAGE LOW - HIGH VACUUM)
ALL VEHICLES WITHOUT 4L80-E TRANSMISSION

1. ENGINE IDLING.
   DOES TECH 1 DISPLAY MAP VOLTAGE BELOW .25 VOLT?

   - **YES** →
     2. IGNITION "OFF."
        - DISCONNECT SENSOR ELECTRICAL CONNECTOR.
        - JUMPER HARNESS TERMINALS "B" TO "C".
        - IGNITION "ON."
        - MAP VOLTAGE SHOULD READ OVER 4.7 VOLTS. DOES IT?

        - **NO** →
          3. IGNITION "OFF."
             - REMOVE JUMPER WIRE.
             - PROBE TERMINAL "B" (CKT 432) WITH A TEST LIGHT TO BATTERY VOLTAGE.
             - IGNITION "ON."
             - TECH 1 SHOULD READ OVER 4 VOLTS. DOES IT?

             - **YES** → 5 VOLT REFERENCE CIRCUIT OPEN OR SHORTED TO GROUND OR FAULTY ECM.
             - **NO** → CKT 432 OPEN OR CKT 432 SHORTED TO GROUND OR CKT 432 SHORTED TO SENSOR GROUND OR FAULTY ECM.

        - **YES** → FAULTY CONNECTION OR SENSOR.

   - **NO** → CODE 34 IS INTERMITTENT. IF NO ADDITIONAL CODES WERE STORED, REFER TO "DIAGNOSTIC AIDS" ON FACING PAGE.

"AFTER REPAIRS," REFER TO CODE CRITERIA ON FACING PAGE AND CONFIRM CODE DOES NOT RESET.

4.3L (VIN W) ENGINE — DIAGNOSTIC TROUBLE CODE CHARTS — MID SIZE VAN (M/L) AND COMPACT PICKUP (S/T)

CODE 42
ELECTRONIC SPARK TIMING (EST)
ALL VEHICLES WITHOUT 4L80-E TRANSMISSION

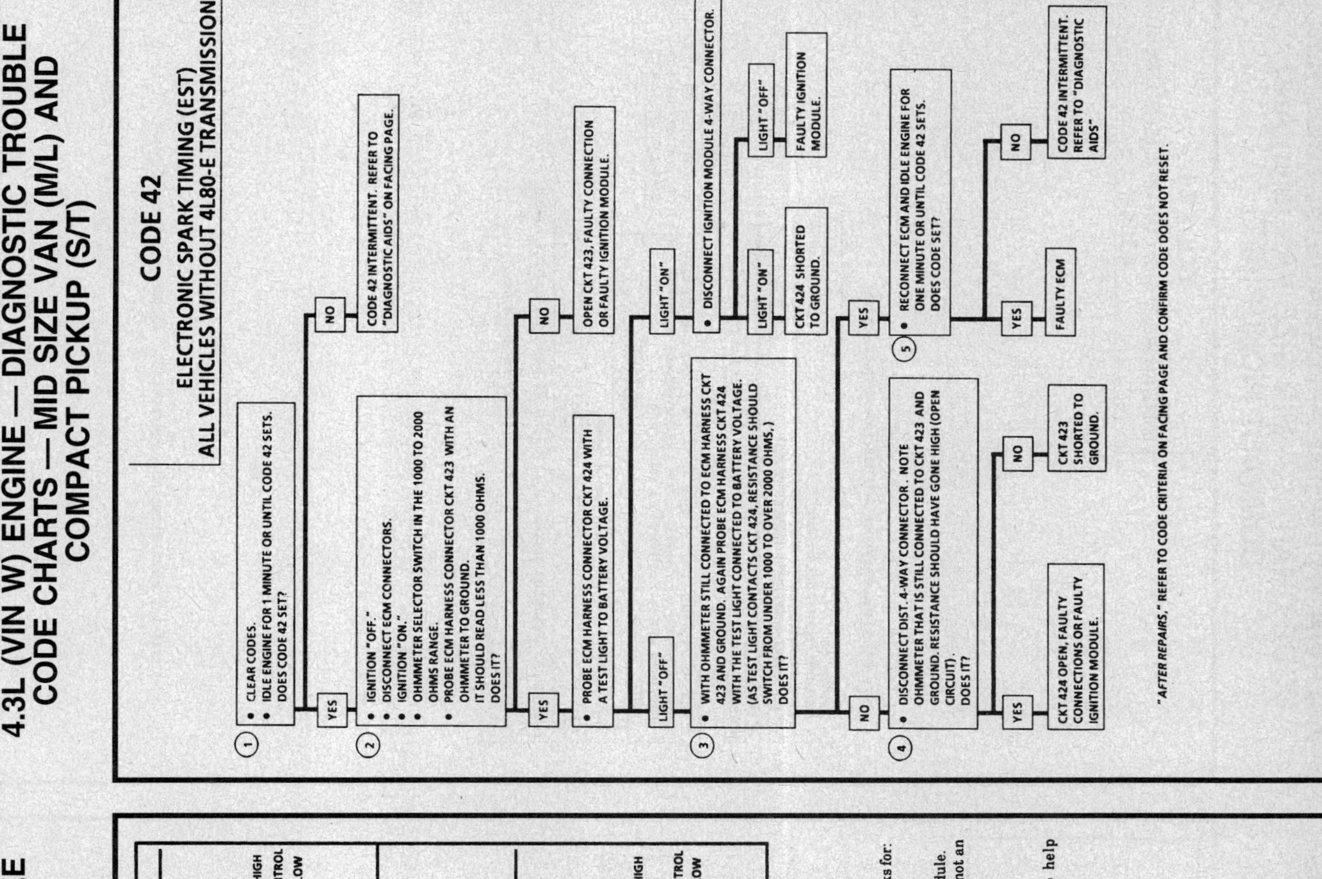

4.3L (VIN W) ENGINE — DIAGNOSTIC TROUBLE CODE CHARTS — MID SIZE VAN (M/L) AND COMPACT PICKUP (S/T)

CODE 42
ELECTRONIC SPARK TIMING (EST)
ALL VEHICLES WITHOUT 4L80-E TRANSMISSION

Test Description: Number(s) below refer to circled number(s) on the diagnostic chart.

1. Code 42 means the ECM has seen an open or short to ground in the EST or bypass circuits. This test confirms Code 42 and that the fault causing the code is present.

2. Checks for a normal EST ground path through the ignition module. An EST CKT 423 shorted to ground will also read less than 500 ohms; however, this will be checked later.

3. As the test light voltage touches CKT 424, the module should switch causing the ohmmeter to "overrange" if the meter is in the 100-200 ohm position. Selecting the 10-20,000 ohms position will indicate above 5000 ohms. The important thing is that the module "switched."

4. The module did not switch and this step checks for:
   - EST CKT 423 shorted to ground.
   - Bypass CKT 424 open.
   - Faulty ignition module connection or module.

5. Confirms that Code 42 is a faulty ECM and not an intermittent in CKTs 423 or 424.

Diagnostic Aids:

The Tech 1 does not have any ability to help diagnose a Code 42 problem.

4.3L (VIN W) ENGINE — DIAGNOSTIC TROUBLE CODE CHARTS — MID SIZE VAN (M/L) AND COMPACT PICKUP (S/T)

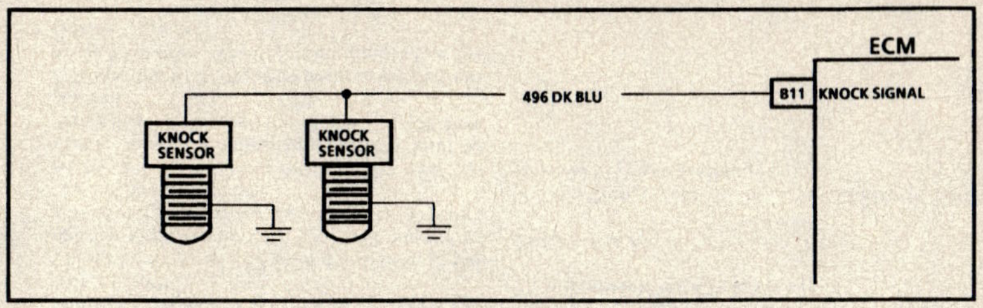

CODE 43
ELECTRONIC SPARK CONTROL (ESC) CIRCUIT
4.3L CPI

Circuit Description:
The Code 43 circuit consists of two knock sensors with one wire that is spliced together and goes directly to the ECM. There are two Code 43 checks performed by the ECM. One check consists on monitoring CKT 496 for a voltage that is above .63 volts and below 4.38 volts.

If voltage is either too high or too low, for 2 seconds, Code 43 will set. Once engine temperature reaches 20°C increase from start up or 85°C, MAP reading above 83 kPa and engine speed below 3800 rpm, the ECM will perform a self check by advancing the timing incrementally up to additional 23° advance while anticipating a knock signal. If no knock signal is received during the test, Code 43 will set.

Test Description: Number(s) below refer to circled number(s) on the diagnostic chart.
1. If an audible knock is heard form the engine, repair the internal engine problem as normally no knock should be detected at idle.
The ECM applies 5 volts on CKT 496 which should be present at the knock sensor terminals when the sensors are disconnected.
2. This test determines if the wiring or if the ESC portion of the MEM-CAL is faulty.
3. An improperly installed knock sensor can prevent the knock sensor from grounding to the block.

Diagnostic Aids:
Check CKT 496 for a potential open or short to ground.
Also, check for proper installation of MEM-CAL.

4.3L (VIN W) ENGINE — DIAGNOSTIC TROUBLE CODE CHARTS — MID SIZE VAN (M/L) AND COMPACT PICKUP (S/T)

CODE 43
ELECTRONIC SPARK CONTROL (ESC) CIRCUIT
4.3L CPI

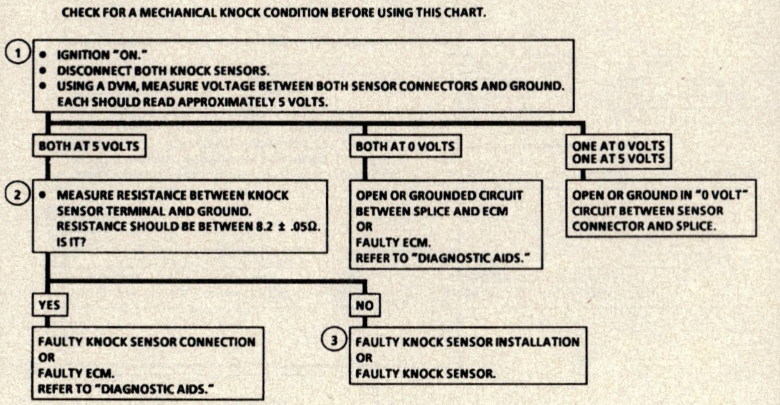

4.3L (VIN W) ENGINE — DIAGNOSTIC TROUBLE CODE CHARTS — MID SIZE VAN (M/L) AND COMPACT PICKUP (S/T)

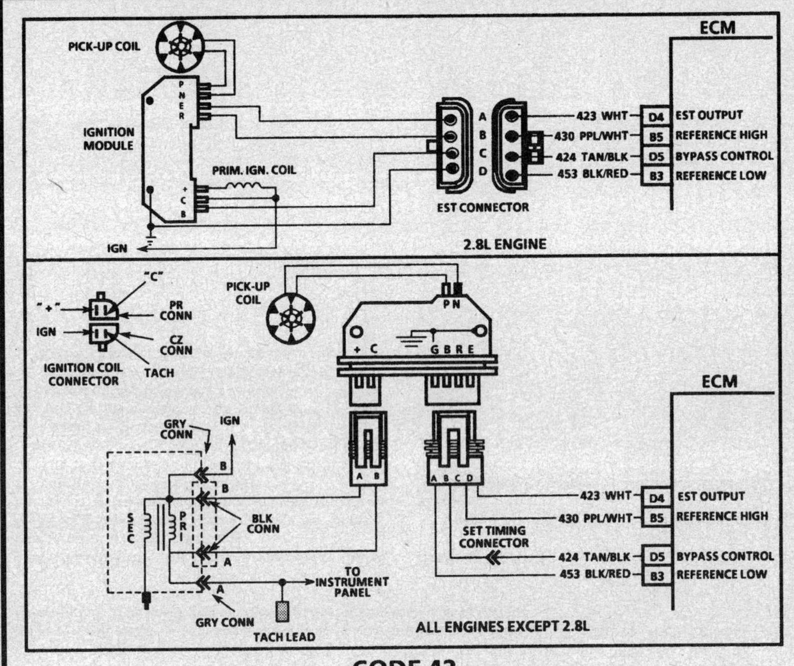

CODE 42
ELECTRONIC SPARK TIMING (EST)
ALL VEHICLES WITHOUT 4L80-E TRANSMISSION

Test Description: Number(s) below refer to circled number(s) on the diagnostic chart.

1. Code 42 means the ECM has seen an open or short to ground in the EST or bypass circuits. This test confirms Code 42 and that the fault causing the code is present.
2. Checks for a normal EST ground path through the ignition module. An EST CKT 423 shorted to ground will also read less than 500 ohms; however, this will be checked later.
3. As the test light voltage touches CKT 424, the module should switch causing the ohmmeter to "overrange" if the meter is in the 100-200 ohm position.
   Selecting the 10-20,000 ohms position will indicate above 5000 ohms. The important thing is that the module "switched."
4. The module did not switch and this step checks for:
   - EST CKT 423 shorted to ground.
   - Bypass CKT 424 open.
   - Faulty ignition module connection or module.
5. Confirms that Code 42 is a faulty ECM and not an intermittent in CKTs 423 or 424.

Diagnostic Aids:

The Tech 1 does not have any ability to help diagnose a Code 42 problem.

4.3L (VIN W) ENGINE — DIAGNOSTIC TROUBLE CODE CHARTS — MID SIZE VAN (M/L) AND COMPACT PICKUP (S/T)

CODE 42
ELECTRONIC SPARK TIMING (EST)
ALL VEHICLES WITHOUT 4L80-E TRANSMISSION

1. - CLEAR CODES.
   - IDLE ENGINE FOR 1 MINUTE OR UNTIL CODE 42 SETS. DOES CODE 42 SET?
   - **YES** → (go to 2)
   - **NO** → CODE 42 INTERMITTENT. REFER TO "DIAGNOSTIC AIDS" ON FACING PAGE.

2. - IGNITION "OFF."
   - DISCONNECT ECM CONNECTORS.
   - IGNITION "ON."
   - OHMMETER SELECTOR SWITCH IN THE 1000 TO 2000 OHMS RANGE.
   - PROBE ECM HARNESS CONNECTOR CKT 423 WITH AN OHMMETER TO GROUND. IT SHOULD READ LESS THAN 1000 OHMS. DOES IT?
   - **YES** → PROBE ECM HARNESS CONNECTOR CKT 424 WITH A TEST LIGHT TO BATTERY VOLTAGE.
     - **LIGHT "OFF"** → (go to 3)
     - **LIGHT "ON"** → DISCONNECT IGNITION MODULE 4-WAY CONNECTOR.
       - **LIGHT "ON"** → CKT 424 SHORTED TO GROUND.
       - **LIGHT "OFF"** → FAULTY IGNITION MODULE.
   - **NO** → OPEN CKT 423, FAULTY CONNECTION OR FAULTY IGNITION MODULE.

3. - WITH OHMMETER STILL CONNECTED TO ECM HARNESS CKT 423 AND GROUND. AGAIN PROBE ECM HARNESS CKT 424 WITH THE TEST LIGHT CONNECTED TO BATTERY VOLTAGE. (AS TEST LIGHT CONTACTS CKT 424, RESISTANCE SHOULD SWITCH FROM UNDER 1000 TO OVER 2000 OHMS,) DOES IT?
   - **NO** → (go to 4)
   - **YES** → (go to 5)

4. - DISCONNECT DIST. 4-WAY CONNECTOR. NOTE OHMMETER THAT IS STILL CONNECTED TO CKT 423 AND GROUND. RESISTANCE SHOULD HAVE GONE HIGH (OPEN CIRCUIT). DOES IT?
   - **YES** → CKT 424 OPEN, FAULTY CONNECTIONS OR FAULTY IGNITION MODULE.
   - **NO** → CKT 423 SHORTED TO GROUND.

5. - RECONNECT ECM AND IDLE ENGINE FOR ONE MINUTE OR UNTIL CODE 42 SETS. DOES CODE SET?
   - **YES** → FAULTY ECM.
   - **NO** → CODE 42 INTERMITTENT. REFER TO "DIAGNOSTIC AIDS"

"AFTER REPAIRS," REFER TO CODE CRITERIA ON FACING PAGE AND CONFIRM CODE DOES NOT RESET.

4.3L (VIN W) ENGINE — DIAGNOSTIC TROUBLE CODE CHARTS — MID SIZE VAN (M/L) AND COMPACT PICKUP (S/T)

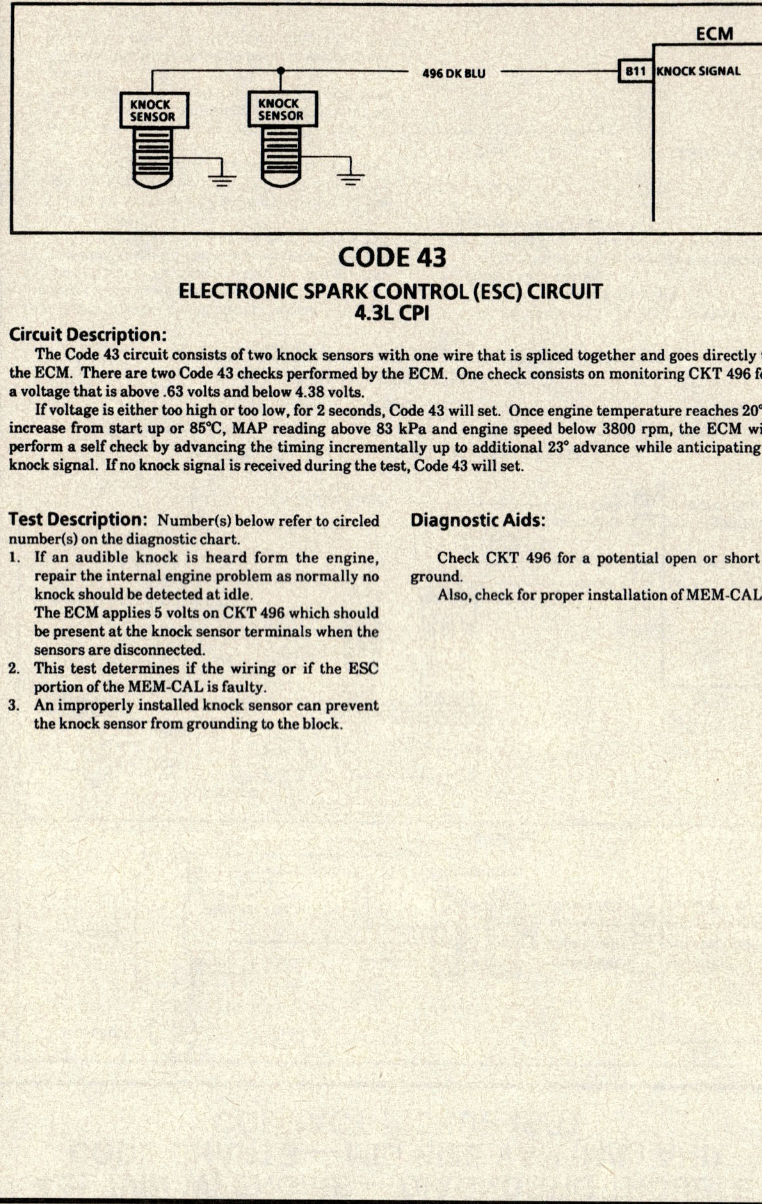

CODE 43
ELECTRONIC SPARK CONTROL (ESC) CIRCUIT
4.3L CPI

Circuit Description:

The Code 43 circuit consists of two knock sensors with one wire that is spliced together and goes directly to the ECM. There are two Code 43 checks performed by the ECM. One check consists on monitoring CKT 496 for a voltage that is above .63 volts and below 4.38 volts.

If voltage is either too high or too low, for 2 seconds, Code 43 will set. Once engine temperature reaches 20°C increase from start up or 85°C, MAP reading above 83 kPa and engine speed below 3800 rpm, the ECM will perform a self check by advancing the timing incrementally up to additional 23° advance while anticipating a knock signal. If no knock signal is received during the test, Code 43 will set.

Test Description: Number(s) below refer to circled number(s) on the diagnostic chart.

1. If an audible knock is heard form the engine, repair the internal engine problem as normally no knock should be detected at idle.
   The ECM applies 5 volts on CKT 496 which should be present at the knock sensor terminals when the sensors are disconnected.
2. This test determines if the wiring or if the ESC portion of the MEM-CAL is faulty.
3. An improperly installed knock sensor can prevent the knock sensor from grounding to the block.

Diagnostic Aids:

Check CKT 496 for a potential open or short to ground.
Also, check for proper installation of MEM-CAL.

4.3L (VIN W) ENGINE — DIAGNOSTIC TROUBLE CODE CHARTS — MID SIZE VAN (M/L) AND COMPACT PICKUP (S/T)

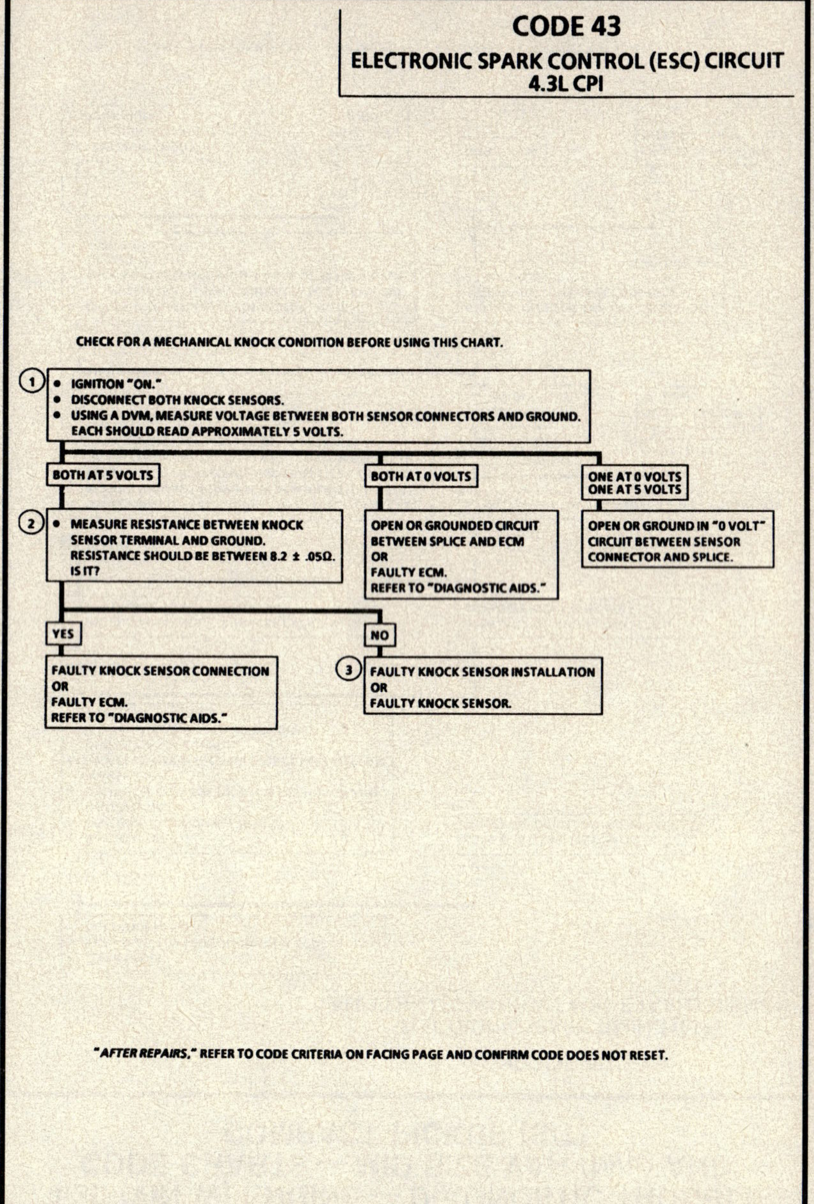

4.3L (VIN W) ENGINE — DIAGNOSTIC TROUBLE CODE CHARTS — MID SIZE VAN (M/L) AND COMPACT PICKUP (S/T)

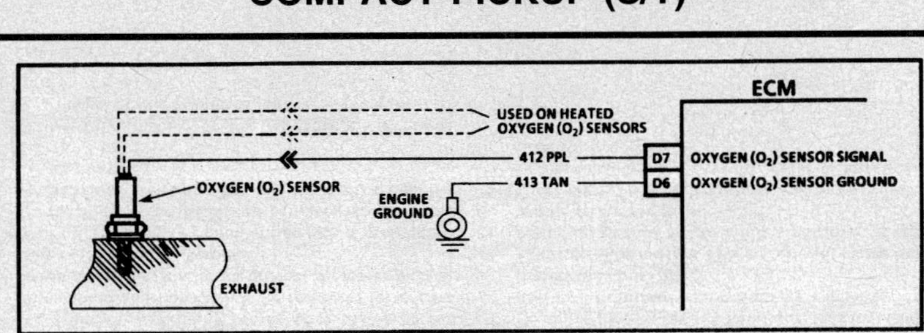

CODE 44
OXYGEN (O₂) SENSOR CIRCUIT
(LEAN EXHAUST INDICATED)
VEHICLES WITHOUT 4L80-E TRANSMISSION

Circuit Description:

The ECM supplies a voltage of about .45 volt between terminals "D6" and "D7". (If measured with a 10 megohm digital voltmeter, this may read as low as .32 volt.) The Oxygen (O₂) sensor varies the voltage within a range of about 1 volt if the exhaust is rich, down through about .10 volt if exhaust is lean.

The sensor is like an open circuit and produces no voltage when it is below about 315°C (600°F). An open sensor circuit or cold sensor causes "Open Loop" operation.

Test Description: Number(s) below refer to circled number(s) on the diagnostic chart.

1. Code 44 is set when the Oxygen (O₂) sensor signal voltage on CKT 412:
   - Remains below .2 volt from 60 seconds to 4 minutes.
   - And the system is operating in "Closed Loop."

Diagnostic Aids:

Using the Tech 1, observe the block learn values at different rpm and air flow conditions. The Tech 1 also displays the block cells, so the block learn values can be checked in each of the cells to determine when the Code 44 may have been set. If the conditions for Code 44 exists, the block learn values will be around 150.

- Oxygen (O₂) Sensor Wire - Sensor pigtail may be mispositioned and contacting the exhaust manifold.
- Heated O₂ Sensor Wire - An oxygen supply inside the O₂ sensor operation. This supply of oxygen is supplied through the O₂ sensor wires. All O₂ sensor wires and connections should be inspected for breaks or contamination that could prevent reference oxygen from reaching the O₂ sensor.
- Lean Injector Poppet Nozzles - Perform "poppet nozzle diagnosis" in "Computer Command Control,"

- Check for intermittent ground in wire between connector and sensor.
- Fuel Contamination - Water, even in small amounts, near the in-tank fuel pump inlet, can be delivered to the injectors. The water causes a lean exhaust and can set a Code 44.
- Fuel Pressure - System will be lean if pressure is too low. It may be necessary to monitor fuel pressure while driving the vehicle at various road speeds and/or loads to confirm. Refer to "Fuel System Diagnosis," CHART A-6.
- Exhaust Leaks - If there is an exhaust leak, the engine can cause outside air to be pulled into the exhaust and past the sensor. Vacuum or crankcase leaks can cause a lean condition.
- AIR System - Be sure air is not being directed to the exhaust ports while in "Closed Loop." If the block learn value goes down while squeezing air hose to left side of exhaust ports, refer to "Air Management,"
- CKT 413 - If CKT 413 is open, the voltage at terminal "D7" will be about .45 volt. This may also cause Code 13 to set.
- If all check OK, the Oxygen (O₂) sensor is faulty.

4.3L (VIN W) ENGINE — DIAGNOSTIC TROUBLE CODE CHARTS — MID SIZE VAN (M/L) AND COMPACT PICKUP (S/T)

CODE 44
OXYGEN (O₂) SENSOR CIRCUIT
(LEAN EXHAUST INDICATED)
VEHICLES WITHOUT 4L80-E TRANSMISSION

(1)
- RUN WARM ENGINE (75°C/167°F TO 95°C/203°F) AT 1200 RPM.
- DOES TECH 1 INDICATE O₂ SENSOR VOLTAGE FIXED BELOW .35 VOLT (350 mV)?

YES
- DISCONNECT O₂ SENSOR.
- WITH ENGINE IDLING, TECH 1 SHOULD DISPLAY O₂ SENSOR VOLTAGE BETWEEN .35 VOLT AND .55 VOLT (350 mV AND 550 mV). DOES IT?

NO
CODE 44 IS INTERMITTENT. IF NO ADDITIONAL CODES WERE STORED, REFER TO "DIAGNOSTIC AIDS" ON FACING PAGE.

YES
REFER TO "DIAGNOSTIC AIDS" ON FACING PAGE.

NO
CKT 412 SHORTED TO GROUND OR FAULTY ECM.

"AFTER REPAIRS," REFER TO CODE CRITERIA ON FACING PAGE AND CONFIRM CODE DOES NOT RESET.

4.3L (VIN W) ENGINE — DIAGNOSTIC TROUBLE CODE CHARTS — MID SIZE VAN (M/L) AND COMPACT PICKUP (S/T)

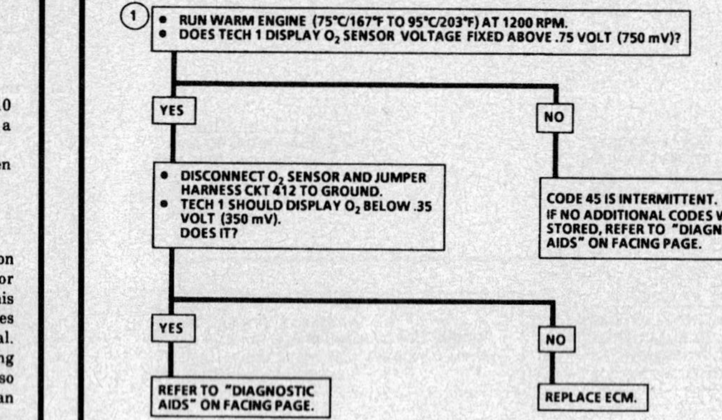

CODE 45
OXYGEN (O₂) SENSOR CIRCUIT
(RICH EXHAUST INDICATED)
VEHICLES WITHOUT 4L80-E TRANSMISSION

Circuit Description:

The ECM supplies a voltage of about .45 volt between terminals "D6" and "D7". (If measured with a 10 megohm digital voltmeter, this may read as low as .32 volt.) The Oxygen (O₂) sensor varies the voltage within a range of about 1 volt if the exhaust is rich, down through about .10 volt if exhaust is lean.

The sensor is like an open circuit and produces no voltage when it is below about 315°C (600°F). An open sensor circuit or cold sensor causes "Open Loop" operation.

Test Description: Number(s) below refer to circled number(s) on the diagnostic chart.

1. Code 45 is set when the Oxygen (O₂)sensor signal voltage or CKT 412:
   - Remains above .7 volt for 60 seconds, and in "Closed Loop."
   - Engine time after start is 1 minute or more.
   - Throttle angle greater than 5% (about .2 volt above idle voltage).

Diagnostic Aids:

Using the "Scan," observe the block learn values at different rpm and air flow conditions to determine when the Code 45 may have been set. If the conditions for Code 45 exist, the block learn values will be around 115. Code 45 will be a hard code on 4.3L CPI engine. The "SES" light will remain "ON" until the ignition is turned "OFF," due to possible high fuel pressure.

An oxygen supply inside the O₂ sensor is necessary for proper O₂ sensor operation. This supply of oxygen is supplied through the O₂ sensor wires. All O₂ sensor wires and connections should be inspected for breaks or contamination that could prevent reference oxygen from reaching the O₂ sensor.

- **Fuel Pressure** - System will go rich if pressure is too high. The ECM can compensate for some increase. However, if it gets too high, a Code 45 may be set.

- <u>Leaking Injector or Poppet Nozzle</u> - Perform "Poppet Nozzle Diagnosis" on 4.3L CPI engine.

- <u>HEI Shielding</u> - An open ground CKT 453 (ignition system reference low) may result in EMI, or induced electrical "noise." The ECM looks at this "noise" as reference pulses. The additional pulses result in a higher than actual engine speed signal. The ECM then delivers too much fuel, causing system to go rich. Engine tachometer will also show higher than actual engine speed which can help in diagnosing this problem.

- <u>Canister Purge</u> - Check for fuel saturation. If full of fuel, check canister control and hoses. Refer to "Evaporative Emission Control,"

- <u>MAP Sensor</u> - An output that causes the ECM to sensor a higher than normal manifold pressure (low vacuum) can cause the system to go rich. Disconnecting the MAP sensor will allow the ECM to set a fixed value for the MAP sensor. Substitute a different MAP sensor if the rich condition is gone while the sensor is disconnected.

- <u>Pressure Regulator</u> - Check for leaking fuel pressure regulator diaphragm by checking for presence of liquid fuel.

- <u>TPS</u> - An intermittent TPS output will cause the system to go rich, due to a false indication of the engine accelerating.

- <u>CTS</u> - Check for a shifted sensor that could cause a rich exhaust but set Code 15. Refer to chart for Code 15.

4.3L (VIN W) ENGINE — DIAGNOSTIC TROUBLE CODE CHARTS — MID SIZE VAN (M/L) AND COMPACT PICKUP (S/T)

CODE 45
OXYGEN (O₂) SENSOR CIRCUIT
(RICH EXHAUST INDICATED)
VEHICLES WITHOUT 4L80-E TRANSMISSION

①
- RUN WARM ENGINE (75°C/167°F TO 95°C/203°F) AT 1200 RPM.
- DOES TECH 1 DISPLAY O₂ SENSOR VOLTAGE FIXED ABOVE .75 VOLT (750 mV)?

YES
- DISCONNECT O₂ SENSOR AND JUMPER HARNESS CKT 412 TO GROUND.
- TECH 1 SHOULD DISPLAY O₂ BELOW .35 VOLT (350 mV). DOES IT?

NO
- CODE 45 IS INTERMITTENT. IF NO ADDITIONAL CODES WERE STORED, REFER TO "DIAGNOSTIC AIDS" ON FACING PAGE.

YES → REFER TO "DIAGNOSTIC AIDS" ON FACING PAGE.

NO → REPLACE ECM.

"AFTER REPAIRS," REFER TO CODE CRITERIA ON FACING PAGE AND CONFIRM CODE DOES NOT RESET.

4.3L (VIN W) ENGINE — DIAGNOSTIC TROUBLE CODE CHARTS — MID SIZE VAN (M/L) AND COMPACT PICKUP (S/T)

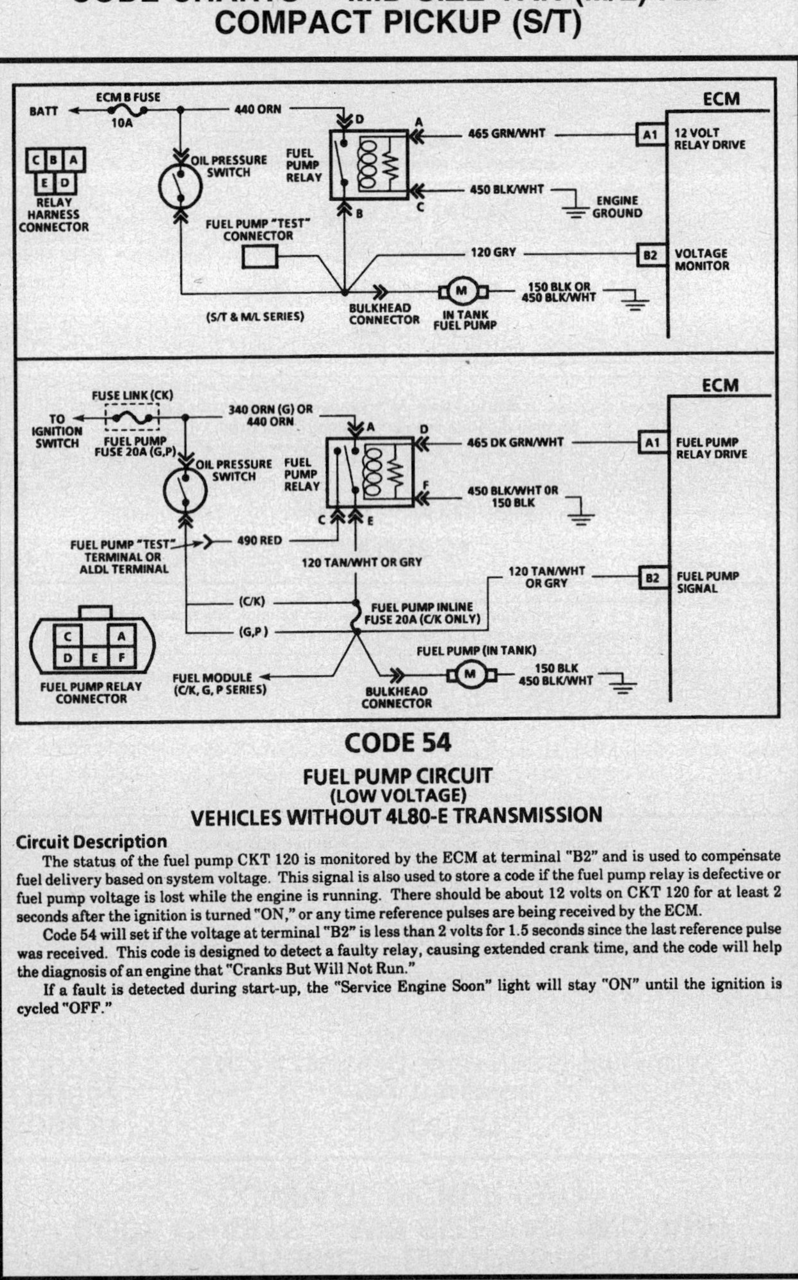

CODE 54
FUEL PUMP CIRCUIT
(LOW VOLTAGE)
VEHICLES WITHOUT 4L80-E TRANSMISSION

Circuit Description

The status of the fuel pump CKT 120 is monitored by the ECM at terminal "B2" and is used to compensate fuel delivery based on system voltage. This signal is also used to store a code if the fuel pump relay is defective or fuel pump voltage is lost while the engine is running. There should be about 12 volts on CKT 120 for at least 2 seconds after the ignition is turned "ON," or any time reference pulses are being received by the ECM.

Code 54 will set if the voltage at terminal "B2" is less than 2 volts for 1.5 seconds since the last reference pulse was received. This code is designed to detect a faulty relay, causing extended crank time, and the code will help the diagnosis of an engine that "Cranks But Will Not Run."

If a fault is detected during start-up, the "Service Engine Soon" light will stay "ON" until the ignition is cycled "OFF."

4.3L (VIN W) ENGINE — DIAGNOSTIC TROUBLE CODE CHARTS — MID SIZE VAN (M/L) AND COMPACT PICKUP (S/T)

CODE 51
CODE 52
CODE 53
CODE 55

CODE 51

FAULTY MEM-CAL

(2.5L, 4.3L CPI ENGINES AND VEHICLES WITH 4L80-E TRANSMISSION)

OR

PROM PROBLEM

(EXCEPT 2.5L, 4.3L CPI ENGINES AND VEHICLES WITHOUT 4L80-E TRANSMISSON)

CHECK THAT ALL PINS ARE FULLY INSERTED IN THE SOCKET. IF OK, REPLACE PROM, CLEAR MEMORY, AND RECHECK. IF CODE 51 REAPPEARS, REPLACE ECM/PCM.

CODE 52

FUEL CAL-PAK MISSING

CHECK FOR MISSING CAL-PAK AND THAT ALL PINS ARE FULLY INSERTED IN THE SOCKET. IF OK, REPLACE ECM. NOTE: ON SOME VEHICLES THE CAL-PAK IS SOLDERED IN.

CODE 53

SYSTEM OVER VOLTAGE

THIS CODE INDICATES THERE IS A BASIC GENERATOR PROBLEM.
- CODE 53 WILL SET IF VOLTAGE AT ECM/PCM TERMINAL "B1" IS GREATER THAN 17.1 VOLTS FOR 2 SECONDS.
- CHECK AND REPAIR CHARGING SYSTEM.

CODE 55

FAULTY ECM/PCM

ALL ENGINES

EXCEPT 2.5L ENGINE

BE SURE ECM/PCM GROUNDS ARE OK AND THAT MEM-CAL IS PROPERLY LATCHED. IF OK, REPLACE ECM/PCM.

4.3L (VIN W) ENGINE — SYSTEM CHECK — MID SIZE VAN (M/L) AND COMPACT PICKUP (S/T)

RESTRICTED EXHAUST SYSTEM CHECK
ALL ENGINES

Proper diagnosis for a restricted exhaust system is essential before any components are replaced. Either of the following procedures may be used for diagnosis, depending upon engine or tool used:

CHECK AT A. I. R. PIPE: **OR**   **CHECK AT O_2 SENSOR:**

CHECK AT A. I. R. PIPE:
1. Remove the rubber hose at the exhaust manifold A.I.R. pipe check valve. Remove check valve.
2. Connect a fuel pump pressure gauge to a hose and nipple from a Propane Enrichment Device (J 26911) (see illustration).
3. Insert the nipple into the exhaust manifold A.I.R. pipe.

CHECK AT O_2 SENSOR:
1. Carefully remove O_2 sensor.
2. Install Borroughs exhaust backpressure tester (BT 8515 or BT 8603) or equivalent in place of O_2 sensor (see illustration).
3. After completing test described below, be sure to coat threads of O_2 sensor with anti-seize compound P/N 5613695 or equivalent prior to re-installation.

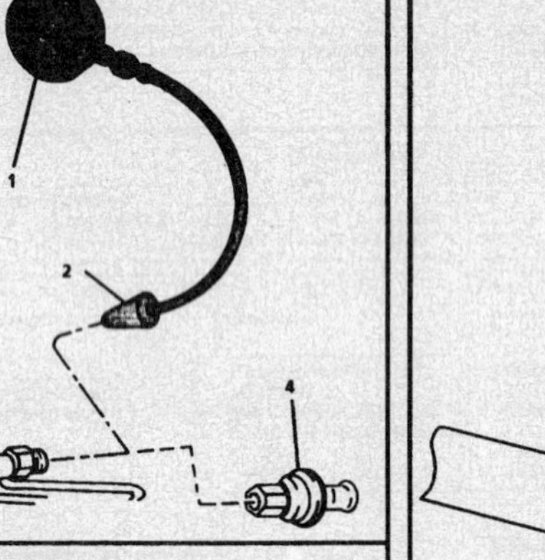

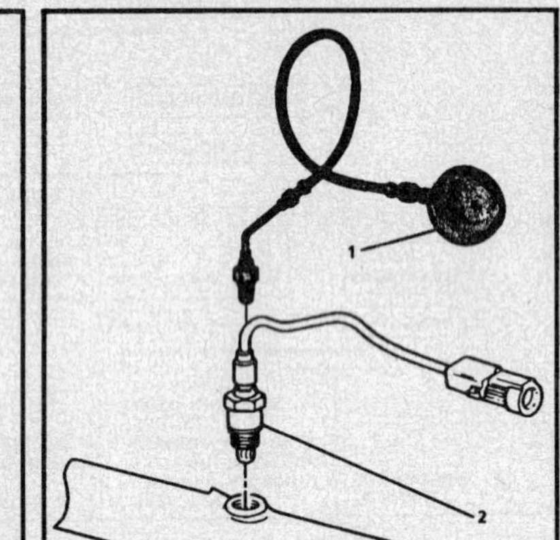

DIAGNOSIS:
1. With the engine idling at normal operating temperature, observe the exhaust system backpressure reading on the gage. Reading should not exceed 8.6 kPa (1.25 psi).
2. Increase engine speed to 2000 rpm and observe gage. Reading should not exceed 20.7 kPa (3 psi).
3. If the backpressure at either speed exceeds specification, a restricted exhaust system is indicated.
4. Inspect the entire exhaust system for a collapsed pipe, heat distress, or possible internal muffler failure.
5. If there are no obvious reasons for the excessive backpressure, the catalytic converter is suspected to be restricted and should be replaced using current recommended procedures.

4.3L (VIN W) ENGINE — SYSTEM CHECK — MID SIZE VAN (M/L) AND COMPACT PICKUP (S/T)

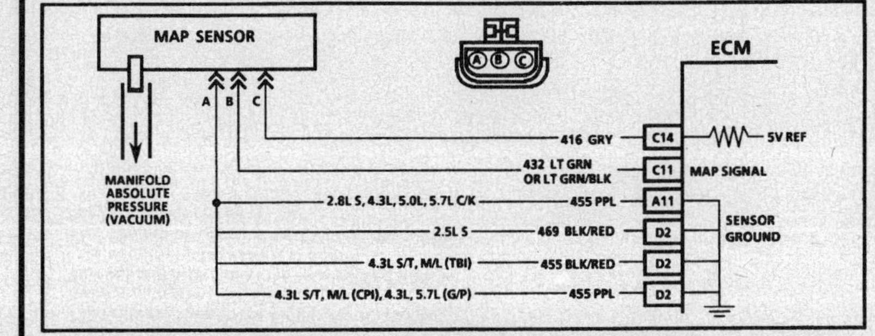

MANIFOLD ABSOLUTE PRESSURE (MAP) OUTPUT CHECK
VEHICLES WITHOUT 4L80-E TRANSMISSION

Circuit Description:

The Manifold Absolute Pressure (MAP) sensor measures the changes in the intake manifold pressure which result from engine load (intake manifold vacuum) and rpm changes; and converts these into a voltage output. The ECM sends a 5 volt reference voltage to the MAP sensor. As the manifold pressure changes, the output voltage of the sensor also changes. By monitoring the sensor output voltage, the ECM knows the manifold pressure. At lower pressure output voltage will be about 1 to 2 volts at idle. While at higher pressure or at Wide Open Throttle (WOT) output voltage will be about 4 to 4.8 volts. The MAP sensor is also used, under certain conditions, to measure barometric pressure, allowing the ECM to make adjustments for different altitudes. The ECM uses the MAP sensor to control fuel delivery and ignition timing.

Test Description: Number(s) below refer to circled number(s) on the diagnostic chart.

? Important

- Be sure to use the same Diagnostic Test Equipment for all measurements.

1. Checks MAP sensor output voltage to the ECM. This voltage, without engine running, represents a barometer reading to the ECM.
   - When comparing Tech 1 "Scan" readings to a known good vehicle, it is important to compare vehicles that use a MAP sensor having the same color insert or having the same "Hot Stamped" number. Refer to figures on facing page.
2. Applying 34 kPa (10" Hg) vacuum to the MAP sensor should cause the voltage to change. Subtract second reading from the first. Voltage value should be greater than 1.5 volts. Upon applying vacuum to the sensor, the change in voltage should be instantaneous. A slow voltage change indicates a faulty sensor.

3. Check vacuum hose to sensor for leaking or restriction. Be sure that no other vacuum devices are connected to the MAP hose.

   NOTICE: Make sure electrical connector remains securely fastened.

4. Disconnect sensor from bracket and twist sensor by hand (only) to check for intermittent connection. Output changes greater than .1 volt indicate a bad connector or connection. If OK, replace sensor.

4.3L (VIN W) ENGINE — SYSTEM CHECK — MID SIZE VAN (M/L) AND COMPACT PICKUP (S/T)

MANIFOLD ABSOLUTE PRESSURE (MAP) OUTPUT CHECK
VEHICLES WITHOUT 4L80-E TRANSMISSION

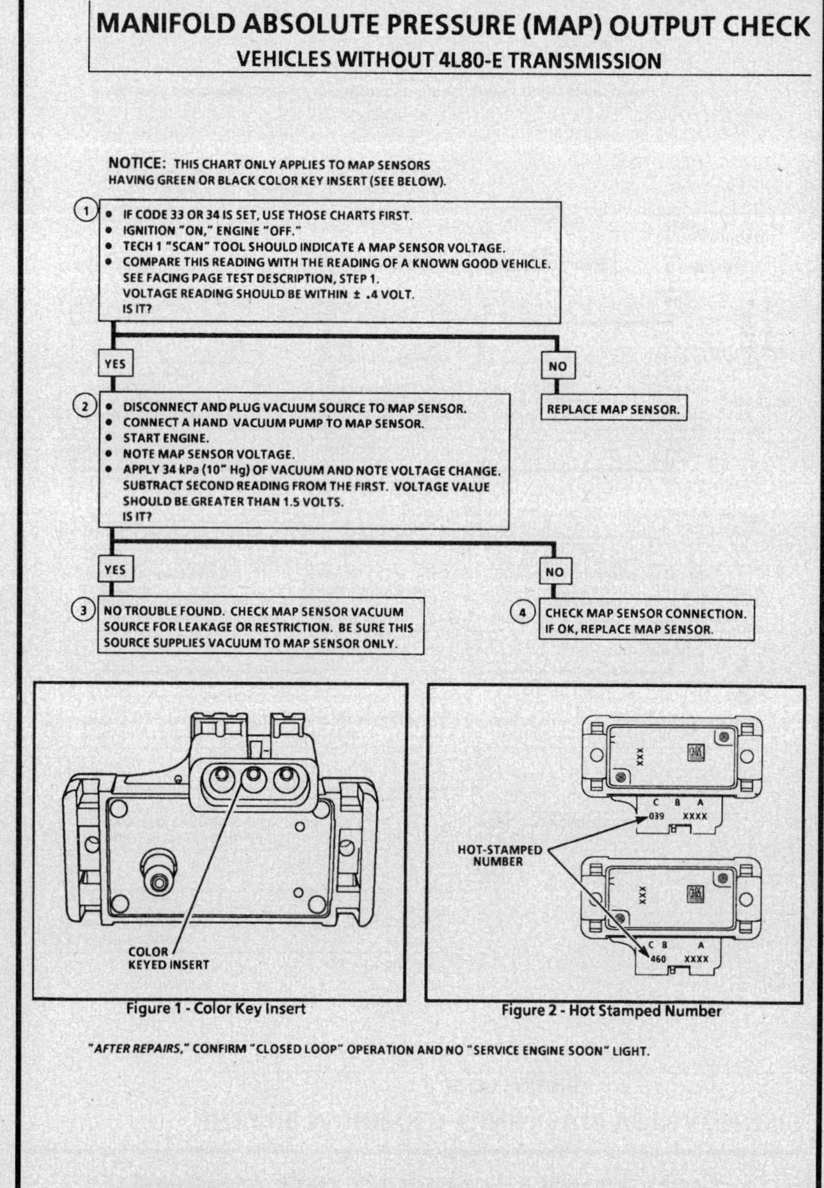

NOTICE: THIS CHART ONLY APPLIES TO MAP SENSORS HAVING GREEN OR BLACK COLOR KEY INSERT (SEE BELOW).

(1)
- IF CODE 33 OR 34 IS SET, USE THOSE CHARTS FIRST.
- IGNITION "ON," ENGINE "OFF."
- TECH 1 "SCAN" TOOL SHOULD INDICATE A MAP SENSOR VOLTAGE.
- COMPARE THIS READING WITH THE READING OF A KNOWN GOOD VEHICLE. SEE FACING PAGE TEST DESCRIPTION, STEP 1. VOLTAGE READING SHOULD BE WITHIN ± .4 VOLT. IS IT?

YES → (2)

NO → REPLACE MAP SENSOR.

(2)
- DISCONNECT AND PLUG VACUUM SOURCE TO MAP SENSOR.
- CONNECT A HAND VACUUM PUMP TO MAP SENSOR.
- START ENGINE.
- NOTE MAP SENSOR VOLTAGE.
- APPLY 34 kPa (10" Hg) OF VACUUM AND NOTE VOLTAGE CHANGE. SUBTRACT SECOND READING FROM THE FIRST. VOLTAGE VALUE SHOULD BE GREATER THAN 1.5 VOLTS. IS IT?

YES → (3) NO TROUBLE FOUND. CHECK MAP SENSOR VACUUM SOURCE FOR LEAKAGE OR RESTRICTION. BE SURE THIS SOURCE SUPPLIES VACUUM TO MAP SENSOR ONLY.

NO → (4) CHECK MAP SENSOR CONNECTION. IF OK, REPLACE MAP SENSOR.

Figure 1 - Color Key Insert

Figure 2 - Hot Stamped Number

"AFTER REPAIRS," CONFIRM "CLOSED LOOP" OPERATION AND NO "SERVICE ENGINE SOON" LIGHT.

4.3L (VIN W) ENGINE — SYSTEM CHECK — MID SIZE VAN (M/L) AND COMPACT PICKUP (S/T)

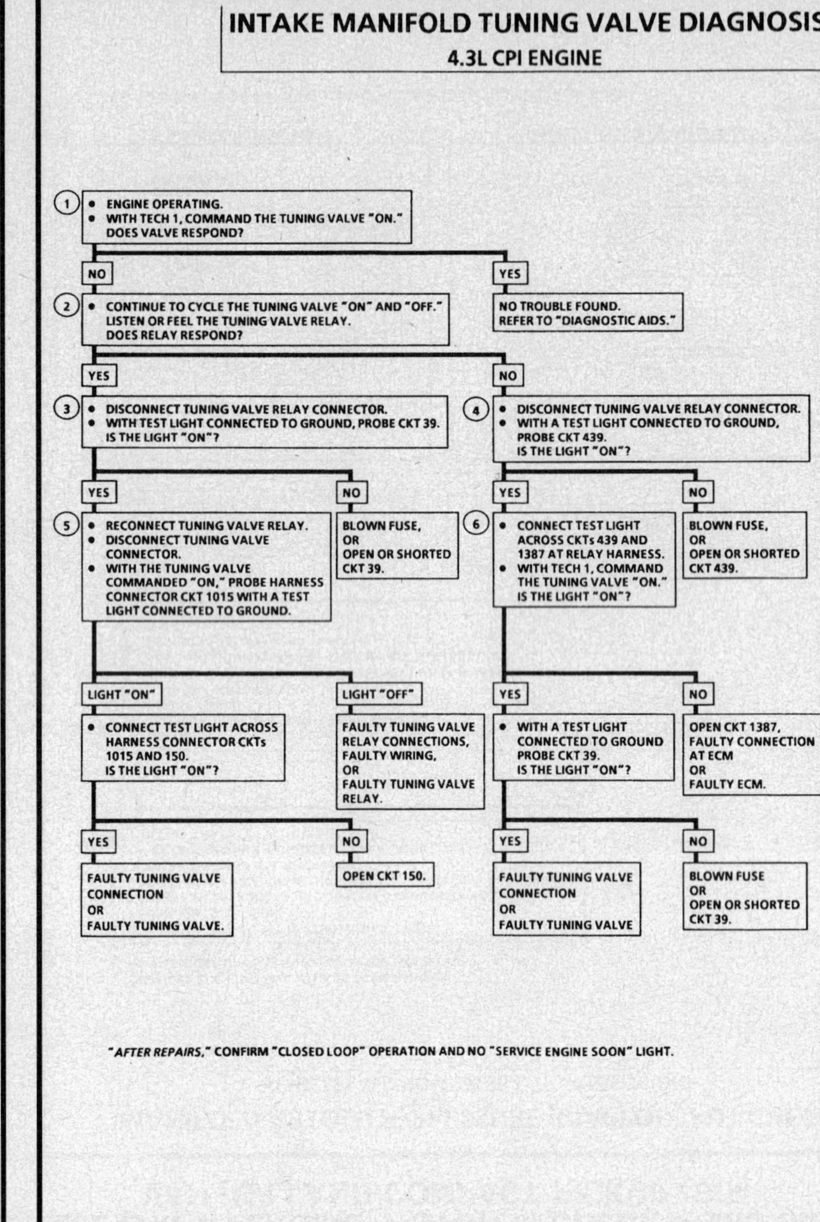

INTAKE MANIFOLD TUNING VALVE DIAGNOSIS
4.3L CPI ENGINE

Circuit Description:

The tuning valve is activated through a relay, by the ECM, during periods of acceleration where engine rpm is over 3000 and throttle position is over 34%.

The tuning valve relay is powered by CKT 439 and turned "ON" and "OFF" when the ECM grounds CKT 1387. When the relay is activated, it applies voltage from CKT 39 to CKT 1015, thus, opening the tuning valve.

Test Description: Number(s) below refer to circled number(s) on the diagnostic chart.

1. This step determines if the system is functioning at the present time.
2. By cycling the tuning valve "ON" and "OFF" will determine if the tuning valve relay is operating.
3. This will determine if the tuning valve relay has ignition voltage on CKT 39.
4. This will determine if the tuning valve relay has ignition voltage on CKT 439.
5. This step will determine if the relay related wiring to valve or the tuning valve is at fault.
6. This step will determine if the ECM, tuning valve relay or related wiring is at fault.

Diagnostic Aids:

- TPS - An intermittent TPS output will cause the ECM to receive a false input.

4.3L (VIN W) ENGINE — SYSTEM CHECK — MID SIZE VAN (M/L) AND COMPACT PICKUP (S/T)

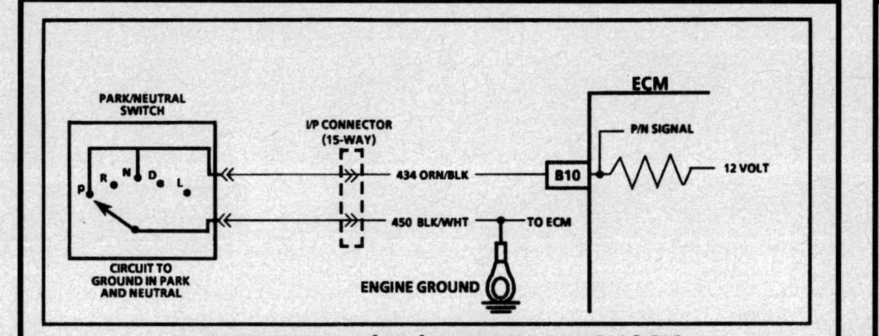

PARK/NEUTRAL (P/N) SWITCH DIAGNOSIS
**ALL VEHICLES WITHOUT 4L80-E TRANSMISSION
AUTO TRANSMISSION ONLY**

Circuit Description:

The Park/Neutral (P/N) switch contacts are closed to ground in park or neutral and open in drive ranges.

The ECM supplies ignition voltage, through a current limiting resistor, to CKT 434 and senses a closed switch, when the voltage on CKT 434 drops to less than one volt.

The ECM uses the P/N signal as one of the inputs to control:
- Idle Air Control (IAC).
- Vehicle Speed Sensor (VSS) Diagnostics.

Test Description: Number(s) below refer to circled number(s) on the diagnostic chart.
1. Checks for a closed switch to ground in park position. Different makes of "Scan" tools will read P/N differently. Refer to operators manual for type of display used for a specific tool.
2. Checks for an open switch in drive or reverse range.
3. Be sure "Scan" indicates drive, even while wiggling shifter to test for an intermittent or misadjusted switch in drive range.

Diagnostic Aids:

If CKT 434 always indicates drive (open), a drop in the idle may exist when the gear selector is moved into drive range.

4.3L (VIN W) ENGINE — SYSTEM CHECK — MID SIZE VAN (M/L) AND COMPACT PICKUP (S/T)

PARK/NEUTRAL (P/N) SWITCH DIAGNOSIS
**ALL VEHICLES WITHOUT 4L80-E TRANSMISSION
AUTOMATIC TRANSMISSION ONLY**

1. WITH TRANSMISSION IN PARK, TECH 1 "SCAN" TOOL SHOULD INDICATE PARK OR NEUTRAL. DOES IT?

— YES →

3. SHIFT TRANSMISSION INTO DRIVE. "SCAN" TOOL SHOULD DISPLAY A CHANGE TO INDICATE DRIVE. DOES IT?

— NO → DISCONNECT P/N SWITCH. THIS SHOULD CAUSE "SCAN" TOOL TO DISPLAY DRIVE RANGE. DOES IT?
- YES → FAULTY P/N SWITCH CONNECTION OR P/N SWITCH MISADJUSTED OR FAULTY P/N SWITCH.
- NO → CKT 434 SHORTED TO GROUND OR FAULTY ECM.

— YES → NO TROUBLE FOUND.

— NO →

2. DISCONNECT PARK/NEUTRAL SWITCH CONNECTOR. JUMPER HARNESS CONNECTOR TERMINALS "A" AND "B". "SCAN" TOOL SHOULD INDICATE PARK OR NEUTRAL. DOES IT?

— NO → JUMPER HARNESS CONNECTOR (CKT 434) TO ENGINE GROUND. "SCAN" TOOL SHOULD INDICATE PARK OR NEUTRAL. DOES IT?
- YES → OPEN GROUND CIRCUIT.
- NO → CKT 434 OPEN OR FAULTY ECM CONNECTION OR ECM.

— YES → FAULTY P/N SWITCH CONNECTION OR P/N SWITCH MISADJUSTED OR FAULTY P/N SWITCH.

"AFTER REPAIRS." CONFIRM "CLOSED LOOP" OPERATION AND NO "SERVICE ENGINE SOON" LIGHT.

4.3L (VIN W) ENGINE — SYSTEM CHECK — MID SIZE VAN (M/L) AND COMPACT PICKUP (S/T)

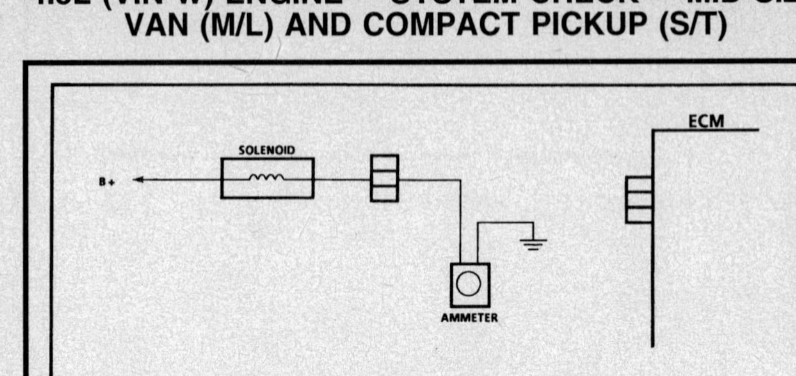

A/C REQUEST SIGNAL DIAGNOSIS

VEHICLES WITHOUT 4L80-E TRANSMISSION

Circuit And Test Description:

Turning "ON" the air conditioning supplies CKT 59 battery voltage to the A/C compressor clutch and to terminal "B8" of the ECM connector to increase and maintain idle speed.

The ECM does not control the A/C compressor clutch, therefore, if A/C does not function, refer to the A/C section of the service manual for diagnosis of the system.

If A/C is operating properly and idle speed dips too low when the A/C compressor turns "ON" or flares too high when the A/C compressor turns "OFF," check for an open CKT 59 to the ECM. If circuits are OK, it is a faulty ECM connector terminal "B8" or ECM.

4.3L (VIN W) ENGINE — SYSTEM CHECK — MID SIZE VAN (M/L) AND COMPACT PICKUP (S/T)

ECM QDR CHECK PROCEDURE

2.5L

QDR NUMBER	ECM OUTPUT TERMINAL	CIRCUIT
1	A5	SES LIGHT
	A4	AC CONTROL
	A2	T.C.C. OR SHIFT LIGHT
	C1	NOT USED
2	A3	EGR CONTROL
	D12	NOT USED
	C2	NOT USED
	A7	NOT USED

2.8L

QDR NUMBER	ECM OUTPUT TERMINAL	CIRCUIT
1	A2	A/C CONTROL
	A3	NOT USED
	C1	NOT USED
	C2	EAC CONTROL
2	A4	EGR CONTROL *
	A5	SES LIGHT
	A7	SHIFT LIGHT
		NOT USED

4.3L CPI

QDR NUMBER	ECM OUTPUT TERMINAL	CIRCUIT
1	A5	SES LIGHT
	A4	NOT USED
	A2	T.C.C.
	C1	NOT USED
2	A3	EGR CONTROL *
	D12	NOT USED
	C2	NOT USED
	A7	INTAKE TUNING VALVE

4.3L TBI, 5.0L, 5.7L, 7.4L

QDR NUMBER	ECM OUTPUT TERMINAL	CIRCUIT
1	A2	NOT USED
	A3	NOT USED
	C1	NOT USED
	C2	EAC CONTROL
2	A4	EGR OR EVRV CONTROL *
	A5	SES LIGHT
	A7	TCC SOLENOID OR SHIFT LIGHT
	B4	NOT USED

* Solid state circuitry cannot be diagnosed by resistance check. Use DVM J 34029-A or equivalent.

4.3L (VIN W) ENGINE — SYSTEM CHECK — MID SIZE VAN (M/L) AND COMPACT PICKUP (S/T)

ECM QDR CHECK PROCEDURE

USE THIS CHECK PROCEDURE ONLY AFTER OTHER DIAGNOSTIC CHARTS IN THE SERVICE MANUAL HAVE DETERMINED THAT THERE WAS AN ECM FAILURE.

- REMOVE THE ECM FROM THE VEHICLE.

- REFER TO LIST BELOW OF THE ECM TERMINALS WHICH ARE QDR OUTPUTS.
- USING THE 100/200K OHM SCALE ON DVM*, MEASURE RESISTANCE BETWEEN THE ECM CASE AND EACH ECM TERMINAL LISTED, BLACK (NEG.) LEAD TO CASE AND RED (POS.) LEAD TO ECM TERMINAL.
- ALL TERMINALS LISTED SHOULD HAVE RESISTANCE OF 50K OHMS OR MORE. DO THEY?

NO

THE PRIOR TEST HAS DETERMINED THAT A QDR IN THE ECM HAS BEEN DAMAGED. IT IS MOST IMPORTANT TO LOCATE AND REPAIR THE CIRCUIT OR COMPONENT THAT CAUSED THE DAMAGE. FAILURE TO DO SO WILL RESULT IN ANOTHER FAILURE OF THE NEWLY REPLACED ECM.
ANY TERMINAL WITH LESS THAN 50K OHMS RESISTANCE IS CONNECTED TO A DEFECTIVE QDR. THE ECM TERMINAL WITH THE LOWEST RESISTANCE WAS CONNECTED TO THE VEHICLE CIRCUIT MOST LIKELY TO HAVE CAUSED THE QDR FAILURE.

- DISCONNECT THE COMPONENT IN THAT VEHICLE CIRCUIT AND CHECK FOR A SHORT TO VOLTAGE. IF THE CIRCUIT IS NOT SHORTED TO VOLTAGE, REPLACE THE COMPONENT IN THAT CIRCUIT AND THE ECM.

YES

- KEY "ON," ENGINE NOT RUNNING.
- USE A FUSED AMMETER CAPABLE OF MEASURING AT LEAST 2 AMPS (J 34029-A OR EQUIVALENT).
- CONNECT ONE LEAD OF THE AMMETER TO CHASSIS GROUND.
- CONNECT THE REMAINING LEAD TO EACH VEHICLE CIRCUIT WHICH WAS LISTED ABOVE.
- MEASURE SUSTAINED CURRENT FLOW THROUGH EACH CIRCUIT FOR 2 MINUTES EACH (IN MOST CASES, THE TCC SOLENOID CANNOT BE EASILY TESTED FOR CURRENT DRAW).
- NOTE AMPERAGE.

IF CIRCUIT(S) HAS MORE THAN 0.75 AMPS CURRENT DRAW.

- CHECK FOR A SHORT TO VOLTAGE IN EXCESSIVE CURRENT DRAW CIRCUIT.
- IF NO SHORT TO VOLTAGE, REPLACE RELATED SOLENOID OR RELAY.

IF NO CIRCUIT HAS MORE THAN 0.75 AMPS CURRENT DRAW.

- REPLACE ECM.

*** USE DVM J 34029-A OR EQUIVALENT**

4.3L (VIN W) ENGINE — SYSTEM CHECK — MID SIZE VAN (M/L) AND COMPACT PICKUP (S/T)

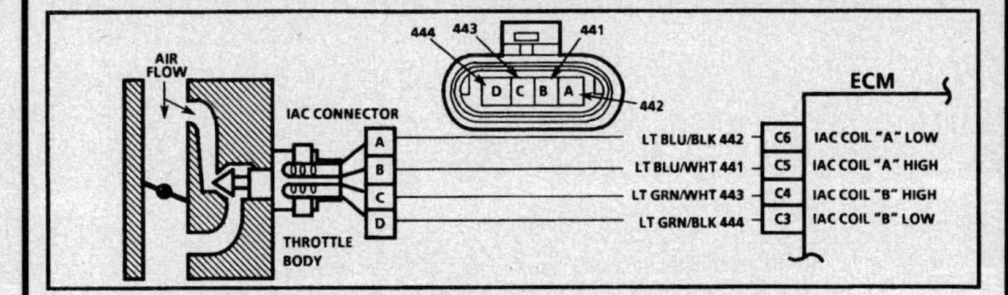

IDLE AIR CONTROL (IAC) SYSTEM CHECK
4.3L CPI ENGINE

Circuit Description:

The ECM controls engine idle speed with the IAC valve. To increase idle speed, the ECM retracts the IAC valve pintle away from the seat, allowing more air to pass by the throttle bore. To decrease idle speed, it extends the IAC valve pintle towards the seat, reducing bypass air flow. A Tech 1 "Scan" tool will read the ECM commands to the IAC valve in counts. Higher the counts indicate more air bypass (higher idle). The lower the counts indicate less air is allowed to bypass (lower idle).

Test Description: Number(s) below refer to circled number(s) on the diagnostic chart.

1. The IAC tester is used to extend and retract the IAC valve. Valve movement is verified by an engine speed change. If no change in engine speed occurs, the valve can be retested when removed from the throttle body.
2. This step checks the quality of the IAC movement in Step 1. Between 700 rpm and about 1500 rpm the engine speed should change smoothly with each flash of the tester light in both extend and retract. If the IAC valve is retracted beyond the control range (about 1500 rpm), it may take many flashes in the extend position before engine speed will begin to drop. This is normal on certain engines, fully extending IAC may cause engine stall. This may be normal.
3. Steps 1 and 2 verified proper IAC valve operation while this step checks the IAC circuits. each lamp on the node light should flash Red and Green while the IAC valve is cycled. While the sequence of color is not important if either light is "OFF" or does not flash Red and Green, check the circuits for faults, beginning with poor terminal contacts.

IAC Valve Reset Procedure

- Ignition "OFF" for 10 seconds.
- Start and run engine for 5 seconds.
- Ignition "OFF" for 10 seconds.

Diagnostic Aids:

A slow, unstable, or fast idle may be caused by a non-IAC system problem that cannot be overcome by the IAC valve. Out of control range, IAC Tech 1 "Scan" tool counts will be above 60 if idle is too low, and zero counts if idle is too high. The following checks should be made to repair a non-IAC system problem.

- **Vacuum Leak (High Idle)** - If idle is too high, stop the engine. Fully extend (low) IAC with tester. Start engine. If idle speed is above 800 rpm, locate and correct vacuum leak including PCV system. Also, check for binding of throttle blade or linkage.
- **System too rich (Low Air/Fuel Ratio)** - The idle speed will be too low. Tech 1 "Scan" tool IAC counts will usually be above 80. System is obviously rich and may exhibit black smoke in exhaust.
  Tech 1 "Scan" tool O_2 voltage will be fixed above 800 mV (.8 volt).
  Check for high fuel pressure, leaking or sticking injector, silicone contaminated O_2 sensors, "Scan" voltage will be slow to respond.
- **Throttle body** - Remove IAC valve and inspect bore for foreign material.
- **IAC Valve Electrical Connections** - IAC valve connections should be carefully checked for proper contact.
- **PCV Valve** - An incorrect or faulty PCV valve may result in an incorrect idle speed.

- **A/C Compressor** - Refer to A/C diagnosis if circuit is shorted to ground. If the relay is faulty, idle problem may exist.

- If intermittent poor drivability or idle symptoms are resolved by disconnecting the IAC, carefully recheck connections, valve terminal resistance or replace IAC.

4.3L (VIN W) ENGINE — SYSTEM CHECK — MID SIZE VAN (M/L) AND COMPACT PICKUP (S/T)

IDLE AIR CONTROL (IAC) SYSTEM CHECK
4.3L CPI ENGINE

NOTICE: IF A REPAIR HAS BEEN MADE, REFER TO THE "IAC RESET PROCEDURE" ON THE FACING PAGE BEFORE RETESTING.

(1)
- IGNITION "OFF," CONNECT IAC DRIVER *TO IAC VALVE.
- SET PARK BRAKE, BLOCK WHEELS, A/C "OFF."
- IDLE ENGINE IN PARK.
- INSTALL "SCAN" TOOL AND DISPLAY RPM.
- WITH IAC DRIVER, EXTEND AND RETRACT IAC VALVE.
- ENGINE RPM SHOULD DECREASE AND INCREASE AS IAC IS CYCLED.

[RPM CHANGES] → (2)

[NO RPM CHANGES] → CHECK IAC PASSAGES. IF OK, REPLACE IAC.

(2) RPM SHOULD CHANGE SMOOTHLY WITH EACH FLASH OF THE IAC DRIVER LIGHT FROM 700 RPM TO ABOUT 1500 RPM. DOES IT?

[YES] → (3)

[NO] → CHECK IAC PASSAGES. IF OK, REPLACE IAC.

(3)
- INSTALL APPROPRIATE IAC NODE LIGHT * IN ECM HARNESS.
- CYCLE IAC DRIVER AND NOTE LIGHTS.
- BOTH LIGHTS SHOULD CYCLE RED AND GREEN BUT NEVER "OFF" AS RPM IS CHANGED OVER ITS RANGE. DO THEY?

[NO] → IF CIRCUIT(S) DID NOT TEST GREEN AND RED, CHECK FOR:
- FAULTY CONNECTOR TERMINAL CONTACTS.
- OPEN CIRCUITS INCLUDING CONNECTIONS.
- CIRCUITS SHORTENED TO GROUND OR VOLTAGE.
- FAULTY ECM CONNECTION OR ECM. REPAIR AS NECESSARY AND RETEST.

[YES] →
- USING THE OTHER CONNECTOR ON THE IAC DRIVER PIGTAIL, CHECK RESISTANCE ACROSS IAC COILS. SHOULD BE 40 TO 80 OHMS BETWEEN IAC TERMINALS "A" TO "B" AND "C" TO "D".

[OK] → CHECK RESISTANCE BETWEEN IAC TERMINALS "B" AND "C" AND "A" AND "D". SHOULD BE INFINITE.

[NOT OK] → REPLACE VALVE AND RETEST.

[OK] → IDLE AIR CONTROL VALVE AND CIRCUIT OK. REFER TO "DIAGNOSTIC AIDS"

[NOT OK] → REPLACE VALVE AND RETEST.

* IAC DRIVER AND NODE LIGHT REQUIRED KIT 222-L FROM: CONCEPT TECHNOLOGY, INC. J 37027 FROM: KENT-MOORE, INC.

4.3L (VIN W) ENGINE — SYSTEM CHECK — MID SIZE VAN (M/L) AND COMPACT PICKUP (S/T)

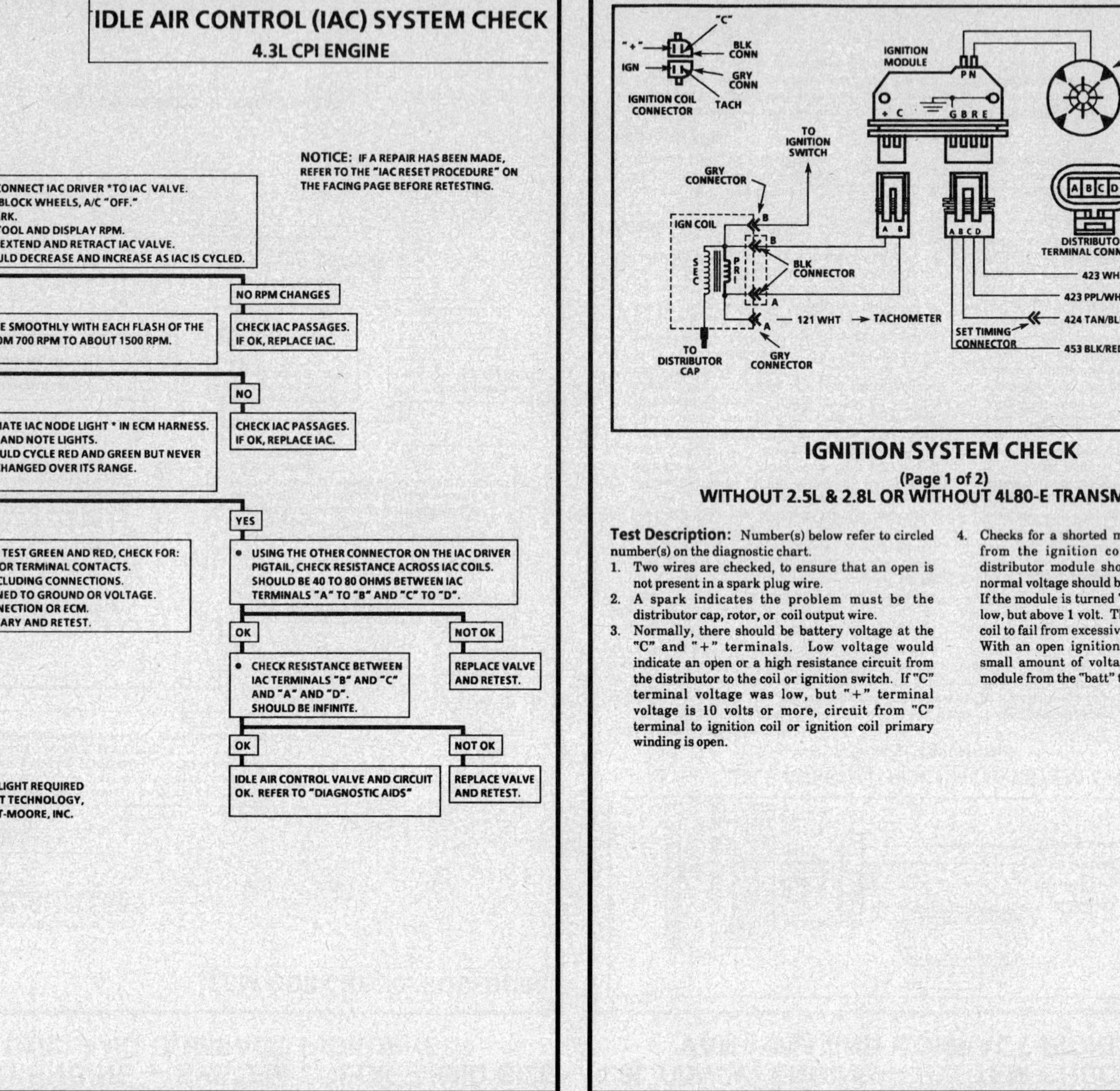

IGNITION SYSTEM CHECK
(Page 1 of 2)
WITHOUT 2.5L & 2.8L OR WITHOUT 4L80-E TRANSMISSION

Test Description: Number(s) below refer to circled number(s) on the diagnostic chart.

1. Two wires are checked, to ensure that an open is not present in a spark plug wire.
2. A spark indicates the problem must be the distributor cap, rotor, or coil output wire.
3. Normally, there should be battery voltage at the "C" and "+" terminals. Low voltage would indicate an open or a high resistance circuit from the distributor to the coil or ignition switch. If "C" terminal voltage was low, but "+" terminal voltage is 10 volts or more, circuit from "C" terminal to ignition coil or ignition coil primary winding is open.
4. Checks for a shorted module or grounded circuit from the ignition coil to the module. The distributor module should be turned "OFF," so normal voltage should be about 12 volts. If the module is turned "ON," the voltage would be low, but above 1 volt. This could cause the ignition coil to fail from excessive heat. With an open ignition coil primary winding, a small amount of voltage will leak through the module from the "batt" to the "tach" terminal.

4.3L (VIN W) ENGINE — SYSTEM CHECK — MID SIZE VAN (M/L) AND COMPACT PICKUP (S/T)

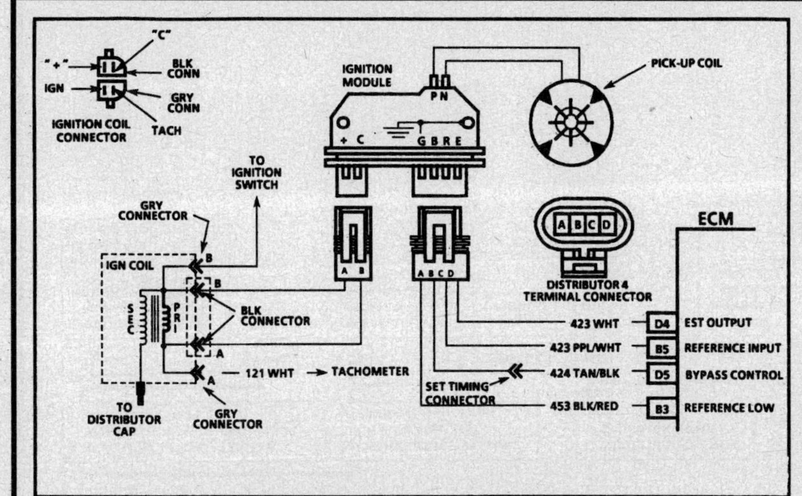

IGNITION SYSTEM CHECK
(Page 2 of 2)
WITHOUT 2.5L & 2.8L OR WITHOUT 4L80-E TRANSMISSION

Test Description: Number(s) below refer to circled number(s) on the diagnostic chart.

5. Applying a voltage (1.35 to 1.50 volts) to module terminal "P" should turn the module "ON" and the "Tach" terminal voltage should drop to about 7-9 volts. This test will determine whether the module or coil is faulty or if the pick-up coil is not generating the proper signal to turn the module "ON." This test can be performed by using a DC test battery with a rating of 1.5 volts. (Such as AA, C, or D cell.) The battery must be a known good battery with a voltage of over 1.35 volts.

6. This should turn "OFF" the module and cause a spark. If no spark occurs, the fault is most likely in the ignition coil because most module problems would have been found before this point in the procedure. A module tester (J 24642) could determine which is at fault.

4.3L (VIN W) ENGINE — SYSTEM CHECK — MID SIZE VAN (M/L) AND COMPACT PICKUP (S/T)

IGNITION SYSTEM CHECK
(Page 1 of 2)
WITHOUT 2.5L & 2.8L OR WITHOUT 4L80-E TRANSMISSION

1.
- CHECK SPARK PLUG WIRES FOR OPEN CIRCUITS, CRACKS IN INSULATION OR IMPROPER SEATING OF TERMINALS AT SPARK PLUGS, DISTRIBUTOR, AND IGNITION COIL BEFORE PROCEEDING WITH THIS TEST. IF A TACHOMETER IS CONNECTED TO THE TACH TERMINAL ON THE IGNITION COIL, DISCONNECT IT BEFORE PROCEEDING WITH TEST.
- CHECK SPARK AT PLUG WITH SPARK TESTER J 26792 OR EQUIVALENT ST 125 WHILE CRANKING (IF NO SPARK ON ONE WIRE, CHECK A SECOND WIRE) A FEW SPARKS AND THEN NOTHING IS CONSIDERED NO SPARK.

NO SPARK →

SPARK → CHECK FUEL, SPARK PLUGS, ETC.

2.
- DISCONNECT 4 TERMINAL DISTRIBUTOR CONNECTOR. CHECK FOR SPARK AT IGNITION COIL TOWER WITH TESTER J 26792 OR EQUIVALENT ST 125 WHILE CRANKING. (LEAVE TESTER CONNECTED TO COIL TOWER FOR STEPS 3-6.)

NO SPARK →

SPARK → CHECK CAP, ROTOR AND COIL WIRES FOR OPENS, DAMAGE, OR WEAR. REPLACE AS NECESSARY.

3.
- DISCONNECT DISTRIBUTOR 2 TERMINAL "C/+" CONNECTOR.
- IGNITION SWITCH "ON" ENGINE STOPPED.
- CHECK VOLTS AT "+" AND "C" TERMINALS OF DISTRIBUTOR HARNESS CONNECTOR.

BOTH TERMINALS 10 VOLTS OR MORE | BOTH TERMINALS UNDER 10 VOLTS. | UNDER 10 VOLTS "C" TERMINAL ONLY.

4.
- RECONNECT DISTRIBUTOR 2 TERMINAL CONNECTOR.
- WITH IGNITION "ON," CHECK VOLTAGE FROM TACH TERMINAL TO GROUND. (TERMINAL MAY BE TAPED BACK IN HARNESS.)

REPAIR WIRE FROM MODULE "+" TERMINAL TO "B" TERMINAL OF BLACK IGNITION COIL CONNECTOR OR PRIMARY CIRCUIT TO IGNITION SWITCH.

CHECK FOR OPEN OR GROUND IN CIRCUIT FROM "C" TERMINAL TO IGNITION COIL. IF CIRCUIT IS OK, CHECK IGNITION COIL OR CONNECTION.

GREATER THAN 10 VOLTS | LESS THAN 1 VOLT | 1 TO 10 VOLTS

- CONNECT TEST LIGHT FROM TACHOMETER TERMINAL TO GROUND.
- CRANK ENGINE AND OBSERVE LIGHT.

REPAIR OPEN TACHOMETER LEAD OR CONNECTOR. REPEAT TEST #4.

REPLACE MODULE AND CHECK FOR SPARK FROM COIL AS IN STEP 6.

REFER TO IGNITION SYSTEM CHECK (2 OF 2).

SPARK → SYSTEM OK

NO SPARK → REPLACE IGNITION COIL.

4.3L (VIN W) ENGINE — SYSTEM CHECK — MID SIZE VAN (M/L) AND COMPACT PICKUP (S/T)

IGNITION SYSTEM CHECK

(Page 2 of 2)
WITHOUT 2.5L & 2.8L OR WITHOUT 4L80-E TRANSMISSION

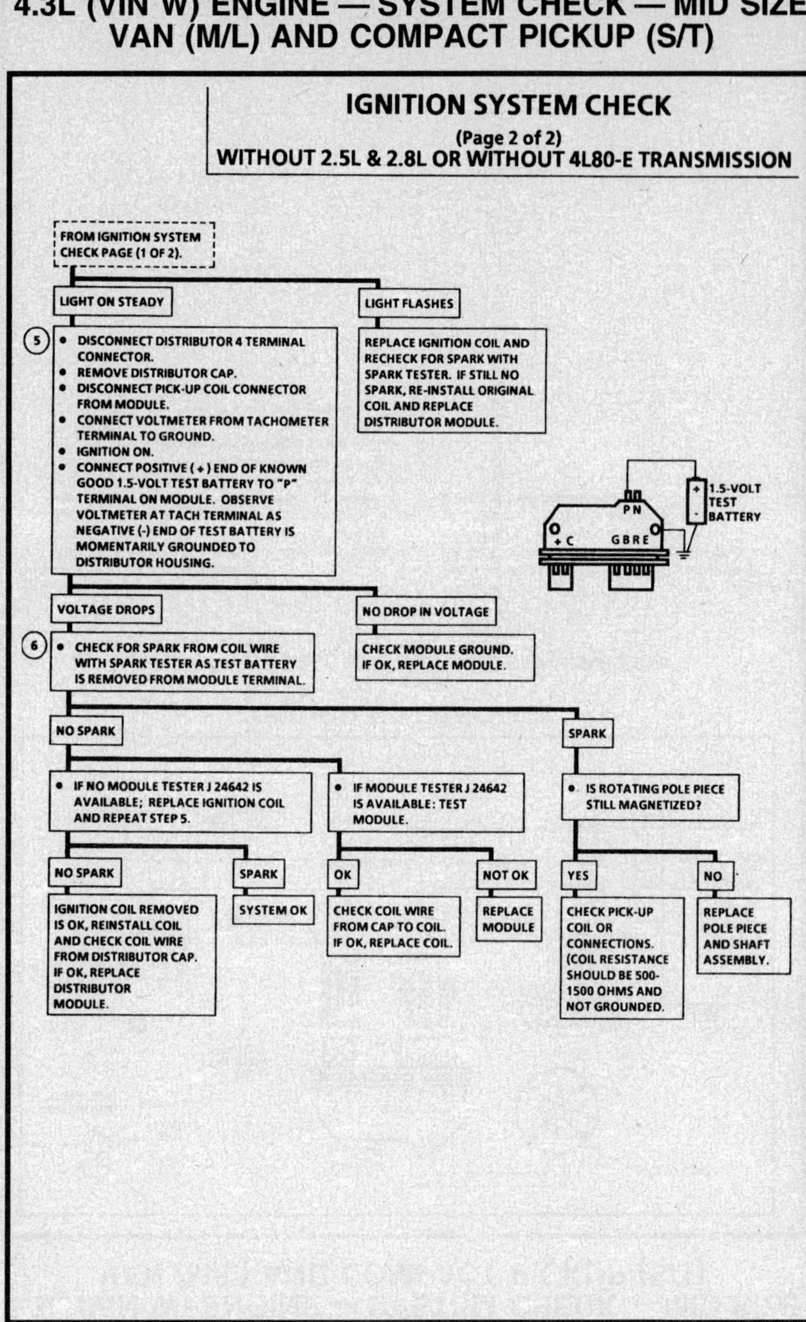

4.3L (VIN W) ENGINE — SYSTEM CHECK — MID SIZE VAN (M/L) AND COMPACT PICKUP (S/T)

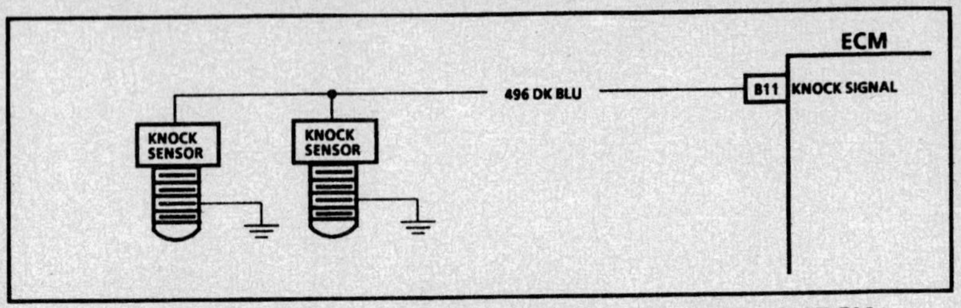

ELECTRONIC SPARK CONTROL (ESC) SYSTEM CHECK

WITH 4.3L CPI

Circuit Description:

The Code 43 circuit consists of two knock sensors with one wire that goes directly to the ECM. There are two Code 43 checks performed by the ECM. One check consists of monitoring CKT 496 for a voltage that is more than .63 volt and less than 4.4 volts.

If voltage is either too high or too low for 2 or more seconds, Code 43 will set. Once engine temperature reaches 85°C, MAP is over 83 kPa, and engine speed is less than 3800 rpm, the ECM will perform a self check. This self check will advance the timing until it receives a knock signal. If no knock signal is received, Code 43 will set.

Diagnostic Aids:

The ECM applies 5 volts to CKT 496. A 8200 ohm resistor in the knock sensors reduces the voltage to about 2.5 volts. When knock occurs, the knock sensor produces a small AC voltage that rides on top of the 2.5 volts already applied. An AC voltage monitor, in the ECM, is able to read this signal as knock and incrementally retard spark.

4.3L (VIN W) ENGINE — SYSTEM CHECK — MID SIZE VAN (M/L) AND COMPACT PICKUP (S/T)

ELECTRONIC SPARK CONTROL (ESC) SYSTEM CHECK
WITH 4.3L CPI

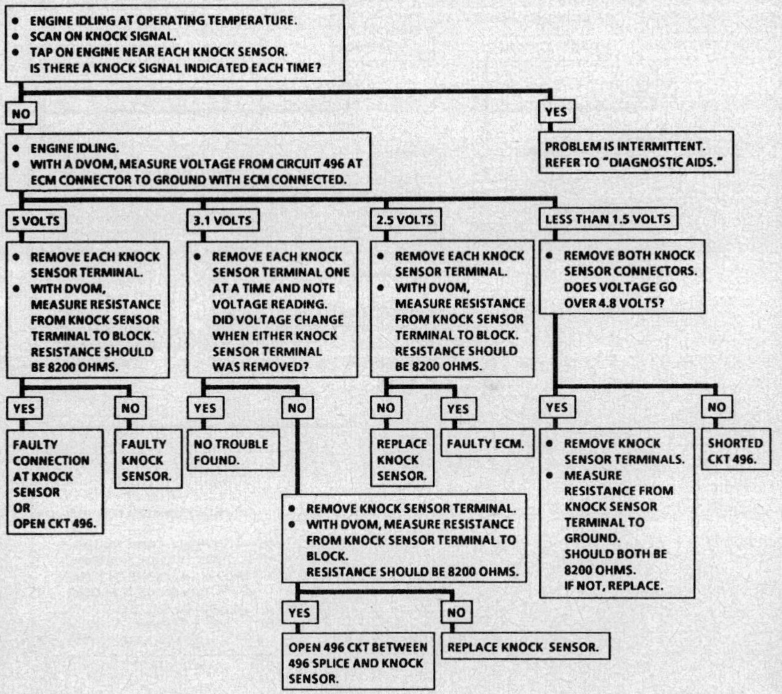

- ENGINE IDLING AT OPERATING TEMPERATURE.
- SCAN ON KNOCK SIGNAL.
- TAP ON ENGINE NEAR EACH KNOCK SENSOR.
  IS THERE A KNOCK SIGNAL INDICATED EACH TIME?

NO

- ENGINE IDLING.
- WITH A DVOM, MEASURE VOLTAGE FROM CIRCUIT 496 AT ECM CONNECTOR TO GROUND WITH ECM CONNECTED.

YES

PROBLEM IS INTERMITTENT. REFER TO "DIAGNOSTIC AIDS."

5 VOLTS

- REMOVE EACH KNOCK SENSOR TERMINAL.
- WITH DVOM, MEASURE RESISTANCE FROM KNOCK SENSOR TERMINAL TO BLOCK. RESISTANCE SHOULD BE 8200 OHMS.

YES — FAULTY CONNECTION AT KNOCK SENSOR OR OPEN CKT 496.

NO — FAULTY KNOCK SENSOR.

3.1 VOLTS

- REMOVE EACH KNOCK SENSOR TERMINAL ONE AT A TIME AND NOTE VOLTAGE READING. DID VOLTAGE CHANGE WHEN EITHER KNOCK SENSOR TERMINAL WAS REMOVED?

YES — NO TROUBLE FOUND.

NO

- REMOVE KNOCK SENSOR TERMINAL.
- WITH DVOM, MEASURE RESISTANCE FROM KNOCK SENSOR TERMINAL TO BLOCK. RESISTANCE SHOULD BE 8200 OHMS.

YES — OPEN 496 CKT BETWEEN 496 SPLICE AND KNOCK SENSOR.

NO — REPLACE KNOCK SENSOR.

2.5 VOLTS

- REMOVE EACH KNOCK SENSOR TERMINAL.
- WITH DVOM, MEASURE RESISTANCE FROM KNOCK SENSOR TERMINAL TO BLOCK. RESISTANCE SHOULD BE 8200 OHMS.

NO — REPLACE KNOCK SENSOR.

YES — FAULTY ECM.

LESS THAN 1.5 VOLTS

- REMOVE BOTH KNOCK SENSOR CONNECTORS. DOES VOLTAGE GO OVER 4.8 VOLTS?

YES

- REMOVE KNOCK SENSOR TERMINALS.
- MEASURE RESISTANCE FROM KNOCK SENSOR TERMINAL TO GROUND. SHOULD BOTH BE 8200 OHMS. IF NOT, REPLACE.

NO — SHORTED CKT 496.

4.3L (VIN W) ENGINE — SYSTEM CHECK — MID SIZE VAN (M/L) AND COMPACT PICKUP (S/T)

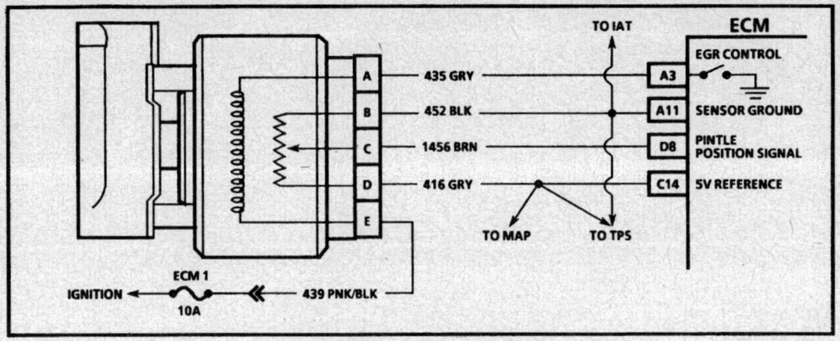

EGR SYSTEM CHECK
WITH LINEAR EGR VALVE

Circuit Description:

To control Exhaust Gas Recirculation (EGR), the ECM operates the linear EGR valve. The linear EGR valve is a small motor with a pintle that is normally closed. By providing a ground path, the ECM will open the pintle and allow exhaust gasses to pass through the valve. The ECM control of the EGR is based on the following inputs:

- Engine coolant temperature - above 25°C
- TPS - "OFF" idle
- MAP
- IAT
- BARO
- Park/Neutral Switch
- Pintle Position Sensor

If Code 23 is stored, refer to Code 23 in COMPUTER COMMAND CONTROL

If Code 32 is stored, refer to Code 32 in COMPUTER COMMAND CONTROL. This chart is used for faulty valve, plugged EGR passages, or improper wiring connections.

Test Description: Number(s) below refer to circled number(s) on the diagnostic chart.

1. At idle, the EGR valve should not be open greater than 3%. A misadjusted Park/Neutral (P/N) Switch or the EGR valve stuck open will allow exhaust gasses to enter the combustion chamber and create a rough idle condition.
2. Checks the pintle's ability to be manually commanded to the desired positions.
3. Checks the electrical circuit of the EGR valve and connecting components.
4. Checks for plugged EGR passages or a faulty EGR valve.

4.3L (VIN W) ENGINE — SYSTEM CHECK — MID SIZE VAN (M/L) AND COMPACT PICKUP (S/T)

EGR SYSTEM CHECK
WITH LINEAR EGR VALVE

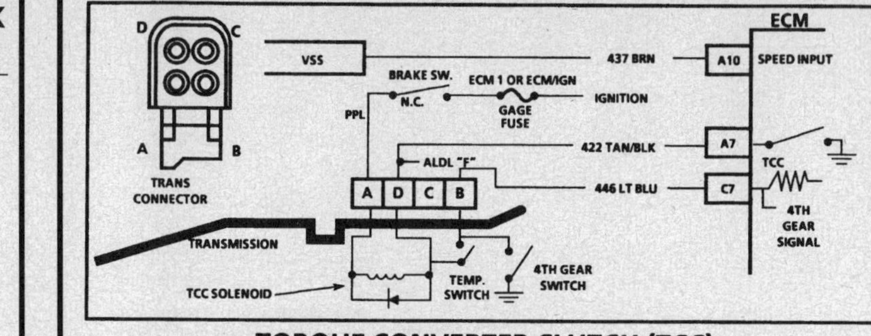

4.3L (VIN W) ENGINE — SYSTEM CHECK — MID SIZE VAN (M/L) AND COMPACT PICKUP (S/T)

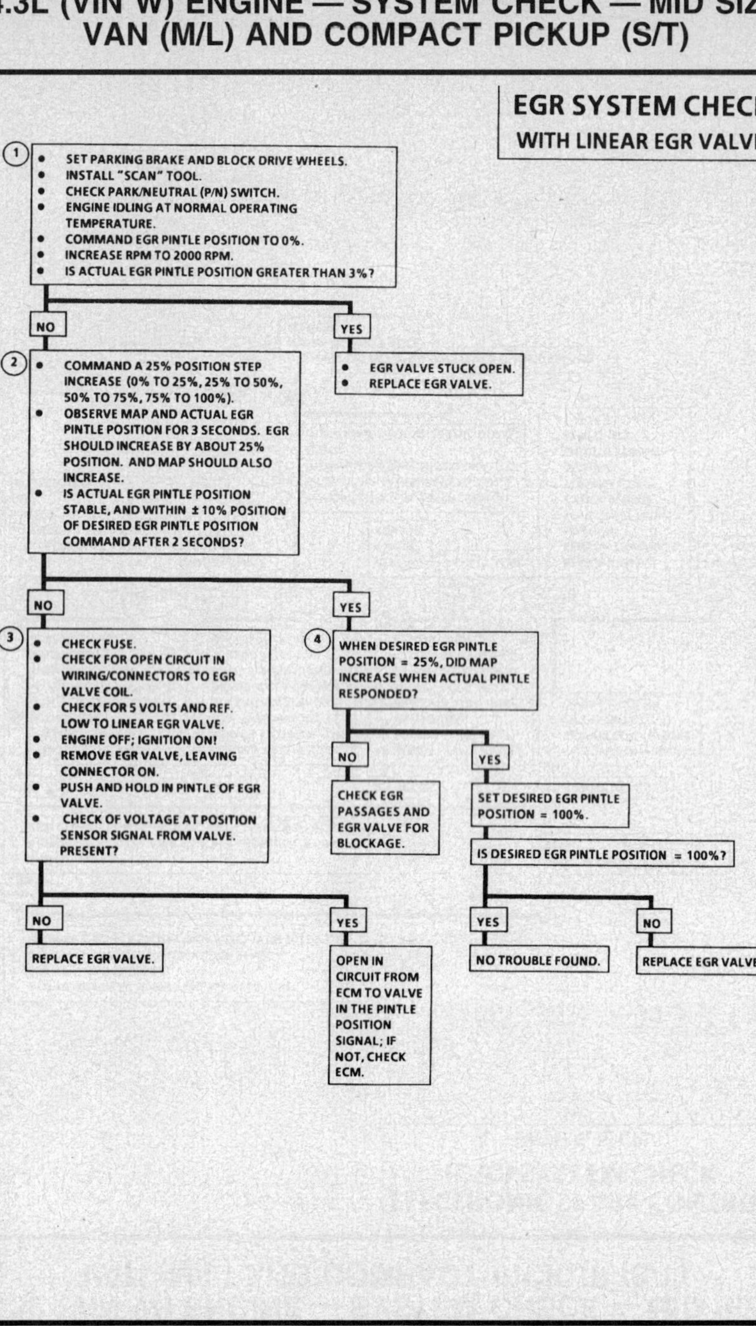

TORQUE CONVERTER CLUTCH (TCC)
(ELECTRICAL DIAGNOSIS)
WITH 4.3L, V8, AND WITHOUT 4L80-E

Circuit Description:

The purpose of the automatic transmission Torque Converter Clutch (TCC) feature is to eliminate the power loss of the torque converter stage when the vehicle is in a cruise condition. This allows for the convenience of the automatic transmission, and the fuel economy of a manual transmission.

Fused battery ignition is supplied to the TCC solenoid through the TCC brake switch. The ECM will engage TCC by grounding CKT 422 to energize the solenoid. TCC will engage when:

- Vehicle speed above 30 mph (48 km/h).
- Engine at normal operating temperature (above 65°C) (149°F).
- Throttle position sensor output not changing, indicating a steady road speed.
- Brake switch closed.
- 3rd or 4th gears.

Test Description: Number(s) below refer to circled number(s) on the diagnostic chart.

1. A test light "ON" indicates battery voltage and continuity through the TCC solenoid is OK.
2. Checks for vehicle speed sensor signal to ECM using a "Scan" tool.
3. Checks for 4th gear signal to ECM. This signal will not prevent TCC engagement, but could cause a change in the engage and disengage speed points.

Diagnostic Aids:

Solenoid coil resistance must measure more than 20 ohms. Less resistance will cause early failure of the ECM "driver."

With an ohmmeter, check the solenoid coil resistance of all ECM controlled solenoids and relays-before installing a replacement ECM. Replace any solenoid or relay that measures less than 20 ohms resistance.

To prevent TCC overheat condition, TCC temperature closes at 279°F ± 7° and reopens at 259°F ± 9°.

4.3L (VIN W) ENGINE — SYSTEM CHECK — MID SIZE VAN (M/L) AND COMPACT PICKUP (S/T)

TORQUE CONVERTER CLUTCH (TCC)

(ELECTRICAL DIAGNOSIS)
WITH 4.3L, V8, AND WITHOUT 4L80-E

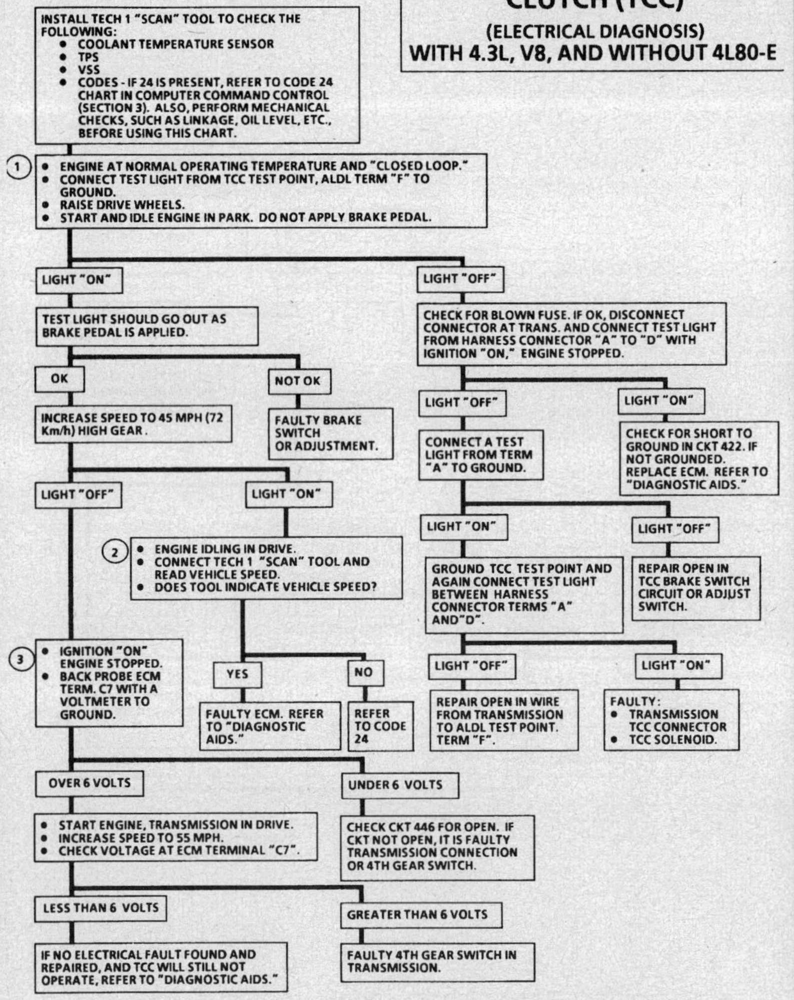
INSTALL TECH 1 "SCAN" TOOL TO CHECK THE FOLLOWING:
• COOLANT TEMPERATURE SENSOR
• TPS
• VSS
• CODES - IF 24 IS PRESENT, REFER TO CODE 24 CHART IN COMPUTER COMMAND CONTROL (SECTION 3). ALSO, PERFORM MECHANICAL CHECKS, SUCH AS LINKAGE, OIL LEVEL, ETC., BEFORE USING THIS CHART.
1
• ENGINE AT NORMAL OPERATING TEMPERATURE AND "CLOSED LOOP."
• CONNECT TEST LIGHT FROM TCC TEST POINT, ALDL TERM "F" TO GROUND.
• RAISE DRIVE WHEELS.
• START AND IDLE ENGINE IN PARK. DO NOT APPLY BRAKE PEDAL.
LIGHT "ON"
TEST LIGHT SHOULD GO OUT AS BRAKE PEDAL IS APPLIED.
OK
INCREASE SPEED TO 45 MPH (72 Km/h) HIGH GEAR.
NOT OK
FAULTY BRAKE SWITCH OR ADJUSTMENT.
LIGHT "OFF"
LIGHT "ON"
2
• ENGINE IDLING IN DRIVE.
• CONNECT TECH 1 "SCAN" TOOL AND READ VEHICLE SPEED.
• DOES TOOL INDICATE VEHICLE SPEED?
YES
FAULTY ECM. REFER TO "DIAGNOSTIC AIDS."
NO
REFER TO CODE 24
3
• IGNITION "ON" ENGINE STOPPED.
• BACK PROBE ECM TERM. C7 WITH A VOLTMETER TO GROUND.
OVER 6 VOLTS
• START ENGINE, TRANSMISSION IN DRIVE.
• INCREASE SPEED TO 55 MPH.
• CHECK VOLTAGE AT ECM TERMINAL "C7".
LESS THAN 6 VOLTS
IF NO ELECTRICAL FAULT FOUND AND REPAIRED, AND TCC WILL STILL NOT OPERATE, REFER TO "DIAGNOSTIC AIDS."
GREATER THAN 6 VOLTS
FAULTY 4TH GEAR SWITCH IN TRANSMISSION.
UNDER 6 VOLTS
CHECK CKT 446 FOR OPEN. IF CKT NOT OPEN, IT IS FAULTY TRANSMISSION CONNECTION OR 4TH GEAR SWITCH.
LIGHT "OFF"
CHECK FOR BLOWN FUSE. IF OK, DISCONNECT CONNECTOR AT TRANS. AND CONNECT TEST LIGHT FROM HARNESS CONNECTOR "A" TO "D" WITH IGNITION "ON," ENGINE STOPPED.
LIGHT "OFF"
CONNECT A TEST LIGHT FROM TERM "A" TO GROUND.
LIGHT "ON"
CHECK FOR SHORT TO GROUND IN CKT 422. IF NOT GROUNDED. REPLACE ECM. REFER TO "DIAGNOSTIC AIDS."
LIGHT "ON"
GROUND TCC TEST POINT AND AGAIN CONNECT TEST LIGHT BETWEEN HARNESS CONNECTOR TERMS "A" AND "D".
LIGHT "OFF"
REPAIR OPEN IN TCC BRAKE SWITCH CIRCUIT OR ADJUST SWITCH.
LIGHT "OFF"
REPAIR OPEN IN WIRE FROM TRANSMISSION TO ALDL TEST POINT. TERM "F".
LIGHT "ON"
FAULTY:
• TRANSMISSION
• TCC CONNECTOR
• TCC SOLENOID.

"AFTER REPAIRS." CONFIRM "CLOSED LOOP" OPERATION AND NO "SERVICE ENGINE SOON" LIGHT.

4.3L (VIN W) ENGINE — SYSTEM CHECK — MID SIZE VAN (M/L) AND COMPACT PICKUP (S/T)

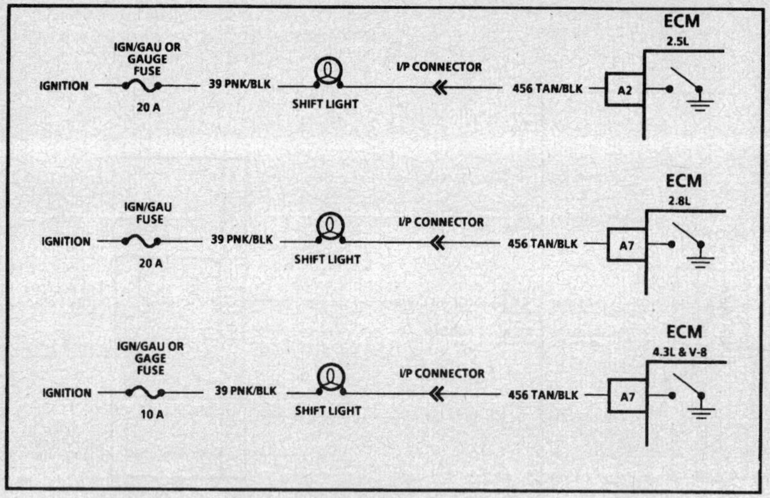
IGNITION
IGN/GAU OR GAUGE FUSE
20 A
39 PNK/BLK
SHIFT LIGHT
I/P CONNECTOR
456 TAN/BLK
A2
ECM
2.5L
IGNITION
IGN/GAU FUSE
20 A
39 PNK/BLK
SHIFT LIGHT
I/P CONNECTOR
456 TAN/BLK
A7
ECM
2.8L
IGNITION
IGN/GAU OR GAGE FUSE
10 A
39 PNK/BLK
SHIFT LIGHT
I/P CONNECTOR
456 TAN/BLK
A7
ECM
4.3L & V-8

MANUAL TRANSMISSION SHIFT LIGHT CHECK

VEHICLES BELOW 8500 GVW ONLY

Circuit Description:

The ECM uses information from the following inputs to control the shift light:

- Coolant Temperature Sensor (CTS).
- Throttle Position Sensor (TPS).
- Vehicle Speed Sensor (VSS).
- RPM.

The ECM uses the measured rpm and the vehicle speed to calculate what gear the vehicle is in. It's this calculation that determines when the shift light should be turned "ON".

Test Description: Number(s) below refer to circled number(s) on the diagnostic chart.

1. This should not turn "ON" the shift light. If the light is "ON," there is a short to ground in CKT 456 wiring, or a fault in the ECM.
2. This should turn "ON" the shift light.
3. This checks for an open in the shift light circuit, or a faulty ECM.

4.3L (VIN W) ENGINE — ECM WIRING — MID SIZE VAN (M/L)

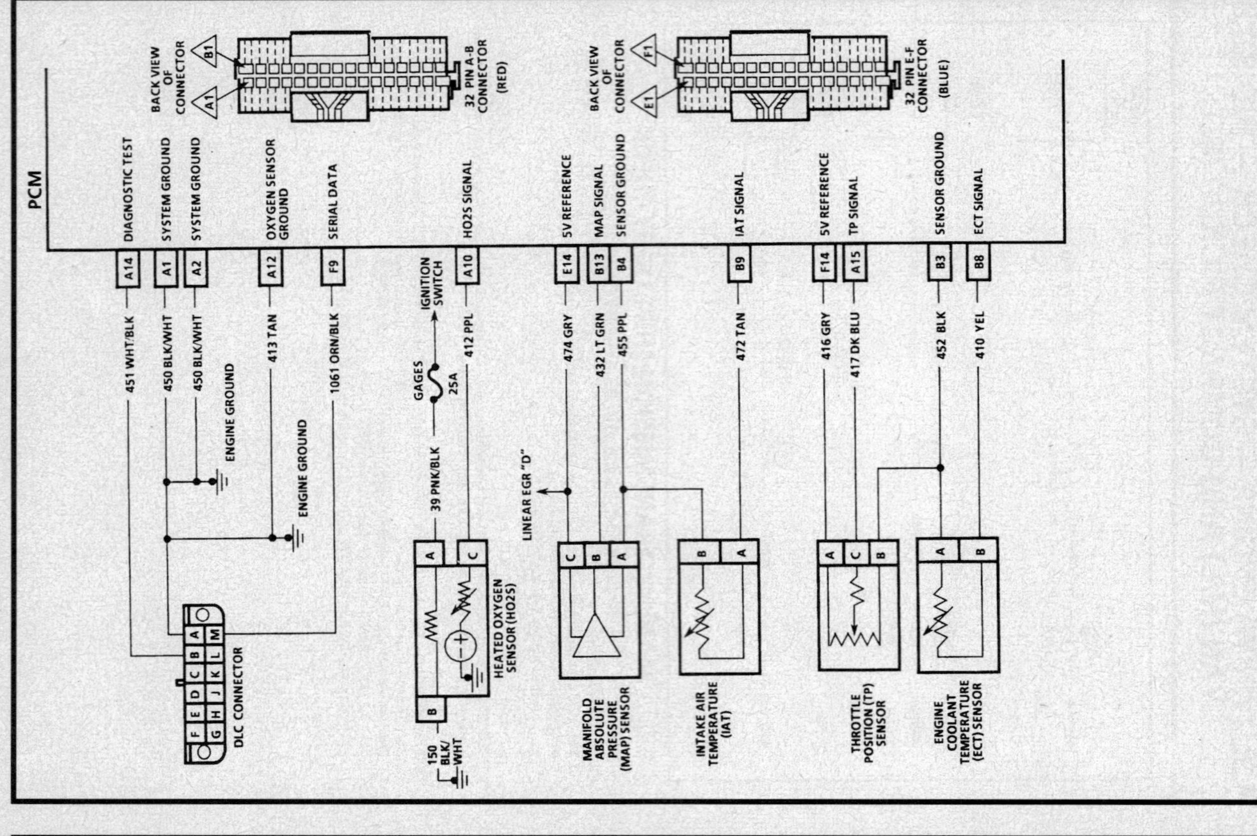

4.3L (VIN W) ENGINE — SYSTEM CHECK — MID SIZE VAN (M/L) AND COMPACT PICKUP (S/T)

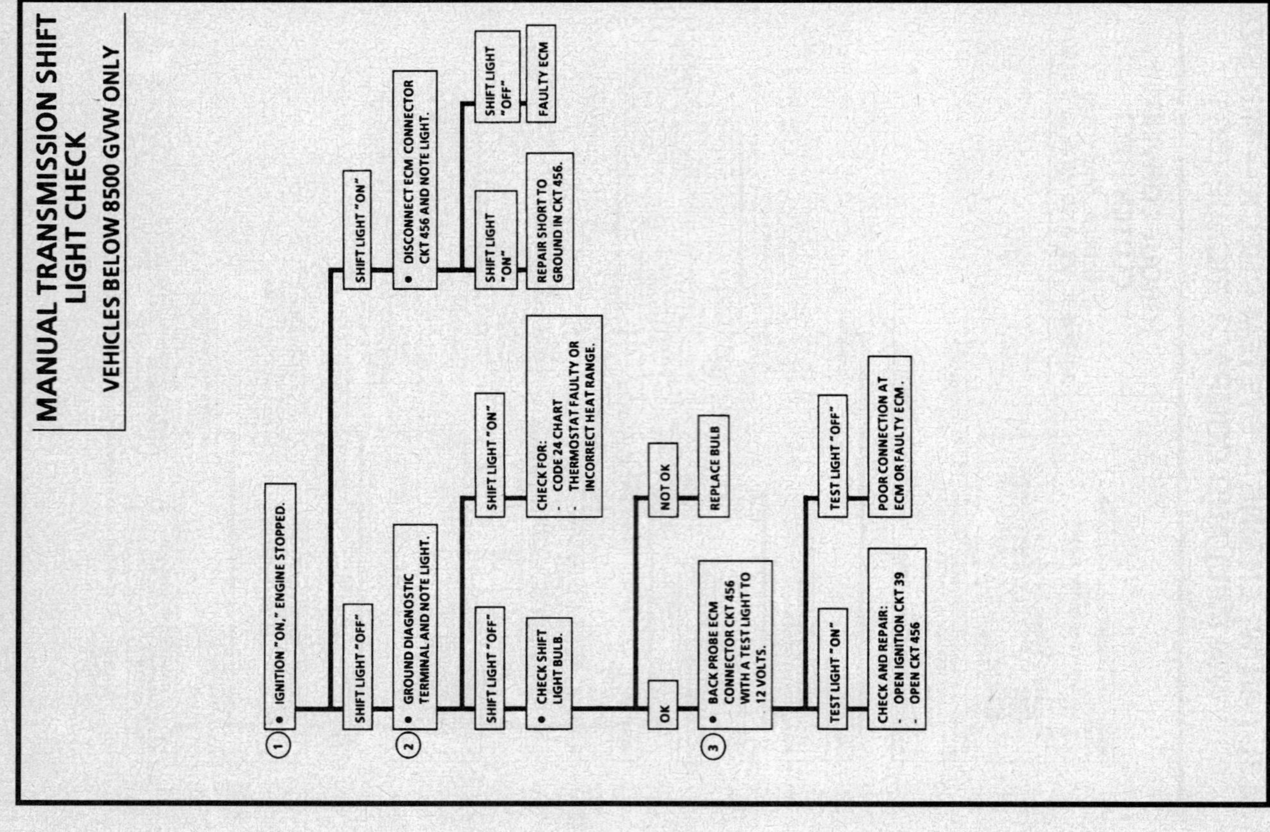

4.3L (VIN W) ENGINE — ECM WIRING — MID SIZE VAN (M/L)

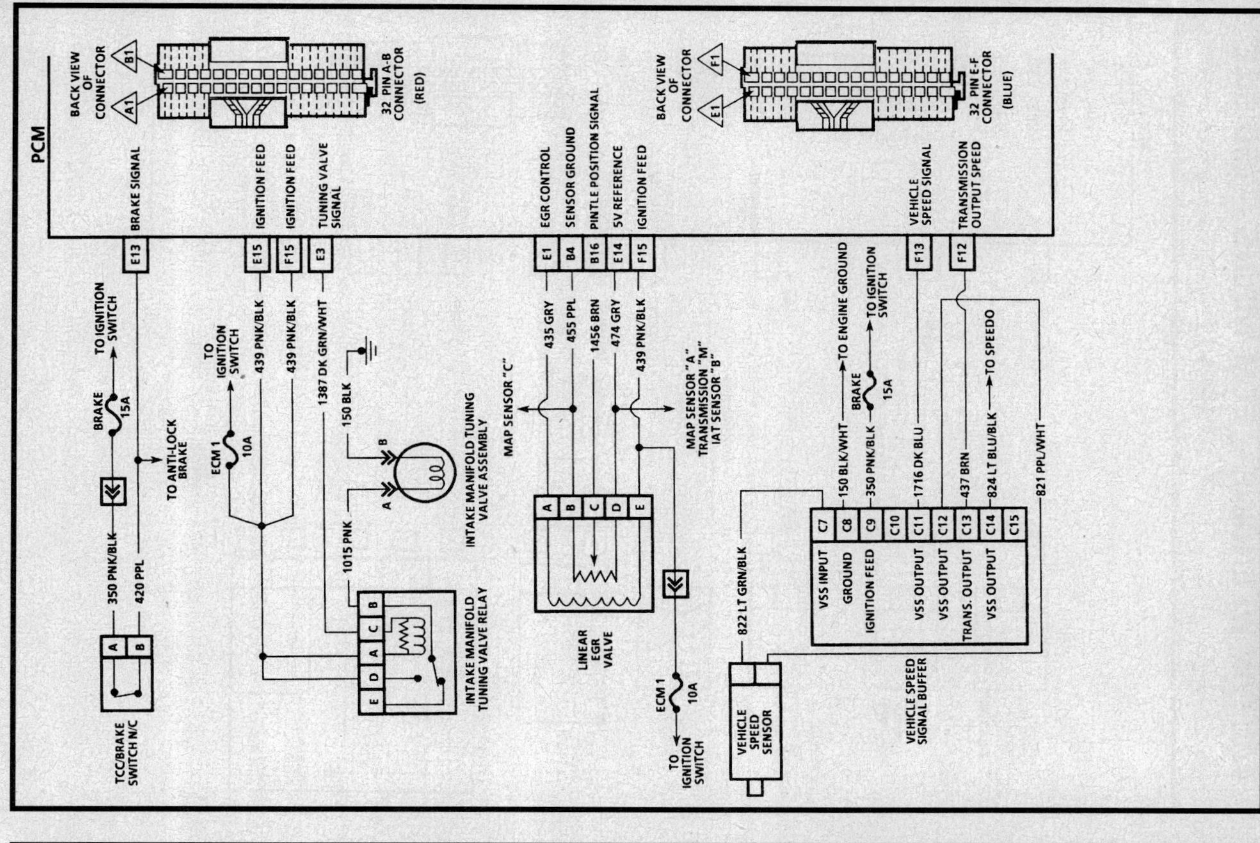

4.3L (VIN W) ENGINE — ECM WIRING — MID SIZE VAN (M/L)

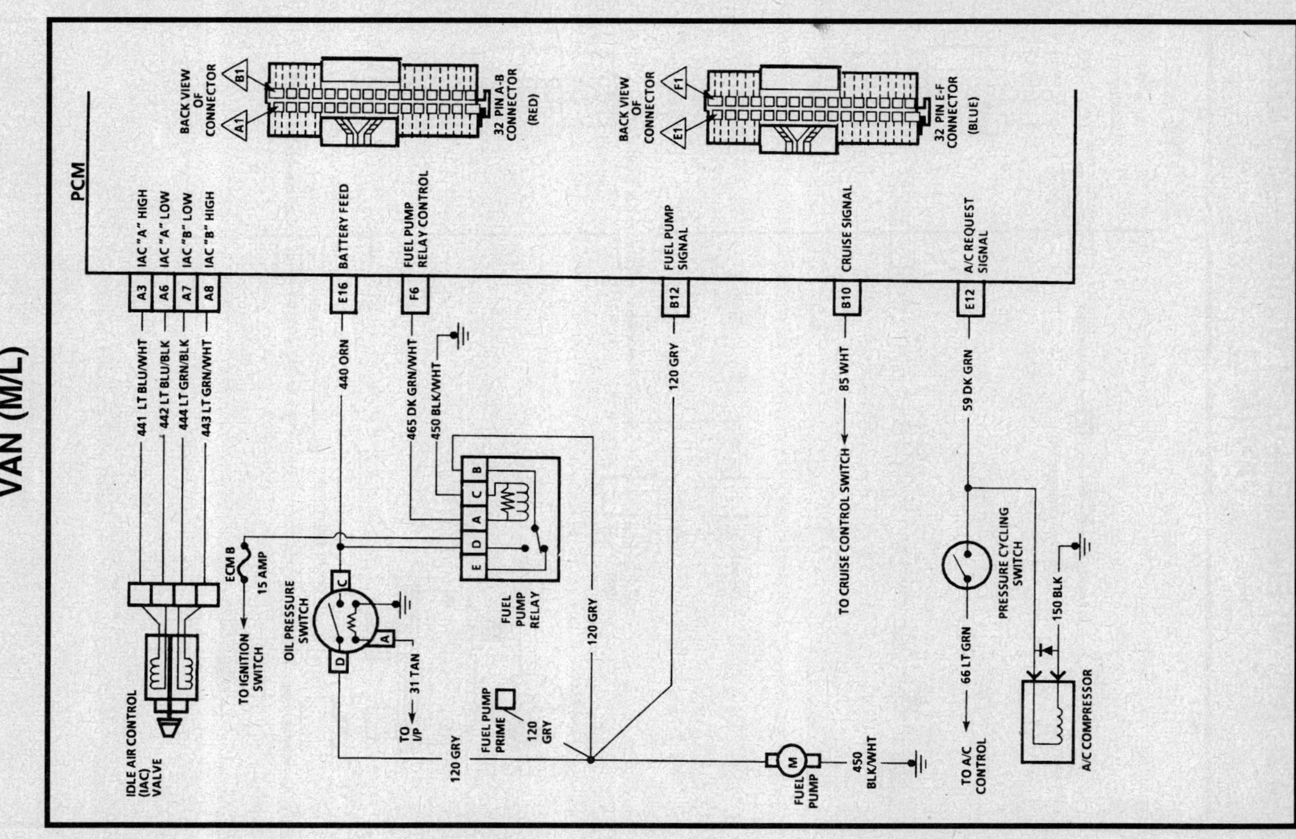

4.3L (VIN W) ENGINE — ECM WIRING — MID SIZE VAN (M/L)

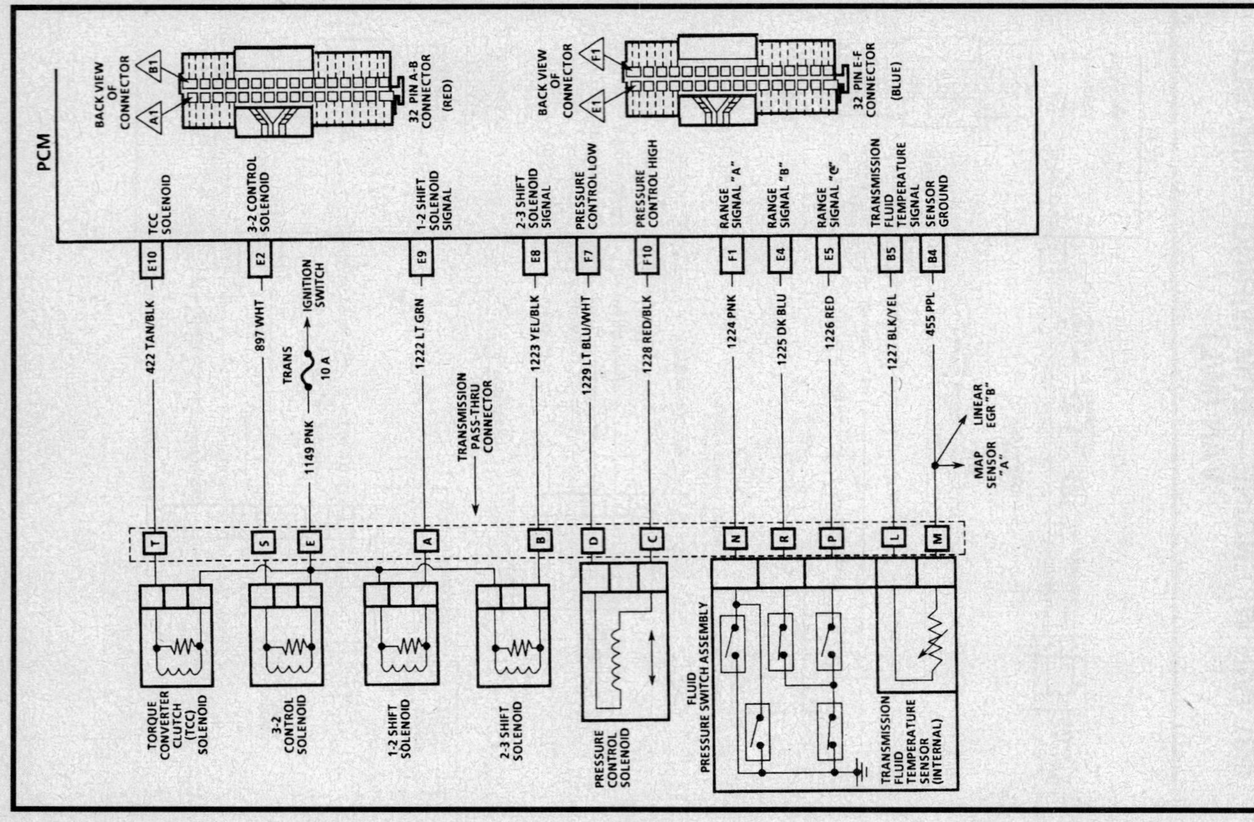

4.3L (VIN W) ENGINE — ECM WIRING — MID SIZE VAN (M/L)

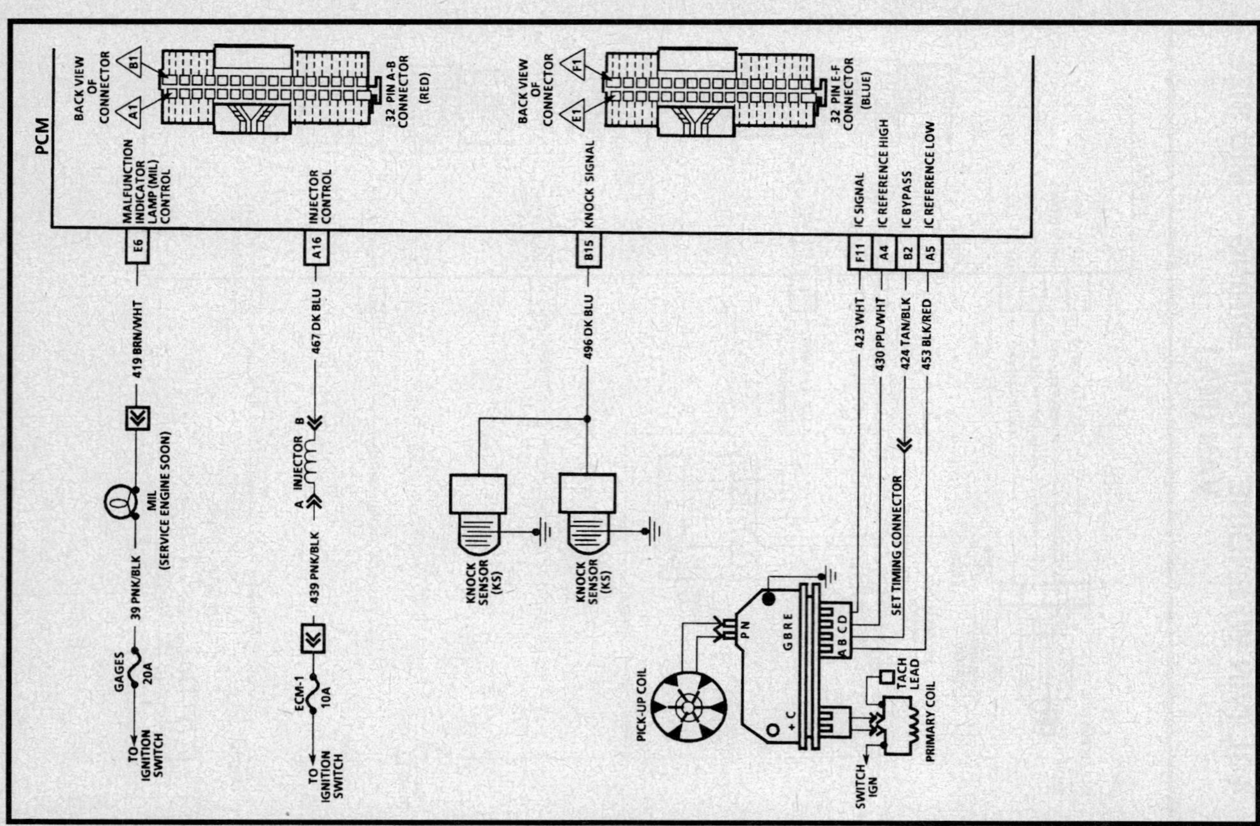

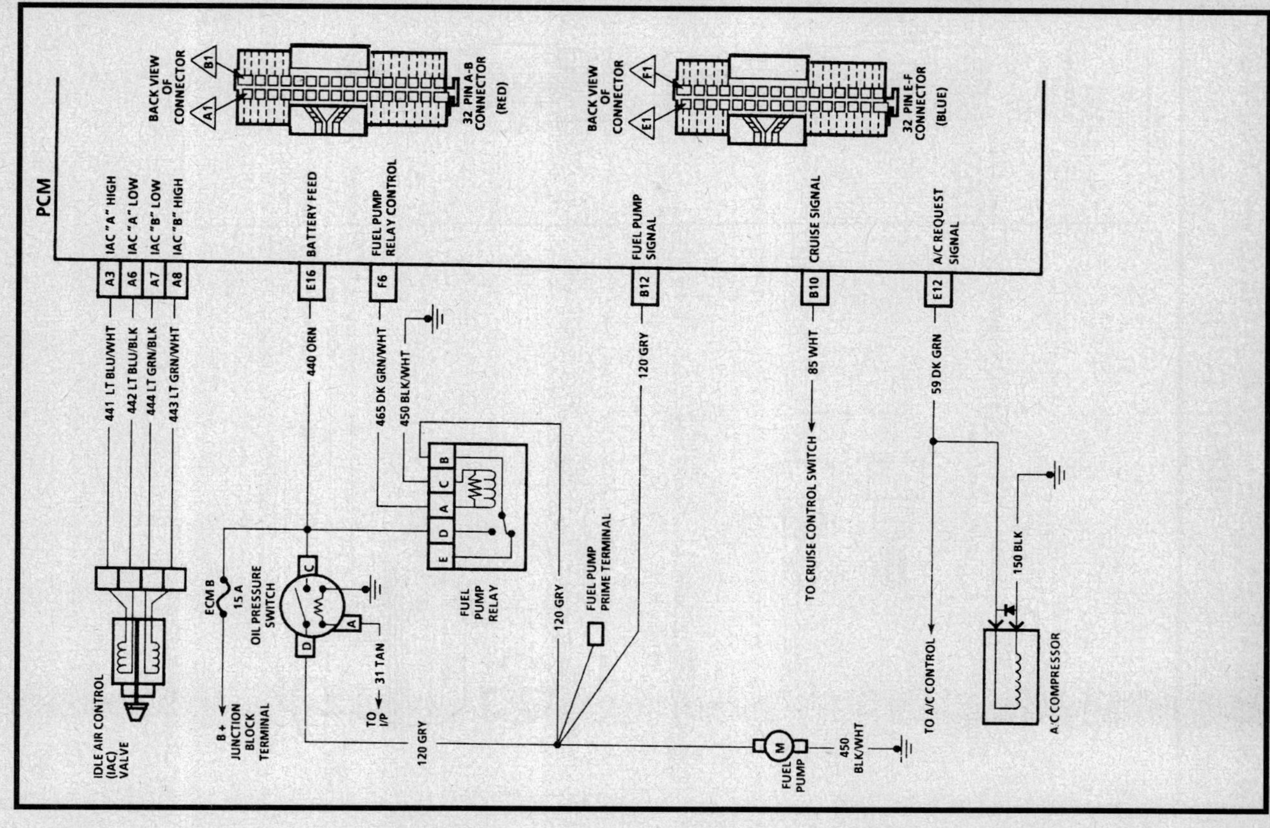
4.3L (VIN W) ENGINE — ECM WIRING — COMPACT PICKUP (S/T)
PCM
BACK VIEW OF CONNECTOR
32 PIN A-B CONNECTOR (RED)
BACK VIEW OF CONNECTOR
32 PIN E-F CONNECTOR (BLUE)
IAC "A" HIGH
IAC "A" LOW
IAC "B" LOW
IAC "B" HIGH
BATTERY FEED
FUEL PUMP RELAY CONTROL
FUEL PUMP SIGNAL
CRUISE SIGNAL
A/C REQUEST SIGNAL
A3
A6
A7
A8
E16
F6
B12
B10
E12
441 LT BLU/WHT
442 LT BLU/BLK
444 LT GRN/BLK
443 LT GRN/WHT
440 ORN
465 DK GRN/WHT
450 BLK/WHT
120 GRY
85 WHT
59 DK GRN
IDLE AIR CONTROL (IAC) VALVE
ECM B 15 A
OIL PRESSURE SWITCH
B+ JUNCTION BLOCK TERMINAL
TO 31 TAN
TO I/P
FUEL PUMP RELAY
FUEL PUMP PRIME TERMINAL
120 GRY
120 GRY
FUEL PUMP
450 BLK/WHT
M
TO CRUISE CONTROL SWITCH
TO A/C CONTROL
150 BLK
A/C COMPRESSOR

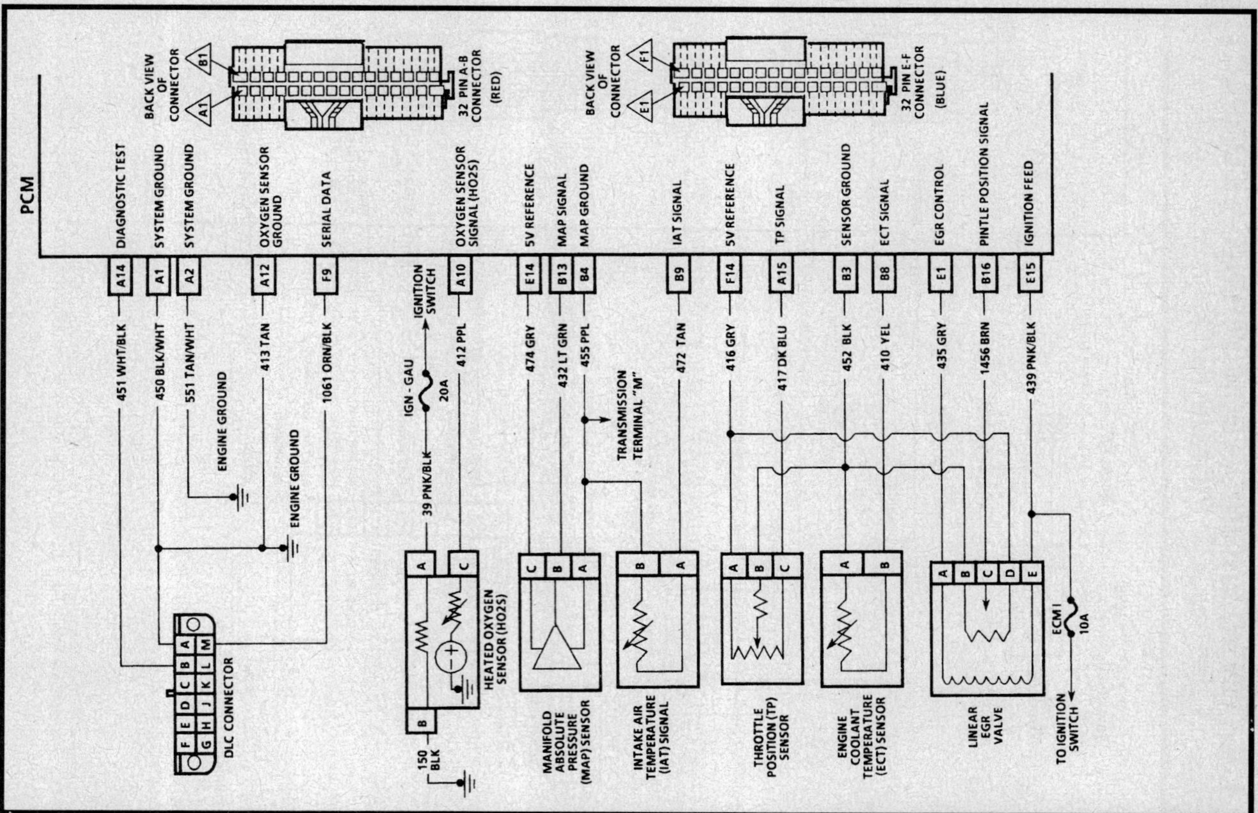
4.3L (VIN W) ENGINE — ECM WIRING — COMPACT PICKUP (S/T)
PCM
BACK VIEW OF CONNECTOR
32 PIN A-B CONNECTOR (RED)
BACK VIEW OF CONNECTOR
32 PIN E-F CONNECTOR (BLUE)
DIAGNOSTIC TEST
SYSTEM GROUND
SYSTEM GROUND
OXYGEN SENSOR GROUND
SERIAL DATA
OXYGEN SENSOR SIGNAL (HO2S)
5V REFERENCE
MAP SIGNAL
MAP GROUND
IAT SIGNAL
5V REFERENCE
TP SIGNAL
SENSOR GROUND
ECT SIGNAL
EGR CONTROL
PINTLE POSITION SIGNAL
IGNITION FEED
A14
A1
A2
A12
F9
A10
E14
B13
B4
B9
F14
A15
B3
B8
E1
B16
E15
451 WHT/BLK
450 BLK/WHT
551 TAN/WHT
413 TAN
1061 ORN/BLK
412 PPL
474 GRY
432 LT GRN
455 PPL
472 TAN
416 GRY
417 DK BLU
452 BLK
410 YEL
435 GRY
1456 BRN
439 PNK/BLK
ENGINE GROUND
ENGINE GROUND
DLC CONNECTOR
F E D C B A
G H J K L M
IGN - GAU
IGNITION SWITCH
20A
39 PNK/BLK
150 BLK
HEATED OXYGEN SENSOR (HO2S)
MANIFOLD ABSOLUTE PRESSURE (MAP) SENSOR
INTAKE AIR TEMPERATURE (IAT) SIGNAL
THROTTLE POSITION (TP) SENSOR
ENGINE COOLANT TEMPERATURE (ECT) SENSOR
TRANSMISSION TERMINAL "M"
LINEAR EGR VALVE
ECM I 10A
TO IGNITION SWITCH

4.3L (VIN W) ENGINE — ECM WIRING — COMPACT PICKUP (S/T)

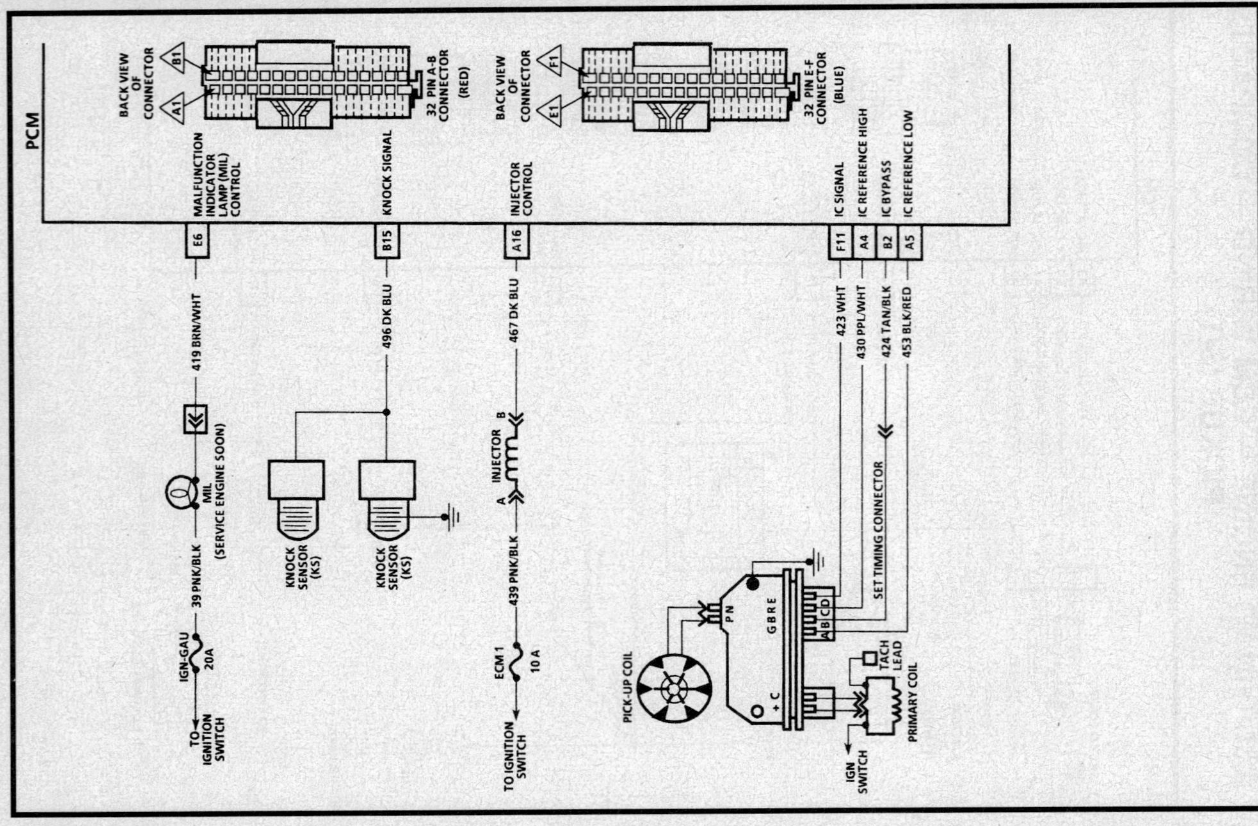

4.3L (VIN W) ENGINE — ECM WIRING — COMPACT PICKUP (S/T)

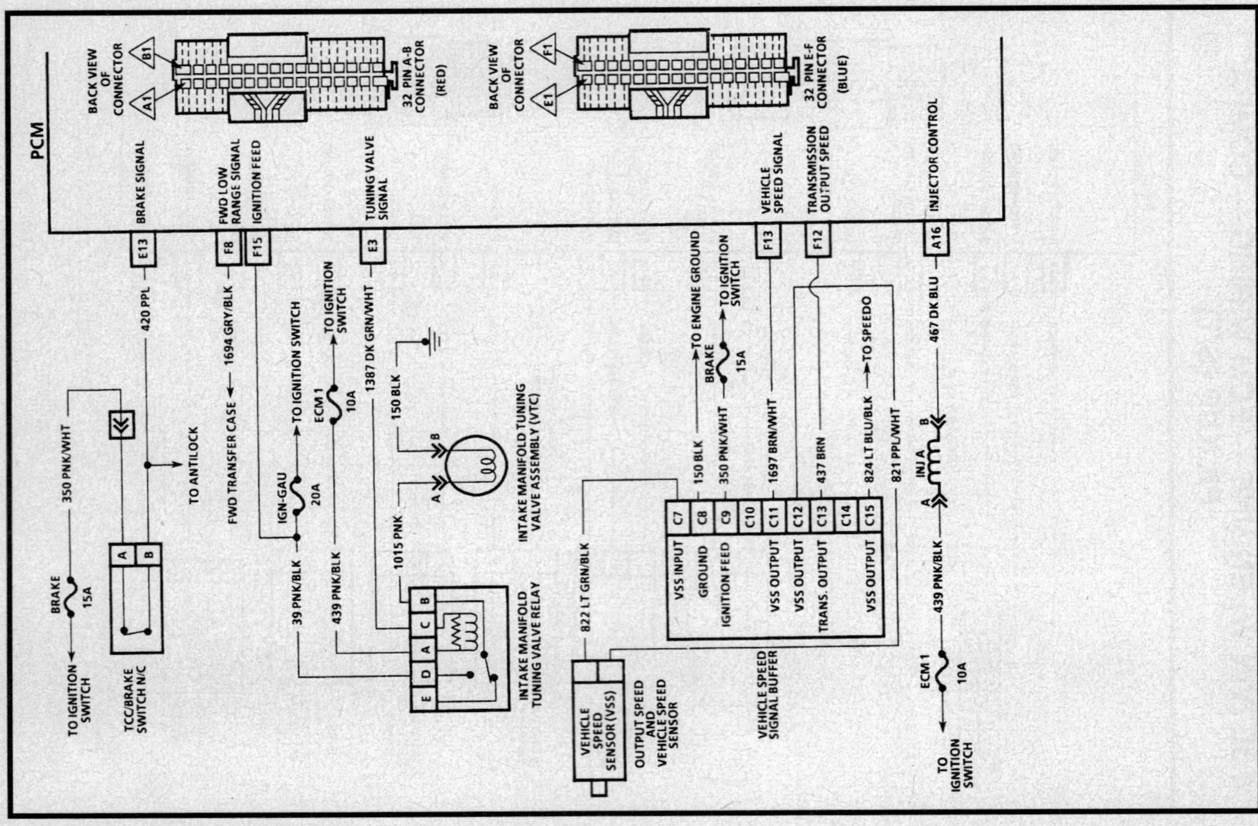

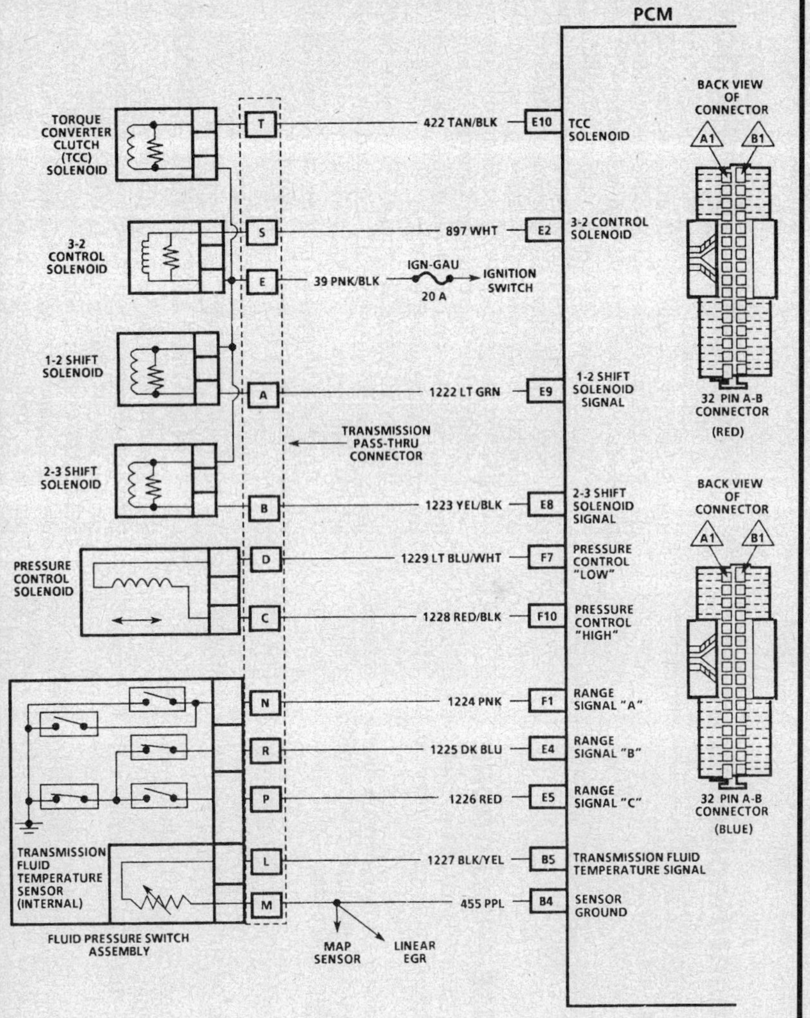

4.3L (VIN W) ENGINE — ECM CONNECTOR — MID SIZE VAN (M/L) AND COMPACT PICKUP (S/T)

ECM CONNECTOR IDENTIFICATION

This ECM voltage chart is for use with a digital voltmeter to further aid in diagnosis. The voltages you get may vary due to low battery charge or other reasons, but they should be very close. The "B+" symbol indicates a nominal system voltage of 12-14 volts.

THE FOLLOWING CONDITIONS MUST BE MET BEFORE TESTING:

- Engine at operating temperature ● Closed Loop ● Engine idling (for "Engine Run" column)
- Test terminal not grounded ● Tech 1 not installed

PIN	CIRCUIT	CKT #	WIRE COLOR	KEY "ON"	ENG. RUN
A1	FUEL PUMP RELAY DRIVER	465	DK GRN WHT	(3)	B+
A2	TCC/CONTROL	422	TAN/BLK	B+	B+
A3	EGR CONTROL	435	GRY	B+	B+
A4	NOT USED			-	-
A5	S.E.S.	419	BRN/WHT	*0	B+
A6	IGNITION	439	PNK/BLK	B+	B+
A7	TUNING VALVE	1387	DK GRN WHT	B+	B+
A8	SERIAL DATA	461 1061	ORN ORN/BLK	(1)	(1)
A9	DIAG. TERMINAL	451	WHT/BLK	5	5
A10	DRAC/VSS	437	BRN	(2)	(2)
A11	TPS/IAT GND	452	BLK	*0	*0
A12	SYSTEM GND	450	BLK/WHT	*0	*0

PIN	CIRCUIT	CKT #	WIRE COLOR	KEY "ON"	ENG. RUN
B1	BATTERY	440	ORN	B+	B+
B2	FUEL PUMP SIGNAL	120	GRY	(3)	B+
B3	DIST. REF. LOW	453	BLK/RED	*0	*0
B4	IGNITION FEED	439	PNK/BLK	B+	B+
B5	DIST. REF. HIGH	430	PPL/WHT	*0	1.3
B6	NOT USED			-	-
B7	NOT USED			-	-
B8	A/C SIGNAL	59	DK GRN	(4)	(4)
B9	HIGH GEAR	446	LT BLU	B+	B+
B10	P/N SW	434	ORN/BLK	(5)	(5)
B11	ESC SIGNAL	496	DK BLU	2.4	2.4
B12	NOT USED			-	-

(1) Varies from 2 to 5 volts.
(2) Varies from 0 volts to battery voltage depending on position of drive wheels.
(3) Battery voltage for 2 seconds after ignition "ON."
(4) 0 volts A/C "OFF," battery voltage A/C "ON."
(5) 0 volts in P/N battery voltage in gear. Less than .5 volts.

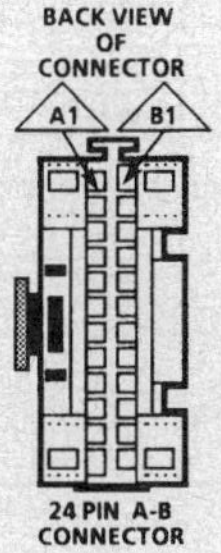

ENGINE: 4.3L (L35) CPI

4.3L (VIN W) ENGINE — ECM WIRING — COMPACT PICKUP (S/T)

4.3L (VIN W) ENGINE — ECM CONNECTOR — MID SIZE VAN (M/L) AND COMPACT PICKUP (S/T)

ECM CONNECTOR IDENTIFICATION

This ECM voltage chart is for use with a digital voltmeter to further aid in diagnosis. The voltages you get may vary due to low battery charge or other reasons, but they should be very close. The "B +" symbol indicates a nominal system voltage of 12-14 volts.

THE FOLLOWING CONDITIONS MUST BE MET BEFORE TESTING:
- Engine at operating temperature • "Closed Loop" • Engine idling (for "Engine Run" column)
- Test terminal not grounded • Tech 1 not installed

PIN	PIN FUNCTION	CKT #	WIRE COLOR	NORMAL VOLTAGE KEY "ON"	NORMAL VOLTAGE ENG RUN
C1	NOT USED	–	–	–	–
C2	NOT USED	–	–	–	–
C3	IAC "B" LOW	444	GRN/BLK	(2)	(2)
C4	IAC "B" HIGH	443	GRN/WHT	(2)	(2)
C5	IAC "A" HIGH	441	BLU/WHT	(2)	(2)
C6	IAC "A" LOW	442	BLU/BLK	(2)	(2)
C7	NOT USED	–	–	–	–
C8	NOT USED	–	–	–	–
C9	NOT USED	–	–	–	–
C10	CTS SIGNAL	410	YEL	(4)	(4)
C11	MAP SIGNAL	432	LT GRN	4.7	1.4 †
C12	IAT SIGNAL	472	TAN	(4)	(4)
C13	TPS SIGNAL	417	DK BLU	(5)	.6
C14	5V REF. (MAP, TPS)	416	GRY	5	5
C15	NOT USED	–	–	–	–
C16	BATTERY +	440	ORN	B +	B +

PIN	PIN FUNCTION	CKT #	WIRE COLOR	NORMAL VOLTAGE KEY "ON"	NORMAL VOLTAGE ENG RUN
D1	SYSTEM GROUND	450 551	BLK/WHT TAN/WHT	*0	*0
D2	SENSOR GROUND (MAP, CTS)	455	PPL	*0	*0
D3	NOT USED	–	–	–	–
D4	EST SIGNAL	423	WHT	*0	1.0
D5	EST BY-PASS	424	TAN/BLK	*0	4.6
D6	O_2 GROUND	413	TAN	*0	*0
D7	O_2 SIGNAL	412	PPL	(1)	(6)
D8	EGR (PINTLE SIGNAL)	1456	BRN	.73	.73
D9	DIRECT GROUND	151	BLK	*0	*0
D10	DIRECT GROUND	151	BLK	*0	*0
D11	NOT USED	–	–	–	–
D12	NOT USED	–	–	–	–
D13	NOT USED	–	–	–	–
D14	NOT USED	–	–	–	–
D15	NOT USED	–	–	–	–
D16	INJECTOR DRIVER	467	DK BLU	B +	B +

(1) .26 to .46 volt within 10 seconds the voltage goes near 0 volt as O_2 heater warms up.
(2) Not usable.
(3) Reads battery voltage for 2 seconds after ignition "ON," then should read 0 volt.
(4) Varies depending on temperature, about 1.3 volt.
(5) .6 volt to about 4.8 volts at Wide Open Throttle (WOT).
(6) Varies from .1 volt to .9 volt.
* Less than .5 volts.
† Varies with manifold pressure.

BACK VIEW OF CONNECTOR
C1 D1
32 PIN C-D CONNECTOR

ENGINE: 4.3L (L35) CPI

Component Replacement

— CAUTION —

Before attempting to remove or service any fuel system component, it is necessary to properly relieve the fuel system pressure.

SERVICE PRECAUTIONS

When working around any part of the fuel system, take the following precautionary steps to prevent fire and/or explosion:
- Disconnect negative terminal from battery (except when testing with battery voltage is required).
- Whenever possible, use a flashlight instead of a drop light.
- Keep all open flame and smoking material out of the area.
- Use a shop cloth or similar to catch fuel when opening a fuel system.
- Relieve fuel system pressure before servicing.
- Use eye protection.
- Always keep a dry chemical (class B) fire extinguisher near the area.

NOTE: Due to the amount of fuel pressure in the fuel lines, prior to performing any work to the fuel system, the fuel system should be de-pressurized.

CLEANING AND INSPECTION

All CPI component parts, with the exception of those noted below, should be cleaned in a cold immersion cleaner such as Carbon X (X–55) or equivalent. The throttle position sensor, idle air control valve, CPI assembly or other components containing rubber should not be placed in a solvent or cleaner bath. A chemical reaction will cause these parts to swell, harden or distort. Do not soak the throttle body with the above parts attached. If the throttle body assembly requires cleaning, soak time in the cleaner should be kept to a minimum. Some vehicles have hidden throttle shaft dust seals that could lose their effectiveness by extended soaking.

1. Clean all parts thoroughly and blow dry with compressed air. Be sure that all fuel and air passages are free of dirt and burrs.

2. Inspect the gasket mating surfaces for damage that could affect proper gasket sealing.

THREAD LOCKING COMPOUND

Service repair kits are supplied with a small vial of thread locking compound with directions for use. If material is not available, use Loctite® 262 or GM part No. 10522624 or equivalent. In precoating the screws, do not use a higher strength locking compound than recommended, thus making removal of the screw extremely difficult, or result in damaging the screw head.

RELIEVING FUEL SYSTEM PRESSURE

Procedure

1. Disconnect the negative battery cable to avoid possible fuel discharge if an accidental attempt is made to start the engine.
2. Loosen the fuel filler cap to relieve fuel tank pressure.
3. Connect fuel gauge J–34730–1, or equivalent, to intake manifold fuel pressure connection tap. Wrap a rag around the pressure tap to absorb any leakage that may occur when installing the gauge.
4. Install bleed hose into a suitable container and open the valve to bleed off the fuel pressure in the fuel system.

ENGINE COOLANT TEMPERATURE (ECT) SENSOR

Removal and Installation

The coolant temperature sensor is usually located on the intake manifold water jacket or near (or in) the thermostat housing.

1. Disconnect the negative battery cable and disconnect the electrical connector at the sensor.
2. Partially drain the cooling system.
3. Remove the threaded temperature sensor from the engine.
4. Apply some pipe tape to the threaded sensor and install the senor. Torque the sensor to 22 ft. lbs. (30 Nm).
5. Reconnect the sensor and the negative battery cable. Refill the cooling system.

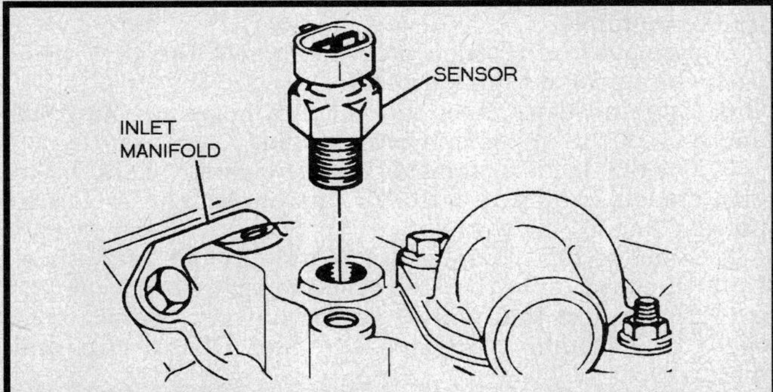

ECT sensor removal

ELECTRONIC CONTROL MODULE (ECM)

To prevent possible electrostatic discharge damage to the ECM, do not touch the connector pins or soldered components on the circuit board. Service of the ECM should normally consists of either replacement of the ECM or a PROM change.

When replacing the ECM, the MEMCAL or PROM and the ECM should be checked first to see if they are the correct parts. If they are, remove the MEMCAL or PROM from the faulty ECM and install it in the new service ECM. The service ECM will not contain a new MEMCAL or PROM.

Removal and Installation

NOTE: When replacing the production ECM with a service ECM (controller), it is important to transfer the broadcast code and production ECM number to the service ECM label. Do not record the number on the ECM cover. This will allow positive identification of the ECM parts throughout the service life of the vehicle.

1. Disconnect the negative battery cable.
2. Remove the right side kick panel to gain access to the ECM. Remove the hush panel, if equipped.
3. Disconnect the connectors from the ECM. On S/T models, remove the VSS buffer connector and rotate the ECM to ease removal.
4. Remove the ECM mounting hardware and remove the ECM from the passenger compartment.

NOTE: Before replacement of a defective ECM, first check the resistance of each ECM controlled solenoid. This can be done at the ECM connector, using an ohmmeter and the ECM connector wiring diagram. Any ECM controlled device with low resistance will damage the replacement ECM due to high current flow through the ECM internal circuits.

5. Installation is the reverse order of the removal procedure. Be sure to remove the new ECM from its packaging and check the service number to make sure it is the same at the defective ECM.
6. After installation, perform the functioal check as follows:
7. Turn the ignition switch to the **ON** position.
8. Enter the diagnostic mode.
9. Allow a Code 12 to flash 4 times to verify that no other codes are present. This indicates that the PROM is installed properly.
10. If trouble Code 51 occurs or if the SERVICE ENGINE SOON light in ON constantly with no codes, the PROM is not fully seated, installed backwards, has bent pins or is defective.
11. If not fully seated, turn the ignition switch **OFF**, then press down firmly on the PROM carrier. Repeat steps 7–11.
12. If installed backwards, replace the PROM.

— WARNING —

Any time the PROM is installed backwards and the ignition switch is turned ON, the PROM is destroyed.

13. If the pins are bent, remove the PROM straighten the pins and reinstall the PROM. If the bent pins break or crack during straightening, discard the PROM and replace with a new PROM. Repeat steps 7–13.

NOTE: To prevent possible electrostatic discharge damage to the PROM or CALPAC, do not touch the component leads and do not remove the integrated circuit from the carrier.

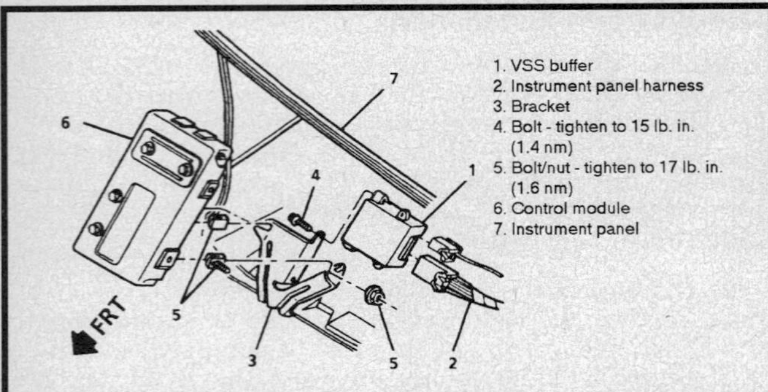

ECM and PROM assembly — Compact truck (S/T)

PROM

Removal and Installation

1. Disconnect the negative battery cable.
2. Remove the ECM.
3. Remove the PROM access panel.
4. Using 2 fingers, push both retaining clips back from the PROM. At the same time, grasp the PROM at both ends and lift up and out of the socket. Do not remove the cover of the PROM.

To install:

5. Before installing a new PROM, be sure the new PROM part number is the same as the old one or the same as the updated number per a service bulletin.
6. Install the PROM with the small notch of the carrier aligned with the small notch in the socket. Press on the PROM carrier until the PROM is firmly seated, do not press on the PROM itself only on the PROM carrier. Press inward on the clips until they snap into place.
7. Install the PROM access cover and reinstall the ECM.
8. Connect the negative battery cable.
9. Perform the functional check as follows:
10. Turn the ignition switch to the **ON** position.
11. Enter the diagnostic mode.
12. Allow a Code 12 to flash 4 times to verify that no other codes are present. This indicates that the PROM is installed properly.
13. If trouble Code 51 occurs or if the SERVICE ENGINE SOON light in ON constantly with no codes, the PROM is not fully seated, installed backwards, has bent pins or is defective.
14. If not fully seated, turn the ignition switch **OFF**, then press down firmly on the PROM carrier. Repeat steps 10–14.
15. If installed backwards, replace the PROM.

── WARNING ──
Any time the PROM is installed backwards and the ignition switch is turned ON, the PROM is destroyed.

16. If the pins are bent, remove the PROM straighten the pins and reinstall the PROM. If the bent pins break or

crack during straightening, discard the PROM and replace with a new PROM. Repeat steps 10–16.

NOTE: To prevent possible electrostatic discharge damage to the PROM or CALPAC, do not touch the component leads and do not remove the integrated circuit from the carrier.

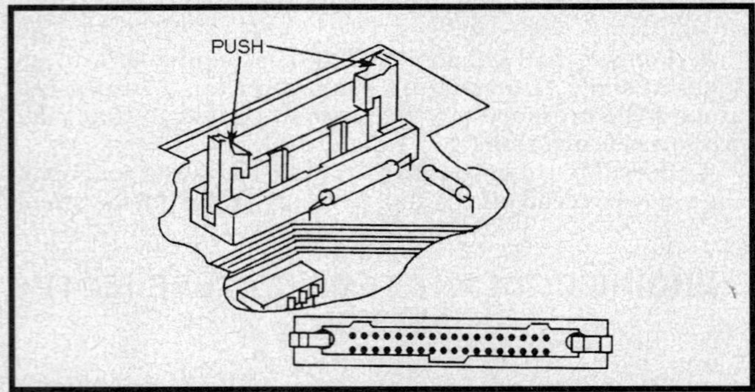

Installing the PROM

CENTRAL PORT FUEL INJECTION (CPI) ASSEMBLY

Removal and Installation

1. Disconnect the negative battery cable.
2. On S/T models, remove the plastic cover. On M/L models, remove the engine cover and the air cleaner box and intake duct.
3. Disconnect the TPS, IAC, MAP and IMTV wire harnesses.
4. Remove the throttle and cruise cable from the upper intake manifold.
5. Remove the ignition coil, disconnect the PCV hose from the upper intake manifold.
6. Tag and disconnect the vacuum hoses at the front and rear of the upper intake manifold.
7. On S/T models, remove the air cleaner snorkel. On M/L models equipped with A/C, disconnect the A/C compressor line.
8. Remove the upper intake manifold bolts and studs. Note the location of the studs for proper installation.
9. Disconnect the injector wiring harness connector at the CPI assembly. Remove the fuel fitting clip and discard.
10. Disconnect the fuel inlet and return tube and fitting assembly. Discard the O-rings.
11. Squeeze the poppet nozzle locking tabs together while lifting the nozzle out of the casting. After removing the 6 poppet nozzles, lift the CPI assembly out of the casting.

To install:

12. Align the CPI assembly grommet with the casting grommet slots and push down until it is seated in the bottom guide hole.
13. Push the poppet nozzles into the castings.

── CAUTION ──
Make sure the poppet nozzles are firmly seated and locked in their casting sockets. An unlocked poppet nozzle could work loose from its socket and result in a fuel leak.

14. Coat new O-rings with clean engine oil and connect the fuel inlet and return tube and fitting assembly.

15. Install a new fuel fitting clip.

16. Pressurize the fuel system and verify there are no fuel leaks.

17. Install the upper intake manifold gasket with the green sealing lines facing up. Be careful not to pinch the injector wires between the upper and lower manifolds.

18. Install the upper intake manifold and retaining bolts and studs. Tighten, in sequence, to 88 inch lbs. (10 Nm).

19. Connect the PCV hose and the vacuum hoses.

20. Connect the throttle and cruise cable to the upper intake manifold. Install the ignition coil.

21. On S/T models, install the air cleaner snorkel. On M/L models equipped with A/C, connect the A/C compressor line.

22. Connect the sensor connectors.

23. On S/T models, install the plastic cover. On M/L models, install the engine cover and recharge the A/C, if equipped.

24. Connect the negative battery cable.

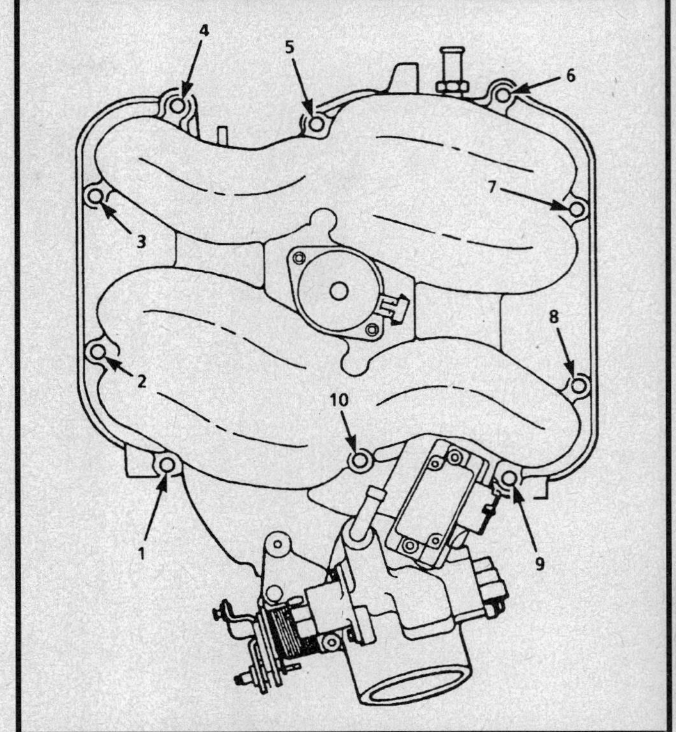

Intake manifold torque sequence

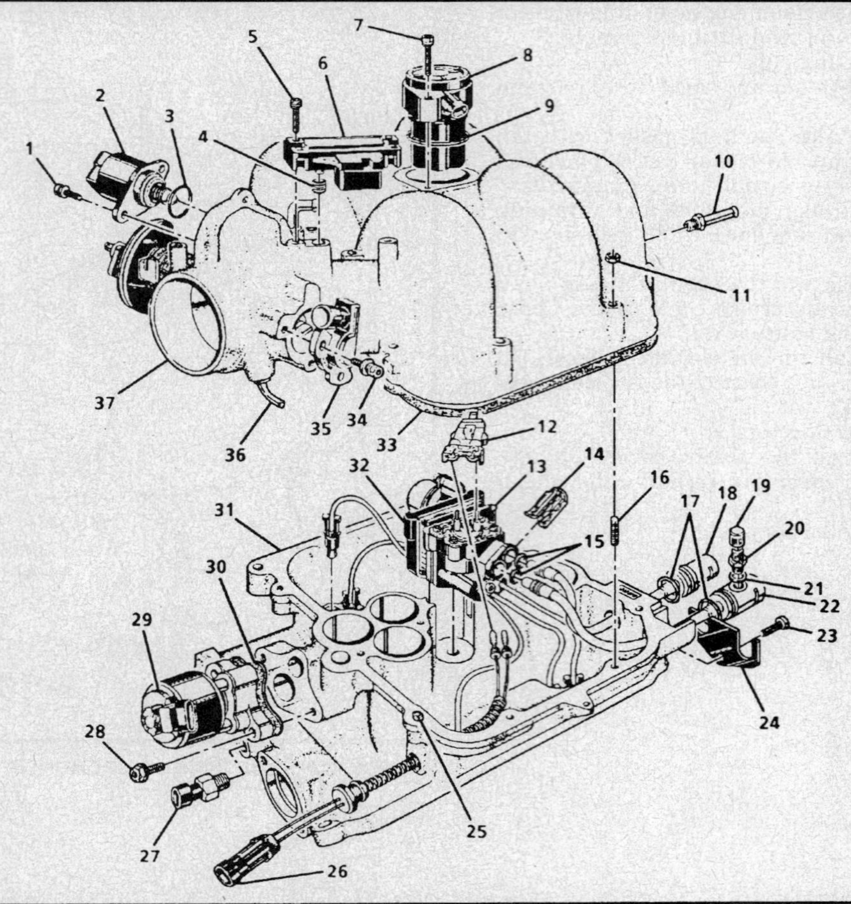

1	BOLT/SCREW - IDLE AIR CONTROL VALVE	14	CLIP - FUEL INJECTION FUEL FEED AND RETURN PIPE
2	VALVE ASSEMBLY - IDLE AIR CONTROL (IAC)	15	SEAL - FUEL INJECTION FUEL FEED AND RETURN PIPE (O-RING)
3	SEAL - IDLE AIR CONTROL VALVE (O-RING)	16	STUD - UPPER INTAKE MANIFOLD
4	SEAL - MAP SENSOR	17	SEAL - LOWER INTAKE MANIFOLD FUEL FEED AND RETURN PIPE (O-RING)
5	BOLT/SCREW - MAP SENSOR	18	PIPE ASSEMBLY - FUEL INJECTION FUEL FEED
6	SENSOR ASSEMBLY - MANIFOLD ABSOLUTE PRESSURE (MAP)	19	CAP - FUEL PRESSURE CONNECTION
7	BOLT/SCREW - INTAKE MANIFOLD TUNING VALVE	20	FUEL PRESSURE CONNECTION ASSEMBLY
8	VALVE ASSEMBLY - INTAKE MANIFOLD TUNING	21	SEAL - FUEL PRESSURE CONNECTION
9	SEAL - INTAKE MANIFOLD VALVE (O-RING)	22	PIPE ASSEMBLY - FUEL INJECTION FUEL RETURN
10	FITTING - POWER BRAKE BOOSTER VACUUM	23	BOLT/SCREW - FUEL INJECTION FUEL FEED AND RETURN PIPE RETAINER
11	NUT - UPPER INTAKE MANIFOLD	24	RETAINER - FUEL INJECTION FUEL FEED AND RETURN PIPE
12	CONNECTOR ASSEMBLY - CENTRAL MULTIPORT FUEL INJECTOR WIRING HARNESS	25	PIN - UPPER INTAKE MANIFOLD LOCATING
13	INJECTOR ASSEMBLY - CENTRAL MULTIPORT FUEL INJECTOR (CMFI)		

26	HARNESS ASSEMBLY - CENTRAL MULTIPORT FUEL INJECTOR WIRING
27	SENSOR ASSEMBLY - ENGINE COOLANT TEMPERATURE (ECT)
28	BOLT/SCREW - EGR VALVE
29	VALVE ASSEMBLY - EGR
30	GASKET - EGR VALVE
31	MANIFOLD ASSEMBLY - LOWER INTAKE
32	SEAL - CENTRAL MULTIPORT FUEL INJECTOR (CMFI)
33	GASKET - UPPER INTAKE MANIFOLD
34	BOLT/SCREW - THROTTLE POSITION SENSOR
35	SENSOR ASSEMBLY - THROTTLE POSITION (TP) SENSOR
36	TUBE - FUEL VAPOR CANISTER PURGE
37	MANIFOLD ASSEMBLY - UPPER INTAKE (WITH THROTTLE BODY)

Fuel injection system components

IDLE AIR CONTROL VALVE

Removal and Installation

1. Disconnect the negative battery cable.
2. Disconnect electrical connector from idle air control valve.
3. Remove the screws retaining the idle air control valve. Remove the idle air control valve from mounting position.

To install:

4. Prior to installing a new idle air control valve, measure the distance the valve plunger is extended. Measurement should be made from the edge of the valves mounting flange to the end of the cone. The distance should not exceed 1⅛ in. (28mm), or damage to the valve may occur when installed. If measuring distance is greater than specified above and the IAC valve is new, press on the valve firmly to retract it, using a slight side to side motion to help it retract easier. Do not attempt to push the pintle in on a used valve, otherwise the valve may become damaged.
5. Use a new gasket and install the idle air control valve in mounting position.
6. Install the retaining screws. Torque the retaining screws to 27 inch lbs. (3.0 Nm). Connect the electrical connector.
7. Reset the IAC valve position:
   a. Turn the ignition switch **ON** for 5 seconds.
   b. Turn the ignition switch **OFF** for 10 seconds.
   c. Start the engine and check for proper idle operation.

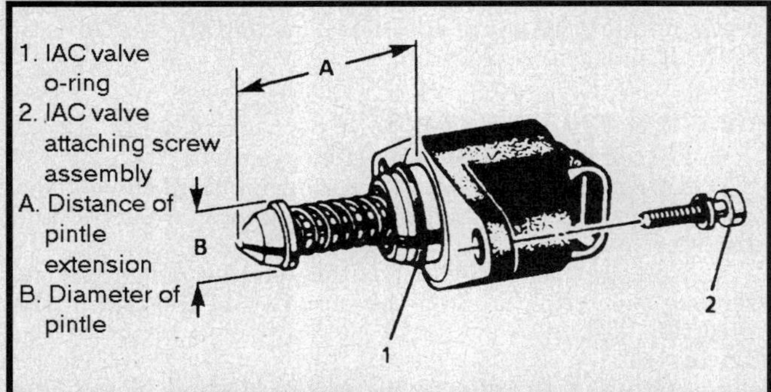

Measuring a new IAC valve pintle length

MANIFOLD ABSOLUTE PRESSURE (MAP) SENSOR

Removal and Installation

1. Disconnect the negative battery cable and remove the vacuum hose from the MAP sensor.
2. Disconnect the electrical connector to the MAP sensor.
3. Remove the MAP sensor retaining screws and remove the MAP sensor.

4. Installation is the reverse order of the removal procedure. Tighten the retaining screws to 27 inch lbs. (3.0 Nm).

INTAKE AIR TEMPERATURE (IAT) SENSOR

Removal and Installation

1. Disconnect the negative battery cable.
2. Disconnect the electrical connector to the IAT sensor.
3. Unscrew the IAT sensor from the intake manifold.
4. Installation is the reverse order of the removal procedure.

OXYGEN SENSOR

When working with the oxygen sensor, the following precautions should be observed:

- The oxygen sensor uses a permanently attached pigtail and connector which should not be removed from the oxygen sensor.
- Damage or removal of the pigtail connector could effect the proper operation of the oxygen sensor.
- The in-line electrical connector and louvered end must be kept free of grease, dirt or other contaminants.
- Avoid using cleaner solvent of any type.
- Do not drop or roughly handle the oxygen sensor.
- It is not recommended to clean or wire brush an oxygen sensor when in question, it is recommended that a new oxygen sensor be used.
- The oxygen sensor may be difficult to remove when the engine temperature is below 120°F (48°C). Excessive force may damage the threads in the exhaust manifold of exhaust pipe.

Removal and Installation

1. Start the engine and let it warm up to 120°F (48°C), stop the engine and disconnect the negative battery cable.
2. Disconnect the electrical connector from the oxygen sensor.

NOTE: Special anti-seize compound is used on the oxygen sensor threads. The compound consists of a liquid graphite and glass beads. The graphite will burn away, but the glass beads will remain, making the sensor easier to remove. New or service sensors will already have the compound applied to the threads. If a oxygen sensor is removed from the engine and if for any reason it is to be reinstalled, the threads must have anti-seize compound applied before installation.

3. Using a special oxygen sensor socket, remove the sensor from the exhaust manifold or exhaust pipe.

To install:

4. Coat the threads of the oxygen sensor with anti-seize compound 5613695 or equivalent, if necessary.
5. Install the sensor and torque it to 30 ft. lbs (41 Nm).
6. Reconnect the electrical connector to the sensor and connect the negative battery cable.

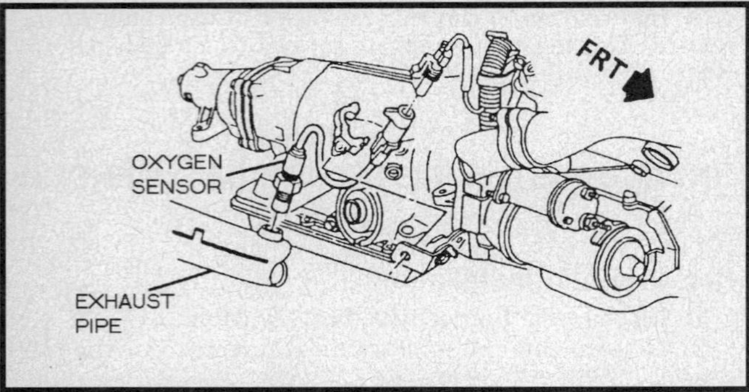

Oxygen sensor removal and installation

VEHICLE SPEED SENSOR

Removal and Installation

1. Disconnect the negative battery cable.
2. Remove the vehicle speed sensor electrical lead from the transmission.
3. Remove the VSS sensor bolt and screw retainer.
4. Remove the vehicle speed sensor assembly and O-ring.
5. Installation is the reverse order of the removal procedure. Use a new O-ring and coat with transmission fluid.

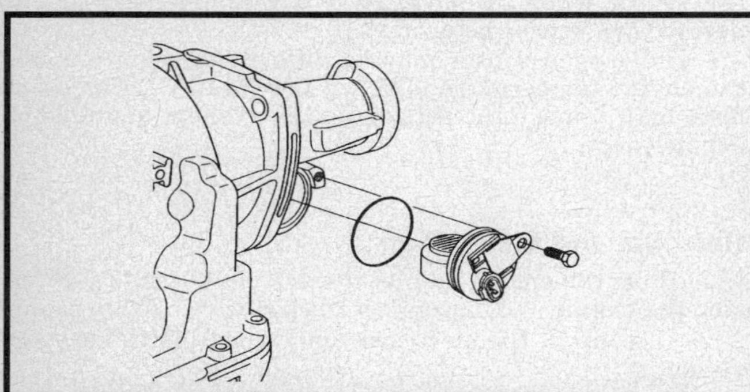

VSS with automatic transmission

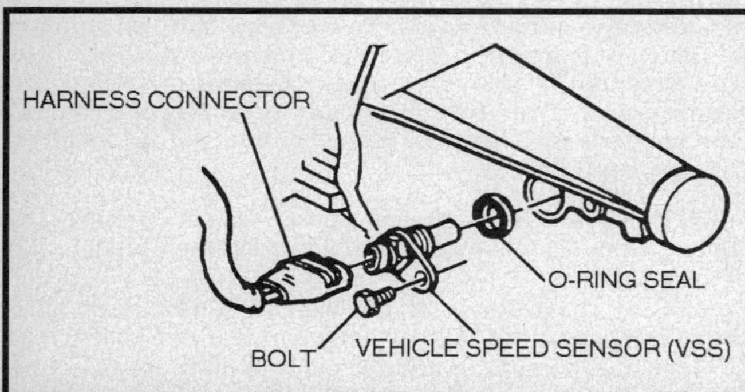

VSS with manual transmission

INTAKE MANIFOLD TUNING VALVE (IMTV)

Removal and Installation

1. Disconnect the negative battery cable.
2. Disconnect the IMTV electrical connector.
3. Remove the IMTV retaining bolts. Remove the IMTV and seal.

To install:

4. Lubricate new O-ring with clean engine oil and install on IMTV.
5. Install the IMTV in the manifold.
6. Install the retaining screws and alternately tighten until they engage the mounting ear surface, then tighten to 18 inch lbs. (2.0 Nm).
7. Connect the electrical connector and the negative battery cable.

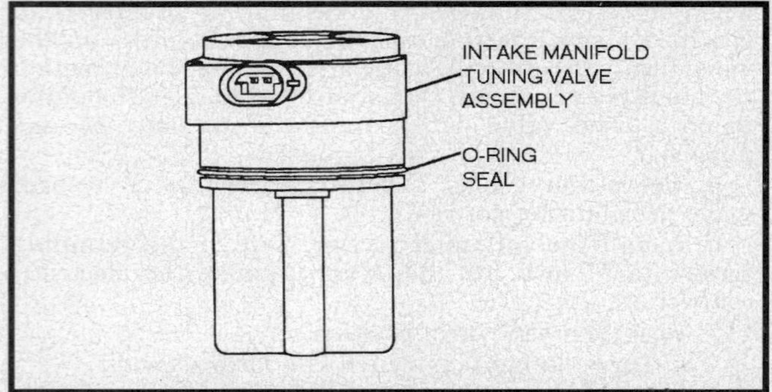

Intake manifold tuning valve

THROTTLE POSITION SENSOR

The throttle position sensor is non-adjustable. If the sensor is found to be out of specifications and the sensor is at fault, it must be replaced.

Removal and Installation

1. Disconnect the negative battery cable.
2. Disconnect the electrical connector from the sensor.
3. Remove the attaching screws, lock washers and retainers.
4. Remove the throttle position sensor. If necessary, remove the screw holding the actuator to the end of the throttle shaft.

To install:

5. With the throttle valve in the normal closed idle position, install the throttle position sensor on the throttle shaft and rotate counterclockwise to align the TPS mounting holes with casting mounting screw holes.
6. Install the retainers, screws and lock washers using a thread locking compound. Tighten to 18 inch lbs. (2.0 Nm).
7. Connect the electrical connector and negative battery cable.

LINEAR EGR VALVE

Removal and Installation

1. Remove the negative battery cable.
2. Disconnect the EGR electrical connector..
3. Remove the retaining bolts and remove the valve and gasket.

DIESEL FUEL SYSTEMS

5 SECTION

LIGHT TRUCKS AND VANS

APPLICATION CHART — LIGHT TRUCKS AND VANS

		APPLICATION CHART				
Model	Body VIN	Engine Liter	Engine VIN	Ignition Type	Fuel Type
Light Truck	C,D,K	6.2	C	—	Diesel
Light Truck	C,D,K	6.2	C	—	Diesel
Light Truck	C,D,K	6.2	C	—	Diesel

General Description

Both the 6.2L Diesel and the 6.5L Turbocharged Diesel engines are available in GM light duty trucks and other chassis up to 16,000 lbs. GVWR. The 6.2L heavy duty engine is equipped with the LL4 emission control package and the 6.5L turbo-diesel engine is equipped with the L65 emission control package. On both of these models there is a crankcase breather valve operated directly by intake manifold vacuum. There are no other sensors or electronic controls. The glow plug control unit is operated with a coolant temperature switch.

The LH6 emission control package on VIN C light duty 6.2L engines includes the crankcase breather and the Diesel Electronic Control System (DECS) for operating emission control equipment, cold starting equipment and the Torque Converter Clutch (TCC) on automatic transmissions. The system is operated by the Engine Control Module (ECM) with a manual transmission or the Powertrain Control Module (PCM) with an automatic transmission. The control module is equipped with a diagnostic program that can detect certain system faults, store trouble codes and notify the driver with a warning light on the instrument panel. Codes and other diagnostic functions can be accessed through the Data Link Connector (DLC) under the dashboard.

With some minor differences in the fuel supply system, the diesel fuel injection system is the same on all engines. The main components of the supply system are a fuel supply pump, fuel filter and water separator, an optional fuel line heater and an excess fuel return system. Some models are also equipped with a secondary filter and a separate water drain on the fuel tank. The main components of the high pressure system are the Stanadyne DB2 distributor type injection pump, Bosch KCA type injector assemblies and the high pressure lines connecting them.

SYSTEM OPERATION

The fuel supply system must complete five functions: transfer fuel from the tank to the injection pump, filter out minute particles that may damage the injection equipment, separate any water from the fuel, warn the driver of a clogged filter or water in the fuel and keep the fuel warm enough to flow through the filter in cold weather. Depending on the equipment package ordered with the vehicle, on most models the diaphragm type mechanical fuel pump is replaced with a solenoid type electric fuel pump.

The distributor type injection pump supplies a metered quantity of fuel to each injector assembly in the correct engine firing order. The pump also controls injection timing and duration depending on engine speed and load. Under all engine loads, the pump governor has ultimate control over the metered fuel quantity and decreases fuel flow to the injectors as engine speed climbs above the maximum rated speed.

The Diesel Emission Control System controls exhaust gas recirculation and cold start timing advance, starts glow plug operation and operates the torque converter clutch on automatic transmissions. The system is controlled by the ECM (or PCM) mounted inside the vehicle. The ECM accepts inputs from the various engine sensors and generates the appropriate output voltages to operate five solenoid valves and activate the cold start timing advance/glow plug system.

Two solenoid valves are used to control vacuum to the EGR valve and a third solenoid valve controls vacuum to the Exhaust Pressure Regulator valve (EPR) (exhaust back pressure must be increased at idle for correct EGR flow). Vacuum for all of these valves is supplied by a belt driven vacuum pump.

When the engine is cold, the ECM operates a cold start system. This includes a solenoid valve on the injection pump to decrease pump housing pressure, allowing a slight injection timing advance. At the same time, an activation signal is supplied to the glow plug control unit. Once the glow plug operation cycle is begun, the separate glow plug control unit will operate the glow plugs through a timed cycle to aid engine warm-up. After starter operation, the cycle is completed even if the engine does not start.

The ECM is equipped with a self diagnostic program to monitor the DEC system. If a malfunction is detected it will alert the driver through the Malfunction Indicator Light (MIL) on the instrument panel.

SYSTEM COMPONENTS

Injection Pump

The Stanadyne DB2 fuel injection pump is a distributor type pump that uses a single high pressure pumping chamber to supply fuel to all of the injectors. The main rotating components are the drive shaft, distributor rotor, internal transfer pump and the governor fly weights. As the rotor spins inside the high pressure hydraulic head,

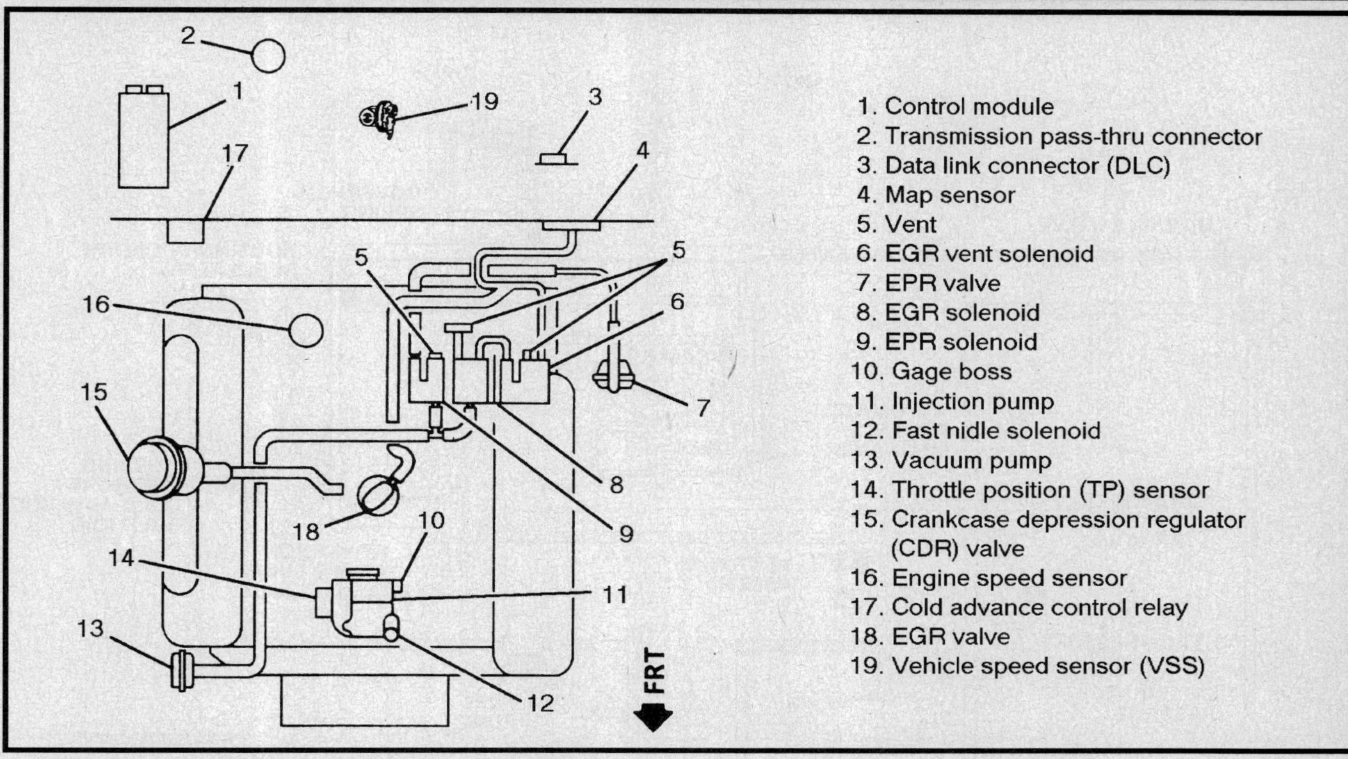

1. Control module
2. Transmission pass-thru connector
3. Data link connector (DLC)
4. Map sensor
5. Vent
6. EGR vent solenoid
7. EPR valve
8. EGR solenoid
9. EPR solenoid
10. Gage boss
11. Injection pump
12. Fast nidle solenoid
13. Vacuum pump
14. Throttle position (TP) sensor
15. Crankcase depression regulator (CDR) valve
16. Engine speed sensor
17. Cold advance control relay
18. EGR valve
19. Vehicle speed sensor (VSS)

DEC system components on light duty 6.2L LH6 engine

the inlet port aligns with a charging passage port in the head and a metered amount of fuel is supplied to the pumping chamber. As rotation continues, the inlet port is closed and the rotor outlet port aligns with a discharge port in the head. At the same time the two plungers in the pumping chamber are forced towards each other by a cam ring around the rotor and high fuel pressure is generated. The spring loaded delivery valve in the rotor is forced open and the fuel is pumped out to the injection line. Fuel delivery ends when the plunger rollers move past the top of the cam lobe, the plungers stop moving towards each other and the pressure drops enough for the delivery valve to close. The delivery valve assures sharp closing of the nozzle while still maintaining residual pressure in the injection line.

Fuel injection timing is controlled by engine speed and load. The amount of fuel metered into the pumping chamber determines how far the plungers are forced apart. Under light loads, less fuel is metered into the pumping chamber and the plungers move only part of the way up their bores. The rollers contact the cam ring later in the rotor rotation cycle and injection begins later in the engine cycle. Injection timing can be controlled by rotating the cam ring slightly to change the time that plunger movement occurs in the cycle.

A timing advance piston in the bottom of the pump housing pushes against the advance pin to control movement of the cam ring. Piston movement is partially controlled by transfer pump pressure that changes with engine speed and load. Additional timing advance control is provided with a spring loaded servo valve built into the advance piston. By controlling the spring pressure on the servo valve, the transfer pump pressure on the high pressure side of the timing advance piston can be held steady

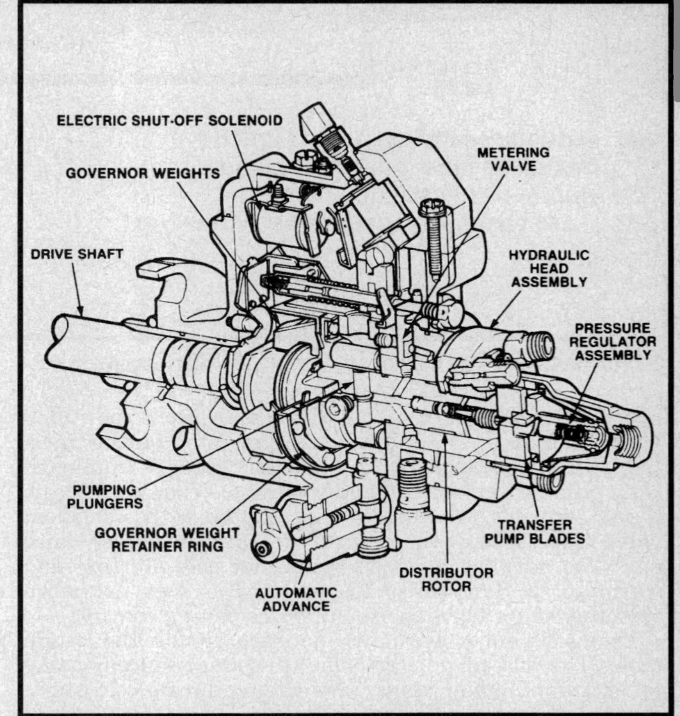

Cutaway drawing of the Stanadyne DB2 diesel fuel injection pump

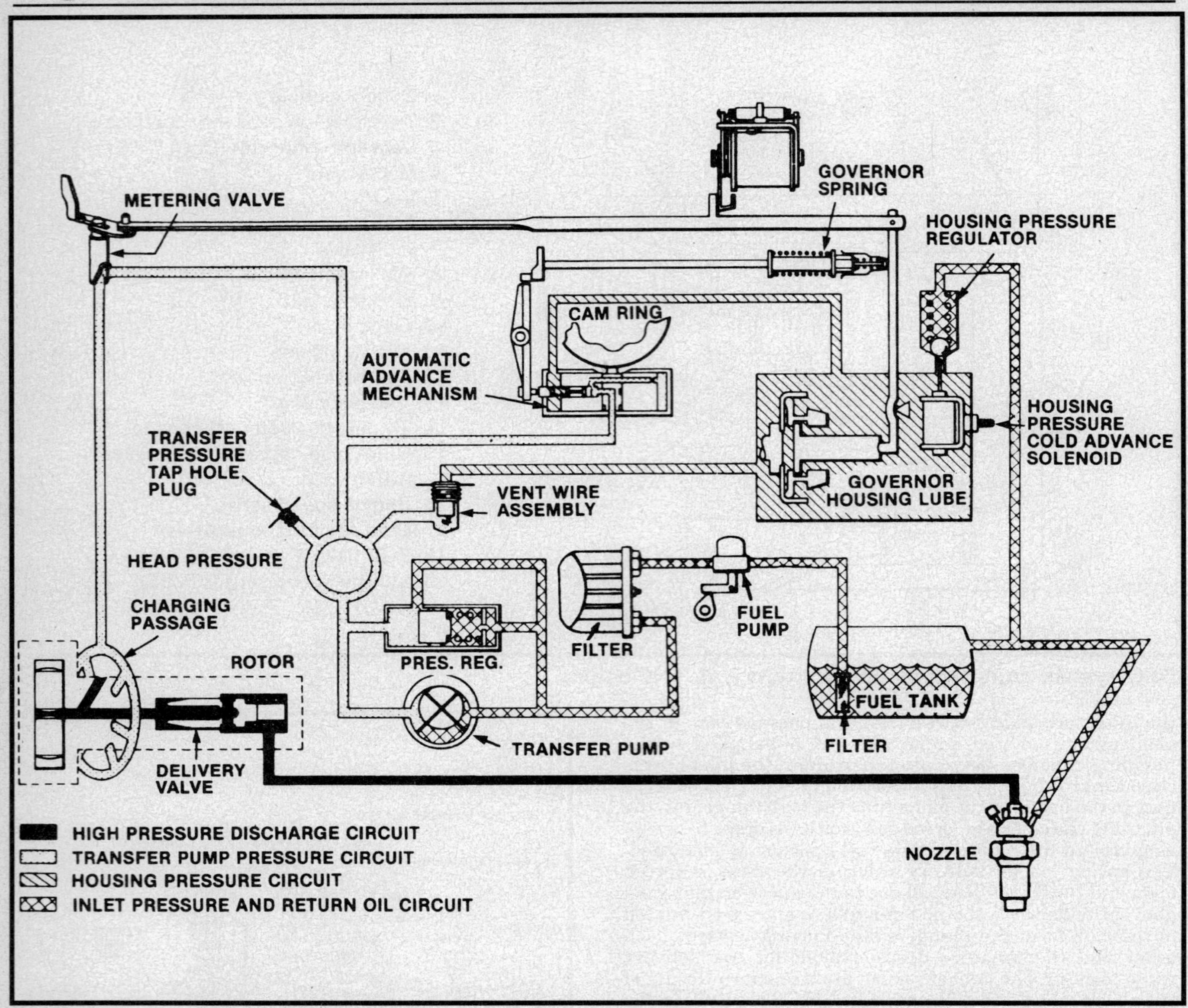

Schematic of fuel supply and injection systems

or increased. The servo valve spring pressure is controlled with the rocker arm and face cam mounted on the injection pump throttle shaft, so injection timing is automatically adjusted for all throttle positions. When the engine is cold, housing pressure can be reduced with a solenoid valve. This lowers the return pressure on the advance piston to advance injection timing for cold starting and warm-up. If the timing advance piston does not move smoothly in its bore, there will be no timing control.

Two solenoid valves are mounted inside the pump cover. The cold advance solenoid is used to reduce housing pressure and advance injection timing during cold starting and warm-up. The shut-solenoid acts on the governor to close the fuel metering valve and shut-down the engine. On some models, an additional solenoid is mounted externally to increase idle speed when engine load increases, such as with an air conditioner compressor.

Shut-Down Solenoid

The shut-down solenoid is mounted inside the pump under the top cover. When voltage is removed from the solenoid by turning the ignition switch **OFF**, it acts directly on the governor linkage to push the metering valve closed and prevent fuel from entering the high pressure pumping chamber in the rotor. If the solenoid malfunctions, it can prevent the engine from starting. If it is not properly positioned when installing the top cover, it can hold the metering valve wide open, causing uncontrolled engine acceleration at the first engine start. If the cover is removed, a special adjustment tool is available for correct installation. The tool is not required but does make the job easier.

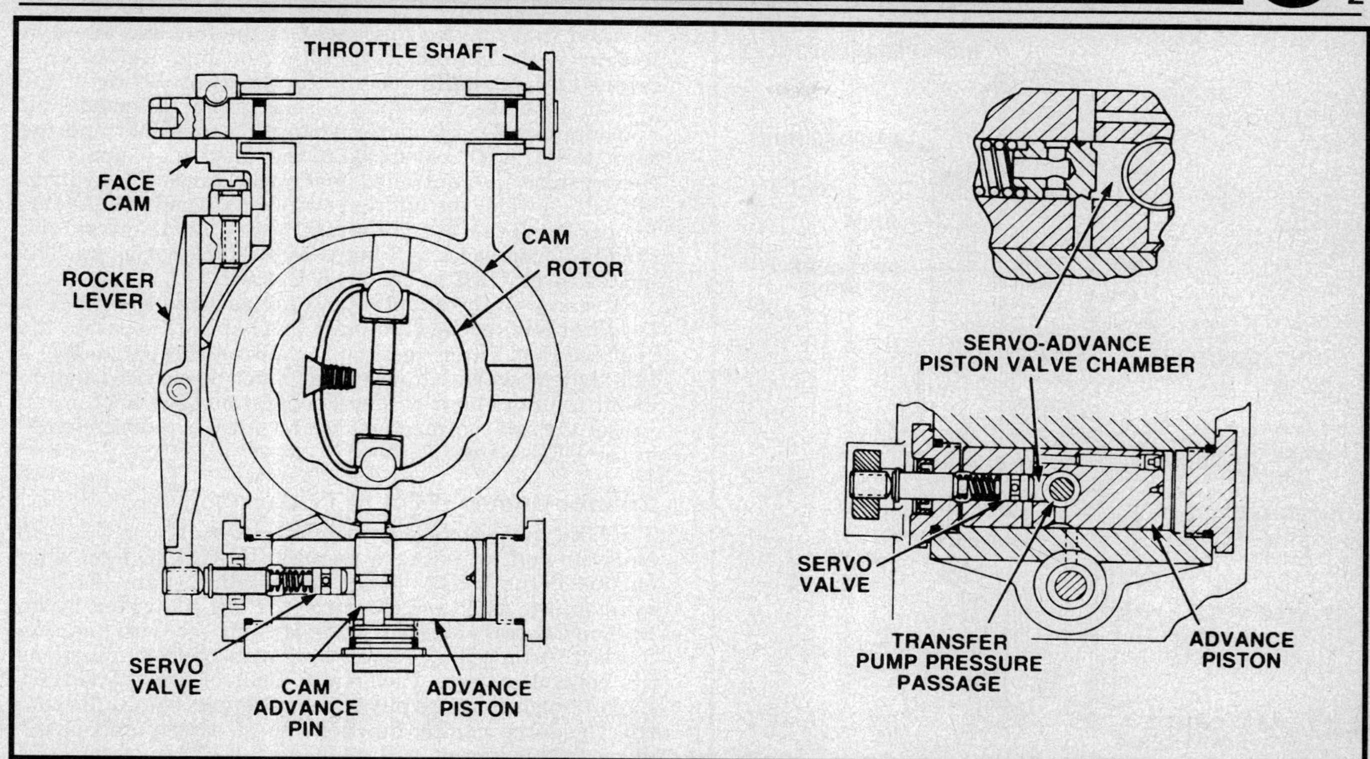

Advance piston moves cam ring to advance injection timing

Cold Advance Solenoid

The cold advance solenoid valve mounted inside the injection pump cover is used to advance the injection timing during cold starting and warm-up. On heavy duty engines, the solenoid is activated by a coolant temperature switch on the rear of the right cylinder head. On the 6.2L engine with the DEC system, the circuit is activated by the ECM through a Cold Advance Control relay that also activates the glow plug control unit. When coolant temperature is below about 80°F (27°C), the relay is energized when the ignition switch is turned **ON**. During starting and warm-up, injection pump housing pressure will be reduced from 10 psi (60 kPa) to zero. This removes the fuel pressure on the return side of the timing advance piston, allowing the transfer pump pressure to move the piston and advance injection timing about 4 degrees. When coolant temperature raises above 95°F (35°C), the control relay is deactivated, the cold advance solenoid is turned **OFF** and injection timing returns to normal control.

The ECM does not monitor the status of this system and a rough cold idle and smoke problem has been noted on several vehicles. These problems may be caused by a faulty oil pressure sending unit that supplies power for the fuel pump or by incorrect operation of the cold advance solenoid. To modify the cold advance solenoid duty cycle, the manufacturer has developed an updated PROM and initial timing specification. The new PROM and specification are available through the OEM parts network. These parts are not available for all vehicles and the vehicle and pump serial numbers must be known when ordering parts.

Fuel Injection Lines and Nozzles

Fuel from the injection pump moves through the high pressure lines to the injector assemblies. The lines are subjected to pressure pulses above 1800 psi (125 BAR) and the rapid increase and decrease in pressure causes gradual erosion of the line from the inside. The lines usually do not fail due to this cavitation erosion but the volume of fuel contained in the lines becomes uneven. The quantity of fuel reaching the injector and the injection timing are affected. Generally when an injection pump is replaced due to high milage, the lines should also be replaced as a set. The major cause of actual line failure is frequent removal and installation or over tightening of the fittings. The flare at each end is relatively soft and takes the shape of the nozzle or pump fitting when the nut is tightened. With repeated installations the flare deforms and can no longer make a tight seal against the fitting. Since equal line length is critical to injection timing, the lines cannot be repaired.

The fuel injector assemblies are Bosch KCA type holders with a single hole pintle type nozzle. Spring pressure holds the nozzle closed and fuel pressure from the injection pump forces the needle off the seat to open the nozzle. The nozzle can be removed from the holder for inspection and replacement. Nozzle opening pressure can be adjusted with a shim above the spring and is normally set at 1810 psi (125 BAR). The assembly is lubricated by the fuel and the holder includes a leak-off port to return excess fuel to the tank. The nozzle holder threads into the cylinder head so the spray is directed into the pre-chamber. A heat shield in the head at the bottom of the nozzle provides a tight seal and must be replaced when ever the holder is removed.

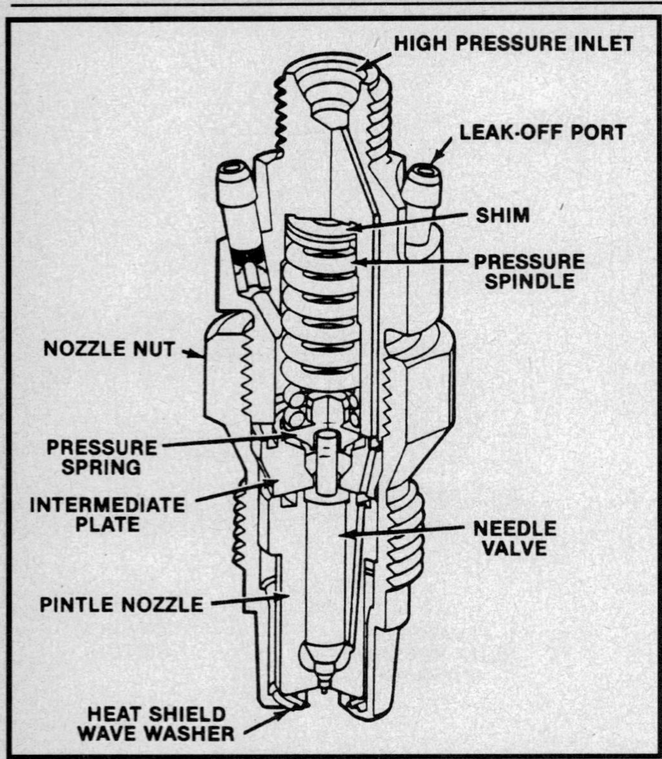

Fuel injector nozzle and holder assembly

Fuel Pump

This pump moves fuel from the tank to the injection pump. Two types of pump have been used depending on model year and equipment on the vehicle; mechanical and electric. The mechanical pump is similar to the diaphragm fuel pump used on other GM engines for many years. It is mounted on the timing chain cover and operated by an eccentric on the front of the camshaft gear.

The electric fuel pump is a recent design and contains no diaphragm to wear out. It is mounted on the left frame rail and consists of a solenoid coil, a plunger with two check valves and a return spring. The solenoid coil is operated by a transistor and control circuit built into the pump housing but isolated from the fuel. The circuit is designed to operate between 8–24 volts, is protected against incorrect polarity and outside interference and will not generate radio interference on AM and FM bands. The pump can handle a wide variety of fuels and can prime itself even with the fuel lines empty. It relies on fuel for cooling so it must not be operated for long periods without fuel flowing.

Power for the pump is supplied by the ignition switch during starting and by the oil pressure sending unit when the engine is running. This assures that the pump will not continue to run if the engine is not running. A faulty oil pressure sending unit can cause intermittent fuel pump operation, which would cause smoke and drivability problems.

Fuel Filter

The two stage fuel filter assembly includes a water separator, water detector, water drain, a fuel heater and a fine particle filter. The first stage collects small droplets of water that may be suspended in the fuel and allows it to drop to the bottom of the filter housing. The water is drained by manually unscrewing the drain fitting at the bottom of the filter adapter assembly. The filter adapter housing includes the water separation container and the filter element. Also mounted to the adapter assembly is a thermostatically controlled heater that turns **ON** at about 46°F (8°C) when the ignition switch is turned **ON**. At this temperature the current draw begins at 13 amps and gradually decreases to 7 amps as the fuel warms up. The heater turns **OFF** at about 85°F (29°C).

A sensor on the adapter determines when the water in the filter has reached a certain level and will operate the "Service Fuel Filter" light on the instrument panel. When this light is activated, the filter must be serviced within about the next hour of engine operation. The heater and sensor are self contained units and operate independently of the DEC system.

Control Module (ECM or PCM)

The DEC system control module is used only on the LH6 emission control package on the VIN C 6.2L light duty engine. If the vehicle is equipped with a manual transmission, the DEC system functions are controlled by an Engine Control Module (ECM). If equipped with an automatic transmission, engine and transmission functions are controlled by a Powertrain Control Module (PCM). Even though they are physically different, both units control the same engine functions so all charts and procedures in this section will refer to the ECM unless a PCM reference is required.

The ECM is mounted under the dashboard on the right side. It supplies a 5 volt reference signal to and accepts return signals from the following sensors:
- Engine Coolant Temperature sensor (ETC)
- Manifold Absolute Pressure sensor (MAP)
- Throttle Position Sensor (TPS)
- Engine Speed Sensor (ESS)
- Vehicle Speed Sensor (VSS)
- Automatic transmission fluid pressure and temperature sensors

Output control voltages are sent to:
- EGR solenoid valve
- EGR solenoid vent valve
- Exhaust Pressure Regulator (EPR) solenoid valve
- Cold timing advance and glow plug control unit relay
- Automatic transmission Torque Converter Clutch (TCC)
- Automatic transmission shift control solenoids
- Malfunction Indicator Light (MIL)

The ECM is equipped with a self diagnostic program that can detect sensor faults, store Diagnostic Trouble Codes (DTC) and alert the driver with the MIL on the instrument panel. By grounding the correct Data Link Connector (DLC) terminal, the on-board diagnostic codes can be read directly from the MIL. For complete diagnosis and testing, a scan tool and test light are required.

If the fault is determined to be in the ECM, it can be serviced in two parts. Inside the ECM is the Programmable Read Only Memory (PROM) chip which stores the specific calibrations required for a particular vehicle/engine combination. Just as in gasoline engine control systems, some repairs or changes in DEC system can be made by simply installing an updated PROM. If the ECM is damaged and must be replaced, it is supplied without a PROM and referred to as the controller. The PROM is referred to as the calibrator. If a faulty PROM is detected,

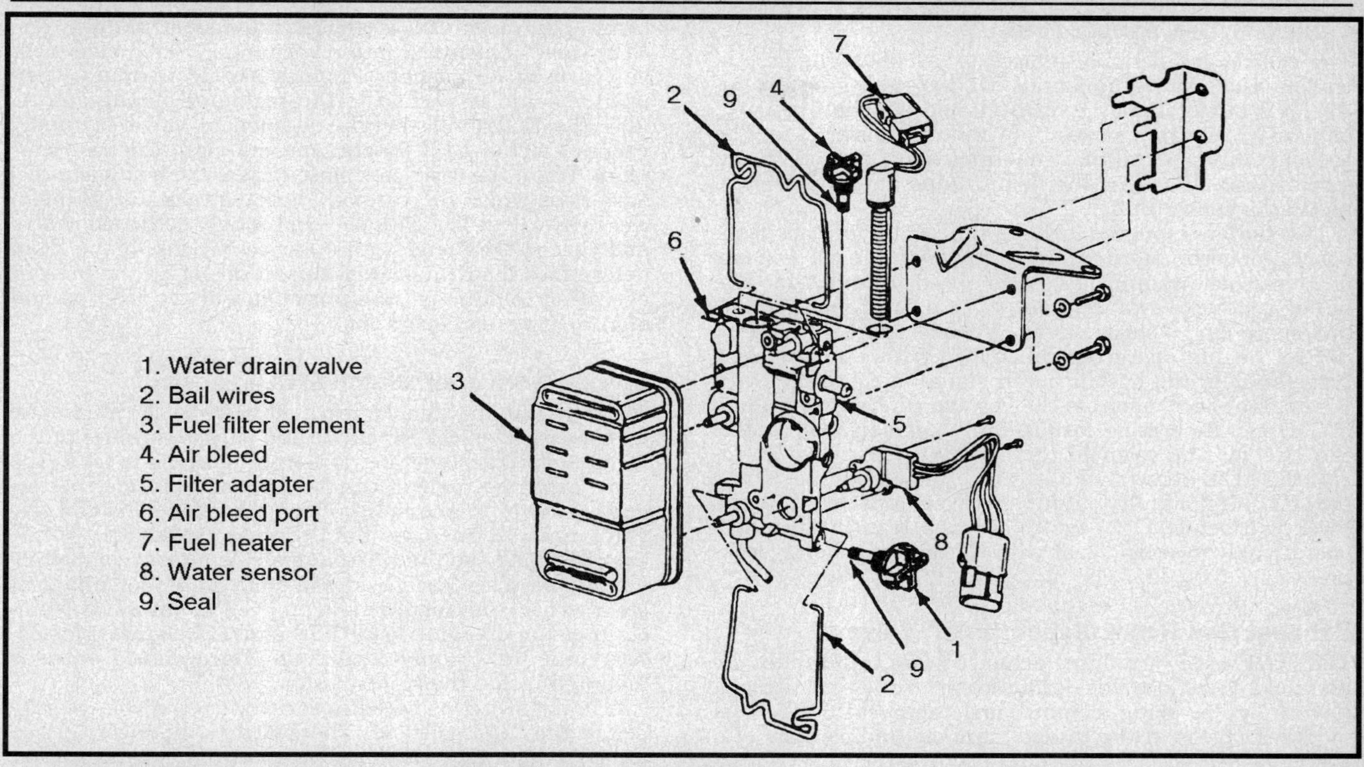

Fuel filter assembly includes a water drain and air bleeder

the ECM will set DTC 51. If a fault in the ECM is detected, it will set DTC 52 or 53. Like all electronic components, the ECM is affected by heat and must not be exposed to high temperatures such as those encountered in a paint baking oven. It should also be removed before using any electric welding equipment on the vehicle.

Engine Coolant Temperature Sensor (ETC)

The ETC sensor is threaded into a coolant passage near the front of the left cylinder head. It is a negative temperature co-efficient thermistor; as temperature rises the resistance of the sensor decreases and as temperature decreases, the resistance of the sensor increases. Resistance changes almost linearly from over 100,000 ohms at -40°F (-40°C) to approximately 70 ohms at 266°F (130°C). The ECM supplies a 5 volt reference signal to the sensor and measures the voltage drop to determine coolant temperature. When coolant temperature is below about 80°F (27°C), the ECM will activate the cold start relay for glow plug control and timing advance. If the ECM detects a faulty ETC signal, it will set DTC 14 or 15.

Manifold Absolute Pressure Sensor (MAP)

The MAP sensor is mounted on the cowl near the wiper motor or on a bracket in front of the engine. It is used to measure vacuum supplied to the EGR valve. It is a pressure transducer that changes the voltage drop across a circuit as the pressure applied to its diaphragm changes. The ECM supplies a 5 volt reference signal to the sensor and measures the voltage return. When vacuum is high (absolute pressure below atmospheric pressure), the voltage signal to the ECM will be low, and when vacuum is low (pressure approaching atmospheric), the voltage signal to the ECM will be high. This information is used to compare actual EGR vacuum signals to ECM duty cycle calculations. If a difference exists between the ECM calculation and what is actually being sensed by the MAP sensor, the ECM will make minor EGR vacuum adjustments with the solenoid valves. If a major difference is found, the ECM will go to a default setting and shut down the EGR system. If the ECM detects a faulty MAP sensor signal, it will set DTC 31 or 33. DTC 32 indicates an incorrect vacuum signal above about 30 mph.

Throttle Position Sensor (TPS)

The TPS is a variable resistance potentiometer connected to the throttle valve shaft on the injection pump. The ECM supplies 5 volts to the sensor and measures the return voltage. At idle position, the voltage signal to the ECM is low, approximately 0.5V. As the shaft moves toward full load position, the return signal increases and final output will be close to 5 volts. The ECM uses this information to calculate engine load for operating the EGR system and the TCC on automatic transmissions. If the ECM detects a faulty TPS signal, it will set DTCs 21, 22 or 23.

Engine Speed Sensor (ESS)

The ESS is mounted between the cylinder heads at the rear of the engine and driven by the camshaft. It is a simple pulse reference sensor that returns the full 5 volt reference signal to the ECM 4 times per camshaft revolution. If the ECM detects a faulty ESS signal, it will set DTC 12. This DTC is not stored in memory and will only be displayed if the ignition switch is **ON** but the engine speed signal is not present.

Vehicle Speed Sensor (VSS)

The vehicle speed signal is used to calculate engine and transmission control functions. On two-wheel drive vehicles, it is mounted on the transmission tail shaft housing; on four-wheel drive models it is on the transfer case. The sensor contains a coil that generates a magnetic field and a rotor that interrupts the field to send an AC pulse signal to the sensor buffer.

The buffer, mounted behind the center of the dashboard, converts and distributes the signal to all systems on the vehicle requiring vehicle speed information. The buffer generates two different signals, transmission output speed and vehicle speed. Vehicle speed is used to control engine operating functions. With an automatic transmission, the output speed signal is compared with the engine speed signal so the Powertrain Control Module (PCM) can determine torque converter slip and control the TCC lock-up solenoid.

If the ECM detects a faulty vehicle speed signal, it will set DTC 16. If a faulty output speed signal is detected, the PCM will set DTC 24 or 72 and the transmission will operate only in second gear with no torque converter lock-up.

Exhaust Gas Recirculation (EGR) Valve

The EGR valve introduces exhaust gases into the intake manifold to reduce the formation of oxides of nitrogen (NOx) by lowering combustion temperatures. It is mounted on top of the intake manifold and consists of a spring loaded poppet valve that is opened with a vacuum diaphragm. The ECM controls the vacuum to the EGR valve with a pair of solenoid valves. The EGR solenoid valve supplies vacuum to the EGR valve and the EGR vent solenoid releases the vacuum.

The signal used to operate the EGR solenoid is not just a simple on-off voltage but a pulse width modulated duty cycle. This allows the ECM to use the solenoid valve as a throttle to provide very fine control over vacuum to the EGR valve and therefore fine, rapid adjustments to EGR valve opening. The EGR vent solenoid allows rapid closing of the EGR valve. Unlike the EGR solenoid, the vent solenoid is not pulse width modulated by the ECM but is either turned **ON** for no EGR, or **OFF** the rest of the time. A vent filter is attached to the solenoid and is a serviceable component. A restricted vent filter will cause the EGR valve to remain open even though the ECM has activated the vent solenoid. The vent filter should be checked every 25,000 miles as part of a complete EGR system inspection.

The ECM makes EGR control decisions based on throttle position and engine RPM and uses MAP sensor readings to determine the EGR duty cycle. Together these sensor and control signals are known as the EGR loop. The solenoid valves are mounted with the EPR solenoid valve as an assembly towards the rear of the engine. If the ECM determines the amount of vacuum for EGR operation is incorrect for the present operating conditions (loop error), it will set DTC 32 and stop EGR system operation.

Exhaust Pressure Regulator (EPR) Valve

The EPR valve is mounted between the exhaust manifold and the exhaust pipe. Since a diesel engine has no throttle, there is little or no intake manifold vacuum at idle. What little vacuum that is available is just enough to operate the crankcase ventilation system, making exhaust gas recirculation almost impossible at idle. The EPR valve is identical to the vacuum operated heat riser valves used on gasoline engines and is normally open until vacuum is applied to the regulator diaphragm. At idle, the ECM will energize a solenoid valve to supply vacuum to the EPR diaphragm and close the restrictor plate. When the restrictor plate closes, exhaust backpressure rises which increases the exhaust gas flow through the EGR valve. The EPR solenoid is only activated at idle and turned **OFF** under all other conditions. If the ECM determines the EPR operation is incorrect for the present operating conditions (loop error), it will set DTC 32 and stop EGR system operation.

Torque Converter Clutch (TCC)

The TCC uses a solenoid operated valve in the automatic transmission to lock the input and output turbines of the torque converter together. When the solenoid is **OFF**, the valve exhausts some of the fluid pressure inside the torque converter to allow slip. When the solenoid is **ON**, the valve closes to increase fluid pressure inside the torque converter and the fluid coupling slip is almost completely eliminated. This feature is used to increase fuel milage at steady operating conditions. The PCM monitors the vehicle speed and engine load (VSS and TPS signals) for determining TCC apply conditions. The solenoid valve is mounted inside the transmission.

A brake switch is wired in series with the voltage supply to the TCC solenoid. The switch is normally closed unless the brake pedal is depressed. When the driver applies the brakes, the switch will open to interrupt the voltage to the TCC solenoid and release the torque converter clutch. If the PCM detects faulty TCC operation, it will set DTC 67 or 69. If a brake switch fault is detected, it will set DTC 37 or 38. The PCM also monitors several other solenoids, fluid pressures and fluid temperature in the automatic transmission and will set DTCs for any faults detected.

Glow Plug System

The glow plugs are six volt heaters that are operated at 12 volts to provide rapid heating inside the pre-combustion chamber. The cold start system on the 6.2L LH6 engine consists of a glow plug control unit, eight glow plugs, the cold advance relay, the cold timing advance solenoid, the ECM and a dash mounted WAIT lamp. The cold advance relay is located at the junction block on the right side of the engine compartment cowl. The glow plug control unit is mounted on a bracket towards the rear of the engine.

If coolant temperature is below about 80°F (27°C) when the ignition switch is turned **ON**, the cold advance relay will be activated by the ECM and power is supplied to the glow plug control unit. Power is supplied to the glow plugs for approximately 5 seconds, then turned **OFF** for about 4.5 seconds. At this point the control unit will begin to cycle the glow plugs **ON** for 1.5 seconds, and then **OFF** for 4.5 seconds. This cycle will repeat for a total duration of approximately 20 seconds, which includes the initial 5 second **ON** period. If the engine is cranked during or after this sequence, the glow plugs will continue to cycle **ON/OFF** for a total time of 25 seconds after the ignition switch is returned to the **RUN** position, whether or not the engine starts. The engine does not have to be running to initiate or terminate glow plug cycling. All times are approximate and will vary with temperature as well as system voltage. The WAIT lamp is illuminated as long as the glow plugs are energized.

Data Link Connector (DLC)

The DLC is the new name for the ALDL used on all GM vehicles. It is used to retrieve serial data and/or Diagnostic Trouble Codes and is mounted under the left side of the dashboard. A scan tool may be used to monitor input information as well as some output data. Stored DTCs can be read with a scan tool connected to the DLC or flashed by the MIL when a jumper wire is connected between terminals A and B with the ignition switch **ON**.

Malfunction Indicator Light (MIL)

This light is on the instrument panel and appears as SERVICE ENGINE SOON. When the ignition switch is first turned **ON**, it will light up as a bulb check, then go out. If the ECM detects a malfunction with the engine operating, the MIL will turn **ON** to warn the driver as long as the malfunction exists. If the malfunction is intermittent, the light will turn **OFF** after about 10 seconds and store a DTC in memory.

DTCs can be read by viewing the flashing MIL on the instrument panel. Each flashing DTC is repeated three times before proceeding to the next code, if one exists. DTC 12 is always displayed when the ignition switch is **ON** and the engine is not running. If only DTC 12 is displayed, no other codes are stored in memory.

Crankcase Depression Regulator (CDR) Valve

All engines are equipped with a CDR valve mounted on the right valve cover. It is used to reduce crankcase pressure at idle and control the flow of crankcase fumes to the intake manifold at higher engine speeds. As intake vacuum increases with engine speed, it pulls the diaphragm partially closed against a spring to control the flow of gasses and limit the vacuum in the crankcase. Too much vacuum creates an oil consumption problem and too little vacuum causes oil leaks. The valve is designed to maintain about 3–4 inches of water (0.75–1.0 kPa) vacuum in the crankcase.

Diagnosis and Testing

SERVICE PRECAUTIONS

- When working with the fuel system, make sure the work area is properly ventilated and take appropriate fire safety precautions.
- When working with diesel fuel systems, cleanliness is critical. Before loosening any fittings on any part of the fuel system, clean away all dirt, oil or fuel residue. Cap all open lines or fittings after removal.
- When working with the DEC system, make sure the ignition switch is **OFF** before disconnecting or connecting any sensor wiring or the ECM connector unless otherwise instructed. If a test requires disconnecting a solenoid wire with the ignition switch **ON**, make sure the wire does not become grounded.
- Do not operate the electric fuel pump any longer than necessary without fuel flowing. The pump relys on fuel flow for cooling.
- The ECM must not be exposed to high current electricity or temperatures in excess of 185°F (85°C) at any

time. The ECM must be disconnected before electric welding on the vehicle or removed before placing the vehicle in a paint baking oven.

- The ECM can be damaged by electrostatic discharge. Touch a vehicle ground or wear a static strap before handling the ECM. Do not touch the ECM connector pins, the PROM pins or any soldered component on the circuit board.

READING CODES

If a malfunction is detected by the ECM, it will store a Diagnostic Trouble Code (DTC) in memory and turn the Malfunction Indicator Light (MIL) **ON**. If the malfunction is intermittent and corrects itself while the engine is running, the MIL will go out after 10 seconds but the DTC will remain in memory for the next 50 engine starts or until the DTC memory is cleared. DTCs can be read with a scan tool or on the MIL on the instrument panel. Some scan tools may also be able to access other ECM data such as VIN code and PROM number. If a scan tool is connected to the Data Link Connector (DLC) and there is no display or if "No DATA" is displayed, refer to the appropriate MIL/SCAN troubleshooting procedure. There are separate procedures for manual and automatic transmissions.

> **NOTE: Begin all troubleshooting procedures with the ON-BOARD DIAGNOSTIC SYSTEM CHECK to make sure the data link or MIL are functioning.**

With Scan Tool

1. Make sure the ignition switch is **OFF** and connect the tool to the DLC under the left side of the dashboard.
2. Connect power to the tool and follow the instructions provided with the tool. It will probably be necessary to turn the ignition switch **ON** to begin data transfer.
3. Codes are reported in numerical order from low to high. Repair malfunctions in that order.
4. If only DTC 12 appears with the engine not running, no DTCs are stored in memory.
5. The scan tool should be able to read sensor input signals and report temperature, throttle position (sensor voltage or percent travel) or sensor output voltage. It will be necessary to run the engine and/or drive the vehicle to read some sensor inputs.

Without Scan Tool

1. Make sure the ignition switch is **OFF** and locate the Diagnostic Connector Link (DLC) under the left side of the dashboard.
2. Connect a jumper wire between terminals A and B, all the way to the right on the top row.
3. Turn the ignition switch **ON** and count the flashes on the "SERVICE ENGINE SOON" light on the instrument panel. It should display DTC 12 three times as follows: flash, pause, flash flash, long pause, etc.
4. If no DTCs are displayed at all or if the lights stays lit continuously, the diagnostic system is malfunctioning. Refer to the appropriate MIL/SCAN troubleshooting procedure.
5. If the light continues to display DTC 12, there are no other DTCs in memory. If there are other DTCs in memory, they will be displayed in numerical order from low to high three times each. Repair malfunctions in that order.

CLEARING CODES

The DTC memory can be cleared by disconnecting the ECM or the battery for at least 30 seconds. If a malfunctions has been repaired but the DTC is not cleared, the DTC will automatically be cleared after 50 engine starts. Some of the troubleshooting procedures require clearing DTCs as part of the diagnosis.

DIAGNOSTIC TROUBLE CODES

Manual Transmission

DTC **12** — ESS signal lost; normal with ignition **ON**, engine not running

DTC **14** — CTS signal high; sets if temperature reading is high or wire is grounded for 5 minutes

DTC **15** — CTS signal low; sets if signal is lost for 5 minutes

DTC **21** — TPS signal high; circuit open or improper adjustment, sets after 30 seconds at idle

DTC **22** — TPS signal low; circuit grounded, sets after 2 minutes above 1250 RPM

DTC **23** — TPS calibration; signal not between 0.25–1.3 volts after 30 seconds at idle

DTC **24** — VSS signal lost; circuit open or grounded, sets after 10 seconds at road speed

DTC **31** — MAP sensor signal low; sets after 10 seconds at idle

DTC **32** — EGR loop error; EGR vacuum incorrect, sets after 10 seconds above 30 MPH

DTC **33** — MAP sensor signal high; sets after 10 seconds at idle

DTC **51** — PROM is faulty or incorrectly installed, sets after 10 seconds

DTC **52** — ECM fault; sets after 10 seconds

DTC **53** — Reference voltage overload; 5 volt power to sensors shorted to ground, sets after 10 seconds; this DTC may also be set by jump starting the engine

Automatic Transmission

DTC **12** — ESS signal lost; normal with ignition **ON**, engine not running

DTC **14** — CTS signal high; sets if temperature reading is high or wire is grounded for 0.5 seconds

DTC **15** — CTS signal low; sets if signal is lost for 0.5 seconds

DTC **16** — Transmission output speed signal low; open circuit to VSS or buffer module

DTC **21** — TPS signal high; circuit open or improper adjustment, sets after 30 seconds at idle

DTC **22** — TPS signal low; circuit grounded, sets after 2 minutes above 1250 RPM

DTC **23** — TPS calibration; signal not between 0.25–1.3 volts after 30 seconds at idle

DTC **24** — VSS signal lost; circuit open or grounded, sets after 10 seconds at road speed

DTC **28** — Fluid pressure switch; invalid combination of transmission fluid pressure switch signals

DTC **31** — MAP sensor signal low; sets after 10 seconds at idle

DTC **32** — EGR loop error; EGR vacuum incorrect, sets after 10 seconds above 30 MPH

DTC **33** — MAP sensor signal high; sets after 10 seconds at idle

DTC **37** — Brake switch stuck **ON**; no brake switch signal when vehicle accelerates from 4 — 21 MPH, must be below 4 and above 20 MPH for 6 seconds, must occur 7 times or more

DTC **38** — Brake switch stuck **OFF**; switch signal remains when speed is above 20 MPH for 6 seconds, then 5 — 20 MPH for 6 seconds, must occur 7 times

DTC **51** — PROM is faulty or incorrectly installed, sets after 10 seconds

DTC **52** — Long system high voltage; charging system voltage above 16 volts for 109 minutes

DTC **53** — System voltage high; system voltage above 19.5 volts for 2 seconds; this DTC may also be set by jump starting the engine

DTC **55** — PCM malfunction

DTC **67** — TCC solenoid; **ON** or **OFF** command voltage exists for more than 2 seconds

DTC **69** — TCC stuck **ON**; no torque converter slip, transmission in 2, 3 or 4 and TPS more than 25% for 4 seconds

DTC **72** — VSS signal; buffer output speed signal changes more than 2050 RPM in P or N or more than 1000 RPM in gear for more than 2 seconds

ON-BOARD DIAGNOSTIC SYSTEM CHECK

Testing

MANUAL TRANSMISSION

NOTE: Although the circuits are similar, terminal designations on the ECM are different on manual and automatic transmission vehicles. Be sure to use the correct wiring schematic.

1. Check the engine compartment for faulty electrical connections and pinched, damaged or disconnected vacuum hoses. Use a hand vacuum pump to check the EGR valve for free movement and diaphragm leakage. Check the EGR vent filter.

2. Run the engine to bring it to normal operating temperature.

3. With the ignition switch **ON** but the engine not running, the MIL on the instrument panel should be **ON** steadily. If the light is **OFF**, go to the MIL/SCAN MALFUNCTION test. If the light is flashing and the wire between terminal B on the DLC and terminal A6 on the ECM is not grounded, the ECM is faulty.

4. With the ignition switch and MIL both **ON**, connect a scan tool or jumper DLC terminals A and B. The scan tool or the MIL should display DTC 12 and/or any other DTC stored in memory. If so, the on-board diagnostic system is functioning properly. If the MIL does not flash or there is no data transfer, go to the MIL/SCAN MALFUNCTION test.

5. Remove the scan tool or DLC jumper and run the engine for at least 2 minutes. If the MIL turns **ON**, there is a system malfunction. Install the scan tool or jumper wire to check the DTC memory. If DTCs 51, 52 or 53 are displayed, address those problems first. For any other DTCs, start with the lowest number.

6. If the MIL stays **OFF** with the engine running, turn the ignition switch **OFF** to avoid setting a false DTC and connect a vacuum gauge in place of the EGR valve.

7. With the engine running at idle and in neutral, the vacuum reading should be steady and above 12 in.Hg (41 kPa). When the throttle is quickly pressed and released, the reading should drop to almost zero. If vacuum is not correct at all conditions, look for a leak or restriction in a hose or a clogged vent filter.

ENGINE SPEED SENSOR CIRCUIT VOLTAGE PARAMETER CHART

ENGINE SPEED SENSOR

Engine	Circuit Voltage
6.2L (VIN C) (LH6)	About 5 volts supplied to sensor connector

ENGINE COOLANT TEMPERATURE CIRCUIT VOLTAGE PARAMETER CHART

COOLANT TEMPERATURE SENSOR

Engine	Circuit Voltage
6.2L (VIN C) (LH6)	About 1.5–2.0 volts returned to ECM at normal operating temperature

ENGINE COOLANT TEMPERATURE SENSOR RESISTANCE PARAMETER CHART

COOLANT, AIR AND OIL TEMPERATURE SENSOR RESISTANCE

Engine	Degrees F	Degrees C	Ohms
All	212	100	177
	194	90	241
	176	80	332
	158	70	467
	140	60	667
	122	50	973
	113	45	1188
	104	40	1459
	95	35	1802
	86	30	2238
	77	25	2796
	68	20	3520
	59	15	4450
	50	10	5670
	41	5	7280
	32	0	9420
	23	−5	12300
	14	−10	16180
	5	−15	21450
	−4	−20	28680
	−22	−30	52700
	−40	−40	100700

THROTTLE POSITION SENSOR RETURN SIGNAL VOLTAGE PARAMETER CHART

THROTTLE POSITION SENSOR

Engine	Signal Voltage	
	Idle	WOT
6.2L (VIN C) (LH6)	0.5–1.25	More than 4.0

VEHICLE SPEED SENSOR PARAMETER CHART

VEHICLE SPEED SENSOR

Engine	Output Signal
6.2L (VIN C) (LH6)	Scan tool reading must match speedometer reading

TRANSMISSION OUTPUT SPEED SIGNAL PARAMETER CHART

TRANSMISSION OUTPUT SPEED SIGNAL

Engine	Output Signal
6.2L (VIN C) (LH6)	Up to 5.0 volts as vehicle speed increases

8. If vacuum is correct, turn the ignition switch **OFF** and reconnect the vacuum hose to the EGR valve. Connect the gauge in place of the EPR valve and start the engine.

9. Vacuum should be above 15 in. Hg (50 kPa) at idle. When the EPR solenoid wiring is disconnected, the vacuum should drop to zero. If not, test the EPR and its solenoid valve.

10. If all tests yield correct results so far but a drivability problem exists, the DEC system is operating correctly and the problem is in the diesel fuel injection system.

AUTOMATIC TRANSMISSION

NOTE: Although the circuits are similar, terminal designations on the ECM are different on manual and automatic transmission vehicles. Be sure to use the correct wiring schematic.

1. Check the engine compartment for faulty electrical connections and pinched, damaged or disconnected vacuum hoses. Use a hand vacuum pump to check the EGR valve for free movement and diaphragm leakage. Check the EGR vent filter.

2. With the ignition switch **ON** but the engine not running, the MIL on the instrument panel should be **ON** steadily. If the light is **OFF**, go to the MIL/SCAN MALFUNCTION test. If the light is flashing and the wire

MAP SENSOR SIGNAL VOLTAGEPARAMETER CHART

MAP SENSOR

Engine	Output Signal
6.2L (VIN C) (LH6)	Signal wire voltage must change as vacuum changes

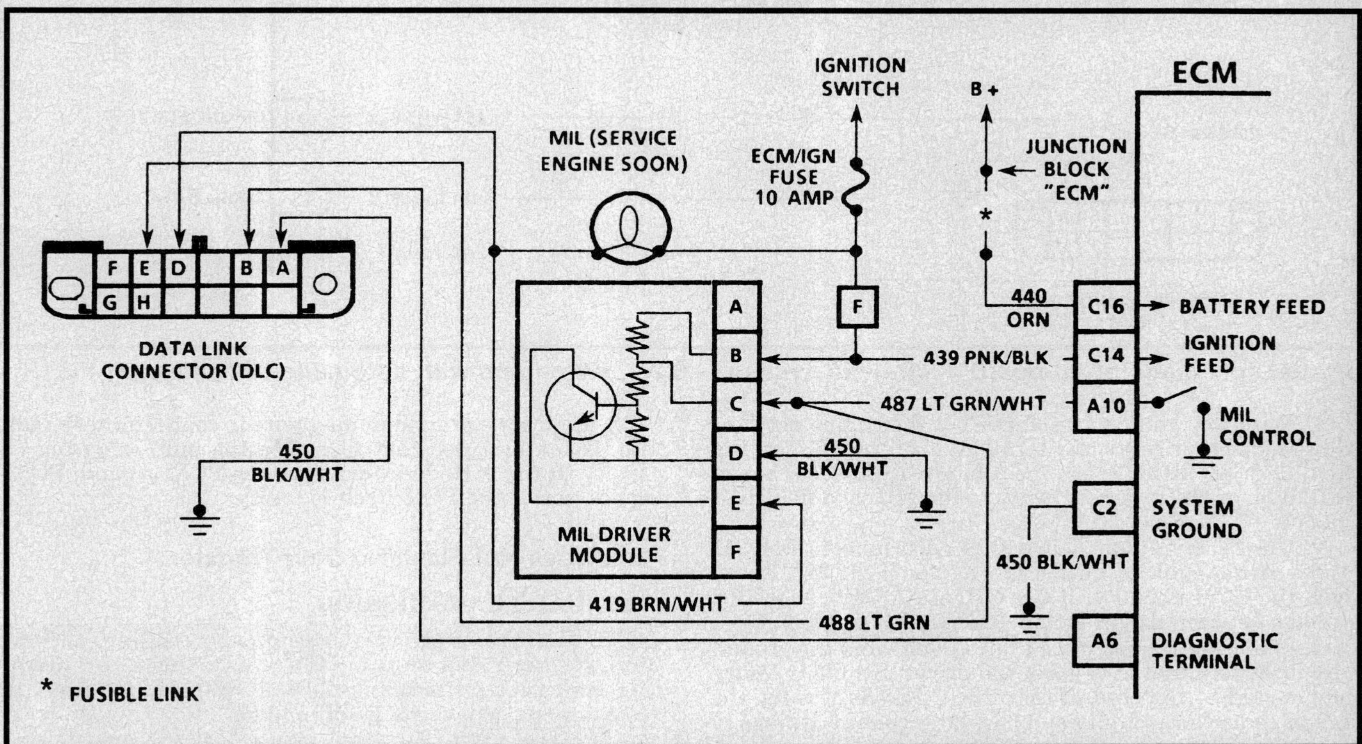

Partial schematic of on-board diagnostic system — 6.2L LH6 engine with manual transmission

between terminal B on the DLC and terminal A14 on the ECM is not grounded, the PCM is faulty.

3. With the ignition switch and MIL both **ON**, connect a scan tool or jumper DLC terminals A and B or ground terminal B. The scan tool or the MIL should display DTC 12 or any other DTC stored in memory. Other engine data should also be available to the scan tool. If the MIL does not flash or there is no data transfer, go to the MIL/SCAN MALFUNCTION test.

4. If the scan tool or MIL are operating correctly, the on-board diagnostic system is operating correctly. If DTC 51, 52 or 53 are displayed, address those problems first. For any other DTCs, start with the lowest number.

MIL/SCAN MALFUNCTION

There are two basic failure modes for the MIL/SCAN functions. If the MIL stays **OFF** or if there is no response when the scan tool is activated, the diagnostic system is not functioning. If the MIL turns **ON** but does not flash or if the scan tool displays NO DATA TRANSFER, the diagnostic system is functioning but cannot report and DTC data.

MIL Stays OFF

MANUAL TRANSMISSION

NOTE: Although the circuits are similar, terminal designations on the ECM are different on manual and automatic transmission vehicles. Be sure to use the correct wiring schematic.

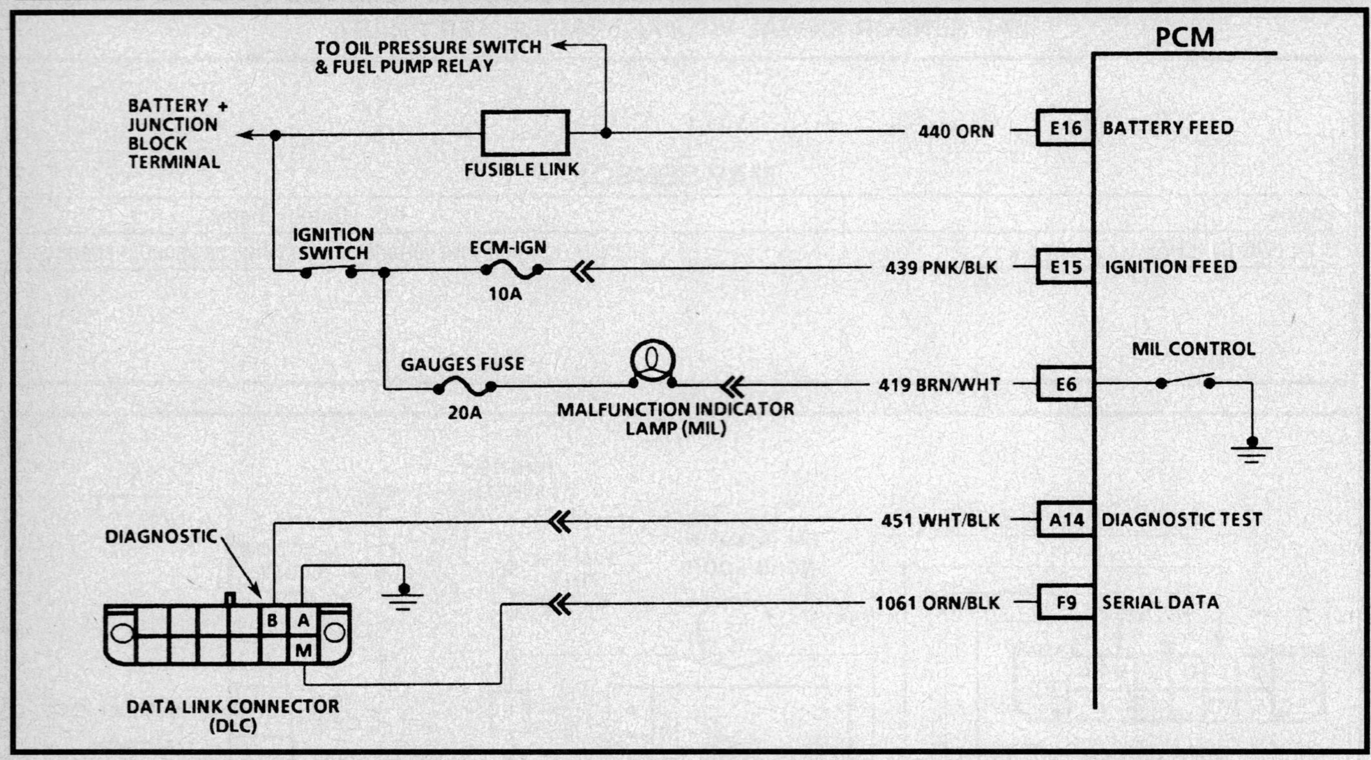

Partial schematic of on-board diagnostic system — 6.2L LH6 engine with automatic transmission

1. With the ignition switch **ON** but the engine not running, momentarily ground terminal E of the DLC. If the MIL does not light, check for a blown ECM fuse, blown MIL bulb or faulty wiring between the MIL and its driver module.

2. Turn the ignition switch **OFF**, disconnect the ECM and turn the ignition switch **ON** again. If the MIL turns **ON**, the ECM is faulty. If the MIL stays **OFF**, check for voltage between driver module terminal C to ground.

3. If terminal C has more than 11 volts and the ground circuit at terminal D is good, the driver module is faulty and should be replaced. If terminal C has 6–11 volts, the driver connector is faulty and may have caused damage to the driver also.

4. If terminal C has less than 6 volts, check for power from the ignition switch to terminal B of the driver module and terminal C14 of the ECM.

5. If the supply voltage is correct, remove the wire from terminal C of the driver connector and reconnect the remaining wires. If the MIL turns **OFF**, the driver module is faulty. If the MIL stays **ON**, the circuit from driver terminal C to ECM terminal A10 is faulty.

AUTOMATIC TRANSMISSION

NOTE: Although the circuits are similar, terminal designations on the ECM are different on manual and automatic transmission vehicles. Be sure to use the correct wiring schematic.

1. Check the fusible link between the battery and the PCM. Check the PCM ignition fuse and the gauges fuse.

2. With the ignition switch **OFF**, disconnect the PCM wiring and turn the ignition **ON** again. Check for power to the PCM at connector terminals E15 and E16.

3. Connect a ground jumper wire to connector terminal E6. If the MIL does not turn **ON**, the bulb or wiring is faulty. If the MIL does turn **ON**, the PCM ground, PCM connector or the PCM itself is faulty.

MIL Does Not Flash/No Data Transfer

MANUAL TRANSMISSION

NOTE: Although the circuits are similar, terminal designations on the ECM are different on manual and automatic transmission vehicles. Be sure to use the correct wiring schematic.

1. Make sure the battery is supplying a full 12 volts and check all fuses and fusible links. If the MIL does not function properly when performing the On–Board Diagnostic Check, turn the ignition switch **OFF** and remove the jumper wire from the DLC.

2. Turn the ignition switch **ON** and check for voltage between MIL driver module terminal C and ground. If there is more than 11 volts, either the circuit from terminal C to terminal A10 on the ECM is shorted to power or the driver module is faulty.

3. If driver terminal C has less than 11 volts, connect the terminal to ground. If the MIL is still **ON**, unplug the driver connector. If the MIL is still **ON**, the brown/white wire from connector terminal E is shorted to ground. If it turns **OFF**, there is a bad connection at terminal C or the driver module is faulty.

4. If the MIL turns **OFF** when terminal C is grounded, remove the jumper wire and connect it to terminal A10 on the ECM. If the MIL turns **ON**, the circuit between terminals C and A10 is open. If the MIL turns **OFF**, ground terminal A6 of the ECM and watch the MIL.

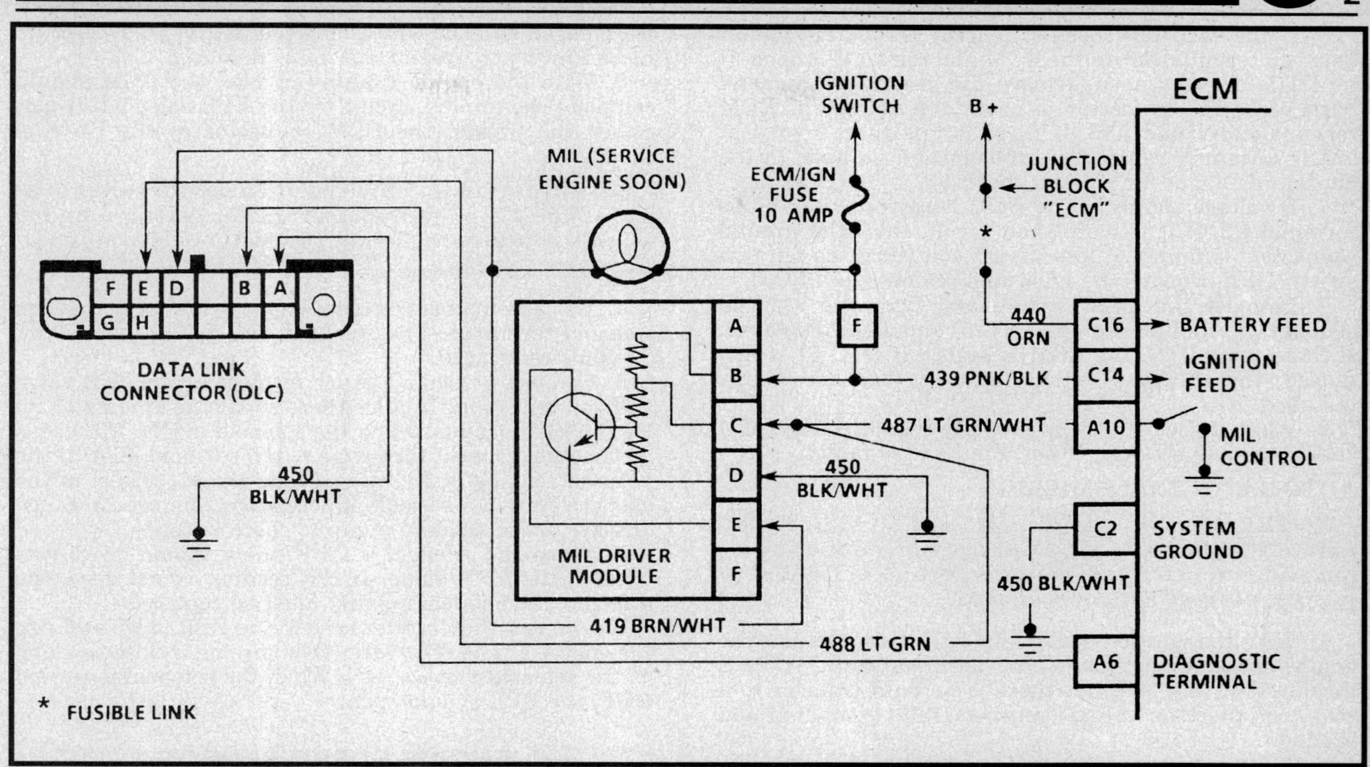

MIL driver module schematic — 6.2L LH6 engine with manual transmission

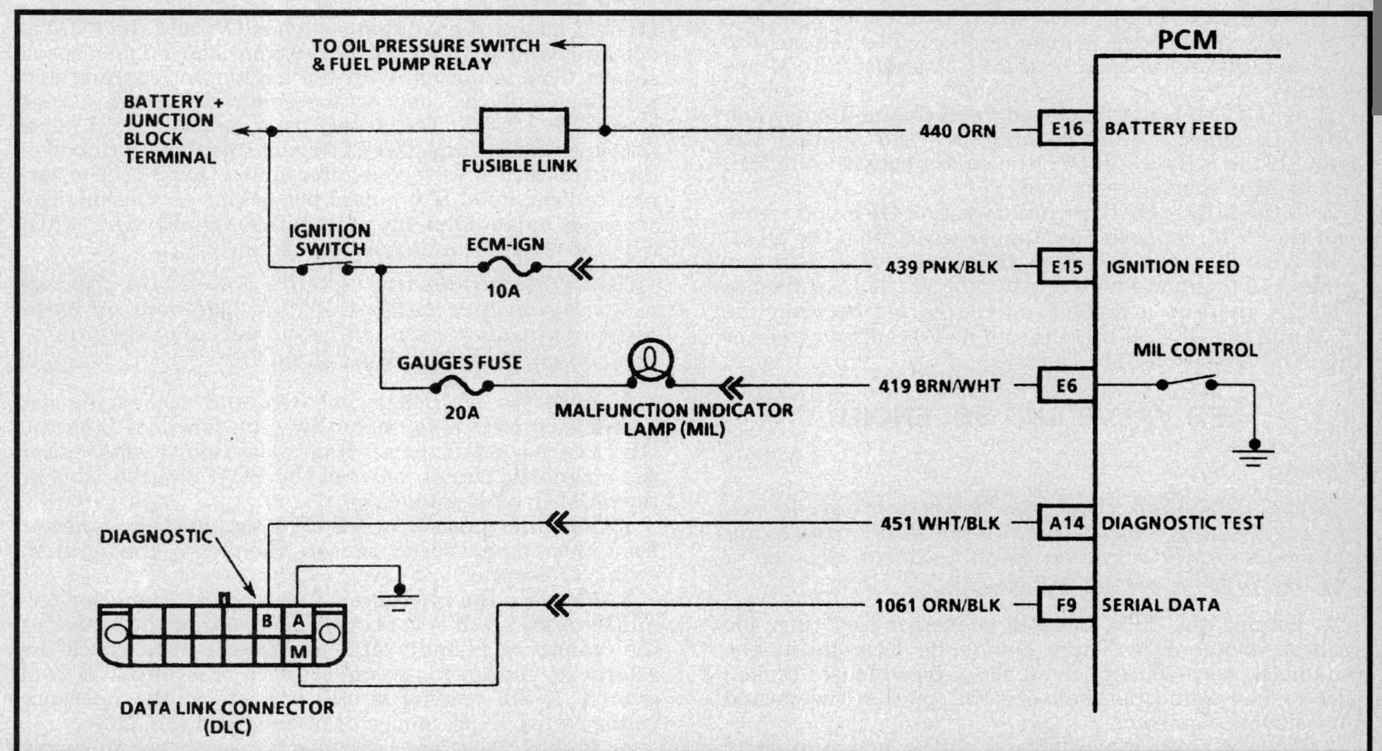

MIL circuit — 6.2L LH6 engine with automatic transmission

5. If the MIL now begins to flash DTCs, the circuit between terminal A6 on the ECM and the DLC is open. If no DTCs are displayed, remove the ground jumper and back probe the connector to check the voltage at ECM terminals A14 and A16. If there is less than 9 volts at either terminal, repair the problem and go back to the On-Board Diagnostic System check.

6. If voltage supply to the ECM is correct, back probe terminal C2. If it is more than 1 volt, the ECM ground connection is faulty. If less than 1 volt, turn the ignition switch **OFF**, remove the ECM and remove the PROM.

7. Connect the ECM wiring and turn the ignition switch **ON**. After 30 seconds, DTC 51 should be displayed. If there is no DTC, the ECM is faulty. If DTC 51 is displayed, the PROM is either faulty or was incorrectly installed.

8. When a good PROM is properly installed, the ECM should display DTC 12. If not, the ECM is faulty.

AUTOMATIC TRANSMISSION

1. With the ignition switch and MIL both **ON** but the engine not running, ground DLC terminal B. If DTC 12 is displayed on the MIL but there is no data transfer to a scan tool, check for voltage between DLC terminal M and ground.

   a. The correct reading is 2.5–5.0 volts. If the reading is correct, the fault is intermittent and may be caused by faulty wiring or connections at the DCL or the ECM.

   b. If the reading is below 2.5 volts or above 5.5 volts, the data link circuit is open or shorted to voltage, the system ground is open or the ECM and/or PROM are faulty.

2. If DTC 12 is not displayed, turn the ignition switch **OFF**, disconnect the PCM and turn the ignition **ON** again. If the MIL is still **ON**, the circuit between the MIL and ECM is shorted to ground.

3. If the MIL is **OFF**, turn the ignition **OFF** and reconnect the PCM. Remove any jumper wires from the DLC.

4. With the ignition switch **ON**, back probe PCM terminal A14 to ground with a test light. If the light flashes DTC 12, there is a problem in the wiring between the PCM and the DLC. If there is still no DTC display, either the PCM or the PROM is faulty.

EPR VALVE AND SOLENOID

Testing

1. Unplug the EPR solenoid connector and turn the ignition switch **ON**. There should be 12 volts at the pink/black wire. If not, check for a blown fuse, broken wire or bad connection between the ignition switch and the solenoid connector.

2. The dark green wire should not have continuity to ground with the engine not running. If it does, turn the ignition switch **OFF**, unplug the ECM connector and check again for continuity to ground at the solenoid connector. If the green wire is still grounded, the insulation

is damaged and the wire is shorted to ground. If there is no continuity to ground, the ECM is faulty.

3. With the engine running at idle, the ECM should complete the ground circuit for the EPR solenoid. If not, either the wiring, the ECM connector or the ECM is faulty.

4. Use an ohmmeter to check the resistance across the solenoid terminals. The solenoid coil should have about 20 ohms resistance.

5. Connect a vacuum gauge in place of the EPR valve and run the engine at idle. There should be at least 15 in. Hg (50 kPa) vacuum when the solenoid is **ON**. If there is no vacuum, connect the gauge to the solenoid inlet. If the vacuum reading is still not correct, check the rest of the vacuum system for a leak. If necessary, connect the gauge directly to the vacuum pump to check output.

6. When the solenoid is **OFF**, there should be no vacuum to the EPR valve. If the reading is not zero, the solenoid valve is leaking and must be replaced.

7. Connect the vacuum hose to the EPR valve and run the engine at idle. The valve should remain closed as long as the solenoid valve is **ON**. When the solenoid is turned **OFF**, the EPR should open.

ENGINE SPEED SENSOR

Testing

DTC 12 means the ignition switch is **ON** and the ECM is not receiving a reference pulse from the engine speed sensor. This is normal with the engine not running and will flash only as long as the condition exists, not be stored as a DTC. The sensor is supplied 5 volts as a reference signal and the ECM measures the number of times the return voltage is interrupted (pulsed). The correct return signal is 4 pulses per engine revolution. The sensor is mounted at the rear of the engine where a distributor would be on a gasoline engine.

1. With the ignition switch **ON** and the engine not running, connect a voltmeter between terminal B on the DLC and ground. If the reading is less than 4 volts, either the diagnostic circuit between the ECM and the DLC or the ECM itself is faulty.

2. Turn the ignition switch **OFF**, unplug the connector from the engine speed sensor, then turn the ignition switch **ON** again.

3. Measure the reference voltage across connector terminals A and B. If it is about 5 volts, either the sensor or the connection is faulty. If it less than 4 volts, check the reference voltage between connector terminal B and ground. If the reading is still low, either the reference voltage wire, ECM connector or the ECM is faulty.

4. If all voltages are correct so far, reconnect all wiring and check the voltage between the sensor ground wire and chassis ground at the back of the ECM connector with the ignition switch **ON**. If the reading is still less than 4 volts, the ground circuit between the sensor and

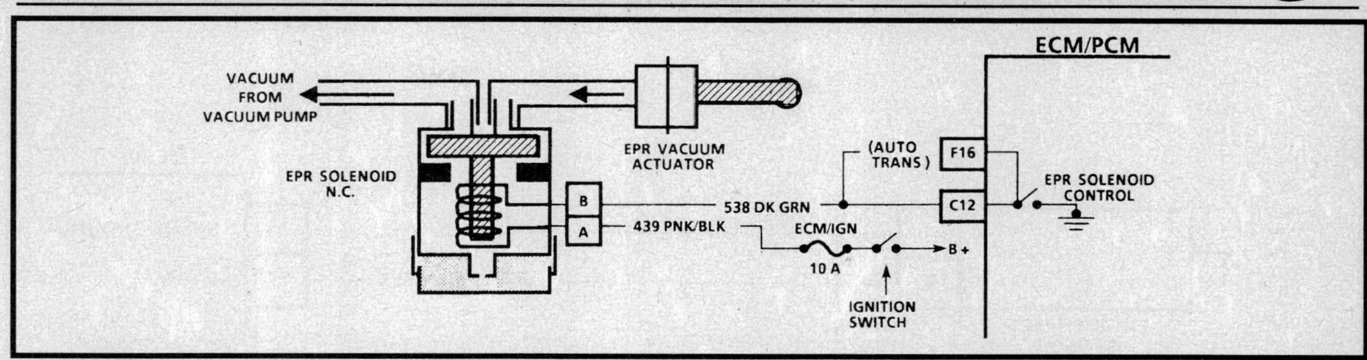

EPR solenoid circuit — 6.2L LH6 engine

the ECM is faulty. If the reading is about 5 volts, either the connector or the ECM is faulty.

ENGINE COOLANT TEMPERATURE SENSOR

Testing

DTC 14 is displayed if the ECT signal indicates a temperature above 275°F (135°C) for more than 3 minutes with a manual transmission or more than 294°F (145°C) for 0.5 seconds with an automatic transmission. It may indicate the 5 volt reference signal is shorted to ground.

DTC 15 is displayed if the signal indicates less than 22°F (-30°C) for more than 5 minutes with a manual transmission or -27°F (-33°C) for 0.5 seconds with an automatic transmission. It may indicate an open circuit.

If DTC 21 also is displayed, check for a faulty ground circuit for all sensors. The circuits and sensor can be checked with a volt/ohmmeter but a scan tool is required for complete testing.

NOTE: Although the circuits are similar, terminal designations on the ECM are different on manual and automatic transmission vehicles. Be sure to use the correct wiring schematic.

NOTE: Most scan tools will display temperature in degrees C.

1. Unplug the sensor connector and turn the ignition switch **ON**. There should be 5 volts to the sensor at the yellow wire terminal. If not, either the wiring, a connector or the ECM is faulty.

2. Turn the ignition switch **OFF**, connect a scan tool to the DLC and connect the yellow sensor wire to ground.

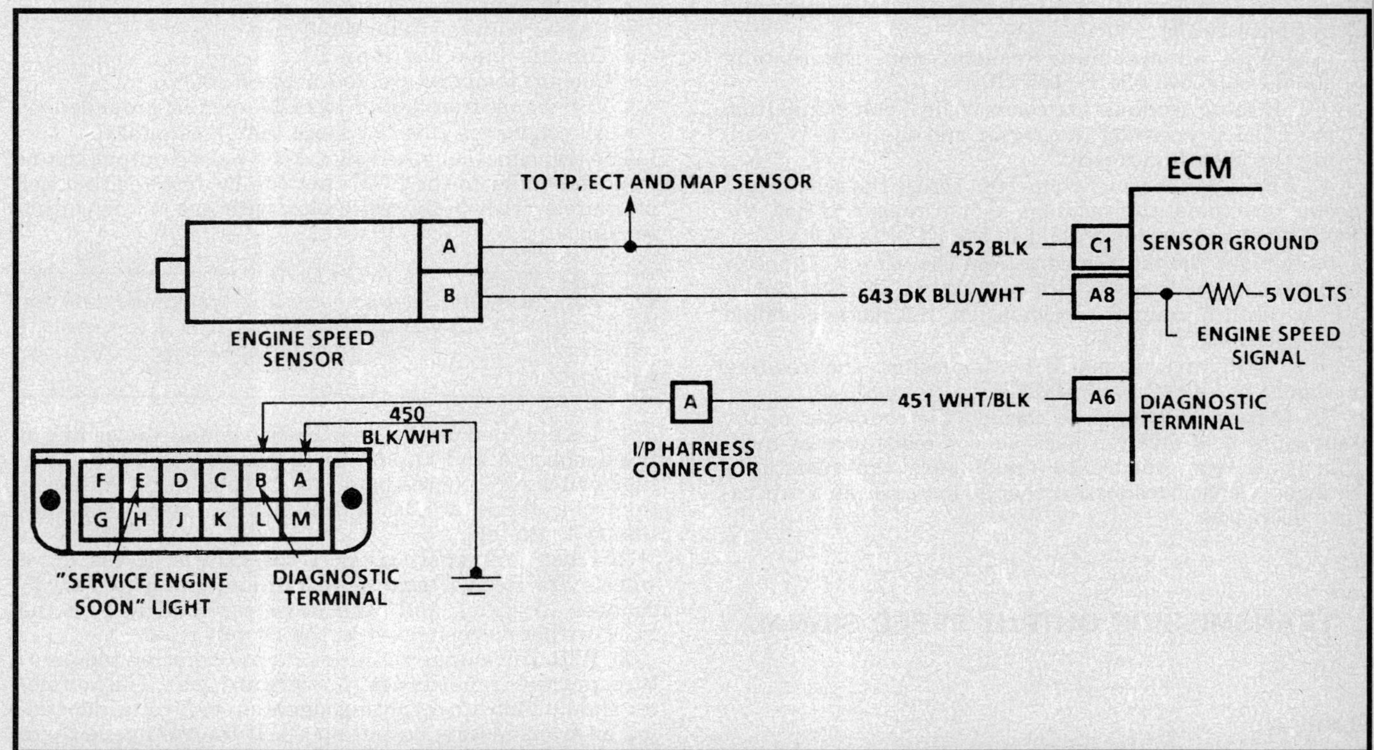

Engine speed sensor wiring schematic — 6.2L LH6 engine with manual transmission

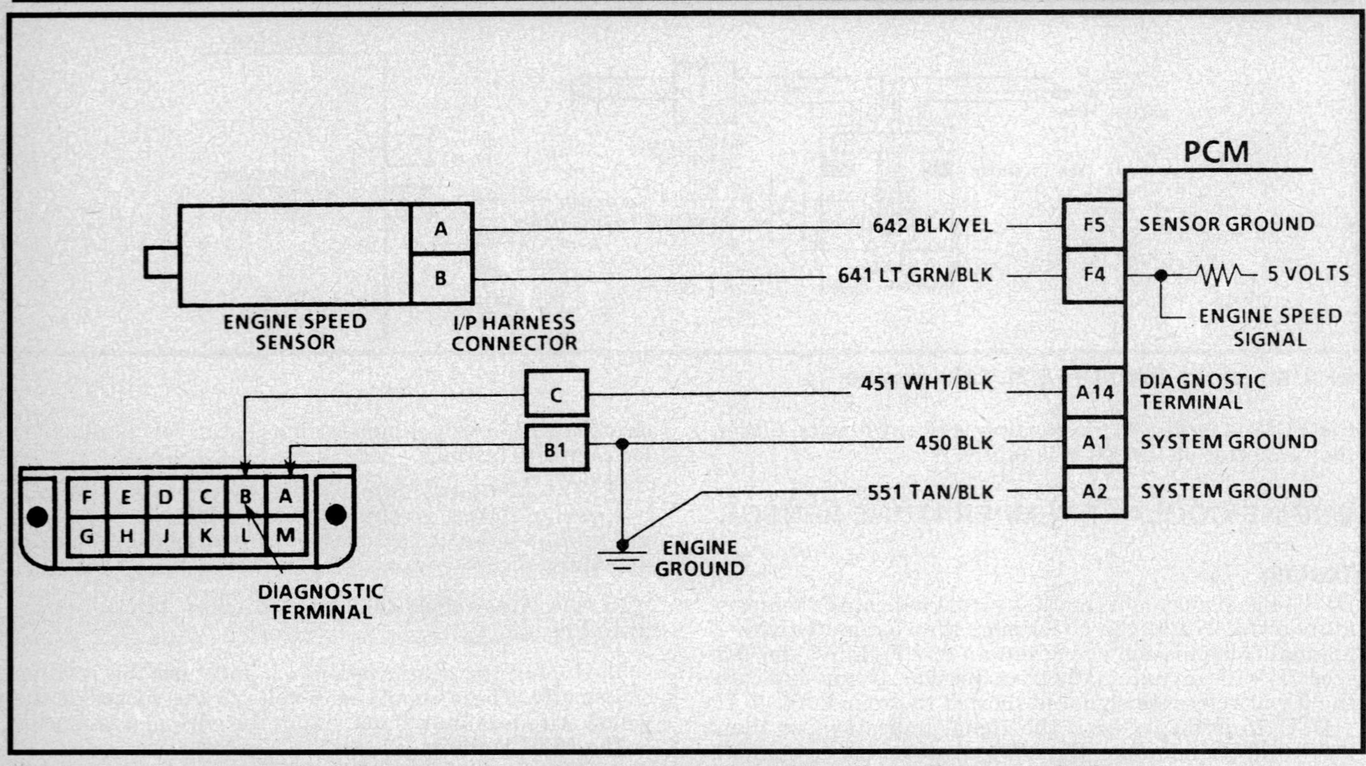

Engine speed sensor wiring schematic — 6.2L LH6 engine with automatic transmission

3. Turn the ignition switch **ON** and observe the temperature readings on the scan tool:

a. With a manual transmission, the reading should be above 266°F (130°C).

b. With an automatic transmission, the reading should be above 294°F (145°C).

c. If these readings are correct, the 5 volt signal from the ECM is reaching the sensor and the ECM is reading the signals correctly.

4. With a jumper wire connected across the sensor connector terminals, the readings should repeat. If not, the ground wire or sensor ground in the ECM is faulty.

5. Remove the jumper wire from the sensor connector and observe the temperature readings on the scan tool:

a. With a manual transmission, the reading should be below -22°F (-30°C).

b. With an automatic transmission, the reading should be below -27°F (-33°C).

6. If these readings are correct, the connector or the sensor itself is faulty. Check sensor resistance at more than one temperature to make sure the resistance changes. Sensor resistance should increase as temperature decreases.

TRANSMISSION OUTPUT SPEED SIGNAL

Testing

DTC 16 indicates the PCM is not receiving a transmission speed signal from the speed sensor buffer. This DTC ap-

plies only to vehicles with automatic transmissions and will be set only under very specific conditions:

- Signal indicates less than 2 mph
- Engine between 1000–4400 rpm
- Throttle angle less than 2%
- Coolant temperature above 140°F (60°C)
- VSS signal from buffer to PCM open or grounded
- All conditions exist for more than 3 seconds

The transmission speed signal is a pulsed output signal from the buffer to the PCM that can be detected (but not measured) with a DC voltmeter with the drive wheels turning.

> **CAUTION**
>
> *This test procedure requires operating the vehicle with the drive wheels turning. Support the rear axle properly or, if road testing, have an assistant in the vehicle to read the scan tool.*

1. Locate the speed sensor buffer behind the center of the dashboard and unplug the connector. With the ignition switch **ON**, there should be 12 volts to the buffer at connector terminal C9. Terminal C8 should have continuity to ground.

2. Raise and safely support the vehicle so the drive wheels are free to turn. Turn the ignition switch **OFF**, connect all wiring and back probe terminal F13 on the PCM with a voltmeter set to the 20 volt scale.

3. With the engine running and at operating temperature, put the transmission in a forward gear. The voltmeter should indicate a varying signal up to 5 volts, depending on transmission output speed. If the voltage is there but the DTC resets after being cleared, the PCM is faulty and will not read the transmission output signal. If there

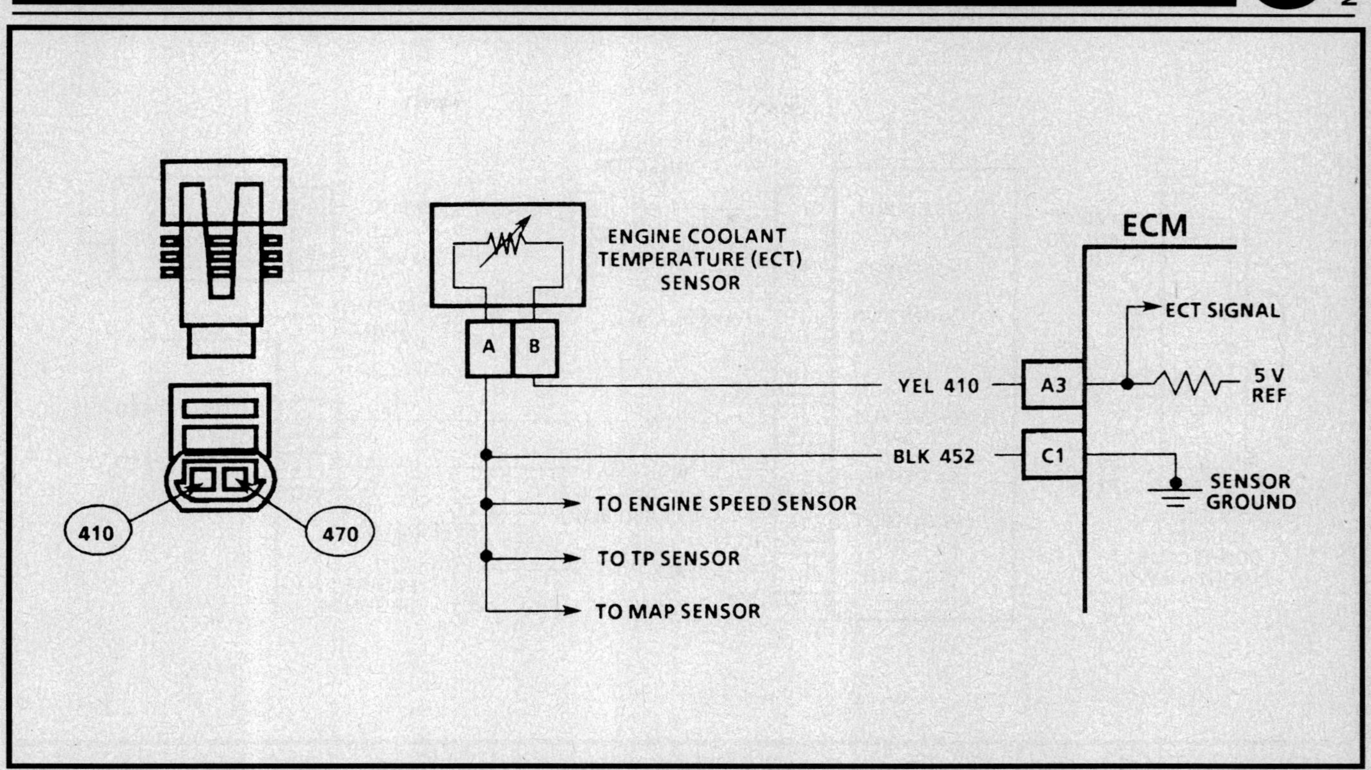

Engine coolant temperature sensor circuit — 6.2L LH6 engine

is no signal, check for a faulty connection or wire between the PCM and the sensor buffer.

4. If the PCM wiring and connector are good but there is no signal from the buffer, back probe terminal C11 on the speed sensor buffer. If there is still no signal, unplug the buffer connector and test again. If there is still no signal, the buffer is faulty.

THROTTLE POSITION SENSOR

Testing

MANUAL TRANSMISSION

DTC 21 usually indicates a short circuit in the sensor wiring and will set only under the following conditions:
- High voltage at ECM terminal A2
- Engine speed less than 1120 rpm
- Conditions exist for more than 2 minutes

DTC 22 usually indicates an open circuit in the sensor wiring or loss of the 5 volt reference and will set only under the following conditions:
- Low voltage at ECM terminal A2
- Engine speed more than 1250 rpm
- Conditions exist for more than 2 minutes

DTC 23 indicates the TPS is not properly calibrated or loss of the 5 volt reference and will set only under the following conditions:
- Signal at ECM terminal not between 0.25–1.35 volts
- Engine speed between 550–650 rpm
- Conditions exist for more than 30 seconds.

If the TPS does not function or if the signal is incorrect, the EGR system will not operate and other DTCs will be set in memory. Always start troubleshooting with the lowest number DTC first and complete the test procedure before replacing any parts.

NOTE: Although the circuits are similar, terminal designations on the ECM are different on manual and automatic transmission vehicles. Be sure to use the correct wiring schematic.

1. Warm the engine to normal operating temperature. Turn the ignition switch **OFF** and disconnect the ECM or the battery for more than 30 seconds to clear the DTC memory. Make sure the ground jumper is removed from the DLC before reconnecting the power.

2. Reconnect the wiring and run the engine at idle with all electrical consumers **OFF** for about 3 minutes or until the MIL turns **ON**. If the MIL does not turn **ON** or if no DTCs are displayed when the diagnostic terminal is grounded, the problem is intermittent. Check for faulty connectors or grounds or problems with the charging system.

3. If DTC 21 or 22 is displayed again, stop the engine and clear the DTC memory again. Reconnect the power and unplug the TPS connector.

4. Turn the ignition switch **ON** and measure the 5 volt reference between connector terminal A (grey wire) and ground. If the voltage is not correct, either the ECM is not supplying the reference voltage or a connector or wiring is faulty. Check the reference voltage at terminal A12 on the ECM and check wiring continuity.

5. Check for continuity to ground at connector terminal C (black wire). If there is no ground, either the ECM is not supplying the sensor ground or there is an open ground circuit between the sensor and the ECM. Check for ground at terminal C1 on the ECM and check wiring continuity.

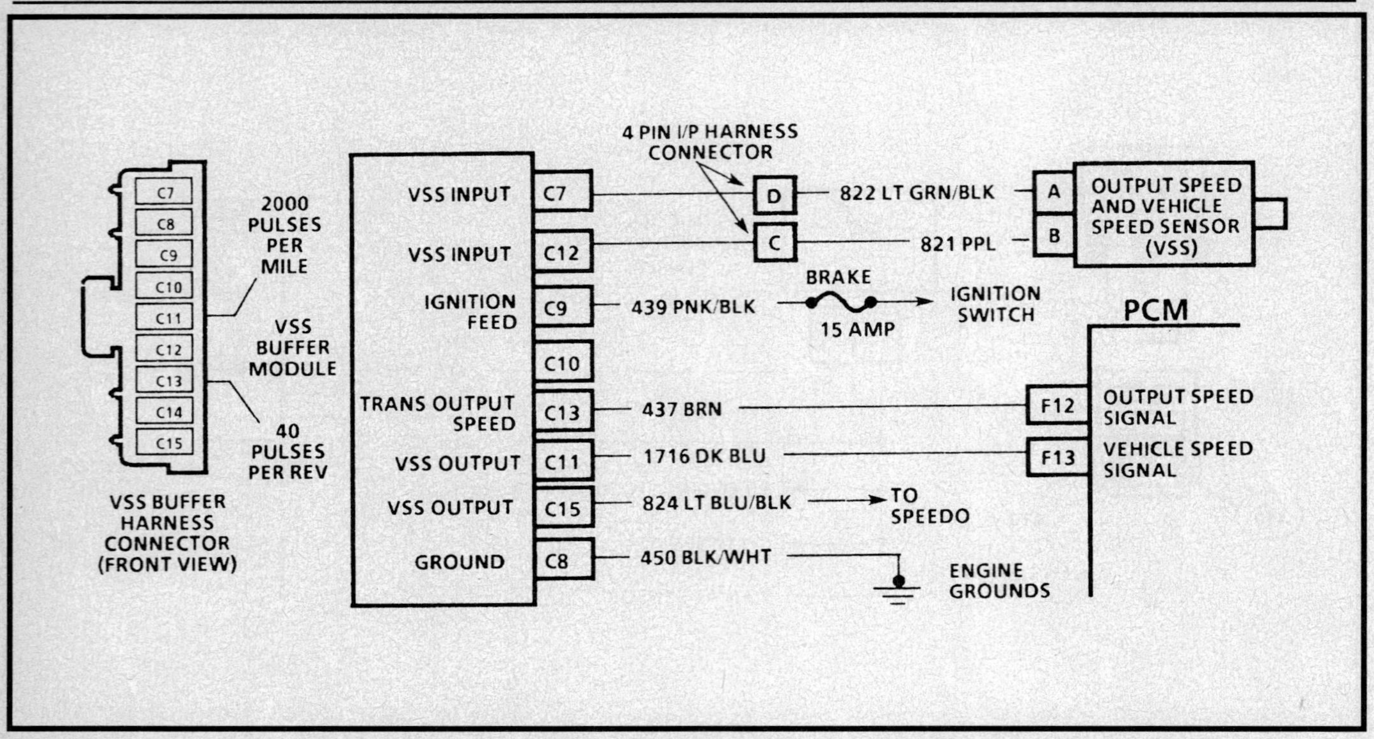

Vehicle speed and transmission output speed buffer circuit — 6.2L LH6 engine with automatic transmission

6. If the reference voltage and ground circuits are good, connect a jumper wire between connector terminals A and B (grey and blue wires). Run the engine as in Step 2 to generate DTCs, then stop the engine and read them.

7. DTC 21 should be displayed. If so, the sensor is faulty or out of adjustment. Turn the ignition switch **OFF**, disconnect the sensor wiring and connect an ohmmeter across sensor terminals A and B. Resistance should change smoothly as the sensor (accelerator pedal) is moved. If the resistance does not change or does not change smoothly, the sensor is faulty and must be replaced.

8. If DTC 23 is displayed at any time, check all connectors, wiring, ground and voltage circuits before adjusting the sensor. Sensor adjustment is required if the pump or sensor have been removed and generally does not change with use.

AUTOMATIC TRANSMISSION

A scan tool displays throttle position in volts or percent of throttle opening. DTC 21 will set only if the signal voltage is greater than about 4.9 volts at full throttle for more than 4 seconds. DTC 22 will set only if signal voltage is below 60mV for more than 4 seconds at any throttle position. DTC 23 will set if the engine is at idle speed (550–650 rpm) for 30 seconds and the sensor signal is above 60mV but not between 0.25–1.35 volts.

If the TPS is not working correctly, the EGR system and the torque converter clutch will not operate and the transmission will have incorrect line pressures, harsh shifts and no fourth gear in hot mode.

NOTE: Although the circuits are similar, terminal designations on the ECM are different on manual and automatic transmission vehicles. Be sure to use the correct wiring schematic.

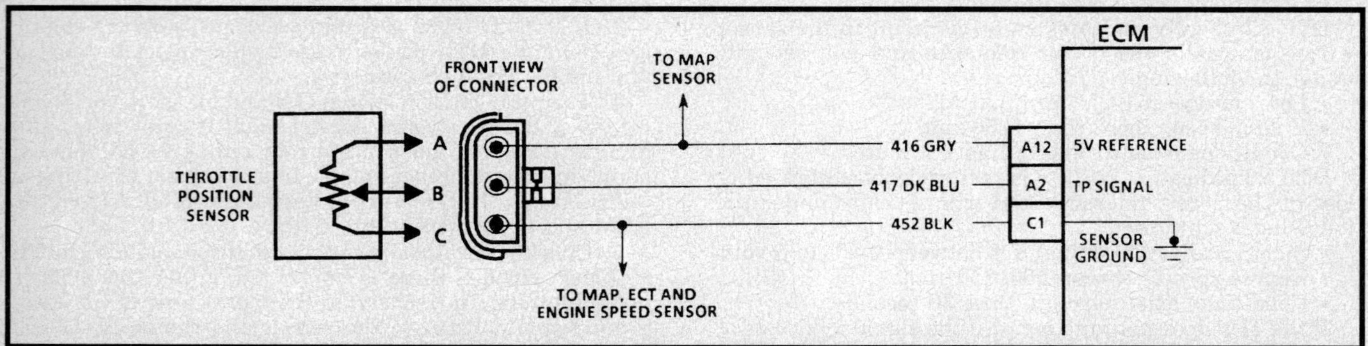

Throttle position sensor wiring schematic — 6.2L LH6 engine with manual transmission

1. Warm the engine to operating temperature, then stop the engine. Connect the scan tool to the DLC, turn the ignition switch **ON** and operate the scan tool to read the TPS.

2. Move the accelerator pedal and watch the scan tool. The TPS voltage should increase smoothly as the pedal is moved. Signal voltage should be about 0.5–1.25 at idle position and increase steadily to more than 4.0 volts as the accelerator pedal is pressed to the floor. If the reading is between 0.25–1.35 volts at idle position, the fault is intermittent. Check for faulty connectors or grounds or problems with the charging system.

3. If DTC 21 is displayed or if the TPS reading is above 1.25 volts at idle position, unplug the sensor connector.

   a. If the voltage stays above 200mV, the blue signal wire between the sensor and the PCM is shorted to voltage or the PCM is faulty.

   b. If the voltage decreases to less than 200mV, check for continuity to ground at the black sensor wire. If the sensor ground circuit is good, either the sensor or the connector is faulty.

4. If DTC 22 is displayed or if the TPS reading is less than 200mV at idle position, unplug the sensor connector and connect a jumper between the grey and blue wire terminals.

   a. If the reading increases to 4 volts or more, the sensor is either faulty or out of adjustment or the sensor connector is faulty.

   b. If the reading does not increase to at least 4 volts, either the PCM is not supplying the 5 volt reference or the wiring and connectors between the sensor and the PCM are faulty. Use a volt/ohmmeter to check the 5 volt reference and sensor ground at the PCM.

5. If DTC 23 is displayed, clear the DTC memory. Reconnect the wiring and run the engine at idle with all electrical consumers **OFF** for about 1 minute or until the MIL turns **ON**. If the MIL does not turn **ON** or if no DTCs are set, the problem is intermittent and the DTC can be ignored.

6. If DTC 23 is set again, turn the ignition switch **OFF** and disconnect the wiring from the sensor and from the PCM. Use an ohmmeter or powered test light to check continuity of the wiring, especially the blue signal wire. If the wiring is good, the sensor is out of adjustment. Sensor adjustment should not change with use and is generally required only if the pump or sensor have been removed.

VEHICLE SPEED SENSOR

Testing

DTC 24 will be set only under specific conditions:
- Speed signal output from the sensor buffer is constant (not pulsed)
- Engine speed is above 2000 rpm
- Vehicle speed signal indicates less than 5 mph
- All conditions exists for more than 40 seconds

The sensor buffer module accepts the speed sensor input and provides a processed signal to the anti-lock brake control unit, the speedometer and the ECM on separate circuits. A scan tool reading of the speed sensor signal should closely match the vehicle speedometer reading. The buffer module is mounted behind the dashboard near the center.

> ### — CAUTION —
> *This test procedure requires operating the vehicle with the drive wheels turning. Support the rear axle properly or, if road testing, have an assistant in the vehicle to read the scan tool.*

NOTE: Although the circuits are similar, terminal designations on the ECM are different on manual and automatic transmission vehicles. Be sure to use the correct wiring schematic.

1. On some models, the brake lights and vehicle speed buffer module are serviced by the same fuse. If the brake lights do not work, repair that problem first.

2. Raise and safely support the rear axle on jackstands and chock the front wheels. If working on a lift, support the axle to prevent excess movement that may effect test data. On 4-WD vehicles, make sure the selector is in 2 wheel drive.

3. Connect a scan tool to the DLC and start the engine. With the engine at idle and the transmission in gear, the speed reading on the scan tool should match the reading on the speedometer. If the readings match, the fault is intermittent. If no other DTCs are set, clear the memory and check for poor connections or ground circuits.

4. If the speedometer functions but the scan tool readings do not agree, backprobe terminal C11 on the sensor buffer while the drive wheels are turning slowly.

   a. If the voltage remains steady, the circuit to the ECM is shorted to ground or the buffer module is faulty.

   b. If the voltage fluctuates up and down, the circuit to the ECM is open or the ECM itself is faulty.

Throttle position sensor circuit — 6.2L LH6 engine with automatic transmission

5. If the speedometer does not function, turn the ignition switch **OFF** and disconnect the speed sensor buffer. Check for continuity to ground at connector terminal C8 and with the ignition **ON**, check for 12 volts at terminal C9.

6. Reconnect the wiring and backprobe a voltmeter across buffer terminals C7 and C12. Turn the drive wheels slowly.

    a. If the voltage fluctuates up and down, check the wiring between the sensor and buffer module. If there are no shorts or poor connections, the speed sensor is faulty.

    b. If the voltage remains steady and the sensor wiring is good, the buffer module is faulty.

MAP SENSOR

Testing

DTC 31 indicates the MAP sensor signal voltage is below the lowest limit used for more than 10 seconds at idle. It can be caused by sensor failure, a short or open circuit or loss of the reference voltage from the ECM.

DTC 32 indicates the MAP sensor signal is signficantly different from the amount of EGR vacuum commanded by the ECM. Together, the vacuum command/vacuum measured signals are known as the EGR loop and the most likely cause for a loop error is a vacuum leak. If the condition exists for more than 10 seconds at more than about 30 mph, the EGR system will be shut down and the DTC will be set in memory.

DTC 33 indicates the MAP sensor signal is above the highest limit used for more than 10 seconds at idle. It can be caused by a stuck solenoid, a clogged vent solenoid filter or a pinched vacuum hose.

NOTE: Although the circuits are similar, terminal designations on the ECM are different on manual and automatic transmission vehicles. Be sure to use the correct wiring schematic.

DTC 31

1. Check the EGR system for vacuum leaks or pinched vacuum hoses.

2. With the engine at operating temperature and at idle, MAP sensor output signal should be above 200mV on a scan tool or 0.25 volts on a meter. If so, the fault is intermittent. If no other DTCs are displayed, check for poor connections at the sensor, clear the memory and test again.

3. If sensor signal voltage is low, turn the ignition switch **OFF** and unplug the sensor connector.

    a. Turn the ignition switch **ON** and check for the 5 volt reference at connector terminal C. If there is no voltage, the wiring from the ECM to the sensor or the ECM itself is faulty.

    b. Check for voltge at connector terminal A. If there is voltage, the sensor wiring is shorted together. If there is no voltage, make sure terminal A has continuity to ground through the ECM sensor ground.

    c. Turn the ignition switch **OFF** and install a jumper wire between connector terminals B and C. Turn the ignition switch **ON**. If the sensor signal now reads the 5 volt reference, the ECM and wiring are good but the connector or the sensor is faulty.

4. Connect an ohmmeter across sensor terminals B and C, connect a hand vacuum pump to the sensor and draw a

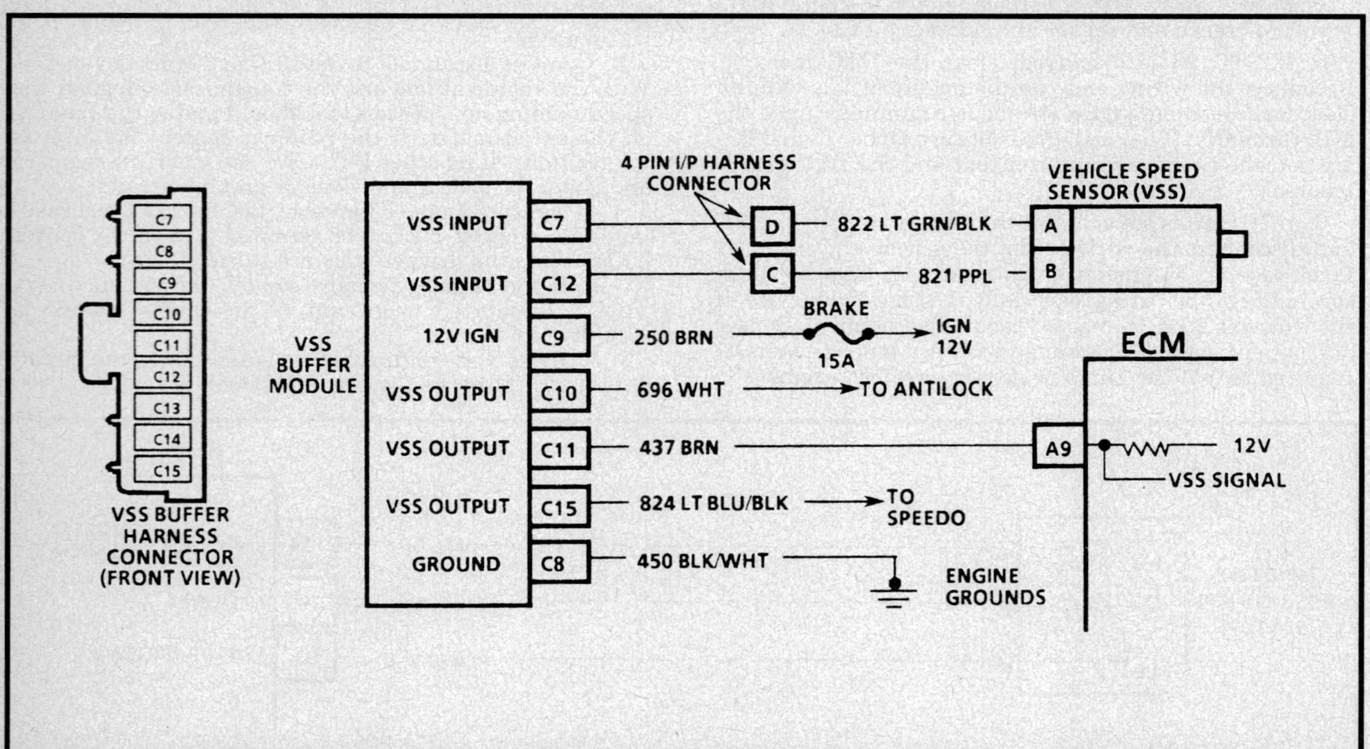

Vehicle speed sensor and buffer schematic — 6.2L LH6 engine with manual transmission

vacuum. If sensor resistance does not change as vacuum changes, the sensor is faulty.

5. After repairing the sensor or wiring, clear the memory and drive the vehicle to see if any other faults still exist that may set DTCs.

DTC 32

1. Turn the ignition switch **OFF** and disconnect the battery or ECM wiring to clear the memory. Connect the wiring and drive the vehicle at more than 30 mph for more than 30 seconds or until the MIL turns **ON**.

2. If the MIL does not turn **ON** or if DTC 32 is not in memory, the fault is intermittent. If no other DTCs are stored, check for a vacuum leak or pinched hose in the EGR system and the air conditioning and cruise control systems, if equipped.

3. If DTC 32 is set again, stop the engine and disconnect the EGR solenoid valve. Use an ohmmeter to check the resistance of the solenoid coil. If resistance is not 16–24 ohms, the solenoid valve is faulty.

4. With the ignition switch **ON** and terminal B of the DLC connected to ground, connect a test light between the terminals of the solenoid connector. The light should flicker faintly. If it glows brightly, the grey wire is shorted to ground. If it does not turn **ON** at all, the solenoid wiring or the ECM is faulty.

5. Connect the solenoid wiring, disconnect the vacuum hoses and turn the ignition switch **ON**.

6. Connect a hand vacuum pump to the vacuum source side of the solenoid and draw a vacuum. If the vacuum does not hold, the solenoid is faulty and the whole assembly must be replaced.

7. If the solenoid holds vacuum, unplug the solenoid connector. If the vacuum does not fall, remove the vent solenoid filter. If vacuum still does not fall, the solenoid is faulty and the whole assembly must be replaced.

8. If all electrical and vacuum tests produce the correct results, there is most likely a vacuum leak or a pinched vacuum hose. Check the vacuum components of the cruise control and air conditioning systems.

DTC 33

1. With the engine running, use a scan tool to read the MAP sensor signal or back probe ECM terminal A1 if equipped with a manual transmission or terminal B13 with an automatic transmission.

2. If the MAP signal is less than 2.5 volts, the fault is intermittent. Check for a bad connection or a leaking or pinched vacuum hose.

3. If the signal is above 2.5 volts, unplug the EGR solenoid and vent solenoid connectors and install a vacuum gauge in place of the EGR valve.

4. If there is no vacuum with the engine running, a valve may be stuck closed or the vacuum pump is faulty.

5. If there is vacuum, connect a test light or voltmeter across the solenoid connector terminals. The light should stay **OFF**. If the light turns **ON** at either connector, check for a short to chassis ground on the solenoid ground circuits. If the wiring is good, the ECM is faulty.

6. If the test light stays **OFF** at both solenoid connectors, the solenoids and ECM are operating properly. The DTC may be caused by a faulty MAP sensor ground circuit or the sensor itself is faulty.

ECM, PROM AND REFERENCE VOLTAGE

Testing

NOTE: Although the circuits are similar, terminal designations on the ECM are different on manual and automatic transmission vehicles. Be sure to use the correct wiring schematic.

1. DTC 51 indicates a failure of the PROM inside the ECM. Remove the PROM, check for bent pins or signs of electrical damage and install the PROM carefully. If DTC 51 is set again, replace the PROM. If the problem is still not corrected, replace the ECM.

2. DTC 52 (manual transmission) or DTC 55 (automatic transmission) indicates a problem with the ECM. It can be caused by poor connections, electrical damage or over heating such as in a paint baking oven. Unplug the ECM connector, carefully check for corrosion or damage and reconnect the wiring. If DTC 52 or 55 is set again, replace the ECM.

3. DTC 53 indicates a reference voltage overload, meaning the 5 volt reference to the sensors is incorrect. This DTC may also be set if the vehicle charging system is not working properly or when jump starting the engine. If other DTCs are also in memory, repair those items first and clear the memory.

4. If DTC 53 is set again, unplug the 24 pin connector from the ECM and check the reference voltage at ECM terminal A 12 with the ignition switch **ON**. If the voltage is correct, the sensor wiring is faulty. If the voltage is not a steady 5 volts, the ECM is faulty and must be replaced.

TORQUE CONVERTER CLUTCH

Testing

DTC 67 indicates the torque converter clutch solenoid is not operating or is operating at the wrong time. DTC 69 indicates the TCC solenoid is stuck **ON** in all gears and may even stall the engine at idle. These problems can be caused by a blown transmission fuse, a faulty brake pedal switch, faulty wiring or connectors or by control problems in the PCM. The TCC solenoid is normally **OFF** and the PCM completes the ground circuit to turn the solenoid **ON**. The circuits between the PCM and the TCC can be checked with a volt/ohmmeter but complete testing requires a scan tool. If DTC 69 is stored in memory but DTC 67 is not, the malfunction is probably an internal mechanical problem in the transmission.

1. Check all fuses and make sure the brake switch works properly. With the ignition switch **ON** and the brake pedal released, there should be voltage to PCM terminal B4 (purple wire). When the pedal is pressed, the brake switch interrupts the voltage to signal the PCM that the brakes have been applied.

2. Connect the scan tool to the DCL and turn the ignition switch **ON**. Command the TCC **ON** and **OFF** three times. If the TCC solenoid changes state as commanded, the control system is functioning properly. If the TCC will not lock-up or unlock, there is a mechanical problem in the transmission.

3. If the TCC does not turn **ON** when commanded, turn the ignition switch **OFF** and disconnect the transmission pass-through connector at the lower right side of

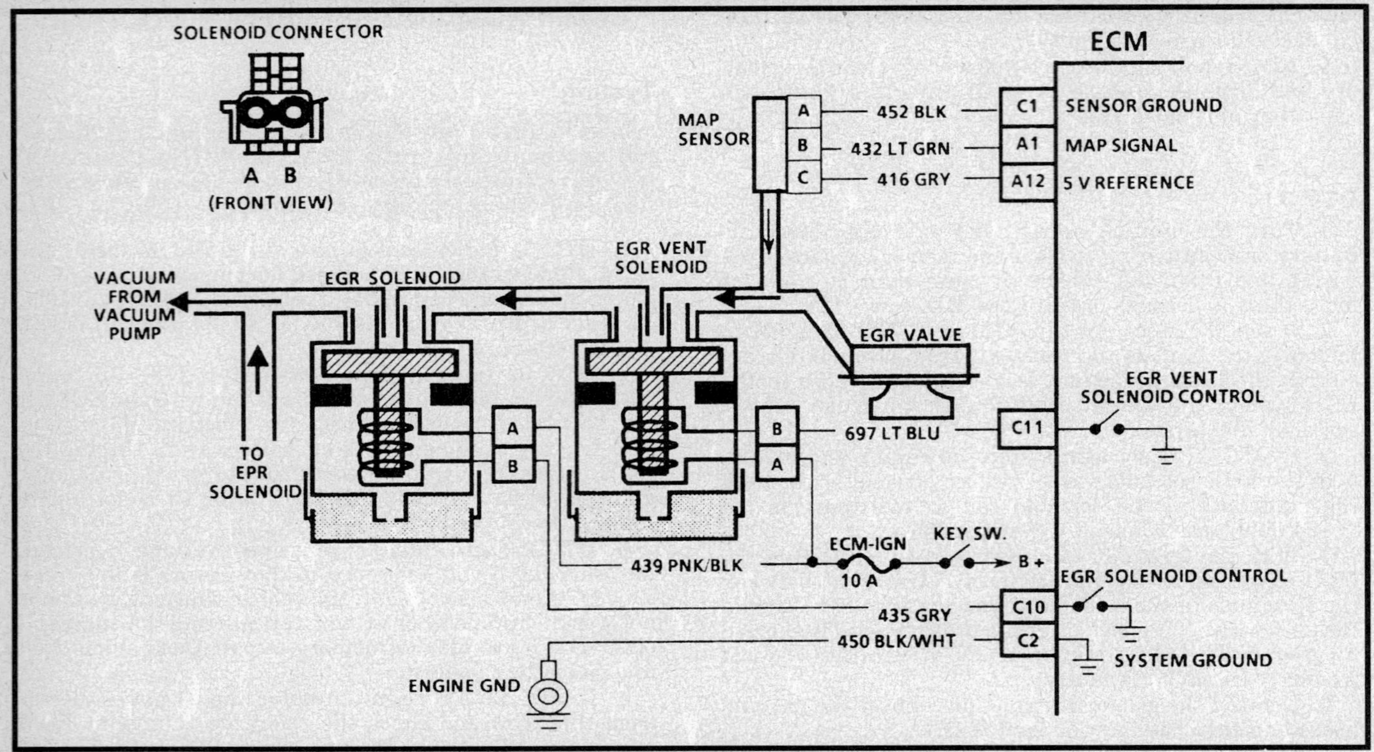

EGR system wiring schematic — 6.2L LH6 engine with manual transmission

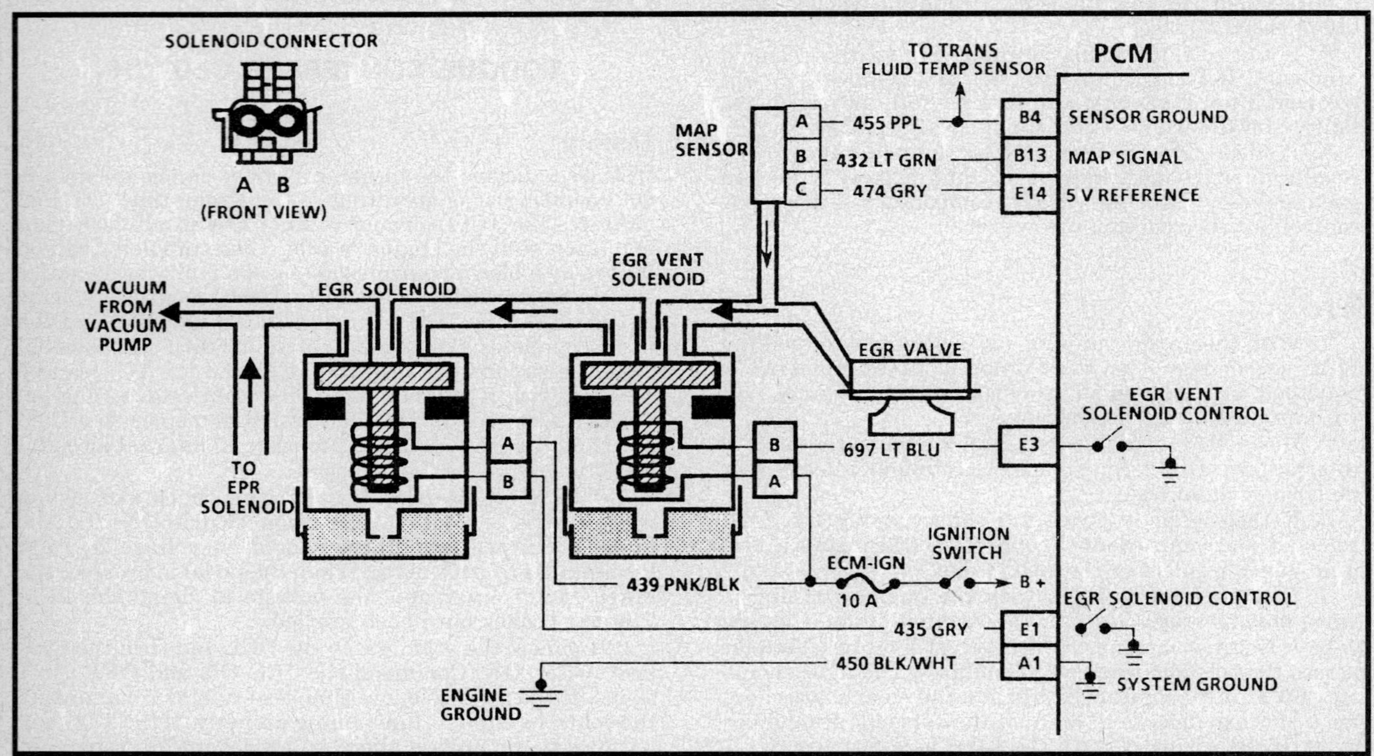

EGR system wiring schematic — 6.2L LH6 engine with automatic transmission

the firewall in the engine compartment. Turn the ignition switch **ON** and observe the scan tool.

    a. If the scan tool indicates the TCC is **OFF**, check for voltage at transmission connector terminal T (tan/black wire). If there is voltage, the wiring is faulty. If there is no voltage, the PCM is faulty.

    b. If the scan tool indicates the TCC is **ON**, measure the TCC solenoid resistance across transmission connector terminals T (tan/black wire) and E (pink wire). If resistance is 20–40 ohms, the solenoid is good and the PCM is faulty. If resistance is not correct, the solenoid or wiring inside the transmission is faulty.

4. If the TCC does not turn **OFF** when commanded, back probe PCM terminal E10.

    a. If there is voltage, the PCM is not opening the TCC solenoid ground. Either the PCM connector or the PCM is faulty.

    b. If there is no voltage, turn the ignition switch **OFF** and disconnect the transmission pass-through connector at the lower right side of the firewall in the engine compartment. Turn the ignition switch **ON** and check for voltage at terminal E (pink wire) on the vehicle side of the connector. If there is no voltage, the TCC power wire or fuse is faulty.

5. If there is voltage to the TCC at the pass-through connector, turn the ignition switch **OFF** and measure the resistance from terminal T (tan/black wire) to ground. The circuit should have high or infinite resistance. If the resistance is low, the tan/black wire is shorted to ground.

6. Connect the voltmeter to terminals E and T on the vehicle side of the pass-through connector and operate the scan tool to command the TCC solenoid.

    a. If the display indicates the solenoid will not turn **OFF** when commanded, the ground circuit between the connector and the PCM or the PCM itself is faulty.

    b. If the display indicates the solenoid is following commands, measure the resistance between terminals T and E on the transmission side of the pass-through connector. If it is not 20–40 ohms, the solenoid or the wiring inside the transmission is faulty.

COLD START ADVANCE SYSTEM

Testing

    NOTE: Although the circuits are similar, terminal designations on the ECM are different on manual and automatic transmission vehicles. Be sure to use the correct wiring schematic.

1. The coolant temperature sensor must function properly. If DTC 14 or 15 is stored in memory, repair that problem first and clear the memory. Locate the cold advance relay mounted on the cowl in the engine compartment next to the fuel pump relay and the main junction block.

2. With the engine coolant above 158°F (70°C), stop the engine and turn the ignition switch **ON**. With all wiring connected, back probe relay terminals E and F (pink/black wires) to make sure there is power to the cold advance relay.

3. Back probe relay terminal A (green/black wire). If there is voltage at terminal A, either terminal D (blue/white wire) is shorted to ground or the ECM is supplying a ground circuit for the relay with the engine warm. Before replacing the ECM, check all solenoids and relays operated by the ECM. A faulty coil could create the false ground.

4. If there is no voltage at terminal A, unplug the coolant temperature sensor and check again. If there is voltage now, the control system is working properly. If there is no timing advance when the engine is cold, either the advance solenoid or the timing advance piston in the pump is faulty.

5. If there is no voltage at terminal A with the coolant temperature sensor disconnected, unplug the connector from the relay and check for continuity to ground at terminal D (blue/white wire). If there is no continuity, either the wire is broken or the ECM is not providing the ground circuit for the relay.

6. With the relay wiring disconnected, check for a circuit to ground at connector terminal A. If the circuit is open, the wire from the advance relay to the advance solenoid is broken or the advance solenoid coil is open.

7. If all voltage and ground circuits are correct, connect 12 volts to terminal F on the relay and ground terminal D. If there is no continuity across terminals E and A, the relay is faulty.

GLOW PLUG SYSTEM

Testing

When engine coolant temperature is below about 150°F (65°C), the ECM will activate the cold advance relay to supply power to the glow plug control unit mounted behind the intake manifold on the engine. When the ignition switch is turned **ON**, the control unit will cycle the glow plugs **ON** and **OFF** for about 20 seconds. The control unit is equipped with a voltage sense circuit to know that power is reaching the glow plugs. After the ignition

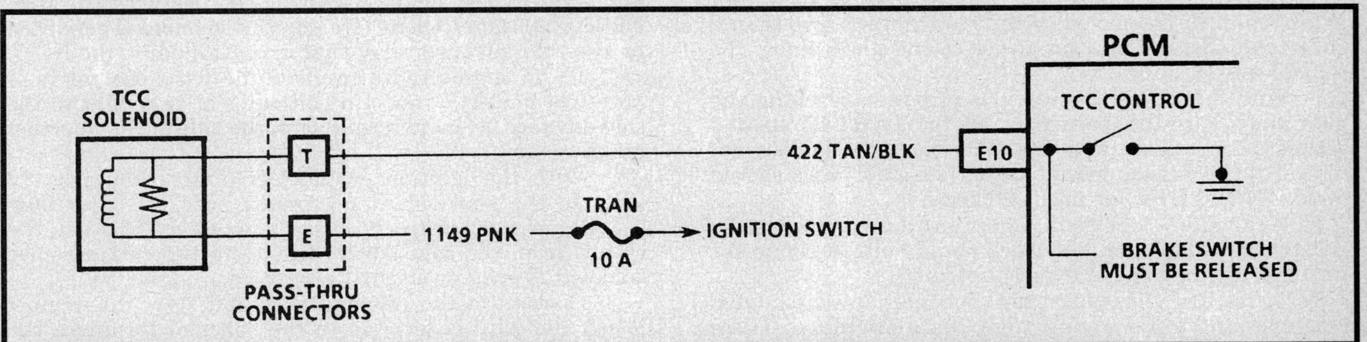

Torque converter clutch solenoid schematic — 6.2L LH6 engine with automatic transmission

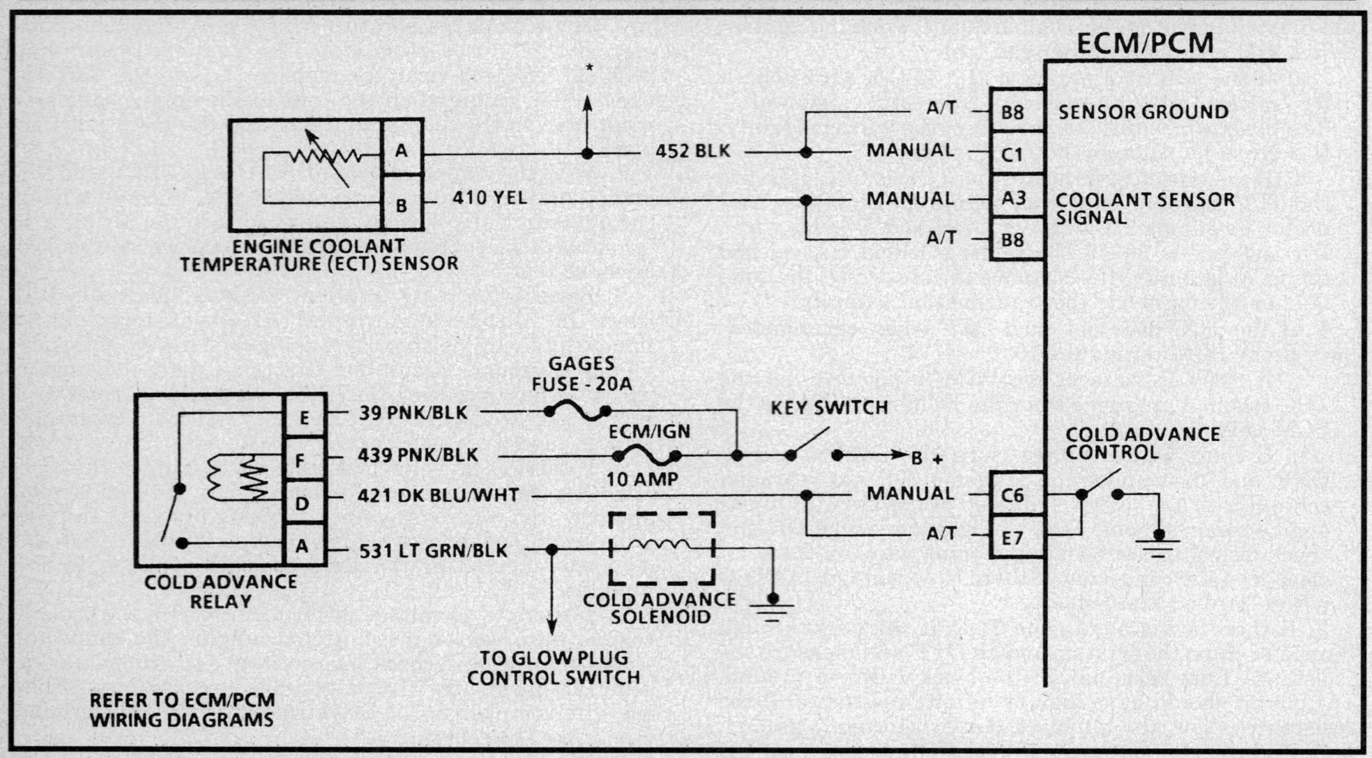

Cold advance relay schematic — 6.2L LH6 engine

switch is turned from **START** to **RUN**, the glow plugs are cycled **ON** and **OFF** again while the engine is warming up. If the after-start feature is not functioning, the engine will still start but may make white smoke and idle poorly until warmed up.

1. Connect a voltmeter or test light to one of the glow plug wires or the output terminal on the glow plug control unit. If the engine is warm, disconnect the coolant temperature sensor.

2. When the ignition switch is turned **ON**, there should be power to the glow plugs for 4–6 seconds. The power should then cycle **ON** and **OFF** for about 20 seconds.

3. If there is no power to the glow plugs, unplug the control unit connector and check for 12 volts from the cold advance relay at connector terminal D. Terminal E should have continuity to ground.

4. Check the resistance between connector terminals C and E. If resistance is more than 2 ohms, the voltage sense wire from the glow plug harness to the control unit is broken or disconnected. If the resistance is less than 2 ohms and there is still no power to the glow plugs, the control unit is faulty.

5. With all wiring connected and power reaching the glow plugs, turn the ignition switch to **START** for about 1 second, then release it to the **RUN** position. It does not matter if the engine doesn't start. The glow plugs should cycle **ON** and **OFF** for about 20 seconds.

6. If the after-start cycle is not functioning, unplug the control unit connector and check for 12 volts at connector terminal B when the starter is operated.

7. To test the glow plugs, turn the ignition switch **OFF** and disconnect the wiring from the glow plugs. There should be 2 ohms resistance between the connector terminal on each glow plug and the engine block. Any glow plug

with higher resistance is faulty and places and extra current burden on the remaining plugs.

NOTE: Do not apply battery voltage directly to the glow plug output stud on the control unit. The glow plugs and control unit will be damaged.

COLD ADVANCE SOLENOID

Testing

When the coolant temperature is below 95°F (35°C), the cold advance solenoid is activated to decrease injection pump housing pressure to advance injection timing. This decreases engine noise and smoke during warm up. The solenoid is mounted inside the pump cover with the shutdown solenoid. On heavy duty engines without electronic engine control systems, the solenoid is activated by a coolant temperature switch. On 6.2L engines with the LH6 Diesel Emission Control System, the solenoid is activated by the cold advance relay that is controlled by the ECM.

1. If the engine is warm, disconnect the coolant temperature sensor. Connect a voltmeter or test light to the cold advance solenoid terminal on the side of the injection pump cover.

2. With the ignition switch **ON**, there should be 12 volts to the solenoid. If not, make sure the glow plug system operates properly. If the glow plugs function, the circuit from the cold advance relay to the cold advance solenoid is open or shorted to ground.

3. Disconnect the solenoid wire and start the engine. When the wire is touched to the solenoid terminal, the timing should advance and the sound of the engine will change.

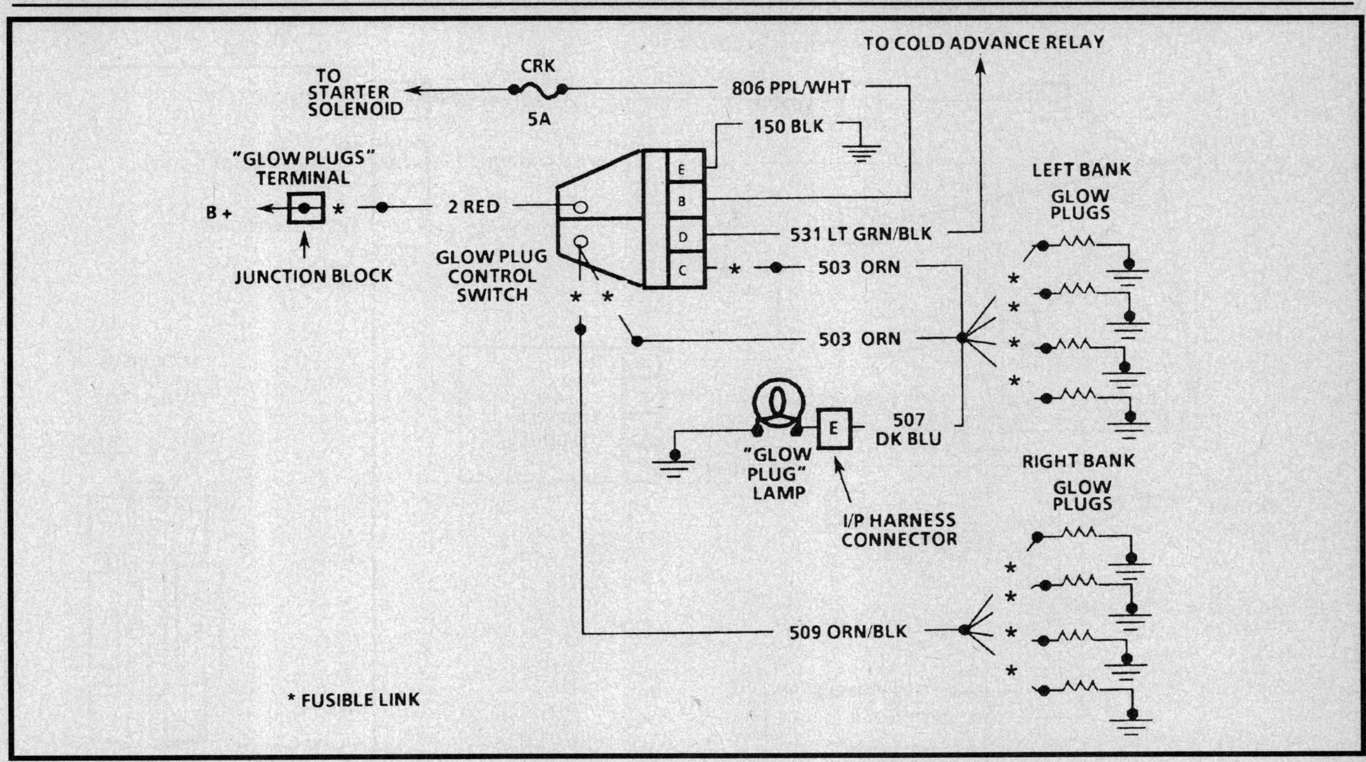

Glow plug system wiring schematic — 6.2L LH6 engine

4. If the engine sound does not change, disconnect the wire again and use an appropriate tool to gently push in on the timing advance servo valve at the bottom of the rocker arm. This should retard injection timing and decrease idle speed.

   a. If the engine sound changes, the timing piston is functioning but the cold advance solenoid is not functioning. The pump cover must be removed to remove the solenoid.

   b. If the sound still does not change, the timing piston is not functioning and the pump must be removed for disassembly.

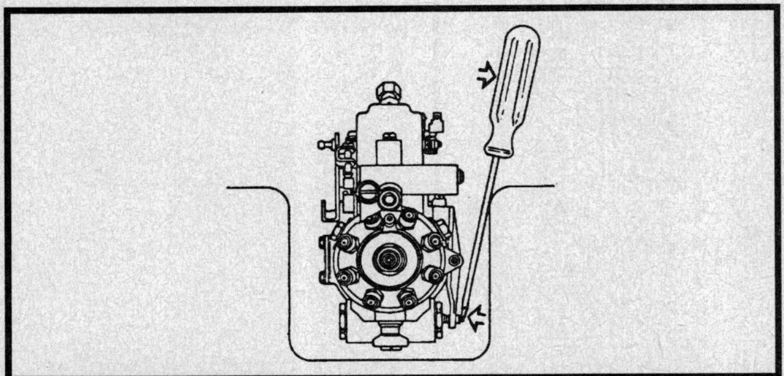

Push in very carefully on the servo valve to retard injection timing at idle — 6.2L engine

Component Replacement

SERVICE PRECAUTIONS

When working around any part of the fuel system, take the following precautionary steps to prevent fire and/or explosion and the resulting property damage or personal injury or death:
- Disconnect negative terminal from battery (except when testing with battery voltage is required).
- Whenever possible, use a flashlight instead of a drop light.
- Keep all open flame and smoking material out of the area.
- Use a shop cloth or similar to catch fuel when opening a fuel system.
- Use eye protection.
- Always keep a dry chemical (class B) fire extinguisher near the area.

INJECTION PUMP

Removal and Installation

1. Disconnect the negative battery cable and remove the air cleaner and fuel filter.
2. Remove the rear alternator bracket.
3. Remove the EGR/EPR solenoid valve assembly from the valve cover.
4. Remove the hoses and fuel lines as required to remove the intake manifold. Cap the fuel line fittings to keep dirt out.
5. Remove the bolts and remove the intake manifold and gasket.

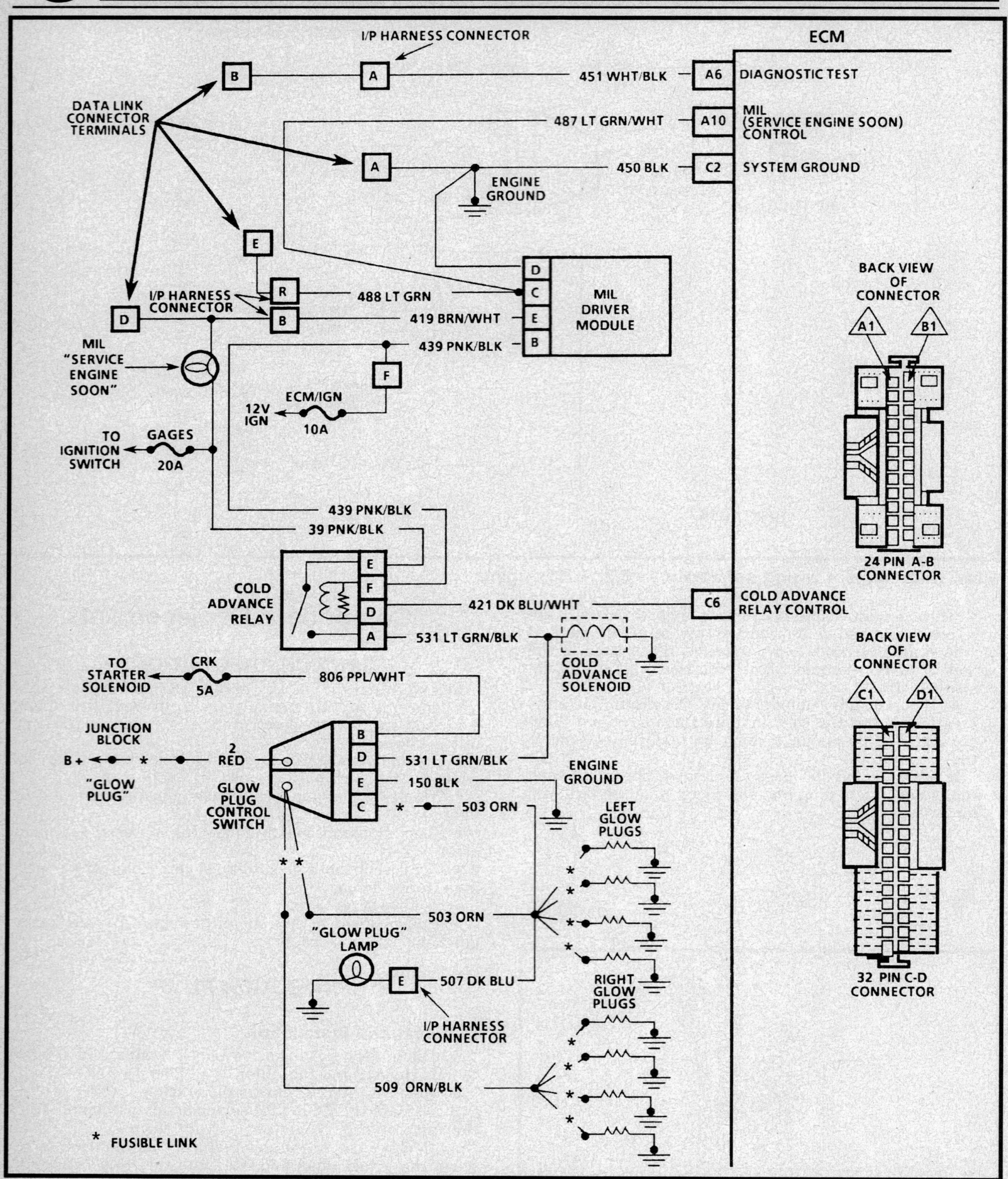

ECM wiring schematic — 6.2L LH6 engine with manual transmission

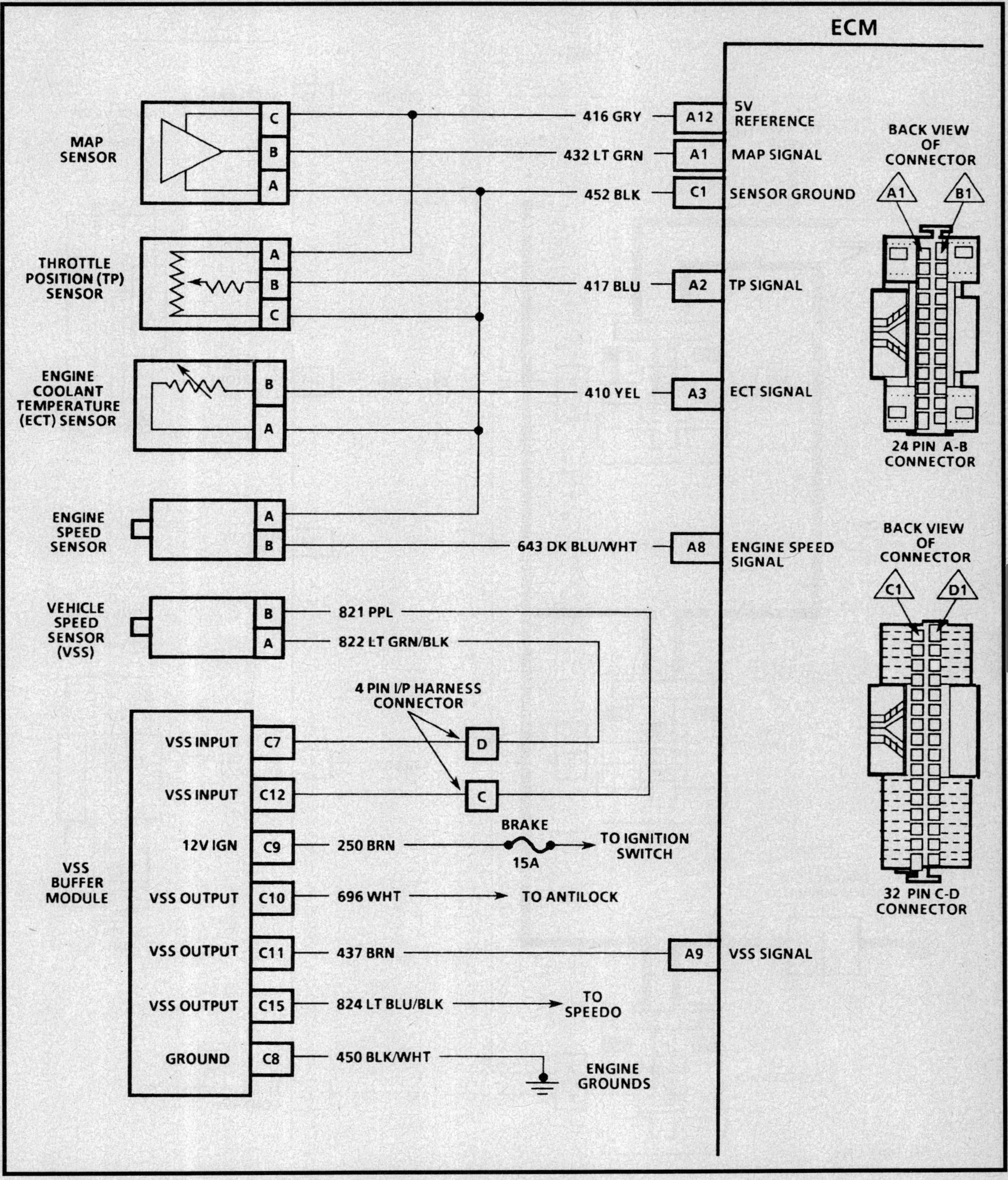

ECM wiring schematic — 6.2L LH6 engine with manual transmission

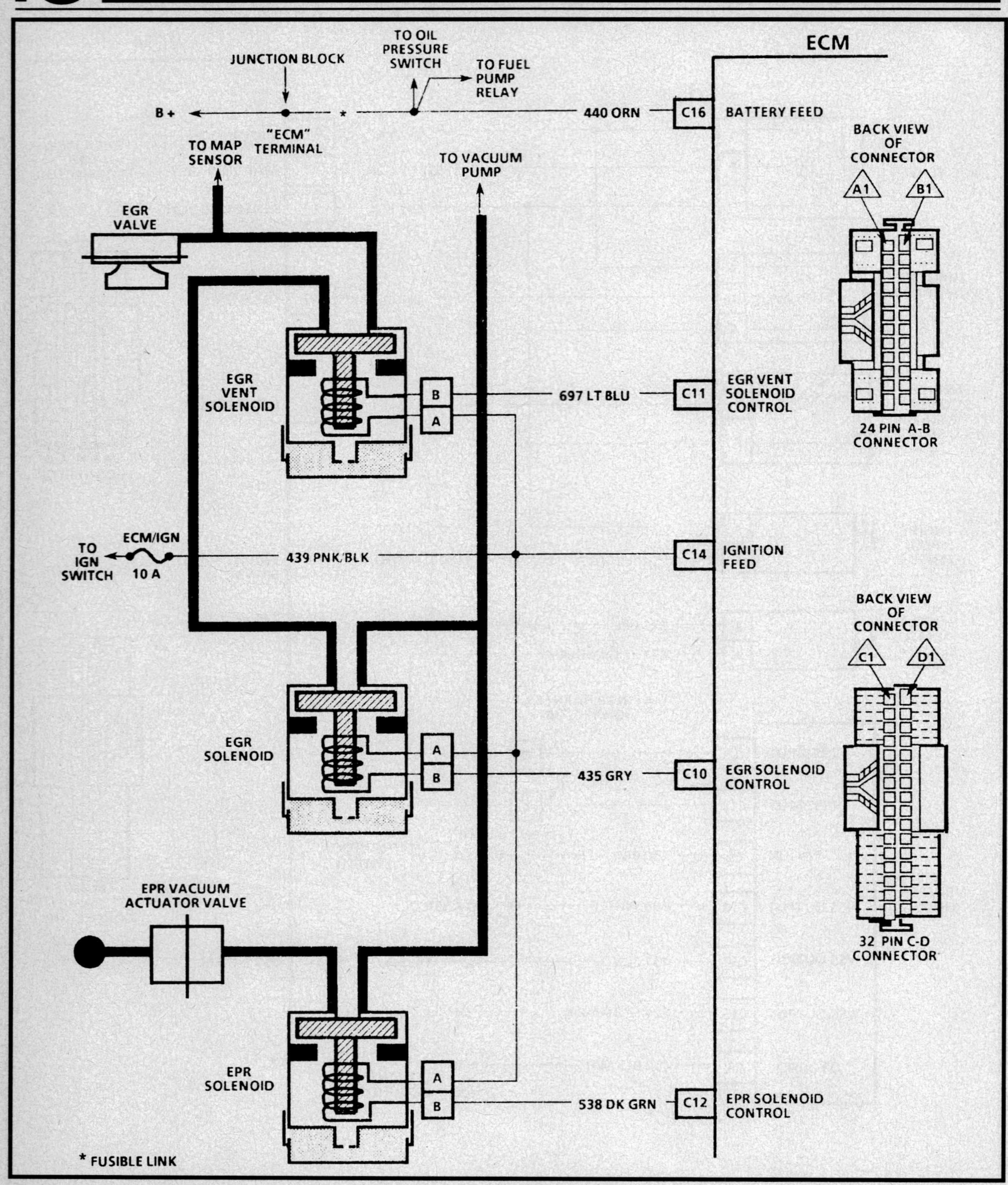

ECM wiring schematic — 6.2L LH6 engine with manual transmission

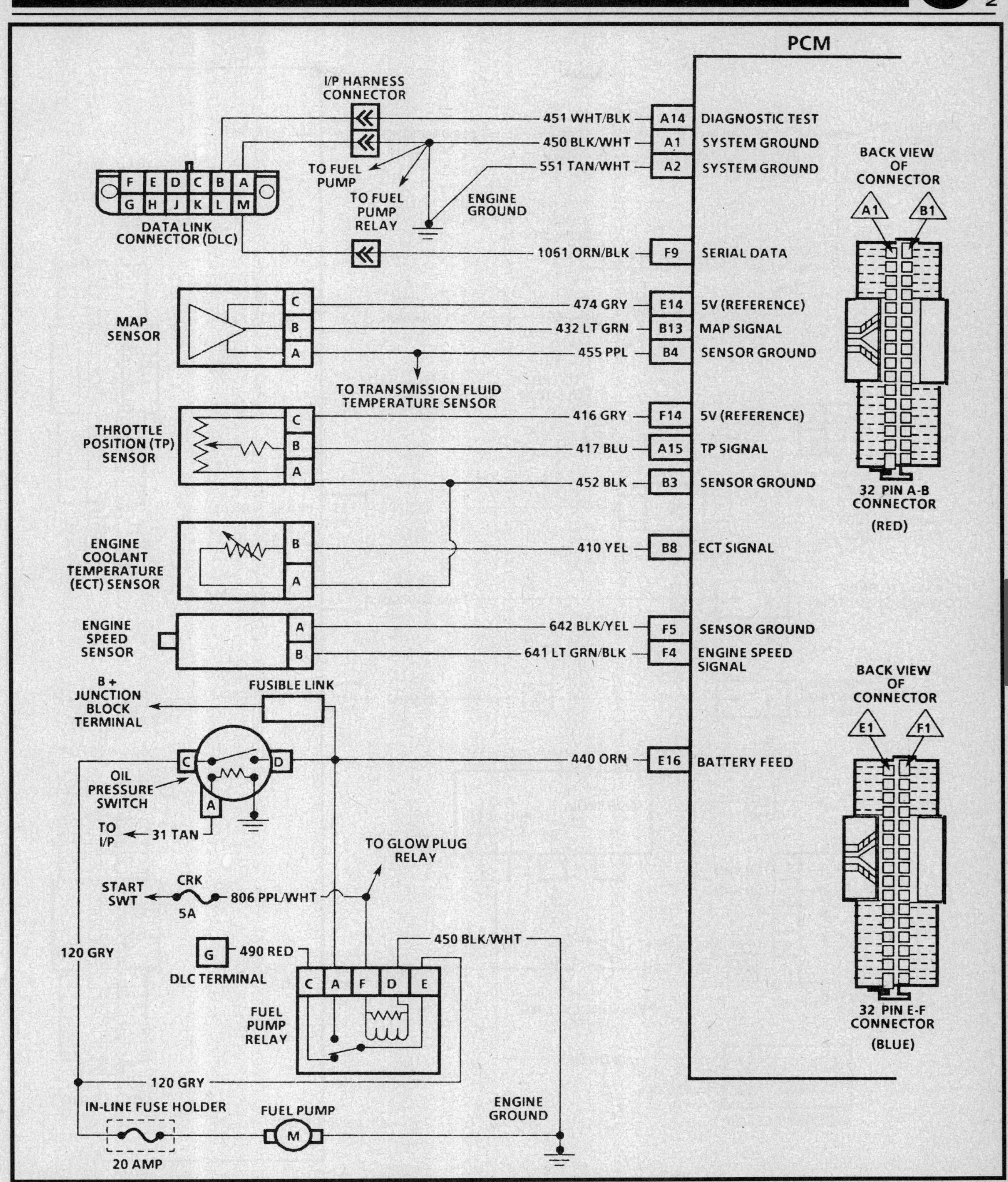

PCM wiring schematic — 6.2L LH6 engine with automatic transmission

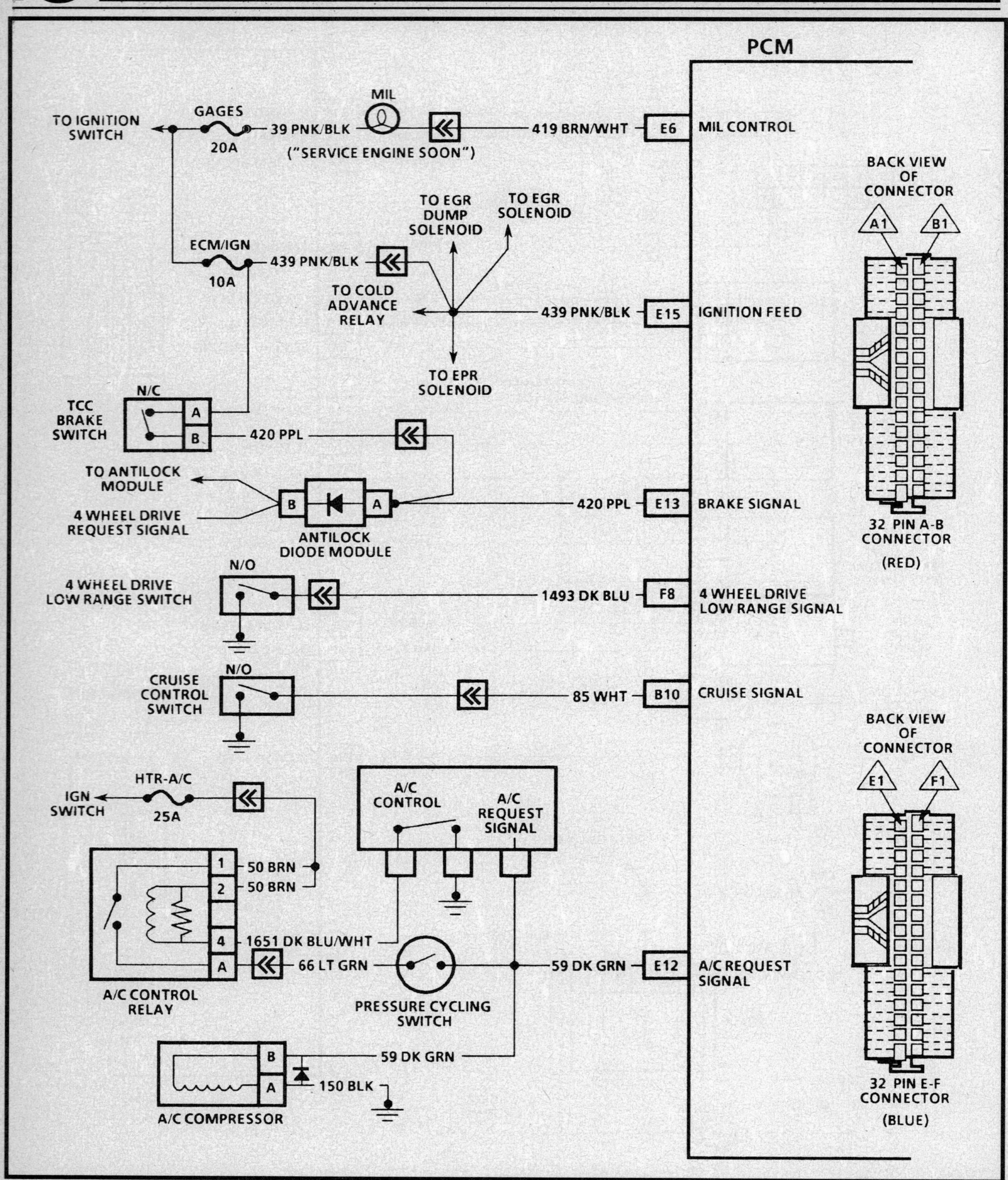

PCM wiring schematic — 6.2L LH6 engine with automatic transmission

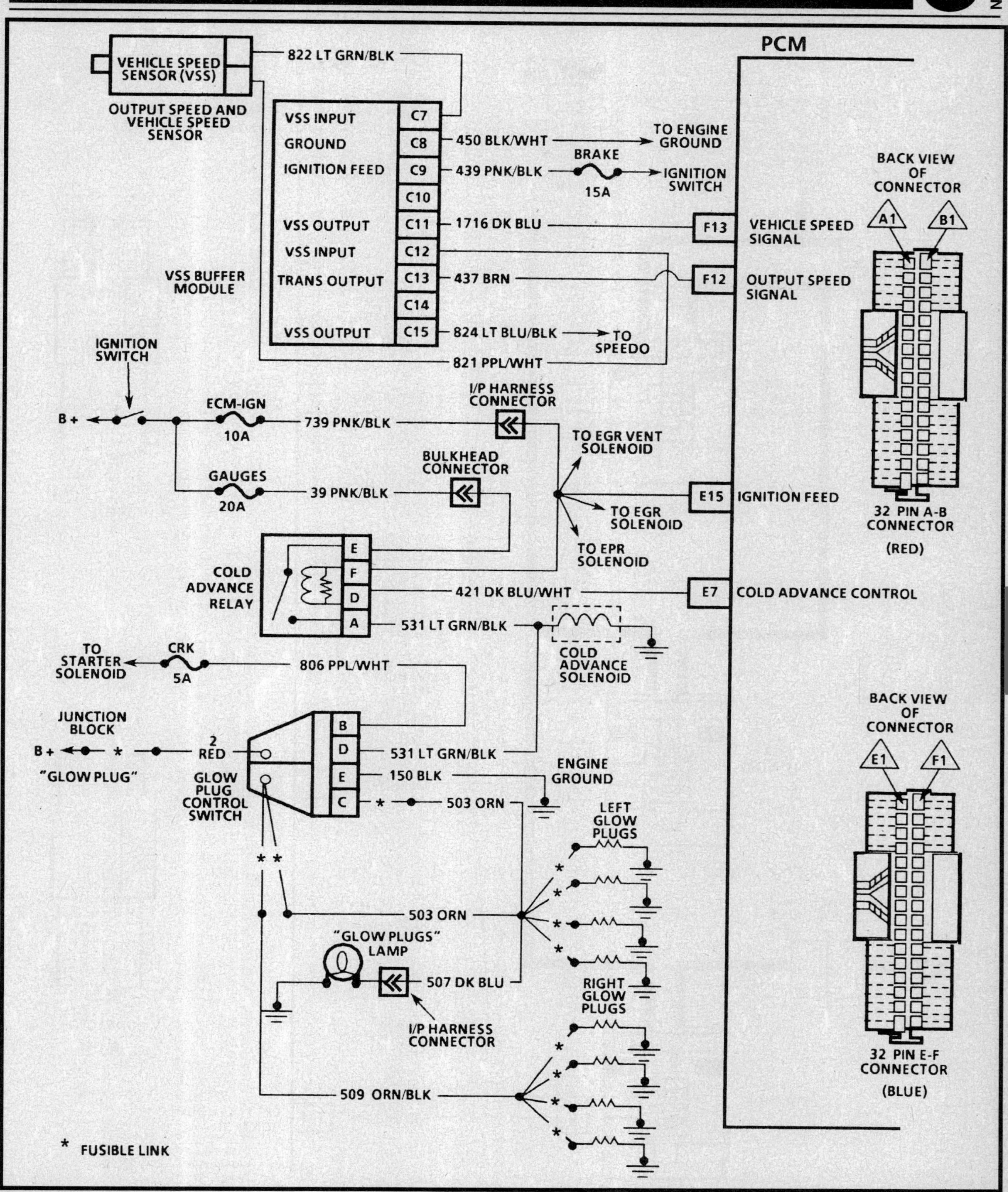

PCM wiring schematic — 6.2L LH6 engine with automatic transmission

DIESEL FUEL SYSTEMS
LIGHT TRUCKS AND VANS

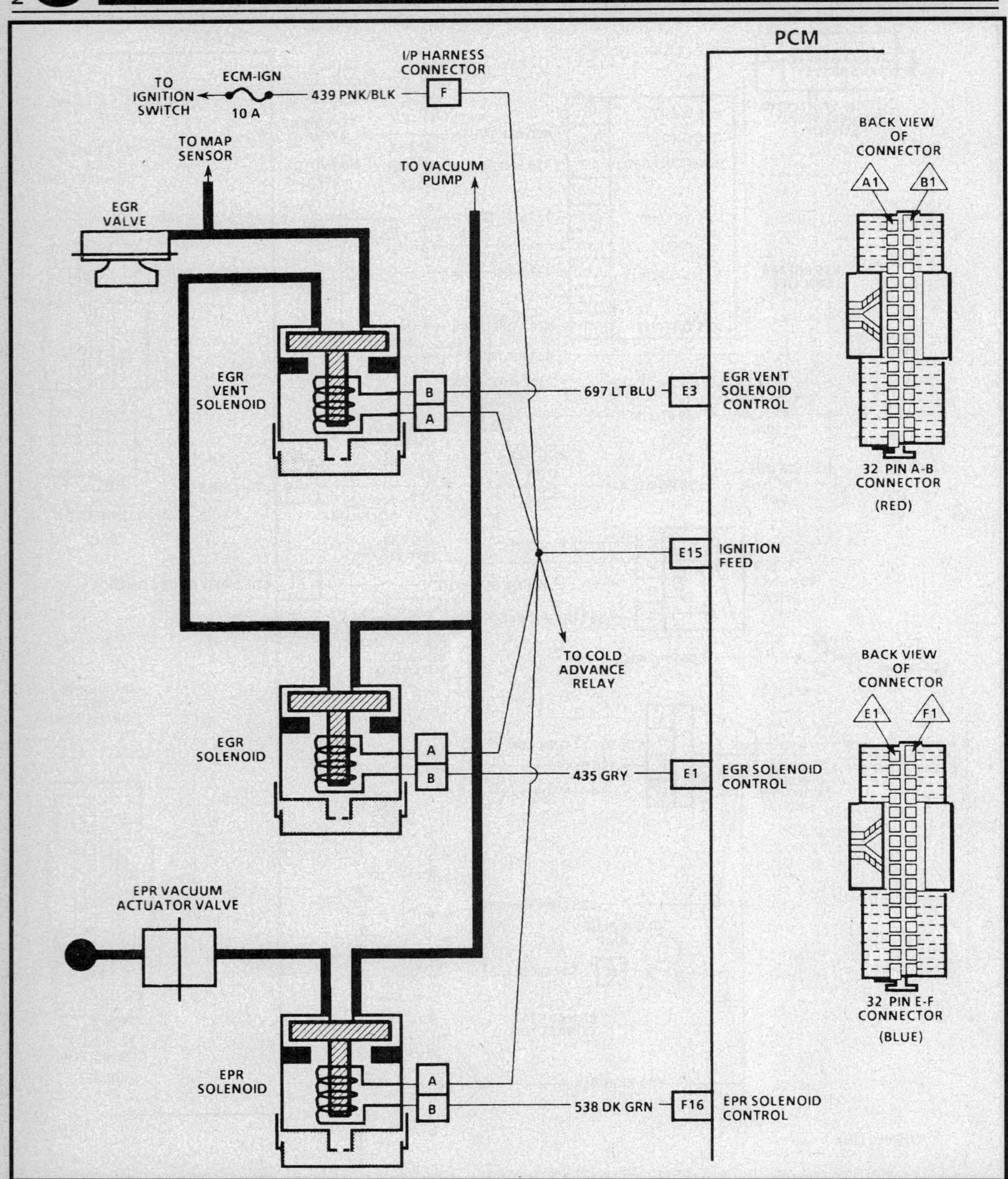

PCM wiring schematic — 6.2L LH6 engine with automatic transmission

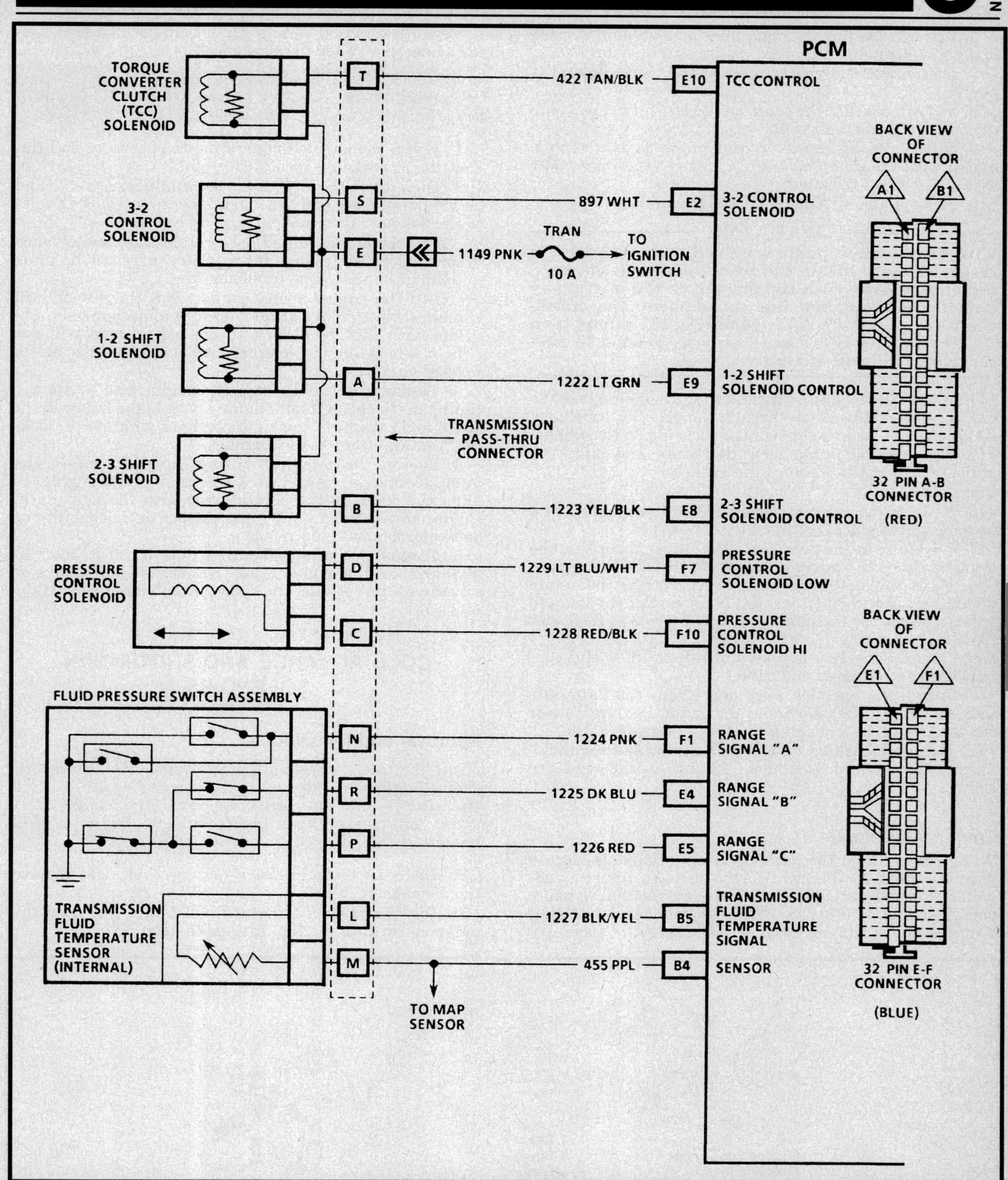

PCM wiring schematic — 6.2L LH6 engine with automatic transmission

6. Remove the fuel injection lines as a set and cap the fittings to keep dirt out of the pump and injectors.

7. Disconnect the control cables and wires from the pump.

8. Disconnect the fuel supply and return lines and cap the fittings to keep dirt out.

9. If the timing mark on the pump is not visible, matchmark the pump mounting flange to the front cover with paint or a scribe tool.

10. Remove the oil filler tube and grommet.

———————— WARNING ————————

With the injection pump removed, the pump gear is resting loosely inside the front cover. Do not turn the engine over with the starter or the gears may jam in the front housing and damage the crankshaft and camshaft. After removing the pump, turn the engine slowly by hand only as needed to preserve correct gear timing.

11. Rotate the engine by hand as required to remove the pump gear bolts working through the oil filler tube hole. Be careful not to drop the bolts or tools inside the engine.

12. Remove the pump mounting nuts and slide the pump off the studs. Remove the gasket.

To install:

13. Make sure the mounting surfaces are clean and fit a new gasket in place.

14. Carefully fit the pump onto the studs. Make sure the locating pin on the pump hub fits into the slot in the gear.

15. With the matchmarks aligned, torque the pump mounting nuts to 30 ft. lbs. (40 Nm). Install the bolts to secure the gear to the pump and torque them to 20 ft. lbs. (25 Nm).

16. Connect the fuel supply and return lines and connect the control wires and cables.

17. Install the injection lines and torque the flare nuts to 20 ft. lbs. (25 Nm). Do not over tighten the flare nuts or the flares will distort and leak.

18. Use new gaskets and install the intake manifold. Torque the bolts to 32 ft. lbs. (42 Nm) in the sequence shown.

Timing Adjustment

When the mark on the pump mounting flange is aligned with the mark on the front cover, injection timing is adjusted to factory specification. If the engine makes white smoke at idle or if a new PROM is being installed that requires a new timing specification, it may be necessary to advance injection timing. Do not adjust injection timing under any other circumstances.

NOTE: Do not run the engine with the pump mounting nuts loose to adjust timing. The pump drive gears could be damaged.

1. Warm the engine to operating temperature and run it at the correct idle speed (650 rpm).

2. Disconnect the wire from the cold advance solenoid on the pump cover and use a jumper wire to apply 12 volts to the solenoid terminal.

   a. If the engine sound does not change, the advance solenoid or the pump timing piston may not be functioning. Repair these problems first.

   b. If the engine sound changes but the engine still makes white smoke at idle, there is a mechanical problem with the engine and timing adjustment will not help. Check engine compression and look for a blown head gasket.

3. To advance timing, stop the engine and locate the timing mark on the front engine cover. Make a new mark 0.039 in. (1.0mm) to the right of the original mark, looking from the front of the engine.

4. Loosen the pump mounting nuts and rotate the pump clockwise (viewed from the front of the engine) to align the mark on the pump flange with the new mark. Do not advance the timing any further or driveability or smoke problems could result.

5. Torque the pump mounting nuts to 30 ft. lbs. (40 Nm) and run the engine again. If there is black smoke or excessive noise at idle, the timing has been advanced too much.

COLD ADVANCE AND SHUT-DOWN SOLENOIDS

Removal and Installation

NOTE: Both solenoids are inside the injection pump cover. Before removing the cover, make sure the pump and surrounding area are clean. Dirt must not enter the pump when the cover is removed.

1. Disconnect the negative battery cable and remove the air cleaner.

2. Disconnect the wiring and fuel return line from the pump cover. Cap the fuel fittings to keep dirt out.

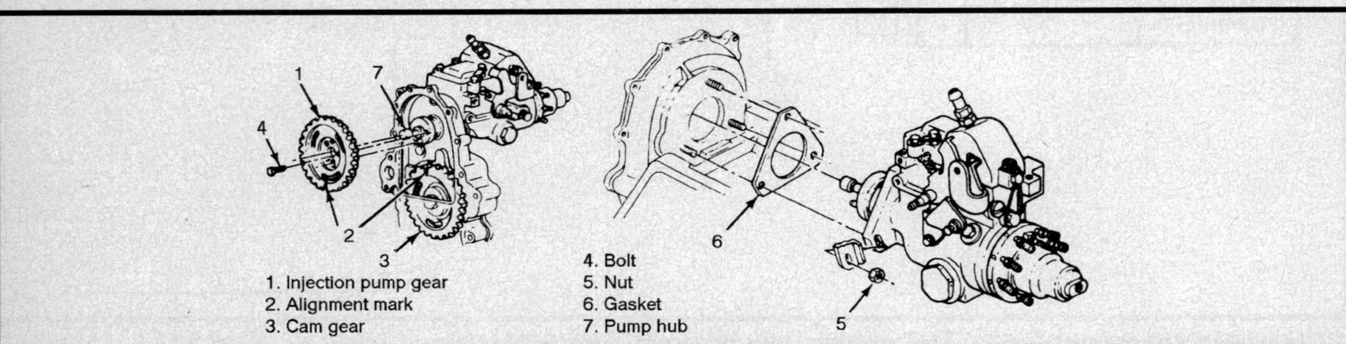

With the pump removed, avoid turning the engine so the gears remain properly aligned — 6.2L engine

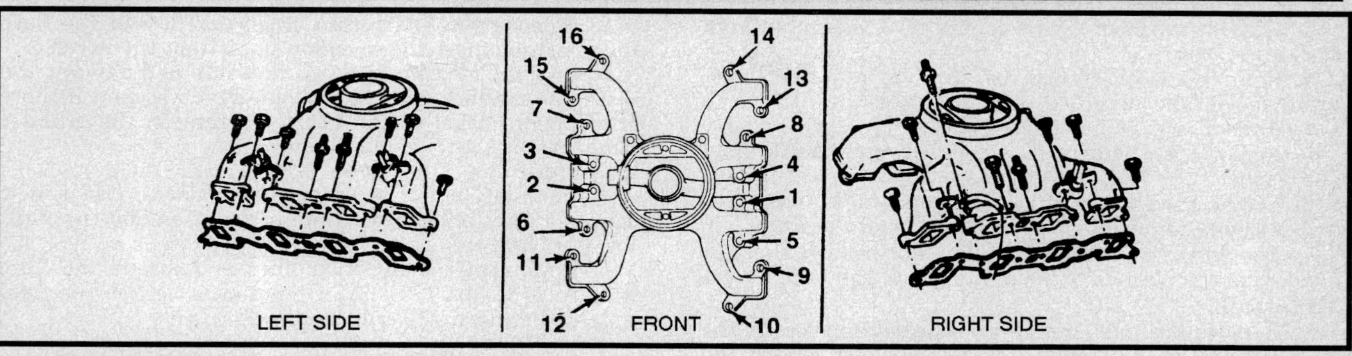

Intake manifold torque sequence — 6.2L engine

3. Remove the screws and lift the pump cover straight off. Remove the nuts to remove the solenoids from the cover.

To install:

4. Make sure the cover and solenoids are clean and fit the solenoids into place. The shut-down solenoid linkage must move freely and the advance solenoid plunger must contact the check ball in the fuel fitting. Install the insulating washers and nuts.

5. Fit a new gasket to the cover.

6. A special Stanadyne tool No. 26528 or equivalent, is used to hold the shut-down solenoid linkage in place for installing the cover. If the tool is not available, use the following procedure to install the cover:

   a. The screws must not be in the pump cover.

   b. Hold the cover about ¼ inch (6mm) towards the front of the engine and about ⅛ inch (3mm) above the pump.

   c. Hold the control lever in idle position and place the cover onto the pump.

   d. Install the screws and washers and connect the battery to test the shut-down solenoid.

> **CAUTION**
>
> *An incorrectly installed cover or shut-down solenoid can cause the engine to "run away" in an uncontrolled acceleration with no way to turn the engine off. This would cause serious engine damage and personal injury. After the cover is installed, test the shut-down solenoid before starting the engine.*

7. To test the solenoid, turn the ignition switch **ON** and touch the pink wire to the solenoid terminal. The solenoid should click as the wire is touched and removed from the connector. If there is no clicking sound, the solenoid pivot arm is on the wrong side of the linkage or the solenoid is not functioning. Remove the cover again and repair the problem.

8. Turn the ignition switch **OFF** and connect the remaining wiring, control cables and fuel lines.

9. Start the engine and check for fuel leaks. The idle will be rough until all air is purged from the pump. Stopping the engine for about 10 minutes will allow the air to raise to the top for faster purging.

FUEL INJECTION LINES

Removal and Installation

1. Disconnect the negative battery cable and remove the air cleaner and fuel filter.

2. Remove the rear alternator bracket.

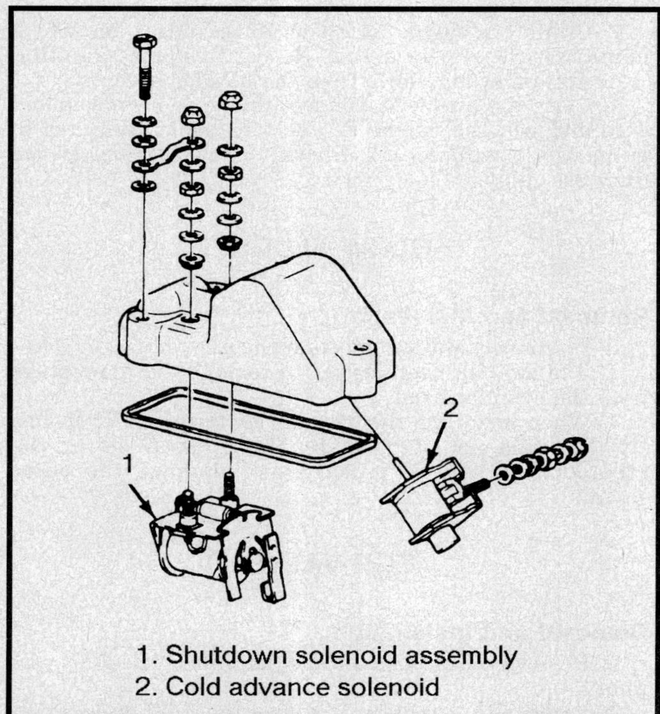

1. Shutdown solenoid assembly
2. Cold advance solenoid

Cold advance solenoid and shut-down solenoid inside pump cover — 6.2L engine

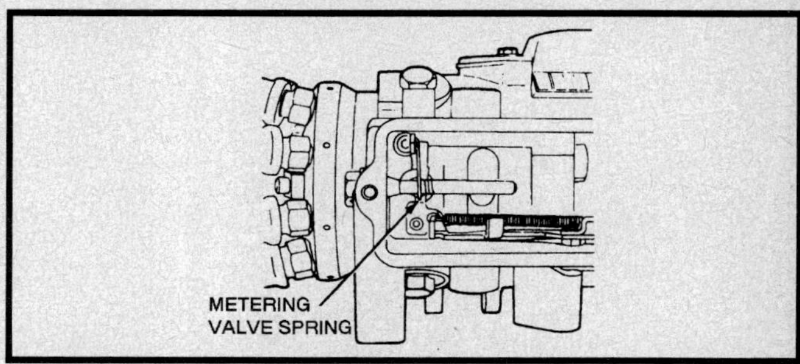

If the shut-down solenoid pivot arm engages the wrong side of the linkage, the engine will run away — 6.2L engine

3. Remove the EGR/EPR solenoid valve assembly from the valve cover.

4. Remove the hoses and fuel lines as required to remove the intake manifold. Cap the fuel line fittings to keep dirt out.

5. Remove the bolts and remove the intake manifold and gasket.

6. Loosen the fittings at the injectors and cap the lines and injectors to keep dirt out. Do not bend the lines.

7. Loosen the lines from the pump and remove them as a set. Cap the fittings to keep dirt out of the pump.

To install:

8. Remove the caps, install the line set and torque the flare nuts to 20 ft. lbs. (25 Nm). Do not overtighten the flare nuts or the flare will split and leak fuel.

9. Use new gaskets and install the intake manifold. Torque the bolts to 32 ft. lbs. (42 Nm) in the correct sequence.

10. Install the solenoid valves and alternator bracket.

11. Install the air cleaner and run the engine to check for leaks.

FUEL INJECTORS

Removal and Installation

NOTE: A special 30mm socket with space for the fuel return hose nipples, tool J 29873, is required to remove and install the injectors. If the nozzle socket is not used, the nipples could be damaged.

1. Disconnect the negative battery cable and remove the fuel line clips.

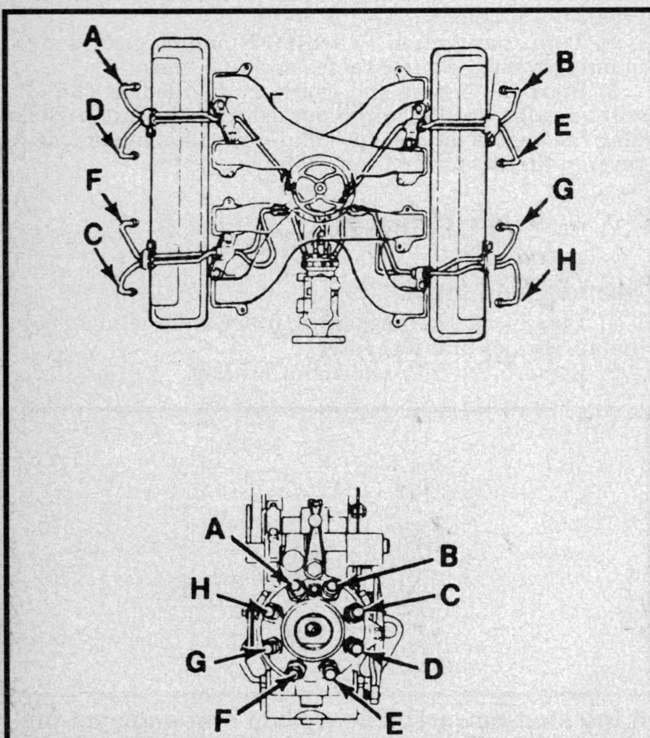

Fuel injection line routing — 6.2L engine

2. Remove the fuel return hoses and loosen the flare nuts to disconnect the injection lines from the nozzles.

3. Cap the nozzles to keep dirt out and remove the nozzle assemblies with the special 30mm socket. If there is a copper gasket in the nozzle opening in the cylinder head, remove it for replacement.

To install:

4. Installation is the reverse of removal. Use a new copper gasket and torque the nozzle assemblies to 50 ft. lbs. (70 Nm).

5. Reconnect the injection lines and torque the flare nuts to 20 ft. lbs. (25 Nm). Do not over tighten the flare nuts or the flare will split and leak.

FUEL FILTER

Removal and Installation

1. Arrange plenty of rags under the filter assembly to catch spilled fuel. Remove the fuel filler cap to relieve any pressure in the tank.

2. Open the bleed screw and water drain valve and drain the filter into a container. There may be more than one quart of fuel to drain out.

3. Pry both bail wires off the filter element and remove the element from the filter base.

To install:

4. Clean any fuel or dirt away from the base and fit a new filter into place. Secure the filter with the bail wires.

5. Close the water drain valve and fit a 1/8 inch ID hose to the air bleed on top of the filter base.

6. Disconnect the shut-down solenoid wire from the injection pump. If the air cleaner is removed, install it now to keep tools or other objects from being sucked into the engine.

7. Crank the engine for about 15 seconds, then wait 1 minute for the starter to cool. Repeat this process until a clear stream of fuel flows from the bleeder hose.

8. Close the air bleed, connect the shut-down solenoid wire and run the engine at idle for about 5 minutes to purge any remaining air. Check for leaking fuel at the filter assembly.

GLOW PLUGS

Removal and Installation

1. Disconnect the wiring from the glow plugs.

2. Use a 3/8 in. deep socket to remove the glow plugs from the cylinder head.

3. When installing the glow plugs, torque to 13 ft. lbs. (17 Nm). Do not over tighten the glow plugs or the tapered seat will be crushed and damage the outer sheath.

EGR VALVE

Removal and Installation

1. Remove the air cleaner and disconnect the EGR vacuum hose.

2. Remove the studs and remove the EGR valve.

3. Installation is the reverse of removal. Use a new gasket and make sure the EGR valve is clean and moves freely.

EPR VALVE

Removal and Installation

1. Disconnect the EPR valve vacuum hose.
2. Remove the nuts to lower the exhaust pipe.
3. Remove the studs from the manifold to remove the EPR valve.
4. Installation is the reverse of removal.

ENGINE COOLANT TEMPERATURE SENSOR

Removal and Installation

1. The sensor is threaded into the intake manifold near the thermostat housing or the left cylinder head near the front. Drain the coolant to bring the level below the sensor.
2. Disconnect the wiring from the sensor and unscrew the sensor from the coolant passage.
3. Installation is the reverse of removal. Use a silicone or non-hardening sealer on the threads and torque to 17 ft. lbs. (23 Nm).

VEHICLE SPEED SENSOR

Removal and Installation

1. Place a pan under the transmission to catch any oil. Disconnect the sensor wiring.
2. Remove the bolt and remove the sensor assembly.
3. Installation is the reverse of removal. Torque the bolt to 97 inch lbs. (11 Nm) and refill the transmission as required.

ENGINE SPEED SENSOR

Removal and Installation

1. Remove the air cleaner and cover the intake manifold opening.
2. Behind the intake manifold, disconnect the wiring and remove the bolt and sensor hold-down clamp.
3. Slide the sensor straight up and remove the gasket.
4. Installation is the reverse of removal. Turn the sensor as needed to engage the camshaft gear and drop into place.

ECM/PCM AND SPEED SENSOR BUFFER

Removal and Installation

NOTE: Electronic control units can be damaged by electrostatic discharge or static electricity. Do not touch control unit connector pins and take appropriate ESD precautions.

1. Disconnect the negative battery cable.
2. To remove the speed sensor buffer, remove the glove compartment inner panel. Disconnect the wiring and remove the buffer.
3. The ECM can be reached under the right side of the dashboard. Disconnect the wiring and remove the ECM.

To install:

4. New ECM and PCM units are sold without a PROM. Be sure to install a PROM before installing the control unit.
5. When installing the control module or buffer, make sure the connectors are clean and the wires are fully pushed into the connector.

PROM

Removal and Installation

NOTE: The PROM can be destroyed by electrostatic discharge. When a PROM is removed, do not touch the pins or allow them to be shorted together.

1. Remove the ECM/PCM. Be sure to work in a clean area and wear a static strap.
2. Remove the PROM access cover.
3. Before removing the PROM, note the reference notch on the end of the PROM carrier. The new PROM must be installed with the notch in the same position. To remove the PROM:

    a. On a PCM (automatic transmission), push out on the clips at each end of the PROM. This will push the PROM up so it can be lifted out of the socket with fingers.

    b. On an ECM, use an appropriate PROM removal tool to carefully pry up each end of the PROM a little at a time until the pins are free of the socket.

To install:

4. Note the reference notch on the end of the PROM. Make sure the new PROM is installed with the notch pointing the same direction.
5. Carefully press the PROM into the socket. Press down on the ends of the PROM until it is fully seated in the socket.
6. Install the access cover and install the ECM/PCM.

THROTTLE POSITION SENSOR

Removal and Installation

1. Remove the air cleaner and cover the opening to keep out dirt and tools.
2. Unplug the TPS connector.
3. Remove the screws and remove the TPS.
4. Installation is the reverse of removal. Adjust the sensor.

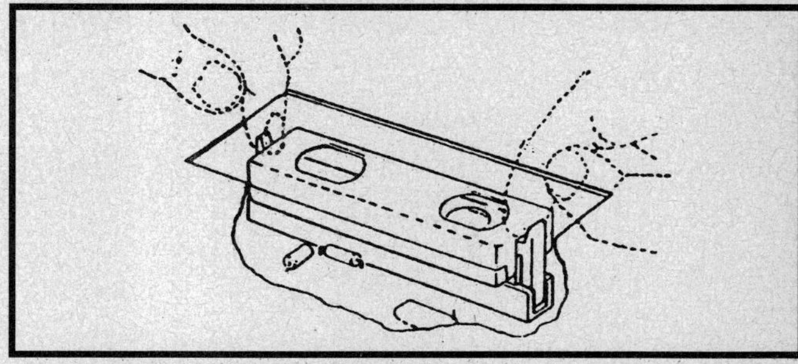

PCM PROM is removed with the retaining clips that fit into the reference notches — 6.2L LH6 engine with automatic transmission

Adjustment

MANUAL TRANSMISSION

NOTE: A special gauge block, tool number J33043-2, is required for correct sensor adjustment. The procedure also requires use of a voltmeter that reads millivolts (mV). Do not attempt sensor adjustment without these tools.

1. Remove the air cleaner assembly and related hoses.
2. Disconnect the TPS wiring and install jumper wires between the connector halves so a voltmeter can be used to read the sensor voltages with the wiring connected.
3. Turn the ignition switch **ON**.
4. Install the gage block, using the .646 side of the block. Position the tool between the gage boss on the injection pump and the full load stop screw on the throttle lever.
5. Rotate the throttle lever and hold the stop screw against the gage block.
6. Measure the voltage across connector terminals A and C (grey and black wires). This reading is the reference voltage; record the reading.
7. Measure the voltage across terminals B and C (blue and black wires). This is the TPS signal voltage.
8. Divide the signal voltage by the reference voltage. The resulting number is called the voltage ratio. It must be within 0.03 volts of the voltage ratio specification that appears on the Vehicle Emission Control Information label. For commercial vehicles, the voltage ratio is 0.69 volts.
9. If adjustment is required, loosen the two attaching screws and rotate the TPS until the correct reading is obtained, then tighten the TPS screws to 53 lb. in. (6 Nm).
10. Let the throttle lever return to the idle stop position and measure voltage across terminals B and C. Voltage should be less than 2.2 volts at closed throttle. If necessary, repeat the adjustment procedure to get the idle and gauge block lever positions correct. If adjustment cannot make both voltages correct, the TPS is faulty and must be replaced.
11. After adjustment is complete, turn the ignition switch **OFF** and reconnect the sensor wiring. Clear the DTC memory and test drive the vehicle.

AUTOMATIC TRANSMISSION

1. Remove the air cleaner assembly and related hoses.
2. Disconnect the TPS wiring and install jumper wires between the connector halves so a voltmeter can be used to read the sensor voltages with the wiring connected.
3. Turn the ignition switch **ON**.
4. Measure the voltage across connector terminals A and C (grey and black wires). This reading is the reference voltage; record the reading.
5. Multiply the reference voltage by 0.33 to obtain the the desired TPS voltage setting. If reference voltage is 5.00 volts, the desired setting is 5.00 x 0.33 = 1.65 volts.
6. Install the gage block, using the .646 side of the block. Position the tool between the gage boss on the injection pump and the full load stop screw on the throttle shaft.
7. With the throttle lever held against the gauge block, measure the voltage across terminals B and C (blue and black wires). This is the TPS signal voltage and must be within 1 percent of the setting calculated in Step 5. Multiply the reading by 0.01 to calculate 1 percent.
8. If adjustment is required, loosen the two attaching screws and rotate the TPS until the correct reading is obtained, then tighten the TPS screws to 53 lb. in. (6 Nm).
9. Let the throttle lever return to the idle stop position, then move the lever against the gauge block again and make sure the reading repeats.
10. Remove the gauge block and hold the lever against the stop screw. The voltage across terminals B and C must be 85–95 percent of the reference voltage.
11. If necessary, repeat the adjustment procedure to get the idle, gauge block and full load lever positions correct. If adjustment cannot make idle and full load voltages correct, the TPS is faulty and must be replaced.
12. After adjustment is complete, turn the ignition switch **OFF** and reconnect the sensor wiring. Clear the DTC memory and test drive the vehicle.

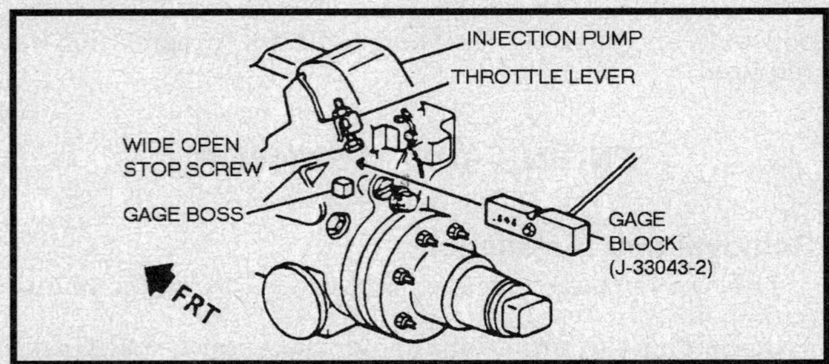

Insert the gauge block between the gauge boss and the full throttle stop screw — 6.2L LH6 engine with manual or automatic transmission

CHILTON PROFESSIONAL AUTOMOTIVE

Dear Valued Customer:

Please use this form to advise us of any questions, comments and/or recommendations that you may have regarding this manual.

Your feedback is extremely important to us so please respond today.

Thank you.

CHILTON PROFESSIONAL AUTOMOTIVE
CUSTOMER COMMENT CARD

Complete Service Manual Title _______________________

Service Manual Part Number _______________________

Section No. _______ Page No. _______ Vehicle Model & Year _______

Comments _______________________

Name _______________________

Company _______________________

Address _______________________

City _______ State _______ Zip _______

Phone ( ) _______ Date _______

CHILTON PROFESSIONAL AUTOMOTIVE

Dear Valued Customer:

Please use this form to advise us of any questions, comments and/or recommendations that you may have regarding this manual.

Your feedback is extremely important to us so please respond today.

Thank you.

CHILTON PROFESSIONAL AUTOMOTIVE
CUSTOMER COMMENT CARD

Complete Service Manual Title _______________________

Service Manual Part Number _______________________

Section No. _______ Page No. _______ Vehicle Model & Year _______

Comments _______________________

Name _______________________

Company _______________________

Address _______________________

City _______ State _______ Zip _______

Phone ( ) _______ Date _______

CHILTON PROFESSIONAL AUTOMOTIVE

Dear Valued Customer:

Please use this form to advise us of any questions, comments and/or recommendations that you may have regarding this manual.

Your feedback is extremely important to us so please respond today.

Thank you.

CHILTON PROFESSIONAL AUTOMOTIVE
CUSTOMER COMMENT CARD

Complete Service Manual Title _______________________

Service Manual Part Number _______________________

Section No. _______ Page No. _______ Vehicle Model & Year _______

Comments _______________________

Name _______________________

Company _______________________

Address _______________________

City _______ State _______ Zip _______

Phone ( ) _______ Date _______

BUSINESS REPLY MAIL
FIRST CLASS   PERMIT NO. 39   WAYNE, PA

POSTAGE WILL BE PAID BY ADDRESSEE

CHILTON BOOK COMPANY
Attn: Assistant Managing Editor—Professional
One Chilton Way
Radnor, PA   19080-9153

NO POSTAGE
NECESSARY
IF MAILED
IN THE
UNITED STATES

BUSINESS REPLY MAIL
FIRST CLASS   PERMIT NO. 39   WAYNE, PA

POSTAGE WILL BE PAID BY ADDRESSEE

CHILTON BOOK COMPANY
Attn: Assistant Managing Editor—Professional
One Chilton Way
Radnor, PA   19080-9153

NO POSTAGE
NECESSARY
IF MAILED
IN THE
UNITED STATES

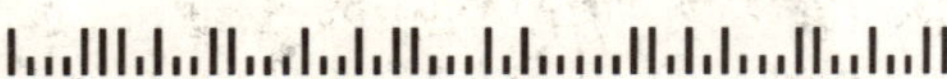

BUSINESS REPLY MAIL
FIRST CLASS   PERMIT NO. 39   WAYNE, PA

POSTAGE WILL BE PAID BY ADDRESSEE

CHILTON BOOK COMPANY
Attn: Assistant Managing Editor—Professional
One Chilton Way
Radnor, PA   19080-9153

NO POSTAGE
NECESSARY
IF MAILED
IN THE
UNITED STATES